Mechanical
Engineers'
Handbook

Mechanical Engineers' Handbook

Second Edition

Edited by
MYER KUTZ
Myer Kutz Associates, Inc.

A Wiley-Interscience Publication
JOHN WILEY & SONS, INC.
New York · Chichester · Weinheim · Brisbane · Singapore · Toronto

For Arlene

Library of Congress Cataloging-in-Publication Data:
Mechanical engineers' handbook / edited by Myer Kutz. — 2nd ed.
    p.   cm.
   Includes index.
    ISBN 0-471-13007-9 (cloth : alk. paper)
    1. Mechanical engineering—Handbooks, manuals, etc. I. Kutz, Myer.
TJ151.M395 1998
621—dc21   97-20306

Printed in the United States of America.

10 9 8 7 6 5 4 3 2

CONTRIBUTORS

Dell K. Allen
Retired from Manufacturing
  Engineering Department
Brigham Young University
Provo, Utah

T. H. Bassford
Inco Alloys International, Inc.
Huntington, West Virginia

Anne Marie Becka, Editor
Metaphase Design Group, Inc.
St. Louis, Missouri

Adrian Bejan
Department of Mechanical Engineering
  and Materials Science
Duke University
Durham, North Carolina

William E. Biles
Industrial Engineering Department
University of Louisville
Louisville, Kentucky

Peter D. Blair
Sigma Xi
The Scientific Research Society
Research Triangle Park, North Carolina

Carl Blumstein
Universitywide Energy Research Group
University of California
Berkeley, California

William Brett
New York, New York

Carl A. Brunner
Retired from U.S. Environmental
  Protection Agency

Benjamin D. Burge
Chips & Technologies, Inc.
San Jose, California

David A. Burge
David A. Burge Co., L.P.A.
Cleveland, Ohio

Robert S. Busk
Hilton Head, South Carolina

Martin S. Chizek
Weinstein Associates
Brunswick, Maine

K. J. Cleetus
Concurrent Engineering Research
  Center
West Virginia University
Morgantown, West Virginia

Jack Collins
Department of Mechanical Engineering
Ohio State University
Columbus, Ohio

Carroll Cone
Toledo, Ohio

K. W. Cooper
Borg Warner Corporation
York, Pennsylvania

Robert L. Crane
Air Force Wright Laboratory
Materials Directorate
Nondestructive Evaluation Branch
  WL/MLLP
Wright Patterson Air Force Base
Dayton, Ohio

Tushar H. Dani
Department of Mechanical Engineering
University of Wisconsin—Madison
Madison, Wisconsin

Steve Daniewicz
Department of Mechanical Engineering
Mississippi State University
Starkville, Mississippi

B. S. Dhillon
Department of Mechanical Engineering
University of Ottawa
Ottawa, Ontario, Canada

George M. Diehl
Consulting Engineer
Machinery Acoustics
Phillipsburg, New Jersey

Charles H. Drummond III
Department of Materials Science and
   Engineering
Ohio State University
Columbus, Ohio

Fritz Dusold
Retired from Mid-Manhattan Library
Science and Business Department
New York, New York

Seymour G. Epstein
The Aluminum Association, Inc.
Washington, D.C.

Warren C. Fackler
Telesis Systems, Inc.
Cedar Rapids, Iowa

Franklin E. Fisher
Mechanical Engineering Department
Loyola Marymount University
Los Angeles, California
and
Senior Staff Engineer
Hughes Aircraft Company (Retired)

Joseph L. Foszcz
Senior Editor, Plant Engineering
   Magazine
Des Plaines, Illinois

Rajit Gadh
Department of Mechanical Engineering
University of Wisconsin—Madison
Madison, Wisconsin

Bernard J. Hamrock
Department of Mechanical Engineering
Ohio State University
Columbus, Ohio

K. E. Hickman
Borg Warner Corporation
York, Pennsylvania

E. L. Hixson
University of Texas
Austin, Texas

Jim Hosier
Inco Alloys International, Inc.
Huntington, West Virginia

V. Jagannathan
Concurrent Engineering Research
   Center
West Virginia University
Morgantown, West Virginia

Byron W. Jones
Kansas State University
Manhattan, Kansas

R. Karinthi
Concurrent Engineering Research
   Center
West Virginia University
Morgantown, West Virginia

J. G. Kaufman
The Aluminum Association, Inc.
Washington, D.C.

Michelle Kazmer
Ford Motor Co.
Dearborn, Michigan

R. Alan Kemerling
Staff Quality Systems Engineer—New
   Product Development
Ethicon Endo-Surgery, Inc.
Cincinnati, Ohio

James G. Keppeler
Progress Energy Corporation
St. Petersburg, Florida

William Kerr
Department of Nuclear Engineering
University of Michigan
Ann Arbor, Michigan

Robert J. King
U.S. Steel Group, USX Corporation
Pittsburgh, Pennsylvania

Calvin J. Kirby
Vice-President Hughes Electronics and
  Chief Executive Officer Hughes
  Avicom International
Tampa, Florida

Donald Knittel
Cabot Corporation
Kokomo, Indiana

Allan Kraus
Allan D. Kraus Associates
Aurora, Ohio

Jan F. Kreider
Jan F. Kreider and Associates, Inc.
Boulder, Colorado

Peter Kuhn
Kuhn and Kuhn, Industrial Energy
  Consultants
Golden Gate Energy Center
Sausalito, California

Myer Kutz
Myer Kutz Associates, Inc.
New York, New York

Gordon Lewis
Digital Equipment Corporation
Maynard, Massachusetts

Peter E. Liley
School of Mechanical Engineering
Purdue University
West Lafayette, Indiana

Joseph A. Maciariello
Horton Professor of Management
Peter F. Drucker Graduate Management
  Center, Claremont Graduate School,
  and Claremont McKenna College
Claremont, California

Hugh R. Martin
University of Waterloo
Waterloo, Ontario, Canada

Theodore E. Matikas
Air Force Wright Laboratory
Materials Directorate
Nondestructive Evaluation Branch
  WL/MLLP
Wright Patterson Air Force Base
Dayton, Ohio

Ronald Douglas Matthews
General Motors Foundation Combustion
  Sciences and Automotive Research
  Laboratories
The University of Texas at Austin
Austin, Texas

F. C. McQuiston
Professor Emeritus
Oklahoma State University
Stillwater, Oklahoma

Howard Mendenhall
Olin Brass
East Alton, Illinois

C. A. Miller
United States Environmental Protection
  Agency
Research Triangle Park, North Carolina

Harold Miller
GE Power Systems
Schenectady, New York

Reuben M. Olson
College of Engineering and Technology
Ohio University
Athens, Ohio

Dennis L. O'Neal
Texas A & M University
College Station, Texas

Joseph W. Palen
Heat Transfer Research, Inc.
College Station, Texas

William J. Palm III
Mechanical Engineering Department
University of Rhode Island
Kingston, Rhode Island

Jerald D. Parker
Professor Emeritus
Oklahoma State University
Stillwater, Oklahoma

Edward N. Peters
General Electric Co.
Selkirk, New York

G. P. "Bud" Peterson
Executive Associate Dean and Associate
  Vice Chancellor of Engineering
Texas A & M University
College Station, Texas

Peter Pollak
The Aluminum Association, Inc.
Washington, D.C.

A. Ravindran
Department of Industrial and
  Manufacturing Engineering
Pennsylvania State University
University Park, Pennsylvania

Thomas G. Ray
Department of Industrial and
  Manufacturing Systems Engineering
Louisiana State University
Baton Rouge, Louisiana

Jack B. ReVelle
Hughes Missile Systems Co.
Tucson, Arizona

Y. V. Reddy
Concurrent Engineering Research
  Center
West Virginia University
Morgantown, West Virginia

Richard J. Reed
North American Manufacturing
  Company
Cleveland, Ohio

G. V. Reklaitis
School of Chemical Engineering
Purdue University
West Lafayette, Indiana

E. A. Ripperger
University of Texas
Austin, Texas

Murray J. Roblin
Chemical and Materials Engineering
  Department
California State Polytechnic University
Pomona, California

Bryce G. Rutter, Ph.D., Principal
Metaphase Design Group, Inc.
St. Louis, Missouri

Andrew P. Sage
School of Information Technology and
  Engineering
George Mason University
Fairfax, Virginia

Robert F. Schmidt
Colonial Metals
Columbia, Pennsylvania

Robert N. Schwarzwalder, Jr.
Ford Motor Co.
Dearborn, Michigan

Cynthia M. Scribner
Hughes Missile Systems Co.
Tucson, Arizona

William A. Smith
College of Engineering
University of South Florida
Tampa, Florida

K. Srinivas
Concurrent Engineering Research
  Center
West Virginia University
Morgantown, West Virginia

Bruce M. Steinetz
NASA Lewis Research Center
Cleveland, Ohio

William G. Steltz
Turboflow International Inc.
Orlando, Florida

Rodney D. Stewart
Mobile Data Services
Huntsville, Alabama

Hans J. Thamhain
Department of Management
Bentley College
Waltham, Massachusetts

Wayne Tustin
Equipment Reliability Institute
Santa Barbara, California

Dennis B. Webster
Department of Industrial and
  Manufacturing Systems Engineering
Louisiana State University
Baton Rouge, Louisiana

Alvin S. Weinstein
Weinstein Associates
Brunswick, Maine

Leonard A. Wenzel
Lehigh University
Bethlehem, Pennsylvania

K. Preston White, Jr.
Department of Systems Engineering
University of Virginia
Charlottesville, Virginia

Mickey R. Wilhelm
Industrial Engineering Department
University of Louisville
Louisville, Kentucky

James B. C. Wu
Cabot Corporation
Kokomo, Indiana

Emory W. Zimmers, Jr.
Enterprise Systems Center
Lehigh University
Bethlehem, Pennsylvania

Magd E. Zohdi
Department of Industrial and
    Manufacturing Engineering
Lousiana State University
Baton Rouge, Louisiana

Carl Zweben
Lockheed Martin Missiles and
    Space—Valley Forge Operations
King of Prussia, Pennsylvania

PREFACE TO THE SECOND EDITION

The two editions of this *Mechanical Engineers' Handbook* are separated by a dozen years. This length of time, especially when measured in cyber years, has dramatic consequences for a technical work that encompasses 78 chapters, each on a different topic.

This second edition of the *Handbook* has 26 entirely new chapters—a third of the book. Thirty-three chapters are revisions (42 percent of the book). Some revisions are by new authors and are so different from their predecessors that they could count as new chapters. Less than one-quarter of the book (19 chapters) is unchanged.

The new edition is the same length overall as the old. Twenty-five chapters from the first edition were deleted in order to make way for new material and because they were either obsolete or could be subsumed into new or revised chapters.

The new edition starts in a very different way than the first edition did. The separate section on digital computers is gone. There does not need to be such a section any longer, inasmuch as information on computers has been integrated into the contents of chapters where computers play an important role. The second edition opens with Chuck Drummond's revision of his chapter on the structure of solid materials. Other revisions in the materials section of the *Handbook* include chapters on aluminum (Seymour Epstein and colleagues at the Aluminum Association), nickel (Jim Hosier), magnesium (Bob Busk), and plastics and elastomers (Edward Peters). In addition, Carl Zweben provided a new version of the chapter on the important topic of composite materials.

In the years from 1986 to 1998, mechanical design entered a new era in which computers have played a larger and larger role. Virtual reality has become a new technology for mechanical engineers. The second edition of the *Handbook* has chapters on ergonomic design and electronic packaging, which the first edition did not have. New teamwork-based methods of product development have evolved, so the topic of concurrent engineering has a prominent place in this section of the *Handbook*.

All told, there are nine new chapters in the mechanical design section. A group of authors at the Concurrent Engineering Research Center—Cleetus, Jagannathan, Reddy, Srinivas, and Karinthi—contributed two chapters. Other new chapters include "Computer-Aided Design" by Emory Zimmers, "Virtual Reality" by Rajit Gadh and Tuskar Dani, "Ergonomic Factors in Design" by Bryce Rutter and Anne Becka, "Electronic Packaging" by Warren Fackler, and "Seal Technology" by Bruce Steinetz. In addition, Balbir Dhillon, who contributed to the first edition of the handbook, wrote two chapters for this edition—"Total Quality Management in Mechanical Design" and "Reliability in Mechanical Design."

Many of the chapters covering mechanical design fundamentals have been revised. Frank Fisher revised his own "Stress Analysis" chapter. The Ravindran-Reklaitis duo revised their own "Design Optimization" chapter. Steve Daniewicz has updated Jack Collins' monumental "Failure Considerations" chapter. Bob Crane, with a new co-author, Theodore E. Matikis, updated his "Nondestructive Testing" chapter.

The authors of the chapters on systems and controls—Andy Sage, Preston White, and Bill Palm—all revised their chapters. Elmer Hixson and Eugene Ripperger revised their "Measurements" chapter.

A key section of the *Handbook* deals with manufacturing engineering. The nine chapters that make up this section in the second edition include four entirely new chapters and five revised chapters. This level of change is a result of new requirements in manufacturing. For example, just-in-time scheduling and a continuous flow in manufacturing processes require well-planned and well-designed manufacturing facilities. The increased use of composite materials has required more sophisticated manufacturing processes. In order to provide greater levels of quality, information systems must be linked directly to production processes. Cutting down on waste when production of a new part starts up requires off-line adjustments in tooling.

The organizers and authors of the majority of the chapters in the manufacturing engineering section—Magd Zohdi and Bill Biles—have again made a significant contribution to the *Handbook*. This time they have been joined by colleagues Dennis Webster and Thomas Ray. In addition, this section features new chapters by Gordon Lewis ("Engineering Design for Economic Production") and Murray Roblin ("Fastening and Joining"), as well as a revision by Dell Allen of his "Classification Systems" chapter.

The section with the least amount of change involves energy and power. But even here there have been numerous major revisions and several important additions. Adrian Bejan revised his "Thermodynamics" chapter and provided a new chapter on "Exergy Analysis and Entropy Generation Minimization." These authors revised their own chapters: Jim Keppler ("Coals, Lignite, Peat"), Peter Blair ("Geothermal Resources"), Joe Palen ("Heat Exchangers, Vaporizers, Condensers"), Allan Kraus ("Cooling Electronic Equipment"), Ronald Douglas Matthews ("Internal Combustion Engines"), and Len Wenzel ("Cryogenic Equipment"). These authors provided new versions of chapters written by others in the first edition: Bud Peterson ("Heat Transfer Fundamentals"), Harold Miller ("Gas Turbines"), and Dennis O'Neal ("Refrigeration"). Joe Fozscz contributed a new chapter on air compressors. I have also included two chapters borrowed from another Wiley handbook: "Steam Turbines" (William G. Steltz) and "Hydraulic Systems" (Hugh R. Martin).

In the pollution control technology area, there are two entirely new chapters—one on air, by Andy Miller, the other on water, by Carl Brunner.

The final section of the *Handbook* deals with a variety of topics that concern mechanical engineers as their careers grow and mature. Few of these topics are static, and the section contains seven new chapters and three revisions. Among the new chapters are "Management Control of Projects" (Joe Maciariello and Calvin Kirby), "Managing People" (Hans Thamhain), "Detailed Cost Estimating" (Rod Stewart), "Total Quality Management and the Mechanical Engineer" (R. Alan Kemmerling and Jack ReVelle), "Registrations, Certifications, and Awards" (Jack ReVelle and Cynthia Scribner), and "What Engineers Need to Know about the Law" (Al Weinstein and Martin Chizek). Revisions include "Finance and the Engineering Function" (Bill Brett), "Patents" (David Burge), and "The Internet and Online Databases" (Bob Schwarzwalder).

I am grateful to all of these authors. What I said in the Preface to the First Edition of the *Handbook* bears repeating: "The wonder is that I was able to obtain . . . articles from such a wide variety of authors. . . . All of them . . . have exhibited great dedication. Writing an article for a handbook is its own reward. A desire to impart information, knowledge, and experience to others for little personal gain is a truly wonderful thing." I also want to thank my editor at Wiley, Bob Argentieri, who has been with this project from the beginning.

MYER KUTZ

New York, New York
December 1997

PREFACE TO THE FIRST EDITION

A handbook's contents portray the interests of the editor. As I pass back and forth over these chapters I find that they reflect my own career.

As an undergraduate in mechanical engineering at MIT I concentrated on mechanical design and actually thought, as I once told my father when he asked me how I intended to make a living, that I would be happy if someone would pay me to solve design problems. I was very young. While still an undergraduate I had a part-time job at the MIT Instrumentation Laboratory, where I did my Bachelor's thesis, *Design of a Bearing Torque Tester.* This was in the late 1950s, in the early days of such new and alternative materials as titanium and composites, covered in detail in this handbook, and when testing and failure analysis were much less sophisticated than they are today. (See the chapters on failure analysis and nondestructive testing.) As things turned out, however, when I went to work at the laboratory full-time, I found myself in a thermal design group working on temperature control for the guidance system of the Polaris missile. Most of my career as a mechanical engineer was focused on thermal design and temperature control. In fact, the title of my 1967 book was just that, *Temperature Control.* So the sections of this handbook on systems analysis, automatic control, thermodynamics, and heat transfer cover disciplines in which I was very much involved professionally throughout the 1960s. (I managed to employ systems engineering terminology in an article in *How Things Work in the Home,* published by Time-Life Books.)

Perhaps it was the Rockefellers' past identification with oil that suggested them as the topic for a book I wrote in the early 1970s (*Rockefeller Power,* published by Simon and Schuster in 1974), and possibly it is stretching the point to say that the chapters in this handbook on energy resources and power reflect what turned out to be a career transition. In 1976 I turned my professional activities to publishing as an acquisition editor for professional and reference books in mechanical engineering and related disciplines at Wiley. Among the most successful titles in my editorial program were books on manufacturing, products liability, patents, and occupational safety and health—all areas where I had no prior professional experience. Sales of these books demonstrated the needs of the marketplace, of course. I would be neglecting the concerns of many mechanical engineers if I did not include these topics in the handbook. Hence, the major sections on manufacturing and management.

More recently my professional life has focused on two major areas—computers and management. For several years I ran the Electronic Publishing Division at Wiley, with responsibility for distributing publications online via telecommunications. The chapters on computers, sources of mechanical engineering information, and online searching all reflect my concerns of this period.

As chairman of the Publications Committee of the American Society of Mechanical Engineers and, currently, as Executive Publisher of the Scientific and Technical

Division at Wiley, I have focused on financial matters. Several chapters of the handbook directly address this work. Worth noting here is that the ASME Board of Governors has identified the interaction of management with finance (and marketing) as being of key strategic importance.

In the past several decades, while my career was undergoing change, mechanical engineering was growing and evolving into a variety of different professions. This handbook, with its wide-ranging emphases, reflects this evolution—just as it must reflect the career growth and change of many mechanical engineers. It was Dick Zeldin, then head of Wiley's handbook program, who suggested that I undertake the editing of this handbook. At the outset, I intended it to be a one-volume updating of the old two-volume Kent's *Mechanical Engineers' Handbook,* which was first copyrighted in 1895. The most recent edition, the 12th, was published in 1950. It still continues to sell despite the fact that some of it is outdated and many topics that have become important over the past 35 years are not included. Not surprisingly, the 1950 edition of Kent contains little or nothing on computers and microprocessors, modern techniques of failure analysis, titanium, plastics and composites, modern techniques of nondestructive testing, group technology in manufacturing, computer integrated manufacturing, control systems design, modern financial techniques, safety engineering, online searching, energy, cogeneration, heat recovery, and nuclear power. The world of Kent, 12th edition, was a different world, a world in which engineering could afford large safety factors. Who cared how much anything weighed, or how much energy was consumed? So it is interesting to look back at the 12th edition and note the emphasis on steel. Steel must have been a relatively happy industry in the immediate postwar period—much happier than today. Obtaining an article on steel was an agonizing process. The industry is so demoralized, it seems, that when the Iron and Steel Institute moved from New York to Washington, it did not take along its library. I finally resolved the difficulty by borrowing part of the article that Wiley published in the Kirk-Othmer *Encyclopedia of Chemical Technology.*

Much of the 12th edition of Kent seemed to defy updating. How could you improve on the section on the efficiency of splices and knots? A revision of the chapter on woodworking (which was combined with a chapter on plastics molding) seemed better left in other hands—and in other publications.

In fact, the break between the 12th edition of Kent and a handbook edited in the 1980s was so severe that it seemed inappropriate to (1) attempt an updating, and (2) call a new mechanical engineers' handbook the 13th edition of Kent. I recommend, therefore, that you not clear space on your shelf for this volume by throwing out your old Kent, but that you keep both.

One thing you may not need the old Kent for, however, is its mathematical tables. With an inexpensive scientific calculator, most numbers you need are at your fingertips. You will not find mathematical tables in these pages. Why destroy forests to make paper for printing tables when ubiquitous microelectronics are so cheap and powerful?

They are changing the profession of mechanical engineering. In fact, CAD/CAM or CIM (computer integrated manufacturing), and robotics, for example, have made mechanical engineering a "hot field." So this handbook begins with chapters on computers and microprocessors (and, later, has a chapter on robots and computer-aided manufacturing).

This order was the recommendation of the handbook's editorial board. I am indebted to them for this wisdom—and for their good sense in confirming the subject matter of the handbook. The final decision on what actually went into the handbook was mine alone—as were the choices of contributors. As it turned out, this was a massive undertaking. Although I have many connections within the mechanical engineering fraternity, not all of the 78 articles came easily. Some did not come forth at all, and I have had to borrow about 10 percent of the articles from other sources. These borrowings have been edited for this handbook, of course, often by the original authors.

The wonder is that I was able to obtain original articles from such a wide variety of authors. Some of them I knew because I had edited their books. Others were colleagues on ASME boards and committees. Others I found after exhaustive correspondence and telephone calls. All of them, no matter what their prior relationship to me, have exhibited great dedication. Writing an article for a handbook is its own reward. A desire to impart information, knowledge, and experience to others for little personal gain is a truly wonderful thing.

My thanks to colleagues Martin Grayson, Thurman Poston, Wiley's excellent production staff, particularly Ed Cantillon, Margaret Comaskey, Douglas Elam, and the unknown copyeditor and proofreaders. Thanks, too, to my secretary, Meryl Weiner, who put up with an awesome amount of last minute typing and organizational work. At this point, authors and editors generally mention spouses and other family members, thanking them for their patience and understanding. I have read hundreds of these citations to late nights in offices and Sunday afternoons spent with galleys spread over the dining room table. I know from personal experience that these statements are not pro forma; they are true, and they come from the heart. So does this: My wife Mandy deserves special thanks, particularly because this handbook was so long in the making.

MYER KUTZ

New York, New York
February 1986

CONTENTS

PART 1   MATERIALS AND MECHANICAL DESIGN

1. Structure of Solids   3
    Charles H. Drummond III

2. Steel   17
    Robert J. King

3. Aluminum and Its Alloys   45
    Seymour G. Epstein, J. G. Kaufman, Peter Pollak

4. Copper and Its Alloys   59
    Howard Mendenhall, Robert F. Schmidt

5. Nickel and Its Alloys   71
    T. H. Bassford, Jim Hosier

6. Titanium and Its Alloys   91
    Donald Knittel, James B. C. Wu

7. Magnesium and Its Alloys   109
    Robert S. Busk

8. Plastics and Elastomers   115
    Edward N. Peters

9. Composite Materials and Mechanical Design   131
    Carl Zweben

10. Stress Analysis   191
    Franklin E. Fisher

11. Concurrent Engineering Revisited: How Far Have We Come?   249
    K. J. Cleetus

12. Concurrent Engineering Technologies   261
    V. Jagannathan, Y. V. Reddy, K. J. Cleetus, K. Srinivas,
    R. Karinthi

13. Computer-Aided Design   275
    Emory Zimmers, Jr.

14. Virtual Reality—A New Technology for the Mechanical
    Engineer   319
    Tushar H. Dani, Rajit Gadh

15. Ergonomic Factors in Design 329
    Bryce G. Rutter, Anne Marie Becka

16. Electronic Packaging 339
    Warren C. Fackler

17. Design Optimization—An Overview 353
    A. Ravindran, G. V. Reklaitis

18. Failure Considerations 377
    Jack Collins, Steve Daniewicz

19. Total Quality Management in Mechanical Design 475
    B. S. Dhillon

20. Reliability in Mechanical Design 487
    B. S. Dhillon

21. Lubrication of Machine Elements 507
    Bernard J. Hamrock

22. Seal Technology 629
    Bruce M. Steinetz

23. Vibration and Shock 661
    Wayne Tustin

24. Noise Measurement and Control 711
    George M. Diehl

25. Nondestructive Testing 729
    Robert L. Crane, Theodore E. Matikas

PART 2   SYSTEMS AND CONTROLS

26. Systems Engineering: Analysis, Design, and Information
    Processing for Analysis and Design 763
    Andrew P. Sage

27. Mathematical Models of Dynamic Physical Systems 795
    K. Preston White, Jr.

28. Basic Control Systems Design 867
    William J. Palm III

29. Measurements 917
    E. L. Hixson, E. A. Ripperger

PART 3   MANUFACTURING ENGINEERING

30. Product Design for Manufacturing and Assembly (DFM&A) 935
    Gordon Lewis

31. Classification Systems 951
    Dell K. Allen

32. Production Planning 987
   Dennis B. Webster, Thomas G. Ray

33. Production Processes and Equipment 1035
   Magd E. Zohdi, William E. Biles, Dennis B. Webster

34. Metal Forming, Shaping, and Casting 1101
   Magd E. Zohdi, Dennis B. Webster, William E. Biles

35. Mechanical Fasteners 1135
   Murray J. Roblin

36. Statistical Quality Control 1175
   Magd E. Zohdi

37. Computer-Integrated Manufacturing 1187
   William E. Biles, Magd E. Zohdi

38. Material Handling 1205
   William E. Biles, Mickey R. Wilhelm, Magd E. Zohdi

PART 4 ENERGY, POWER, AND POLLUTION CONTROL TECHNOLOGY

39. Thermophysical Properties of Fluids 1245
   Peter E. Liley

40. Fluid Mechanics 1289
   Reuben M. Olson

41. Thermodynamics Fundamentals 1331
   Adrian Bejan

42. Exergy Analysis and Entropy Generation Minimization 1351
   Adrian Bejan

43. Heat Transfer Fundamentals 1367
   G. P. "Bud" Peterson

44. Combustion 1431
   Richard J. Reed

45. Furnaces 1449
   Carroll Cone

46. Gaseous Fuels 1505
   Richard J. Reed

47. Liquid Fossil Fuels from Petroleum 1517
   Richard J. Reed

48. Coals, Lignite, Peat 1535
   James G. Keppeler

49. Solar Energy Applications 1549
   Jan F. Kreider

50. Geothermal Resources: An Introduction   1583
    Peter D. Blair

51. Energy Auditing   1591
    Carl Blumstein, Peter Kuhn

52. Heat Exchangers, Vaporizers, Condensers   1607
    Joseph W. Palen

53. Air Heating   1641
    Richard J. Reed

54. Cooling Electronic Equipment   1649
    Allan Kraus

55. Pumps and Fans   1681
    William A. Smith

56. Nuclear Power   1699
    William Kerr

57. Gas Turbines   1723
    Harold Miller

58. Steam Turbines   1765
    William G. Steltz

59. Internal Combustion Engines   1801
    Ronald Douglas Matthews

60. Hydraulic Systems   1831
    Hugh R. Martin

61. Air Compressors   1865
    Joseph L. Foszcz

62. Refrigeration   1879
    Dennis L. O'Neal, K. W. Cooper, K. E. Hickman

63. Cryogenic Systems   1915
    Leonard A. Wenzel

64. Indoor Environmental Control   1973
    Jerald D. Parker, F. C. McQuiston

65. Air Pollution–Control Technologies   2011
    C. A. Miller

66. Water Pollution–Control Technology   2031
    Carl A. Brunner

PART 5   MANAGEMENT, FINANCE, QUALITY, LAW, AND RESEARCH

67. Management Control of Projects   2047
    Joseph A. Maciariello, Calvin J. Kirby

68. Managing People   2085
    Hans J. Thamhain

69. Finance and the Engineering Function 2097
William Brett

70. Detailed Cost Estimating 2117
Rodney D. Stewart

71. Investment Analysis 2143
Byron W. Jones

72. Total Quality Management and the Mechanical Engineer 2159
R. Alan Kemerling, Jack B. ReVelle

73. Registrations, Certifications, and Awards 2177
Jack B. ReVelle, Cynthia M. Scribner

74. Safety Engineering 2193
Jack B. ReVelle

75. What the Law Requires of the Engineer 2229
Alvin S. Weinstein, Martin S. Chizek

76. Patents 2247
David A. Burge, Benjamin D. Burge

77. Electronic Information Resources: Your On-Line Bookshelf 2269
Robert N. Schwarzwalder, Jr., Michelle Kazmer

78. Sources of Mechanical Engineering Information 2287
Fritz Dusold, Myer Kutz

INDEX **2293**

PART 1

MATERIALS AND
MECHANICAL DESIGN

CHAPTER 1

STRUCTURE OF SOLIDS

Charles H. Drummond III
Department of Materials Science and Engineering
Ohio State University
Columbus, Ohio

1.1	**INTRODUCTION**	**3**		1.2.2 Alloys	13
	1.1.1 Effects of Structure on			1.2.3 Noncrystalline Metals	13
	Properties	3			
	1.1.2 Atomic Structure	3	**1.3**	**CERAMICS**	**14**
	1.1.3 Bonding	4		1.3.1 Crystalline Ceramics	14
	1.1.4 Simple Structures	4		1.3.2 Noncrystalline Ceramics	14
	1.1.5 Crystallography	5		1.3.3 Glass–Ceramics	15
	1.1.6 States of Matter	7			
	1.1.7 Polymorphism	8	**1.4**	**POLYMERS**	**15**
	1.1.8 Defects	8			
			1.5	**COMPOSITES AND COATINGS**	**15**
1.2	**METALS**	**12**		1.5.1 Fiberglass	15
	1.2.1 Structures	12		1.5.2 Coatings	15

1.1 INTRODUCTION

1.1.1 Effects of Structure on Properties

Physical properties of metals, ceramics, and polymers, such as ductility, thermal expansion, heat capacity, elastic modulus, electrical conductivity, and dielectric and magnetic properties, are a direct result of the structure and bonding of the atoms and ions in the material. An understanding of the origin of the differences in these properties is of great engineering importance.

In single crystals, a physical property such as thermal expansion varies with direction, reflecting the crystal structure; whereas in polycrystalline and amorphous materials, a property does not vary with direction, reflecting the average property of the individual crystals or the randomness of the amorphous structure. Most engineering materials are polycrystalline, composed of many grains, and thus an understanding of the properties requires not only a knowledge of the structure of the single grains but also a knowledge of grain size and orientation, grain boundaries, and other phases present; that is, a knowledge of the microstructure of this material.

1.1.2 Atomic Structure

Atoms consist of electrons, protons, and neutrons. The central nucleus consists of positively charged protons and electrically neutral neutrons. Negatively charged electrons are in orbits about the nucleus in different energy levels, occupying a much larger volume than the nucleus.

In an atom, the number of electrons equals the number of protons and, hence, an atom is neutral. The atomic number of an element is given by the number of protons, and the atomic weight is given by the total number of protons and neutrons. (The weight of the electrons is negligible.) Thus, hydrogen, H, with one proton and one electron, has an atomic number of 1 and an atomic weight of 1 and is the first element in the periodic chart. Oxygen, O, with atomic number 8, has eight protons and eight neutrons and, hence, an atomic weight of 16.

Completed electronic shells have a lower energy than partially filled orbitals when bonded to other atoms. As a result of this energy reduction, atoms share electrons to complete the shells, or gain or lose electrons to form completed shells. In the latter case, ions are formed in which the

Mechanical Engineers' Handbook, 2nd ed., Edited by Myer Kutz.
ISBN 0-471-13007-9 © 1998 John Wiley & Sons, Inc.

number of electrons is not equal to the number of protons. Thus, O by gaining two electrons, has a charge of -2 and forms the oxygen ion O^{2-}.

The periodic chart arranges elements in columns of the same electronic configuration. The first column consists of the alkalies Li, Na, K, Cs, Rb; each has one electron in the outer shell that can be lost. Similarly, the second column of alkaline-earths can form Mg^{2+}, Ca^{2+}, Sr^{2+}, Ba^{2+} by losing two electrons. The seventh column consists of the halogens Fl, Cl, Br, I, which by gaining one electron become the halides, all with a charge of -1. The eighth column consists of the inert gases He, Ne, Ar, K, Xe, with completed shells. The bonding of the elements and ions with similar electronic configurations is similar. Moving down a column increases the number of electrons and, hence, the atom's size increases even though the outer electronic configuration remains the same.

The outer electrons that are lost, gained, or shared are called valence electrons, and the inner electrons are called core electrons. For the most part, the valence electrons are important in determining the nature of the bonding and, hence, the structure and properties of the materials.

1.1.3 Bonding

When two atoms or ions are within atomic distances of each other, distances of 0.5–3.0Å, bonding may occur between the atoms or ions. The resulting reduction in energy due to an attractive force leads to the formation of polyatomic gas molecules, liquids, and solids. If the energy of the bonds is large (75–275 kcal/mol), primary bonds are formed—metallic, ionic, or covalent. If the energy of the bond is smaller (1–10 kcal/mol), secondary bonds are formed—van der Waals and hydrogen. In addition, combinations of bond types, such as a mixture of ionic and covalent bonds, may occur.

Metallic Bonding

In a metallic crystal, an ordered arrangement of nuclei and their electrons is embedded in a cloud of valence electrons, which are shared throughout the lattice. The resulting bonding is a nondirectional primary bond. Since the binding energy of the valence electrons is relatively small, the mobility of these electrons is high and creates high electrical and thermal conductivity. The atoms are approximately spherical in shape as a result of the shape of completed inner shell. Examples of metals are Cu, Au, Ag, and Na.

Ionic Bonding

The strongest type of bonding between two oppositely charged particles is called ionic bonding. The positively charged ions (cations) attract as many negatively charged ions (anions) as they can and form ionic bonds. The primary bond formed is nondirectional if the bonding is purely ionic. Li^+ and F^- in LiF form predominately ionic bonds. In general, since the electrons are strongly bonded, electrical and thermal conductivities are much smaller than in metals and, thus, ionic bonded materials are classified as insulators or dielectrics.

Covalent Bonding

Covalent bonding results from an overlap or sharing, not from gain or loss of valence electrons. A net reduction of energy as a result of each atom's completing the other's orbital also results in a primary bond, but it is directional. The directionality is a result of the shape of the orbitals involved in the bonding. When C is covalently bonded to four other C's in diamond, the bonding is purely covalent and the configuration of these four bonds is tetrahedral. When B, however, is bonded to three other B's, a triangular configuration is formed. Organic polymers and diatomic gases such as Cl_2 are typical examples of covalent bonding. As a result of the strong bonding of the valence electrons, these materials, for the most part, have low electrical and thermal conductivity.

Van der Waals and Hydrogen Bonding

Van der Waals bonds are secondary bonds, the result of fluctuating dipoles, due to the fact that at an instant of time the centers of positive and negative charge do not coincide. An example is an inert gas such as Ar, which below $-190°C$ forms a solid as a result of these weak attractive forces. Similar weak forces exist in molecules and solids. Hydrogen bonds are also secondary bonds, but they are the result of permanent dipoles. For example, the water molecule, H_2O, is nonlinear and the bonding between H and an adjacent O in water results in H_2O being a liquid above 0°C a 1 atm pressure rather than a gas, as is the case for other molecules of comparable molecular weight.

1.1.4 Simple Structures

If atoms or ions are considered to be spheres, then the most efficient packing of the spheres in space will form their most stable structure. However, the type of bonding—in particular, directional bonding—may affect the structure formed. In two dimensions, there is only one configuration that most efficiently fills space, the close-packed layer (see Fig. 1.1). If similar layers are stacked to form a three-dimensional structure, an infinite number of configurations is possible. Two are important. In

Fig. 1.1 Close-packed layer.

both, the first two layers are the same. In the first layer (*A*), the point at the center of three spheres provides a hollow for a fourth sphere to rest. A second close-packed layer (*B*) then can be placed on the first layer, with each sphere occupying the hollow. With the addition of a third layer to these two layers, two choices are possible. A sphere in the third layer can be placed above a sphere in the first layer in the spaces marked (•) in Fig. 1.2 or above a hollow not occupied by a sphere spaces marked (x) in the second layer. If the first stacking arrangement is continued, that is, the first and third layers in registry with each other (denoted *ABABA* . . .), the hexagonal close-packed (hcp) structure is generated, so called because of the hexagonal symmetry of the structure. If the second stacking arrangement is continued, that is, the first and third layers are not on top of each other (denoted *ABCABC* . . .), the cubic close-packed or face-centered cubic (fcc) structure is generated, so called because the structure formed is a face-centered cube. Both structures are shown in Fig. 1.3. In both structures, 74% of the volume is occupied and each sphere is contacted by 12 spheres (or 12 nearest neighbors), although the arrangement is different. Another common structure is the body-centered cubic (bcc) structure shown in Fig. 1.3. Here, each sphere has eight nearest neighbors, with another six at a slightly greater distance. The volume fraction occupied is 68%. In the hcp and fcc structures, the stacking of a fourth sphere on top of three in any close-packed layer generates a tetrahedral site or void, as shown in Fig. 1.4. Into such a site a smaller sphere with a coordination number of four could fit. Three spheres from each of two layers generate an octahedral site or void, as shown in Fig. 1.4. Into such a site a smaller sphere with a coordination number of six could fit. In the hcp and fcc structures, there are two tetrahedral and one octahedral sites per packing sphere; however, the arrangement of these sites is different.

1.1.5 Crystallography

All possible crystallographic structures are described in terms of 14 Bravais space lattices—only 14 different ways of periodically arranging points in space. These are shown in Fig. 1.5. Each of the

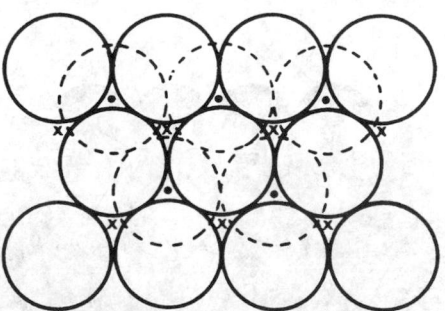

Fig. 1.2 Two possible sites for sphere in fcc and hcp structures: x and • (from D. M. Adams, *Inorganic Solids,* Wiley, New York, 1974).

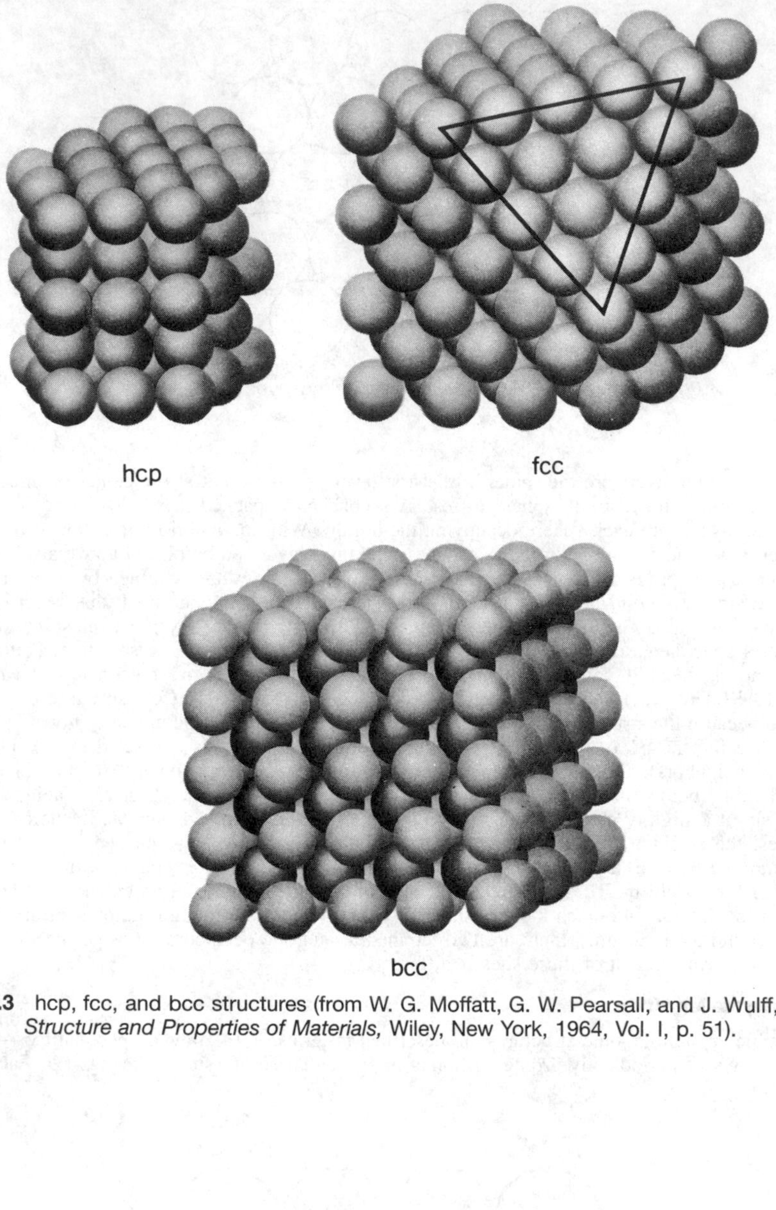

Fig. 1.3 hcp, fcc, and bcc structures (from W. G. Moffatt, G. W. Pearsall, and J. Wulff, *The Structure and Properties of Materials,* Wiley, New York, 1964, Vol. I, p. 51).

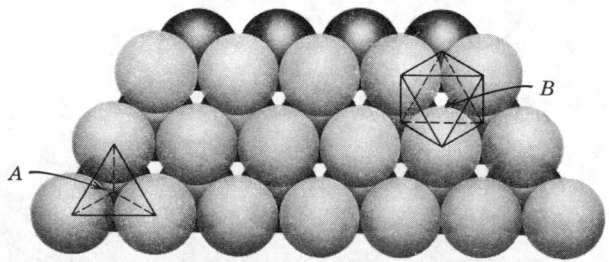

Fig. 1.4 Tetrahedral and octahedral sites (from G. W. Moffatt, G. W. Pearsall, and J. Wulff, *The Structure and Properties of Materials,* Wiley, New York, 1964, Vol. I, p. 58).

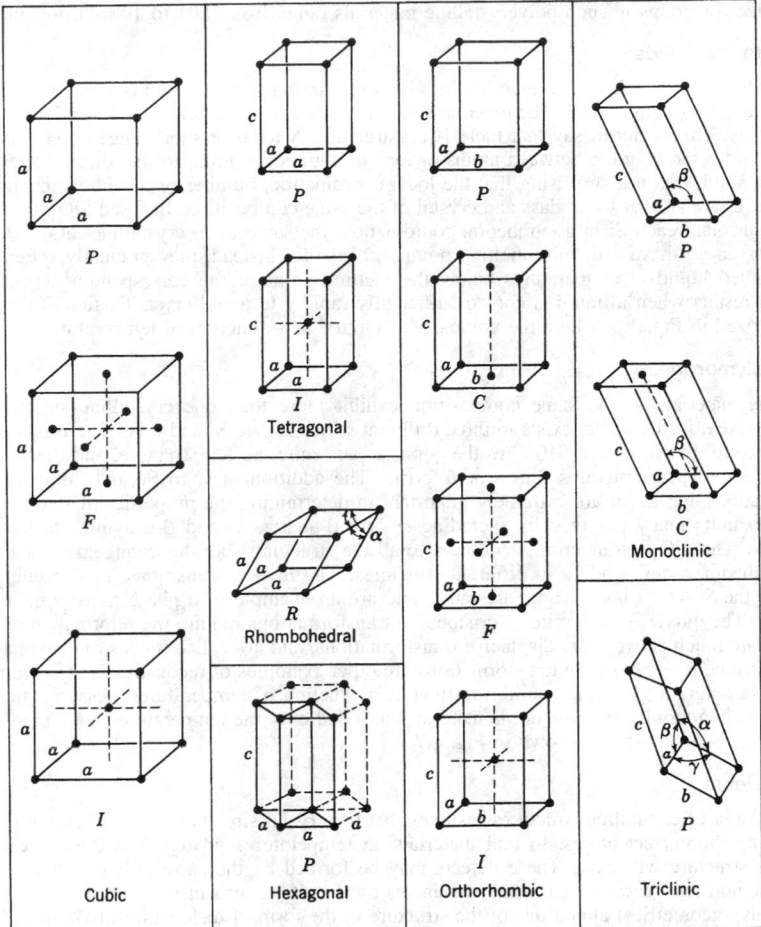

Fig. 1.5 Bravais lattices (from W. G. Moffatt, G. W. Pearsall, and J. Wulff, *The Structure and Properties of Materials,* Wiley, New York, 1964, Vol. I, p. 47).

positions in a given space lattice is equivalent and an atom or ion or group of atoms or ions can be centered on each position. Each of the lattices is described by a unit cell, as shown in Fig. 1.5. The seven crystallographic systems are also shown in Fig. 1.5.

1.1.6 States of Matter

Matter can be divided into gases, liquids, and solids. In gases and liquids, the positions of the atoms are not fixed with time, whereas in solids they are. Distances between atoms in gases are an order of magnitude or greater than the size of the atoms, whereas in solids and liquids closest distances between atoms are only approximately the size of the atoms. Almost all engineering materials are solids, either crystalline or noncrystalline.

Crystalline Solids

In crystalline solids, the atoms or ions occupy fixed positions and vibrate about these equilibrium positions. The arrangement of the positions is some periodic array, as discussed in Section 1.1.5. At 0°K, except for a small zero-point vibration, the oscillation of the atoms is zero. With increasing temperature the amplitude and frequency of vibration increase up to the melting point. At the melting point, the crystalline structure is destroyed, and the material melts to form a liquid. For a particular single crystal the external shape is determined by the symmetry of the crystal class to which it belongs. Most engineering materials are not single crystals but polycrystalline, consisting of many small crystals. These crystals are often randomly oriented and may be of the same composition or of different composition or of different structures. There may be small voids between these grains. Typical sizes of grains in such polycrystalline materials range from 0.01 to 10 mm in diameter.

Noncrystalline Solids

Noncrystalline solids (glasses) are solids in which the arrangement of atoms is periodic (random) and lacks any long-range order. The external shape is without form and has no defined external faces like a crystal. This is not to say that there is no structure. A local or short-range order exists in the structure. Since the bonding between atoms or ions in a glass is similar to that of the corresponding crystalline solid, it is not surprising that the local coordination, number of neighbors, configuration, and distances are similar for a glass and crystal of the same composition. In fused SiO_2, for example, four O's surround each Si in a tetrahedral coordination, the same as in crystalline SiO_2.

Glasses do not have a definite melting point, crystals do. Instead, they gradually soften to form a supercooled liquid at temperatures below the melting point of the corresponding crystal. Glass formation results when a liquid is cooled sufficiently rapidly to avoid crystallization. This behavior is summarized in Fig. 1.6, where the volume V is plotted as a function of temperature T.

1.1.7 Polymorphism

Crystalline materials of the same composition exhibit more than one crystalline structure called polymorphs. Fe, for example, exists in three different structures: α, γ, and δ Fe. The α phase, ferrite, a bcc structure, transforms at 910°C to the γ phase, austenite, an fcc structure, and then at 1400°C changes back to bcc structures δ-iron or δ-ferrite. The addition of C to Fe and the reactions and transformations that occur are extremely important in determining the properties of steel.

SiO_2 exhibits many polymorphs, including α- and β-quartz, α- and β-tridymite, and α- and β-cristobalite. The SiO_4 tetrahedron is common to all the structures, but the arrangement or linking of these tetrahedra varies, leading to different structures. The $\alpha \rightarrow \beta$ transitions involve only a slight change in the Si–O–Si bond angle, are rapid, and are an example of a phase transformation called displacive. The quartz $\rightarrow$ tridymite $\rightarrow$ cristobalite transformations require the reformation of the new structure, are much slower than displacive transformations, and are called reconstructive phase transformations. The $\alpha \rightarrow \gamma \rightarrow \delta$ Fe transformations are other examples of reconstructive transformations.

A phase diagram gives the equilibrium phases a function of temperature, pressure, and composition. More commonly, the pressure is fixed at 1 atm and only the temperature and composition are varied. The Fe–C diagram is shown in Fig. 1.7.

1.1.8 Defects

The discussion of crystalline structures assumes that the crystal structures are perfect, with each site occupied by the correct atoms. In real materials, at temperatures greater than 0°K, defects in the crystalline structure will exist. These defects may be formed by the substitution of atoms different from those normally occupying the site, vacancies on the site, atoms in sites not normally occupied (interstitials), geometrical alterations of the structure in the form of dislocations, twin boundaries, or grain boundaries.

Solid Solution

When atoms or ions are approximately the same size, they may substitute for another in the structure. For example, Cu and Au have similar radii and at high temperature form a complete solid solution,

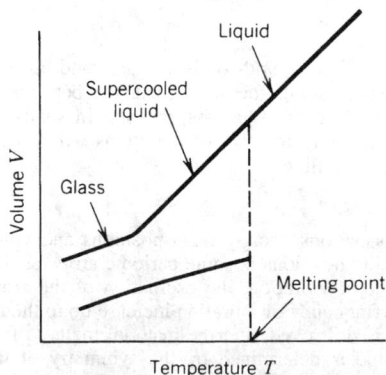

Fig. 1.6 Glass formation.

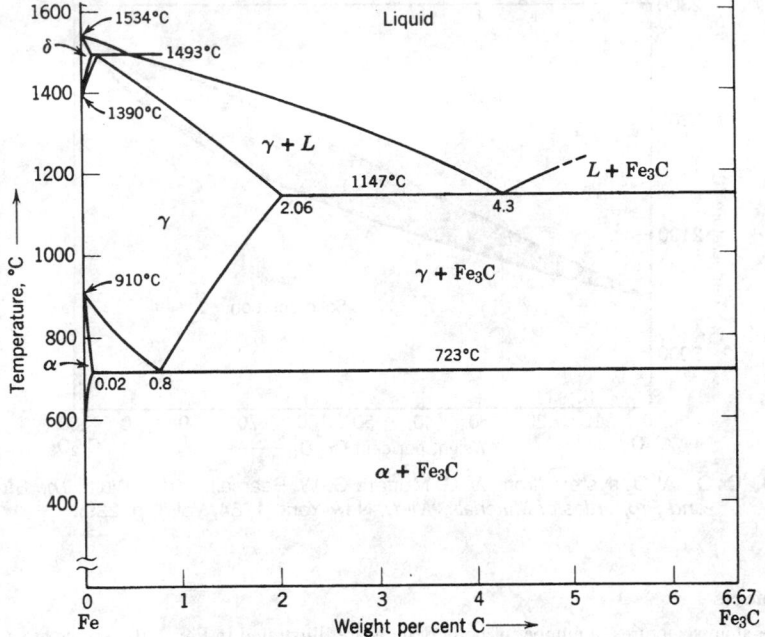

Fig. 1.7 Fe–C phase diagram (from W. G. Moffatt, G. W. Pearsall, and J. Wulff, *The Structure and Properties of Materials,* Wiley, New York, 1964, Vol. I, p. 185).

as shown in Fig. 1.8. A ceramic example is the Cr_2O_3–Al_2O_3 system shown in Fig. 1.9, where Cr and Al substitute for each other. Cr^{3+} has a radius of 0.76 Å and Al has a radius of 0.67 Å. Complete solid solution is not possible if the size difference between atoms or ions is too large, if the structures of the end members are different, or if there are charge differences between ions being substituted. In the last case, substitution is possible only if the charge is compensated for by the creation of vacancies or by oxidation or reduction of ions.

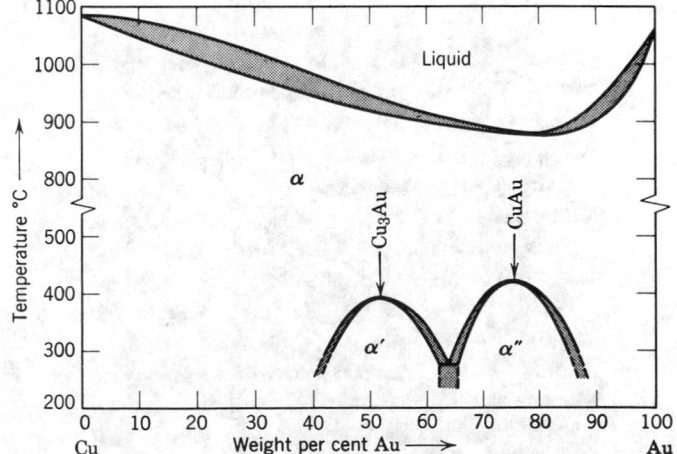

Fig. 1.8 Cu–Au system (from W. G. Moffatt, G. W. Pearsall, and J. Wulff, *The Structure and Properties of Materials,* Wiley, New York, 1964, Vol. I, p. 230).

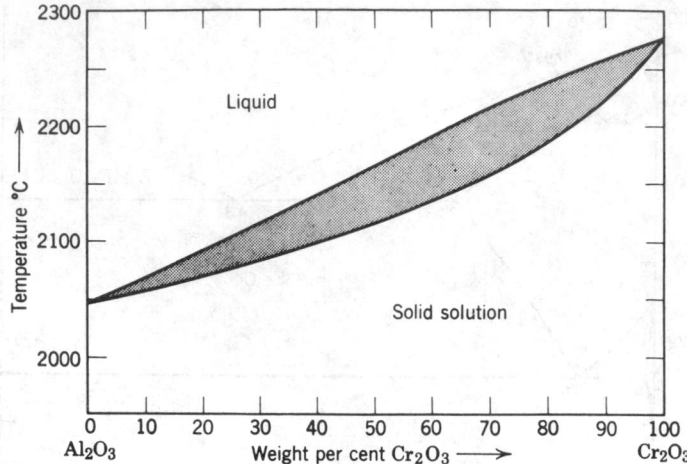

Fig. 1.9 Cr_2O_3–Al_2O_3 system (from W. G. Moffatt, G. W. Pearsall, and J. Wulff, *The Structure and Properties of Materials,* Wiley, New York, 1964, Vol. I, p. 229).

Point Defects

For single-atom structures, a number of point defects are illustrated in Fig. 1.10. Shown are a vacancy (an absent atom); an interstitial atom, occupying a normally unoccupied site; and two types of impurities, one in an interstitial site and the other substituting for an atom. In Fig. 1.11 a number of point defects are shown for an ionic compound AB. Substitutional ions, vacancies, and impurity ions are shown. In ionic compounds, because charges must be balanced, when a cation is removed, an anion is also removed. The resulting vacancy and interstitial point defects are called a Schottky pair. A Frenkel defect occurs when an ion is removed from its normal site and is placed in an interstitial site. The presence of defects—interstitials and vacancies—is necessary for diffusion to occur in many crystalline solids.

Dislocations

Two basic types of dislocations exist in solids—edge and screw dislocations. An edge dislocation consists of an extra plane of atoms, as shown in Fig. 1.12. It is represented by the symbol ⊥ and has associated compression and tension. A screw dislocation is formed by the atom planes spiraling

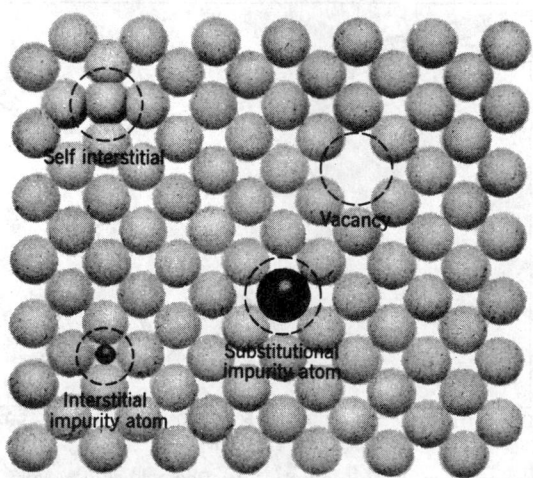

Fig. 1.10 Point defects (from W. G. Moffatt, G. W. Pearsall, and J. Wulff, *The Structure and Properties of Materials,* Wiley, New York, 1964, Vol. I, p. 77).

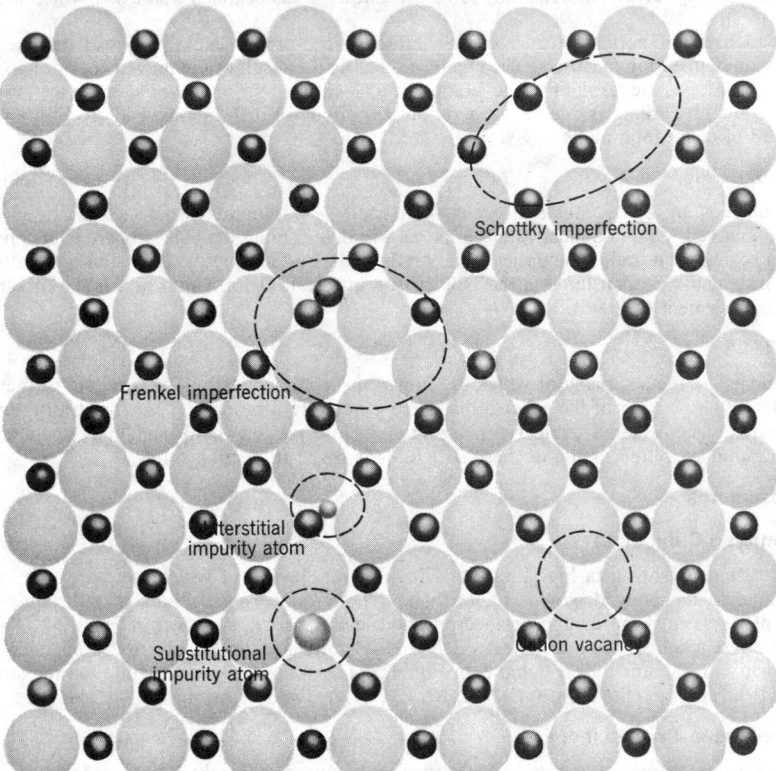

Fig. 1.11 Point defects in a compound *AB* (from W. G. Moffatt, G. W. Pearsall, and J. Wulff, *The Structure and Properties of Materials,* Wiley, New York, 1964, Vol. I, p. 78).

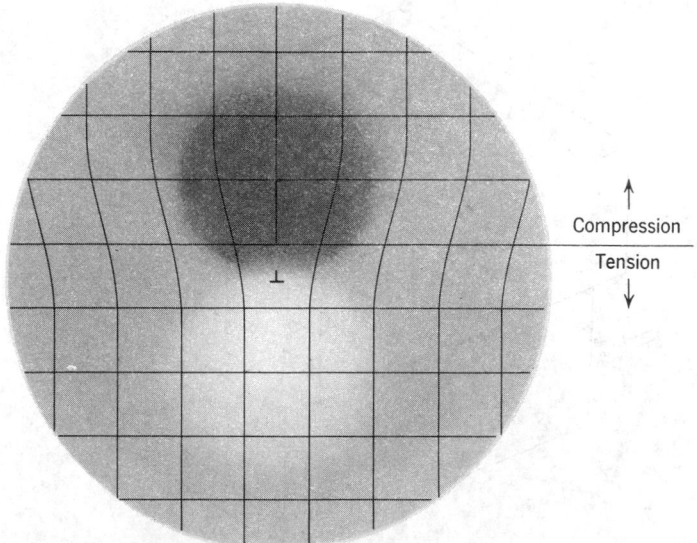

Fig. 1.12 Edge dislocation (from W. G. Moffatt, G. W. Pearsall, and J. Wulff, *The Structure and Properties of Materials,* Wiley, New York, 1964, Vol. I, p. 85).

and is shown in Fig. 1.13. Combinations of screw and edge dislocations also exist, which are called mixed dislocations.

Dislocations are important because of their effect on the properties, in particular the mechanical properties, of engineering materials. The slip of a metal is the result of the movement of dislocations; plastic deformation is the result of the generation of dislocations; the increased strength and brittleness as a result of cold working is due to a generation and pileup of dislocations; and creep in a material is the result of dislocation climb.

Grain Boundaries

Grain boundaries are the regions that occur when there is no alignment between grains in a poly-crystalline material. Grain boundaries are important in determining the bulk properties of a material. Impurities segregate at grain boundaries if they reduce the surface energy. Diffusion is usually faster along grain boundaries than through the bulk of the material. Deformation of a material can occur by relative movement of grains.

1.2 METALS

Most elements are metals, many of which are important technologically. The structure of metals can be considered the packing of the spheres that most efficiently fills space. Three basic structures will be considered: face-centered cubic (fcc), hexagonal close-packed (hcp), and body-centered cubic (bcc). An introductory discussion of these structures is given in Section 1.1.4.

1.2.1 Structures

Face-Centered Cubic (fcc)

The fcc structure is shown in Fig. 1.3. The *ABCABC* . . . layers, of which there are four sets, are perpendicular to the body diagonals of the cube. The 12 nearest neighbors at a distance D (the diameter of a sphere) form a cubo-octahedron about each sphere, as shown in Fig. 1.14. There are six next nearest neighbors at a distance $\sqrt{2}\,D$ and 24 third-nearest neighbors at a distance $\sqrt{3}\,D$. The symmetry of the structure is cubic F in Fig. 1.5. The following metals adopt the fcc structures as one of their polymorphs: Al, Ca, Fe, Co, Ni, Cu, Sr, Y, Rh, Pd, Ag, Ir, Pt, Au, and Pb.

Hexagonal Close-Packed (hcp)

The hcp structure is shown in Fig. 1.3. There is only one close-packed direction with a packing sequence *ABAB* . . . The hexagonal symmetry is shown in Fig. 1.5. As in the fcc structure, there are 12 nearest neighbors, but their configuration is different in the form of a twinned cubooctahedron, as shown in Fig. 1.14. There are six next nearest neighbors, as in the fcc structure, but only two third-nearest neighbors at a distance $\sqrt{8/3}\,D$ or $1.633D$, the distance from one sphere to the spheres

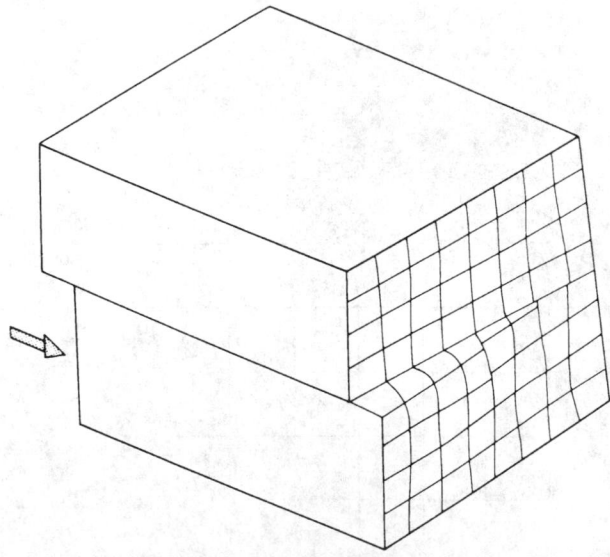

Fig. 1.13 Screw dislocation (from W. D. Kingery, H. K. Bowen, and D. R. Uhlmann, *Introduction to Ceramics,* Wiley, New York, 1976).

hcp   fcc

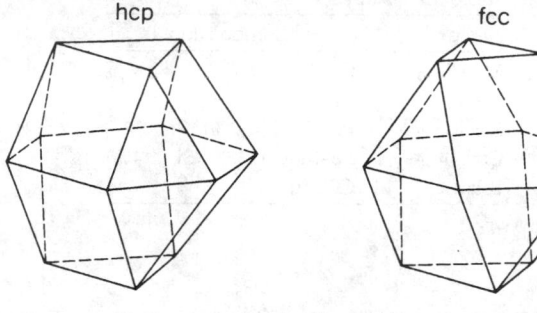

Twinned Cubo-Octahedron   Cubo-Octahedron

Fig. 1.14 Configuration of nearest neighbors in hcp and fcc structures (from W. G. Moffatt, G. W. Pearsall, and J. Wulff, *The Structure and Properties of Materials,* Wiley, New York, 1964.

in the second layer above or below the given sphere. The c/a ratio $= 1.633$ is defined in Fig. 1.5 for the hexagonal lattice. If the shape of the atoms is ellipsoidal rather than spherical, then the c/a ratio deviates from the 1.633 value. Metals with the hcp structure and their c/a ratio are given in Table 1.1.

Body-Centered Cubic (bcc)

The bcc structure is shown in Fig. 1.3. There, the distance of the next nearest neighbors is close to the nearest-neighbor distance. Thus, the effective coordination number is 14, comparable to the fcc and hcp structures. Metals that have the bcc structure are Li, Na, K, Ca, V, Ti, Cr, Fe, Rb, Sr, Nb, Mo, Cs, Ba, Hf, Ta, and W.

The structure of a particular metal adopts cannot be explained only in terms of volume occupied or number of nearest neighbors. The energy differences between fcc, hcp, and bcc structures are very small. The nature of the bonding and the electronic configuration also play important roles.

1.2.2 Alloys

The addition of a second element (or more) to a metal results in an alloy, which may have improved engineering properties. Some examples of alloys are given in Table 1.2. The extent of solid solution and phases formed is given by the appropriate phase diagram. The extent of solid solution is determined by the relative sizes of the atoms.

1.2.3 Noncrystalline Metals

Noncrystalline or amorphous metals can be prepared in the form of ribbon or film by rapid quenching techniques (cooling rate $> 10^5$ deg/sec), such as splat cooling or vapor, electrolytic, or chemical deposition. Compositions include a metal and metalloid, such as Si, Ge, P, Sb, or C, with generally 80 wt% metal. Typical compositions are Ni_3P, Au_3Si and Pd–Fe–Si. The structure of these materials consists of a dense, random packing of the metal in which approximately 64% of the volume is occupied and in which the metalloid occupies irregularly shaped tetrahedra, octahedra, and other sites and stabilizes the structure. Improved mechanical properties, including higher strengths, greater ductility, improved corrosion resistance, and interesting magnetic properties, make these promising engineering materials.

Table 1.1   hcp Metals

Metal	c/a	Metal	c/a
Be	1.568	Y	1.571
Na	1.63	Zr	1.593
Mg	1.623	Ru	1.582
Sc	1.594	Cd	1.886
Co	1.623	Hf	1.581
Zn	1.856	Re	1.615
Sr	1.63	Os	1.579

Table 1.2   Alloys

Name	Composition
Monel	Cu–Ni
Bronze	Cu–Sn
Steel	Fe–C (C < 2%)
Cast iron	Fe–Si–C (2% < Si < 4%)
Brass	Cu–Zn

1.3 CERAMICS

Ceramics are nonmetallic inorganic materials. Thus, the oxides of all metals such as Fe_2O_3, TiO_2, Al_2O_3, and SiO_2 and materials such as diamond, SiN, SiC, and Si are considered to be ceramics.

1.3.1 Crystalline Ceramics

Most oxides can be considered close packings of oxygen ions with the cations occupying the tetrahedral and/or octahedral sites in the structure. As an example, α-alumina (α-Al_2O_3) consists of an hcp packing of O^{2-} with two thirds of the octahedral sites occupied by Al^{3+} in an orderly fashion. Since for each O^{2-} there exist one octahedral and two tetrahedral sites, in Al_2O_3 there would be three octahedral sites in which two Al^{3+} are placed thus two-thirds of the octahedral and none of the tetrahedral sites are filled. The compound is electrically neutral, since $2 \times (3+)$ (Al) $= 3 \times (2-)$ (O). If the Al is shared by six O's, then $3/6 = 1/2$ of its charge is contributed to each O. For the charge on each O to be satisfied, four Al's need to be coordinated to each O, since $4(1/2) = 2$. A notation to indicate the coordination scheme for α-Al_2O_3 is 6:4—each Al is coordinated to six O's and each oxygen is coordinated to four Al's.

A summary of common ceramic structures is given in Table 1.3. The structure of silicates is complicated, but the basic unit is the SiO_4 tetrahedron. The three polymorphs of SiO_2—quartz, tridymite and cristobalite—have different arrangements for the linking of all four vertices of the tetrahedron. Each Si is bonded to four O's and each O is bonded to two Si's. In the layer silicates such as micas, clays, and talc, only three of the vertices are linked. The result is a laminar structure in which the bonding between layers is a weaker ionic bonding, hydrogen bonding, or van der Waals bonding, respectively, for mica, clay, and talc.

Of particular importance in semiconductors is the diamond structure. In this structure, each atom is tetrahedrally coordinated to four other atoms. The predominant covalent bonding of the structure is manifested by the high degree of directionality in the bonding. In addition to diamond, Si and Ge have this structure, as do other semiconductors that have been doped with other elements.

1.3.2 Noncrystalline Ceramics

Common glass compositions are fused SiO_2, soda–lime silica, soda–borosilicate, and alkali–lead silicate. Glass formers such as SiO_2 and B_2O_3 are characterized by a high viscosity at the melting point and readily form glasses when cooled. Network modifiers such as Na_2O and CaO do not form glasses unless quenched at extremely high rates. Intermediaries such as Al_2O_3 and PbO, while not readily forming glasses by themselves, can be present in high concentrations when combined with glass formers.

Amorphous or fused SiO_2 has Si tetrahedrally coordinated to four O's, with each O bonded to two Si's. Thus, SiO_4 tetrahedra are linked, sharing all four vertices in a continuous three-dimensional network. The structure has short-range order but no long-range order. The introduction of network modifiers results in the formation of nonbridging oxygen—oxygen bonded to only one Si and thus

Table 1.3   Common Ceramic Structures

Structure	Examples	Coordination	Packing of Anion
Rock salt	MgO, CaO, SrO, FeO	6:6	fcc
Zincblende	SiC	4:4	ccp
Rutile	TiO_2, GeO_2	6:3	Distorted hcp
Perovskite	$SrTiO_3$, $BaTiO_3$	12:6:6	fcc
Spinel	$MgAl_2O_4$, $FeAl_2O_4$	4:6:4	fcc
Corundum	α-Al_2O_3, Fe_2O_3	6:4	hcp
Fluorite	UO_2, ZrO_2	8:4	Simple cube

negatively charged. The cation, such as Na^+, is in the interstitial sites balancing the charge. The result is an increase in density, a large decrease in viscosity, and a decrease in thermal expansion with increasing alkali content. The alkaline earths behave in a similar manner. Commercial soda–lime–silica glass (72 wt% SiO_2, 12–15 wt% Na_2O, 10–15 wt% CaO) has a broken up silica network with Na and Ca ions in large interstitials.

1.3.3 Glass–Ceramics

Glass–ceramics are materials that have been fabricated as glasses and then crystallized as a result of controlled nucleation and growth. In most cases, nucleating agents such as TiO_2, P_2O_5, Pt, or ZrO_2 are added to aid in crystallization. The microstructure of many glass–ceramics consists of 95–98% crystalline phase with a grain size < 1 μm embedded in a small amount of a pore-free glassy phase. Typical composition systems are Li_2O–Al_2O_3–SiO_2, and Na_2O–BaO–Al_2O_3–SiO_2. Some of the desirable properties of various glass–ceramic systems are zero or very low thermal expansion, high mechanical strength, high electrical resistivity, and machinability.

1.4 POLYMERS

Polymers are organic materials that consist of chains of C and H. The intrachain bonding is covalent, while the interchain bonding is van der Waals. The repeating structural units, monomers, are linked together to form the polymer.

Isomers are organic compounds of the same composition but with a different arrangement of the atoms. Copolymerization is the process of linking different polymers together. Many polymers are noncrystalline because the long chains become entangled or because of side groups attached to the chain, particularly if they are large or irregularly placed. Both of these factors make it difficult to crystallize the chains. The addition of plasticizers—low-molecular-weight compounds that separate the chains—also help prevent crystallization.

The manner in which the polymers are formed affects the final structure. Bifunctional monomers result in two bonds that form linear chains, whereas trifunctional or tetrafunctional monomers result in network or framework polymers. This results in cross linking and in increased structural rigidity and less elasticity. The shape of the linear polymers can be altered by the addition of side groups; not only does the packing become less ordered, but the interbonding becomes stronger. Branching, the splitting of the polymer chain, is another way to introduce three dimensions into the polymer structure.

Since polymers are organic materials when compared to metals and ceramics, they tend to have low melting or softening temperatures and are flammable. The elastic moduli are lower by several orders of magnitude and they serve as electrical insulators.

1.5 COMPOSITES AND COATINGS

Many modern engineering materials have been developed by combining two or more materials into a single material, a composite, or by coating one material with another. The structure of such composites and coating will be discussed in general, but the specifics will not be covered.

1.5.1 Fiberglass

Fiberglass is formed from glass fibers impregnated in an order or random manner in a plastic material. The fibers are usually of a composition known as E-glass (SiO_2, 54 wt%; Al_2O_3, 14 wt%; B_2O_3, 10 wt%; MgO, 4.5 wt%; and CaO, 17.5 wt%), are typically 0.00023–0.00053 in. in diameter, and woven together to form continuous fibers or to form cloth.

1.5.2 Coatings

Various coatings used in engineering applications are summarized in Table 1.4. Coatings can serve as a protective layer for the substrate and/or alter the appearance of the surface. The structures of the coating and the substrate have previously been discussed. Of great importance is the bonding structure at the interface between the coating and the substrate.

In general, the bonding will be affected by atomistic and microscopic considerations of the surfaces. The bonding in the material, whether metallic, ionic, or covalent, may be continued or altered

Table 1.4 Coatings

Coating	Composition	Substrate
Enamel	Inorganic glass	Metal
Glaze	Inorganic glass	Ceramic
Paint	Organic	Metal, polymer
Galvanized and plating	Metal	Metal

in the interface. Such factors as surface roughness, porosity, and oxidation/reduction, and the presence of impurities, will affect the bonding at the interface.

Enamels are used on metals to protect the surface from oxidation and to change the color and appearance of the surface. The vitreous enamel is fused to the surface of the metal. The bonding changes from metallic to ionic–covalent on the enamel. The thermal expansion of the enamel is usually less than that of the metal substrate, so that the enamel surface is in compression, thus improving the mechanical properties of the enamel. Glazes are used to decrease the porosity of the ceramic substrate and to alter the appearance of the surface.

BIBLIOGRAPHY

Adams, D. M., *Inorganic Solids,* Wiley, New York, 1974.

Barrett, C. S. and T. B. Massalski, *Structure of Metals,* 3rd ed., Pergamon Press, Oxford, 1980.

Barrett, C. R., W. D. Nix, and A. S. Tetelman, *The Principles of Engineering Materials,* Prentice-Hall, Englewood Cliffs, NJ, 1973.

Flinn, R. A., and P. K. Trojan, *Engineering Materials and Their Applications,* 4th ed., Houghton Mifflin, Boston, MA, 1990.

Guy, A. G., *Essentials of Materials Science,* McGraw-Hill, New York, 1976.

Kingery, W. D., H. K. Bowen, and D. R. Uhlmann, *Introduction to Ceramics,* 2nd ed., Wiley, New York, 1976.

Moffatt, W. G., G. W. Pearsall, and J. Wulff, *The Structure and Properties of Materials,* Wiley, New York, 1964, Vol. 1.

The Structure and Properties of Materials, 4 Vols., Wiley, New York: Vol. 1, *Structures,* W. G. Moffatt, G. W. Pearsall, and J. Wulff (eds.), 1964; Vol. 2, *Thermodynamics of Structure,* H. H. Brophy, R. M. Rose, and J. Wulff (eds.), 1964; Vol. 3, *Mechanical Behavior,* H. W. Hayden, W. G. Moffatt, and J. Wulff (eds.), 1965; Vol. 4, *Electronic Properties,* R. M. Rose, L. A. Shepard, and J. Wulff (eds.), 1966.

Van Vlack, L. H., *Elements of Materials Science,* 6th ed., Addison-Wesley, Reading, MA, 1989.

CHAPTER 2

STEEL

Robert J. King
U.S. Steel Group, USX Corporation
Pittsburgh, Pennsylvania

2.1	**METALLOGRAPHY AND HEAT**			2.4.5	Austempering	28
	TREATMENT	**18**		2.4.6	Normalizing	28
				2.4.7	Annealing	29
2.2	**IRON-IRON CARBIDE PHASE**			2.4.8	Isothermal Annealing	29
	DIAGRAM	**19**		2.4.9	Spheroidization Annealing	31
	2.2.1 Changes on Heating and			2.2.10	Process Annealing	31
	Cooling Pure Iron	19		2.4.11	Carburizing	31
	2.2.2 Changes on Heating and			2.4.12	Nitriding	31
	Cooling Eutectoid Steel	19				
	2.2.3 Changes on Heating and		**2.5**	**CARBON STEELS**		**31**
	Cooling Hypoeutectoid Steels	20		2.5.1	Properties	32
	2.2.4 Changes on Heating and			2.5.2	Microstructure and Grain	
	Cooling Hypereutectoid Steels	20			Size	32
	2.2.5 Effect on Alloys on the			2.5.3	Microstructure of Cast Steels	33
	Equilibrium Diagram	20		2.5.4	Hot Working	33
	2.2.6 Grain Size—Austenite	20		2.5.5	Cold Working	34
	2.2.7 Microscopic-Grain-Size			2.5.6	Heat Treatment	34
	Determination	21		2.5.7	Residual Elements	35
	2.2.8 Fine- and Coarse-Grain					
	Steels	21	**2.6**	**DUAL-PHASE SHEET STEELS**		**35**
	2.2.9 Phase Transformations—					
	Austenite	21	**2.7**	**ALLOY STEELS**		**36**
	2.2.10 Isothermal Transformation			2.7.1	Functions of Alloying	
	Diagram	21			Elements	36
	2.2.11 Pearlite	23		2.7.2	Thermomechanical	
	2.2.12 Bainite	23			Treatment	36
	2.2.13 Martensite	23		2.7.3	High-Strength Low-Alloy	
	2.2.14 Phase Properties—Pearlite	23			(HSLA) Steels	36
	2.2.15 Phase Properties—Bainite	23		2.7.4	AISI Alloy Steels	36
	2.2.16 Phase Properties—Martensite	23		2.7.5	Alloy Tool Steels	37
	2.2.17 Tempered Martensite	23		2.7.6	Stainless Steels	37
	2.2.18 Transformation Rates	23		2.7.7	Martensitic Stainless Steels	37
	2.2.19 Continuous Cooling	24		2.7.8	Ferrite Stainless Steels	39
				2.7.9	Austenitic Stainless Steels	39
2.3	**HARDENABILITY**	**25**		2.7.10	High-Temperature Service,	
					Heat-Resisting Steels	40
2.4	**HEAT-TREATING PROCESSES**	**26**		2.7.11	Quenched and Tempered	
	2.4.1 Austenitization	26			Low-Carbon Constructional	
	2.4.2 Quenching	27			Alloy Steels	41
	2.4.3 Tempering	27		2.7.12	Maraging Steels	41
	2.4.4 Martempering	28		2.7.13	Silicon-Steel Electrical	
					Sheets	41

Reprinted from *Kirk-Othmer Encyclopedia of Chemical Technology*, 3rd ed., Wiley, New York, 1983, Vol. 21, by permission of the publisher.

Mechanical Engineers' Handbook, 2nd ed., Edited by Myer Kutz.
ISBN 0-471-13007-9   © 1998 John Wiley & Sons, Inc.

2.1 METALLOGRAPHY AND HEAT TREATMENT

The great advantage of steel as an engineering material is its versatility, which arises from the fact that its properties can be controlled and changed by heat treatment.[1-3] Thus, if steel is to be formed into some intricate shape, it can be made very soft and ductile by heat treatment; on the other hand, heat treatment can also impart high strength.

The physical and mechanical properties of steel depend on its constitution, that is, the nature, distribution, and amounts of its metallographic constituents as distinct from its chemical composition. The amount and distribution of iron and iron carbide determine the properties, although most plain carbon steels also contain manganese, silicon, phosphorus, sulfur, oxygen, and traces of nitrogen, hydrogen, and other chemical elements such as aluminum and copper. These elements may modify, to a certain extent, the main effects of iron and iron carbide, but the influence of iron carbide always predominates. This is true even of medium-alloy steels, which may contain considerable amounts of nickel, chromium, and molybdenum.

The iron in steel is called ferrite. In pure iron-carbon alloys, the ferrite consists of iron with a trace of carbon in solution, but in steels it may also contain alloying elements such as manganese, silicon, or nickel. The atomic arrangement in crystals of the allotrophic forms of iron is shown in Fig. 2.1.

Cementite, the term for iron carbide in steel, is the form in which carbon appears in steels. It has the formula Fe_3C, and consists of 6.67% carbon and 93.33% iron. Little is known about its properties, except that it is very hard and brittle. As the hardest constituent of plain carbon steel, it scratches glass and feldspar but not quartz. It exhibits about two-thirds the induction of pure iron in a strong magnetic field.

Austenite is the high-temperature phase of steel. Upon cooling, it gives ferrite and cementite. Austenite is a homogeneous phase, consisting of a solid solution of carbon in the γ form of iron. It forms when steel is heated above 790°C. The limiting temperatures for its formation vary with composition and are discussed below. The atomic structure of austenite is that of γ iron, fcc; the atomic spacing varies with the carbon content.

When a plain carbon steel of ~ 0.80% carbon content is cooled slowly from the temperature range at which austenite is stable, ferrite and cementite precipitate together in a characteristically lamellar structure known as pearlite. It is similar in its characteristics to a eutectic structure but, since it is formed from a solid solution rather than from a liquid phase, it is known as a eutectoid structure. At carbon contents above and below 0.80%, pearlite of ~ 0.80% carbon is likewise formed on slow cooling, but excess ferrite or cementite precipitates first, usually as a grain-boundary network, but occasionally also along the cleavage planes of austenite. The excess ferrite or cementite rejected by the cooling austenite is known as a proeutectoid constituent. The carbon content of a slowly cooled steel can be estimated from the relative amounts of pearlite and proeutectoid constituents in the microstructure.

Bainite is a decomposition product of austenite consisting of an aggregate of ferrite and cementite. It forms at temperatures lower than those where very fine pearlite forms and higher than those at which martensite begins to form on cooling. Metallographically, its appearance is feathery if formed

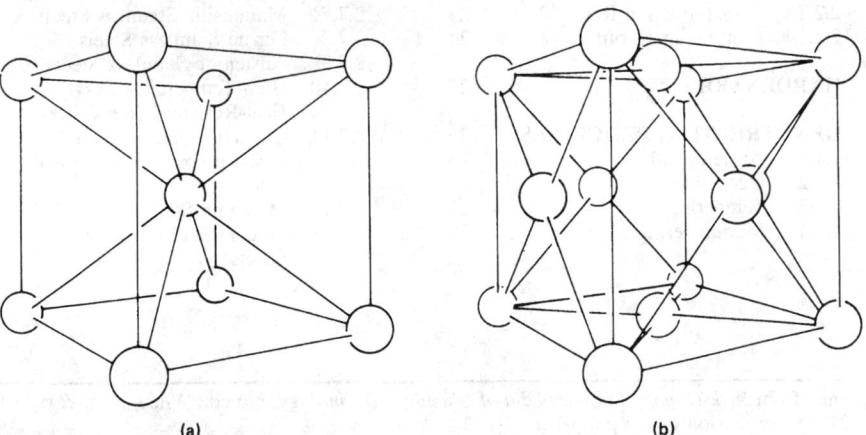

(a)   (b)

Fig. 2.1 Crystalline structure of allotropic forms of iron. Each white sphere represents an atom of (a) α and δ iron in bcc form, and (b) γ iron, in fcc (from Ref. 1).

in the upper part of the temperature range, or acicular (needlelike) and resembling tempered martensite if formed in the lower part.

Martensite in steel is a metastable phase formed by the transformation of austenite below the temperature called the M_s temperature, where martensite begins to form as austenite is cooled continuously. Martensite is an interstitial supersaturated solid solution of carbon in iron with a body-centred tetragonal lattice. Its microstructure is acicular.

2.2   IRON–IRON CARBIDE PHASE DIAGRAM

The iron–iron carbide phase diagram (Fig. 2.2) furnishes a map showing the ranges of compositions and temperatures in which the various phases such as austenite, ferrite, and cementite are present in slowly cooled steels. The diagram covers the temperature range from 600°C to the melting point of iron, and carbon contents from 0 to 5%. In steels and cast irons, carbon can be present either as iron carbide (cementite) or as graphite. Under equilibrium conditions, only graphite is present because iron carbide is unstable with respect to iron and graphite. However, in commercial steels, iron carbide is present instead of graphite. When a steel containing carbon solidifies, the carbon in the steel usually solidifies as iron carbide. Although the iron carbide in a steel can change to graphite and iron when the steel is held at ~ 900°C for several days or weeks, iron carbide in steel under normal conditions is quite stable.

The portion of the iron–iron carbide diagram of interest here is that part extending from 0 to 2.01% carbon. Its application to heat treatment can be illustrated by considering the changes occurring on heating and cooling steels of selected carbon contents.

Iron occurs in two allotropic forms, α or δ (the latter at a very high temperature) and γ (see Fig. 2.1.) The temperatures at which these phase changes occur are known as the critical temperatures, and the boundaries in Fig. 2.2 show how these temperatures are affected by composition. For pure iron, these temperatures are 910°C for the α–γ phase change and 1390° for the γ–δ phase change.

2.2.1   Changes on Heating and Cooling Pure Iron

The only changes occurring on heating or cooling pure iron are the reversible changes at ~910°C from bcc α iron to fcc γ iron and from the fcc δ iron to bcc γ iron at ~1390°C.

2.2.2   Changes on Heating and Cooling Eutectoid Steel

Eutectoid steels are those that contain 0.8% carbon. The diagram shows that at and below 727°C the constituents are α–ferrite and cementite. At 600°C, the α-ferrite may dissolve as much as 0.007% carbon. Up to 727°C, the solubility of carbon in the ferrite increases until, at this temperature, the

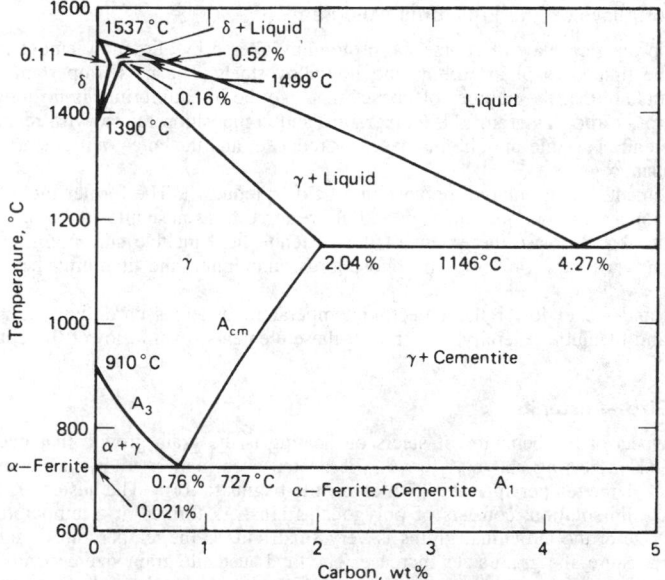

Fig. 2.2   Iron–iron carbide phase diagram (from Ref. 1).

ferrite contains about 0.02% carbon. The phase change on heating an 0.8% carbon steel occurs at 727°C which is designated as A_1, as the eutectoid or lower critical temperature. On heating just above this temperature, all ferrite and cementite transform to austenite, and on slow cooling the reverse change occurs.

When a eutectoid steel is slowly cooled from ~738°C, the ferrite and cementite form in alternate layers of microscopic thickness. Under the microscope at low magnification, this mixture of ferrite and cementite has an appearance similar to that of a pearl and is therefore called pearlite.

2.2.3 Changes on Heating and Cooling Hypoeutectoid Steels

Hypoeutectoid steels are those that contain less carbon than the eutectoid steels. If the steel contains more than 0.02% carbon, the constituents present at and below 727°C are usually ferrite and pearlite; the relative amounts depend on the carbon content. As the carbon content increases, the amount of ferrite decreases and the amount of pearlite increases.

The first phase change on heating, if the steel contains more than 0.02% carbon, occurs at 727°C. On heating just above this temperature, the pearlite changes to austenite. The excess ferrite, called proeutectoid ferrite, remains unchanged. As the temperature rises further above A_1, the austenite dissolves more and more of the surrounding proeutectoid ferrite, becoming lower and lower in carbon content until all the proeutectoid ferrite is dissolved in the austenite, which now has the same average carbon content as the steel.

On slow cooling the reverse changes occur. Ferrite precipitates, generally at the grain boundaries of the austenite, which becomes progressively richer in carbon. Just above A_1, the austenite is substantially of eutectoid composition, 0.8% carbon.

2.2.4 Changes on Heating and Cooling Hypereutectoid Steels

The behavior on heating and cooling hypereutectoid steels (steels containing >0.80% carbon) is similar to that of hypoeutectoid steels, except that the excess constituent is cementite rather than ferrite. Thus, on heating above A_1, the austentie gradually dissolves the excess cementite until at the A_{cm} temperature the proeutectoid cementite has been completely dissolved and austenite of the same carbon content as the steel is formed. Similarly, on cooling below A_{cm}, cementite precipitates and the carbon content of the austenite approaches the eutectoid composition. On cooling below A_1, this eutectoid austenite changes to pearlite and the room-temperature composition is, therefore, pearlite and proeutectoid cementite.

Early iron–carbon equilibrium diagrams indicated a critical temperature at ~768°C. It has since been found that there is no true phase change at this point. However, between ~768 and 790°C there is a gradual magnetic change, since ferrite is magnetic below this range and paramagnetic above it. This change, occurring at what formerly was called the A_2 change, is of little or no significance with regard to the heat treatment of steel.

2.2.5 Effect of Alloys on the Equilibrium Diagram

The iron–carbon diagram may, of course, be profoundly altered by alloying elements, and its application should be limited to plain carbon and low-alloy steels. The most important effects of the alloying elements are that the number of phases that may be in equilibrium is no longer limited to two as in the iron–carbon diagram; the temperature and composition range, with respect to carbon, over which austenite is stable may be increased or reduced; and the eutectoid temperature and composition may change.

Alloying elements either enlarge the austenite field or reduce it. The former include manganese, nickel, cobalt, copper, carbon, and nitrogen and are referred to as austenite formers.

The elements that decrease the extent of the austenite field include chromium, silicon, molybdenum, tungsten, vanadium, tin, niobium, phosphorus, aluminum, and titanium; they are known as ferrite formers.

Manganese and nickel lower the eutectoid temperature, whereas chromium, tungsten, silicon, molybdenum, and titanium generally raise it. All these elements seem to lower the eutectoid carbon content.

2.2.6 Grain Size—Austenite

A significant aspect of the behavior of steels on heating is the grain growth that occurs when the austenite, formed on heating above A_3 or A_{cm}, is heated even higher; A_3 is the upper critical temperature and A_{cm} is the temperature at which cementite begins to form. The austenite, like any metal composed of a solid solution, consists of polygonal grains. As formed at a temperature just above A_3 or A_{cm}, the size of the individual grains is very small but, as the temperature is increased above the critical temperature, the grain sizes increase. The final austenite grain size depends, therefore, on the temperature above the critical temperature to which the steel is heated. The grain size of the austenite has a marked influence on transformation behavior during cooling and on the grain size of the constituents of the final microstructure. Grain growth may be inhibited by carbides that dissolve slowly or by dispersion of nonmetallic inclusions. Hot working refines the coarse grain formed by reheating steel to the relatively high temperatures used in forging or rolling, and the grain size of hot-worked steel is determined largely by the temperature at which the final stage of the hot-working process is carried out. The general effects of austenite grain size on the properties of heat-treated steel are summarized in Table 2.1.

2.2.7   Microscopic-Grain-Size Determination

The microscopic grain size of steel is customarily determined from a polished plane section prepared in such a way as to delineate the grain boundaries. The grain size can be estimated by several methods. The results can be expressed as diameter of average grain in millimeters (reciprocal of the square root of the number of grains per mm^2), number of grains per unit area, number of grains per unit volume, or a micrograin-size number obtained by comparing the microstructure of the sample with a series of standard charts.

2.2.8   Fine- and Coarse-Grain Steels

As mentioned previously, austenite-grain growth may be inhibited by undissolved carbides or non-metallic inclusions. Steels of this type are commonly referred to as fine-grained steels, whereas steels that are free from grain-growth inhibitors are known as coarse-grained steels.

The general pattern of grain coarsening when steel is heated above the critical temperature is as follows: Coarse-grained steel coarsens gradually and consistently as the temperature is increased, whereas fine-grained steel coarsens only slightly, if at all, until a certain temperature known as the coarsening temperature is reached, after which abrupt coarsening occurs. Heat treatment can make any type of steel either fine or coarse grained; as a matter of fact, at temperatures above its coarsening temperature, the fine-grained steel usually exhibits a coarser grain size than the coarse-grained steel at the same temperature.

Making steels that remain fine grained above 925°C involves the judicious use of deoxidation with aluminum. The inhibiting agent in such steels is generally conjectured to be a submicroscopic dispersion of aluminum nitride or, perhaps at times, aluminum oxide.

2.2.9   Phase Transformations—Austenite

At equilibrium, that is, with very slow cooling, austenite transforms to pearlite when cooled below the A_1 temperature. When austenite is cooled more rapidly, this transformation is depressed and occurs at a lower temperature. The faster the cooling rate, the lower the temperature at which transformation occurs. Furthermore, the nature of the ferrite-carbide aggregate formed when the austenite transforms varies markedly with the transformation temperature, and the properites are found to vary correspondingly. Thus, heat treatment involves a controlled supercooling of austenite, and in order to take full advantage of the wide range of structures and properties that this treatment permits, a knowledge of the transformation behavior of austenite and the properties of the resulting aggregates is essential.

2.2.10   Isothermal Transformation Diagram

The transformation behavior of austenite is best studied by observing the isothermal transformation at a series of temperatures below A_1. The transformation progress is ordinarily followed metallo-graphically in such a way that both the time-temperature relationships and the manner in which the microstructure changes are established. The times at which transformation begins and ends at a given temperature are plotted, and curves depicting the transformation behavior as a function of temperature are obtained by joining these points (Fig. 2.3) Such a diagram is referred to as an isothermal trans-formation (IT) diagram, a time-temperature-transformation (TTT) diagram, or, an S curve.[4]

Table 2.1   Trends in Heat-Treated Products

Property	Coarse-grain Austenite	Fine-grain Austenite
Quenched and Tempered Products		
Hardenability	Increasing	Decreasing
Toughness	Decreasing	Increasing
Distortion	More	Less
Quench cracking	More	Less
Internal stress	Higher	Lower
Annealed or Normalized Products		
Machinability		
Rough finish	Better	Inferior
Fine finish	Inferior	Better

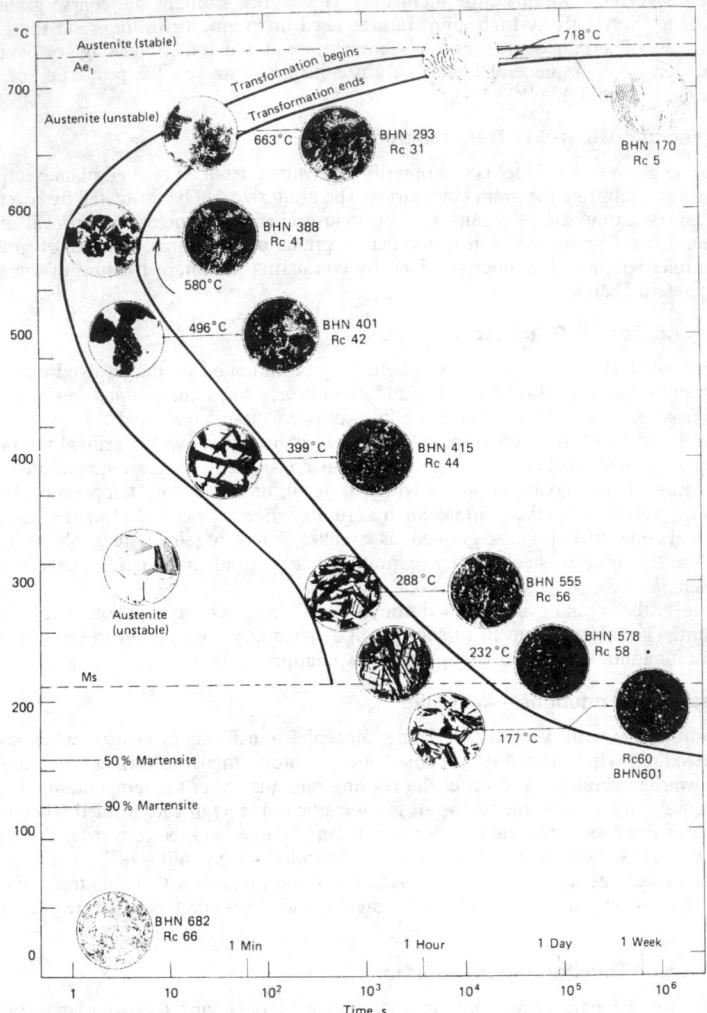

Fig. 2.3 Isothermal transformation diagram for a plain carbon eutectoid steel; $Ae_1 = A_1$ temperature at equilibrium; BHN = Brinell hardness number; Rc = Rockwell hardness scale C. C,0.89%; Mn, 0.29% austenitized at 885°C; grain size, 4–5; photomicrographs originally ×2500.

The IT diagram for a cutectoid carbon steel is shown in Fig. 2.3 In addition to the lines depicting the transformation, the diagram shows microstructures at various stages of transformation and hardness values. Thus, the diagram illustrates the characteristic subcritical austenite transformation behavior, the manner in which microstructure changes with transformation temperature, and the general relationship between these microstructural changes and hardness.

As the diagram indicates, the characteristic isothermal transformation behavior at any temperature above the temperature at which transformation to martensite begins (the M_s temperature) takes place over a period of time, known as the incubation period, in which no transformation occurs, followed by a period of time during which the transformation proceeds until the austenite has been transformed completely. The transformation is relatively slow at the beginning and toward the end, but much more rapid during the intermediate period in which ~25–75% of the austenite is transformed. Both the incubation period and the time required for completion of the transformation depend on the temperature.

The behavior depicted in this program is typical of plain carbon steels, with the shortest incubation period occurring at ~540°C. Much longer times are required for transformation as the temperature approaches either the Ae_1 or the M_s temperature. This A_1 temperature is lowered slightly during cooling and increased slightly during heating. The 540°C temperature, at which the transformation begins in the shortest time period is commonly referred to as the nose of the IT diagram. If complete transformation is to occur at temperatures below this nose, the steel must be cooled rapidly enough to prevent transformation at the nose temperature. Microstructures resulting from transformation at these lower temperatures exhibit superior strength and toughness.

2.2.11 Pearlite

In carbon and low-alloy steels, transformation over the temperature range of ~700–540°C gives pearlite microstructures of the characteristic lamellar type. As the transformation temperature falls, the lamellae move closer and the hardness increases.

2.2.12 Bainite

Transformation to bainite occurs over the temperature range of ~540–230°C. The acicular bainite microstructures differ markedly from the pearlite microstructures. Here again, the hardness increases as the transformation temperature decreases, although the bainite formed at the highest possible temperature is often softer than pearlite formed at a still higher temperature.

2.2.13 Martensite

Transformation to martensite, which in the steel illustrated in Fig. 2.3 begins at ~230°C, differs from transformation to pearlite or bainite because it is not time dependent, but occurs almost instantly during cooling. The degree of transformation depends only on the temperature to which it is cooled. Thus, in this steel of Fig. 2.3, transformation to martensite starts on cooling to 230°C (designated as the M_s temperature). The martensite is 50% transformed on cooling to ~150°C, and the transformation is essentially completed at ~90°C (designated as the M_f temperature). The microstructure of martensite is acicular. It is the hardest austenite transformation product but brittle; this brittleness can be reduced by tempering as discussed below.

2.2.14 Phase Properties—Pearlite

Pearlites are softer than bainites or martensites. However, they are less ductile than the lower-temperature bainites and, for a given hardness, far less ductile than tempered martensite. As the transformation temperature decreases within the pearlite range, the interlamellar spacing decreases, and these fine pearlites, formed near the nose of the isothermal diagram, are both harder and more ductile than the coarse pearlites formed at higher temperatures. Thus, although as a class pearlite tends to be soft and not very ductile, its hardness and toughness both increase markedly with decreasing transformation temperatures.

2.2.15 Phase Properties—Bainite

In a given steel, bainite microstructures are generally found to be both harder and tougher than pearlite, although less hard than martensite. Bainite properites generally improve as the transformation temperature decreases and lower bainite compares favorably with tempered martensite at the same hardness or exceeds it in toughness. Upper bainite, on the other hand, may be somewhat deficient in toughness as compared with fine pearlite of the same hardness.[4]

2.2.16 Phase Properties—Martensite

Martensite is the hardest and most brittle microstructure obtainable in a given steel. The hardness of martensite increases with increasing carbon content up to the eutectoid composition, and, at a given carbon content, varies with the cooling rate.

Although for some applications, particularly those involving wear resistance, the hardness of martensite is desirable in spite of the accompanying brittleness, this microstructure is mainly important as starting material for tempered martensite structures, which have definitely superior properties.

2.2.17 Tempered Martensite

Martensite is tempered by heating to a temperature ranging from 170 to 700°C for 30 min to several hours. This treatment causes the martensite to transform to ferrite interspersed with small particles of cementite. Higher temperatures and longer tempering periods cause the cementite particles to increase in size and the steel to become more ductile and lose strength. Tempered martensitic structures are, as a class, characterized by toughness at any strength. The diagram of Fig. 2.4 describes, within ± 10%, the mechanical properties of tempered martensite, regardless of composition. For example, a steel consisting of tempered martensite, with an ultimate strength of 1035 MPa (150,000 psi), might be expected to exhibit elongation of 16–20%, reduction of area of between 54 and 64%, yield point of 860–980 MPa (125,000–142,000 psi), and Brinell hardness of about 295–320. Because of its high ductility at a given hardness, this is the structure that is preferred.

2.2.18 Transformation Rates

The main factors affecting transformation rates of austenite are composition, grain size, and homogeneity. In general, increasing carbon and alloy content as well as increasing grain size tend to lower

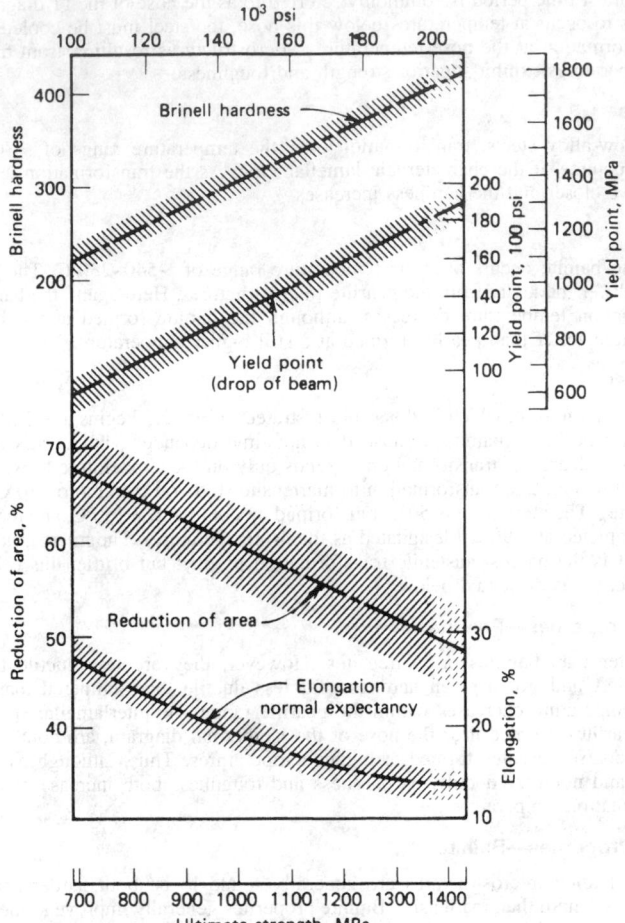

Fig. 2.4 Properties of tempered martensite (from Ref. 1). Fully heat-treated miscellaneous analyses, low-alloy steels; 0.30–0.50% C.

transformation rates. These effects are reflected in the isothermal transformation curve for a given steel.

2.2.19 Continuous Cooling

The basic information depicted by an IT diagram illustrates the structure formed if the cooling is interrupted and the reaction is completed at a given temperature. The information is also useful for interpreting behavior when the cooling proceeds directly without interruption, as in the case of annealing, normalizing, and quenching. In these processes, the residence time at a single temperature is generally insufficient for the reaction to go to completion; instead, the final structure consists of an association of microstructures which were formed individually at successivley lower temperatures as the piece cooled. However, the tendency to form seveal structures is still explained by the isothermal diagram.[5,6]

The final microstructure after continuous cooling depends on the times spent at the various transformation-temperature ranges through which a piece is cooled. The transformation behavior on continuous cooling thus represents an integration of these times by constructing a continuous-cooling diagram at constant rates similar to the isothermal transformation diagram (see Fig. 2.5). This diagram lies below and to the right of the corresponding IT diagram if plotted on the same coordinates; that is, transformation on continuous cooling starts at a lower temperature and after a longer time than the intersection of the cooling curve and the isothermal diagram would predict. This displacement is a function of the cooling rate, and increases with increasing cooling rate.

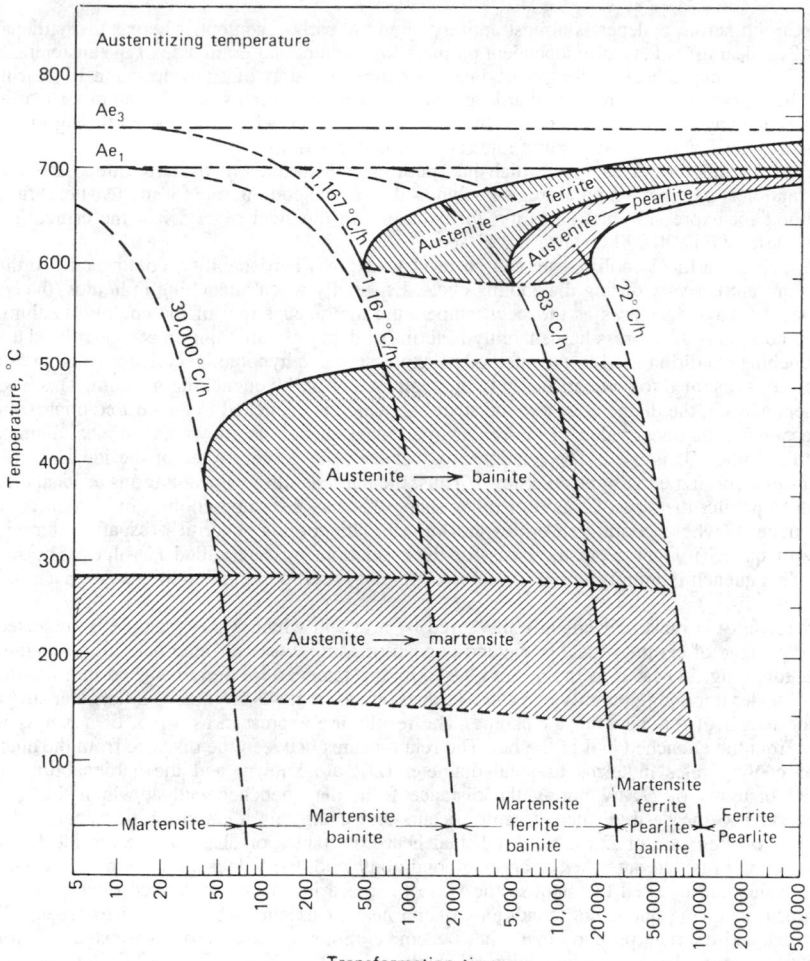

Fig. 2.5 Continuous-cooling transformation diagram for a type 4340 alloy steel, with superimposed cooling curves illustrating the manner in which transformation behavior during continuous cooling governs final microstructure (from Ref. 1). Ae_3 = critical temperature at equilibrium.

Several cooling-rate curves have been superimposed on Fig. 2.5. The changes occurring during these cooling cycles illustrate the manner in which diagrams of this nature can be correlated with heat-treating processes and used to predict the resulting microstructure.

Considering, first, the relatively low cooling rate ($< 22°C/hr$), the steel is cooled through the regions in which transformations to ferrite and pearlite occur which constitute the final microstructure. This cooling rate corresponds to a slow cooling in the furnace such as might be used in annealing.

At a higher cooling rate ($22–83°C/hr$), such as might be obtained on normalizing a large forging, the ferrite, pearlite, bainite, and martensite fields are traversed and the final microstructure contains all these constituents.

At cooling rates of $1167–30,000°C/hr$, the microstructure is free of proeutectoid ferrite and consists largely of bainite and a small amount of martensite. A cooling rate of at least $30,000°C/hr$ is necessary to obtain the fully martensitic structure desired as a starting point for tempered martensite.

Thus, the final microstructure, and therefore the properties of the steel, depend upon the transformation behavior of the austenite and the cooling conditions, and can be predicted if these factors are known.

2.3 HARDENABILITY

Hardenability refers to the depth of hardening or to the size of a piece that can be hardened under given cooling conditions, and not to the maximum hardness that can be obtained in a given steel.[7,8]

The maximum hardness depends almost entirely upon the carbon content, whereas the hardenability (depth of hardening) is far more dependent on the alloy content and grain size of the austenite. Steels whose IT diagrams indicate a long time interval before the start of transformation to pearlite are useful when large sections are to be hardened, since if steel is to transform to bainite or martensite, it must escape any transformation to pearlite. Therefore, the steel must be cooled through the high-temperature transformation ranges at a rate rapid enough for transformation not to occur even at the nose of the IT diagram. This rate, which just permits transformation to martensite without earlier transformation at a higher temperature, is known as the critical cooling rate for martensite. It furnishes one method for expressing hardenability; for example, in the steel of Fig. 2.5, the critical cooling rate for martensite is 30,000°C/hr or 8.3°C/sec.

Although the critical cooling rate can be used to express hardenability, cooling rates ordinarily are not constant but vary during the cooling cycle. Especially when quenching in liquids, the cooling rate of steel always decreases as the steel temperature approaches that of the cooling medium. It is therefore customary to express hardenability in terms of depth of hardening in a standardized quench. The quenching condition used in this method of expression is a hypothetical one in which the surface of the piece is assumed to come instantly to the temperature of the quenching medium. This is known as an ideal quench; the diameter of a round steel bar, which is quenched to the desired microstructure, or corresponding hardness value, at the center in an ideal quench, is known as the ideal diameter for which the symbol D_I is used. The relationships between the cooling rates of the ideal quench and those of other cooling conditions are known. Thus, the hardenability values in terms of ideal diameter are used to predict the size of round or other shape that has the same cooling rate when cooled in actual quenches whose cooling severities are known. The cooling severities (usually referred to as severity of quench) which form the basis for these relationships are called H values. The H value for the ideal quench is infinity; those for some commonly used cooling conditions are given in Table 2.2.

Hardenability is most conveniently measured by a test in which a steel sample is subjected to a continuous range of cooling rates. In the end-quench or Jominy test, a round bar, 25 mm in diameter and 102 mm long, is heated to the desired austenitizing temperature and quenched in a fixture by a stream of water impinging on only one end. Hardness measurements are made on flats that are ground along the length of the bar after quenching. The results are expressed as a plot of hardness versus distance from the quenched end of the bar. The relationships between the distance from the quenched end and cooling rates in terms of ideal diameter (D_I) are known, and the hardenability can be evaluated in terms of D_I by noting the distance from the quenched end at which the hardness corresponding to the desired microstructure occurs and using this relationship to establish the corresponding cooling rate or D_I value. Published heat-flow tables or charts relate the ideal-diameter value to cooling rates in quenches or cooling conditions whose H values are known. Thus, the ideal-diameter value can be used to establish the size of a piece in which the desired microstructure can be obtained under the quenching conditions of the heat treatment to be used. The hardenability of steel is such an important property that it has become common practice to purchase steels to specified hardenability limits. Such steels are called H steels.

2.4 HEAT-TREATING PROCESSES

In heat-treating processes, steel is usually heated above the A_3 point and then cooled at a rate that results in the microstructure that gives the desired properties.[9,10]

2.4.1 Austentization

The steel is first heated above the temperature at which austenite is formed. The actual austenitizing temperature should be high enough to dissolve the carbides completely and take advantage of the hardening effects of the alloying elements. In some cases, such as tool steels or high-carbon steels,

Table 2.2 *H* Values Designating Severity of Quench for Commonly Used Cooling Conditions[a]

Degree of Agitation of Medium	Quenching Medium		
	Oil	Water	Brine
None	0.25-0.30	0.9-1.0	2
Mild	0.30-0.35	1.0-1.1	2.0-2.2
Moderate	0.35-0.40	1.2-1.3	
Good	0.40-0.50	1.4-1.5	
Strong	0.50-0.80	1.6-2.0	
Violent	0.80-1.1	4.0	5.0

[a]H values are proportional to the heat-extracting capacity of the medium.

undissolved carbides may be retained for wear resistance. The temperature should not be high enough to produce pronounced grain growth. The piece should be heated long enough for complete solution; for low-alloy steels in a normally loaded furnace, 1.8 min/mm of diameter or thickness usually suffices.

Excessive heating rates may create high stresses, resulting in distortion or cracking. Certain types of continuous furnaces, salt baths, and radiant-heating furnaces provide very rapid heating, but pre-heating of the steel may be necessary to avoid distortion or cracking, and sufficient time must be allowed for uniform heating throughout. Unless special precautions are taken, heating causes scaling or oxidation, and may result in decarburization; controlled-atmosphere furnaces or salt baths minimize these effects.

2.4.2 Quenching

The primary purpose of quenching is to cool rapidly enough to suppress all transformation at temperatures above the M_s temperature. The cooling rate required depends on the size of the piece and the hardenability of the steel. The preferred quenching media are water, oils, and brine. The temperature gradients set up by quenching create high thermal and transformational stresses which may lead to cracking and distortion; a quenching rate no faster than necessary should be employed to minimize these stresses. Agitation of the cooling medium accelerates cooling and improves uniformity. Cooling should be long enough to permit complete transformation to martensite. Then, in order to minimize cracking from quenching stresses, the article should be transferred immediately to the tempering furnace (Fig. 2.6).

2.4.3 Tempering

Quenching forms very hard, brittle martensite with high residual stresses. Tempering relieves these stresses and improves ductility, although at some expense of strength and hardness. The operation consists of heating at temperatures below the lower critical temperature (A_1).

Measurements of stress relaxation on tempering indicate that, in a plain carbon steel, residual stresses are significantly lowered by heating to temperatures as low as 150°C, but that temperatures of 480°C and above are required to reduce these stresses to very low values. The times and temperatures required for stress relief depend on the high-temperature yield strength of the steel, since stress relief results from the localized plastic flow that occurs when the steel is heated to a temperature at which yield strength decreases. This phenomenon may be affected markedly by composition, and particularly by alloy additions. The toughness of quenched steel, as measured by the notch impact test, first increases on tempering up to 200°C, then decreases on tempering between 200 and 310°C,

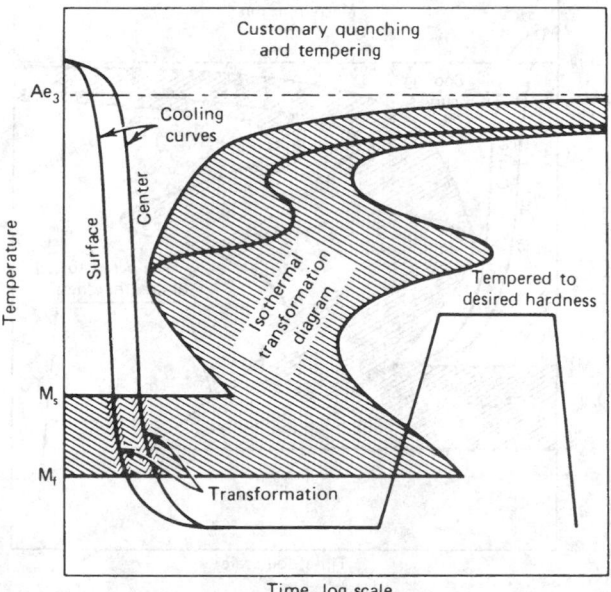

Fig. 2.6 Transformation diagram for quenching and tempering martensite; the product is tempered martensite (from Ref. 1).

and finally increases rapidly on tempering at 425°C and above. This behavior is characteristic and, in general, temperatures of 230-310°C should be avoided.

In order to minimize cracking, tempering should follow quenching immediately. Any appreciable delay may promote cracking.

The tempering of martensite results in a contraction, and if the heating is not uniform, stresses result, Similarly, heating too rapidly may be dangerous because of the sharp temperature gradient set up between the surface and the interior. Recirculating-air furnaces can be used to obtain uniform heating. Oil or salt baths are commonly used for low-temperature tempering; lead or salt baths are used at higher temperatures.

Some steels lose toughness on slow cooling from ~540°C and above, a phenomenon known as temper brittleness; rapid cooling after tempering is desirable in these cases.

2.4.4 Martempering

A modified quenching procedure known as martempering minimizes the high stresses created by the transformation to martensite during the rapid cooling characteristic of ordinary quenching (see Fig. 2.7). In practice, it is ordinarily carried out by quenching in a molten-salt bath just above the M_s temperature. Transformation to martensite does not begin until the piece reaches the temperature of the salt bath and is removed to cool relatively slowly in air. Since the temperature gradient characteristic of conventional quenching is absent, the stresses produced by the transformation are much lower and a greater freedom from distortion and cracking is obtained. After martempering, the piece may be tempered to the desired strength.

2.4.5 Austempering

As discussed earlier, lower bainite is generally as strong as and somewhat more ductile than tempered martensite. Austempering, which is an isothermal heat treatment that results in lower bainite, offers an alternative heat treatment for obtaining optimum strength and ductility.

In austempering the article is quenched to the desired temperature in the lower bainite region, usually in molten salt, and kept at this temperature until transformation is complete (see Fig. 2.8). Usually, it is held twice as long as the period indicated by the IT diagram. The article may be quenched or air cooled to room temperature after transformation is complete, and may be tempered to lower hardness if desired.

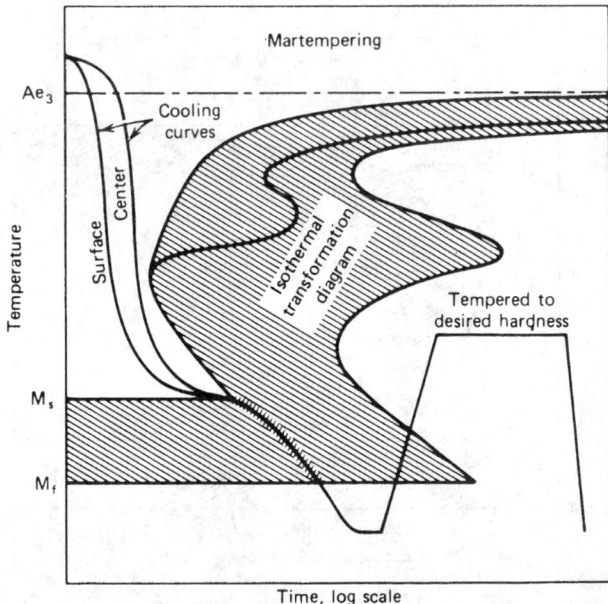

Fig. 2.7 Transformation diagram for martempering; the product is tempered martensite (from Ref. 1).

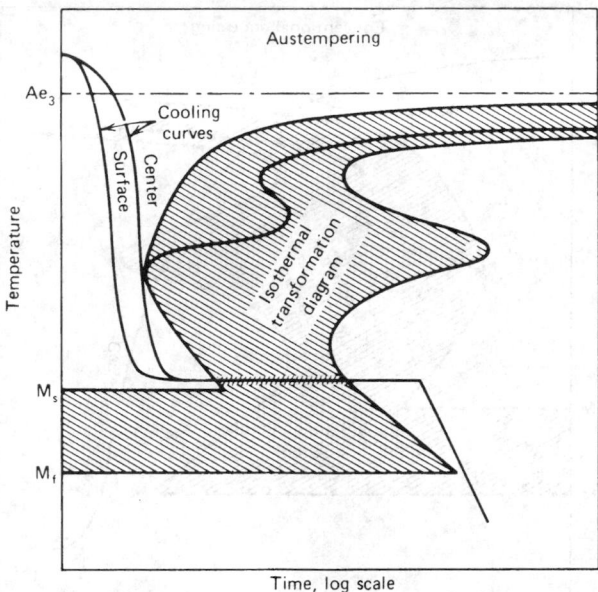

Fig. 2.8 Transformation diagram for austempering; the product is bainite (from Ref. 1).

2.4.6 Normalizing

In this operation, steel is heated above its upper critical temperature (A_3) and cooled in air. The purpose of this treatment is to refine the grain and to obtain a carbide size and distribution that is more favorable for carbide solution on subsequent heat treatment than the earlier as-rolled structure.

The as-rolled grain size, depending principally on the finishing temperature in the rolling operation, is subject to wide variations. The coarse grain size resulting from a high finishing temperature can be refined by normalizing to establish a uniform, relatively fine-grained microstructure.

In alloy steels, particularly if they have been slowly cooled after rolling, the carbides in the as-rolled condition tend to be massive and are difficult to dissolve on subsequent austenitization. The carbide size is subject to wide variations, depending on the rolling and slow cooling. Here again, normalizing tends to establish a more uniform and finer carbide particle size, which facilitates subsequent heat treatment.

The usual practice is to normalize at 50–80°C above the upper critical temperature; however, for some alloy steels considerably higher temperatures may be used. Heating may be carried out in any type of furnace that permits uniform heating and good temperature control.

2.4.7 Annealing

Annealing relieves cooling stresses induced by hot- or cold-working and softens the steel to improve its machinability or formability. It may involve only a subcritical heating to relieve stresses, recrystallize cold-worked material, or spheroidize carbides; it may involve heating above the upper critical temperature (A_3) with subsequent transformation to pearlite or directly to a spheroidized structure on cooling.

The most favorable microstructure for machinability in the low- or medium-carbon steels is coarse pearlite. The customary heat treatment to develop this microstructure is a full annealing, illustrated in Fig. 2.9. It consists of austenitizing at a relatively high temperature to obtain full carbide solution, followed by slow cooling to give transformation exclusively in the high-temperature end of the pearlite range. This simple heat treatment is reliable for most steels. It is, however, rather time-consuming since it involves slow cooling over the entire temperature range from the austenitizing temperature to a temperature well below that at which transformation is complete.

2.4.8 Isothermal Annealing

Annealing to coarse pearlite can be carried out isothermally by cooling to the proper temperature for transformation to coarse pearlite and holding until transformation is complete. This method, called isothermal annealing, is illustrated in Fig. 2.10. It may save considerable time over the full-annealing

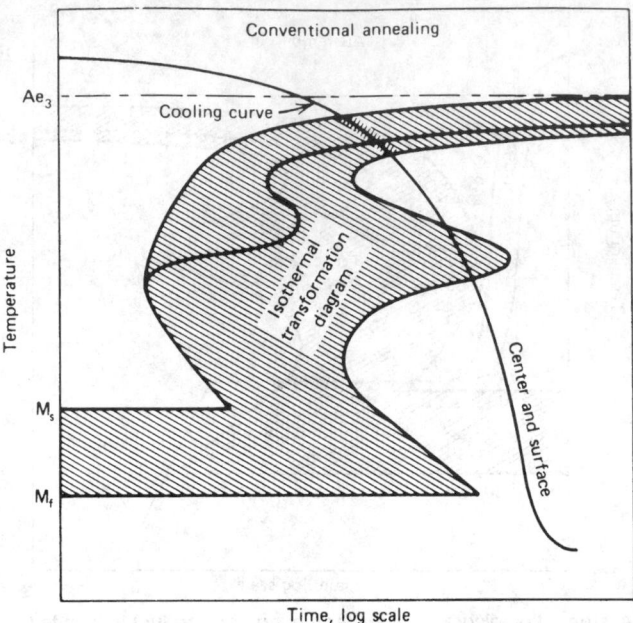

Fig. 2.9 Transformation diagram for full annealing; the product is ferrite and pearlite (from Ref. 1).

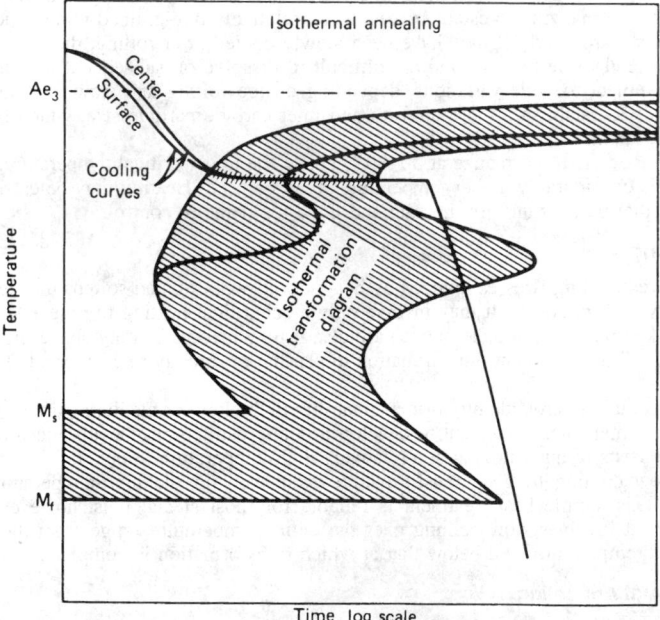

Fig. 2.10 Transformation diagram for isothermal annealing; the product is ferrite and pearlite (from Ref. 1).

process described previously, since neither the time from the austenitizing temperature to the transformation temperature, nor from the transformation temperature to room temperature, is critical; these may be shortened as desired. If extreme softness of the coarsest pearlite is not necessary, the transformation may be carried out at the nose of the IT curve, where the transformation is completed rapidly and the operation further expedited: the pearlite in this case is much finer and harder.

Isothermal annealing can be conveniently adapted to continuous annealing, usually in specially designed furnaces, when it is commonly referred to as cycle annealing.

2.4.9 Spheroidization Annealing

Coarse pearlite microstructures are too hard for optimum machinability in the higher carbon steels. Such steels are customarily annealed to develop spheroidized microstructures by tempering the as-rolled, slowly cooled, or normalized materials just below the lower critical temperature range. Such an operation is known as subcritical annealing. Full spheroidization may require long holding times at the subcritical temperature and the method may be slow, but it is simple and may be more convenient than annealing above the critical temperature.

The annealing procedures described above to produce pearlite can, with some modifications, give spheroidized microstructures. If free carbide remains after austenitizing, transformation in the temperature range where coarse pearlite ordinarily would form proceeds to spheroidized rather than pearlite microstructures. Thus, heat treatment to form spheroidized microstructures can be carried out like heat treatment for pearlite, except for the lower austenitizing temperatures. Spheroidization annealing may thus involve a slow cooling similar to the full-annealing treatment used for pearlite, or it may be a treatment similar to isothermal annealing. An austenitizing temperature not more than 55°C above the lower critical temperature is customarily used for this supercritical annealing.

2.4.10 Process Annealing

Process annealing is the term used for subcritical annealing of cold-worked materials. It customarily involves heating at a temperature high enough to cause recrystallization of the cold-worked material and to soften the steel. The most important example of process annealing is the box annealing of cold-rolled low-carbon sheet steel. The sheets are enclosed in a large box that can be sealed to permit the use of a controlled atmosphere to prevent oxidation. Annealing is usually carried out between 590 and 700°C. The operation usually takes ~24 hr, after which the charge is cooled slowly within the box; the entire process takes ~40 hr.

2.4.11 Carburizing

In carburizing, low-carbon steel acquires a high-carbon surface layer by heating in contact with carbonaceous materials. On quenching after carburizing, the high-carbon skin hardens, whereas the low-carbon core remains comparatively soft. The result is a highly wear-resistant exterior over a very tough interior. This material is particularly suitable for gears, camshafts, etc. Carburizing is most commonly carried out by packing the steel in boxes with carbonaceous solids, sealing to exclude the atmosphere, and heating to about 925°C for a period of time depending on the depth desired; this method is called pack carburizing. Alternatively, the steel may be heated in contact with carburizing gases in which case the process is called gas carburizing; or, least commonly, in liquid baths of carburizing salts, in which case it is known as liquid carburizing.

2.4.12 Nitriding

The nitrogen case-hardening process, termed nitriding, consists of subjecting machined and (preferably) heat-treated parts to the action of a nitrogenous medium, commonly ammonia gas, under conditions whereby surface hardness is imparted without requiring any further treatment. Wear resistance, retention of hardness at high temperatures, and resistance to certain types of corrosion are also imparted by nitriding.

2.5 CARBON STEELS

The plain carbon steels represent by far the largest volume produced, with the most diverse applications of any engineering material, including castings, forgings, tubular products, plates, sheet and strip, wire and wire products, structural shapes, bars, and railway materials (rails, wheels, and axles). Carbon steels are made by all modern steelmaking processes and, depending on their carbon content and intended purpose, may be rimmed, semikilled, or fully killed.[11–15]

The American Iron and Steel Institute has published standard composition ranges for plain carbon steels, which in each composition range are assigned an identifying number according to a method of classification (see Table 2.3). In this system, carbon steels are assigned to one of three series: 10xx (nonresulfurized), 11xx (resulfurized), and 12xx (rephosphorized and resulfurized). The 10xx steels are made with low phosphorus and sulfur contents, 0.04% max and 0.050% max, respectively. Sulfur in amounts as high as 0.33% max may be added to the 11xx and as high as 0.35% max to

Table 2.3 Standard Numerical Designations of Plain Carbon and Constructional Alloy Steels (AISI-SAE Designations)[1]

Series Designation[a]	Types	Series Designation[a]	Types
10xx	Nonresulfurized carbon-steel grades	47xx	1.05% Ni–0.45% Cr–0.20% Mo
11xx	Resulfurized carbon-steel grades	48xx	3.5% Ni–0.25% Mo
12xx	Rephosphorized and resulfurized Carbon-steel grades	50xx	0.28 or 0.40% Cr
13xx	1.75% Mn	51xx	0.80, 0.90, 0.95, 1.00, or 1.05% Cr
23xx	3.50% Ni	5xxxx	1.00% C–0.50, 1.00, or 1,45% Cr
25xx	5.00% Ni	61xx	0.80 or 0.95% Cr–0.10 or 0.15% V
31xx	1.25% Ni–0.65% Cr	86xx	0.55% Ni–0.50 or 0.65% Cr–0.20% Mo
33xx	3.5% Ni–1.55% Cr	87xx	0.55% Ni–0.50% Cr–0.25% Mo
40xx	0.25% Mo	92xx	0.85% Mn–2.00% Si
41xx	0.50 or 0.95% Cr–0.12 or 0.20% Mo	93xx	3.25% Ni–1.20% Cr–0.12% Mo
43xx	1.80% Ni–0.50 or 0.80% Cr–0.25% Mo	98xx	1.00% Ni–0.80% Cr–0.25% Mo
46xx	1.55 or 1.80% Ni–0.20 or 0.25% Mo		

[a]The first figure indicates the class to which the steel belongs; 1xxx indicates a carbon steel, 2xxx a nickel steel, and 3xxx a nickel-chromium steel. In the case of alloy steels, the second figure generally indicates the approximate percentage of the principal alloying element. Usually, the last two or three figures (represented in the table by x) indicate the average carbon content in points or hundredths of 1 wt %. Thus, a nickel steel containing a 3.5% nickel and 0.30% carbon would be designated as 2330.

the 12xx steels to improve machinability. In addition, phosphorus up to 0.12% max may be added to the 12xx steels to increase stiffness.

In identifying a particular steel, the letters x are replaced by two digits representing average carbon content; for example, an AISI No. 1040 steel would have an average carbon content of 0.40%, with a tolerance of $\pm 0.03\%$, giving a range of 0.37 to 0.43% carbon.

2.5.1 Properties

The properties of plain carbon steels are governed principally by carbon content and microstructure. The fact that properties can be controlled by heat treatment has been discussed in Section 2.1. Most plain carbon steels, however, are used without heat treatment.

The properties of plain carbon steels may be modified by residual elements other than the carbon, manganese, silicon, phosphorus, and sulfur that are always present, as well as gases, especially oxygen, nitrogen, and hydrogen, and their reaction products. These incidental elements are usually acquired from scrap, deoxidizers, or the furnace atmosphere. The gas content depends mostly on melting, deoxidizing, and pouring procedures; consequently, the properties of plain carbon steels depend heavily on the manufacturing techniques.

The average mechanical peoperties of as-rolled 2.5-cm bars of carbon steels as a function of carbon contents are shown in Fig. 2.11. This diagram is an illustration of the effect of carbon content when microstructure and grain size are held approximately constant.

2.5.2 Microstructure and Grain Size

The carbon steels with relatively low hardenability are predominantly pearlitic in the cast, rolled, or forged state. The constituents of the hypoeutectoid steels are, therefore, ferrite and pearlite, and of the hypereutectoid steels, cementite and pearlite. As discussed earlier, the properties of such pearlitic steels depend primarily on the interlamellar spacing of the pearlite and the grain size. Both hardness and ductility increase as the interlamellar spacing or the pearlite-transformation temperature decreases, whereas the ductility increases with decreasing grain size. The austenite-transformation behavior in carbon steel is determined almost entirely by carbon and manganese content; the effects of

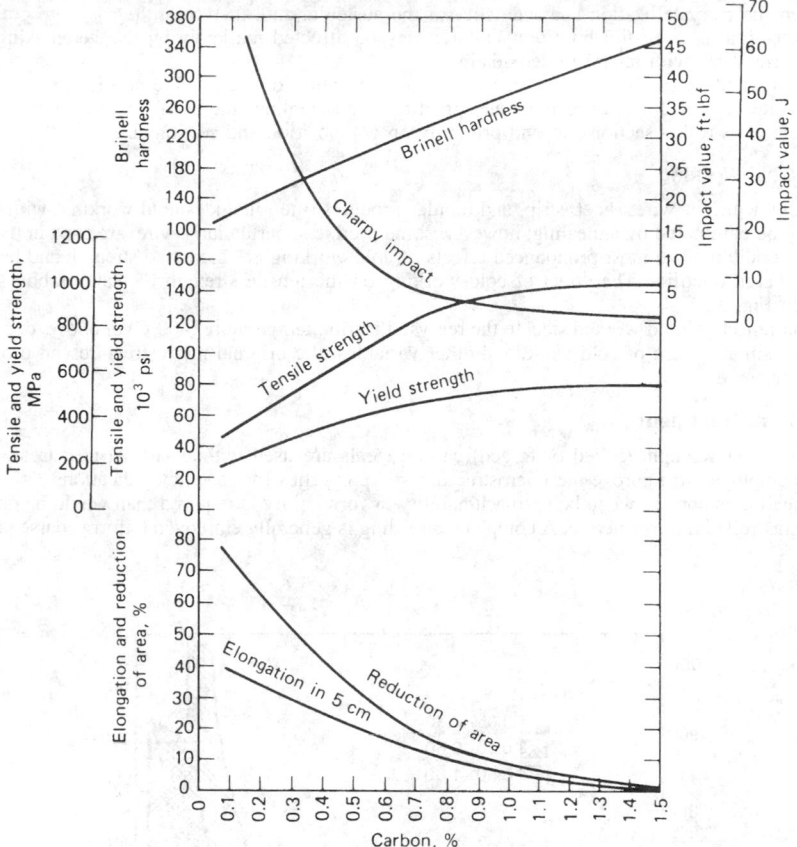

Fig. 2.11   Variations in average mechanical peoperties of as-rolled 2.5-cm bars of plain carbon steels, as a function of carbon content (from Ref. 1).

phosphorus and sulfur are almost negligible; and the silicon content is normally so low as to have no influence. The carbon content is ordinarily chosen in accordance with the strength desired, and the manganese content selected to produce suitable microstructure and properties at that carbon level under the given cooling conditions.

2.5.3   Microstructure of Cast Steels

Cast steel is generally coarse grained, since austenite forms at high temperature and the pearlite is usually coarse, in as much as cooling through the ciritical range is slow, particularly if the casting is cooled in the mold. In hypoeutectoid steels, ferrite ordinarily precipitates at the original austenite grain boundaries during cooling. In hypereutectoid steels, cementite is similarly precipitated. Such mixtures of ferrite or cementite and coarse pearlite have poor strength and ductility properties, and heat treatment is usually necessary to obtain suitable microstructures and properties in cast steels.

2.5.4   Hot Working

Many carbon steels are used in the form of as-rolled finished sections. The microstructure and properties of these sections are determined largely by composition, rolling procedures, and cooling conditions. The rolling or hot working of these sections is ordinarily carried out in the temperature range in which the steel is austenic, with four principal effects: Considerable homogenization occurs during the heating for rolling, tending to eliminate dendrite segregation present in the ingot; the dendritic structure is broken up during rolling; recrystallization takes place during rolling, with final austenitic grain size determined by the temperature at which the last passes are made (the finishing temperature); and dendrites and inclusions are reoriented, with markedly improved ductility, in the rolling direction.

Thus, homogeneity and grain size of the austenite is largely determined by the rolling technique. However, the recrystallization characteristics of the austenite and, therefore, the austenite grain size characteristic at a given finishing temperature, may be affected markedly by the steelmaking technique, particularly with regard to deoxidation.

The distribution of the ferrite or cementite and the nature of the pearlite are determined by the cooling rate after rolling. Since the usual practice is air cooling, the final microstructure and the properties of as-rolled sections depend primarily on composition and section size.

2.5.5 Cold Working

The manufacture of wire, sheet, strip, and tubular products often includes cold working, with effects that may be eliminated by annealing; however, some products, particularly wire, are used in the cold-worked condition. The most pronounced effects of cold working are increased strength and hardness and decreased ductility. The effect of cold working on the tensile strength of plain carbon steel is shown in Fig. 2.12.

Upon reheating cold-worked steel to the recrystallization temperature (400°C) or above, depending on composition, extent of cold work, and other variables, the original microstructure and properties may be restored.

2.5.6 Heat Treatment

Although most wrought (rolled or forged) carbon steels are used without a final heat treatment, it may be employed to improve the microstructure and properties for specific applications.

Annealing is applied when better machinability or formability is required than would be obtained with the as-rolled microstructure. A complete annealing is generally employed to form coarse pearlite,

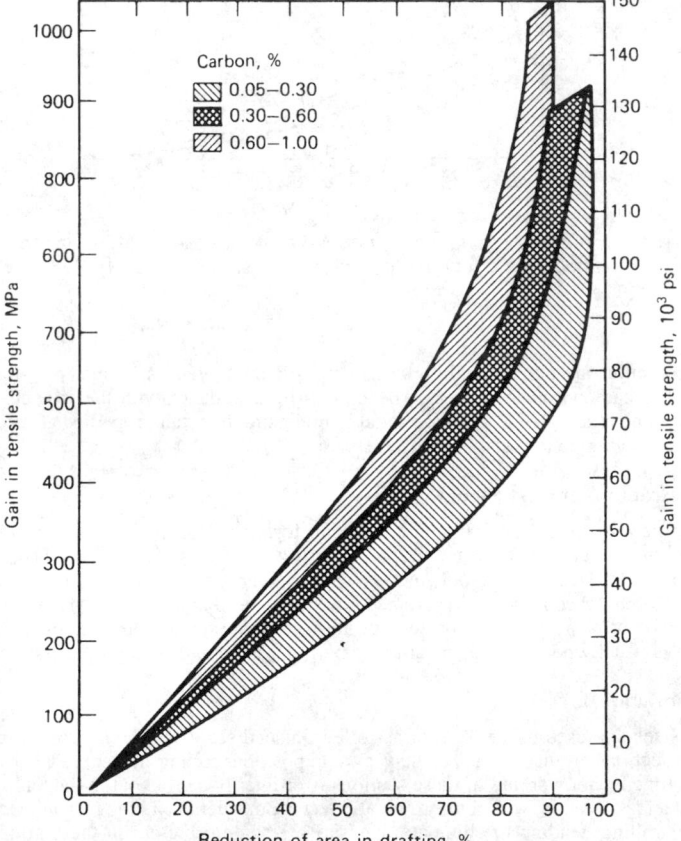

Fig. 2.12 Increase of tensile strength of plain carbon steel with increased cold working (from Ref. 1).

although a subcritical annealing or spheroidizing treatment is occasionally used. Process annealing for optimum formability is universal with cold-rolled strip and sheet and cold-worked tubing.

The grain size of as-rolled products depends largely on the finishing temperature but is difficult to control. A final normalizing treatment from a relatively low temperature may establish a fine, uniform grain size for applications in which ductility and toughness are critical.

Quenching and tempering of plain carbon steels are being more frequently applied. In one type of treatment, the steel is heat treated to produce tempered martensite, but because of relatively low hardenability, the operation is limited to section sizes of not more than 10-13 mm. In the other type, large sections of plain carbon steels are quenched and tempered to produce fine pearlite microstructures with much better strength and ductility than those of the coarse pearlite microstructures in as-rolled or normalized products.

Thin sections of carbon steels ($\leq$ 5 mm) are particularly suitable for the production of parts requiring toughness at high hardness by austempering.

2.5.7 Residual Elements

In addition to the carbon, manganese, phosphorus, sulfur, and silicon that are always present, carbon steels may contain small amounts of gases, such as hydrogen, oxygen, or nitrogen, introduced during the steelmaking process; nickel, copper, molybdenum, chromium, and tin, which may be present in the scrap, and aluminum, titanium, vanadium, or zirconium, which may be introduced during deoxidation.

Oxygen and nitrogen cause the phoenomenon called aging, mainfested as a spontaneous increase in hardness at room temperature and believed to be a precipitation effect.

An embrittling effect, the mechanism of which is not completely understood, is caused by a hydrogen content of more than ~3 ppm. As discussed earlier, the content of hydrogen and other gases can be reduced by vacuum degassing.

Alloying elements such as nickel, chromium, molybdenum, and copper, which may be introduced with scrap, do, of course, increase the hardenability although only slightly since the concentrations are ordinarily low. However, the heat-treating characteristics may change, and for applications in which ductility is important, as in low-carbon steels for deep drawing, the increased hardness imparted by these elements may be harmful.

Tin, even in low amounts, is harmful in steels for deep drawing; for most applications, however, the effect of tin in the quantities ordinarily present is negligible.

Aluminum is generally desirable as a grain refiner and tends to reduce the susceptibility of carbon steel to aging associated with strain. Unfortunately, it tends to promote graphitization and is, therefore, undesirable in steels used at high temperatures. The other elements that may be introduced as deoxidizers, such as titanium, vanadium, or zirconium, are ordinarily present in such small amounts as to be ineffective.

2.6 DUAL-PHASE SHEET STEELS

Dual-phase steels derive their name from their unique microstructure of a mixture of ferrite and martensite phases. This microstructure is developed in hot- and cold-rolled sheet by using a combination of steel composition and heat treatment that changes an initial microstructure of ferrite and pearlite (or iron carbide) to ferrite and martensite.[16-22]

Normally, high-strength hot-rolled sheets are manufactured by hot rolling and cooling on a hot-strip mill, which produces a microstructure of ferrite and pearlite. On heating to ~750–850°C, a microstructure of ferrite and austenite is produced, and by cooling at an appropriate rate (which depends on steel composition or hardenability), the austenite is transformed to a very hard martensite phase contained within the soft, ductile ferrite matrix. The final ferrite–martensite microstructure, which may contain 5–30% martensite (increasing amounts increase the strength), may be considered as a composite; the strength may therefore be estimated, according to a simple law of mixtures, from the strengths and volume fractions of the individual phases.

The properties of a steel (0.11% C, 1.6%, Mn, 0.60% Si, 0.04% V) in the hot-rolled (ferrite and pearlite) and in the heat-treated (ferrite-martensite double-phase) state are given below (to convert MPa to psi, multiply by 145):

	Yield Strength (MPa)	Tensile Strength (MPa)	Total Elongation (%)
Hot rolled	480	4275	24
Dual phase	345	4516	32

Although the tensile (ultimate) strength of the steel is little affected by heat treating, the yield strength is substantially reduced and the ductility markedly improved. The low yield strength allows for the easy initiation of plastic deformation during press forming of dual-phase sheet material.

However, dual-phase steels have the unique capacity to strain harder rapidly so that after a few percent deformation (3–5%) the yield strength exceeds 550 MPa (80,000 psi).

Dual-phase steels have found application in automotive bumpers and wheels where high ductility is requried to form the complex shapes. The development of very high strength of ~550 MPa (80,000 psi) allows thinner, lighter weight sheet to be used, instead of steels having strengths of only 200–350 MPa (30,000–50,000 psi). However, the heat-treating step increases production costs.

2.7 ALLOY STEELS

As a class, alloy steels may be defined as steels having enhanced properties owing to the presence of one or more special elements or larger proportions of elements (such as silicon and manganese) than are ordinarily present in carbon steel. Steels containing alloying elements are classified into high-strength low-alloy (HSLA) steels; AISI alloy steels; alloy tool steels; stainless steels; heat-resistant steels; and electrical steels (silicon steels). In addition, there are numerous steels, some with proprietary compositions, with exceptional properties developed to meet unusually severe requirements. The relatively small production of such steels does not reflect their engineering importance.[23,24]

2.7.1 Functions of Alloying Elements

In the broadest sense, alloy steels may contain up to ~50% of alloying elements which directly enhance properties. Examples are the increased corrosion resistance of high chromium steels, the enhanced electric properties of silicon steels, the improved strength of the HSLA steels, and the improved hardenability and tempering characteristics of the AISI alloy steels.

2.7.2 Thermomechanical Treatment

The conventional method of producing high-strength steels has been to add alloy elements such as Cr, Ni, and Mo to the liquid steel. The resulting alloy steels are often heat treated after rolling to develop the desired strength without excessive loss of toughness (resistance to cracking upon impact). In the 1970s, a less expensive method was developed to produce HSLA steels with improved toughness and yield strength ranging from 400 to 600 MPa (60,000 to 85,000 psi). In this thermomechanical treatment, the working of the steel is controlled while its temperature is changing and it is being hot rolled between 1300 and 750°C to its final thickness.[25–29] The HSLA steels that are commonly strengthened by thermomechanical treatment, also called controlled rolling, generally contain 0.05–0.20% carbon, 0.40–1.60% manganese, 0.05–0.50% silicon, plus 0.01–0.30% of one or more of the following elements: aluminum, molybdenum, niobium, titanium, and vanadium. Thermomechanical treatment usually involves a substantial degree of rolling, such as a 50–75% decrease in thickness in the last rolling passes; temperature maintained between 750 and 950°C; and a controlled rate of cooling after hot rolling. This procedure gives a very fine steel grain size and imparts strength and toughness. Steels so treated are increasingly used in automobiles and oil and gas pipelines.

2.7.3 High-Strength Low-Alloy (HSLA) Steels

HSLA steels are categorized according to mechanical properties, particularly the yield point; for example, within certain thickness limits they have yield points ranging from 310 to 450 MPa (45,000 to 65,000 psi) as compared with 225 to 250 MPa (33,000 to 36,000 psi) for structural carbon steel. This classification is in contrast to the usual classification into plain carbon or structural-carbon steels, alloy steels, and stainless steels on the basis of alloying elements.

The superior mechanical properites of HSLA steels are obtained by the addition of alloying elements (other than carbon), singly and in combination. Each steel must meet similar minimum mechanical requirements. They are available for structural use as sheets, strips, bars, and plates, and in various other shapes. They are not to be considered as special-purpose steels or requiring heat treatment.

To be of commercial interest, HSLA steels must offer economic advantages. They should be much stronger and often tougher than structural carbon steel. In addition, they must have sufficient ductility, formability, and weldability to be fabricated by customary techniques. Improved resistance to corrosion is often required. The abrasion resistance of these steels is somewhat higher than that of structural carbon steel containing 0.15–0.20% carbon. Superior mechanical properties permits the use of HSLA steels in structures with a higher unit working stress; this generally permits reduced section thickness with corresponding decrease in weight. Thus, HSLA steels may be substituted for structural carbon steel without change in section, resulting in a stronger and more duable structure without weight increase.

2.7.4 AISI Alloy Steels

The American Iron and Steel Institute defines alloy steels as follows: "By common custom steel is considered to be alloy steel when the maximum of the range given for the content of alloying elements exceeds one or more of the following limits: manganese, 1.65%; silicon, 0.60%; copper, 0.60%; or in which a definite range or a definite minimum quantity of any of the following elements is specified or required within the limits of the recognized field of constructional alloy steels: aluminum, boron, chromium up to 3.99%, cobalt, columbium (niobium), molybdenum, nickel, titanium, tungsten, vanadium, zirconium, or any other alloying element added to obtain a desired alloying effect."[30] Steels that contain 4.00% or more of chromium are included by convention among the special types of alloy steels known as stainless steels subsequently discussed.[31-37]

Steels that fall within the AISI definition have been standardized and classified jointly by AISI and SAE as shown in Table 2.3. They represent by far the largest alloy steel production and are generally known as AISI alloy steels. They are also commonly referred to as constructional alloy steels.

The effect of the alloying elements on AISI steels is indirect since alloying elements control microstructure through their effect on hardenability. They permit the attainment of desirable microstructures and properties over a much wider range of sizes and sections than is possible with carbon steels.

2.7.5 Alloy Tool Steels

Alloy tool steels are classified roughly into three groups: Low-alloy tool steels, to which alloying elements have been added to impart hardenability higher than that of plain carbon tool steels; accordingly, they may be hardened in heavier sections or with less drastic quenches to minimize distortion; intermediate-alloy tool steels usually contain elements such as tungsten, molybdenum, or vanadium, which form hard, wear-resistant carbides; high-speed tool steels contain large amounts of carbide-forming elements that serve not only to furnish wear-resisting carbides, but also promote the phenomenon known as secondary hardening and thereby increase resistance to softening at elevated temperatures.

2.7.6 Stainless Steels

Stainless steels are more resistant to rusting and staining than plain carbon and low-alloy steels.[38-46] This superior corrosion resistance is due to the addition of chromium. Although other elements, such as copper, aluminum, silicon, nickel, and molybdenum, also increase corrosion resistance, they are limited in their usefulness.

No single nation can claim credit for the development of the stainless steels; Germany, the United Kingdom, and the United States share alike in their developoment. In the United Kingdom in 1912, during the search for steel that would resist fouling in gun barrels, a corrosion-resistant composition was reported of 12.8% chromium and 0.24% carbon. It was suggested that this composition be used for cutlery. In fact, the composition of AISI type 420 steel (12–14% chromium, 0.15% carbon) is similar to that of the first corrosion-resistant steel.

The higher chromium–iron alloys were developed in the United States from the early 20th century on, when the effect of chromium on oxidation resistance at 1090°C was fist noticed. Oxidation resistance increased markedly as the chromium content was raised above 20%. Even now and with steels containing appreciable quantities of nickel, 20% chromium seems to be the minimum amount necessary for oxidation resistance at 1090°C.

The austenitic iron–chromium–nickel alloys were developed in Germany around 1910 in a search for materials for use in pyrometer tubes. Further work led to the versatile 18% chromium–8% nickel steels, so-called 18–8, which are widely used today.

The chromium content seems to be the controlling factor and its effect may be enhanced by additions of molybdenum, nickel, and other elements. The mechanical properties of the stainless steels, like those of the plain carbon and lower-alloy steels, are functions of structure and composition. Thus, austenitic steels possess the best impact properties at low temperatures and the highest strength at high temperatures, whereas martensitic steels are the hardest at room temperature. Thus, stainless steels, which are available in a variety of structures, exhibit a range of mechanical properties which, combined with their excellent corrosion resistance, makes these steels highly versatile from the standpoint of design.

The standard AISI and SAE types are identified in Table 2.4.

2.7.7 Martensitic Stainless Steels

Martensitic stainless steels are iron–chromium alloys that are hardenable by heat treatment. They include types 403, 410, 414, 416, 420, 431, 440A, 440B, 440C, 501, and 502 (see Table 2.4). The most widely used is type 410, containing 11.50–13.50% chromium and <0.15% carbon. In the annealed condition, this grade may be drawn or formed. It is an air-hardening steel, affording a wide range of properties by heat treatment. In sheet or strip form, type 410 is used extensively in the petroleum industry for ballast trays and liners. It is also used for parts of furnaces operating below 650°C, and for blades and buckets in steam turbines.

Type 420, with ~0.35% carbon and a resultant increased hardness, is used for cutlery. In bar form, it is used for valves, valve stems, valve seats, and shafting where corrosion and wear resistance are needed. Type 440 may be employed for surgical instruments, especially those requiring a durable

Table 2.4   Standard Stainless and Heat-Resisting Steel Products[1]

AISI Type Number	SAE Type[a] Number	Chemical Composition, %			
		Carbon	Chromium	Nickel	Other
201	30201	0.15 max	16.00–18.00	3.50–5.50	Mn 5.50–7.50[b] P 0.06 max[c] N 0.25 max
202	30202	0.15 max	17.00–19.00	4.00–6.00	Mn 7.50–10.00 P 0.06 max N 0.25 max
301	30301	0.15 max	16.00–18.00	6.00–8.00	
302	30302	0.15 max	17.00–19.00	8.00–10.00	
302B	30302B	0.15 max	17.00–19.00	8.00–10.00	Si 2.00–3.00[d] P 0.20 max S 0.15 min[e]
303	30303	0.15 max	17.00–19.00	8.00–10.00	Mo 0.60 max
303Se	30303Se	0.15 max	17.00–19.00	8.00–10.00	P 0.20 max S 0.06 max Se 0.15 min
303SeA		0.08 max	17.25–18.75	11.50–13.00	Se 0.15–0.35
304	30304	0.08 max	18.00–20.00	8.00–10.00	
304L		0.030 max	18.00–20.00	8.00–10.00	
305	30305	0.12 max	17.00–19.00	10.00–13.00	
307		0.07–0.15	19.50–21.50	9.00–10.50	Mo residual only
308	30308	0.08 max	1900–21.00	10.00–12.00	
308 Mod		0.07–0.15	19.50–21.50	9.00–10.50	Mo residual only
309	30309	0.20 max	22.00–24.00	12.00–15.00	
309S	30309S	0.08 max	22.00–24.00	12.00–15.00	
309SCb		0.08 max	22.00–24.00	12.00–15.00	NbTa min. 10 times carbon Ta 0.10 max
309SCbTa		0.08 max	22.00–24.00	12.00–15.00	NbTa min. 10 times carbon
310	30310	0.25 max	24.00–26.00	19.00–22.00	
314	30314	0.25 max	23.00–26.00	19.00–22.00	
316	30316	0.08 max	16.00–18.00	10.00–14.00	Mo 2.00–3.00
316L	30316L	0.030 max	16.00–18.00	10.00–14.00	Mo 2.00–3.00
317	30317	0.08 max	18.00–20.00	11.00–15.00	Mo 3.00–4.00
318		0.10 max	16.00–18.00	10.00–14.00	Mo 2.00–3.00 NbTa min. 10 times carbon
D319		0.07 max	17.50–19.50	11.00–15.00	Mn 2.00 max Si 1.00 max Mo 2.25–3.00
321	30321	0.08 max	17.00–19.00	9.00–12.00	Ti min. 5 times carbon
330		0.25 max	14.00–16.00	33.00–36.00	
347	30347	0.08 max	17.00–19.00	9.00–13.00	NbTa min. 10 times carbon
348	30348	0.08 max	17.00–19.00	9.00–13.00	NbTa min, 10 times carbon Ta 0.10 max Co 0.20 max
403	51403	0.15 max	11.50–13.00		
405	51405	0.08 max	11.50–14.50		Al 0.10–0.30
410	51410	0.15 max	11.50–13.50		
410Mo		0.15 max	11.50–13.50		Mo 0.40–0.60
414	51414	0.15 max	11.50–13.50	1.25–2.50	
416	51416	0.15 max	12.00–14.00		P 0.06 max S 0.15 min Mo 0.60 max
410Se	51410Se	0.15 max	12.00–14.00		P 0.06 mas S 0.06 max Se 0.15 min
420	51420	>0.15	12.00–14.00		

Table 2.4 *(Continued)*

AISI Type Number	SAE Type[a] Number	Chemical Composition, %			
		Carbon	Chromium	Nickel	Other
420F	51420F	>0.15	12.00–14.00		S[f]
430	51430	0.12 max	14.00–18.00		
430F	51430F	0.12 max	14.00–18.00		P 0.06 max
					S 0.15 min
					Mo 0.60 max
430Ti		0.10 max	16.00–18.00		Ti 0.30–0.70
431	51431	0.20 max	15.00–17.00	1.25–2.50	
434A		0.05–0.10	15.00–17.00		Cu 0.75–1.10
442	51442	0.25 max	18.00–23.00		
446	51446	0.20 max	23.00–27.00		N 0.25 max
501	51501	>0.10	4.00–6.00		Mo 0.40–0.65
502	51502	0.10 max	4.00–6.00		Mo 0.40–0.65

[a] SAE chemical composition (ladle) ranges may differ slightly in certain elements from AISI limits.
[b] Manganese: All steels of AISI Type 300 series—2.00% max. All steels of AISI Type 400 and 500 series—1.00% max except 416, 416Se, 430F, and 430Se (1.25% max) and Type 446 (1.5% max).
[c] Phosphorus: All steels of AISI Type 200 series—0.060% max. All steels of AISI Type 300 series—0.045% max except Types 303 and 303Se (0.20% max). All steels of AISI Type 400 and 500 series—0.040% max except Types 416, 416Se, 430F, and 430FSe (0.060% max).
[d] Silicon: All steels of AISI Type 200, 300, 400 and 500 series—1.00% max except where otherwise indicated.
[e] Sulfur: All steels of AISI Type 200, 300, 400, and 500 series—0.30% max except Types 303, 416, and 430F (0.15% min) and Types 303Se, 416Se, and 430FSe (0.060% max).
[f] No restriction.

cutting edge. The necessary hardness for different applications can be obtained by selecting grade A, B, or C, with increasing carbon content in that order.

Other martensitic grades are types 501 and 502, the former with > 0.10% and the latter < 0.10% carbon; both contain 4.6% chromium. These grades are also air hardening, but do not have the corrosion resistance of the 12% chromium grades. Type 501 and 502 have wide application in the petroleum industry for hot lines, bubble towers, valves, and plates.

2.7.8 Ferrite Stainless Steels

These steels are iron–chromium alloys that are largely ferritic and not hardenable by heat treatment (ignoring the 475°C embrittlement). They include types 405, 430, 430F, and 446 (see Table 2.4).

The most common ferritic grade is type 430, containing 0.12% carbon or less and 14–18% chromium. Because of its higher chromium content, the corrosion resistance of type 430 is superior to that of the martensitic grades. Furthermore, type 430 may be drawn, formed, and, with proper techniques, welded. It is widely used for automotive and architectural trim. It is employed in equipment for the manufacture and handling of nitric acid to which it is resistant. Type 430 does not have high creep strength but is suitable for some types of service up to 815°C and thus has application in combustion chambers for domestic heating furnaces.

The high chromium content of type 446 (23–27% chromium) imparts excellent heat resistance, although its high-temperature strength is only slightly better than that of carbon steel. Type 446 is used in sheet or strip form up to 1150°C. This grade does not have the good drawing characteristics of type 430, but it may be formed. Accordingly, it is widely used for furnace parts such as muffles, burner sleeves, and annealing baskets. Its resistance to nitric and other oxidizing acids makes it suitable for chemical-processing equipment.

2.7.9 Austenitic Stainless Steels

These steels are iron-chromium–nickel alloys not hardenable by heat treatment and predominantly austenitic. They include types 301, 302, 302B, 303, 304, 304L, 305, 308, 309, 310, 314, 316, 316L, 317, 321, and 347. In some recently developed austenitic stainless steels, all or part of the nickel is replaced by manganese and nitrogen in proper amounts, as in one proprietary steel and types 201 and 202 (see Table 2.4).

The most widely used austenitic stainless steel is type 302, known as 18–8; it has excellent corrosion resistance and, because of its austenitic structure, excellent ductility. It may be deep drawn or strongly formed. It can be readily welded, but carbide precipitation must be avoided in and near the weld by cooling rapidly enough after welding. Where carbide precipitation presents problems, types 321, 347, or 304L may be used. The applications of type 302 are wide and varied, including kitchen equipment and utensils; dairy installations; transportation equipment; and oil-, chemical-, paper-, and food-processing machinery.

The low nickel content of type 301 causes it to harden faster than type 302 because of reduced austenite stability. Accordingly, although type 301 can be drawn successfully, its drawing properties are not as good as those of type 302. For the same reason, type 301 can be cold rolled to very high strength.

Type 301, because of its lower carbon content, is not as prone as type 302 to give carbide precipitation problems in welding. In addition, its somewhat higher chromium content makes it slightly more resistant to corrosion. It is used to withstand severe corrosive conditions in the paper, chemical, and other industries.

The austenitic stainless steels are widely used for high-temperature service.

Types 321 and 347, with additions of titanium and niobium, respectively, are used in welding applications and high-temperature service under corrosive conditions. Type 304L may be used as an alternative for types 321 and 347 in welding and stress-relieving applications below 426°C.

The addition of 2–4% molybdenum to the basic 18–8 composition produces types 316 and 317 with improved corrosion resistance. These grades are employed in the textile, paper, and chemical industries where strong sulfates, chlorides, and phosphates and reducing acids such as sulfuric, sulfurous, acetic, and hydrochloric acids are used in such concentrations that the use of corrosion-resistant alloys is mandatory. Types 316 and 317 have the highest creep and rupture strengths of any commercial stainless steels.

The austenitic stainless steels most resistant to oxidation are types 309 and 310. Because of their high chromium and nickel contents, these steels resist scaling at temperatures up to 1090 and 1150°C and, consequently, are used for furnace parts and heat exchangers. They are somewhat harder and not as ductile as the 18–8 types, but they may be drawn and formed. They can be welded readily and have increasing use in the manufacture of jet-propulsion motors and industrial-furnace equipment.

For applications requiring good machinability, type 303 containing sulfur or selenium may be used.

2.7.10 High-Temperature Service, Heat-Resisting Steels

The term high-temperature service comprises many types of operations in many industries. Conventional high-temperature equipment includes steam boilers and turbines, gas turbines, cracking stills, tar stills, hydrogenation vessels, heat-treating furnaces, and fittings for diesel and other internal-combustion engines. Numerous steels are available from which to select the one best suited for each of the foregoing applications. Where unusual conditions occur, modification of the chemical composition may adapt an existing steel grade to service conditions. In some cases, however, entirely new alloy combinations must be developed to meet service requirements. For example, the aircraft and missile industries have encountered design problems of increased complexity, requiring metals of great high-temperature strength for both power plants and structures, and new steels are constantly under development to meet these requirements.[47,48]

A number of steels suitable for high-temperature service are given in Table 2.5.

The design of load-bearing structures for service at room temperature is generally based on the yield strength or for some applications on the tensile strength. The metal behaves essentially in an elastic manner, that is, the structure undergoes an elastic deformation immediately upon load appli-

Table 2.5 Alloy Composition of High-Temperature Steels[1]

Ferritic Steels	Austenitic Steels	AISI Type
0.5% Mo	18% Cr-8% Ni	304
0.5% Cr-0.5% Mo	18% Cr-8% Ni with Mo	316
1% Cr-0.5% Mo	18% Cr-8% Ni with Ti	321
2% Cr-0.5% Mo	18% Cr-8% Ni with Nb	347
2.25% Cr-1% Mo	25% Cr-12% Ni	309
3% Cr-0.5% Mo-1.5% Si	25% Cr-20% Ni	310
5% Cr-0.5% Mo-1.5% Si		
5% Cr-0.5% Mo. with Nb added		
5% Cr-0.5% Mo, with Ti added		
9% Cr-1% Mo		
12% Cr		410
17% Cr		430
27% Cr		446

cation and no further deformation occurs with time; when the load is removed, the structure returns to its original dimensions.

At high temperature, the behavior is different. A structure designed according to the principles employed for room-temperature service continues to deform with time after load application, even though the design data may have been based on tension tests at the temperature of interest. This deformation with time is called creep, since at the design stresses at which it first was recognized it occurred at a relatively low rate.

In spite of the fact that plain carbon steel has lower resistance to creep than high-temperature alloy steels, it is widely used in such applications up to 540°C, where rapid oxidation commences and a chromium-bearing steel must be employed. Low-alloy steels containing small amounts of chromium and molybdenum have higher creep strengths than carbon steel and are employed where materials of higher strength are needed. Above ~540°C, the amount of chromium required to impart oxidation resistance increases rapidly. The 2% chromium steels containing molybdenum are useful up to ~620°C, whereas 10–14% chromium steels may be employed up to ~700–760°C. Above this temperature, the austenitic 18–8 steels are commonly used; their oxidation resistance is considered adequate up to ~815°C. For service between 815 and 1090°C, steels containing 25% chromium and 20% nickel, or 27% chromium are used.

The behavior of steels at high temperature is quite complex, and only a few design considerations have been mentioned here.

2.7.11 Quenched and Tempered Low-Carbon Constructional Alloy Steels

A class of quenched and tempered low-carbon constructional alloy steels has been very extensively used in a wide variety of applications such as pressure vessels, mining and earth-moving equipment, and large steel structures.[49–51]

As a general class, these steels are referred to as low-carbon martensites to differentiate them from constructional alloy steels of higher carbon content, such as AISI alloy steels, that develop high-carbon martensite upon quenching. They are characterized by a relatively high strength, with minimum yield strengths of 690 MPa (100,000 psi), toughness down to –45°C, and weldability with joints showing full joint efficiency when welded with low-hydrogen electrodes. They are most commonly used in the form of plates, but also sheet products, bars, structural shapes, forgings, or semi-finished products.

Several steel-producing companies manufacture such steels under various tradenames; their compositions are proprietary.

2.7.12 Maraging Steels

A group of high-nickel martensitic steels called maraging steels contain so little carbon that they are referred to as carbon-free iron–nickel martensites.[52,53]

Iron–carbon martensite is hard and brittle in the as-quenched condition and becomes softer and more ductile when tempered. Carbon-free iron–nickel martensite, on the other hand, is relatively soft and ductile and becomes hard, strong, and tough when subjected to an aging treatment at 480°C.

The first iron–nickel martensitic alloys contained ~0.01% carbon, 20 or 25% nickel, and 1.5–2.5% aluminum and titanium. Later an 18% nickel steel containing cobalt, molybdenum, and titanium was developed, and still more recently a series of 12% nickel steels containing chromium and molybdenum came on the market.

By adjusting the content of cobalt, molybdenum, and titanium, the 18% nickel steel can attain yield strengths of 1380–2070 MPa (200,000–300,000 psi) after the aging treatment. Similarly, yield strengths fo 12% nickel steel in the range of 1035–1380 MPa (150,000–200,000 psi) can be developed by adjusting its composition.

2.7.13 Silicon-Steel Electrical Sheets

The silicon steels are characterized by relatively high permeability, high electrical resistance, and low hysteresis loss when used in magnetic circuits. First patented in the United Kingdom around 1900, the silicon steels permitted the development of more powerful electrical equipment and have furthered the rapid growth of the electrical power industry. Steels containing 0.5–5% silicon are produced in sheet form for the laminated magnetic cores of electrical equipment and are referred to as electrical sheets.[54–56]

The grain-oriented steels, containing ~3.25% silicon, are used in the highest efficiency distribution and power transformers and in large turbine generators. They are processed in a special way to give them directional properties related to orientation of the crystals making up the structure of the steel in a preferred direction.

The nonoriented steels are subdivided into low-silicon steels, containing ~0.5–1.5% silicon, used mainly in rotors and stators of motors and generators. Steels containing ~1% silicon are used for reactors, relays, and small intermittent-duty transformers.

Intermediate-silicon steels (2.5–3.5% Si) are used in motors and generators of average to high efficiency and in small- to medium-size intermittent-duty transformers, reactors, and motors.

High-silicon steels (~3.75–5.00% Si) are used in power transformers and high-efficiency motors, generators, and transformers, and in communications equipment.

REFERENCES

1. H. E. McGannon, (ed.), *The Making, Shaping and Treating of Steel*, 9th ed., U.S. Steel Corporation, Pittsburgh, PA, 1971.
2. *Applications of Modern Metallographic Techniques*, STP 480, American Society for Testing Materials, Philadelphia, PA, 1970.
3. W. C. Leslie, *The Physical Metallurgy of Steels*, Hemisphere Publishing, McGraw-Hill, New York, 1981
4. E. C. Bain and H. W. Paxton, *Alloying Elements in Steel*, American Society for Metals, Metals Park, OH, 1961.
5. *Heat Treatment '79*, Metals Society Publication 261, Metals Society, London, 1980.
6. K. E. Thelning, *Steel and Its Heat Treatment: Bofors Handbook*, Butterworths, Boston, 1975.
7. D. V. Doane and J. S. Kirkaldy, *Hardenability Concepts with Applications to Steel*, The Metallurgical Society–AIME, Warrendale, PA, 1978.
8. C. A. Siebert, D. V. Doane, and D. H. Breen, *The Hardenability of Steels*, American Society for Metals, Metals Park, OH, 1977.
9. G. Krauss, *Principles of Heat Treatment of Steel*, American Society for Metals, Metals Park, OH, 1980.
10. M. Atkins, *Atlas of Continuous Cooling Transformations Diagrams for Engineering Steels*, British Steel Corp., Sheffield, UK, 1978.
11. J. S. Blair, *The Profitable Way: Carbon Sheet Steel Specifying and Purchasing Handbook* General Electric Technology Marketing, Schenectady, NY, 1978.
12. J. S. Blair, *The Profitable Way: Carbon Strip Steel Specifying and Purchasing Handbook*, General Electric Technology Marketing, Schenectady, NY, 1978.
13 J. D. Jevons, *The Metallurgy of Deep Drawing and Pressing*, Wiley, New York, 1942.
14. *Low Carbon Structural Steels for the Eighties*, Institution of Metallurgists, London, 1977.
15. J. S. Blair, *The Profitable Way: Carbon Plate Steel Specifying and Purchasing Handbook*, General Electric Technology Marketing, Schenectady, NY, 1978.
16. R. A. Kot and J. W. Morris (eds.), *Structure and Properties of Highly Formable Dual-Phase Steels*, The Metallurgical Society–AIME, Warrendale, PA, 1979.
17. A. T. Davenport (ed.), *Formable HSLA and Dual-Phase Steels*, The Metallurgical Society–AIME, Warrendale, PA, 1979.
18. A. B. Rothwell and J. M. Gray (eds.), *Welding of HSLA (Microalloyed) Structural Steels*, American Society for Metals, Metals Park, OH, 1978.
19. R. A. Kot and B. L. Bramfitt, *Fundamentals of Dual-Phase Steels*, The Metallurgical Society–AIME, Warrendale, PA, 1981.
20. P. E. Repas, *Iron Steelmaker* **7**, 12 (1980).
21. M. D. Baughman, K. L. Fetters, G. Perrault, Jr., and K. Toda, *Iron Steel Eng.* **56**, 52 (1979).
22. A. P. Coldren and G. T. Eldis, *J. Met.* **32**, 41 (1980).
23. B. P. Bardes (ed.), *Metals Handbook*, 9th ed., American Society for Metals, Metals Park, OH, 1978, Vol. 1, p. 127.
24. *Nickel Alloy Steels Data Book*, International Nickel Co., Inc., New York, 1967.
25. *Micro Alloying 75: Proceedings of an International Symposium on High-Strength Low-Alloy Steels*, Union Carbide Corp., New York, 1977.
26. F. B. Pickering, *Physical Metallurgy and the Design of Steels*, Applied Science Publishers, London, 1978.
27. G. R. Speich and D. S. Dabkowski, in *The Hot Deformation of Austenite*, J. B. Ballance (ed.), The Metallurgical Society–AIME, Warrendale, PA, 1977.
28. *Iron Age* **214**, MP9 (Dec. 9, 1974).
29. A. T. Davenport, D. R. DiMicco, and D. W. Dickinson, *J. Met.* **32**, 28 (1980).
30. *Steel Products Manual: Strip Steel*, American Iron and Steel Institute, Washington, DC, 1978.
31. *Alloy Cross Index*, Mechanical Properties Data Center, Battelle's Columbus Laboratories, Columbus, OH, 1981.
32. P. M. Unterweiser, *Worldwide Guide to Equivalent Irons and Steels*, American Society for Metals, Metals Park, OH, 1979.
33. *Unified Numbering System for Metals and Alloys*, Society of Automotive Engineers, Warrendale, PA, 1977.

34. *Handbook of Comparative World Steel Standards*, International Technical Information Institute, Tokyo, Japan, 1980.

35. R. B. Ross, *Metallic Materials Specification Handbook*, E. and F. N. Spon Ltd., New York, 1980.

36. C. W. Wegst, *Key to Steel (Stahlschluessel)*, Verlag Stahlschluessel Wegst KG, Marbach/Neckar, Federal Republic of Germany, 1974.

37. M. J. Wahll and R. F. Frontani, *Handbook of Soviet Alloy Compositions*, Metals and Ceramics Information Center, Battelle's Columbus Laboratories, Columbus, OH, 1976.

38. *Source Book on Stainless Steels*, American Society for Metals, Metals Park, OH, 1976.

39. K. G. Brickner and co-workers, *Selection of Stainless Steels*, American Society for Metals, Metals Park, OH, 1968.

40. D. Peckner and I. M. Bernstein, *Handbook of Stainless Steels*, McGraw-Hill, New York, 1977.

41. W. F. Simmons and R. B. Gunia, *Compilation and Index of Trade Names, Specifications, and Producers of Stainless Alloys and Superalloys*, Data Series, DS45A, American Society for Testing Materials, Philadelphia, PA, 1972.

42. G. E. Rowan and co-workers, *Forming of Stainless Steels*, American Society for Metals, Metals Park, OH, 1968.

43. R. A. Lula (ed.), *Toughness of Ferritic Stainless Steels*, STP 706, American Society for Testing Materials, Philadelphia, PA, 1980.

44. C. R. Brinkman and H. W. Garvin (eds.), *Properties of Austenitic Stainless Steels and Their Weld Metals*, STP 679, American Society for Testing Materials, Philadelphia, PA, 1978.

45. J. J. Demo (ed.), *Structure, Constitution, and General Characteristics of Wrought Ferritic Stainless Steels*, STP 619, American Society for Testing Materials, Philadelphia, PA, 1977.

46. F. B. Pickering (ed.), *The Metallurgical Evolution of Stainless Steels*, American Society for Metals, Metals Park, OH, 1979.

47. E. F. Bradley (ed.), *Source Book on Materials for Elevated-Temperature Application*, American Society for Metals, Metals Park, OH, 1979.

48. G. V. Smith (ed.), *Ductility and Toughness Considerations in Elevated Temperature Service*, American Society of Mechanical Engineers, New York, 1978.

49. R. L. Brockenbrough and B. G. Johnston, *USS Steel Design Manual*, ADUSS 27-3400-03, U.S. Steel Corp., Pittsburgh, PA, 1974.

50. J. H. Gross, *Trans. ASME: J. Pressure Vessel Technology* **96**, 9 (1974).

51. *Annual Book of ASTM Standards, Part 4—Steel*, American Society for Testing Materials, Philadelphia, PA, 1982, p. 465.

52. S. Floreen and G. R. Speich, *Trans. ASM* **57**, 714 (1964).

53. *Third Maraging Steel Project Review*, AD 425 299, Air Force Systems Command Technical Documentary Report No. RTD-TDR-63-4048, Nov. 1963, available from National Technical Information Service of the United States Department of Commerce, Springfield, VA.

54. A. E. DeBarr, *Soft Magnetic Materials Used in Industry*, Reinhold, New York, 1953.

55. F. Brailsford, *Magnetic Materials*, Wiley, New York, 1960.

56. R. M. Bozorth, *Ferromagnetism*, Van Nostrand, New York, 1951.

57. R. J. King, "Steel," in *Kirk-Othmer Encyclopedia of Chemical Technology*, 3rd ed., Wiley, New York, 1983, Vol. 21.

CHAPTER 3

ALUMINUM AND ITS ALLOYS

Seymour G. Epstein
J. G. Kaufman
Peter Pollak
The Aluminum Association, Inc.
Washington, D.C.

3.1	INTRODUCTION	45
3.2	PROPERTIES OF ALUMINUM	45
3.3	ALUMINUM ALLOYS	46
3.4	ALLOY DESIGNATION SYSTEMS	46
3.5	MECHANICAL PROPERTIES OF ALUMINUM ALLOYS	48
3.6	WORKING STRESSES	49
3.7	CHARACTERISTICS	51
3.7.1	Resistance to General Corrosion	51
3.7.2	Workability	51
3.7.3	Weldability and Brazeability	51
3.8	TYPICAL APPLICATIONS	52
3.9	MACHINING ALUMINUM	53

3.9.1	Cutting Tools	53
3.9.2	Single-Point Tool Operations	53
3.9.3	Multipoint Tool Operations	54
3.10	CORROSION BEHAVIOR	54
3.10.1	General Corrosion	55
3.10.2	Pitting Corrosion	55
3.10.3	Galvanic Corrosion	56
3.11	FINISHING ALUMINUM	56
3.11.1	Mechanical Finishes	56
3.11.2	Chemical Finishes	56
3.11.3	Electrochemical Finishes	56
3.11.4	Clear Anodizing	57
3.11.5	Color Anodizing	57
3.11.6	Integral Color Anodizing	57
3.11.7	Electrolytically Deposited Coloring	57
3.11.8	Hard Anodizing	57
3.11.9	Electroplating	57
3.11.10	Applied Coatings	57
3.12	SUMMARY	57

3.1 INTRODUCTION

Aluminum is the most abundant metal and the third most abundant chemical element in the earth's crust, comprising over 8% of its weight. Only oxygen and silicon are more prevalent. Yet, until about 150 years ago aluminum in its metallic form was unknown to man. The reason for this is that aluminum, unlike iron or copper, does not exist as a metal in nature. Because of its chemical activity and its affinity for oxygen, aluminum is always found combined with other elements, mainly as aluminum oxide. As such it is found in nearly all clays and many minerals. Rubies and sapphires are aluminum oxide colored by trace impurities, and corundum, also aluminum oxide, is the second hardest naturally occurring substance on earth—only a diamond is harder.

It was not until 1886 that scientists learned how to economically extract aluminum from aluminum oxide via electrolytic reduction. Yet in the more than 100 years since that time, aluminum has become the second most widely used of the approximately 60 naturally occurring metals, behind only iron.

3.2 PROPERTIES OF ALUMINUM

Let us consider the properties of aluminum that lead to its wide use.

One property of aluminum that everyone is familiar with is its light weight or, technically, its low specific gravity. The specific gravity of aluminum is only 2.7 times that of water, and roughly one-

Mechanical Engineers' Handbook, 2nd ed., Edited by Myer Kutz.
ISBN 0-471-13007-9 © 1998 John Wiley & Sons, Inc.

third that of steel or copper. An easy number to remember is that 1 in.^{3} of aluminum weighs 0.1 lb; 1 ft^3 weighs 170 lb compared to 62 lb for water and 490 lb for steel. The following are some other properties of aluminum and its alloys that will be examined in more detail in later sections:

Formability. Aluminum can be formed by every process in use today and in more ways than any other metal. Its relatively low melting point, 1220°F, while restricting high-temperature applications to about 500–600°F, does make it easy to cast, and there are over 1000 foundries casting aluminum in this country.

Mechanical Properties. Through alloying, naturally soft aluminum can attain strengths twice that of mild steel.

Strength-to-Weight Ratio. Some aluminum alloys are among the highest strength to weight materials in use today, in a class with titanium and superalloy steels. This is why aluminum alloys are the principal structural metal for commercial and military aircraft.

Cryogenic Properties. Unlike most steels, which tend to become brittle at cryogenic temperatures, aluminum alloys actually get tougher at low temperatures and hence enjoy many cryogenic applications.

Corrosion Resistance. Aluminum possesses excellent resistance to corrosion by natural atmospheres and by many foods and chemicals.

High Electrical and Thermal Conductivity. On a volume basis the electrical conductivity of pure aluminum is roughly 60% of the International Annealed Copper Standard, but pound for pound aluminum is a better conductor of heat and electricity than copper and is surpassed only by sodium, which is a difficult metal to use in everyday situations.

Reflectivity. Aluminum can accept surface treatment to become an excellent reflector and it does not dull from normal oxidation.

Finishability. Aluminum can be finished in more ways than any other metal used today.

3.3 ALUMINUM ALLOYS

While commercially pure aluminum (defined as at least 99% aluminum) does find application in electrical conductors, chemical equipment, and sheet metal work, it is a relatively weak material, and its use is restricted to applications where strength is not an important factor. Some strengthening of the pure metal can be achieved through cold working, called strain hardening. However, much greater strengthening is obtained through alloying with other metals, and the alloys themselves can be further strengthened through strain hardening or heat treating. Other properties, such as castability and machinability, are also improved by alloying. Thus, aluminum alloys are much more widely used than is the pure metal, and in many cases, when aluminum is mentioned, the reference is actually to one of the many commercial alloys of aluminum.

The principal alloying additions to aluminum are copper, manganese, silicon, magnesium, and zinc; other elements are also added in smaller amounts for metallurgical purposes. Since there have been literally hundreds of aluminum alloys developed for commercial use, the Aluminum Association formulated and administers special alloy designation systems to distinguish and classify the alloys in a meaningful manner.

3.4 ALLOY DESIGNATION SYSTEMS

Aluminum alloys are divided into two classes according to how they are produced: wrought and cast. The wrought category is a broad one, since aluminum alloys may be shaped by virtually every known process, including rolling, extruding, drawing, forging, and a number of other, more specialized processes. Cast alloys are those that are poured molten into sand (sand casting) or high-strength steel (permanent mold or die casting) molds, and are allowed to solidify to produce the desired shape. The wrought and cast alloys are quite different in composition; wrought alloys must be ductile for fabrication, while cast alloys must be fluid for castability.

In 1974, the Association published a designation system for wrought aluminum alloys that classifies the alloys by major alloying additions. This system is now recognized worldwide under the International Accord for Aluminum Alloy Designations, administered by the Aluminum Association, and is published as American Standards Institute (ANSI) Standard H35.1. More recently, a similar system for casting alloys was introduced.

Each wrought or cast aluminum alloy is designated by a number to distinguish it as a wrought or cast alloy and to categorize the alloy. A wrought alloy is given a four-digit number. The first digit classifies the alloy by alloy series, or principal alloying element. The second digit, if different than 0, denotes a modification in the basic alloy. The third and fourth digits form an arbitrary number

Table 3.1 Designation System for Wrought Aluminum Alloys

Alloy Series	Description or Major Alloying Element
1xxx	99.00% minimum aluminum
2xxx	Copper
3xxx	Manganese
4xxx	Silicon
5xxx	Magnesium
6xxx	Magnesium and silicon
7xxx	Zinc
8xxx	Other element
9xxx	Unused series

which identifies the specific alloy in the series.* A cast alloy is assigned a three-digit number followed by a decimal. Here again the first digit signifies the alloy series or principal addition; the second and third digits identify the specific alloy; the decimal indicates whether the alloy composition is for the final casting (0.0) or for ingot (0.1 or 0.2). A capital letter prefix (A, B, C, etc.) indicates a modification of the basic alloy.

The designation systems for wrought and cast aluminum alloys are shown in Tables 3.1 and 3.2, respectively.

Specification of an aluminum alloy is not complete without designating the metallurgical condition, or temper, of the alloy. A temper designation system, unique for aluminum alloys, was developed by the Aluminum Association and is used for all wrought and cast alloys. The temper designation follows the alloy designation, the two being separated by a hyphen. Basic temper designations consist of letters; subdivisions, where required, are indicated by one or more digits following the letter. The basic tempers are:

F—As-Fabricated. Applies to the products of shaping processes in which no special control over thermal conditions or strain hardening is employed. For wrought products, there are no mechanical property limits.

O—Annealed. Applies to wrought products that are annealed to obtain the lowest strength temper, and to cast products that are annealed to improve ductility and dimensional stability. The O may be followed by a digit other than zero.

Table 3.2 Designation System for Cast Aluminum Alloys

Alloy Series	Description or Major Alloying Element
1xx.x	99.00% minimum aluminum
2xx.x	Copper
3xx.x	Silicon plus copper and/or magnesium
4xx.x	Silicon
5xx.x	Magnesium
6xx.x	Unused series
7xx.x	Zinc
8xx.x	Tin
9xx.x	Other element

*An exception is for the 1xxx series alloys, where the last two digits indicate the minimum aluminum percentage. For example, alloy 1060 contains a minimum of 99.60% aluminum.

Table 3.3   Subdivisions of H Temper: Strain Hardened

First digit indicates basic operations:

H1—Strain hardened only

H2—Strain hardened and partially annealed

H3—Strain hardened and stabilized

H4—Strain hardened, lacquered, or painted

Second digit indicates degree of strain hardening:

HX2—Quarter hard

HX4—Half hard

HX8—Full hard

HX9—Extra hard

Third digit indicates variation of two-digit temper.

H—Strain-Hardened (Wrought Products Only). Applies to products that have their strength increased by strain hardening, with or without supplementary thermal treatments to produce some reduction in strength. The H is always followed by two or more digits. (See Table 3.3.)

W—Solution Heat Treated. An unstable temper applicable only to alloys that spontaneously age at room temperature after solution heat treatment. This designation is specific only when the period of natural aging is indicated; for example: W ½ hr.

T—Thermally Treated to Produce Stable Tempers Other than F, O, or H. Applies to products that are thermally treated, with or without supplementary strain hardening, to produce stable tempers. The T is always followed by one or more digits. (See Table 3.4.)

3.5   MECHANICAL PROPERTIES OF ALUMINUM ALLOYS

Wrought aluminum alloys are generally thought of in two categories: nonheat-treatable and heat-treatable. Nonheat-treatable alloys are those that derive their strength from the hardening effect of elements such as manganese, iron, silicon, and magnesium, and are further strengthened by strain hardening. They include the 1xxx, 3xxx, 4xxx, and 5xxx series alloys. Heat-treatable alloys are

Table 3.4   Subdivions of T Temper: Thermally Treated

First digit indicates specific sequence of treatments:

T1—Cooled from an elevated-temperature shaping process and naturally aged to a substantially stable condition

T2—Cooled from an elevated-temperature shaping process, cold worked, and naturally aged to a substantially stable condition

T3—Solution heat-treated, cold worked, and naturally aged to a substantially stable condition

T4—Solution heat-treated and naturally aged to a substantially stable condition

T5—Cooled from an elevated-temperature shaping process and then artifically aged

T6—Solution heat-treated and then artifically aged

T7—Solution heat-treated and overaged/stabilized

T8—Solution heat-treated, cold worked, and then artificially aged

T9—Solution heat-treated, artificially aged, and then cold worked

T10—Cooled from an elevated-temperature shaping process, cold worked, and then artificially aged

Second digit indicates variation in basic treatment:

Examples:

T42 or T62—Heat treated to temper by user

Additional digits indicate stress relief:

Examples:

TX51 or TXX51—Stress relieved by stretching

TX52 or TXX52—Stress relieved by compressing

TX54 or TXX54—Stress relieved by combination of stretching and compressing strengthened by a combination of solution heat treatment and natural or controlled aging for precipitation hardening, and include the 2xxx, some 4xxx, 6xxx, and 7xxx series alloys. Castings are not normally strain hardened, but many are solution heat-treated and aged for added strength.

In Table 3.5 typical mechanical properties are shown for several representative nonheat-treatable alloys in the annealed, half-hard and full-hard tempers; values for super purity aluminum (99.99%) are included for comparison. Typical properties are usually higher than minimum, or guaranteed, properties and are not meant for design purposes but are useful for comparisons. It should be noted that pure aluminum can be substantially strain hardened, but a mere 1% alloying addition produces a comparable tensile strength to that of fully hardened pure aluminum with much greater ductility in the alloy. And the alloys can then be strain hardened to produce even greater strengths. Thus, the alloying effect is compounded. Note also that, while strain hardening increases both tensile and yield strengths, the effect is more pronounced for the yield strength so that it approaches the tensile strength in the fully hardened temper. Ductility and workability are reduced as the material is strain hardened, and most alloys have limited formability in the fully hardened tempers.

Table 3.6 lists typical mechanical properties and nominal compositions of some representative heat-treatable aluminum alloys. One can readily see that the strengthening effect of the alloying ingredients in these alloys is not reflected in the annealed condition to the same extent as in the nonheat-treatable alloys, but the true value of the additions can be seen in the aged condition. Presently, heat-treatable alloys are available with tensile strengths approaching 100,000 psi.

Again, casting alloys cannot be work hardened and are either used in as-cast or heat-treated conditions. Typical mechanical properties for commonly used casting alloys range from 20 to 50 ksi for ultimate tensile strength, from 15 to 50 ksi tensile yield strength and up to 20% elongation. The range of strengths available with wrought aluminum alloys is shown graphically in Fig. 3.1.

3.6 WORKING STRESSES

Aluminum is used in a wide variety of structural applications. These range from curtain walls on buildings to tanks and piping for handling cryogenic liquids, and even bridges and major buildings and roof structures. In establishing appropriate working stresses the factors of safety applied to the ultimate strength and yield strength of the aluminum alloy vary with the specific application. For building and similar type structures a factor of safety of 1.95 is applied to the tensile ultimate strength

Table 3.5 Typical Mechanical Properties of Representative Nonheat-Treatable Aluminum Alloys (Not for Design Purposes)

Alloy	Nominal Composition	Temper	Tensile Strength (ksi)	Yield Strength (ksi)	Elongation (% in 2 in)	Hardness (BHN)
1199	99.9+% Al	O	6.5	1.5	50	—
		H18	17	16	5	—
1100	99+% Al	O	13	5	35	23
		H14	18	17	9	32
		H18	24	22	5	44
3003	1.2% Mn	O	16	6	30	28
		H14	22	21	8	40
		H18	29	27	4	55
3004	1.2% Mn	O	26	10	20	43
	1.0% Mg	H34	35	29	9	63
		H38	41	36	5	77
5005	0.8% Mg	O	13	6	25	28
		H14	23	22	6	41
		H18	29	28	4	51
5052	2.5% Mg	O	28	13	25	47
		H34	38	31	10	68
		H38	42	37	7	77
5456	5.1% Mg	O	45	23	24	70
	0.8% Mn	H321, H116	51	37	16	90
B443.0	5.0% Si	F[a]	19	8	8	40
		F[b]	23	9	10	45
514.0	4.0% Mg	F[a]	25	12	9	50

[a]Sand cast.

[b]Permanent mold cast.

Table 3.6 Typical Mechanical Properties of Representative Heat-Treatable Aluminum Alloys (Not for Design Purposes)

Alloy	Nominal Composition	Temper	Tensile Strength (ksi)	Yield Strength (ksi)	Elongation (% in 2 in)	Hardness (BHN)
2024	4.4% Cu	O	27	11	20	47
	1.5% Mg	T4	68	47	20	120
	0.6% Mn	T6	69	57	10	125
		T86	75	71	6	135
2219	6.3% Cu	T62	60	42	10	—
6061	1.0% Mg	O	18	8	25	30
	0.6% Si	T4	35	21	22	65
		T6	45	40	12	95
6063	0.40 Si	O	13	7	—	25
	0.70 Mg	T6	35	31	12	73
7075	5.6% Zn	O	33	15	17	60
	2.5% Mg	T6	83	73	11	150
	1.6% Cu	T73	73	63	13	—
356.0	7.0% Si	T6[a]	33	24	3.5	70
	0.3% Mg	F[b]	26	18	5	—
		T6[b]	37	27	5	80

[a]Sand cast.

[b]Permanent mold cast.

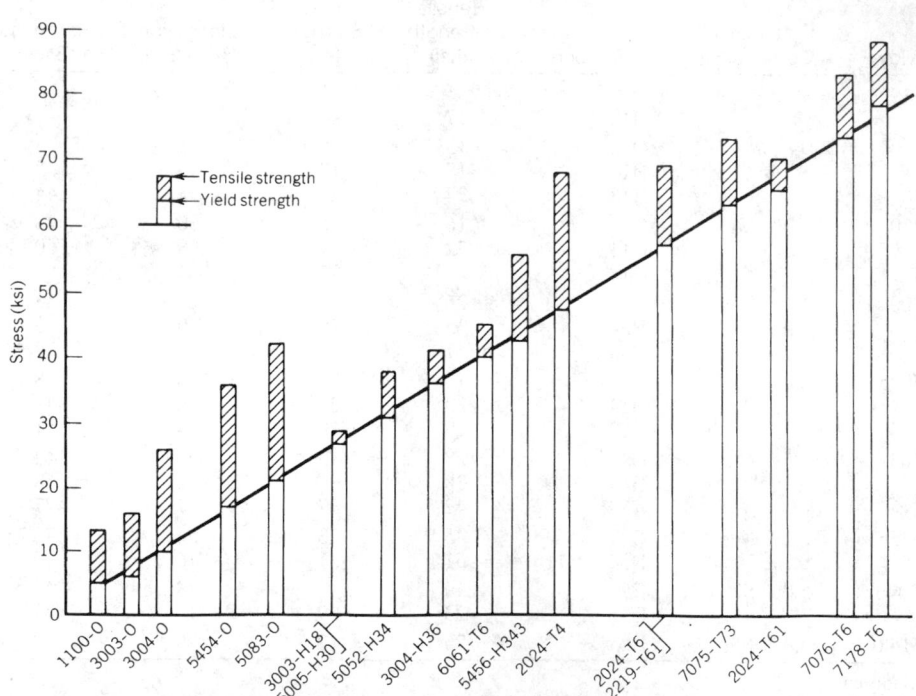

Fig. 3.1 Comparison of strengths of wrought aluminum alloys.

and 1.65 on the yield strength. For bridges and similar type structures the factors of safety are 2.20 on tensile ultimate strength and 1.85 on yield strength. For other types of applications the factors of safety may differ.

Selection of the working stresses and safety factors for a particular application should be based on codes, specifications, and standards covering that application published by agencies of government or nationally recognized trade and professional organizations.

For building and bridge design, reference should be made to the *Aluminum Design Manual*, published by the Aluminum Association. For boiler and pressure vessel design, reference should be made to the *Boiler and Pressure Vessel Code* published by the American Society of Mechanical Engineers.

For information on available codes, standards and specifications for other applications, the Aluminum Association may be consulted at 900 19th Street, NW, Washington, DC 20006.

3.7 CHARACTERISTICS

In addition to strength, the combination of alloy and temper determine other characteristics such as corrosion resistance, workability, machinability, etc. Some of the more important characteristics of representative aluminum alloys are compared in Table 3.7. The ratings A through E are relative ratings to compare wrought and cast aluminum alloys *within each category* and are explained below. Where a range of ratings is given, the first rating applies to the alloy in the annealed condition and the second rating is for the alloy when fully hardened. Alloys shown are representative and other alloys of the same type generally have comparable ratings.

3.7.1 Resistance to General Corrosion

Ratings are based on exposures to sodium chloride solution by intermittent spraying or immersion. In general, alloys with A and B ratings can be used in industrial and seacoast atmospheres and in many applications without protection. Alloys with C, D, and E ratings generally should be protected, at least on faying surfaces.

3.7.2 Workability

Ratings A through D for workability (cold) are relative ratings in decreasing order of merit.

3.7.3 Weldability and Brazeability

Aluminum alloys can be joined by most fusion and solid-state welding processes as well as by brazing and soldering. Fusion welding is commonly done by gas metal-arc welding (GMAW) and gas tungsten-arc welding (GTAW).

The relative weldability and brazeability of representative aluminum alloys is covered in Table 3.7, where ratings A through D are defined as follows:

A = Generally weldable by all commercial procedures and methods.

Table 3.7   Comparative Characteristics of Representative Aluminum Alloys

Alloy	Resistance to General Corrosion	Workability[c]	Machinability	Brazeability	Weldability (Arc)
1100	A	A–C	E–D	A	A
2024	D[a]	C–D	B	D	B–C
3003	A	A–C	E–D	A	A
3004	A	A–C	D–C	B	A
5005	A	A–C	E–D	B	A
5052	A	A–C	D–C	C	A
5456	A[b]	B–C	D–C	D	A
6061	B	A–C	D–C	A	A
7075	C[a]	D	B	D	C
356.0	B	A	C	—	A
B443.0	B	A	E	—	A
514.0	A	D	A	—	C

[a]E in thick sections.

[b]May differ if material heated for long periods.

[c]Castability for casting alloys.

Table 3.8   Practical Aluminum Thickness Ranges for Various Joining Processes

	Thickness (in) [or Area (in²)]	
Joining Process	Minimum	Maximum
Gas metal-arc welding	0.12	No limit
Gas tungsten-arc welding	0.02	1
Resistance spot welding	Foil	0.18
Resistance seam welding	0.01	0.18
Flash welding	0.05	(12)
Stud welding	0.02	No limit
Cold welding—butt joint	(0.0005)	(0.2)
Cold welding—lap joint	Foil	0.015
Ultrasonic welding	Foil	0.12
Electron beam welding	0.02	6
Brazing	0.006	No limit

[a]Reprinted from the American Welding Society, *Welding Handbook*, 7th ed., Miami, FL, 1982.

B = Weldable with special techniques or for specific applications that justify preliminary trials or testing to develop welding procedures and weld performance.

C = Limited weldability because of crack sensitivity or loss in resistance to corrosion and mechanical properties.

D = No commonly used welding methods have been developed.

Table 3.8 gives practical thickness or cross-sectional areas that can be joined by various processes.

3.8   TYPICAL APPLICATIONS

Typical applications of commonly used wrought aluminum alloys are listed in Tables 3.9 and 3.10. By comparing these with Tables 3.5, 3.6, and 3.7, one can readily see that application is based on properties such as strength, corrosion resistance, weldability, etc. Where one desired property, such as high strength, is the prime requisite, then steps must be taken to overcome a possible undesirable characteristic, such as relatively poor corrosion resistance. In this case, the high-strength alloy would be protected by a protective coating such as cladding, which will be described in a later section. Conversely, where resistance to attack is the prime requisite, then one of the more corrosion-resistant

Table 3.9   Typical Applications of Wrought Nonheat-Treatable Aluminum Alloys

Alloy Series	Typical Alloys	Typical Applications
1xxx	1350	Electrical conductor
	1060	Chemical equipment, tank cars
	1100	Sheet metal work, cooking utensils, decorative
3xxx	3003, 3004	Sheet metal work, chemical equipment, storage tanks, beverage cans, heat exchangers
4xxx	4043	Welding electrodes
	4343	Brazing alloy
5xxx	5005, 5050, 5052, 5657	Decorative and automotive trim, architectural and anodized, sheet meal work, appliances bridge and building structures, beverage can ends
5xxx (>2.5% Mg)	5083, 5086, 5182, 5454, 5456	Marine, welded structures, storage tanks, pressure vessels, armor plate, cryogenics, beverage can easy open ends, automotive structures

Table 3.10 Typical Applications of Wrought Heat-Treatable Alloys

Alloy Series	Typical Alloys	Typical Applications
2xxx	2011	Screw machine products
(Al-Cu)	2219	Structural, high temperature
2xxx (Al-Cu-Mg)	2014, 2024, 2618	Aircraft structures and engines, truck frames and wheels, automotive structures
6xxx	6061, 6063	Marine, truck frames and bodies, structures, architectural, furniture, bridge decks, automotive structures
7xxx (Al-Zn-Mg)	7004, 7005	Structural, cryogenic, missile
7xxx (Al-Zn-Mg-Cu)	7001, 7075, 7178	High-strength structural and aircraft

alloys would be employed and assurance of adequate strengths would be met through proper design. The best combination of strength and corrosion resistance for consumer applications in wrought products is found among the 5xxx and 6xxx series alloys. Several casting alloys have good corrosion resistance, and aluminum castings are widely used as cooking utensils and components of food processing equipment as well as for valves, fittings, and other components in various chemical applications.

3.9 MACHINING ALUMINUM

Aluminum alloys are readily machined and offer such advantages as almost unlimited cutting speed, good dimensional control, low cutting force, and excellent life. Relative machinability of commonly used alloys are classified as A, B, C, D, or E (see Table 3.7).

3.9.1 Cutting Tools

Cutting tool geometry is described by seven elements: top or back rake angle, side rake angle, end relief angle, side relief angle, end cutting edge angle, and nose radius.

The depth of cut may be in the range of $\frac{1}{16}$–$\frac{1}{4}$ in. for small work up to $\frac{1}{2}$–$1\frac{1}{2}$ in. for large work. The feed depends on finish. Rough cuts vary from 0.006 to 0.080 in. and finishing cuts from 0.002 to 0.006 in. Speed should be as high as possible, up to 15,000 fpm.

Cutting forces for an alloy such as 6061-T651 are 0.30–0.50 hp/in.^{3}/min for a 0° rake angle and 0.25–0.35 hp/in.^{3}/min for a 20° rake angle.

Lubrication such as light mineral or soluble oil is desirable for high production. Alloys with a machinability rating of A or B may not need lubrication.

The main types of cutting tool materials include water-hardening steels, high-speed steels, hard-cast alloys, sintered carbides and diamonds:

1. Water-hardening steels (plain carbon or with additions of chromium, vanadium, or tungsten) are lowest in first cost. They soften if cutting edge temperatures exceed 300–400°F; have low resistance to edge wear; and are suitable for low cutting speeds and limited production runs.

2. High-speed steels are available in a number of forms, are heat treatable, permit machining at rapid rates, allow cutting edge temperatures of over 1000°F, and resist shock better than hard-cast or sintered carbides.

3. Hard-cast alloys are cast closely to finish size, are not heat treated, and lie between high-speed steels and carbides in terms of heat resistance, wear, and initial cost. They will not take severe shock loads.

4. Sintered carbide tools are available in solid form or as inserts. They permit speeds 10–30 times faster than for high-speed steels. They can be used for most machining operations. They should be used only when they can be supported rigidly and when there is sufficient power and speed. Many types are available.

5. Mounted diamonds are used for finishing cuts where an extremely high-quality surface is required.

3.9.2 Single-Point Tool Operations

1. *Turning.* Aluminum alloys should be turned at high speeds with the work held rigidly and supported adequately to minimize distortion.

2. *Boring.* All types of tooling are suitable. Much higher speeds can be employed than for boring ferrous materials. Carbide tips are normally used in high-speed boring in vertical or horizontal boring machines.

3. *Planing and Shaping.* Aluminum permits maximum table speeds and high metal removal rates. Tools should not strike the work on the return stroke.

3.9.3 Multipoint Tool Operations

Milling

Removal rate is high with correct cutter design, speed and feed, machine rigidity, and power. When cutting speeds are high, the heat developed is retained mostly in the chips, with the balance absorbed by the coolant. Speeds are high with cutters of high-speed and cast alloys, and very high with sintered carbide cutters.

All common types of solid-tooth, high-carbon, or high-speed steel cutters can be employed. High-carbon cutters operating at a maximum edge temperature of 400°F are preferred for short run production. For long runs, high-speed steel or inserted-tooth cutters are used.

Speeds of 15,000 fpm are not uncommon for carbide cutters. Maximum speeds for high-speed and high-carbon-steel cutters are around 5000 fpm and 600 fpm, respectively.

Drilling

General-purpose drills with bright finishes are satisfactory for use on aluminum. Better results may be obtained with drills having a high helix angle. Flute areas should be large; the point angle should be 118° (130°–140° for deeper holes). Cutting lips should be equal in size. Lip relief angles are between 12° and 20°, increasing toward the center to hold the chisel angle between 130° and 145°.

No set rule can be given for achieving the correct web thickness. Generally, for aluminum, it may be thinner at the point without tool breakage.

A ⅛-in. drill at 6000 rpm has a peripheral speed of 2000 fpm. For drilling aluminum, machines are available with speeds up to 80,000 rpm.

If excessive heat is generated, hold diameter may be reduced even below drill size. With proper drills, feeds, speeds, and lubrication, no heat problem should occur.

For a feed of 0.008 ipr, and a depth to diameter ratio of 4:1, the thrust value is 170 lb and the torque value is 10 lb-in. for a ¼-in. drill with alloy 6061-T651. Aluminum alloys can be counterbored, tapped, threaded by cutting or rolling, and broached. Machining fluid should be used copiously.

Grinding

Resin-bounded silicon carbide wheels of medium hardness are used for rough grinding of aluminum. Finish grinding requires softer, vitrified-bonded wheels. Wheels speeds can vary from 5500 to 6000 fpm. Abrasive belt grinding employs belt speeds from 4600 to 5000 sfpm. Grain size of silicon carbide abrasive varies from 36 to 80 for rough cuts and from 120 to 180 for finishing cuts. For contact wheel abrasive belt grinding, speeds are 4500–6500 sfpm. Silicon carbide or aluminum oxide belts (24–80 grit) are used for rough cuts.

Sawing, Shearing, Routing, and Arc Cutting Aluminum

Correct tooth contour is most important in *circular sawing.* The preferred saw blade has an alternate hollow ground side—rake teeth at about 15°. Operating speeds are 4000–15,000 fpm. Lower speeds are recommended for semi-high-speed steel, intermediate speeds for high-speed inserted-tooth steel blades, and high speeds for carbide-tipped blades.

Band sawing speeds should be between 2000 and 5000 fpm. Spring-tempered blades are recommended for sheet and soft blades with hardened teeth for plate. Tooth pitch should not exceed material thickness: four to five teeth to the inch for spring tempered, six to eight teeth to the inch for flexible backed. Contour sawing is readily carried out. Lubricant should be applied to the back of the blade.

Shearing of sheet may be done on guillotine shears. The clearance between blades is generally 10–12% of sheet thickness down to 5–6% for light gauge soft alloy sheet. Hold-down pads, shear beds, and tables should be covered to prevent marring. Routing can also be used with 0.188–0.50 in. material routed at feeds of 10–30 ipm. Plates of 3-in.-thick heat-treated material can be routed at feeds up to 10 ipm.

Chipless machining of aluminum can be carried out using shear spinning rotary swaging, internal swaging, thread rolling, and flame cutting.

3.10 CORROSION BEHAVIOR

Although aluminum is a chemically active metal, its resistance to corrosion is attributable to an invisible oxide film that forms naturally and is always present unless it is deliberately prevented from forming. Scratch the oxide from the surface and, in air, the oxide immediately reforms. Once formed, the oxide effectively protects the metal from chemical attack and also from further oxidation. Some properties of this natural oxide are:

1. It is very thin—200–400 billionths of an inch thick.
2. It is tenacious. Unlike iron oxide or rust which spalls from the surface leaving a fresh surface to oxidize, aluminum oxide adheres tightly to aluminum.
3. It is hard. Aluminum oxide is one of the hardest substances known.
4. It is relatively stable and chemically inert.
5. It is transparent and does not detract from the metal's appearance.

3.10.1 General Corrosion

The general corrosion behavior of aluminum alloys depends basically on three factors: (1) the stability of the oxide film, (2) the environment, and (3) the alloying elements; these factors are not independent of one another. The oxide film is considered stable between pH 4.5 and 9.0; however, aluminum can be attacked by certain anions and cations in neutral solutions, and it is resistant to some acids and alkalies.

In general, aluminum alloys have good corrosion resistance in the following environments: atmosphere, most fresh waters, seawater, most soils, most foods, and many chemicals. Since "good corrosion resistance" is intended to mean that the material will give long service life without surface protection, in support of this rating is the following list of established applications of aluminum in various environments:

In Atmosphere. Roofing and siding, truck and aircraft skin, architectural.

With Most Fresh Waters. Storage tanks, pipelines, heat exchangers, pleasure boats.

In Seawater. Ship hulls and superstructures, buoys, pipelines.

In Soils. Pipelines and drainage pipes.

With Foods. Cooking utensils, tanks and equipment, cans and packaging.

With Chemicals. Storage tanks, processing and transporting equipment.

It is generally true that the higher the aluminum purity, the greater is its corrosion resistance. However, certain elements can be alloyed with aluminum without reducing its corrosion resistance and in some cases an improvement actually results. Those elements having little or no effect include Mn, Mg, Zn, Si, Sb, Bi, Pb, Ti; those having a detrimental effect include Cu, Fe, Ni:

Al–Mn Alloys. Al–Mn alloys (3xxx series) have good corrosion resistance and may possibly be better than 1100 alloy in marine environments and for cooking utensils because of a reduced effect by Fe in these alloys.

Al–Mg Alloys. Al–Mg alloys (5xxx series) are as corrosion resistant as 1xxx alloys and even more resistant to salt water and some alkaline solutions. In general, they offer the best combination of strength and corrosion resistance of all aluminum alloys.

Al–Mg–Si Alloys. Al–Mg–Si alloys (6xxx series) have good resistance to atmosphere corrosion, but generally slightly lower resistance than Al–Mg alloys. They can be used unprotected in most atmospheres and waters.

Alclad Alloys. Alclad alloys are composite wrought products comprised of an aluminum alloy core with a thin layer of corrosion—protective pure aluminum or aluminum alloy metallurgically bonded to one or both surfaces of the core. As a class, alclad alloys have a very high resistance to corrosion. The cladding is anodic to the core and thus protects the core.

3.10.2 Pitting Corrosion

Pitting is the most common corrosive attack on aluminum alloy products. Pits form at localized discontinuities in the oxide film on aluminum exposed to atmosphere, fresh water, or saltwater, or other neutral electrolytes. Since in highly acidic or alkaline solutions the oxide film is usually unstable pitting generally occurs in a pH range of about 4.5–9.0. The pits can be minute and concentrated and can vary in size and be widely scattered, depending on alloy composition, oxide film quality, and the nature of the corrodent.

The resistance of aluminum to pitting depends significantly on its purity; the purest metal is the most resistant. The presence of other elements in aluminum, except Mn, Mg, and Zn, increases in susceptibility to pitting. Copper and iron have the greatest effect on susceptibility. Alclad alloys have greatest resistance to penetration since any pitting is confined to the more anodic cladding until the cladding is consumed.

3.10.3 Galvanic Corrosion

Aluminum in contact with a dissimilar metal in the presence of an electrolyte tends to corrode more rapidly than if exposed by itself to the same environment; this is referred to as galvanic corrosion.

The tendency of one metal to cause galvanic corrosion of another can be predicted from a "galvanic series," which depends on environments. Such a series is listed below; the anodic metal is usually corroded by contact with a more cathodic one:

Anodic	Magnesium and zinc	Protect aluminum
	Aluminum, cadmium, and chromium	Neutral and safe in most environments
	Steel and iron	Cause slow action on aluminum except in marine environments
	Lead	Safe except for severe marine or industrial atmospheres
	Copper and nickel	Tend to corrode aluminum
Cathodic	Stainless steel	Safe in most atmospheres and fresh water; tends to corrode aluminum in severe marine atmospheres

Since galvanic corrosion is akin to a battery and depends on current flow, several factors determine the severity of attack. These are:

Electrolyte Conductivity. The higher the electrical conductivity, the greater the corrosive effect.

Polarization. Some couples polarize strongly to reduce the current flow appreciably. For example, stainless steel is highly cathodic to aluminum, but because of polarization the two can safely be used together in many environments.

Anode/Cathode Area Ratios. A high ratio minimizes galvanic attack; a low ratio tends to cause severe galvanic corrosion.

3.11   FINISHING ALUMINUM

The aluminum surface can be finished in more different ways than any other metal. The normal finishing operations fall into four categories: mechanical, chemical, electrochemical, and applied. They are usually performed in the order listed although one or more of the processes can be eliminated depending on the final effect desired.

3.11.1   Mechanical Finishes

This is an important starting point since even with subsequent finishing operations, a rough, smooth, or textured surface may be retained and observed. For many applications, the as-fabricated finish may be good enough. Or this surface can be changed by grinding, polishing, buffing, abrasive or shot blasting, tumbling or burnishing, or even hammering for special effects. Rolled surfaces of sheet or foil can be made highly specular by use of polishing rolls; one side or two sides bright and textures can be obtained by using textured rolls.

3.11.2   Chemical Finishes

A chemical finish is often applied after the mechanical finish. The most widely used chemical finishes include caustic etching for a matte finish, design etching, chemical brightening, conversion coatings, and immersion coatings. Conversion coatings chemically convert the natural oxide coating on aluminum to a chromate, a phosphate, or a combination chromate-phosphate coating, and they are principally used and are the recommended ways to prepare the aluminum surface for painting.

3.11.3   Electrochemical Finishes

These include electrobrightening for maximum specularity, electroplating of another metal such as nickel or chromium for hardness and wear resistance, and, most importantly, anodizing. Anodizing is an electrochemical process whereby the natural oxide layer is increased in thickness over a thousand times and made more dense for increased resistance to corrosion and abrasion resistance.

The anodic oxide forms by the growth of cells, each cell containing a central pore. The pores are sealed by immersing the metal into very hot or boiling water. Sealing is an important step and will affect the appearance and properties of the anodized coating. There are several varieties of anodized coatings.

3.11.4   Clear Anodizing

On many aluminum alloys a thick, transparent oxide layer can be obtained by anodizing in a sulfuric acid solution—this is called clear anodizing. The thickness of the layer depends on the current density and the time in solution, and is usually between 0.1 and 1 mil in thickness.

3.11.5 Color Anodizing

Color can be added to the film simply by immersing the metal immediately after anodizing and before sealing into a vat containing a dye or metallic coloring agents and then sealing the film. A wide range of colors have been imparted to aluminum in this fashion for many years. However, the colors imparted in this manner tend to fade from prolonged exposure to sunlight.

3.11.6 Integral Color Anodizing

More lightfast colors for outdoor use are achieved through integral color anodizing. These are proprietary processes utilizing electrolytes containing organic acids and, in some cases, small amounts of impurities are added to the metal itself to bring about the desired colors. This is usually a one-stage process and the color forms as an integral part of the anodize. Colors are for the most part limited to golds, bronzes, grays, and blacks.

3.11.7 Electrolytically Deposited Coloring

Electrolytically deposited coloring is another means of imparting lightfast colors. Following sulfuric acid anodize, the parts are transferred to a second solution containing metallic pigments that are driven into the coating by an electric current.

3.11.8 Hard Anodizing

Hard anodizing, or hardcoating as it is sometimes called, usually involves anodizing in a combination of acids and produces a very dense coating, often 1–5 mils thick. It is very resistant to wear and is normally intended for engineering applications rather than appearance.

3.11.9 Electroplating

In electroplating, a metal such as chromium or nickel is deposited on the aluminum surface from a solution containing that metal. This usually is done for appearance or to improve the hardness or abrasion resistance of the surface. Electroplating has a "smoothing out" effect, whereas anodized coatings follow the contours of the base metal surface thus preserving a matte or a polished surface as well as any other patterns applied prior to the anodize.

3.11.10 Applied Coatings

Applied coatings include porcelain enamel, paints and organic coatings, and laminates such as plastic, paper, or wood veneers. Probably as much aluminum produced today is painted as is anodized. Adhesion can be excellent when the surface has been prepared properly. For best results paint should be applied over a clean conversion-coated or anodized surface.

3.12 SUMMARY

Listed below are some of the characteristics of aluminum and its alloys that lead to their widespread application in nearly every segment of the economy. It is safe to say that no other material offers the same combination of properties.

- *Lightweight.* Very few metals have a lower density than aluminum, and they are not in common usage. Iron and copper are roughly three times as dense, titanium over 60% more dense than aluminum.
- *Good Formability.* Aluminum can be fabricated or shaped by just about every known method and is consequently available in a wide variety of forms.
- *Wide Range of Mechanical Properties.* Aluminum can be utilized as a weak, highly ductile material or, through alloying, as a material with a tensile strength approaching 100,000 psi.
- *High Strength-to-Weight Ratio.* Because of the combination of low density and high tensile strength, some aluminum alloys possess superior strength to weight-ratios, equalled or surpassed only by highly alloyed and strengthened steels and titanium.
- *Good Cryogenic Properties.* Aluminum does not become brittle at very low temperatures; indeed, mechanical properties of most aluminum alloys actually improve with decreasing temperature.
- *Good Weatherability and General Corrosion Resistance.* Aluminum does not rust away in the atmosphere and usually requires no surface protection. It is highly resistant to attack from a number of chemicals.
- *High Electrical and Thermal Conductivity.* Pound for pound, aluminum conducts electricity and heat better than any other material except sodium, which can only be used under very special conditions. On a volume basis, only copper, silver, and gold are better conductors.
- *High Reflectivity.* Normally, aluminum reflects 80% of white light, and this value can be increased with special processing.

- *Finishability.* Aluminum is unique among the architectural metals in respect to the variety of finishes employed.

BIBLIOGRAPHY

Aluminum and Aluminum Alloys, American Society for Metals, Metals Park, OH.

Aluminum Brazing Handbook, Aluminum Association, Washington, DC.

Aluminum Casting Technology, American Foundrymen's Society, Des Plains, IL.

Aluminum Design Manual, Aluminum Association, Washington, DC.

Aluminum Electrical Conductor Handbook, Aluminum Association, Washington DC.

Aluminum: Properties and Physical Metallurgy, ASM International, Metals Park, OH.

Aluminum Soldering Handbook, Aluminum Association, Washington, DC.

Aluminum Standards and Data, Aluminum Association, Washington, DC.

Forming and Machining Aluminum, Aluminum Association, Washington, DC.

Goddard, H. P., et al., *The Corrosion of Light Metals*, Wiley, New York.

Guidelines for the Use of Aluminum with Food and Chemicals, Aluminum Association, Washington, DC.

Handbook of Corrosion Data, ASM, Metals Park, OH.

Manford, J. Dean, *Handbook of Aluminum Bonding Technology and Data*, Marcel Dekker, New York.

Metals Handbook Series, ASM, Metals Park, OH.

Standards for Aluminum Sand and Permanent Mold Castings, Aluminum Association, Washington, DC.

The Surface Treatment and Finishing of Aluminum and Its Alloys, American Society for Metals, Metals Park, OH.

Welding Aluminum: Theory and Practice, Aluminum Association, Washington, DC.

CHAPTER 4

COPPER AND ITS ALLOYS

Howard Mendenhall
Olin Brass
East Alton, Illinois

Robert F. Schmidt
Colonial Metals
Columbia, Pennsylvania

4.1	**COPPER**	**59**	4.2.1 Introduction	60
	4.1.1 Composition of Commercial		4.2.2 Selection of Alloy	62
	Copper	59	4.2.3 Fabrication	62
	4.1.2 Hardening Copper	60	4.2.4 Mechanical and Physical	
	4.1.3 Corrosion	60	Properties	68
	4.1.4 Fabrication	60	4.2.5 Special Alloys	68
4.2	**SAND-CAST COPPER-BASE**			
	ALLOYS	**60**		

4.1 COPPER

Howard Mendenhall

4.1.1 Composition of Commercial Copper

Specifications for copper, generally accepted by industry, are the ASTM standard specifications. These also cover silver-bearing copper. (See Table 1)

Low-resistance copper, used for electrical purposes, may be electrolytically or fire refined. It is required to have a content of copper plus silver not less than 99.90%. Maximum permissible resistivities in international ohms (meter, gram) are: copper wire bars, 0.15328; ingots and ingot bars, 0.15694.

	Mechanical Properties of Copper		
	Annealed	Cold Rolled or Drawn	Cast
Tensile strength			
psi	30,000–40,000	32,000–60,000	20,000–30,000
MPa	210–280	220–400	140–210
Elongation in 2 in.	25–40%	2–35%	25–45%
Reduction of area	40–60%	2–4%	—
Rockwell F hardness	65 max	54–100	—
Rockwell 30T hardness	31 max	18–70	—

Mechanical Engineers' Handbook, 2nd ed., Edited by Myer Kutz.
ISBN 0-471-13007-9   © 1998 John Wiley & Sons, Inc.

Physical Properties of Copper

Density	0.323 lb/in.³			8.94 g/cm³	
Melting point	1981°F			1083°C	
Coefficient of linear thermal expansion	0.0000094/°F	(68–212°F)		0.0000170/°C	(20–100°C)
	0.0000097/°F	(68–392°F)		0.0000174/°C	(20–200°C)
	0.0000099/°F	(68–572°F)		0.0000178/°C	(20–300°C)
Pattern shrinkage	¼ in./ft			2%	
Thermal conductivity	226 Btu/ft²/ft/hr/°F at 68°F			398 W/m/°C at 27°C	
Electric resistivity	10.3 ohms (circular mil/ft) at 68°F			1.71 microhm/cm at 20°C	
Temperature coefficient of electric resistivity	0.023 ohms/°F at 68°F			0.0068/°C at 20°C	
Specific heat	—			0.386 J/g/°C at 20°C	
Magnetic property				Diamagnetic	
Optional property				Selectively reflecting	
Young's modulus	17,300,000 psi			119,300 MPa	

ASTM Specification B216-78, *Fire-Refined Copper for Wrought Products and Alloys*, calls for the following analysis: Cu + Ag, min 99.88%; As, max 0.012%; Sb, max 0.003%; Se + Te, max 0.025%; Ni, max 0.05%; Bi, max 0.003%; Pb, max 0.004%.

Oxygen-free high-conductivity copper is a highly ductile material, made under conditions that prevent the entrance of oxygen and the formation of copper oxide. It is utilized in deep-drawing, spinning, and edge-bending operations, and in welding, brazing, and other hot-working operations where embrittlement must be avoided. It has the same conductivity and tensile properties as tough pitch electrolytic copper.

Deoxidized copper containing silver has been utilized to increase softening resistance of copper. It does not affect oxygen level. A number of elements that reduce oxygen in copper, such as Zr, Cr, B, P, can also provide some softening resistance.

4.1.2 Hardening Copper

There are three methods for hardening copper: grain-size control, cold working, or alloying. When copper is hardened with tin, silicon, or aluminum, it generally is called bronze; when hardened with zinc, it is called brass.

4.1.3 Corrosion

Copper is resistant to the action of seawater and to atmospheric corrosion. It is not resistant to the common acids, and is unsatisfactory in service with ammonia and with most compounds of sulfur. Manufacturers should be consulted in regard to its use under corrosive conditions.

4.1.4 Fabrication

Copper may be hot forged, hot or cold rolled, hot extruded, hot pierced, and drawn, stamped, or spun cold. It can be silver-soldered, brazed, and welded. For brazing in reducing atmosphere or for welding by the oxyacetylene torch or electric arc, deoxidized copper will give more satisfactory joints than electrolytic or silver bearing copper. High-temperature exposure of copper containing oxygen, in reducing atmosphere, leads to decomposition of copper oxide and formation of steam with resulting embrittlement. Copper is annealed from 480 to 1400°F, depending on the properties desired. Ordinary commercial annealing is done in the neighborhood of 1100°F. Inert or reducing atmospheres give best surface quality; however, high temperature annealing of oxygen-containing coppers in reducing atmosphere can cause embrittlement. Copper may be electrodeposited from the alkaline cyanide solution, or from the acid sulfate solution.

4.2 SAND-CAST COPPER-BASE ALLOYS

Robert F. Schmidt

4.2.1 Introduction

The information required for selection of cast copper-base alloys for various types of applications can be found in Table 4.1. The principal data required by engineers and designers for castings made of copper-base alloys are given in Table 4.2. A cross-reference chart is shown in Table 4.3 for quick reference in locating the specifications applying to these alloys. Additional information in regard to

Table 4.1 Application for Copper-Base Alloys

Uses	Types of Alloys	Alloy Number
Andirons	Leaded yellow brass	C85200
Architectural trim	Leaded red brass	C83600
	Leaded yellow brass	C85400
	Leaded nickel silver	C97400
Ball bearing races	Manganese bronze	C86200
	Aluminum bronze	C95400
	Leaded yellow brass	C85200
Bearings, high speed, low load	High-leaded tin bronze	C93200
		C93800
		C93700
Bearings, low speed, heavy load	Tin bronze	C91300
		C91000
	Manganese bronze	C86300
	Aluminum bronze	C95400
Bearings, medium speed	High-leaded tin bronze	C93700
		C93800
Bells	Tin bronze	C91300
	Silicon bronze	C87200
Carburetors	Leaded red brass	C83600
	Leaded tin bronze	C92200
Cocks and faucets	Leaded semired brass	C84400
		C84800
	Leaded yellow brass	C85200
Corrosion resistance to acids	Aluminum bronze	C95400
	Leaded nickel bronze	C97600
alkalies	Silicon bronze	C87200
seawater	Nickel aluminum bronze	C95800
water	Leaded red brass	C83600
	Leaded semired brass	C84400
Electrical hardware	Leaded red brass	C83300
	Silicon bronze	C87200
	Aluminum bronze	C95400
Fittings	Leaded semired brass	C84400
Food-handling equipment	Leaded nickel bronze	C97600
		C97800
Gears	Tin bronze	C90700
		C91600
	Aluminum bronze	C95400
General hardware	Leaded red brass	C83600
Gun mounts	Manganese bronze	C86200
	Aluminum bronze	C95300
High-strength alloy	Manganese bronze	C86300
Impellers	Tin bronze	C90300
	Leaded red brass	C83600
	Aluminum bronze	C95400
	Silicon brass	C87200
Landing gear parts	Aluminum bronze	C95400
Lever arms	Manganese bronze	C86500
Marine castings and fittings	Manganese bronze	C86500
		C86200
	Aluminum bronze	C95800
Marine propellers	Aluminum bronze	C95800
	Manganese bronze	C86500
Musical instruments	Leaded nickel bronze	C97800
Ornamental bronze	Leaded yellow brass	C85200

Table 4.1   (Continued)

Uses	Types of Alloys	Alloy Number
Pickling baskets	Aluminum bronze	C95300
Piston rings	Tin bronze	C90500
		C91300
Plumbing fixtures	Leaded semired brass	C84400
		C84800
Pump bodies	Tin bronze	C90300
	Leaded tin bronze	C93800
	Aluminum bronze	C95800
Steam fittings and valves	Leaded tin bronze	C92200
		C92300
Valves, high pressure	Leaded tin bronze	C92200
		C92600
Valves, low pressures	Leaded red brass	C83600
	Leaded semired brass	C84400
Valve seats for elevated temperature	Leaded nickel bronze	C97800
Valve stems	Silicon brass	C87500
	Silicon bronze	C87200
Wear parts	High-leaded tin bronze	C93700
		C93800
Weldability	Tin bronze	C90700
	Manganese bronze	C86500
	Aluminum bronze	All grades
	Silicon bronze	C87200
Welding jaws	Aluminum bronze	C95300
Wormwheels	Aluminum bronze	C95500

special alloys, such as high conductivity copper, chromium-copper, and beryllium copper, is covered in Section 4.2.5.

4.2.2   Selection of Alloy

Table 4.1 is an outline of the various types of allows generally used for the purposes shown. When specifying a specific alloy for a new application, the foundry or ingot maker should be consulted. This is particularly important where corrosion resistance is involved or specific mechanical properties are required. While all copper-base alloys have good general corrosion resistance, specific environments, especially chemical, can cause corrosive attack or stress corrosion cracking. An example of this is the stress corrosion cracking that occurs when a manganese bronze alloy (high-strength yellow brass) is placed under load in certain environments.

The typical and minimum properties shown in Table 4.2 for the various alloys are for room temperature. The effect of elevated temperature on mechanical properties should be considered for any given application. The ingot maker or foundry should be consulted for this information.

Since copper-base alloy castings are often used for pressure-tight value and pump parts, caution should be exercised in alloy selection. In general, when small-sized, thin-wall castings are used, such as valve bodies with up to 3-in. openings, with all sections up to 1 in., the leaded red brass and leaded tin bronze alloys should be specified. When heavy-wall valves and pump bodies over 1-in. thickness are used, the castings should be made of nickel aluminum bronze or 70/30 cupronickel. These alloy preferences are based on differences in solidification behavior.

4.2.3   Fabrication

All sand-cast copper-base alloys can be machined, although some are far more machinable than others. The alloys containing lead, such as the leaded red brasses, leaded tin bronzes, and high-leaded tin bronzes, are very easily machined. On the other hand, aluminum and manganese bronzes do not machine easily. However, use of carbide tooling, proper tool angles, and coolants permit successful machining. In regard to weldability, no leaded alloys should be welded. In general, the aluminum bronzes, silicon bronzes, and α-β manganese bronzes can be welded successfully. This also applies

Table 4.2 Sand-Cast Copper-Base Alloys

Composition columns: for C83600–C85700 and C90300–C93800 the columns are **Cu, Sn, Pb, Zn, Others**; for C86200–C87500 the columns are **Cu, Zn, Fe, Al, Mn, Others** (shown below as "Sn / Zn", "Pb / Fe", "Zn / Al", "Mn").

UNS Number	Ingot Number	Cu	Sn / Zn	Pb / Fe	Zn / Al	Mn	Others	Yield Strength[a] ksi (MPa)	Tensile Strength[a] ksi (MPa)	Elongation[a] (%)	Brinell Hardness (500 kg)	Impact Strength (Izod) (ft-lb)	Electrical Conductivity (%, IACS)	Pattern Shrinkage (in./ft)
C83600	115	85	5	5	5			14 (97)	30 (207)	20	65	9	15	11/64
C83800	120	83	4	6	7			13 (90)	30 (207)	20	60	8	15.2	11/64
C84400	123	81	3	7	9			13 (90)	29 (200)	18	55	8	16.7	11/64
C84800	130	76	3	6	15			12 (83)	28 (193)	16	55	12[b]	16.6	11/64
C85200	400	72	1	3	24			11 (76)	35 (241)	25	46		18.6	3/16
C85400	403	67	1	3	29			14 (97)	30 (207)	20	53		19.6	3/16
C85700	405.2	61	1	1	37.3		0.3 Al	18 (124)	40 (276)	15	76		21.8	7/32
C86200	423	64	26	3	4	3		45 (310)	90 (621)	18	180[c]	12	7.4	1/4
C86300	424	62	26	3	6	3	0.75 Pb	60 (414)	110 (758)	12	225[c]	15	8.0	9/32
C86400	420	58	38	1	0.75	0.25	1 Pb	20 (138)	60 (414)	15	105[c]	30	19.3	1/4
C86500	421	58	39	1	1	1		25 (172)	65 (448)	20	130[c]	32[b]	20.6	1/4
C87200	500	92	4				4 Si	18 (124)	45 (310)	20	87	33	6.1	1/4
C87500	500	95					1 Mn, 4 Si	18 (124)	45 (310)	20	88	33	5.9	1/4
C87500	500	82	14		3		4 Si	24 (165)	60 (414)	16	115[c]	32[b]	6.1	15/64
C90300	225	88	8	0	4			18 (124)	40 (276)	20	70	14[b]	12.4	3/16
C90500	210	88	10	0	2			18 (124)	40 (276)	20	75	10	10.9	3/16
C92200	245	86	6	1½	4½			16 (110)	34 (234)	24	64	19[b]	14.3	3/16
C92300	230	87	8	1	4			16 (110)	36 (248)	18	70	14	12.3	3/16
C92600	215	87	10	1	2			18 (124)	40 (276)	20	72	7	10.0	3/16
C93200	315	83	7	7	3			14 (97)	30 (207)	15	67	5	12.4	7/32
C93700	305	80	10	10				12 (83)	30 (207)	15	67	5	10.1	1/8
C93800	319	78	7	15				14 (97)	26 (179)	12	58	5	11.6	5/32

Table 4.2 (Continued)

UNS Number	Ingot Number	Nominal Composition (% by Weight)	Yield Strength[a] ksi (MPa)	Tensile Strength[a] ksi (MPa)	Elongation[a] (%)	Brinell Hardness (500 kg)	Impact Strength (Izod) (ft-lb)	Electrical Conductivity (%, IACS)	Pattern Shrinkage (in./ft)
C95200	415	Cu 88, Fe 3, Al 9	25 (172) / 29 (200)	65 (448) / 80 (552)	20 / 38	120[c]	35	12.2	7/32
C95300	415	Cu 89, Fe 1, Al 10	25 (172) / 27 (186)	65 (448) / 75 (517)	20 / 25	140[c]	30	15.3	7/32
C95400	415	Cu 86, Fe 3½, Al 10½	30 (207) / 36 (248)	75 (517) / 92 (634)	12 / 18	156[c]	15	13	9/32
C95410	415	Cu 84, Fe 4, Ni 2, Al 10	30 (207) / 36 (248)	75 (517) / 96 (662)	12 / 15	176[c]	15	13	9/32
C95500	415	Cu 81, Fe 4, Ni 4, Al 11	40 (276) / 44 (303)	90 (621) / 102 (703)	6 / 12	200[c]	13	8.8	3/16
C95800	415	Cu 81½, Fe 4, Ni 4½, Al 9, 1 Mn	35 (241) / 37 (255)	85 (586) / 96 (662)	15 / 25	160[c]	20	7.0	3/16
C96400	415	Cu 68, Fe 1, Ni 30, 1 Nb	32 (221) / 37 (255)	60 (414) / 68 (469)	20 / 28	140[c]	78[b]	5.0	3/16
C97300	410	Cu 57, Sn 2, Pb 9, Zn 20, 12 Ni	15 (103) / 17 (117)	30 (207) / 36 (248)	8 / 25	60		5.9	1/8
C97400	411	Cu 60, Sn 3, Pb 5, Zn 16, 16 Ni	16 (110) / 17 (117)	30 (207) / 38 (262)	8 / 20	70		5.5	1/8
C97600	412	Cu 64, Sn 4, Pb 4, Zn 8, 20 Ni	17 (117) / 25 (172)	40 (276) / 47 (324)	22 / 22	85	11[e]	4.8	1/8
C97800	413	Cu 66, Sn 5, Pb 2, Zn 2, 25 Ni	22 (151) / 30 (207)	50 (345) / 55 (379)	10 / 15	130[b]		4.5	3/16
C81100	—	Cu 99.7	6 (41)	20 (138)	50	40		92	1/4
C81400	—	Cu 99, 1 Cr	36 (248)	53 (365[d])	11[d]	B69[e]		60	1/4
C82500	—	Cu 97½, 2 Be, 0.5 Cr, 0.25 Si	45 (310)	80 (551) / 160[d,e]	20	B82.5[e] / C40[d,e]	84	20[d]	3/16
C83300	131	Cu 93, Sn 1, Pb 2, Zn 4	14 (97)	30 (207)	25 / 35	35		32	3/16
C83450	—	Cu 88, Sn 2½, Pb 2, Zn 6½, 1 Ni	18 (124)	35 (241)	10 / 34	55		20	3/16
C91100	205	Cu 89, Sn 11, Pb 2		44 (303)	2	80		9.6	3/16
C91300	—	Cu 84, Sn 16		35 (241)	0.5	135[c]		8.5	3/16
C91600	194	Cu 81, Sn 19, 1½ Ni	17 (117)	44 (303)	16	170[c]		10.0	3/16
C92900	205A, 206A	Cu 84, Sn 10, Pb 2½, 3½ Ni	25 (172)	47 (324)	8 / 20	85	12	9.2	3/16
C99400	—	Cu 90.5, Fe 2, Ni 2.2, Al 1.2, 3 Zn, 1.2 Si	30 (207) / 34 (234)	60 (414) / 66 (445)	20 / 25	125[c]		16.6	3/16
C99500	—	Cu 88, Fe 4, Ni 4.5, Al 1.2, 1.2 Si	40 (276)	70 (483)	12	145[c]		13.7	3/16
C99700	—	Cu 58, Ni 5, Al 1, 13 Mn, 23 Zn	25 (172)	55 (379)	25	110[c]		3.0	1/4
C99750	—	Cu 58, Al 1, 20 Mn, 20 Zn, 1 Pb	32 (220)	65 (448)	30	110[c]		2.0	1/4

[a] Left column is minimum; right column is typical; yield strength is 0.5% extension under load.
[b] Impact strength, Charpy (ft-lb).
[c] Brinell hardness (3000 kg).
[d] Heat treated.
[e] Rockwell.

64

Table 4.3 Copper-Base Alloy Casting Specifications

Alloy Number	American Society for Testing Materials			Federal		Military	Society of Automotive Engineers	
	Commercial Designation	Specification Number	Alloy Number	QQ-C-390A Alloy Designation	Former Specification		Current	Former
C83600	85–5–5–5	B62,B584 B271,B505	C83600	836	QQ-L-225(2)	MIL-C-11866(25) MIL-C-15345(1) MIL-C-22087(2) MIL-C-22229(836)	836	40
C83800	83–4–6–7	B271,B584 B505	C83800	838	QQ-L-225(17			
C84400	81–3–7–9	B271,B584 B505	C84400	844	QQ-L-225(11)	MIL-B-11553(11) MIL-B-18343		
C84800	76–2½–6½–15	B271,B584 B505	C84800					
C85200	72–1–3–24	B271 B584	C85200	852	QQ-B-621(C)			
C85400	67–1–3–29	B271 B584	C85400	854	QQ-B-621(B)		854	41
C85700	61–1–1–37	B271 B584	C85700	857	QQ-B-621(A)	MIL-C-15345(3) MIL-C-11866(27)		
C86200	90,000 tensile manganese bronze	B271,B584 B505	C86200	862	QQ-B-726(B)	MIL-C-11866(20) MIL-C-22087(7) MIL-C-22229(862)	862	430A
C86300	110,000 tensile manganese bronze	B22,B505 B271,B584	C86300	863	QQ-B-726(C)	MIL-C-11866(21) MIL-C-15345(6) MIL-C-22087(9) MIL-C-22229(863)	863	430B
C86400	60,000 tensile manganese bronze	B271 B584	C86400	864	QQ-B-726(D) QQ-B-726(D)			
C86500	65,000 tensile manganese bronze	B271,B584 B505	C86500	865	QQ-B-726(A)	MIL-C-15345(4) MIL-C-22087(5) MIL-C-22229(865)	865	43

Table 4.3 (Continued)

Alloy Number	Commercial Designation	American Society for Testing Materials		Federal		Military	Society of Automotive Engineers	
		Specification Number	Alloy Number	QQ-C-390A Alloy Designation	Former Specification		Current	Former
C87200	5% zinc max silicon bronze	B271 B584	C87200	872	QQ-593(B)	MIL-C-11866(19) MIL-C-22229(872)		
C87500	82–14–4 silicon brass	B271 B584	C87500		QQ-593(A)			
C90300	88–8–0–4	B271,B584 B505	C90300	903	QQ-L-225(5)	MIL-C-11866(26) MIL-C-15345(8) MIL-C-22087(3) MIL-C-22229(903)	903	620
C90500	88–10–0–2	B22,B505 B271,B584	C90500	905	QQ-L-225(16)		905	62
C92200	88–6–½–4½	B61,B505 B271,B584	C92200	922	QQ-L-225(1)	MIL-C-15345(9) MIL-B-16541	922	622
C92300	87–8–1–4	B271, B505,B584	C92300	923	QQ-L-225(6-6X)	MIL-C-15345(10)	923	621
C93200	83–7–7–3	B271,B584 B505	C93200	932	QQ-L-225(12)	MIL-B-11553(12) MIL-B-16261(6)	932	660
C93500	85–5–9–1	B271,B584 B505	C93500	935	QQ-L-225(14)		935	66
C93700	80–10–10	B22,B505	C93700	937		MIL-B-13506(792,797)	937	64
C93800	78–7–15	B271,B584 B66,B271, B144,B505, B584	C93800	938	QQ-L-225(7)		938	67
C95200	88–3–9 aluminum bronze	B148,B505 B271	C95200	952	QQ-B-671(1)	MIL-C-22087(6) MIL-C-22229(952)	952	68A

C95300	89-1-10 aluminum bronze	B148,B505 B271	C95300	953	QQ-B-671(2)	MIL-C-11866(22)	953	68B
C95400	85-4-11 aluminum bronze	B148,B505 B271	C95400	954	QQ-B-671(3)	MIL-C-11866(23) MIL-C-15345(13)		
C95500	81-4-11-4 aluminum bronze	B148,B505 B271	C95500	955	QQ-B-671(4)	MIL-C-11866(24) MIL-C-15345(14) MIL-C-22087(8) MIL-C-22229(955)		
C95800	81-4-9-5-1mN aluminum bronze	B148 b271	C95800	958		MIL-C-15345(38) MIL-B-21230(1) MIL-B-24480 MIL-B-22229(958)		
C96400	70-30 cupronickel	B369 B505	C96400	964		MIL-C-15345(24) MIL-C-20159(1) MIL-C-15345(7)		
C97300	12% nickel silver	B271 B584	C97300					
C94700	16% nickel silver							
C97600	20% nickel bronze	B271 B584	C97600					
C97800	25% nickel bronze	B271 B584	C97800					

to tin bronzes and 70/30 cupronickel. These alloys not only can be joined to other materials by welding, but can also be repaired by welding if exhibiting casting defects such as shrinkage porosity. All copper-base alloys can be joined by brazing.

4.2.4 Mechanical and Physical Properties

The mechanical and physical properties of the most widely used copper-base casting alloys are given in Table 4.2. Alloy numbers used are the UNS numbers developed by the Copper Development Association (CDA) and now adopted by the American Society for Testing Materials (ASTM), Society for Automotive Engineers (SAE), and the U.S. Government. Also shown for reference purposes are the ingot numbers still used by the ingot makers. Much of the data shown in Table 4.2 were taken from *Standards Handbook*, Part 7, Alloy Data, published by CDA. Table 4.2 not only shows the typical properties that can be attained, but also the minimum values called for in the various specifications listed in Table 4.3. These properties, of course, can only be attained when care is taken toward proper melting, gating, feeding, and venting of casting molds.

The CDA *Standards Handbook*, Part 7, contains a very complete list of physical properties on not only the alloys shown in Table 4.2, but also other alloys less widely used.

4.2.5 Special Alloys

There are a number of alloys shown in Table 4.2 that are used for special purposes and amount to much less tonnage than the red brasses, leaded red brasses, tin bronzes, manganese bronzes, and aluminum bronzes. The following sections mention the more widely used of the special alloy families.

Gear Bronzes

High-tin alloys such as C90700 (89% copper, 11% tin), C91600 (88% copper, 10% tin, 2% nickel), and C92900 (84% copper, 10% tin, 2½% lead, 3½% nickel) are widely used for cast bronze gears. In addition to these tin bronze alloys, aluminum bronze, such as C95400 (86% copper, 4% iron, and 10% aluminum) is also used for gear applications.

Bridge Bearing Plates

These castings are made almost entirely to ASTM B22 specification and are generally made from copper-tin alloys like C91300 (81% copper, 19% tin) and C91100 (86% copper, 14% tin). Three other alloys, specified under ASTM B22 are C86300 high-tensile manganese bronze, C90500 tin bronze, and C93700 high-leaded tin bronze.

Piston Rings

Tin bronzes, such as C91300 and C91100, are commonly used for piston rings. These castings are usually made by the centrifugal castings process.

High Conductivity

When the electrical conductivity of pure copper is required, it can be melted and deoxidized and poured into casting molds. Care must be taken to avoid contamination by elements usually present in cast copper-base alloys, such as phosphorous, iron, zinc, tin, and nickel. Electrical conductivity values of 85% to 90% IACS can be attained with low level impurities present. This alloy is C81100.

Moderate Conductivity, High Strength. All of the alloys shown in Table 4.2 have electrical conductivity less than 25% IACS. However, there are additional copper-base alloys available with higher electrical conductivity. Beryllium copper and low-tin bronzes are examples of alloys in the 25-35% IACS range. C83300, which has 32% IACS, has a composition of 93% copper, 1% tin, 2% lead, and 4% zinc. A typical beryllium copper casting alloy with around 25% IACS is C82500, which has as-cast typical properties of 80,000 psi tensile strength and 20% elongation in 2 in., and after heat treatment has a tensile strength of 155,000 psi and elongation of 1% in 2 in. Hardness of this alloy is typically Rockwell C40 in the heat-treated condition and Rockwell B82 when as-cast. This alloy has a composition of 2% beryllium, 0.5% cobalt, 0.25% silicon, and 97.20% copper.

When some strength is required in addition to high electrical conductivity, the best casting alloy is chromium copper, alloy C81400. This alloy is made up of 0.9% chromium, 0.1% silicon, and 99% copper. It is heat treatable and maintains an electrical conductivity of 85% IACS, a tensile strength of 51,000 psi, a yield strength of 40,000 psi, and an elongation of 17%. The hardness value for this alloy is 105 under a 500-kg load.

BIBLIOGRAPHY

Books

ASTM Book of Standards, Part 2.01, American Society for Testing Materials, Philadelphia, PA, 1983, Table 11-3.

Copper-Base Alloys Foundry Practice, 3rd ed., American Foundrymen's Society, Des Plaines, IL, 1965, Section 11.3.

Metals Handbook, 9th ed., American Society for Metals, Metals Park, OH, 1979, Vol. 2, Sections 11.1 and 11.2.

SAE Handbook, Society of Automotive Engineers, Warrendale, PA, 1982, Table 11-3.

Standards Handbook, Part 7, Cast Products, Copper Development Association, Greenwich, CT, 1978, Table 11-2 and Section 11.4.

Standards Handbook, Part 6, Specifications Index, Copper Development Association, Greenwich, CT, 1983, Table 11-3.

Periodicals

Foundry, Penton/IPC, Cleveland, OH.

Modern Castings, American Foundrymen's Society, Des Plaines, IL.

Transaction

Transaction, American Foundrymen's Society, Des Plaines, IL.

CHAPTER 5

NICKEL AND ITS ALLOYS

T. H. Bassford
Jim Hosier
Inco Alloys International, Inc.
Huntington, West Virginia

5.1	INTRODUCTION	71	5.5 HEAT TREATMENT	84
			5.5.1 Reducing Atmosphere	84
5.2	NICKEL ALLOYS	72	5.5.2 Prepared Atmosphere	85
	5.2.1 Classification of Alloys	72		
	5.2.2 Discussion and Applications	72	5.6 WELDING	86
5.3	CORROSION	80	5.7 MACHINING	86
5.4	FABRICATION	82	5.8 CLOSURE	88
	5.4.1 Resistance to Deformation	82		
	5.4.2 Strain Hardening	82		

5.1 INTRODUCTION

Nickel, the 24th element in abundance, has an average content of 0.016% in the outer 10 miles of the earth's crust. This is greater than the total for copper, zinc, and lead. However, few of these deposits scattered throughout the world are of commercial importance. Oxide ores commonly called laterites are largely distributed in the tropics. The igneous rocks contain high magnesium contents and have been concentrated by weathering. Of the total known ore deposits, more than 80% is contained in laterite ores. The sulfide ores found in the northern hemispheres do not easily concentrate by weathering. The sulfide ores in the Sudbury district of Ontario, which contain important by-products such as copper, cobalt, iron, and precious metals are the world's greatest single source of nickel.[1]

Nickel has an atomic number of 28 and is one of the transition elements in the fourth series in the periodic table. The atomic weight is 58.71 and density is 8.902 g/cm^3. Useful properties of the element are the modulus of elasticity and its magnetic and magnetostrictive properties, and high thermal and electrical conductivity. Hydrogen is readily adsorbed on the surface of nickel. Nickel will also adsorb other gases such as carbon monoxide, carbon dioxide, and ethylene. It is this capability of surface adsorption of certain gases without forming stable compounds that makes nickel an important catalyst.[2]

As an alloying element, nickel is used in hardenable steels, stainless steels, special corrosion-resistant and high-temperature alloys, copper–nickel, "nickel–silvers," and aluminum–nickel. Nickel imparts ductility and toughness to cast iron.

Approximately 10% of the total annual production of nickel is consumed by electroplating processes. Nickel can be electrodeposited to develop mechanical properties of the same order as wrought nickel; however, special plating baths are available that will yield nickel deposits possessing a hardness as high as 450 Vickers (425 BHN). The most extensive use of nickel plate is for corrosion protection of iron and steel parts and zinc-base die castings used in the automotive field. For these applications, a layer of nickel, 0.0015–0.003 in. thick, is used. This nickel plate is then finished or covered with a chromium plate consisting in thickness of about 1% of the underlying nickel plate thickness in order to maintain a brilliant, tarnish-free, hard exterior surface.

Mechanical Engineers' Handbook, 2nd ed., Edited by Myer Kutz.
ISBN 0-471-13007-9 © 1998 John Wiley & Sons, Inc.

5.2 NICKEL ALLOYS

Most of the alloys listed and discussed are in commercial production. However, producers from time to time introduce improved modifications that make previous alloys obsolete. For this reason, or economic reasons, they may remove certain alloys from their commercial product line. Some of these alloys have been included to show how a particular composition compares with the strength or corrosion resistance of currently produced commercial alloys.

5.2.1 Classification of Alloys

Nickel and its alloys can be classified into the following groups on the basis of chemical composition.[3]

Nickel

(1) Pure nickel, electrolytic (99.56% Ni), carbonyl nickel powder and pellet (99.95% Ni); (2) commercially pure wrought nickel (99.6–99.97% nickel); and (3) anodes (99.3% Ni).

Nickel and Copper

(1) Low-nickel alloys (2–13% Ni); (2) cupronickels (10–30% Ni); (3) coinage alloy (25% Ni); (4) electrical resistance alloy (45% Ni); (5) nonmagnetic alloys (up to 60% Ni); and (6) high-nickel alloys, Monel (over 50% Ni).

Nickel and Iron

Wrought alloy steels (0.5–9% Ni); (2) cast alloy steels (0.5–9% Ni); (3) alloy cast irons (1–6 and 14–36% Ni); (4) magnetic alloys (20–90% Ni): (a) controlled coefficient of expansion (COE) alloys (29.5–32.5% Ni) and (b) high-permeability alloys (49–80% Ni); (5) nonmagnetic alloys (10–20% Ni); (6) clad steels (5–40% Ni); (7) thermal expansion alloys: (a) low expansion (36–50% Ni) and (b) selected expansion (22–50% Ni).

Iron, Nickel, and Chromium

(1) Heat-resisting alloys (40–85% Ni); (2) electrical resistance alloys (35–60% Ni); (3) iron-base superalloys (9–26% Ni); (4) stainless steels (2–25% Ni); (5) valve steels (2–13% Ni); (6) iron-base superalloys (0.2–9% Ni); (7) maraging steels (18% Ni).

Nickel, Chromium, Molybdenum, and Iron

(1) Nickel-base solution-strengthened alloys (40–70% Ni); (2) nickel-base precipitation-strengthened alloys (40–80% Ni).

Powder-Metallurgy Alloys

(1) Nickel-base dispersion strengthened (78–98% Ni); (2) nickel-base mechanically alloyed oxide-dispersion-strengthened (ODS) alloys (69–80% Ni).

The nominal chemical composition of nickel-base alloys is given in Table 5.1. This table does not include alloys with less than 30% Ni, cast alloys, or welding products. For these and those alloys not listed, the chemical composition and applicable specifications can be found in the *Unified Numbering System for Metals and Alloys*, published by the Society of Automotive Engineers, Inc.

5.2.2 Discussion and Applications

The same grouping of alloys used in Tables 5.1, 5.2, and 5.3, which give chemical composition and mechanical properties, will be used for discussion of the various attributes and uses of the alloys as a group. Many of the alloy designations are registered trademarks of producer companies.

Nickel Alloys

The corrosion resistance of nickel makes it particularly useful for maintaining product purity in the handling of foods, synthetic fibers, and caustic alkalies, and also in structural applications where resistance to corrosion is a prime consideration. It is a general-purpose material used when the special properties of the other nickel alloys are not required. Other useful features of the alloy are its magnetic and magnetostrictive properties; high thermal and electrical conductivity; low gas content; and low vapor pressure.[4]

Typical *nickel 200* applications are food-processing equipment, chemical shipping drums, electrical and electronic parts, aerospace and missile components, caustic handling equipment and piping, and transducers.

Nickel 201 is preferred to nickel 200 for applications involving exposure to temperatures above 316°C (600°F). Nickel 201 is used as coinage, plater bars, and combustion boats in addition to some of the applications for Nickel 200.

Permanickel alloy 300 by virtue of the magnesium content is age-hardenable. But, because of its low alloy content, alloy 300 retains many of the characteristics of nickel. Typical applications are

Table 5.1. Nonimal Chemical Composition (wt%)

Material	Ni	Cu	Fe	Cr	Mo	Al	Ti	Nb	Mn	Si	C	Other Elements
Nickel												
Nickel 200	99.6	—	—	—	—	—	—	—	0.23	0.03	0.07	—
Nickel 201	99.7	—	—	—	—	—	—	—	0.23	0.03	0.01	—
Permanickel alloy 300	98.7	—	0.02	—	—	—	0.49	—	0.11	0.04	0.29	0.38 Mg
Duranickel alloy 301	94.3	—	0.08	—	—	4.44	0.44	—	0.25	0.50	0.16	—
Nickel–Copper												
Monel alloy 400	65.4	32	1.00	—	—	—	—	—	1.0	0.10	0.12	—
Monel alloy 404	54.6	45.3	0.03	—	—	—	—	—	0.01	0.04	0.07	—
Monel alloy R-405	65.3	31.6	1.25	—	—	0.1	—	—	1.0	0.17	0.15	0.04 S
Monel alloy K-500	65.0	30	0.64	—	—	2.94	0.48	—	0.70	0.12	0.17	—
Nickel–Chromium–Iron												
Inconel alloy 600	76	0.25	8.0	15.5	—	—	—	—	0.5	0.25	0.08	—
Inconel alloy 601	60.5	0.50	14.1	23.0	—	1.35	—	—	0.5	0.25	0.05	—
Inconel alloy 690	60	—	9.0	30	—	—	—	—	—	—	0.01	—
Inconel alloy 706	41.5	0.15	40	16	—	0.20	1.8	3	0.18	0.18	0.03	—
Inconel alloy 718	53.5	0.15	18.5	19	3.0	0.5	0.9	5.1	0.18	0.18	0.04	—
Inconel alloy X-750	73	0.25	7	15.5	—	0.70	2.5	1	0.50	0.25	0.04	—
Nickel–Iron–Chromium												
Incoloy alloy 800	31	0.38	46	20	—	0.38	0.38	—	0.75	0.50	0.05	—
Incoloy alloy 800H	31	0.38	46	20	—	0.38	0.38	—	0.75	0.50	0.07	—
Incoloy alloy 825	42	1.75	30	22.5	3	0.10	0.90	—	0.50	0.25	0.01	—
Incoloy alloy 925	43.2	1.8	28	21	3	0.35	2.10	—	0.60	0.22	0.03	—
Pyromet 860	44	—	Bal	13	6	1.0	3.0	—	0.25	0.10	0.05	4.0 Co
Refractaloy 26	38	—	Bal	18	3.2	0.2	2.6	—	0.8	1.0	0.03	20 Co
Nickel–Iron												
Nilo alloy 36	36	—	61.5	—	—	—	—	—	0.5	0.09	0.03	—
Nilo alloy 42	41.6	—	57.4	—	—	—	—	—	0.5	0.06	0.03	—
Ni–Span–C alloy 902	42.3	0.05	48.5	5.33	—	0.55	2.6	—	0.40	0.50	0.03	—
Incoloy alloy 903	38	—	41.5	—	—	0.90	1.40	2.9	0.09	0.17	0.02	14 Co
Incoloy alloy 907	37.6	0.10	41.9	—	—	1.5	—	4.70	0.05	0.08	0.02	14 Co

Table 5.1 *(Continued)*

Material	Ni	Cu	Fe	Cr	Mo	Al	Ti	Nb	Mn	Si	C	Other Elements
Nickel–Chromium–Molybdenum												
Hastelloy alloy X	Bal[c]	—	19	22	9	—	—	—	<1	1.5	0.10	—
Hastelloy alloy G	Bal	2	19.5	22	6.5	—	—	2.1	<1	<1	<0.05	<1 W, <2.5 Co
Hastelloy alloy C-276	Bal	—	5.5	15.5	16	—	—	—	<0.08	<1	<0.01	2.5 Co, 4 W, 0.35 V
Hastelloy alloy C	Bal	—	<3	16	15.5	—	<0.7	—	<0.08	<1	<0.01	<2 Co
Inconel alloy 617	54	—	—	22	9	1	—	—	—	—	0.07	12.5 Co
Inconel alloy 625	Bal	—	2.5	21.5	9	<0.4	<0.4	3.6	<0.5	<0.5	0.03	—
MAR-M-252	Bal	—	—	19	10	1	2.6	—	—	—	0.15	10 Co, 0.005 B
Rene' 41	Bal	—	—	19	10	1.5	3.1	—	—	—	0.09	11 Co, <0.010 B
Rene' 95	Bal	—	—	14	3.5	3.5	2.5	3.5	—	—	0.15	8 Co, 3.5 W, 0.01 B, 0.05 Zr
Astroloy	Bal	—	—	15	5.3	4.4	3.5	—	—	—	0.06	15 Co
Udimet 500	Bal	—	<0.5	19	4	3.0	3.0	—	—	—	0.08	18 Co 0.007 B
Udimet 520	Bal	—	—	19	6	2.0	3.0	—	—	—	0.05	12 Co. 1 W, 0.005 B
Udimet 600	Bal	—	<4	17	4	4.2	2.9	—	—	—	0.04	16 Co. 0.02 B
Udimet 700	Bal	—	—	15	5.0	4.4	3.5	—	—	—	0.07	18.5 Co, 0.025 B
Udimet 1753	Bal	—	9.5	16.3	1.6	1.9	3.2	—	0.1	0.05	0.24	7.2 Co, 8.4 W, 0.008 B, 0.06 Zr
Waspaloy	Bal	<0.1	<2	19	4.3	1.5	3	—	—	—	0.08	14 Co, 0.006 B, 0.05 Zr
Nickel-Powder Alloys (Dispersion Strengthened)												
TD–nickel	98	—	—	—	—	—	—	—	—	—	—	2 ThO$_2$
TD–NiCr	Bal	—	—	20	—	—	—	—	—	—	—	1.7 ThO$_2$
Nickel-Powder Alloys (Mechanically Alloyed)												
Inconel alloy MA 754	78	—	1.0	20	—	0.3	0.5	—	—	—	0.05	0.6 Y$_2$O$_3$
Inconel alloy MA 6000	69	—	—	15	2	4.5	2.5	—	—	—	0.05	4 W, 2 Ta, 1.1 Y$_2$O$_3$

[a] Minimum.　[b] Maximum.　[c] Balance.

Table 5.2 Mechanical Properties of Nickel Alloys

Material	0.2% Yield Strength (ksi)[a]	Tensile Strength (ksi)[a]	Elongation (%)	Rockwell Hardness
Nickel				
Nickel 200	21.5	67	47	55 Rb
Nickel 201	15	58.5	50	45 Rb
Permanickel alloy 300	38	95	30	79 Rb
Duranickel alloy 301	132	185	28	36 Rc
Nickel–Copper				
Monel alloy 400	31	79	52	73 Rb
Monel alloy 404	31	69	40	68 Rb
Monel alloy R-405	56	91	35	86 Rb
Monel alloy K-500	111	160	24	25 Rc
Nickel–Chromium–Iron				
Inconel alloy 600	50	112	41	90 Rb
Inconel alloy 601	35	102	49	81 Rb
Inconel alloy 690	53	106	41	97 Rb
Inconel alloy 706	158	193	21	40 Rc
Inconel alloy 718	168	205	20	46 Rc
Inconel alloy X-750	102	174	25	33 Rc
Nickel–Iron–Chromium				
Incoloy alloy 800	48	88	43	84 Rb
Incoloy alloy 800H	29	81	52	72 Rb
Incoloy alloy 825	44	97	53	84 Rb
Incoloy alloy 925	119	176	24	34 Rc
Pyromet 860	115	180	21	37 Rc
Refractaloy 26	100	170	18	—
Nickel–Iron				
Nilo alloy 42	37	72	43	80 Rb
Ni–Span–C alloy 902	137	150	12	33 Rc
Incoloy alloy 903	174	198	14	39 Rc
Incoloy alloy 907	163	195	15	42 Rc
Nickel–Chromium–Molybdenum				
Hastelloy alloy X	52	114	43	—
Hastelloy alloy G	56	103	48.3	86 Rb
Hastelloy alloy C-276	51	109	65	—
Inconel alloy 617	43	107	70	81 Rb
Inconel alloy 625	63	140	51	96 Rb
MAR-M-252	122	180	16	—
Rene' 41	120	160	18	—
Rene' 95	190	235	15	—
Astroloy	152	205	16	—
Udimet 500	122	190	32	—
Udimet 520	125	190	21	—
Udimet 600	132	190	13	—
Udimet 700	140	204	17	—
Udimet 1753	130	194	20	39 Rc
Waspaloy	115	185	25	—
Nickel-Powder Alloys (Dispersion Strengthened)				
TD–Nickel	45	65	15	—
TD–NiCr	89	137	20	—
Nickel-Powder Alloys (Mechanically Alloyed)				
Inconel alloy MA 754	85	140	21	—
Inconel alloy MA 6000	187	189	3.5	—

[a] MPa = ksi × 6.895.

Table 5.3   1000-hr Rupture Stress (ksi)[a]

	1200°F	1500°F	1800°F	2000°F
Nickel–Chromium–Iron				
Inconel alloy 600	14.5	3.7	1.5	—
Inconel alloy 601	28	6.2	2.2	1.0
Inconel alloy 690	16	—	—	—
Inconel alloy 706	85	—	—	—
Inconel alloy 718	85	—	—	—
Inconel alloy X-750	68	17	—	—
Nickel–Iron–Chromium				
Incoloy alloy 800	20	—	—	—
Incoloy alloy 800H	23	6.8	1.9	0.9
Incoloy alloy 825	26	6.0	1.3	—
Pyromet 860	81	17	—	—
Refractaloy 26	65	15.5	—	—
Nickel–Chromium–Moloybdenum				
Hastelloy alloy X	31	9.5	—	—
Inconel alloy 617	52	14	3.8	1.5
Inconel alloy 625	60	7.5	—	—
MAR-M-252	79	22.5	—	—
Rene' 41	102	29	—	—
Rene' 95	125	—	—	—
Astroloy	112	42	8	—
Udimet 500	110	30	—	—
Udimet 520	85	33	—	—
Udimet 600	—	37	—	—
Udimet 700	102	43	7.5	—
Udimet 1753	98	34	6.5	—
Waspaloy	89	26	—	—
Nickel-Powder Alloys (Dispersion Strengthened)				
TD–Nickel	21	15	10	7
TD–NiCr	—	—	8	5
Nickel-Powder Alloys (Mechanically Alloyed)				
Inconel alloy MA 754	38	—	19	14
Inconel alloy MA 6000	—	—	22	15

[a] MPa ksi × 6.895.

grid lateral winding wires, magnetostriction devices, thermostat contact arms, solid-state capacitors, grid side rods, diaphragms, springs, clips, and fuel cells.

Duranickel alloy 301 is another age-hardenable high nickel alloy, but is made heat treatable by aluminum and titanium additions. The important features of alloy 301 are high strength and hardness, good corrosion resistance, and good spring properties up to 316°C (600°F); and it is on these mechanical considerations that selection of the alloy is usually based. Typical applications are extrusion press parts, molds used in the glass industry, clips, diaphragms, and springs.

Nickel–Copper Alloys

Nickel–copper alloys are characterized by high strength, weldability, excellent corrosion resistance, and toughness over a wide temperature range. They have excellent service in seawater or brackish water under high-velocity conditions, as in propellers, propeller shafts, pump shafts, and impellers and condenser tubes, where resistance to the effects of cavitation and erosion are important. Corrosion rates in strongly agitated and aerated seawater usually do not exceed 1 mil/year.

Monel alloy 400 has low corrosion rates in chlorinated solvents, glass-etching agents, sulfuric and many other acids, and practically all alkalies, and it is resistant to stress-corrosion cracking. Alloy 400 is useful up to 538°C (1000°F) in oxidizing atmospheres, and even higher temperatures may be used if the environment is reducing. Springs of this material are used in corrosive environments up to 232°C (450°F). Typical applications are valves and pumps; pump and propeller shafts; marine fixtures and fasteners; electrical and electronic components; chemical processing equipment; gasoline and freshwater tanks; crude petroleum stills, process vessels, and piping; boiler feedwater heaters and other heat exchangers; and deaerating heaters.

Monel alloy 404 is characterized by low magnetic permeability and excellent brazing characteristics. Residual elements are controlled at low levels to provide a clean, wettable surface even after prolonged firing in wet hydrogen. Alloy 404 has a low Curie temperature and its magnetic properties are not appreciably affected by processing or fabrication. This magnetic stability makes alloy 404 particularly suitable for electronic applications. Much of the strength of alloy 404 is retained at outgassing temperatures. Thermal expansion of alloy 404 is sufficiently close to that of many other alloys as to permit the firing of composite metal tubes with negligible distortion. Typical applications are waveguides, metal-to-ceramic seals, transistor capsules, and power tubes.

Monel alloy R-405 is a free-machining material intended almost exclusively for use as stock for automatic screw machines. It is similar to alloy 400 except that a controlled amount of sulfur is added for improved machining characteristics. The corrosion resistance of alloy R-405 is essentially the same as that of alloy 400, but the range of mechanical properties differs slightly. Typical applications are water meter parts, screw machine products, fasteners for nuclear applications, and valve seat inserts.

Monel alloy K-500 is an age-hardenable alloy that combines the excellent corrosion resistance characteristics of the Monel nickel–copper alloys with the added advantage of increased strength and hardness. Age hardening increases its strength and hardness. Still better properties are achieved when the alloy is cold-worked prior to the aging treatment. Alloy K-500 has good mechanical properties over a wide temperature range. Strength is maintained up to about 649°C (1200°F), and the alloy is strong, tough, and ductile at temperatures as low as −253°C (−423°F). It also has low permeability and is nonmagnetic to −134°C (−210°F). Alloy K-500 has low corrosion rates in a wide variety of environments. Typical applications are pump shafts and impellers, doctor blades and scrapers, oil-well drill collars and instruments, electronic components, and springs.

Nickel–Chromium–Iron Alloys

This family of alloys was developed for high-temperature oxidizing environments. These alloys typically contain 50–80% nickel, which permits the addition of other alloying elements to improve strength and corrosion resistance while maintaining toughness.

Inconel alloy 600 is a standard engineering material for use in severely corrosive environments at elevated temperatures. It is resistant to oxidation at temperatures up to 1177°C (2150°F). In addition to corrosion and oxidation resistance, alloy 600 presents a desirable combination of high strength and workability, and is hardened and strengthened by cold-working. This alloy maintains strength, ductility, and toughness at cryogenic as well as elevated temperatures. Because of its resistance to chloride-ion stress-corrosion cracking and corrosion by high-purity water, it is used in nuclear reactors. For this service, the alloy is produced to exacting specifications and is designated Inconel alloy 600T. Typical applications are furnace muffles, electronic components, heat-exchanger tubing, chemical- and food-processing equipment, carburizing baskets, fixtures and rotors, reactor control rods, nuclear reactor components, primary heat-exchanger tubing, springs, and primary water piping. Alloy 600, being one of the early high-temperature, corrosion-resistant alloys, can be thought of as being the basis of many of our present day special-purpose high-nickel alloys, as illustrated in Fig. 5.1.

Inconel alloy 601 has shown very low rates of oxidation and scaling at temperatures as high as 1093°C (2000°F). The high chromium content (nominally 23%) gives alloy 601 resistance to oxidizing, carburizing, and sulfur-containing environments. Oxidation resistance is further enhanced by the aluminum content. Typical applications are heat-treating baskets and fixtures, radiant furnace tubes, strand-annealing tubes, thermocouple protection tubes, and furnace muffles and retorts.

Inconel alloy 690 is a high-chromium nickel alloy having very low corrosion rates in many corrosive aqueous media and high-temperature atmospheres. In various types of high-temperature water, alloy 690 also displays low corrosion rates and excellent resistance to stress-corrosion cracking—desirable attributes for nuclear steam-generator tubing. In addition, the alloy's resistance to sulfur-containing gases makes it a useful material for such applications as coal-gasification units, burners and ducts for processing sulfuric acid, furnaces for petrochemical processing, and recuperators and incinerators.

Inconel alloy 706 is a precipitation-hardenable alloy with characteristics similar to alloy 718, except that alloy 706 has considerably improved machinability. It also has good resistance to oxidation and corrosion over a broad range of temperatures and environments. Like alloy 718, alloy 706 has excellent resistance to postweld strain-age cracking. Typical applications are gas-turbine components and other parts that must have high strength combined with good machinability and weldability.

Inconel alloy 718 is an age-hardenable high-strength alloy suitable for service at temperatures from −253°C (−423°F) to 704°C (1300°F). The fatigue strength of alloy 718 is high, and the alloy exhibits high stress-rupture strength up to 704°C (1300°F) as well as oxidation resistance up to 982°C (1800°F). It also offers good corrosion resistance to a wide variety of environments. The outstanding characteristic of alloy 718 is its slow response to age hardening. The slow response enables the material to be welded and annealed with no spontaneous hardening unless it is cooled slowly. Alloy 718 can also be repair-welded in the fully aged condition. Typical applications are jet engine components, pump bodies and parts, rocket motors and thrust reversers, and spacecraft.

Inconel alloy X-750 is an age-hardenable nickel–chromium–iron alloy used for its corrosion and oxidation resistance and high creep-rupture strength up to 816°C (1500°F). The alloy is made age-hardenable by the addition of aluminum, columbium, and titanium, which combine with nickel, during

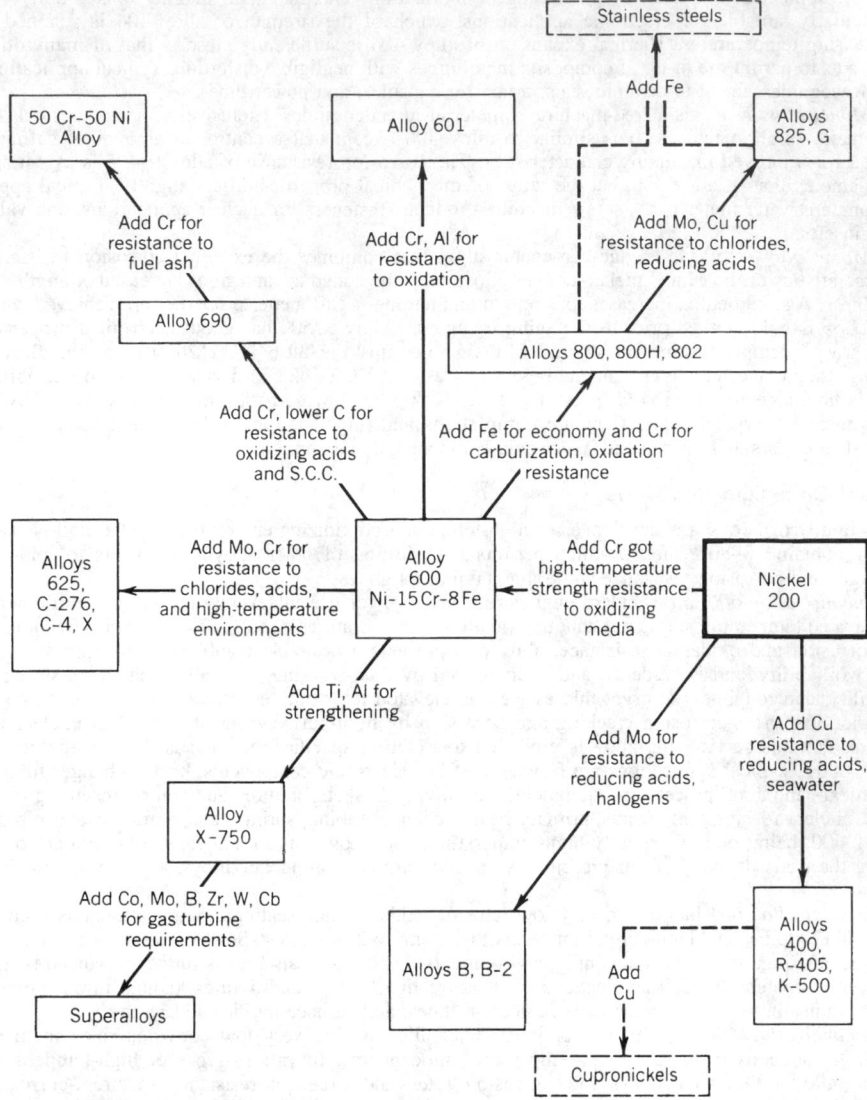

Fig. 5.1 Some compositional modifications of nickel and its alloys to produce special properties.

proper heat treatment, to form the intermetallic compound $Ni_3(Al, Ti)$. Alloy X-750, originally developed for gas turbines and jet engines, has been adopted for a wide variety of other uses because of its favorable combination of properties. Excellent relaxation resistance makes alloy X-750 suitable for springs operating at temperatures up to about 649°C (1200°F). The material also exhibits good strength and ductility at temperatures as low as −253°C (−423°F). Alloy X-750 also exhibits high resistance to chloride-ion stress-corrosion cracking even in the fully age-hardened condition. Typical applications are gas-turbine parts (aviation and industrial), springs (steam service), nuclear reactors, bolts, vacuum envelopes, heat-treating fixtures, extrusion dies, aircraft sheet, bellows, and forming tools.

Nickel–Iron–Chromium Alloys

This series of alloys typically contains 30−45% Ni and is used in elevated- or high-temperature environments where resistance to oxidation or corrosion is required.

Incoloy alloy 800 is a widely used material of construction for equipment that must resist corrosion, have high strength, or resist oxidation and carburization. The chromium in the alloy imparts resistance to high-temperature oxidation and general corrosion. Nickel maintains an austenitic structure so that the alloy remains ductile after elevated-temperature exposure. The nickel content also contributes resistance to scaling, general corrosion, and stress-corrosion cracking. Typical applications are heat-treating equipment and heat exchangers in the chemical, petrochemical, and nuclear industries, especially where resistance to stress-corrosion cracking is required. Considerable quantities are used for sheathing on electric heating elements.

Incoloy alloy 800H is a version of Incoloy alloy 800 having significantly higher creep and rupture strength. The two alloys have the same chemical composition with the exception that the carbon content of alloy 800H is restricted to the upper portion of the standard range for alloy 800. In addition to a controlled carbon content, alloy 800H receives an annealing treatment that produces a coarse grain size—an ASTM number of 5 or coarser. The annealing treatment and carbon content are responsible for the alloy's greater creep and rupture strength.

Alloy 800H is useful for many applications involving long-term exposure to elevated temperatures or corrosive atmospheres. In chemical and petrochemical processing, the alloy is used in steam/hydrocarbon reforming for catalyst tubing, convection tubing, pigtails, outlet manifolds, quenching-system piping, and transfer piping; in ethylene production for both convection and cracking tubes; in oxo-alcohol production for tubing in hydrogenation heaters; in hydrodealkylation units for heater tubing; and in production of vinyl chloride monomer for cracking tubes, return bends, and inlet and outlet flanges.

Industrial heating is another area of wide usage for alloy 800H. In various types of heat-treating furnaces, the alloy is used for radiant tubes, muffles, retorts, and assorted furnace fixtures. Alloy 800H is also used in power generation for steam superheater tubing and high-temperature heat exchangers in gas-cooled nuclear reactors.

Incoloy alloy 825 was developed for use in aggressively corrosive environments. The nickel content of the alloy is sufficient to make it resistant to chloride-ion stress-corrosion cracking, and, with molybdenum and copper, alloy 825 has resistance to reducing acids. Chromium confers resistance to oxidizing chemicals. The alloy also resists pitting and intergranular attack when heated in the critical sensitization temperature range. Alloy 825 offers exceptional resistance to corrosion by sulfuric acid solutions, phosphoric acid solutions, and seawater. Typical applications are phosphoric acid evaporators, pickling-tank heaters, pickling hooks and equipment, chemical-process equipment, spent nuclear fuel element recovery, propeller shafts, tank trucks, and oil-country cold-worked tubulars.

Incoloy alloy 925 was developed for severe conditions found in corrosive wells containing H_2S, CO_2, and brine at high pressures. Alloy 925 is a weldable, age-hardenable alloy having corrosion and stress-corrosion resistance similar to Incoloy alloy 825. It is recommended for applications where alloy 825 does not have adequate yield or tensile strength for service in the production of oil and gas, such as valve bodies, hanger bars, flow lines, casing, and other tools and equipment.

Pyromet 860 and *Refractaloy 26* are high-temperature precipitation-hardenable alloys with lower nickel content than Inconel alloy X-750 but with additions of cobalt and molybdenum. The precipitation-hardening elements are the same except the Al/Ti ratio is reversed with titanium content being greater than aluminum. Typical applications of both alloys are critical components of gas turbines, bolts, and structural members.[8]

Nickel–Iron

The nickel–iron alloys listed in Table 5.1 as a group have a low coefficient of expansion that remains virtually constant to a temperature below the Curie temperature for each alloy. A major application for *Nilo alloy 36* is tooling for curing composite airframe components. The thermal expansion characteristics of *Nilo alloy 42* are particularly useful for semiconductor lead frames and glass-sealing applications.

Ni-Span-C alloy 902 and *Incoloy alloys 903 and 907* are precipitation-hardenable alloys with similar thermal expansion characteristics to Nilo alloy 42 but having different constant coefficient of expansion temperature range. Alloy 902 is frequently used in precision apparatus where elastic members must maintain a constant frequency when subjected to temperature fluctuations. Alloys 903 and 907 are being used in aircraft jet engines for members requiring high-temperature strengths to 649°C (1200°F) with thermal expansion controlled to maintain low clearance.

Nickel–Chromium–Molybdenum Alloys

This group of alloys contains 45–60% Ni and was developed for severe corrosion environments. Many of these alloys also have good oxidation resistance and some have useful strength to 1093°C (2000°F).

Hastelloy alloy X is a non-age-hardenable nickel–chromium–iron–molybdenum alloy developed for high-temperature service up to 1204°C (2200°F). Typical applications are furnace hardware subjected to oxidizing, reducing, and neutral atmospheres; aircraft jet engine tail pipes; and combustion cans and afterburner components.[5,6]

Hastelloy alloy C is a mildly age-hardenable alloy similar in composition to alloy X except nearly all the iron is replaced with molybdenum and nickel. It is highly resistant to strongly oxidizing acids, salts, and chlorine. It has good high-temperature strength. Typical applications are chemical, petrochemical, and oil refinery equipment; aircraft jet engines; and heat-treating equipment.[6,7]

Hastelloy alloy C-276 is a modification of Hastelloy alloy C where the carbon and silicon content is reduced to very low levels to diminish carbide precipitation in the heat-affected zone of weldments. Alloy C-276 is non-age-hardenable and is used in the solution-treated condition. No postwelding heat treatment is necessary for chemical-process equipment. Typical applications are chemical- and petrochemical-process equipment, aircraft jet engines, and deep sour gas wells.[6,7]

Hastelloy alloy G is a non-age-hardenable alloy similar to the composition of alloy X but with 2% copper and 2% columbium and lower carbon content. Alloy G is resistant to pitting and stress-corrosion cracking. Typical applications are paper and pulp equipment, phosphate fertilizer, and synthetic fiber processing.[6,7]

Inconel alloy 617 is a solid-solution-strengthened alloy containing cobalt that has an exceptional combination of high-temperature strength and oxidation resistance which makes alloy 617 a useful material for gas-turbine aircraft engines and other applications involving exposure to extreme temperatures, such as, steam generator tubing and pressure vessels for advanced high-temperature gas-cooled nuclear reactors.

Inconel alloy 625, like alloy 617, is a solid-solution-strengthened alloy but containing columbium instead of cobalt. This combination of elements is responsible for superior resistance to a wide range of corrosive environments of unusual severity as well as to high-temperature effects such as oxidation and carburization. The properties of alloy 625 that make it attractive for seawater applications are freedom from pitting and crevice corrosion, high corrosion fatigue strength, high tensile strength, and resistance to chloride-ion stress-corrosion cracking. Typical applications are wire rope for mooring cables; propeller blades; submarine propeller sleeves and seals; submarine snorkel tubes; aircraft ducting, exhausts, thrust-reverser, and spray bars; and power plant scrubbers, stack liners, and bellows.

MAR-M-252, Rene' 41, Rene' 95, and *Astroloy* are a group of age-hardenable nickel-base alloys containing 10–15% cobalt designed for highly stressed parts operating at temperatures from 871 to 982°C (1600 to 1800°F) in jet engines. MAR-M-252 and Rene' 41 have nearly the same composition but Rene' 41 contains more of the age-hardening elements allowing higher strengths to be obtained. Rene' 95, of similar base composition but in addition containing 3.5% columbium and 3.5% tungsten, is used at temperatures between 371 and 649°C (700 and 1200°F). Its primary use is as disks, shaft retaining rings, and other rotating parts in aircraft engines of various types.[6–8]

Udimet 500, 520, 600, and 700 and *Unitemp 1753* are age-hardenable, nickel-base alloys having high strength at temperatures up to 982°C (1800°F). All contain a significant amount of cobalt. Applications include jet engine gas-turbine blades, combustion chambers, rotor disks, and other high-temperature components.[6–8]

Waspaloy is an age-hardenable nickel-base alloy developed to have high strength up to 760°C (1400°F) combined with oxidation resistance to 871°C (1600°F). Applications are jet engine turbine buckets and disks, air frame assemblies, missile systems, and high-temperature bolts and fasteners.[6–8]

Nickel Powder Alloys (Dispersion Strengthened)

These oxide dispersion strengthened (ODS) alloys are produced by a proprietary powder metallurgical process using thoria as the dispersoid. The mechanical properties to a large extent are determined by the processing history. The preferred thermomechanical processing results in an oriented texture with grain aspect ratios of about 3:1 to 6:1.

TD–nickel and *TD–NiCr* are dispersion-hardened nickel alloys developing useful strengths up to 1204°C (2200°F). These alloys are difficult to fusion weld without reducing the high-temperature strength. Brazing is used in the manufacture of jet engine hardware. Applications are jet engine parts, rocket nozzles, and afterburner liners.[6–8]

Nickel Powder Alloys (Mechanically Alloyed)

Inconel alloy MA 754 and *Inconel alloy MA 6000* are ODS nickel-base alloys produced by mechanical alloying.[9,10] An yttrium oxide dispersoid imparts high creep-rupture strength up to 1149°C (2100°F). MA 6000 is also age-hardenable, which increases strength at low temperatures up to 760°C (1400°F) These mechanical alloys like the thoria-strengthened alloys described are difficult to fusion weld without reducing high-temperature strength. Useful strength is obtained by brazing. MA 754 is being used as aircraft gas-turbine vanes and bands. Applications for MA 6000 are aircraft gas turbine buckets and test grips.

5.3 CORROSION

It is well recognized that the potential saving is very great by utilizing available and economic practices to improve corrosion prevention and control. Not only should the designer consider initial cost of materials, but he or she should also include the cost of maintenance, length of service, downtime cost, and replacement costs. This type of cost analysis can frequently show that more highly alloyed, corrosion-resistant materials are more cost effective. The National Commission on

Materials Policy concluded that one of the "most obvious opportunities for material economy is control of corrosion."

Studies have shown that the total cost of corrosion is astonishing. The overall cost of corrosion in the United States was estimated by the National Bureau of Standards in 1978 and updated by Battelle scientists in 1995. According to a report released in April, metallic corrosion costs the United States about $300 billion a year. The report, released by Battelle (Columbus, Ohio) and Specialty Steel Industry of North America (SSINA, Washington, DC), claims that about one-third of the costs of corrosion ($100 billion) is avoidable and could be saved by broader use of corrosion-resistant materials and the application of best anticorrosion technology from design through maintenance.

Since becoming commercially available shortly after the turn of the century, nickel has become very important in combating corrosion. It is a major constituent in the plated coatings and claddings applied to steel, corrosion-resistant stainless steels, copper–nickel and nickel–copper alloys, high-nickel alloys, and commercially pure nickel alloys. Not only is nickel a corrosion-resistant element in its own right, but, owing to its high tolerance for alloying, it has been possible to develop many metallurgically stable, special-purpose alloys.[11]

Figure 5.1 shows the relationship of these alloys and the major effect of alloying elements. Alloy 600 with 15% chromium, one of the earliest of the nickel–chromium alloys, can be thought of as the base for other alloys. Chromium imparts resistance to oxidizing environments and high-temperature strength. Increasing chromium to 30%, as in alloy 690, increases resistance to stress-corrosion cracking, nitric acid, steam, and oxidizing gases. Increasing chromium to 50% increases resistance to melting sulfates and vanadates found in fuel ash. High-temperature oxidation resistance is also improved by alloying with aluminum in conjunction with high chromium (e.g., alloy 601).

Without chromium, nickel by itself is used as a corrosion-resistant material in food processing and in high-temperature caustic and gaseous chlorine or chloride environments.

Of importance for aqueous reducing acids, oxidizing chloride environments, and seawater are alloy 625 and alloy C-276, which contain 9% and 16% molybdenum, respectively, and are among the most resistant alloys currently available. Low-level titanium and aluminum additions provide γ' strengthening while retaining good corrosion resistance, as in alloy X-750. Cobalt and other alloying element additions provide jet engine materials (superalloys) that combine high-temperature strength with resistance to gaseous oxidation and sulfidation.

Another technologically important group of materials are the higher-iron alloys, which were originally developed to conserve nickel and are often regarded as intermediate in performance and cost between nickel alloys and stainless steels. The prototype, alloy 800 (Fe/33% Ni/21% Cr), is a general purpose alloy with good high-temperature strength and resistance to steam and oxidizing or carburizing gases. Alloying with molybdenum and chromium, as in alloy 825 and alloy G, improves resistance to reducing acids and localized corrosion in chlorides.

Another important category is the nickel–copper alloys. At the higher-nickel end are the Monel alloys (30–45% Cu, balance Ni) used for corrosive chemicals such as hydrofluoric acid, and severe marine environments. At the higher-copper end are the cupronickels (10–30% Ni, balance Cu), which are widely used for marine applications because of their fouling resistance.

Nickel alloys exhibit high resistance to attack under nitriding conditions (e.g., in dissociated ammonia) and in chlorine or chloride gases. Corrosion in the latter at elevated temperatures proceeds by the formation and volatilization of chloride scales, and high-nickel contents are beneficial since nickel forms one of the least volatile chlorides. Conversely, in sulfidizing environments, high-nickel alloys without chromium can exhibit attack due to the formation of a low-melting-point $Ni-Ni_3Si_2$ eutectic. However high chromium contents appear to limit this form of attack.[5]

Friend explains corrosion reactions as wet or dry:[11]

The term wet corrosion usually refers to all forms of corrosive attack by aqueous solutions of electrolytes, which can range from pure water (a weak electrolyte) to aqueous solutions of acids or bases or of their salts, including neutral salts. It also includes natural environments such as the atmosphere, natural waters, soils, and others, irrespective or whether the metal is in contact with a condensed film or droplets of moisture or is completely immersed. Corrosion by aqueous environments is electrochemical in nature, assuming the presence of anodic and cathodic areas on the surface of the metal even though these areas may be so small as to be indistinguishable by experimental methods and the distance between them may be only of atomic dimensions.

The term dry corrosion implies the absence of water or an aqueous solution. It generally is applied to metal/gas or metal/vapor reactions involving gases such as oxygen, halogens, hydrogen sulfide, and sulfur vapor and even to "dry" steam at elevated temperatures. . . . High-temperature oxidation of metals has been considered to be an electrochemical phenomenon since it involves the diffusion of metal ions outward, or of reactant ions inward, through the corrosion product film, accompanied by a flow of electrons.

The decision to use a particular alloy in a commercial application is usually based on past corrosion experience and laboratory or field testing using test spools of candidate alloys. Most often weight loss is measured to rank various alloys; however, many service failures are due to localized attack such as pitting, crevice corrosion, intergranular corrosion, and stress-corrosion cracking, which must be measured by other means.

A number of investigations have shown the effect of nickel on the different forms of corrosion. Figure 5.2 shows the galvanic series of many alloys in flowing seawater. This series gives an indication of the rate of corrosion between different metals or alloys when they are electrically coupled in an electrolyte. The metal close to the active end of the chart will behave as an anode and corrode, and the metal closer to the noble end will act as a cathode and be protected. Increasing the nickel content will move an alloy more to the noble end of the series. There are galvanic series for other corrosive environments, and the film-forming characteristics of each material may change this series somewhat. Seawater is normally used as a rough guide to the relative positions of alloys in solution of good electrical conductivity such as mineral acids or salts.

Residual stresses from cold rolling or forming do not have any significant effect on the general corrosion rate. However, many low-nickel-containing steels are subject to stress-corrosion cracking in chloride-containing environments. Figure 5.3 from work by LaQue and Copson[12] shows that nickel–chromium and nickel–chromium–iron alloys containing about 45% Ni or more are immune from stress-corrosion cracking in boiling 42% magnesium chloride.[11]

When localized corrosion occurs in well-defined areas, such corrosion is commonly called *pitting attack*. This type of corrosion typically occurs when the protective film is broken or is penetrated by a chloride–iron and the film is unable to repair itself quickly. The addition of chromium and particularly molybdenum makes nickel-base alloys less susceptible to pitting attack, as shown in Fig. 5.4, which shows a very good relationship between critical[11] pitting temperature in a salt solution. Along with significant increases in chromium and/or molybdenum, the iron content must be replaced with more nickel in wrought alloys to resist the formation of embrittling phases.[12,13]

Air *oxidation* at moderately high temperatures will form an intermediate subsurface layer between the alloy and gas quickly. Alloying of the base alloy can affect this subscale oxide and, therefore, control the rate of oxidation. At constant temperature, the resistance to oxidation is largely a function of chromium content. Early work by Eiselstein and Skinner has shown that nickel content is very beneficial under cyclic temperature conditions as shown in Fig. 5.5.[14]

5.4 FABRICATION

The excellent ductility and malleability of nickel and nickel-base alloys in the annealed condition make them adaptable to virtually all methods of cold fabrication. As other engineering properties vary within this group of alloys, formability ranges from moderately easy to difficult in relation to other materials.

5.4.1 Resistance to Deformation

Resistance to deformation, usually expressed in terms of hardness or yield strength, is a primary consideration in cold forming. Deformation resistance is moderately low for the nickel and nickel–copper systems and moderately high for the nickel–chromium and nickel–iron–chromium systems. However, when properly annealed, even the high-strength alloys have a substantial range between yield and ultimate tensile strength. This range is the plastic region of the material and all cold forming is accomplished within the limits of this region. Hence, the high-strength alloys require only stronger tooling and more powerful equipment for successful cold forming. Nominal tensile properties and hardnesses are given in Table 5.2.

5.4.2 Strain Hardening

A universal characteristic of the high-nickel alloys is that they have face-centered-cubic crystallographic structures, and, consequently, are subject to rapid strain hardening. This characteristic is used to advantage in increasing the room-temperature tensile properties and hardness of alloys that otherwise would have low mechanical strength, or in adding strength to those alloys that are hardened by a precipitation heat treatment. Because of this increased strength, large reductions can be made without rupture of the material. However, the number of reductions in a forming sequence will be limited before annealing is required, and the percentage reduction in each successive operation must be reduced.

Since strain hardening is related to the solid-solution strengthening of alloying elements, the strain-hardening rate generally increases with the complexity of the alloy. Accordingly, strain-hardening rates range from moderately low for nickel and nickel–copper alloys to moderately high for nickel–chromium and nickel–iron–chromium alloys. Similarly, the age-hardenable alloys have higher strain-hardening rates than their solid-solution equivalents. Figure 5.6 compares the strain-hardening rates of some nickel alloys with those of other materials as shown by the increase in hardness with increasing cold reduction.

Laboratory tests have indicated that the shear strength of the high-nickel alloys in double shear averages about 65% of the ultimate tensile strength (see Table 5.4). These values, however, were obtained under essentially static conditions using laboratory testing equipment having sharp edges

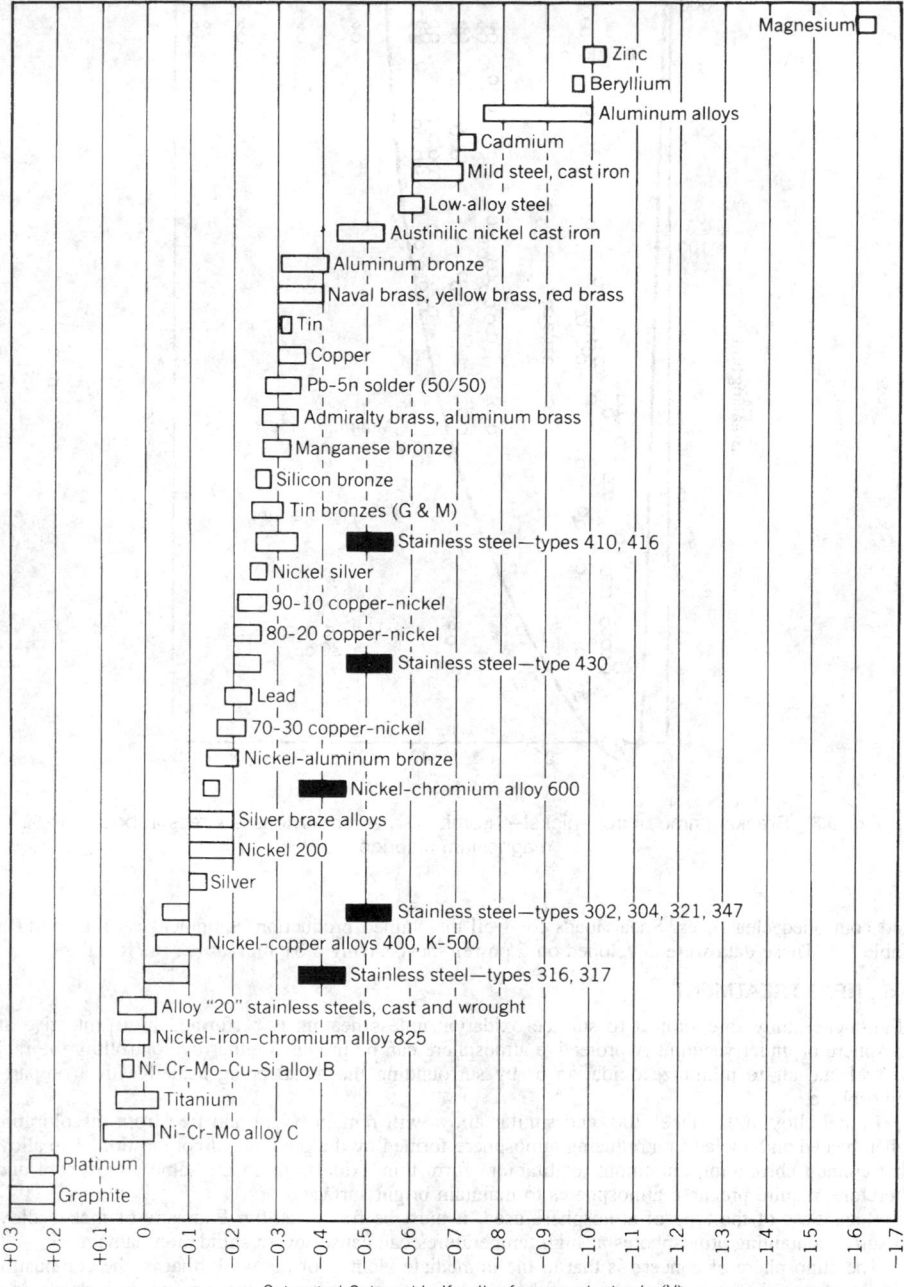

Fig. 5.2 Corrosion potentials in flowing seawater (8-13 ft/sec), temperature range 50-80°F. Alloys are listed in the order of the potential they exhibit in flowing seawater. Certain alloys, indicated by solid boxes, in low velocity or poorly aerated water, and at shielded areas, may become active and exhibit a potential near −0.5 V.

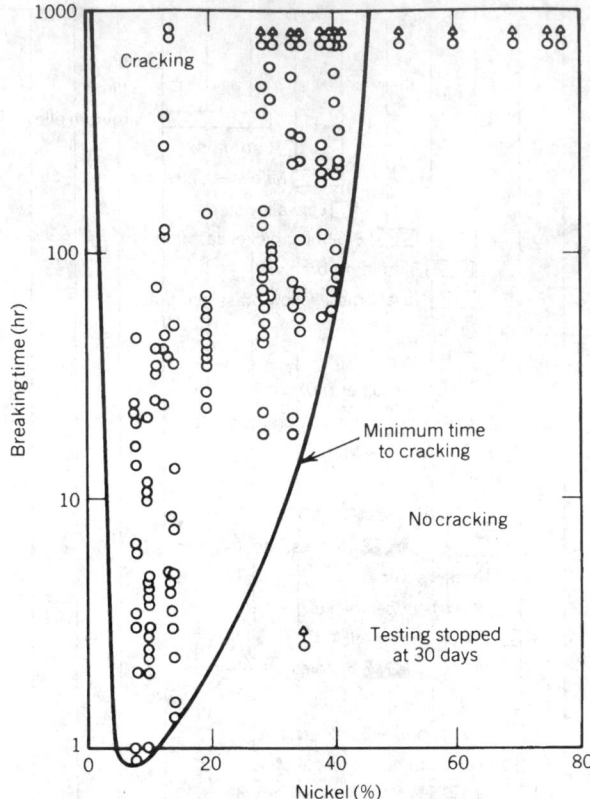

Fig. 5.3 Breaking time of iron–nickel–chromium wires under tensile stress in boiling 42% magnesium chloride.

and controlled clearances. Shear loads for well-maintained production equipment can be found in Table 5.5. These data were developed on a power shear having a 31 mm/m (⅜ in./ft) rake.

5.5 HEAT TREATMENT

High-nickel alloys are subject to surface oxidation unless heating is performed in a protective atmosphere or under vacuum. A protective atmosphere can be provided either by controlling the ratio of fuel and air to minimize oxidation or by surrounding the metal being heated with a prepared atmosphere.

Monel alloy 400, Nickel 200, and similar alloys will remain bright and free from discoloration when heated and cooled in a reducing atmosphere formed by the products of combustion. The alloys that contain chromium, aluminum, or titanium form thin oxide films in the same atmosphere and, therefore, require prepared atmospheres to maintain bright surfaces.

Regardless of the type of atmosphere used, it must be free of sulfur. Exposure of nickel alloys to sulfur-containing atmospheres at high temperatures can cause severe sulfidation damage.

The atmosphere of concern is that in the immediate vicinity of the work, that is, the combustion gases that actually contact the surface of the metal. The true condition of the atmosphere is determined by analyzing gas samples taken at various points about the metal surface.

Furnace atmospheres can be checked for excessive sulfur by heating a small test piece of the material, for example, 13 mm (½ in.) diameter rod or 13 mm × 25 mm (½ in. × 1 in.) flat bar, to the required temperature and holding it at temperature for 10–15 min. The piece is then air cooled or water quenched and bent through 180° flat on itself. If heating conditions are correct, there will be no evidence of cracking.

5.5.1 Reducing Atmosphere

The most common protective atmosphere used in heating the nickel alloys is that provided by controlling the ratio between the fuel and air supplied to the burners. A suitable reducing condition can

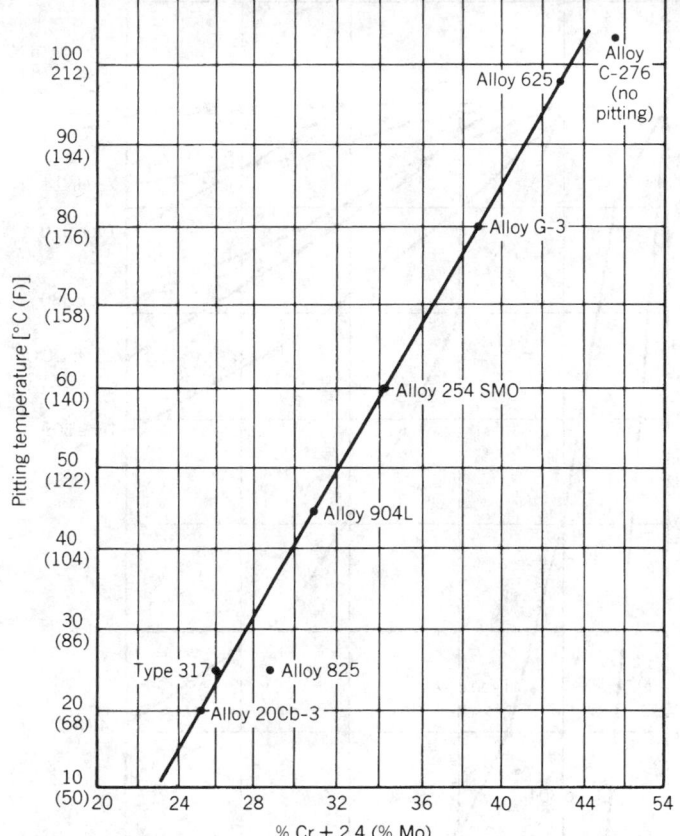

Fig. 5.4 Critical temperature for pitting in 4% NaCl + 1% $Fe_2 (SO_4)_3$ + 0.01 M HCl versus composition for Fe–Ni–Cr–Mo alloys.

be obtained by using a slight excess of fuel so that the products of combustion contain at least 4%, preferably 6%, of carbon monoxide plus hydrogen. The atmosphere should not be permitted to alternate from reducing to oxidizing; only a slight excess of fuel over air is needed.

It is important that combustion take place before the mixture of fuel and air comes into contact with the work, otherwise the metal may be embrittled. To ensure proper combustion, ample space should be provided to burn the fuel completely before the hot gases contact the work. Direct impingement of the flame can cause cracking.

5.5.2 Prepared Atmosphere

Various prepared atmospheres can be introduced into the heating and cooling chambers of furnaces to prevent oxidation of nickel alloys. Although these atmospheres can be added to the products of combustion in a directly fired furnace, they are more commonly used with indirectly heated equipment. Prepared protective atmospheres suitable for use with the nickel alloys include dried hydrogen, dried nitrogen, dried argon or any other inert gas, dissociated ammonia, and cracked or partially reacted natural gas. For the protection of pure nickel and nickel–copper alloys, cracked natural gas should be limited to a dew point of −1 to 4°C (30 to 40°F).

Figure 5.7 indicates that at a temperature of 1093°C (2000°F), a hydrogen dew point of less than −30°C (−20°F) is required to reduce chromium oxide to chromium; at 815°C (1500°F) the dew point must be below −50°C (−60°F). The values were derived from the thermodynamic relationships of pure metals with their oxides at equilibrium, and should be used only as a guide to the behavior of complex alloys under nonequilibrium conditions. However, these curves have shown a close correlation with practical experience. For example, Inconel alloy 600 and Incoloy alloy 800 are successfully bright-annealed in hydrogen having a dew point of −35 to −40°C (−30 to −40°F).

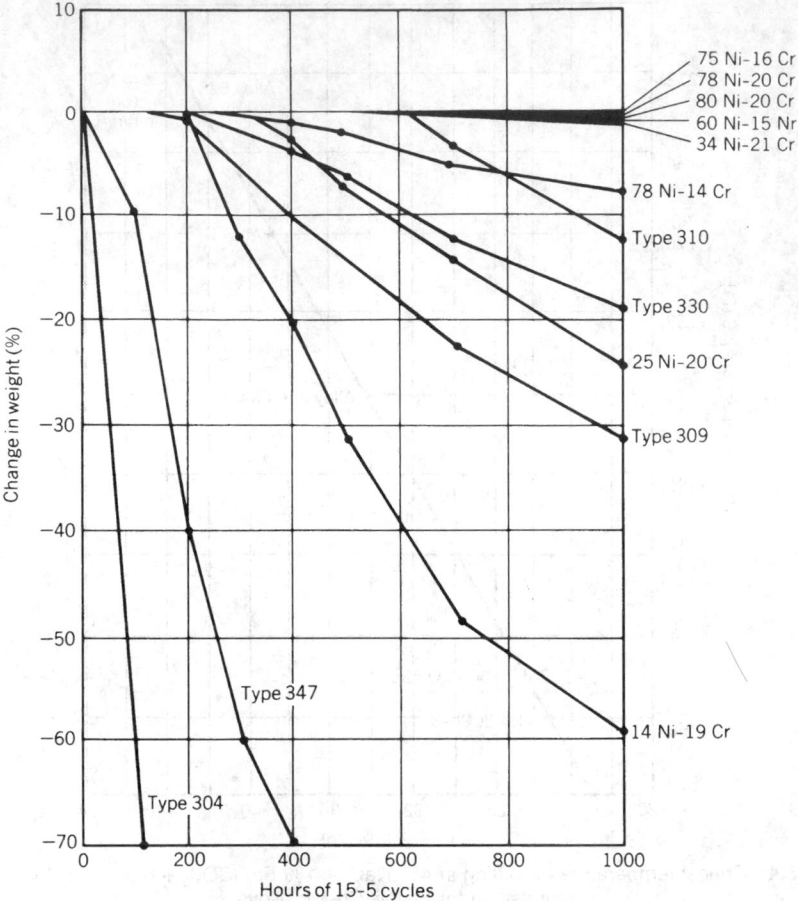

Fig. 5.5   Effect of nickel content on air oxidation of alloys. Each cycle consisted of 15 min at 1800°F followed by a 5-min air cooling.

As indicated in Fig. 5.7, lower dew points are required as the temperature is lowered. To minimize oxidation during cooling, the chromium-containing alloys must be cooled rapidly in a protective atmosphere.

5.6   WELDING

Cleanliness is the single most important requirement for successful welded joints in nickel alloys. At high temperatures, nickel and its alloys are susceptible to embrittlement by sulfur, phosphorus, lead, and other low-melting-point substances. Such substances are often present in materials used in normal manufacturing/fabrication processes; some examples are grease, oil, paint, cutting fluids, marking crayons and inks, processing chemicals, machine lubricants, and temperature-indicating sticks, pellets, or lacquers. Since it is frequently impractical to avoid the use of these materials during processing and fabrication of the alloys, it is mandatory that the metal be thoroughly cleaned prior to any welding operation or other high-temperature exposure.

Before maintenance welding is done on high-nickel alloys that have been in service, products of corrosion and other foreign materials must be removed from the vicinity of the weld. Clean, bright base metal should extend 50–75 mm (2–3 in.) from the joint on both sides of the material. This prevents embrittlement by alloying of corrosion products during the welding process. Cleaning can be done mechanically by grinding with a fine grit wheel or disk, or chemically by pickling.

5.7   MACHINING

Nickel and nickel-base alloys can be machined by the same techniques used for iron-base alloys. However, higher loads will be imparted to the tooling requiring heavy-duty equipment to withstand

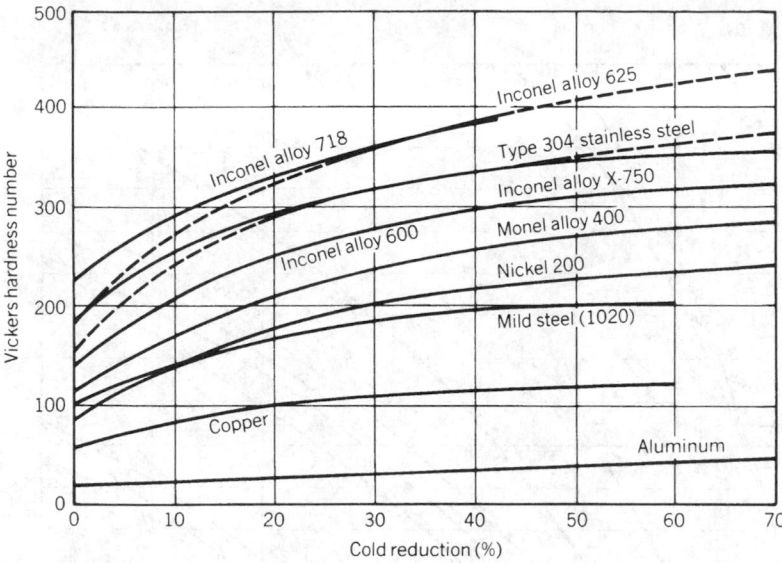

Fig. 5.6 Effect of cold work on hardness.

Table 5.4 Strength in Double Shear of Nickel and Nickel Alloys

Alloy	Condition	Shear Strength (ksi)[a]	Tensile Strength (ksi)	Hardness
Nickel 200	Annealed	52	68	46 Rb
	Half-hard	58	79	84 Rb
	Full-hard	75	121	100 Rb
Monel alloy 400	Hot-rolled, annealed	48	73	65 Rb
	Cold-rolled, annealed	49	76	60 Rb
Inconel alloy 600	Annealed	60	85	71 Rb
	Half-hard	66	98	98 Rb
	Full-Hard	82	152	31 Rc
Inconel alloy X-750	Age-hardened[b]	112	171	36 Re

[a] MPa = ksi × 6.895.
[b] Mill-annealed and aged 1300°F (750°C)/20 hr.

Table 5.5 Shear Load for Power Shearing of 6.35-mm (0.250-in.) Guage Annealed Nickel Alloys at 31 mm/m ($^3/_8$ in./ft.) Rake as Compared with Mild Steel

Alloy	Tensile Strength (ksi)[a]	Hardness (Rb)	Shear Load (lb)[b]	Shear Load in Percent of Same Gauge of Mild Steel
Nickel 200	60	60	61,000	200
Monel alloy 400	77	75	66,000	210
Inconel alloy 600	92	79	51,000	160
Inconel alloy 625	124	95	55,000	180
Inconel alloy 718	121	98	50,000	160
Inconel alloy X-750	111	88	57,000	180
Mild steel	50	60	31,000	100

[a] MPa = ksi × 6.895.
[b] kg = lb × 0.4536.

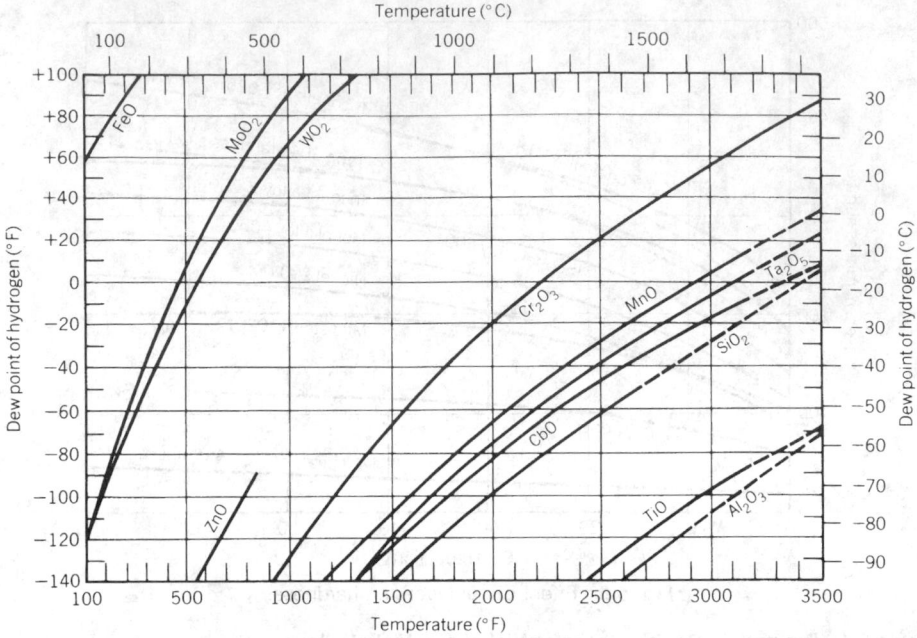

Fig. 5.7 Metal/metal oxide equilibria in hydrogen atmospheres.

Table 5.6   Registered Trademarks of Producer Company

Trademark	Owner
Duranickel	Inco family of companies
Hastelloy	Haynes International, Inc.
Incoloy	Inco family of companies
Inconel	Inco family of companies
MAR-M	Martin Marietta Corp.
Monel	Inco family of companies
Nilo	Inco family of companies
Ni–Span–C	Inco family of companies
Permanickel	Inco family of companies
Pyromet	Carpenter Technology Corp.
Rene	General Electric Co.
Rene' 41	Allvac Metals Corp.
Udimet	Special Metals Corp.
Waspaloy	United Aircraft Corp.

the load and coolants to dissipate the heat generated. The cutting tool edge must be maintained sharp and have the proper geometry.

5.8   CLOSURE

There has been a vast amount of nickel-alloy developments since the 1950 edition of *Kent's Mechanical Engineer's Handbook*. It has not been possible to give the composition and discuss each commercial alloy and, therefore, one should refer to publications like Refs. 6–8 for alloy listings, which are revised periodically to include the latest alloys available. (See Table 5.6 for the producer companies of some of the alloys mentioned in this chapter.)

REFERENCES

1. Joseph R. Boldt, Jr., *The Winning of Nickel*, Van Nostrand, New York, 1967.
2. *Nickel and Its Alloys*, NBS Monograph 106, May, 1968.
3. *Kent's Mechanical Engineer's Handbook*, 1950 edition, pp. 4–50 to 4–60.
4. Huntington Alloys, Inc., *Alloy Handbook*, and *Bulletins*.
5. Inco internal communication by A.J. Sedriks.
6. *Alloy Digest*, Engineering Alloy Digest, Inc., 1983.
7. *Aerospace Structural Metals Handbook*, 1983.
8. *Materials and Processing Databook*, 1983 Metals Progress.
9. J. S. Benjamin, *Met. Trans. AIME* **1**, 2943 (1970).
10. J. P. Morse and J. S. Benjamin, *J. Met.* **29** (12), 9 (1977).
11. Wayne Z. Friend, *Corrosion of Nickel and Nickel-Base Alloy*, Wiley, New York, 1980.
12. F. L. LaQue and H. R. Copson, *Corrosion Resistance of Metals and Alloys*, 2nd ed., Reinhold, New York, 1963.
13. J. Kolts et al., "Highly Alloyed Austenitic Materials for Corrosion Service," *Metal Prog.*, 25–36 (September, 1983).
14. *High Temperature Corrosion in Refinery and Petrochemical Service*, Inco Publication, 1960.

CHAPTER 6

TITANIUM AND ITS ALLOYS

Donald Knittel
James B. C. Wu
Cabot Corporation
Kokomo, Indiana

6.1	INTRODUCTION	91		6.5.2 Drawing	100
				6.5.3 Bending	104
6.2	ALLOYS	92		6.5.4 Cutting and Grinding	104
	6.2.1 Aerospace Alloys	94		6.5.5 Welding	104
	6.2.2 Nonaerospace Alloys	95			
	6.2.3 Other Alloys	96	6.6	SPECIFICATIONS, STANDARDS, AND QUALITY CONTROL	105
6.3	PHYSICAL PROPERTIES	96			
			6.7	HEALTH AND SAFETY FACTORS	107
6.4	CORROSION RESISTANCE	97			
6.5	FABRICATION	98	6.8	USES	107
	6.5.1 Boiler Code	98			

6.1 INTRODUCTION

Titanium was first identified as a constituent of the earth's crust in the late 1700s. In 1790, William Gregor, an English clergyman and mineralogist, discovered a black magnetic sand (ilmenite), which he called menaccanite after his local parish. In 1795, a German chemist found that a Hungarian mineral, rutile, was the oxide of a new element he called titan, after the mythical Titans of ancient Greece. In the early 1900s, a sulfate purification process was developed to commercially obtain high-purity TiO_2 for the pigment industry, and titanium pigment became available in both the United States and Europe. During this period, titanium was also used as an alloying element in irons and steels. In 1910, 99.5% pure titanium metal was produced at General Electric from titanium tetrachloride and sodium in an evacuated steel container. Since the metal did not have the desired properties, further work was discouraged. However, this reaction formed the basis for the commercial sodium reduction process. In the 1920s, ductile titanium was prepared with an iodide dissociation method combined with Hunter's sodium reduction process.

In the early 1930s, a magnesium vacuum reduction process was developed for reduction of titanium tetrachloride to metal. Based on this process, the U.S. Bureau of Mines (BOM) initiated a program in 1940 to develop commercial production. Some years later, the BOM publicized its work on titanium and made samples available to the industrial community. By 1948, the BOM produced batch sizes of 104 kg. In the same year, E. I. du Pont de Nemours & Co., Inc., announced commercial availability of titanium, and the modern titanium metals industry began.[1]

By the mid-1950s, this new metals industry had become well established, with six producers, two other companies with tentative production plans, and more than 25 institutions engaged in research projects. Titanium, termed the wonder metal, was billed as the successor to aluminum and stainless steels. When, in the 1950s, the DOD (titanium's most staunch supporter) shifted emphasis from aircraft to missiles, the demand for titanium sharply declined. Only two of the original titanium metal plants are still in use, the Titanium Metals Corporation of America's (TMCA) plant in Henderson,

Reprinted with additions from *Kirk–Othmer Encyclopedia of Chemical Technology*, 3rd ed., Wiley, New York, 1983, Vol. 23, by permission of the publisher.

Mechanical Engineers' Handbook, 2nd ed., Edited by Myer Kutz.
ISBN 0-471-13007-9 © 1998 John Wiley & Sons, Inc.

Nevada, and National Distillers & Chemical Corporation's two-stage sodium reduction plant built in the late 1950s at Ashtabula, Ohio, which now houses the sponge production facility for RMI Corporation (formerly Reactor Metals, Inc.).

Overoptimism followed by disappointment has characterized the titanium-metals industry. In the late 1960s, the future again appeared bright. Supersonic transports and desalination plants were intended to use large amounts of titanium. Oregon Metallurgical Corporation, a titanium melter, decided at that time to become a fully integrated producer (i.e., from raw material to mill products). However, the supersonic transports and the desalination industry did not grow as expected. Nevertheless, in the late 1970s and early 1980s, the titanium-metal demand again exceeded capacity and both the United States and Japan expanded capacities. This growth was stimulated by greater acceptance of titanium in the chemical-process industry, power-industry requirements for seawater cooling, and commercial and military aircraft demands. However, with the economic recession of 1981–1983, the demand dropped well below capacity and the industry was again faced with hard times.

6.2 ALLOYS

Titanium alloy systems have been studied extensively. A single company evaluated over 3000 compositions in 8 years. Alloy development has been aimed at elevated-temperature aerospace applications, strength for structural applications, and aqueous corosion resistance. The principal effort has been in aerospace applications to replace nickel- and cobalt-base alloys in the 500–900°C ranges. To date, titanium alloys have replaced steel in the 200–500°C range. The useful strength and corrosion-resistance temperature limit is ~550°C.

The addition of alloying elements alters the α-β transformation temperature. Elements that raise the transformation temperature are called α stabilizers; elements that depress the transformation temperature are called β stabilizers; the latter are divided into β-isomorphous and β-eutectoid types. The β-isomorphous elements have limited α solubility, and increasing additions of these elements progressively depresses the transformation temperature. The β-eutectoid elements have restricted beta solubility and form intermetallic compounds by eutectoid decomposition of the β phase. The binary phase diagram illustrating these three types of alloy systems is shown in Fig. 6.1

The important α-stabilizing alloying elements include aluminum, tin, zirconium, and the interstitial alloying elements (i.e., elements that do not occupy lattice positions) oxygen, nitrogen, and carbon. Small quantities of interstitial alloying elements, generally considered to be impurities, have a very great effect on strength and ultimately embrittle the titanium at room temperature.[3] The effects of oxygen, nitrogen and carbon on the ultimate tensile properties and elongation are shown in Table 6.1. These elements are always present and are difficult to control. Nitrogen has the greatest effect, and commercial alloys specify its limit to be less than 0.05 wt %. It may also be present as nitride (TiN) inclusions, which are detrimental to critical aerospace structural applications. Oxygen additions increase strength and serve to identify several commercial grades. This strengthening effect diminishes at elevated temperatures and under creep conditions at room temperature. For cryogenic service, low oxygen content is specified (<1300 ppm) because high concentrations of interstitial impurities increase sensitivity to cracking, cold brittleness, and fracture temperatures. Alloys with low interstitial

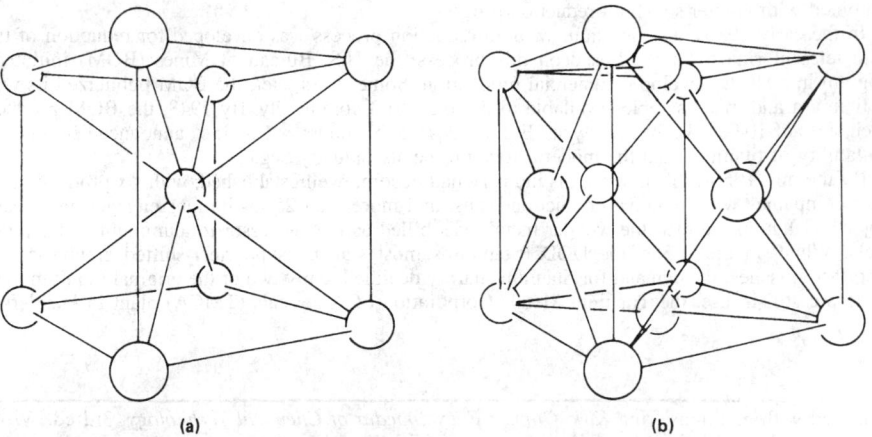

(a)   (b)

Fig. 6.1 The effect of alloying elements on the phase diagram of titanium: (a) α-stabilized system, (b) β-isomorphous system, and (c) β-eutectoid system.[2]

Table 6.1   Effects of O, N, and C on the Ultimate Tensile Strength[2,3]

Concentration of Impurity, wt%	Oxygen[b,c]		Nitrogen[b,c]		Carbon[b,c]	
	UT MPa[d]	Elong., %	UT MPa[d]	Elong., %	UT MPa[d]	Elong., %
0.025	330	37	380	35	310	40
0.05	365	35	460	28	330	39
0.1	440	30	550	20	370	36
0.15	490	27	630	15	415	32
0.2	545	25	700	13	450	26
0.3	640	23	embrittles		500	21
0.5	790	18			520	18
0.7	930	8			525	17

[a] Tests were conducted using titanium produced by the iodide process.
[b] UT = ultimate tensile stress.
[c] Elongation on 2.54 cm.
[d] To convert MPa to psi, multiply by 145.

content are identified as ELI (extra-low interstitials) after the alloy name. Carbon does not affect strength at concentration above 0.25 wt % because carbides (TiC) are formed. Carbon content is usually specified at 0.08 wt % max.[4]

The most important alloying element is aluminum, an α stabilizer. It is not expensive, and its atomic weight is less than that of titanium; hence, aluminum additions lower the density. The mechanical strength of titanium can be increased considerably by aluminum additions. Even though the solubility range of aluminum extends to 27 wt %, above 7.5 wt % the alloy becomes too difficult to fabricate and embrittles. The embrittlement is caused by a coherently ordered phase based on Ti_3Al. Other α-stabilizing elements also cause phase ordering. An empirical relationship below which ordering does not occur is[5]

$$\text{wt \% Al} + \frac{\text{wt \% Sn}}{3} + \frac{\text{wt \% Zr}}{6} + 10 \times \text{wt \% O} \le 9$$

The important β-stabilizing alloying elements are the bcc elements vanadium, molybdenum, tantalum, and niobium of the β-isomorphous type and manganese, iron, chromium, cobalt, nickel, copper, and silicon of the β-eutectoid type. The β-eutectoid elements arranged in order of increasing tendency to form compounds are shown in Table 6.2. The elements copper, silicon, nickel, and cobalt are termed active eutectoid forms because of a rapid decomposition of β to α and a compound. The other elements in Table 6.2 are sluggish in their eutectoid reactions.

Alloys of the β type respond to heat treatment, are characterized by higher density than pure titanium, and are easily fabricated. The purpose of β alloying is to form an all-β-phase alloy with commercially useful qualities, form alloys with duplex α and β structure to enhance heat-treatment

Table 6.2 β-Eutectoid Elements in Order of Increasing Tendency to Form Compounds[2,6]

Element	Eutectoid Composition, wt %	Eutectoid Temperature, °C	Composition for β Retention on Quenching, wt %
manganese	20	550	6.5
iron	15	600	4.0
chromium	15	675	8.0
cobalt	9	685	7.0
nickel	7	770	8.0
copper	7	790	13.0
silicon	0.9	860	

response (i.e., changing the α and β volume ratio), or use β-eutectoid elements for intermetallic hardnening. The most important commercial β-alloying element is vanadium.

6.2.1 Aerospace Alloys

The alloys of titanium for aerospace use can be divided into three categories: an all-α structure, a mixed α-β structure, and an all-β structure. The α-β structure alloys are further divided into near-α alloys (<2% β stabilizers). Most of the approximately 100 commercially available alloys (approximately 30 in the United States, 40 in the USSR, and 10 in Europe and Japan) are of the α-β structure type.[7] Some of these, produced in the United States, are given in Table 6.3 along with some wrought properties.[8–10] The most important commercial alloy is Ti–6 Al–4 V, an α-β alloy with a good combination of strength and ductility. It can be age-hardened and has moderate ductility, and an excellent record of successful applications. It is mostly used for compressor blades and disks in aircraft gas-turbine engines, and also in lower-temperature engine applications such as rotating disks and fans. It is also used for rocket-motor cases, structural forgings, steam-turbine blades, and cryogenic parts for which ELI grades are usually specified.

Other commercially important α-β alloys are Ti–3 Al–2.5 V, Ti–6 Al–6 V–2 Sn, and Ti–10 V–2 Fe–3 Al (see Table 6.3). As a group, these alloys have good strength, moderate ductility, and can be age-hardened.[10,11] Weldability becomes more difficult with increasing β constituents, and fabrication of strip, foil, sheet, and tubing may be difficult. Temperature tolerances are lower than those of the α or near-α alloys. The alloy Ti–3 Al–2.5 V (called one-half Ti–6 Al–4 V) is easier to fabricate than Ti–6 Al–4 V and is used primarily as seamless aircraft-hydraulic tubing. The alloy Ti–6 Al–6 V–2 Sn is used for some aircraft forgings because it has a higher strength than Ti–6 Al–4 V. The alloy Ti–10 V–2 Fe–3 Al is easier to forge at lower temperatures than Ti–6 Al–4 V because it contains more β-alloying constituents and has good fracture toughness. This alloy can be hardened to high strengths [1.24–1.38 GPa or $(1.8-2) \times 10^5$ psi] and is expected to be used as forgings for airframe structures to replace steel below temperatures of 300°C[12],

Table 6.3 Properties, Specifications and Applications of Wrought Titanium Alloys[2,9,10]

Nominal Composition, wt %	CAS Registry No.	ASTM B-265	CLTE[a], μm/(m · K) 21-100°C	CLTE[a], μm/(m · K) 21-538°C	Average Physical Properties Modulus of Elasticity[b], GPa[c]	Modulus of Rigidity[b], GPa[c]	Poisson's[b] Ratio	Density, g/cm³	Condition
commercially pure									
99.5 Ti		grade 1	8.7	9.8	102	39	0.34	4.5	annealed
99.2 Ti		grade 2	8.7	9.8	102	39	0.34	4.5	annealed
99.1 Ti		grade 3	8.7	9.8	103	39	0.34	4.5	annealed
99.0 Ti		grade 4	8.7	9.8	104	39	0.34	4.5	annealed
99.2 Ti[g]		grade 7	8.7	9.8	102	39	0.34	4.5	annealed
98.9 Ti[h]								4.5	annealed
Ti–5 Al–2.5 Sn[i]	[11109-19-6]	grade 6	9.4	9.6	110			4.5	annealed
Ti–8 Al–1 Mo, 1 V[i]	[39303-55-4]		8.5	10.1	124	47	0.32	4.4	duplex annealed
Ti–6 Al–2 Sn 4 Zr–2 Mo[i]	[11109-15-2]		7.8	8.1	114			4.5	annealed
Ti–3 Al–2.5 V[i]	[11109-23-2]		9.6	9.9	107			4.5	annealed
Ti–6 Al–4 V[i]	[12743-70-3]	grade 5	8.7	9.6	114	42	0.342	4.4	annealed
Ti–6 Al–6 V, 2 Sn[i]	[12606-77-8]		9.0	9.6	110			4.5	annealed
Ti–10 V–2 Fe, 3 Al[i]	[51809-47-3]				112			4.6	solution and age

[a] CLTE = coefficient of linear thermal expansion.
[b] Room temperature.
[c] To convert GPa to psi, multiply by 145,000.
[d] To convert MPa to psi, multiply by 145.

The only α alloy of commercial importance is Ti–5 Al–2.5 Sn. It is weldable, has good elevated-temperature stability, and good oxidation resistance to about 600°C. It is used for forgings and sheet-metal parts such as aircraft-engine compressor cases because of weldability.

The commercially important near-α alloys are Ti–8 Al–1 Mo–1 V and Ti–6 Al–2 Sn–4 Zr–2 Mo. They exhibit good creep resistance and the excellent weldability and high strength of α alloys; the temperature limit is ~500°C. Alloy Ti–8 Al–1 Mo–1 V is used for compressor blades because of its high elastic modules and creep resistance; however, it may suffer from ordering embrittlement. Alloy Ti–6 Al–2 Sn–4 Zr–2 Mo is also used for blades and disks in aircraft engines. The service temperature limit of 470 °C is ~70°C higher than that of Ti–8 Al–1 Mo–1 V.[5]

Commercialization of β alloys has not been very successful. Even though alloys with high strength [up to 1.5 GPa (217,500 psi)] were made, they suffered from intermetallic and ω-phase embrittlement. These alloys are metallurgically unstable and have little practical use above 250°C. They are fabricable but welds are not ductile. This alloy type is used in the cold-drawn or cold-rolled condition and finds application in spring manufacture (alloy Ti–13 V–11 Cr–3 Al).[13] There is one commercially available alloy of the β-eutectoid type (Ti–2.5 Cu) that uses a true precipitation-hardening mechanism to increase strength. The precipitate is Ti_2Cu. This alloy is only slightly heat treatable; it is used in engine castings and flanges.[5]

6.2.2 Nonaerospace Alloys

The nonaerospace alloys are used primarily in industrial applications. The four grades (ASTM grade 1 through grade 4) differ primarily in oxygen and iron content (see Table 6.4). ASTM grade 1 has the highest purity and the lowest strength (strength is controlled by impurities). The two other alloys of this group are ASTM grade 7, Ti–0.2 Pd, and ASTM grade 12, Ti–0.8 Ni–0.3 Mo. The alloys in this group are distinguished by excellent weldability, formability, and corrosion resistance. The strength, however, is not maintained at elevated temperatures (see Table 6.3). The primary use of alloys in this group is in industrial-processing equipment (i.e., tanks, heat exchangers, pumps, elec-

Average Mechanical Properties										
Room Temperature					Extreme Temperatures				Charpy	
Tensile Strength, MPa[d]	Yield Strength, MPa[d]	Elonga-tion, %	Reduction in Area, %	Test Temperature, °C	Tensile Strength, MPa[d]	Yield Strength, MPa[d]	Elonga-tion, %	Reduction in Area, %	Impact Strength, J/m[e]	Hardness[f]
331	241	30	55	315	152	97	32	80		HB 120
434	346	28	50	315	193	117	35	75	43	HB 200
517	448	25	45	315	234	138	34	75	38	HB 225
662	586	20	40	315	310	172	25	70	20	HB 265
434	346	28	50	315	186	110	37	75	43	HB 200
517	448	25	42	315	324	207	32			
862	807	16	40	315	565	448	18	45	26	HRC 36
1000	952	15	28	540	621	517	25	55	33	HRC 35
979	896	15	35	540	648	490	26	60		HRC 32
690	586	20		315	483	345	25			
993	924	14	30	540	531	427	35	50	19	HRC 36
1069	1000	14	30	315	931	807	18	42	18	HRC 38
1276	1200	10	19	315	1103	979	13	42		

[e] To convert J/m to ft-lb/in., divide by 53.38.
[f] HB = Brinnell, HRC = Rockwell (C-scale).
[g] Also contains 0.2 Pd.
[h] Also contains 0.8 Ni and 0.3 Mo.
[i] Numerical designations = wt % of element.

Table 6.4   ASTM Requirements for Different Titanium Grades[2,4]

Element	Grade 1	Grade 2	Grade 3	Grade 4	Grade 7	Grade 12
nitrogen, max	0.03	0.03	0.05	0.05	0.03	0.03
carbon, max	0.10	0.10	0.10	0.10	0.10	0.08
hydrogen, max	0.015	0.015	0.015	0.015	0.015	0.015
iron, max	0.20	0.30	0.30	0.50	0.30	0.30
oxygen, max	0.18	0.25	0.35	0.40	0.25	0.25
palladium					0.21–0.25	
molybdenum						0.2–0.4
nickel						0.6–0.9
residuals, max						
each	0.1	0.1	0.1	0.1	0.1	0.1
total	0.4	0.4	0.4	0.4	0.4	0.4
titanium	remainder	remainder	remainder	remainder	remainder	remainder

trodes, etc.), even though there is some use in airframes and aircraft engines. The ASTM grade 1 is used where higher purity is desired, for example, as weld wire for grade 2 fabrication and as sheet for explosive bonding to steel. Grade 1 is manufactured from high-purity sponge. The ASTM grade 2 is the most commonly used grade of commercially pure titanium. The chemistry for this grade is easy to meet with most sponge. The ASTM grades 3 and 4 are higher strength versions of grade 2; grades 7 and 12 have better corrosion resistance than grade 2 in reducing acids and acid chlorides. However, grade 7 is expensive and grade 12 is not readily available.

6.2.3   Other Alloys

Other alloying ranges include the aluminides (TiAl and Ti_3Al), the superconducting alloys (Ti–Nb type), the shape-memory alloys (Ni–Ti type), and the hydrogen-storage alloys (Fe–Ti). The aluminides TiAl and Ti_3Al have excellent high-temperature strengths, comparable to those of nickel- and cobalt-base alloys, with less than half the density. These alloys exhibit ultimate strengths of 1 GPa (145,000 psi), and 800 MPa (116,000 psi) yield, respectively, 4–5% elongation, and 7% reduction in area. Strengths are maintained to 800–900°C. The modulus of elasticity is high [125–165 GPa, (18–24) $\times$ 10^6 psi], and oxidation resistance is good.[8] The aluminides are intended for both static and rotating parts in the turbine section of gas-turbine aircraft engines.

Titanium alloyed with niobium exhibits superconductivity, and a lack of electrical resistance below 10K. Composition ranges from 25 to 50 wt % Ti. These alloys are β-phase alloys with superconducting transitional temperatures at $\sim$10 K. Their use is of interest for power generation, propulsion devices, fusion research, and electronic devices.[14]

Titanium alloyed with nickel exhibits a memory effect, that is, the metal form switches from one specific shape to another in response to temperature changes. The group of Ti–Ni alloys (nitinol) was developed by the Navy in the early 1960s for F-14 fighter jets. The compositions are typically Ti with 55 wt % Ni. The transition temperature ranges from –100°C to >100°C and is controlled by additional alloying elements. These alloys are of interest for thermostats, recapture of waste heat, pipe joining, etc. The nitinols have not been extensively used because of high price and fabrication difficulties.[15]

Titanium alloyed with iron is a leading candidate for solid-hybride energy-storage material for automotive fuel. The hydride $FeTiH_2$ absorbs and releases hydrogen at low temperatures. This hydride stores 0.9 kW·hr/kg. To provide the energy equivalent to a tank of gasoline would require about 800 kg $FeTiH_2$.[8]

6.3   PHYSICAL PROPERTIES

The physical properties of titanium are given in Table 6.5. The most important physical property of titanium from a commercial viewpoint is the ratio of its strength [ultimate strength > 690 MPa (100,000 psi)] at a density of 4.507 g/cm^3. Titanium alloys have a higher yield strength-to-density rating between –200 and 540°C than either aluminum alloys or steel.[6,16] Titanium alloys can be made with strength equivalent to high-strength steel, yet with density $\sim$60% that of iron alloys. At ambient temperatures, titanium's strength-to-weight ratio is equal to that of magnesium, one and one-half times greater than that of aluminum, two times greater than that of stainless steel, and three times greater than that of nickel. Alloys of titanium have much higher strength-to-weight ratios than alloys

Table 6.5 Physical Properties of Titanium[2]

Property	Value
melting point, °C	1668 ± 5
boiling point, °C	3260
density, g/cm³	
α phase at 20°C	4.507
β phase at 885°C	4.35
allotropic transformation, °C	882.5
latent heat of fusion, kJ/kg[a]	440
latent heat of transition, kJ/kg[a]	91.8
latent heat of vaporization, MJ/kg[a]	9.83
entropy at 25°C, J/mol[a]	30.3
thermal expansion coefficient at 20°C per °C	8.41×10^{-6}
thermal conductivity at 25°C, W/(m·K)	21.9
emissivity	9.43
electrical resistivity at 20°C, nΩ·m	420
magnetic susceptibility, mks	180×10^{-6}
modulus of elasticity, GPa[b]	
tension	ca 101
compression	103
shear	44
Poisson's ratio	~0.41
lattice constants, nm	$a_0 = 0.29503$
α, 25°C	$c_0 = 0.46531$
β, 900°C	$a_0 = 0.332$
vapor pressure, kPa[c]	$\log P_{kPa} = 5.7904 - 24644/T - 0.000227\, T$
specific heat, J/(kg·K)[d]	$C_p = 669.0 - 0.037188\, T - 1.080 \times 10^7/T^2$

[a] To convert J to cal, divide by 4.184.
[b] To convert GPa to psi, multiply by 145,000.
[c] To convert $\log P_{kPa}$ to $\log P_{atm}$, add 2.0056 to the constant.
[d] $T > 298$ K.

of nickel, aluminum, or magnesium, and stainless steel. Because of its high melting point, titanium can be alloyed to maintain strength well above the useful limits of magnesium and aluminum alloys. This property gives titanium a unique position in applications between 150 and 550°C where the strength-to-weight ratio is the sole criterion.

Solid titanium exists in two allotropic crystalline forms. The α phase, stable below 882.5°C, is a hexagonal closed-packed structure, whereas the β phase, a bcc crystalline structure, is stable between 882.5°C and the melting point of 1668°C. The high-temperature β phase can be found at room temperature when β-stabilizing elements are present as impurities or additions (see Section 6.2). The α and β phases can be distinguished by examining an unetched polished mount with polarized light. The α is optically active and changes from light to dark as the microscope stage is rotated. The microstructure of titanium is difficult to interpret without knowledge of the alloy content, working temperature, and thermal treatment.[6,17,18]

The heat-transfer qualities of titanium are characterized by the coefficient of thermal conductivity. Even though this is low, heat transfer in service approaches that of admiralty brass (thermal conductivity seven times greater) because titanium's greater strength permits thinner-walled equipment, relative absence of corrosion scale, erosion-corrosion resistance permitting higher operating velocities, and inherently passive film.

6.4 CORROSION RESISTANCE

Titanium is immune to corrosion in all naturally occurring environments. It does not corrode in air, even if polluted or moist with ocean spray. It does not corrode in soil or even the deep salt-mine-type environments where nuclear waste might be buried. It does not corrode in any naturally occurring water and most industrial wastewater streams. For these reasons, titanium has been termed the metal for the earth, and 20–30% of consumption is used in corrosion-resistance applications.

Even though titanium is an active metal, it resists decomposition because of a tenacious protective oxide film. This film is insoluble, repairable, and nonporous in many chemical media and provides excellent corrosion resistance. However, where this oxide film is broken, the corrosion rate is very rapid. However, usually the presence of a small amount of water is sufficient to repair the damaged oxide film. In a seawater solution, this film is maintained in the passive region from ~ −0.2 to 10 V versus the saturated calomel electrode.[19,20]

Titanium is resistant to corrosion attack in oxidizing, neutral, and inhibited reducing conditions. Examples of oxidizing environments are nitric acid, oxidizing chloride ($FeCl_3$ and $CuCl_2$) solutions, and wet chlorine gas. Neutral conditions include all neutral waters (fresh, salt, and brackish), neutral salt solutions, and natural soil environments. Examples of inhibited reducing conditions are in hydrochloric or sulfuric acids with oxidizing inhibitors and in organic acids inhibited with small amounts of water. Corrosion resistances to a variety of media are given in Table 6.6.[22] Titanium resistance to aqueous chloride solutions and chlorine account for most of its use in corrosion-resistant applications.

Titanium corrodes very rapidly in acid fluoride environments. The degree of attack generally increases with the acidity and the fluoride content. It is attacked in boiling HCl or H_2SO_4 at acid concentrations >1% or in ~10 wt % acid concentration at room temperature. Titanium is also attacked by hot caustic solutions, phosphoric acid solutions (concentrations above 25 wt %), boiling $AlCl_3$ (concentrations >15 wt %), dry chlorine gas, anhydrous ammonia above 150°C, and dry hydrogen–dihydrogen sulfide above 150°C.

Titanium is susceptible to pitting and crevice corrosion in aqueous chloride environments. The area of susceptibility is shown in Fig 6.2 as a function of temperature and sodium chloride content.[22] The susceptibility also depends on pH. The susceptibility temperature increases parabolically from 65°C as pH is increased from zero. With ASTM grades 7 or 12, crevice-corrosion attack is not observed above pH 2 until ~270°C. Noble alloying elements shift the equilibrium potential into the passive region where a protective film is formed and maintained.

Titanium does not stress crack in environments that cause stress cracking of other metal alloys (i.e., boiling 42% $MgCl_2$, NaOH, sulfides, etc.). Some of the alloys are susceptible to hot-salt stress cracking; however, this is a laboratory observation and has not been confirmed in service. Titanium stress cracks in methanol containing acid chlorides or sulfates, red fuming nitric acid, nitrogen tetroxide, and trichloroethylene.

Titanium is susceptible to failure by hydrogen embrittlement. Hydrogen attack initiates at sites of surface iron contamination or when titanium is galvanically coupled with iron.[23] In hydrogen-containing environments, titanium absorbs hydrogen above 80°C or in areas of high stress. If the surface oxide is removed by vacuum annealing or abrasion, pure dry hydrogen reacts at lower temperatures. Small amounts of oxygen or water vapor repair the oxide film and prevent this occurence. Molybdenum-containing alloys are less susceptible to hydrogen attack. Titanium resists this oxidation in air up to 650°C. Noticeable scale forms and embrittlement occurs at higher temperatures. Surface contaminants accelerate oxidation. In the presence of oxygen, the metal does not react significantly with nitrogen. Spontaneous ignition occurs in gas mixtures containing more than 40% oxygen under impact loading or abrasion. Ignition also occurs in dry halogen gases.

Titanium resists erosion–corrosion by fast-moving sand-laden water. In a high-velocity sand-laden seawater test (8.2 m/sec) for a 60-day period, titanium performed more than 100 times better than 18 Cr-8 Ni stainless steel, Monel, or 70 Cu–30 Ni. Resistance to cavitation (i.e., corrosion on surfaces exposed to high-velocity liquids) is better than by most other structural metals.[21,22]

In galvanic coupling, titanium is usually the cathode metal and, consequently, is not attacked. The galvanic potential in flowing seawater in relation to other metals is shown in Table 6.7.[21] Since titanium is the cathode metal, hydrogen attack may be of concern, as it occurs with titanium coupled to iron.

6.5 FABRICATION

Titanium can be fabricated similarly to nickel-base alloys and stainless steels. However, the characteristics of titanium have to be taken into account. Compared to these materials, titanium has:

1. Lower modulus of elasticity.
2. Lower ductility.
3. Higher melting point.
4. Lower thermal conductivity.
5. Smaller strain-hardening coefficient, thereby, lower uniform elongation.
6. Greater tendency to cold weld, thereby, greater tendency to gall or seize.
7. Greater tendency to be contaminated by oxygen, nitrogen, hydrogen, and carbon.

6.5.1 Boiler Code

The allowable stress values as determined by the Boiler and Pressure Vessel Committee of the American Society of Mechanical Engineers are listed in Tables 6.8 and 6.9 for various titanium grades and product forms.

Table 6.6   Corrosion Data for ASTM Grade 2 Titanium[2,16,21]

Media	Conc, wt %	Temperature, °C	Corrosion Rate, mm/yr
acetaldehyde	100	149	0.0
acetic acid	5–99.7	124	0.0
adipic acid	67	232	0.0
aluminum chloride, aerated	10	100	0.002
	10	150	0.03
	20	149	16
	25	20	0.001
	25	100	6.6
	40	121	109
ammonia + 28% urea + 20.5% H_2O + 19% CO_2 + 0.3% inerts + air	32.2	182	0.08
ammonia carbamate	50	100	0.0
ammonium perchlorate aerated	20	88	0.0
aniline hydrochloride	20	100	0.0
aqua regia	3:1	RT	0.0
	3:1	79	0.9
barium chloride, aerated	5–20	100	<0.003
bromine–water solution		RT	0.0
calcium chloride		RT	0.0
	5	100	0.005
	10	100	0.007
	20	100	0.02
	55	104	0.0005
	60	149	<0.003
	62	154	0.05–0.4
	73	177	2.1
calcium hypochlorite	6	100	0.001
chlorine gas, wet	>0.7 H_2O	RT	0.0
	>0.5 H_2O	200	0.0
chlorine gas, dry	<0.5 H_2O	RT	may react
chlorine dioxide in steam	5	99	0.0
chloracetic acid	100	189	<0.1
chromic acid	50	24	0.01
citric acid	25	100	0.0009
copper sulfate + 2% H_2SO_4	saturated	RT	0.02
cupric chloride, aerated	1–20	100	<0.01
cyclohexane (plus traces of formic acid)		150	0.003
ethylene dichloride	100	boiling	0.005–0.1
ferric chloride	10–30	100	<0.1
formic acid, nonaerated	10	100	2.4
hydrochloric acid, aerated	5	35	0.04
	20	35	4.4
HCl, chlorine saturated	5	190	<0.03
HCl + 10% HNO_3	5	38	0.0
HCl + 1% CrO_3	5	93	0.03
hydrofluoric acid	1–48	RT	rapid
hydrogen peroxide	3	RT	<0.1
hydrogen sulfide, steam and 0.077% mercaptans	7.65	93–110	0.0
hypochlorous acid + Cl_2O and Cl_2	17	38	0.00003

Table 6.6 *(Continued)*

Media	Conc, wt %	Temperature, °C	Corrosion Rate, mm/yr
lactic acid	10	boiling	<0.1
manganous chloride, aerated	5–20	100	0.0
magnesium chloride	5–40	boiling	0.0
mercuric chloride, aerated	1	100	0.0003
	5	100	0.01
	10	100	0.001
	55	102	0.0
mercury	100	RT	0.0
nickel chloride, aerated	5–20	100	0.0004
nitric acid	17	boiling	0.08–0.1
	70	boiling	0.05–0.9
nitric acid, red fuming	<about 2% H_2O	RT	ignition sensitive
	>about 2% H_2O	RT	nonignition sensitive
oxalic acid	1	37	0.3
oxygen, pure			ignition sensitive
phenol	saturated	21	0.1
phosphoric acid	10–30	RT	0.02–0.05
	10	boiling	10
potassium chloride	saturated	60	<0.0002
potassium dichromate			0.0
potassium hydroxide	50	27	0.01
	50	boiling	2.7
seawater, ten year test			0.0
sodium chlorate	saturated	boiling	0.0
sodium chloride	saturated	boiling	0.0
sodium chloride, titanium in contact with Teflon	23	boiling	crevice attack
sodium dichromate	saturated	RT	0.0
sodium hypochlorite + 12–15% sodium chloride + 1% sodium hydroxide + 1–2% sodium carbonate	1.5–4	66–93	0.03
stannic chloride	5	100	0.003
	24	boiling	0.04
sulfuric acid	1	boiling	2.5
sulfuric acid + 0.25% $CuSO_4$	5	93	0.0
terephthalic acid	77	218	0.0
urea–ammonia reaction mass		elevated temperature and pressure	no attack
zinc chloride	20	104	0.0
	50	150	0.0
	75	200	0.5
	80	200	203

6.5.2 Drawing

Commercially pure titanium can be cold drawn by tools required for austenitic stainless steels. Alpha-beta alloys, such as Ti–6 Al–4 V, are difficult to draw at room temperature. The following considerations should be given to drawing of titanium:

1. Slow drawing speeds are recommended.

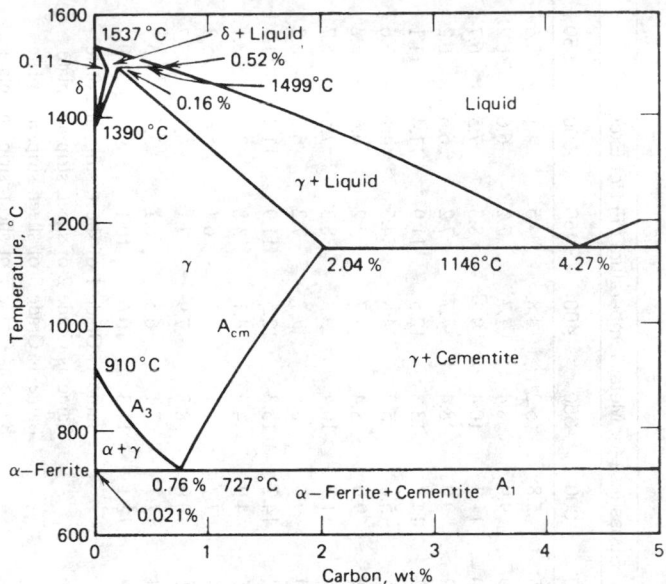

Fig. 6.2 Corrosion characteristics of titanium in aqueous NaCl solution.[23]

Table 6.7 Galvanic Series in Flowing Seawater 4 m/sec at 24°C[2,23]

Metal	Potential, V[a]
T304 stainless steel, passive	0.08
Monel alloy	0.08
Hastelloy alloy C	0.08
unalloyed titanium	0.10
silver	0.13
T410 stainless steel, passive	0.15
nickel	0.20
T430 stainless steel, passive	0.22
70–30 copper–nickel	0.25
90–10 copper–nickel	0.28
admiralty brass	0.29
G bronze	0.31
aluminum brass	0.32
copper	0.36
naval brass	0.40
T410 stainless steel, active	0.52
T304 stainless steel, active	0.53
T430 stainless steel, active	0.57
carbon steel	0.61
cast iron	0.61
aluminum	0.79
zinc	1.03

[a] Steady-state potential, negative to saturated calomel half-cell.

Table 6.8 Maximum Allowable Stress Values in Tension for Titanium and Its Alloys[24]

Form and Specification Number	Grade	Condition	Maximum Allowable Stress (ksi) for Metal Temperature (°F) Not Exceeding										
			100	150	200	250	300	350	400	450	500	550	600
Sheet, strip, plate, SB-265	1		8.8	8.1	7.3	6.5	5.8	5.2	4.8	4.5	4.1	3.6	3.1
	2	Annealed	12.5	12.0	10.9	9.9	9.0	8.4	7.7	7.2	6.6	6.2	5.7
	3	Annealed	16.3	15.6	14.3	13.0	11.7	10.4	9.3	8.3	7.5	6.7	6.0
	7		12.5	12.0	10.9	9.9	9.0	8.4	7.7	7.2	6.6	6.2	5.7
Bar, billet, SB-348	12		17.5	17.5	16.4	15.2	14.2	13.3	12.5	11.9	11.4	11.1	10.8
Pipe, SB-337	1	Seamless annealed	8.8	8.1	7.3	6.5	5.8	5.2	4.8	4.5	4.1	3.6	3.1
	2	Seamless annealed	12.5	12.0	10.9	9.9	9.0	8.4	7.7	7.2	6.6	6.2	5.7
Tubing, SB-338	3	Seamless annealed	16.3	15.6	14.3	13.0	11.7	10.4	9.3	8.3	7.5	6.7	6.0
	7	Seamless annealed	12.5	12.0	10.9	9.9	9.0	8.4	7.7	7.2	6.6	6.2	5.7
	12	Seamless annealed	17.5	17.5	16.4	15.2	14.2	13.3	12.5	11.9	11.4	11.1	10.8
Pipe, SB-337	1	Weld, annealed[a]	7.5	6.9	6.2	5.5	4.9	4.4	4.1	3.8	3.5	3.1	2.6
	2	Weld, annealed[a]	10.6	10.2	9.3	8.4	7.7	7.1	6.5	6.1	5.6	5.3	4.8
Tubing, SB-338	3	Weld, annealed[a]	13.9	13.3	12.2	11.1	10.0	8.8	7.9	7.1	6.4	5.7	5.1
	7	Weld, annealed[a]	10.6	10.2	9.3	8.4	7.7	7.1	6.5	6.1	5.6	5.3	4.8
	12		14.9	14.9	13.9	12.9	12.1	11.3	10.6	10.1	9.7	9.4	9.2
Forgings, SB-381	1		Same as Grade 1 of sheet, strip, and plate										
	F2	Annealed	Same as Grade 2 of sheet, strip and plate										
	F3	Annealed	Same as Grade 3 of sheet, strip, and plate										
	F7		Same as Grade 7 of sheet, strip and plate										
	F12		Same as Grade 12 of sheet, strip, and plate										

[a] 85% joint efficiency has been used in determining the allowable stress values for welded pipe and tube. Filler metal shall not be used in the manufacture of welded tubing or pipe.

Table 6.9 Design Stress Intensity Values in Tension for Titanium and Its Alloys[25]

Form and Specification Number	Grade	Condition	Design Stress Intensity (ksi) for Metal Temperature (°F) Not Exceeding										
			100	150	200	250	300	350	400	450	500	550	600
Sheet, strip, plate, SB-265	1	Annealed	11.7	10.8	9.7	8.6	7.7	6.9	6.4	6.0	5.3	4.7	4.2
	2	Annealed	16.7	16.7	16.7	13.7	12.3	10.9	9.8	8.8	8.0	7.5	7.3
	3	Annealed	21.7	20.8	19.0	17.3	15.6	13.9	12.3	11.1	9.9	8.9	8.0
	7	Annealed	16.7	16.7	16.7	13.7	12.3	10.9	9.8	8.8	8.0	7.5	7.3
Bar, billet, SB-348													
Pipe, SB-337	1	Seamless annealed	11.7	10.8	9.7	8.6	7.7	6.9	6.4	6.0	5.3	4.7	4.2
	2	Seamless annealed	16.7	16.7	16.7	13.7	12.3	10.9	9.8	8.8	8.0	7.5	7.3
Tubing, SB-338	3	Seamless annealed	21.7	20.8	19.0	17.3	15.6	13.9	12.3	11.1	9.9	8.9	8.0
	7	Seamless annealed	16.7	16.7	16.7	13.7	12.3	10.9	9.8	8.8	8.0	7.5	7.3
	1	Weld, annealed[a]	9.9	9.2	8.3	7.3	6.5	5.9	5.4	5.1	4.5	4.0	3.6
	2	Weld, annealed[a]	14.2	14.2	14.2	11.6	10.5	9.3	8.3	7.5	6.8	7.4	6.2
	3	Weld, annealed[a]	18.4	17.7	16.2	14.7	13.3	11.8	10.5	9.4	8.4	7.6	6.8
	7	Weld, annealed[a]	14.2	14.2	14.2	11.6	10.5	9.3	8.3	7.5	6.8	6.4	6.2
Forgings, SB-381	F1	Annealed	11.7	10.8	9.7	8.6	7.7	6.9	6.4	6.0	5.3	4.7	4.2
	F2	Annealed	16.7	16.7	16.7	13.7	12.3	10.9	9.8	8.8	8.0	7.5	7.3
	F3	Annealed	21.7	20.8	19.0	17.3	15.6	13.9	12.9	11.1	9.9	8.9	8.0
	F7	Annealed	16.7	16.7	16.7	13.7	12.3	10.9	9.8	8.8	8.0	7.5	7.3

[a] A quality factor of 0.85 has been applied in arriving at the design intensity values for this material. Filler metal shall not be used in the manufacture of welded tubing or pipe.

2. Blanks should be profiled and cleaned.

3. Cleanliness should be maintained on the dies and blanks.

4. Room-temperature deformations should be held to 8% maximum on a single draw before annealing.

5. Proper lubrication, preferably using dry-film types with antigalling constituents, should be applied to blanks.

6. The large amount of springback should be considered.

Hot drawing of titanium is capable of resulting in deeper draws, lower loads, and less distortion. The recommended drawing temperature ranges are 200–300°C for commercially pure titanium and 500–650°C for titanium alloys, such as Ti–6 Al–4 V.

6.5.3 Bending

Titanium can be bent with press brake equipment used for cold-forming stainless steels. The minimum bend radii for bending various titanium alloys through an angle of 105° are given in Table 6.10. More severe bends can be accomplished at 200°C or higher temperatures depending on the alloy and the bend required. The amount of springback decreases with increasing temperature. For bending operations above 550°C, a descaling operation may be necessary to remove the surface oxide layer.

Titanium tubes (<25 mm outside diameter) can be bent at a radius equal to two to three times the outside diameter of the tube. For tubes with an outside diameter larger than 25 mm, larger bend radii are recommended for room-temperature bending. Tighter bends can be obtained by heating the tube above 200°C. Owing to the low modulus, titanium has a tendency to buckle under compressive stress. Therefore, it is recommended that both the inside and outside surfaces of the bend be subjected to tension to avoid buckling.

6.5.4 Cutting and Grinding

Titanium can be sheared, flame cut, saw cut, and abrasive cut. Sheared edges should be examined for cracks for plates over 9.5 mm thick. Flame-cut edges are recommended with oxygen and carbon. It is recommended that at least 1.6 mm below the surface (measured from the lowest point of the cut roughness) be removed by grinding or machining. Thick plates may require removal of additional thickness. Generous amounts of coolant should be used in saw cutting and abrasive cutting to keep the workpiece cool and to minimize sparking. Water or water-soluble oil is recommended.

Abrasive grinding can be used on titanium, but care should be taken to avoid excessive heat buildup and contamination. The temperature of the grinding sparks is very high and precautionary measures should be taken accordingly.

6.5.5 Welding

Commercially pure titanium and most titanium alloys can be readily welded using the gas metal-arc (GMA) or gas tungsten-arc (GTA) welding process. Owing to titanium's highly reactive nature, the welding processes involving a noninert gas or a flux, such as, oxyacetylene-shielded metal arc, flux-cored arc, and submerged arc welding, are not suitable. For the same reason, welding titanium requires a clear environment and good inert-gas shielding in either GMA or GTA welding.

Table 6.10 Minimum Bend Radius for Annealed Titanium Sheet (ASTM B-265)

| Grade | Minimum Bend Radius[a] | |
	Under 1.8 mm Thick	1.8–4.75 mm Thick
1	1.5t	2.0t
2	2.0t	2.5t
3	2.0t	2.5t
4	2.5t	3.0t
5	4.5t	5.0t
6	4.0t	4.5t
7	2.0t	2.5t
10	3.0t	3.0t
11	1.5t	2.0t
12	2.0t	2.5t

[a] In multiples of sheet thickness, t, for room-temperature bending through and angle of 105°.

Common joint designs can be used for welding titanium as long as the design allows proper inert-gas shielding. The joint surfaces must be clean and free of grease, oil, moisture, visible oxides, and other contaminants. The oxides can be removed by grinding, brushing with a stainless-steel wire brush, or pickling in a room-temperature solution containing 30% nitric acid and 3% hydrofluoric acid, by weight.

A primary shield is needed for the molten weld puddle and a trailing secondary shield for the solidified weld deposit and the heat-affected zone. In addition, a backup shield is also needed for the backside of the weld and the heat-affected zone. Argon is generally preferred to helium for primary shielding because of better arc stability. Argon-helium mixtures can be used in some conditions where high voltage and deep penetration are desired. Either argon or helium can be used in the secondary and backup shielding.

The mechanical properties of the welds depend on the alloying elements. The welds generally have higher strength but less ductility than the parent metals as shown in Table 6.11.

Other than the GTA and GMA welding processes, titanium can also be welded by electron beam, resistance, plasma arc, and friction welding. In general, titanium cannot be welded with a dissimilar metal owing to the fact that it forms brittle intermetallic compounds with most other metals. Mechanical joining is recommended when joining titanium with a dissimilar metal.

6.6   SPECIFICATIONS, STANDARDS, AND QUALITY CONTROL

The alloys of titanium have compositional specifications tabulated by ASTM. The ASTM specification number is given in Table 6.3 for the commercially important alloys. Military specifications are found under MIL-T-9046 and MIL-T-9047, and aerospace material specifications for bar, sheet, tubing, and wire under specification numbers 4900–4980. Each aircraft company has its own set of alloy specifications.

The alloy name in the United States usually includes a company name or trademark in conjunction with the composition for alloyed titanium or the strength (ultimate tensile strength for TMCA and yield strength for other U.S. producers) for unalloyed titanium. The common alloys and company designations are shown in Table 6.12.

Table 6.11   Comparison in Tensile Properties Between Weld and Parent Metal of Titanium Alloys[26]

Alloy	Condition	Yield Strength (MPa)	Tensile Strength (MPa)	Elongation (%)
Grade 1	Parent metal	215	315	50.4
	Single-bead metal	255	345	37.5
	Multiple-bead weld	270	365	37.7
Grade 2	Parent metal	325	460	26.2
	Single-bead weld	380	505	18.3
	Multiple-bead weld	385	510	13.3
Grade 3	Parent metal	395	545	25.9
	Single-bead weld	475	605	15.5
	Multiple-bead weld	480	615	14.7
Grade 4	Parent metal	530	660	22.3
	Single-bead weld	580	695	16.4
	Multiple-bead weld	585	710	16.0
Grade 5	Parent metal	945	1000	11.0
	Single-bead weld	920	1060	3.5
	Multiple-bead weld	945	1090	3.2
Grade 6	Parent metal	805	850	15.7
	Single-bead weld	770	920	9.8
	Multiple-bead weld	820	935	7.5
Grade 9	Parent metal	670	705	15.2
	Single-bead weld	600	705	12.7
	Multiple-bead weld	625	745	11.2
Ti–8 Al–1 Mo–1 V	Parent metal	1020	1060	15.0
	Single-bead weld	930	1085	5.5
	Multiple-bead weld	960	1115	3.2
Ti–6 Al–6 V–2 Sn	Parent metal	1005	1060	9.8
	Single-bead weld	1255	1295	0.3
	Multiple-bead weld	—	1280	0.1

Table 6.12 Company Names of Common Titanium Alloys[2,7,9]

Alloys	ASTM	Cabot	IMI[a]	RMI	Timet	USSR
99.5 Ti	grade 2	CABOT Ti 40	IMI-125	RMI 40	Ti-50 A	VT1-0
99.2 Ti	grade 3	CABOT Ti 55	IMI-130	RMI 55	Ti-65 A	VT1
99.0 Ti	grade 4	CABOT Ti 70	IMI-155	RMI 70	Ti-75 A	VT1-1
Ti-5 Al-2.5 Sn	grade 6		IMI-315	RMI 5 Al-2.5 Sn	Ti-5 Al-2.5 Sn	VT5-1
Ti-6 Al-4 V	grade 5	CABOT Ti-6 Al-4 V	IMI-317	RMI 6 Al-4V	Ti-6 Al-4 V	VT6

[a] IMI = IMI Limited, Witton, Birmingham, UK.

Since titanium alloys are used in a variety of applications, several different material and quality standards are specified. Among them are ASTM, ASME, ASM, U.S. military, and a number of proprietary sources. The correct chemistry is basic to obtaining mechanical and other properties required for a given application. Minor elements controlled by specification include carbon, iron, hydrogen, nitrogen, and oxygen. For more stringent applications, yttrium may also be specified. In addition, control of the thermomechanical processing and subsequent heat treatment is vital to obtaining desired properties. For extremely critical applications, such as rotating parts in aircraft gas turbines, raw materials, melting parameters, chemistry, thermomechaical processing, heat treatment, test, and finishing operations must be carefully and closely controlled at each step to ensure that requried characteristics are present in the products supplied.

6.7 HEALTH AND SAFETY FACTORS

Titanium and its corrosion products are nontoxic. A safety problem does exist with titanium grindings, turnings, and some corrosion products which are pyrophoric. Grindings and turnings should be stored in a closed container and not left on the floor. Smoking must be prohibited in areas where titanium is ground or turned and, if a fire occurs, it must be extinguished with a class D extinguisher (for use against metal fires). The larger the surface area, the more pyrophoric the titanium fines. When titanium equipment is being worked on, all flammable products and corrosive products must be removed, and the area must be well ventilated. A pyrophoric corrosion product has been observed in environments or dry Cl_2 gas and in dry red fuming nitric acids.

6.8 USES

Titanium is primarily used in the form of high-purity titanium oxide. Although the principal application of high-purity (pigment-grade) TiO_2 is in paint pigments, other important uses are in plastics (for color in floor-covering products and to help protect plastic products and foodstuffs contained in plastic bags from ultraviolet radiation deterioration), in paper (as a filler and whitener), and in rubber. Future application areas include TiO_2 single-crystal electrodes for water decomposition for the production of hydrogen fuel, flue-gas denitrification catalysts, and high-purity TiO_2 to make barium titanate thermistors.

Titanium metal was first established as a material for aerospace, "metal-for-air" applications. In the late 1970s, it was developed as "metal-for-sea" uses. The metal-for-air and metal-for-sea designations characterize Japanese market development goals. In terms of volume, the U.S. titanium industry is still in the metal-for-air development stage; the statements about metal-for-earth and -sea reflect an optimistic outlook.

In the United States, the high strength-to-weight ratio of titanium accounts for approximately 70% of its uses. Before 1970, the high strength-to-weight ratio was the basis of over 90% of applications, such as engines, where the advantage of light weight is translated to higher flying, faster planes. Aerospace applications have shaped and controlled the titanium-metal industry.

The use of titanium in aircraft is divided about equally between engines and airframes. For engine components, titanium is limited because of temperature constraints at the compressor area where it is used as blades, casings, and disks. In the frame, it is used in bulkheads, firewall, flap tracks, landing-gear parts, wiring pivot structures, fasteners, rotor hubs, and hot-area skins. In the F-15, titanium accounts for about 32% of the structural weight. Design changes and weight savings owing to the use of titanium in Pratt and Whitney's JT3D engine, employed to power Boeing 707 and Douglas DC 8 aircraft, resulted in 42% more takeoff thrust, 13% lower specific fuel consumption, and 18% less weight than the prior JT3C engine.[3]

The other outstanding property of titanium metal is its corrosion resistance, although its use in corrosion-resistance applications in 1980 in the United States was a mere 5000 tons or ~0.001% of the metal used in corrosion-resistance markets. The largest application was heat-exchanger pipes and tubing (~800 μm or 22 gauge welded) for the power industry, and marine and desalination applications, where titanium provides protection against corrosion by seawater, brackish water, and other estuary waters containing high concentrations of chlorides and industrial wastes.

Titanium metal is especially utilized in environments of wet chlorine gas and bleaching solutions, that is, in the chlor-alkali industry and the pulp and paper industries. Here, titanium is used as anodes for chlorine production, chlorine-caustic scrubbers, pulp washers, and Cl_2, ClO_2, and $HClO_4$ storage and piping equipment.

In the chemical industry, titanium is used in heat-exchanger tubing for salt production, in the production of ethylene glycol, ethylene oxide, propylene oxide, and terephthalic acid, and in industrial wastewater treatment. Titanium is used in environments of aqueous chloride salts ($ZnCl_2$, NH_4Cl, $CaCl_2$, $MgCl_2$, etc.), chlorine gas, chlorinated hydrocarbons, and nitric acid.

In metal recovery, titanium is used for ore-leaching solutions and as racks for metal plating. The leaching solutions contain HCl or H_2SO_4 with enough ferric or cupric ions to inhibit the corrosion of titanium. In metal-plating applications, titanium is cathodically protected against H_2SO_4 and chrome-plating solution corrosion. An important factor in using titanium for metal-plating applications is that the minute amount of dissolved titanium ions does not plate out as an impurity in the coatings.

In oil and gas refinery applications, titanium is used as protection in environments of H_2S, SO_2, CO_2, NH_3, caustic solutions, steam and cooling water. It is used in heat-exchanger condensers for the fractional condensation of crude hydrocarbons, NH_3, propane, and desulfurization products using seawater or brackish water for cooling.

Other application areas include nuclear-waste storage canisters, pacemaker castings, implantations, geothermal equipment, automotive connection rods, and ordnance.

REFERENCES

1. S. C. Williams, *Report on Titanium*, J. W. Edwards, Inc. Ann Arbor, MI, 1965.
2. D. Knittel, "Titanium and Titanium Alloys," in *Kirk-Othmer Encyclopedia of Chemical Technology*, 3rd ed., Wiley, New York, 1983, Vol. 23.
3. A. D. McQuillan and M. K. McQuillan, in *Metallurgy of the Rarer Metals*, H. M. Finniston (ed.), Academic, New York, 1956, p. 335.
4. *ASTM Standard Specification for Titanium and Titanium Alloy Strip, Sheet, and Plate*. ANSI-ASTM B265-79, American Society for Testing and Materials, Philadelphia, PA, Oct. 1980.
5. R. M. Duncan, P. A. Blenkinsop, and R. E. Goosey, in *The Development of Gas Turbine Materials*, G. W. Meethan (ed.), Wiley, New York, 1981, p. 63.
6. *Facts About the Metallography of Titanium*, RMI Company, Niles, OH, 1975.
7. R. A. Wood, *The Titanium Industry in the Mid-1970's*, Battelle Report MCIC-75-26, Battelle Memorial Institute, Columbus, OH, June 1975.
8. R. I. Jaffee, in *Titanium '80 Science and Technology*, H. Kimura and O. Izumi (eds.), The Metallurgical Society/American Institute of Mining, Metallurgical and Petroleum Engineers, Warrendale, PA, 1980, p. 53.
9. H. Hucek and M. Wahll, *Handbook of International Alloy Compositions and Designations*, Battelle Report MCIC-HB-09, Battelle Memorial Institute, Columbus, OH, Nov. 1976, Vol. 1.
10. *Metals Prog. Databook* **110**, 94 (June 1976).
11. S. G. Glazunov, in *Titanium Alloys for Modern Technology*, N. P. Sazhin and co-workers (eds.), NASA TT F-596, National Aeronautics and Space Administration, Washington, DC, March 1970, p. 11.
12. C. C. Chen and R. R. Boyer, *J. Met.* **31**, 33 (1979).
13. E. L. Hayman, D. W. Greenwood, and B. G. Martin, *Exp. Mech.* **17**, 161 (May 1977).
14. E. M. Savitskiy, M. I. Bychkova, and V. V. Baron, Ref. 8 p. 735.
15. C. M. Wayman, *J. Met.* **32**, 129 (1980).
16. *How to Use Titanium Properties and Fabrication of Titanium Mill Products*, Titanium Metals Corporation of America, Pittsburgh, PA, 1975.
17. H. R. Ogden and F. C. Holden, *Metallography of Titanium Alloys*, TML Report No. 103, Battelle Memorial Institute, Columbus, OH, May 29, 1958.
18. *Metals Handbook*, American Society for Metals, Metals Park, OH, 1972, Vol. 7.
19. T. R. Beck, in *Localized Corrosion*, R. W. Staehle, B. F. Brown, J. Kruger, and A. Agarwal (eds.), National Association of Corrosion Engineers, Houston, TX, 1974, Vol. Nace-3, p. 644.
20. E. E. Millaway, *Mater. Prot.* **4**, 16 (1965).
21. L. C. Covington, R. W. Schultz, and I. A. Fronson, *Chem. Eng. Prog.* **74**, 67 (1978).
22. *Titanium for Industrial Brine and Sea Water Service*, Titanium Metal Corporation of America, Pittsburgh, PA, 1968.
23. L. C. Covington and R. W. Schultz, in *Industrial Applications of Titanium and Zirconium*, STP 728, E. W. Kleefisch (ed.), American Society for Testing and Materials, Philadelphia, PA, 1981, p. 163.
24. Boiler and Pressure Vessel Code, Section VIII—Division I.
25. Boiler and Pressure Vessel Code, Section VIII—Division 2.
26. *Metals Handbook*, 9th ed., American Society for Metals, Metals Park, OH, 1980, Vol. 3.

CHAPTER 7

MAGNESIUM AND ITS ALLOYS

Robert S. Busk
Hilton Head, South Carolina

7.1	INTRODUCTION	109	7.4	FABRICATION		110
				7.4.1 Machining		110
7.2	USES	109		7.4.2 Joining		110
	7.2.1 Nonstructural Applications	109		7.4.3 Forming		112
	7.2.2 Structural Applications	109				
			7.5	CORRROSION AND FINISHING		113
7.3	ALLOYS AND PROPERTIES	110		7.5.1 Chemical-Conversion		
	7.3.1 Mechanical Properties of			Coatings		113
	Castings	110		7.5.2 Anodic Coatings		113
	7.3.2 Mechanical Properties of			7.5.3 Pointing		113
	Wrought Products	110		7.5.4 Electroplating		113
	7.3.3 Physical Properties	110				

7.1 INTRODUCTION

Magnesium, with a specific gravity of only 1.74, is the lowest-density metal available for engineering use. It is produced either by electrolytic reduction of $MgCl_2$ or by chemical reduction of MgO by Si in the form of ferrosilicon. $MgCl_2$ is obtained from seawater, brine deposits, or salt lakes. MgO is obtained principally from seawater or dolomite. Because of the widespread, easy availability of magnesium ores (e.g., from the ocean), the ore supply is, in human terms, inexhaustible.

7.2 USES

Magnesium is used both as a structural, load-bearing material and in applications that exploit its chemical and metallurgical properties.

7.2.1 Nonstructural Applications

Because of its high place in the electromotive series, magnesium is used as a sacrificial anode to protect steel from corrosion; some examples are the protection of buried pipelines and the prolongation of the life of household hot-water tanks. Alloys used for this purpose are produced by permanent-mold castings and by extrusion.

Magnesium in powder form is added to gray cast iron to produce ductile, or nodular, iron, an alloy that has many of the producibility advantages of cast iron but is ductile and strong.

A significant use for magnesium powder is its addition to the iron tapped from blast furnaces to remove sulfur prior to converting to steel, thereby increasing the efficiency of the blast furnace and improving the toughness of the steel.

Magnesium powder is also used to produce the Grignard reagent, an organic intermediate used in turn to produce fine chemicals and pharmaceuticals.

Magnesium sheet and extrusions are used to produce photoengravings.

Magnesium in ingot form is one of the principal alloying additions to aluminum, imparting improved strength and corrosion resistance to that metal.

7.2.2 Structural Applications

Magnesium structures are made from sand, permanent-mold, investment, and die casting, and from sheet, plate, extrusions, and forgings. The base forms produced in these ways are fabricated into

Mechanical Engineers' Handbook, 2nd ed., Edited by Myer Kutz.
ISBN 0-471-13007-9 © 1998 John Wiley & Sons, Inc.

finished products by machining, forming, and joining. Finishing for protective or decorative purposes is by chemical-conversion coatings, painting, or electroplating.

The most rapidly growing method of producing structural parts is die casting. This method is frequently the most economical to produce a given part and is especially effective in producing parts with very thin sections. A stimulus for the recent very high growth rate has been the development of a high-purity corrosion-resistant alloy that makes unnecessary the protective finishing of many parts. See alloy AZ91D in Table 7.1. Die castings are produced by cold chamber, by hot chamber, and by a recently developed method analogous to the injection molding of plastic parts. The latter technique, known as Thixomolding,[1,2,3,4] uses a machine that advances the alloy in a semisolid state by means of a screw and then injects an accumulated amount into the die. The melting step is eliminated, production rates are at least as high as for hot-chamber die casting, and metal quality is superior to that produced by either cold- or hot-chamber die casting. Two major fields dominate the die-casting markets: automotive (e.g., housings, brake pedals, transmissions, instrument panels) and computers (e.g., housings, disc readers).

Those properties mainly significant for structural applications are density (automotive and aerospace vehicle parts; portable tools such as chain saws; containers such as for computers, cameras, briefcases; sports equipment such as catcher's masks, archery bows); high damping capacity (antivibration platforms for electronic equipment; walls for sound attenuation); excellent machinability (jigs and fixtures for manufacturing processes); high corrosion-resistance in an alkaline environment (cement tools).

7.3 ALLOYS AND PROPERTIES

Many alloys have been developed to provide a range of properties and characteristics to meet the needs of a wide variety of applications. The most frequently used are given in Table 7.1. There are two major classes—one containing aluminum as the principal alloying ingredient, the other containing zirconium. Those containing aluminum are strong and ductile, and have excellent resistance to atmospheric corrosion. Since zirconium is a potent grain refiner for magnesium alloys but is incompatible with the presence of aluminum in magnesium, it is added to all alloys not containing aluminum. Within this class, those alloys containing rare earth or yttrium are especially suited to applications at temperatures ranging to as high as 300°C. Those not containing rare-earth or yttrium have zinc as a principal alloying element and are strong, ductile, and tough.

Recently, the high-purity casting alloys, AZ91E for sand and permanent mold castings and AZ91D, AM60B, AM50A, and AS41B for die castings, have been developed. The high-purity die casting alloys are superior in corrosion resistance to the commonly used aluminum die casting alloy. These alloys have been largely responsible for the large expansion in magnesium automotive applications.

7.3.1 Mechanical Properties of Castings

Magnesium castings are produced in sand, permanent, investment, pressure die-casting molds.

Castings produced in sand molds range in size from a few pounds to a few thousand pounds and can be very simple to extremely complex in shape. If production runs are large enough to justify higher tooling costs, then permanent instead of sand molds are used. The use of low pressure to fill a permanent mold is a low-cost method that is also used. Investment casting is a specialized technique that permits the casting of very thin and intricate sections with excellent surface and high mechanical properties. Die casting is a process for the production of castings with good dimensional tolerances, good surface, and acceptable properties at quite low cost.

Mechanical properties of cast alloys are given in Table 7.2.

7.3.2 Mechanical Properties of Wrought Products

Wrought products are produced as forgings, extrusions, sheet, and plate. Mechanical properties are given in Table 7.3.

7.3.3 Physical Properties

A selection of physical properties of pure magnesium is given in Table 7.4. Most of these are insensitive to alloy addition, but melting point, density, and electrical resistivity vary enough that these properties are listed for alloys in Table 7.5.

7.4 FABRICATION

7.4.1 Machining

Magnesium is the easiest of all metals to machine: it requires only low power and produces clean, broken chips, resulting in good surfaces even with heavy cuts.

7.4.2 Joining

All standard methods of joining can be used, including welding, riveting, brazing, and adhesive bonding.

Table 7.1 Magnesium Alloys in Common Use

ASTM Designation	Ag	Al	Fe max	Mn	Ni max	Rare Earth	Si	Zn	Zr	Forms
AM50A		4.9	0.004	0.32	0.002			0.22		DC
AM60B		6.0	0.005	0.42	0.002			0.22max		DC
AS41B		4.2	0.0035	0.52	0.002		1.0	0.12		DC
AZ31B		3	0.005	0.6	0.005			1		S, P, F, E
AZ61A		6.5	0.005	0.33	0.005			0.9		F, E
AZ80A		8.5	0.005	0.31	0.005			0.5		F ,E
AZ81A		7.6		0.24				0.7		SC, PM, IC
AZ91D		9	0.005	0.33	0.002			0.7		DC
AZ91E		9	0.005	0.26	0.0010			0.7		SC. PM
EZ33A						3.2		2.5	0.7	SC, PM
K1A									0.7	SC, PM
M1A				1.6						E
QE22A	2.5					2.2			0.7	S, PM, IC
WE43A			0.01	0.15	0.005	A		0.20	0.7	S, PM, IC
WE54A				0.15	0.005	B			0.7	S, PM, IC
ZE41A				0.15		1.2		4.2	0.7	S, PM, IC
ZE63A						2.6		5.8	0.7	S, PM, IC
ZK40A								4	0.7	E
ZK60A								5.5	0.7	F, E

A = 4 Yttrium; 3 RE
B = 5.1 Yttrium; 4 R.E.
DC = die casting; E = extrusion; F = forging; IC = investment casting; P = plate; PM = permanent mold; S = sheet; SC = sand casting

Table 7.2 Typical Mechanical Properties for Castings

Alloy	Temper	Tensile Strength (MPa)	Yield Strength (MPa)	Elongation in 2 in. (%)
Sand and Permanent Mold Castings				
AZ81A	T4	276	85	15
AZ91E	F	165	95	3
	T4	275	85	14
	T6	275	195	6
EZ33A	T5	160	105	3
K1A	F	185	51	20
QE22A	T6	275	205	4
WE43A	T6	235	190	4
WE54A	T6	270	195	4
ZE63A	T6	295	190	7
Investment Castings				
AZ81A	T4	275	100	12
AZ91E	F	165	100	2
	T4	275	100	12
	T5	180	100	3
	T7	275	140	5
EZ33A	T5	255	110	4
K1A	F	175	60	20
QE22A	T6	260	185	4
Die Castings				
AM50A	F	200	110	10
AM60B	F	220	130	8
AS41B	F	210	140	6
AZ91D	F	230	160	3

Table 7.3 Typical Mechanical Properties of Wrought Products

Alloy	Temper	Tensile Strength (MPa)	Yield Strength (MPa)		Elongation in 2 in. (%)
			Tensile	Compressive	
Sheet and Plate					
AZ31B	O	255	150	110	21
	H24	290	220	180	15
Extrusions					
AZ31B	F	260	200	95	15
AZ61A	F	310	230	130	16
AZ80A	F	340	250	140	11
	T5	380	275	240	7
M1A	F	255	180	125	12
ZK40A	T5	275	255	140	4
ZK60A	F	340	250	185	14
	T5	365	305	250	11
Forgings					
AZ31B	F	260	195	85	9
AZ61A	F	195	180	115	12
AZ80A	F	315	215	170	8
	T5	345	235	195	6
	T6	345	250	185	5
ZK60A	T5	305	205	195	16
	T6	325	270	170	11

Welding is by inert-gas-shielded processes using either helium or argon, and either MIG or TIG. Alloys containing more than 1.5% aluminum should be stress-relieved after welding in order to prevent stress-corrosion cracking due to residual stresses associated with the weld joint. Rivets for magnesium are of aluminum rather than magnesium. Galvanic attack is minimized or eliminated by using aluminum rivets made of an alloy high in magnesium, such as 5056. Brazing is used, but not extensively, since it can be done only on alloys with a high melting point, such as AZ31B or K1A. Adhesive bonding is straightforward, and no special problems related to magnesium are encountered.

7.4.3 Forming

Magnesium alloys are formed by all the usual techniques, such as deep drawing, bending, spinning, rubber forming, stretch forming, and dimpling.

In general, it is preferable to form magnesium in the temperature range of 150–300°C. While this requires more elaborate tooling, there is some compensation in the ability to produce deeper draws (thus fewer tools) and in the elimination or minimizing of springback. Hydraulic rather than mechanical presses are preferred.

Table 7.4 Physical Properties of Pure Magnesium

Density	1.718 g/cm³ (Ref. 5)
Melting point	650°C (Ref. 6)
Boiling point	1107°C (Ref. 6)
Thermal expansion	25.2×10^{-6}/K (Ref. 7)
Specific heat	1.025 kJ/kg·K at 20°C (Ref. 8)
Latent heat of fusion	360–377 kJ/kg (Ref. 8)
Latent heat of sublimation	6113–6238 kJ/kg Ref. 6)
Latent heat of vaporization	5150–5400 kJ/kg (Ref. 6)
Heat of combustion	25,020 kJ/kg (Ref. 10)
Electrical resistivity	4.45 ohm meter $\times 10^{-8}$
Crystal structure	Close-packed hexagonal: $a + 0.32087$ nm; $c = 0.5209$ nm; $c/a = 1.6236$ (Ref. 9)
Young's modulus	45 Gpa
Modulus of rigidity	16.5 Gpa
Poisson's ratio	0.35

Table 7.5 Physical Properties of Alloys[10]

Alloy	Density (g/cm³)	Melting Point (°C) Liquidus	Melting Point (°C) Solidus	Electrical Resistivity (ohm-metres × 10^{-8})
AM60B	1.79	615	540	
AS41B	1.77	620	565	13.0
AZ31B	1.77	632	605	9.2
AZ61A	1.8	620	525	12.5
AZ80A	1.8	610	490	15.6
AZ81A	1.80	610	490	13.0
AZ91D	1.81	595	470	17.0
EZ33A	1.83	645	545	7.0
K1A	1.74	649	648	5.7
M1A	1.76	649	648	5.4
QE22A	1.81	645	545	6.8
ZK60A	1.83	635	520	5.7

7.5 CORROSION AND FINISHING

Magnesium is highly resistant to alkalies and to chromic and hydrofluoric acids. In these environments, no protection is usually necessary. On the other hand, magnesium is less resistant to other acidic or salt-laden environments. While most magnesium alloys can be exposed without protection to dry atmosphere, it is generally desirable to provide a protective finish.

Magnesium is anodic to any other structural metal and will be preferentially attacked in the presence of an electrolyte. Therefore, galvanic contact must be avoided by separating magnesium from other metals by the use of films and tapes. These precautions do not apply in the case of 5056 aluminum alloy, since the galvanic attack in this case is minimal.

Because magnesium is not resistant to acid attack, standing water (which will become acidic by absorption of CO_2 from the atmosphere) must be avoided by providing drain holes.

7.5.1 Chemical-Conversion Coatings

There are a large number of chemical-conversion processes based on chromates, fluorides, or phosphates. These are simple to apply and provide good protection themselves, in addition to being a good paint base.

7.5.2 Anodic Coatings

There are a number of good anodic coatings that offer excellent corrosion protection and also provide a good paint base.

7.5.3 Painting

If a good chemical-conversion or anodic coating is present, any paint will provide protection. Best protection results from the use of baked, alkaline-resistant paints.

7.5.4 Electroplating

Once a zinc coating is deposited chemically, followed by a copper strike, standard electroplating procedures can be applied to magnesium to give decorative and protective finishes.

REFERENCES

1. M. C. Flemings, "A History of the Development of Rheocasting," in *Proceedings of the Work Shop on Rheocasting*, Army Materials and Mechanics Research Center, Feb. 3–4, 1977, pp. 3–10.
2. S. C. Erickson, "A Process for the Thixotropic Casting of Magnesium Alloy Parts," in *Proceedings of the International Magnesium Association*, May 17–20, 1987, p. 39.
3. R. D. Carnahan, R. Kilbert and L. Pasternak, "Advances in Thixomolding," in *Proceedings of the International Magnesium Association*, May 17–18, 1994, p. 21.
4. K. Saito, "Thixomolding of Magnesium Alloys," in *Proceedings of the International Magnesium Association*, June 2–4, 1996.
5. R. S. Busk, *Trans. AIME* **194**, 207 (1952).
6. D. R. Stull and G. C. Sinke, *Thermodynamic Properties of the Elements*, Vol. 18, *Advances in Chemistry*, American Chemical Society, Washington, DC, 1956.
7. P. Hidnert and W. T. Sweeney, *J. Res. Nat. Bur. St.* **1**, 771 (1955).
8. R. A. McDonald and D. R. Stull, *J. Am. Chem. Soc.* **77**, 529 (1955).

9. R. S. Busk, *Trans. AIME*, **188**, 1460 (1950).

10. J. W. Frederickson, "Pure Magnesium," in *Metals Handbook*, 8th ed., American Society for Metals, Metals Park, OH, 1961, Vol. 1.

11. *Physical Properties of Magnesium and Magnesium Alloys*, Dow Chemical Company, 1967.

BIBLIOGRAPHY

Bothwell, M. R., *The Corrosion of Light Metals*, Wiley, New York, 1967.

Busk, R. S., *Magnesium Products Design*, Marcel Dekker, New York, 1987.

Emley, E. F., *Principles of Magnesium Technology*, Pergamon Press, New York, 1966.

Fabricating with Magnesium, Dow Chemical Company.

Machining Magnesium, Dow Chemical Company.

"Nonferrous Metal Products," in *Annual Book of ASTM Standards*, **02.02**, ASTM, 1995.

Operations in Magnesium Finishing, Dow Chemical Company.

"Properties of Magnesium Alloys," in *Metals Handbook*, 10th ed., American Society for Metals, Metals Park, OH, 1990, Vol. 2.

Roberts, C. S., *Magnesium and Its Alloys*, Wiley, New York, 1960.

"Selection and Application of Magnesium and Magnesium Alloys," in *Metals Handbook*, 10th ed., American Society for Metals, Metals Park, OH, 1990, Vol. 2.

CHAPTER 8

PLASTICS AND ELASTOMERS

Edward N. Peters
General Electric Company
Selkirk, New York

8.1 INTRODUCTION 115

8.2 COMMODITY
THERMOPLASTICS 116
  8.2.1 Polyethylene 116
  8.2.2 Polypropylene 116
  8.2.3 Polystyrene 117
  8.2.4 Impact Polystyrene 117
  8.2.5 SAN (Styrene/Acrylonitrile
        Copolymer) 117
  8.2.6 ABS 118
  8.2.7 Polyvinyl Chloride 118
  8.2.8 Poly(vinylidine chloride) 119
  8.2.9 Poly(methyl Methacrylate) 119
  8.2.10 Poly(ethylene Terephthalate) 119

8.3 ENGINEERING
THERMOPLASTICS 120
  8.3.1 Polyesters (Thermoplastic) 120
  8.3.2 Polyamides (Nylon) 120
  8.3.3 Polyacetals 121
  8.3.4 Polyphenylene Sulfide 121
  8.3.5 Polycarbonates 122
  8.3.6 Polysulfone 122
  8.3.7 Modified Polyphenylene Ether 123
  8.3.8 Polyimides 123

8.4 FLUORINATED
THERMOPLASTICS 124
  8.4.1 Poly(tetrafluoroethylene) 124
  8.4.2 Poly(chlorotrifluoroethylene) 124
  8.4.3 Fluorinated Ethylene-
        Propylene 125
  8.4.4 Polyvinylidine Fluoride 125
  8.4.5 Poly(ethylene
        chlorotrifluoroethylene) 128
  8.4.6 Poly(vinyl fluoride) 128

8.5 THERMOSETS 128
  8.5.1 Phenolic Resins 128
  8.5.2 Epoxy Resins 128
  8.5.3 Unsaturated Polyesters 128
  8.5.4 Alkyd Resins 129
  8.5.5 Diallyl Phthalate 129
  8.5.6 Amino Resins 129

8.6 GENERAL-PURPOSE
ELASTOMERS 129

8.7 SPECIALTY ELASTOMERS 129

8.1 INTRODUCTION

The use of plastics has increased almost 20-fold in the last 30 years. Plastics have come on the scene as the result of a continual search for man-made substances that can perform better or can be produced at a lower cost than natural materials such as wood, glass, and metal, which require mining, refining, processing, milling, and machining. Plastics can also increase productivity by producing finished parts and consolidating parts. Thus, an item made from several metal parts that require separate fabrication and assembly can often be consolidated into one or two plastic parts. Such increases in productivity have led to fantastic growth.

Plastics can be classified in several ways. The two major classifications are thermosetting materials and thermoplastic materials. As the name implies, thermosetting plastics or thermosets are set, cured, or hardened into a permanent shape. The curing that usually occurs rapidly under heat or UV light leads to an irreversible cross-linking of the polymer. Thermoplastics differ from thermosetting materials in that they do not set or cure under heat. When heated, thermoplastics merely soften to a mobile, flowable state where they can be shaped into useful objects. Upon cooling, the thermoplastics harden and hold their shape. Thermoplastics can be repeatedly softened by heat and shaped.

Mechanical Engineers' Handbook, 2nd ed., Edited by Myer Kutz.
ISBN 0-471-13007-9 © 1998 John Wiley & Sons, Inc.

Thermoplastics can be classified as amorphous or semicrystalline plastics. Most polymers are either completely amorphous or have an amorphous component even if they are crystalline. Amorphous polymers are hard, rigid glasses below a fairly sharply defined temperature, which is known as the glass transition temperature. Above the glass transition temperature, the amorphous polymer becomes soft and flexible and can be shaped. Mechanical properties show profound changes near the glass transition temperature. Many polymers are not completely amorphous but are semicrystalline. Semicrystalline polymers have melting points that are above their glass transition temperatures. The degree of crystallinity and the morphology of the crystalline phase have an important effect on mechanical properties. Crystalline plastics will become less rigid near their glass transition temperature but will not flow until the temperature is above the crystalline melting point. At ambient temperatures, crystalline/semicrystalline plastics have greater rigidity, hardness, density, lubricity, creep resistance, and solvent resistance than amorphous plastics.

From a cost and performance standpoint, polymers can be classified as either commodity or engineering plastics.

Another important class of polymeric resins are elastomers. Elastomers have glass transition temperatures below room temperature. Thus, elastomeric materials are rubber-like polymers at room temperatures, but below their glass transition temperature they will become rigid and lose their rubbery characteristics.

8.2 COMMODITY THERMOPLASTICS

The commodity thermoplastics include polyolefins and side-chain-substituted vinyl polymers.

8.2.1 Polyethylene

Polyethylenes (PEs) have the largest volume use of any plastic. They are prepared by the catalytic polymerization of ethylene. Depending on the mode of polymerization, one can obtain a high-density (HDPE) or a low-density (LDPE) polyethylene polymer. LDPE is prepared under more vigorous conditions, resulting in short-chain branching. Linear low-density polyethylene (LLDPE) is prepared by introducing short-branching via copolymerization with a small amount of long-chain olefin.

Polyethylenes are crystalline thermoplastics that exhibit toughness, near-zero moisture absorption, excellent chemical resistance, excellent electrical insulating properties, low coefficient of friction, and ease of processing. Their heat deflection temperatures are reasonable but not high. The branching in LLDPE and LDPE decreases the crystallinity. HDPE exhibits greater stiffness, rigidity, improved heat resistance, and increased resistance to permeability than LDPE and LLDPE. Some typical properties of PEs are listed in Table 8.1.

Uses. HDPE's major use is in blow-molded bottles, drums, carboys automotive gas tanks; injection-molded material-handling pallets, trash and garbage containers, and household and automotive parts; and extruded pipe.

LDPE/LLDPEs find major applications in film form for food packaging, as a vapor barrier film, plastic bags; for extruded wire and cable insulation; and for bottles, closures and toys.

8.2.2 Polypropylene

Polypropylene (PP) is prepared by the catalyzed polymerization of propylene. PP is a highly crystalline thermoplastic that exhibits low density, rigidity, excellent chemical resistance, negligible water absorption, and excellent electrical properties. Its properties appear in Table 8.2.

Table 8.1 Typical Property Values for Polyethylenes

Property	HDPE	LLDPE/LDPE
Density (Mg/m³)	0.96–0.97	0.90–0.93
Tensile modulus (GPa)	0.76–1.0	—
Tensile strength (MPa)	25–32	4–20
Elongation at break (%)	500–700	275–600
Flexural modulus (GPa)	0.8–1.0	0.2–0.4
Vicat soft point (°C)	120–129	80–98
Brittle temperature (°C)	−100 to −70	−85 to −35
Hardness (Shore)	D60–D69	D45–D55
Dielectric constant (10^6 Hz)	—	2.3
Dielectric strength (MV/m)	—	9–21
Dissipation factor (10^6 Hz)	—	0.0002
Linear mold shrinkage (in./in.)	0.007–0.009	0.015–0.035

Table 8.2   Typical Property Values for Polypropylenes

Density (Mg/m³)	0.09–0.93
Tensile modulus (GPa)	1.8
Tensile strength (MPa)	37
Elongation at break (%)	10–60
Heat deflection at 0.45 MPa (°C)	100–105
Heat deflection at 1.81 MPa (°C)	60–65
Vicat soft point (°C)	130–148
Linear thermal expansion (mm/mm·K)	3.8×10^{-5}
Hardness (Shore)	D76
Volume resistivity (Ω·cm)	1.0×10^{17}
Linear mold shrinkage (in./in.)	0.01–0.02

Uses. End uses for PP are in blow-molding bottles and automotive parts; injection-molding closures, appliances, housewares, automotive parts, and toys. PP can be extruded into fibers and filaments for use in carpets, rugs, and cordage.

8.2.3   Polystyrene

Catalytic polymerization of styrene yields polystyrene (PS), a clear, amorphous polymer with a moderately high heat deflection temperature. PS has excellent electrical insulating properties, but, it is brittle under impact and exhibits very poor resistance to surfactants and solvents. Its properties appear in Table 8.3.

Uses. Ease of processing, rigidity, clarity, and low cost combine to support applications in toys, displays, and housewares. PS foams can readily be prepared and are characterized by excellent low thermal conductivity, high strength-to-weight ratio, low water absorption, and excellent energy absorption. These attributes have made PS foam of special interest as insulation boards for construction, protective packaging materials, insulated drinking cups, and flotation devices.

8.2.4   Impact Polystyrene

Copolymerization of styrene with a rubber, polybutadiene, can reduce brittleness of PS, but only at the expense of rigidity and heat deflection temperature. Impact polystyrene (IPS) or high-impact polystyrene (HIPS) can be prepared, depending on the levels of rubber. These materials are translucent to opaque and generally exhibit poor weathering characteristics. Typical properties appear in Table 8.3.

8.2.5   SAN (Styrene/Acrylonitrile Copolymer)

Copolymerization of styrene with a moderate amount of acrylonitrile provides a clear, amorphous polymer (SAN) with increased heat deflection temperature and chemical resistance compared to polystyrene. However, impact resistance is still poor. Typical properties appear in Table 8.3

Uses. SAN is utilized in typical PS-type applications where a slight increase in heat deflection temperature and/or chemical resistance is needed, such as housewares and appliances.

Table 8.3   Typical Properties of Styrene Thermoplastics

Property	PS	SAN	IPS/HIPS	ABS
Density (Mg/m³)	1.050	1.080	1.02–1.04	1.05–1.07
Tensile modulus (GPa)	2.76–3.1	3.4–3.9	2.0–2.4	2.5–2.7
Tensile strength (MPa)	41–52	65–76	26–40	36–40
Elongation at break (%)	1.5–2.5	—	—	15–25
Heat deflection temperature at 1.81 MPa (°C)	82–93	100–105	80–87	80–95
Vicat soft point (°C)	98–107	110	88–101	90–100
Notched Izod (kJ/m)	0.02	0.02	0.1–0.3	0.1–0.5
Linear thermal expansion (10^{-5} mm/mm·K)	5–7	6.4–6.7	7.0–7.5	7.5–9.5
Hardness (Rockwell)	M60–M75	M80–M83	M45, L55	R69–R115
Linear mold shrinkage (in./in.)	0.007	0.003–0.004	0.007	0.0055

8.2.6 ABS

ABS is a terpolymer prepared from the combination of acrylonitrile, butadiene (as polybutadiene), and styrene monomers. Compared to PS, ABS exhibits good impact strength, improved chemical resistance, and similar heat deflection temperature. ABS is also opaque. Properties are a function of the ratio of the three monomers. Typical properites are shown in Table 8.3.

Uses. The previously mentioned properties of ABS make it suitable for tough consumer products; automotive parts; business machine housings; telephones; appliances; luggage; and pipe, fittings, and consuits.

8.2.7 Polyvinyl Chloride

The catalytic polymerization of vinyl chloride yields polyvinyl chloride. It is commonly referred to as PVC or vinyl and is second only to polyethylene in volume use. Normally, PVC has a low degree of crystallinity and good transparency. The high chlorine content of the polymer produces advantages in flame resistance, fair heat deflection temperature, good electrical properties, and good chemical resistance. However, the chlorine also makes PVC difficult to process. The chlorine atoms have a tendency to split out under the influence of heat during processing and heat and light during end use in finished products, producing discoloration and embrittlement. Therefore, special stabilizer systems are often used with PVC to retard degradation.

There are two major sub-classifications of PVC: rigid and flexible (plasticized). In addition, there are also foamed PVC and PVC copolymers. Typical properties of PVC resins appear in Table 8.4.

Rigid PVC

PVC alone is a fairly good rigid polymer, but it is difficult to process and has low impact strength. Both of these properties are improved by the addition of elastomers or impact modified graft copolymers, such as ABS and impact acrylic polymers. These improve the melt flow during processing and improve the impact strength without seriously lowering the rigidity or the heat deflection temperature.

Uses. With this improved balance of properties, rigid PVCs are used in such applications as door and window frames; pipe, fittings, and conduit; building panels and siding; rainwater gutters and down spouts; credit cards; and flooring.

Plasticized PVC

Flexible PVC is a plasticized material. The PVC is softened by the addition of compatible, nonvolatile, liquid plasticizers. The plasticizers, which are usually used in > 20 parts per hundred resins, lower the crystallinity in PVC and act as internal lubricants to give a clear, flexible plastic. Plasticized PVC is also available in liquid formulations known as plastisols or organosols.

Uses. Plasticized PVC is used for wire and cable insulation, outdoor apparel, rainwear, flooring, interior wall covering, upholstery, automotive seat covers, garden hose, toys, clear tubing, shoes, tablecloths, and shower curtains. Plastisols are used in coating fabric, paper, and metal; and rotationally cast into balls, dolls, and so on.

Table 8.4 Typical Property Values for Polyvinyl Chloride Materials

Property	General Purpose	Rigid	Rigid Foam	Plasticized	Copolymer
Density (Mg/m³)	1.40	1.34–1.39	0.75	1.29–1.34	1.37
Tensile modulus (GPa)	3.45	2.41–2.45	—	—	3.15
Tensile strength (MPa)	8.7	37.2–42.4	> 13.8	14–26	52–55
Elongation at break (%)	113	—	> 40	250–400	—
Notched Izod (kJ/m)	0.53	0.74–1.12	> 0.06	—	0.02
Heat deflection temperature at 1.81 MPa (°C)	77	73–77	65	—	65
Brittle temperature (°C)	—	—	—	−60 to −30	—
Hardness	D85 (Shore)	R107–R122 (Rockwell)	D55 (Shore)	A71–A96 (Shore)	—
Linear thermal expansion (10⁻⁵ mm/mm·K)	7.00	5.94	5.58	—	—
Linear mold shrinkage (in./in.)	0.003	—	—	—	—

Foamed PVC

Rigid PVC can be foamed to a low-density cellular material that is used for decorative moldings and trim.

Uses. Foamed plastisols add greatly to the softness and energy absorption already inherent in plasticized PVC, giving richness and warmth to leather-like upholstery, clothing, shoe fabrics, handbags, luggage, and auto door panels; and energy absorption for quiet and comfort in flooring, carpet backing, auto headliners, and so on.

PVC Copolymers

Copolymerization of vinyl chloride with 10–15% vinyl acetate gives a vinyl polymer with improved flexibility and less crystallinity than PVC, making such copolymers easier to process without detracting seriously from the rigidity and heat deflection temperature. These copolymers find primary applications in flooring and solution coatings.

8.2.8 Poly(vinylidene chloride)

Poly(vinylidene chloride) is prepared by the catalytic polymerization of 1,1-dichloroethylene. This crystalline polymer exhibits high strength, abrasion resistance, high melting point, better than ordinary heat resistance (100°C maximum service temperature), and outstanding impermeability to oil, grease, water vapor, oxygen, and carbon dioxide. It is used for packaging films, coatings, and monofilaments.

When the polymer is extruded into film, quenched, and oriented, the crystallinity is fine enough to produce high clarity and flexibility. These properties contribute to widespread use in packaging film, especially for food products that require impermeable barrier protection.

Poly(vinylidene chloride) and/or copolymers with vinyl chloride, alkyl acrylate, or acrylonitrile are used in coating paper, paperboard, or other films to provide more economical, impermeable materials.

A small amount of poly(vinylidene chloride) is extruded into monofilament and tape that is used in outdoor furniture upholstery.

8.2.9 Poly(methyl Methacrylate)

The catalytic polymerization of methylmethacrylate yields poly(methyl methacrylate) (PMMA), a strong, rigid, clear, amorphous polymer. PMMA has excellent resistance to weathering, low water absorption, and good electrical resistivity. PMMA properties appear in Table 8.5.

Uses. PMMA is used for glazing, lighting diffusers, skylights, outdoor signs, and automobile taillights.

8.2.10 Poly(ethylene Terephthalate)

Poly(ethylene terephthalate) (PET) is prepared from the condensation polymerization of dimethyl terephthalate and ethylene glycol. PET is a crystalline polymer that exhibits high modulus, high strength, high melting point, good electrical properties, and moisture and solvent resistance. PET crystallizes slowly, hence blow-molded and extruded objects are clear. Injection-molding grades are nucleated to facilitate crystallization and shorten the molding cycle. Nucleated PET resins are opaque.

Uses. Primary applications of PET include blow-molded beverage bottles; fibers for wash and wear, wrinkle-resistant fabrics; and films that are used in food packaging, electrical applications (capacitors, etc.), magnetic recording tape, and graphic arts.

Table 8.5 Typical Properties of Poly(methyl Methacrylate)

Property	PMMA
Density (Mg/m³)	1.18–1.19
Tensile modulus (GPa)	3.10
Tensile strength (MPa)	72
Elongation at break (%)	5
Notched Izod (kJ/m)	0.4
Heat deflection temperature at 1.81 MPa (°C)	96
Continuous service temperature (°C)	88
Hardness (Rockwell)	M90–M100
Linear thermal expansion (10^{-5} mm/mm·K)	6.3
Linear mold shrinkage (in./in.)	0.002–0.008

8.3 ENGINEERING THERMOPLASTICS

Engineering thermoplastics comprise a special high-performance segment of synthetic plastic materials that offer premium properties. When properly formulated, they may be shaped into mechanically functional, semiprecision parts or structural components. "Mechanically functional" implies that the parts may be subjected to mechanical stress, impact, flexure, vibration, sliding friction, temperature extremes, hostile environments, etc., and continue to function.

As substitutes for metal in the construction of mechanical apparatus, engineering plastics offer advantages such as transparency, light weight, self-lubrication, and economy in fabrication and decorating. Replacement of metals by plastic is favored as the physical properties and operating temperature ranges of plastics improve and as the cost of metals and their fabrication increases.

8.3.1 Polyesters (Thermoplastic)

Poly(butylene terephthalate) (PBT) is prepared from the condensation polymerization of butanediol with dimethyl terephthalate. PBT is a crystalline polymer that has a fast rate of crystallization, which facilitates rapid molding cycles. It seems to have a unique and favorable balance of properties between polyamides and polyacetals. PBT has low moisture absorption, extremely good self-lubricity, fatigue resistance, solvent resistance, and good maintenance of mechanical properties at elevated temperatures. PBT resins are often used with reinforcing materials like glass fiber to enhance strength, modulus and heat deflection temperature. Properties appear in Table 8.6.

Uses. Applications of PBT include gears, rollers, bearing, housings for pumps, and appliances, impellers, pulleys, switch parts, automotive components, and electrical/electronic components. A high-density PBT is used in countertops and sinks.

8.3.2 Polyamides (Nylon)

The two major types of polyamides (PA) are nylon 6 (PA6) and nylon 66 (PA66). Polycaprolactam or nylon 6 is prepared by the polymerization of caprolactam. Poly(hexamethylene adipamide) or nylon 66 is derived from the condensation polymerization of hexamethylene diamine with adipic acid. Polyamides are crystalline polymers. Nylon's key features include a high degree of solvent resistance, toughness, and fatigue resistance. Nylons do exhibit a tendency to creep under applied load. Glass fibers or mineral fillers are often used to enhance the properties of polyamides. In addition, the properties of nylon are greatly affected by moisture, which acts as a plasticizer. Properties of nylon 6 and 66 with and without glass fiber appear in Table 8.7.

Uses. The largest application of nylons is in fibers. Molded applications include automotive components, related machine parts (gears, cams, pulleys, rollers, boat propellers, etc.), appliance parts, and electrical insulation.

Modified Polyamides

Moisture has a profound effect on the properties of polyamides. Water acts as a plasticizer in polyamides, lowering their rigidity and strength while increasing their ductility. Moreover, an increase in moisture has a negative effect on dimensional stability. Polyamides have been modified by blending with poly(phenylene ether) (PPE) in order to minimize the effect of moisture. In PA/PPE alloys, the polyamide is the continuous phase and imparts good solvent resistance. The PPE is a dispersed phase and acts as a reinforcement of the crystalline nylon matrix, giving improved stiffness and toughness versus the unfilled nylon resin. Since PPE does not absorb any significant amount of moisture, the

Table 8.6 Typical Properties of Poly(butylene Terephthalate)

Property	PBT	PBT + 40% Glass Fiber
Density (Mg/m^3)	1.300	1.600
Flexural modulus (GPa)	2.4	9.0
Flexural strength (MPa)	88	207
Elongation at break (%)	300	3
Notched Izod (kJ/m)	0.06	0.12
Heat deflection temperature at 0.45 MPa (°C)	154	232
Heat deflection temperature at 1.81 MPa (°C)	54	232
Hardness (Rockwell)	R117	M86
Linear thermal expansion (10^{-5} mm/mm·K)	9.54	1.89
Linear mold shrinkage (in./in.)	0.020	< 0.007

Table 8.7 Typical Properties of Polyamides

Property	PA6	PA6 + 40% Glass Fiber	PA66	PA66 + 40% Glass Fiber
Density (Mg/m³)	1.130	1.460	1.140	1.440
Flexural modulus (GPa)	2.8	10.3	2.8	9.3
Flexural strength (MPa)	113	248	—	219
Elongation at break (%)	150	3	60	4
Notched Izod (kJ/m)	0.06	0.16	0.05	0.14
Heat deflection temperature at 0.45 MPa (°C)	170	218	235	260
Heat deflection temperature at 1.81 MPa (°C)	64	216	90	250
Hardness (Rockwell)	R119	M92	R121	M119
Linear thermal expansion (10^{-5} mm/mm·K)	8.28	2.16	8.10	3.42
Linear mold shrinkage (in./in.)	0.013	0.003	0.0150	0.0025

effect of moisture on properties and dimensional stability is reduced in PPE/PA blends versus polyamides. In addition, heat deflection temperatures are enhanced. Properties are shown in Table 8.8.

Uses. PA/PPE alloys are used in automotive body panels (fenders and quarter panels), automotive wheel covers, exterior truck parts, under-the-hood automotive parts (air intake resonators, electrical junction boxes and connectors), fluid handling applications (pumps, etc.).

8.3.3 Polyacetals

Polyacetals are prepared via the polymerization of formaldehyde or the copolymerization of formaldehyde with ethylene oxide. Polyacetals are crystalline polymers that exhibit rigidity, high strength, solvent resistance, fatique resistance, toughness, self-lubricity, and cold-flow resistance. They also exhibit a tendency to thermally depolymerize and, hence are difficult to flame-retard. Properties are enhanced by the addition of glass fiber or mineral fillers. Typical properties appear in Table 8.9.

Uses. Applications of polyacetals include moving parts in appliances and machines (gears, bearings, bushings, etc.), in automobiles (door handles, etc.), and in plumbing (valves, pumps, faucets, etc.).

8.3.4 Polyphenylene Sulfide

The condensation polymerization of dichlorobenzene and sodium sulfide yields a crystalline polymer, polyphenylene sulfide (PPS). It is characterized by high heat resistance, rigidity, excellent chemical resistance, low friction coefficient, good abrasion resistance, and electrical properties. PPS is somewhat difficult to process due to the very high melting temperature, relatively poor flow characteristics, and some tendency for slight cross linking during processing. PPS resins normally contain glass fibers for mineral fillers. Properties appear in Table 8.10.

Uses. The unreinforced resin is used only in coatings. The reinforced materials are used in aerospace applications, pump components, electrical/electronic components, appliance parts, and in automotive applications.

Table 8.8 Typical Properties of PPE/Polyamide 66 Alloys

Property	Unfilled PA	Unfilled PPE/PA	10% Glass Fiber PA	10% Glass Fiber PPE/PA	30% Glass Fiber PA	30% Glass Fiber PPE/PA
Density (Mg/m³)	1.14	1.10	1.204	1.163	1.37	1.33
Flexural modulus (GPa)						
Dry as molded	2.8	2.2	4.5	3.8	8.3	8.1
100% relative humidity	0.48	0.63	2.3	2.6	4.1	5.8
at 150°C	0.21	0.70	0.9	1.6	3.2	4.3
Flexural strength (MPa)						
Dry as molded	96	92	151	146	275	251
100% relative humidity	26	60	93	109	200	210
at 150°C	14	28	55	60	122	128

Table 8.9   Typical Properties of Polyacetals

Property	Polyacetal	Polyacetal + 40% Glass Fiber
Density (Mg/m³)	1.420	1.740
Flexural modulus (GPa)	2.7	11.0
Flexural strength (MPa	107	117
Elongation at break (%)	75	1.5
Notched Izod (kJ/m)	0.12	0.05
Heat deflection temperature at 0.45 MPa (°C)	170	167
Heat deflection temperature at 1.81 MPa (°C)	124	164
Hardness (Rockwell)	M94	R118
Linear thermal expansion (10^{-5} mm/mm·K)	10.4	3.2
Linear mold shrinkage (in./in.)	0.02	0.003

8.3.5   Polycarbonates

Most commercial polycarbonates are derived from the reaction of bisphenol A and phosgene. Polycarbonates (PCs) are transparent amorphous polymers. PCs are among the stronger, tougher, and more rigid thermoplastics. Polycarbonates also show resistance to creep and excellent electrical insulating characteristics. Polycarbonate properties are shown in Table 8.11.

Uses. Applications of PC include safety glazing, safety shields, non-breakable windows, automotive taillights, lenses, electrical relay covers, various appliance parts and housings, power tool housings, automotive exterior parts, and blow-molded bottles.

Polycarbonate/ABS Alloys

PC/ABS blends are prepared by extruder blending of PC and ABS resins and offer a unique balance of properties. The addition of ABS improves the melt processing of the blend, which facilitates filling large, thin-walled parts. The toughness (especially at low temperatures) of PC is enhanced by the blending with ABS while maintaining the high strength and rigidity. The properties are a function of the ratio of ABS to polycarbonate. Properties appear in Table 8.12.

Uses. PC/ABS is used in automotive body panels (doors), housewares (small appliances). PC/ABS has become the resin of choice for business equipment because of the combination of processing ease and toughness.

8.3.6   Polysulfone

Polysulfone is prepared from the condensation polymerization of bisphenol A and dichlorodiphenyl sulfone. The transparent, amorphous resin is characterized by excellent thermo-oxidative stability, hydrolytic stability, and creep resistance. Properties appear in Table 8.13.

Uses. Typical applications of polysulfones include microwave cookware, medical equipment where sterilization by steam is required, coffee makers, and electrical/electronic components.

Table 8.10   Typical Properties of Poly(phenylene Sulfide)

Property	PPS + 40% Glass Fiber
Density (Mg/m³)	1.640
Tensile modulus (GPa)	7.7
Tensile strength (MPa)	135
Elongation at break (%)	1.3
Flexural modulus (GPa)	11.7
Flexural strength (MPa)	200
Notched Izod (kJ/m)	0.08
Heat deflection temperature at 1.81 MPa (°C)	> 260
Constant service temperature (°C)	232
Hardness (Rockwell)	R123
Linear thermal expansion (10^{-5} mm/mm·K)	4.0
Linear mold shrinkage (in./in.)	0.004

Table 8.11 Typical Properties of Polycarbonates

Property	PC	PC + 40% Glass Fiber
Density (Mg/m³)	1.200	1.520
Tensile modulus (GPa)	2.4	11.6
Tensile strength (MPa)	65	158
Elongation at break (%)	110	4
Flexural modulus (GPa)	2.3	9.7
Flexural strength (MPa)	93	186
Notched Izod (kJ/m)	0.86	0.13
Heat deflection temperature at 0.45 MPa (°C)	138	154
Heat deflection temperature at 1.81 MPa (°C)	132	146
Constant service temperature (°C)	121	135
Hardness (Rockwell)	M70	M93
Linear thermal expansion (10^{-5} mm/mm·K)	6.74	1.67
Linear mold shrinkage (in./in.)	0.006	0.0015

8.3.7 Modified Polyphenylene Ether

Poly(2,6-dimethyl phenylene ether) (PPE) is prepared by the polymerization of 2,6-dimethylphenol. This amorphous polymer has a very high glass transition temperature, high heat deflection temperature, and no hydrolizable bonds. PPE is usually blended with styrenics (i.e., HIPS, ABS, etc.) to form a family of modified polyphenylene ether-based resins (and with polyamides, as described earlier). These amorphous blends cover a wide range of heat deflection temperatures, depending on the ratio of PPE to HIPS. They are characterized by high toughness, outstanding dimensional stability at elevated temperatures, outstanding hydrolytic stability, long-term stability under load, and excellent dielectric properties over a wide range of frequencies and temperatures. Their properties appear in Table 8.14.

Uses. Applications include automotive (instrument panels, trim, etc.), TV cabinets, electrical connectors, pumps, plumbing fixtures, and small appliances.

8.3.8 Polyimides

Polyimides are a class of polymers prepared from the condensation reaction of a dicarboxylic acid anhydride with a diamine. Thermoplastic and thermoset grades of polyimides are available. The thermoset polyimides are among the most heat-resistant polymers; they can withstand temperatures up to 250°C. Thermoplastic polyimides, which can be processed by standard techniques, fall into two main categories — polyetherimides (PEI) and polyamideimides (PAI).

In general, polyimides have high heat resistance, high deflection temperatures, very good electrical properties, very good wear resistance, superior dimensional stability, outstanding flame resistance, and very high strength and rigidity. Polyimide properties appear in Table 8.15.

Uses. Polyimide applications include gears, bushings, bearings, seals, insulators, electrical/electronic components (printed wiring boards, connectors, etc.), cooking utensils, microwave oven components, and structural components.

Table 8.12 Typical Properties of Polycarbonates/ABS Blends

Properties	PA/ABS Ratio (wt/wt)			
	0/100	50/50	80/20	100/00
Density (Mg/m³)	1.06	1.13	1.17	1.20
Tensile modulus (GPa)	1.8	1.9	2.5	2.4
Tensile strength (MPa)	40	57	60	65
Elongation at break (%)	20	70	150	110
Notched Izod:				
at 25°C (kJ/m)	0.30	0.69	0.75	0.86
at −20°C (kJ/m)	0.11	0.32	0.64	0.15
Heat deflection temperature at 1.81 MPa (°C)	80	100	113	132

Table 8.13 Typical Properties of Polysulfone

Property	Polysulfone
Density (Mg/m³)	1.240
Tensile modulus (GPa)	2.48
Tensile strength (MPa)	70
Elongation at break (%)	75
Flexural modulus (GPa)	2.69
Flexural strength (MPa)	106
Notched Izod (kJ/m)	0.07
Heat deflection temperature at 1.81 MPa (°C)	174
Constant service temperature (°C)	150
Hardness (Rockwell)	M69
Linear thermal expansion (10^{-5} mm/mm·K)	5.6
Linear mold shrinkage (in./in.)	0.007

8.4 FLUORINATED THERMOPLASTICS

In general, fluoropolymers or fluoroplastics are a family of fluorine-containing thermoplastics that exhibit some unusual properties. These properties include inertness to most chemicals, resistance to high temperatures, extremely low coefficient of friction, and excellent dielectric properties. Mechanical properties are normally low, but can be enhanced with glass or carbon fiber or molybdenum disulfide fillers. Properties are shown in Table 8.16.

8.4.1 Poly(tetrafluoroethylene)

Poly(tetrafluoroethylene) (PTFE) is a crystalline, very heat-resistant (up to 250°C) chemical-resistant polymer. PTFE has the lowest coefficient of friction of any polymer. It does not soften like other thermoplastics, and has to be processed by unconventional techniques (PTFE powder is compacted to the desired shape and sintered).

Uses. PTFE applications include non-stick coatings on cookware; non-lubricated bearings; chemical-resistant pipe, fittings, valves, and pump parts; high-temperature electrical parts; and gaskets, seals, and packings.

8.4.2 Poly(chlorotrifluoroethylene)

Poly(chlorotrifluoroethylene) (CTFE) is less crystalline and exhibits higher rigidity and strength than PTFE. Poly(chlorotrifluoroethylene) has excellent chemical resistance and heat resistance up to 200°C. Unlike PTFE, CTFE can be molded and extruded by conventional processing techniques.

Table 8.14 Typical Properties of Modified Polyphenylene Ether Resins

Property	190 Grade	225 Grade	300 Grade
Density (Mg/m³)	1.080	1.090	1.060
Tensile modulus (GPa)	2.5	2.4	—
Tensile strength (MPa)	48	55	76
Elongation at break (%)	35	35	—
Flexural modulus (GPa)	2.2	2.4	2.4
Flexural strength (MPa)	57	76	104
Notched Izod (kJ/m)	0.37	0.32	0.53
Heat deflection temperature at 0.45 MPa (°C)	96	118	157
Heat deflection temperature at 1.81 MPa (°C)	88	107	149
Constant service temperature (°C)	—	95	—
Hardness (Rockwell)	R115	R116	R119
Linear thermal expansion (10^{-5} mm/mm·K)	—	—	5.9
Linear mold shrinkage (in./in.)	0.006	0.006	0.006

Table 8.15 Typical Properties of Polyimides

Property	Polyimide	Polyetherimide		Polyamideimide	
		Unfilled	30% Glass Fiber	Unfilled	30% Glass Fiber
Density (Mg/m³)	—	1.27	1.51	1.38	1.57
Tensile modulus (GPa)	2.65	2.97	10.3	4.83	10.7
Tensile strength (MPa)	195	97	193	117	205
Elongation at break (%)	90	60	3	10	5
Notched Izod (kJ/m)	—	0.6	0.11	0.13	0.11
Heat deflection temperature at 0.45 MPa (°C)	—	410	414	—	—
Heat deflection temperature at 1.81 MPa (°C)	—	392	410	260	274
Constant service temperature (°C)					
Hardness (Rockwell)	—	R109	M125	E78	E94
Linear thermal expansion (10⁻⁵ mm/mm·K)	—	5.6	2.0	3.60	1.80
Linear mold shrinkage (in./in.)	—	0.5	0.2	—	0.25

Uses. CTFE applications include electrical insulation, cable jacketing, electrical and electronic coil forms, pipe and pump parts, valve diaphragms, and coatings for corrosive process equipment and other industrial parts.

8.4.3 Fluorinated Ethylene-Propylene

Copolymerization of tetrafluoroethylene with some hexafluoropropylene produces fluorinated ethylene–propylene polymer (FEP), which has less crystallinity, lower melting point, and improved impact strength than PTFE. FEP can be molded by normal thermoplastic techniques.

Uses. FEP applications include wire insulation and jacketing, high-frequency connectors, coils, gaskets, and tube sockets.

8.4.4 Polyvinylidene Fluoride

Polyvinylidene fluoride (PVDF) has high tensile strength and better ability to be processed but less thermal and chemical resistance than FEP, CTFE, and PTFE.

Table 8.16 Typical Properties of Fluoropolymers

Property	PTFE	CTFE	FEP	ETFE	ECTFE
Density (Mg/m³)	2.160	2.100	2.150	1.700	1.680
Tensile modulus (GPa)	—	14.3	—	—	—
Tensile strength (MPa)	27.6	39.4	20.7	44.8	48.3
Elongation at break (%)	~ 275	~ 150	~ 300	100–300	200
Notched Izod (kJ/m)	—	0.27	0.15	—	—
Heat deflection temperature at 0.45 MPa (°C)	—	126	—	104	116
Heat deflection temperature at 1.81 MPa (°C)	—	75	—	71	77
Constant service temperature (°C)	260	199	204	—	150–170
Hardness	D55–65 (Shore)	D75–80 (Shore)	D55 (Shore)	D75 (Shore)	R93 (Rockwell)
Dielectric strength (MV/m)	23.6	19.7	82.7	7.9	19.3
Dielectric constant at 10² Hz	2.1	3.0	2.1	2.6	2.5
Dielectric constant at 10³ Hz	2.1	2.7	—	2.6	2.5
Linear thermal expansion (10⁻⁵ mm/mm·K)	9.9	4.8	9.3	13.68	—
Linear mold shrinkage (in./in.)	0.033–0.053	0.008	–	–	< 0.025

Table 8.17 Properties of General Purpose Elastomers

Rubber	ASTM Nomenclature	Outstanding Characteristic	Property Deficiency	Temperature Use Range (°C)
Butadiene rubber	BR	Very flexible; resistance to wear	Sensitive to oxidation; poor resistance to fuels and oil	−100 to 90
Natural rubber	NR	Similar to BR but less resilient	Similar to BR	−50 to 80
Isoprene rubber	IR	Similar to BR but less resilient	Similar to BR	−50 to 80
Isobutylene–isoprene rubber (butyl rubber)	IIR	High flexibility; low permeability to air		−45 to 150
Chloroprene	CR	Flame resistant; fair fuel and oil resistance; increased resistance toward oxygen, ozone, heat, light	Poor low temperature flexibility	−40 to 115
Nitrile-butadiene	NBR	Good resistance to fuels, oils, and solvents; improved abrasion resistance	Lower resilience; higher hysteresis; poor electrical properties; poorer low temperature flexibility	−50 to 80
Styrene–butadiene rubber	SBR	Relatively low cost	Less resilience; higher hysteresis; limited low temperature flexibility	−50 to 80
Ethylene-propylene copolymer	EPDM	Resistance to ozone and weathering	Poor hydrocarbon and oil resistance	−50 to < 175
Polysulfide	T	Chemical resistance; resistance to ozone and weathering	Creep; low resilience	−45 to 120

126

Table 8.18 Properties of Specialty Elastomers

Elastomer	ASTM Nomenclature	Temperature Use Range (°C)	Outstanding Characteristic	Typical Applications
Silicones (polydimethylsiloxane)	MQ	−100 to 300	Wide temperature range; resistance to aging, ozone, sunlight; very high gas permeability	Seals, molded and extruded goods; adhesives, sealants; biomedical; personal care products
Fluoroelastomers	CFM	−40 to 200	Resistance to heat, oils, chemical	Seals such as O-rings, corrosion resistant coatings
Acrylic	AR	−40 to 200	Oil, oxygen, ozone, and sunlight resistance	Seals, hose
Epichlorohydrin	ECO	−18 to 150	Resistance to oil, fuels; some flame resistance; low gas permeability	Hose, tubing, coated fabrics, vibration isolators
Chlorosulfonated	CSM	−40 to 150	Resistance to oil, ozone weathering, oxidizing chemicals	Automotive hose, wire and cable, linings for reservoirs
Chlorinated polyethylene	CM	−40 to 150	Resistance to oils, ozone, chemicals	Impact modifier, automotive applications
Ethylene–acrylic		−40 to 175	Resistance to ozone, weathering	Seals, insulation, vibration damping
Propylene oxide		−6– to 150	Low temperature properties	Motor mounts

127

Uses. Polyvinylidene fluoride applications include seals and gaskets, diaphragms, and piping.

8.4.5 Poly(ethylene chlorotrifluoroethylene)

The copolymer of ethylene and chlorotrifluoroethylene is poly(ethylene chlorotrifluoroethylene) (ECTFE). It has high strength and chemical and impact resistance. ECTFE can be processed by conventional techniques.

Uses. Poly(ethylene chlorotrifluoroethylene) applications include wire and cable coatings, chemical resistant coatings and linings, molded lab ware, and medical packing.

8.4.6 Poly(vinyl fluoride)

Poly(vinyl fluoride) films exhibit excellent outdoor durability. It is the least chemical-resistant fluoropolymer.

Uses. Poly(vinyl fluoride) uses include glazing, lighting, and coatings on presurfaced exterior building panels.

8.5 THERMOSETS

Thermosetting resins are used in molded and laminated plastics. They are first polymerized into a low molecular weight, linear or slightly branched polymer or oligomer that is still soluble, fusible, and highly reactive during final processing. Thermoset resins are generally highly filled with mineral fillers and glass fibers. Thermosets are generally catalyzed and/or heated to finish the polymerization reaction, which cross links them to almost infinite molecular weight. This step is often referred to as curing. Such cured polymers cannot be reprocessed or reshaped.

The high filler loading and the high cross-link density of thermoset resins result in very high densities and very low ductility, but very high rigidity and good chemical resistance.

8.5.1 Phenolic Resins

Phonolic resins combine the high reactivity of phenol and formaldehyde to form pre-polymers and oligomers called resoles and novolaks. These materials are combined with fibrous fillers to give a phenolic resin, which when heated provides rapid, complete cross linking to cured structures. The highly cross-linked aromatic structure has high hardness, rigidity, strength, heat resistance, chemical resistance, and good electrical properties.

Uses. Phenolic applications include automotive uses (distributor caps, rotors, brake linings), appliance parts (pot handles, knobs, bases, electrical/electronic components (connectors, circuit breakers, switches), and adhesive in laminates (e.g., plywood).

8.5.2 Epoxy Resins

The most common epoxy resins are prepared from the reaction of bisphenol A and epichlorohydrin to yield low molecular weight resins that are liquid either at room temperature or on warming. Each polymer chain usually contains two or more epoxy groups. The high reactivity of the epoxides with amines, anhydrides, and other curing agents provides facile conversion into highly cross-linked materials. Cured epoxy resins exhibit hardness, strength, heat resistance, electrical resistance, and broad chemical resistance.

Uses. Epoxy resins are used in glass-reinforced, high-strength composites in aerospace, pipes, tanks, pressure vessels; encapsulation or casting of various electrical and electronic components (printed wiring boards, etc.); adhesives; protective coatings in appliances, flooring, and industrial equipment; and sealants.

8.5.3 Unsaturated Polyesters

Unsaturated polyesters are prepared by the condensation polymerization of various diols and maleic anhydride to give a very viscous liquid that is then dissolved in styrene monomer. The addition of styrene lowers the viscosity to a level suitable for impregnation and lamination of glass fibers. The low molecular weight polyester has numerous fumarate ester units that provide easy reactivity with styrene monomer. In combination with reinforcing materials like glass fiber, cured resins offer outstanding strength, high rigidity, high strength-to-weight ratio, impact strength, and chemical resistance. Properly formulated, reinforced unsaturated polyesters are commonly referred to as sheet molding compound (SMC) or reinforced plastics. SMC typically is formulated with 50% calcium carbonate filler, 25% long glass fiber (> 1 in.), and 25% unsaturated polyester. The highly filled nature of SMC results in high density and a brittle, easily pitted surface.

Bulk molding compound (BMC) is formulated similarly to SMC except that 1/4-inch chopped glass is used. The shorter glass length gives easier process but lower strength and impact.

Uses. The prime use of unsaturated polyesters is in combination with glass fibers in high-strength composites and in SMC and BMC materials. The applications include transportation markets (large body parts for automobiles, trucks, trailers, buses, and aircraft), marine markets (small to medium-sized boat hulls and associated marine equipment), building panels, housing and bathroom components (bathtub and shower stalls), appliances, and electrical/electronic components.

8.5.4 Alkyd Resins

Alkyd resins are based on branched pre-polymers from glycerol, phthalic anhydride, and glyceryl esters of fatty acids. Alkyds have excellent heat resistance, are dimensionally stable at high temperatures, and have excellent dielectric strength (> 14 MV/m), high resistance to electrical leakage, and excellent arc resistance.

Uses. Alkyd resin applications include drying oils in enamel paints, lacquers for automobiles and appliances; and molding compounds when formulated with reinforcing fillers for electrical applications (circuit breaker insulation, encapsulation of capacitors and resistors, and coil forms).

8.5.5 Diallyl Phthalate

Diallyl Phthalate (DAP) is the most widely used compound in the allylic family of thermosets. The neat resin is a medium viscosity liquid. These low molecular weight pre-polymers can be reinforced and compression molded into highly cross-linked, completely cured products.

The most outstanding properties of cured DAP are excellent dimensional stability and high insulation resistance. In addition, DAP materials have high dielectric strength and excellent arc resistance and chemical resistance.

Uses. DAP applications include electronic parts, electrical connectors, bases, and housings. DAP is also used as a coating and impregnating material.

8.5.6 Amino Resins

The two main members of the amino family of thermosets are the melamine- and urea-based resins. They are prepared from the reaction of melamine and urea with formaldehyde. In general, these materials exhibit extreme hardness, scratch resistance, electrical resistance, and chemical resistance.

Uses. Melamine resins find use in colorful, rugged dinnerware; decorative laminates (countertops, tabletops, and furniture surfacing); electrical applications (switchboard panels, circuit breaker parts, arc barriers, and armature and slot wedges); and adhesives and coatings.

Urea resins are used in particle board binders, decorative housings, closures, elecrical parts, coatings, and paper and textile treatment.

8.6 GENERAL-PURPOSE ELASTOMERS

Elastomers are polymers that can be stretched substantially beyond their original length and will retract rapidly and forcibly to essentially their original dimensions upon release of the force.

The optimum properties and/or economics of many rubbers are obtained through formulating with reinforcing agents, fillers, extending oils, vulcanizing agents, antioxidants, pigments, and so on. End-use markets for formulated rubbers include automotive tire products (including tubes, retread applications, valve stems, and inner liners), adhesives, cements, caulks, sealants, latex foam products, hose (automotive, industrial, and consumer applications), belting (V-conveyor and trimming), footwear (heels, soles, slab stock, boots, and canvas), and molded, extruded, and calendered products (athletic goods, flooring, gaskets, household products, O-rings, blown sponge, thread, and rubber sundries). A list of general-purpose elastomers and properties is summarized in Table 8.17.

8.7 SPECIALTY ELASTOMERS

Specialty rubbers offer higher performance than general-purpose rubbers and find use in more demanding applications. They are more costly and hence are produced in smaller volumes. Properties and uses are summarized in Table 8.18.

CHAPTER 9

COMPOSITE MATERIALS AND MECHANICAL DESIGN

Carl Zweben
Lockheed Martin Missiles and Space—Valley Forge Operations
King of Prussia, Pennsylvania

9.1	**INTRODUCTION**	**131**
	9.1.1 Classes and Characteristics of Composite Materials	132
	9.1.2 Comparative Properties of Composite Materials	133
	9.1.3 Manufacturing Considerations	136
9.2	**REINFORCEMENTS AND MATRIX MATERIALS**	**136**
	9.2.1 Reinforcements	137
	9.2.2 Matrix Materials	139
9.3	**PROPERTIES OF COMPOSITE MATERIALS**	**143**
	9.3.1 Mechanical Properties of Composite Materials	144
	9.3.2 Physical Properties of Composite Materials	153
9.4	**PROCESSES**	**161**
	9.4.1 Polymer Matrix Composites	163
	9.4.2 Metal Matrix Composites	163
	9.4.3 Ceramic Matrix Composites	163
	9.4.4 Carbon/Carbon Composites	163
9.5	**APPLICATIONS**	**163**
	9.5.1 Aerospace and Defense	164
	9.5.2 Machine Components	166
	9.5.3 Electronic Packaging and Thermal Control	168
	9.5.4 Internal Combustion Engines	168
	9.5.5 Transportation	170
	9.5.6 Process Industries, High-Temperature Applications, and Wear-, Corrosion-, and Oxidation-Resistant Equipment	176
	9.5.7 Offshore and Onshore Oil Exploration and Production Equipment	178
	9.5.8 Dimensionally Stable Devices	178
	9.5.9 Biomedical Applications	179
	9.5.10 Sports and Leisure Equipment	180
	9.5.11 Marine Structures	182
	9.5.12 Miscellaneous Applications	182
9.6	**DESIGN AND ANALYSIS**	**184**
	9.6.1 Polymer Matrix Composites	185
	9.6.2 Metal Matrix Composites	187
	9.6.3 Ceramic Matrix Composites	187
	9.6.4 Carbon/Carbon Composites	187

9.1 INTRODUCTION

The development of composite materials and related design and manufacturing technologies is one of the most important advances in the history of materials. Composites are multifunctional materials having unprecedented mechanical and physical properties that can be tailored to meet the requirements of a particular application. Many composites also exhibit great resistance to high-temperature corrosion and oxidation and wear. These unique characteristics provide the mechanical engineer with design opportunities not possible with conventional monolithic (unreinforced) materials. Composites technology also makes possible the use of an entire class of solid materials, ceramics, in applications for which monolithic versions are unsuited because of their great strength scatter and poor resistance to mechanical and thermal shock. Further, many manufacturing processes for composites are well adapted to the fabrication of large, complex structures, which allows consolidation of parts, reducing manufacturing costs.

Mechanical Engineers' Handbook, 2nd ed., Edited by Myer Kutz.
ISBN 0-471-13007-9   © 1998 John Wiley & Sons, Inc.

Composites are important materials that are now used widely, not only in the aerospace industry, but also in a large and increasing number of commercial mechanical engineering applications, such as internal combustion engines; machine components; thermal control and electronic packaging; automobile, train, and aircraft structures and mechanical components, such as brakes, drive shafts, flywheels, tanks, and pressure vessels; dimensionally stable components; process industries equipment requiring resistance to high-temperature corrosion, oxidation, and wear; offshore and onshore oil exploration and production; marine structures; sports and leisure equipment; and biomedical devices.

It should be noted that biological structural materials occurring in nature are typically some type of composite. Common examples are wood, bamboo, bone, teeth, and shell. Further, use of artificial composite materials is not new. Straw-reinforced mud bricks were employed in biblical times. Using modern terminology, discussed later, this material would be classified as an organic fiber-reinforced ceramic matrix composite.

In this chapter, we consider the properties of reinforcements and matrix materials (Section 9.2), properties of composites (Section 9.3), how they are made (Section 9.4), their use in mechanical engineering applications (Section 9.5), and special design considerations for composites (Section 9.6).

9.1.1 Classes and Characteristics of Composite Materials

There is no universally accepted definition of a composite material. For the purpose of this work, we consider a composite to be a material consisting of two or more distinct phases, bonded together.[1]

Solid materials can be divided into four categories: polymers, metals, ceramics, and carbon, which we consider as a separate class because of its unique characteristics. We find both reinforcements and matrix materials in all four categories. This gives us the ability to create a limitless number of new material systems with unique properties that cannot be obtained with any single monolithic material. Table 9.1 shows the types of material combinations now in use.

Composites are usually classified by the type of material used for the matrix. The four primary categories of composites are polymer matrix composites (PMCs), metal matrix composites (MMCs), ceramic matrix composites (CMCs), and carbon/carbon composites (CCCs). At this time, PMCs are the most widely used class of composites. However, there are important applications of the other types, which are indicative of their great potential in mechanical engineering applications.

Figure 9.1 shows the main types of reinforcements used in composite materials: aligned continuous fibers, discontinuous fibers, whiskers (elongated single crystals), particles, and numerous forms of fibrous architectures produced by textile technology, such as fabrics and braids. Increasingly, designers are using hybrid composites that combine different types of reinforcements to achieve more efficiency and to reduce cost.

A common way to represent fiber-reinforced composites is to show the fiber and matrix separated by a slash. For example, carbon fiber-reinforced epoxy is typically written "carbon/epoxy," or, "C/Ep." We represent particle reinforcements by enclosing them in parentheses followed by "p"; thus, silicon carbide (SiC) particle-reinforced aluminum appears as "(SiC)p/Al."

Composites are strongly heterogeneous materials; that is, the properties of a composite vary considerably from point to point in the material, depending on which material phase the point is located in. Monolithic ceramics and metallic alloys are usually considered to be homogeneous materials, to a first approximation.

Many artificial composites, especially those reinforced with fibers, are anisotropic, which means their properties vary with direction (the properties of isotropic materials are the same in every direction). This is a characteristic they share with a widely used natural fibrous composite, wood. As for wood, when structures made from artificial fibrous composites are required to carry load in more than one direction, they are used in laminated form.

Many fiber-reinforced composites, especially PMCs, MMCs, and CCCs, do not display plastic behavior as metals do, which makes them more sensitive to stress concentrations. However, the absence of plastic deformation does not mean that composites are brittle materials like monolithic ceramics. The heterogeneous nature of composites results in complex failure mechanisms that impart toughness. Fiber-reinforced materials have been found to produce durable, reliable structural components in countless applications. The unique characteristics of composite materials, especially anisotropy, require the use of special design methods, which are discussed in Section 9.6.

Table 9.1 Types of Composite Materials

Reinforcement	Matrix			
	Polymer	Metal	Ceramic	Carbon
Polymer	X	X	X	X
Metal	X	X	X	X
Ceramic	X	X	X	X
Carbon	X	X	X	X

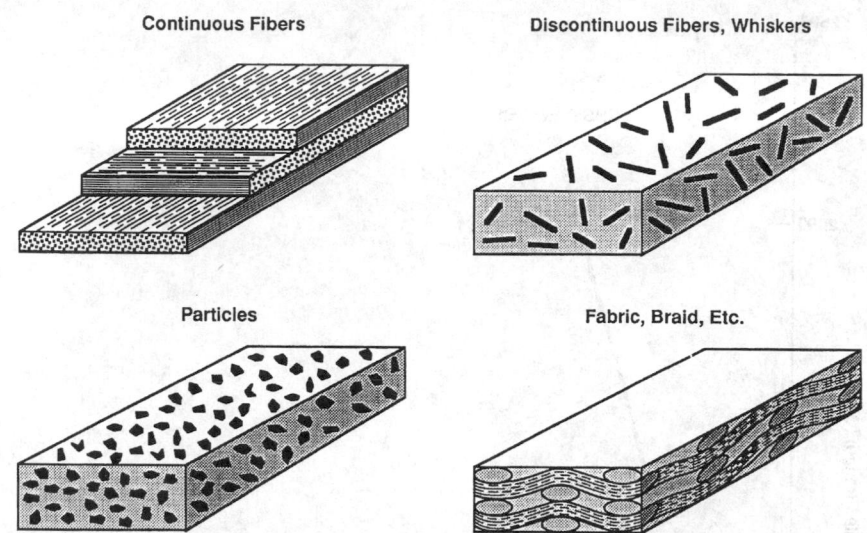

Fig. 9.1 Reinforcement forms.

9.1.2 Comparative Properties of Composite Materials

There are a large and increasing number of materials that fall in each of the four types of composites, making generalization difficult. However, as a class of materials, composites tend to have the following characteristics: high strength; high modulus; low density; excellent resistance to fatigue, creep, creep rupture, corrosion, and wear; and low coefficient of thermal expansion (CTE). As for monolithic materials, each of the four classes of composites has its own particular attributes. For example, CMCs tend to have particularly good resistance to corrosion, oxidation, and wear, along with high-temperature capability.

For applications in which both mechanical properties and low weight are important, useful figures of merit are specific strength (strength divided by specific gravity or density) and specific stiffness (stiffness divided by specific gravity or density). Figure 9.2 presents specific stiffness and specific tensile strength of conventional structural metals (steel, titanium, aluminum, magnesium, and beryllium), two engineering ceramics (silicon nitride and alumina), and selected composite materials. The composites are PMCs reinforced with selected continuous fibers—carbon, aramid, E-glass, and boron—and an MMC, aluminum containing silicon carbide particles. Also shown is beryllium–aluminum, which can be considered a type of metal matrix composite, rather than an alloy, because the mutual solubility of the constituents at room temperature is low.

The carbon fibers represented in Figure 9.2 are made from several types of precursor materials: polyacrilonitrile (PAN), petroleum pitch, and coal tar pitch. Characteristics of the two types of pitch-based fibers tend to be similar but very different from those made from PAN. Several types of carbon fibers are represented: standard-modulus (SM) PAN, ultrahigh-strength (UHS) PAN, ultrahigh-modulus (UHM) PAN, and ultrahigh-modulus (UHM) pitch. These fibers are discussed in Section 9.2. It should be noted that there are dozens of different kinds of commercial carbon fibers, and new ones are continually being developed.

Because the properties of fiber-reinforced composites depend strongly on fiber orientation, fiber-reinforced polymers are represented by lines. The upper end corresponds to the axial properties of a unidirectional laminate, in which all the fibers are aligned in one direction. The lower end represents a quasi-isotropic laminate having equal stiffness and approximately equal strength characteristics in all directions in the plane of the fibers.

As Figure 9.2 shows, composites offer order-of-magnitude improvements over metals in both specific strength and stiffness. It has been observed that order-of-magnitude improvements in key properties typically produce revolutionary effects in a technology. Consequently, it is not surprising that composites are having such a dramatic influence in engineering applications.

In addition to their exceptional static strength properties, fiber-reinforced polymers also have excellent resistance to fatigue loading. Figure 9.3 shows how the number of cycles to failure (N) varies with maximum stress (S) for aluminum and selected unidirectional PMCs subjected to tension-tension fatigue. The ratio of minimum stress to maximum stress (R) is 0.1. The composites consist of epoxy matrices reinforced with key fibers: aramid, boron, SM carbon, high-strength (HS) glass, and E-glass. Because of their excellent fatigue resistance, composites have largely replaced metals

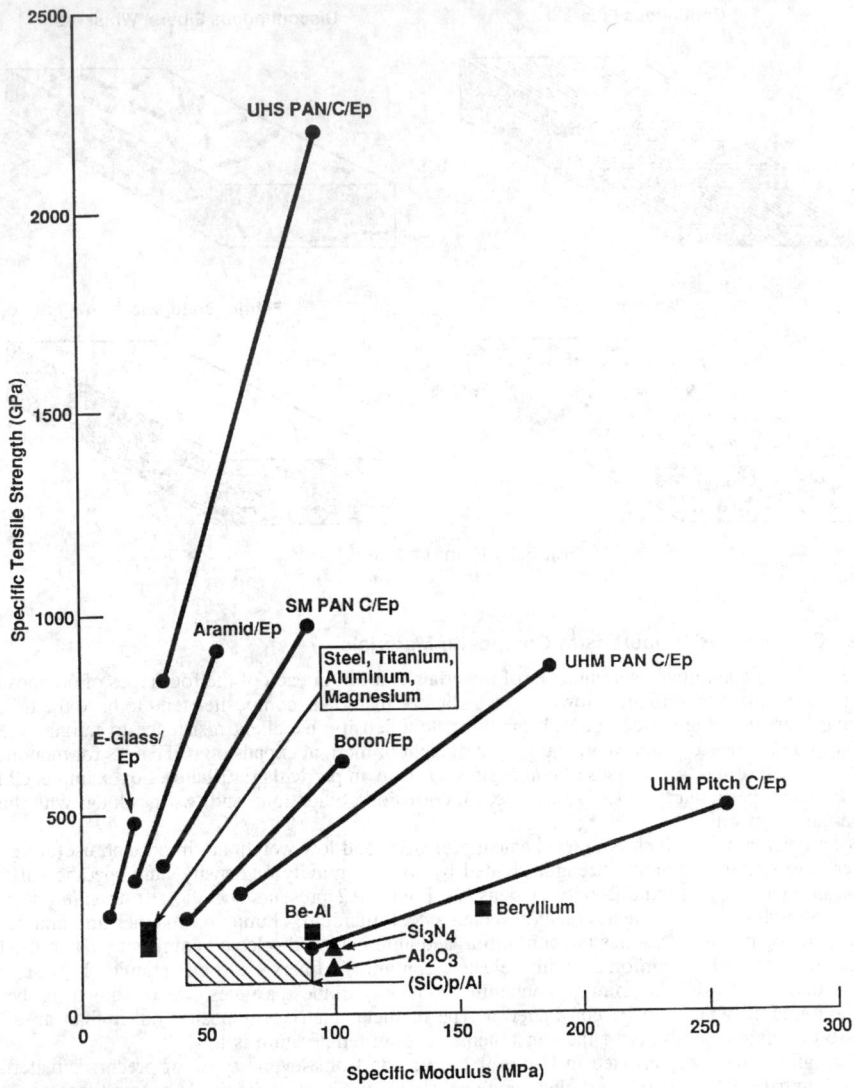

Fig. 9.2 Specific tensile strength (tensile strength divided by density) as a function of specific modulus (modulus divided by density) of composite materials and monolithic metals and ceramics.

in fatigue-critical aerospace applications, such as helicopter rotor blades. Composites also are being used in commercial fatigue-critical applications, such as automobile springs (see Section 9.5).

The outstanding mechanical properties of composite materials have been a key reason for their extensive use in structures. However, composites also have important physical properties, especially low, tailorable CTE and high-thermal conductivity, that are key reasons for their selection in an increasing number of applications.

Many composites, such as PMCs reinforced with carbon and aramid fibers, and silicon carbide particle-reinforced aluminum, have low CTEs, which are advantageous in applications requiring dimensional stability. By appropriate selection of reinforcements and matrix materials, it is possible to produce composites with near-zero CTEs.

Coefficient of thermal expansion tailorability provides a way to minimize thermal stresses and distortions that often arise when dissimilar materials are joined. For example, Figure 9.4 shows how the CTE of silicon carbide particle-reinforced aluminum varies with particle content. By varying the

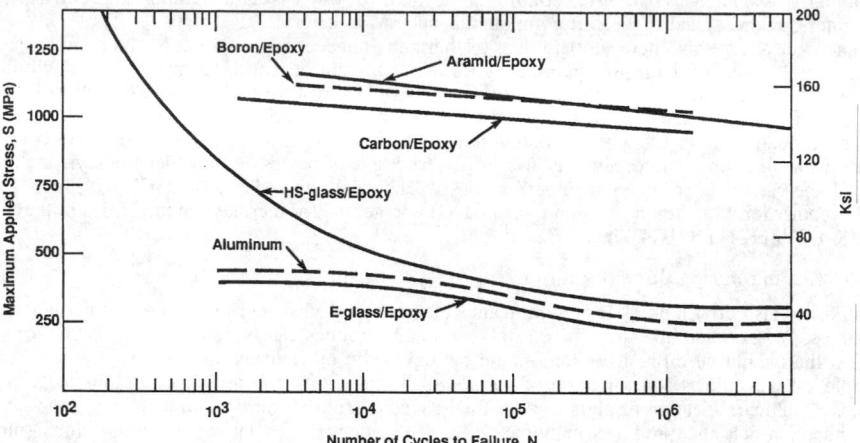

Fig. 9.3 Number of cycles to failure as a function of maximum stress for aluminum and unidirectional polymer matrix composites subjected to tension-tension fatigue with a stress ratio, R = 0.1 (from Ref. 2).

amount of reinforcement, it is possible to match the CTEs of a variety of key engineering materials, such as steel, titanium, and alumina (aluminum oxide).

The ability to tailor CTE is particularly important in applications such as electronic packaging, where thermal stresses can cause failure of ceramic substrates, semiconductors, and solder joints.

Another unique and increasingly important property of some composites is their exceptionally high-thermal conductivity. This is leading to increasing use of composites in applications for which heat dissipation is a key design consideration. In addition, the low densities of composites make them

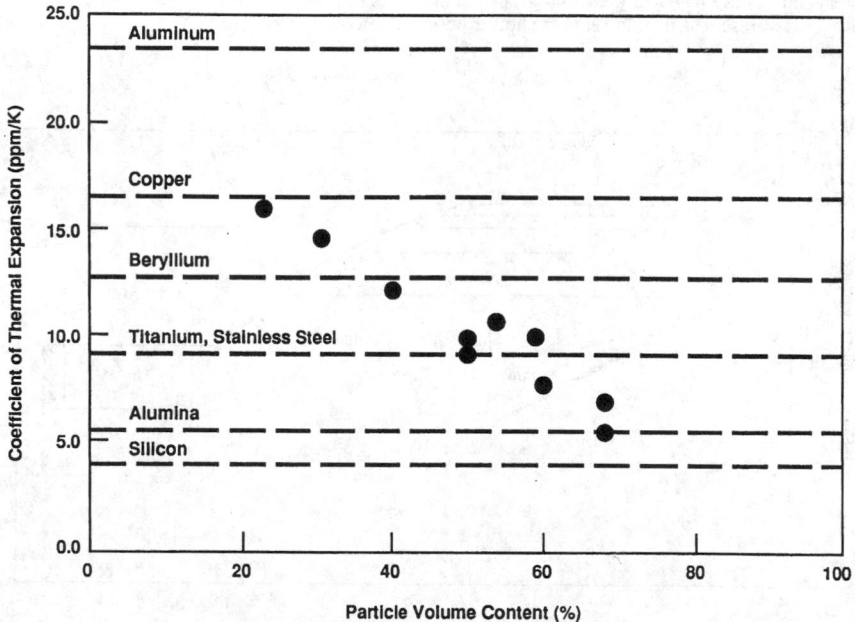

Fig. 9.4 Variation of coefficient of thermal expansion with particle volume fraction for silicon carbide particle-reinforced aluminum (from Ref. 3).

particularly advantageous in thermal control applications for which weight is important, such as laptop computers, avionics, and spacecraft components, such as radiators.

There are a large and increasing number of thermally conductive composites, which are discussed in Section 9.3. One of the most important types of reinforcements for these materials is pitch fibers. Figure 9.5 shows how thermal conductivity varies with electrical resistivity for conventional metals and carbon fibers. It can be seen that PAN-based fibers have relatively low thermal conductivities. However, pitch-based fibers with thermal conductivities more than twice that of copper are commercially available. These reinforcements also have very high-stiffnesses and low densities. At the upper end of the carbon fiber curve are fibers made by chemical vapor deposition (CVD). Fibers made from another form of carbon, diamond, also have the potential for thermal conductivities in the range of 2000 W/m K (1160 BTU/h · ft · F).

9.1.3 Manufacturing Considerations

Composites also offer a number of significant manufacturing advantages over monolithic metals and ceramics. For example, fiber-reinforced polymers and ceramics can be fabricated in large, complex shapes that would be difficult or impossible to make with other materials. The ability to fabricate complex shapes allows consolidation of parts, which reduces machining and assembly costs. Some processes allow fabrication of parts to their final shape (net shape) or close to their final shape (near-net shape), which also produces manufacturing cost savings. The relative ease with which smooth shapes can be made is a significant factor in the use of composites in aircraft and other applications for which aerodynamic considerations are important. Manufacturing processes for composites are covered in Section 9.4.

9.2 REINFORCEMENTS AND MATRIX MATERIALS

As discussed in Section 9.1, we divide solid materials into four classes: polymers, metals, ceramics, and carbon. There are reinforcements and matrix materials in each category. In this section, we consider the characteristics of key reinforcements and matrices.

There are important issues that must be discussed before we present constituent properties. The conventional materials used in mechanical engineering applications are primarily structural metals, for most of which there are industry and government specifications. The situation is very different for composites. Most reinforcements and matrices are proprietary materials for which there are no industry standards. This is similar to the current status of ceramics. The situation is further complicated by the fact that there are many test methods in use to measure mechanical and physical properties of reinforcements and matrix materials. As a result, there are often conflicting material property data in the usual sources, published papers, and manufacturers' literature. The data presented in this article represent a carefully evaluated distillation of information from many sources. The principal sources are listed in the bibliography and references. In view of the uncertainties discussed, the properties presented in this section should be considered approximate values.

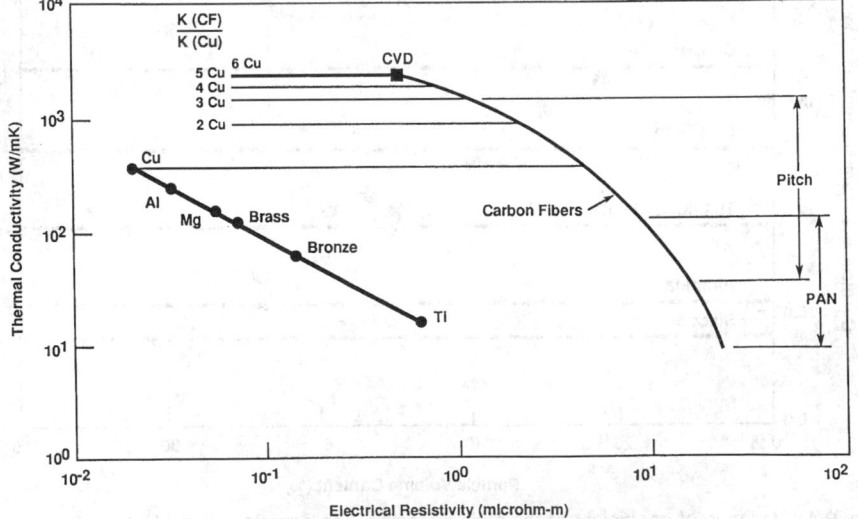

Fig. 9.5 Thermal conductivity as a function of electrical resistivity of metals and carbon fibers (adapted from one of Amoco Performance Products).

Because of the large number of matrix materials and reinforcements, we are forced to be selective. Further, space limitations prevent presentation of a complete set of properties. Consequently, properties cited are room temperature values, unless otherwise stated.

9.2.1 Reinforcements

The four key types of reinforcements used in composites are continuous fibers, discontinuous fibers, whiskers (elongated single crystals), and particles (Fig. 9.1). Continuous, aligned fibers are the most efficient reinforcement form and are widely used, especially in high-performance applications. However, for ease of fabrication and to achieve specific properties, such as improved through-thickness strength, continuous fibers are converted into a wide variety of reinforcement forms using textile technology. Key among them at this time are two-dimensional and three-dimensional fabrics and braids.

Fibers

The development of fibers with unprecedented properties has been largely responsible for the great importance of composites and the revolutionary improvements in properties compared to conventional materials that they offer. The key fibers for mechanical engineering applications are glasses, carbons (also called graphites), several types of ceramics, and high-modulus organics. Most fibers are produced in the form of multifilament bundles called strands or ends in their untwisted forms, and yarns when twisted. Some fibers are produced as monofilaments, which generally have much larger diameters than strand filaments. Table 9.2 presents properties of key fibers, which are discussed in the following subsections.

Fiber strength requires some discussion. Most of the key fibrous reinforcements are made of brittle ceramics or carbon. It is well known that the strengths of monolithic ceramics decrease with increasing material volume because of the increasing probability of finding strength-limiting flaws. This is called size effect. As a result of size effect, fiber strength typically decreases monotonically with increasing gage length (and diameter). Flaw sensitivity also results in considerable strength scatter at a fixed test length. Consequently, there is no single value that characterizes fiber strength. This is also true of key organic reinforcements, such as aramid fibers. Consequently, the values presented in Table 9.2 should be considered approximate values and are useful primarily for comparative purposes. Note that, because unsupported fibers buckle under very low stresses, it is very difficult to measure their inherent compression strength, and these properties are almost never reported. Instead, composite compression strength is measured directly.

Glass Fibers. Glass fibers are used primarily to reinforce polymers. The leading types of glass fibers for mechanical engineering applications are E-glass and high-strength (HS) glass. E-glass fibers, the first major composite reinforcement, were originally developed for electrical insulation applica-

Table 9.2 Properties of Key Reinforcing Fibers

Fiber	Density g/cm³ (Pci)	Axial Modulus GPa (Msi)	Tensile Strength MPa (Ksi)	Axial Coefficient of Thermal Expansion ppm/K (ppm/F)		Axial Thermal Conductivity W/mK
E-glass	2.6 (0.094)	70 (10)	2000 (300)	5	(2.8)	0.9
HS glass	2.5 (0.090)	83 (12)	4200 (650)	4.1	(2.3)	0.9
Aramid	1.4 (0.052)	124 (18)	3200 (500)	−5.2	(−2.9)	0.04
Boron	2.6 (0.094)	400 (58)	3600 (520)	4.5	(2.5)	—
SM carbon (PAN)	1.7 (0.061)	235 (34)	3200 (500)	−0.5	(−0.3)	9
UHM carbon (PAN)	1.9 (0.069)	590 (86)	3800 (550)	−1	(−0.6)	18
UHS carbon (PAN)	1.8 (0.065)	290 (42)	7000 (1000)	−1.5	(−0.8)	160
UHM carbon (pitch)	2.2 (0.079)	895 (130)	2200 (320)	−1.6	(−0.9)	640
UHK carbon (pitch)	2.2 (0.079)	830 (120)	2200 (320)	−1.6	(−0.9)	1100
SiC monofilament	3.0 (0.11)	400 (58)	3600 (520)	4.9	(2.7)	—
SiC multifilament	3.0 (0.11)	400 (58)	3100 (450)	—		—
Si-C-O	2.6 (0.094)	190 (28)	2900 (430)	3.9	(2.2)	1.4
Si-Ti-C-O	2.4 (0.087)	190 (27)	3300 (470)	3.1	(1.7)	—
Aluminum oxide	3.9 (0.14)	370 (54)	1900 (280)	7.9	(4.4)	—
High-density Polyethylene	0.97 (0.035)	172 (25)	3000 (440)	—		—

tions (that is the origin of the "E"). E-glass is, by many orders of magnitude, the most widely used of all fibrous reinforcements. The primary reasons for this are its low cost and early development compared to other fibers. Glass fibers are produced as multifilament bundles. Filament diameters range from 3–20 micrometers (118–787 microinches). Table 9.2 presents representative properties of E-glass and HS glass fibers.

E-glass fibers have relatively low elastic moduli compared to other reinforcements. In addition, E-glass fibers are susceptible to creep and creep (stress) rupture. HS glass is stiffer and stronger than E-glass, and has better resistance to fatigue and creep.

The thermal and electrical conductivities of glass fibers are low, and glass fiber-reinforced PMCs are often used as thermal and electrical insulators. The CTE of glass fibers is also low compared to most metals.

Carbon (Graphite) Fibers. Carbon fibers, commonly called graphite fibers in the United States, are used as reinforcements for polymers, metals, ceramics, and carbon. There are dozens of commercial carbon fibers, with a wide range of strengths and moduli. As a class of reinforcements, carbon fibers are characterized by high-stiffness and strength, and low density and CTE. Fibers with tensile moduli as high as 895 GPa (130 Msi) and with tensile strengths of 7000 MPa (1000 Ksi) are commercially available. Carbon fibers have excellent resistance to creep, stress rupture, fatigue, and corrosive environments, although they oxidize at high-temperatures. Some carbon fibers also have extremely high-thermal conductivities—many times that of copper. This characteristic is of considerable interest in electronic packaging and other applications where thermal control is important. Carbon fibers are the workhorse reinforcements in high-performance aerospace and commercial PMCs and some CMCs. Of course, as the name suggests, carbon fibers are also the reinforcements in carbon/carbon composites.

Most carbon fibers are highly anisotropic. Axial stiffness, tension and compression strength, and thermal conductivity are typically much greater than the corresponding properties in the radial direction. Carbon fibers generally have small, negative axial CTEs (which means that they get shorter when heated) and positive radial CTEs. Diameters of common reinforcing fibers, which are produced in the form of multifilament bundles, range from 4–10 micrometers (160–390 microinches). Carbon fiber stress–strain curves tend to be nonlinear. Modulus increases under increasing tensile stress and decreases under increasing compressive stress.

Carbon fibers are made primarily from three key precursor materials: polyacrylonitrile (PAN), petroleum pitch, and coal tar pitch. Rayon-based fibers, once the primary CCC reinforcement, are far less common in new applications. Experimental fibers also have been made by chemical vapor deposition. Some of these have reported axial thermal conductivities as high as 2000 W/m K, five times that of copper.

PAN-based materials are the most widely used carbon fibers. There are dozens on the market. Fiber axial moduli range from 235 GPa (34 Msi) to 590 GPa (85 Msi). They generally provide composites with excellent tensile and compressive strength properties, although compressive strength tends to drop off as modulus increases. Fibers having tensile strengths as high as 7 GPa (1 Msi) are available. Table 9.2 presents properties of three types of PAN-based carbon fibers and two types of pitch-based carbon fibers. The PAN-based fibers are standard modulus (SM), ultrahigh-strength (UHS) and ultrahigh-modulus (UHM). SM PAN fibers are the most widely used type of carbon fiber reinforcement. They are one of the first types commercialized and tend to be the least expensive. UHS PAN carbon fibers are the strongest type of another widely used class of carbon fiber, usually called intermediate modulus (IM) because the axial modulus of these fibers falls between those of SM and modulus carbon fibers.

A key advantage of pitch-based fibers is that they can be produced with much higher axial moduli than those made from PAN precursors. For example, UHM pitch fibers with moduli as high as 895 GPa (130 Msi) are available. In addition, some pitch fibers, which we designate UHK, have extremely high-axial thermal conductivities. There are commercial UHK fibers with a nominal axial thermal conductivity of 1100 W/m K, almost three times that of copper. However, composites made from pitch-based carbon fibers generally are somewhat weaker in tension and shear, and much weaker in compression, than those using PAN-based reinforcements.

Boron Fibers. Boron fibers are primarily used to reinforce polymers and metals. Boron fibers are produced as monofilaments (single filaments) by chemical vapor deposition of boron on a tungsten wire or carbon filament, the former being the most widely used. They have relatively large diameters, 100–140 micrometers (4000–5600 microinches), compared to most other reinforcements. Table 9.2 presents representative properties of boron fibers having a tungsten core and diameter of 140 micrometers. The properties of boron fibers are influenced by the ratio of overall fiber diameter to that of the tungsten core. For example, fiber specific gravity is 2.57 for 100-micrometer fibers and 2.49 for 140-micrometer fibers.

Fibers Based on Silicon Carbide. Silicon carbide-based fibers are primarily used to reinforce metals and ceramics. There are a number of commercial fibers based on silicon carbide. One type, a monofilament, is produced by chemical vapor deposition of high-purity silicon carbide on a carbon monofilament core. Some versions use a carbon-rich surface layer that serves as a reaction barrier. There are a number of multifilament silicon carbide-based fibers which are made by pyrolysis of polymers. Some of these contain varying amounts of silicon, carbon and oxygen, titanium, nitrogen, zirconium, and hydrogen. Table 9.2 presents properties of selected silicon carbide-based fibers.

Fibers Based on Alumina. Alumina-based fibers are primarily used to reinforce metals and ceramics. Like silicon–carbide-based fibers, they have a number of different chemical formulations. The primary constituents, in addition to alumina, are boria, silica, and zirconia. Table 9.2 presents properties of high-purity alumina fibers.

Aramid Fibers. Aramid, or aromatic, polyamide fibers are high-modulus organic reinforcements primarily used to reinforce polymers and for ballistic protection. There are a number of commercial aramid fibers produced by several manufacturers. Like other reinforcements, they are proprietary materials with different properties. Table 9.2 presents properties of one of the most widely used aramid fibers.

High-Density Polyethylene Fibers. High-density polyethylene fibers are primarily used to reinforce polymers and for ballistic protection. Table 9.2 presents properties of a common reinforcing fiber. The properties of high-density polyethylene tend to decrease significantly with increasing temperature, and they tend to creep significantly under load, even at low temperatures.

9.2.2 Matrix Materials

The four classes of matrix materials are polymers, metals, ceramics, and carbon. Table 9.3 presents representative properties of selected matrix materials in each category. As the table shows, the properties of the four types differ substantially. These differences have profound effects on the properties of the composites using them. In this section, we examine characteristics of key materials in each class.

Polymer Matrix Materials

There are two major classes of polymers used as matrix materials: thermosets and thermoplastics. Thermosets are materials that undergo a curing process during part fabrication, after which they are rigid and cannot be reformed. Thermoplastics, on the other hand, can be repeatedly softened and reformed by application of heat. Thermoplastics are often subdivided into several types: amorphous, crystalline, and liquid crystal. There are numerous types of polymers in both classes. Thermosets tend to be more resistant to solvents and corrosive environments than thermoplastics, but there are exceptions to this rule. Resin selection is based on design requirements, as well as manufacturing and cost considerations. Table 9.4 presents representative properties of common matrix polymers.

Polymer matrices generally are relatively weak, low-stiffness, viscoelastic materials. The strength and stiffness of PMCs come primarily from the fiber phase. One of the key issues in matrix selection is maximum service temperature. The properties of polymers decrease with increasing temperature. A widely used measure of comparative temperature resistance of polymers is glass transition temperature (Tg), which is the approximate temperature at which a polymer transitions from a relatively rigid material to a rubbery one. Polymers typically suffer significant losses in both strength and stiffness above their glass transition temperatures. New polymers with increasing temperature capability are continually being developed, allowing them to compete with a wider range of metals. For example, carbon fiber-reinforced polyimides have replaced titanium in some aircraft gas turbine engine parts.

An important consideration in selection of polymer matrices is their moisture sensitivity. Resins tend to absorb water, which causes dimensional changes and reduction of elevated temperature strength and stiffness. The amount of moisture absorption, typically measured as percent weight gain, depends on the polymer and relative humidity. Resins also desorb moisture when placed in a drier atmosphere. The rate of absorption and desorption depends strongly on temperature. The moisture sensitivity of resins varies widely; some are very resistant.

In a vacuum, resins outgas water and organic and inorganic chemicals, which can condense on surfaces with which they come in contact. This can be a problem in optical systems and can affect surface properties critical for thermal control, such as absorptivity and emissivity. Outgassing can be controlled by resin selection and baking out the component.

Thermosetting Resins. The key types of thermosetting resins used in composites are epoxies, bismaleimides, thermosetting polyimides, cyanate esters, thermosetting polyesters, vinyl esters, and phenolics.

Epoxies are the workhorse materials for airframe structures and other aerospace applications, with decades of successful flight experience to their credit. They produce composites with excellent structural properties. Epoxies tend to be rather brittle materials, but toughened formulations with greatly improved impact resistance are available. The maximum service temperature is affected by reduced elevated temperature structural properties resulting from water absorption. A typical airframe limit is about 120°C (250°F).

Table 9.3 Properties of Selected Matrix Materials

Material	Class	Density g/cm³ (Pci)	Modulus GPa (Msi)	Tensile Strength MPa (Ksi)	Tensile Failure Strain %	Thermal Conductivity W/mK (BTU/h·ft·F)	Coefficient of Thermal Expansion ppm/K (ppm/F)
Epoxy	Polymer	1.8 (0.065)	3.5 (0.5)	70 (10)	3	0.1 (0.06)	60 (33)
Aluminum (6061)	Metal	2.7 (0.098)	69 (10)	300 (43)	10	180 (104)	23 (13)
Titanium (6Al-4V)	Metal	4.4 (0.16)	105 (15.2)	1100 (160)	10	16 (9.5)	9.5 (5.3)
Silicon Carbide	Ceramic	2.9 (0.106)	520 (75)	—	< 0.1	81 (47)	4.9 (2.7)
Alumina	Ceramic	3.9 (0.141)	380 (55)	—	< 0.1	20 (120)	6.7 (3.7)
Glass (borosilicate)	Ceramic	2.2 (0.079)	63 (9)	—	< 0.1	2 (1)	5 (3)
Carbon	Carbon	1.8 (0.065)	20 (3)	—	< 0.1	5–90 (3–50)	2 (1)

Table 9.4 Properties of Selected Thermosetting and Thermoplastic Matrices

	Density g/cm³ (Pci)	Modulus GPa (Msi)	Tensile Strength MPa (Ksi)	Elongation to Break (%)	Thermal Conductivity W/mK	Coefficient of Thermal Expansion ppm/K (ppm/F)
Epoxy (1)	1.1–1.4 (0.040–0.050)	3–6 (0.43–0.88)	35–100 (5–15)	1–6	0.1	60 (33)
Thermosetting polyester (1)	1.2–1.5 (0.043–0.054)	2–4.5 (0.29–0.65)	40–90 (6–13)	2	0.2	100–200 (56–110)
Polypropylene (2)	0.90 (0.032)	1–4 (0.15–0.58)	25–38 (4–6)	> 300	0.2	110 (61)
Nylon 6-6 (2)	1.14 (0.041)	1.4–2.8 (0.20–0.41)	60–75 (9–11)	40–80	0.2	90 (50)
Polycarbonate (2)	1.06–1.20 (0.038–0.043)	2.2–2.4 (0.32–0.35)	45–70 (7–10)	50–100	0.2	70 (39)
Polysulfone (2)	1.25 (0.045)	2.2 (0.32)	76 (11)	50–100	—	56 (31)
Polyetherimide (2)	1.27 (0.046)	3.3 (0.48)	110 (16)	60	—	62 (34)
Polyamideimide (2)	1.4 (0.050)	4.8 (0.7)	190 (28)	17	—	63 (35)
Polyphenylene sulfide (2)	1.36 (0.049)	3.8 (0.55)	65 (10)	4	—	54 (30)
Polyether etherketone (2)	1.26–1.32 (0.046–0.048)	3.6 (0.52)	93 (13)	50	—	47 (26)

(1) Thermoset, (2) Thermoplastic.

Bismaleimide resins are used for aerospace applications requiring higher temperature capabilities than can be achieved by epoxies. They are employed for temperatures of up to about 200°C (390°F).

Thermosetting polyimides are used for applications with temperatures as high as 250°C to 290°C (500°F to 550°F).

Cyanate ester resins are not as moisture sensitive as epoxies and tend to outgas much less. Formulations with operating temperatures as high as 205°C (400°F) are available.

Thermosetting polyesters are the workhorse resins in commercial applications. They are relatively inexpensive, easy to process, and corrosion resistant.

Vinyl esters are also widely used in commercial applications. They have better corrosion resistance than polyesters, but are somewhat more expensive.

Phenolic resins have good high-temperature resistance and produce less smoke and toxic products than most resins when burned. They are used in applications such as aircraft interiors and offshore oil platform structures, for which fire resistance is a key design requirement.

Thermoplastic Resins. Thermoplastics are divided into three main classes: amorphous, crystalline, and liquid crystal. Polycarbonate, acrylonitrile–butadiene–styrene (ABS), polystyrene, polysulfone, and polyetherimide are amorphous materials. Crystalline thermoplastics include nylon, polyethylene, polyphenylene sulfide, polypropylene, acetal, polyethersulfone, and polyether etherketone (PEEK). Amorphous thermoplastics tend to have poor solvent resistance. Crystalline materials tend to be better in this respect. Relatively inexpensive thermoplastics such as nylon are extensively used with chopped E-glass fiber reinforcements in countless injection-molded parts. There are an increasing number of applications using continuous fiber-reinforced thermoplastics.

Metals

The metals initially used for MMC matrix materials generally were conventional alloys. Over time, however, many special matrix materials tailored for use in composites have been developed. The key metallic matrix materials used for structural MMCs are alloys of aluminum, titanium, iron, and intermetallic compounds, such as titanium aluminides. However, many other metals have been used as matrix materials, such as copper, lead, magnesium, cobalt, silver, and superalloys. The *in situ* properties of metals in a composite depend on the manufacturing process and, because metals are elastic–plastic materials, the history of mechanical stresses and temperature changes to which they are subjected.

Ceramic Matrix Materials

The key ceramics used as CMC matrices are silicon carbide, alumina, silicon nitride, mullite, and various cements. The properties of ceramics, especially strength, are even more process-sensitive than those of metals. In practice, it is very difficult to determine the *in situ* properties of ceramic matrix materials in a composite.

As discussed earlier, in the section on fiber properties, ceramics are very flaw-sensitive, resulting in a decrease in strength with increasing material volume, a phenomenon called "size effect." As a result, there is no single value that describes the tensile strength of ceramics. In fact, because of the very brittle nature of ceramics, it is difficult to measure tensile strength, and flexural strength (often called modulus of rupture) is typically reported. It should be noted that flexural strength is also dependent on specimen size and is generally much higher than that of a tensile coupon of the same dimensions. In view of the great difficulty in measuring a simple property like tensile strength, which arises from their flaw sensitivity, it is not surprising that monolithic ceramics have had limited success in applications where they are subjected to significant tensile stresses.

The fracture toughness of ceramics is typically in the range of 3–6 MPa $\cdot$ m$^{1/2}$. Those of transformation-toughened materials are somewhat higher. For comparison, the fracture toughnesses of structural metals are generally greater than 20 MPa $\cdot$ m$^{1/2}$.

Carbon Matrix Materials

Carbon is a remarkable material. It includes materials ranging from lubricants to diamonds and structural fibers. The forms of carbon matrices resulting from the various carbon/carbon manufacturing processes tend to be rather weak, brittle materials. Some forms have very high-thermal conductivities. As for ceramics, *in situ* matrix properties are difficult to measure.

9.3   PROPERTIES OF COMPOSITE MATERIALS

There are a large and increasing number of materials in all four classes of composites: polymer matrix composites (PMCs), metal matrix composites (MMCs), ceramic matrix composites (CMCs), and carbon/carbon composites (CCCs). In this section, we present mechanical and physical properties of some of the key materials in each class.

Initially, the excellent mechanical properties of composites was the main reason for their use. However, there are an increasing number of applications for which the unique and tailorable physical properties of composites are key considerations. For example, the extremely high-thermal conductivity and tailorable coefficient of thermal expansion (CTE) of some composite material systems are leading to their increasing use in electronic packaging. Similarly, the extremely high-stiffness, near-zero CTE, and low density of carbon fiber-reinforced polymers have made these composites the materials of choice in spacecraft structures.

Composites are complex, heterogeneous, and often anisotropic material systems. Their properties are affected by many variables, including *in situ* constituent properties; reinforcement form, volume fraction and geometry; properties of the interphase, the region where the reinforcement and matrix are joined (also called the interface); and void content. The process by which the composite is made affects many of these variables. The same matrix material and reinforcements, when combined by different processes, may result in composites with very different properties.

Several other important things must be kept in mind when considering composite properties. For one, most composites are proprietary material systems made by proprietary processes. There are few industry or government specifications for composites, as there are for many monolithic structural metals. However, this is also the case for many monolithic ceramics and polymers, which are widely used engineering materials. Despite their inherently proprietary nature, some widely used composite materials made by a number of manufacturers have similar properties. A notable example is standard-modulus (SM) carbon fiber-reinforced epoxy.

Another critical issue is that properties are sensitive to the test methods by which they are measured, and there are many different test methods used throughout the industry. Further, test results are very sensitive to the skill of the technician performing the test. Because of these factors, it is very common to find significant differences in reported properties of what is nominally the same composite material.

In Section 9.2, we discussed the issue of size effect, which is the decrease in strength with increasing material volume that is observed in monolithic ceramics key reinforcing fibers. There is some evidence, suggestive but not conclusive, of size effects in composite strength properties, as well. However, if composite strength size effects exist at all, they are much less severe than for fibers by themselves. The reason is that the presence of a matrix results in very different failure mechanisms. However, until the issues are resolved definitively, caution should be used in extrapolating strength data from small coupons to large structures, which may have volumes many orders of magnitude greater.

As mentioned earlier, the properties of composites are very sensitive to reinforcement form, volume fraction, and geometry. This is illustrated in Table 9.5, which presents the properties of several common types of E-glass fiber-reinforced polyester composites. The reinforcement forms are discontinuous fibers, woven roving (a heavy fabric), and straight, parallel continuous fibers. As we shall see, discontinuous reinforcement is not as efficient as continuous. However, discontinuous fibers allow the composite material to flow during processing, facilitating fabrication of complex molded parts.

The composites using discontinuous fibers are divided into three categories. One is bulk molding compound (BMC), also called dough molding compound, in which fibers are relatively short, about 3–12 mm, and are nominally randomly oriented in three dimensions. BMC also has a very high loading of mineral particles, such as calcium carbonate, which are added for a variety of reasons: to reduce dimensional changes from resin shrinkage, to obtain a smooth surface, and to reduce cost, among others. Because it contains both particulate and fibrous reinforcement, BMC can be considered a type of hybrid composite.

The second type of composite is chopped strand mat (CSM), which contains discontinuous fibers, typically about 25 mm long, nominally randomly oriented in two directions. The third material is sheet molding compound (SMC), which contains chopped fibers 25–50 mm in length, also nominally randomly oriented in two dimensions. Like BMC, SMC also contains particulate mineral fillers, such as calcium carbonate and clay.

Table 9.5 Effect of Fiber Form and Volume Fraction on Mechanical Properties of E-Glass-Reinforced Polyester[4]

	Bulk Molding Compound	Sheet Molding Compound	Chopped Strand Mat	Woven Roving	Unidirectional Axial	Unidirectional Transverse
Glass content (wt %)	20	30	30	50	70	70
Tensile modulus GPa (Msi)	9 (1.3)	13 (1.9)	7.7 (1.1)	16 (2.3)	42 (6.1)	12 (1.7)
Tensile strength MPa (Ksi)	45 (6.5)	85 (12)	95 (14)	250 (36)	750 (110)	50 (7)

The first thing to note in comparing the materials in Table 9.5 is that fiber content, here presented in the form of weight percent, differs considerably for the four materials. This is significant, because, as discussed in Section 9.2, the strength and stiffness of polyester and most polymer matrices is considerably lower than those of E-glass, carbon, and other reinforcing fibers. Composites reinforced with randomly oriented fibers tend to have lower volume fractions than those made with aligned fibers or fabrics. There is a notable exception to this. Some composites with discontinuous-fiber reinforcement are made by chopping up composites reinforced with aligned continuous fibers or fabrics that have high-fiber contents.

Examination of Table 9.5 shows that the modulus of SMC is considerably greater than that of CSM, even though both have the same fiber content. This is because SMC also has particulate reinforcement. Note, however, that although the particles improve modulus, they do not increase strength. This is generally the case for particle-reinforced polymers, but, as we will see later, particles often do enhance the strengths of MMCs and CMCs, as well as their moduli.

We observe that the modulus of the BMC composite is greater than that of CSM and SMC, even though the former has a much lower fiber content. Most likely, this results from the high-mineral content and also the possibility that the fibers are oriented in the direction of test, and are not truly random. Many processes, especially those involving material flow, tend to orient fibers in one or more preferred directions. If so, then one would find the modulus of the BMC to be much lower than the one presented in the table if measured in other directions. This illustrates one of the limitations of using discontinuous fiber reinforcement: it is often difficult to control fiber orientation.

The moduli and strengths of the composites reinforced with fabrics and aligned fibers are much higher than those with discontinuous fibers, when the former two types of materials are tested parallel to fiber directions. For example, the tensile strength of woven roving is more than twice that of CSM. The properties presented are measured parallel to the warp direction of the fabric (the warp direction is the lengthwise direction of the fabric). The elastic and strength properties in the fill direction, perpendicular to the warp, typically are similar to, but somewhat lower than, those in the warp direction. Here, we assume that the fabric is "balanced," which means that the number of fibers in the warp and fill directions per unit length are approximately equal. Note, however, that the elastic modulus, tensile strength, and compressive strength at 45° to the warp and fill directions of a fabric are much lower than the corresponding values in the warp and fill directions. This is discussed further in the sections that cover design.

As Table 9.5 shows, the axial modulus and tensile strength of the unidirectional composite are much greater than those of the fabric. However, the modulus and strength of the unidirectional composite in the transverse direction are considerably lower than the corresponding axial properties. Further, the transverse strength is considerably lower than that of SMC and CSM. In general, the strength of PMCs is weak in directions for which there are no fibers. The low transverse moduli and strengths of unidirectional PMCs are commonly overcome by use of laminates with fibers in several directions. Low through-thickness strength can be improved by use of three-dimensional reinforcement forms. Often, the designer simply assures that through-thickness stresses are within the capability of the material.

In this section, we present representative mechanical and physical properties of key composite materials of interest for a broad range of mechanical engineering applications. The properties represent a distillation of values from many sources. Because of space limitations, it is necessary to be selective in our choice of materials and properties presented. It is simply not possible to present a complete set of data that will cover every possible application. As discussed earlier, there are many textile forms, such as woven fabrics, used as reinforcements. However, we concentrate on aligned, continuous fibers because they produce the highest strength and stiffness. To do a thorough evaluation of composites, the design engineer should consider alternative reinforcement forms. Unless otherwise stated, room temperature property values are presented. We consider mechanical properties in Section 9.3.1 and physical in Section 9.3.2.

9.3.1 Mechanical Properties of Composite Materials

In this section, we consider mechanical properties of key PMCs, MMCs, CMCs, and CCCs that are of greatest interest for mechanical engineering applications.

Mechanical Properties of Polymer Matrix Composites

As discussed earlier, polymers are relatively weak, low-stiffness materials. In order to obtain materials with mechanical properties that are acceptable for structural applications, it is necessary to reinforce them with continuous or discontinuous fibers. The addition of ceramic or metallic particles to polymers results in materials which have increased modulus, but, as a rule, strength typically does not increase significantly, and may actually decrease. However, there are many particle-reinforced polymers used in electronic packaging, primarily because of their physical properties. For these applications, ceramic particles, such as alumina, aluminum nitride, boron nitride, and even diamond, are added to obtain an electrically insulating material with higher thermal conductivity and lower CTE than the monolithic base polymer. Metallic particles such as silver and aluminum are added to create materials which are both electrically and thermally conductive. These materials have replaced lead-based solders in many applications. There are also magnetic composites made by incorporating ferrous or permanent magnet particles in various polymers. A common example is magnetic tape used to record audio and video.

We focus on composites reinforced with continuous fibers because they are the most efficient structural materials. Table 9.6 presents room temperature mechanical properties of unidirectional polymer matrix composites reinforced with key fibers: E-glass, aramid, boron, standard-modulus (SM) PAN (polyacrilonitrile) carbon, ultrahigh-strength (UHS) PAN carbon, ultrahigh-modulus (UHM) PAN carbon, ultrahigh-modulus (UHM) pitch carbon, and ultrahigh-thermal conductivity (UHK) pitch carbon. We assume that the fiber volume fraction is 60%, a typical value. As discussed in Section 9.2, UHS PAN carbon is the strongest type of intermediate-modulus (IM) carbon fiber.

The properties presented in Table 9.6 are representative of what can be obtained at room temperature with a well-made PMC employing an epoxy matrix. Epoxies are widely used, provide good mechanical properties, and can be considered a reference matrix material. Properties of composites using other resins may differ from these, and have to be examined on a case-by-case basis.

The properties of PMCs, especially strengths, depend strongly on temperature. The temperature dependence of polymer properties differs considerably. This is also true for different epoxy formulations, which have different cure and glass transition temperatures. Some polymers, such as polyimides, have good elevated temperature properties that allow them to compete with titanium. There are aircraft gas turbine engine components employing polyimide matrices that see service temperatures as high as 290°C (550°F). Here again, the effect of temperature on composite properties has to be considered on a case-by-case basis.

The properties shown in Table 9.6 are axial, transverse and shear moduli, Poisson's ratio, tensile and compressive strengths in the axial and transverse directions, and inplane shear strength. The Poisson's ratio presented is called the major Poisson's ratio. It is defined as the ratio of the magnitude of transverse strain divided by axial strain when the composite is loaded in the axial direction. Note that transverse moduli and strengths are much lower than corresponding axial values.

As discussed in Section 9.2, carbon fibers display nonlinear stress–strain behavior. Their moduli increase under increasing tensile stress and decrease under increasing compressive stress. This makes the method of calculating modulus critical. Various tangent and secant definitions are used throughout the industry, contributing to the confusion in reported properties. The values presented in Table 9.6, which are approximate, are based on tangents to the stress–strain curves at the origin. Using this definition, tensile and compressive moduli are usually very similar. However, this is not the case for moduli using various secant definitions. Using these definitions typically produces compression moduli that are significantly lower than tension moduli.

Because of the low transverse strengths of unidirectional laminates, they are rarely used in structural applications. The design engineer uses laminates with layers in several directions to meet requirements for strength, stiffness, buckling, and so on. There are an infinite number of laminate geometries that can be selected. For comparative purposes, it is useful to consider quasi-isotropic laminates, which have the same elastic properties in all directions in the plane. Laminates are quasi-isotropic when they have the same percentage of layers every $180/n°$, where $n \geq 3$. The most common quasi-isotropic laminates have layers which repeat every 60, 45, or 30°. We note, however, that strength properties in the plane are not isotropic for these laminates, although they tend to become more uniform as the angle of repetition becomes smaller.

Table 9.7 presents the mechanical properties of quasi-isotropic laminates. Note that the moduli and strengths are much lower than the axial properties of unidirectional laminates made of the same material. In most applications, laminate geometry is such that the maximum axial modulus and tensile and compressive strengths fall somewhere between axial unidirectional and quasi-isotropic values.

The tension-tension fatigue behavior of unidirectional composites, discussed in Section 9.1, is one of their great advantages over metals (Fig. 9.6). In general the tension-tension S–N curves (curves of maximum stress plotted as a function of cycles to failure) of PMCs reinforced with carbon, boron, and aramid fibers are relatively flat. Glass fiber-reinforced composites show a greater reduction in strength with increasing number of cycles. Still, PMCs reinforced with HS glass are widely used in applications for which fatigue resistance is a critical design consideration, such as helicopter rotors.

Metals are more likely to fail in fatigue when subjected to fluctuating tensile rather than compressive load. This is because they tend to fail by crack propagation under fatigue loading. However, the failure modes in composites are very different and more complex. One consequence is that composites tend to be more susceptible to fatigue failure when loaded in compression. Figure 9.6 shows the cycles to failure as a function of maximum stress for carbon fiber-reinforced epoxy laminates subjected to tension–tension and compression–compression fatigue. The laminates have 60% of their layers oriented at 0°, 20% at +45° and 20% at −45°. They are subjected to a fluctuating load in the 0° direction. The ratios of minimum stress-to-maximum stress (R) for tensile and compressive fatigue are 0.1 and 10, respectively. We observe that the reduction in strength is much greater for compression–compression fatigue. However, the composite compressive fatigue strength at 10^7 cycles is still considerably greater than the corresponding tensile value for aluminum.

Table 9.6 Mechanical Properties of Selected Unidirectional Polymer Matrix Composites

Fiber	Axial Modulus GPa (Msi)	Transverse Modulus GPa (Msi)	Inplane Shear Modulus GPa (Msi)	Poisson's Ratio	Axial Tensile Strength MPa (Ksi)	Transverse Tensile Strength MPa (Ksi)	Axial Compressive Strength MPa (Ksi)	Transverse Compressive Strength MPa (Ksi)	Inplane Shear Strength MPa (Ksi)
E-glass	45 (6.5)	12 (1.8)	5.5 (0.8)	0.28	1020 (150)	40 (7)	620 (90)	140 (20)	70 (10)
Aramid	76 (11)	5.5 (0.8)	2.1 (0.3)	0.34	1240 (180)	30 (4.3)	280 (40)	140 (20)	60 (9)
Boron	210 (30)	19 (2.7)	4.8 (0.7)	0.25	1240 (180)	70 (10)	3310 (480)	280 (40)	90 (13)
SM carbon (PAN)	145 (21)	10 (1.5)	4.1 (0.6)	0.25	1520 (220)	41 (6)	1380 (200)	170 (25)	80 (12)
UHS carbon (PAN)	170 (25)	10 (1.5)	4.1 (0.6)	0.25	3530 (510)	41 (6)	1380 (200)	170 (25)	80 (12)
UHM carbon (PAN)	310 (45)	9 (1.3)	4.1 (0.6)	0.20	1380 (200)	41 (6)	760 (110)	170 (25)	80 (12)
UHM carbon (pitch)	480 (70)	9 (1.3)	4.1 (0.6)	0.25	900 (130)	20 (3)	280 (40)	100 (15)	41 (6)
UHK carbon (pitch)	480 (70)	9 (1.3)	4.1 (0.6)	0.25	900 (130)	20 (3)	280 (40)	100 (15)	41 (6)

Table 9.7 Mechanical Properties of Selected Quasi-Isotropic Polymer Matrix Composites

Fiber	Axial Modulus GPa (Msi)	Transverse Modulus GPa (Msi)	Inplane Shear Modulus GPa (Msi)	Poisson's Ratio	Axial Tensile Strength MPa (Ksi)	Transverse Tensile Strength MPa (Ksi)	Axial Compressive Strength MPa (Ksi)	Transverse Compressive Strength MPa (Ksi)	Inplane Shear Strength MPa (Ksi)
E-glass	23 (3.4)	23 (3.4)	9.0 (1.3)	0.28	550 (80)	550 (80)	330 (48)	330 (48)	250 (37)
Aramid	29 (4.2)	29 (4.2)	11 (1.6)	0.32	460 (67)	460 (67)	190 (28)	190 (28)	65 (9.4)
Boron	80 (11.6)	80 (11.6)	30 (4.3)	0.33	480 (69)	480 (69)	1100 (160)	1100 (160)	360 (52)
SM carbon (PAN)	54 (7.8)	54 (7.8)	21 (3.0)	0.31	580 (84)	580 (84)	580 (84)	580 (84)	410 (59)
UHS carbon (PAN)	63 (9.1)	63 (9.1)	21 (3.0)	0.31	1350 (200)	1350 (200)	580 (84)	580 (84)	410 (59)
UHM carbon (PAN)	110 (16)	110 (16)	41 (6.0)	0.32	490 (71)	490 (71)	270 (39)	70 (39)	205 (30)
UHM carbon (pitch)	165 (24)	165 (24)	63 (9.2)	0.32	310 (45)	310 (45)	96 (14)	96 (14)	73 (11)
UHK carbon (pitch)	165 (24)	165 (24)	63 (9.2)	0.32	310 (45)	310 (45)	96 (14)	96 (14)	73 (11)

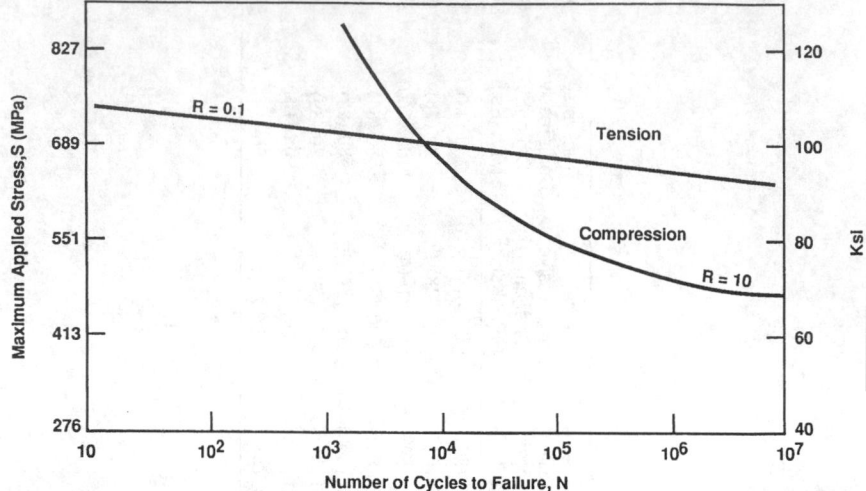

Fig. 9.6 Cycles to failure as a function of maximum stress for carbon fiber-reinforced epoxy laminates loaded in tension-tension (R = 0.1) and compression-compression (R = −10) fatigue (after Ref. 5.).

Polymer matrix composites reinforced with carbon and boron are very resistant to deformation and failure under sustained static load when they are loaded in a fiber-dominated direction. (These phenomena are called creep and creep rupture, respectively.) The creep and creep rupture behavior of aramid is not quite as good. Glass fibers display significant creep, and creep rupture is an important design consideration. Polymers are viscoelastic materials that typically display significant creep when they are not constrained with fibers. Therefore, creep should be considered when composites are subjected to significant stresses in matrix-dominated directions, such as the laminate through-thickness direction.

Mechanical Properties of Metal Matrix Composites

Monolithic metallic alloys are the most widely used materials in mechanical engineering applications. By reinforcing them with continuous fibers, discontinuous fibers, whiskers and particles, we create new materials with enhanced or modified properties, such as higher strength and stiffness, better wear resistance, lower CTE, and so on. In some cases, the improvements are dramatic.

The greatest increases in strength and modulus are achieved with continuous fibers. However, the relatively high-cost of many continuous reinforcing fibers used in MMCs has limited the application of these materials. The most widely used MMCs are reinforced with discontinuous fibers or particles. This may change as new, lower-cost continuous fibers and processes are developed and as cost drops with increasing production volume.

Continuous Fiber-Reinforced MMCs. One of the major advantages of MMCs reinforced with continuous fibers over PMCs is that many, if not most, unidirectional MMCs have much greater transverse strengths, which allow them to be used in a unidirectional configuration. Table 9.8 presents representative mechanical properties of selected unidirectional MMCs reinforced with continuous fibers corresponding to a nominal fiber volume fraction of 50%. The values represent a distillation obtained from numerous sources. In general, the axial moduli of the composites are much greater than those of the monolithic base metals used for the matrices. However, MMC transverse strengths are typically lower than those of the parent matrix materials.

Mechanical Properties of Discontinuous Fiber-Reinforced MMCs. One of the primary mechanical engineering applications of discontinuous fiber-reinforced MMCs is in internal combustion engine components (see Section 9.5.4). Fibers are added primarily to improve the wear resistance and elevated temperature strength and fatigue properties of aluminum. The improvement in wear resistance eliminates the need for cast iron sleeves in engine blocks and cast iron insert rings in pistons. Fiber-reinforced aluminum composites also have higher thermal conductivities than cast iron and, when fiber volume fractions are relatively low, their CTEs are closer to that of unreinforced aluminum, reducing thermal stresses.

The key reinforcements used in internal combustion engine components to increase wear resistance are discontinuous alumina and alumina–silica fibers. In one application, Honda Prelude engine

Table 9.8 Mechanical Properties of Selected Unidirectional Continuous Fiber-Reinforced Metal Matrix Composites

Fiber	Matrix	Density g/cm³ (Pci)	Axial Modulus GPa (Msi)	Transverse Modulus GPa (Msi)	Axial Tensile Strength MPa (Ksi)	Transverse Tensile Strength MPa (Ksi)	Axial Compressive Strength MPa (Ksi)
UHM carbon (pitch)	Aluminum	2.4 (0.090)	450 (65)	15 (5)	690 (100)	15 (5)	340 (50)
Boron	Aluminum	2.6 (0.095)	210 (30)	140 (20)	1240 (180)	140 (20)	1720 (250)
Alumina	Aluminum	3.2 (0.12)	240 (35)	130 (19)	1700 (250)	120 (17)	1800 (260)
Silicon carbide	Titanium	3.6 (0.13)	260 (38)	170 (25)	1700 (250)	340 (50)	2760 (400)

blocks, carbon fibers are combined with alumina to tailor both wear resistance and coefficient of friction of cylinder walls. Wear resistance is not an inherent property, so that there is no single value that characterizes a material. However, in engine tests, it was found that ring groove wear for an alumina fiber-reinforced aluminum piston was significantly less than that for one with a cast iron insert.

Mechanical Properties of Particle-Reinforced MMCs. Particle-reinforced metals are a particularly important class of MMCs for engineering applications. A wide range of materials fall into this category, and a number of them have been used for many years. An important example is a material consisting of tungsten carbide particles embedded in a cobalt matrix that is used extensively in cutting tools and dies. This composite, often referred to as a cermet, cemented carbide, or simply, but incorrectly, "tungsten carbide," has much better fracture toughness than monolithic tungsten carbide, which is a brittle ceramic material. Another interesting MMC, tungsten carbide particle-reinforced silver, is a key circuit breaker contact pad material. Here, the composite provides good electrical conductivity and much greater hardness and wear resistance than monolithic silver, which is too soft to be used in this application. Ferrous alloys reinforced with titanium carbide particles, discussed in the next subsection, have been used for many years in commercial applications. Compared to the monolithic base metals, they offer greater wear resistance and stiffness and lower density.

Mechanical Properties of Titanium Carbide Particle-Reinforced Steel. A number of ferrous alloys reinforced with titanium carbide particles have been used in mechanical system applications for many years. To illustrate the effect of the particulate reinforcements, we consider a particular composite consisting of austenitic stainless steel reinforced with 45% by volume of titanium carbide particles. The modulus of the composite is 304 GPa (44 Msi) compared to 193 GPa (28 Msi) for the monolithic base metal. The specific gravity of the composite is 6.45, about 20% lower than that of monolithic matrix, 8.03. The specific stiffness of the composite is almost double that of the unreinforced metal.

Mechanical Properties of Silicon Carbide Particle-Reinforced Aluminum. Aluminum reinforced with silicon carbide particles is one of the most important of the newer types of MMCs. A wide range of materials fall into this category. They are made by a variety of processes, which are discussed in Section 9.4. Properties depend on the type of particle, particle volume fraction, matrix alloy, and the process used to make them. Table 9.9 shows how representative composite properties vary with particle volume fraction. In general, as particle volume fraction increases, modulus and yield strength increase and fracture toughness and tensile ultimate strain decrease. Particle reinforcement also improves short-term elevated temperature strength properties and fatigue resistance.

Mechanical Properties of Alumina Particle-Reinforced Aluminum. Alumina particles are used to reinforce aluminum as an alternative to silicon carbide particles because they do not react as readily with the matrix at high temperatures and are less expensive. Consequently, alumina-reinforced composites can be used in a wider range of processes and applications. However, the stiffness and thermal conductivity of alumina are lower than the corresponding properties of silicon carbide and these characteristics are reflected in somewhat lower values for composite properties.

Mechanical Properties of Ceramic Matrix Composites

Ceramics, in general, are characterized by high stiffness and hardness, resistance to wear, corrosion and oxidation, and high-temperature operational capability. However, they also have serious deficiencies that have severely limited their use in applications that are subjected to significant tensile stresses. Ceramics have very low fracture toughness, which makes them very sensitive to the presence of small flaws. This results in great strength scatter and poor resistance to thermal and mechanical shock. Civil engineers recognized this deficiency long ago and, in construction, ceramic materials like stone and concrete are rarely used to carry tensile loads. In concrete, this function has been relegated to reinforcing bars made of steel or, more recently, PMCs. An important exception has been in lightly loaded structures where dispersed reinforcing fibers of asbestos, steel, glass and carbon allow modest tensile stresses to be supported.

In CMCs, fibers, whiskers, and particles are combined with ceramic matrices to improve fracture toughness, which reduces strength scatter and improves thermal and mechanical shock resistance. By a wide margin, the greatest increases in fracture resistance result from the use of continuous fibers. Table 9.10 compares fracture toughnesses of structural metallic alloys with those of monolithic ceramics and CMCs reinforced with whiskers and continuous fibers. The low fracture toughness of monolithic ceramics gives rise to very small critical flaw sizes. For example, the critical flaw sizes for monolithic ceramics corresponding to a failure stress of 700 MPa (about 100 Ksi) are in the range of 20–80 micrometers. Flaws of this size are difficult to detect with conventional nondestructive techniques.

The addition of continuous fibers to ceramics can, if done properly, significantly increase the effective fracture toughness of ceramics. For example, as Table 9.10 shows, addition of silicon carbide fibers to a silicon carbide matrix results in a CMC having a fracture toughness in the range of aluminum alloys.

Table 9.9 Mechanical Properties of Silicon Carbide Particle-Reinforced Aluminum

Property	Aluminum (6061-T6)	Titanium (6Al-4V)	Steel (4340)	Composite Particle Volume Fraction		
				25	55	70
Modulus, GPa (Msi)	69 (10)	113 (16.5)	200 (29)	114 (17)	186 (27)	265 (38)
Tensile yield strength, MPa (Ksi)	275 (40)	1000 (145)	1480 (215)	400 (58)	495 (72)	225 (33)
Tensile ultimate strength, MPa (Ksi)	310 (45)	1100 (160)	1790 (260)	485 (70)	530 (77)	225 (33)
Elongation (%)	15	5	10	3.8	0.6	0.1
Density, g/cm³ (lb/in.³)	2.77 (0.10)	4.43 (0.16)	7.76 (0.28)	2.88 (0.104)	2.96 (0.107)	3.00 (0.108)
Specific modulus, GPa	5	26	26	40	63	88

Table 9.10   Fracture Toughness of Structural Alloys, Monolithic Ceramics, and Ceramic Matrix Composites

Matrix	Reinforcement	Fracture Toughness MPa m$^{1/2}$
Aluminum	none	30–45
Steel	none	40–65[a]
Alumina	none	3–5
Silicon carbide	none	3–4
Alumina	Zirconia particles[b]	6–15
Alumina	Silicon carbide whiskers	5–10
Silicon carbide	Continuous silicon carbide fibers	25–30

[a]The toughness of some alloys can be much higher.
[b]Transformation-toughened.

The addition of continuous fibers to a ceramic matrix also changes the failure mode. Figure 9.7 compares the tensile stress-strain curves for a typical monolithic ceramic and a conceptual continuous fiber-reinforced CMC. The monolithic material has a linear stress-strain curve and fails catastrophically at a low strain level. However, the CMC displays a nonlinear stress-strain curve with much more area under the curve, indicating that more energy is absorbed during failure and that the material has a less catastrophic failure mode. The fiber-matrix interphase properties must be carefully tailored and maintained over the life of the composite to obtain this desirable behavior.

Although the CMC stress-strain curve looks, at first, like that of an elastic-plastic metal, this is deceiving. The departure from linearity in the CMC results from internal damage mechanisms, such as the formation of microcracks in the matrix. The fibers bridge the cracks, preventing them from propagating. However, the internal damage is irreversible. As the figure shows, the slope of the stress-strain curve during unloading and subsequent reloading is much lower than that representing initial loading. For an elastic–plastic material, the slopes of the unloading and reloading curves are parallel to the initial elastic slope.

There are numerous CMCs at various stages of development. One of the most mature types consists of a silicon carbide matrix reinforced with fabric woven of silicon carbide-based fibers. These composites are commonly referred to as SiC/SiC. We consider one version. Because the modulus of the particular silicon carbide-based fibers used in this material is lower than that of pure silicon carbide, the modulus of the composite, about 210 GPa (30 Msi), is lower than that of monolithic silicon carbide, 440 GPa (64 Msi). The flexural strength of the composite parallel to the fabric warp direction, about 300 MPa (44 Ksi), is maintained to a temperature of at least 1100°C for short

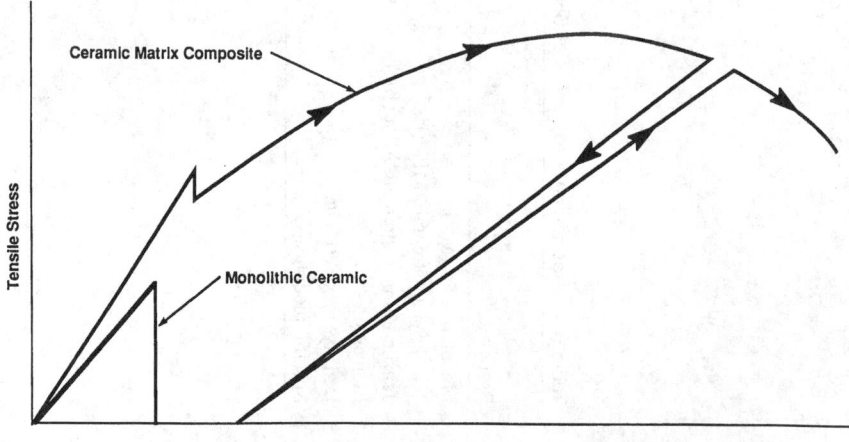

Fig. 9.7   Stress-strain curves for a monolithic ceramic and ceramic matrix composite reinforced with continuous fibers.

times. Long-term strength behavior depends on degradation of the fibers, matrix, and interphase. Because of the continuous fiber reinforcement, SiC/SiC displays excellent resistance to severe thermal shock.

Mechanical Properties of Carbon/Carbon Composites

Carbon/carbon composites consist of continuous and discontinuous carbon fibers embedded in carbon matrices. As for other composites, there are a wide range of materials that fall in this category. The variables affecting properties include type of fiber, reinforcement form, and volume fraction and matrix characteristics.

Historically, CCCs were first used because of their excellent resistance to high-temperature ablation. Initially, strengths and stiffnesses were low, but these properties have steadily increased over the years. As discussed in Section 9.5, CCCs are an important class of materials in high-temperature applications such as aircraft brakes, rocket nozzles, racing car brakes and clutches, glass-making equipment, and electronic packaging, among others.

One of the most significant limitations of CCCs is oxidation, which begins at a temperature threshold of approximately 370°C (700°F) for unprotected materials. Addition of oxidation inhibitors raises the threshold substantially. In inert atmospheres, CCCs retain their properties to temperatures as high as 2800°C (5000°F).

Carbon matrices are typically weak, brittle, low-stiffness materials. As a result, transverse and through-thickness elastic moduli and strength properties of unidirectional CCCs are low. Because of this, two-dimensional and three-dimensional reinforcement forms are commonly used. In the direction of fibrous reinforcement, it is possible to obtain moduli as high as 340 GPa (50 Msi), tensile strengths as high as 700 MPa (100 Ksi), and compressive strengths as high as 800 MPa (110 Ksi). In directions orthogonal to fiber directions, elastic moduli are in the range of 10 MPa (1.5 Ksi), tensile strengths 14 MPa (2 Ksi), and compressive strengths 34 MPa (5 Ksi).

9.3.2 Physical Properties of Composite Materials

Material physical properties are critical for many applications. In this category, we include, among others, density, CTE, thermal conductivity, and electromagnetic characteristics. In this section, we concentrate on the properties of most general interest to mechanical engineers: density, CTE, and thermal conductivity.

Thermal control is a particularly important consideration in electronic packaging because failure rates of semiconductors increase exponentially with temperature. Since conduction is an important method of heat removal, thermal conductivity is a key material property. For many applications, such as spacecraft, aircraft, and portable systems, weight is also an important factor, and consequently, material density is also significant. A useful figure of merit is specific thermal conductivity, defined as thermal conductivity divided by density. Specific thermal conductivity is analogous to specific modulus and specific strength.

In addition to thermal conductivity and density, CTE is also of great significance in many applications. For example, semiconductors and ceramic substrates used in electronics are brittle materials with coefficients of expansion in the range of about 3–7 ppm/K. Semiconductors and ceramic substrates are typically attached to supporting components, such as packages, printed circuit boards (PCBs), and heat sinks with solder or an adhesive. If the CTE of the supporting material is significantly different from that of the ceramic or semiconductor, thermal stresses arise when the assembly is subjected to a change in temperature. These stresses can result in failure of the components or the joint between them.

A great advantage of composites is that there are an increasing number of material systems that combine high thermal conductivity with tailorable CTE, low density, and excellent mechanical properties. Composites can truly be called multifunctional materials.

The key composite materials of interest for thermal control are PMCs, MMCs, and CCCs reinforced with ultrahigh-thermal conductivity (UHK) carbon fibers, which, as discussed in Section 9.2, are made from pitch; silicon carbide particle-reinforced aluminum; beryllium oxide particle-reinforced beryllium; and diamond particle-reinforced aluminum and copper. There also are a number of other special CCCs developed specifically for thermal control applications.

Table 9.11 presents physical properties of a variety of unidirectional composites reinforced with UHK carbon fibers, along with those of monolithic copper and 6063 aluminum for comparison. Unidirectional composites are useful for directing heat in a particular direction. The particular fibers represented have a nominal axial thermal conductivity of 1100 W/m K. Predicted properties are shown for four matrices: epoxy, aluminum, copper, and carbon. Typical reinforcement volume fractions (V/O) are assumed. As Table 9.11 shows, the specific axial thermal conductivities of the composites are significantly greater than those of aluminum and copper.

Figure 9.8 presents thermal conductivity as a function of CTE for various materials used in electronic packaging. Materials shown include silicon (Si) and gallium arsenide (GaAs) semiconductors; alumina (Al_2O_3), beryllium oxide (BeO), and aluminum nitride (AlN) ceramic substrates; and monolithic aluminum, beryllium, copper, silver, and Kovar™, a nickel-iron alloy. Other monolithic

Table 9.11 Physical Properties of Selected Unidirectional Composites and Monolithic Metals

Matrix	Reinforcement	V/O %	Density g/cm³ (Pci)	Axial Coefficient of Thermal Expansion ppm/K (ppm/F)	Axial Thermal Conductivity W/m K (BTU/h·ft·F)	Transverse Thermal Conductivity W/m K (BTU/h·ft·F)	Specific Axial Thermal Conductivity W/m K (BTU/h·ft·F)
Aluminum (6063)	—	—	2.7 (0.098)	23 (13)	218 (126)	218 (126)	81
Copper	—	—	8.9 (0.32)	17 (9.8)	400 (230)	400 (230)	45
Epoxy	UHK carbon fibers	60	1.8 (0.065)	−1.2 (−0.7)	660 (380)	2 (1.1)	370
Aluminum	UHK carbon fibers	50	2.45 (0.088)	−0.5 (−0.3)	660 (380)	50 (29)	110
Copper	UHK carbon fibers	50	5.55 (0.20)	−0.5 (−0.3)	745 (430)	140 (81)	130
Carbon	UHK carbon fibers	40	1.85 (0.067)	−1.5 (−0.8)	740 (430)	45 (26)	400

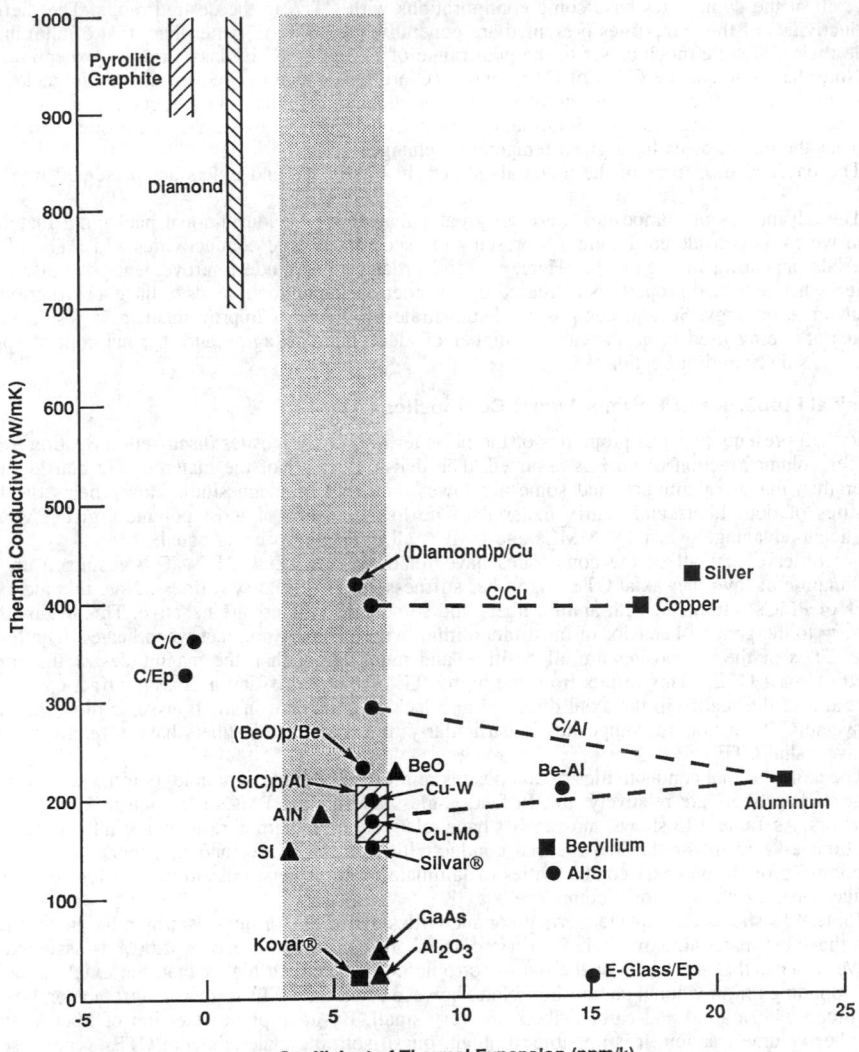

Fig. 9.8 Thermal conductivity as a function of coefficient of thermal expansion for selected monolithic materials and composites used in electronic packaging.

materials included are diamond and pyrolitic graphite, which have very high thermal conductivities in some forms. The figure also presents metal–metal composites, such as copper–tungsten (Cu–W), copper–molybdenum (Cu–Mo), beryllium–aluminum (Be–Al), aluminum–silicon (Al–Si), and Silvar®, which contains silver and a nickel iron alloy. The latter materials can be considered as composites rather than true alloys because the two components have low solubility and appear as distinct phases at room temperature.

As Figure 9.8 shows, aluminum, copper, and silver have relatively high thermal conductivities but have CTEs much greater than desirable for most electronic packaging applications. By combining these metals with various reinforcements, it is possible to create new materials having CTEs isotropic in two dimensions (quasi-isotropic) or three dimensions in the desired range. The figure shows a number of composites: copper reinforced with UHK carbon fibers (C/Cu), aluminum reinforced with UHK carbon fibers (C/Al), carbon reinforced with UHK carbon fibers (C/C), epoxy reinforced with UHK carbon fibers (C/Ep), aluminum reinforced with silicon carbide particles [(SiC)p/Al], beryllium oxide particle-reinforced beryllium [(BeO)p/Be], diamond particle-reinforced copper [(Diamond)p/Cu], and E-glass fiber-reinforced epoxy (E-glass/Ep). With the exception of E-glass/Ep, C/Ep, and

C/C, all of the composites have some configurations with CTEs in the desired range. The thermal conductivities of the composites presented are generally similar to, or better than, that of aluminum, while their CTEs are much closer to the goal range of 3–7 ppm/K. E-glass/Ep is an exception.

Note that although the CTEs of C/Ep and C/C are lower than desired for electronic packaging applications, the differences between their CTEs and those of ceramics and semiconductors are much less than the differences for aluminum and copper. Consequently, use of the composites can result in lower thermal stresses for a given temperature change.

The physical properties of the materials shown in Figure 9.8 and others are presented in Table 9.12.

The advantages of composites are even greater than those of conventional packaging materials when weight is considered. Figure 9.9 presents the specific thermal conductivities and CTEs of the materials appearing in Figure 9.8. Here, we find order-of-magnitude improvements. As discussed earlier, when a critical property is increased by an order of magnitude it tends to have a revolutionary effect on technology. Several composites demonstrate this level of improvement; as a result, composites are being used in an increasing number of electronic packaging and thermal control applications, as discussed in Section 9.5.

Physical Properties of Polymer Matrix Composites

Table 9.13 presents physical properties of the polymer matrix composites discussed in Section 9.3.1. A fiber volume fraction of 60% is assumed. The densities of all of the materials are considerably lower than that of aluminum, and some are lower than that of magnesium. This reflects the low densities of both fibers and matrix materials. The low densities of most polymers give PMCs a significant advantage over most MMCs and CMCs, all other things being equal.

We observe that all of the composites have relatively low-axial CTEs. This results from the combination of low-fiber-axial CTE, high fiber stiffness, and low matrix stiffness. Note that the axial CTEs of PMCs reinforced with aramid fibers and some carbon fibers are negative. This means that, contrary to the general behavior of most monolithic materials, they contract when heated. The transverse CTEs of the composites are all positive, and much larger than the magnitudes of the corresponding axial CTEs. This results from the high CTE of the matrix and a Poisson effect caused by constraint of the matrix in the axial direction and lack of constraint in the transverse direction. The transverse CTE of aramid composites is particularly high because the fibers have a relatively high positive radial CTE.

The axial thermal conductivities of composites reinforced with glass, aramid, boron, and a number of the carbon fibers are relatively low. In fact, E-glass and aramid PMCs are often used as thermal insulators. As Table 9.13 shows, most PMCs have relatively high thermal resistivities in the transverse direction, as a result of the low thermal conductivities of the matrix and the fibers in the radial direction. Through-thickness conductivities of laminates tend to be similar to the transverse thermal conductivities of unidirectional composites.

Table 9.14 shows the inplane thermal conductivities and CTEs of quasi-isotropic laminates made from the same material as in Table 9.13. Here again, a fiber volume fraction of 60% is assumed.

We observe that the CTEs of the quasi-isotropic composites are higher than the axial values of corresponding unidirectional composites. Note, however, that the CTEs of quasi-isotropic composites reinforced with aramid and carbon fibers are very small. By appropriate selection of fiber, matrix, and fiber volume fraction, it is possible to obtain quasi-isotropic materials with CTEs very close to zero. Note that through-thickness CTEs for these laminates typically will be positive and relatively large. However, this is not a significant issue for many applications.

Turning to thermal conductivity, we find that quasi-isotropic laminates reinforced with UHM pitch carbon fibers have an inplane thermal conductivity similar to that of aluminum alloys, while UHK pitch carbon fibers provide laminates with a conductivity more than 50% higher. Both materials have densities 35% lower than aluminum.

As mentioned above, through-thickness thermal conductivities of laminates tend to be similar to the transverse thermal conductivities of unidirectional composites, which are relatively low. However, if laminate thickness is small, this may not be a significant limitation.

Physical Properties of Metal Matrix Composites

In this section, we consider physical properties of selected unidirectional fiber-reinforced MMCs and of silicon carbide particle-reinforced aluminum MMCs.

Physical Properties of Continuous Fiber-Reinforced Metal Matrix Composites. Table 9.11 presents physical properties of unidirectional composites consisting of UHK pitch carbon fibers in aluminum and copper matrices. These materials both have very low, slightly negative axial CTEs for the assumed fiber volume fraction of 50%. As the table shows, the axial thermal conductivities for MMCs with aluminum and copper matrices are substantially greater than that of monolithic copper. A major advantage of having thermally conductive matrix materials is that the resulting composite

Table 9.12 Physical Properties of Isotropic and Quasi-Isotropic Composites and Monolithic Materials Used in Electronic Packaging

Matrix	Reinforcement	V/O %	Density g/cm³ (Pci)	Coefficient of Thermal Expansion ppm/K (ppm/F)	Thermal Conductivity W/m K (BTU/h·ft·F)	Specific Thermal Conductivity W/m K
Aluminum (6063)	—	—	2.7 (0.098)	23 (13)	218 (126)	81
Copper	—	—	8.9 (0.32)	17 (9.8)	400 (230)	45
Beryllium	—	—	1.86 (0.067)	13 (7.2)	150 (87)	81
Magnesium	—	—	1.80 (0.065)	25 (14)	54 (31)	12
Titanium	—	—	4.4 (0.16)	9.5 (5.3)	16 (9.5)	4
Stainless steel (304)	—	—	8.0 (0.29)	17 (9.6)	16 (9.4)	2
Molybdenum	—	—	10.2 (0.37)	5.0 (2.8)	140 (80)	14
Tungsten	—	—	19.3 (0.695)	4.5 (2.5)	180 (104)	9
Invar®	—	—	8.0 (0.29)	1.6 (0.9)	10 (6)	1
Kovar®	—	—	8.3 (0.30)	5.9 (3.2)	17 (10)	2
Alumina (99% pure)	—	—	3.9 (0.141)	6.7 (3.7)	20 (12)	5
Beryllia	—	—	2.9 (0.105)	6.7 (3.7)	250 (145)	86
Aluminum nitride	—	—	3.2 (0.116)	4.5 (2.5)	250 (145)	78
Silicon	—	—	2.3 (0.084)	4.1 (2.3)	150 (87)	65
Gallium arsenide	—	—	5.3 (0.19)	5.8 (3.2)	44 (25)	8
Diamond	—	—	3.5 (0.13)	1.0 (0.6)	2000 (1160)	570
Pyrolitic graphite	—	—	2.3 (0.083)	−1 (−0.6)	1700 (980)	750
Aluminum-silicon	—	—	2.5 (0.091)	13.5 (7.5)	126 (73)	50
Beryllium-aluminum	—	—	2.1 (0.076)	13.9 (7.7)	210 (121)	100
Copper-tungsten (10/90)	—	—	17 (0.61)	6.5 (3.6)	209 (121)	12
Copper-molybdenum (15/85)	—	—	10 (0.36)	6.6 (3.7)	184 (106)	18
Aluminum	SiC particles	70	3.0 (0.108)	6.5 (3.6)	190 (110)	63
Beryllium	BeO particles	60	2.6 (0.094)	6.1 (3.4)	240 (139)	92
Copper	Diamond particles	55	5.9 (0.21)	5.8 (3.2)	420 (243)	71
Epoxy	UHK carbon fibers	60	1.8 (0.065)	−0.7 (−0.4)	330 (191)	183
Aluminum	UHK carbon fibers	26	2.6 (0.094)	6.5 (3.6)	290 (168)	112
Copper	UHK carbon fibers	26	7.2 (0.26)	6.5 (3.6)	400 (230)	56
Carbon	UHK carbon fibers	40	1.8 (0.065)	−1 (−0.6)	360 (208)	195

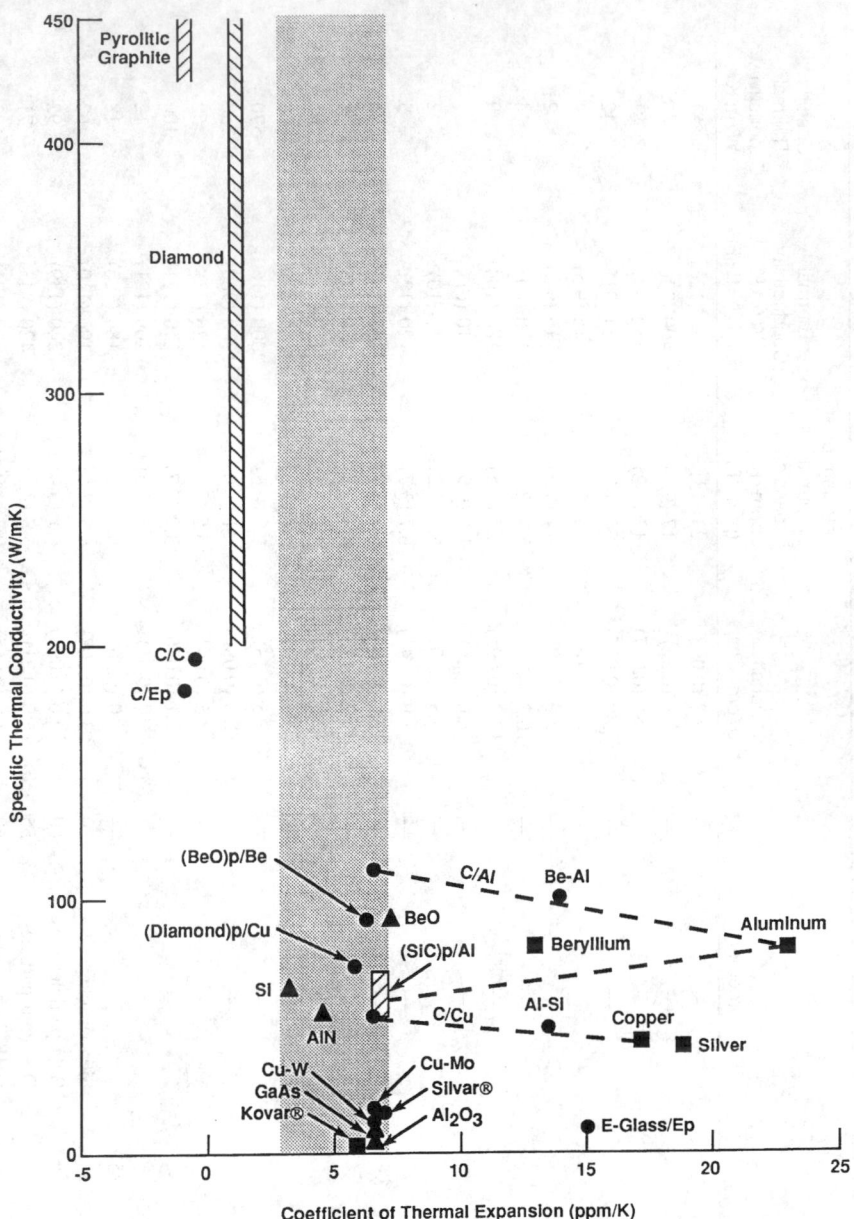

Fig. 9.9 Specific thermal conductivity (thermal conductivity divided by specific gravity) for selected monolithic materials and composites used in electronic packaging.

transverse and through-thickness thermal conductivities are more than an order of magnitude higher
than those of an epoxy-matrix composite.

Table 9.12 presents the properties of quasi-isotropic composites composed of aluminum and copper matrices reinforced with UHK pitch carbon fibers. Here, the fiber volume fraction of about 26%
has been chosen to achieve an inplane similar to that of aluminum oxide, 6.5 ppm/K (3.6 ppm/F).
The inplane thermal conductivities of the aluminum- and copper-matrix composites are 290 W/m K
(168 BTU/h · ft · F) and 400 W/m K (230 BTU/h · ft · F), respectively. These values are considerably
greater than those of any material with a similar CTE, with the exception of diamond particle-
reinforced copper, which is discussed later. Because of the lower fiber volume fractions, the through-

Table 9.13 Physical Properties of Selected Unidirectional Polymer Matrix Composites

Fiber	Density g/cm³ (Pci)	Axial CTE 10⁻⁶/K (10⁻⁶/F)	Transverse CTE 10⁻⁶/K (10⁻⁶/F)	Axial Thermal Conductivity W/m K (BTU/h·ft·F)	Transverse Thermal Conductivity W/m K (BTU/h·ft·F)
E-glass	2.1 (0.075)	6.3 (3.5)	22 (12)	1.2 (0.7)	0.6 (0.3)
Aramid	1.38 (0.050)	−4.0 (−2.2)	58 (32)	1.7 (1.0)	0.1 (0.08)
Boron	2.0 (0.073)	4.5 (2.5)	23 (13)	2.2 (1.3)	0.7 (0.4)
SM carbon (PAN)	1.58 (0.057)	0.9 (0.5)	27 (15)	5 (3)	0.5 (0.3)
UHS carbon (PAN)	1.61 (0.058)	0.5 (0.3)	27 (15)	10 (6)	0.5 (0.3)
UHM carbon (PAN)	1.66 (0.060)	−0.9 (−0.5)	40 (22)	45 (26)	0.5 (0.3)
UHM carbon (pitch)	1.80 (0.065)	−1.1 (−0.6)	27 (15)	380 (220)	10 (6)
UHK carbon (pitch)	1.80 (0.065)	−1.1 (−0.6)	27 (15)	660 (380)	10 (6)

Table 9.14 Physical Properties of Selected Quasi-Isotropic Polymer Matrix Composites

Fiber	Density g/cm³ (Pci)	Axial CTE 10⁻⁶/K (10⁻⁶/F)	Transverse CTE 10⁻⁶/K (10⁻⁶/F)	Axial Thermal Conductivity W/m K (BTU/h·ft·F)	Transverse Thermal Conductivity W/m K (BTU/h·ft·F)
E-glass	2.1 (0.075)	10 (5.6)	10 (5.6)	0.9 (0.5)	0.9 (0.5)
Aramid	1.38 (0.050)	1.4 (0.8)	1.4 (0.8)	0.9 (0.5)	0.9 (0.5)
Boron	2.0 (0.073)	6.5 (3.6)	6.5 (3.6)	1.4 (0.8)	1.4 (0.8)
SM carbon (PAN)	1.58 (0.057)	3.1 (1.7)	3.1 (1.7)	2.8 (1.6)	2.8 (1.6)
UHS carbon (PAN)	1.61 (0.058)	2.3 (1.3)	2.3 (1.3)	6 (3)	6 (3)
UHM carbon (PAN)	1.66 (0.060)	0.4 (0.2)	0.4 (0.2)	23 (13)	23 (13)
UHM carbon (pitch)	1.80 (0.065)	−0.4 (−0.2)	−0.4 (−0.2)	195 (113)	195 (113)
UHK carbon (pitch)	1.80 (0.065)	−0.4 (−0.2)	−0.4 (−0.2)	335 (195)	335 (195)

thickness thermal conductivities of these composites will be somewhat higher than those of the unidirectional composites presented in Table 9.11.

Physical Properties of Particle-Reinforced Metal Matrix Composites. In this section, we consider the physical properties of silicon carbide particle-reinforced aluminum and diamond particle-reinforced copper.

Physical Properties of Silicon Carbide Particle-Reinforced Metal Matrix Composites. The physical properties of particle-reinforced composites tend to be isotropic (in three dimensions). As for all composites, the physical properties of silicon carbide particle-reinforced aluminum depend on constituent properties and reinforcement volume fraction. Figure 9.4 shows how the CTE of (SiC)p/Al varies with particle volume fraction for typical commercial materials. Table 9.15 presents density and CTE for several specific volume fractions, along with data for monolithic aluminum, titanium, and steel.

Thermal conductivity depends strongly on the corresponding properties of the matrix, reinforcement, and particle volume fraction. The thermal conductivity of very pure silicon carbide is slightly higher than that of copper. However, those of commercial particles are much lower. The thermal conductivities of silicon carbide particle-reinforced aluminum used in electronic packaging applications tend to be in the range of monolithic aluminum alloys, about 160–218 W/m K (92–126 Btu/h·ft·F).

Physical Properties of Diamond Particle-Reinforced Copper Metal Matrix Composites. Table 9.12 presents the physical properties of diamond particle-reinforced copper composites, which are developmental materials. As for other particle-reinforced composites, the properties can be expected to be relatively isotropic. This material has a thermal conductivity somewhat higher than that of monolithic copper, a much lower density, and a CTE in the range of semiconductors and ceramic substrates. This unique combination of properties makes this composite an attractive candidate for electronic packaging applications.

Physical Properties of Ceramic Matrix Composites

As discussed in Section 9.1, there are many CMCs and they are at various stages of development. One of the more mature systems is silicon carbide fiber-reinforced silicon carbide (SiC/SiC). For a fabric-reinforced composite with a fiber volume fraction of 40%, the density is 2.5 g/cm³ (0.090 Pci), the CTE is 3 ppm/K (1.7 ppm/F), the inplane thermal conductivity is 19 W/m K (11 BTU/h·ft·F) and the through-thickness value is 9.5 W/m K (5.5 BTU/h·ft·F).

Physical Properties of Carbon/Carbon Composites

The CTE of CCCs depends on fiber type, volume fraction, and geometry and matrix characteristics. In the fiber direction, CTE tends to be negative with a small absolute value. Perpendicular to the fiber direction, composite CTE is dominated by matrix properties. As a rule, the magnitude of transverse CTE is small. Both positive and negative values have been reported.

It is well known that in some forms carbon has exceptionally high thermal conductivities. For example, pyrolitic graphite can have a thermal conductivity as high as 2000 W/m K (1160 BTU/h·ft·F), five times that of copper. The conductivity of some types of diamond is much higher. Some CCCs also have very high thermal conductivities. Values have been reported as high as 400 W/m K (230 BTU/h·ft·F) for quasi-isotropic composites and 700 W/m K (400 BTU/h·ft·F) for unidirectional materials.

9.4 PROCESSES

Selection of the processes by which a system will be fabricated is a key part of design. Processes for making composites have a critical influence on material properties, reliability, cost and schedule. One of the big advantages of composites is that there are processes that allow consolidation of parts, which can reduce overall complexity and cost and improve performance. Many processes allow fabrication of parts to their final shape (net shape processes) or close to their final shape (near-net shape processes). This can eliminate or reduce the need for, and cost associated with, machining and joining. This section describes key fabrication processes for the four classes of composites.

Conceptually, fabrication of composite components consists of four parts: combination of constituents, material consolidation, shaping, and joining. Some processes combine two or more of these steps. There are a number of processes, especially those based on infiltration of liquid matrices, that have analogues in all four classes of composites. Infiltration processes typically use a preform, in which the reinforcement is shaped and held together by a temporary (fugitive) or permanent binder. In the case of fibrous reinforcements, the preform must maintain the correct fiber orientation and volume fraction during processing to make sure that the strength and stiffness properties of the finished part meet design requirements.

An important consideration in all processes is minimization of voids, or porosity, which typically has a deleterious effect on properties.

Table 9.15 Physical Properties of Silicon Carbide Particle-Reinforced Aluminum

Property	Aluminum (6061-T6)	Titanium (6Al-4V)	Steel (4340)	Composite Particle Volume Fraction		
				25	55	70
CTE, 10^{-6}/K (10^{-6}/F)	23 (13)	9.5 (5.3)	12 (6.6)	16.4 (9.1)	10.4 (5.8)	6.2 (3.4)
Thermal Conductivity W/m·K (BTU/h·ft·F)	218 (126)	16 (9.5)	17 (9.4)	160–220 (92–126)	160–220 (92–126)	160–220 (92–126)
Density, g/cm³ (Pci)	2.77 (0.10)	4.43 (0.16)	7.76 (0.28)	2.88 (0.104)	2.96 (0.107)	3.00 (0.108)

9.4.1 Polymer Matrix Composites

There are a large and increasing number of processes for making PMC parts. Many are not very labor-intensive and can make near-net shape components. For thermoplastic matrices reinforced with discontinuous fibers, one of the most widely used processes is injection molding. However, as discussed in Section 9.3, the stiffness and strength of resulting parts are relatively low. This section focuses on processes for making composites with continuous fibers.

Many PMC processes combine fibers and matrices directly. However, a number use an intermediate material called a prepreg, which stands for preimpregnated material, consisting of fibers embedded in a thermoplastic or partially cured thermoset matrix. The most common forms of prepreg are unidirectional tapes and impregnated tows and fabrics.

Material consolidation is commonly achieved by application of heat and pressure. For thermosetting resins, consolidation involves a complex physical–chemical process, which is accelerated by subjecting the material to elevated temperature. However, some resins undergo cure at room temperature. Another way to cure resins without temperature is by use of electron bombardment. As part of the consolidation process, uncured laminates are often placed in an evacuated bag, called a vacuum bag, which applies atmospheric pressure when evacuated. The vacuum-bagged assembly is typically cured in an oven or autoclave. The latter also applies pressure significantly above the atmospheric level.

PMC parts are usually shaped by use of molds made from a variety of materials: steel, aluminum, bulk graphite, and also PMCs reinforced with E-glass and carbon fibers. Sometimes molds with embedded heaters are used.

The key processes for making PMC parts are filament winding, fiber placement, compression molding, pultrusion, prepreg lay-up, resin film infusion and resin transfer molding. The latter process uses a fiber preform which is placed in a mold.

9.4.2 Metal Matrix Composites

An important consideration in selection of manufacturing processes for MMCs is that reinforcements and matrices can react at elevated temperatures, degrading material properties. To overcome this problem, reinforcements are often coated with barrier materials. Many of the processes for making MMCs with continuous fiber reinforcements are very expensive. However, considerable effort has been devoted to development of relatively inexpensive processes that can make net shape or near-net shape parts that require little or no machining to achieve their final configuration.

Manufacturing processes for MMCs are based on a variety of approaches for combining constituents and consolidating the resulting material: powder metallurgy, ingot metallurgy, plasma spraying, chemical vapor deposition, physical vapor deposition, electrochemical plating, diffusion bonding, hot pressing, remelt casting, pressureless casting, and pressure casting. The last two processes use preforms.

Some MMCs are made by *in situ* reaction. For example, a composite consisting of aluminum reinforced with titanium carbide particles has been made by introducing a gas containing carbon into a molten alloy containing aluminum and titanium.

9.4.3 Ceramic Matrix Composites

As for MMCs, an important consideration in fabrication of CMCs is that reinforcements and matrices can react at high temperatures. An additional issue is that ceramics are very difficult to machine, so that it is desirable to fabricate parts that are close to their final shape. A number of CMC processes have this feature. In addition, some processes make it possible to fabricate CMC parts that would be difficult or impossible to create out of monolithic ceramics.

Key processes for CMCs include chemical vapor infiltration (CVI); infiltration of preforms with slurries, sol-gels, and molten ceramics; *in situ* chemical reaction; sintering; hot pressing; and hot isostatic processing. Another process infiltrates preforms with selected polymers that are then pyrolyzed to form a ceramic material.

9.4.4 Carbon/Carbon Composites

CCCs are primarily made by chemical vapor infiltration (CVI), also called chemical vapor deposition (CVD), and by infiltration of pitch or various resins. Following infiltration, the material is pyrolyzed, which removes most non-carbonaceous elements. This process is repeated several times until the desired material density is achieved.

9.5 APPLICATIONS

Composites are now being used in a large and increasing number of important mechanical engineering applications. In this section, we discuss some of the more significant current and emerging applications.

It is generally known that glass fiber-reinforced polymer (GFRP) composites have been used extensively as engineering materials for decades. The most widely recognized applications are probably boats, electrical equipment, and automobile and truck body components. It is generally known, for example, that the Corvette body is made of fiberglass and has been for many years. However, many materials that are actually composites, but are not recognized as such, also have been used for a long time in mechanical engineering applications. One example is cermets, which are ceramic particles bound together with metals; hence the name. These materials fall in the category of metal matrix composites. Cemented carbides are one type of cermet. What are commonly called "tungsten carbide" cutting tools and dies are, in most cases, not made of monolithic tungsten carbide, which is too brittle for many applications. Instead, they are actually MMCs consisting of tungsten carbide particles embedded in a high-temperature metallic matrix such as cobalt. The composite has a much higher fracture toughness than monolithic tungsten carbide.

Another example of unrecognized composites are industrial circuit breaker contact pads, made of silver reinforced with tungsten carbide particles, which impart hardness and wear resistance (Fig. 9.10). The silver provides electrical conductivity. This MMC is a good illustration of an application for which a new multifunctional material was developed to meet requirements for a combination of physical and mechanical properties.

In this section, we consider representative examples of composite usage in mechanical engineering applications, including aerospace and defense; electronic packaging and thermal control; machine components; internal combustion engines; transportation; process industries, high temperature and wear, corrosion and oxidation-resistant equipment; offshore and onshore oil exploration and production equipment; dimensionally stable components; biomedical applications; sports and leisure equipment; marine structures and miscellaneous applications. Use of composites is now so extensive that it is impossible to present a complete list. Instead, we have selected applications that, for the most part, are commercially successful and illustrate the potential for composite materials in various aspects of mechanical engineering.

9.5.1 Aerospace and Defense

Composites are baseline materials in a wide range of aerospace and defense structural applications, including military and commercial aircraft, spacecraft, and missiles. They are also used in aircraft gas turbine engine components, propellers, and helicopter rotors. Aircraft brakes are covered in another subsection.

PMCs are the workhorse materials for most aerospace and defense applications. Standard modulus and intermediate modulus carbon fibers are the leading reinforcements, followed by aramid and glass. Boron fibers are used in some of the original composite aircraft structures and special applications requiring high compressive strength. For low-temperature airframe and other applications, epoxies are the key matrix resin. For higher temperatures, bismaleimides, polyimides, and phenolics are employed. Thermoplastic resins increasingly are finding their way into new applications.

The key properties of composites that have led to their use in aircraft structures are high specific stiffness and strength and excellent fatigue resistance. For example, composites have largely replaced

Fig. 9.10 Commercial circuit breaker uses tungsten carbide particle-reinforced silver contact pads.

monolithic aluminum in helicopter rotors because they extend fatigue life by factors of up to six times those of metallic designs.

The amount of composites used in aircraft structures varies by type of aircraft and the time at which they were developed. The B-2 "Stealth" Bomber makes extensive use of carbon fiber-reinforced PMCs (Fig. 9.11).

In general, aircraft that take off and land vertically (VTOL aircraft), such as helicopters and tilt wing vehicles, use the highest percentage of composites in their structures. For all practical purposes, most new VTOL aircraft have all-composite structures. The V-22 Osprey uses PMCs reinforced with carbon, aramid, and glass fibers in the fuselage, wings, empennage (tail section) and rotors (Fig. 9.12).

Use of composites in commercial passenger aircraft is limited by practical manufacturing problems in making very large structures and by cost. Still, use of composites has increased steadily. For example, the Boeing 777 has an all-composite empennage.

Fig. 9.11 The B-2 "Stealth" Bomber airframe makes extensive use of carbon fiber-reinforced polymer matrix composites (courtesy Northrop Grumman).

Fig. 9.12 The V-22 Osprey uses polymer matrix composites in the fuselage, empennage, and rotors (courtesy Boeing).

Thrust-to-weight ratio is an important figure of merit for aircraft gas turbine engines and other propulsion systems. Because of this, there has been considerable work devoted to the development of a variety of composite components. Production applications include carbon fiber-reinforced polymer fan blades, exit guide vanes, and nacelle components; silicon carbide particle-reinforced aluminum exit guide vanes; and CMC engine flaps made of silicon carbide reinforced with carbon and with silicon carbide fibers.

There has been extensive development of MMCs with titanium and titanium aluminide matrices reinforced with silicon carbide fibers aimed at high-temperature engine and fuselage structures. Composites using intermetallic materials, such as titanium aluminide, are often called intermetallic matrix composites (IMCs).

The key design requirements for spacecraft structures are high specific stiffness and low thermal distortion, along with high specific strength for those components that see high loads during launch. The key reinforcements are high-stiffness PAN- and pitch-based carbon fibers. Figure 9.13 shows the NASA Upper Atmosphere Research Satellite structure, which is made of high-modulus PAN carbon/epoxy. For most spacecraft, thermal control is also an important design consideration, due in large part to the absence of convection as a cooling mechanism in space. Because of this, there is increasing interest in thermally conductive materials, including PMCs reinforced with ultrahigh-modulus pitch-based carbon fibers for structural components such as radiators, and for electronic packaging. MMCs are also being used for thermal control and electronic packaging applications. See Section 9.5.3 for a more detailed discussion of these applications.

The Space Shuttle Orbiters use boron fiber-reinforced aluminum struts in their center fuselage sections and CCC nose caps and wing leading edges.

The Hubble Space Telescope high-gain antenna masts, which also function as wave guides, are made of an MMC consisting of ultrahigh-modulus pitch-based carbon fibers in an aluminum matrix.

Missiles, especially those with solid rocket motors, have used PMCs for many years. In fact, high-strength glass was originally developed for this application. As for most aerospace applications, epoxies are the most common matrix resins. Over the years, new fibers with increasingly higher specific strengths—first aramid, then ultrahigh-strength carbon—have displaced glass in high-performance applications. However, high-strength glass is still used in a wide variety of related applications, such as launch tubes for shoulder-fired anti-tank rockets.

Carbon/carbon composites are widely used in rocket nozzle throat inserts.

9.5.2 Machine Components

Composites increasingly are being used in machine components because they reduce mass and thermal distortion and have excellent resistance to corrosion and fatigue.

Fig. 9.13 The Upper Atmosphere Research Satellite structure is composed of lightweight high-modulus carbon fiber-reinforced epoxy struts, which provide high stiffness and strength and low coefficient of thermal expansion.

One of the most successful applications has been in rollers and shafts used in machines that handle rolls of paper, thin plastic film, fiber products, and audio tape. Figure 9.14 shows a chromium-plated carbon fiber-reinforced epoxy roller used in production of audio tape. The low rotary inertia of the composite part allows it to start and stop more quickly than the baseline metal design. This reduces the amount of defective tape resulting from differential slippage between roller and tape.

Rollers as long as 10.7 m (35 ft) and 0.43 m (17 in.) in diameter have been produced. In these applications, use of carbon fiber-reinforced polymers has resulted in reported mass reductions of 30% to 60%. This enables some shafts to be handled by one person instead of two (Fig. 9.15). It also reduces shaft rotary inertia, which, as for the audio machine roller discussed in the previous paragraph, allows machines to be stopped more quickly without damaging the plastic or paper. The higher critical speeds of composite shafts also allow them to be operated at higher speeds. In addition, the high stiffness of composite shafts reduces lateral displacement under load. PMC rollers can be coated with a variety of materials, including metals and elastomers.

PMCs also have been used in translating parts, such as tubes used to remove plastic parts from injection molding machines. In another application, use of a carbon fiber-reinforced epoxy robotic arm in a computer cartridge-retrieval system doubled the cartridge-exchange rate compared to the original aluminum design.

Specific strength is an important figure of merit for materials used in flywheels. Composites have received considerable attention for this reason (Fig. 9.16). Another advantage of composites is that their modes of failure tend to be less catastrophic than for metal designs. The latter, when they fail, often liberate large pieces of high-velocity, shrapnel-like jagged metal that are dangerous and difficult to contain.

The high specific stiffness and low coefficient of thermal expansion (CTE) of silicon carbide particle-reinforced aluminum has led to its use in machine parts for which low vibration, mass, and thermal distortion are important, such as photolithography stages (Fig. 9.17). The absence of out-gassing is another advantage of MMC components.

Figure 9.18 shows a developmental actuator housing made of silicon carbide particle-reinforced aluminum. Properties of interest here are high specific stiffness and yield strength. In addition, compared to monolithic aluminum, the composite offers a closer CTE match to steel than monolithic aluminum, and better wear resistance.

The excellent hardness, wear resistance, and smooth surface of a silicon carbide whisker-reinforced alumina CMC resulted in the adoption of this material for use in beverage can-forming equipment. Here, we find a CMC replacing what is in fact a metal matrix composite; a cemented carbide or cermet, consisting of tungsten carbide particles in a cobalt binder.

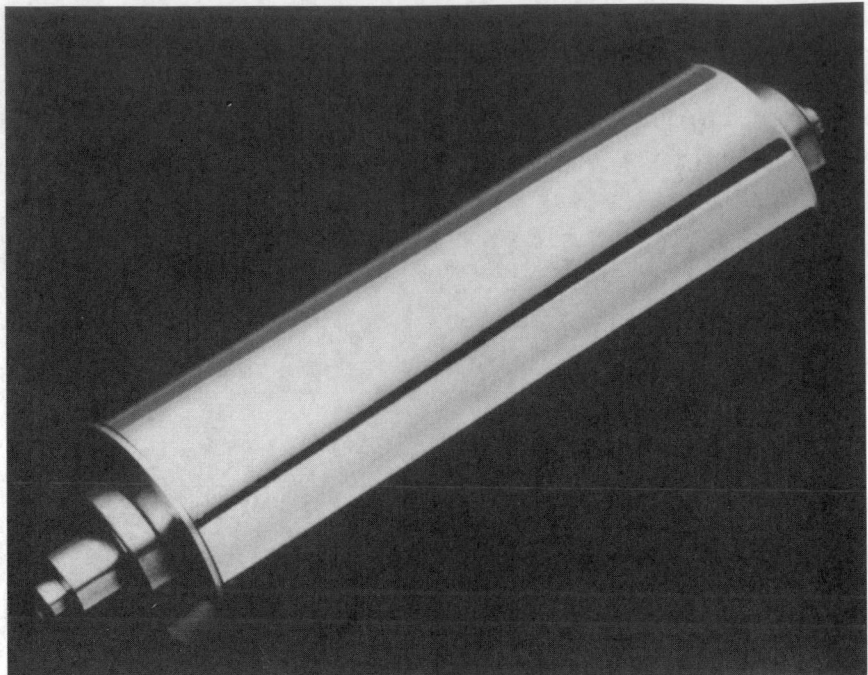

Fig. 9.14 Metal plated carbon/epoxy roller used in production of audio tape has a much lower rotary inertia than a metal roller, decreasing smearing during startup and shutdown (courtesy Tonen).

9.5.3 Electronic Packaging and Thermal Control

Composites increasingly are being used in thermal control and electronic packaging applications because of their high thermal conductivities, low densities, tailorable CTEs, and availability of net shape and near-net shape fabrication processes. The materials of interest are PMCs, MMCs, and CCCs.

Electronic Packaging

Electronic packaging is commonly divided into various levels, starting at the level of the integrated circuit and progressing upwards to the enclosure and support structure. Composites are used in all of these levels. Components made of composites include carriers, packages, heat sinks, enclosures, and support structures. Key production materials include silicon carbide particle-reinforced aluminum, beryllium oxide particle-reinforced beryllium, ultrahigh-thermal-conductivity (UHK) pitch-based carbon fiber-reinforced polymers, metals, and CCCs. Various types of composite components are used in electronic devices for cellular telephone ground telephone stations, electrical vehicles, aircraft, spacecraft, and missiles. Figure 9.19 shows a spacecraft electronics module housing made of beryllium oxide particle-reinforced beryllium. MMCs also have been successfully used in many aircraft electronic systems. For example, Figure 9.20 shows a printed circuit board heat sink (also called a cold plate or thermal plane) made of silicon carbide particle-reinforced aluminum.

Thermal Control

The key composite materials used in thermal control applications are UHK carbon fiber-reinforced polymers. For the most part, the applications include components that have structural as well as thermal control applications. Examples include the Boeing 777 aircraft engine nacelle honeycomb cores and spacecraft radiator panels and battery sleeves.

9.5.4 Internal Combustion Engines

There have been a number of historic uses of MMCs in automobile internal combustion engines. In the early 1980s, Toyota introduced an MMC diesel engine piston consisting of aluminum locally reinforced in the top ring groove region with discontinuous alumina-silica fibers and with discontin-

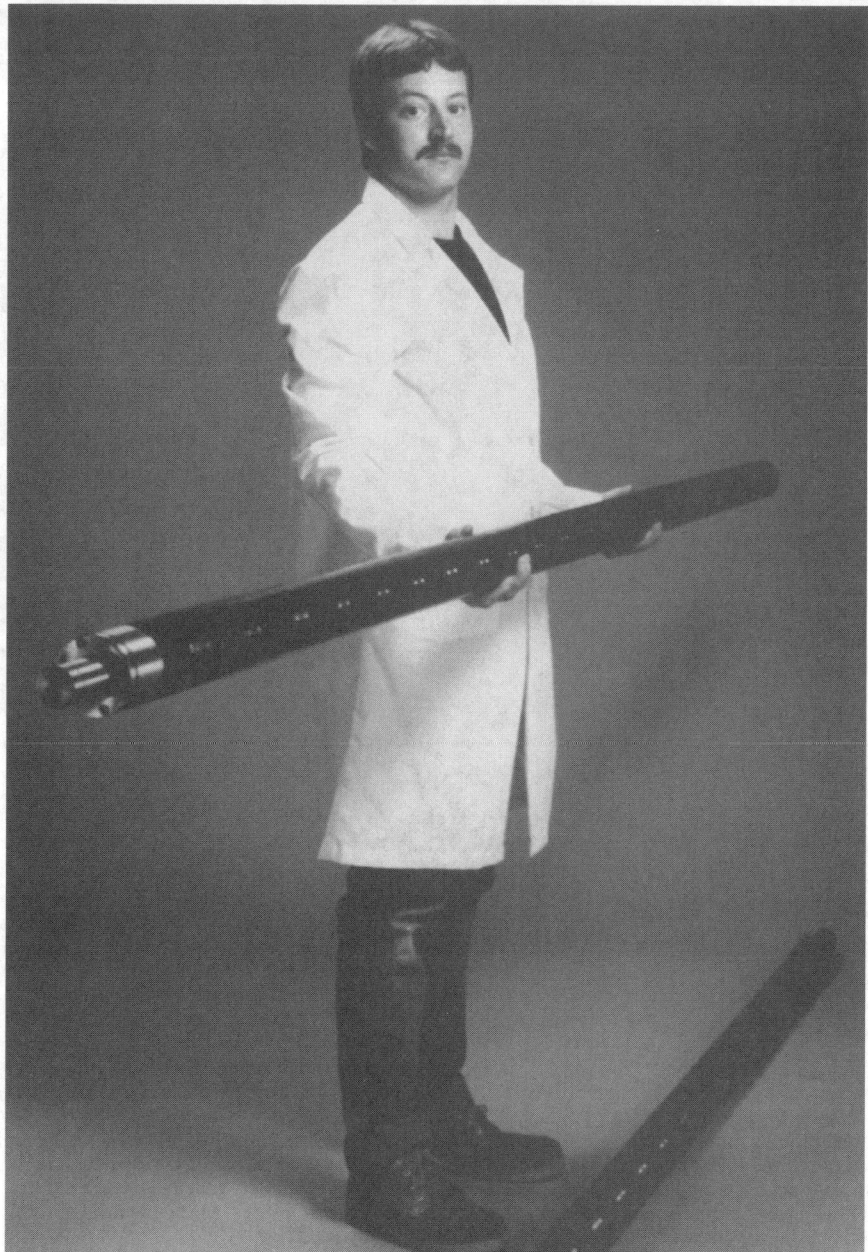

Fig. 9.15 The lower weight of carbon/epoxy rollers used in printing, paper, and conversion equipment facilitates handling. Lower rotary inertia results in reduced tendency to tear paper and plastic film during startup and shutdown (courtesy Du Pont).

uous alumina fibers. The pistons are made by pressure infiltration of a preform. Here, the ceramic fibers provide increased wear resistance, replacing a heavier nickel cast iron insert that was used with the original monolithic aluminum piston.

In the early 1990s, Honda began production of aluminum engine blocks reinforced in the cylinder wall regions with a combination of carbon and alumina fibers. Use of fiber reinforcement allowed the removal of cast iron cylinder liners that had been required because of the poor wear resistance

Fig. 9.16 Developmental flywheel for automobile energy storage combines a carbon/epoxy rim and a high-strength glass/epoxy disk.

of monolithic aluminum. As for the Toyota pistons, the engine blocks are made by a pressure infiltration process. The Honda engine uses hybrid fiber preforms consisting of discontinuous alumina and carbon fibers with a ceramic binder. The advantages of the composite design are greater bore diameter with no increase in overall engine size, higher thermal conductivity in the cylinder walls, and reduced weight. Figure 9.21 shows one of the engine blocks with a section cut away. The fiber-reinforced regions are clearly visible in a close-up view of the cylinder walls (Fig. 9.22).

Other engine components under evaluation are carbon/carbon pistons; MMC connecting rods and piston wrist pins; and CMC diesel engine exhaust valve guides.

9.5.5 Transportation

Composites are used in a wide variety of transportation applications, including automobile, truck, and train bodies; drive shafts; brakes; springs; and natural gas vehicle cylinders. There is also considerable interest in composite flywheels as a source of energy storage in vehicles. This subject is covered in Section 9.5.2.

Automobile, Truck, and Train Bodies

As mentioned in the introduction to this section, it is widely known that for many years, the GM Corvette has had a PMC body consisting of chopped glass fiber-reinforced thermosetting polyester. However, the body is semi-structural and primary loads are supported by a steel frame. A key reason for use of PMCs reinforced with chopped glass fibers in automotive components is that these materials

Fig. 9.17 Silicon carbide particle-reinforced aluminum photolithography stage has the same stiffness as the cast iron baseline, but is 60% lighter and has a much higher thermal conductivity, reducing thermal gradients and resulting distortion (courtesy Lanxide).

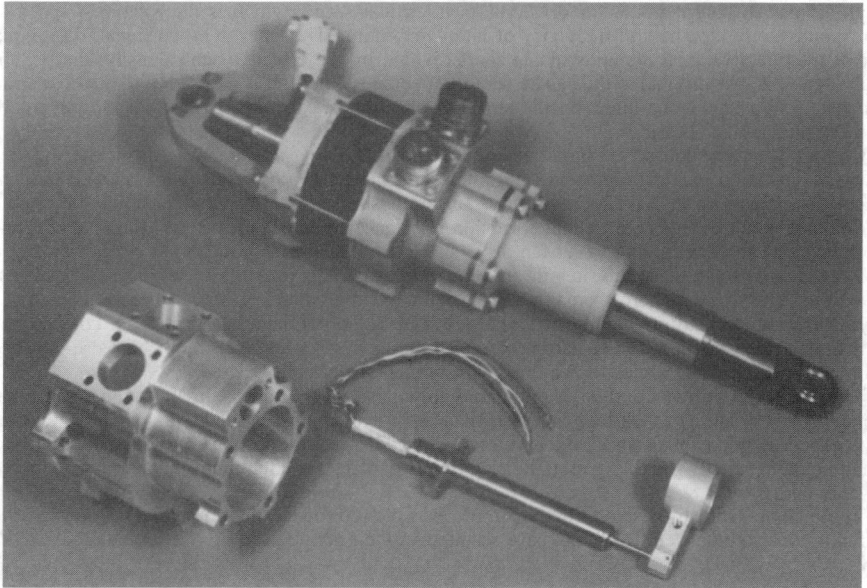

Fig. 9.18 Silicon carbide particle-reinforced aluminum actuator housings provide higher stiffness and wear resistance and lower coefficient of thermal expansion than aluminum (courtesy DWA Aluminum Composites).

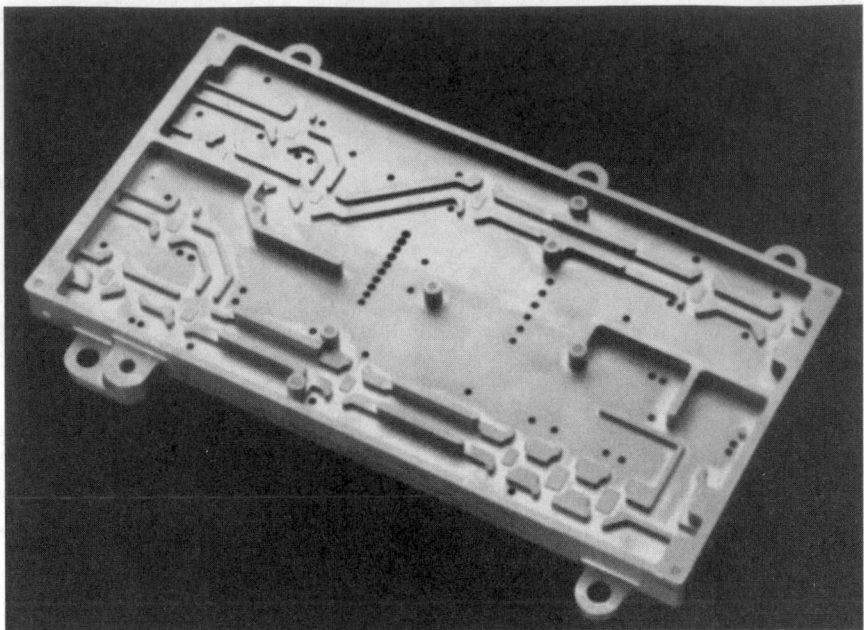

Fig. 9.19 Beryllium oxide particle-reinforced beryllium RF electronic housing provides reduced mass, high thermal conductivity, and coefficient of thermal expansion in the range of ceramic substrates and semiconductors (courtesy Brush Wellman).

allow complex shapes to be made in one piece, replacing numerous steel stampings that must be joined by welding or mechanical fastening, thereby reducing labor costs.

Drive Shafts

A critical design consideration for drive shafts is critical speed, which is the rotational speed that corresponds to the first natural frequency of lateral vibration. The latter is proportional to the square root of the effective axial modulus of the shaft divided by the effective shaft density; that is, shaft critical speed is proportional to the square root of specific stiffness. It has been found that in a variety of mechanical systems, the high specific stiffness of composites makes it possible to eliminate the need for intermediate bearings.

Composite production drive shafts are used in boats, cooling tower fans, and pickup trucks. In the last application, use of composites eliminates the need for universal joints, as well as center support bearings (Fig. 9.23). The lower mass of composite shafts also reduces vibrational loads on bearings, reducing wear. The excellent corrosion resistance of composites is an additional advantage in applications such as cooling tower fan drive shafts (see Section 9.5.6).

Another advantage of composites in drive shafts is that it is possible to vary the ratio of axial-to-torsional stiffness far more than is possible with metallic shafts. This can be accomplished by varying the number and orientation of the layers, and by appropriate use of material combinations. For example, it is possible to use carbon fibers in the axial direction to achieve high critical speed, and glass fibers at other angles to achieve low torsional stiffness, if desired.

The number of different designs and material combinations is limitless. In almost all cases, carbon fibers are used because of their high specific stiffness. Often, E-glass is used as an outer layer because of its excellent impact resistance and lower cost. In one case, carbon fibers are applied axially to a thin aluminum shaft. E-glass is used to electrically isolate the aluminum and carbon to prevent galvanic corrosion.

The high specific stiffness of silicon carbide particle-reinforced aluminum and the low cost and weldability of some material systems have resulted in their adoption in production automobile drive shafts.

Brakes for Automobiles, Trains, Aircraft, and Special Applications

Volumetric constraints and the need to reduce weight have led to the use of a variety of composites for automobile, train, aircraft, and special application brake components.

Fig. 9.20 Silicon carbide particle-reinforced aluminum printed circuit board heat sink is much lighter and has a higher specific stiffness than the copper–molybdenum baseline, and provides similar thermal performance (courtesy Lanxide Electronic Products).

Carbon/carbon composites have been used for some years in aircraft brakes in place of steel, resulting in a substantial weight reduction. Carbon/carbon has also been used in racing car and racing motorcycle brakes.

The wear resistance of monolithic aluminum generally is not good enough for brake rotors. However, introduction of ceramic particles, such as silicon carbide and alumina, results in materials with greatly improved resistance to wear. Ceramic particle-reinforced aluminum MMCs are being used in both automobile and railway car brake rotors in place of cast iron. In these applications, the high thermal conductivity of the composite is an advantage. However, the relatively low melting point of aluminum prevents the use of composites employing this metal as a matrix in rotors which see very high temperatures. The high specific stiffness and wear resistance of silicon carbide particle-reinforced aluminum have led to the evaluation of these MMCs in brake calipers. Figure 9.24 shows ceramic particle-reinforced aluminum brake rotors and caliper components.

Another interesting application for ceramic particle-reinforced aluminum MMCs is in amusement car rail brakes (see Section 9.5.12).

Automobile Springs

The Corvette uses structural GFRP leaf springs that are reinforced with continuous glass fibers. These have been used successfully for many years in what is a very demanding, cost-sensitive application.

Natural Gas Vehicle Cylinders

There is considerable interest in use of natural gas as a fuel for automobiles and trucks. Pressure vessels to contain the natural gas are required for the vehicles, refueling stations, and trucks to transport the fuel. The weight and cost of vehicle fuel tanks are major issues. A variety of composite designs that demonstrate weight savings over steel have been developed. They use steel, aluminum, or polymeric liners overwrapped with carbon fiber, glass fiber, or a combination of the two, embedded

Fig. 9.21   Honda Prelude engine block has cylinder walls that are reinforced with a combination of alumina and carbon fibers, eliminating the need for cast iron sleeves. The result is an engine with better thermal performance and a higher power-to-weight ratio (courtesy Honda).

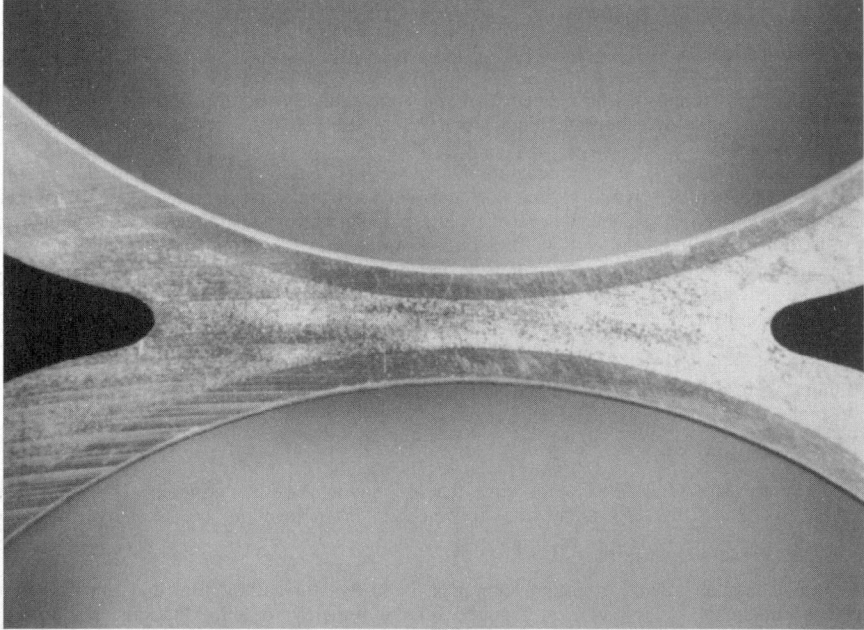

Fig. 9.22   Close-up of Honda Prelude cylinder walls showing region of fibrous reinforcement (courtesy Honda).

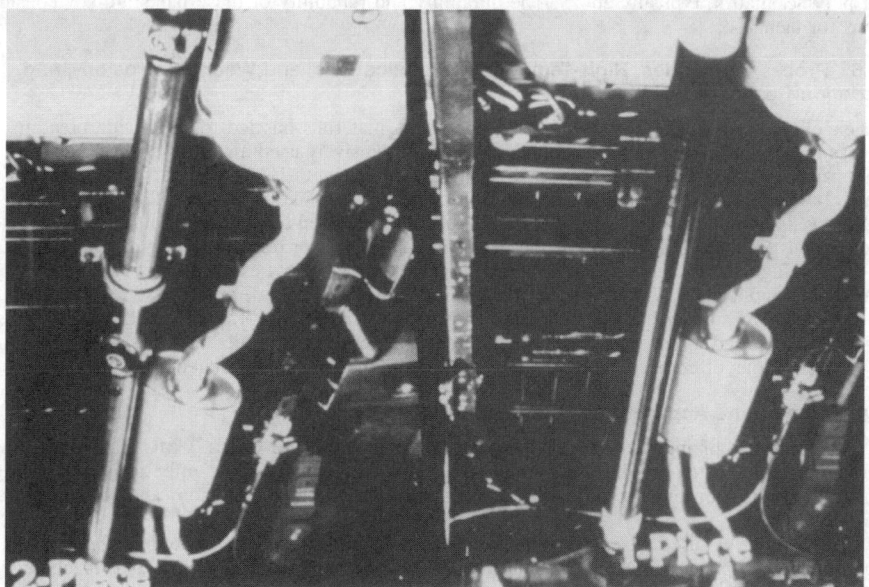

Fig. 9.23 One-piece pickup truck drive shaft consists of outer layers of carbon- and glass fiber-reinforced polymer that are pultruded over an inner aluminum tube. The composite drive shaft replaces a two-piece steel shaft that requires an intermediate support bearing and universal joint (courtesy MMFG).

Fig. 9.24 Silicon carbide particle-reinforced aluminum brake rotors, calipers, and other parts provide higher specific stiffness and better wear resistance than monolithic aluminum and are lighter than cast iron (courtesy Lanxide).

in a polymer matrix, typically epoxy. The durability and reliability of these tanks are key considerations for their use.

9.5.6 Process Industries, High-Temperature Applications, and Wear-, Corrosion-, and Oxidation-Resistant Equipment

The excellent corrosion resistance of many composite materials has led to their widespread use in process industries equipment. Undoubtedly, the most extensively used materials are PMCs consisting of thermosetting polyester and vinyl ester resins reinforced with E-glass fiber. These materials are relatively inexpensive and easily formed into products such as pipes, tanks, and flue liners. However, GFRP has its limitations. E-glass is susceptible to creep and creep rupture and is attacked by a variety of chemicals, including alkalies. For these reasons, E-glass fiber-reinforced polymers are typically not used in high-stress components. In addition, polyesters and vinyl esters are not suitable for high-temperature applications. Other types of composite materials overcome the limitations of GFRP and are finding increasing use in applications for which resistance to corrosion, oxidation, wear, and erosion are required, often in high-temperature environments. In this section, we consider representative applications of composites in a variety of process industries and related equipment.

High-Temperature Applications

The key materials of interest for high-temperature applications are CCCs, CMCs, and PMCs with high-temperature matrices. These materials, especially CMCs and CCCs, offer resistance to high-temperature corrosion and oxidation, as well as resistance to wear, erosion, and mechanical and thermal shock.

CCCs are being used in equipment to make glass products, such as bottles. Production and experimental components include GOB distributors, interceptors, pads, and conveyor machine wear guides. Use of carbon/carbon eliminates the need for water cooling, coatings, and lubricants required for steel parts. In some applications, the CCC parts have shown significant reduction in wear.

Carbon fiber-reinforced high-temperature thermoplastic composites are also being used in glass-handling equipment. The key advantages of this material are its low thermal conductivity, which reduces glass checking (microcracking), and its wear resistance, which reduces down time for part replacement.

A wide variety of ceramic matrix composites are being used in production and developmental high-temperature applications, including industrial gas turbine combustor liners and turbine rotor tip shrouds; radiant burner and immersion tubes; high-temperature gas filters; reverberatory screens for porous radiant burners; heat exchanger tubes and tube headers; and tube hangers for crude oil preheat furnaces. Figure 9.25 shows a number of developmental continuous fiber CMC parts made by polymer impregnation and pyrolysis: combustor liners, chemical pump components, high-temperature pipe hangers, and turbine seals. Figure 9.26 shows a CMC hot gas candle filter composed of alumina–boria–silica fibers in a silicon carbide matrix made by chemical vapor deposition.

In another high-temperature application, silicon carbide whisker-reinforced silicon nitride ladles are being used for casting molten aluminum.

Wear- and Erosion-Resistant Applications

PMCs, MMCs, CMCs, and CCCs are all being used in a variety of applications for which wear and erosion resistance is an important consideration in material selection.

Polymers are reinforced with a variety of materials to reduce coefficient of friction and wear and improve strength characteristics: carbon particles, molybdenum disulfide particles, carbon fibers, glass fibers, and aramid fibers.

As discussed in Sections 9.5.4 and 9.5.5, addition of ceramic reinforcements, such as aluminum oxide fibers, to aluminum significantly increases its wear resistance, allowing it to be used in wear-critical applications such as pistons and brake rotors and internal combustion engine blocks.

However, CMCs probably offer the greatest potential for applications requiring resistance to severe wear and erosion. One of the most important composites for these applications is silicon carbide particle-reinforced alumina $[(SiC)p/Al_2O_3]$. The material also contains some residual metal alloy. A significant benefit of this material is that the process used to make it, directed metal oxidation, allows the fabrication of large, complex components that are difficult to make out of monolithic ceramics.

CMCs are now being used in industries such as mining, mineral processing, metalworking, and chemical processing. Figure 9.27 shows components made of $(SiC)p/Al_2O_3$, including impellers, pipeline chokes and liners for pumps, chutes, and valves, and hydrocyclones.

Corrosion-Resistant Applications

As discussed earlier, E-glass-reinforced polyester and vinyl ester PMCs have been extensively used for decades in corrosion-resistant applications, such as chemical industry tanks, flue liners, pumps, and pipes. However, there are applications for which GFRP is not well suited. For example, carbon fibers are much more resistant than glass fibers to chemical attack, creep, and creep rupture, and

Fig. 9.25 Continuous fiber-reinforced ceramic matrix composite pipe hangers, combustor liners, chemical pump components, and other parts provide better thermal and mechanical shock resistance than monolithic ceramics and better oxidation and corrosion resistance than baseline metal designs (courtesy Dow Corning).

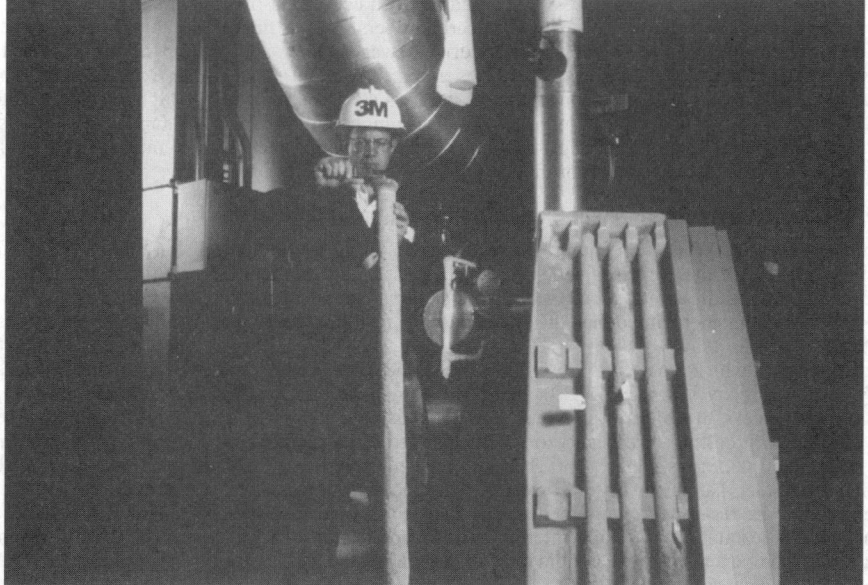

Fig. 9.26 Alumina–boria–silica fiber-reinforced silicon carbide ceramic matrix composite hot gas candle filter has better thermal and mechanical shock resistance than monolithic ceramics and is more resistant to corrosion and oxidation than metal filters (courtesy 3M).

Fig. 9.27 Silicon carbide particle-reinforced alumina ceramic matrix composite parts for wear-resistant applications, including impellers, pipeline chokes and liners for pumps, chutes, valves, and hydrocyclones (courtesy Lanxide).

have much higher specific stiffness. Carbon fiber-reinforced vinyl ester rods have been used in place of titanium in printed circuit production systems, where they are subjected to a variety of corrosive etchant materials. The high specific stiffness of the PMC rods results in less deflection than for titanium. Glass fiber-reinforced rods would deflect much more. Thermoplastics, such as polyether etherketone reinforced with carbon fibers, are being used in pump parts. In this application, carbon fibers provide increased corrosion resistance and reduced coefficient of friction compared to glass.

Epoxy-matrix drive shafts reinforced with carbon fibers, E-glass fibers, or a combination of these, are being used in corrosive environments to drive sewage pumps and cooling tower fans used in power plants, chemical manufacturing facilities and refineries. In some of these applications, composite shafts up to 6.1 m (20 ft) long replace stainless steel. Because of the high specific stiffness and strength of carbon fibers, the composite shafts have higher critical speeds and much lower masses, reducing static and vibratory bearing loads and often eliminating the need for intermediate support bearings. Figure 9.28 shows a carbon fiber-reinforced epoxy cooling tower drive shaft.

9.5.7 Offshore and Onshore Oil Exploration and Production Equipment

Oil exploration and production equipment requirements place severe demands on materials. To function successfully in these environments, materials must be durable and have good resistance to corrosion and fatigue. In addition, as offshore oil exploration moves to increasing depths, equipment mass is becoming more important. These needs are resulting in increasing interest in composite materials.

Sucker rods, which are used to raise oil to the surface, have been made of E-glass fiber-reinforced vinyl ester for many years (Fig. 9.29). Here, the composite offers corrosion resistance and weight savings over steel. Oil well drill pipe has been made using a combination of carbon and glass fibers.

The excellent corrosion resistance of GFRP has led to its successful use in gratings and railings for offshore oil platforms. Figure 9.30 shows E-glass fiber-reinforced phenolic grating, which is 80% lighter than steel, has much better corrosion resistance and lower thermal conductivity, and meets strength and fire-resistance requirements. The increasing water depth at which these platforms are being used is leading to increasing interest in other applications, such as mooring lines, drill pipes, and risers. Components using a combination of carbon fibers and glass fibers in vinyl ester and other resins are candidates to replace steel.

9.5.8 Dimensionally Stable Devices

The low CTE and low density of composite materials make them attractive for applications in which dimensional stability and mass are important. Examples include countless spacecraft optical and RF

Fig. 9.28 Corrosion-resistant carbon fiber-reinforced epoxy cooling tower drive shaft eliminates requirement for intermediate support bearings (courtesy Addax).

systems, such as the Hubble Space Telescope metering truss, wave guides, antenna reflectors, electro-optical systems, and laser devices. Composites also have been used in commercial measuring equipment, such as coordinate measuring machines.

The key composites in these applications are carbon fiber-reinforced PMCs and silicon carbide particle-reinforced aluminum MMCs. Often, CFRPs are used in place of Invar℗, a nickel–iron alloy that has a low CTE but a relatively high density, 8.0 g/cm³ (0.29 Pci). Epoxies have been the traditional matrix materials, but they are being replaced with cyanate esters, which are less susceptible to moisture distortion and have less outgassing. Figure 9.31 shows a developmental electro-optical system gimbal composed of parts made from two types of carbon fiber-reinforced epoxy and from silicon carbide particle-reinforced aluminum. The MMC was used for parts that have complex shapes and are not well suited for carbon/epoxy. Use of composites substantially reduces mass and thermal distortion compared to the aluminum baseline.

A limited number of production mirrors have been made of silicon carbide particle-reinforced aluminum. Metal-coated carbon fiber-reinforced PMCs also are being investigated for lightweight, dimensionally stable mirrors.

9.5.9 Biomedical Applications

Composites are being used for an increasing number of biomedical applications, including x-ray equipment, prosthetics, orthotics, implants, dental restorative materials and wheelchairs. In addition to the usual requirements for stiffness, strength, and so on, materials used for implants must be compatible with the human body.

Carbon fiber-reinforced epoxy is widely used in x-ray film cassettes and tables and stretchers used to support patients in x-ray devices, such as tomography machines. Here, the high specific stiffness and strength of carbon/epoxy reduces the mass of the support equipment and cassettes, allowing the radiologist to lower the x-ray dosages to which patients are exposed.

Carbon fiber-reinforced polymers are extensively used in artificial fingers, arms, legs, hips and feet. They are also used in leg braces and wheelchairs. In all of these applications, the devices are lighter than metallic designs.

PMCs have been used for many years as dental restorative materials. Here, the reinforcements are glasses and fumed silica particles, which provide hardness, wear resistance, and esthetic qualities, and reduce overall composite shrinkage during cure. Compositions with particle loadings as high as 80% are used. In recent years, titanium posts used to attach artificial replacement teeth to the jaw have been replaced by ones made of carbon fiber-reinforced epoxy.

Fig. 9.29 Corrosion-resistant E-glass fiber-reinforced vinyl ester sucker rods used to pump oil (courtesy MMFG).

There is considerable research into development of PMC and CCC implant materials. One potential application is joint replacement. Here, work is under way to improve the resistance to wear and creep of ultrahigh-weight polyethylene, which has been used in a monolithic form for many years.

Another goal is to replace titanium and chromium alloys used for bone reinforcement and replacement. In these applications, the objective is to obtain materials with lower modulus than the incumbents. The reason for this is that the high stiffness of metals reduces stress in the adjacent bone, leading to mass loss. Candidate replacement materials are carbon fiber-reinforced polymers and CCCs.

9.5.10 Sports and Leisure Equipment

PMCs have been used successfully in sports equipment for many years. The key reinforcements are E-glass and, for high-performance products, carbon. The amount of carbon fiber used in golf club shafts alone rivals that used in the airframe industry. Boron and aramid fibers are used in specialized applications. Figure 9.32 shows an array of equipment made from carbon fibers, including golf club shafts, skis, tennis and other rackets, fishing rods, and others. PMCs also have been very successful in high-performance bicycle frames and wheels. There are numerous other PMC sports and leisure equipment applications, including surfboards, water skis, snowmobiles, and many others.

Fig. 9.30 Corrosion-resistant E-glass fiber-reinforced phenolic grating is 80% lighter than steel, has lower thermal conductivity, and meets strength and fire resistance requirements (courtesy MMFG).

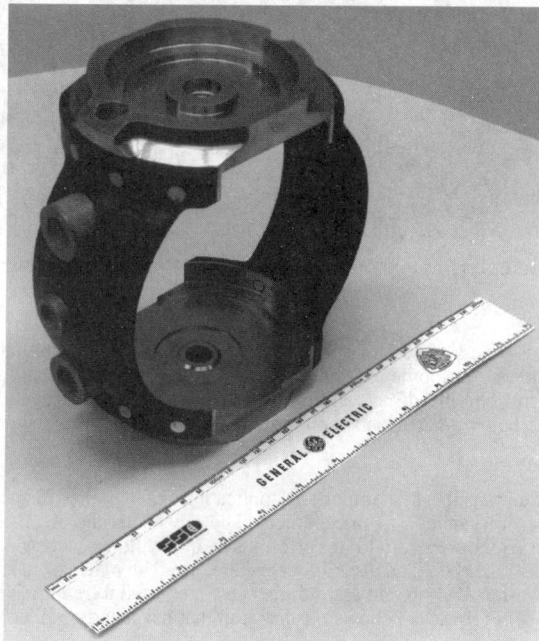

Fig. 9.31 Developmental lightweight, dimensionally stable electro-optical system gimbal composed of parts made from two types of carbon fiber-reinforced epoxy and from silicon carbide particle-reinforced aluminum.

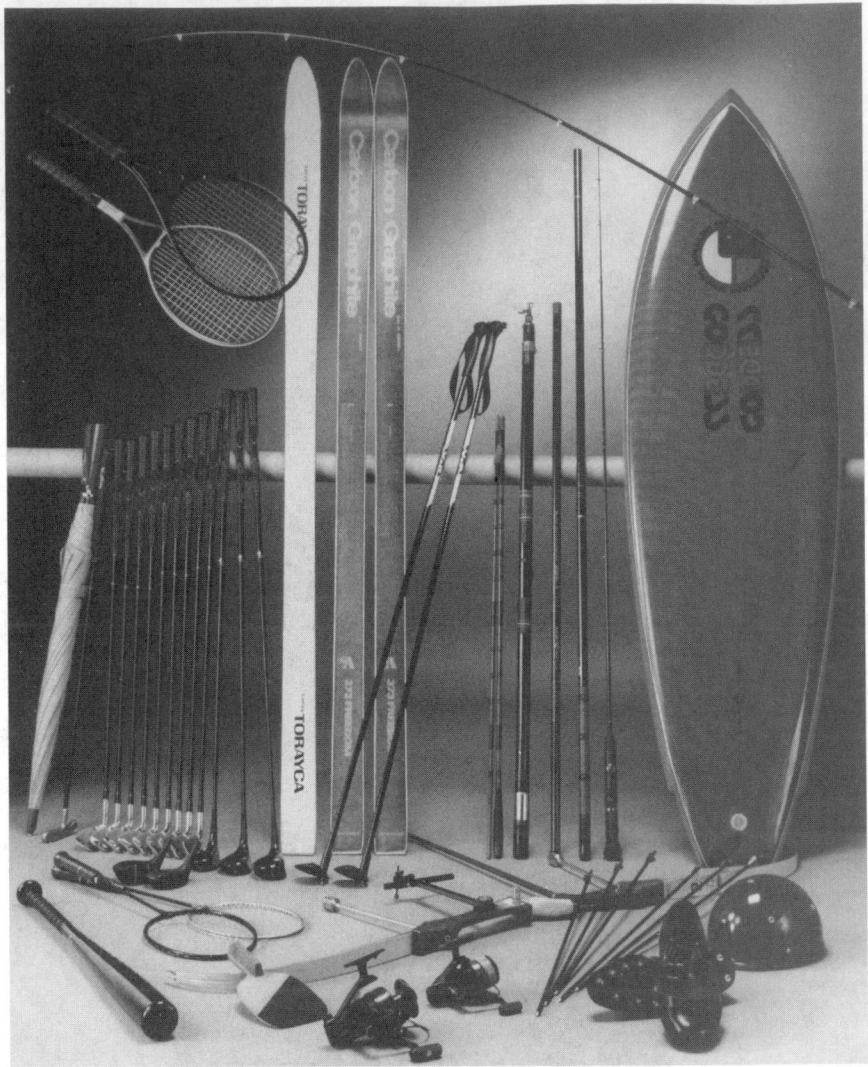

Fig. 9.32 Carbon fiber-reinforced polymer sports equipment (courtesy Toray).

MMCs have been used in a variety of specialized applications, such as mountain bike frames and wheels. Figure 9.33 shows developmental sports equipment using titanium carbide particle-reinforced titanium, including a golf club head, bat, and ice skate blade. In the latter application, the composite offers light weight and better wear resistance than monolithic titanium.

9.5.11 Marine Structures

Boats and ships were among the first important applications of polymer matrix composites. Applications range in size from canoes to mine hunters. The key materials are E-glass fibers and thermosetting polyester resins. However, in high-performance applications, such as Americas Cup sailboat hulls, booms, and masts, carbon and aramid fibers are used in place of glass, and epoxy resins frequently replace polyester. Carbon and aramid fibers are also used to reinforce sails to help maintain their aerodynamic shape. Figure 9.34 shows a catamaran that has a carbon fiber-reinforced PMC hull.

9.5.12 Miscellaneous Applications

In addition to the applications cited earlier in this section, there are countless other products using composite materials. We consider a few of these, including wind turbine blades, musical instruments, audio speakers, pressure vessels, and one other unique application.

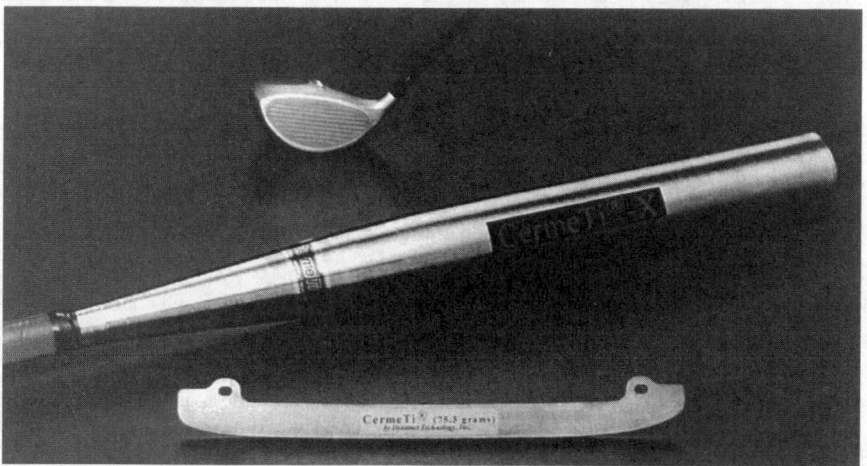

Fig. 9.33 Developmental sports equipment using titanium carbide particle-reinforced titanium, including a golf club head, bat, and ice skate blade (courtesy Dynamet Technology).

Wind turbines (i.e., windmills) have been used as a source of power for centuries. In the last few decades, there has been significant interest in use of wind turbines as renewable, nonpolluting source of electric power. Numerous devices have been installed in regions with high average annual wind speeds. Blades for many of these systems have been made of various polymers, notably epoxies and polyesters reinforced primarily with glass fibers, and in some instances carbon fibers. The reasons for use of composites are good fatigue and corrosion resistance, relative ease of fabrication, and cost-effectiveness.

Fig. 9.34 Catamaran with carbon fiber-reinforced hull (courtesy Toray).

Wood, an anisotropic fibrous material, has been used from time immemorial as a material of construction for musical instruments. In recent years, glass and carbon fiber-reinforced polymers, primarily epoxy, have been introduced in a wide range of instruments, including guitars, electric basses, banjos, mandolins, and violins. Carbon fiber reinforced plastics also have been used in violin bows. PMCs reinforced with aramid fibers are used in drum sticks and heads.

Audio speakers made of carbon fiber- and aramid fiber-reinforced epoxy have been used in a number of production applications (Fig. 9.35).

Pressure vessels used in natural gas vehicles were mentioned in Section 9.5.5. There have been many other applications of PMC pressure vessels, including firefighter breathing tanks, pressurization tanks for aircraft escape slides, and spacecraft pressure tanks. Reinforcements include high-strength glass, carbon, and aramid fibers. Epoxies are the leading matrix materials.

Figure 9.36 shows one of the more unusual applications for composites, a silicon carbide particle-reinforced aluminum brake rail mounted on a vehicle used in a theme park ride. Here the composite provides much better wear resistance than monolithic aluminum, with a negligible increase in weight.

9.6 DESIGN AND ANALYSIS

The most widely used materials of construction in mechanical engineering applications, monolithic metals and ceramics, are typically considered to be isotropic for purposes of design and analysis. Particle-reinforced composites also tend to be relatively isotropic. However, composites reinforced with fibers, especially continuous fibers, are typically strongly anisotropic and require special design and analysis methods. In this section, we consider how the special characteristics of composites influence the design process. We concentrate on composites reinforced with continuous fibers, which includes reinforcement forms such as fabrics and braids, because these are the most efficient materials.

As discussed in earlier sections, there are four key classes of composites: polymer matrix composites (PMCs), metal matrix composites (MMCs), ceramic matrix composites (CMCs), and carbon/carbon composites (CCCs). We consider all four, but focus on PMCs, which are the most widely used class of composites at this time and are likely to remain so for some time to come. Much of the discussion of PMC design and analysis, especially that dealing with consideration of the importance of elastic property anisotropy, applies to all fiber-reinforced composites.

The design process is an iterative one. After the critical step of establishing requirements, the engineer develops a preliminary design, which is then analyzed to determine whether it meets requirements. If analysis shows that safety margins are too large or too small, the design is refined, and the process repeated.

For a composite component, the designer selects the overall configuration; reinforcement types, forms, and volume fractions; matrix material; and the number of layers, along with their thicknesses and orientations.

Fig. 9.35 Carbon fiber-reinforced epoxy audio speaker (courtesy Tonen).

Fig. 9.36 Theme park ride vehicle uses a silicon carbide particle-reinforced aluminum brake rail (courtesy DWA Aluminum Composites).

An important consideration is selection of the manufacturing process, which, as discussed in previous section, has critical effects on material properties and cost. Experience has shown that in developing composite components, it is particularly important to involve manufacturing, quality assurance, and procurement personnel from the start.

In the next section, we discuss the design process for a PMC component. We then examine special considerations for MMCs, CMCs, and CCCs. Because design and analysis of composite components is very complex, it is not possible to cover the subject in detail.

9.6.1 Polymer Matrix Composites

As discussed in Sections 9.1 and 9.3, PMCs, which derive their strength and stiffness from the fibrous reinforcement phase, are, like wood, typically strongly anisotropic. PMCs are weak and have low stiffness in directions that are perpendicular to fiber directions and planes which are not intersected by fibers. We call these matrix-dominated directions and properties. Examples are transverse directions in unidirectional composites and interlaminar planes in laminates. As a consequence of the low transverse and through-thickness strengths of PMCs, unidirectional laminates are rarely used in structural applications.

Because PMC laminates are strongly anisotropic, isotropic analytical methods generally cannot be used. Anisotropy affects virtually all aspects of design and analysis, including deflections, natural frequencies, buckling loads, and failure modes. Fortunately, analytical methods for anisotropic structures are well developed. This is true for both closed-form anisotropic solutions and finite elements methods.

As a simple illustration of the differences between isotropic plates and anisotropic laminates, consider elastic constants. For isotropic materials, there are only two independent elastic constants. For example, the extensional modulus (E), shear modulus (G), and Poisson's ratio, (v) are related by the formula E = 2G(1 + v). This is generally not valid for anisotropic laminates, for which there are four independent inplane elastic constants. In common engineering, they are usually Ex, Ey, Gxy, and vxy. Here, Ex and Ey are the extensional moduli in the x- and y-directions, Gxy is the inplane shear modulus, and vxy is the major inplane Poisson's ratio. The latter is defined as the ratio of the magnitude of the strain in the y-direction divided by the magnitude of the strain in the x-direction when an extensional stress is applied in the x-direction.

When a tensile stress is applied to an isotropic material, it produces an extensional strain in the direction of the applied load and a lateral, Poisson, contraction in the perpendicular direction. There is no shear distortion. Conversely, when a shear stress is applied to an isotropic material, it produces only shear strain, and not extensional strain. However, for an arbitrary anisotropic material, application of an extensional load produces not only extensional strains in the directions parallel and perpendicular to the applied load, but shear strains, as well. This is called tension-shear coupling. When a shear stress is applied, it produces extensional as well as shear strains.

When anisotropic materials are laminated and subjected to an inplane tensile stress, the general laminate response is much more complex than that of a plate made of an isotropic material. In the most general case of an arbitrary laminate, the tensile stress will produce not only extension in the direction of the load and lateral contraction, but also bending, twisting, and inplane shear deformation.

To minimize coupling, laminates are designed to be balanced and symmetric. A balanced laminate is one for which the directions of the layers above the mid-plane are a mirror image of those below it. A symmetric laminate is one for which for every layer having an orientation of $+\theta$ direction with respect to a reference axis, there exists an identical laminate in the $-\theta$ direction.

Although coupling is undesirable in most cases, there have been a few designs where selected coupling has been used to advantage. Examples are aircraft with forward swept wings and bicycle cranks.

It is important to note that the properties of laminates are very sensitive to laminate geometry. Further, anisotropic laminates can have characteristics very different from those of monolithic materials. Often, these properties are counter-intuitive. For example, the Poisson's ratio of laminates having fibers in the $+45°$ and $-45°$ directions can be much greater than 0.5, compared to about 0.3 for most metals. Addition of fibers at $90°$ can reduce this value significantly.

Theoretically the designer can select from an infinite number of laminate geometries to meet requirements for a particular component. In practice, however, it is common to choose laminates from discrete families with fibers in a few directions. The most common family has fibers in four directions: $0°$, $+45°$, $-45°$, and $90°$. The designer selects laminates having various percentages of layers in the four directions, usually making sure that they are balanced and symmetric. To assure adequate strength in all directions, many organizations use the "10% Rule." Using this convention, at least 10% of the layers in a laminate are placed in each of the four key directions.

A critical design consideration for laminated PMCs is minimization of through-thickness stresses. This is very different from the situation for monolithic structures, for which these stresses are typically considered to be of secondary importance and are ignored. Interlaminar stresses arise from a variety of sources: out-of-plane loads; curvature; stress waves from impact loads; and free-edge effects. Interlaminar stresses in curved regions are caused by mechanical loads, and by differential thermal and moisture expansion in the inplane and through-thickness directions.

Computer programs based on laminated plate theory are widely used in design and analysis. These programs are used to predict laminate properties and to define laminate response and layer stresses and strains resulting from applied loads, moments, and changes in temperature and moisture level. Laminated plate analysis is also used to generate carpet plots for properties of laminated plates which are used in preliminary design.

The stress-strain curves for PMCs are essentially linear to failure, although, as discussed in Sections 9.2 and 9.3, composites reinforced with carbon and aramid fibers do display some nonlinearity. As a consequence of the lack of plastic deformation, under static loading, PMC laminates are sensitive to stress concentrations, such as those that arise at joints and cutouts. However, composite stress concentrations are relatively insensitive to fatigue loading. In fact, fatigue loading often results in local microdamage that reduces the effect of the stress concentration. This is the opposite of monolithic metals, which are relatively insensitive to static stress concentrations because of plasticity, but sensitive to stress concentrations under fatigue loading, which causes propagation of through-thickness cracks.

Prediction of laminate failure under applied load is commonly based on a variety of failure theories that are applied to stresses or strains on a layer-by-layer basis. Layer stresses and strains are determined using finite element analysis combined with laminated plate theory. Failure theories are based on maximum stress, maximum strain, or numerous interaction formulas for stress or strain components.

The joining of composites is a critical design issue because of their sensitivity to stress concentrations. Joining is accomplished by adhesive bonding, mechanical fasteners, or a combination of these. As a rule, adhesive joints are the most efficient structurally, but are sensitive to manufacturing processes and environmental degradation. Mechanically fastened joints are used for very highly loaded structures, especially those subjected to fatigue loading and for which environmental degradation is a concern. However, because mechanical joints are less efficient, there is typically a weight penalty associated with their use. Stresses arising from differences in Poisson's ratios and coefficients of thermal expansion (CTEs) are important considerations when composites are joined to metals or other laminates. Stresses caused by moisture expansion also should be considered.

Galvanic corrosion is an important issue whenever dissimilar materials are joined. This is especially true for carbon and aluminum. The problem can be overcome by electrically isolating the two materials or by using compatible materials.

As for all materials, design allowables for PMCs should take into account the loading conditions and environment, including temperature to which they will be subjected.

9.6.2 Metal Matrix Composites

As discussed in Sections 9.3 and 9.5, the leading types of reinforcements for MMCs are continuous fibers, discontinuous fibers, and particles. Continuous fibers provide materials with the highest strength and stiffnesses. Discontinuous fibers are primarily used to increase wear resistance and elevated temperature static and fatigue strengths. Particles provide isotropic materials with high specific modulus and yield strength, improved elevated temperature strength properties and wear resistance, and reduced CTE.

For all MMCs, an important consideration is degradation of properties resulting from interactions between the reinforcement and matrix at elevated temperature, which can occur during manufacture or in service. This is defined experimentally.

In contrast to PMCs, the high transverse strength of many MMCs allows use of unidirectional laminates in structures. A good example are the Space Shuttle Orbiter boron–aluminum struts cited in Section 9.5. An important consideration for MMCs reinforced with continuous fibers is that they display elastic-plastic characteristics.

As discussed in Section 9.5, fiber orientation has a critical influence on composite properties. A critical consideration for MMCs reinforced with discontinuous fibers is to assure that the manufacturing process results in components which have fiber volume fractions and orientations that meet design requirements.

The isotropic nature of particle-reinforced MMCs significantly simplifies design. Major considerations for these materials are that they tend to have lower elongations and fracture toughnesses than the base metal.

9.6.3 Ceramic Matrix Composites

As for MMCs, interactions between matrix and reinforcement at elevated temperatures is an important consideration. In addition, formation of matrix cracking exposes reinforcements to the environment, which can degrade the properties of the interphase or the fiber itself, resulting in embrittlement and weakening of the material.

Another design consideration for CMCs is that, like PMCs, they have relatively weak properties in fiber-dominated directions. They also are sensitive to stress concentrations that arise at joints and cutouts.

9.6.4 Carbon/Carbon Composites

The comments for CMCs generally apply to CCCs, although the problem of fiber-matrix interaction is not a serious consideration for the latter. A major consideration for CCCs is that they typically have even weaker matrix-dominated properties than PMCs and CMCs.

A critical issue for CMCs is elevated temperature oxidation, which, as discussed in Section 9.3, can be reduced by use of coatings and oxidation inhibitors in the matrix.

REFERENCES

1. A. Kelly (ed.), *Concise Encyclopedia of Composite Materials*, rev. ed., Pergamon Press, Oxford, 1994.
2. Z. L. H. Miner, R. A. Wolffe, and C. Zweben, "Fatigue, Creep and Impact Resistance of Kevlar® 49 Reinforced Composites," in *Composite Reliability*, ASTM STP 580, American Society for Testing and Materials, Philadelphia, PA, 1975.
3. C. Zweben, "Overview of Metal Matrix Composites for Electronic Packaging and Thermal Management," *JOM* (July 1992).
4. A. F. Johnson, "Glass-Reinforced Plastics: Thermosetting Resins," in *Concise Encyclopedia of Composite Materials*, A. Kelly (ed.), rev. ed., Pergamon Press, Oxford, 1994.
5. J. Halpin, Lecture Notes, UCLA short course "Fiber Composites: Design, Evaluation, and Quality Assurance."

BIBLIOGRAPHY

Advanced Materials by Design, OTA-E-351, U.S. Congress Office of Technology Assessment, U.S. Government Printing Office, Washington, DC, June 1988.

Agarwal, B. D., and L. J. Broutman, *Analysis and Performance of Fiber Composites*, Wiley, New York, 1981.

Allen, H. G., *Analysis and Design of Structural Sandwich Panels*, Pergamon Press, Oxford, 1969.

Ambartsumyan, S. A., *Theory of Anisotropic Plates*, Technomic, Lancaster, PA, 1970.

Ashby, M. F., *Material Selection in Mechanical Design*, Pergamon Press, Oxford, 1992.

Ashton, J. E., and J. M. Whitney, *Theory of Laminated Plates*, Technomic, Lancaster, PA, 1970.

Bulletin 54, Chromalloy Metal Tectonics Company.

Calcote, L. R., *The Analysis of Laminated Composite Structures*, Van Nostrand Reinhold, New York, 1969.

Chou, T.-W., *Microstructural Design of Composite Materials*, Cambridge University Press, Cambridge, 1992.

——— (ed.), *Materials Science Handbook*, Vol. 13, *Structural Properties of Composites*, VCH, Weinheim, Federal Republic of Germany, 1993.

Design Guide for Advanced Composite Applications, Advanstar Communications, Duluth, MN, 1993.

Deve, H. E., and C. McCullough, "Continuous-Fiber Reinforced Al Composites: A New Generation," *JOM*, 33–37 (July 1995).

Donomoto, T., et al., "Ceramic Fiber Reinforced Piston for High Performance Diesel Engines," SAE Technical Paper No. 830252, 1983.

Dvorak, G. J. (ed.), *Inelastic Deformation of Composite Materials*, Proceedings of the 19901 IUTAM Symposium, Springer-Verlag, New York, 1991.

Engineered Materials Handbook, Vol. 1, *Composites*, American Society for Metals, Materials Park, OH, 1987.

Fisher, K., "Industrial Applications," *High-Performance Composites*, 46–49 (May/June 1995).

Grimes, G. G., et al., "Tape Composite Material Allowables Application in Airframe Design/Analysis," *Composites Engineering* **3**(7/8), 777–804 (1993).

Halpin, J. C., *Primer on Composite Materials: Analysis*, Technomic, Lancaster, PA, 1984.

Hayashi, T., H. Ushio, and M. Ebisawa, "The Properties of Hybrid Fiber Reinforced Metal and Its Application for Engine Block," SAE Technical Paper No. 890557, 1989.

Hoskin, B., and A. A. Baker, *Composite Materials for Aircraft Structures*, American Institute of Aeronautics and Astronautics, New York, 1986.

Hull, D., *An Introduction to Composite Materials*, Cambridge University Press, Cambridge, 1981.

Jones, R. M., *Mechanics of Composite Materials*, McGraw-Hill, New York, 1975.

Kedward, K., "Designing with Composites," Lecture Notes for Short Course "Composite Materials: Selection, Design and Manufacture for Engineering Applications," UCLA Extension, University of California, Los Angeles, CA, 1996.

Kennedy, C. R., "Reinforced Ceramics Via Oxidation of Molten Metals," *Ceramic Industry*, 26–30 (December 1994).

Kerns, J. A., et al., "Dymalloy, A Composite Substrate for High Power Density Electronic Components," in *Proceedings of the 1995 International Symposium on Microelectronics*, Los Angeles, CA, 1995.

Kliger, H. S., and E. R. Barker, "A Comparative Study of the Corrosion Resistance of Carbon and Glass Fibers," in 39th Annual Conference, Reinforced Plastics/Composites Institute, The Society of the Plastics Industry, Inc., January 1984.

Ko, F. K., "Advanced Textile Structural Composites," in *Advanced Topics in Materials Science and Engineering*, J. I. Moran-Lopez and J. M. Sanchez (eds.), Plenum Press, New York, 1993.

———, "Three-Dimensional Fabrics for Composites," in *Textile Structural Composites*, T.-W. Chou and F. K. Ko (eds.), Elsevier Science Publishers, B. V., Amsterdam, 1989, pp. 129-171.

Kulkarni, S. V., and C. Zweben (eds.), *Composites in Pressure Vessels and Piping*, American Society of Mechanical Engineering, New York, 1977.

Kwarteng K., and C. Stark, "Carbon Fiber Reinforced PEEK (APC-2/AS-4) Composites for Orthopaedic Implants," *SAMPE Quarterly*, 10–14 (Oct. 1990).

Lekhnitskii, S. G., *Anisotropic Plates*, Gordon and Breach, 1968.

Mallick, P. K., *Fiber-Reinforced Composites: Materials, Manufacturing and Design,* 2nd ed., Marcel Dekker, New York, 1993.

Marshall, A. C., *Composite Basics*, 4th ed., Marshall Consulting, Walnut Creek, CA., 1994.

Marshall, D. B., and A. G. Evans, "Failure Mechanisms in Ceramic-Fiber/Ceramic-Matrix Composites," *Journal of the American Ceramic Society* **68**(5), 225–231 (May 1985).

McConnell, V. P., "Fail-Safe Ceramics," *High-Performance Composites*, 27–31 (March/April 1996).

———, "Industrial Applications," *Advanced Composites*, 31–38 (March/April 1992).

Meyers, M. A., and O. T. Inal (eds.), *Frontiers in Materials Technologies*, Materials Science Monographs, 26, Elsevier, Oxford, 1985.

Military Handbook—5F, *Metallic Materials and Elements for Aerospace Vehicle Structures*, U.S. Department of Defense, Washington, D.C., December 1992.

Morrell, R., *Handbook of Properties of Technical and Engineering Ceramics*, Her Majesty's Stationery Office, London, 1985, Part 1: An Introduction for the Engineer and Designer.

Norman, J. C., and C. Zweben, "Kevlar® 49/Thornel® 300 Hybrid Fabric Composites for Aerospace Applications," *SAMPE Quarterly* **7**(4), 1–10 (July 1976).

Pfeifer, W. H., et al., "High Conductivity Carbon-Carbon Composites for SEM-E Heat Sinks," in Sixth International SAMPE Electronic Materials and Processes Conference, Baltimore, MD, June 22–25, 1992.

Premkumar, M. K., W. H. Hunt, Jr., and R. R. Sawtell, "Aluminum Composite Materials for Multichip Modules," *JOM*, 22–28 (July 1992).

"Properties of SiC/SiC Laminates (0/90 Fabric Layup)," Preliminary Engineering Data, Publication H-28488, Du Pont Composites.

Rawal, S. P., M. S. Misra, and R. G. Wendt, "Composite Materials for Space Applications," NASA CR-187472, National Aeronautics and Space Administration, Hampton, VA, 1990.

Savage, G., *Carbon–Carbon Composites*, Chapman and Hall, London, 1993.

Schmidt, K. A., and C. Zweben, "Advanced Composite Packaging Materials," in *Electronic Materials Handbook*, Vol. 1, *Packaging*, American Society for Metals, Materials Park, OH, 1989.

Schwartz, M. M., *Fabrication of Composite Materials*, American Society for Metals, Metals Park, OH, 1985.

Shih, W. T., F. H. Ho, and B. B. Burkett, "Carbon–Carbon (C–C) Composites for Thermal Plane Applications," in Seventh International SAMPE Electronics Conference, June 20–23, 1994, Parsippany, NJ.

Shtessel, V. E., and M. J. Koczak, "The Production of Metal Matrix Composites by Reactive Processes," *Materials Technology* **9**(7/8), 154–158 (1994).

Smith, D. L., K. E. Davidson, and L. S. Thiebert, "Carbon–Carbon Composites (CCC): A Historical Perspective," in *Proceedings*, 41st International SAMPE Symposium, March 24-28, 1996.

Smith, W. S., M. W. Wardle, and C. Zweben, "Test Methods for Fiber Tensile Strength, Composite Flexural Modulus and Properties of Fabric-Reinforced Laminates," in *Composite Materials: Testing and Design (Fifth Conference)*, ASTM STP 674, S. W. Tsai (ed.), American Society for Testing and Materials, PA, 1979, pp. 228-262.

Strong, A. B., *Fundamentals of Composites Manufacturing: Materials, Methods, and Applications*, Society of Manufacturing Engineers, Dearborn, MI, 1989.

"Top Twenty Awards," *Materials Engineering*, 28–33 (Nov. 1985).

Tsai, S. W., *Composites Design*, 4th ed., Think Composites, Dayton, OH, 1988.

Tsai, S. W., & H. T. Hahn, *Introduction to Composite Materials*, Technomic, Lancaster, PA, 1980.

Vasiliev, V. V., *Mechanics of Composite Structures*, R. M. Jones, English Edition Editor, Taylor and Francis, Washington, DC, 1988.

Vinson, J. R., and R. L. Sierakowski, *The Behavior of Structures Composed of Composite Materials*, Kluwer Academic Publishers, Dordrecht, 1993.

Warren, R. (ed.), *Ceramic–Matrix Composites*, Chapman and Hall, New York, 1992.

Zweben, C., "Advanced Composite Materials for Process Industries and Corrosion Resistant Applications," in *Proceedings of the Conference on Advances in Materials Technology for Process Industries' Needs*, National Association of Corrosion Engineers, Houston, TX, 1985, pp. 163–172.

———, "Advanced Composites—A Revolution for the Designer," in AIAA 50th Anniversary Annual Meeting and Technical Display, "Learn from the Masters" Lecture Series, Paper No. 81-0894, Long Beach, May 1981.

———"Is There a Size Effect in Composite Materials and Structures?," *Composites* **25**, 451–454 (1994).

———"Mechanical and Thermal Properties of Silicon Carbide Particle-Reinforced Aluminum," in *Thermal and Mechanical Behavior of Metal Matrix and Ceramic Matrix Composites*, ASTM STP 1080, L. M. Kennedy, H. H. Moeller, and W. S. Johnson (eds.), American Society for Testing and Materials, Philadelphia, PA, 1989.

———, "Metal Matrix Composites: Aerospace Applications," in *Encyclopedia of Advanced Materials*, M. C. Flemings et al. (eds.), Pergamon Press, Oxford, 1994.

———, "Simple, Design-Oriented Composite Failure Criteria Incorporating Size Effects," in *Proceedings*, Tenth International Conference on Composite Materials, ICCM-10, Whistler, BC, Canada, August 1995.

———, "Tensile Strength of Fiber-Reinforced Composites: Basic Concepts and Recent Developments," in *Composite Materials: Testing and Design*, ASTM STP 460, American Society for Testing and Materials, Philadelphia, PA, 1970, pp. 528–539.

————, "The Future of Advanced Composite Electronic Packaging," in *Materials for Electronic Packaging*, D. D. L. Chung (ed.), Butterworth-Heinemann, Oxford, 1995.

————, "Thermomechanical Properties of Fibrous Composite Materials: Theory," in *Encyclopedia of Materials Science and Engineering*, M. B. Bever (ed.), Pergamon Press, Oxford, 1986.

CHAPTER 10

STRESS ANALYSIS

Franklin E. Fisher
Mechanical Engineering Department
Loyola Marymount University
Los Angeles, California
and
Senior Staff Engineer
Hughes Aircraft Company (Retired)

10.1	**STRESSES, STRAINS, STRESS INTENSITY**	**191**
	10.1.1 Fundamental Definitions	191
	10.1.2 Work and Resilience	197
10.2	**DISCONTINUITIES, STRESS CONCENTRATION**	**199**
10.3	**COMBINED STRESSES**	**199**
10.4	**CREEP**	**203**
10.5	**FATIGUE**	**205**
	10.5.1 Modes of Failure	206
10.6	**BEAMS**	**207**
	10.6.1 Theory of Flexure	207
	10.6.2 Design of Beams	212
	10.6.3 Continuous Beams	217
	10.6.4 Curved Beams	220
	10.6.5 Impact Stresses in Bars and Beams	220
	10.6.6 Steady and Impulsive Vibratory Stresses	224
10.7	**SHAFTS, BENDING, AND TORSION**	**224**
	10.7.1 Definitions	224
	10.7.2 Determination of Torsional Stresses in Shafts	225
	10.7.3 Bending and Torsional Stresses	229
10.8	**COLUMNS**	**229**
	10.8.1 Definitions	229
	10.8.2 Theory	230
	10.8.3 Wooden Columns	232
	10.8.4 Steel Columns	232
10.9	**CYLINDERS, SPHERES, AND PLATES**	**235**
	10.9.1 Thin Cylinders and Spheres under Internal Pressure	235
	10.9.2 Thick Cylinders and Spheres	235
	10.9.3 Plates	237
	10.9.4 Trunnion	237
	10.9.5 Socket Action	237
10.10	**CONTACT STRESSES**	**242**
10.11	**ROTATING ELEMENTS**	**244**
	10.11.1 Shafts	244
	10.11.2 Disks	244
	10.11.3 Blades	244
10.12	**DESIGN SOLUTION SOURCES AND GUIDELINES**	**244**
	10.12.1 Computers	244
	10.12.2 Testing	245

10.1 STRESSES, STRAINS, STRESS INTENSITY

10.1.1 Fundamental Definitions

Static Stresses

TOTAL STRESS on a section mn through a loaded body is the resultant force S exerted by one part of the body on the other part in order to maintain in equilibrium the external loads acting on the

Revised from Chapter 8, *Kent's Mechanical Engineer's Handbook,* 12th ed., by John M. Lessells and G. S. Cherniak.

Mechanical Engineers' Handbook, 2nd ed., Edited by Myer Kutz.
ISBN 0-471-13007-9 © 1998 John Wiley & Sons, Inc.

part. Thus, in Figs. 10.1, 10.2, and 10.3 the total stress on section mn due to the external load P is S. The units in which it is expressed are those of load, that is, pounds, tons, etc.

UNIT STRESS more commonly called stress σ, is the total stress per unit of area at section mn. In general it varies from point to point over the section. Its value at any point of a section is the total stress on an elementary part of the area, including the point divided by the elementary total stress on an elementary part of the area, including the point divided by the elementary area. If in Figs 10.1, 10,2, and 10.3 the loaded bodies are one unit thick and four units wide, then when the total stress S is uniformly distributed over the area, $\sigma = P/A = P/4$. Unit stresses are expressed in pounds per square inch, tons per square foot, etc.

TENSILE STRESS OR TENSION is the internal total stress S exerted by the material fibers to resist the action of an external force P (Fig. 10.1), tending to separate the material into two parts along the line mn. For equilibrium conditions to exist, the tensile stress at any cross section will be equal and opposite in direction to the external force P. If the internal total stress S is distributed uniformly over the area, the stress can be considered as unit tensile stress $\sigma = S/A$.

COMPRESSIVE STRESS OR COMPRESSION is the internal total stress S exerted by the fibers to resist the action of an external force P (Fig. 10.2) tending to decrease the length of the material. For equilibrium conditions to exist, the compressive stress at any cross section will be equal and opposite in direction to the external force P. If the internal total stress S is distributed uniformly over the area, the unit compressive stress $\sigma = S/A$.

SHEAR STRESS is the internal total stress S exerted by the material fibers along the plane mn (Fig. 10.3) to resist the action of the external forces, tending to slide the adjacent parts in opposite directions. For equilibrium conditions to exist, the shear stress at any cross section will be equal and opposite in direction to the external force P. If the internal total stress S is uniformly distributed over the area, the unit shear stress $\tau = S/A$.

NORMAL STRESS is the component of the resultant stress that acts normal to the area considered (Fig. 10.4).

AXIAL STRESS is a special case of normal stress and may be either tensile or compressive. It is the stress existing in a straight homogeneous bar when the resultant of the applied loads coincides with the axis of the bar.

SIMPLE STRESS exists when either tension, compression, or shear is considered to operate singly on a body.

TOTAL STRAIN on a loaded body is the total elongation produced by the influence of an external load. Thus, in Fig. 10.4, the total strain is equal to δ. It is expressed in units of length, that is, inches, feet, etc.

UNIT STRAIN or deformation per unit length is the total amount of deformation divided by the original length of the body before the load causing the strain was applied. Thus, if the total elongation is δ in an original gage length l, the unit strain $e = \delta/l$. Unit strains are expressed in inches per inch and feet per foot.

TENSILE STRAIN is the strain produced in a specimen by tensile stresses, which in turn are caused by external forces.

COMPRESSIVE STRAIN is the strain produced in a bar by compressive stresses, which in turn are caused by external forces.

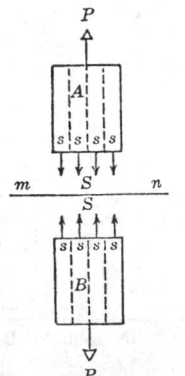

Fig. 10.1 Tensile stress.

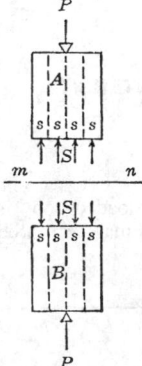

Fig. 10.2 Compressive stress.

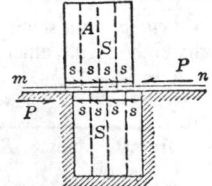

Fig. 10.3 Shear stress.

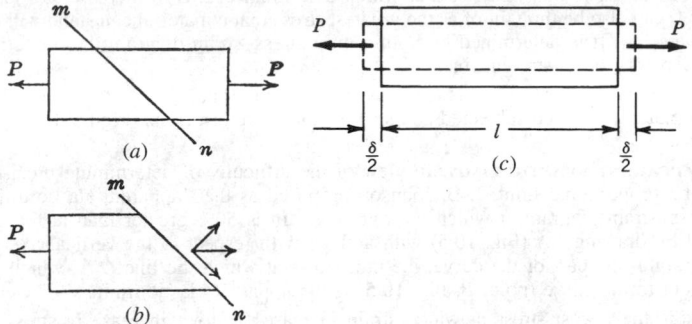

Fig. 10.4 Normal and shear stress components of resultant stress on section *mn* and strain due to tension.

SHEAR STRAIN is a strain produced in a bar by the external shearing forces.

POISSON'S RATIO is the ratio of lateral unit strain to longitudinal unit strain under the conditions of uniform and uniaxial longitudinal stress within the proportional limit. It serves as a measure of lateral stiffness. Average values of Poisson's ratio for the usual materials of construction are:

Material	Steel	Wrought Iron	Cast Iron	Brass	Concrete
Poisson's ratio	0.300	0.280	0.270	0.340	0.100

ELASTICITY is that property of a material that enables it to deform or undergo strain and return to its original shape upon the removal of the load.

HOOKE'S LAW states that within certain limits (not to exceed the proportional limit) the elongation of a bar produced by an external force is proportional to the tensile stress developed. Hooke's law gives the simplest relation between stress and strain.

PLASTICITY is that state of matter where permanent deformations or strains may occur without fracture. A material is plastic if the smallest load increment produces a permanent deformation. A perfectly plastic material is nonelastic and has no ultimate strength in the ordinary meaning of that term. Lead is a plastic material. A prism tested in compression will deform permanently under a small load and will continue to deform as the load is increased, until it flattens to a thin sheet. Wrought iron and steel are plastic when stressed beyond the elastic limit in compression. When stressed beyond the elastic limit in tension, they are partly elastic and partly plastic, the degree of plasticity increasing as the ultimate strength is approached.

STRESS–STRAIN RELATIONSHIP gives the relation between unit stress and unit strain when plotted on a stress–strain diagram in which the ordinate represents unit stress and the abscissa represents unit strain. Figure 10.5 shows a typical tension stress–strain curve for medium steel. The form of the curve obtained will vary according to the material, and the curve for compression will be different from the one for tension. For some materials like cast iron, concrete, and timber, no part of the curve is a straight line.

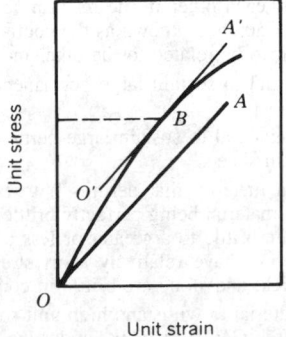

Fig. 10.5 Stress–strain relationship showing determination of apparent elastic limit.

PROPORTIONAL LIMIT is that unit stress at which unit strain begins to increase at a faster rate than unit stress. It can also be thought of as the greatest stress that a material can stand without deviating from Hooke's law. It is determined by noting on a stress–strain diagram the unit stress at which the curve departs from a straight line.

ELASTIC LIMIT is the least stress that will cause permanent strain, that is, the maximum unit stress to which a material may be subjected and still be able to return to its original form upon removal of the stress.

JOHNSON'S APPARENT ELASTIC LIMIT. In view of the difficulty of determining precisely for some materials the proportional limit, J. B. Johnson proposed as the "apparent elastic limit" the point on the stress–strain diagram at which the rate of strain is 50% greater than at the original. It is determined by drawing OA (Fig. 10.5) with a slope with respect to the vertical axis 50% greater than the straight-line part of the curve; the unit stress at which the line $O'A'$ which is parallel to OA is tangent to the curve (point B, Fig. 10.5) is the apparent elastic limit.

YIELD POINT is the lowest stress at which strain increases without increase in stress. Only a few materials exhibit a true yield point. For other materials the term is sometimes used as synonymous with yield strength.

YIELD STRENGTH is the unit stress at which a material exhibits a specified permanent deformation or state. It is a measure of the useful limit of materials, particularly of those whose stress–strain curve in the region of yield is smooth and gradually curved.

ULTIMATE STRENGTH is the highest unit stress a material can sustain in tension, compression, or shear before rupturing.

RUPTURE STRENGTH OR BREAKING STRENGTH is the unit stress at which a material breaks or ruptures. It is observed in tests on steel to be slightly less than the ultimate strength because of a large reduction in area before rupture.

MODULUS OF ELASTICITY (Young's modulus) in tension and compression is the rate of change of unit stress with respect to unit strain for the condition of uniaxial stress within the proportional limit. For most materials the modulus of elasticity is the same for tension and compression.

MODULUS OF RIGIDITY (modulus of elasticity in shear) is the rate of change of unit shear stress with respect to unit shear strain for the condition of pure shear within the proportional limit. For metals it is equal to approximately 0.4 of the modulus of elasticity.

TRUE STRESS is defined as a ratio of applied axial load to the corresponding cross-sectional area. The units of true stress may be expressed in pounds per square inch, pounds per square foot, etc.,

$$\sigma = \frac{P}{A}$$

where σ is the true stress, pounds per square inch, P is the axial load, pounds, and A is the smallest value of cross-sectional area existing under the applied load P, square inches.

TRUE STRAIN is defined as a function of the original diameter to the instantaneous diameter of the test specimen:

$$q = 2 \log_e \frac{d_0}{d} \text{ in./in.}$$

where q = true strain, inches per inch, d_0 = original diameter of test specimen, inches, and d = instantaneous diameter of test specimen, inches.

TRUE STRESS–STRAIN RELATIONSHIP is obtained when the values of true stress and the corresponding true strain are plotted against each other in the resulting curve (Fig. 10.6). The slope of the nearly straight line leading up to fracture is known as the coefficient of strain hardening. It as well as the true tensile strength appear to be related to the other mechanical properties.

DUCTILITY is the ability of a material to sustain large permanent deformations in tension, such as drawing into a wire.

MALLEABILITY is the ability of a material to sustain large permanent deformations in compression, such as beating or rolling into thin sheets.

BRITTLENESS is that property of a material that permits it to be only slightly deformed without rupture. Brittleness is relative, no material being perfectly brittle, that is, capable of no deformation before rupture. Many materials are brittle to a greater or less degree, glass being one of the most brittle of materials. Brittle materials have relatively short stress–strain curves. Of the common structural materials, cast iron, brick, and stone are brittle in comparison with steel.

TOUGHNESS is the ability of the material to withstand high unit stress together with great unit strain, without complete fracture. The area $OAGH$, or OJK, under the curve of the stress–strain diagram

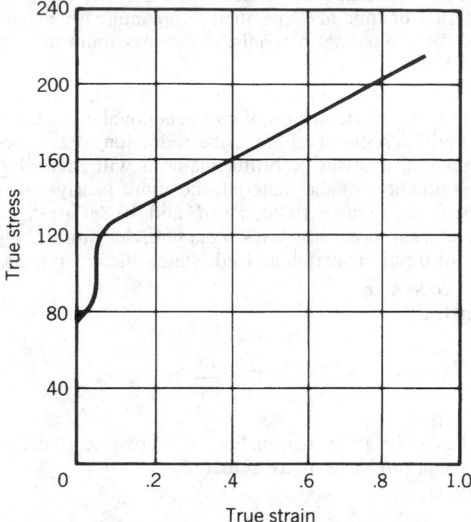

Fig. 10.6 True stress–strain relationship.

(Fig. 10.7), is a measure of the toughness of the material. The distinction between ductility and toughness is that ductility deals only with the ability to deform, whereas toughness considers both the ability to deform and the stress developed during deformation.

STIFFNESS is the ability to resist deformation under stress. The modulus of elasticity is the criterion of the stiffness of a material.

HARDNESS is the ability to resist very small indentations, abrasion, and plastic deformation. There is no single measure of hardness, as it is not a single property but a combination of several properties.

CREEP or flow of metals is a phase of plastic or inelastic action. Some solids, as asphalt or paraffin, flow appreciably at room temperatures under extremely small stresses; zinc, plastics, fiber-reinforced plastics, lead, and tin show signs of creep at room temperature under moderate stresses. At sufficiently high temperatures, practically all metals creep under stresses that vary with temperature, the higher the temperature the lower being the stress at which creep takes place. The deformation due to creep continues to increase indefinitely and becomes of extreme importance in members subjected to high temperatures, as parts in turbines, boilers, super-heaters, etc.

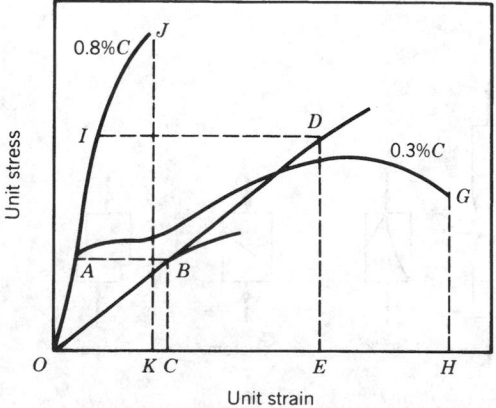

Fig. 10.7 Toughness comparison.

Creep limit is the maximum unit stress under which unit distortion will not exceed a specified value during a given period of time at a specified temperature. A value much used in tests, and suggested as a standard for comparing materials; is the maximum unit stress at which creep does not exceed 1% in 100,000 hours.

TYPES OF FRACTURE. A bar of brittle material, such as cast iron, will rupture in a tension test in a clean sharp fracture with very little reduction of cross-sectional area and very little elongation (Fig. 10.8a). In a ductile material, as structural steel, the reduction of area and elongation are greater (Fig. 10.8b). In compression, a prism of brittle material will break by shearing along oblique planes; the greater the brittleness of the material, the more nearly will these planes parallel the direction of the applied force. Figures 10.8c, 10.8d, and 10.8e, arranged in order of brittleness, illustrate the type of fracture in prisms of brick, concrete, and timber. Figure 10.8f represents the deformation of a prism of plastic material, as lead, which flattens out under load without failure.

RELATIONS OF ELASTIC CONSTANTS

Modulus of elasticity, E:

$$E = \frac{Pl}{Ae}$$

where P = load, pounds, l = length of bar, inches, A = cross-sectional area acted on by the axial load, P, and e = total strain produced by axial load P.

Modulus of rigidity, G:

$$G = \frac{E}{2(1 + \nu)}$$

where E = modulus of elasticity and ν = Poisson's ratio.

Bulk modulus, K, is the ratio of normal stress to the change in volume.

Relationships. The following relationships exist between the modulus of elasticity E, the modulus of rigidity G, the bulk modulus of elasticity K, and Poisson's ratio ν:

$$E = 2G(1 + \nu); \qquad G = \frac{E}{2(1 + \nu)}; \qquad \nu = \frac{E - 2G}{2G}$$

$$K = \frac{E}{3(1 - 2\nu)}; \qquad \nu = \frac{3K - E}{6K}$$

ALLOWABLE UNIT STRESS, also called allowable working unit stress, allowable stress, or working stress, is the maximum unit stress to which it is considered safe to subject a member in service. The term allowable stress is preferable to working stress, since the latter often is used to indicate the actual stress in a material when in service. Allowable unit stresses for different materials for various conditions of service are specified by different authorities on the basis of test or experience. In general, for ductile materials, allowable stress is considerably less than the yield point.

FACTOR OF SAFETY is the ratio of ultimate strength of the material to allowable stress. The term was originated for determining allowable stress. The ultimate strength of a given material divided by an arbitrary factor of safety, dependent on material and the use to which it is to be put, gives

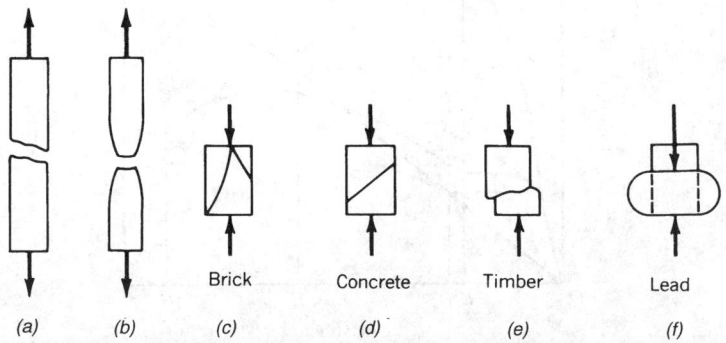

| Brick | Concrete | Timber | Lead |

| (a) | (b) | (c) | (d) | (e) | (f) |

Fig. 10.8 (a) Brittle and (b) ductile fractures in tension and compression fractures.

the allowable stress. In present design practice, it is customary to use allowable stress as specified by recognized authorities or building codes rather than an arbitrary factor of safety. One reason for this is that the factor of safety is misleading, in that it implies a greater degree of safety than actually exists. For example, a factor of safety of 4 does not mean that a member can carry a load four times as great as that for which it was designed. It also should be clearly understood that, even though each part of a machine is designed with the same factor of safety, the machine as a whole does not have that factor of safety. When one part is stressed beyond the proportional limit, or particularly the yield point, the load or stress distribution may be completely changed throughout the entire machine or structure, and its ability to function thus may be changed, even though no part has ruptured.

Although no definite rules can be given, if a factor of safety is to be used, the following circumstances should be taken into account in its selection:

1. When the ultimate strength of the material is known within narrow limits, as for structural steel for which tests of samples have been made, when the load is entirely a steady one of a known amount and there is no reason to fear the deterioration of the metal by corrosion, the lowest factor that should be adopted is 3.
2. When the circumstances of (1) are modified by a portion of the load being variable, as in floors of warehouses, the factor should not be less than 4.
3. When the whole load, or nearly the whole, is likely to be alternately put on and taken off, as in suspension rods of floors of bridges, the factor should be 5 or 6.
4. When the stresses are reversed in direction from tension to compression, as in some bridge diagonals and parts of machines, the factor should be not less than 6.
5. When the piece is subjected to repeated shocks, the factor should be not less than 10.
6. When the piece is subjected to deterioration from corrosion, the section should be sufficiently increased to allow for a definite amount of corrosion before the piece is so far weakened by it as to require removal.
7. When the strength of the material or the amount of the load or both are uncertain, the factor should be increased by an allowance sufficient to cover the amount of the uncertainty.
8. When the strains are complex and of uncertain amount, such as those in the crankshaft of a reversing engine, a very high factor is necessary, possibly even as high as 40.
9. If the property loss caused by failure of the part may be large or if loss of life may result, as in a derrick hoisting materials over a crowded street, the factor should be large.

Dynamic Stresses

DYNAMIC STRESSES occur where the dimension of time is necessary in defining the loads. They include creep, fatigue, and impact stresses.

CREEP STRESSES occur when either the load or deformation progressively vary with time. They are usually associated with noncyclic phenomena.

FATIGUE STRESSES occur when type cyclic variation of either load or strain is coincident with respect to time.

IMPACT STRESSES occur from loads which are transient with time. The duration of the load application is of the same order of magnitude as the natural period of vibration of the specimen.

10.1.2 Work and Resilience

EXTERNAL WORK. Let P = axial load, pounds, on a bar, producing an internal stress not exceeding the elastic limit; σ = unit stress produced by P, pounds per square inch; A = cross-sectional area, square inches; l = length of bar, inches; e = deformation, inches; E = modulus of elasticity; W = external work performed on bar, inch-pounds = $\frac{1}{2}Pe$. Then

$$W = \frac{1}{2} A\sigma \left(\frac{\sigma l}{E} \right) = \frac{1}{2} \left(\frac{\sigma^2}{E} \right) Al \qquad (10.1)$$

The factor $\frac{1}{2}(\sigma^2/E)$ is the work required per unit volume, the volume being Al. It is represented on the stress–strain diagram by the area ODE or area OBC (Fig. 10.9), which DE and BC are ordinates representing the unit stresses considered.

RESILIENCE is the strain energy that may be recovered from a deformed body when the load causing the stress is removed. Within the proportional limit, the resilience is equal to the external work performed in deforming the bar, and may be determined by Eq. (10.1). When σ is equal to the proportional limit, the factor $\frac{1}{2}(\sigma^2/E)$ is the *modulus of resilience,* that is, the measure of capacity of a unit volume of material to store strain energy up to the proportional limit. Average values of

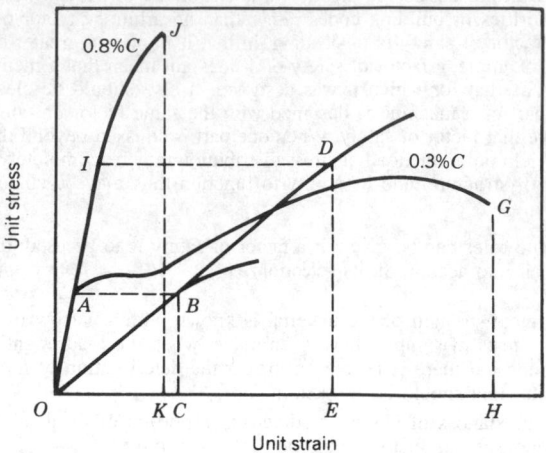

Fig. 10.9 Work areas on stress–strain diagram.

the modulus of resilience under tensile stress are given in Table 10.1.

The total resilience of a bar is the product of its volume and the modulus of resilience. These formulas for work performed on a bar, and its resilience, do not apply if the unit stress is greater than the proportional limit.

WORK REQUIRED FOR RUPTURE. Since beyond the proportional limit the strains are not proportional to the stresses, $\frac{1}{2}P$ does not express the mean value of the force acting. Equation (10.1), therefore, does not express the work required for strain after the proportional limit of the material has been passed, and cannot express the work required for rupture. The work required per unit volume to produce strains beyond the proportional limit or to cause rupture may be determined from the stress–strain diagram as it is measured by the area under the stress–strain curve up to the strain in question, as *OAGH* or *OJK* (Fig. 10.9). This area, however, does not represent the resilience, since part of the work done on the bar is present in the form of hysteresis losses and cannot be recovered.

DAMPING CAPACITY (HYSTERESIS). Observations show that when a tensile load is applied to a bar, it does not produce the complete elongation immediately, but there is a definite time lapse which

Table 10.1   Modulus of Resilience and Relative Toughness under Tensile Stress (Avg. Values)

Material	Modulus of Resilience (in.-lb/in.^{3})	Relative Toughness (Area under Curve of Stress-Deformation Diagram)
Gray cast iron	1.2	70
Malleable cast iron	17.4	−3,800
Wrought iron	11.6	11,000
Low-carbon steel	15.0	15,700
Medium-carbon steel	34.0	16,300
High-carbon steel	94.0	5,000
Ni-Cr steel, hot-rolled	94.0	44,000
Vanadium steel, 0.98% C, 0.2% V, heat-treated	260.0	22,000
Duralumin, 17 ST	45.0	10,000
Rolled bronze	57.0	15,500
Rolled brass	40.0	10,000
Oak	2.3[a]	13[a]

[a]Bending.

depends on the nature of the material and the magnitude of the stresses involved. In parallel with this it is also noted that, upon unloading, complete recovery of energy does not occur. This phenomenon is variously termed *elastic hysteresis* or, for vibratory stresses, damping. Figure 10.10 shows a typical hysteresis loop obtained for one cycle of loading. The area of this hysteresis loop, representing the energy dissipated per cycle, is a measure of the damping properties of the material. While the exact mechanism of damping has not been fully investigated, it has been found that under vibratory conditions the energy dissipated in this manner varies approximately as the cube of the stress.

10.2   DISCONTINUITIES, STRESS CONCENTRATION

The direct design procedure assumes no abrupt changes in cross-section, discontinuities in the surface, or holes, through the member. In most structural parts this is not the case. The stresses produced at these discontinuities are different in magnitude from those calculated by various design methods. The effect of the localized increase in stress, such as that caused by a notch, fillet, hole, or similar *stress raiser*, depends mainly on the type of loading, the geometry of the part, and the material. As a result, it is necessary to consider a stress-concentration factor K_t, which is defined by the relationship

$$K_t = \frac{\sigma_{max}}{\sigma_{nominal}} \tag{10.2}$$

In general σ_{max} will have to be determined by the methods of experimental stress analysis or the theory of elasticity, and $\sigma_{nominal}$ by a simple theory such as $\sigma = P/A$, $\sigma = Mc/I$, $\tau = Tc/J$ without taking into account the variations in stress conditions caused by geometrical discontinuities such as holes, grooves, and fillets. For ductile materials it is not customary to apply stress-concentration factors to members under static loading. For brittle materials, however, stress concentration is serious and should be considered.

Stress-Concentration Factors for Fillets, Keyways, Holes, and Shafts

In Table 10.2 selected stress-concentration factors have been given from a complete table in Refs. 1, 2, and 4.

10.3   COMBINED STRESSES

Under certain circumstances of loading a body is subjected to a combination of tensile, compressive, and/or shear stresses. For example, a shaft that is simultaneously bent and twisted is subjected to combined stresses, namely, longitudinal tension and compression and torsional shear. For the purposes of analysis it is convenient to reduce such systems of combined stresses to a basic system of stress coordinates known as principal stresses. These stresses act on axes that differ in general from the axes along which the applied stresses are acting and represent the maximum and minimum values of the normal stresses for the particular point considered.

Determination of Principal Stresses

The expressions for the principal stresses in terms of the stresses along the x and y axes are

$$\sigma_1 = \frac{\sigma_x + \sigma_y}{2} + \sqrt{\left(\frac{\sigma_x - \sigma_y}{2}\right)^2 + \tau_{xy}^2} \tag{10.3}$$

$$\sigma_2 = \frac{\sigma_x + \sigma_y}{2} - \sqrt{\left(\frac{\sigma_x - \sigma_y}{2}\right)^2 + \tau_{xy}^2} \tag{10.4}$$

$$\tau_1 = \pm \sqrt{\left(\frac{\sigma_x - \sigma_y}{2}\right)^2 + \tau_{xy}^2} \tag{10.5}$$

where σ_1, σ_2, and τ_1 are the principal stress components and σ_x, σ_y, and τ_{xy} are the calculated stress components, all of which are determined at any particular point (Fig. 10.11).

Graphical Method of Principal Stress Determination—Mohr's Circle

Let the axes x and y be chosen to represent the directions of the applied normal and shearing stresses, respectively (Fig. 10.12). Lay off to suitable scale distances $OA = \sigma_x$, $OB = \sigma_y$, and $BC = AD = \tau_{xy}$. With point E as a center construct the circle DFC. Then OF and OG are the principal stresses σ_1 and σ_2, respectively, and EC is the maximum shear stress τ_1. The inverse also holds—that is, given the principal stresses, σ_x and σ_y can be determined on any plane passing through the point.

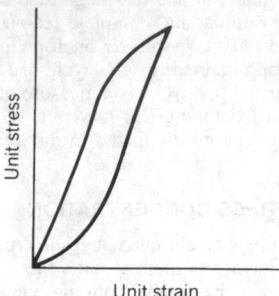

Fig. 10.10   Hysteresis loop for loading and unloading.

Stress–Strain Relations

The linear relation between components of stress and strain is known as *Hooke's law*. This relation for the two-dimensional case can be expressed as

$$e_x = \frac{1}{E} (\sigma_x - \nu\sigma_y) \tag{10.6}$$

$$e_y = \frac{1}{E} (\sigma_y - \nu\sigma_x) \tag{10.7}$$

$$\gamma_{xy} = \frac{1}{G} \tau_{xy} \tag{10.8}$$

where σ_x, σ_y, and τ_{xy} are the stress components of a particular point, ν is Poisson's ratio, E is modulus of elasticity, G is modulus of rigidity, and e_x, e_y, and γ_{xy} are strain components.

The determination of the magnitudes and directions of the principal stresses and strains and of the maximum shearing stresses is carried out for the purpose of establishing criteria of failure within the material under the anticipated loading conditions. To this end several theories have been advanced to elucidate these criteria. The more noteworthy ones are listed below. The theories are based on the assumption that the principal stresses do not change with time, an assumption that is justified since the applied loads in most cases are synchronous.

Maximum-Stress Theory (Rankine's Theory)

This theory is based on the assumption that failure will occur when the maximum value of the greatest principal stress reaches the value of the maximum stress σ_{max} at failure in the case of simple axial loading. Failure is then defined as

Table 10.2   Stress-Concentration Factors[a]

Type		K_t Factors					
Circular hole in plate or rectangular bar	$\dfrac{h}{a} = 0.67$	0.77	0.91	1.07	1.29	1.56	
	$k = 4.37$	3.92	3.61	3.40	3.25	3.16	
tSquare shoulder with fillet for rectangular and circular cross sections in bending	$\dfrac{h}{r} \Big/ \dfrac{r}{d}$	0.05	0.10	0.20	0.27	0.50	1.0
	0.5	1.61	1.49	1.39	1.34	1.22	1.07
	1.0	1.91	1.70	1.48	1.38	1.22	1.08
	1.5	2.00	1.73	1.50	1.39	1.23	1.08
	2.0		1.74	1.52	1.39	1.23	1.09
	3.5		1.76	1.54	1.40	1.23	1.10

[a]Adapted by permission from R. J. Roark and W. C. Young, *Formulas for Stress and Strain,* 6th ed., McGraw-Hill, New York, 1989.

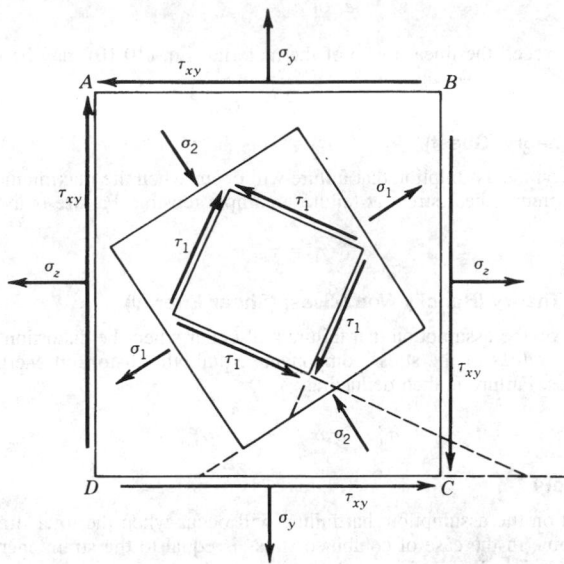

Fig. 10.11 Diagram showing relative orientation of stresses. (Reproduced by permission from J. Marin, *Mechanical Properties of Materials and Design,* McGraw-Hill, New York, 1942.)

$$\sigma_1 \text{ or } \sigma_2 = \sigma_{max} \tag{10.9}$$

Maximum-Strain Theory (Saint Venant)

This theory is based on the assumption that failure will occur when the maximum value of the greatest principal strain reaches the value of the maximum strain e_{max} at failure in the case of simple axial loading. Failure is then defined as

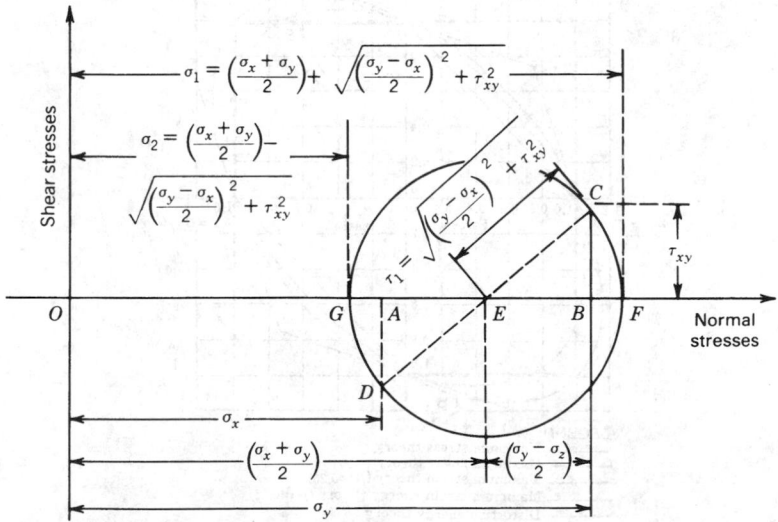

Fig. 10.12 Mohr's circle used for the determination of the principal stresses. (Reproduced by permission from J. Marin, *Mechanical Properties of Materials and Design,* McGraw-Hill, New York, 1942.)

$$e_1 \text{ or } e_2 = e_{max} \tag{10.10}$$

If e_{max} does not exceed the linear range of the material, Eq. (10.10) may be written as

$$\sigma_1 - \nu\sigma_2 = \sigma_{max} \tag{10.11}$$

Maximum-Shear Theory (Guest)

This theory is based on the assumption that failure will occur when the maximum shear stress reaches the value of the maximum shear stress at failure in simple tension. Failure is then defined as

$$\tau_1 = \tau_{max} \tag{10.12}$$

Distortion-Energy Theory (Hencky–Von Mises) (Shear Energy)

This theory is based on the assumption that failure will occur when the distortion energy corresponding to the maximum values of the stress components equals the distortion energy at failure for the maximum axial stress. Failure is then defined as

$$\sigma_1^2 - \sigma_1\sigma_2 + \sigma_2^2 = \sigma_{max}^2 \tag{10.13}$$

Strain-Energy Theory

This theory is based on the assumption that failure will occur when the total strain energy of deformation per unit volume in the case of combined stress is equal to the strain energy per unit volume at failure in simple tension. Failure is then defined as

$$\sigma_1^2 - 2\nu\sigma_1\sigma_2 + \sigma_2^2 = \sigma_{max}^2 \tag{10.14}$$

Comparison of Theories

Figure 10.13 compares the five foregoing theories. In general the distortion-energy theory is the most satisfactory for ductile materials and the maximum-stress theory is the most satisfactory for brittle materials. The maximum-shear theory gives conservative results for both ductile and brittle materials. The conditions for yielding, according to the various theories, are given in Table 10.3, taking $\nu = 0.300$ as for steel.

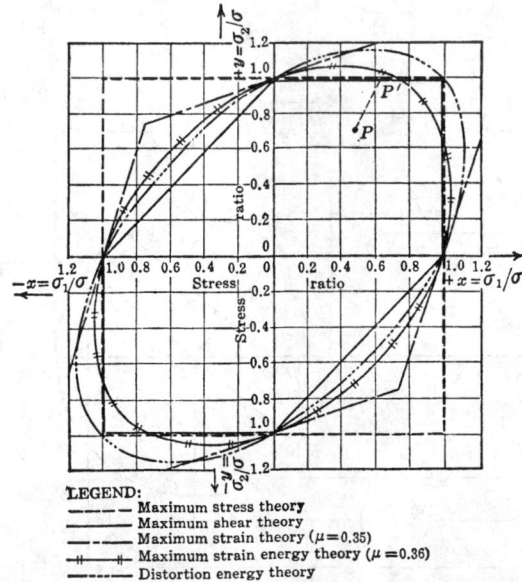

LEGEND:
- — — — Maximum stress theory
- ———— Maximum shear theory
- — · · — Maximum strain theory ($\mu = 0.35$)
- ─╫──╫─ Maximum strain energy theory ($\mu = 0.36$)
- ———— Distortion energy theory

Fig. 10.13 Comparison of five theories of failure. (Reproduced by permission from J. Marin, *Mechanical Properties of Materials and Design*, McGraw-Hill, New York, 1942.)

Table 10.3 Comparison of Stress Theories

$\tau = \sigma_{yp}$	(from the maximum-stress theory
$\tau = 0.77\sigma_{yp}$	(from the maximum-strain theory)
$\tau = 0.50\sigma_{yp}$	(from the maximum-shear theory)
$\tau = 0.62\sigma_{yp}$	(from the maximum-strain-energy theory)

Static Working Stresses

Ductile Materials. For ductile materials the criteria for working stresses are

$$\sigma_w = \frac{\sigma_{yp}}{n} \quad \text{(tension and compression)} \tag{10.15}$$

$$\tau_w = \frac{1}{2}\frac{\sigma_{yp}}{n} \tag{10.16}$$

Brittle Materials. For brittle materials the criteria for working stresses are

$$\sigma_w = \frac{\sigma_{ultimate}}{K_t \times n} \quad \text{(tension)} \tag{10.17}$$

$$\sigma_w = \frac{\sigma_{compressive}}{K_t \times n} \quad \text{(compression)} \tag{10.18}$$

where K_t is the stress-concentration factor, n is the factor of safety, σ_w and τ_w are working stresses, and σ_{yp} is stress at the yield point.

Working-Stress Equations for the Various Theories.
Stress Theory

$$\sigma_w = \frac{\sigma_x + \sigma_y}{2} \pm \sqrt{\left(\frac{\sigma_x - \sigma_y}{2}\right)^2 + \tau_{xy}^2} \tag{10.19}$$

Shear Theory

$$\sigma_w = 2\sqrt{\left(\frac{\sigma_x - \sigma_y}{2}\right)^2 + \tau_{xy}^2} \tag{10.20}$$

Strain Theory

$$\sigma_w = (1 - \nu)\left(\frac{\sigma_x + \sigma_y}{2}\right) + (1 + \nu)\sqrt{\left(\frac{\sigma_x - \sigma_y}{2}\right)^2 + \tau_{xy}^2} \tag{10.21}$$

Distortion-Energy Theory

$$\sigma_w = \sqrt{\sigma_x^2 - \sigma_x\sigma_y + \sigma_y^2 + 3\tau_{xy}^2} \tag{10.22}$$

Strain-Energy Theory

$$\sigma_w = \sqrt{\sigma_x^2 - 2\nu\sigma_x\sigma_y + \sigma_y^2 + 2(1 + \nu)\tau_{xy}^2} \tag{10.23}$$

where σ_x, σ_y, τ_{xy} are the stress components of a particular point, ν is Poisson's ratio, and σ_w is working stress.

10.4 CREEP

Introduction

Materials subjected to a constant stress at elevated temperatures deform continuously with time, and the behavior under these conditions is different from the behavior at normal temperatures. This continuous deformation with time is called creep. In some applications the permissible creep deformations are critical, in others of no significance. But the existence of creep necessitates information on the creep deformations that may occur during the expected life of the machine. Plastic, zinc, tin, and fiber-reinforced plastics creep at room temperature. Aluminum and magnesium alloys start to creep at around 300°F. Steels above 650°F must be checked for creep.

Mechanism of Creep Failure

There are generally four distinct phases distinguishable during the course of creep failure. The elapsed time per stage depends on the material, temperature, and stress condition. They are: (1) Initial phase—where the total deformation is partially elastic and partially plastic. (2) Second phase—where the creep rate decreases with time, indicating the effect of strain hardening. (3) Third phase—where the effect of strain hardening is counteracted by the annealing influence of the high temperature which produces a constant or minimum creep rate. (4) Final phase—where the creep rate increases until fracture occurs owing to the decrease in cross-sectional area of the specimen.

Creep Equations

In conducting a conventional creep test, curves of strain as a function of time are obtained for groups of specimens; each specimen in one group is subjected to a different constant stress, while all of the specimens in the group are tested at one temperature.

In this manner families of curves like those shown in Fig. 10.14 are obtained. Several methods have been proposed for the interpretation of such data. (See Refs. 1 and 3.) Two frequently used expressions of the creep properties of a material can be derived from the data in the following form:

$$C = B\sigma^m \tag{10.24}$$
$$\epsilon = \epsilon_0 + Ct$$

where C = creep rate, B, m = experimental constants, σ = stress, ϵ = creep strain at any time t, ϵ_0 = zero-time strain intercept, and t = time. See Fig. 10.15.

Stress Relaxation

Various types of bolted joints and shrink or press fit assemblies and springs are applications of creep taking place with diminishing stress. This deformation tends to loosen the joint and produce a stress reduction or stress relaxation. The performance of a material to be used under diminishing creep-stress condition is determined by a tensile stress-relaxation test.

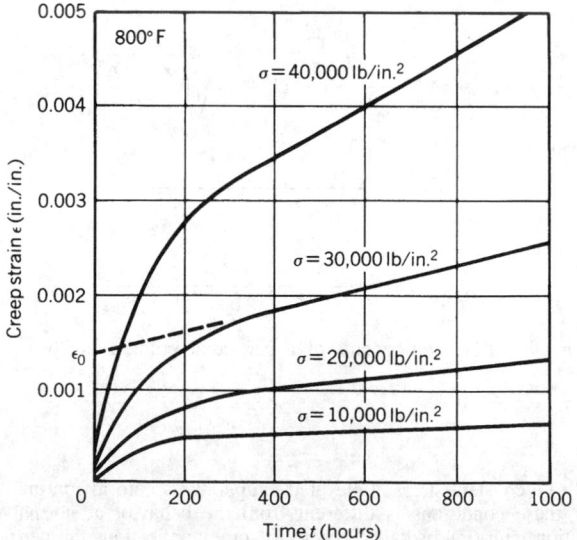

Fig. 10.14 Curves of creep strain for various stress levels.

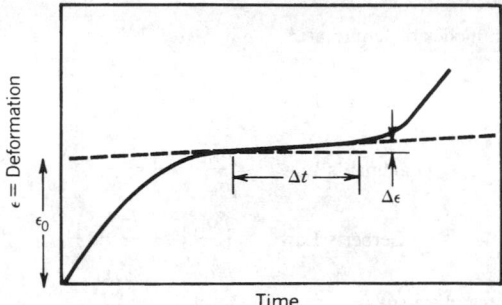

Fig. 10.15 Method of determining creep rate.

10.5 FATIGUE

Definitions

STRESS CYCLE. A stress cycle is the smallest section of the stress-time function that is repeated identically and periodically, as shown in Fig. 10.16.

MAXIMUM STRESS. σ_{max} is the largest algebraic value of the stress in the stress cycle, being positive for a tensile stress and negative for a compressive stress.

MINIMUM STRESS. σ_{min} is the smallest algebraic value of the stress in the stress cycle, being positive for a tensile stress and negative for a compressive stress.

RANGE OF STRESS. σ_r is the algebraic difference between the maximum and minimum stress in one cycle:

$$\sigma_r = \sigma_{max} - \sigma_{min} \tag{10.25}$$

For most cases of fatigue testing the stress varies about zero stress, but other types of variation may be experienced.

ALTERNATING-STRESS AMPLITUDE (VARIABLE STRESS COMPONENT). σ_a is one-half the range of stress, $\sigma_a = \sigma_r/2$.

MEAN STRESS (STEADY STRESS COMPONENT). σ_m is the algebraic mean of the maximum and minimum stress in one cycle:

$$\sigma_m = \frac{\sigma_{max} + \sigma_{min}}{2} \tag{10.26}$$

STRESS RATIO. R is the algebraic ratio of the minimum stress and the maximum stress in one cycle.

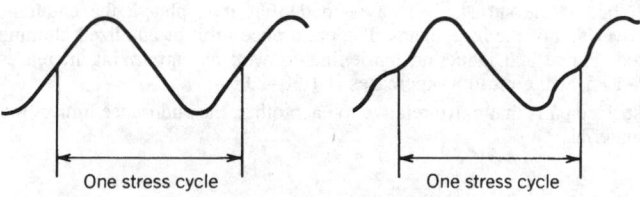

One stress cycle   One stress cycle

Fig. 10.16 Definition of one stress cycle.

10.5.1 Modes of Failure

The three most common modes of failure are*

$$\text{Soderberg's Law} \qquad \frac{\sigma_m}{\sigma_y} + \frac{\sigma_a}{\sigma_e} = \frac{1}{N} \tag{10.27}$$

$$\text{Goodman's Law} \qquad \frac{\sigma_m}{\sigma_u} + \frac{\sigma_a}{\sigma_e} = \frac{1}{N} \tag{10.28}$$

$$\text{Gerber's Law} \qquad \left(\frac{\sigma_m}{\sigma_u}\right)^2 + \frac{\sigma_a}{\sigma_e} = \frac{1}{N} \tag{10.29}$$

From distortion energy for plane stress

$$\sigma_m = \sqrt{\sigma_{xm}^2 - \sigma_{xm}\sigma_{ym} + \sigma_{ym}^2 + 3\tau_{xym}^2} \tag{10.30}$$

$$\sigma_a = \sqrt{\sigma_{xa}^2 - \sigma_{xa}\sigma_{ya} + \sigma_{ya}^2 + 3\tau_{xya}^2} \tag{10.31}$$

The stress concentration factor,[4] K_t or K_f, is applied to the individual stress for both σ_a and σ_m for brittle materials and only to σ_a for ductile materials. N is a reasonable factor of safety. σ_u is the ultimate tensile strength, and σ_y is the yield strength. σ_e is developed from the endurance limit σ_e' and reduced or increased depending on conditions and manufacturing procedures and to keep σ_e less than the yield strength:

$$\sigma_e = k_a k_b \cdots k_n \, \sigma_e'$$

where σ_e' (Ref. 1) for various materials is:

Steel	$0.5\sigma_u$ and never greater than 100 kpsi at 10^6 cycles
Magnesium	$0.35\sigma_u$ at 10^8 cycles
Nonferrous alloys	$0.35\sigma_u$ at 10^8 cycles
Aluminum alloys	$(0.16–0.3)\sigma_u$ at 5×10^8 cycles (see Military Handbook 5D)

and where the other k factors are affected as follows:

Surface Condition. For surfaces that are from machined to ground, the k_a varies from 0.7 to 1.0. When surface finish is known, k_a can be found[1] more accurately.

Size and Shape. If the size of the part is 0.30 in. or larger, the reduction is 0.85 or less, depending on the size.

Reliability. The endurance limit and material properties are averages and both should be corrected. A reliability of 90% reduces values 0.897, while one of 99% reduces 0.814.

Temperature. The endurance limit at $-190°C$ increases 1.54–2.57 for steels, 1.14 for aluminums, and 1.4 for titaniums. The endurance limit is reduced approximately 0.68 for some steels at 1382°F, 0.24 for aluminum around 662°F, and 0.4 for magnesium alloys at 572°F.

Residual Stresses. For steel, shot peening increases the endurance limit 1.04–1.22 for polished surfaces, 1.25 for machined surfaces, 1.25–1.5 for rolled surfaces, and 2–3 for forged surfaces. The shot-peening effect disappears above 500°F for steels and above 250°F for aluminum. Surface rolling affects the steel endurance limit approximately the same as shot peening, while the endurance limit is increased 1.2–1.3 in aluminum, 1.5 in magnesium, and 1.2–2.93 in cast iron.

Corrosion. A corrosive environment decreases the endurance limit of anodized aluminum and magnesium 0.76–1.00, while nitrided steel and most materials are reduced 0.6–0.8.

Surface Treatments. Nickel plating reduces the endurance limit of 1008 steel 0.01 and of 1063 steel 0.77, but, if the surface is shot peened after it is plated, the endurance limit can be increased over that of the base metal. The endurance limit of anodized aluminum is in general not affected. Flame and induction hardening as well as carburizing increases the endurance limit 1.62–1.85, while nitriding increases it 1.30–2.00.

Fretting. In surface pairs that move relative to each other, the endurance limit is reduced 0.70–0.90 for each material.

---

*This section is condensed from Ref. 1, Chap. 12.

Radiation. Radiation tends to increase tensile strength but to decrease ductility.

In discussions on fatigue it should be emphasized that most designs must pass vibration testing. When sizing parts so that they can be modeled on a computer, the designer needs a starting point until feedback is received from the modeling. A helpful starting point is to estimate the static load to be carried, to find the level of vibration testing in G levels, to assume that the part vibrates with a magnification of 10, and to multiply these together to get an equivalent static load. The stress level should be $\sigma_u/4$, which should be less than the yield strength. When the design is modeled, changes can be made to bring the design within the required limits.

10.6 BEAMS

10.6.1 Theory of Flexure

Types of Beams

A beam is a bar or structural member subjected to transverse loads that tend to bend it. Any structural members acts as a beam if bending is induced by external transverse forces.

A **simple beam** (Fig. 10.17a) is a horizontal member that rests on two supports at the ends of the beam. All parts between the supports have free movement in a vertical plane under the influence of vertical loads.

A **fixed beam, constrained beam,** or **restrained beam** (Fig. 10.17b) is rigidly fixed at both ends or rigidly fixed at one end and simply supported at the other.

A **continuous beam** (Fig. 10.17c) is a member resting on more than two supports.

A **cantilever beam** (Fig. 10.17d) is a member with one end projecting beyond the point of support, free to move in a vertical plane under the influence of vertical loads placed between the free end and the support.

Phenomena of Flexure

When a simple beam bends under its own weight, the fibers on the upper or concave side are shortened, and the stress acting on them is compression; the fibers on the under or convex side are lengthened, and the stress acting on them is tension. In addition, shear exists along each cross section, the intensity of which is greatest along the sections at the two supports and zero at the middle section.

When a cantilever beam bends under its own weight, the fibers on the upper or convex side are lengthened under tensile stresses; the fibers on the under or concave side are shortened under compressive stresses, the shear is greatest along the section at the support, and zero at the free end.

The **neutral surface** is that horizontal section between the concave and convex surfaces of a loaded beam, where there is no change in the length of the fibers and no tensile or compressive stresses acting upon them.

The **neutral axis** is the trace of the neutral surface on any cross section of a beam. (See Fig. 10.18).

The **elastic curve** of a beam is the curve formed by the intersection of the neutral surface with the side of the beam, it being assumed that the longitudinal stresses on the fibers are within the elastic limit.

Reactions at Supports

The reactions, or upward pressures at the points of support, are computed by applying the following conditions necessary for equilibrium of a system of vertical forces in the same plane: (1) The algebraic sum of all vertical forces must equal zero; that is, the sum of the reactions equals the sum of the downward loads. (2) The algebraic sum of the moments of all the vertical forces must equal zero.

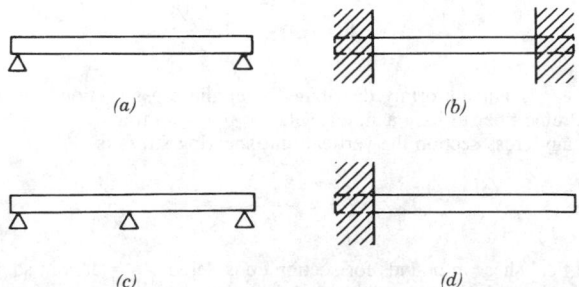

Fig. 10.17 (a) Simple, (b) constrained, (c) continuous, and (d) cantilever beams.

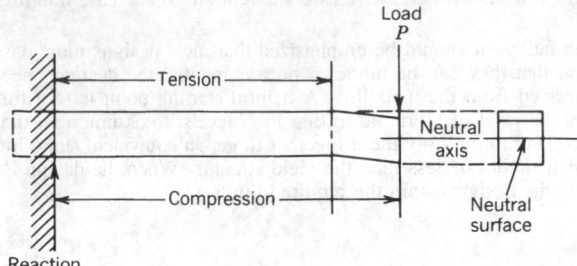

Fig. 10.18   Loads and stress conditions in a cantilever beam.

Condition (1) applies to cantilever beams and to simple beams uniformly loaded, or with equal concentrated loads placed at equal distances from the center of the beam. In the cantilever beam, the reaction is the sum of all the vertical forces acting downward, comprising the weight of the beam and the superposed loads. In the simple beam each reaction is equal to one-half the total load, consisting of the weight of the beam and the superposed loads. Condition (2) applies to a simple beam not uniformly loaded. The reactions are computed separately, by determining the moment of the several loads about each support. The sum of the moments of the load around one support is equal to the moment of the reaction of the other support around the first support.

Conditions of Equilibrium

The fundamental laws for the stresses at any cross section of a beam in equilibrium are: (1) Sum of horizontal tensile stresses = sum of horizontal compressive stresses. (2) Resisting shear = vertical shear. (3) Resisting moment = bending moment.

Vertical Shear.   At any cross section of a beam the resultant of the external vertical forces acting on one side of the section is equal and opposite to the resultant of the external vertical forces acting on the other side of the section. These forces tend to cause the beam to shear vertically along the section. The value of either resultant is known as the vertical shear at the section considered. It is computed by finding the algebraic sum of the vertical forces to the left of the section; that is, it is equal to the left reaction minus the sum of the vertical downward forces acting between the left support and the section.

A **shear diagram** is a graphic representation of the vertical shear at all cross sections of the beam. Thus in the uniformly loaded simple beam (Table 10.5) the ordinates to the line represent to scale the intensity of the vertical shear at the corresponding sections of the beam. The vertical shear is greatest at the supports, where it is equal to the reactions, and it is zero at the center of the span. In the cantilever beam (Table 10.5) the vertical shear is greatest at the point of support, where it is equal to the reaction, and it is zero at the free end. Table 10.5 shows graphically the vertical shear on all sections of a simple beam carrying two concentrated loads at equal distances from the supports, the weight of the beam being neglected.

Resisting Shear.   The tendency of a beam to shear vertically along any cross section, due to the vertical shear, is opposed by an internal shearing stress at that cross section known as the resisting shear; it is equal to the algebraic sum of the vertical components of all the internal stresses acting on the cross section.

If V = vertical shear, pounds; V_r = resisting shear, pounds; τ = average unit shearing stress, pounds per square inch; and A = area of the section, square inches, then at any cross section

$$V_r = V = \tau A; \qquad \tau = \frac{V}{A} \tag{10.32}$$

The resisting shear is not uniformly distributed over the cross section, but the intensity varies from zero at the extreme fiber to its maximum value at the neutral axis.

At any point in any cross section the vertical unit shearing stress is

$$\tau = \frac{VA'c'}{It} \tag{10.33}$$

where V = total vertical shear in pounds for section considered; A' = area in square inches of cross section between a horizontal plane through the point where shear is being found and the extreme fiber on the same side of the neutral axis; c' = distance in inches from neutral axis to center of gravity of area A'; I = moment of inertia of the section, inches^4; t = width of section at plane of shear, inches. Maximum value of the unit shearing stress, where A = total area, square inches, of cross section of the beam, is

$$\text{For a solid rectangular beam:} \quad \tau = \frac{3V}{2A} \tag{10.34}$$

$$\text{For a solid circular beam:} \quad \tau = \frac{4V}{3A} \tag{10.35}$$

Horizontal Shear. In a beam, at any cross section where there is a vertical shearing force, there must be resultant unit shearing stresses acting on the vertical faces of particles that lie at that section. On a horizontal surface of such a particle, there is a unit shearing stress equal to the unit shearing stress on a vertical surface of the particle. Equation (10.33) therefore, also gives the horizontal unit shearing stress at any point on the cross section of a beam.

Bending moment, at any cross section of a beam, is the algebraic sum of the moments of the external forces acting on either side of the section. It is positive when it causes the beam to bend convex downward, hence causing compression in upper fibers and tension in lower fibers of the beam. When the bending moment is determined from the forces that lie to the left of the section, it is positive if they act in a clockwise direction; if determined from forces on the right side, it is positive if they act in a counterclockwise direction. If the moments of upward forces are given positive signs, and the moments of downward forces are given negative signs, the bending moment will always have the correct sign, whether determined from the right or left side. The bending moment should be determined for the side for which the calculation will be simplest.

In Table 10.5 let M be the bending moment, pound-inches, at a section of a simple beam at a distance x, inches, from the left support; w = weight of beam per 1 in. of length; l = length of the beam, inches. Then the reactions are $\frac{1}{2}wl$, and $M = \frac{1}{2}wlx - \frac{1}{2}xwx$. For the sections at the supports, $x = 0$ or l and $M = 0$. For the section at the center of the span $x = \frac{1}{2}l$ and $M = \frac{1}{8}wl^2 = \frac{1}{8}Wl$, where W = total weight.

A **moment diagram** Table 10.5 shows the bending moment at all cross sections of a beam. Ordinates to the curve represent to scale the moments at the corresponding cross sections. The curve for a simple beam uniformly loaded is a parabola, showing $M = 0$ at the supports and $M = \frac{1}{8}wl^2 = \frac{1}{8}Wl$ at the center, M being in pound-inches.

The dangerous section is the cross section of a beam where the bending moment is greatest. In a cantilever beam it is at the point of support, regardless of the disposition of the loads. In a simple beam it is that section where the vertical shear changes from positive to negative, and it may be located graphically by constructing a shear diagram or numerically by taking the left reaction and subtracting the loads in order from the left until a point is reached where the sum of the loads subtracted equals the reaction. For a simple beam, uniformly loaded, the dangerous section is at the center of the span.

The tendency to rotate about a point in any cross section of a beam is due to the bending moment at that section. This tendency is resisted by the **resisting moment,** which is the algebraic sum of the moments of all the horizontal stresses with reference to the same point.

Formula for Flexure

Let M = bending moment; M_r = resisting moment of the horizontal fiber stresses; σ = unit stress (tensile or compressive) on any fiber, usually that one most remote from the neutral surface; c = distance of that fiber from the neutral surface. Then

$$M = M_r = \frac{\sigma I}{c} \tag{10.36}$$

$$\sigma = \frac{Mc}{I} \tag{10.37}$$

where I = moment of inertia of the cross section with respect to its neutral axis. If σ is in pounds per square inch, M must be in pound-inches, I in inches^4 and c in inches.

Equation (10.37) is the basis of the design and investigation of beams. It is true only when the maximum horizontal fiber stress σ does not exceed the proportional limit of the material.

Moment of inertia is the sum of the products of each elementary area of the cross section multiplied by the square of the distance of that area from the assumed axis of rotation, or

$$I = \Sigma r^2 \, \Delta A = \int r^2 \, dA \tag{10.38}$$

where Σ is the sign of summation, ΔA is an elementary area of the section, and r is the distance of ΔA from the axis. The moment of inertia is greatest in those sections (such as I-beams) having much of the area concentrated at a distance from the axis. Unless otherwise stated, the neutral axis is the axis of rotation considered. I usually is expressed in inches[4]. See Table 10.4 for values of moments of inertia of various sections.

Modulus of rupture is the term applied to the value of σ as found by Eq. (10.37), when a beam is loaded to the point of rupture. Since Eq. (10.37) is true only for stresses within the proportional limit, the value σ of the rupture strength so found is incorrect. However, the equation is used, as a measure of the ultimate load-carrying capacity of a beam. The modulus of rupture does not show

Table 10.4　Elements of Sections

A = area of section

I = moment of inertia about axis $I\text{-}I$

c = distance from axis $I\text{-}I$ to remotest point of section

I/c = section modulus

r = radius of gyration

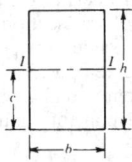 **RECTANGLE** Axis through center $A = bh$ $c = h/2$ $I = bh^3/12$ $I/c = bh^2/6$ $r = h/\sqrt{12} = 0.289h$	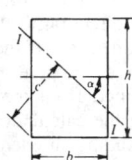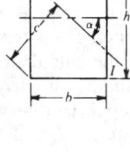 **RECTANGLE** Axis any line through center of gravity $A = bh$ $c = (b \sin \alpha + h \cos \alpha)/2$ $I =$ $bh(b^2 \sin^2 \alpha + h^2 \cos^2 \alpha)/12$ $I/c =$ $\dfrac{bh(b^2 \sin^2 \alpha + h^2 \cos^2 \alpha)}{6(b \sin \alpha + h \cos \alpha)}$ $r =$ $\sqrt{(b^2 \sin^2 \alpha + h^2 \cos^2 \alpha)/12}$
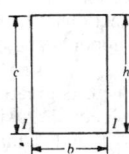 **RECTANGLE** Axis on base $A = bh$ $c = h$ $I = bh^3/3$ $I/c = bh^2/3$ $r = h/\sqrt{3} = 0.577h$	**TRIANGLE** Axis through center of gravity $A = bh/2$ $c = 2/3\ h$ $I = bh^3/36$ $I/c = bh^2/24$ $r = h/\sqrt{18} = 0.236h$
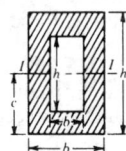 **HOLLOW RECTANGLE** Axis through center $A = bh - b_1 h_1$ $c = h/2$ $I = (bh^3 - b_1 h_1^3)/12$ $I/c = (bh^3 - b_1 h_1^3)/6h$ $r = \sqrt{\dfrac{bh^3 - b_1 h_1^3}{12(bh - b_1 h_1)}}$	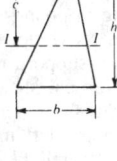 **TRIANGLE** Axis through base $A = bh/2$ $c = h$ $I = bh^3/12$ $I/c = bh^2/12$ $r = h/\sqrt{6} = 0.408h$
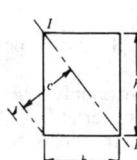 **RECTANGLE** Axis on diagonal $A = bh$ $c = bh/\sqrt{b^2 + h^2}$ $I = b^3 h^3/6(b^2 + h^2)$ $I/c = b^2 h^2/6\sqrt{(b^2 + h^2)}$ $r = bh/\sqrt{6(b^2 + h^2)}$	**TRIANGLE** Axis through apex $A = bh/2$ $c = h$ $I = bh^3/4$ $I/c = bh^2/4$ $r = h/\sqrt{2} = 0.707h$

Table 10.4 *(Continued)*

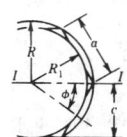

EQUILATERAL POLYGON

Axis through center, parallel to one side. n = number of sides

$$A = nR_1^2 \tan \phi$$

$$c = a/2 \tan \phi = R_1$$

$$I = \{A(12R_1^2 + a^2)\}/48$$

$$I/c = \{A(12R_1^2 + a^2)\}/48R_1$$

$$r = \sqrt{(12R_1^2 + a^2)/48}$$

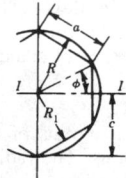

EQUILATERAL POLYGON

Axis through center, normal to side. n = number of sides

$$A = nR_1^2 \tan \phi$$

$$c = a/(2 \sin \phi) = R$$

$$I = \{A(6R^2 - a^2)\}/24$$

$$I/c = \{A(6R^2 - a^2)\}/24R$$

$$r = \sqrt{(6R^2 - a^2)/24}$$

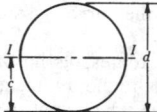

CIRCLE

Axis through center

$$A = \pi d^2/4 = 0.7854d^2$$

$$c = d/2$$

$$I = \pi d^4/64 = 0.0491d^4$$

$$I/c = \pi d^3/32 = 0.0982d^3$$

$$r = d/4$$

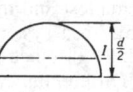

HALF CIRCLE

Axis through center of gravity

$$A = \pi d^2/8 = 0.3927d^2$$

$$c = \{d(3\pi - 4)\}/6\pi = 0.2878d.$$

$$I = \{d^4(9\pi^2 - 64)\}/1152\pi = 0.0068d^4$$

$$I/c = \frac{\{d^3(9\pi^2 - 64)\}}{\{192(3\pi - 4)\}} = 0.0238d^3$$

$$r = \{d\sqrt{(9\pi^2 - 64)}\}/12\pi = 0.1322d$$

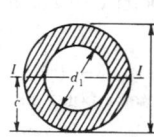

HOLLOW CIRCLE

Axis through center

$$A = \pi(d^2 - d_1^2)/4 = 0.7854(d^2 - d_1^2)$$

$$c = d/2$$

$$I = \pi(d^4 - d_1^4)/64 = 0.0491(d^4 - d_1^4)$$

$$I/c = \pi(d^4 - d_1^4)/32d = 0.0982(d^4 - d_1^4)/d$$

$$r = \sqrt{(d^2 + d_1^2)/4}$$

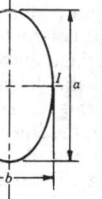

ELLIPSE

Axis through center

$$A = \pi ab/4 = 0.7854ab$$

$$c = a/2$$

$$I = \pi a^3 b/64 = 0.0491a^3 b$$

$$I/c = \pi a^2 b/32 = 0.0982a^2 b$$

$$r = a/4$$

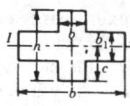

CROSSED RECTANGLES

Axis through center

$$A = th + t_1(b - t)$$

$$c = h/2$$

$$I = \{th^3 + t_1^3(b - t)\}/12$$

$$I/c = \{th^3 + t_1^3(b - t)\}/6h$$

$$r = \sqrt{\frac{th^3 + t_1^3(b - t)}{12\{th + t_1(b - t)\}}}$$

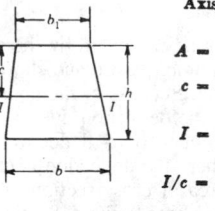

TRAPEZOID

Axis through center of gravity

$$A = \{(b + b_1)h\}/2$$

$$c = \frac{\{(b_1 + 2b)h\}}{\{3(b + b_1)\}}$$

$$I = \frac{h^3(b^2 + 4bb_1 + b_1^2)}{36(b + b_1)}$$

$$I/c = \frac{h^2(b^2 + 4bb_1 + b_1^2)}{12(b_1 + 2b)}$$

$$r = \frac{h}{6(b + b_1)} \sqrt{2(b^2 + 4bb_1 + b_1^2)}$$

the actual stress in the extreme fiber of a beam; it is useful only as a basis of comparison. If the strength of a beam in tension differs from its strength in compression, the modulus of rupture is intermediate between the two.

Section modulus, the factor I/c in flexure [Eq. (10.36)], is expressed in inches³. It is the measure of a capacity of a section to resist a bending moment. For values of I/c for simple shapes, see Table 10.4. See Refs. 6 and 17 for properties of standard steel and aluminum structural shapes.

Elastic Deflection of Beams

When a beam bends under load, all points of the elastic curve except those over the supports are deflected from their original positions. The radius of curvature ρ of the elastic curve at any section is expressed as

$$\rho = \frac{EI}{M} \tag{10.39}$$

where E = modulus of elasticity of the material, pounds per square inch; I = moment of inertia, inches⁴, of the cross section with reference to its neutral axis; M = bending moment, pound-inches, at the section considered. Where there is no bending moment, ρ is infinity and the curve is a straight line; where M is greatest, ρ is smallest and the curvature, therefore, is greatest.

If the elastic curve is referred to a system of coordinate axes in which x represents horizontal distances, y vertical distances, and l distances along the curve, the value of ρ is found, by the aid of the calculus, to be $d^3l/dx \cdot d^2y$. Differential equation (10.40) of the elastic curve which applies to all beams when the elastic limit of the material is not exceeded is obtained by substituting this value in the expression $\rho = EI/M$ and assuming that dx and dl are practically equal:

$$EI \frac{d^2y}{dx^2} = M \tag{10.40}$$

Equation (10.40) is used to determine the deflection of any point of the elastic curve, by regarding the point of support as the origin of the coordinate axis, taking y as the vertical deflection at any point on the curve and x as the horizontal distance from the support to the point considered. The values of E, I, and M are substituted and the expression is integrated twice, giving proper values to the constants of integration, and the deflection y is determined for any point. See Table 10.5.

For example, a cantilever beam in Table 10.5 has a length = l, inches, and carries a load, P, pounds, at the free end. It is required to find the deflection of the elastic curve at a point distant x, inches, from the support, the weight of the beam being neglected.

The moment $M = -P(l - x)$. By substitution in Eq. (10.40), the equation for the elastic curve becomes $EI(d^2y/dx^2) = -Pl + Px$. By integrating and determining the constant of integration by the condition that $dy/dx = 0$ when $x = 0$, $EI(dy/dx) = -Plx + \frac{1}{2}Px^2$ results. By integrating a second time and determining the constant by the condition that $x = 0$ when $y = 0$, $EIy = -\frac{1}{2}Plx^2 + \frac{1}{6}Px^3$, which is the equation of the elastic curve, results. When $x = l$, the value of y, or the deflection in inches at the free end, is found to be $-Pl^3/3EI$.

Deflection due to Shear

The deflection of a beam as computed by the ordinary formulas is that due to flexural stresses only. The deflection in honeycomb, plastic and short beams due to vertical shear can be considerable, and should always be checked. Because of the nonuniform distribution of the shear over the cross section of the beam, computing the deflection due to shear by exact methods is difficult. It may be approximated by $y_s = M/AE_s$, where y_s = deflection, inches, due to shear; M = bending moment, pound-inches, at the section where the deflection is calculated; E_s = modulus of elasticity in shear, pounds per square inch; A = area of cross section of beam, square inches.⁷ For a rectangular section, the ratio of deflection due to shear to the deflection due to bending, will be less than 5% if the depth of the beam is less than one-eighth of the length.

10.6.2 Design of Beams

Design Procedure

In designing a beam the procedure is: (1) Compute reactions. (2) Determine position of the dangerous section and the bending moment at that section. (3) Divide the maximum bending moment (expressed in pound-inches) by the allowable unit stress (expressed in pounds per square inch) to obtain the minimum value of the section modulus. (4) Select a beam section with a section modulus equal to or slightly greater than the section modulus required.

Table 10.5 Bending Moment, Vertical Shear, and Deflection of Beams of Uniform Cross Section under Various Conditions of Loading

P = concentrated loads, lb
R_1, R_2 = reactions, lb
w = uniform load per unit of length, lb per in.
W = total uniform load on beam, lb
l = length of beam, in
z = distance from support to any section, in
E = modulus of elasticity, psi

I = moment of inertia, in.[4]
V_z = vertical shear at any section, lb
V = maximum vertical shear, lb
M_z = bending moment at any section, lb-in.
M = maximum bending moment, lb-in.
y = maximum deflection, in.

SIMPLE BEAM—UNIFORM LOAD

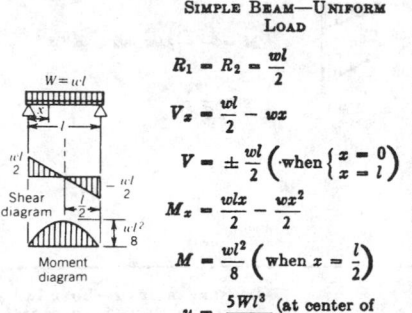

$$R_1 = R_2 = \frac{wl}{2}$$

$$V_z = \frac{wl}{2} - wx$$

$$V = \pm \frac{wl}{2} \left(\text{when} \begin{cases} x = 0 \\ x = l \end{cases} \right)$$

$$M_z = \frac{wlx}{2} - \frac{wx^2}{2}$$

$$M = \frac{wl^2}{8} \left(\text{when } x = \frac{l}{2} \right)$$

$$y = \frac{5Wl^3}{384EI} \text{ (at center of span)}$$

SIMPLE BEAM—CONCENTRATED LOAD AT ANY POINT

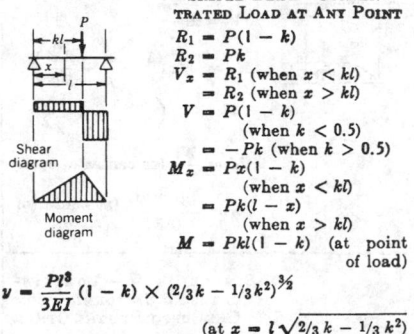

$R_1 = P(1 - k)$
$R_2 = Pk$
$V_z = R_1$ (when $x < kl$)
   = R_2 (when $x > kl$)
$V = P(1 - k)$
   (when $k < 0.5$)
   = $-Pk$ (when $k > 0.5$)
$M_z = Px(1 - k)$
   (when $x < kl$)
   = $Pk(l - x)$
   (when $x > kl$)
$M = Pkl(1 - k)$ (at point of load)

$$y = \frac{Pl^3}{3EI}(1 - k) \times (2/3 k - 1/3 k^2)^{3/2}$$

$$(\text{at } x = l \sqrt{2/3 k - 1/3 k^2})$$

SIMPLE BEAM—CONCENTRATED LOAD AT CENTER

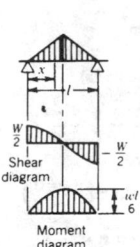

$$R_1 = R_2 = \frac{P}{2}$$

$$V_z = V = \pm \frac{P}{2}$$

$$M_z = \frac{Px}{2}$$

$$M = \frac{Pl}{4} \left(\text{when } x = \frac{l}{2} \right)$$

$$y = \frac{Pl^3}{48EI} \text{ (at center of span)}$$

SIMPLE BEAM—TWO EQUAL CONCENTRATED LOADS AT EQUAL DISTANCES FROM SUPPORTS

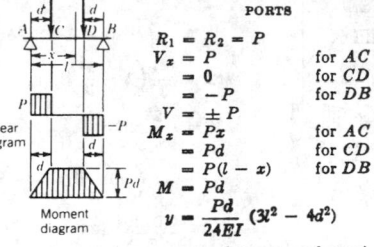

$R_1 = R_2 = P$
$V_z = P$ for AC
   = 0   for CD
   = $-P$ for DB
$V = \pm P$
$M_z = Px$ for AC
   = Pd for CD
   = $P(l - x)$ for DB
$M = Pd$

$$y = \frac{Pd}{24EI}(3l^2 - 4d^2)$$

$$(\text{at center of span})$$

SIMPLE BEAM—LOAD INCREASING UNIFORMLY FROM SUPPORTS TO CENTER OF SPAN

$$R_1 = R_2 = \frac{W}{2}$$

$$V_z = W \left(\frac{1}{2} - \frac{2x^2}{l^2} \right)$$

$$\left(\text{when } x < \frac{l}{2} \right)$$

$$V = \pm \frac{W}{2} \text{ (at supports)}$$

$$M_z = Wx \left(\frac{1}{2} - \frac{2x^2}{3l^2} \right)$$

$$M = \frac{Wl}{6} \text{ (at center of span)}$$

$$y = \frac{Wl^3}{60EI} \text{ (at center of span)}$$

CANTILEVER BEAM—LOAD CONCENTRATED AT FREE END

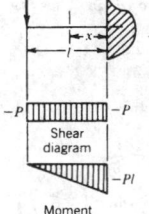

$$R = P$$

$$V_z = V = -P$$

$$M_z = -P(l - x)$$

$$M = -Pl (\text{when } x = 0)$$

$$y = \frac{Pl^3}{3EI}$$

Table 10.5 *(Continued)*

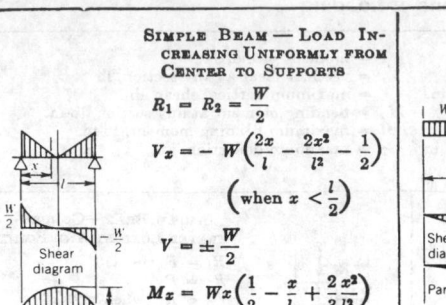

SIMPLE BEAM — LOAD INCREASING UNIFORMLY FROM CENTER TO SUPPORTS

$$R_1 = R_2 = \frac{W}{2}$$

$$V_x = -W\left(\frac{2x}{l} - \frac{2x^2}{l^2} - \frac{1}{2}\right)$$

$$\left(\text{when } x < \frac{l}{2}\right)$$

$$V = \pm \frac{W}{2}$$

$$M_x = Wx\left(\frac{1}{2} - \frac{x}{l} + \frac{2}{3}\frac{x^2}{l^2}\right)$$

$$\left(\text{when } x < \frac{l}{2}\right)$$

$$M = \frac{Wl}{12} \text{ (at center of span)}$$

$$y = \frac{3}{320}\frac{Wl^3}{EI} \text{ (at center of span)}$$

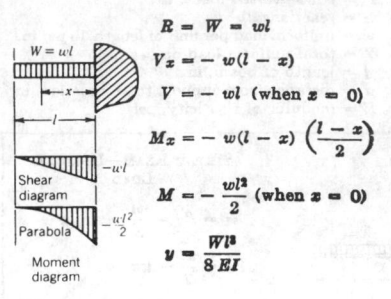

CANTILEVER BEAM—UNIFORM LOAD

$$R = W = wl$$

$$V_x = -w(l - x)$$

$$V = -wl \text{ (when } x = 0)$$

$$M_x = -w(l - x)\left(\frac{l - x}{2}\right)$$

$$M = -\frac{wl^2}{2} \text{ (when } x = 0)$$

$$y = \frac{Wl^3}{8EI}$$

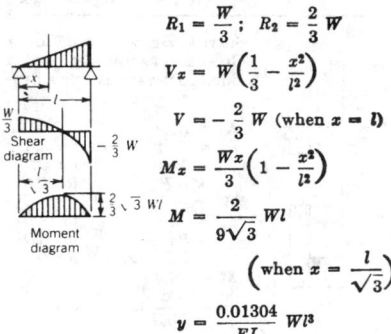

SIMPLE BEAM — LOAD INCREASING UNIFORMLY FROM ONE SUPPORT TO THE OTHER

$$R_1 = \frac{W}{3}; \quad R_2 = \frac{2}{3}W$$

$$V_x = W\left(\frac{1}{3} - \frac{x^2}{l^2}\right)$$

$$V = -\frac{2}{3}W \text{ (when } x = l)$$

$$M_x = \frac{Wx}{3}\left(1 - \frac{x^2}{l^2}\right)$$

$$M = \frac{2}{9\sqrt{3}}Wl$$

$$\left(\text{when } x = \frac{l}{\sqrt{3}}\right)$$

$$y = \frac{0.01304}{EI}Wl^3$$

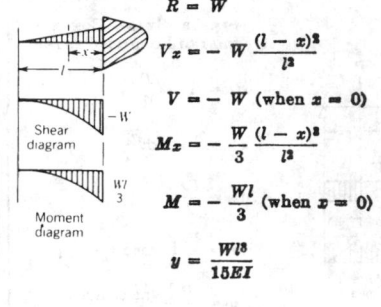

CANTILEVER BEAM—LOAD INCREASING UNIFORMLY FROM FREE END TO SUPPORT

$$R = W$$

$$V_x = -W\frac{(l - x)^2}{l^2}$$

$$V = -W \text{ (when } x = 0)$$

$$M_x = -\frac{W}{3}\frac{(l - x)^3}{l^2}$$

$$M = -\frac{Wl}{3} \text{ (when } x = 0)$$

$$y = \frac{Wl^3}{15EI}$$

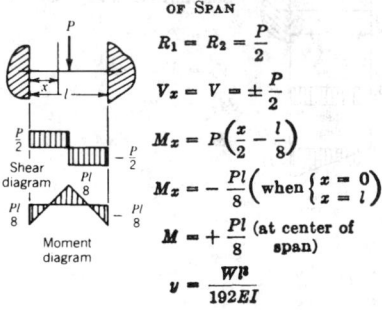

FIXED BEAM — CONCENTRATED LOAD AT CENTER OF SPAN

$$R_1 = R_2 = \frac{P}{2}$$

$$V_x = V = \pm\frac{P}{2}$$

$$M_x = P\left(\frac{x}{2} - \frac{l}{8}\right)$$

$$M_x = -\frac{Pl}{8}\left(\text{when } \begin{cases} x = 0 \\ x = l \end{cases}\right)$$

$$M = +\frac{Pl}{8} \text{ (at center of span)}$$

$$y = \frac{Wl^3}{192EI}$$

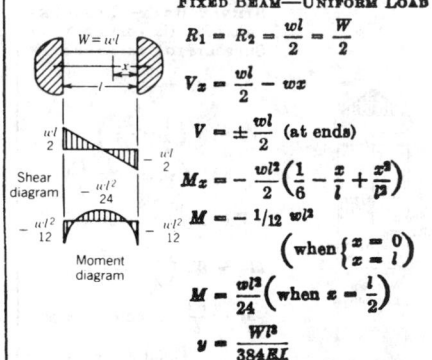

FIXED BEAM—UNIFORM LOAD

$$R_1 = R_2 = \frac{wl}{2} = \frac{W}{2}$$

$$V_x = \frac{wl}{2} - wx$$

$$V = \pm\frac{wl}{2} \text{ (at ends)}$$

$$M_x = -\frac{wl^2}{2}\left(\frac{1}{6} - \frac{x}{l} + \frac{x^2}{l^2}\right)$$

$$M = -1/12 \, wl^2$$

$$\left(\text{when } \begin{cases} x = 0 \\ x = l \end{cases}\right)$$

$$M = \frac{wl^2}{24}\left(\text{when } x = \frac{l}{2}\right)$$

$$y = \frac{Wl^3}{384EI}$$

Table 10.5 *(Continued)*

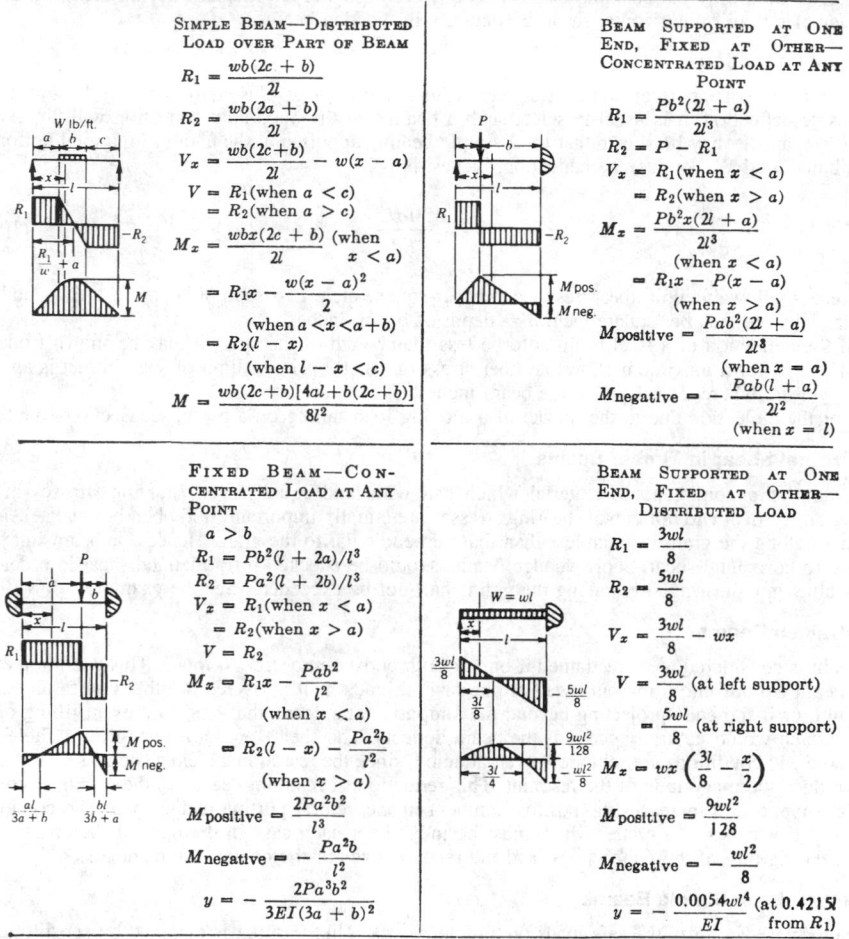

SIMPLE BEAM—DISTRIBUTED
LOAD OVER PART OF BEAM

$$R_1 = \frac{wb(2c + b)}{2l}$$

$$R_2 = \frac{wb(2a + b)}{2l}$$

$$V_x = \frac{wb(2c+b)}{2l} - w(x - a)$$

$$V = R_1 (\text{when } a < c)$$
$$= R_2 (\text{when } a > c)$$

$$M_x = \frac{wbx(2c + b)}{2l} \text{ (when } x < a)$$

$$= R_1 x - \frac{w(x - a)^2}{2}$$
$$(\text{when } a < x < a + b)$$

$$= R_2 (l - x)$$
$$(\text{when } l - x < c)$$

$$M = \frac{wb(2c+b)[4al+b(2c+b)]}{8l^2}$$

BEAM SUPPORTED AT ONE
END, FIXED AT OTHER—
CONCENTRATED LOAD AT ANY
POINT

$$R_1 = \frac{Pb^2(2l + a)}{2l^3}$$

$$R_2 = P - R_1$$

$$V_x = R_1 (\text{when } x < a)$$
$$= R_2 (\text{when } x > a)$$

$$M_x = \frac{Pb^2 x(2l + a)}{2l^3}$$
$$(\text{when } x < a)$$

$$= R_1 x - P(x - a)$$
$$(\text{when } x > a)$$

$$M_{\text{positive}} = \frac{Pab^2(2l + a)}{2l^3}$$
$$(\text{when } x = a)$$

$$M_{\text{negative}} = -\frac{Pab(l + a)}{2l^2}$$
$$(\text{when } x = l)$$

FIXED BEAM—CON-
CENTRATED LOAD AT ANY
POINT

$$a > b$$

$$R_1 = Pb^2(l + 2a)/l^3$$
$$R_2 = Pa^2(l + 2b)/l^3$$
$$V_x = R_1 (\text{when } x < a)$$
$$= R_2 (\text{when } x > a)$$
$$V = R_2$$

$$M_x = R_1 x - \frac{Pab^2}{l^2}$$
$$(\text{when } x < a)$$

$$= R_2 (l - x) - \frac{Pa^2 b}{l^2}$$
$$(\text{when } x > a)$$

$$M_{\text{positive}} = \frac{2Pa^2 b^2}{l^3}$$

$$M_{\text{negative}} = -\frac{Pa^2 b}{l^2}$$

$$y = -\frac{2Pa^3 b^2}{3EI(3a + b)^2}$$

BEAM SUPPORTED AT ONE
END, FIXED AT OTHER—
DISTRIBUTED LOAD

$$R_1 = \frac{3wl}{8}$$

$$R_2 = \frac{5wl}{8}$$

$$V_x = \frac{3wl}{8} - wx$$

$$V = \frac{3wl}{8} \text{ (at left support)}$$

$$= \frac{5wl}{8} \text{ (at right support)}$$

$$M_x = wx \left(\frac{3l}{8} - \frac{x}{2} \right)$$

$$M_{\text{positive}} = \frac{9wl^2}{128}$$

$$M_{\text{negative}} = -\frac{wl^2}{8}$$

$$y = -\frac{0.0054wl^4}{EI} \text{ (at } 0.4215l \text{ from } R_1)$$

Web Shear

A beam designed in the foregoing manner is safe against rupture of the extreme fibers due to bending in a vertical plane, and usually the cross section will have sufficient area to sustain the shearing stresses with safety. For short beams carrying heavy loads, however, the vertical shear at the supports is large, and it may be necessary to increase the area of the section to keep the unit shearing stress within the limit allowed. For steel beams, the average unit shearing stress is computed by $\tau = V/A$, where V = total vertical shear, pounds; A = area of web, square inches.

Shear Center

Closed or solid cross sections with two axes of symmetry will have a shear center at the origin. If the loads are applied here, then the bending moment can be used to calculate the deflections and bending stress, which means there are no torsional stresses. The open section or unsymmetrical section generally has a shear center that is offset on one axis of symmetry and must be calculated.[2,8,9] The load applied at this location will develop bending stresses and deflections. If any sizable torsion is developed, then torsional stresses and rotations must be accounted for.

Miscellaneous Considerations

Other considerations which will influence the choice of section under certain conditions of loading are: (1) Maximum vertical deflection that may be permitted in beams coming in contact with plaster.

(2) Danger of failure by sidewise bending in long beams, unbraced against lateral deflection. (3) Danger of failure by the buckling of the web of steel beams of short span carrying heavy loads. (4) Danger of failure by horizontal shear, particularly in wooden beams.

Vertical Deflection

If a beam is to support or come in contact with materials like plaster, which may be broken by excessive deflection, it is usual to select such a beam that the maximum deflection will not exceed ($\frac{1}{360}$ × span). It may be shown that for a simple beam, supported at the ends, with a total uniformly distributed load W, pounds, the deflection, inches, is

$$y = \frac{30\sigma L^2}{Ed} \tag{10.41}$$

where σ = allowable unit fiber stress, pounds per square inch; L = span of beam, feet; E = modulus of elasticity, pounds per square inch; d = depth of beam, inches.

If the deflection of a steel beam is to be less than $\frac{1}{360}$th of the span, it may be shown from Eq. (10.41) that, for a maximum allowable fiber stress of 18,000 psi, the limit of span in feet is approximately $1.8d$, where d = depth of the beam inches.

For the deflection due to the impact of a moving load falling on a beam, see Section 10.6.6.

Horizontal Shear in Timber Beams

In beams of a homogeneous material which can withstand equally well shearing stresses in any direction, vertical and horizontal shearing stresses are equally important. In timber, however, shearing strength along the grain is much less than that perpendicular to the grain. Hence, the beams may fail owing to horizontal shear. Short wooden beams should be checked for horizontal shear in order that allowable unit shearing stress along the grain shall not be exceeded. (See the example below.)

Restrained Beams

A beam is considered to be restrained if one or both ends are not free to rotate. This condition exists if a beam is built into a masonry wall at one or both ends, if it is riveted or otherwise fastened to a column, or if the ends projecting beyond the supports carry loads that tend to prevent tilting of the ends which would naturally occur as the beam deflects. The shears and moments give in Table 10.5 for fixed end conditions are seldom, if ever, attained, since the restraining elements themselves deform and reduce the magnitude of the restraint. This reduction of restraint decreases the negative moment at the support and increases the positive moment in the central portion of the span. The amount of restraint that exists is a matter which must be judged for each case in the light of the construction used, the rigidity of the connections, and the relative sizes of the connecting members.

Safe Loads on Simple Beams

Equation (10.42) gives the safe loads on simple beams. This formula is obtained by substituting in the flexure equation (10.36), the value of M for a simple beam, uniformly loaded, as given in Table 10.5. Let W = total load, pounds; σ = extreme fiber unit stress, pounds per square inch; S = section modulus, inches^3; L = length of span, feet. Then

$$W = \frac{2}{3} \sigma \frac{S}{L} \tag{10.42}$$

If σ is taken as a maximum allowable unit fiber stress, this equation gives the maximum allowable load on the beam. Most building codes permit a value of σ = 18,000 psi for quiescent loads on steel. For this value of σ, Eq. (10.42) becomes

$$W = \frac{12,000 \, S}{L} \tag{10.43}$$

If the load is concentrated at the center of the span, the safe load is one-half the value given by Eq. (10.43). If the load is neither uniformly distributed nor concentrated at the center of the span, the maximum bending moment must be used. The foregoing equations are for beams laterally supported and are for flexure only. The other factors which influence the strength of the beam, as shearing, buckling, etc., must also be considered.

Use of Tables in Design

The following is an example in the use of tables for the design of a wooden beam.

Example 10.1

Design a southern pine girder, of common structural grade, to carry a load of 9600 lb distributed uniformly over a 16-ft span in the interior of a building, the beam being a simple beam, freely supported at each end.

Solution. From Table 10.5, the bending moment of a simple beam uniformly loaded is $M = wl^2/8$. Since $W = wl$ and $l = 12L$,

$$M = 9600 \times 16 \times {}^{12}\!/_8 = 230,400 \text{ lb-in.}$$

If the allowable unit stress on yellow pine is 1200 psi,

$$\frac{I}{c} = \frac{230,400}{1200} = 192 \text{ in.}^3$$

From Table 10.4, the section modulus of a rectangular section is $bd^2/6$. Assume $b = 8$ in. Then $8d^2/6 = 192$, and $d = \sqrt{144} = 12.0$ in. A beam 8 by 12 in. is selected tentatively, and checked for shear.

Maximum shearing stress (horizontal and vertical) is at the neutral surface over the supports. Equation (10.34) for horizontal shear in a solid rectangular beam is $\tau = 3V/2A$; $V = 9600/2 = 4800$, and $A = 8 \times 12 = 96$, whence $\tau = (3 \times 4800)/(2 \times 96) = 75$ psi.

If the safe horizontal unit shearing stress for common-grade southern yellow pine is 88 psi, and since the actual horizontal unit shearing stress is less than 88 lb, the 8 by 12 in. beam will be satisfactory.

A **beam of uniform strength** is one in which the dimensions are such that the maximum fiber stress σ is the same throughout the length of the beam. The form of the beam is determined by finding the areas of various cross sections from the flexure formula $M = \sigma I/c$, keeping σ constant and making I/c vary with M. For a rectangular section of width b and depth d, the section modulus $I/c = \frac{1}{6}bd^2$, and, therefore, $M = \frac{1}{6}\sigma bd^2$. By making bd^2 vary with M, the dimensions of the various sections are obtained. Table 10.6 gives the dimensions b and d, at any section, the maximum unit fiber stress σ and the maximum deflection y, of some rectangular beams of uniform strength. In this table, the bending moment has been assumed to be the controlling factor. On account of the vertical shear near the ends of the beams, the area of the sections must be increased over that given by an amount necessary to keep the unit shearing stress within the allowable unit shearing stress. The discussion of beams of uniform strength, although of considerable theoretical interest, is of little practical value since the cost of fabrication will offset any economy in the use of the material. A plate girder in a bridge or a building is an approximation in practice to a steel beam of uniform strength.

10.6.3 Continuous Beams

As in simple beams, the expressions $M = \sigma I/c$ and $\tau = V/A$ govern the design and investigation of beams resting on more than two supports. In the case of continuous beams, however, the reactions cannot be obtained in the manner described for simple beams. Instead, the bending moments at the various sections must be determined, and from these values the vertical shears at the sections and the reactions at the supports may be derived.

Consider the second span of length l_2, inches, of the continuous beam (Fig. 10.19). Vertical shear V_x at any section distant x, inches, from the left support of the span is equal to the algebraic sum of all the vertical forces on one side of the section. Thus, if V_2 = vertical shear at a section to the right of, but infinitely close to, the left support, $w_2 x$ = uniform load, and ΣP_2 = sum of the concentrated loads along the distance x, applied at a distance kl_2 from the left support, k being a fraction less than unity, then

$$V_x = V_2 - w_2 x - \Sigma P_2 \qquad (10.44)$$

At any section, distant x from the left support, the bending moment is equal to the algebraic sum of the moments of all forces on one side of the section. If M_2 is the moment, pound-inches, at the support to the left,

$$M_x = M_2 + V_2 x - \frac{w_2 x^2}{2} - \Sigma P_2(x - kl_2) \qquad (10.45)$$

Assume that $x = l_2$. Then M_x becomes the moment M_3 at the next support to the right, and the expression may be written

Table 10.6 Rectangular Beams of Uniform Strength*

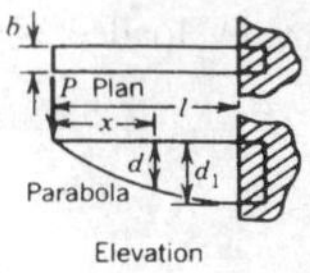 **I. CANTILEVER BEAM LOADED AT FREE END** Width is constant. Depth varies. $$d = d_1\sqrt{x/l}$$ $$\sigma = 6Pl/bd_1{}^2$$ $$y = 8Pl^3/Ebd_1{}^3$$ Elevation is formed by a straight line and a parabola with its vertex at the loaded end.	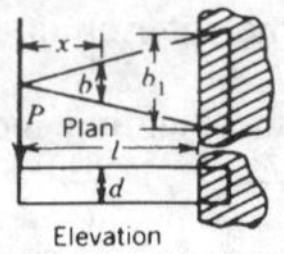 **II. CANTILEVER BEAM LOADED AT FREE END** Depth constant. Width varies. $$b = b_1 x/l$$ $$\sigma = 6Pl/b_1 d^2$$ $$y = 6Pl^3/Eb_1 d^3$$
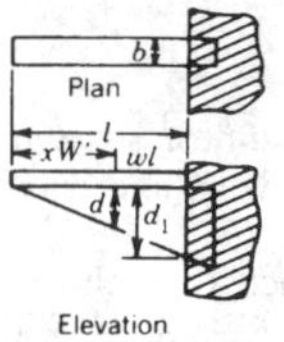 **III. CANTILEVER BEAM UNIFORMLY LOADED** Width is constant. Depth varies. $$d = (x/l)d_1$$ $$\sigma = 3wl^2/bd_1{}^2$$ $$y = 6wl^4/bEd_1{}^3$$	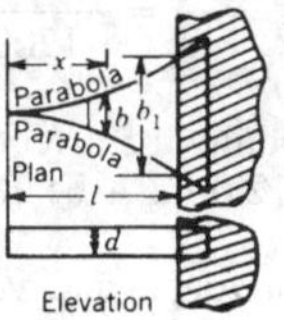 **IV. CANTILEVER BEAM UNIFORMLY LOADED** Depth is constant. Width varies. $$b = b_1 x^2/l^2$$ $$\sigma = 3wl^2/b_1 d^2$$ $$y = 3wl^4/b_1 Ed^3$$
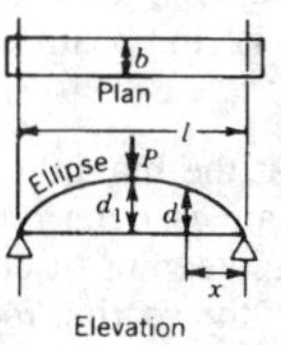 **V. SIMPLE BEAM UNIFORMLY LOADED** Width is constant. Depth varies. $$d = \sqrt{\dfrac{4d_1{}^2(lx - x^2)}{l^2}}$$ $$\sigma = \dfrac{3wl^2}{4bd_1{}^2}$$ Elevation is formed by a straight line and an ellipse.	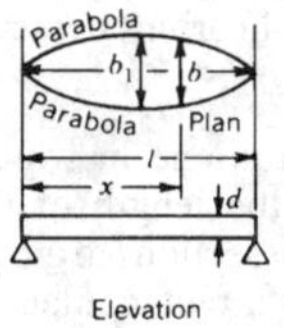 **VI. SIMPLE BEAM UNIFORMLY LOADED** Depth is constant. Width varies. $$b = \dfrac{4b_1}{l^2}(lx - x^2)$$ $$\sigma = \dfrac{3}{4}\dfrac{wl^2}{b_1 d^2}$$ Plan is two parabolas, with vertices at center of span.
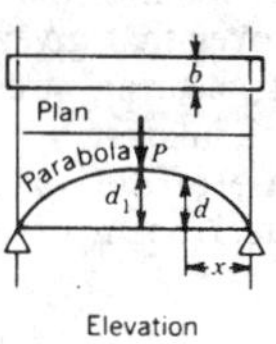 **VII. SIMPLE BEAM LOADED AT CENTER OF SPAN** Width is constant. Depth varies. $$d = d_1\sqrt{2x/l}$$ $$\sigma = \dfrac{3}{2}\dfrac{Pl}{bd_1{}^2}$$ $$y = \dfrac{1}{2}\dfrac{Pl^3}{Ebd_1{}^3}$$ Elevation is a parabola with vertices at points of support.	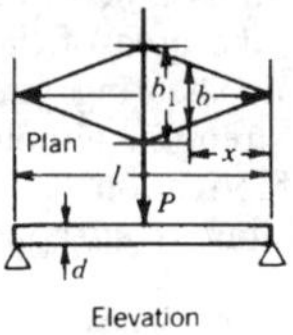 **VIII. SIMPLE BEAM LOADED AT CENTER OF SPAN** Depth is constant. Width varies. $$b = 2b_1 x/l$$ $$\sigma = \dfrac{3}{2}\dfrac{Pl}{b_1 d^2}$$ $$y = \dfrac{3}{8}\dfrac{Pl^3}{Eb_1 d^3}$$ Plan is two triangles with vertices at points of support.

* The sections of the beams near the ends must be increased over the amounts shown to resist the vertical shear expressed by the formula $\tau = 3/2 \; V/A$.

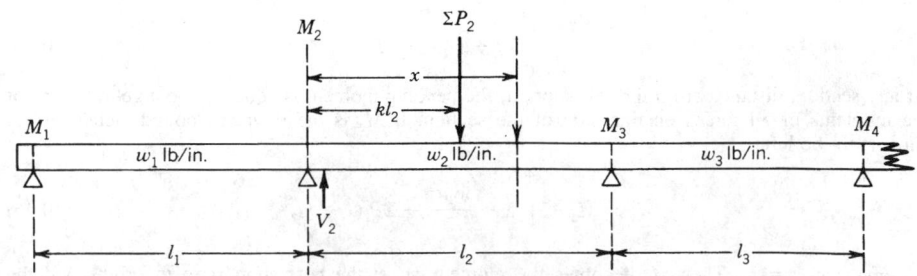

Fig. 10.19 Continuous beam.

$$V_2 l_2 = M_3 - M_2 + \frac{w_2 l_2^2}{2} + \Sigma P_2 (l_2 - k l_2) \tag{10.46}$$

From Eqs. (10.44), (10.45), and (10.46) it is evident that the bending moment M_x and the shear V_x at any section between two consecutive supports may be determined if the bending moments M_2 and M_3 at those supports are known.

To determine bending moments at the supports an expression known as the *theorem of three moments* is used. This gives the relation between the moments at any three consecutive supports of a beam. For beams with the supports on the same level, and uniformly loaded over each span, the formula is

$$M_1 l_1 + 2M_2 (l_1 + l_2) + M_3 l_2 = -\tfrac{1}{4} w_1 l_1^3 - \tfrac{1}{4} w_2 l_2^3 \tag{10.47}$$

where M_1, M_2, and M_3 = moments of three consecutive supports; l_1 = length between first and second support; l_2 = length between second and third support; w_1 = uniform load per lineal unit over the first span; w_2 = uniform load per lineal unit over the second span. When both spans are of equal length and when the load on each span is the same, $l_1 = l_2$, $w_1 = w_2$, and Eq. (10.47) reduces to

$$M_1 + 4M_2 + M_3 = -\tfrac{1}{2} w l^2 \tag{10.48}$$

which applies to most cases in practice.

Equations (10.47) and (10.48) are used as follows: For any continuous beam of n spans there are $(n + 1)$ supports. Assuming the ends of the beam to be simply supported without any overhang, the moments at the end supports are zero, and there are, therefore, to be determined $(n - 1)$ moments at the other supports. This may be done by writing $(n - 1)$ equations of the form of Eqs. (10.47) and (10.48) for each support. These equations will contain $(n - 1)$ unknown moments, and their solution will give values of M_1, M_2, M_3, etc., expressed as coefficients of $w l^2$. The shear V_1 at any support may be determined by substituting values of M_1 and M_2 in Eq. (10.46), and the bending moment at any point in any span may be obtained by Eq. (10.45). The shear at any point in any span may be determined from Eq. (10.44).

Figure 10.20 gives values and diagrams for the reactions, shears, and moments at all sections of continuous beams uniformly loaded up to five spans. Note that the reaction at any support is equal to the sum of the shears to the right and to the left of that support.

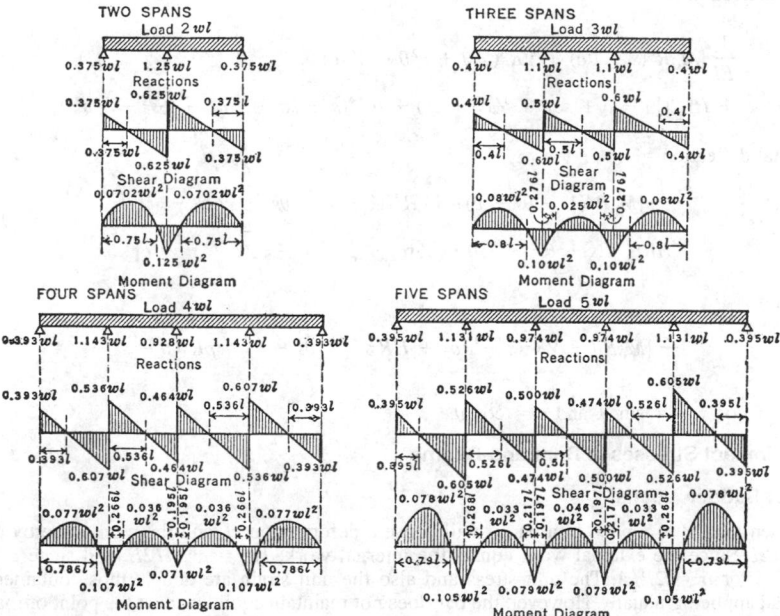

Fig. 10.20 Shear and moment diagrams of continuous beams.

10.6.4   Curved Beams

The derivation of the flexure formula, $\sigma = Mc/I$, assumes that the beam is initially straight; therefore, any deviation from this condition introduces an error in the value of the stress. If the curvature is slight, the error involved is not large, but in beams with a large amount of curvature, as hooks, chain links, and frames of punch presses, the error involved in the use of the ordinary flexure formula is considerable. The effect of the curvature is to increase the stress in the inside and to decrease it on the outside fibers of the beam and to shift the position of the neutral axis from the centroidal axis toward the concave or inner side.

The correct value for the unit fiber stress may be found by introducing a correction factor in the flexure formula, $\sigma = K(P/A \pm Mc/I)$; the factor K depends on the shape of the beam and on the ratio R/c, where R = distance, inches, from the centroidal axis of the section to the center of curvature of the central axis of the unstressed beam; and c = instance, inches, of centroidal axis from the extreme fiber on the inner or concave side. Reference 8 has an analysis of curved beams, as does Table 10.7, which gives values of K for a number of shapes and ratios of R/c. For slightly different shapes or proportions K may be found by interpolation.

Deflection of Curved and Slender Curved Beams

The deflection of curved beams,[8,9] Fig. 10.21, in the curved portion can be found by

$$U = \int \frac{1}{2} \frac{P^2}{EA} \, ds + \int \frac{\phi V^2}{GA} \, ds + \int \frac{1}{2} \frac{M^2}{EAy_oR} \, ds + \int \frac{MP}{EAR} \, ds \tag{10.49}$$

$$\frac{\partial U}{\partial Q} = \delta_Q \tag{10.50}$$

where Q is a fictitious load of a couple where the deflection or rotation is desired or can be thought of as a 1 lb load or 1 in.-lb couple. y_o is from Table 10.7, ϕ is a shape factor[2] often taken as 1, and ds is $R \, d\theta$. When $R/c > 4$, the last two terms condense to the integral of $(M^2/2EI) \, ds$. When the length of the curved portion to the depth of the beam is greater than 10, the second term of Eq. (10.49) can be dropped. When in doubt, include all terms.

When beams are not curved (Fig. 10.22), such as some clamps, the following equations (used by permission of McGraw-Hill from the 4th ed. of Ref. 2) are useful:

$$M = M_0 + HR\left[\sin(\theta - x) - x\right] - VR\left[\cos(\theta - x) - c\right] + pR^2(1 - u) \tag{10.51}$$

Vertical deflection =

$$\frac{1}{EI} [M_0 R^2(s - \theta c) + VR^3(\tfrac{1}{2}\theta + c^2\theta - \tfrac{3}{2}sc) \\ + HR^3(\tfrac{1}{2} - c + sc\theta + \tfrac{1}{2}c^2 - s^2) + pR^4(s + sc - \tfrac{3}{2}\theta c - \tfrac{1}{2}s^3 - \tfrac{1}{2}c^2s)] \tag{10.52}$$

Horizontal deflection =

$$\frac{1}{EI} [M_0 R^2(1 - \theta s - c) + VR^3(\tfrac{1}{2} - c + \theta sc + \tfrac{1}{2}c^2 - s^2) + \\ HR^3(-2s + \theta s^2 + \tfrac{1}{2}\theta + \tfrac{3}{2}sc) + pR^4(1 - \tfrac{3}{2}\theta s + s^2 - c] \tag{10.53}$$

Rotation =

$$\frac{1}{EI} [M_0 R\theta + VR^2(s - \theta c) + HR^2(1 - \theta s - c) + pR^3(\theta - s)] \tag{10.54}$$

where $u = \cos x$, $s = \sin \theta$, and $c = \cos \theta$.

10.6.5   Impact Stresses in Bars and Beams

Effect of Sudden Loads

If a sudden load P is applied to a bar, it will cause a deformation el, and the work done by the load will be Pel. Since the external work equals the internal work, $Pel = \sigma^2Al/2E$, and since $e = \sigma/E$, $P = \sigma A/2$, or $\sigma = 2P/A$. The unit stress and also the unit strain are double those obtained by an equal load applied gradually. However, the bar does not maintain equilibrium at the point of maximum stress and strain. After a series of oscillations, however, in which the surplus energy is dissipated in damping, the bar finally comes to rest with the same strain and stress as that due to the equal static load.

Table 10.7 Values of Constant K for Curved Beams

Section	R/c	Values of K Inside Fiber	Values of K Outside Fiber	Y_0*/R	Section	R/c	Values of K Inside Fiber	Values of K Outside Fiber	Y_0*/R
	1.2	3.41	.54	.224		1.2	2.89	.57	.305
	1.4	2.40	.60	.151		1.4	2.13	.63	.204
	1.6	1.96	.65	.108		1.6	1.79	.67	.149
	1.8	1.75	.68	.084		1.8	1.63	.70	.112
	2.0	1.62	.71	.069		2.0	1.52	.73	.090
	3.0	1.33	.79	.030		3.0	1.30	.81	.041
	4.0	1.23	.84	.016		4.0	1.20	.85	.021
	6.0	1.14	.89	.0070		6.0	1.12	.90	.0093
	8.0	1.10	.91	.0039		8.0	1.09	.92	.0052
	10.0	1.08	.93	.0025		10.0	1.07	.94	.0033
	1.2	3.01	.54	.336		1.2	3.09	.56	.336
	1.4	2.18	.60	.229		1.4	2.25	.62	.229
	1.6	1.87	.65	.168		1.6	1.91	.66	.168
	1.8	1.69	.68	.128		1.8	1.73	.70	.128
	2.0	1.58	.71	.102		2.0	1.61	.73	.102
	3.0	1.33	.80	.046		3.0	1.37	.81	.046
	4.0	1.23	.84	.024		4.0	1.26	.86	.024
	6.0	1.13	.88	.011		6.0	1.17	.91	.011
	8.0	1.10	.91	.0060		8.0	1.13	.94	.0060
	10.0	1.08	.93	.0039		10.0	1.11	.95	.0039
	1.2	3.14	.52	.352		1.2	3.26	.44	.361
	1.4	2.29	.54	.243		1.4	2.39	.50	.251
	1.6	1.93	.62	.179		1.6	1.99	.54	.186
	1.8	1.74	.65	.138		1.8	1.78	.57	.144
	2.0	1.61	.68	.110		2.0	1.66	.60	.116
	3.0	1.34	.76	.050		3.0	1.37	.70	.052
	4.0	1.24	.82	.028		4.0	1.27	.75	.029
	6.0	1.15	.87	.012		6.0	1.16	.82	.013
	8.0	1.12	.91	.0060		8.0	1.12	.86	.0060
	10.0	1.11	.93	.0039		10.0	1.09	.88	.0039
	1.2	3.63	.58	.418		1.2	3.55	.67	.409
	1.4	2.54	.63	.299		1.4	2.48	.72	.292
	1.6	2.14	.67	.229		1.6	2.07	.76	.224
	1.8	1.89	.70	.183		1.8	1.83	.78	.178
	2.0	1.73	.72	.149		2.0	1.69	.80	.144
	3.0	1.41	.79	.069		3.0	1.38	.86	.067
	4.0	1.29	.83	.040		4.0	1.26	.89	.038
	6.0	1.18	.88	.018		6.0	1.15	.92	.018
	8.0	1.13	.91	.010		8.0	1.10	.94	.010
	10.0	1.10	.92	.0065		10.0	1.08	.95	.0065
	1.2	2.52	.67	.408		1.2	2.37	.73	.453
	1.4	1.90	.71	.285		1.4	1.79	.77	.319
	1.6	1.63	.75	.208		1.6	1.56	.79	.236
	1.8	1.50	.77	.160		1.8	1.44	.81	.183
	2.0	1.41	.79	.127		2.0	1.36	.83	.147
	3.0	1.23	.86	.058		3.0	1.19	.88	.067
	4.0	1.16	.89	.030		4.0	1.13	.91	.036
	6.0	1.10	.92	.013		6.0	1.08	.94	.016
	8.0	1.07	.94	.0076		8.0	1.06	.95	.0089
	10.0	1.05	.95	.0048		10.0	1.05	.96	.0057
	1.2	3.28	.58	.269		1.2	2.63	.68	.399
	1.4	2.31	.64	.182		1.4	1.97	.73	.280
	1.6	1.89	.68	.134		1.6	1.66	.76	.205
	1.8	1.70	.71	.104		1.8	1.51	.78	.159
	2.0	1.57	.73	.083		2.0	1.43	.80	.127
	3.0	1.31	.81	.038		3.0	1.23	.86	.058
	4.0	1.21	.85	.020		4.0	1.15	.89	.031
	6.0	1.13	.90	.0087		6.0	1.09	.92	.014
	8.0	1.10	.92	.0049		8.0	1.07	.94	.0076
	10.0	1.07	.93	.0031		10.0	1.06	.95	.0048

* Y_0 is distance from centroidal axis to neutral axis, where beam is subjected to pure bending.

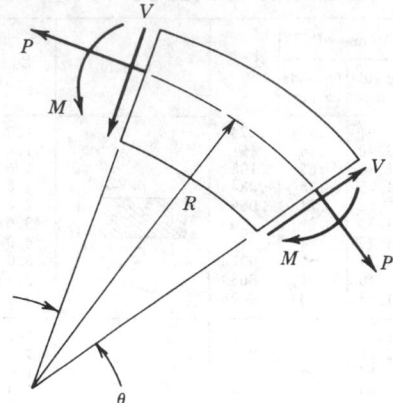

Fig. 10.21 Positive sign convention for curved beams.

Stress Due to Live Loads

In structural design two loads are considered, the dead load or weight of the structure and the live load or superimposed loads to be carried. The stresses due to the dead load and to the live load are computed separately, each being regarded as a static load. It is obvious that the stress due to the live load may be greatly increased, depending on the suddenness with which the load is applied. It has been shown above that the stress due to a suddenly applied load is double the stress caused by a static load. The term *coefficient of impact* is used extensively in structural engineering to denote the number by which the computed static stress is multiplied to obtain the value of the increased stress assumed to be caused by the suddenness of the application of the live load. If σ = static unit stress computed from the live load, and i = coefficient of impact, then the increase of unit stress due to sudden loading is $i\sigma$, and the total unit stress due to live load is $\sigma + i\sigma$. The value of i has been determined by empirical methods and varies according to different conditions.

In the building codes of most cities, specified floor loadings for buildings include the impact allowance, and no increase is needed for live loads except for special cases of vibration or other unusual conditions. For railroad bridges, the value of i depends upon the proportion of the length of the bridge which is loaded. No increase in the static stress is needed when the mass of the structure, as in monolithic concrete, is great. For machinery and for unusual conditions, such as elevator machinery and its supports, each structure should be considered by itself and the coefficient assumed accordingly. It should be noted that the meaning of the word impact used above differs somewhat from its strict theoretical meaning and as it is used in the next paragraph. The use of the terms impact and coefficient of impact in connection with live load stresses is, however, very general.

Axial Impact on Bars

A load P dropped from a height h onto the end of a vertical bar of cross-sectional area A, rigidly secured at the bottom end, produces in the bar a unit stress which increases from 0 up to σ', with a corresponding total strain increasing from 0 up to e_1. The work done on the bar is $P(h + e_1)$, which, provided no energy is expended in hysteresis losses or in giving velocity to the bar, is equal to the energy $\frac{1}{2}\sigma' A e_1$ stored in the bar; that is,

$$P(h + e_1) = \frac{1}{2}\sigma' A e_1 \tag{10.55}$$

If e = strain produced by a static load P, within the proportional limit

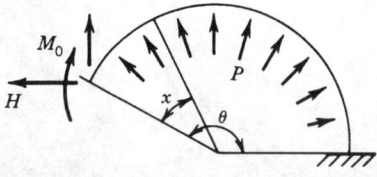

Fig. 10.22 Circular cantilever with end loading and uniform radial pressure p lb/linear in.

$$\frac{e}{e_1} = \frac{P/A}{\sigma'} \tag{10.56}$$

Combining this with Eq. (10.55) gives

$$\sigma' = \sigma + \sigma \sqrt{1 + 2\frac{h}{e}} \tag{10.57}$$

$$e_1 = e + e \sqrt{1 + 2\frac{h}{e}} \tag{10.58}$$

A wrought-iron bar 1 in. square and 5 ft long under a static load of 5000 lb will be shortened about 0.012 in., assuming no lateral flexure to occur; but, if a weight of 5000 lb drops on its end from a height of 0.048 in., a stress of 20,000 lb will be produced.

Equations (10.57) and (10.58) give values of stress and strain that are somewhat high because part of the energy of the applied force is not effective in producing stress, but is expended in overcoming the inertia of the bar and in producing local stresses. For light bars they give approximately correct results.

If the bar is horizontal and is struck at one end by a weight P, moving with a velocity V, the strain produced is e_1. Then, as before, $\frac{1}{2}\sigma'Ae_1 = Ph$. In this case $h = V^2/2g$ = height from which P would have to fall to acquire velocity V (g = acceleration due to gravity = 32.16 ft/sec^2). Combining with Eq. (10.56),

$$\sigma' = \sigma \sqrt{2\frac{h}{e}} \tag{10.59}$$

$$e_1 = e \sqrt{2\frac{h}{e}} \tag{10.60}$$

Impact on Beams

If a weight P falls on a horizontal beam from a height h, producing a maximum deflection y and a maximum unit stress σ' in the extreme fiber, the values of σ' and y are given by

$$\sigma' = \sigma + \sigma \sqrt{1 + 2\frac{h}{y}} \tag{10.61}$$

$$y_1 = y + y \sqrt{1 + 2\frac{h}{y}} \tag{10.62}$$

where σ = extreme fiber unit stress and y = deflection due to P, considered as a static load. The value of σ may be obtained from the flexure formula [Eq. (10.37)]; that of y from the proper formula for deflection under static load.

If a weight P moving horizontally with a velocity V strikes a beam (the ends of which are secured against horizontal movement), the maximum fiber unit stress and the maximum lateral deflection are given by

$$\sigma' = \sigma \sqrt{2\frac{h}{y}} \tag{10.63}$$

$$y_1 = y \sqrt{2\frac{h}{y}} \tag{10.64}$$

where σ and y are as before and h is height through which P would have to fall to acquire the velocity V. These formulas, like those for axial impact on bars, give results higher than those observed in tests, particularly if the weight of the beam is great. For further discussion, see Ref. 7.

Rupture from Impact

Rupture may be caused by impact provided the load has the requisite velocity. The above formulas, however, do not apply since they are valid only for stresses within the proportional limit. It has been found that the dynamic properties of a material are dependent on volume, velocity of the applied load, and material condition. If the velocity of the applied load is kept within certain limiting values, the total energy values for static and dynamic conditions are identical. If the velocity is increased, the impact values are considerably reduced. For further information, see Ref. 10.

10.6.6 Steady and Impulsive Vibratory Stresses

For steady vibratory stresses of a weight, W, supported by a beam or rod, the deflection of the bar, or beam, will be increased by the dynamic magnification factor. The relation is given by

$$\delta_{\text{dynamic}} = \delta_{\text{static}} \times \text{dynamic magnification factor}$$

An example of the calculating procedure for the case of no damping losses is

$$\delta_{\text{dynamic}} = \delta_{\text{static}} \times \frac{1}{1 - (\omega/\omega_n)^2} \tag{10.65}$$

where ω is the frequency of oscillation of the load and ω_n is the natural frequency of oscillation of a weight on the bar.

For the same beam excited by a single sine pulse of magnitude A in./sec^2 and a sec duration, then for $t < a$ a good approximation is

$$\sigma_{\text{dynamic}} = \frac{\delta_{\text{static}}(A/g)}{1 - \left(\dfrac{\omega}{4\pi\omega_n}\right)^2}\left[\sin \omega t - \frac{1}{4\pi^2}\left(\frac{\omega}{\omega_n}\right)\sin \omega_n t\right] \tag{10.66}$$

where A/g is the number of g's and ω is π/a.

10.7 SHAFTS, BENDING, AND TORSION

10.7.1 Definitions

TORSIONAL STRESS. A bar is under torsional stress when it is held fast at one end, and a force acts at the other end to twist the bar. In a round bar (Fig. 10.23) with a constant force acting, the straight line ab becomes the helix ad, and a radial line in the cross section, ob, moves to the position od. The angle bad remains constant while the angle bod increases with the length of the bar. Each cross section of the bar tends to shear off the one adjacent to it, and in any cross section the shearing stress at any point is normal to a radial line drawn through the point. Within the shearing proportional limit, a radial line of the cross section remains straight after the twisting force has been applied, and the unit shearing stress at any point is proportional to its distance from the axis.

TWISTING MOMENT, T, is equal to the product of the resultant, P, of the twisting forces, multiplied by its distance from the axis, p.

RESISTING MOMENT, T_r, in torsion, is equal to the sum of the moments of the unit shearing stresses acting along a cross section with respect to the axis of the bar. If dA is an elementary area of the section at a distance of z units from the axis of a circular shaft (Fig. 10.23b), and c is the distance from the axis to the outside of the cross section where the unit shearing stress is τ, then the unit shearing stress acting on dA is $(\tau z/c)\,dA$, its moment with respect to the axis is $(\tau z^2/c)\,dA$, an the sum of all the moments of the unit shearing stresses on the cross section is $\int (\tau z^2/c)\,dA$. In

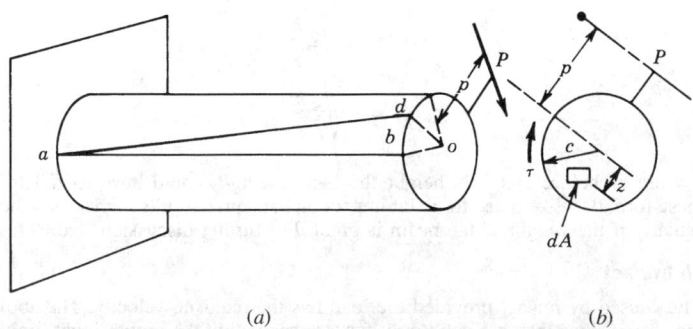

(a)   (b)

Fig. 10.23 Round bar subject to torsional stress.

this expression the factor $\int z^2\, dA$ is the polar moment of inertia of the section with respect to the axis. Denoting this by J, the resisting moment may be written $\tau J/c$.

THE POLAR MOMENT OF INERTIA of a surface about an axis through its center of gravity and perpendicular to the surface is the sum of the products obtained by multiplying each elementary area by the square of its distance from the center of gravity of its surface; it is equal to the sum of the moments of inertia taken with respect to two axes in the plane of the surface at right angles to each other passing through the center of gravity. It is represented by J, inches^4. For the cross section of a round shaft,

$$J = \tfrac{1}{32}\pi d^4 \quad \text{or} \quad \tfrac{1}{2}\pi r^4 \tag{10.67}$$

For a hollow shaft,

$$J = \tfrac{1}{32}\pi(d^4 - d_1^4) \tag{10.68}$$

where d is the outside and d_1 is the inside diameter, inches, or

$$J = \tfrac{1}{2}\pi(r^4 - r_1^4) \tag{10.69}$$

where r is the outside and r_1 the inside radius, inches.

THE POLAR RADIUS OF GYRATION, k_p, sometimes is used in formulas; it is defined as the radius of a circumference along which the entire area of a surface might be concentrated and have the same polar moment of inertia as the distributed area. For a solid circular section,

$$k_p^2 = \tfrac{1}{8}d^2 \tag{10.70}$$

For a hollow circular section,

$$k_p^2 = \tfrac{1}{8}(d^2 - d_1^2) \tag{10.71}$$

10.7.2 Determination of Torsional Stresses in Shafts

Torsion Formula for Round Shafts

The conditions of equilibrium require that the twisting moment, T, be opposed by an equal resisting moment, T_r, so that for the values of the maximum unit shearing stress, τ, within the proportional limit, the torsion formula for round shafts becomes

$$T_r = T = \tau\frac{J}{c} \tag{10.72}$$

if τ is in pounds per square inch, then T_r and T must be in pound-inches, J is in inches^4, and c is in inches. For solid round shafts having a diameter, d, inches,

$$J = \tfrac{1}{32}\pi d^4 \quad \text{and} \quad c = \tfrac{1}{2}d \tag{10.73}$$

and

$$T = \tfrac{1}{16}\pi d^3 \tau \quad \text{or} \quad \tau = \frac{16T}{\pi d^3} \tag{10.74}$$

For hollow round shafts,

$$J = \frac{\pi(d^4 - d_1^4)}{32} \quad \text{and} \quad c = \tfrac{1}{2}d \tag{10.75}$$

and the formula becomes

$$T = \frac{\tau\pi(d^4 - d_1^4)}{16d} \quad \text{or} \quad \tau = \frac{16Td}{\pi(d^4 - d_1^4)} \tag{10.76}$$

The torsion formula applies only to solid circular shafts or hollow circular shafts, and then only when the load is applied in a plane perpendicular to the axis of the shaft and when the shearing proportional limit of the material is not exceeded.

Shearing Stress in Terms of Horsepower

If the shaft is to be used for the transmission of power, the value of T, pound-inches, in the above formulas becomes $63,030H/N$, where H = horsepower to be transmitted and N = revolutions per minute. The maximum unit shearing stress, pounds per square inch, then is

$$\text{For solid round shafts:} \qquad \tau = \frac{321,000H}{Nd^3} \tag{10.77}$$

$$\text{For hollow round shafts:} \qquad \tau = \frac{321,000Hd}{N(d^4 - d_1^4)} \tag{10.78}$$

If τ is taken as the allowable unit shearing stress, the diameter, d, inches, necessary to transmit a given horsepower at a given shaft speed can then be determined. These formulas give the stress due to torsion only, and allowance must be made for any other loads, as the weight of shaft and pulley, and tension in belts.

Angle of Twist

When the unit shearing stress τ does not exceed the proportional limit, the angle bod (Fig. 10.23) for a solid round shaft may be computed from the formula

$$\theta = \frac{Tl}{GJ} \tag{10.79}$$

where θ = angle in radians; l = length of shaft in inches; G = shearing modulus of elasticity of the material; T = twisting moment, pound-inches. Values of G for different materials are steel, 12,000,000; wrought iron, 10,000,000; and cast iron, 6,000,000.

When the angle of twist on a section begins to increase in a greater ratio than the twisting moment, it may be assumed that the shearing stress on the outside of the section has reached the proportional limit. The shearing stress at this point may be determined by substituting the twisting moment at this instant in the torsion formula.

Torsion of Noncircular Cross Sections

The analysis of shearing stress distribution along noncircular cross sections of bars under torsion is complex. By drawing two lines at right angles through the center of gravity of a section before twisting, and observing the angular distortion after twisting, it as been found from many experiments that in noncircular sections the shearing unit stresses are not proportional to their distances from the axis. Thus in a rectangular bar there is no shearing stress at the corners of the sections, and the stress at the middle of the wide side is greater than at the middle of the narrow side. In an elliptical bar the shearing stress is greater along the flat side than at the round side.

It has been found by tests[5,11] as well as by mathematical analysis that the torsional resistance of a section, made up of a number of rectangular parts, is approximately equal to the sum of the resistances of the separate parts. It is on this basis that nearly all the formulas for noncircular sections have been developed. For example, the torsional resistance of an I-beam is approximately equal to the sum of the torsional resistances of the web and the outstanding flanges. In an I-beam in torsion the maximum shearing stress will occur at the middle of the side of the web, except where the flanges are thicker than the web, and then the maximum stress will be at the midpoint of the width of the flange. Reentrant angles, as those in I-beams and channels, are always a source of weakness in members subjected to torsion. Table 10.8 gives values of the maximum unit shearing stress τ and the angle of twist θ induced by twisting bars of various cross sections, it being assumed that τ is not greater than the proportional limit.

Torsion of thin-wall closed sections, Fig. 10.24,

$$T = 2qA \tag{10.80}$$

$$q = \tau t \tag{10.81}$$

$$\theta_i = \frac{\theta}{L} = \frac{T}{2A} \frac{1}{2AG} \frac{S}{t} = \frac{T}{GJ} \tag{10.82}$$

where S is the arc length around area A over which τ acts for a thin-wall section; shear buckling should be checked. When more than one cell is used[1,12] or if section is not constructed of a single material,[12] the calculations become more involved:

$$J = \frac{4A^2}{\oint ds/t} \tag{10.83}$$

Table 10.8 Formulas for Torsional Deformation and Stress

General formulas: $\theta = \dfrac{TL}{KG}$, $\tau = \dfrac{T}{Q}$, where θ = angle of twist, radians; T = twisting moment, in.-lb; L = length, in.; τ = unit shear stress, psi; G = modulus of rigidity, psi; K, in.^{4}; and Q, in.^{3} are functions of the cross section.

Shape	Formula for K in $\theta = \dfrac{TL}{KG}$	Formula for Shear Stress
	$K = \dfrac{\pi d^4}{32}$	$\tau = \dfrac{16T}{\pi d^3}$
	$K = \frac{1}{32}\pi(d^4 - d_1{}^4)$	$\tau = \dfrac{16Td}{\pi(d^4 - d_1{}^4)}$
	$K = \frac{2}{3}\,\pi r t^3$	$\tau = \dfrac{3T}{2\pi r t^2}$
	$K = \dfrac{\pi a^3 b^3}{a^2 + b^2}$	$\tau = \dfrac{2T}{\pi a b^2}$
	$K = \dfrac{\pi a_1{}^3 b_1{}^3}{a_1{}^2 + b_1{}^2}[(1+q)^4 - 1]$ $q = \dfrac{a - a_1}{a_1}$ $q = \dfrac{b - b_1}{b_1}$	$\tau = \dfrac{2T}{\pi a_1 b_1{}^2[(1+q)^4 - 1]}$
	$K = \dfrac{b^4\sqrt{3}}{80}$	$\tau = \dfrac{20T}{b^3}$
	$K = 2.69b^4$	$\tau = \dfrac{1.09T}{b^3}$
	$K = \dfrac{ab^3}{16}\left[\dfrac{16}{3} - 3.36\dfrac{b}{a}\left(1 - \dfrac{b^4}{12a^4}\right)\right]$	$\tau = \dfrac{(3a + 1.8b)\,T}{a^2 b^2}$
	$K = \dfrac{2t_1 t_2(a - t_2)^2(b - t_1)^2}{at_2 + bt_1 - t_2{}^2 - t_1{}^2}$	$\tau = \dfrac{T}{2t_2(a - t_2)(b - t_1)}$
	$K = 0.1406b^4$	$\tau = \dfrac{4.8T}{b^3}$

Table 10.8 *(Continued)*

General formulas: $\theta = \dfrac{TL}{KG}$, $\tau = \dfrac{T}{Q}$, where θ = angle of twist, radians; T = twisting moment, in.-lb; L = length, in.; τ = unit shear stress, psi; G = modulus of rigidity, psi; K, in.^{4}; and Q, in.^{3} are functions of the cross section.

Shape	Formula for K in $\theta = \dfrac{TL}{KG}$	Formula for Shear Stress
	r = fillet radius D = diameter largest inscribed circle $K = 2K_1 + K_2 + 2\alpha D^4$ $K_1 = ab^3\left[\dfrac{1}{3} - 0.21\dfrac{b}{a}\left(1 - \dfrac{b^4}{12a^4}\right)\right]$ $K_2 = cd^3\left[\dfrac{1}{3} - 0.105\dfrac{d}{c}\left(1 - \dfrac{d^4}{192c^4}\right)\right]$ $\alpha = \dfrac{b}{d}\left(0.07 + 0.076\dfrac{r}{b}\right)$	For all solid sections of irregular form the maximum shear stress occurs at or very near one of the points where the largest inscribed circle touches the boundary, and of these, at the one where the curvature of the boundary is algebraically least. (Convexity represents positive, concavity negative, curvature of the boundary.) At a point where the curvature is positive (boundary of section straight or convex) this maximum stress is given approximately by:
	$K = 2K_1 + K_2 + 2\alpha D^4$ $K_1 = ab^3\left[\dfrac{1}{3} - 0.21\dfrac{b}{a}\left(1 - \dfrac{b^4}{12a^4}\right)\right]$ $K_2 = 1/3\,cd^3$ $\alpha = \dfrac{t}{t_1}\left(0.15 + 0.1\dfrac{r}{b}\right)$ $t = b$ if $b < d$ $t = d$ if $d < b$ $t_1 = b$ if $b > d$ $t_1 = d$ if $d > b$	$\tau = G\dfrac{\theta}{L}c$ or $\tau = \dfrac{T}{K}c$ where $c = \dfrac{D}{1 + \dfrac{\pi^2 D^4}{16A^2}} \times$ $\left[1 + 0.15\left(\dfrac{\pi^2 D^4}{16A^2} - \dfrac{D}{2r}\right)\right]$
	$K = K_1 + K_2 + \alpha D^4$ $K_1 = ab^3\left[\dfrac{1}{3} - 0.21\dfrac{b}{a}\left(1 - \dfrac{b^4}{12a^4}\right)\right]$ $K_2 = cd^3\left[\dfrac{1}{3} - 0.105\dfrac{d}{c}\left(1 - \dfrac{d^4}{192c^4}\right)\right]$ $\alpha = \dfrac{b}{d}\left(0.07 + 0.076\dfrac{r}{b}\right)$	where D = diameter of largest inscribed circle, r = radius of curvature of boundary at the point (positive for this case), A = area of the section.

Ultimate Strength in Torsion

In a torsion failure, the outer fibers of a section are the first to shear, and the rupture extends toward the axis as the twisting is continued. The torsion formula for round shafts has no theoretical basis after the shearing stresses on the outer fibers exceed the proportional limit, as the stresses along the section then are no longer proportional to their distances from the axis. It is convenient, however, to compare the torsional strength of various materials by using the formula to compute values of τ at which rupture takes place. These computed values of the maximum stress sustained before rupture are somewhat higher for iron and steel than the ultimate strength of the materials in direct shear. Computed values of the ultimate strength in torsion are found by experiment to be: cast iron, 30,000 psi; wrought iron, 55,000 psi; medium steel, 65,000 psi; timber, 2000 psi. These computed values of twisting strength may be used in the torsion formula to determine the probable twisting moment that will cause rupture of a given round bar or to determine the size of a bar that will be ruptured by a given twisting moment. In design, large factors of safety should be taken, especially when the stress is reversed as in reversing engines and when the torsional stress is combined with other stresses as in shafting.

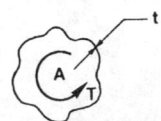

Fig. 10.24 Thin-walled tube.

10.7.3 Bending and Torsional Stresses

The stress for combined bending and torsion can be found from Eqs. (10.20), shear theory, and (10.22), distortion energy, with $\sigma_y = 0$:

$$\frac{\sigma_w}{2} = \sqrt{\left(\frac{Mc}{2I}\right)^2 + \left(\frac{Tr}{J}\right)^2} \tag{10.84}$$

For solid round rods, this equation reduces to

$$\frac{\sigma_w}{2} = \frac{16}{\pi d^3} \sqrt{M^2 + T^2} \tag{10.85}$$

From distortion energy

$$\sigma = \sqrt{\left(\frac{Mc}{I}\right)^2 + 3\left(\frac{Tr}{J}\right)^2} \tag{10.86}$$

For solid round rods, the equation yields

$$\sigma = \frac{32}{\pi d^3} \sqrt{M^2 + \tfrac{3}{4}T^2} \tag{10.87}$$

10.8 COLUMNS

10.8.1 Definitions

A COLUMN OR STRUT is a bar or structural member under axial compression, which has an unbraced length greater than about eight or ten times the least dimension of its cross section. On account of its length, it is impossible to hold a column in a straight line under a load; a slight sidewise bending always occurs, causing flexural stresses in addition to the compressive stresses induced directly by the load. The lateral deflection will be in a direction perpendicular to that axis of the cross section about which the moment of inertia is the least. Thus in Fig. 10.25a the column will bend in a direction perpendicular to aa, in Fig. 10.25b it will bend perpendicular to aa or bb, and in Fig. 10.25c it is likely to bend in any direction.

RADIUS OF GYRATION of a section with respect to a given axis is equal to the square root of the quotient of the moment of inertia with respect to that axis, divided by the area of the section, that is

$$k = \sqrt{\frac{I}{A}}; \qquad \frac{I}{A} = k^2 \tag{10.88}$$

where I is the moment of inertia and A is the sectional area. Unless otherwise mentioned, an axis

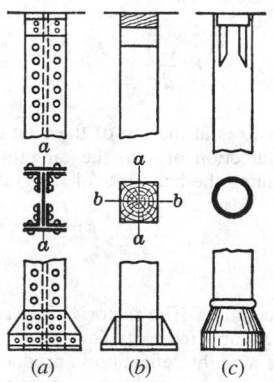

(a)   (b)   (c)

Fig. 10.25 Column end designs.

through the center of gravity of the section is the axis considered. As in beams, the moment of inertia is an important factor in the ability of the column to resist bending, but for purposes of computation it is more convenient to use the radius of gyration.

LENGTH OF A COLUMN is the distance between points unsupported against lateral deflection.

SLENDERNESS RATIO is the length l divided by the least radius of gyration k, both in inches. For steel, a *short column* is one in which $l/k < 20$ or 30, and its failure under load is due mainly to direct compression; in a *medium-length column*, $l/k =$ about 30–175, failure is by a combination of direct compression and bending; in a *long column*, $l/k >$ about 175–200, failure is mainly by bending. For timber columns these ratios are about 0–30, 30–90, and above 90 respectively. The load which will cause a column to fail decreases as l/k increases. The above ratios apply to round-end columns, If the ends are fixed (see below), the effective slenderness ratio is one-half that for round-end columns, as the distance between the points of inflection is one-half of the total length of the column. For flat ends it is intermediate between the two.

CONDITIONS OF ENDS. The various conditions which may exist at the ends of columns usually are divided into four classes: (1) Columns with round ends; the bearing at either end has perfect freedom of motion, as there would be with a ball-and-socket joint at each end. (2) Columns with hinged ends; they have perfect freedom of motion at the ends in one plane, as in compression members in bridge trusses where loads are transmitted through end pins. (3) Columns with flat ends; the bearing surface is normal to the axis of the column and of sufficient area to give at least partial fixity to the ends of the columns against lateral deflection. (4) Columns with fixed ends; the ends are rigidly secured, so that under any load the tangent to the elastic curve at the ends will be parallel to the axis in its original position.

Experiments prove that columns with fixed ends are stronger than columns with flat, hinged, or round ends, and that columns with round ends are weaker than any of the other types. Columns with hinged ends are equivalent to those with round ends in the plane in which they have free movement; columns with flat ends have a value intermediate between those with fixed ends and those with round ends. If often happens that columns have one end fixed and one end hinged, or some other combination. Their relative values may be taken as intermediate between those represented by the condition at either end. The extent to which strength is increased by fixing the ends depends on the length of column, fixed ends having a greater effect on long columns than on short ones.

10.8.2 Theory

There is no exact theoretical formula that gives the strength of a column of any length under an axial load. Formulas involving the use of empirical coefficients have been deduced, however, and they give results that are consistent with the results of tests.

Euler's Formula

Euler's formula assumes that the failure of a column is due solely to the stresses induced by sidewise bending. This assumption is not true for short columns, which fail mainly by direct compression, nor is it true for columns of medium length. The failure in such cases is by a combination of direct compression and bending. For columns in which $l/k > 200$, Euler's formula is approximately correct and agrees closely with the results of tests.

Let $P =$ axial load, pounds; $l =$ length of column, inches; $I =$ least moment of inertia, inches^4; $k =$ least radius of gyration, inches; $E =$ modulus of elasticity; $y =$ lateral deflection, inches, at any point along the column, that is caused by load P. If a column has round ends, so that the bending is not restrained, the equation of its elastic curve is

$$EI \frac{d^2y}{dx^2} = -Py \tag{10.89}$$

when the origin of the coordinate axes is at the top of the column, the positive direction of x being taken downward and the positive direction of y in the direction of the deflection. Integrating the above expression twice and determining the constants of integration give

$$P = \Omega \pi^2 \frac{EI}{l^2} \tag{10.90}$$

which is Euler's formula for long columns. The factor Ω is a constant depending on the condition of the ends. For round ends $\Omega = 1$; for fixed ends $\Omega = 4$; for one end round and the other fixed $\Omega = 2.05$. P is the load at which, if a slight deflection is produced, the column will not return to its original position. If P is decreased, the column will approach its original position, but if P is increased, the deflection will increase until the column fails by bending.

For columns with value of l/k less than about 150, Euler's formula gives results distinctly higher than those observed in tests. Euler's formula is now little used except for long members and as a basis for the analysis of the stresses in some types of structural and machine parts. It always gives an *ultimate* and never an allowable load.

Secant Formula

The deflection of the column is used in the derivation of the Euler formula, but if the load were truly axial it would be impossible to compute the deflection. If the column is assumed to have an initial eccentricity of load of e in. (see Ref. 7, for suggested values of e), the equation for the deflection y becomes

$$y_{max} = e \left(\sec \frac{l}{2} \sqrt{\frac{P}{EI}} - 1 \right)$$ (10.91)

The maximum unit compressive stress becomes

$$\sigma = \frac{P}{A} \left(1 + \frac{ec}{k^2} \sec \frac{l}{2} \sqrt{\frac{P}{EI}} \right)$$ (10.92)

where l = length of column, inches; P = total load, pounds; A = area, square inches; I = moment of inertia, inches⁴; k = radius of gyration, inches; c = distance from neutral axis to the most compressed fiber, inches; E = modulus of elasticity; both I and k are taken with respect to the axis about which bending takes place. The ASCE indicates $ec/k^2 = 0.25$ for central loading. Because the formula contains the secant of the angle $(l/2) \sqrt{P/EI}$, it is sometimes called the *secant formula*. It has been suggested by the Committee on Steel-Column Research[13,14] that the best rational column formula can be constructed on the secant type, although of course it must contain experimental constants.

The secant formula can be used also for columns that are eccentrically loaded, if e is taken as the actual eccentricity plus the assumed initial eccentricity.

Eccentric Loads on Short Compression Members

Where a direct push acting on a member does not pass through the centroid but at a distance e, inches, from it, both direct and bending stresses are produced. For short compression members in which column action may be neglected, the direct unit stress is P/A, where P = total load, pounds, and A = area of cross section, square inches. The bending unit stress is Mc/I, where $M = Pe$ = bending moment, pound-inches; c = distance, inches, from the centroid to the fiber in which the stress is desired; I = moment of inertia, inches⁴. The total unit stress at any point in the section is $\sigma = P/A + Pec/I$, or $\sigma = (P/A)(1 + ec/k^2)$, since $I = Ak^2$, where k = radius of gyration, inches.

Eccentric Loads on Columns

Various column formulas must be modified when the loads are not balanced, that is, when the resultant of the loads is not in line with the axis of the column. If P = load, pounds, applied at a distance e in. from the axis, bending moment $M = Pe$. Maximum unit stress σ, pounds per square inch, due to this bending moment alone, is $\sigma = Mc/I = Pec/Ak^2$, where c = distance, inches, from the axis to the most remote fiber on the concave side; A = sectional area in square inches; k = radius of gyration in the direction of the bending, inches. This unit stress must be added to the unit stress that would be induced if the resultant load were applied in line with the axis of the column.

The secant formula, Eq. (10.92), also can be used for columns that are eccentrically loaded if e is taken as the actual eccentricity plus the assumed initial eccentricity.

Column Subjected to Transverse or Cross-Bending Loads

A compression member that is subjected to cross-bending loads may be considered to be (1) a beam subjected to end thrust or (2) a column subjected to cross-bending loads, depending on the relative magnitude of the end thrust and cross-bending loads, and on the dimensions of the member. The various column formulas may be modified so as to include the effect of cross-bending loads. In this form the modified secant formula for transverse loads is

$$\sigma = \frac{P}{A} \left[1 + (e + y) \frac{c}{k^2} \sec \frac{l}{2k} \sqrt{\frac{P}{AE}} \right] + \frac{Mc}{Ak^2}$$ (10.93)

In the formula, σ = maximum unit stress on concave side, pounds per square inch; P = axial end load, pounds; A = cross-sectional area, square inches; M = moment due to cross-bending load, pound-inches; y = deflection due to cross-bending load, inches; k = radius of gyration, inches; l = length of column, inches; e = assumed initial eccentricity, inches; c = distance, inches, from axis to the most remote fiber on the concave side.

10.8.3 Wooden Columns

Wooden Column Formulas

One of the principal formulas is that formerly used by the AREA, $P/A = \sigma_1(1 - l/60d)$, where P/A = allowable unit load, pounds per square inch; σ_1 = allowable unit stress in direct compression on short blocks, pounds per square inch; l = length, inches; d = least dimension, inches. This formula is being replaced rapidly by formulas recommended by the ASTM and AREA. Committees of these societies, working with the U.S. Forest Products Laboratory, classified timber columns in three groups (ASTM Standards, 1937, D245-37):

1. *Short Columns.* The ratio of unsupported length to least dimension does not exceed 11. For these columns, the allowable unit stress should not be greater than the values given in Table 10.9 under compression parallel to the grain.

2. *Intermediate-Length Columns.* Where the ratio of unsupported length to least dimension is greater than 10, Eq. (10.94), of the fourth power parabolic type, shall be used to determine allowable unit stress, until this allowable unit stress is equal to two-thirds of the allowable unit stress for short columns.

$$\frac{P}{A} = \sigma_1 \left[1 - \frac{1}{3} \left(\frac{l}{Kd} \right)^4 \right] \tag{10.94}$$

where P = total load, pounds; A = area, square inches; σ_1 = allowable unit compressive stress parallel to grain, pounds per square inch (see Table 10.9); l = unsupported length, inches; d = least dimension, inches; $K = l/d$ at the point of tangency of the parabolic and Euler curves, at which $P/A = \frac{2}{3}\sigma_1$. The value of K for any species and grade is $\pi/2\sqrt{E/6\sigma_1}$, where E = modulus of elasticity.

3. *Long Columns.* Where P/A as computed by Eq. (10.94) is less than $\frac{2}{3}\sigma_1$, Eq. (10.95) of the Euler type, which includes a factor of safety of 3, shall be used:

$$\frac{P}{A} = \frac{1}{36} \left[\frac{\pi^2 E}{(l/d)^2} \right] \tag{10.95}$$

Timber columns should be limited to a ratio of l/d equal to 50. No higher loads are allowed for square-ended columns. The strength of round columns may be considered the same as that of square columns of the same cross-sectional area.

Use of Timber Column Formulas

The values of E (modulus of elasticity) and σ_1 (compression parallel to grain) in the above formulas are given in Table 10.9. Table 10.10 gives the computed values of K for some common types of timbers. These may be substituted directly in Eq. (10.94) for intermediate-length columns, or may be used in conjunction with Table 10.11, which gives the strength of columns of intermediate length, expressed as a percentage of strength (σ_1) of short columns. In the tables, the term "continuously dry" refers to interior construction where there is no excessive dampness or humidity; "occasionally wet but quickly dry" refers to bridges, trestles, bleachers, and grandstands; "usually wet" refers to timber in contact with the earth or exposed to waves or tidewater.

10.8.4 Steel Columns

Types

Two general types of steel columns are in use: (1) rolled shapes and (2) built-up sections. The rolled shapes are easily fabricated, accessible for painting, neat in appearance where they are not covered, and convenient in making connections. A disadvantage is the probability that thick sections are of lower-strength material than thin sections because of the difficulty of adequately rolling the thick material. For the effect of thickness of material on yield point, see Ref. 14, p. 1377.

General Principles in Design

The design of steel columns is always a cut-and-try method, as no law governs the relation between area and radius of gyration of the section. A column of given area is selected, and the amount of load that it will carry is computed by the proper formula. If the allowable load so computed is less than that to be carried, a larger column is selected and the load for it is computed, the process being repeated until a proper section is found.

Table 10.9 Basic Stresses for Clear Material*

Species	Extreme Fiber in Bending or Tension Parallel to Grain	Maximum Horizontal Shear	Compression Perpendicular to Grain	Compression Parallel to Grain $L/d = 11$ or Less	Modulus of Elasticity in Bending
Softwoods					
Baldcypress (Southern cypress)	1900	150	300	1450	1,200,000
Cedars					
Redcedar, Western	1300	120	200	950	1,000,000
White-cedar, Atlantic (Southern white-cedar) and northern	1100	100	180	750	800,000
White-cedar, Port Orford	1600	130	250	1200	1,500,000
Yellow-cedar, Alaska (Alaska cedar)	1600	130	250	1050	1,200,000
Douglas-fir, coast region	2200	130	320	1450	1,600,000
Douglas-fir, coast region, close-grained	2350	130	340	1550	1,600,000
Douglas-fir, Rocky Mountain region	1600	120	280	1050	1,200,000
Douglas-fir, dense, all regions	2550	150	380	1700	1,600,000
Fir, California red, grand, noble, and white	1600	100	300	950	1,100,000
Fir, balsam	1300	100	150	950	1,000,000
Hemlock, Eastern	1600	100	300	950	1,100,000
Hemlock, Western (West Coast hemlock)	1900	110	300	1200	1,400,000
Larch, Western	2200	130	320	1450	1,500,000
Pine, Eastern white (Northern white), ponderosa, sugar, and Western white (Idaho white)	1300	120	250	1000	1,000,000
Pine, jack	1600	120	220	1050	1,100,000
Pine, lodgepole	1300	90	220	950	1,000,000
Pine, red (Norway pine)	1600	120	220	1050	1,200,000
Pine, southern yellow	2200	160	320	1450	1,600,000
Pine, southern yellow, dense	2550	190	380	1700	1,600,000
Redwood	1750	100	250	1350	1,200,000
Redwood, close-grained	1900	100	270	1450	1,200,000
Spruce, Engelmann	1100	100	180	800	800,000
Spruce, red, white, and Sitka	1600	120	250	1050	1,200,000
Tamarack	1750	140	300	1350	1,300,000
Hardwoods					
Ash, black	1450	130	300	850	1,100,000
Ash, commercial white	2050	185	500	1450	1,500,000
Beech, American	2200	185	500	1600	1,600,000
Birch, sweet and yellow	2200	185	500	1600	1,600,000
Cottonwood, Eastern	1100	90	150	800	1,000,000
Elm, American and slippery (white or soft elm)	1600	150	250	1050	1,200,000
Elm, rock	2200	185	500	1600	1,300,000
Gums, blackgum, sweetgum (red or sap gum)	1600	150	300	1050	1,200,000
Hickory, true and pecan	2800	205	600	2000	1,800,000
Maple, black and sugar (hard maple)	2200	185	500	1600	1,600,000
Oak, commercial red and white	2050	185	500	1350	1,500,000
Tupelo	1600	150	300	1050	1,200,000
Yellow poplar	1300	120	220	950	1,100,000

*These stresses are applicable with certain adjustments to material of any degree of seasoning.

(For use in determining working stresses according to the grade of timber and other applicable factors. All values are in pounds per square inch. U.S. Forest Products Laboratory.)

Table 10.10 Values of K for Columns of Intermediate Length

	ASTM Standards, 1937, D245–37					
	Continuously Dry		Occasionally Wet		Usually Wet	
Species	Select	Common	Select	Common	Select	Common
Cedar, western red	24.2	27.1	24.2	27.1	25.1	28.1
Cedar, Port Orford	23.4	26.2	24.6	27.4	25.6	28.7
Douglas fir, coast region	23.7	27.3	24.9	28.6	27.0	31.1
Douglas fir, dense	22.6	25.3	23.8	26.5	25.8	28.8
Douglas fir, Rocky Mountain region	24.8	27.8	24.8	27.8	26.5	29.7
Hemlock, west coast	25.3	28.3	25.3	28.3	26.8	30.0
Larch, western	22.0	24.6	23.1	25.8	25.8	28.8
Oak, red and white	24.8	27.8	26.1	29.3	27.7	31.1
Pine, southern	.....	27.3	.....	28.6	.....	31.1
Pine, dense	22.6	25.3	23.8	26.5	25.8	28.8
Redwood	22.2	24.8	23.4	26.1	25.6	28.6
Spruce, red, white, Sitka	24.8	27.8	25.6	28.7	27.5	30.8

A few general principles should guide in proportioning columns. The radius of gyration should be approximately the same in the two directions at right angles to each other; the slenderness ratio of the separate parts of the column should not be greater than that of the column as a whole; the different parts should be adequately connected in order that the column may function as a single unit; the material should be distributed as far as possible from the centerline in order to increase the radius of gyration.

Steel Column Formulas

A variety of steel column formulas are in use, differing mostly in the value of unit stress allowed with various values of l/k. See Ref. 15, for a summary of the formulas.

Test on Steel Columns

After the collapse of the Quebec Bridge in 1907 as a result of a column failure, the ASCE, the AREA, and the U.S. Bureau of Standards cooperated in tests of full-sized steel columns. The results of these tests are reported in Ref. 16, pp. 1583–1688. The tests showed that, for columns of the proportions commonly used, the effect of variation in the steel, kinks, initial stresses, and similar

Table 10.11 Strength of Columns of Intermediate Length, Expressed as a Percentage of Strength of Short Columns

ASTM Standards, 1937, D245–37
Values for expression $\{1 - \frac{1}{3}(l/Kd)^4\}$ in eq. 33

Ratio of Length to Least Dimension in Rectangular Timbers, l/d

K	12	13	14	15	16	17	18	19	20	21	22	23	24	25	26	27	28	29	30	31
22	97	96	95	93	91	88	85	81	77	72	67	.....								
23	98	97	95	94	92	90	87	84	81	77	72	67	.....							
24	98	97	96	95	93	92	89	87	84	80	76	72	67	.....						
25	98	98	97	96	94	93	91	89	86	83	80	76	72	67	.....					
26	99	98	97	96	95	93	92	91	89	86	83	80	76	72	67	.....				
27	99	98	98	97	96	95	93	92	90	88	85	82	79	74	71	67	.....			
28	99	98	98	97	96	95	94	93	91	89	87	85	82	79	75	71	67	.....		
29	99	99	98	98	97	96	95	94	92	91	89	87	84	82	79	75	71	67	.....	
30	99	99	98	98	97	97	96	95	94	92	90	88	86	84	81	78	75	71	67	..
31	99	99	99	98	98	97	96	95	94	93	92	90	88	86	84	81	78	75	71	67

Note. This table can also be used for columns not rectangular, the l/d being equivalent to $0.289l/k$, where k is the least radius of gyration of the section.

defects in the column was more important than the effect of length. They also showed that the thin metal gave definitely higher strength, per unit area, than the thicker metal of the same type of section.

10.9 CYLINDERS, SPHERES, AND PLATES

10.9.1 Thin Cylinders and Spheres under Internal Pressure

A cylinder is regarded as thin when the thickness of the wall is small compared with the mean diameter, or $d/t > 20$. There are only tensile membrane stresses in the wall developed by the internal pressure p

$$\frac{\sigma_1}{R_1} + \frac{\sigma_2}{R_2} = \frac{p}{t} \qquad (10.96)$$

In the case of a cylinder where R_1, the curvature, is R and R_2 is infinite, and the hoop stress is

$$\sigma_1 = \sigma_h = \frac{pR}{t} \qquad (10.97)$$

If the two equations are compared, it is seen that the resistance to rupture by circumferential stress [Eq. (10.97)] is one-half the resistance to rupture by longitudinal stress [Eq. (10.98)]. For this reason cylindrical boilers are single riveted in the circumferential seams and double or triple riveted in the longitudinal seams.

From the equations of equilibrium, the longitudinal stress is

$$\sigma_2 = \sigma_L = \frac{pR}{2t} \qquad (10.98)$$

For a sphere, using Eq. (10.96), $R_1 = R_2 = R$ and $\sigma_1 = \sigma_2$, making

$$\sigma_1 = \sigma_2 = \frac{pR}{2t} \qquad (10.99)$$

In using the foregoing formulas to design cylindrical shells or piping, thickness t must be increased to compensate for rivet holes in the joints. Water pipes, particularly those of cast iron, require a high factor of safety, which results in increased thickness to provide security against shocks caused by water hammer or rough handling before they are laid. Equation (10.98) applies also to the stresses in the walls of a thin hollow sphere, hemisphere, or dome. When holes are cut, the tensile stresses must be found by the method used in riveted joints.

Thin Cylinders under External Pressure

Equations (10.97) and (10.98) apply equally well to cases of external pressure if P is given a negative sign, but the stresses so found are significant only if the pressure and dimensions are such that no buckling can occur.

10.9.2 Thick Cylinders and Spheres

Cylinders

When the thickness of the shell or wall is relatively large, as in guns, hydraulic machinery piping, and similar installations, the variation in stress from the inner surface to the outer surface is relatively large, and the ordinary formulas for thin wall cylinders are no longer applicable. In Fig. 10.26 the stresses, strains, and deflections are related[1,18,19] by

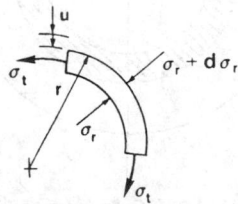

Fig. 10.26 Cylindrical element.

$$\sigma_t = \frac{E}{1 - \nu^2}(\epsilon_t + \nu\epsilon_r) = \frac{E}{1 - \nu^2}\left[\frac{u}{r} + \nu\frac{\partial u}{\partial r}\right] \tag{10.100}$$

$$\sigma_r = \frac{E}{1 - \nu^2}(\epsilon_r + \nu\epsilon_t) = \frac{E}{1 - \nu^2}\left[\frac{\partial u}{\partial r} + \nu\frac{u}{r}\right] \tag{10.101}$$

where E is the modulus and ν is Poisson's ratio. In a cylinder (Fig. 10.27) that has internal and external pressures, p_i and p_o; internal and external radii, a and b; $K = b/a$; the stresses are

$$\sigma_t = \frac{p_i}{K^2 - 1}\left(1 + \frac{b^2}{r^2}\right) - \frac{p_o K^2}{K^2 - 1}\left(1 + \frac{a^2}{r^2}\right) \tag{10.102}$$

$$\sigma_r = \frac{p_i}{K^2 - 1}\left(1 - \frac{b^2}{r^2}\right) - \frac{p_o K^2}{K^2 - 1}\left(1 - \frac{a^2}{r^2}\right) \tag{10.103}$$

if $p_o = 0$, and σ_t, σ_r are maximum at $r = a$; if $p_i = 0$, σ_t is maximum at $r = a$; and σ_r is maximum at $r = b$.

In shrinkage fits, Fig. 10.27, a hollow cylinder is pressed over a cylinder with a radial interference δ at $r = b$. p_f, the pressure between the cylinders, can be found from

$$\delta = \frac{bp_f}{E_o}\left(\frac{c^2 + b^2}{c^2 - b^2} + \nu_o\right) + \frac{bp_f}{E_i}\left(\frac{a^2 + b^2}{b^2 - a^2} - \nu_i\right) \tag{10.104}$$

The radial deflection can be found at a which shrinks and c which expands by knowing σ_r is zero and using Eqs. (10.100) and (10.101):

$$u_a = \frac{\sigma_t}{E_i}a, \qquad u_c = \frac{\sigma_t}{E_o}c \tag{10.105}$$

Spheres

The stress, strain, and deflections[19,20] are related by

$$\sigma_t = \frac{E}{1 - \nu - 2\nu^2}[\epsilon_t + \nu\epsilon_r] = \frac{E}{1 - \nu - 2\nu^2}\left[\frac{u}{r} + \nu\frac{\partial u}{\partial r}\right] \tag{10.106}$$

$$\sigma_r = \frac{E}{1 - \nu - 2\nu^2}[2\nu\epsilon_t + (1 - \nu)\epsilon_r] = \frac{E}{1 - \nu - 2\nu^2}\left[2\nu\frac{u}{r} + (1 - \nu)\frac{\partial u}{\partial r}\right] \tag{10.107}$$

The stresses for a thick wall sphere with internal and external pressure, p_i and p_o, and $K = b/a$ are

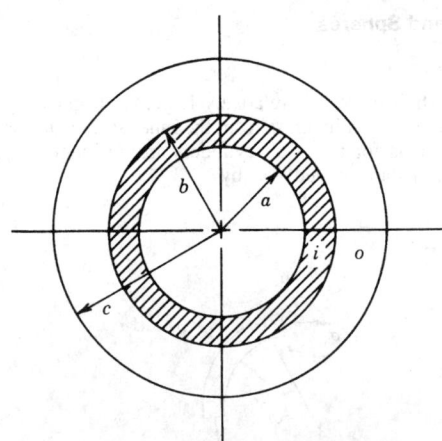

Fig. 10.27 Cylinder press fit.

$$\sigma_t = \frac{p_i(1 + b^3/2r^3)}{K^3 - 1} - \frac{p_oK^3(1 + a^3/2r^3)}{K^3 - 1} \qquad (10.108)$$

$$\sigma_r = \frac{p_i(1 - b^3/r^3)}{K^3 - 1} - \frac{p_oK^3(1 - a^3/r^3)}{K^3 - 1} \qquad (10.109)$$

If $p_i = 0$, $\sigma_r = 0$ at $r = a$, then

$$u_a = (1 - \nu)\frac{\sigma_t}{E}a \qquad (10.110)$$

Conversely, if $p_o = 0$, $\sigma_r = 0$ at $r = b$, then

$$u_b = (1 - \nu)\frac{\sigma_t}{E}b \qquad (10.111)$$

10.9.3   Plates

The formulas that apply for plates are based on the assumptions that the plate is flat, of uniform thickness, and of homogeneous isotropic material, thickness is not greater than one-fourth the least transverse dimension, maximum deflection is not more than one-half the thickness, all forces are normal to the plane of the plate, and the plate is nowhere stressed beyond the elastic limit. In Table 10.12 are formulas for deflection and stress for various shapes, forms of load and edge conditions. For further information see Refs. 12 and 21.

10.9.4   Trunnion

A solid shaft (Fig. 10.28) on a round or rectangular plate loaded with a bending moment is called a trunnion. The loading generally is developed from a bearing mounted on the solid shaft. For a round, simply supported plate

$$\sigma_r = \frac{\beta M}{at^2} \qquad (10.112)$$

$$\theta = \frac{\gamma M}{Et^3} \qquad (10.113)$$

$$\left.\begin{array}{l} \beta = 10^{(0.7634 - 1.252x)} \\ \log \gamma = 0.248 - \pi x^{1.5} \end{array}\right\} \; 0 < x = \frac{b}{a} < 1 \qquad (10.114)$$

For the fixed-end plate

$$\left.\begin{array}{l} \beta = 10^{(1 - 1.959x)} \\ \log \gamma = 0.179 - 3.75x^{1.5} \end{array}\right\} \; 0 < x = \frac{b}{a} < 1 \qquad (10.115)$$

The equations for β, γ are derived from curve fitting of data (see, for example, Refs. 2, 4th ed., and 21).

10.9.5   Socket Action

In Fig. 10.29a, summation of moments in the middle of the wall yields

$$2\left[\left(\frac{\omega''}{2}\frac{l}{2}\right)\left(\frac{2}{3}\frac{l}{2}\right)\right] = F\left(a + \frac{l}{2}\right)$$

$$\omega'' = \frac{6}{l^2}\left[F\left(a + \frac{l}{2}\right)\right] \qquad (10.116)$$

Summation of forces in the horizontal gives

$$\omega' = \frac{F}{l} \qquad (10.117)$$

At B, the bearing pressure in Fig. 10.29c is

Table 10.12   Formulas for Flat Plates[a]

Notation: W = total applied load, lb; w = unit applied load, psi; t = thickness of plate, in.; σ = stress at surface of plate, psi; y = vertical deflection of plate from original position, in.; E = modulus of elasticity; m = reciprocal of ν, Poisson's ratio. q denotes any given point on the surface of plate; r denotes the distance of q from the center of a circular plate. Other dimensions and corresponding symbols are indicated on figures. Positive sign for σ indicates tension at upper surface and equal compression at lower surface; negative sign indicates reverse condition. Positive sign for y indicates upward deflection, negative sign downward deflection. Subscripts $r, t, a,$ and b used with σ denote, respectively, radial direction, tangential direction, direction of dimension a, and direction of dimension b. All dimensions are in inches. All logarithms are to the base e ($\log_e x = 2.3026 \log_{10} x$).

TYPE OF LOAD AND SUPPORT	FORMULAS FOR STRESS AND DEFLECTION
	CIRCULAR FLAT PLATES

Outer edges supported. Uniform load over entire surface.

$w \pi a^2 = W$

At center:

$$\max \sigma_r = \sigma_t = \frac{-3W}{8\pi m t^2}(3m+1) \qquad \max y = -\frac{3W(m-1)(5m+1)a^2}{16\pi E m^2 t^3}$$

At q:

$$\sigma_r = -\frac{3W}{8\pi m t^2}\left[(3m+1)\left(1-\frac{r^2}{a^2}\right)\right] \qquad \sigma_t = -\frac{3W}{8\pi m t^2}\left[(3m+1)-(m+3)\frac{r^2}{a^2}\right]$$

$$y = -\frac{3W(m^2-1)}{8\pi E m^2 t^3}\left[\frac{(5m+1)a^2}{2(m+1)}+\frac{r^4}{2a^2}-\frac{(3m+1)r^2}{m+1}\right]$$

Outer edges fixed. Uniform load over entire surface.

$W = w\pi a^2$

At center:

$$\sigma_r = \sigma_t = -\frac{3W(m+1)}{8\pi m t^2} \qquad \max y = -\frac{3W(m^2-1)a^2}{16\pi E m^2 t^3}$$

At q:

$$\sigma_r = \frac{3W}{8\pi m t^2}\left[(3m+1)\frac{r^2}{a^2}-(m+1)\right] \qquad \sigma_t = \frac{3W}{8\pi m t^2}\left[(m+3)\frac{r^2}{a^2}-(m+1)\right]$$

$$y = \frac{-3W(m^2-1)}{16\pi E m^2 t^3}\left[\frac{(a^2-r^2)^2}{a^2}\right]$$

Outer edges supported. Uniform load over concentric circular area of radius r_0.

At $q, r < r_0$:

$$\sigma_r = -\frac{3W}{2\pi m t^2}\left[m+(m+1)\log\frac{a}{r_0}-(m-1)\frac{r_0^2}{4a^2}-(3m+1)\frac{r^2}{4r_0^2}\right]$$

$$\sigma_t = -\frac{3W}{2\pi m t^2}\left[m+(m+1)\log\frac{a}{r_0}-(m-1)\frac{r_0^2}{4a^2}-(m+3)\frac{r^2}{4r_0^2}\right]$$

$$y = -\frac{3W(m^2-1)}{16\pi E m^2 t^3}\left[4a^2-5r_0^2+\frac{r^4}{r_0^2}-(8r^2+4r_0^2)\log\frac{a}{r_0}-\frac{2(m-1)r_0^2(a^2-r^2)}{(m+1)a^2}+\frac{8m(a^2-r^2)}{m+1}\right]$$

$W = w\pi r_0^2$

At $q, r > r_0$:

$$\sigma_r = -\frac{3W}{2\pi m t^2}\left[(m+1)\log\frac{a}{r}-(m-1)\frac{r_0^2}{4a^2}+(m-1)\frac{r_0^2}{4r^2}\right]$$

$$\sigma_t = -\frac{3W}{2\pi m t^2}\left[(m-1)+(m+1)\log\frac{a}{r}-(m-1)\frac{r_0^2}{4a^2}-(m-1)\frac{r_0^2}{4r^2}\right]$$

$$y = -\frac{3W(m^2-1)}{16\pi E m^2 t^3}\left[\frac{(12m+4)(a^2-r^2)}{m+1}-\frac{2(m-1)r_0^2(a^2-r^2)}{(m+1)a^2}-(8r^2+4r_0^2)\log\frac{a}{r}\right]$$

At center:

$$\max \sigma_r = \sigma_t = -\frac{3W}{2\pi m t^2}\left[m+(m+1)\log\frac{a}{r_0}-(m-1)\frac{r_0^2}{4a^2}\right]$$

$$\max y = -\frac{3W(m^2-1)}{16\pi E m^2 t^3}\left[\frac{(12m+4)a^2}{m+1}-4r_0^2\log\frac{a}{r_0}-\frac{(7m+3)r_0^2}{m+1}\right]$$

[a] By permission from Ref. 22.

Table 10.12 (*Continued*)

Type of Load and Support	Formulas for Stress and Deflection

CIRCULAR FLAT PLATES

Outer edges supported. Uniform load on concentric circular ring of radius r_0.

At q, $r < r_0$:

$$\max \sigma_r = \sigma_t = -\frac{3W}{2\pi mt^2}\left[\frac{1}{2}(m-1) + (m+1)\log\frac{a}{r_0} - (m-1)\frac{r_0^2}{2a^2}\right]$$

$$y = -\frac{3W(m^2-1)}{2\pi Em^2t^3}\left[\frac{(3m+1)(a^2-r^2)}{2(m+1)} - (r^2+r_0^2)\log\frac{a}{r_0} + (r^2-r_0^2)\right.$$
$$\left. - \frac{(m-1)r_0^2(a^2-r^2)}{2(m+1)a^2}\right]$$

At q, $r > r_0$:

$$\sigma_r = -\frac{3W}{2\pi mt^2}\left[(m+1)\log\frac{a}{r} + (m-1)\frac{r_0^2}{2r^2} - (m-1)\frac{r_0^2}{2a^2}\right]$$

$$\sigma_t = -\frac{3W}{2\pi mt^2}\left[(m-1) + (m+1)\log\frac{a}{r} - (m-1)\frac{r_0^2}{2r^2} - (m-1)\frac{r_0^2}{2a^2}\right]$$

$$y = -\frac{3W(m^2-1)}{2\pi Em^2t^3}\left[\frac{(3m+1)(a^2-r^2)}{2(m+1)} - (r^2+r_0^2)\log\frac{a}{r} - \frac{(m-1)r_0^2(a^2-r^2)}{2(m+1)a^2}\right]$$

Outer edges fixed. Uniform load over concentric circular area of radius r_0.

At q, $r < r_0$:

$$\sigma_r = -\frac{3W}{2\pi mt^2}\left[(m+1)\log\frac{a}{r_0} + (m+1)\frac{r_0^2}{4a^2} - (3m+1)\frac{r^2}{4r_0^2}\right]$$

$$\sigma_t = -\frac{3W}{2\pi mt^2}\left[(m+1)\log\frac{a}{r_0} + (m+1)\frac{r_0^2}{4a^2} - (m+3)\frac{r^2}{4r_0^2}\right]$$

$$y = -\frac{3W(m^2-1)}{16\pi Em^2t^3}\left[4a^2 - (8r^2+4r_0^2)\log\frac{a}{r_0} - \frac{2r^2r_0^2}{a^2} + \frac{r^4}{r_0^2} - 3r_0^2\right]$$

At q, $r > r_0$:

$$\sigma_r = -\frac{3W}{2\pi mt^2}\left[(m+1)\log\frac{a}{r} + (m+1)\frac{r_0^2}{4a^2} + (m-1)\frac{r_0^2}{4r^2} - m\right]$$

$$\sigma_t = -\frac{3W}{2\pi mt^2}\left[(m+1)\log\frac{a}{r} + (m+1)\frac{r_0^2}{4a^2} - (m-1)\frac{r_0^2}{4r^2} - 1\right]$$

$$y = -\frac{3W(m^2-1)}{16\pi Em^2t^3}\left[4a^2 - (8r^2+4r_0^2)\log\frac{a}{r} - \frac{2r^2r_0^2}{a^2} - 4r^2 + 2r_0^2\right]$$

At center:

$$\sigma_r = \sigma_t = -\frac{3W}{2\pi mt^2}\left[(m+1)\log\frac{a}{r_0} + (m+1)\frac{r_0^2}{4a^2}\right] = \max\sigma_r \text{ when } r_0 < 0.588a$$

$$\max y = -\frac{3W(m^2-1)}{16\pi Em^2t^3}\left[4a^2 - 4r_0^2\log\frac{a}{r_0} - 3r_0^2\right]$$

$W = w\pi r_0^2$

Outer edges fixed. Uniform load on concentric circular ring of radius r_0.

At q, $r < r_0$:

$$\sigma_r = \sigma_t = -\frac{3W}{4\pi mt^2}\left[(m+1)\left(2\log\frac{a}{r_0} + \frac{r_0^2}{a^2} - 1\right)\right] = \max\sigma \text{ when } r < 0.31a$$

$$y = -\frac{3W(m^2-1)}{2\pi Em^2t^3}\left[\frac{1}{2}\left(1 + \frac{r_0^2}{a^2}\right)(a^2-r^2) - (r^2+r_0^2)\log\frac{a}{r_0} + (r^2-r_0^2)\right]$$

At q, $r > r_0$:

$$\sigma_r = -\frac{3W}{4\pi mt^2}\left[(m+1)\left(2\log\frac{a}{r} + \frac{r_0^2}{a^2}\right) + (m-1)\frac{r_0^2}{r^2} - 2m\right]$$

$$\sigma_t = -\frac{3W}{4\pi mt^2}\left[(m+1)\left(2\log\frac{a}{r} + \frac{r_0^2}{a^2}\right) - (m-1)\frac{r_0^2}{r^2} - 2\right]$$

$$y = -\frac{3W(m^2-1)}{2\pi Em^2t^3}\left[\frac{1}{2}\left(1 + \frac{r_0^2}{a^2}\right)(a^2-r^2) - (r^2+r_0^2)\log\frac{a}{r}\right]$$

At center:

$$\max y = -\frac{3W(m^2-1)}{2\pi Em^2t^3}\left[\frac{1}{2}(a^2-r_0^2) - r_0^2\log\frac{a}{r_0}\right]$$

Table 10.12 *(Continued)*

Type of Load and Support	Formulas for Stress and Deflection
	CIRCULAR FLAT PLATES WITH CONCENTRIC CIRCULAR HOLE
Outer edge supported. Uniform load over entire surface. 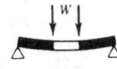 $W = w\pi(a^2 - b^2)$	At inner edge: $$\max \sigma = \sigma_t = -\frac{3w}{4mt^2(a^2 - b^2)}\left[a^4(3m+1) + b^4(m-1) - 4ma^2b^2 - 4(m+1)a^2b^2 \log\frac{a}{b}\right]$$ $$\max y = -\frac{3w(m^2-1)}{2m^2Et^3}\left[\frac{a^4(5m+1)}{8(m+1)} + \frac{b^4(7m+3)}{8(m+1)} - \frac{a^2b^2(3m+1)}{2(m+1)} \right.$$ $$\left. + \frac{a^2b^2(3m+1)}{2(m-1)}\log\frac{a}{b} - \frac{2a^2b^4(m+1)}{(a^2-b^2)(m-1)}\left(\log\frac{a}{b}\right)^2\right]$$
Outer edge supported. Uniform load along inner edge. 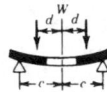 W	At inner edge: $$\max \sigma = \sigma_t = -\frac{3W}{2\pi mt^2}\left[\frac{2a^2(m+1)}{a^2-b^2}\log\frac{a}{b} + (m-1)\right]$$ $$\max y = -\frac{3W(m^2-1)}{4\pi Em^2 t^3}\left[\frac{(a^2-b^2)(3m+1)}{(m+1)} + \frac{4a^2b^2(m+1)}{(m-1)(a^2-b^2)}\left(\log\frac{a}{b}\right)^2\right]$$
Supported along concentric circle near outer edge. Uniform load along concentric circle near inner edge. 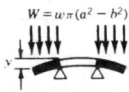	At inner edge: $$\max \sigma = \sigma_t = -\frac{3W}{2\pi mt^2}\left[\frac{2a^2(m+1)}{a^2-b^2}\log\frac{c}{d} + (m-1)\frac{c^2-d^2}{a^2-b^2}\right]$$
Inner edge supported. Uniform load over entire surface. 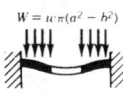 $W = w\pi(a^2 - b^2)$	At inner edge: $$\max \sigma = \sigma_t = \frac{3w}{4mt^2(a^2-b^2)}\left[4a^4(m+1)\log\frac{a}{b} + 4a^2b^2 + b^4(m-1) - a^4(m+3)\right]$$ At outer edge: $$\max y = \frac{3w(m-1)}{16Em^2t^3}\left[a^4(7m+3) + b^4(5m+1) - a^2b^2(12m+4)\right.$$ $$\left. - \frac{4a^2b^2(3m+1)(m+1)}{(m-1)}\log\frac{a}{b} + \frac{16a^4b^2(m+1)^2}{(a^2-b^2)(m-1)}\left(\log\frac{a}{b}\right)^2\right]$$
Outer edge fixed and supported. Uniform load over entire surface. 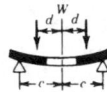 $W = w\pi(a^2 - b^2)$	At outer edge: $$\max \sigma_r = \frac{3w}{4t^2}\left[a^2 - 2b^2 + \frac{b^4(m-1) - 4b^4(m+1)\log\frac{a}{b} + a^2b^2(m+1)}{a^2(m-1) + b^2(m+1)}\right] = \nu \max \sigma$$ At inner edge: $$\max \sigma_t = -\frac{3w(m^2-1)}{4mt^2}\left[\frac{a^4 - b^4 - 4a^2b^2\log\frac{a}{b}}{a^2(m-1) + b^2(m+1)}\right]$$ $$\max y = -\frac{3w(m^2-1)}{16m^2Et^3}\left[a^4 + 5b^4 - 6a^2b^2 + 8b^4\log\frac{a}{b}\right.$$ $$\left. + \frac{\left\{[-8b^6(m+1) + 4a^2b^4(3m+1) + 4a^4b^2(m+1)]\log\frac{a}{b} - 16a^2b^4(m+1)\left(\log\frac{a}{b}\right)^2\right\}}{a^2(m-1) + b^2(m+1)} \right.$$ $$\left. + \frac{+ 4a^4b^4 - 2a^4b^2(m+1) + 2b^6(m-1)}{a^2(m-1) + b^2(m+1)}\right]$$

Table 10.12   (Continued)

Type of Load and Support	Formulas for Stress and Deflection
	CIRCULAR FLAT PLATES WITH CONCENTRIC CIRCULAR HOLE

Outer edge fixed and supported. Uniform load along inner edge.

At outer edge:

$$\max \sigma_r = \frac{3W}{2\pi t^2}\left[1 - \frac{2mb^2 - 2b^2(m+1)\log\frac{a}{b}}{a^2(m-1)+b^2(m+1)}\right] = \max \sigma \text{ when } \frac{a}{b} < 2.4$$

At inner edge:

$$\max \sigma_t = \frac{3W}{2\pi m t^2}\left[1 + \frac{ma^2(m-1) - mb^2(m+1) - 2(m^2-1)a^2\log\frac{a}{b}}{a^2(m-1)+b^2(m+1)}\right]$$

$$= \max \sigma \text{ when } \frac{a}{b} > 2.4$$

$$\max y = -\frac{3W(m^2-1)}{4\pi m^2 E t^3} \times$$

$$\left[a^2 - b^2 + \frac{2mb^2(a^2-b^2) - 8ma^2b^2\log\frac{a}{b} + 4a^2b^2(m+1)\left(\log\frac{a}{b}\right)^2}{a^2(m-1)+b^2(m+1)}\right]$$

Outer edge fixed. Uniform moment along inner edge.

At inner edge:

$$\max \sigma_r = \frac{6M}{t^2}$$

$$\max y = \frac{6M(m^2-1)}{mEt^3}\left[\frac{a^2b^2 - b^4 - 2a^2b^2\log\frac{a}{b}}{a^2(m-1)+b^2(m+1)}\right]$$

At outer edge:

$$\sigma_r = -\frac{6M}{t^2}\left[\frac{2mb^2}{(m+1)b^2+(m-1)a^2}\right]$$

Outer edge supported. Unequal uniform moments along edges.

$M_a \; M_b \; M_b \; M_a$

At q:

$$\sigma_r = \frac{6}{t^2(a^2-b^2)}\left[a^2M_a - b^2M_b - \frac{a^2b^2}{r^2}(M_a - M_b)\right]$$

$$\sigma_t = \frac{6}{t^2(a^2-b^2)}\left[a^2M_a - b^2M_b + \frac{a^2b^2}{r^2}(M_a - M_b)\right]$$

From outer edge level:

$$y = \frac{12(m^2-1)}{mEt^3(a^2-b^2)}\left[\frac{a^2-r^2}{2}\left(\frac{a^2M_a - b^2M_b}{m+1}\right) + \log\frac{a}{r}\left(\frac{a^2b^2(M_a - M_b)}{m-1}\right)\right]$$

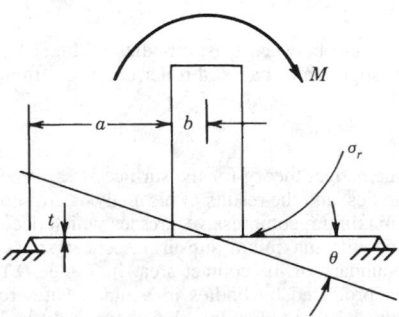

Fig. 10.28   Simply supported trunnion.

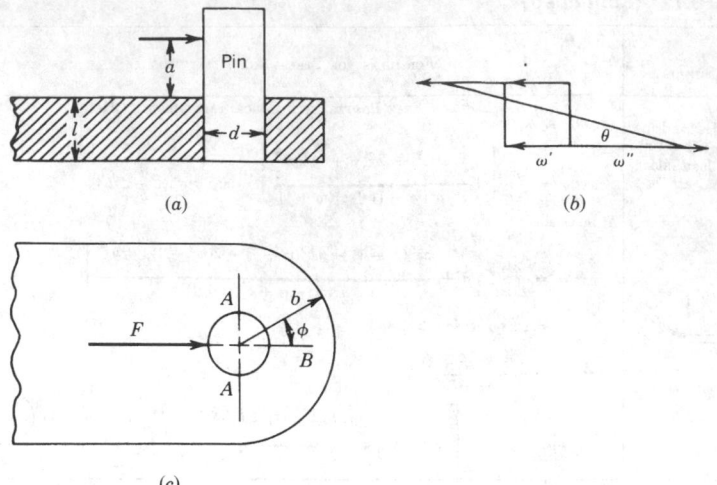

(a)   (b)

(c)

Fig. 10.29   Socket action near an edge.

$$p_i = \frac{\omega' + \omega''}{d} \text{ psi} \tag{10.118}$$

In Eq. (10.102) $p_o = 0$ and

$$\sigma_t = \frac{p_i}{R^2 - 1}\left[1 + \left(\frac{b}{d/2}\right)^2\right]$$

At A in Fig. 10.29c

$$\sigma = \frac{\phi 8F}{\pi^2 bl}$$

where $2b/d = 2, 4$ and $\phi = 4.3, 4.4$;

$$F = (\omega' + \omega'')l$$

If a pin is pressed into the frame hole, σ_t created by p_f [Eq. (10.104)] must be added. Furthermore, if the pin and frame are different metals, additional σ_t will be created by temperature changes that vary p_f.

The stress in the pin can be found from the maximum moment developed by ω' and ω'', and then calculating the bending stress.

10.10   CONTACT STRESSES

The stresses caused by the pressure between elastic bodies (Table 10.13) are of importance in connection with the design or investigation of ball and roller bearings, trunnions, expansion rollers, track stresses, gear teeth, etc.

Contact Stress Theory

H. Hertz[23] developed the mathematical theory for the surface stresses and the deformations produced by pressure between curved bodies, and the results of his analysis are supported by research. Formulas based on this theory give the maximum compressive stresses which occur at the center of the surfaces of contact, but do not consider the maximum subsurface shear stresses nor the maximum tensile stresses which occur at the boundary of the contact area. In Table 10.13 formulas are given for the elastic stress and deformation produced by bodies in contact. Numerous tests have been made to determine the bearing strength of balls and rollers, but there is difficulty in interpreting the results

Table 10.13   Areas of Contact and Pressures with Two Surfaces in Contact

Poisson's ratio $= 0.3$; $P =$ load, lb; $P_1 =$ load per in. of length, lb; $E =$ modulus of elasticity.

Character of Surfaces	Maximum Pressure, s, at Center of Contact, psi	Radius, r, or Width, b, of Contact Area, in.
Two spheres	$s = 0.616 \sqrt[3]{PE^2 \left(\dfrac{d_1 + d_2}{d_1 d_2}\right)^2}$	$r = 0.881 \sqrt[3]{\dfrac{P}{E}\left(\dfrac{d_1 d_2}{d_1 + d_2}\right)}$
Sphere and plane	$s = 0.616 \sqrt[3]{\dfrac{PE^2}{d^2}}$	$r = 0.881 \sqrt[3]{\dfrac{Pd}{E}}$
Sphere and hollow sphere	$s = 0.616 \sqrt[3]{PE^2 \left(\dfrac{d_2 - d_1}{d_1 d_2}\right)^2}$	$r = 0.881 \sqrt[3]{\dfrac{P}{E}\left(\dfrac{d_1 d_2}{d_2 - d_1}\right)}$
Cylinder and plane	$s = 0.591 \sqrt{\dfrac{P_1 E}{d}}$	$b = 2.15 \sqrt{\dfrac{P_1 d}{E}}$
Two cylinders	$s = 0.591 \sqrt{P_1 E \left(\dfrac{d_1 + d_2}{d_1 d_2}\right)}$	$b = 2.15 \sqrt{\dfrac{P_1}{E}\left(\dfrac{d_1 d_2}{d_1 + d_2}\right)}$
General case of two bodies in contact	$s = \dfrac{1.5P}{\pi c d}$	$c = \alpha \sqrt[3]{\dfrac{P\delta}{K}}$ $d = \beta \sqrt[3]{\dfrac{P\delta}{K}}$ $\delta = \dfrac{4}{\dfrac{1}{R_1} + \dfrac{1}{R_2} + \dfrac{1}{R_1'} + \dfrac{1}{R_2'}}$ $K = \dfrac{8}{3}\dfrac{E_1 E_2}{E_2(1-\nu_1^2) + E_1(1-\nu_2^2)}$

$$\theta = \text{arc cos } \frac{1}{4}\delta \sqrt{\left[\left(\frac{1}{R_1} - \frac{1}{R_1'}\right)^2 + \left(\frac{1}{R_2} - \frac{1}{R_2'}\right)^2 + 2\left(\frac{1}{R_1} - \frac{1}{R_1'}\right)\left(\frac{1}{R_2} - \frac{1}{R_2'}\right)\cos 2\phi\right]}$$

θ	0°	10°	20°	30°	40°	50°	60°	70°	80°	90°
α	∞	6.612	3.778	2.731	2.136	1.754	1.486	1.284	1.128	1.00
β	0	0.319	0.408	0.493	0.567	0.641	0.717	0.802	0.893	1.00

for lack of a satisfactory criterion of failure. One arbitrary criterion of failure is the amount of allowable plastic yielding. For further information on contact stresses see Refs. 2, 24, and 25.

10.11   ROTATING ELEMENTS

10.11.1   Shafts

The stress[1] in the center of a rotating shaft or solid cylinder is

$$\sigma_r = \sigma_h = \frac{3 - 2\nu}{8(1 - \nu)} \left(\frac{\gamma\omega^2}{g} \right) r_o^2 \tag{10.119}$$

$$\sigma_z = \frac{\nu\gamma\omega^2}{4g(1 - \nu)} r_o^2 \tag{10.120}$$

where ν is Poisson's ratio, ω is in rad/sec, γ is the density in lb/in.[3], and g is 386 in./sec[2]. The limiting ω can be found by using distortion energy; however, most shafts support loads and are limited by critical speeds from torsional or bending modes of vibration. Holzer's method and Dunkerley's equation are used.

10.11.2   Disks

A rotating disk[1,9,19] of inside radius a and outside radius b has $\sigma_r = 0$ at a and b, while σ_t is

$$\sigma_{ta} = \frac{3 + \nu}{4g} \gamma\omega^2 \left(b^2 + \frac{1 - \nu}{3 + \nu} a^2 \right) \tag{10.121}$$

$$\sigma_{tb} = \frac{3 + \nu}{4g} \gamma\omega^2 \left(a^2 + \frac{1 - \nu}{3 + \nu} b^2 \right) \tag{10.122}$$

Substitution in Eq. (10.105) gives the outside and inside radial expansions.
   The solid disk of radius b has stresses at the center

$$\sigma_t = \sigma_r = \frac{3 + \nu}{8g} \gamma\omega^2 b^2 \tag{10.123}$$

Substitution into the distortion energy [Eq. (10.22)] can give one the limiting speed.

10.11.3   Blades

Blades attached to a rotating shaft will experience a tensile force at the attachment to the shaft. These can be found from dynamics of machinery texts; however, the forces developed from a fluid driven by the blades develop more problems. The blades, if not in the plane, will develop additional forces and moments from the driving force plus vibration of the blades on the shaft.

10.12   DESIGN SOLUTION SOURCES AND GUIDELINES

Designs are composed of simple elements, as discussed here. These elements are subjected to temperature extremes, vibrations, and environmental effects that cause them to creep, buckle, yield, and corrode. Finding solutions to model these cases, as elements, can be difficult and when found the solutions are complex to follow, let alone to calculate. See Refs. 2, 21, and 26–32 Handbooks cataloging known solutions. Always cross-check with another reference. The Handbooks of Roark & Young and Blevins have been computerized using a TK solver and are distributed by UTS software. These closed form solutions would ease some of the more complicated calculations and checks finite element solutions using a computer.

10.12.1   Computers

Most computer set-ups use linear elastic solutions where the analyst supplies mechanical properties of materials such as yield and ultimate strengths and cross-sectional properties like area and area moments of inertia. When solving more complex problems, some concerns to keep in mind:

Questions to Be Asked

1. Will I know if this model buckles?
2. Can one use a non-linear stress-strain curve?
3. Is there any provision for creep and buckling?

4. How large and complex a structure can be solved? Look at a solved problem and relate it to future problems.

Things to Watch and Note

1. Press fit joints, flanges, pins, bolts, welds and bonds, and any connection interface present modeling problems. The stress analysis of a single loaded weld is not a simple task. The stress solution for a trunnion with more complexity, such as seal grooves in the plate, requires many small finite elements to converge to a closed form solution (Eq. 10.112).

2. Vibration solutions with connection interfaces can give frequency solutions with 50% error with many connections and still have 10–15% error with no connections. The computer solution appears to be always on the high side.

3. Detailed fatigue stresses on elements can be derived out of the loads by printing out the force variation.

4. Materials. The materials have good operating range[33] and limitations for spring stress relaxation at higher temperatures or lower limits can be applicable to structural members.

Nickel Alloys, Inconels and similar materials.	$-300°F \leq T \leq 1020°F$
300, 400, 17-4, 17-7 stainless or austenitic, martensite, and precipitation-hardening stainless steels.	$-110°F \leq T \leq 570°F$
Spring steels	$-5°F \leq T \leq 430°F$
Patented cold drawn carbon steels	$-110°F \leq T \leq 300°F$
Copper Beryllium	$-330°F \leq T \leq 260°F$
Titanium Alloys	
Bronzes	$-40°F \leq T \leq 175°F$
Aluminum	$-300°F \leq T \leq 400°F$
Magnesium	$-300°F \leq T \leq 350°F$

The high temperatures are for the onset of creep and stress relaxation and lower mechanical properties with higher temperature. The low temperatures show higher mechanical properties but are shock-sensitive. Always examine for the mechanical properties for the temperature range and thermal expansion.[34-37] The mechanical properties at room temperature have predictable distributions with ample sample sizes, but if the temperature is varied, similar published results are not readily available.

Rubber, plastics, and elastomers have glassy transition temperatures below which the material is putty-like and above which the material is rock-like and brittle. All material mechanical properties vary a great deal due to temperature. This makes computer solutions much more complex. Testing is the final reliable check.

10.12.2 Testing

Most designs must pass some sets of vibration, environmental, and screen testing before delivery to a customer. It is at this time that design flaws show up and frequencies, stresses, and so on are verified. Some preliminary testing might help:

1. Compare impact hammer frequency test of part of or an entire system to the computer and hand calculations. The physical testing includes the boundary values sometimes difficult to simulate on a computer.

2. Spot bond optical parts to dissimilar metal structural frame, which must be hot and cold soak tested to see if the bonding fractures the optical parts. Computers cannot predict a failure of this type well.

3. Check testing of joints and seal surfaces with pressure-sensitive gaskets to see if the developed pressures are sufficient to maintain the design to proper requirements. Then use operational testing to check for thermal warping of these critical surfaces.

4. Pressurize or load brazed, welded, or soldered part to check the process and its calculations for the pressures and loads.

5. Rapid Prototyping.[38] This method could be used to check a photoelastic model by vibrating it or freezing stresses in the model from static loads. It also could define areas of high stress for a smaller grid finite element modeling. Stress coating on a regular plastic model could also point out areas of high stress.

REFERENCES

1. J. H. Faupel and F. E. Fisher, *Engineering Design*, 2nd ed., Wiley, New York, 1981.
2. R. J. Roark and W. C. Young, *Formulas for Stress and Strain*, 6th ed., McGraw-Hill, New York, 1989.
3. J. Marin, *Mechanical Properties of Materials and Design*, McGraw-Hill, New York, 1942.
4. R. E. Peterson, *Stress Concentration Factors*, 2nd ed., Wiley, New York, 1974.
5. Young, *Bulletin 4*, School of Engineering Research, University of Toronto.
6. *Aluminum Standards and Data*, 3rd ed., Aluminum Association, New York, 1972.
7. F. B. Seely and J. O. Smith, *Resistance of Materials*, 4th ed., Wiley, New York, 1957.
8. A. P. Boresi, O. Sidebottom, F. B. Seely, and J. O. Smith, *Advanced Mechanics of Materials*, 3rd ed., Wiley, New York, 1978.
9. S. P. Timoshenko, *Strength of Materials*, 3rd ed., Krieger, Melbourne, FL, 1958, Vols. I and II.
10. H. C. Mann, *Proc. Am. Soc. Testing Materials*, 1935, 1936, and 1937.
11. Bach, Elastizität u. Festigkeit.
12. R. M. Rivello, *Theory Analysis of Flight Structures*, McGraw-Hill, New York, 1969.
13. B. G. Johnston (ed.), *Structural Research Council, Stability Design Criteria for Metal Structures*, 3rd ed., Wiley, New York, 1976.
14. *Trans. Am. Soc. Civil Engr.*, **xcviii** (1933).
15. S. P. Timoshenko and J. M. Gere, *Theory of Elastic Stability*, 2nd ed., McGraw-Hill, New York, 1961.
16. *Trans. Am. Soc. Civil Engr.*, **lxxxiii** (1919–20).
17. *AISC Handbook*, American Institute of Steel Construction, New York.
18. R. C. Juvinall, *Stress, Strain and Strength*, McGraw-Hill, New York, 1967.
19. S. P. Timoshenko and J. N. Goodier, *Theory of Elastic Stability*, 3rd ed., McGraw-Hill, New York, 1970.
20. M. Hetényi, *Handbook of Experimental Stress Analysis*, Wiley, New York, 1950.
21. W. Griffel, *Handbook of Formulas for Stress and Strain*, Frederick Ungar, New York, 1966.
22. R. J. Roark, *Formulas for Stress and Strain*, 2nd ed., McGraw-Hill, New York, 1943.
23. H. Hertz, *Gesammelte Werke*, Vol. 1, Leipzig, 1895.
24. R. K. Allen, *Rolling Bearings*, Pitman and Sons, London, 1945.
25. A. Palmgren, *Ball and Roller Bearing Engineering*, SKF Industries, Philadelphia, PA, 1945.
26. R. D. Blevins, *Formulas for Natural Frequency and Mode Shapes*, Krieger, Melbourne, FL, 1993.
27. R. D. Blevins, *Flow-Induced Vibration*, 2nd ed., Krieger, Melbourne, FL, 1994.
28. W. Flügge (ed.), *Handbook of Engineering Mechanics*, 1st ed., McGraw-Hill, New York, 1962.
29. A. W. Leissa, *Vibration of Plates NASA SP-160 (N70-18461) NTIS*, Springfield, VA.
30. A. W. Leissa, *Vibration of Shells NASA SP-288 (N73-26924) NTIS*, Springfield, VA.
31. A. Kleinlogel, *Rigid Frame Formulas*, 12th ed., Frederick Ungar, New York, 1958.
32. V. Leontovich, *Frames and Arches*, McGraw-Hill, New York, 1959.
33. M. O'Malley, "The Effect of Extreme Temperature on Spring Performance," *Springs* (May 1986).
34. Mil HDBK 5F, *Metallic Materials for Aerospace Structures*, 2 Vols., Department of Defense, 1990.
35. *Aerospace Structural Metals Handbook*, Five Vols., CINDAS/USAF CRDA, Purdue University, West Lafayette, IN, 1993.
36. *Structural Alloys Handbook*, 3 Vols., CINDAS, Purdue University, West Lafayette, IN, 1993.
37. *Thermophysical Properties of Matter*, Vol. 12, *Metallic Expansion*, 1995; Vol. 13, *Non-Metallic Thermoexpansion*, 1977, IFI/Phenium, New York.
38. S. Ashley, "Rapid Prototyping Is Coming of Age," *Mechanical Engineering* (July 1995).

BIBLIOGRAPHY

Almen, J. O., and P. H. Black, *Residual Stresses and Fatigue in Metals*, McGraw-Hill, New York, 1963.

Di Giovanni, M., *Flat and Corrugated Diaphragm Design Handbook*, Marcel Dekker, New York, 1982.

Osgood, W. R. (ed.), *Residual Stresses in Metals and Metal Construction*, Reinhold, New York, 1954.

Proceedings of the Society for Experimental Stress Analysis.

Symposium on Internal Stresses in Metals and Alloys, Institute of Metals, London, 1948.

Vande Walle, L. J., *Residual Stress for Designers and Metallurgists*, 1980 American Society for Metals Conference, American Society for Metals, Metals Park, OH, 1981.

CHAPTER 11

CONCURRENT ENGINEERING REVISITED: HOW FAR HAVE WE COME?

K. J. Cleetus
Concurrent Engineering Research Center
West Virginia University
Morgantown, West Virginia

11.1 ORIGIN OF CE	249	11.10 CE AND INNOVATION	253
11.2 ADOPTION OF CE	250	11.11 CE LESSONS	253
11.3 DEFINITION OF CE	250	11.12 CONCURRENT ENGINEERING TECHNOLOGIES	253
11.4 THE CE TEAM	250	11.12.1 Communication	253
		11.12.2 Task Coordination	254
11.5 THE ESSENCE OF CE	250	11.12.3 Negotiation/Tradeoff	255
		11.12.4 Data-Sharing	256
11.6 BARRIERS TO CE	251	11.12.5 Electronic Design Notebooks	257
11.7 APPLICABILITY OF CE	252	11.12.6 Process Libraries	257
11.8 CE AND THE INDIVIDUAL	252	11.13 APPLICATION OF CE PRINCIPLES	258
11.9 TEAMWORK CAN LEAD TO CHAOS	252		

11.1 ORIGIN OF CE

Concurrent engineering (CE) was a phenomenon of the 1980s. It arose in the Department of Defense (DoD) when it was realized that a number of new defense products were being designed without any thought given at the time of design to whether the design was manufacturable. Lack of consideration of manufacturability led to many revisions at a late stage when shortcomings of the design were discovered by the factory.

From this came the notable study of projects in 13 companies that led to a report of the Institute of Defense Analyses,[1] in which the first definition of concurrent engineering was given:

> Concurrent engineering is a systematic approach to the integrated, concurrent design of products and their related processes, including manufacture and support. This approach is intended to cause the developers, from the outset, to consider all elements of the product life cycle from conception through disposal, including quality, cost, schedule, and user requirements.

Thereafter, the concept was given currency in many ways. DoD projects came to insist on adherence to CE principles; contractors had to demonstrate how they would take into account the concerns of

This work was funded in part by DARPA grant number MDA-972-91-J-1022 awarded to the Concurring Research Center at West Virginia University.

Mechanical Engineers' Handbook, 2nd ed., Edited by Myer Kutz.
ISBN 0-471-13007-9 © 1998 John Wiley & Sons, Inc.

manufacturability at the time of design, and in general, how activities usually undertaken late in a project would be given early consideration. One of the barriers was the need for additional expense early in the project to form a larger group of people representing downstream perspectives of the design (manufacturing, maintenance, disposal, etc.). The phasing of DoD development budgets at the time was not compatible with incurring a higher expenditure early, in the hope that it would be more than offset by lower development costs later on, because there would be fewer glitches in manufacturing, fewer engineering changes, and so on.

11.2 ADOPTION OF CE

The adoption of CE was even more spirited in civilian manufacturing companies. Automotive companies, the civilian aircraft industry, electronics, and other significant sectors of the economy espoused concurrent engineering with vigor. Product development was the focus of concurrent engineering when it appeared, and it was noteworthy that the new lessons of CE were fully incorporated into the then 10-year-old efforts to achieve higher quality in American manufacturing. CE was seen as adding the time dimension to achieving quality, showing that not only was quality enhanced (the design and the manufactured reality would be compatible), but the time taken to achieve it could be decreased by getting all the relevant perspectives to work in concert from the very beginning.

The way in which civilian industry went about CE was not the same everywhere. Some chose to set up product development under one large roof so that engineers from all disciplines could interact with each other. The Chrysler Technology Center[2] is a well-known embodiment of this idea. Another approach to CE emphasized the need to share documents electronically among many people in the course of a very large project, so that the time-consuming and voluminous documentation could be developed faster and placed in the hands of the right people as soon as possible.

11.3 DEFINITION OF CE

Quite early in the development of CE, the Concurrent Engineering Research Center (CERC) was set up by ARPA, the Advanced Research Projects Agency (then DARPA, with the D for Defense). When CERC began examining CE a new, more general definition[3] was put forward:

> *CE is a systematic approach to integrated product development that emphasizes response to customer expectations and embodies team values of cooperation, trust and sharing in such a manner that decision making proceeds with large intervals of parallel working by all life-cycle perspectives early in the process, synchronized by comparatively brief exchanges to produce consensus.*

According to this definition, CE applies not just to engineering or product development, but to the general problem of decision-making in any domain; for that purpose, a new process was advocated, conforming to certain principles. The goal was set as response to customer expectations, a goal that was expected to pervade the actions of all perspectives at all times, though who the customer was still needed to be defined for each perspective. Most important of all, the essential means of achieving CE were set forth: teams of people working together with a shared goal and a set of values that included openness to sharing information at all levels, especially in a horizontal manner within the team, trusting that the exploitation of early information would not be to the disadvantage of the information donor.

11.4 THE CE TEAM

The mental picture is of a team composed of all the perspectives of the product being developed. The team meets together throughout the project; in a sense, the whole process of product development can be viewed as a series of meetings to share ideas and technical data and plan work, punctuated by long intervals of individual task accomplishment by members of the team.

11.5 THE ESSENCE OF CE

At the heart of CE lay some very simple ideas:

- *Focusing on Customers*
  - Formulating goals that seek out the customer's input early
  - Using metrics to evaluate tasks that are customer-driven
  - Tracking changes in customer requirements
- *Teaming*
  - Devolution of decision-making on a team selected for the purpose
  - Inclusion of all perspectives in the team
  - Joint problem-solving
  - Willingness to compromise

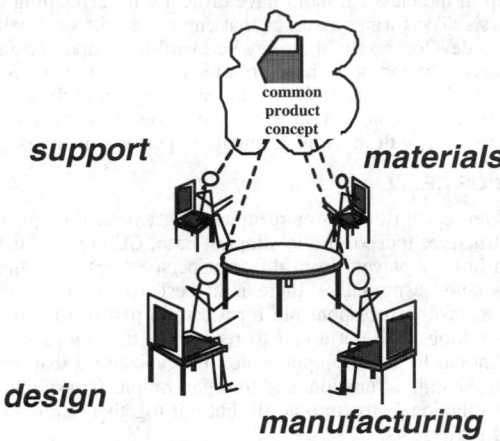

support   materials

design
                        manufacturing

Fig. 11.1 A CE team.

- *Working Cooperatively*
    - Attaining consensus
    - Planning early
    - Managing tradeoffs
    - Communicating regularly
    - Sharing knowledge early
    - Propagating the influence of one decision on others
    - Reducing risk
- *Improving Processes*
    - Planning a concurrent style of working and task breakdown among perspectives
    - Ability to work without complete data
    - Willingness to brook periods of inconsistency
- *Systems Engineering*
    - Integrating and automating

No single idea of CE was new in the late 1980s, when the rush began, but putting it all together resulted in new insights and inspiration for a new approach. The uniqueness of CE lay not so much in the fundamental insights as in some practical consequences of those insights:

- It defines the right composition of a team by an operational test
- It tries to resolve the conflict between:
    - Parallel working and consistency
    - Individual empowerment and teamwork
    - Early propagation of influence and rigorous system engineering
- It puts emphasis on the process rather than the organization structure
- It posits that all three conflicting goals can be achieved simultaneously:
    - Cost reduction
    - Quality improvement
    - Time speed-up

11.6 BARRIERS TO CE

The contradictions in CE are very real and some of the barriers to adopting it arise from the inability to reconcile them in a practical way for the routine working of the team. A new set of values has to be understood and observed by team members. Organizations previously given to a command style of work performance have to realize first the greater fruits of empowering individuals to exercise their own initiative, with goals set only at the most general levels possible. From there they have to learn the additional imperatives implicit in the accountability of individuals to teams.

Engineers brought up in the classical mold have difficulty in reconciling the demands of parallel work and rigorous analysis. The former assumes that engineers will work with tentative and incomplete data and proceed to develop results that may be overthrown upon critique at the next meeting when the results are shared. On the other hand, rigorous analysis works sequentially, step by step, with full information. It is discomfiting to have the usual sequential mode of working set aside in the interest of faster discovery of conflicts among perspectives. It will even be necessary to develop new tools and methods to cope with incomplete inputs to perform analyses.

11.7 APPLICABILITY OF CE

CE has a point of confluence with the newer reengineering tenets: it emphasizes the Process rather than the Organization structure. Indeed, in the simplest case, CE presents the organization structure as a team leader with a host of players from different perspectives — a single flat entity. In more complex cases (such as defense products), there is at best a two-level structure of a team leader supervising several sub-teams, each responsible for a certain part of the product. CE, in any event, is a particular way of working; it is not a call to reorganize the company or change the reporting structure. Therefore, it should be equally applicable to organizations that have the classic structure of departments, each representing a function, and to organizations (consulting companies are typical) who frequently have no functional structure at all, but put together teams to service a client drawn from several specialized individuals.

Though CE was invented in the context of ensuring that the design of a product is manufacturable, it applies as much to service companies as it does to companies that design and manufacture products. In services, too, a full range of perspectives should be brought to bear so that every aspect that can affect the service is allowed to contribute to the solution being developed, for the client. The software industry is a case in point. Clearly, the users, the system maintenance people, the designers, the programmers, the test staff, the technical documentation staff, and sometimes even hardware designers have to work together, in parallel.

Only thus is time saved and quality improved simultaneously. Quality can, of course, be improved by spending more time, and therefore more money, but how to accomplish it in less time and at less expense is what CE is all about. In that respect, it takes further and gives new meaning to Deming's statement "Quality costs less." He was alluding to the consequences of poor quality for the customer who has to put up with the problem and for the supplier who has to fix it. But CE says that you can achieve better quality in less time if you start by involving all the perspectives and share information among them continually so that they can react immediately when any decision seems to have poor consequences for later stages.

11.8 CE AND THE INDIVIDUAL

How desirable is CE? In spite of the emphasis on the team, one can agree that the quality of the company's work is determined to a great extent by the qualities of the individual, such as:

- Diligence in work
- Dedication to quality
- Level of skill
- Repertoire of tools
- Inventiveness

The team is the proper enabler for this individual competence to be raised to a collective level for tasks that require multiple people. The majority of modern artifacts, no matter how seemingly simple, involve multiple design and manufacturing technologies and several areas of competence to deliver. We should look upon the team as the way to deliver a consistent product when the required competence exceeds what any one or two individuals may possess.

11.9 TEAMWORK CAN LEAD TO CHAOS

Teams, though, result in chaos more often than not. There are many reasons:

- Teams are often a sham, never destined by their originators to coalesce.
- Teams rarely invest enough in achieving a common vision before setting out on the detailed work.
- Customer focus is more easily stated than subscribed to in practice.
- Teams do not keep practicing and working on team processes.
- Coordination is not given sufficient importance.
- Motivating factors are geared to recognize individual work, rather than team achievement.
- The leader of the team is not up to the job.

It is not easy or mechanical to work in teams and achieve a team identity. Achieving it overnight by decree is impossible, but without it the CE vision will remain a mirage.

11.10 CE AND INNOVATION

There is a mistaken idea that inventiveness is not encouraged in a team. It is thought that all invention or discovery results from quiet meditation by an individual who thinks long and hard about a subject and plays around with many ideas until a happy thought strikes him or her. This view is promoted by the long history of science and technology, where such, has been the mode of operation. Hence, people will be tempted to assume that the new mode of teamwork might result in product harmony and speed, but it will not break new ground or result in anything like a recognizable new invention, a patent, or a revolutionary product. This view has been overtaken by the sheer complexity of modern products and the pervasiveness of multiple technologies in the life cycle of single products. It has become commonplace even in science to have large teams of people working on experiments, whether it be in genetics, space science, or particle physics.

Wherein lies the source of discovery in such cases? One may recognize that the clash of ideas can spark new insights when people of like interests but different viewpoints congregate to discuss and explore a new field. This happens every day at scientific conferences; researchers return from the best of these gatherings to work at their research with quite new motivations and insights, garnered from discussions at a conference.

The same thing happens in a team at work continually. Complementary strengths occur in the persons forming the group, providing the foil needed to make new ideas emerge. The group milieu also provides the critique that "new" ideas must endure to survive and become "good" ideas. The cut and thrust of group debate allows ideas to be sifted more quickly.

11.11 CE LESSONS

CE is becoming more widespread. The principles are recognized and companies value it. But they have discovered:

- CE does not work without a mandate from above.
- CE does not work without a strong team leader.
- CE works best with physical collocation of the entire team.
- Technology-based CE is best achieved by integrating the tools employed by several perspectives.

For the computer scientist, the most interesting aspect of CE is that a number of new and old information technologies can be exploited to improve the efficiency of the CE process. As may be expected, these are the technologies to support collaboration within a group.

11.12 CONCURRENT ENGINEERING TECHNOLOGIES

11.12.1 Communication

The simplest of these are the technologies of communication, which have undergone a revolution. E-mail, in use for a quarter of a century, has now become the accustomed medium of communication among people at technical companies, even between those only a door or two from their working colleagues. This has been supplemented over the last few years by the appearance of multimedia mail, allowing complex documents to be exchanged. Simple team communication, ranging from notices and instructions to requests for information and action items, can be communicated very effectively by multimedia mail, especially if (as in Lotus Notes,[4]) the e-mail is given the structure of a database with templates to display the characteristic formats of different types of communication within the company.

Multimedia desktop conferencing software[5] has further augmented the possibilities for groups that need to meet from time to time. Team meetings can now be held over the network, with live graphics, video if need be, and interaction capabilities for everyone, quite akin to what they would have if everyone were in the same room. The blackboard at which technical people like to discuss has been replaced by a virtual "whiteboard" on which anyone can place a drawing and have it made visible to persons half a continent away on their own screens, who can immediately respond—by voice with speech packets conveyed by the data network, by modifying the drawing and annotating it for others to see, by playing a video for the sake of the audience, and so on. The possibilities are limitless, and even if body warmth is not communicated on wires, all the important details formerly missing in fax, phone, and so on are now richly present.

The lesson is, however, that if some of this is not captured systematically and made part of a meeting record, it will be lost for future decision-making. A structured way (decisions, action items, etc.) of capturing the record will enable building indexes for future reference. Storing the record and indexing it not only saves the corporate memory of salient events, but also provides a mine of

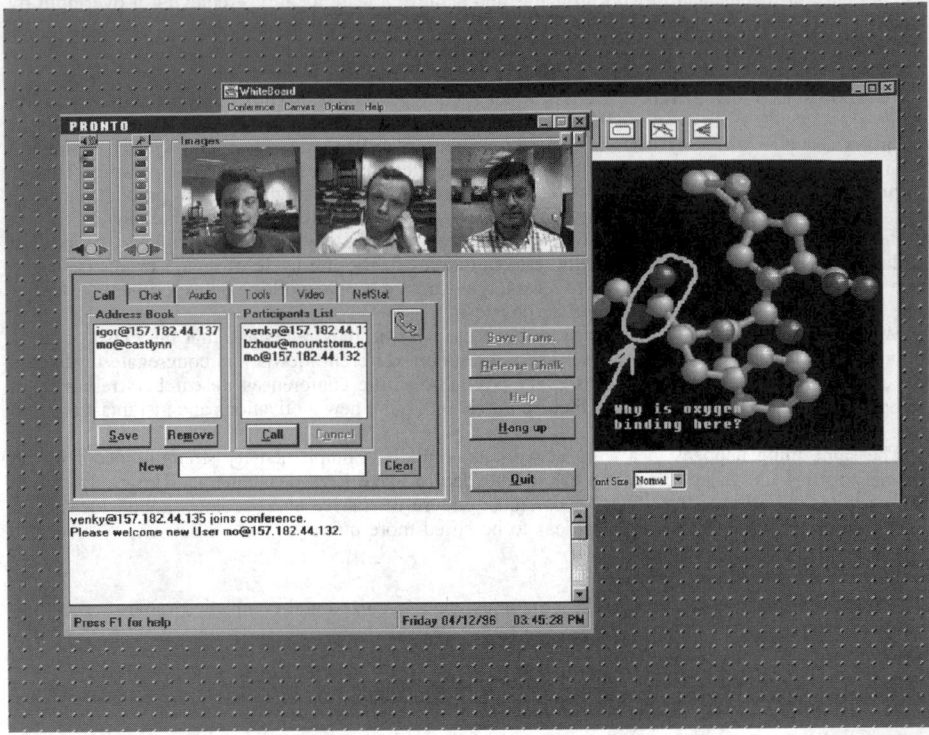

Fig. 11.2 A screen from the MONET desktop conferencing software.

information for turning past project history into a set of episodes that can be used as lessons for designers.

11.12.2 Task Coordination

When multiple persons work on a project, there is an essential need for coordinating their work over the long haul, so that their mutual interdependence is recognized:

1. Their dependence on each other for completion of prior assigned tasks, and

2. Their dependence on each other for the input required by assigned tasks and the output resulting from task performance.

When the tasks are performed in a sequence by different people in a wholly predictable manner, the technology of *workflow* is applicable.[6] Workflow packages allow a company to design a sequence of tasks to create a process that can be repetitively used. The workflow takes care of many things: notifying a person when a prior task is completed and he or she needs to perform the next task in sequence, making the package of data required to perform the tasks flow automatically to the desktop of the person performing it, so that all supporting paper documents may be eliminated, and storing the results of the tasks under stringent database security so that the final outputs and the intermediate stages of processing are all readily accessible from a database long after the process is over. Notification, routing, and database storage are important to coordination, and modern workflow packages have extensive and flexible support for all phases. However, workflow is not very good for non-repeatable processes, such as the classic CE case of product development.

For a one-off project, such as product development, there are no repeatable processes, except on a microscale. The task network for the whole project will be unique. In such cases, workflow has little to offer for collaboration support and one must appeal to standard project management—with a collaboration twist:

- The task network should be visible to all.
- Progress should be reportable by each individual from his or her workplace via computer.
- The data sharing should be via network databases.
- The flow of instructions, drawings, and work authorization should take place without paper.

- Questions should be handled electronically.
- The whole project history should be stored.

Such project-management packages are scarce today, but undoubtedly it is the wave of the future to conduct group work over computer networks. The difficulty is that you have to combine classic PERT techniques with interactive, distributed, task, and data management. We already see the glimmerings of a such new generation of desktop project-coordination tools with a significant collaboration flavor.[7] Such packages go beyond the standard metrics of time and cost to provide unique project assessment tools. All the metrics can be evaluated for single tasks, for sub-projects containing several tasks, or for the entire project. The display of these metrics as colors in Gantt charts provides a very useful qualitative view for managers. For instance, if they wish to assess the understandability metric of the project tasks, it can be shown on a Gantt chart in which the perfectly understandable tasks (all outstanding questions satisfactorily answered) are shown as green, those with over half the questions answered are shown as yellow, and those with less are shown as red.

11.12.3 Negotiation/Tradeoff

All design is compromise. During the collaboration, when alternatives are considered to solve a problem or different concepts are being evaluated to realize a product, it is necessary to have quantitative analyses of the cost/benefit. Simple spreadsheets are a way of capturing design goals, criteria, and design variables that influence the product performance from a customer standpoint.

It is impossible, however, to develop general support for this because the computation of product performance criteria is often a tedious numerical process that needs many custom-designed computer programs. Human judgment will intervene, too. Thus tools to support tradeoff among alternatives must perforce be custom and proprietary in nature.

However, when the weighing of ideas and concepts can be done in a less technical way and the collective judgment of the team can be made to bear on a decision, some computer support can be afforded for the process. Tools can be designed to support brainstorming and the evaluation of ideas among a team of distributed individuals over the network. One such is the software called Group Systems V from Ventana Corp;[8] another is the Group Decision System (GDS) of CyberMarché,[9] itself modeled on CM/1 from Corporate Memory Systems.[10] The latter software allows the team to start a discussion focused on a problem under a leader. Members of the team can supply solutions, supported by arguments. The whole decision grows as a tree from the discussion node to the solution nodes, and thence to the argument nodes. The leader can put in motion a decision-pruning process at the end that involves voting, first on the arguments and then on the solutions, having in mind the customer criteria stated in advance.

Group Systems V is a capable package that is being used in many organizations to aid the group decision-making process while preserving anonymity. It has a very flexible set of voting protocols

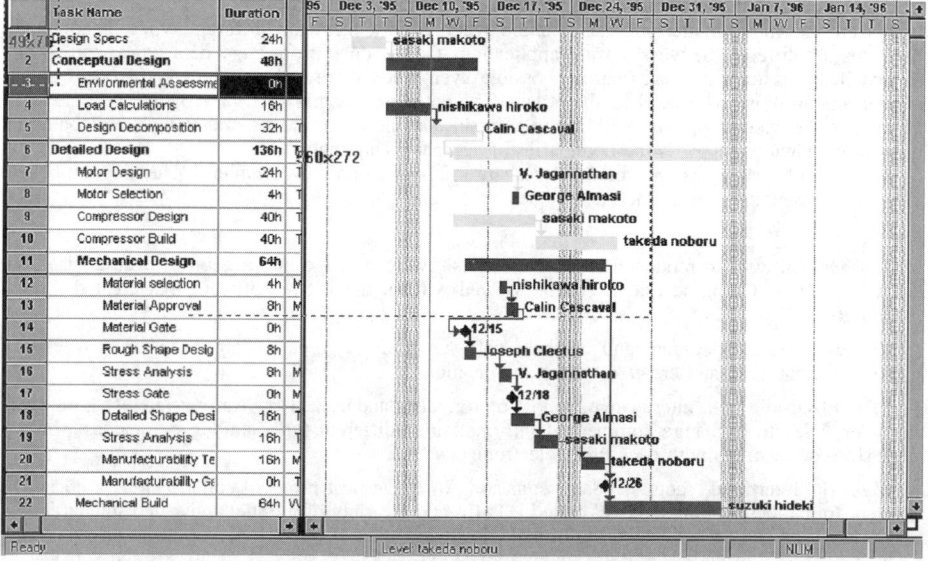

Fig. 11.3 A distributed project assessment chart.

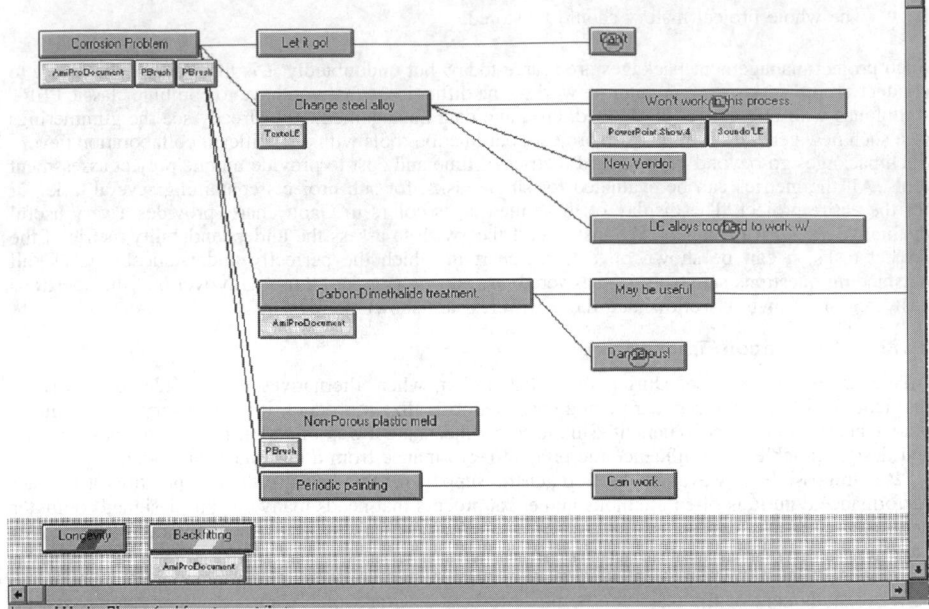

Fig. 11.4　The main screen of a well-developed discussion in GDS.

and automatically produces the minutes of group sessions held under its aegis. A good team facilitator is a must, and the process is usually carried out in a decision room equipped with networked personal computers, though there is nothing in the software itself to prevent its use by remotely connected team members. It is the team leader function that chiefly demands face-to-face interaction to drive the meeting from one stage to the next. In this it falls short of the kind of remote communication that is preferred, in general, for collaboration tools.

11.12.4　Data-Sharing

The most important among technologies to support the collaboration required by concurrent engineering are those that enable the storing and subsequent sharing of information. The technical barriers to this are enormous in the general situation when the users are on different types of computer platforms, on different networks, using applications that produce proprietary data types, and so on. The challenge is to make data-sharing possible even when there is heterogeneity of one kind or another among those who need to share data. The consensus is that such heterogeneity is the rule, rather than the exception, even within one company or one department—and today's collaboration must be worldwide and cross the boundaries of companies and networks.

A variety of techniques and software to share information are now available. The diagram below shows a representative set of such techniques.

Acrobat: Software to render the output of any software package into a neutral form, devised by the Adobe Corp., so that it can be viewed without the native software that created the output.

Notes: Forms-based e-mail and database software useful to create shared databases over a wide area network and create workflow applications.[4]

EDI: Electronic data interchange, a host of formats standardized by various interest groups under ANSI to exchange commercial information of different types among companies who wish to do business with each other electronically.[11]

Web: The Internet de facto standard, consisting of a transport protocol called HTTP, a document format description standard called HTML, and a variety of graphic and video standards, so that users can access linked multimedia documents placed on Web servers from anywhere in the world. Currently, there are many application packages, such as databases and spreadsheets, that are readily accessible using the Web interface, without extra work.[12]

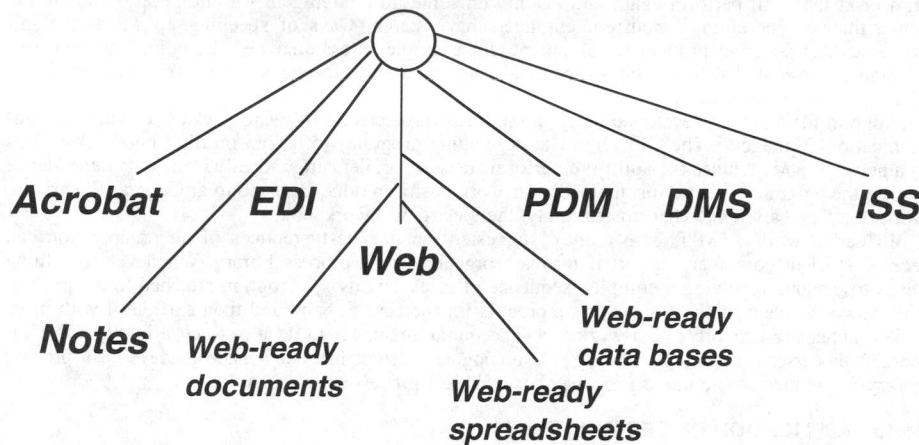

Fig. 11.5 Data-sharing technologies.

PDM: Product Data Management Systems that store information about products, their decomposition structure, and revision history, in order to support a company's manufacturing, inventory control, maintenance, and other technical activities.[13]

DMS: Document-management systems that perform much the same as PDMs, for the more general class of documents occurring in a company. Also enables fast retrieval by indexes, keywords, document types, and even text search.[14]

ISS: An information-sharing system that constructs a model of the disparate information lying in different databases, builds gateways to each of them, and provides a single system image to external users who wish to access data from any of the constituent databases. A virtual unification of databases.[15]

11.12.5 Electronic Design Notebooks

The support for design rationale capture is quite poor in engineering organizations. The best that is done in a systematic fashion is to annotate drawings when they are revised to state on the drawing the nature of the revision and its cause. However, this has many weaknesses in a complex project.

When it is first made, the drawing does not contain a justification for each of its constituent parts, and that situation will persist through all its revisions. Much of the design rationale and original design intent will never be captured. Further, when a change is made to one part of the product, consequential changes may need to be made to other parts. Unless the link between the parts is maintained and examined whenever changes are made, it is possible that a revision will be made that is inconsistent, unsafe, or inefficient. When variant products are designed, what can be changed and should be changed to suit the new requirements, and what may not be changed without severe disadvantage and redesign, is not clear. Indeed, a design process involves alternatives that were considered and rejected, and the memory of the unselected alternatives is lost in conventional records, since it may not even have been recorded.

The idea was mooted long ago that these things could be cured if only the designers captured their daily work electronically. Many so-called electronic design notebooks (EDN) were prototyped[16] so that the steps of the design work done by individual engineers could be captured in documents with CAD images, analytic results, annotations, design criteria, and so on recorded, along with indexes by which they can be retrieved in future. Since a great deal of product-development work is done on the computer, it might be possible to add the additional support to capture the important results and annotate them without too much added effort for the working engineers, whose aversion to documentation is itself a difficult barrier.

EDNs have not come into vogue, probably because they are still clumsy to use and demand a lot of work for little payoff—a payoff, moreover, that accrues not to the documentor, but to some future engineers on another project. It remains one of the aspects of collaboration that has the least satisfactory support, perhaps because the goal is to encourage collaboration between people over unknown expanses of space and time and with unspecified projects lying in the future.

11.12.6 Process Libraries

Though processes in product development are not strictly repeatable on a macro-level, when you zoom in and look at the individual tasks performed by development engineers, there are many standard tasks they will perform again and again with some little intelligent variation appropriate to the new situation. Therefore, concurrent engineers have sought ways of speeding up standard mini-processes that occur in product development (for example, stress analysis). Usually, a sequence of programs is executed with the piping of data from one step to the next, with some conversion and filtering in the general case.

Support for this was developed early in the DICE project in a software package called the Communications Manager,[17] The CM allows the executing programs to be run on different workstations in a network, and includes the ability to perform steps in parallel automatically if the data dependence among the programs will permit that. Similar work has been done by a group at General Electric, in work that they call, somewhat misleadingly, the electronic "workbook."[16]

In earlier work at MIT, T. Malone[18] suggested that the best practices of an organization can receive efficient computer support if they are organized as a process library. Whenever something needs to be done involving a complex sequence of steps, an adviser program attached to the process library would select the best appropriate process for the case at hand and then execute it with more or less automatic support, like a workflow. The initial attempt is to develop such a library for commercial processes, such as purchasing. No analogous attempt has been made to develop a process library of product development processes, or technical processes, in general.

11.13 APPLICATION OF CE PRINCIPLES

The principles of concurrent engineering are applicable to every field of human endeavor. There is no problem of any dimension that does not require far more knowledge and competence than a single human can possibly possess today. It seems a commonplace observation that a problem can have a satisfactory solution only if all aspects of it are considered and every potential perspective satisfied. But for one reason or another—sometimes from a desire to exercise autocratic power, sometimes from unreasoned haste, often from a lack of mental rigor to see the problem whole, and perhaps from a lack of appreciation of what concurrent engineering entails—problem-solving is oversimplified.

Ironically, achieving a satisfactory solution takes more time when shortcuts are taken that violate CE principles. This is the enduring lesson, often learned and forgotten. Though some of the technology that can support CE has been described, this compendium must not give the impression that the technology is primary. It is not. What is of primary importance is to act according to CE principles, by seeing the problem as a whole, involving all pertinent perspectives from the start, instilling a common vision, and having the will to exchange openly and pursue a path of collaboration.

Of course, much of the supporting information technology is just as universal as the CE principles. Therefore, the underlying collaboration technology, no less than the CE principles that it attempts to realize, can be applied equally, for example, to planning and executing an economic development project and to a technical engineering project.

REFERENCES

1. R. I. Winner, J. P. Pennell, H. E. Bertrand, and M. M. G. Slusarzuk, *The Role of Concurrent Engineering in Weapon Systems Acquisition*, Institute of Defense Analyses Report R-338, December 1988.

2. http://www.chryslercorp.com/

3. K. J. Cleetus, *Definition of CE*, CERC Technical Report CERC-TR-RN-92-003, March 1992.

4. Lotus Notes Spec Sheet. Part No. 34717, Lotus Development Corporation, Cambridge, MA. Also found at http://www.lotus.com/notesdoc/21d6.htm

5. Kankanahalli et al., *MONET: A Multimedia Conferencing System for Collocating People and Programs*, CALS and CE Conference, June 1991.

6. K. Center and S. Henry, "A New Paradigm for Business Processes," in Workflow Conference on Business Process Technology, The Workflow Institute, San Jose, CA, 1993.

7. K. J. Cleetus, C. G. Cascaval and K. Matsuzaki, *PACT—A Software Package to Manage Projects and Coordinate People* (to be published).

8. J. Nunamaker, A. R. Dennis, J. S. Valacich, D. R. Vogel and J. F. George, "Electronic Meeting Systems to Support Group Work," in *Communications of the ACM* **34**(7), 40 (July 19).

9. K. J. Cleetus and G. Almasi, *GDS—A Group Decision System for Teams* (to be published).

10. E. J. Conklin, "Capturing Organizational Memory," in *Proceedings of Groupware '92*, D. Coleman (ed.), Morgan Kaufmann, San Francisco.

11. B. Orlando, *Electronic Data Interchange (EDI) in a CALS/CE Environment: A User's View*, CE and CALS Washington, 1993.

12. What Is the Web?, a presentation found at http://www.cern.ch/CERN/WorldWideWeb/WWWandCERN.html

13. Hewlett-Packard Co., White Paper, *Understanding Product Data Management*, found at http://www.ideal.com/pdmic/undrstnd.html

14. Novell Inc., White Paper, *SoftSolutions Document Management Architecture,* found at http://wp.novell.com/groupwar/softsol.htm

15. V. Jagannathan, R. Karinthi, M. Sobolewski, and G. Almasi, "Model Based Information Access, in *Proceedings of the Second Workshop on Enabling Technologies: Infrastructure for Collaborative Enterprises,* IEEE Computer Society Press, Los Alamitos, CA, 1993.

16. J. W. Lewis and K. J. Singh, "Electronic Design Notebooks (EDN): Technical Issues," in *Proceedings of Concurrent Engineering: Research and Applications Conference,* Concurrent Technologies Corporation, Johnstown, PA, 1995.

17. R. Kannan, K. J. Cleetus and R. Reddy, "The Local Concurrency Manager in Distributed Computing," in *Proceedings of the Second National Symposium in Concurrent Engineering,* Concurrent Engineering Research Center, Morgantown, WV, February 1990.

18. T. W. Malone, K. Crowston, J. Lee, and B. Pentland, "Tools for Inventing Organizations: Toward a Handbook of Organizational Processes," in *Proceedings of the Second Workshop on Enabling Technologies: Infrastructure for Collaborative Enterprises,* IEEE Computer Society Press, Los Alamitos, CA, 1993.

CHAPTER 12

CONCURRENT ENGINEERING TECHNOLOGIES

V. Jagannathan
Y. V. Reddy
K. J. Cleetus
K. Srinivas
R. Karinthi
Concurrent Engineering Research Center
West Virginia University
Morgantown, West Virginia

12.1	INTRODUCTION	261
12.2	COLLOCATION SERVICES	262
	12.2.1 Technology Overview	262
12.3	COORDINATION SERVICES	264
	12.3.1 Technology Overview	265
12.4	INFORMATION SHARING	267
	12.4.1 Technology Overview	267
	12.4.2 Related Research	270
12.5	CORPORATE HISTORY MANAGEMENT SERVICES	271
	12.5.1 Issues in Product History	271
12.6	CONCLUSION	274

12.1 INTRODUCTION

Effective collaboration among members of a team is the key to success, whether the team is a group of engineers designing a new engine or a group of physicians planning a medical procedure. This is now widely recognized, as can be seen by the numerous national initiatives emphasizing teamwork such as Concurrent Engineering (CE)[1], Total Quality Management (TQM), Integrated Product Development (IPD), Open System Architecture for CIM, and the Virtual Enterprise (VE).

Advances in database and networking technology, Internet technologies, groupware, multimedia, and graphical user interfaces, as well as a steep drop in the cost of computing, make possible the creation of a truly collaborative environment that transcends the barriers of distance, time, and heterogeneity of computer equipment. The ideal collaborative environment will enable any member of a team to communicate spontaneously, and thereby collaborate, with any other member of the team. This chapter provides an overview of technologies that facilitate geographically distributed teams to work together. Four primary categories of infrastructural services are needed to support collaboration: collocation services; coordination services; information-sharing and integration services; and corporate history management services. These are discussed in the remainder of the chapter.

Numerous people over the past eight years contributed to the development of the material presented here. In particular, Ravi Raman, Dan Nichols, and Felix Londono have contributed to various versions of the material.

This work was funded in part by DARPA grant number MDA972-91-J-1022 and NASA grant number NAG 5-2129, awarded to the Concurrent Engineering Research Center at West Virginia University.

Mechanical Engineers' Handbook, 2nd ed., Edited by Myer Kutz.
ISBN 0-471-13007-9 © 1998 John Wiley & Sons, Inc.

12.2   COLLOCATION SERVICES

Informal meetings and scheduled conferences are essential for teamwork, since they provide opportunities to inform others of ongoing work, consider cross-functional issues, and negotiate to harmonize viewpoints across multiple perspectives. Studies have shown that knowledge workers spend a large percentage (20–70%) of their time attending meetings and conferences. The parallelism implicit in concurrent engineering requires that team members perform simultaneous work activities without waiting for each subteam to report its results; the penalty for this method is a lack of consistency. Therefore, the basic concurrent engineering process itself envisages periodic meetings to bring about convergence.

When people who are geographically dispersed are required to meet at the same location, however, significant travel time, money, and energy are spent, leading to decreased productivity. Moreover, in most meetings and conferences, the participants are not equipped with all the information they might need to function effectively. In other words, they are dislocated from their ideal work environment. Furthermore, some meetings are totally unstructured and free-form, which diminishes their effectiveness. And many of these meetings have no effective mechanism for archiving all of the events that occur for future use.

The solution to these problems is to use existing computer and communications technology to overcome the distance barrier. The use of technology cannot only cut travel costs and time, it can also increase productivity by enabling individuals to, in a manner of speaking, bring their offices to their meetings.

This section describes the technologies available to support meetings involving distributed members of a virtual team.

12.2.1   Technology Overview

Several computer-based services are now deployed to support group communication. These services can be classified according to time and distance (Table 12.1).

Electronic Messaging

The simplest group communication service is electronic mail (e-mail), now widely used in industry, government research labs, and universities, and available to the general public through long distance carriers. Electronic mail is a useful facility for keeping members of a team in contact during a project. Minor structuring of the messages can provide a convenient way of performing daily work: disseminating task assignments, receiving notices of various kinds, and requesting information. A slight enhancement of electronic mail is the electronic bulletin board, which can serve as a discussion forum for recording and gathering views, ideas, analyses, and other information. An electronic bulletin board not only enables the rapid development of ideas and consensus, but it also generates an automatic corporate memory. A further enhancement indexes the discussion messages so that anyone who wishes to benefit from earlier knowledge can rapidly retrieve the archives.

Recent products, such as Lotus Notes, carry this structuring much further in two respects: first, in generating structured databases to serve the messages belonging to different categories (for example, engineering change notices, task assignments, customer complaints, meeting announcements, etc.) using organization-specific indexes; second, by making it possible for messages to contain attached graphic files produced by any application so that the recipient can view them as long as the application can be executed on his or her computer.

Computer-Supported Meetings

Computer-supported meetings come in several varieties.

Xerox Colab started with the idea of holding a problem-solving meeting around a "chalkboard" with several participants. They constructed a meeting room in which each participant had a computer in a connected network, and everyone could view and manipulate the contents of an electronic chalkboard on the computer screen via a "what you see is what I see" (WYSIWIS) chalkboard. The participants can also converse face to face. At the front of the room is a large electronic chalkboard—a

Table 12.1   Communication Services

	Same Time	Different Time
Same place	Computer support for face-to-face meetings	E-mail, computer supported asynchronous meetings
Different place	Desktop multimedia conferencing from the workplace, computer supported meetings	E-mail, computer supported asynchronous meetings

larger-scale replica of what is displayed on each screen—and a podium from which a speaker can manipulate that chalkboard directly.

The chalkboard provides a shared memory, enabling meetings to be focused and allowing direct and simultaneous participation. Later, Colab evolved in two directions.

The first direction involves a structured and recorded meeting. The meeting is organized into distinct phases—brainstorming for ideas, organizing by categorizing and ordering the ideas, and finally, evaluating the ideas and agreeing upon conclusions. The basic tool is a word processor with an outlining capability and a list manager. Once again, the participants operate from their own workstations. The team can determine whether the sessions are done synchronously with all the participants simultaneously present, at different times, or as soon as possible. With this kind of meeting facility, the entire group advances to the next stage only when the previous stage is complete.

This structured decision support capability is carried much further in some recent commercial software, compressing the time taken to arrive at decisions, particularly in the synchronous mode of operation, by a factor of 10! Group Systems V (from Ventana Corporation) is a Group Decision Support System (GDSS) allowing synchronous or asynchronous meetings to take place with some structured phases of decision-making; that is electronic brainstorming. Other options include Categorizer, Voting (seven kinds), Topic Commenter, Group Dictionary, Alternative Evaluation (multiple criteria voting), Policy Formation (group writing to devise a short mission or policy), Idea Organization (powerful list-building and organization, e.g., nominal group technique), Group Outliner, Questionnaire (on-line, fill-in-the-blanks, survey tool), Stakeholder Identification, Group Writer (everybody works on different sections), and Group Matrix (two-way analysis of agreement between values for different criteria for different alternatives). VisionQuest (from Collaborative Tech Corp.) is much like Group Systems V and is in regular use at an electronic meeting room for rent at a Marriott Hotel in Washington, D.C.

The second direction Xerox Colab took involved the addition of an argumentation facility, the Argnoter. With this facility, someone proposes an idea, someone else then raises an argument for or against the idea, others ask questions regarding the proposal, and yet others raise issues that come up in the consideration of the proposal. This sequence of argumentation is structured and made commonly visible to all the participants, whether or not they are currently signed on to the discussion or arrive later and wish to take part.

Many incarnations of Argnoter now exist. IBIS (Issue Based Information System) and gIBIS (Graphical IBIS) are two examples. CM/1, a commercialization of gIBIS designed at the Microelectronics Computer Corporation (MCC), is a groupware system for qualitative decision-making support, shared issue exploration, decision mapping by a group, and documentation of decision rationale. The anticipated results include organizational learning, better decision-making, and greatly enhanced productivity for collaborative work groups.

Desktop Conferencing

Another type of tool for virtual meetings is the *desktop conferencing system*. A desktop conferencing system consists of hardware and software that enable real-time, full-motion video and real-time audio conferencing. Some systems also enable application sharing and the archival of meeting minutes.

Desktop conferencing systems provide a significant advantage to professionals who need to frequently consult and cooperate with team members at other sites. Users can conveniently and effectively communicate in face-to-face meetings because they can see each other, notice each other's facial expressions, hear each other's voices clearly, and use whiteboards and other media to draw pictures, take notes, and point to items on the screen.

There are currently a number of desktop conferencing tools available commercially, including Intel's ProShare and CU-SeeMe (Cornell University and White Pine Software). Research prototypes include the Meeting on the Network (MONET) system at the Concurrent Engineering Research Center, West Virginia University.

Application Sharing

Application-sharing technology makes the information displayed on one computer simultaneously available on multiple computers. This is a very powerful technology for collaborative work and it has innumerable applications. This technology has been used to develop group editing tools, whereby a number of people can jointly work on developing a document. It can also allow people to present their data, viewgraphs, spreadsheets, design documents, and other materials to other people, all from their own workstations. It also addresses how multiple participants can interact with an application program (such as a finite element modeler), make changes, and see the effects of the changes. Example of application-sharing tools include the COMIX system (West Virginia University), XTV (Old Dominion University) and Shared-X (Hewlett-Packard).

Conferencing and application-sharing technology are rapidly maturing and show great potential in supporting collaboration over the network. Flexible support for latecomers to such technology-assisted meetings still raises some hurdles. Some interesting solutions are suggested by Abdel-Wahab of Old Dominion University. For instance, how can such a person be rapidly briefed on what has transpired up to that point? There is also the issue of managing multiple applications simultaneously. For instance, in a meeting that involves marketing, design, and customer support, marketing may want to share a document and design the output from a CAD tool, and customer support may want to open a spreadsheet or database with failure rate information. Finally, the most significant challenge in deploying this technology is how to deal with the heterogeneity in hardware and software. Building tools that work on different hardware platforms and can cooperate with the software that exists on all of these platforms is still a difficult problem.

Audio Technology

Advances in the field of digital audio are opening new doors in improving productivity in our group work environments. From one's desktop computer, it is now possible to participate in a conference call, send and receive voice mail, annotate documents with voice clips, and give remote viewgraph presentations. Advances in voice synthesis have resulted in more realistic synthetic speech. There are some programs (e.g., at MIT's Media Lab) that can even provide the speech inflections associated with laughter and other emotions. Advances in speech recognition will one day make possible the conversion of voice annotations and speech to text. That day is not far off!

Some of the remaining challenges include managing audio quality over the network, delay and jitter control, and satisfying the hard real-time constraints posed by the nature of audio data.

Video Technology

Rapid advances in video and compression hardware technologies are making it feasible to develop and deploy multimedia applications. These include video conferencing over a computer network, support for multimedia mail, and use of this technology for a variety of other collaborative work.

Transmission Technology

One method for transmitting data over networks efficiently to a large number of computers is *multicasting*. Like radio and television broadcasts, computers can tune to specific frequencies to intercept messages destined for multiple host machines located worldwide. Ideally, only a single message is needed to contact all host machines. For unreliable delivery, the sender can simply send the message continuously, as in the case of audio and video data. For reliable delivery, the sender can request positive acknowledgments from a specific number of recipients.

Multicasting is currently being used experimentally to send audio, video, and shared data over *wide area networks* (WANs). The experiment, known as the *Multicast Backbone* (MBONE), has been using multicasting successfully since 1993 via Level 2 IP (Internet protocol) packets. The bandwidth required is at least T1 for a limited number of conferences. The same software and protocols, however, should be usable given larger-capacity networks (i.e., T3). Indeed, many MBONE sites (Xerox, Bellcore, Lawrence Berkeley Laboratory) are currently running on an experimental Gigabit network known as Xunet. The IP protocols are also evolving and the future versions, such as IP.v6, have better support for multicasting.

Asynchronous Transfer Mode (ATM) technology is also evolving rapidly and is inherently better suited for the high-bandwidth and real-time needs of high-quality conferencing and video-on-demand applications.

On the other end of the bandwidth spectrum are mobile and wireless links.

12.3 COORDINATION SERVICES

Traditionally, task coordination has been largely a human process. With the significant growth in the employment of multidisciplinary tiger teams, however, computer support is critical for group decision-making and negotiation, especially over a geographically dispersed network. Particular features of task coordination systems include common visibility of activities and data, planning and scheduling of activities, change notification, and constraint management across multiple perspectives.

Some systems, such as bulletin boards and electronic mail, provide an initial underpinning to support group working, but are very limited and informal. Fundamentally, they only allow for the exchange of messages, although they are being adapted to support brainstorming and group discussions. However, they do not support structured decision group working.

The team structure must be expressible in computer structures which mirror the organization of the project into a number of teams spanning many functional areas and ultimately many organizations. This imposes a substantial requirement that a project-coordination intelligence be pervasive in the network, so that wherever a person is located, that person can be deputed to belong to several teams at once. The team's membership profiles, constraints, common workspace, and tasks thus become visible, making it possible for a person to belong to any project, serve any role, and participate in all the team interactions at once, without leaving the workstation.

Coordination theory and technology are topics of high interest in current and recent research activities.

As part of the DARPA Initiative in Concurrent Engineering at the Concurrent Engineering Research Center, a system known as the Project Coordination Board was developed to support coordination of product development activities.

Coordination is being considered as part of the Computer-Supported Cooperative Work (CSCW) research efforts. The National Science Foundation has launched the Coordination Theory and Collaboration Technology Initiative under the Computer and Information Science and Engineering Directorate (CISE).

The Center for Coordination Science at the Massachusetts Institute of Technology is an interdisciplinary team studying new ways to organize human activity and developing new technologies to help people work together more effectively. In their view, coordination technology will provide benefits to humanity equivalent to those provided by the economies of production and transportation during the Industrial Revolution.

The technical approach of the Coordination Theory and Technology Project at MCC uses distributed systems technology to support the flexible automation necessary for coordinating people, tasks, and resources involved in organizational activities.

12.3.1 Technology Overview

Particular features of task-coordination systems include common visibility of activities and data, planning and scheduling of activities, change notification, and constraint management across multiple perspectives.

Common Visibility and Change Notification

Concurrent work involving many functional areas must be coordinated via a common workspace in which the actual work of product developers is made visible to assure structured group working. Conceptually, the common workspace is equivalent to the meeting table around which product developers gather to discuss and reach consensus in traditional engineering environments.

The common workspace must provide constant visibility of a unified cross-functional product model that provides directives for the information needed by product developers. Product developers can view and access components of the product structure within each domain of specialization. The common workspace, through nodes in the product structure, provides access to design information required during the product development cycle. It is also through iteration with this product model structure that product developers assert their design decisions onto the common workspace: a product developer locates some information required for analysis, executes a tool, and obtains the results by selecting appropriate nodes in the product structure. In principle, the common workspace provides immediate access to the information required by product developers to do their work. It also allows product developers to share the results of their work with their peer team members.

Product developers are concerned about how design decisions made by others influence their work. The product-development effort is driven, in part, by the existence of customer requirements, rules of design, policies, constitutive equations of engineering relating variables in different domains, and so on. They help guide and shape the product-development process, but they also raise conflicts and inconsistencies across perspectives. The common workspace must provide product developers with visibility of conflicts and inconsistencies across perspectives that affect their work. This aspect of visibility is supported by functionality that manages all types of dependencies, relationships, and constraints that exist between components of the product model. Notification mechanisms are implanted to support visibility of design decisions as they affect work of the virtual team members. This is key in assuring that everyone affected by a design decision participates in the final decision-making process.

Visibility of work concerns visibility of the activities performed by the group. Management of the relationship between the product and the process models must result in a dynamic approach to process management in which tasks are planned, scheduled, and monitored over the computer network. The common workspace is the place where the network of activities becomes visible. Product developers use the common workspace to view and respond to "work units" assigned to them by a project leader during the product development life cycle.

Finally, being the medium for interaction among product developers, the common workspace is also the place where consensus about conflicting design decisions is reached by product developers. This negotiation framework in the common workspace allows for the resolution of conflicts while exploiting trade-off analysis information to assure that the best design alternative is selected from the various design decisions posted into the common workspace. Functionality to support negotiation to reach consensus has two requirements. First, the framework for negotiation must provide means for human interaction across the computer network. This requirement was covered above in Section 12.2. Second, the framework for negotiation must exploit information available from the system to facilitate trade-off and multiobjective analysis in the decision-making process.

Managing Workflow

Coordination of the virtual team involves management of the ongoing concurrent work in many functional areas. It involves managing (over the computer network) the workflow of activities performed by the virtual team members.

Existing project-management packages are all oriented to repetitive and completely foreseen sequences of tasks from beginning to end. A concurrent engineering approach to product development calls for exploration, opportunistic contributions, and joint planning of work. Therefore, dynamic workflow management is required to support the planning, scheduling, and monitoring of tasks.

In principle, workflow management should extend the meaning of "manufacturing process" to all the processes occurring in the product development cycle. Also, workflow management must provide for the management of the critical relationship between the product model and the process model.

Workflow management involves the ability to reuse process models from previous projects. Each process or activity must be linked to the part or sub-part that is the focus of the activity, in order to provide for management of the relationship between the product model and the process model. Process models must be refined into atomic activities woven into a network of task scheduling units, providing for the initialization, dissemination, retrieval, monitoring, and overall management of these task units over the computer network. It means that team members receive electronic work orders via the computer screen and respond to them appropriately. This is core functionality to provide for dynamic management of the workflow in a CE environment

Tracking Design Progress

One aspect of assessing progress in a design is the quality measure attached to the results of the tasks. If assistance to track the results against product performance goals is not present, task leaders will be able to see much activity, but they will be unable to determine whether the activity is converging to the customer-desired product. A system to evaluate product performance metrics whenever the leader desires or when sufficient data are available would lessen the burden of tracking progress.

Assessment involves the ability to judge the quality of the evolving design based on initial specifications such as customer requirements, safety regulations, and standards. Evaluating these criteria requires that they be known, and this information can come from the standards and guidelines adopted and constraints created by specific customer requirements. Performance metrics, used to measure the performance of various components, also offer assessment capabilities. This can involve optimization of parameter values and requires the ability to specify the desired criteria.

Another area of assessment involves such "ility" components as manufacturability and maintainability, which are not explicit aspects of customer requirements but are determined by engineering domains and are an integral part of the design. Finally, product development performance measurements, such as adherence to the task schedule, prove a key area of assessment. These assessment capabilities partly depend on the availability of domain specific tools to evaluate performance requirements, especially with "ility" components and performance metrics.

The key requirement for monitoring the progress of a design is passing the "ACID" test: the ability to ACCESS information about the current design and past, similar designs; the ability to CHOOSE the parameters of the design to monitor and the ability to choose analysis routines to further analyze the information; the ability to INTERPRET the information that are collected; and the ability to DISPLAY the information gathered in an intuitive fashion.

Assessments can be valuable only with accurate, complete information. Therefore, to monitor the progress of the design, any source of information that can contribute to the assessment must be accessible. This includes information about the current design as well as past designs of a similar nature. Often, past designs can lend themselves to comparisons of current design situations.

Access stands not only for access to information, but also for access to analysis routines. These routines can provide more than just a filtered view of the data. Quality evaluation methods can provide valuable insight into the current state of the design based on the intended goals.

Once access is provided to the information and the analysis tools, users must be able to choose what information to monitor and which routines to invoke to analyze the data. Choosing the information to monitor implies determination of the key parameters crucial to the design. Techniques that can aid a designer in choosing the key parameters include Quality Function Deployment (QFD) and constraint management.

Constraint Management

QFD indicates customer requirements; by weighting those requirements for importance, the key ones can be determined. Through successive iterations of QFD, these can be translated into lower-level constraints on product parameters which can be maintained by a constraint manager. Through this translation, low-level product parameters can be monitored and their values propagated to determine their effect on customer requirements. In this manner, the progress of the design can be determined based on customer requirements.

The constraints maintained for a design come from a variety of sources—not just customer requirements. In fact, one method of tracking design progress (and design "goodness") is by monitoring constraint satisfaction. Through the constraints, allowable/optimal values for parameters can be determined. Therefore, by using both QFD and constraint management, a designer can determine the key parameters to monitor, and, in some cases, the measurements to determine the direction of progress made on those parameters.

Once the user has determined the key parameters to monitor and indicated the tools and methods that should be invoked to evaluate those parameters, the collection and interpretation of those parameters should occur automatically. As sufficient real-time data are collected for the methods to interpret the data, they should be invoked.

The information that is being monitored should be displayed in a manner intuitive to the decision-making process. That is, the user should see the results in a natural manner without the need to sift through the reports/information the tool generates. To accomplish this, the tool should have access to graphical (plots, charts, histograms, etc.) and textual (tables, text, etc.) libraries to compose reports in a variety of manners. In addition, these results should be displayed in a timely manner. Reports concerning the assessments should be available as needed.

Another feature of display is the ability to "navigate" through the results to view aspects of the monitored design to the required level of detail. That is, users should see a high-level report of the design progress, but should be able to view aspects of the monitored information in sufficient detail to make quality decisions about the design.

As an example, consider an assessment which monitors constraint activity, collecting all of the constraint violations. At the end of the week the Project Leader (PL) notices in the report provided that a constraint on the blade's strength has been violated ten times whereas no other constraint has been violated more than twice. The PL may want to know which designers have violated that constraint and what tasks caused the constraint violations. The desired information should be available through the report through requesting more information about the constraint's history.

Once this information is displayed, the PL notices that the aerodynamics engineer has violated the constraint seven of the ten times when attempting to determine the optimal flow path for the blade. This information may cause the PL to request a meeting between the engineer who set the constraint (in this case, the mechanical engineer) and the aerodynamics engineer to resolve the problem so that the strength constraint can be satisfied and the flow path can be optimized.

12.4 INFORMATION SHARING

In addition to communicating with each other and coordinating teamwork, team members must have ready access to the information necessary and appropriate for their tasks. Because corporate information exists in a variety of computer information repositories, such as databases, documents, drawings, and data files, it is imperative that an information-sharing system be utilized to provide a single interface to these sources of information.

The information must also be indexed and accessible via a type of electronic card catalog. This is a recognized need for which an ANSI standard solution exists—Information Resource Dictionary System (IRDS). An IRDS contains metadata—data describing other data in the organization. It consists of a dictionary which identifies various information resources and describes their logical structures, and a directory that describes the location and protocol by which such information may be accessed.

Some of the major criteria for information-sharing technologies include:

- Ease of use
- Performance objectives such as response times
- Integration with existing/fielded systems/environments
- Cost-effectiveness, economic viability, scalability: some of the major factors relating to scalability are the number and types of information, the number (total and simultaneous) of users (providers), and the geographic distribution of the elements of the system
- Ability to address security, integrity, consistency, and proprietary and intellectual property rights concerns
- Need for legislative mandate—some of the technological solutions such as digital signatures have not been put to test in a court of law and may require legislative support for implementation

12.4.1 Technology Overview

A variety of techniques have been developed to enable wide-area access to information. Increasingly, organizations are beginning to rely on the Internet-based information technologies to communicate and share information. Particularly effective are tools that provide access to enterprise databases over the Internet, especially over the emerging World Wide Web.

Current and Emerging Approaches and Systems

Organizations have typically employed centralized information servers based on large mainframe computers. These solutions have migrated to smaller machines that effectively employ the client–server approach to provide information to distributed users.

Organizations that do not have to deal with a significant number of fielded/deployed systems, organizations that have complete control over their entire operations, and small businesses can adopt this strategy.

Centralized solutions are also available for unstructured information, and document-imaging solutions are available commercially that can store and serve gigabits of information. Examples of research systems include the object-oriented database management system—OMEGA, the multimedia object presentation manager MINOS, and the multimedia office server MULTOS.

Distributed DBMS solutions have now been available commercially for several years. Organizations that can dictate a homogenous solution have fielded such ways of accessing information successfully. For instance, a company can dictate that their entire distributed databases should be based on Oracle™ or Sybase™ products. Using commercial off-the-shelf (COTS) products does involve a significant amount of customization and front-end application work, however.

Security in current-day systems is handled primarily through private networks. The use of such private networks will eventually become obsolete as the Internet becomes more prevalent and cost-effective. Solutions for addressing security concerns on the Internet are beginning to appear in the marketplace.

In the context of integrating multiple databases and information repositories, several approaches have emerged over the last decade. The *federated* or *multidatabase* approaches (see Ref. 2) present the user with a collection of local schemas along with tools for information sharing among the databases. In this case, the user integrates only the necessary portions of the databases. There are several advantages to this approach, such as increased security, easier maintenance, and the ability to deal with inconsistent databases. Such an approach is suitable when the different databases in the federation contain similar data, but not when a wide variety of information repositories (not all of which are databases) need to be integrated. In such cases, the user needs to be guided through the information available via a model.

The World Wide Web

The World Wide Web (WWW or Web) is accessible through commercial on-line services and through popular Web browsers, such as Netscape and Mosaic (available in the public domain). Web browsers provide a point-and-click metaphor for accessing a hyperlinked collection of documents written using the Hypertext Markup Language (HTML) and available on the Internet. HTML is one of the family of languages that conform to the Standard Generalized Markup Language (SGML), an international standard for specifying neutral-format documents. HTML documents are served by servers that adhere to the Hypertext Transfer Protocol (HTTP), which was designed to efficiently support multiple independent requests for documents. These servers do not maintain any state information; each request for a document is an independent transaction. To support the dynamic creation of HTML documents, the HTTP servers support a Common Gateway Interface (CGI). Typically, the HTTP servers invoke CGI programs—frequently called CGI scripts—when requested to serve specific documents. CGI scripts can be written to provide access to, and present information coming from, a variety of sources.

CORBA Standards

The Common Object Request Broker Architecture (CORBA) standard was developed out of the need for interoperable solutions that work across multiple hardware and software platforms. This standard is promoted by the Object Management Group (OMG), whose membership includes over 500 hardware and software vendors. The CORBA architecture, and particularly the Version 2.0 standard, promotes interoperability to a hitherto unprecedented level: it promotes independence in hardware architecture, language, and location. For instance, by complying with CORBA, software services can be written in any language (e.g., C, C++, Scheme, or even Fortran), run on any machine (e.g., Sparc, Silicon Graphics, Macintosh, PC), use any operating system (e.g., Windows NT, Unix), and be accessed by client software, which could be in turn be written in any language. Of course, the success of this standard—and it already has achieved a fair measure of success—depends upon the availability of support for developing Object Request Broker (ORB) services in the respective languages and operating environments. This support is currently commercially available for all of the major hardware platforms.

The CORBA standard defines services that are based on object-orientation principles and requires the definition of services using the Interface Definition Language (IDL).

The major components of CORBA are:

Object Request Broker (ORB) core and interface

Interface Definition Language (IDL)

Dynamic Invocation Interface (DII)

Object Adapters

The architecture defined by the CORBA standard is shown in Fig. 12.1.

Web*

Web* is a software that is part of the Information Sharing System in West Virginia University. The goal of the Information Sharing System (ISS) is to provide the means for an organization to effectively disseminate information, thus enabling effective work in collaborative endeavors. Because corporate information exists in a variety of computer information repositories, such as databases, documents, drawings, and data files, it is imperative that these sources be integrated with an information-sharing system to enable wider use. CERC's Web* (pronounced "WebStar") software, currently released into the public domain, can be used to integrate multiple information sources. Web* allows the exploitation of the World Wide Web and the CORBA environment.

The Web* software allows the linking of any information source to a World Wide Web client, such as Mosaic or Netscape, by allowing a person to specify HTML or other ASCII-based templates which are dynamically filled in upon the request of a user. The templates contain embedded TCL commands, which are interpreted and can be used to retrieve and dynamically fill the templates with information. Web* comes with interfaces to call CORBA-compliant services. One of the key features of Web* is that it provides mechanisms to deal with the stateless nature of the HTTP protocol. The architecture of Web* is shown in Fig. 12.2.

Scripting Languages

Scripting languages have long been used for routine chores that do not need the full power of a programming language such as C or C++. Some examples of scripting languages include shell scripts in Unix, the Practical Extraction and Report Language (Perl), and the Tool Command Language (Tcl). One characteristic that most scripting languages have in common is that they are usually interpreted. Because of this fact, interpreted programming languages such as Basic, Common Lisp, or Scheme can also be used as scripting languages.

Mediators

In future information systems, mediators and mediator-like systems will play an important role in providing enterprise-wide information sharing. This emerging subfield of computer science, originally pioneered by Distributed Artificial Intelligence (DAI) researchers, advocates the notion of intelligent

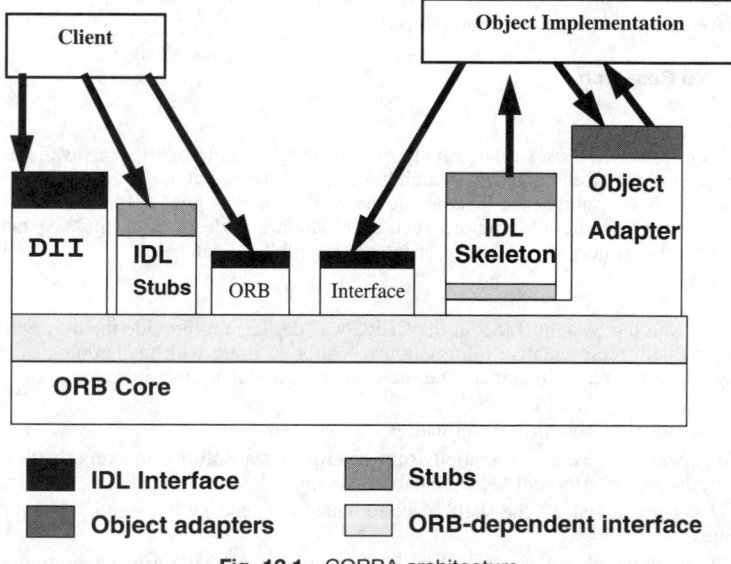

Fig. 12.1 CORBA architecture.

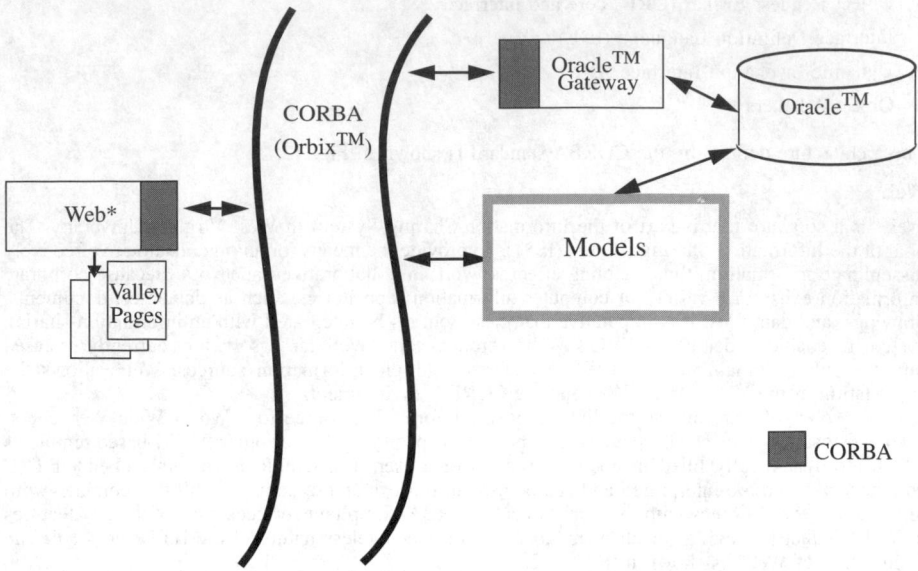

Fig. 12.2 Web* architecture.

computing agents that cooperate over a network to accomplish a set of objectives. Each agent is independent and maintains its own view of the world (in DAI terminology, a model of itself and others). Recently, agent-based architectures have come to the forefront of enterprise information systems and are currently referred to as *mediators*, based on the term used by Wiederhold.[3] The notion of mediators in the context of developing an information system hinges on the premise that intelligent assistants can be built that

- Filter and forward information
- Maintain a model of what information is available and in what format
- Facilitate the specification of user profiles on what information they are interested in and what they should be notified about
- Anticipate the needs of the user it is supporting
- Retrieve and provide the information needed

12.4.2 Related Research

Carnot

The Carnot project[4] at MCC uses a large body of knowledge or ontology that is available from the Cyc knowledge base (see Refs. 5 and 6), which has a rich representational structure. The bulk of the knowledge in Cyc is common sense facts about the world (such as how tall humans are) and does not include enterprise-specific information. Such information is added to Cyc, thus expanding Cyc's ontology. The Carnot project is also aimed at integrating multiple databases.

ISS

The Information Sharing System (ISS)[7] at the CERC was designed to provide the user with a system that is integrated with representative information repositories along with methodologies that enable the implementation in a phased manner. The major themes in this project include:

- Use of a model of enterprise information
- Uniform way to access information from heterogeneous information repositories including multimedia repositories and legacy database systems
- Use of a commercial off-the-shelf object-oriented database systems and relational database systems
- Use of model-based and case-based retrieval techniques for enterprise integration

12.5   CORPORATE HISTORY MANAGEMENT SERVICES

In any product development environment, there is a need to electronically capture the design intent and the evolution of the product from conceptual design to retirement. Corporate history is useful for future product design as well as for documenting existing products. Indexing, linking, and storing various types of documents (design, manufacturing, specifications, etc.) and archiving decisions reached in meetings among team members are some of the features of corporate history management systems.

Traditionally, the history of a product has been captured in notebooks and the minds of the designers and other team members involved in product development at various stages. They documented the tasks and activities and the decisions taken at each step. Often, such information is in the form of the daily notes of the designer, memos, pictures or a piece of conversation. The following are some of the reasons for capturing product history.

- In order to be able to produce good-quality products efficiently, a virtual team member must be able to look up the history of related products. This way, the team member gets to know the good decisions, the rationale behind the decisions, the lessons learned in the earlier ventures, why a particular decision was made in the context of several alternatives, etc. Access to such information is vital for developing good-quality products in a competitive world.
- Product history is vital in improving product quality. For example, if the designer of a new engine of a car had the maintenance records on the previous versions of the car readily available, then he or she would be able to evaluate how the previous designs of the engine fared and use this information in the new design.
- Product history is important even in the context of a single product. A product goes through an entire life cycle from conceptual design to maintenance and retirement. There will be several versions of the product during its lifetime. Unless the rationale behind each decision is documented and understood, team members with various perspectives will not be able to respond to these decisions. When problems arise, it becomes especially important to know why a certain design decision was made. Knowing the rationale helps in altering the decisions if need be.
- Product history is also important, for legal reasons. Even beyond the useful life of a product, a corporation needs to keep a record of the events in the product development, such as the studies conducted and decisions made. This is to provide some immunity against liability lawsuits that may come much later. Documented product history will also be useful in patent law in proving the originality and precedence of a particular work.

With the above motivation for product history, let us examine how it is currently being done, and why we want a computer-supported tool for doing this. Even though computers have become commonplace in every office, they are still not actively being used for capturing design history. The current mechanism for capturing decisions is the engineering notebook, into which an engineer writes down all the design decisions and sometimes pastes additional material. Below are some of the virtues of capturing design history electronically.

- Engineering notebooks cannot handle multimedia information; voice and video especially cannot be archived effectively. This is a significant drawback of the notebooks, because voice annotations and graphical illustrations that support design decisions are never recorded formally and are lost after a meeting has ended. With the development of multimedia editors and high-resolution display workstations, however, capturing graphics and video is becoming easy.
- Notebooks cannot be shared and accessed easily, particularly in a geographically scattered organization. Current computer networks are making information easily accessible over long distances.
- Notebooks cannot aid in retrieving items of interest to a team member. Computer-supported tools can aid in retrieving and traversing information. Such searching capabilities are one of the key advantages of computer tools.

12.5.1   Issues in Product History

Process Modeling

Understanding the product-development process is the first step in building a tool to capture product history. A design progresses through several stages before it is developed into a product, and the design history must provide support for capturing all the significant events in this process. Therefore, it is important to understand what the customer (virtual team member) needs to support his work in its various stages. In this regard, providing the functionality of a design notebook is only a first step.

Books have several limitations when it comes to retrieval, and one cannot pose intelligent queries to them. Some level of organization can be achieved through the use of tables of contents, indices, and such, but the human still has to do most of the work. This could be cumbersome if one has to track through a whole pile of books, even if they are all physically co-located. A *to be* scenario must be created based on what the team members want given the state of the art of the computing technology.

Representation

A rich representation structure is necessary for capturing the complexity of product history. This includes representations for product history as well as the data types in product history. One of the key aspects of a product history data system is heterogeneity, which means that it is capable of handling a variety of data formats.

Representing product history involves the use of a variety of data formats. Plain text, text with special fonts, 2D and 3D drawings, IGES files, schemas in EXPRESS, program code, solid models, graphic images, and voice are all examples. In addition, some enterprises have document formats and forms that are specific to the enterprise, such as the Design Study Summary (DSS) and Item of Information (IOI) used by General Electric.

In addition to the above data formats, a product history tool should be capable of representing various conceptual entities such as a design, design rationale, project, team, experimental results, analytic results, constraints, and so on. One of the important notions to be captured is the rationale behind decisions in the course of product development. Design rationale is an explanation of why an artifact is designed the way it is. A product can be thought of as a physical realization of a set of goals. The product represents an optimization of the goals subject to resource and other constraints. For example, a manufacturer might be making a computer with the goals of large memory, high CPU performance, fast disk access, high-resolution monitor, and so on. However, all of these have to be achieved subject to cost constraints and application demands. Thus, within a spectrum of choices, the final product represents the best alternative. Product history can be decomposed along several axes, with each axis providing a view. The above decomposition of a product into a set of goals is a *structural* view. Thus, the design of a CAD workstation involves a high-resolution monitor, a processor board, frame buffers, graphics accelerators, track ball, and so on. Since products evolve, they go from one version to the next. For example, Version 2 of a chip design could point to Version 3 of the chip, and in order for one to fully appreciate the design decisions of Version 3 one must examine the previous version of the design. Thus, there is a *temporal* view: a slice through the time line. There is a third view, the *logical* or rational view, which shows the supporting facts behind each decision and, very often, a causal chain. For example, we might want to provide a high-resolution monitor because we want to display 3D solids accurately, which is in turn necessary because we want to develop a CAD workstation for mechanical designers, and mechanical designers deal with 3D solids. Any representation scheme must support these three axes of product history.

In fact, the logical view could also present several issues[8,9] that are relevant in a particular design. An issue could be argued by taking one of several positions, and each position in turn has an argument, or a sequence of logical assertions, that support that position.

Retrieval

A product's history data collection can become very large, and, unless efficient ways of retrieving it are provided, it will be quite useless. For this reason, computer-based tools have a distinct advantage over paper. When a team has a large collection of notebooks containing designs, it is not easy to search them for a specific item of interest. This is particularly true if only some constraints on the item are known. For example, a user might want to retrieve the design of a turbine blade for which John or Pete was in the design team and that was drawn in the late 1980s. Product history could be organized by time, projects, teams, leaders, events, and so on, and retrieval by any of these indices is required. In addition to standard index-based retrieval, users need querying capabilities as well as associative retrieval.

A design will have a number of attributes, such as the date, designer, team members, project lead, key words, and rationale. It should be possible to retrieve a design based on any one of them. In addition, if old designs are to be reused and refitted by people who are not aware of the specific design attributes, associative retrieval techniques must be provided. For this purpose, case-based reasoning[10] seems to offer a lot of promise. Case-based reasoning is a method of reasoning from analogies—that is, reasoning from old cases or experiences in an effort to solve problems, critique solutions, explain anomalous situations, or interpret situations. Normally, people tend to solve problems by analogy, but they are not good at retrieving the relevant cases, especially when dealing with a large number of them. In particular, medical-expertise and legal education are case-oriented. Since computers are good at searching large databases, augmenting their capabilities with case-based reasoning will make the relevant cases available to a human. The case-based approach is not intended to supplant humans in decision-making. Instead, it supports human decision-making by providing the cases based on analogies and cues rather than specific indices. One of the key issues here is to figure the set of abstract cases that a specific case typifies; otherwise, an archived case will be applicable only when the new case matches it very closely. The design rationale can serve as a way of capturing the abstract cases. Once the cases relevant to a particular situation have been retrieved, the user can utilize conflict-resolution strategies to find the most appropriate case. Adapting the case to the current situation is the task of the human.

Navigation

Retrieval can enable one to access only one particular item. However, once the user has retrieved an item, he or she often needs to examine a related or consequential item. For example, while examining one version of a design, a team member might want to examine the next version or perhaps the components of the design. While examining a particular design decision, he or she might want to look at the alternative designs that were considered but not chosen. In a book, the table of contents, lists of figures and tables, glossary, index, and bibliography provide ways to navigate through the book. The sequence of pages, chapters, sections, and so on, in a book provide a simple linear order of navigation. When it comes to navigating information stored on the computer, however, hyperlinks provide a way of navigating complex documents.

Hyperlinks can provide all the linear navigational capabilities available in a book, and much more. A link can include information about its type structure, which identifies different types of links. For example, *immediate predecessor* is a relation between designs, which is a special case of the *predecessor* relation. If a design is represented as a node, then other menu items associated with the node can take the user to various other nodes, depending on the relation chosen. Effective ways of visualizing these hyperlinks must be provided.

Creation and Update

In general, the product history is meant to be read-only. However, this applies mostly to past history; the current events need to be updated into the product history. This could be at the level of conceptual design or in the later stages of product development. In each of these cases, the user must be able to manipulate heterogeneous data formats—for example, cutting and pasting text, voice, graphics, and so on—freely into the document being created or modified.

Even with the increasing use of computer tools, there is a reason why the traditional notebook is still the medium of choice in design. For the creation of new designs, no other medium offers the flexibility and agility offered by a pad of paper. During the conceptual design stage, the designer must be able to give free reign to his/her thoughts and visions. Many of the current editors do not allow multiple media and a wide variety of data formats. More importantly, they constrain the user to a set of entities depending on whether they are *text mode, graphics mode*, and so on. Such a scheme can be quite constraining for a designer who needs an environment that can capture his/her thoughts. Ideally, the designer would like a sketchpad-like facility in which he/she can draw freehand and the computer will interpret the drawings, i.e., a tool that acts both as a computer and as a pencil. Currently, tools such as the PenPoint system by GO Corporation are beginning to address this problem.

User Interface

In addition to sketchpad-like interface, any tool for capturing product history needs to include a multimedia editor, which enables communication and editing of audio, video, text, and graphics. The user-interface must be programmable to accept a variety of documents, such as process diagrams, design reports, and memos. One of the important features needed is the ability to establish as well as visualize hyperlinks. Say the link from one version of a design to the next is called the *next-version* hyperlink, and the link from a component to its parent is called the *part-of*. The user needs to be able to visualize these hyperlinks and to distinguish between the two kinds. For example, the two links could be shown in different colors. Such capabilities are required so the user can visualize a complex document at various levels. For example, he or she might want to look at the rationale for a particular design, or at the successive versions of a design at a coarse grain. Thus, there should be ways for a user to visualize a design along specific axes, such as the structural and logical axes, and ways to visualize a design at a more detailed level.

Security

The product data history will be organized at several levels. Certain kinds of information may be personal to an individual team member, and certain data may be exchanged among all the members of a team or within a project. As people move in and out of various projects, their access must be enabled and disabled accordingly.

A concept that addresses the issues of logical partitioning as well as selective access is that of a notebook. A notebook need not be a continuous document such as a single word-processed document. It is, rather, a collection of separate documents linked together in assorted ways. An example of such is a *lab notebook*, which contains the versioned documents of a team member and his/her notes. This workspace belongs exclusively to the owner. A different kind of notebook, such as a Project/

Patent notebook, is a read-only document, to which team members may submit entries and also read but cannot alter.

The Environment of a VTM

A tool for representing product history is an essential part of the environment of a Virtual Team Member. It should not be an isolated tool but should be integrated as part of the CE environment. A CE environment provides a team member with a single *shell* or working environment; within this environment, he or she can move freely from one CE tool to another and can drag the constraints from one design and apply them in another context. He/she should be able to communicate to this tool from the other tools in the environment. For example, the product history tool must communicate with a tool for multimedia conferencing over the network. This means that the minutes of a multimedia conference could be part of the archived documents and could become part of the product history. Likewise, in the middle of a conference, a team member should be able to access information about the product history and be able to include it as part of a conference session.

User Acceptance and Validation

As with any computer tool, the final benefits of a product history tool depend on the acceptance and use of, and feedback from, its end users. The tool must not alter the existing practices and protocols of recording design history, but must support the same practices in electronic form. It must not require team members to learn and use a lot of computer jargon, thereby distracting them from their main tasks. The tool also must be programmable to produce specialized documents that are specific to certain corporations. This way, the knowledge is captured in the electronic form, enabling easy subsequent access, and at the same time the user can make the transition smoothly to the computer tools.

12.6 CONCLUSION

This chapter has provided an overview of the technology elements needed to support concurrent engineering teams. A number of commercial tools now on the marketplace address some of the needs and requirements presented in this chapter. The emergence of high-speed networks, the explosion in the adoption of World Wide Web technology, and the maturation of integration frameworks now make it possible to support the notion of a virtual team—a geographically distributed team of engineers.

REFERENCES

1. K. J. Cleetus, "Definition of Concurrent Engineering," in *CERC Technical Report Series*, CERC-TR-RN-92-003, Concurrent Engineering Research Center, West Virginia University, Morgantown WV, May 1992.
2. W. Litwin, L. Mark, and N. Roussopoulos, "Interoperability of Multiple Autonomous Databases," *ACM Computing Surveys* 22, 267–293 (1990).
3. G. Wiederhold, "Mediators in the Architecture of Future Information Systems," *IEEE Computer* 25 (3), 50–62 (March 1992).
4. C. Collet, M. Huhns, and W.-M. Shen, "Resource Integration Using a Large Knowledge Base in Carnot," *IEEE Computer*, 24 (12), 55–62 (December 1991).
5. D. Lenat and R. V. Guha, *Building Large Knowledge-Based Systems: Representation and Inference in the Cyc Project*, Addison-Wesley, Reading, MH, 1990.
6. R. V. Guha and D. Lenat, "Cyc: A Midterm Report," *AI Magazine* 11 (3), 32 (Fall 1990).
7. V. Jagannathan et al., "Model-Based Information Access," in *Proceedings of the Second Workshop on Enabling Technologies: Infrastructure for Collaborative Enterprises*, IEEE Computer Press, Los Alamitos, CA, Summer 1993, pp. 198–212.
8. W. Kunz and H. Rittel, *Issues as Elements of Information Systems*, Technical Report Working Paper No. 131, Institute of Urban and Regional Development, University of California, Berkeley, 1970.
9. H. Rittel, *Apis: A Concept for an Argumentative Planning Information System*, Technical Report Working Paper No. 324, Institute of Urban and Regional Development, University of California, 1980.
10. E. Rich and K. Knight, "Common Sense," in *Artificial Intelligence*, R. R. Donnelley & Sons, 1991.

CHAPTER 13

COMPUTER-AIDED DESIGN

Dr. Emory W. Zimmers, Jr., & Technical Staff
Enterprise Systems Center
Lehigh University
Bethlehem, PA

13.1	**INTRODUCTION TO COMPUTER-AIDED DESIGN (CAD)**	**275**
	13.1.1 A Historical Perspective of CAD	276
	13.1.2 The Design Process	276
	13.1.3 Applying Computers to Design	278
13.2	**HARDWARE**	**282**
	13.2.1 Input/Output and Central Processing Unit (CPU)	282
13.3	**THE COMPUTER**	**283**
	13.3.1 Computer Evolution	284
	13.3.2 Categories of Computers	284
	13.3.3 Central Processing Unit (CPU)	285
	13.3.4 RISC and CISC Computers	285
	13.3.5 Parallel Processing	287
13.4	**MEMORY SYSTEMS**	**287**
	13.4.1 Organizational Methods	287
	13.4.2 Internal Memory and Related Techniques	288
	13.4.3 External Memory	289
	13.4.4 Magnetic Disks	289
	13.4.5 Magnetic Tape	290
	13.4.6 Optical Data Storage	290
13.5	**INPUT DEVICES**	**290**
	13.5.1 Keyboard	290
	13.5.2 Touch Pad	291

	13.5.3 Mouse	291
	13.5.4 Trackball	291
	13.5.5 Light Pen	291
	13.5.6 Digitizer	292
	13.5.7 Scanner	293
13.6	**OUTPUT DEVICES**	**293**
	13.6.1 Electronic Displays	293
	13.6.2 Hard Copy Devices	294
13.7	**SOFTWARE**	**296**
	13.7.1 Operating Systems	296
	13.7.2 Graphical User Interface (GUI) and the X Window System	298
	13.7.3 Computer Languages	299
13.8	**CAD SOFTWARE**	**301**
	13.8.1 Graphics Software	301
	13.8.2 Solid Modeling	302
13.9	**CAD STANDARDS AND TRANSLATORS**	**309**
	13.9.1 Analysis Software	311
13.10	**APPLICATIONS OF CAD**	**314**
	13.10.1 Optimization Applications	314
	13.10.2 Virtual Prototyping	315
	13.10.3 Rapid Prototyping	316
	13.10.4 Computer-Aided Manufacturing (CAM)	317

13.1 INTRODUCTION TO CAD

Computer-aided design (CAD) uses the mathematical and graphic-processing power of the computer to assist the engineer in the creation, modification, analysis, and display of designs. Many factors have contributed to CAD technology becoming a necessary tool in the engineering world, such as the computer's speed at processing complex equations and managing technical databases. CAD combines the characteristics of designer and computer that are best applicable to the design process.

The combination of human creativity with computer technology provides the design efficiency that has made CAD such a popular design tool. CAD is often thought of simply as computer-aided

Mechanical Engineers' Handbook, 2nd ed., Edited by Myer Kutz.
ISBN 0-471-13007-9 © 1998 John Wiley & Sons, Inc.

drafting, and its use as an electronic drawing board is a powerful tool in itself. The functions of a CAD system extend far beyond its ability to represent and manipulate graphics. Geometric modeling, engineering analysis, simulation, and the communication of the design information can also be performed using CAD.

13.1.1 A Historical Perspective of CAD

Graphical representation of data, in many ways, forms the basis of CAD. An early application of computer graphics was used in the SAGE (Semi-Automatic Ground Environment) Air Defense Command and Control System in the 1950s. SAGE converted radar information into computer-generated images on a cathode ray tube (CRT) display. It also used an input device, the light pen, to select information directly from the CRT screen.

Another significant advancement in computer graphics technology occurred in 1963, when Ivan Sutherland, in his doctoral thesis at MIT, described the SKETCHPAD system. The SKETCHPAD system was driven by a Lincoln TX-2 computer. With SKETCHPAD, images could be created and manipulated using the light pen. Graphical manipulations such as translation, rotation, and scaling could all be accomplished on-screen using SKETCHPAD. Computer applications based on Sutherland's approach have become known as interactive computer graphics (ICG). The graphical capabilities of SKETCHPAD showed the potential for computerized drawing in design. The high cost of computer hardware in the 1960s limited the use of ICG systems to large corporations, such as those in the automotive and aerospace industries, which could justify the initial investment. With the rapid development of computer technology, computers became more powerful, using faster processors and greater data storage capabilities. Their physical size and cost decreased, and computers became affordable to smaller companies and personal users. Today it is rare to find an engineering, design, or architectural firm of any size without a working CAD system running on a personal computer or a workstation.

13.1.2 The Design Process

Before any discussion of computer-aided design, it is necessary to understand the design process in general. What is the series of events that leads to the beginning of a design project? How does the engineer go about the process of designing something? How does one arrive at the conclusion that the design has been completed? We address these questions by defining the process (Fig. 13.1) in terms of six distinct stages:

1. Customer input and perception of need
2. Problem definition
3. Synthesis
4. Analysis and optimization
5. Evaluation
6. Final design and specification

A need is usually perceived in one of two ways. Someone must recognize either a problem in an existing design or a customer-driven opportunity in the marketplace for a new product. In either case, a need exists which can be addressed by modifying an existing design or developing an entirely new design. Because the need for change may only be indicated by subtle circumstances, such as noise, marginal performance characteristics, or deviations from quality standards, the design engineer who identifies the need has taken a first step in correcting the problem. That step sets in motion processes that may allow others to see the need more readily and possibly enroll them in the solution process.

Once the decision has been made to take corrective action to the need at hand, the problem must be defined as a particular problem to be solved such that all significant parameters in the problem are defined. These parameters often include cost limits, quality standards, size and weight characteristics, and functional characteristics. Often, specifications may be defined by the capabilities of the manufacturing process. Anything that will influence the engineer in choosing design features must be included in the definition of the problem. Careful planning in this stage can lead to fewer iterations in subsequent design stages.

Once the problem has been fully defined in this way, the designer moves on to the synthesis stage, where knowledge and creativity can be applied to conceptualize an initial design. Teamwork can make the design more successful and effective at this stage. That design is then subjected to various forms of analysis, which may reveal specific problems in the initial design. The designer then takes the analytical results and applies them in an iteration of the synthesis stage. These iterations may continue through several cycles of synthesis and analysis until the design is optimized.

The design is then evaluated according to the parameters set forth in the problem definition. A scale prototype is often fabricated to perform further analysis and to assess operating performance, quality, reliability, and other criteria. If a design flaw is revealed during this stage, the design moves back to the synthesis/analysis stages for reoptimization, and the process moves in this circular manner until the design clears the evaluative stage and is ready for presentation.

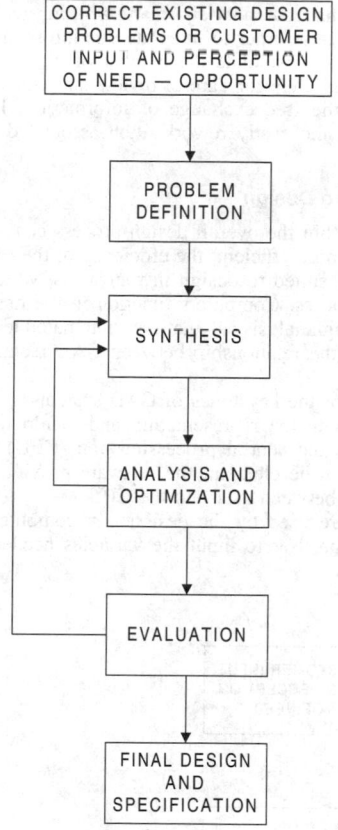

Fig. 13.1   The general design process.

Final design and specification represents the last stage of the design process. Communicating the design to others in such a way that its manufacture and marketing are seen as vital to the organization is essential. When the design has been fully approved, detailed engineering drawings are produced, complete with specifications for components, subassemblies, and the tools and fixtures required to manufacture the product and the associated costs of production. These can then be transferred manually or digitally, using CAD data, to the various departments responsible for manufacture.

In every branch of engineering, prior to the implementation of CAD, design has traditionally been accomplished manually on the drawing board. The resulting drawing, complete with significant details, was then subjected to analysis using complex mathematical formulae and then sent back to the drawing board with suggestions for improving the design. The same iterative procedure was followed and, because of the manual nature of the drawing and the subsequent analysis, the whole procedure was time-consuming and labor-intensive. CAD has allowed the designer to bypass much of the manual drafting and analysis that was previously required, making the design process flow more smoothly and much more efficiently.

It is helpful to understand the general product development process as a step-wise process. However, in today's engineering environment, the steps outlined above have become consolidated into a more streamlined approach called *concurrent engineering.* This approach enables teams to work concurrently by providing common ground for interrelated product development tasks. Product information can be easily communicated among all development processes: design, manufacturing, marketing, management, and supplier networks. Concurrent engineering recognizes that fewer iterations result in less time and money spent in moving from design concept to manufacture and from manufacturing to market. The related processes of Design for Manufacturing (DFM) and Design for Assembly (DFA) have become integral parts of the concurrent engineering approach.

Design for Manufacturing and Design for Assembly methods use cross-disciplinary input from a variety of sources (e.g., design engineers, manufacturing engineers, suppliers, and shop-floor representatives) to facilitate the efficient design of a product that can be manufactured, assembled, and marketed in the shortest possible period of time. Products designed using DFM and DFA are often simpler, cost less, and reach the marketplace in far less time than traditionally designed products. DFM focuses on determining what materials and manufacturing techniques will result in the most efficient use of available resources in order to integrate this information early in the design process. The DFA methodology strives to consolidate the number of parts wherever possible, uses gravity-assisted assembly techniques, and calls for careful review and consensus approval of designs early in the process. By facilitating the free exchange of information, DFM and DFA methods allow engineering companies to avoid the costly rework often associated with repeated iterations of the design process.

13.1.3 Applying Computers to Design

Many of the individual tasks within the overall design process can be performed using a computer. As each of these tasks is made more efficient, the efficiency of the overall process increases as well. The computer is especially well suited to design in four areas, which correspond to the latter four stages of the general design process. Computers function in the design process through geometric modeling capabilities, engineering analysis calculations, automated testing procedures, and automated drafting. Figure 13.2 illustrates the relationship between CAD technology and the final four stages of the design process.

Geometric modeling is one of the keystones of CAD systems. It uses mathematical descriptions of geometric elements to facilitate the representation and manipulation of graphical images on a computer display screen. While the central processing unit (CPU) provides the ability to quickly make the calculations specific to the element, the software provides the instructions necessary for efficient transfer of information between user and the CPU.

Three types of commands are used by the designer in computerized geometric modeling. The first type of command allows the user to input the variables needed by the computer to represent

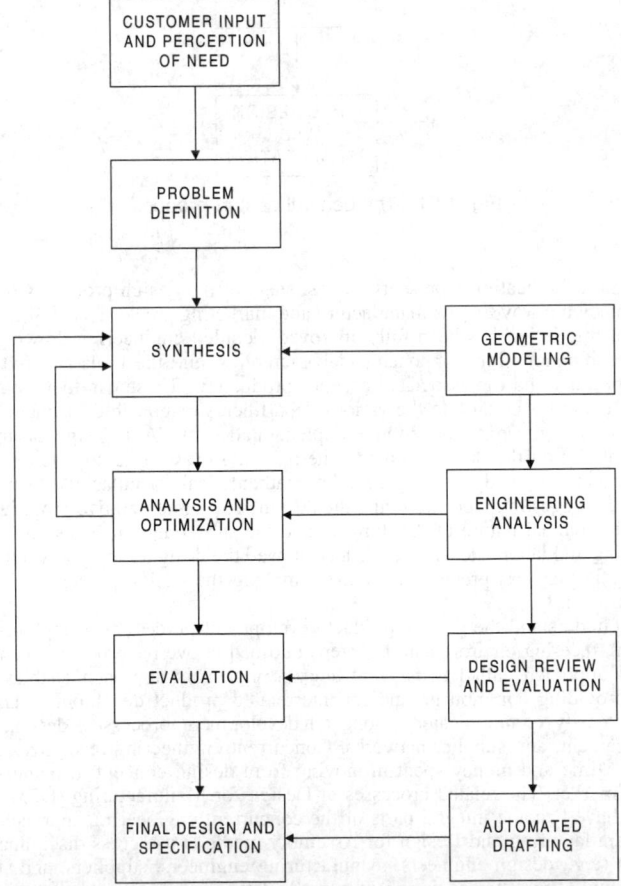

Fig. 13.2 Application of computers to the design process.

basic geometric elements such as points, lines, arcs, circles, splines, and ellipses. The second type of command is used to transform these elements. Commonly performed transformations in CAD include scaling, rotation, and translation. The third type of command allows the various elements previously created by the first two commands to be joined into a desired shape.

During the whole geometric modeling process, mathematical operations are at work that can be easily stored as computerized data and retrieved as needed for review, analysis, and modification. There are different ways of displaying the same data on the CRT screen, depending on the needs or preferences of the designer. One method is to display the design as a two-dimensional representation of a flat object formed by interconnecting lines. Another method displays the design as a three-dimensional representation of objects. In three-dimensional representations, there are four types of modeling approaches:

- Wireframe modeling
- Surface modeling
- Solid modeling
- Hybrid solid modeling

A *wireframe model* is a skeletal description of a three-dimensional object. It consists only of points, lines, and curves that describe the boundaries of the object. There are no surfaces in a wireframe model. Three-dimensional wireframe representations can cause the viewer some confusion because all of the lines defining the object appear on the two-dimensional display screen. This makes it hard for the viewer to tell whether the model is being viewed from above or below, inside or outside.

Surface modeling defines not only the edge of the three-dimensional object, but also its surface. In surface modeling, two different types of surfaces can be generated: faceted surfaces using a polygon mesh and true curve surfaces. NURBS (Non-Uniform Rational B-Spline) is a B-spline curve or surface defined by a series of weighted control points and one or more knot vectors. It can exactly represent a wide range of curves such as arcs and conics. The greater flexibility for controlling continuity is one advantage of NURBS. NURBS can precisely model nearly all kinds of surfaces more robustly than the polynomial-based curves that were used in earlier surface models. The surface modeling is more sophisticated than wireframe modeling. Here, the computer still defines the object in terms of a wireframe but can generate a surface "skin" to cover the frame, thus giving the illusion of a "real" object. However, because the computer has the image stored in its data as a wireframe representation having no mass, physical properties cannot be calculated directly from the image data. Surface models are very advantageous due to point-to-point data collections usually required for Numerical Control (NC) programs in computer-aided manufacturing (CAM) applications. Most surface modeling systems also produce the stereolithographic data required for rapid prototyping systems.

Solid modeling defines the surfaces of an object, with the added attributes of volume and mass. This allows image data to be used in calculating the physical properties of the final product. Solid modeling software uses one of two methods: constructive solid geometry (CSG) or boundary representation (B-rep). The CSG method uses Boolean operations (union, subtraction, intersection) on two sets of objects to define composite models. For example, a cylinder can be subtracted from a cube. B-rep is a representation of a solid model that defines an object in terms of its surface boundaries: faces, edges, and vertices.

Hybrid solid modeling allows the user to represent a part with a mixture of wireframe, surface modeling, and solid geometry. The I-DEAS Master Modeler offers this representation feature.

In CAD software, the hidden-line command can remove the background lines of the object in a model. Certain features have been developed to minimize the ambiguity of wireframe representations. These features include using dashed lines to represent the background of a view, or removing those background lines altogether. The latter method is appropriately referred to as *hidden-line removal*. The hidden-line removal feature makes it easier to visualize the model because the back faces are not displayed. Shading removes hidden lines and assigns flat colors to visible surfaces. Rendering adds and adjusts lights and materials to surfaces to produce realistic effects. Shading and rendering can greatly enhance the realism of the 3D image. Figures 13.3(*a*) and (*b*) show the same object, represented as a pure wireframe and a wireframe with hidden-line removal.

Engineering analysis can be performed using one of two approaches: analytical or experimental. Using the analytical method, the design is subjected to simulated conditions, using any number of analytical formulae. By contrast, the experimental approach to analysis requires that a prototype be constructed and subsequently subjected to various experiments to yield data that might not be available through purely analytical methods.

There are various analytical methods available to the designer using a CAD system. Finite element analysis and static and dynamic analysis are all commonly performed analytical methods available in CAD.

Finite element analysis (FEA) is a computer numerical analysis program (Fig. 13.4) used to solve the complex problems in many engineering and scientific fields, such as structural analysis (stress,

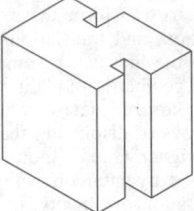

Fig. 13.3 (*a*) Pure wireframe model. (*b*) Wireframe model with hidden-line removal feature.

deflection, vibration), thermal analysis (steady state and transient), and fluid dynamics analysis (laminar and turbulent flow).

The finite element method divides a given physical or mathematical model into smaller and simpler elements, performs analysis on each individual element, using the required mathematics. It then assembles the individual solutions of the elements to reach a global solution for the model. FEA software programs usually consist of three parts: the preprocessor, the solver, and the postprocessor.

The program inputs are prepared in the preprocessor. Model geometry can be defined or imported from CAD software. Meshes are generated on a surface or solid model to form the elements. Element properties and material descriptions can be assigned to the model. Finally, the boundary conditions

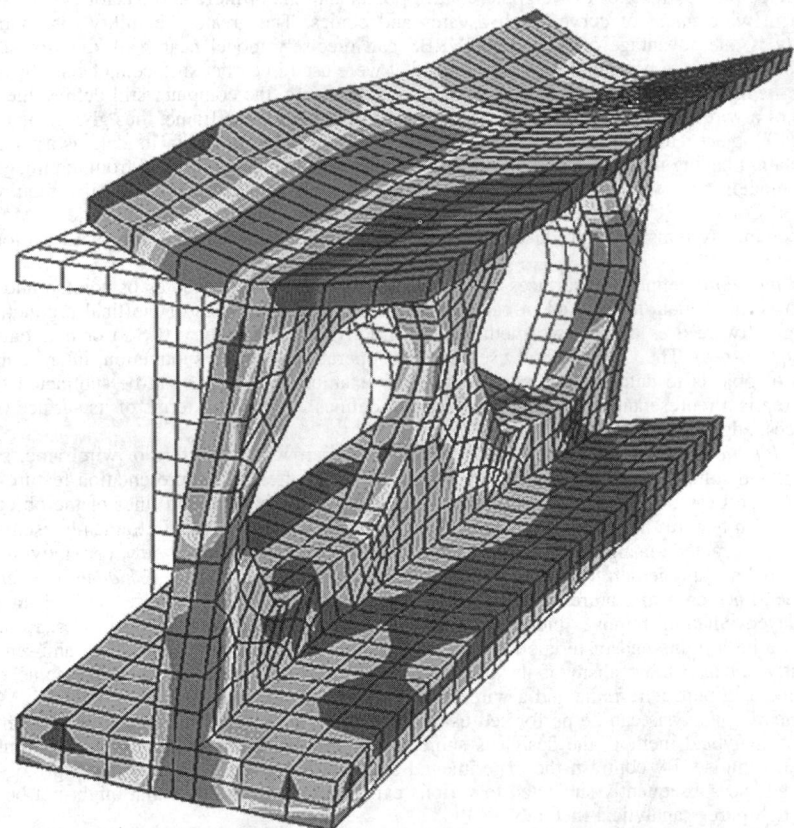

Fig. 13.4 Finite element analysis of random vibration in a beam. Colors or gray scales are often used to show degrees of stress and deflection. The original shape is also outlined without shading for reference (courtesy of Algor, Inc.).

and loads are applied to the elements and their nodes. Certain checks must be completed before the analysis calculation. These include checking for duplication of nodes and elements and verifying the element connectivity of the surface elements so that the surface normals are all in the same direction. In order to optimize disk space and running time, the nodes and elements should usually be renumbered and sequenced. Many analysis options are available in the analysis solver to execute the model. The element stiffness matrices can be formulated and solved to form a global stiffness value for the model solution. The results of the analysis data are then interpreted by the postprocessor in an orderly manner. The postprocessor in most FEA applications offers graphical output and animation displays. Many vendors of CAD software are also developing pre- and post processors that allow the user to visualize their input and output graphically. FEA is a powerful tool in effectively synthesizing a design into an optimized product.

Kinematic analysis and synthesis (Fig. 13.5) studies the motion or position of a set of rigid bodies in a system without reference to the forces causing that motion or the mass of the bodies. It allows engineers to see how the mechanisms they design will function in motion. This luxury enables the designer to avoid faulty designs and also to apply the design to a variety of scenarios without constructing a physical prototype. Synthesis of the data extracted from kinematic analysis in numerous iterations of the process leads to optimization of the design. The increased number of trials that kinematic analysis allows the engineer to perform may have profound results in optimizing the behavior of the resulting mechanism before actual production.

Static analysis determines reaction forces at the joint positions of resting mechanisms when a constant load is applied. As long as zero velocity is assumed, static analysis can be performed on mechanisms at different points of their range of motion. Static analysis allows the designer to determine the reaction forces on whole mechanical systems as well as interconnection forces transmitted to their individual joints. The data extracted from static analysis can be useful in determining compatibility with the various criteria set out in the problem definition. These criteria may include reliability, fatigue, and performance considerations to be analyzed through stress analysis methods.

Dynamic analysis combines motion with forces in a mechanical system to calculate positions, velocities, accelerations, and reaction forces on parts in the system. The analysis is performed stepwise within a given interval of time. Each degree of freedom is associated with a specific coordinate for which initial position and velocity must be supplied. The computer model from which the design

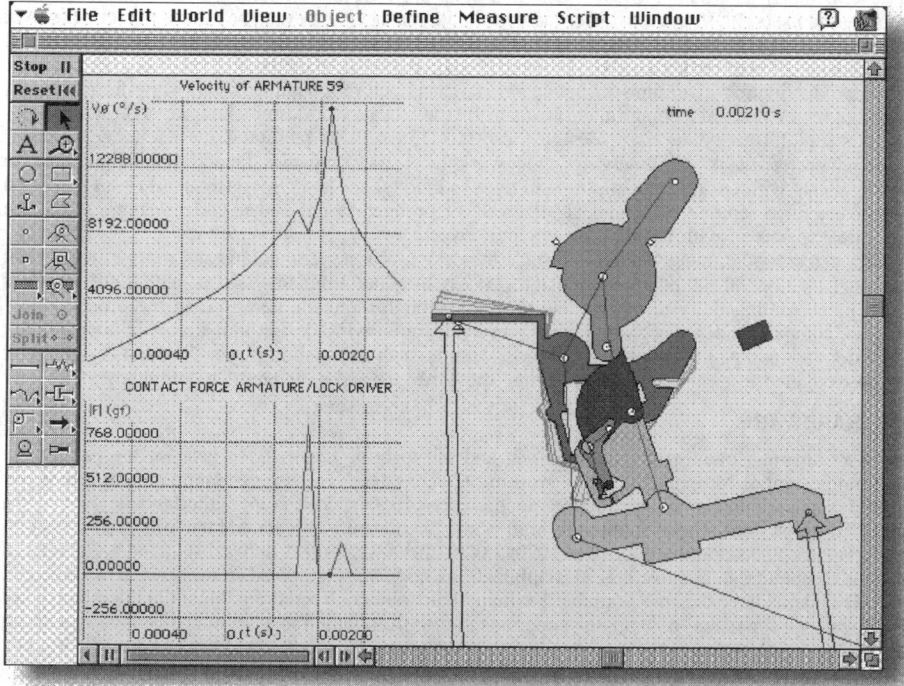

Fig. 13.5 Kinematic analysis of a switch mechanism (image courtesy of Knowledge Solutions, Inc.).

is analyzed is created by defining the system in various ways. Generally, data relating to individual parts, joints, forces, and overall system coordination must be supplied by the user, either directly or through a manipulation of data within the software.

The results of all of these types of analyses are typically available in many forms, depending on the needs of the designer. All of these analytical methods will be discussed in greater detail in Section 13.8.

Experimental analysis involves fabricating a prototype and subjecting it to various experimental methods. Although this usually takes place in the later stages of design, CAD systems enable the designer to make more effective use of experimental data, especially where analytical methods are thought to be unreliable for the given model. CAD also provides a useful platform for incorporating experimental results into the design process when experimental analysis is performed in earlier iterations of the process.

Design review can be easily accomplished using CAD. The accuracy of the design can be checked using automated tolerancing and dimensioning routines to reduce the possibility of error. Layering is a technique which allows the designer to superimpose images upon one another. This can be quite useful during the evaluative stage of the design process by allowing the designer to check the dimensions of a final design visually against the dimensions of stages of the design's proposed manufacture, ensuring that sufficient material is present in preliminary stages for correct manufacture.

Interference checking can also be performed using CAD. This procedure involves making sure that no two parts of a design occupy the same space at the same time.

Automated drafting capabilities in CAD systems facilitate presentation, which is the final stage of the design process. CAD data, stored in computer memory, can be sent to a pen plotter or other hard-copy device (see Section 13.6.2) to produce a detailed drawing quickly and easily. In the early days of CAD, this feature was the primary rationale for investing in a CAD system. Drafting conventions, including but not limited to dimensioning, crosshatching, scaling of the design, and enlarged views of parts or other design areas, can be included automatically in nearly all CAD systems. Detail and assembly drawings, bills of materials (BOM), and cross-sectioned views of design parts are also automated and simplified through CAD. In addition, most systems are capable of presenting as many as six views of the design automatically. Drafting standards defined by a company can be programmed into the system such that all final drafts will comply with the standard.

Documentation of the design is also simplified using CAD. Product Data Management (PDM) has become an important application associated with CAD. PDM allows companies to make CAD data available interdepartmentally on a computer network. This approach holds significant advantages over conventional data management. PDM is not simply a database holding CAD data as a library for interested users. PDM systems offer increased data management efficiency through a client-server relationship among individual computers and a networked server. Benefits of implementing a PDM system include faster retrieval of CAD files through keyword searches and other search features; automated distribution of designs to management, manufacturing engineers, and shop-floor workers for design review; recordkeeping functions that provide a history of design changes; and data security functions limiting access levels to design files (Fig. 13.6). PDM facilitates the exchange of information characteristic of the emerging agile workplace. As companies face increased pressure to provide clients with customized solutions to their individual needs, PDM systems allow an increased level of teamwork among personnel at all levels of product design and manufacturing, cutting the costs often associated with information lag and rework.

Although computer-aided design has made the design process less tedious and more efficient than traditional methods, the fundamental design process in general remains unchanged. It still requires human input and ingenuity to initiate and proceed through the many iterations of the process. Nevertheless, computer-aided design is such a powerful, time-saving design tool that it is now difficult to function in a competitive engineering world without such a system in place. The CAD system will now be examined in terms of its components: the hardware and software of a computer.

13.2 HARDWARE

Just as a draftsman traditionally requires pen and ink to bring creativity to bear on the page, there are certain essential components to any working CAD system. The use of computers for interactive graphics applications can be traced back to the early 1960s, when Ivan Sutherland developed the SKETCHPAD system. The prohibitively high cost of hardware made general use of interactive computer graphics uneconomical until the 1970s. With the development and subsequent popularity of personal computers, interactive graphics applications now are widespread in homes and workplaces.

CAD systems have become available for many hardware configurations. Most CAD systems have been developed for standard computer systems, ranging from mainframes to microcomputers. Others, like turnkey CAD systems, come with all of the hardware and software required to run a particular CAD application, and are supplied by specialized vendors.

13.2.1 Input/Output and Central Processing Unit (CPU)

The above systems all share a dependence on components that allow the actual interaction between computer and users. These electronic components are categorized under two general headings: input

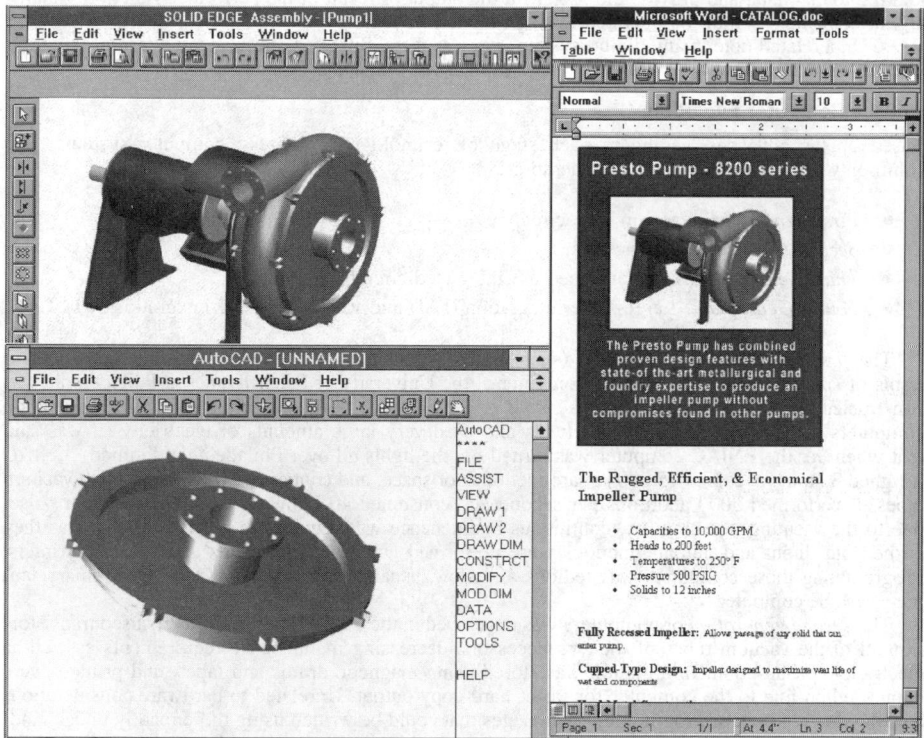

Fig. 13.6 CAD files can be used in conjunction with other applications. The above illustration shows Integraph Corporation's Solid Edge software operating in conjunction with AutoCAD from Autodesk, Inc. and Microsoft Word (image courtesy of Integraph Corporation).

devices and output devices. Input devices transfer information from the designer into the computer's Central Processing Unit (CPU) so that the data, encoded in binary sequencing, may be manipulated and analyzed efficiently. Output devices do exactly the opposite. They transfer binary data from the CPU back to the user in a usable (usually visual) format. Both types of devices are required in a CAD system. Without an input device, no information can be transferred to the CPU for processing, and without an output device, any information in the CPU is of little use to the designer because binary code is lengthy and tedious.

13.3 THE COMPUTER

Although the influence of computer technology is a somewhat recent phenomenon due to the reduced cost of computers over the last two decades, the philosophical basis for the construction and employment of computing systems has a longer history than 20 years.

Charles Babbage, a nineteenth-century mathematician at Cambridge University in England, is often cited as a pioneer in the computing field. Babbage designed an "analytical engine," the capabilities of which would have surprisingly foreshadowed the same basic functions of today's computers had his design not been limited by the manufacturing capabilities of his time. The analytical engine was designed with considerations for input, storage, mathematical calculation, grouping results, and printing results in typeface. Other, less complex mechanical forms of computers include the slide rule and even the abacus.

The vast majority of contemporary computers are digital, although some analog computers do exist. This latter type has been relegated almost to a footnote in contemporary computing due to the overwhelming advances made in digital technology. The difference between digital and analog systems lies in the binary code. Digital computers use a system of switches with two settings, "on" or "off." These settings are typically represented as "0" for "off" and "1" for "on."

Although digital computers vary in size, shape, price, and capabilities, all digital computers have four common features. First, the circuits used can exist in one of two states, either "on" or "off." This characteristic yields the basis for binary logic. Second, all share the ability to store data in binary form. Third, all digital computers can receive external input data, perform various functions relating to that data, and provide the user with the output or result of the performed function. Finally, digital computers can all be operated through the use of instructions organized into sets of separate steps. On a related note, many digital systems possess the ability to perform many different functions at the same time, using a technique known as *parallel processing*.

13.3.1 Computer Evolution

Based on the advances leading to each stage of technological progress, computer systems have commonly been grouped into four generations:

- *First Generation:* Vacuum tube circuitry
- *Second Generation:* Transistors
- *Third Generation:* Small and medium integrated circuits
- *Fourth Generation:* Large-scale integration (LSI) and very large-scale integration (VLSI)

The *first generation* of computers (such as ENIAC in the 1940s) were huge machines both in terms of size and mass. The ENIAC computer at the University of Pennsylvania in Philadelphia was constructed during World War II to calculate projectile trajectories. The circuitry of first-generation computers was composed of vacuum tubes and used very large amounts of electricity (it was said that whenever the ENIAC computer was turned on, the lights all over Philadelphia dimmed). ENIAC weighed 30 tons, occupied 15,000 square feet of floor space, and contained more than 18,000 vacuum tubes. It performed 5000 additions per second and consumed 40 kilowatts of power per hour. Also, due to the vacuum tube circuitry, continuous maintenance was required to change the tubes as they burned out. Input and output functions were performed using punched cards and separate printers. Programming these computers was tedious and slow, usually performed directly in the binary language of the computer.

The *second generation* of computers was developed in the 1950s. These computers used transistors instead of the vacuum tubes of their predecessors, decreasing maintenance requirements as well as electricity consumption. Information was stored using magnetic drums and tapes, and printers were connected on-line to the computer for faster hard-copy output. Unrelated to hardware considerations was the development of programming languages that could be written using more readily understandable commands and then separately converted into the binary data required by the computer.

Third-generation computers were distinguished by the advent of the integrated circuit in the late 1960s, which made computers faster and more compact. Storage, input, and output capabilities also increased dramatically. High-level software languages, such as COBOL, FORTRAN, and BASIC, were developed and gained popularity. These languages were written in a way that the programmer could more readily understand and assembled automatically into a set of instructions for the computer to follow. The most significant development of this period was a downward cost spiral that precipitated the popularity of minicomputers—smaller computers designed for use by one user or a small number of users at a time, as opposed to the larger mainframes of previous generations.

In the *fourth generation* of digital computers, the steady decrease in processing times and cost for computer technology has continued with a corresponding increase in memory and computational capabilities. With large-scale integration (LSI), more than 1000 components can be placed on a single integrated-circuit chip. Very large-scale integration (VLSI) chips contain more than 10,000 components; current VLSI chips have 100,000 or more components on each chip. The semiconductor technology developed in the 1970s condensed whole computers into the size of a single chip, known as a microprocessor. Semiconductors were responsible for the arrival of "personal computers" in the late 1970s and early 1980s.

13.3.2 Categories of Computers

Computers can be divided into categories, depending on their size and capabilities. Traditionally, computers are grouped under the following headings:

- Supercomputers
- Mainframes
- Minicomputers
- Microcomputers

Supercomputers are the world's most powerful computers, often with processing speeds in excess of 20 million computations per second. The performance of the CRAY-2 supercomputer was rated at 100 million floating point operations per second (MFLOPS). Supercomputers are often used to calculate extensive mathematical problems for scientific research purposes. These problems are characterized by the need for high precision and repetitive performance of floating-point arithmetic operations on large arrays of numbers.

Mainframes have large memory capabilities coupled with extremely fast processing speeds. These computers are less powerful systems than supercomputers, but they are used in large CAD systems where a significant amount of highly accurate analysis must occur. Mainframes are highly applicable to analytical methods and are often used in dynamic analysis, stress analysis, heat transfer analysis, and other analytical methods. This type of computer system is used most often in large engineering corporations, such as those in the automotive or aerospace industries, where centralized computing and data storage are essential. Mainframes support multiple users (some over 500) at terminals, giving them access almost instantaneously to the data required to design and share information among the project team. Because of their extensive memory capabilities, mainframes are also used for large database maintenance. Mainframe computers usually require a specialized support staff for maintenance and programming. The typical configuration of a mainframe system is a processor with 32-bit and 64-bit word addressing, 64 megabytes (MB) to 2 gigabytes (GB) of memory, and several gigabytes of storage space.

Minicomputers are somewhat smaller and less powerful than a mainframe, but they nevertheless offer a powerful, less expensive alternative to mainframe systems where a centralized computing environment is desired. They introduced the concept of distributed data processing. A typical minicomputer is available with 16- to 32-bit word addressing, several megabytes of memory, and multiple disk drives amounting to several megabytes to gigabytes of storage space. Turnkey CAD systems were offered as minicomputer systems in the late 1970s and early 1980s. A number of display terminals can be supported by minicomputer systems, and on-line printers for minicomputer systems are capable of delivering between two and three thousand lines of text per minute.

Microcomputers, which include the personal computer and engineering workstations, are desktop-size or smaller computers. These computers have seen the greatest growth in the number of systems being sold and used since the early 1980s. There are various reasons for this trend. Microcomputers are quickly becoming more powerful, with greater memory capabilities. A wide range of microcomputers are available with 8- to 32-bit word addressing, several megabytes of memory, and built-in hard disk, floppy disk, CD-ROM, and tape backup systems.

Many companies operate best using a decentralized approach to computing; however, networks have become increasingly common in microcomputer environments in order to provide some of the advantages of centralized computing when desirable. Powerful servers that support massive client-server networks have largely replaced the huge mainframe computers. Even the power of a contemporary PC exceeds that of a mainframe from the 1960s and 1970s. Furthermore, the computational capability of engineering workstations today exceeds that of most minicomputers. The latest trend is to classify computers as supercomputers, servers, workstations, large PCs, and small PCs.

One common differential between types of computers is the word length. The term *word length* does not refer to words in human language; rather, it signifies the number of places in the base-2 units of the machine language in the various types of computer. Mainframes have traditionally run using 32-bit words, with minicomputers typically having 16-bit capabilities and microcomputers 8-bit capabilities. Word length influences processing speeds and memory-addressing capabilities in computer systems. Longer word length means that more information can be operated on or transferred to a different part of the system in fewer steps, thereby taking less time. The word length also influences memory capabilities by making virtual memory techniques available. While formerly applicable as a general rule for distinguishing the capabilities of the various types, various word lengths are now available in all types of computers. Even some home electronic game systems now employ 32-bit technology in a system about the size of a large textbook.

13.3.3 Central Processing Unit (CPU)

The computer's central processing unit (CPU) is the portion of a computer that retrieves and executes instructions. The CPU is essentially the brain of a CAD system. It consists of an arithmetic and logic unit (ALU), a control unit, and various registers. The CPU is often simply referred to as the *processor.*

The ALU performs arithmetic operations, logic operations, and related operations, according to the program instructions. The control unit controls all CPU operations, including ALU operations, the movement of data within the CPU, and the exchange of data and control signals across external interfaces (e.g., the system bus). Registers are high-speed internal memory-storage units within the CPU. Some registers are user-visible; that is, available to the programmer via the machine instruction set. Other registers are dedicated strictly to the CPU for control purposes. An internal clock synchronizes all CPU components. The clock speed (the number of clock pulses per second) is measured in megahertz (MHz) or millions of clock pulses per second. The clock speed essentially measures how fast an instruction is processed by the CPU.

13.3.4 RISC and CISC Computers

Computers can be divided into two categories, depending on their method of using instructions:

- Reduced Instruction Set Computers (RISC)

- Complex Instruction Set Computers (CISC)

The reasons for designing CISC computers are to simplify compilers and to improve performance. Underlying both of these reasons was the shift to high-level languages (HLLs) in computer programming. Computer architects attempted to design machines that provided better support for HLLs. CISC was expected to yield smaller programs that would execute faster.

RISC technology is very new by comparison. RISC computers use fewer and simpler instructions than conventional CISC computers. Simpler instructions reduce the complexity of the circuits required to implement an instruction, thereby allowing individual instructions to execute quickly. RISC machines generally show a higher level of performance than a comparably complex CISC system, despite the fact that a RISC processor executes more instructions to accomplish a given task than does a CISC processor. The following characteristics are common on all RISC-architecture computers: one instruction per machine cycle, unique register-to-register operations, and an instruction pipeline. RISC architecture often includes more general-purpose registers to maximize the number of operations that take place on the CPU. CISC computers, in contrast, employ more memory referencing. Studies have shown that the compilers on CISC machines tend to favor simpler instructions, such that the conciseness of the complex instruction sets seldom comes into play. The expectation that a CISC computer would produce smaller programs may not be realized because the more complex the instruction set, the more processor time is required to decode and execute each instruction. Longer opcodes required in CISC architecture produce longer instructions. In computationally intensive applications, such as FEA, where calculation times are often measured in hours, RISC processors are generally more efficient performing floating point operations than CISC processors.

Most RISC architecture can be found in workstations that run on the UNIX operating system, such as Sparc MIPS, DEC Alpha, PA-RISC, and PowerPC. The Pentium architecture, however, is an excellent example of CISC design. It represents the result of decades of CISC architecture evolution. Pentium architecture incorporates the sophisticated design principles once found only on mainframes, supercomputers, and servers.

The RISC versus CISC debate continues to drive technology in new directions. There is a growing realization in the industry that RISC and CISC may benefit from each other. Notably, more recent designs, like the PowerPC line of the Apple Macintosh, are no longer "pure" RISC, and the more recent CISC designs, like the Pentium P7, have incorporated some RISC-like features.

Engineering PCs

Computer-aided design projects often range from simple 2D drawings to graphics-intensive engineering applications. Computationally intensive number crunching in 3D surface and solid modeling, photorealistic rendering, and finite element analysis applications demand a great deal from a personal computer. Careful selection of a PC for these applications requires an examination of the capabilities of the CPU, RAM capacity, disk space, operating system, network features, and graphics capabilities. The industry advances quickly, especially in microprocessor capabilities. The following PC configurations list minimum requirements for various CAD applications. It should be noted that because of rapid advances in the industry, this listing may be dated by the time of publication.

For 2D drafting applications, a low-end PC is sufficient. This denotes a PC equipped with a 486 processor running on a 16-bit DOS operating system with 16 MB of RAM, or Microsoft Windows with 32 MB of RAM. (The operating system needs 4 MB, most drafting applications require at least 8 MB of RAM, and Windows holds as much data as possible in RAM.) Eight K of on-board cache memory and 256 K to 512 K of external cache for faster response is recommended. A 500-MB, fast SCSI (Smaller Computer System Interface) hard drive is minimum. SCSI is a type of bus used to support local disk drives and other peripherals. Five hundred MB is considered minimum because CAD and SWAP files require approximately 150 MB, and the operating system itself usually requires about 100 MB. A 16-inch high-resolution (1024 × 768), 256-color monitor should also be considered a minimum requirement. CD-ROM drives and fast modems with transfer rates of 28,800 baud are essential for non-networked tasks.

For 3D modeling and FEA applications, a Pentium 100 MHz processor running on a 32-bit Windows NT operating system works significantly better. The minimum memory requirement for Windows NT is 32 MB and the operating system requires 160 MB of hard-drive space. The system should have a 16-K internal cache with 256 K to 512 K of external cache for increased performance. The monitor for these applications should be 21 in. with .25 ultrafine dot pitch, high resolution (1600 × 1280), and 65,536 colors. Peripheral Component Interconnect (PCI) local bus (a high-bandwidth, processor-independent bus) and a minimum 1-GB SCSI-2 hard drive are also required.

Engineering Workstations

The Intel Pentium CPU microprocessor reduced the performance gap between PCs and workstations. A current trend is the merging of PCs and workstations into "personal workstations." Pentium Pro and RISC processors, such as DEC Alpha, Sparc MIPS, PowerPC, PA-RISC, are all powerful personal workstations with high-performance graphics accelerators.

Until recently, operating systems were the main distinction between a low-end workstation and a high-end PC. The UNIX operating system, which supports multitasking and networking, usually ran on a workstation. DOS, Windows, and the Macintosh operating systems, which perform single tasks, usually ran on PCs. That distinction is beginning to disappear because of the birth of Microsoft Windows NT and IBM's OS/2. Both of these operating systems support multitasking and networking, as well. Other operating systems designed for workstations are being modified for use on a PC, such as IBM's AIX. UNIX now runs on laptop computers with the PowerPC microprocessor.

Many CAD and FEA software applications are traditionally UNIX-based applications. Since there are significant differences in price between UNIX and Windows NT, more and more CAD and FEA vendors have released versions of their software for Windows NT. Windows NT is now being offered on workstations with processors such as DEC Alpha and MIPS R4000. Memory capacity for a personal workstation can be up to 256 MB, with up to 2 GB of disk space, such as those from Silicon Graphics, Inc.'s Indy Systems. Thanks to 64-bit technology and a scalable modular platform, high-end workstation performance can now boast supercomputer-like performance at a fraction of the cost of a mainframe or supercomputer. These systems include the HP Series 700, IBM RS/6000, Sun Sparcstation 10, Silicon Graphics Indigo, and certain models of DEC Alpha/AXP. The noted performance gain is a result of using more powerful 64-bit word addressing and up to 350 MHz clock speed. Dual processors available in some engineering workstations with the DEC Alpha system, Intel Pentium chip, or Motorola 68060 chip allow some degree of scalable parallel processing within current engineering workstations.

13.3.5 Parallel Processing

To reach higher levels of productivity in the analysis of complex structures with thousands of components, such as in combustion engines or crash simulations, the application of parallel processing was introduced. Two parallel processing methods used in engineering applications are:

- Massively parallel processing (MPP)
- Scalable parallel processing (SPP)

MPP machines combine a number of processors into one machine and boast large amounts of processing power. While these machines were expected to revolutionize computerized engineering analysis, users could not simply upload existing applications to these half-billion dollar computers. On the contrary, programming MPP machines proved to be difficult even for experienced programmers, and few applications were ever developed for general use.

SPP, however, has shown extraordinary potential for use in general engineering analysis. SPP techniques essentially link PCs or workstations, each with existing memory and disk storage capacities and CPU capabilities, allowing the attributes of each individual machine to be applied to the computations. In addition, there is no lack of application software for SPP users. Applications for thermal, dynamic, stress (both linear and non-linear), and fluid analyses are available for use with SPP computers.

13.4 MEMORY SYSTEMS

Memory systems store program information and other data in a manner that facilitates efficient access. In order to accomplish this task, various technologies and organizational methods are employed. The typical computer system is equipped with a hierarchy of memory subsystems: some (internal to the system) can be directly accessed by the processor and some can be accessed by the processor via input/output devices (external). Internal memory consists of main memory, cache, and registers (the CPU's own local memory). External memory consists of secondary storage devices, such as magnetic disk and tape, and optical memory storage on CD-ROM.

13.4.1 Organizational Methods

Memory systems organize binary data into addressable words where data can be stored or retrieved. Some systems use a method called "interleaved memory," which refers to an ability to access more than one word at a time. Interleaved memory is an advantage in situations where it is likely that the second word will be required by the CPU soon after the first.

The memory circuitry is a separate entity from the processing circuitry of the CPU. Therefore, in order for the two circuit systems to communicate, a memory controller is required. The controller deciphers the requests for memory information or storage access from the CPU and initiates the proper sequence of events. Because a controller can only control a certain amount of memory, multiple controllers are often used within a single system. The implementation of multiple memory controllers allows for interleaved memory transfer. Each controller, with its own memory domain within a system, increases efficient data transfer by allowing sequential data to be stored in different domains such that before one controller is finished transferring one part of the sequence, the next controller is already beginning its operation sequence. In some systems, memory can be shared among different CPUs of the same computer. The controllers used in conjunction with a multiprocessor system in this case would have access ports for each of the individual processors.

13.4.2 Internal Memory and Related Techniques

Registers

Memory within the CPU for data required to perform a specific task, such as the operand or result of a mathematical calculation, is stored in memory devices called *registers*. These registers are accessible by specific commands from the CPU. Other registers in the CPU are inaccessible for memory storage but are included as a part of the working system as a whole. Generally, registers hold the same number of bits, or binary digits, as the word length of the CPU. The number of registers in a CPU is variable, ranging from 1 to 16 or more.

Metal Oxide Semiconductor (MOS) Random Access Memory (RAM)

This type of storage is semiconductor-based memory, with wide-ranging applications in nearly all computer systems. RAM can either be dynamic or static. Dynamic RAM (DRAM) uses circuitry that must be periodically refreshed by rewriting the data in each block of memory. Static RAM (SRAM) does not require the refreshing but is usually more expensive than dynamic RAM. In general, static RAMs are faster than dynamic RAMs. Both types of RAM are volatile; they lose their memory when power is turned off. Thus, RAM can be used only as temporary storage. To compensate, programs can be stored on magnetic disks or tapes or on other forms of solid-state memory, or a battery can be used to supply the necessary power to maintain semiconductor memory when the CPU is not in use.

Often, it is useful to separate memory into two types: Random Access Memory (RAM) and Read Only Memory (ROM). RAM is used in memory blocks from which the CPU must be able to "read" (access stored information) and "write" (store new data). Because information in RAM can be accessed, subsequently modified, and possibly erased, it does not provide the needed security for important programs. For those applications, such as system programs, function tables, and library subroutines, data are best stored in ROM memory. Because access is usually limited to retrieval, data are not easily altered and the integrity of key programs is ensured. Some ROMs can actually be erased and rewritten under certain conditions. These types are useful when a program needs periodic alterations but should be protected from general access and possible accidental erasure.

Cache Memory

Cache memory is designed to hold information relating to frequently used applications and subroutines in active memory. Cache memory, however, is usually a separate piece of hardware between the CPU and main memory. It provides faster data transfer to the CPU than does main memory, but usually at a higher cost as well. This cost is usually well justified, especially if the computer is to be used with repetitive programs, where the cache can dramatically increase the speed with which programs run, and increase the efficiency of the user. Cache sizes of between 1 K and 512 K are usually offered for lower average cost per bit and faster average access time.

Virtual Memory

This technique addresses the problem of very large programs that use extensive address space and operate within a limited memory capacity. Programs use the registers of the CPU to keep the most active applications of the program available quickly. Other, less active applications are stored on magnetic disk space until needed. If needed, the application called for will be directed to occupy a less active register and the data formerly in that register will be saved onto the disk in order to maintain any changes made to the data during execution of the program.

Memory Addressing

Some instructions from the CPU may require an operation involving one or more operands stored in memory. Operands of this type include those for logic, mathematical operations, and so forth. For example, an addressing mode might supply the ability to take operands from various locations in memory and store the result in a separate location. Most CPUs employ a variety of addressing modes, depending on the operation. The variety of addressing modes for different tasks is a benefit in most CPUs. It provides a degree of flexibility in data management that increases efficiency and processing ability. Furthermore, memory addressing is made even more efficient in some systems by the ability to operate on single bits within an 8-digit byte, on the byte itself, and sometimes on words of 16 and 32 bits. A technique known as *extended addressing* can further increase the memory capabilities of the computer. Using this technique, the program running extended addressing functions considers memory as a number of pages. Each page is assigned a relocation constant that is combined with the other addresses on that page to form a longer address than would normally be used. In an 8-bit computer, extended addresses might take the form of 10-bit words, which increases the number of addresses possible within the system.

13.4.3 External Memory

Most computer systems use some sort of magnetic storage system to store data after semiconductor memory has been erased due to a loss in power. Although computer systems of the 1990s have main memories much larger than those commonly used in previous decades, contemporary computer programs have many more capabilities, and hence take up more memory space, than programs of preceding decades. This leaves the main memory and cache memory still requiring external (with respect to the CPU) storage capabilities. Magnetic storage systems are one answer to this problem. These storage technologies provide the added benefit of an ability to be copy data an almost unlimited number of times. Some copies can be stored away from the computer for security, while others may be kept nearby, perhaps on-line to facilitate access to the stored data. The two main magnetic storage configurations are disks and tapes.

13.4.4 Magnetic Disks

Magnetic disks are connected to the CPU through a controller attached to an input/output port. The controller can often control a number of disk drives and is usually programmed with a significant amount of data relating to error detection, data transfer, and other pertinent information.

Disk drives use a drive motor and one or more "heads" to read and write data on the disk. The drive motor turns a spindle on which the disk rests, rotating the disk at a controlled speed. A standard speed allows data written on one disk drive to be read on another drive. The head mechanism holds a read/write head for each recording surface on the disk. The heads are normally held away from the surface of the disk but placed extremely close to the surface or directly on the surface of the disk, depending on the type of drive, during data transfer. The magnetic disk itself is coated with a magnetic oxide, which forms the actual storage medium. The oxide must be organized or "formatted" into closely spaced, concentric circular tracks. The tracks must be positioned accurately and consistently on all disks in order for the head mechanism to position itself accurately over a specific track. Data are recorded and subsequently read as analog variations in the magnetic field of the oxide medium. These data are transferred to the disk drive controller, which converts the analog signal to a digital one for processing by the CPU. Disks are further formatted into blocks, or "sectors," of equal area. On most disks, these blocks are established in the medium by the manufacturer. The seek time is the time needed to position the head at the desired track. Rotation latency time is the time elapsed for an appropriate sector to rotate and line up with the head. The sum of the seek time and the latency time gives the overall access time. This time consideration can be useful during the purchase of a disk drive, since the access time will affect the efficiency of the overall system if frequent storage access is required. Soft-sectoring, a technique in which the disk drive will establish blocks of unequal area on the disk, can also be used. Finding a particular track on a magnetic disk, where the distance between tracks can be as little as 0.01 mm, is no easy task. All hard disk drives use a servo-control mechanism to ensure accurate positioning over a desired track. Many floppy drives, where the distance between adjacent tracks is not as small, use a stepping motor system.

Floppy Disks

The floppy disk was designed as a cheap and simple device to provide quick, reliable access to information stored as a back-up to computer memory. Physically, the floppy disk is a flexible diskette coated with magnetic oxide and contained within a square plastic housing. The housing has openings which allow the spindle of the drive motor to turn the disk and provide space for the head mechanism to make contact with the disk's surface. Floppy disks have continued to shrink in physical size and grow in terms of storage limits since their inception. The earliest floppy diskettes were 8 in. in diameter and capable of storing 256 K of data. These disks also typically used only one side of the diskette. The most commonly used diskettes today are 3.5 in. in diameter, and are capable of storing 1.44 MB of data using both sides of the disk. Access times for floppy disks are 100–500 msec and data transfer rates are generally lower than 300 K/sec. Floppy disk drive systems provide reliable data storage on a medium which can be removed from one computer and used on another. They are among the most common magnetic data storage systems currently in use.

Hard Disks

This term is a general heading for a category of disk drives where the disk remains within the drive. Because the disk and drive are housed within the same sealed unit, hard disks provide some significant advantages over floppy disks. The disk medium remains free of contaminants, resulting in greater reliability and data accuracy. The heads of a hard disk are lightweight and designed to hover aerodynamically and extremely close to the disk surface without touching. This virtually eliminates wear on either the disk or the drive heads. Shorter access times are also seen using hard disk systems because two heads are normally associated with each surface.

Portable hard disks supplying 1 GB of storage space are now available. These disks provide extra security for important files and provide a convenient medium for storing large amounts of data for back-up or other purposes.

13.4.5 Magnetic Tape

Magnetic tape was the first application of magnetic data storage employed with computer systems. This medium is effective, to a large extent, due to existing standards for data format. These standards, like those for magnetic disks, allow data to be used on different types of computers. Various forms of magnetic tape storage systems have been developed and satisfy specific user needs.

Standard industrial tape drives use reels of 12.7-mm-wide tape at lengths of 731 or 365 m, coated with an oxide medium similar to that used in magnetic disks. The tape moves past read/write heads which provide data transfer at 800–6250 bits per linear inch. The tape motion is servo-controlled for a high degree of accuracy, with the tape winding more than 180° around a capstan to provide sufficient physical control. The motors driving the tape reels must also be carefully controlled to ensure proper tape tension. Two mechanisms are typically used to ensure this control. In older tape drives, the tape moves over fixed- and variable-position pulleys. The variable pulleys are attached to a spring-loaded tension arm drawing the tape in a "W" form between the pulleys. The position of the tension arm gives the motor control mechanism the necessary information regarding tape winding and release to ensure the proper tension. Most current tape drive designs have abandoned the tension arm in favor of the following vacuum chamber technique. Between each reel and capstan, a vacuum chamber draws the tape into a loop 1–2 m long. The length of the tape in the chamber is detected using photoelectric sensors and this information is used by the motor controller to govern the movement of the reels. Smaller magnetic tape drives are available for smaller systems such as microcomputers. In these drives, the tape is normally housed in a plastic cartridge that protects the tape medium from contamination. This provides excellent data integrity on a much smaller scale than industrial-type systems. An 8-mm tape is usually used with a workstation system, while PC-based tape drives are usually on a 4-mm format. Because magnetic tape moves in a linear fashion, it is inefficient for applications requiring rapid random access to stored data. It does, however, provide an excellent means for back-up protection of important data.

13.4.6 Optical Data Storage

While magnetic systems are a popular and reliable method for storing large amounts of data, they can become quite cumbersome in terms of physical size as the amount of data increases. There is also the limitation on speed imposed by the need to convert between the digital signal of the CPU and an analog signal. Digital data storage addresses both of these problems. First, the data storage capability of a compact disk (CD), essentially identical to that used in musical recordings, is 680 K, and the technology for downloading the information stored on CDs keeps advancing. The initial transfer speed of CD drives was 150 K/sec. Currently, quad-speed CD drives, capable of data transfer at 600 K/sec, have become standard, and 6X speed drives delivering data at 900 K/sec are already on the market. Second, because the storage medium and the CPU use the same digital data format, there is no need for a controller to convert the signal from an analog signal. Often, compact disk storage in computers is referred to as CD-ROM, because the technology to write information in digital form on the compact disk is still out of reach in terms of price for most computer users. The technology does exist, however, and with time the price will surely make the ability to write data to a CD more generally available. Most CAD software packages are now available on CD-ROM.

13.5 INPUT DEVICES

Commonly used input devices in CAD systems include the alphanumeric keyboard, the mouse, the light pen, and the digitizer. All of these allow information transfer from the device to the CPU. The information being transferred can be alphanumeric or functional (in order to use command paths in the software) or graphic in nature. In either case, the devices allow an interface between the designer's thoughts and the machine that will assist in the design process.

13.5.1 Keyboard

The alphanumeric keyboard is one of the most recognizable computer input devices. Rows of letters and numbers (typically laid out like a typewriter keyboard) are used with other functional keys. These keys are either dedicated to tasks such as control of cursor placement on a display screen or definable by the user, transfer bits of information to the CPU in one of several ways. Key depression can be detected through a simple mechanical switch, a change in magnetic coupling, or a change in capacitance. The alphanumeric keyboard is dedicated to the input of alphanumeric information and special commands via function keys.

Special programmed-function keyboards with 16–32 buttons can also be used in conjunction with a CAD system. These keyboards are often separate from the alphanumeric keyboard, but the keys can similarly be dedicated or definable to specific CAD tasks. Some keyboards will employ cardboard or plastic overlays that show the function of each key. In the case where the keyboard is applicable to several tasks within the general CAD techniques, the overlays show which functions the keys will command different techniques.

13.5.2   Touch Pad

The touch pad is a device that allows command inputs and data manipulation to take place directly on the screen. The touch pad is mounted over the screen of the display terminal, and the user can select areas or on-screen commands by touching a finger to the pad. Various techniques are employed in touch pads to detect the position of the user's finger. Low-resolution pads employ a series of light-emitting diodes (LEDs) and photodiodes in the x- and y-axes of the pad. When the user's finger touches the pad, a beam of light is broken between an LED and a photodiode, which determines a position. Pads of this type generally supply 10–50 resolvable positions in each axis. A high-resolution panel design generates high-frequency shock waves traveling orthogonally through the glass. When the user touches the panel, part of the waves in both directions is deflected back to the source. The coordinates of this input can then be calculated by determining how long after the wave was generated it was reflected back to the source. Panels of this type can supply resolution of up to 500 positions in each direction. A different high-resolution panel design uses two transparent panel layers. One layer is conductive while the other is resistive. The pressure of a finger on the panel causes the voltage to drop in the resistive layer, and the measurement of the drop can be used to calculate the coordinates of the input.

The input of graphical data is somewhat clumsy using a keyboard or touch pad. For this reason, various input devices that are specialized for graphics input have been created and are widely used in CAD.

13.5.3   Mouse

The mouse is used for graphical cursor control. A mouse conveys cursor-placement information in the x–y coordinate plane to the CPU. A spherical roller is housed within the mouse such that the roller touches the plane upon which the mouse is resting. When the mouse is moved along a flat surface, the spherical roller simultaneously contacts two orthogonal potentiometers, each of which is connected to an analog–to–digital converter. The orthogonal potentiometers send x- and y-axis vector information via a connecting wire to the CPU, which performs the necessary vector additions to allow cursor control in any direction on a 2D display screen. Often, a mouse will be equipped with one to three pressure-sensitive buttons that assist in the selection of on-screen command paths. Since the mouse is inexpensive and simple to use, it has become a standard computer input device.

13.5.4   Trackball

This device operates much like a mouse in reverse. The main components of the mouse are also present in the trackball. Like the mouse, the trackball uses a spherical roller that comes into contact with two orthogonally placed potentiometers, sending x- and y-axis vector information to the CPU via a connecting wire. The difference between the mouse and the trackball lies in the placement of its spherical roller. In a trackball, the spherical roller rests on a base and is controlled directly by manual manipulation. As in a mouse, buttons may be present to facilitate the use of on-screen commands.

13.5.5   Light Pen

Another computer input device that can be used in CAD systems is the light pen. Somewhat mis-named, the light pen does not project light; rather, it detects light from a raster-scan cathode ray tube (CRT) screen (see Section 13.6). Light pens used in the SAGE Radar system in the 1950s resembled guns that were pointed at the screen, with input delivered through a trigger-like device. Contemporary light pens are hand-held cylindrical instruments approximately the size of an ink pen. At one end of the cylinder is a lens and a photo-optical sensor. The other end of the cylinder is connected to the computer by a cable. The pen detects the timing of the screen's repeated illuminations (a process so fast that the constant flicker one would expect is absent and the screen appears to maintain a constant 2D image) by detecting the light pulse at the desired location when the screen is illuminated. The pulse is then transferred through the cable to the CPU, which uses the pulse to determine where the light pen is in contact with the screen and uses this information to track its position continuously. The location is determined by correlating the pulse with the graphical display data in the CPU to identify what graphical information was being displayed at the given time and location on the screen. The device can also send a second type of signal to the CPU, indicating the selection of a point on the screen when a button on the pen is depressed. Light pens can thus be used to create lines and shapes that appear instantaneously to the human eye on the computer display. Light pens also select on-screen command paths in a manner similar to the mouse and trackball. Small graphical areas, sometimes called icons, on the screen can be associated with programmed software commands. If the user points the light pen at an icon and depresses the light pen button, the desired command will be executed on-screen, bypassing the need for keyboard inputs.

13.5.6 Digitizer

A digitizer is an input device consisting of a large, flat surface coupled with an electronic tracking device, or cursor. The cursor is tracked by the tablet underneath it and buttons on the cursor act as switches to allow the user to input position data and commands. Digitizing tablets apply different technologies to sense and track cursor position (Fig. 13.7). The three most common techniques use electromagnetic, electrostatic, and magnetostrictive methods to track the cursor. Electromagnetic tablets have a grid of wires underlying the tablet surface. Either the cursor or the tablet generates an alternating current that is detected by a magnetic receiver in the complementary device. The receiver generates and sends a digital signal to the CPU, giving the cursor's position. Despite their use of electromagnetism, these types of digitizers are not compromised by magnetic or conductive materials on their surface. Electrostatic digitizers generate a variable electric field that is detected by the tracking device. The frequency of the field variations and the time at which the field is sensed provide the information necessary to give accurate coordinates. Electrostatic digitizers function accurately in contact with paper, plastic, or any other material with a small dielectric constant. They do lose accuracy, however, when even partially conductive materials are in close proximity to the tablet. Magnetostrictive tablets use an underlying wire grid similar to that used in electromagnetic tablets. These tablets, however, use magnetostrictive wires (i.e., wires which change dimension depending on a magnetic field) in the grid. A magnetic pulse initiated at one end of a wire propagates through the wire as a wave. The cursor senses the wave using a loop of wire and relays a signal to the CPU, which then couples the time of the cursor signal with the time elapsed since the wave originated to give the position. These tablets require periodic remagnetization and recalibration to maintain their functional ability.

Digitizing tablets can usually employ various modes of operation. One mode allows the input of individual points. Other modes allow a continuous stream of points to be tracked into the CPU, either with or without one of the cursor buttons depressed, depending on the needs of the user. A digitizing-rate function, which enables the user to specify the number of points to be tracked in a given period of time, is also often present in CAD systems with digitizers. The rate can be adjusted as necessary to facilitate the accurate input of curves.

Whatever the type, digitizers are highly accurate graphical input devices and strongly suited to drafting original designs and to tracing existing designs from a hard-copy drawing. Resolution can be up to 1000 lines per linear inch. Tablet sizes typically range from 10×11 in. to 44×60 in.

Fig. 13.7 Digitizing Tablet and Cursor (courtesy of Summagraphics, Inc.).

Often, plastic sheets with areas for command functions, such as switching between modes, as discussed above, are laid over the tablet to allow the designer to access software commands directly through the tablet using one of the buttons on the tracking device. Many of these commands deal with the generation of graphical elements, such as lines, circles, and other geometries.

13.5.7 Scanner

Scanners have been used as computer input devices for some time, but their employment in CAD systems has been somewhat limited until recently due to resolution limits. A scanner uses a photo-optical sensor to detect areas of light and dark on the page when illuminated by a bright light source. The information from the sensors is digitized and sent to the CPU through a connecting cable. Special software that supplies optical character recognition (OCR) is capable of recognizing alphanumeric characters on the scanned image so that engineering notes and other text can be maintained with the image once the scanning is complete.

13.6 OUTPUT DEVICES

Just as a CAD system requires input devices to transfer information from the user to the CPU, output devices are also necessary to transfer data in visual form back to the user. Electronic displays provide real-time feedback to the user, enabling visualization and modification of information without hard-copy production. Often, however, a hard-copy is required for presentation or evaluation, and the devices that use the data in the CPU or stored in memory to create the desired copy are a second category of output devices.

13.6.1 Electronic Displays

Contemporary computer graphics displays use a cathode ray tube (CRT) (Fig. 13.8) to generate an image on the display screen. The CRT heats a cathode to project a beam of electrons onto a phosphor-coated glass screen. The electron beam energizes the phosphor coating at the point of contact, causing the phosphor to glow.

CRTs employ two different techniques, stroke-writing and raster-scan, to direct the beam onto the screen (Fig. 13.9). A CRT using the stroke-writing technique directs the beam only along the vectors given by the graphics data in the CPU. In a raster-scan CRT, the electron beam sweeps systematically from left to right and top to bottom at a continuous rate, employing what is known as *rasterization*. The image is created by turning the electron beam on or off at various points along the sweep, depending on whether a light or a dark dot is required at those points to create a recognizable image.

Regardless of the technique used to create an image on a CRT screen, the phosphor glows for a very short period of time after being energized by the electron beam. Three types of graphics terminals make up the vast majority of those used in CAD systems. Each employs a different approach to create a continuous image on the screen.

Vector Refresh Terminals

These early graphics terminals (circa 1960s) use the stroke-writing approach in the CRT. The electron beam continuously refreshes the image at speeds around 40–50 cycles per second to avoid a noticeable flicker. Refresh terminals permit a high degree of movement in the displayed image, as well as high resolution. Selective erasing or editing is possible at any time without erasing and repainting the entire image.

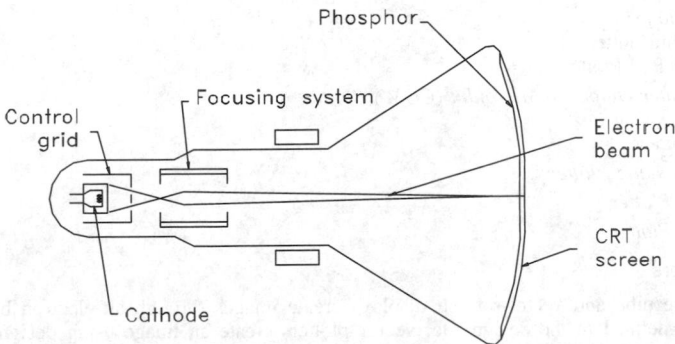

Fig. 13.8 Diagrammatic representation of a cathode ray tube (CRT).

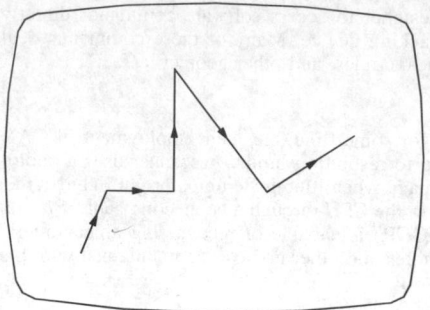 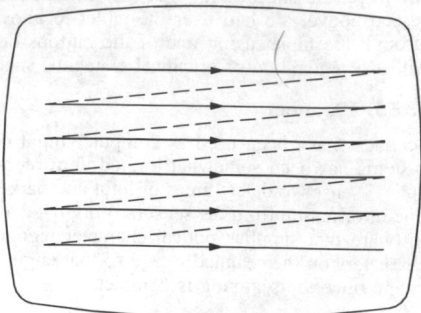

Fig. 13.9 (a) Electron beam direction in a stroke-writing CRT. (b) Electron beam pattern in a raster-scan CRT.

Direct View Storage Tube (DVST)

DVST was one of the most widely used graphics display devices in the mid- to late 1970s. In a DVST terminal, the CRT uses the stroke-writing technique to generate the desired image on the screen. One or more electron flood-guns continuously supply the necessary energy to maintain the image on the screen. DVST generates a long-lasting, flicker-free image with high resolution and no refreshing. It handles an almost unlimited amount of data. However, display dynamics are limited, since DVST terminals do not permit selective erasure. The image also does not appear as bright as with a vector refresh or raster-scan terminal.

Raster Scan Terminals

As their name suggests, these displays use rasterization to create the image on the CRT screen. Much like a television set (the difference being that a TV input signal is analog while that of a computer terminal is digital), these terminals refresh the image by continuously sweeping over the screen in a constant pattern. This allows real-time changes to be made to the image. Raster-scan is currently a dominant technology in CAD graphics display. Raster-scan terminal features include brightness, accuracy, selective erasure, dynamic motion capabilities, and the potential for unlimited color. The raster-scan terminal can display a large amount of information without flicker, although the resolution is not as high as with a DVST display. Raster-scan technology has moved rapidly since 1982 and, with the introduction of graphic accelerators and additional add-on frame buffers to workstations and PCs, affordable, high-quality graphics have finally reached the end-user.

13.6.2   Hard Copy Devices

Despite the ease with which design files can be managed using computer technology, a hard copy of the work is often required for recordkeeping and presentation. Output devices have been developed to interface with the computer and produce a hard copy of the requisite design or file. The processes used in hard copy devices are analogous to those used in CRT displays. The various types of hard-copy devices are shown in the following outline:

Vector Plotters
1. *Pen Plotter*
      Drum Plotter
      Flat-Bed Plotter
2. *Computer Output to Microfilm (COM) Plotter*

Raster Plotters
1. *Electrostatic Plotter*
2. *Inkjet Plotter*
3. *Laser Plotter*

Vector Plotters

Just as storage tube and vector-refresh displays create images through an electron beam directed along vectors defined in the design file, vector plotters create an image using designated vectors. Vector plotters produce very high-resolution hard copies. Two common kinds of vector plotters are the pen plotter and the COM plotter.

Pen Plotter. Pen plotters use mechanical ink pens, directed along design vectors to create images on paper or similar media. Pen plotters are further divided into drum- and flat-bed types.

Drum Plotter. The drum plotter (Fig. 13.10) consists of a cylindrical drum, most often mounted horizontally, and a pen mounted on a slide parallel to the surface of the drum. The plotter matches the motion of the drum and that of the pen along the slide to draw the appropriate vectors and create the desired image. The drum plotter is able to produce hard copies that are limited in length only by the amount of paper on a roll.

Flat-Bed Plotter. Flat-bed plotters act on the same vector-plotting concept, but the paper is attached to a flat surface. The writing tool is again movable along a metal slide. The slide provides one of the two coordinate axes upon which the design will be printed. The parallel tracks along which the slide itself moves provides the other axis. As the pen moves along the slide, the slide moves along the other axis, enabling the two-dimensional vectors to be drawn.

Computer-Output to Microfilm (COM) Plotter. COM plotters constitute a third type of vector plotter. These plotters produce images on film rather than on paper, using light instead of ink. These expensive units facilitate efficient archiving of designs. Designs can be stored using a fraction of the space required for hard-copy filing and enlarged to original size when needed. The cost of producing a plot using a COM plotter can be significantly less than the cost associated with a flat-bed or drum plot, but the quality upon enlargement is generally poorer than that obtained using a pen plotter.

Raster Plotters

Many plotters require rasterization, breaking up the image into a series of dots that will then be reconstituted to recreate the image in much the same way that a raster-scan CRT screen uses these dots to produce a recognizable image on the display terminal. Common raster plotters include the electrostatic plotter, inkjet plotter, and laser plotter. These types of plotters generally produce lower-resolution images than those generated on a pen plotter. However, because the time to produce a rasterized plot is independent of complexity, these have the advantage of being significantly faster and are often used to create preliminary hard copies before a higher-resolution plot is performed.

Electrostatic Plotter. Electrostatic or direct printing forms images for thermosensitive media by application of heat from nibs on thermal printing heads. The paper turns black at the precise points where heat is applied. Electrostatic plotters have few moving parts and thus are reliable and quiet.

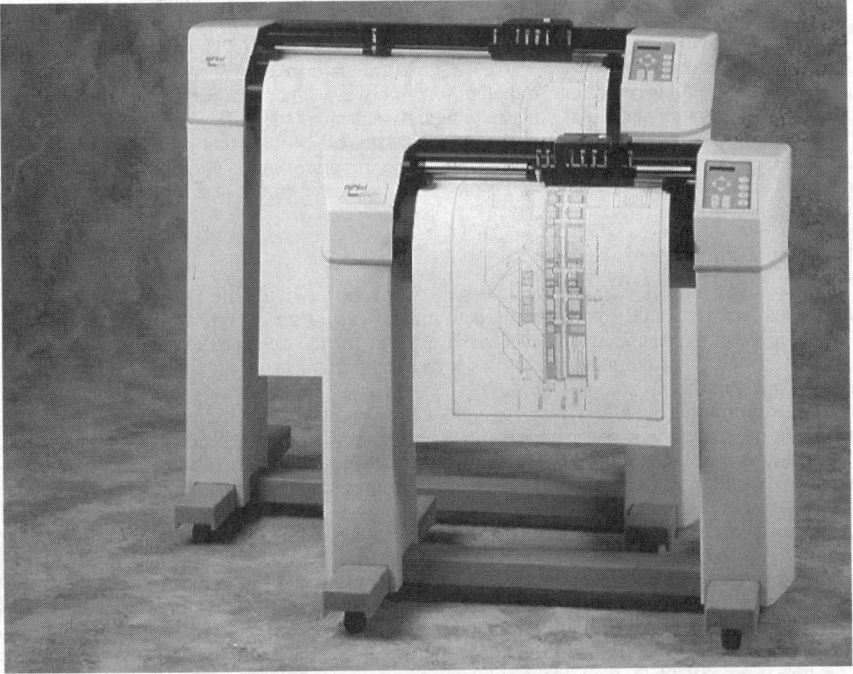

Fig. 13.10 Two drum-type plotters (courtesy of Summagraphics, Inc.).

Resolution can be as high as 400×800 dpi, with gray scales ranging from 16–128 values. These are medium- to high-throughput devices, producing complex images in about a minute. On-board computing facilities, such as RISC processors and fast hard disk storage mechanisms, contribute to rapid drawing and processing speeds. Expansion slots accommodate interface cards for LANs or parallel ports.

Inkjet Plotter. Inkjet plotters and printers fire tiny ink droplets at paper or a similar medium from minute nozzles in the printing head. Heat generated by a separate heating element almost instantaneously vaporizes the ink. The resulting bubble generates a pressure wave that ejects an ink droplet from the nozzle. Once the pressure pulse passes, ink vapor condenses and the negative pressure produced as the bubble contracts draws fresh ink into the nozzle. These plotters do not require special paper and can also be used for preliminary drafts. Inkjet plotters are available both as desktop units for 8.5×11-in. graphics and in wide format for engineering CAD drawings. Typical full-color resolution is 360 dpi, with black-and-white resolution rising to 700×720 dpi. These devices handle both roll-feed and cut sheet media in widths ranging from 8.5–36 in. Also, ink capacity in recently developed plotters has increased, allowing these devices to handle large rolls of paper without depleting any one ink color. Inkjet plotters are very user-friendly, often including sensors for the ink supply and ink flow that warn users of an empty cartridge or of ink stoppage, allowing replacement without losing a print. Other sensors eliminate printing voids and unwanted marks caused by bubbles in the ink lines. Special print modes typically handle high-resolution printing by repeatedly going over image areas to smooth image lines. In addition, inkjet plotters typically contain 6–64 megabytes of image memory and options such as hard drives, an Ethernet interface for networking, and built-in Postscript interpreters for faster processing. Inkjet plotters and printers are increasingly dominating other output technologies, such as pen plotters, in the design laboratory.

Laser Plotter. Laser plotters produce fairly high-quality hard copies in a shorter period of time than pen plotters. A laser housed within the plotter projects rasterized image data in the form of light onto a photostatic drum. As the drum rotates further about its axis, it is dusted with an electrically charged powder known as toner. The toner adheres to the drum wherever the drum has been charged by the laser light. The paper is brought into contact with the drum and the toner is released onto the paper, where it is fixed by a heat source close to the exit point. Laser plotters can quickly produce images in black and white or in color, and resolution is high.

13.7 SOFTWARE

Software is the collection of executable computer programs including operating systems, languages, and application programs. All of the hardware described above can do nothing without software to support it. In its broadest definition, software is a group of stored commands, sometimes known as a program, that provides an interface between the binary code of the CPU and the thought processes of the user. The commands provide the CPU with the information necessary to drive graphical displays and other output devices, to establish links between input devices and the CPU. The commands also define paths that enable other command sequences to operate. Software operates at all levels of computer function. Operating systems are a type of software that provides a platform upon which other programs may run. Likewise, individual programs often provide a platform for the operation of subroutines, which are smaller programs dedicated to the performance of specific tasks within the context of the larger program.

13.7.1 Operating Systems

Operating systems have developed over the past 50 years for two main purposes. First, operating systems attempt to schedule computational activities to ensure good performance of the computing system. Second, they provide a convenient environment for the development and execution of programs. An operating system may function as a single program or as a collection of programs that interact with each other in a variety of ways.

An operating system has four major components: process management, memory management, input/output operations, and file management. The operating system schedules and performs input/output, allocates resources and memory space and provides monitoring and security functions. It governs the execution and operation of various system programs and applications such as compilers, databases, and CAD software.

Operating systems that serve several users simultaneously (e.g., UNIX) are more complicated than those serving only a single user (e.g., MS-DOS, Macintosh Operating System). The two main themes in operating systems for multiple users are multiprogramming and multitasking.

Multiprogramming provides for the interleaved execution of two or more computer programs (jobs) by a single processor. In multiprogramming, while the current job is waiting for the input/output (I/O) to complete, the CPU is simply switched to execute another job. When that job is waiting for I/O to complete, the CPU is switched to another job, and so on. Eventually, the first job completes its I/O functions and is serviced by the CPU again. As long as there is some job to complete, the CPU remains active. Holding multiple jobs in memory at one time requires special hardware to protect each job, some form of memory management, and CPU scheduling. Multiprogramming increases CPU use and decreases the total time needed to execute the jobs, resulting in greater throughput.

The techniques that use multiprogramming to handle multiple interactive jobs are referred to as *multitasking* or *time-sharing*. Multitasking or time-sharing is a logical extension of multiprogramming for situations where an interactive mode is essential. The processor's time is shared among multiple users. Time-sharing was developed in the 1960s, when most computers were large, costly mainframes. The requirement for an interactive computing facility could not be met by the use of a dedicated computer. An interactive system is used when a short response time is required. Time-sharing operating systems are very sophisticated, requiring extra disk management facilities and an on-line file system having protective mechanisms as well.

The following sections discuss the two most widely used operating systems for CAD applications, UNIX and Windows NT. It should be noted that both of these operating systems can run on the same hardware architecture.

UNIX

The first version of UNIX was developed in 1969 by Ken Thompson and Dennis Ritchie of the Research Group of Bell Laboratories to run on a PDP-7 minicomputer. The first two versions of UNIX were created using assembly language, while the third version was written using the C programming language. As UNIX evolved, it became widely used at universities, research and government institutions, and eventually in the commercial world. UNIX quickly became the most portable of operating systems, operable on almost all general-purpose computers. It runs on personal computers, workstations, minicomputers, mainframes, and supercomputers. UNIX has become the preferred program-development platform for many applications, such as graphics, networking, and databases. A proliferation of new versions of UNIX has led to a strong demand for UNIX standards. Most existing versions can be traced back to one of two sources: AT&T System V or 4.3 BSD (Berkeley UNIX) from the University of California, Berkeley (one of the most influential versions).

UNIX was designed to be a time-sharing, multi-user operating system. UNIX supports multiple processes (multiprogramming). A process can easily create new processes with the fork system call. Processes can communicate with pipes or sockets. CPU scheduling is a simple priority algorithm. Memory management is a variable-region algorithm with swapping supported by paging. The file system is a multilevel tree that allows users to create their own subdirectories. In UNIX, I/O devices such as printers, tape drives, keyboards, and terminal screens are all treated as ordinary files (file metaphor) by both programmers and users. This simplifies many routine tasks and is a key component in extensibility of the systems. Certifiable security that protect users' data and network support are also two important features.

UNIX consists of two separable parts: the kernel and the system programs. The kernel is the collection of software that provides the basic capabilities of the operating system. In UNIX, the kernel provides the file system, CPU scheduling, memory management, and other operating system functions (I/O devices, signals) through system calls. System calls can be grouped into three categories: file manipulation, process control, and information manipulation. Systems programs use the kernel-supported system calls to provide useful functions, such as compilation and file manipulation. Programs, both system and user-written, are normally executed by a command interpreter. The command interpreter is a user process called a *shell*. Users can write their own shell. There are, however, several shells in general use. The Bourne shell, written by Steve Bourne, is the most widely available. The C shell, mostly by Bill Joy, is the most popular on BSD systems. The Korn Shell, by David Korn, has also become quite popular in recent years.

Windows NT

The development effort for the new high-end operating system in the Microsoft Windows family, Windows NT (New Technology), has been led by David Culter since 1988. Market requirements and sound design characteristics shaped the Windows NT development. The architects of "NT," as it is popularly known, capitalized on the strengths of UNIX while avoiding its pitfalls. Windows NT and UNIX share striking similarities. There are also marked differences between the two systems. UNIX was designed for host-based terminal computing (multi-user) in 1969, while Windows NT was designed for client/server distributed computing in 1990. The users on single-user general-purpose workstations (clients) can connect to multi-user general-purpose servers with the processing load shared between them. There are two Windows NT-based operating systems: Windows NT Server and Windows NT Workstation. The Windows NT Workstation is simply a scaled-down version of Windows NT Server in terms of hardware and software. Windows NT is a microkernel-based operating system. The operating system runs in privileged processor mode (kernel mode) and has access to system data and hardware. Applications run on a non-privileged processor mode (user mode) and have limited access to system data and hardware through a set of digitally controlled application programming interfaces (APIs). Windows NT also supports both single-processor and symmetric multiprocessing (SMP) operations. Multiprocessing refers to computers with more than one processor. A multiprocessing computer is able to execute multiple threads simultaneously, one for each processor in the computer. In SMP, any processor can run any type of thread. The processors communicate with each other through shared memory. SMP provides better load-balancing and fault-tolerance. The Win32 subsystem is the most critical of the Windows NT environment subsystems. It provides the graphical user interface and controls all user input and application output.

Windows NT is a fully 32-bit operating system with all 32-bit device drivers, paving the way for future development. It makes administration easy by providing more flexible built-in utilities and removes diagnostic tools. Windows NT Workstation provides full crash protection to maximize up-time and reduce support costs. Windows NT is a complete operating system with fully integrated networking, including built-in support for multiple network protocols. Security is pervasive in Windows NT to protect system files from error and tampering. The NT file system (NTFS) provides security for multiple users on a machine.

Windows NT, like UNIX, is a portable operating system. It runs on many different hardware platforms and supports a multitude of peripheral devices. It integrates preemptive multitasking for both 16- and 32-bit applications into the operating system, so it transparently shares the CPUs among the running applications. More usable memory is available due to advanced memory features of Windows NT. There are more than 1400 32-bit applications available for Windows NT today, including all major CAD and FEA software applications.

Hardware requirements for the Windows NT operating system fall into three main categories: processor, memory, and disk space. In general, Windows NT Server requires more in each of the three categories than does its sister operating system, the Windows NT Workstation. The minimum processor requirements are a 32-bit x86-based microprocessor (Intel 80386/25 or higher), Intel Pentium, Apple Power-PC, or other supported RISC-based processor, such as the MIPS R4000 or Digital Alpha AXP. The minimum memory requirement is 16 MB. The minimum disk space requirements for just the operating system are in the 100-MB range. NT Workstation requires 75 MB for x86 and 97 MB for RISC. For the NT Server, 90 MB for x86 and 110 MB for RISC are required. There is no need to add additional disk space for any application that is run on the NT operating system.

13.7.2 Graphical User Interface (GUI) and the X Window System

DOS, UNIX, and other command-line operating systems have long been criticized for the complexity of their user interface. For this reason, GUI is one of the most important and exciting developments of this decade. The emergence of GUI revolutionized the methods of man-machine interaction used in the modern computer. GUIs are available for almost every type of computer and operating system on the market. A GUI is distinguished by its appearance and by the way an operator's actions and input options are handled. There are over a dozen GUIs. They may look slightly different, but they all share certain basic similarities. These include the following: a pointing device (mouse or digitizer), a bit-mapped display, windows, on-screen menus, icons, dialog boxes, buttons, sliders, check boxes, and an object-action paradigm. Simplicity, ease of use, and enhanced productivity are all benefits of a GUI. GUIs have fast become important features of CAD software.

Graphical user interface systems were first envisioned by Vannevar Bush in a 1945 journal article. Xerox was researching graphical user interface tools at the Palo Alto Research Center throughout the 1970s. By 1983, every major workstation vendor had a proprietary window system. It was not until 1984, however, when Apple introduced the Macintosh computer, that a truly robust window environment reached the average consumer. In 1984, a project called Athena at MIT gave rise to the X Window system. Athena investigated the use of networked graphics workstations as a teaching aid for students in various disciplines. The research showed that people could learn to use applications with a GUI much more quickly than by learning commands.

The X Window system is a non-vendor-specific window system. It was specifically developed to provide a common window system across networks connecting machines from different vendors. Typically, the communication is via Transmission Control Protocal/Internet Protocal (TCP/IP) over an Ethernet network. The X Window system (X-Windows or X) is not a GUI. It is a portable, network-transparent window system that acts as a foundation upon which to build GUIs (such as AT&T's OpenLook, OSF/Motif, and DEC Windows). The X Window system provides a standard means of communicating between dissimilar machines on a network and can be viewed in a window. The unique benefit provided by a window system is the ability to have multiple views showing different processes on different networks. Since the X Window system is in the public domain and not specific to any platform or operating system, it has become the de facto window system in heterogeneous environments from PCs to mainframes.

Unfortunately, a window environment does not come without a price. Extra layers of software separate the user and the operating system, such as window system, GUI, and an Application Programming Interface (ToolKit) in a UNIX operating environment. GUIs also place extra demands on hardware. All visualization workstations require more powerful processing capabilities (> 6 MIPS), large CPU memory and disk subsystems, built-in network Input/Output (I/O) with typically Ethernet high-speed internal bus structures ($\geq$ 32 MB/sec)—high-resolution monitors ($\geq$ 1024 $\times$ 768), more colors ($>$ 256), and so on.

For PCs, both operating systems and GUIs are in a tremendous state of flux. Microsoft Windows, Windows NT, and Windows 95 are expected to dominate the market, followed by the Macintosh. For workstations, the OSF/Motif interface on an X-Windows system seems to have the best potential to become an industry-wide graphical user interface standard.

13.7.3 Computer Languages

The computer must be able to understand the commands it is given in order to perform desired tasks at hand. The binary code used by the computer circuitry is very easy for the computer to understand, but can be tedious and almost indecipherable to the human programmer. Languages for computer programming have developed to facilitate the programmer's job. Languages are often categorized as low- or high-level languages.

Low-Level Languages

The term *low-level* refers to languages that are easy for the computer to understand. These languages are often specific to a particular type of computer, so that programs created on one type of computer must be modified to run on another type. Machine language (ML) and assembly language (AL) are both considered low-level languages.

Machine language is the binary code that the computer understands. ML uses an operator command coupled with one or more operands. The operator command is the binary code for a specific function, such as addition. The numbers to be added, in this example, are operands. Operators are also binary codes, arbitrary with respect to the machine used. For a hypothetical computer, all operator codes are established to be eight digits, with the operator command appearing after the two operands. If the operator code for addition then were 01100110, the binary (base 2) representation of the two numbers added would be followed by the code for addition. A command line to perform the addition of 21 and 14 would then be written as follows:

$$000101010000111001100110$$

The two operands are written in their 8 bit binary forms (21_{10} as 00010101_2 and 14_{10} as 00001110_2) and are followed by the operator command (01100110 for addition). The binary nature of this language makes programming difficult and error-correction even more so.

AL operates in a similar manner to ML but substitutes words for machine codes. The program is written using these one-to-one relationships between words and binary codes and separately assembled through software into binary sequences. Both ML and AL are time-intensive for the programmer and, because of the differences in logic circuitry between types of computers, the languages are specific to the computer being used. High-level languages address the problems presented by these low-level languages in various ways.

High-Level Languages (HLLs)

High-level languages give the programmer the ability to bypass much of the tediousness of programming involved in low-level languages. Often many ML commands will be combined within one HLL statement. The programming statements in HLL are converted to ML using a compiler. The compiler uses a low-level language to translate the HLL commands into ML and check for errors. The net gain in terms of programming time and accuracy far outweighs the extra time required to compile the code. Because of their programming advantages, HLLs are far more popular and widely used than low-level languages. The following commonly used programming languages are described below:

- FORTRAN
- Pascal
- BASIC
- C
- C++

FORTRAN (FORmula TRANslation). Developed at IBM between 1954 and 1957 to perform complex calculations, this language employs a hierarchical structure similar to that used by mathematicians to perform operations. The programmer uses formulas and operations in the order that would be used to perform the calculation manually. This makes the language very easy to use. FORTRAN can perform simple as well as complex calculations. FORTRAN is used primarily for scientific or engineering applications. CFP95 Suite, a software benchmarking product by Standard

Performance Evaluation Corp. (SPEC) is written in FORTRAN. It contains 10 CPU-intensive floating point benchmarks.

The programming field in FORTRAN is composed of 80 columns, arranged in groups relating to a programming function. The label or statement number occupies columns 1–5. If a statement extends beyond the statement field, a continuation symbol is entered in column 6 of the next line, allowing the statement to continue on that line. The programming statements in FORTRAN are entered in columns 7–72. The maximum number of lines in a FORTRAN statement is 20. Columns 73–80 are used for identification purposes. Information in these columns is ignored by the compiler, as are any statements with a C entered in column 1.

Despite its abilities, there are several inherent disadvantages to FORTRAN. Text is difficult to read, write, and manipulate. Commands for program flow are complicated and a subroutine cannot go back to itself to perform the same function.

Pascal. Pascal is a programming language with many different applications. It was developed by Niklaus Wirth in Switzerland during the early 1970s and named after the French mathematician Blaise Pascal. Pascal can be used in programs relating to mathematical calculations, file processing and manipulation, and other general-purpose applications.

A program written in Pascal has three main sections: the program name, the variable declaration, and the body of the program. The program name is typically the word *PROGRAM* followed by its title. The variable declaration includes defining the names and types of variables to be used. Pascal can use various types of data and the user can also define new data types, depending on the requirements for the program. Defined data types used in Pascal include strings, arrays, sets, records, files, and pointers. Strings consist of collections of characters to be treated as a single unit. Arrays are sequential tables of data. Sets define a data set collected with regard to sequence. Records are mixed data types organized into a hierarchical structure. Files refer to collections of records outside of the program itself, and pointers provide flexible referencing to data. The body of the program uses commands to execute the desired functions. The commands in Pascal are based on English and are arranged in terms of separate procedures and functions, both of which must have a defined beginning and end. A function can be used to execute an equation and a procedure is used to perform sets of equations in a defined order. Variables can be either "global" or "local," depending on whether they are to be used throughout the program or within a particular procedure. Pascal is somewhat similar to FORTRAN in its logical operation, except that Pascal uses symbolic operators while FORTRAN operates using commands. The structure of Pascal allows it to be applicable to areas other than mathematical computation.

BASIC (Beginners All-Purpose Symbolic Interactive Code). BASIC was developed at Dartmouth College by John Kemeny and Thomas Kurtz in the mid-1960s. BASIC uses mathematical programming techniques similar to FORTRAN and the simplified format and data manipulation capabilities similar to Pascal. As in FORTRAN, BASIC programs are written using line numbers to facilitate program organization and flow. Because of its simplicity, BASIC is an ideal language for the beginning programmer. BASIC runs in either direct or programming modes. In the direct mode, the program allows the user to perform a simple command directly, yielding an instantaneous result. The programming mode is distinguished by the use of line numbers that establish the sequence of the programming steps. For example, if the user wishes to see the words *PLEASE ENTER DIAMETER* displayed on the screen immediately, he would execute the command *PRINT "PLEASE ENTER DIAMETER."* If, however, that phrase were to appear in a program, the above command would be preceded by the appropriate line number.

The compiler used in the BASIC language is unlike the compiler used for either FORTRAN or Pascal. Whereas other HLL compilers check for errors and execute the program as a whole unit, a BASIC program is checked and compiled line by line during program execution. BASIC is often referred to as an "interpreted" language as opposed to a compiled one, since it interprets the program into ML line by line. This condition allows for simplified error debugging. In BASIC, if an error is detected, it can be corrected immediately, while in FORTRAN and Pascal, the programmer must go back to the source program in order to correct the problem and then recompile the program as a separate step. The interpretive nature of BASIC does cause programs to run significantly more slowly than in either Pascal or FORTRAN.

C. C was developed from the B language by Dennis Ritchie in 1972. C was standardized by the late 1970s when B. W. Kernighan and Ritchie's book *The C Programming Language* was published. C was developed specifically as a tool for the writing of operating systems and compilers. It originally became most widely known as the development language for the UNIX operating systems. C expanded at a tremendous rate over many hardware platforms. This led to many variations and a lot of confusion and, while these variations were similar, there were notable differences. This was a problem for developers that wanted to write programs that ran on several platforms. In 1989, the American National Standards Committee on Computers and Information Processing approved a standard version of C. This version is known as *ANSI C* and it includes a definition of a set of library routines for file operations, memory allocation, and string manipulation.

A program written in C appears similar to Pascal. C, however, is not as rigidly structured as Pascal. There are sections for the declaration of the main body of the program and the declaration of variables. C, like Pascal, can use various types of data and the programmer can also define new data types. C has a rich set of data types, including arrays, sets, records, files, and pointers. C allows for far more flexibility than Pascal in the creation of new data types and the implementation of existing data types. Pointers in C are more powerful than they are in Pascal. Pointers are variables that point not to data but to the memory location of data. Pointers also keep track of what type of data is stored there. A pointer can be defined as a pointer to an integer or a pointer to a character. CINT95 Suite, a software benchmarking product, is written in C. It contains eight CPU-intensive integer benchmarks.

C++. C++ is a superset of the C language developed by Bjarne Stroustrup in 1986. C++'s most important addition to the C language is the ability to do object-oriented programming. Object-oriented programming places more emphasis on the data of a program. Programs are structured around objects. An object is a combination of the program's data and code. Like a traditional variable, an object stores data, but unlike traditional languages, objects can also do things. For example, an object called *triangle* might store both the dimensions of the triangle and the instructions on how to draw the triangle. Object-oriented programming has led to a major increase in productivity in the development of applications over traditional programming techniques.

A program written in C++ no longer resembles C or Pascal. More emphasis is placed on a modular design around objects. The main section of a C++ code should be very small and may only call one or two functions, and the declaration of variables in the main function should be avoided. Global variables and functions are avoided at all costs and the use of variables in local objects is stressed. The avoidance of global variables and functions that do large amounts of work is intended to increase security and make programs easier to develop, debug, and modify.

Some computer languages have been developed or modified for use with software applications for the Windows NT operating system. These include languages such as Ada, COBOL, Forth, LISP, Prolog, Visual BASIC, and Visual C++.

13.8   CAD SOFTWARE

Contemporary CAD software is often sold in "packages" that feature all of the programs needed for CAD applications. These fall into two categories: graphics software and analysis software. Graphics software makes use of the CPU and its peripheral input/output devices to generate a design and represent it on-screen. Analysis software makes use of the stored data relating to the design and applies them to dimensional modeling and various analytical methods using the computational speed of the CPU.

13.8.1   Graphics Software

Traditional drafting has consisted of the creation of two-dimensional technical drawings that operated in the synthesis stage of the general design process. However, contemporary computer graphics software, including that used in CAD systems, enables designs to be represented pictorially on the screen such that the human mind may create perspective, thus giving the illusion of three dimensions on a 2D screen. Regardless of the design representation, the drafting itself only involves taking the conceptual solution for the previously recognized and defined problem and representing it pictorially. It has been asserted above that this "electronic drawing-board" feature is one of the advantages of computer-aided design. But how does that drawing board operate?

The drawing board available through CAD systems is largely a result of the supporting graphics software. That software facilitates graphical representation of a design on-screen by converting graphical input into Cartesian coordinates along x-, y-, and sometimes z-axes. Design elements such as geometric shapes are often programmed directly into the software for simplified geometric representation. The coordinates of the lines and shapes created by the user can then be organized into a matrix and manipulated through matrix multiplication, and the resulting points, lines, and shapes are relayed back to the graphics software and, finally, the display screen for simplified editing of designs. Because the whole process can take as little as a few nanoseconds, the user sees the results almost instantaneously. Some basic graphical techniques that can be used in CAD systems include scaling, rotation, and translation. All are accomplished through an application of matrix manipulation to the image coordinates. While matrix mathematics provides the basis for the movement and manipulation of a drawing, much of CAD software is dedicated to simplifying the process of drafting itself because creating the drawing line by line, shape by shape is a lengthy and tedious process in itself. CAD systems offer users various techniques that can shorten the initial drafting time.

Geometric Definition

All CAD systems offer defined geometric elements that can be called into the drawing by the execution of a software command. The user must usually specify the variables specific to the desired element. For example, the CAD software might have, stored in the program, the mathematical definition of a circle. In the x–y coordinate plane, that definition is the following equation:

$$(x - m)^2 + (y - n)^2 = r^2$$

Here, the radius of the circle with its center at (m, n) is r. If the user specifies m, n, and r, a circle of the specified size will be represented on-screen at the given coordinates. A similar process can be applied to many other graphical elements. Once defined and stored as an equation, the variables of size and location can be applied to create the shape on-screen quickly and easily. This is not to imply that a user must input the necessary data in numerical form. Often, a graphical input device such as a mouse, trackball, digitizer, or light pen can be used to specify a point from which a line (sometimes referred to as a *rubber-band line* due to the variable length of the line as the cursor is moved toward or away from the given point) can be extended until the desired length is reached. A second input specifies that the desired endpoint has been reached, and variables can be calculated from the line itself. For a rectangle or square, the line might represent a diagonal from which the lengths of the sides could be extrapolated. In the example of the circle above, the user would specify that a circle was to be drawn using a screen command or other input method. The first point could be established on-screen as the center. Then the line extending away from the center would define the radius. Often the software will show the shape changing size as the line lengthens or shortens. When the radial line corresponds to the circle of desired size, the second point is defined. The coordinates of the two defined points give the variables needed for the program to draw the circle. The center is given by the coordinates of the first point and the radius is easily calculated by determining the length of the line between points 1 and 2. Most engineering designs are much more complex than simple, whole shapes, and CAD systems are capable of combining shapes in various ways to create the desired design.

The combination of defined geometric elements enables the designer to create many unique geometries quickly and easily on a CAD system. The concepts involved in two-dimensional combinations are illustrated before moving on to three-dimensional combinations.

Once the desired geometric elements have been called into the program, they can be defined as *cells,* individual design elements within the program. These cells can then be added as well as subtracted in any number of ways to create the desired image. For example, a rectangle might be defined as cell "A" and a circle might be defined as cell "B." When these designations have been made, the designer can add the two geometries or subtract one from the other, using Boolean logic commands such as union, intersection, and difference. The concept for two dimensions is illustrated by Fig. 13.11. The new shape can also be defined as a cell and combined in a similar manner to other primitives or conglomerate shapes. Cell definition, therefore, is recognized as a very powerful tool in CAD.

13.8.2 Solid Modeling

Three-dimensional geometric or solid modeling capabilities follow the same basic concept illustrated above, but with some other important considerations. First, there are various approaches to creating the design in three dimensions (Fig. 13.12). Second, different operators in solid-modeling software may be at work in constructing the 3D geometry.

In CAD solid-modeling software, there are various approaches that define the way in which the user creates the model. Since the introduction of solid-modeling capabilities into the CAD main-

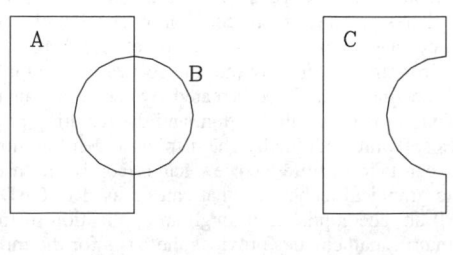

$$A \; - \; B \; = \; C$$

Fig. 13.11 Two-dimensional example of Boolean difference.

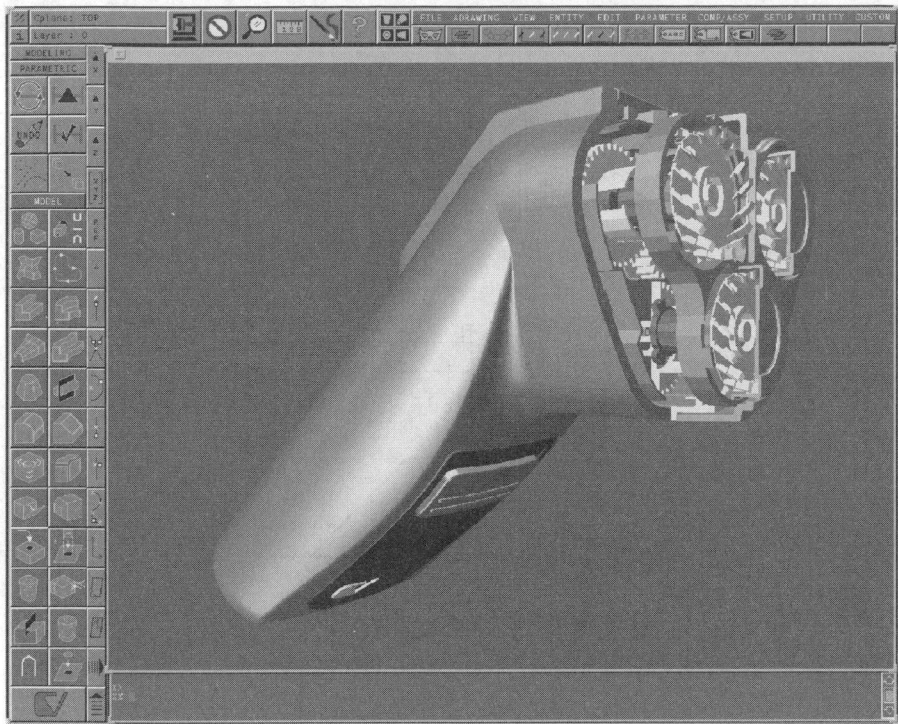

Fig. 13.12 Solid model of an electric shaver design (courtesy of ComputerVision, Inc.).

stream, various functional approaches to solid modeling have been developed. Many CAD software packages today support dimension-driven solid-modeling capabilities, which include variational design, parametric design, and feature-based modeling.

Dimension-driven design denotes a system whereby the model is defined as sets of equations that are solved sequentially. These equations allow the designer to specify constraints, such as that one plane must always be parallel to another. If the orientation of the first plane is changed, the angle of the second plane will likewise be changed to maintain the parallel relationship. This approach gets its name from the fact that the equations used often define the distances between data points.

The *variational modeling* method describes the design in terms of a sketch that can later be readily converted to a 3D mathematical model with set dimensions. If the designer changes the design, the model must then be completely recalculated. This approach is quite flexible because it takes the dimension-driven approach of handling equations sequentially and makes it nonsequential. Dimensions can then be modified in any order, making it well suited for use early in the design process when the design geometry might change dramatically. Variational modeling also saves computational time (thus increasing the run-speed of the program) by eliminating the need to solve any irrelevant equations. Variational sketching (Fig. 13.13) involves creating two-dimensional profiles of the design that can represent end views and cross sections. Using this approach, the designer typically focuses on creating the desired shape with little regard for dimensional parameters. Once the design shape has been created, a separate dimensioning capability can scale the design to the desired dimensions.

Parametric modeling solves engineering equations between sets of parameters such as size parameters and geometric parameters. Size parameters are dimensions such as the diameter and depth of a hole. Geometric parameters are constraints such as tangential, perpendicular, or concentric relationships. Parametric modeling approaches keep a record of operations performed on the design such that relationships between design elements can be inferred and incorporated into later changes in the design, thus making the change with a certain degree of acquired knowledge about the relationships between parts and design elements. For example, using the parametric approach, if a recessed area in the surface of a design should always have a blind hole in the exact center of the area, and the recessed portion of the surface is moved, the parametric modeling software will also move the blind hole to the new center. In Fig. 13.14, if a bolt circle (BC) is concentric with a bored hole and the bored hole is moved, the bolt circle will also move and remain concentric with the bored

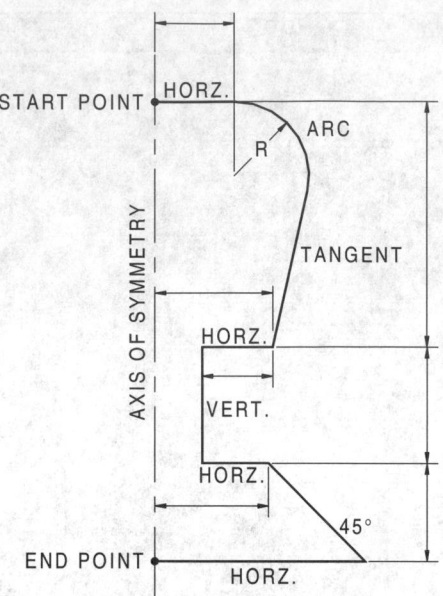

Fig. 13.13 Variational sketch.

hole. The dimensions of the parameters may also be modified using parametric modeling. The design is modified through a change in these parameters, either internally, within the program, or from an external data source, such as a separate database.

Feature-based modeling allows the designer to construct solid models from geometric features, which are industrial standard objects such as holes, slots, shells, and grooves (Fig. 13.15). For example, a hole can be defined using a "through-hole" feature. Whenever this feature is used, independent on the thickness of the material through which the hole passes, the hole will always be open at both sides. In variational modeling, by contrast, if a hole were created in a plane of specified thickness and the thickness were increased, the hole would be a blind hole until the designer adjusted the dimensions of the hole to provide an opening at both ends. The major advantage of feature-based modeling is the maintenance of design intent regardless of dimensional changes in design. Another significant advantage in using a feature-based approach is the capability to change many design elements relating to a change in a certain part. For example, if the threading of a bolt is changed, the threading of the associated nut would be changed automatically, and if that bolt design were used more than once in the design, all bolts and nuts could similarly be altered in one step. A knowledge base and inference engine make feature-based modeling more intelligent in some feature-based CAD systems.

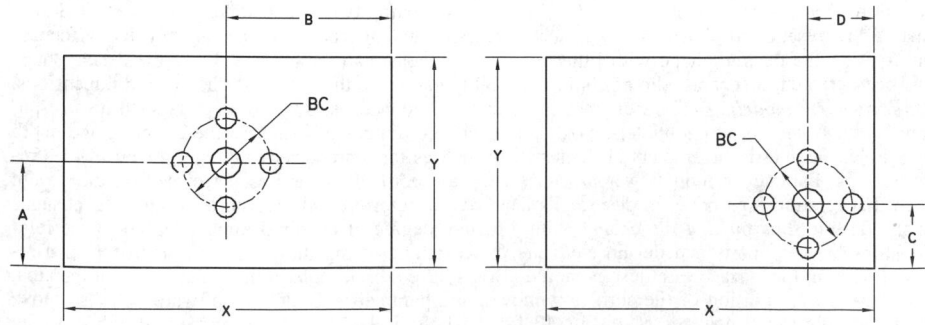

Fig. 13.14 Parametric modeling.

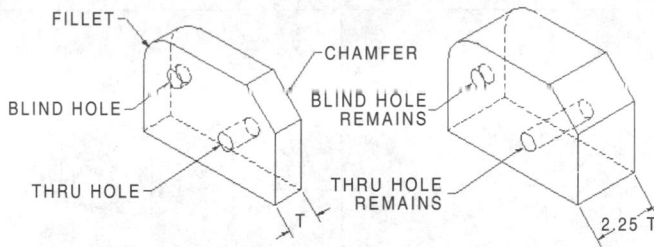

Fig. 13.15 Feature-based modeling.

Regardless of the modeling approach employed by a software package, there are usually two basic methods for creating 3D solid models: constructive solid geometry (CSG) and boundary representation (B-rep). Most CAD applications offer both methods.

With the CSG method, using defined solid geometries, such as those for a cube, sphere, cylinder, prism, and so on, the user can combine them by subsequently employing a Boolean logic operator, such as union, subtraction, difference, and intersection, to generate a more complex part. In three dimensions, the Boolean difference between a cylinder and a torus might appear as in Fig. 13.16.

The boundary representation method is a modeling feature in 3D representation. Using this technique, the designer first creates a 2D profile of the part. Then, using a linear, rotational, or compound sweep, the designer extends the profile along a line, about an axis, or along an arbitrary curved path, respectively, to define a 3D image with volume. Figure 13.17 illustrates the linear, rotational, and compound sweep methods.

Software manufacturers approach solid modeling differently. Nevertheless, every comprehensive solid modeler should have five basic functional capabilities: interactive drawing, a solid modeler, a dimensional constraint engine, a feature manager, and assembly managers.

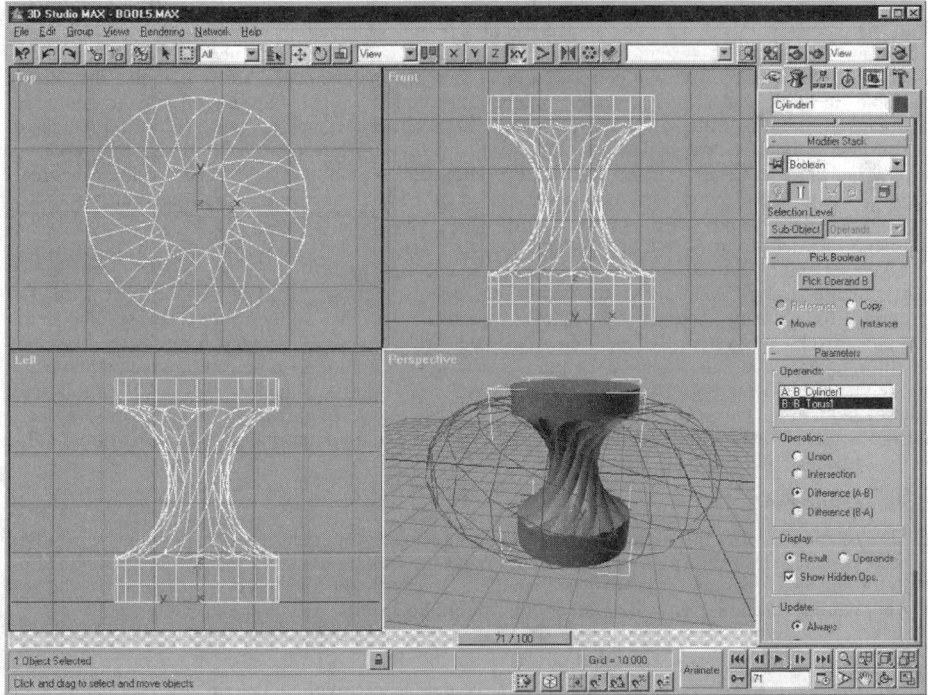

Fig. 13.16 Boolean difference between a cylinder and a torus using Autodesk 3-D StudioMax software (courtesy of Autodesk, Inc.).

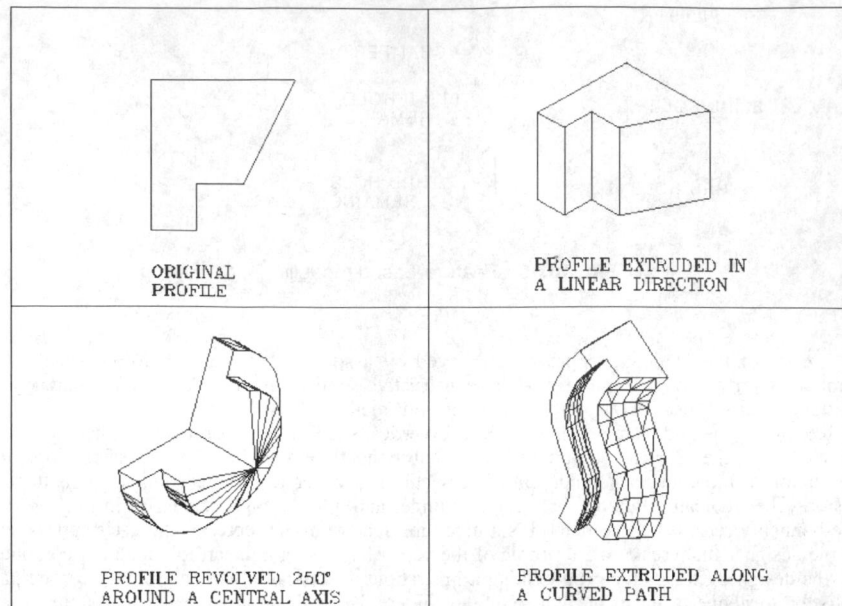

Fig. 13.17   Various common sweep methods in CAD software.

The drawing capabilities should indicate shapes and profiles quickly and easily, usually with one or two mouse clicks. The purpose of drawing interactively on the screen should be to capture the basic concept information in the computer as efficiently as possible. The solid modeler should be able to combine geometric elements using Boolean logic commands and transform 2D cross sections into 3D models with volume using linear, rotational, and compound sweep methods. A dimensional constraint engine controls relational variables associated with the model such that when the model is changed, the variables change correspondingly. It is the dimensional constraint engine that allows variables to be defined in terms of their relatedness to other variables instead of as fixed geometric elements in the design file. The feature manager allows features such as holes, slots, and flanges to be introduced into the design. These features can save time in later iterations of design and represent a major advance in CAD system software in recent years. Assembly management involves the treatment of design units as conglomerate entities, often called *cells,* as a single functional unit. Assembly cells make management of the design a fairly easy task, since the user can essentially group any elements of interest into a cell, and perform selective tasks on the cell as a whole.

Another significant advance in solid modeling over the past 20 years has been the creation of parts libraries using CAD data files. In early systems, geometries had to be created from within the program. Today, many systems will accept geometries from other systems and software. The Initial Graphics Exchange System (IGES) is an ANSI standard that defines a neutral form for the exchange of information among dissimilar CAD and CAM systems. Significant time can be saved when using models from differing sources. Often, corporations will supply magnetic disks or CD-ROMs with catalogued listings of various parts and products. In this way, the engineer can focus on the major design considerations without constantly redesigning small, common parts such as bearings, bolts, cogs, sprockets, and so on.

Editing Features

CAD systems also offer the engineer powerful editing features that reduce the design time by avoiding all the manual redrawing that was traditionally required. Common editing features are performed on cells of single or conglomerate geometric shape elements. Most CAD systems offer all of the following editing functions, as well as others that might be specific to a program being used:

- *Movement.* Allows a cell to be moved to another location on the display screen
- *Duplication.* Allows a cell to appear at a second location without deleting the original location
- *Rotation.* Rotates a cell a given angle about an axis
- *Mirroring.* Displays a mirror image of the cell about a plane

- *Deletion.* Removes the cell from the display and the design data file
- *Removal.* Erases the cell from the display, but maintains it in the design data file
- *Trim.* Removes any part of the cell extending beyond a defined point, line, or plane
- *Scaling.* Enlarges or reduces the cell by a specified factor along x-, y-, and z-axes
- *Offsetting.* Creates a new object that is similar to a selected object at a specified distance
- *Chamfering.* Connects two nonparallel objects by extending or trimming them to intersect or join with a beveled line
- *Filleting.* Connects two objects with a smoothly fitted arc of a specified radius
- *Hatching.* User can edit both hatch boundaries and hatch patterns

Most of the editing features offered in CAD are transformations performed using algebraic matrix manipulations.

Transformations

Transformation in general refers to the movement or other manipulation of graphical data. Two-dimensional transformations are considered first in order to illustrate the basic concepts. Later, these concepts are applied to geometries with three dimensions.

Two-Dimensional Transformations. To locate a point in a two-axis Cartesian coordinate system, x and y values are specified. This two-dimensional point can be modeled as a 1×2 matrix: (x,y). For example, the matrix $p = (3,2)$ would be interpreted to be a point that is 3 units from the origin in the x-direction and 2 units from the origin in the y-direction.

This method of representation can be conveniently extended to define a line segment as a 2×2 matrix by giving the x and y coordinates of the two end points of the line. The notation would be

$$l = \begin{bmatrix} x_1 & y_1 \\ x_2 & y_2 \end{bmatrix}$$

Using the rules of matrix algebra, a point or line (or other geometric element represented in matrix notation) can be operated on by a transformation matrix to yield a new element.

There are several common transformations: translation, scaling, and rotation.

Translation. Translation involves moving the element from one location to another. In the case of a line segment, the operation would be

$$\begin{cases} x_1' = x_1 + \Delta x & y_1' = y_1 + \Delta y \\ x_2' = x_2 + \Delta x & y_2' = y_2 + \Delta y \end{cases}$$

where x', y' are the coordinates of the translated line segment,
$\quad x$, y are the coordinates of the original line segment,
$\quad \Delta x$ and Δy are the movements in the x and y directions, respectively.

In the matrix notation, this can be represented as

$$l' = l + T$$

where
$$T = \begin{bmatrix} \Delta x & \Delta y \\ \Delta x & \Delta y \end{bmatrix}$$
is the translation matrix.

Any other geometric element can be translated in space by adding Δx to the current x value and Δy to the current y value of each point that defines the element.

Scaling. The scaling transformation enlarges or reduces the size of elements. Scaling of an element is used to enlarge it or reduce its size. The scaling need not necessarily be done equally in the x and y directions. For example, a circle could be transformed into an ellipse by using unequal x and y scaling factors.

A line segment can be scaled by the scaling matrix as follows:

$$l' = l \times S$$

where
$$S = \begin{bmatrix} \alpha & 0 \\ 0 & \beta \end{bmatrix}$$
is the scaling matrix.

Note that the x scaling factor α and y scaling factor β are not necessarily the same. This would produce an alteration in the size of the element by the factor α in the x-direction and by the factor β in the y-direction. It also has the effect of repositioning the element with respect to the Cartesian system origin. If the scaling factors are less than one, the size of the element is reduced and it is moved closer to the origin. If the scaling factors are larger than one, the element is enlarged and removed farther from the origin. Scaling can also occur without moving the relative position of the element with respect to the origin. In this case, the element could be translated to the origin, scaled, and translated back to the origin location.

Rotation. In this transformation, the geometric element is rotated about the origin by an angle θ. For a positive angle, the rotation is in the counterclockwise direction. This accomplishes rotation of the element by the same angle, but it also moves the element. In matrix notation, the procedure would be as follows:

$$l' = l \times R$$

where $R = \begin{bmatrix} \cos\theta & \sin\theta \\ -\sin\theta & \cos\theta \end{bmatrix}$ is the rotation matrix.

Besides rotating about the origin point (0, 0), it might be important in some instances to rotate the given geometry about an arbitrary point in space. This is achieved by first moving the center of the geometry to the desired point and then rotating the object. Once the rotation is performed, the transformed geometry is translated back to its original position.

Concatenation. The previous single transformations can be combined as a sequence of transformations. This is called *concatenation,* and the combined transformations are called *concatenated transformations.*

During the editing process, when a graphic model is being developed, the use of concatenated transformations is quite common. It would be unusual that only a single transformation would be needed to accomplish a desired manipulation of the image. One example in which combinations of transformations would be required would be to uniformly scale a line l and then rotate the scaled geometry by an angle θ about the origin. The resulting new line is then

$$l' = l \times R \times S$$

where R is the rotation matrix and S is the scaling matrix. A concatenation matrix can then be defined as

$$C = RS$$

Concatenation is a unique feature used in many CAD functions in which a number of transformations are applied to a geometry. The advantage is in the amount of multiplication performed to get the desired picture. In the concatenation procedure, the transformation matrix C is first evaluated and then stored for future use. This eliminates the need of premultiplying the individual matrix to yield the desired transformed geometry.

The above concatenation matrix cannot be used in the example of rotating geometry about an arbitrary point. In this case, the sequence would be translation to the origin, rotation about the origin, then translation back to the original location. Note that the translation has to be done separately.

Three-Dimensional Transformations. Transformations by matrix methods can be extended to three-dimensional space. The same three general categories defined in the preceding section are considered.

Translation. The translation matrix for a point defined in three dimensions would be

$$T = (\Delta x, \quad \Delta y, \quad \Delta z)$$

An element would be translated by adding the increments Δx, Δy, and Δz to the respective coordinates of each of the points defining the three-dimensional geometry element.

Scaling. The scaling transformation is given by

$$S = \begin{bmatrix} \alpha & 0 & 0 \\ 0 & \beta & 0 \\ 0 & 0 & \gamma \end{bmatrix}$$

Equal values of α, β, γ, produce a uniform scaling in all three directions.

Rotation. Rotation in three dimensions can be defined for each of the axes. Rotation about the z axis by an angle θ_z is accomplished by the matrix

$$R_z = \begin{bmatrix} \cos\theta_z & -\sin\theta_z & 0 \\ \sin\theta_z & \cos\theta_z & 0 \\ 0 & 0 & 1 \end{bmatrix}$$

Rotation about the y axis by the angle θ_y is accomplished similarly

$$R_y = \begin{bmatrix} \cos\theta_y & 0 & \sin\theta_y \\ 0 & 1 & 0 \\ -\sin\theta_y & 0 & \cos\theta_y \end{bmatrix}$$

Rotation about the x axis by the angle θ_x is performed with an analogous transformation matrix

$$R_x = \begin{bmatrix} 1 & 0 & 0 \\ 0 & \cos\theta_x & -\sin\theta_x \\ 0 & \sin\theta_x & \cos\theta_x \end{bmatrix}$$

All the three rotations about x, y, and z axes can be concatenated to form a rotation about an arbitrary axis.

Graphical Representation of Image Data

As discussed in the introduction to this section, one of the major advantages of CAD is its ability to display the design interactively on the computer display screen. Wireframe representations, whether in 2D or 3D, can be ambiguous and difficult to understand. Because mechanical and other engineering designs often involve three-dimensional parts and systems, CAD systems that offer 3D representation capabilities have quickly become the most popular in engineering design.

In order to generate a 2D view from a 3D model, the CAD software must be given information describing the viewpoint of the user. With this information, the computer can calculate angles of view and determine which surfaces of the design would be visible from the given point. The software typically uses surfaces that are closest to the viewer to block out surfaces that would be hidden from view. Then, applying this technique and working in a direction away from the viewer, the software determines which surfaces are visible. The next step determines the virtual distance between viewer and model, allowing those areas outside the boundaries of the screen to be excluded from consideration. Then the colors displayed on each surface must be determined by combining considerations of the user's preferences for light source and surface color. The simulated light source can play a very important role in realistically displaying the image by influencing the values of colors chosen for the design and by determining reflection and shadow placements (see Fig. 13.18). Current high-end CAD software can simulate a variety of light sources, including spot-lighting and sunlight, either direct or through some opening such as a door or window.

Some systems will even allow surface textures to be chosen and displayed. Once these determinations have been made, the software calculates the color and value for each pixel in the raster-scan display terminal. Since these calculations are computationally intensive, the choice of hardware is often just as important as the software used when employing solid-modeling programs with advanced surface representation features.

13.9 CAD STANDARDS AND TRANSLATORS

In order for CAD applications to run across systems from various vendors, four main formats facilitate this data exchange:

- IGES
- STEP
- DXF
- ACIS (American Committee for Interoperable Systems)

IGES (Initial Graphics Exchange Specification)

IGES is an ANSI standard for the digital representation and exchange of information between CAD/CAM systems. 2D geometry and 3D constructive solid geometry (CSG) can be translated into IGES format. New versions of IGES also support boundary representation (B-rep) solid modeling capabilities. Common translators (IGES-in and IGES-out) functions available in the IGES library

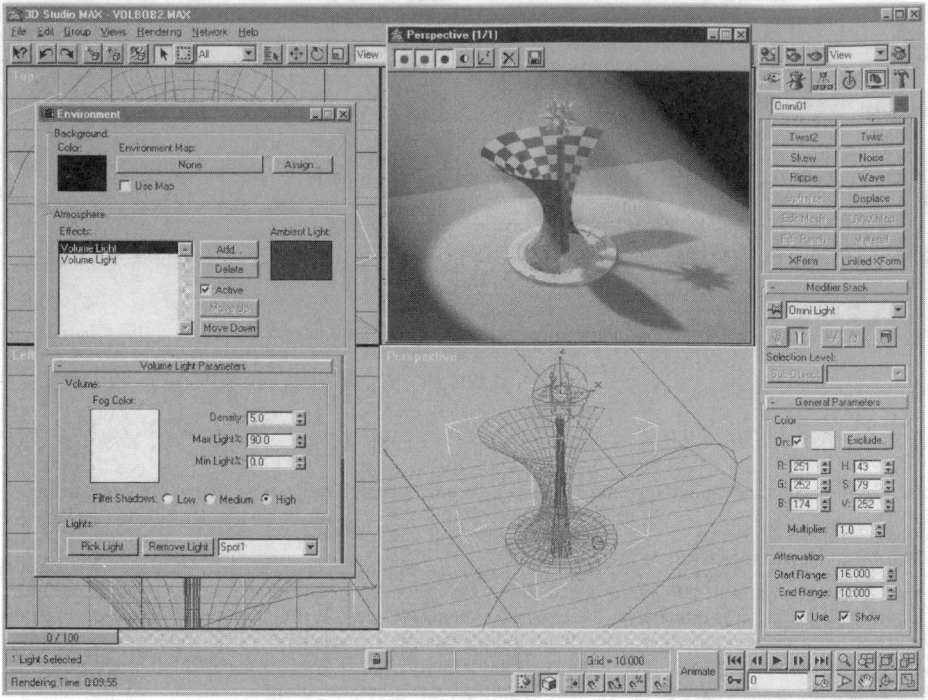

Fig. 13.18   Vase design rendered with directed spot-lighting and shadows in Autodesk 3-D StudioMax software (courtesy of Autodesk Inc.).

include IGES file parsing and formatting, general entity manipulation routines, common math utilities for matrix, vector, and other applications, and a robust set of geometry-conversion routines and linear approximation facilities.

STEP (Standard for the Exchange of Product Model Data)

STEP is an international standard. It provides one natural format that can apply to CAD data throughout the life cycle of a product. STEP offers features and benefits that are absent from IGES. STEP is a collection of standards. The user can pull out an IGES specification and get all the data required in one document. STEP can also transfer B-rep solids between CAD systems. STEP differs from IGES in how it defines data. In IGES, the user pulls out the spec, reads it, and implements what it says. In STEP, the implementor takes the definition and runs it through a special compiler that then delivers the code. This process assures that there is no ambiguous understanding of data among implementors. The conformance testing for STEP will eventually be built into the standard.

DXF (Drawing Exchange Format)

DXF, developed by Autodesk, Inc. for AutoCAD software, is the de facto standard for exchanging CAD/CAM data on a PC-based system. Only 2D drawing information can be converted into DXF, either in ASCII or in binary format.

ACIS (American Committee for Interoperable Systems)

The ACIS modeling kernel is a set of software algorithms used for creating solid-modeling packages. Software developers license ACIS routines from the developer, Spatial Technology Corp., to simplify the task of writing new solid modelers. The key benefit of this approach is that models created using software based on ACIS should run unchanged with other brands of ACIS-based modelers. This eliminates the need to use IGES translators for transferring model data back and forth among applications. ACIS-based packages have become commercially available for CAD/CAM and FEA software packages. Output files from ACIS have the suffix "*.SAT."

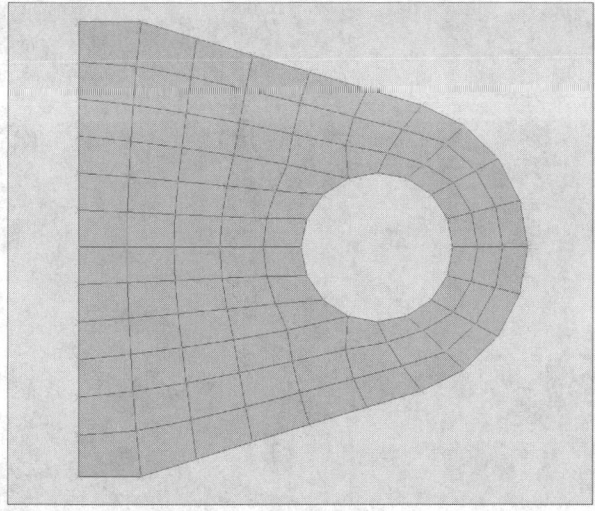

Fig. 13.19

13.9.1 Analysis Software

An important part of the design process is the simulation of the performance of a designed device. A fastener is designed to work under certain static or dynamic loads. The temperature distribution in a CPU chip may need to be calculated to determine the heat transfer behavior and possible thermal stress. Turbulent flow over a turbine blade controls cooling but may induce vibration. Whatever the device being designed, there are many possible influences on the device's performance.

The load-types listed above can be calculated using finite element analysis (FEA). The analysis divides a given domain into smaller, discrete fundamental parts called elements. An analysis of each element is then conducted using the required mathematics. Finally, the solution to the problem as a whole is determined through an aggregation of the individual solutions of the elements. In this manner, complex problems can be solved by dividing the problem into smaller and simpler problems upon which approximate solutions can be obtained. General-purpose FEA software programs have been generalized such that users do not need to have detailed knowledge of FEA.

A finite element model can be thought of as a system of solid blocks (elements) assembled together (Fig. 13.19). Several types of elements that are available in the finite-element library are

Fig. 13.20

Fig. 13.21   Rapid Prototype (R) and Part cast from Prototype (L) (courtesy of ProtoCam, Inc.).

given below. Well-known general purpose FEA packages, such as NASTRAN and ANSYS, provide an element library.

To demonstrate the concept of FEA, a two-dimensional bracket is shown divided into quadrilateral elements, each having four nodes. Elements are joined to each other at nodal points. When a load is applied to the structure, all elements deform until all forces balance. For each element in the model, equations can be written that relate displacement and forces at the nodes. Each node has a potential of displacement in x and y directions under F_x and F_y (x and y components of the nodal force) so that one element needs eight equations to express its displacement. The displacements and forces are identified by a coordinate numbering system for recognition by the computer program. For example, d_{xi}^I is the displacement in the x direction for element I at node i, while d_{yi}^I is the displacement in the y direction for the same node in element I. Forces are identified in a similar manner, so that F_{xi}^I is the force in the x direction for element I at the node i.

A set of equations relating displacements and forces for the element I should take the form of the basic spring equation, $F = kd$.

$$k_{11}d_{xi}^I + k_{12}d_{yi}^I + k_{13}d_{xj}^I + k_{14}d_{yj}^I + k_{15}d_{xk}^I + k_{16}d_{yk}^I + k_{17}d_{xl}^I + k_{18}d_{yl}^I = F_{xi}^I$$
$$k_{21}d_{xi}^I + k_{22}d_{yi}^I + k_{23}d_{xj}^I + k_{24}d_{yj}^I + k_{25}d_{xk}^I + k_{26}d_{yk}^I + k_{27}d_{xl}^I + k_{28}d_{yl}^I = F_{yi}^I$$
$$k_{31}d_{xi}^I + k_{32}d_{yi}^I + k_{33}d_{xj}^I + k_{34}d_{yj}^I + k_{35}d_{xk}^I + k_{36}d_{yk}^I + k_{37}d_{xl}^I + k_{38}d_{yl}^I = F_{xj}^I$$
$$k_{41}d_{xi}^I + k_{42}d_{yi}^I + k_{43}d_{xj}^I + k_{44}d_{yj}^I + k_{45}d_{xk}^I + k_{46}d_{yk}^I + k_{47}d_{xl}^I + k_{48}d_{yl}^I = F_{yj}^I$$
$$k_{51}d_{xi}^I + k_{52}d_{yi}^I + k_{53}d_{xj}^I + k_{54}d_{yj}^I + k_{55}d_{xk}^I + k_{56}d_{yk}^I + k_{57}d_{xl}^I + k_{58}d_{yl}^I = F_{xk}^I$$
$$k_{61}d_{xi}^I + k_{62}d_{yi}^I + k_{63}d_{xj}^I + k_{64}d_{yj}^I + k_{65}d_{xk}^I + k_{66}d_{yk}^I + k_{67}d_{xl}^I + k_{68}d_{yl}^I = F_{yk}^I$$
$$k_{71}d_{xi}^I + k_{72}d_{yi}^I + k_{73}d_{xj}^I + k_{74}d_{yj}^I + k_{75}d_{xk}^I + k_{76}d_{yk}^I + k_{77}d_{xl}^I + k_{78}d_{yl}^I = F_{xl}^I$$
$$k_{81}d_{xi}^I + k_{82}d_{yi}^I + k_{83}d_{xj}^I + k_{84}d_{yj}^I + k_{85}d_{xk}^I + k_{86}d_{yk}^I + k_{87}d_{yl}^I + k_{88}d_{yi}^I = F_{yl}^I$$

The k parameters are stiffness coefficients that relate the nodal deflections and forces. They are determined by the governing equations of the problem using given material properties such as Young's modulus and Poisson's ratio and from element geometry.

The set of equations can be written in matrix form for ease of operation as follows:

$$
\begin{pmatrix}
k_{11} & k_{12} & k_{13} & k_{14} & k_{15} & k_{16} & k_{17} & k_{18} \\
k_{21} & k_{22} & k_{23} & k_{24} & k_{25} & k_{26} & k_{27} & k_{28} \\
k_{31} & k_{32} & k_{33} & k_{34} & k_{35} & k_{36} & k_{37} & k_{38} \\
k_{41} & k_{42} & k_{43} & k_{44} & k_{45} & k_{46} & k_{47} & k_{48} \\
k_{51} & k_{52} & k_{53} & k_{54} & k_{55} & k_{56} & k_{57} & k_{58} \\
k_{61} & k_{62} & k_{63} & k_{64} & k_{65} & k_{66} & k_{67} & k_{68} \\
k_{71} & k_{72} & k_{73} & k_{74} & k_{75} & k_{76} & k_{77} & k_{78} \\
k_{81} & k_{82} & k_{83} & k_{84} & k_{85} & k_{86} & k_{87} & k_{88}
\end{pmatrix}
\times
\begin{Bmatrix}
d_{xi}^I \\ d_{yi}^I \\ d_{xj}^I \\ d_{yj}^I \\ d_{xk}^I \\ d_{yk}^I \\ d_{xl}^I \\ d_{yl}^I
\end{Bmatrix}
=
\begin{Bmatrix}
F_{xi}^I \\ F_{yi}^I \\ F_{xj}^I \\ F_{yj}^I \\ F_{xk}^I \\ F_{yk}^I \\ F_{xl}^I \\ F_{yl}^I
\end{Bmatrix}
$$

When a structure is modeled, individual sets of matrix equations are automatically generated for each element. The elements in the model share common nodes so that individual sets of matrix equations can be combined into a global set of matrix equations. This global set relates all of the nodal deflections to the nodal forces. Nodal deflections are solved simultaneously from the global matrix. When displacements for all nodes are known, the state of deformation of each element is known and stress can be determined through stress-strain relations.

For a two-dimensional structure problem, each node displacement has three degrees of freedom, one translational in each of x and y directions and a rotational in the (x,y) plane. In a three-dimensional structure problem, the displacement vector can have up to six degrees of freedom for each nodal point. Each degree of freedom at a nodal point may be unconstrained (unknown) or constrained. The nodal constraint can be given as a fixed value or a defined relation with its adjacent nodes. One or more constraints must be given prior to solving a structure problem. These constraints are referred to as *boundary conditions*.

Finite element analysis obtains stresses, temperatures, velocity potentials, and other desired unknown variables in the analyzed model by minimizing an energy function. The law of conservation of energy is a well-known principle of physics. It states that, unless atomic energy is involved, the total energy of a system must be zero. Thus, the finite element energy functional must equal zero.

The finite element method obtains the correct solution for any analyzed model by minimizing the energy functional. Thus, the obtained solution satisfies the law of conservation of energy.

The minimum of the functional is found by setting to zero the derivative of the functional with respect to the unknown nodal point potential. It is known from calculus that the minimum of any function has a slope or derivative equal to zero. Thus, the basic equation for finite element analysis is

$$\frac{dF}{dp} = 0$$

where F is the functional and p is the unknown nodal point potential to be calculated. The finite element method can be applied to many different problem types. In each case, F and p vary with the type of problem.

Problem Types

Linear Statics.   Linear static analysis represents the most basic type of analysis. The term *linear* means that the stress is proportional to strain (i.e., the materials follow Hooke's law). The term *static* implies that forces do not vary with time, or that time variation is insignificant and can therefore be safely ignored.

Assuming the stress is within the linear stress-strain range, a beam under constant load can be analyzed as a linear static problem. Another example of a linear statics is a steady-state temperature distribution within a constant material property structure. The temperature differences cause thermal expansion, which in turn induces thermal stress.

Buckling.   In linear static analysis, a structure is assumed to be in a state of stable equilibrium. As the applied load is removed, the structure is assumed to return to its original, undeformed position. Under certain combinations of loadings, however, the structure continues to deform without an increase in the magnitude of loading. In this case, the structure has buckled or become unstable. For elastic, or linear, buckling analysis, it is assumed that there is no yielding of the structure and that the direction of applied forces does not change.

Elastic buckling incorporates the effect of differential stiffness, which includes higher-order strain displacement relationships, that are functions of the geometry, element type, and applied loads. From a physical standpoint, the differential stiffness represents a linear approximation of softening (reducing) the stiffness matrix for a compressive axial load and stiffening (increasing) the stiffness matrix for a tensile axial load.

In buckling analysis, eigenvalues are solved. These are scaling factors used to multiply the applied load in order to produce the critical buckling load. In general, only the lowest buckling load is of interest, since the structure will fail before reaching any of the higher-order buckling loads. Therefore, usually only the lowest eigenvalue needs to be computed.

Normal Modes.   Normal modes analysis computes the natural frequencies and mode shapes of a structure. Natural frequencies are the frequencies at which a structure will tend to vibrate if subjected to a disturbance. For example, the strings of a piano are each tuned to vibrate at a specific frequency. The deformed shape at a specific natural frequency is called the *mode shape*. Normal modes analysis is also called *real eigenvalue analysis*.

Normal modes analysis forms the foundation for a thorough understanding of the dynamic characteristics of the structure. In static analysis, the displacements are the true physical displacements due to the applied loads. In normal modes analysis, because there is no applied load, the mode shape components can all be scaled by an arbitrary factor for each mode.

Nonlinear Statics.   Nonlinear structural analysis must be considered if large displacements occur with linear materials (geometric nonlinearity), or if structural materials behave in a nonlinear stress-strain relationship (material nonlinearity), or a combination of large displacements and nonlinear stress-strain effects occurs. An example of geometric nonlinear statics is shown when a structure is loaded above its yield point. The structure will then tend to be less stiff, permanent deformation will occur, and Hooke's law will not be applicable anymore. In material nonlinear analysis, the material stiffness matrix will change during the computation. Another example of nonlinear analysis includes the contacting problem, where a gap may appear and/or sliding may occur between mating components during load application or removal.

Dynamic Response.   Dynamic response in general consists of frequency response and transient response. Frequency response analysis computes structural response to steady-state oscillatory excitation. Examples of oscillatory excitation include rotating machinery, unbalanced tires, and helicopter blades. In frequency response, excitation is explicitly defined in the frequency domain. All of the applied forces are known at each forcing frequency. Forces can be in the form of applied forces and/or enforced motions. The most common engineering problem is to apply steady-state sinusoidally varying loads at several points on a structure and determine its response over a frequency range of interest. Transient response analysis is the most general method for computing forced dynamic response. The purpose of a transient response analysis is to compute the behavior of a structure subjected to time-varying excitation. All of the forces applied to the structure are known at each instant in time. The important results obtained from a transient analysis are typically displacements, velocities, and accelerations of the structure, as well as the corresponding stresses.

13.10   APPLICATIONS OF CAD

Computer-aided design has been presented in terms of its applicability to design, the hardware and software used, and its capabilities as an entity unto itself. The use of CAD data in conjunction with specialized applications is now reviewed. These applications fall outside the realm of CAD software in a strict sense; however, they provide opportunities for the designer to use the data generated through CAD in new and innovative techniques that can similarly affect design efficiency. Many of the items discussed in this section apply to the evaluative stage of design. Some of the basic analytical methods that can be used in CAD to optimize designs have already been presented. Options open to the design engineer using information from a CAD database and special applications are now presented.

13.10.1   Optimization Applications

As designs become more complex, engineers need fast, reliable tools. Over the last 20 years, finite element analysis has become the major tool used to identify and solve design problems. Increased design efficiency provided by CAD has been augmented by the application of finite element methods to analysis, but engineers still often use a trial-and-error method for correcting the problems identified through FEA. This method inevitably increases the time and effort associated with design because it increases the time needed for interaction with the computer. As well, solution possibilities are often limited by the designer's personal experiences.

Design optimization seeks to eliminate much of this extra time by applying a logical mathematical method to facilitate modification of complex designs. Optimization strives to minimize or maximize a characteristic, such as weight or physical size, that is subjected to constraints on one or more parameters. Either the size, shape, or both determines the approach used to optimize a design. Optimizing the size is usually easier than optimizing the shape of a design. Optimizing the thickness of a plate does not significantly change its geometry. On the other hand, optimizing a design parameter, such as the radius of a hole, does change the geometry during shape optimization.

Optimization approaches were difficult to implement in the engineering environment because the process was somewhat academic in nature and not viewed as easily applicable to design practices. However, if viewed as a part of the process itself, optimization techniques can be readily understood and implemented in the design process. Iterations of the design procedure occur as they normally do in design up to a point. At that point, the designer implements the optimization program. The objectives and constraints upon the optimization must first be defined. The optimization program then evaluates the design with respect to the objectives and constraints and makes automated adjustments in the design. Because the process is automatic, engineers should have the ability to monitor the progress of the design during optimization, stop the program if necessary, and begin again.

The power of optimization programs is largely a function of the capabilities of the design software used in earlier stages. Two- and three-dimensional applications require automatic and parametric meshing capabilities. Linear static, natural frequencies, mode shapes, linearized buckling, and steady-state analyses are required for other applications. Because the design geometry and mesh can change during optimization iterations, error estimate and adaptive control must be included in the optimization program. Also, when separate parts are to be assembled and analyzed as a whole, it is often helpful to the program to connect different meshes and element types without regard to nodal or elemental interface matches.

Preliminary design data used to meet the desired design goals through evaluation, remeshing, and revision. Acceptable tolerances must then be entered along with imposed constraints on the optimization. The engineer should be able to choose from a large selection of design objectives and behavior constraints and use these with ease. Also, constraints from a variety of analytical procedures should be supported so that optimization routines can use the data from previously performed analyses.

Although designers usually find optimization of shape more difficult to perform than of size, the use of parametric modeling capabilities in some CAD software minimizes this difficulty. Shape optimization is an important tool in many industries, including shipbuilding, aerospace, and automotive manufacturing. The shape of a model can be designed using any number of parameters, but as few as possible should be used, for the sake of simplicity. If the designer cannot define the parameters, neither design nor optimization can take place. Often, the designer will hold a mental note on the significance of each parameter. Therefore, designer input is crucial during an optimization run.

13.10.2   Virtual Prototyping

The creation of physical models for evaluation can often be time-consuming and provide limited productivity. By employing kinematic and dynamic analyses on a design within the computer environment, time is saved and often the result of the analysis is more useful than experimental results from physical prototypes. Physical prototyping often requires a great deal of manual work, not only to create the parts of the model, but to assemble them and apply the instrumentation needed as well. Virtual prototyping uses kinematic and dynamic analytical methods to perform many of the same tests on a design model. The inherent advantage of virtual prototyping is that it allows the engineer to fine-tune the design before a physical prototype is created. When the prototype is eventually fabricated, the designer is likely to have better information with which to create and test the model.

Physical models can provide the engineer with valuable design data, but the time required to create a physical prototype is long and must be repeated often through iterations of the process. A second disadvantage is that through repeated iterations, the design is usually changed, so that time is lost in the process when parts are reconstructed as a working model. Too often, the time invested in prototype construction and testing reveals less useful data than expected.

Virtual prototyping of a design is one possible solution to the problems of physical prototyping. Virtual prototyping employs computer-based testing so that progressive design changes can be incorporated quickly and efficiently into the prototype model. Also, with virtual prototyping, tests can be performed on the system or its parts in a way that might not be possible in a laboratory setting. For example, the instrumentation required to test the performance of a small part in a system might disrupt the system itself, thus denying the engineer the accurate information needed to optimize the design. Virtual prototyping can also apply forces to the design that would be impossible to apply in the laboratory. For example, if a satellite is to be constructed, the design should be exposed to zero gravity in order to simulate its performance properly.

Prototyping and testing capabilities have been enhanced by rapid prototyping systems with the ability to convert CAD data quickly into solid full-scale models that can be examined and tested. The major advantage of rapid prototyping is in the ability of the design to be seen and felt by the designer and less technically adept personnel, especially when esthetic considerations must come into play. While rapid prototyping will be discussed further below, even this technology is somewhat limited in testing operations. For example, in systems with moving parts, joining rapid prototype models can be difficult and time-consuming. With a virtual prototyping system, connections between parts can be made with one or two simple inputs. Since the goal is to provide as much data in as little time as possible, use of virtual prototyping before a prototype is fabricated can strongly benefit the design project.

Engineers increasingly perform kinematic and dynamic analyses on a virtual prototype because a well-designed simulation leads to information that can be used to modify design parameters and characteristics that might not have otherwise been considered. Kinematic and dynamic analysis methods apply the laws of physics to a computerized model in order to analyze the motion of parts within the system and evaluate the overall interaction and performance of the system as a whole. At one time a mainframe computer was required to perform the necessary calculations to provide a realistic motion simulation. Today, microcomputers have the computational speed and memory capabilities necessary to perform such simulations on the desktop.

One advantage of kinematic/dynamic analysis software is that it allows the engineer to overload forces on the model deliberately. Because the model can be reconstructed in an instant, the engineer can take advantage of the destructive testing data. Physical prototypes would have to be fabricated and reconstructed every time the test was repeated. There are many situations in which physical prototypes must be constructed, but those situations can often be made more efficient and informative by the application of virtual prototyping analyses.

13.10.3 Rapid Prototyping

One of the most recent applications of CAD technology has been in the area of rapid prototyping. Physical models traditionally have the characteristic of being one of the best evaluative tools for influencing the design process. Unfortunately, they have also represented the most time-consuming and costly stage of the design process. Rapid prototyping addresses this problem, combining CAD data with sintering, layering, or deposition techniques to create a solid physical model of the design or part. The rapid prototyping industry is currently developing technology to enable the small-scale production of real parts, as well as molds and dies that can then be used in subsequent traditional manufacturing methods. These two goals are causing the industry to become specialized into two major sectors. The first sector aims to create small rapid prototyping machines that one day might become as common in the design office as printers and plotters are today. The second branch of the rapid prototyping industry is specializing in the production of highly accurate, structurally sound parts to be used in the manufacturing process.

Stereolithography

A stereolithography machine divides a 3D CAD model into slices as thin as 0.0025 in. and sends the information to an ultraviolet laser beam. The laser traces the slice onto a container of photocurable liquid polymer, crosslinking (solidifying) the polymer into a layer of resin. The first layer is then lowered by the height of the next slice. The process is then repeated and, with each repetition, the solid resin is lowered by an increment equal to the height of the next slice until the prototype has been completed. The workspace on one large stereolithography machine has a workspace of 20 $\times$ 20 $\times$ 20 in.

Early stereolithographed parts were made of acrylate resins using the *Tri-Hatch* build style. This resulted in fabricated parts being very brittle, with rough surface finishes, and significantly less accuracy than provided by today's stereolithography machines. Since 1990, advances in hardware, software, polymers, and processing methods have resulted in improved accuracy. The standard measurement for accuracy used in stereolithography applications is the $\varepsilon(90)$ value, which indicates the degree of 90th percentile error. Early machines were capable of an $\varepsilon(90)$ accuracy of about 400 microns. In 1990, a new technique called the *Weave* build style increased the $\varepsilon(90)$ accuracy to 300 microns. The increased accuracy led the application of the models for verifying designs in checking for interference, tolerance build-ups, and other potential design flaws. The next advance in stereolithography at 3-D Systems was the release of the "Star-Weave" structure, when $\varepsilon(90)$ reached 200 microns. At this accuracy, engineers could begin to use stereolithographed parts in iterations of the design process and for optimization routines. There are some applications for which computational simulations are simply not accurate enough; for example, airflow through an inlet manifold on an automobile engine. Chrysler Corporation, for example, has used stereolithography to create manifold parts and subsequently tested them in the laboratory to identify the optimal design.

All of the benefits realized from the fabrication of plastic parts could potentially be greater with the ability to use the technology to create metal parts. The initial focus at 3-D Systems was on using the stereolithographed part in an investment casting system. The problem was that early stereolithographs were solid, causing thermal expansion to place stress on the ceramic shell, breaking the mold. The problem was addressed with the QuickCast build geometry. The structure of the stereolithograph is not solid, but about two-thirds hollow, with an open lattice structure. The internal lattice provides the structural strength and the somewhat hollow properties of the model generate a smooth surface definition. The key to investment casting using this approach is to ensure that any liquid polymer left within the lattice structure of the pattern has an escape route to avoid its solidification to a thickness that would cause enough thermal expansion to break the ceramic mold. Therefore, a drain hole is usually provided to allow whatever resin is left in the mold to escape before the final UV-curing process. A vent hole prevents a complete vacuum in the internal lattice.

A further advance in stereolithography technology was developed by Thomas Pang, an organic chemist at 3-D Systems, who developed an epoxy resin with significant advantages over the acrylate plastics originally used. The acrylates have high viscosity but not very high green strength. The lattice triangles in acrylates could not be too large, or they would sag, and the high viscosity of the liquid meant that the resin flowed very slowly through the drain holes in the part. Epoxy resin offers one-tenth the viscosity of acrylates coupled with a fourfold increase in green strength. Also, the linear shrinkage effective upon hardening is decreased with the epoxy technology. Acrylates show 0.6–1.0% shrinkage, while epoxy resin offers only 0.06% linear shrinkage. With the epoxy resin, $\varepsilon(90)$ values began to approach $100\mu m$. For these reasons, epoxy resin is viewed as a major improvement over acrylate systems.

The QuickCast system, using first acrylate plastics and then epoxy resin, has opened up the market to rapid manufacturing of accurate functional prototypes in aluminum, stainless steel, carbon steel, and others. The accuracy of the prototypes keeps growing greater, as shown by an $\varepsilon(90)$ value of

91μm achieved in December 1993 with the SLA–500/30 machine. Current research focus is on using RP technology to create investment-cast steel tool molds for low-level manufacturing purposes. Ford Motor Co. implemented such technology in the production of a wiper-blade cover for its mid-year 1994 Ford Explorer. Other RP companies have also jumped into the rapid tooling market. DTM Corp. of Austin, Texas, has introduced a laser sintering process capable of sintering precursor powders into nylon, polycarbonate, and casting wax parts. The biggest advantage at DTM is the use of laser sintering to create metal prototypes. In the process, called *RapidTool* at DTM, the powder used in their Sinterstation machines is a combination low-carbon steel and thermoplastic binder. The laser uses the data from a CAD file to trace the incremental slices of a part onto the powder, causing the plastic to bind with the metal, holding the shape of the part. A low-temperature furnace burns off the plastic binder and then the temperature is raised to lightly fuse the steel particles, leaving an internal steel skeleton. The steel skeleton is subsequently infused with copper to provide a composite metal part slightly more than 50% iron by weight. The tools created using RapidTool technology use similar processes as those created using aluminum tooling. Accuracy using RapidTool is projected to reach 0.003 in. on features and 0.010 in. on any dimension.

Other materials for use with RP are currently being tested and implemented by various other companies involved in the growing rapid-prototyping industry. The technology presents not only the opportunity, but the realistic opportunity, for further automatization in the design and implementation environments. One major factor impeding the implementation of such technology on a large scale is cost. Currently, even desktop systems, such as those from 3-D Systems, range from about $100,000 to $450,000, thus limiting their use in small and mid-size corporations. The cost associated with materials is also high, meaning that virtual prototyping, at least for the time being, can cut the cost of the rapid prototyping and testing process before investing in the fabrication of even a rapid prototype. As with most technologies, however, prices are expected to drop with time, and when they do, it is expected that RP technology will become as much of a fixture in the design engineering environment as CAD itself has become.

13.10.4 Computer-Aided Manufacturing (CAM)

Although this chapter is primarily about CAD, we would be terribly remiss not to mention CAM in terms of our applications discussion. The two techniques are so integrally related in today's manufacturing environment that often, one is not mentioned without the other. Acronyms such as CAD/CAM, CADAM (computer-aided design and manufacturing), and CIM (computer-integrated manufacturing) are often used to describe the marriage between the two. In essence, CAM uses data prepared through CAD to streamline the manufacturing process through the use of tools such as computer numerical control and robotics. CAM will be discussed in much further detail in another chapter of this handbook. We strongly refer the reader to that chapter to foster some basic understanding of this related subject.

BIBLIOGRAPHY

Ali, H., "Optimization for Finite Element Applications," *Mechanical Engineering*, 68–70 (December 1994).

Amirouche, F. M. L., *Computer-Aided Design and Manufacturing*, Prentice-Hall, Englewood Cliffs, NJ, 1993.

Ashley, S., "Prototyping with Advanced Tools," *Mechanical Engineering*, 48–55 (June 1994).

"Basics of Design Engineering," *Machine Design*, 47–83 (February 8, 1996).

"Basics of Design Engineering," *Machine Design*, 83–126 (July 1995).

"CAD/CAM Industry Report 1994," *Machine Design*, 36–98 (May 23, 1994).

Deitz, D., "PowerPC: The New Chip on the Block," *Mechanical Engineering*, 58–62 (January 1996).

Dvorak, P. (ed), "Engineering on the Other Personal Computer," *Machine Design*, 42–52 (October 26, 1995).

———, "Windows NT Makes CAD Hum," *Machine Design*, 46–52 (January 10, 1994).

"Engineering Drives Document Management," Special Editorial Supplement, *Machine Design*, 77–84 (June 15, 1995).

Foley, J. D., et al., *Computer Graphics: Principles and Practice*, 2nd ed., Addison-Wesley, New York, 1990.

Groover, M. P., and E. W. Zimmers, Jr., *CAD/CAM: Computer-Aided Design and Manufacturing*, Prentice-Hall, Englewood Cliffs, NJ, 1984.

Hanratty, P. J., "Making Solid Modeling Easier to Use," *Mechanical Engineering*, 112–114 (March 1994).

Hodson, W. K. (ed), *Maynard's Industrial Engineering Handbook*, 4th ed., McGraw-Hill, New York, 1992.

Hordeski, M. F., *CAD/CAM Techniques*, Reston, Reston, VA, 1986.

Krouse, J. K., *What Every Engineer Should Know About Computer-Aided Design and Computer-Aided Manufacturing,* Marcel Dekker, New York, 1982.

Lee, G., "Virtual Prototyping on Personal Computers," *Mechanical Engineering,* 70–73 (July 1995).

Masson, R., "Parallel and Almost Personal," *Machine Design,* 70–76 (April 20, 1995).

Microsoft Windows NT from a UNIX Point of View, Business Systems Technology Series, Microsoft Corp.

Norton, R. L., Jr., "Push Information, Not Paper," *Machine Design,* 105–109 (December 12, 1994).

Puttre, M., "Taking Control of the Desktop," *Mechanical Engineering,* 62–66 (September 1994).

Shigley, J. E., and C. R. Mischke, *Mechanical Engineering Design,* 5th ed., McGraw-Hill, New York, 1989.

Stallings, W., *Computer Organization and Architecture: Designing for Performance,* 4th ed., Prentice-Hall, Englewood Cliffs, NJ, 1996.

Teschler, L. (ed.), "Why PDM Projects Go Astray," *Machine Design,* 78–82 (February 22, 1996).

Wallach, S., and J. Swanson, "Higher Productivity with Scalable Parallel Processing," *Mechanical Engineering,* 72–74 (December 1994).

CHAPTER 14

VIRTUAL REALITY—A NEW TECHNOLOGY FOR THE MECHANICAL ENGINEER

Tushar H. Dani
Rajit Gadh
Department of Mechanical Engineering
University of Wisconsin—Madison
Madison, Wisconsin

14.1 INTRODUCTION	319	
14.2 VIRTUAL REALITY	319	
14.3 VR TECHNOLOGY	320	
14.3.1 VR Hardware	320	
14.3.2 VR Software	322	
14.4 VR SYSTEM ARCHITECTURE	323	
14.5 THREE-DIMENSIONAL COMPUTER GRAPHICS vs. VR	324	
14.5.1 Immersive VR System	324	

14.5.2 Desktop VR Systems 325
14.5.3 Hybrid Systems 325

14.6 VR FOR MECHANICAL ENGINEERING 325
   14.6.1 Enhanced Visualization 325
   14.6.2 VR–CAD 325

14.7 VIRTUAL PROTOTYPING/ MANUFACTURING AND VR 326

14.1 INTRODUCTION

In recent times, the term *virtual* has seen increasing usage in the mechanical engineering discipline as a qualifier to describe a broad range of technologies. Examples of usage include *"virtual reality," "virtual prototyping,"* and *"virtual manufacturing."* In this chapter, the meaning of the term *virtual reality* (VR) is explained and the associated hardware and software technology is described. Next, the role of virtual reality as a tool for the mechanical engineer in the design and manufacturing process is highlighted. Finally, the terms *virtual prototyping* and *virtual manufacturing* are discussed.

14.2 VIRTUAL REALITY

The term *virtual reality* is an oxymoron, as it translates to "reality that does not exist." In practice, however, it refers to a broad range of technologies that have become available in recent years to allow generation of synthetic computer-generated (and hence virtual) environments within which a person can interact with objects as if he or she were in the real world (reality).[1] In other instances, it is used as a qualifier to describe some computer applications, such as a virtual reality system for concept shape design or a virtual reality system for robot path planning.

Hence, the term by itself has no meaning unless it is used in the context of some technology or application. Keeping in mind this association of VR with technology, the next section deals with various elements of VR technology that have developed over the last few years. Note that even though the concept of VR has existed since the late 1980s, only in the last two to three years has it gained a lot of exposure in industry and the media. The main reason for this is that the VR technology has become available at an affordable price so as to be considered a viable tool for interactive design and analysis.

Mechanical Engineers' Handbook, 2nd ed., Edited by Myer Kutz.
ISBN 0-471-13007-9 © 1998 John Wiley & Sons, Inc.

Later, we will focus on VR applications, which allow such VR technology to be put to good use. In particular, a VR-based application is compared to a typical three-dimensional (3D) computer-aided-design (CAD) application to highlight the similarities and differences between them.

14.3 VR TECHNOLOGY

Typically, in the print media or television, images of VR include glove-type devices and/or so-called head mounted displays (HMDs). Though the glove and HMD are not the only devices that can be used in a virtual environment (VE), they do convey to the viewer the essential features associated with a VE: a high degree of immersion, and interactivity.

Immersion refers to the ability of the synthetic environment to cause the user to feel as if he or she is in a computer-generated virtual world. The immersive capabilities can be judged, for example, by the quality of graphics presented (how real does the scene look?) or by the types of devices used (HMD, for example). All VEs need not be immersive, as will become clearer from later sections.

Interactivity is determined by the extent to which the user can interact with the virtual world being presented and the ways he or she can interact with the virtual world: for example, how the user can interact with the VE (using the glove) and the speed with which the scene is updated in response to user actions. This display update rate becomes an important ergonomic factor, especially in immersive systems, where a lag between the user's actions and the scene displayed can cause nausea.

With reference to the typical glove/HMD combination, the glove-type device is used to replace the mouse/keyboard input and provides the interactivity, while the HMD is used to provide the immersion. Though the glove and head-mounted display combination are the most visible elements of a VR system, there are other components of a VR that must be considered. First, the glove and HMD are not the only devices that can be used in a VE. There are many other devices in the market that can be used for providing the 3D interactions capabilities. These are discussed in Section 14.3.1.

Second, the software in a VR system plays an equally important role in determining the behavior of the system, is discussed. A wide variety of software tools for VR system are described in Section 14.3.2.

Third, the need for real-time performance, combined with the need to interface with a wide range of devices, requires that special attention be paid to the architecture of a VR system. An example of a typical VR system architecture is provided in Section 14.3.1.

14.3.1 VR Hardware

The hardware in a VE consists of three components: the main processor, input devices, and output devices (Fig. 14.1). In the initial stages of VR technology development, in the 1990s, there was a limited choice of computer systems that could be used for VR applications. Currently, all major UNIX workstation vendors have specific platforms targeted to the VR market. These workstations usually have a enhanced graphics performance and specific hardware to support VR-type activity. However, with improvements in the processing speeds, of PCs, they are also becoming viable alternatives to more expensive UNIX-based systems. With prices much lower than their workstation counterparts, these are popular with VR enthusiasts and researchers (with limited budgets) alike. The popularity of the PC-based VR systems has spawned a whole range of affordable PC-based VR interaction devices, some examples of which are provided in this section.

Main Processor

The main processor or virtual environment generator[2] creates the virtual environment and handles the interactions with the user. It provides the computing power to run the various aspects of the virtual world simulation.

The first task of the virtual environment generator is to display the virtual world. An important factor to consider in the display process is the number of frames per second of the scene that can be displayed. Since the goal of a VE is to look and feel like a real environment, the main processor must be sufficiently powerful (computationally) to be able to render the scene at an acceptable frame rate. A measure of the speed of such a processor is the number of shaded polygons it can render per second. Typical speeds for UNIX-based Silicon Graphics machines range from 60,000 Tmesh/sec (Triangular Mesh) for an Indigo2XL to 1.6 million Tmesh/sec for a Power Onyx/12.[3]

The second task of the main processor is to interface with the different input and output devices that are so important in providing the interactiveness in the VE. Depending on the platform used, a wide range of input and output devices are available. A brief summary of such devices is provided in the next two sections. Detailed description of such devices and hardware can be found in Ref. 4.

Input Devices

Input devices provide the means for the user to interact with the virtual world. The virtual world, in turn, responds to the user's actions by sending feedback through various output devices, such as a visual display. Since the principal objective of a VE is to provide realistic interaction with the virtual

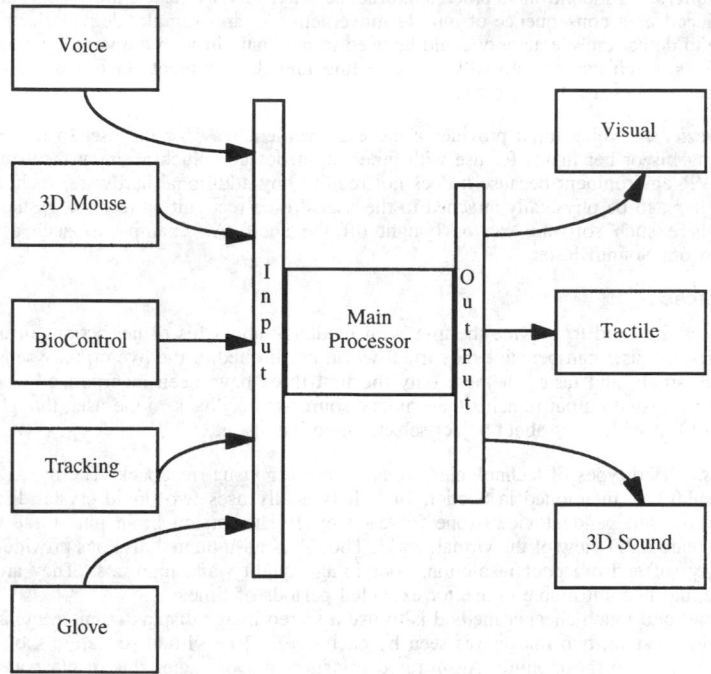

Fig. 14.1 Hardware in a VR system.

world, input devices play an important role in a VR system. The mouse/keyboard interaction is still used in some VR environments, but the new generation of 3D devices that provide the tools to reach into the 3D virtual world.

Based on their usage, input devices can be grouped into five categories: tracking, pointing, hand-input, voice-based, and devices based on bio-sensors. Of these, the first four types are typically used in VR systems. Note that of the devices described below, only the devices in the first three categories are used in VEs.

Tracking Devices.   These devices are used in position and orientation tracking of a user's head and/or hand. These data are then used to update the virtual world scene. The tracker is sometimes also used to track the user's hand position (usually wearing a glove; see below) in space so that interactions with objects in the 3D world are possible. Tracking sensors based on mechanical, ultrasonic, magnetic, and optical systems are available. One example of such a device is the Ascension tracker.[5]

Point Input Devices.   These devices have been adapted from the mouse/trackball technology to provide a more advanced form of data input. Included in this category is the 6-degree of freedom (6-dof) mouse and force ball. The 6-dof mouse functions like a normal mouse on the desktop but as a 6-dof device once lifted off the desktop. A force ball uses mechanical strains developed to measure the forces and torques the user applies in each of the possible three directions. An example of force ball-type technology is the SpaceBall. Another device that behaves like a 6-dof mouse is the Logitech Flying Mouse, which looks like a mouse but uses ultrasonic waves for tracking position in 3D space.

Glove-Type Devices.   These consist of a wired cloth glove that is worn over the hand like a normal glove. Fiber-optical, electrical, or resistive sensors are used to measure the position of the joints of the fingers. The glove is used as a gestural input device in the VE. This usually requires the development of gesture-recognition software to interpret the gestures and translate them into commands the VR software can understand. The glove is typically used along with a tracking device that measures the position and orientation of the glove in 3D space. Note that some gloves do provide some rudimentary form of tracking and hence do not require the use of a separate tracking device. One example of such a glove is the PowerGlove[6] which is quite popular with VR home enthusiasts since it is very affordable. Other costlier and more sophisticated versions, such as the CyberGlove, are also available.

Biocontrollers. Biocontrollers process indirect activity, such as muscle movements and electrical signals produced as a consequence of muscle movement. As an example, dermal electrodes placed near the eye to detect muscle activity could be used to navigate through the virtual worlds by simple eye movements. Such devices are still in the testing and development stage and are not quite as popular as the devices mentioned earlier.

Audio Devices. Voice input provides a more convenient way for the user to interact with the VE by freeing his or her hands for use with other input devices. Such an input mechanism is very useful in a VR environment because it does not require any additional hardware, such as the glove or biocontrollers, to be physically attached to the user. Voice-recognition technology has evolved to the point where such software can be bought off the shelf. An example of such a software is VoiceAssist from SoundBlaster.

Output Devices

Output devices are used to provide the user with feedback about his or her actions in the VE. The ways in which the user can perceive the virtual world are limited to the five primary senses of sight, sound, touch, smell, and taste. Of these only the first three have been incorporated in commercial output devices. Visual output remains the primary source of feedback to the user, though sound can also be used to provide cues about object selection, collisions, etc.

Graphics. Two types of technologies are available for visual feedback. The first, HMD (head-mounted display), is mentioned in Section 14.3. It typically uses two liquid crystal display (LCD) screens to show independent views (one for each eye). The human brain puts these two images together to create a 3D view of the virtual world. Though head-mounted displays provide immersion, they currently suffer from poor resolution, poor image quality, and high cost. They are also quite cumbersome and uncomfortable to use for extended periods of time.

The second and much cheaper method is to use a stereo image display monitor and LCD shutter glasses. In this system, two images (as seen by each eye) of the virtual scene are show alternately at a very high rate on the monitor. An infrared transmitter coordinates this display rate to the frequency with which each of the glasses is blacked out. A 3D image is thus perceived by the user. One such popular device is the StereoGraphics EyeGlasses system.[7]

Audio. After sight, sound is the most important sensory channel for virtual experiences. It has the advantage of being a channel of communication that can be processed in parallel with visual information. The most apparent use is to provide auditory feedback to the user about his or her actions in the virtual world. An example is to provide audio cues if a collision occurs or an object is successfully selected. Three-dimensional sound, in which the different sounds would appear to come from separate locations, can be used to provide a more realistic VR experience. Since most workstations and PCs nowadays are equipped with sound cards, incorporating sound into the VE is thus not a difficult task.

Contact. This type of feedback could either be touch or force.[8] Such tactile feedback devices allow a user to feel forces and resistance of objects in the virtual environment. One method of simulating different textures for tactile feedback is to use electrical signals on the fingertips. Another approach has been to use inflatable air pockets in a glove to provide touch feedback. For force feedback, some kind of mechanical device (arm) is used to provide resistance as the user tries to manipulate objects in the virtual world. An example of such a device is the PHANToM haptic interface, which allows a user to "feel" virtual objects.[9]

14.3.2   VR Software

As should be clear from the preceding discussion, VR technology provides the tools for an enhanced level of interaction in three dimensions with the computer. The need for real-time performance while depicting complex virtual environments and the ability to interface to a wide variety of specialized devices require VR software to have features that are clearly not needed in typical computer applications. Existing approaches to VR content creation have typically taken the following approaches[10]: virtual world authoring tools and VR toolkits. A third category is the Virtual Reality Modeling Language (VRML) and the associated "viewers" which are rapidly becoming a standard way for users to share "virtual worlds" across the World Wide Web.

Virtual World Authoring and Playback Tools

One approach to designing VR applications is first to create the virtual world that the user will experience (including ascribing behavior to objects in that world) and then to use this as an input to a separate "playback" application. The "playback" is not strictly a playback in the sense that users are still allowed to move about and interact in the virtual world. An example of this would be a walk-through kind of application, where a static model of a house can be created (authored) and the user can then visualize and interact with it using VR devices (the playback application).

Authoring tools usually allow creation of virtual worlds using the mouse and keyboard and without requiring programs in C or C++. However, this ease of use comes at the cost of flexibility, in the sense that the user may not have complete control over the virtual world being played back. Yet such systems are popular when a high degree of user interaction, such as allowing the user to change the virtual environment on the fly, is not important to the application being developed and when programming in C or C++ is not desired. Examples of such tools are the SuperScape,[11] Virtus,[12] and VREAM[13] systems.

VR Toolkits

VR Toolkits usually consist of programming libraries in C or C++ that provide a set of functions that handle several aspects of the interaction within the virtual environment. They are usually used to develop custom VR applications with a higher degree of user interaction than the walk-through applications mentioned above. An example of this would be a VR-based driver training system, where in addition to the visual rendering, vehicle kinematics and dynamics must also be simulated.

In general, VR toolkits provide functions that include the handling of input/output devices and geometry creation facilities. The toolkits typically provide built-in device drivers for interfacing with a wide range of commercial input and output devices, thus saving the need for the programmer to be familiar with the characteristics of each device. They also provide rendering functions such as shading and texturing. In addition, the toolkits may also provide functions to create new types of objects or geometry interactively in the virtual environment. Examples of such toolkits include the dVise library[14] the WorldToolKit library,[15] and Autodesk's Cyberspace Development Kit.[16]

VRML

The Virtual Reality Modeling Language (VRML) is a relative newcomer in the field of VR software. It was originally conceptualized as a language for Internet-based VR applications but is gaining popularity as a possible tool for distributed design over the Internet and World Wide Web.

VRML is the language used to describe a virtual scene. The description thus created is then fed into a VRML viewer (or VRML browser) to view and interact with the scene. In some respects, VRML can be thought of as fitting into the category of virtual world authoring tools and playback discussed above. Though the attempt to integrate CAD into VRML is still in the initial phase, it certainly offers new and interesting possibilities. For example, different components of a product may be designed in physically different locations. All of these could be linked together (using the Internet) and viewed through a VRML viewer (with all the advantages of a 3D interactive environment), and any changes could be directed to the person in charge of designing that particular component. Further details on VRML can be found at the VRML site.[17]

14.4   VR SYSTEM ARCHITECTURE

To understand the architectural requirements of a VR system, it will be instructive to compare it with a standard 3D CAD application. A typical CAD software program consists of three basic components: the user input processing component, the application component, and the output component. The input processing component captures and processes the user input (typically from the mouse/keyboard) and provides these data to the application component. The application component allows the user to model and edit the geometry being designed until a satisfactory result is obtained. The output component provides a graphical representation of the model the user is creating (typically on a computer screen).

For a VR system, components similar to those in CAD software can be identified. One major difference between a traditional CAD system and a VR-based application system is obviously the input and output devices provided. Keeping in mind the need for realism, it is imperative to maintain a reasonable performance for the VR application. Here "performance" refers to the response of the virtual environment to the user's actions. For example, if there is too much lag between the time a person moves his or her hand and the time the image of the hand is updated on the display, the user will get disoriented very quickly.

One way to overcome this difficulty is to maintain a high frame rate (i.e., number of screen updates per second) for providing the graphical output. This can be achieved by distributing the input processing, geometric modeling, and output processing tasks amongst different processors. The reason for distributing the tasks is to reduce the computational load on the main processor (Fig. 14.2).

Typical approaches adopted are to run the input and output processing component on another processor (Windows-based PC or a Macintosh) while doing the display on the main processor. In addition to reducing the computational workload on the main processor, another benefit of running the input component on a PC is that there are a wide variety of devices available for the PC platform, as opposed to the UNIX platform. This also has an important practical advantage in that a much

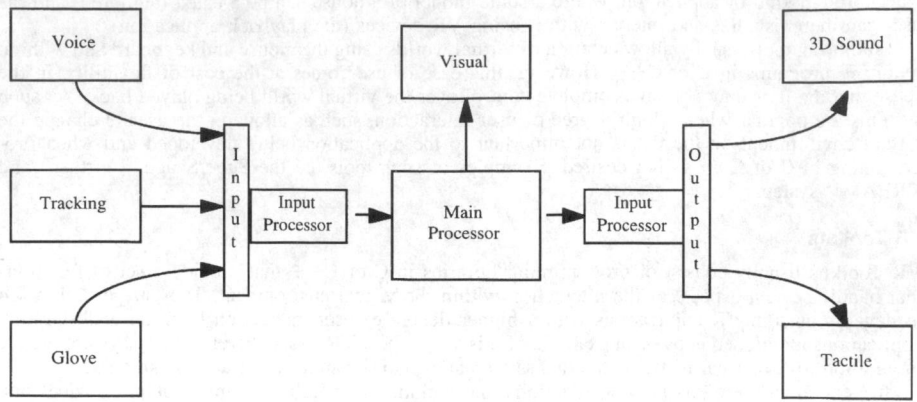

Fig. 14.2 VR system architecture.

wider (and cheaper) range of devices is available for a PC or Macintosh than for its workstation counterparts.

14.5   THREE-DIMENSIONAL COMPUTER GRAPHICS vs. VR

We will now consider virtual reality in the context of applications. So far it has been stressed that VR applications must provide interactive and immersive environments. However, a typical CAD application is interactive (although based on using a 2D mouse) and can be "used" with StereoGlasses (to provide immersion), and thus can be considered to meet the requirements for a VR system. Yet such CAD systems are not referred to as VR systems, despite providing a "virtual" world where the designer can in effect create objects in 3D.

Thus, there seems to be some kind of basic level of interaction and immersion that must be met before a system can be classified as a VR system (Fig. 14.3). The boundaries between a 3D application and VR are not very clear, but in general, a VR application will require 3D input devices (as opposed to a mouse device) and will also provide enhanced feedback, either sound- or contact-type, in addition to the display (typically stereoscopic). On the basis of the level of realism intended (often proportional to cost) and hardware used, VR systems can be classified as *immersive* or *desktop*.[18]

14.5.1   Immersive VR Systems

In an immersive system, the user's field of view is completely surrounded by a synthetic, computer-generated 3D environment. Such a system is useful in an application in which it is important that the user perceives that he or she is part of the virtual environment; for example, an application that allows a student driver to obtain training in a virtual environment. VR devices such as head-position trackers, head-mounted displays, and data gloves are commonly used in such systems to give a feeling of immersion. As the aim of these systems is to provide realism to the user, they require the use of very high-speed computers and other expensive hardware. Typical examples of such applications include virtual walk-throughs of buildings or driving a virtual vehicle.[2]

Fig. 14.3 CAD vs. Desktop vs. Immersive Systems.

14.5.2 Desktop VR Systems

Desktop VR systems are typically more economical than immersive systems. Desktop systems let users view and interact with entities in a 3D environment using a stereo display monitor and stereo glasses. Such systems are adequate for tasks where immersion is not essential, such as CAD. For interaction with the 3D world, devices like the Spaceball and/or gloves can be used. Since desktop-based VR environments do not need devices such as head-mounted displays, they are simpler and cheaper to implement than immersive systems. Additional features such as voice recognition capability and sound output can further enhance the usability of a desktop system without requiring the use of significant additional hardware.

14.5.3 Hybrid Systems

A new category of VR systems, which can be classified as hybrid systems, attempts to preserve the benefits of HMD-based systems, such as higher degree of immersion, with the comfort of desktop VR systems. Since HMD's can be cumbersome to use, hybrid systems provide immersion by using projectors to display the computer images (usually stereoscopic) on a large screen. These can be in either a vertical (wall) or horizontal (table) configuration. As in desktop systems, they typically require the user to wear the lightweight LCD glasses and use standard position trackers to track the user's head and hand position. Examples of this approach are the CAVE[19] system developed at the University of Chicago, Virtual Workbench[20] developed by Fakespace Inc., and the Virtual Design Studio project at University of Wisconsin-Madison.[21] A comprehensive list of projection-based VR systems can be found at the Projected VR Systems site.[22]

14.6 VR FOR MECHANICAL ENGINEERING

Until recently, the usability of CAD systems has been constrained by the lack of appropriate hardware devices to interact with computer models in three dimensions and the lack of software that exploits the advantages of this new generation of devices. Thus, the user interface for CAD programs has remained essentially the mouse/keyboard/menu paradigm.

The availability of VR technology has allowed the user interface to expand beyond the realm of the mouse and keyboard. Consequently, new software has evolved that allows the usage of such devices in various tasks. The VR systems for mechanical engineering can be divided into two categories, depending on the amount of interactivity possible between the user and the design environment: (1) those that support visualization and (2) those that allow design activity of some type. A summary of how VR is being used in other CAD/CAM applications can be found at the National Institute of Standards and Technology (NIST) World Wide Web site.[23]

14.6.1 Enhanced Visualization

Enhanced visualization systems allow the user to view the CAD model in a 3D environment to get a better idea of the shape features of the parts and their relationships. Models created in existing CAD systems are "imported," after an appropriate translation process, into a VR environment. Once the part is imported into the VR environment, 3D interaction devices such as gloves and 3D display monitors can be used to examine the models in a "true" 3D environment.

Enhanced visualization systems typically use 3D navigational devices such as Spaceball, flying mouse, etc., and stereo monitors with shutter eyeglasses, to allow enhanced visualization of a product or prototype. The Mitre corporation[24] has developed several virtual environments, including the Microdesigner, which enable a designer to review 3D designs. Researchers at Sun Microsystems[25] have developed a Virtual Lathe with which a user can view the action of a cutting tool and control the tool in 3D. Other examples of such work include the VENUS project[26] and research at Clemson University.[27]

14.6.2 VR–CAD

The second category of software allows design activity in the VR environment. The advantage of design activity (as opposed to just visualization) in a VR environment is that the designer is no longer limited to a traditional 2D interface when making 3D designs. Such systems use a variety of input devices (gloves, 3D navigation devices, etc.) to provide a 3D interface for design and interaction. In addition, they also support alternative methods of user input, such as voice and gestures.

Examples of VR–CAD systems include the DesignSpace[28] system, currently under development at Stanford University. It allows conceptual design and assembly, using voice and gestures in a networked virtual environment. Another system that allows design is the Virtual Workshop[29] developed at MIT, which allows parts to be created in a virtual metal and woodworking shop. Other systems include the 3-Draw system[30] and JDCAD system.[31] The 3-Draw system uses a 3D input device to let the designer sketch out ideas in three dimensions. The JDCAD system uses a pair of 3D input devices and 3D user interface menus to allow design of components.

The authors are currently developing a system called Conceptual Virtual Design System (COVIRDS), a VR system that allows the designer to create concept shape designs in a 3D environment. COVIRDS[32] is designed to solve some of the limitations of existing CAD systems. It has an intuitive interface so that designers without CAD system expertise can use the computer to create concept shapes using natural interaction mechanisms, such as voice commands and gestures.

14.7 VIRTUAL PROTOTYPING/MANUFACTURING AND VR

The terms *virtual prototyping* and *virtual manufacturing* are commonly used in academia and industry and can be easily confused with virtual reality (technology or applications). *Virtual,* as used in virtual prototyping or virtual manufacturing, refers to the use of a computer to make a prototype or aid in manufacturing a product. The discussion below applies both to virtual prototyping and virtual manufacturing.

Virtual prototyping refers to the design and analysis of a product without actually making a physical prototype of the part. *Virtual* here refers to the fact that the result of the design is not yet created in its final form, only a visual representation of the object that is presented to the user for observation, analysis, and manipulation. This prototype does not necessarily have all the features of the final product, but has enough of the key features to allow testing of the product design against the product requirements.

The simplest example of virtual prototyping tool is a 3D CAD system that allows a user to design, create, and analyze a part. However, since a 3D model is difficult to visualize on a 2D screen, one approach that has developed is to use a VR-based design and visualization system. The VR-based CAD system (as discussed in Section 14.6.2) allows changes to the "virtual prototype" to be made instantaneously, thus allowing the designer to experiment with different shapes in a short period of time. The importance of getting an optimum design lies in the fact that once the concept design is decided 60–70% of the costs of a product are committed. A poor design decision may result in increasing the downstream (committed) cost significantly. Hence, VR can be used as a tool to facilitate virtual prototyping and manufacturing. However, note that virtual prototyping or manufacturing does not require the use of VR. For more details on virtual manufacturing/prototyping see Ref. 33.

REFERENCES

1. N. I. Durlach and A. S. Mavor (eds.), "Virtual Reality: Scientific and Technological Challenges," in *National Research Council*, National Academic Press, Washington, DC, 1995.
2. R. S. Kalawsky, *The Science of Virtual Reality and Virtual Environments*, Addison-Wesley, New York, 1993.
3. Silicon Graphics Computer Systems, Periodic Table, WWW URL: ftp://ftp.sgi.com/sgi/Periodic Table.ps.Z.
4. K. Pimental and K. Teixeira, *Virtual Reality: Through the New Looking Glass*, Intel/Windcrest/McGraw-Hill, New York, 1993.
5. Ascension Technology Home Page, WWW URL: http://world.std.com/~ascen
6. Mattel PowerGlove Home Page, WWW URL: http://www.spies.com/jet/vr/vr.html
7. StereoGraphics Corp. Home Page, WWW URL: http://www.stereographics.com
8. M. A. Gigante, "Virtual Reality: Enabling Technologies," in *Virtual Reality Systems*, E. A. Earnshaw (ed.), Academic Press, 1993.
9. T. H. Massie and J. K. Salisbury, "The PHANToM Haptic Interface: A Device for Probing Virtual Objects," in *Proceedings of the ASME Winter Annual Meeting, Symposium on Haptic Interface for Virtual Environment and Teleoperator Systems*, Chicago, November 1994, WWW URL: http://www.mit.edu:8001/people/proven/Phantom/
10. J. Isdale, "What Is Virtual Reality," On-line Document, WWW URL: ftp://ftp.hitl.washington.edu/pub/scivw/WWW/scivw.html
11. Superscape Home Page, WWW URL: http://www.superscape.com/index.html
12. Virtus Corp. Home Page, WWW URL: http://www.virtus.com
13. VREAM V1.1 Webview and VR Creator (VREAM Inc.) Home Page, WWW URL: http://www.vream.com/vream/vream.html
14. Division Ltd. Home Page, WWW URL: http//www.division.com
15. Sense8 Corp. Home Page, WWW URL: http://www.sense8.com
16. Autodesk CDK Home Page, WWW URL: http://www.autodesk.com/prod/mm/cyber.htm
17. Virtual Reality Modeling Language, WWW URL: http://www.wired.com/vrml
18. L. Jacobson, "Virtual Reality: A Status Report," *AI Expert* **6** (8), 26–33 (August 1991).
19. The CAVE System—http://evlweb.eecs.uic.edu/pape/CAVE.
20. The Virtual WorkBench—http://www.fakespace.com/new_pro.html

21. Virtual Design Studio—http://icarve.me.wisc.edu/groups/virtual
22. Projected VR—http://evlweb.eecs.uic.edu/pape/CAVE/projectedVR.html
23. S. Ressler, "Virtual Reality for Manufacturing—Case Studies," WWW URL: http://nemo.ncsl.nist.gov/sressler/projects/mfg/
24. P. T. Breen Jr., "The Reality of Fantasy: Real Applications for Virtual Environments," *Information Display,* **11** (8), 15–18 (1992).
25. T. Studt, "REALITY: From Toys to Research Tools," *R&D Magazine* (March 1993).
26. VENUS, Virtual Prototyping Project, CERN, Switzerland, WWW URL: http://sgvenus.cern.ch/VENUS/vr_project.html
27. D. Fadel et al., "A Link between Virtual and Physical Prototyping," in *Proceeedings of the SME Rapid Prototyping and Manufacturing Conference,* Detroit, May 2–4, 1995, WWW URL: http://fantasia.eng.clemson.edu/vr/research.html
28. W. L. Chapin, T. A. Lacey, and L. Leifer, "DesignSpace — A Manual Interaction Environment for Computer-Aided Design," in *Proceedings of the ACM SIGCHI 1994 Conference: CHI'94 Human Factors in Computing Systems,* Boston, WWW URL: http://gummo.stanford.edu/html/DesignSpace/home.html
29. J. W. Barrus and W. Flowers, "The Virtual Workshop: A Simulated Environment for Mechanical Design," in *SIGGRAPH '94 Proceedings.*
30. E. Sachs, A. Roberts, and D. Stoops, "3-Draw: A Tool for Designing 3D Shapes," *IEEE Computer Graphics and Applications* **11** (11), 18–26 (1991).
31. J. Liang, "JDCAD: A Highly Interactive 3D Modeling System," *Computers and Graphics* **18**, 499–506 (1994).
32. T. H. Dani and R. Gadh, "COVIRDS : A Conceptual Virtual Design System," in *Proceedings of the Computers in Engineering Conference and the Engineering Database Symposium of the ASME,* Boston, 1995, WWW URL: http://icarve.me.wisc.edu/groups/virtual/
33. W. E. Alzheimer, M. Shahinpoor, and S. L. Stanton (eds.), "Virtual Manufacturing," in *Proceedings of the 2nd Agile Manufacturing Conference (AMC '95),* Albuquerque, NM, March 16–17, 1995.

BIBLIOGRAPHY

Human Interface Technology Lab Home Page (T. Emerson, Librarian), University of Washington, Seattle, WWW URL: http://www.hitl.washington.edu/people/diderot

Virtual IO Home Page, WWW URL: http://www.vio.com

K. Warwick, J. Gray, and D. Roberts (eds.), *Virtual Reality in Engineering,* Institution of Electrical Engineers, London, 1993.

Bowman, D., Conceptual Design Space Project, WWW URL: http://www.cc.gatech.edu/gvu/virtual/CDS/

Deitz, D., "Educating Engineers for the Digital Age," *Mechanical Engineering* **117** (9), 77–80 (September 1995).

CHAPTER 15

ERGONOMIC FACTORS IN DESIGN

Bryce G. Rutter, Ph.D., Principal
Anne Marie Becka, Editor
Metaphase Design Group, Inc.
St. Louis, Missouri

15.1	ERGONOMICS	329
15.2	HUMAN PERFORMANCE	329
	15.2.1 Physical Ergonomics	330
	15.2.2 Perceptual and Cognitive Ergonomics	330
15.3	THE DESIGN PROCESS	330
15.4	DESIGN RESEARCH	331
15.5	ERGONOMIC ANALYSES	332
	15.5.1 Anatomical Analysis	332
	15.5.2 Biomechanical Analysis	333
	15.5.3 Task Analysis	335
	15.5.4 Link Analysis	335

	15.5.5 Motion Analysis	335
	15.5.6 Thermographic Imprint Analysis	336
	15.5.7 Low-Speed Cine Analysis	336
15.6	DESIGN RESEARCH METHODS	336
	15.6.1 Competitive Product Analysis	336
	15.6.2 Product Performance Analysis	336
	15.6.3 Usability Studies	336
15.7	COST–BENEFIT ANALYSIS OF ERGONOMICS AND DESIGN RESEARCH	337

15.1 ERGONOMICS

A widespread increase in the availability of technology in the second half of the twentieth century has meant that more and more people come in contact with a variety of product designs on a daily basis. Regardless of this increase in the number and types of human users, many engineers still concentrate their design efforts on the machine or system alone, forcing the user to adjust to fit the product. Such readjustments on the part of the user can lead to discomfort and dissatisfaction with the design, as well as more serious effects, such as safety hazards and personal injury.

Ergonomics (also called human factors) is an applied science that makes the user central to design by improving the fit between that user and his or her tools, equipment, and environment. Key here is that designs are developed to fit both the physiological and psychological needs of the user. Ergonomists examine all ranges of the human interface, from static anthropometric measures and movement ranges to users' perceptions of a product. This interface involves both software (displays, electronic controls, etc.) and hardware (knobs, grips, physical configurations, etc.) issues.

Ergonomics grew into a distinct scientific discipline during the second world war. What began as a form of engineering (human engineering or human factors engineering) has come to encompass a wide range of interdisciplinary professions, including psychology, industrial design, medicine, and computer science. Its practitioners' range in focus includes concept modeling and product design, job performance analysis, functional analysis, workspace and equipment design, computer interfaces, environment design, and so forth.

15.2 HUMAN PERFORMANCE

The true basis of ergonomics is understanding the limitations of human performance capabilities relative to product interaction. These limitations are either physical or cognitive/perceptual in nature, but all address how people respond to man-made designs. Such interface analysis is crucial to establishing a safe and effective system of operation or environment for the user.

Mechanical Engineers' Handbook, 2nd ed., Edited by Myer Kutz.
ISBN 0-471-13007-9 © 1998 John Wiley & Sons, Inc.

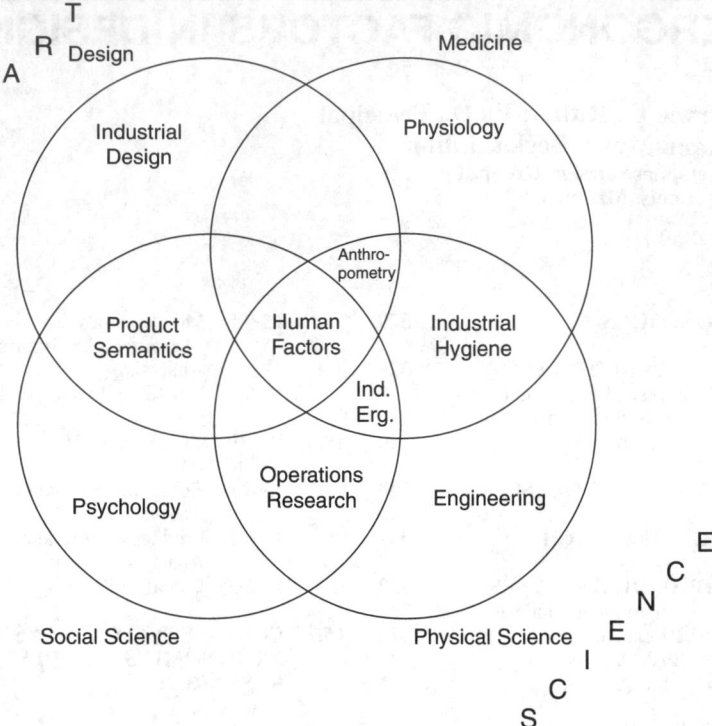

Fig. 15.1 General interdisciplinary nature of human factors with selected examples (from Ref. 1, p. 90. Reprinted by permission).

15.2.1 Physical Ergonomics

A thorough understanding of the physical characteristics of a wide range of people is essential to any product that is designed for human use. When analyzing design relative to human performance, ergonomists study static anthropometric data, which includes sizing percentiles (e.g., lengths, measurements) of a wide range of populations, including gender, age, race, and other such factors. Ranges of joint motions, strengths, and grips for varying human percentiles are also reviewed. These data serve as valuable information to designers and help ensure the final product will physically fit the targeted end-user, be it a child, the aged, or a particular racial population, and so forth.

15.2.2 Perceptual and Cognitive Ergonomics

Proper fit of a product to a user does not end with the physical interface. The perceptual and cognitive demands a product places on a user must also be examined. Note that a great misconception underlying these capabilities is that they address emotive responses of the user. However, neither are qualitative findings; both types offer fact-based, quantitative data to be used in product development.

Perceptual responses are those filtered through one or more of the five senses, such as tactile and auditory feedback of controls. Cognitive responses are based on logic, reason, and how users process information. Cognitive issues include intuitiveness of control features and functions as well as icon representation and label comprehension.

15.3 THE DESIGN PROCESS

Implementing an ergonomics program can help ensure a product's successful transition from the drawing board to the end-user. However, human factors cannot be examined in a vacuum. Ergonomists must work directly with designers and engineers throughout the entire design development process, each providing feedback to the other during concept development and testing. In addition to standard ergonomic analyses, design research should be conducted with targeted end-users to identify design problems that are often overlooked by the engineer, who examines the product only within the design environment. Such end-user research serves to measure a design's overall efficacy on a wide range

Table 15.1   Body Dimensions of U.S. Civilian Adults, Female/Male, in cm[a]

	Percentiles			
	5th	50th	95th	SD
Heights				
(f above floor, s above seat)				
Stature ("height") [f]	152.78/ 164.69	162.94/ 175.58	173.73/ 186.65	6.36/6.68
Eye height [f]	141.52/ 152.82	151.61/ 163.39	162.13/ 174.29	6.25/6.57
Shoulder (acromial) height [f]	124.09/ 134.16	133.36/ 144.25	143.20/ 154.56	5.79/6.20
Elbow height [f]	92.63/ 99.52	99.79/ 107.25	107.40/ 115.28	4.48/4.81
Wrist height [f]	72.79/ 77.79	79.03/ 84.65	85.51/ 91.52	3.86/4.15
Crotch height [f]	70.02/ 76.44	77.14/ 83.72	84.58/ 91.64	4.41/4.62
Height (sitting) [s]	79.53/ 85.45	85.20/ 91.39	91.02/ 97.19	3.49/3.56
Eye height (sitting) [s]	68.46/ 73.50	73.87/ 79.20	79.43/ 84.80	3.32/3.42
Shoulder (acromial) height (sitting) [f]	50.91/ 54.85	55.55/ 59.78	60.36/ 64.63	2.86/2.96
Elbow height (sitting) [s]	17.57/ 18.41	22.05/ 23.06	26.44/ 27.37	2.68/2.72
Thigh height (sitting) [s]	14.04/ 14.86	15.89/ 16.82	18.02/ 18.99	1.21/1.26
Knee height (sitting) [f]	47.40/ 51.44	51.54/ 55.88	56.02/ 60.57	2.63/2.79
Popliteal height (sitting) [f]	35.13/ 39.46	38.94/ 43.41	42.94/ 47.63	2.37/2.49
Depths				
Forward (thumbtip) reach	67.67/ 73.92	73.46/ 80.08	79.67/ 86.70	3.64/3.92
Buttock–knee distance (sitting)	54.21/ 56.90	58.89/ 61.64	63.98/ 66.74	2.96/2.99
Buttock–popliteal distance (sitting)	44.00/ 45.81	48.17/ 50.04	52.77/ 54.55	2.66/2.66
Elbow–fingertip distance	40.62/ 44.79	44.29/ 48.40	48.25/ 52.42	2.34/2.33
Chest depth	20.86/ 20.96	23.94/ 24.32	27.78/ 28.04	2.11/2.15
Breadths				
Forearm–forearm breadth	41.47/ 47.74	46.85/ 54.61	52.84/ 62.06	3.47/4.36
Hip breadth (sitting)	34.25/ 32.87	38.45/ 36.68	43.22/ 41.16	2.72/2.52
Head Dimensions				
Head circumference	52.25/ 54.27	54.62/ 56.77	57.05/ 59.35	1.46/1.54
Head breadth	13.66/ 14.31	14.44/ 15.17	15.27/ 16.08	0.49/0.54
Interpupillary breadth	5.66/ 5.88	6.23/ 6.47	6.85/ 7.10	0.36/0.37
Foot Dimensions				
Foot length	22.44/ 24.88	24.44/ 26.97	26.46/ 29.20	1.22/1.31
Foot breadth	8.16/ 9.23	8.97/ 10.06	9.78/ 10.95	0.49/0.53
Lateral malleolus height [f]	5.23/ 5.84	6.06/ 6.71	6.97/ 7.64	0.53/0.55
Hand Dimensions				
Circumference, metacarpale	17.25/ 19.85	18.62/ 21.38	20.03/ 23.03	0.85/0.97
Hand length	16.50/ 17.87	18.05/ 19.38	19.69/ 21.06	0.97/0.98
Hand breadth, metacarpale	7.34/ 8.36	7.94/ 9.04	8.56/ 9.76	0.38/0.42
Thumb breadth, interphalangeal	1.86/ 2.19	2.07/ 2.41	2.29/ 2.65	0.13/0.14
Weight (in kg)	39.2[b]/ 57.7[b]	62.01/ 78.49	84.8[b]/ 99.3[b]	13.8[b]/12.6[b]

[a]Adapted from U.S. Army data reported by Gordon et al. (1989) (from K. Kroemer, H. Kroemer, and K. Kroemer-Elbert, *Ergonomics: How to Design for Ease and Efficiency,* p. 30. ©1994. Reprinted by permission of Prentice-Hall, Englewood Cliffs, NJ).
[b] Estimated.
Note: In this table, the entries in the 50th percentile column are actually "mean" (average) values. The 5th and 95th percentile values are from measured data, not calculated (except for weight). Thus, the values given may be slightly different from those obtained by subtracting 1.65 SD from the mean (50th) percentile, or by adding 1.65 SD to it.

of user perception and knowledge levels. Resulting data can provide a tangible starting point upon which design revisions or new product concepts can be made.

15.4   DESIGN RESEARCH

A core component of a successful product design is understanding the wants and needs of the product's end-users. Therefore, talking with target customers to gain insight into their requirements is a logical step in concept development. Unfortunately, most manufacturers and engineers approach this issue through "gut-feeling" guesswork — fabricating a list of items or issues based on the premonitions of the development team or head of manufacturing. This method of design development is doomed from its inception, as engineers and manufacturers are often so far removed from their customer base that the resulting products never meet users' requirements or expectations.

Other times, manufacturers circumvent actual end-user research in lieu of product assessment by their marketing department. This form of "research" is extremely qualitative and often unsubstantiated by end-user feedback. Worse yet is when manufacturers base product design requirements on results of a survey of sales personnel. It is generally believed that because sales personnel are on the floor daily with customers, they have insight into customers' wants and needs. However, such methods can be disastrous, as sales representatives are not trained to observe and categorize human behavior, as many human factors specialists are.

15.5 ERGONOMIC ANALYSES

Ergonomic assessments successfully define special requirements of unique user groups by providing a comprehensive assessment of the degree of compatibility between the user, the product, and the user's workspace. Data collected include empirical measures of workspace envelopes, task and link analyses (used to identify inefficiencies in the conduct of work, illogical procedures, and hazards), and definitions of anthropometric requirements (the dimensions of the human body). Several types of ergonomic analyses are listed below.

15.5.1 Anatomical Analysis

An anatomical analysis is the study of the interaction between a product and various anatomical features of the user's body (e.g., the musculoskeletal system, nerves, veins and arteries, joints, etc.). The goal of this analysis is to identify biological constraints for design that, if exceeded, may lead to user discomfort, stress, strain, pain, or occupational disability. Typically, a product's effect on the muscular, skeletal, nervous, and circulatory systems is explored.

Design programs in which this type of analysis is especially important are those that involve large forces being exerted, rapidly repeating body motions, and/or high pressure on a portion of the user's anatomy. An anatomical analysis allows ergonomists to identify potentially harmful effects of the use

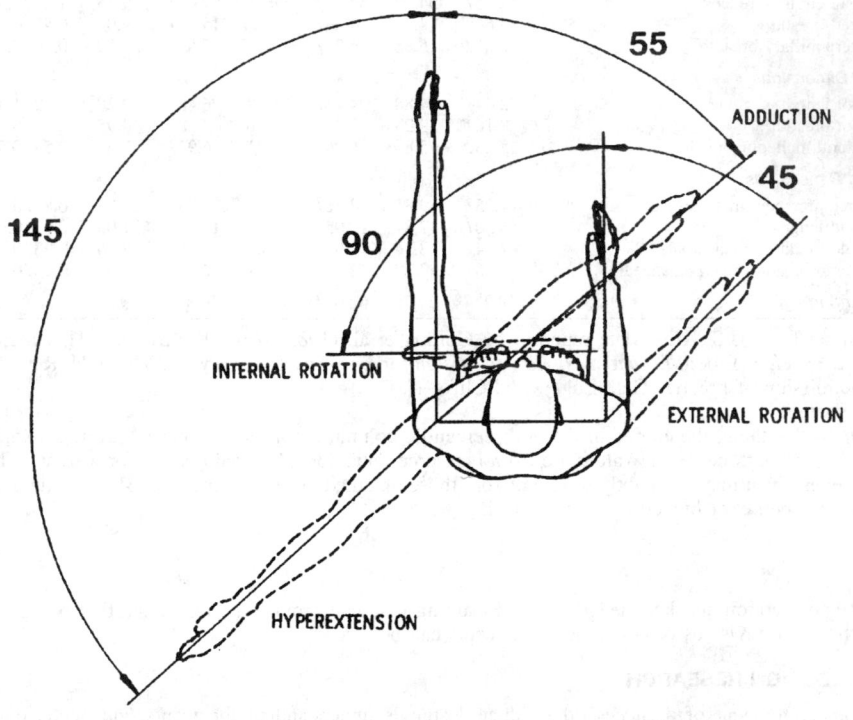

SHOULDER

Fig. 15.2 Selected examples of range of joint motions: upper extremities (from B. G. Rutter, *Dynamic Anatomical Anthropometry*. ©1981. Reprinted by permission).

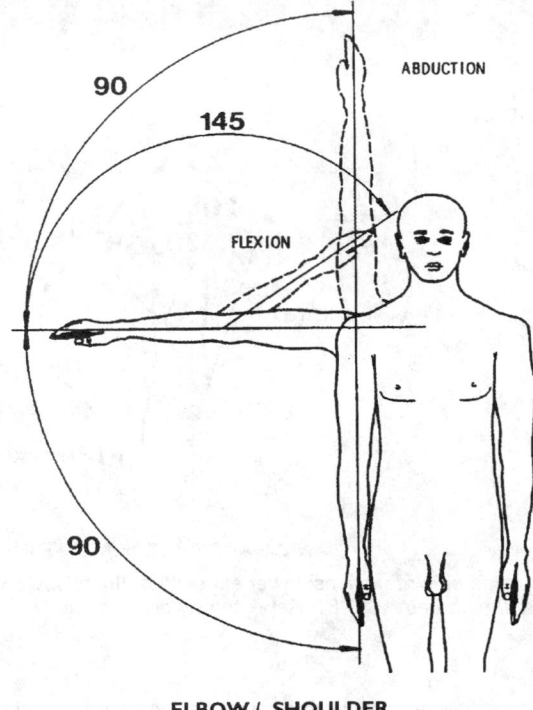

ELBOW / SHOULDER

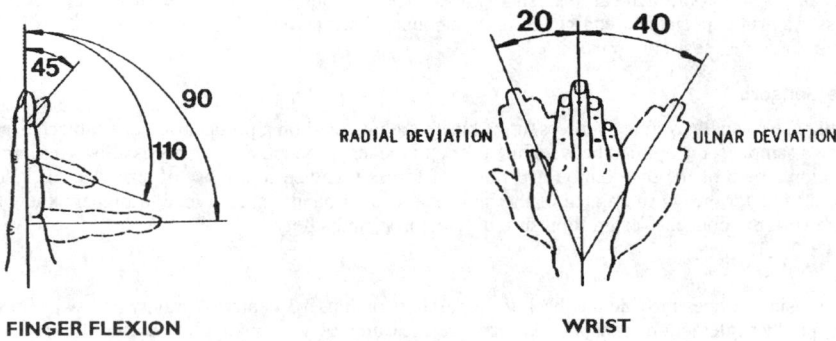

FINGER FLEXION   WRIST

Fig. 15.2 (Continued).

of a product on its users. It also provides design guidelines in the form of constraints on the user interface. The various anatomical systems affect the level of anatomical analyses. In addition, the type of product being designed and the nature of the interaction between the user and the product determine what anatomical features need to be considered in the analysis. Such analysis is best when performed by someone trained in kinesiology (the study of human movement).

15.5.2 Biomechanical Analysis

Biomechanical analysis involves modeling the human body as a mechanical system. The various measurement tools used in biomechanical analysis all provide information about the mechanics of the user's body when interacting with a product or performing a task. Such analysis is appropriate when the goal is to quantitatively assess or validate the efficiency and/or safety of one or more

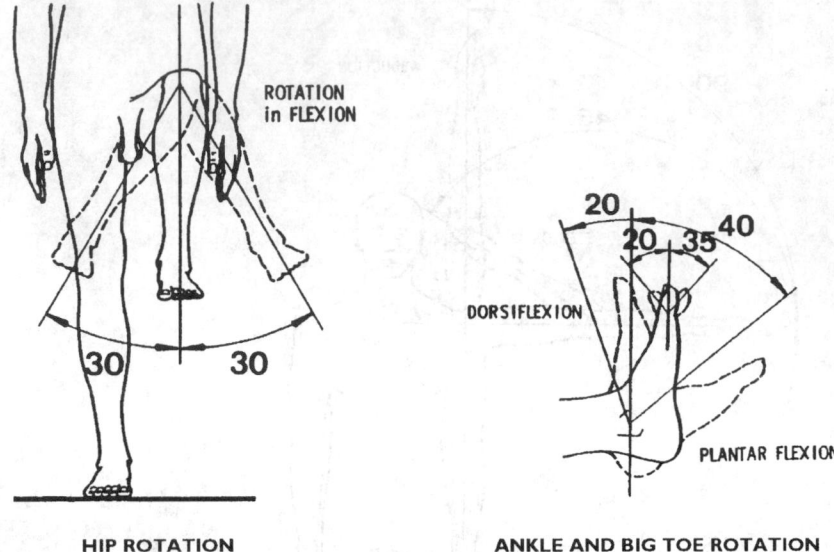

Fig. 15.3 Selected examples of range of joint motions: lower extremities. (from B. G. Rutter, *Dynamic Anatomical Anthropometry*. ©1981. Reprinted by permission).

products. When precise measurements of the human interaction are required, a biomechanical analysis is essential. It provides quantitative measures of the patterns of muscular exertion and/or body position during actuation. This information provides an indication of the biomechanical efficiency and safety of the product tested.

Performing a biomechanical analysis using any of the four tools discussed below is a complex process requiring specialized equipment and personnel. Various other biomechanical tools exist. The following are the most commonly used.

Force Sensors

For this type of analysis, force sensors/transducers are mounted on a product or a test subject. Signals provide a sample force applied between the user and the test product. Such analysis allows researchers to develop a map of the distribution and range of forces involved in the use of a product. If loading is found to be too heavy in an area of the body that cannot handle such a load, designers know they must rework the concept design to ensure user comfort and safety.

Force Plates

These sensing devices provide feedback to researchers on a user's center of gravity and sway/motion during product interaction. Sensors takes sample measures of weights applied during different positions of user activity. These measures allow researchers to determine the activity's affect on a body in order to determine possible stress, strain, fatigue, and injury to the user.

Accelerometers

These devices measure the rate of movement change over time in order to determine user velocity during product use. Sensors sample the range of acceleration of different parts of the user's body in order to determine overall movement rates. Such data are critical in that it tells researchers how using different products affects users' movement (i.e., level of fatigue) over time.

Data Glove

This research tool has sensors that measure the movement of a user's hand and all related digits during product operation. The data collected allows researchers to track grip extents, various grip architects, grasping strategies, and the range of movement of the entire arm during product use. Researchers analyze this information to determine whether a product will cause overextension, thereby causing pain and injury.

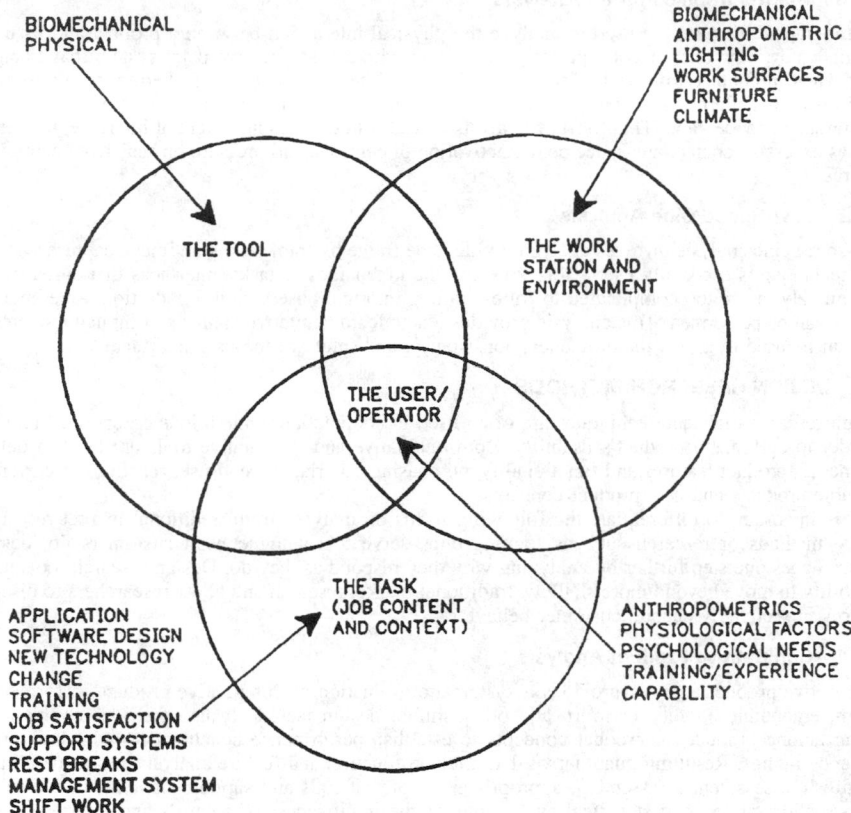

BIOMECHANICAL
PHYSICAL

BIOMECHANICAL
ANTHROPOMETRIC
LIGHTING
WORK SURFACES
FURNITURE
CLIMATE

THE TOOL

THE WORK
STATION AND
ENVIRONMENT

THE USER/
OPERATOR

THE TASK
(JOB CONTENT
AND CONTEXT)

APPLICATION
SOFTWARE DESIGN
NEW TECHNOLOGY
CHANGE
TRAINING
JOB SATISFACTION
SUPPORT SYSTEMS
REST BREAKS
MANAGEMENT SYSTEM
SHIFT WORK

ANTHROPOMETRICS
PHYSIOLOGICAL FACTORS
PSYCHOLOGICAL NEEDS
TRAINING/EXPERIENCE
CAPABILITY

Fig. 15.4 Model of user interface issues: factors involved. (from Ref. 2. Reprinted by permission).

15.5.3 Task Analysis

Task analysis involves breaking a job function down into its constituent parts, assessing human resources and time requirements, then using the information to redesign the task to optimize user output. The systematic breakdown of the individual tasks into sub-tasks allows a thorough review and the subsequent improvement of a product or system. Task analysis has been applied to the assessment and redesign of products, industrial worksites, information displays, product control panels, architectural layouts, and so on. It is most effective in the review of multisequential and/or complicated activities.

15.5.4 Link Analysis

Link analysis is used to identify inefficiencies in time/motion paths of a user performing a task and details frequencies of such paths. Time/motion analysis traces can be performed directly on photographs of a task/product and can be recorded with a motion-detection system. Results graphically illustrate the human interface and allow for the identification of inefficiencies and repetitive motions that are nonproductive or may lead to diminished productivity or injury. Link analysis provides a graphic measure of the user interface and contributes a relatively quick yet precise evaluation of the path of human interaction with a product or system.

15.5.5 Motion Analysis

This assessment method determines the kinematics (measurements of the space/time attributes of human movement) of the user interface. Motion analysis provides a detailed quantitative profile of a movement required for a particular product-related activity. Such analysis provides a detailed quantitative assessment of a product's efficiency, consistency, and safety.

15.5.6 Thermographic Imprint Analysis

Thermographic imprints are used to analyze the physical interaction between a product and its user. For this study, a product or concept design is treated with a heat-sensitive paint system that changes color when in contact with heat from the user's body. The result is a visual thermographic imprint that illustrates patterns of contact between the user and product that are not readily available from photographs or videotape. This form of analysis is useful in diagnosing potential interface problems, such as excessive contact areas, accidental activation of controls, and pressure on sensitive anatomical features.

15.5.7 Low-Speed Cine Analysis

Low-speed cine analysis involves analyzing videotape frame by frame using an image capture system. This technique is especially useful for exploring the kinematics of tasks/interfaces that either occur very quickly or are too complicated to follow through normal observation. In addition, time/motion studies can be performed. This analysis provides a detailed quantitative analysis of human movement that can be used to assess the efficiency, consistency, and safety of the user interface.

15.6 DESIGN RESEARCH METHODS

Design research tools and techniques are often used in conjunction with various ergonomic analyses in order to optimize a product's usability. Both qualitative and quantitative tools are used to define user needs, product features and functionality, purchasing criteria, and end-user reactions to currently available products and new product concepts.

It is important to differentiate the following forms of analyses from traditional market research. Where methods of research such as focus groups serve to catalogue what customers do, design research goes one step further by analyzing why they respond as they do. Design research tools have the ability to move beyond more shallow, traditional market research and allow researchers to discern patterns in seemingly chaotic customer behavior.

15.6.1 Competitive Product Analysis

Competitive product analysis provides a systematic evaluation of competitive product performance, design, ergonomics, safety, comfort, and other similar design factors. It also provides comparative testing among products or product concepts to establish performance benchmarks and relative performance ratings. Resulting quantitative data include function and feature analysis, assessment of fit and finish, assessment of assembly, appropriateness of materials and support manuals, effectiveness of instruction guides, and statistical evaluations of the intuitiveness of controls and interface logic. Competitive product analysis can range from an internalized approach to a user-based treatise whose scope is determined by demographic guidelines.

The exploratory nature of this form of research dictates that it be executed early in the product development process. An abbreviated competitive product analysis with users can serve as a form of design validation after a product has been developed.

15.6.2 Product Performance Analysis

This form of design research involves the quantitative testing and evaluation of a product's performance attributes. Techniques include on-site and laboratory measures relative to use efficiency, product efficacy, and safety, using various types of sensing technologies. Depending upon the scope of inquiry and product/system being assessed, it may include motion analysis, biomechanical analysis, low-speed cine analysis, and so on. Quantitative data resulting from these analyses include such measurements as error rates, reaction and response times, motion-velocity analysis, acceleration, jerk, and movement paths of body parts. Qualitative evaluations, such as surveys, interviews, and observation, are also utilized.

Product performance analysis provides the clearest and most quantified and qualified measure in the process of existing product, competitive product, mock-ups, or prototype assessment. This design-research method can be used to establish industry benchmarks by ascertaining performance levels on a currently marketed product or in the concept development phase to assist in the selection of the optimal design.

15.6.3 Usability Studies

Usability studies provide both quantitative and qualitative information relative to the user's physical, cognitive, and perceptual relationship with the product. Test subjects are allowed to interact with the product for a period of time in the environment in which the product would normally be used. Researchers then interview users for a detailed understanding of the product's feature and functions, its ease of use, intuitiveness of operation, and so forth. Perceptual and actual responses are measured at various stages of product contact for an understanding of both the psychological and physiological interface.

15.7 COST–BENEFIT ANALYSIS OF ERGONOMICS AND DESIGN RESEARCH

Numerous industry studies have clearly illustrated positive cost–benefit advantages of implementing ergonomic programs. After all, what manufacturers cannot attest to some number of "No Problem Found" (NPF) returns of products? A buyer returns a product simply because it "does not work." Put simply, the product did not fit the user in some manner: perhaps it caused the user discomfort, perhaps he or she just could not figure out how to get the product to work, or perhaps the buyer thought the product was just too complicated or difficult to even try to use. In all of these scenarios, no fault can be found with the product or design except that it was developed without the user in mind.

Bear in mind that the cost of corrections to a poorly designed product geometrically increases throughout the development process. Therefore, human factors specialists should begin working with engineers and designers in the early stages of product development. When ergonomists are called in to fix a product that has already been sent to market and failed, costs will escalate.

A manufacturer's decision to adopt an ergonomic orientation will serve to reposition its products from a commodity-based supplier to a supplier of high-value products. Integrating ergonomics into a design program ensures more comfortable, safe, and productive design solutions and a better overall product for the end-user.

REFERENCES

1. N. M. Simonelli, "Product Design and Human Factors Diversity: What You See Is What You Get," in *Ergonomics: Harness the Power of Human Factors in Your Business*, E. T. Klemmer (ed.), Ablex, Norwood, NJ, 1989.
2. W. E. Baker, "Human Factors, Ergonomics, and Usability: Principals and Practice," in *Ergonomics: Harness the Power of Human Factors in Your Business*, E. T. Klemmer (ed.), Ablex, Norwood, NJ, 1989.

BIBLIOGRAPHY

Anderson, J. E., *Grant's Atlas of Anatomy*, 10th ed., Williams and Wilkins, Baltimore, 1991.

Anthropology Research Project, Webb Associates, *Anthropometric Source Book, NASA Reference Publication 1024*, 3 Vols., National Aeronautics and Space Administration Scientific and Technical Office, Yellow Springs, OH, 1978.

Boff, K. R., and J. E. Lincoln, *Engineering Data Compendium: Human Perception and Performance*, 3 Vols. Harry G. Armstrong Aerospace Medical Research Laboratory, Wright-Patterson Air Force Base, OH, 1988.

Croney, J., *Anthropometry for Designers*, Van Nostrand Reinhold, New York, 1981.

Eastman Kodak Company, Human Factors Section, Health, Safety and Human Factors Laboratory, *Ergonomic Design for People at Work*, 2 Vols., Van Nostrand Reinhold, New York, 1986.

Grandjean, E., *Fitting the Task to the Man: An Ergonomic Approach*, Taylor and Francis, London, 1988.

Kroemer, K., H. Kroemer, and K. Kroemer-Elbert, *Engineering Physiology: Bases of Human Factors/Ergonomics*, 2nd ed., Van Nostrand Reinhold, New York, 1990.

Kroemer, K., H. Kroemer, and K. Kroemer-Elbert, *Ergonomics: How to Design for Ease and Efficiency*, Prentice-Hall, Englewood Cliffs, NJ, 1994.

McCormick, E. J., *Human Factors in Engineering and Design*, McGraw-Hill, New York, 1982.

Panero, J., and M. Zelnick, *Human Dimension and Interior Space*, Whitney Library of Design; Watson-Guptil, New York, 1979.

Pheasant, S., *Bodyspace: Anthropometry, Ergonomics and Design*, Taylor and Francis, London, 1986.

Rutter, B. G., *Dynamic Anatomical Anthropometry*, University of Illinois Press, Urbana–Champaign, Illinois, 1981.

Salvendy, G. (ed.), *Handbook of Human Factors*, Wiley, New York, 1987.

Tilley, A. R., *The Measure of Man and Woman: Human Factors in Design*, Whitney Library of Design; Watson-Guptil, New York, 1993.

Wickens, C. D., *Engineering Psychology and Human Performance*, 2nd ed., HarperCollins, New York, 1991.

Woodson, W. E., B. Tillman, and P. Tillman, *Human Factors Design Handbook*, 2nd ed., McGraw-Hill, New York, 1992.

CHAPTER 16

ELECTRONIC PACKAGING

Warren C. Fackler, P.E.
Telesis Systems, Inc.
Cedar Rapids, Iowa

16.1	INTRODUCTION	339
	16.1.1 Scope	339
	16.1.2 Overview	340
	16.1.3 Design Techniques	340
16.2	COMPONENT MOUNTING	341
	16.2.1 General	341
	16.2.2 Specific Components	341
	16.2.3 Discrete Components	341
	16.2.4 Printed Circuit Board Components	341
16.3	FASTENING AND JOINING	342
	16.3.1 General	342
	16.3.2 Mechanical Fastening	342
	16.3.3 Welding and Soldering	343
	16.3.4 Adhesives	344
16.4	INTERCONNECTION	344
	16.4.1 General	344
	16.4.2 Discrete Wiring	344
	16.4.3 Board Level	344
	16.4.4 Intramodule	344
	16.4.5 Intermodule	344
	16.4.6 Interequipment	344
	16.4.7 Fiber-Optic Connections	344
16.5	MATERIALS SELECTION	345
	16.5.1 General	345
	16.5.2 Materials	345
	16.5.3 Metals	345
	16.5.4 Plastics and Adhesives	345
	16.5.5 Ceramics and Glasses	345
	16.5.6 Corrosion	345
16.6	SHOCK AND VIBRATION	345
	16.6.1 General	345
	16.6.2 Environmental Loads	345
	16.6.3 Life	346
	16.6.4 Shock	346
	16.6.5 Vibration	346
	16.6.6 Testing	347
16.7	STRUCTURAL DESIGN	347
	16.7.1 General	347
	16.7.2 Strength	347
	16.7.3 Complexity	347
	16.7.4 Degree of Enclosure	347
	16.7.5 Thermal Expansion and Stresses	348
16.8	THERMAL DESIGN	348
	16.8.1 General	348
	16.8.2 Heat Transfer Modes	348
16.9	MANUFACTURABILITY	350
	16.9.1 General	350
	16.9.2 Assembly Considerations	350
	16.9.3 Design to Process	350
	16.9.4 Concurrent Engineering	350
16.10	PROTECTIVE PACKAGING	350
	16.10.1 General	350
	16.10.2 Storage Environment Protection	350
	16.10.3 Shipping Environment Protection	351

16.1 INTRODUCTION

16.1.1 Scope

Electronic packaging is a multidisciplinary process consisting of the physical design, product development, manufacture, and field support required to transform an electronic circuit into functional electronic equipment.

The categories of technical knowledge and design emphasis applicable to a given electronic product vary significantly in priority, depending on the intended product application (e.g., aerospace,

Mechanical Engineers' Handbook, 2nd ed., Edited by Myer Kutz.
ISBN 0-471-13007-9 © 1998 John Wiley & Sons, Inc.

automotive, computers, consumer goods, industrial equipment, marine equipment, medical equipment, military equipment, telephony, test equipment, etc.).

The key to successful electronic packaging is the ability to identify the applicable field of technology most likely to offer a solution to a design problem and then to apply that technology correctly in association with related technologies.

The focus in this chapter is on identification and categorization of the type of problem to be solved and selection of the most appropriate approach to that problem. For development of analytical solutions, detailed material properties, and appropriate manufacturing and assembly processes, the reader is referred to other chapters in this handbook, references at the end of this chapter, [1-3] or appropriate resources in each field.

The most important technical design considerations are listed below.

Component mounting techniques include mechanical, metallurgical, and adhesive techniques, which vary as a function of physical, thermal, and electrical interconnection requirements.

Fastening and joining techniques include threaded fasteners, rivets, welding, soldering, brazing, and adhesives utilized to mount and interconnect parts of an equipment, providing protection for contained circuit elements.

Interconnection techniques address the methods used to electrically interconnect passive and active circuit elements, including bonding, deposition, soldering, wiring, and connector systems.

Material selection techniques are used to identify and employ the most appropriate, cost-effective, and durable combination of materials for the intended product application.

Shock and vibration design practices offer methods to avoid product degradation from critical dynamic loads imposed in service.

Structural design of a system, enclosure, module, or bracket involves analytical, empirical, and experimental techniques to predict mechanical stresses. Included are thermal stresses, deformation under load, degree of enclosure, and RFI/EMI protection.

Thermal design includes the methods employed to control component temperatures to achieve satisfactory product reliability. Conduction, free and forced convection, radiation, liquid and evaporative cooling may be utilized within an equipment or between the equipment and the local environment.

Manufacturability of an electronic equipment depends on techniques to achieve ease of assembly, component selection, utilization of design rules consistent with manufacturing processes, and application of concurrent engineering and total quality management techniques.

Protective packaging includes the techniques employed to ensure that a product will survive handling, shipping, and storage environments without damage.

16.1.2 Overview

The design and analysis of electronic equipment consists of a hierarchical continuum, each level with similar yet varying characteristics. The goal is to provide the most favorable conditions for reliable operation of every component within an equipment. Working from the external environment inward:

1. *Exterior Conditions.* Service and storage environments define overall outer structural attachments, environmental conditions, electrical interconnection, power source, heat rejection, and ergonometric human factors requirements.

2. *Internal Conditions.* The equipment enclosure and structure provide mounting, thermal, and electrical interfaces between the outer environment and the internal environment, which contains electronic modules, subassemblies, and components.

3. *Component Environments.* Module and subassembly structures define the interface between individual electronic components and the equipment's internal environment.

4. *Component Requirements.* Components reside in modules and subassemblies and possess physical and operational characteristics defined by the component manufacturer and verified by test (temperature sensitivity, heat generation, mechanical stresses, shock and vibration fragility, operational life, assembly loads, reliability, mounting criteria, etc.).

Ongoing reductions in component sizes and power-dissipation levels will continue to compress equipment size and per-component thermal dissipation. Design emphasis will continue regarding addition of operational features, increased reliability, reduced maintenance, higher component density per unit volume, routing and wiring for very high-speed circuit operation, and reduction of production costs and schedules.

16.1.3 Design Techniques

Computer-based analysis and design environments and programs exist to aid with electronic packaging design and development tasks.[4] Most computer algorithms are adaptations of codes generated for other but related purposes (e.g., finite element techniques for structures and thermal analysis, fluid flow analysis and visualization, solid modeling and drafting programs, printed circuit board design programs, and others). In many instances, the underlying computational assumptions inherent in the program are not thoroughly documented.

There are several opportunities to introduce errors when building a predictive model of a product. The electronics packaging engineer must possess a basic understanding of each physical phenomenon and the underlying assumptions implicit in each type of analytical model as applied to the specific equipment under analysis.

16.2 COMPONENT MOUNTING

16.2.1 General

Components consist of any active (transistor, integrated circuit, display, disk drive, other) or passive (connector, wire, resistor, switch, heat sink, other) element mounted on or within electronic equipment. Components may require specific mounting techniques, such as socket-mounted relays, power transformers, heat sink or chassis-mounted semiconductor devices (Triacs, silicon-controlled rectifiers (SCRs), power transistors, other). Smaller electronic components (resistors, capacitors, integrated circuits, other) may be mounted to a rigid or flexible printed circuit board. Discrete components may be mounted either to a structure or to a printed circuit board.

16.2.2 Specific Components

Specific components include components and subassemblies that are not appropriate to printed circuit board mounting due to component size, special mounting needs, interconnection, serviceability, cost, or accessibility requirements.

Component-mounting techniques must be consistent with the requirements of each specific component. Component specifications provided by the manufacturer usually provide a guide to mounting requirements. Examples of specific components include disk drives (may require vibration-absorption mounting), liquid crystal displays (may require temperature control and avoidance of mechanical twist), power relays (vibration isolation and mechanical retention), panel-mounted switches and controls (environmental suitability and ergonometric considerations), connectors (strength, keying, and accessibility), and devices that generate significant amounts of heat.

16.2.3 Discrete Components

Discrete components are circuit elements not incorporated into an integrated circuit. Discrete components are mechanically attached to a structure, lead-soldered to electrical terminals, or soldered to a printed circuit board. Examples of discrete components include resistors and capacitors in leaded packages, individual transistors, rectifiers, bridges, relays, and light-emitting diodes (LEDs).

For the non-printed circuit board mounting of discrete components, a variety of mounting techniques are employed, depending on the detailed configuration of the discrete component to be mounted.

16.2.4 Printed Circuit Board Components

A printed circuit board consists of a substrate (usually FR4 glass epoxy) with a conductive layer (usually copper) that has been etched to reproduce a pattern of component mounting pads and interconnecting traces. A printed circuit board may be constructed of other substrates and circuit conductive materials to improve dissipation of heat and reductions in stresses due to thermal expansion between components and the substrate.

The printed circuit board[5] may have the etched circuit pattern on one side only or on both sides, with or without plated through-holes connecting the traces on either side of the board. Multilayer printed circuit boards offer additional planes of circuit trace patterns, with or without buried vias, to interconnect closely spaced multileaded components.

The use of increasingly smaller components and integrated circuits with greater internal complexity and high connection point counts of beyond 400 for an individual device forces ever-decreasing trace widths. Trace widths and spaces between traces of 0.010 in. or wider are common, as are fine-line board traces and spaces of from 0.010 to 0.006 in. Very fine-line boards from 0.006 in. to 0.001 traces and 0.002 spaces or smaller are difficult to achieve in production.

Flexible printed circuit boards constructed from thin polyester film substrates with copper conductors are fabricated in single, double, or multilayer format. With or without components attached, the flexible circuit board permits the shaping of a circuit to fit within an enclosure without the mechanical restrictions applicable to rigid circuit assemblies. Flexible printed circuits may be combined with rigid printed circuit boards to eliminate connectors and wiring harnesses by using the flexible circuits as interconnection between rigid board assemblies.

The various types of components mounted to a printed circuit board may be classified as either leaded components or surface-mounted components.

Leaded components are mounted by inserting component leads through holes in the printed circuit board and soldering the leads into place. Lead-trimming and board-cleaning operations follow. This technology is mature. Leaded components consist of discrete components and leaded integrated circuit (dual in-line package (DIP) with two rows of pins, and single in-line package (SIP)) packages.

A variation of leaded components is the pin grid array (PGA) package, where the integrated circuit is housed in a plastic or ceramic carrier and a matrix of pins extends from the bottom of the matrix for insertion into a printed circuit board. Such packages have pin counts up to 168 and higher.

Very large-scale integrated (VLSI) circuits[6] combine a multiplicity of circuit functions on an often custom-designed integrated circuit.

Surface-mount technology (SMT)[7] consists of attaching non-leaded packages to the printed circuit board by placing the components on patterns of conductors that have been coated with solder paste. Following placement, the assembly is heated to reflow the solder paste and bond the components to the printed circuit board.

Converting from a through-hole design to an SMT design usually reduces the printed circuit board area to about 40% of the original size. The area reduction is highly dependent on the specific components employed, interconnection, and mechanical considerations.

SMT components include:

- Small outline integrated circuits (SOIC), similar in appearance to DIP packages, except that the body of the component is smaller and the pins are replaced by gull wing or j-type lead configurations.

- Common SMT discrete package sizes known as 1206, 0805, 0603, and 0402 for resistors, capacitors and diodes; with EIA A, B, C, and D; and MELF packages for various types of capacitors.

- Plastic or ceramic leaded chip carriers (PLCC or CLCC), rectangular carriers with j-leads around all four edges. These components may be directly soldered to the printed circuit board or installed into a socket that in turn is soldered to the printed circuit board.

- Chip on board (COB), which consists of adhesive-bonding a basic silicon chip die to the printed circuit board, beam-welding leads from the die to the printed circuit board, and encapsulating the die and leads in a drop of adhesive potting compound.

- Ball grid array (BGA) packages, much like PGA packages, except that instead of an array of pins protruding from the bottom of the component, there is an array of solder balls, each attached to a pad on the component. The component may be either a plastic (PBGA) or a ceramic (CBGA) package. The BGA is placed onto a corresponding artwork pattern on the printed circuit board and the assembly is subjected to heat to reflow the solder balls, thus attaching the BGA to the printed circuit board.

- Flip chip package, a component package manufactured with small solder balls placed directly on the circuit substrate where electric connections are required. The substrate is then "flipped" or turned over so that the solder balls may be fused by reflow directly to pads on a printed circuit board.

- Multichip module (MCM), a component package, houses more than one interconnected silicon die within a subassembly. The subassembly is then attached to a printed circuit board as a through-hole or SMT component. In one manifestation, the MCM is a SIP circuit board mounted to the main printed circuit board assembly.

- Silicon on silicon (SOS), a component package consisting of silicon die attached to a silicon substrate to create a custom integrated circuit assembly. The subassembly is attached to the printed circuit board like a conventional component.

16.3 FASTENING AND JOINING

16.3.1 General

Fastening and joining techniques are used to achieve mechanical assembly of the electronic equipment. Fastening may involve attachment of the electronic product into its use environment, fabrication of the product mechanical structure, attachment of subassemblies, modules, or printed circuit boards into the equipment, attachment of a specific component, or attachment of a discrete component to a structure or printed circuit board. In each case, the fastening requirements are different and must be evaluated for each specific application.

16.3.2 Mechanical Fastening

Conventional machine design techniques apply to the design of mechanical joints employing threaded fasteners, rivets, and pins. These techniques are employed when strength and deflection are the design criterion; for example, attachment of an electronic equipment to its host structure and attachment of specific components to the structure of the electronic equipment. Dynamic loads (shock and vibration) require additional consideration, as does selection of fastener materials to avoid corrosion.

In many mechanical fastening applications within an electronics equipment, strength is not an issue and fastener size is selected based on the need to reduce the number of screw sizes (cost issue) and the space available for mechanical fastening. In these cases, for commercial applications where corrosive environments are not a significant issue, cadmium-plated fasteners are employed. For instances where dissimilar metal fastener and component parts are exposed to moisture or corrosive environments, stainless steel fasteners are advised.

Screw head selection is important in electronic equipment applications. Phillips-head screws are preferred over slotted head screws due to their ability to gain increased tightening torque. Pan-head screws are preferred over round-head screws due to their absence of sharp edges. Flat-head screws are used to hide the screw head within the material thickness of one of the structural elements; however, there is no allowance for tolerances that exist between flat-head screws in a multifastener joint.

When threaded fasteners are used, there is concern that the joint will loosen and become ineffective over time. Such loosening may be caused by thermal cycling or vibration. It is necessary to ensure that the threaded joint maintains strength. Techniques to prolong threaded joint integrity include:

- Using a lock washer between the nut and the base material, or under the screw head if the nut is part of or pressed into the base material. If a nut is used, place a flat washer between the lock washer and the base material to avoid damage to the base material.
- If an electrical bond must be established through the threaded joint, a tooth-type lock washer without a flat washer must be employed.
- Using a compression nut (formed to cause friction between the nut and the screw threads). This device loses effectiveness if frequently removed and may require replacement.
- Using a nut or screw with a compressible insert. This applies to screw sizes of #6 and larger. The same warning on reuse applies as for the compression nut.
- Using a screw-retention adhesive material on the threads prior to making the joint. The adhesive must be reapplied each time the joint is disassembled. Various degrees of hold are available.
- Using anti-rotation wire through a hole in the nut or in the head of the screw. Applicable to larger bolts only.
- Tooth-type lock washers should not be used in contact with printed circuit board or other non-metallic materials.
- Joints where one or more elements are capable of cold flow, e.g., nylon, plastics, and soft metal, require a retention method other than compression-type lock washers.

Rivets used in electronics assembly may be solid or tubular. Do not depend on a riveted joint to provide long-term electrical connectivity. Cold flow will lead to joint looseness when plastic materials are involved. Rivet material must be compatible with other materials in the joint to avoid corrosion.

Pins pressed into holes in mating parts are sometimes used to make permanent joints. Pin joints may be disassembled, but a larger-diameter pin may be required to achieve full joint strength upon reassembly. Materials selection is important to avoid corrosion.

16.3.3 Welding and Soldering

Conventional spot welding, inert gas welding, torch welding, and brazing[8] are used in the construction of metal chassis and other structural components. Such joints have consistent electrical conductivity. Material properties in the heat-affected zone are often altered and may cause mechanical failure. Lap joints must be cleaned and protected from ingestion of contaminates, which may eventually cause corrosion, loss of electrical conductivity, and mechanical failure of the joint.

Lead–tin solder is used to make electrical joints[9–11] and is the material that binds components to printed circuit boards. Eutectic 63% lead/37% tin solder has a relatively low melting point and is used for attachment of components to circuit boards. Sixty percent lead/40% tin solder is commonly used for cable and connector applications. Special alloy solders contain other metals, such as silver, for applications where standard solder may leach away material from electroplated contacts.

In applications where a soldered electrical joint is needed and mechanical stresses will be present, the joint must be designed to accept the mechanical stresses without the solder present. Under load, a solder joint will creep until the loads are eliminated or the joint fails. As a result, solder is generally used only for electrical connection purposes and not for carrying mechanical loads.

Solder is the only means of mechanical and electrical support for surface-mounted parts on a circuit board assembly. Successful surface-mount design requires that the mass of the individual parts be very small and that the circuit board be protected from bending stresses so that attachment points will not eventually fail due to creep or fatigue fracture. Due to variances in the coefficient of thermal expansion between the circuit board substrate and the component materials, solder joints will be subjected to thermal cycling-induced stresses caused by environmental or operationally generated temperature changes.

16.3.4 Adhesives

Adhesives[12] are used in electronic equipment for a variety of purposes, such as component attachment to circuit boards in preparation for wave soldering, encapsulants used to encase and protect components and circuits, and adhesives used to seal mechanical joints to avoid liquid and gas leakage.

Adhesive joints withstand shear loads, but are much weaker when subjected to peeling loads. The load-bearing properties of cured adhesive joints (creep, stiffness, modulus of elasticity, and shear stresses) may vary significantly over temperature ranges often experienced in service. Successful joints using adhesives are designed to bear mechanical loads without the adhesive present, with the adhesive applied to achieve seal.

Adhesives may release chemicals and gases that are corrosive to materials used in construction of electronic components. Such adhesives must be avoided or fully cured prior to introduction into a sealed electronic enclosure.

16.4 INTERCONNECTION

16.4.1 General

Interconnection techniques are used to electrically connect circuit elements and electronic assemblies. Different design criteria apply to the various levels of interconnection. The categories of interconnection are as follows.

16.4.2 Discrete Wiring

Discrete wiring involves the connection from one component to another by use of electronic hookup wire, which may either be insulated or uninsulated. In either case, the individual connections are made by mechanically by forming the component leads to fit the support terminals prior to applying soldering to the connection. Care is taken to route wires away from sharp objects and to avoid placing mechanical stresses on the electrical joints.

16.4.3 Board Level

Board-level interconnection is accomplished by soldering components to a conductive pattern etched into the printed circuit board. Panel- or bracket-mounted parts may require discrete wiring between the component and the printed circuit board. Board assemblies sometimes consist of two or more individual circuit boards where a smaller board assembly is soldered directly to a host circuit board.

Socket-type connectors may be soldered to a circuit board to receive integrated circuits, relays, memory chips, and other discrete components. Care is exercised to ensure that the socket provides mechanical retention of the part to prevent the part from being dislodged by transportation and service environments.

16.4.4 Intramodule

Discrete components and circuit board assemblies located within an electronic subassembly, or module, are interconnected within the module. In addition, the module circuits and components are presented to an interface, such as one or more connectors, to facilitate interconnection with other modules or cable assemblies.

16.4.5 Intermodule

Individual modules are interconnected to achieve system-level functions required of the equipment of which they are a part. Modules may plug together directly using connectors mounted to each module, be interconnected by cable and wiring harness assemblies, or plug into connectors arrayed on a common interconnection circuit board sometimes called a "mother" board.

16.4.6 Interequipment

System-level interconnection between electronic equipment may consist of wiring harness assemblies, fiber-optic cables, or wireless interconnection.

16.4.7 Fiber-Optic Connections

Fiber-optic[13] links are sometimes employed instead of conventional metallic conductors to interconnect electronic systems. Fiber-optic communications consists of transmitting a modulated light beam through a small-diameter (100 micrometers) glass fiber to a receiver, where the modulated light signal is transformed to an electrical signal. Used extensively in communications, fiber-optic links are valuable for transmitting information but cannot carry electrical current. Design is centered on methods to provide connectors and splices without inducing signal reflection and attenuation. The design must accommodate minimum bend radii, which are a function of the number of fibers in a cable, and the fibers must be supported to avoid excessive mechanical stresses.

16.5 MATERIALS SELECTION

16.5.1 General

Electronic equipment enclosures, structure, and internal mounting brackets and devices are fabricated from a variety of materials. Materials selection consists of employing the materials that have the required physical properties, are suitable when used in combination with other materials, and may be fabricated.

16.5.2 Materials

A wide variety of materials are used in electronic packaging. Key considerations are strength, electrical conductivity, thermal conductivity, thermal coefficient of expansion, and manufacturability. Materials range from electrically conductive (used to conduct signals) to non-conductive (electrical insulators), and include ferrous (iron-bearing) metals, non-ferrous metals, plastics, ceramics, and glasses. Materials are selected based on the requirements of the intended application. The electronics packaging engineer is often required to use components where the component materials selection was determined by the component manufacturer. Such component materials must be identified and often require protection to assure maximum component life.

16.5.3 Metals

A variety of metals[14] are used in electronic equipment. Their properties are well documented in printed and electronic database files. Metals commonly encountered and used in electronic packaging include both non-ferrous[15] and ferrous[14] alloys.

16.5.4 Plastics and Adhesives

Several families of plastics[16,17] are used in electronic equipment, with family member variations formulated to solve very specific problems. Adhesives used in electronic packaging[12] are often found as subsets of plastic family members (e.g., epoxy adhesives). Individual manufacturers of plastics sometimes focus on a given family. The properties of plastic family members are found in lists and databases that address the family of plastics under consideration.

16.5.5 Ceramics and Glasses

Ceramic materials[18,19] are commonly employed in electronic components, less commonly in design of electronic equipment due to brittleness and sensitivity to mechanical bending and shock loads. Ceramics are used as incompressible electrical insulators,[20] which may be formulated to conduct heat away from critical components. Glass applications include semiconductor manufacturing (silicon die) and as a sealing material between metal and ceramic parts.

16.5.6 Corrosion

Corrosion is the result of an electrochemical reaction where metals ranking at different levels on the electrogalvanic chart are in the presence of an electrolyte. This situation is similar to that in a storage cell, where the anodic element suffers sacrificial deterioration. Corrosion failures may manifest themselves as loss of electrical conductivity or loss of strength in a joint. In some metals, corrosion leaches elements from grain boundaries and leads to weakened structural properties. Corrosion may occur at interruptions in the plating that expose the base metal to which the plating is applied.

Methods to control corrosion[16] include selection of materials with least offset in the galvanic series, use of electrical insulators between metals to break the current path from anode to cathode, and protection against the introduction of electrolytes.

16.6 SHOCK AND VIBRATION

16.6.1 General

Shock and vibration[21] loads consist of implusive and repetitive mechanical forces acting on an equipment.

16.6.2 Environmental Loads

Sources of shock loads include objects striking an equipment, structural-borne stress waves such as those caused by gunfire recoil, the equipment falling and striking other objects, and forces induced by handling and shipment.

Vibration sources include motion induced by rotating machinery, aerodynamic or hydrodynamic buffeting, and motion caused by usage and transportation.

16.6.3 Life

Equipment life is reduced by shock loads, which fracture components and cause catastrophic breakage or deformation. Fatigue and wear failures result from vibration-induced or other repetitive stresses that produce incremental damage that accumulates until failure occurs.

16.6.4 Shock

Shock is a sudden change in momentum of a body. A shock pulse may range from a simple step function or haversine pulse to a brief but complex waveform composed of several frequencies. The shock pulse may result in bending displacement and subsequent (ringing) vibration of the equipment or elements thereof. Shock pulses of a duration near the fundamental or harmonic of the resonance of the structure often cause greatly magnified and destructive responses. Shock failures include:

1. Permanent localized deformation at point of impact
2. Permanent deformation within an equipment if structural elements such as mounting brackets are deformed or fractured
3. Secondary impact failures within an equipment should structural deformations cause components to strike adjacent surfaces
4. Temporary or permanent malperformance of an operating equipment
5. Failure of fasteners, structural joints, and mounting attachment points
6. Breakage of fragile components and structural elements

Design techniques employed to avoid shock-induced[7,22] damage include:

1. Characterization of the shock-producing event in terms of impulse waveform, energy, and point of application
2. Computation or empirical determination of equipment responses to the shock pulse in terms of acceleration (or "g" level) vs. time
3. Modification of the equipment structure to avoid resonant frequencies that coincide with the frequency content of the shock pulse
4. Assuring that the strength of structural elements is adequate to withstand the dynamic "g" loading without either permanent deformation or harmful displacements due to bending
5. Selecting and using components that are known to withstand the internal shock environment to which they are subjected when the local mounting structure responds to the shock pulse
6. Employing protective measures such as energy absorbing or resonance modifying materials between the equipment and the point of shock application, or within the equipment to mount fragile components

16.6.5 Vibration

The response of an equipment to vibration can be damaging if the equipment or elements thereof are resonant within the pass band of the excitation spectra. Vibration failures include:

1. Fretting, wear, and loosening of mechanical joints, thermal joints and fasteners; and within components such as connectors, switches, and potentiometers
2. Fatigue-induced structural failure of brackets, circuit boards, and components
3. Physical and operational failures should individual structural element bending displacements produce impact with adjacent objects
4. Deviations in the performance of electronic components caused by relative motion of elements within the component or by the relative motion between a component and other objects

Design techniques employed to avoid vibration-induced damage include:

1. Characterization of the energy and frequency content of the source of vibration excitation
2. Analytical and empirical determination of equipment primary, secondary structural responses, and component sensitivity to vibration excitation in the pass band of the source vibration
3. Control of individual resonance response frequencies of an equipment structure and internal elements to avoid coincidence of resonance frequencies
4. Employment of materials that have adequate fatigue life to withstand the cumulative damage predicted to occur over the life of the equipment
5. Use of energy-absorbing materials between the equipment and the excitation source, and within the equipment for the mounting of sensitive components

16.6.6 Testing

The primary purposes of testing related to shock and vibration are to verify and characterize the dynamic response of the equipment and components thereof to a dynamic environment and to demonstrate that the final equipment design will withstand the test environment specified for the equipment under evaluation.

Basic characterization testing is usually performed on an electrodynamic vibration machine with the unit under test hard-mounted to a vibration fixture that has no resonance in the pass band of the excitation spectrum. The test input is a low-displacement-level sinusoid that is slowly varied in frequency (swept) over the frequency range of interest. Sine sweep testing produces a history of the response (displacement or acceleration) of selected points on the equipment to sinusoidal excitation over the tested excitation frequencies and displacements.

Caution is advised when using a hard-mount vibration fixture, as the fixture is very stiff and capable of injecting more energy into a test specimen at specimen resonance than would be experienced in service. For this reason, the test input signal should be of low amplitude. In service, the reaction of a less stiff mounting structure to the specimen at specimen resonance would significantly reduce the energy injected into the specimen. If a specimen response history is known prior to testing, the test system may be set to control input levels to reproduce the response history as measured by a control accelerometer placed at the location on the test specimen where the field vibration history was measured.

Vibration-test information is used to aid in adjusting the equipment design to avoid unfavorable responses to the service excitation, such as the occurrence of coupled resonance (e.g., a component having a resonance frequency coincident with the resonance frequency of its supporting structure; or structure having a significant resonance which coincides with the frequency of an input shock spectrum). Individual components are often tested to determine and document the excitation levels and frequencies at which they malperform. This type of testing is fundamental to both shock and vibration design.

For more complex vibration-service input spectra, such as multiple sinusoidal or random vibration spectra, additional testing is performed, using the more complex input waveform on product elements to gain assurance that the responses thereof are predictable. The final test exposes the equipment to specified vibration frequencies, levels, and duration, which may vary by axis of excitation and may be combined with other variables such as temperature, humidity, and altitude environments.

16.7 STRUCTURAL DESIGN

16.7.1 General

Structural design of a system,[22-24] equipment structure, module structure, or bracket involves analytical, empirical, and experimental techniques to predict and thus control mechanical stresses.

16.7.2 Strength

Strength is the ability of a material to bear both static (sustained) and dynamic (time-varying) loads without significant permanent deformation. Many non-ferrous materials suffer permanent deformation under sustained loads (creep). Ductile materials withstand dynamic loads better than brittle materials, which may fracture under sudden load application. Materials such as plastics often exhibit significant changes in material properties over the temperature range encountered by a product.

Many equipment require control of deflection or deformation during service. Such structural elements are designed for stiffness to control deflection but must be checked to assure that strength criteria are achieved.

16.7.3 Complexity

An equipment is viewed as a collection of individual elements interconnected to achieve an overall systems function. Each element may be individually modeled; however, the equipment model becomes complex when the elements are interconnected. The static or dynamic response of one element becomes the input or forcing function for elements mounted to it.

The concept of mechanical impedance[25] applies to dynamic environments and refers to the reaction between a structural element or component and its mounting points over a range of excitation frequencies. The reaction force at the structural interface or mounting point is a function of the resonance response of an element and may have an amplifying or damping effect on the mounting structure, depending on the spectrum of the excitation. Mechanical impedance design involves control of element resonance and structure resonance, providing compatible impedance for interconnected structural and component elements.

16.7.4 Degree of Enclosure

Degree of enclosure is the extent to which the components within an electronic equipment are isolated from the surrounding environment.

For vented enclosures, the design must provide drain holes to facilitate elimination of induced liquids and condensation. Convection-cooled equipment used in environments with airborne particles may require filtration. Equipment cooled by forced air usually require filtration on air inlets.

Completely (hermetically) sealed equipment enclosures using metal or glass seals permit the internal humidity and pressure to be defined when the unit is sealed. It is necessary to control the dryness of internal gases to protect from condensation, induced corrosion and to assure that internal pressures due to heating in combination with external ambient pressures (e.g., due to altitude changes) do not exceed structural deformation limitations and stress capabilities of the enclosure.

Partially sealed enclosures using permeable sealing materials (e.g., adhesives and plastics, etc.) are vulnerable to penetration by water vapor and other gases. Pachen's law states that the total pressure inside an enclosure is the sum of the partial pressures of the constituent gases. When the external partial pressure of a constituent gas is higher than the internal partial pressure of that gas, regardless of the total pressure inside the equipment, the gas will permeate the seal until the internal and external partial pressures are equalized. When the gas is water vapor and is ingested into an equipment, condensation will occur during temperature cycles, resulting in corrosion and perhaps interruption of electrical signals. Permeable seals do not protect from internal moisture damage and corrosion.

Equipment that operate in the presence of explosive gases must incorporate components that cannot cause ignition, and exposed circuits must operate at low voltage and current conditions so that short-circuit heating is controlled or eliminated. Vented equipment require use of flame-propagation barriers, such as screen mesh, that demonstrate under test that should ignition occur inside the unit, the flame front will not propagate into the outer environment.

16.7.5 Thermal Expansion and Stresses

The coefficient of thermal expansion is a material property and varies widely among the materials used in the construction of an electronic equipment. When bonded or fastened together and subjected to temperature changes, materials with different coefficients of thermal expansion cause bending and shear stresses that may be detrimental to the operation or life of an equipment.

Thermal cycling of bonded elements leads to failure, such as loss of electrical contact between bolted joints, cracking and breaking of ceramic parts bonded to plastic or metal surfaces, and solder joint failure. Thermal stresses are reduced by selecting adjoining materials with the least difference in coefficient of thermal expansion.

16.8 THERMAL DESIGN

16.8.1 General

The object of thermal design[26,27] is to control component temperatures to achieve satisfactory product reliability.[28] Component-fabrication techniques, such as complementary metal oxide semiconductor (CMOS), greatly reduce power requirements and component heat generation. Continuing reductions in equipment size lead to increased component density and power generated per unit volume. Thus, even when an equipment employs low-power components, thermal design practices must be applied.

Thermal design hierarchy includes:

1. Equipment total heat generation and how that heat will be dissipated to the local external environment
2. Equipment internal environment, which is the environment experienced by modules, subassemblies, and components
3. Control of critical component temperatures

Thermal design also includes consideration of temperature sensitivity of materials, finishes, adhesives, and lubricants.

Heat flow and temperature are analogous to current and voltage. *Thermal resistance* (in °C per watt) relate temperature to the flow of thermal energy in the same manner as Ohm's law relates voltage to the flow of electrical current.

Thermal resistances are used to characterize heat flow through a material, components such as heat sinks, interfaces between components and mounting surfaces, interfaces between structural elements and joints, and interfaces between the equipment and the local external environment.

Thermal design includes definition of heat flow paths from the component to the ultimate heat sink and, for each heat flow path, the identification and selection of thermal resistances that ensure that component temperatures are maintained at acceptable levels.

16.8.2 Heat Transfer Modes

Conduction

Conduction is the transfer of thermal energy through a material medium, which may be solid, liquid, or gas. Conduction of heat from a source to the ultimate heat sink includes, as appropriate:

1. Component internal heat transfer from heat source to the component interface with local air and mounting surfaces. Component specifications usually include a thermal resistance, which relates an internal critical point temperature (such as a semiconductor junction) to a specified location on the component package.
2. Contact resistance between the component and its mounting surface. Contact resistance depends on contact area, pressure (over time), and presence of thermal grease or other materials used to lower contact resistance. Contact resistances are often defined experimentally.
3. Thermal conduction through structural elements is defined in tables of material properties as the coefficient of thermal conductivity.
4. Interface resistances due to structural joints, often defined experimentally.
5. Thermal resistance of heat sinks[29] used to dissipate energy to gas or liquid coolants. Conduction heat sinks include liquid-cooled cold plates, convection heat sinks, evaporative devices such as heat pipes, and thermoelectric (Peltier effect) cooling devices.

Free Convection

Free convection involves the rejection of thermal energy to air or gases surrounding a component or equipment in the absence of mechanically or environmentally induced motion of the gases. Free convection is a continuing process where the warming of gases in immediate contact with a warm body produces natural buoyancy and movement of the heated gases away from the warm object. Cooler gases are drawn toward the warm body, replacing the escaping warm gases.

Free convection thermal resistances depend on orientation of warm surfaces relative to gravity, surface area, and surface finish. Free convection is enhanced by extended surfaces,[29] such as fins, to increase the effective contact area between the warm surface and local gases. Manufacturers of commercial heat sinks provide thermal-resistance information.

Equipment cooled by free convection require venting to permit the escape of warm gases and entrance of cooler gases. Warm components must be in the flow path of the cooling gases and internal obstructions must be avoided that would impede flow of the cooling gases.

Natural convection cooling is a mass rate of flow process, and cooling effectiveness is decreased by reduction in ambient air pressure (thus lowering the density of the cooling gas) such as occurs at higher altitudes.

Forced Convection

Forced convection may be produced mechanically with fans and blowers, by the result of equipment movement during use, or by naturally occurring air movement (wind) over unsheltered equipment. Forced convection thermal resistances are much lower than thermal resistances effected by natural convection cooling.

Forced-convection cooling air is first ducted to the most temperature-sensitive components, with less sensitive components located downstream, based on temperature sensitivity.

Forced-convection heat sinks[30] are often used to cool primary heat dissipation components and assemblies.

Forced-convection thermal resistance is sensitive to the mass rate of flow and velocity of the cooling air. Mass rate of flow is pressure-dependent and decreases with reductions in ambient pressure. Decreased mass rate of flow causes increases in thermal resistance. Increased air velocity past the cooled object reduces the thickness of the boundary layer and decreases the thermal resistance, creating improved cooling.

Radiation

Radiation is the transfer of energy from a warmer object to a cooler object by infrared radiation. Unlike convection cooling, radiation heat transfer occurs in vacuum and is not dependent upon the presence of gaseous media between the objects.

The effectiveness of radiation heat transfer is dependent on the temperature differential between the objects, the distance, the projected area, and the emissivity of the emitting and receiving surfaces. Radiation cooling (or heating) is reduced if smoke or other particulate matter is suspended in intervening gases.

Radiation is the primary mode of heat transfer in outer space. However, unless large temperature differences and short separation distances are experienced between earthbound objects, radiation involves a small fraction of the total heat flow.

Solar radiation causes heat buildup when ultraviolet rays (radiated from a high-temperature source) strike the surface of a closed equipment case, thus adding to the heat load within the equipment. Should solar radiation pass into a device, as through plastic or glass that is transparent to ultraviolet rays, the radiation will heat internal surfaces, which will then radiate at infrared frequencies to which the transparent material is usually opaque. In this way, thermal energy is trapped inside the enclosure and may lead to excessive internal temperatures. Solar panels make use of this phenomenon to heat water and living spaces. Solar-induced heat loading must be considered whenever electronic equipment is exposed to the sun.

Evaporation

Evaporation cooling and condensation techniques utilize the latent heat of vaporization of a heat transfer liquid, such as water, alcohol, freon, or other liquid, to effect temperature control. Thus, when an object is submerged within a liquid, the temperature of the object is controlled to the boiling point of the liquid (100 °C for water) within which it is submerged.

Heat pipe and evaporation chamber devices utilize evaporation by containing a liquid and providing an internal capillary structure to return condensed liquid from the cool end of the device to the heated end of the device.

16.9 MANUFACTURABILITY

16.9.1 General

The manufacturability of an electronic equipment[31,2,32] depends on the techniques used to achieve ease of assembly, component selection, and utilization of design rules consistent with manufacturing processes.

16.9.2 Assembly Considerations

The ease of assembly of an electronic equipment is dependent upon careful design of the product, with produceability and maintainability as major considerations. Products that involve materials requiring few, if any, secondary operations (such as pressing, welding, soldering, drilling, bending, and the need for critical mechanical alignment procedures) are easier to assemble, with fewer quality errors, than products involving many such operations.

Design of components and structures that may be correctly assembled in only one manner, and that may serve more than one function within the product, tend to reduce assembly time and errors. The equipment design must provide adequate physical space for the tools needed to accomplish assembly. Labor-intensive operations such as specifying a joint to be sealed using adhesive may be avoided by specifying an easily installed gasket that may cost less when the cost of the assembly labor plus the cost of the gasket is considered.

16.9.3 Design to Process

The electronic equipment designer must consider the manufacturing processes to be employed in the fabrication of the components and structural elements specified for an equipment. Component part design must be consistent with the available machinery and processes that will be utilized to fabricate the component. This includes the proper selection of materials, component dimensions, edge distances, tooling clearances, process limitations, processing temperatures, application of component finishes, and related issues.

16.9.4 Concurrent Engineering

In the quest for improving product design schedules, equipment quality, and the reduction of project costs, concurrent engineering practices are becoming increasingly common.

Successful concurrent engineering teams work closely together and are required to share product-development information on a nearly real-time basis. This approach permits a variety of team members to work simultaneously on product-design aspects that in the past were handled in a sequential manner. In this way, process engineers, manufacturing engineers, quality control personnel, purchasing personnel, and other team members may influence the design earlier in the product-development process.

16.10 PROTECTIVE PACKAGING

16.10.1 General

Protective packaging includes the techniques employed to ensure that a product will survive handling, shipping, and storage environments without degradation.

16.10.2 Storage Environment Protection

Electronic equipment may be subjected to storage for extended periods of time. The storage environments may include exposure to more severe temperature, pressure, and humidity variations than the product will experience after being placed into service.

Protective packaging selected for storage must withstand the storage environments and offer protection to the enclosed product.

The materials from which storage containers and fillers are selected must be chemically inert and not introduce detrimental effects of the stored equipment. For example, some paper products contain sulfur, the fumes of which accelerate tarnishing, thus increasing contact resistance of silver-plated contacts.

16.10.3 Shipping Environment Protection

Protective packaging must withstand transportation environments and the handling associated with movement of the equipment to, from, and between carriers. The transportation and handling environments may include exposure to more severe shock and vibration than the product will experience after being placed into service.

The shipping containers must tolerate stacking and handling by mechanical lifting devices. When it is predicted that the transportation environment may be more severe than the protected product will withstand, the container may include packing materials to cushion the products from vibration and shock loads. Packaging protection and service life are usually verified by testing prior to use.

REFERENCES

1. C. A. Harper, *Electronic Packaging and Interconnection Handbook*, McGraw-Hill, New York, 1991.
2. B. S. Matisoff, *Handbook of Electronics Packaging Design and Engineering*, Van Nostrand Reinhold, New York, 1982.
3. M. Pecht, *Handbook of Electronic Packaging Design*, Marcel Dekker, New York, 1991.
4. D. Agonafer and R. E. Fulton, *Computer Aided Design in Electronic Packaging*, EEP, Vol. 3, ASME Press, New York, 1992.
5. G. L. Ginsberg, *Printed Circuits Design*, McGraw-Hill, New York, 1990.
6. R. J. Hannemann, A. D. Kraus, and M. G. Pecht, *Physical Design of VLSI Systems*, Wiley-Interscience, New York, 1994.
7. C. Capillpo, *Surface Mount Technology*, McGraw-Hill, New York, 1990.
8. R. K. Wassink, *Soldering in Electronics*, 2nd ed., Electrochemical Publications, Ayr, Scotland, 1989.
9. A. Rahn, *The Basics of Soldering*, Wiley-Interscience, New York, 1993.
10. R. W. Woodgate, *Handbook of Machine Soldering*, Wiley-Interscience, New York, 1988.
11. M. G. Pecht, *Soldering Processes and Equipment*, Wiley-Interscience, New York, 1993.
12. A. J. Kinloch, *Adhesion and Adhesives*, Chapman and Hall, New York, 1987.
13. F. C. Allard, *Fiber Optics Handbook*, McGraw-Hill, New York, 1990.
14. *Metals Handbook*, Desk ed., American Society for Metals, Materials Park, OH, 1985.
15. *Electronic Materials Handbook*, Vol. 1, American Society for Metals, Materials Park, OH, 1990.
16. E. A. Muccio, *Plastics Processing Technology*, American Society for Metals, Materials Park, OH, 1994.
17. *Engineered Materials Handbook*, American Society for Metals, Materials Park, OH, 1995.
18. R. C. Buchanan, *Ceramic Materials for Electronics*, Marcel Dekker, New York, 1986.
19. J. B. Wachtman, *Mechanical Properties of Ceramics*, Wiley-Interscience, New York, 1996.
20. W. T. Shugg, *Handbook of Electrical and Electronic Insulating Materials*, Van Nostrand Reinhold, New York, 1986.
21. C. M. Harris and C. E. Crede, *Shock and Vibration Handbook*, 3rd ed., McGraw-Hill, New York, 1988.
22. J. H. Williams, *Fundamentals of Applied Dynamics*, MIT Press, Cambridge, MA, 1995.
23. E. Suhir, *Structural Analysis in Microelectronics*, Van Nostrand Reinhold, New York, 1992.
24. P. A. Engel, *Structural Analysis of Printed Circuit Board Systems*, Springer-Verlag, New York, 1993.
25. W. C. Fackler, *Equivalent Techniques for Vibration Testing*, SVM-9, Naval Research Laboratory, Washington, DC, 1972.
26. A. D. Kraus and A. Bar-Cohen, *Thermal Analysis and Control of Electronic Equipment*, McGraw-Hill, New York, 1993.
27. D. S. Steinberg, *Cooling Techniques for Electronic Equipment*, 2nd ed., Wiley-Interscience, New York, 1991.
28. F. Jensen, *Electronic Component Reliability*, Wiley-Interscience, New York, 1995.
29. A. D. Kraus and A. Bar-Cohen, *Design and Analysis of Heat Sinks*, Wiley, New York, 1995.
30. W. T. Kays and A. L. London, *Compact Heat Exchangers*, McGraw-Hill, New York, 1984.
31. Conference proceedings, several authors, *Design for Manufacturability*, ASME Press, New York, 1994.
32. Y. C. Lee and T. J. Bennett, *Manufacturing Aspects in Electronic Packaging*, EEP, Vol. 2/PED, Vol. 60, ASME Press, New York, 1992.

BIBLIOGRAPHY

Marcus, P., and J. Oudar, *Corrosion Mechanisms in Theory and Practice*, Marcel Dekker, New York, 1995.

McConnell, K. G., *Vibration Testing: Theory and Practice*, Wiley-Interscience, New York, 1995.

Seraphim, D. P., R. Lansk, and C.-Y. Li, *Principles of Electronic Packaging*, McGraw-Hill, New York, 1989.

Sherwani, Y., and B. Sandeep, *Introduction to Multichip Modules*, Wiley-Interscience, New York, 1995.

Steinberg, D. S., *Vibrational Analysis for Electronic Equipment*, 2nd ed., Wiley-Interscience, New York, 1988.

CHAPTER 17

DESIGN OPTIMIZATION— AN OVERVIEW

A. Ravindran
Department of Industrial and Manufacturing Engineering
Pennsylvania State University
University Park, Pennsylvania

G. V. Reklaitis
School of Chemical Engineering
Purdue University
West Lafayette, Indiana

17.1	**INTRODUCTION**	**353**	**17.4**	**STRUCTURE OF OPTIMIZATION PROBLEMS** 366
17.2	**REQUIREMENTS FOR THE APPLICATION OF OPTIMIZATION METHODS**	**354**	**17.5**	**OVERVIEW OF OPTIMIZATION METHODS** 368
	17.2.1 Defining the System Boundaries	354		17.5.1 Unconstrained Optimization Methods 368
	17.2.2 The Performance Criterion	354		17.5.2 Constrained Optimization
	17.2.3 The Independent Variables	355		Methods 369
	17.2.4 The System Model	355		17.5.3 Code Availability 372
17.3	**APPLICATIONS OF OPTIMIZATION IN ENGINEERING**	**356**	**17.6**	**SUMMARY** 373
	17.3.1 Design Applications	357		
	17.3.2 Operations and Planning Applications	362		
	17.3.3 Analysis and Data Reduction Applications	364		

17.1 INTRODUCTION

This chapter presents an overview of optimization theory and its application to problems arising in engineering. In the most general terms, optimization theory is a body of mathematical results and numerical methods for finding and identifying the best candidate from a collection of alternatives without having to enumerate and evaluate explicitly all possible alternatives. The process of optimization lies at the root of engineering, since the classical function of the engineer is to design new, better, more efficient, and less expensive systems, as well as to devise plans and procedures for the improved operation of existing systems. The power of optimization methods to determie the best case without actually testing all possible cases comes through the use of a modest level of mathematics and at the cost of performing iterative numerical calculations using clearly defined logical procedures or algorithms implemented on computing machines. Because of the scope of most engineering applications and the tedium of the numerical calculations involved in optimization algorithms, the techniques of optimization are intended primarily for computer implementation.

Mechanical Engineers' Handbook, 2nd ed., Edited by Myer Kutz.
ISBN 0-471-13007-9   © 1998 John Wiley & Sons, Inc.

17.2 REQUIREMENTS FOR THE APPLICATION OF OPTIMIZATION METHODS

In order to apply the mathematical results and numerical techniques of optimization theory to concrete engineering problems it is necessary to delineate clearly the boundaries of the engineering system to be optimized, to define the quantitative criterion on the basis of which candidates will be ranked to determine the "best," to select the system variables that will be used to characterize or identify candidates, and to define a model that will express the manner in which the variables are related. This composite activity constitutes the process of *formulating* the engineering optimization problem. Good problem formulation is the key to the success of an optimization study and is to a large degree an art. It is learned through practice and the study of successful applications and is based on the knowledge of the strengths, weaknesses, and peculiarities of the techniques provided by optimization theory.

17.2.1 Defining the System Boundaries

Before undertaking any optimization study it is important to define clearly the boundaries of the system under investigation. In this context a system is the restricted portion of the universe under consideration. The system boundaries are simply the limits that separate the system from the remainder of the universe. They serve to isolate the system from its surroundings, because, for purposes of analysis, all interactions between the system and its surroundings are assumed to be frozen at selected, representative levels. Since interactions, nonetheless, always exist, the act of defining the system boundaries is the first step in the process of approximating the real system.

In many situations it may turn out that the initial choice of system boundary is too restrictive. In order to analyze a given engineering system fully it may be necessary to expand the system boundaries to include other subsystems that strongly affect the operation of the system under study. For instance, suppose a manufacturing operation has a point shop in which finished parts are mounted on an assembly line and painted in different colors. In an initial study of the paint shop we may consider it in isolation from the rest of the plant. However, we may find that the optimal batch size and color sequence we deduce for this system are strongly influenced by the operation of the fabrication department that produces the finished parts. A decision thus has to be made whether to expand the system boundaries to include the fabrication department. An expansion of the system boundaries certainly increases the size and complexity of the composite system and thus may make the study much more difficult. Clearly, in order to make our work as engineers more manageable, we would prefer as much as possible to break down large complex systems into smaller subsystems that can be dealt with individually. However, we must recognize that this decomposition is in itself a potentially serious approximation of reality.

17.2.2 The Performance Criterion

Given that we have selected the system of interest and have defined its boundaries, we next need to select a criterion on the basis of which the performance or design of the system can be evaluated so that the "best" design or set of operating conditions can be identified. In many engineering applications, an economic criterion is selected. However, there is a considerable choice in the precise definition of such a criterion: total capital cost, annual cost, annual net profit, return on investment, cost to benefit ratio, or net present worth. In other applications a criterion may involve some technology factors, for instance, minimum production time, maximum production rate, minimum energy utilization, maximum torque, and minimum weight. Regardless of the criterion selected, in the context of optimization the "best" will always mean the candidate system with either the *minimum* or the *maximum* value of the performance index.

It is important to note that within the context of the optimization methods, only *one* critrion or performance measure is used to define the optimum. It is not possible to find a solution that, say, simultaneously minimizes cost and maximizes reliability and minimizes energy utilization. This again is an important simplification of reality, because in many practical situations it would be desirable to achieve a solution that is "best" with respect to a number of different criteria. One way of treating multiple competing objectives is to select one criterion as primary and the remaining criteria as secondary. The primary criterion is then used as an optimization performance measure, while the secondary criteria are assigned acceptable minimum or maximum values and are treated as problem constraints. However, if careful considerations were not given while selecting the acceptable levels, a feasible design that satisfies all the constraints may not exist. This problem is overcome by a technique called *goal programming,* which is fast becoming a practical method for handling multiple criteria. In this method, all the objectives are assigned target levels for achievement and a relative priority on achieving these levels. Goal programming treats these targets as goals to aspire for and not as absolute constraints. It then attempts to find an optimal solution that comes as "close as possible" to the targets in the order of specified priorities. Readers interested in multiple criteria optimizations are directed to recent specialized texts.[1,2]

17.2.3 The Independent Variables

The third key element in formulating a problem for optimization is the selection of the independent variables that are adequate to characterize the possible candidate designs or operating conditions of the system. There are several factors that must be considered in selecting the independent variables.

First, it is necessary to distinguish between variables whose values are amenable to change and variables whose values are fixed by external factors, lying outside the boundaries selected for the system in question. For instance, in the case of the paint shop, the types of parts and the colors to be used are clearly fixed by product specifications or customer orders. These are specified system parameters. On the other hand, the order in which the colors are sequenced is, within constraints imposed by the types of parts available and inventory requirements, an independent variable that can be varied in establishing a production plan.

Furthermore, it is important to differentiate between system parameters that can be treated as fixed and those that are subject to fluctuations which are influenced by external and uncontrollable factors. For instance, in the case of the paint shop, equipment breakdown and worker absenteeism may be sufficiently high to influence the shop operations seriously. Clearly, variations in these key system parameters must be taken into account in the production planning problem formulation if the resulting optimal plan is to be realistic and operable.

Second, it is important to include in the formulation all of the important variables that influence the operation of the system or affect the design definition. For instance, if in the design of a gas storage system we include the height, diameter, and wall thickness of a cylindrical tank as independent variables, but exclude the possibility of using a compressor to raise the storage pressure, we may well obtain a very poor design. For the selected fixed pressure we would certainly find the least cost tank dimensions. However, by including the storage pressure as an independent variable and adding the compressor cost to our performance criterion, we could obtain a design that has a lower overall cost because of a reduction in the required tank volume. Thus, the independent variables must be selected so that all important alternatives are included in the formulation. Exclusion of possible alternatives, in general, will lead to suboptimal solutions.

Finally, a third consideration in the selection of variables is the level of detail to which the system is considered. While it is important to treat all of the key independent variables, it is equally important not to obscure the problem by the inclusion of a large number of fine details of subordinate importance. For instance, in the preliminary design of a process involving a number of different pieces of equipment—pressure vessels, towers, pumps, compressors, and heat exchangers—one would normally not explicitly consider all of the fine details of the design of each individual unit. A heat exchanger may well be characterized by a heat-transfer surface area as well as shell-side and tube-side pressure drops. Detailed design variables such as number and size of tubes, number of tube and shell passes, baffle spacing, header type, and shell dimensions would normally be considered in a separate design study involving that unit by itself. In selecting the independent variables a good rule to follow is to include only those variables that have a significant impact on the composite system performance criterion.

17.2.4 The System Model

Once the performance criterion and the independent variables have been selected, then the next step in problem formulation is the assembly of the model that describes the manner in which the problem variables are related and the performance criterion is influenced by the independent variables. In principle, optimization studies may be performed by experimenting directly with the system. Thus, the independent variables of the system or process may be set to selected values, the system operated under those conditions, and the system performance index evaluated using the observed performance. The optimization methodology would then be used to predict improved choices of the independent variable values and the experiments continued in this fashion. In practice most optimization studies are carried out with the help of a *model*, a simplified mathematical representation of the real system. Models are used because it is too expensive or time consuming or risky to use the real system to carry out the study. Models are typically used in engineering design because they offer the cheapest and fastest way of studying the effects of changes in key design variables on system performance.

In general, the model will be composed of the basic material and energy balance equations, engineering design relations, and physical property equations that describe the physical phenomena taking place in the system. These equations will normally be supplemented by inequalities that define allowable operating ranges, specify minimum or maximum performance requirements, or set bounds on resource availabilities. In sum, the model consists of all of the elements that normally must be considered in calculating a design or in predicting the performance of an engineering system. Quite clearly the assembly of a model is a very time-consuming activity, and it is one that requires a thorough understanding of the system being considered. In simple terms, a model is a collection of equations and inequalities that define how the system variables are related and that constrain the variables to take on acceptable values.

From the preceding discussion, we observe that a problem suitable for the application of optimization methodology consists of a performance measure, a set of independent variables, and a model relating the variables. Given these rather general and abstract requirements, it is evident that the methods of optimization can be applied to a very wide variety of applications. We shall illustrate next a few engineering design applications and their model formulations.

17.3 APPLICATIONS OF OPTIMIZATION IN ENGINEERING

Optimization theory finds ready application in all branches of engineering in four primary areas:

1. Design of components of entire systems.
2. Planning and analysis of existing operations.
3. Engineering analysis and data reduction.
4. Control of dynamic systems.

In this section we briefly consider representative applications from the first three areas.

In considering the application of optimization methods in design and operations, the reader should keep in mind that the optimization step is but one step in the overall process of arriving at an optimal design or an efficient operation. Generally, that overall process will, as shown in Fig. 17.1, consist of an iterative cycle involving synthesis or definition of the structure of the system, model formulation, model parameter optimization, and analysis of the resulting solution. The final optimal design or new operating plan will be obtained only after solving a series of optimization problems, the solution to each of which will have served to generate new ideas for further system structures. In the interest of brevity, the examples in this section show only one pass of this iterative cycle and focus mainly on preparations for the optimization step. This focus should not be interpreted as an indication of the

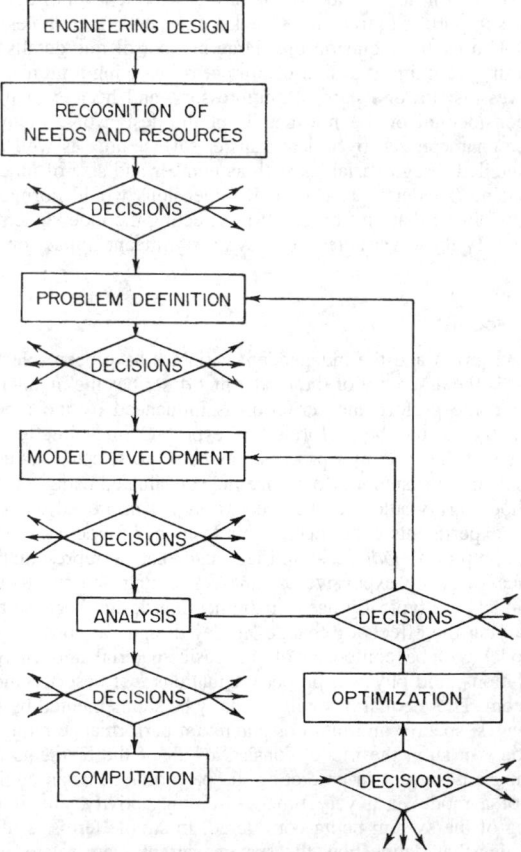

Fig. 17.1 Optimal design process.

dominant role of optimization methods in the engineering design and systems analysis process. Optimization theory is but a very powerful tool that, to be effective, must be used skillfully and intelligently by an engineer who thoroughly understands the system under study. The primary objective of the following example is simply to illustrate the wide variety but common form of the optimization problems that arise in the design and analysis process.

17.3.1 Design Applications

Applications in engineering design range from the design of individual structural members to the design of separate pieces of equipment to the preliminary design of entire production facilities. For purposes of optimization the shape or structure of the system is assumed known and optimization problem reduces to the selection of values of the unit dimensions and operating variables that will yield the best value of the selected performance criterion.

Example 17.1 Design of an Oxygen Supply System

Description. The basic oxygen furnace (BOF) used in the production of steel is a large fed-batch chemical reactor that employs pure oxygen. The furnace is operated in a cyclic fashion: ore and flux are charged to the unit, are treated for a specified time period, and then are discharged. This cyclic operation gives rise to a cyclically varying demand rate for oxygen. As shown in Fig. 17.2, over each cycle there is a time interval of length t_1 of low demand rate, D_0, and a time interval $(t_2 - t_1)$ of high demand rate, D_1. The oxygen used in the BOF is produced in an oxygen plant. Oxygen plants are standard process plants in which oxygen is separated from air using a combination of refrigeration and distillation. These are highly automated plants, which are designed to deliver a fixed oxygen rate. In order to mesh the continuous oxygen plant with the cyclically operating BOF, a simple inventory system shown in Fig. 17.3 and consisting of a compressor and a storage tank must be designed. A number of design possibilities can be considered. In the simplest case, one could select the oxygen plant capacity to be equal to D_1, the high demand rate. During the low-demand interval the excess oxygen could just be vented to the air. At the other extreme, one could also select the oxygen plant capacity to be just enough to produce the amount of oxygen required by the BOF over a cycle. During the low-demand interval, the excess oxygen production would then be compressed and stored for use during the high-demand interval of the cycle. Intermediate designs could involve some combination of venting and storage of oxygen. The problem is to select the optimal design.

Formulation. The system of concern will consist of the O_2 plant, the compressor, and the storage tank. The BOF and its demand cycle are assumed fixed by external factors. A reasonable performance index for the design is the total annual cost, which consists of the oxygen production cost (fixed and variable), the compressor operating cost, and the fixed costs of the compressor and of the storage

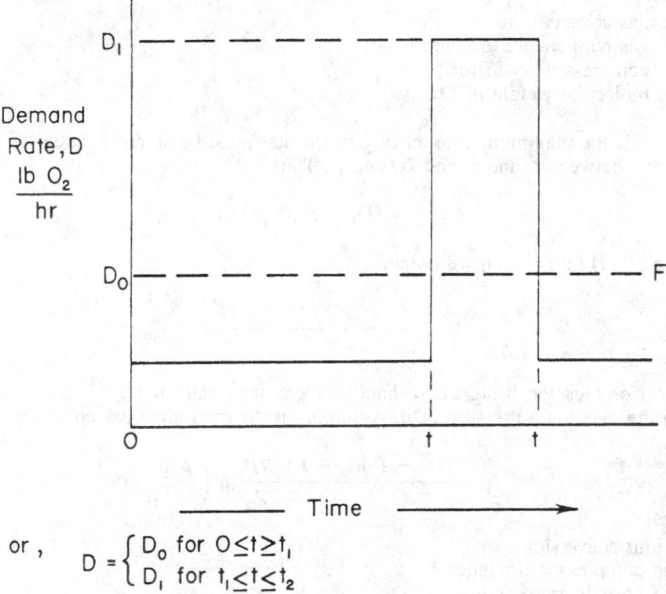

$$D = \begin{cases} D_0 & \text{for } 0 \leq t \geq t_1 \\ D_1 & \text{for } t_1 \leq t \leq t_2 \end{cases}$$

Fig. 17.2 Oxygen demand cycle.

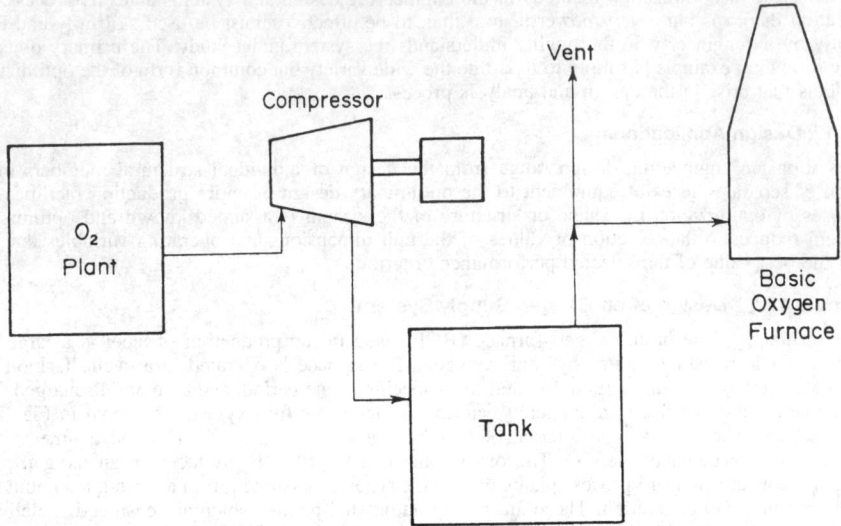

Fig. 17.3 Design of oxygen production system.

vessel. The key independent variables are the oxygen plant production rate F (lb O_2/hr), the compressor and storage tank design capacities, H (hp) and V (ft³), respectively, and the maximum tank pressure, p (psia). Presumably the oxygen plant design is standard, so that the production rate fully characterizes the plant. Similarly, we assume that the storage tank will be of a standard design approved for O_2 service.

The model will consist of the basic design equations that relate the key independent variables.

If I_{max} is the maximum amount of oxygen that must be stored, then using the corrected gas law we have

$$V = \frac{I_{max}}{M} \frac{RT}{p} z \tag{17.1}$$

where R = the gas constant
$\quad T$ = the gas temperature (assume fixed)
$\quad z$ = the compressibility factor
$\quad M$ = the molecular weight of O_2

From Fig. 17.1, the maximum amount of oxygen that must be stored is equal to the area under the demand curve between t_1 and t_2 and D_1 and F. Thus,

$$I_{max} = (D_1 - F)(t_2 - t_1) \tag{17.2}$$

Substituting (17.2) into (17.1), we obtain

$$V = \frac{(D_1 - F)(t_2 - t_1)}{M} \frac{RT}{p} z \tag{17.3}$$

The compressor must be designed to handle a gas flow rate of $(D_1 - F)(t_2 - t_1)/t_1$ and to compress it to the maximum pressure of p. Assuming isothermal ideal gas compression,[3]

$$H = \frac{(D_1 - F)(t_2 - t_1)}{t_1} \frac{RT}{k_1 k_2} \ln\left(\frac{p}{p_0}\right) \tag{17.4}$$

where k_1 = a unit conversion factor
$\quad k_2$ = the compressor efficiency
$\quad p_0$ = the O_2 delivery pressure

In addition to (17.3) and (17.4), the O_2 plant rate F must be adequate to supply the total oxygen demand, or

$$F \geq \frac{D_0 t + D_1(t_2 - t_1)}{t_2} \tag{17.5}$$

Moreover, the maximum tank pressure must be greater than the O_2 delivery pressure,

$$p \geq p_0 \tag{17.6}$$

The performance criterion will consist of the oxygen plant annual cost,

$$C_1(\$/\text{yr}) = a_1 + a_2 F \tag{17.7}$$

where a_1 and a_2 are empirical constants for plants of this general type and include fuel, water, and labor costs.

The capital cost of storage vessels is given by a power-law correlation,

$$C_2(\$) = b_1 V^{b_2} \tag{17.8}$$

where b_1 and b_2 are empirical constants appropriate for vessels of a specific construction.

The capital cost of compressors is similarly obtained from a correlation,

$$C_3(\$) = b_3 H^{b_4} \tag{17.9}$$

The compressor power cost will, as an approximation, be given by

$$b_5 t_1 H$$

where b_5 is the cost of power.

The total cost function will thus be of the form,

$$\text{Annual cost} = a_1 + a_2 F + d\{b_1 V^{b_2} + b_3 H^{b_4}\} + N b_5 t_1 H \tag{17.10}$$

where N = the number of cycles per year
d = an appropriate annual cost factor

The complete design optimization problem thus consists of the problem of minimizing (17.10), by the appropriate choice of F, V, H, and p, subject to Eqs. (17.3) and (17.4) as well as inequalities (17.5) and (17.6).

The solution of this problem will clearly be affected by the choice of the cycle parameters (N, D_0, D_1, t_1, and t_2), the cost parameters (a_1, a_2, b_1–b_5, and d), as well as the physical parameters (T, p_0, k_2, z, and M).

In principle, we could solve this problem by eliminating V and H from (17.10) using (17.3) and (17.4), thus obtaining a two-variable problem. We could then plot the contours of the cost function (17.10) in the plane of the two variables F and p, impose the inequalities (17.5) and (17.6), and determine the minimum point from the plot. However, the methods discussed in subsequent chapters allow us to obtain the solution with much less work. For further details and a study of solutions for various parameter values the reader is invited to consult Ref. 4.

The preceding example presented a preliminary design problem formulation for a system consisting of several pieces of equipment. The next example illustrates a detailed design of a single structural element.

Example 17.2 Design of a Welded Beam

Description. A beam A is to be welded to a rigid support member B. The welded beam is to consist of 1010 steel and is to support a force F of 6000 lb. The dimensions of the beam are to be selected so that the system cost is minimized. A schematic of the system is shown in Fig. 17.4.

Formulation. The appropriate system boundaries are quite self-evident. The system consists of the beam A and the weld required to secure it to B. The independent or design variables in this case are the dimensions h, l, t, and b as shown in Fig. 17.4. The length L is assumed to be specified at 14 in. For notational convenience we redefine these four variables in terms of the vector of unknowns $\mathbf{x}$,

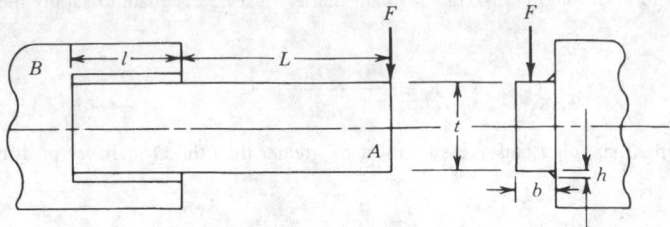

Fig. 17.4 Welded beam.

$$\mathbf{x} = [x_1, x_2, x_3, x_4]^T = [h, l, t, b]^T$$

The performance index appropriate to this design is the cost of a weld assembly. The major cost components of such an assembly are (a) set-up labor cost, (b) welding labor cost, and (c) material cost:

$$F(x) = c_0 + c_1 + c_2 \tag{17.11}$$

where $F(x)$ = cost function
$\quad c_0$ = set-up cost
$\quad c_1$ = welding labor cost
$\quad c_2$ = material cost

Set-Up Cost: c_0. The company has chosen to make this component a weldment, because of the existence of a welding assembly line. Furthermore, assume that fixtures for set-up and holding of the bar during welding are readily available. The cost c_0 can, therefore, be ignored in this particular total cost model.

Welding Labor Cost: c_1. Assume that the welding will be done by machine at a total cost of $10 per hour (including operating and maintenance expense). Furthermore, suppose that the machine can lay down 1 in.³ of weld in 6 min. Therefore, the labor cost is

$$c_1 = \left(10 \,\frac{\$}{\mathrm{hr}}\right)\left(\frac{1 \text{ hr}}{60 \text{ min}}\right)\left(6 \,\frac{\text{min}}{\text{in.}^3}\right) V_w = 1 \left(\frac{\$}{\text{in.}^3}\right) V_w$$

where V_w = weld volume, in.³
    Material Cost: c_2.

$$c_2 = c_3 V_w + c_4 V_B$$

where c_3 = $/volume of weld material = $(0.37)(0.283)(\$/\text{in.}^3)$
$\quad c_4$ = $/volume of bar stock = $(0.17)(0.283)(\$/\text{in.}^3)$
$\quad V_B$ = volume of bar A (in.³)

From the geometry,

$$V_w = 2(\tfrac{1}{2}h^2 l) - h^2 l \quad\text{and}\quad V_B = tb(L + l)$$

so

$$c_2 = c_3 h^2 l + c_4 tb(L + l)$$

Therefore, the cost function becomes

$$F(x) = h^2 l + c_3 h^2 l + c_4 tb(L + l) \tag{17.12}$$

or, in terms of the x variables

$$F(x) = (l + c_3)x_1^2 x_2 + c_4 x_3 x_4 (L + x_2) \tag{17.13}$$

Note all combinations of x_1, x_2, x_3, and x_4 can be allowed if the structure is to support the load required. Several functional relationships between the design variables that delimit the region of feasibility must certainly be defined. These relationships, expressed in the form of inequalities, represent the design model. Let us first define the inequalities and then discuss their interpretation.

The inequities are:

$$g_1(x) = \tau_d - \tau(x) \geq 0 \tag{17.14}$$

$$g_2(x) = \sigma_d - \sigma(x) \geq 0 \tag{17.15}$$

$$g_3(x) = x_4 - x_1 \geq 0 \tag{17.16}$$

$$g_4(x) = x_2 \geq 0 \tag{17.17}$$

$$g_5(x) = x_3 \geq 0 \tag{17.18}$$

$$g_6(x) = P_c(x) - F \geq 0 \tag{17.19}$$

$$g_7(x) = x_1 - 0.125 \geq 0 \tag{17.20}$$

$$g_8(x) = 0.25 - DEL(x) \geq 0 \tag{17.21}$$

where τ_d = design shear stress of weld
$\tau(x)$ = maximum shear stress in weld; a function of x
σ_d = design normal stress for beam material
$\sigma(x)$ = maximum normal stress in beam; a function of x
$P_c(x)$ = bar buckling load; a function of x
$DEL(x)$ = bar end deflection; a function of x

In order to complete the model it is necessary to define the important stress states.

Weld stress: $\tau(x)$. After Shigley,[5] the weld shear stress has two components, τ' and τ'', where τ' is the primary stress acting over the weld throat area and τ'' is a secondary torsional stress:

$$\tau' = F/\sqrt{2}x_1x_2 \quad \text{and} \quad \tau'' = MR/J$$

with $M = F[L + (x_2/2)]$
$R = \{(x_2^2/4) + [(x_3 + x_1)/2]^2\}^{1/2}$
$J = 2\{0.707x_1x_2[x_2^2/12 + (x_3 + x_1)/2)^2]\}$

where M = moment of F about the center of gravity of the weld group
J = polar moment of inertia of the weld group

Therefore, the weld stress τ becomes

$$\tau(x) = [(\tau')^2 + 2\tau'\tau'' \cos \theta + (\tau'')^2]^{1/2}$$

where $\cos \theta = x_2/2R$.

Bar Bending Stress: $\sigma(x)$. The maximum bending stress can be shown to be equal to

$$\sigma(x) = 6FL/x_4x_3^2$$

Bar Buckling Load: $P_c(x)$. If the ratio $t/b = x_3/x_4$ grows large, there is a tendency for the bar to buckle. Those combinations of x_3 and x_4 that will cause this buckling to occur must be disallowed. It has been shown[6] that for narrow rectangular bars, a good approximation to the buckling load is

$$P_c(x) = \frac{4.013\sqrt{EI\alpha}}{L^2}\left[1 - \frac{x_3}{2L}\sqrt{\frac{EI}{\alpha}}\right]$$

where E = Young's modulus = 30×10^6 psi
$I = \frac{1}{12}x_3x_4^3$
$\alpha = \frac{1}{3}Gx_3x_4^3$
G = shearing modulus = 12×10^6 psi

Bar deflection: $DEL(x)$. To calculate the deflection assume the bar to be a cantilever of length L. Thus,

$$DEL(x) = 4FL^3 / Ex_3^3 x_4$$

The remaining inequalities are interpreted as follows.

g_3 states that it is not practical to have the weld thickness greater than the bar thickness. g_4 and g_5 are nonnegativity restrictions on x_2 and x_3. Note that the nonnegativity of x_1 and x_4 are implied by g_3 and g_7. Constraint g_6 ensures that the buckling load is not exceeded. Inequality g_7 specifies that it is not physically possible to produce an extremely small weld.

Finally, the two parameters τ_d and σ_d in g_1 and g_2 depend on the material of construction. For 1010 steel $\tau_d = 13,600$ psi and $\sigma_d = 30,000$ psi are appropriate.

The complete design optimization problem thus consists of the cost function (17.13) and the complex system of inequalities that results when the stress formulas are substituted into (17.14) through (17.21). All of these functions are expressed in terms of four independent variables.

This problem is sufficiently complex that graphical solution is patently infeasible. However, the optimum design can readily be obtained numerically using the methods of subsequent sections. For a further discussion of this problem and its solution the reader is directed to Ref. 7.

17.3.2   Operations and Planning Applications

The second major area of engineering application of optimization is found in the tuning of existing operations. We shall discuss an application of goal programming model for machinability data optimization in metal cutting.[8]

Example 17.3   An Economic Machining Problem with Two Competing Objectives

Consider a single-point, single-pass turning operation in metal cutting wherein an optimum set of cutting speed and feed rate is to be chosen which balances the conflict between metal removal rate and tool life as well as being within the restrictions of horsepower, surface finish, and other cutting conditions. In developing the mathematical model of this problem, the following constraints will be considered for the machining parameters:

Constraint 1: Maximum Permissible Feed.

$$f \leq f_{max} \tag{17.22}$$

where f is the feed in inches per revolution. f_{max} is usually determined by a cutting force restriction or by surface finish requirements.[9]

Constraint 2: Maximum Cutting Speed Possible.   If v is the cutting speed in surface feet per minute, then

$$v \leq v_{max} \tag{17.23}$$

where

$$v_{max} = \frac{\pi D N_{max}}{12}$$

and

$$N_{max} = \text{maximum spindle speed available on the machine}$$

Constraint 3: Maximum Horsepower Available.   If P_{max} is the maximum horsepower available at the spindle, then

$$vf^\alpha \leq \frac{P_{max}(33,000)}{c_t d_c^\beta}$$

where α, β, and c_t are constants.[9] d_c is the depth of cut in inches, which is fixed at a given value. For a given P_{max}, c_t, β, and d_c, the right-hand side of the above constraint will be a constant. Hence, the horsepower constraint can be written simply as

$$vf^\alpha \leq \text{constant} \tag{17.24}$$

Constraint 4: Nonnegativity Restrictions on Feed Rate and Speed.

$$v, f \geq 0 \tag{17.25}$$

In optimizing metal cutting there are a number of optimality criteria that can be used. Suppose we consider the following objectives in our optimization: (i) maximize metal removal rate (MRR), (ii) maximize tool life (TL). The expression for MRR is

$$\text{MRR} = 12vfd_c \text{ in.}^3/\text{min} \tag{17.26}$$

TL for a given depth of cut is given by

$$\text{TL} = \frac{A}{v^{1/n}f^{1/n_1}} \tag{17.27}$$

where A, n, and n_1 are constants. We note that the MRR objective is directly proportional to feed and speed, while the TL objective is inversely proportional to feed and speed. In general, there is no single solution to a problem formulated in this way, since MRR and TL are competing objectives and their respective maxima must include some compromise between the maximum of MRR and the maximum of TL.

A Goal Programming Model

Goal programming is a technique specifically designed to solve problems involving complex, usually conflicting multiple objectives. Goal programming requires the user to select a set of goals (which may or may not be realistic) that ought to be achieved (if possible) for the various objectives. It then uses preemptive weights or priority factors to rank the different goals and tries to obtain an optimal solution satisfying as many goals as possible. For this, it creates a single objective function that minimizes the deviations from the stated goals according to their relative importance.

Before we discuss the goal programming formulation of the machining problem, we should discuss the difference between the terms "real constraint" and "goal constraint" (or simply "goal") as used in goal programming models. The real constraints are absolute restrictions placed on the behavior of the design variables, while the goal constraints are conditions one would like to achieve but are not mandatory. For instance, a real constraint given by

$$x_1 + x_2 = 3$$

requires all possible values of $x_1 + x_2$ to always equal 3. As opposed to this, if we simply had a goal requiring $x_1 + x_2 = 3$, then this is not mandatory and we can choose values of x_1, x_2 such that $x_1 + x_2 \geq 3$ as well as $x_1 + x_2 \leq 3$. In a goal constraint positive and negative deviational variables are introduced as follows:

$$x_1 + x_2 + d_1^- - d_1^+ = 3, \qquad d_1^-, d_1^+ \geq 0$$

Note that if $d_1^- > 0$, then $x_1 + x_2 < 3$, and if $d_1^+ > 0$, then $x_1 + x_2 > 3$. By assigning suitable preemptive weights on d_1^- and d_1^+, the model will try to achieve the sum $x_1 + x_2$ as close as possible to 3.

Returning to the machining problem with competing objectives, suppose that management considers that a given single-point, single-pass turning operation will be operating at an acceptable efficiency level if the following goals are met as closely as possible.

1. The MRR must be greater than or equal to a given rate M_1 (in.^{3}/min).
2. The tool life must equal T_1 (min).

In addition, management requires that a higher priority be given to achieving the first goal than the second.

The goal programming approach may be illustrated by expressing each of the goals as goal constraints as shown below. Taking the MRR goal first,

$$12vfd_c + d_1^- - d_1^+ = M_1$$

where d_1^- represents the amount by which the MRR goal is underachieved, and d_1^+ represents any overachievement of the MRR goal. Similarly, the TL goal can be expressed as

$$\frac{A}{v^{1/n}f^{1/n_1}} + d_2^- - d_2^+ = T_1$$

Since the objective is to have an MRR of at least M_1, the objective function must be set up so that a high penalty will be assigned to the underachievement variable d_1^-. No penalty will be assigned to d_1^+. In order to achieve a tool life of T_1, penalties must be associated with both d_2^- and d_2^+ so that both of these variables are minimized to their fullest extent. The relative magnitudes of these penalties must reflect the fact that the first goal is considered to be more important that the second. Accordingly, the goal programming objective function for this problem is

$$\text{Minimize } z = P_1 d_2^- + P_2(d_2^- + d_2^+)$$

where P_1 and P_2 are nonnumerical preemptive priority factors such that $P_1 >>> P_2$ (i.e., P_1 is infinitely larger than P_2). With this objective function every effort will be made to satisfy completely the first goal before any attempt is made to satisfy the second.

In order to express the problem as a linear goal programming problem, M_1 is replaced by M_2, where

$$M_2 = \frac{M_1}{12 d_c}$$

The goal T_1 is replaced by T_2, where

$$T_2 = \frac{A}{T_1}$$

and logarithms are taken of the goals and constraints. The problem can then be stated as follows:

$$\text{Minimize } z = P_1 d_1^- + P_2(d_2^- + d_2^+)$$

Subject to

(MRR goal)	$\log v + \log f + d_1^- - d_1^+ = \log M_2$
(TL goal)	$(1/n) \log v + (1/n_1) \log f + d_2^- - d_2^+ = \log T_2$
(f_{max} constraint)	$\log f \leq \log f_{max}$
(V_{max} constraint)	$\log v \leq \log v_{max}$
(Horsepower constraint)	$\log v + \alpha \log f \leq \log \text{constant}$

$$\log v, \log f, d_1^-, d_1^+, d_2^-, d_2^+ \geq 0$$

We would like to reemphasize here that the last three inequalities are real constraints on feed, speed, and horsepower that must be satisfied at all times, while the equations for MRR and TL are simply goal constraints. For a further discussion of this problem and its solution, see Ref. 8. An efficient algorithm and a computer code for solving linear goal programming problems is given in Ref. 10. Readers interested in other optimization models in metal cutting should see Ref. 11. The textbook by Lee[12] contains a good discussion of goal programming theory and its applications.

17.3.3 Analysis and Data Reduction Applications

A further fertile area for the application of optimization techniques in engineering can be found in nonlinear regression problems as well as in many analysis problems arising in engineering science. A very common problem arising in engineering model development is the need to determine the parameters of some semitheoretical model given a set of experimental data. This data reduction or regression problem inherently transforms to an optimization problem, because the model parameters must be selected so that the model fits the data as closely as possible.

Suppose some variable y is assumed to be dependent on an independent variable x and related to x through a postulated equation $y = f(x, \theta_1, \theta_2)$, which depends on two parameters θ_1 and θ_2. To establish the appropriate values of θ_1 and θ_2, we run a series of experiments in which we adjust the independent variable x and measure the resulting y. As a result of a series of N experiments covering the range of x of interest, a set of y and x values (y_i, x_i), $i = 1, \ldots, N$, is available. Using these data we now try to "fit" our function to the data by adjusting θ_1 and θ_2 until we get a "good fit." The most commonly used measure of a "good fit" is the *least squares criterion,*

$$L(\theta_1, \theta_2) = \sum_{i=1}^{N} [y_i - f(x_i, \theta_1, \theta_2)]^2 \tag{17.28}$$

The difference $y_i - f(x_i, \theta_1, \theta_2)$ between the experimental value y_i and the predicted value $f(x_i, \theta_1, \theta_2)$ measures how close our model prediction is to the data and is called the *residual*. The sum of the squares of the residuals at all the experimental points gives an indication of goodness of fit. Clearly, if $L(\theta_1, \theta_2)$ is equal to zero, then the choice of θ_1, θ_2 has led to a perfect fit; the data points fall exactly on the predicted curve. The data-fitting problem can thus be viewed as an optimization problem in which $L(\theta_1, \theta_2)$ is minimized by appropriate choice of θ_1 and θ_2.

Example 17.4   Nonlinear Curve Fitting

Description.   The pressure-molar-volume-temperature relationship of real gases is known to deviate from that predicted by the ideal gas relationship

$$Pv = RT$$

where P = pressure (atm)
   v = molar volume (cm^3/g · mol)
   T = temperature (K)
   R = gas constant (82.06 atm · cm^3/g · mol · K)

The semiempirical Redlich–Kwong equation

$$P = \frac{RT}{v - b} - \frac{a}{T^{1/2}v(v + b)} \tag{17.29}$$

is intended to direct for the departure from ideality but involves two empirical constants a and b whose values are best determined from experimental data. A series of PvT measurements listed in Table 17.1 are made for CO_2, from which a and b are to be estimated using nonlinear regression.

Formulation.   Parameters a and b will be determined by minimizing the least squares function (17.28). In the present case, the function will take the form

$$\sum_{i=1}^{\delta} \left[P_i - \frac{RT_i}{v_i - b} + \frac{a}{T^{1/2}v_i(v_i + b)} \right]^2 \tag{17.30}$$

where P_i is the experimental value at experiment i, and the remaining two terms correspond to the value of P predicted from Eq. (17.29) for the conditions of experiment i for some selected value of the parameters a and b. For instance, the term corresponding to the first experimental point will be

$$\left(33 - \frac{82.06(273)}{500 - b} + \frac{a}{(273)^{1/2}(500)(500 + b)} \right)^2$$

Function (17.30) is thus a two-variable function whose value is to be minimized by appropriate choice of the independent variables a and b. If the Redlich–Kwong equation were to precisely match the data, then at the optimum the function (17.30) would be exactly equal to zero. In general, because of experimental error and because the equation is too simple to accurately model the CO_2 nonidealities, Eq. (17.30) will not be equal to zero at the optimum. For instance, the optimal values of $a = 6.377 \times 10^7$ and $b = 29.7$ still yield a squared residual of 9.7×10^{-2}.

Table 17.1 PvT Data for CO_2

Experiment Number	P (atm)	v (cm^3/g · mol)	$T°$(K)
1	33	500	273
2	43	500	323
3	45	600	373
4	26	700	273
5	37	600	323
6	39	700	373
7	38	400	273
8	63.6	400	373

17.4 STRUCTURE OF OPTIMIZATION PROBLEMS

Although the application problems discussed in the previous section originate from radically different sources and involve different systems, at root they have a remarkably similar form. All four can be expressed as problems requiring the minimization of a real-valued function $f(x)$ of an N-component vector argument $x = (x_1, x_2, \ldots, x_N)$ whose values are restricted to satisfy a number of real-valued equations $h_k(x) = 0$, a set of inequalities $g_j(x) \geq 0$, and the variable bounds $x_i^{(U)} \geq x_i \geq x_i^{(L)}$. In subsequent discussions we will refer to the function $f(x)$ as the *objective function,* to the equations $h_k(x) = 0$ as the *equality constraints,* and to the inequalities $g_j(x) \geq 0$ as the *inequality constraints.* For our purposes, these problem functions will always be assumed to be real valued, and their number will always be finite.

The general problem,

$$\text{Minimize } f(x)$$
$$\text{Subject to } h_k(x) = 0 \qquad k = 1, \ldots, K$$
$$g_j(x) \geq 0 \qquad j = 1, \ldots, J$$
$$x_i^{(L)} \geq x_i \geq x_i^{(L)} \qquad i = 1, \ldots, N$$

is called the *constrained* optimization problem. For instance, Examples 17.1, 17.2, and 17.3 are all constrained problems. The problem in which there are no constraints, that is,

$$J = K = 0$$

and

$$x_i^{(U)} = -x_i^{(L)} = \infty, \qquad i = 1, \ldots, N$$

is called the *unconstrained* optimization problem. Example 17.4 is an unconstrained problem. Optimization problems can be classified further based on the structure of the functions f, h_k, and g_j and on the dimensionality of x. Figure 17.5 illustrates one such classification. The basic subdivision is between unconstrained and constrained problems. There are two important classes of methods for solving the unconstrained problems. The direct search methods require only that the objective function be evaluated at different points, at least through experimentation. Gradient-based methods require the analytical form of the objective function and its derivatives.

An important class of constrained optimization problems is *linear programming,* which requires both the objective function and the constraints to be linear functions. Out of all optimization models, linear programming models are the most widely used and accepted in practiced. Professionally written software programs are available from all major computer manufacturers for solving very large linear programming problems. Unlike the other optimization problems that require special solution methods based on the problem structure, linear programming has just one common algorithm, known as the "simplex method," for solving all types of linear programming problems. This essentially has contributed to the successful applications of linear programming models in practice. In 1984, Narendra Karmarkar,[13] an AT&T researcher, developed an interior point algorithm, which was claimed to be 50 times faster than the simplex method for solving linear programming problems. By 1990, Karmarkar's seminal work had spawned hundreds of research papers and a large class of interior point methods. It has become clear that while the initial claims are somewhat exaggerated, interior point methods do become competitive for very large problems. For a discussion of interior point methods, see Refs. 14 and 15.

Integer programming (**IP**) is another important class of linearly constrained problems where some or all of the design variables are restricted to be integers. But solutions of IP problems are generally difficult, time-consuming, and expensive. Hence, a practical approach is to treat all the integer variables as continuous, solve the associated LP problem, and round off the fractional values to the nearest integers such that the constraints are not violated. This generally produces a good integer solution close to the optimal integer solution, particularly when the values of the variables are large. However, such an approach would fail when the values of the variables are small or binary valued (0 or 1). A good rule of thumb is to treat any integer variable whose value will be at least 20 as continuous and use special purpose IP algorithms for the rest. For a complete discussion of integer programming applications and algorithms, see Refs. 16 and 17.

The next class of optimization problems involves nonlinear objective functions and linear constraints. Under this class we have the following:

1. Quadratic programming, whose objective is a quadratic function.
2. Convex programming, whose objective is a special nonlinear function satisfying an important mathematical property called "convexity."

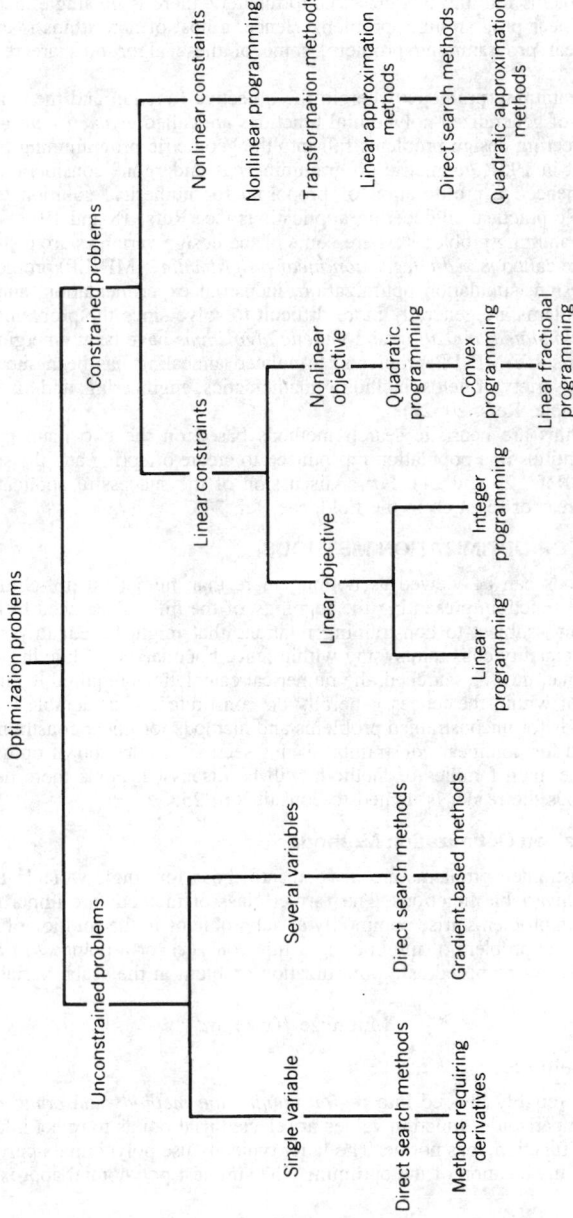

Fig. 17.5 Classification of optimization problems.

3. Linear fractional programming, whose objective is the ratio of two linear functions.

Special-purpose algorithms that take advantage of the particular form of the objective functions are available for solving these problems.

The most general optimization problems involve nonlinear objective functions and nonlinear constraints and are generally grouped under the term "nonlinear programming." The majority of engineering design problems fall into this class. Unfortunately, there is no single method that is best for solving every nonlinear programming problem. Hence, a host of algorithms is available for solving the general nonlinear programming problem, some of these algorithms are reviewed in the next section.

Nonlinear programming problems wherein the objective function and the constraints can be expressed as the sum of generalized polynomial functions are called *geometric programming* problems. A number of engineering design problems fall into the geometric programming framework. Since its earlier development in 1961, geometric programming has undergone considerable theoretical development, has experienced a proliferation of proposals for numerical solution techniques, and has enjoyed considerable practical engineering applications (see Refs. 18 and 19).

Nonlinear programming problems where some of the design variables are restricted to be discrete or integer valued are called *mixed integer nonlinear programming* (MINLP) problems. Such problems arise in process design, simulation optimization, industrial experimentation, and reliability optimization. MINLP problems are generally more difficult to solve since the problems have several local optima. Recently, *simulated annealing* and *genetic algorithms* have been emerging as powerful heuristic algorithms to solve MINLP problems. Simulated annealing has been successfully applied to solve problems in a variety of fields, including mathematics, engineering, and mathematical programming (see, for example, Refs. 20–22).

Genetic algorithms are heuristic search methods based on the two main principles of natural genetics, namely, entities in a population reproduced to create offspring and the survival of the fittest (see, for example, Refs. 23 and 21). For a discussion of the successful applications of genetic algorithms and the areas of research in the field, see Ref. 24.

17.5 OVERVIEW OF OPTIMIZATION METHODS

Optimization methods can be viewed as nothing more than numerical hill-climbing procedures in which the objective function, presenting the topology of the hill, is searched to identify the highest point—or maximum—subject to constraining relations that might be equality constraints (stay on winding path) or inequality constraints (stay within fence boundaries). While the constraints do serve to reduce the area that must be searched, the numerical calculations required to ensure that the search stays on the path or within the fences generally do constitute a considerable burden. Accordingly, optimization methods for unconstrained problems and methods for linear constraints are less complex than those designed for nonlinear constraints. In this section, a selection of optimization techniques representative of the main families of methods will be discussed. For a more detailed presentation of individual methods the reader is invited to consult Ref. 25.

17.5.1 Unconstrained Optimization Methods

Methods for unconstrained problems are divided into those for single-variable functions and those appropriate for multivariable functions. The former class of methods are important because single-variable optimization problems arise commonly as subproblems in the solution of multivariable problems. For instance, the problem of minimizing a function $f(x)$ for a point x^0 in a direction d (often called a *line search*) can be posed as a minimization problem in the scalar variable α:

$$\text{Minimize } f(x^0 + \alpha d)$$

Single Variable Methods

These methods are roughly divided into *region elimination methods* and *point estimation methods*. The former use comparison of function values at selected trial points to reject intervals within which the optimum of the function does not lie. The latter typically use polynomial approximating functions to estimate directly the location of the optimum. The simplest polynomial approximating function is the quadratic

$$\tilde{f}(x) = ax^2 + bx + c$$

whose coefficients a, b, c can be evaluated readily from those trial values of the actual function. The point at which the derivative of $\tilde{f}$ is zero is used readily to predict the location of the optimum of the true function

$$\bar{x} = -b/2a$$

The process is repeated using successively improved trial values until the differences between successive estimates $\bar{x}$ become sufficiently small.

Multivariable Unconstrained Methods

These algorithms can be divided into *direct search methods* and *gradient-based methods*. The former methods only use direct function values to guide the search, while the latter also require the computation of function gradient and, in some cases, second derivative values. Direct search methods in widespread use in engineering applications include the simplex search, the pattern search method of Hooke and Jeeves, random-sampling-based methods, and the conjugate directions method of Powell (see Chap. 3 of Ref. 25). All but the last of these methods make no assumptions about the smoothness of the function contours and hence can be applied to both discontinuous and discrete-valued objective functions.

Gradient-based methods can be grouped into the classical methods of steepest descent (Cauchy) and Newton's method and the modern quasi-Newton methods such as the conjugate gradient, Davidson–Fletcher–Powell, and Broyden–Fletcher–Shanno algorithms. All gradient-based methods employ the first derivative or gradient of the function at the current best solution estimate $\bar{x}$ to compute a direction in which the objective function value is guaranteed to decrease (a *descent* direction). For instance, Cauchy's classical method used the direction.

$$d = -\nabla f(\bar{x})$$

followed by a line search from $\bar{x}$ in this direction. In Newton's method the gradient vector is premultiplied by the matrix of second derivatives to obtain an improved direction vector

$$d = -(\nabla^2 f(\bar{x}))^{-1} \nabla f(\bar{x})$$

which in theory at least yields very good convergence behavior. However, the computation of $\nabla^2 f$ is often too burdensome for engineering applications. Instead, in recent years quasi-Newton methods have found increased application. In these methods, the direction vector is computed as

$$d = -H \nabla f(\bar{x})$$

where H is a matrix whose elements are updated as the iterations proceed using only values of gradient and function value difference from successive estimates. Quasi-Newton methods differ in the details of H updating, but all use the general form

$$H^{n+1} = H^n + C^n$$

where H^n is the previous value of H and C^n is a suitable correction matrix. The attractive feature of this family of methods is that convergence rates approaching those of Newton's method are attained without the need for computing $\nabla^2 f$ or solving the linear equation set

$$\nabla^2 f(\bar{x}) \cdot d = -f(\bar{x})$$

to obtain d. Recent developments in these methods have focused on strategies for eliminating the need for detailed line searching along the direction vectors and on enhancements for solving very large problems. For a detailed discussion of quasi-Newton methods the reader is directed to Refs. 25 and 26.

17.5.2 Constrained Optimization Methods

Constrained optimization methods can be classified into those applicable to totally linear or at least linearly constrained problems and those applicable to general nonlinear problems. The linear or linearly constrained problems can be well solved using methods of linear programming and extensions, as discussed earlier. The algorithms suitable for general nonlinear problems comprise four broad categories of methods:

1. Direct search methods that use only objective and constraint function values.
2. Transformation methods that use constructions that aggregate constraints with the original objective function to form a single composite unconstrained function.
3. Linearization methods that use linear approximations of the nonlinear problem functions to produce efficient search directions.
4. Successive quadratic programming methods that use quasi-Newton constructions to solve the general problem via a series of subproblems with quadratic objective function and linear constraints.

Direct Search

The direct search methods essentially consist of extensions of unconstrained direct search procedures to accommodate constraints. These extensions are generally only possible with inequality constraints or linear equality constraints. Nonlinear equalities must be treated by implicit or explicit variable elimination. That is, each equality constraint is either explicitly solved for a selected variable and used to eliminate that variable from the search or the equality constraints are numerically solved for values of the dependent variables for each trial point in the space of the independent variables.

For example, the problem

$$\text{Minimize} \quad f(\mathbf{x}) = x_1 x_2 x_3$$
$$\text{Subject to} \quad h_1(\mathbf{x}) = x_1 + x_2 + x_3 - 1 = 0$$
$$h_2(\mathbf{x}) = x_1^2 x_3 + x_2 x_3^2 + x_2^{-1} - 2 = 0$$
$$\mathbf{0} \le (x_1, x_3) \le \tfrac{1}{2}$$

involves two equality constraints and hence can be viewed as consisting of two dependent variables and one independent variable. Clearly, h_1 can be solved for x_1 to yield

$$x_1 = 1 - x_2 - x_3$$

Thus, on substitution the problem reduces to

$$\text{Minimize} \quad (1 - x_2 - x_3) x_2 x_3$$
$$\text{Subject to} \quad (1 - x_2 - x_3)^2 x_3 + x_2 x_3^2 + x_2^{-1}(1 - x_2 - x_3)\, 2 = 0$$
$$0 \le 1 - x_2 - x_3 \le \tfrac{1}{2}$$
$$0 \le x_3 \le \tfrac{1}{2}$$

Solution of the remaining equality constraint for one variable, say x_3, in terms of the other is very difficult. Instead, for each value of the independent variable x_2, the corresponding value of x_3 would have to be calculated numerically using some root-finding method.

Some of the more widely used direct search methods include the adaptation of the simplex search due to Box (called the *complex method*), various direct random-sampling-type methods, and combined random sampling/heuristic procedures such as the combinatorial heuristic meethod[27] advanced for the solution of complex optimal mechanism design problems.

A typical direct sampling procedure is given by the formula,

$$x_i p = \bar{x}_i \times z_i (2r - 1)^k, \qquad \text{for each variable } x_i, \quad i = 1, \ldots, n$$

where $\bar{x}_i$ = the current best value of variable i
  z_i = the allowable range of the variable i
  r = a random variable uniformly distributed on the interval 0–1
  k = an adaptive parameter whose value is adjusted based on past successes or failures in the search

For given $\bar{x}$, z, and k, r is sampled N times and the new point x^p evaluated. If x^p satisfies all constraints, it is retained; if it is infeasible, it is rejected and a new set of N r values is generated. If x^p is feasible, $f(x^p)$ is compared to $f(\bar{x})$, and if improvement is found, x^p replaces $\bar{x}$. Otherwise x^p is rejected. The parameter k is an adaptive parameter whose value will regulate the contraction or expansion of the sampling region. A typical adjustment procedure for k might be to increase k by 2 whenever a specified number of improved points is found or to decrease it by 2 when no improvement is found after a certain number of trials.

The general experience with direct search and especially random-sampling-based methods for constrained problems is that they can be quite effective for severely nonlinear problems that involve multiple local minima but are of low dimensionality.

Transformation Methods

This family consists of strategies for converting the general constrained problem to a parametrized unconstrained problem that is solved repeatedly for successive values of the parameters. The approaches can be grouped into the penalty/barrier function constructions, exact penalty methods, and augmented Lagrangian methods. The classical penalty function approach is to transform the general constrained problem to the form

$$P(x, R) = f(x) + \Omega(R, g(x), h(x))$$

where R = the penalty parameter
Ω = the penalty term

The ideal penalty function will have the property that

$$P(x, R) = \begin{cases} f(x), & \text{if } x \text{ is feasible} \\ \infty, & \text{if } x \text{ is infeasible} \end{cases}$$

Given this idealized construction, $P(x, R)$ could be minimized using any unconstrained optimization method, and, hence, the underlying constrained problem would have been solved. In practice such radical discontinuities cannot be tolerated from a numerical point of view, and, hence, practical penalty functions use penalty terms of the form

$$\Omega(R, g, h) = R \left(\sum_k h_k(x) \right)^2 + R(\Sigma(\min(0, g_j(x)))^2)$$

A series of unconstrained minimizations of $P(x, R)$ with different values of R are carried out beginning with a low value of R (say $R = 1$) and progressing to very large values of R. For low values of R, the unconstrained minima of $P(x, R)$ obtained will involve considerable constraint violations. As R increases, the violations decrease until in the limit as $R \to \infty$, the violations will approach zero. A large number of different forms of the Ω function have been proposed; however, all forms share the common feature that a sequence of problems must be solved and that, as the penalty parameter R becomes large, the penalty function becomes increasingly distorted and thus its minimization becomes increasingly more difficult. As a result the penalty function approach is best used for modestly sized problems (2–10 variables), few nonlinear equalities (2–5), and a modest number of inequalities. In engineering applications, the unconstrained subproblems are most commonly minimized using direct search methods, although successful use of quasi-Newton methods is also reported.

The exact penalty function and augmented Lagrangian approaches have been developed in an attempt to circumvent the need to force convergence by using increasing values of the penalty parameter. One typical representative of this type of method is the so-called meethod of multipliers.[28] In this method, once a sufficiently large value of R is reached, further increases are not required. However, the method does involve additional finite parameters that must be updated between subproblem solutions. Computational evidence reported to date suggests that, while augmented Lagrangian approaches are more reliable than penalty-function methods, they, as a class, are not suitable for larger dimensionality problems.

Linearization Methods

The common characteristic of this family of methods, is the use of local linear approximations to the nonlinear problem functions to define suitable, preferably feasible, directions for search. Well-known members of this family include the method of feasible directions, the gradient projection meethod, and the generalized reduced gradient (GRG) method. Of these, the GRG method has seen the widest engineering application.

The key constructions of the GRG method are the following:

1. The calculation of the reduced objective function gradient $\nabla \bar{f}$.
2. The use of the reduced gradient to determine a direction vector in the space of the independent variables.
3. The adjustment of the dependent variable values using Newton's method so as to achieve constraint satisfaction.

Given a feasible point x^0, the gradients of the equality constraints are evaluated and used to form the constraint Jacobian matrix A. This matrix is partitioned into a square submatrix J and the residual rectangular matrix C where the variable associated with the columns of J are the dependent variables and those associated with C are the independent variables.

If J is selected to have nonzero determinant, then the reduced gradient is defined as

$$\nabla \bar{f}(x^0) = \nabla \bar{f} - \nabla \hat{f} J^{-1} C$$

where $\nabla \bar{f}$ is the subvector of objective function partial derivatives corresponding to the (dependent) variables and $\hat{f}$ is the corresponding subvector whose components correspond to the independent variables. The reduced gradient $\nabla \hat{f}$ provides an estimate of the rate of change of $f(x)$ with respect to the independent variables when the dependent variables are adjusted to satisfy the linear approximations to the constraints.

Given $\nabla \bar{f}$, in the simplest version of GRF algorithm, the direction subvector for the independent variables $\bar{d}$ is selected to be the reduced gradient descent direction

$$\bar{d} = -\nabla \bar{f}$$

For a given step α in that direction, the constraints are solved iteratively to determine the value of the dependent variables $\hat{x}$ that will lead to a feasible point. Thus, the system

$$h_k(\bar{x}^0 + \alpha\bar{d}, \hat{x}) = 0, \qquad k = 1, \ldots, K$$

is solved for the K unknown variables $\hat{x}$. The new feasible point is checked to determine whether an improved objective value has been obtained and, if not, α is reduced and the solution for $\hat{x}$ repeated. The overall algorithm terminates when a point is reached at which the reduced gradient is sufficiently close to zero.

The GRG algorithm has been extended to accommodate inequality constraints as well as variable bounds. Moreover, the use of efficient equation solving procedures, line search procedures for α, and quasi-Newton formulas to generate improved direction vectors $\bar{d}$ have been investigated. A commercial quality GRG code will incorporate such developments and thus will constitute a reasonably complex software package. Computational testing using such codes indicates that GRG implementations are among the most robust and efficient general purpose nonlinear optimization methods currently available.[29] One of the particular advantages of this algorithm, which can be critical in engineering applications, is that it generates *feasible* intermediate points; hence, it can be interrupted prior to final convergence to yield a feasible solution. Of course, this attractive feature and the general efficiency of the method are attained at the price of providing (analytically or numerically) the values of the partial derivatives of all of the model functions.

Successive Quadratic Programming (SQP) Methods

This family of methods seeks to attain superior convergence rates by employing subproblems constructed using higher-order approximating functions than those employed by the linearization methods. The SQP methods are still the subject of active research; hence, developments and enhancements are proceeding apace. However, the basic form of the algorithm is well established and can be sketched out as follows.

At a given point x^0, a direction finding subproblem is constructed, which takes the form of a quadratic programming problem:

$$\text{Minimize} \quad \nabla^T f \cdot d + \tfrac{1}{2} d^T H d$$
$$\text{Subject to} \quad h_k(x^0) + \nabla^T h_k(x^0) \, d = 0$$
$$g_j(x^0) + \nabla^T g_j(x^0) \, d \geq 0$$

The symmetric matrix H is a quasi-Newton approximation of the matrix of second derivatives of a composite function (the Lagrangian) containing terms corresponding to all of the functions f, h_k, and g_j. H is updated using only gradient differences as in the unconstrained case. The direction vector d is used to conduct a line search, which seeks to minimize a penalty function of the type discussed earlier. The penalty function is required because, in general, the intermediate points produced in this method will be infeasible. Use of the penalty function ensures that improvements are achieved in either the objective function values or the constraint violations or both. One major advantage of the method is that very efficient methods are available for solving large quadratic programming problems and, hence, that the method is suitable for large scale applications. Recent computational testing indicates that the SQP approach is very efficient, outperforming even the best GRG codes.[30] However, it is restricted to models in which infeasibilities can be tolerated and will produce feasible solutions only when the algorithm has converged.

17.5.3   Code Availability

With the exception of the direct search methods and the transformation-type methods, the development of computer programs implementing state-of-the-art optimization algorithms is a major effort requiring expertise in numerical methods in general and numerical linear algebra in particular. For that reason, it is generally recommended that engineers involved in design optimization studies take advantage of the number of good quality implementations now available through various public sources.

Commercial computer codes for solving LP/IP/NLP problems are available from many computer manufacturers and private companies who specialize in marketing software for major computer systems. Depending on their capabilities, these codes vary in their complexity, ease of use, and cost (see, for example, Ref. 34). LP models with a few hundred constraints can now be solved on personal computers (PCs). There are now at least a hundred small companies marketing LP software for PCs. For a 1995 survey of LP software for personal computers, see Ref. 35.

Nash[36] presents a 1995 survey of nonlinear programming software that will run on PC compatibles, Macintosh systems, and UNIX-based workstations. Detailed product descriptions, prices, and capabilities of 30 NLP software are included in the survey. There are now LP/IP/NLP solvers that can be invoked directly from inside spreadsheet packages. For example, Microsoft Excel and Microsoft Office for Windows and Macintosh contain a general-purpose optimizer for solving small-scale linear, integer, and nonlinear programming problems. Borland's Quattro-Pro also has a built-in solver for optimization. In both spreadsheet programs, the LP optimizer is based on the simplex algorithm, while the NLP optimizer is based on the GRG algorithm.

There are now modeling languages that allow the user to express a model in a very compact algebraic form, with whole classes of constraints and variables defined over index sets. Models with thousands of constraints and variables can be defined in a couple of pages, in a syntax that is very close to standard algebraic notation. The algebraic form of the model is kept separate from the actual data for any particular instance of the model. The computer takes over the responsibility of transforming the abstract form of the model and the specific data into a specific constraint matrix. This has greatly simplified the building, and even more the changing, of optimization models. There are several modeling languages available for PCs. The two high-end products are GAMS (General Algebraic Modeling System) and AMPL (A Mathematical Programming Language). For a reference on GAMS, see Ref. 37. For a general introduction to modeling languages, see Refs. 34 and 38, and for an excellent discussion of AMPL, see Ref. 39.

Readers with access to the Internet can get a complete list of optimization software available for LP, IP, and NLP problems at the following NEOS web site:

http://www.mcs.anl.gov/home/otc

This site provides access not only to the software guide but also to the other optimization-related sites that are continually updated. The NEOS guide on optimization software is based on the textbook by More and Wright,[40] an excellent resource for those interested in a broad review of the various optimization methods and their computer codes. The book is divided into two parts. Part I has an overview of algorithms for different optimization problems, categorized as unconstrained optimization, nonlinear least squares, nonlinear equations, linear programming, quadratic programming, bound-constrained optimization, network optimization, and integer programming. Part II includes product descriptions of 75 software packages that implement the algorithms described in Part I. Much of the software described in this book is in the public domain and can be obtained through the Internet.

17.6 SUMMARY

In this chapter an overview was given of the elements and methods comprising design optimization methodology. The key element in the overall process of design optimization was seen to be the engineering model of the system constructed for this purpose. The assumptions and formulation details of the model govern the quality and relevance of the optimal design obtained. Hence, it is clear that design optimization studies cannot be relegated to optimization software specialists but are the proper domain of the well-informed design engineer.

The chapter also gave a structural classification of optimization problems and a broad brush review of the main families of optimization methods. Clearly this review can only hope to serve as entry point to this broad field. For a more complete discussion of optimization techniques with emphasis on engineering applications, guidelines for model formulation, practical solution strategies, and available computer software, the readers are referred to the text by Reklaitis, Ravindran, and Ragsdell.[25]

The Design Automation Committee of the Design Engineering Division of ASME has been sponsoring conferences devoted to engineering design optimization. Several of these presentations have subsequently appeared in the *Journal of Mechanical Design, ASME Transactions.* Ragsdell[31] presents a review of the papers published up to 1977 in the areas of machine design applications and numerical methods in design optimization. ASME published, in 1981, a special volume entitled *Progress in Engineering Optimization,* edited by Mayne and Ragsdell.[32] It contains several articles pertaining to advances in optimization methods and their engineering applications in the areas of mechanism design, structural design, optimization of hydraulic networks, design of helical springs, optimization of hydrostatic journal bearing, and others. Finally, the persistent and mathematically oriented reader may wish to pursue the fine exposition given by Avrial,[33] which explores the theoretical properties and issues of nonlinear programming methods.

REFERENCES

1. M. Zeleny, *Multiple Criteria Decision Making,* McGraw-Hill, New York, 1982.
2. T. L. Vincent and W. J. Grantham, *Optimality in Parametric Systems,* Wiley, New York, 1981.

3. K. E. Bett, J. S. Rowlinson, and G. Saville, *Thermodynamics for Chemical Engineers,* MIT Press, Cambridge, MA, 1975.

4. F. C. Jen, C. C. Pegels, and T. M. Dupuis, "Optimal Capacities of Production Facilities," *Management Science* **14B,** 570–580 (1968).

5. J. E. Shigley, *Mechanical Engineering Design,* McGraw-Hill, New York, 1973, p. 271.

6. S. Timoshenko and J. Gere, *Theory of Elastic Stability,* McGraw-Hill, New York, 1961, p. 257.

7. K. M. Ragsdell and D. T. Phillips, "Optimal Design of a Class of Welded Structures Using Geometric Programming," *ASME J. Eng. Ind. Ser. B* **98**(3), 1021–1025 (1975).

8. R. H. Philipson and A. Ravindran, "Application of Goal Programming to Machinability Data Optimization," *Journal of Mechanical Design, Trans. of ASME* **100,** 286–291 (1978).

9. E. J. A. Armarego and R. H. Brown, *The Machining of Metals,* Prentice-Hall, Englewood Cliffs, NJ, 1969.

10. J. L. Arthur and A. Ravindran, "PAGP-Partitioning Algorithm for (Linear) Goal Programming Problems," *ACM Transactions on Mathematical Software* **6,** 378–386 (1980).

11. R. H. Philipson and A. Ravindran, "Application of Mathematical Programming to Metal Cutting," *Mathematical Programming Study* **11,** 116–134 (1979).

12. S. M. Lee, *Goal Programming for Decision Analysis,* Auerbach Publishers, Philadelphia, PA, 1972.

13. N. K. Karmarkar, "A New Polynomial Time Algorithm for Linear Programming," *Combinatorica* **4,** 373–395 (1984).

14. A. Arbel, *Exploring Interior Point Linear Programming: Algorithms and Software,* MIT Press, Cambridge, MA, 1993.

15. S.-C. Fang and S. Puthenpura, *Linear Optimization and Extensions,* Prentice-Hall, NJ, 1993.

16. K. G. Murty, *Operations Research: Deterministic Optimization Models,* Prentice-Hall, Englewood Cliffs, 1995.

17. G. L. Nemhauser and L. A. Wolsey, *Integer and Combinatorial Optimization,* Wiley, New York, 1988.

18. C. S. Beightler and D. T. Phillips, *Applied Geometric Programming,* Wiley, New York, 1976.

19. M. J. Rijckaert, "Engineering Applications of Geometric Programming," in *Optimization and Design,* M. Avriel, M. J. Rijckaert, and D. J. Wilde (eds.), Prentice-Hall, Englewood Cliffs, NJ, 1974.

20. I. O. Bohachevsky, M. E. Johnson, and M. L. Stein, "Generalized Simulated Annealing for Function Optimization," *Technometrics* **28,** 209–217 (1986).

21. L. Davis (ed.), *Genetic Algorithms and Simulated Annealing,* Pitman, London, 1987.

22. S. Kirkpatrick, C. D. Gelatt, and M. P. Vecchi, "Optimization by Simulated Annealing," *Science,* **220,** 670–680 (1983).

23. D. E. Goldberg, *Genetic Algorithm in Search, Optimization, and Machine Learning,* Addison-Wesley, Reading, MA, 1989.

24. A. Maria, "Genetic Algorithms for Multimodal Continuous Optimization Problems," Ph.D. Diss., University of Oklahoma, Norman, OK, 1995.

25. G. V. Reklaitis, A. Ravindran, and K. M. Ragsdell, *Engineering Optimization: Methods and Applications,* Wiley, New York, 1983.

26. R. Fletcher, *Practical Methods of Optimization,* 2nd ed., Wiley, New York, 1987.

27. T. W. Lee and F. Fruedenstein, "Heuristic Combinatorial Optimization in the Kinematic Design of Mechanisms: Part 1: Theory," *J. Eng. Ind. Trans. ASME,* 1277–1280 (1976).

28. S. B. Schuldt, G. A. Gabriele, R. R. Root, E. Sandgren, and K. M. Ragsdell, "Application of a New Penalty Function Method to Design Optimization," *J. Eng. Ind. Trans. ASME,* 31–36 (1977).

29. E. Sandgren and K. M. Ragsdell, "The Utility of Nonlinear Programming Algorithms: A Comparative Study—Parts 1 and 2," *Journal of Mechanical Design, Trans. of ASME* **102,** 540–541 (1980).

30. K. Schittkowski, *Nonlinear Programming Codes: Information, Tests, Performance,* Lecture Notes in Economics and Mathematical Systems, Vol. 183, Springer-Verlag, New York, 1980.

31. K. M. Ragsdell, "Design and Automation," *Journal of Mechanical Design, Trans. of ASME* **102,** 424–429 (1980).

32. R. W. Mayne and K. M. Ragsdell (eds.), *Progress in Engineering Optimization,* ASME, New York, 1981.

33. M. Avriel, *Nonlinear Programming: Analysis and Methods,* Prentice-Hall, Englewood Cliffs, NJ, 1976.

34. R. Sharda, *Linear and Discrete Optimization and Modeling Software: A Resource Handbook,* Lionheart, Atlanta, GA, 1993.

35. R. Sharda, "Linear Programming Solver Software for Personal Computers: 1995 Report," *OR/MS Today* **22,** 49–57 (1995).

36. S. G. Nash, "Software Survey NLP," *OR/MS Today* **22,** 60–71 (1995).

37. A. Brooke, D. Kendrick, and A. Meeraus, *GAMS: A User's Guide,* Scientific Press, Redwood City, CA, 1988.

38. R. Sharda and G. Rampal, "Algebraic Modeling Languages on PC's," *OR/MS Today* **22,** 58–63, 1995.

39. R. Fourer, D. M. Gay, and B. W. Kernighan, "A Modelling Language for Mathematical Programming," *Management Science* **36,** 519–554 (1990).

40. J. J. More and S. J. Wright, *Optimization Software Guide,* SIAM Publications, Philadelphia, PA, 1993.

CHAPTER 18

FAILURE CONSIDERATIONS

Jack Collins
Department of Mechanical Engineering
Ohio State University
Columbus, Ohio

Steve Daniewicz
Department of Mechanical Engineering
Mississippi State University
Starkville, Mississippi

18.1	**CRITERIA OF FAILURE**	**377**		18.5.10	Damage Tolerance and Fracture Control	436
18.2	**FAILURE MODES**	**378**	**18.6**	**CREEP AND STRESS RUPTURE**	**437**
18.3	**ELASTIC DEFORMATION AND YIELDING**	**382**		18.6.1	Prediction of Long-Term Creep Behavior	439
				18.6.2	Creep under Uniaxial State of Stress	440
18.4	**FRACTURE MECHANICS AND UNSTABLE CRACK GROWTH**	**383**		18.6.3	Creep under Multiaxial State of Stress	442
				18.6.4	Cumulative Creep	442
18.5	**FATIGUE AND STRESS CONCENTRATION**	**396**				
	18.5.1 Fatigue Loading and Laboratory Testing	397	**18.7**	**COMBINED CREEP AND FATIGUE**	**443**
	18.5.2 The S–N–P Curves— A Basic Design Tool	401	**18.8**	**FRETTING AND WEAR**	**449**
	18.5.3 Factors That Affect S–N–P Curves	402		18.8.1 Fretting Phenomena	450
	18.5.4 Nonzero Mean and Multiaxial Fatigue Stresses	402		18.8.2 Wear Phenomena	456
	18.5.5 Spectrum Loading and Cumulative Damage	410	**18.9**	**CORROSION AND STRESS CORROSION**	**462**
	18.5.6 Stress Concentration	414		18.9.1 Types of Corrosion	463
	18.5.7 Low-Cycle Fatigue	420		18.9.2 Stress Corrosion Cracking	467
	18.5.8 Three-Phase Approach for Fatigue Life Prediction	429	**18.10**	**FAILURE ANALYSIS AND RETROSPECTIVE DESIGN**	**468**
	18.5.9 Service Spectrum Simulation and Full-Scale Testing	435				

18.1 CRITERIA OF FAILURE

Any change in the size, shape, or material properties of a structure, machine, or machine part that renders it incapable of performing its intended function must be regarded as a mechanical failure of the device. It should be carefully noted that the key concept here is that *improper functioning* of a machine part constitutes failure. Thus, a shear pin that does *not* separate into two or more pieces upon the application of a preselected overload must be regarded as having failed as surely as a drive shaft has failed if it *does* separate into two pieces under normal expected operating loads.

Mechanical Engineers' Handbook, 2nd ed., Edited by Myer Kutz.
ISBN 0-471-13007-9 © 1998 John Wiley & Sons, Inc.

Failure of a device or structure to function properly might be brought about by any one or a combination of many different responses to loads and environments while in service. For example, too much or too little elastic deformation might produce failure. A fractured load-carrying structural member or a shear pin that does not shear under overload conditions each would constitute failure. Progression of a crack due to fluctuating loads or aggressive environment might lead to failure after a period of time if resulting excessive deflection or fracture interferes with proper machine function.

A primary responsibility of any mechanical designer is to ensure that his or her design functions as intended for the prescribed design lifetime and, at the same time, that it be competitive in the marketplace. Success in designing competitive products while averting premature mechanical failures can be achieved consistently only by recognizing and evaluating all potential modes of failure that might govern the design. To recognize potential failure modes a designer must be acquainted with the array of failure modes observed in practice, and with the conditions leading to these failures. The following section summarizes the mechanical failure modes most commonly observed in practice, followed by a brief description of each one.

18.2 FAILURE MODES

A failure mode may be defined as the physical process or processes that take place or that combine their effects to produce a failure, as just discussed. In the following list of commonly observed failure modes it may be noted that some failure modes are unilateral phenomena, whereas others are combined phenomena. For example, fatigue is listed as a failure mode, corrosion is listed as a failure mode, and corrosion fatigue is listed as still another failure mode. Such combinations are included because they are commonly observed, important, and often *synergistic*. In the case of corrosion fatigue, for example, the presence of active corrosion aggravates the fatigue process and at the same time the presence of a fluctuating load accelerates the corrosion process.

The following list is not presented in any special order but it includes all commonly observed modes of mechanical failure:[1]

1. Force and/or temperature-induced elastic deformation.
2. Yielding.
3. Brinnelling.
4. Ductile rupture.
5. Brittle fracture.
6. Fatigue:
    a. High-cycle fatigue
    b. Low-cycle fatigue
    c. Thermal fatigue
    d. Surface fatigue
    e. Impact fatigue
    f. Corrosion fatigue
    g. Fretting fatigue
7. Corrosion:
    a. Direct chemical attack
    b. Galvanic corrosion
    c. Crevice corrosion
    d. Pitting corrosion
    e. Intergranular corrosion
    f. Selective leaching
    g. Erosion corrosion
    h. Cavitation corrosion
    i. Hydrogen damage
    j. Biological corrosion
    k. Stress corrosion
8. Wear:
    a. Adhesive wear
    b. Abrasive wear
    c. Corrosive wear
    d. Surface fatigue wear
    e. Deformation wear
    f. Impact wear

       g. Fretting wear
  9. Impact:
       a. Impact fracture
       b. Impact deformation
       c. Impact wear
       d. Impact fretting
       e. Impact fatigue
 10. Fretting:
       a. Fretting fatigue
       b. Fretting wear
       c. Fretting corrosion
 11. Creep.
 12. Thermal relaxation.
 13. Stress rupture.
 14. Thermal shock.
 15. Galling and seizure.
 16. Spalling.
 17. Radiation damage.
 18. Buckling.
 19. Creep buckling.
 20. Stress corrosion.
 21. Corrosion wear.
 22. Corrosion fatigue.
 23. Combined creep and fatigue.

As commonly used in engineering practice, the failure modes just listed may be defined and described briefly as follows. It should be emphasized that these failure modes only produce failure when they generate a set of circumstances that interferes with the proper functioning of a machine or device.

Force and/or temperature-induced elastic deformation failure occurs whenever the elastic (recoverable) deformation in a machine member, brought about by the imposed operational loads or temperatures, becomes large enough to interfere with the ability of the machine to perform its intended function satisfactorily.

Yielding failure occurs when the plastic (unrecoverable) deformation in a ductile machine member, brought about by the imposed operational loads or motions, becomes large enough to interfere with the ability of the machine to perform its intended function satisfactorily.

Brinnelling failure occurs when the static forces between two curved surfaces in contact result in local yielding of one or both mating members to produce a permanent surface discontinuity of significant size. For example, if a ball bearing is statically loaded so that a ball is forced to indent permanently the race through local plastic flow, the race is brinnelled. Subsequent operation of the bearing might result in intolerably increased vibration, noise, and heating; and, therefore, failure would have occurred.

Ductile rupture failure occurs when the plastic deformation, in a machine part that exhibits ductile behavior, is carried to the extreme so that the member separates into two pieces. Initiation and coalescence of internal voids slowly propagate to failure, leaving a dull, fibrous rupture surface.

Brittle fracture failure occurs when the elastic deformation, in a machine part that exhibits brittle behavior, is carried to the extreme so that the primary interatomic bonds are broken and the member separates into two or more pieces. Preexisting flaws or growing cracks form initiation sites for very rapid crack propagation to catastrophic failure, leaving a granular, multifaceted fracture surface.

Fatigue failure is a general term given to the sudden and catastrophic separation of a machine part into two or more pieces as a result of the application of fluctuating loads or deformations over a period of time. Failure takes place by the initiation and propagation of a crack until it becomes unstable and propagates suddenly to failure. The loads and deformations that typically cause failure by fatigue are far below the static failure levels. When loads or deformations are of such magnitude that more than about 10,000 cycles are required to produce failure, the phenomenon is usually termed *high-cycle fatigue.* When loads or deformations are of such magnitude that less than about 10,000 cycles are required to produce failure, the phenomenon is usually termed *low-cycle fatigue.* When load or strain cycling is produced by a fluctuating temperature field in the machine part, the process is usually termed *thermal fatigue. Surface fatigue* failure, usually associated with rolling surfaces in contact, manifests itself as pitting, cracking, and spalling of the contacting surfaces as a result of the cyclic Hertz contact stresses that result in maximum values of cyclic shear stresses slightly below the surface. The cyclic subsurface shear stresses generate cracks that propagate to the contacting surface, dislodging particles in the process to produce surface pitting. This phenomenon is often viewed as a type of wear. Impact fatigue, corrosion fatigue, and fretting fatigue are described later.

Corrosion failure, a very broad term, implies that a machine part is rendered incapable of performing its intended function because of the undesired deterioration of the material as a result of chemical or electrochemical interaction with the environment. Corrosion often interacts with other failure modes such as wear or fatigue. The many forms of corrosion include the following. *Direct chemical attack,* perhaps the most common type of corrosion, involves corrosive attack of the surface of the machine part exposed to the corrosive media, more or less uniformly over the entire exposed surface. *Galvanic corrosion* is an accelerated electrochemical corrosion that occurs when two dissimilar metals in electrical contact are made part of a circuit completed by a connecting pool or film of electrolyte or corrosive medium, leading to current flow and ensuing corrosion. *Crevice corrosion* is the accelerated corrosion process highly localized within crevices, cracks, or joints where small volume regions of stagnant solution are trapped in contact with the corroding metal. *Pitting corrosion* is a very localized attack that leads to the development of an array of holes or pits that penetrate the metal. *Intergranular corrosion* is the localized attack occurring at grain boundaries of certain copper, chromium, nickel, aluminum, magnesium, and zinc alloys when they are improperly heat treated or welded. Formation of local galvanic cells that precipitate corrosion products at the grain boundaries seriously degrades the material strength because of the intergranular corrosive process. *Selective leaching* is a corrosion process in which one element of a solid alloy is removed, such as in dezincification of brass alloys or graphitization of gray cast irons. *Erosion corrosion* is the accelerated chemical attack that results when abrasive or viscid material flows past a containing surface, continuously baring fresh, unprotected material to the corrosive medium. *Cavitation corrosion* is the accelerated chemical corrosion that results when, because of differences in vapor pressure, certain bubbles and cavities within a fluid collapse adjacent to the pressure-vessel walls, causing particles of the surface to be expelled, baring fresh, unprotected surface to the corrosive medium. *Hydrogen damage,* while not considered to be a form of direct corrosion, is induced by corrosion. Hydrogen damage includes hydrogen blistering, hydrogen embrittlement, hydrogen attack, and decarburization. *Biological corrosion* is a corrosion process that results from the activity of living organisms, usually by virtue of their processes of food ingestion and waste elimination, in which the waste products are corrosive acids or hydroxides. *Stress corrosion,* an extremely important type of corrosion, is described separately later.

Wear is the undesired cumulative change in dimensions brought about by the gradual removal of discrete particles from contacting surfaces in motion, usually sliding, predominantly as a result of mechanical action. Wear is not a single process, but a number of different processes that can take place by themselves or in combination, resulting in material removal from contacting surfaces through a complex combination of local shearing, plowing, gouging, welding, tearing, and others. *Adhesive wear* takes place because of high local pressure and welding at asperity contact sites, followed by motion-induced plastic deformation and rupture of asperity functions, with resulting metal removal or transfer. *Abrasive wear* takes place when the wear particles are removed from the surface by the plowing, gouging, and cutting action of the asperities of a harder mating surface or by hard particles entrapped between the mating surfaces. When the conditions for either adhesive wear or abrasive wear coexist with conditions that lead to corrosion, the processes interact synergistically to produce *corrosive wear.* As described earlier, *surface fatigue wear* is a wear phenomenon associated with curved surfaces in rolling or sliding contact, in which subsurface cyclic shear stresses initiate microcracks that propagate to the surface to spall out macroscopic particles and form wear pits. *Deformation wear* arises as a result of repeated *plastic* deformation at the wearing surfaces, producing a matrix of cracks that grow and coalesce to form wear particles. Deformation wear is often caused by severe impact loading. *Impact wear* is impact-induced repeated *elastic* deformation at the wearing surface that produces a matrix of cracks that grows in accordance with the surface fatigue description just given. Fretting wear is described later.

Impact failure results when a machine member is subjected to nonstatic loads that produce in the part stresses or deformations of such magnitude that the member no longer is capable of performing its function. The failure is brought about by the interaction of stress or strain waves generated by dynamic or suddenly applied loads, which may induce local stresses and strains many times greater than would be induced by the static application of the same loads. If the magnitudes of the stresses and strains are sufficiently high to cause separation into two or more parts, the failure is called *impact fracture.* If the impact produces intolerable elastic or plastic deformation, the resulting failure is called *impact deformation.* If repeated impacts induce cyclic elastic strains that lead to initiation of a matrix of fatigue cracks, which grows to failure by the surface fatigue phenomenon described earlier, the process is called *impact wear.* If fretting action, as described in the next paragraph, is induced by the small lateral relative displacements between two surfaces as they impact together, where the small displacements are caused by Poisson strains or small tangential "glancing" velocity components, the phenomenon is called *impact fretting. Impact fatigue* failure occurs when impact loading is applied repetitively to a machine member until failure occurs by the nucleation and propagation of a fatigue crack.

Fretting action may occur at the interface between any two solid bodies whenever they are pressed together by a normal force and subjected to small-amplitude cyclic relative motion with respect to each other. Fretting usually takes place in joints that are not intended to move but, because of vibrational loads or deformations, experience minute cyclic relative motions. Typically, debris produced by fretting action is trapped between the surfaces because of the small motions involved. *Fretting fatigue* failure is the premature fatigue fracture of a machine member subjected to fluctuating loads or strains together with conditions that simultaneously produce fretting action. The surface discontinuities and microcracks generated by the fretting action act as fatigue crack nuclei that propagate to failure under conditions of fatigue loading that would otherwise be acceptable. Fretting fatigue failure is an insidious failure mode because the fretting action is usually hidden within a joint where it cannot be seen and leads to premature, or even unexpected, fatigue failure of a sudden and catastrophic nature. *Fretting wear* failure results when the changes in dimensions of the mating parts, because of the presence of fretting action, become large enough to interfere with proper design function or large enough to produce geometrical stress concentration of such magnitude that failure ensues as a result of excessive local stress levels. *Fretting corrosion* failure occurs when a machine part is rendered incapable of performing its intended function because of the surface degradation of the material from which the part is made, as a result of fretting action.

Creep failure results whenever the plastic deformation in a machine member accrues over a period of time under the influence of stress and temperature until the accumulated dimensional changes interfere with the ability of the machine part to perform satisfactorily its intended function. Three stages of creep are often observed: (1) transient or primary creep during which time the rate of strain decreases, (2) steady-state or secondary creep during which time the rate of strain is virtually constant, and (3) tertiary creep during which time the creep strain rate increases, often rapidly, until rupture occurs. This terminal rupture is often called creep rupture and may or may not occur, depending on the stress–time–temperature conditions.

Thermal relaxation failure occurs when the dimensional changes due to the creep process result in the relaxation of a prestrained or prestressed member until it no longer is able to perform its intended function. For example, if the prestressed flange bolts of a high-temperature pressure vessel relax over a period of time because of creep in the bolts, so that, finally, the peak pressure surges exceed the bolt preload to violate the flange seal, the bolts will have failed because of thermal relaxation.

Stress rupture failure is intimately related to the creep process except that the combination of stress, time, and temperature is such that rupture into two parts is ensured. In stress rupture failures the combination of stress and temperature is often such that the period of steady-state creep is short or nonexistent.

Thermal shock failure occurs when the thermal gradients generated in a machine part are so pronounced that differential thermal strains exceed the ability of the material to sustain them without yielding or fracture.

Galling failure occurs when two sliding surfaces are subjected to such a combination of loads, sliding velocities, temperatures, environments, and lubricants that massive surface destruction is caused by welding and tearing, plowing, gouging, significant plastic deformation of surface asperities, and metal transfer between the two surfaces. Galling may be thought of as a severe extension of the adhesive wear process. When such action results in significant impairment to intended surface sliding or in seizure, the joint is said to have failed by galling. *Seizure* is an extension of the galling process to such severity that the two parts are virtually welded together so that relative motion is no longer possible.

Spalling failure occurs whenever a particle is spontaneously dislodged from the surface of a machine part so as to prevent the proper function of the member. Armor plate fails by spalling, for example, when a striking missile on the exposed side of an armor shield generates a stress wave that propagates across the plate in such a way as to dislodge or spall a secondary missile of lethal potential on the protected side. Another example of spalling failure is manifested in rolling contact bearings and gear teeth because of the action of surface fatigue as described earlier.

Radiation damage failure occurs when the changes in material properties induced by exposure to a nuclear radiation field are of such a type and magnitude that the machine part is no longer able to perform its intended function, usually as a result of the triggering of some other failure mode, and often related to loss in ductility associated with radiation exposure. Elastomers and polymers are typically more susceptible to radiation damage than are metals, whose strength properties are sometimes enhanced rather than damaged by exposure to a radiation field, although ductility is usually decreased.

Buckling failure occurs when, because of a critical combination of magnitude and/or point of load application, together with the geometrical configuration of a machine member, the deflection of the member suddenly increases greatly with only a slight change in load. This nonlinear response results in a buckling failure if the buckled member is no longer capable of performing its design function.

Creep buckling failure occurs when, after a period of time, the creep process results in an unstable combination of the loading and geometry of a machine part so that the critical buckling limit is exceeded and failure ensues.

Stress corrosion failure occurs when the applied stresses on a machine part in a corrosive environment generate a field of localized surface cracks, usually along grain boundaries, that render the part incapable of performing its function, often because of triggering some other failure mode. Stress corrosion is a very important type of corrosion failure mode because so many different metals are susceptible to it. For example, a variety of iron, steel, stainless-steel, copper, and aluminum alloys are subject to stress corrosion cracking if placed in certain adverse corrosive media.

Corrosion wear failure is a combination failure mode in which corrosion and wear combine their deleterious effects to incapacitate a machine part. The corrosion process often produces a hard, abrasive corrosion product that accelerates the wear, while the wear process constantly removes the protective corrosion layer from the surface, baring fresh metal to the corrosive medium and thus accelerating the corrosion. The two modes combine to make the result more serious than either of the modes would have been otherwise.

Corrosion fatigue is a combination failure mode in which corrosion and fatigue combine their deleterious effects to cause failure of a machine part. The corrosion process often forms pits and surface discontinuities that act as stress raisers which in turn accelerate fatigue failure. Furthermore, cracks in the usually brittle corrosion layer also act as fatigue crack nuclei that propagate into the base material. On the other hand, the cyclic loads or strains cause cracking and flaking of the corrosion layer, which bares fresh metal to the corrosive medium. Thus, each process accelerates the other, often making the result disproportionately serious.

Combined creep and fatigue failure is a combination failure mode in which all of the conditions for both creep failure and fatigue exist simultaneously, each process influencing the other to produce failure. The interaction of creep and fatigue is probably synergistic but is not well understood.

18.3   ELASTIC DEFORMATION AND YIELDING

Small changes in the interatomic spacing of a material, brought about by applied forces or changing temperatures, are manifested macroscopically as elastic strain. Although the maximum elastic strain in crystalline solids, including engineering metals, is typically very small, the force required to produce the small strain is usually large; hence, the accompanying stress is large. On the other hand, certain other noncrystalline materials such as elastomers may exhibit recoverable (but not necessarily linear) strains of several hundred percent. For uniaxial loading of a machine or structural element, the total elastic deformation of the member may be found by integrating the elastic strain over the length of the element. Thus, for a uniform bar subjected to uniaxial loading the total deformation of the bar in the axial direction is

$$\Delta l = l\epsilon \tag{18.1}$$

where Δl is total axial deformation of the bar, l is the original bar length, and ϵ is the axial elastic strain. If Δl exceeds the design-allowable axial deformation, failure will occur. For example, if the axial deformation of an aircraft gas-turbine blade, due to the centrifugal force field, exceeds the tip clearance gap, failure will occur because of force-induced elastic deformation. Likewise, if thermal expansion of the blade produces a blade-axial deformation that exceeds the tip clearance gap, failure will occur because of temperature-induced elastic deformation.

When the state of stress is more complicated, it becomes necessary to calculate the elastic strains induced by the multiaxial states of stress in three mutually perpendicular directions through the use of the generalized Hooke's law equations given by

$$\epsilon_x = \frac{1}{E}\left[\sigma_x - \nu(\sigma_y + \sigma_z)\right]$$

$$\epsilon_y = \frac{1}{E}\left[\sigma_y - \nu(\sigma_x + \sigma_z)\right] \tag{18.2}$$

$$\epsilon_z = \frac{1}{E}\left[\sigma_z - \nu(\sigma_x + \sigma_y)\right]$$

where σ_x, σ_y, and σ_z are the normal stresses in the three coordinate directions, E and ν are Young's modulus and Poisson's ratio, respectively, and ϵ_x, ϵ_y, and ϵ_z are the elastic strains in the three coordinate directions. Again, total elastic deformation of a member in any of the coordinate directions may be found by integrating the strain over the member's length in that direction. If the change in length of the member in any direction exceeds the design-allowable deformation in that direction, failure will occur. The use of commercial finite element analysis software packages is one commonly used means of determining both the elastic strains produced in a structural element and the subsequent elastic deformations produced.

If applied loads reach certain critical levels, the atoms within the microstructure may be moved into new equilibrium positions and the induced strains are not fully recovered upon release of the loads. Such permanent strains, usually the result of slip, are called plastic strains, and the macroscopic permanent deformation due to plastic strain is called yielding. If applied loads are increased even more, the plastic deformation process may be carried to the point of instability where *necking* begins: internal voids form and slowly coalesce to finally produce a ductile rupture of the loaded member.

After plastic deformation has been initiated, the Hooke's law equations (18.2) are no longer valid and the predictions of plastic strains and deformations under multiaxial states of stress are more difficult. If a designer can tolerate a prescribed plastic deformation without experiencing failure, these plastic deformations may be determined using plasticity theory. Many commercial finite element analysis software packages now posses the capability to compute both plastic strains and deformations for a prescribed nonlinear elastic-plastic constitutive relation.

For the case of simple uniaxial loading, the onset of yielding may be accurately predicted to occur when the uniaxial maximum normal stress reaches a value equal to the yield point strength of the material read from an engineering stress–strain curve. If the loading is more complicated, and a multiaxial state of stress is produced by the loads, the onset of yielding may no longer be predicted by comparing any one of the normal stress components with uniaxial material yield strength, not even the maximum principal normal stress. Onset of yielding for multiaxially stressed critical points in a machine or structure is more accurately predicated through the use of a *combined stress theory of failure,* which has experimentally been validated for the prediction of yielding. The two most widely accepted theories for predicting the onset of yielding are the distortion energy theory (also called the octahedral shear stress theory or the Huber–von Mises–Hencky theory) and the maximum shearing stress theory. The distortion energy theory is somewhat more accurate while the maximum shearing stress theory may be slightly easier to use.

In words, the distortion energy theory may be expressed as follows:

Failure is predicted to occur in the multiaxial state of stress when the distortion energy per unit volume becomes equal to or exceeds the distortion energy per unit volume at the time of failure in a simple uniaxial stress test using a specimen of the same material.

Mathematically, the distortion energy theory may be formulated as

Failure is predicted by the distortion energy theory to occur if

$$\tfrac{1}{2}[(\sigma_1 - \sigma_2)^2 + (\sigma_2 - \sigma_3)^2 + (\sigma_3 - \sigma_1)^2] \geq \sigma_f^2 \tag{18.3}$$

The maximum shearing stress theory may be stated in words as:

Failure is predicted to occur in the multiaxial state of stress when the maximum shearing stress magnitude becomes equal to or exceeds the maximum shearing stress magnitude at the time of failure in a simple uniaxial stress test using a specimen of the same material.

Mathematically, the maximum shearing stress theory becomes:

Failure is predicted by the maximum shearing stress theory to occur if

$$\sigma_1 - \sigma_3 \geq \sigma_f \tag{18.4}$$

where σ_1, σ_2, and σ_3 are the principal stresses at a point, ordered such that $\sigma_1 \geq \sigma_2 \geq \sigma_3$, and σ_f is the uniaxial failure strength in tension.

Comparisons of these two failure theories with experimental data on yielding are shown in Fig. 18.1 for a variety of materials and different biaxial states of stress.

18.4 FRACTURE MECHANICS AND UNSTABLE CRACK GROWTH

When the material behavior is brittle rather than ductile, the mechanics of the failure process are much different. Instead of the slow coalescence of voids associated with ductile rupture, brittle fracture proceeds by the high-velocity propagation of a crack across the loaded member. If the material behavior is clearly brittle, fracture may be predicted with reasonable accuracy through use of the maximum normal stress theory of failure. In words, the maximum normal stress theory may be expressed as follows:

Failure is predicted to occur in the multiaxial state of stress when the maximum principal normal stress becomes equal to or exceeds the maximum normal stress at the time of failure in a simple uniaxial stress test using a specimen of the same material.

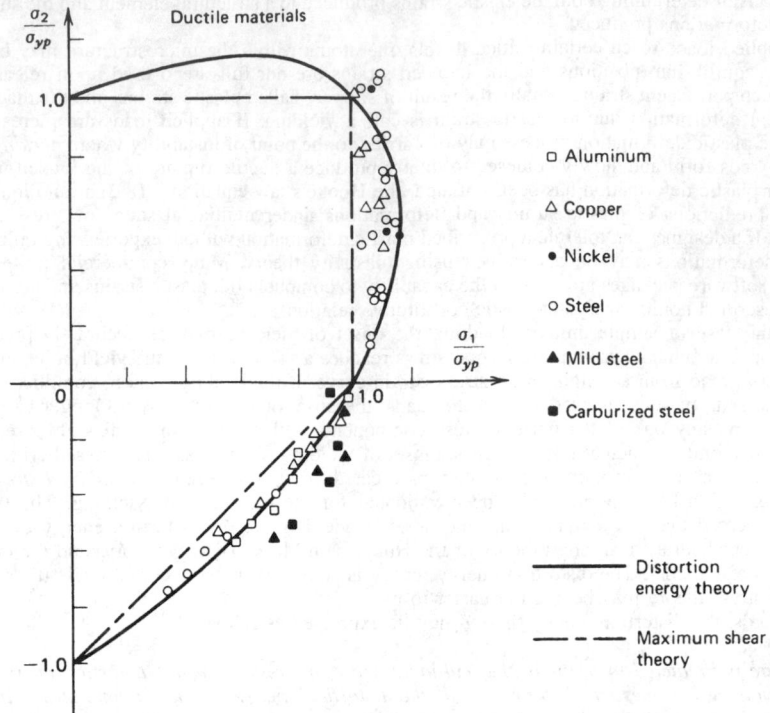

Fig. 18.1 Comparison of biaxial yield strength data with theories of failure for a variety of ductile materials.

Mathematically, the maximum normal stress theory becomes:

Failure is predicted by the maximum normal stress theory to occur if

$$\sigma_1 > \sigma_t \qquad \sigma_3 \le \sigma_c \qquad (18.5)$$

where σ_1, σ_2, and σ_3 are the principal stresses at a point, ordered such that $\sigma_1 \ge \sigma_2 \ge \sigma_3$, σ_t is the uniaxial failure strength in tension, and σ_c is the uniaxial failure strength in compression. Comparison of this failure theory with experimental data on brittle fracture for different biaxial states of stress is shown in Fig. 18.2.

On the other hand, more recent experience has led to the understanding that nominally ductile materials may also fail by a brittle fracture response in the presence of cracks or flaws if the combination of crack size, geometry of the part, temperature, and/or loading rate lies within certain critical regions. Furthermore, the development of higher-strength structural alloys, the wider use of welding, and the use of thicker sections in some cases have combined their influence to reduce toward a critical level the capacity of some structural members to accommodate local plastic strain without fracture. At the same time, fabrication by welding, residual stresses due to machining, and assembly mismatch in production have increased the need for accommodating local plastic strain to prevent failure. Fluctuating service loads of greater severity and more aggressive environments have also contributed to unexpected fractures. From the study of all these factors the basic concepts of *fracture control* were conceived and developed. Fracture control consists, simply, of controlling the nominal stress and crack size so that the combination always lies below a critical level for the material being used in a given design application.

An important observation in studying fracture behavior is that the magnitude of the nominal applied stress that causes fracture is related to the size of the crack or cracklike flaw within the structure.[2] For example, observations of the behavior of central through-the-thickness cracks, oriented normal to the applied tensile stress, in steel and aluminum plates, yielded the results shown in Figs. 18.3 and 18.4. In these tests, as the tensile loading on the precracked plates was slowly increased, the crack extension slowly increased for a time and then abruptly extended to failure by rapid crack

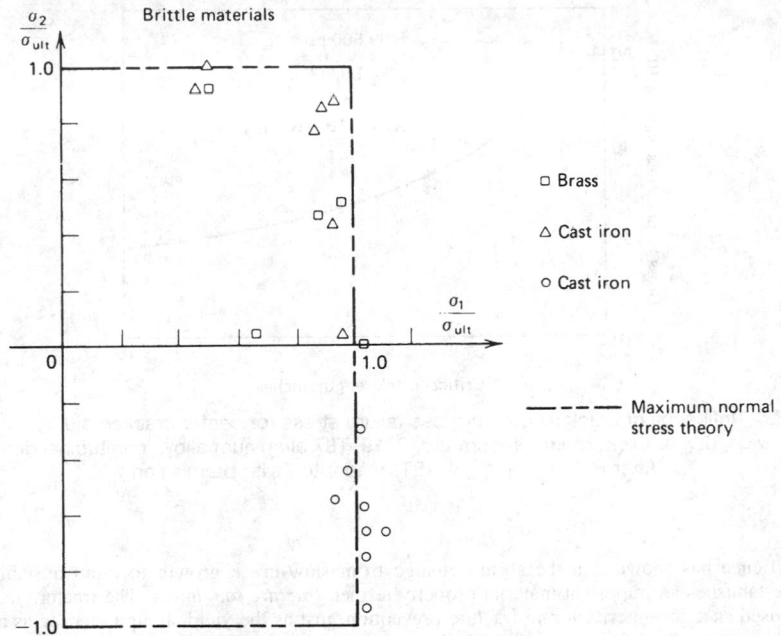

Fig. 18.2   Comparison of biaxial brittle fracture strength data with maximum normal stress theory for several brittle materials.

propagation. Slow stable crack growth was characterized by speeds of the order of fractions of an inch per minute. Rapid crack propagation was characterized by speeds of the order of hundreds of feet per second. The data of Figs. 18.3 and 18.4 indicate that for longer initial crack length the fracture stress, that is, the stress corresponding to the onset of rapid crack extension, was lower. For the aluminum alloy the fracture stress was less than the yield strength for cracks longer than about 0.75 in. For the steel alloy the fracture stress was less than the yield strength for cracks longer than about 0.5 in. In both cases, for shorter cracks the fracture stress approaches the ultimate strength of the material determined from a conventional uniaxial tension test.

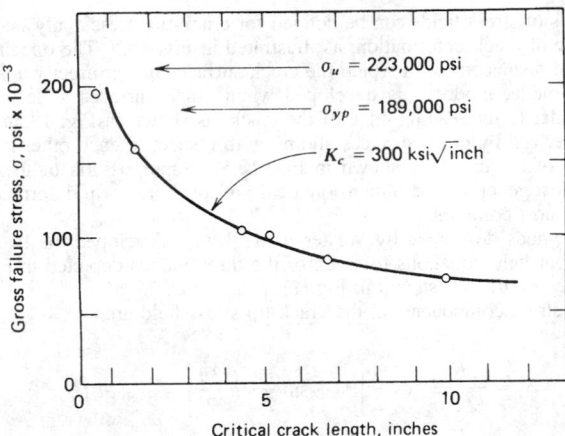

Fig. 18.3   Influence of crack length on gross failure stress for center cracked steel plate, 36 in. wide, 0.14 in. thick, room temperature, 4330 M steel, longitudinal direction. (After Ref. 2, copyright ASTM: adapted with permission.)

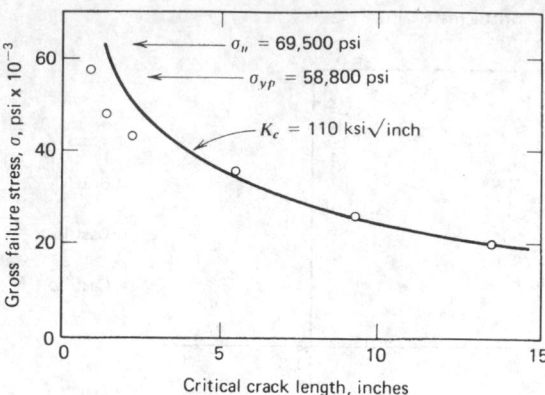

Fig. 18.4 Influence of crack length on gross failure stress for center cracked aluminum plate, 24 in. wide, 0.1 in. thick, room temperature, 2219–T87 aluminum alloy, longitudinal direction. (After Ref. 2, copyright ASTM, adapted with permission.)

Experience has shown that the abrupt change from slow crack growth to rapid unstable crack growth establishes an important material property termed *fracture toughness*. The fracture toughness may be used as a design criterion in fracture prevention, just as the yield strength is used as a design criterion in prevention of yielding of a ductile material under static loading.

In many cases slow crack propagation is also of interest, especially under conditions of fluctuating loads and/or aggressive environments. In analyses and predictions involving fatigue failure phenomena, characterization of the rate of slow crack extension and the initial flaw size, together with critical crack size, are used to determine the useful life of a component or structure subjected to fluctuating loads. The topic of fatigue crack propagation is discussed further in Section 18.5.

The simplest useful model for stress at the tip of a crack is based on the assumptions of linear elastic material behavior and a two-dimensional analysis; thus, the procedure is often referred to as linear elastic fracture mechanics. Although the validity of the linear elastic assumption may be questioned in view of plastic zone formation at the tip of a crack in any real engineering material, as long as "small-scale yielding" occurs, that is, as long as the plastic zone size remains small compared to the dimensions of the crack, the linear elastic model gives good engineering results. Thus, the small-scale yielding concept implies that the small plastic zone is confined within a linear elastic field surrounding the crack tip. If the material properties, section size, loading conditions, and environment combine in such a way that "large-scale" plastic zones are formed, the basic assumptions of linear elastic fracture mechanics are violated, and elastic–plastic fracture mechanics methods must be employed.

Three basic types of stress fields can be defined for crack-tip stress analysis, each one associated with a distinct mode of crack deformation, as illustrated in Fig. 18.5. The opening mode, mode I, is associated with local displacement in which the crack surfaces move directly apart, as shown in Fig. 18.5a. The sliding mode, mode II, is developed when crack surfaces slide over each other in a direction perpendicular to the leading edge of the crack, as shown in Fig. 18.5b. The tearing mode, mode III, is characterized by crack surfaces sliding with respect to each other in a direction parallel to the leading edge of the crack, as shown in Fig. 18.5c. Superposition of these three modes will fully describe the most general three-dimensional case of local crack-tip deformation and stress field, although mode I is most common.

Based on the methods developed by Westergaard,[3] Irwin[4] developed the two-dimensional stress field and displacement field equations for each of the three modes depicted in Fig. 18.5, expressing them in terms of the coordinates shown in Fig. 18.6.

For mode I, the stress components in the crack-tip stress field are

$$\sigma_x = \frac{K_I}{\sqrt{2\pi r}} \cos \frac{\theta}{2} \left[1 - \sin \frac{\theta}{2} \sin \frac{3\theta}{2} \right] + \sigma_{x0} + [O]r^{1/2} \tag{18.6}$$

$$\sigma_y = \frac{K_I}{\sqrt{2\pi r}} \cos \frac{\theta}{2} \left[1 + \sin \frac{\theta}{2} \sin \frac{3\theta}{2} \right] + [O]r^{1/2} \tag{18.7}$$

$$\tau_{xy} = \frac{K_I}{\sqrt{2\pi r}} \sin \frac{\theta}{2} \cos \frac{\theta}{2} \cos \frac{3\theta}{2} + [O]r^{1/2} \tag{18.8}$$

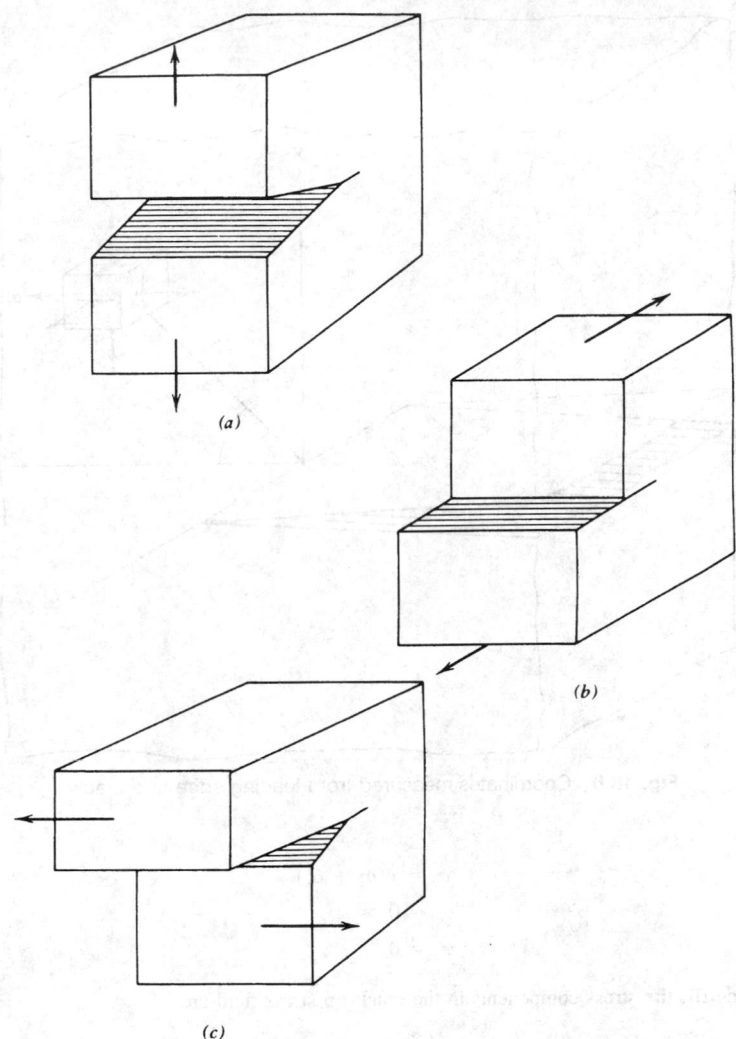

Fig. 18.5 Basic modes of crack displacement: (a) mode I; (b) model II; (c) mode III.

For conditions of plane strain, where displacements in the z direction are constrained to be zero (thick members), the remaining three stress components are

$$\sigma_z = \nu(\sigma_x + \sigma_y) \tag{18.9}$$

$$\tau_{xz} = 0 \tag{18.10}$$

$$\tau_{yz} = 0 \tag{18.11}$$

For mode II, the stress components in the crack-tip stress field are

$$\sigma_x = \frac{-K_{\mathrm{II}}}{\sqrt{2\pi r}} \sin\frac{\theta}{2}\left[2 + \cos\frac{\theta}{2}\cos\frac{3\theta}{2}\right] + \sigma_{x0} + [O]r^{1/2} \tag{18.12}$$

$$\sigma_y = \frac{K_{\mathrm{II}}}{\sqrt{2\pi r}} \sin\frac{\theta}{2}\cos\frac{\theta}{2}\cos\frac{3\theta}{2} + [O]r^{1/2} \tag{18.13}$$

$$\tau_{xy} = \frac{K_{\mathrm{II}}}{\sqrt{2\pi r}} \cos\frac{\theta}{2}\left[1 - \sin\frac{\theta}{2}\sin\frac{3\theta}{2}\right] + [O]r^{1/2} \tag{18.14}$$

and, for plane strain conditions,

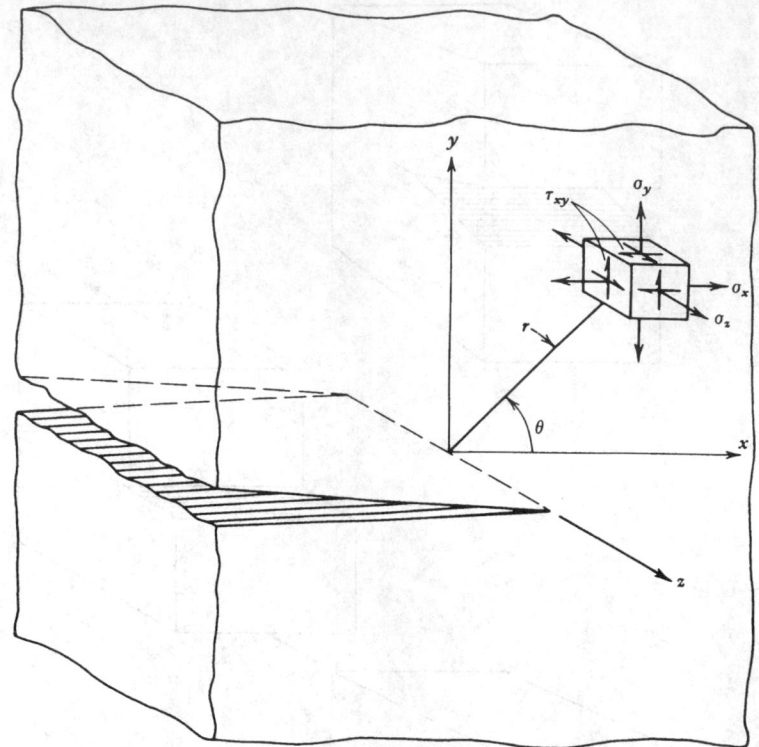

Fig. 18.6 Coordinates measured from leading edge of a crack.

$$\sigma_z = \nu(\sigma_x + \sigma_y) \tag{18.15}$$

$$\tau_{xy} = 0 \tag{18.16}$$

$$\tau_{yz} = 0 \tag{18.17}$$

For mode III, the stress components in the crack-tip stress field are

$$\tau_{xz} = \frac{-K_{\mathrm{III}}}{\sqrt{2\pi r}} \sin\frac{\theta}{2} + \tau_{xz0} + [O]r^{1/2} \tag{18.18}$$

$$\tau_{yz} = \frac{K_{\mathrm{III}}}{\sqrt{2\pi r}} \cos\frac{\theta}{2} + [O]r^{1/2} \tag{18.19}$$

$$\sigma_x = \sigma_y = \sigma_z = \tau_{xy} = 0 \tag{18.20}$$

In expressions (18.6) through (18.20), higher-order terms such as the uniform stresses parallel to cracks σ_{x0} and τ_{xz0}, and terms of the order of $r^{1/2}$, that is, $[O]r^{1/2}$, are indicated. These terms are usually neglected as higher order compared to the leading $1/\sqrt{r}$ term. The parameters K_{I}, K_{II}, and K_{III} are called crack-tip stress field intensity factors, or simply *stress intensity factors*. They represent the *strength* of the stress field surrounding the tip of the crack. Physically, K_{I}, K_{II}, and K_{III} may be interpreted as the intensity of load transmittal through the crack-tip region, induced by the introduction of a crack into a flaw-free body. Since fracture is induced by the crack-tip stress field, the stress intensity factors are primary correlation parameters in current practice.

For remote loadings, in general, the expressions for the stress intensity factor are of the form

$$K = C\sigma\sqrt{\pi a} \tag{18.21}$$

where C is dependent on the type of loading and the geometry away from the crack. Much work has been completed in determining values of C for a wide variety of conditions. (See, for example, Ref. 5.)

Many commercial finite element analysis software packages possess special crack tip elements allowing the numerical computation of stress intensity factors. A discussion of some of the techniques employed within these software packages is given by Anderson.[6] Through the use of weight functions,[7] stress intensity factors may be computed using numerical integration. This analysis technique is ideally suited for use with personal computers and will be discussed in more detail in the following paragraphs.

For a given cracked plate, for example, Fig. 18.6, the factor K increases proportionally with gross nominal stress and also is a function of the instantaneous crack length. Thus, K is a single-parameter measure of the stress field around the crack tip. The value of K associated with the onset of rapid crack extension has been designated the *critical stress intensity, K_c*. As noted earlier in Figs. 18.3 and 18.4, the onset of rapid crack propagation for specimens with different initial crack lengths occurs at different values of gross-section stress, but at a constant value of K_c. Thus, K_c provides a single-parameter fracture criterion that allows the prediction of fracture based on (18.21). That is, *fracture is predicted to occur if*

$$C\sigma\sqrt{\pi a} \geq K_c \tag{18.22}$$

In studying material behavior, one finds that for a given material, depending on the state of stress at the crack tip, the critical stress intensity K_c decreases to a lower limiting value as the state of strain approaches the condition of plane strain. This lower limiting value defines a basic material property K_{Ic}, the *plane strain fracture toughness* for the material. Standard test methods have been established for the determination of K_{Ic} values.[8] A few data are shown in Table 18.1. Useful compilations of fracture toughness values have been prepared by several organizations and individuals. These include Refs. 10–16.

For the plane strain fracture toughness K_{Ic} to be a valid failure prediction criterion for a specimen or a machine part, plane strain conditions must exist at the crack tip; that is, the material must be *thick* enough to ensure plane strain conditions. It has been estimated empirically that for plane strain conditions the minimum material thickness B must be

$$B \geq 2.5 \left(\frac{K_{Ic}}{\sigma_{yp}}\right)^2 \tag{18.23}$$

where σ_{yp} is the material yield strength.

If the material is not thick enough to meet the criterion of (18.23), plane stress is a more likely state of stress at the crack tip; and K_c, the critical stress intensity factor for failure prediction under plane stress conditions, may be estimated using a semi-empirical relationship for K_c.[17,18]

$$K_C = K_{IC} \left[1 + \frac{1.4}{B^2} \left(\frac{K_{IC}}{\sigma_{yp}}\right)^4 \right]^{1/2} \tag{18.24}$$

As long as the crack-tip plastic zone remains in the regime of small-scale yielding, this estimation

Table 18.1 Yield Strength and Plane Strain Fracture Toughness Data for Selected Engineering Alloys[9,10]

Alloy	Form	Test Temperature °F	Test Temperature °C	σ_{yp} ksi	σ_{yp} MPa	K_{Ic} ksi$\sqrt{\text{in}}$	K_{Ic} MPa$\sqrt{\text{m}}$
4340 (500°F temper) steel	Plate	70	21	217–238	1495–1640	45–57	50–63
4340 (800°F temper) steel	Forged	70	21	197–211	1360–1455	72–83	79–91
D6AC (1000°F temper) steel	Plate	70	21	217	1495	93	102
D6AC (1000°F temper) steel	Plate	−65	−54	228	1570	56	62
A 538 steel				250	1722	100	111
2014-T6 aluminum	Forged	75	24	64	440	28	31
2024-T351 aluminum	Plate	80	27	54–56	370–385	28–40	31–44
7075-T6 aluminum				85	585	30	33
7075-T651 aluminum	Plate	70	21	75–81	515–560	25–28	27–31
7075-T7351 aluminum	Plate	70	21	58–66	400–455	28–32	31–35
Ti-6Al-4V titanium	Plate	74	23	119	820	96	106

procedure provides a good design approach. For conditions that result in large crack-tip plastic zones (large applied stresses, large crack lengths), performing a failure assessment using linear elastic fracture mechanics (LEFM) is invalid and potentially nonconservative. A general rule of thumb is that plasticity effects become significant when the applied stresses approach 50% of the yield stress, but this is by no means a universal rule.[6] When small-scale yielding is not generated at the crack tip, a better design approach would involve the implementation of an appropriate elastic-plastic fracture mechanics procedure, such as a *failure assessment diagram* (FAD). This methodology will be discussed in more detail in the following paragraphs.

A more useful estimate defining the limits of LEFM applicability may be formulated using the fracture toughness test specimen size requirements set forth by the ASTM.[8] For a valid plane strain fracture toughness test, one implying that LEFM is valid, the crack size a must satisfy $a > 2.5$ $(K_{IC}/\sigma_{yp})^2$. From Eq. (18.7), with $\theta = 0$, the stress σ_y will equal the yield stress σ_{yp} when

$$r_y = \alpha \left(\frac{K_I}{\sigma_{yp}}\right)^2 \tag{18.25}$$

where $\alpha = 1/2\pi$. This equation is an initial estimate of the plastic zone size in a nonhardening material under plane stress conditions. For plane strain conditions $\alpha = 1/6\pi$.[19] A subsequent force balance may be used to show that a more reasonable estimate is $2r_y$.[19] However, strain hardening effects will reduce the size of the plastic zone; consequently Eq. (18.25) may be considered a first order estimate of the actual plastic zone size.

With $\alpha = 1/6\pi$ and $K_I = K_{IC}$, Eq. (18.25) gives the maximum possible plastic zone size under plane strain conditions. Substituting this result into the ASTM crack size requirement gives $a > 50\,r_y$ (approximately). Defining the distance d as the smallest in-plane dimension from the crack tip to the nearest free surface as shown in Fig. 18.7, this result suggests that, in general, small-scale yielding conditions may be expected if[6]

$$r_y < d/50 \tag{18.26}$$

where r_y is the maximum plastic zone size, computed from Eq. (18.25) with $\alpha = 1/6\pi$, $K_I = K_{IC}$ for plane strain and $\alpha = 1/2\pi$, $K_I = K_C$ for plane stress.

In predicting failure for designing a part so that failure will not occur, a designer must, at an early stage, identify the probable mode of failure, employ a suitable "modulus" by which severity of loading and environment may be represented analytically, select a material and geometry for the proposed part, and obtain pertinent critical material strength properties related to the probable failure mode. He must next calculate the magnitude of the selected "modulus" under applicable loading and environmental conditions and compare the calculated magnitude of the modulus with the proper critical material strength property. Failure is predicted to occur if the magnitude of the selected modulus equals or exceeds the critical material strength parameter.

For example, if a designer determines yielding to be a potential failure mode for his or her part, he or she would probably select stress (σ) as his or her "modulus" and the uniaxial yield point strength (σ_{yp}) as the critical material strength parameter. The designer would then assess the quality of his or her design by asserting that *failure is predicted to occur if*

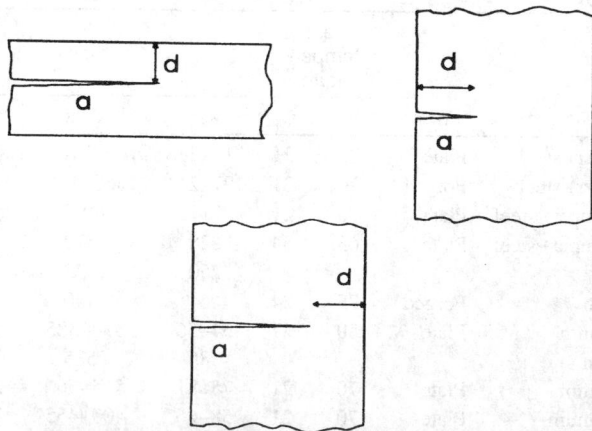

Fig. 18.7   Smallest in-plane distance d from crack tip to nearest free surface.

$$\sigma \geq \sigma_{yp} \tag{18.27}$$

The fracture mechanics approach is useful to the designer in precisely the same way when brittle fracture is a possible failure mode. The designer would select stress intensity factor K as his or her "modulus" and fracture toughness K_c as the appropriate critical strength parameter and assert that failure is predicted to occur if

$$K \geq K_c \tag{18.28}$$

Although the details of calculating K and determining K_c for some cases may be difficult, the basic concept of predicting failure by brittle fracture is no more complicated than this. It is worth noting that in most cases a designer would be well advised to consider both the possibility of failure by brittle fracture and also the possibility of failure by yielding.

To utilize (18.28) as a design or failure prediction tool, the stress intensity factor must be determined for the particular loading and geometry of the part or structure under investigation. To illustrate the procedure, several configurations are mentioned here, with many more solutions available in the literature. (See, for example, Refs. 5, 20, and 21.)

For *central through-the-thickness cracks* and *single edge through-the-thickness cracks* under *direct tension loading* or *shear loading,* the form of the stress intensity factor is

$$K = C\sigma_t\sqrt{\pi a} \tag{18.29}$$

or

$$K = C\tau\sqrt{\pi a} \tag{18.30}$$

where C is a function of geometry and crack displacement mode, as given in Figs. 18.8 and 18.9.

For a *beam* with a *single through-the-thickness edge crack* under a *pure bending moment,* the form of the stress intensity factor is

$$K_1 = C_1\sigma_b\sqrt{\pi a} \tag{18.31}$$

where C_1 is a function of geometry, as given in Fig. 18.10, and the gross section bending stress σ_b is

$$\sigma_b = \frac{6M}{tb^2} \tag{18.32}$$

For a *through-the-thickness crack emanating from a circular hole* in an infinite plate under *biaxial tension,* the form of the stress intensity factor is

$$K_1 = C_1\sigma\sqrt{\pi a} \tag{18.33}$$

where C_1 is a function of geometry and the ratio of biaxial stress components, as shown in Fig. 18.11.

For a *part-through thumbnail surface crack* in a plate subjected to *uniform tension loading,* the form of the stress intensity factor is

$$K_1 = \frac{1.12}{\sqrt{Q}}\,\sigma_t\sqrt{\pi a} \tag{18.34}$$

where Q is a surface flaw shape parameter that depends on the ratio of crack depth to length and the ratio of nominal applied stress to yield strength of the material, as shown in Fig.18.12.

Stress intensity factors may also be determined using weight functions.[6,7,23] Generally, this technique involves the use of numerical integration. It is ideally suited for use with personal computers. The primary advantage of the weight function technique is its capability to consider arbitrary applied stress distributions, as opposed to the pure bending or uniform tension stress distributions illustrated in Figs. 18.8, 18.9, 18.10, and 18.11.

Weight functions are unique for a given cracked geometry, with many weight functions available in the literature (see, for example, Refs. 7 and 23). Given the weight function $m(x, a)$ for a given geometry exhibiting a through-thickness crack of length a, and the applied stress distribution $\sigma(x)$ along the crack line x in the uncracked body, the stress intensity factor may be written

$$K_I(a) = \int_o^a \sigma(x)m(x, a)\,dx \tag{18.35}$$

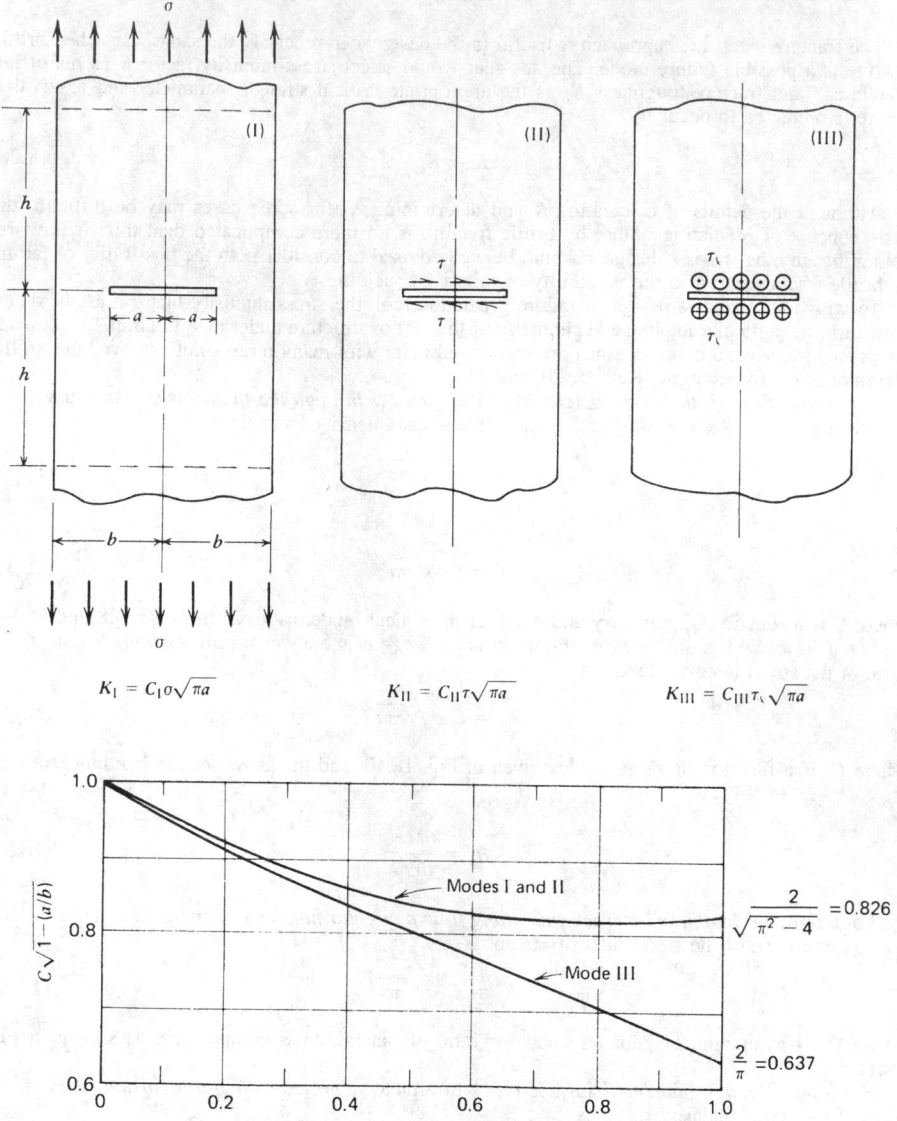

$$K_I = C_I \sigma \sqrt{\pi a} \qquad K_{II} = C_{II} \tau \sqrt{\pi a} \qquad K_{III} = C_{III} \tau_\backslash \sqrt{\pi a}$$

Fig. 18.8 Stress intensity factors K_I, K_{II}, and K_{III} for center cracked test specimen. (From Ref. 5, copyright Del Research Corp.; adapted with permission.)

For a single through-thickness crack of length a in a strip with width b (see Fig. 18.9), an approximate weight function is given by[23]

$$m(x, a) = \sqrt{\frac{2}{\pi (a - x)}} \left[1 + m_1 \left(1 - \frac{x}{a} \right) + m_2 \left(1 - \frac{x}{a} \right)^2 \right] \qquad (18.36)$$

where m_1 and m_2 are functions of the crack length to width ratio (a/b). For $0 \le a/b \le 1/2$

$$
\begin{aligned}
m_1 &= A_1 + B_1(a/b)^2 + C_1(a/b)^6 \\
m_2 &= A_2 + B_2(a/b)^2 + C_2(a/b)^6 \\
A_1 &= 0.6147, \ B_1 = 17.1844, \ C_1 = 8.7822 \\
A_2 &= 0.2502, \ B_2 = 3.2899, \ C_2 = 70.0444
\end{aligned}
\qquad (18.37)
$$

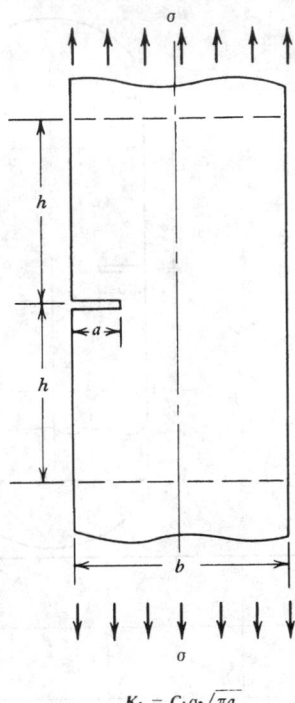

$$K_1 = C_1 \sigma \sqrt{\pi a}$$

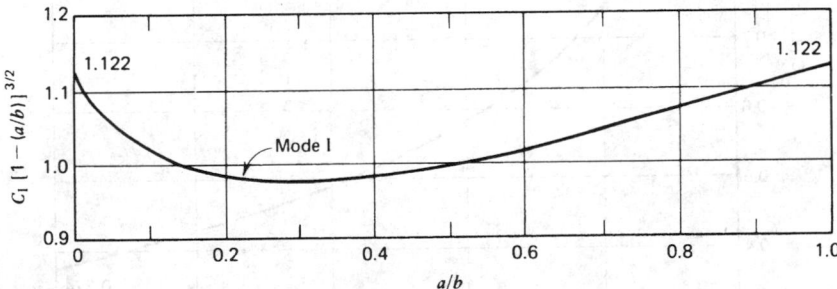

Fig. 18.9 Stress intensity factor K_1 for single edge notch test specimen. (From Ref. 5, copyright Del Research Corp.; reprinted with permission.)

This weight function, in conjunction with Eq. (18.35), may be used to compute stress intensity factors for arbitrary applied stress distributions $\sigma(x)$ determined through an analysis of the uncracked body.

Using the stress intensity factor, together with fracture toughness properties for the material of interest, a designer may utilize (18.28) to predict failure or, more important, to design a part so that failure will not occur under service loading. It should be reiterated that fracture toughness is not only a function of metallurgical factors such as alloy composition and heat treatment, but a function of service temperature, loading rate, and state of stress in the vicinity of the crack tip as well. In many practical applications the plastic zone size ahead of the crack tip becomes so large that the assumption of small-scale yielding is no longer valid, and elastic analyses using the stress intensity factor are no longer appropriate. When small-scale yielding conditions are not satisfied, a better design approach would be to implement an appropriate elastic-plastic fracture mechanics (EPFM) methodology. One such methodology involves the use of a *failure assessment diagram* (FAD).[6,18,24]

Under small-scale yielding conditions, fracture under predominantly elastic conditions is predicted to occur when the stress intensity factor equals or exceeds the material fracture toughness. Alternately, failure by plastic collapse may occur if the plastic zone becomes sufficiently large such that it

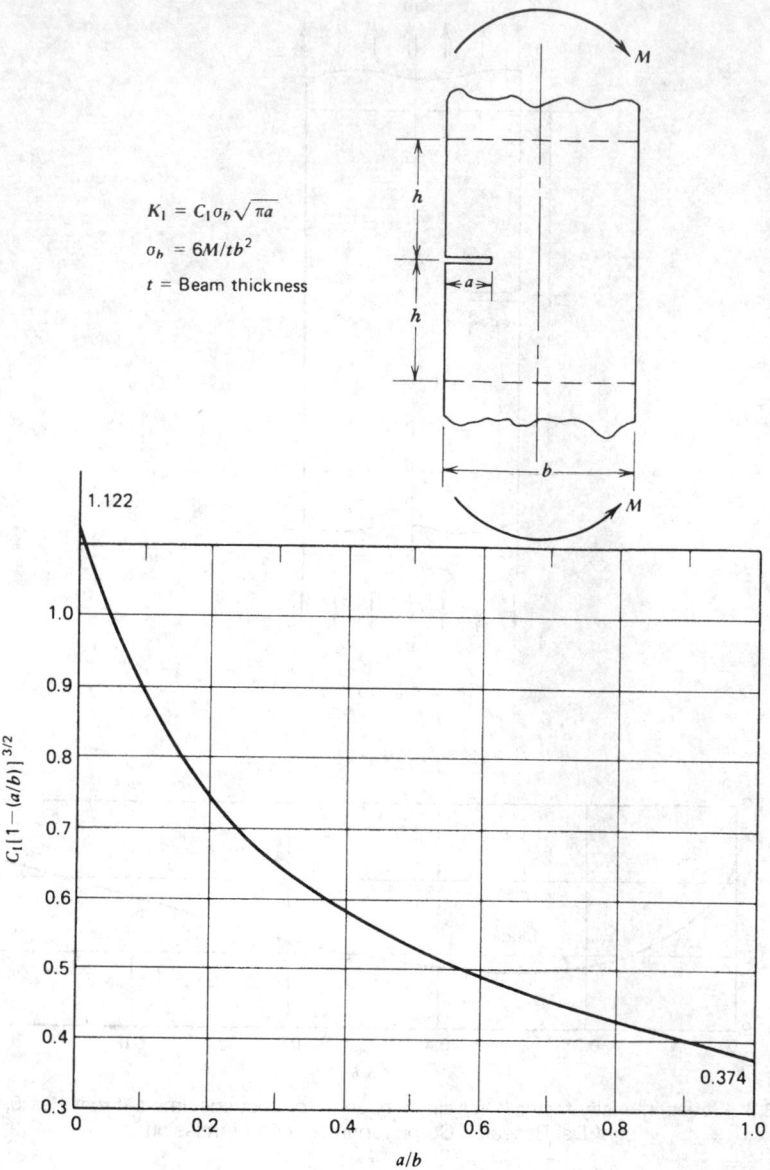

Fig. 18.10 Stress intensity factor K_I for single through-the-thickness edge crack under pure bending moment. (From Ref. 5, copyright Del Research Corp.; adapted with permission.)

encompasses the entire remaining ligament ahead of the crack. Plastic collapse is predicted to occur when the applied stress equals the plastic collapse stress σ_c. This applied stress corresponds to a stress in the uncracked ligament equal to the yield stress. Under intermediate conditions, in which the crack tip plastic zone does not encompass the entire remaining ligament and yet is not small, an interaction between elastic fracture and plastic collapse defines the governing failure mode. The FAD allows an approximate assessment of this interaction.

Defining $K_r = K_I/K_c$ and $S_r = \sigma/\sigma_c$, where σ is the stress used to compute K_I, failure is predicted to occur when

$$K_r = S_r \left\{ \frac{8}{\pi^2} \ln \left[\sec \left(\frac{\pi}{2} S_r \right) \right] \right\}^{-1/2} \tag{18.38}$$

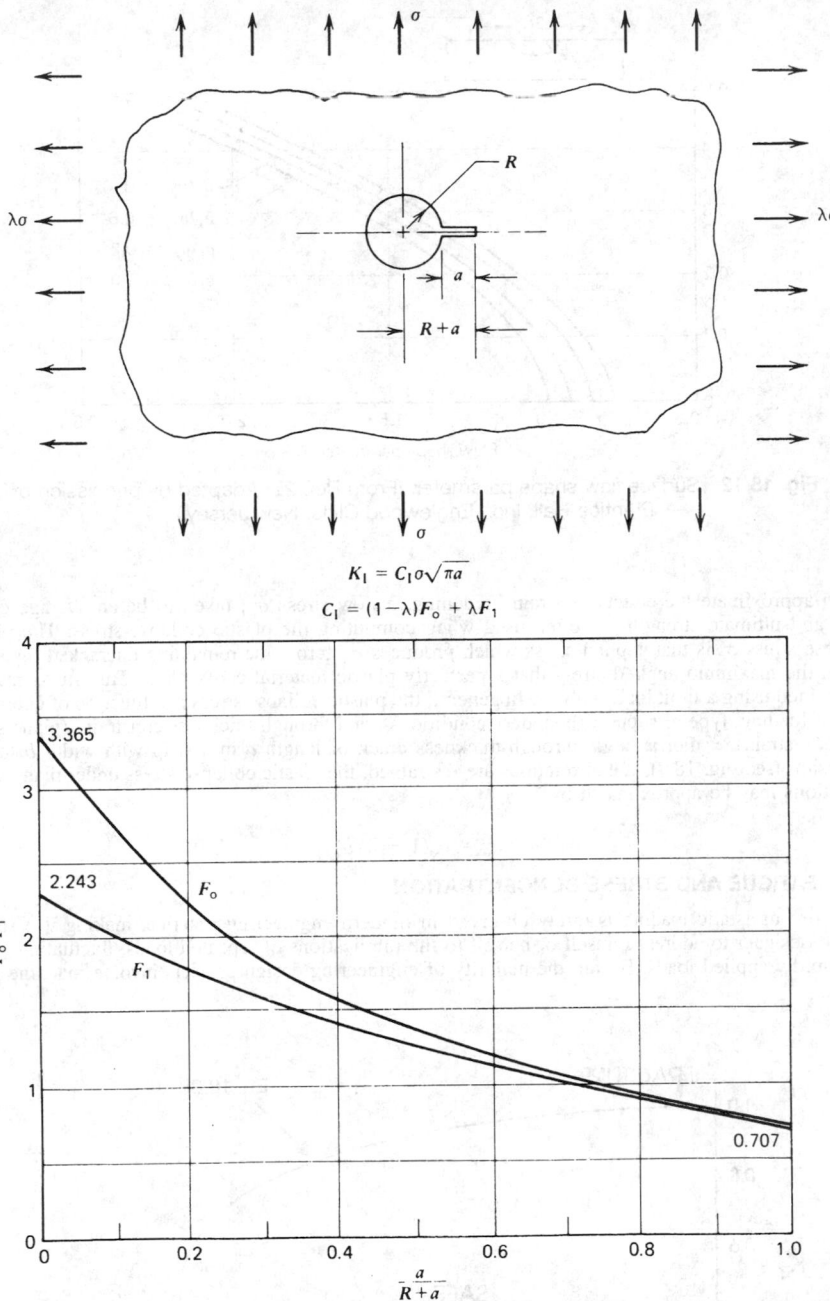

Fig. 18.11   Stress intensity factor K_1 for a through-the-thickness crack emanating from a circular hole in an infinite plate under biaxial tension. (From Ref. 5, copyright Del Research Corp.; adapted with permission.)

Equation (18.38) represents a failure curve in the K_r–S_r plane. This curve is illustrated in Fig. 18.13 and is known as the failure assessment diagram or the R6 curve. The integrity of a flawed structure may be assessed by computing S_r and K_r and plotting this point on the FAD. For a point falling within the curve, no failure is predicted. For points falling on or outside the curve, failure is predicted to occur.

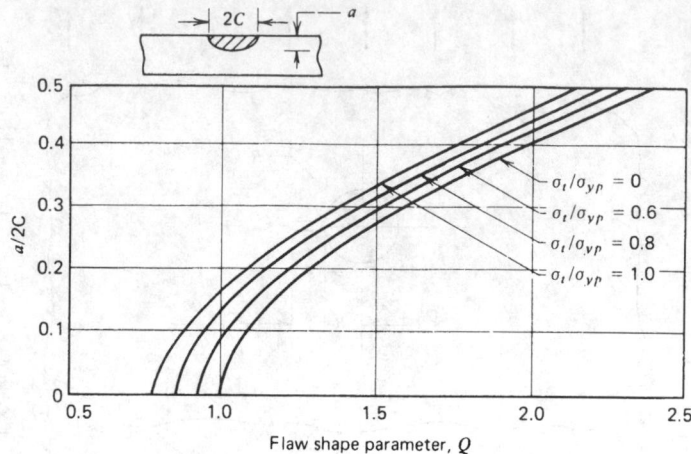

Fig. 18.12 Surface flaw shape parameter. (From Ref. 22. Adapted by permission of Prentice-Hall, Inc., Englewood Cliffs, New Jersey.)

To approximate the effects of strain hardening, a flow stress σ_o, taken to be an average of the yield and ultimate strengths, is often used when computing the plastic collapse stress. The plastic collapse stress σ_c is that applied stress which produces σ_o across the remaining uncracked ligament, and is the maximum applied stress that a perfectly plastic material can sustain. This stress may be determined using a limit load analysis. In general, the plastic collapse stress is a function of geometry, type of loading, type of support (boundary conditions), and through-thickness constraint (plane stress or plane strain).[6,25] For a single through-thickness crack of length a in a strip with width b loaded in tension (see Fig. 18.9), if end rotations are restrained, the plastic collapse stress under plane stress conditions may be approximated by[25]

$$\sigma_c = \sigma_o(1 - a/b) \tag{18.39}$$

18.5 FATIGUE AND STRESS CONCENTRATION

Static or quasistatic loading is rarely observed in modern engineering practice, making it essential for the designer to address himself or herself to the implications of repeated loads, fluctuating loads, and rapidly applied loads. By far, the majority of engineering design projects involve machine parts

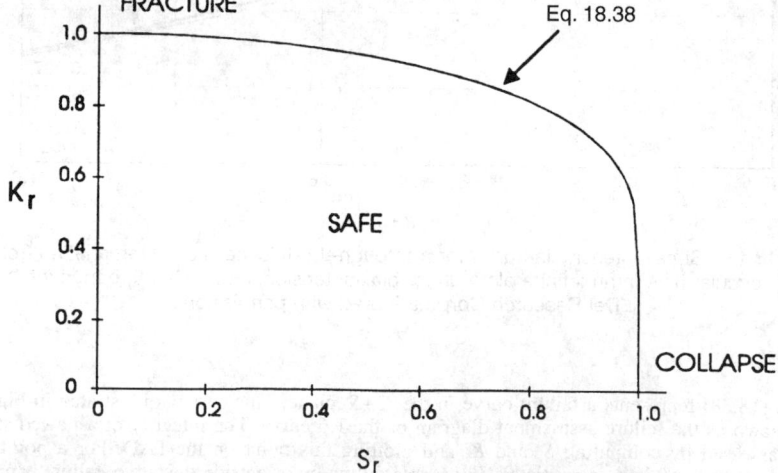

Fig. 18.13 Failure assessment diagram.

subjected to fluctuating or cyclic loads. Such loading induces fluctuating or cyclic stresses that often result in failure by fatigue.

Fatigue failure investigations over the years have led to the observation that the fatigue process actually embraces two domains of cyclic stressing or straining that are significantly different in character, and in each of which failure is probably produced by different physical mechanisms. One domain of cyclic loading is that for which significant plastic strain occurs during each cycle. This domain is associated with high loads and short lives, or low numbers of cycles to produce fatigue failure, and is commonly referred to as *low-cycle fatigue*. The other domain of cyclic loading is that for which the strain cycles are largely confined to the elastic range. This domain is associated with lower loads and long lives, or high numbers of cycles to produce fatigue failure, and is commonly referred to as *high-cycle fatigue*. Low-cycle fatigue is typically associated with cycle lives from 1 up to about 10^4 or 10^5 cycles. Fatigue may be characterized as a progressive failure phenomenon that proceeds by the *initiation* and *propagation* of cracks to an unstable size. Although there is not complete agreement on the microscopic details of the initiation and propagation of the cracks, processes of reversed slip and dislocation interaction appear to produce fatigue nuclei from which cracks may grow. Finally, the crack length reaches a critical dimension and one additional cycle then causes complete failure. The final failure region will typically show evidence of plastic deformation produced just prior to final separation. For ductile materials the final fracture area often appears as a shear lip produced by crack propagation along the planes of maximum shear.

Although designers find these basic observations of great interest, they must be even more interested in the macroscopic phenomenological aspects of fatigue failure and in avoiding fatigue failure during the design life. Some of the macroscopic effects and basic data requiring consideration in designing under fatigue loading include:

1. The effects of a simple, completely reversed alternating stress on the strength and properties of engineering materials.
2. The effects of a steady stress with superposed alternating component, that is, the effects of cyclic stresses with a nonzero mean.
3. The effects of alternating stresses in a multiaxial state of stress.
4. The effects of stress gradients and residual stresses, such as imposed by shot peening or cold rolling, for example.
5. The effects of stress raisers, such as notches, fillets, holes, threads, riveted joints, and welds.
6. The effects of surface finish, including the effects of machining, cladding, electroplating, and coating.
7. The effects of temperature on fatigue behavior of engineering materials.
8. The effects of size of the structural element.
9. The effects of accumulating cycles at various stress levels and the permanence of the effect.
10. The extent of the variation in fatigue properties to be expected for a given material.
11. The effects of humidity, corrosive media, and other environmental factors.
12. The effects of interaction between fatigue and other modes of failure, such as creep, corrosion, and fretting.

18.5.1 Fatigue Loading and Laboratory Testing

Faced with the design of a fatigue-sensitive element in a machine or structure, a designer is very interested in the fatigue response of engineering materials to various loadings that might occur throughout the design life of the machine under consideration. That is, the designer is interested in the effects of various *loading spectra* and associated *stress spectra*, which will in general be a function of the design configuration and the operational use of the machine.

Perhaps the simplest fatigue stress spectrum to which an element may be subjected is a zero-mean sinusoidal stress-time pattern of constant amplitude and fixed frequency, applied for a specified number of cycles. Such a stress–time pattern, often referred to as a completely reversed cyclic stress, is illustrated in Fig. 18.14a. Utilizing the sketch of Fig. 18.14, we can conveniently define several useful terms and symbols; these include:

$$\sigma_{max} = \text{maximum stress in the cycle}$$

$$\sigma_m = \text{mean stress} = (\sigma_{max} + \sigma_{min})/2$$

$$\sigma_{min} = \text{minimum stress in the cycle}$$

$$\sigma_a = \text{alternating stress amplitude} = (\sigma_{max} - \sigma_{min})/2$$

$$\Delta\sigma = \text{range of stress} = \sigma_{max} - \sigma_{min}$$

$$R = \text{stress ratio} = \sigma_{min}/\sigma_{max}$$

$$A = \text{amplitude ratio} = \sigma_a/\sigma_m = (1 - R)/(1 + R)$$

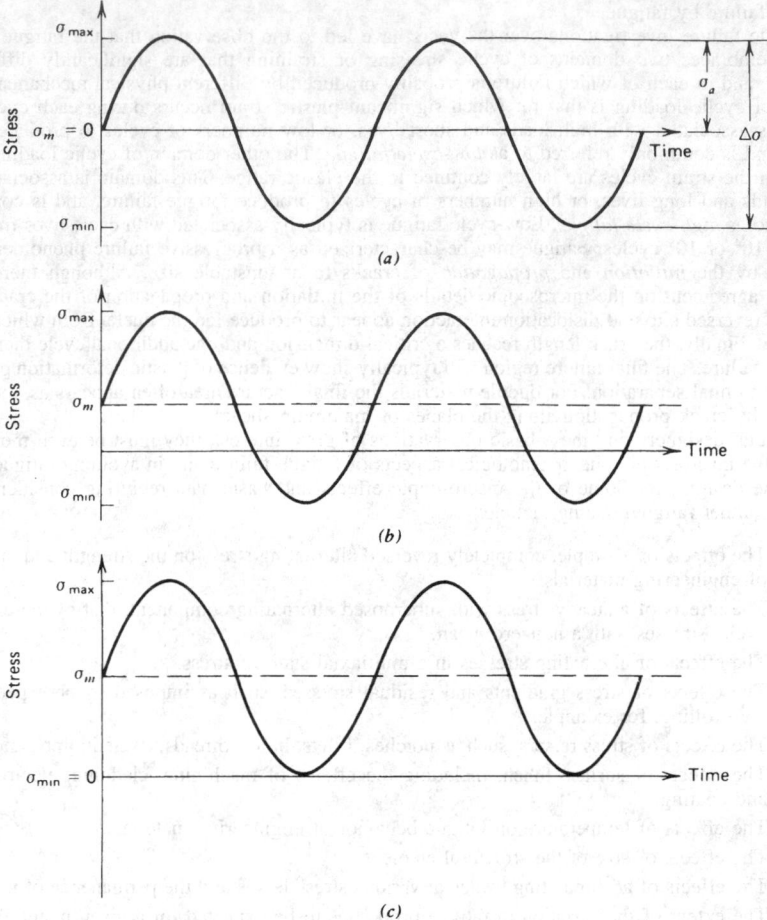

Fig. 18.14 Several constant-amplitude stress–time patterns of interest: (a) completely reversed, $R = -1$; (b) nonzero mean stress; (c) released tension, $R = 0$.

Any two of the quantities just defined, except the combinations σ_a and $\Delta\sigma$ or the combination A and R, are sufficient to describe completely the stress–time pattern above.

More complicated stress–time patterns are produced when the mean stress, or stress amplitude, or both mean and stress amplitude change during the operational cycle, as illustrated in Fig. 18.15. It may be noted that this stress–time spectrum is beginning to approach a degree of realism. Finally, in Fig. 18.16 a sketch of a realistic stress spectrum is given. This type of quasirandom stress–time pattern might be encountered in an airframe structural member during a typical mission including refueling, taxi, takeoff, gusts, maneuvers, and landing. The obtaining of useful, realistic data is a challenging task in itself. Instrumentation of existing machines, such as operational aircraft, provide some useful information to the designer if his or her mission is similar to the one performed by the instrumented machine. Recorded data from accelerometers, strain gauges, and other transducers may in any event provide a basis from which a statistical representation can be developed and extrapolated to future needs if the fatigue processes are understood.

Basic data for evaluating the response of materials, parts, or structures are obtained from carefully controlled laboratory tests. Various types of testing machines and systems commonly used include:

1. Rotating-bending machines:
   a. Constant bending moment type
   b. Cantilever bending type
2. Reciprocating-bending machines.

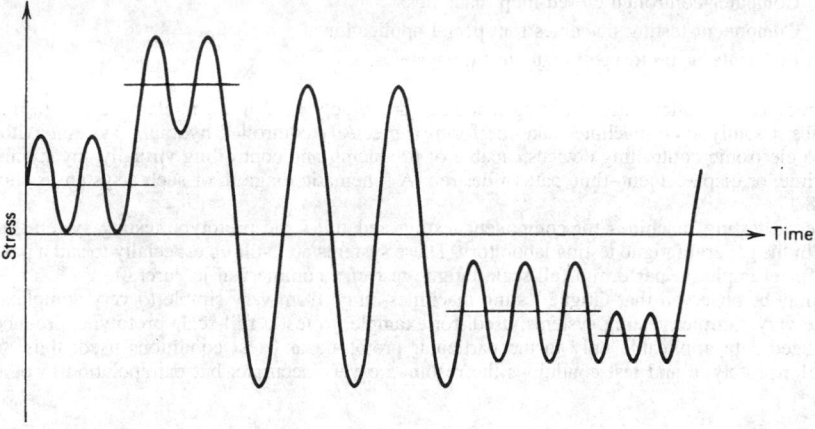

Fig. 18.15 Stress–time pattern in which both mean and amplitude change to produce a more complicated stress spectrum.

3. Axial direct-stress machines:
   a. Brute-force type
   b. Resonant type
4. Vibrating shaker machines:
   a. Mechanical type
   b. Electromagnetic type
5. Repeated torsion machines.
6. Multiaxial stress machines.

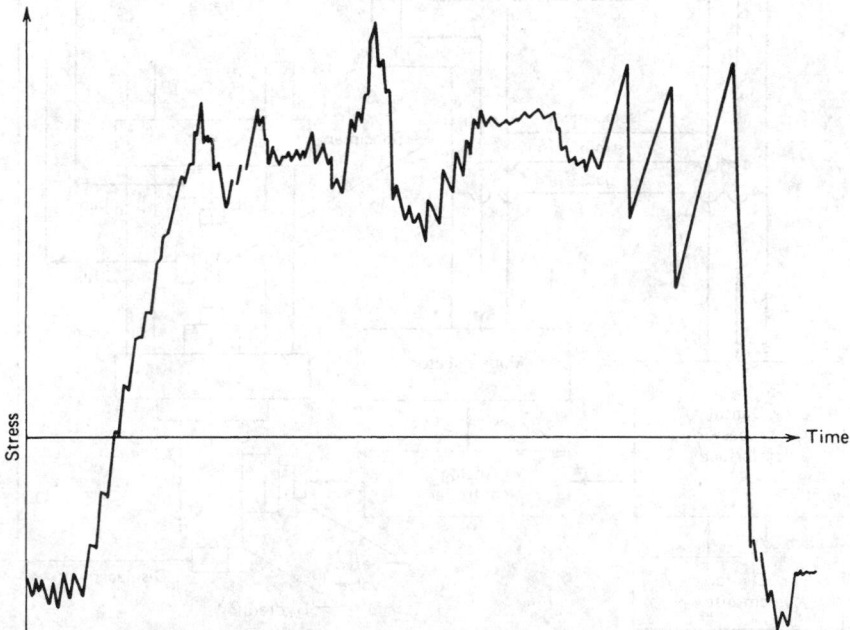

Fig. 18.16 A quasirandom stress–time pattern that might be typical of an operational aircraft during any given mission.

7. Computer-controlled closed-loop machines.
8. Component testing machines for special applications.
9. Full-scale or prototype fatigue testing systems.

Computer-controlled fatigue testing machines are widely used in all modern fatigue testing laboratories. Usually such machines take the form of precisely controlled hydraulic systems with feedback to electronic controlling devices capable of producing and controlling virtually any strain–time, load–time, or displacement–time pattern desired. A schematic diagram of such a system is shown in Fig. 18.17.

Special testing machines for component testing and full-scale prototype testing systems are not found in the general fatigue testing laboratory. These systems are built up especially to suit a particular need, for example, to perform a full-scale fatigue test of a commercial jet aircraft.

It may be observed that fatigue testing machines range from very simple to very complex.

The very complex testing systems, used, for example, to test a full-scale prototype, produce very specialized data applicable only to the particular prototype and test conditions used; thus, for the particular prototype and test conditions the results are very accurate, but extrapolation to other test

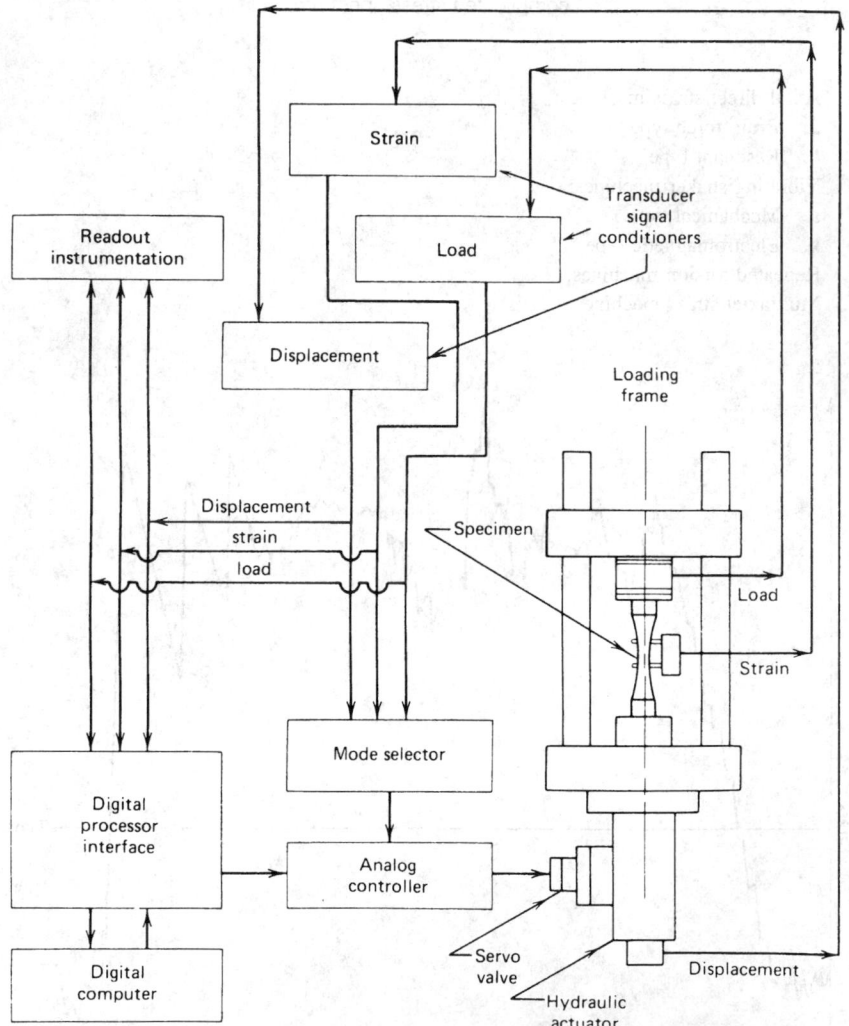

Fig. 18.17 Schematic diagram of a computer-controlled closed-loop fatigue testing machine.

conditions and other pieces of hardware is difficult, if not impossible. On the other hand, simple smooth-specimen laboratory fatigue data are very general and can be utilized in designing virtually any piece of hardware made of the specimen material. However, to use such data in practice requires a quantitative knowledge of many pertinent differences between the laboratory and the application, including the effects of nonzero mean stress, varying stress amplitude, environment, size, temperature, surface finish, residual stress pattern, and others. Fatigue testing is performed at the extremely simple level of smooth specimen testing, the extremely complex level of full-scale prototype testing, and everywhere in the spectrum between. Valid arguments can be made for testing at all levels.

18.5.2 The *S–N–P* Curves—A Basic Design Tool

Basic fatigue data in the high-cycle life range can be conveniently displayed on a plot of cyclic stress level versus the logarithm of life, or alternatively, on a log–log plot of stress versus life. These plots, called *S–N* curves, constitute design information of fundamental importance for machine parts subjected to repeated loading. Because of the scatter of fatigue life data at any given stress level, it must be recognized that there is not only one *S–N* curve for a given material, but a family of *S–N* curves with probability of failure as the parameter. These curves are called the *S–N–P* curves, or curves of constant probability of failure on a stress-versus-life plot. A representative family of *S–N–P* curves is illustrated in Fig. 18.18. It should also be noted that references to the "*S–N* curve" in the literature generally refer to the mean curve unless otherwise specified. Details regarding fatigue testing and the experimental generation of *S–N–P* curves may be found in Ref. 1.

The mean *S–N* curves sketched in Fig. 18.19 distinguish two types of material response to cyclic loading commonly observed. The ferrous alloys and titanium exhibit a steep branch in the relatively short life range, leveling off to approach a stress asymptote at longer lives. This stress asymptote is called the *fatigue limit* (formerly called endurance limit) and is the stress level below which an infinite number of cycles can be sustained without failure. The nonferrous alloys do not exhibit an asymptote, and the curve of stress versus life continues to drop off indefinitely. For such alloys there is no fatigue limit, and failure as a result of cyclic load is only a matter of applying enough cycles. All materials, however, exhibit a relatively flat curve in the long-life range.

To characterize the failure response of nonferrous materials, and of ferrous alloys in the finite-life range, the term *fatigue strength at a specified life, S_N,* is used. The term fatigue strength identifies the stress level at which failure will occur at the specified life. The specification of *fatigue strength* without specifying the corresponding life is meaningless. The specification of a *fatigue limit* always implies infinite life.

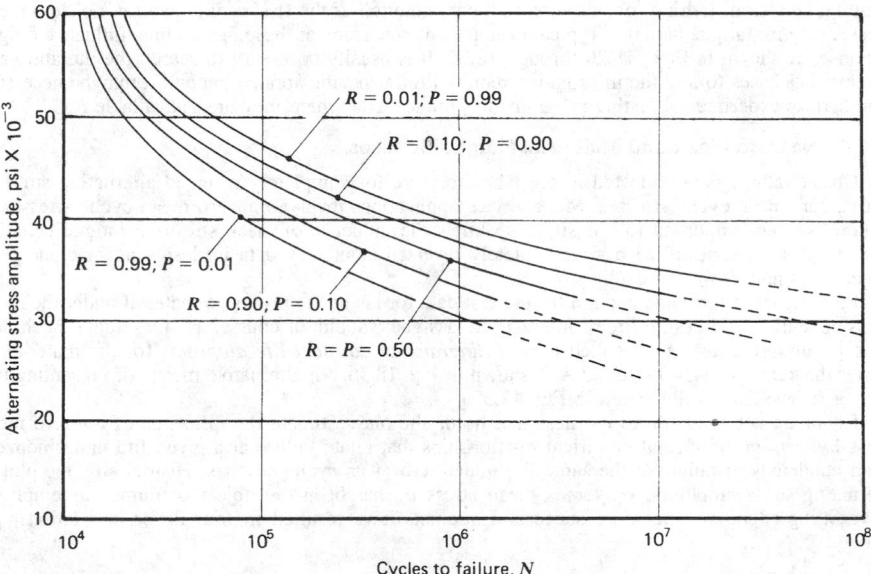

Fig. 18.18 Family of *S–N–P* curves, or *R–S–N* curves, for 7075-T6 aluminum alloy. Note: *P* = probability of failure; *R* = reliability = 1 − *P*. (Adapted from Ref. 31, p. 117; with permission from John Wiley & Sons, Inc.)

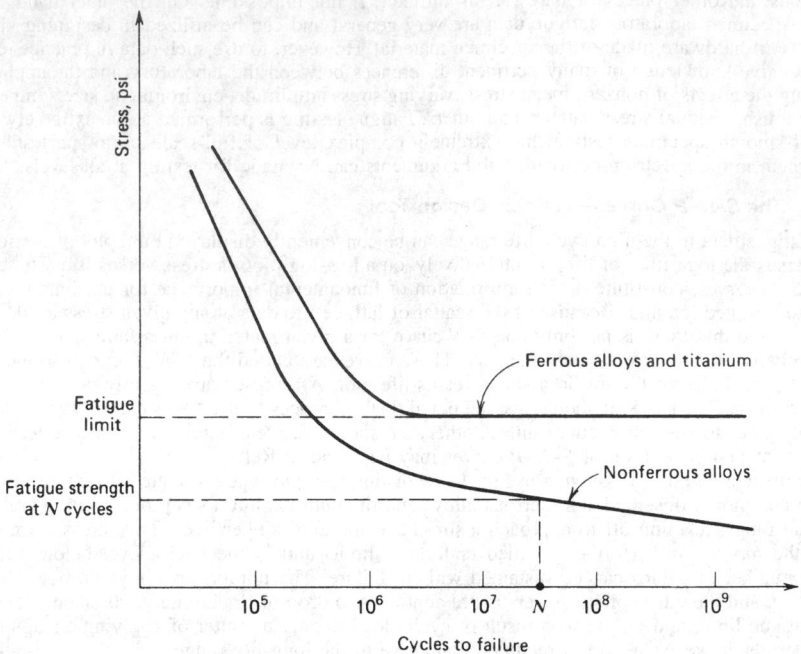

Fig. 18.19   Two types of material response to cyclic loading.

18.5.3   Factors That Affect *S–N–P* Curves

There are many factors that may influence the fatigue failure response of machine parts or laboratory specimens, including material composition, grain size and grain direction, heat treatment, welding, geometrical discontinuities, size effects, surface conditions, residual surface stresses, operating temperature, corrosion, fretting, operating speed, configuration of the stress–time pattern, nonzero mean stress, and prior fatigue damage. Typical examples of how some of these factors may influence fatigue response are shown in Figs. 18.20 through 18.35. It is usually necessary to search the literature and existing data bases to find the information required for a specific application and it may be necessary to undertake experimental testing programs to produce data where they are unavailable.

18.5.4   Nonzero Mean and Multiaxial Fatigue Stresses

Most basic fatigue data collected in the laboratory are for completely reversed alternating stresses, that is, zero mean cyclic stresses. Most service applications involve nonzero mean cyclic stresses. It is therefore very important to a designer to know the influence of mean stress on fatigue behavior so that he or she can utilize basic completely reversed laboratory data in designing machine parts subjected to nonzero mean cyclic stresses.

If a designer is fortunate enough to find test data for his or her proposed material under the mean stress conditions and design life of interest, the designer should, of course, use these data. Such data are typically presented on so-called *master diagrams* or *constant life diagrams* for the material. A master diagram for a 4340 steel alloy is shown in Fig. 18.36. An alternative means of presenting this type of fatigue data is illustrated in Fig. 18.37.

If data are not available to the designer, he or she may estimate the influence of nonzero mean stress by any one of several empirical relationships that relate failure at a given life under nonzero mean conditions to failure at the same life under zero mean cyclic stresses. Historically, the plot of alternating stress amplitude σ_a versus mean stress σ_m has been the object of numerous empirical curve-fitting attempts. The more successful attempts have resulted in four different relationships, namely:

1. Goodman's linear relationship.
2. Gerber's parabolic relationship.
3. Soderberg's linear relationship.
4. The elliptic relationship.

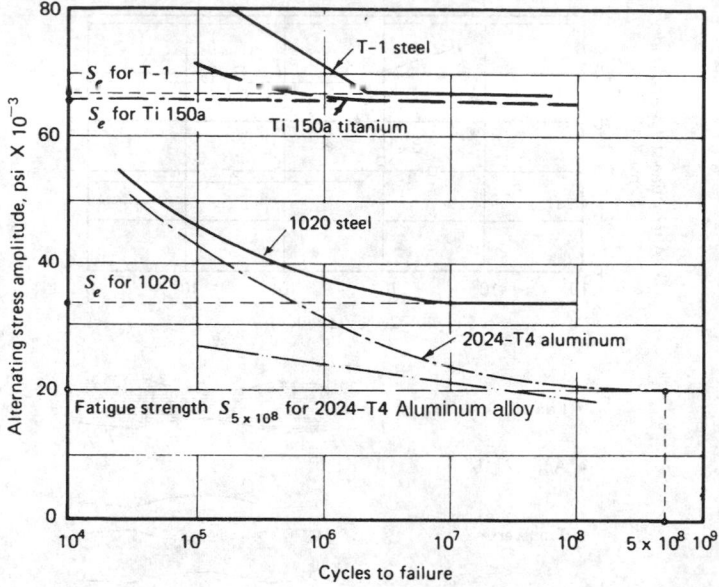

Fig. 18.20 Effect of material composition on the *S–N* curve. Note that ferrous and titanium alloys exhibit a well-defined fatigue limit, whereas other alloy compositions do not. (Data from Refs. 26 and 27.)

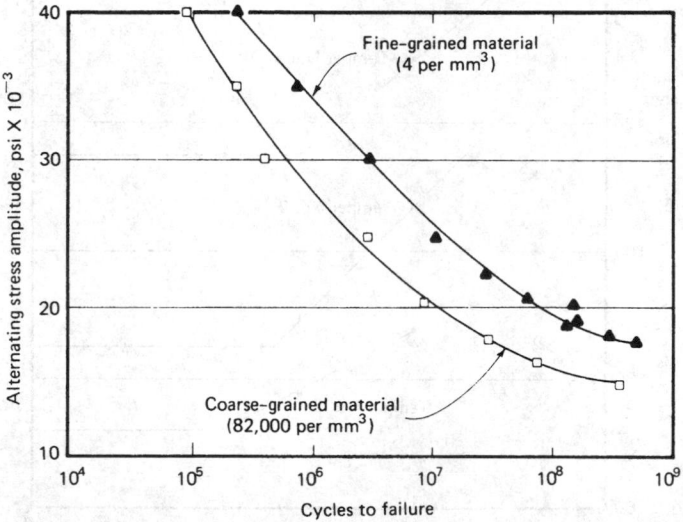

Fig. 18.21 Effect of grain size on the *S–N* curve for 18S aluminum alloy. Average diameter ratio of coarse to fine grains is approximately 27 to 1. Nominal composition: 4.0% copper, 2.0% nickel, 0.6% magnesium. Note that at a life of 10^8 cycles of the mean fatigue strength of the coarse-grained material is about 3000 psi lower than for fine-grained material. (Data from Ref. 28; adapted from *Fatigue and Fracture of Metals,* by W. M. Murray, by permission of the MIT Press, Cambridge, Massachusetts, copyright, 1952.)

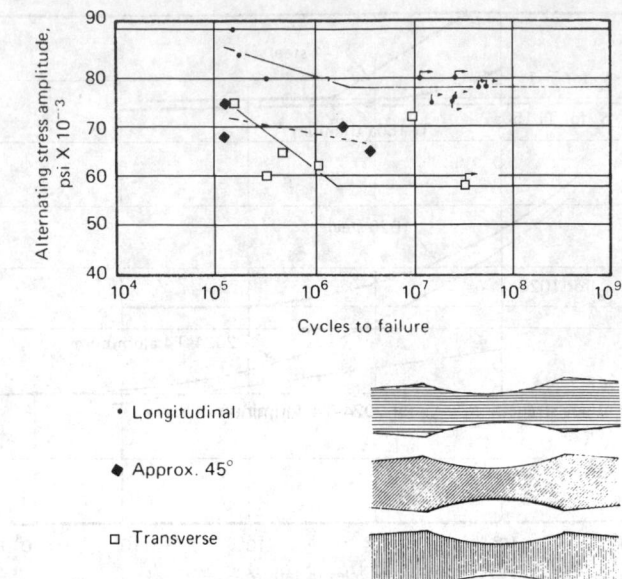

Fig. 18.22 Effect on the *S–N* curve of grain flow direction relative to longitudinal loading direction for specimens machined from crankshaft forgings. Nominal composition: 0.41% carbon, 0.47% manganese, 0.01% silicon, 0.04% phosphorous, 1.8% nickel. S_u = 139,000 psi, S_{yp} = 115,000 psi, e (2.0 in.) = 20%. (Data from Ref. 29.)

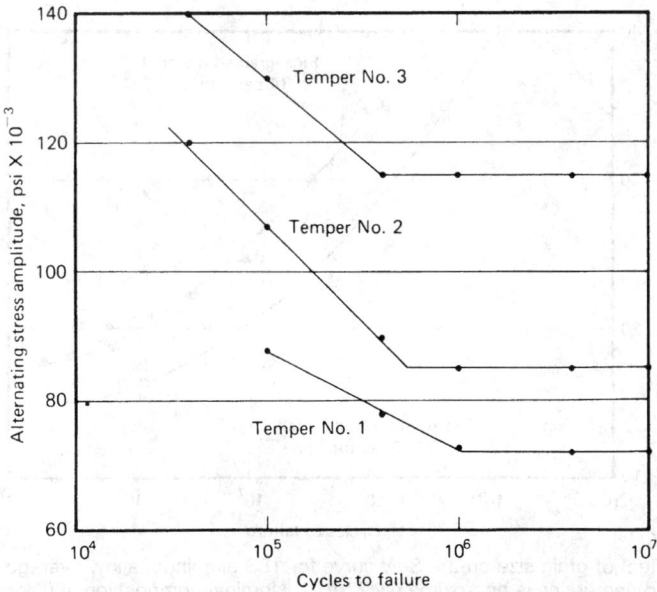

Fig. 18.23 Effects of heat treatment on the *S–N* curve of SAE 4130 steel, using 0.19-in.-diameter rotating bending specimens cut from ⅜-in. plate, 1625°F, oil quenched, followed by three different tempers. Temper No. 1: S_u = 129,000 psi; S_{yp} = 118,000 psi. Temper No. 2: S_u = 150,000 psi; S_{yp} = 143,000 psi. Temper No. 3: S_u = 206,000 psi; S_{yp} = 194,000 psi.

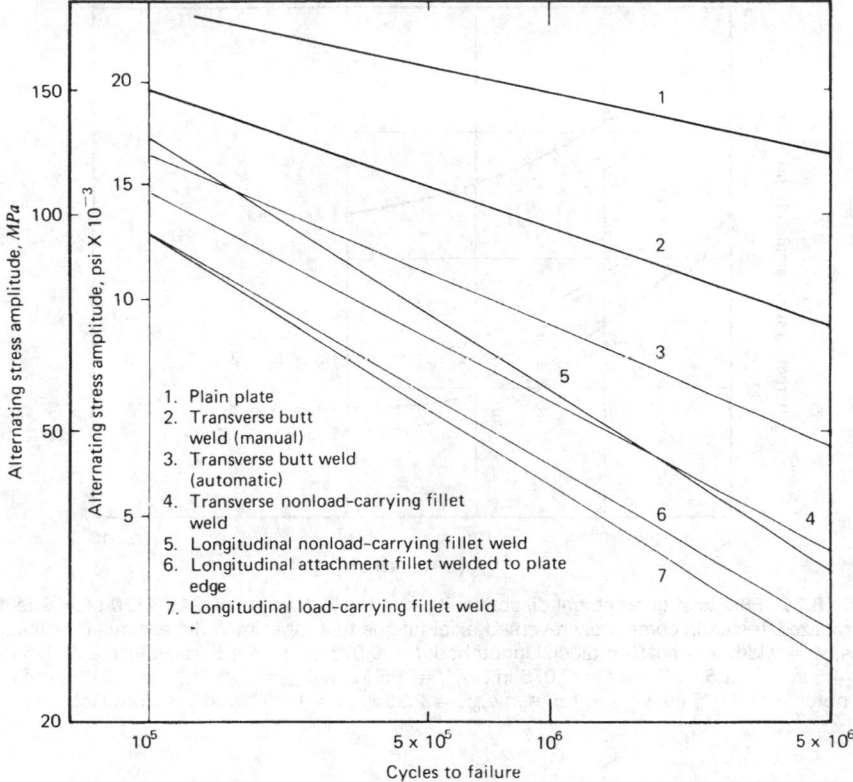

Fig. 18.24   Effects of welding detail on the *S–N* curve of structural steel, with yield strength in the range 30,000–52,000 psi. Tests were released tension ($\sigma_{min} = 0$). (Data from Ref. 30.)

A modified form of the Goodman relationship is recommended for general use under conditions of high-cycle fatigue. For tensile mean stress ($\sigma_m > 0$), this relationship may be written

$$\frac{\sigma_a}{\sigma_N} + \frac{\sigma_m}{\sigma_u} = 1 \tag{18.40}$$

where σ_u is the material ultimate strength and σ_N is the zero mean stress fatigue strength for a given number of cycles N. For a given alternating stress, compressive mean stresses ($\sigma_m < 0$) have been empirically observed to exert no influence on fatigue life. Thus, for $\sigma_m < 0$, the fatigue response is identical to that for $\sigma_m = 0$ with $\sigma_a = \sigma_N$.

The modified Goodman relationship is illustrated in Fig. 18.38. This curve is a failure locus for the case of *uniaxial* fatigue stressing. Any cyclic loading that produces an alternating stress and mean stress that exceeds the bounds of the locus will cause failure in fewer than N cycles. Any alternating stress–mean stress combination that lies within the locus will result in more than N cycles without failure. Combinations that just touch the locus produce failure in exactly N cycles. The modified Goodman relationship shown in Fig. 18.38 considers fatigue failure exclusively. The reader is cautioned to insure that the maximum and minimum stresses produced by the cyclic loading do not exceed the material yield strength σ_{yp} such that failure by yielding would be predicted to occur.

For a given design life N, Eq. (18.40) may be used to estimate whether fatigue failure will occur under any nonzero mean stress condition if the ultimate strength σ_u and the completely reversed ($\sigma_m = 0$) fatigue strength σ_N for the material are known. These material properties are usually available.

If the machine part under consideration is subjected not only to nonzero mean stress, but also to a multiaxial state of stress, then multiaxial fatigue must be considered. Historically, the majority of fatigue-related research has been focused on uniaxial loading conditions, and consequently multiaxial fatigue is not as well characterized.

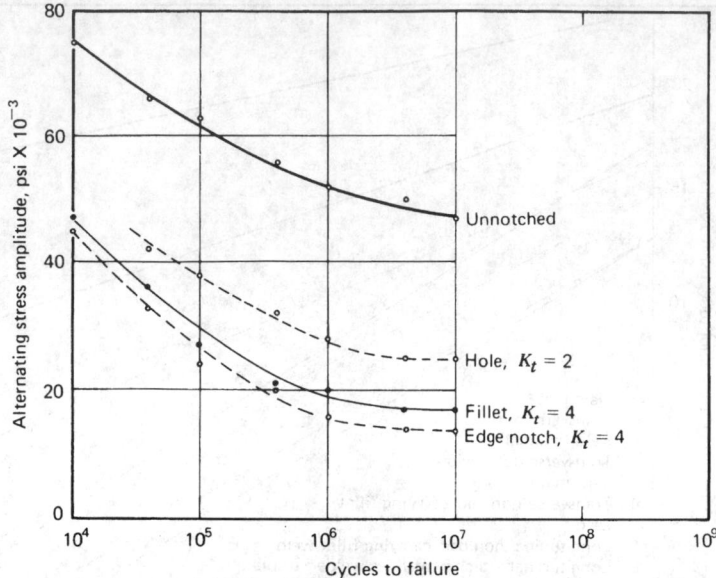

Fig. 18.25 Effects of geometrical discontinuities on the *S–N* curve of SAE 4130 steel sheet, normalized, tested in completely reversed axial fatigue test. Specimen dimensions (t = thickness, w = width, r = notch radius): Unnotched: t = 0.075 in., w = 1.5 in. Hole: t = 0.075 in., w = 4.5 in., r = 1.5 in. Fillet: t = 0.075 in., w_{net} = 1.5 in., w_{gross} = 2.25 in., r = 0.0195 in. Edge notch: t = 0.075 in., w_{net} = 1.5 in., w_{gross} = 2.25 in., r = 0.057 in. (Data from Ref. 26.)

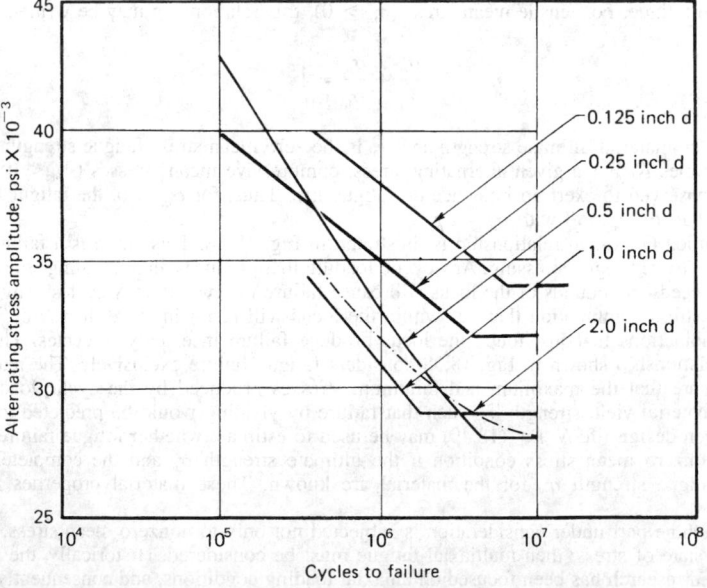

Fig. 18.26 Size effects on the *S–N* curve of SAE 1020 steel specimens cut from a 3½-in.-diameter hot-rolled bar, testing in rotating bending. (Data from Ref. 32.)

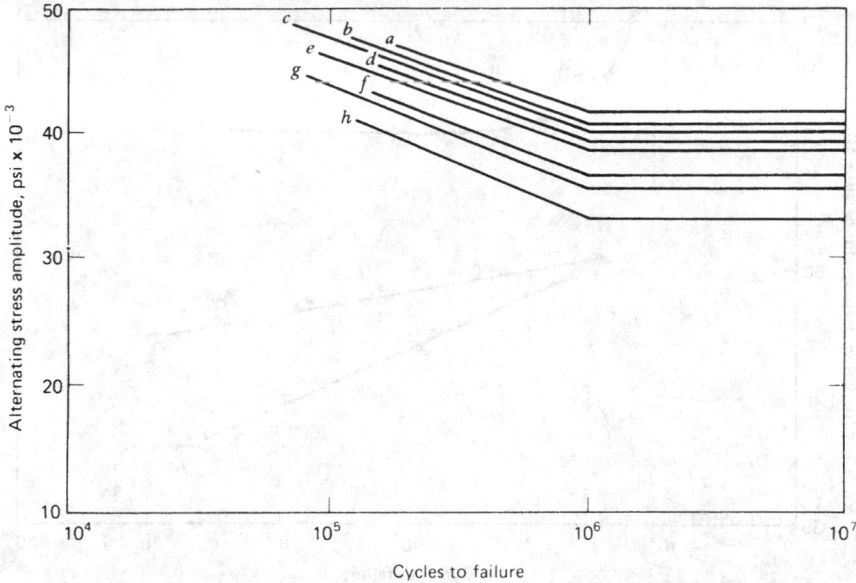

Fig. 18.27 Effect of surface finish on the *S–N* curve of 0.33% carbon steel specimens, testing in a rotating cantilever beam machine: (*a*) high polish, longitudinal direction; (*b*) FF emery finish; (*c*) No. 1 emery finish; (*d*) coarse emery finish; (*e*) smooth file; (*f*) as-turned; (*g*) bastard file; (*h*) coarse file. (Data from Ref. 33.)

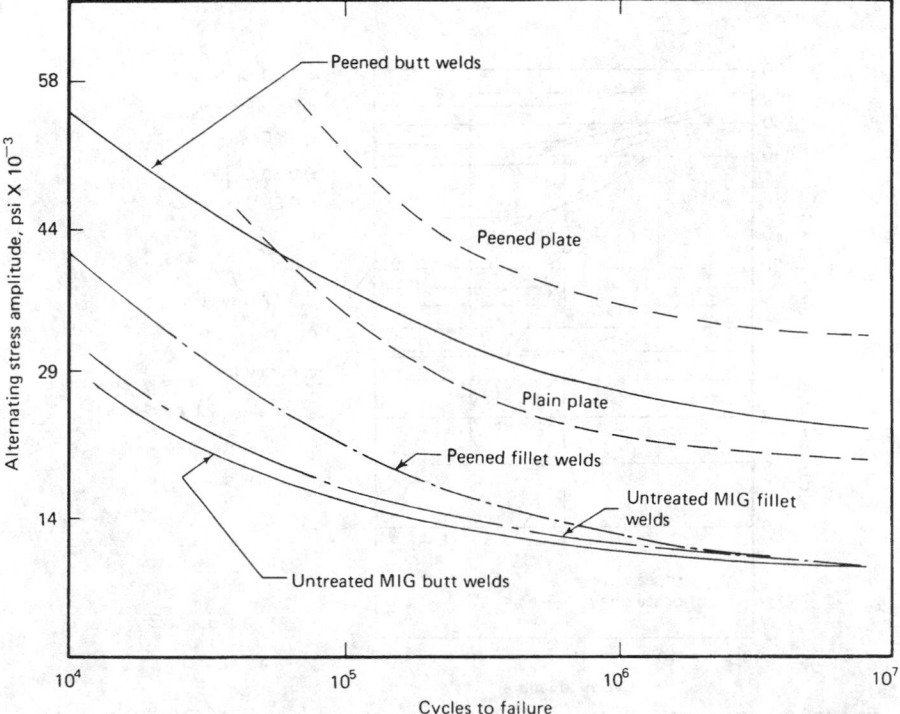

Fig. 18.28 Effect of shot peening on the *S–N* curves for welded and unwelded steel plate. (Data from Ref. 30.)

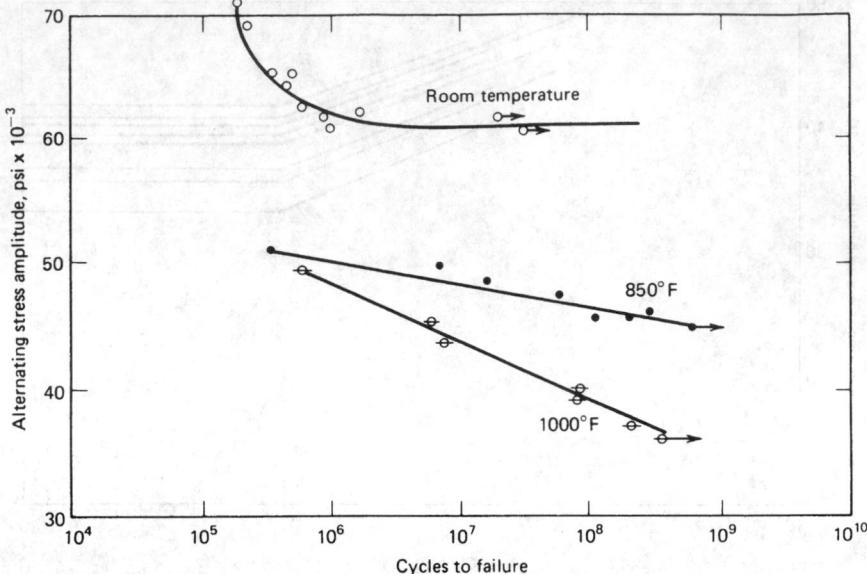

Fig. 18.29   Effect of operating temperature on the *S–N* curve of a 12% chromium steel alloy.
Alloy composition = 0.10% C, 0.45% Mn, 0.21% Ni, 12.3% Cr, and 0.38% Mo.
(Data from Ref. 34.)

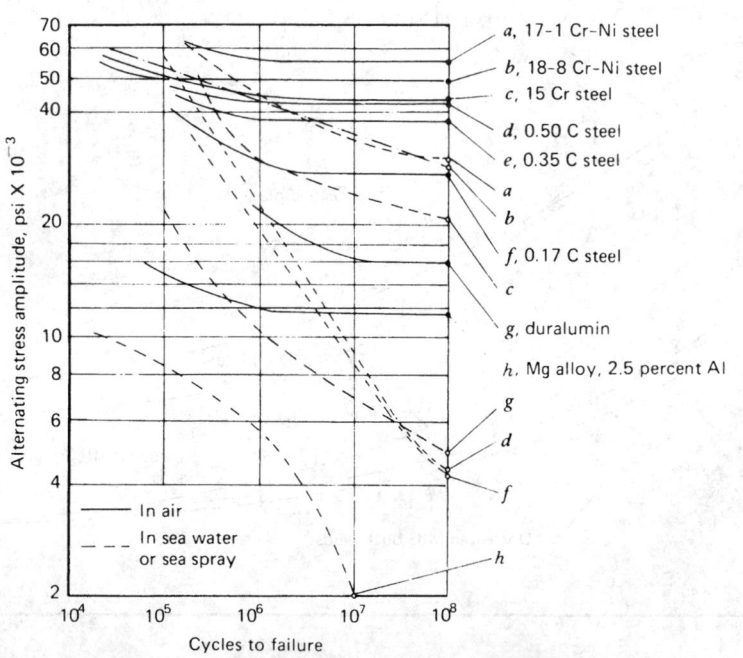

Fig. 18.30   Effects of corrosion on the *S–N* curves of various aircraft materials tested in
push–pull loading in seawater or sea spray. (Data from Ref. 35.)

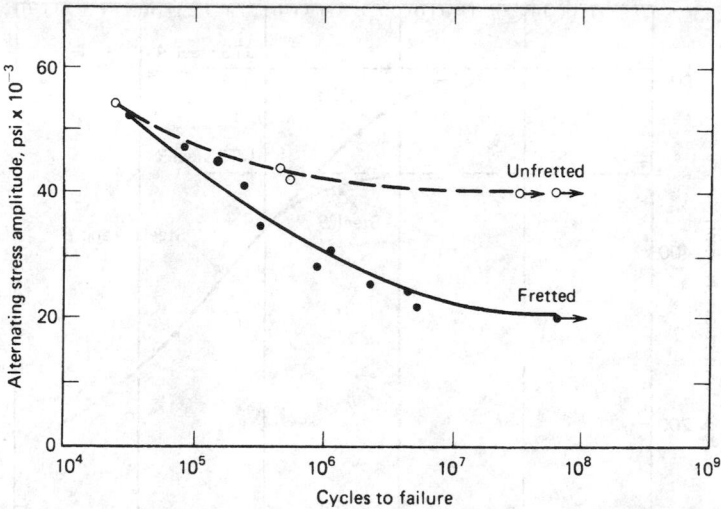

Fig. 18.31   Effect of fretting on the *S–N* curve of a forged 0.24% steel. (Data from Ref. 36; reprinted with permission from McGraw-Hill Book Company.)

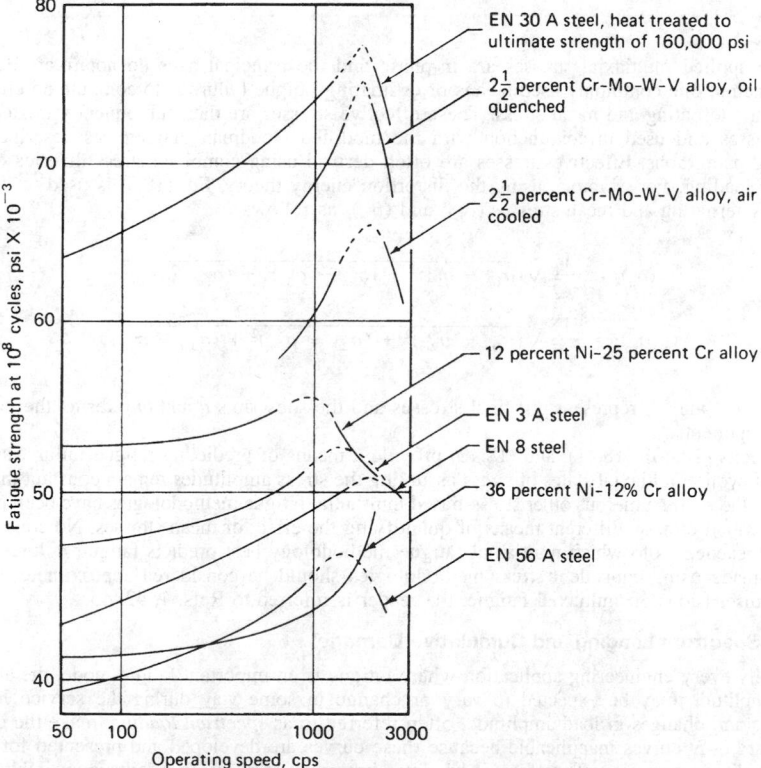

Fig. 18.32   Effect of operating speed on the fatigue strength at 10^8 cycles for several different ferrous alloys. (Data from Ref. 37, p. 381.)

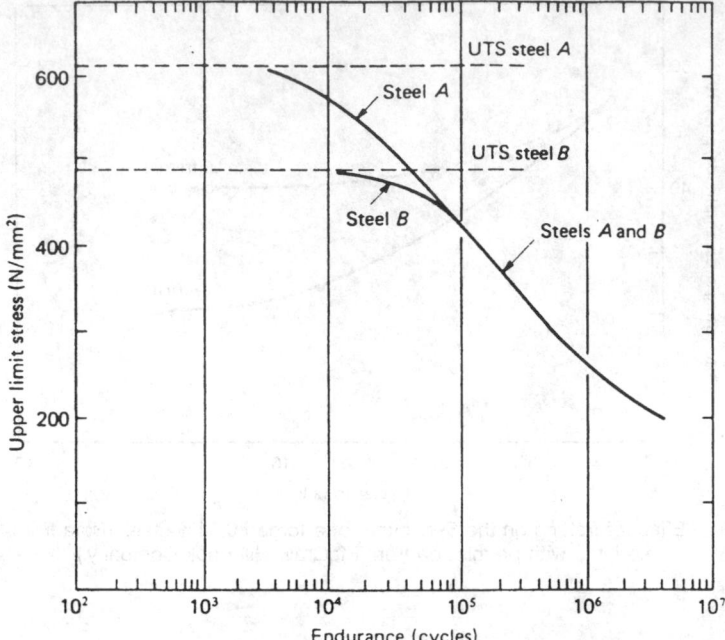

Fig. 18.33 Effect of ultimate strength on the S–N curve for transverse butt welds in two steels. (Data from Ref. 38.)

If the applied multiaxial stresses are in-phase and the principal axes do not rotate during the cyclic loading, one commonly used means of estimating fatigue failure is to compute an effective or equivalent alternating and mean stress. These effective stresses are then subsequently treated as uni-axial stresses and used in conjunction with the modified Goodman diagram, as described in the preceding paragraphs. Effective stresses are often derived using combined stress theories of failure for static loading. For example, using the distortion energy theory, Eq. (18.3) is used to define the effective alternating and mean stresses $(\sigma_a)_e$ and $(\sigma_m)_e$ as follows

$$(\sigma_a)_e = \frac{1}{\sqrt{2}} \sqrt{(\sigma_{1a} - \sigma_{2a})^2 + (\sigma_{2a} - \sigma_{3a})^2 + (\sigma_{3a} - \sigma_{1a})^2} \qquad (18.41)$$

$$(\sigma_m)_e = \frac{1}{\sqrt{2}} \sqrt{(\sigma_{1m} - \sigma_{2m})^2 + (\sigma_{2m} - \sigma_{3m})^2 + (\sigma_{3m} - \sigma_{1m})^2} \qquad (18.42)$$

where σ_1, σ_2, and σ_3 represent principal stresses and the subscripts a and m refer to alternating and mean components.

Equations (18.40), (18.41), and (18.42) provide a means of predicting fatigue failure under con-ditions of cyclic multiaxial states of stress assuming the stress amplitudes remain constant throughout the life of the part. Numerous other stress-based multiaxial fatigue methodologies have been proposed, some of which employ different means of quantifying the effect of mean stresses. No consensus has yet been reached as to which multiaxial fatigue methodology best predicts fatigue failure, and pre-dictions made using equivalent stress methodologies should be considered approximate. For more detailed discussions of multiaxial fatigue, the reader is referred to Refs. 1, 42–45.

18.5.5 Spectrum Loading and Cumulative Damage

In virtually every engineering application where fatigue is an important failure mode, the alternating stress amplitude may be expected to vary or change in some way during the service life. Such variations and changes in load amplitude, often referred to as *spectrum loading,* make the direct use of standard S–N curves inapplicable because these curves are developed and presented for constant stress amplitude operation. Therefore, it becomes important to a designer to have available a theory or hypothesis, verified by experimental observations, that will permit good design estimates to be made for operation under conditions of spectrum loading using the standard constant amplitude S–N curves.

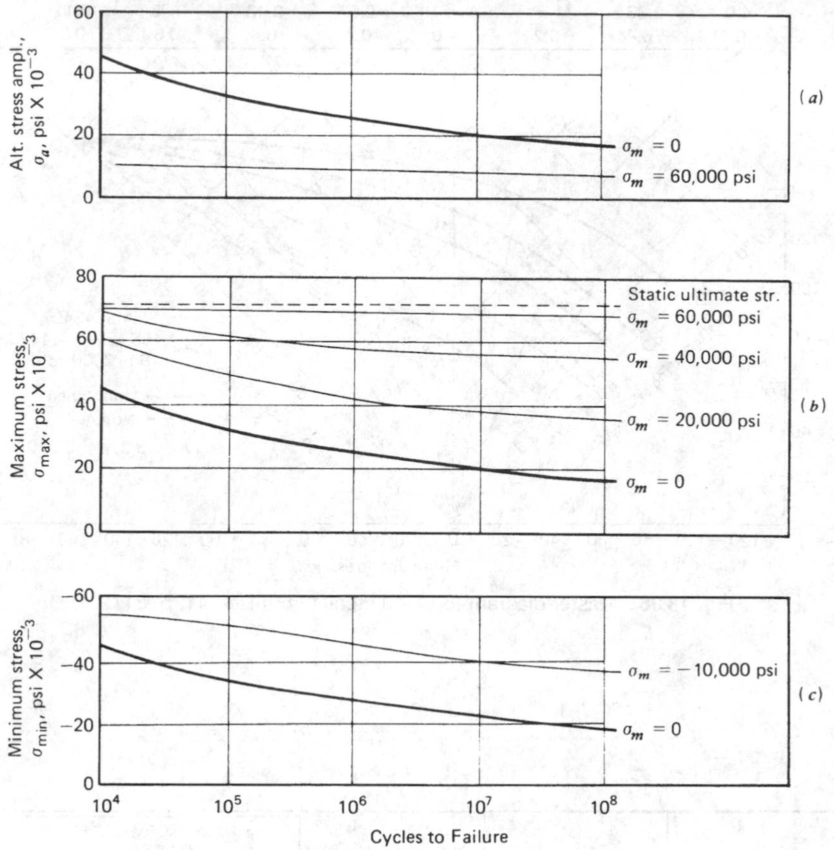

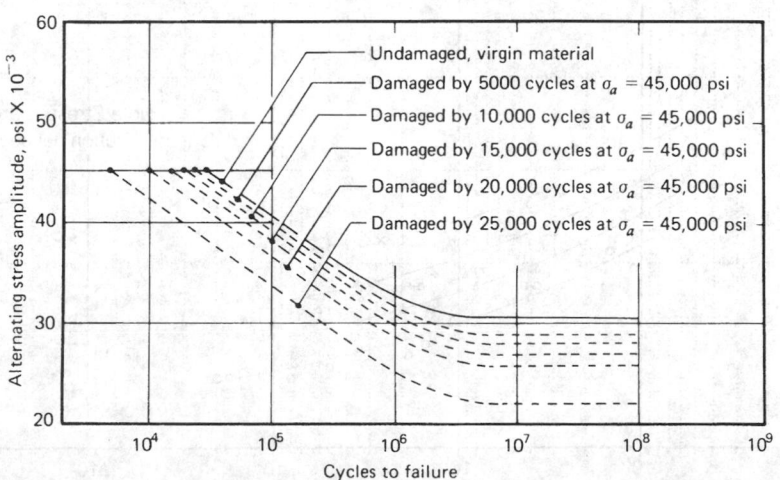

Fig. 18.34   Various ways of presenting the influence of nonzero mean stress on the fatigue behavior of 2014-T6 aluminum alloy. (Adapted from Ref. 39.)

Fig. 18.35   Illustration of the influence of accumulated fatigue damage on subsequent fatigue behavior of carbon steel. Note: Life of virgin material at σ_a = 45,000 psi is approximately 30,000 cycles. (Data from Ref. 40; reprinted with permission from John Wiley & Sons, Inc.)

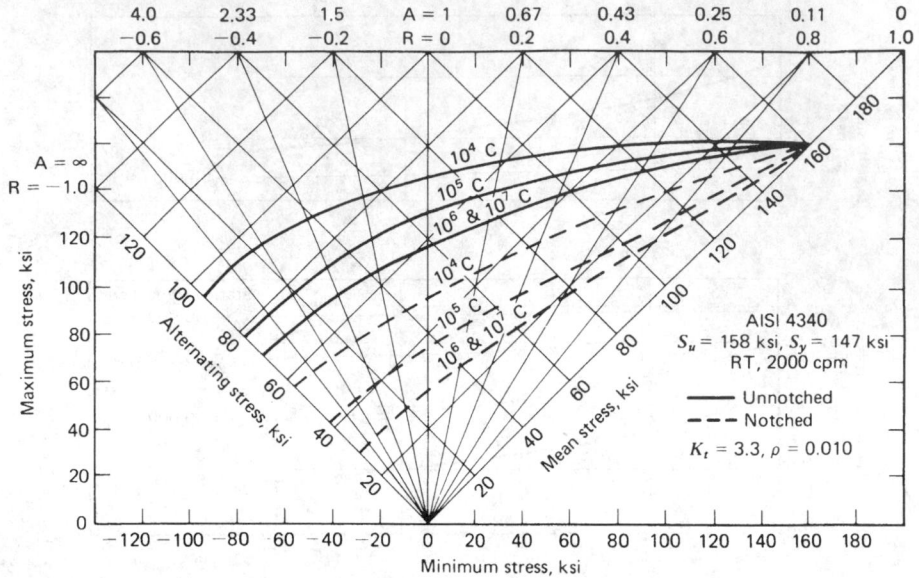

Fig. 18.36 Master diagram for 4340 steel. (From Ref. 41, p. 317.)

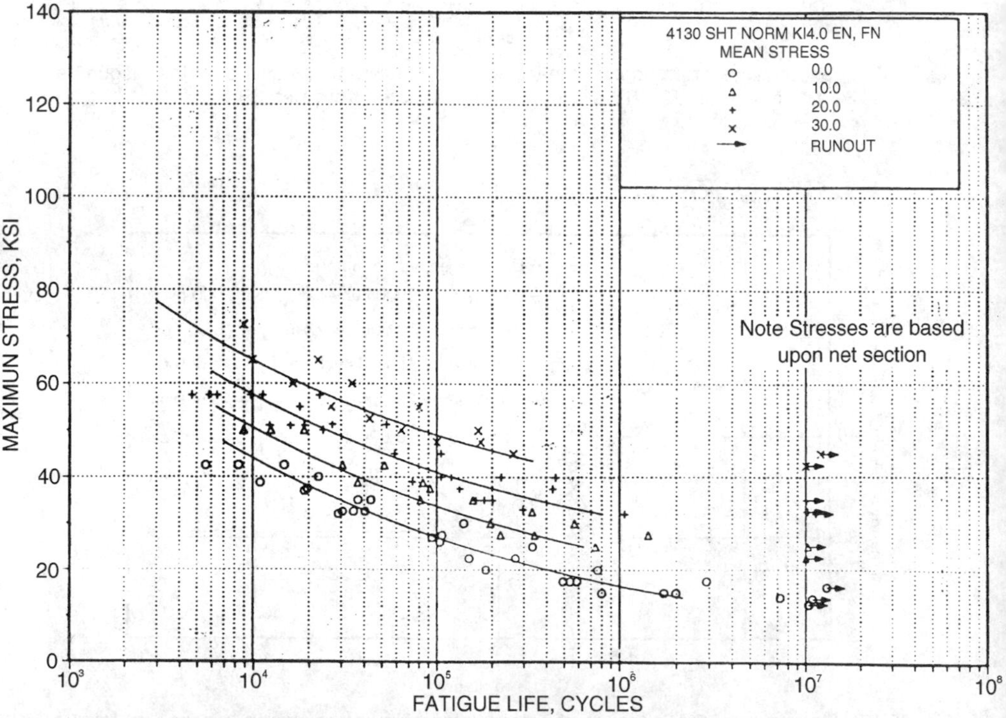

Fig. 18.37 Best-fit S–N curves for notched 4130 alloy steel sheet, $K_t = 4.0$. (From Ref. 11.)

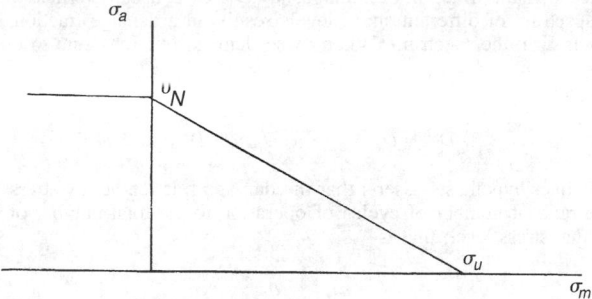

Fig. 18.38 Modified Goodman relationship.

The basic postulate adopted by all fatigue investigators working with spectrum loading is that operation at any given cyclic stress amplitude will produce *fatigue damage,* the seriousness of which will be related to the number of cycles of operation at that stress amplitude and also related to the total number of cycles that would be required to produce failure of an undamaged specimen at that stress amplitude. It is further postulated that the damage incurred is permanent and operation at several different stress amplitudes in sequence will result in an accumulation of total damage equal to the sum of the damage increments accrued at each individual stress level. When the total accumulated damage reaches a critical value, fatigue failure occurs. Although the concept is simple in principle, much difficulty is encountered in practice because the proper assessment of the amount of damage incurred by operation at any given stress level S_i for a specified number of cycles n_i is not straightforward. Many different *cumulative damage* theories have been proposed for the purposes of assessing fatigue damage caused by operation at any given stress level and the addition of damage increments to properly predict failure under conditions of spectrum loading. The first cumulative damage theory was proposed by Palmgren in 1924 and later developed by Miner in 1945. This linear theory, which is still widely used, is referred to as the *Palmgren–Miner hypothesis* or the *linear damage rule.* The theory may be described using the *S–N* plot shown in Fig. 18.39.

By definition of the *S–N* curve, operation at a constant stress amplitude S_1 will produce complete damage, or failure, in N_1 cycles. Operation at stress amplitude S_1 for a number of cycles n_1 smaller

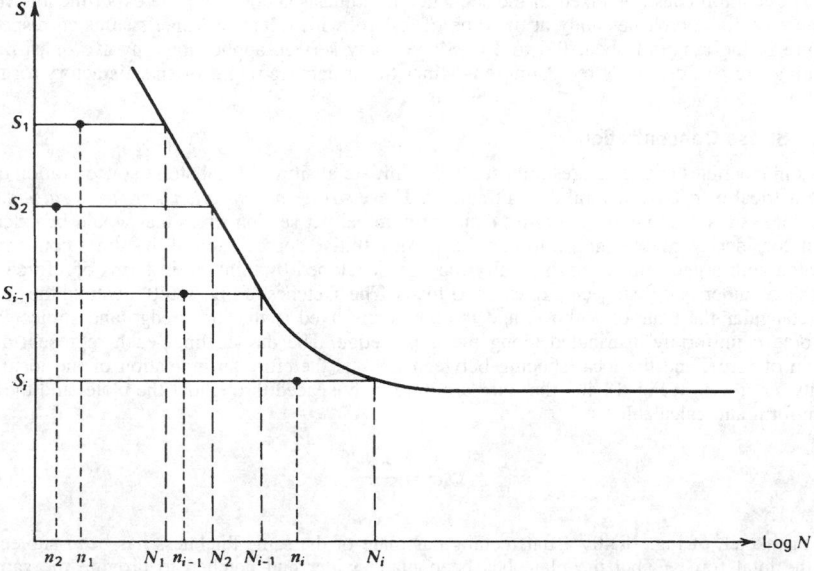

Fig. 18.39 Illustration of spectrum loading where n_i cycles of operation are accrued at each of the different corresponding stress levels S_i, and the N_i are cycles to failure at each S_i.

than N_1 will produce a smaller fraction of damage, say D_1. D_1 is usually termed the damage fraction. Operation over a spectrum of different stress levels results in a damage fraction D_i for each of the different stress levels S_i in the spectrum. When these damage fractions sum to unity, failure is predicted; that is,

Failure is predicted to occur if:

$$D_1 + D_2 + \cdots + D_{i-1} + D_i \geq 1 \qquad (18.43)$$

The Palmgren–Miner hypothesis asserts that the damage fraction at any stress level S_i is linearly proportional to the ratio of number of cycles of operation to the total number of cycles that would produce failure at that stress level; that is

$$D_i = \frac{n_i}{N_i} \qquad (18.44)$$

By the Palmgren–Miner hypothesis, then, utilizing (18.44), we may write (18.43) as

Failure is predicted to occur if:

$$\frac{n_1}{N_1} + \frac{n_2}{N_2} + \cdots + \frac{n_{i-1}}{N_{i-1}} + \frac{n_i}{N_i} \geq 1 \qquad (18.45)$$

or

Failure is predicted to occur if:

$$\sum_{j=1}^{i} \frac{n_j}{N_j} \geq 1 \qquad (18.46)$$

This is a complete statement of the Palmgren–Miner hypothesis or the linear damage rule. It has one important virtue, namely, *simplicity;* and for this reason it is widely used. It must be recognized, however, that in its simplicity certain significant influences are unaccounted for, and failure prediction errors may therefore be expected. Perhaps the most significant shortcomings of the linear theory are that no influence of the order of application of various stress levels is recognized, and damage is assumed to accumulate at the same rate at a given stress level without regard to past history. Experimental data indicate that the order in which various stress levels are applied does have a significant influence and also that damage rate at a given stress level is a function of prior cyclic stress history. Experimental values for the Miner's sum at the time of failure often range from about ¼ to about 4, depending on the type of decreasing or increasing cyclic stress amplitudes used. If the various cyclic stress amplitudes are mixed in the sequence in a quasi-random way, the experimental Miner's sum more nearly approaches unity at the time of failure, with values of Miner's sums corresponding to failure in the range of about 0.6 to 1.6. Since many service applications involve quasi-random fluctuating stresses, the use of the Palmgren–Miner linear damage rule is often satisfactory for failure protection.

18.5.6 Stress Concentration

Failures in machines and structures almost always initiate at sites of local stress concentration caused by geometrical or microstructural discontinuities. These *stress concentrations,* or *stress raisers,* often lead to local stresses many times higher than the nominal net section stress that would be calculated without considering stress concentration effects. An intuitive appreciation of the stress concentration associated with a geometrical discontinuity may be developed by thinking in terms of "force flow" through a member as it is subjected to external loads. The sketches of Fig. 18.40 illustrate the concept. The rectangular flat plate of width w and thickness t is fixed at the lower edge and subjected to a total force F uniformly distributed along the upper edge. The dashed lines each represent a fixed quantum of force, and the local spacing between lines is therefore an indication of the local force intensity, or stress. In Fig. 18.40a the lines are uniformly spaced throughout the plate, and the stress σ is uniform and calculable as

$$\sigma = \frac{F}{wt} \qquad (18.47)$$

In the sketch of Fig. 18.40b a flat rectangular plate of the same thickness has been subjected to the same total force F, but the plate has been made wider and notched to provide the same net section width w at the site of the notch. The lines of force flow may be visualized in very much the same way that streamlines would be visualized in the steady flow of a fluid through a channel with

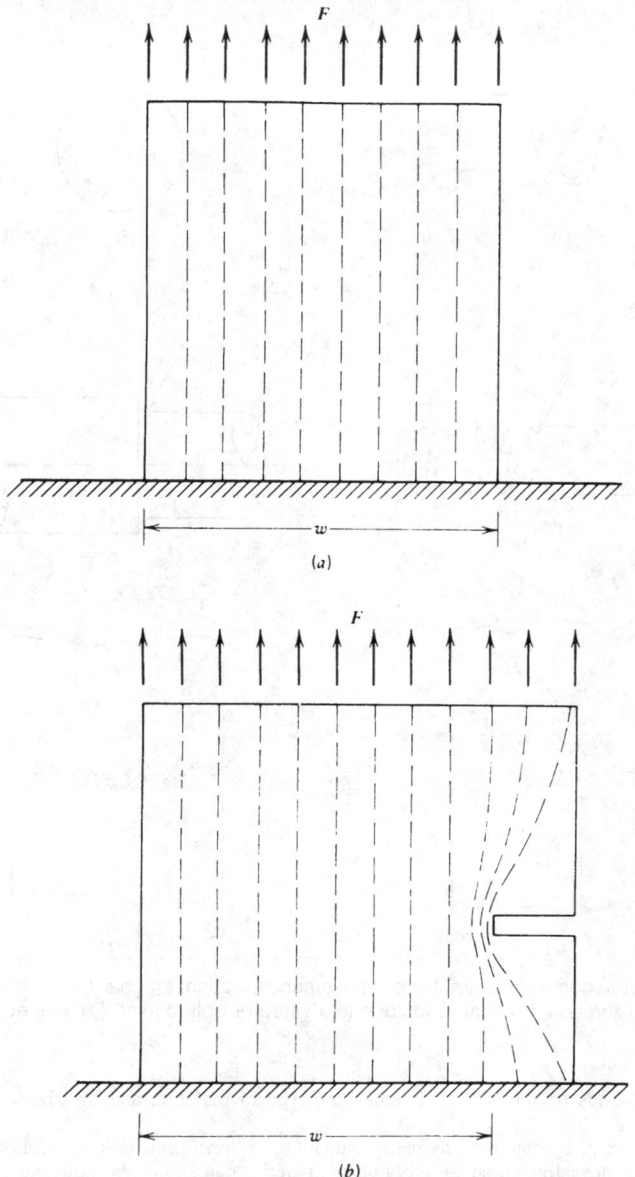

Fig. 18.40   Intuitive concept of stress concentration: (a) without stress concentration; (b) with stress concentration.

the same shape as the plate cross section. No force can be supported across the notch, and therefore the lines of force flow must pass around the root of the notch. In so doing, force flow lines crowd together locally near the root of the notch, producing a higher force intensity, or stress, at the notch root. Thus, the local stress is raised or concentrated near the notch root, and even though the net section nominal stress is still properly calculated by (18.47), the actual local stress at the root of the notch may be many times higher than the calculated nominal stress. Many common examples of stress concentration may be cited, some of which are illustrated in Fig. 18.41. Discontinuities at the roots of gear teeth, at the corners of keyways in shafting, at the roots of screw threads, at the fillets of shaft shoulders, around rivet holes and bolt holes, and in the neighborhood of welded joints all constitute stress raisers that usually must be considered by a designer. The seriousness of the stress

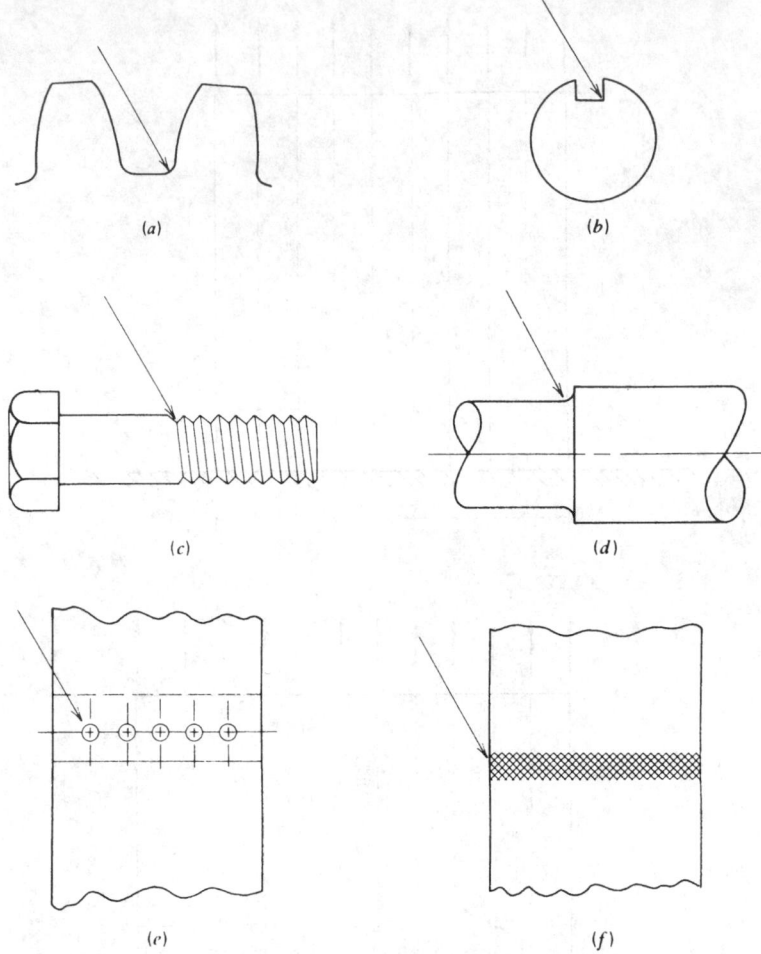

Fig. 18.41 Some common examples of stress concentration: (a) gear teeth; (b) shaft keyway; (c) bolt threads; (d) shaft-shoulder; (e) riveted or bolted joint; (f) welded joint.

concentration depends on the type of loading, the type of material, and the size and shape of the discontinuity.

Stress raisers may be classified as being either *highly local* or *widely distributed*. Highly local stress raisers are those for which the volume of material containing the concentration of stress is negligibly small compared to the overall volume of the stressed member. Widely distributed stress raisers are those for which the volume of material containing the concentration of stress is a significant portion of the overall volume of the stressed member.

The theoretical elastic stress concentration factor, K_t, is defined to be the ratio of the actual maximum local stress in the region of the discontinuity to the nominal net section stress calculated by simple theory as if the discontinuity exerted no stress concentration effect; that is,

$$K_t = \frac{\text{actual maximum stress}}{\text{nominal stress}} \tag{18.48}$$

and the magnitude of K_t is found to be a function of geometry and type of loading, but not a function of material. It should be noted that this definition of K_t is valid only for stress levels within the elastic range, and must be suitably modified if stresses are in the plastic range.

The fatigue stress concentration factor, K_f, is defined to be the ratio of the effective fatigue stress at the root of the discontinuity to the nominal fatigue stress calculated as if the notch has no stress concentrating effect. This definition is made for the high-cycle fatigue range and must be suitably modified for the low-cycle fatigue range. Thus, the fatigue stress concentration factor may be defined as

$$K_f = \frac{\text{effective fatigue stress}}{\text{nominal fatigue stress}} \qquad (18.49)$$

Stress concentration factors are determined in a variety of different ways, including direct measurement of strain, utilization of photoelastic techniques, application of the principles of the theory of elasticity, and finite-element analysis. Numerical values for a wide variety of geometries and types of loading are presented in Ref. 46. Examples of typical charts for selection of stress concentration factor K_t are shown in Fig. 18.42. With a sharply notched specimen or machine part, it is clear that even a moderate load may produce actual stresses at the root of the notch that exceed the yield point of the material locally. The local yielding causes a redistribution of stresses, and the theoretical elastic stress concentration factor K_t no longer describes the ratio of actual to nominal stresses accurately, since the actual maximum stress is relatively lower compared to the nominal stress than it would be if the material remained elastic. That is, the stress concentration factor is diminished in magnitude by local plastic flow, whereas the local strain is made larger than would be predicted by elastic theory.

Mathematical solutions of elastic–plastic stress and strain distributions around notches are relatively difficult to obtain, even using numerical solutions and digital computer techniques. One of the more successful approximations for stress concentration due to a circular hole in a very wide plate under tension has been given as[47]

$$K = 1 + 2\frac{E_s}{E} \qquad (18.50)$$

where E = Young's modulus
E_s = secant modulus
K = stress concentration factor

Figure 18.43 illustrates values of stress concentration factor and strain concentration factor computed for a specific case by (18.50) and compared with values measured using very small strain gages. The agreement between calculated and measured values is good. Unlike the theoretical stress concentration factor K_t, the fatigue stress concentration factor K_f is a *function of the material* as well as geometry and type of loading. To account for the influence of material characteristics, a *notch sensitivity index q* has been defined to relate the actual effect of a notch on fatigue strength of a material to the effect that might be predicted solely on the basis of elastic theory. The definition of notch sensitivity index q is given by

$$q = \frac{K_f - 1}{K_t - 1} \qquad (18.51)$$

where K_f = fatigue stress concentration factor
K_t = theoretical stress concentration factor
q = notch sensitivity index valid for high-cycle fatigue range

The reason for subtracting unity from the numerator and denominator in this definition is to provide a scale for q that ranges from zero for no notch effect to unity for full notch effect. That is, for full notch effect K_f is equal to K_t. The notch sensitivity index has been found to be a function of both material and notch radius. Scatter in the experimental data is a serious problem in evaluating notch sensitivity index, as may be seen in the experimental results shown in Fig. 18.44. The notch sensitivity index for a range of steels and an aluminum alloy are shown in Fig. 18.45 for axial, bending, and torsional loading. These curves provide sufficient accuracy for most design applications and clearly demonstrate that the notch sensitivity index is a function of both the material and the notch root radius. An expression for fatigue stress concentration factor may be written from (18.51) as

$$K_f = q(K_t - 1) + 1 \qquad (18.52)$$

where the theoretical elastic stress concentration factor K_t may be determined on the basis of geometry and loading from handbook charts such as those depicted in Fig. 18.42. The notch sensitivity index q may be read from charts, such as the one shown in Fig. 18.45.

For uniaxial states of cyclic stress it is sometimes convenient to use K_f as a "strength reduction factor" rather than as a "stress concentration factor." That is, for uniaxial stressing only, a designer may choose to divide the fatigue limit by K_f rather than multiplying the applied nominal cyclic stress

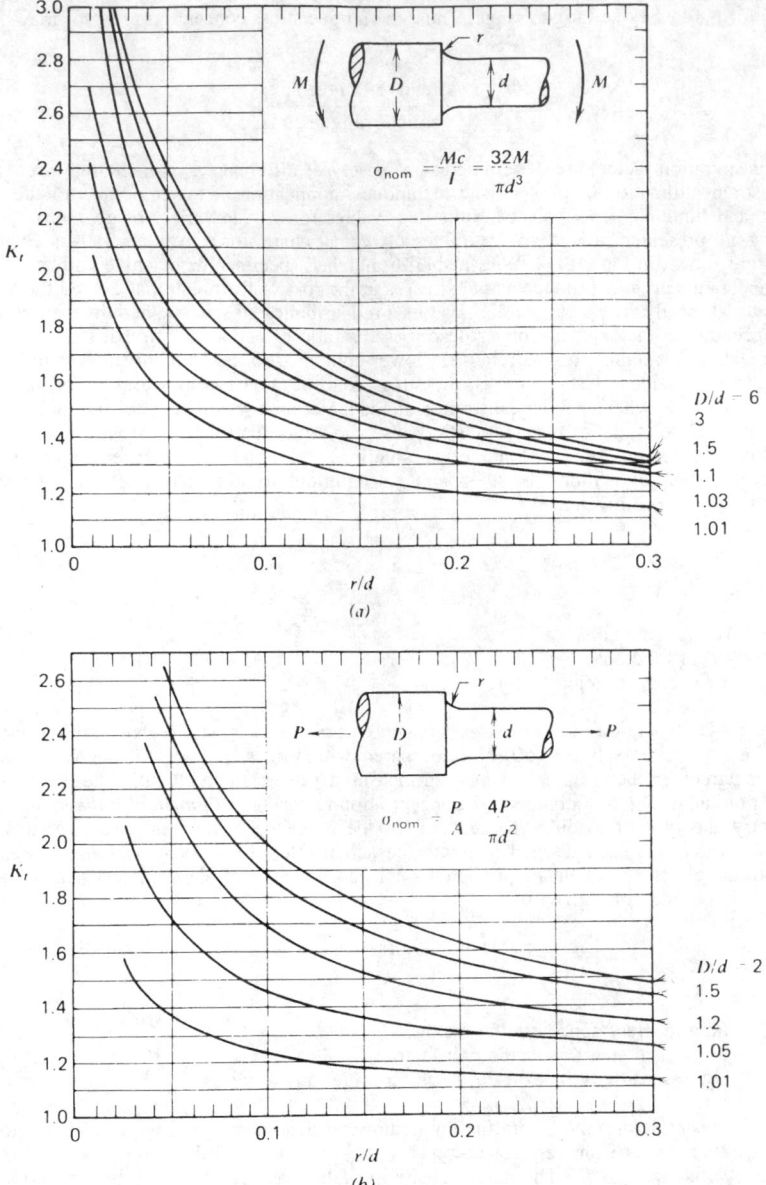

Fig. 18.42 Stress concentration factors for a shaft with a fillet subjected to (a) bending, (b) axial load, or (c) torsion. (From Ref. 46, adapted with permission from John Wiley & Sons, Inc.)

by K_f. Although, conceptually, it is clearly more correct to think of K_f as a stress concentration factor, computationally, it is equivalent, and often simpler, to use K_f as a strength reduction factor. For multiaxial states of stress, however, K_f must be used as a stress concentration factor, since an appropriate value of K_f used as a strength reduction factor would be undefined.

The fatigue stress concentration factor (or strength reduction factor) determined from (18.52) is strictly applicable only in the high-cycle fatigue range, that is, for cycle lives of 10^5–10^6 cycles and larger. For ductile materials and static loads, effects of stress concentration may usually be neglected. Thus, in the intermediate- and low-cycle life range from a quarter cycle (static load) up to about 10^5–10^6 cycles, the stress concentration factor changes from unity to K_f. As shown in Fig. 18.46,

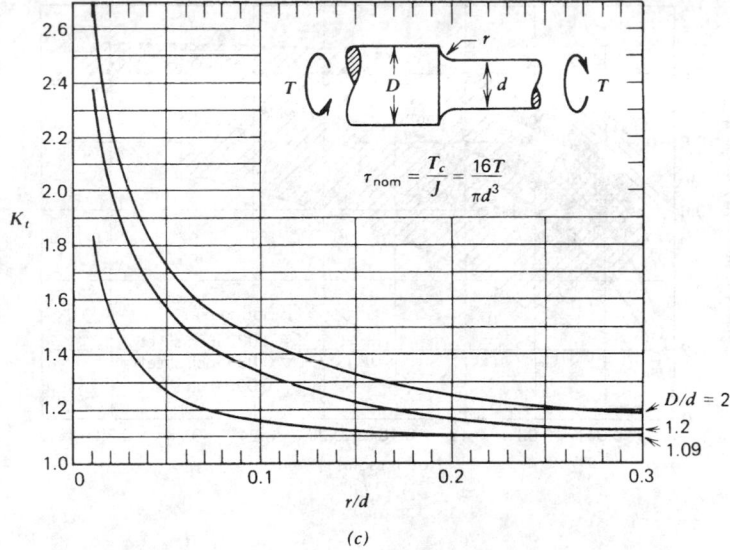

$$\tau_{nom} = \frac{T_c}{J} = \frac{16T}{\pi d^3}$$

$D/d = 2$
1.2
1.09

r/d

(c)

Fig. 18.42   (Continued)

the notched and unnotched S–N curves converge as they approach the low-cycle end of the range and coincide at the quarter cycle point A. Many materials exhibit fatigue stress concentration factors very near unity for lives less than 1000 cycles. Estimates of fatigue stress concentration factor are often made by constructing a straight line on a semilogarithmic S–N plot from the ultimate strength at a life of 1 cycle to the unnotched fatigue strength divided by K_f at a life of 10^6 cycles. Such a straight line construction is shown in Fig. 18.46. The ratio of unnotched to notched fatigue strength values read at a specific life then becomes the estimate of fatigue stress concentration factor to be used for that life.

Finally, it should be noted that experimental investigations have indicated that for fatigue of *ductile materials* the fatigue stress concentration factor should be applied *only* to the *alternating component* of stress and not to the steady component of stress that exists in any nonzero mean cyclic stress. For

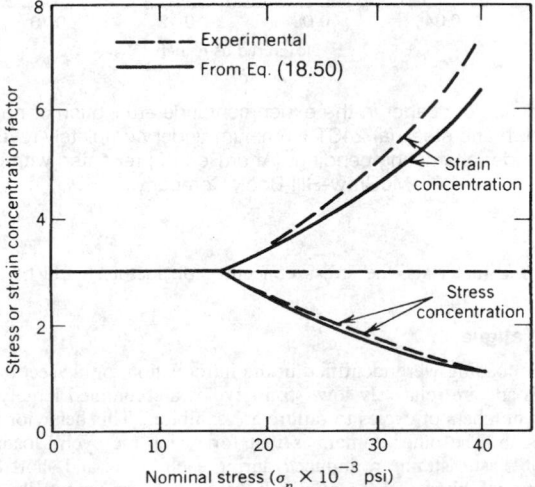

Fig. 18.43 Effect of plasticity on stress and strain concentration. (After Ref. 41.)

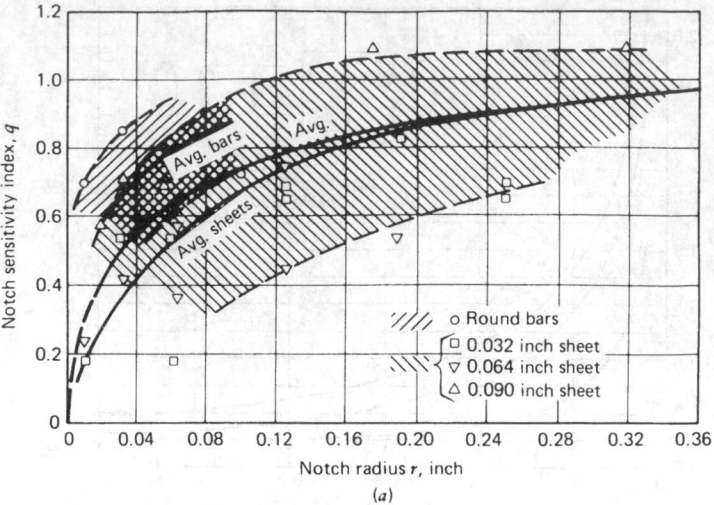

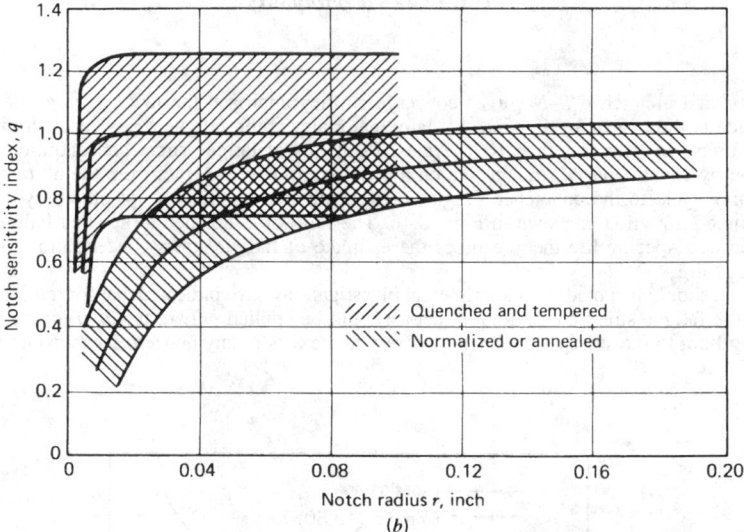

Fig. 18.44 An indication of scatter in the experimental determination of notch sensitivity index for alloys of aluminum and steel. (*a*) 24ST aluminum under completely reversed axial loading. (*b*) Steel alloys under alternating bending. (After Ref. 48; reprinted with permission from McGraw-Hill Book Company.)

fatigue loading of *brittle* materials, the stress concentration factor should be applied to the steady component as well.

18.5.7 Low-Cycle Fatigue

Two domains of cyclic loading were identified in the introduction to this section. One domain is that for which the cyclic loads are relatively low, strain cycles are confined largely to the elastic range, and long lives or high numbers of cycles to failure are exhibited. This behavior has traditionally been called high-cycle fatigue. The other domain is that for which the cyclic loads are relatively high, significant amounts of plastic strain are induced during each cycle, and short lives or low numbers of cycles to failure are exhibited if these relatively high loads are repeatedly applied. This type of behavior has been commonly called *low-cycle fatigue* or, more recently, cyclic strain-controlled fa-

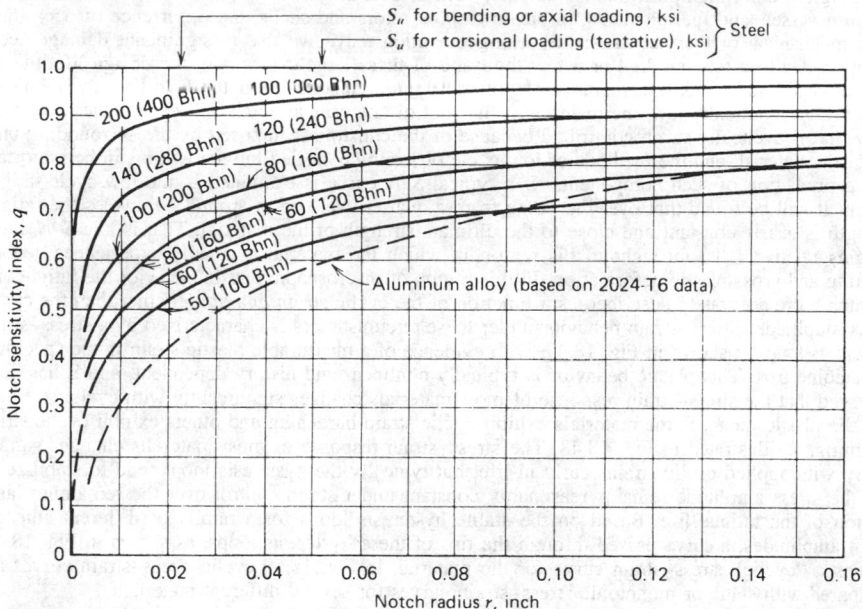

Fig. 18.45 Curves of notch sensitivity index versus notch radius for a range of steels and an aluminum alloy subjected to axial, bending, and torsional loading. (After Ref. 49; reprinted with permission from McGraw-Hill Book Company.)

tigue. The transition from low-cycle-fatigue behavior to high-cycle fatigue behavior generally occurs in the range from about 10^4 to 10^5 cycles, and some investigators define the low-cycle-fatigue range to be failure in 50,000 cycles or less.[50] Although the usual objective of an engineering designer is to provide long life, there are several circumstances in which the low-cycle fatigue or strain-controlled life response is of great importance. For example, in the design of high-performance devices such as missiles and rockets, the total design lifetime may be only a few hundred or a few thousand cycles from launch to delivery, and low-cycle-fatigue analysis and design methods are of direct interest. In

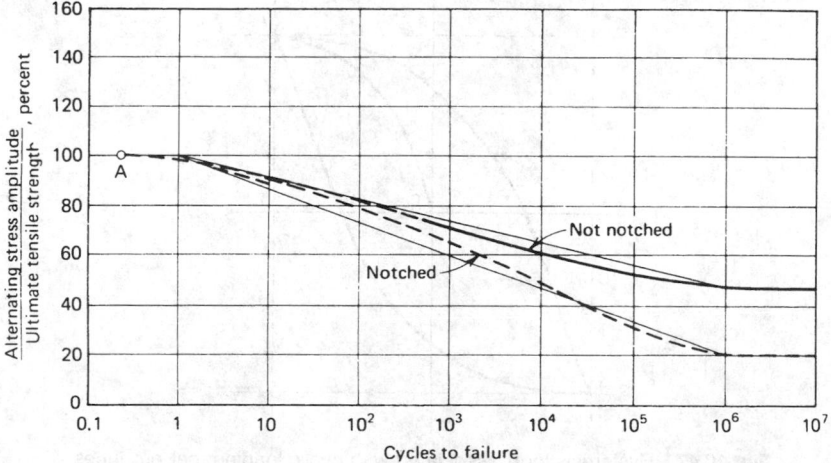

Fig. 18.46 S–N curves for notched and unnotched specimens subjected to completely reversed axial loading. (After Ref. 28, by permission of The MIT Press.)

the design of other high-performance devices, such as aircraft gas-turbine blades and wheels, nuclear pressure vessels and fuel elements, or steam turbine rotors and shells, the occurrence of occasional large mechanical or thermal transients during operation may give rise to significant damage accumulation due to a few hundred or a few thousand of these large cycles over the design lifetime, so that low-cycle-fatigue design methods are of great importance. Even if the loads on a machine or structure are nominally low, the material at the root of any critical notch will experience local plasticity that is cyclically strain controlled because of the constraints imposed by the surrounding bulk of elastic material, and the methods of low-cycle or strain-controlled fatigue will again be important in life prediction of such components. If a typical S–N curve is examined in the low-cycle-fatigue region, it will be found that over the range from a quarter cycle up to around 10^3 cycles the fatigue strength is nearly constant and close to the ultimate strength of the material. That is, the S–N curve remains relatively flat throughout this region in which the material cyclically experiences general yielding and gross plastic deformation. In this region of macroscopic plastic behavior the fatigue life is much more accurately described as a function of the cyclic *strain amplitude* rather than the cyclic stress amplitude. Stress–strain behavior under these circumstances is characterized by a stress–strain hysteresis loop, as shown in Fig. 18.47, with evidence of a measurable plastic strain in the specimen or machine part. This plastic behavior is typically nonlinear and history dependent, and it has been observed that the stress–strain response of most materials changes significantly with cyclic straining into the plastic range. Some materials exhibit cyclic strain hardening and others exhibit cyclic strain softening, as illustrated in Fig. 18.48. The stress–strain response of most materials changes significantly with applied cyclic strains early in life, but typically the hysteresis loops tend to stabilize so that the stress amplitude remains reasonably constant under strain control over the remaining large portion of the fatigue life. Based on the stable hysteresis loops for a family of different constant strain amplitudes, a curve passed through the tips of these hysteresis loops, as shown in Fig. 18.49, defines a "cyclic" stress–strain curve for the material. In Fig. 18.50 cyclic stress–strain curves are compared with static or monotonic stress–strain curves for several different materials.

The usual method of displaying the results of low-cycle fatigue tests is to plot the logarithm of strain amplitude or strain range versus the logarithm of number of cycles (or reversals) to failure. Sometimes the plastic strain amplitude or strain range is plotted, and sometimes the total strain amplitude or strain range is plotted as the ordinate. Early experimental investigations had indicated that if the plastic strain amplitude were plotted versus cycles to failure on a log–log plot, the data would approximate a straight line with a slope of about −0.5. Subsequent investigations have indicated that the slope ranges from about −0.7 to −0.5. Such plots seem to be remarkably similar for a wide range of materials,[51] as indicated in Fig. 18.51. Experimental evidence accumulated by various investigators in recent years seems to indicate that the cyclic life is better related to total strain than to plastic strain, especially at the longer life end of the low-cyclic range. An example of a plot of strain amplitude versus life is shown in Fig. 18.52, separately showing the plastic strain amplitude and the total strain amplitude for a nickel–steel alloy.

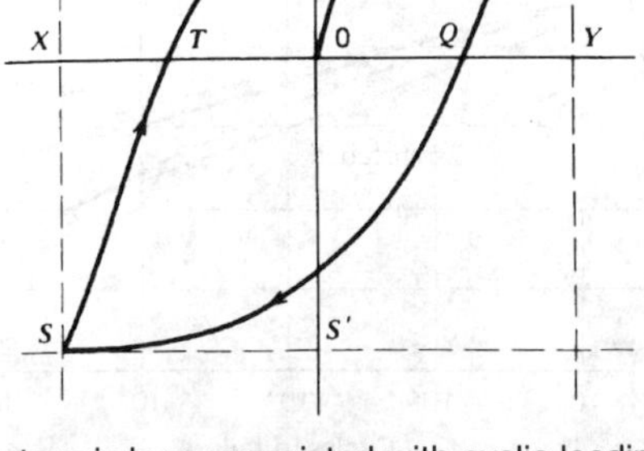

Fig. 18.47 Hysteresis loop associated with cyclic loading that produces low-cycle-fatigue damage.

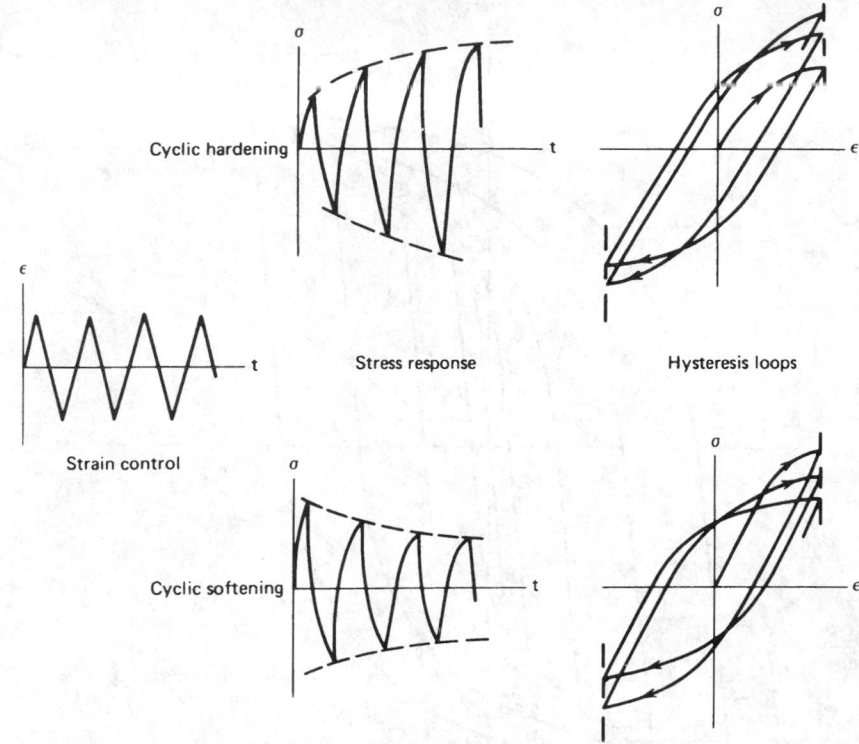

Fig. 18.48 Illustration of cyclic strain hardening and cyclic strain softening phenomena under strain control. (From Ref. 55, copyright ASTM; reprinted with permission.)

Data such as those shown in Fig. 18.51 led to the proposal of an empirical equation relating plastic strain $\Delta\epsilon_p$ to failure life N_f for completely reversed strain cycling under uniaxial stress conditions in the low-cycle-fatigue regime. The relationship, independently proposed by Manson[52] and Coffin,[53] may be expressed as

$$\frac{\Delta\epsilon_p}{2} = \epsilon'_f(2N_f)^c \tag{18.53}$$

where $\Delta\epsilon_p/2$ = plastic strain amplitude
ϵ'_f = fatigue ductility coefficient, defined as strain intercept at one load reversal, that is, at $2N_f = 1$ (see Fig. 18.53)
$2N_f$ = total reversals to failure
c = fatigue ductility exponent defined as slope of plastic strain amplitude versus reversals to failure curve on log–log plot (see Fig. 18.53)

Later work by many investigators, capitalizing on the Manson–Coffin equation of (18.53), has indicated that total strain amplitude, the sum of elastic strain amplitude plus plastic strain amplitude, may be better correlated to life. As shown in Fig. 18.53, the total strain amplitude is the sum of elastic plus plastic components. This has been modeled mathematically by Morrow et al.[54] as

$$\frac{\Delta\epsilon}{2} = \frac{\sigma'_f}{E}(2N_f)^b + \epsilon'_f(2N_f)^c \tag{18.54}$$

where constants b and σ'_f/E are the slope and one-reversal intercept of the plastic curve in Fig. 18.53, and constants c and ϵ'_f are the slope and one-reversal intercept of the plastic curve in Fig. 18.53. Typically,[55] b ranges from about -0.05 to -0.15 and c ranges from about -0.5 to -0.8.

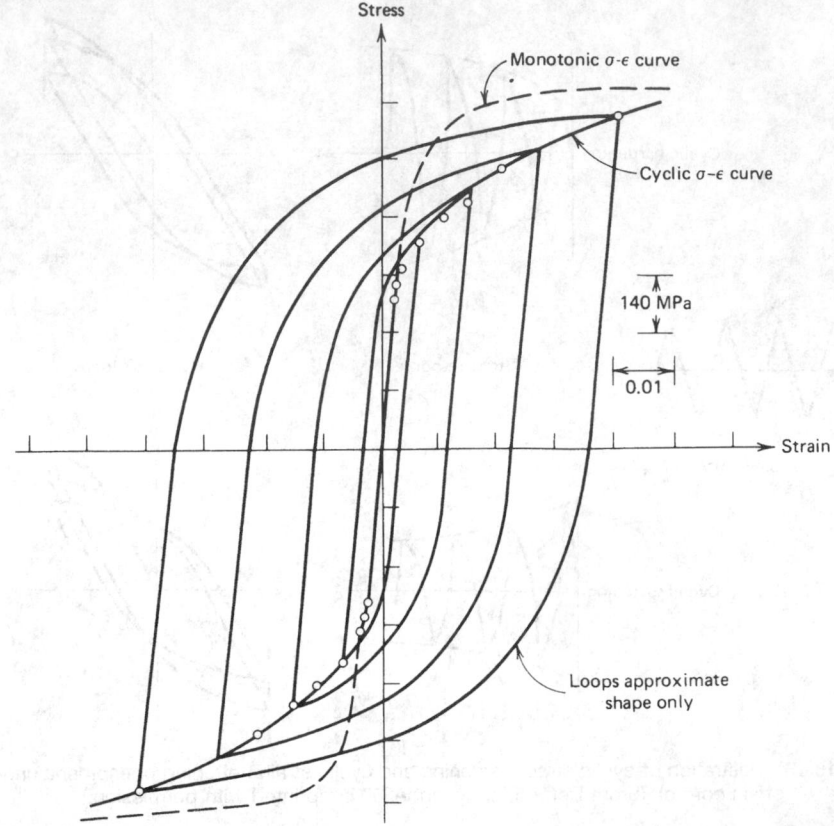

Fig. 18.49 Cyclic stress–strain curve compared to monotonic stress–strain curve for SAE 4340 steel. (From Ref. 55, copyright ASTM; reprinted with permission.)

Although these constants are best evaluated from cyclic testing, they may be approximated from static properties, if fatigue data are unavailable. This may be done by taking σ_f' equal to true fracture strength σ_f, ϵ_f' equal to true fracture ductility ϵ_f, c equal to -0.6 and b equal to $-0.16 \log 2\sigma_f/\sigma_u$. However, actual fatigue data should be used where available.

From the schematic representation of (18.54) in Fig. 18.53, it may be noted that at short lives the plastic strain amplitude component dominates, whereas at longer lives the elastic strain amplitude component dominates. The point at which the elastic and plastic curves intersect has been called the "transition life." A thoughtful consideration of (18.54) and Fig. 18.53 leads to the observation that for design lives less than the transition life, materials with high fatigue ductility (high fracture ductility) are superior, whereas design lives greater than the transition life demand materials with high values of true fracture strength. This contradictory set of requirements, namely, high strength with high ductility, requires careful consideration on the part of the designer to match the appropriate material to the application, with careful attention to the magnitude of operating strain amplitude. This point is emphasized in Fig. 18.54 where three idealized materials are shown, one very high strength material (strong), one very ductile material (ductile), and one whose properties lie between the two extremes (tough). These curves cross at about 10^3 cycles (2×10^3 reversals) at a total cyclic strain amplitude of about 0.01. Thus, one would select the "strong" material for design life requirements greater than about 10^3 cycles, pick the "ductile" material for design life requirements shorter than about 10^3 cycles, and "optimize" with the "tough" material for spectrum loading of a more complicated nature. It is interesting to note that all types of materials seem to have about the same fatigue resistance for total strain amplitude around 0.01, corresponding to a failure life of about 10^3 cycles. Presently attainable combinations of true fracture strength and true fracture ductility are illustrated in Fig. 18.55.

The effects of nonzero mean *strain* under low-cycle fatigue conditions have been studied by relatively few investigators, principally for the case of tensile mean strain. The experimental results

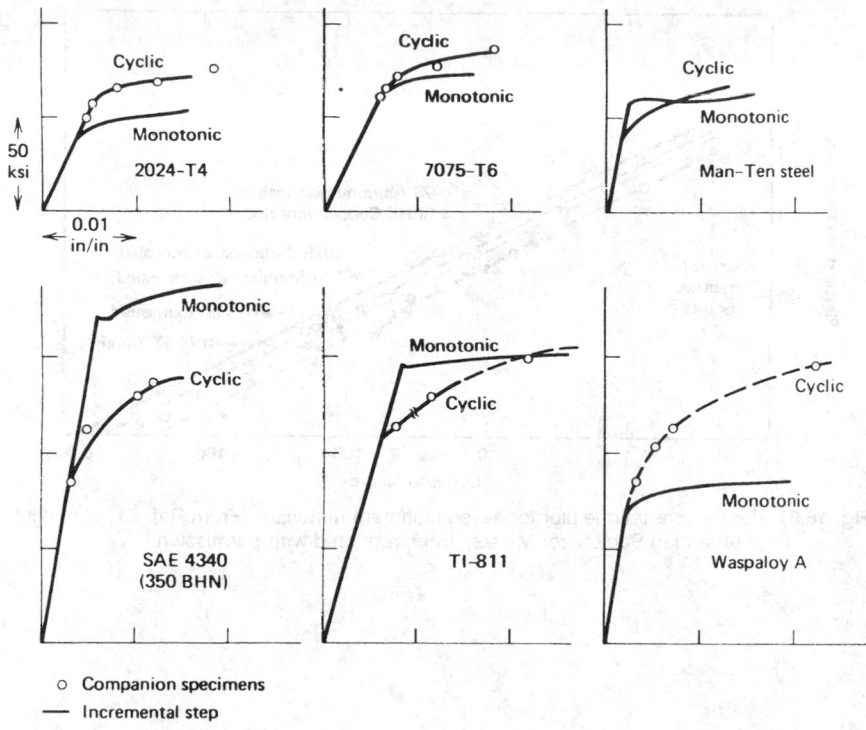

o Companion specimens

— Incremental step

Material	Condition	0.2 percent yield strength, monotonic σ_{yp}/cyclic σ_{yp} ksi	Strain-hardening exponent n(monotonic)/n'(cyclic)	Cyclic behavior
OFHC copper	Annealed	3/20	0.40/0.15	Hardens
	Partial annealed	37/29	0.13/0.16	Stable
	Cold worked	50/34	0.10/0.12	Softens
2024 aluminum alloy	T4	44/65	0.20/0.11	Hardens
7075 aluminum alloy	T6	68/75	0.11/0.11	Hardens
Man-Ten steel	As-received	55/50	0.15/0.16	Softens and hardens
SAE 4340 steel	Quenched and tempered, 350 BHN	170/110	0.066/0.14	Softens
Ti-8Al-1Mo-1V	Duplex annealed	145/115	0.078/0.14	Softens and hardens
Waspaloy		79/102	0.11/0.17	Hardens
SAE 1045 steel	Quenched and tempered, 595 BHN	270/250	0.071/0.14	Stable
	Quenched and tempered, 500 BHN	245/185	0.047/0.12	Softens
	Quenched and tempered, 450 BHN	220/140	0.041/0.15	Softens
	Quenched and tempered, 390 BHN	185/110	0.044/0.17	Softens
SAE 4142 steel	As-quenched, 670 BHN	235/...	0.14/...	Hardens
	Quenched and tempered, 560 BHN	245/250	0.092/0.13	Stable
	Quenched and tempered, 475 BHN	250/195	0.048/0.12	Softens
	Quenched and tempered, 450 BHN	230/155	0.040/0.17	Softens
	Quenched and tempered, 380 BHN	200/120	0.051/0.18	Softens

Fig. 18.50 Cyclic stress–strain behavior of several materials. (From Ref. 56, copyright ASTM; reprinted with permission.)

of these few investigations indicate that the effect of a compressive mean strain on low-cycle fatigue life is essentially the same as the effect of a tensile mean strain if their magnitudes are the same. These results also indicate that mean strain effects are of primary importance only in the operating range where the *plastic* strain component dominates, that is, at design lives less than the transition life for the material.

The effects of nonzero mean *stress* are of primary importance only in the operating range where the *elastic* strain component dominates, that is, at design lives greater than the transition life of the material.

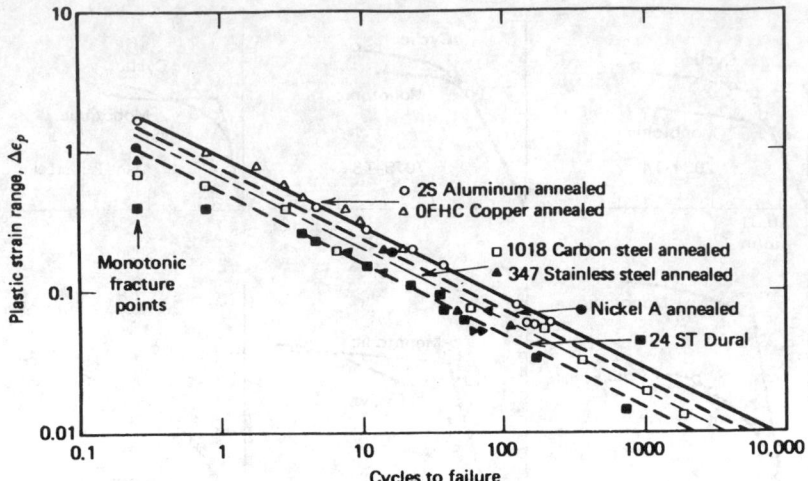

Fig. 18.51   Low-cycle-fatigue plot for several different materials. (From Ref. 51, copyright
American Society for Metals, 1959; reprinted with permission.)

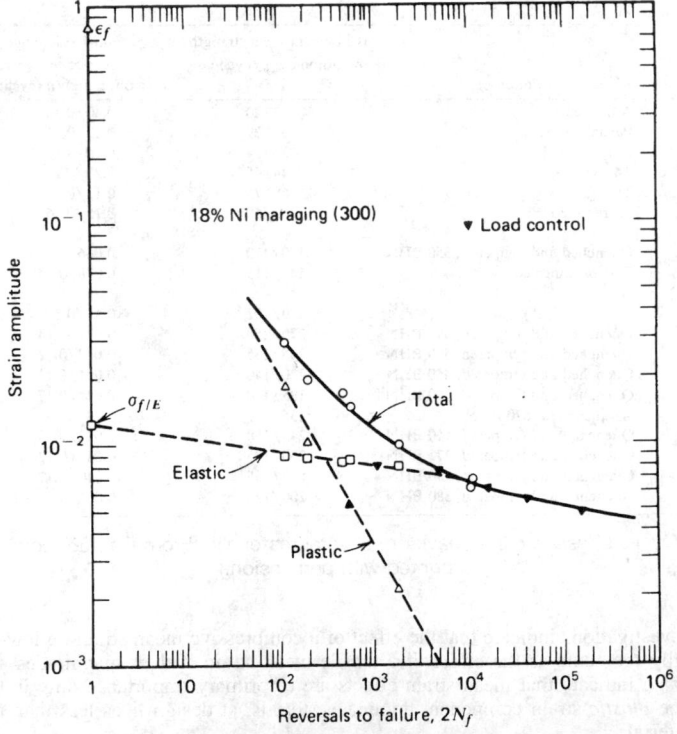

Fig. 18.52   Strain amplitude versus life for 18% Ni maraging steel, separately showing elastic,
plastic, and total components of the strain. (From Ref. 55, copyright ASTM;
reprinted with permission.)

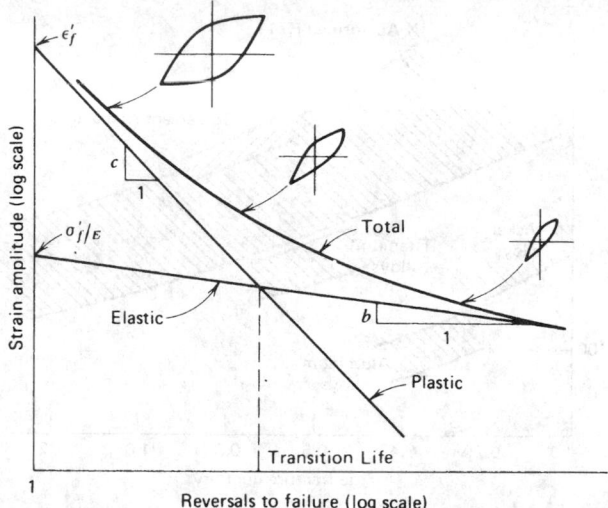

Fig. 18.53 Schematic representation of elastic, plastic, and total strain amplitude versus fatigue life. (From Ref. 55, copyright ASTM; reprinted with permission.)

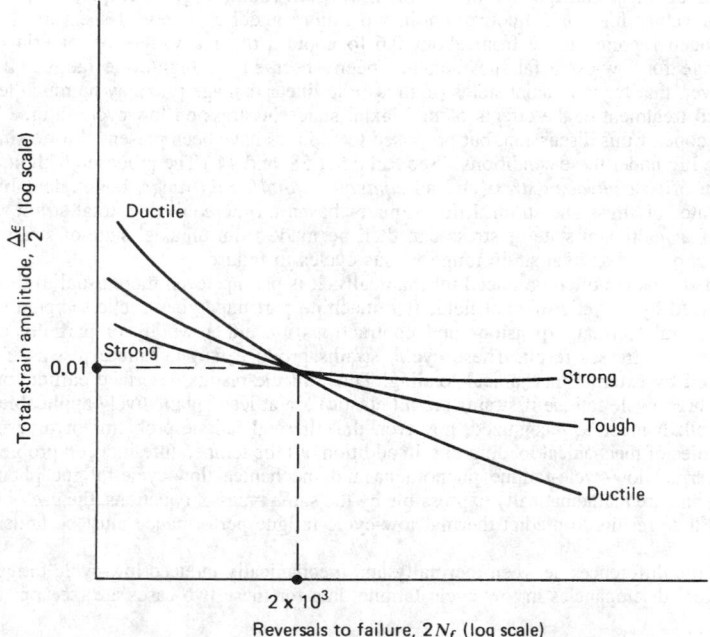

Fig. 18.54 Idealized representation of cyclic strain failure resistance of various types of materials. (After Ref. 55, copyright ASTM; reprinted with permission.)

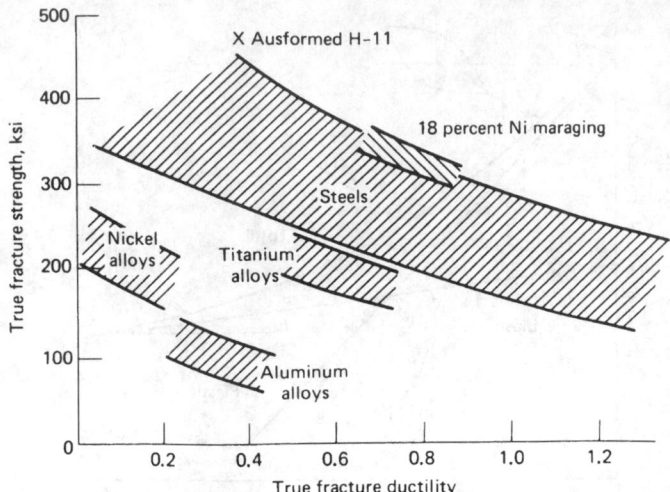

Fig. 18.55 Monotonic fracture strength–ductility combinations presently attainable for various alloy classes. (From Ref. 55, copyright ASTM; reprinted with permission.)

Just as in the case of high-cycle fatigue, the assessment of low-cycle-fatigue damage under conditions where the cyclic strain amplitude ranges over a spectrum of values requires the use of a cumulative damage theory. Many investigators have studied cumulative damage effects in low-cycle fatigue with the general conclusion that a linear damage rule of the Palmgren–Miner type yields acceptable results if local stress–strain behavior can be accurately determined as a function of applied loads and if cycle counting is properly conducted.

If, for example, the fatigue-modified Neuber rule described in Section 18.5.8 is utilized to determine local stress and strain amplitude history and if the rain-flow cycle counting method, also described in Section 18.5.8, is properly applied to the local strain–time spectrum, results of using the Palmgren–Miner linear damage rule of (18.46) have been found to give acceptable life predictions. The range of values for $\Sigma(n/N)$ corresponding to failure under a variety of spectrum loading conditions has been reported to be from about 0.6 to about 1.6 for a variety of materials. This is a narrower range for low-cycle fatigue than has been observed for high-cycle fatigue. It should be noted, however, that for multiaxial states of stress, the linear damage rule may be much less reliable.

A detailed treatment of the effects of multiaxial states of stress on low-cycle-fatigue behavior is beyond the scope of this discussion, but proposed techniques have been presented for estimating low-cycle-fatigue life under these conditions. (See Refs. 57, 58, and 44.) The proposed technique involves the definition of an *equivalent stress* and an *equivalent total strain range,* both calculable from the multiaxial states of stress and strain. Life estimates based on the equivalent total strain range under conditions of a multiaxial state of stress can then be made from uniaxial state-of-stress low-cycle-fatigue data expressed as total strain range versus cycles to failure.

Although strains are often produced mechanically, it is perhaps even more usual to find the cyclic strains produced by a cyclic thermal field. If a machine part undergoes cyclic temperature changes and if the natural thermal expansions and contractions are either wholly or partially constrained, cyclic strains and stresses result. These cyclic strains produce fatigue failure just as if the strains were produced by external mechanical loading. Thus, all the results described earlier for low-cycle fatigue (and high-cycle fatigue if strains are all elastic) are at least qualitatively applicable to thermal fatigue as well. It must be recognized, however, that thermal fatigue problems involve not only all the complexities of mechanical loading but, in addition, all the temperature-induced problems as well.

While thermal low-cycle-fatigue phenomena and mechanical low-cycle-fatigue phenomena are very similar, and are mathematically expressible by the same types of equations, the use of mechanical low-cycle-fatigue results to predict thermal low-cycle-fatigue performance must be undertaken with care.

Some of the differences between thermally and mechanically induced low-cycle fatigue that give rise to apparent discrepancies in low-cycle-fatigue data for these two cases are (see pp. 270–272 of Ref. 58):

1. In thermal fatigue the plastic strain tends to become concentrated in the hottest regions of the body, since the yield point is locally reduced at these hottest regions.

2. In thermal fatigue there is often a localized region of strain developed by virtue of plastic flow during the compressive branch of the strain cycle to produce a bulging at the hottest region. This is followed by a necking tendency adjacent to the bulge caused by plastic flow during the tensile branch of the strain cycle upon cooling.

3. Cyclic variations in temperature may, in and of themselves, have important effects on the material properties and ability to resist low-cycle-fatigue failure.

4. There may be interaction effects caused by superposition of simultaneous variations in temperatures and strains.

5. Rates at which the strain cycling is induced may have an important effect, since the testing speeds in thermal fatigue tests are often greatly different from the rates used in mechanical low-cycle–fatigue tests.

For these reasons caution is necessary in the prediction of thermal low-cycle-fatigue behavior from mechanical low-cycle-fatigue results or vice versa.

18.5.8 Three-Phase Approach for Fatigue Life Prediction

In recent years it has been recognized that the fatigue failure process involves three phases. A crack initiation phase occurs first, followed by a crack propagation phase; finally, when the crack reaches a critical size, the final phase of unstable rapid crack growth to fracture completes the failure process. The modeling of each of these phases has been under intense scrutiny. For the crack initiation phase, the most promising approach seems to be the "local stress–strain method."

The basic premise of the local stress–strain approach is that the local fatigue response of the material at the critical point, that is, the site of crack initiation, is analogous to the fatigue response of a small, smooth specimen subjected to the same cyclic strains and stresses.[44,45,54,61] This concept is illustrated schematically in Fig. 18.56 for a simple notched plate under cyclic loading. The cyclic stress–strain response of the critical material may be determined from the characterizing smooth specimen through appropriate laboratory testing. To properly perform such laboratory tests, the local cyclic stress–strain history at the critical point in the structure must be determined, by either analytical or experimental means. Thus, valid stress analysis procedures, finite-element modeling, or experimental strain measurements are necessary, and the ability to properly account for plastic behavior must be included. In performing smooth specimen tests of this type it must be recognized that the phenomena of cyclic hardening, cyclic softening, and cycle-dependent stress relaxation, as well as sequential loading effects and residual stress effects, may be experienced by the specimen as it accumulates fatigue damage presumed to be the same as at the critical point in the structural member being simulated. Some data have been accumulated to support the validity of this postulate.[59]

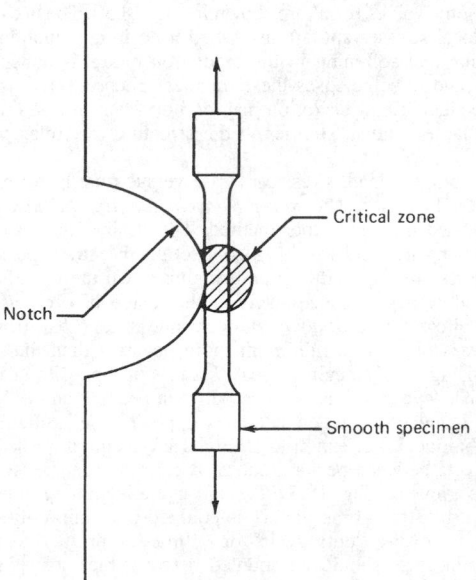

Fig. 18.56 Smooth specimen analog of material at critical point in the structure. (See Ref. 54.)

Computer simulation of the smooth specimen simulation has also been shown to be feasible.[44,60,61] To successfully utilize this simulation technique, it is necessary to have access to both monotonic and cyclic material properties, since the stress–strain response of most materials changes significantly with cyclic straining into the plastic range as already noted in Fig. 18.50. If cyclic materials response data are not available from the literature or an accessible data bank, it is necessary to perform enough smooth specimen testing to characterize the cyclic stress–strain response and fracture resistance of the material. With cyclic materials properties available, the computer simulation model for prediction of crack initiation must contain the following abilities:

1. To compute local stresses and strains, including means and ranges, from the applied loads and geometry of the structure.
2. To count cycles and associate mean and range values of stress and strain with each cycle.
3. To convert nonzero mean cycles to equivalent completely reversed cycles.
4. To compute fatigue damage in each cycle from stress and/or strain amplitudes and cyclic materials properties.
5. To compute damage cycle by cycle and sum the damage to give desired prediction of crack initiation.

To compute local stresses and strains from the external loading and geometry, Neuber's rule is often used in conjunction with the cyclic stress–strain properties and fatigue stress concentration factor. Neuber's rule may be modified for application to fatigue loading[44,45,61,62] by utilizing the fatigue stress concentration factor K_f together with nominal stress *range* ΔS, local stress *range* $\Delta\sigma$, and local strain *range* $\Delta\epsilon$ to develop the expression

$$\frac{(K_f\,\Delta S)^2}{E} = \Delta\sigma\,\Delta\epsilon \tag{18.55}$$

All the terms on the left-hand side of (18.55) are known from the geometry and from loading and material properties of the structure. To resolve the right-hand term, an empirical expression for the cyclic stress–strain curve that is satisfactory for most engineering metals may be obtained by separating the cyclic strain amplitude $\Delta\epsilon/2$ into elastic and plastic components to yield

$$\frac{\Delta\epsilon}{2} = \frac{\Delta\epsilon_e}{2} + \frac{\Delta\epsilon_p}{2} = \frac{\Delta\sigma}{2E} + \left[\frac{\Delta\sigma}{2k'}\right]^{1/n'} \tag{18.56}$$

where k' and n' are the *cyclic* strength coefficient and *cyclic* strain-hardening exponent, respectively, determined from the intercept and slope of a log–log plot of cyclic stress amplitude versus cyclic plastic strain amplitude. Some values for n' are shown in Fig. 18.50. The use of (18.56), then, provides a means for computing local stresses and strains, when used in conjunction with (18.55). The need to include cyclic hardening and softening in the prediction model is dependent on the accuracy of other parts of the model, and in some cases these transient phenomena may be regarded as second-order effects. Cycle-dependent stress relaxation may be important in cases in which occasional large overload cycles produce large residual stresses in the structure that relax with additional cycles of local plastic strain.

To properly interpret complex load, stress, or strain versus time histories requires that an appropriate cycle counting method be used. The *rain-flow cycle counting method,* illustrated in Fig. 18.57, is probably more widely used than any other method. The strain–time history is plotted so that the time axis is vertically downward, and the lines connecting the strain peaks are imagined to be a series of roofs. Several rules are imposed on rain dripping down these roofs so that cycles and half-cycles are defined. Rain flow begins successively at the inside of each strain peak. The rain flow initiating at each peak is allowed to drip down and continue except that, if it initiates at a minimum, it must stop when it comes opposite a minimum more negative than the minimum from which it initiated. For example, in Fig. 18.57 begin at peak 1 and stop opposite peak 9, peak 9 being more negative than peak 1. A half-cycle is thus counted between peaks 1 and 8. Similarly, if the rain flow initiates at a maximum, it must stop when it comes opposite a maximum more positive than the maximum from which it initiated. For example, in Fig. 18.57 begin at peak 2 and stop opposite peak 4, thus counting a half-cycle between peaks 2 and 3. A rain flow must also stop if it meets the rain from a roof above. For example, in Fig. 18.57, the half-cycle beginning at peak 3 ends beneath peak 2. Note that every part of the strain–time history is counted once and only once.

If cycles are to be counted for a "duty cycle" or a "mission profile" that is repeated until failure occurs, one complete strain cycle should be counted between the most positive and most negative peaks in the sequence, and other smaller complete cycles that are interruptions of this largest cycle should also be counted. This will be accomplished by the rain-flow method if the cycle counting is

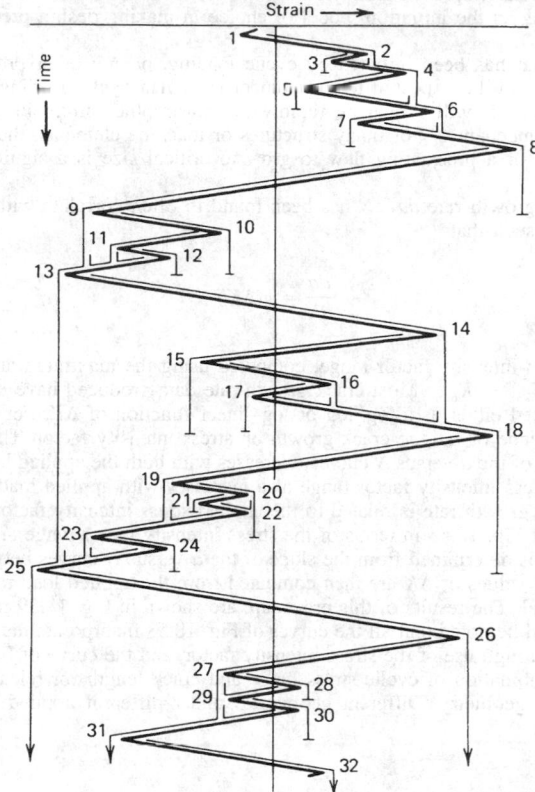

Fig. 18.57 Example of rain-flow cycle counting method. (After Ref. 63.)

started at either the most positive or most negative peak in the sequence. Using this procedure to identify the maximum and minimum strain (or stress), the range and mean may be tabulated for each cycle in the stress–strain history.

The nonzero mean stress cycles may be converted to equivalent completely reversed cycles by utilizing either the modified Goodman equation or some empirical expression based on specific material data. Using Eq. (18.40), for a tensile mean stress the equivalent completely reversed stress σ_{eqC-R} is

$$\sigma_{eqC-R} = \frac{\sigma_a}{1 - \sigma_m/\sigma_u}, \qquad \sigma_m \geq 0 \qquad (18.57)$$

and for a compressive mean stress

$$\sigma_{eqC-R} = \sigma_a, \qquad \sigma_m \leq 0 \qquad (18.58)$$

To compute the fatigue damage in each cycle associated with the equivalent completely reversed stress and strain range, it is necessary to have available data for strain amplitude versus cycles to failure, N_f (or reversals to failure, $2N_f$), as illustrated in Fig. 18.53. An expression for total strain amplitude has already been given in Eq. (18.54) as a function of total number of cycles to failure.

Damage is summed by utilizing an appropriate cumulative damage theory. Using the procedure described in this section for prediction of crack initiation, the Palmgren–Miner linear damage hypothesis of (18.46) gives results that are as good as any other proposed technique. Thus, when the sum of cycle ratios becomes equal to unity, it is predicted that crack initiation has occurred. The prediction techniques described in this section are practical only with the help of a computer program designed to perform the tedious cycle-by-cycle analyses involved. Another practical difficulty lies in the definition and detection of an "initiated" crack. Nevertheless, the state of the art in prediction of fatigue crack initiation has improved greatly during the past decade. The local stress–strain approach appears to be emerging as the initiation model of choice in making design predictions of initiation life.

A fatigue crack that has been initiated by cyclic loading, or any other preexisting flaw in the structure or material, may be expected to grow under sustained cyclic loading until it reaches the critical size from which it will propagate rapidly to catastrophic failure in accordance with the principles of fracture mechanics. For many structures or machine elements, the time required for a fatigue-initiated crack or a preexisting flaw to grow to critical size is a significant portion of the useful life.

The fatigue crack growth rate da/dN has been found to often correlate with the crack-tip stress intensity factor range such that

$$\frac{da}{dN} = g(\Delta K) \tag{18.59}$$

where ΔK is the stress intensity factor range, computed using the maximum and minimum applied stresses with $\Delta K = K_{max} - K_{min}$. Most crack growth rate data produced have been characterized in terms of ΔK and plotted either as a log–log or log–linear function of ΔK. For example, Fig. 18.58 illustrates the dependence of fatigue crack growth on stress intensity factor. The crack growth rate, indicated by the slope of the a versus N curves, increases with both the applied load and crack length. Since the crack-tip stress intensity factor range also increases with applied load and crack length, it is clear that the crack growth rate is related to the applied stress intensity factor range.

To plot the data of Fig. 18.58 in terms of the stress intensity factor range and crack growth rate, the crack growth rate is determined from the slope of the a versus N curves between successive data points. Corresponding values of ΔK are then computed from the applied load range and mean crack length for each interval. The results of this procedure are shown in Fig. 18.59 for the data presented in Fig. 18.58. It should be noted that all the curves of Fig. 18.58 incorporate themselves into a single curve in Fig. 18.59 through use of the stress intensity factor, and the curve of Fig. 18.59 is therefore applicable to any combination of cyclic stress range and crack length for released loading ($R = 0$) on specimens of this geometry. Different geometries under different applied stresses will exhibit

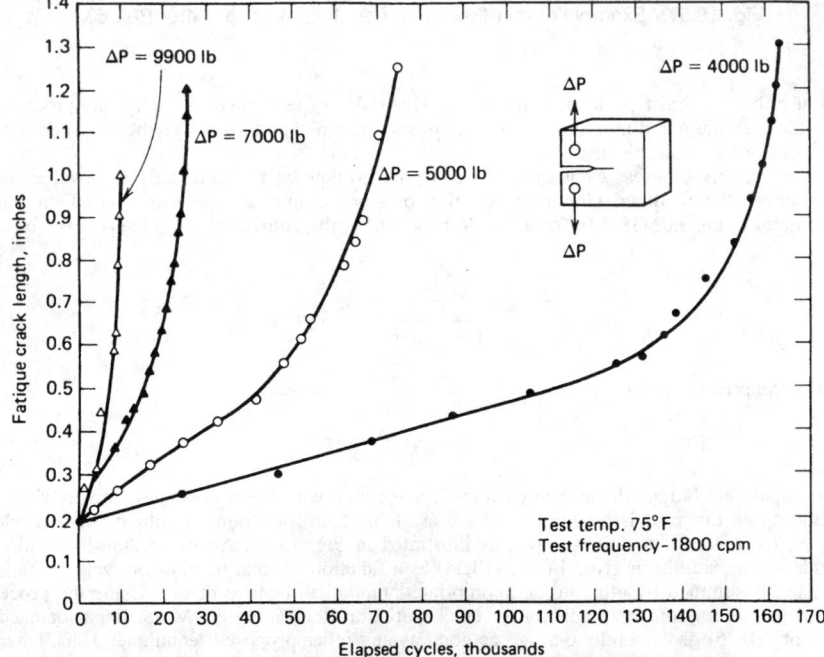

Fig. 18.58 Effect of cyclic-load range on crack growth in Ni–Mo–V alloy steel for released tension loading. (From Ref. 73, copyright Society for Experimental Stress Analysis, 1971; reprinted with permission.)

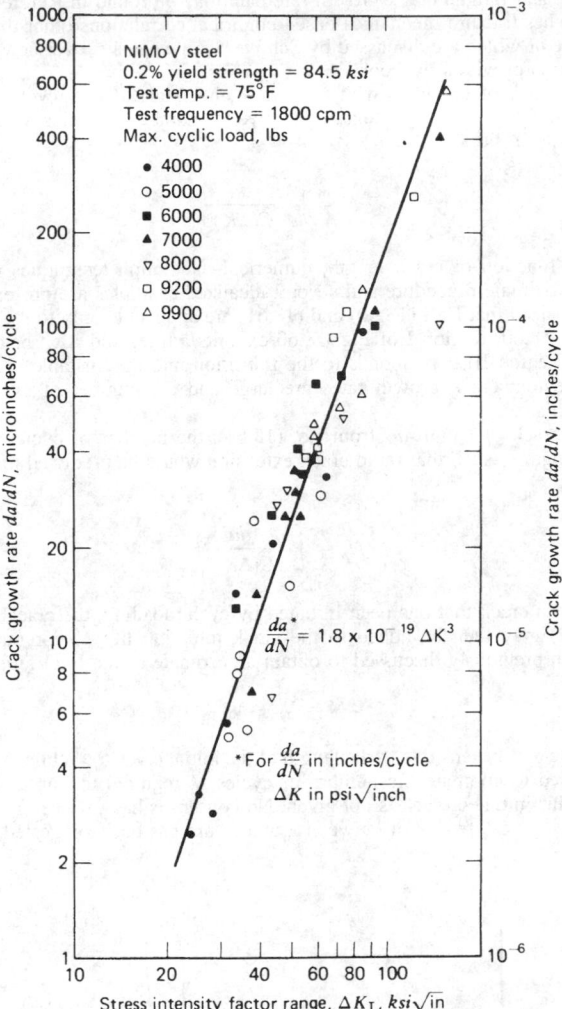

Fig. 18.59   Crack growth rate as a function of stress-intensity range for Ni–Mo–V steel. (From Ref. 73, copyright Society for Experimental Stress Analysis, 1971; reprinted with permission.)

identical crack-tip stress fields if the stress intensity factors are equal. Thus, because the stress intensity factor characterizes the state of stress near the crack tip, the fatigue crack growth rate correlation shown in Fig. 18.59 is applicable to any cyclically loaded component with $R = 0$ manufactured using the same material. This allows crack growth data generated from simple laboratory specimens to be utilized for approximate crack growth predictions in more complex geometries.

Fatigue crack growth rate data similar to that shown in Fig. 18.59 have been reported for a wide variety of engineering metals. The linear behavior observed using log–log coordinates suggests that Eq. (18.59) may be generalized as follows

$$\frac{da}{dN} = C(\Delta K)^n \qquad (18.60)$$

where n is the slope of the log da/dN versus log ΔK plot and C is the da/dN value found by extending the straight line to a ΔK value of unity. This relationship was first proposed by Paris.[66] The empirical parameters C and n are a function of material type, R ratio, thickness, temperature, environment, and loading frequency. Standard methods have been established for conducting fatigue crack growth tests,[64] and fatigue crack growth rate data may be found in References 10, 11, 13, 14, 15, and 16. Many other fracture mechanics-based empirical correlations other than Eq. (18.60) have been proposed, some of which are discussed by Schijve.[65] An extensive overview of the fatigue crack propagation problem is provided by Pook.[71]

Given an initial crack of length a_i with this crack either initiated by cyclic loading or initially present as a flaw, Eq. (18.60) may be integrated to give the number of cycles N required to propagate a crack to a size a_N such that

$$N = \int_{a_i}^{a_N} \frac{da}{C(\Delta K)^n} \tag{18.61}$$

Given that ΔK is a function of crack length, numerical integration techniques may be required to compute N. An approximate procedure and several idealized examples are presented by Parker.[23]

It must be emphasized that Eqs. (18.60) and (18.61) are applicable only to region II crack growth, as illustrated in Fig. 18.60. Region I of Fig. 18.60 exhibits a threshold ΔK_{th} below which the crack will not propagate. Region III corresponds to the transition into the unstable regime of rapid crack extension. In this region, crack growth rates are large and the number of cycles associated with growth in this region small.

Given an initial crack of length a_i, from Eq. (18.61), the number of cycles required to grow a crack to a critical length a_c such that rapid crack extension would be predicted may be approximated as

$$N_p = \int_{a_i}^{a_c} \frac{da}{C(\Delta K)^n} \tag{18.62}$$

Assuming an initial crack that has been initiated by cyclic loading, the crack propagation life N_p given by Eq. (18.62) may then be added to the crack initiation life N_i computed using the local stress-strain approach previously discussed to obtain an estimate of the total fatigue life N with

$$N = N_i + N_p \tag{18.63}$$

Such estimates are highly sensitive to the length of the initial crack a_i. While the local stress–strain approach may be used to compute the number of cycles N_i required to initiate a crack, the corresponding length of this initiated crack is not given. No consensus has yet been reached regarding the length of this initiated crack. A size of between 2 and 3 mm has been suggested,[61] as cracks of this

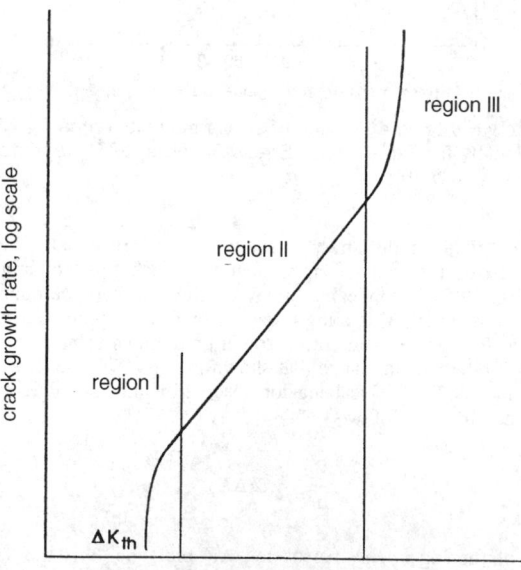

Fig. 18.60 Schematic representation of fatigue crack growth rate.

size normally exist at fracture in the small laboratory specimens used to generate material property data for the local stress–strain approach. An alternative approach would involve the assumption of a preexisting material or manufacturing defect such that $N_i \cong 0$. For example, such an assumption is often made during the analysis of welded joints.[67] If nondestructive techniques are employed, a reasonable assumption for the size of this initial defect would be the smallest flaw that could avoid detection.

Crack growth rates determined from constant amplitude cyclic loading tests are approximately the same as for random loading tests in which the maximum stress is held constant but mean and range of stress vary randomly. However, in random loading tests where the maximum stress is also allowed to vary, the sequence of loading cycles may have a marked effect on crack growth rate, with the overall crack growth being significantly higher for random loading spectra.

Many investigations have shown a significant delay in crack propagation following intermittent application of high stresses. That is, fatigue damage and crack extension are dependent on preceding cyclic load history. This dependence of crack extension on preceding history and the effects upon future damage increments are referred to as *interaction* effects. Most of the interaction studies conducted have dealt with *retardation* of crack growth as a result of the application of occasional tensile overload cycles. Retardation may be characterized as a period of reduced crack growth rate following the application of a peak load or loads higher and in the same direction as those peaks that follow.

The modeling of interaction effects requires consideration of crack tip plasticity and its subsequent influence. In metals of all types, cracks will remain closed or partially closed for a portion of the applied cyclic load as a consequence of plastically deformed material left in the wake of the growing crack. Under cyclic loading, crack growth will occur during the loading portion of the cycle. Given that a plastic zone exists at the crack tip prior to crack extension, as the material at the crack tip separates, the newly formed crack surfaces will exhibit a layer of plastically deformed material along the newly formed crack faces. Subsequent unloading will compress this plastically deformed material, closing the crack while the applied stress remains tensile. This phenomenon is known as plasticity-induced fatigue crack closure and was first discussed by Elber.[69] Upon reloading during the following cycle, crack growth will not continue unless the applied load is sufficiently large such that the compressive stresses acting along the crack surfaces are overcome and the crack is fully opened. This load is known as the crack opening load and has been demonstrated to be a key parameter in determining fatigue crack growth rates under both constant amplitude and spectrum loading. Further information regarding crack closure may be found in Refs. 1, 6, 18, 42, 43, 61, 65, and 70.

Discussion to this point has been limited to the growth of through-thickness cracks under mode I loading. While mode I loading is often dominant, under the most general circumstances the applied cyclic loads will generate stress intensity factor ranges ΔK_I, ΔK_{II}, and ΔK_{III} at the crack tip, and *mixed mode* fatigue crack growth must be considered. Modeling methodologies for mixed mode fatigue crack growth are discussed in Refs. 19 and 23. In addition, fatigue cracks in machine elements and structures are often not through-thickness cracks but rather surface cracks that extend partially through the thickness. Such surface cracks are often semi-elliptical in shape and the analysis of these cracks is considerably more complicated. Information regarding surface cracks may be found in Refs. 6 and 68.

Recent research has suggested that when fatigue cracks are small, crack growth rates are larger than would be predicted using Eq. (18.60) for a given ΔK.[72] Small-crack behavior is often important, as a significant portion of the fatigue life may be spent in the small-crack regime. The fatigue crack propagation life N_p will also be influenced by the presence of residual stresses such as might exist as a consequence of welding, heat treatment, carburizing, grinding, or shot-peening. Compressive residual stresses are beneficial, decreasing the rate of fatigue crack growth and increasing propagation life. While approximate methodologies exist for incorporating the effects of residual stress within fatigue crack growth predictions,[61] residual stress distributions are often difficult to characterize.

Reasonable design estimates for fatigue life may be obtained using Eq. (18.63). However, the many uncertainties typically associated with fatigue life predictions emphasize the essential requirement to conduct full-scale fatigue tests to provide acceptable reliability.

18.5.9 Service Spectrum Simulation and Full-Scale Testing

To achieve a reliable fatigue-resistant design configuration, a realistic service loading simulation test or a full-scale fatigue test will generally be required. For service load simulation testing it is essential that both the test specimen and the loading spectrum, including sequence, be representative of actual service conditions. This means that the specimen should be an actual component, a complete structural subassembly, or a complete full-scale machine or structure. It also means that an exact simulation of the service load–time history would be best, if known. Usually, however, an estimated load–time history must be generated on the basis of a statistical representation of similar mission loading spectra obtained from similar instrumented articles already in actual use.

In designing service loading simulation spectra, special attention must be given to the highest load levels to be incorporated because they may exert a major influence on crack propagation and fatigue life. As noted in Section 18.5.8, the life-enhancing retardation phenomenon associated with large load peaks can be extremely important. It should be recognized that the largest peak loads in service will vary from article to article on a statistical basis, so that some articles will experience the maximum load peak more than once, whereas others will never see this load level. For these reasons it is usual to truncate the design loading spectrum, discarding all load levels that occur fewer than 10 times during the projected design lifetime. Although it may seem intuitively wrong to discard the high load peaks to achieve a more critical simulation test spectrum, it must be realized that the crack retardation effect of the occasional high loads gives a longer test life than if the high loads were omitted. Thus, it is conservative to omit the 10 highest loads from the simulation test spectrum. Likewise, truncation is utilized in analyses for establishing suitable inspection intervals for aircraft so that propagating cracks will have a higher probability of being detected before they approach critical size.

The application of failsafe or limit loads at regular intervals during a service spectrum simulation or full-scale fatigue test *must be avoided,* since they may contribute to crack growth retardation that is not typical of the actual flight spectrum. Limit loading should be applied only at the end of the fatigue test.

Low-amplitude cycles are often omitted from service simulation testing to save time. However, it must be recognized that these cycles may contribute to fatigue crack nucleation through the fretting process, and omitting these low-amplitude cycles, and the fretting they would normally produce, may result in unsafe simulation test predictions of service fatigue life.

Full-scale fatigue testing of an article such as a newly designed aircraft is extremely expensive. Such tests generally would be regarded as accelerated tests, where flight simulation testing over a 6- to 12-month testing period is designed to represent 10 or more years of actual service history. Flight simulation testing of an aircraft component on a modern closed-loop fatigue testing machine may be accomplished in one or two weeks or less. The benefits of full-scale testing include (1) discovery of fatigue-critical elements and design deficiencies, (2) determination of times to detectable cracking, (3) obtaining data on crack propagation, (4) determination of remaining safe life with cracks, (5) determination of residual strength, (6) establishment of proper inspection intervals, and (7) development of repair methods. Factors that may influence the full-scale fatigue simulation test results include loading rate, environmental factors, statistical scatter, and loading spectrum deviations. Actual service life is usually shorter than the simulation test life, sometimes by factors of as much as 2 to 4. However, in recent years the agreement between full-scale flight simulation testing and in-service experience has improved significantly. Full-scale flight simulation testing is often continued over the long term so that fatigue failures in the test will lead the fleet experience by enough time to redesign and install whatever modifications are required to prevent catastrophic fleet failures in service before they occur.

18.5.10 Damage Tolerance and Fracture Control

The concept of "damage-tolerant" structure, which has developed primarily within the aerospace industry, is characterized by structural configurations that are designed to minimize the loss of aircraft because of the propagation of undetected flaws, cracks, or other similar damage. There are two major design objectives that must be met to produce a damage-tolerant structure. These objectives are controlled safe flaw growth, or safe life with cracks, and positive damage containment, which implies a safe remaining or residual strength. These objectives should not be considered as separate or distinct requirements, however, because it is only by their judicious combination that effective fracture control can be achieved. Furthermore, it must be emphasized that damage-tolerant design is not a substitute for a careful fatigue analysis and design as discussed earlier, because the achievement of "fatigue quality" through careful stress analysis, geometry selection, detail design, material selection, surface finish, and workmanship is a necessary prerequisite to effective damage-tolerant design and fracture control.

The general goals of damage-tolerant design and fracture control include the selection of fracture-resistant materials and manufacturing processes, the design for inspectability, and the use of damage-tolerant structural configurations such as multiple load paths or crack stoppers, in addition to the usual rules of good design practice.

In the application of the fracture control philosophy, the basic assumption is made that flaws do exist even in new structures and that they may go undetected. The first major requirement for damage tolerance, therefore, is that any member in the structure, including each element of a redundant load path group, must have a safe life with assumed cracks present. For any specific application the primary factors influencing the design include the type or class of structure, the quality of the nondestructive inspection (NDI) techniques used in production assembly, the accessibility of the structure to inspection, the assurance that the member will be inspected on schedule when in service, and the probability that a flaw of subcritical size will go undetected even though periodic in-service inspections are made on schedule.

Most structural arrangements may be classified according to load path as class 1, single load path; class 2, single primary load path with auxiliary crack arrest features; or class 3, multiple or redundant load path. Each of these structural classes is illustrated in Fig. 18.61. Clearly, for the class 1 structure it is essential to satisfy the safe-life-with-cracks requirement, because failure is catastrophic. For class

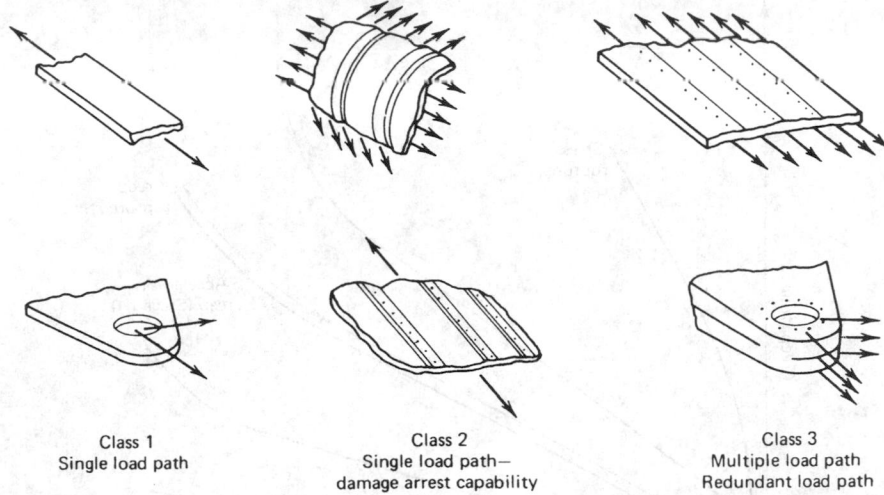

<table>
<tr><td align="center">Class 1
Single load path</td><td align="center">Class 2
Single load path—
damage arrest capability</td><td align="center">Class 3
Multiple load path
Redundant load path</td></tr>
</table>

Fig. 18.61 Structural arrangements. (After Ref. 74.)

2 structures, including pressurized cabins and pressure vessels, relatively large amounts of damage may be contained by providing tear straps or stiffeners. There is usually a high probability of damage detection for a class 2 structure because of fuel or pressure leakage, that is, "leak-before-break" design is characteristic of class 2 structures. Class 3 structures are usually designed to provide a specified percentage of the original strength, that is, a specified residual strength, during and subsequent to the failure of one element. This is often called "failsafe" type of structure. However, the preexisting flaw concept requires that all members, including every member of a multiple load path structure, be assumed to contain flaws. It is usual to assume a smaller initial flaw size for class 3 structures because it is appropriate to take a larger risk of operating with cracks if multiple load paths are available.

The development of inspection procedures is an important part of any fracture control program. Appropriate inspection procedures must be established for each structural element, and regions within elements may be classified with respect to required NDI sensitivity. Inspection intervals are established on the basis of crack growth information assuming a specified initial flaw size and a "detectable" flaw size that depends on the NDI procedure. Inspection intervals are established to ensure that an undetected flaw will not grow to critical size before the next inspection, with a comfortable margin of safety. The intervals are usually picked so that two inspections will occur before any crack will reach critical size.

A good fracture-control program should encompass and interact with design, materials selection, fabrication, inspection, and operational phases in the development of any high-performance engineering system.

18.6 CREEP AND STRESS RUPTURE

Creep in its simplest form is the progressive accumulation of plastic strain in a specimen or machine part under stress at elevated temperature over a period of time. Creep failure occurs when the accumulated creep strain results in a deformation of the machine part that exceeds the design limits. *Creep rupture* is an extension of the creep process to the limiting condition where the stressed member actually separates into two parts. *Stress rupture* is a term used interchangeably by many with creep rupture; however, others reserve the term stress rupture for the rupture termination of a creep process in which steady-state creep is never reached, and use the term creep rupture for the rupture termination of a creep process in which a period of steady-state creep has persisted. Figure 18.62 illustrates these differences. The interaction of creep and stress rupture with cyclic stressing and the fatigue process has not yet been clearly understood but is of great importance in many modern high-performance engineering systems.

Creep strains of engineering significance are not usually encountered until the operating temperatures reach a range of approximately 35–70% of the melting point on a scale of absolute temperature. The approximate melting temperature for several substances is shown in Table 18.2.

Not only is excessive deformation due to creep an important consideration, but other consequences of the creep process may also be important. These might include creep rupture, thermal relaxation, dynamic creep under cyclic loads or cyclic temperatures, creep and rupture under multiaxial states of stress, cumulative creep effects, and effects of combined creep and fatigue.

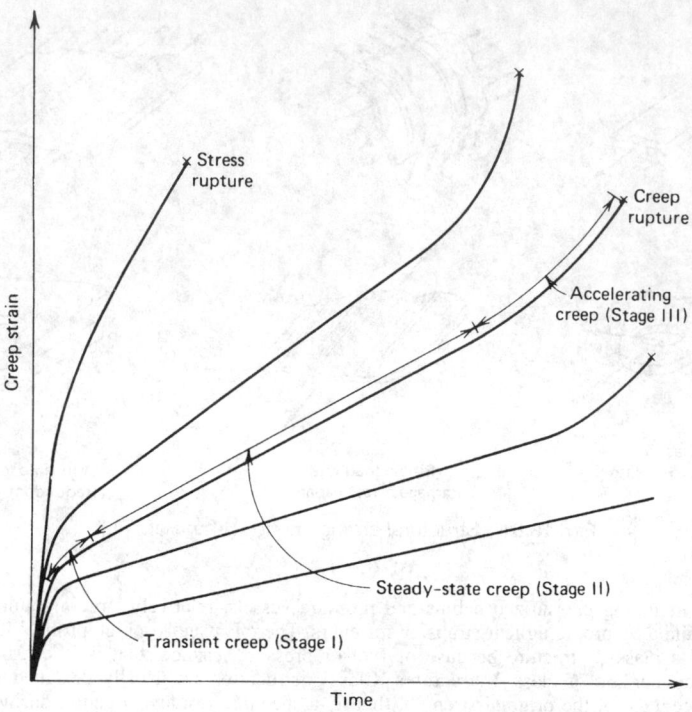

Fig. 18.62 Illustration of creep and stress rupture.

Table 18.2   Melting Temperatures[49]

Material	°F	°C
Hafnium carbide	7030	3887
Graphite (sublimes)	6330	3500
Tungsten	6100	3370
Tungsten carbide	5190	2867
Magnesia	5070	2800
Molybdenum	4740	2620
Boron	4170	2300
Titanium	3260	1795
Platinum	3180	1750
Silica	3140	1728
Chromium	3000	1650
Iron	2800	1540
Stainless steels	2640	1450
Steel	2550	1400
Aluminum alloys	1220	660
Magnesium alloys	1200	650
Lead alloys	605	320

Creep deformation and rupture are initiated in the grain boundaries and proceed by sliding and separation. Thus, creep rupture failures are intercrystalline, in contrast, for example, to the transcrystalline failure surface exhibited by room-temperature fatigue failures. Although creep is a plastic flow phenomenon, the intercrystalline failure path gives a rupture surface that has the appearance of brittle fracture. Creep rupture typically occurs without necking and without warning. Current state-of-the-art knowledge does not permit a reliable prediction of creep or stress rupture properties on a theoretical basis. Furthermore, there seems to be little or no correlation between the creep properties of a material and its room-temperature mechanical properties. Therefore, test data and empirical methods of extending these data are relied on heavily for prediction of creep behavior under anticipated service conditions.

Metallurgical stability under long-time exposure to elevated temperatures is mandatory for good creep-resistant alloys. Prolonged time at elevated temperatures acts as a tempering process, and any improvement in properties originally gained by quenching may be lost. Resistance to oxidation and other corrosive media are also usually important attributes for a good creep-resistant alloy. Larger grain size may also be advantageous since this reduces the length of grain boundary, where much of the creep process resides.

18.6.1 Prediction of Long-Term Creep Behavior

Much time and effort has been expended in attempting to device good short-time creep tests for accurate and reliable prediction of long-term creep and stress rupture behavior. It appears, however, that really reliable creep data can be obtained only by conducting long-term creep tests that duplicate actual service loading and temperature conditions as nearly as possible. Unfortunately, designers are unable to wait for years to obtain design data needed in creep failure analysis. Therefore, certain useful techniques have been developed for approximating long-term creep behavior based on a series of short-term tests. Data from creep testing may be cross plotted in a variety of different ways. The basic variables involved are stress, strain, time, temperature, and, perhaps, strain rate. Any two of these basic variables may be selected as plotting coordinates, with the remaining variables treated as parametric constants for a given curve. Three commonly used methods for extrapolating short-time creep data to long-term applications are the abridged method, the mechanical acceleration method, and the thermal acceleration method. In the abridged method of creep testing the tests are conducted at several different stress levels and at the contemplated operating temperature. The data are plotted as creep strain versus time for a family of stress levels, all run at constant temperature. The curves are plotted out to the laboratory test duration and then extrapolated to the required design life. In the mechanical acceleration method of creep testing, the stress levels used in the laboratory tests are significantly higher than the contemplated design stress levels, so the limiting design strains are reached in a much shorter time than in actual service. The data taken in the mechanical acceleration method are plotted as stress level versus time for a family of constant strain curves all run at a constant temperature. The thermal acceleration method involves laboratory testing at temperatures much higher than the actual service temperature expected. The data are plotted as stress versus time for a family of constant temperatures where the creep strain produced is constant for the whole plot.

It is important to recognize that such extrapolations are not able to predict the potential of failure by creep rupture prior to reaching the creep design life. In any testing method it should be noted that creep testing guidelines usually dictate that test periods of less than 1% of the expected life are not deemed to give significant results. Tests extending to at least 10% of the expected life are preferred where feasible.

Several different theories have been proposed in recent years to correlate the results of short-time elevated-temperature tests with long-term service performance at more moderate temperatures. The more accurate and useful of these proposals to date are the Larson–Miller theory and the Manson–Haferd theory.

The Larson–Miller theory[75] postulates that for each combination of material and stress level there exists a unique value of a parameter P that is related to temperature and time by the equation

$$P = (\theta + 460)(C + \log_{10} t) \qquad (18.64)$$

where P = Larson–Miller parameter, constant for a given material and stress level
$\quad \theta$ = temperature, °F
$\quad C$ = constant, usually assumed to be 20
$\quad t$ = time in hours to rupture or to reach a specified value of creep strain

This equation was investigated for both creep and rupture for some 28 different materials by Larson and Miller with good success. By using (18.64) it is a simple matter to find a short-term combination of temperature and time that is equivalent to any desired long-term service requirement. For example, for any given material at a specified stress level the test conditions listed in Table 18.3 should be equivalent to the operating conditions.

Table 18.3 Equivalent Conditions Based on Larson–Miller Parameter

Operating Condition	Equivalent Test Condition
10,000 hours at 1000°F	13 hours at 1200°F
1,000 hours at 1200°F	12 hours at 1350°F
1,000 hours at 1350°F	12 hours at 1500°F
1,000 hours at 300°F	2.2 hours at 400°F

The Manson–Haferd[76] theory postulates that for a given material and stress level there exists a unique value of a parameter P' that is related to temperature and time by the equation

$$P' = \frac{\theta - \theta_a}{\log_{10} t - \log_{10} t_a} \tag{18.65}$$

where P' = Manson–Haferd parameter, constant for a given material and stress level
$\quad\theta$ = temperature, °F
$\quad t$ = time in hours to rupture or to reach a specified value of creep strain
$\quad\theta_a, t_a$ = material constants

In the Manson–Haferd equation values of the constants for several materials are shown in Table 18.4.

18.6.2 Creep under Uniaxial State of Stress

Many relationships have been proposed to relate stress, strain, time, and temperature in the creep process. If one investigates experimental creep strain versus time data, it will be observed that the data are close to linear for a wide variety of materials when plotted on log strain versus log time coordinates. Such a plot is shown, for example, in Fig. 18.63 for three different materials. An equation describing this type of behavior is

$$\delta = At^a \tag{18.66}$$

where δ = true creep strain
$\quad t$ = time
$\quad A, a$ = empirical constants

Differentiating (18.66) with respect to time gives

$$\dot{\delta} = aAt^{(a-1)} \tag{18.67}$$

or, setting $aA = b$ and $(1 - a) = n$,

$$\dot{\delta} = bt^{-n} \tag{18.68}$$

This equation represents a variety of different types of creep strain versus time curves, depending on the magnitude of the exponent n. If n is zero, the behavior, characteristic of high temperatures, is termed *constant creep rate,* and the creep strain is given as

Table 18.4 Constants for Manson–Haferd Equation[76]

Material	Creep or Rupture	θ_a	$\log_{10} t_a$
25–20 stainless steel	Rupture	100	14
18–8 stainless steel	Rupture	100	15
S-590 alloy	Rupture	0	21
DM steel	Rupture	100	22
Inconel X	Rupture	100	24
Nimonic 80	Rupture	100	17
Nimonic 80	0.2 percent plastic strain	100	17
Nimonic 80	0.1 percent plastic strain	100	17

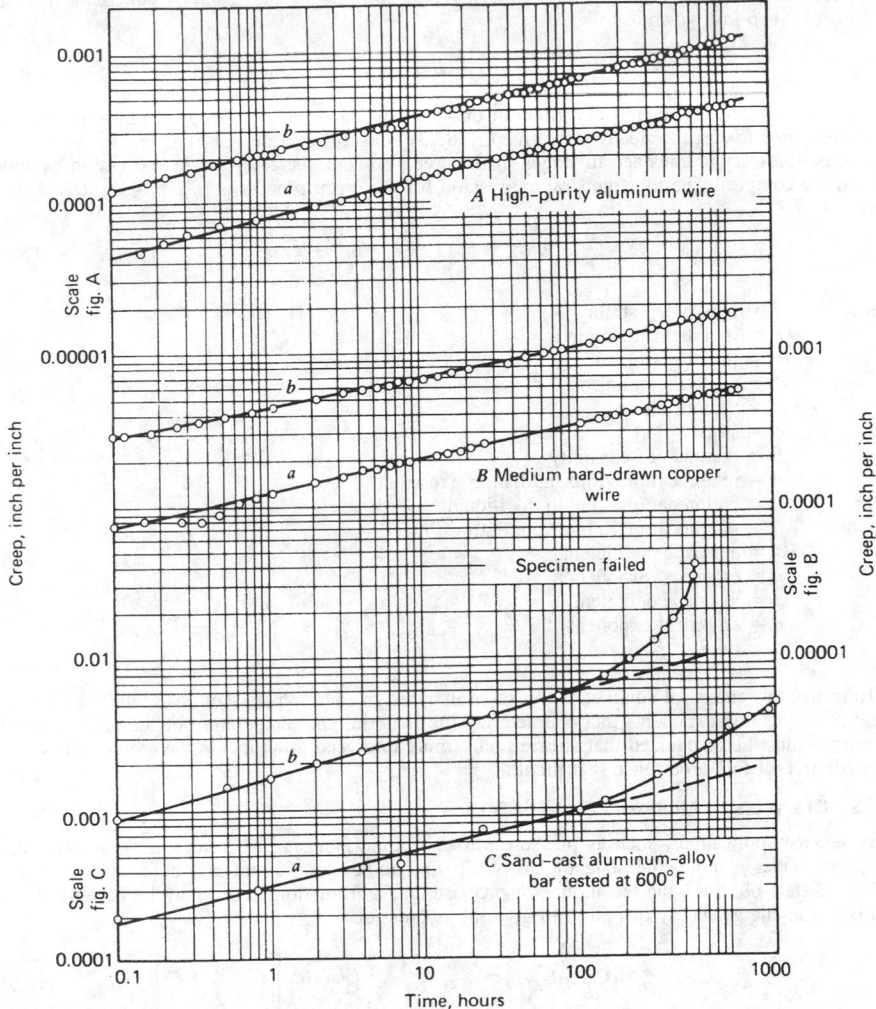

Fig. 18.63 Creep curves for three materials plotted on log–log coordinates. (From Ref. 77.)

$$\delta = b_1 t + C_1 \tag{18.69}$$

If n lies between 0 and 1, the behavior is termed *parabolic creep,* and the creep strain is given by

$$\delta = b_3 t^m + C_3 \tag{18.70}$$

This type of creep behavior occurs at intermediate and high temperatures. The coefficient b_3 increases exponentially with stress and temperature, and the exponent m decreases with stress and increases with temperature. The influence of stress level σ on creep rate can often be represented by the empirical expression

$$\dot{\delta} = B\sigma^N \tag{18.71}$$

Assuming the stress σ to be independent of time, we may integrate (18.71) to yield the creep strain

$$\delta = Bt\sigma^N + C' \tag{18.72}$$

If the constant C' is small compared with $Bt\sigma^N$, as it often is, the result is called the *log–log stress–time creep law*, given as

$$\delta = Bt\sigma^N \tag{18.73}$$

As long as the instantaneous deformation on load application and the stage I transient creep are small compared to stage II steady-state creep, (18.73) is useful as a design tool.

If it is necessary to consider all stages of the creep process, the creep strain expression becomes much more complex. The most general expression for the creep process is (see p. 438 of Ref. 78)

$$\delta = \frac{\sigma}{E} + k_1\sigma^m + k_2(1 - e^{-qt})\sigma^n + k_3t\sigma^p \tag{18.74}$$

where
$$\delta = \text{total creep strain}$$
$$\sigma/E = \text{initial elastic strain}$$
$$k_1\sigma^m = \text{initial plastic strain}$$
$$k_2(1 - e^{-qt})\sigma^n = \text{anelastic strain}$$
$$k_3t\sigma^p = \text{viscous strain}$$
$$\sigma = \text{stress}$$
$$E = \text{modulus of elasticity}$$
$$m = \text{reciprocal of strain-hardening exponent}$$
$$k_1 = \text{reciprocal of strength coefficient}$$
$$q = \text{reciprocal of Kelvin retardation time}$$
$$k_2 = \text{anelastic coefficient}$$
$$n = \text{empirical exponent}$$
$$k_3 = \text{viscous coefficient}$$
$$p = \text{empirical exponent}$$
$$t = \text{time}$$

To utilize this empirical nonlinear expression in a design environment requires specific knowledge of the constants and exponents that characterize the material and temperature of the application. In all cases it must be recognized that stress rupture may intervene to terminate the creep process, and the prediction of this occurrence is difficult.

18.6.3 Creep under Multiaxial State of Stress

Many service applications, such as pressure vessels, piping, and turbine rotors, may involve creep conditions under a multiaxial state of stress. To determine creep strain and deformation under a multiaxial state of stress, the techniques of proportional deformation theory may be combined with the distortion energy theory of failure to give the expressions

$$\delta_1 = Bt(\sigma_1')^N[\alpha^2 + \beta^2 - \alpha\beta - \alpha - \beta + 1]^{(N-1)/2}\left[1 - \frac{\alpha}{2} - \frac{\beta}{2}\right] \tag{18.75}$$

$$\delta_2 = Bt(\sigma_1')^N[\alpha^2 + \beta^2 - \alpha\beta - \alpha - \beta + 1]^{(N-1)/2}\left[\alpha - \frac{\beta}{2} - \frac{1}{2}\right] \tag{18.76}$$

$$\delta_3 = Bt(\sigma_1')^N[\alpha^2 + \beta^2 - \alpha\beta - \alpha - \beta + 1]^{(N-1)/2}\left[\beta - \frac{\alpha}{2} - \frac{1}{2}\right] \tag{18.77}$$

where $\delta_1, \delta_2, \delta_3 = \text{principal true strains}$
$$\sigma_1', \sigma_2', \sigma_3' = \text{principal true stresses}$$
$$\alpha = \sigma_2'/\sigma_1'$$
$$\beta = \sigma_3'/\sigma_1'$$
$$B, N = \text{experimentally determined uniaxial creep parameters}$$

These three equations completely define the principal creep strains in terms of the principal creep stresses and the experimentally determined uniaxial tensile creep parameters B and N. Predictions of creep behavior in any multiaxial state of stress can be made by these equations, based only on the results of a simple uniaxial creep test.

18.6.4 Cumulative Creep

There is at the present time no universally accepted method for estimating the creep strain accumulated as a result of exposure for various periods of time at different temperatures and stress levels. However, several different techniques for making such estimates have been proposed. The simplest of these is a linear hypothesis suggested by Robinson.[79] A generalized version of the Robinson hypothesis may be written as follows: If a design limit of creep strain δ_D is specified, it is predicted that the creep strain δ_D will be reached when

$$\sum_{i=1}^{k} \frac{t_i}{L_i} = 1 \qquad (18.78)$$

where t_i = time of exposure at the ith combination of stress level and temperature
$\quad\quad L_i$ = time required to produce creep strain δ_D if entire exposure were held constant at the ith combination of stress level and temperature

Stress rupture may also be predicted by (18.78) if the L_i values correspond to stress rupture. This prediction technique gives relatively accurate results if the creep deformation is dominated by stage II steady-state creep behavior. Under other circumstances the method may yield predictions that are seriously in error.

Other cumulative creep prediction techniques that have been proposed include the time-hardening rule, the strain-hardening rule, and the life-fraction rule. The time-hardening rule is based on the assumption that the major factor governing the creep rate is the length of exposure at a given temperature and stress level, no matter what the past history of exposure has been. The strain-hardening rule is based on the assumption that the major factor governing the creep rate is the amount of prior strain, no matter what the past history of exposure has been. The life-fraction rule is a compromise between the time-hardening rule and the strain-hardening rule which accounts for influence of both time history and strain history. The life-fraction rule is probably the most accurate of these prediction techniques.

18.7 COMBINED CREEP AND FATIGUE

There are several important high-performance applications of current interest in which conditions persist that lead to combined creep and fatigue. For example, aircraft gas turbines and nuclear power reactors are subjected to this combination of failure modes. To make matters worse, the duty cycle in these applications might include a sequence of events including fluctuating stress levels at constant temperature, fluctuating temperature levels at constant stress, and periods during which both stress and temperature are simultaneously fluctuating. Furthermore, there is evidence to indicate that the fatigue and creep processes interact to produce a synergistic response.

It has been observed that interrupted stressing may accelerate, retard, or leave unaffected the time under stress required to produce stress rupture. The same observation has also been made with respect to creep rate. Temperature cycling at constant stress level may also produce a variety of responses, depending on material properties and the details of the temperature cycle.

No general law has been found by which cumulative creep and stress rupture response under temperature cycling at constant stress or stress cycling at constant temperature in the creep range can be accurately predicted. However, some recent progress has been made in developing life prediction techniques for combined creep and fatigue. For example, a procedure sometimes used to predict failure under combined creep and fatigue conditions for isothermal cyclic stressing is to assume that the creep behavior is controlled by the mean stress σ_m and that the fatigue behavior is controlled by the stress amplitude σ_a, with the two processes combining linearly to produce failure. This approach is similar to the development of the Goodman diagram described in Section 18.5.4 except that instead of an intercept of σ_u on the σ_m axis, as shown in Fig. 18.38, the intercept used is the *creep-limited static stress* σ_{cr}, as shown in Fig. 18.64. The creep-limited static stress corresponds either to the design limit on creep strain at the design life or to creep rupture at the design life, depending on which failure mode governs. The linear prediction rule then may be stated as

Failure is predicted to occur under combined isothermal creep and fatigue if

$$\frac{\sigma_a}{\sigma_N} + \frac{\sigma_m}{\sigma_{cr}} \geq 1 \qquad (18.79)$$

An elliptic relationship is also shown in Fig. 18.64, which may be written as

Failure is predicted to occur under combined isothermal creep and fatigue if

$$\left(\frac{\sigma_a}{\sigma_N}\right)^2 + \left(\frac{\sigma_m}{\sigma_{cr}}\right)^2 \geq 1 \qquad (18.80)$$

The linear rule is usually (but not always) conservative. In the higher-temperature portion of the creep range the elliptic relationship usually gives better agreement with data. For example, in Fig. 18.65a actual data for combined isothermal creep and fatigue tests are shown for several different

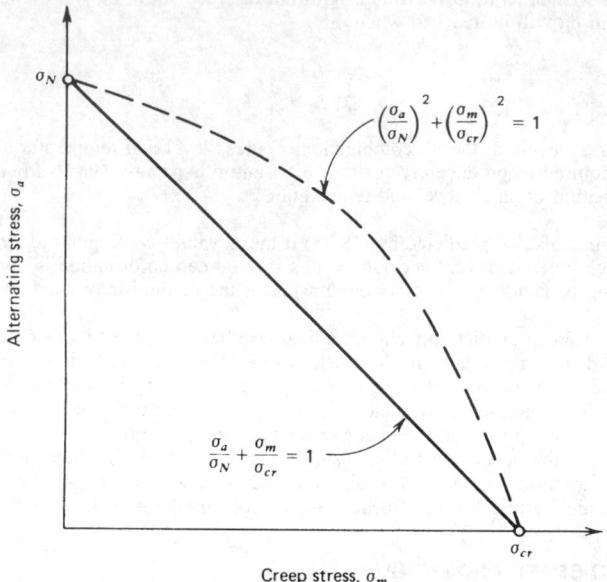

Fig. 18.64 Failure prediction diagram for combined creep and fatigue under constant-temperature conditions.

temperatures using a cobalt-base S-816 alloy. The elliptic approximation is clearly better at higher temperatures for this alloy. Similar data are shown in Fig. 18.65*b* for 2024 aluminum alloy. Detailed studies of the relationships among creep strain, strain at rupture, mean stress, and alternating stress amplitude over a range of stresses and constant temperatures involve extensive, complex testing programs. The results of one study of this type[82] are shown in Fig. 18.66 for S-816 alloy at two different temperatures.

Several other empirical methods have recently been proposed for the purpose of making life predictions under more general conditions of combined creep and low-cycle fatigue. These methods include:

1. Frequency-modified stress and strain-range method.[83]
2. Total time to fracture versus time-of-one-cycle method.[84]
3. Total time to fracture versus number of cycles to fracture method.[85]
4. Summation of damage fractions using interspersed fatigue with creep method.[86]
5. Strain-range partitioning method.[87]

The frequency-modified strain-range approach of Coffin was developed by including frequency-dependent terms in the basic Manson–Coffin–Morrow equation, cited earlier as (18.54). The resulting equation can be expressed as

$$\Delta\epsilon = AN_f^a\nu^b + BN_f^c\nu^d \tag{18.81}$$

where the first term on the right-hand side of the equation represents the elastic component of strain range, and the second term represents the plastic component. The constants A and B are the intercepts, respectively, of the elastic and plastic strain components at $N_f = 1$ cycle and $\nu = 1$ cycle/min. The exponents a, b, c, and d are constants for a particular material at a given temperature. When the constants are experimentally evaluated, this expression provides a relationship between total strain range $\Delta\epsilon$ and cycles to failure N_f.

The total time to fracture versus time-of-one-cycle method is based on the expression

$$t_f = \frac{N_f}{\nu} = Ct_c^k \tag{18.82}$$

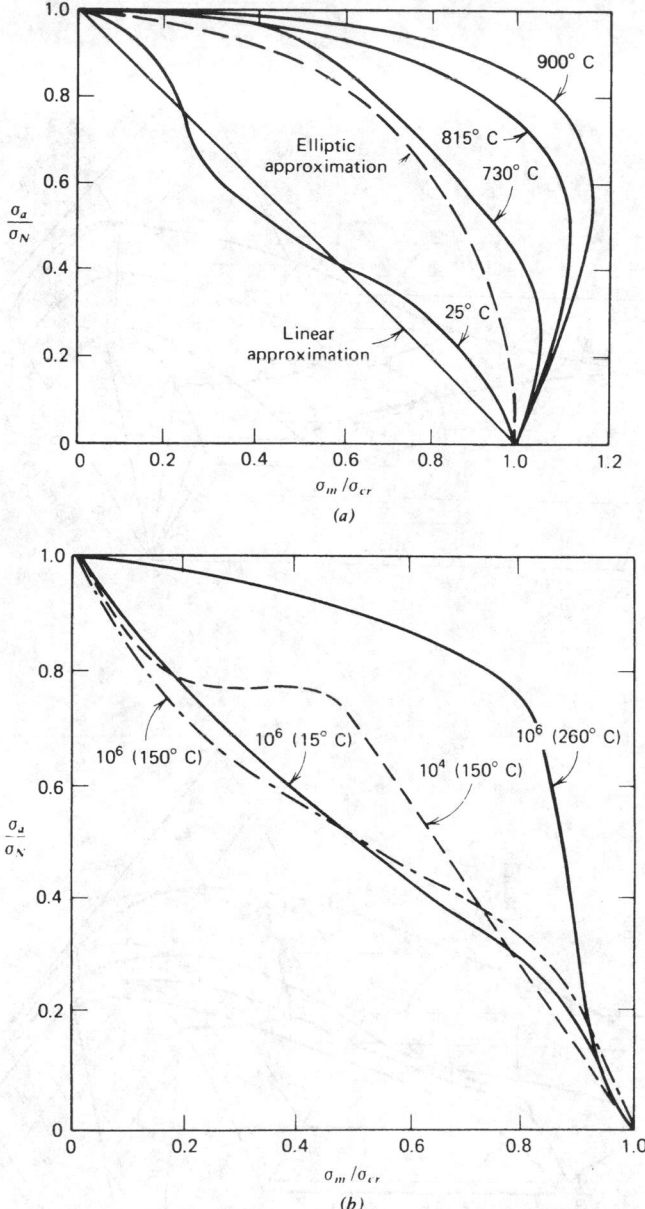

Fig. 18.65 Combined isothermal creep and fatigue data plotted on coordinates suggested in Figure 18.64. (a) Data for S-816 alloy for 100-hr life, where σ_N is fatigue strength for 100-hr life and σ_{cr} is creep rupture stress for 100-hr life. (From Refs. 80 and 81.) (b) Data for 2024 aluminum alloy, where σ_N is fatigue strength for life indicated on curves and σ_{cr} is creep stress for corresponding time to rupture. (From Refs. 80 and 82.)

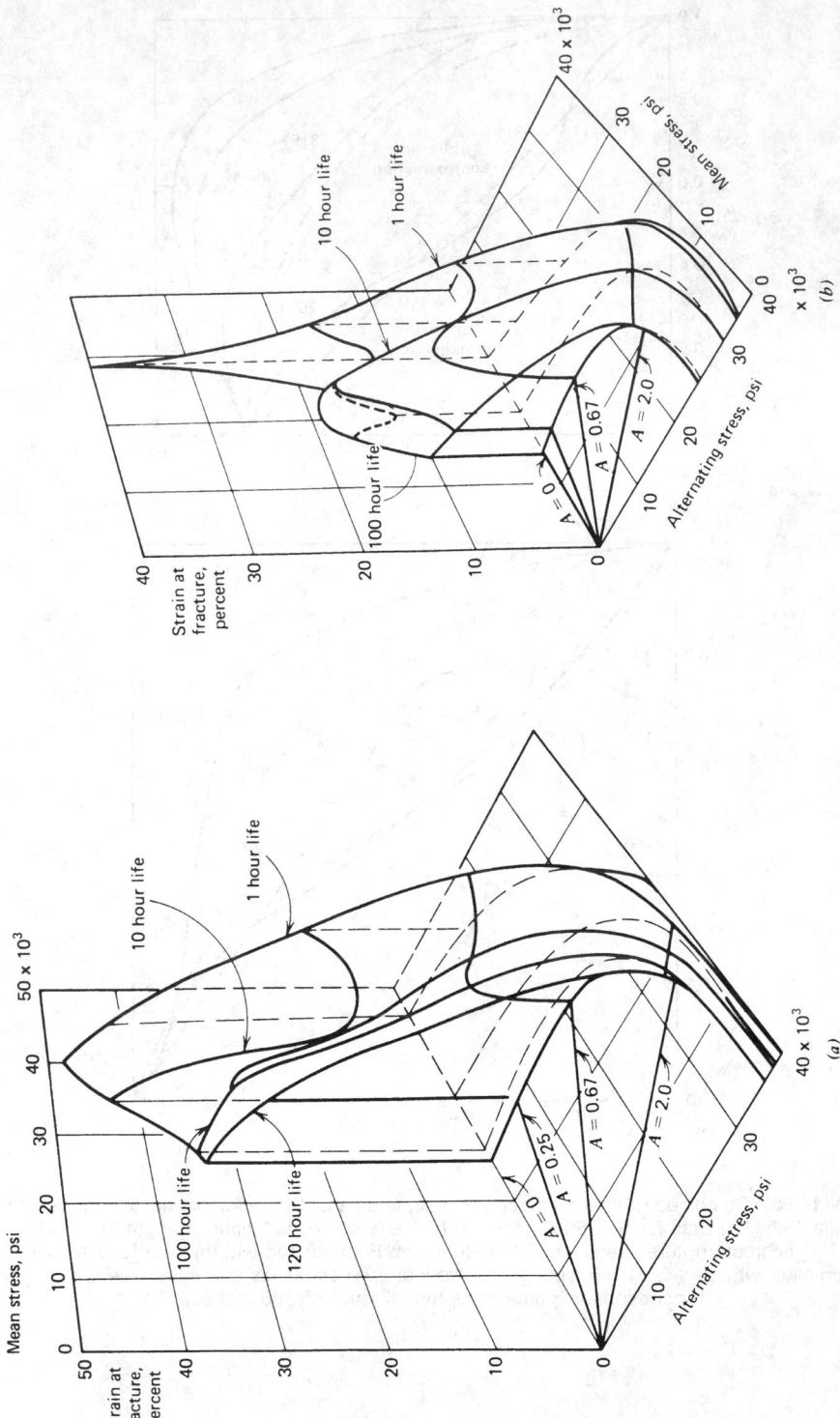

Fig. 18.66 Strain at fracture for various combinations of mean and alternating stresses in unnotched specimens of S-816 alloy. (a) Data taken at 900°C. (b) Data taken at 816°C. (From Refs. 80 and 81.)

where t_f is the total time to fracture in minutes, ν is frequency expressed in cycles per minute, N_f is total cycles to failure, $t_c = 1/\nu$ is the time for one cycle in minutes, and C and k are constants for a particular material at a particular temperature for a particular total strain range.

The total time to fracture versus number-of-cycles method characterizes the fatigue–creep interaction as

$$t_f = DN_f^{-m} \tag{18.83}$$

which is identical to (18.82) if $D = C^{1/(1-k)}$ and $m = k/(1 - k)$. However, it has been postulated that there are three different sets of constants D and m: one set for continuous cycling at varying strain rates, a second set for cyclic relaxation, and a third set for cyclic creep.

The interspersed fatigue and creep analysis proposed by the Metal Properties Council involves the use of a specified combined test cycle on unnotched bars. The test cycle consists of a specified period at constant tensile load followed by various numbers of fully reversed strain-controlled fatigue cycles. The specified test cycle is repeated until failure occurs. For example, in one investigation the specified combined test cycle consisted of 23 hr at constant tensile load followed by either 1.5, 2.5, 5.5, or 22.5 fully reversed strain-controlled fatigue cycles. The failure data are then plotted as fatigue damage fraction versus creep damage fraction, as illustrated in Fig. 18.67.

The fatigue damage fraction is the ratio of total number of fatigue cycles N_f' included in the combined test cycle divided by the number of fatigue cycles N_f to cause failure if no creep time were interspersed. The creep damage fraction is the ratio of total creep time t_{cr} included in the combined test cycle divided by the total creep life to failure t_f if no fatigue cycles were interspersed. A "best-fit" curve through the data provides the basis for making a graphical estimate of life under combined creep and fatigue conditions, as shown in Fig. 18.67.

The strain-range partitioning method is based on the concept that any cycle of completely reversed inelastic strain may be partitioned into the following strain-range components: completely reversed plasticity, $\Delta\epsilon_{pp}$; tensile plasticity reversed by compressive creep, $\Delta\epsilon_{pc}$; tensile creep reversed by compressive plasticity, $\Delta\epsilon_{cp}$; and completely reversed creep, $\Delta\epsilon_{cc}$. The first letter of each subscript

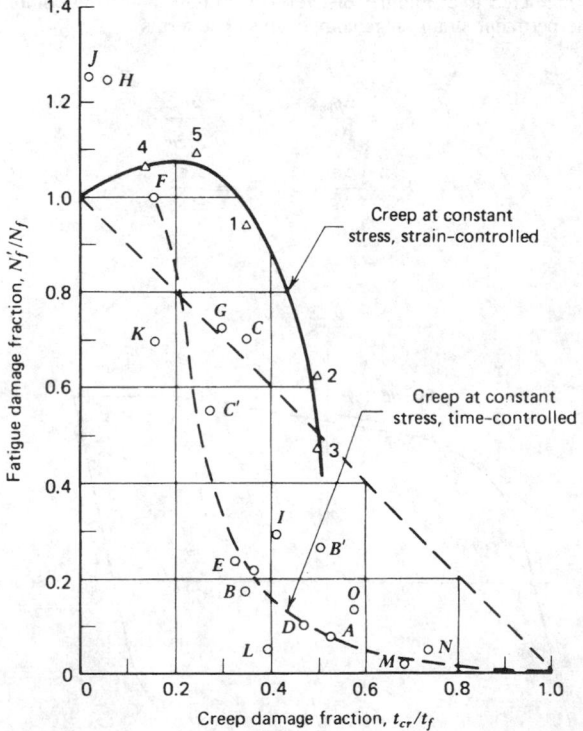

Fig. 18.67 Plot of fatigue damage fraction versus creep damage fraction for 1 Cr–1 Mo–¼ V rotor steel at 1000°F in air, using the method of the Metal Properties Council. (After Ref. 88, copyright Society for Experimental Stress Analysis, 1973; reprinted with permission.)

in the notation, c for creep or p for plastic deformation, refers to the type of strain imposed during the tensile portion of the cycle, and the second letter refers to the type of strain imposed during the compressive portion of the cycle. The term *plastic deformation* or *plastic flow* in this context refers to *time-independent* plastic strain that occurs by crystallographic slip within the crystal grains. The term *creep* refers to *time-dependent* plastic deformation that occurs by a combination of diffusion within the grains together with grain boundary sliding between the grains. The concept is illustrated in Fig. 18.68.

It may be noted in Fig. 18.68 that tensile inelastic strain, represented as $\overline{AD}$ is the sum of plastic strain $\overline{AC}$ plus creep strain $\overline{CD}$. Also, compressive inelastic strain $\overline{DA}$ is the sum of plastic strain $\overline{DB}$ plus creep strain $\overline{BA}$. In general, $\overline{AC}$ will not be equal to $\overline{DB}$, nor will $\overline{CD}$ be equal to $\overline{BA}$. However, since we are dealing with a closed hysteresis loop, $\overline{AD}$ does equal $\overline{DA}$. The partitioned strain ranges are obtained in the following manner.[89] The completely reversed portion of the plastic strain range, $\Delta\epsilon_{pp}$, is the smaller of the two plastic flow components, which in Fig. 18.68 is equal to $\overline{DB}$. Likewise, the completely reversed portion of the creep strain range, $\Delta\epsilon_{cc}$, is the smaller of the two creep components, which in Fig. 18.68 is equal to $\overline{CD}$. As can be seen graphically, the difference between the two plastic components must be equal to the difference between the two creep components, or $\overline{AC} - \overline{DB}$ must equal $\overline{BA} - \overline{CD}$. This difference then is either $\Delta\epsilon_{pc}$ or $\Delta\epsilon_{cp}$, in accordance with the notation just defined. For the case illustrated in Fig. 18.68, the difference is $\Delta\epsilon_{pc}$, since the tensile plastic strain component is greater than the compressive plastic strain component. It follows from this discussion that the sum of the partitioned strain ranges will necessarily be equal to the total inelastic strain range, or the width of the hysteresis loop.

It is next assumed that a unique relationship exists between cyclic life to failure and each of the four strain-range components listed. Available data indicate that these relationships are of the form of the basic Manson–Coffin–Morrow expression (18.54), as indicated, for example, in Fig. 18.69 for a type 316 stainless-steel alloy at 1300°F. The governing life prediction equation, or "interaction damage rule," is then postulated to be

$$\frac{1}{N_{\text{pred}}} = \frac{F_{pp}}{N_{pp}} + \frac{F_{pc}}{N_{pc}} + \frac{F_{cp}}{N_{cp}} + \frac{F_{cc}}{N_{cc}} \tag{18.84}$$

where N_{pred} is the predicted total number of cycles to failure under the combined *straining* cycle containing all of the pertinent strain range components. The terms F_{pp}, F_{pc}, F_{cp}, and F_{cc} are defined as

$$F_{pp} = \frac{\Delta\epsilon_{pp}}{\Delta\epsilon_p}, \qquad F_{pc} = \frac{\Delta\epsilon_{pc}}{\Delta\epsilon_p}$$

$$F_{cp} = \frac{\Delta\epsilon_{cp}}{\Delta\epsilon_p}, \qquad F_{cc} = \frac{\Delta\epsilon_{cc}}{\Delta\epsilon_p} \tag{18.85}$$

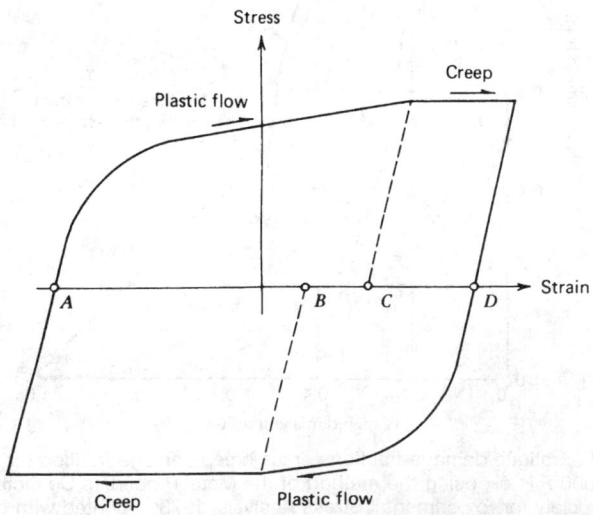

Fig. 18.68 Typical hysteresis loop.

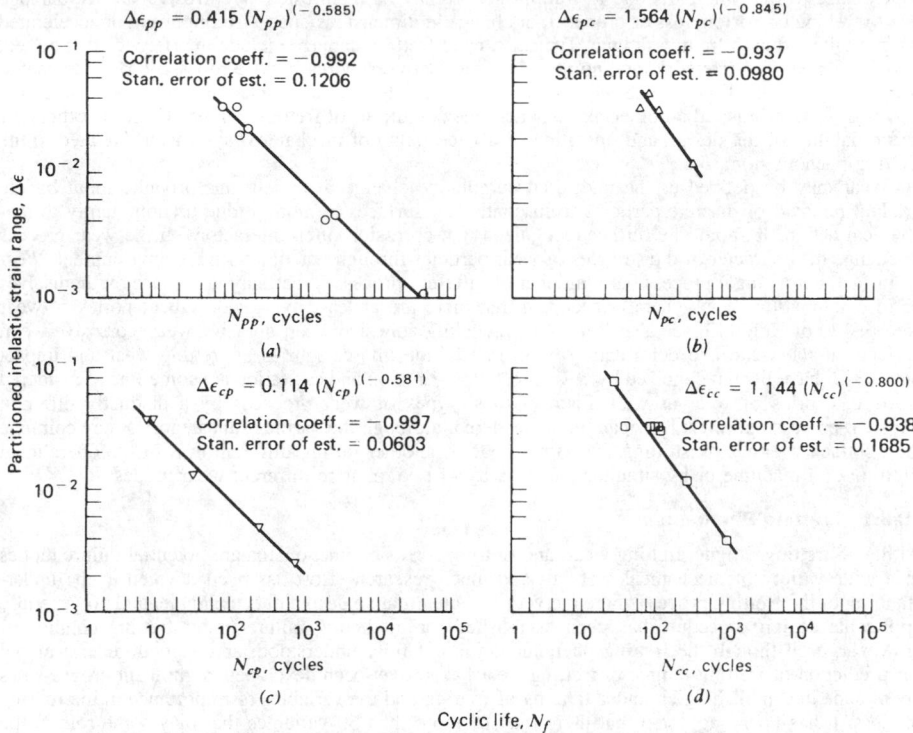

$$\Delta\epsilon_{pp} = 0.415 \, (N_{pp})^{(-0.585)}$$

Correlation coeff. = −0.992
Stan. error of est. = 0.1206

N_{pp}, cycles

(a)

$$\Delta\epsilon_{pc} = 1.564 \, (N_{pc})^{(-0.845)}$$

Correlation coeff. = −0.937
Stan. error of est. = 0.0980

N_{pc}, cycles

(b)

$$\Delta\epsilon_{cp} = 0.114 \, (N_{cp})^{(-0.581)}$$

Correlation coeff. = −0.997
Stan. error of est. = 0.0603

N_{cp}, cycles

(c)

$$\Delta\epsilon_{cc} = 1.144 \, (N_{cc})^{(-0.800)}$$

Correlation coeff. = −0.938
Stan. error of est. = 0.1685

N_{cc}, cycles

(d)

Partitioned inelastic strain range, $\Delta\epsilon$

Cyclic life, N_f

Fig. 18.69 Summary of partitioned strain-life relations for type 316 stainless steel at 1300°F (After Ref. 90): (a) pp-type strain range; (b) pc-type strain range; (c) cp-type strain range; (d) cc-type strain range.

for any selected inelastic strain range $\Delta\epsilon_p$, using information from a plot of experimental data such as that shown in Fig. 18.69. The partitioned failure lives N_{pp}, N_{pc}, N_{cp}, and N_{cc} are also obtained from Fig. 18.69. The use of (18.84) has, in several investigations,[90–95] shown the predicted lives to be acceptably accurate, with most experimental results falling with a scatter band of $\pm 2N_f$ of the predicted value.

More recent investigations have indicated that improvements in predictions by the strain-range partitioning method may be achieved by using the "creep" ductility and "plastic" ductility of a material determined in the actual service environment, to "normalize" the strain versus life equations prior to using (18.85). Procedures for using the strain-range partitioning method under conditions of multiaxial loading have also been proposed[94] but remain to be verified more fully.

18.8 FRETTING AND WEAR

Fretting and wear share many common characteristics but, at the same time, are distinctly different in several ways. Basically, fretting action has, for many years, been defined as a combined mechanical and chemical action in which contacting surfaces of two solid bodies are pressed together by a normal force and are caused to execute oscillatory sliding relative motion, wherein the magnitude of normal force is great enough and the amplitude of the oscillatory sliding motion is small enough to significantly restrict the flow of fretting debris away from the originating site.[96] More recent definitions of fretting action have been broadened to include cases in which contacting surfaces periodically separate and then reengage, as well as cases in which the fluctuating friction-induced surface tractions produce stress fields that may ultimately result in failure. The complexities of fretting action have been discussed by numerous investigators, who have postulated the combination of many mechanical, chemical, thermal, and other phenomena that interact to produce fretting. Among the postulated phenomena are plastic deformation caused by surface asperities plowing through each other, welding and tearing of contacting asperities, shear and rupture of asperities, friction-generated subsurface shearing stresses, dislodging of particles and corrosion products at the surfaces, chemical reactions, debris accumulation and entrapment, abrasive action, microcrack initiation, and surface delamination.[97–112]

Damage to machine parts due to fretting action may be manifested as corrosive surface damage due to fretting corrosion, loss of proper fit or change in dimensions due to fretting wear, or accelerated fatigue failure due to fretting fatigue. Typical sites of fretting damage include interference fits; bolted, keyed, splined, and riveted joints; points of contact between wires in wire ropes and flexible shafts; friction clamps; small-amplitude-of-oscillation bearings of all kinds; contacting surfaces between the leaves of leaf springs; ad all other places where the conditions of fretting persist. Thus, the efficiency and reliability of the design and operation of a wide range of mechanical systems are related to the fretting phenomenon.

Wear may be defined as the undesired cumulative change in dimensions brought about by the gradual removal of discrete particles from contacting surfaces in motion, due predominantly to mechanical action. It should be further recognized that corrosion often interacts with the wear process to change the character of the surfaces of wear particles through reaction with the environment. Wear is, in fact, not a single process but a number of different processes that may take place by themselves or in combination. It is generally accepted that there are at least five major subcategories of wear (see p. 120 of Ref. 113, see also Ref. 114), including adhesive wear, abrasive wear, corrosive wear, surface fatigue wear, and deformation wear. In addition, the categories of fretting wear and impact wear[115-117] have been recognized by wear specialists. Erosion and cavitation are sometimes considered to be categories of wear as well. Each of these types of wear proceeds by a distinctly different physical process and must be separately considered, although the various subcategories may combine their influence either by shifting from one mode to another during different eras in the operational lifetime of a machine or by simultaneous activity of two or more different wear modes.

18.8.1   Fretting Phenomena

Although fretting fatigue, fretting wear, and fretting corrosion phenomena are potential failure modes in a wide variety of mechanical systems, and much research effort has been devoted to the understanding of the fretting process, there are very few quantitative design data available, and no generally applicable design procedure has been established for predicting failure under fretting conditions. However, even though the fretting phenomenon is not fully understood, and a good general model for prediction of fretting fatigue or fretting wear has not yet been developed, significant progress has been made in establishing an understanding of fretting and the variables of importance in the fretting process. It has been suggested that there may be more than 50 variables that play some role in the fretting process.[118] Of these, however, there are probably only eight that are of major importance; they are:

1. The magnitude of relative motion between the fretting surfaces.
2. The magnitude and distribution of pressure between the surfaces at the fretting interface.
3. The state of stress, including magnitude, direction, and variation with respect to time in the region of the fretting surfaces.
4. The number of fretting cycles accumulated.
5. The material, and surface condition, from which each of the fretting members is fabricated.
6. Cyclic frequency of relative motion between the two members being fretted.
7. Temperature in the region of the two surfaces being fretted.
8. Atmospheric environment surrounding the surfaces being fretted.

These variables interact so that a quantitative prediction of the influence of any given variable is very dependent on all the other variables in any specific application or test. Also, the combination of variables that produce a very serious consequence in terms of fretting fatigue damage may be quite different from the combinations of variables that produce serious fretting wear damage. No general techniques yet exist for quantitatively predicting the influence of the important variables of fretting fatigue and fretting wear damage, although many special cases have been investigated. However, it has been observed that certain trends usually exist when the variables just listed are changed. For example, fretting damage tends to increase with increasing contact pressure until a nominal pressure of a few thousand pounds per square inch is reached, and further increases in pressure seem to have relatively little direct effect. The state of stress is important, especially in fretting fatigue. Fretting damage accumulates with increasing numbers of cycles at widely different rates, depending on specific operating conditions. Fretting damage is strongly influenced by the material properties of the fretting pair—surface hardness, roughness, and finish. No clear trends have been established regarding frequency effects on fretting damage, and although both temperature and atmospheric environment are important influencing factors, their influences have not been clearly established. A clear presentation of the current state of knowledge relative to these various parameters is given, however, in Ref. 109.

Fretting fatigue is fatigue damage directly attributable to fretting action. It has been suggested that premature fatigue nuclei may be generated by fretting through either abrasive pit-digging action, asperity-contact microcrack initiation,[119] friction-generated cyclic stresses that lead to the formation of microcracks,[120] or subsurface cyclic shear stresses that lead to surface delamination in the fretting zone.[112] Under the abrasive pit-digging hypothesis, it is conjectured that tiny grooves or elongated pits are produced at the fretting interface by the asperities and abrasive debris particles moving under the influence of oscillatory relative motion. A pattern of tiny grooves would be produced in the fretted region with their longitudinal axes all approximately parallel and in the direction of fretting motion, as shown schematically in Fig. 18.70.

The asperity-contact microcrack initiation mechanism is postulated to proceed due to the contact force between the tip of an asperity on one surface and another asperity on the mating surface as the surfaces move back and forth. If the initial contact does not shear one or the other asperity from its base, the repeated contacts at the tips of the asperities give rise to cyclic or fatigue stresses in the region at the base of each asperity. It has been estimated[105] that under such conditions the region at the base of each asperity is subjected to large local stresses that probably lead to the nucleation of fatigue microcracks at these sites. As shown schematically in Fig. 18.71, it would be expected that the asperity-contact mechanism would produce an array of microcracks whose longitudinal axes would be generally perpendicular to the direction of fretting motion.

The friction-generated cyclic stress fretting hypothesis[107] is based on the observation that when one member is pressed against the other and caused to undergo fretting motion, the tractive friction force induces a compressive tangential stress component in a volume of material that lies ahead of the fretting motion, and a tensile tangential stress component in a volume of material that lies behind the fretting motion, as shown in Fig. 18.72a. When the fretting direction is reversed, the tensile and compressive regions change places. Thus, the volume of material adjacent to the contact zone is subjected to a cyclic stress that is postulated to generate a field of microcracks at these sites. Furthermore, the geometrical stress concentration associated with the clamped joint may contribute to microcrack generation at these sites.[108] As shown in Fig. 18.72c, it would be expected that the friction-generated microcrack mechanism would produce an array of microcracks whose longitudinal axes would be generally perpendicular to the direction of fretting motion. These cracks would lie in a region adjacent to the fretting contact zone.

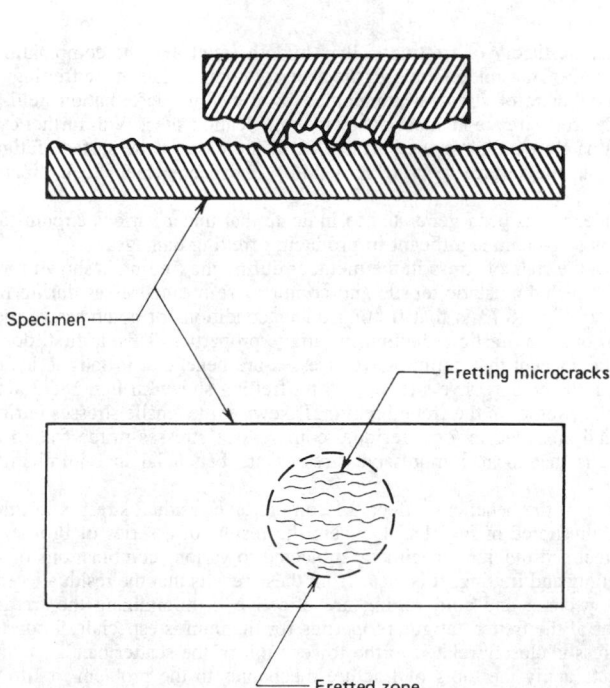

Fig. 18.70 Idealized schematic illustration of the stress concentrations produced by the abrasive pit-digging mechanism.

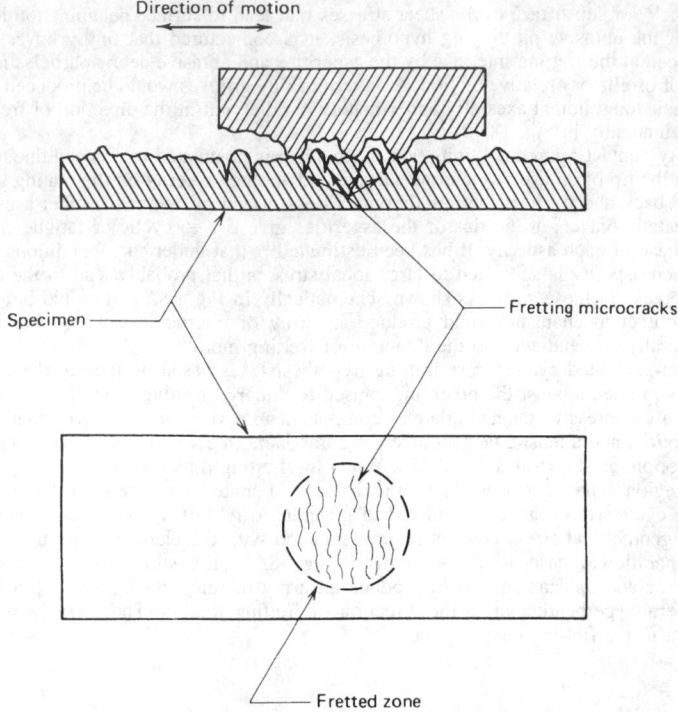

Fig. 18.71 Idealized schematic illustration of the stress concentrations produced by the asperity-contact microcrack initiation mechanism.

In the delamination theory of fretting[112] it is hypothesized that the combination of normal and tangential tractive forces transmitted through the asperity-contact sites at the fretting interface produce a complex multiaxial state of stress, accompanied by a cycling deformation field, which produces subsurface peak shearing stress and subsurface crack nucleation sites. With further cycling, the cracks propagate approximately parallel to the surface, as in the case of the surface fatigue phenomenon, finally propagating to the surface to produce a thin wear sheet, which "delaminates" to become a particle of debris.

Supporting evidence has been generated to indicate that under various circumstances each of the four mechanisms is active and significant in producing fretting damage.

The influence of the state of stress in the member during the fretting is shown for several different cases in Fig. 18.73, including static tensile and compressive mean stresses during fretting. An interesting observation in Fig. 18.73 is that fretting under conditions of compressive mean stress, either static or cyclic, produces a drastic reduction in fatigue properties. This, at first, does not seem to be in keeping with the concept that compressive stresses are beneficial in fatigue loading. However, it was deduced[121] that the compressive stresses during fretting shown in Fig. 18.73 actually resulted in local residual tensile stresses in the fretted region. Likewise, the tensile stresses during fretting shown in Fig. 18.73 actually resulted in local residual compressive stresses in the fretted region. The conclusion, therefore, is that local compressive stresses are beneficial in minimizing fretting fatigue damage.

Further evidence of the beneficial effects of compressive residual stresses in minimizing fretting fatigue damage is illustrated in Fig. 18.74, where the results of a series of Prot (fatigue limit) tests are reported for steel and titanium specimens subjected to various combinations of shot peening and fretting or cold rolling and fretting. It is clear from these results that the residual compressive stresses produced by shot peening and cold rolling are effective in minimizing the fretting damage. The reduction in scatter of the fretted fatigue properties for titanium is especially important to a designer because design stress is closely related to the lower limit of the scatter band.

Recent efforts to apply the tools of fracture mechanics to the problem of life prediction under fretting fatigue conditions have produced encouraging preliminary results that may ultimately provide designers with a viable quantitative approach.[122] These studies emphasize that the principal effect of fretting in the fatigue failure process is to accelerate crack initiation and the early stages of crack growth, and they suggest that when cracks have reached a sufficient length, the fretting no longer

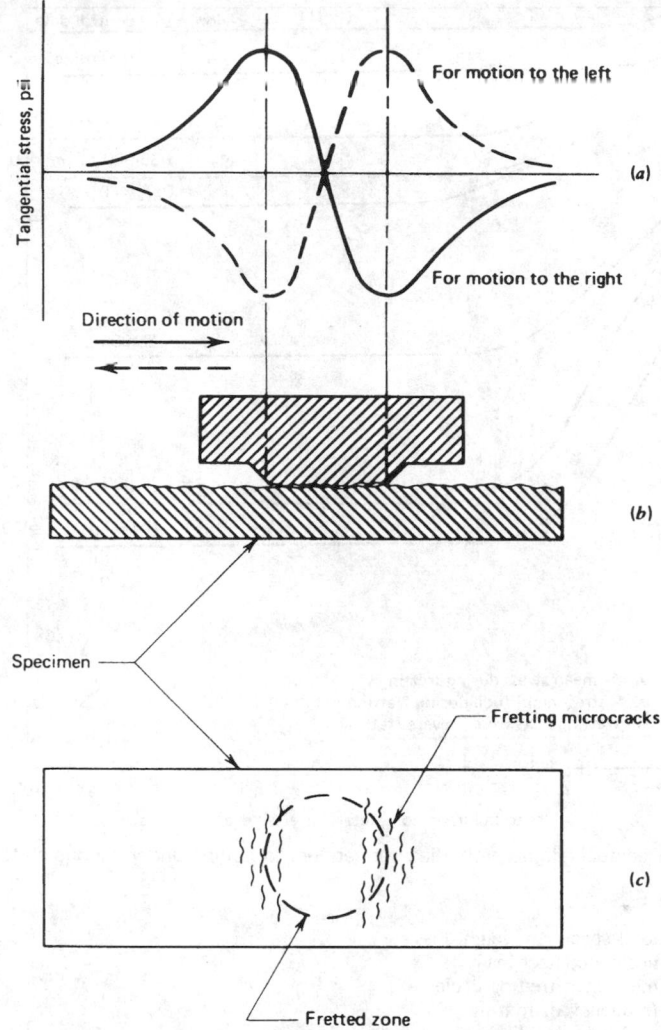

Fig. 18.72 Idealized schematic illustration of the tangential stress components and micro-cracks produced by the friction-generated microcrack initiation mechanism.

has a significant influence on crack propagation. At this point the fracture mechanics description of crack propagation described in Section 18.5.8 becomes valid.

In the final analysis, it is necessary to evaluate the seriousness of fretting fatigue damage in any specific design by running simulated service tests on specimens or components. Within the current state-of-the-art knowledge in the area of fretting fatigue, there is no other safe course of action open to the designer.

Fretting wear is a change in dimensions through wear directly attributable to the fretting process between two mating surfaces. It is thought that the abrasive pit-digging mechanism, the asperity-contact microcrack initiation mechanism, and the wear-sheet delamination mechanism may all be important in most fretting wear failures. As in the case of fretting fatigue, there has been no good model developed to describe the fretting wear phenomenon in a way useful for design. An expression for weight loss due to fretting has been proposed[102] as

$$W_{\text{total}} = (k_0 L^{1/2} - k_1 L) \frac{C}{F} + k_2 SLC \qquad (18.86)$$

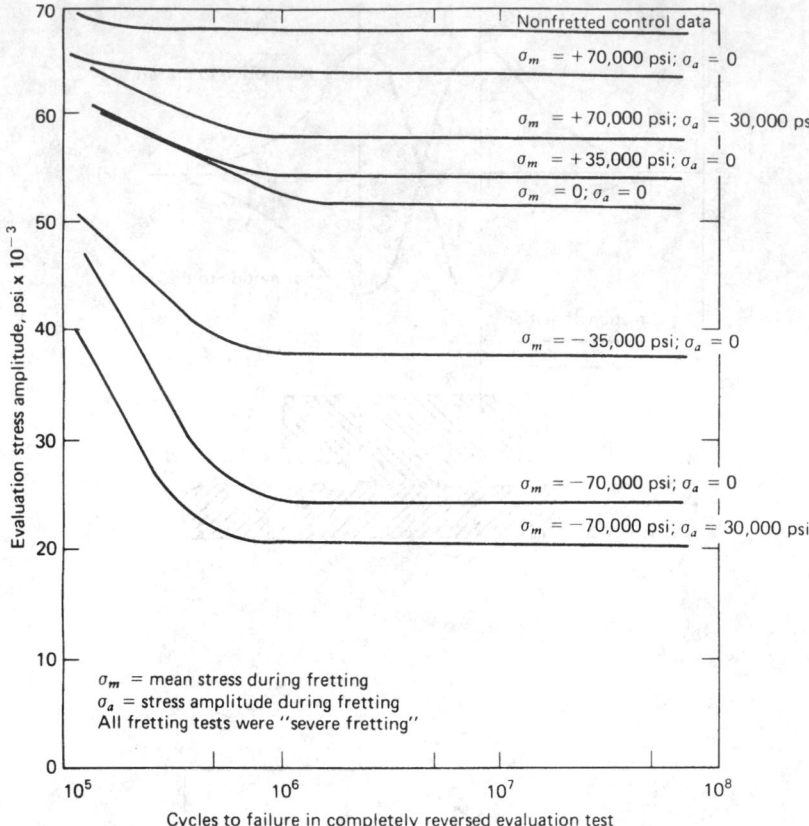

Fig. 18.73 Residual fatigue properties subsequent to fretting under various states of stress.

where W_{total} = total specimen weight loss
        L = normal contact load
        C = number of fretting cycles
        F = frequency of fretting
        S = peak-to-peak slip between fretting surfaces
k_0, k_1, k_2 = constants to be empirically determined

This equation has been shown to give relatively good agreement with experimental data over a range of fretting conditions using mild steel specimens.[102] However, weight loss is not of direct use to a designer. Wear depth is of more interest. Prediction of wear depth in an actual design application must in general be based on simulated service testing.

Some investigators have suggested that estimates of fretting wear depth may be based on the classical adhesive or abrasive wear equations, in which wear depth is proportional to load and total distance slid, where the total distance slid is calculated by multiplying relative motion per cycle times number of cycles. Although there are some supporting data for such a procedure,[123] more investigation is required before it could be recommended as an acceptable approach for general application.

If fretting wear at a support interface, such as between tubes and support plates of a steam generator or heat exchanger or between fuel pins and support grids of a reactor core, produces loss of fit at a support site, impact fretting may occur. Impact fretting is fretting action induced by the small lateral relative displacements between two surfaces when they impact together, where the small displacements are caused by Poisson strains or small tangential "glancing" velocity components. Impact fretting has only recently been addressed in the literature,[124] but it should be noted that under certain circumstances impact fretting may be a potential failure mode of great importance.

Fretting corrosion may be defined as any corrosive surface involvement resulting as a direct result of fretting action. The consequences of fretting corrosion are generally much less severe than for either fretting wear or fretting fatigue. Note that the term *fretting corrosion* is not being used here

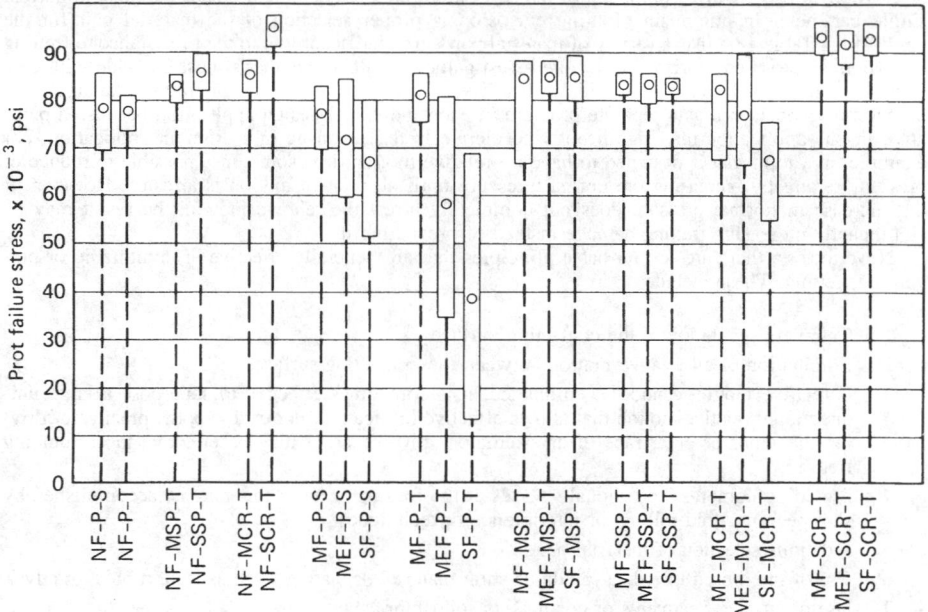

Test conditions used (see table for key symbols)

Test Condition Used	Code Designation	Sample Size	Mean Prot Failure Stress, psi	Unbiased Standard Deviation, psi
Nonfretted, polished, SAE 4340 steel	NF-P-S	15	78,200	5,456
Nonfretted, polished, Ti-140-A titanium	NF-P-T	15	77,800	2,454
Nonfretted, mildly shot-peened, Ti-140-A titanium	NF-MSP-T	15	83,100	1,637
Nonfretted, severely shot-peened, Ti-140-A titanium	NF-SSP-T	15	85,700	2,398
Nonfretted, mildly cold-rolled, Ti-140-A titanium	NF-MCR-T	15	85,430	1,924
Nonfretted, severely cold-rolled, Ti-140-A titanium	NF-SCR-T	15	95,400	2,120
Mildly fretted, polished, SAE 4340 steel	MF-P-S	15	77,280	4,155
Medium fretted, polished, SAE 4340 steel	MeF-P-S	15	71,850	5,492
Severely fretted, polished, SAE 4340 steel	SF-P-S	15	67,700	6,532
Mildly fretted, polished, Ti-140-A titanium	MF-P-T	15	81,050	3,733
Medium fretted, polished, Ti-140-A titanium	MeF-P-T	15	58,140	15,715
Severely fretted, polished, Ti-140-A titanium	SF-P-T	15	38,660	19,342
Mildly fretted, mildly shot-peened, Ti-140-A titanium	MF-MSP-T	15	84,520	5,239
Medium fretted, mildly shot-peened, Ti-140-A titanium	MeF-MSP-T	15	84,930	2,446
Severely fretted, mildly shot-peened, Ti-140-A titanium	SF-MSP-T	15	84,870	2,647
Mildly fretted, severely shot-peened, Ti-140-A titanium	MF-SSP-T	15	83,600	1,474
Medium fretted, severely shot-peened, Ti-140-A titanium	MeF-SSP-T	15	83,240	1,332
Severely fretted, severely shot-peened, Ti-140-A titanium	SF-SSP-T	15	83,110	1,280
Mildly fretted, mildly cold-rolled, Ti-140-A titanium	MF-MCR-T	15	82,050	4,313
Medium fretted, mildly cold-rolled, Ti-140-A titanium	MeF-MCR-T	15	76,930	8,305
Severely fretted, mildly cold-rolled, Ti-140-A titanium	SF-MCR-T	15	67,960	5,682
Mildly fretted, severely cold-rolled, Ti-140-A titanium	MF-SCR-T	15	93,690	1,858
Medium fretted, severely cold-rolled, Ti-140-A titanium	MeF-SCR-T	15	91,950	2,098
Severely fretted, severely cold-rolled, Ti-140-A titanium	SF-SCR-T	15	93,150	1,365

Fig. 18.74 Fatigue properties of fretted steel and titanium specimens with various degrees of shot peening and cold rolling. (See Ref. 106.)

as a synonym for fretting, as in much of the early literature on this topic. Perhaps the most important single parameter in minimizing fretting corrosion is proper selection of the material pair for the application. Table 18.5 lists a variety of material pairs grouped according to their resistance to fretting corrosion.[125] Cross comparisons from one investigator's results to another's must be made with care because testing conditions varied widely. The minimization or prevention of fretting damage must be carefully considered as a separate problem in each individual design application because a palliative in one application may significantly accelerate fretting damage in a different application. For example, in a joint that is designed to have no relative motion, it is sometimes possible to reduce or prevent fretting by increasing the normal pressure until all relative motion is arrested. However, if the increase in normal pressure does not completely arrest the relative motion, the result may be significantly increasing fretting damage instead of preventing it.

Nevertheless, there are several basic principles that are generally effective in minimizing or preventing fretting. These include:

1. Complete separation of the contacting surfaces.
2. Elimination of all relative motion between the contacting surfaces.
3. If relative motion cannot be eliminated, it is sometimes effective to superpose a large unidirectional relative motion that allows effective lubrication. For example, the practice of driving the inner or outer race of an oscillatory pivot bearing may be effective in eliminating fretting.
4. Providing compressive residual stresses at the fretting surface; this may be accomplished by shot peening, cold rolling, or interference fit techniques.
5. Judicious selection of material pairs.
6. Use of interposed low-shear-modulus shim material or plating, such as lead, rubber, or silver.
7. Use of surface treatments or coatings as solid lubricants.
8. Use of surface grooving or roughening to provide debris escape routes and differential strain matching through elastic action.

Of all these techniques, only the first two are completely effective in preventing fretting. The remaining concepts, however, may often be used to minimize fretting damage and yield an acceptable design.

18.8.2 Wear Phenomena

The complexity of the wear process may be better appreciated by recognizing that many variables are involved, including the hardness, toughness, ductility, modulus of elasticity, yield strength, fatigue properties, and structure and composition of the mating surfaces, as well as geometry, contact pressure, temperature, state of stress, stress distribution, coefficient of friction, sliding distance, relative velocity, surface finish, lubricants, contaminants, and ambient atmosphere at the wearing interface. Clearance versus contact-time history of the wearing surfaces may also be an important factor in some cases. Although the wear processes are complex, progress has been made in recent years toward development of quantitative empirical relationships for the various subcategories of wear under specified operating conditions. Adhesive wear is often characterized as the most basic or fundamental subcategory of wear since it occurs to some degree whenever two solid surfaces are in rubbing contact and remains active even when all other modes of wear have been eliminated. The phenomenon of adhesive wear may be best understood by recalling that all real surfaces, no matter how carefully prepared and polished, exhibit a general waviness upon which is superposed a distribution of local protuberances or asperities. As two surfaces are brought into contact, therefore, only a relatively few asperities actually touch, and the *real* area of contact is only a small fraction of the *apparent* contact area. (See Chap. 1 of Ref. 126 and Chap. 2 of Ref. 127.) Thus, even under very small applied loads the local pressures at the contact sites become high enough to exceed the yield strength of one or both surfaces, and local plastic flow ensues. If the contacting surfaces are clean and uncorroded, the very intimate contact generated by this local plastic flow brings the atoms of the two contacting surfaces close enough together to call into play strong adhesive forces. This process is sometimes called *cold welding*. Then if the surfaces are subjected to relative sliding motion, the cold-welded junctions must be broken. Whether they break at the original interface or elsewhere within the asperity depends on surface conditions, temperature distribution, strain-hardening characteristics, local geometry, and stress distribution. If the junction is broken away from the original interface, a particle of one surface is transferred to the other surface, marking one event in the adhesive wear process. Later sliding interactions may dislodge the transferred particles as loose wear particles, or they may remain attached. If this adhesive wear process becomes severe and large-scale metal transfer takes place, the phenomenon is called *galling*. If the galling becomes so severe that two surfaces adhere over a large region so that the actuating forces can no longer produce relative motion between them, the phenomenon is called *seizure*. If properly controlled, however, the adhesive wear rate may be

Table 18.5 Fretting Corrosion Resistance of Various Material Pairs[125]

	Material Pairs Having Good Fretting Corrosion Resistance		
Sakmann and Rightmire	Lead	on	Steel
	Silver plate	on	Steel
	Silver plate	on	Silver plate
	'Parco-lubrized' steel	on	Steel
Gray and Jenny	Grit blasted steel plus lead plate	on	Steel (very good)
	1/16 in. nylon insert	on	Steel (very good)
	Zinc and iron phosphated (Bonderizing) steel	on	Steel (good with thick coat)
McDowell	Laminated plastic	on	Gold plate
	Hard tool steel	on	Tool steel
	Cold-rolled steel	on	Cold-rolled steel
	Cast iron	on	Cast iron with phosphate coating
	Cast iron	on	Cast iron with rubber cement
	Cast iron	on	Cast iron with tungsten sulphide coating
	Cast iron	on	Cast iron with rubber insert
	Cast iron	on	Cast iron with Molykote lubricant
	Cast iron	on	Stainless steel with Molykote lubricant
	Material Pairs Having Intermediate Fretting Corrosion Resistance		
Sakmann and Rightmire	Cadmium	on	Steel
	Zinc	on	Steel
	Copper alloy	on	Steel
	Zinc	on	Aluminum
	Copper plate	on	Aluminum
	Nickel plate	on	Aluminum
	Silver plate	on	Aluminum
	Iron plate	on	Aluminum
Gray and Jenny	Sulphide coated bronze	on	Steel
	Cast bronze	on	"Parco-lubrized" steel
	Magnesium	on	"Parco-lubrized" steel
	Grit-blasted steel	on	Steel
McDowell	Cast iron	on	Cast iron (rough or smooth surface)
	Copper	on	Cast iron
	Brass	on	Cast iron
	Zinc	on	Cast iron
	Cast iron	on	Silver plate
	Cast iron	on	Copper plate
	Magnesium	on	Copper plate
	Zirconium	on	Zirconium
Sakmann and Rightmire	Steel	on	Steel
	Nickel	on	Steel
	Aluminum	on	Steel
	Al-Si alloy	on	Steel
	Antimony plate	on	Steel
	Tin	on	Steel
	Aluminium	on	Aluminum
	Zinc plate	on	Aluminum
Gray and Jenny	Grit blast plus silver plate	on	Steel*
	Steel	on	Steel
	Grit blast plus copper plate	on	Steel
	Grit blast plus tin plate	on	Steel
	Grit blast and aluminium foil	on	Steel
	Be-Cu insert	on	Steel
	Magnesium	on	Steel
	Nitrided steel	on	Chromium plated steel†

Table 18.5 *(Continued)*

	Material Pairs Having Poor Fretting Corrosion Resistance		
McDowell	Aluminium	on	Cast iron
	Aluminum	on	Stainless steel
	Magnesium	on	Cast iron
	Cast iron	on	Chromium plate
	Laminated plastic	on	Cast iron
	Bakelite	on	Cast iron
	Hard tool steel	on	Stainless steel
	Chromium plate	on	Chromium plate
	Cast iron	on	Tin plate
	Gold plate	on	Gold plate

*Possibly effective with light loads and thick (0.005 inch) silver plate.

†Some improvement by heating chromium plated steel to 538°C for 1 hour.

low and self-limiting, often being exploited in the "wearing-in" process to improve mating surfaces such as bearings or cylinders so that full film lubrication may be effectively used.

One quantitative estimate of the amount of adhesive wear is given as follows (see Ref. 113 and Chaps. 2 and 6 of Ref. 128):

$$d_{adh} = \frac{V_{adh}}{A_a} = \left(\frac{k}{9\sigma_{yp}}\right)\left(\frac{W}{A_a}\right) L_s \qquad (18.87)$$

or

$$d_{adh} = k_{adh} p_m L_s \qquad (18.88)$$

where d_{adh} is the average wear depth, A_a is the apparent contact area, L_s is the total sliding distance, V_{adh} is the wear volume, W is the applied load, $p_m = W/A_a$ is the mean nominal contact pressure between bearing surfaces, and $k_{adh} = k/9\sigma_{yp}$ is a wear coefficient that depends on the probability of formation of a transferred fragment and the yield strength (or hardness) of the softer material. Typical values of the wear constant k for several material pairs are shown in Table 18.6, and the influence of lubrication on the wear constant k is indicated in Table 18.7.

Noting from (18.88) that

$$k_{adh} = \frac{d_{adh}}{p_m L_s} \qquad (18.89)$$

it may be observed that if the ratio $d_{adh}/p_m L_s$ is experimentally found to be constant, (18.88) should be valid. Experimental evidence has been accumulated (see pp. 124–125 of Ref. 113) to confirm that for a given material pair this ratio is constant up to mean nominal contact pressures approximately equal to the uniaxial yield strength. Above this level the adhesive wear coefficient increases rapidly, with attendant severe galling and seizure.

Table 18.6 Archard Adhesive Wear Constant k for Various Unlubricated Material Pairs in Sliding Contact[a]

Material Pair	Wear Constant k
Zinc on zinc	160×10^{-3}
Low-carbon steel on low-carbon steel	45×10^{-3}
Copper on copper	32×10^{-3}
Stainless steel on stainless steel	21×10^{-3}
Copper (on low-carbon steel)	1.5×10^{-3}
Low-carbon steel (on copper)	0.5×10^{-3}
Bakelite on bakelite	0.02×10^{-3}

[a]From Chap. 6 of Ref. 128, with permission of John Wiley & Sons.

Table 18.7 Order of Magnitude Values for Adhesive Wear Constant k Under Various Conditions of Lubrication[a]

| | Metal (on Metal) | | Nonmetal |
Lubrication Condition	Like	Unlike	(on Metal)
Unlubricated	5×10^{-3}	2×10^{-4}	5×10^{-6}
Poorly lubricated	2×10^{-4}	2×10^{-4}	5×10^{-6}
Average lubrication	2×10^{-5}	2×10^{-5}	5×10^{-6}
Excellent lubrication	2×10^{-6} to 10^{-7}	2×10^{-6} to 10^{-7}	2×10^{-6}

[a]From Chap. 6 of Ref. 128, with permission of John Wiley & Sons.

In the selection of metal combinations to provide resistance to adhesive wear, it has been found that the sliding pair should be composed of mutually insoluble metals and that at least one of the metals should be from the B subgroup of the periodic table. (See p. 31 of Ref. 129.) The reasons for these observations are that the number of cold-weld junctions formed is a function of the mutual solubility, and the strength of the junction bonds is a function of the bonding characteristics of the metals involved. The metals in the B subgroup of the periodic table are characterized by weak, brittle covalent bonds. These criteria have been verified experimentally, as shown in Table 18.8, where 114 of 123 pairs tested substantiated the criteria.

In the case of abrasive wear, the wear particles are removed from the surface by the plowing and gouging action of the asperities of a harder mating surface or by hard particles trapped between the rubbing surfaces. This type of wear is manifested by a system of surface grooves and scratches, often called *scoring*. The abrasive wear condition in which the hard asperities of one surface wear away the mating surface is commonly called *two-body wear,* and the condition in which hard abrasive particles between the two surfaces cause the wear is called *three-body wear.*

An average abrasive wear depth d_{abr} may then be estimated as

$$d_{abr} = \frac{V_{abr}}{A_a} = \frac{(\tan \theta)_m}{3 \pi \sigma_{yp}} \left(\frac{W}{A_a} \right) L_s \tag{18.90}$$

or

$$d_{abr} = k_{abr} p_m L_s \tag{18.91}$$

where W is total applied load, $(\tan \theta)_m$ is a weighted mean value for all asperities, L_s is a total distance of sliding, σ_{yp} is the uniaxial yield point strength for the softer material, V_{abr} is abrasive wear volume, $p_m = W/A_a$ is mean nominal contact pressure between bearing surfaces, and $k_{abr} = (\tan \theta)_m / 3 \pi \sigma_{yp}$ is an abrasive wear coefficient that depends on the roughness characteristics of the surface and the yield strength (or hardness) of the softer material.

Comparing (18.90) for abrasive wear volume with (18.87) for adhesive wear volume, we note that they are formally the same except the constant $k/3$ in the adhesive wear equation is replaced by $(\tan \theta)_m / \pi$ in the abrasive wear equation. Typical values of the wear constant $3(\tan \theta)_m / \pi$ for several materials are shown in Table 18.9. As indicated in Table 18.9, experimental evidence shows that k_{abr} for three-body wear is typically about an order of magnitude smaller than for the two-body case, probably because the trapped particles tend to roll much of the time and cut only a small part of the time.

In selecting materials for abrasive wear resistance, it has been established that both hardness and modulus of elasticity are key properties. Increasing wear resistance is associated with higher hardness and lower modulus of elasticity since both the amount of elastic deformation and the amount of elastic energy that can be stored at the surface are increased by higher hardness and lower modulus of elasticity.

Table 18.10 tabulates several materials in order of descending values of (hardness)/(modulus of elasticity). Well-controlled experimental data are not yet available, but general experience would provide an ordering of materials for decreasing wear resistance compatible with the array of Table 18.10. When the conditions for adhesive or abrasive wear exist together with conditions that lead to corrosion, the two processes persist together and often interact synergistically. If the corrosion product is hard and abrasive, dislodged corrosion particles trapped between contacting surfaces will accelerate the abrasive wear process. In turn, the wear process may remove the "protective" surface layer of corrosion product to bare new metal to the corrosive atmosphere, thereby accelerating the corrosion process. Thus, the corrosion wear process may be self-accelerating and may lead to high rates of wear.

Table 18.8 Adhesive Wear Behavior of Various Pairs[a]

Description of Metal Pair	Material Combination				Remarks
	Al Disk	Steel Disk	Cu Disk	Ag Disk	
Soluble pairs with poor adhesive wear resistance	Be	Be	Be	Be	These pairs substantiate the criteria of solubility and B subgroup metals
	Mg	—	Mg	Mg	
	Al	Al	Al	—	
	Si	Si	Si	Si	
	Ca	—	Ca	—	
	Ti	Ti	Ti	—	
	Cr	Cr	—	—	
	—	Mn	—	—	
	Fe	Fe	—	—	
	Co	Co	Co	—	
	Ni	Ni	Ni	—	
	Cu	—	Cu	—	
	—	Zn	Zn	—	
	Zr	Zr	Zr	Zr	
	Nb	Nb	Nb	—	
	Mo	Mo	Mo	—	
	Rh	Rh	Rh	—	
	—	Pd	—	—	
	Ag	—	Ag	—	
	—	—	Cd	Cd	
	—	—	In	In	
	Sn	—	Sn	—	
	Ce	Ce	Ce	—	
	Ta	Ta	Ta	—	
	W	W	W	—	
	—	Ir	—	—	
	Pt	Pt	Pt	—	
	Au	Au	Au	Au	
	Th	Th	Th	Th	
	U	U	U	U	
Soluble pairs with fair or good adhesive wear resistance. (F) = Fair	—	Cu(F)	—		These pairs do not substantiate the stated criteria
	Zn(F)	—	—		
	—	—	Sb(F)		
Insoluble pairs, neither from the B subgroup, with poor adhesive wear resistance		Li			These pairs substantiate the stated criteria
		Mg			
		Ca			
		Ba			
Insoluble pairs, one from the B subgroup, with fair or good adhesive wear resistance. (F) = Fair	—	C(F)	—	—	These pairs substantiate the stated criteria
	—	—	—	Ti(F)	
	—	—	Cr(F)	Cr(F)	
	—	—	—	Fe(F)	
	—	—	—	Co(F)	
	—	—	Ge(F)	—	
	—	Se(F)	Se(F)	—	
	—	—	—	Nb(F)	
	—	Ag	—	—	
	Cd	Cd	—	—	
	In	In	—	—	
	—	Sn(F)	—	—	
	—	Sb(F)	Sb	—	
	Te(F)	Te(F)	Te(F)	—	
	Tl	Tl	Tl	—	
	Pb(F)	Pb	Pb	—	
	Bi(F)	Bi	Bi(F)	—	
Insoluble pairs, one from the B subgroup, with poor adhesive wear resistance	C	—	C	C	These pairs do not substantiate the stated criteria
	—	—	—	Ni	
	Se	—	—	—	
	—	—	—	Mo	

[a]See pp. 34–35 of Ref. 129.

Table 18.9 Abrasive Wear Constant 3(tan θ)$_m$/π for Various Materials in Sliding Contact as Reported by Different Investigators[a]

Materials	Wear Type	Particle Size, μ	3(tan θ)$_m$/π
Many	Two-body	—	180×10^{-3}
Many	Two-body	110	150×10^{-3}
Many	Two-body	40–150	120×10^{-3}
Steel	Two-body	260	80×10^{-3}
Many	Two-body	80	24×10^{-3}
Brass	Two-body	70	16×10^{-3}
Steel	Three-body	150	6×10^{-3}
Steel	Three-body	80	4.5×15^{-3}
Many	Three-body	40	2×10^{-3}

[a]See p. 169 of Ref. 128. Reprinted with permission from John Wiley & Sons.

On the other hand, some corrosion products, for example, metallic phosphates, sulfides, and chlorides, form as soft lubricative films that actually improve the wear rate markedly, especially if adhesive wear is the dominant phenomenon.

Three major wear control methods have been defined, as follows (see p. 36 of Ref. 129): *principle of protective layers,* including protection by lubricant, surface film, paint, plating, phosphate, chemical, flame-sprayed, or other types of interfacial layers: *principle of conversion,* in which wear is converted from destructive to permissible levels through better choice of metal pairs, hardness, surface finish, or contact pressure: and *principle of diversion,* in which the wear is diverted to an economical replaceable wear element that is periodically discarded and replaced as "wear out" occurs. When two surfaces operate in rolling contact, the wear phenomenon is quite different from the wear of sliding surfaces just described, although the "delamination" theory[130] is very similar to the mechanism of wear between rolling surfaces in contact as described here. Rolling surfaces in contact result

Table 18.10 Values of (Hardness/Modulus of Elasticity) for Various Materials[113] †

Material	Condition	BHN*/($E \times 10^{-6}$) (in mixed units)
Alundum (Al_2O_3)	Bonded	143
Chrome plate	Bright	83
Gray iron	Hard	33
Tungsten carbide	9% Co	22
Steel	Hard	21
Titanium	Hard	17
Aluminum alloy	Hard	11
Gray iron	As cast	10
Structural steel	Soft	5
Malleable iron	Soft	5
Wrought iron	Soft	3.5
Chromium metal	As cast	3.5
Copper	Soft	2.5
Silver	Pure	2.3
Aluminum	Pure	2.0
Lead	Pure	2.0
Tin	Pure	0.7

†Reprinted from copyrighted work with permission; courtesy of Elsevier Publishing Company.
*Brinell hardness number.

in Hertz contact stresses that produce maximum values of shear stress slightly below the surface. (See, for example, p. 389 of Ref. 131.) As the rolling contact zone moves past a given location on the surface, the subsurface peak shear stress cycles from zero to a maximum value and back to zero, thus producing a cyclic stress field. Such conditions may lead to fatigue failure by the initiation of a subsurface crack that propagates under repeated cyclic loading and that may ultimately propagate to the surface to spall out a macroscopic surface particle to form a wear pit. This action, called *surface fatigue wear*, is a common failure mode in antifriction bearings, gears, and cams, and all machine parts that involve rolling surfaces in contact. Deformation wear arises as a result of repeated plastic deformations at the wearing surfaces; this wear may induce a matrix of cracks that grow and coalesce to form wear particles or may produce cumulative permanent plastic deformations that finally grow into an unacceptable surface indentation or wear scar. Deformation wear is generally caused by conditions that lead to impact loading between the two wearing surfaces. Although some progress has been made in deformation wear analysis, the techniques are highly specialized. Fretting wear, which has received renewed attention in the recent literature (see p. 55 of Ref. 132 and p. 75 of Ref. 128), has already been discussed. Impact wear is a term reserved for impact-induced repeated elastic deformations at the wearing surfaces that produce a matrix of cracks that grow in accordance with surface fatigue phenomena. Under some circumstances impact wear may be generated by purely normal impacts, and under other circumstances the impact may contain elements of rolling and/or sliding as well. The severity of the impact is generally measured or expressed in terms of the kinetic energy of the striking mass. The geometry of the impacting surfaces and the material properties of the two contacting surfaces play a major role in the determination of severity of impact wear damage. The objective of a designer faced with impact wear as a potential failure mode is to predict the size of the wear scar, or its depth, as a function of the number of repetitive load cycles.

An *empirical* approach to the prediction of sliding wear has been developed,[133] and the pertinent empirical constants have been evaluated for a wide variety of materials and lubricant combinations for various operating conditions. This empirical development permits the designer to specify a design configuration to ensure "zero wear" during the specified design lifetime. *Zero wear* is defined to be wear of such small magnitude that the surface finish is not significantly altered by the wear process. That is, the wear depth for zero wear is of the order of one-half the peak-to-peak surface finish dimension.

If a *pass* is defined to be a distance of sliding W equal to the dimension of the contact area in the direction of sliding, N is the number of passes, τ_{max} is the maximum shearing stress in the vicinity of the surface, τ_{yp} is the shear yield point of the specified material, and γ_r is a constant for the particular combination of materials and lubricant, then the empirical model asserts that there will be "zero wear" for N passes if

$$\tau_{max} \leq \left[\frac{2 \times 10^3}{N} \right]^{1/9} \gamma_r \tau_{yp} \tag{18.92}$$

or, to interpret it differently, the number of passes that can be accommodated without exceeding the zero wear level is given by

$$N = 2 \times 10^3 \left[\frac{\gamma_r \tau_{yp}}{\tau_{max}} \right]^9 \tag{18.93}$$

It may be noted that the constant γ_r is referred to 2000 passes and must be experimentally determined. For quasihydrodynamic lubrication, γ_r ranges between 0.54 and 1. For dry or boundary lubrication, γ_r is 0.54 for materials with low susceptibility to adhesive wear and 0.20 for materials with high susceptibility to adhesive wear.

Calculation of the maximum shear stress τ_{max} in the vicinity of the contacting surface must include both the normal force and the friction force. Thus, for conforming geometries, such as a flat surface on a flat surface or a shaft in a journal bearing, a critical point at the contacting interface may be analyzed by the maximum shear stress theory to determine τ_{max}.

The number of passes will usually require expression as a function of the number of cycles, strokes, oscillations, or hours of operation in the design lifetime.

Utilizing these definitions and a proper stress analysis at the wear interface allows one to design for "zero wear" through use of Eqs. (18.92) or (18.93).

18.9 CORROSION AND STRESS CORROSION

Corrosion may be defined as the undesired deterioration of a material through chemical or electrochemical interaction with the environment, or destruction of materials by means other than purely mechanical action. Failure by corrosion occurs when the corrosive action renders the corroded device incapable of performing its design function. Corrosion often interacts synergistically with another failure mode, such as wear or fatigue, to produce the even more serious combined failure modes, such as corrosion wear or corrosion fatigue. Failure by corrosion and protection against failure by corrosion has been estimated to cost in excess of 8 billion dollars annually in the United States alone. (See p. 1 of Ref. 134.)

The complexity of the corrosion process may be better appreciated by recognizing that many variables are involved, including environmental, electrochemical, and metallurgical aspects. For example, anodic reactions and rate of oxidation; cathodic reactions and rate of reduction; corrosion inhibition, polarization, or retardation; passivity phenomena; effect of oxidizers; effect of velocity; temperature; corrosive concentration; galvanic coupling; and metallurgical structure all influence the type and rate of the corrosion process.

Corrosion processes have been categorized in many different ways. One convenient classification divides corrosion phenomena into the following types (see p. 28 of Ref. 134 and p. 85 of Ref. 135): direct chemical attack, galvanic corrosion, crevice corrosion, pitting corrosion, intergranular corrosion, selective leaching, erosion corrosion, cavitation corrosion, hydrogen damage, biological corrosion, and stress corrosion cracking. Depending on the types of environment, loading, and mechanical function of the machine parts involved, any of the types of corrosion may combine their influence with other failure modes to produce premature failures. Of particular concern are interactions that lead to failure by corrosion wear, corrosion fatigue, fretting fatigue, and corrosion-induced brittle fracture.

18.9.1 Types of Corrosion

Direct chemical attack is probably the most common type of corrosion. Under this type of corrosive attack the surface of the machine part exposed to the corrosive media is attacked more or less uniformly over its entire surface, resulting in a progressive deterioration and dimensional reduction of sound load-carrying net cross section. The rate of corrosion due to direct attack can usually be estimated from relatively simple laboratory tests in which small specimens of the selected material are exposed to a well-simulated actual environment, with frequent weight change and dimensional measurements carefully taken. The corrosion rate is usually expressed in mils per year (mpy) and may be calculated as (see p. 133 of Ref. 134)

$$R = \frac{534W}{\gamma At} \tag{18.94}$$

where R is rate of corrosion penetration in mils (1 mil = 0.001 in.) per year (mpy), W is weight loss in milligrams, A is exposed area of the specimen in square inches, γ is density of the specimen in grams per cubic centimeter, and t is exposure time in hours. Use of this corrosion rate expression in predicting corrosion penetration in actual service is usually successful if the environment has been properly simulated in the laboratory. Corrosion rate data for many different combinations of materials and environments are available in the literature.[136-138] Figure 18.75 illustrates one presentation of such data.

Direct chemical attack may be reduced in severity or prevented by any one or a combination of several means, including selecting proper materials to suit the environment; using plating, flame spraying, cladding, hot dipping, vapor deposition, conversion coatings, and organic coatings or paint to protect the base material; changing the environment by using lower temperature or lower velocity, removing oxygen, changing corrosive concentration, or adding corrosion inhibitors; using cathodic protection in which electrons are supplied to the metal surface to be protected either by galvanic coupling to a sacrificial anode or by an external power supply; or adopting other suitable design modifications.

Galvanic corrosion is an accelerated electrochemical corrosion that occurs when two dissimilar metals in electrical contact are made part of a circuit completed by a connecting pool or film of electrolyte or corrosive medium. Under these circumstances, the potential difference between the dissimilar metals produces a current flow through the connecting electrolyte, which leads to corrosion, concentrated primarily in the more anodic or less noble metal of the pair. This type of action is completely analogous to a simple battery cell. Current must flow to produce galvanic corrosion, and, in general, more current flow means more serious corrosion. The relative tendencies of various metals to form galvanic cells, and the probable direction of the galvanic action, are illustrated for several commercial metals and alloys in seawater in Table 18.11. (See p. 32 of Ref. 134 or p. 86 of Ref. 135.)

Ideally, tests in the actual service environment should be conducted; but, if such data are unavailable, the data of Table 18.11 should give a good indication of possible galvanic action. The farther apart the two dissimilar metals are in the galvanic series, the more serious the galvanic corrosion problem may be. Material pairs within any bracketed group exhibit little or no galvanic action. It should be noted, however, that there are sometimes exceptions to the galvanic series of Table 18.11, so wherever possible corrosion tests should be performed with actual materials in the actual service environment.

The accelerated galvanic corrosion is usually most severe near the junction between the two metals, decreasing in severity at locations farther from the junction. The ratio of cathodic area to anodic area exposed to the electrolyte has a significant effect on corrosion rate. It is *desirable* to

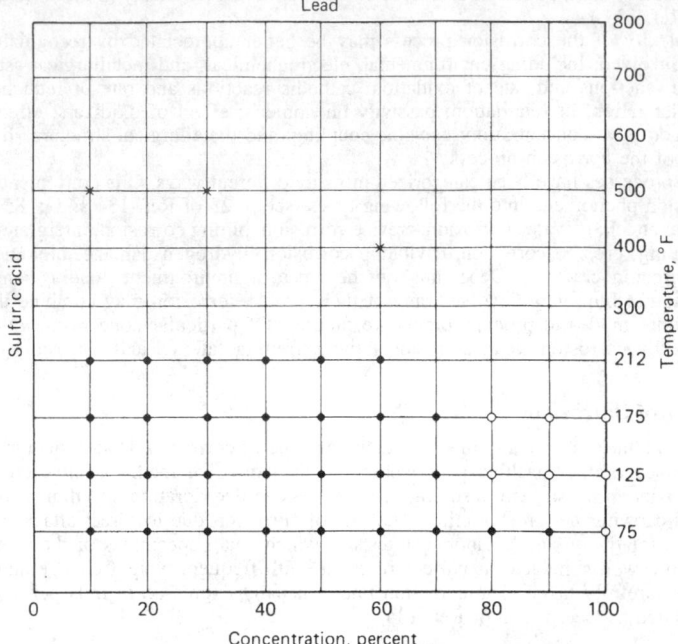

Fig. 18.75 Nelson's method for summarizing corrosion rate data for lead in sulfuric acid environment as a function of concentration and temperature. (See Ref. 136; reprinted with permission of McGraw-Hill Book Company.)

have a *small ratio* of cathode area to anode area. For this reason, if only *one* of two dissimilar metals in electrical contact is to be coated for corrosion protection, the *more* noble or more corrosion-resistant metal should be coated. Although this at first may seem the wrong metal to coat, the area effect, which produces anodic corrosion rate of 10^2–10^3 times cathodic corrosion rates for equal areas, provides the logic for this assertion.

Galvanic corrosion may be reduced in severity or prevented by one or a combination of several steps, including the selection of material pairs as close together as possible in the galvanic series, preferably in the same bracketed group; electrical insulation of one dissimilar metal from the other as completely as possible; maintaining as small a ratio of cathode area to anode area as possible; proper use and maintenance of coatings; the use of inhibitors to decrease the aggressiveness of the corroding medium; and the use of cathodic protection in which a third metal element anodic to both members of the operating pair is used as a sacrificial anode that may require periodic replacement.

Crevice corrosion is an accelerated corrosion process highly localized within crevices, cracks, and other small-volume regions of stagnant solution in contact with the corroding metal. For example, crevice corrosion may be expected in gasketed joints; clamped interfaces; lap joints; rolled joints; under bolt and rivet heads; and under foreign deposits of dirt, sand, scale, or corrosion product. Until recently, crevice corrosion was thought to result from differences in either oxygen concentration or metal ion concentration in the crevice compared to its surroundings. More recent studies seem to indicate, however, that the local oxidation and reduction reactions result in oxygen depletion in the stagnant crevice region, which leads to an excess positive charge in the crevice due to increased metal ion concentration. This, in turn, leads to a flow of chloride and hydrogen ions into the crevice, both of which accelerate the corrosion rate within the crevice. Such accelerated crevice corrosion is highly localized and often requires a lengthy incubation period of perhaps many months before it gets under way. Once started, the rate of corrosion accelerates to become a serious problem. To be

Table 18.11 Galvanic Series of Several Commercial Metals and Alloys in Seawater[a]

↑	Platnium
	Gold
Noble or	Graphite
cathodic	Titanium
(protected	Silver
end)	⎡ Chlorimet 3 (62 Ni, 18 Cr, 18 Mo) ⎤
	⎣ Hastelloy C (62 Ni, 17 C, 15 Mo) ⎦
	⎡ 18-8 Mo stainless steel (passive) ⎤
	⎢ 18-8 stainless steel (passive) ⎥
	⎣ Chromium stainless steel 11-30% Cr (passive) ⎦
	⎡ Inconel (passive)(80 Ni, 13 Cr, 7 Fe) ⎤
	⎣ Nickel (passive) ⎦
	Silver solder
	⎡ Monel (70 Ni, 30 Cu) ⎤
	⎢ Cupronickels (60-90 Cu, 40-10 Ni) ⎥
	⎢ Bronzes (Cu-Sn) ⎥
	⎢ Copper ⎥
	⎣ Brasses (Cu-Zn) ⎦
	⎡ Chlorimet 2 (66 Ni, 32 Mo, 1 Fe) ⎤
	⎣ Hastelloy B (60 Ni, 30 Mo, 6 Fe, 1 Mn) ⎦
	⎡ Inconel (active) ⎤
	⎣ Nickel (active) ⎦
	Tin
	Lead
	Lead-tin solders
	⎡ 18-8 Mo stainless steel (active) ⎤
	⎣ 18-8 stainless steel (active) ⎦
	Ni-Resist (high Ni cast iron)
	Chromium stainless steel, 13% Cr (active)
	⎡ Cast iron ⎤
	⎣ Steel or iron ⎦
Active or	2024 aluminum (4.5 Cu, 1.5 Mg, 0.6 Mn)
anodic	Cadmium
(corroded	Commercially pure aluminum (1100)
end)	Zinc
↓	Magnesium and magnesium alloys

[a]See p. 32 of Ref. 134. Reprinted with permission of McGraw-Hill Book Company.

susceptible to crevice corrosion attack, the stagnant region must be wide enough to allow the liquid to enter but narrow enough to maintain stagnation. This usually implies cracks and crevices of a few thousandths to a few hundredths of an inch in width.

To reduce the severity of crevice corrosion, or prevent it, it is necessary to eliminate the cracks and crevices. This may involve caulking or seal welding existing lap joints; redesign to replace riveted or bolted joints by sound, welded joints; filtering foreign material from the working fluid; inspection and removal of corrosion deposits; or using nonabsorbent gasket materials. Pitting corrosion is a very localized attack that leads to the development of an array of holes or pits that penetrate the metal. The pits, which typically are about as deep as they are across, may be widely scattered or so heavily concentrated that they simply appear as a rough surface. The mechanism of pit growth is virtually identical to that of crevice corrosion described, except that an existing crevice is not required to initiate pitting corrosion. The pit is probably initiated by a momentary attack due to a random variation in fluid concentration or a tiny surface scratch or defect. Some pits may become inactive because of a stray convective current, whereas others may grow large enough to provide a stagnant region of stable size, which then continues to grow over a long period of time at an accelerating rate. Pits usually grow in the direction of the gravity force field since the dense concentrated solution in a pit is required for it to grow actively. Most pits, therefore, grow downward from horizontal surfaces to ultimately perforate the wall. Fewer pits are formed on vertical walls, and very few pits grow upward from the bottom surface.

Measurement and assessment of pitting corrosion damage is difficult because of its highly local nature. Pit depth varies widely and, as in the case of fatigue damage, a statistical approach must be taken in which the probability of a pit of specified depth may be established in laboratory testing. Unfortunately, a significant size effect influences depth of pitting, and this must be taken into account when predicting service life of a machine part based on laboratory pitting corrosion data.

The control or prevention of pitting corrosion consists primarily of the wise selection of material to resist pitting or, since pitting is usually the result of stagnant conditions, imparting velocity to the fluid. Increasing its velocity may also decrease pitting corrosion attack.

Because of the atomic mismatch at the grain boundaries of polycrystalline metals, the stored strain energy is higher in the grain boundary regions than in the grains themselves. These high-energy grain boundaries are more chemically reactive than the grains. Under certain conditions depletion or en-richment of an alloying element or impurity concentration at the grain boundaries may locally change the composition of a corrosion-resistant metal, making it susceptible to corrosive attack. Localized attack of this vulnerable region near the grain boundaries is called intergranular corrosion. In partic-ular, the austenitic stainless steels are vulnerable to intergranular corrosion if *sensitized* by heating into the temperature range from 950° to 1450°F, which causes depletion of the chromium near the grain boundaries as chromium carbide is precipitated at the boundaries. The chromium-poor regions then corrode because of local galvanic cell action, and the grains literally fall out of the matrix. A special case of intergranular corrosion, called "weld decay," is generated in the portion of the weld-affected zone, which is heated into the sensitizing temperature range.

To minimize the susceptibility of austenitic stainless steels to intergranular corrosion, the carbon content may be lowered to below 0.03%, stabilizers may be added to prevent depletion of the chro-mium near the grain boundaries, or a high-temperature solution heat treatment, called quench-annealing, may be employed to produce a more homogeneous alloy.

Other alloys susceptible to intergranular corrosion include certain aluminum alloys, magnesium alloys, copper-based alloys, and die-cast zinc alloys in unfavorable environments.

The corrosion phenomenon in which one element of a solid alloy is removed is termed selective leaching. Although the selective leaching process may occur in any of several alloy systems, the more common examples are *dezincification* of brass alloys and *graphitization* of gray cast iron. Dezincification may occur as either a highly local "plug-type" or a broadly distributed layer-type attack. In either case, the dezincified region is porous, brittle, and weak. Dezincification may be minimized by adding inhibitors such as arsenic, antimony, or phosphorous to the alloy; by lowering oxygen in the environment; or by using cathodic protection.

In the case of graphitization of gray cast iron, the environment selectively leaches the iron matrix to leave the graphite network intact to form an active galvanic cell. Corrosion then proceeds to destroy the machine part. Use of other alloys, such as nodular or malleable cast iron, mitigates the problem because there is no graphite network in these alloys to support the corrosion residue. Other alloy systems in adverse environments that may experience selective leaching include aluminum bronzes, silicon bronzes, and cobalt alloys.

Erosion corrosion is an accelerated, direct chemical attack of a metal surface due to the action of a moving corrosive medium. Because of the abrasive wear action of the moving fluid, the formation of a protective layer of corrosion product is inhibited or prevented, and the corroding medium has direct access to bare, unprotected metal. Erosion corrosion is usually characterized by a pattern of grooves or peaks and valleys generated by the flow pattern of the corrosive medium. Most alloys are susceptible to erosion corrosion, and many different types of corrosive media may induce erosion corrosion, including flowing gases, liquids, and solid aggregates. Erosion corrosion may become a problem in such machine parts as valves, pumps, blowers, turbine blades and nozzles, conveyors, and piping and ducting systems, especially in the regions of bends and elbows.

Erosion corrosion is influenced by the velocity of the flowing corrosive medium, turbulence of the flow, impingement characteristics, concentration of abrasive solids, and characteristics of the metal alloy surface exposed to the flow. Methods of minimizing or preventing erosion corrosion include reducing the velocity, eliminating or reducing turbulence, avoiding sudden changes in the direction of flow, eliminating direct impingement where possible, filtering out abrasive particles, using harder and more corrosion-resistant alloys, reducing the temperature, using appropriate surface coatings, and using cathodic protection techniques.

Cavitation often occurs in hydraulic systems, such as turbines, pumps, and piping, when pressure changes in a flowing liquid give rise to the formation and collapse of vapor bubbles at or near the containing metal surface. The impact associated with vapor bubble collapse may produce high-pressure shock waves that may plastically deform the metal locally or destroy any protective surface film of corrosion product and locally accelerate the corrosion process. Furthermore, the tiny depressions so formed act as a nucleus for subsequent vapor bubbles, which continue to form and collapse at the same site to produce deep pits and pockmarks by the combined action of mechanical deformation and accelerated chemical corrosion. This phenomenon is called cavitation corrosion. Cavitation corrosion may be reduced or prevented by eliminating the cavitation through appropriate design changes. Smoothing the surfaces, coating the walls, using corrosion-resistant materials, minimizing pressure differences in the cycle, and using cathodic protection are design changes that may be effective.

Hydrogen damage, although not considered to be a form of direct corrosion, is often induced by corrosion. Any damage caused in a metal by the presence of hydrogen or the interaction with hydrogen is called hydrogen damage. Hydrogen damage includes hydrogen blistering, hydrogen embrittlement, hydrogen attack, and decarburization.

Hydrogen blistering is caused by the diffusion of hydrogen atoms into a void within a metallic structure where they combined to form molecular hydrogen. The hydrogen pressure builds to a high level that, in some cases, causes blistering, yielding, and rupture. Hydrogen blistering may be minimized by using materials without voids, by using corrosion inhibitors, or by using hydrogen-impervious coatings.

Hydrogen embrittlement is also caused by the penetration of hydrogen into the metallic structure to form brittle hydrides and pin dislocation movement to reduce slip, but the exact mechanism is not yet fully understood. Hydrogen embrittlement is more serious at the higher-strength levels of susceptible alloys, which include most of the high-strength steels. Reduction and prevention of hydrogen embrittlement may be accomplished by "baking out" the hydrogen at relatively low temperatures for several hours, use of corrosion inhibitors, or use of less susceptible alloys.

Decarburization and hydrogen attack are both high-temperature phenomena. At high temperatures hydrogen removes carbon from an alloy, often reducing its tensile strength and increasing its creep rate. This carbon-removing process is called *decarburization*. It is also possible that the hydrogen may lead to the formation of methane in the metal voids, which may expand to form cracks, another form of hydrogen attack. Proper selection of alloys and coatings is helpful in prevention of these corrosion-related problems.

Biological corrosion is a corrosion process or processes that results from the activity of living organisms. These organisms may be microorganisms, such as aerobic or anaerobic bacteria, or they may be macroorganisms, such as fungi, mold, algae, or barnacles. The organisms may influence or produce corrosion by virtue of their processes of food ingestion and waste elimination. There are, for example, sulfate-reducing anaerobic bacteria, which produce iron sulfide when in contact with buried steel structures, and aerobic sulfur-oxidizing bacteria, which produce localized concentrations of sulfuric acid and serious corrosive attack on buried steel and concrete pipe lines. There are also iron bacteria, which ingest ferrous iron and precipitate ferrous hydroxide to produce local crevice corrosion attack. Other bacteria oxidize ammonia to nitric acid, which attacks most metals, and most bacteria produce carbon dioxide, which may form the corrosive agent carbonic acid. Fungi and mold assimilate organic matter and produce organic acids. Simply by their presence, fungi may provide the site for crevice corrosion attacks, as does the presence of attached barnacles and algae. Prevention or minimization of biological corrosion may be accomplished by altering the environment or by using proper coatings, corrosion inhibitors, bactericides or fungicides, or cathodic protection.

18.9.2 Stress Corrosion Cracking

Stress corrosion cracking is an extremely important failure mode because it occurs in a wide variety of different alloys. This type of failure results from a field of cracks produced in a metal alloy under the combined influence of tensile stress and a corrosive environment. The metal alloy is not attacked over most of its surface, but a system of intergranular or transgranular cracks propagates through the matrix over a period of time.

Stress levels that produce stress corrosion cracking are well below the yield strength of the material, and residual stresses as well as applied stresses may produce failure. The lower the stress level, the longer is the time required to produce cracking, and there appears to be a threshold stress level below which stress corrosion cracking does not occur. (See p. 96 of Ref. 134.)

The chemical compositions of the environments that lead to stress corrosion cracking are highly specific and peculiar to the alloy system, and no general patterns have been observed. For example, austenitic stainless steels are susceptible to stress corrosion cracking in chloride environments but not in ammonia environments, whereas brasses are susceptible to stress corrosion cracking in ammonia environments but not in chloride environments. Thus, the "season cracking" of brass cartridge cases in the crimped zones was found to be stress corrosion cracking due to the ammonia resulting from decomposition of organic matter. Likewise, "caustic embrittlement" of steel boilers, which resulted in many explosive failures, was found to be stress corrosion cracking due to sodium hydroxide in the boiler water.

Stress corrosion cracking is influenced by stress level, alloy composition, type of environment, and temperature. Crack propagation seems to be intermittent, and the crack grows to a critical size, after which a sudden and catastrophic failure ensues in accordance with the laws of fracture mechanics. Stress corrosion crack growth in a statically loaded machine part takes place through the interaction of mechanical strains and chemical corrosion processes at the crack tip. The largest value of plane strain stress intensity factor for which crack growth does not take place in a corrosive environment is designated K_{Iscc}. In many cases, corrosion fatigue behavior is also related to the magnitude of K_{Iscc}.[9]

Prevention of stress corrosion cracking may be attempted by lowering the stress below the critical threshold level, choice of a better alloy for the environment, changing the environment to eliminate the critical corrosive element, use of corrosion inhibitors, or use of cathodic protection. Before cathodic protection is implemented care must be taken to ensure that the phenomenon is indeed stress corrosion cracking because hydrogen embrittlement is accelerated by cathodic protection techniques.

18.10 FAILURE ANALYSIS AND RETROSPECTIVE DESIGN

In spite of all efforts to design and manufacture machines and structures to function properly without failure, failures do occur. Whether the failure consequences simply represent an annoying inconvenience, such as a "binding" support on the sliding patio screen, or a catastrophic loss of life and property, as in the crash of a jumbo jet, it is the responsibility of the designer to glean all of the information possible from the failure event so that similar events can be avoided in the future. Effective assessment of service failures usually requires the intense interactive scrutiny of a team of specialists, including at least a mechanical designer and a materials engineer trained in failure analysis techniques. The team might often include a manufacturing engineer and a field service engineer as well. The mission of the failure analysis team is to discover the initiating cause of failure, identify the best solution, and redesign the product to prevent future failures. Although the results of failure analysis investigations may often be closely related to product liability litigation, the legal issues will not be addressed in this discussion.

Techniques utilized in the failure analysis effort include the inspection and documentation of the event through direct examination, photographs and eyewitness reports; preservation of all parts, especially failed parts; and pertinent calculations, analyses, and examinations that may help establish and validate the cause of failure. The materials engineer may utilize macroscopic examination, low-power magnification, microscopic examination, transmission or scanning electron microscopic techniques, energy-dispersive X-ray techniques, hardness tests, spectrographic analysis, metallographic examination, or other techniques of determining the failure type, failure location, material abnormalities, and potential causes of failure. The designer may perform stress and deflection analyses, examine geometry, assess service loading and environmental influences, reexamine the kinematics and dynamics of the application, and attempt to reconstruct the failure scenario. Other team members may examine the quality of manufacture, the quality of maintenance, the possibility of unusual or unconventional usage by the operator, or other factors that may have played a role in the service failure. Piecing all of this information together, it is the objective of the failure analysis team to identify as accurately as possible the probable cause of failure.

As undesirable as service failures may be, the results of a well-executed failure analysis may be transformed directly into improved product reliability by a designer who capitalizes on service failure data and failure analysis results. These techniques of retrospective design have become important working tools of the profession and are likely to continue to grow in importance.

REFERENCES

1. J. A. Collins, *Failure of Materials in Mechanical Design; Analysis, Prediction, Prevention,* 2nd ed., Wiley, New York, 1993.

2. "Progress in Measuring Fracture Toughness and Using Fracture Mechanics," *Materials Research and Standards,* 103–119 (March 1964).

3. H. M. Westergaard, "Bearing Pressures and Cracks," *Trans. ASME* **61**, A49 (1939).

4. G. R. Irwin, "Analysis of Stresses and Strains Near the End of a Crack Traversing a Plate," *Journal of Applied Mechanics* **24**, 361 (1957).

5. H. Tada, P. C. Paris, and G. R. Irwin, *The Stress Analysis of Cracks Handbook,* 2nd ed., Paris Productions, St. Louis, 1985.

6. T. L. Anderson, *Fracture Mechanics, Fundamentals and Applications,* 2nd ed., CRC Press, Boca Raton, FL, 1995.

7. X. Wu and A. J. Carlsson, *Weight Functions and Stress Intensity Factor Solutions,* Pergamon Press, Oxford, 1991.

8. E 399-90, "Standard Test Method for Plane-Strain Fracture Toughness of Metallic Materials," Annual Book of ASTM Standards, Vol. 03.01, American Society for Testing and Materials, Philadelphia, PA, 1991.

9. R. W. Hertzberg, *Deformation and Fracture Mechanics of Engineering Materials,* 3rd ed., Wiley, New York, 1989.

10. J. P. Gallagher (ed.), *Damage Tolerant Design Handbook,* 4 Vols., Metals and Ceramics Information Center, Battelle Columbus Labs., Columbus, OH, 1983.

11. *Metallic Materials and Elements for Aerospace Vehicle Structures,* MIL-HDBK-5F, 2 Vols., U. S. Dept. of Defense, Naval Publications and Forms Center, Philadelphia, PA, 1987.

12. W. T. Matthews, *Plane Strain Fracture Toughness (K_{IC}) Data Handbook for Metals,* Report No. AMMRC MS 73-6, U. S. Army Materiel Command, NTIS, Springfield, VA, 1973.

13. C. M. Hudson and S. K. Seward, "A Compendium of Sources of Fracture Toughness and Fatigue Crack Growth Data for Metallic Alloys," *International Journal of Fracture* **14**, R151 (1978).

14. C. M. Hudson and S. K. Seward, "A Compendium of Sources of Fracture Toughness and Fatigue Crack Growth Data for Metallic Alloys-Part II," *International Journal of Fracture* **20**, R59 (1982).

15. C. M. Hudson and S. K. Seward, "A Compendium of Sources of Fracture Toughness and Fatigue Crack Growth Data for Metallic Alloys-Part III," *International Journal of Fracture* **39**, R43 (1989).

16. C. M. Hudson and J. J. Ferrainolo, "A Compendium of Sources of Fracture Toughness and Fatigue Crack Growth Data for Metallic Alloys-Part IV," *International Journal of Fracture* **48**, R19 (1991).

17. G. R. Irwin, "Fracture Mode Transition for a Crack Traversing a Plate," *Journal of Basic Engineering* **82**, 417 (1960).

18. M. F. Kanninen and C. H. Popelar, *Advanced Fracture Mechanics,* Oxford University Press, New York, 1985.

19. D. Broek, *Elementary Engineering Fracture Mechanics,* 4th ed., Kluwer, London, 1986.

20. G. C. Sih, *Handbook of Stress Intensity Factors for Researchers and Engineers,* Institute of Fracture and Solid Mechanics, Lehigh University, Bethlehem, PA, 1973.

21. D. P. Rooke and D. J. Cartwright, *Compendium of Stress Intensity Factors,* Her Majesty's Stationery Office, London, 1976.

22. J. M. Barsom and S. T. Rolfe *Fracture and Fatigue Control in Structures,* 2nd ed., Prentice-Hall, Englewood Cliffs, NJ, 1987.

23. A. P. Parker, *The Mechanics of Fracture and Fatigue,* E. & F. N. Spon Ltd., New York, 1981.

24. R. P. Harrison, K. Loosemore, and I. Milne, "Assessment of the Integrity of Structures Containing Cracks," CEGB Report R/H/R6, Central Electricity Generating Board, United Kingdom (1976).

25. A. G. Miller, "Review of Limit Loads of Structures Containing Defects," *International Journal of Pressure Vessels and Piping* **32**, 197 (1988).

26. H. J. Grover, S. A. Gordon, and L. R. Jackson, *Fatigue of Metals and Structures,* Government Printing Office, Washington, DC, 1954.

27. A. Higdon, E. H. Ohlsen, W. B. Stiles, J. A. Weese, and W. F. Riley, *Mechanics of Materials,* 4th ed., Wiley, New York, 1985.

28. W. M. Murray (ed.), *Symposium on Fatigue and Fracture of Metals,* Wiley, New York, 1952.

29. J. B. Johnson, "Aircraft Engine Material," *SAE Journal* **40**, 153–162 (1937).

30. *Proceedings of the Conference on Welded Structures,* The Welding Institute, Cambridge, England, 1971, Vols. I and II.

31. A. F. Madayag, *Metal Fatigue, Theory and Design,* Wiley, New York, 1969.

32. H. F. Moore, "A Study of Size Effect and Notch Sensitivity in Fatigue Tests of Steel," *ASTM Proceedings* **45**, 507 (1945).

33. W. N. Thomas, "Effect of Scratches and Various Workshop Finishes Upon the Fatigue Strength of Steel," *Engineering* **116**, 449ff (1923).

34. G. V. Smith, *Properties of Metals at Elevated Temperatures,* McGraw-Hill, New York, 1950.

35. H. J. Gough and D. G. Sopwith, *Journal of Iron and Steel Institute* **127**, 301 (1933).

36. O. J. Horger, *Metals Engineering Design* (ASME Handbook), American Society of Mechanical Engineers, McGraw-Hill, New York, 1953.

37. *Proceedings of International Conference on Fatigue,* American Society of Mechanical Engineers (jointly with Institution of Mechanical Engineers), New York, 1956.

38. N. E. Frost and K. Denton, "The Fatigue Strength of Butt Welded Joints in Low Alloy Structural Steels," *British Welding Journal* **14**(4) (1967).

39. F. M. Howell and J. L. Miller, "Axial Stress Fatigue Strengths of Several Structural Aluminum Alloys," *ASTM Proceedings* **55**, 955 (1955).

40. Battelle Memorial Institute, *Prevention of Fatigue in Metals*, Wiley, New York, 1941.

41. H. J. Grover, *Fatigue of Aircraft Structures*, Government Printing Office, Washington, DC, 1966.

42. J. A. Bannantine, J. J. Comer, and J. L. Handrock, *Fundamentals of Metal Fatigue Analysis*, Prentice-Hall, Englewood Cliffs, NJ, 1990.

43. S. Suresh, *Fatigue of Materials*, Cambridge University Press, 1991.

44. N. E. Dowling, *Mechanical Behavior of Materials*, Prentice-Hall, Englewood Cliffs, NJ, 1993.

45. H. O. Fuchs and R. I. Stephens, *Metal Fatigue in Engineering*, Wiley, New York, 1980.

46. R. E. Peterson, *Stress Concentration Factors*, Wiley, New York, 1974.

47. E. Z. Stowell, "Stress and Strain Concentration at a Circular Hole in an Infinite Plate," NACA TN2073, National Advisory Committee for Aeronautics, Cleveland, OH, April 1950.

48. G. Sines and J. L. Waisman, *Metal Fatigue*, McGraw-Hill, New York, 1959.

49. R. C. Juvinall, *Engineering Considerations of Stress, Strain, and Strength*, McGraw-Hill, New York, 1967.

50. *Manual on Low Cycle Fatigue Testing*, STP 465, American Society for Testing and Materials, Philadelphia, PA, 1969.

51. J. F. Tavernelli and L. F. Coffin, Jr., "A Compilation and Interpretation of Cyclic Strain Fatigue Tests on Metals," *ASM Transactions of American Society for Metals* 51, 438–453 (1959).

52. S. S. Manson, "Behavior of Materials Under Conditions of Thermal Stress," NACA TN-2933, National Advisory Committee for Aeronautics, Cleveland, OH, 1954.

53. L. F. Coffin, Jr., "Design Aspects of High Temperature Fatigue with Particular Reference to Thermal Stresses," *ASME Transactions* 78, 527–532 (1955).

54. J. Morrow, J. F. Martin, and N. E. Dowling, "Local Stress–Strain Approach to Cumulative Fatigue Damage Analysis," Final Report, T. & A. M. Report No. 379, Department of Theoretical and Applied Mechanics, University of Illinois, Urbana, IL, January 1974.

55. R. W. Landgraf, "The Resistance of Metal to Cyclic Deformation," *Achievement of High Fatigue Resistance in Metals Alloys*, STP-467, American Society for Testing and Materials, Philadelphia, PA, 1970.

56. B. M. Wundt, *Effects of Notches on Low Cycle Fatigue*, STP-490, American Society for Testing and Materials, Philadelphia, PA, 1972.

57. E. Krempl, *The Influence of State of Stress on Low-Cycle Fatigue of Structural Materials: A Literature Survey and Interpretation Report*, STP-549, American Society for Testing and Materials, Philadelphia, PA, 1974.

58. S. S. Manson, *Thermal Stress and Low Cycle Fatigue*, McGraw-Hill, New York, 1966.

59. S. J. Stadnick and J. Morrow, "Techniques for Smooth Specimen Simulation of the Fatigue Behavior of Notched Members," *Testing for Prediction of Material Performance in Structures and Components*, STP-515, American Society for Testing and Materials, Philadelphia, PA, 1972.

60. J. F. Martin, T. H. Topper, and G. M. Sinclair, "Computer Based Simulation of Cyclic Stress–Strain Behavior with Applications to Fatigue," *Materials Research and Standards* 11(2), 23 (February 1971).

61. R. C. Rice (ed.), *Fatigue Design Handbook*, 2nd ed., SAE Pub. No. AE-10, Society of Automotive Engineers, Warrendale, PA, 1988.

62. T. H. Topper, R. M. Wetzel, and J. Morrow, "Neuber's Rule Applied to Fatigue of Notched Specimens," *Journal of Materials* 4(1), 200–209 (1969).

63. N. E. Dowling, "Fatigue Failure Predictions for Complicated Stress-Strain Histories," *Journal of Materials* 7 (1), 71–87 (1972).

64. E 647-88a, "Standard Test Method for Measurement of Fatigue Crack Growth Rates," American Society for Testing and Materials, Philadelphia, PA, 1988.

65. J. Schijve, "Four Lectures on Fatigue Crack Growth," *Engineering Fracture Mechanics* 11, 167–221 (1979).

66. P. C. Paris and F. Erdogan, "A Critical Analysis of Crack Propagation Laws," *Trans. ASME, Journal of Basic Engineering* 85, 528–534 (1963).

67. S. J. Maddox, *Fatigue Strength of Welded Structures*, 2nd ed., Abington, 1991.

68. W. G. Reuter, J. H. Underwood, and J. C. Newman (eds.), *Surface-Crack Growth: Models, Experiments and Structures*, STP-1060, American Society for Testing and Materials, Philadelphia, PA, 1990.

69. W. Elber, *Eng. Fracture Mech.* 2, 37–45 (1970).

70. J. C. Newman and W. Elber (eds.), *Mechanics of Fatigue Crack Closure*, STP-982, American Society for Testing and Materials, Philadelphia, PA, 1988.

71. L. P. Pook, *The Role of Crack Growth in Metal Fatigue,* The Metals Society, London, 1983.
72. R. D. Ritchie and J. Lankford (eds.), *Small Fatigue Cracks,* The Metallurgical Society, Inc., 1986.
73. W. G. Clark, Jr., "Fracture Mechanics in Fatigue," *Experimental Mechanics* (September 1971).
74. H. A. Wood, "Fracture Control Procedures for Aircraft Structural Integrity," AFFDL Report TR-21-89, Wright-Patterson Air Force Base, OH, July 1971.
75. F. R. Larson and J. Miller, "Time–Temperature Relationships for Rupture and Creep Stresses," *ASME Transactions* **74**, 765 (1952).
76. S. S. Manson and A. M. Haferd, "A Linear Time–Temperature Relation for Extrapolation of Creep and Stress Rupture Data," NACA Technical Note 2890, National Advisory Committee for Aeronautics, Cleveland, OH, March 1953.
77. R. G. Sturm, C. Dumont, and F. M. Howell, "A Method of Analyzing Creep Data," *ASME Transactions* **58**, A62 (1936).
78. N. H. Polakowski and E. J. Ripling, *Strength and Structure of Engineering Materials.* Prentice-Hall, Englewood Cliffs, NJ, 1966.
79. E. L. Robinson, "Effect of Temperature Variation on the Long-Time Rupture Strength of Steels," *ASME Transactions* **74**, 777–781 (1952).
80. A. J. Kennedy, *Processes of Creep and Fatigue in Metals,* Wiley, New York, 1963.
81. F. W. Demoney and B. J. Lazan, WADC Tech Report 53-510, Wright-Patterson Air Force Base, OH, 1954.
82. F. Vitovec and B. J. Lazan, *Symposium on Metallic Materials for Service at Temperatures Above 1600°F,* STP-174, American Society for Testing and Materials, Philadelphia, PA, 1956.
83. L. F. Coffin, Jr., "The Effect of Frequency on the Cyclic Strain and Low Cycle Fatigue Behavior of Cast Udimet 500 at Elevated Temperature," *Metallurgical Transactions* **12**, 3105–3113 (November 1971).
84. J. B. Conway and J. T. Berling, "A New Correlation of Low-Cycle Fatigue Data Involving Hold Periods," *Metallurgical Transactions* **1**(1), 324–325 (January 1970).
85. J. R. Ellis and E. P. Esztergar, "Considerations of Creep-Fatigue Interaction in Design Analysis," *Symposium on Design for Elevated Temperature Environment,* ASME, New York, 1971, pp. 29–33.
86. R. M. Curran and B. M. Wundt, "A Program to Study Low-Cycle Fatigue and Creep Interaction in Steels at Elevated Temperatures," *Current Evaluation of 2¼ Chrome 1 Molybdenum Steel in Pressure Vessels and Piping,* ASME, New York, 1972, pp. 49–82.
87. S. S. Manson, G. R. Halford, and M. H. Hirschberg, "Creep-Fatigue Analysis by Strain-Range Partitioning," *Symposium on Design for Elevated Temperature Environment,* ASME, New York, 1971, pp. 12–24.
88. M. M. Leven, "The Interaction of Creep and Fatigue for a Rotor Steel," *Experimental Mechanics* **13**(9), 353–372 (September 1973).
89. S. S. Manson, G. R. Halford, and M. H. Hirschberg, NASA Technical Memo TMX-67838, Lewis Research Center, Cleveland, OH, May 1971.
90. J. F. Saltsman and G. R. Halford, "Application of Strain-Range Partitioning to the Prediction of Creep-Fatigue Lives of AISI Types 304 and 316 Stainless Steel," NASA Technical Memo TMX-71898, Lewis Research Center, Cleveland, OH, September 1976.
91. C. G. Annis, M. C. Van Wanderham, and R. M. Wallace, "Strain-Range Partitioning Behavior of an Automotive Turbine Alloy," Final Report NASA TR 134974, February 1976.
92. M. H. Hirschberg and G. R. Halford, "Use of Strain-Range Partitioning to Predict High-Temperature Low Cycle Fatigue Life," NASA TN D-8072, January 1976.
93. J. F. Saltsman and G. R. Halford, "Application of Strain-Range Partitioning to the Prediction of MPC Creep-Fatigue Data for 2¼ Cr-1 Mo Steel," NASA TMX-73474, December 1976.
94. S. S. Manson and G. R. Halford, "Treatment of Multiaxial Creep-Fatigue by Strain-Range Partitioning," NASA TMX-73488, December 1976.
95. *Characterization of Low Cycle High Temperature Fatigue by the Strain-Range Partitioning Method,* AGARD Conference Proceedings No. 243, distributed by NASA, Langley Field, VA, April 1978.
96. J. A. Collins, "Fretting-Fatigue Damage-Factor Determination," *Journal of Engineering for Industry* **87**(8), 298–302 (August 1965).
97. D. Godfrey, "Investigation of Fretting by Microscopic Observation," NACA Report 1009, Cleveland, OH, 1951 (formerly TN-2039, February 1950).

98. F. P. Bowden and D. Tabor, *The Friction and Lubrication of Solids,* Oxford University Press, Amen House, London, 1950.

99. D. Godfrey and J. M. Baily, "Coefficient of Friction and Damage to Contact Area During the Early Stages of Fretting; I—Glass, Copper, or Steel Against Copper," NACA TN-3011, Cleveland, OH, September 1953.

100. M. E. Merchant, "The Mechanism of Static Friction," *Journal of Applied Physics* **11**(3), 232 (1940).

101. E. E. Bisson, R. L. Johnson, M. A. Swikert, and D. Godfrey, "Friction, Wear, and Surface Damage of Metals as Affected by Solid Surface Films," NACA TN-3444, Cleveland, OH, May 1955.

102. H. H. Uhlig, "Mechanisms of Fretting Corrosion," *Journal of Applied Mechanics* **76**, 401–407 (1954).

103. I. M. Feng and B. G. Rightmire, "The Mechanism of Fretting," *Lubrication Engineering* **9**, 134ff (June 1953).

104. I. M. Feng, "Fundamental Study of the Mechanism of Fretting," Final Report, Lubrication Laboratory, Massachusetts Institute of Technology, Cambridge, 1955.

105. H. T. Corten, "Factors Influencing Fretting Fatigue Strength," T. & A. M. Report No. 88, Department of Theoretical and Applied Mechanics, University of Illinois, Urbana, IL, June 1955.

106. W. L. Starkey, S. M. Marco, and J. A. Collins, "Effects of Fretting on Fatigue Characteristics of Titanium–Steel and Steel–Steel Joints," ASME Paper 57-A-113, New York, 1957.

107. W. D. Milestone, "Fretting and Fretting-Fatigue in Metal-to-Metal Contacts," ASME Paper 71-DE-38, New York, 1971.

108. G. P. Wright and J. J. O'Connor, "The Influence of Fretting and Geometric Stress Concentrations on the Fatigue Strength of Clamped Joints," *Proceedings, Institution of Mechanical Engineers,* 186 (1972).

109. R. B. Waterhouse, *Fretting Corrosion,* Pergamon Press, New York, 1972.

110. "Fretting in Aircraft Systems," AGARD Conference Proceedings CP161, distributed through NASA, Langley Field, VA, 1974.

111. "Control of Fretting Fatigue," Report No. NMAB-333, National Academy of Sciences, National Materials Advisory Board, Washington, DC, 1977.

112. N. P. Suh, S. Jahanmir, J. Fleming, and E. P. Abrahamson, "The Delamination Theory of Wear—II," Progress Report, Materials Processing Lab, Mechanical Engineering Dept., MIT Press, Cambridge, MA, September 1975.

113. J. T. Burwell, Jr., "Survey of Possible Wear Mechanisms," *Wear* **1**, 119–141 (1957).

114. M. B. Peterson, M. K. Gabel, and M. J. Derine, "Understanding Wear"; K. C. Ludema, "A Perspective on Wear Models"; E. Rabinowicz, "The Physics and Chemistry of Surfaces"; J. McGrew, "Design for Wear of Sliding Bearings"; R. G. Bayer, "Design for Wear of Lightly Loaded Surfaces," *ASTM Standardization News* **2**(9), 9–32 (September 1974).

115. R. B. Waterhouse, *Fretting Corrosion,* Pergamon Press, New York, 1972.

116. P. A. Engel, "Predicting Impact Wear," *Machine Design,* 100–105 (May 1977).

117. P. A. Engel, *Impact Wear of Materials,* Elsevier, New York, 1976.

118. J. A. Collins, "A Study of the Phenomenon of Fretting-Fatigue with Emphasis on Stress-Field Effects," Dissertation, Ohio State University, Columbus, 1963.

119. J. A. Collins and F. M. Tovey, "Fretting Fatigue Mechanisms and the Effect of Direction of Fretting Motion on Fatigue Strength," *Journal of Materials* **7**(4) (December 1972).

120. W. D. Milestone, "An Investigation of the Basic Mechanism of Mechanical Fretting and Fretting-Fatigue at Metal-to-Metal Joints, with Emphasis on the Effects of Friction and Friction-Induced Stresses," Dissertation, Ohio State University, Columbus, 1966.

121. J. A. Collins and S. M. Marco, "The Effect of Stress Direction During Fretting on Subsequent Fatigue Life," *ASTM Proceedings* **64**, 547 (1964).

122. J. A. Alic, "Fretting Fatigue in the Presence of Periodic High Tensile or Compressive Loads," Final Scientific Report, Grant No. AFOSR77-3422, Wright-Patterson AFB, Ohio, April 1979.

123. H. Lyons, "An Investigation of the Phenomenon of Fretting-Wear and Attendant Parametric Effects Towards Development of Failure Prediction Criteria," Ph.D. Dissertation, Ohio State University, Columbus, 1978.

124. P. L. Ko, "Experimental Studies of Tube Fretting in Steam Generators and Heat Exchangers," ASME/CSME Pressure Vessels and Piping Conference, Nuclear and Materials Division, Montreal, Canada, June 1978.

125. R. B. Heywood, *Designing Against Fatigue of Metals,* Reinhold, New York, 1962.

126. F. P. Bowden and D. Tabor, *Friction and Lubrication of Solids,* Oxford University Press, London, 1950.

127. F. P. Bowden and D. Tabor, *Friction and Lubrication,* Methuen, London, 1967.

128. E. Rabinowicz, *Friction and Wear of Materials,* Wiley, New York, 1966.

129. C. Lipson, *Wear Considerations in Design,* Prentice-Hall, Englewood Cliffs, NJ, 1967.

130. N. P. Suh, "The Delamination Theory of Wear," *Wear* **25**, 111–124 (1973).

131. J. E. Shigley, *Mechanical Engineering Design,* 2nd ed., McGraw-Hill, New York, 1972.

132. W. A. Glaeser, K. C. Ludema, and J. K. Rhee (eds.), *Wear Materials,* American Society of Mechanical Engineers, New York, April 25–28, 1977.

133. C. W. MacGregor (ed.), *Handbook of Analytical Design for Wear,* Plenum Press, New York, 1964.

134. M. G. Fontana and N. D. Greene, *Corrosion Engineering,* McGraw-Hill, New York, 1967.

135. L. S. Seabright and R. J. Fabian, "The Many Faces of Corrosion," *Materials in Design Engineering* **57**(1) (January 1963).

136. G. Nelson, *Corrosion Data Survey,* National Association of Corrosion Engineers, Houston, TX, 1972.

137. H. H. Uhlig (ed.), *Corrosion Handbook,* Wiley, New York, 1948.

138. E. Rabald, *Corrosion Guide,* Elsevier, New York, 1951.

CHAPTER 19

TOTAL QUALITY MANAGEMENT IN MECHANICAL DESIGN

B. S. Dhillon
Department of Mechanical Engineering
University of Ottawa
Ottawa, Ontario, Canada

19.1	**INTRODUCTION**	**475**	
19.2	**TQM IN GENERAL**	**476**	
	19.2.1 Total	476	
	19.2.2 Quality	476	
	19.2.3 Management	476	
19.3	**DEMING'S APPROACH TO TQM**	**477**	
19.4	**QUALITY IN THE DESIGN PHASE**	**477**	
	19.4.1 Product Design Review	477	
	19.4.2 Process Design Review	478	
	19.4.3 Plans for Acquisition and Process Control	479	
	19.4.4 Guidelines for Improving Design Quality	479	

	19.4.5 Taguchi's Quality Philosophy Summary and Kume's Approach for Process Improvement	480	
19.5	**QUALITY TOOLS AND METHODS**	**480**	
	19.5.1 Fishbone Diagram	480	
	19.5.2 Pareto Diagram	481	
	19.5.3 Kaizen Method	481	
	19.5.4 Force Field Analysis	481	
	19.5.5 Customer Needs Mapping Method	482	
	19.5.6 Control Charts	482	
	19.5.7 Poka-Yoke Method	482	
	19.5.8 Benchmarking	483	
	19.5.9 Hoshin Planning Method	484	
	19.5.10 Gap Analysis Method	484	

19.1 INTRODUCTION

In today's competitive environment, the age-old belief of many companies that "the customer is always right" has a new twist. In order to survive, companies are focusing their entire organization on customer satisfaction. The approach followed for ensuring customer satisfaction is known as Total Quality Management (TQM). The challenge is to "manage" so that the "total" and the "quality" are experienced in an effective manner.[1]

Though modern quality control dates back to 1916, the real beginning of TQM can be considered the late 1940s, when such figures as W. E. Deming, J. M. Juran, and A. V. Feigenbaum played an instrumental role.[2] In subsequent years, the TQM approach was more widely practiced in Japan than anywhere else. In 1951, the Japanese Union of Scientists and Engineers introduced a prize, named after W. E. Deming, for the organization that implemented the most successful quality policies. On similar lines, in 1987, the U. S. government introduced the Malcolm Baldrige Award.

Quality cannot be inspected out of a product; it must be built in. The consideration of quality in design begins during the specification-writing phase. Many factors contribute to the success of the quality consideration in engineering or mechanical design. TQM is a useful tool for application during the design phase. It should be noted that the material presented in this section does not specifically deal with mechanical design, but with the design in general. The same material is equally applicable to the design of mechanical items. This chapter presents topics such as TQM in general, Deming's approach to TQM, quality in design, quality tools and techniques, and selected references on TQM and design quality.

Mechanical Engineers' Handbook, 2nd ed., Edited by Myer Kutz.
ISBN 0-471-13007-9 © 1998 John Wiley & Sons, Inc.

19.2 TQM IN GENERAL

The term *quality* may simply be defined as providing customers with products and services that meet their needs in an effective manner. TQM focuses on customer satisfaction. The three words that make up this concept—"total," "quality," and "management"—are discussed separately below.[1]

19.2.1 Total

This calls for the involvement of all the aspects of the organization in satisfying the customer, a goal that can only be accomplished if the usefulness is recognized of having partnership environment at each stage of the business process both within and outside the organization, as applicable. With respect to the outside stage of the business process, the important critical factors for a successful supplier–customer relationship are

1. Development of a customer–supplier relationship based on mutual trust, respect, and benefit
2. Development of in-house requirements by customers
3. Customers making suppliers clearly understand their requirements
4. Customers selecting their potential suppliers with mechanisms in place to achieve zero defects
5. Regular monitoring of suppliers' processes and products by the customers

19.2.2 Quality

Any company or organization in pursuit of TQM must define the term *quality* clearly and precisely. It may be said that quality is deceptively simple but endlessly complicated, and numerous definitions have been proposed, such as "quality = people + attitude"; "providing error-free products and services to customers on time"; and "satisfying the requirements and expectations of customers". Another definition is offered here: "quality means providing both external and internal customers with innovative goods and services that meet their needs effectively."

This definition has three important dimensions:

1. It focuses on satisfying the needs of customers
2. Organizations using this definition provide both products and services, which jointly determine the customer's perception of the company in question
3. The concerned companies have both external and internal customers

According to a survey reported in Ref. 1, 82% of the definitions indicated that quality is defined by the customer, not by the supplier. The top five quality measures identified by the respondents were customer feedback (22.92%), customer complaints (16.67%), net profits (10.42%), returning customers (10.42%), and product defects (8.33%).

19.2.3 Management

The approach to management is instrumental in determining companies' ability to attain corporate goals and allocate resources effectively. TQM calls for a radical change in involving employees in company decision-making, as their contribution and participation are vital to orienting all areas of business in providing quality products to customers. It must be remembered that over the years the *Fortune* 1000 companies in the United States have reported such benefits of employee-involvement as increased employee trust in management, improved product quality, improved employee safety/ health, increase in productivity, improved management decision-making, increased worker satisfaction, improvement in employee quality of work life, improved union–management relations, improved implementation of technology, improved organization processes, eliminated layers of management, and better customer service.

Companies considering the introduction of TQM will have to see their employees in a new way, for the change in management philosophy needed to truly manage total quality is nothing short of dramatic. Furthermore, it is important that the managment infrastructure lay the foundation for involving the entire workforce in the pursuit of customer satisfaction.

The Senior Management Role

Senior management must show enthusiasm for improving product quality if employees are to seriously consider its importance. The following steps by management are useful in gaining commitment to total quality:[3]

- Announce absolutely clear quality policies and goals and ensure that these are explained to everyone involved.
- Regularly show management support through action.

- Ensure that everyone in the organization understands his or her necessary input in making quality happen.
- Eradicate any opportunity for compromising conformance.
- Make it clearly known to everyone concerned, including suppliers, that they are an important element in contributing to the quality of the end product.

19.3 DEMING'S APPROACH TO TQM

One of the pioneers of the TQM concept has expressed his views on improving quality. His fourteen-point approach is as follows:[4]

1. Establish consistency of purpose for improving services.
2. Adopt the new philosophy for making the accepted levels of defects, delays, or mistakes unwanted.
3. Stop reliance on mass inspection as it neither improves nor guarantees quality. Remember that teamwork between the firm and its suppliers is the way for the process of improvement.
4. Stop awarding business with respect to the price.
5. Discover problems. Management must work continually to improve the system.
6. Take advantage of modern methods used for training. In developing a training program, take into consideration such items as
   - Identification of company objectives
   - Identification of the training goals
   - Understanding of goals by everyone involved
   - Orientation of new employees
   - Training of supervisors in statistical thinking
   - Team-building
   - Analysis of the teaching need
7. Institute modern supervision approaches.
8. Eradicate fear so that everyone involved may work to his or her full capacity.
9. Tear down department barriers so that everyone can work as a team member.
10. Eliminate items such as goals, posters, and slogans that call for new productivity levels without the improvement of methods.
11. Make your organization free of work standards prescribing numeric quotas.
12. Eliminate factors that inhibit employee workmanship pride.
13. Establish an effective education and training program.
14. Develop a program that will push the above 13 points every day for never-ending improvement.

19.4 QUALITY IN THE DESIGN PHASE

Although TQM will help generally to improve design quality, specific quality-related steps are also necessary during the design phase. These additional steps will further enhance the product design.

An informal review during specification writing may be regarded as the beginning of quality assurance in the design phase. As soon as the first draft of the specification is complete, the detailed analysis begins.

Some of the important areas assocated with quality in design are discussed separately below.[5]

19.4.1 Product Design Review

Various types of design reviews are conducted during the product-design phase. One reason for performing these reviews is to improve quality. Design reviews conducted during the design phase include preliminary design review, detailed design reviews (the number of which may vary from one project to another), critical design review (the purpose of this review is to approve the final design), preproduction design review (this review is performed after the prototype tests), postproduction design review, and operations and support design review.

The consideration of quality begins at the preliminary design review and becomes stronger as the design develops. The role of quality assurance in preliminary design review is to ensure that the new design is free of quality problems of similar existing designs. This requires a good knowledge of the strengths and weaknesses of the competing products. The following approaches are quite useful in ensuring quality during the design phase.

Quality Function Deployment (QFD), Quality Loss Function, and Benchmarking

Quality function deployment is a value-analysis tool used during product and process development. It is an extremely useful concept for developing test strategies and translating needs to specification.

QFD was developed in Japan. In the case of new product development, it is simply a matrix of consumer/customer requirements versus design requirements. Some of the sources for the input are market surveys, interviews, and brainstorming. To use the example of an automobile, customer needs include price, expectations at delivery (safety, perceived quality, service ability, performance, workmanship, etc.), and expectations over time (including customer support, durability, reliability, performance, repair part availability, low preventive maintenance and maintenance cost, mean time between failures within prediction, etc.).

Finally, QFD helps to turn needs into design engineering requirements.

The basis for the quality loss function is that if all parts are produced close to their specified values, then it is fair to expect best product performance and lower cost to society. According to Taguchi,[6] quality cost goes up not only when the finished product is outside given specifications, but also when it deviates from the set target value within the specifications.

One important point to note, using Taguchi's philosophy, is that a product's final quality and cost are determined to a large extent by its design and manufacturing processes. It may be said that the loss function concept is simply the application of a life cycle cost model to quality assurance. Taguchi expresses the loss function as follows:

$$L(x) = c(x - T_v)^2 \tag{19.1}$$

where x = the variable
$L(x)$ = the loss at x
T_v = the targeted value of the variable at which the product is expected to show its best performance
c = the proportionality constant
$(x - T_v)$ = the deviation from the target value

In the formulation of the loss function, assumptions are made, such as zero loss at the target value and that the dissatisfaction of customer is proportional only to the deviation from the target value. The value of the proportionality constant, c, can be determined by estimating the loss value for an unacceptable deviation, such as the tolerance limit. Thus, the following relationship can be used to estimate the value of c:

$$c = \frac{L_a}{\Delta^2} \tag{19.2}$$

where L_a = the amount of loss expressed in dollars
Δ = the deviation amount from the target value T_v

Example 19.1
Assume that the estimated loss for Rockwell hardness number beyond 56 is $150 and the targeted value of the hardness number is 52. Estimate the value of the proportionality constant. Substituting the given data into Eq. (19.2), we get

$$c = \frac{150}{(56 - 52)^2}$$
$$= 9.375$$

Thus, the value of the proportionality constant is 9.375.

Benchmarking is a process of comparing in-house products and processes with the most effective in the field and setting objectives for gaining a competitive advantage. The following steps are associated with benchmarking:[7]

- Identify items and their associated key features to benchmark during product planning.
- Select companies, industries, or technologies to benchmark against. Determine existing strengths of the items to benchmark against.
- Determine the best-in-class target from each selected benchmark item.
- Evaluate, as appropriate, in-house processes and technologies with respect to benchmarks.
- Set improvement targets remembering that the best-in-class target is always a moving target.

19.4.2 Process Design Review

Soon after the approval of a preliminary design, a process flowchart is prepared. In order to assume the proper consideration being given to quality, the quality engineer works along with process and reliability engineers.

For the correct functioning of the process, the quality engineer's expertise in variation control provides important input.

Lack of integration between quality assurance and manufacturing is one of the main reasons for the failure of the team effort. The performance of process failure mode and effect analysis (FMEA) helps this integration to take place early. The consideration of the total manufacturing process performance by the FMEA concept, rather than that of the mere equipment, is also a useful step in this regard. For FMEA to produce promising results, the quality and manufacturing engineers have to work as a team. Nevertheless, FMEA is a useful tool for performing analysis of a new process, including analysis of receiving, handling, and storing materials and tools. Also, the participation of suppliers in FMEA studies enhances FMEA's value. The following steps are associated with the process of FMEA:[8,9]

- Develop process flowchart that includes all process inputs: materials, storage and handling, transportation, etc.
- List all components/elements of the process.
- Write down each component/element description and identify all possible failure modes.
- Assign failure rate/probability to each component/element failure mode.
- Describe each failure mode cause and effect.
- Enter remarks for each failure mode.
- Review each critical failure mode and take corrective measures.

19.4.3 Plans for Acquisition and Process Control

The development of quality assurance plans for procurement and process control during the design phase is useful for improving product quality. One immediate advantage is the smooth transition from design to production. The equipment-procurement plan should include such items as equipment-performance verification, statistical tolerance analysis, testing for part interchangeability, and pilot runs. Similarly, the component-procurement quality plans should address concerns and cooperation on areas including component qualification, closed-loop failure managment, process control implementation throughout the production lines, and standard and special screening tests.

Prior to embarking on product manufacturing, there is a definite need for the identification of the critical points where the probability of occurrence of a serious defect is quite high. Thus, the process control plans should be developed by applying the quality function deployment and process FMEA. These plans should include items such as

- Acceptance of standard definitions
- Procedures to monitor defects
- Approaches for controlling critical process points

19.4.4 Guidelines for Improving Design Quality

During the product design phase, there are various measures concerned professionals should take to improve quality. These include[10,11] designing for effective testing, simplifying assembly and making it foolproof, designing for robustness, minimizing the number of parts, reducing the number of different parts, using well-understood and repeatable processes, minimizing engineering changes to released products, eliminating adjustments, selecting components that can withstand process operations, and laying out components for reliable process completion. These factors may be taken into consideration during designing and/or during design reviews.

Past experience has shown that guidelines such as those listed above lead to many benefits, including

- Increase in part yield
- Decrease in degradation of performance with time
- Improvement in product reliability
- Reduction in part damage
- Better serviceability
- Improvement in consistency in part quality
- Reduction in the volume of drawings and instructions to control
- Lower assembly error rate

Today, many engineering systems use computer technology, to varying degrees. This means that it is important not only to have good-quality hardware, but also good-quality software. The software-development environment possesses certain characteristics that may adversely affect its quality, including outdated support tools, cost and time constraints, complex hardware, variations in programmer skill, poorly defined customer objectives, a small project staff, high programmer turnover, and software-naive customers.

Directly or indirectly, from the TQM perspective, some basic rules must be followed:[12]

- Do not leave management to managers alone. Remember that everyone in an organization is a manager of tasks or processes and, in fact, those closest to the task or process should play a leading role in its management.
- Quality through inspection is no longer a competitive option today. A sensible approach for companies to increase their market share is to design quality into products and processes.
- Random variability exists in all processes. Failure to take this into account by engineering design and control methods will lead to high production costs and out-of-specification products.
- Today's customers want reliable, safe, and low-cost products to satisfy their needs. Remember that use of the latest technology will not alone hold the market share; the manufacturer must also champion its customers' concerns.
- Remember that experimentation belongs on the manufacturing floor, not just in the research laboratory. It is impossible to reproduce the exact production environment in the laboratory. Further, do not overlook teaching the methods of experimentation to production people.

Additional information on improving design quality may be found in Refs. 13–15.

19.4.5 Taguchi's Quality Philosophy Summary and Kume's Approach for Process Improvement

Taguchi's approach was discussed earlier, but because of its importance, this section summarizes his quality philosophy again, in seven basic steps:[12,16]

1. A critical element of a manufactured item's quality is the total loss generated by that item to society as whole.
2. In today's market, continuous cost reduction and quality improvement are critical for companies to stay in business.
3. Design and its associated manufacturing processes determine, to a large extent, the ultimate quality and cost of a manufactured product.
4. Unceasing reduction in a product-performance characteristic's variation from its target value is part of a continuous quality-improvement effort.
5. The loss of customers due to variation in an item's performance is frequently almost proportional to the square of the performance characteristic's deviation from its target value.
6. The identification of the product and process parameter settings that reduce performance variation can be accomplished through statistically designed experiments.
7. Reduction in the performance variation of a product of process can be achieved by exploiting the product or process parameter nonlinear effects on performance characteristics.

To improve process, Kume[17] outlined a seven-step approach:

1. Select project.
2. Observe the process under consideration.
3. Perform process analysis.
4. Take corrective measures.
5. Evaluate effectiveness of corrective measures.
6. Standardize the change.
7. Review and make appropriate modifications, if applicable, in future plans.

19.5 QUALITY TOOLS AND METHODS

Over the years, many quality-improvement tools and methods have been developed by researchers and others. Effective application of these tools and techniques becomes a vital element in the success of the TQM concept during product design. Examples of these tools and techniques are control charts, fishbone or cause-and-effect diagram, Pareto diagram, Poka-yoke, force field analysis, benchmarking, Kaizen, customer needs mapping, Hoshin planning technique, and gap analysis.[4] These approaches are described below.[4]

19.5.1 Fishbone Diagram

This approach, also known as the cause-and-effect or Ishikawa diagram, was originally developed by K. Ishikawa in Japan. The diagram serves as a useful tool in quality-related studies to perform cause-

and-effect analysis for generating ideas and finding the root cause of a problem for investigation. The diagram somewhat resembles a fishbone; thus the name. A typical fishbone diagram is shown in Fig. 19.1. The diagram depicts the "effect" on the right hand (i.e., in the "fish head"). The boxes in the main area are for writing in possible causes. All of these boxes (i.e., in the main area) are connected to the central "fish spine" or the main line. For example, in the case of total quality management, the "fish head" or "effect" box will become "customer satisfaction" and the boxes in the main area will represent people, methods, machines, materials, and so on.

Major steps for developing a fishbone diagram are as follows:[4]

- Establish problem statement.
- Brainstorm to highlight possible causes.
- Categorize major causes into natural grouping and stratify by steps of the process.
- Insert the problem or effect in the "fish head" box on the right-hand side. Develop the diagram by unifying the causes through following the necessary process steps.
- Refine categories by asking questions such as "What causes this?" and "Why does this condition exist?"

19.5.2   Pareto Diagram

An Italian economist, Vilfredo Pareto (1848–1923) developed a formula in 1897 to show that the distribution of income is uneven.[14,15] In 1907, a similar theory was put forward in a diagram by M. C. Lorenz, a U.S. economist. In later years, J. M. Juran[19] applied Lorenz's diagram to quality problems and called it *Pareto analysis*.

In quality-control work, Pareto analysis simply means, for example, that there are always a few kinds of defects in the hardware manufacture that loom large in occurrence frequency and severity. Economically, these defects are costly, and thus of great significance. Alternatively, it may simply be stated that on the average about 80% of the costs occur due to 20% of the defects.

The Pareto diagram, derived from the above reasoning, is helpful in identifying the spot for concerted effort. The Pareto diagram is a type of frequency chart with bars arranged in descending order from left to right, visually highlighting the major problem areas. The Pareto principle can be quite instrumental in TQM effort, particularly in improving quality of product designs.

19.5.3   Kaizen Method

"Kaizen"[4] means improvement in Japanese, and the Kaizen philosophy maintains that the current way of life deserves to be improved on a continuous basis. This philosophy is broader than TQM because it calls for ongoing improvement as workers, leaders, managers, and so on. Thus, Kaizen includes TQM, quality circles, zero defects, new product design, continuous quality improvement, customer service agreements, and so on.

Kaizen is often referred to as "the improvement movement" because it encompasses constant improvement in social life, working life, and the home life of everyone.

19.5.4   Force Field Analysis

This method was developed by Kurk Lewin[4] to identify forces existing in a situation. It calls first for clear understanding of the driving and restraining forces and then for developing plans to implement change.

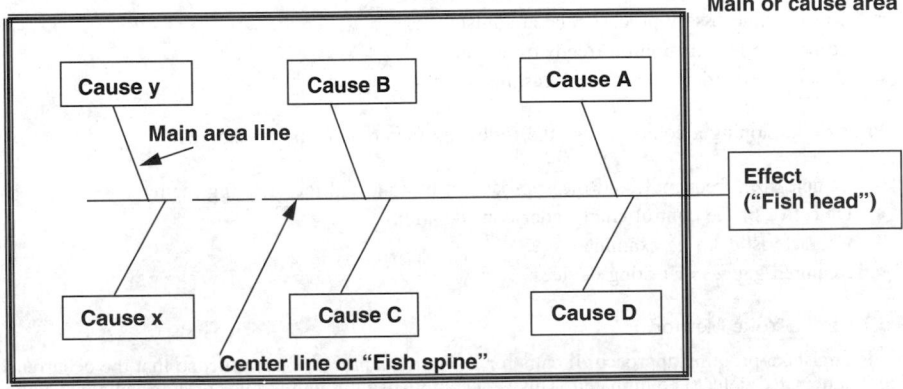

Fig. 19.1 Fishbone diagram layout.

The change is considered as a dynamic process and as the result of a struggle between driving forces (i.e., those forces seeking to upset the status quo) and restraining forces (i.e., those forces attempting to maintain the status quo). A change occurs only when the driving forces are stronger than the restraining forces.

The group brainstorming method[20] serves as a useful tool to identify the driving forces and restraining forces in a given situation.

With respect to improving engineering design quality, the force field analysis facilitates changes in the following way:

- It forces the concerned personnel to identify and to think through the certain facets of an acquired change.
- It highlights the priority order of the involved driving and restraining forces.
- It leads to the establishment of a priorty action plan.

An example of using the force field analysis technique is given in Ref. 4.

19.5.5 Customer Needs Mapping Method

This approach is used to identify consumer requirements and then to identify in-house processes' ability to meet those requirements statisfactorily. A process's two major customers are the external customer (the purchaser of the product or service) and the internal customer (the next step in the process of receiving the output). Past experience has shown that the internal customer is often overlooked by groups such as inventory control, accounting, facilities, and computer.

Both the external and internal customer's wants or requirements can be identified through brainstorming, customer interviews, and so on.

Some of the advantages of customer needs mapping are as follows:

- It enhances the understanding of the customer background.
- It highlights customer wants.
- It translates customer needs into design features.
- It focuses attention on process steps important to customers.
- It highlights overlooked customer needs.

19.5.6 Control Charts

Control charts were developed by Walter A. Shewhart[21] of Bell Telephone Laboratories in 1924 for analyzing discrete or continuous data collected over a period of time. A control chart is a graphical tool used for assessing the states of control of a given process. The variations or changes are inherent in all processes, but their magnitude may be large or very small. Minimizing or eliminating such variations will help to improve quality of a product or service.

Physically, a control chart is composed of a center line (CL), upper control limit (UCL), and lower control limit (LCL). In most cases, a control chart uses control limits of plus or minus three standard deviations from the mean of the quality or other characteristic under study.

Following are some of the reasons for using control charts for improving product design quality:[20]

- To provide a visual display of a process
- To determine if a process is in statistical control
- To stop unnecessary process-related adjustments
- To provide information on trends over time
- To take appropriate corrective measures

Prior to developing a control chart, the following factors must be considered:

- Sample size, frequency, and the approach to be followed for selecting them
- Objective of the control chart under consideration
- Characteristics to be examined
- Required gauges or testing devices

19.5.7 Poka-Yoke Method

This is a mistake-proofing approach. It calls for designing a process or product so that the occurrence of any anticipated defect is eliminated. This is accomplished through the use of automatic test equipment that "inspects" the operations performed in manufacturing a product and then allows the product to proceed only when everything is correct. Poka-Yoke makes it possible to achieve the goal of zero defects in production process.

19.5.8 Benchmarking*

This may be described as a strategy of duplicating the best practice of an organization excelling in a specified business function. The benchmarking may be grouped into the following five categories:

1. *Competitive benchmarking* is concerned with identifying the important competitive characteristics of a competitive product or service and then comparing them to your own.
2. *Internal benchmarking* is concerned with identifying and comparing internal repetitive operational functions among divisions and/or branches.
3. *Industrial or functional benchmarking* is concerned with comparing functions within the same industry.
4. *Shadow benchmarking* is concerned with monitoring important product and service attributes of a dominant competitor in the field as well as meeting changes as they happen.
5. *World-class benchmarking* is concerned with comparing processes across diverse industries.

In order to perform benchmarking, two teams are formed: one for need assessment and one for actual benchmarking. The following factors should be identified for the need assessment:

- Key strategically important success factors for the organization
- The firm's differentiating factors from the point of view of the customer
- Factors that significantly impact quality, costs, or cycle times
- The most important area for improvement
- The data requirement for determining the critical success factor's effectiveness

The benchmarking team conducts its mission in six steps:

1. Developing operational definitions for the key success factors identified
2. Baselining operations by making use of the operational definitions developed
3. Identifying the best-in-class ideas through brainstorming
4. Collecting data through appropriate means
5. Performing analysis and communicating findings
6. Developing strategies by implementing procedures to lower the cycle time

In conducting benchmarking studies, it must be remembered that some of the common benchmarking characteristics are data-collection difficulty, cost and speed of performing benchmarking, risk of adoption, expected benefits, and difficulty in convincing management to adopt improvement ideas.

The following index could be used to measure the gap between the company performance and the benchmark performance:

$$G = \left(\frac{BP}{CP} - 1\right)(100) \tag{19.3}$$

where G = the gap factor expressed in percentage
CP = the (your) company performance
BP = the benchmark performance

Example 19.2

Assume that the benchmark response time to process a customer is 6 minutes and your firm's response time for the same process is 10 minutes. Calculate the value of the gap factor.

Substituting the given data into Eq. (19.3), we get

$$G = \left(\frac{6}{10} - 1\right)(100) = -40\%$$

There is a 40% gap in performance.

---

*This approach was briefly discussed earlier and is described again here because of its importance.

19.5.9 Hoshin Planning Method

This method,[22] also known as the "seven management tools," helps to tie quality-improvement activities to the long-term plans of the organization. Hoshin planning focuses on policy-development issues: planning objective identification, management and employee action identification, and so on. The following are the three basic Hoshin planning processes:

1. *General planning* begins with the study of consumers for the purpose of focusing the organization's attention on satisfying their needs.
2. *Intermediate planning* begins after the general planning is over. It breaks down the general planning premises into various segments for the purpose of addressing them individually.
3. *Detailed planning* begins after the completion of the intermediate planning and is assisted by the arrow diagram and by the process decision program chart.

The seven management tools related to or used in each of the above three areas are as follows:

General

1. Interrelationship diagram
2. Affinity chart

Intermediate

3. Matrix data analysis
4. Tree diagram
5. Matrix diagram

Detailed

6. Arrow diagram
7. Process decision program chart

Each of the above management tools is discussed below.

The *interrelationship diagram* is used to identify cause-and-effect links among ideas produced. It is particularly useful in situations where there is a requirement to identify root causes. One important limitation of the interrelationship diagram is the overwhelming attempts to identify linkages between all generated ideas.

The *affinity chart* is used to sort related ideas into groups and then label each similar group. The affinity chart is extremely useful in handling large volumes of ideas, including the requirement to identify broad issues.

The *matrix data analysis* is used to show linkages between two variables, particularly when there is a requirement to show visually the strength of their relationships. The main drawback of this approach is that only two relationships can be compared at a time.

The *tree diagram* is used to map out required tasks into detailed groupings. This method is extremely useful when there is a need to divide broad tasks or general objectives into subtasks.

The *matrix diagram* is used to show relationships between activities, such as tasks and people. It is an extremely useful tool for showing relationships clearly.

The *arrow diagram* is used as a detailed planning and scheduling tool and helps to identify time requirements and relationships among activities. The arrow diagram is an extremely powerful tool in situations requiring detailed planning and control of complex tasks with many interrelationships.

The *process decision program chart* is used to map out contingencies along with countermeasures. The process decision program chart is an advantage in implementing a new plan with potential problems so that the countermeasures can be thought through.

19.5.10 Gap Analysis Method

This method is used to understand services offered from different perspectives. The method considers five major gaps that are evaluated so that when differences are highlighted between perceptions, corrective measures can be initiated to narrow the gap or difference.

1. *Consumer expectation and management perception gap*
2. *Management perception of consumer expectation and service quality specification gap*
3. *Service quality specifications and service delivery gap*
4. *External communication and service delivery gap*
5. *Consumer expectation concerning the service and the actual service received gap*

REFERENCES

1. C. R. Farquhar and C. G. Johnston, *Total Quality Management: A Competitive Imperative*, Report No. 60-90-E, Conference Board of Canada, Ottawa, Ont., 1990.
2. C. D. Gevirtz, *Developing New Products with TQM*, McGraw-Hill, New York, 1994.
3. P. B. Crosby, *The Eternally Successful Organization*, McGraw-Hill, New York, 1988.
4. P. Mears, *Quality Improvement Tools and Techniques*, McGraw-Hill, New York, 1995.
5. D. G. Raheja, *Assurance Technologies*, McGraw-Hill, New York, 1991.
6. G. Taguchi, E. A. Elsayed, and T. C. Hsiang, *Quality Engineering in Production Systems*, McGraw-Hill, New York, 1989.
7. *Total Quality Management: A Guide for Implementation*, Document No. DOD 5000.51.6 (draft), U.S. Department of Defense, Washington, DC, March 23, 1989.
8. B. S. Dhillon, *Systems Reliability, Maintainability, and Management*, Petrocelli Books, New York, 1983.
9. B. S. Dhillon, and C. Singh, *Engineering Reliability: New Techniques and Applications*, Wiley, New York, 1981.
10. D. Daetz, "The Effect of Product Design on Product Quality and Product Cost," *Quality Progress* **20**, 63–67 (June 1987).
11. J. R. Evans and W. M. Lindsay, *The Management and Control of Quality*, West, New York, 1989.
12. R. H. Lochner, and J. E. Matar, *Designing for Quality*, ASQC Quality Press, Milwaukee, WI, 1990.
13. J. A. Burgess, "Assuring the Quality of Design," *Machine Design*, 65–69 (February 1982).
14. J. A. Burgess, *Design Assurance for Engineers and Managers*, Marcel Dekker, New York, 1984.
15. B. S. Dhillon, *Quality Control, Reliability and Engineering Design*, Marcel Dekker, New York, 1985.
16. R. N. Kackar, "Taguchi's Quality Philosophy: Analysis and Commentary," *Quality Progress* **19**, 21–29 (1986).
17. H. Kume, *Statistical Methods for Quality Improvement*, Japanese Quality Press, Tokyo, 1987.
18. K. Ishikawa, *Quality Control Circles at Work*, Asian Productivity Organization, Tokyo, 1984.
19. J. M. Juran, F. M. Gryna, and R. S. Bingham (eds.), *Quality Control Handbook*, McGraw-Hill, New York, 1979.
20. B. S. Dhillon, *Engineering Design: A Modern Approach*, Richard D. Irwin, Burr Ridge, IL, 1996.
21. *Statistical Quality Control Handbook*, AT&T Technologies, Indianapolis, 1956.
22. B. King, *Hoshin Planning: The Developmental Approach*, Methuen, Boston, MA, 1969.

BIBLIOGRAPHY

TQM

Baker, W., *TQM: A Philosophy and Style of Managing*, Faculty of Administration, University of Ottawa, Ottawa, Ont., 1992.

Farquhar, C. R., and C. G. Johnston, *Total Quality Management: A Competitive Imperative*, Report No. 60-90-E, Conference Board of Canada, Ottawa, Ont., 1990.

Feigenbaum, A. V., *Total Quality Control*, McGraw-Hill, New York, 1983.

Gevirtz, C. D., *Developing New Products with TQM*, McGraw-Hill, New York, 1994.

Mears, P., "TQM Contributors," in *Quality Improvement Tools and Techniques*, McGraw-Hill, New York, 1995, pp. 229–246.

Oakland, J. S. (ed.), *Total Quality Management*, Proceedings of the Second International Conference, IFS Publications, Kempston, Bedford, UK, 1989.

Shores, A. R., *Survival of the Fittest: Total Quality Control and Management*, ASQC Quality Press, Milwaukee, WI, 1988.

Stein, R. E., *The Next Phase of Total Quality Management*, Marcel Dekker, New York, 1994.

Tenner, R. R., and I. J. Detoro, *Total Quality Management: Three Steps to Continuous Improvement*, Addison-Wesley, Reading, MA, 1992.

Design Quality

Burgess, J. A., "Assuring the Quality of Design," *Machine Design*, 65–69 (February 1982).

Chaparian, A. P., "Teammates: Design and Quality Engineers," *Quality Progress* **10** (4), 16–17 (April 1977).

Colloquium on Management of Design Quality Assurance, IEE Colloquium Digest No. 1988/6, Institution of Electrical Engineers, London, 1988.

Daetz, D., "The Effect of Product Design on Product Quality and Product Cost," *Quality Progress* **20,** 63–67 (June 1987).

Evans, J. R., and W. M. Lindsay, "Quality and Product Design," in *The Management and Control of Quality*, West, New York, 1982, pp. 188–221.

Lockner, R. H., and J. E. Matar, *Designing for Quality*, ASQC Quality Press, Milwaukee, WI, 1990.

Michalek, J. M., and R. K. Holmes, "Quality Engineering Techniques in Product Design/Process," in *Quality Control in Manufacturing*, Society of Automotive Engineers, SP-483, pp. 17–22.

Phadke, M. S., *Quality Engineering Using Robust Design*, Prentice-Hall, Englewood Cliffs, NJ, 1986.

Pignatiello, J. J., and J. S. Ramberg, "Discussion on Off-line Quality Control, Parameter Design, and the Taguchi Method," *Journal of Quality Technology* **17,** 198–206 (1985).

Quality Assurance in the Design of Nuclear Power Plants: A Safety Guide, Report No. 50-SG-QA6, International Atomic Energy Agency, Vienna, 1981.

Quality Assurance in the Procurement, Design, and Manufacture of Nuclear Fuel Assemblies: A Safety Guide, Report No. 50-SG-QA 11, International Atomic Agency, Vienna, 1983.

Ross, P. J., *Taguchi Techniques for Quality Engineering*, McGraw-Hill, New York, 1988.

CHAPTER 20

RELIABILITY IN MECHANICAL DESIGN

B. S. Dhillon
Department of Mechanical Engineering
University of Ottawa
Ottawa, Ontario, Canada

20.1 INTRODUCTION	**487**	
20.2 BASIC RELIABILITY NETWORKS	**488**	
20.2.1 Series Network	488	
20.2.2 Parallel Network	488	
20.2.3 *k*-out-of-*n* Unit Network	489	
20.2.4 Standby System	490	
20.3 MECHANICAL FAILURE MODES AND CAUSES	**491**	
20.4 RELIABILITY-BASED DESIGN	**491**	
20.5 DESIGN-RELIABILITY TOOLS	**492**	
20.5.1 Failure Modes and Effects Analysis (FMEA)	492	
20.5.2 Fault Tree	494	

20.5.3 Failure Rate Modeling and Parts Count Method	496	
20.5.4 Stress-Strength Interference Theory Approach	497	
20.5.5 Network Reduction Method	498	
20.5.6 Markov Modeling	498	
20.5.7 Safety Factors	500	
20.6 DESIGN LIFE-CYCLE COSTING	**501**	
20.7 RISK ASSESSMENT	**501**	
20.7.1 Risk-Analysis Process and Its Application Benefits	502	
20.7.2 Risk Analysis Techniques	502	
20.8 FAILURE DATA	**504**	

20.1 INTRODUCTION

The history of the application of probability concepts to electric power systems goes back to the 1930s.[1-6] However, the beginning of the reliability field is generally regarded as World War II, when Germans applied basic reliability concept to improve reliability of their V1 and V2 rockets.

During the period from 1945–1950 the U.S. Army, Navy, and Air Force conducted various studies that revealed a definite need to improve equipment reliability. As a result of this effort, the Department of Defense, in 1950, established an ad hoc committee on reliability. In 1952, this committee was transformed to a group called the Advisory Group on the Reliability of Electronic Equipment (AGREE). In 1957, this group's report, known as the AGREE Report, was published, and it subsequently led to a specification on the reliability of military electronic equipment.

The first issue of a journal on reliability appeared in 1952, published by the Institute of Electrical and Electronic Engineers (IEEE). The first symposium on reliability and quality control was held in 1954. Since those days, the field of reliability has developed into many specialized areas: mechanical reliability, software reliability, power system reliability, and so on. Most of the published literature on the field is listed in Refs. 7, 8.

The history of mechanical reliability in particular goes back to 1951, when W. Weibull[9] developed a statistical distribution, now known as the Weibull distribution, for material strength and life length. The work of A. M. Freudenthal[10,11] in the 1950s is also regarded as an important milestone in the history of mechanical reliability.

The efforts of the National Aeronautics and Space Administration (NASA) in the early 1960s also played a pivotal role in the development of the mechanical reliability field,[12] due primarily to two factors: the loss of Syncom I in space in 1963, due to a bursting high-pressure gas tank, and the loss of Mariner III in 1964, due to mechanical failure. Many projects concerning mechanical relia-

Mechanical Engineers' Handbook, 2nd ed., Edited by Myer Kutz.
ISBN 0-471-13007-9 © 1998 John Wiley & Sons, Inc.

bility were initiated and completed by NASA. A comprehensive list of publications on mechanical reliability is given in Ref. 13.

20.2 BASIC RELIABILITY NETWORKS

A system component may form various different configurations: series, parallel, k-out-of-n, standby, and so on. In the published reliability literature, these configurations are known as the standard configurations. During the mechanical design process, it might be desirable to evaluate the reliability or the values of other related parameters of systems forming such configurations. These networks are described in the following pages.

20.2.1 Series Network

The block diagram of an "n" unit series network is shown in Fig. 20.1. Each block represents a system unit or component. If any one of the components fails, the system fails; thus, all of the series units must work successfully for the system to succeed.

For independent units, the reliability of the network shown in Fig. 20.1 is

$$R_s = R_1 R_2 R_3 \cdots R_n \tag{20.1}$$

where R_s = the series system reliability
$\quad n$ = the number of units
$\quad R_i$ = the reliability of unit i; for $i = 1, 2, 3, \cdots, n$

For units' constant failure rates, Eq. (20.1) becomes [14]

$$R_s(t) = e^{-\lambda_1 t} \cdot e^{-\lambda_2 t} \cdot e^{-\lambda_3 t} \cdots e^{-\lambda_n t}$$
$$= e^{-\sum_{i=1}^{n} \lambda_i t} \tag{20.2}$$

where $R_s(t)$ = the series system reliability at time t
$\quad \lambda_i$ = the unit i constant failure rate, for $i = 1, 2, 3, \cdots, n$

The system hazard rate or the total failure rate is given by [14]

$$\lambda_s(t) = -\frac{1}{R_s(t)} \frac{dR_s(t)}{dt} = \sum_{i=1}^{n} \lambda_i \tag{20.3}$$

where $\lambda_s(t)$ = the series system total failure rate or the hazard rate

Note that the series system failure rate is the sum of the unit failure rates. In mechanical or in other design analysis, when the failure rates are added, it is automatically assumed that the units are acting in series. This is the worst-case design assumption—if any one unit fails, the system fails. In engineering design specifications, the adding up of all system component failure rates is often specified.

The system mean time to failure is expressed by[13]

$$MTTF_s = \lim_{s \to 0} R_s(s) = \frac{1}{\sum_{i=1}^{n} \lambda_i} \tag{20.4}$$

where $MTTF_s$ = the series system mean time to failure
s (in brackets) = the Laplace transform variable
$\quad R_s(s)$ = the Laplace transform of the series system reliability

20.2.2 Parallel Network

The block diagram of an "n" unit parallel network is shown in Fig. 20.2. As in the case of the series network, each block represents a system unit or component. All of the system units are assumed to

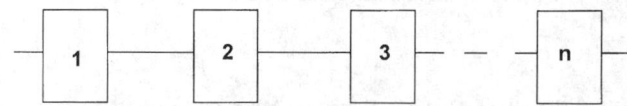

Fig. 20.1 Block diagram representing a series system.

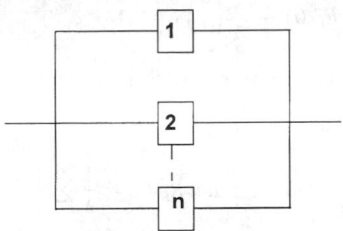

Fig. 20.2 Parallel network block diagram.

be active and at least one unit must function normally for the system to succeed, meaning that this type of configuration may be used to improve a mechanical system's reliability during the design phase.

For independent units, the reliability of the parallel network shown in Fig. 20.2 is given by[13]

$$R_p = 1 - (1 - R_1)(1 - R_2) \cdots (1 - R_n) \tag{20.5}$$

where R_p = the parallel network reliability

For constant failure rates of the units, Eq. (20.5) becomes

$$R_p(t) = 1 - (1 - e^{-\lambda_1 t})(1 - e^{-\lambda_2 t}) \cdots (1 - e^{-\lambda_n t}) \tag{20.6}$$

where $R_p(t)$ = the parallel network reliability at time t

Obviously, Eqs. (20.5) and (20.6) indicate that system reliability increases with the increasing values of n.

For identical units, the system mean time to failure is given by[14]

$$MTTF_p = \lim_{s \to 0} R_p(s) = \frac{1}{\lambda} \sum_{i=1}^{n} \frac{1}{i} \tag{20.7}$$

where $MTTF_p$ = the parallel network mean time to failure
$R_p(s)$ = the Laplace transform of the parallel network reliability
λ = the constant failure rate of a unit

20.2.3 k-out-of-n Unit Network

This arrangement is basically a parallel network with a condition that at least k units out of the total of n units must function normally for the system to succeed. This network is sometimes referred to as partially redundant network. An example might be a Jumbo 747. If a condition is imposed that at least three out of four of its engines must operate normally for the aircraft to fly successfully, then this system becomes a special case of the k-out-of-n unit network. Thus, in this case, $k = 3$ and $n = 4$.

For independent and identical units, the k-out-of-n unit network reliability is[14]

$$R_{k/n} = \sum_{i=k}^{n} \binom{n}{i} R^i (1 - R)^{n-i} \tag{20.8}$$

where

$$\binom{n}{i} = \frac{n!}{i! \, (n - i)!}$$

R = the unit reliability
$R_{k/n}$ = the k-out-of-n unit network reliability

Note that at $k = 1$, the k-out-of-n unit network reduces to a parallel network and at $k = n$, it becomes a series system.

For constant unit failure rates, Eq. (20.8) is rewritten to the following form:[13]

$$R_{k/n}(t) = \sum_{i=k}^{n} \binom{n}{i} e^{-i\lambda t} (1 - e^{-\lambda t})^{n-t}$$ (20.9)

where $R_{k/n}(t)$ = is the k-out-of-n unit network reliability at time t

The system mean time to failure is given by[13]

$$MTTF_{k/n} = \lim_{s \to 0} R_{k/n}(s) = \frac{1}{\lambda} \sum_{i=k}^{n} \frac{1}{i}$$ (20.10)

where $MTTF_{k/n}$ = the mean time to failure of the k-out-of-n unit network
$\quad R_{k/n}(s)$ = the Laplace transform of the k-out-of-n unit network reliability.

20.2.4 Standby System

The block diagram of an $(n + 1)$ unit standby system is shown in Fig. 20.3. Each block represents a unit or a component of the system. In the standby system case, as shown in Fig. 20.3, one unit operates and n units are kept on standby.

During the mechanical design process, this type of redundancy is sometimes adopted to improve system reliability.

If we assume independent and identical units, perfect switching, and standby units as good as new, then the standby system reliability is given by[14]

$$R_{ss}(t) = \sum_{i=0}^{n} \left\{ \int_{0}^{t} \lambda(t)dt \right\}^{i} e^{-\int_{0}^{t}\lambda(t)dt} \Big/ i!$$ (20.11)

where $R_{ss}(t)$ = the standby system reliability at time t
$\quad n$ = the number of standbys
$\quad \lambda(t)$ = the unit hazard rate or time-dependent failure rate

For two non-identical units (i.e., one operating, the other on standby), the system reliability is expressed by[15]

$$R_{ss}(t) = R_{o}(t) + \int_{o}^{t} f_{o}(t_1)R_{su}(t - t_1)\, dt_1$$ (20.12)

where $R_{o}(t)$ = the operating unit reliability at time t
$\quad R_{su}(t)$ = the standby unit reliability at time t
$\quad f_{o}(t_1)$ = the operating unit failure density function

For known reliability of the switching mechanism, Eq. (20.12) is modified to

$$R_{ss}(t) = R_{o}(t) + R_{sw} \int_{o}^{t} f_{o}(t_1)R_{su}(t - t_1)\, dt_1$$ (20.13)

where R_{sw} = the reliability of the switching mechanism

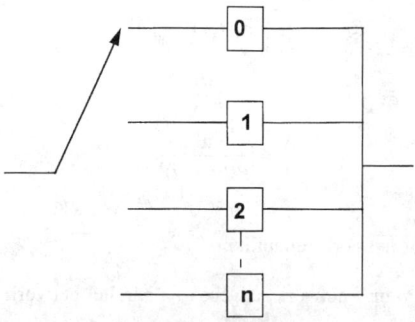

Fig. 20.3 An $(n + 1)$ unit standby system block diagram.

For identical units and constant unit failure rates, Eq. (20.13) simplifies to

$$R_{ss}(t) = e^{-\lambda t}(1 + R_{sw}\lambda t) \tag{20.14}$$

where λ = the unit constant failure rate

20.3 MECHANICAL FAILURE MODES AND CAUSES

There are certain failure modes and causes associated with mechanical products. The proper identification of relevant failure modes and their causes during the design process would certainly help to improve the reliability of design under consideration.

Mechanical and structural parts function adequately within specific useful lives. Beyond those lives, they cannot be used for effective mission, safe mission, and so on. A mechanical failure may be defined as any change in the shape, size, or material properties of a structure, piece of equipment, or equipment part that renders it unfit to perform its specified mission satisfactorily.[13] One of the factors for the failure of a mechanical part is the specified magnitude and type of load. The basic types of loads are dynamic, cyclic, and static. There are many types of failures that result from different types of loads: tearing, spalling, buckling, abrading, wear, crushing, fracture, and creep.[16] In fact, there are many different modes of mechanical failures.[17]

- Brinelling
- Thermal shock
- Ductile rupture
- Fatigue
- Creep
- Corrosion
- Fretting
- Stress rupture
- Brittle fracture
- Radiation damage
- Galling and seizure
- Thermal relaxation
- Temperature-induced elastic deformation
- Force-induced elastic deformation
- Impact

Field experience has shown that there are various causes of mechanical failures, including[18] defective design, wear-out, manufacturing defects, incorrect installation, gradual deterioration in performance, and failure of other parts.

Some of the important failure modes and their associated characteristics are presented below.[19]

- *Creep.* This may be described as the steady flow of metal under a sustained load. The cause of a failure is the continuing creep deformation in situations when either a rupture occurs or a limiting acceptable level of distortion is exceeded.
- *Corrosion.* This may be described as the degradation of metal surfaces under service or storage conditions because of direct chemical or electrochemical reaction with its environment. Usually, stress accelerates the corrosion damage. In hydrogen embrittlement, the metal ductility increases due to hydrogen absorption, leading either to fracture or to brittle failure under impact loads at high-strain rates or under static loads at low-strain rates, respectively.
- *Static failure.* Many of the materials fail by fracture due to the application of static loads beyond the ultimate strength.
- *Wear.* This occurs in contacts such as sliding, rolling, or impact, due to gradual destruction of a metal surface through contact with another metal or non-metal surface.
- *Fatigue failure.* In the presence of cyclic loads, materials can fail by fracture even when the maximum cyclic stress magnitude is well below the yield strength.

20.4 RELIABILITY-BASED DESIGN

It would be unwise to expect a system to perform to a desired level of reliability unless it is specifically designed for that reliability. The specification of desired system/equipment/part reliability in the design specification due to factors such as well-publicized failures (e.g., the space shuttle *Challenger* disaster and the Chernobyl nuclear accident) has increased the importance of reliability-based design. The starting point for the reliability-based design is during the writing of the design specification. In this phase, all reliability needs and specifications are entrenched into the design specification. Examples of these requirements might include item mean time to failure (MTBF), mean time to repair (MTTR), test or demonstration procedures to be used, and applicable document.

The U.S. Department of Defense, over the years, has developed various reliability documents for use during the design and development of an engineering item. Many times, such documents are entrenched into the item design specification document. Table 20.1 presents some of these documents. Many professional bodies and other organizations have also developed documents on various aspects of reliability.[7,8,14–16] References 15 and 20 provide descriptions of documents developed by the U.S. Department of Defense.

Reliability is an important consideration during the design phase. According to Ref. 21, as many as 60% of failures can be eliminated through design changes. There are many strategies the designer could follow to improve design:

1. Eliminate failure modes.
2. Focus design for fault tolerance.
3. Focus design for fail safe.
4. Focus design to include mechanism for early warnings of failure through fault diagnosis.

During the design phase of a product, various types of reliability and maintainability analyses can be performed, including reliability evaluation and modeling, reliability allocation, maintainability evaluation, human factors/reliability evaluation, reliability testing, reliability growth modeling, and life-cycle costing. In addition, some of the design improvement strategies are zero-failure design, fault-tolerant design, built-in testing, derating, design for damage detection, modular design, design for fault isolation, and maintenance-free design. During design reviews, reliability and maintainability-related actions recommended/taken are to be thoroughly reviewed from desirable aspects.

20.5 DESIGN-RELIABILITY TOOLS

There are many reliability analysis techniques and methods available to design professionals during the design phase. These include failure modes and effects analysis (FMEA), stress-strength modeling, fault tree analysis, network reduction, Markov modeling, and safety factors. All of these techniques are applicable in evaluating mechanical designs.

20.5.1 Failure Modes and Effects Analysis (FMEA)

FMEA is a vital tool for evaluating system design from the point of view of reliability. It was developed in the early 1950s to evaluate the design of various flight control systems.[22]

The difference between the FMEA and failure modes, effects, and criticality analysis (FMECA) is that FMEA is a qualitative technique used to evaluate a design, whereas FMECA is composed of

Table 20.1 Selected Reliability Documents Developed by the U.S. Department of Defense[20]

No.	Document No.	Document Title
1	M1L-HDBK-217	Reliability prediction of electronic equipment
2	M1L-STD-781	Reliability design qualification and production-acceptance tests: exponential distribution
3	M1L-HDBK-472	Maintainability prediction
4	RADC-TR-83-72	Evolution and practical application of failure modes and effects analysis (FMEA)
5	NPRD-2	Nonelectronic parts reliability data
6	RADC-TR-75-22	Nonelectronic reliability notebook
7	M1L-STD-1629	Procedures for performing a failure mode, effect, and criticality analysis (FMECA)
8	M1L-STD-1635 (EC)	Reliability growth testing
9	M1L-STD-721	Definition of terms for reliability and maintainability
10	M1L-STD-785	Reliability program for systems and equipment development and production
11	M1L-STD-965	Parts control program
12	M1L-STD-756	Reliability modeling and prediction
13	M1L-STD-2084	General requirements for maintainability
14	M1L-STD-882	System safety program requirements
15	M1L-STD-2155	Failure-reporting analysis and corrective action system

FMEA and criticality analysis (CA). Criticality analysis is a quantitative method used to rank critical failure mode effects by taking into consideration their occurrence probabilities.

As FMEA is a widely used method in industry, there are many standards/documents written on it. In fact, Ref. 23 collected and evaluated 45 of such publications prepared by organizations such as the U.S. Department of Defense (DOD), National Aeronautics and Space Administration (NASA), Institute of Electrical and Electronic Engineers (IEEE), and so on. These documents include:[24]

- *DOD:* M1L-STD-785A (1969), M1L-STD-1629 (draft) (1980), M1L-STD-2070(AS) (1977), M1L-STD-1543 (1974), AMCP-706-196 (1976)
- *NASA:* NHB 5300.4 (1A) (1970), ARAC Proj. 79-7 (1976)
- *IEEE:* ANSI N 41.4 (1976)

Details of the above documents as well as a list of publications on FMEA are given in Ref. 24. There can be many reasons for conducting FMEA, including:[25]

- To identify design weaknesses
- To help in choosing design alternatives during the initial design stages
- To help in recommending design changes
- To help in understanding all conceivable failure modes and their associated effects
- To help in establishing corrective action priorities
- To help in recommending test programs

In performing FMEA, the analyst seeks answers to various questions for each component of the concerned system, such as, How can the component fail and what are the possible failure modes? What are all the possible effects associated with each failure mode? How can the failure be detected? What is the criticality of the failure effects? Are there any safeguards against the possible failure?

Procedure for Performing FMEA

This procedure is composed of four steps:

1. Establishing analysis scope
2. Collecting data
3. Preparing the component list
4. Preparing FMEA sheets

Establishing Analysis Scope. This is concerned with establishing system boundaries and the extent of the analysis. The analysis may encompass information on various areas concerning each potential component failure: failure frequency, underlying causes of the failure, safeguards, possible failure effects, detection of failure, and failure effect criticality. Furthermore, the extent of FMEA depends on the timing of performance of FMEA; for example, conceptual design stage and detailed design stage. In this case, the extent of FMEA may be broader for the detailed design analysis stage than for the conceptual design stage. In any case, the extent of the analysis should be decided on the merits of each case.

Collecting Data. Because performing FMEA requires various kinds of data, professionals conducting FMEA should have access to documents concerning specifications, operating procedures, system configurations, and so on. In addition, the FMEA team, as applicable, should collect desired information by interviewing design professionals, operation/maintenance engineers, component suppliers, and external experts for collecting desirable information.

Preparing the Component List. The preparation of the component list is absolutely necessary prior to embarking on performing FMEA. In the past, it has proven useful to include operating conditions, environmental conditions, and functions in the component list.

Preparing FMEA Sheet. FMEA is conducted using FMEA sheets. These sheets include areas on which information is desirable, such as part, function, failure mode, cause of failure, failure effect, failure detection, safety feature, frequency of failure, effect criticality, and remarks.

- *Part* is concerned with the identification and description of the part/component in question.
- *Function* is concerned with describing the function of the part in various different operational modes.
- *Failure mode* is concerned with the determination of all possible failure modes associated with a part, e.g., open, short, close, premature, and degraded.
- *Cause of failure* is concerned with the identification of all possible causes of a failure.

- *Failure effect* is concerned with the identification of all possible failure effects.
- *Failure detection* is concerned with the identification of all possible ways and means of detecting a failure.
- *Safety feature* is concerned with the identification of built-in safety provisions associated with a failure.
- *Frequency of failure* is concerned with determination of failure occurrence frequency.
- *Effect criticality* is concerned with ranking the failure according to its criticality, e.g., critical (i.e., potentially hazardous), major (i.e., reliability and availability will be affected significantly but it is not a safety hazard), minor (i.e., reliability and availability will be affected somewhat but it is not a safety hazard), insignificant (i.e., little effect on reliability and availability and it will not be a safety hazard).
- *Remarks* is concerned with listing any remark concerning the failure in question, as well as possible recommendations.

One of the major advantages of FMEA is that it helps to identify system weaknesses at the early design stage. Thus, remedial measures may be taken immediately during the design phase.

The major drawback of FMEA is that it is a "single failure analysis." In other words, FMEA is not well suited for determining the combined effects of multiple failures.

20.5.2 Fault Tree

This method, so called because it arranges fault events in a tree-shaped diagram, is one of the most widely used techniques for performing system reliability analysis. In particular, it is probably the most widely used method in the nuclear power industry. The technique is well suited for determining the combined effects of multiple failures.

The fault tree technique is more costly to use than the FMEA approach. It was developed in the early 1960s in Bell Telephone Laboratories to evaluate the reliability of the Minuteman Launch Control System. Since that time, hundreds of publications on the method have appeared. References 26–27 describe it in detail.

The fault tree analysis begins by identifying an undesirable event, called the "top event," associated with a system. The fault events that could cause the occurrence of the top event are generated and connected by logic gates known as *AND*, *OR*, and so on. The construction of a fault tree proceeds by generation of fault events (by asking the question "How could this event occur?") in a successive manner until the fault events need not be developed further. These events are known as primary or elementary events. In simple terms, the fault tree may be described as the logic structure relating the top event to the primary events.

Fig. 20.4 presents four basic symbols associated with the fault tree method.

- *Circle* is used to represent a basic fault event, i.e., the failure of an elementary component. The component failure parameters, such as probability, failure, and repair rates, are obtained from field data or other sources.
- *Rectangle* is used to represent an event resulting from the combination of fault events through the input of a logic gate.

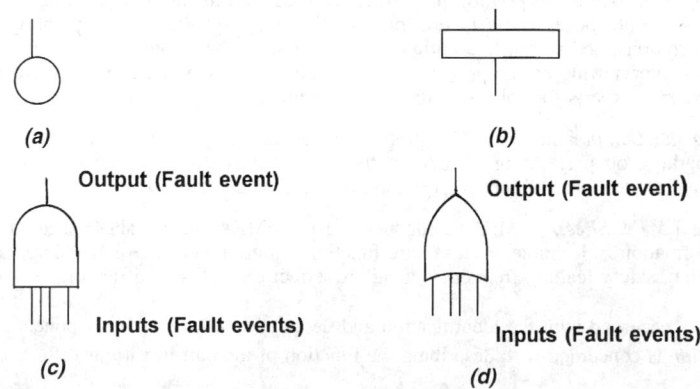

Fig. 20.4 Basic fault tree symbols (a) basic fault event, (b) resultant event, (c) AND gate, (d) OR gate.

- *AND gate* is used to denote a situation that an output event occurs if all the input fault events occur.
- *OR gate* is used to denote a situation that an output event occurs if any one or more of the input fault events occur.

The construction of fault trees using the symbols shown in Fig. 20.4 is demonstrated through the following example.

Example 20.1
Construct a fault tree of a simple system concerning hot water supply to the kitchen of a house. Assume that the hot water faucet only fails to open and the top event is kitchen without hot water. In addition, gas is used to heat water.

A simplified fault tree of a kitchen without hot water is shown in Fig. 20.5. This fault tree indicates that if any one of the E_i, for $i = 1, 2, 3, 4, 5$, fault event (i.e., fault events denoted by circles) occurs, there will be no hot water in kitchen.

The probability of occurrence of the top event Z_o (i.e., no hot water in kitchen) can be estimated, if the occurrence probabilities of the fault events E_1, E_2, E_3, E_4, and E_5 are known, using the formula given below.

The probability of occurrence of the OR gate output fault event, say x, is given by

$$P_{OR}(x) = 1 - \prod_{i=1}^{n} \left[1 - P(E_i) \right] \tag{20.15}$$

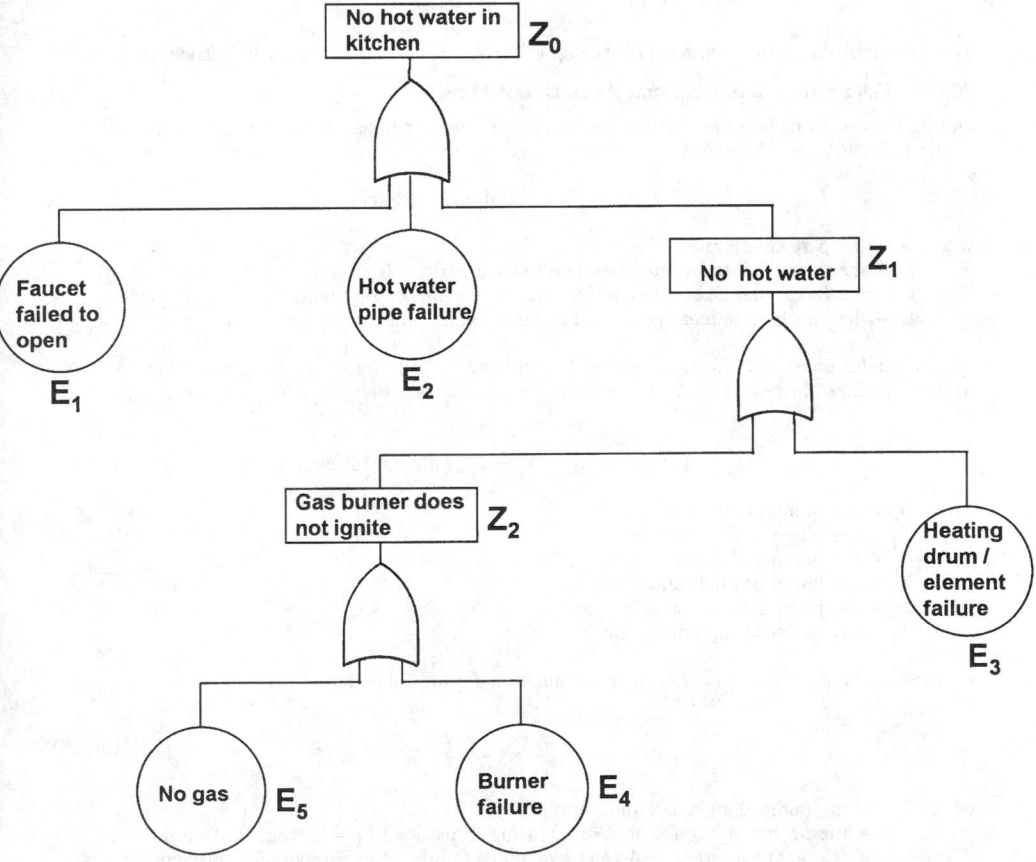

Fig. 20.5   Fault tree for kitchen without hot water.

where n = the number of independent input fault events
$P(E_i)$ = the probability of occurrence of the input fault event E_i, for i = 1, 2, 3, 4, and 5

Similarly, the probability of occurrence of the AND gate output fault event, say y, is given by

$$P_{AND}(y) = \prod_{i=1}^{n} P(E_i) \tag{20.16}$$

Example 20.2
Assume that the probability of occurrence of fault events E_1, E_2, E_3, E_4, and E_5 shown in Fig. 20.5 are 0.01, 0.02, 0.03, 0.04, and 0.05, respectively. Calculate the probability of occurrence of top event Z_0.

Substituting the specified data into Eq. (20.15), we get the probabilities of occurrence of events Z_2, Z_1, Z_0, respectively

$$P(Z_2) = P(E_4) + P(E_5) - P(E_4)\,P(E_5)$$
$$= (0.04) + (0.05) - (0.04)\,(0.05)$$
$$= 0.088$$
$$P(Z_1) = P(Z_2) + P(E_3) - P(Z_2) \cdot P(E_3)$$
$$= (0.088) + (0.03) - (0.088)\,(0.03)$$
$$= 0.11536$$
$$P(Z_0) = 1 - [1 - P(E_1)]\,[1 - P(E_2)]\,[1 - P(Z_1)]$$
$$= 1 - (1 - 0.01)\,(1 - 0.02)\,(1 - 0.11536)$$
$$= 0.14172$$

Thus, the probability of occurrence of the top event Z_0, that is, no hot water in kitchen, is 0.14172.

20.5.3 Failure Rate Modeling and Parts Count Method

During the design phase to predict the failure rate of a large number of electronic parts, the equation of the following form is used:[28]

$$\lambda = \lambda_b f_1 f_2 \cdots \text{(failures}/10^6 \text{ hr)} \tag{20.17}$$

where λ = the part failure rate
f_1 = the factor that takes into consideration the part quality level
f_2 = the factor that takes into consideration the influence of environment on the part
λ_b = the part base failure rate related to temperature and electrical stresses

On similar lines, Ref. 29 has proposed to estimate the failure rates of various mechanical parts, devices, and so on. For example, to estimate the failure rate of pumps, the following equation is proposed:

$$\lambda_p = \lambda_1 + \lambda_2 + \lambda_3 + \lambda_4 + \lambda_5, \text{ failures}/10^6 \text{ cycles} \tag{20.18}$$

where λ_p = the pump failure rate
λ_1 = the pump shaft failure rate
λ_2 = the pump seal failure rate
λ_3 = the pump bearing failure rate
λ_4 = the pump fluid driver failure rate
λ_5 = the pump casing failure rate

In turn, the pump shaft failure rate is obtained using the following relationship:

$$\lambda_1 = \lambda_{psb} \prod_{i=1}^{6} \theta_i \tag{20.19}$$

where λ_{psb} = the pump shaft base failure rate
θ_i = the ith modifying factor; i = 1 (casing thrust load), i = 2 (shaft surface finish), i = 3 (Contamination), i = 4 (material temperature), i = 5 (pump displacement), i = 6 (material endurance limit)

The values of the above factors are tabulated under the varying conditions in Ref. 29. Reference 29 also provides similar formulas for obtaining failure rates of pump bearings, seals, fluid driver, and casing.

The parts count method is used to estimate system/equipment failure rate during early design stages as well as during bid proposal. The following expression is used to estimate system/equipment failure rate:

$$\lambda_s = \sum_{i=1}^{n} N_i (\lambda_c Q_c)_i \text{ failures}/10^6 \text{ hour} \tag{20.20}$$

where λ_s = the system/equipment failure rate
$\quad N_i$ = the number of ith generic component
$\quad \lambda_c$ = the ith generic component failure rate expressed in failures/10^6 hour
$\quad Q_c$ = the quality factor associated with ith generic component
$\quad n$ = the number of different generic component categories

The values of λ_c and Q_c are given in Ref. 28. It is to be noted that Eq. (20.20) is based on the assumption that the operational environment of the entire equipment/system is the same.

20.5.4 Stress–Strength Interference Theory Approach

This is a useful approach to determine reliability of a mechanical item when its associated stress and strength probability density functions are known. In this case, the item reliability may be defined as the probability that the failure-governing stress will not exceed the failure-governing strength. Thus, mathematically, the item reliability is expressed by

$$R(x < y) = P(y > x) \tag{20.21}$$

where x = the stress variable
$\quad y$ = the strength variable
$\quad P$ = the probability
$\quad R$ = item reliability

Equation (20.21) is rewritten in the following form:[13,26]

$$R(x < y) = \int_{-\infty}^{\infty} f(y) \left[\int_{-\infty}^{y} f(x) \, dx \right] dy \tag{20.22}$$

where $f(x)$ = the probability density function of the stress
$\quad f(y)$ = the probability density function of the strength

Several alternative forms of Eq. (20.22) are given in Ref. 13. In order to demonstrate the applicability of Eq. (20.22), we assume that the item stress and strength are defined by the following probability density functions:[13]

$$f(x) = \alpha e^{-\alpha x}, \, x > 0 \tag{20.23}$$

$$f(y) = \frac{1}{\sigma \sqrt{2\pi}} \exp\left[-\frac{1}{2} \left(\frac{y-\mu}{\sigma} \right)^2 \right], \, -\infty < y < \infty \tag{20.24}$$

where α = the reciprocal of the mean stress
$\quad \mu$ = the mean strength
$\quad \sigma$ = the strength standard deviation

Substituting Eqs. (20.23) and (20.24) into Eq. (20.22) yields[13,30]

$$R = \int_{-\infty}^{\infty} \frac{1}{\sqrt{2\pi}} \exp\left\{ -\frac{1}{2} \left(\frac{y-\mu}{\sigma} \right)^2 \right\} \left[\int_{-\infty}^{\infty} \alpha e^{-\alpha x} \, dx \right] dy$$

$$= 1 - \exp\left[-\frac{1}{2} (2\mu\alpha - \sigma^2\alpha^2) \right] \tag{20.25}$$

Reliability expressions for various other combinations of stress and strength probability density functions are given in Ref. 13. This reference also provides a graphical approach based on Mellin transforms to estimate mechanical item reliability.

20.5.5 Network Reduction Method

This is probably the simplest and the most straightforward approach to determine the reliability of systems composed of configurations such as series, parallel, and so on. The approach is concerned with sequentially reducing the series and parallel configurations to equivalent hypothetical components until the whole system becomes a single hypothetical component or unit. The approach is demonstrated through the following example.

Example 20.3
Evaluate the reliability of Fig. 20.6 block diagram given each unit's reliability between zero and one.
Using Eq. (20.1), the reliability of Fig. 20.6 subsystem A is

$$R_A = R_1 R_2 R_3$$
$$= (0.9)\,(0.8)\,(0.85)$$
$$= 0.612$$

The above result allows us to reduce subsystem A to a single hypothetical component/unit with reliability $R_A = 0.612$, as shown in Fig. 20.7.
Using Eq. (20.5), the reliability of Fig. 20.7 subsystem B is given by

$$R_B = 1 - (1 - R_4)\,(1 - R_A)$$
$$= 1 - (0.3)\,(0.388)$$
$$= 0.8836$$

Using the above result, we have reduced the Fig. 20.7 subsystem B to a single hypothetical component/unit with reliability $R_B = 0.8836$, as shown in Fig. 20.8.
With the aid of Eq. (20.1), the Fig. 20.8 reliability is

$$R_s = R_B R_5 = (0.8836)\,(0.95)$$
$$= 0.8394$$

Thus, the Fig. 20.8 network is reduced to a single hypothetical unit with reliability $R_s = 0.8394$.

20.5.6 Markov Modeling

This method is probably used more widely than any other reliability prediction method. It is extremely useful in performing reliability and availability analysis of systems with dependent failure and repair

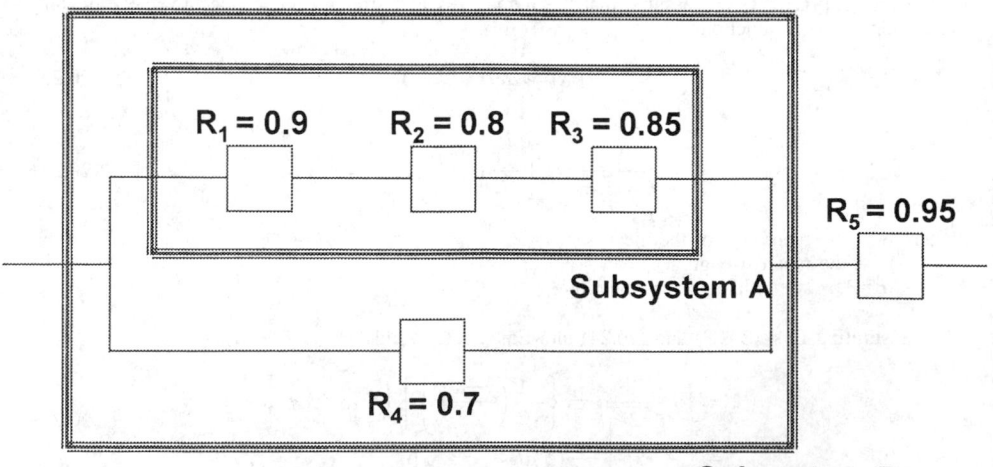

Fig. 20.6 Block diagram of a system.

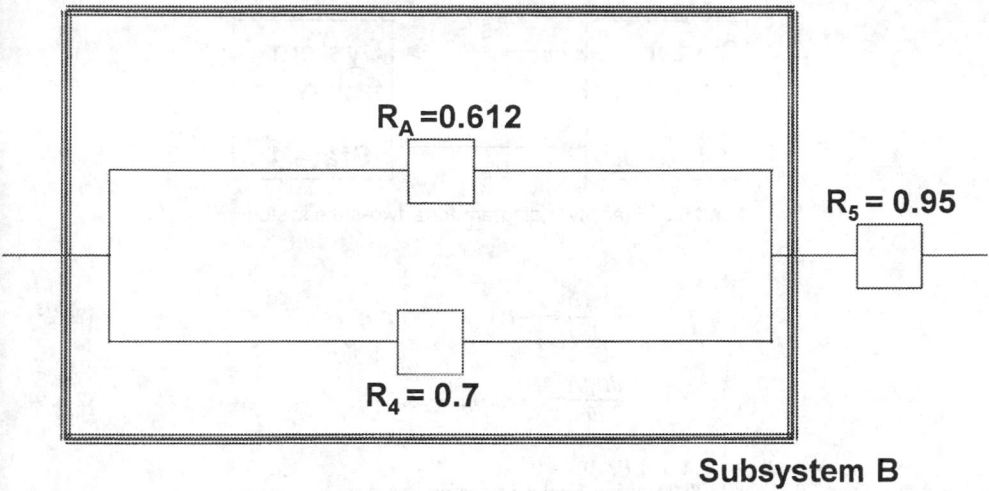

Fig. 20.7 Reduced Fig. 20.6 network.

modes as well as constant failure and repair rates. However, the method breaks down for a system with non-constant failure and repair rates. The following assumptions are made to formulate Markov state equations:[31]

- All occurrences are independent of each other.
- The probability of more than one transition occurrence from one state to the next state in finite time interval, Δt, is negligible.
- The probability of occurrence from one state to another in the finite time interval Δt is given by $\alpha \Delta t$, where the α is the constant transition rate from one state to another.

This method is demonstrated through the following example.

Example 20.4
Develop state probability expressions for a two-state system whose state–space diagram is shown in Fig. 20.9.
The Markov equations associated with Fig. 20.9 are as follows:

$$P_0\,(t + \Delta\,t) = P_0(t)\,(1 - \lambda_s\Delta t) + P_1(t)\,\mu_s\Delta t \tag{20.26}$$

$$P_1\,(t + \Delta\,t) = P_1(t)\,(1 - \mu_s\Delta t) + P_0(t)\,\lambda_s\Delta t \tag{20.27}$$

where $P_0(t)$ = the probability that the system is in state 0 at time t
$P_1(t)$ = the probability that the system is in state 1 at time t
$\lambda_s\Delta t$ = the transition probability that the system has failed in time Δt
$\mu_s\Delta t$ = the transition probability that the system is repaired in time Δt
$(1 - \lambda_s\Delta t)$ = the probability of no failure transition from state 0 to state 1 in time Δt
$(1 - \mu_s\Delta t)$ = the probability of no repair transition from state 1 to state 0 in time Δt

Rearranging Eqs. (20.26) and (20.27), we get the following differential equations:

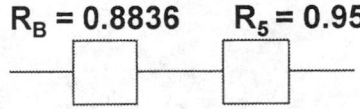

Fig. 20.8 Reduced Fig. 20.7 network.

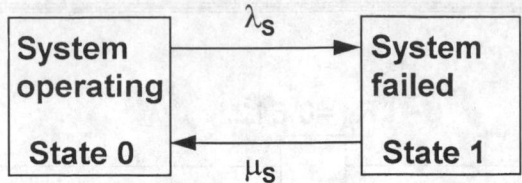

Fig. 20.9 Transition diagram for a two-state system.

$$\frac{dP_0(t)}{dt} = -P_0(t)\lambda_s + P_1(t)\mu_s \qquad (20.28)$$

$$\frac{dP_1(t)}{dt} = -P_1(t)\mu_s + P_0(t)\lambda_s \qquad (20.29)$$

At time $t = 0$, $P_0(0) = 1$ and $P_1(0) = 0$
Solving Eqs. (20.28) and (20.29) using Laplace transforms, we get

$$P_0(s) = \frac{(s + \mu_s)}{s^2 + (\lambda_s + \mu_s)\, s} \qquad (20.30)$$

$$P_1(s) = \frac{\lambda}{s^2 + (\lambda + \mu)\, s} \qquad (20.31)$$

The inverse Laplace transforms of Eqs. (20.30) and (20.31) are as follows:

$$P_0(t) = \frac{\mu_s}{\lambda_s + \mu_s} + \frac{\mu_s}{\lambda_s + \mu_s}\, e^{-(\lambda_s + \mu_s)t} \qquad (20.32)$$

$$P_1(t) = \frac{\lambda_s}{\lambda_s + \mu_s} - \frac{\lambda_s}{\lambda_s + \mu_s}\, e^{-(\lambda_s + \mu_s)t} \qquad (20.33)$$

For the given values of λ_s and μ_s, we can obtain the availability and unavailability of the system at any time t using Eqs. (20.32) and (20.33), respectively.

20.5.7 Safety Factors

Safety factors are often used to design reliable mechanical systems, equipment, and devices. The factor used to determine the safeness of a member is known as the factor of safety. This approach can provide satisfactory design in situations where the safety factors are established from the previous experience. Otherwise, design solely based on such factors could be misleading. There are various definitions used to define a safety factor.[13] Two examples of such definitions are presented below.

Definition I
According to Refs. 31 and 32, the safety factor is expressed as follows:

$$S_f = \frac{S_u}{S_w} \qquad (20.34)$$

where S_f = the safety factor
S_u = the ultimate strength expressed in pounds per square inch (psi)
S_w = the working stress expressed in psi

Definition II
The safety factor is defined by[33]

$$S_f = \frac{S_m}{S} \qquad (20.35)$$

where S_f = the safety factor
      S_m = the mean strength
      S = the mean stress

20.6   DESIGN LIFE-CYCLE COSTING

The life-cycle costing concept plays an important role during the design phase of an engineering product, as design decisions may directly or indirectly relate to the product cost. For example, the design simplification may reduce the operational cost of the product. One important application of the life-cycle costing concept during the design phase is in making decisions concerning alternative designs.

The term *life-cycle costing* was first coined in 1965.[34] Life-cycle cost (LCC) is defined as the sum of all costs incurred during the life time of an item; that is, the sum of procurement and ownership costs. This concept is applicable not only to engineering products, but also to buildings, other civil engineering structures, and so on. Most of the published literature on LCC is listed in Ref. 35.

Over the years, many different mathematical models have been developed to estimate product life-cycle cost. Some of these models are presented below.

Life-Cycle Cost Model I
The life-cycle cost of a product is expressed by[35]

$$LCC = RK + NRK \tag{20.36}$$

where RK = the recurring cost, composed of such elements as maintenance cost, labour cost, operating cost, inventory cost, and support cost
   NRK = the non-recurring cost, with elements such as training cost, research and development cost, procurement cost, reliability and maintainability improvement cost, support cost, qualification approval cost, installation cost, transportation cost, test equipment cost, and the cost of life-cycle cost management

Life-Cycle Cost Model II
The life-cycle cost is composed of three components:

$$LCC = PK + ILK + RK \tag{20.37}$$

where PK = the procurement cost representing the total of the unit prices
   ILK = the initial logistic cost, made up of the one-time costs, such as acquisition of new support equipment, not accounted for in the life-cycle costing of solicitation and training, and existing support equipment modifications and initial technical data-management cost
   RK = the recurring cost, composed of elements such as maintenance cost, operating cost, and management cost.

Life-Cycle Cost Model III
This model is specifically concerned with estimating life-cycle cost of switching power supplies,[36] which is expressed by

$$LCC = IK + FK \tag{20.38}$$

where IK = the initial cost and FK the failure cost, expressed by

$$FK = \lambda(EL)\,(RK + SK) \tag{20.39}$$

where λ = the switching power supply failure rate
   EL = the expected life of the switching power supply
   RK = the repair cost
   SK = the cost of the spares

20.7   RISK ASSESSMENT

Risk is present in all human activity. It can be health and safety-related or it can be economic (e.g., loss of equipment and production due to accidents involving fires, explosions, etc.). Risk may be described as a measure of the probability and security of a negative effect to health, equipment/property, or the environment.[37] Two important terms related to risk are described separately below.

Risk assessment is the process of risk analysis and risk evaluation. Risk analysis uses available data to determine risk to humans, environment, or equipment/property from hazards. It is usually composed of three steps: scope definition, hazard identification, and risk determination. Risk evaluation is the stage at which values and judgments enter the decision process.

Risk management is the total process of risk assessment and risk control. In turn, risk control is the decision-making process concerned with managing risk, and the implementations, enforcement, and reevaluation of its effectiveness from time to time, using risk assessment final results or conclusions as one of the inputs.

20.7.1 Risk-Analysis Process and Its Application Benefits

The risk-analysis process is made up of six steps:

1. Scope definition
2. Hazard identification
3. Risk estimation
4. Documentation
5. Verification
6. Analysis update

In establishing overall plan of risk analysis involves describing problems and formulating the objective, defining the system under study, highlighting assumptions and constraints associated with the analysis, identifying the decisions to be made, and documenting the risk-analysis plan.

Hazard identification involves identifying the hazards that generate risk in the system. Risk estimation is accomplished in the following steps:

- Hazard source investigation
- Performance of pathway analysis to trace the hazard from its source to its potential receptors
- Selection of methods/models to estimate the risk
- Evaluation of data needs
- Outlining the rationales and assumptions associated with methods, models, and data
- Estimation of risk for determining the impact on the concerned receptor
- Risk-estimation documentation

Documentation involves the documentation of the risk-analysis plan, preliminary evaluation, and risk estimation, in order to verify the integrity and correctiveness of the analysis process. It includes reviewing scope appropriateness, critical assumptions, appropriateness of methods, models and data used, analysis performed, and analysis insensitiveness.

Analysis update calls for revision of the analysis as new information becomes available.

Some of the advantages of risk-analysis applications are potential hazards and failure modes identification, better understanding of the system, risk comparisons to similar system/equipment/devices, better decisions regarding safety-improvement expenditures, and quantitative risk statements.

20.7.2 Risk-Analysis Techniques

There are various methods used to perform risk analysis.[37–40] However, the relevance and suitability of these methods prior to their applications must be carefully considered. Factors to be considered include a given method's appropriateness to the system, its scientific defensibility, whether it generates results in a form that enhances understanding of the risk occurrence, and how simple it is to use.

After the objectives and scope of the risk analysis have been defined, the methods should be selected, based on such factors as the objectives of the study, the phase of development, system and hazard types under study, the level of risk, the required levels of manpower, and resources, information and data needs, and capability for updating analysis.

Methods for performing risk analysis of engineering systems may be divided into two categories:

- *Hazard identification.* This requires that the system under consideration be systematically reviewed to identify inherent hazards and their type. The hazard-identification process makes use of experiences gained from previous risk-analysis studies and historical data. The methods under the hazard identification category are failure modes and effects analysis (FMEA), hazard and operability studies (HAZOP), fault tree analysis, and event tree analysis (ETA).
- *Risk estimation.* This is concerned with the risk quantitative analysis. It requires estimates of the frequency and consequences of hazardous events, system failure, and human error. Two methods under the risk-estimation category are frequency analysis and consequence analysis.

All of the above-mentioned methods are described below.

Hazard and Operability Study (HAZOP)

This is a form of FMEA originally developed for applications in process industries. HAZOP is a systematic approach for identifying hazards and operational problems throughout a facility. It has three objectives: to develop full facility description; to review systematically each facility or process element to identify how deviations from the design intentions can happen; and to judge whether such deviations can result in hazards or operating problems.

HAZOP can be applied during various stages of design or to process plants in operation. Its application during the early phase of design can often lead to safer detailed design. HAZOP involves the following steps:

- Establishing study objectives and scope
- Forming the HAZOP team, composed of suitable members from design and operation areas
- Obtaining necessary drawings, process description, and other relevant documentation (e.g., process flow sheets, equipment specification, layout drawings, and operating and maintenance procedures)
- Performing analysis of all major pieces of equipment, system, etc.
- Documenting consequences concerning deviation from the normal state and highlighting those deviations considered hazardous and credible

Failure Modes and Effects Analysis (FMEA)

This method is widely used in system reliability and safety analyses, and is equally applicable in risk-analysis studies. The technique is described above.

Fault Tree Analysis (FTA)

This technique is widely used in safety and reliability analyses of engineering systems—in particular, nuclear power-generation systems. Its applications in risk analysis are equally effective. The approach is discussed above.

Event Tree Analysis (ETA)

This is a "bottom-up" technique used to identify the possible outcomes where the occurrence of an initiating event is known. ETA is often used to analyze more complex systems than the ones handled by FMEA.[37,38,41,42] ETA is useful in analyzing facilities having engineered accident-mitigating factors to identify the event sequence that follows the initiating event and to generate given consequences. Generally, it is assumed that each sequence event is either a success or a failure.

Because of the inductive nature of ETA, the fundamental question asked is, "What happens if . . . ?" ETA studies highlight the relationship between the success or failure of various mitigating systems as well as the hazardous events that follow the single initiating event. Some of the additional points associated with ETA follow:

- It is a good idea to identify events that require further investigation using FTA.
- It is absolutely necessary to identify all possible initiating events in order to carry out a comprehensive risk assessment.
- ETA application always leaves the possibility of overlooking some important initiating events.
- It is difficult for ETA to incorporate delayed success or recovery events, as event trees cover only the success and failure states of a system.

Consequence Analysis

This is concerned with determining the impact of the undesired event on adjacent people, property, or the environment. Generally, for risk calculations concerning safety, it consists of determining the probability that people at different distances and environments from the event source will suffer illness or injury. Some examples of the undesired event are fires, explosions, release of toxic materials, and projection of debris. More specifically, the consequence analysis or models are required to predict probability of casualties. Consequence analysis should also consider the following:

- Basing analysis on selected undesirable events
- Corrective measures to eradicate consequences
- Describing series of consequences from undesirable events
- Conditions or situations having effects on the series of consequences
- Existence of the criteria used for accomplishing the identification of consequences
- Immediate and aftermath consequences

Table 20.2   Failure Rates for Selected Mechanical Parts

No.	Part	Failure Rate per 10^6 hr
1	Hair spring	1.0
2	Seal, O-ring	0.2
3	Bearing, roller	0.139–7.31
4	Mechanical joint	0.2
5	Compressor	0.84–198.0
7	Nut or bolt	0.02
8	Pipe	0.2
9	Piston	1.0
10	Gasket	0.5

Frequency Analysis

This is concerned with estimating the occurrence frequency of undesired events or accident scenarios (identified at the hazard-identification stage). Two commonly used approaches in performing frequency analysis are as follows:

- Making use of the past frequency data concerning the events under consideration to predict the frequency of their future occurrence
- Employing methods such as ETA and FTA to estimate event-occurrence frequencies

The approaches are complementary. Each has strengths where the other has weaknesses. All in all, whenever it is feasible, each approach should be employed to serve as a check on the other one.

20.8   FAILURE DATA

Failure data provide invaluable information to reliability engineers, design engineers, management, and so on concerning the product performance. These data are the final proof of the success or failure of the effort expended during the design and manufacture of a product used under designed conditions. During the design phase of a product, past information concerning its failures plays a critical role in reliability analysis of that product. Some of the uses of the failure data are estimating item failure rate, performing effective design reviews, predicting reliability and maintainability of redundant systems, conducting tradeoff and life cycle cost studies, and performing preventive maintenance and replacement studies.

There are various ways and means of collecting failure data. For example, during the equipment life cycle, there are eight identifiable data sources:[43]

- Repair facility reports
- Previous experience with similar or identical items
- Warranty claims
- Tests conducted during field demonstration, environmental qualification approval, and field installation
- Customer's failure-reporting systems
- Factory acceptance testing
- Developmental testing of the item
- Inspection records generated by quality control and manufacturing groups

See Refs. 28, 44–48 for some sources of obtaining failure data on mechanical parts. Reference 43 lists over 350 sources for obtaining various types of failure data. Table 20.2 presents failure rates for selected mechanical parts.

REFERENCES

1. W. J. Lyman, "Fundamental Considerations in Preparing a Master System Plan," *Electrical World* **101**, 778–792 (1933).
2. P. E. Benner, "The Use of the Theory of Probability to Determine Spare Capacity," *General Electric Review* **37**, 345–348.
3. S. A. Smith, "Service Reliability Measured by Probabilities of Outage," *Electrical World*, **103**, 222–225 (1934).
4. S. M. Dean, "Considerations Involved in Making System Investments for Improved Service Reliability," *Edison Electric Inst. Bull.* **6**, 491–496 (1938).

5. S. A. Smith, "Probability Theory and Spare Equipment," *Edison Electric Inst. Bull.* (March 1934).

6. S. A. Smith, "Spare Capacity Fixed by Probabilities of Outage," *Electrical World* **103**, 222–225 (1934).

7. B. S. Dhillon, *Reliability and Quality Control: Bibliography on General and Specialized Areas*, Beta, 1992.

8. B. S. Dhillon, *Reliability Engineering Applications: Bibliography on Important Application Areas*, Beta, 1992.

9. W. Weibull, "A Statistical Distribution Function of Wide Applicability," *Journal of Applied Mechanics* **18**, 293–297 (1951).

10. A. M. Freudenthal and E. J. Gumbel, "Failure and Survival in Fatigue," *Journal of Applied Physics* **25**, 110–120 (1954).

11. A. M. Freudenthal, "Safety and the Probability of Structural Failure," *Trans. Am. Society of Civil Engineers* **121**, 1337–1397 (1956).

12. W. M. Redler, "Mechanical Reliability Research in the National Aeronautics and Space Administration," in *Proceedings of the Reliability and Maintainability Conference*, 1966, pp. 763–768.

13. B. S. Dhillon, *Mechanical Reliability: Theory, Models, and Applications*, American Institute of Aeronautics and Astronautics, Washington, DC, 1988.

14. B. S. Dhillon, *Reliability Engineering in Systems Design and Operation*, Van Nostrand Reinhold, New York, 1983.

15. W. Grant-Ireson and C. F. Coombs (eds.), *Handbook of Reliability Engineering and Management*, McGraw-Hill, New York, 1988.

16. S. S. Rao, *Reliability-Based Design*, McGraw-Hill, New York, 1992.

17. J. A. Coolins, *Failure of Materials in Mechanical Design*, Wiley, New York, 1981.

18. C. Lipson, *Analysis and Prevention of Mechanical Failures*, Course Notes No. 8007, University of Michigan, Ann Arbor, June 1980.

19. N. A. Tiner, Failure Analysis with the Electron Microscope, *Fox-Mathis*, Los Angeles, 1973.

20. J. W. Wilbur and N. B. Fuqua, *A Primer for DOD Reliability*, Maintainability and Safety Standards Document No. PRIM 1, 1988, Rome Air Development Center, Griffiss Air Force Base, Rome, NY, 1988.

21. D. G. Raheja, *Assurance Technologies*, McGraw-Hill, New York, 1991.

22. J. S. Countinho, "Failure Effect Analysis," *Trans. N.Y.* Academy of Sciences **26**, 564–584 (1964).

23. *Procedures for Performing a Failure Modes and Effects and Criticality Analysis*, MIL-STD-1629, Department of Defense, Washington, DC, 1980.

24. B. S. Dhillon, "Failure Modes and Effects Analysis—Bibliography," *Microelectronics and Reliability* **32**, 719–732 (1992).

25. C. Sundararajan, *Guide to Reliability Engineering*, Van Nostrand Reinhold, New York, 1991.

26. B. S. Dhillon and C. Singh, *Engineering Reliability: New Techniques and Applications*, Wiley, New York, 1981.

27. B. S. Dhillon, "Fault Tree Analysis," in *Mechanical Engineers Handbook*, 1st ed., M. Kutz (ed.), Wiley, New York, 1986, pp. 354–369.

28. *Reliability Prediction of Electronic Equipment*, MIL-HDBK-217, U.S. Department of Defense, Washington, DC, 1992. (Available from Rome Air Development Center, Griffiss Air Force Base, Rome, NY, 13441. This document also includes electromechanical devices.)

29. J. D. Raze et al., "Reliability Models for Mechanical Equipment," *Proceedings of the Annual Reliability and Maintainability Symposium*, 1987, pp. 130–134.

30. D. Kececioglu and D. Li, "Exact Solutions for the Prediction of the Reliability of Mechanical Components and Structural Members," in *Proceedings of the Failure Prevention and Reliability Conference*, American Society of Mechanical Engineers, New York, 1985, pp. 115–122.

31. V. M. Faires, *Design of Machine Elements*, Macmillan, New York, 1955.

32. G. M. Howell, "Factors of Safety," *Machine Design*, 76–81, (July 1956).

33. R. B. McCalley, "Nomogram for Selection of Safety Factors," *Design News*, 138–141, (Sept. 1957).

34. *Life Cycle Costing in Equipment Procurement*, Report No. LMI Task 4C-5, Logistics Management Institute (LMI), Washington, DC, April 1965.

35. B. S. Dhillon, *Life Cycle Costing: Techniques, Models, and Applications*, Gordon and Breach Science Publishers, New York, 1989.

36. D. Monteith and B. Shaw, "Improved R, M, and LCC for Switching Power Supplies," in *Proceedings of the Annual Reliability and Maintainability Symposium*, 1979, pp. 262–265.
37. *Risk Analysis Requirements and Guidelines*, CAN/CSA-Q634-91, Canadian Standards Association, 1991. (Available from the Canadian Standards Association, 178 Rexdale Boulevard, Rexdale, Ont., Canada, M9W 1R3.)
38. W. E. Wesley, "Engineering Risk Analysis," in *Technological Risk Assessment*, P. F. Rice, L. A. Sagan, and C. G. Whipple, (eds.), Martinus Nijhoff, The Hague, 1984, pp. 49–84.
39. V. Covello and M. Merkhofer, *Risk Assessment and Risk Assessment Methods: The State of the Art*, NSF Report, National Science Foundation (NSF), Washington, DC, 1984.
40. B. S. Dhillon and S. N. Rayapati, "Chemical Systems Reliability: A Survey," *IEEE Trans. on Reliability*, **37**, 199–208 (1988).
41. S. J. Cox and N. R. S. Tait, *Reliability, Safety and Risk Management*, Butterworth-Heinemann, Oxford, 1991.
42. R. Ramakumar, *Engineering Reliability: Fundamentals and Applications*, Prentice-Hall, Englewood Cliffs, New Jersey, 1993.
43. B. S. Dhillon and H. C. Viswanath, "Bibliography of Literature on Failure Data," *Microelectronics and Reliability* **30**, 723–750 (1990).
44. R. E. Schafer et al., *RADC Non-Electronic Reliability Notebook*, Rept. RADC-TR-85-194, Reliability Analysis Center, Rome Air Development Center (RADC), Griffiss Air Force Base, Rome, NY, 1985.
45. IEEE Nuclear Reliability Data Manual, IEEE Std. 500, Wiley, New York, 1977.
46. H. P. Bloch and F. K. Geitner, *Practical Machinery Management for Process Plants: Machinery Failure Analysis and Troubleshooting*, Gulf, Houston, 1983, pp. 628–630.
47. T. Anderson and M. Misund, "Pipe Reliability: An Investigation of Pipeline Failure Characteristics and Analysis of Pipeline Failure Rates for Submarine and Cross-Country Pipelines," *Journal of Petroleum Technology*, 709–717 (April 1983).
48. S. O. Nilsson, "Reliability Data on Automotive Components," in *Proceedings of the Annual Reliability and Maintainability Symposium*, 1975, pp. 276–279.

BIBLIOGRAPHY*

Bompas-Smith, J. H., *Mechanical Survival*, McGraw-Hill, London, 1973.

Carter, A. D. S., *Mechanical Reliability*, Macmillan Education, London, 1986.

Dhillon, B. S., *Robot Reliability and Safety*, Springer-Verlag, New York, 1991.

Frankel, E. G., *Systems Reliability and Risk Analysis*, Martinus Nijhoff, The Hague, 1984.

Haugen, E. B., *Probabilistic Mechanical Design*, Wiley, New York, 1980.

Kapur, K. C., and L. R. Lamberson, *Reliability in Engineering Design*, Wiley, New York, 1977.

Kivenson, G., *Durability and Reliability in Engineering Design*, Hayden, New York, 1971.

Little, A., *Reliability of Shell Buckling Predictions*, MIT Press, Cambridge, MA, 1964.

Mechanical Reliability Concepts, ASME, New York, 1965.

Middendorf, W. H., *Design of Devices and Systems*, Marcel Dekker, New York, 1990.

Milestone, W. D. (ed.), *Reliability, Stress Analysis and Failure Prevention Methods in Mechanical Design*, ASME, New York, 1980.

Shooman, M. L., *Probabilistic Reliability: An Engineering Approach*, R. E. Krieger, Melbourne, FL, 1990.

Siddell, J. N., *Probabilistic Engineering Design*, Marcel Dekker, New York, 1983.

*Additional publications on mechanical design reliability may be found in Refs. 7 and 13.

CHAPTER 21

LUBRICATION OF MACHINE ELEMENTS

Bernard J. Hamrock
Department of Mechanical Engineering
Ohio State University
Columbus, Ohio

SYMBOLS		**508**	**21.3**	**ELASTOHYDRODYNAMIC**		
				LUBRICATION		**556**
21.1	**LUBRICATION**			21.3.1	Contact Stresses and	
	FUNDAMENTALS	**512**			Deformations	558
	21.1.1 Conformal and			21.3.2	Dimensionless Grouping	566
	Nonconformal Surfaces	512		21.3.3	Hard-EHL Results	568
	21.1.2 Bearing Selection	513		21.3.4	Soft-EHL Results	572
	21.1.3 Lubricants	516		21.3.5	Film Thickness for Different	
	21.1.4 Lubrication Regimes	518			Regimes of Fluid-Film	
	21.1.5 Relevant Equations	520			Lubrication	573
				21.3.6	Rolling-Element Bearings	576
21.2	**HYDRODYNAMIC AND**					
	HYDROSTATIC		**21.4**	**BOUNDARY LUBRICATION**		**616**
	LUBRICATION	**523**		21.4.1	Formation of Films	618
	21.2.1 Liquid-Lubricated			21.4.2	Physical Properties of	
	Hydrodynamic Journal				Boundary Films	619
	Bearings	524		21.4.3	Film Thickness	621
	21.2.2 Liquid-Lubricated			21.4.4	Effect of Operating	
	Hydrodynamic Thrust				Variables	621
	Bearings	530		21.4.5	Extreme-Pressure (EP)	
	21.2.3 Hydrostatic Bearings	536			Lubricants	623
	21.2.4 Gas-Lubricated					
	Hydrodynamic Bearings	545				

By the middle of this century two distinct regimes of lubrication were generally recognized. The first of these was hydrodynamic lubrication. The development of the understanding of this lubrication regime began with the classical experiments of Tower,[1] in which the existence of a film was detected from measurements of pressure within the lubricant, and of Petrov,[2] who reached the same conclusion from friction measurements. This work was closely followed by Reynolds' celebrated analytical paper[3] in which he used a reduced form of the Navier–Stokes equations in association with the continuity equation to generate a second-order differential equation for the pressure in the narrow, converging gap of a bearing contact. Such a pressure enables a load to be transmitted between the surfaces with very low friction since the surfaces are completely separated by a film of fluid. In such a situation it is the physical properties of the lubricant, notably the dynamic viscosity, that dictate the behavior of the contact.

The second lubrication regime clearly recognized by 1950 was boundary lubrication. The understanding of this lubrication regime is normally attributed to Hardy and Doubleday,[4,5] who found that very thin films adhering to surfaces were often sufficient to assist relative sliding. They concluded that under such circumstances the chemical composition of the fluid is important, and they introduced the term "boundary lubrication." Boundary lubrication is at the opposite end of the lubrication

Mechanical Engineers' Handbook, 2nd ed., Edited by Myer Kutz.
ISBN 0-471-13007-9 © 1998 John Wiley & Sons, Inc.

spectrum from hydrodynamic lubrication. In boundary lubrication it is the physical and chemical properties of thin films of molecular proportions and the surfaces to which they are attached that determine contact behavior. The lubricant viscosity is not an influential parameter.

In the last 30 years research has been devoted to a better understanding and more precise definition of other lubrication regimes between these extremes. One such lubrication regime occurs in nonconformal contacts, where the pressures are high and the bearing surfaces deform elastically. In this situation the viscosity of the lubricant may rise considerably, and this further assists the formation of an effective fluid film. A lubricated contact in which such effects are to be found is said to be operating elastohydrodynamically. Significant progress has been made in our understanding of the mechanism of elastohydrodynamic lubrication, generally viewed as reaching maturity.

This chapter describes briefly the science of these three lubrication regimes (hydrodynamic, elastohydrodynamic, and boundary) and then demonstrates how this science is used in the design of machine elements.

SYMBOLS

A_p total projected pad area, m^2

a_b groove width ratio

a_f bearing-pad load coefficient

B total conformity of ball bearing

b semiminor axis of contact, m; width of pad, m

$\bar{b}$ length ratio, b_s/b_r

b_g length of feed groove region, m

b_r length of ridge region, m

b_s length of step region, m

C dynamic load capacity, N

C_l load coefficient, $F/p_a Rl$

c radial clearance of journal bearing, m

c' pivot circle clearance, m

c_b bearing clearance at pad minimum film thickness (Fig. 21.16), m

c_d orifice discharge coefficient

D distance between race curvature centers, m

$\tilde{D}$ material factor

D_x diameter of contact ellipse along x axis, m

D_y diameter of contact ellipse along y axis, m

d diameter of rolling element or diameter of journal, m

d_a overall diameter of ball bearing (Fig. 21.76), m

d_b bore diameter of ball bearing, m

d_c diameter of capillary tube, m

d_i inner-race diameter of ball bearing, m

d_o outer-race diameter of ball bearing, m

$\bar{d}_o$ diameter of orifice, m

E modulus of elasticity, N/m^2

E' effective elastic modulus, $2 \left(\dfrac{1 - \nu_a^2}{E_a} + \dfrac{1 - \nu_b^2}{E_b} \right)^{-1}$, N/m^2

$\tilde{E}$ metallurgical processing factor

$\mathscr{E}$ elliptic integral of second kind

e eccentricity of journal bearing, m

F applied normal load, N

F' load per unit length, N/m

$\tilde{F}$ lubrication factor

$\mathscr{F}$ elliptic integral of first kind

F_c pad load component along line of centers (Fig. 21.41), N

F_e rolling-element-bearing equivalent load, N

F_r applied radial load, N

F_s pad load component normal to line of centers (Fig. 21.41), N

F_t applied thrust load, N

f	race conformity ratio
f_c	coefficient dependent on materials and rolling-element bearing type (Table 21.19)
G	dimensionless materials parameter
$\tilde{G}$	speed effect factor
G_f	groove factor
g_e	dimensionless elasticity parameter, $W^{8/3}/U^2$
g_v	dimensionless viscosity parameter, GW^3/U^2
H	dimensionless film thickness, h/R_x
$\tilde{H}$	misalignment factor
H_a	dimensionless film thickness ratio, h_s/h_r
H_b	pad pumping power, N m/sec
H_c	power consumed in friction per pad, W
H_f	pad power coefficient
H_{min}	dimensionless minimum film thickness, h_{min}/R_x
$\hat{H}_{min}$	dimensionless minimum film thickness, $H_{min}(W/U)^2$
H_p	dimensionless pivot film thickness, h_p/c
H_t	dimensionless trailing-edge film thickness, h_t/c
h	film thickness, m
$\bar{h}_i$	film thickness ratio, h_i/h_o
h_i	inlet film thickness, m
h_l	leading-edge film thickness, m
h_{min}	minimum film thickness, m
h_o	outlet film thickness, m
h_p	film thickness at pivot, m
h_r	film thickness in ridge region, m
h_s	film thickness in step region, m
h_t	film thickness at trailing edge, m
h_0	film constant, m
J	number of stress cycles
K	load deflection constant
$\bar{K}$	dimensionless stiffness coefficient, cK_p/p_aRl
K_a	dimensionless stiffness, $-c\,\partial\bar{W}/\partial c$
K_p	film stiffness, N/m
K_1	load-deflection constant for a roller bearing
$K_{1.5}$	load-deflection constant for a ball bearing
$\bar{K}_\infty$	dimensionless stiffness, cK_p/p_aRl
k	ellipticity parameter, D_y/D_x
k_c	capillary tube constant, m³
k_o	orifice constant, $m^4/N^{1/2}$ sec
L	fatigue life
L_a	adjusted fatigue life
L_{10}	fatigue life where 90% of bearing population will endure
L_{50}	fatigue life where 50% of bearing population will endure
l	bearing length, m
l_c	length of capillary tube, m
l_r	roller effective length, m
l_t	roller length, m
l_v	length dimension in stress volume, m
l_1	total axial length of groove, m
M	probability of failure
$\bar{M}$	stability parameter, $\overline{m}p_a h_r^5/2R^5/\eta^2$
m	number of rows of rolling elements
$\bar{m}$	mass supported by bearing, N sec²/m

m_p preload factor

N rotational speed, rps

N_R Reynolds number

n number of rolling elements or number of pads or grooves

P dimensionless pressure, p/E'

P_d diametral clearance, m

P_e free endplay, m

p pressure, N/m^2

p_a ambient pressure, N/m^2

p_l lift pressure, N/m^2

p_{max} maximum pressure, N/m^2

p_r recess pressure, N/m^2

p_s bearing supply pressure, N/m^2

Q volume flow of lubricant, m^3/sec

$\overline{Q}$ dimensionless flow, $3\eta Q/\pi p_a h_r^3$

Q_c volume flow of lubricant in capillary, m^3/sec

Q_o volume flow of lubricant in orifice, m^3/sec

Q_s volume side flow of lubricant, m^3/sec

q constant, $\pi/2 - 1$

q_f bearing-pad flow coefficient

R curvature sum on shaft or bearing radius, m

$\overline{R}$ groove length fraction, $(R_o - R_g)/(R_o - R_i)$

R_g groove radius (Fig. 21.60), m

R_o orifice radius, m

R_x effective radius in x direction, m

R_y effective radius in y direction, m

R_1 outer radius of sector thrust bearing, m

R_2 inner radius of sector thrust bearing, m

r race curvature radius, m

r_c roller corner radius, m

S probability of survival

Sm Sommerfeld number for journal bearings, $\eta N d^3 l/2Fc^2$

Sm_t Sommerfeld number for thrust bearings, $\eta ubl^2/Fh_0^2$

s shoulder height, m

T tangential force, N

$\overline{T}$ dimensionless torque, $6\,T_r/\pi p_a/(R_1^2 + R_2^2)\,h_r\Lambda_c$

T_c critical temperature

T_r torque, N m

U dimensionless speed parameter, $u\eta_0/E'R_x$

u mean surface velocity in direction of motion, m/sec

v elementary volume, m^3

N dimensionless load parameter, $F/E'R_x^2$

$\overline{W}$ dimensionless load capacity, $F/p_a l(b_r + b_s + b_g)$

$\overline{W}_\infty$ dimensionless load, $1.5G_f F/\pi p_a(R_1^2 - R_2^2)$

X, Y factors for calculation of equivalent load

x,y,z coordinate system

$\overline{x}$ distance from inlet edge of pad to pivot, m

α radius ratio, R_y/R_x

α_a offset factor

α_b groove width ratio, $b_s/(b_r + b_s)$

α_p angular extent of pad, deg

α_r radius ratio, R_2/R_1

β contact angle, deg

β'	iterated value of contact angle, deg
β_a	groove angle, deg
β_f	free or initial contact angle, deg
β_p	angle between load direction and pivot, deg
Γ	curvature difference
γ	groove length ratio, l_1/l
Δ	rms surface finish, m
δ	total elastic deformation, m
ϵ	eccentricity ratio, e/c
η	absolute viscosity of lubricant, N sec/m^2
η_k	kinematic viscosity, ν/ρ, m^2/sec
η_0	viscosity at atmospheric pressure, N sec/m^2
θ	angle used to define shoulder height, deg
$\overline{\theta}$	dimensionless step location, $\theta_i/(\theta_i + \theta_o)$
θ_g	angular extent of lubrication feed groove, deg
θ_i	angular extent of ridge region, deg
θ_o	angular extent of step region, deg
Λ	film parameter (ratio of minimum film thickness to composite surface roughness)
Λ_c	dimensionless bearing number, $3\eta\omega(R_1^2 - R_2^2)/p_a h_r^2$
Λ_j	dimensionless bearing number, $6\eta\omega R^2/p_a c^2$
Λ_t	dimensionless bearing number, $6\eta u l/p_a h_r^2$
λ	length-to-width ratio
λ_a	length ratio, $(b_r + b_s + b_g)/l$
λ_b	$(1 + 2/3\alpha)^{-1}$
μ	coefficient of friction, T/F
ν	Poisson's ratio
ξ	pressure–viscosity coefficient of lubricant, m^2/N
ξ_p	angle between line of centers and pad leading edge, deg
ρ	lubricant density, N sec^2/m^4
ρ_0	density at atmospheric pressure, N sec^2/m^4
$\sigma_{\max}$	maximum Hertzian stress, N/m^2
τ	shear stress, N/m^2
τ_0	maximum shear stress, N/m^2
ϕ	attitude angle in journal bearings, deg
ϕ_p	angle between pad leading edge and pivot, deg
ψ	angular location, deg
ψ_t	angular limit of ψ, deg
ψ_s	step location parameter, $b_s/(b_r + b_s + b_g)$
ω	angular velocity, rad/sec
ω_B	angular velocity of rolling-element race contact, rad/sec
ω_b	angular velocity of rolling element about its own center, rad/sec
ω_c	angular velocity of rolling element about shaft center, rad/sec
ω_d	rotor whirl frequency, rad/sec
$\overline{\omega}_d$	whirl frequency ratio, ω_d/ω_j
ω_j	journal rotational speed, rad/sec

Sub-scripts

a	solid a
b	solid b
EHL	elastohydrodynamic lubrication
e	elastic
HL	hydrodynamic lubrication
i	inner

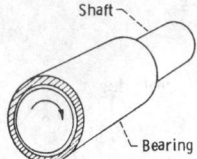

Fig. 21.1 Conformal surfaces. (From Ref. 6.)

iv	isoviscous
o	outer
pv	piezoviscous
r	rigid
x,y,z	coordinate system

21.1 LUBRICATION FUNDAMENTALS

A lubricant is any substance that is used to reduce friction and wear and to provide smooth running and a satisfactory life for machine elements. Most lubricants are liquids (like minerals oils, the synthetic esters and silicone fluids, and water), but they may be solids (such as polytetrafluorethylene) for use in dry bearings, or gases (such as air) for use in gas bearings. An understanding of the physical and chemical interactions between the lubricant and the tribological surfaces is necessary if the machine elements are to be provided with satisfactory life. To help in the understanding of this tribological behavior, the first section describes some lubrication fundamentals.

21.1.1 Conformal and Nonconformal Surfaces

Hydrodynamic lubrication is generally characterized by surfaces that are conformal; that is, the surfaces fit snugly into each other with a high degree of geometrical conformity (as shown in Fig. 21.1), so that the load is carried over a relatively large area. Furthermore, the load-carrying surface remains essentially constant while the load is increased. Fluid-film journal bearings (as shown in Fig. 21.1) and slider bearings exhibit conformal surfaces. In journal bearings the radial clearance between the shaft and bearing is typically one-thousandth of the shaft diameter; in slider bearings the inclination of the bearing surface to the runner is typically one part in a thousand. These converging surfaces, coupled with the fact that there is relative motion and a viscous fluid separating the surfaces, enable a positive pressure to be developed and exhibit a capacity to support a normal applied load. The magnitude of the pressure developed *is not* generally large enough to cause significant elastic deformation of the surfaces. The minimum film thickness in a hydrodynamically lubricated bearing is a function of applied load, speed, lubricant viscosity, and geometry. The relationship between the minimum film thickness h_{min} and the speed u and applied normal load F is given as

$$(h_{min})_{HL} \propto \left(\frac{u}{F}\right)^{1/2} \tag{21.1}$$

More coverage of hydrodynamic lubrication can be found in Section 21.2.

Many machine elements have contacting surfaces that *do not* conform to each other very well, as shown in Fig. 21.2 for a rolling-element bearing. The full burden of the load must then be carried by a very small contact area. In general, the contact areas between nonconformal surfaces enlarge considerably with increasing load, but they are still smaller than the contact areas between conformal

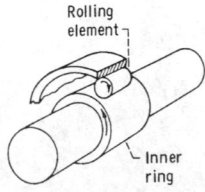

Fig. 21.2 Nonconformal surfaces. (From Ref. 6.)

surfaces. Some examples of nonconformal surfaces are mating gear teeth, cams and followers, and rolling-element bearings (as shown in Fig. 21.2). The mode of lubrication normally found in these nonconformal contacts is elastohydrodynamic lubrication. The requirements necessary for hydrodynamic lubrication (converging surfaces, relative motion, and viscous fluid) are also required for elastohydrodynamic lubrication.

The relationship between the minimum film thickness and normal applied load and speed for an elastohydrodynamically lubricated contact is

$$(h_{min})_{EHL} \propto F^{-0.073} \tag{21.2}$$

$$(h_{min})_{EHL} \propto u^{0.68} \tag{21.3}$$

Comparing the results of Eqs. (21.2) and (21.3) with that obtained for hydrodynamic lubrication expressed in Eq. (21.1) indicates that:

1. The exponent on the normal applied load is nearly seven times larger for hydrodynamic lubrication than for elastohydrodynamic lubrication. This implies that in elastohydrodynamic lubrication the film thickness is only slightly affected by load while in hydrodynamic lubrication it is significantly affected by load.
2. The exponent on mean velocity is slightly higher for elastohydrodynamic lubrication than that found for hydrodynamic lubrication.

More discussion of elastohydrodynamic lubrication can be found in Section 21.3.

The load per unit area in conformal bearings is relatively low, typically averaging only 1 MN/m^2 and seldom over 7 MN/m^2. By contrast, the load per unit area in nonconformal contacts will generally exceed 700 MN/m^2 even at modest applied loads. These high pressures result in elastic deformation of the bearing materials such that elliptical contact areas are formed for oil-film generation and load support. The significance of the high contact pressures is that they result in a considerable increase in fluid viscosity. Inasmuch as viscosity is a measure of a fluid's resistance to flow, this increase greatly enhances the lubricant's ability to support load without being squeezed out of the contact zone. The high contact pressures in nonconforming surfaces therefore result in both an elastic deformation of the surfaces and large increases in the fluid's viscosity. The minimum film thickness is a function of the parameters found for hydrodynamic lubrication with the addition of an effective modulus of elasticity parameter for the bearing materials and a pressure–viscosity coefficient for the lubricant.

21.1.2 Bearing Selection

Ball bearings are used in many kinds of machines and devices with rotating parts. The designer is often confronted with decisions on whether a nonconformal bearing such as a rolling-element bearing or a conformal bearing such as a hydrodynamic bearing should be used in a particular application. The following characteristics make rolling-element bearings *more desirable* than hydrodynamic bearings in many situations:

1. Low starting and good operating friction
2. The ability to support combined radial and thrust loads
3. Less sensitivity to interruptions in lubrication
4. No self-excited instabilities
5. Good low-temperature starting

Within reasonable limits changes in load, speed, and operating temperature have but little effect on the satisfactory performance of rolling-element bearings.

The following characteristics make nonconformal bearings such as rolling-element bearings *less desirable* than conformal (hydrodynamic) bearings:

1. Finite fatigue life subject to wide fluctuations
2. Large space required in the radial direction
3. Low damping capacity
4. High noise level
5. More severe alignment requirements
6. Higher cost

Each type of bearing has its particular strong points, and care should be taken in choosing the most appropriate type of bearing for a given application.

The Engineering Services Data Unit documents[7,8] provide an excellent guide to the selection of the type of journal or thrust bearing most likely to give the required performance when considering the load, speed, and geometry of the bearing. The following types of bearings were considered:

1. **Rubbing bearings**, where the two bearing surfaces rub together (e.g., unlubricated bushings made from materials based on nylon, polytetrafluoroethylene, also known as PTFE, and carbon).

2. **Oil-impregnated porous metal bearings**, where a porous metal bushing is impregnated with lubricant and thus gives a self-lubricating effect (as in sintered-iron and sintered-bronze bearings).

3. **Rolling-element bearings**, where relative motion is facilitated by interposing rolling elements between stationary and moving components (as in ball, roller, and needle bearings).

4. **Hydrodynamic film bearings**, where the surfaces in relative motion are kept apart by pressures generated hydrodynamically in the lubricant film.

Figure 21.3, reproduced from the Engineering Sciences Data Unit publication,[7] gives a guide to the typical load that can be carried at various speeds, for a nominal life of 10,000 hr at room temperature, by journal bearings of various types on shafts of the diameters quoted. The heavy curves

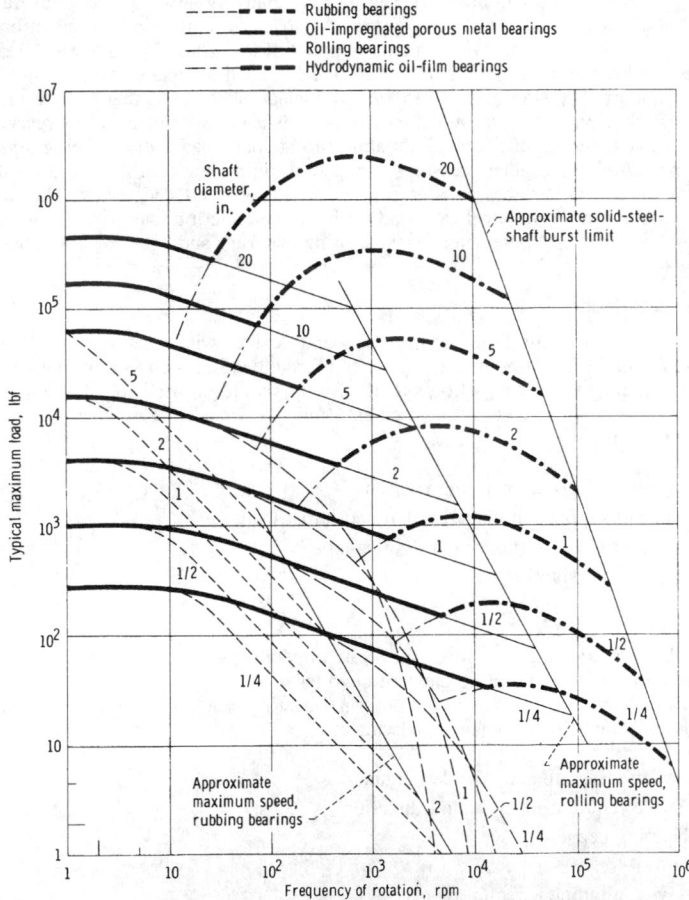

Fig. 21.3 General guide to journal bearing type. (Except for roller bearings, curves are drawn for bearings with width equal to diameter. A medium-viscosity mineral oil lubricant is assumed for hydrodynamic bearings.) (From Ref. 7.)

indicate the preferred type of journal bearing for a particular load, speed, and diameter and thus divide the graph into distinct regions. From Fig. 21.3 it is observed that rolling-element bearings are preferred at lower speeds and hydrodynamic oil film bearings are preferred at higher speeds. Rubbing bearings and oil-impregnated porous metal bearings are not preferred for any of the speeds, loads, or shaft diameters considered. Also, as the shaft diameter is increased, the transitional point at which hydrodynamic bearings are preferred over rolling-element bearings moves to the left.

The applied load and speed are usually known, and this enables a preliminary assessment to be made of the type of journal bearing most likely to be suitable for a particular application. In many cases the shaft diameter will have been determined by other considerations, and Fig. 21.3 can be used to find the type of journal bearing that will give adequate load capacity at the required speed. These curves are based upon good engineering practice and commercially available parts. Higher loads and speeds or smaller shaft diameters are possible with exceptionally high engineering standards or specially produced materials. Except for rolling-element bearings the curves are drawn for bearings with a width equal to the diameter. A medium-viscosity mineral oil lubricant is assumed for the hydrodynamic bearings.

Similarly, Fig. 21.4, reproduced from the Engineering Sciences Data Unit publication,[8] gives a guide to the typical maximum load that can be carried at various speeds for a nominal life of 10,000 hr at room temperature by thrust bearings of the various diameters quoted. The heavy curves again indicate the preferred type of bearing for a particular load, speed, and diameter and thus divide the graph into major regions. As with the journal bearing results (Fig. 21.3) at the hydrodynamic bearing is preferred at lower speeds. A difference between Figs. 21.3 and 21.4 is that at very low speeds

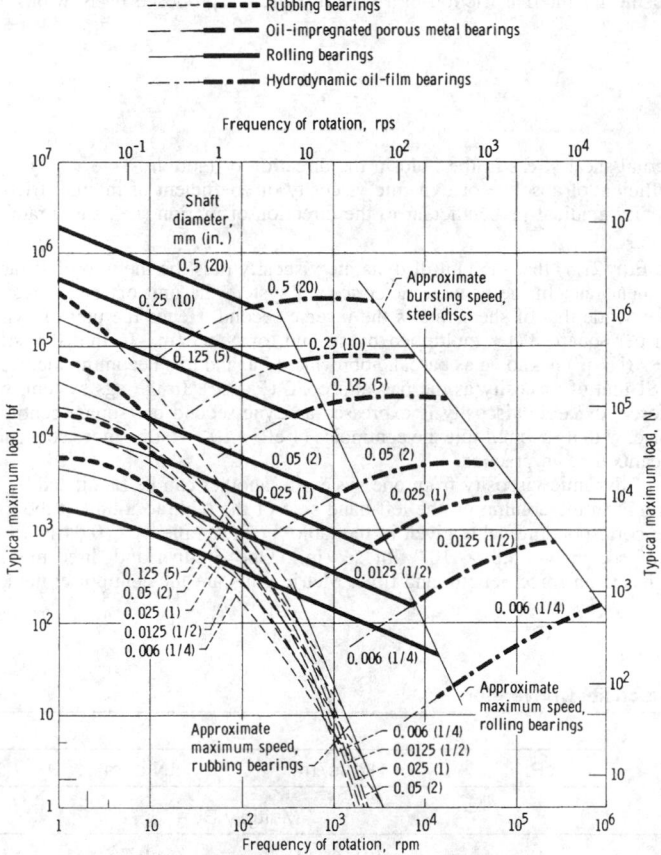

Fig. 21.4 General guide to thrust bearing type. (Except for roller bearings, curves are drawn for typical ratios of inside to outside diameter. A medium-viscosity mineral oil lubricant is assumed for hydrodynamic bearings.) (From Ref. 8.)

there is a portion of the latter figure in which the rubbing bearing is preferred. Also, as the shaft diameter is increased, the transitional point at which hydrodynamic bearings are preferred over rolling-element bearings moves to the left. Note also from this figure that oil-impregnated porous metal bearings are not preferred for any of the speeds, loads, or shaft diameters considered.

21.1.3 Lubricants

Both oils and greases are extensively used as lubricants for all types of machine elements over wide range of speeds, pressures, and operating temperatures. Frequently, the choice is determined by considerations other than lubrication requirements. The requirements of the lubricant for successful operation of nonconformal contacts such as in rolling-element bearings and gears are considerably more stringent than those for conformal bearings and therefore will be the primary concern in this section.

Because of its fluidity oil has several advantages over grease: It can enter the loaded conjunction most readily to flush away contaminants, such as water and dirt, and, particularly, to transfer heat from heavily loaded machine elements. Grease, however, is extensively used because it permits simplified designs of housings and enclosures, which require less maintenance, and because it is more effective in sealing against dirt and contaminants.

Viscosity

In hydrodynamic and elastohydrodynamic lubrication the most important physical property of a lubricant is its viscosity. The viscosity of a fluid may be associated with its resistance to flow, that is, with the resistance arising from intermolecular forces and internal friction as the molecules move past each other. Thick fluids, like molasses, have relatively high viscosity; they do not flow easily. Thinner fluids, like water, have lower viscosity; they flow very easily.

The relationship for internal friction in a viscous fluid (as proposed by Newton)[9] can be written as

$$\tau = \eta \frac{du}{dz} \tag{21.4}$$

where τ = internal shear stress in the fluid in the direction of motion
  η = coefficient of absolute or dynamic viscosity or coefficient of internal friction
  du/dz = velocity gradient perpendicular to the direction of motion (i.e., shear rate)

It follows from Eq. (21.4) that the unit of dynamic viscosity must be the unit of shear stress divided by the unit of shear rate. In the newton-meter-second system the unit of shear stress is the newton per square meter while that of shear rate is the inverse second. Hence the unit of dynamic viscosity will be newton per square meter multiplied by second, or N sec/m². In the SI system the unit of pressure or stress (N/m²) is known as pascal, abbreviated Pa, and it is becoming increasingly common to refer to the SI unit of viscosity as the pascal-second (Pa sec). In the cgs system, where the dyne is the unit of force, dynamic viscosity is expressed as dyne-second per square centimeter. This unit is called the poise, with its submultiple the centipoise ($1 \text{ cP} = 10^{-2} \text{ P}$) of a more convenient magnitude for many lubricants used in practice.

Conversion of dynamic viscosity from one system to another can be facilitated by Table 21.1. To convert from a unit in the column on the left-hand side of the table to a unit at the top of the table, multiply by the corresponding value given in the table. For example, $\eta = 0.04$ N sec/m² = $0.04 \times 1.45 \times 10^{-4}$ lbf sec/in.² = 5.8×10^{-6} lbf sec/in.². One English and three metric systems are presented—all based on force, length, and time. Metric units are the centipoise, the kilogram force-

Table 21.1 Viscosity Conversion

	To—			
	cP	kgf s/m²	N s/m²	lbf s/in²
To Convert From—	Multiply By—			
cP	1	1.02×10^{-4}	10^{-3}	1.45×10^{-7}
kgf s/m²	9.807×10^3	1	9.807	1.422×10^{-3}
N s/m²	10^3	1.02×10^{-1}	1	1.45×10^{-4}
lbf s/in²	6.9×10^6	7.034×10^2	6.9×10^3	1

second per square meter, and the newton-second per square meter (or Pa sec). The English unit is pound force-second per square inch, or reyn, in honor of Osborne Reynolds.

In many situations it is convenient to use the *kinematic viscosity* rather than the dynamic viscosity. The kinematic viscosity η_k is equal to the dynamic viscosity η divided by the density ρ of the fluid ($\eta_k = \eta/\rho$). The ratio is literally kinematic, all trace of force or mass cancelling out. The unit of kinematic viscosity may be written in SI units as square meters per second or in English units as square inches per second or, in cgs units, as square centimeters per second. The name stoke, in honor of Sir George Gabriel Stokes, was proposed for the cgs unit by Max Jakob in 1928. The centistoke, or one-hundredth part, is an everyday unit of more convenient size, corresponding to the centipoise.

The viscosity of a given lubricant varies within a given machine element as a result of the nonuniformity of pressure or temperature prevailing in the lubricant film. Indeed, many lubricated machine elements operate over ranges of pressure or temperature so extensive that the consequent variations in the viscosity of the lubricant may become substantial and, in turn, may dominate the operating characteristics of machine elements. Consequently, an adequate knowledge of the viscosity–pressure and viscosity–pressure–temperature relationships of lubricants is indispensable.

Oil Lubrication

Except for a few special requirements, petroleum oils satisfy most operating conditions in machine elements. High-quality products, free from adulterants that can have an abrasive or lapping action, are recommended. Animal or vegetable oils or petroleum oils of poor quality tend to oxidize, to develop acids, and to form sludge or resinlike deposits on the bearing surfaces. They thus penalize bearing performance or endurance.

A composite of recommended lubricant kinematic viscosities at 38°C (100°F) is shown in Fig. 21.5. The ordinate of this figure is the speed factor, which is bearing bore size measured in millimeters multiplied by the speed in revolutions per minute. In many rolling-element-bearing applications an

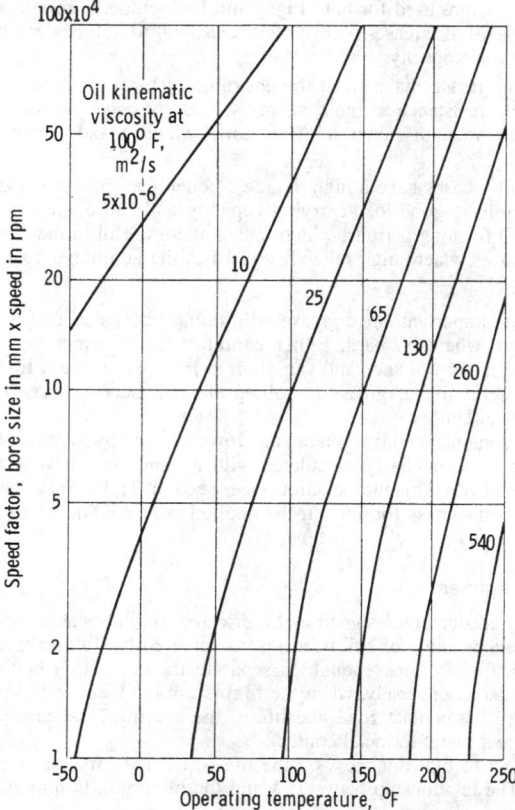

Fig. 21.5 Recommended lubricant viscosities for ball bearings. (From Ref. 10.)

oil equivalent to an SAE-10 motor oil [4×10^{-6} m²/sec, or 40 cS, at 38°C (100°F)] or a light turbine oil is the most frequent choice.

For a number of military applications where the operational requirements span the temperature range −54 to 204°C (−65 to 400°F), synthetic oils are used. Ester lubricants are most frequently employed in this temperature range. In applications where temperatures exceed 260°C (500°F), most synthetics will quickly break down, and either a solid lubricant (e.g., MoS_2) or a polyphenyl ether is recommended. A more detailed discussion of synthetic lubricants can be found in Bisson and Anderson.[11]

Grease Lubrication

The simplest method of lubricating a bearing is to apply grease, because of its relatively nonfluid characteristics. The danger of leakage is reduced, and the housing and enclosure can be simpler and less costly than those used with oil. Grease can be packed into bearings and retained with inexpensive enclosures, but packing should not be excessive and the manufacturer's recommendations should be closely adhered to.

The major limitation of grease lubrication is that it is not particularly useful in high-speed applications. In general, it is not employed for speed factors over 200,000, although selected greases have been used successfully for higher speed factors with special designs.

Greases vary widely in properties depending on the type and grade or consistency. For this reason few specific recommendations can be made. Greases used for most bearing operating conditions consist of petroleum, diester, polyester, or silicone oils thickened with sodium or lithium soaps or with more recently developed nonsoap thickeners. General characteristics of greases are as follows:

1. Petroleum oil greases are best for general-purpose operation from −34 to 149°C (−30 to 300°F).

2. Diester oil greases are designed for low-temperature service down to −54°C (−65°F).

3. Ester-based greases are similar to diester oil greases but have better high-temperature characteristics, covering the range from −73 to 177°C (−100 to 350°F).

4. Silicone oil greases are used for both high- and low-temperature operation, over the widest temperature range of all greases [−73 to 232°C (−100 to 450°F)], but have the disadvantage of low load-carrying capacity.

5. Fluorosilicone oil greases have all of the desirable features of silicone oil greases plus good load capacity and resistance to fuels, solvents, and corrosive substances. They have a very low volatility in vacuum down to 10^{-7} torr, which makes them useful in aerospace applications.

6. Perfluorinated oil greases have a high degree of chemical inertness and are completely non-flammable. They have good load-carrying capacity and can operate at temperatures as high as 280°C (550°F) for long periods, which makes them useful in the chemical processing and aerospace industries, where high reliability justifies the additional cost.

Grease consistency is important since grease will slump badly and churn excessively when too soft and fail to lubricate when too hard. Either condition causes improper lubrication, excessive temperature rise, and poor performance and can shorten machine element life. A valuable guide to the estimation of the useful life of grease in rolling-element bearings has been published by the Engineering Sciences Data Unit.[12]

It has recently been demonstrated by Aihara and Dowson[13] and by Wilson[14] that the film thickness in grease-lubricated components can be calculated with adquate accuracy by using the viscosity of the base oil in the elastohydrodynamic equation (see Section 21.3). This enables the elastohydrodynamic lubrication film thickness formulas to be applied with confidence to grease-lubricated machine elements.

21.1.4 Lubrication Regimes

If a machine element is adequately designed and lubricated, the lubricated surfaces are separated by a lubricant film. Endurance testing of ball bearings, as reported by Tallian et al.,[15] has demonstrated that when the lubricant film is thick enough to separate the contacting bodies, fatigue life of the bearing is greatly extended. Conversely, when the film is not thick enough to provide full separation between the asperities in the contact zone, the life of the bearing is adversely affected by the high shear resulting from direct metal-to-metal contact.

To establish the effect of film thickness on the life of the machine element, we first introduce a relevant parameter Λ. The relationship between Λ and the minimum film thickness $h_{\min}$ is defined to be

$$\Lambda = \frac{h_{\min}}{(\Delta_a^2 + \Delta_b^2)^{1/2}} \tag{21.5}$$

where Δ_a = rms surface finish of surface a
$\quad\quad \Delta_b$ = rms surface finish of surface b

Hence Λ is just the minimum film thickness in units of the composite roughness of the two bearing surfaces.

Hydrodynamic Lubrication Regime

Hydrodynamic lubrication occurs when the lubricant film is sufficiently thick to prevent the opposite solids from coming into contact. This condition is often referred to as the ideal form of lubrication since it provides low friction and a high resistance to wear. The lubrication of the contact is governed by the bulk physical properties of the lubricant, notably viscosity, and the frictional characteristics arise purely from the shearing of the viscous lubricant. The pressure developed in the oil film of hydrodynamically lubricated bearings is due to two factors:

1. The geometry of the moving surfaces produces a convergent film shape.
2. The viscosity of the liquid results in a resistance to flow.

The lubricant films are normally many times thicker than the surface roughness so that the physical properties of the lubricant dictate contact behavior. The film thickness normally exceeds 10^{-6} m. For hydrodynamic lubrication the film parameter Λ, defined in Eq. (21.5), is an excess of 10 and may even rise to 100. Films of this thickness are clearly also insensitive to chemical action in surface layers of molecular proportions.

For normal load support to occur in bearings, positive pressure profiles must develop over the length of the bearing. Three different forms of hydrodynamic lubrication are presented in Fig. 21.6.

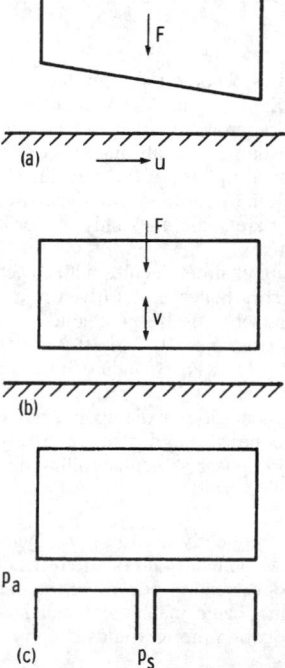

Fig. 21.6 Mechanisms of load support for hydrodynamic lubrication. (*a*) Slider bearing. (*b*) Squeeze film bearing. (*c*) Externally pressurized bearing.

Figure 21.6a shows a slider bearing. For a positive load to be developed in the slider bearing shown in Fig. 21.6a the lubricant film thickness must be decreasing in the direction of sliding.

A squeeze film bearing is another mechanism of load support of hydrodynamic lubrication, and it is illustrated in Fig. 21.6b. The squeeze action is the normal approach of the bearing surfaces. The squeeze mechanism of pressure generation provides a valuable cushioning effect when the bearing surfaces tend to be pressed together. Positive pressures will be generated when the film thickness is diminishing.

An externally pressurized bearing is yet a third mechanism of load support of hydrodynamic lubrication, and it is illustrated in Fig. 21.6c. The pressure drop across the bearing is used to support the load. The load capacity is independent of the motion of the bearing and the viscosity of the lubricant. There is no problem of contact at starting and stopping as with the other hydrodynamically lubricated bearings because pressure is applied before starting and is maintained until after stopping.

Hydrodynamically lubricated bearings are discussed further in Section 21.2.

Elastohydrodynamic Lubrication Regime

Elastohydrodynamic lubrication is a form of hydrodynamic lubrication where elastic deformation of the bearing surfaces becomes significant. It is usually associated with highly stressed machine components of low conformity. There are two distinct forms of elastohydrodynamic lubrication (EHL).

Hard EHL.   Hard EHL relates to materials of *high* elastic modulus, such as metals. In this form of lubrication both the elastic deformation and the pressure–viscosity effects are equally important. Engineering applications in which elastohydrodynamic lubrication are important for high-elastic-modulus materials include gears and rolling-element bearings.

Soft EHL.   Soft EHL relates to materials of *low* elastic modulus, such as rubber. For these materials the elastic distortions are large, even with light loads. Another feature of the elastohydrodynamics of low-elastic-modulus materials is the negligible effect of the relatively low pressures on the viscosity of the lubricating fluid. Engineering applications in which elastohydrodynamic lubrication are important for low-elastic-modulus materials include seals, human joints, tires, and a number of lubricated elastomeric material machine elements.

The common factors in hard and soft EHL are that the local elastic deformation of the solids provides coherent fluid films and that asperity interaction is largely prevented. Elastohydrodynamic lubrication normally occurs in contacts where the minimum film thickness is in the range 0.1 μm < h_{min} ≤ 10 μm and the film parameter Λ is in the range 3 ≤ Λ < 10. Elastohydrodynamic lubrication is discussed further in Section 21.3.

Boundary Lubrication Regime

If in a lubricated contact the pressures become too high, the running speeds too low, or the surface roughness too great, penetration of the lubricant film will occur. Contact will take place between the asperities. The friction will rise and approach that encountered in dry friction between solids. More importantly, wear will take place. Adding a small quantity of certain active organic compounds to the lubricating oil can, however, extend the life of the machine elements. These additives are present in small quantities (< 1%) and function by forming low-shear-strength surface films strongly attached to the metal surfaces. Although they are sometimes only one or two molecules thick, such films are able to prevent metal-to-metal contact.

Some boundary lubricants are long-chain molecules with an active end group, typically an alcohol, an amine, or a fatty acid. When such a material, dissolved in a mineral oil, meets a metal or other solid surface, the active end group attaches itself to the solid and gradually builds up a surface layer. The surface films vary in thickness from 5×10^{-9} to 10^{-8} m depending on molecular size, and the film parameter Λ is less than unity ($\Lambda < 1$). Boundary lubrication is discussed further in Section 21.4.

Figure 21.7 illustrates the film conditions existing in hydrodynamic, elastohydrodynamic, and boundary lubrication. The surface slopes in this figure are greatly distorted for the purpose of illustration. To scale, real surfaces would appear as gently rolling hills rather than sharp peaks.

21.1.5   Relevant Equations

This section presents the equations frequently used in hydrodynamic and elastohydrodynamic lubrication theory. They are not relevant to boundary lubrication since in this lubrication regime bulk fluid effects are negligible. The differential equation governing the pressure distribution in hydrodynamically and elastohydrodynamically lubricated machine elements is known as the Reynolds equation. For steady-state hydrodynamic lubrication the Reynolds equation normally appears as

$$\frac{\partial}{\partial x}\left(h^3 \frac{\partial p}{\partial x}\right) + \frac{\partial}{\partial y}\left(h^3 \frac{\partial p}{\partial y}\right) = 12\eta u \frac{\partial h}{\partial x} \tag{21.6}$$

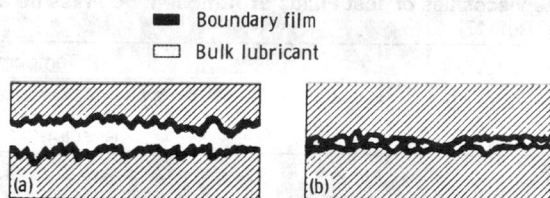

■ Boundary film

▭ Bulk lubricant

Fig. 21.7 Film conditions of lubrication regimes, (a) Hydrodynamic and elastohydrodynamic lubrication—surfaces separated by bulk lubricant film. (b) Boundary lubrication—performance essentially dependent on boundary film.

where h = film shape measured in the z direction, m
  p = pressure, N/m^2
  η = lubricant viscosity, N sec/m^2
  u = mean velocity, $(u_a + u_b)/2$, m/sec

Solutions of Eq. (21.6) are rarely achieved analytically, and approximate numerical solutions are sought.

For elastohydrodynamic lubrication the steady-state form of the Reynolds equation normally appears as

$$\frac{\partial}{\partial x}\left(\frac{\rho h^3}{\eta}\frac{\partial p}{\partial x}\right) + \frac{\partial}{\partial y}\left(\frac{\rho h^3}{\eta}\frac{\partial p}{\partial y}\right) = 12u\frac{\partial(\rho h)}{\partial x} \qquad (21.7)$$

where ρ is lubricant density in N sec^2/m^2. The essential difference between Eqs. (21.6) and (21.7) is that Eq. (21.7) allows for variation of viscosity and density in the x and y directions. Equations (21.6) and (21.7) allow for the bearing surfaces to be of finite length in the y direction. Side leakage, or flow in the y direction, is associated with the second term in Eqs. (21.6) and (21.7). The solution of Eq. (21.7) is considerably more difficult than that of Eq. (21.6); therefore, only numerical solutions are available.

The viscosity of a fluid may be associated with the resistance to flow, with the resistance arising from the intermolecular forces and internal friction as the molecules move past each other. Because of the much larger pressure variation in the lubricant conjunction, the viscosity of the lubricant for elastohydrodynamic lubrication does not remain constant as is approximately true for hydrodynamic lubrication.

As long ago as 1893, Barus[16] proposed the following formula for the isothermal viscosity–pressure dependence of liquids:

$$\eta = \eta_0 e^{\xi p} \qquad (21.8)$$

where η_0 = viscosity at atmospheric pressure
  ξ = pressure–viscosity coefficient of lubricant

The pressure–viscosity coefficient ξ characterizes the liquid considered and depends in most cases only on temperature, not on pressure.

Table 21.2 lists the absolute viscosities of 12 lubricants at atmospheric pressure and three temperatures as obtained from Jones et al.[17] These values would correspond to η_0 to be used in Eq. (21.8) for the particular fluid and temperature to be used. The 12 fluids with manufacturer and manufacturer's designation are shown in Table 21.3. The pressure–viscosity coefficients ξ, expressed in square meters per newton, for these 12 fluids at three different temperatures are shown in Table 21.4.

For a comparable change in pressure the relative density change is smaller than the viscosity change. However, very high pressures exist in elastohydrodynamic films, and the liquid can no longer be considered as an incompressible medium. From Dowson and Higginson[18] the density can be written as

$$\rho = \rho_0\left(1 + \frac{0.6p}{1 + 1.7p}\right) \qquad (21.9)$$

where p is given in gigapascals.

Table 21.2   Absolute Viscosities of Test Fluids at Atmospheric Pressure and Three Temperatures (From Ref. 17)

	Temperature, °C		
	38	99	149
Test Fluid	Absolute Viscosity, η, cP		
Advanced ester	25.3	4.75	2.06
Formulated advanced ester	27.6	4.96	2.15
Polyalkyl aromatic	25.5	4.08	1.80
Polyalkyl aromatic + 10 wt % heavy resin	32.2	4.97	2.03
Synthetic paraffinic oil (lot 3)	414	34.3	10.9
Synthetic paraffinic oil (lot 4)	375	34.7	10.1
Synthetic paraffinic oil (lot 4) + antiwear additive	375	34.7	10.1
Synthetic paraffinic oil (lot 2) + antiwear additive	370	32.0	9.93
C-ether	29.5	4.67	2.20
Superrefined naphthenic mineral oil	68.1	6.86	2.74
Synthetic hydrocarbon (traction fluid)	34.3	3.53	1.62
Fluorinated polyether	181	20.2	6.68

The film shape appearing in Eq. (21.7) can be written with sufficient accuracy as

$$h = h_0 + \frac{x^2}{2R_x} + \frac{y^2}{2R_y} + \delta(x,y) \qquad (21.10)$$

where h_0 = constant, m
$\delta(x,y)$ = total elastic deformation, m
R_x = effective radius in x direction, m
R_y = effective radius in y direction, m

The elastic deformation can be written, from standard elasticity theory, in the form

Table 21.3   Fluids with Manufacturer and Manufacturer's Designation (From Ref. 17)

Test Fluid	Manufacturer	Designation
Advanced ester	Shell Oil Co.	Aeroshell® turbine oil 555 (base oil)
Formulated advanced ester	Shell Oil Co.	Aeroshell® turbine oil 555 (WRGL-358)
Polyalkyl aromatic	Continental Oil Co.	DN-600
Synthetic paraffinic oil (lot 3)	Mobil Oil Corp.	XRM 109F3
Synthetic paraffinic oil (lot 4)		XRM 109F4
Synthetic paraffinic oil + antiwear additive (lot 2)		XRM 177F2
Synthetic paraffinic oil + antiwear additive (lot 4)		XRM 177F4
C-ether	Monsanto Co.	MCS-418
Superrefined naphthenic mineral oil	Humble Oil and Refining Co.	FN 2961
Synthetic hydrocarbon (traction fluid)	Monsanto Co.	MCS-460
Fluorinated polyether	DuPont Co.	PR 143 AB (Lot 10)

Table 21.4 Pressure–Viscosity Coefficients for Test Fluids at Three Temperatures (From Ref. 17)

Test Fluid	Temperature, °C		
	38	99	149
	Pressure-viscosity Coefficient, ξ, m²/N		
Advanced ester	1.28×10^{-8}	0.987×10^{-8}	0.851×10^{-8}
Formulated advanced ester	1.37	1.00	.874
Polyalkyl aromatic	1.58	1.25	1.01
Polyalkyl aromatic + 10 wt % heavy resin	1.70	1.28	1.06
Synthetic paraffinic oil (lot 3)	1.77	1.51	1.09
Synthetic paraffinic oil (lot 4)	1.99	1.51	1.29
Synthetic paraffinic oil (lot 4) + antiwear additive	1.96	1.55	1.25
Synthetic paraffinic oil (lot 2) + antiwear additive	1.81	1.37	1.13
C-ether	1.80	.980	.795
Superrefined naphthenic mineral oil	2.51	1.54	1.27
Synthetic hydrocarbon (traction fluid)	3.12	1.71	.939
Fluorinated polyether	4.17	3.24	3.02

$$\delta(x,y) = \frac{2}{\pi E'} \iint_\Lambda \frac{p(x,y)\, dx_1 dy_1}{[(x - x_1)^2 + (y - y_1)^2]^{1/2}} \tag{21.11}$$

where

$$E' = 2 \left(\frac{1 - \nu_a^2}{E_a} + \frac{1 - \nu_b^2}{E_b} \right)^{-1} \tag{21.12}$$

and ν = Poisson's ratio
E = modulus of elasticity, N/m²

Therefore, Eq. (21.6) is normally involved in hydrodynamic lubrication situations, while Eqs. (21.7)–(21.11) are normally involved in elastohydrodynamic lubrication situations.

21.2 HYDRODYNAMIC AND HYDROSTATIC LUBRICATION

Surfaces lubricated hydrodynamically are normally conformal as pointed out in Section 21.1.1. The conformal nature of the surfaces can take its form either as a thrust bearing or as a journal bearing, both of which will be considered in this section. Three features must exist for hydrodynamic lubrication to occur:

1. A viscous fluid must separate the lubricated surfaces.
2. There must be relative motion between the surfaces.
3. The geometry of the film shape must be larger in the inlet than at the outlet so that a convergent wedge of lubricant is formed.

If feature 2 is absent, lubrication can still be achieved by establishing relative motion between the fluid and the surfaces through external pressurization. This is discussed further in Section 21.2.3.

In hydrodynamic lubrication the entire friction arises from the shearing of the lubricant film so that it is determined by the viscosity of the oil: the thinner (or less viscous) the oil, the lower the friction. The great advantages of hydrodynamic lubrication are that the friction can be very low ($\mu \simeq 0.001$) and, in the ideal case, there is no wear of the moving parts. The main problems in hydrodynamic lubrication are associated with starting or stopping since the oil film thickness theoretically is zero when the speed is zero.

The emphasis in this section is on hydrodynamic and hydrostatic lubrication. This section is not intended to be all inclusive but rather to typify the situations existing in hydrodynamic and hydrostatic lubrication. For additional information the reader is recommended to investigate Gross et al.,[19] Reiger,[20] Pinkus and Sternlicht,[21] and Rippel.[22]

21.2.1 Liquid-Lubricated Hydrodynamic Journal Bearings

Journal bearings, as shown in Fig. 21.8, are used to support shafts and to carry radial loads with minimum power loss and minimum wear. The bearing can be represented by a plain cylindrical bush wrapped around the shaft, but practical bearings can adopt a variety of forms. The lubricant is supplied at some convenient point through a hole or a groove. If the bearing extends around the full 360° of the shaft, the bearing is described as a full journal bearing. If the angle of wrap is less than 360°, the term "partial journal bearing" is employed.

Plain

Journal bearings rely on the motion of the shaft to generate the load-supporting pressures in the lubricant film. The shaft does not normally run concentric with the bearing center. The distance between the shaft center and the bearing center is known as the eccentricity. This eccentric position within the bearing clearance is influenced by the load that it carries. The amount of eccentricity adjusts itself until the load is balanced by the pressure generated in the converging portion of the bearing. The pressure generated, and therefore the load capacity of the bearing, depends on the shaft eccentricity e, the frequency of rotation N, and the effective viscosity of the lubricant η in the converging film, as well as the bearing dimensions l and d and the clearance c. The three dimensionless groupings normally used for journal bearings are:

1. The eccentricity ratio, $\epsilon = e/c$
2. The length-to-diameter ratio, $\lambda = l/d$
3. The Sommerfeld number, $Sm = \eta N d^3 l / 2Fc^2$

When designing a journal bearing, the first requirement to be met is that it should operate with an adequate minimum film thickness, which is directly related to the eccentricity ($h_{min} = c - e$). Figures 21.9, 21.10, and 21.11 show the eccentricity ratio, the dimensionless minimum film thickness, and the dimensionless Sommerfeld number for, respectively, a full journal bearing and partial journal bearings of 180° and 120°. In these figures a recommended operating eccentricity ratio is indicated as well as a preferred operational area. The left boundary of the shaded zone defines the optimum eccentricity ratio for minimum coefficient of friction, and the right boundary is the optimum eccentricity ratio for maximum load. In these figures it can be observed that the shaded area is significantly reduced for the partial bearings as compared with the full journal bearing. These plots were adapted from results given in Raimondi and Boyd.[23]

Figures 21.12, 21.13, and 21.14 show a plot of attitude angle ϕ (angle between the direction of the load and a line drawn through the centers of the bearing and the journal) and the bearing characteristic number for various length-to-diameter ratios for, respectively, a full journal bearing and partial journal bearings of 180° and 120°. This angle establishes where the minimum and maximum film thicknesses are located within the bearing. These plots were also adapted from results given in Raimondi and Boyd,[23] where additional information about the coefficient of friction, the flow variable, the temperature rise, and the maximum film pressure ratio for a complete range of length-to-diameter ratios as well as for full or partial journal bearings can be found.

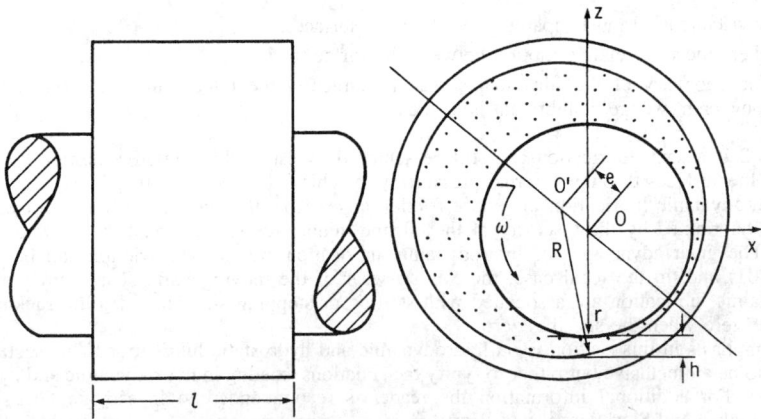

Fig. 21.8　Journal bearing.

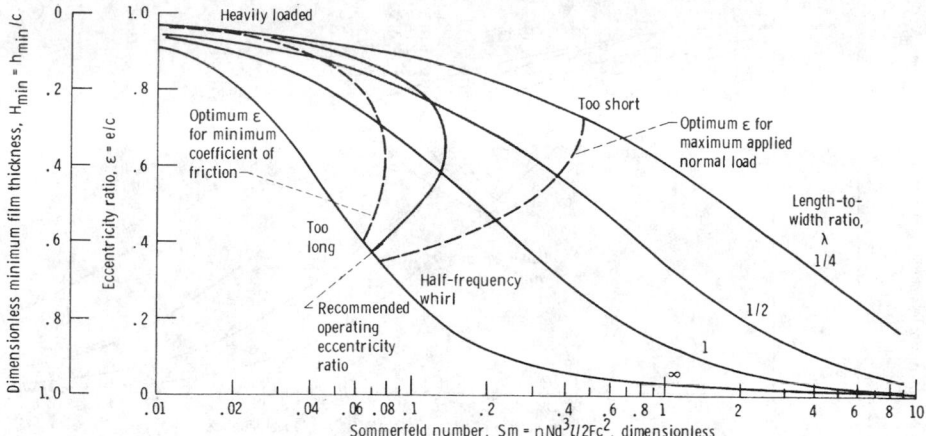

Fig. 21.9 Design figure showing eccentricity ratio, dimensionless minimum film thickness, and Sommerfeld number for full journal bearings. (Adapted from Ref. 23.)

Nonplain

As applications have demanded higher speeds, vibration problems due to critical speeds, imbalance, and instability have created a need for journal bearing geometries other than plain journal bearings. These geometries have various patterns of variable clearance so as to create pad film thicknesses that have more strongly converging and diverging regions. Figure 21.15 shows elliptical, offset half, three-lobe, and four-lobe bearings—bearings different from the plain journal bearing. An excellent discussion of the performance of these bearings is provided in Allaire and Flack,[24] and some of their conclusions are presented here. In Fig. 21.15, each pad is moved in toward the center of the bearing some fraction of the pad clearance in order to make the fluid-film thickness more converging and diverging than that which occurs in a plain journal bearing. The pad center of curvature is indicated by a cross. Generally, these bearings give good suppression of instabilities in the system but can be subject to subsynchronous vibration at high speeds. Accurate manufacturing of these bearings is not always easy to obtain.

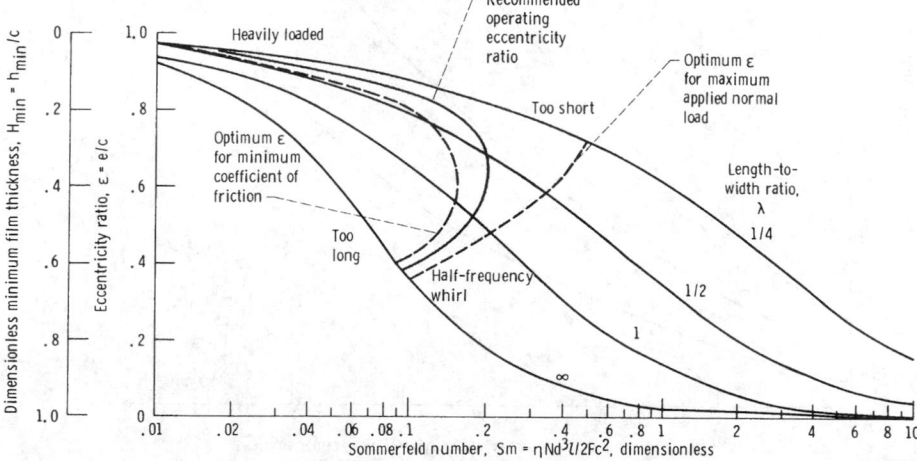

Fig. 21.10 Design figure showing eccentricity ratio, dimensionless minimum film thickness, and Sommerfeld number for 180° partial journal bearings, centrally loaded. (Adapted from Ref. 23.)

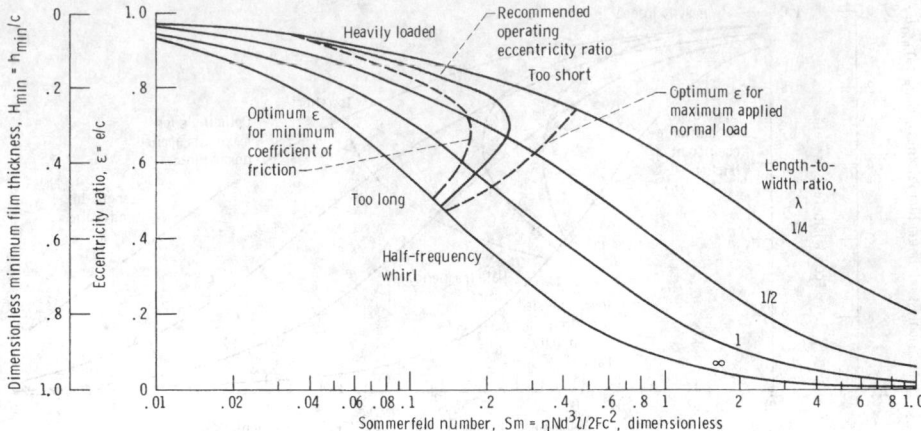

Fig. 21.11 Design figure showing eccentricity ratio, dimensionless minimum film thickness, and Sommerfeld number for 120° partial journal bearings, centrally loaded. (Adapted from Ref. 23.)

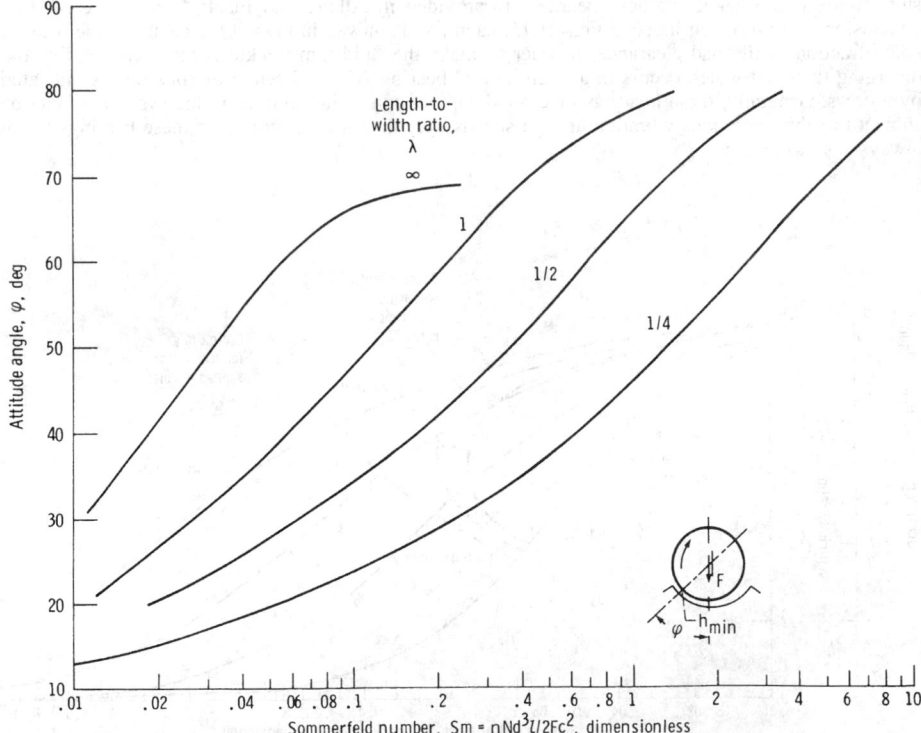

Fig. 21.12 Design figure showing attitude angle (position of minimum film thickness) and Sommerfeld number for full journal bearings, centrally loaded. (Adapted from Ref. 23.)

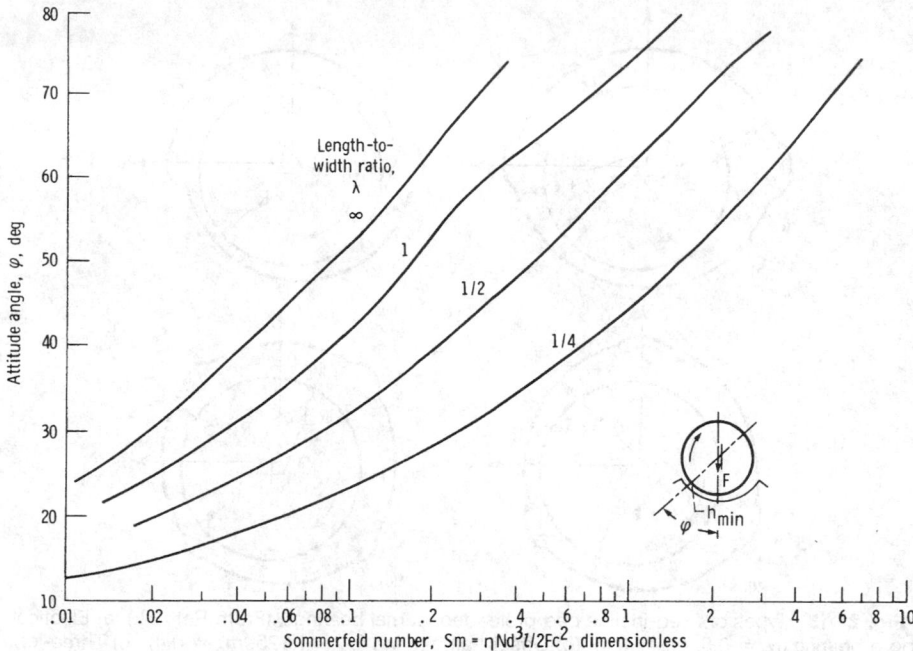

Fig. 21.13 Design figure showing attitude angle (position of minimum film thickness) and Sommerfeld number for 180° partial journal bearings, centrally loaded. (Adapted from Ref. 23.)

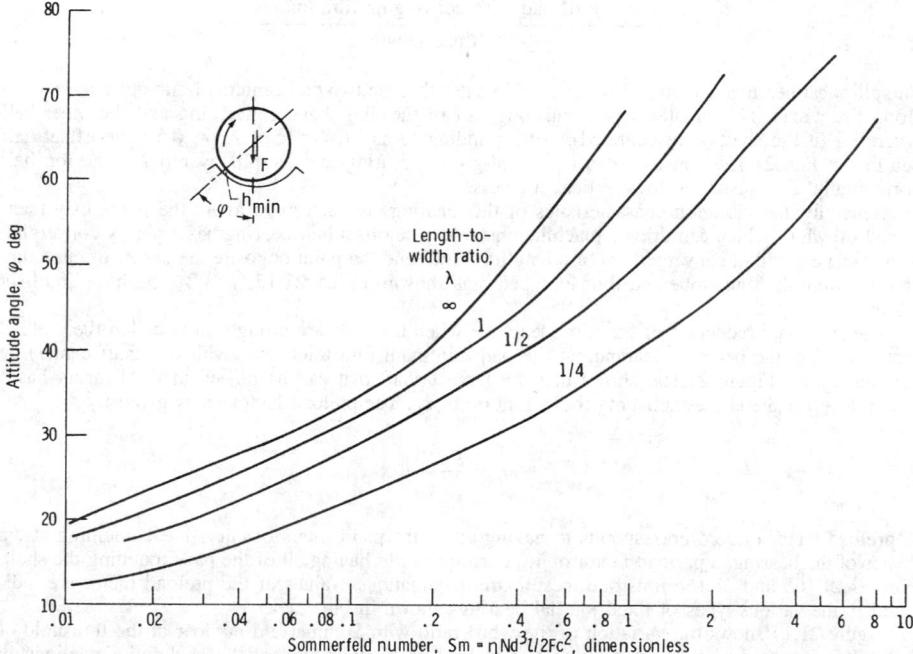

Fig. 21.14 Design figure showing attitude angle (position of minimum film thickness) and Sommerfeld number for 120° partial journal bearings, centrally loaded. (Adapted from Ref. 23.)

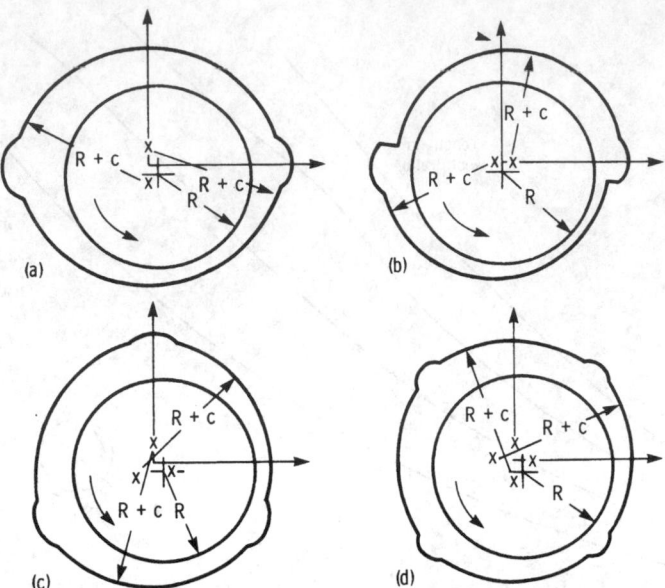

Fig. 21.15 Types of fixed-incline pad preloaded journal bearings. (From Ref. 24.) (a) Elliptical bore bearing ($\alpha_a = 0.5$, $m_p = 0.4$). (b) Offset half bearing ($\alpha_a = 1.125$, $m_p = 0.4$). (c) Three-lobe bearing ($\alpha_a = 0.5$, $m_p = 0.4$). (d) Four-lobe bearing ($\alpha_a = 0.5$, $m_p = 0.4$).

A key parameter used in describing these bearings is the fraction of length in which the film thickness is converging to the full pad length, called the offset factor and defined as

$$\alpha_a = \frac{\text{length of pad with converging film thickness}}{\text{full pad length}}$$

The elliptical bearing, shown in Fig. 21.15, indicates that the two pad centers of curvature are moved along the y axis. This creates a pad with one-half of the film shape converging and the other half diverging (if the shaft were centered), corresponding to an offset factor $\alpha_a = 0.5$. The offset half bearing in Fig. 21.15b consists of a two-axial-groove bearing that is split by moving the top half horizontally. This results in low vertical stiffness.

Generally, the vibration characteristics of this bearing are such as to avoid the previously mentioned oil whirl, which can drive a machine unstable. The offset half bearing has a purely converging film thickness with a converged pad arc length of 160° and the point opposite the center of curvature at 180°. Both the three-lobe and four-lobe bearings shown in Figs. 21.15c and 21.15d have an offset factor of $\alpha_a = 0.5$.

The fractional reduction of the film clearance when the pads are brought in is called the preload factor m_p. Let the bearing clearance at the pad minimum film thickness (with the shaft center) be denoted by c_b. Figure 21.16a shows that the largest shaft that can be placed in the bearing has a radius $R + c_b$, thereby establishing the definition of c_b. The preload factor m_p is given by

$$m_p = \frac{c - c_b}{c}$$

A preload factor of zero corresponds to having all of the pad centers of curvature coinciding at the center of the bearing; a preload factor of 1.0 corresponds to having all of the pads touching the shaft. Figures 21.16b and 21.16c illustrate these extreme situations. Values of the preload factor are indicated in the various types of fixed journal bearings shown in Fig. 21.15.

Figure 21.17 shows the variation of the whirl ratio with Sommerfeld number at the threshold of instability for the four bearing types shown in Fig. 21.15. It is evident that a definite relationship exists between the stability and whirl ratio such that the more stable bearing distinctly whirls at a lower speed ratio. With the exception of the elliptical bearing, all bearings whirl at speeds less than

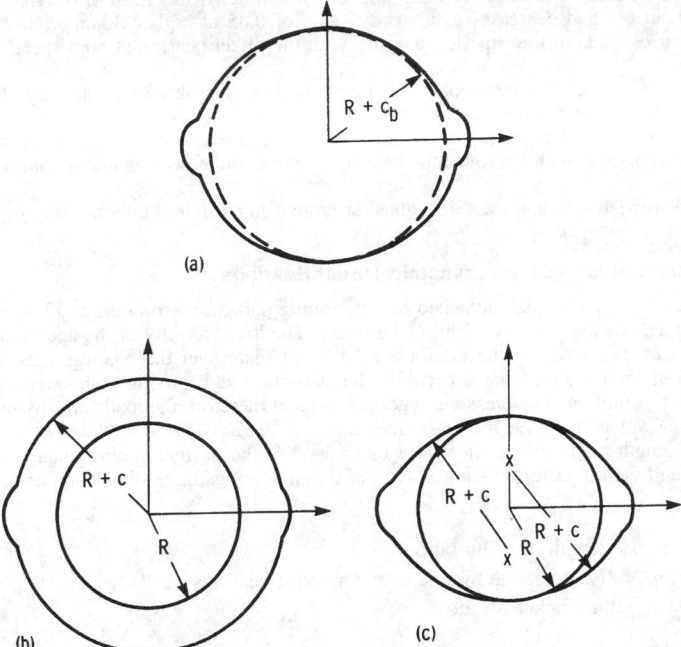

Fig. 21.16 Effect of preload on two-lobe bearings. (From Ref. 24.) (a) Largest shaft that fits in bearing. (b) $m = 0$, largest shaft $= R + c$, bearing clearance $c_b = $ (c). (c) $m = 1.0$, largest shaft $= R$, bearing clearance $c_b = 0$.

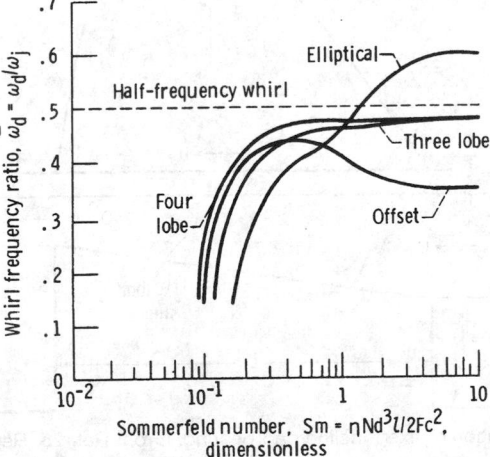

Fig. 21.17 Chart for determining whirl frequency ratio. (From Ref. 24.)

0.48 of the rotor speed. The offset bearing attains a maximum whirl ratio of 0.44 at a Sommerfeld number of about 0.4 and decreases to a steady value of 0.35 at higher Sommerfeld numbers. This observation corresponds to the superior stability with the offset bearing at high-speed and light-load operations.

The whirl ratios with the three-lobe and four-lobe bearings share similar characteristics. They both rise sharply at low Sommerfeld numbers and remain fairly constant for most portions of the curves. Asymptotic whirl ratios of 0.47 and 0.48, respectively, are reached at high Sommerfeld numbers. In comparison with the four-lobe bearing, the three-lobe bearing always has the lower whirl ratio.

The elliptical bearing is the least desirable for large Sommerfeld numbers. At $Sm > 1.3$ the ratio exceeds 0.5.

21.2.2 Liquid-Lubricated Hydrodynamic Thrust Bearings

In a thrust bearing, a thrust plate attached to, or forming part of, the rotating shaft is separated from the sector-shaped bearing pads by a film of lubricant. The load capacity of the bearing arises entirely from the pressure generated by the motion of the thrust plate over the bearing pads. This action is achieved only if the clearance space between the stationary and moving components is convergent in the direction of motion. The pressure generated in, and therefore the load capacity of, the bearing, depends on the velocity of the moving slider $u = (R_1 + R_2)\omega/2 = \pi(R_1 + R_2)N$, the effective viscosity, the length of the pad l, the width of the pad b, the normal applied load F, the inlet film thickness h_i, and the outlet film thickness h_o. For thrust bearings three dimensionless parameters are used:

1. $\lambda = l/b$, pad length-to-width ratio
2. $Sm_t = \eta u b l^2 / F h_o^2$, Sommerfeld number for thrust bearings
3. $\overline{h}_i = h_i/h_o$, film thickness ratio

It is important to recognize that the total thrust load F is equal to nF, where n is the number of pads in a thrust bearing. In this section three different thrust bearings will be investigated. Two fixed-pad types, a fixed incline and a step sector, and a pivoted-pad type will be discussed.

Fixed-Incline Pad

The simplest form of fixed-pad thrust bearing provides only straight-line motion and consists of a flat surface sliding over a fixed pad or land having a profile similar to that shown in Fig. 21.18. The fixed-pad bearing depends for its operation on the lubricant being drawn into a wedge-shaped space

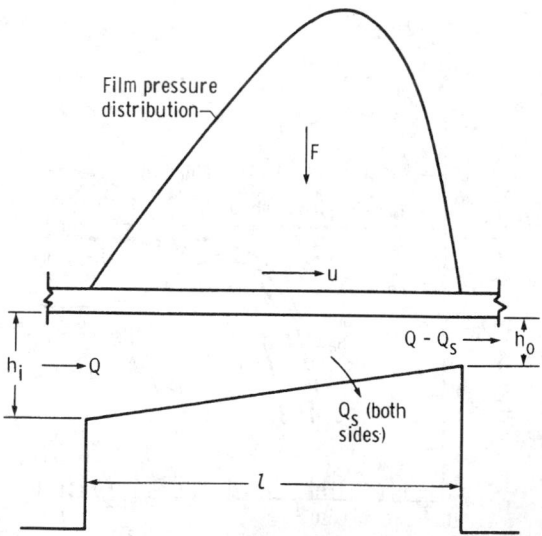

Fig. 21.18 Configuration of fixed-incline pad bearing. (From Ref. 25. Reprinted by permission of ASME.)

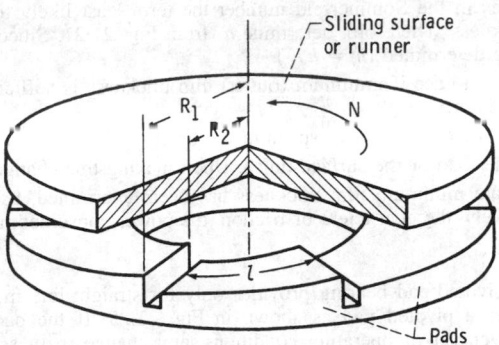

Fig. 21.19 Configuration of fixed-incline pad thrust bearing. (From Ref. 25.)

and thus producing pressure that counteracts the load and prevents contact between the sliding parts. Since the wedge action only takes place when the sliding surface moves in the direction in which the lubricant film converges, the fixed-incline bearing, shown in Fig. 21.18, can only carry load for this direction of operation. If reversibility is desired, a combination of two or more pads with their surfaces sloped in opposite direction is required. Fixed-incline pads are used in multiples as in the thrust bearing shown in Fig. 21.19.

The following procedure assists in the design of a fixed-incline pad thrust bearing:

1. Choose a pad width-to-length ratio. A square pad ($\lambda = 1$) is generally felt to give good performance. From Fig. 21.20, if it is known whether maximum load or minimum power is most important in the particular application, a value of the film thickness ratio can be determined.

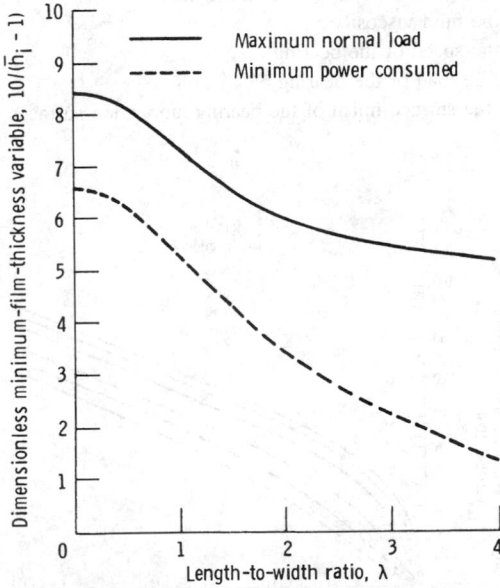

Fig. 21.20 Chart for determining minimum film thickness corresponding to maximum load or minimum power less for various pad proportions—fixed-incline pad bearings. (From Ref. 25. Reprinted by permission of ASME.)

2. Within the terms in the Sommerfeld number the term least likely to be preassigned is the outlet film thickness. Therefore, determine h_o from Fig. 21.21. Since $\bar{h}_i$ is known from Fig. 21.20, h_i can be determined ($h_i = \bar{h}_i h_o$).

3. Check Table 21.5 to see if minimum (outlet) film thickness is sufficient for the preassigned surface finish. If not:

   a. Increase the fluid viscosity or speed of the bearing.

   b. Decrease the load or the surface finish. Upon making this change return to step 1.

4. Once an adequate minimum film thickness has been determined, use Figs. 21.22–21.24 to obtain, respectively, the coefficient of friction, the power consumed, and the flow.

Pivoted Pad

The simplest form of pivoted-pad bearing provides only for straight-line motion and consists of a flat surface sliding over a pivoted pad as shown in Fig. 21.25. If the pad is assumed to be in equilibrium under a given set of operating conditions, any change in these conditions, such as a change in load, speed, or viscosity, will alter the pressure distribution and thus momentarily shift the center of pressure and create a moment that causes the pad to change its inclination until a new position of equilibrium is established. It can be shown that if the position of that pivot, as defined by the distance $\bar{x}$, is fixed by choosing $\bar{x}/l$, the ratio of the inlet film thickness to the outlet film thickness, h_i/h_o, also becomes fixed and is independent of load, speed, and viscosity. Thus the pad will automatically alter its inclination so as to maintain a constant value of h_i/h_o.

Pivoted pads are sometimes used in multiples as pivoted-pad thrust bearings, shown in Fig. 21.26. Calculations are carried through for a single pad, and the properties for the complete bearing are found by combining these calculations in the proper manner.

Normally, a pivoted pad, will only carry load if the pivot is placed somewhere between the center of the pad and the outlet edge ($0.5 < \bar{x}/l \leq 1.0$). With the pivot so placed, the pad therefore can only carry load for one direction of rotation.

The following procedure helps in the design of pivoted-pad thrust bearings:

1. Having established if minimum power or maximum load is more critical in the particular application and chosen a pad length-to-width ratio, establish the pivot position from Fig. 21.27.

2. In the Sommerfeld number for thrust bearings the unknown parameter is usually the outlet or minimum film thickness. Therefore, establish the value of h_o from Fig. 21.28.

3. Check Table 21.5 to see if the outlet film thickness is sufficient for the preassigned surface finish. If sufficient, go on to step 4. If not, consider:

   a. Increasing the fluid viscosity

   b. Increasing the speed of the bearing

   c. Decreasing the load of the bearing

   d. Decreasing the surface finish of the bearing lubrication surfaces

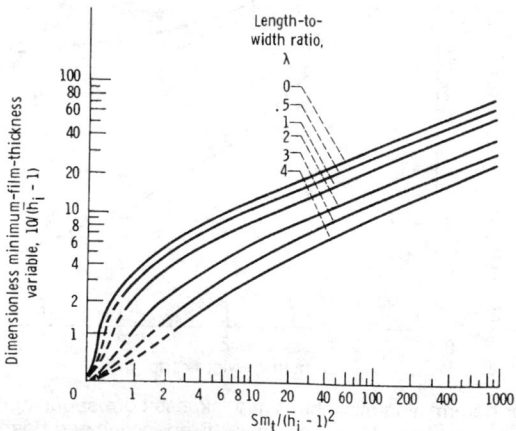

Fig. 21.21 Chart for determining minimum film thickness for fixed-incline pad thrust bearings. (From Ref. 25. Reprinted by permission of ASME.)

Table 21.5 Allowable Minimum Outlet Film Thickness for a Given Surface Finish (From Ref. 8)

Surface Finish					Allowable Minimum Outlet Film Thickness[a], h_0	
Familiar British Units, μin, CLA[b]	SI Units, μin[c] CLA	Description of Surface	Examples of Manufacturing Methods	Approximate Relative Costs	Familiar British Units, in.	SI Units, m
4–8	0.1–0.2	Mirror-like surface without toolmarks, close tolerances	Grind, lap, and superfinish	17–20	0.00010	0.0000025
8–16	0.2–0.4	Smooth surface without scratches, close tolerances	Grind and lap	17–20	.00025	.0000062
16–32	0.4–0.8	Smooth surface, close tolerances	Grind, file, and lap	10	.00050	.0000125
32–63	0.8–1.6	Accurate bearing surface without toolmarks	Grind, precision mill, and file	7	.00100	.000025
63–125	1.6–3.2	Smooth surface without objectionable toolmarks, moderate tolerances	Shape, mill, grind, and turn	5	.00200	.000050

[a]The values of film thickness are given only for guidance. They indicate the film thickness required to avoid metal-to-metal contact under clean oil conditions with no misalignment. It may be necessary to take a larger film thickness than that indicated (e.g., to obtain an acceptable temperature rise). It has been assumed that the average surface finish of the pads is the same as that of the runner.

[b]CLA = centerline average.

[c]μm = micrometer; 40 μin. (microinch) = 1 μm.

Upon making this change return to step 1.

4. Once an adequate outlet film thickness is established, determine the film thickness ratio, power loss, coefficient of friction, and flow from Figs. 21.29–21.32.

Step Sector

The configuration of a step-sector thrust bearing is shown in Fig. 21.33. The parameters used to define the dimensionless load and stiffness are:

1. $\bar{h}_i = h_i/h_o$, film thickness ratio.
2. $\bar{\theta} = \theta_i/(\theta_i + \theta_o)$, dimensionless step location.
3. n, number of sectors.
4. $\alpha_r = R_2 R_1$, radius ratio.
5. θ_g, angular extent of lubrication feed groove.

Note that the first four parameters are dimensionless and the fifth is dimensional and expressed in radians.

The optimum parallel step-sector bearing for *maximum load capacity* for a given α_r and θ_g is

$$\bar{\theta}_{\text{opt}} = 0.558, \quad (\bar{h}_i)_{\text{opt}} = 1.668, \quad \text{and} \quad n_{\text{opt}} = \frac{2\pi}{\theta_g + \dfrac{2.24(1 - \alpha_r)}{1 + \alpha_r}}$$

where n_{opt} is rounded off to the nearest integer and its minimum value is 3. For *maximum stiffness*, results are identical to the above with the exception that $(\bar{h}_i)_{\text{opt}} = 1.467$. These results are obtained from Hamrock.[26]

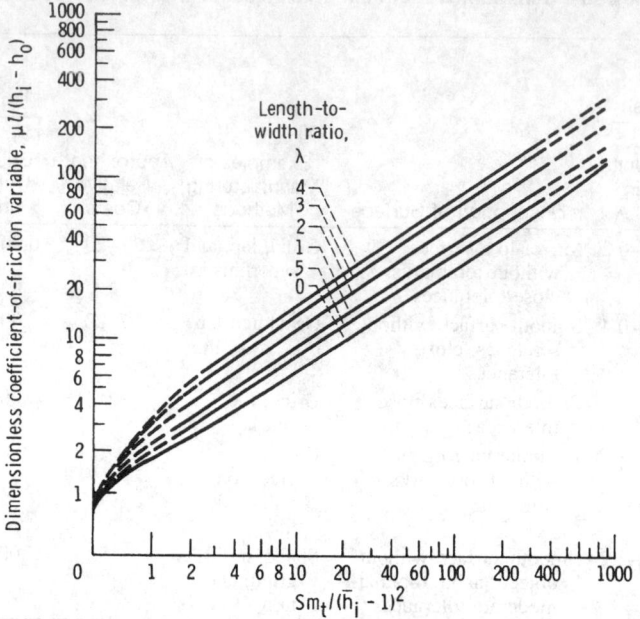

Fig. 21.22 Chart for determining coefficient of friction for fixed-incline pad thrust bearings. (From Ref. 25. Reprinted by permission of ASME.)

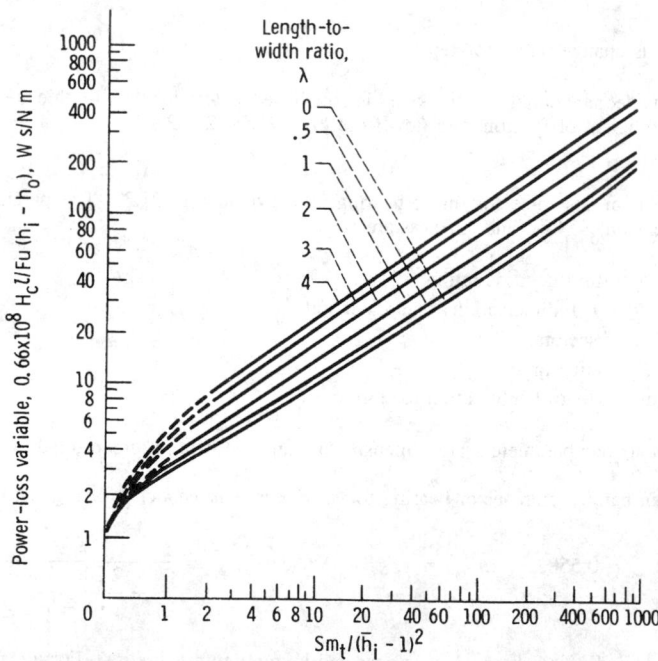

Fig. 21.23 Chart for determining power loss for fixed-incline pad thrust bearings. (From Ref. 25. Reprinted by permission of ASME.)

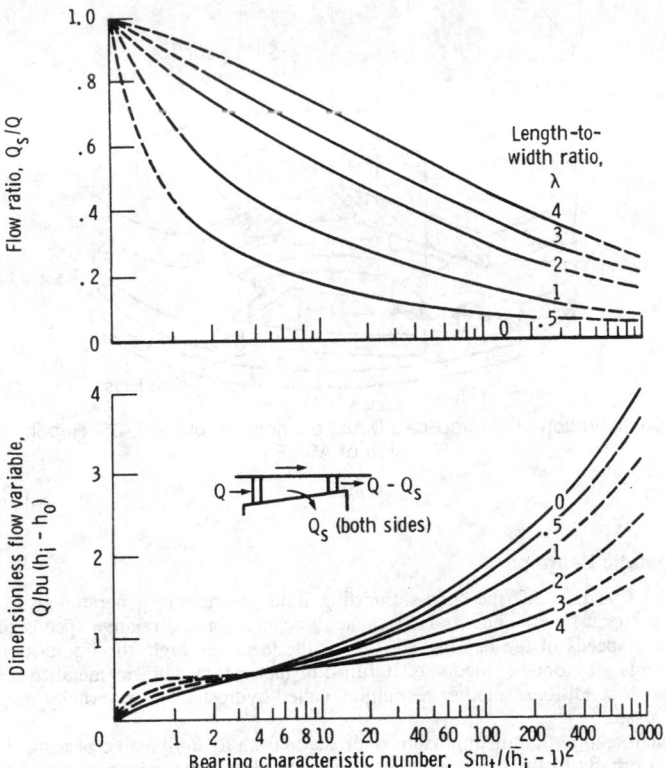

Fig. 21.24 Charts for determining lubricant flow for fixed-incline pad thrust bearings. (From Ref. 25. Reprinted by permission of ASME.)

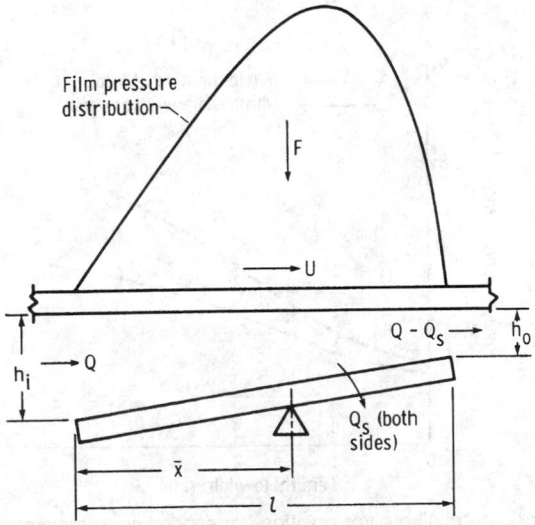

Fig. 21.25 Configuration of pivoted-pad bearings. (From Ref. 25. Reprinted by permission of ASME.)

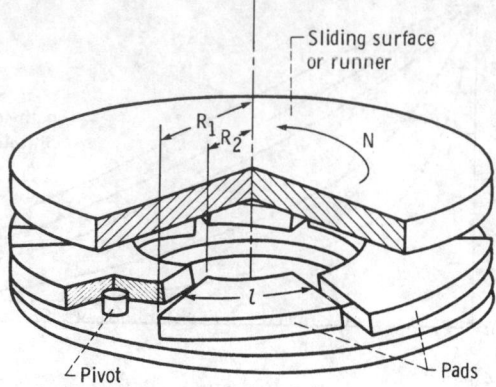

Fig. 21.26 Configuration of pivoted-pad thrust bearings. (From Ref. 25. Reprinted by permission of ASME.)

21.2.3 Hydrostatic Bearings

In Sections 21.2.1 and 21.2.2 the load-supporting fluid pressure is generated by relative motion between the bearing surfaces. Thus its load capacity depends on the relative speeds of the surfaces. When the relative speeds of the bearing are low or the loads are high, the liquid-lubricated journal and thrust bearings may not be adequate. If full-film lubrication with no metal-to-metal contact is desired under such conditions, another technique, called hydrostatic or externally pressurized lubrication, may be used.

The one salient feature that distinguishes hydrostatic from hydrodynamic bearings is that the fluid is pressurized externally to the bearings and the pressure drop across the bearing is used to support the load. The load capacity is independent of the motion of bearing surfaces or the fluid viscosity. There is no problem of contact of the surfaces at starting and stopping as with conventional hydrodynamically lubricated bearings because pressure is applied before starting and maintained until after stopping. Hydrostatic bearings can be very useful under conditions of little or no relative motion and under extreme conditions of temperature or corrosivity, where it may be necessary to use bearing materials with poor boundary lubricating properties. Surface contact can be avoided completely, so

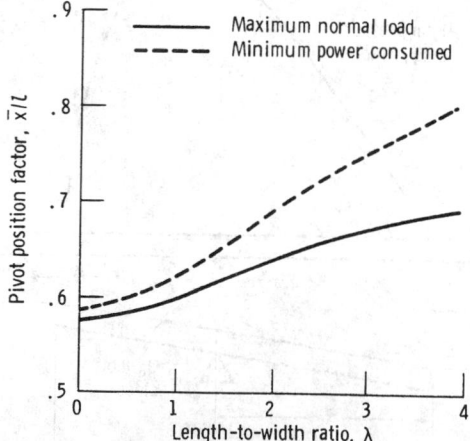

Fig. 21.27 Chart for determining pivot position corresponding to maximum load or minimum power loss for various pad proportions—pivoted-pad bearings. (From Ref. 25. Reprinted by permission of ASME.)

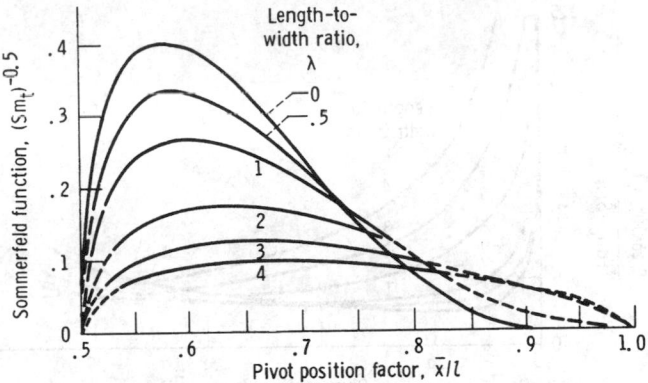

Fig. 21.28 Chart for determining outlet film thickness for pivoted-pad thrust bearings. (From Ref. 25. Reprinted by permission of ASME.)

material properties are much less important than in hydrodynamic bearings. The load capacity of a hydrostatic bearing is proportional to the available pressure.

Hydrostatic bearings do, however, require an external source of pressurization such as a pump. This represents an additional system complication and cost.

The chief advantage of hydrostatic bearings is their ability to support extremely heavy loads at slow speeds with a minimum of driving force. For this reason they have been successfully applied in rolling mills, machine tools, radio and optical telescopes, large radar antennas, and other heavily loaded, slowly moving equipment.

The formation of a fluid film in a hydrostatic bearing system is shown in Fig. 21.34. A simple bearing system with the pressure source at zero pressure is shown in Fig. 21.34a. The runner under the influence of a load F is seated on the bearing pad. As the source pressure builds up, Fig. 21.34b, the pressure in the pad recess also increases. The pressure in the recess is built up to a point, Fig. 21.34c, where the pressure on the runner over an area equal to the pad recess area is just sufficient to lift the load. This is commonly called the lift pressure. Just after the runner separates from the bearing pad, Fig. 21.34d, the pressure in the recess is less than that required to lift the bearing runner $(p_r < p_l)$. After lift, flow commences through the system. Therefore, a pressure drop exists between

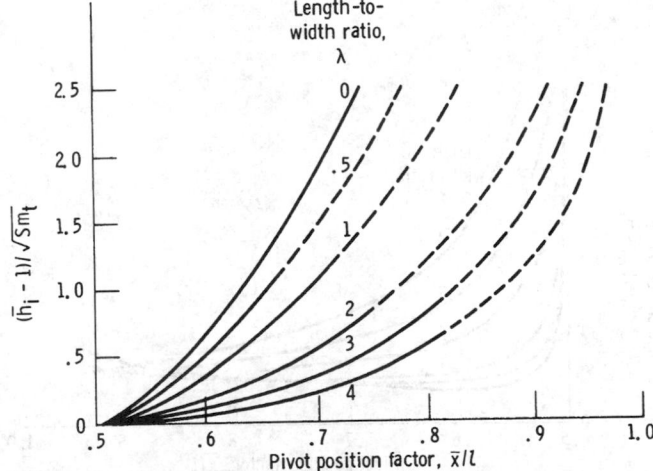

Fig. 21.29 Chart for determining film thickness ratio $\bar{h}_i$ for pivoted-pad thrust bearings. (From Ref. 25. Reprinted by permission of ASME.)

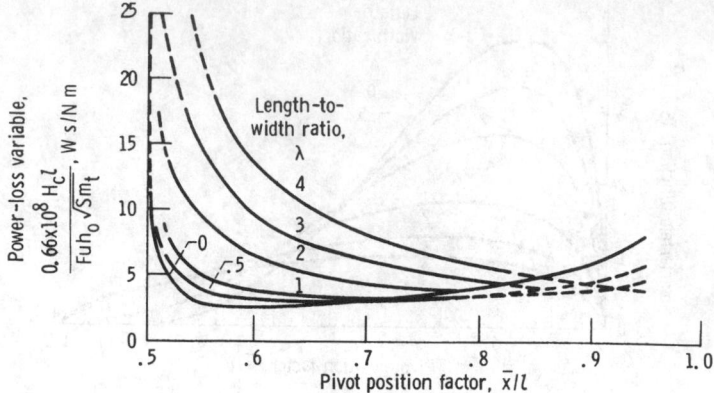

Fig. 21.30 Chart for determining power loss for pivoted-pad thrust bearings. (From Ref. 25. Reprinted by permission of ASME.)

the pressure source and the bearing (across the restrictor) and from the recess to the exit of the bearing.

If more load is added to the bearing, Fig. 21.34e, the film thickness will decrease and the recess pressure will rise until pressure within the bearing clearance and the recess is sufficient to carry the increased load. If the load is now decreased to less than the original, Fig. 21.34f, the film thickness will increase to some higher value and the recess pressure will decrease accordingly. The maximum load that can be supported by the pad will be reached, theoretically, when the pressure in the recess is equal to the pressure at the source. If a load greater than this is applied, the bearing will seat and remain seated until the load is reduced and can again be supported by the supply pressure.

Pad Coefficients

To find the load-carrying capacity and flow requirements of any given hydrostatic bearing pad, it is necessary to determine certain pad coefficients. Since the selection of pad and recess geometries is up to the designer, the major design problem is the determination of particular bearing coefficients for particular geometries.

The load-carrying capacity of a bearing pad, regardless of its shape or size, can be expressed as

$$F = a_f A_p p_r \tag{21.13}$$

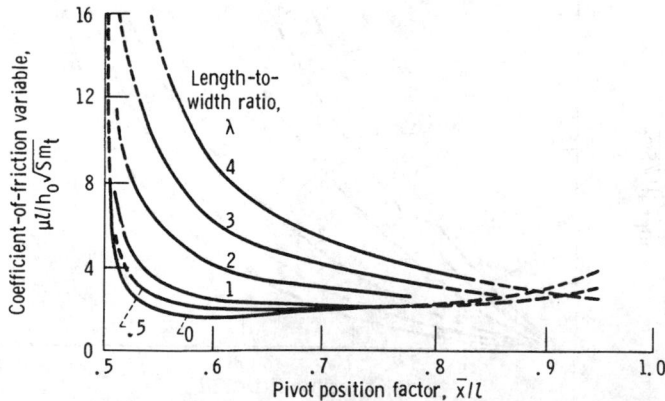

Fig. 21.31 Chart for determining coefficient of friction for pivot-pad thrust bearings. (From Ref. 25. Reprinted by permission of ASME.)

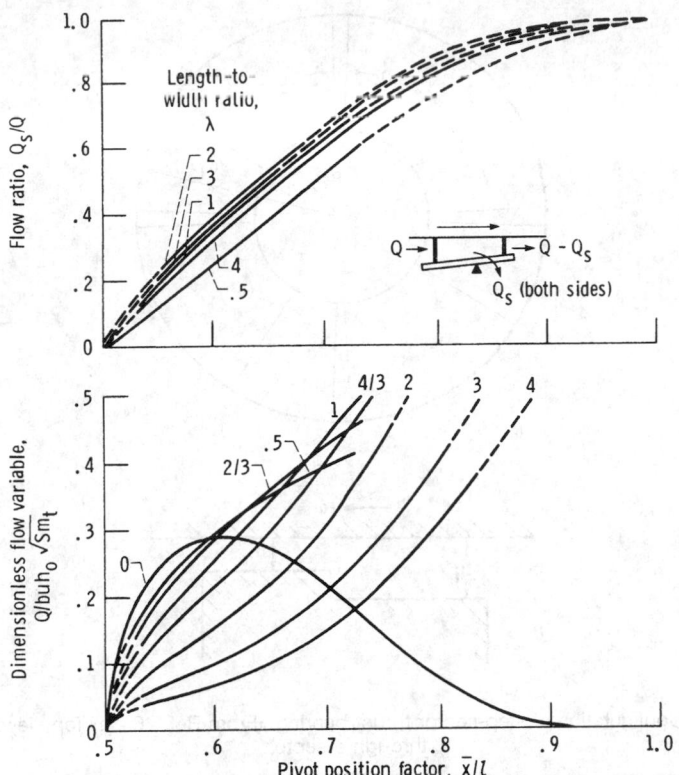

Fig. 21.32 Chart for determining lubricant flow for pivot-pad thrust bearings. (From Ref. 25. Reprinted by permission of ASME.)

where a_f = bearing pad load coefficients
  A_p = total projected pad area, m^2
  p_r = recess pressure, N/m^2

The amount of lubricant flow across a pad and through the bearing clearance is

$$Q = q_f \frac{F}{A_p} \frac{h^3}{\eta} \tag{21.14}$$

where q_f = pad flow coefficient
  h = film thickness, m
  η = lubricant absolute viscosity, N sec/m^2

The pumping power required by the hydrostatic pad can be evaluated by determining the product of recess pressure and flow:

$$H_b = p_r Q = H_f \left(\frac{F}{A_p}\right)^2 \frac{h^3}{\eta} \tag{21.15}$$

where $H_f = q_f/a_f$ is the bearing pad power coefficient. Therefore, in designing hydrostatic bearings the designer is primarily concerned with the bearing coefficients (a_f, q_f, and H_f) expressed in Eqs. (21.13)–(21.15).

Bearing coefficients are dimensionless quantities that relate performance characteristics of load, flow, and power to physical parameters. The bearing coefficients for two types of bearing pads will be considered, both of which exhibit pure radial flow and are flat, thrust-loaded types of bearings. For other types of hydrostatic bearings the reader is referred to Rippel.[22]

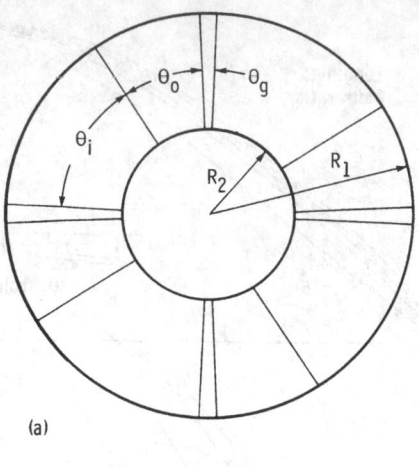

(a)

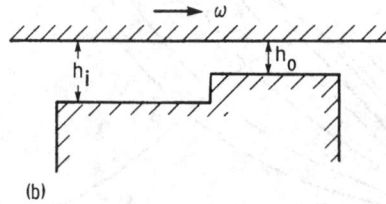

(b)

Fig. 21.33 Configuration of step-sector thrust bearing. (From Ref. 26.) (a) Top view. (b) Section through a sector.

Cicular Step Bearing Pad. The bearing coefficients for this type of pad are expressed as

$$a_f = \frac{1}{2}\left[\frac{1 - (R_o/R)^2}{\log_e (R/R_o)}\right] \tag{21.16}$$

$$q_f = \frac{\pi}{3}\left[\frac{1}{1 - (R_o/R)^2}\right] \tag{21.17}$$

$$H_f = \frac{2\pi \log_e(R/R_o)}{3[1 - (R_o/R)^2]^2} \tag{21.18}$$

For this type of pad the total projected bearing pad area A_p is equal to πR^2.

Figure 21.35 shows the three bearing pad coefficients for various ratios of recess radius to bearing radius for a circular step thrust bearing. The bearing-pad load coefficient a_f varies from zero for extremely small recesses to unity for bearings having large recesses with respect to pad dimensions. In a sense, a_f is a measure of how efficiently the bearing uses the recess pressure to support the applied load.

In Fig. 21.35 we see that the pad flow coefficient q_f varies from unity for pads with relatively small recesses to a value approaching infinity for bearings with extremely large recesses. Physically, as the recess becomes larger with respect to the bearing, the hydraulic resistance to fluid flow decreases, and thus flow increases.

From Fig. 21.35, the power coefficient H_f approaches infinity for very small recesses, decreases to a minimum value as the recess size increases, and approaches infinity again for very large recesses. For this particular bearing the minimum value of H_f occurs at a ratio of recess radius to bearing radius R_o/R of 0.53. All bearing-pad configurations exhibit minimum values of H_f when their ratios of recess length to bearing length are approximately 0.4 to 0.6.

Annular Thrust Bearing. Figure 21.36 shows an annular thrust pad bearing. In this bearing the lubricant flows from the annular recess over the inner and outer sills. For this type of bearing the pad coefficients are

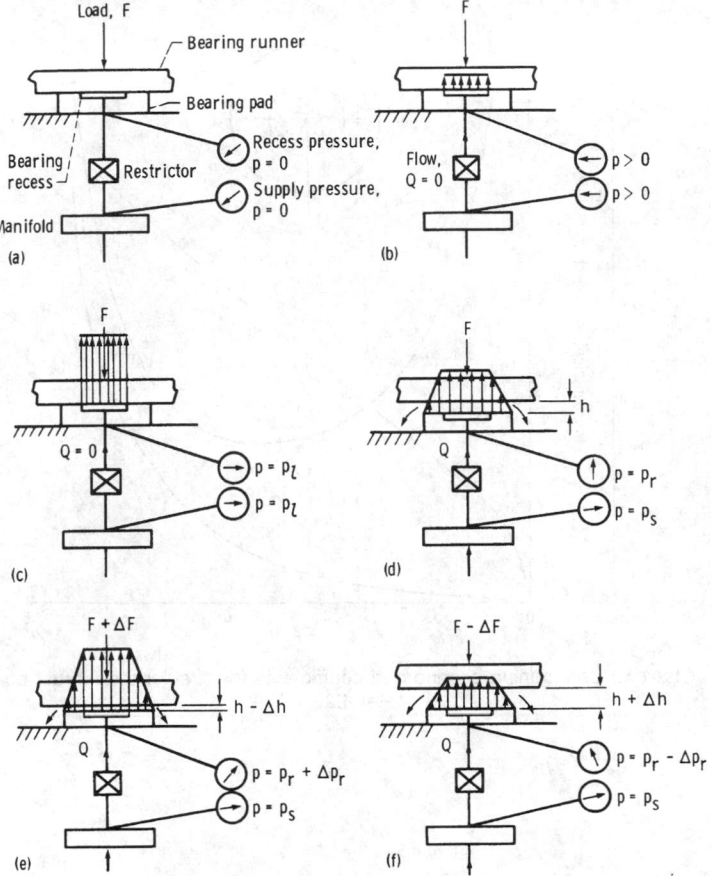

Fig. 21.34 Formation of fluid film in hydrostatic bearing system. (From Ref. 22.) (a) Pump off. (b) Pressure building up. (c) Pressure × recess area = F. (d) Bearing operating. (e) Increased load. (f) Decreased load.

$$a_f = \frac{1}{2(R_4^2 - R_1^2)} \left[\frac{R_4^2 - R_3^2}{\log_e(R_4/R_3)} - \frac{R_2^2 - R_1^2}{\log_e(R_2/R_1)} \right] \qquad (21.19)$$

$$q_f = \frac{\pi}{6q_f} \left[\frac{1}{\log_e(R_4/R_3)} - \frac{1}{\log_e(R_2/R_1)} \right] \qquad (21.20)$$

$$H_f = \frac{q_f}{a_f} \qquad (21.21)$$

For this bearing the total projected bearing-pad area is

$$A_p = \pi(R_4^2 - R_1^2) \qquad (21.22)$$

Figure 21.37 shows the bearing-pad load coefficient for an annular thrust pad bearing as obtained from Eqs. (21.19)–(21.21). For this figure it is assumed that the annular recess is centrally located within the bearing width; this therefore implies that $R_1 + R_4 = R_2 + R_3$. The curve for a_f applies for all R_1/R_4 ratios.

The hydrostatic bearings considered in this section have been limited to flat thrust-loaded bearings. Design information about other pad configurations can be obtained from Rippel.[22] The approach used for the simple, flat, thrust-loaded pad configuration is helpful in considering the more complex geometries covered by Rippel.[22]

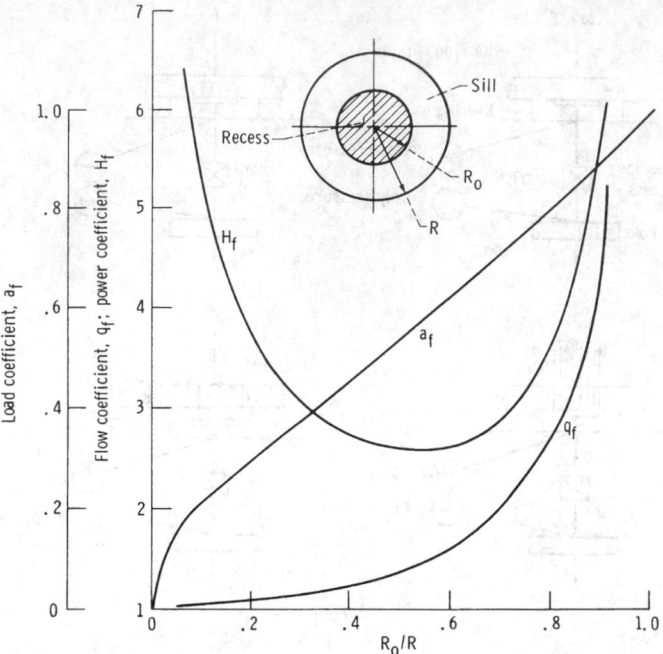

Fig. 21.35 Chart for determining bearing pad coefficients for circular step thrust bearing. (From Ref. 22.)

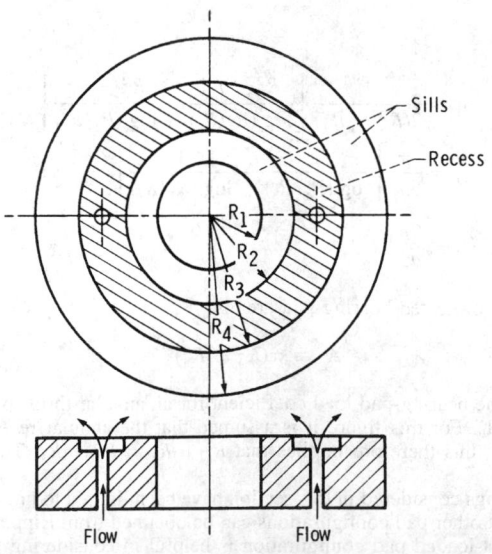

Fig. 21.36 Configuration of annular thrust pad bearing. (From Ref. 22.)

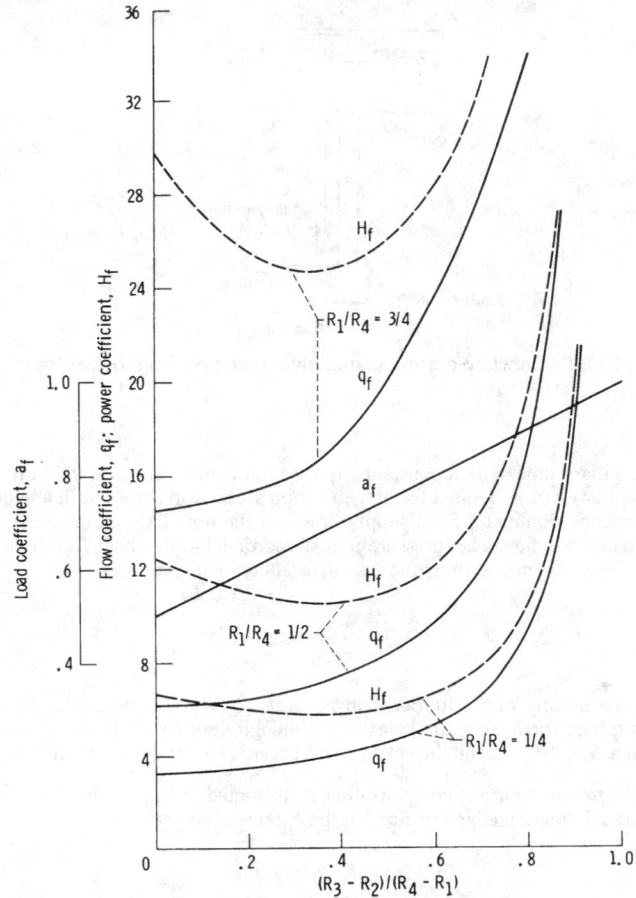

Fig. 21.37 Chart for determining bearing pad coefficients for annular thrust pad bearings. (From Ref. 22.)

Compensating Elements

As compared with common bearing types, hydrostatic bearings are relatively complex systems. In addition to the bearing pad, the system includes a pump and compensating elements. Three common types of compensating elements for hydrostatic bearings are the capillary tube, the sharp-edged orifice, and constant-flow-valve compensation.

Capillary Compensation. Figure 21.38 shows a capillary-compensated hydrostatic bearing as obtained from Rippel.[22] The small diameter of the capillary tube provides restriction and resultant pressure drop in the bearing pad. The characteristic feature of capillary compensation is a long tube or a relatively small diameter ($l_c > 20d_c$). The laminar flow of fluid through such a tube while neglecting entrance and exit effects and viscosity changes due to temperature and pressure effects can be expressed as

$$Q_c = \frac{k_c(p_s - p_r)}{\eta} \tag{21.23}$$

where

$$k_c = \frac{\pi d_c^4}{128 l_c} \tag{21.24}$$

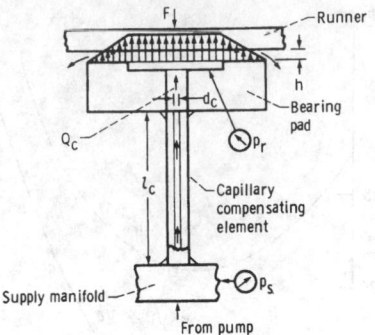

Fig. 21.38 Capillary-compensated hydrostatic bearing. (From Ref. 22.)

For a given capillary tube, k_c is a constant expressed in cubic meters. Thus, from Eq. (21.23) the flow through a capillary tube is related linearly to the pressure drop across it. In a hydrostatic bearing with capillary compensation and a fixed supply pressure, the flow through the bearing will decrease with increasing load since the pocket pressure p_r is proportional to the load. To satisfy the assumption of laminar flow, Reynolds number must be less than 2000 when expressed as

$$N_R = \frac{4\rho Q_c}{\pi d_c \eta} < 2000 \qquad (21.25)$$

where ρ is the mass density of the lubricant in N $\sec^2/\text{m}^4$. Hypodermic needle tubing serves quite well as capillary tubing for hydrostatic bearings. Although very small diameter tubing is available, diameters less than 6×10^{-4} m should not be used because of their tendency to clog.

Orifice Compensation.   Orifice compensation is illustrated in Fig. 21.39. The flow of an incompressible fluid through a sharp-edged orifice can be expressed as

$$Q_o = k_o(p_s - p_r)^{1/2} \qquad (21.26)$$

where

$$k_o = \frac{\pi c_d d_o^2}{\sqrt{8\rho}}$$

and c_d is the orifice discharge coefficient. For a given orifice size and given lubricant, k_o is a constant

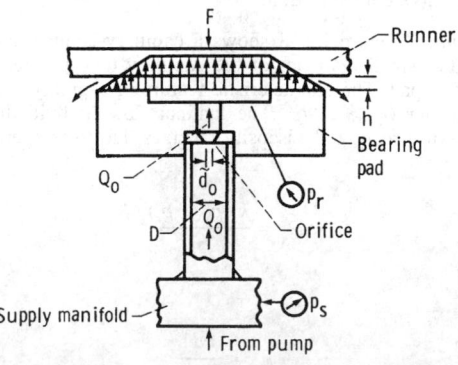

Fig. 21.39 Orifice-compensated hydrostatic bearing. (From Ref. 22.)

expressed in $m^4/sec\ N^{1/2}$. Thus, from Eq. (21.26) flow through an orifice is proportional to the square root of the pressure difference across the orifice.

The discharge coefficient c_d is a function of Reynolds number. For an orifice the Reynolds number is

$$N_R = \frac{d_o}{\eta}\,[2\rho(p_s - p_r)]^{1/2} \tag{21.27}$$

For a Reynolds number greater than approximately 15, which is the usual case in orifice-compensated hydrostatic bearings, c_d is about 0.6 for $d_o/D < 0.1$. For a Reynolds number less than 15, the discharge coefficient is approximately

$$c_d = 0.20\sqrt{N_R} \tag{21.28}$$

The pipe diameter D at the orifice should be at least 10 times the orifice diameter d_o. Sharp-edged orifices, depending on their diameters, have a tendency to clog, therefore orifice diameters d_o less than 5×10^{-4} m should be avoided.

Constant-Flow-Valve Compensation. Constant-flow-valve compensation is illustrated in Fig. 21.40. This type of restrictor has a constant flow regardless of the pressure difference across the valve. Hence, the flow is independent of recess pressure.

The relative ranking of the three types of compensating elements with regard to a number of considerations is given in Table 21.6. A rating of 1 in this table indicates best or most desirable. This table should help in deciding which type of compensation is most desirable in a particular application.

Basically, any type of compensating element can be designed into a hydrostatic bearing system if loads on the bearing never change. But if stiffness, load, or flow vary, the choice of the proper compensating element becomes more difficult and the reader is again referred to Rippel.[22]

21.2.4 Gas-Lubricated Hydrodynamic Bearings

A relatively recent (within the last 30 years) extension of hydrodynamic lubrication that is of growing importance is gas lubrication. It consists of using air or some other gas as a lubricant rather than a mineral oil. The viscosity of air is 1000 times smaller than that of very thin mineral oils. Consequently, the viscous resistance is very much less. However, the distance of nearest approach (i.e., the closest distance between the shaft and the bearing) is also correspondingly smaller, so that special precautions must be taken. To obtain full benefits from gas lubrication, the following should be observed:

1. Surfaces must have a very fine finish.
2. Alignment must be very good.
3. Dimensions and clearances must be very accurate.

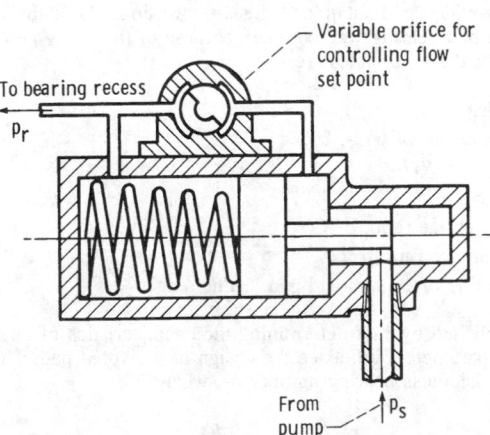

Fig. 21.40 Constant-flow-valve compensation in hydrostatic bearing. (From Ref. 22.)

Table 21.6 Compensating-Element Considerations[a] (From Ref. 22)

	Compensating Element		
Consideration	Capillary	Orifice	Constant-flow Valve
Initial cost	2	1	3
Cost to fabricate and install	2	3	1
Space required	2	1	3
Reliability	1	2	3
Useful life	1	2	3
Commercial availability	2	3	1
Tendency to clog	1	2	3
Serviceability	2	1	3
Adjustability	3	2	1

[a]Rating of 1 is best or most desirable.

4.   Speeds must be high.

5.   Loading must be relatively low.

Another main difference between the behavior of similar gas and liquid films besides that of viscosity is the compressibility of the gas. At low relative speeds it is reasonable to expect the gas-film density to remain nearly constant and the film therefore to behave as if it were incompressible. At high speeds, however, the density change is likely to become of primary importance so that such gas-film properties must differ appreciably from those of similar liquid films.

Gas-lubricated bearings can also operate at very high temperatures since the lubricant will not degrade chemically. Furthermore, if air is used as the lubricant, it costs nothing. Gas bearings are finding increasing use in gas-cycle machinery where the cycle gas is used in the bearings, thus eliminating the need for a conventional lubrication system; in gyros, where precision and constancy of torque are critical; in food and textile processing machinery, where cleanliness and absence of contaminants are critical; and also in the magnetic recording tape industry.

Journal Bearings

Plain gas-lubricated journal bearings are of little interest because of their poor stability characteristics. Lightly loaded bearings that operate at low eccentricity ratios are subjected to fractional frequency whirl, which can result in bearing destruction. Two types of gas-lubricated journal bearings find widespread use, namely, the pivoted pad and the herringbone groove.

Pivoted Pad.   Pivoted-pad journal bearings are most frequently used as shaft supports in gas-bearing machinery because of their excellent stability characteristics. An individual pivot pad and shaft are shown in Fig. 21.41, and a three-pad pivoted-pad bearing assembly is shown in Fig. 21.42. Generally, each pad provides pad rotation degrees of freedom about three orthogonal axes (pitch, roll, and yaw). Pivoted-pad bearings are complex because of the many geometric variables involved in their design. Some of these variables are:

1.   Number of pads.
2.   Circumferential extent of pads, α_p.
3.   Aspect ratio of pad, R/l.
4.   Pivot location, ϕ_p/α_p.
5.   Machined-in clearance ratio, c/R.
6.   Pivot circle clearance ratio, c'/R.
7.   Angle between line of centers and pad leading edge, ξ_p.

Analysis is accomplished by first determining the characteristics of an individual pad. Both geometric and operating parameters influence the design of a pivoted pad. The operating parameter of importance is the dimensionless bearing number Λ_j, where

$$\Lambda_j = \frac{6\eta\omega R^2}{p_a c^2}$$

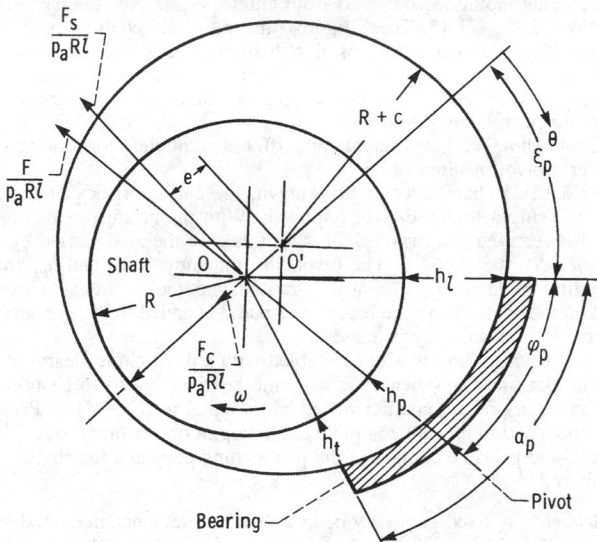

Fig. 21.41 Geometry of individual shoe-shaft bearing. (From Ref. 27.)

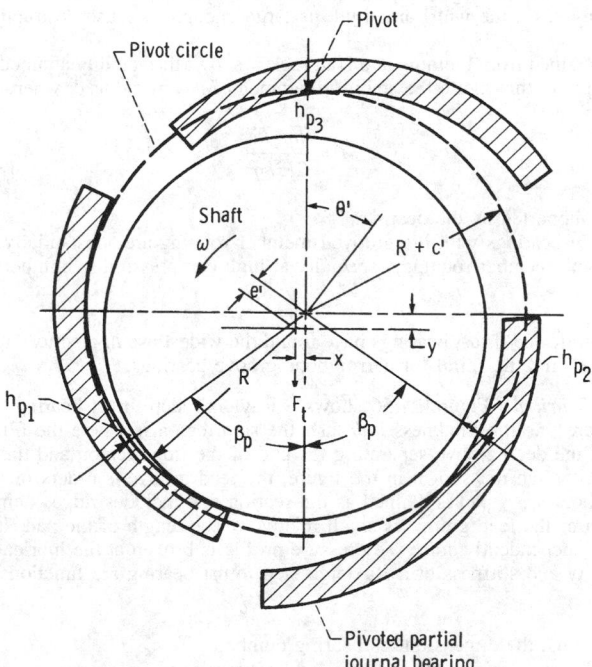

Fig. 21.42 Geometry of pivoted-pad journal bearing with three shoes. (From Ref. 27.)

The results of computer solutions obtained from Gunter et al.[27] for the performance of a single pad are shown in Figs. 21.43–21.45. These figures illustrate load coefficient, pivot film thickness, and trailing-edge film thickness as functions of pivot location and eccentricity ratio. These field maps apply for a pad with a radius-to-width ratio of 0.606, a circumferential extent of 94.5° (an aspect ratio of 1), and $\Lambda_j = 3.5$. For other geometries and Λ values similar maps must be generated. Additional maps are given in Gunter et al.[27]

Figures 21.46–21.48 show load coefficient and stiffness coefficient for a range of Λ_j values up to 4. These plots are for a pivot position of $\frac{2}{3}$.

When the individual pad characteristics are known, the characteristics of the multipad bearing can be determined by using a trial-and-error approach. With the arrangement shown in Fig. 21.42, the load is directed between the two lower pivots. For this case the load carried by each of the lower pads is initially assumed to be $F \cos \beta$. The pivot film thicknesses h_{p_1} and h_{p_2} are then calculated. The upper-pad pivot film thickness h_{p_3}, eccentricity ratio ϵ, and load coefficient C_{l_3} can be determined. The additional load on the shaft due to the reaction of pad 3 is added to the system load. Calculations are repeated until the desired accuracy is achieved.

Pivoted-pad journal bearings are usually assembled with a pivot circle clearance c' somewhat less than the machined-in clearance c. When $c'/c < 1$, the bearing is said to be preloaded. Preload is usually given in terms of a preload coefficient, which is equal to $(c - c')/c$. Preloading is used to increase bearing stiffness and to prevent complete unloading of one or more pads. The latter condition can lead to pad flutter and possible contact of the pad leading edge and the shaft, which, in turn, can result in bearing failure.

Herringbone Groove. A fixed-geometry bearing that has demonstrated good stability characteristics and thus promise for use in high-speed gas bearings is the herringbone bearing. It consists of a circular journal and bearing sleeve with shallow, herringbone-shaped grooves cut into either member. Figure 21.49 illustrates a partially grooved herringbone journal bearing. In this figure the groove and bearing parameters are also indicated. Figures 21.50–21.54 were obtained from Hamrock and Fleming[28] and are design charts that present curves for optimizing the design parameters for herringbone journal bearings for maximum radial load. The (*a*) portion of these figures is for the grooved member rotating and the (*b*) portion is for the smooth member rotating. The only groove parameter not represented in these figures is the number of grooves to be used. From Hamrock and Fleming[28] it was found that the *minimum* number of grooves to be placed around the journal can be represented by $n \geq \Lambda_j/5$.

More than any other factors, self-excited whirl instability and low-load capacity limit the usefulness of gas-lubricated journal bearings. The whirl problem is the tendency of the journal center to orbit the bearing center at an angular speed less than or equal to half that of the journal about its own center. In many cases the whirl amplitude is large enough to cause destructive contact of the bearing surfaces.

Figure 21.55, obtained from Fleming and Hamrock,[29] shows the stability attained by the optimized herringbone bearings. In this figure the stability parameter $\overline{M}$ is introduced, where

$$\overline{M} = \frac{\overline{m} p_a h_r^5}{2R^5 l \eta^2}$$

and $\overline{m}$ is the mass supported by the bearing.

In Fig. 21.55, the bearings with the grooved member rotating are substantially more stable than those with the smooth member rotating, especially at high compressibility numbers.

Thrust Bearings

Two types of gas-lubricated thrust bearings have found the widest use in practical applications. These are the Rayleigh step and the spiral- or herringbone-groove bearings.

Rayleigh Step Bearing. Figure 21.56 shows a Rayleigh step thrust bearing. In this figure the ridge region is where the film thickness is h_r and the step region is where the film thickness is h_s. The feed groove is the deep groove separating the end of the ridge region and the beginning of the next step region. Although not shown in the figure, the feed groove is orders of magnitude deeper than the film thickness h_r. A pad is defined as the section that includes ridge, step, and feed groove regions. The length of the feed groove is small relative to the length of the pad. It should be noted that each pad acts independently since the pressure profile is broken at the lubrication feed groove.

The load capacity and stiffness of a Rayleigh step thrust bearing are functions of the following parameters:

1. $\Lambda_t = 6 \eta u l / p_a h_r^2$, the dimensionless bearing number.
2. $\lambda_a = (b_r + b_s + b_g)/l$, length ratio.
3. $H_a = h_s/h_r$, film thickness ratio.

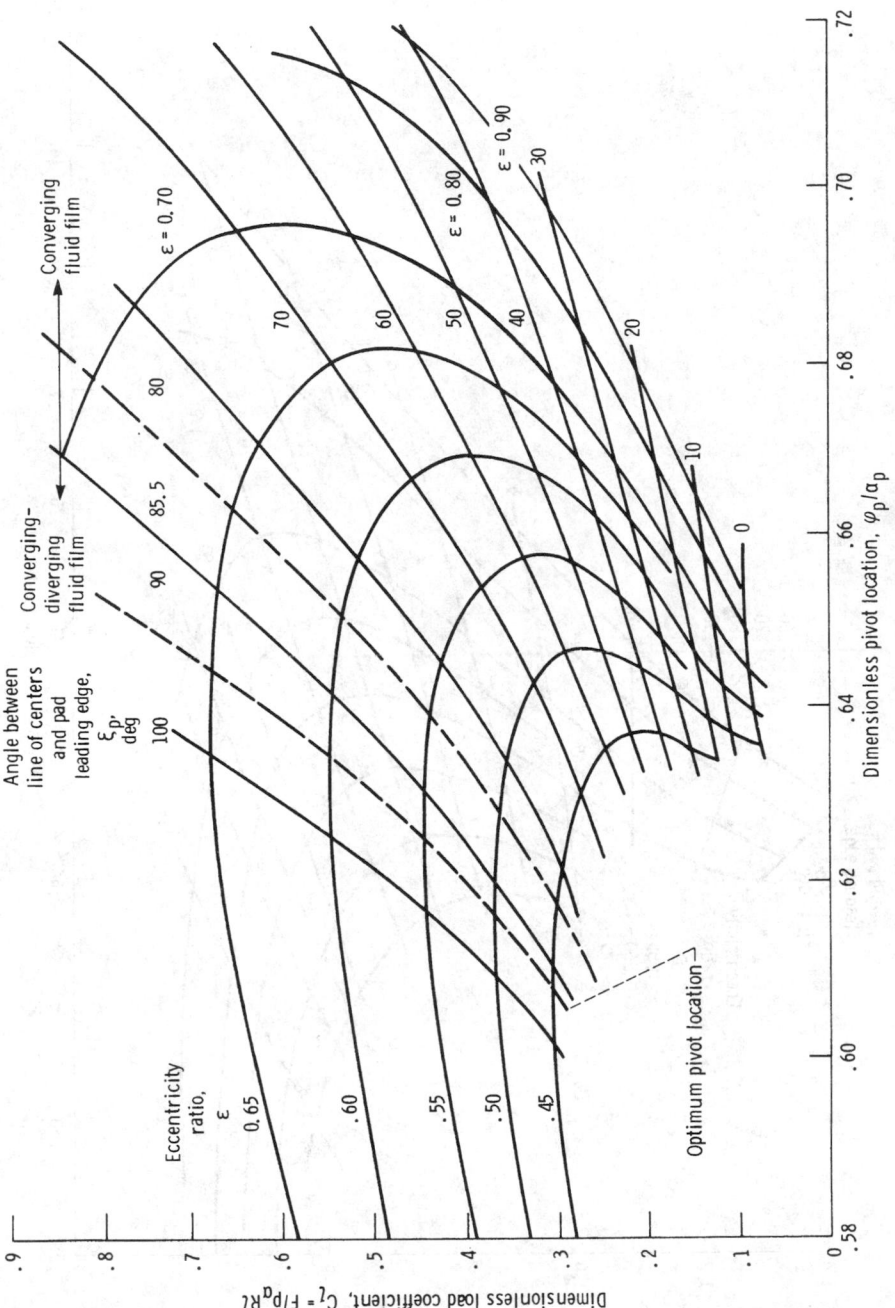

Fig. 21.43 Chart for determining load coefficient. Bearing radius-to-length ratio, R/l, 0.6061; angular extent of pad, α_p, 94.5°; dimensionless bearing number, A_j, 3.5. (From Ref. 27.)

549

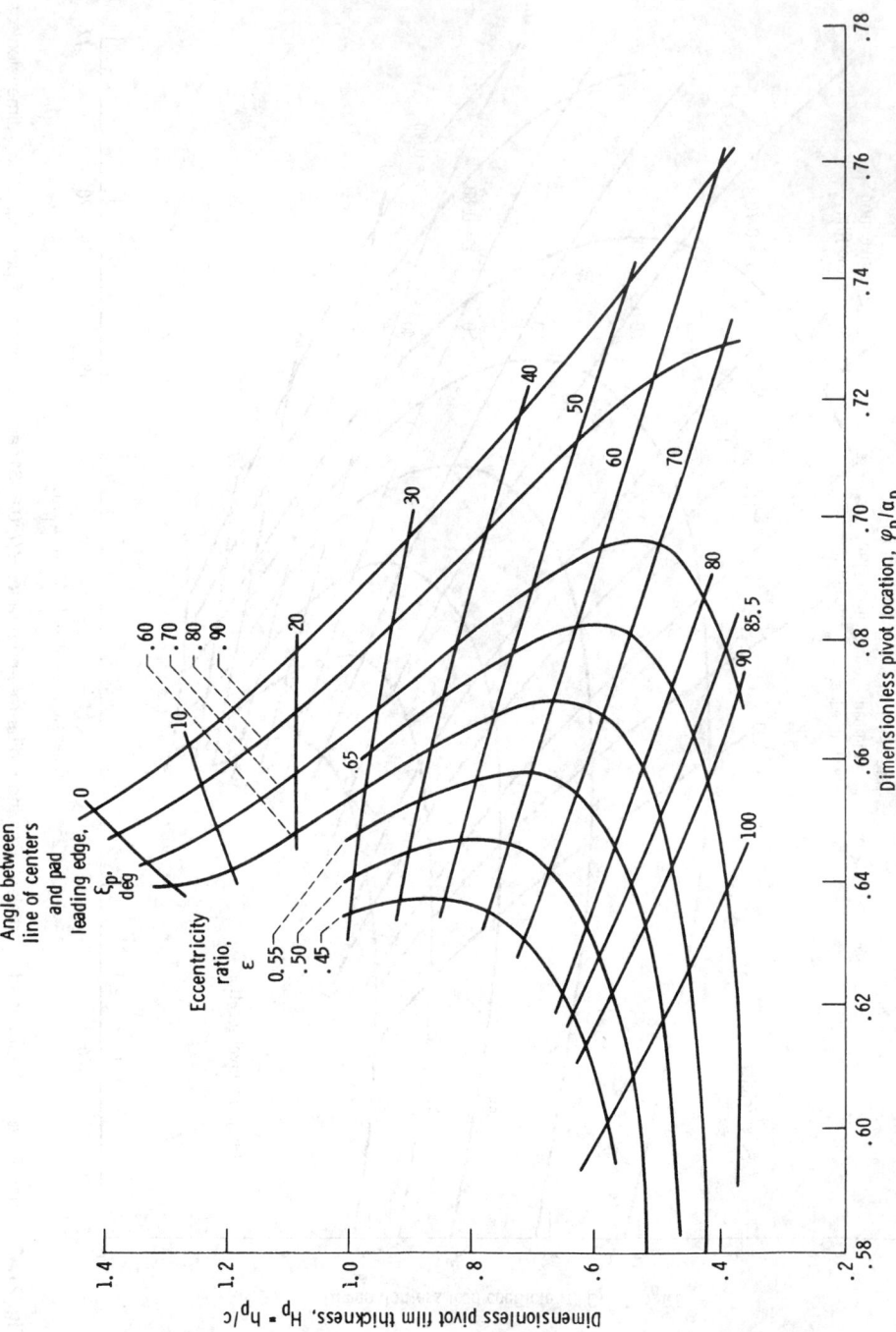

Fig. 21.44 Chart for determining pivot film thickness. Bearing radius-to-length ratio, R/l, 0.6061; angular extent of pad, α_p, 94.5°; dimensionless bearing number Λ, 3.5. (From Ref. 27.)

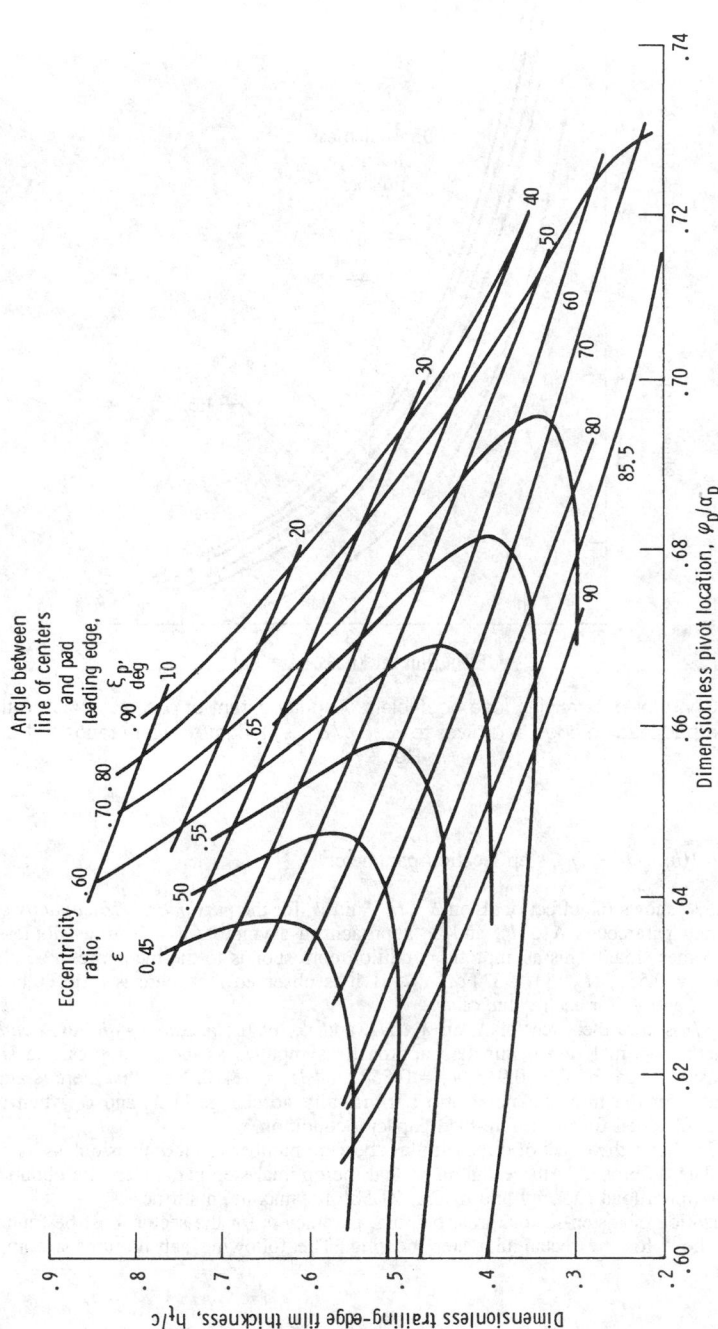

Fig. 21.45 Chart for determining trailing-edge film thickness. Bearing radius-to-length ratio, R/l, 0.6061; angular extent of pad, α_p, 94.5°; dimensionless bearing number, A_j, 3.5. (From Ref. 27.)

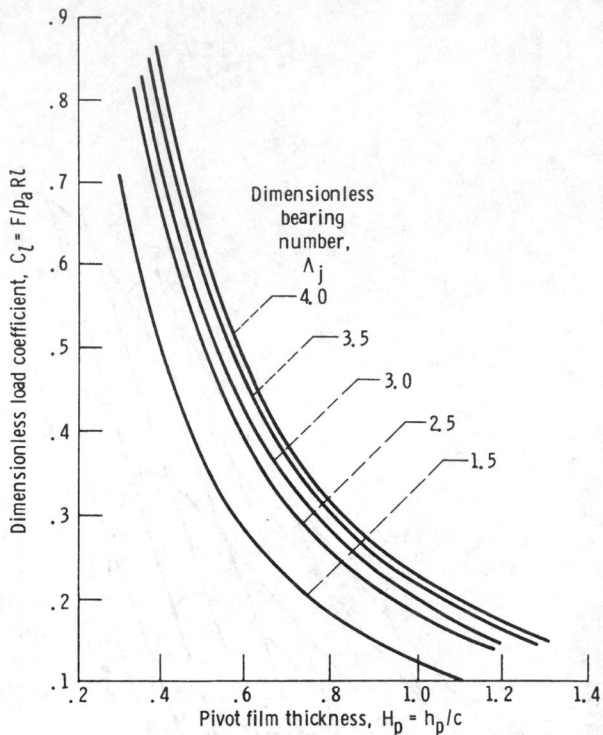

Fig. 21.46 Chart for determining load coefficient. Angular extent of pad, α_p, 94.5°; ratio of angle between pad leading edge and pivot to α_p, φ_p/α_p, $\frac{2}{3}$; length-to-width ratio, λ, 1.0. (From Ref. 27.)

4. $\psi_s = b_s/(b_s + b_r + b_g)$, step location parameter.

Figure 21.57a shows the effect of Λ_t on λ_a, H_a, and ψ_s for the *maximum-load-capacity condition*. The optimal step parameters Λ_a, H_a, and ψ_s approach an asymptote as the dimensionless bearing number Λ_t becomes small. This asymptotic condition corresponds to the incompressible solution or $\lambda_a = 0.918$, $\psi_s = 0.555$, $H_a = 1.693$. For $\Lambda_t > 1$ it is observed that there is a different optimum value of λ_a, H_a, and ψ_s for each value of Λ_t.

Figure 21.57b shows the effect of Λ_t on λ_a, H_a, and ψ_s for the *maximum-stiffness condition*. As in Fig. 21.57a the optimal step parameters approach asymptotes as the incompressible solution is reached. The asymptotes are $\lambda_a = 0.915$, $\psi_s = 0.557$, and $H_a = 1.470$. Note that there is a difference in the asymptote for the film thickness ratio but virtually no change in λ_a and ψ_s when compared with the results obtained for maximum-load-capacity condition.

Figure 21.58 shows the effect of dimensionless bearing number Λ_t on dimensionless load capacity and stiffness. The difference in these figures is that the optimal step parameters are obtained in Fig. 21.58a for maximum load capacity and in Fig. 21.58b for maximum stiffness.

For optimization of a step-sector thrust bearing, parameters for the sector must be found that are analogous to those for the rectangular step bearing. The following substitutions accomplish this transformation:

$$l \rightarrow R_1 - R_2$$
$$n(b_s + b_r + b_g) \rightarrow \pi(R_1 + R_2)$$
$$u \rightarrow \frac{\omega}{2}(R_1 + R_2)$$

where n is the number of pads placed in the step sector. By making use of these equations, the dimensionless bearing number can be rewritten as

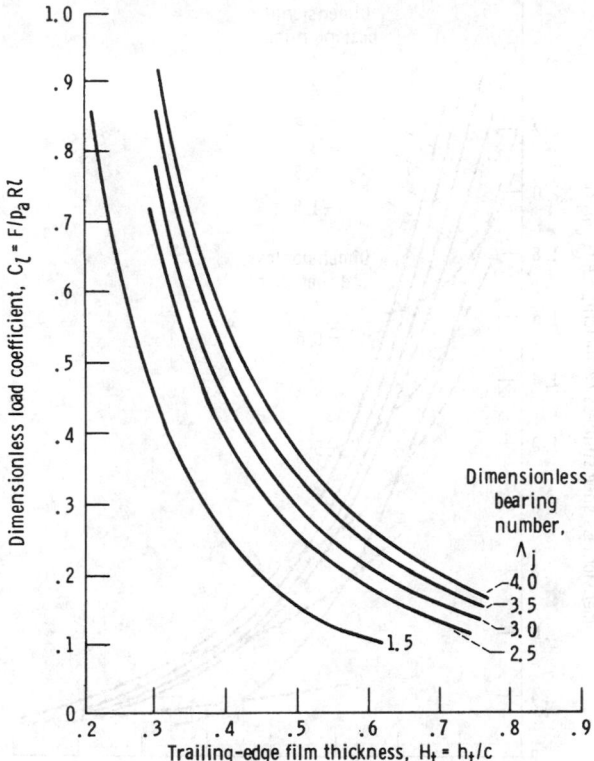

Fig. 21.47 Chart for determining load coefficient. Angular extent of pad, α_p, 94.5°; ratio of angle between pad loading edge and pivot to α_p, φ_p/α_p, $\frac{2}{3}$; length-to-width ratio, λ, 1.0. (From Ref. 27.)

$$\Lambda_c = \frac{3\eta\omega(R_1^2 - R_2^2)}{p_a h_r^2}$$

The optimal number of pads to be placed in the sector is obtained from the formula

$$n = \frac{\pi(R_1 + R_2)}{(\lambda_a)_{opt}(R_1 - R_2)}$$

where $(\lambda_a)_{opt}$ is obtained from Fig. 21.57a or 21.57b for a given dimensionless bearing number Λ_r. Since n will not normally be an integer, rounding it to the nearest integer is required. Therefore, through the parameter transformation discussed above, the results presented in Figs. 21.57 and 21.58 are directly usable in designing optimal step-sector gas-lubricated thrust bearings.

Spiral-Groove Thrust Bearings. An inward-pumping spiral-groove thrust bearing is shown in Fig. 21.59. An inward-pumping thrust bearing is somewhat more efficient than an outward-pumping thrust bearing and therefore is the only type considered here.

The dimensionless parameters normally associated with a spiral-groove thrust bearing are:

1. Angle of inclination, β_a.
2. Width ratio, $\bar{b} = b_s/b_r$.
3. Film ratio, $H_a = h_s/h_r$.
4. Radius ratio, $\alpha_r = R_2/R_1$.
5. Groove length fraction, $\bar{R} = (R_1 - R_g)/(R_1 - R_2)$.
6. Number of grooves, n.

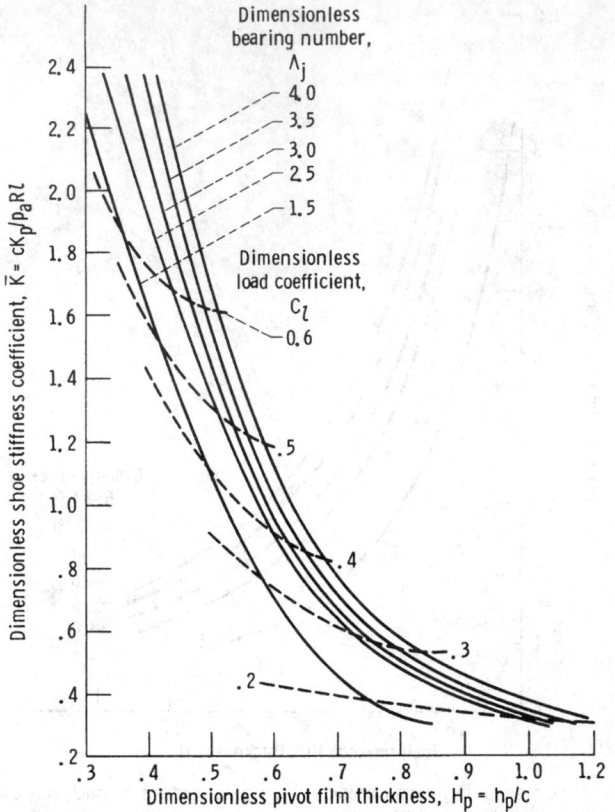

Fig. 21.48 Chart for determining shoe stiffness coefficient. (From Ref. 27.)

7. Dimensionless bearing number, $\Lambda_c = 3\eta\omega(R_1^2 - R_2^2)/p_a h_r^2$.

The first six parameters are geometrical parameters and the last parameter is an operating parameter.

The performance of spiral-groove thrust bearings is represented by the following dimensionless parameters:

Load

$$\overline{W}_\infty = \frac{1.5G_f F}{\pi p_a(R_1^2 - R_2^2)} \qquad (21.29)$$

Stiffness

$$\overline{K}_\infty = \frac{1.5h_r G_f K_p}{\pi p_a(R_1^2 - R_2^2)} \qquad (21.30)$$

Flow

$$\overline{Q} = \frac{3\eta Q}{\pi p_a h_r^3} \qquad (21.31)$$

Torque

$$\overline{T} = \frac{6T_r}{\pi p_a(R_1^2 + R_2^2)h_r\Lambda_c} \qquad (21.32)$$

When the geometrical and operating parameters are specified, the load, stiffness, flow, and torsion can be obtained.

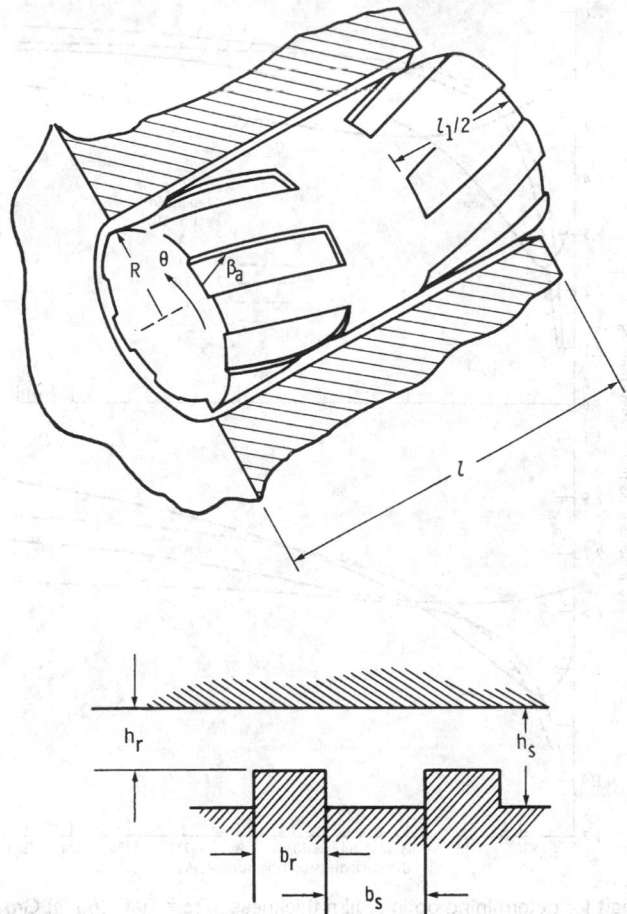

Fig. 21.49 Configuration of concentric herringbone-groove journal bearing. Bearing parameters; $\lambda = l/2R$; $\Lambda_j = 6\mu UR/p_a h_r^2$. Groove parameters; $H_a = h_s/h_r$; $\alpha_b = b_s/(b_r + b_s)$; β_a, $\gamma = l_1/l$; n. (From Ref. 28.)

The design charts of Reiger[20] are reproduced as Figs. 21.60–21.66. Figure 21.60 shows the dimensionless load for various radius ratios as a function of dimensionless bearing number Λ_c. This figure can be used to calculate the dimensionless load for a finite number of grooves; Fig. 21.61 can be used to determine the value of the groove factor. Figure 21.62 shows curves of dimensionless stiffness; Fig. 21.63 shows curves of dimensionless flow; and Fig. 21.64 shows curves of dimensionless torque. Optimized groove geometry parameters can be obtained from Fig. 21.65. Finally, Fig. 21.66 is used to calculate groove radius R_g (shown in Fig. 21.59). Figure 21.66 shows the required groove length fraction $\bar{R} = (R_o - R_g)/(R_o - R_i)$ to ensure stability from self-excited oscillations.

In a typical design problem the given factors are load, speed, bearing envelope, gas viscosity, ambient pressure, and an allowable radius-to-clearance ratio. The maximum value of the radius-to-clearance ratio is usually dictated by the distortion likely to occur to the bearing surfaces. Typical values are 5000–10,000. The procedure normally followed in designing a spiral-groove thrust bearing while using the design curves given in Figs. 21.60–21.66 is as follows:

1. Select the number of grooves n.
2. From Fig. 21.61 determine the groove factor G_f for given $\alpha_r = R_i/R_o$ and n.
3. Calculate $\bar{W}_\infty = 1.5 G_f F/\pi p_a(R_1^2 - R_2^2)$.
4. If $\bar{W}_\infty < 0.8$, R_1 must be increased. Return to step 2.
5. From Fig. 21.60, given $\bar{W}_\infty$ and α_r establish Λ_c.

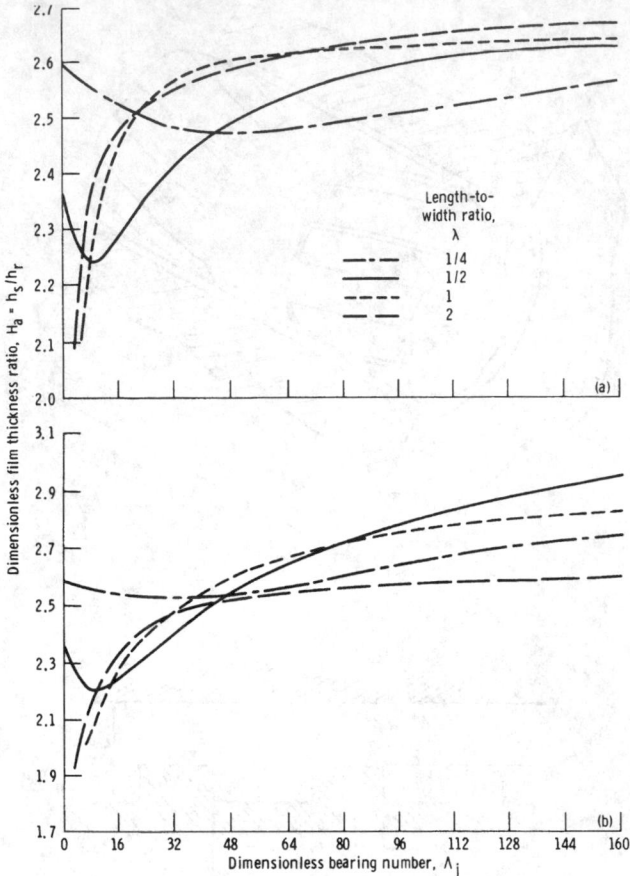

Fig. 21.50   Chart for determining optimal film thickness. (From Ref. 28.) (a) Grooved member rotating. (b) Smooth member rotating.

6. Calculate

$$\frac{R_1}{h_r} = \left\{ \frac{\Lambda_c p_a}{3\eta(\omega_h - \omega_o)[1 - (R_2/R_1)^2]} \right\}^{1/2}$$

If $R_1/h_r > 10{,}000$ (or whatever preassigned radius-to-clearance ratio), a larger bearing or higher speed is required. Return to step 2. If these changes cannot be made, an externally pressurized bearing must be used.

7. Having established what α_r and Λ_c should be, obtain values of $\overline{K}_\infty$, $\overline{Q}$, and $\overline{T}$ from Figs. 21.62, 21.63, and 21.64, respectively. From Eqs. (21.29), (21.30), and (21.31) calculate K_p, Q, and T_r.

8. From Fig. 21.65 obtain groove geometry (b, β_a, and H_a) and from Fig. 21.66 obtain R_g.

21.3   ELASTOHYDRODYNAMIC LUBRICATION

Downson[31] defines elastohydrodynamic lubrication (EHL) as "the study of situations in which elastic deformation of the surrounding solids plays a significant role in the hydrodynamic lubrication process." Elastohydrodynamic lubrication implies complete fluid-film lubrication and no asperity inter-action of the surfaces. There are two distinct forms of elastohydrodynamic lubrication.

1. *Hard EHL.*   Hard EHL relates to materials of high elastic modulus, such as metals. In this form of lubrication not only are the elastic deformation effects important, but the pressure–viscosity

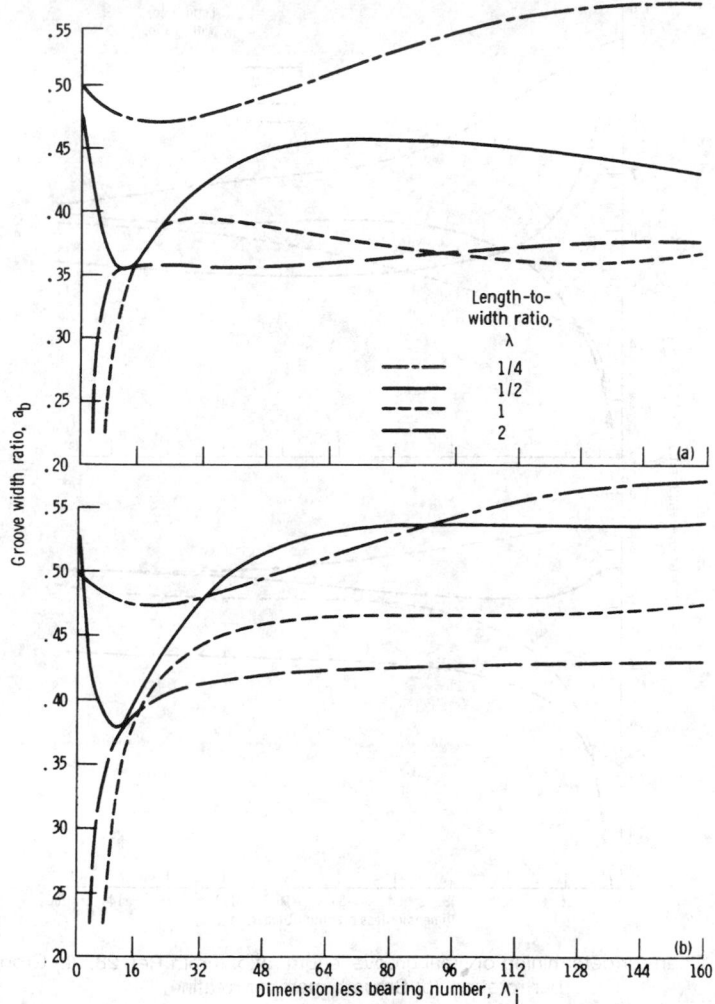

Fig. 21.51   Chart for determining optimal groove width ratio. (From Ref. 28.) (*a*) Grooved member rotating. (*b*) Smooth member rotating.

effects are equally as important. Engineering applications in which this form of lubrication is dominant include gears and rolling-element bearings.

2. *Soft EHL.* Soft EHL relates to materials of low elastic modulus, such as rubber. For these materials that elastic distortions are large, even with light loads. Another feature is the negligible pressure–viscosity effect on the lubricating film. Engineering applications in which soft EHL is important include seals, human joints, tires, and a number of lubricated elastomeric material machine elements.

The recognition and understanding of elastohydrodynamic lubrication presents one of the major developments in the field of tribology in this century. The revelation of a previously unsuspected regime of lubrication is clearly an event of importance in tribology. Elastohydrodynamic lubrication not only explained the remarkable physical action responsible for the effective lubrication of many machine elements, but it also brought order to the understanding of the complete spectrum of lubrication regimes, ranging from boundary to hydrodynamic.

A way of coming to an understanding of elastohydrodynamic lubrication is to compare it to hydrodynamic lubrication. The major developments that have led to our present understanding of hydrodynamic lubrication[1,3] predate the major developments of elastohydrodynamic lubrication[32,33]

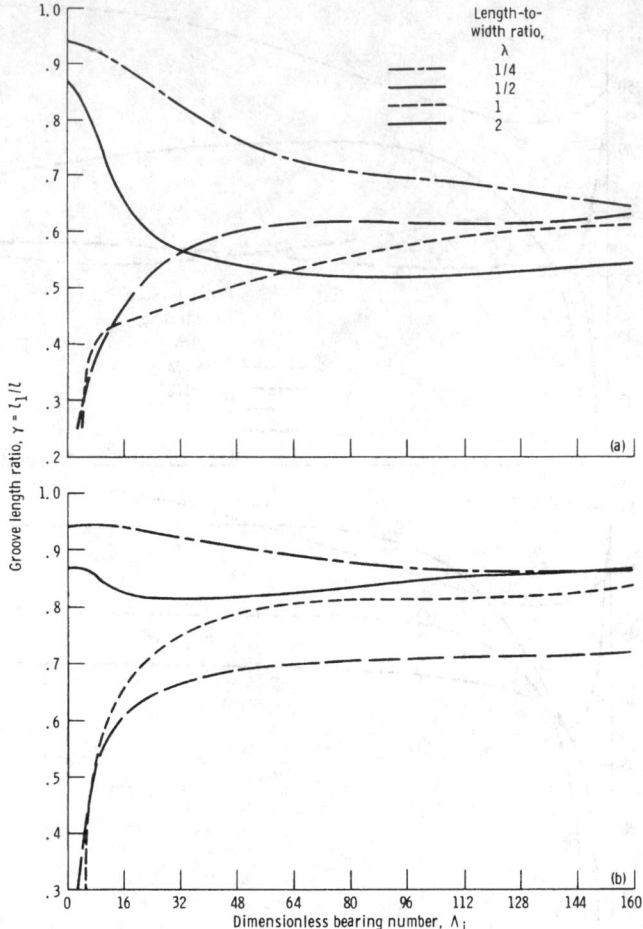

Fig. 21.52 Chart for determining optimal groove length ratio. (From Ref. 28.) (*a*) Grooved member rotating. (*b*) Smooth member rotating.

by 65 years. Both hydrodynamic and elastohydrodynamic lubrication are considered as fluid-film lubrication in that the lubricant film is sufficiently thick to prevent the opposing solids from coming into contact. Fluid-film lubrication is often referred to as the ideal form of lubrication since it provides low friction and high resistance to wear.

This section highlights some of the important aspects of elastohydrodynamic lubrication while illustrating its use in a number of applications. It is not intended to be exhaustive but to point out the significant features of this important regime of lubrication. For more details the reader is referred to Hamrock and Dowson.[10]

21.3.1 Contact Stresses and Deformations

As was pointed out in Section 21.1.1, elastohydrodynamic lubrication is the mode of lubrication normally found in nonconformal contacts such as rolling-element bearings. A load–deflection relationship for nonconformal contacts is developed in this section. The deformation within the contact is calculated from, among other things, the ellipticity parameter and the elliptic integrals of the first and second kinds. Simplified expressions that allow quick calculations of the stresses and deformations to be made easily from a knowledge of the applied load, the material properties, and the geometry of the contacting elements are presented in this section.

Elliptical Contacts

The undeformed geometry of contacting solids in a nonconformal contact can be represented by two ellipsoids. The two solids with different radii of curvature in a pair of principal planes (*x* and *y*)

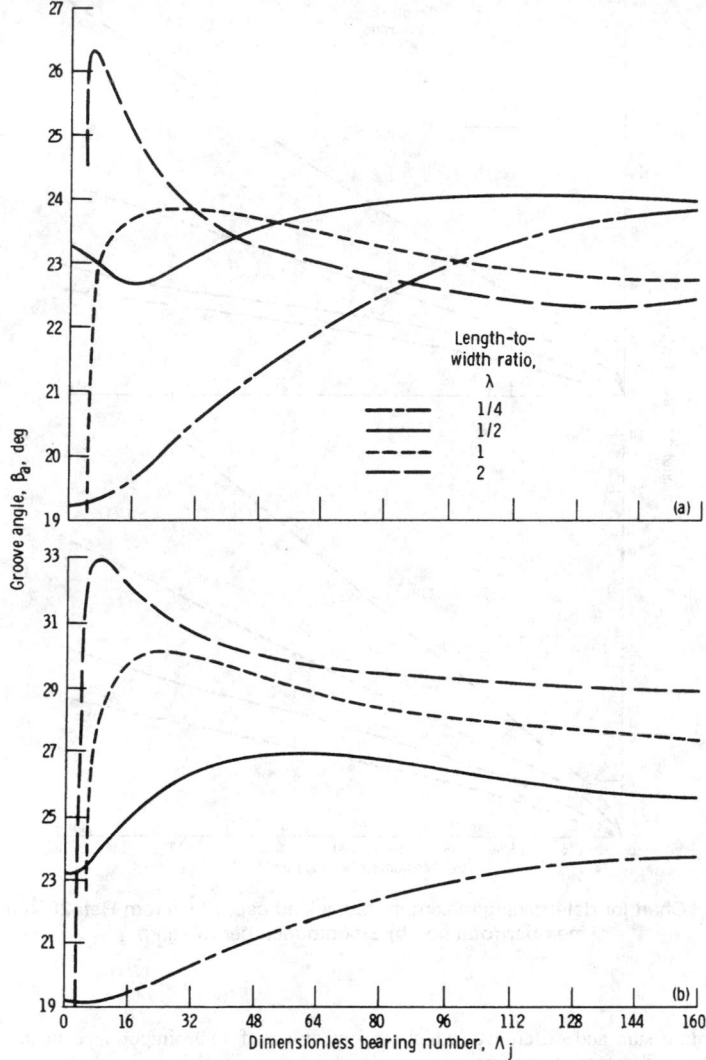

Fig. 21.53 Chart for determining optimal groove angle. (From Ref. 28.) (a) Grooved member rotating. (b) Smooth member rotating.

passing through the contact between the solids make contact at a single point under the condition of zero applied load. Such a condition is called point contact and is shown in Fig. 21.67, where the radii of curvature are denoted by r's. It is assumed that convex surfaces, as shown in Fig. 21.67, exhibit positive curvature and concave surfaces exhibit negative curvature. Therefore if the center of curvature lies within the solids, the radius of curvature is positive; if the center of curvature lies outside the solids, the radius of curvature is negative. It is important to note that if coordinates x and y are chosen such that

$$\frac{1}{r_{ax}} + \frac{1}{r_{bx}} > \frac{1}{r_{ay}} + \frac{1}{r_{by}}$$

(21.33)

coordinate x then determines the direction of the semiminor axis of the contact area when a load is applied and y determines the direction of the semimajor axis. The direction of motion is always considered to be along the x axis.

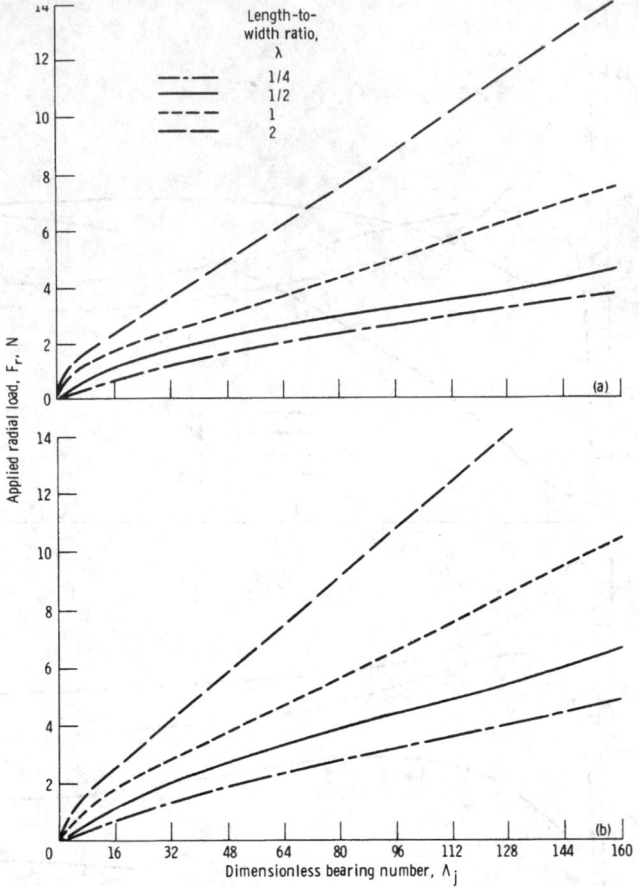

Fig. 21.54 Chart for determining maximum radial load capacity. (From Ref. 28.) (a) Grooved member rotating. (b) Smooth member rotating.

The curvature sum and difference, which are quantities of some importance in the analysis of contact stresses and deformations, are

$$\frac{1}{R} = \frac{1}{R_x} + \frac{1}{R_y} \tag{21.34}$$

$$\Gamma = R\left(\frac{1}{R_x} - \frac{1}{R_y}\right) \tag{21.35}$$

where

$$\frac{1}{R_x} = \frac{1}{r_{ax}} + \frac{1}{r_{bx}} \tag{21.36}$$

$$\frac{1}{R_y} = \frac{1}{r_{ay}} + \frac{1}{r_{by}} \tag{21.37}$$

$$\alpha = \frac{R_y}{R_x} \tag{21.38}$$

Equations (21.36) and (21.37) effectively redefine the problem of two ellipsoidal solids approaching one another in terms of an equivalent ellipsoidal solid of radii R_x and R_y approaching a plane.

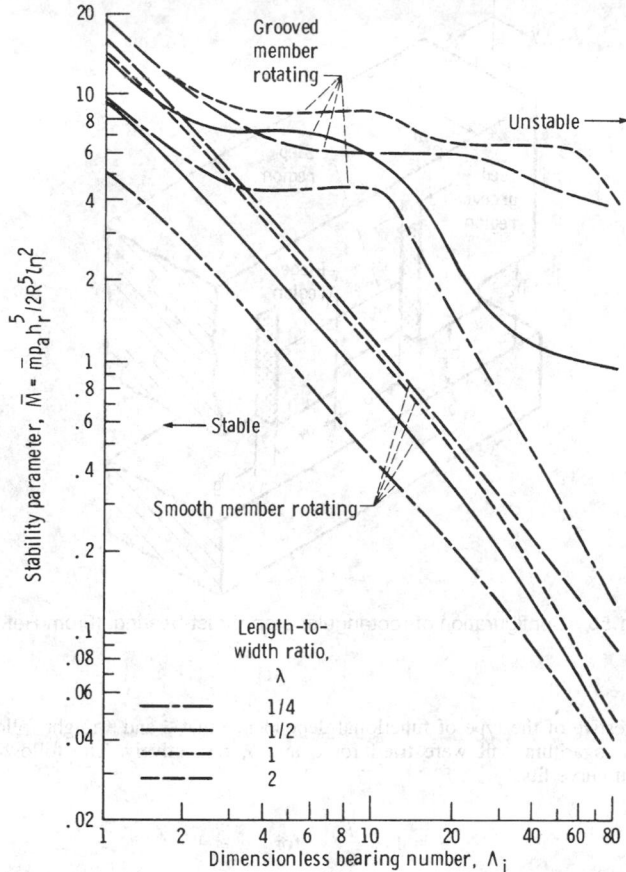

Fig. 21.55 Chart for determining maximum stability of herringbone-groove bearings. (From Ref. 29.)

The ellipticity parameter k is defined as the elliptical-contact diameter in the y direction (transverse direction) divided by the elliptical-contact diameter in the x direction (direction of motion) or $k = D_y/D_x$. If Eq. (21.33) is satisfied and $\alpha \geq 1$, the contact ellipse will be oriented so that its major diameter will be transverse to the direction of motion, and, consequently, $k \geq 1$. Otherwise, the major diameter would lie along the direction of motion with both $\alpha \leq 1$ and $k \leq 1$. Figure 21.68 shows the ellipticity parameter and the elliptic integrals of the first and second kinds for a range of curvature ratios ($\alpha = R_y/R_x$) usually encountered in concentrated contacts.

Simplified Solutions for $\alpha > 1$. The classical Hertzian solution requires the calculation of the ellipticity parameter k and the complete elliptic integrals of the first and second kinds $\mathcal{F}$ and $\mathcal{E}$. This entails finding a solution to a transcendental equation relating k, $\mathcal{F}$, and $\mathcal{E}$ to the geometry of the contacting solids. Possible approaches include an iterative numerical procedure, as described, for example, by Hamrock and Anderson,[35] or the use of charts, as shown by Jones.[36] Hamrock and Brewe[34] provide a shortcut to the classical Hertzian solution for the local stress and deformation of two elastic bodies in contact. The shortcut is accomplished by using simplified forms of the ellipticity parameter and the complete elliptic integrals, expressing them as functions of the geometry. The results of Hamrock and Brewe's work[34] are summarized here.

A power fit using linear regression by the method of least squares resulted in the following expression for the ellipticity parameter:

$$k = \alpha^{2/\pi}, \quad \text{for} \quad \alpha \geq 1 \tag{21.39}$$

The asymptotic behavior of $\mathcal{E}$ and $\mathcal{F}$ ($\alpha \rightarrow 1$ implies $\mathcal{E} \rightarrow \mathcal{F} \rightarrow \pi/2$, and $\alpha \rightarrow \infty$ implies $\mathcal{F} \rightarrow \infty$ and

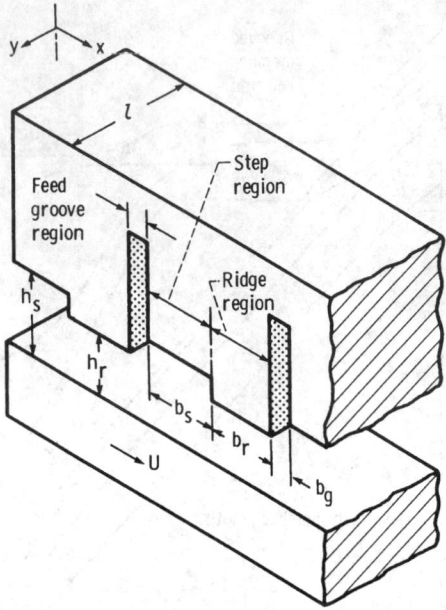

Fig. 21.56 Configuration of rectangular step thrust bearing. (From Ref. 30.)

$\mathscr{E} \to 1$) was suggestive of the type of functional dependence that $\mathscr{E}$ and $\mathscr{F}$ might follow. As a result, an inverse and a logarithmic fit were tried for $\mathscr{E}$ and $\mathscr{F}$, respectively. The following expressions provided excellent curve fits:

$$\mathscr{E} = 1 + \frac{q}{\alpha} \quad \text{for} \quad \alpha \ge 1 \tag{21.40}$$

$$\mathscr{F} = \frac{\pi}{2} + q \ln \alpha \quad \text{for} \quad \alpha \ge 1 \tag{21.41}$$

where

$$q = \frac{\pi}{2} - 1 \tag{21.42}$$

When the ellipticity parameter k [Eq. (21.39)], the elliptic integrals of the first and second kinds [Eqs. (21.40) and (21.41)], the normal applied load F, Poisson's ratio ν, and the modulus of elasticity E of the contacting solids are known, we can write the major and minor axes of the contact ellipse and the maximum deformation at the center of the contact, from the analysis of Hertz,[37] as

$$D_y = 2 \left(\frac{6k^2 \mathscr{E} FR}{\pi E'} \right)^{1/3} \tag{21.43}$$

$$D_x = 2 \left(\frac{6 \mathscr{E} FR}{\pi k E'} \right)^{1/3} \tag{21.44}$$

$$\delta = F \left[\left(\frac{9}{2 \mathscr{E} R} \right) \left(\frac{F}{\pi k E'} \right)^2 \right]^{1/3} \tag{21.45}$$

where [as in Eq. (21.12)]

$$E' = 2 \left(\frac{1 - \nu_a^2}{E_a} + \frac{1 - \nu_b^2}{E_b} \right)^{-1} \tag{21.46}$$

In these equations D_y and D_x are proportional to $F^{1/3}$ and δ is proportional to $F^{2/3}$.

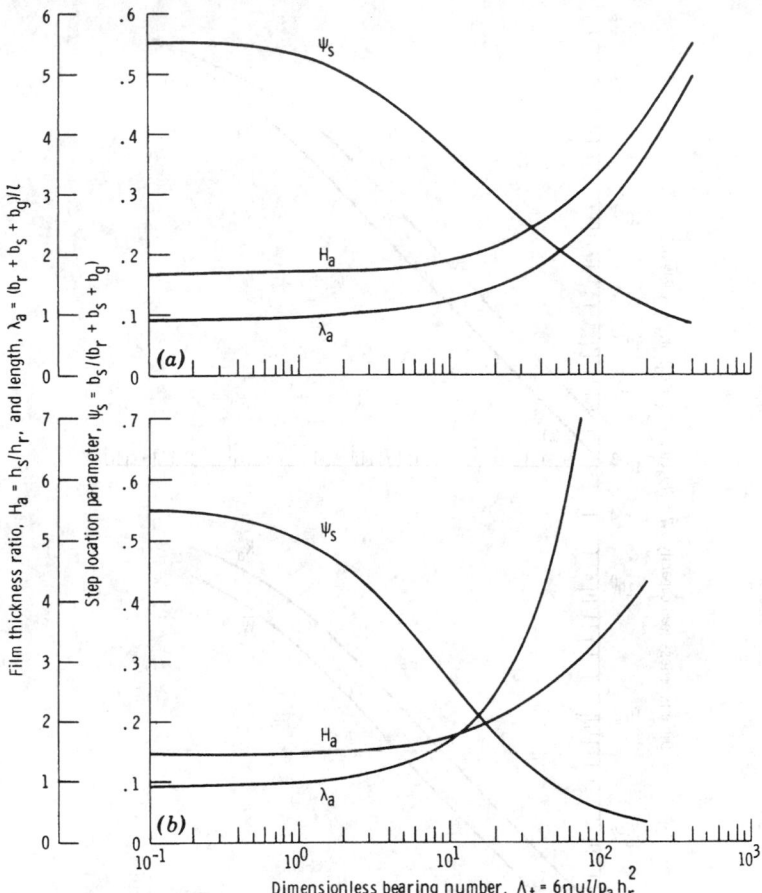

Fig. 21.57 Chart for determining optimal step parameters. (From Ref. 30.) (a) Maximum dimensionless load. (b) Maximum dimensionless stiffness.

The maximum Hertzian stress at the center of the contact can also be determined by using Eqs. (21.42) and (21.44)

$$\sigma_{max} = \frac{6F}{\pi D_x D_y} \tag{21.47}$$

Simplified Solutions for $\alpha \leq 1$. Table 21.7 gives the simplified equations for $\alpha < 1$ as well as for $\alpha \geq 1$. Recall that $\alpha \geq 1$ implies $k \geq 1$ and Eq. (21.33) is satisfied, and $\alpha < 1$ implies $k < 1$ and Eq. (21.33) is not satisfied. It is important to make the proper evaluation of α, since it has a great significance in the outcome of the simplified equations.

Figure 21.69 shows three diverse situations in which the simplified equations can be usefully applied. The locomotive wheel on a rail (Fig. 21.69a) illustrates an example in which the ellipticity parameter k and the radius ratio α are less than 1. The ball rolling against a flat plate (Fig. 21.69b) provides pure circular contact (i.e., $\alpha = k = 1.0$). Figure 21.69c shows how the contact ellipse is formed in the ball–outer-race contact of a ball bearing. Here the semimajor axis is normal to the direction of rolling and, consequently, α and k are greater than 1. Table 21.8 shows how the degree of conformity affects the contact parameters for the various cases illustrated in Fig. 21.69.

Rectangular Contacts

For this situation the contact ellipse discussed in the preceding section is of infinite length in the transverse direction ($D_y \rightarrow \infty$). This type of contact is exemplified by a cylinder loaded against a

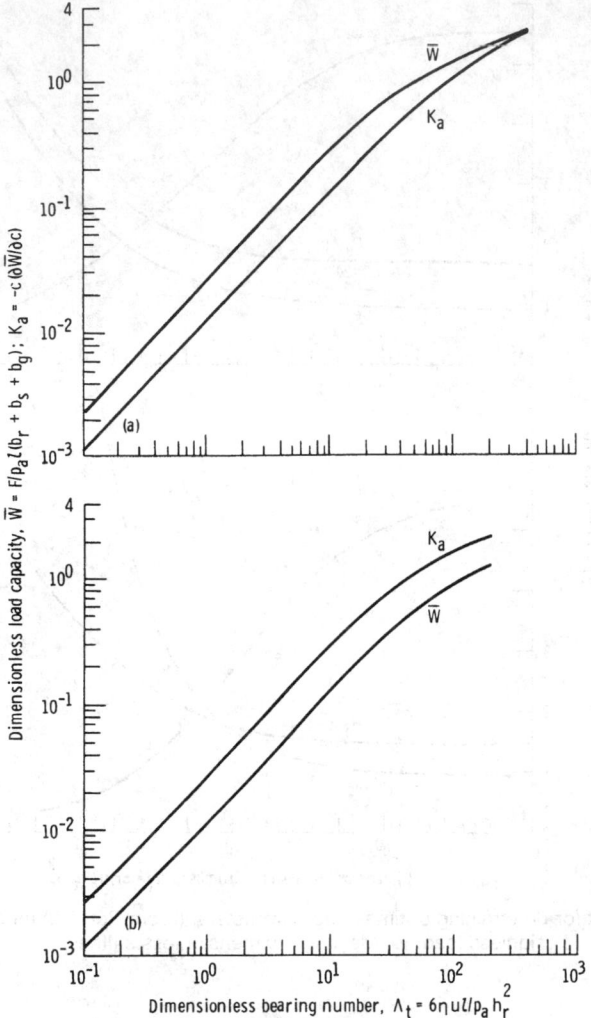

Fig. 21.58 Chart for determining dimensionless load capacity and stiffness. (From Ref. 30.) (a) Maximum dimensionless load capacity. (b) Maximum stiffness.

plate, a groove, or another parallel cylinder or by a roller loaded against an inner or outer ring. In these situations the contact semiwidth is given by

$$b = R_x \left(\frac{8W}{\pi}\right)^{1/2} \tag{21.48}$$

where

$$W = \frac{F'}{E'R_x} \tag{21.49}$$

and F' is the load per unit length along the contact.

The maximum deformation due to the approach of centers of two cylinders can be written as[12]

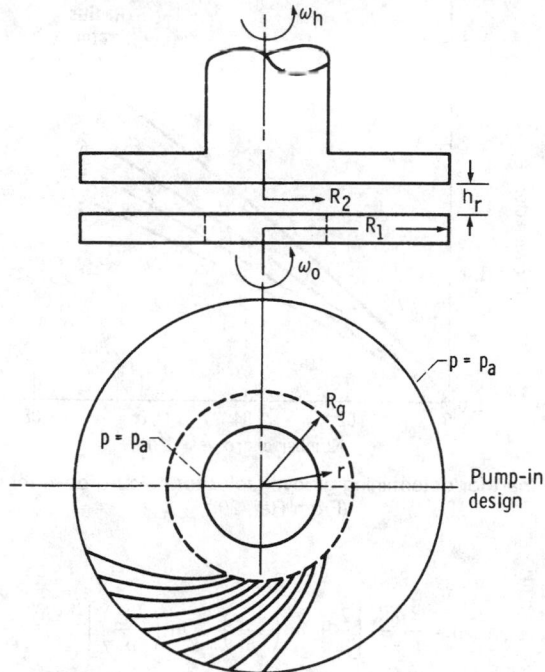

Fig. 21.59 Configuration of spiral-groove thrust bearing. (From Ref. 20.)

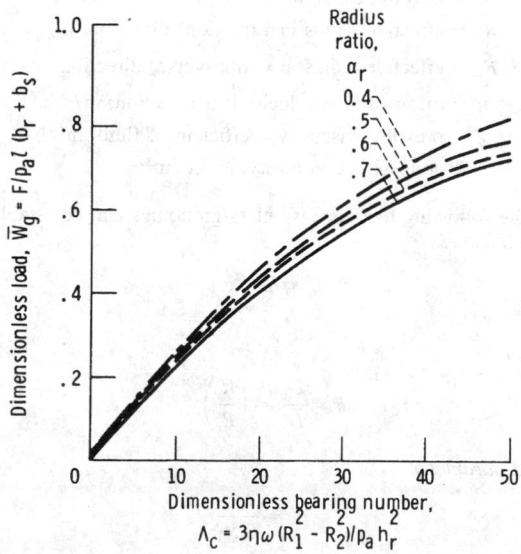

Fig. 21.60 Chart for determining load for spiral-groove thrust bearings. (From Ref. 20.)

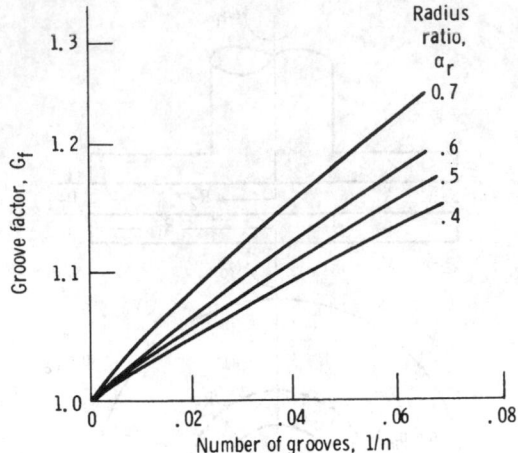

Fig. 21.61 Chart for determining groove factor for spiral-groove thrust bearings. (From Ref. 20.)

$$\delta = \frac{2WR_x}{\pi}\left[\frac{2}{3} + \ln\left(\frac{2r_{ax}}{b}\right) + \ln\left(\frac{2r_{bx}}{b}\right)\right] \tag{21.50}$$

The maximum Hertzian stress in a rectangular contact can be written as

$$\sigma_{\max} = E'\left(\frac{W}{2\pi}\right)^{1/2} \tag{21.51}$$

21.3.2 Dimensionless Grouping

The variables appearing in elastohydrodynamic lubrication theory are

E' = effective elastic modulus, N/m^2

F = normal applied load, N

h = film thickness, m

R_x = effective radius in x (motion) direction, m

R_y = effective radius in y (transverse) direction, m

u = mean surface velocity in x direction, m/sec

ξ = pressure–viscosity coefficient of fluid, m^2/N

η_0 = atmospheric viscosity, $N\ sec/m^2$;

From these variables the following five dimensionless groupings can be established.
Dimensionless film thickness

$$H = \frac{h}{R_x} \tag{21.52}$$

Ellipticity parameter

$$k = \frac{D_y}{D_x} = \left(\frac{R_y}{R_x}\right)^{2/\pi} \tag{21.53}$$

Dimensionless load parameter

$$W = \frac{F}{E'R_x^2} \tag{21.54}$$

Dimensionless speed parameter

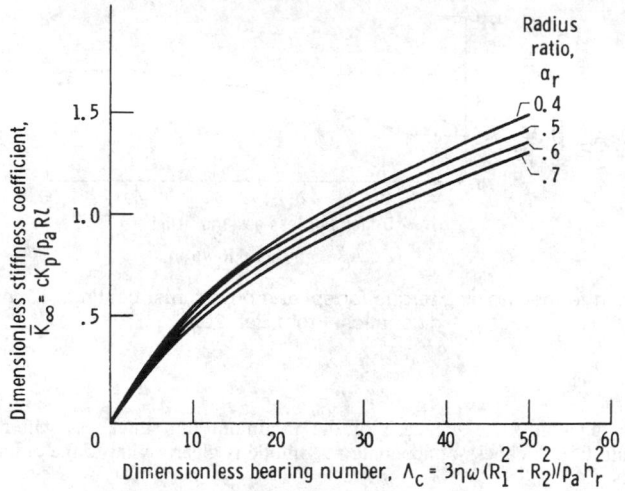

Fig. 21.62 Chart for determining stiffness for spiral-groove thrust bearings. (From Ref. 20.)

$$U = \frac{\eta_0 u}{E' R_x} \tag{21.55}$$

Dimensionless materials parameter

$$G = \xi E' \tag{21.56}$$

The dimensionless minimum film thickness can now be written as a function of the other parameters involved:

$$H = f(k, U, W, G)$$

The most important practical aspect of elastohydrodynamic lubrication theory becomes the deter-

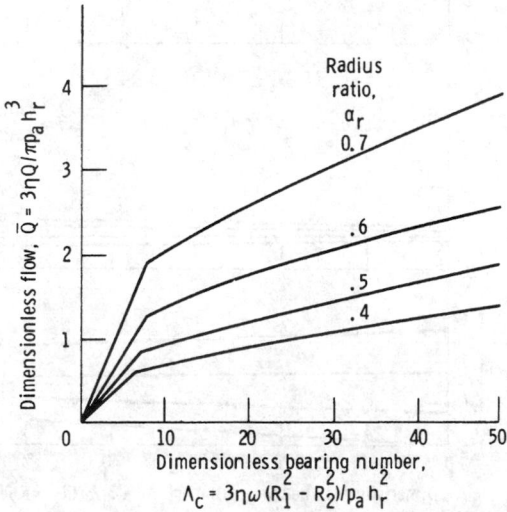

Fig. 21.63 Chart for determining flow for spiral-groove thrust bearings. (From Ref. 20.)

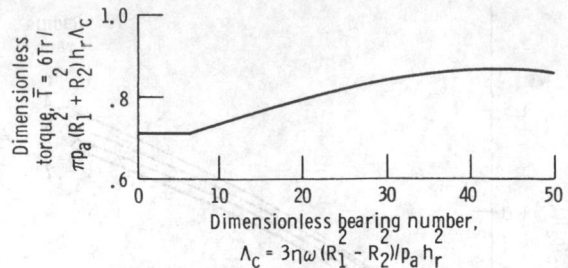

Fig. 21.64 Chart for determining torque for spiral-groove thrust bearings. (Curve is for all radius ratios. From Ref. 20.)

mination of this function f for the case of the minimum film thickness within a conjunction. Maintaining a fluid-film thickness of adequate magnitude is clearly vital to the efficient operation of machine elements.

21.3.3 Hard-EHL Results

By using the numerical procedures outlined in Hamrock and Dowson,[38] the influence of the ellipticity parameter and the dimensionless speed, load, and materials parameters on minimum film thickness was investigated by Hamrock and Dowson.[39] The ellipticity parameter k was varied from 1 (a ball-on-plate configuration) to 8 (a configuration approaching a rectangular contact). The dimensionless speed parameter U was varied over a range of nearly two orders of magnitude, and the dimensionless load parameter W over a range of one order of magnitude. Situations equivalent to using materials of bronze, steel, and silicon nitride and lubricants of paraffinic and naphthenic oils were considered in the investigation of the role of the dimensionless materials parameter G. Thirty-four cases were used in generating the minimum-film-thickness formula for hard EHL given here:

$$H_{\min} = 3.63 \, U^{0.68} G^{0.49} W^{-0.073} (1 - e^{-0.68k}) \qquad (21.57)$$

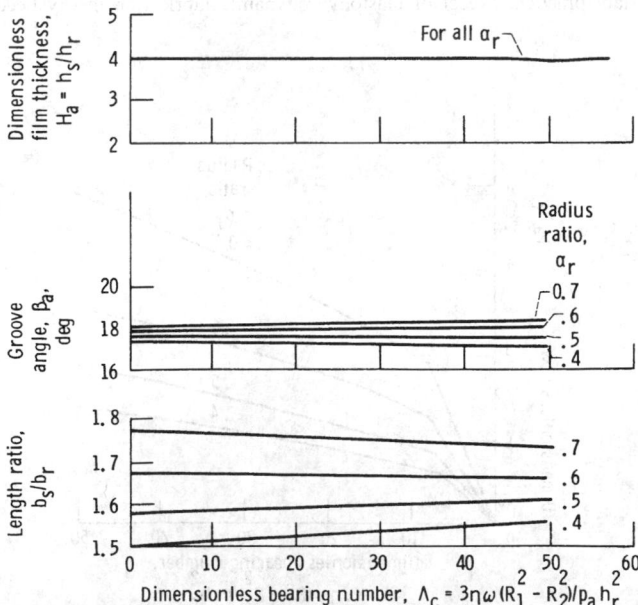

Fig. 21.65 Chart for determining optimal groove geometry for spiral-groove thrust bearings. (from Ref. 20.)

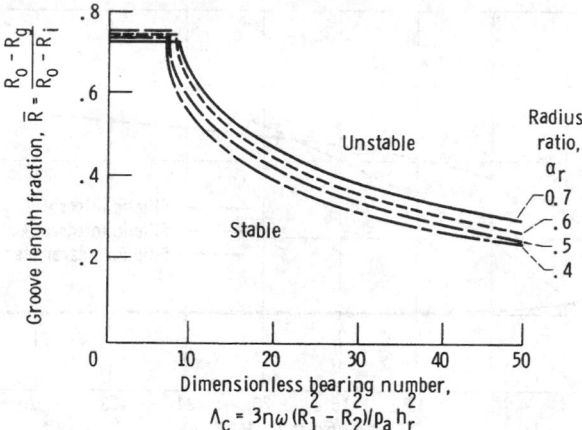

Fig. 21.66 Chart for determining groove length fraction for spiral-groove thrust bearings. (From Ref. 20.)

In this equation the dominant exponent occurs on the speed parameters, while the exponent on the load parameter is very small and negative. The materials parameter also carries a significant exponent, although the range of this variable in engineering situations is limited.

In addition to the minimum-film-thickness formula, contour plots of pressure and film thickness throughout the entire conjunction can be obtained from the numerical results. A representative contour plot of dimensionless pressure is shown in Fig. 21.70 for $k = 1.25$, $U = 0.168 \times 10^{-11}$, and $G = 4522$. In this figure and in Fig. 21.71, the $+$ symbol indicates the center of the Hertzian contact zone. The dimensionless representation of the X and Y coordinates causes the actual Hertzian contact ellipse to be a circle regardless of the value of the ellipticity parameter. The Hertzian contact circle is shown by asterisks. On this figure is a key showing the contour labels and each corresponding value of dimensionless pressure. The inlet region is to the left and the exit region is to the right. The pressure gradient at the exit end of the conjunction is much larger than that in the inlet region. In Fig. 21.70 a pressure spike is visible at the exit of the contact.

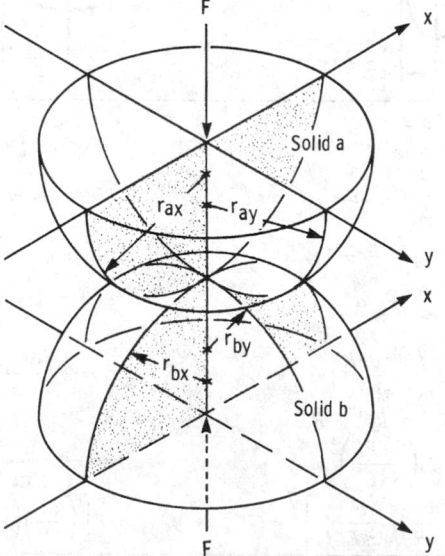

Fig. 21.67 Geometry of contacting elastic solids. (From Ref. 10.)

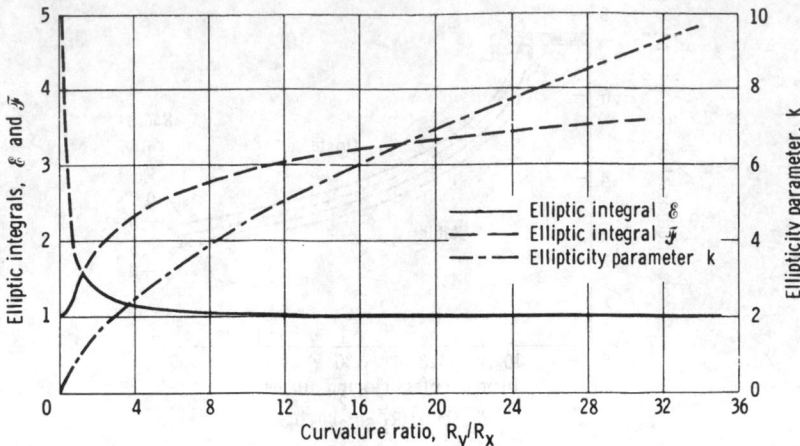

Fig. 21.68 Chart for determining ellipticity parameter and elliptic integrals of first and second kinds. (From Ref. 34.)

Contour plots of the film thickness are shown in Fig. 21.71 for the same case as Fig. 21.70. In this figure two minimum regions occur in well-defined lobes that follow, and are close to, the edge of the Hertzian contact circle. These results contain all of the essential features of available experimental observations based on optical interferometry.[40]

Table 21.7   Simplified Equations (From Ref. 6)

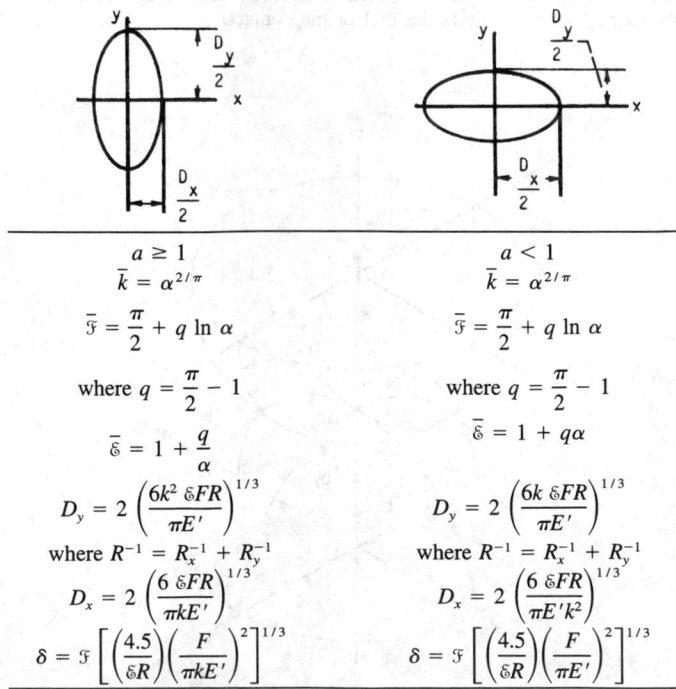

$a \geq 1$	$a < 1$
$\bar{k} = \alpha^{2/\pi}$	$\bar{k} = \alpha^{2/\pi}$
$\bar{\mathcal{F}} = \dfrac{\pi}{2} + q \ln \alpha$	$\bar{\mathcal{F}} = \dfrac{\pi}{2} + q \ln \alpha$
where $q = \dfrac{\pi}{2} - 1$	where $q = \dfrac{\pi}{2} - 1$
$\bar{\mathcal{E}} = 1 + \dfrac{q}{\alpha}$	$\bar{\mathcal{E}} = 1 + q\alpha$
$D_y = 2\left(\dfrac{6k^2\,\mathcal{E}FR}{\pi E'}\right)^{1/3}$	$D_y = 2\left(\dfrac{6k\,\mathcal{E}FR}{\pi E'}\right)^{1/3}$
where $R^{-1} = R_x^{-1} + R_y^{-1}$	where $R^{-1} = R_x^{-1} + R_y^{-1}$
$D_x = 2\left(\dfrac{6\,\mathcal{E}FR}{\pi kE'}\right)^{1/3}$	$D_x = 2\left(\dfrac{6\,\mathcal{E}FR}{\pi E'k^2}\right)^{1/3}$
$\delta = \mathcal{F}\left[\left(\dfrac{4.5}{\mathcal{E}R}\right)\left(\dfrac{F}{\pi kE'}\right)^2\right]^{1/3}$	$\delta = \mathcal{F}\left[\left(\dfrac{4.5}{\mathcal{E}R}\right)\left(\dfrac{F}{\pi E'}\right)^2\right]^{1/3}$

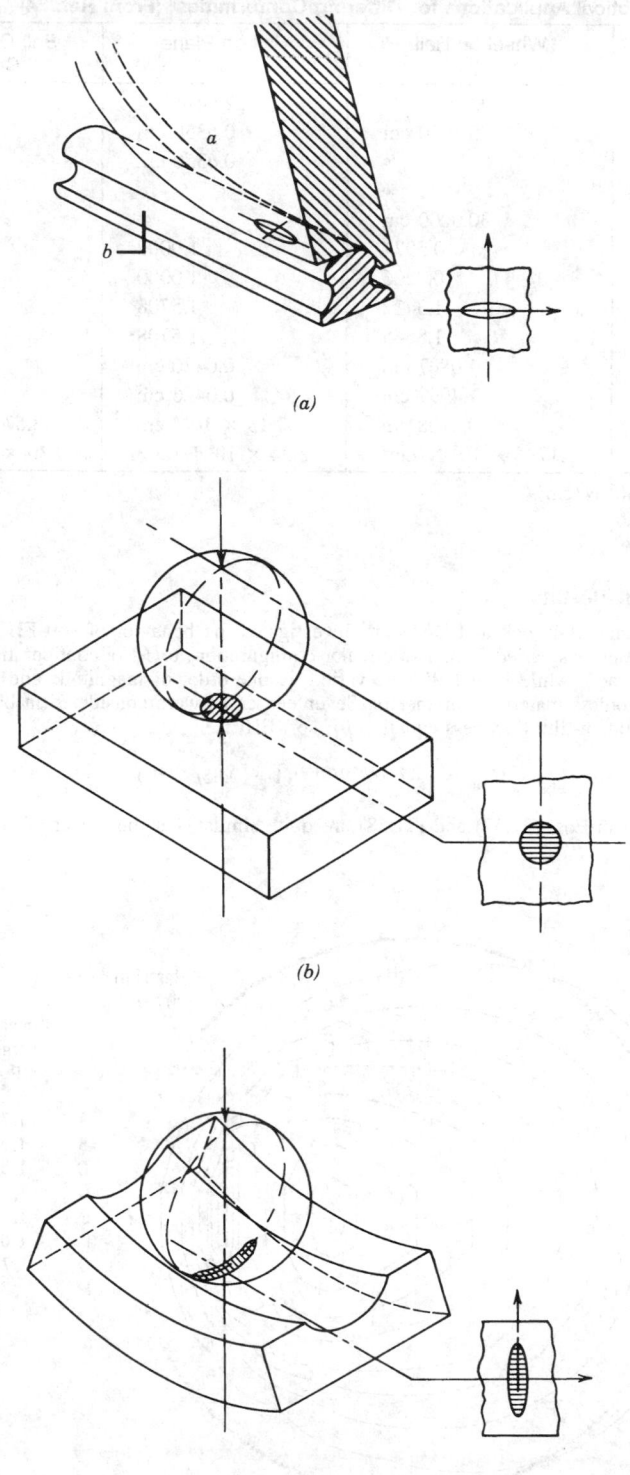

Fig. 21.69 Three degrees of conformity. (From Ref. 34.) (*a*) Wheel on rail. (*b*) Ball on plane.
(*c*) Ball–outer-race contact.

Table 21.8   Practical Applications for Differing Conformities[a] (From Ref. 34)

Contact Parameters	Wheel on Rail	Ball on Plane	Ball-Outer-Race Contact
F	1.00×10^5 N	222.4111 N	222.4111 N
r_{ax}	50.1900 cm	0.6350 cm	0.6350 cm
r_{ay}	∞	0.6350 cm	0.6350 cm
r_{bx}	∞	∞	-3.8900 cm
r_{by}	30.0000 cm	∞	-0.6600 cm
u	0.5977	1.0000	22.0905
k	0.7206	1.0000	7.1738
$\mathcal{E}$	1.3412	1.5708	1.0258
$\mathcal{F}$	1.8645	1.5708	3.3375
D_y	1.0807 cm	0.0426 cm	0.1810 cm
D_x	1.4997 cm	0.0426 cm	0.0252 cm
δ	0.0108 cm	7.13×10^{-4} cm	3.57×10^{-4} cm
σ_{max}	1.1784×10^5 N/cm^2	2.34×10^5 N/cm^2	9.30×10^4 N/cm^2

$^aE' = 2.197 \times 10^7$ N/cm^2.

21.3.4   Soft-EHL Results

In a similar manner, Hamrock and Dowson[41] investigated the behavior of soft-EHL contacts. The ellipticity parameter was varied from 1 (a circular configuration) to 12 (a configuration approaching a rectangular contact), while U and W were varied by one order of magnitude and there were two different dimensionless materials parameters. Seventeen cases were considered in obtaining the dimensionless minimum-film-thickness equation for soft EHL:

$$H_{min} = 7.43 U^{0.65} W^{-0.21}(1 - 0.85 e^{-0.31k}) \tag{21.58}$$

The powers of U in Eqs. (21.57) and (21.58) are quite similar, but the power of W is much more

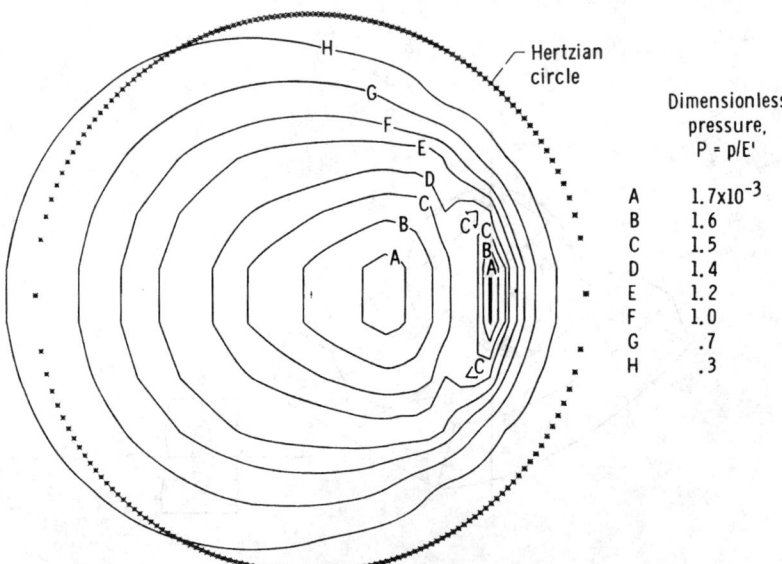

	Dimensionless pressure, $P = p/E'$
A	1.7×10^{-3}
B	1.6
C	1.5
D	1.4
E	1.2
F	1.0
G	.7
H	.3

Fig. 21.70   Contour plot of dimensionless pressure. $k = 1.25$; $U = 0.168 \times 10^{-11}$; $W = 0.111 \times 10^{-6}$; $G = 4522$. (From Ref. 39.)

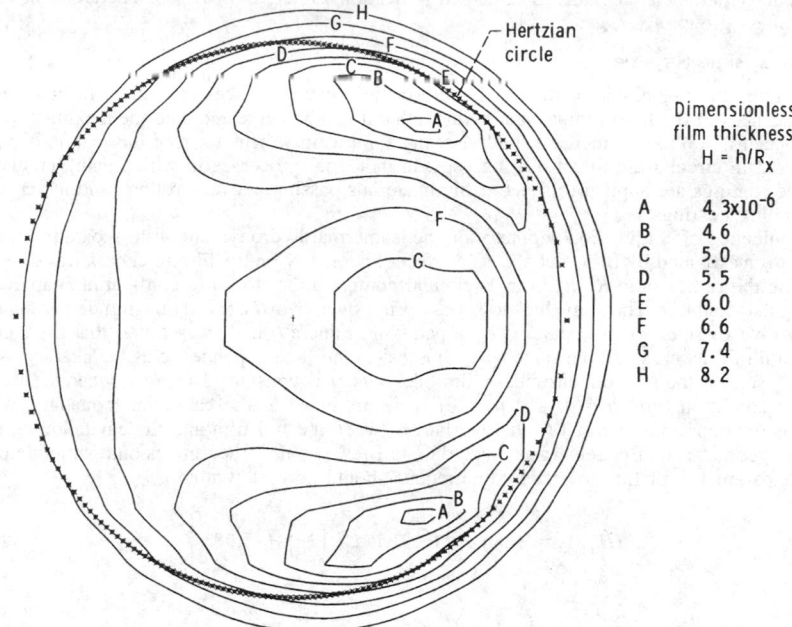

Dimensionless
film thickness,
$H = h/R_x$

A	4.3×10^{-6}
B	4.6
C	5.0
D	5.5
E	6.0
F	6.6
G	7.4
H	8.2

Fig. 21.71 Contour plot of dimensionless film thickness. $k = 1.25$; $U = 0.168 \times 10^{-11}$; $W = 0.111 \times 10^{-6}$; $G = 4522$. (From Ref. 39.)

significant for soft-EHL results. The expression showing the effect of the ellipticity parameter is of exponential form in both equations, but with quite different constants.

A major difference between Eqs. (21.57) and (21.58) is the absence of the materials parameter in the expression for soft EHL. There are two reasons for this: one is the negligible effect of the relatively low pressures on the viscosity of the lubricating fluid, and the other is the way in which the role of elasticity is automatically incorporated into the prediction of conjunction behavior through the parameters U and W. Apparently the chief effect of elasticity is to allow the Hertzian contact zone to grow in response to increases in load.

21.3.5 Film Thickness for Different Regimes of Fluid-Film Lubrication

The types of lubrication that exist within nonconformal contacts like that shown in Fig. 21.70 are influenced by two major physical effects: the elastic deformation of the solids under an applied load and the increase in fluid viscosity with pressure. Therefore, it is possible to have four regimes of fluid-film lubrication, depending on the magnitude of these effects and on their relative importance. In this section because of the need to represent the four fluid-film lubrication regimes graphically, the dimensionless grouping presented in Section 21.3.2 will need to be recast. That is, the set of dimensionless parameters given in Section 21.3.2 {H, U, W, G, and k}—will be reduced by one parameter without any loss of generality. Thus the dimensionless groupings to be used here are:

Dimensionless film parameter

$$\hat{H} = H \left(\frac{W}{U}\right)^2 \qquad (21.59)$$

Dimensionless viscosity parameter

$$g_v = \frac{GW^3}{U^2} \qquad (21.60)$$

Dimensionless elasticity parameter

$$g_e = \frac{W^{8/3}}{U^2}$$

The ellipticity parameter remains as discussed in Section 21.3.1, Eq. (21.39). Therefore the reduced dimensionless group is $\{\hat{H}, g_v, g_e, k\}$.

Isoviscous-Rigid Regime

In this regime the magnitude of the elastic deformation of the surfaces is such an insignificant part of the thickness of the fluid film separating them that it can be neglected, and the maximum pressure in the contact is too low to increase fluid viscosity significantly. This form of lubrication is typically encountered in circular-arc thrust bearing pads; in industrial processes in which paint, emulsion, or protective coatings are applied to sheet or film materials passing between rollers; and in very lightly loaded rolling bearings.

The influence of conjunction geometry on the isothermal hydrodynamic film separating two rigid solids was investigated by Brewe et al.[42] The effect of geometry on the film thickness was determined by varying the radius ratio R_y/R_x from 1 (circular configuration) to 36 (a configuration approaching a rectangular contact). The film thickness was varied over two orders of magnitude for conditions representative of steel solids separated by a paraffinic mineral oil. It was found that the computed minimum film thickness had the same speed, viscosity, and load dependence as the classical Kapitza solution,[43] so that the new dimensionless film thickness H is constant. However, when the Reynolds cavitation condition ($\partial p/\partial n = 0$ and $p = 0$) was introduced at the cavitation boundary, where n represents the coordinate normal to the interface between the full film and the cavitation region, an additional geometrical effect emerged. According to Brewe et al.,[42] the dimensionless minimum-film-thickness parameter for the isoviscous-rigid regime should now be written as

$$(\hat{H}_{\min})_{ir} = 128\alpha\lambda_b^2 \left[0.131 \tan^{-1}\left(\frac{\alpha}{2}\right) + 1.683 \right]^2 \tag{21.62}$$

where

$$\alpha = \frac{R_y}{R_x} \approx (k)^{\pi/2} \tag{21.63}$$

and

$$\lambda_b = \left(1 + \frac{2}{3\alpha} \right)^{-1} \tag{21.64}$$

In Eq. (21.62) the dimensionless film thickness parameter $\hat{H}$ is shown to be strictly a function only of the geometry of the contact described by the ratio $\alpha = R_y/R_x$.

Piezoviscous-Rigid Regime

If the pressure within the contact is sufficiently high to increase the fluid viscosity within the conjunction significantly, it may be necessary to consider the pressure–viscosity characteristics of the lubricant while assuming that the solids remain rigid. For the latter part of this assumption to be valid, it is necessary that the deformation of the surfaces remain an insignificant part of the fluid-film thickness. This form of lubrication may be encountered on roller end-guide flanges, in contacts in moderately loaded cylindrical tapered rollers, and between some piston rings and cylinder liners.

From Hamrock and Dowson[44] the minimum-film-thickness parameter for the piezoviscous-rigid regime can be written as

$$(\hat{H}_{\min})_{pvr} = 1.66 \, g_v^{2/3} (1 - e^{-0.68k}) \tag{21.65}$$

Note the absence of the dimensionless elasticity parameter g_e from Eq. (21.65).

Isoviscous-Elastic (Soft-EHL) Regime

In this regime the elastic deformation of the solids is a significant part of the thickness of the fluid film separating them, but the pressure within the contact is quite low and insufficient to cause any substantial increase in viscosity. This situation arises with materials of low elastic modulus (such as rubber), and it is a form of lubrication that may be encountered in seals, human joints, tires, and elastomeric material machine elements.

If the film thickness equation for soft EHL [Eq. (21.58)] is rewritten in terms of the reduced dimensionless grouping, the minimum-film-thickness parameter for the isoviscous-elastic regime can be written as

$$(\hat{H}_{\min})_{ie} = 8.70 \, g_e^{0.67} (1 - 0.85e^{-0.31k}) \tag{21.66}$$

Note the absence of the dimensionless viscosity parameter g_v from Eq. (21.66).

Piezoviscous-Elastic (Hard-EHL) Regime

In fully developed elastohydrodynamic lubrication the elastic deformation of the solids is often a significant part of the thickness of the fluid film separating them, and the pressure within the contact is high enough to cause a significant increase in the viscosity of the lubricant. This form of lubrication is typically encountered in ball and roller bearings, gears, and cams.

Once the film thickness equation [Eq. (21.57)] has been rewritten in terms of the reduced dimensionless grouping, the minimum film parameter for the piezoviscous-elastic regime can be written as

$$(\hat{H}_{min})_{pve} = 3.42 \, g_v^{0.49} g_e^{0.17} (1 - e^{-0.68k}) \tag{21.67}$$

An interesting observation to make in comparing Eqs. (21.65) through (21.67) is that in each case the sum of the exponents on g_v and g_e is close to the value of $\frac{2}{3}$ required for complete dimensional representation of these three lubrication regimes: piezoviscous-rigid, isoviscous-elastic, and piezoviscous-elastic.

Contour Plots

Having expressed the dimensionless minimum-film-thickness parameter for the four fluid-film regimes in Eqs. (21.62) to (21.67), Hamrock and Dowson[44] used these relationships to develop a map of the lubrication regimes in the form of dimensionless minimum-film-thickness parameter contours. Some of these maps are shown in Figs. 21.72–21.74 on a log-log grid of the dimensionless viscosity and elasticity parameters for ellipticity parameters of 1, 3, and 6, respectively. The procedure used to obtain these figures can be found in Ref. 44. The four lubrication regimes are clearly shown in Figs. 21.72–21.74. By using these figures for given values of the parameters k, g_v, and g_e, the fluid-film lubrication regime in which any elliptical conjunction is operating can be ascertained and the approximate value of $\hat{H}_{min}$ can be determined. When the lubrication regime is known, a more accurate value of $\hat{H}_{min}$ can be obtained by using the appropriate dimensionless minimum-film-thickness equation. These results are particularly useful in initial investigations of many practical lubrication problems involving elliptical conjunctions.

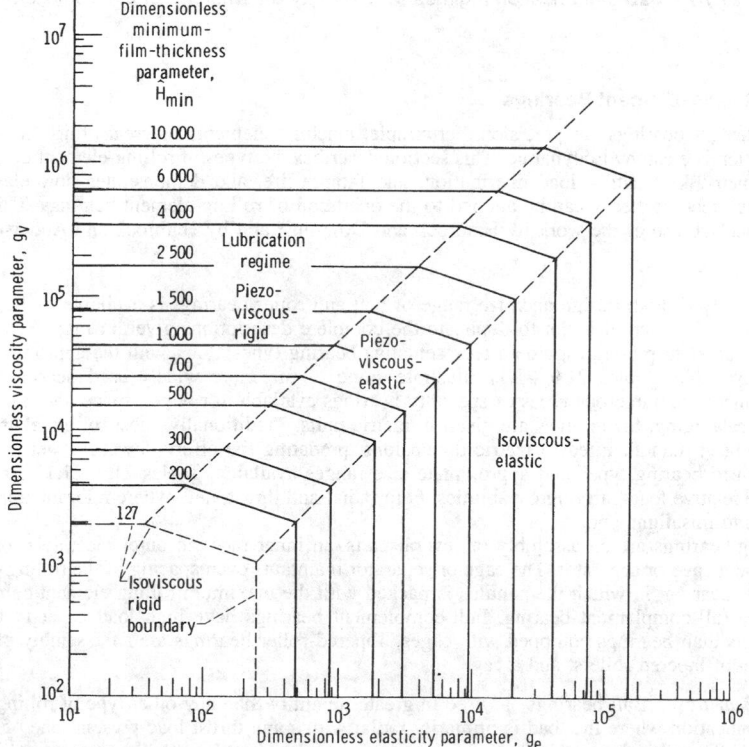

Fig. 21.72 Map of lubrication regimes for ellipticity parameter k of 1. (From Ref. 44.)

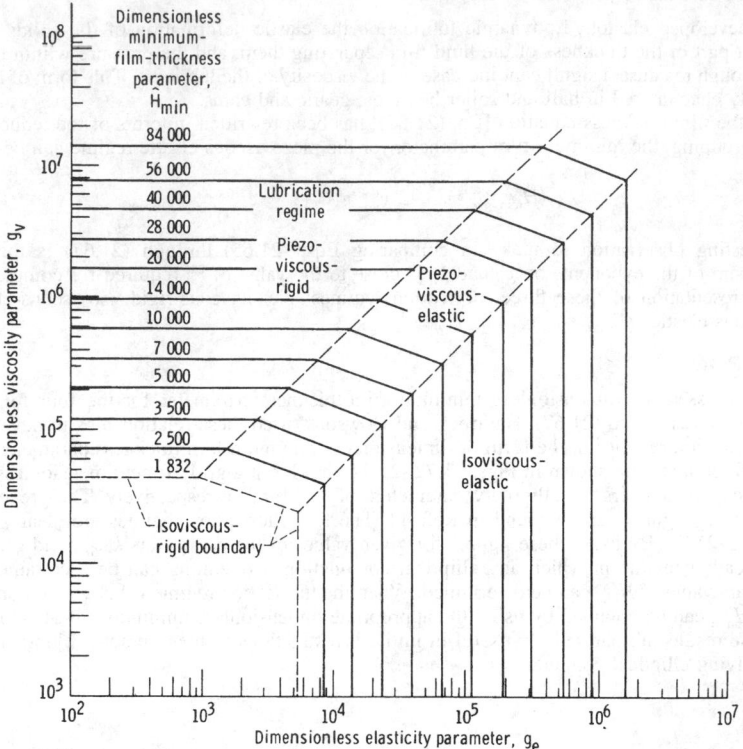

Fig. 21.73 Map of lubrication regimes for ellipticity parameter k of 3. (From Ref. 44.)

21.3.6 Rolling-Element Bearings

Rolling-element bearings are precision, yet simple, machine elements of great utility, whose mode of lubrication is elastohydrodynamic. This section describes the types of rolling-element bearings and their geometry, kinematics, load distribution, and fatigue life, and demonstrates how elastohydrodynamic lubrication theory can be applied to the operation of rolling-element bearings. This section makes extensive use of the work by Hamrock and Dowson[10] and by Hamrock and Anderson.[6]

Bearing Types

A great variety of both design and size range of ball and roller bearings is available to the designer. The intent of this section is not to duplicate the complete descriptions given in manufacturers' catalogs, but rather to present a guide a representative bearing types along with the approximate range of sizes available. Tables 21.9–21.17 illustrate some of the more widely used bearing types. In addition, there are numerous types of specialty bearings available for which space does not permit a complete cataloging. Size ranges are given in metric units. Traditionally, most rolling-element bearings have been manufactured to metric dimensions, predating the efforts toward a metric standard. In addition to bearing types and approximate size ranges available, Tables 21.9–21.17 also list approximate relative load-carrying capabilities, both radial and thrust, and, where relevant, approximate tolerances to misalignment.

Rolling bearings are an assembly of several parts—an inner race, an outer race, a set of balls or rollers, and a cage or separator. The cage or separator maintains even spacing of the rolling elements. A cageless bearing, in which the annulus is packed with the maximum rolling-element complement, is called a full-complement bearing. Full-complement bearings have high load capacity but lower speed limits than bearings equipped with cages. Tapered-roller bearings are an assembly of a cup, a cone, a set of tapered rollers, and a cage.

Ball Bearings.   Ball bearings are used in greater quantity than any other type of rolling bearing. For an application where the load is primarily radial with some thrust load present, one of the types in Table 21.9 can be chosen. A Conrad, or deep-groove, bearing has a ball complement limited by

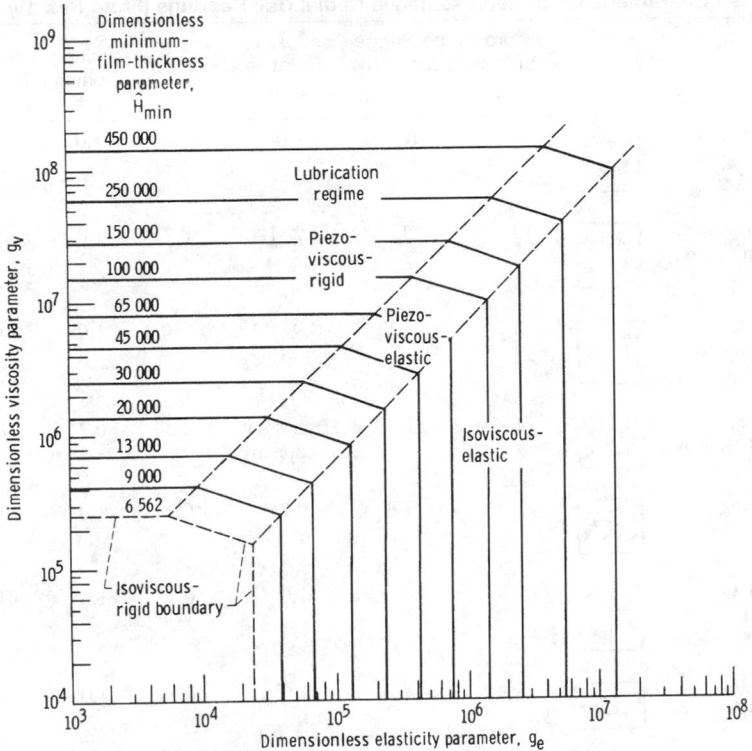

Fig. 21.74 Map of lubrication regimes for ellipticity parameter k of 6. (From Ref. 44.)

the number of balls that can be packed into the annulus between the inner and outer races with the inner race resting against the inside diameter of the outer race. A stamped and riveted two-piece cage, piloted on the ball set, or a machined two-piece cage, ball piloted or race piloted, is almost always used in a Conrad bearing. The only exception is a one-piece cage with open-sided pockets that is snapped into place. A filling-notch bearing has both inner and outer races notched so that a ball complement limited only by the annular space between the races can be used. It has low thrust capacity because of the filling notch.

The self-aligning internal bearing shown in Table 21.9 has an outer-race ball path ground in a spherical shape so that it can accept high levels of misalignment. The self-aligning external bearing has a multipiece outer race with a spherical interface. It too can accept high misalignment and has higher capacity than the self-aligning internal bearing. However, the external self-aligning bearing is somewhat less self-aligning than its internal counterpart because of friction in the multipiece outer race.

Representative angular-contact ball bearings are illustrated in Table 21.10. An angular-contact ball bearing has a two-shouldered ball groove in one race and a single-shouldered ball groove in the other race. Thus it is capable of supporting only a unidirectional thrust load. The cutaway shoulder allows assembly of the bearing by snapping over the ball set after it is positioned in the cage and outer race. This also permits use of a one-piece, machined, race-piloted cage that can be balanced for high-speed operation. Typical contact angles vary from 15° to 25°.

Angular-contact ball bearings are used in duplex pairs mounted either back to back or face to face as shown in Table 21.10. Duplex bearing pairs are manufactured so that they "preload" each other when clamped together in the housing and on the shaft. The use of preloading provides stiffer shaft support and helps prevent bearing skidding at light loads. Proper levels of preload can be obtained from the manufacturer. A duplex pair can support bidirectional thrust load. The back-to-back arrangement offers more resistance to moment or overturning loads than does the face-to-face arrangement.

Where thrust loads exceed the capability of a simple bearing, two bearings can be used in tandem, with both bearings supporting part of the thrust load. Three or more bearings are occasionally used

Table 21.9 Characteristics of Representative Radial Ball Bearings (From Ref. 10)

Type		Approximate Range of Bore Sizes, mm		Relative Capacity		Limiting Speed Factor	Tolerance to Misalignment
		Minimum	Maximum	Radial	Thrust		
Conrad or deep groove		3	1060	1.00	[a]0.7	1.0	±0°15′
Maximum capacity or filling notch		10	130	1.2–1.4	[a]0.2	1.0	±0°3′
Magneto or counterbored outer		3	200	0.9–1.3	[b]0.5–0.9	1.0	±0°5′
Airframe or aircraft control		4.826	31.75	High static capacity	[a]0.5	0.2	0°
Self-aligning, internal		5	120	0.7	[b]0.2	1.0	±2°30′
Self-aligning, external		—	—	1.0	[a]0.7	1.0	High
Double row, maximum		6	110	1.5	[a]0.2	1.0	±0°3′
Double row, deep groove		6	110	1.5	[a]1.4	1.0	0°

[a]Two directions.
[b]One direction.

in tandem, but this is discouraged because of the difficulty in achieving good load sharing. Even slight differences in operating temperature will cause a maldistribution of load sharing.

The split-ring bearing shown in Table 21.10 offers several advantages. The split ring (usually the inner) has its ball groove ground as a circular arc with a shim between the ring halves. The shim is then removed when the bearing is assembled so that the split-ring ball groove has the shape of a gothic arch. This reduces the axial play for a given radial play and results in more accurate axial positioning of the shaft. The bearing can support bidirectional thrust loads but must not be operated for prolonged periods of time at predominantly radial loads. This results in three-point ball–race contact and relatively high frictional losses. As with the conventional angular-contact bearing, a one-piece precision-machined cage is used.

Ball thrust bearings (90° contact angle), Table 21.11, are used almost exclusively for machinery with vertical oriented shafts. The flat-race bearing allows eccentricity of the fixed and rotating members. An additional bearing must be used for radial positioning. It has low load capacity because of the very small ball–race contacts and consequent high Hertzian stress. Grooved-race bearings have higher load capacities and are capable of supporting low-magnitude radial loads. All of the pure thrust ball bearings have modest speed capability because of the 90° contact angle and the consequent high level of ball spinning and frictional losses.

Roller Bearings. Cylindrical roller bearings, Table 21.12, provide purely radial load support in most applications. An N or U type of bearing will allow free axial movement of the shaft relative to the housing to accommodate differences in thermal growth. An F or J type of bearing will support a light thrust load in one direction; and a T type of bearing will support a light bidirectional thrust load.

Table 21.10 Characteristics of Representative Angular-Contact Ball Bearings (From Ref. 10)

Type		Approximate Range of Bore Sizes, mm		Relative Capacity		Limiting Speed Factor	Tolerance to Misalignment
		Minimum	Maximum	Radial	Thrust		
One-directional thrust		10	320	[b]1.00–1.15	[a,b]1.5–2.3	[b]1.1–3.0	±0°2'
Duplex, back to back		10	320	1.85	[c]1.5	3.0	0°
Duplex, face to face		10	320	1.85	[c]1.5	3.0	0°
Duplex, tandem		10	320	1.85	[a]2.4	3.0	0°
Two-directional or split ring		10	110	1.15	[c]1.5	3.0	±0°2'
Double row		10	140	1.5	[c]1.85	0.8	0°
Double row, maximum		10	110	1.65	[a]0.5 [d]1.5	0.7	0°

[a]One direction.
[b]Depends on contact angle.
[c]Two directions.
[d]In other direction.

Table 21.11 Characteristics of Representative Thrust Ball Bearings (From Ref. 10)

Type		Approximate Range of Bore Sizes, mm		Relative Capacity		Limiting Speed Factor	Tolerance to Misalignment
		Minimum	Maximum	Radial	Thrust		
One directional, flat race		6.45	88.9	0	[a]0.7	0.10	[b]0°
One directional, grooved race		6.45	1180	0	[a]1.5	0.30	0°
Two directional, grooved race		15	220	0	[c]1.5	0.30	0°

[a]One direction.
[b]Accepts eccentricity.
[c]Two directions.

Table 21.12   Characteristics of Representative Cylindrical Roller Bearings (From Ref. 10)

Type		Approximate Range of Bore Sizes, mm		Relative Capacity		Limiting Speed Factor	Tolerance to Misalignment
		Minimum	Maximum	Radial	Thrust		
Separable outer ring, nonlocating (RN, RIN)		10	320	1.55	0	1.20	±0°5′
Separable inner ring, nonlocating (RU, RIU)		12	500	1.55	0	1.20	±0°5′
Separable outer ring, one-direction locating (RF, RIF)		40	177.8	1.55	[a]Locating	1.15	±0°5′
Separable inner ring, one-direction locating (RJ, RIJ)		12	320	1.55	[a]Locating	1.15	±0°5′
Self-contained, two-direction locating		12	100	1.35	[b]Locating	1.15	±0°5′
Separable inner ring, two-direction locating (RT, RIT)		20	320	1.55	[b]Locating	1.15	±0°5′
Nonlocating, full complement (RK, RIK)		17	75	2.10	0	0.20	±0°5′
Double row, separable outer ring, nonlocating (RD)		30	1060	1.85	0	1.00	0°
Double row, separable inner ring, nonlocating		70	1060	1.85	0	1.00	0°

[a]One direction.
[b]Two directions.

Cylindrical roller bearings have moderately high radial load capacity as well as high-speed capability. Their speed capability exceeds that of either spherical or tapered-roller bearings. A commonly used bearing combination for support of a high-speed rotor is an angular-contact ball bearing or duplex pair and a cylindrical roller bearing.

As explained in the following section on bearing geometry, the rollers in cylindrical roller bearings are seldom pure cylinders. They are crowned or made slightly barrel shaped to relieve stress concentrations of the roller ends when any misalignment of the shaft and housing is present.

Cylindrical roller bearings may be equipped with one- or two-piece cages, usually race piloted. For greater load capacity, full-complement bearings can be used, but at a significant sacrifice in speed capability.

Table 21.13 Characteristics of Representative Spherical Roller Bearings (From Ref. 10)

Type		Approximate Range of Bore Sizes, mm		Relative Capacity		Limiting Speed Factor	Tolerance to Misalignment
		Minimum	Maximum	Radial	Thrust		
Single row, barrel or convex		20	320	2.10	0.20	0.50	±2°
Double row, barrel or convex		25	1250	2.40	0.70	0.50	±1°30′
Thrust		85	360	[a]0.10 [b]0.10	[a]1.80 [b]2.40	0.35–0.50	±3°
Double row, concave		50	130	2.40	0.70	0.50	±1°30′

[a]Symmetric rollers.
[b]Asymmetric rollers.

Spherical roller bearings, Tables 21.13–21.15, are made as either single- or double-row bearings. The more popular bearing design uses barrel-shaped rollers. An alternative design employs hourglass-shaped rollers. Spherical roller bearings combine very high radial load capacity with modest thrust load capacity (with the exception of the thrust type) and excellent tolerance to misalignment. They find widespread use in heavy-duty rolling mill and industrial gear drives, where all of these bearing characteristics are requisite.

Tapered-roller bearings, Table 21.16, are also made as single- or double-row bearings with combinations of one- or two-piece cups and cones. A four-row bearing assembly with two- or three-piece cups and cones is also available. Bearings are made with either a standard angle for applications in which moderate thrust loads are present or with a steep angle for high thrust capacity. Standard and special cages are available to suit the application requirements.

Single-row tapered-roller bearings must be used in pairs because a radially loaded bearing generates a thrust reaction that must be taken by a second bearing. Tapered-roller bearings are normally set up with spacers designed so that they operate with some internal play. Manufacturers' engineering journals should be consulted for proper setup procedures.

Needle roller bearings, Table 21.17, are characterized by compactness in the radial direction and are frequently used without an inner race. In the latter case the shaft is hardened and ground to serve

Table 21.14 Characteristics of Standardized Double-Row, Spherical Roller Bearings (From Ref. 10)

Type		Roller Design	Retainer Design	Roller Guidance	Roller-race Contact
SLB		Symmetric	Machined, roller piloted	Retainer pockets	Modified line, both races
SC		Symmetric	Stamped, race piloted	Floating guide ring	Modified line, both races
SD		Asymmetric	Machined, race piloted	Inner-ring center rib	Line contact, outer; point contact, inner

Table 21.15 Characteristics of Spherical Roller Bearings (From Ref. 10)

Series	Types	Approximate Range of Bore Sizes, mm		Approximate Relative Capacity[a]		Limiting Speed Factor
		Minimum	Maximum	Radial	Thrust	
202	Single-row barrel	20	320	1.0	0.11	0.5
203	Single-row barrel	20	240	1.7	.18	.5
204	Single-row barrel	25	110	2.1	.22	.4
212	SLB	35	75	1.0	.26	.6
213	SLB	30	70	1.7	.53	
22, 22K	SLB, SC, SD	30	320	1.7	.46	
23, 23K	SLB, SC, SD	40	280	2.7	1.0	
30, 30K	SLB, SC, SD	120	1250	1.2	.29	.7
31, 31K	SLB, SC, SD	110	1250	1.7	.54	.6
32, 32K	SLB, SC, SD	100	850	2.1	.78	.6
39, 39K	SD	120	1250	.7	.18	.7
40, 40K	SD	180	250	1.5	—	.7

[a]Load capacities are comparative within the various series of spherical roller bearings only. For a given envelope size, a spherical roller bearing has a radial capacity approximately equal to that of a cylindrical roller bearing.

as the inner race. Drawn cups, both open and closed end, are frequently used for grease retention. Drawn cups are thin walled and require substantial support from the housing. Heavy-duty roller bearings have relatively rigid races and are more akin to cylindrical roller bearings with long-length-to-diameter-ratio rollers.

Needle roller bearings are more speed limited than cylindrical roller bearings because of roller skewing at high speeds. A high percentage of needle roller bearings are full-complement bearings. Relative to a caged needle bearing, these have higher load capacity but lower speed capability.

There are many types of specialty bearings available other than those discussed here. Aircraft bearings for control systems, thin-section bearings, and fractured-ring bearings are some of the more widely used bearings among the many types manufactured. A complete coverage of all bearing types is beyond the scope of this chapter.

Angular-contact ball bearings and cylindrical roller bearings are generally considered to have the highest speed capabilities. Speed limits of roller bearings are discussed in conjunction with lubrication methods. The lubrication system employed has as great an influence on limiting bearing speed as does the bearing design.

Geometry

The operating characteristics of a rolling-element bearing depend greatly on the diametral clearance of the bearing. This clearance varies for the different types of bearings discussed in the preceding section. In this section, the principal geometrical relationships governing the operation of unloaded rolling-element bearings are developed. This information will be of vital interest when such quantities as stress, deflection, load capacity, and life are considered in subsequent sections. Although bearings rarely operate in the unloaded state, an understanding of this section is vital to the appreciation of the remaining sections.

Geometry of Ball Bearings

Pitch Diameter and Clearance.   The cross section through a radial, single-row ball bearing shown in Fig. 21.75 depicts the radial clearance and various diameters. The pitch diameter d_e is the mean of the inner- and outer-race contact diameters and is given by

$$d_e = d_i + \tfrac{1}{2}(d_o - d_i) \quad \text{or} \quad d_e = \tfrac{1}{2}(d_o + d_i) \tag{21.68}$$

Also from Fig. 21.75, the diametral clearance denoted by P_d can be written as

$$P_d = d_o - d_i - 2d \tag{21.69}$$

Diametral clearance may therefore be thought of as the maximum distance that one race can move diametrally with respect to the other when no measurable force is applied and both races lie in the

Table 21.16   Characteristics of Representative Tapered Roller Bearings (From Ref. 10)

Type		Subtype	Approximate Range of Bore Sizes, mm	
			Minimum	Maximum
Single row (TS)		TST—Tapered bore	8	1690
		TSS—Steep angle	24	430
		TS—Pin cage	16	1270
		TSE, TSK—keyway cones	—	
		TSF, TSSF—flanged cup	12	380
		TSG—steering gear (without cone)	8	1070
			—	
Two row, double cone, single cups (TDI)		TDIK, TDIT,	30	1200
		TDITP—tapered bore	30	860
		TDIE, TDIKE—slotted	24	690
		double cone	55	520
		TDIS—steep angle		
Two row, double cup, single cones, adjustable (TDO)		TDO	8	1830
		TDOS—steep angle	20	1430
Two row, double cup, single cones, nonadjustabe (TNA)		TNA	20	60
		TNASW—slotted cones	30	260
		TNASWE—extended cone rib	20	305
			8	70
		TNASWH—slotted cones, sealed	—	—
		TNADA, TNHDADX—self-aligning cup AD		
Four row, cup adjusted (TQO)		TQO, TQOT—tapered bore	70	1500
			250	1500
Four row cup adjusted (TQI)		TQIT—tapered bore	—	—

same plane. Although diametral clearance is generally used in connection with single-row radial bearings, Eq. (21.69) is also applicable to angular-contact bearings.

Race Conformity.   Race conformity is a measure of the geometrical conformity of the race and the ball in a plane passing through the bearing axis, which is a line passing through the center of the bearing perpendicular to its plane and transverse to the race. Figure 21.76 is a cross section of a ball bearing showing race conformity, expressed as

$$f = \frac{r}{d} \tag{21.70}$$

For perfect conformity, where the radius of the race is equal to the ball radius, f is equal to ½. The closer the race conforms to the ball, the greater the frictional heat within the contact. On the other hand, open-race curvature and reduced geometrical conformity, which reduce friction, also increase the maximum contact stresses and, consequently, reduce the bearing fatigue life. For this reason, most ball bearings made today have race conformity ratios in the range $0.51 \le f \le 0.54$, with $f = 0.52$ being the most common value. The race conformity ratio for the outer race is usually made slightly larger than that for the inner race to compensate for the closer conformity in the plane of the bearing between the outer race and ball than between the inner race and ball. This tends to equalize the contact stresses at the inner- and outer-race contacts. The difference in race conformities does not normally exceed 0.02.

Contact Angle.   Radial bearings have some axial play since they are generally designed to have a diametral clearance, as shown in Fig. 21.77. This implies a free-contact angle different from zero. Angular-contact bearings are specifically designed to operate under thrust loads. The clearance built

Table 21.17   Characteristics of Representative Needle Roller Bearings (From Ref. 10)

Type		Bore Sizes, mm		Relative Load Capacity		Limiting Speed Factor	Misalignment Tolerance
		Minimum	Maximum	Dynamic	Static		
Drawn cup, needle	Open end / Closed end	3	185	High	Moderate	0.3	Low
Drawn cup, needle, grease retained		4	25	High	Moderate	0.3	Low
Drawn cup, roller	Open end / Closed end	5	70	Moderate	Moderate	0.9	Moderate
Heavy-duty roller		16	235	Very high	Moderate	1.0	Moderate
Caged roller		12	100	Very high	High	1.0	Moderate
Cam follower		12	150	Moderate to high	Moderate to high	0.3–0.9	Low
Needle thrust		6	105	Very high	Very high	0.7	Low

into the unloaded bearing, along with the race conformity ratio, determines the bearing free-contact angle. Figure 21.77 shows a radial bearing with contact due to the axial shift of the inner and outer races when no measurable force is applied.

Before the free-contact angle is discussed, it is important to define the distance between the centers of curvature of the two races in line with the center of the ball in both Figs. 21.77a and 21.77b. This distance—denoted by x in Fig. 21.77a and by D in Fig. 21.77b—depends on race radius and ball diameter. Denoting quantities referred to the inner and outer races by subscripts i and o, respectively, we see from Figs. 21.77a and 21.77b that

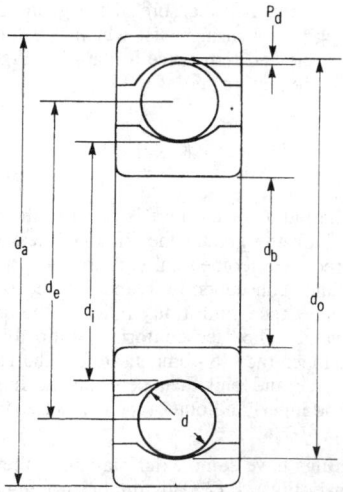

Fig. 21.75   Cross section through radial, single-row ball bearing. (From Ref. 10.)

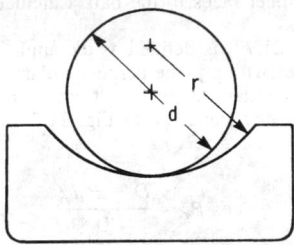

Fig. 21.76 Cross section of ball and outer race, showing race conformity. (From Ref. 10.)

$$\frac{P_d}{4} + d + \frac{P_d}{4} = r_o - x + r_i$$

or

$$x = r_o + r_i - d - \frac{P_d}{2}$$

and

$$d = r_o - D + r_i$$

or

$$D = r_o + r_i - d \tag{21.71}$$

From these equations, we can write

$$x = D - \frac{P_d}{2}$$

This distance, shown in Fig. 21.77, will be useful in defining the contact angle.
By using Eq. (21.70), we can write Eq. (21.71) as

$$D = Bd \tag{21.72}$$

where

$$B = f_o + f_i - 1 \tag{21.73}$$

The quantity B in Eq. (21.72) is known as the total conformity ratio and is a measure of the combined

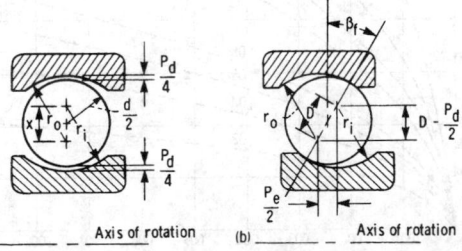

Fig. 21.77 Cross section of radial ball bearing, showing ball–race contact due to axial shift of inner and outer rings. (From Ref. 10.) (a) Initial position. (b) Shifted position.

conformity of both the outer and inner races to the ball. Calculations of bearing deflection in later sections depend on the quantity B.

The free-contact angle β_f (Fig. 21.77) is defined as the angle made by a line through the points of contact of the ball and both races with a plane perpendicular to the bearing axis of rotation when no measurable force is applied. Note that the centers of curvature of both the outer and inner races lie on the line defining the free-contact angle. From Fig. 21.77, the expression for the free-contact angle can be written as

$$\cos \beta_f = \frac{D - P_d/2}{D} \tag{21.74}$$

By using Eqs. (21.69) and (21.71), we can write Eq. (21.74) as

$$\beta_f = \cos^{-1}\left[\frac{r_o + r_i - \frac{1}{2}(d_o - d_i)}{r_o + r_i - d}\right] \tag{21.75}$$

Equation (21.75) shows that if the size of the balls is increased and everything else remains constant, the free-contact angle is decreased. Similarly, if the ball size is decreased, the free-contact angle is increased.

From Eq. (21.74) the diametral clearance P_d can be written as

$$P_d = 2D(1 - \cos \beta_f) \tag{21.76}$$

This is an alternative definition of the diametral clearance given in Eq. (21.69).

Endplay. Free endplay P_e is the maximum axial movement of the inner race with respect to the outer race when both races are coaxially centered and no measurable force is applied. Free endplay depends on total curvature and contact angle, as shown in Fig. 21.77, and can be written as

$$P_e = 2D \sin \beta_f \tag{21.77}$$

The variation of free-contact angle and free endplay with the ratio $P_d/2d$ is shown in Fig. 21.78 for

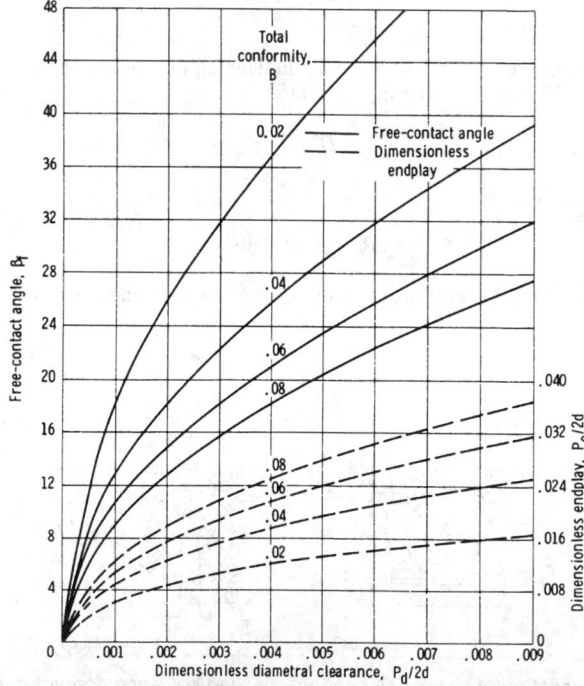

Fig. 21.78 Chart for determining free-contact angle and endplay. (From Ref. 10.)

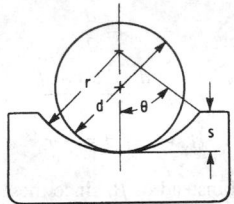

Fig. 21.79 Shoulder height in ball bearing. (From Ref. 10.)

four values of total conformity normally found in single-row ball bearings. Eliminating β_f in Eqs. (21.76) and (21.77) enables the establishment of the following relationships between free endplay and diametral clearance:

$$P_d = 2D - [(2D)^2 - P_c^2]^{1/2}$$

$$P_e = (4DP_d - P_d^2)^{1/2}$$

Shoulder Height. The shoulder height of ball bearings is illustrated in Fig. 21.79. Shoulder height, or race depth, is the depth of the race groove measured from the shoulder to the bottom of the groove and is denoted by s in Fig. 21.79. From this figure the equation defining the shoulder height can be written as

$$s = r(1 - \cos \theta) \tag{21.78}$$

The maximum possible diametral clearance for complete retention of the ball–race contact within the race under zero thrust load is given by

$$(P_d)_{\max} = \frac{2Ds}{r}$$

Curvature Sum and Difference. A cross section of a ball bearing operating at a contact angle β is shown in Fig. 21.80. Equivalent radii of curvature for both inner- and outer-race contacts in, and normal to, the direction of rolling can be calculated from this figure. The radii of curvature for the ball–inner-race contact are

$$r_{ax} = r_{ay} = \frac{d}{2} \tag{21.79}$$

$$r_{bx} = \frac{d_c - d \cos \beta}{2 \cos \beta} \tag{21.80}$$

$$r_{by} = -f_i d = -r_i \tag{21.81}$$

The radii of curvature for the ball–outer-race contact are

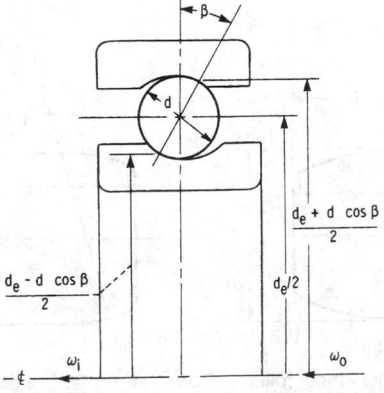

Fig. 21.80 Cross section of ball bearing. (From Ref. 10.)

$$r_{ax} = r_{ay} = \frac{d}{2} \tag{21.82}$$

$$r_{bx} = \frac{d_c + d \cos \beta}{2 \cos \beta} \tag{21.83}$$

$$r_{by} = -f_o d = -r_o \tag{21.84}$$

In Eqs. (21.80) and (21.81), β is used instead of β_f since these equations are also valid when a load is applied to the contact. By setting $\beta = 0°$, Eqs. (21.79)–(21.84) are equally valid for radial ball bearings. For thrust ball bearings, $r_{bx} = \infty$ and the other radii are defined as given in the preceding equations.

Equations (21.36) and (21.37) effectively redefine the problem of two ellipsoidal solids approaching one another in terms of an equivalent ellipsoidal solid of radii R_x and R_y approaching a plane. From the radius-of-curvature expressions, the radii R_x and R_y for the contact example discussed earlier can be written for the ball–inner-race contact as

$$R_x = \frac{d(d_e - d \cos \beta)}{2d_e} \tag{21.85}$$

$$R_y = \frac{f_i d}{2f_i - 1} \tag{21.86}$$

and for the ball–outer-race contact as

$$R_x = \frac{d(d_e + d \cos \beta)}{2d_e} \tag{21.87}$$

$$R_y = \frac{f_o d}{2f_o - 1} \tag{21.88}$$

Roller Bearings. The equations developed for the pitch diameter d_e and diametral clearance P_d for ball bearings in Eqs. (21.68) and (21.69), respectively, are directly applicable for roller bearings.

Crowning. To prevent high stresses at the edges of the rollers in cylindrical roller bearings, the rollers are usually crowned as shown in Fig. 21.81. A fully crowned roller is shown in Fig. 21.81a and a partially crowned roller in Fig. 21.81b. In this figure the crown curvature is greatly exaggerated for clarity. The crowning of rollers also gives the bearing protection against the effects of slight misalignment. For cylindrical rollers, $r_{ay}/d \approx 10^2$. In contrast, for spherical rollers in spherical roller bearings, as shown in Fig. 21.81, $r_{ay}/d \approx 4$. In Fig. 21.81 it is observed that the roller effective length l_r is the length presumed to be in contact with the races under loading. Generally, the roller effective length can be written as

$$l_r = l_t - 2r_c$$

where r_c is the roller corner radius or the grinding undercut, whichever is larger.

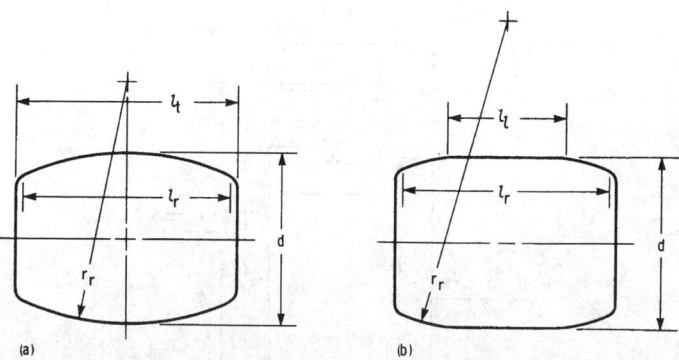

Fig. 21.81 Spherical and cylindrical rollers. (From Ref. 6.) (a) Spherical roller (fully crowned). (b) Cylindrical roller (partially crowned).

Race Conformity. Race conformity applies to roller bearings much as it applies to ball bearings. It is a measure of the geometrical conformity of the race and the roller. Figure 21.82 shows a cross section of a spherical roller bearing. From this figure the race conformity can be written as

$$f = \frac{r}{2r_{ay}}$$

In this equation if subscripts i or o are added to f and r, we obtain the values for the race conformity for the inner- and outer-race contacts.

Free Endplay and Contact Angle. Cylindrical roller bearings have a contact angle of zero and may take thrust load only by virtue of axial flanges. Tapered-roller bearings must be subjected to a thrust load or the inner and outer races (the cone and cup) will not remain assembled; therefore, tapered-roller bearings do not exhibit free diametral play. Radial spherical roller bearings are, however, normally assembled with free diametral play and, hence, exhibit free endplay. The diametral play P_d for a spherical roller bearing is the same as that obtained for ball bearings as expressed in Eq. (21.69). This diametral play as well as endplay is shown in Fig. 21.83 for a spherical roller bearing. From this figure we can write that

$$r_o \cos \beta = \left(r_o - \frac{P_d}{2} \right) \cos \gamma$$

or

$$\beta = \cos^{-1} \left[\left(1 - \frac{P_d}{2r_o} \right) \cos \gamma \right]$$

Also from Fig. 21.83 the free endplay can be written as

$$P_c = 2r_o(\sin \beta - \sin \gamma) + P_d \sin \gamma$$

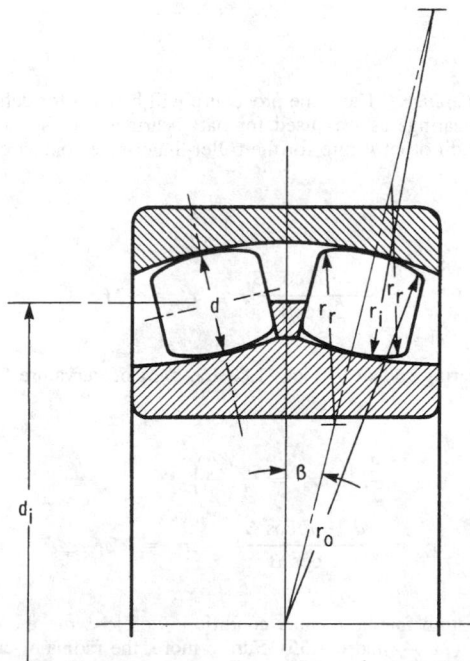

Fig. 21.82 Spherical roller bearing geometry. (From Ref. 6.)

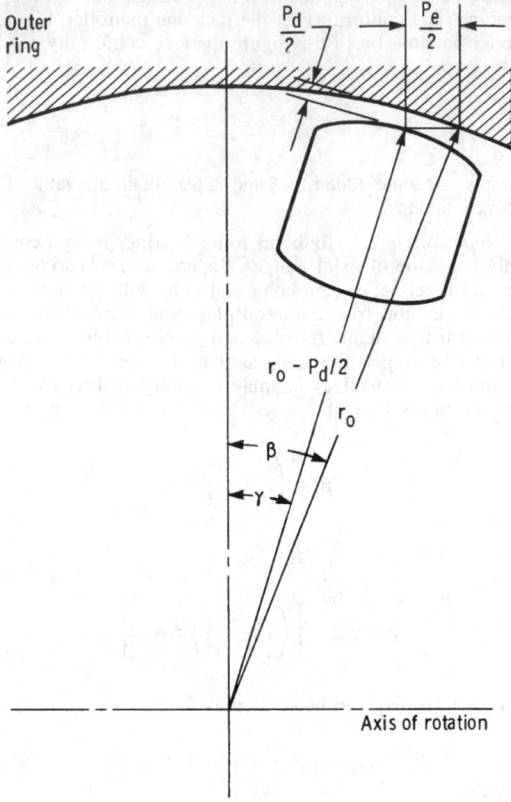

Fig. 21.83 Schematic diagram of spherical roller bearing, showing diametral play and endplay. (From Ref. 6.)

Curvature Sum and Difference. The same procedure will be used for defining the curvature sum and difference for roller bearings as was used for ball bearings. For spherical roller bearings, as shown in Fig. 21.82, the radii of curvature for the roller–inner-race contact can be written as

$$r_{ax} = \frac{d}{2}, \qquad r_{ay} = f_i\left(\frac{r_i}{2}\right)$$

$$r_{bx} = \frac{d_e - d\cos\beta}{2\cos\beta}, \qquad r_{by} = -2f_i r_{ay}$$

For the spherical roller bearing shown in Fig. 21.82 the radii of curvature for the roller–outer-race contact can be written as

$$r_{ax} = \frac{d}{2}, \qquad r_{ay} = f_o\left(\frac{r_o}{2}\right)$$

$$r_{bx} = -\frac{d_e + d\cos\beta}{2\cos\beta}, \qquad r_{by} = -2f_o r_{ay}$$

Knowing the radii of curvature for the contact condition, we can write the curvature sum and difference directly from Eqs. (21.34) and (21.35). Furthermore, the radius-of-curvature expressions R_x and R_y for spherical roller bearings can be written for the roller–inner-race contact as

$$R_x = \frac{d(d_c - d \cos \beta)}{2d_c} \tag{21.89}$$

$$R_y = \frac{2r_{ay}f_i}{2f_i - 1} \tag{21.90}$$

and for the roller–outer-race contact as

$$R_x = \frac{d(d_c + d \cos \beta)}{2d_c} \tag{21.91}$$

$$R_y = \frac{2r_{ay}f_o}{2f_o - 1} \tag{21.92}$$

Kinematics

The relative motions of the separator, the balls or rollers, and the races of rolling-element bearings are important to understanding their performance. The relative velocities in a ball bearing are somewhat more complex than those in roller bearings, the latter being analogous to the specialized case of a zero- or fixed-value contact-angle ball bearing. For that reason the ball bearing is used as an example here to develop approximate expressions for relative velocities. These are useful for rapid but reasonably accurate calculation of elastohydrodynamic film thickness, which can be used with surface roughnesses to calculate the lubrication life factor.

When a ball bearing operates at high speeds, the centrifugal force acting on the ball creates a difference between the inner- and outer-race contact angles, as shown in Fig. 21.84, in order to maintain force equilibrium on the ball. For the most general case of rolling and spinning at both inner- and outer-race contacts, the rolling and spinning velocities of the ball are as shown in Fig. 21.85.

The equations for ball and separator angular velocity for all combinations of inner- and outer-race rotation were developed by Jones.[45] Without introducing additional relationships to describe the elastohydrodynamic conditions at both ball–race contacts, however, the ball–spin-axis orientation angle ϕ cannot be obtained. As mentioned, this requires a long numerical solution except for the two extreme cases of outer- or inner-race control. These are illustrated in Fig. 21.86.

Race control assumes that pure rolling occurs at the controlling race, with all of the ball spin occurring at the other race contact. The orientation of the ball rotational axis is then easily determinable from bearing geometry. Race control probably occurs only in dry bearings or dry-film-lubricated bearings where Coulomb friction conditions exist in the ball–race contact ellipses. Pure rolling will occur at the race contact with the higher magnitude spin-opposing moment. This is usually the inner race at low speeds and the outer race at high speeds.

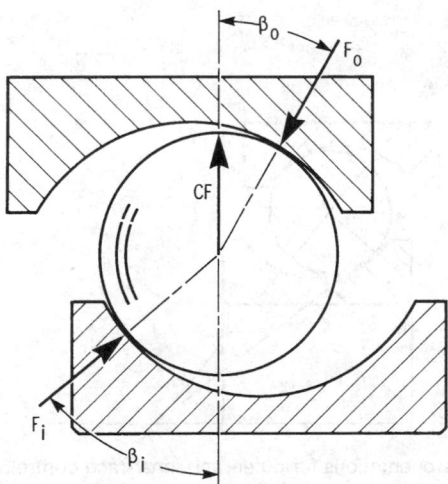

Fig. 21.84 Contact angles in a ball bearing at appreciable speeds. (From Ref. 6.)

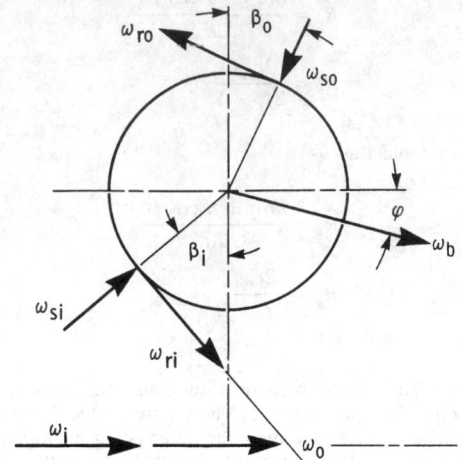

Fig. 21.85 Angular velocities of a ball. (From Ref. 6.)

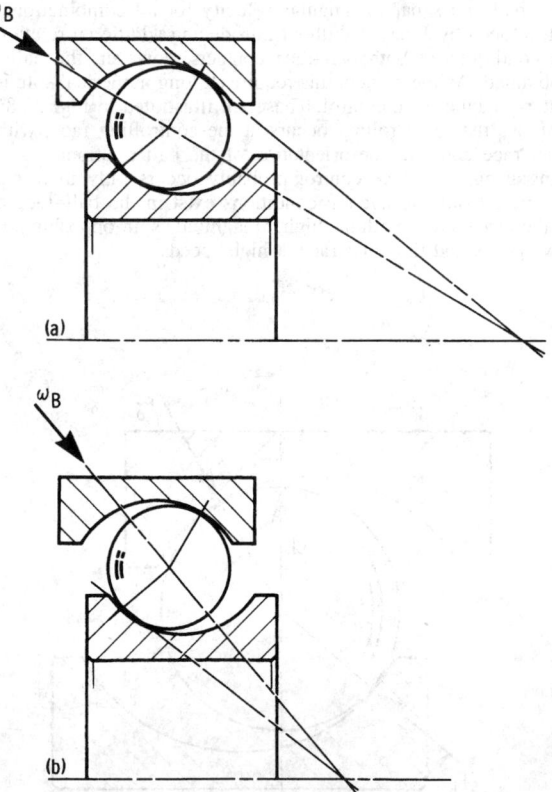

Fig. 21.86 Ball spin axis orientations for outer- and inner-race control. (From Ref. 6.) (a) Outer-race control. (b) Inner-race control.

In oil-lubricated bearings in which elastohydrodynamic films exist in both ball–race contacts, rolling with spin occurs at both contacts. Therefore, precise ball motions can only be determined through use of a computer analysis. We can approximate the situation with a reasonable degree of accuracy, however, by assuming that the ball rolling axis is normal to the line drawn through the centers of the two ball–race contacts. This is shown in Fig. 21.80.

The angular velocity of the separator or ball set ω_c about the shaft axis can be shown to be

$$\omega_c = \frac{(v_i + v_o)/2}{d_e/2}$$

$$= \frac{1}{2}\left[\omega_i\left(1 - \frac{d\cos\beta}{d_e}\right) + \omega_o\left(1 + \frac{d\cos\beta}{d_e}\right)\right] \tag{21.93}$$

where v_i and v_o are the linear velocities of the inner and outer contacts. The angular velocity of a ball ω_b about its own axis is

$$\omega_b = \frac{v_i - v_o}{d_e/2}$$

$$= \frac{d_e}{2d}\left[\omega_i\left(1 - \frac{d\cos\beta}{d_e}\right) - \omega_o\left(1 + \frac{d\cos\beta}{d_e}\right)\right] \tag{21.94}$$

To calculate the velocities of the ball–race contacts, which are required for calculating elastohydrodynamic film thicknesses, it is convenient to use a coordinate system that rotates at ω_c. This fixes the ball–race contacts relative to the observer. In the rotating coordinate system the angular velocities of the inner and outer races become

$$\omega_{ir} = \omega_i - \omega_c = \left(\frac{\omega_i - \omega_o}{2}\right)\left(1 + \frac{d\cos\beta}{d_e}\right)$$

$$\omega_{or} = \omega_o - \omega_c = \left(\frac{\omega_o - \omega_i}{2}\right)\left(1 - \frac{d\cos\beta}{d_e}\right)$$

The surface velocities entering the ball–inner-race contact for pure rolling are

$$u_{ai} = u_{bi} = \left(\frac{d_e - d\cos\beta}{2}\right)\omega_{ir} \tag{21.95}$$

or

$$u_{ai} = u_{bi} = \frac{d_e(\omega_i - \omega_o)}{4}\left(1 - \frac{d^2\cos^2\beta}{d_e^2}\right) \tag{21.96}$$

and those at the ball–outer-race contact are

$$u_{ao} = u_{bo} = \left(\frac{d_e + d\cos\beta}{2}\right)\omega_{or} \tag{21.97}$$

or

$$u_{ao} = u_{bo} = \frac{d_e(\omega_o - \omega_i)}{4}\left(1 - \frac{d^2\cos^2\beta}{d_e^2}\right) \tag{21.97}$$

For a cylindrical roller bearing $\beta = 0°$ and Eqs. (21.92), (21.94), (21.96), and (21.97) become, if d is roller diameter,

$$\omega_c = \frac{1}{2}\left[\omega_i\left(1 - \frac{d}{d_e}\right) + \omega_o\left(1 + \frac{d}{d_e}\right)\right]$$

$$\omega_R = \frac{d_e}{2d}\left[\omega_i\left(1 - \frac{d}{d_e}\right) + \omega_o\left(1 + \frac{d}{d_e}\right)\right]$$

$$u_{ai} = u_{bi} = \frac{d_e(\omega_i - \omega_o)}{4}\left(1 - \frac{d^2}{d_e^2}\right)$$

$$u_{ao} = u_{bo} = \frac{d_e(\omega_o - \omega_i)}{4}\left(1 - \frac{d^2}{d_e^2}\right)$$

(21.98)

For a tapered-roller bearing, equations directly analogous to those for a ball bearing can be used if d is the average diameter of the tapered roller, d_e is the diameter at which the geometric center of the rollers is located, and ω is the angle as shown in Fig. 21.87.

Static Load Distribution

Having defined a simple analytical expression for the deformation in terms of load in Section 21.3.1, it is possible to consider how the bearing load is distributed among the rolling elements. Most rolling-element bearing applications involve steady-state rotation of either the inner or outer race or both; however, the speeds of rotation are usually not so great as to cause ball or roller centrifugal forces or gyroscopic moments of significant magnitudes. In analyzing the loading distribution on the rolling elements, it is usually satisfactory to ignore these effects in most applications. In this section the load–deflection relationships for ball and roller bearings are given, along with radial and thrust load distributions of statically loaded rolling elements.

Load–Deflection Relationships. For an elliptical contact the load–deflection relationship given in Eq. (21.45) can be written as

$$F = K_{1.5}\delta^{3/2}$$

(21.99)

where

$$K_{1.5} = \pi k E'\left(\frac{2\delta R}{9\mathscr{F}^3}\right)^{1/2}$$

(21.100)

Similarly for a rectangular contact, Eq. (21.50) gives

$$F = K_1\delta$$

where

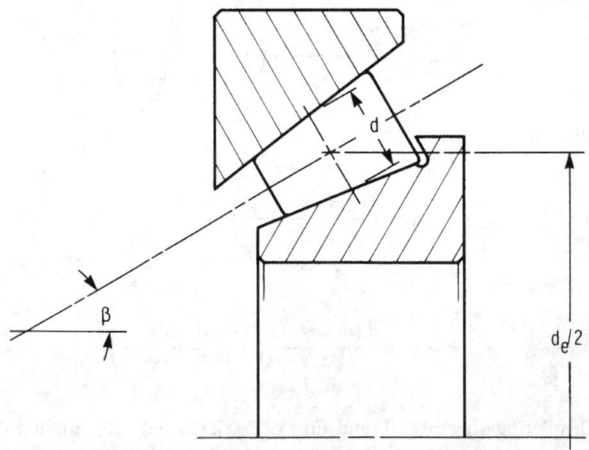

Fig. 21.87 Simplified geometry for tapered-roller bearing. (From Ref. 6.)

$$K_1 = \left(\frac{\pi l E'}{2}\right)\left[\frac{1}{\frac{2}{3} + \ln(2r_{ax}/b) + \ln(2r_{bx}/b)}\right] \tag{21.101}$$

In general, then,

$$F = K_j \delta^j \tag{21.102}$$

in which $j = 1.5$ for ball bearings and 1.0 for roller bearings. The total normal approach between two races separated by a rolling element is the sum of the deformations under load between the rolling element and both races. Therefore

$$\delta = \delta_o + \delta_i \tag{21.103}$$

where

$$\delta_o = \left[\frac{F}{(K_j)_o}\right]^{1/j} \tag{21.104}$$

$$\delta_i = \left[\frac{F}{(K_j)_i}\right]^{1/j} \tag{21.105}$$

Substituting Eqs. (21.103)–(21.105) into Eq. (21.102) gives

$$K_j = \frac{1}{\{[1/(K_j)_o]^{1/j} + [1/(K_j)_i]^{1/j}\}^j}$$

Recall that $(K_j)_o$ and $(K_j)_i$ are defined by Eq. (21.100) or (21.101) for an elliptical or rectangular contact, respectively. From these equations we observe that $(K_j)_o$ and $(K_j)_i$ are functions of only the geometry of the contact and the material properties. The radial and thrust load analyses are presented in the following two sections and are directly applicable for radially loaded ball and roller bearings and thrust-loaded ball bearings.

Radially Loaded Ball and Roller Bearings. A radially loaded rolling element with radial clearance P_d is shown in Fig. 21.88. In the concentric position shown in Fig. 21.88a, a uniform radial clearance between the rolling element and the races of $P_d/2$ is evident. The application of a small radial load to the shaft causes the inner race to move a distance $P_d/2$ before contact is made between a rolling element located on the load line and the inner and outer races. At any angle there will still be a radial clearance c that, if P_d is small compared with the radius of the tracks, can be expressed with adequate accuracy by

$$c = (1 - \cos \psi) \, P_d/2$$

On the load line where $\psi = 0$ the clearance is zero, but when $\psi = 90°$, the clearance retains its initial value of $P_d/2$.

The application of further load will cause elastic deformation of the balls and the elimination of clearance around an arc $2\psi_c$. If the interference or total elastic compression on the load is $\delta_{\max}$, the corresponding elastic compression of the ball δ_ψ along a radius at angle ψ to the load line will be given by

$$\delta_\psi = (\delta_{\max} \cos \psi - c) = (\delta_{\max} + P_d/2) \cos \psi - P_d/2$$

This assumes that the races are rigid. Now, it is clear from Fig. 21.88 that $(\delta_{\max} + P_d/2)$ represents the total relative radial displacement of the inner and outer races. Hence,

$$\delta_\psi = \delta \cos \psi - P_d/2 \tag{21.106}$$

The relationship between load and elastic compression along the radius at angle ψ to the load vector is given by Eq. (21.102) as

$$F_\psi = K_j \delta_\psi^j$$

Substituting Eq. (21.106) into this equation gives

$$F_\psi = K_j(\delta \cos \psi - P_d/2)^j$$

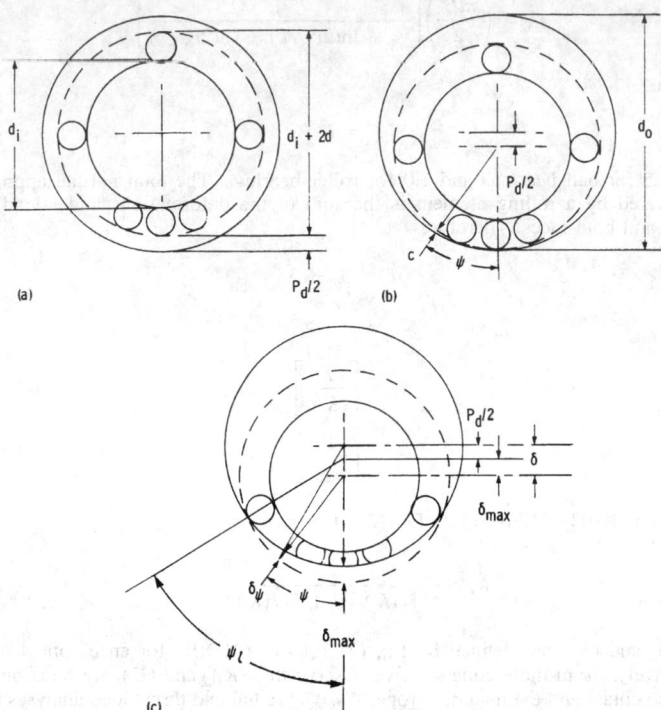

Fig. 21.88 Radially loaded rolling-element bearing. (From Ref. 10.) (a) Concentric arrangement. (b) Initial contact. (c) Interference.

For static equilibrium the applied load must equal the sum of the components of the rolling-element loads parallel to the direction of the applied load:

$$F_r = \sum F_\psi \cos \psi$$

Therefore

$$F_r = K_j \sum \left(\delta \cos \psi - \frac{P_d}{2} \right)^j \cos\psi \qquad (21.107)$$

The angular extent of the bearing arc $2\psi_l$ in which the rolling elements are loaded is obtained by setting the root expression in Eq. (21.107) equal to zero and solving for ψ:

$$\psi_l = \cos^{-1} \left(\frac{P_d}{2\delta} \right)$$

The summation in Eq. (21.107) applies only to the angular extent of the loaded region. This equation can be written for a roller bearing as

$$F_r = \left(\psi_l - \frac{P_d}{2\delta} \sin \psi_l \right) \frac{n K_1 \delta}{2\pi} \qquad (21.108)$$

and similarly in integral form for a ball bearing as

$$F_r = \frac{n}{\pi} K_{1.5} \delta^{3/2} \int_0^{\psi_l} \left(\cos \psi - \frac{P_d}{2\delta} \right)^{3/2} \cos \psi \, d\psi$$

The integral in the equation can be reduced to a standard elliptic integral by the hypergeometric series and the beta function. If the integral is numerically evaluated directly, the following approximate expression is derived:

$$\int_0^{\psi_l} \left(\cos \psi - \frac{P_d}{2\delta} \right)^{3/2} \cos \psi \, d\psi = 2.491 \left\{ \left[1 + \left(\frac{P_d/2\delta - 1}{1.23} \right)^2 \right]^{1/2} - 1 \right\}$$

This approximate expression fits the exact numerical solution to within $\pm 2\%$ for a complete range of $P_d/2\delta$.

The load carried by the most heavily loaded ball is obtained by substituting $\psi = 0°$ in Eq. (21.107) and dropping the summation sign:

$$F_{\max} = K_j \delta^j \left(1 - \frac{P_d}{2\delta} \right)^j$$

Dividing the maximum ball load [Eq. (21.109)] by the total radial load for a roller bearing [Eq. (21.108)] gives

$$F_r = \frac{[\psi_l - (P_d/2\delta) \sin \psi_l] \, nF_{\max}}{2\pi(1 - P_d/2\delta)} \tag{21.109}$$

and similarly for a ball bearing

$$F_r = \frac{nF_{\max}}{Z} \tag{21.110}$$

where

$$Z = \frac{\pi(1 - P_d/2\delta)^{3/2}}{2.491 \left\{ \left[1 + \left(\frac{1 - P_d/2\delta}{1.23} \right)^2 \right]^{1/2} - 1 \right\}} \tag{21.111}$$

For *roller bearings* when the diametral clearance P_d is zero, Eq. (21.105) gives

$$F_r = \frac{nF_{\max}}{4} \tag{21.112}$$

For *ball bearings* when the diametral clearance P_d is zero, the value of Z in Eq. (21.110) becomes 4.37. This is the value derived by Stribeck[46] for ball bearings of zero diametral clearance. The approach used by Stribeck was to evaluate the finite summation for various numbers of balls. He then derived the celebrated Stribeck equation for static load-carrying capacity by writing the more conservative value of 5 for the theoretical value of 4.37:

$$F_r = \frac{nF_{\max}}{5} \tag{21.113}$$

In using Eq. (21.113), it should be remembered that Z was considered to be a constant and that the effects of clearance and applied load on load distribution were not taken into account. However, these effects were considered in obtaining Eq. (21.110).

Thrust-Loaded Ball Bearings.   The static-thrust-load capacity of a ball bearing may be defined as the maximum thrust load that the bearing can endure before the contact ellipse approaches a race shoulder, as shown in Fig. 21.89, or the load at which the allowable mean compressive stress is reached, whichever is smaller. Both the limiting shoulder height and the mean compressive stress must be calculated to find the static-thrust-load capacity.

The contact ellipse in a bearing race under a load is shown in Fig. 21.89. Each ball is subjected to an identical thrust component F_t/n, where F_t is the total thrust load. The initial contact angle before the application of a thrust load is denoted by β_f. Under load, the normal ball thrust load F acts at the contact angle β and is written as

$$F = \frac{F_t}{n} \sin \beta \tag{21.114}$$

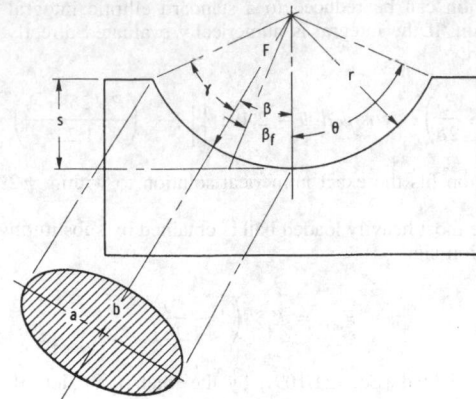

Fig. 21.89 Contact ellipse in bearing race. (From Ref. 10.)

A cross section through an angular-contact bearing under a thrust load F_t is shown in Fig. 21.90. From this figure the contact angle after the thrust load has been applied can be written as

$$\beta = \cos^{-1}\left(\frac{D - P_d/2}{D + \delta}\right) \tag{21.115}$$

The initial contact angle was given in Eq. (21.74). Using that equation and rearranging terms in Eq. (21.115) give, solely from geometry (Fig. 21.90),

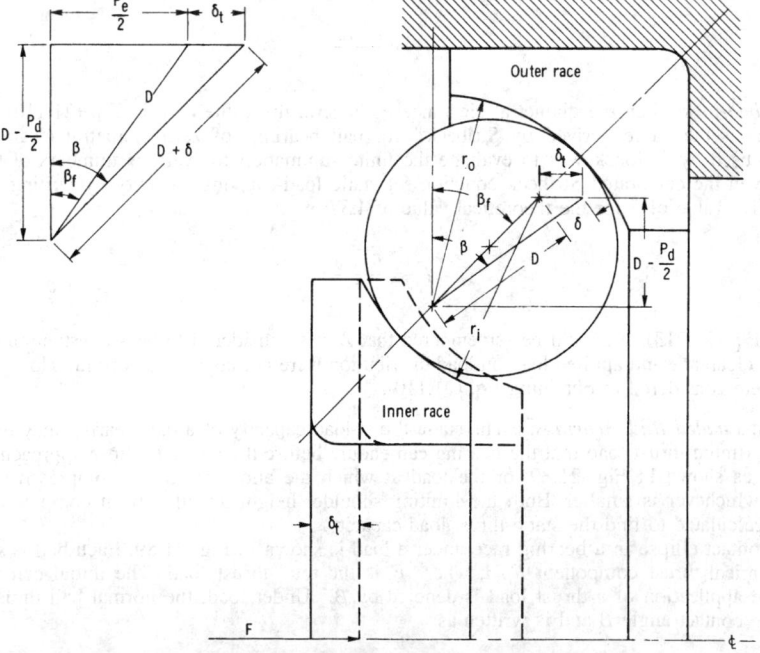

Fig. 21.90 Angular-contact ball bearing under thrust load. (From Ref. 10.)

$$\delta = D \left(\frac{\cos \beta_f}{\cos \beta} - 1 \right) = \delta_o + \delta_i$$

$$= \left[\frac{F}{(K_j)_o} \right]^{1/j} + \left[\frac{F}{(K_j)_i} \right]^{1/j}$$

$$K_j = 1 \bigg/ \left\{ \left[\frac{1}{(K_j)_o} \right]^{1/j} + \left[\frac{1}{(K_j)_i} \right]^{1/j} \right\}^j$$

$$K_j = 1 \bigg/ \left\{ \left[\frac{4.5 \, \mathfrak{F}_o^3}{\pi k_o E_o' (R_o \, \mathcal{E}_o)^{1/2}} \right]^{2/3} + \left[\frac{4.5 \, \mathfrak{F}_i^3}{\pi k_i E_i (R_i \, \mathcal{E}_i)^{1/2}} \right]^{2/3} \right\} \tag{21.116}$$

$$F = K_j D^{3/2} \left(\frac{\cos \beta_f}{\cos \beta} - 1 \right)^{3/2} \tag{21.117}$$

where

$$K_{1.5} = \pi k E' \left(\frac{R \, \mathcal{E}}{4.5 \, \mathfrak{F}^3} \right)^{1/2} \tag{21.118}$$

and k, $\mathcal{E}$, and $\mathfrak{F}$ are given by Eqs. (21.39), (21.40), and (21.41), respectively.
    From Eqs. (21.114) and (21.117), we can write

$$\frac{F_t}{n \sin \beta} = F \tag{21.119}$$

$$\frac{F_t}{n K_j D^{3/2}} = \sin \beta \left(\frac{\cos \beta_f}{\cos \beta} - 1 \right)^{3/2}$$

This equation can be solved numerically by the Newton–Raphson method. The iterative equation to be satisfied is

$$\beta' - \beta = \frac{\dfrac{F_t}{n K_{1.5} D^{3/2}} - \sin \beta \left(\dfrac{\cos \beta_f}{\cos \beta} - 1 \right)^{3/2}}{\cos \beta \left(\dfrac{\cos \beta_f}{\cos \beta} - 1 \right)^{3/2} + \dfrac{3}{2} \cos \beta_f \tan^2 \beta \left(\dfrac{\cos \beta_f}{\cos \beta} - 1 \right)^{1/2}} \tag{21.120}$$

In this equation convergence is satisfied when $\beta' - \beta$ becomes essentially zero.
    When a thrust load is applied, the shoulder height is limited to the distance by which the pressure–contact ellipse can approach the shoulder. As long as the following inequality is satisfied, the pressure–contact ellipse will not exceed the shoulder height limit:

$$\theta > \beta + \sin^{-1} \left(\frac{D_y}{fd} \right)$$

From Fig. 21.79 and Eq. (21.68), the angle used to define the shoulder height θ can be written as

$$\theta = \cos^{-1} \left(\frac{1 - s}{fd} \right)$$

From Fig. 21.77 the axial deflection δ_t corresponding to a thrust load can be written as

$$\delta_t = (D + \delta) \sin \beta - D \sin \beta_f \tag{21.121}$$

Substituting Eq. (21.116) into Eq. (21.121) gives

$$\delta_t = \frac{D \sin(\beta - \beta_f)}{\cos \beta}$$

Having determined β from Eq. (21.120) and β_f from Eq. (21.103), we can easily evaluate the relationship for δ_t.

Preloading. The use of angular-contact bearings as duplex pairs preloaded against each other is discussed in the first subsection in Section 21.3.6. As shown in Table 21.10 duplex bearing pairs are used in either back to back or face-to-face arrangements. Such bearings are usually preloaded against each other by providing what is called "stickout" in the manufacture of the bearing. This is illustrated in Fig. 21.91 for a bearing pair used in a back-to-back arrangement. The magnitude of the stickout and the bearing design determine the level of preload on each bearing when the bearings are clamped together as in Fig. 21.91. The magnitude of preload and the load–deflection characteristics for a given bearing pair can be calculated by using Eqs. (21.74), (21.99), (21.114), and (21.116)–(21.119).

The relationship of initial preload, system load, and final load for bearings a and b is shown in Fig. 21.92. The load–deflection curve follows the relationship $\delta = KF^{2/3}$. When a system thrust load F_t is imposed on the bearing pairs, the magnitude of load on bearing b increases while that on bearing a decreases until the difference equals the system load. The physical situation demands that the change in each bearing deflection be the same ($\Delta a = \Delta b$ in Fig. 21.92). The increments in bearing load, however, are not the same. This is important because it always requires a system thrust load far greater than twice the preload before one bearing becomes unloaded. Prevention of bearing unloading, which can result in skidding and early failure, is an objective of preloading.

Rolling Bearing Fatigue Life

Contact Fatigue Theory. Rolling fatigue is a material failure caused by the application of repeated stresses to a small volume of material. It is a unique failure type. It is essentially a process of seeking out the weakest point at which the first failure will occur. A typical spall is shown in Fig. 21.93. We can surmise that on a microscale there will be a wide dispersion in material strength or resistance to fatigue because of inhomogeneities in the material. Because bearing materials are complex alloys, we would not expect them to be homogeneous nor equally resistant to failure at all points. Therefore, the fatigue process can be expected to be one in which a group of supposedly identical specimens exhibit wide variations in failure time when stressed in the same way. For this reason it is necessary to treat the fatigue process statistically.

To be able to predict how long a typical bearing will run under a specific load, we must have the following two essential pieces of information:

1. An accurate, quantitative estimate of the life dispersion or scatter.

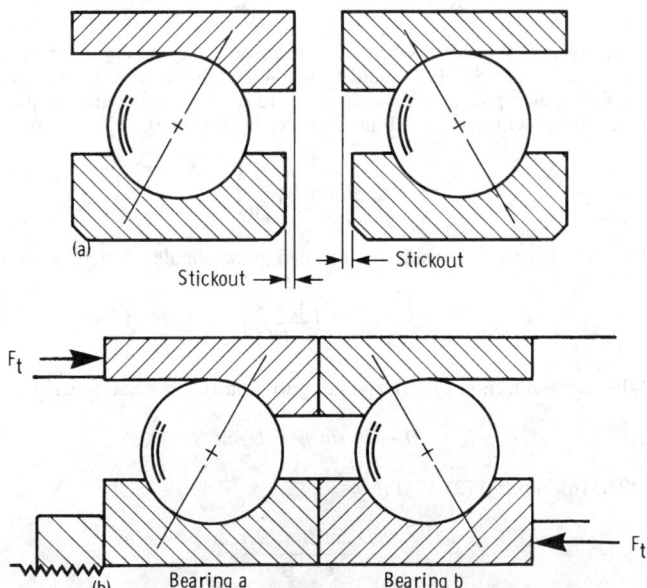

Fig. 21.91 Angular-contact bearings in back-to-back arrangement, shown individually as manufactured and as mounted with preload. (From Ref. 6.) (*a*) Separated. (*b*) Mounted and preloaded.

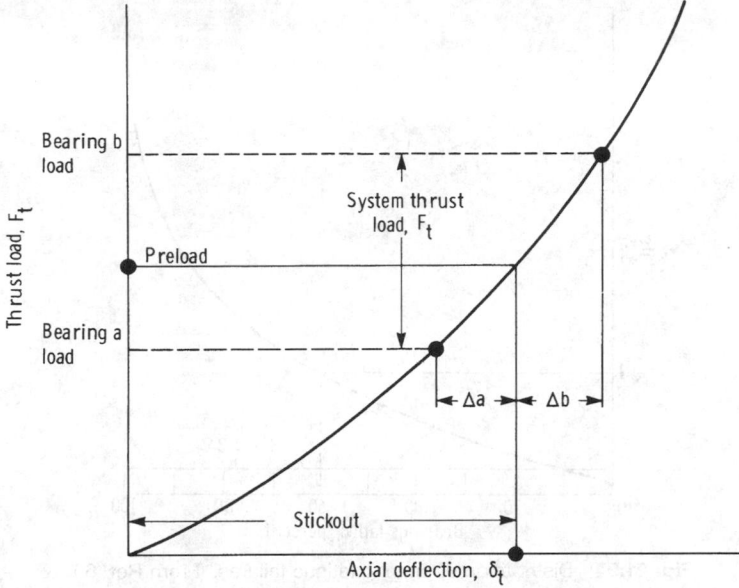

Fig. 21.92 Thrust-load-axial-deflection curve for a typical ball bearing. (From Ref. 6.)

2. The life at a given survival rate or reliability level. This translates into an expression for the "load capacity," or the ability of the bearing to endure a given load for a stipulated number of stress cycles or revolutions. If a group of supposedly identical bearings is tested at a specific load and speed, there will be a wide scatter in bearing lives, as shown in Fig. 21.94.

The Weibull Distribution. Weibull[47] postulates that the fatigue lives of a homogeneous group of rolling-element bearings are dispersed according to the following relation:

$$\ln \ln \frac{1}{S} = e_1 \ln L/A$$

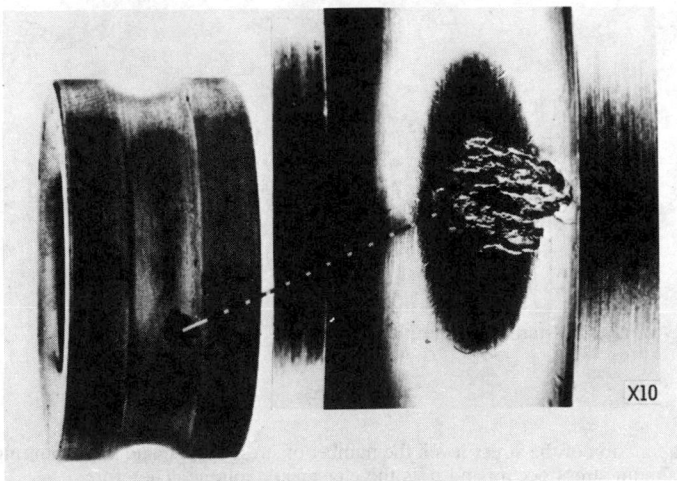

Fig. 21.93 Typical fatigue spall.

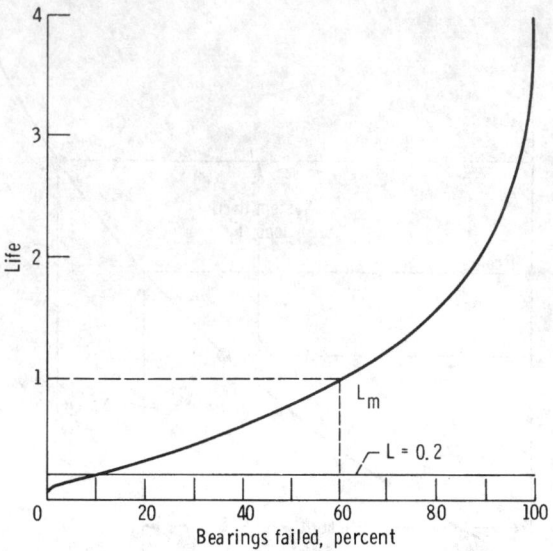

Fig. 21.94 Distribution of bearing fatigue failures. (From Ref. 6.)

where S is the probability of survival, L is the fatigue life, and e_1 and A are constants. The Weibull distribution results from a statistical theory of strength based on probability theory, where the dependence of strength on volume is explained by the dispersion in material strength. This is the "weakest link" theory.

Consider a volume being stressed that is broken up into m similar volumes:

$$S_1 = 1 - M_1 \qquad S_2 = 1 - M_2 \qquad S_3 = 1 - M_3 \qquad \cdots \qquad S_m = 1 - M_m$$

The M's represent the probability of failure and the S's represent the probability of survival. For the entire volume we can write

$$S = S_1 \cdot S_2 \cdot S_3 \cdot \cdots \cdot S_m$$

Then

$$1 - M = (1 - M_1)(1 - M_2)(1 - M_3) \cdots (1 - M_m)$$

$$1 - M = \prod_{i=1}^{m} (1 - M_i)$$

$$S = \prod_{i=1}^{m} (1 - M_i)$$

The probability of a crack starting in the ith volume is

$$M_i = f(x)v_i$$

where $f(x)$ is a function of the stress level, the number of stress cycles, and the depth into the material where the maximum stress occurs and v_i is the elementary volume. Therefore,

$$S = \prod_{i=1}^{m} [1 - f(x)v_i]$$

$$\ln S = \sum_{i=1}^{m} \ln[1 - f(x)v_i]$$

Now if $f(x)v_i \ll 1$, then $\ln[1 - f(x)v_i] = -f(x)v_i$ and

$$\ln S = -\sum_{i=1}^{m} f(x)v_i$$

Let $v_i \rightarrow 0$; then

$$\sum_{i=1}^{m} f(x)v_i = \int (x)\, dv = f(x)V$$

Lundberg and Palmgren[48] assume that $f(x)$ could be expressed as a power function of shear stress τ_0, number of stress cycles, J, and depth to the maximum shear stress Z_0:

$$f(x) = \frac{\tau_0^{c_1} J^{c_2}}{Z_0^{c_3}} \tag{21.122}$$

They also choose as the stressed volume

$$V = D_y Z_0 l_v$$

Then

$$\ln S = -\frac{\tau_0^{c_1} J^{c_2} D_y l_v}{Z_0^{c_3 - 1}}$$

or

$$\ln \frac{1}{S} = \frac{\tau_0^{c_1} J^{c_2} D_y l_v}{Z_0^{c_3 - 1}}$$

For a specific bearing and load (e.g., stress) τ_0, D_y, l_v, and Z_0 are all constant, so that

$$\ln \frac{1}{S} \approx J^{c_2}$$

Designating J as life L in stress cycles gives

$$\ln \frac{1}{S} = \left(\frac{L}{A}\right)^{c_2}$$

or

$$\ln \ln \frac{1}{S} = c_2 \ln \left(\frac{L}{A}\right) \tag{21.123}$$

This is the Weibull distribution, which relates probability of survival and life. It has two principal functions. First, bearing fatigue lives plot as a straight line on Weibull coordinates (log log vs log), so that the life at any reliability level can be determined. Of most interest are the L_{10} life ($S = 0.9$) and the L_{50} life ($S = 0.5$). Bearing load ratings are based on the L_{10} life. Second, Eq. (21.123) can be used to determine what the L_{10} life must be to obtain a required life at any reliability level. The L_{10} life is calculated, from the load on the bearing and the bearing dynamic capacity or load rating given in manufacturers' catalogs and engineering journals, by using the equation

$$L = \left(\frac{C}{F_e}\right)^m$$

where C = basic dynamic capacity or load rating
$\quad\ \ F_e$ = equivalent bearing load
$\quad\ \ m$ = 3 for elliptical contacts and 10/3 for rectangular contacts

A typical Weibull plot is shown in Fig. 21.95.

Lundberg–Palmgren Theory. The Lundberg–Palmgren theory, on which bearing ratings are based, is expressed by Eq. (21.122). The exponents in this equation are determined experimentally from the dispersion of bearing lives and the dependence of life on load, geometry, and bearing size. As a standard of reference, all bearing load ratings are expressed in terms of the specific dynamic capacity C, which, by definition, is the load that a bearing can carry for 10^6 inner-race revolutions with a 90% chance of survival.

Factors on which specific dynamic capacity and bearing life depend are:

1. Size of rolling element.
2. Number of rolling elements per row.
3. Number of rows of rolling elements.
4. Conformity between rolling elements and races.
5. Contact angle under load.
6. Material properties.
7. Lubricant properties.
8. Operating temperature.
9. Operating speed.

Only factors 1–5 are incorporated in bearing dynamic capacities developed from the Lundberg–Palmgren theory. The remaining factors must be taken into account in the life adjustment factors discussed later.

The formulas for specific dynamic capacity as developed by Lundberg and Palmgren[48,49] are as follows:

For radial ball bearings with $d \le 25$ mm,

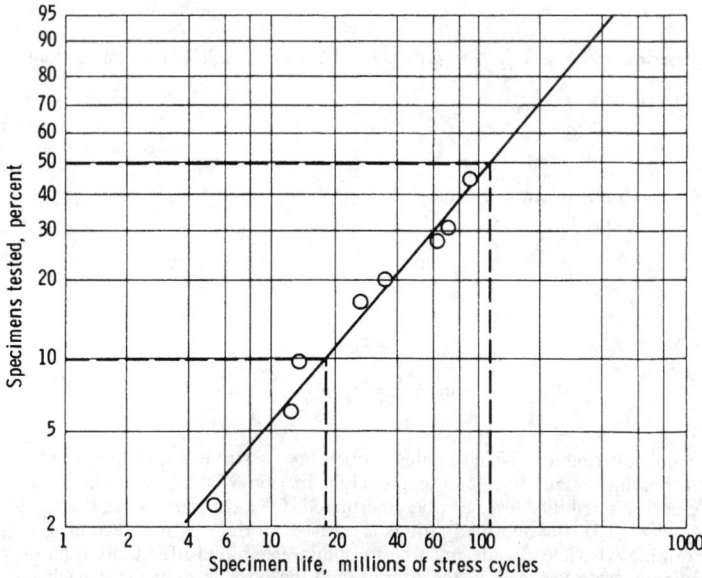

Fig. 21.95 Typical Weibull plot of bearing fatigue failures. (From Ref. 10.)

$$C = f_c (i \cos \beta)^{0.7} n^{2/3} \left(\frac{d}{0.0254} \right)^{1.8}$$

where d = diameter of rolling element, m
i = number of rows of rolling elements
n = number of rolling elements per row
β = contact angle
f_c = coefficient dependent on material and bearing type

For radial ball bearings with $d \geq 25$ mm,

$$C = f_c (i \cos \beta)^{0.7} n^{2/3} \left(\frac{d}{0.0254} \right)^{1.4}$$

For radial roller bearings,

$$C = f_c (i \cos \beta)^{0.78} n^{3/4} \left(\frac{d}{0.0254} \right)^{1.07} \left(\frac{l_t}{0.0254} \right)^{0.78}$$

where l_t is roller length in meters.
For thrust ball bearings with $\beta \neq 90°$,

$$C = f_c (i \cos \beta)^{0.7} (\tan \beta) n^{2/3} \left(\frac{d}{0.0254} \right)^{1.8}$$

For thrust roller bearings with $\beta \neq 90°$,

$$C = f_c (i \cos \beta)^{0.78} (\tan \beta) n^{3/4} \left(\frac{l_t}{0.0254} \right)^{0.78}$$

For thrust ball bearings with $\beta = 90°$,

$$C = f_c i^{0.7} n^{2/3} \left(\frac{d}{0.0254} \right)^{1.8}$$

For thrust roller bearings with $\beta = 90°$,

$$C = f_c i^{0.78} n^{3/4} \left(\frac{d}{0.0254} \right)^{1.07} \left(\frac{l_t}{0.0254} \right)^{0.78}$$

For ordinary bearing steels such as SAE 52100 with mineral oil lubrication, f_c can be evaluated by using Tables 21.18 and 21.19, but a more convenient method is to use tabulated values from the most recent Antifriction Bearing Manufacturers Association (AFBMA) documents on dynamic load ratings and life.[50] The value of C is calculated or determined from bearing manufacturers' catalogs. The equivalent load F_e can be calculated from the equation

$$F_e = XF_r + YF_t$$

Factors X and Y are given in bearing manufacturers' catalogs for specific bearings.
In addition to specific dynamic capacity C, every bearing has a specific static capacity, usually designated as C_0. Specific static capacity is defined as the load that, under static conditions, will result in a permanent deformation of 0.0001 times the rolling-element diameter. For some bearings C_0 is less than C, so it is important to avoid exposing a bearing to a static load that exceeds C_0. Values of C_0 are also given in bearing manufacturers' catalogs.

The AFBMA Method. Shortly after publication of the Lundberg–Palmgren theory, the AFBMA began efforts to standardize methods for establishing bearing load ratings and making life predictions. Standardized methods of establishing load ratings for ball bearings[51] and roller bearings[52] were devised, based essentially on the Lundberg–Palmgren theory. These early standards are published in their entirety in Jones.[45] In recent years significant advances have been made in rolling-element bearing material quality and in our understanding of the role of lubrication in bearing life through the development of elastohydrodynamic theory. Therefore the original AFBMA standards in

Table 21.18 Capacity Formulas for Rectangular and Elliptic Contacts[a] (From Ref. 6)

Function	Elliptical Contact of Ball Bearings			Rectangular Contact of Roller Bearings		
c	$f_c f_a i^{0.7} N^{2/3} d^{1.8}$			$f_c f_a i^{7/9} N^{3/4} d^{29/27} l_{t,i}^{7/9}$		
f_c	$g_c f_1 f_2 \left(\dfrac{d_i}{d_i - d}\right)^{0.41}$			$g_c f_1 f_2$		
g_c	$\left[1 + \left(\dfrac{c_i}{c_o}\right)^{10/8}\right]^{-0.8}$			$\left[1 + \left(\dfrac{c_i}{c_o}\right)^{9/2}\right]^{-2/9}$		
c_i/c_o	$f_3 \left[\dfrac{d_i(d_o - d)}{d_o(d_i - d)}\right]^{0.41}$			$f_3 \left(\dfrac{l_{t,i}}{l_{t,o}}\right)^{7/9}$		
	Radial	**Thrust**		**Radial**	**Thrust**	
		$\beta \neq 90°$	$\beta = 90°$		$\beta \neq 90°$	$\beta = 90°$
γ	$\dfrac{d \cos \beta}{d_e}$		$\dfrac{d}{d_e}$	$\dfrac{d \cos \beta}{d_e}$		$\dfrac{d}{d_e}$
f_a	$(\cos \beta)^{0.7}$	$(\cos \beta)^{0.7} \tan \beta$	1	$(\cos \beta)^{7/9}$	$(\cos \beta)^{7/9} \tan \beta$	1
f_1	3.7–4.1	6–10		18–25	36–60	
f_2	$\dfrac{\gamma^{0.3}(1 - \gamma)^{1.39}}{(1 + \gamma)^{1/3}}$	$\gamma^{0.3}$		$\dfrac{\gamma^{2/9}(1 - \gamma)^{29/27}}{(1 + \gamma)^{1/3}}$	$\gamma^{2/9}$	
f_3	$104 f_4$	f_4	1	$1.14 f_4$	f_4	1
f_4	$\left(\dfrac{1 - \gamma}{1 + \gamma}\right)^{1.72}$			$\left(\dfrac{1 - \gamma}{1 + \gamma}\right)^{38/37}$		

[a] Units in kg and mm.

AFBMA[51,52] have been updated with life adjustment factors. These factors have been incorporated into ISO,[50] which is discussed in the following section.

Life Adjustment Factors. A comprehensive study of the factors affecting the fatigue life of bearings, which were not taken account of in the Lundberg–Palmgren theory, is reported in Bamberger et al.[53] In that reference it was assumed that the various environmental or bearing design factors are multiplicative in their effect on bearing life. The following equation results:

$$L_A = (\tilde{D})(\tilde{E})(\tilde{F})(\tilde{G})(\tilde{H})L_{10}$$

or

$$L_A = (\tilde{D})(\tilde{E})(\tilde{F})(\tilde{G})(\tilde{H})(C/F_e)^m$$

where $\tilde{D}$ = materials factor
$\tilde{E}$ = metallurgical processing factor
$\tilde{F}$ = lubrication factor
$\tilde{G}$ = speed effect factor
$\tilde{H}$ = misalignment factor
F_e = bearing equivalent load
m = load-life exponent; either 3 for ball bearings or 10/3 for roller bearings

Factors, $\tilde{D}$, $\tilde{E}$, and $\tilde{F}$ are briefly reviewed here. The reader is referred to Bamberger et al.[53] for a complete discussion of all five life adjustment factors.

Materials Factors $\tilde{D}$ and $\tilde{E}$. For over a century, AISI 52100 steel has been the predominant material for rolling-element bearings. In fact, the basic dynamic capacity as defined by AFBMA in 1949 is based on an air-melted 52100 steel, hardened to at least Rockwell C 58. Since that time, better control of air-melting processes and the introduction of vacuum remelting processes have resulted in more homogeneous steels with fewer impurities. Such steels have extended rolling-element bearing fatigue lives to several times the AFBMA or catalog life. Life improvements of 3–8 times are not uncommon. Other steel compositions, such as AISI M-1 and AISI M-50, chosen for their

Table 21.19 Capacity Formulas for Mixed Rectangular and Elliptical Contacts[a] (From Ref. 6)

Function	Radial Bearing	Thrust Bearing $\beta \neq 90°$	Thrust Bearing $\beta = 90°$	Radial Bearing	Thrust Bearing $\beta \neq 90°$	Thrust Bearing $\beta = 90°$
	Inner Race			Outer Race		
γ	$\dfrac{d\cos\beta}{d_e}$		$\dfrac{d}{d_e}$	$\dfrac{d\cos\beta}{d_e}$		$\dfrac{d}{d_e}$
	Rectangular Contact c_i			Elliptical Contact c_o		
c_i or c_o	$f_1 f_2 f_a i^{7/9} N^{3/4} D^{29/27} l_{t,i}^{7/9}$			$f_1 f_2 f_a \left(\dfrac{2R}{D}\dfrac{r_o}{r_o - R}\right)^{0.41} i^{0.7} N^{2/3} D^{1.8}$		
f_a	$(\cos\beta)^{7/9}$	$(\cos\beta)^{7/9}\tan\beta$	1	$(\cos\beta)^{0.7}$	$(\cos\beta)^{0.7}\tan\beta$	1
f_1	18–25	36–60		3.5–3.9	6–10	
f_2	$\dfrac{\gamma^{2/9}(1-\gamma)^{29/27}}{(1+\gamma)^{1/3}}$		$\gamma^{3/9}$	$\dfrac{\gamma^{0.3}(1+\gamma)^{1.39}}{(1-\gamma)^{1/3}}$		$\gamma^{0.3}$
	Point Contact c_i			Line Contact c_o		
c_i or c_o	$f_1 f_2 f_a \left(\dfrac{2R}{D}\dfrac{r_i}{r_i - R}\right)^{0.41} i^{0.7} n^{2/3} d^{1.8}$			$f_1 f_2 f_a i^{7/9} n^{3/4} d^{29/27} l_{t,o}^{7.9}$		
f_a	$(\cos\alpha)^{0.7}$	$(\cos\alpha)^{0.7}\tan\alpha$	1	$(\cos\alpha)^{7/9}$	$(\cos\alpha)^{7/9}\tan\alpha$	1
f_1	3.7–4.1	6–10		15–22	36–60	
f_2	$\dfrac{\gamma^{0.3}(1-\gamma)^{1.39}}{(1+\gamma)^{1/3}}$		$\gamma^{0.3}$	$\dfrac{\gamma^{2/9}(1+\gamma)^{29/27}}{(1-\gamma)^{1/3}}$		$\gamma^{2/9}$

[a] $C = C_i [1 + (C_i/C_o)^4]^{1/4}$ units in kg and mm.

higher-temperature capabilities and resistance to corrosion, also have shown greater resistance to fatigue pitting when vacuum melting techniques are employed. Case-hardened materials, such as AISI 4620, AISI 4118, and AISI 8620, used primarily for roller bearings, have the advantage of a tough, ductile steel core with a hard, fatigue-resistant surface.

The recommended $\tilde{D}$ factors for various alloys processed by air melting are shown in Table 21.20. Insufficient definitive life data were found for case-hardened materials to recommended $\tilde{D}$ factors for

Table 21.20 Material Factor for Through-Hardened Bearing Materials[a] (From Ref. 53)

Material	$\tilde{D}$-Factor
52100	2.0
M-1	.6
M-2	.6
M-10	2.0
M-50	2.0
T-1	.6
Halmo	2.0
M-42	.2
WB 49	.6
440C	0.6–0.8

[a] Air-melted materials assumed.

them. It is recommended that the user refer to the bearing manufacturer for the choice of a specific case-hardened material.

The metallurgical processing variables considered in the development of the $\tilde{E}$ factor included melting practice (air and vacuum melting) and metal working (thermomechanical working). Thermomechanical working of M-50 has also been shown to result in improved life, but it is costly and still not fully developed as a processing technique. Bamberger et al.[53] recommended an $\tilde{E}$ factor of 3 for consumable-electrode-vacuum-melted materials.

The translation of factors into a standard[50] is discussed later.

Lubrication Factor $\tilde{F}$. Until approximately 1960 the role of the lubricant between surfaces in rolling contact was not fully appreciated. Metal-to-metal contact was presumed to occur in all applications with attendant required boundary lubrication. The development of elastohydrodynamic lubrication theory showed that lubricant films of thickness of the order of microinches and tens of microinches occur in rolling contact. Since surface finishes are of the same order of magnitude as the lubricant film thicknesses, the significance of rolling-element bearing surface roughnesses to bearing performance became apparent. Tallian[54] first reported on the importance on bearing life of the ratio of elastohydrodynamic lubrication film thickness to surface roughness. Figure 21.96 shows life as a percentage of calculated L_{10} life as a function of Λ, where

$$\Lambda = \frac{h_{min}}{(\Delta_a^2 + \Delta_b^2)^{1/2}}$$

Figure 21.97, from Bamberger et al.,[53] presents a curve of the recommended $\tilde{F}$ factor as a function of the Λ parameter. A mean of the curves presented in Tallian[54] for ball bearings and in Skurka[55] for roller bearings is recommended for use. A formula for calculating the minimum film thickness h_{min} in the hard-EHL regime is given in Eq. (21.57).

The results of Bamberger et al.[53] have not been fully accepted into the current AFBMA standard represented by ISO.[50] The standard presents the following:

1. Life and dynamic load rating formulas for radial and thrust ball bearings and radial and thrust roller bearings.

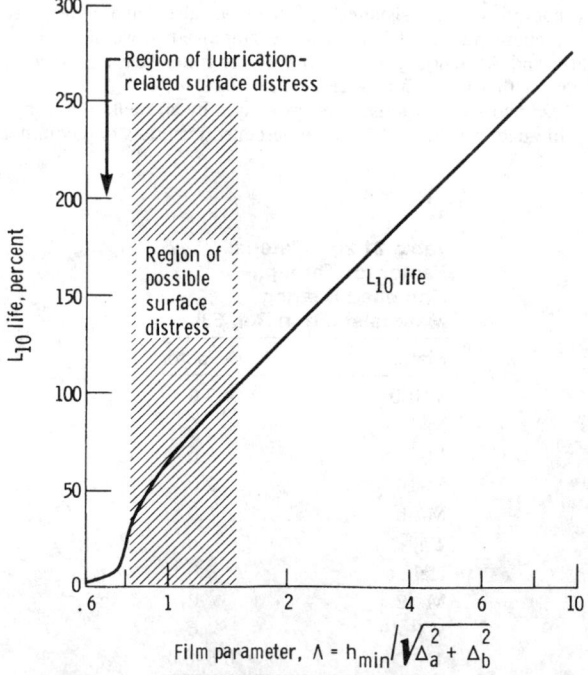

Fig. 21.96   Chart for determining group fatigue life L_{10}. (From Ref. 54.)

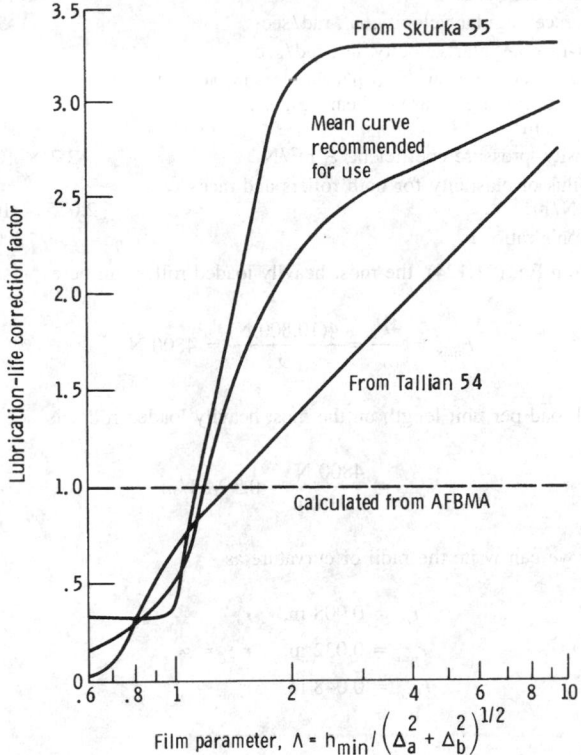

Fig. 21.97 Chart for determining lubrication-life correction factor. (From Ref. 53.)

2. Tables of f_c for all cases.
3. Tables of X and Y factors for calculating equivalent loads.
4. Load rating formulas for multirow bearings.
5. Life correction factors for high-reliability levels a_1, materials a_2, and lubrication or operating conditions a_3.

Procedures for calculating a_2 and a_3 are less than definitive, reflecting the need for additional research, life data, and operating experience.

Applications

In this section two applications of the film thickness equations developed throughout this chapter are presented to illustrate how the fluid-film lubrication conditions in machine elements can be analyzed. Specifically, a typical roller bearing and a typical ball bearing problem are considered.

    Cylindrical-Roller-Bearing Problem. The equations for elastohydrodynamic film thickness that have been developed earlier relate primarily to elliptical contacts, but they are sufficiently general to allow them to be used with adequate accuracy in line-contact problems, as would be found in a cylindrical roller bearing. Therefore, the minimum elastohydrodynamic film thicknesses on the inner and outer races of a cylindrical roller bearing with the following dimensions are calculated:

Inner-race diameter, d_i, mm (m)	65 (0.064)
Outer-race diameter, d_o, mm (m)	96 (0.096)
Diameter of cylindrical rollers, d, mm (m)	16 (0.016)
Axial length of cylindrical rollers, l, mm (m)	16 (0.016)
Number of rollers in complete bearing, n	9

A bearing of this kind might well experience the following operating conditions:

Radial load, F_r, N   10,800

Inner-race angular velocity, ω_i, rad/sec   524

Outer-race angular velocity, ω_o, rad/sec   0

Lubricant viscosity at atmospheric pressure at
  operating temperature of bearings, η_o,
  N sec/m²   0.01

Viscosity–pressure coefficient, ξ, m²/N 2.2×10^{-8}

Modulus of elasticity for both rollers and races,
  E, N/m² 2.075×10^{11}

Poisson's ratio, ν 0.3

Calculation. From Eq. (21.124), the most heavily loaded roller can be expressed as

$$F_{max} = \frac{4F_r}{n} = \frac{4(10{,}800 \text{ N})}{9} = 4800 \text{ N} \tag{21.124}$$

Therefore, the radial load per unit length on the most heavily loaded roller is

$$F'_{max} = \frac{4800 \text{ N}}{0.016 \text{ m}} = 0.3 \text{ MN/m} \tag{21.125}$$

From Fig. 21.98 we can write the radii of curvature as

$$r_{ax} = 0.008 \text{ m}, \quad r_{ay} = \infty$$

$$r_{bx,i} = 0.032 \text{ m}, \quad r_{by,i} = \infty$$

$$r_{bx,o} = 0.048 \text{ m}, \quad r_{by,o} = \infty$$

Then

$$\frac{1}{R_{x,i}} = \frac{1}{0.008} + \frac{1}{0.032} = \frac{5}{0.032}$$

giving $R_{x,i} = 0.0064$ m,

$$\frac{1}{R_{x,o}} = \frac{1}{0.008} - \frac{1}{0.048} = \frac{5}{0.048} \tag{21.126}$$

giving $R_{x,o} = 0.0096$ m, and

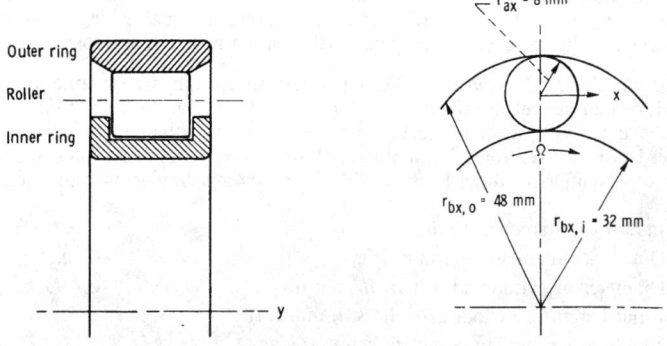

Fig. 21.98 Roller bearing example: $r_{ay} = r_{by,f} = r_{by,o} = \infty$.

$$\frac{1}{R_{y,i}} = \frac{1}{R_{y,o}} = \frac{1}{\infty} + \frac{1}{\infty} = 0 \tag{21.127}$$

giving $R_{y,i} = R_{y,o} = \infty$

From the input information, the effective modulus of elasticity can be written as

$$E' = 2\left(\frac{1 - \nu_a^2}{E_a} + \frac{1 - \nu_b^2}{E_b}\right) = 2.28 \times 10^{11} \text{ N/m}^2 \tag{21.128}$$

For pure rolling, the surface velocity u relative to the lubricated conjunctions for a cylindrical roller is

$$u = |\omega_i - \omega_o| \frac{d_e^2 - d^2}{4d_e} \tag{21.129}$$

where d_e is the pitch diameter and d is the roller diameter.

$$d_e = \frac{d_o + d_i}{2} = \frac{0.096 + 0.064}{2} = 0.08 \text{ m} \tag{21.130}$$

Hence,

$$u = \frac{0.08^2 - 0.16^2}{4 \times 0.08} |524 - 0| = 10.061 \text{ m/sec} \tag{21.131}$$

The dimensionless speed, materials, and load parameters for the inner- and outer-race conjunctions thus become

$$U_i = \frac{\eta_0 u}{E' R_{x,i}} = \frac{0.01 \times 10.061}{2.28 \times 10^{11} \times 0.0064} = 6.895 \times 10^{-11} \tag{21.132}$$

$$G_i = \xi E' = 5016 \tag{21.133}$$

$$W_i = \frac{F}{E'(R_{x,i})^2} = \frac{4800}{2.28 \times 10^{11} \times (0.0064)^2} = 5.140 \times 10^{-4} \tag{21.134}$$

$$U_o = \frac{\eta_0 u}{E' R_{x,o}} = \frac{0.01 \times 10.061}{2.28 \times 10^{11} \times 0.0096} = 4.597 \times 10^{-11} \tag{21.135}$$

$$G_o = \xi E' = 5016 \tag{21.136}$$

$$W_o = \frac{F}{E'(R_{x,o})^2} = \frac{4800}{2.28 \times 10^{11} \times (0.0096)^2} = 2.284 \times 10^{-4} \tag{21.137}$$

The appropriate elliptical-contact elastohydrodynamic film thickness equation for a fully flooded conjunction is developed in Section 21.3.3 and recorded as Eq. (21.138):

$$H_{min} = \frac{h_{min}}{R_x} = 3.63 \, U^{0.68} G^{0.49} W^{-0.073} (1 - e^{-0.68k}) \tag{21.138}$$

For a roller bearing, $k = \infty$ and this equation reduces to

$$H_{min} = 3.63 U^{0.68} G^{0.49} W^{-0.073}$$

The dimensionless film thickness for the roller–inner-race conjunction is

$$H_{min} = \frac{h_{min}}{R_{x,i}} = 3.63 \times 1.231 \times 10^{-7} \times 65.04 \times 1.783 = 50.5 \times 10^{-6}$$

and hence

$$h_{min} = 0.0064 \times 50.5 \times 10^{-6} = 0.32 \ \mu m$$

The dimensionless film thickness for the roller–outer-race conjunction is

$$H_{min} = \frac{h_{min}}{R_{x,o}} = 3.63 \times 9.343 \times 10^{-8} \times 65.04 \times 1.844 = 40.7 \times 10^{-6}$$

and hence

$$h_{min} = 0.0096 \times 40.7 \times 10^{-6} = 0.39 \ \mu m$$

It is clear from these calculations that the smaller minimum film thickness in the bearing occurs at the roller–inner-race conjunction, where the geometrical conformity is less favorable. It was found that if the ratio of minimum film thickness to composite surface roughness is greater than 3, an adequate elastohydrodynamic film is maintained. This implies that a composite surface roughness of $< 0.1 \ \mu m$ is needed to ensure that an elastohydrodynamic film is maintained.

Radial Ball Bearing Problem.   Consider a single-row, radial, deep-groove ball bearing with the following dimensions:

Inner-race diameter, d_i, m	0.052291
Outer-race diameter, d_o, m	0.077706
Ball diameter, d, m	0.012700
Number of balls in complete bearing, n	9
Inner-groove radius, r_i, m	0.006604
Outer-groove radius, r_o, m	0.006604
Contact angle, β, deg	0
rms surface finish of balls, Δ_b, μm	0.0625
rms surface finish of races, Δ_a, μm	0.175

A bearing of this kind might well experience the following operating conditions:

Radial load, F_r, N	8900
Inner-race angular velocity, ω_i, rad/sec	400
Outer-race angular velocity, ω_o, rad/sec	0
Lubricant viscosity at atmospheric pressure and effective operating temperature of bearing, η_0, N sec/m^2	0.04
Viscosity–pressure coefficient, ξ, m^2/N	2.3×10^{-8}
Modulus of elasticity for both balls and races, E, N/m^2	2×10^{11}
Poisson's ratio for both balls and races, ν	0.3

The essential features of the geometry of the inner and outer conjunctions (Figs. 21.75 and 21.76) can be ascertained as follows:

Pitch diameter [Eq. (21.68)]:

$$d_e = 0.5(d_o + d_i) = 0.065 \ m$$

Diametral clearance [Eq. (21.69)]:

$$P_d = d_o - d_i - 2d = 1.5 \times 10^{-5} \ m$$

Race conformity [Eq. (21.70)]:

$$f_i = f_o = \frac{r}{d} = 0.52$$

Equivalent radius [Eq. (21.85)]:

$$R_{x,i} = \frac{d(d_e - d)}{2d_e} = 0.00511 \text{ m}$$

Equivalent radius [Eq. (21.87)]:

$$R_{x,o} = \frac{d(d_e + d)}{2d_e} = 0.00759 \text{ m}$$

Equivalent radius [Eq. (21.86)]:

$$R_{y,i} = \frac{f_i d}{2f_i - 1} = 0.165 \text{ m}$$

Equivalent radius [Eq. (21.88)]:

$$R_{y,o} = \frac{f_o d}{2f_o - 1} = 0.165 \text{ m}$$

The curvature sum

$$\frac{1}{R_i} = \frac{1}{R_{x,i}} + \frac{1}{R_{y,i}} = 201.76 \qquad (21.139)$$

gives $R_i = 4.956 \times 10^{-3}$ m, and the curvature sum

$$\frac{1}{R_o} = \frac{1}{R_{x,o}} + \frac{1}{R_{y,o}} = 137.81 \qquad (21.140)$$

gives $R_o = 7.256 \times 10^{-3}$ m. Also, $\alpha_i = R_{y,i}/R_{x,i} = 32.35$ and $\alpha_o = R_{y,o}/R_{x,o} = 21.74$.
The nature of the Hertzian contact conditions can now be assessed.
Ellipticity parameters:

$$k_i = \alpha_i^{2/\pi} = 9.42, \qquad k_o = \alpha_o^{2/\pi} = 7.09 \qquad (21.141)$$

Elliptic integrals:

$$q = \frac{\pi}{2} - 1$$

$$\mathcal{E}_i = 1 + \frac{q}{\alpha_i} = 1.0188, \qquad \mathcal{E}_o = 1 + \frac{q}{\alpha_o} = 1.0278 \qquad (21.142)$$

$$\mathcal{F}_i = \frac{\pi}{2} + q \ln \alpha_i = 3.6205, \qquad \mathcal{F}_o = \frac{\pi}{2} + q \ln \alpha_o = 3.3823 \qquad (21.143)$$

The effective elastic modulus E' is given by

$$E' = 2 \left(\frac{1 - v_a^2}{E_a} + \frac{1 - v_b^2}{E_b} \right)^{-1} = 2.198 \times 10^{11} \text{ N/m}^2$$

To determine the load carried by the most heavily loaded ball in the bearing, it is necessary to adopt an iterative procedure based on the calculation of local static compression and the analysis presented in the fourth subsection in Section 21.3.6. Stribeck[46] found that the value of Z was about 4.37 in the expression

$$F_{\max} = \frac{ZF_r}{n}$$

where $F_{\max}$ = load on most heavily loaded ball
F_r = radial load on bearing
n = number of balls

However, it is customary to adopt a value of $Z = 5$ in simple calculations in order to produce a conservative design, and this value will be used to begin the iterative procedure.

Stage 1.   Assume $Z = 5$. Then

$$F_{max} = \frac{5F_r}{9} = \frac{5}{9} \times 8900 = 4944 \text{ N} \qquad (21.144)$$

The maximum local elastic compression is

$$\delta_i = \mathfrak{I}_i \left[\left(\frac{9}{2\delta_i R_i} \right) \left(\frac{F_{max}}{\pi k_i E'} \right)^2 \right]^{1/3} = 2.902 \times 10^{-5} \text{ m} \qquad (21.145)$$

$$\delta_o = \mathfrak{I}_o \left[\left(\frac{9}{2\delta_o R_o} \right) \left(\frac{F_{max}}{\pi k_o E'} \right)^2 \right]^{1/3} = 2.877 \times 10^{-5} \text{ m}$$

The sum of the local compressions on the inner and outer races is

$$\delta = \delta_i + \delta_o = 5.799 \times 10^{-5} \text{ m}$$

A better value for Z can now be obtained from

$$Z = \frac{\pi(1 - P_d/2\delta)^{3/2}}{2.491 \left\{ \left[1 + \left(\frac{1 - P_d/2\delta}{1.23} \right)^2 \right]^{1/2} - 1 \right\}}$$

since $P_d/2\delta = (1.5 \times 10^{-5})/(5.779 \times 10^{-5}) = 0.1298$. Thus

$$Z = 4.551$$

Stage 2.

$$Z = 4.551$$
$$F_{max} = (4.551 \times 8900)/9 = 4500 \text{ N}$$
$$\delta_i = 2.725 \times 10^{-5} \text{ m}, \qquad \delta_o = 2.702 \times 10^{-5} \text{ m}$$
$$\delta = 5.427 \times 10^{-5} \text{ m}$$
$$\frac{P_d}{2\delta} = 0.1382$$

Thus

$$Z = 4.565$$

Stage 3.

$$Z = 4.565$$
$$F_{max} = \frac{4.565 \times 8900}{9} = 4514 \text{ N}$$
$$\delta_i = 2.731 \times 10^{-5} \text{ m}, \qquad \delta_o = 2.708 \times 10^{-5} \text{ m}$$
$$\delta = 5.439 \times 10^{-5} \text{ m}$$
$$\frac{P_d}{2\delta} = 0.1379$$

and hence

$$Z = 4.564$$

This value is very close to the previous value from stage 2 of 4.565, and a further iteration confirms its accuracy.

Stage 4.

$$Z = 4.564$$

$$F_{max} = \frac{4.564 \times 8900}{9} = 4513 \text{ N}$$

$$\delta_i = 2.731 \times 10^{-5} \text{ m}, \qquad \delta_o = 2.707 \times 10^{-5} \text{ m}$$

$$\delta = 5.438 \times 10^{-5} \text{ m}$$

$$\frac{P_d}{2\delta} = 0.1379$$

and hence

$$Z = 4.564$$

The load on the most heavily loaded ball is thus 4513 N.

Elastohydrodynamic Minimum Film Thickness. For pure rolling

$$u = |\omega_o - \omega_i| \frac{d_e^2 - d^2}{4d_e} = 6.252 \text{ m/sec} \tag{21.146}$$

The dimensionless speed, materials, and load parameters for the inner- and outer-race conjunctions thus become

$$U_i = \frac{\eta_0 u}{E' R_{x,i}} = \frac{0.04 \times 6.252}{2.198 \times 10^{11} \times 5.11 \times 10^{-3}} = 2.227 \times 10^{-10} \tag{21.147}$$

$$G_i = \xi E' = 2.3 \times 10^{-8} \times 2.198 \times 10^{11} = 5055 \tag{21.148}$$

$$W_i = \frac{F}{E'(R_{x,i})^2} = \frac{4513}{2.198 \times 10^{11} \times (5.11)^2 \times 10^{-6}} = 7.863 \times 10^{-4} \tag{21.149}$$

$$U_o = \frac{\eta_0 u}{E' R_{x,o}} = \frac{0.04 \times 6.252}{2.198 \times 10^{11} \times 7.59 \times 10^{-3}} = 1.499 \times 10^{-10} \tag{21.150}$$

$$G_o = \xi E' = 2.3 \times 10^{-8} \times 2.198 \times 10^{11} = 5055 \tag{21.151}$$

$$W_o = \frac{F}{E'(R_{x,o})^2} = \frac{4513}{2.198 \times 10^{11} \times (7.59)^2 \times 10^{-6}} = 3.564 \times 10^{-4} \tag{21.152}$$

The dimensionless minimum elastohydrodynamic film thickness in a fully flooded elliptical contact is given by

$$H_{min} = \frac{h_{min}}{R_x} = 3.63 U^{0.68} G^{0.49} W^{-0.073} (1 - e^{-0.68k}) \tag{21.153}$$

For the ball-inner-race conjunction it is

$$\begin{aligned}(H_{min})_i &= 3.63 \times 2.732 \times 10^{-7} \times 65.29 \times 1.685 \times 0.9983 \\ &= 1.09 \times 10^{-4}\end{aligned} \tag{21.154}$$

Thus

$$(h_{min})_i = 1.09 \times 10^{-4} R_{x,i} = 0.557 \ \mu\text{m}$$

The lubrication factor Λ discussed in the fifth subsection of Section 21.3.6 was found to play a significant role in determining the fatigue life of rolling-element bearings. In this case

$$\Lambda_i = \frac{(h_{min})_i}{(\Delta_a^2 + \Delta_b^2)^{1/2}} = \frac{0.557 \times 10^{-6}}{[(0.175)^2 + (0.06225)^2]^{1/2} \times 10^{-6}} = 3.00 \tag{21.155}$$

Ball–outer-race conjunction is given by

$$(H_{\min})_o = \frac{(h_{\min})_o}{R_{x,o}} = 3.63 U_o^{0.68} G^{0.49} W^{-0.073} \left(1 - e^{-0.68 k_o}\right)$$

$$= 3.63 \times 2.087 \times 10^{-7} \times 65.29 \times 1.785 \times 0.9919 \qquad (21.156)$$

$$= 0.876 \times 10^{-4}$$

Thus

$$(h_{\min})_o = 0.876 \times 10^{-4} R_{x,o} = 0.665 \ \mu\text{m}$$

In this case, the lubrication factor Λ is given by

$$\Lambda_o = \frac{0.665 \times 10^{-6}}{[(0.175)^2 + (0.0625)^2]^{1/2} \times 10^{-6}} = 3.58 \qquad (21.157)$$

Once again, it is evident that the smaller minimum film thickness occurs between the most heavily loaded ball and the inner race. However, in this case the minimum elastohydrodynamic film thickness is about three times the composite surface roughness, and the bearing lubrication can be deemed to be entirely satisfactory. Indeed, it is clear from Fig. 21.97 that very little improvement in the lubrication factor $\tilde{F}$ and thus in the fatigue life of the bearing could be achieved by further improving the minimum film thickness and hence Λ.

21.4   BOUNDARY LUBRICATION

If the pressures in fluid-film-lubricated machine elements are too high, the running speeds are too low, or the surface roughness is too great, penetration of the lubricant film will occur. Contact will take place between asperities, leading to a rise in friction and wear rate. Figure 21.99 (obtained from Bowden and Tabor[56]) shows the behavior of the coefficient of friction in the different lubrication regimes. It is to be noted in this figure that in boundary lubrication, although the friction is much higher than in the hydrodynamic regime, it is still much lower than for unlubricated surfaces. As the running conditions are made more severe, the amount of lubricant breakdown increases, until the system scores or seizes so badly that the machine element can no longer operate successfully.

Figure 21.100 shows the wear rate in the different lubrication regimes as determined by the operating load. In the hydrodynamic and elastohydrodynamic lubrication regimes, since there is no asperity contact, there is little or no wear. In the boundary lubrication regime the degree of asperity interaction and wear rate increases as the load increases. The transition from boundary lubrication to an unlubricated condition is marked by a drastic change in wear rate. Machine elements cannot operate successfully in the unlubricated region. Together Figs. 21.99 and 21.100 show that both friction and wear can be greatly decreased by providing a boundary lubricant to unlubricated surfaces.

Understanding boundary lubrication depends first on recognizing that bearing surfaces have asperities that are large compared with molecular dimensions. On the smoothest machined surfaces these asperities may be 25 nm (0.025 μm) high; on rougher surfaces they may be ten to several hundred times higher. Figure 21.101 illustrates typical surface roughness as a random distribution of

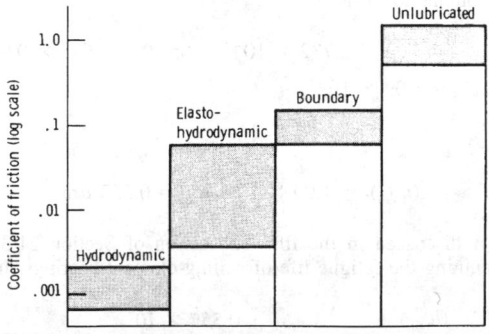

Fig. 21.99 Schematic drawing showing how type of lubrication shifts from hydrodynamic to elastohydrodynamic to boundary lubrication as the severity of running conditions is increased. (From Ref. 56.)

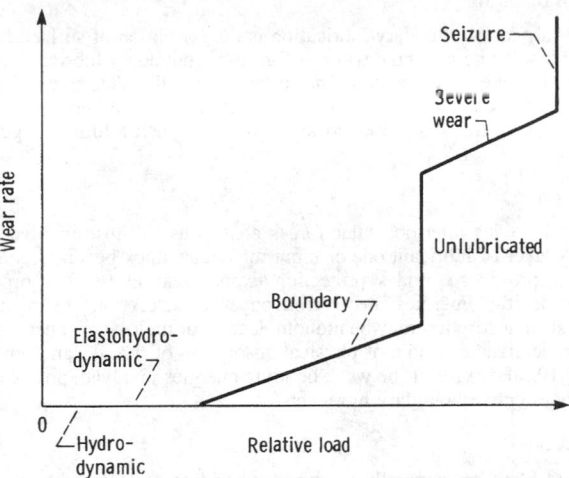

Fig. 21.100 Chart for determining wear rate for various lubrication regimes. (From Ref. 57.)

hills and valleys with varying heights, spacing, and slopes. In the absence of hydrodynamic or elastohydrodynamic pressures these hills or asperities must support all of the load between the bearing surfaces. Understanding boundary lubrication also depends on recognizing that bearing surfaces are often covered by boundary lubricant films such as are idealized in Fig. 21.101. These films separate the bearing materials and, by shearing preferentially, provide some control of friction, wear, and surface damage.

Many mechanism, such as door hinges, operate totally under conditions (high load, low speed) of boundary lubrication. Others are designed to operate under full hydrodynamic or elastohydrodynamic lubrication. However, as the oil film thickness is a function of speed, the film will be unable to provide complete separation of the surfaces during startup and rundown, and the condition of boundary lubrication will exist. The problem from the boundary lubrication standpoint is to provide a boundary film with the proper physical characteristics to control friction and wear. The work of Bowden and Tabor,[56] Godfrey,[59] and Jones[60] was relied upon in writing the sections that follow.

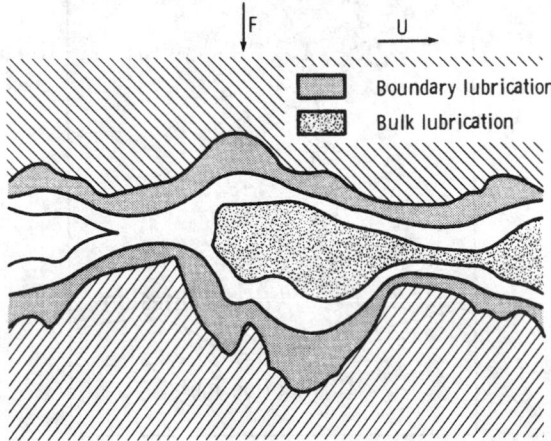

Fig. 21.101 Lubricated bearing surfaces. (From Ref. 58.)

21.4.1 Formation of Films

The most important aspect of boundary lubrication is the formation of surface films that will protect the contacting surfaces. There are three ways of forming a boundary lubricant film; physical adsorption, chemisorption, and chemical reaction. The surface action that determines the behavior of boundary lubricant films is the energy binding the film molecules to the surface, a measure of the film strength. The formation of films is presented in the order of such a film strength, the weakest being presented first.

Physical Adsorption

Physical adsorption involves intermolecular forces analogous to those involved in condensation of vapors to liquids. A layer of lubricant one or more molecules thick becomes attached to the surfaces of the solids, and this provides a modest protection against wear. Physical adsorption is usually rapid, reversible, and nonspecific. Energies involved in physical adsorption are in the range of heats of condensations. Physical adsorption may be monomolecular or multilayer. There is no electron transfer in this process. An idealized example of physical adsorption of hexadecanol on an unreactive metal is shown in Fig. 21.102. Because of the weak bonding energies involved, physically adsorbed species are usually not very effective boundary lubricants.

Chemical Adsorption

Chemically adsorbed films are generally produced by adding animal and vegetable fats and oils to the base oils. These additives contain long-chain fatty acid molecules, which exhibit great affinity for metals at their active ends. The usual configuration of these polar molecules resembles that of a carpet pile with the molecules standing perpendicular to the surface. Such fatty acid molecules form metal soaps that are low-shear-strength materials with coefficients of friction in the range 0.10–0.15. The soap film is dense because of the preferred orientation of the molecules. For example, on a steel surface stearic acid will form a monomolecular layer of iron stearate, a soap containing 10^{14} molecules/cm^2 of surface. The effectiveness of these layers is limited by the melting point of the soap (180°C for iron stearate). It is clearly essential to choose an additive that will react with the bearing metals, so that less reactive, inert metals like gold and platinum are not effectively lubricated by fatty acids.

Examples of fatty acid additives are stearic, oleic, and lauric acid. The soap films formed by these acids might reduce the coefficient of friction to 50% of that obtained by a straight mineral oil. They

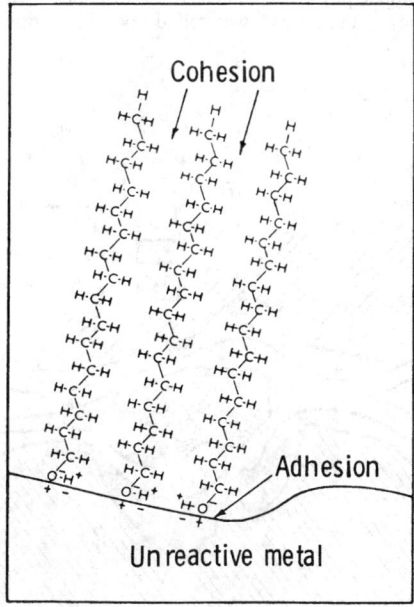

Fig. 21.102 Physical adsorption of hexadecanol. (From Ref. 59.)

provide satisfactory boundary lubrication at moderate loads, temperatures, and speeds and are often successful in situations showing evidence of mild surface distress.

Chemisorption of a film on a surface is usually specific, may be rapid or slow, and is not always reversible. Energies involved are large enough to imply that a chemical bond has formed (i.e., electron transfer has taken place). In contrast to physical adsorption, chemisorption may require an activation energy. A film may be physically adsorbed at low temperatures and chemisorbed at higher temperatures. In addition, physical adsorption may occur on top of a chemisorbed film. An example of a film of stearic acid chemisorbed on an iron oxide surface to form iron stearate is shown in Fig. 21.103.

Chemical Reaction

Films formed by chemical reaction provide the greatest film strength and are used in the most severe operating conditions. If the load and sliding speeds are high, significant contact temperatures will be developed. It has already been noted that films formed by physical and chemical adsorption cease to be effective above certain transition temperatures, but some additives start to react and form new high-melting-point inorganic solids at high temperatures. For example, sulfur will start to react at about 100°C to form sulfides with melting points of over 1000°C. Lubricants containing additives like sulfur, chlorine, phosphorous, and zinc are often referred to as extreme-pressure (EP) lubricants, since they are effective in the most arduous conditions.

The formation of a chemical reaction film is specific; may be rapid or slow (depending on temperature, reactivity, and other conditions); and is irreversible. An idealized example of a reacted film of iron sulfide on an iron surface is shown in Fig. 21.104.

21.4.2 Physical Properties of Boundary Films

The two physical properties of boundary films that are most important in determining their effectiveness in protecting surfaces are melting point and shear strength. It is assumed that the film thicknesses involved are sufficient to allow these properties to be well defined.

Melting Point

The melting point of a surface film appears to be one discriminating physical property governing failure temperature for a wide range of materials including inorganic salts. It is based on the observation that only a surface film that is solid can properly interfere with potentially damaging asperity contacts. Conversely, a liquid film allows high friction and wear. Under practical conditions, physically adsorbed additives are known to be effective only at low temperatures, and chemisorbed addi-

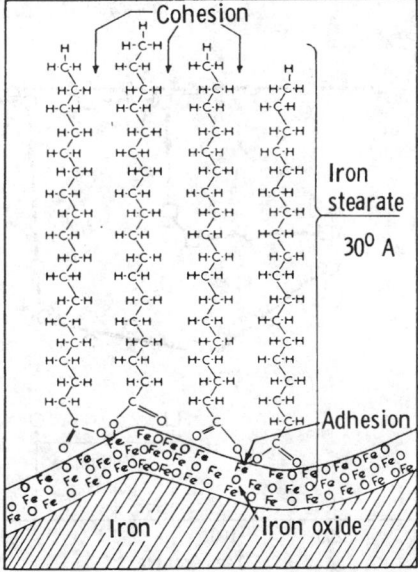

Fig. 21.103 Chemisorption of stearic acid on iron surface to form iron stearate. (From Ref. 59.)

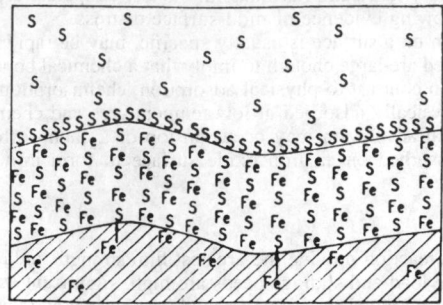

Fig. 21.104 Formation of inorganic film by reaction of sulfur with iron to form iron sulfide. (From Ref. 59.)

tives at moderate temperatures. High-melting-point inorganic materials are used for high-temperature lubricants.

The correlation of melting point with failure temperature has been established for a variety of organic films. An illustration is given in Fig. 21.105 (obtained from Russell et al.[61]) showing the friction transition for copper lubricated with pure hydrocarbons. Friction data for two hydrocarbons (mesitylene and dotriacontane) are given in Fig. 21.105 as a function of temperature. In this figure the boundary film failure occurs at the melting point of each hydrocarbon.

In contrast, chemisorption of fatty acids on reactive metals yields failure temperature based on the softening point of the soap rather than the melting point of the parent fatty acid.

Shear Strength

The shear strength of a boundary lubricating film should be directly reflected in the friction coefficient. In general, this is true with low-shear-strength soaps yielding low friction and high-shear-

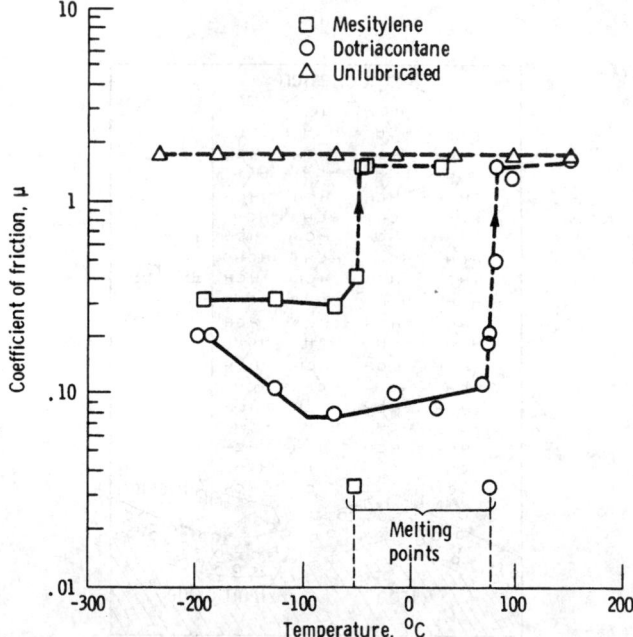

Fig. 21.105 Chart for determining friction of copper lubricated with hydrocarbons in dry helium. (From Ref. 61.)

strength salts yielding high friction. However, the important parameter in boundary friction is the ratio of shear strength of the film to that of the substrate. This relationship is shown in Fig. 21.106, where the ratio is plotted on the horizontal axis with a value of 1 at the left and zero at the right. These results are in agreement with experience. For example, on steel an MoS_2 film gives low friction and Fe_2O_3 gives high friction. The results from Fig. 21.106 also indicate how the same friction value can be obtained with various combinations provided that the ratio is the same. It is important to recognize that shear strength is also affected by pressure and temperature.

21.4.3 Film Thickness

Boundary film thickness can vary from a few angstroms (adsorbed gas) to thousands of angstroms (chemical reaction films). In general, as the thickness of a boundary film increases, the coefficient of friction decreases. This effect is shown in Fig. 21.107*a*, which shows the coefficient of friction plotted against oxide film thickness formed on a copper surface. However, continued increases in thickness may result in an increase in friction. This effect is shown in Fig. 21.107*b*, which shows the coefficient of friction plotted against indium film thickness on copper surface. It should also be pointed out that the shear strengths of all boundary films decrease as their thicknesses increase, which may be related to the effect seen in Fig. 21.107*b*.

For physically adsorbed or chemisorbed films, surface protection is usually enhanced by increasing film thickness. The frictional transition temperature of multilayers also increases with increasing number of layers.

For thick chemically reacted films there is an optimum thickness for minimum wear that depends on temperature, concentration, or load conditions. The relationship between wear and lubricant (or additive) reactivity is shown in Fig. 21.108. Here, if reactivity is not great enough to produce a thick enough film, adhesion wear occurs. On the other hand, if the material is too reactive, very thick films are formed and corrosive wear ensues.

21.4.4 Effect of Operating Variables

The effect of load, speed, temperature, and atmosphere can be important for the friction and wear of boundary lubrication films. Such effects are considered in this section.

On Friction

Load. The coefficient of friction is essentially constant with increasing load.

Speed. In general, in the absence of viscosity effects, friction changes little with speed over a sliding speed range of 0.005 to 1.0 cm/sec. When viscosity effects do come into play, two types of behavior are observed, as shown in Fig. 21.109. In this figure relatively nonpolar materials such as mineral oils show a decrease in friction with increasing speed, while polar fatty acids show the opposite trend. At higher speeds viscous effects will be present, and increases in friction are normally observed.

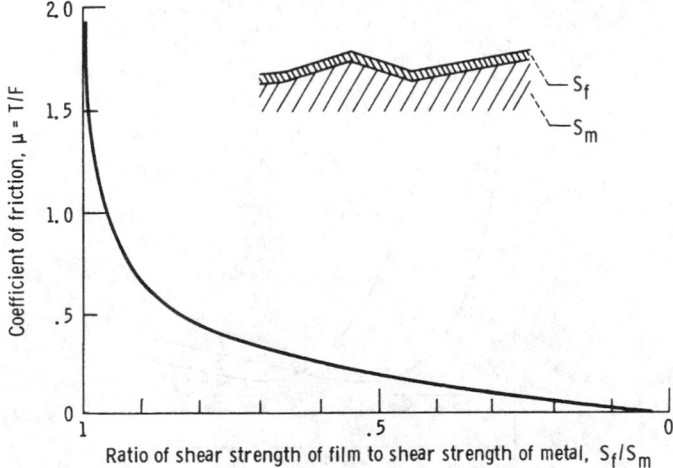

Fig. 21.106 Chart for determining friction as function of shear strength ratio. (From Ref. 59.)

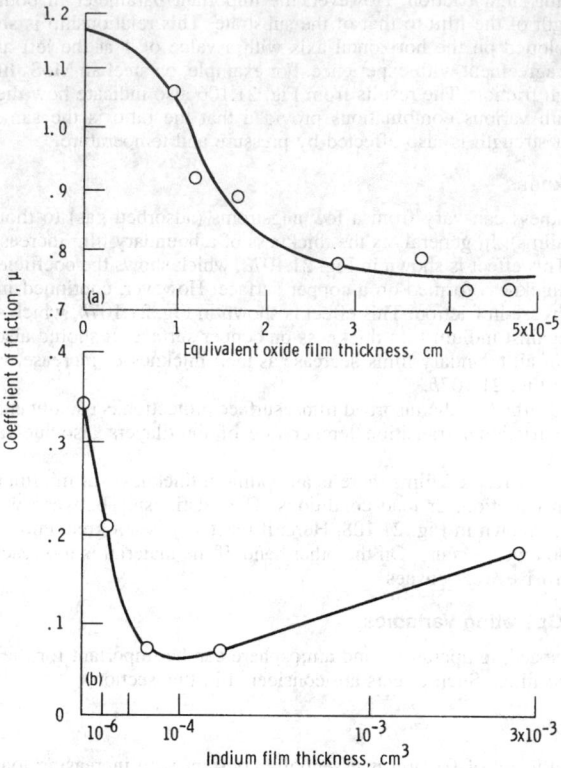

Fig. 21.107   Chart for determining relationship of friction and thickness of films on copper surfaces. (From Ref. 62.)

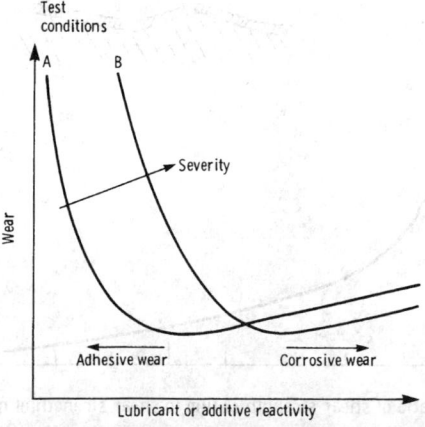

Fig. 21.108   Relationship between wear and lubricant reactivity. (From Ref. 63.)

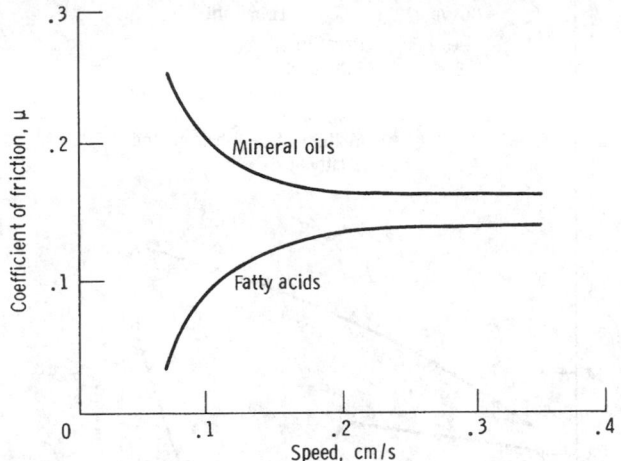

Fig. 21.109 Effect of speed on coefficient of friction. (From Ref. 64.)

Temperature. It is difficult to make general comments on the effect of temperature on boundary friction since so much depends on the other conditions and the type of materials present. Temperature can cause disruption, desorption, or decomposition of boundary films. It can also provide activation energy for chemisorption or chemical reactions.

Atmosphere. The presence of oxygen and water vapor in the atmosphere can greatly affect the chemical processes that occur in the boundary layer. These processes can, in turn, affect the friction coefficient.

On Wear

Load. It is generally agreed that wear increases with increasing load, but no simple relationship seems to exist, at least before the transition to severe wear occurs. At this point a discontinuity of wear versus load is often like that illustrated in Fig. 21.100.

Speed. For practical purposes, wear rate in a boundary lubrication regime is essentially independent of speed. This assumes no boundary film failure due to contact temperature rise.

Temperature. As was the case for friction, there is no way to generalize the effect of temperature on wear. The statement that pertains to friction also pertains to wear.

Atmosphere. Oxygen has been shown to be an important ingredient in boundary lubrication experiments involving load-carrying additives. The presence of oxygen or moisture in the test atmosphere has a great effect on the wear properties of lubricants containing aromatic species.

21.4.5 Extreme-Pressure (EP) Lubricants

The best boundary lubricant films cease to be effective above 200–250°C. At these high temperatures the lubricant film may iodize. For operation under more severe conditions, EP lubricants might be considered.

Extreme-pressure lubricants usually consist of a small quantity of an EP additive dissolved in a lubricating oil, usually referred to as the base oil. The most common additives used for this purpose contain phosphorus, chlorine, or sulfur. In general, these materials function by reacting with the surface to form a surface film that prevents metal-to-metal contact. If, in addition, the surface film formed has a low shear strength, it will not only protect the surface, but it will also give a low coefficient of friction. Chloride films give a lower coefficient of friction ($\mu = 0.2$) than sulfide films ($\mu = 0.5$). Sulfide films, however, are more stable, are unaffected by moisture, and retain their lubricating properties to very high temperatures.

Although EP additives function by reacting with the surface, they must not be too reactive, otherwise chemical corrosion may be more troublesome than frictional wear. They should only react when there is a danger of seizure, usually noted by a sharp rise in local or global temperature. For this reason it is often an advantage to incorporate in a lubricant a small quantity of a fatty acid that can provide effective lubrication at temperatures below those at which the additive becomes reactive.

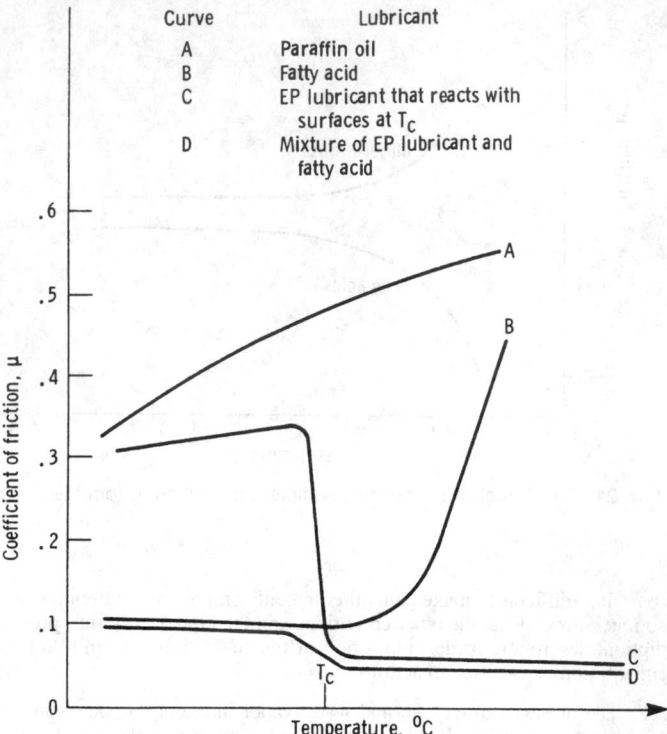

Fig. 21.110   Graph showing frictional behavior of metal surfaces with various lubricants. (From Ref. 56.)

Bowden and Tabor[56] describe this behavior in Fig. 21.110, where the coefficient of friction is plotted against temperature. Curve A is for paraffin oil (the base oil) and shows that the friction is initially high and increases as the temperature is raised. Curve B is for a fatty acid dissolved in the base oil: it reacts with the surface to form a metallic soap, which provides good lubrication from room temperature up to the temperature at which the soap begins to soften. Curve C is for a typical EP additive in the base oil; this reacts very slowly below the temperature T_c, so that in this range the lubrication is poor, while above T_c the protective film is formed and effective lubrication is provided to a very high temperature. Curve D is the result obtained when the fatty acid is added to the EP solution. Good lubrication is provided by the fatty acid below T_c, while above this temperature the greater part of the lubrication is due to the additive. At still higher temperatures, a deterioration of lubricating properties will also occur for both curves C and D.

REFERENCES

1. B. Tower, "First Report on Friction Experiments (Friction of Lubricated Bearings)," *Proc. Inst. Mech. Eng., London,* 632–659 (1883).

2. N. P. Petrov, "Friction in Machines and the Effect of the Lubricant," *Inzh. Zh. St-Petreb.* **1**, 71–140 (1883); **2**, 227–279 (1883); **3**, 377–436 (1883); **4**, 535–564 (1883).

3. O. Reynolds, "On the Theory of Lubrication and Its Application to Mr. Beauchamp Tower's Experiments, Including an Experimental Determination of the Viscosity of Olive Oil," *Philos. Trans. R. Soc. London* **177**, 157–234 (1886).

4. W. B. Hardy and I. Doubleday, "Boundary Lubrication—The Temperature Coefficient," *Proc. R. Soc.* A101, 487–492 (1922).

5. W. B. Hardy and I. Doubleday, "Boundary Lubrication—The Paraffin Series," *Proc. R. Soc.* **A104**, 25–39 (1922).

6. B. J. Hamrock and W. J. Anderson, *Rolling-Element Bearings,* NASA RP-1105, 1983.

7. ESDU, "General Guide to the Choice of Journal Bearing Type," Engineering Sciences Data Unit, Item 65007, Institution of Mechanical Engineers, London, 1965.

8. ESDU, "General Guide to the Choice of Thrust Bearing Type," Engineering Sciences Data Unit, Item 67033, Institution of Mechanical Engineers, London, 1967.

9. I. Newton, *Philosophiae Naturalis Principia Mathematica.* 1687. Imprimature S. Pepys. Reg. Soc. Praess. 5 Julii 1866. Revised and supplied with a historical and explanatory appendix by F. Cajori, edited by R. T. Crawford, 1934. Published by the University of California Press, Berkeley and Los Angeles, 1966.

10. B. J. Hamrock and D. Dowson, *Ball Bearing Lubrication—The Elastohydrodynamics of Elliptical Contacts,* Wiley, New York, 1981.

11. E. E. Bisson and W. J. Anderson, *Advanced Bearing Technology,* NASA SP-38, 1964.

12. ESDU, "Contact Stresses," Engineering Sciences Data Unit, Item 78035, Institution of Mechanical Engineers, London, 1978.

13. S. Aihara and D. Dowson, "A Study of Film Thickness in Grease Lubricated Elastohydrodynamic Contacts," in *Proceedings of Fifth Leeds–Lyon Symposium on Tribology on "Elastohydrodynamics and Related Topics,"* D. Dowson, C. M. Taylor, M. Godet, and D. Berthe (eds.), Mechanical Engineering Publications, Bury St. Edmunds, Suffolk, 1979, pp. 104–115.

14. A. R. Wilson, "The Relative Thickness of Grease and Oil Films in Rolling Bearings," *Proc. Inst. Mech. Eng., London* **193**(17), 185–192 (1979).

15. T. Tallian, L. Sibley, and R. Valori, "Elastohydrodynamic Film Effect on the Load-Life Behavior of Rolling Contacts," ASME Paper 65-LUB-11, 1965.

16. C. Barus, "Isotherms, Isopeistics and Isometrics Relative to Viscosity," *Am. J. Sci.* **45**, 87–96 (1893).

17. W. R. Jones, R. L. Johnson, W. O. Winer, and D. M. Sanborn, "Pressure-Viscosity Measurements for Several Lubricants to 5.5×10^8 Newtons Per Square Meter (8×10^4 psi) and 149°C (300°F)," *ASLE Trans.* **18**(4), 249–262 (1975).

18. D. Dowson and G. R. Higginson, *Elastohydrodynamic Lubrication, the Fundamentals of Roller and Gear Lubrication,* Pergamon, Oxford, 1966.

19. W. A. Gross, L. A. Matsch, V. Castelli, A. Eshel, and M. Wildmann, *Fluid Film Lubrication,* Wiley, New York, 1980.

20. N. F. Reiger, *Design of Gas Bearings,* Mechanical Technology, Inc., Latham, New York, 1967.

21. O. Pinkus and B. Sternlicht, *Theory of Hydrodynamic Lubrication,* McGraw-Hill, New York, 1961.

22. H. C. Rippel, *Cast Bronze Hydrostatic Bearing Design Manual,* Cast Bronze Bearing Institute, Inc., Cleveland, OH, 1963.

23. A. A. Raimondi and J. Boyd, "A Solution for the Finite Journal Bearing and Its Application to Analysis and Design; III," *Trans. ASLE* **1**(1), 194–209 (1959).

24. P. E. Allaire and R. D. Flack, "Journal Bearing Design for High Speed Turbomachinery," *Bearing Design—Historical Aspects, Present Technology and Future Problems,* W. J. Anderson (ed.), American Society of Mechanical Engineers, New York, 1980, pp. 111–160.

25. A. A. Raimondi and J. Boyd, "Applying Bearing Theory to the Analysis and Design of Pad-type Bearings," *Trans. ASME,* 287–309 (April 1955).

26. B. J. Hamrock, "Optimum Parallel Step-Sector Bearing Lubricated with an Incompressible Fluid," NASA TM-83356, 1983.

27. E. J. Gunter, J. G. Hinkle, and D. D. Fuller, "Design Guide for Gas-Lubricated Tilting-Pad Journal and Thrust Bearings with Special Reference to High Speed Rotors," Franklin Institute Research Laboratories Report I-A2392-3-1, 1964.

28. B. J. Hamrock and D. P. Fleming, "Optimization of Self-Acting Herringbone Grooved Journal Bearings for Minimum Radial Load," in *Proceedings of Fifth International Gas Bearing Symposium,* University of Southampton, Southampton, England, 1971, Paper 13.

29. D. P. Fleming and B. J. Hamrock, "Optimization of Self-Acting Herringbone Journal Bearings for Maximum Stability," *6th International Gas Bearing Symposium,* University of Southampton, Southampton, England, 1974, Paper c1, pp. 1–11.

30. B. J. Hamrock, "Optimization of Self-Acting Step Thrust Bearings for Load Capacity and Stiffness," *ASLE Trans.* **15**(3), 159–170 (1972).

31. D. Dowson, "Elastohydrodynamic Lubrication—An Introduction and a Review of Theoretical Studies," *Institution of Mechanical Engineers, London, Proceedings,* Vol. 180, Pt. 3B, 1965, pp. 7–16.

32. A. N. Grubin, "Fundamentals of the Hydrodynamic Theory of Lubrication of Heavily Loaded Cylindrical Surfaces," in *Investigation of the Contact Machine Components,* Kh. F. Ketova (ed.), translation of Russian Book No. 30, Central Scientific Institute of Technology and Mechanical Engineering, Moscow, 1949, Chap. 2. (Available from Dept. of Scientific and Industrial Research,

Great Britain, Transl. CTS-235, and from Special Libraries Association, Chicago, Transl. R-3554.)

33. A. J. Petrusevich, "Fundamental Conclusion from the Contact-Hydrodynamic Theory of Lubrication," *Zv. Akad, Nauk, SSSR (OTN)* **2,** 209 (1951).

34. B. J. Hamrock and D. Brewe, "Simplified Solution for Stresses and Deformation," *J. Lubr. Technol.* **105**(2), 171–177 (1983).

35. B. J. Hamrock and W. J. Anderson, "Analysis of an Arched Outer-Race Ball Bearing Considering Centrifugal Forces," *J. Lubr. Techn.* **95**(3), 265–276 (1973).

36. A. B. Jones, "Analysis of Stresses and Deflections," New Departure Engineering Data, General Motors Corporation, Bristol, CT, 1946.

37. H. Hertz, "The Contact of Elastic Solids," *J. Reine Angew. Math.* **92,** 156–171 (1881).

38. B. J. Hamrock and D. Dowson, "Isothermal Elastohydrodynamic Lubrication of Point Contacts, Part I—Theoretical Formulation," *J. Lubr. Technol.* **98**(2), 223–229 (1976).

39. B. J. Hamrock and D. Dowson, "Isothermal Elastohydrodynamic Lubrication of Point Contacts, Part III—Fully Flooded Results," *J. Lubr. Technol.* **99**(2), 264–276 (1977).

40. A. Cameron and R. Gohar, "Theoretical and Experimental Studies of the Oil Film in Lubricated Point Contacts," *Proc. R. Soc. London, Ser. A,* **291,** 520–536 (1966).

41. B. J. Hamrock and D. Dowson, "Elastohydrodynamic Lubrication of Elliptical Contacts for Materials of Low Elastic Modulus, Part I—Fully Flooded Conjunction," *J. Lubr. Technol.* **100**(2), 236–245 (1978).

42. D. E. Brewe, B. J. Hamrock, and C. M. Taylor, "Effect of Geometry on Hydrodynamic Film Thickness," *J. Lubr. Technol.,* **101**(2), 231–239 (1979).

43. P. L. Kapitza, "Hydrodynamic Theory of Lubrication During Rolling," *Zh. Tekh. Fig.* **25**(4), 747–762 (1955).

44. B. J. Hamrock and D. Dowson, "Minimum Film Thickness in Elliptical Contacts for Different Regimes of Fluid-Film Lubrication," in *Proceedings of Fifth Leeds–Lyon Symposium on Tribology on Elastohydrodynamics and Related Topics,* D. Dowson, C. M. Taylor, M. Godet, and D. Berthe (eds.), Mechanical Engineering Publications, Bury St. Edmunds, Suffolk, 1979, pp. 22–27.

45. A. B. Jones, "The Mathematical Theory of Rolling Element Bearings," in *Mechanical Design and Systems Handbook,* H. A. Rothbart (ed.), McGraw-Hill, New York, 1964, pp. 13-1–13-76.

46. R. Stribeck, "Kugellager fur beliebige Belastungen," *Z. VDI-Zeitschrift* **45**(3), 73–125 (1901).

47. W. Weibull, "A Statistical Representation of Fatigue Failures in Solids," *Trans. Roy. Inst. Tech., Stockholm* **27** (1949).

48. G. Lundberg and A. Palmgren, "Dynamic Capacity of Rolling Bearings," *Acta Polytechnica,* Mechanical Engineering Series, Vol. I, No. 3, 1947.

49. G. Lundberg and A. Palmgren, "Dynamic Capacity of Rolling Bearings," *Acta Polytechnica,* Mechanical Engineering Series, Vol. II, No. 4, 1952.

50. ISO, "Rolling Bearings, Dynamic Load Ratings and Rating Life," ISO/TC4/JC8, Revision of ISOR281. Issued by International Organization for Standardization, Technical Committee ISO/TC4, 1976.

51. AFBMA, *Method of Evaluating Load Ratings for Ball Bearings,* AFBMA Standard System No. 9, Revision No. 4, Anti-Friction Bearing Manufacturers Association, Inc., Arlington, VA, 1960.

52. AFBMA, *Method of Evaluating Load Ratings for Roller Bearings,* AFBMA Standard System No. 11, Anti-Friction Bearing Manufacturers Association, Inc., Arlington, VA, 1960.

53. E. N. Bamberger, T. A. Harris, W. M. Kacmarsky, C. A. Moyer, R. J. Parker, J. J. Sherlock, and E. V. Zaretsky, *Life Adjustment Factors for Ball and Roller Bearings—An Engineering Design Guide,* American Society for Mechanical Engineers, New York, 1971.

54. T. E. Tallian, "On Competing Failure Modes in Rolling Contact," *Trans. ASLE* **10,** 418–439 (1967).

55. J. C. Skurka, "Elastohydrodynamic Lubrication of Roller Bearings," *J. Lubr. Technol.* **92**(2), 281–291 (1970).

56. F. P. Bowden and D. Tabor, *Friction—An Introduction to Tribology,* Heinemann, London, 1973.

57. A. Beerbower, "Boundary Lubrication," *Scientific and Technical Application Forecasts,* Department of the Army, DAC-19-69-C-0033, 1972.

58. R. S. Fein and F. J. Villforth, "Lubrication Fundamentals," *Lubrication,* Texaco Inc., New York, 1973, pp. 77–88.

59. D. Godfrey, "Boundary Lubrication," in *Interdisciplinary Approach to Friction and Wear,* P. M. Ku (ed.), NASA SP-181, 1968, pp. 335–384.

60. W. R. Jones, *Boundary Lubrication—Revisited,* NASA TM-82858, 1982.

61. J. A. Russell, W. C. Campbell, R. A. Burton, and P. M. Ku, "Boundary Lubrication Behavior of Organic Films at Low Temperatures," *ASLE Trans.* **8**(1), 48 (1965).

62. I. V. Kragelski, *Friction and Wear,* Butterworth, London, 1965, pp. 158–163.

63. C. N. Rowe, "Wear Corrosion and Erosion," *Interdisciplinary Approach to Liquid Lubricant Technology,* P. M. Ku (ed.), NASA SP-237, 1973, pp. 469–527.

64. D. Clayton, "An Introduction to Boundary and Extreme Pressure Lubrication," *Physics of Lubrication, Br. J. Appl. Phys.* **2,** Suppl. 1, 25 (1951).

CHAPTER 22

SEAL TECHNOLOGY

Bruce M. Steinetz
NASA Lewis Research Center
Cleveland, Ohio

22.1	**INTRODUCTION**	**629**	**22.3**	**DYNAMIC SEALS**	**638**
				22.3.1 Initial Seal Selection	638
22.2	**STATIC SEALS**	**629**		22.3.2 Mechanical Face Seals	642
	22.2.1 Gaskets	629		22.3.3 Emission Concerns	644
	22.2.2 O-Rings	634		22.3.4 Noncontacting Seals for High-Speed/Aerospace Applications	646
	22.2.3 Packings and Braided Rope Seals	637		22.3.5 Labyrinth Seals	650
				22.3.6 Honeycomb Seals	653
				22.3.7 Brush Seals	654

22.1 INTRODUCTION

Seals are required to fulfill critical needs in meeting the ever-increasing system-performance requirements of modern machinery. Approaching a seal design, one has a wide range of available seal choices. This chapter aids the practicing engineer in making an initial seal selection and provides current reference material to aid in the final design and application.

This chapter provides design insight and application for both static and dynamic seals. Static seals reviewed include gaskets, O-rings, and selected packings. Dynamic seals reviewed include mechanical face, labyrinth, honeycomb, and brush seals. For each of these seals, typical configurations, materials, and applications are covered. Where applicable, seal flow models are presented.

22.2 STATIC SEALS

22.2.1 Gaskets

Gaskets are used to effect a seal between two mating surfaces subjected to differential pressures. Gasket types and materials are limited only by one's imagination. Table 22.1 lists some common gasket materials and Table 22.2[1] lists common elastomer properties. The following gasket characteristics are considered important for good sealing performance.[2] Selecting the gasket material that has the best balance of the following properties will result in the best practical gasket design.

- Chemical compatibility
- Heat resistance
- Compressibility
- Microconformability (asperity sealing)
- Recovery
- Creep relaxation
- Erosion resistance
- Compressive strength (crush resistance)
- Tensile strength (blowout resistance)
- Shear strength (flange shearing movement)
- Removal or "Z" strength

Mechanical Engineers' Handbook, 2nd ed., Edited by Myer Kutz.
ISBN 0-471-13007-9   © 1998 John Wiley & Sons, Inc.

Table 22.1   Common Gasket Materials, Gasket Factors (*m*) and Minimum Design Seating Stress (*y*) (Table 2-5.1 ASME Code for Pressure Vessels, 1995)

Gasket Material	Gasket Factor m	Min. Design Seating Stress y, psi	Sketches
Self-energizing types (O-rings, metallic, elastomer, other gasket types considered as self-sealing)	0	0	...
Elastomers without fabric or high percent of asbestos fiber:			
Below 75A Shore Durometer	0.50	0	
75A or higher Shore Durometer	1.00	200	
Asbestos with suitable binder for operating conditions:			
⅛ in. thick	2.00	1600	
1⁄16 in. thick	2.75	3700	
1⁄32 in. thick	3.50	6500	
Elastomers with cotton fabric insertion	1.25	400	
Elastomers with asbestos fabric insertion (with or without wire reinforcement):			
3-ply	2.25	2200	
2-ply	2.50	2900	
1-ply	2.75	3700	
Vegetable fiber	1.75	1100	
Spiral-wound metal, asbestos filled:			
Carbon	2.50	10,000	
Stainless, Monel, and nickel-base alloys	3.00	10,000	
Corrugated metal, asbestos inserted, or corrugated metal, jacketed asbestos filled:			
Soft aluminum	2.50	2900	
Soft copper or brass	2.75	3700	
Iron or soft steel	3.00	4500	
Monel or 4%–6% chrome	3.25	5500	
Stainless steels and nickel-base alloys	3.50	6500	
Corrugated metal:			
Soft aluminum	2.75	3700	
Soft copper or brass	3.00	4500	
Iron or soft steel	3.25	5500	
Monel or 4%–6% chrome	3.50	6500	
Stainless steels and nickel-base alloys	3.75	7600	

Table 22.1 *(Continued)*

Gasket Material	Gasket Factor m	Min. Design Seating Stress y, psi	Sketches
Flat metal, jacketed asbestos filled:			
Soft aluminum	3.25	5500	
Soft copper or brass	3.50	6500	
Iron or soft steel	3.75	7600	
Monel	3.50	8000	
4%–6% chrome	3.75	9000	
Stainless steels and nickel-base alloys	3.75	9000	
Grooved metal:			
Soft aluminum	3.25	5500	
Soft copper or brass	3.50	6500	
Iron or soft steel	3.75	7600	
Monel or 4%–6% chrome	3.75	9000	
Stainless steels and nickel-base alloys	4.25	10,100	
Solid flat metal:			
Soft aluminum	4.00	8800	
Soft copper or brass	4.75	13,000	
Iron or soft steel	5.50	18,000	
Monel or 4%–6% chrome	6.00	21,800	
Stainless steels and nickel-base alloys	6.50	26,000	
Ring joint:			
Iron or soft steel	5.50	18,000	
Monel or 4%–6% chrome	6.00	21,800	
Stainless steels and nickel-base alloys	6.50	26,000	

- Antistick
- Heat conductivity
- Acoustic isolation
- Dimensional stability

Nonmetallic Gaskets. Most *nonmetallic gaskets* consist of a fibrous base held together with some form of an elastomeric binder. A gasket is formulated to provide the best load-bearing properties while being compatible with the fluid being sealed.

Nonmetallic gaskets are often reinforced to improve torque retention and blowout resistance for more severe service requirements. Some types of reinforcements include perforated cores, solid cores, perforated skins, and solid skins, each suited for specific applications. After a gasket material has been reinforced by either material additions or laminating, manufacturers can emboss the gasket raising a sealing lip, which increases localized pressures, thereby increasing sealability.

Metallic Gaskets. *Metallic gaskets* are generally used where either the joint temperature or load is extreme or in applications where the joint might be exposed to particularly caustic chemicals. A good seal capable of withstanding very high temperature is possible if the joint is designed to yield locally over a narrow location with application of bolt load. Some of the most common metallic gaskets range from soft varieties, such as copper, aluminum, brass, and nickel, to highly alloyed steels. Noble metals, such as platinum, silver, and gold, also have been used in difficult locations.

Metallic gaskets are available in both standard and custom designs. Since there is such a wide variety of designs and materials used, it is recommended that the reader directly contact metallic gasket suppliers for design and sealing information.

Required Bolt Load

ASME Method. The ASME Code for Pressure Vessels, Section VIII, Div. 1, App. 2, is the most commonly used design method for gasketed joints where important joint properties, including flange thickness, bolt size and pattern, are specified. Because of the absence of leakage considerations, it should be noted that the ASME is currently evaluating the Pressure Vessel Research Council's method for gasket design. It is likely that a nonmandatory appendix to the Code will appear first (see discussion in Ref. 3).

An integral part of the AMSE Code revolves around two gasket factors:

1. An m factor, often called the gasket-maintenance factor, is associated with the hydrostatic end force and the operation of the joint.
2. The y factor is a rough measure of the minimum seating stress associated with a particular gasket material. The y factor pertains only to the initial assembly of the joint.

The ASME Code makes use of two basic equations to calculate bolt load, with the larger calculated load being used for design:

$$W_{m1} = H + H_p = \frac{\pi}{4} G^2 P + 2\pi b G m P$$

$$W_{m2} = H_y = \pi b G y$$

where W_{m1} = minimum required bolt load from maximum operating or working conditions, lb
W_{m2} = minimum required initial bolt load for gasket seating (atmospheric-temperature conditions) without internal pressure, lb
H = total hydrostatic end force, lb $[(\pi/4)G^2 P]$
H_p = total joint–contact–surface compression load, lb
H_y = total joint–contact–surface seating load, lb
G = diameter at location of gasket load reaction; generally defined as follows: When $b_0 <$ ¼ in., G = mean diameter of gasket contact face, in.; When $b_0 >$ ¼ in., G = outside diameter of gasket contact face less $2b$, in.
P = maximum internal design pressure, psi
b = effective gasket or joint–contact–surface *seating* width, in.
b = b_0 when $b_0 \leq$ ¼ in.
b = $0.5\sqrt{b_0}$ when $b_0 >$ ¼ in.
$2b$ = effective gasket or joint–contact–surface *pressure* width, in.
b_0 = basic gasket seating width per ASME Table 2-5.2. The table defines b_0 in terms of flange finish and type of gasket, usually from one-half to one-fourth gasket contact width
m = gasket factor per ASME Table 2-5.1 (repeated here as Table 22.1).
y = gasket or joint–contact–surface unit seating load, per ASME Table 2-5.1 (repeated here as Table 22.1), psi

The factor m provides a margin of safety to be applied when the hydrostatic end force becomes a determining factor. Unfortunately, this value is difficult to obtain experimentally since it is not a constant. The equation for W_{m2} assumes that a certain unit stress is required on a gasket to make it conform to the sealing surfaces and be effective. The second empirical constant y represents the gasket yield-stress value and is very difficult to obtain experimentally.

Practical Considerations

Flange Surfaces. Preparing the flange surfaces is paramount for effecting a good gasket seal. Surface finish affects the degree of sealability. The rougher the surface, the more bolt load required to provide an adequate seal. Extremely smooth finishes can cause problems for high operating pressures, as lower frictional resistance leads to a higher tendency for blowout. Surface finish lay is important in certain applications to mitigate leakage. Orienting finish marks transverse to the normal leakage path will generally improve sealability.

Flange Thickness. Flange thickness must also be sized correctly to transmit bolt clamping load to the area between the bolts. Maintaining seal loads at the midpoint between the bolts must be kept constantly in mind. Adequate thickness is also required to minimize the bowing of the flange. If the flange is too thin, the bowing will become excessive and no bolt load will be carried to the midpoint, preventing sealing.

Bolt Pattern. Bolt pattern and frequency are critical in effecting a good seal. The best bolt clamping pattern is invariably a combination of the maximum practical number of bolts, optimum spacing, and positioning.

One can envision the bolt loading pattern as a series of straight lines drawn from bolt to adjacent bolt until the circuit is completed. If the sealing areas lie on either side of this pattern, it will likely be a potential leakage location. Figure 22.1 shows an example of the various conditions.[2] If bolts

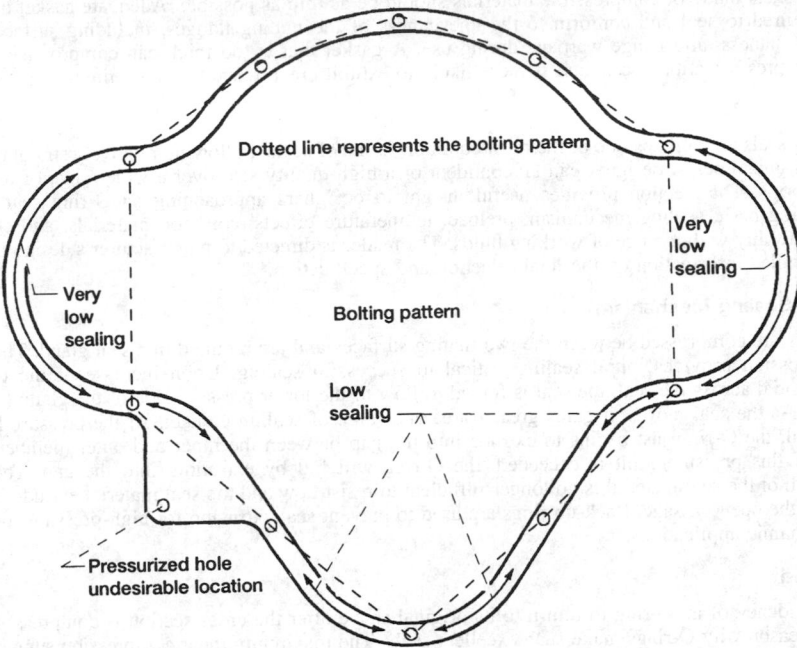

Fig. 22.1 Bolting pattern indicating poor sealing areas. (From Ref. 2.)

cannot be easily repositioned on a problematic flange, Fig. 22.2 illustrates techniques to improve gasket effectiveness through reducing gasket face width where bolt load is minimum. Note that gasket width is retained in the vicinity of the bolt to support local bolt loads and minimize gasket tearing.

Gasket Thickness and Compressibility. Gasket thickness and compressibility must be matched to the rigidity, roughness, and unevenness of the mating flanges. An effective gasket seal is achieved only if the stress level imposed on the gasket at installation is adequate for the specific gasket and joint requirements.

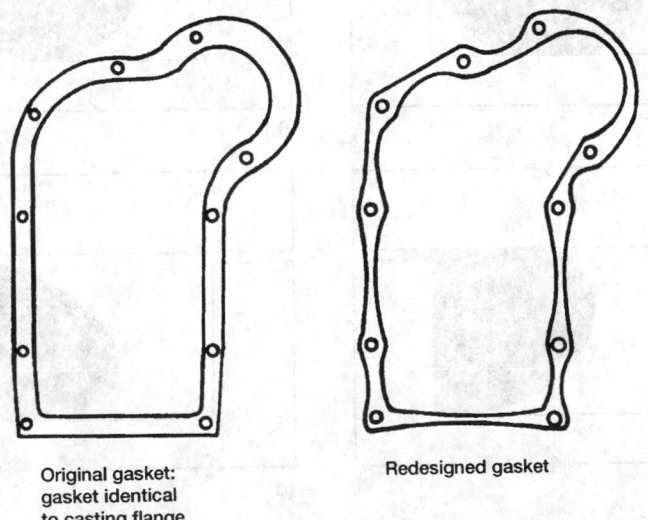

Original gasket:
gasket identical
to casting flange

Redesigned gasket

Fig. 22.2 Original vs. redesigned gasket for improved sealing. (From Ref. 2.)

Gaskets made of compressible materials should be as thin as possible. Adequate gasket thickness is required to seal and conform to the unevenness of the mating flanges, including surface finish, flange flatness, and flange warpage during use. A gasket that is too thick can compromise the seal during pressurization cycles and is more likely to exhibit creep relaxation over time.

22.2.2 O-Rings

O-ring seals are perhaps one of the most common forms of seals. Following relatively straightforward design guidelines, a designer can be confident of a high-quality seal over a wide range of operating conditions. This section provides useful insight to designers approaching an O-ring seal design, including basic sealing mechanism, preload, temperature effects, common materials, and chemical compatibility with a range of working fluids. The reader is directed to manufacturer's design manuals for detailed information on the final selection and specification.[4]

Basic Sealing Mechanism

O-rings are compressed between the two mating surfaces and are retained in a seal gland. The initial compression provides initial sealing critical to successful sealing. Upon increase of the pressure differential across the seal, the seal is forced to flow to the lower pressure side of the gland (see Fig. 22.3). As the seal moves, it gains greater area and force of sealing contact. At the pressure limit of the seal, the O-ring just begins to extrude into the gap between the inner and outer member of the gap. If this pressure limit is exceeded, the O-ring will fail by extruding into the gap. The shear strength of the seal material is no longer sufficient to resist flow and the seal material extrudes (flows) out of the open passage. Back-up rings are used to prevent seal extrusion for high-pressure static and for dynamic applications.

Preload

The tendency of an O-ring to return to its original shape after the cross section is compressed is the basic reason why O-rings make such excellent seals. The maximum linear compression suggested by manufacturers is 30% for static applications and 16% for dynamic seals (up to 25% for small cross-sectional diameters). Compression less than these values is acceptable, within reason, if assembly

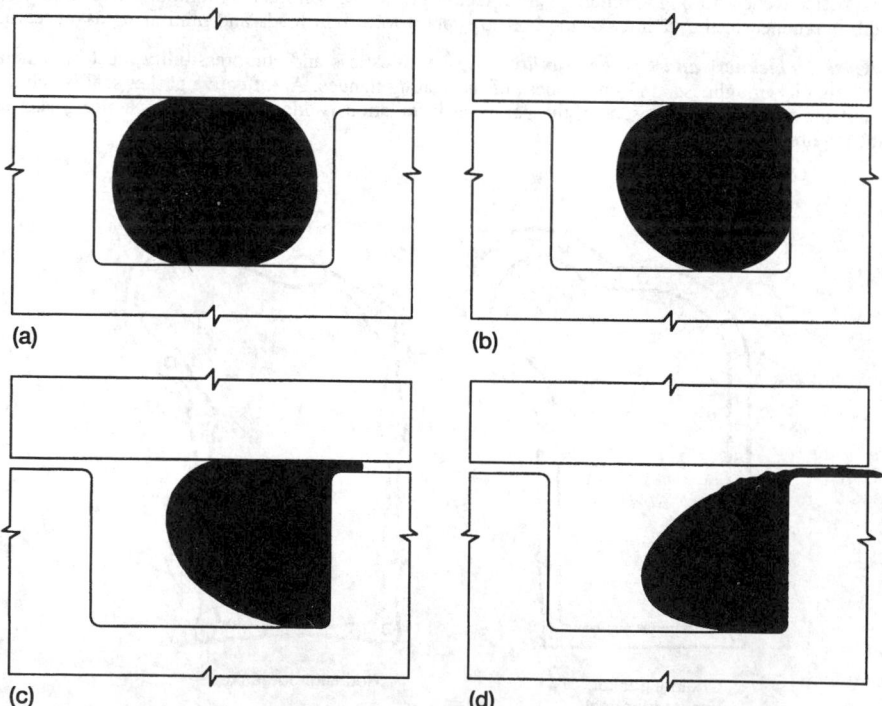

Fig. 22.3 Basic O-ring sealing mechanism. (a) O-ring installed; (b) O-ring under pressure; (c) O-ring extruding; (d) O-ring failure. (From Ref. 4.)

problems are an issue. Manufacturers recommend[4] a minimum amount of initial linear compression to overcome compression set that O-rings exhibit.

O-ring compression force depends principally on the hardness of the O-ring, its cross-sectional dimension, and the amount of compression. Figure 22.4 illustrates the range of compressive force per linear inch of seal for typical linear percent compressions (0.139 in. cross-section diameter) and compound hardness (Shore A hardness scale). Softer compounds provide better sealing ability, as the rubber flows more easily into the grooves. Harder compounds are specified for high pressures, to limit chance of extruding into the groove, and to improve wear life for dynamic service. For most applications, compounds having a Type A durometer hardness from 70–80 are the most suitable compromise.[4]

Thermal Effects

O-ring seals respond to temperature changes. Therefore, it is critical to ensure the correct material and hardness is selected for the application. High temperatures soften compounds. This softening can negatively affect the seal's extrusion resistance at temperature. Over long periods of time at high temperature, chemical changes occur. These generally cause an increase in hardness, along with volume and compression-set changes.

O-ring compounds harden and contract at cold temperatures. These effects can both lead to a loss of seal if initial compression is not set properly. Because the compound is harder, it does not flow into the mating surface irregularities as well. Just as important, the more common O-ring materials have a coefficient of thermal expansion 10 times greater than that of steel (i.e., nitrile CTE is 6.2×10^{-5}°F).

Groove dimensions must be sized correctly to account for this dimensional change. Manufacturers design charts[4] are devised such that proper O-ring sealing is ensured for the temperature ranges for standard elastomeric materials. However, the designer may want to modify gland dimensions for a given application that experiences only high or low temperatures in order to maintain a particular squeeze on the O-ring. Martini[5] gives several practical examples showing how to tailor groove dimensions to maintain a given squeeze for the operating temperature.

Material Selection / Chemical Compatibility

Seal compounds must work properly over the required temperature range, have the proper hardness to resist extrusion while effectively sealing, and must also resist chemical attack and resultant swelling caused by the operating fluids. Table 22.2 summarizes the most important elastomers, their working temperature range, and their resistance to a range of common working fluids.

Rotary Applications

O-rings are also used to seal rotary shafts where surface speeds and pressures are relatively low. One factor that must be carefully considered when applying O-ring seals to rotary applications is the Gow-

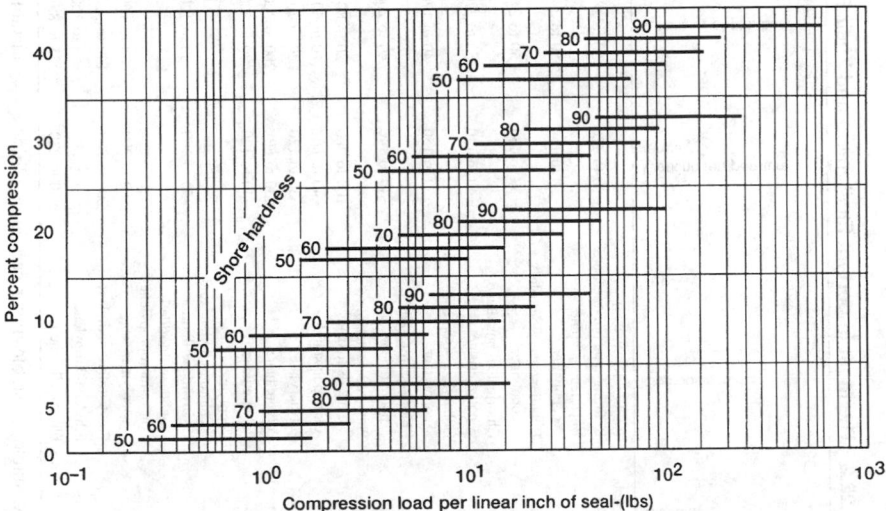

Fig. 22.4 Effect of percent compression and material Shore hardness on seal compression load, 0.139-in. cross section. (From Ref. 4.)

Table 22.2 The Most Important Elastomers and Their Properties[1]

Elastomer	Composition	Working temperature range, °C	Tensile strength, bar	Elongation, %	Hardness, °Shore	Water	Steam	Hydraulic fluids, non-flammable (ester-based)	Mineral fats and oils	Vegetable and animal fats and oils	Ozone	Hydrocarbons Aliphatic	Hydrocarbons Aromatic	Hydrocarbons Halogenated	Alcohols	Ketones	Esters	Dilute acids	Concentrated acids	Dilute alkalis	Concentrated alkalis	Saline solutions
Natural rubber	Rubber, K. W. Coil Refining-type polymerisate	−30 to 120	50 to 280	1000	30 to 98	x	x	—	—	—	—	—	—	—	x	x	x	o	—	x	o	x
S.B.R.	Butadiene–styrene copolymer	−30 to 130	50 to 240	700	40 to 95	x	x	—	—	—	o	—	—	—	x	x	—	x	o	x	x	x
Nitrile N	Butadiene–acrylonitrile copolymer	−30 to 130	50 to 240	700	40 to 95	x	o	—	x	x	o	x	o	—	x	—	—	x	—	o	o	x
Neoprene	Chlorinated–butadiene polymerisate	−40 to 140	50 to 270	800	40 to 95	x	x	—	o	o	x	o	—	—	x	o	o	x	o	x	x	x
Butyl	Isobutylene–isoprene copolymer	−50 to 150	40 to 170	900	40 to 90	x	x	o	o	o	x	—	o	—	x	o	—	x	o	x	x	x
Hypalon	Chloro–sulfonated polyethylene	−40 to 140	40 to 200	600	40 to 95	x	o	—	o	o	x	—	o	—	x	o	—	x	o	x	x	x
Silicone rubber	Polycondensates of dialkylsiloxanes	−100 to 200	20 to 80	500	40 to 80	o	—	o	x	x	x	o	o	o	x	o	x	x	o	x	o	o
Thiokol	Alkylopolysulfide	−40 to 80	10 to 60	200	65 to 80	x	—	x	x	x	x	x	o	o	x	o	x	x	o	x	x	x
Polyacrylic	Polyacrylate	−30 to 120	20 to 70	700	70 to 85	o	—	x	o	x	x	x	o	o	o	—	x	x	o	x	o	o
Vulcollan	Polyurethane	−30 to 80	200 to 320	600	70 to 95	o	—	x	x	x	x	x	o	o	o	x	x	—	o	x	—	x
Adiprene	Polyurethane	−40 to 120	80 to 300	700	70 to 95	x	o	x	x	x	x	x	o	o	o	x	x	—	o	o	—	o
Kel-F	Copolymer of chlorotriethylene and vinylidene fluoride	−50 to 180	30 to 120	700	60 to 90	—	—	—	—	—	—	—	—	—	o	—	—	—	—	—	—	—
Viton	Vinylidene fluoride–hexafluoropropylene copolymer	−60 to 200	80 to 160	300	60 to 95	x	x	x	x	x	x	x	x	x	x	—	—	x	x	x	x	x
PTFE	Polytetrafluoroethylene	−200 to 280	140 to 310	200	55D	x	x	x	x	x	x	x	x	o	x	x	—	x	x	—	—	x
E.P.R.	Ethylene–propylene	−55 to 200	50 to 160	400	70 to 95	x	x	x	x	x	x	x	x	o	x	x	o	x	x	x	x	x
F.S.R.	Fluoro–silicone rubber	−60 to 230	55 to 85	400	40 to 80	o	o	—	—	x	x	x	o	o	x	—	o	o	—	x	o	x

Note: x, stable; o, stable under certain conditions; —, unstable.

Joule effect.[5] When a rubber O-ring is stretched slightly around a rotating shaft, (e.g. put in tension) friction between the ring and shaft generates heat causing the ring to contract, exhibiting a negative expansion coefficient. As the ring contracts friction forces increase generating additional heat and further contraction. This positive-feedback cycle causes rapid seal failures. Similar failures in reciprocating applications and static applications are unusual because surface speeds are too low to initiate the cycle. Further, in reciprocating applications the seal is moved into contact with cooler adjacent material. To prevent the failure cycle, O-rings are not stretched over shafts but are oversized slightly (circumferentially) and compressed into the sealing groove. The pre-compression of the cross-section results in O-ring stresses that oppose the contraction stress preventing the failure cycle described. Martini[5] provides guidelines for specifying the O-ring seal. Following appropriate techniques O-ring seals have run for significant periods of time at speeds up to 750 fpm and pressures up to 200 psi.

22.2.3 Packings and Braided Rope Seals

Rope packings used to seal stuffing boxes and valves and prevent excessive leakage can be traced back to the early days of the Industrial Revolution. An excellent summary of types of rope seal packings is given in Ref. 6. Novel adaptations of these seal packings have been required as temperatures have continued to rise to meet modern system requirements. New ceramic materials are being investigated to replace asbestos in a variety of gasket and rope-packing constructions.

Materials

Packing materials are selected for intended-temperature and chemical environment. Graphite-based packing/gaskets are rated for up to 1000°F for oxidizing environments and up to 5400°F for reducing environments.[7] Used within its recommended temperature, graphite will provide a good seal with acceptable ability to track joint movement during temperature/pressure excursions. Graphite can be laminated with itself to increase thickness or with metal/plastic to improve handling and mechanical strength. Table 22.2 provides working temperatures for conventional (e.g., nitrile, PTFE, neoprene, amongst others) gasket/packings. Table 22.3 provides typical maximum working temperatures for high temperature gasket/packing materials.

Packings and Braided Rope Seals for High-Temperature Service

High-temperature packings and rope seals are required for a variety of applications, including sealing: furnace joints, locations within continuous casting units (gate seals, mold seals, runners, spouts, etc.), amongst others. High-temperature packings are used for numerous aerospace applications, including turbine casing and turbine engine locations, Space Shuttle thermal protection systems, and nozzle joint seals.

Aircraft engine turbine inlet temperatures and industrial system temperatures continue to climb to meet aggressive cycle thermal efficiency goals. Advanced material systems, including monolithic/composite ceramics, intermetallic alloys (i.e., nickel aluminide), and carbon–carbon composites, are

Table 22.3 Gasket/Rope Seal Materials

Fiber Material	Maximum Working Temperature °F
Graphite	
Oxidizing environment	1000
Reducing	5400
Fiberglass (glass dependent)	1000
Superalloy metals (depending on alloy)	1300–1600
Oxide Ceramics (Ref. Tompkins 1995)*	
62% Al_2O_3 24% SiO_2 14% B_2O_3 (Nextel 312)	1800†
70% Al_2O_3 28% SiO_2 2% B_2O_3 (Nextel 440)	2000†
73% Al_2O_3 27% SiO_2 (Nextel 550)	2100†

*Tompkins, T. L. "Ceramic Oxide Fibers: Building Blocks for New Applications," Ceramic Industry Publ, Business News Publishing, April, 1995.
†Temperature at which fiber retains 50% (nominal) room temperature strength.

being explored to meet aggressive temperature, durability, and weight requirements. Incorporating these materials in the high-temperature locations in the system, designers must overcome materials issues, such as differences in thermal expansion rates and lack of material ductility.

Designers are finding that one way to avoid cracking and buckling of the high-temperature brittle components rigidly mounted in their support structures is to allow relative motion between the primary and supporting components.[8] Often this joint occurs in a location where differential pressures exist, requiring high-temperature seals. These seals or packings must exhibit the following important properties: operate hot ($\geq 1300°F$); exhibit low leakage; resist mechanical scrubbing caused by differential thermal growth and acoustic loads; seal complex geometries; retain resilience after cycling; and support structural loads.

In an industrial seal application, a high-temperature all-ceramic seal is being used to seal the interface between a low-expansion rate primary structure and the surrounding support structure. The seal consists of a dense uniaxial fiber core overbraided with two two-dimensional braided sheath layers.[8] Both core and sheath are composed of 8 μm alumina–silica fibers (Nextel 550) capable of withstanding 2000+°F temperatures. In this application over a heat/cool cycle, the support structure moves 0.3 in. relative to the primary structure, precluding normal fixed-attachment techniques. Leakage flows for the all-ceramic seal are shown in Fig. 22.5 for three temperatures after simulated scrubbing[8] (10 cycles × 0.3-in. at 1300°F).

In a turbine vane application, the conventional braze joint is replaced with a floating seal arrangement incorporating a small-diameter ($\frac{1}{16}$-in.) rope seal (Fig. 22.6). The seal is designed to serve as a seal and a compliant mount, allowing relative thermal growth between the high-temperature turbine vane and the lower-temperature support structure, preventing thermal strains and stresses. A hybrid seal consisting of a dense uniaxial ceramic core (8 μm alumina–silica Nextel 550 fibers) overbraided with a superalloy wire (0.0016-in. diameter Haynes 188 alloy) abrasion-resistant sheath has proven successful for this application.[9] Leakage flows for the hybrid seal are shown in Fig. 22.7 for two temperatures, and pressures under two preload conditions after simulated scrubbing (10 cycles × 0.3-in. at 1300°F).

Recent studies[8] have shown the benefits of high sheath braid angle and double-stage seals for reducing leakage. Increasing hybrid seal sheath braid angle and increasing core coverage led to increased compressive force (for the same linear seal compression) and one-third the leakage of the conventional hybrid design. Adding a second seal stage reduced seal leakage 30% relative to a single stage.

22.3 DYNAMIC SEALS

22.3.1 Initial Seal Selection

An engineer approaching a dynamic seal design has a wide range of seals to choose from. A partial list of seals available ranges from the mechanical face seal through the labyrinth and brush seal, as

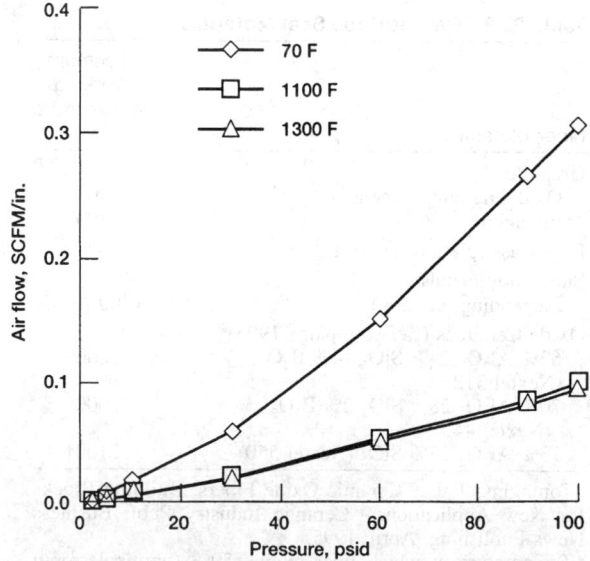

Fig. 22.5 Flow vs. pressure data for 3 temperatures, $\frac{1}{16}$ in. diameter all-ceramic seal, 0.022 in. seal compression, after scrubbing. (From Ref. 8.)

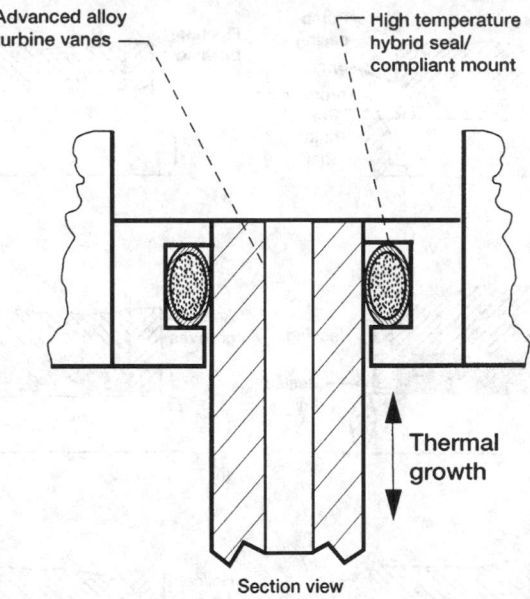

Fig. 22.6 Schematic of turbine vane seal. (From Ref. 9.)

indicated in Fig. 22.8. To aid in the initial seal selection, a "decision tree" has been proposed by Fern and Nau.[10] The decision tree (see Fig. 22.9) has been updated for the current work to account for the emergence of brush seals. In this chart, a majority of answers either "yes" or "no" to the questions at each stage leads the designer to an appropriate seal starting point. If answers are equally divided, both alternatives should be explored using other design criteria, such as performance, size, and cost.

The scope of this chapter does not permit treatment of every entry in the decision tree. However, several examples are given below to aid in understanding its use.

Radial lip seals are used to prevent fluids, normally lubricated, from leaking around shafts and their housings. They are also used to prevent dust, dirt, and foreign contaminants from entering the

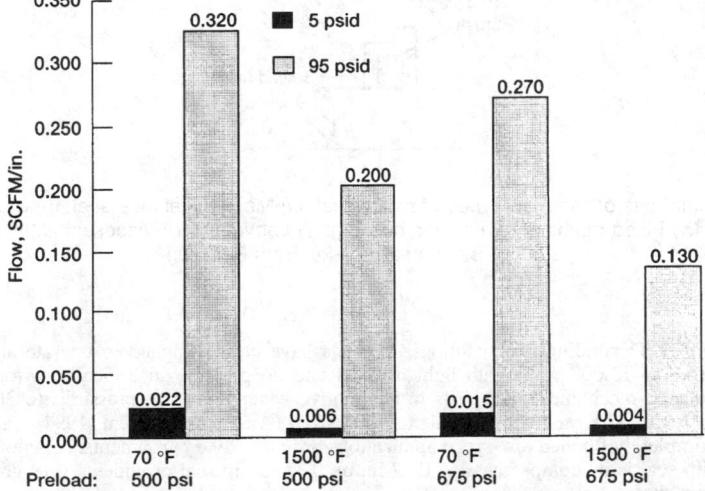

Fig. 22.7 The effect of temperature, pressure, and representative compression on seal flow after cycling for 0.060-in. hybrid vane seal. (From Ref. 9.)

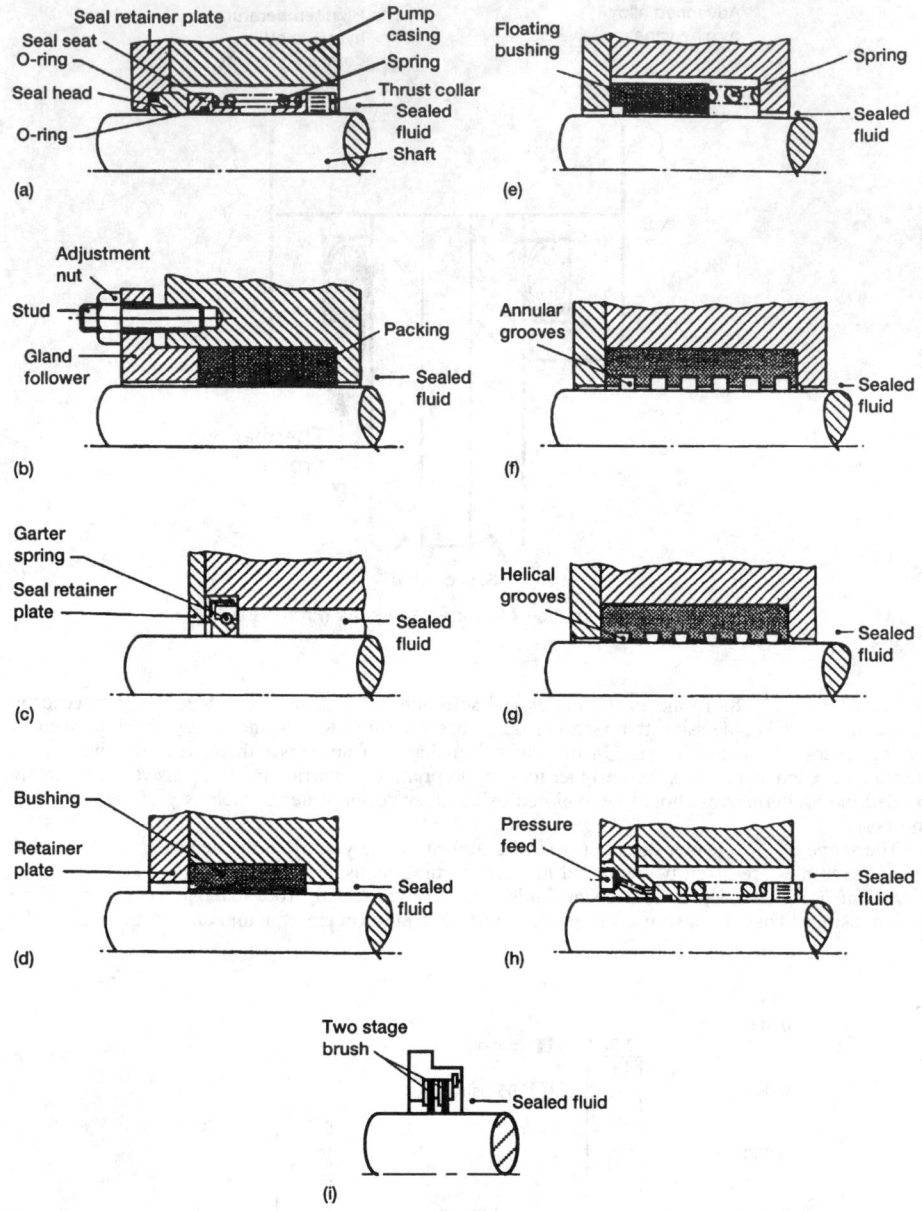

Fig. 22.8 Examples of the main types of rotary seal. (*a*) Mechanical face seal; (*b*) Stuffing box; (*c*) Lip seal; (*d*) Fixed bushing; (*e*) Floating bushing; (*f*) Labyrinth; (*g*) Viscoseal; (*h*) Hydrostatic seal; (*i*) Brush seal. ((*a*)–(*h*) From Ref. 10.)

lubricant chamber. Depending on conditions, lip seals have been designed to operate at very high shaft speeds (6,000–12,000 rpm) with light oil mist and no pressure in a clean environment. Lip seals have replaced mechanical face seals in automotive water pumps at pressures to 30 psi, temperatures −45°F to 350°F, and shaft speeds to 8000 sfpm (American Variseal, 1994). Lip seals are also used in completely flooded low-speed applications or in muddy environments. A major advantage of the radial lip seal is its compactness. A 0.32-in. by 0.32-in. lip seal provides a very good seal for a 2-in. diameter shaft.

Mechanical face seals are capable of handling much higher pressures and a wider range of fluids. Mechanical face seals are recommended over brush seals where very high pressures must be sealed

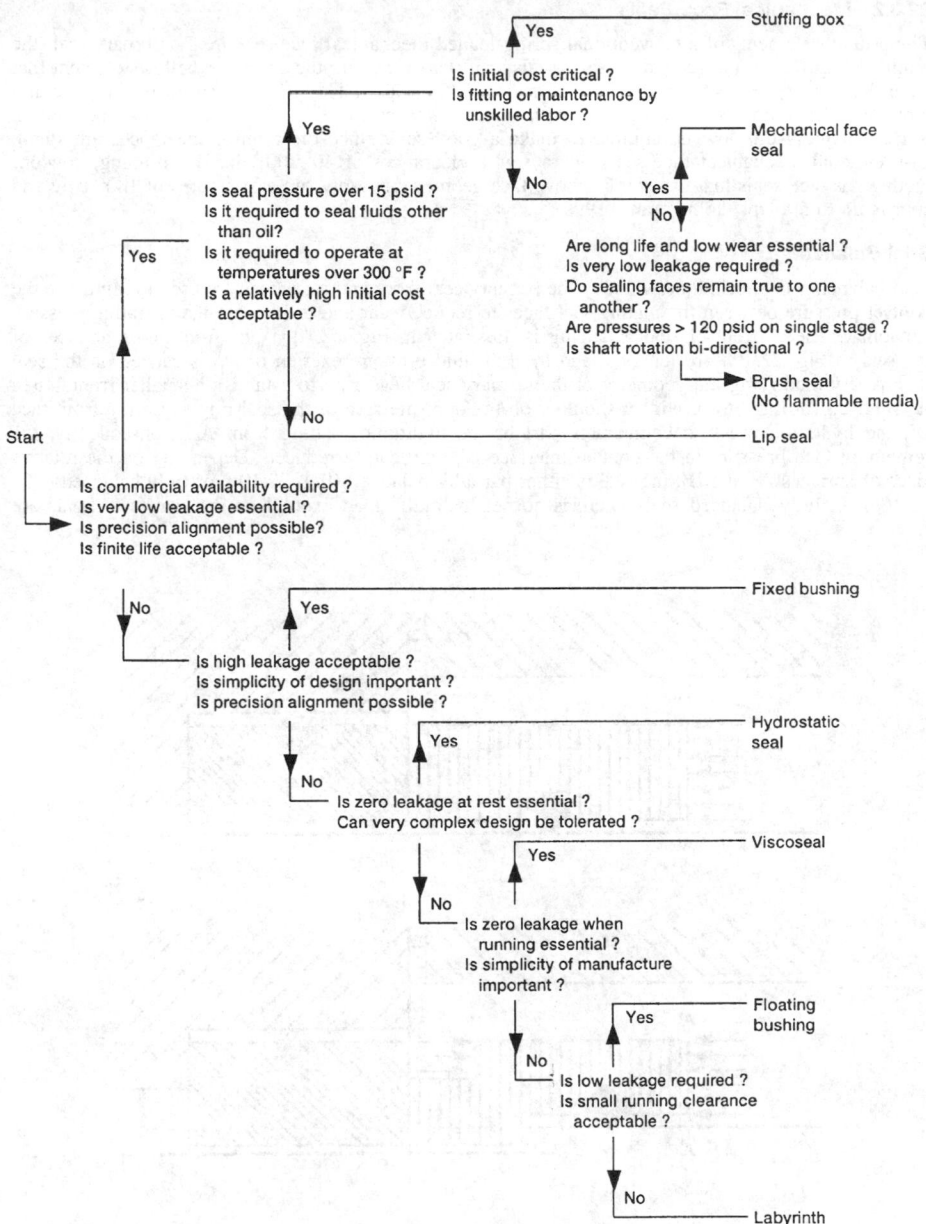

Fig. 22.9 Seal selection chart (a majority answer of "yes" or "no" to the question at each stage leads the reader to the appropriate decision; if answers are equally divided both alternatives should be explored). (Adapted from Ref. 10.)

in a single stage. Mechanical face seals have a lower leakage than brush seals because their effective clearances are several times smaller. However, the mechanical face seal requires much better control of dimensions and tolerates less shaft misalignment and runout, thereby increasing costs.

Turbine Engine Seals. Readers interested particularly in turbine engine seals are referred to Steinetz and Hendricks,[11] (1997) which reviews in greater depth the tradeoffs in selecting seals for turbine engine applications. Technical factors increasing seal design complexity for aircraft engines include high temperatures ($\geq 1000°F$), high surface speeds (up to 1500 fps), rapid thermal/structural transients, maneuver and landing loads, and the requirement to be lightweight.

22.3.2 Mechanical Face Seals

The primary elements of a conventional spring-loaded mechanical face seal are the primary seal (the main sealing faces), the secondary seal (seals shaft leakage), and the spring or bellows element that keep the primary seal surfaces in contact, shown in Fig. 22.8. The primary seal faces are generally lapped to demanding surface flatness, with surface flatness of 40 μin (1 micron) not uncommon. Surface flatness this low is required to make a good seal, since the running clearances are small. Conventional mechanical face seals operate with clearances of 40–200 μin. Dry-running, noncontacting gas face seals that use spiral groove face geometry reliably run at pressures of 1800 psig and speeds up to 590 fps (John Crane, 1993).

Seal Balance

Seal balancing is a technique whereby the primary seal front and rear areas are used to minimize the contact pressure between the mating seal faces to reduce wear and to increase the operating pressure capability. The concept of seal balancing is illustrated in Fig. 22.10.[12] The front and rear faces of the seal in Fig. 22.10a are identical and the full fluid pressure exerted on A′ is carried on the seal face A. By modifying the geometry of the primary seal head ring to establish a smaller frontal area A′ (Fig. 22.10b) and to provide a shoulder on the opposite side of the seal ring to form a front face B′, the hydraulic pressure counteracts part of the hydraulic loading from A′. Consequently, the remaining face pressure in the contact interface is significantly reduced. Depending on the relative sizes of surfaces A′ and B′, the seal is either partially balanced (Fig. 22.10b) or fully balanced (Fig. 22.10c). In fully balanced seals, there is no net hydraulic load exerted on the seal face. Seals are

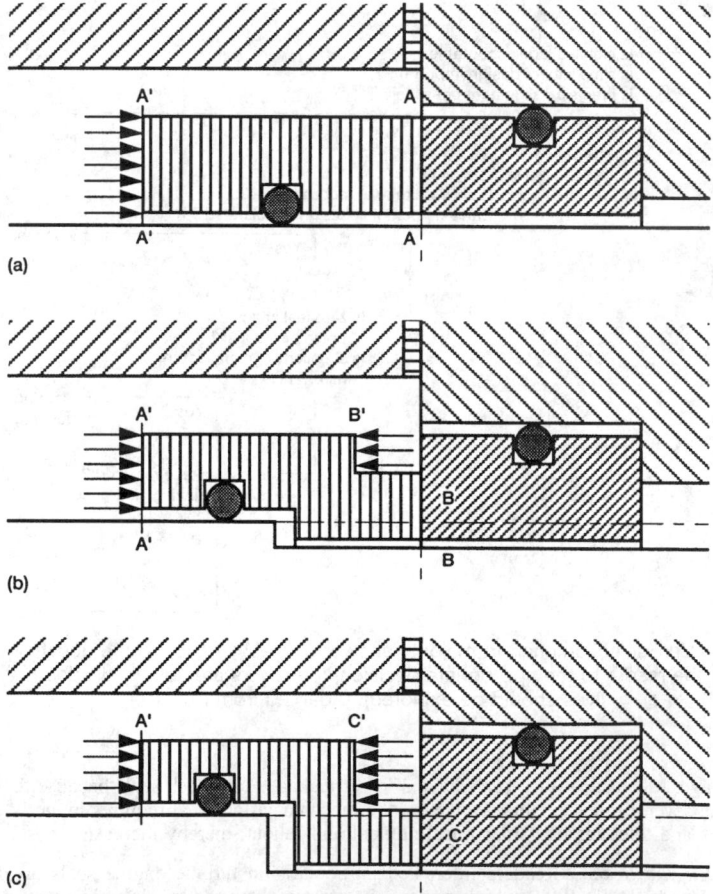

Fig. 22.10 Illustration of face seal balance conditions. (a) Unbalanced; (b) Partially balanced; (c) Fully balanced. (From Ref. 12.)

generally run with a partial balance, however, to minimize face loads and wear while keeping the seal closed during possible transient overpressure conditions. Partially balanced seals can run at pressures greater than six times unbalanced seals can for the same speed and temperature conditions.

Mechanical Face Seal Leakage

Liquid Flow. Minimizing leakage between seal faces is possible only through maintaining small clearances. Volumetric flow (Q) can be determined for the following two conditions (Lebeck, 1991).[13]

For Coned Faces:

$$Q = \frac{\pi \phi r_m}{3\mu} \left(\frac{P_o - P_i}{1/h_o^2 - 1/h_i^2} \right)$$

For Parallel Faces:

$$Q = \frac{-\pi r_m h^3}{6\mu} \frac{(P_o - P_i)}{(r_o - r_i)} \qquad h_o = h_i \quad \text{and} \quad ((r_o - r_i)/r_m < 0.1)$$

where ϕ (radians) is the cone angle (positive if faces are convergent travelling inward radially); r_o, r_i (in.) outer and inner radii; r_m (in.) mean radius (in.); h_o, h_i (in.) outer and inner film thicknesses; P_o, P_i (psi) outer and inner pressures; μ (lbf · s/in.²) viscosity. The need for small clearances is demonstrated by noting that doubling the film clearance, h, increases the leakage flow eight-fold.

Gas Flow. Closed-form equations for gas flow through parallel faces can be written only for conditions of laminar flow (Reynolds No. < 2300). For laminar flow with a parabolic pressure distribution across the seal faces, the mass flow is given as (Lebeck, 1991):[13]

$$\dot{m} = \frac{\pi r_m h^3}{12\mu RT} \frac{(P_i^2 - P_o^2)}{(r_o - r_i)} \qquad (r_o - r_i)/r_m < 0.1$$

where R is the gas constant (53.3 lbf · ft/lb$_m$ · °R for air), and T (°R) is the gas temperature (isothermal throughout).

In cases where flow is both laminar and turbulent, iterative schemes must be employed. See Refs. 13 and 14 for numerical algorithms to use in solving for the seal leakage rates. Reference 15 treats the most general case of two-phase flow through the seal faces.

Seal Face Flatness

In addition to lapping faces to the 40 μin. flatness, there are several other points to consider. The lapped rings should be mounted on surfaces that are themselves flat. The ring must be stiff enough to resist distortions caused either by thermal or fluid pressure stresses.

The primary mode of distortion of a mechanical seal face under combined fluid and thermal stresses is solid body rotation about the seal's neutral axis.[10] If the sum of the moments M (in.-lb/in.) per unit of circumference around the neutral axis can be calculated, then the angular deflection θ (radians) of the sealing face, can be obtained from

$$\theta = Mr_m^2/EI$$

where E (psi) = Young's modulus
$\quad I$ (in.⁴) = the second moment of areas about the neutral axis
$\quad r_m$ (in.) = the mean radius of the seal ring

Face Seal Materials

Selecting the correct materials for a given seal application is critical to ensuring desired performance and durability. Seal components for which material selection is important from a tribology standpoint are the stationary nosepiece (or primary seal ring) and the mating ring (or seal seat). Properties considered ideal for the primary seal ring are shown below.[16]

1. Mechanical:
    (a) High modulus of elasticity
    (b) High tensile strength
    (c) Low coefficient of friction
    (d) Excellent wear characteristics and hardness
    (e) Self-lubrication

2. Thermal:
    (a) Low coefficient of expansion
    (b) High thermal conductivity
    (c) Thermal shock resistance
    (d) Thermal stability
3. Chemical:
    (a) Corrosion resistance
    (b) Good wetability
4. Miscellaneous:
    (a) Dimensional stability
    (b) Good machinability and ease of manufacture
    (c) Low cost and ready availability

Carbon–graphite is often the first choice for one of the running seal surfaces because of its superior dry-running (i.e., start-up) behavior. It can run against itself, metals, or ceramics without galling or seizing. Carbon–graphite is generally impregnated with resin or with a metal to increase thermal conductivity and bearing characteristics. In cases where the seal will see considerable abrasives, carbon may wear excessively and then it is desirable to select very hard seal-face materials. A preferred combination for very long wear (subject to other constraints) is tungsten carbide running on tungsten carbide. For a comprehensive coverage of face seal material selection, including chemical compatibility, see Ref. 17.

Secondary seals are either O-rings or bellows. Temperature ranges and chemical compatibility for common O-ring secondary seals such as nitrile, fluorocarbon (Viton), and PTFE (Teflon) are provided in Table 22.2.

22.3.3 Emission Concerns

Mechanical face seals have played and will continue to play a major role for many years in minimizing emissions to the atmosphere. New federal, state, and local environmental regulations have intensified the focus on mechanical face seal performance in terms of emissions. Within a short time, regulators have gone from little or no concern about fugitive hazardous emissions to a position of severely restricting all hazardous emissions. For instance, under the authority of Title III of the 1990 Clean Air Act Amendment (CAAA), the U.S. Environmental Protection Agency (EPA) adopted the National Emission Standards for Hazardous Air Pollutants (NESHAP) for the control of emissions of volatile hazardous air pollutants (Ref. STLE, 1994).[18] Leak definition per the regulation (EPA HON Subpart H (5)) are defined as follows:

Phase I: 10,000 parts per million volumetric (ppmv), beginning on compliance date

Phase II: 5000 ppmv, 1 year after compliance date

Phase III: 1000–5000 ppmv, depending on application, 2½ years after compliance date

The Clean Air Act regulations require U.S. plants to reduce emissions of 189 hazardous air pollutants by 80% in the next several years.[19] The American Petroleum Industry (API) has responded with a standard of its own, known as API 682, that seeks to reduce maintenance costs and control volatile organic compounds (VOC) emissions on centrifugal and rotary pumps in heavy service. API 682, a pump shaft sealing standard, is designed to help refinery pump operators and similar users comply with environmental emissions regulations. These regulations will continue to have a major impact on users of valves, pumps, compressors and other processing devices. Seal users are cautioned to check with their state and local air quality control authorities for specific information.

Sealing Approaches for Emissions Controls

The Society of Tribologists and Lubrication Engineers published a guideline of mechanical seals for meeting the fugitive emissions requirements.[18] Seal technology available meets approximately 95% of current and anticipated federal, state, and local emission regulations. Applications not falling within the guidelines include food, pharmaceutical, and monomer-type products where dual seals cannot be used because of product purity requirements and chemical reaction of dual seal buffer fluids with the sealed product.

Three sealing approaches for meeting the new regulatory requirements are discussed below: single seals, tandem seals, and double seals.[18]

Single Seals.   The most economical approach available is the single seal mounted inside a stuffing box (Fig. 22.11). Generally, this type of seal uses the pumped product for lubrication. Due to some finite clearance between the faces, there is a small amount of leakage to atmosphere. Using current technology in the design of a single seal, emissions can be controlled to 500 ppm based on

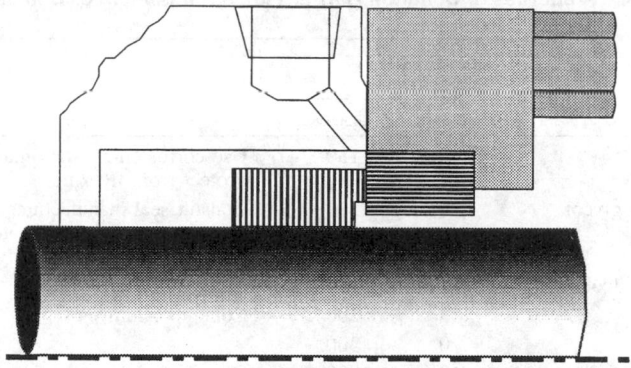

Fig. 22.11   Single seal. (From Ref. 18.)

both laboratory and field test data. Emission to atmosphere can be eliminated by venting the atmospheric side to a vapor recovery or disposal system. Using this approach, emissions readings approaching zero can be achieved. Since single seals have a minimum of contacting parts and normally require minimum support systems, they are considered highly reliable.

Tandem Seals.   Tandem seals consist of two seal assemblies between which a barrier fluid operates at a pressure less than the pumped process pressure. The inboard primary seal seals the full pumped product pressure, and the outboard seal typically seals a nonpressurized barrier fluid (Fig. 22.12). Tandem seal system designs are available that provide zero emission of the pumped product to the environment, provided the vapor pressure of the product is higher than that of the barrier fluid and the product is immiscible in the barrier fluid. The barrier fluid isolates the pumped product from the atmosphere and is maintained by a support system. This supply system generally includes a supply tank assembly and optional cooling system and means for drawing off the volatile component (generally at the top of the supply tank). Examples of common barrier fluids are found in Table 22.4.

Tandem seal systems also provide a high level of sealing and reliability, and are simple systems to maintain, due to the typical use of nonpressurized barrier fluid. Pumped product contamination by the barrier fluid is avoided since the barrier fluid is at a lower pressure than the pumped product.

Double Seals.   Double seals differ from tandem seals in that the barrier fluid between the primary and outboard seal is pressurized (Fig. 22.13). Double seals can be either externally or internally pressurized. An externally pressurized system requires a lubrication unit to pressurize the barrier fluid above the pumped product pressure and to provide cooling. An internally pressurized double seal refers to a system that internally pressurizes the fluid film at the inboard faces as the shaft rotates. In this case, the barrier fluid in the seal chamber is normally at atmospheric pressure. This results in less heat generation from the system.

Application Guide.   The areas of application based on emissions to atmosphere for the three types of seals discussed are illustrated in Fig. 22.14. The scope of this chart is for seals less than 6 in. in diameter, for pressures 600 psig and less, and for surface speeds up to 5600 fpm. Waterbury[19]

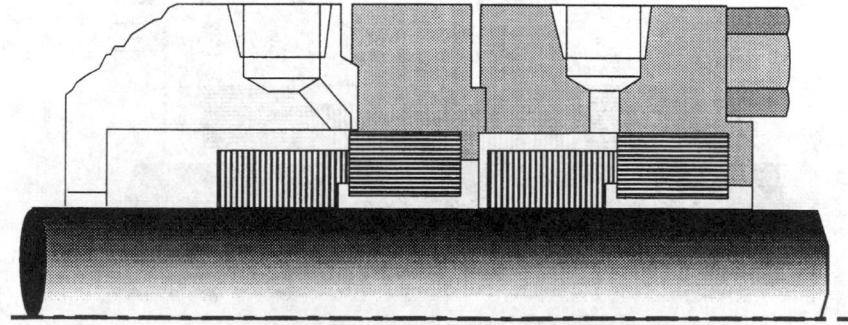

Fig. 22.12   Tandem seal. (From Ref. 18.)

Table 22.4 Properties of Common Barrier Fluids for Tandem or Double Seals[a]

Barrier Fluid	Temperature Limits °F		Comments
	Lower	Upper	
Water	40	180	Use corrosion-resistant materials Protect from freezing
Propylene glycol	−76	368	Consult seal manufacturer for proper mixture with water to avoid excessive viscosity
n–Propyl alcohol	−147	157	
ATF	55	200	Contains additives
Kerosene	0	300	
No. 2 diesel fuel	10	300	Contains additives

[a]STLE Society of Tribologists and Lubrication Engineers, "Guidelines for Meeting Emission Regulations for Rotating Machinery with Mechanical Seals," Special Publication SP-30, 1990.

provides a modern overview of several commercial products aimed at achieving zero leakage or leak-free operation in compliance with current regulations.

22.3.4 Noncontacting Seals for High-Speed/Aerospace Applications

For very high-speed turbomachinery, including gas turbines, seal runner speeds may reach speeds greater than 1300 fps, requiring novel seal arrangements to overcome wear and pressure limitations of conventional face seals. Two classes of seals are used that rely on a thin film of air to separate the seal faces. Hydrostatic face seals port high pressure fluid to the sealing face to induce opening force and maintain controlled face separation (see Fig. 22.15). The fluid pressure developed between the faces is dependent upon the gap dimension and the pressure varies between the lower and upper limits shown in the figure. Any change in the design clearance results in an increase or decrease of the opening force in a stabilizing sense. Of the four configurations shown, the coned seal configuration is the most popular. Converging faces are used to provide seal stability. Hydrostatic face seals suffer from contact during startup. To overcome this, the seals can be externally pressurized, but this adds cost and complexity.

The aspirating hydrostatic face seal (Fig. 22.15d) under development by GE and Stein Seal for turbine engine applications provides a unique failsafe feature.[20–22] The seal is designed to be open during initial rotation and after system shutdown—the two periods during which potentially damaging rubs are most common. Upon system pressurization, the aspirating teeth set up an initial pressure drop across the seal (6 psi nominal) that generates a closing force to overcome the retraction spring force F_s causing the seal to close to its operating clearance (nominal 0.0015–0.0025 in.). System pressure is ported to the face seal to prevent touchdown and provide good film stiffness during

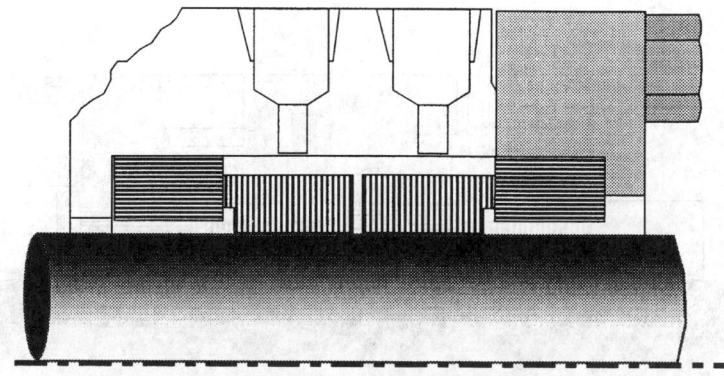

Fig. 22.13 Double seal. (From Ref. 18.)

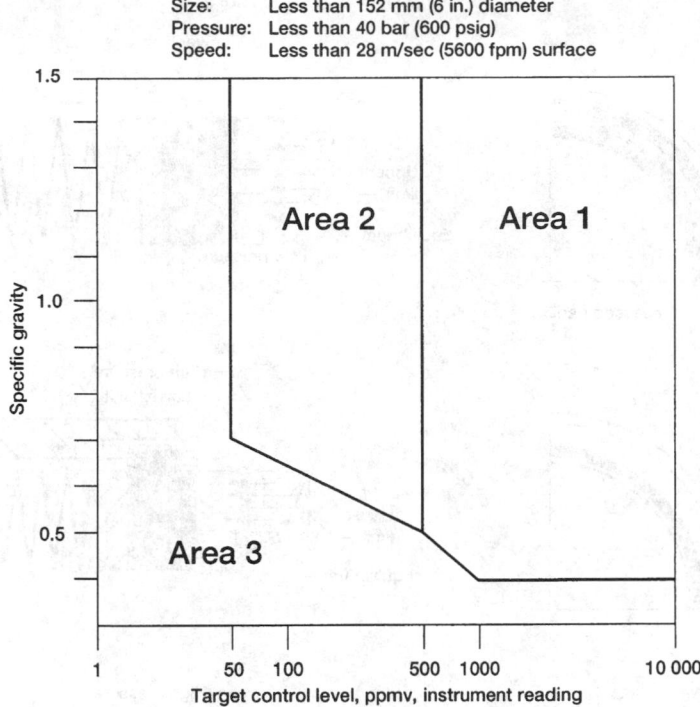

Size: Less than 152 mm (6 in.) diameter
Pressure: Less than 40 bar (600 psig)
Speed: Less than 28 m/sec (5600 fpm) surface

Chart area	Recommended technology
1	General purpose single seals, or dual (double and tandem) seals
2	Special purpose single seals, or dual (double and tandem) seals
3	Dual pressurized (double) seals Single or dual non-pressurized (tandem) seals vented to a closed vent system, above 0.4 specific gravity

Fig. 22.14 Application guide to control emissions. (From Ref. 18.)

operation. At engine shutdown, the pressure across the seal drops and the springs retract the seal away from the rotor, preventing contact.

Hydrodynamic or self-acting face seals incorporate lift pockets to generate a hydrodynamic film between the two faces to prevent seal contact. A number of lift pocket configurations are employed, including shrouded Rayleigh step, spiral groove, circular groove, and annular groove (Fig. 22.16). In these designs, hydrodynamic lift is independent of the seal pressure; it is proportional to the rotation speed and to the fluid viscosity. Therefore a minimum speed is required to develop sufficient lift force for face separation. Hydrodynamic seals operate on small (≤0.0005 in. nominal) clearances, resulting in very low leakage compared to labyrinth or brush seals, as shown in Fig. 22.17.[23] Because rubbing occurs during startup and shutdown, seal face materials must be selected for good rubbing characteristics for low wear (see Face Seal Materials, above).

Computer Analysis Tools: Face/Annular Seals

To aid aerospace and industrial seal designers alike, NASA sponsored the development of computer codes to predict the seal performance under a variety of conditions.[24] Codes were developed to treat both incompressible (e.g., liquid) and compressible (e.g., gas) flow conditions. In general, the codes assess seal performance characteristics, including load capacity, leakage flow, power requirements,

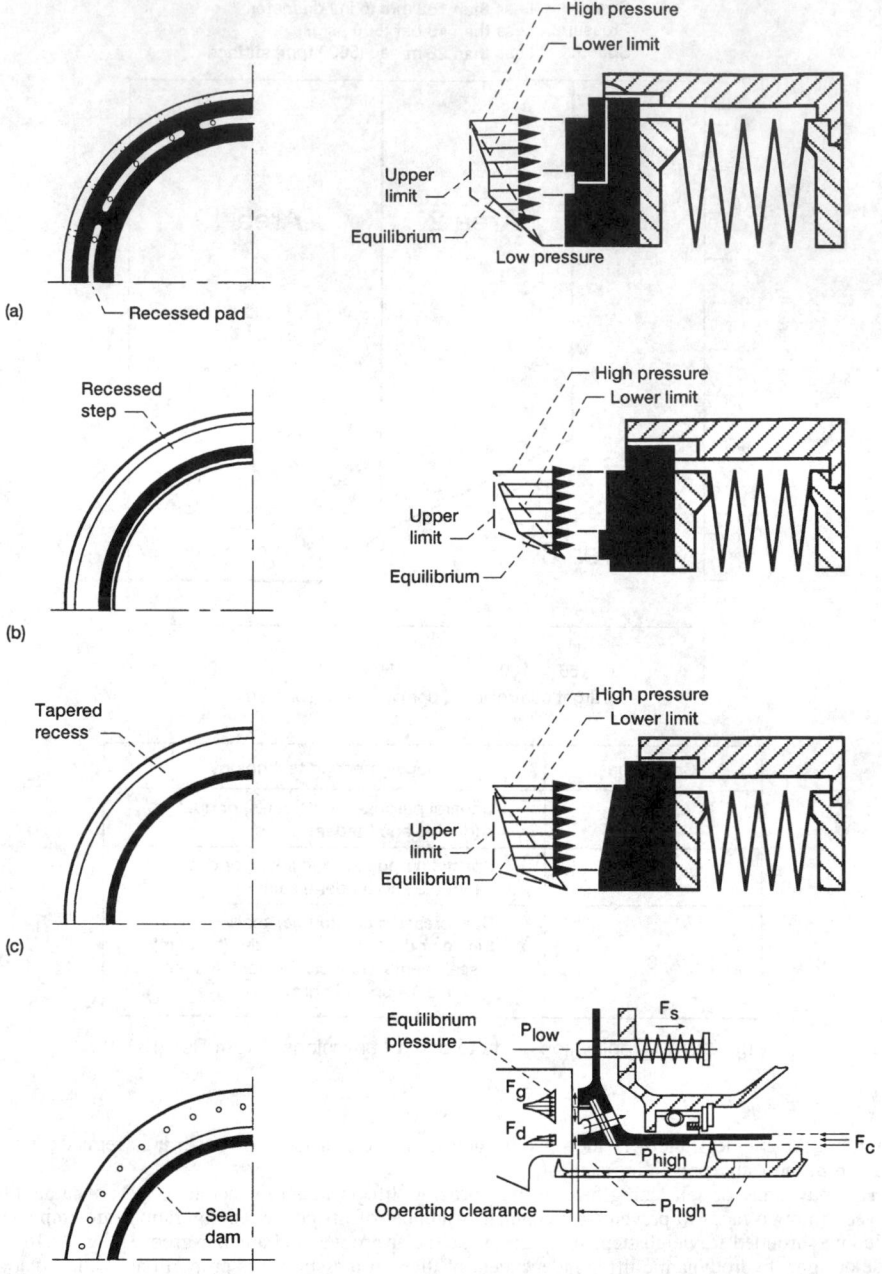

Fig. 22.15 Self-energized hydrostatic noncontacting mechanical face seals. (a) recessed pads with orifice compensation; (b) recessed step; (c) convergent tapered face; (d) aspirating seal. ((a)–(c) from Ref. 1; (d) from Ref. 20.)

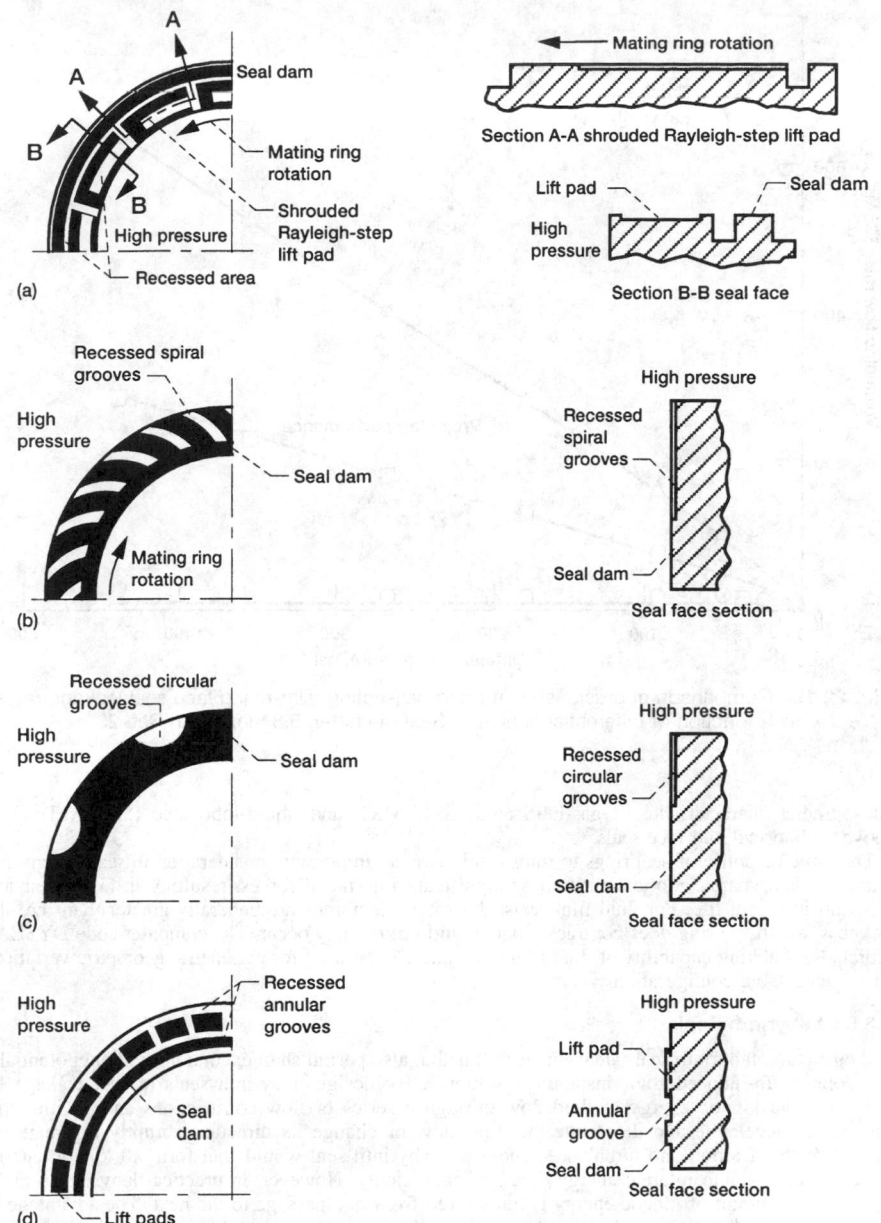

Fig. 22.16 Various types of hydrodynamic noncontacting mechanical face seals. (a) Shrouded Rayleigh step; (b) Spiral groove; (c) Circular groove; (d) Annular groove. (From Ref. 1.)

and dynamic characteristics in the form of stiffness and damping coefficients. These performance characteristics are computed as functions of seal and groove geometry, loads or film thicknesses, running speed, fluid viscosity, and boundary pressures. The GFACE code predicts performance for the following face seal geometries: hydrostatic, hydrostatic recess, radial and circumferential Rayleigh step, and radial and circumferential tapered land. The GCYLT code predicts performance for both hydrodynamic and hydrostatic cylindrical seals, including the following geometries: circumferential multilobe and Rayleigh step, Rayleigh step in direction of flow, tapered and self-energized hydrostatic. A description of these codes and their validation is given by Shapiro.[25] The SPIRALG/SPIRALI

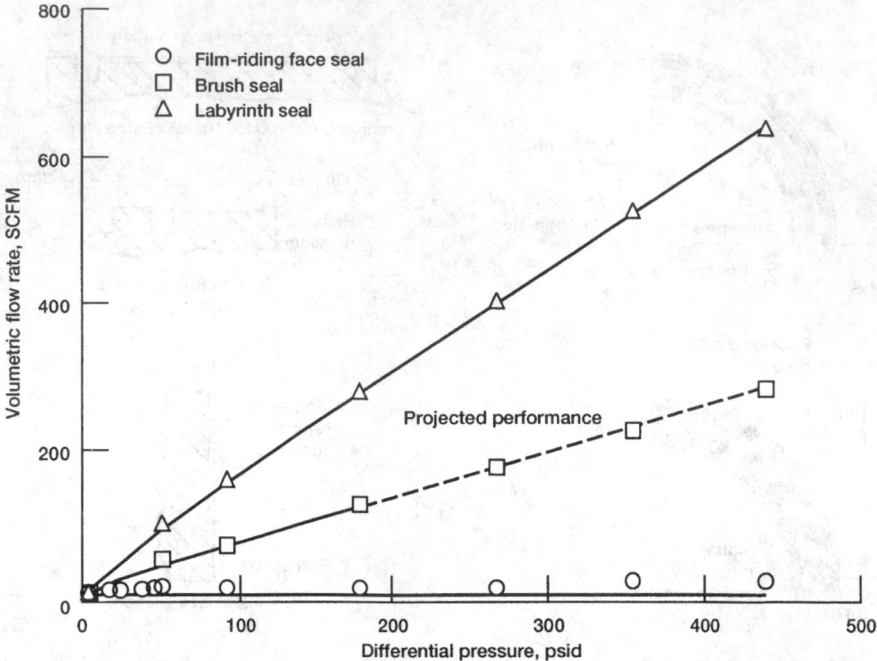

Fig. 22.17   Comparison of brush, labyrinth and self-acting, film-riding face seal leakage rates as a function of differential pressure. Seal diameter, 5.84 in. (From Ref. 23.)

codes predict characteristics of gas-lubricated (SPIRALG) and liquid-lubricated (SPIRALI) spiral groove, cylindrical and face seals.[26]

Dynamic response of seal rings to rotor motions is an important consideration in seal design. For contact seals, dynamic motion can impose significant interfacial forces, resulting in high wear and reduction in useful life. For fluid film seals, the rotor excursions are generally greater than the film thickness, and if the ring does not track, contact and failure may occur. The computer code DYSEAL predicts the tracking capability of fluid film seals and can be used for parametric geometric variations to find acceptable configurations.[27]

22.3.5   Labyrinth Seals

By their nature, labyrinth seals are clearance seals that also permit shaft excursions without potentially catastrophic rub-induced rotor instability problems. By design, labyrinth seals restrict leakage by dissipating the kinetic energy of fluid flow through a series of flow constrictions and cavities that sequentially accelerate and decelerate the fluid flow or change its direction abruptly to create the maximum flow friction and turbulence. The ideal labyrinth seal would transform all kinetic energy at each throttling into internal energy (heat) in each cavity. However, in practical labyrinth seals, a considerable amount of kinetic energy is transferred from one passage to the next. The advantage of labyrinth seals is that the speed and pressure capability is limited only by the structural design. One disadvantage, however, is a relatively high leakage rate. Labyrinth seals are used in so many gas sealing applications because of their very high running speed (1500 ft/s), pressure (250 psi), and temperature (≥1300°F), and the need to accommodate shaft excursions caused by transient loads. Labyrinth seal leakage rates have been reduced over the years through novel design concepts, but are still higher than desired because labyrinth seal leakage is clearance-dependent and this clearance opens due to periodic transient rubs.

Seal Configurations

Labyrinth seals can be configured in many ways (Fig. 22.18). The labyrinth seal configurations typically used are straight, angled-teeth straight, stepped, staggered, and abradable or wear-in. Optimizing labyrinth seal geometry depends on the given application and greatly affects the labyrinth seal leakage. Stepped labyrinth seals have been used extensively as turbine interstage air seals. Leak-

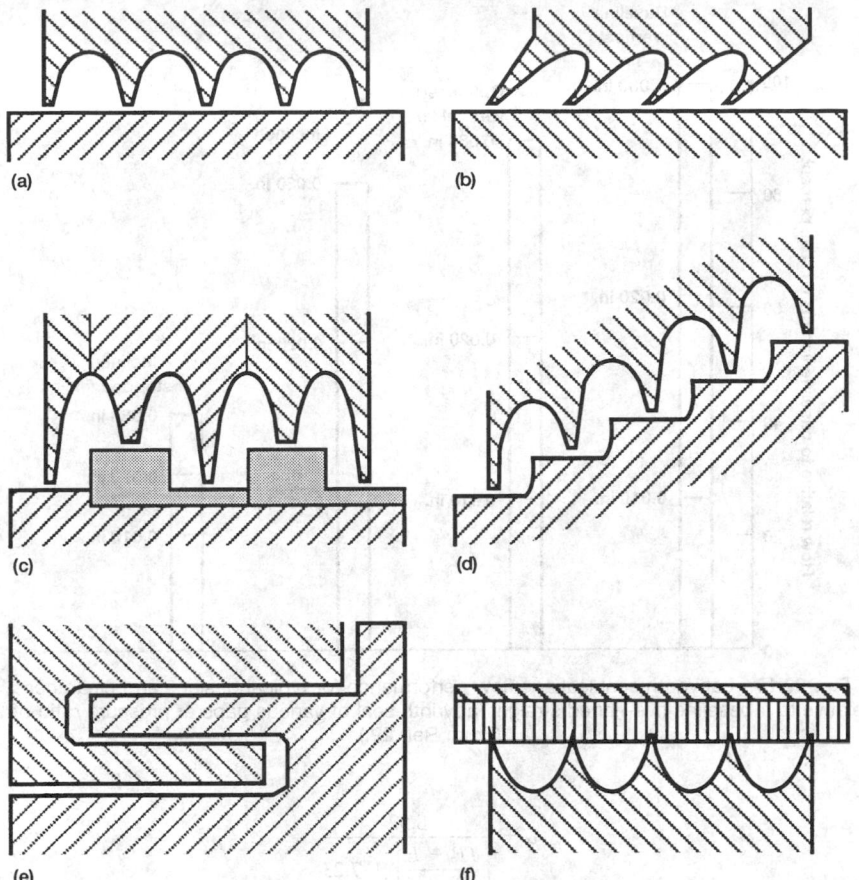

Fig. 22.18 Labyrinth seal configurations. (*a*) Straight labyrinth; (*b*) Inclined- or angled-teeth straight labyrinth; (*c*) Staggered labyrinth; (*d*) Stepped labyrinth; (*e*) Interlocking labyrinth; (*f*) Abradable (wear-in) labyrinth. (From Ref. 28.)

age flow through inclined, stepped labyrinths is about 40% that of straight labyrinths for similar conditions (Fig. 22.19). Performance benefits of stepped labyrinths must be balanced with other design issues. They require more radial space, are more difficult to manufacture, and may produce an undesirable thrust load because of the stepped area.

Leakage Flow Modeling

Leakage flow through labyrinth seals is generally modeled as a sequential series of throttlings through the narrow blade tip clearances. Ideally, the kinetic energy increase across each annular orifice would be completely dissipated in the cavity. However dissipation is not complete. Various authors handle this in different ways: Egli[30] introduced the concept of "carryover" to account for the incomplete dissipation of kinetic energy in straight labyrinth seals. Vermes[31] introduces the residual energy factor, α, to account for the residual energy in the flow as it passes from one stage to the next:

$$W = 5.76K \frac{A_g}{[RT_o]^{1/2}} \frac{P_o}{[1-\alpha]^{1/2}} \beta \quad \text{where} \quad \beta = \left[\frac{1 - \left[\dfrac{P_N}{P_o}\right]^2}{N - \ln\left[\dfrac{P_N}{P_o}\right]} \right]^{1/2}$$

and the residual energy factor

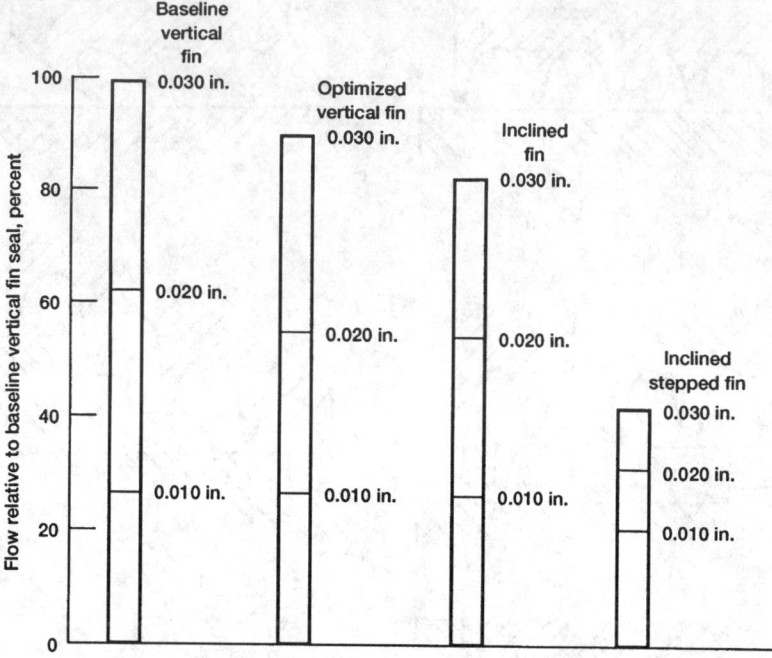

Fig. 22.19 Labyrinth seal leakage flow performance for typical designs and clearances, relative to a baseline five-finned straight labyrinth seal of various gaps at pressure ratio of 2. (From Ref. 29.)

$$\alpha = \cfrac{8.52}{\left[\cfrac{TP - L}{c}\right] + 7.23}$$

where A_g = flow area of single annular orifice (sq. in.)
c = clearance (in.)
g_c = gravitational constant (32.2 ft/s^2)
G = mass flux (lb$_m$/ft$^2 \cdot$ s)
K = clearance factor for annular orifice (see Fig. 20.20)
L = tooth width at sealing point (in.)
N = number of teeth (in.)
N_{Re} = Reynolds number, defined as $G(c/12)/\mu g_c$
TP = tooth pitch (in.)
P_o, P_N = inlet pressure, pressure at tooth N
R = gas constant (lbf $\cdot$ ft/lbm $\cdot$ °R)
T_o = gas inlet temperature (°R)
W = weight flow lb/s
μ = gas viscosity (lb$_f \cdot$ s/ft^2)

The clearance factor is plotted in Fig. 22.20 for a range of Reynolds numbers and tooth width-to-clearance ratios. Since K is a function of N_{Re} and since N_{Re} is a function of the unknown mass flow, the necessary first approximation can be made with $K = 0.67$. Vermes[31] also presents methods for calculating mass flow for stepped labyrinth seals and for off-design conditions (e.g., the stepped seal teeth are offset from their natural lands). Tooth shape also plays a role in leakage resistance. Mahler[32] showed that sharp corners provide the highest leakage resistance.

Applications

There are innumerable applications of labyrinth seals in the field. They are used to seal rolling element bearings, machine spindles, and other applications where some leakage can be tolerated. Since the

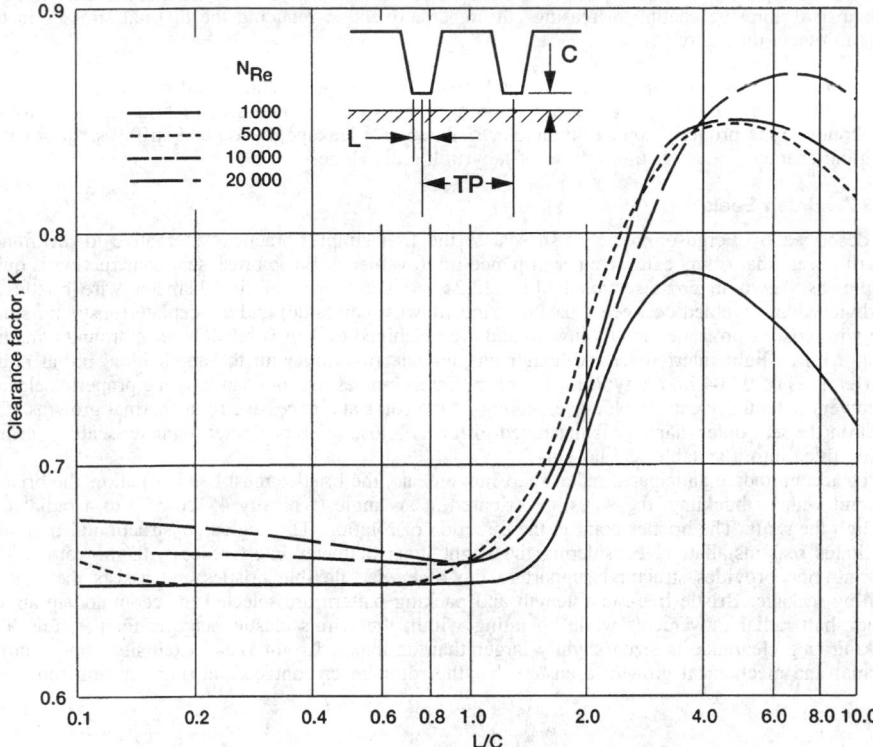

Fig. 22.20 Clearance factor (K) versus ratio of tooth width (L) to tooth clearance (C). (For various Reynolds Numbers (N_{Re}).) (From Ref. 31.)

development of the gas turbine engine, the labyrinth seal has been perhaps the most common seal applied to sealing both primary and secondary airflow.[11] Its combined pressure–speed–life limits have for many years exceeded those of its rubbing-contact seal competitors. Labyrinth seals are also used extensively in cryogenic rocket turbopump applications.

Computer Analysis Tools: Labyrinth Seals

The computer code KTK calculates the leakage and pressure distribution through labyrinth seal based on a detailed knife-to-knife (KTK) analysis. This code was developed by Allison Gas Turbines for the Air Force[33] and is also documented in Shapiro et al.[27] Rhode and Nail[34] present recent work in the application of a Reynolds-averaged computer code to generic labyrinth seals operating in the compressible region Mach number ≥ 0.3.

22.3.6 Honeycomb Seals

Honeycomb seals are used extensively in mating contact with labyrinth knife-edges machined onto the rotor in applications where there are significant shaft movements. After brazing the honeycomb material to the case, the inner diameter is machined to seal tolerance requirements. Properly designed honeycomb seals, in extensive tests performed by Stocker et al.[35] under NASA contract, showed dramatic leakage reductions under select gap and honeycomb cell-size combinations.

For applications where low leakage is paramount, designers will specify a small radial clearance between the labyrinth teeth and abradable surface (honeycomb or sprayed abradable). Designers will take advantage of normal centrifugal growth of the rotor to reduce this clearance to line-to-line and often to a wear-in condition, making an effective labyrinth seal. A "green" slow speed-ramp wear-in cycle is recommended.

Materials. Honeycomb elements are often fabricated of Hastelloy X,[36] a nickel-base alloy. Honeycomb seals provide for low-energy rubs when transient conditions cause the labyrinth knife-edges to wear into the surface (low-energy rubs minimize potentially damaging shaft vibrations). In very high surface speed applications and where temperatures are high the labyrinth teeth are "tipped" with a hard abrasive coating increasing cutting effectiveness, reducing the thermal stresses in the labyrinth teeth during rubs.

Honeycomb Annular Seals.   Honeycomb seals are also being considered now as annular seals to greatly improve damping over either smooth surfaces or labyrinth seals. Childs et al.[37] showed that honeycombs properly applied in annular seals control leakage, have good stiffness, and exhibit damping characteristics six times those of labyrinth seals alone.

22.3.7 Brush Seals

As described by Ferguson,[29] the brush seal is the first simple, practical alternative to the finned labyrinth seal that offers extensive performance improvements. Basic brush seal construction is quite simple, as shown in cross section in Fig. 22.21. A dense pack of fine-diameter wire bristles is sandwiched and welded between a backing ring (downstream side) and a sideplate (upstream side). The wire bristles protrude radially inward and are machined to form a brush bore fit around a mating rotor, with a slight interference. Although interference fits vary with the application, initial radial interferences of 0.004 in. are typical. Brush seal interferences and preload must be properly selected to prevent potentially catastrophic overheating of the rotor and excessive rotor thermal growths. The weld on the seal outer diameter is machined to form a close-tolerance outer diameter sealing surface that is fitted into a suitable seal housing.

To accommodate anticipated radial shaft movements, the bristles must bend. To allow the bristles to bend without buckling, the wires are oriented at an angle (typically 45° to 55°) to a radial line through the shaft. The bristles point in the direction of rotation. The angled construction also greatly facilitates seal installation, considering the slight inner-diameter interference with the rotors. The backing ring provides structural support to the otherwise flexible bristles and assists the seal in limiting leakage. Bristle free-radial-length and packing pattern are selected to accommodate anticipated shaft radial movements while operating within the wire's elastic range at temperature. The backing ring clearance is sized slightly larger than anticipated rotor radial excursions and relative thermal and mechanical growth to ensure that the rotor never contacts the ring, causing rotor and

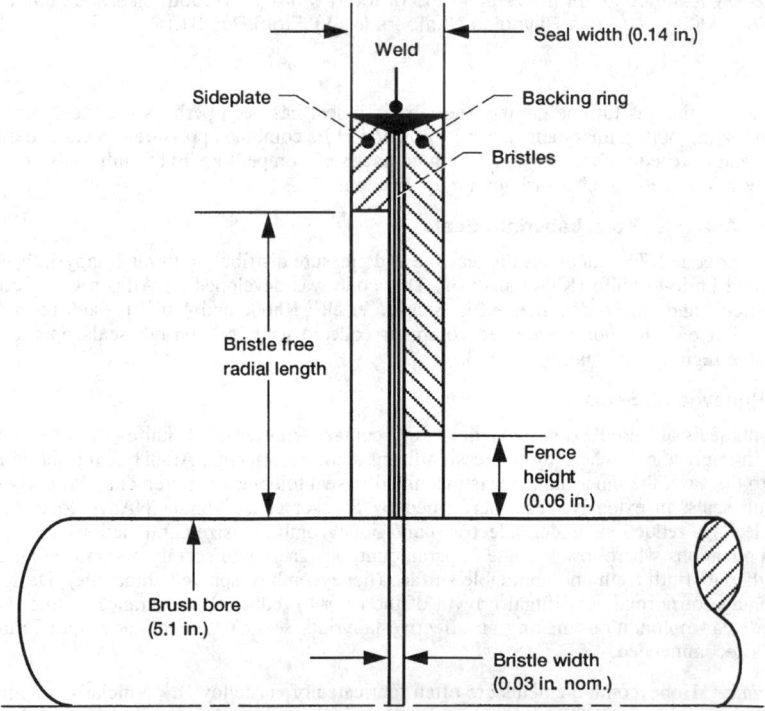

Fig. 22.21   Brush seal cross section with typical dimensions. (From Ref. 11.)

casing damage. An abradable rub surface added to the backing ring has been proposed to mitigate this problem by allowing tighter backing-plate clearances.

Standard single-stage brush seals typically are manufactured using 0.0028-in.-diameter bristles. Bristle pack widths are usually maintained around 0.030 in. and the backplate is in contact with the last row of downstream bristles. This basic design is limited to gas pressures below 70 psid.

Brush seal designs for higher-pressure applications require bristle packs that have higher axial stiffness to prevent the bristles from blowing under the backing ring. Short et al.,[38] amongst others, have developed brush seals rated for pressure differentials of 120 psid and above using 0.006-in.-diameter bristles and a thicker brush pack width (0.05 in.).

Multiple brush seals are generally used where large pressure drops ($\geq$145 psid) must be accommodated. The primary reason for using multiple seals is not to improve sealing but to reduce pressure-induced distortions in the brush pack, namely axial brush distortions under the backing ring, that cause wear. Researchers have noticed greater wear on the downstream brush if the flow jet coming from the upstream brush is not deflected away from the downstream brush–rotor contact.

Leakage Performance Comparisons

Ferguson[29] compared brush seal leakage with that of traditional five-finned labyrinth seals of various configurations. The results of this study (Fig. 22.22) indicate that the flow of a *new* brush seal is only 4% that of a vertical finned seal with a 0.03-in. radial gap and one-fifth that of an inclined-fin labyrinth seal with a step up and a 0.01-in. gap.

Addy et al.[39] showed similar large reductions in leakage testing a 5.1-in. bore seal across a wide temperature and speed range. Table 22.5 compares air leakage between a new brush seal and similarly sized labyrinth seals.

Effects of Speed. Addy et al.[39] found that brush seal flow parameter did not appreciably change when tested at speeds to 30,000 rpm and temperatures to 700°F. Stocker[35] found that rotation reduced straight labyrinth seal leakage by up to 10% for smooth and abradable lands but speed had negligible effect when run with honeycomb lands.

Brush seals are more robust than labyrinth seals. Bristle flexibility allows the brush to return to its normal operating position after the pressure has been removed, even after large excursions. Labyrinth seals incur permanent clearance increases under such conditions, degrading seal and engine performance.

Installed Performance. Mahler and Boyes[40] have made leakage comparisons of new and aircraft engine-tested brush seals. They concluded that performance did not deteriorate significantly for periods approaching one engine overhaul cycle (3000 hrs). Of the three brush seals examined, the

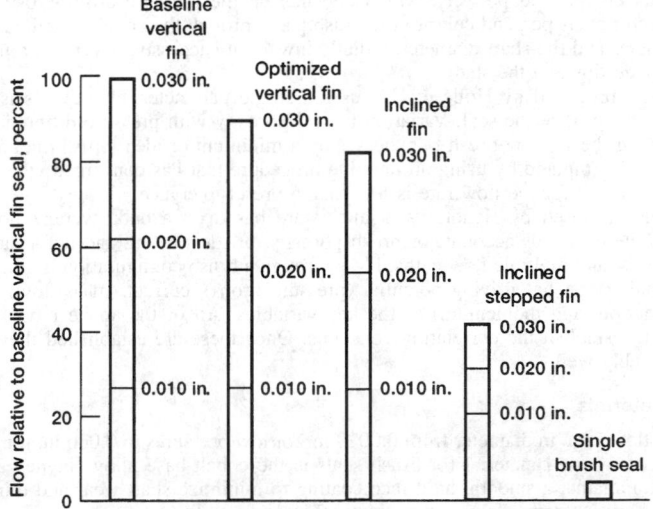

Fig. 22.22 Sealing performance of new brush seals relative to baseline five-finned labyrinth seals of various radial gaps at pressure ratio of 2. (From Ref. 29.)

Table 22.5 Comparison of New Labyrinth Seal (Smooth and Honeycomb Lands) vs. New Brush Seal Leakage Rates for Comparable Conditions

Seal	Rotor Dia. (in.)	Seal Clearance (in.)	Brush Seal Interference (in.)	Pressure Ratio P_i/P_o	ϕ-Flow Parameter $\dfrac{lb_m \cdot R^{1/2}}{lb_f \cdot s}$	Mass Flow $\dfrac{lb_m}{s}$
4-Tooth labyrinth vs. smooth land[a]	6.0	0.010	–	3.0	0.36	0.141
4-Tooth labyrinth vs. honeycomb land 0.062-in. cell size[a]	6.0	0.010	–	3.0	0.35	0.137
Brush seal[b]	5.1	–	0.004	3.0	0.0053	0.0099

Labyrinth seal: 4-tooth labyrinth; 0.11-in. pitch; 0.11-in. knife height.
Brush seal: 0.028-in. brush width; 0.0028-in. diameter bristles; 0.06-in. fence clearance.
Static, 0 rpm

Flow parameter, $\phi = \dfrac{m\sqrt{T_i}}{P_i A}$

[a]Ref. 35.
[b]Ref. 39.

"worst-case" brush seal's leakage rates doubled compared to a new brush seal. Even so, brush seal leakage was still less than half the leakage of the labyrinth seal.

Brush seals are not a solution for all seal problems. However, when they are applied within design limits, brush seal leakage will be lower than that of competing labyrinth seals and remain closer to design goals even after transient rub conditions.

In order to properly design and specify brush seals for an application, many design factors must be considered and traded-off. A comprehensive brush seal design algorithm was proposed by Holle and Krishan.[41] An iterative process must be followed to satisfy seal basic geometry, stress, thermal (especially during transient rub conditions), leakage, and life constraints to arrive at an acceptable design.

Brush Seal Flow Modeling

Brush seal flow modeling is complicated by several factors unique to porous structures, in that the leakage depends on the seal porosity, which depends on the pressure drop across the seal. Flow through the brush travels perpendicular to the brush pack through the annulus formed by the backing ring inner diameter and the shaft diameter, radially inward at successive layers within the brush, and between the bristle tips and the shaft.

A flow model proposed by Holle et al.[42] uses a single parameter, effective brush thickness, to correlate the flows through the seal. Variation in seal porosity with pressure difference is accounted for by normalizing the varying brush thicknesses by a minimum or ideal brush thickness. Maximum seal flow rates are computed by using an iterative procedure that has converged when the difference in successive iterations for the flow rate is less than a preset tolerance.

Flow models proposed by Hendricks et al.[43,44] are based on a bulk average flow through the porous media. These models account for brush porosity, bristle loading and deformation, brush geometry parameters, and multiple flow paths. Flow through a brush configuration is simulated by using an electrical analog that has driving potential (pressure drops), current (mass flow), and resistance (flow losses, friction, and momentum) as the key variables. All of the above models require some empirical data to establish the correlating constants. Once these are established the models predict seal flow reasonably well.

Brush Seal Materials

Brush wire bristles range in diameter from 0.0028 in. for low pressures to 0.006 in. for high pressures. The most commonly used material for brush seals is the cobalt-base alloy Haynes 25. Brush seals are generally run against a smooth, hard-face coating to minimize shaft wear and minimize chances of wear-induced cracks from affecting the structural integrity of the rotor. The usual coatings selected are ceramic, including chromium carbide and aluminum oxide. Selecting the correct mating wire and shaft surface finish for a given application can reduce friction heating and extend seal life through reduced oxidation and wear. For extreme operating temperatures to above 1300°F, Derby[45] has shown low wear and friction for the nickel-based superalloy Haynes 214 (heat treated for high strength) running against a solid-film lubricated hard-face coating Triboglide. Fellenstein et al.[46,47] investigated a number of bristle/rotor coating material pairs corroborating the benefits of Haynes 25 wires run against chrome-carbide but observed Haynes 214 bristle flairing when run against chrome-carbide and zirconia coatings.

Applications

Brush seals are seeing extensive service in both commercial and military turbine engines. Allison Engines has implemented brush seals in engines for the Saab 2000, Cessna Citation-X, and V-22 Osprey. GE has implemented a number of brush seals in the balance piston region of the GE90 engine for the Boeing 777 aircraft. PW has entered revenue service with brush seals in three locations[40] on the PW4168 for Airbus aircraft and on the PW4084 for the Boeing 777. Brush seals are also finding their way into industrial applications. Chupp et al.[48,49] are evaluating brush seals for large utility combined-cycle industrial turbines to exploit brush seal's low leakage to help meet plant efficiency goals (approaching 60%) and to reduce life cycle costs over a minimum life of 24,000 hr.

Ongoing Developments

Brush seals exhibit three other phenomena, seal "hysteresis," bristle "stiffening," and "pressure closing," that are beginning to receive attention. As described by Short et al.[38] and Basu et al.,[50] after the rotor moves into the bristle pack (due to radial excursions or thermal growths), the displaced bristles do not immediately recover against the frictional forces between them and the backing ring. As a result, a significant leakage increase (more than double) was observed following rotor movement.[50] This leakage hysteresis exists until after the pressure load is removed (e.g., after the engine is shut down). Furthermore, the bristle pack exhibits a considerable stiffening effect with application of pressure. This phenomenon results from interbristle friction loads making it more difficult for the brush bristles to flex during shaft excursions. Air leaking through the seal also exerts a radially inward force on the bristles, resulting in what has been termed "pressure-closing" or bristle "blow-down." This extra contact load especially on the upstream side of the brush affects the life of the seal (upstream bristles are worn in either a scalloped or coned configuration) and higher interface contact pressure.

Addressing these problems, Short et al.[38] demonstrated that relieving the brush back plate to minimize frictional contact, using larger (0.006-in. diameter) bristles and incorporating a flow deflector to mitigate pressure-closing produces a stable, low-wear, high-pressure (120 psid) brush seal.

Long life and durability under very high-temperature ($\geq$1300°F) conditions are hurdles to overcome to meet goals of advanced turbine engines under development for next-generation commercial subsonic, supersonic, and military fighter engine requirements. The tribology phenomena are complex and installation specific. In order to extend engine life and bring down maintenance costs, research and development are continuing in this area. To extend brush seal lives at high temperature, Addy et al.,[39] Hendricks et al.,[51] and Howe[52] are investigating approaches to replace metallic bristles with ceramic fibers. Ceramic fibers offer the potential for operating above 815°C (1500°F) and for reducing bristle wear rates and increasing seal lives while maintaining good flow resistance. Though early results indicate rotor coating wear, ceramic brush leakage rates were less than half those of labyrinth seal (0.007-in. clearance) and bristle wear was low.[39]

REFERENCES

1. I. E. Etsion and B. M. Steinetz "Seals," in *Mechanical Design Handbook,* H. A. Rothbart, (ed.), McGraw-Hill, New York, 1996, Section 17.

2. R. V. Brink, D. E. Czernik, and L. A. Horve, *Handbook of Fluid Sealing,* McGraw-Hill, New York, 1993.

3. J. F. Payne, A. Bazergui, and G. F. Leon, "Getting New Gasket Design Constants from Gasket Tightness Data," Special Supplement, *Experimental Techniques,* 22–27 (November 1988).

4. *Parker O-ring Handbook,* Cleveland, OH, 1992.

5. Martini, L. J. *Practical Seal Design,* Marcel Dekker, 1984.

6. A. Mathews and G. R. McKillop, "Compression Packings," in *Machine Design Seals Reference Issue,* Penton, March 1967, Chap. 8.

7. R. A. Howard, P. S. Petrunich, and K. C. Schmidt, *Grafoil Engine Design Manual,* Vol. 1, Union Carbide Corp., 1987.

8. B. M. Steinetz and M. Adams, "Effects of Compression, Staging and Braid Angle on Braided Rope Seal Performance," NASA TM-107504, AIAA-97-2872; 1997 AIAA Joint Propulsion Conference, Seattle WA, July 7–9, 1997.

9. B. M. Steinetz et al., *High Temperature Braided Rope Seals for Static Sealing Applications,* NASA TM-107233; also AIAA J. of Propulsion and Power, Vol. 13 No. 5, 1997.

10. A. G. Fern and B. S. Nau, *Seals,* Engineering Design Guide 15, published for Design Council, British Standards Institution and Council of Engineering Institutions, Oxford University Press, 1976.

11. B. M. Steinetz and R. C. Hendricks, "Aircraft Engine Seals," Chapter 9 of *Tribology for Aerospace Applications,* STLE Special Publication SP-37, 1997.

12. H. Buchter, *Industrial Sealing Technology,* Wiley, New York, 1979.

13. A. O. Lebeck, *Principles and Design of Mechanical Face Seals,* Wiley, New York, 1991.

14. J. Zuk and P. J. Smith, *Quasi-One Dimensional Compressible Flow Across Face Seals and Narrow Slots—II. Computer Program,* NASA TN D-6787, 1972.

15. W. F. Hughes et al., *Dynamics of Face and Annular Seals With Two-Phase Flow,* NASA CR-4256, 1989.

16. P. F. Brown, "Status of Understanding for Seal Materials," *Tribology in the 80's,* NASA CP-23000-VOL-2, Vol. 2, 1984, pp. 811–829.

17. J. C. Dahlheimer, *Mechanical Face Seal Handbook,* Chilton Book Co, Philadelphia, 1972.

18. Society of Tribologists and Lubrication Engineers, *Guidelines for Meeting Emission Regulations for Rotating Machinery with Mechanical Seals,* Special Publication SP-30, revised 1994.

19. R. C. Waterbury, "Zero-Leak Seals Cut Emissions," *Pumps and Systems Magazine,* AES Marketing, Fort Collins, CO, July, 1996.

20. H. Hwang, T. Tseng, B. Shucktis, and B. Steinetz, *Advanced Seals for Engine Secondary Flowpath,* AIAA-95-2618. Presented at the 1995 AIAA/ASME/SAE/ASEE Joint Propulsion Conference, San Diego, CA, 1995.

21. C. E. Wolfe et al., *Full Scale Testing and Analytical Validation of an Aspirating Face Seal,* AIAA Paper 96-2802, 1996.

22. B. Bagepalli et al., *Dynamic Analysis of an Aspirating Face Seal for Aircraft-Engine Applications,* AIAA Paper 96-2803, 1996.

23. J. Munson, *Testing of a High Performance Compressor Discharge Seal,* AIAA Paper 93-1997, 1993.

24. R. C. Hendricks, *Seals Code Development—'95,* NASA CP-10181, 1995.

25. W. Shapiro, *Numerical, Analytical, Experimental Study of Fluid Dynamic Forces in Seals, Volume 2–Description of Gas Seal Codes GCYLT and GFACE,* NASA Contract Report for Contract NAS3-25644, September 1995.

26. J. Walowit and W. Shapiro, *Numerical, Analytical, Experimental Study of Fluid Dynamic Forces in Seals, Volume 3—Description of Spiral-Groove Codes SPIRALG and SPIRALI,* NASA Contract Report for Contract NAS3-25644, September 1995.

27. W. Shapiro et al., *Numerical, Analytical, Experimental Study of Fluid Dynamic Forces in Seals, Volume 5—Description of Seal Dynamics Code DYSEAL and Labyrinth Seals Code KTK,* NASA Contract Report for Contract NAS3-25644, September 1995.

28. R. E. Burcham and R. B. Keller, Jr., *Liquid Rocket Engine Turbopump Rotating-Shaft Seals,* NASA SP-8121, 1979.

29. J. G. Ferguson, *Brushes as High Performance Gas Turbine Seals,* ASME Paper 88-GT-182, 1988.

30. A. Egli, "The Leakage of Steam Through Labyrinth Seals," *ASME Transactions* **57**(3), 115–122 (1935).

31. G. Vermes, "A Fluid Mechanics Approach to the Labyrinth Seal Leakage Problem," *Journal of Engineering for Power* **83**(2), 161–169, 1961.

32. F. H. Mahler, *Advanced Seal Technology,* Report PWA-4372, Contract F33615-71-C-1534, Pratt and Whitney Aircraft Co., East Hartford, CT, 1972.

33. D. L. Tipton, T. E. Scott, and R. E. Vogel, *Labyrinth Seal Analysis: Volume III—Analytical and Experimental Development of a Design Model for Labyrinth Seals,* AFWAL TR-85-2103, Allison Gas Turbine Division, General Motors, Corp., 1986.

34. D. L. Rhode and G. H. Nail, "Computation of Cavity-by-Cavity Flow Development in Generic Labyrinth Seals," *Journal of Tribology* **14**, 47–51, 1992.

35. H. L. Stocker, D. M. Cox, and G. F. Holle, *Aerodynamic Performance of Conventional and Advanced Design Labyrinth Seals with Solid Smooth, Abradable, and Honeycomb Lands—Gas Turbine Engines.* NASA CR-135307, 1977.

36. Z. Galel, F. Brindisi, and D. Norstrom, *Chemical Stripping of Honeycomb Airseals, Overview and Update,* ASME Paper 90-GT-318, 1990.

37. D. W. Childs, D. Elrod, and K. Hale, "Annular Honeycomb Seals: Test Results for Leakage and Rotordynamic Coefficients—Comparison to Labyrinth and Smooth Configurations," in

Rotordynamic Instability Problems in High-Performance Turbomachinery, NASA CP-3026, pp. 143–159, 1989.

38. J. F. Short et al., *Advanced Brush Seal Development,* AIAA Paper 96-2907, 1996.

39. H. E. Addy et al., *Preliminary Results of Silicon Carbide Brush Seal Testing at NASA Lewis Research Center,* AIAA Paper 95-2763, 1995.

40. F. Mahler and E. Boyes, *The Application of Brush Seals in Large Commercial Jet Engines,* AIAA Paper 95-2617, 1995.

41. G. F. Holle and M. R. Krishnan, *Gas Turbine Engine Brush Seal Applications,* AIAA Paper 90-2142, 1990.

42. G. F. Holle, R. E. Chupp, and C. A. Dowler, "Brush Seal Leakage Correlations Based on Effective Thickness," *Fourth International Symposium on Transport Phenomena and Dynamics of Rotating Machinery,* preprint Vol. A., 1992, pp. 296–304.

43. R. C. Hendricks et al., *A Bulk Flow Model of a Brush Seal System,* ASME Paper 91-GT-325, 1991.

44. R. C. Hendricks et al., "Investigation of Flows in Bristle and Fiberglass Brush Seal Configurations," in *Fourth International Symposium on Transport Phenomena and Dynamics of Rotating Machinery,* preprint Vol. A, 1992, pp. 315–325.

45. J. Derby and R. England, *Tribopair Evaluation of Brush Seal Applications,* AIAA Paper 92-3715, 1992.

46. J. Fellenstein, C. Della Corte, K. D. Moore, and E. Boyes, *High Temperature Brush Seal Tuft Testing of Metallic Bristles vs Chrome Carbide,* NASA TM-107238, AIAA-96-2908, 1996.

47. J. A. Fellenstein, C. Della Corte, K. A. Moore, and E. Boyes, *High Temperature Brush Seal Tuft Testing of Selected Nickel–Chrome and Cobalt–Chrome Superalloys,* NASA TM-107497, AIAA-97-2634, 1997.

48. R. E. Chupp, R. P. Johnson, and R. G. Loewenthal, *Brush Seal Development for Large Industrial Gas Turbines,* AIAA Paper 95-3146, 1995.

49. R. E. Chupp, R. J. Prior, and R. G. Loewenthal, *Update on Brush Seal Development for Large Industrial Gas Turbines,* AIAA Paper 96-3306, 1996.

50. P. Basu et al., *Hysteresis and Bristle Stiffening Effects of Conventional Brush Seals,* AIAA Paper 93-1996, 1993.

51. R. C. Hendricks, R. Flower, and H. Howe, "Development of a Brush Seals Program Leading to Ceramic Brush Seals," in *Seals Flow Code Development—'93,* NASA CP-10136, pp. 99–117, 1994.

52. H. Howe, "Ceramic Brush Seals Development," in *Seals Flow Code Development—'93,* NASA CP-10136, 1994, pp. 133–150.

BIBLIOGRAPHY

American Society of Mechanical Engineers, *Code for Pressure Vessels,* Sec. VIII, Div. 1, App. 2, 1995.

American Variseal, "Variseal™ Design Guide," AVDG394 American Variseal Co., Broomfield, CO, 1994.

Crane, J., "Dry Running Noncontacting Gas Seal" Bulletin No. S-3030, 1993.

Howard, R. A., *Grafoil Engineering Design Manual,* Union Carbide Corp., Cleveland, OH, 1987.

CHAPTER 23

VIBRATION AND SHOCK

Wayne Tustin
Equipment Reliability Institute
Santa Barbara, California

23.1	VIBRATION	661
23.2	ROTATIONAL IMBALANCE	668
23.3	VIBRATION MEASUREMENT	673
23.4	ACCELERATION MEASUREMENT	681
23.5	SHOCK MEASUREMENT AND ANALYSIS	692
23.6	SHOCK TESTING	695
23.7	SHAKE TESTS FOR ELECTRONIC ASSEMBLIES	705

23.1 VIBRATION

In any structure or assembly, certain whole-body motions and certain deformations are more common than others; the most likely (easiest to excite) motions will occur at certain natural frequencies. Certain exciting or forcing frequencies may coincide with the natural frequencies (resonance) and give relatively severe vibration responses.

We will now discuss the much-simplified system shown in Fig. 23.1. It includes a weight W (it is technically preferred to use mass M here, but weight W is what people tend to think about), a spring of stiffness K, and a viscous damper of damping constant C. K is usually called the spring rate; a static force of K newtons will statically deflect the spring by δ mm, so that spring length l becomes $\delta + l$. (In "English" units, a force of K lb will statically deflect the spring by 1 in.) This simplified system is constrained to just one motion—vertical translation of the mass. Such single-degree-of-freedom (SDF) systems are not found in the real world, but the dynamic behavior of many real systems approximate the behavior of SDF systems over small ranges of frequency.

Suppose that we pull weight W down a short distance further and then let it go. The system will oscillate with W moving up-and-down at natural frequency f_N, expressed in cycles per second (cps) or in hertz (Hz); this condition is called "free vibration." Let us here ignore the effect of the damper, which acts like the "shock absorbers" or dampers on your automobile's suspension—using up vibratory energy so that oscillations die out. f_N may be calculated by

$$f_N = \frac{1}{2\pi} \sqrt{\frac{Kg}{W}} \tag{23.1}$$

It is often convenient to relate f_N to the static deflection δ due to the force caused by earth's gravity, $F = W = Mg$, where $g = 386$ in./sec^2 = 9807 mm/sec^2, opposed by spring stiffness K expressed in either lb/in. or N/mm. On the moon, both g and W would be considerably less (about one-sixth as large as on earth). Yet f_N will be the same. Classical texts show Eq. (23.1) as

$$f_N = \frac{1}{2\pi} \sqrt{\frac{K}{M}}$$

In the "English" System:

$$\delta = \frac{F}{K} = \frac{W}{K}$$

Mechanical Engineers' Handbook, 2nd ed., Edited by Myer Kutz.
ISBN 0-471-13007-9 © 1998 John Wiley & Sons, Inc.

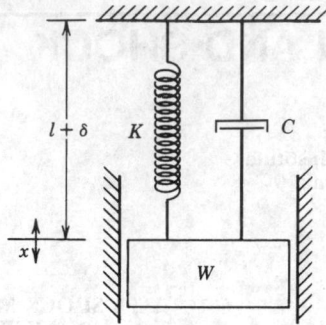

Fig. 23.1   Single-degree-of-freedom system.

Then

$$f_N = \frac{1}{2\pi} \sqrt{\frac{g}{\delta}} = \frac{1}{2\pi} \sqrt{\frac{386}{\delta}}$$
$$= \frac{19.7}{2\pi\sqrt{\delta}} = \frac{3.13}{\sqrt{\delta}}$$

(23.2a)

In the International System:

$$\delta = \frac{F}{K} = \frac{Mg}{K}$$

Then

$$f_N = \frac{1}{2\pi} \sqrt{\frac{g}{\delta}} = \frac{1}{2\pi} \sqrt{\frac{9807}{\delta}}$$
$$= \frac{99.1}{2\pi\sqrt{\delta}} = \frac{15.76}{\sqrt{\delta}}$$

(23.2b)

Relationships (23.2a) and (23.2b) often appear on specialized "vibration calculators." As increasingly large mass is supported by a spring; δ becomes larger and f_N drops.

Let $K = 1000$ lb/in. and vary W:

W (lb)	$\delta = \dfrac{W}{K}$ (in.)	f_N (Hz)
0.001	0.000 001	3 130
0.01	0.000 01	990
0.1	0.000 1	313
1.	0.001	99
10.	0.01	31.3
100.	0.1	9.9
1 000.	1.	3.13
10 000.	10.	0.99

Let $K = 1000$ N/mm and vary M:

M (kg)	$\delta = \dfrac{9.81M}{K}$	f_N (Hz)
0.00102	10 nm	4 980
0.0102	100 nm	1 576
0.102	1 μm	498
1.02	10 μm	157.6
10.2	100 μm	49.8
102.	1 mm	15.76
1 020.	10 mm	4.98
102 000.	1 m	0.498

Note, from Eqs. (23.2a) and (23.2b), that f_N depends on δ, and thus on both M and K (or W and K). As long as both load and stiffness change proportionately, f_N does not change.

The peak-to-peak or double displacement amplitude D will remain constant if there is no damping to use up energy. The potential energy we put into the spring becomes zero each time the mass passes through the original position and becomes maximum at each extreme. Kinetic energy becomes maximum as the mass passes through zero (greatest velocity) and becomes zero at each extreme (zero velocity). Without damping, energy is continually transferred back and forth from potential to kinetic energy. But with damping, motion gradually decreases; energy is converted to heat. A vibration pickup on the weight would give oscilloscope time history patterns like Fig. 23.2; more damping was present for the lower pattern and motion decreased more rapidly.

Assume that the "support" at the top of Fig. 23.1 is vibrating with a constant D of, say, 1 in. Its frequency may be varied. How much vibration will occur at weight W? The answer will depend on

1. The frequency of "input" vibration.
2. The natural frequency and damping of the system.

Let us assume that this system has an f_N of 1 Hz while the forcing frequency is 0.1 Hz, one-tenth the natural frequency, Fig. 23.3. We will find that weight W has about the same motion as does the input, around 1 in D. Find this at the left edge of Fig. 23.3; transmissibility, the ratio of response vibration divided by input vibration, is $1/1 = 1$. As we increase the forcing frequency, we find that the response increases. How much? It depends on the amount of damping in the system. Let us assume that our system is lightly damped, that it has a ratio C/C_c of 0.05 (ratio of actual damping to "critical" damping is 0.05). When our forcing frequency reaches 1 Hz (exactly f_N), weight W has a response D of about 10 in., 10 times as great as the input D. At this "maximum response" frequency, we have the condition of "resonance"; the exciting frequency is the same as the f_N of the load. As we further increase the forcing frequency (see Fig. 23.3), we find that response decreases. At 1,414 times f_N, the response has dropped so that D is again 1 in. As we further increase the forcing frequency, the response decreases further. At a forcing frequency of 2 Hz, the response D will be about 0.3 in. and at 3 Hz it will be about 0.1 in.

Note that the abscissa of Fig. 23.3 is "normalized"; that is, the transmissibility values of the preceding paragraph would be found for another system whose natural frequency is 10 Hz, when the forcing frequency is, respectively, 1, 10, 14.14, 20, and 30 Hz. Note also that the vertical scale of Fig. 23.3 can represent (in addition to ratios of motion) ratios of force, where force can be measured in pounds or newtons.

The region above 1.414 times f_N (where transmissibility is less than 1) is called the region of "isolation." That is, weight W has less vibration than the input; it is *isolated*. This illustrates the use of vibration isolators—rubber elements or springs that reduce the vibration input to delicate units in aircraft, missiles, ships, and other vehicles, and on certain machines. We normally try to set f_N (by selecting isolators) considerably below the expected forcing frequency of vibration. Thus, if our

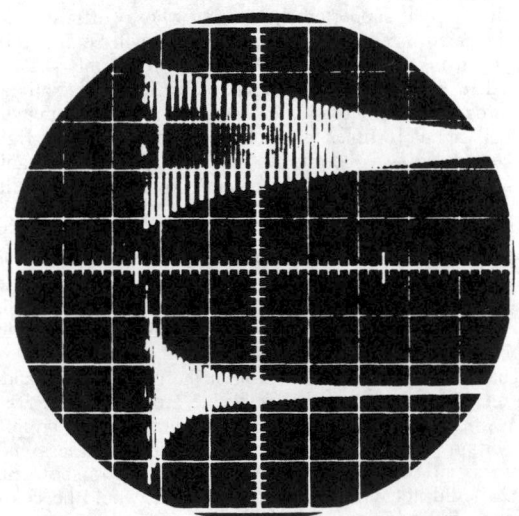

Fig. 23.2 Oscilloscope time history patterns of damped vibration.

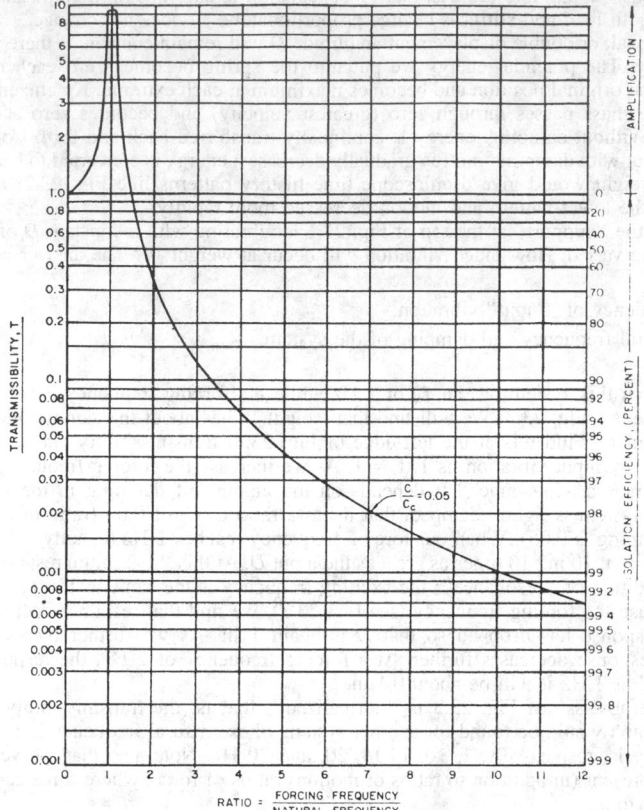

Fig. 23.3 Transmissibility of a lightly damped system.

support vibrates at 50 Hz, we might select isolators whose K makes f_N 25 Hz or less. According to Fig. 23.3, we will have best isolation at 50 Hz if f_N is as low as possible. (However, we should not use too-soft isolators because of instabilities that could arise from too-large static deflections, and because of need for excessive clearance to any nearby structures.)

Imagine a system with a weight supported by a spring whose stiffness K is sufficient that $f_N = 10$ Hz. At an exciting frequency of 50 Hz, the frequency ratio will be 50/10 or 5, and we can read transmissibility = 0.042 from Fig. 23.3. The weight would "feel" only 4.2% as much vibration as if it were rigidly mounted to the support. We might also read the "isolation efficiency" as being 96%. However, as the source of 50-Hz vibration comes up to speed (passing slowly through 10 Hz), the isolated item will "feel" about 10 times as much vibration as if it were rigidly attached, without any isolators. Here is where damping is helpful: to limit the "Q" or "mechanical buildup" at resonance. Observe Fig. 23.4, plotted for several different values of damping. With little damping present, there is much resonant magnification of the input vibration. With more damping, maximum transmissibility is not so high. For instance, when C/C_c is 0.01, "Q" is about 40. Even higher Q values are found with certain structures having little damping; Q's over 1000 are sometimes found.

Most structures (ships, aircraft, missiles, etc.) have Q's ranging from 10 to 40. Bonded rubber vibration isolating systems often have Q's around 10; if additional damping is needed (to keep Q lower), dashpots or rubbing elements may be used. Note that there is less buildup at resonance, but that isolation is not as effective when damping is present.

Figure 23.1 shows guides or constraints that restrict the motion to up-and-down translation. A single measurement on an SDF system will describe the arrangement of its parts at any instant. Another SDF system is a wheel attached to a shaft. If the wheel is given an initial twist, that system will also oscillate at a certain f_N, which is determined by shaft stiffness and wheel inertia. This imagined rotational system is an exact counterpart of the SDF system shown in Fig. 23.1.

If weight W in Fig. 23.1 did not have the guides shown, it would be possible for weight W to move in five other motions—five additional degrees of freedom. Visualize the six possibilities:

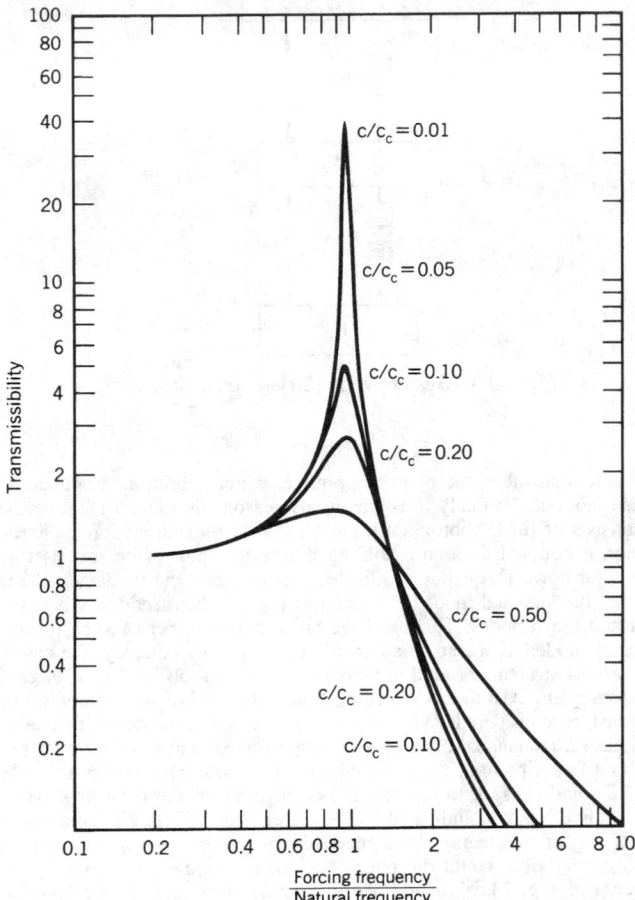

Fig. 23.4   Transmissibility for several different values of damping.

Vertical translation.
North–south translation.
East–west translation.
Rotation about the vertical axis.
Rotation about the north–south axis.
Rotation about the east–west axis.

Now this solid body has six degrees of freedom—six measurements would be required in order to describe the various whole-body motions that may be occurring.

Suppose now that the system of Fig. 23.1 were attached to another mass, which in turn is supported by another spring and damper, as shown in Fig. 23.5. The reader will recognize that this is more typical of many actual systems than is Fig. 23.1. Machine tools, for example, are seldom attached directly to bedrock, but rather to other structures that have their own vibration characteristics. Weight W_2 will introduce six additional degrees of freedom, making a total of 12 for the system of Fig. 23.5. That is, in order to describe all of the possible solid-body motions of W_1 and W_2, it would be necessary to consider all 12 motions and to describe the instantaneous positions of the two masses.

The reader can extend that reasoning to include additional masses, springs, and dampers, and additional degrees of freedom—possible motions. Finally, consider a continuous beam or plate, where mass, spring, and damping are distributed rather than being concentrated as in Fig. 23.1 and 23.5. Now we have an infinite number of possible motions, depending on the exciting frequency, the

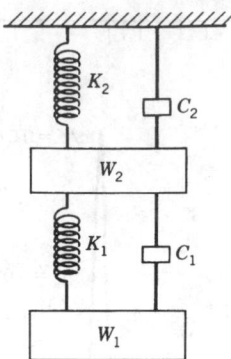

Fig. 23.5 System with 12 degrees of freedom.

distribution of mass and stiffness, the point or points at which vibration is applied, etc. Fortunately, only a few of these motions are likely to occur in any reasonable range of frequencies.

Figure 23.6 consists of three photographs, each taken at a different forcing frequency. Up-and-down shaker motion is coupled through a holding fixture to a pair of beams. (Let us only consider the left-hand beams for now.) Frequency is adjusted first to excite the fundamental shape (also called the "first mode" or "fundamental mode") of beam response. Then frequency is readjusted higher to excite the second and third modes. There will be an infinite number of such resonant frequencies, an infinite number of modes, if we go to an infinitely high test frequency. We can mentally extend this reasoning to various structures found in vehicles, machine tools, and appliances, etc.: an infinite number of resonances could exist in any structure. Fortunately, there are usually limits to the exciting frequencies that must be considered. Whenever the frequency coincides with one of the f_N's of a structure, we will have a resonance; high stresses, large forces, and large motions result.

All remarks about free vibration, f_N, and damping apply to continuous systems. We can "pluck" the beam of Fig. 23.6 and cause it to respond in one or more of the patterns shown. We know that vibration will gradually die out, as indicated by the upper trace of Fig. 23.2, because there is internal (or hysteresis) damping of the beam. Stress reversals create heat, which uses up vibratory energy. With more damping, vibration would die out faster, as in the lower trace of Fig. 23.2. The right-hand cantilever beam of Fig. 23.6 is made of thin metal layers joined by a layer of a viscoelastic damping material, as in Fig. 23.7. Shear forces between the layers use up vibratory energy faster and free vibration dies out more quickly.

Statements about forced vibration and resonance also apply to continuous systems. At very low frequencies, the motion is the same (transmissibility = 1) at all points along the beams. At certain f_N's, large motion results.

The points of minimum D are called "nodes" and the points having maximum D are called "anti-nodes." With a strobe light and/or vibration sensors we could show that (in a pure mode) all points along the beam are moving either in-phase or out-of-phase and that phase reverses from one side of a node to the other. Also bending occurs at the attachment point and at the antinodes; these are the locations where fatigue failures usually occur.

It is possible for several modes to occur at once, if several natural frequencies are present as in complex or in broad-band random vibration. It is also possible for several modes to simultaneously be excited by a shock pulse (and then to die out). Certain modes are most likely to cause failure in a particular installation; the frequencies causing such critical modes are called "critical frequencies."

Intense sounds can cause modes to be excited, especially in thin panels; stress levels in aircraft and missile skins may cause fatigue failures. Damping treatments on the skin are often very effective in reducing such vibration.

Figure 23.6 shows forced vibration of two beams whose length is adjusted until their f_N's are identical. The conventional solid beam responds more violently than the damped laminated beam. We say that the maximum transmissibility (often called mechanical "Q") of the solid beam is much greater than that of the damped beam. Assuming that the vibration continues indefinitely, which beam will probably fail in fatigue first? Here is one reason for using damping.

Resonance is sometimes helpful and desirable; at other times it is harmful. Some readers will be familiar with deliberate applications of vibration to move bulk materials, to compact materials, to remove entrapped gases, or to perform fatigue tests. Maximum vibration is achieved (assuming the vibratory force is limited) by operating the system at resonance.

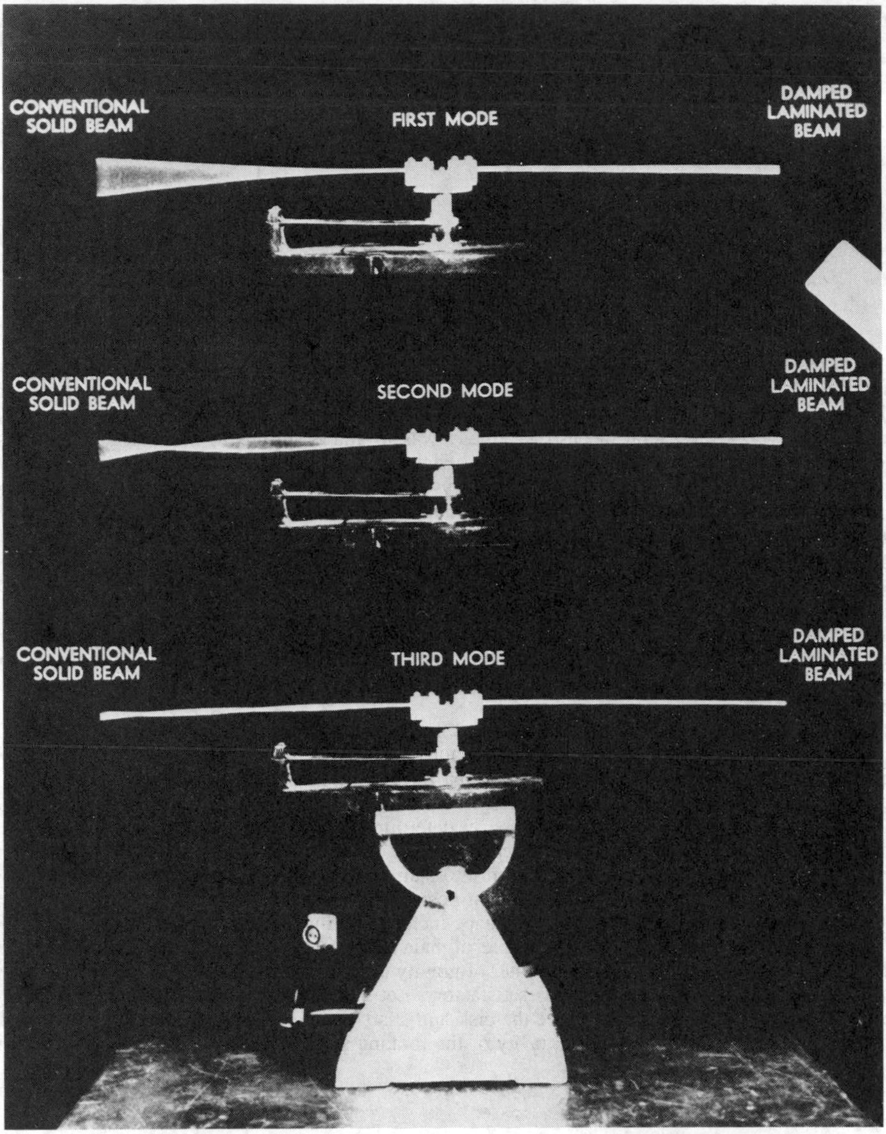

Fig. 23.6 Pair of beams excited at three different forcing frequencies.

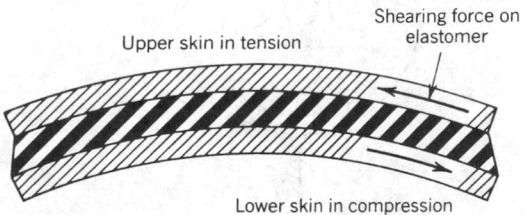

Upper skin in tension

Shearing force on elastomer

Lower skin in compression

Fig. 23.7 Detail of laminated beam. (Courtesy of Lord Manufacturing Co.)

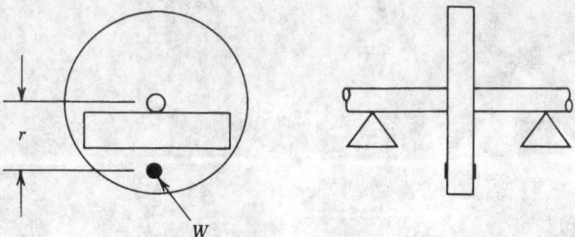

Fig. 23.8   Static balancing of a disk.

A resonant vibration absorber can sometimes reduce motion, if the vibration input to a structure is at a fixed frequency. Imagine, for example, weaving machines in a relatively soft, multistory factory building. They happen to excite an up-and-down resonant motion of their floor. (Some "old timers" claim they saw D's of several inches.) A remedy was to attach springs to the undersides of those floors, directly beneath the offending machines. Each spring supported a pail which was gradually filled with sand until the f_N of the spring/pail was equal to the exciting frequency of the weaving machine above it. A dramatic reduction in floor motion told maintenance people that the spring was correctly loaded. Similar methods (using tanks filled with water) have been used on ships. However, any change in exciting frequency necessitates a bothersome readjustment.

23.2   ROTATIONAL IMBALANCE

Where rotating engines are used in ships, automobiles, aircraft, or other vehicles; where turbines are used in vehicles or electrical power generating stations; where propellers are used in ships and aircraft—in all of these and many other varied applications, imbalance of the rotating members causes vibration.

Consider first the simple disk shown in Fig. 23.8. This disk has some extra material on one side so that the center of gravity is not at the rotational center. If we attach this disk to a shaft and allow the shaft to rotate on knife-edges, we observe that the system comes to rest with the heavy side of the disk downward. This type of imbalance is called *static* imbalance, since it can be detected statically. It can be measured statically, also, by determining some weight W at some radius r that must be attached to the side opposite the heavy side, in order to restore the center of gravity to the rotational center and thus to bring the system into static balance; that is, so that the disk will have no preferred position and will rest in any angular position. The product Wr is the value of the original imbalance. It is often expressed in units of ounce-inches, gram-millimeters, etc.

Static balancing is the simplest technique of balancing, and is often used for the wheels on automobiles, for instance. It locates the center of gravity at the center of the wheel. But we will show that this compensation is not completely satisfactory. Let us now support the disc and shaft of Fig. 23.8 by a bearing at each end and cause the disk and shaft to spin. A rotating vector force of $Mr\omega^2$ lb results, in phase with the center of gravity of the rotating system, as shown in Fig. 23.9. W is the

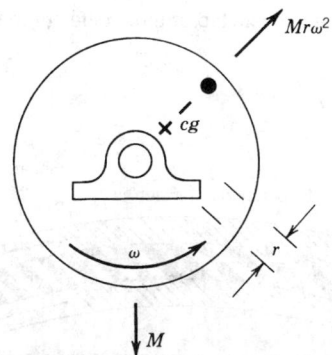

Fig. 23.9   Unbalanced disk in rotation.

total weight, ω is the angular velocity in radians per second, and r is the radial distance (inches or mm) from the shaft center to the center of gravity.

The shaft and bearings must absorb and transmit not only the weight of the rotor but also a new force, one which rotates; one which, at high rotational speeds, may be greater than the weight of the rotor.

Though your automotive mechanic may only statically balance the wheels of your car by adding wheel weights on the light side of each wheel, this static balancing results in noticeable improvement in car ride, in passenger comfort, and in tire wear, because the force $Mr\omega^2$ is greatly reduced.

A numerical example may interest the reader. Imbalance can be measured in ounce-inches (or, in metric units, gram-millimeters). One ounce-inch means that an excess or deficiency of weight of one ounce exists at a radius of 1 in. How big is an ounce-inch? It sounds quite small, but at high rotational speeds (since force is proportional to the square of rotational speed) this "small" imbalance can cause very high forces. You will recall that centrifugal force may be calculated by

$$F = \frac{Mv^2}{r}$$

where M is the mass in kilograms (or weight in pounds divided by 386 in./sec², the acceleration due to the earth's gravity); r is the radius in inches; and v is the tangential velocity in inches per second. We can calculate $v = 2\pi fr$, where f is the frequency of rotation in hertz and r is the radius in inches or mm. Then

$$F = \frac{M}{r}(2\pi fr)^2 = 4\pi^2 Mf^2 r = \frac{4\pi^2}{386} Wrf^2 = 0.1023 Wrf^2$$

Let us calculate the force that results from 1 oz-in. of imbalance on a member rotating at 8000 rpm or 133 rps:

$$1 \text{ oz} = \frac{1}{16} \text{ lb}$$

$$r = 1 \text{ in.}$$

Then

$$F = 0.1023(\tfrac{1}{16})(1)(133)^2 = 114 \text{ lb}$$

If this centrifugal force of 114 lb occurred in an electrical motor, for instance, whose weight was less than 114 lb, the imbalance force acting through the bearings would lift the motor off its supposed once each revolution, or 8000 times a minute. If the motor were fastened to some framework, vibratory force would be apparent.

In most rotating elements, such as motor armatures or engine crankshafts, the mass of the rotor is distributed along the shaft rather than being concentrated in a disk as shown in Figs. 23.8 and 23.9. If we test such a rotor as we tested in Fig. 23.8, we may find that we have static balance, then the rotating element has no preferred angular position, and that the center of gravity coincides with the shaft center. But when we spin such a unit, we may find severe forces being transmitted by shaft and bearings. Obviously we are not truly balanced; since this new imbalance is apparent only when the system is rotated, we call it *dynamic imbalance*.

As a simplified example of such a system, consider Fig. 23.10. If the two imbalances P and Q are exactly equal, if they are exactly 180° apart, and if the two disks are otherwise uniform and identical, this system will be statically balanced. But if we rotate the shaft, each disk will have rotating centrifugal force similar to Fig. 23.9. These two forces are out-of-phase with each other. The result is dynamic imbalance forces in our simple two-disk system; they must be countered by two rotating forces rather than by one rotating force as before, in static balancing.

If we again consider one wheel of our automobile, having spent our money for static balancing only, we may still have unbalanced forces at high speeds. We may find it necessary to both statically and dynamically balance the wheel to reduce the forces to zero. Few automotive mechanics, doing this work every day, are aware of this. You will find it quite difficult to find repair shops that both statically and dynamically balance the wheels of your automobile, but the results, in increased comfort and tire wear, often repay one for the effort and expense.

Imagine that we have a perfectly homogeneous and balanced rotor. Now we add a weight on one side of the midpoint. If we now spin this rotor, but do not rigidly restrain its movement, the motion will resemble the left sketch in Fig. 23.11; the centerline of the shaft will trace out a cylinder. On the other hand, suppose that we had added two equal weights on opposite sides, equidistant from the center, so that we have static balance. If we spin the rotor, its centerline will trace out two cones,

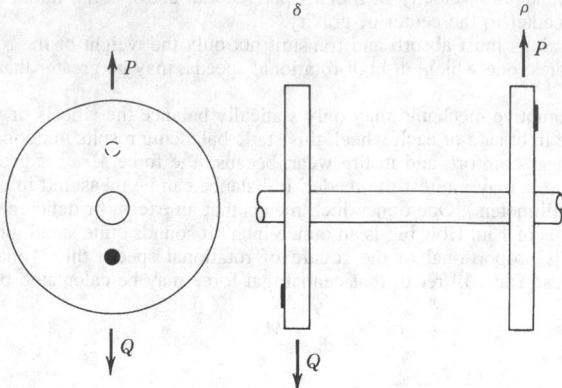

Fig. 23.10 Schematic of unbalanced shaft.

as shown in the right sketch; the apex of each cone will be at the center of gravity of the rotor. In practical unbalanced rotors, the motion will be some complex combination of these two movements.

Imbalance in machinery rotors can come from a number of sources. One is lack of symmetry; the configuration of the rotor may not be symmetrical in design, or a core may have shifted in casting, or a rough cast or forged area may not be machined. Another source is lack of homogeneity in the material due, perhaps, to blowholes in a casting or to some other variation in density. The rotor (a fan blade, for example) may distort at operating rpm. The bearings may not be aligned properly.

Generally, manufacturing processes are the major source of imbalance; this includes manufacturing tolerances and processes that permit any unmachined portions, any eccentricity or lack of squareness with the shaft, or any tolerances that permit parts of the rotor to shift during assembly. When possible, rotors should be designed for inherent balance. If operating speeds are low, balancing may not be necessary; today's trends are all toward higher rpm and toward lighter-weight assemblies; balancing is more required than it was formerly.

Figure 23.12 shows two unbalanced disks on a shaft; they represent the general case of any rotor, but the explanation is simpler if the weight is concentrated into two disks. The shaft is supported by two bearings, a distance l apart. At a given rotational speed, one disk generates a centrifugal force P, while the other disk generates a centrifugal force Q. In the plane of bearing 1, forces P and Q may be resolved into two forces by setting the sum of the moments about plane 2 equal to zero; the force diagram is shown in Fig. 23.12. Similarly, in the plane of bearing 2, forces P and Q may be resolved into two forces by setting the sum of the moments about plane 1 equal to zero; this force diagram is also shown in Fig. 23.12. Both force diagrams represent rotating vectors with angular velocity ω radians per second. Force P has been replaced by two forces, one at each bearing plane:

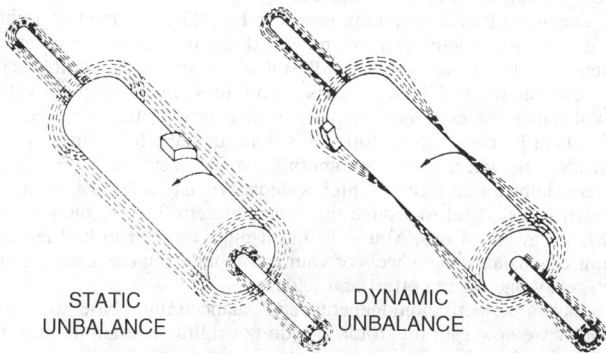

STATIC
UNBALANCE

DYNAMIC
UNBALANCE

Fig. 23.11 Imbalance in a rotor; the rotor on the left has a weight on one side at the midpoint; the rotor on the right has two equal weights on opposite sides, equidistant from the center.

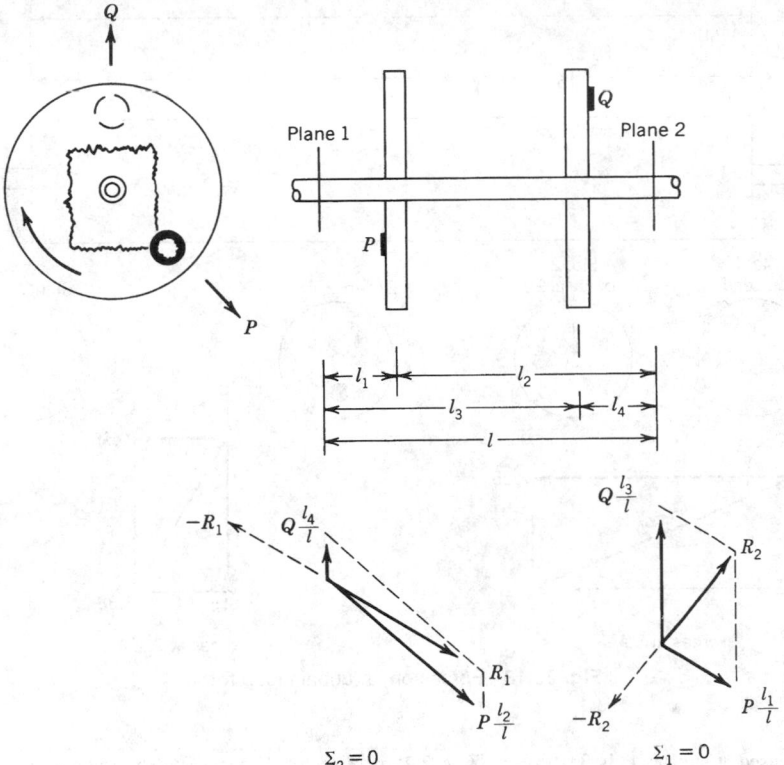

Fig. 23.12 Representation of the general case of an unbalanced rotor.

$P(l_1/l)$ and $P(l_2/l)$. Similarly, force Q may be replaced by two forces, one at each bearing plane: $Q(l_3/l)$ and $Q(l_4/l)$.

We may combine the two forces at bearing plane 1 into one resultant force R_1. We may combine the two forces at bearing plane 2 into one resultant force R_2. If we can somehow apply rotating counterforces at the bearings, namely, $-R_1$ and $-R_2$, we will achieve complete static and dynamic balance. No force will be transmitted from our rotor to its bearings.

In practice, of course, we do not actually use forces at the bearings. We add balancing weights at two locations along the shaft and at the proper angles around the shaft. These weights each generate a centrifugal force, which is also proportional to the square of shaft speed. Or we may remove weight. Either way, we must achieve forces $-R_1$ and $-R_2$. The amount of each weight added (or subtracted) depends on where along the shaft we can conveniently perform the physical operation.

When we have our automobile wheels balanced, to use this simple example for the last time, only the weight and angular position of the counterweights may be selected. The weights are always attached to either the inner or outer rim of the wheel. In early automobile wheels, with their large diameter/thickness ratio, static balancing was usually sufficient. Modern automobiles have smaller, thicker wheels (turning at higher speeds) and they approach Fig. 23.11. Dynamic balancing is definitely better.

In the case of a rotating armature we may subtract weight at a convenient point along the length of the armature (usually at the end) and at the proper angular position by simply drilling out a bit of the armature material. This operation is repeated at the other end. Then the armature is both statically and dynamically balanced.

As an example, consider the rotor shown in Fig. 23.13. This rotor has 3 oz-in. of imbalance at station 2, located 3 in. from the left end, and at an angular position of 90° from an arbitrary reference. Another imbalance of 2 oz-in. exists at station 3, located 5 in. from the right end, and at an angle of 180° from the same reference. We want to statically and dynamically balance the rotor, by means of corrections at the two ends, at stations 1 and 4.

Let us now draw a vector diagram of the forces at station 1, as shown in Fig. 23.13. First we take a summation of forces about station 4, just as we did in Fig. 23.12. The imbalance at station 2,

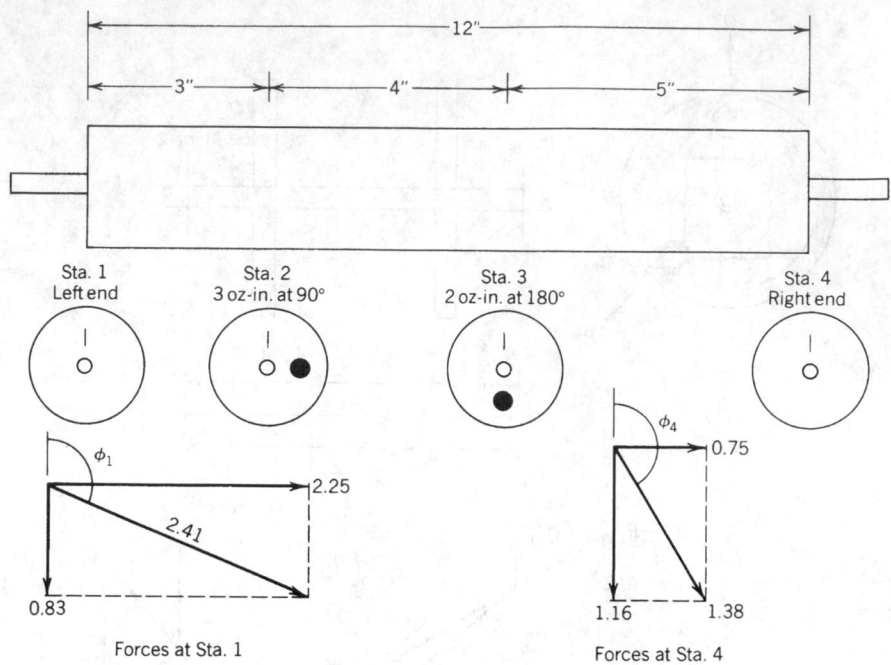

Fig. 23.13 Forces on an unbalanced rotor.

when sensed at station 1, is $3 \times \frac{9}{12} = \frac{9}{4} = 2.25$ or a vector 2.25 at 90°. The imbalance at station 3, when sensed at station 1, is $2 \times \frac{5}{12} = \frac{5}{6} = 0.83$ or a vector 0.83 at 180°. By taking the square root of 2.25^2 plus 0.83^2, we find that we must remove 2.41 oz-in. Now at what angle shall we remove weight?

$$\phi_1 = 90° + \tan^{-1} \frac{0.83}{2.25}$$

$$= 90° + 20.3° = 110.3°$$

So we see that we must remove 2.41 oz-in. at an angle of 110.3° from our reference. We may do this by removing 2.41 oz at a radius of 1 in., or by removing 1 oz at a radius of 2.41 in. or any convenient combination of weight and radius.

Let us now draw a vector diagram of forces at station 4, summing forces about station 1. The imbalance at station 2, when sensed at station 4, is $3 \times \frac{3}{12} = \frac{3}{4} = 0.75$ or a vector 0.75 at 90°. The imbalance at station 3, when sensed at station 4, is $2 \times \frac{7}{12} = \frac{7}{6} = 1.16$ or a vector of 1.16 at 180°. By taking the square root of 0.75^2 plus 1.16^2, we get a resultant of 1.38 oz-in. Now to get the proper angle:

$$\phi_4 = 180° - \tan^{-1} \frac{0.75}{1.16}$$

$$= 180° - 32.8° = 147.2°$$

So we see that we must remove 1.38 oz-in. at an angle of 147.2° from our reference. We may again select any convenient combination of weight and radius along this 147.2° radius line.

The reader must be cautioned that elastic bodies cannot be balanced in this manner; imbalance must be removed in the plane where it exists, rather than in some plane that happens to be convenient.

We have reviewed the meanings of several terms and shown how balancing may be accomplished. Now we will discuss the test equipment used in learning the amount and location of imbalance.

The simplest balancing machines make use of gravity. A rotor, on its shaft, may rest on hard knife-edges, which are straight and level. The heavy side of the disk will seek out its lowest level and thus automatically indicate the angle at which a balancing weight must be added. Various weights are added until the disk has no preferred position. Or the disk may be horizontal with its center pivoted on a point, or attached to a string. The heavy side will seek its lowest level, thus indicating the angle at which a correcting weight should be added. Weight is added until the disk is level. With either of these gravity machines, which accomplish static balancing only (the disk is stationary, not rotating), we recognize that balancing could have been accomplished by removing material from the heavy side, rather than by adding material to the light side.

Static balancing is occasionally helpful as a first step in dynamic balancing. A rotor may be so badly unbalanced that it cannot be brought up to speed in a dynamic balancing machine. It must be given a preliminary static balance.

There are many variations on the basic idea of centrifugal (dynamic) balancing machines. We will not attempt to describe any one unit; rather we will mention rather briefly some of the variations that a reader may encounter.

Bearings may be supported in a soft, flexible manner, to approximate the unrestricted rotor of Fig. 23.11. Bearings are supported by thin rods or wires which are soft in a single plane; the resulting motion is primarily in that plane. Motion of the flexible bearing supports is measured as an indication of imbalance force. The earliest machines measured motion mechanically. Subsequent machines measured motion electrically, which pickups that sensed either displacement or velocity. Newer machines employ force-sensing transducers to generate a signal for the electronic system. A readout informs the operator how much imbalance exists and where, and usually tells this in terms of how much material must be removed (or added) and where. Some machines require that the work-piece be removed to, say, a drill press. Others combine the balancing operation with drilling, welding, soldering, etc., so the part can be balanced and rechecked without removing it from the machine. On some machines the entire operation is automatic.

The reader may be interested to learn that transducers and meters of some balancing machines are capable of measuring peak-to-peak displacements of about 0.08 μin. This extreme sensitivity makes it possible to detect very small imbalances.

Table 23.1 indicates the balancing accuracy of some common commercial items. Gryscopes are held to even closer tolerances, on the order of 0.000 000 25 oz-in., since even slight vibration causes long-time drifting.

In-place dynamic balancing of rotating machinery is often needed when it is impossible or inconvenient to remove a rotor for balancing. A stroboscopic light is synchronized to the rpm of the rotor to be balanced; this is usually done with a vibration pickup on the frame of the machine. It may be necessary to "tune out" interfering vibrations from other machines by means of a filter between pickup and strobe light; when this is accomplished, the offending rotor will appear to be "frozen" and not moving. A reference mark on the rotor and the angle at which the mark appears indicate the angle of imbalance. Readout from the vibration pickup is proportional to the amount of imbalance.

The rotor is stopped, and a compensating weight is attached at some location. The rotor is again spun at operating rpm and any improvement (or degradation) is noted in the vibration level, along with the new reference angle. Different weights and radial and angular locations are tried until satisfactory balance is achieved. The "cut-and-try" process is greatly speeded by drawing vector diagrams representing imbalance forces or from use of a specially programmed calculator. Skill in this work comes from wide experience; a wide range of problems arises, as compared to production-line balancing work.

23.3 VIBRATION MEASUREMENT

Let us first discuss the measurement of sinusoidal vibratory *displacement,* normally the peak-to-peak displacement or double-amplitude D measured in inches or in millimeters. If a structure is steadily

Table 23.1 Selected Balancing Accuracies

Part	Accuracy of Balance (oz-in.)
Vacuum sweeper armature	0.00005–0.0001
Automobile crankshaft	0.0001–0.0002
Automobile flywheel	0.0002–0.0005
Automobile clutch	0.001–0.002
Aircraft crankshaft	0.0001–0.0004
Aircraft supercharger	0.00055–0.0001
Electric motor armatures, 1 hp	
1800 rpm	0.0001–0.0002
3600 rpm	0.00005–0.0001
Small instrument motor armatures	0.000005–0.00001

vibrating with a large enough D, we can estimate motion by merely holding a ruler alongside. The reader is requested to mentally estimate the smallest D which would be read with $\pm 10\%$ accuracy, $\frac{1}{2}$ or $\frac{1}{4}$ in.? 10 or 5 mm? If we could hold the ruler steady, if the vibration remained constant, and if we took several readings and averaged them, we might get an accuracy of $\pm 10\%$ at $\frac{1}{4}$ in. or 10 mm D.

Optical techniques for measuring vibratory displacement D are limited in accuracy, especially at the small displacements we find in vibration measurement and testing. Let us calculate D under the test conditions of $10g$ (peak) acceleration at a frequency of 1000 Hz by using the following relationships:

English unit example:

$$A = 0.0511f^2D$$

or

$$D = \frac{A}{0.0511f^2}$$
$$= \frac{10}{.0511(1000)(1000)}$$
$$= 0.0001957 \text{ in.}$$

International System example

$$A = 0.002\ 02f^2D$$

or

$$D = \frac{A}{0.002\ 02f^2}$$
$$= \frac{10}{0.002\ 02(1000)(1000)}$$
$$= 0.00495 \text{ mm}$$

This is about 200 μin., an example of the extremely small displacements we find in much vibration work. At higher frequencies, D is still smaller; for example, at $10g$ (peak), 2000 Hz, D would be about 50 μin. or 1.3 μm.

A very popular optical technique uses the "optical wedge" device, shown in Fig. 23.14a, which is cemented onto a structure that will be vibrating. This device depends on our eyes' persistence of vision; it works best above 15 Hz. Velocity is zero at the extremes of position. There we get a crisp image. In between, where the pattern is moving, we let a lighter, gray image. The result *appears* to be two images, overlapping each other as in Fig. 23.14b. The ends of the images will be separated by exactly D. D is read by noting where the two crisp images intersect, then by looking directly below this intersection on the ruled scale. In Fig. 23.17b D is 0.50 in. With great care, accuracies of $\pm 10\%$ or even $\pm 5\%$ are possible, particularly at larger amplitudes. This technique is most useful when an "unknown" D is to be measured.

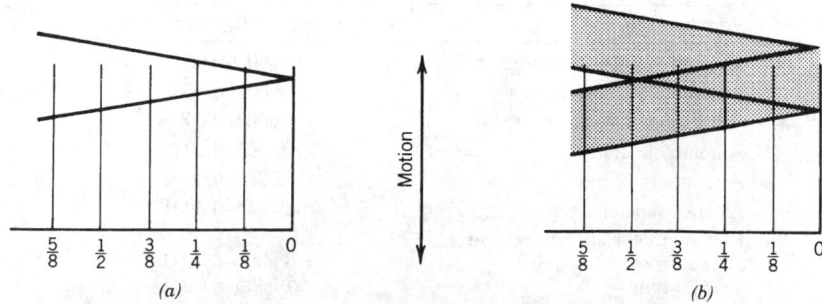

Fig. 23.14 Vibration measurement by "optical wedge" device.

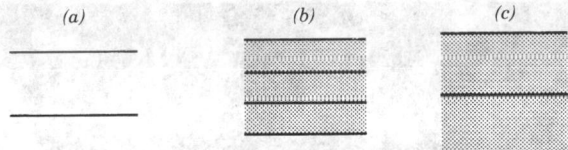

Fig. 23.15 Vibration calibration.

When one wants an optical indication that D has reached a particular value, the scheme of Fig. 23.15 is useful. Two parallel lines, perpendicular to the direction of motion, are drawn ½ in. apart. When the target is stationary, the lines appear as in Fig. 23.15a. To adjust the D of a shaker to, say, ¼ in., increase shaker power until two gray bands separated by a white band, all ¼ in. high, appear as in Fig. 23.15b. For a D of ½ in., increase the power to the shaker until the two gray bands just merge, as in Fig. 23.15c.

Suppose you wish to calibrate an accelerometer/cable/amplifier/meter system at 10g (peak) and 20 Hz. Carefully rule lines 0.49 in. apart, then adjust D until the pattern resembles Fig. 23.15c. Some laboratories use lines scribed through dull paint on the surface of a metal block bolted to the table of a shaker. Lines are scribed with spacings of 0.100, 0.200, 0.036, 0.49, etc., in. These D's, at known frequencies, give useful levels of velocity and acceleration to check velocity pickups and accelerometers. Use a magnifying glass or a low-power microscope to determine when areas merge.

Accuracy statements on these techniques are not conclusive. Factors include accuracy of original construction of the device used, accuracy and judgment of the user, the D being read (greatest accuracy at large D's), etc. Only with all factors aiding the operator do optical pattern techniques give ±5% accuracy.

For measuring D's smaller than 0.1 in., use optical magnification. Use a 40× or 50× microscope with internal "cross hairs" and with cross-feeds that permit moving the entire microscope body. A stroboscopic light is helpful. Focus the microscope on some well-defined point or edge on the vibrating structure (no motion); with the structure vibrating, move the microscope body until the cross hairs line up with the extreme excursion of the point or edge being observed. Read a dial wheel on the cross-feed. Move the microscope until the cross hairs line up with the other extreme position, and take another reading. D equals the distance the microscope body was moved—the difference in readings. In practice, this technique is difficult to use: the "neutral" position of the vibrating structure may shift and lead to large errors. "Filar" elements inside the microscope are helpful.

A better microscope technique, generally used with about a 40× or 50× instrument, requires a calibrated reticle or eyepiece, having regularly spaced lines, usually 0.001 in. apart. Before the structure commences to vibrate, focus the microscope on a bit of Scotchlite tape, fine garnet, or emery paper. You will see the grains of the target; with a strong light from the side, each grain will reflect a bright point of light, as in Fig. 23.16a. When the structure vibrates, each point of light "stretches" into a line as in Fig. 23.16b. The length of any line is estimated by comparing it with the rulings. In Fig. 23.16b, the rulings are 0.001 in. apart, and D is 0.005 in.

Optical (and mechanical) magnification is also used in largely obsolete hand-held instruments typified by Fig. 23.17. A probe maintains contact with a vibrating surface. Motion resulted in a bright pattern appearing on a scale: the pattern width represents D. A similar unit permanently recorded waveforms of instantaneous displacement versus time on a paper strip. As paper speed was known,

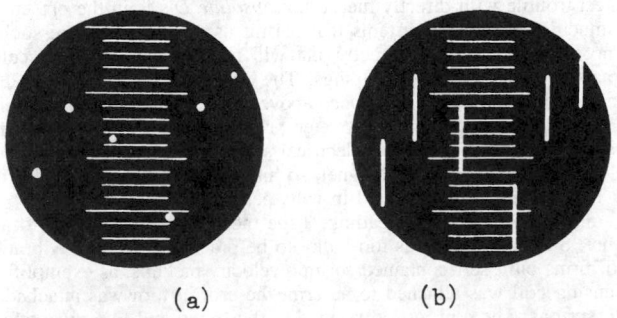

Fig. 23.16 Vibration measurement with a microscope.

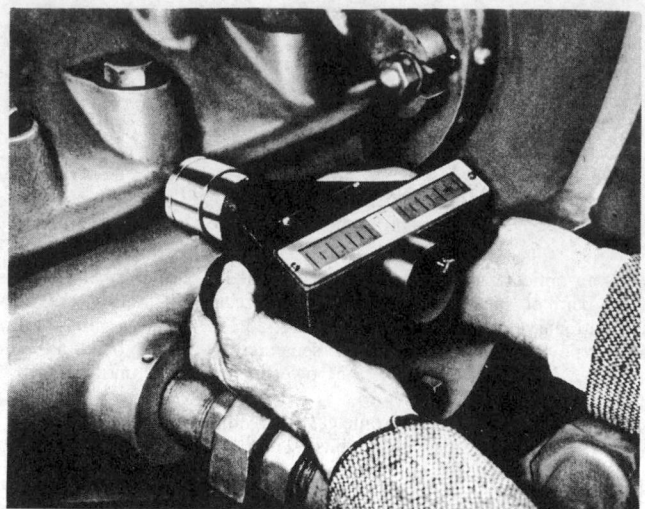

Fig. 23.17 Hand-held vibration measurement instrument.

frequency could be calculated. These instruments were satisfactory when a vibrating machine was so heavy that its motion was not affected by the mass of the probe. The instrument case could not move. Thus the operator stood on a nonvibrating support and held the instrument steady against forces transmitted through it. These limitations were never fully met; consequently, these instruments were seldom used for measuring the tiny displacements found in most vibration situations, originating in faulty bearings, gears, etc. They were mainly used in measuring the result of rotational imbalance upon heavy machines such as engines, pumps, etc.

There are a number of displacement-sensing pickups that send an electrical signal to remote amplifiers, meters, oscilloscopes, oscillographs, tape recorders, analyzers, etc. Several different physical methods can convert changes in position into electrical signals: these include linkages from vibrating structures to the sliders of stationary variable resistances; also noncontacting variations in capacitance, inductance, eddy currents, etc., which are caused by changes in distance. Note that some portion of each instrument must be held stationary, if one is to gain accurate information about the *absolute* motion of a vibrating structure. This is very difficult to do. The "background noise" of vibration in many industrial buildings is 0.005 in. D or more; this will certainly prevent accurate measurement of, say, 0.001 in. D. And many times we want to read D's of only a few microinches. The frame vibration of a gas turbine is normally very small (unless resonances in the support structure are excited). However, there may be large, low-frequency motion due to other machines nearby; this may "mask" the signal of interest.

On the other hand, displacement sensors read *relative* displacements well. One type is inserted into a drilled, threaded hole in the frame of a machine. Its sensing end is close to a rotating shaft, either perpendicular to the shaft axis for measuring eccentric shaft motion or parallel to the shaft for measuring axial shaft motion. If motion is not perfect, a signal proportional to relative motion between shaft and housing will result.

Since the greatest trouble with directly measuring *absolute D*'s is in the presence of background vibration, we are much interested in systems that permit us to largely ignore such background vibration. We will not try to measure D directly, but will measure velocity or acceleration and then convert those signals into displacement readings. The velocity pickup of Fig. 23.18, with minor modifications, eliminated the problems mentioned above. A coil was attached to a vibrating structure; leads were brought out to an electronic voltmeter or a recorder. A permanent magnet was held (without any motion!) close to the coil; the electrical signal generated by the coil was proportional to the relative velocity between coil and magnet, so any motion of the magnet created an error in the measurement. This system was calibrated in volts per unit of velocity.

If the magnet could be held still, the reading of the meter was proportional to actual velocity of the moving structure. Such a pickup was too bulky to be practical for much vibration measurement work. A number of firms built self-contained seismic velocity pickups, as exemplified by Fig. 23.19. In this unit, the sensing coil was attached to an arm; the arm in turn was attached to the frame by pivots and delicate springs. The coil moved relative to the frame and to a magnetic field which was supplied by Alnico permanent magnets. The entire pickup was attached to the structure being inves-

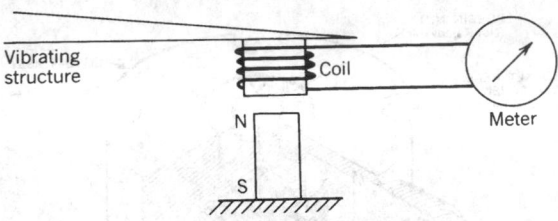

Fig. 23.18 A velocity pickup.

tigated. At the pickup's f_N of about 5 Hz, the arm and coil moved more than did the pickup frame. This unit was only used at frequencies above 10 Hz where the arm and coil were "isolated" from vibration; thus they remained stationary (seismic support). The magnetic field swept up-and-down across the coil; this generated a voltage proportional to the velocity of the pickup.

Functions of the coil and magnet may be reversed. The permanent magnet of Fig. 23.20 was seismically suspended. When the case vibration was well above resonance (operation of this unit starts at 45 Hz), the magnet remained stationary. The coil, attached to the frame, swept through the magnetic field so that a voltage was induced. The frequency range is 45–1500 Hz. Stroke was limited; the maximum D was 0.1 in. Some velocity pickups were built for lower frequencies and longer strokes. Accelerations greater than $50g$ could cause damage.

The velocity pickup shown in Fig. 23.21 was similar to the unit of Fig. 23.19. However, a length of 0.030 in. drill rod, connected to the sensing coil, passed out through the housing. This unit was normally held by hand, with the "probe" firmly touching a vibrating surface. Motion was transmitted to the coil, in which a voltage proportional to velocity was generated. This unit was very useful for hand-scanning over a vibrating surface, to locate points of maximum and minimum motion.

Velocity pickups had many advantages: they were self-contained devices requiring no external source of power—no dc or ac excitation. Because of their low internal impedance, they could be used at great distances from the readout instrument. They were most used in the frequency range 10–500 Hz. A typical sensitivity was 100 millivolts (peak) per inch per second (peak) velocity.

Velocity pickups had certain disadvantages, of course. They were generally quite large and heavy, especially as compared with some accelerometers. They could not be used at very large displacements, because of limited stroke of their moving parts. Since they will sometimes be used close to their natural frequencies, some form of damping (usually oil or eddy current) was needed; this introduced certain problems in measuring any nonsinusoidal motion.

Readout from velocity pickups was very simple, as in Fig. 23.18. The pickup sensitivity was known in terms of millivolts per unit of velocity. An electronic voltmeter reading in terms of millivolts could be interpreted in terms of velocity.

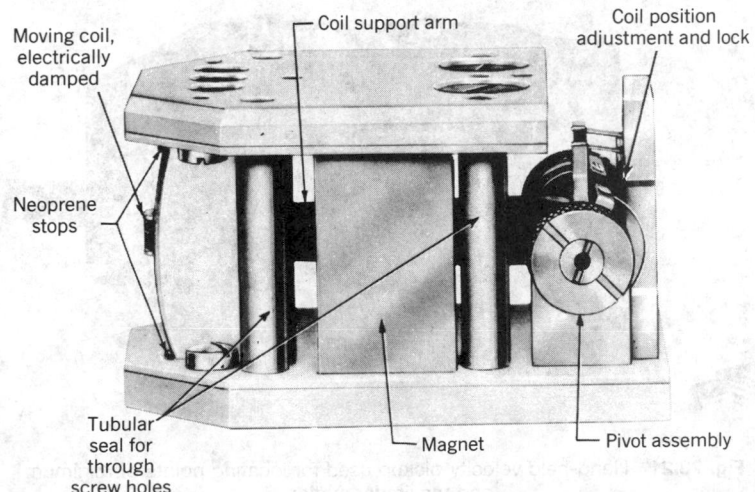

Fig. 23.19 A self-contained seismic velocity pickup. (Courtesy of Vibra-Metrics Inc.)

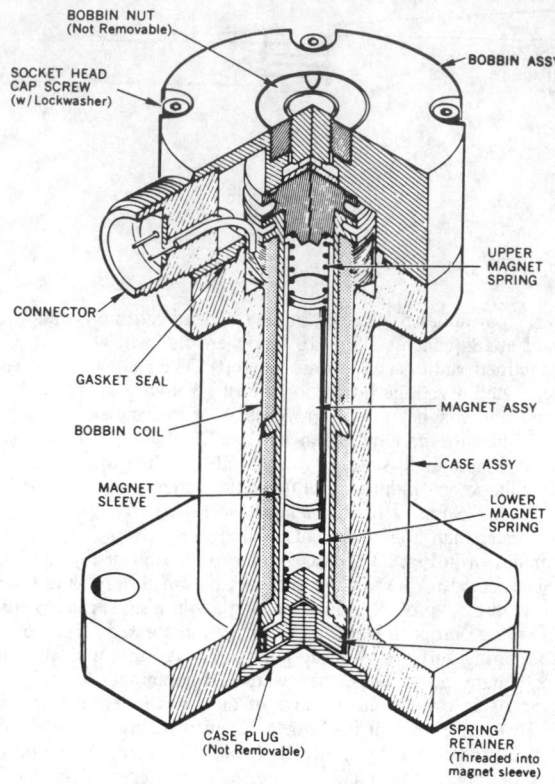

Fig. 23.20 Velocity pickup with permanent magnet suspended seismically. (Courtesy of CEC.)

Fig. 23.21 Hand-held velocity pickup used for locating points of minimum and maximum motion.

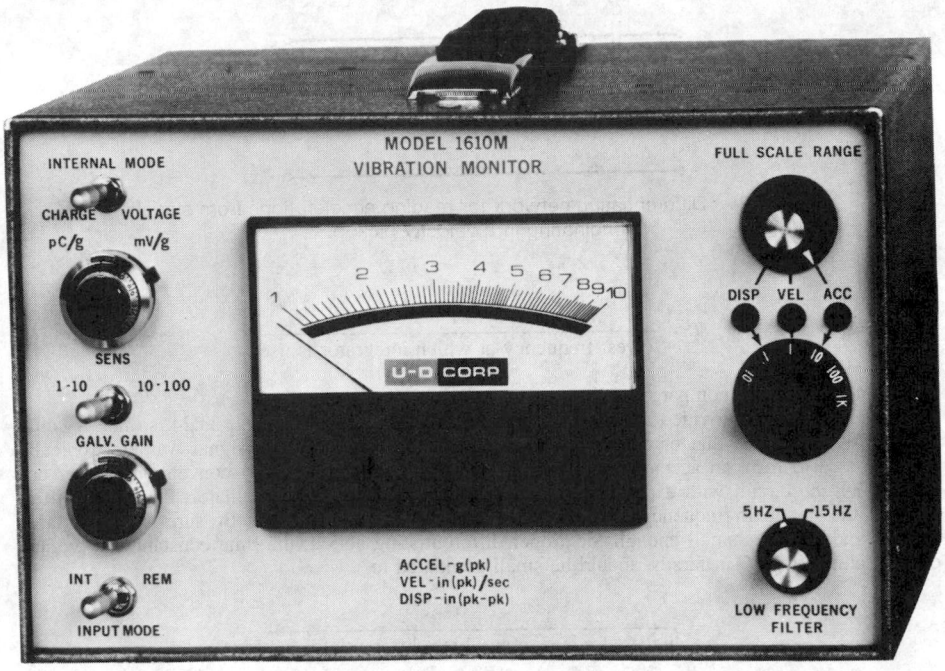

Fig. 23.22 A specialized vibration meter.

If you knew the velocity and the frequency of *sinusoidal* vibration, but wished to know the displacement or acceleration, you could calculate these. If you did not need extreme accuracy, you could use one of the cardboard vibration calculators that are available from some shaker and accelerometer manufacturers. One of the advantages of velocity pickups when used with a common electronic voltmeter: values of velocity, displacement, and/or acceleration were easily determined. This advantage was very important around 1945–1955; funds for purchase of more sophisticated vibration instruments were not readily available.

Many measurements that formerly were made with coil-and-magnet velocity pickups now employ an accelerometer packaged with an integrating network. Power is of course required.

Specialized vibration meters such as the unit in Fig. 23.22 were soon developed. When one wished to read D on a meter whose input signal comes from a velocity pickup, the meter had to contain an integrating network similar to Fig. 23.23. The velocity signal was electronically integrated to form a displacement signal, which was then measured by conventional meter circuitry. Test the circuit with a constant-voltage, variable-frequency test oscillator; the circuit integrates the signal. If frequency doubles, the output voltage should drop to ½, etc. Be sure that R is large enough and that X_c is small enough for proper integrating action; the "time constant," the product of R in ohms times C in farads, should be large compared to the period T, where T is

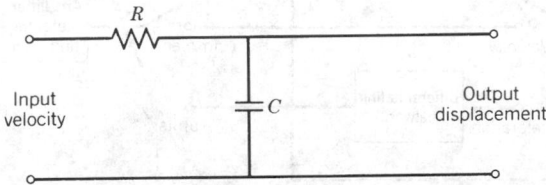

Fig. 23.23 Integrating network for reading displacements when input signals come from a velocity pickup.

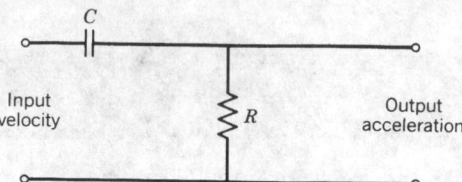

Fig. 23.24 Differentiating network for reading accelerations from signals originating in a velocity pickup.

$$T = \frac{1}{\text{lowest frequency at which integrator is used}}$$

and R should be large compared to X_c.

Most meters also provided a differentiating network, something like Fig. 23.24, so that A could be read from a signal that originated in a velocity pickup. The velocity signal was electronically differentiated to form an acceleration signal, which was then measured by conventional meter circuitry. Test the circuit with a constant-voltage, variable-frequency test oscillator; the circuit differentiated the signal. If frequency doubled, the output voltage doubled, etc. Be sure that X_c is large enough and that R is small enough for proper differentiating action; the time constant, the product of R in ohms times C in farads, should be small compared to

$$T = \frac{1}{\text{highest frequency at which differentiator is used}}$$

and R should be small compared to X_c.

Vibration meters are simply electronic voltmeters with additional features. Most provided for inputs from velocity pickups and from velocity coils built into older model shakers. Most (see Fig. 23.25) had a three-position switch marked Displacement-Velocity-Acceleration. In the Velocity position, the unit was a simple ac voltmeter. The variable-gain "normalizing" control was set for the particular pickup sensitivity being used. When such a meter was used to read displacement, an integrating network was inserted into the signal path, as in Fig. 23.25, by switching to Displacement. For reading acceleration, the signal was switched through a differentiating network, as shown, by switching to Acceleration.

All meters today provide for accelerometer inputs, as in Fig. 23.25. If the incoming signal is twice integrated, it becomes proportional to displacement. Integration and differentiation "waste" much signal; therefore preceding and subsequent stages must have considerable gain. Enough voltage and power must be developed to run the detector, the indicating meter and any recording galvanometer that may be used for a permanent record of vibration.

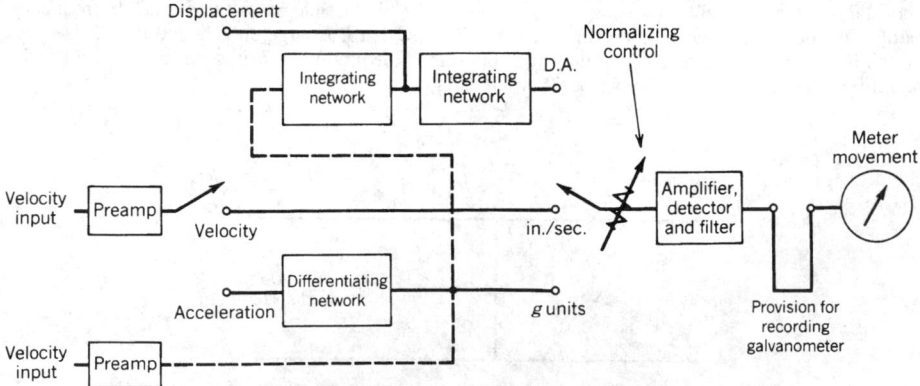

Fig. 23.25 A vibration meter circuit.

Differentiating and integrating networks are not satisfactory over many octaves of frequency. Fortunately, displacement information is usually only needed over a range such as 5–100 Hz, so integrating components are chosen for best accuracy in that region. Acceleration data are usually needed above about 50 Hz; when velocity pickups are used, differentiating components are selected for best operation from 50 to perhaps 2000 Hz. Serious inaccuracies result from use beyond the frequencies indicated in Table 23.2. Table 23.2 gives typical tolerances for the vibration meter only, not including the velocity pickup or accelerometer.

23.4 ACCELERATION MEASUREMENT

Let us now take up the subject of accelerometers—units whose instantaneous output voltage is proportional to the instantaneous value of acceleration. Most vibration and shock measurements are even more true today than in 1986 are made with accelerometers.

At high frequencies, accelerometers generate a larger signal than do velocity pickups. This may be shown by calculating the peak velocity that would exist if the peak acceleration were 1g at a frequency of 2000 Hz:

English units:

$$V = \frac{A}{0.0162f}$$

$$V = 0.031 \text{ in./sec (peak)}$$

SI units:

$$V = \frac{A}{0.000\ 642f}$$

$$V = 0.78 \text{ mm/sec (peak)}$$

Let us assume that a velocity pickup has a sensitivity of 105 mV/in. sec. It will generate about 3.22 mV (peak) or about 2.28 mV (rms). (At 200 Hz, 1g, the velocity would be 10 times greater, of course, and so would be the voltage.) A signal of only 2 or 3 mV rms is difficult to measure under some conditions. We will show that accelerometers built for these higher frequencies have a number of advantages; one advantage is higher output signal at higher test frequencies found in modern vibration tests and field measurements.

A reasonable sensitivity figure for a crystal accelerometer is 10 mV (peak) per g (peak); this sensitivity is relatively constant up to, say, 10,000 Hz. In the situation discussed above (1g at 2000 Hz) the accelerometer would generate 10 mV (peak) or 7.07 mV rms. This is about three times as much signal as the velocity pickup generates. This advantage would double with each doubling of frequency (octave).

Accelerometers always operate *below* their natural frequencies. Thus the natural frequency must be high; the useful "flat" range is to about one-fifth of the natural frequency, where the sensitivity is about 4% higher than it is at low frequencies. A 50-kHz unit thus permits operation to 10 kHz.

How might we build an instrument that would sense acceleration? The automobile enthusiast senses acceleration by "feeling" his or her body sink into the seat cushions. As the car and seat gain a high velocity (accelerate), a deflection proportional to the intensity of acceleration is noted. His or her body has inertia and tends to keep its former velocity. If we could measure the amount of deflection, we would have a crude accelerometer. In Fig. 23.27 we see a crude accelerometer and how acceleration could be measured on a meter. A cantilever beam supports a weight W. On the top and bottom are cemented strain gages, elements that change resistance when stretched or compressed. When the structure being observed accelerates upward, the weight W, having inertia, tends to stay behind; the top side of the beam stretches while the bottom side shortens.

Table 23.2 Vibration Meter Data

Function	Range	Frequency Limits	Tolerance
Velocity	0.01–100 in./sec	5–500 Hz	±2%
Displacement	0.001–10 in. D	5–5 000 Hz	±2%
Acceleration	0.1–1 000g	10–5 000 Hz	±1%
		5–10 000 Hz	±2%
		2–25 000 Hz	±5%

This results in R_1 increasing while R_2 decreases. The electrical signal to the meter (which had been zero before, when $R_1 = R_2$) becomes negative. The larger the acceleration, the larger the voltage and deflection of the meter pointer. On a downward acceleration, all these actions reverse.

For very low frequencies, we can use a dc meter, as shown in Fig. 23.27. The pointer could be adjusted to the center of the scale for zero acceleration; it will swing left for upward acceleration and right for downward acceleration. If we gradually increase the vibration frequency, the meter will not "follow" but will just quiver about the zero position. We will have to switch to an ac meter or use an oscilloscope. The ac meter will probably display the rms value of the alternating voltage input. Its reading will remain constant as long as the acceleration peak or rms value of each cycle of vibration remains the same. The oscilloscope will show a sinusoidal pattern, with peaks representing peak acceleration.

Let us assume that our crude accelerometer has a natural frequency of 20 Hz; this is 1.0 on the frequency scale of Fig. 23.26. As we slowly increase the frequency of vibration, holding this peak acceleration constant, we will observe higher and higher readings as we approach 20 Hz, as indicated by Fig. 23.26. Thus we can use our crude accelerometer only to 4 or 5 Hz; at higher frequencies it gives false readings. We could put some kind of dash pot or other damper on the beam. This would prevent so much buildup and would permit testing to perhaps 10 or 15 Hz. However, this would introduce phase-shift problems if we tried to measure nonsinusoidal waveforms or shock. We could use a stiffer beam (or a lighter weight) in order to get a higher natural frequency. While this would permit testing to higher frequencies, it would reduce the bending and thus the electrical response per g, the sensitivity.

Accelerometers using this principle are used in laboratories and in navigational equipment. But the configuration of Fig. 23.27 would be hard to package. Figure 23.28 shows a practical instrument. A mass is free to slide in the direction shown; it is restrained by guides against other motions. The mass is "sprung" from the case by four wire springs, which are also electrical strain gages; these change resistance when deformed. If the case accelerates to the right (due to acceleration of the structure to which the unit is attached), R_3 and R_4 increase while R_1 and R_2 decrease. Electrical polarity reverses when acceleration reverses. Output voltage is proportional to the intensity of acceleration. In place of a battery or other direct-current supply, as shown in Fig. 23.28, excitation could be an alternating current or "carrier" supply. Variations in resistance would amplitude modulate the carrier; this has some advantages in that electronic amplification is simpler. Strain gage accelerometers generally have low natural frequencies. Sensitivity has to be sacrificed if higher natural frequencies are desired. Typical natural frequencies are 90, 155, and 250 Hz with full-scale ranges of $\pm 1g$, $\pm 2.5g$, and $\pm 5g$, respectively.

One way to get a higher frequency range without great loss of sensitivity is to use piezoresistive units, suggested by Fig. 23.29. Note that the spring rate (and thus the natural frequency) is determined by how much material is cut away from the mass. Semiconductor strain gages (higher gage factor than wire) sense any deflection of the mass that may result from an input acceleration. A typical piezoresistive accelerometer has a natural frequency of 10,000 Hz with a useful frequency range of 0–2000 Hz; its full-scale range is $\pm 250g$, while its sensitivity is about 1 mV/g.

The most sensitive (capable of sensing distant earthquakes in micro-g's) and most precise (used in navigation of spacecraft) accelerometer is the servo or force-rebalance type, suggested by Fig. 23.30. Acceleration of the case results in very slight relative motion between mass and case; this is sensed by an internal position pickup. The resulting signal is amplified, then it passes through a coil

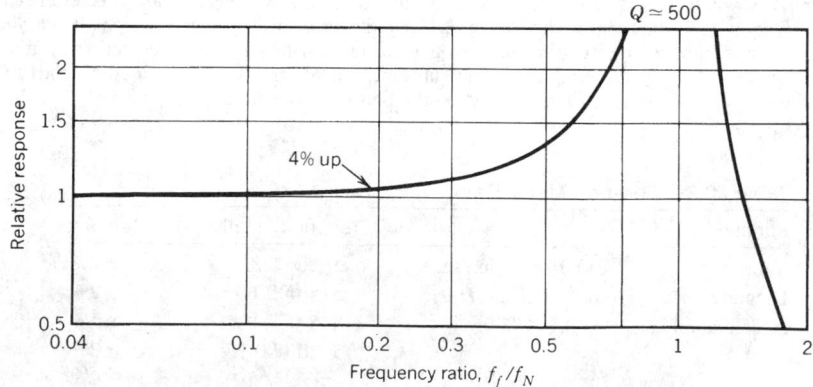

Fig. 23.26 Useful "flat" range for an accelerometer.

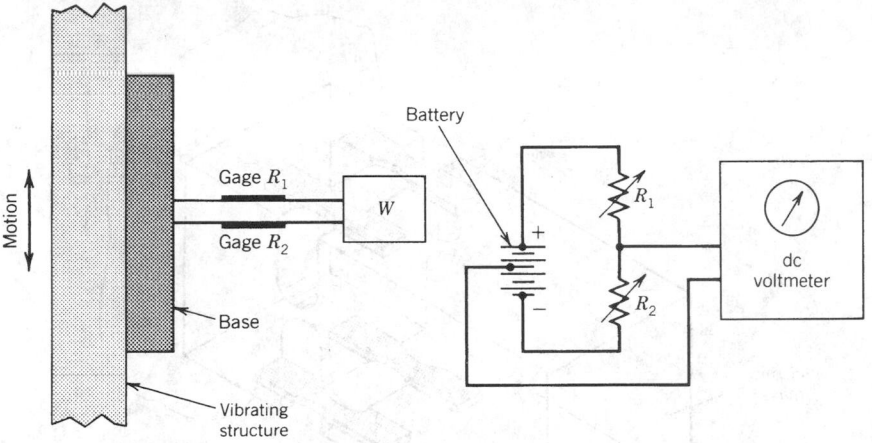

Fig. 23.27 A crude accelerometer.

on the mass, restoring the mass to its original position. The restoring current also acts as the output signal from the accelerometer.

The piezoresistive, servo, and strain-gage types have one important advantage over most piezoelectric crystal types: they can be easily calibrated by merely turning them over; this is a static calibration at $\pm 1g$, which is very useful. (Some piezoelectrics used with some charge amplifiers may also be calibrated in this manner.) Imagine that the accelerometer of Fig. 23.28 were turned onto its left end; then the mass would stretch R_3 and R_4 while it shortens R_1 and R_2, giving a dc output for $1g$. Turning the accelerometer on its other end would reverse the electrical signal. This calibration can only be performed on accelerometers having response down to zero frequency. Piezoresistive and strain-gage types can be used and calibrated on centrifuges where the acceleration is one-directional and constant at $1g$, $10g$, etc. For some vibration measurements, however, it is best to suppress electrical signals due to static accelerations; in missile testing, for example, we may be interested only in vibratory loadings on structures. We can insert a blocking capacitor to suppress zero-frequency signals if they are troublesome.

A different technique of generating a signal proportional to acceleration is shown by Fig. 23.31. This impractical, but possible, device includes a bar made from a natural crystalline material such as a quartz or from a synthetic crystal such as barium titanate. Such a bar will develop an electrical charge Q when it is bent. (This is quite similar to some phonograph pickups.) Upward acceleration of a test structure and the inertia-caused lagging of the mass will cause such bending and charge Q. Downward acceleration will cause reversed polarity. We can electrically measure either charge Q or voltage E. If the measuring system permits, static acceleration will give zero-frequency charge and voltage, but usually there is some low-frequency limitation on the system response.

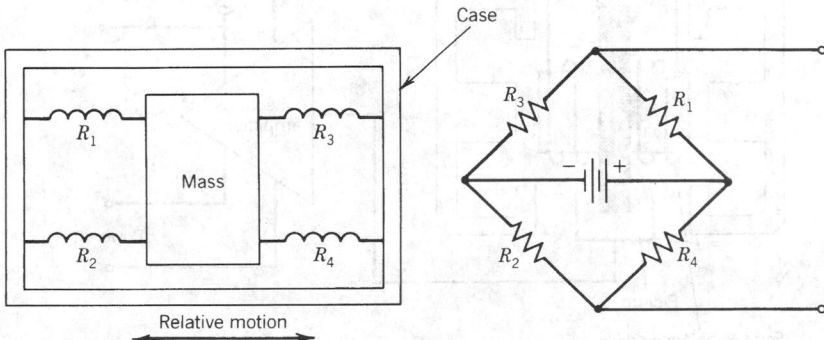

Fig. 23.28 A practical accelerometer.

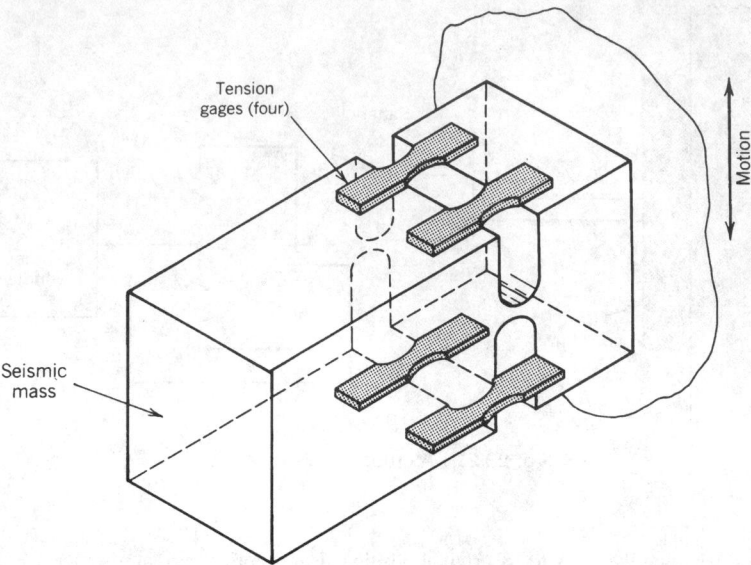

Fig. 23.29 A peizoresistive accelerometer.

The design of Fig. 23.31 would be difficult to package, whereas the basic designs of Fig. 23.32 can give high natural frequencies, have wide dynamic range (fraction of $1g$ to several thousand g's), are small and rugged, and are quite simple to mount mechanically. All crystal accelerometers are self-generating, in comparison with the strain-gage and piezoresistive types which must be supplied with power. The two units at the left, upper row, Fig. 23.32 are no longer in production by any manufacturer; they were too sensitive to method of mounting (attachment torque, for example) and to acoustic excitation (behave something like microphones). The compression designs may show an effect called "base strain sensitivity." If located on a structure that bends, the electrical output may masquerade as vibration.

Severe mechanical shock or very high temperatures can depolarize synthetic crystals, but not natural quartz crystals. "Ground loops" can give electrical signals on the accelerometer leads; these

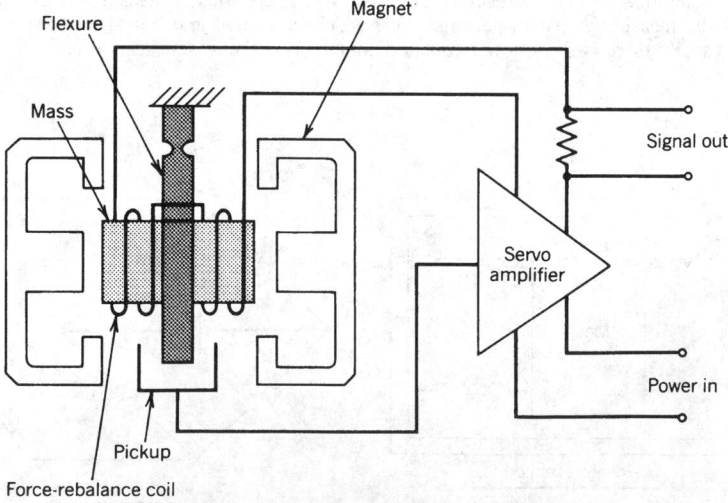

Fig. 23.30 A servo or force-rebalance accelerometer.

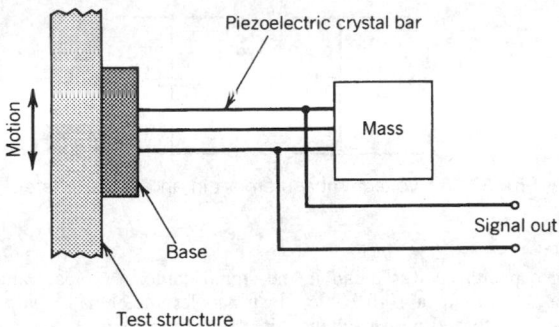

Fig. 23.31   A device for generating a signal proportional to acceleration.

can be avoided by using special designs in which the accelerometer is electrically insulated from its case, or by electrically insulating the case from the vibrating structure. The latter method can change the frequency response of the unit, however, and any insulation should be present at the time of calibration.

The compression types, Fig. 23.32, generate electrical charge when the crystal element is compressed due to upward acceleration or decompressed due to downward acceleration. Similarly, a crystal plate bends in the bender type and a crystal ring or flat sandwich shears in the shear type.

An analogy for readers not inclined toward understanding of electronics: accelerometers behave something like a sponge. When the crystal is squeezed (or bent or sheared), it puts out "juice" (electrons). If we remove the squeeze, the juice returns. If we measure the amount of juice squeezed out (electrical charge), we have a measure of acceleration. Electrically, this requires a *charge amplifier,* a storage-measuring device for electrons.

At one time, voltage was measured. Voltage E was proportional to charge Q divided by shunt capacitance C; see Fig. 23.33. An accelerometer's sensitivity was given in terms of charge (picocoulombs or 10^{-12} C) per g or in terms of voltage per g when used with a specified shunt capacitance or with a specified cable:

$$E = \frac{Q}{C_{total}} = \frac{Q}{C_a + C_c}$$

Part of the shunting capacitance was found in the accelerometer itself; a typical value was 500 pF.

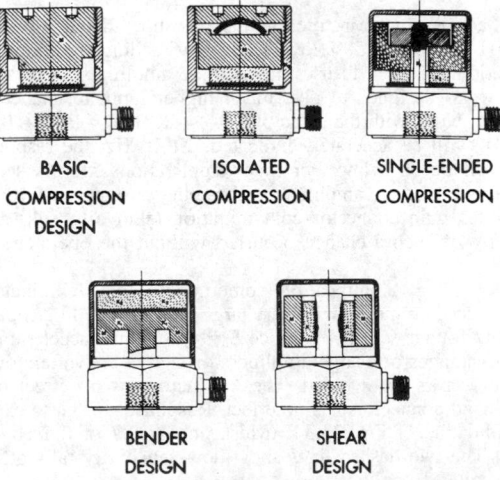

Fig. 23.32   Basic designs for accelerometers.

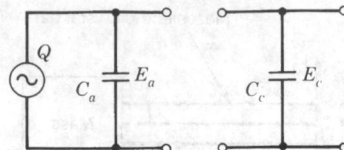

Fig. 23.33   Voltage measurement in an accelerometer.

Part of the shunting capacitance was found in the signal cable; a typical value was 30 pF (30 $\times$ 10^{-12} F) per foot or 300 pF for a 10-ft cable. If an accelerometer had a voltage sensitivity, open circuit, of 10 mV per g with 500 pF capacitance, it would drop to

$$\frac{10 \text{ mV}}{g} \times \frac{500}{500 + 300} = 10(0.625) = 6.24 \text{ mV}/g$$

This was a severe loss of signal for only 10 ft of cable. More important, there was always danger that the user would forget to apply correction factors for the particular cable he or she used. Laboratory calibrations were usually done with a standard value of external capacitance, such as 300 pF. The capacitances of test cables used for environmental or service tests affected the voltage per g; test personnel remembered to correct for the cables used or they would make errors in their measurements. With charge amplifiers (storage devices), the amount of shunting capacitance used is not so important. Sensitivity is essentially the same for any length cable.

We mentioned earlier the difficulty in measuring the dc (static) potential generated when crystal accelerometers receive a static acceleration. The electrical charge Q "leaks off" rather quickly through the approximately 10 mΩ input resistance of a typical meter. If the test structure were vibrating at perhaps 1000 Hz, we could read the output voltage on an ac electronic voltmeter, because successive cycles of vibration would generate new voltage pulsations across the total capacitance before the previous pulsation leaked off.

Here we are describing the effect of input resistance of the first electronic unit in the signal path. The shunting resistance of electronic voltmeters, oscilloscopes, and other instruments affects their readings at low frequencies. The time constant T of the accelerometer-cable-instrument combination is equal to the product of shunting resistance R times the shunting capacitance C: that is, $T = RC$. Typically, R is 10^7 Ω. A typical value of C is 5×10^{-10} F, or 500 pF. In this instance $T = RC = 0.005$ sec. When the vibration is above 1/0.005 sec or 200 Hz, we will get an accurate (error about -1.5% at 200 Hz) indication of our acceleration, since

$$f \approx \frac{1}{RC} \approx 200 \text{ Hz}$$

Graphs published in accelerometer instruction manuals show that at 100 Hz the instrument reading will be 5% low; at 18 Hz, it will be 50% low. These low readings require that data be corrected.

The traditional solution has been to use intermediate amplifiers (sometimes called cathode followers or emitter followers) designed to give much higher input resistance. A typical unit has an input resistance of 10^3 Ω. Now, with the same 500 pF system, $T = RC = 0.05$ sec. Vibration at all frequencies above 20 Hz will be accurately indicated. At 10 Hz, the response will be 5% low. At 1.8 Hz, the response will be 50% low. For many applications, this is satisfactory. For very-low frequency performance, intermediate amplifiers with input resistances of 10^9 Ω have been used, but there are many difficulties with connector contamination (skin oils, salt, moisture, etc.) that may change system sensitivity; if such a change occurred without the operator's knowledge, the results could be most serious!

Another approach to voltage amplification, sometimes used to eliminate need for intermediate amplifiers, has employed accelerometers with very large internal shunting capacitance. However, these generally have a very low voltage sensitivity. Since $E = Q/C$, small accelerations are hard to measure.

One of the major advantages of charge amplifiers or charge converters over voltage amplifiers is that these low-frequency errors are avoided. The RC factor has no effect on sensitivity. Moderate contamination of cables and connectors has no effect. Response *can* go to extremely low frequencies as suggested by the solid line of Fig. 23.34. (Most stop at 0.2 or 0.5 Hz to avoid responding to thermal gradients, etc.) The two dashed lines show how sensitivity falls off with loads of different resistance, when using voltage amplification.

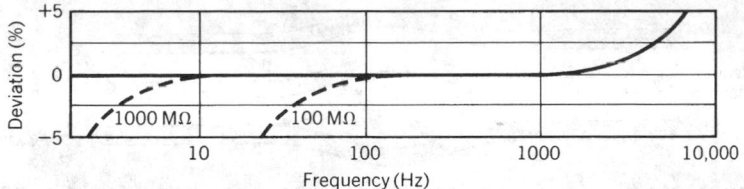

Fig. 23.34 Deviation in frequency measurement at extremely low frequencies.

Ideally, there should be a constant ratio (called sensitivity) between the electrical output from an accelerometer and the mechanical input. Then the electrical signal is an accurate measure of the mechanical input. Whenever that ratio changes, measurement errors can creep in; data must be corrected. Not only is this true for measurement of sinusoidal vibration, but also for measurement of shock (especially long-duration shock pulses) and of random vibration.

If the measurement system (accelerometer, *all* amplifiers in the system *and* the readout device) responds clear down to zero frequency, the electrical signal will be a duplicate of the mechanical input. If the mechanical input is one of the shock pulses shown by solid lines of Fig. 23.35, the electrical signal *should* duplicate it. This requires that RC be much greater than T, where T is the

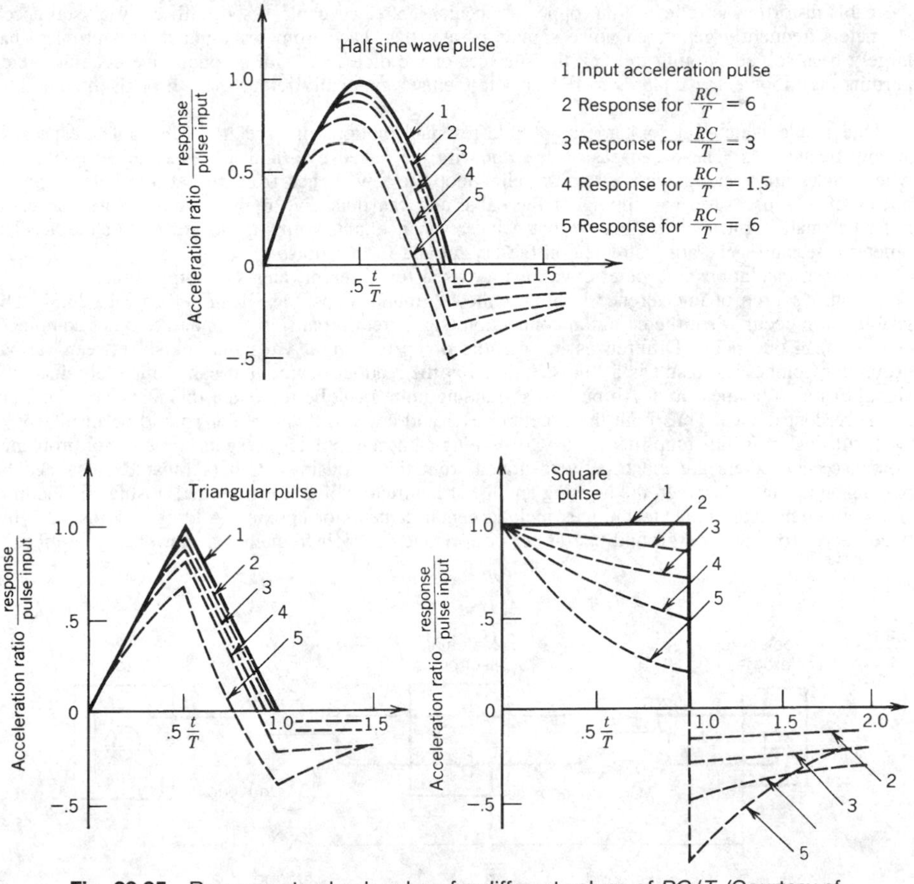

Half sine wave pulse

1 Input acceleration pulse

2 Response for $\frac{RC}{T} = 6$

3 Response for $\frac{RC}{T} = 3$

4 Response for $\frac{RC}{T} = 1.5$

5 Response for $\frac{RC}{T} = .6$

Triangular pulse

Square pulse

Fig. 23.35 Response to shock pulses for different values of RC/T. (Courtesy of Endevco Corp.)

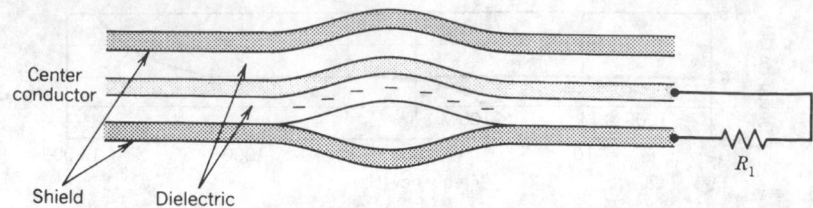

Fig. 23.36 Distortion in accelerometer cable.

time duration of the mechanical shock pulse. For instance, with a rectangular pulse, if $RC = 50T$, there will only be a 2% error at the end of the pulse, whereas if $RC = 20T$, there will be a 5% error. The magnitude of these errors is shown by Fig. 23.35 for three common shock pulses for $RC = 6T$, $3T$, $1.5T$, and $0.6T$.

The cable from an accelerometer to the first electronic unit is important in that it must transmit the signal without any distortion or introducing any "noise." Coaxial cable is used to minimize pickup from strong electrical or magnetic fields. However, when this cable is overly flexed or mechanically distorted, it can generate unwanted electrical signals, or "noise." See Fig. 23.36. If the shield parts from the dielectric, a voltage is generated by "triboelectricity." The excess of electrons on the dielectric is transferred to the center conductor at that point; excess electrons flow along the center conductor to the input resistance of the first electronic unit then back along the shield. The amplifier cannot distinguish between this brief flow of electrons and an accelerometer signal. When the cable distortion is relieved, the opposite occurs. Untreated cables used with early crystal accelerometers frequently generated noise signals greater than accelerometer signals! This problem has largely been solved through treating the surfaces of the dielectric with a conductive coating, which permits the triboelectric charges to redistribute themselves locally rather than through the amplifier input.

One should watch out for large-amplitude resonant motions of accelerometer cables, especially at low frequencies. These can easily be seen with the naked eye, and are especially troublesome when cables are secured with a "strain-relief loop," a few inches from the accelerometer. Shock pulses often cause severe whipping of the cable and can thus give faulty signals. Note that cable must be small, light, and flexible so that cable resonance and whipping do not distort the accelerometer case, thus generating stresses on the crystal and further noise.

Different insulations and outer coverings are used for different ranges of temperature.

Another source of interference is often called "ground loops," as illustrated in Fig. 23.37. The problem can occur when the common connection (signal return path) is grounded to other equipment at more than one point. Differences in potential of up to several volts may exist between various grounding points. The result is a "noise" signal at the readout device that can completely mask the signal that is to be measured. All but one grounding point must be removed; this is usually the input to the readout device. Thus both the accelerometer and any in-line amplifier should be insulated.

Insulating an in-line amplifier is easy. Insulating an accelerometer presents some minor problems. Some accelerometers are electrically insulated from their housings. Others must be insulated by placing an insulator between the housing and the structure to which it is mounted. Insulated mounting studs are often used; other techniques include dental cements or epoxies. Adding any insulator between accelerometer and ground can affect performance at high frequencies, however; the effect of

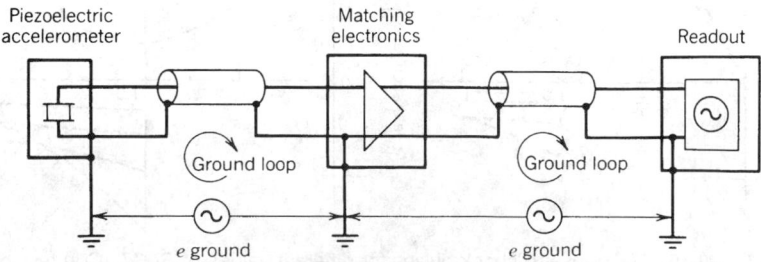

Fig. 23.37 Ground loops.

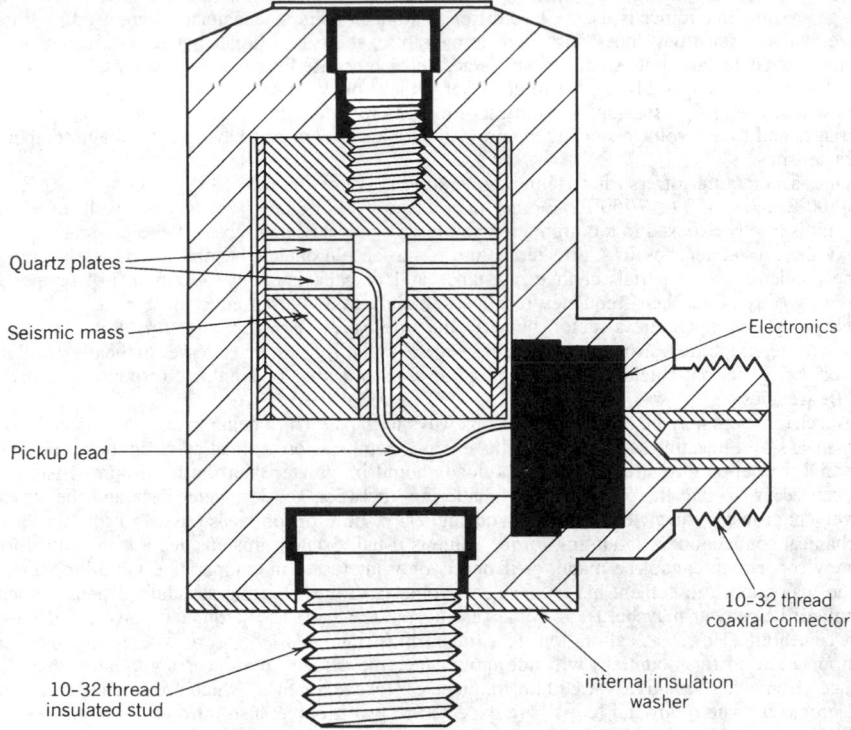

Quartz plates

Seismic mass

Pickup lead

Electronics

10-32 thread
coaxial connector

10-32 thread
insulated stud

internal insulation
washer

Fig. 23.38 Packaging together of accelerometer and solid-state amplifier.

any electrical insulation on the mechanical properties of the system should be checked by the calibration laboratory *before* they are used for vibration or shock tests.

Today's vibration and shock measurements benefit from the packaging together of accelerometer and a solid-state amplifier (Fig. 23.38). The signal lead is thus protected against any contamination and leakage. Also, shunt capacitance cannot change. High electrical output (several volts per g) is obtained if several transistors are used. Some use only one transistor, mainly to give very low output impedance (typically only 100 Ω). Long cables may be used without much electrical interference. Ground loops do not present much difficulty, nor does cable noise. This innovation offers large advantages, though at an increase in cost (partly offset by simpler external electronics). All manufacturers of accelerometers offer them.

We would like vibration pickups such as accelerometers to respond only to vibration, and not at all to other environments such as acoustical fields, pressure fields, temperature, etc. Let us briefly discuss the effect of varying ambient temperature on the sensitivity of an accelerometer. Figure 23.39 shows such an effect upon a particular crystal accelerometer. The solid line shows how sensitivity

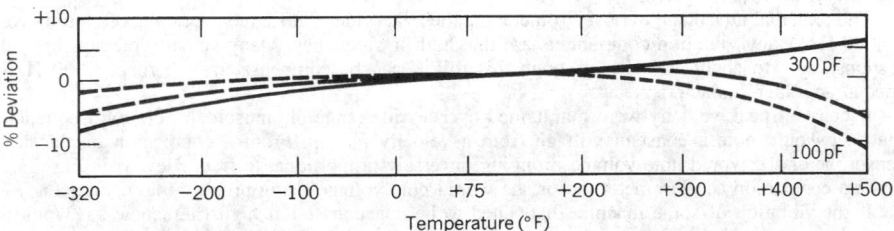

Fig. 23.39 Effect of temperature on sensitivity of an accelerometer.

varies when charge-sensing electronics is used. The dashed lines show how sensitivity varies when voltage-sensing electronics is used. (Two different values of shunt capacitance were used.) Although, as we can see, sensitivity does vary with temperature, there is a broad flat region in which these sensitivity changes are quite small. (Today's accelerometers are far superior to early day units in this regard.) Curves such as Fig. 23.39 often differ widely between voltage and charge response; it is important to use charge-sensing electronics with accelerometers having flat charge-temperature characteristics and to use voltage-sensing electronics with accelerometers having flat voltage-temperature characteristics.

One manufacturer offers four usable temperature ranges: -65 to $+185°F$, -65 to $+350°F$, -320 to $+500°F$, and -452 to $+750°F$. Another manufacturer offers units said to work well to $+1100°F$. If a unit is briefly exposed to too high a temperature, it may still be usable (after a calibration check). But if the crystal reaches its Curie temperature, the accelerometer becomes completely unusable. Some accelerometer materials change resistance and/or capacitance with extremes of temperature; this can greatly affect their frequency responses, mainly at low test frequencies.

Not only is temperature a factor, but also the *rate of change of temperature*. With some accelerometers, temperature transients can produce several volts signal due to stresses in the crystal element and/or case. Thermal shielding is often helpful, as is blocking of amplifier response to extremely low frequencies.

Accelerometers may be attached in various ways to the structure being investigated. The best way is by means of a machine screw or stud; this exerts a compression preload between the accelerometer base and the supporting structure. This preload should be greater than the maximum tension force that can occur, so that there will be no "chattering" between accelerometer base and the structure. At very high test frequencies (5000 Hz and higher), a little oil or grease assists in getting a solid mechanical connection. A mounting torque value is usually stated; this ensures a firm connection. If a screw connection cannot be made, or if one is only interested in comparative vibration values, he or she may choose to cement his or her accelerometer to the vibrating structure. Dental cement or Eastman 910 cement may be used. Some (usually quite small) accelerometers have a flat base for easy cementing. However, calibration data (taken in a standards laboratory where only screw connections are used for mounting) will not apply, since the cement introduces compliance that affects the accelerometer's sensitivity at certain frequencies. Insulated studs, which are often used to mount accelerometers when "ground loops" are expected to give trouble, also introduce compliance.

Some types of accelerometers, particularly the compression types, exhibit an effect called base-strain sensitivity. As suggested by Fig. 23.40, an accelerometer may be located at a point of small motion. Yet the readout instrument may show large motion. Why? Sometimes the difficulty is distortion of the base caused by bending of the structure being measured, as would be shown by strain measurements; this distortion can cause an alterating force on the crystal and a resulting electrical signal. One solution is to select accelerometers that do not have much such sensitivity. Another is to reduce the effect by mounting the accelerometer on an insulated mounting stud that will transmit vibration but not bending force.

Both in industrial and in aerospace applications, displacement and velocity pickups have been losing favor since about 1955, when rugged piezoelectric accelerometers first became available. Let us discuss a few relative advantages of these accelerometers. First, they are far more rugged; some accelerometers for shock measurements are routinely calibrated at levels greater than $10,000g$. In comparison, few velocity pickups will withstand $100g$ accelerations and are easily damaged by rough handling. Noncontacting displacement sensors are no doubt ragged enough to withstand any ordinary industrial application.

We might note that aircraft engine vibrations (in aircraft and test stands) were once measured with velocity sensors mounted close to main bearings; output was integrated and read on a meter as inches D. This practice carried into jet engines; a typical overall displacement limit was 0.003 in. Recently, however, most people are convinced that acceleration measurements are better indications of internal forces due to failed gears, bad bearings, imbalance due to failed turbine blades, etc. The giant Lockheed C5A was the first to carry accelerometers on each engine, with a centralized vibration monitoring and alarm system.

The principal advantage of accelerometers is their far wider flat frequency range, often to beyond 10,000 Hz. Many vibration components are that high in frequency. Many velocity pickups have flat response only to about 100 Hz, although this rolloff can be compensated to perhaps 2000 Hz by special amplifier characteristics.

If constant peak velocity were maintained over a wide range of sinusoidal vibration frequencies, the user would note a constant voltage from a velocity pickup. From a companion displacement sensor, he or she would note voltage dropping, inversely proportional to frequency, as in Fig. 23.41. From a companion accelerometer, he or she would note voltage rising proportional to frequency.

If the vibration of some machine happened to be concentrated at high frequencies, it would be much less observable on the signal from a displacement pickup than on the signal from an accelerometer. Such high-frequency vibration is characteristic of the rubbing of machine parts that are supposed to slide smoothly. High-frequency vibration caused by rough bearings, rough gears, etc.,

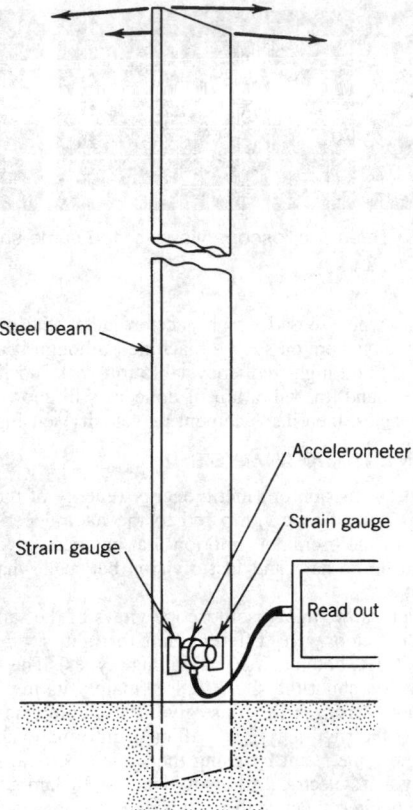

Fig. 23.40 Location of accelerometer at point of small motion.

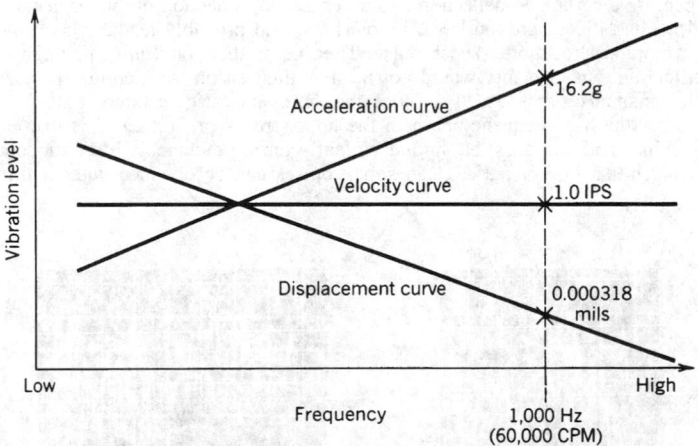

Fig. 23.41 With constant peak velocity maintained over a wide range of sinusoidal vibration frequencies, displacement voltage is inversely proportional to frequency and acceleration voltage is proportional to frequency.

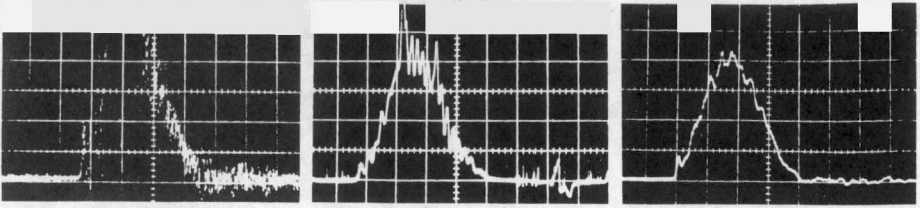

Fig. 23.42 Three oscilloscope views of the same shock pulse.

then, is most easily observed when sensed by an accelerometer, the preferred pickup. An accelerometer can be located nearly anywhere on such a machine (although it is preferable to place it close to a critical bearing, gear, etc.) and high-frequency vibration will find its way through the structure to be sensed. If, on the other hand, the vibration of concern will show up as whole-body vibration at relatively low shaft frequencies, then displacement or velocity sensing is better.

23.5 SHOCK-MEASUREMENT AND ANALYSIS

Shock may be defined as a transmission of kinetic energy (energy of motion) to a system that takes place in a relatively short period of time compared to the natural period of the system. Shock is followed by a natural decay of the oscillatory motion that has been given to the system. The reader will recognize that there can be no hard and fast division between what we casually call vibration and what we casually call shock.

Consider Fig. 23.42, which shows three oscilloscope views of the same shock pulse. They appear different because the accelerometers were different (left to right; $f_N = 35{,}000$ Hz, undamped piezoelectric; $f_N = 2000$ and 850 Hz, both damped strain gage types). The trace at the left of Fig. 23.42 gives us much more information about the shock pulse (mainly its high-frequency content) than do the others. The others are not "wrong"; they just give a less complete picture. However, for loads that are not rigid, the trace at the right may give sufficient information.

We can get very much the same result by using the same electrical pulse (Fig. 23.42 at the left) and by passing it through low-pass electrical filters to decrease the amount of high-frequency "hash." There is a real temptation to filter shock pulses (this temptation should be resisted) and there is considerable past history and existing practice to be overcome. Observe Fig. 23.43, oscilloscope views of shock pulses originating in a piezoelectric accelerometer during a shock test on a very old type of shock test machine. Test personnel were told to expect a shock pulse approximating a half-sine wave. The upper traces contain so much hash that any possible half-sine pulse is obscured. But by a little filtering, the half-sine pulse becomes visible; with more filtering, it becomes easy to see. However, note how the height of the right-hand lower-trace pulse is lowered. Also note how it appears to be stretched out in time. If the operator had previously calibrated the oscilloscope (using sinusoidal excitation) at Y_1 g/in. on the vertical scale, and at X_1 msec/in. on the horizontal scale, he or she would be expecting a certain Y_2 deflection and a certain X_2 deflection on his or her oscilloscope. If Y_2 were too low, and if X_2 were too large, he or she would probably readjust his or her shock test machine for a more severe, shorter shock pulse. Then he or she would *think* he or she was meeting the test specification, but in reality would not be. For this reason, accelerometer manufacturers, in their instruction manuals, often say, "For shock measurements, use *no* filtering at all."

The reason for the high-frequency hash in the upper traces of Fig. 23.43 is that early designed shock test machines had carriages that rattled violently during the shock. More modern types often have cast carriages that do not rattle; clean shock pulses can be obtained without filtering. But to

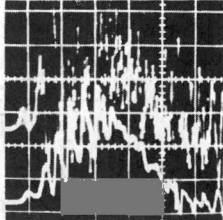
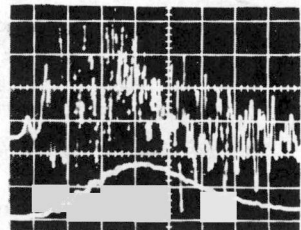

Fig. 23.43 High-frequency hash due to rattling of carriages in shock-testing machines.

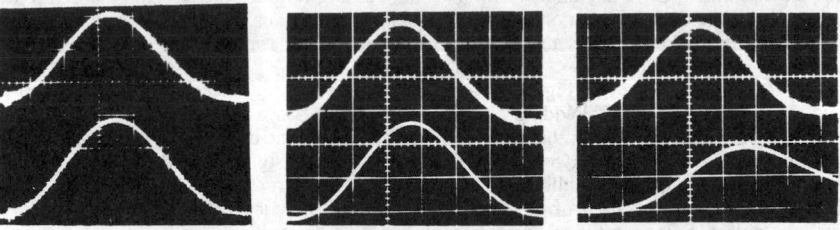

Fig. 23.44 Lower traces filtered; upper traces unfiltered.

illustrate answers to the question "How much filtering *can* I use?," consider Fig. 23.44. In all three cases, taken on the same machine, the upper traces are unfiltered. The frequency ranges for the lower traces were, respectively, 0.06–60,000 Hz, 0.06–600 Hz, and 0.06–60 Hz. Why only 60 Hz? It may be shown that this shock pulse, approximately 0.011 sec in duration (11 msec) contains frequencies that are multiples of 45 Hz. An old "rule of thumb" in such cases was to set the cutoff frequency of the filter at 1.5 × the lowest frequency present; thus 60 Hz. It was felt that this would show how much motion was present at the lowest frequency, but would block motion signals at other frequencies. True enough, but look at the result. A better "rule of thumb" would be 15 × lowest frequency present; in this case 600 Hz.

The low-frequency response must extend nearly to zero. The lower trace of Fig. 23.45 here shows what happens with a low-frequency cutoff of 6 Hz; note the negative undershoot of the trace after the shock pulse was over. There was also a reduction in apparent pulse height, but this is harder to see.

Since so-called half-sine pulses more closely approximate what is called a haversine pulse, there is a little difficulty in establishing, from a record, just when the shock pulse starts and when it stops. One method is illustrated by Fig. 23.46. Some specifications also give the "rise time" and the "fall time" of a shock pulse; these are also shown by Fig. 23.46.

Accelerometer systems used for shock measurement and analysis should be shock calibrated. An accelerometer might be shown to have a sensitivity of, say, 10 mV/g under sinusoidal vibration conditions at vibration intensities of 50g or 100g, the maximum available from most shakers. But what will the sensitivity be at 1000g or 10,000g? Is the accelerometer linear? It is sometimes possible

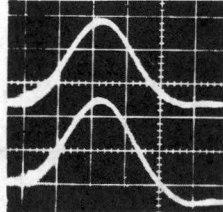

Fig. 23.45 Lower trace filtered with low-frequency cutoff of 6 Hz.

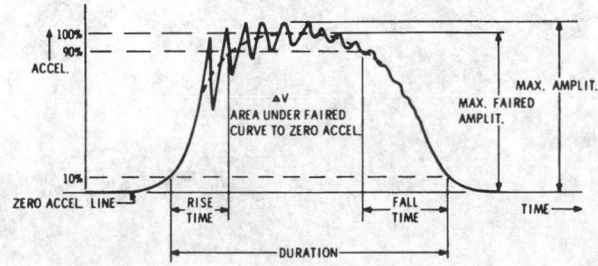

Fig. 23.46 Half-sine pulse.

to sinusoidal vibration-calibrate an accelerometer on a tuning fork or other resonant structure (driven by a shaker) at 500g or 1000g, but only at one frequency per structure. Shock calibration as suggested by Figs. 23.47 and 23.48 gives sensitivity at pulses to perhaps 15,000g and at various pulse time durations.

In the drop-ball calibrator of Fig. 23.47, the accelerometer is at rest, attached to an anvil which is struck by a falling steel ball. Padding on the impact surface varies pulse intensity and duration. In the machine shown in Fig. 23.48, the accelerometer being calibrated is attached to a falling carriage and is stopped by a target, which contains a force gage (previously calibrated by dead weights).

Another advantage of shock calibration (in addition to linearity checks of accelerometer sensitivity) is that the electronic portions of the measurement system are checked for performance with short, severe input pulses. Some amplifiers, detectors, and readout devices work satisfactorily on sinusoidal input signals but fail miserably with pulse inputs.

Readout of shock pulses is most often done on an oscilloscope. Pulse height (acceleration intensity), also pulse shape and duration, can be read if the oscilloscope is a storage-type unit. Alternately, a Polaroid or other camera may be used for a permanent record. Oscillographs running at high speed, with high-frequency galvanometers, are also used. Peak-holding meters in which the maximum pulse height is held until a "reset" button is pushed are sometimes used. However, these tell nothing about the pulse shape or duration.

In our discussion of shock thus far we have concerned ourselves only with converting motion into an electrical signal and with readout. We generally use accelerometers that provide a description of pulse height, duration, and shape. This description tells us very little about the energy content of the shock pulse at various frequencies. We can appreciate that a designer responsible for building a product that will pass a shock test should have frequency information in order that his design will avoid having any resonances where there is much excitation.

How might we analyze a shock pulse in terms of frequency? One method is to employ a mathematical procedure known as the Fourier transform to convert from the time domain (instantaneous force, acceleration, velocity, or displacement versus time) into the frequency domain (force, accel-

Fig. 23.47 Drop-ball calibrator.

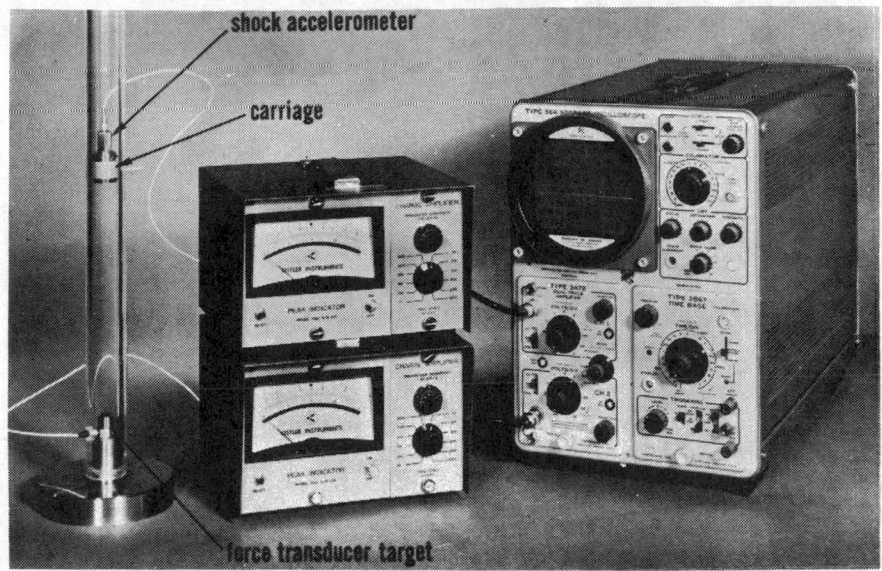

Fig. 23.48 Calibration of accelerometer: accelerometer is attached to a falling carriage and is stopped by a target containing a force gage previously calibrated by dead weights.

eration, velocity, or displacement versus frequency). For all but simple time histories (such as the left-hand column of Fig. 23.49), this procedure entails vast amounts of work. If a digital computer is available, one can use a special program to save that work. Further time savings, on the order of 300:1, are possible by use of the fairly recent fast Fourier transform (FFT) program. Special purpose hybrid computers can also make this transformation. With a frequency description of a shock that is likely to occur in service, an equipment designer knows what the natural frequencies should be, in order to avoid damage.

Another way to analyze a shock pulse in terms of frequency is by means of what is called a *response spectrum*. Imagine that an array of spring-mass-damper systems (such as a group of reeds), as suggested by Fig. 23.50, each with a different natural frequency (but the same mechanical buildup Q) is attached to a structure. A single unit from that array is shown in Fig. 23.51. A shock pulse occurs. Its displacement time history is recorded as x, the shock input. Some method, such as the pen making trace y, is used to measure the maximum response motion of each reed. Some reeds will respond more than others, depending on the amount of energy in the shock motion at different frequencies. A graph of maximum response is made versus reed natural frequency; this graph is called a response spectrum. It gives information about relative amounts of energy at various frequencies. Note that it does not describe the shock pulse itself; it only describes what the pulse does to a series of idealized mechanical systems. The right-hand column of Fig. 23.49 indicates the response spectra of the several common shock time histories shown in the left-hand column. The residual spectrum deals only with events during the decay period. The primary spectrum (not shown) deals only with events during the shock. The maximum spectrum is the larger of the two at any instant of time. The maximum value of the maximum spectrum (maximax value) is often considered.

Note that the symmetrical time histories (half-sine, square, and triangular shock pulses) have wide variations from one frequency to another, as opposed to the sawtooth pulse, which is relatively constant at all frequencies. The outcome of a shock test using a sawtooth pulse is thus much less dependent on the specific natural frequencies of a particular test item. This is the main reason why symmetrical pulses are less often used today for tests, and why sawtooth pulses are today most popular.

23.6 SHOCK TESTING

A shock test is generally an environmental test, used to screen "good" items from "bad" items. Test items which break or which do not operate correctly after the shock, or which do not operate correctly during the shock (depending on how failure is defined) are considered to be "bad." Let us look at three major classifications of shock tests and the specifications that describe the tests.

First let us consider the "big-bang" tests, performed in accordance with specifications such as MIL-S-901. Such specifications reference drawings which tell how the shock test machine is to be built. The smaller Navy machine, Fig. 23.52, tests loads to 400 lb fastened to an 800 lb anvil plate

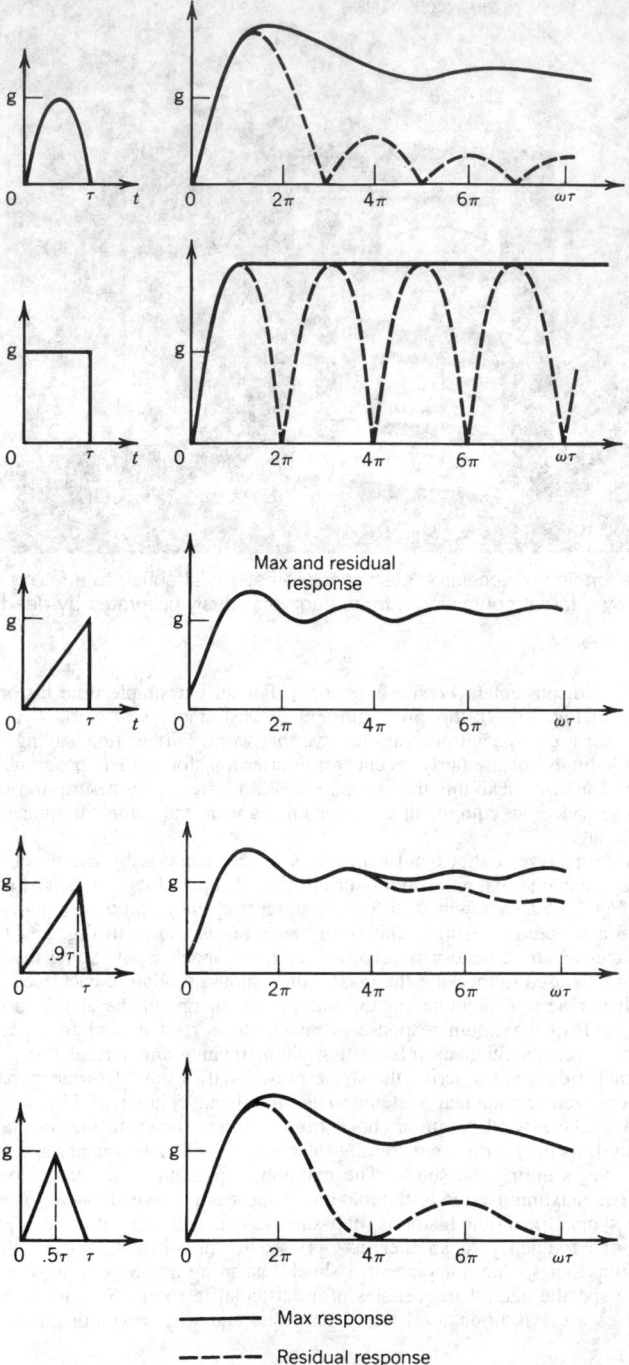

Fig. 23.49 Conversion of shock pulses from time domain to frequency domain.

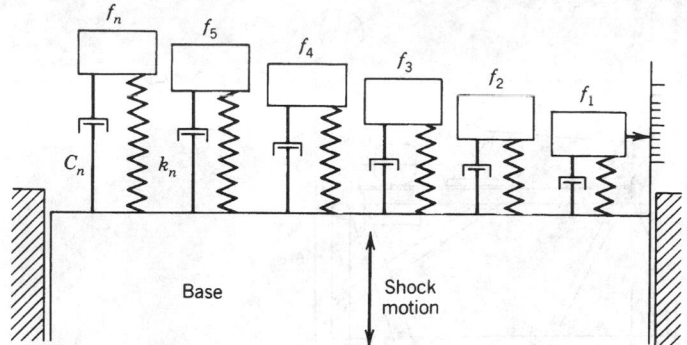

Fig. 23.50 An array of damper systems used to determine response spectrum for a shock pulse.

(vertical) and struck by a swinging 400 lb hammer, which is dropped from a specified height. The larger Navy machine, Fig. 23.53, tests 250–6000 lb loads fastened to a 3500 lb anvil plate (horizontal) and struck by a 3000 lb hammer, which is dropped from a specified height. Still larger Navy loads (to 400,000 lb) are fastened to special steel barges and are tested by setting off underwater explosives nearby.

These test machines and their operation are defined by the test specification. Specifications tell nothing about the shock pulse that results or about the effect of the shock pulse on test items (real or idealized). In fact, with these specifications and machines, there is no need for any instrumentation other than a ruler! The Navy has quite a bit of experience indicating that equipment items which have passed such tests are rugged and capable of withstanding the effects of a ship's own gunfire and of enemy explosions.

Consider Fig. 23.54, which shows several paths for energy transmission from an underwater explosion to the hull of a submarine. Depending on the relative amounts of energy following the various paths, the shock motion of the hull will be different. No one simple specification could duplicate the effects of the variety of shock pulses that the hull could receive.

Consider next Fig. 23.55, which gives the shock spectra at three different levels in a surface ship. Note, curve A, that high-frequency energy is found only at or near the ship's bottom. Energy at B, higher in the hull, shows less high-frequency energy. Energy at C, in the structure higher than the hull, shows still less high-frequency energy but accentuated energy at low frequencies. No one simple

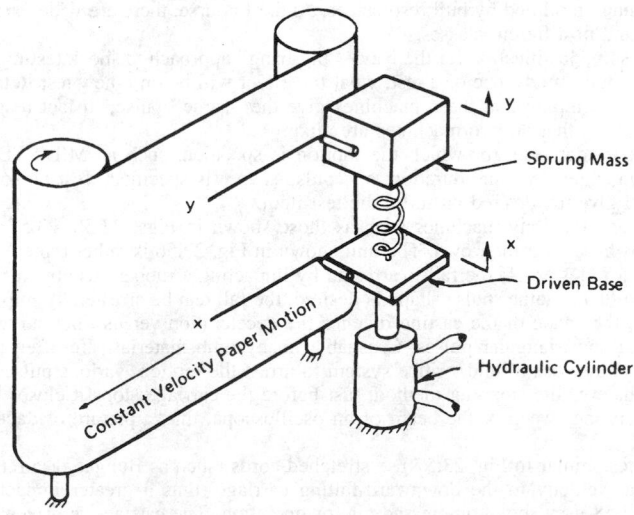

Fig. 23.51 A single-unit damper system.

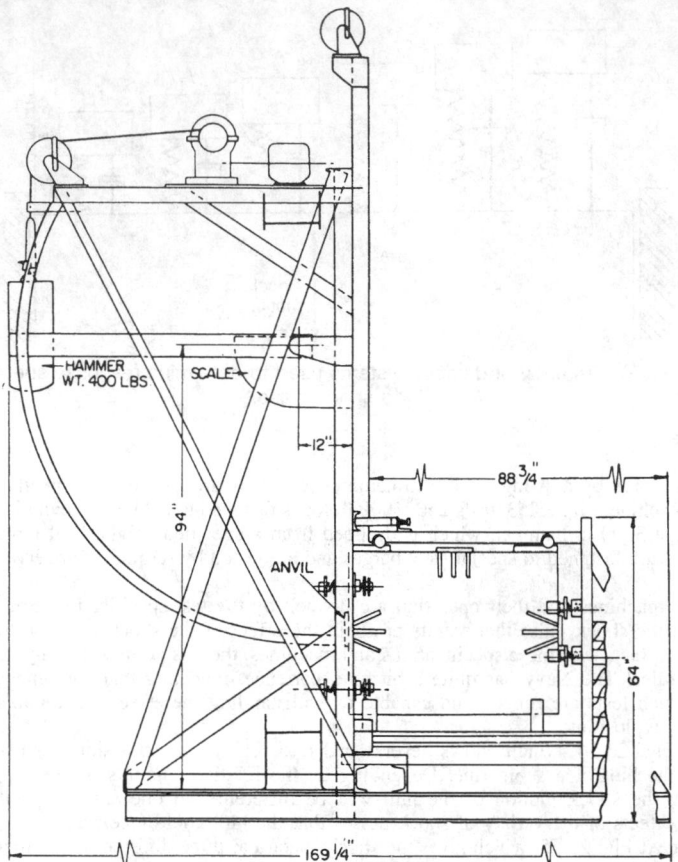

Fig. 23.52   "Smaller" Navy shock test machine.

specification could duplicate the effects of even a simple standard pulse received by the hull, when, that pulse is so much modified by hull resonances. And, of course, there are wide variations between ships of a class and of different classes.

Thus there is some justification for the Navy "big-bang" approach to shock testing. The drawback is that one cannot tell, in advance of a test, what the effect will be on a new test item. Furthermore, there is no way to compare that these machines give the "same" pulse; in fact response spectrum investigations indicate that no two machines are alike.

Let us next consider tests for which the motion is specified, such as MIL-STD-810C. Here a definite waveform (intensity, time duration, and pulse shape) is specified. Test personnel select any machine that will give the desired pulse on the test item.

Most test laboratories buy machines such as those shown in Figs. 23.56, 23.57, and 23.58, although a few labs have built their own. The unit shown in Fig. 23.56 is rather typical of the common free-fall gravity drop testers. If the fall is arrested by impacting a rubber cushion, a haversine deceleration pulse results. If another pulse shape is desired, the fall can be arrested by impacting a shaped soft lead casting; the shape of the casting dictates the deceleration versus time curve. Lead is often used for sawtooth and triangular pulses. Crushable honeycomb materials are often used for square pulses. Some machines use a fluid/orifice system to arrest the motion; various pulse shapes may be programmed. Whatever the stopping method, just before the carriage stops it closes a switch which initiates or triggers the sweep of the beam of an oscilloscope, thus a picture of deceleration versus time is obtained.

Some machines similar to Fig. 23.57 use stretched cords (such as Bungee or aircraft shock cord) to give additional velocity to the downward-falling carriage, thus a greater impact. The machine shown by Fig. 23.58 uses shop compressed air for operation. The carriage is driven downward and impacts an arresting device. The Hyge machine illustrated by Fig. 23.58 uses dry nitrogen under high pressure to accelerate a carriage for test; a braking system later arrests the carriage gently.

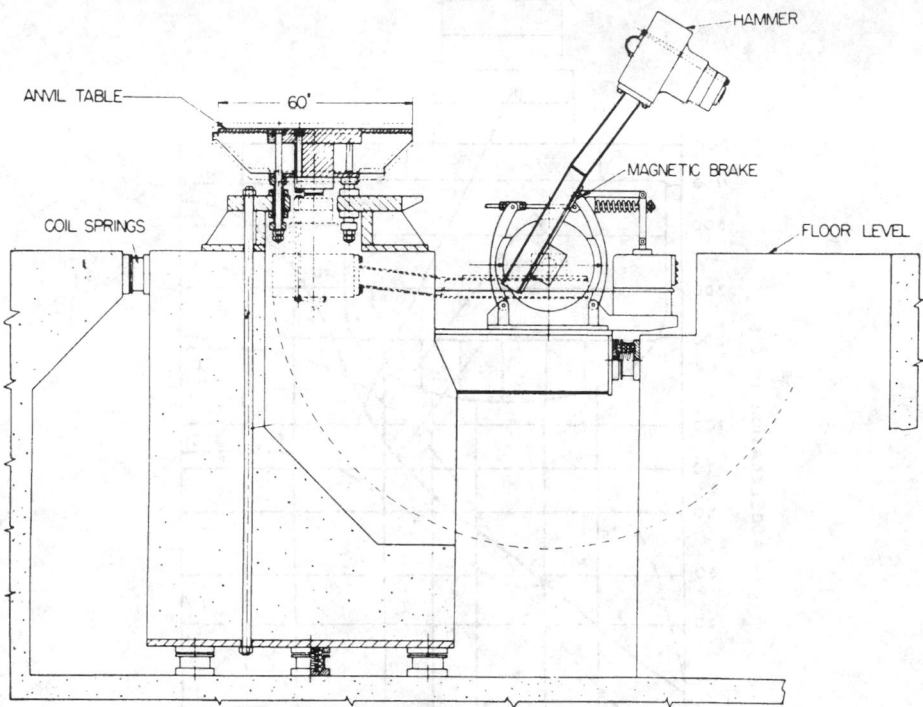

Fig. 23.53 "Larger" Navy shock test machine.

Still another way to generate a specified pulse of acceleration versus time is to use a shaker. Electrohydraulic shakers have been used; one advantage is their long stroke, required for long-duration shock pulses. A few electromagnetic shakers have been built with a long stroke for the same purpose. In general, though, the 1-in. peak-to-peak motion of typical shakers restricts their usefulness for testing with specified (half-sine, haversine, square, triangular, sawtooth, etc.) pulse shapes, especially if pulses are more than a very few milliseconds long.

There are several advantages in using a shaker for shock testing: the cost of a shock test machine may be saved; the cost of an extra test fixture to adapt a test item to a shock test machine may be saved if shock testing may be done on an existing shaker; reversal of the direction of a shock is as

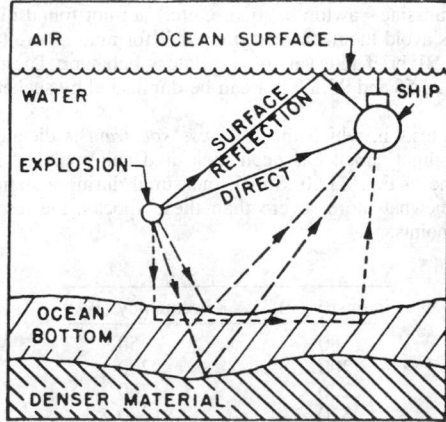

Fig. 23.54 Paths for energy transmission from underwater explosion to hull of submarine.

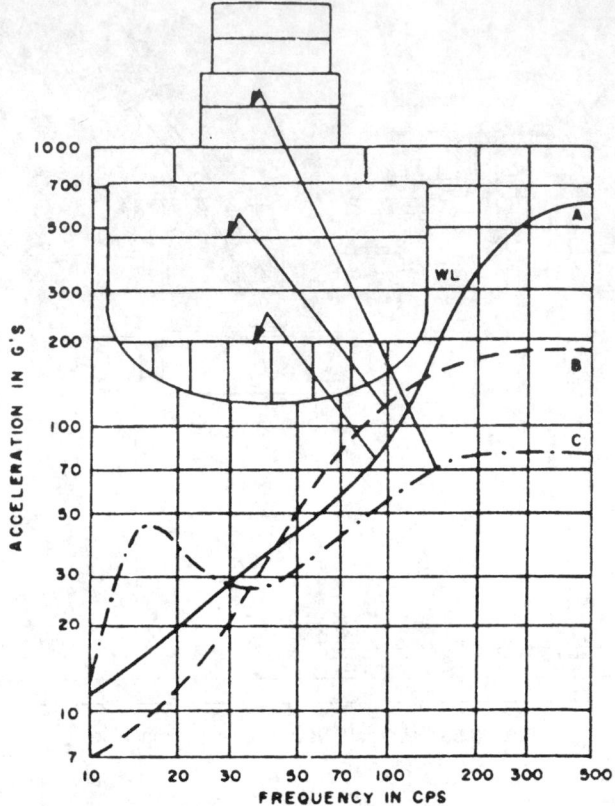

Fig. 23.55 Shock spectra at three different levels in a surface ship.

simple as reversing a switch, whereas a complex attachment to a shock test machine would be required; setup time may be greatly reduced, compared to transferring a load from vibration test to a shock test machine.

Figure 23.59a suggests older analog equipment. Usually, a switch on the pulse generator was closed once for each pulse. Proper shaping of the pulse was done with the gain of the power amplifier reduced; by means of trial and error the pulse was shaped until it met the requirements of the test specification. Then the power amplifier gain was increased for the actual shock test. Figure 23.59b shows that the electrical pulse (when viewed on an oscilloscope) looks much different from the motion pulse. Digital computers control these tests now.

Simple pulse shapes (half-sine, sawtooth, square, etc.) are not found in the real world and many newer testing specifications avoid them. Instead, they call for much more typical time histories such as the one set into Fig. 23.60, brief intervals of oscillatory behavior. These cannot be duplicated by machines typified by Figs. 23.56 and 23.57, but can be duplicated by shakers operating under control of digital computers.

Finally, let us consider tests in which the *response spectrum* is dictated as in MIL-STD-810D. Let us imagine that an in-flight shock has been measured and analyzed in terms of its frequency content, as in the lower line of Fig. 23.60. Shock measured during a static firing is also shown. In order to specify a test somewhat more severe than these shocks, the test requirement envelope is describe by the following points:

Intensity (g)	Frequency (Hz)
80	50
200	200
200	700
320	1000
320	2000

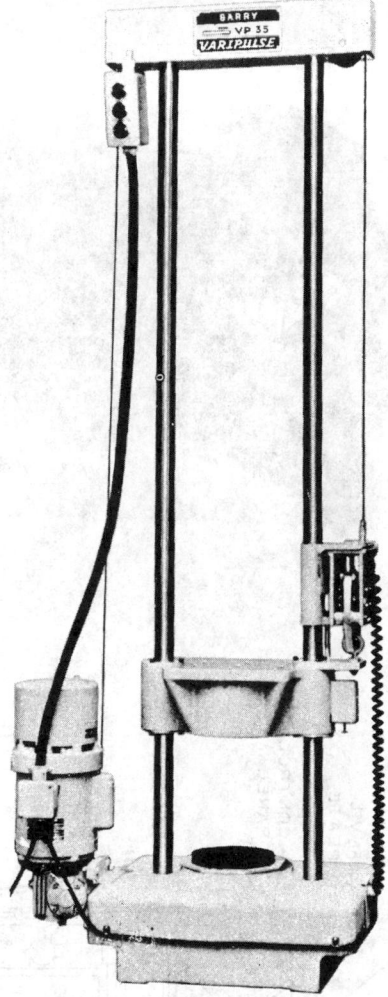

Fig. 23.56 A free-fall gravity drop tester.

These coordinate points are connected by straight lines on a semilog plot, as in the upper line of Fig. 23.60. A shock pulse of mechanical motion must be generated. Its shape and duration are less important than that the response spectrum be as specified. A signal from the control accelerometer will be analyzed by a shock response spectrum analyzer.

Early shock response spectrum testing (1960s and early 1970s) was done as follows: A single pulse (short, with much high-frequency content) was generated, then frequency divided by a bank of filters. Each filter had an adjustable attenuator to adjust the energy content in its frequency "window" of the final pulse. The synthesized pulse then went to the power amplifier and shaker. The response spectrum was analyzed with an identical filter bank. Energy in the various frequency windows could be adjusted so that the response spectrum met requirements.

Since the energy content of the shock pulse at various frequencies was specified, the test operator had to vary the shaker's motion in order to meet the specification. This process closely resembled the process of equalizing a shaker and test load for random vibration testing. Note in Fig. 23.61 the similarity between the block diagrams of equipment needed for random vibration testing and for shock spectrum testing.

With the introduction of digital control techniques, it became feasible to reproduce very complex time histories, not only for long-time vibration tests but also for short-time shock tests.

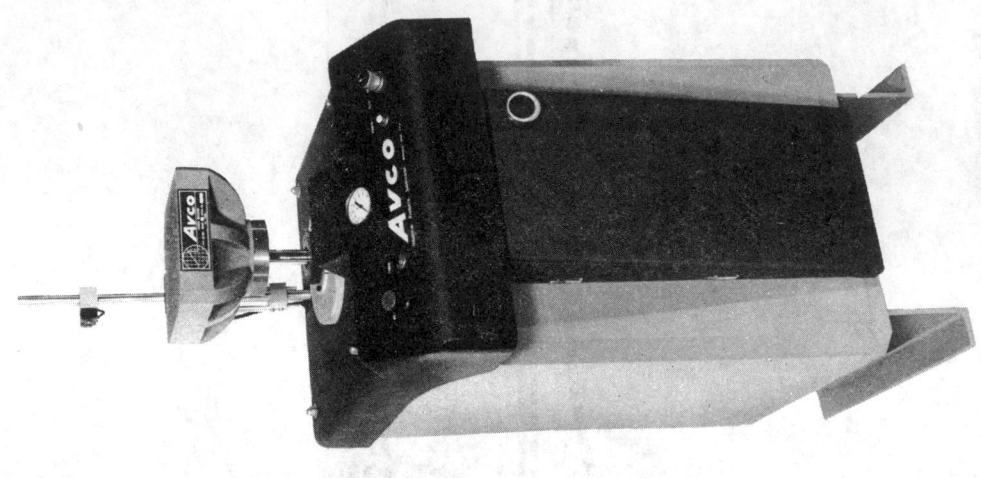

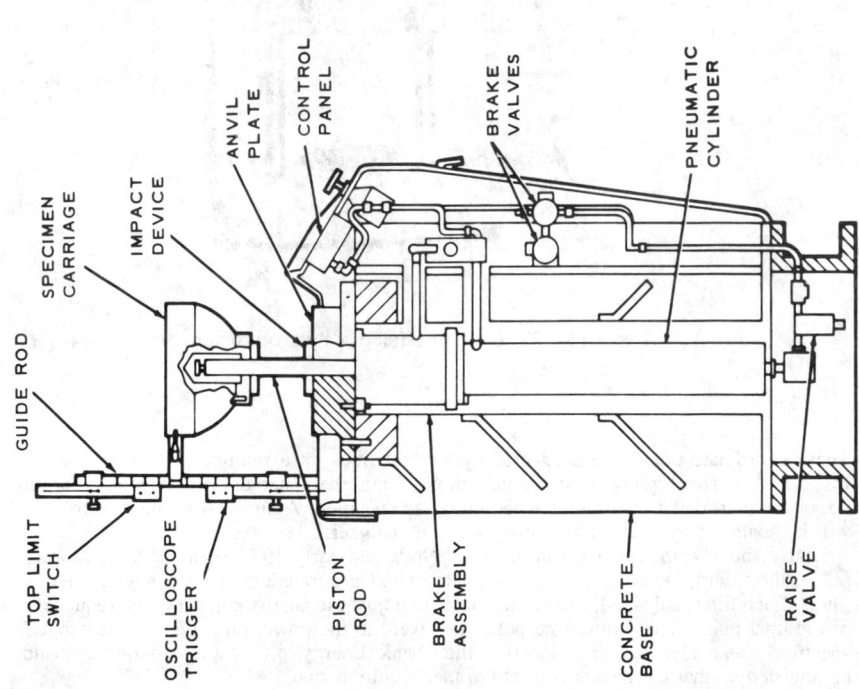

TOP LIMIT
SWITCH

GUIDE ROD

OSCILLOSCOPE
TRIGGER

SPECIMEN
CARRIAGE

IMPACT
DEVICE

PISTON
ROD

ANVIL
PLATE

CONTROL
PANEL

BRAKE
ASSEMBLY

BRAKE
VALVES

CONCRETE
BASE

PNEUMATIC
CYLINDER

RAISE
VALVE

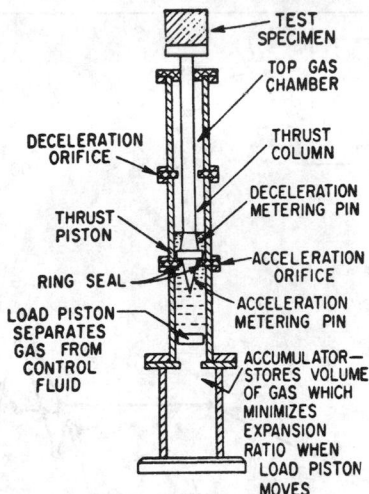

Fig. 23.58 A shock testing machine that uses dry nitrogen under high pressure to accelerate the carriage.

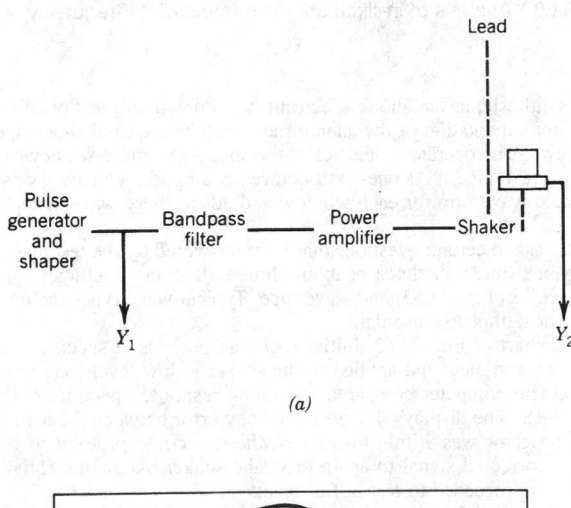

(a)

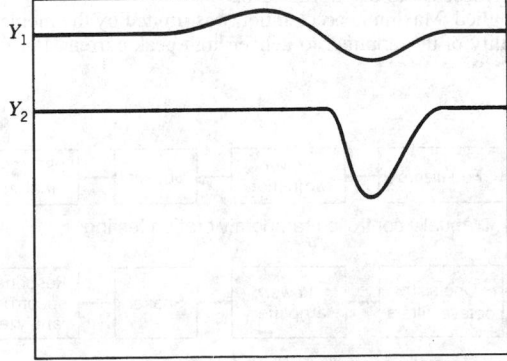

(b)

Fig. 23.59 (a) Equipment needed for a shock test. (b) Y_1: electrical pulse; Y_2: motion pulse.

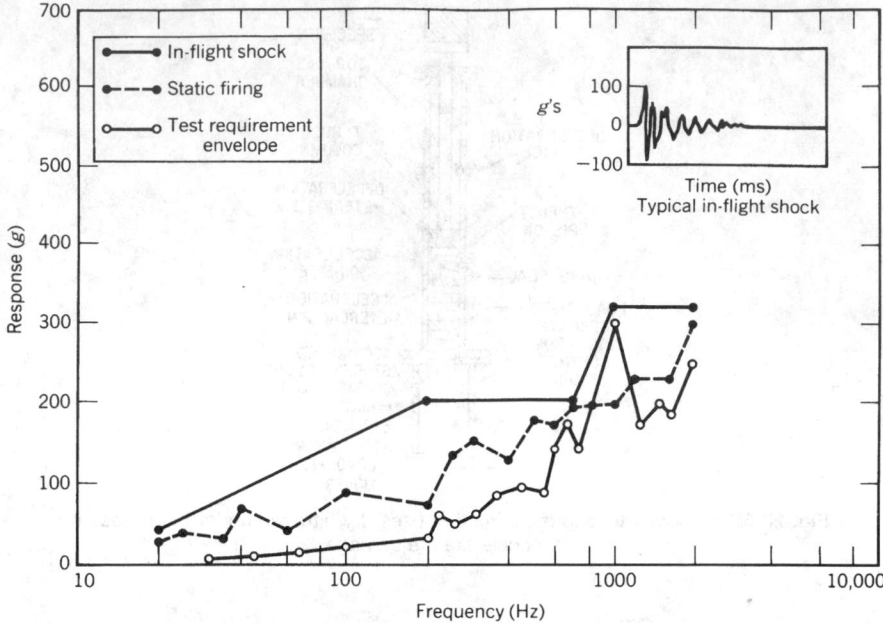

Fig. 23.60 Analysis of in-flight shock in terms of its frequency content.

One method of synthesizing the shock spectrum was known as the "parallel filters" method. It was a direct digital implementation of the analog method. The required shock spectrum was divided into frequency windows. The operator could select the spacing of these windows (one-half, one-third, and one-sixth octave are typical). At one-sixth octave spacing, 53 windows was typical. The minicomputer chose a basic waveform for each window and added these waveforms to form a composite waveform.

The minicomputer asked certain questions that were answered by the test operator. These included the date, test title, shock duration, shock peak amplitude, damping coefficient, number of shocks to be applied (see "repeat" in Fig. 23.62) and wave type. Typical wave types included half-sine, square, sawtooth, triangular, or damped sinusoidal.

Consider the flow chart of Fig. 23.62. Initial shock amplitudes at specific frequencies were generated by the computer, summed, and applied to the shaker at low level (one-fourth or one-eighth of full level is typical). The computer compared the shock response spectrum (SRS) of the low-level shock to the desired SRS. The display CRT presented any error between the actual and desired SRS's as a percentage. If the error was within tolerances, the test could proceed to full level. If not, the computer produced a corrected signal to again drive the shaker. After this equalization process was completed, the operator proceeded to test at full level.

Certain limitations applied. Maximum acceleration was limited by the momentary force capability of the shaker and the ability of the amplifier to deliver high peak current. Pulse duration was limited

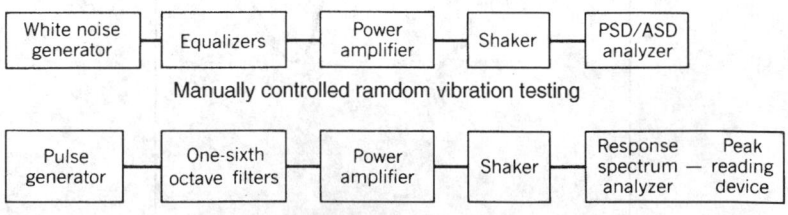

Fig. 23.61 Flow charts for random vibration and shock spectrum testing.

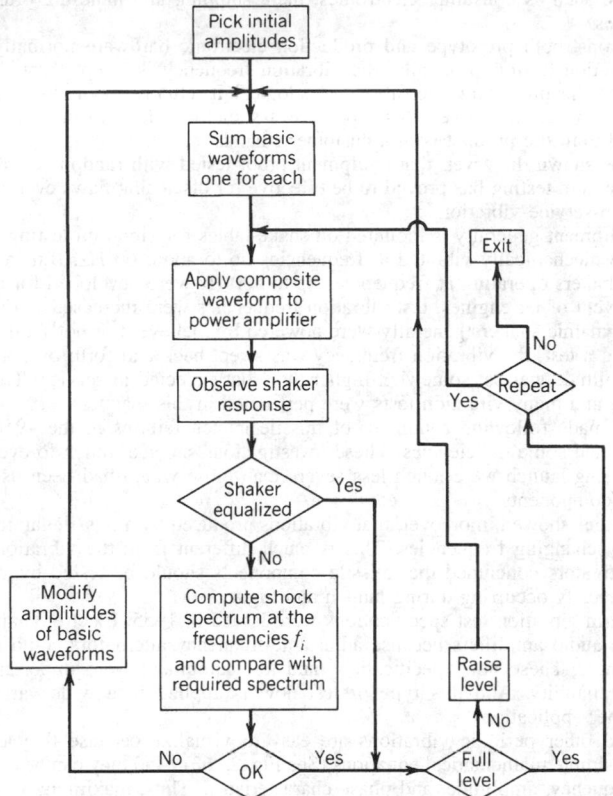

Fig. 23.62 Flow chart for shock spectrum testing to required spectrum.

by several factors, including shaker displacement and power-amplifier low-frequency response. When test parameters exceeded shaker system limits, testing was performed by one of the older methods (Figs. 23.56, 23.57, and 23.58). (Even when a conventional shock machine is used to generate a shock pulse, one can measure and record pulses digitally. Several manufacturers offer units that digitize a shock pulse signal and store it in memory. Most provide readouts that display peak amplitude in g's and pulse duration in milliseconds. Most provide for a plotter that gives the operator a hard copy for the final test report. These units are good for recording time histories but cannot by themselves analyze shock spectra.)

With their FFT capability, shaker-control computers are used for broad-band modal analysis of shock transients. Shock spectrum analysis is used to learn about the dynamic properties of an elastic structure. Once these properties (resonant frequencies, mode shapes, and damping factors) are known, the structure's dynamic behavior can be predicted and/or modified.

Simple pulse shapes (half-sine, sawtooth, square, etc.) are seldom, if ever, found in service or transportation, where typical time histories are brief intervals of oscillatory behavior (see Fig. 23.60). The latter cannot be duplicated by machines such as those shown in Figs. 23.57 and 23.58. Thus, the newer test specifications (Fig. 23.61) must be performed under the control of a digital shaker system.

23.7 SHAKE TESTS FOR ELECTRONIC ASSEMBLIES*

Not all electronic gear enjoys the protected environment of a computer, business machine, or bench instrument. A substantial portion of electronic assemblies ends up in vehicles, aircraft, military hardware, or hand-held devices where service life entails a fair amount of shock and vibration. And in

---

*From *Machine Design*, January 22, 1981, used by permission.

some applications, such as consumer electronics, mere shipping and handling often impose severe mechanical stresses.

For these reasons, both prototype and production electronic hardware normally is subjected to some sort of vibration testing. Typically, the vibration frequency is swept through some specified spectrum. Although this procedure ostensibly seems logical, it actually may induce failures that would not occur in field service. Moreover, this type of test sometimes fails to uncover true weaknesses that are not found until the product is in a customer's hands.

Experience has shown, however, that equipment proof-tested with random vibrations rarely fails in service. And random testing has proved to be effective for disclosing flaws even in equipment that is not subject to in-service vibration.

Electronic equipment generally is mounted on shake-tables for vibration testing. For many years, shake-tables were mechanically vibrated at frequencies up to about 60 Hz. But in the early 1950s, electromagnetic shakers operating at frequencies up to 500 Hz were developed for aircraft vibration. Later with the advent of jet engines, test vibration frequencies were increased to over 2000 Hz.

Early electrodynamic shakers generally were powered by (believe it or not!) adjustable-frequency alternators. During a test, the vibration frequency was swept back and forth over a frequency band, and vibration amplitude was set somewhat higher than that expected in service. This was known as sine-wave testing, and many vibration tests were performed in this manner.

Investigations made following a number of missile launch failures in the 1950s indicated that sine-wave testing had some deficiencies. These investigations showed, much to everyone's surprise, that vibrations during launch were much less severe than those which had been used for sine-wave testing the failed components.

Additional studies showed, moreover, that vibrations produced by a missile launch contain a wide band of randomly changing frequencies. This is much different from the vibrations used for sine-wave tests. Investigators concluded that missile components should be tested by vibrations closely matching those actually occurring during launch and flight.

The first random vibration test specifications were issued in 1955. Electrodynamic shakers were powered by large audio amplifiers because adjustable-frequency alternators could not produce random-vibration power. These new specifications and test equipment were important factors in improving missile reliability. And this type of test now is applied to a wide variety of electronic equipment for other applications.

Sine-wave and other periodic vibrations are easy to visualize because they can be described graphically or by simple mathematical equations (See Fig. 23.63). And they can be completely defined by specifying frequency, amplitude, and phase characteristics. Thus, maximum acceleration, an important parameter for test purposes, is calculable.

Random vibrations, however, are difficult to picture because they have nonperiodic waveforms that change magnitude unpredictably. A random vibration, though often limited to a specific frequency band, contains many frequencies simultaneously. And all frequencies in the band are possible.

Neither instantaneous velocities nor instantaneous displacements can be predicted for random vibrations. Thus, maximum acceleration is indeterminate, and statistical theory must be used to calculate rms acceleration and to predict the probability for any specific instantaneous acceleration.

A random vibration can be defined for test purposes by specifying an acceleration spectral density (ASD) value for all frequencies included in the test. (Acceleration spectral density sometimes is known as power spectral density or PSD.) The concept for acceleration spectral density is derived from statistical theory. It is typically specified in g^2/Hz units and is the square root of the vibrational power contained in a 1-Hz-wide slice of the defined random vibration. The rms acceleration for a 1-Hz-wide slice can be calculated by extracting the square root of its ASD value (Fig. 23.64).

An ASD chart can be produced by plotting ASD values versus frequency on log-log graph paper. This type of chart, by showing upper and lower frequency limits as well as ASD values, enables one to visualize frequency and vibrational power distribution.

Neither instantaneous nor maximum acceleration values can be determined from an ASD chart. That information can be obtained only by using probability techniques. But rms acceleration can be calculated by extracting the square root of the area under the ASD curve.

The simplest type of ASD chart has a constant ASD value for an entire test-frequency band, the ASD value dropping sharply to zero at the maximum and minimum frequencies. This is known as "white" random vibration (Fig. 23.65).

Based on probability theory, instantaneous acceleration rates are less than the rms value 68% of the time, less than twice the rms value 95% of the time, and over three times the rms value only 0.3% of the time. Theoretically, the acceleration rate can occasionally be infinitely high. However, test equipment is not capable of producing infinitely high acceleration rates, even briefly. Thus, maximum acceleration rates for random vibration tests typically are limited to about three times the rms value. Simulation of road and off-road vibration inputs to land vehicles calls for higher peaks and for spending more time at those higher peaks.

Test specifications for products subject to in-service vibrations typically are based on measurements and analysis of vibration that the product will experience. These specifications then may be

Sine-wave vibrations can be represented by a simple equation. And the instantaneous acceleration produced by a sine wave can be determined by taking the second derivative of the sinc-wave equation:

$$a = 4\pi^2 f^2 X \sin 2\pi ft$$

where a = instantaneous acceleration, f = frequency, X = peak amplitude, and t = time. From this, maximum acceleration can be calculated:

$$A = 0.511 f^2 D$$

where A = maximum acceleration in in./sec, f = frequency in Hz, and $D = 2X$ = total displacement in in.

Random vibrations, however, cannot be represented by a simple mathematical expression. Moreover, maximum acceleration for random vibrations is indeterminant. But other useful acceleration values can be obtained from statistical principles. Thus, if the instantaneous acceleration produced by a random vibration is observed over a long period, the mean, the variance, and the standard deviation can be measured and calculated:

$$\bar{a} = \frac{I}{T} \int_0^T a(t)\, dt$$

$$\overline{a^2} = \frac{I}{T} \int_0^T a(t)^2\, dt$$

$$\sigma = \sqrt{\lim_{T\to\infty} \frac{I}{T} \int_0^T a(t)^2\, dt}$$

where $\bar{a}$ = mean acceleration, $\overline{a^2}$ = acceleration standard variance, σ = acceleration deviation, and T = measurement period.

The mean acceleration (usually zero) has little value for test purposes, but variance does. Variance, or mean square acceleration, is proportional to the total vibrational power. And standard deviation is a useful figure because it is equal to rms, or effective, acceleration.

Acceleration variance, or mean square acceleration, for a 1-Hz-wide slice of a random vibration is known as acceleration spectral density (ASD) and is represented by

$$\text{ASD} = \lim_{\Delta f \to 1} \frac{a(f)^2}{\Delta f}$$

Acceleration spectral density is a useful value because it can be used for presenting random vibrations graphically. And ASD can be measured easily.

Fig. 23.63 Statistical acceleration.

suitable only for those particular conditions. However, some test specifications have been widely used for many different products and applications.

Random Vibration is widely used for stress screening electronic products that are not subject to heavy vibration in service. A vibration stress screen can uncover design weaknesses as well as faulty components, nicked wires, cold-soldered joints, and other workmanship flaws. Products that will be stress screened in production generally should be subjected to the same test during development.

Stress screening is a technique for identifying flaws in subassemblies or in complete products before they are placed in service. Stress screening typically includes burn-in procedures and temperature cycling as well as vibration.

Stress screening is not a test in the generally accepted sense of the word. The word test implies that a product must meet certain performance criteria. But in stress screening, the only real failure is one where a flawed item is approved for use.

Sine-wave vibrations were once used for stress screening electronic products. But experience has shown that random vibrations are more effective for the purpose. As in pure performance testing, sine-wave vibrations can cause failures that would not occur in service, and they can leave some flaws hidden. Random vibrations, however, rarely trigger unwarranted failures, and they often expose design and workmanship faults that are undiscovered by sine waves.

When a printed circuit board (PCB), for example, is subjected to sine-wave vibrations, the amplitude of the board vibration at its center typically is far greater than that of the shaker table. This occurs when a test frequency coincides with the board's natural frequency. The ratio of the board amplitude to the shaker table amplitude is known as the Q of the board. The value of Q is a function of an article's damping characteristic.

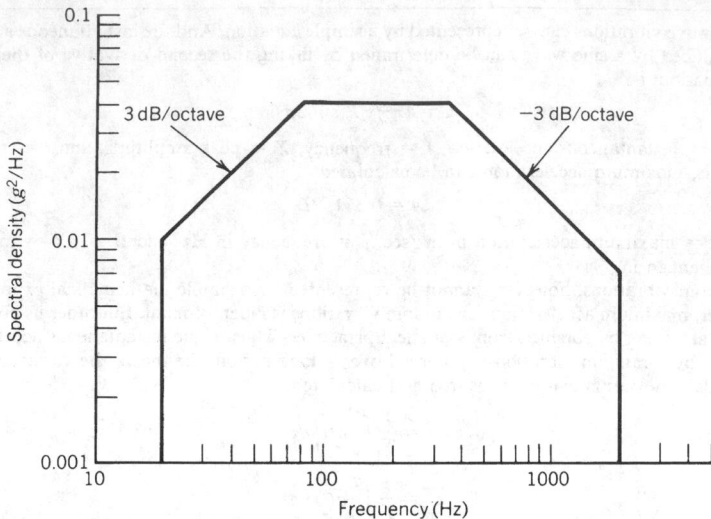

Fig. 23.64 This typical spectral density chart defines a random vibration test that is widely used for stress-screening electronic equipment. Spectral density, measured in g^2/Hz, is proportional to the vibrational power contained in a 1-Hz-wide slice of a random vibration spectrum. The square root of the area under the curve is an rms 6g, the effective acceleration value for the entire spectrum. The use of dB/octave for defining the sloping portions of the diagram is based on filter rolloff terminology.

If a PCB has a Q of 25, its amplitude, when subjected to a sine-wave vibration having a frequency equal to the board's natural frequency, will be 25 times the shaker amplitude. However, if the PCB is subjected to random vibrations, the PCB amplitude will be much lower, proportional to the square root of Q, or in this case, 5.

In sine-wave testing, moreover, all vibrational energy is concentrated in one-frequency. Thus, when that frequency briefly coincides with the natural frequency of an article under test, all the vibrational energy is passed on to the article. And though for sine waves maximum acceleration is only $\sqrt{2}$ times the rms values, instantaneous acceleration is high most of the time.

Random vibrations briefly can produce very high acceleration rates. But instantaneous acceleration rates are less than the rms value most of the time. And vibrational energy is distributed in both time

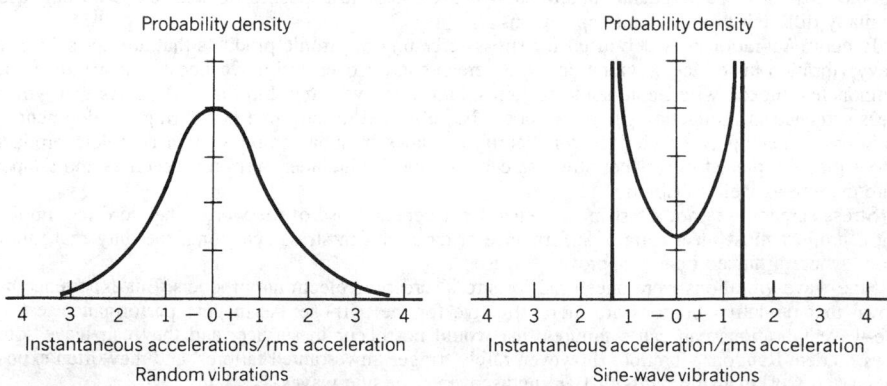

Fig. 23.65 Instantaneous acceleration rates for wide-band random vibrations are defined by a Gaussian probability density curve as shown on the left. This curve indicates that acceleration is close to zero most of the time. High rates do occur, but only rarely. Instantaneous acceleration rates for a sine wave are described by the curve on the right. This curve shows that acceleration is near the rms value most of the time and only rarely close to zero.

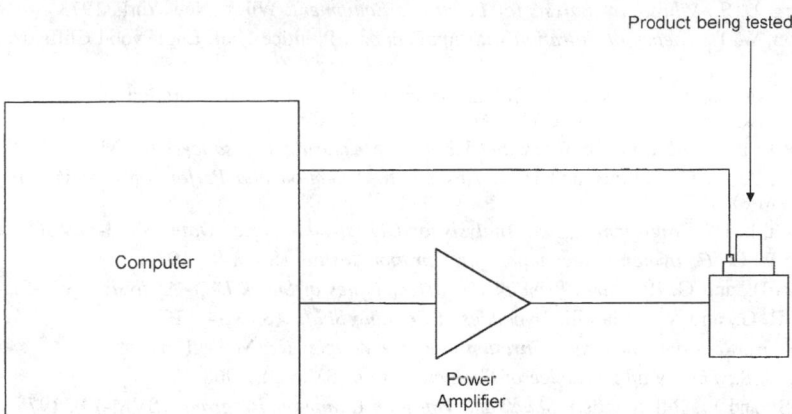

Product being tested

Computer

Power
Amplifier

Fig. 23.66 Input signals for a random vibration shaker system are produced by a computer. The spectrum of shaker motion is compared with the desired spectrum the operator had previously keyed in. The computer quickly corrects the actual spectrum to agree.

and frequency domains. Thus, any one narrow frequency band contains only a small portion of the total energy available. And the narrower the band, the less energy it contains. Items have a high Q are affected only by frequencies quite close to their natural frequency. Thus, random vibrations typically are gentler than sinusoidal vibrations. And random vibration tests are more representative of transportation and in-service vibrations (Fig. 23.66).

Random vibration screens for products that are not subject to vibration in service must, of necessity, be developed empirically. One widely used screen is described in NAVMAT P-9492, a Navy publication on stress screening. This called for rms $6g$ random vibration having a 20–2000-Hz frequency band. The test period was 10 min for products tested on one axis. Test period for additional axes was five minutes each.

An rms $6g$ screen may seem excessively severe for stress screening products that are not subject to severe vibration in service. But screens are not intended for qualification or acceptance. The purpose is to weed out flawed products. And a brief period of an rms $6g$ random vibration will not harm properly designed and manufactured electronic products.

The NAVMAT P-9492 test often is run in conjunction with a temperature cycling test. The temperature range is −65 to 131°F with 40°F/min ramps. Temperature extremes are held until internal parts have stabilized to within 5°F of the specific temperature. Products are cycled 1–10 times through this temperature range, depending on product complexity. Those containing 4000 components are cycled 10 times, but those containing 100 or less are cycled only once.

Sine-wave vibrations should be used for fatigue testing products that are subject to in-service vibration of known frequency. And when a product has been designed to be insensitive to certain frequencies, a sine-wave test is a valid means for demonstrating that the product does not in fact resonate at those frequencies. Sine-wave vibrations also are required for calibrating vibration sensors and systems.

BIBLIOGRAPHY

Bendat, J. S., and A. G. Piersol, *Random Data: Analysis and Measurement Procedures,* Wiley-Interscience, New York, 1971.

Broch, J. T., *Mechanical Vibration and Shock Measurements,* 2nd ed., Bruel & Kjaer Measurement Systems, Denmark, 1979.

Crandall, S. H. (ed.), *Random Vibration*, MIT Press, Cambridge, MA, 1963.

Crandall, S. H., and W. D. Mark, *Random Vibration in Mechanical Systems*, Academic, New York, 1963.

Harris, C. E., and C. E. Crede (eds.), *Shock and Vibration Handbook*, McGraw-Hill, New York, 1961.

———, *Shock and Vibration Handbook*, 2nd ed., McGraw-Hill, New York, 1976.

Keast, D. N., *Measurements in Mechanical Dynamics*, McGraw-Hill, New York, 1967.

Morrison, R., *Grounding and Shielding Techniques in Instrumentation*, Wiley, New York, 1967.

Morrow, C. T., *Shock and Vibration Engineering*, Vol. 1, Wiley, New York, 1963.

Salter, J. P., *Steady-State Vibration*, Kenneth Mason, London, 1969.

Steinberg, D. S., *Vibration Analysis for Electronic Equipment*, Wiley, New York, 1973.

Thomson, W. T., *Theory of Vibration with Applications*, Prentice-Hall, Englewood Cliffs, NJ, 1981.

The following texts are printed in volume form by the Shock and Vibration Information Center, SAVIAC 2231 Crystal Drive 711, Arlington, VA 22202

Bouche, R. R., *Calibration of Shock and Vibration Measuring Transducers* (SVM-11), 1979.

Curtis, A. J., N. G. Tinling, and H. T. Abstein, Jr., *Selection and Performance of Vibration Tests* (SVM-8), 1971.

Enochson, L. D., *Programming and Analysis for Digital Time Series Data* (SVM-3), 1968.

Fackler, W. C., *Equivalence Techniques for Vibration Testing* (SVM-9), 1972.

Kelly, R. D., and G. Richman, *Principles and Techniques of Shock Data Analysis* (SVM-5), 1969.

Loewy, R. G., and V. J. Piarulli, *Dynamics of Rotating Shafts* (SVM-4), 1969.

Lyon, R. H., *Random Noise and Vibration in Space Vehicles* (SVM-1), 1967.

Mustin, G. S., *Theory and Practice of Cushion Design* (SVM-2), 1968.

Pilkey B., and W. Pilkey (eds.), *Shock and Vibration Computer Programs* (SVM-10), 1975.

Ruzicka, J. E., and T. R. Derby, *Influence of Damping in Vibration Isolation* (SVM-7), 1971.

Sevin, E., and W. D. Pilkey, *Optimum Shock and Vibration Isolation* (SVM-6), 1971.

CHAPTER 24

NOISE MEASUREMENT AND CONTROL

George M. Diehl, P.E.
Consulting Engineer
Machinery Acoustics
Phillipsburg, New Jersey

24.1	SOUND CHARACTERISTICS	711	24.14	MACHINES IN SEMIREVERBERANT LOCATIONS	716
24.2	FREQUENCY AND WAVELENGTH	712	24.15	TWO-SURFACE METHOD	717
24.3	VELOCITY OF SOUND	712	24.16	MACHINERY NOISE CONTROL	719
24.4	SOUND POWER AND SOUND PRESSURE	712	24.17	SOUND ABSORPTION	719
24.5	DECIBELS AND LEVELS	712	24.18	NOISE REDUCTION DUE TO INCREASED ABSORPTION IN ROOM	720
24.6	COMBINING DECIBELS	712	24.19	SOUND ISOLATION	720
24.7	SOUND PRODUCED BY SEVERAL MACHINES OF THE SAME TYPE	713	24.20	SINGLE PANEL	721
24.8	AVERAGING DECIBELS	715	24.21	COMPOSITE PANEL	721
24.9	SOUND-LEVEL METER	715	24.22	ACOUSTIC ENCLOSURES	722
24.10	SOUND ANALYZERS	715	24.23	DOUBLE WALLS	723
24.11	CORRECTION FOR BACKGROUND NOISE	715	24.24	VIBRATION ISOLATION	723
24.12	MEASUREMENT OF MACHINE NOISE	716	24.25	VIBRATION DAMPING	725
24.13	SMALL MACHINES IN A FREE FIELD	716	24.26	MUFFLERS	725
			24.27	SOUND CONTROL RECOMMENDATIONS	727

24.1 SOUND CHARACTERISTICS

Sound is a compressional wave. The particles of the medium carrying the wave vibrate longitudinally, or back and forth, in the direction of travel of the wave, producing alternating regions of compression and rarefaction. In the compressed zones the particles move forward in the direction of travel, whereas in the rarefied zones they move opposite to the direction of travel. Sound waves differ from light

Mechanical Engineers' Handbook, 2nd ed., Edited by Myer Kutz.
ISBN 0-471-13007-9 © 1998 John Wiley & Sons, Inc.

waves in that light consists of transverse waves, or waves that vibrate in a plane normal to the direction of propagation.

24.2 FREQUENCY AND WAVELENGTH

Wavelength, the distance from one compressed zone to the next, is the distance the wave travels during one cycle. Frequency is the number of complete waves transmitted per second. Wavelength and frequency are related by the equation

$$v = f\lambda$$

where v = velocity of sound, in meters per second
$\quad f$ = frequency, in cycles per second or hertz
$\quad \lambda$ = wavelength, in meters

24.3 VELOCITY OF SOUND

The velocity of sound in air depends on the temperature, and is equal to

$$v = 20.05 \sqrt{273.2 + C^\circ} \text{ m/sec}$$

where C° is the temperature in degrees Celsius.
The velocity in the air may also be expressed as

$$v = 49.03 \sqrt{459.7 + F^\circ} \text{ ft/sec}$$

where F° is the temperature in degrees Fahrenheit.
The velocity of sound in various materials is shown in Tables 24.1, 24.2, and 24.3.

24.4 SOUND POWER AND SOUND PRESSURE

Sound power is measured in watts. It is independent of distance from the source, and independent of the environment. Sound intensity, or watts per unit area, is dependent on distance. Total radiated sound power may be considered to pass through a spherical surface surrounding the source. Since the radius of the sphere increases with distance, the intensity, or watts per unit area, must also decrease with distance from the source.

Microphones, sound-measuring instruments, and the ear of a listener respond to changing pressures in a sound wave. Sound power, which cannot be measured directly, is proportional to the mean-square sound pressure, p^2, and can be determined from it.

24.5 DECIBELS AND LEVELS

In acoustics, sound is expressed in decibels instead of watts. By definition, a decibel is 10 times the logarithm, to the base 10, of a ratio of two powers, or powerlike quantities. The reference power is 1 pW, or 10^{-12} W. Therefore,

$$L_W = 10 \log \frac{W}{10^{-12}} \tag{24.1}$$

where L_W = sound power level in dB
$\quad W$ = sound power in watts
$\quad \log$ = logarithm to base 10

Sound pressure level is 10 times the logarithm of the pressure ratio squared, or 20 times the logarithm of the pressure ratio. The reference sound pressure is 20 μPa, or 20×10^{-6} Pa. Therefore,

$$L_p = 20 \log \frac{p}{20 \times 10^{-6}} \tag{24.2}$$

where L_p = sound pressure level in dB
$\quad p$ = root-mean-square sound pressure in Pa
$\quad \log$ = logarithm to base 10

24.6 COMBINING DECIBELS

It is often necessary to combine sound levels from several sources. For example, it may be desired to estimate the combined effect of adding another machine in an area where other equipment is operating. The procedure for doing this is to combine the sounds on an energy basis, as follows:

Table 24.1   Velocity of Sound in Solids

Material	Longitudinal Bar Velocity		Plate (Bulk) Velocity	
	cm/sec	fps	cm/sec	fps
Aluminum	5.24×10^5	1.72×10^4	6.4×10^5	2.1×10^4
Antimony	3.40×10^5	1.12×10^4	—	—
Bismuth	1.79×10^5	5.87×10^3	2.18×10^5	7.15×10^3
Brass	3.42×10^5	1.12×10^4	4.25×10^5	1.39×10^4
Cadmium	2.40×10^5	7.87×10^3	2.78×10^5	9.12×10^3
Constantan	4.30×10^5	1.41×10^4	5.24×10^5	1.72×10^4
Copper	3.58×10^5	1.17×10^4	4.60×10^5	1.51×10^4
German silver	3.58×10^5	1.17×10^4	4.76×10^5	1.56×10^4
Gold	2.03×10^5	6.66×10^3	3.24×10^5	1.06×10^4
Iridium	4.79×10^5	1.57×10^4	—	—
Iron	5.17×10^5	1.70×10^4	5.85×10^5	1.92×10^4
Lead	1.25×10^5	4.10×10^3	2.40×10^5	7.87×10^3
Magnesium	4.90×10^5	1.61×10^4	—	—
Manganese	3.83×10^5	1.26×10^4	4.66×10^5	1.53×10^4
Nickel	4.76×10^5	1.56×10^4	5.60×10^5	1.84×10^4
Platinum	2.80×10^5	9.19×10^3	3.96×10^5	1.30×10^4
Silver	2.64×10^5	8.66×10^3	3.60×10^5	1.18×10^4
Steel	5.05×10^5	1.66×10^4	6.10×10^5	2.00×10^4
Tantalum	3.35×10^5	1.10×10^4	—	—
Tin	2.73×10^5	8.96×10^3	3.32×10^5	1.09×10^4
Tungsten	4.31×10^5	1.41×10^4	5.46×10^5	1.79×10^4
Zinc	3.81×10^5	1.25×10^4	4.17×10^5	1.37×10^4
Cork	5.00×10^4	1.64×10^3	—	—
Crystals				
Quartz X cut	5.44×10^5	1.78×10^4	5.72×10^5	1.88×10^4
Rock salt X cut	4.51×10^5	1.48×10^4	4.78×10^5	1.57×10^4
Glass				
Heavy flint	3.49×10^5	1.15×10^4	3.76×10^5	1.23×10^4
Extra heavy flint	4.55×10^5	1.49×10^4	4.80×10^5	1.57×10^4
Heaviest crown	4.71×10^5	1.55×10^4	5.26×10^5	1.73×10^4
Crown	5.30×10^5	1.74×10^4	5.66×10^5	1.86×10^4
Quartz	5.37×10^5	1.76×10^4	5.57×10^5	1.81×10^4
Granite	3.95×10^5	1.30×10^4	—	—
Ivory	3.01×10^5	9.88×10^3	—	—
Marble	3.81×10^5	1.25×10^4	—	—
Slate	4.51×10^5	1.48×10^4	—	—
Wood				
Elm	1.01×10^5	3.31×10^3	—	—
Oak	4.10×10^5	1.35×10^4	—	—

$$L_p = 10 \log [10^{0.1L_1} + 10^{0.1L_2} + \cdots + 10^{0.1L_n}] \qquad (24.3)$$

where L_p = total sound pressure level in dB
  L_1 = sound pressure level of source No. 1
  L_n = sound pressure level of source No. n
  log = logarithm to base 10

24.7   SOUND PRODUCED BY SEVERAL MACHINES OF THE SAME TYPE

The total sound produced by a number of machines of the same type can be determined by adding $10 \log n$ to the sound produced by one machine alone. That is,

Table 24.2 Velocity of Sound in Liquids

Material	Temperature °C	°F	Velocity cm/sec	fps
Alcohol, ethyl	12.5	54.5	1.21×10^5	3.97×10^3
	20	68	1.17×10^5	3.84×10^3
Benzene	20	68	1.32×10^5	4.33×10^3
Carbon bisulfide	20	68	1.16×10^5	3.81×10^3
Chloroform	20	68	1.00×10^5	3.28×10^3
Ether, ethyl	20	68	1.01×10^5	3.31×10^3
Glycerine	20	68	1.92×10^5	6.30×10^3
Mercury	20	68	1.45×10^5	4.76×10^3
Pentane	20	68	1.02×10^5	3.35×10^3
Petroleum	15	59	1.33×10^5	4.36×10^3
Turpentine	3.5	38.3	1.37×10^5	4.49×10^3
	27	80.6	1.28×10^5	4.20×10^3
Water, fresh	17	62.6	1.43×10^5	4.69×10^3
Water, sea	17	62.6	1.51×10^5	4.95×10^3

$$L_p(n) = L_p + 10 \log n$$

where $L_p(n)$ = sound pressure level of n machines
L_p = sound pressure level of one machine
n = number of machines of the same type

In practice, the increase in sound pressure level measured at any location seldom exceeds 6 dB, no matter how many machines are operating. This is because of the necessary spacing between machines, and the fact that sound pressure level decreases with distance.

Table 24.3 Velocity of Sound in Gases

Material	Temperature °C	°F	Velocity cm/sec	fps
Air	0	32	3.31×10^4	1.09×10^3
	20	68	3.43×10^4	1.13×10^3
Ammonia gas	0	32	4.15×10^4	1.48×10^3
Carbon dioxide	0	32	2.59×10^4	8.50×10^2
Carbon monoxide	0	32	3.33×10^4	1.09×10^3
Chlorine	0	32	2.06×10^4	6.76×10^2
Ethane	10	50	3.08×10^4	1.01×10^3
Ethylene	0	32	3.17×10^4	1.04×10^3
Hydrogen	0	32	1.28×10^5	4.20×10^3
Hydrogen chloride	0	32	2.96×10^4	9.71×10^2
Hydrogen sulfide	0	32	2.89×10^4	9.48×10^2
Methane	0	32	4.30×10^4	1.41×10^3
Nitric oxide	10	50	3.24×10^4	1.06×10^3
Nitrogen	0	32	3.34×10^4	1.10×10^3
	20	68	3.51×10^4	1.15×10^3
Nitrous oxide	0	32	2.60×10^4	8.53×10^2
Oxygen	0	32	3.16×10^4	1.04×10^3
	20	68	3.28×10^4	1.08×10^3
Sulfur dioxide	0	32	2.13×10^4	6.99×10^2
Water vapor	0	32	1.01×10^4	3.31×10^2
	100	212	1.05×10^4	3.45×10^2

24.8 AVERAGING DECIBELS

There are many occasions when the average of a number of decibel readings must be calculated. One example is when sound power level is to be determined from a number of sound pressure level readings. In such cases the average may be calculated as follows:

$$\overline{L_p} = 10 \log \left\{ \frac{1}{n} \left[10^{0.1L_1} + 10^{0.1L_2} + \cdots + 10^{0.1L_n} \right] \right\} \tag{24.4}$$

where $\overline{L_p}$ = average sound pressure level in dB
L_1 = sound pressure level at location No. 1
L_n = sound pressure level at location No. n
n = number of locations
log = logarithm to base 10

The calculation may be simplified if the difference between maximum and minimum sound pressure levels is small. In such cases arithmetic averaging may be used instead of logarithmic averaging, as follows:

If the difference between the maximum and minimum of the measured sound pressure levels is 5 dB or less, average the levels arithmetically.

If the difference between maximum and minimum sound pressure levels is between 5 and 10 dB, average the levels arithmetically and add 1 dB.

The results will usually be correct within 1 dB when compared to the average calculated by Eq. (24.4).

24.9 SOUND-LEVEL METER

The basic instrument in all sound measurements is the sound-level meter. It consists of a microphone, a calibrated attenuator, an indicating meter, and weighting networks. The meter reading is in terms of root-mean-square sound pressure level.

The A-weighting network is the one most often used. Its response characteristics approximate the response of the human ear, which is not as sensitive to low-frequency sounds as it is to high-frequency sounds. A-weighted measurements can be used for estimating annoyance caused by noise and for estimating the risk of noise-induced hearing damage. Sound levels read with the A-network are referred to as dBA.

24.10 SOUND ANALYZERS

The octave-band analyzer is the most common analyzer for industrial noise measurements. It separates complex sounds into frequency bands one octave in width, and measures the level in each of the bands.

An octave is the interval between two sounds having a frequency ratio of two. That is, the upper cutoff frequency is twice the lower cutoff frequency. The particular octaves read by the analyzer are identified by the center frequency of the octave. The center frequency of each octave is its geometric mean, or the square root of the product of the lower and upper cutoff frequencies. That is,

$$f_0 = \sqrt{f_1 f_2}$$

where f_0 = the center frequency, in Hz
f_1 = the lower cutoff frequency, in Hz
f_2 = the upper cutoff frequency, in Hz

f_1 and f_2 can be determined from the center frequency. Since $f_2 = 2f_1$ it can be shown that $f_1 = f_0/\sqrt{2}$ and $f_2 = \sqrt{2} f_0$.

Third-octave band analyzers divide the sound into frequency bands one-third octave in width. The upper cutoff frequency is equal to $2^{1/3}$, or 1.26, times the lower cutoff frequency.

When unknown frequency components must be identified for noise control purposes, narrow-band analyzers must be used. They are available with various bandwidths.

24.11 CORRECTION FOR BACKGROUND NOISE

The effect of ambient or background noise should be considered when measuring machine noise. Ambient noise should preferably be at least 10 dB below the machine noise. When the difference is less than 10 dB, adjustments should be made to the measured levels as shown in Table 24.4.

Table 24.4   Correction for Background Sound

Level Increase Due to the Machine (dB)	Value to Be Subtracted from Measured Level (dB)
3	3.0
4	2.2
5	1.7
6	1.3
7	1.0
8	0.8
9	0.6
10	0.5

If the difference between machine octave-band sound pressure levels and background octave-band sound pressure levels is less than 6 dB, the accuracy of the adjusted sound pressure levels will be decreased. Valid measurements cannot be made if the difference is less than 3 dB.

24.12   MEASUREMENT OF MACHINE NOISE

The noise produced by a machine may be evaluated in various ways, depending on the purpose of the measurement and the environmental conditions at the machine. Measurements are usually made in overall A-weighted sound pressure levels, plus either octave-band or third-octave-band sound pressure levels. Sound power levels are calculated from sound pressure level measurements.

24.13   SMALL MACHINES IN A FREE FIELD

A free field is one in which the effects of the boundaries are negligible, such as outdoors, or in a very large room. When small machines are sound tested in such locations, measurements at a single location are often sufficient. Many sound test codes specify measurements at a distance of 1 m from the machine.

Sound power levels, octave band, third-octave band, or A-weighted, may be determined by the following equation:

$$L_W = L_p + 20 \log r + 7.8 \tag{24.5}$$

where L_W = sound power level, in dB
L_p = sound pressure level, in dB
r = distance from source, in m
$\log$ = logarithm to base 10

24.14   MACHINES IN SEMIREVERBERANT LOCATIONS

Machines are almost always installed in semireverberant environments. Sound pressure levels measured in such locations will be greater than they would be in a free field. Before sound power levels are calculated adjustments must be made to the sound pressure level measurements.

There are several methods for determining the effect of the environment. One uses a calibrated reference sound source, with known sound power levels, in octave or third-octave bands. Sound pressure levels are measured on the machine under test, at predetermined microphone locations. The machine under test is then replaced by the reference sound source, and measurements are repeated. Sound power levels can then be calculated as follows:

$$L_{Wx} = \overline{L_{px}} + (L_{Ws} - \overline{L_{ps}}) \tag{24.6}$$

where L_{Wx} = band sound power level of the machine under test
$\overline{L_{px}}$ = average sound pressure level measured on the machine under test
L_{Ws} = band sound power level of the reference source
$\overline{L_{ps}}$ = average sound pressure level on the reference source

Another procedure for qualifying the environment uses a reverberation test. High-speed recording equipment and a special noise source are used to measure the time for the sound pressure level, originally in a steady state, to decrease 60 dB after the special noise source is stopped. This reverberation time must be measured for each frequency, or each frequency band of interest.

Unfortunately, neither of these two laboratory procedures is suitable for sound tests on large machinery, which must be tested where it is installed. This type of machinery usually cannot be shut down while tests are being made on a reference sound source, and reverberation tests cannot be made in many industrial areas because ambient noise and machine noise interfere with reverberation time measurements.

24.15 TWO-SURFACE METHOD

A procedure that can be used in most industrial areas to determine sound pressure levels and sound power levels of large operating machinery is called the two-surface method. It has definite advantages over other laboratory-type tests. The machine under test can continue to operate. Expensive, special instrumentation is not required to measure reverberation time. No calibrated reference source is needed; the machine is its own sound source. The only instrumentation required is a sound level meter and an octave-band analyzer. The procedure consists of measuring sound pressure levels on two imaginary surfaces enclosing the machine under test. The first measurement surface, S_1, is a rectangular parallelepiped 1 m away from a reference surface. The reference surface is the smallest imaginary rectangular parallelepiped that will just enclose the machine, and terminate on the reflecting plane, or floor. The area, in square meters, of the first measurement surface is given by the formula

$$S_1 = ab + 2ac + 2bc \qquad (24.7)$$

where $a = L + 2$
$b = W + 2$
$c = H + 1$

and L, W, and H are the length, width, and height of the reference parallelepiped, in meters.

The second measurement surface, S_2, is a similar but larger, rectangular parallelepiped, located at some greater distance from the reference surface. The area, in square meters, of the second measurement surface is given by the formula

$$S_2 = de + 2df + 2ef \qquad (24.8)$$

where $d = L + 2x$
$e = W + 2x$
$f = H + x$

and x is the distance in meters from the reference surface to S_2.

Microphone locations are usually those shown on Fig. 24.1.

First, the measured sound pressure levels should be corrected for background noise as shown in Table 24.4. Next, the average sound pressure levels, in each octave band of interest, should be calculated as shown in Eq. (24.4).

Octave-band sound pressure levels, corrected for both background noise and for the semireverberant environment, may then be calculated by the equations

$$\overline{L_p} = \overline{L_{p1}} - C \qquad (24.9)$$

$$C = 10 \log \left\{ \left[\frac{K}{K-1} \right] \left[1 - \frac{S_1}{S_2} \right] \right\} \qquad (24.10)$$

$$K = 10^{0.1(\overline{L_{p1}} - \overline{L_{p2}})} \qquad (24.11)$$

where $\overline{L_p}$ = average octave-band sound pressure level over area S_1, corrected for both background sound and environment
$\overline{L_{p1}}$ = average octave-band sound pressure level over area S_1, corrected for background sound only
C = environmental correction
$\overline{L_{p2}}$ = average octave-band sound pressure level over area S_2, corrected for background sound

As an alternative, the environmental correction C may be obtained from Fig. 24.2.

Sound power levels, in each octave band of interest, may be calculated by the equation

$$L_W = \overline{L_p} + 10 \log \left[\frac{S_1}{S_0} \right] \qquad (24.12)$$

where L_W = octave-band sound power level, in dB
$\overline{L_p}$ = average octave-band sound pressure level over area S_1, corrected for both background sound and environment
S_1 = area of measurement surface S_1, in m^2
S_0 = 1 m^2

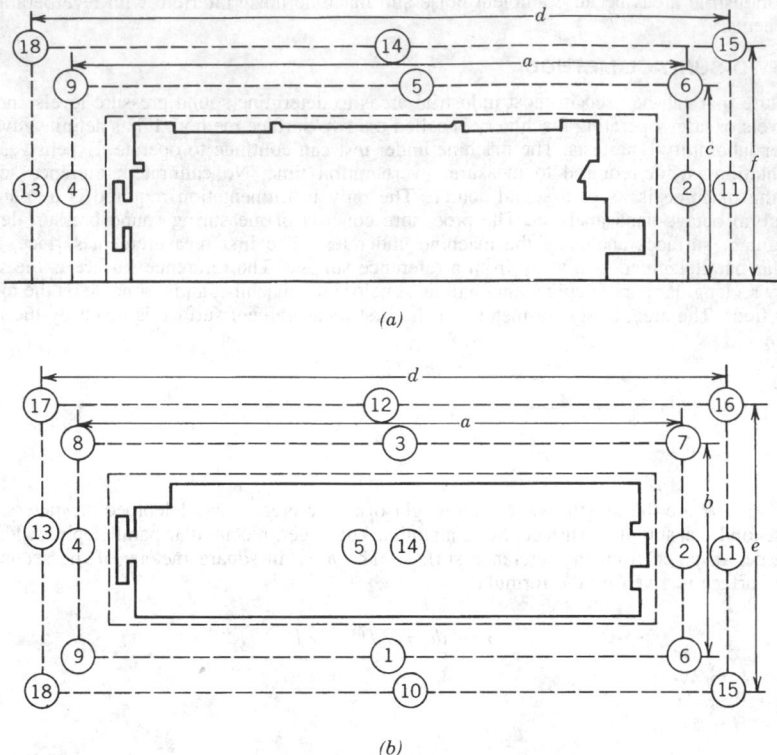

Fig. 24.1 Microphone locations: (*a*) side view; (*b*) plan view.

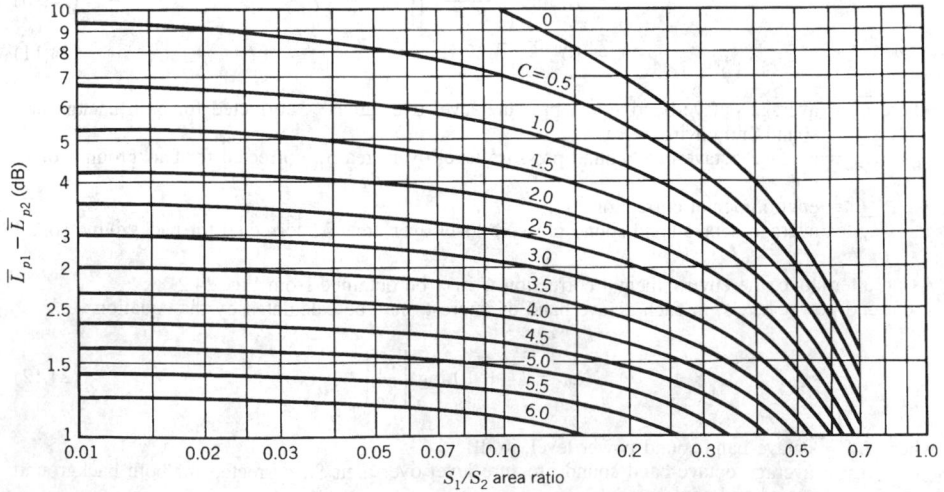

Fig. 24.2 S_1S_2 area ratio.

For simplicity, this equation can be written

$$L_W = \overline{L_p} + 10 \log S_1$$

24.16   MACHINERY NOISE CONTROL

There are five basic methods used to reduce noise: sound absorption, sound isolation, vibration isolation, vibration damping, and mufflers. In most cases several of the available methods are used in combination to achieve a satisfactory solution. Actually, most sound-absorbing materials provide some isolation, although it may be very small; and most sound-isolating materials provide some absorption, even though it may be negligible. Many mufflers rely heavily on absorption, although they are classified as a separate means of sound control.

24.17   SOUND ABSORPTION

The sound-absorbing ability of a material is given in terms of an absorption coefficient, designated by α. Absorption coefficient is defined as the ratio of the energy absorbed by the surface to the energy incident on the surface. Therefore, α can be anywhere between 0 and 1. When $\alpha = 0$, all the incident sound energy is reflected; when $\alpha = 1$, all the energy is absorbed.

The value of the absorption coefficient depends on the frequency. Therefore, when specifying the sound-absorbing qualities of a material, either a table or a curve showing α as a function of frequency is required. Sometimes, for simplicity, the acoustical performance of a material is stated at 500 Hz only, or by a noise reduction coefficient (NRC) that is obtained by averaging, to the nearest multiple of 0.05, the absorption coefficients at 250, 500, 1000, and 2000 Hz.

The absorption coefficient varies somewhat with the angle of incidence of the sound wave. Therefore, for practical use, a statistical average absorption coefficient at each frequency is usually measured and stated by the manufacturer. It is often better to select a sound-absorbing material on the basis of its characteristics for a particular noise rather than by its average sound-absorbing qualities.

Sound absorption is a function of the length of path relative to the wavelength of the sound, and not the absolute length of the path of sound in the material. This means that at low frequencies the thickness of the material becomes important, and absorption increases with thickness. Low-frequency absorption can be improved further by mounting the material at a distance of one-quarter wavelength from a wall, instead of directly on it.

Table 24.5 shows absorption coefficients of various materials used in construction.

The sound absorption of a surface, expressed in either square feet of absorption, or sabins, is equal to the area of the surface, in square feet, times the absorption coefficient of the material on the surface.

Average absorption coefficient, $\overline{\alpha}$, is calculated as follows:

$$\overline{\alpha} = \frac{\alpha_1 S_1 + \alpha_2 S_2 + \cdots + \alpha_n S_n}{S_1 + S_2 + \cdots + S_n} \qquad (24.13)$$

Table 24.5   Absorption Coefficients

Material	125 cps	250 cps	500 cps	1000 cps	2000 cps	4000 cps
Brick, unglazed	0.03	0.03	0.03	0.04	0.05	0.07
Brick, unglazed, painted	0.01	0.01	0.02	0.02	0.02	0.03
Concrete block	0.36	0.44	0.31	0.29	0.39	0.25
Concrete block, painted	0.10	0.05	0.06	0.07	0.09	0.08
Concrete	0.01	0.01	0.015	0.02	0.02	0.02
Wood	0.15	0.11	0.10	0.07	0.06	0.07
Glass, ordinary window	0.35	0.25	0.18	0.12	0.07	0.04
Plaster	0.013	0.015	0.02	0.03	0.04	0.05
Plywood	0.28	0.22	0.17	0.09	0.10	0.11
Tile	0.02	0.03	0.03	0.03	0.03	0.02
6 lb/ft^2 fiberglass	0.48	0.82	0.97	0.99	0.90	0.86

where $\bar{\alpha}$ = the average absorption coefficient
$\alpha_1, \alpha_2, \alpha_n$ = the absorption coefficients of materials on various surfaces
S_1, S_2, S_n = the areas of various surfaces

24.18 NOISE REDUCTION DUE TO INCREASED ABSORPTION IN ROOM

A machine in a large room radiates noise that decreases at a rate inversely proportional to the square of the distance from the source. Soon after the machine is started the sound wave impinges on a wall. Some of the sound energy is absorbed by the wall, and some is reflected. The sound intensity will not be constant throughout the room. Close to the machine the sound field will be dominated by the source, almost as though it were in a free field, while farther away the sound will be dominated by the diffuse field, caused by sound reflections. The distance where the free field and the diffuse field conditions control the sound depends on the average absorption coefficient of the surfaces of the room and the wall area. This critical distance can be calculated by the following equation:

$$r_c = 0.2 \sqrt{R} \tag{24.14}$$

where r_c = distance from source, in m
R = room constant of the room, in m²

Room constant is equal to the product of the average absorption coefficient of the room and the total internal area of the room divided by the quantity one minus the average absorption coefficient. That is,

$$R = \frac{S_t \bar{\alpha}}{1 - \bar{\alpha}} \tag{24.15}$$

where R = the room constant, in m²
$\bar{\alpha}$ = the average absorption coefficient
S_t = the total area of the room, in m²

Essentially free-field conditions exist farther from a machine in a room with a large room constant than they do in a room with a small room constant.

The distance r_c determines where absorption will reduce noise in the room. An operator standing close to a noisy machine will not benefit by adding sound-absorbing material to the walls and ceiling. Most of the noise heard by the operator is radiated directly by the machine, and very little is reflected noise. On the other hand, listeners farther away, at distances greater than r_c, will benefit from the increased absorption.

The noise reduction in those areas can be estimated by the following equation:

$$NR = 10 \log \frac{\bar{\alpha_2} S}{\bar{\alpha_1} S} \tag{24.16}$$

where NR = far field noise reduction, in dB
$\bar{\alpha_1}S$ = room absorption before treatment
$\bar{\alpha_2}S$ = room absorption after treatment

Equation (24.16) shows that doubling the absorption will reduce noise by 3 dB. It requires another doubling of the absorption to get another 3 dB reduction. This is much more difficult than getting the first doubling, and considerably more expensive.

24.19 SOUND ISOLATION

Noise may be reduced by placing a barrier or wall between a noise source and a listener. The effectiveness of such a barrier is described by its transmission coefficient.

Sound transmission coefficient of a partition is defined as the fraction of incident sound transmitted through it.

Sound transmission loss is a measure of sound-isolating ability, and is equal to the number of decibels by which sound energy is reduced in transmission through a partition. By definition, it is 10 times the logarithm to the base 10 of the reciprocal of the sound transmission coefficient. That is,

$$TL = 10 \log \frac{1}{\tau} \tag{24.17}$$

**Table 24.6 Transmission Loss of
Building Materials**

Item	TL
Hollow-core door (³⁄₁₆-in. panels)	15
1¾-in. solid-core oak door	20
2½-in. heavy wood door	25–30
4-in. cinder block	20–25
4-in. cinder block, plastered	40
4-in. cinder slab	40–45
4-in. slab-suspended concrete, plastered	50
Two 4-in. cinder blocks—4-in. air space	55
4-in. brick	45
4-in. brick, plastered	47
8-in. brick, plastered	50
Two 8-in. cinder block—4-in. air space	57

where TL = the transmission loss, in dB
$\quad\tau$ = the transmission coefficient

Transmission of sound through a rigid partition or solid wall is accomplished mainly by the forced vibration of the wall. That is, the partition is forced to vibrate by the pressure variations in the sound wave.

Under certain conditions porous materials can be used to isolate high-frequency sound, and, in general, the loss provided by a uniform porous material is directly proportional to the thickness of the material. For most applications, however, sound absorbing materials are very ineffective sound isolators because they have the wrong characteristics. They are porous, instead of airtight, and they are lightweight, instead of heavy. The transmission loss of nonporous materials is determined by weight per square foot of surface area, and how well all cracks and openings are sealed. Transmission loss is affected also by dynamic bending stiffness and internal damping. Table 24.6 shows the transmission loss of various materials used in construction.

24.20 SINGLE PANEL

The simplest type of sound-isolating barrier is a single, homogeneous, nonporous partition. In general, the transmission loss of a single wall of this type is proportional to the logarithm of the mass. Its isolating ability also increases with frequency, and the approximate relationship is given by the following equation:

$$TL = 20 \log W + 20 \log f - 33 \qquad (24.18)$$

where TL = the transmission loss, in dB
$\quad W$ = the surface weight, in pounds per square foot
$\quad f$ = the frequency, in Hz

This means that the transmission loss increases 6 dB each time the weight is doubled, and 6 dB each time the frequency is doubled. In practice, these numbers are each about 5 dB, instead of 6 dB.

In general, a single partition or barrier should not be counted on to provide noise reduction of more than about 10 dB.

24.21 COMPOSITE PANEL

Many walls or sound barriers are made of several different materials. For example, machinery enclosures are commonly constructed of sheet steel, but they may have a glass window to observe instruments inside the enclosure. The transmission coefficients of the two materials are different. Another example is when there are necessary cracks or openings in the enclosure where it fits around a rotating shaft. In this case, the transmission loss of the opening is zero, and the transmission coefficient is 1.0.

The effectiveness of such a wall is related to both the transmission coefficients of the materials in the wall and the areas of the sections. A large area transmits more noise than a small one made of the same material. Also, more noise can be transmitted by a large area with a relatively small transmission coefficient than by a small one with a comparatively high transmission coefficient. On the other hand, a small area with a high transmission coefficient can ruin the effectiveness of an otherwise excellently designed enclosure. Both transmission coefficients and areas must be controlled carefully.

The average transmission coefficient of a composite panel is

$$\bar{\tau} = \frac{\tau_1 S_1 + \tau_2 S_2 + \tau_3 S_3 + \cdots + \tau_n S_n}{S_1 + S_2 + S_3 + \cdots + S_n} \tag{24.19}$$

where $\bar{\tau}$ = the average transmission coefficient
$\tau_1, \ldots, \tau_n$ = the transmission coefficients of the various areas
$S_1, \ldots, S_n$ = the various areas

Figure 24.3 shows how the transmission loss of a composite wall or panel may be determined from the transmission loss values of its parts. It also shows the damaging effect of small leaks in the enclosure. In this case the transmission loss of the leak opening is zero.

A leak of only 0.1% in an expensive, high-quality door or barrier, constructed of material with a transmission loss of 50 dB, would reduce the transmission-loss by 20 dB, resulting in a *TL* of only 30 dB instead of 50 dB. A much less expensive 30 dB barrier would be reduced by only 3 dB, resulting in a *TL* of 27 dB. This shows that small leaks are more damaging to a high-quality enclosure than to a lower-quality one.

24.22 ACOUSTIC ENCLOSURES

When machinery noise must be reduced by 20 dB or more, it is usually necessary to use complete enclosures. It must be kept in mind that the actual decrease in noise produced by an enclosure depends on other things as well as the transmission loss of the enclosure material. Vibration resonances must be avoided, or their effects must be reduced by damping; structural and mechanical connections must not be permitted to short circuit the enclosure; and the enclosure must be sealed as well as possible to prevent acoustic leaks. In addition, the actual noise reduction depends on the acoustic properties of the room in which the enclosure is located. For this reason, published data on transmission loss of various materials should not be assumed to be the same as the noise reduction that will be obtained when using those materials in enclosures. How materials are used in machinery enclosures is just as important as which materials are used.

A better description of the performance of an acoustic enclosure is given by its noise reduction *NR*, which is defined as the difference in sound pressure level between the enclosure and the receiving room. The relation between noise reduction and transmission loss is given by the following equation:

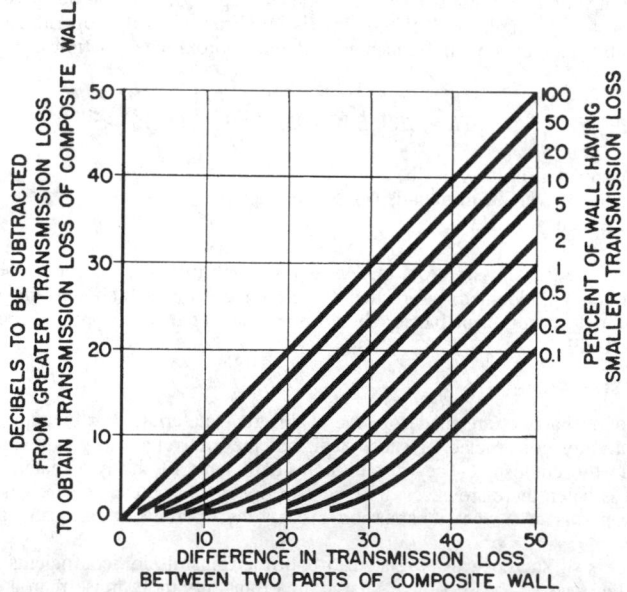

Fig. 24.3 Decibels to be subtracted from greater transmission loss to obtain transmission loss of composite wall. The difference in transmission loss between two parts of composite wall. Percentage of wall having smaller transmission loss.

$$NR = TL - 10 \log \left[\frac{1}{4} + \frac{S_{\text{wall}}}{R_{\text{room}}} \right] \tag{24.20}$$

where NR = the difference in sound pressure level, in dB, between the enclosure and the receiving room
TL = the transmission loss of the enclosure walls, in dB
S_{wall} = the area of the enclosure walls, in square feet
R_{room} = the room constant of the receiving room, in square feet

Equation (24.20) indicates that if the room constant is very large, like it would be outdoors, or in a very large area with sound-absorbing material on the walls and ceiling, the NR could exceed the TL by almost 6 dB. In most industrial areas the NR is approximately equal to, or is several dB less than, the TL. '

If there is no absorption inside the enclosure, and it is highly reverberant, like smooth sheet steel, sound reflects back and forth many times. As the noise source continues to radiate noise, with none of it being absorbed, the noise continues to increase without limit. Theoretically, with zero absorption, the sound level will increase to such a value that no enclosure can contain it.

Practically, this condition cannot exist; there will always be some absorption present, even though it may be very little. However, the sound level inside the enclosure will be greater than it would be without the enclosure. For this reason, the sound level inside the enclosure should be assumed to equal the actual noise source plus 10 dB, unless a calculation shows it to be otherwise.

When absorbing material is added to the inside of the enclosure, sound energy decreases each time it is reflected.

An approximate method for estimating the noise reduction of an enclosure is

$$NR = 10 \log \left[1 + \frac{\overline{\alpha}}{\overline{\tau}} \right] \tag{24.21}$$

where NR = the noise reduction, in dB
$\overline{a}$ = the average absorption coefficient of the inside of the enclosure
$\overline{\tau}$ = the average transmission coefficient of the enclosure

Equation (24.21) shows that in the theoretical case where there is no sound absorption, there is no noise reduction.

24.23 DOUBLE WALLS

A 4-in.-thick brick wall has a transmission loss of about 45 dB. An 8-in.-thick brick wall, with twice as much weight, has a transmission loss of about 50 dB. After a certain point has been reached it is found to be impractical to try to obtain higher isolation values simply by doubling the weight, since both the weight and the cost become excessive, and only a 5 dB improvement is gained for each doubling of weight.

An increase can be obtained, however, by using double-wall construction. That is, two 4-in.-thick walls separated by an air space are better than one 8-in. wall. However, noise radiated by the first panel can excite vibration of the second one and cause it to radiate noise. If there are any mechanical connections between the two panels, vibration of one directly couples to the other, and much of the benefit of double-wall construction is lost.

There is another factor that can reduce the effectiveness of double-wall construction. Each of the walls represents a mass, and the air space between them acts as a spring. This mass–spring–mass combination has a series of resonances that greatly reduce the transmission loss at the corresponding frequencies. The effect of the resonances can be reduced by adding sound-absorbing material in the space between the panels.

24.24 VIBRATION ISOLATION

There are many instances where airborne sound can be reduced substantially by isolating a vibrating part from the rest of the structure. A vibration isolator, in its simplest form is some type of resilient support. The purpose of the isolator may be to reduce the magnitude of force transmitted from a vibrating machine or part of a machine to its supporting structure. Conversely, its purpose may be to reduce the amplitude of motion transmitted from a vibrating support to a part of the system that is radiating noise due to its vibration.

Vibration isolators can be in the form of steel springs, cork, felt, rubber, plastic, or dense fiberglass. Steel springs can be calculated quite accurately and can do an excellent job of vibration isolation. However, they also can have resonances, and high-frequency vibrations can travel through them readily, even though they are effectively isolating the lower frequencies. For this reason, springs are usually used in combination with elastomers or similar materials. Elastomers, plastics, and materials of this type have high internal damping and do not perform well below about 15 Hz. However, this is below the audible range, and, therefore, it does not limit their use in any way for effective sound control.

The noise reduction that can be obtained by installing an isolator depends on the characteristics of the isolator and the associated mechanical structure. For example, the attenuation that can be obtained by spring isolators depends not only on the spring constant, or spring stiffness (the force necessary to stretch or compress the spring one unit of length), but also on the mass load on the spring, the mass and stiffness of the foundation, and the type of excitation.

If the foundation is very massive and rigid, and if the mounted machine vibrates at constant amplitude, the reduction in force on the foundation is independent of frequency. If the machine vibrates at a constant force, the reduction in force depends on the ratio of the exciting frequency to the natural frequency of the system.

When a vibrating machine is mounted on an isolator, the ratio of the force applied to the isolator by the machine, to the force transmitted by the isolator, to the foundation is called the "transmissibility." That is,

$$\text{transmissibility} = \frac{\text{transmitted force}}{\text{impressed force}}$$

Under ideal conditions this ratio would be zero. In practice, the objective is to make it as small as possible. This can be done by designing the system so that the natural frequency of the mounted machine is very low compared to the frequency of the exciting force.

If no damping is present, the transmissibility can be expressed by the following equation:

$$T = \frac{1}{1 - (\omega/\omega_n)^2} \tag{24.22}$$

where T = the transmissibility, expressed as a fraction
ω = the circular frequency of the exciting force, in radians per second
ω_n = the circular frequency of the mounted system, in radians per second

When $\omega/\omega_n = 0$, the transmissibility equals 1.0. That is, there is no benefit obtained from the isolator.

If ω/ω_n is greater than zero but less than 1.41, the isolator actually increases the magnitude of the transmitted force. This is called the "region of amplification." In fact, when ω/ω_n equals 1.0, the theoretical amplitude of the transmitted force goes to infinity, since this is the point where the frequency of the disturbing force equals the system natural frequency.

Equation (24.22) indicates that the transmissibility becomes negative when ω/ω_n is greater than 1.0. The negative number is simply due to the phase relation between force and motion, and it can be disregarded when considering only the amount of transmitted force.

Since vibration isolation is achieved only when ω/ω_n is greater than 1.41, the equation for transmissibility can be written so that T is positive:

$$T = \frac{1}{(\omega/\omega_n)^2 - 1}$$

Also, since $\omega = 2\pi f$,

$$T = \frac{1}{(f/f_n)^2 - 1} \tag{24.23}$$

The static deflection of a spring when stretched or compressed by a weight is related to its natural frequency by the equation

$$f_n = 3.14 \sqrt{\frac{1}{d}} \tag{24.24}$$

where f_n = the natural frequency, in hertz
d = the deflection, in inches

When this is substituted in the equation for transmissibility, Eq. (24.23), it can be shown that

$$d = \left(\frac{3.14}{f}\right)^2 \left(\frac{1}{T} + 1\right) \tag{24.25}$$

This shows that the transmissibility can be determined from the deflection of the isolator due to its supported load.

Equations (24.23) and (24.25) can be plotted, as shown in Fig. 24.4, for convenience in selecting isolator natural frequencies or deflections.

For critical applications, the natural frequency of the isolator should be about one-tenth to one-sixth of the disturbing frequency. That is, the transmissibility should be between 1 and 3%. For less critical conditions, the natural frequency of the isolator should be about one-sixth to one-third of the driving frequency, with transmissibility between 3 and 12%.

24.25 VIBRATION DAMPING

Complex mechanical systems have many resonant frequencies, and whenever an exciting frequency is coincident with one of the resonant frequencies, the amplitude of vibration is limited only by the amount of damping in the system. If the exciting force is wide band, several resonant vibrations can occur simultaneously, thereby compounding the problem. Damping is one of the most important factors in noise and vibration control.

There are three kinds of damping. Viscous damping is the type that is produced by viscous resistance in a fluid, for example, a dashpot. The damping force is proportional to velocity. Dry friction, or Coulomb damping, produces a constant damping force, independent of displacement and velocity. The damping force is produced by dry surfaces rubbing together, and it is opposite in direction to that of the velocity. Hysteresis damping, also called material damping, produces a force that is in phase with the velocity but is proportional to displacement. This is the type of damping found in solid materials, such as elastomers, widely used in sound control.

A large amount of noise radiated from machine parts comes from vibration of large areas or panels. These parts may be integral parts of the machine or attachments to the machine. They can be flat or curved, and vibration can be caused by either mechanical or acoustic excitation. The radiated noise is a maximum when the parts are vibrating in resonance.

When the excitation is mechanical, vibration isolation may be all that is needed. In other instances, the resonant response can be reduced by bonding a layer of energy-dissipating polymeric material to the structure. When the structure bends, the damping material is placed alternately in tension and compression, thus dissipating the energy as heat.

This extensional or free-layer damping is remarkably effective in reducing resonant vibration and noise in relatively thin, lightweight structures such as panels. It becomes less effective as the structure stiffness increases, because of the excessive increase in thickness of the required damping layer.

In a vibrating structure, the amount of energy dissipated is a function of the amount of energy necessary to deflect the structure, compared to that required to deflect the damping material. If 99% of the vibration energy is required to deflect the structure, and 1% is required to deflect the damping layer, then only 1% of the vibration energy is dissipated.

Resonant vibration amplitude in heavier structures can be controlled effectively by using constrained-layer damping. In this method, a relatively thin layer of viscoelastic damping material is constrained between the structure and a stiff cover plate. Vibration energy is removed from the system by the shear motion of the damping layer.

24.26 MUFFLERS

Silencers, or mufflers, are usually divided into two categories: absorptive and reactive. The absorptive type, as the name indicates, removes sound energy by the use of sound-absorbing materials. They have relatively wide-band noise-reduction characteristics, and are usually applied to problems associated with continuous spectra, such as fans, centrifugal compressors, jet engines, and gas turbines. They are also used in cases where a narrow-band noise predominates, but the frequency varies because of a wide range of operating speed.

A variety of sound-absorbing materials are used in many different configurations, determined by the level of the unsilenced noise and its frequency content, the type of gas being used, the allowable pressure drop through the silencer, the gas velocity, gas temperature and pressure, and the noise criterion to be met.

Fiberglass or mineral wool with density approximately 0.5–6.0 lb/ft^3 is frequently used in absorptive silencers. These materials are relatively inexpensive and have good sound-absorbing characteristics. They operate on the principle that sound energy causes the material fibers to move, converting the sound energy into mechanical vibration and heat. The fibers do not become very warm since the sound energy is actually quite low, even at fairly high decibel levels.

The simplest kind of absorptive muffler is a lined duct, where the absorbing material is either added to the inside of the duct walls or the duct walls themselves are made of sound-absorbing material. The attenuation depends on the duct length, thickness of the lining, area of the air passage, type of absorbing material, and frequency of the sound passing through.

The acoustical performance of absorptive mufflers is improved by adding parallel or annular baffles to increase the amount of absorption. This also increases pressure drop through the muffler, so that spacing and area must be carefully controlled.

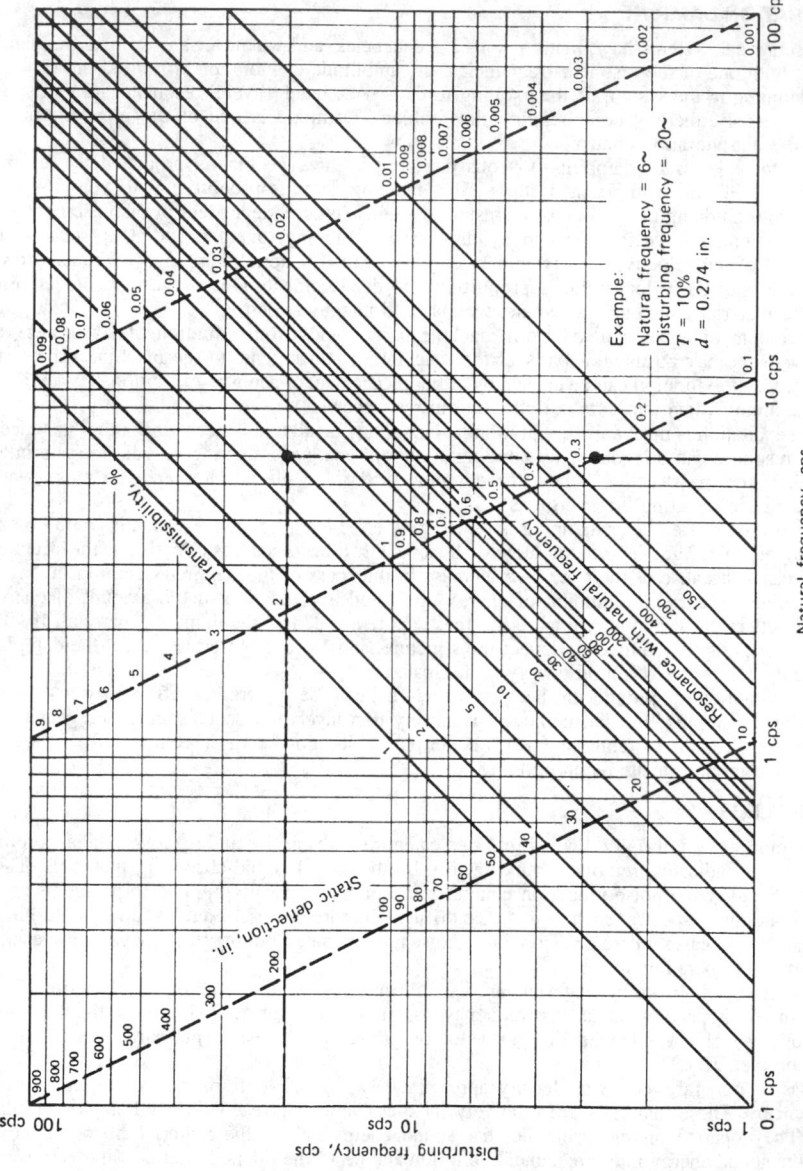

Fig. 24.4 Transmissibility of flexible mountings: $T = 1/[(\omega/\omega_n)^2 - 1]$.

Natural frequency, cps

Disturbing frequency, cps

Static deflection, in.

Transmissibility, %

Resonance with natural frequency

Example:
Natural frequency = 6~
Disturbing frequency = 20~
$T = 10\%$
$d = 0.274$ in.

Reactive mufflers have a characteristic performance that does not depend to any great extent on the presence of sound-absorbing material, but utilizes the reflection characteristics and attenuating properties of conical connectors, expansion chambers, side branch resonators, tail pipes, and so on, to accomplish sound reduction.

Expansion chambers operate most efficiently in applications involving discrete frequencies rather than broad-band noise. The length of the chamber is adjusted so that reflected waves cancel the incident waves, and since wavelength depends on frequency, expansion chambers should be tuned to some particular frequency. When a number of discrete frequencies must be attenuated, several expansion chambers can be placed in series, each tuned to a particular wavelength.

An effective type of reactive muffler, called a Helmholtz resonator, consists of a vessel containing a volume of air, that is connected to a noise source, such as a piping system. When a pure-tone sound wave is propagated along the pipe, the air in the vessel expands and contracts. By proper design of the area and length of the neck, and volume of the chamber, sound wave cancellation can be obtained, thereby reducing the tone. This type of resonator produces maximum noise reduction over a very narrow frequency range, but it is possible to combine several Helmholtz resonators on a piping system so that not only will each cancel out at its own frequency, but they can be made to overlap so that noise is attenuated over a wider range instead of at sharply tuned points.

Helmholtz resonators are normally located in side branches, and for this reason they do not affect flow in the main pipe.

The resonant frequency of these devices can be calculated by the equation

$$f = \frac{C}{2\pi}\sqrt{\frac{A}{LV}} \tag{24.26}$$

where f = the resonant frequency, in hertz
  C = the speed of sound in the fluid, in feet per second
  A = the cross-sectional area of the neck, in square feet
  L = the length of the neck, in feet
  V = the volume of the chamber, in cubic feet

The performance of all types of mufflers can be stated in various ways. Not everyone uses the same terminology, but in general, the following definitions apply:

INSERTION LOSS is defined as the difference between two sound pressure levels measured at the same point in space before and after a muffler is inserted in the system.

DYNAMIC INSERTION LOSS is the same as insertion loss, except that it is measured when the muffler is operating under rated flow conditions. Therefore, the dynamic insertion loss is of more interest than ratings based on no-flow conditions.

TRANSMISSION LOSS is defined as the ratio of sound power incident on the muffler to the sound power transmitted by the muffler. It cannot be measured directly, and it is difficult to calculate analytically. For these reasons the transmission loss of a muffler has little practical application.

ATTENUATION is used to describe the decrease in sound power as sound waves travel through the muffler. It does not convey information about how a muffler performs in a system.

NOISE REDUCTION is defined as the difference between sound pressure levels measured at the inlet of a muffler and those at the outlet.

24.27 SOUND CONTROL RECOMMENDATIONS

Sound control procedures should be applied during the design stages of a machine, whenever possible. A list of recommendations for noise reduction follows:

1. Reduce horsepower. Noise is proportional to horsepower. Therefore, the machine should be matched to the job. Excess horsepower means excess noise.

2. Reduce speed. Slow-speed machinery is quieter than high-speed machinery.

3. Keep impeller tip speeds low. However, it is better to keep the rpm low and the impeller diameter large than to keep the rpm high and the impeller diameter small, even though the tip speeds are the same.

4. Improve dynamic balance. This decreases rotating forces, structure-borne sound, and the excitation of structural resonances.

5. Reduce the ratio of rotating masses to fixed masses.

6. Reduce mechanical run-out of shafts. This improves the initial static and dynamic balance.

7. Avoid structural resonances. These are often responsible for many unidentified components in the radiated sound. In addition to being excited by sinusoidal forcing frequencies, they can be excited by impacting parts and sliding and rubbing contacts.

8. Eliminate or reduce impacts. Either reduce the mass of impacting parts or their striking velocities.

9. Reduce peak acceleration. Reduce the rate of change of velocity of moving parts by using the maximum time possible to produce the required velocity change, and by keeping the acceleration as nearly constant as possible over the available time period.

10. Improve lubrication. Inadequate lubrication is often the cause of bearing noise, structure-borne noise due to friction, and the excitation of structural resonances.

11. Maintain closer tolerances and clearances in bearings and moving parts.

12. Install bearings correctly. Improper installation accounts for approximately half of bearing noise problems.

13. Improve alignment. Improper alignment is a major source of noise and vibration.

14. Use center of gravity mounting whenever feasible. When supports are symmetrical with respect to the center of gravity, translational modes of vibration do not couple to rotational modes.

15. Maintain adequate separation between operating speeds and lateral and torsional resonant speeds.

16. Consider the shape of impeller vanes from an acoustic standpoint. Some configurations are noisier than others.

17. Keep the distance between impeller vanes and cutwater or diffuser vanes as large as possible. Close spacing is a major source of noise.

18. Select combinations of rotating and stationary vanes that are not likely to excite strong vibration and noise.

19. Design turning vanes properly. They are a source of self-generated noise.

20. Keep the areas of inlet passages as large as possible and their length as short as possible.

21. Remove or keep at a minimum any obstructions, bends, or abrupt changes in fluid passages.

22. Pay special attention to inlet design. This is extremely important in noise generation.

23. Item 22 applies also to the discharge, but the inlet is more important than the discharge from an acoustic standpoint.

24. Maintain gradual, not abrupt, transition from one area to the next in all fluid passages.

25. Reduce flow velocities in passages, pipes, and so on. Noise can be reduced substantially by reducing flow velocities.

26. Reduce jet velocities. Jet noise is proportional to the eighth power of the velocity.

27. Reduce large radiating areas. Surfaces radiating certain frequencies can often be divided into smaller areas with less radiating efficiency.

28. Disconnect possible sound radiating parts from other vibrating parts by installing vibration breaks to eliminate metal to metal contact.

29. Provide openings or air leaks in large radiating areas so that air can move through them. This reduces pressure build-up and decreases radiated noise.

30. Reduce clearances, piston weights, and connecting rod weights in reciprocating machinery to reduce piston impacts.

31. Apply additional sound control devices, such as inlet and discharge silencers and acoustic enclosures.

32. When acoustic enclosures are used, make sure that all openings are sealed properly.

33. Install machinery on adequate mountings and foundations to reduce structure-borne sound and vibration.

34. Take advantage of all directivity effects whenever possible by directing inlet and discharge openings away from listeners or critical areas.

35. When a machine must meet a particular sound specification, purchase driving motors, turbines, gears, and auxiliary equipment that produce 3- to 5-dB lower sound levels than the machine alone. This ensures that the combination is in compliance with the specification.

CHAPTER 25

NONDESTRUCTIVE TESTING

Robert L. Crane
Theodore E. Matikas
Air Force Wright Laboratory
Materials Directorate
Nondestructive Evaluation Branch
WL/MLLP
Wright Patterson Air Force Base
Dayton, Ohio

25.1	**INTRODUCTION**	**729**
25.2	**LIQUID PENETRANTS**	**730**
	25.2.1 The Penetrant Process	730
	25.2.2 Categories of Penetrants	730
	25.2.3 Reference Standards	730
	25.2.4 Limitations of Penetrant Inspections	730
25.3	**ULTRASONIC METHODS**	**732**
	25.3.1 Sound Waves	733
	25.3.2 Reflection and Transmission of Sound	733
	25.3.3 Refraction of Sound	735
	25.3.4 The Inspection Process	737
25.4	**RADIOGRAPHY**	**738**
	25.4.1 The Generation and Absorption of X Radiation	739
	25.4.2 Neutron Radiography	740
	25.4.3 Attenuation of X Radiation	741
	25.4.4 Film-Based Radiography	742
	25.4.5 The Penetrameter	743
	25.4.6 Real-Time Radiography	744
	25.4.7 Computed Tomography	744
25.5	**EDDY CURRENT INSPECTION**	**746**
	25.5.1 The Skin Effect	746
	25.5.2 The Impedance Plane	746
	25.5.3 Liftoff of the Inspection Coil from the Specimen	747
25.6	**THERMAL METHODS**	**750**
	25.6.1 Infrared Cameras	750
	25.6.2 Thermal Paints	751
	25.6.3 Thermal Testing	751
25.7	**MAGNETIC PARTICLE METHOD**	**751**
	25.7.1 The Magnetizing Field	751
	25.7.2 Continuous versus Noncontinuous Fields	752
	25.7.3 The Inspection Process	753
	25.7.4 Demagnetizing the Part	753
APPENDIX A	**ULTRASONIC PROPERTIES OF COMMON MATERIALS**	**754**
APPENDIX B	**ELECTRICAL RESISTIVITIES AND CONDUCTIVITIES OF COMMERCIAL METALS AND ALLOYS**	**759**

25.1 INTRODUCTION

Nondestructive evaluation (NDE) encompasses those physical and chemical tests that are used to determine if a component or structure can perform its intended function without the test methods impairing the component's performance. Until recently, NDE was relegated to detecting physical flaws and estimating their dimensions. These data were used to determine if a component should be scrapped or repaired, based on quality-acceptance criteria. Such traditional definitions are being expanded as requirements for high-reliability, cost-effective NDE tests are increasing. In addition, NDE techniques are changing as they become an integral part of the automated manufacturing process.

Mechanical Engineers' Handbook, 2nd ed., Edited by Myer Kutz.
ISBN 0-471-13007-9 © 1998 John Wiley & Sons, Inc.

This chapter is but a brief review of the more commonly used NDE methods. Those who require more detailed information on standard NDE practices should consult Refs. 1–6 at the end of the chapter. For information on recent advances in NDE research the reader is referred to Refs. 7–14.

The NDE methods reviewed here consist of the five classical techniques—penetrants, ultrasonic methods, radiography, magnetic particle tests, and eddy current methods. Additionally, we have briefly covered thermal-inspection methods.

25.2 LIQUID PENETRANTS

Liquid penetrants are used to detect surface-connected discontinuities in solid, nonporous materials. The method uses a brightly colored penetrating liquid that is applied to the surface of a clean part. The liquid in time enters the discontinuity and is later withdrawn to provide a surface indication of the flaw. This process is depicted schematically in Fig. 25.1. A penetrant flaw indication in turbine blade is shown in Fig. 25.2.

25.2.1 The Penetrant Process

Technical societies and military specifications have developed classification systems for penetrants. Society documents (typically ASTM E165) categorize penetrants into two methods (visible and fluorescent) and three types (water washable, post-emulsifiable, and solvent removable). Penetrants, then, are classified by type of dye, rinse process, and sensitivity. See Ref. 1, Vol. 2, for a more detailed discussion of penetrant testing.

The first step in penetrant testing (PT) or inspection is to clean the part (Fig. 25.1*a* and 25.1*b*). Many times this critical step is the most neglected phase of the inspection. Since PT detects only flaws that are open to the surface, the flaw and part surface must, prior to inspection, be free of dirt, grease, oil, water, chemicals, and other foreign materials. Typical cleaning procedures use vapor degreasers, ultrasonic cleaners, alkaline cleaners, or solvents.

After the surface is clean, a penetrant is applied to the part by dipping, spraying, or brushing. Step 2 in Fig. 25.1*c* shows the penetrant on the part surface and in the flaw. In the case of tight surface openings, such as fatigue cracks, the penetrant must be allowed to remain on the part for a minimum of 30 minutes to enhance the probability of complete flaw filling. Fluorescent dye penetrants are used for many inspections where high sensitivity is required.

At the conclusion of the minimum dwell time, the penetrant on the surface of the part is removed by one of three processes, depending on the characteristics of the inspection penetrant. Ideally, only the surface penetrant is removed and the penetrant in the flaw is left undisturbed (Fig. 25.1*c*).

The final step in a basic penetrant inspection is the application of a developer, wet or dry, to the part surface. The developer aids in the withdrawal of penetrant from the flaw and provides a suitable background for flaw detection. The part is then viewed under a suitable light source; either ultraviolet or visible light. White light is used for visible penetrants while ultraviolet light is used for fluorescent penetrants. A typical penetrant indication for a crack in a jet engine turbine blade is shown in Fig. 25.2.

25.2.2 Categories of Penetrants

Once the penetrant material is applied to the surface of the part, it must be removed before an inspection can be carried out. Penetrants are often categorized by their removal method. There are generally three methods of removing the penetrant and thus three categories. Water-washable penetrants contain an emulsifier that permits water to wet the penetrant and carry it from the part, much as a detergent removes stains from clothing during washing. The penetrant is usually removed with a water spray. Post-emulsifiable penetrants require that an emulsifier be applied to the part to permit water to remove the excess penetrant. After a short dwell time, during which the emulsifier mixes with the surface penetrant, a water spray cleans the part. For solvent-removable penetrants, the excess material is usually removed with a solvent spray and wiping. This process is generally used in field applications where water-removal techniques are not applicable.

25.2.3 Reference Standards

Several types of reference standards are used to check the effectiveness of liquid-penetrant systems. One of the oldest and most often-used methods involves chromium-cracked panels, which are available in sets containing fine, medium, and coarse cracks. The panels are capable of classifying penetrant materials by sensitivity and identifying changes in the penetrant process.

25.2.4 Limitations of Penetrant Inspections

The major limitation of liquid-penetrant inspection is that it can only detect flaws that are open to the surface. Other methods are used for detecting subsurface flaws. Another factor that may inhibit the effectiveness of liquid-penetrant inspection is the surface roughness of the part being inspected. Very rough surfaces are likely to produce excessive background or false indications during inspection. Although the liquid-penetrant method is used to inspect some porous parts, such as powder metallurgy

Table 25.1 Capabilities of the Common NDE Methods

Method	Typical Flaws Detected	Typical Application	Advantages	Disadvantages
Radiography	Voids, porosity, inclusions, and cracks	Castings, forgings, weldments, and structural assemblies	Detects internal flaws; useful on a wide variety of geometric shapes; portable; provides a permanent record	High cost; insensitive to thin laminar flaws, such as tight fatigue cracks and delaminations; potential health hazard
Liquid penetrants	Cracks, gouges, porosity, laps, and seams open to a surface	Castings, forgings, weldments, and components subject to fatigue or stress–corrosion cracking	Inexpensive; easy to apply; portable; easily interpreted	Flaw must be open to an accessible surface, level of detectability operator-dependent
Eddy current testing	Cracks, and variations in alloy composition or heat treatment, wall thickness, dimensions	Tubing, local regions of sheet metal, alloy sorting, and coating thickness measurement	Moderate cost, readily automated; portable	Detects flaws that change in conductivity of metals; shallow penetration; geometry-sensitive
Magnetic particles	Cracks, laps, voids, porosity, and inclusions	Castings, forgings, and extrusions	Simple; inexpensive; detects shallow subsurface flaws as well as surface flaws	Useful for ferromagnetic materials only; surface preparation required, irrelevant indications often occur; operator-dependent
Thermal testing	Voids or disbonds in both metallic and nonmetallic materials, location of hot or cold spots in thermally active assemblies	Laminated structures, honeycomb, and electronic circuit boards	Produces a thermal image that is easily interpreted	Difficult to control surface emissivity; poor discrimination
Ultrasonic testing	Cracks, voids, porosity, inclusions and delaminations and lack of bonding between dissimilar materials	Composites, forgings, castings, and weldments and pipes	Excellent depth penetration; good sensitivity and resolution; can provide permanent record	Requires acoustic coupling to component; slow; interpretation is often difficult

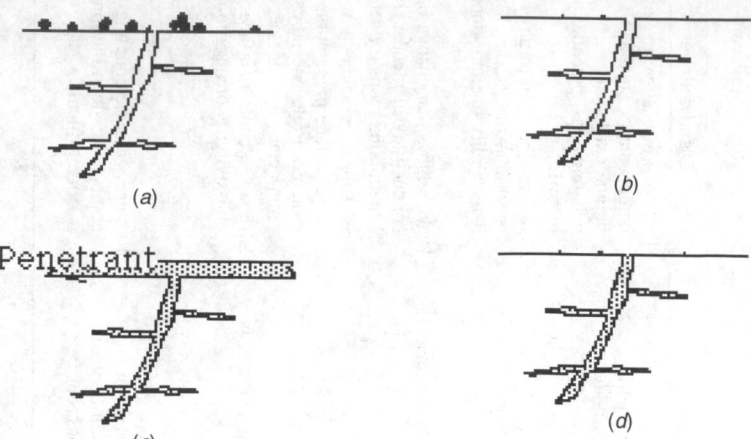

Fig. 25.1 (a) Schematic representation of a part surface before cleaning for penetrant inspection; (b) part surface after cleaning and before penetrant application; (c) part after penetrant application; (d) part after excess penetrant has been removed.

parts, the process generally is not well suited for the inspection of porous materials because the background penetrant from pores obscures flaw indications.

25.3 ULTRASONIC METHODS

Ultrasonic methods utilize sound waves to inspect the interior of materials. Sound waves are mechanical or elastic waves and are composed of oscillations of discrete particles of the material. The process of inspection using sound waves is quite analogous to the use of sonar to detect schools of fish or map the ocean floor. Both government and industry have developed standards to regulate ultrasonic inspections. These include, but are not limited to, the American Society for Testing and Materials Specifications 214-68, 428-71, and 494-75, and military specification MIL-1-8950H. Acoustic and ultrasonic testing takes many forms, from simple coin-tapping to transmission of sonic waves into a material and analyzing the returning echoes for the information they contain about its internal structure. Reference 15 provides an exhaustive treatment of this inspection technique.

Instruments operating in the frequency range between 20 and 500 kHz are usually defined as sonic instruments, while above 500 kHz is the domain of ultrasonic methods. In order to generate and receive the ultrasonic wave, a piezoelectric transducer is usually used to convert electrical signals

Fig. 25.2 Penetrant indication of a crack running along the edge of a jet engine turbine blade. Ultraviolet light causes the extracted penetrant to glow.

to sound wave signals and vice versa. This transducer usually consists of a piezoelectric crystal mounted in a waterproof housing that facilitates its electrical connection to a pulsar (transmitter) receiver. In the transmit mode, a high-voltage, short-duration pulse of electrical energy is applied to the crystal, causing it to change shape rapidly and emit a high-frequency pulse of acoustic energy. In the receive mode, any ultrasonic waves or echoes returning from the acoustic path, which includes the coupling media and part, compress the piezoelectric crystal, producing an electrical signal that is amplified and processed by the receiver.

25.3.1 Sound Waves

Ultrasonic waves have several characteristics, such as wavelength (λ), frequency (f), velocity (v), pressure (P), and amplitude (a). The following relationship between wavelength, frequency, and sound velocity is valid for all types of waves

$$f \times \lambda = v$$

For example, the wavelength of longitudinal ultrasonic waves of frequency 2 MHz propagating in steel is 3 mm and the wavelength of shear waves is 1.6 mm.

The sound pressure is related to the particles' amplitude by the relation, where the terms were defined in the previous paragraph.

$$P = 2\pi f \times \rho \times v \times a$$

Ultrasonic waves are reflected from all interfaces/boundaries that separate media with different acoustic impedances, a phenomenon quite similar to the reflection of electrical signals in transmission lines. The acoustic impedance Z of any medium capable of supporting sound waves is defined by

$$Z = \rho \times v$$

where ρ = the density of the medium in g/cm³
  v = the velocity of sound along the direction of propagation

Materials with high acoustic impedance are called (sonically) hard in contrast with (sonically) soft materials. For example, steel ($Z = 7.7$ g/cm³ $\times$ 5.9 km/sec = 45.4 $\times$ 10⁶ kg/m² sec) is sonically harder than aluminum ($Z = 2.7$ g/cm³ $\times$ 6.3 km/sec = 17 $\times$ 10⁶ kg/m² sec). An extensive list of acoustic properties of many common materials is provided in Appendix A.

25.3.2 Reflection and Transmission of Sound

Since very nearly all the acoustic energy incident on air/solid interfaces is reflected because of the large impedance mismatch of these two media, a coupling medium with an impedance closer to that of the part is needed to transmit ultrasonic energy into the part under examination. A liquid couplant has obvious advantages for components with complex external geometries, and water is the couplant of choice for most inspection situations. The receiver, in addition to amplifying the returning echoes, also time-gates echoes that return between the front surface and rear surfaces of the component. Thus, any unusually occurring echo can either be displayed separately or used to set off an alarm.

A schematic diagram of a typical ultrasonic pulse echo setup is shown in Fig. 25.3. This display of voltage amplitude versus time or depth (if acoustic velocity is known) at a single point of the specimen is known as an A-scan. In the setup shown in Fig. 25.3, the first signal corresponds to the reflection of the ultrasonic wave from the front surface of the sample (FS), the last signal corresponds to the reflection of the ultrasonic wave from the back surface of the sample (BS), and the signal in between corresponds to the defect echo from inside the component.

The portion of sound energy that is reflected from or transmitted through each interface is a function of the impedances of media on each side of the interface. The reflection coefficient R (ratio of the sound pressures or intensities of the reflected and incident waves) and transmission coefficient T (ratio of the sound pressures or intensities of the transmitted and incident waves) for an acoustic wave normally incident onto an interface are

$$R = \frac{p_r}{p_i} = \frac{Z_I - Z_{II}}{Z_I + Z_{II}}$$

$$R_{pwr} = \frac{I_r}{I_i} = \left(\frac{Z_I - Z_{II}}{Z_I + Z_{II}}\right)^2$$

Likewise, the transmission coefficients, T and T_{pwr}, are defined as

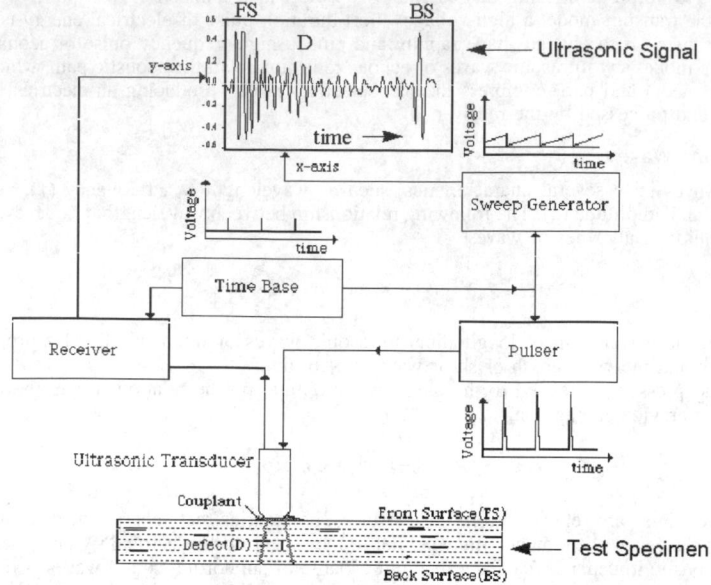

Fig. 25.3   Schematic representation of ultrasonic data collection and display in the A-scan mode.

$$T = \frac{p_t}{p_i} = \frac{2Z_{II}}{Z_I + Z_{II}}$$

$$T_{pwr} = \frac{I_t}{I_i} = \frac{4\dfrac{Z_{II}}{Z_I}}{\left(1 + \dfrac{Z_{II}}{Z_I}\right)^2}$$

where I_i, I_r, and I_t = the incident, reflected, and transmitted acoustic field intensities, respectively
Z_I = the acoustic impedance of the medium from which the sound is incident
Z_{II} = the acoustic impedance into which the wave is transmitted.

From these equations, it is apparent that for a crack like flaw containing air, Z_I = 450 kg/cm² sec, located in, say, a piece of steel, Z_{II} = 45.4 × 10⁶ kg/m² sec, the reflection coefficient for the flaw is practically −1.0. The minus sign indicates a phase change of 180° for the reflected pulse (note that the defect echo signal in Fig. 25.3 is inverted or phase shifted 180° from the front surface signal).

Effectively no acoustic energy is transmitted across an air gap, necessitating the use of water as a coupling medium in ultrasonic testing. The acoustic properties of several common materials are shown in Appendix A. These data are useful for a number of simple, yet informative, calculations.

Thus far, the discussion has involved only longitudinal waves. This type of wave motion is the only type that can travel through fluids such as air and water. This wave motion is quite similar to the motion one would observe in a spring, or a Slinky toy, where the displacement and wave motion are collinear (the oscillations occur in the direction of wave propagation). This wave is also called compressional or dilatational since compressional and dilatational forces are active in it. Audible sound waves transmitting acoustic energy from a source through the air to our ears are compressional waves. This mode of wave propagation is supported in liquids and gases as well as in solids. However, a solid medium can also support other modes of wave propagation, such as shear waves, Rayleigh or surface waves, and so on. Shear or transverse waves have a wave motion that is analogous to the motion one gets by snapping a rope; that is, the displacement of the rope is perpendicular to the direction of wave propagation. The velocity of this wave mode is about one-half that of compressional

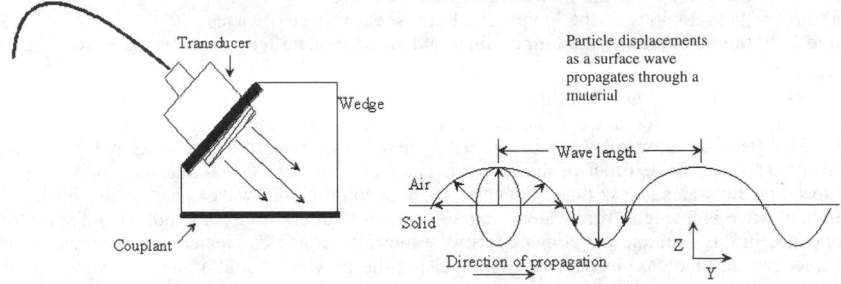

Fig. 25.4 Generation and propagation of surface waves in a material.

waves and is only supported by solid media, as shown in Appendix A. Shear waves can be generated when a longitudinal wave is incident on a fluid/solid interface at angles of incidence other than 0°. Rayleigh or surface waves have elliptical wave motion, as shown in Fig. 25.4, and about one ultrasonic wavelength penetration in the material. Therefore, they are used to detect surface and very near-surface flaws. The velocity of Rayleigh waves is about 90% of the shear wave velocity. Their generation requires a special device, as shown in Fig. 25.4, which enables an incident ultrasonic wave on the sample at a specific angle that is characteristic of the material (Rayleigh angle). See Ref. 15 for more details.

25.3.3 Refraction of Sound

The direction of propagation of acoustic waves is described by the acoustic equivalent of Snell's law. Referring to Fig. 25.5, the directions of propagation are determined with the following equation:

$$\frac{\sin \theta_i}{c_I} = \frac{\sin \theta_r}{c_r} = \frac{\sin \gamma_r}{b_r} = \frac{\sin \theta_t}{c_t} = \frac{\sin \gamma_t}{b_t}$$

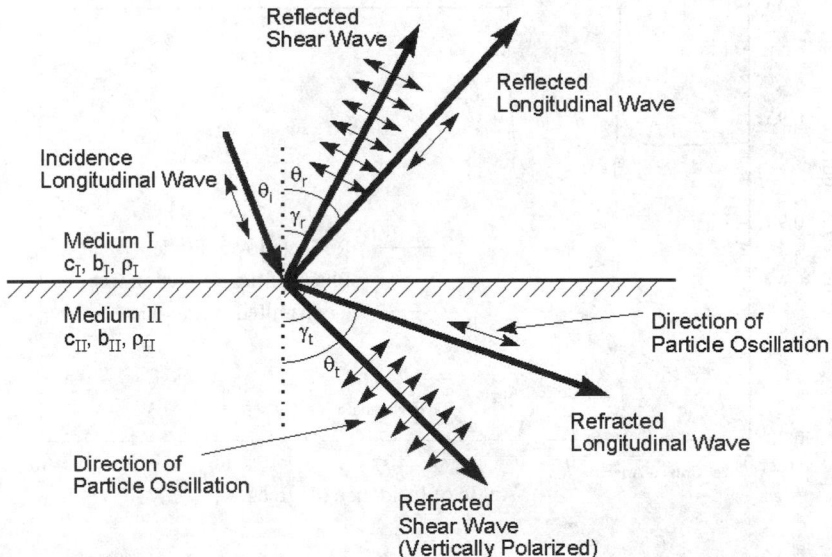

Fig. 25.5 Schematic representation of Snell's law and the mode conversion of a longitudinal wave incident on a solid–solid interface.

where c_I = the velocity of the incident longitudinal wave
  c_r and b_r = the velocities of the longitudinal and shear reflected waves
  c_t and b_t = the velocities of the longitudinal and shear transmitted waves in the solid II

In the case of steel/water interface, there are no shear reflected waves and the above relationship is simplified. Since the water has a lower wave speed than either the compressional or shear wave speeds of the steel, the acoustic waves in the metal are refracted toward from the normal. The situation changes dramatically if the order of the media is changed to a water/steel interface. In this case, the wave speed of the water is less than either the shear or longitudinal wave speed of the steel. With a longitudinal wave is incident from water at an angle other than 90° both longitudinal and shear waves are generated in the steel and travel away from the interface at angles greater than the incident wave. This effect can be predicted using Snell's law. Using the previous equation and the wave speeds of steel from Appendix A the reader will note that the longitudinal wave is refracted further from the normal than the shear wave. As the angle of incidence is increased, there will be an angle where the longitudinal wave is refracted parallel to the surface or simply propagates along the interface. This angle is called the first critical angle. If the angle of incidence is increased further, there will be a point where the shear wave also disappears. This angle is called the second critical angle. A computer-drawn curve is shown in Fig. 25.6, in which the normalized acoustic energy reflected and refracted at a water/steel interface are plotted as a function of angle of incidence. Note that the longitudinal or first critical angle for steel occurs at 14.5°. Likewise, second critical angle occurs at 30°. If the angle of incidence is increased above the first critical angles, then only a shear wave is generated in the metal and propagates at an angle of refraction determined by Snell's law. Angles of incidence above the second critical angle produce a complete reflection of the incident acoustic waves—that is, no acoustic energy enters the solid. At a specific angle of incidence (Rayleigh angle), surface acoustic waves are generated on the material. The Rayleigh angle can be easily calculated from Snell's law when the refraction angle is 90°. The Rayleigh angle for steel occurs at 29.5. Between the two critical angles, only the shear wave is present in the material. In this region, shear wave testing is performed, which has two advantages. First, with only one type of wave present, the

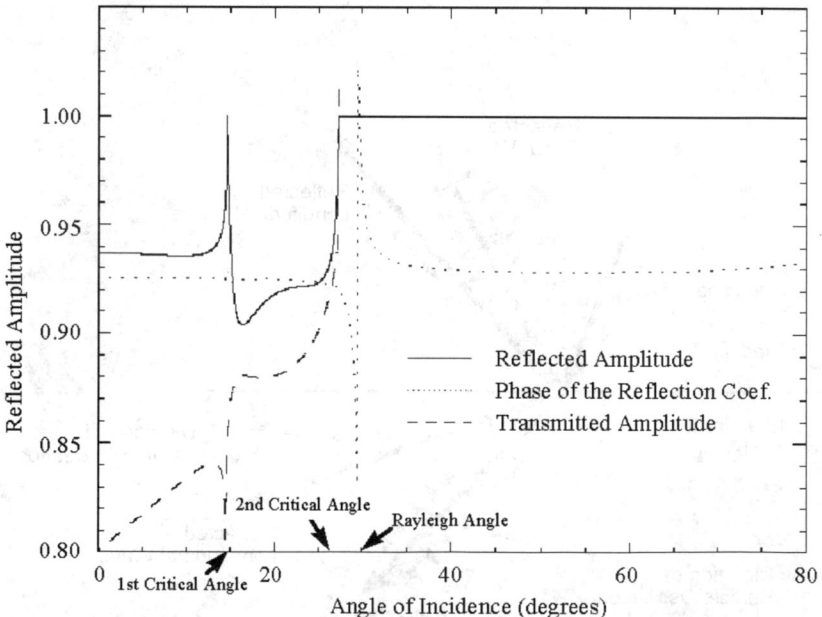

Fig. 25.6 Amplitude (energy flux) and phase of the reflected coefficient and transmitted amplitude vs. angle of incidence for a longitudinal wave incident on a water–steel interface. The arrows indicate the critical angles for the interface.

ambiguity that would exist concerning where a reflected wave originates is not present. Second, the lower wave speed of the shear wave means that about twice the time is available for resolving distances within the part under examination. These advantages mean shear wave inspection is often chosen for inspection of thin metallic structures, such as those in aircraft.

Using only Snell's law and the relationships for the reflection and transmission coefficients, a great deal of information can be deduced about any ultrasonic inspection situation in which the acoustic wave is incident at 90° to the surface. For other angles of incidence or for a component containing one or more thin layers, a computer program is used to analyze the acoustic interactions. For more complicated materials or structures, such as fiber-reinforced composites, any analytical predictions require computer computation for even relatively simple situations. In these cases, more complicated modes of wave propagation occur, such as Lamb waves (plate waves), Stoneley waves (interface waves), Love waves (guided in layers of a solid material coated onto another one), and so on.

25.3.4 The Inspection Process

Once the type of inspection has been chosen and the optimum experimental parameters have been determined, it remains only to choose the mode of presentation of the data. If the size of the flaw is small compared to the transducer, then the A-scan method can be chosen, as shown in Fig. 25.3. The acquisition of a series of A-scans obtained by scanning the transducer in one dimension (line) is called a B-scan. In the A-scan mode, the voltage output of the transducer is displayed versus time or depth in the part. The size of the flaw is often inferred by comparing the size of the defect signal to a set of standard calibration blocks, which have varying sizes of flat bottom holes drilled in one end. For a specific transducer, the magnitude of the signal from the flat bottom hole is an increasing function of hole diameter. In this way, an equivalent flat-bottom-hole size can be given for a flaw signal obtained from a defect in a component. The equivalent size is meaningful only for flaws that are nearly perpendicular to the ultrasonic signal path.

If the flaw size is larger than the transducer or if a number of flaws are expected, then the C-scan mode is usually selected. In this inspection mode, as shown in Fig. 25.7, the transducer is scanned back and forth in two coordinates across the part. When a flaw signal is detected between the front and back surface signals, then a line the size of intersection of the raster-scan with the flaw is left blank on a piece of paper or CRT screen. In this manner, a planar projection of each flaw is viewed and its positional relationships to others and to the part boundaries are easily assessed. Unfortunately, in this mode the depth information for each flaw is frequently lost; therefore, this mode is mostly used for thin, layered aircraft structures. However, if the ultrasonic signals are monitored at a particular time window (time gate), an image can be obtained. This C-scan can represent various characteristics of the ultrasonic signal detected in the particular time gate, such as peak-to-

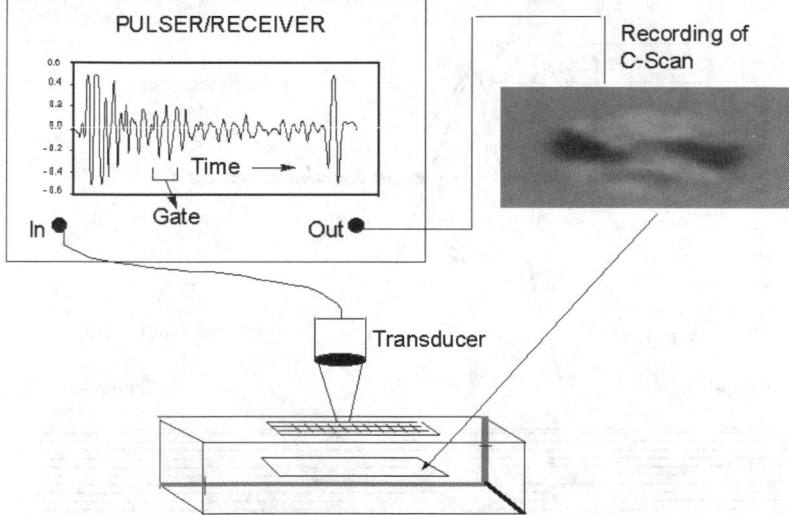

Fig. 25.7 Schematic representation of ultrasonic data collection. The data are displayed using the C-scan mode. The image shows a defect located at a certain depth in the material.

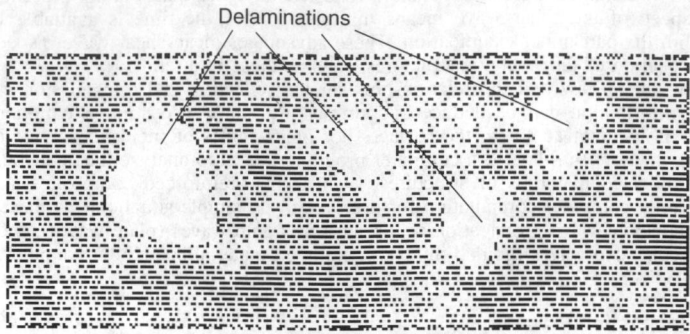

Fig. 25.8 Typical C-scan image of composite specimen, showing delaminations and porosity.

peak amplitude, positive or negative peaks, time of flight, mean value of amplitude, and so on. The C-scan provides a visual representation of a slice of the material at a certain depth and is very useful for nondestructive inspection.

Depending on the structural complexity and the attenuation of the signal by the material and electronic instrumentation, flaws as small as 0.015 in., in one dimension, can be reliably detected and quantified using this ultrasonic method. An example of a typical C-scan printout of an adhesively bonded test panel is shown in Fig. 25.8. While the panel was fabricated with Teflon void-simulating implants, the numerous white areas indicate the presence of a great deal of porosity in the adhesive. For a much more extensive treatment of this inspection technique, see Ref. 1, Vol. 7.

25.4 RADIOGRAPHY

Radiography is an NDE method in which the projected X-ray attenuation for many straight line paths through a specimen are recorded as a two-dimensional image on recording medium. For a more detailed description of radiography testing, see Ref. 1, Vol. 3.

This process, shown schematically in Fig. 25.9, records visually any feature that changes the attenuation of the X-ray beam along the path that the X-ray photons take through the structure. This local change in attenuation produces a change in the density or darkness of the film or electronic recording device at that location. This change in brightness, which is sometimes a mere shadow, is used by the inspector to detect internal anomalies. In this task, the inspector is greatly aided in detecting and quantifying flaws by knowing the geometry of the part and how this relates to the

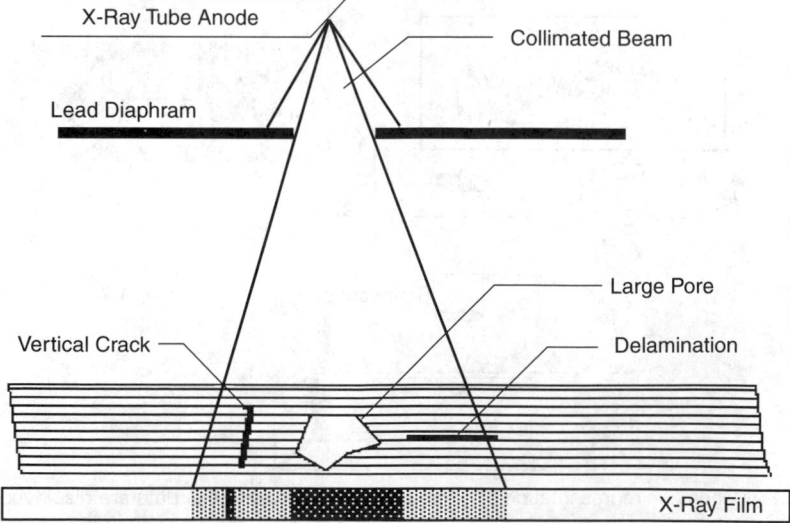

Fig. 25.9 Schematic radiograph of a thin plate with two types of flaws.

image. It should be noted in Fig. 25.9 that only flaws that change the attenuation of the X-ray beam on passage through the part are recorded. For example, a delamination in a composite or laminated material that is not the result of missing material is not visible because there is no change in attenuation of the X-ray beam. Flaws that are oriented perpendicular to the plate do attenuate the X-ray beam as much as adjacent beam paths and therefore are easily detected as a darkened line or area on the film. An example of a crack in the correct orientation to be visible on a radiograph of a piece of tubing is shown in Fig. 25.10.

25.4.1 The Generation and Absorption of X Radiation

X radiation can be produced via a number of processes. The most common way of producing X rays is with an electron tube in which a beam of energetic electrons impact a metal target. As the electrons are rapidly decelerated in this collision, a wide band of X radiation is produced, analogous to white light. This is referred to as Bremsstrahlung or breaking radiation. Higher-energy electrons produce shorter-wavelength or more energetic X rays. The relationship between the shortest-wavelength X radiation produced and the highest voltage applied to the tube is given by

$$\lambda = \frac{12,336}{\text{voltage}}$$

where λ = the shortest wavelength of the X radiation-produced in Ångstroms.

The more energetic the radiation, the more penetrating power it has. Therefore, higher-energy radiation is used on dense materials, such as metals. While it is possible to analytically predict what X-ray energy would provide the best image for a specific material and geometry, a simpler method of arriving at the optimum X-ray energy is to use the curves shown in Fig. 25.11. Note that high-energy radiation is used for dense materials, such as steels, or for thick, less dense materials, such as large composite solid rocket motors. An alternative to Fig. 25.11 is the table of radiographic equivalence factors shown in Table 25.2.[16] Aluminum is the standard material for voltages below 100 Kv, while steel is the standard above this voltage. When radiographing another material its thickness is multiplied by the equivalency factor, of Table 25.2, to obtain the equivalent thickness of the standard material to obtain an acceptable radiograph. For example, if one needed to produce a radiograph of a 0.75-in.-thick piece of brass with a 400-Kv X-ray source, one would multiply the 0.75 by the factor of 1.3 to obtain 0.98. This means that an acceptable radiograph of the brass plates would be obtained with the same exposure parameters as would be used for 0.98 in. (approximately 1 in.) of steel.

Penetrating radiation for radiography can also be obtained from the decay of radioactive sources. This is usually referred to as gamma radiation. These radiation sources have distinct characteristics that distinguish them from X-ray tubes. First, gamma radiation is very nearly monochromatic; that is, the radiation spectrum contains only one or two dominant characteristic energies. Second, the energies of most sources are in the million volt range, making them ideal for inspecting highly attenuated materials and structures. Third, the small size of these sources permits them to be used in situations where an X-ray tube could not fit into a small space. Fourth, since the gamma-ray source is continually decaying, adjustments to the exposure time must be made in order to achieve consistent results over time. Finally, the operator must be cognizant that the source is always on and is therefore

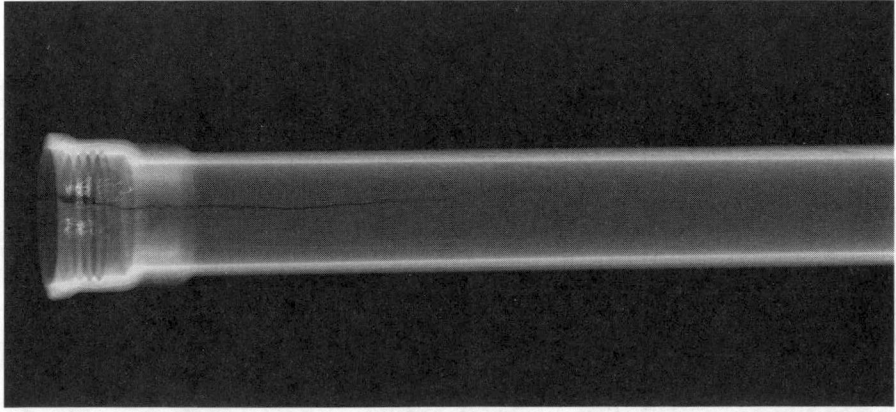

Fig. 25.10 Radiograph of an aluminum tubing with a crack.

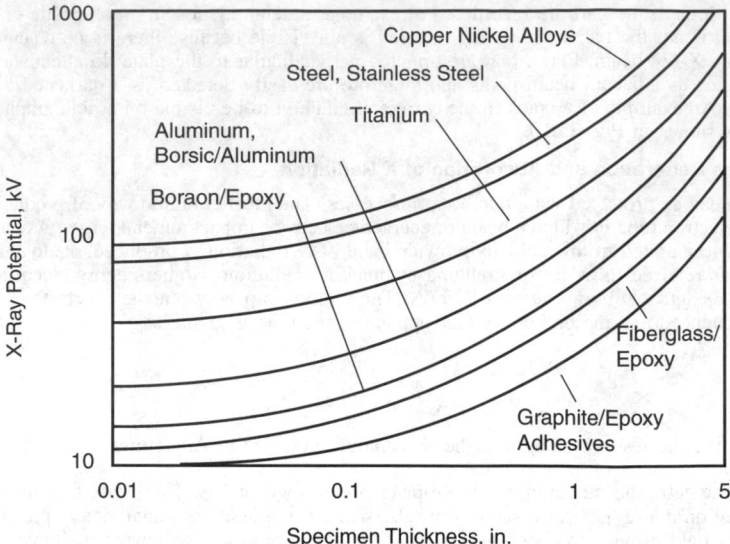

Fig. 25.11 Plot of the X-ray tube voltage vs. thickness for several important industrial materials.

a constant safety hazard. Since, in use, gamma radiography differs little from standard practice with X-ray tubes, we will make no further distinction between the two.

25.4.2 Neutron Radiography

Neutron radiography[17] is useful for inspecting materials and structures in specialized circumstances. The attenuation of neutrons is not related to atomic number in a smooth manner as it is with X-rays. While x-radiation is most heavily absorbed by high atomic number elements, this is not true of neutrons as shown in Fig. 25.12. The reader can easily discern that hydrogen adsorbs neutrons to a greater extent than do most metals. This means that hydrogen containing materials could be detected

Table 25.2 Radiographic Equivalence Factors

Energy Level	100 kV	150 kV	220 kV	250 kV	400 kV	I MeV	2 MeV	4–25 MeV	192Ir	60Co
Metal										
Magnesium	0.05	0.05	0.08							
Aluminum	0.08	0.12	0.18						0.35	0.35
Aluminum alloy	0.10	0.14	0.18						0.35	0.35
Titanium		0.54	0.54		0.71	0.9	0.9	0.9	0.9	0.9
Iron/all steels	1.0	1.0	1.0	1.0	1.0	1.0	1.0	1.0	1.0	1.0
Copper	1.5	1.6	1.4	1.4	1.4	1.1	1.1	1.2	1.1	1.1
Zinc		1.4	1.3		1.3			1.2	1.1	1.0
Brass		1.4	1.3		1.3	1.2	1.1	1.0	1.1	1.0
Inconel X		1.4	1.3		1.3	1.3	1.3	1.3	1.3	1.3
Monel	1.7		1.2							
Zirconium	2.4	2.3	2.0	1.7	1.5	1.0	1.0	1.0	1.2	1.0
Lead	14.0	14.0	12.0			5.0	2.5	2.7	4.0	2.3
Halfnium			14.0	12.0	9.0	3.0				
Uranium			20.0	16.0	12.0	4.0		3.9	12.6	3.4

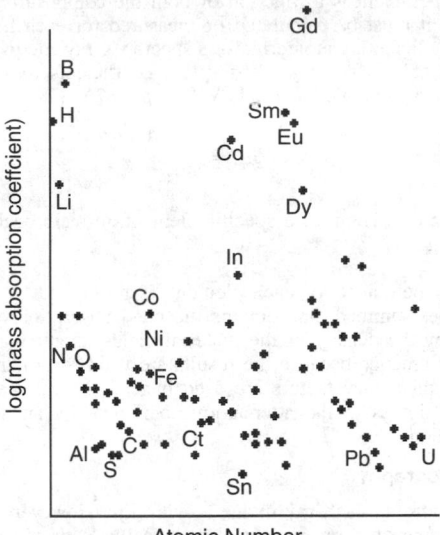

Fig. 25.12 A plot of neutron mass absorption coefficient vs. atomic number.

inside a metal container. Neutron radiography can be used to detect adhesive flaws in bonded metal structures because the hydrogen containing adhesive absorbs more neutrons than the metal structure—see the schematic representation of a neutron radiograph of an adhesively bonded specimen with various flaws shown in Fig. 25.13.

Neutron radiography, however, does have several constraints. First, neutrons do not expose radiographic film and therefore fluorescing materials are often used to produce an image or screens. The image produced in this manner is not as sharp and well-defined as one produced with X rays. Second, there does not exist a suitable portable source of neutrons, which means that a nuclear reactor is used to supply the penetrating radiation. Even with these restrictions, there are times when there is no other alternative to neutron radiography and the utility of the method outweighs its expense and difficulty of usage.

25.4.3 Attenuation of X Radiation

The interpretation of a radiograph requires some fundamental understanding of the X-ray absorption. The basic relationship governing this phenomenon is

$$I = I_0 e^{-\mu x}$$

where I and I_0 = the transmitted and incident X-ray beam intensities, respectively
μ = the attenuation coefficient of the material in cm^{-1}
x = the thickness of the specimen, in cm

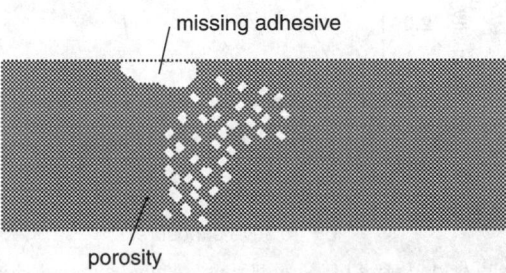

Fig. 25.13 Idealized neutron radiograph of bonded aluminum panel with two types of flaws.

Since the attenuation coefficient is a function of both the composition of the specimen and the wavelength of the X rays, it must be calculated or measured for each inspection. It is possible to calculate the attenuation coefficient of a material for a specific X-ray energy using the mass absorption coefficient, μ_m, as defined below. The mass absorption coefficients for most elements are readily available for a variety of X-ray energies. (Ref. 1, Vol. 3, pp. 836–878)

$$\mu_m = \frac{\mu}{\rho}$$

where μ_m = the attenuation coefficient of a specific element, in reciprocal cm
ρ = its density in g/cm^3

If the mass absorption coefficient for each element, is multiplied by its weight fraction in the material and these quantities summed, one obtains the mass absorption coefficient of the material. Multiplying this quantity by the density of the material yields its attenuation coefficient. This procedure is not often used in practice because the results are valid for a narrow band of wavelengths. In practice, radiographic equivalency factors are used instead. This procedure points out that each element in a material contributes to the attenuation coefficient by an amount proportional to its percentage of the composition.

25.4.4 Film-Based Radiography

The classical method of recording an X-ray image is with film. However, new solid state X-ray area detectors permits the recording of X-ray images electronically. Both of these recording mechanisms utilize similar mathematical relationships to describe their sensitivity to defects. These will be covered next from the standpoint of film.

The relationship between the darkness on an X-ray film and the quantity of radiation falling on it is shown in Fig. 25.14. This is a log-log plot of darkness or film density and relative exposure. Relative exposure may be varied by changing either the time of exposure, intensity of the beam or specimen thickness. The slope of the curve along its linear portion is referred to as the film gamma, γ. Film has characteristics that are quite analogous to electronic devices. For example, the greater the gamma or amplification capability of the film, the smaller its dynamic range—range of exposures over which thickness changes in part will be accurately recorded. If it is desirable to use a high gamma film to detect subtle flaws in a part with several thicknesses, then the radiographer will frequently use two different film types in the same cassette—see Fig. 25.15. In this way, each film will be optimum for the different thicknesses of the part.

It is possible to calculate the minimum detectable flaw size for a specific radiographic inspection. Additionally, a method is available to check the radiographic procedure to determine if the radiograph was taken in a manner that the smallest flaws are detectable. In sensitivity calculations a small flaw is represented by a small change in thickness of the part. Whether a defect is detectable depends on the correct alignment of the flaw with respect to the X-ray source and film. The following calculation

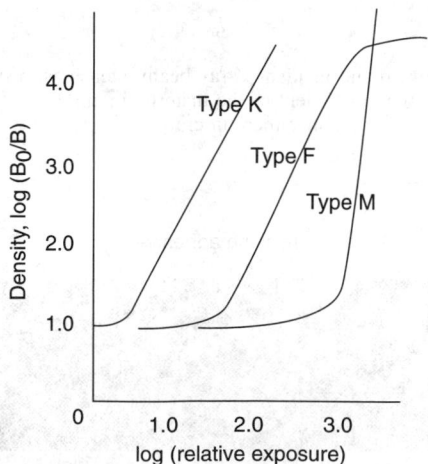

Fig. 25.14 Density or darkness of X-ray film vs. relative exposure for three common radiography films.

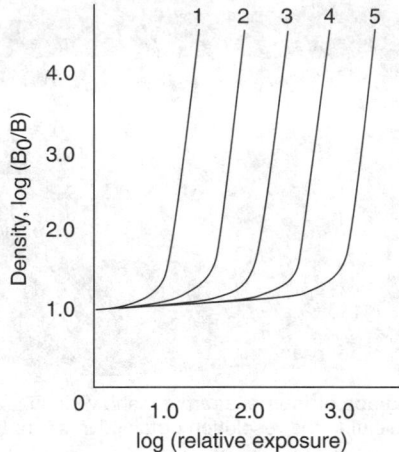

Fig. 25.15 Density vs. relative exposure for films that could be used in a multiple film exposure to obtain optimum flaw detectability in a complex part.

only is an estimation of the best sensitivity of the radiographic process to small changes in part thickness.

Using a knowledge of the minimum density difference that is detectable by the average radiographic inspector, the following relationship can be derived that relates the radiographic sensitivity, S, in percent to thickness changes to film and specimen parameters:

$$S = \frac{2.3}{\gamma \times \mu \times x}$$

where γ = the film gamma for the exposure conditions used
μ = the attenuation coefficient of the specimen
x = the maximum thickness of the part being inspected

25.4.5 The Penetrameter

The radiographic process is usually checked with a small device called a penetrameter that is shown schematically in Fig. 25.16. Its image on a radiograph is shown schematically in Fig. 25.17. The penetrameter is a thin strip of metal in which three holes of varying sizes are machined. It is composed of the same material as the specimen under inspection and has a thickness either 1%, 2%, or 4% of the maximum thickness of the part. The holes in the penetrameter have diameters that are 1, 2, and 4 times the thickness of the penetrameter. The sensitivity achieved for each radiographic can be easily determined by noting the smallest hole just visible in the thinnest penetrameter on a film and referring to Table 25.3. Using the formula for sensitivity and then noting if that level was achieved in practice, the radiographic process can be quantitatively controlled. While this procedure does not offer any guarantee of detecting flaws, it is quite useful in controlling the mechanics of the inspection.

Other variables of the radiographic process may also be easily and rapidly changed with the aid of tables, graphs, and nomograms, which are usually provided by film manufacturers free of charge. For more information in this regard, see the commercial literature.

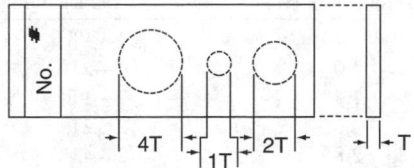

Fig. 25.16 Schematic of typical X-ray penetrameter.

Fig. 25.17   Idealized radiograph of the penetrameter shown in Fig. 25.16. The 1T hole is just visible, indicating the resolution obtained in the radiograph.

25.4.6   Real-Time Radiography

While film radiography represents the bulk of the radiographic NDE performed at this time, new methods of both recording the data and analyzing it are rapidly becoming accepted. For example, filmless radiography (FR) uses solid-state detectors or television detection and image-processing methods instead of film to record the image. These methods have several advantages along with some disadvantages. For example, real time radiography (RTR) or FR permits the viewing of a radiographic image while the specimen is being moved. This often permits the detection of flaws that would normally be missed in conventional film radiography because of the limited number of views or exposures usually taken. The price to be paid for these advantages can be the lower resolution of the FR system when compared to film. Typical resolution capabilities of an FR system are in the range from 4 to perhaps 12 line pairs/mm, while film has resolution capabilities in the range from 10–100 line pairs/mm. This means that some types of very fine flaws are not detectable with FR and the inspector must resort to film. Another factor that must be considered when comparing these two methods is the possibility of digital image processing, which is easily implemented with FR systems. These methods can be of enormous benefit in enhancing flaws that might otherwise be invisible to the inspector. While the images on film can also be enhanced using the same image-processing schemes, they cannot be performed in real or near real time, as can be done with an electronic system.

25.4.7   Computed Tomography

Another advance in industrial radiography that promises to have a major impact on the interaction of NDE with engineering design and analysis is computed tomography (CT). CT produces an image of a thin slice of the specimen under examination. This slice is parallel to the path of the X-ray beam as it passes through the specimen as contrasted to classical radiography, wherein the image is formed on a plane perpendicular to the path of the X-ray beam on passage through the specimen. While the classical radiographic image is difficult to interpret because it represents a collapse of all of the images of the specimen between the source of X rays and the recording media, the CT image appears as a thin slice of the specimen.

Table 25.3

Sensitivity, S (%)	Quality Level (% T—Hole Diameter)
0.7	1—1T
1.0	1—2T
1.4	2—1T
2.0	2—2T
2.8	2—4T
4.0	4—2T

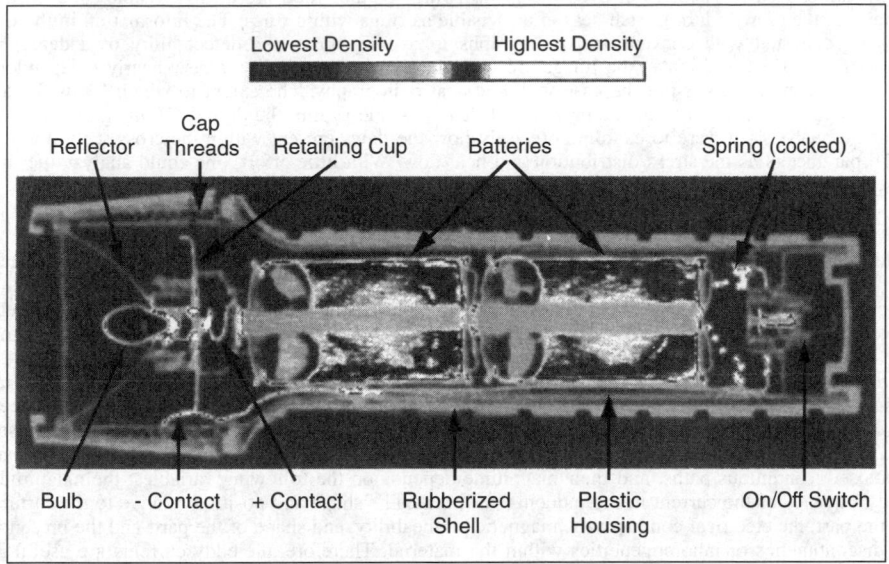

Fig. 25.18 Computed tomograph of flashlight.

This comparison is best explained with actual images from these two modalities. Figure 25.10 shows a typical industrial radiograph where one can easily see the image of the top and bottom surfaces of the tube. Contrast this with the image in Fig. 25.18, which shows CT slice through a flashlight. The individual components of the flashlight are easily visible and defects in its assembly could be easily detected. An image in even finer detail than this that shows the microstructural details of a material is shown in Fig. 25.19. Everyone will recognize that this is a pencil. Not only are the

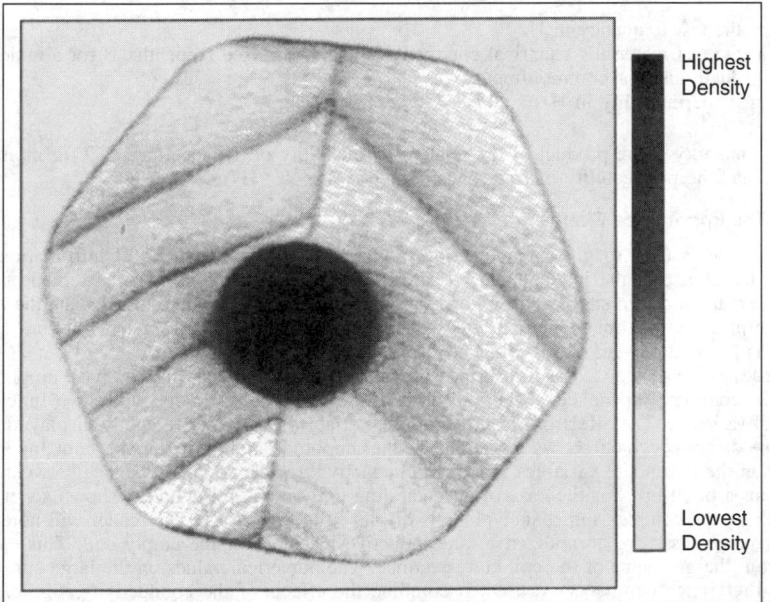

Fig. 25.19 Computed tomograph of a pencil. Note the yearly growth rings in wood.

key features clearly visible, but even the growth rings of the wood are clearly visible. In fact, the details of the growth during each season are visible as rings within rings. The information in the CT image contrasted with conventional radiographs is striking. First, the detectability of a defect is independent of its position in the image. Second, the defect detectability is very nearly independent of its orientation. This is not the case with classical radiography. The extent to which CT will alter radiographic NDE is only now being realized. It is possible to link the digital CT image with finite element analysis software to examine precisely how the flaws present within the cross section affect such parameters as the stress distribution and heat flow. With little effort, one could analyze the full three-dimensional performance of many engineering structures.

25.5 EDDY CURRENT INSPECTION

Eddy current (EC) methods are used to inspect electrically conducting components for flaws, such as surface connected and near-surface cracks and voids; heat-treatment; external dimensions and thickness of thin metals; and thickness of nonconducting coatings on a metal substrate. Quite often, several of these conditions can be monitored simultaneously if instrumentation capable of measuring the phase of the eddy current signal is used.

This NDE method is based on the principle that eddy currents are induced in a conducting material when a coil or array of conductors (probe) with an alternating or pulsated electric current is placed in close proximity to its surface. The induced currents create an electromagnetic field that opposes the field of the inducing coil in accordance with Lenz's law. The eddy currents circulate in the part in closed, continuous paths, and their magnitude depends on the following variables: the magnitude and frequency of the current in the inducing coil; the coil's shape and position relative to the surface of the part; the electrical conductivity, magnetic permeability, and shape of the part; and the presence of discontinuities or inhomogeneities within the material. Therefore, the eddy currents are useful in measuring the properties of materials and detecting discontinuities or variations in geometry of components.

25.5.1 The Skin Effect

Since alternating currents are necessary to perform this type of inspection, information from the inspection is limited to the near-surface region by the skin effect. Within the material, the eddy current density decreases exponentially with the depth. The density of the eddy current field falls off exponentially with depth and diminishes to a value of about 37% of the at-surface value at a depth referred to as the standard depth of penetration (SDP). The SDP, in meters, can be calculated with the simple formula

$$SDP = \frac{1}{\sqrt{\pi f \sigma \mu}}$$

where f = the test frequency in Hz
σ = the test material's electrical conductivity in mho/m (see Appendix B for a table of conductivities for common metals)
μ = its permeability in H/m

This latter quantity is the product of the relative permeability of the specimen, 1.0 for nonmagnetic materials, and the permeability of free space, which is 4×10^7 H/m.

25.5.2 The Impedance Plane

While the SDP is used to give an indication of the depth from which useful information can be obtained, the choice of the independent variables in most test situations is usually made using the impedance plane diagram suggested by Förster.[18] It is theoretically possible to calculate the optimum inspection parameters from numerical codes based on Maxwell's equations, but this is a laborious task that is justified only in special situations.

The eddy currents induced at the surface of a material are time-varying and have amplitude and phase. The complex impedance of the coil used in the inspection of a specimen is a function of a number of variables. The effect of changes in these variables can be conveniently displayed with the impedance diagram, which shows variations in the amplitude and phase of the coil impedance as functions of the dependent variables specimen conductivity, thickness, and distance between the coil and specimen or liftoff. For the case of an encircling coil on a solid cylinder, shown schematically in Fig. 25.20, the complex impedance plane is displayed in Fig. 25.21. The reader will note that the ordinate and abscissa are normalized by the inductive reactance of the empty coil. This eliminates the effect of the geometry of the coil and specimen. The numerical values on the large curve, which are called reference numbers, are used to combine the effects of the conductivity, size of the test specimen, and the frequency of the measurement into a single parameter. This yields a diagram that is useful for most test conditions. The reference numbers shown on the outermost curve are obtained with the following relationship, for nonmagnetic materials.

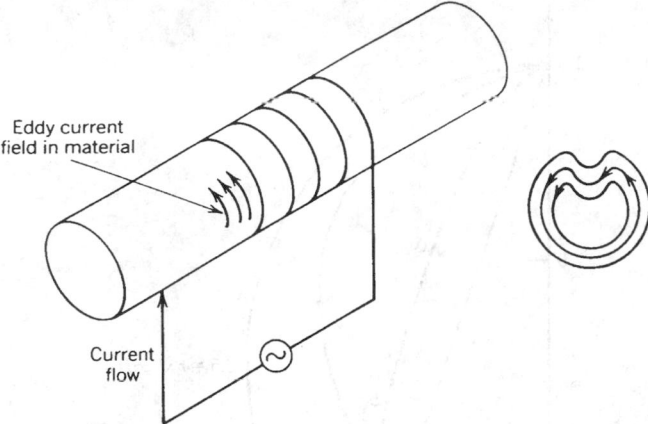

Fig. 25.20 Schematic representation of eddy current inspection of a solid cylinder. Also shown are the eddy current paths within the cross section of the cylinder in the vicinity of a crack.

$$\text{reference \#} = r\sqrt{2\pi f \mu \sigma}$$

where r = the radius of the bar in meters

f = the frequency of the test in Hz

μ = the magnetic permeability of free space (4×10^{-7} H/m)

σ = the conductivity of the specimen in mho/m

The outer curve in Fig. 25.21 is useful only for the case where the coil is the same size as the solid cylinder under test. For those cases where the coil is larger than the test specimen, a coil filling factor must be calculated and the appropriate point on a new curve located. This is easily accomplished using the fill factor N of the coil, which is defined as

$$N = \left(\frac{\text{diameter}_{\text{specimen}}}{\text{diameter}_{\text{coil}}}\right)^2$$

Figure 25.21 shows the impedance plane with a curve for a specimen/coil inspection geometry with a fill factor of 0.75. Note that the reference numbers on the curves representing the different fill factors can be determined by projecting a straight line from 1.0 on the ordinate to the reference number of interest, as is shown for the reference number 5.0. Both the fill factor and the reference number change when the size of the specimen or coil changes. Assume that a reference number of 5.0 is appropriate to a specific test with $N = 1.0$; if the coil diameter is changed so that the fill factor becomes 0.75, then the new reference number will be equal to $5.0 \times \sqrt{0.75} = 4.33$. While the actual change in reference number for this case follows the path indicated by the dotted line of this curve, we have estimated the change along the straight line. This yields a small error in optimizing the test setup, but is sufficient for most purposes. For a more detailed treatment of the impedence plane the reader is referred to Ref. 1, Vol. 4. The inspection geometry discussed thus far has been for a solid cylinder. The other geometry of general interest is the thin-walled tube in this case the skin effect limits the thickness of metal that may be effectively inspected.

For an infinitely thin-walled tube, the impedance plane is shown in Fig. 25.22. Also included in this figure is the curve for a solid cylinder. The dotted lines that connect these two cases are for thin-walled cylinders of varying thicknesses. The semicircular curve for the thin cylinder is used in the same manner as described above for the solid cylinder.

25.5.3 Liftoff of the Inspection Coil from the Specimen

In most inspection situations, the only independent variables are frequency and liftoff. High frequency excitations are frequently used for detecting defects, such as surface connected cracks or corrosion, while low frequencies are used to detect subsurface flaws. It is also possible to change the coil shape and measurement configuration to enhance detectability, but the discussion of these parameters is beyond the scope of this article and the reader is referred to the literature for a discussion of these more complex variables. The relationships discussed thus far may be put into application by examining a small section of Fig. 25.22. This figure is shown in expanded form in Figure 25.23. In this figure, changes in thickness, liftoff, and conductivity are represented by vectors. These vectors all

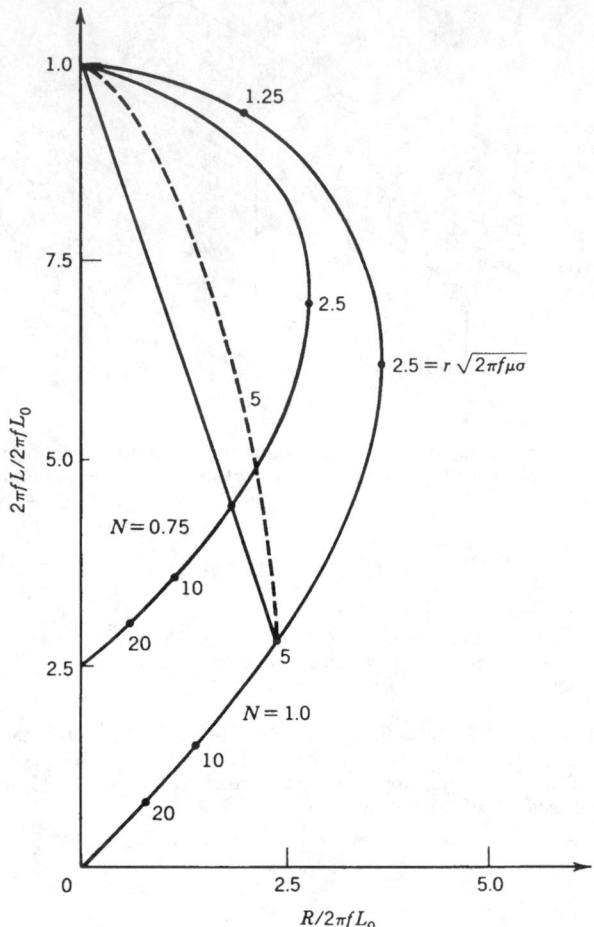

Fig. 25.21 The normalized impedance diagram for a long encircling coil on a solid, nonferromagnetic cylinder. For $N = 1$ the coil and cylinder have the same diameter; while for $N = 0.75$ the coil is approximately 1.155 times larger than the cylinder.

point in different directions representing the phase of the different possible signals. Instrumentation with phase discrimination circuitry can differentiate between these signals and therefore is often capable of detecting two changes in specimen condition at once. Changes in conductivity can arise from several different conditions. For example, aluminum alloys can have different conductivities depending on their heat treatment. Changes in apparent conductivity are also due to the presence of cracks or voids. A crack decreases the apparent conductivity of the specimen because the eddy currents must travel a longer distance to complete their circuit within the material. Liftoff and wall thinning are also shown on Fig. 25.23. Thus, two different flaw conditions can be rapidly detected. There are situations where changes in wall thickness and liftoff result in signals that are very nearly out of phase and therefore the net change is not detectable. If this situation is suspected, then inspection at two different frequencies could permit the detection of this situation. There are other inspection situations that cannot be covered in this brief description. These include the inspection of ferromagnetic alloys, plate and sheet stock, and the measurement of film thicknesses on metal substrates. For a treatment of these and other special applications of eddy current NDE, the reader is referred to Ref. 1, Vol. 4.

There are numerous methods of making eddy current NDE measurements. Two of the more common generic methods are shown schematically in Fig. 25.24. In the absolute coil arrangement, very accurate measurements can be made with the differences between the two samples. In the differential coil method, it is the differences between the two variables at two slightly different

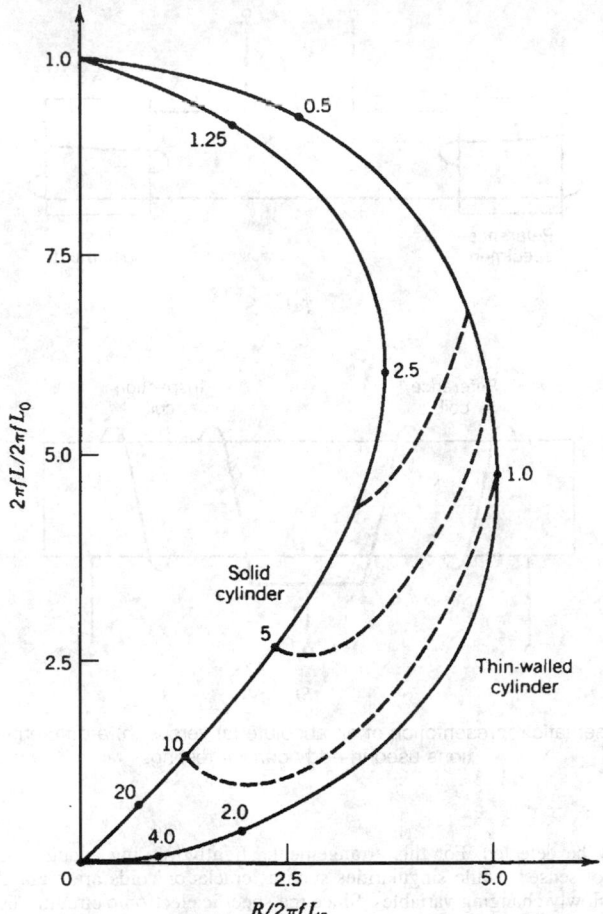

Fig. 25.22 The normalized impedance diagram for a long encircling coil on a solid and thin-walled solid, nonferromagnetic cylinders. The dashed lines represent the effects of varying wall thickness.

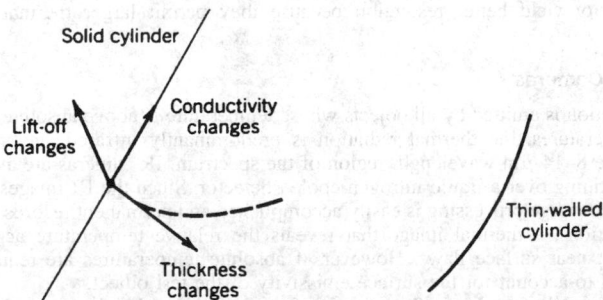

Fig. 25.23 The effects of various changes in inspection conditions on signal changes in the impedance plane of Fig. 25.22. Phase differentiation is relatively easily accomplished with current instrumentation.

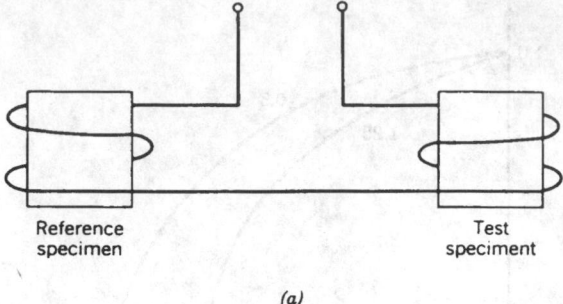

(a)

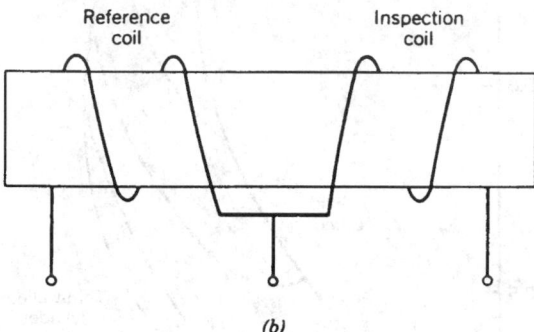

(b)

Fig. 25.24   A schematic representation of an absolute (*a*) versus (*b*) a differential coil configurations used in eddy current testing.

locations that may be detected. For this arrangement, slightly varying changes in dimensions and conductivity are not sensed, while singularities such as cracks or voids are highlighted, even in the presence of other slowly changing variables. Since the specific electronic circuitry used to accomplish this task can vary dramatically, depending on the specific inspection situation, the reader is referred to the current NDE and instrumentation literature listed in the References.

25.6   THERMAL METHODS

Thermal nondestructive test methods involve detecting either the surface or the temperature of a test object. This technique is used to detect the flow of thermal energy either into or out of a specimen, which indicate surface or near-surface defects. Other properties that can influence this NDE test are the specific heat, density, thermal conductivity, and emissivity of the test specimen. Defects that are usually detected include porosity, cracks, and laminations. The sensitivity of any thermal method is greatest for near-surface flaws and degrades rapidly for deeply buried flaws. Materials with lower thermal conductivity yield better resolution because they permit larger thermal gradients to be obtained.

25.6.1   Infrared Cameras

Infrared (IR) radiation is emitted by all objects whose temperature is above absolute zero. For objects at moderate temperatures, the thermal radiation is predominantly infrared and measurements are concentrated in the 8–14 μm wavelength region of the spectrum. IR cameras are available that view large areas by scanning over a liquid nitrogen-cooled detector. Since the IR images can be stored in digital form, further image processing is easily accomplished and permanent records can be produced. For many applications, a thermal image that reveals the relative temperature across an object is sufficient to detect near surface flaws. However, if absolute temperatures are required, the camera must be calibrated to account for the surface emissivity of the test object.

Thermography's ability to detect flaws is significantly affected by the type of flaw and its orientation with respect to the surface of the object. To have a maximum effect on the surface temperatures, the flaw must interrupt heat flow to the surface. Since a flaw can occur at any angle to the surface, the important parameter is the projected area of the flaw in the field of view of the camera.

Subsurface flaws, such as cracks parallel to the surface, porosity, and debonding of a surface layer, are easily detected. Cracks that are perpendicular to the object surface can present an infinitesimal area to the camera and therefore are very difficult to detect using thermography.

Many other NDE methods can provide better spatial resolution than thermal methods, due to spreading of thermal energy as it diffuses to the surface of the specimen. The greatest advantage to thermography is that it is a noncontact, remote technique requiring only line-of-sight access IR camera based an object. Large areas can be viewed rapidly since scan rates for IR cameras run between 16 and 30 frames per second. Temperature differences of 0.02°C or less can be detected in a controlled environment.

25.6.2 Thermal Paints

A number of contact thermal methods are available to provide temperature or temperature-distribution data for surfaces. These methods involve applying a coating to the sample and observing the coating change color as the object is thermally cycled. Several different types of coatings are available that cover a wide temperature range. Temperature-sensitive pigments in the form of paints have been made to cover a temperature range from 40°–1600°C. Thermochromic compounds and liquid crystals change color at a specific surface temperature. The advantages of these materials are the simplicity of the test and the relatively low cost if small areas are involved.

25.6.3 Thermal Testing

Excellent results may be achieved if the IR detection can be performed in a dynamic thermal environment where the transient effects of a heat or work input into the object are monitored. This enhances detection of areas where different heat transfer rates occur. Applications involving steady-state conditions are more limited. Thermography has been successfully used in several different areas of testing. In medicine it is used to detect tumors, in aircraft manufacture or maintenance it is used to detect debonding in layered structures, in the electronics industry it is used to detect poor thermal performance of circuit board components, and it is sometimes used to detect stress-induced thermal gradients around defects in dynamically loaded test samples. For more information on thermal NDE methods, see Ref. 1, Vols. 8 and 9, and Ref. 4.

25.7 MAGNETIC PARTICLE METHOD

The magnetic particle method of nondestructive testing is used to locate surface and subsurface discontinuities in ferromagnetic materials.[1] An excellent short reference for this NDE method is Ref. 2, especially chapters 10–16. This method is based on the principle that magnetic lines of force, when present in a magnetized ferromagnetic material, are distorted by changes in material continuity, such as cracks or inclusions, as shown schematically in Fig. 25.25. If the flaw is open at the surface, the flux lines bulge or escape from the surface at the site of the discontinuity. Even near-surface flaws, such as nonmagnetic inclusions, cause the same bulging of the lines of force above the surface. This distorted field, usually referred to as a leakage field, is used to reveal the presence of the discontinuity when fine magnetic particles are attracted to it. If these particles are fluorescent, then their presence at a flaw will be visible under ultraviolet light, much as penetrant indications. See Fig. 25.7. Magnetic particle inspection is used principally for all the inspection for steel components because it is fast, easily implemented, and has rather simple flaw indications. The part is usually magnetized with an electric current and then a solution containing fluorescent particles is applied. The particles that stick to the part form the indication of the flaw.

25.7.1 The Magnetizing Field

The magnetizing field may be applied in any one of a number of ways. Its function is to generate a magnetic field in the part. The application of a magnetizing force (H) generates a magnetic flux (B) in the part as shown schematically in Fig. 25.26. The magnetic flux density, B, has units of tesla or webers/m^2 and the strength of the magnetic field or magnetic flux intensity, H, has units of amperes/meter. Instrumentation is often calibrated in oersteds for H and gauss for B. Referring to

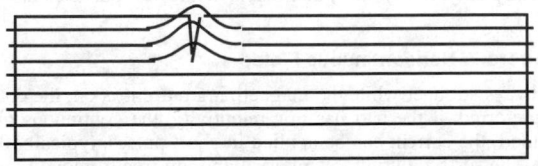

Fig. 25.25 Schematic representation of the magnetic lines of flux in a ferromagnetic metal near a flaw. Small magnetic particles are attracted to the leakage field associated with the flaw.

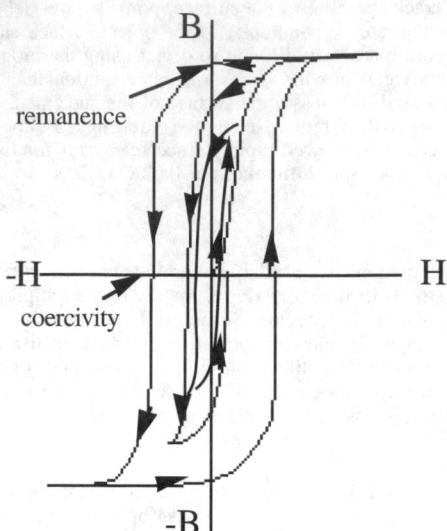

Fig. 25.26 A magnetic flux intensity, *H*, versus magnetic flux density, *B*, hysteresis curve for a
typical steel. Initial magnetization starts at the origin and progresses as shown by the arrows.
Demagnetization follows the arrows of the smaller hysteresis loops.

Fig. 25.26, one starts at the origin as a magnetizing force *H* is applied to a specimen. The magnetic
field *B* internal to the specimen increases in a nonlinear fashion along the path shown by the arrows.
If the force (*H*) is reversed, then the magnetic field (*B*) does not return to zero, but follows the arrows
around the curve as shown. Note that once the magnetizing force is removed, the flux density does
not return to zero, but remains at an elevated value called the material's remanence, B_r. This is where
most magnetic particle inspections are performed. Also note that an appreciable reverse magnetic
force $-H_c$ must be applied before the internal field density is again zero. This point $-H_c$ is referred
to as the coercivity of the material. If the magnetizing force is applied and reversed while decreasing
its strength, in the manner shown, then the material will respond by continually moving around this
hysteresis loop.

Selection of the type of magnetizing current depends primarily on whether the defects are open
to the surface or are wholly below it. Alternating current (ac) magnetizing currents are best for the
detection of surface discontinuities because the current is concentrated in the near-surface region of
the part. Direct current (dc) magnetizing currents are best suited for subsurface discontinuities because
of the current's deeper penetration of the part. While dc can be obtained from batteries or dc generators, it is usually produced by half-wave or full-wave rectification of commercial power. Rectified
current is classified as half-wave direct current (HWDC) or full-wave direct current (FWDC). Alternating current fields are usually obtained from conventional power mains, but are supplied to the part
at reduced voltage, for reasons of safety and the high-current requirements of the magnetizing process.

Two general types of magnetic particles are available. One type of particle is low-carbon steel
with high permeability and low retentivity, which is used dry and consists of different sizes and
shapes to respond to both weak and strong leakage fields. The other type of particle consists of
extremely fine particles of magnetic iron oxide that are suspended in a liquid (either a petroleum
distillate or water). These particles are smaller and have a lower permeability than the dry particles.
Their small mass permits them to be held by the weak leakage fields at very fine surface cracks.
Magnetic particles are available in several colors to increase their contrast against different surfaces.
Dry powders are typically gray, red, yellow, and black, while wet particles are usually red, black, or
fluorescent.

25.7.2 Continuous versus Noncontinuous Fields

Because the field is always stronger while the magnetizing current is on, the continuous magnetizing
method is generally preferred, if the part has low retentivity, the continuous method must be used.
In the continuous method, the current can be applied in short pulses, typically 0.5 sec. The magnetic
particles are applied to the surface during this interval and are free to move to the site of the leakage
fields. In this case, the use of liquid-suspended fluorescent particles yields the most sensitive detection

method. For field inspections, the magnetizing current is usually continuously on during the test to give time for the powder to migrate to the defect site.

In the residual method, the particles are applied after the magnetizing current is removed. This method is particularly suited for production inspection of multiple parts.

The choice of direction of the magnetizing field (B) within the part involves the nature of the flaw and its direction with respect to the surface and the major axis of the part. In circular magnetization, the field runs circumferentially around the part. It is induced into the part by passing current through it between two contacting electrodes. Since only flaws perpendicular to the magnetizing lines are readily detectable, circular magnetization is used to detect flaws that are parallel or less than 45° to the surface of the part. Longitudinal magnetization is usually produced by placing the part in a coil. It creates a field running lengthwise through the part and is used to detect transverse discontinuities to the axis of the part.

25.7.3 The Inspection Process

The surface of the part to be examined should be essentially clean, dry, and free of contaminants, such as oil, grease, loose rust, loose sand, loose scale, lint, thick paint, welding flux, and weld splatter. Cleaning of the test part may be accomplished by detergents, organic solvents, or mechanical means.

Portable and stationary equipment are available. Selection of the specific type depends on the nature and location of testing. Portable equipment is available in lightweight units (35–90 lb), which can be readily taken to the inspection site. Generally, these units operate off 115, 230, or 460 V ac and supply current outputs of 750–1500 A in half-wave or ac.

25.7.4 Demagnetizing the Part

Once the inspection process is complete, the part must be demagnetized. This is done by one of several means, depending on the subsequent usage of the part. A simple way of demagnetizing the part is to do what many electronic technicians do to remove any residual magnetism from small tools. In this case, the tool is placed in the coil of a soldering iron and slowly withdrawn. This has the effect of retracing the hysteresis loop a large number of times, each time with a smaller magnetizing force applied to the tool. The tool will then have a very small remnant magnetic field that is, for all practical purposes, zero. This same process is accomplished with an industrial part by slowly reducing and reversing the magnetizing current till it is essentially zero. This process of reducing the residual magnetic field is shown schematically by following the arrows in Fig. 25.26. Another way to demagnetize the part is to heat it above the materials' Curie temperature, about 550°C for iron, where all residual magnetism disappears. This last process is the best means of removing all magnetism, but it does require the expense and time of an elevated heat treatment.

APPENDIX A
ULTRASONIC PROPERTIES OF COMMON MATERIALS

Ultrasonic Properties of Liquids

Liquid (20°C unless noted)	Longitudinal Wave Speed × 10⁵ cm/sec	Density gm/cm³
Acetic Acid	1.173	1.049
Acetone	1.192	0.792
Amyl Acetate (26°C)	1.168	0.879
Aniline	0.656	1.022
Benzene	1.326	0.879
Blood (Horse) (37°C)	1.571	
Bromoform	0.928	2.890
n-Buytl Alcohol	1.268	0.810
Caprylic Acid	1.331	0.910
Carbon Disulfide	1.158	1.263
Carbon Tetrachloride	0.938	1.595
Cloroform	1.005	1.498
Formaldehyde (25°C)	1.587	0.815
Gasoline (34°C)	1.25	0.803
Glycerin	1.923	1.261
Kerosene (25°C)	1.315	0.82
Mercury	1.451	13.546
Methyl Alcohol	1.123	0.796
Oils		
Campor (25°C)	1.390	
Castor	1.500	0.969
Condenser	1.432	
Olive (22°C)	1.440	0.918
SAE 20	1.74	0.87
Sperm (32°C)	1.411	
Transformer	1.38	0.92
Oleic Acid	1.333	0.873
n-Pentane	1.044	
Silicon Tetrachloride (30°C)	0.766	1.483
Toluene	1.328	0.67
Water (distilled)	1.482	1.00
m-Xylene	1.340	0.864

Ultrasonic Properties of Solids: Metals

Metal (20°C unless noted)	Longitudinal Wave Speed × 10⁵ cm/sec	Transverse Wave Speed × 10⁵ cm/sec	Density gm/cm³
Aluminum			
Al (1100)	6.31	3.08	2.71
Al (2014)	6.37	3.07	2.80
Al (2024-T4)	6.37	3.16	2.77
Al (2117-T4)	6.50	3.12	2.80
Al (6061-T6)	6.31	3.14	2.70
Bearing Babbit	2.30		10.1
Beryllium	12.890	8.880	1.82
Bismuth	2.18	1.10	9.80
Brass (70% Cu & 30% Zn)	4.37	2.10	8.50
Brass (Naval)	4.43	2.12	8.42

Ultrasonic Properties of Solids: Metals (*Continued*)

Metal (20°C unless noted)	Longitudinal Wave Speed × 10⁵ cm/sec	Transverse Wave Speed × 10⁵cm/sec	Density gm/cm³
Bronze (5% P)	3.53	2.32	8.86
Cadmium	2.78	1.50	8.64
Cerium	2.424	1.415	6.77
Chromium	6.608	4.005	7.20
Cobalt	5.88	3.10	8.90
Columbium	4.92	2.10	8.57
Constantan	5.177	2.625	8.88
Copper	4.759	2.325	8.93
Copper (110)	4.70	2.26	8.9
Dysprosium	2.296	1.733	8.53
Erbium	2.064	1.807	9.06
Europium	1.931	1.237	5.17
Gadolinium	2.927	1.677	7.89
Germanium	5.18	3.10	5.47
Gold	3.24	1.20	19.32
Hafnium	2.84		13.3
Hastelloy X	5.79	2.74	8.23
Hastelloy C	5.84	2.90	8.94
Holmium	3.089	1.729	8.80
Indium	2.56	0.74	7.30
Invar	4.657	2.658	
Lanthanium	2.362	1.486	6.16
Lead	2.160	0.700	11.34
Lead (96% Pb & 6% Sb)	2.16	0.81	10.88
Lutetium	2.765	1.574	9.85
Magnesium	5.823	3.163	1.74
AM-35	5.79	3.10	1.74
FS-1	5.47	3.03	1.69
J-1	5.67	3.01	1.70
M1A	5.74	3.10	1.76
O-1	5.80	3.04	1.72
Manganese	4.66	2.35	7.39
Manganin	4.66	2.35	8.40
Molybdenium	6.29	3.35	10.2
Nickel			
Pure	5.63	2.96	8.88
Inconel	5.82	3.02	8.5
Inconel (X-750)	5.94	3.12	8.3
Inconel (wrought)	7.82	3.02	8.25
Monel	5.35	2.72	8.83
Monel (wrought)	6.02	2.72	8.83
Silver–Nickel (18%)	4.62	2.32	8.75
German Silver	4.76	2.16	8.40
Neodynium	2.751	1.502	7.10
Platinum	3.96	1.67	21.4
Potassium	2.47	1.22	0.862
Praseodynium	2.639	1.437	6.75
Samarium	2.875	1.618	7.48
Silver	3.60	1.59	10.5
Sodium	3.03	1.70	0.97

Ultrasonic Properties of Solids: Metals (*Continued*)

Metal (20°C unless noted)	Longitudinal Wave Speed $\times\ 10^5$ cm/sec	Transverse Wave Speed $\times\ 10^5$ cm/sec	Density gm/cm^3
Steel			
1020	5.89	3.24	7.71
1095	5.90	3.19	7.80
4150, Rc14	5.86	2.79	7.84
4150, Rc 18	5.88	3.18	7.82
4150, Rc 43	5.87	3.20	7.81
4150, Rc 64	5.83	2.77	7.80
4340	5.85	3.24	7.80
52100 Annealed	5.99	3.27	7.83
52100 Hardened	5.89	3.20	7.8
D6 Tool Steel Annealed	6.14	3.31	7.7
Stainless Steels			
302	5.66	3.12	7.9
304L	5.64	3.07	7.9
347	5.74	3.10	7.91
410	5.39	2.99	7.67
430	6.01	3.36	7.7
Tantalum	4.10	2.90	16.6
Thorium	2.94	1.56	11.3
Thulium	3.009	1.809	9.29
Tin	3.32	1.67	7.29
Titanium (Ti-6-4)	6.18	3.29	4.50
Tungsten			
Annealed	5.221	2.887	19.25
Drawn	5.410	2.640	19.25
Uranium	3.37	1.98	18.7
Vanadium	6.023	2.774	6.03
Ytterbium	1.946	1.193	6.99
Yttrium	4.10	2.38	4.34
Zinc	4.187	2.421	7.10
Zirconium	4.65	2.25	6.48

Ultrasonic Properties of Solids: Ceramics

Ceramic (20°C unless noted)	Longitudinal Wave Speed $\times\ 10^5$ cm/sec	Transverse Wave Speed $\times\ 10^5$ cm/sec	Density gm/cm^3
Alumium Oxide	10.84	6.36	3.98
Barium Nitrate	4.12	2.28	3.24
Barium Titanate	5.65	3.03	5.5
Bone (Human Tibia)	4.00	1.97	1.7–2.0
Cobalt Oxide	6.56	3.32	6.39
Concrete	4.25–5.25		2.60
Glass			
Crown	5.66	3.42	2.50
Flint	4.26	2.56	3.60
Lead	3.76	2.22	4.6
Plate	5.77	3.43	2.51
Pyrex	5.57	3.44	2.23
Soft	5.40		2.40
Gramite	3.95		2.75
Graphite	4.21	2.03	2.25
Ice (−16°C)	3.83	1.92	0.94
Indium Antimonide	3.59	1.91	
Lead Nitrate	3.28	1.47	4.53
Lithium Fluoride	6.56	3.84	2.64

Ultrasonic Properties of Solids: Ceramics (*Continued*)

Ceramic (20°C unless noted)	Longitudinal Wave Speed $\times 10^5$ cm/sec	Transverse Wave Speed $\times 10^5$ cm/sec	Density gm/cm^3
Magnesium Oxide	9.32	5.76	3.58
Manganese Oxide	6.68	3.59	5.37
Nickel Oxide	6.60	3.68	6.79
Porcelain	5.34	3.12	2.41
Quartz			
Crystalline	5.73		2.65
Fused	5.57	3.52	2.60
Polycrystalline	5.75	3.72	2.65
Rock Salt	4.60	2.71	2.17
Titanium Dioxide (Rutile)	8.72	4.44	4.26
Sandstone	2.92	1.84	2.2–2.4
Sapphire (c-axis)	11.91	7.66	3.97
Slate	4.50		2.6–3.3
Titanium Carbide	8.27	5.16	5.15
Tourmaline (Z-cut)	7.54		3.10
Tungsten Carbide	6.66	3.98	10.15
Ytterium Iron Garnet	7.29	4.41	5.17
Zinc Sulfide	5.17	2.42	4.02
Zinc Oxide	6.00	2.84	5.61

Ultrasonic Properties of Solids: Polymers

Polymer (20°C unless noted)	Longitudinal Wave Speed $\times 10^5$ cm/sec	Transverse Wave Speed $\times 10^5$ cm/sec	Density gm/cm^3
Acrylic Resin	2.67	1.12	1.18
Bakelite	2.59		1.40
Buytl Rubber	1.99		1.13
Cellulose Acetate	2.45		1.30
Cork	0.5		0.2
Delrin (Acetalhomo-Polymer) (0°C)	2.515		1.42
Ebonite	2.50		1.15
Lexan			
(Polycarbonate 0°C)	2.28		1.19
Neoprene	1.730		1.42
Nylon	2.68		
Nylon 6,6	1.68		
Parafin	2.20	0.83	
Perspex	2.70	1.33	1.29
Phenolic	1.42		1.34
Plexiglas			
UVA	2.76		1.27
UVA II	2.73	1.43	1.18
Polyacrylonitrile–butadiene–styrene I	2.16	1.43	1.18
Polyacrylonitrile–butadiene–styrene II	2.20	0.810	1.022
Polybutadiene Rubber	1.57		1.10
Polycaprolactam	2.700	1.12	1.146
Polycarboranesilonane	1.450		1.041
Polydimethylsiloxane	1.020		1.045

Ultrasonic Properties of Solids: Polymers (Continued)

Polymer (20°C unless noted)	Longitudinal Wave Speed $\times 10^5$ cm/sec	Transverse Wave Speed $\times 10^5$ cm/sec	Density gm/cm^3
Polyepoxide + glass spheres I	2.220	1.170	0.691
Polyepoxide + glass spheres II	2.400	1.280	0.718
Polyepoxide + glass spheres III	2.100	1.020	0.793
Polyepoxide + MPDA	2.820	1.230	1.205
Polyester + water	1.840	0.650	1.042
Polythylene	2.67		1.10
Polyhexamethylene adipamide	2.710	1.120	1.147
Polymethacrylate	2.690	1.344	1.191
Polyoxymethylene	2.440	1.000	1.425
Polyproplene	2.650	1.300	0.913
Polystyrene	2.400	1.150	1.052
Polysulfane Resin	2.297		1.24
Polytetrafluoroethylene (Teflon)	1.380		2.177
Polyvinylbutyral	2.350		1.107
Polyvinyl Chloride	2.300		
Polyvinylidene Chloride	2.400		
Polyvinylidene Fluoride	1.930		1.779
Rubber			
India	1.48		0.90
Natural	1.55		1.12
Rubber/Carbon (100/40)	1.68		
Silicon Rubber	0.948		1.48

Calculated Ultrasonic Properties of Composites

Composite (20°C)	Longitudinal Wave Speed $\times 10^5$ cm/sec	Transverse Wave Speed $\times 10^5$ cm/sec	Density gm/cm^3
Glass/Epoxy (parallel to fibers)	5.18	1.63	1.91
Glass/Epoxy (perpendicular to fibers)	3.16	1.72	1.91
Graphite/Epoxy (parallel to fibers)	9.62	1.96	1.57
Graphite/Epoxy (perpendicular to fibers)	2.96	1.96	1.57
Boron/Epoxy (parallel to fibers)	10.60	1.72	1.91
Boron/Epoxy (perpendicular to fibers)	3.34	1.85	1.91

APPENDIX B

ELECTRICAL RESISTIVITIES AND CONDUCTIVITIES OF COMMERCIAL METALS AND ALLOYS

Common Name (Classification)	Resistivity (micro-ohm cm)	Conductivity (% IACS)
Ingot iron (included for comparison)	9	19
Plain carbon steel (AISI-SAE 1020)	10	17
Stainless steel type 304	72	2.4
Cast gray iron (ASTH A48-48. Class 25)	67	2.6
Malleable iron (ASTM A 47)	30	5.7
Ductile cast iron (ASTH A339, A395)	60	2.9
Ni (resist cast iron, type 2)	170	1.0
Cast 28-7 alloy 11D (ASTH A297-63T)	41	4.2
Hastelloy C	139	1.2
Hastelloy X	115	1.5
Haynes Stellite alloy 25	8B	2.0
Inconel X (annealed)	122	1.4
Inconel 600	98	1.7
Aluminum alloy 3003, rolled (ASTH B221)	4	43
Aluminum alloy 2017, annealed (ASTH B221)	4	43
Aluminum alloy 380 (ASTH SC84B)	7.5	23
A# Aluminum alloy		
6061-T-6	4.1	42
7075-T-6	5.3	32
2024-T-4	5.2	30
Copper (ASTH B152, B124, B133, B1, B2, B3)	1.7	1.0×10^2
Yellow brass or high brass (ASTH B36, B134, B135)	7	25
70-30 brass	6.2	28
Aluminum bronze ASTH B 169, alloy A ASTH B124, B130	12	14
Phosphor bronzes	16	11
Nickel silver I B% alloy A wrought (ASTH B 122, No. 2)	29	5.9
Cupronickel 30%	35	4.9
Red brass, cast (ASTH B30, No. 4A)	11	16
Chemical lead	21	8.2
Antimonial lead (hard lead)	23	7.5
Solder 50-50	15	N
Ti-6Al-4V alloy	172	1.0
Magnesium alloy AZ31 B	9	19
K Monel	58	3.0
Nickel (ASTH B160, B161, B162)	10	17
Cupronickel 55-45 (constantan)	49	3.5
Commercial titanium	80	2.2
Waspaloy	123	1.4
Zinc (ASTH B69)	6	29
Zircaloy-2	72	2.4
Zirconium (commercial)	41	4.2

REFERENCES

1. *The Nondestructive Testing Handbooks,* 2nd ed., American Society for Nondestructive Testing, Columbus, OH: Vol. 1, *Leak Testing,* R. C. McMaster (ed.); Vol. 2, *Liquid Penetrant Testing,* R. C. McMaster (ed.); Vol. 3, *Radiography and Radiation Testing,* L. E. Bryant (ed.); Vol. 4, *Electromagnetic Testing,* M. L. Mester (ed.); Vol. 5, *Acoustic Emission,* R. K. Miller (ed.); Vol. 6, *Magnetic Particle Testing,* J. T. Schmidt and K. Skeie (eds.); Vol. 7, *Ultrasonic Testing,* A. S. Birks and R. E. Green, Jr. (eds.); Vol. 8, *Visual and Optical Testing,* M. W. Allgaier and S. Ness (eds.); Vol. 9, *Special Nondestructive Testing Methods,* R. K. Stanley (ed.).

2. D. E. Bray and R. K. Stanley, *Nondestructive Evaluation, A Tool for Design, Manufacturing, and Service,* McGraw-Hill, New York, 1989.

3. R. Halmshaw, *Nondestructive Testing Handbook,* 2nd ed., Chapman & Hall, London, 1991.

4. *Metals Handbook: Nondestructive Evaluation and Quality Control,* Vol. 11, American Society for Metals, Metals Park, OH, 1976.

5. R. A. Kline, *Nondestructive Characterization of Materials,* Technomic. Lancaster, PA, 1992.

6. *Annual Book of ASTM Standards: Part II, Metallography and Nondestructive Testing,* American Society for Testing and Materials, Philadelphia, PA.

7. *Materials Evaluation,* American Society for Nondestructive Testing, Columbus, OH.

8. *British Journal of Nondestructive Testing,* British Institute of Nondestructive Testing, Northampton, UK.

9. *NDT International,* IPC Business Press Ltd., Sussex, UK.

10. *Non-Destructive Testing Journal,* Energiteknik, Japan (c/o Japan Technical Services Corp., 3F Ohkura Bldg., 4-10 Shiba-Daimon 1 Chrome, Mitato-ku Tokyo, 105, JAPAN).

11. *Soviet Journal of Nondestructive Testing,* translated by Consultants Bureau, New York.

12. *Journal of Nondestructive Testing,* Plenum Press, New York.

13. K. Ono (ed.), *Journal of Acoustic Emission,* Plenum Press, New York.

14. *Sensors—The Journal of Machine Perception,* Helmers Publishing, Peterborough, NH.

15. J. Krautkramer and H. Krautkramer, *Ultrasonic Testing of Materials,* 4th ed., Springer-Verlag, New York, 1995.

16. R. A. Quinn, *Industrial Radiology—Theory and Practice,* Eastman Kodak, Rochester, NY, 1980.

17. H. Burger, *Neutron Radiography: Methods, Capabilities and Applications,* Elsevier, New York, 1965.

18. F. Förster, "Theoretische und experimentalle Grundlagen der zerstörungfreien Werkstoffprufung mit Wirbelstromverfahern. I. Das Tastpulverfahern," *Z. Metallk.* **43,** 163–171 (1952).

PART 2
SYSTEMS AND CONTROLS

CHAPTER 26

SYSTEMS ENGINEERING: ANALYSIS, DESIGN, AND INFORMATION PROCESSING FOR ANALYSIS AND DESIGN

Andrew P. Sage
School of Information Technology and Engineering
George Mason University
Fairfax, Virginia

26.1	INTRODUCTION	763	26.5	SYSTEM DESIGN		784
			26.5.1	The Purposes of Systems Design		784
26.2	THE SYSTEM LIFE CYCLE AND FUNCTIONAL ELEMENTS OF SYSTEMS ENGINEERING	765	26.5.2	Operational Environments and Decision Situation Models		785
			26.5.3	The Development of Aids for the Systems Design Process		786
26.3	SYSTEMS ENGINEERING OBJECTIVES	770	26.5.4	Leadership Requirements for Design		789
26.4	SYSTEMS ENGINEERING METHODOLOGY AND		26.5.5	System Evaluation		790
	METHODS	771	26.5.6	Evaluation Test Instruments		791
	26.4.1 Issue Formulation	771				
	26.4.2 Issue Analysis	775	26.6	CONCLUSIONS		792
	26.4.3 Information Processing by Humans and Organizations	779				
	26.4.4 Interpretation	782				
	26.4.5 The Central Role of Information in Systems Engineering	783				

26.1 INTRODUCTION

Systems engineering is a management technology. Technology involves the organization and delivery of science for the (presumed) betterment of humankind. Management involves the interaction of the organization, and the humans in the organization, with the environment. Here, we interpret environment in a very general sense to include the complete external milieu surrounding individuals and organizations. Hence, systems engineering as a management technology involves three ingredients: science, organizations, and their environments. Information, and knowledge, is ubiquitous throughout systems engineering and management efforts and is, in reality, a fourth ingredient. Systems engineering is thus seen to involve science, organizations and humans, environments, technologies, and information and knowledge.

The process of systems engineering involves working with clients in order to assist them in the organization of information and knowledge to aid in judgment and choice of activities. These activities result in the making of decisions and associated resource allocations through enhanced efficiency, effectiveness, equity, and explicability as a result of systems engineering efforts.

Mechanical Engineers' Handbook, 2nd ed., Edited by Myer Kutz.
ISBN 0-471-13007-9   © 1998 John Wiley & Sons, Inc.

This set of action alternatives is selected from a larger set, in accordance with a value system, in order to influence future conditions. Development of a set of rational policy or action alternatives must be based on formation and identification of candidate alternative policies and objectives against which to evaluate the impacts of these proposed activities, such as to enable selection of efficient, effective, and equitable alternatives for implementation.

In this chapter, we are concerned with the engineering of large-scale systems, or *systems engineering*.[1] We are especially concerned with strategic level systems engineering, or *systems management*.[2] We begin by first discussing the need for systems engineering and then providing some definitions of systems engineering. We next present a structure describing the systems engineering process. The result of this is a *life-cycle model* for systems engineering processes. This is used to motivate discussion of the functional levels, or considerations, involved in systems engineering efforts: *systems engineering methods and tools*, *systems methodology or processes*, and *systems management*. Considerably more details are presented in Refs. 1 and 2, which are the sources from which most of this chapter is derived.

Systems engineering is an appropriate combination of mathematical, behavioral, and management theories in a useful setting appropriate for the resolution of complex real world issues of large scale and scope. As such, systems engineering consists of the use of management, behavioral, and mathematical constructs to identify, structure, analyze, evaluate, and interpret generally incomplete, uncertain, imprecise, and otherwise imperfect information. When associated with a value system, this information leads to knowledge to permit decisions that have been evolved with maximum possible understanding of their impacts. A central need, but by no means the only need, in systems engineering is to select an appropriate life cycle, or process, that is explicit, rational, and compatible with the implementation framework extant, and the perspectives and knowledge bases of those responsible for decision activities. When this is accomplished, an appropriate choice of systems engineering methods and tools may be made to enable full implementation of the life-cycle process.

Information is a very important quantity that is assumed to be present in the management technology that is systems engineering. This strongly couples notions of systems engineering with those of technical direction or systems management of technological development, rather than exclusively with one or more of the methods of systems engineering, important as they may be for the ultimate success of a systems engineering effort. It suggests that *systems engineering is the management technology that controls a total system life-cycle process, which involves and which results in the definition, development, and deployment of a system that is of high quality, trustworthy, and cost-effective in meeting user needs.* This process-oriented notion of systems engineering and systems management will be emphasized here.

Among the appropriate conditions for use of systems engineering are the following:

- There are many considerations and interrelations.
- There are far-reaching and controversial value judgments.
- There are multidisciplinary and interdisciplinary considerations.
- The available information is uncertain, imprecise, incomplete, or otherwise flawed.
- Future events are uncertain and difficult to predict.
- Institutional and organizational considerations play an important role.
- There is a need for explicit and explicable consideration of the efficiency, effectiveness, and equity of alternative courses of action.

There are a number of results potentially attainable from use of systems engineering approaches. These include:

- Identification of perceived needs in terms of identified objectives and values of a client group
- Identification or definition of a set of user or client requirements for the product system or service system that will ultimately be fielded
- Enhanced identification of a wide range of proposed alternatives or policies that might satisfy these needs, achieve the objectives of the clients in a high-quality and trustworthy fashion, and fulfill the requirements definition
- Increased understanding of issues that led to the effort, and the impacts of alternative actions upon these issues
- Ranking of these identified alternative courses of action in terms of the utility (benefits and costs) in achieving objectives, satisfying needs, and fulfilling requirements
- A set of alternatives that is selected for implementation, generally by a group of content specialists responsible for detailed design and implementation, and an appropriate plan for action to achieve this implementation

Ultimately the action plans result in a working product or service and are maintained over time in subsequent phases of the post-deployment efforts that also involve systems engineering.

To develop professionals capable of coping satisfactorily with diverse factors involved in wide scope problem-solving is a primary goal of systems engineering and systems engineering education. This does not imply that a single individual or even a small group can, despite its strong motivation, solve all of the problems involved in a systems study. Such a requirement would demand total and absolute intellectual maturity on the part of the systems engineer and such is surely not realistic. It is also unrealistic to believe that issues can be resolved without very close association with a number of people who have stakes, and who thereby become stakeholders, in problem-solution efforts. Consequently, systems engineers must be capable of facilitation and communication of knowledge between the diverse group of professionals, and their publics, that are involved in wide-scope problem-solving. This requires that systems engineers be knowledgeable and able to use not only the technical methods-based tools that are needed for issue and problem resolution, but the behavioral constructs and management abilities that are also needed for resolution of complex, large-scale problems. Intelligence, imagination, and creativity are necessary but not sufficient for proper use of the procedures of systems engineering. Facility in human relations and effectiveness as a broker of information among parties at interest in a systems engineering program are very much needed as well.

It is this blending of the technical, managerial, and behavioral that is a normative goal of success for systems engineering education and for systems engineering professional practice. Thus, systems engineering involves

- The sciences and the various methods, analysis, and measurement perspectives associated with the sciences
- Life-cycle process models for definition, development, and deployment of systems
- The systems management issues associated with choice of an appropriate process
- Organizations and humans, and the understanding of organizational and human behavior
- Environments and understanding of the diverse interactions of organizations of people, technologies, and institutions with their environments
- Information, and the way in which it can and should be processed to facilitate all aspects of systems engineering efforts

Successful systems engineering must be practiced at three levels: systems methods and measurements, systems processes and methodology, and systems management. Systems engineers must be aware of a wide variety of methods that assist in the formulation, analysis, and interpretation of contemporary issues. They must be familiar with systems engineering process life cycles (or methodology, as an open set of problem-solving procedures) in order to be able to select eclectic approaches that are best suited to the task at hand. Finally, a knowledge of systems management is necessary in order to be able to select life-cycle processes that are best matched to behavioral and organizational concerns and realities.

All three of these levels, suggested in Fig. 26.1, are important. To neglect any of them in the practice of systems engineering is to invite failure. It is generally not fully meaningful to talk only of a method or algorithm as a useful system-fielding or life-cycle process. It is ultimately meaningful to talk of a particular process as being useful. A process or product line that is truly useful for the fielding of a system will depend on the methods that are available, the operational environment, and leadership facets associated with use of the system and the system fielding process. Thus systems management, systems engineering processes, and systems engineering methods and measurements do, separately and collectively, play a fundamental role in systems engineering.

26.2 THE SYSTEM LIFE CYCLE AND FUNCTIONAL ELEMENTS OF SYSTEMS ENGINEERING

We have provided one definition of systems engineering thus far. It is primarily a structural and process-oriented definition. A related definition, in terms of purpose, is that "systems engineering is management technology to assist and support policy-making, planning, decision-making, and associated resource allocation or action deployment for the purpose of acquiring a product desired by customers or clients. Systems engineers accomplish this by quantitative and qualitative formulation, analysis, and interpretation of the impacts of action alternatives upon the needs perspectives, the institutional perspectives, and the value perspectives of their clients or customers." Each of these three steps is generally needed in solving systems engineering problems. Issue *formulation* is an effort to identify the needs to be fulfilled and the requirements associated with these in terms of objectives to be satisfied, constraints and alterables that affect issue resolution, and generation of potential alternative courses of action. Issue *analysis* enables us to determine the impacts of the identified alternative courses of action, including possible refinement of these alternatives. Issue

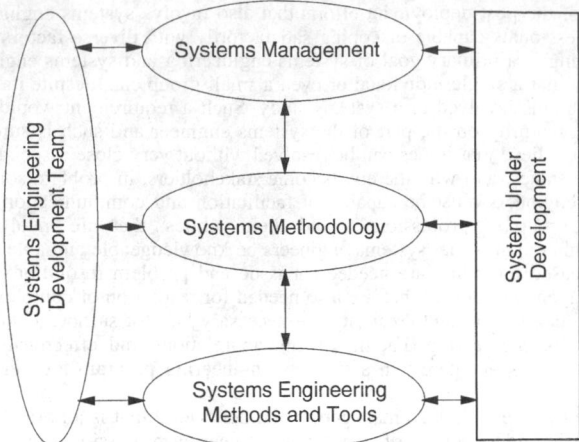

Fig. 26.1 Conceptual illustration of the three levels for systems engineering.

interpretation enables us to rank in order the alternatives in terms of need satisfaction and to select one for implementation or additional study. This particular listing of three systems engineering steps and their descriptions is rather formal. Often, issues are resolved this way. The steps of formulation, analysis, and interpretation may also be accomplished on as "as-if" basis by application of a variety of often useful heuristic approaches. These may well be quite appropriate in situations where the problem-solver is experientially familiar with the task at hand and the environment into which the task is imbedded.[1]

The key words in this definition are "formulation," "analysis," and "interpretation." In fact, all of systems engineering can be thought of as consisting of formulation, analysis, and interpretation efforts, together with the systems management and technical direction efforts necessary to bring this about. We may exercise these in a formal sense throughout each of the several phases of a systems engineering life cycle, or in an "as-if" or experientially based intuitive sense. These formulation, analysis, and interpretation efforts are the step-wise or microlevel components that comprise a part of the structural framework for systems methodology. They are needed for each phase in a systems engineering effort, although the specific formulation methods, analysis methods, and interpretation methods may differ considerably across the phases.

We can also think of a functional definition of systems engineering: "Systems engineering is the art and science of producing a product, based on phased efforts, that satisfies user needs. The system is functional, reliable, of high quality, and trustworthy, and has been developed within cost and time constraints through use of an appropriate set of methods and tools."

Systems engineers are very concerned with the appropriate *definition*, *development*, and *deployment* of product systems and service systems. These comprise a set of phases for a systems engineering life cycle. There are many ways to describe the life-cycle phases of the systems engineering process, and we have described a number of them in Refs. 1 and 2. Each of these basic life-cycle models, and those that are outgrowths of them, is comprised of these three phases of definition, development, and deployment. For pragmatic reasons, a typical life cycle will almost always contain more than three phases. Often, it takes on the "waterfall" pattern illustrated in Fig. 26.2, although there are a number of modifications of the basic waterfall, or "grand-design," life cycles that allow for incremental and evolutionary development of systems life-cycle processes.[2]

A successful approach to systems engineering as an intellectual and action-based approach for increased innovation and productivity and other contemporary challenges must be capable of issue formulation, analysis, and interpretation at the level of institutions and values as well as at the level of symptoms. Systems engineering approaches must allow for the incorporation of need and value perspectives as well as technology perspectives into models and postulates used to evolve and evaluate policies or activities that may result in technological and other innovations.

In actual practice, the steps of the systems process (formulation, analysis, and interpretation) are applied iteratively, across each of the phases of a systems engineering effort, and there is much feedback from one step to the other. This occurs because of the learning that is accomplished in the process of problem-solution. Underlying all of this is the need for a general understanding of the diversity of the many systems engineering methods and algorithms that are available and their role in a systems engineering process. The knowledge taxonomy for systems engineering, which consists

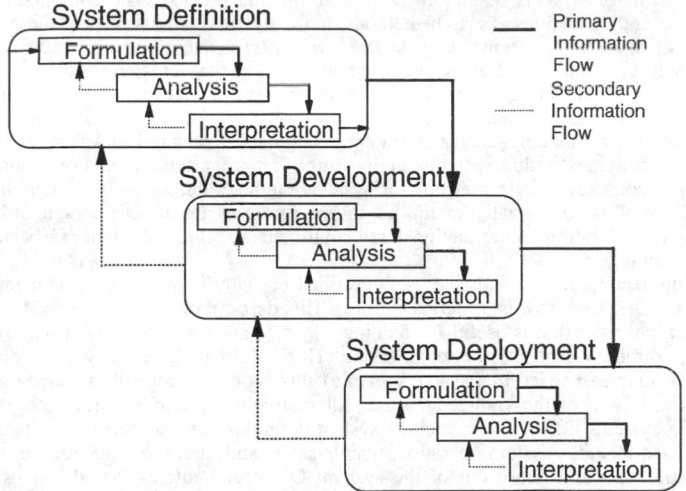

Fig. 26.2 One representation of three systems engineering steps within each of three life-cycle phases.

of the major intellectual categories into which systems efforts may be categorized, is of considerable importance. The categories include systems methods and measurements, systems engineering processes or systems methodology, and systems management. These are used, as suggested in Fig. 26.3, to produce a *systems*, which is a generic term that we use to describe a product or a service.

The methods and metrics associated with systems engineering involve the development and application of concepts that form the basis for problem formulation and solution in systems engineering. Numerous tools for mathematical systems theory have been developed, including operations research (linear programming, nonlinear programming, dynamic programming, graph theory, etc.), decision and control theory, statistical analysis, economic systems analysis, and modeling and simulation. Systems science is also concerned with psychology and human factors concepts, social interaction and human judgment research, nominal group processes, and other behavioral science efforts. Of very special significance for systems engineering is the interaction of the behavioral and the algo-

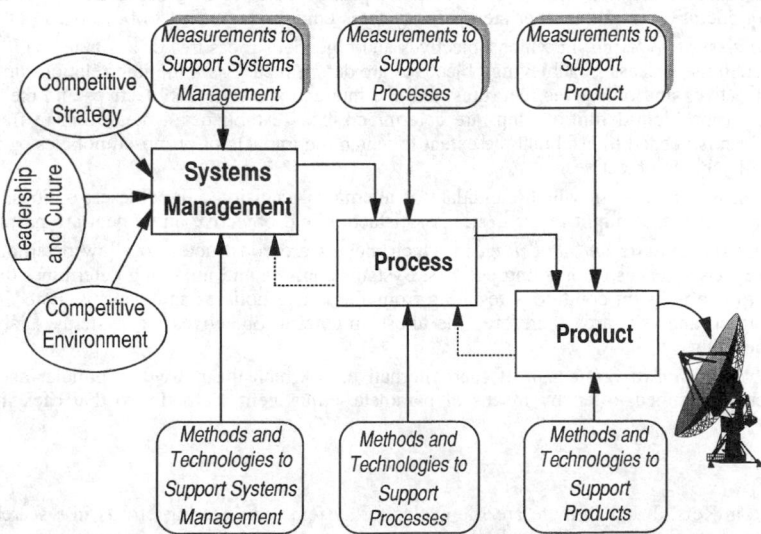

Fig. 26.3 Representation of the structure systems engineering and management functional efforts.

rithmic components of systems science in the choice-making process. The combination of a set of systems science and operations research methods and a set of relations among these methods and activities constitutes what is known as a *methodology*. References 3 and 4 discuss a number of systems engineering methods and associated methodologies for systems engineering.

As we use it here, a methodology is an open set of procedures that provides the means for solving problems. The tools or the content of systems engineering consists of a variety of algorithms and concepts that use words, mathematics, and graphics. These are structured in ways that enable various problem-solving activities within systems engineering. Particular sets of relations among tools and activities, which constitute the framework for systems engineering, are of special importance here. Existence and use of an appropriate systems engineering process are of considerable utility in dealing with the many considerations, interrelations, and controversial value judgments associated with contemporary problems.

Systems engineering can be and has been described in many ways. Of particular importance is a morphological description; that is, in terms of form. This description leads to a specific methodology that results in a process* that is useful for fielding a system and/or issue resolution. We can discuss the knowledge dimension of systems engineering. This would include the various disciplines and professions that may be needed in a systems team to allow it to accomplish intended purposes of the team, such as provision of the knowledge base. Alternatively, we may speak of the phases or time dimension of a systems effort. These include system definition, development, and deployment. The deployment phase includes system operation, maintenance, and finally modification or reengineering or ultimate retirement and phase out of the system. Of special interest are the steps of the logic structure or logic dimension of systems engineering:

- Formulation of issues, or identification of problems or issues in terms of needs and constraints, objectives, or values associated with issue resolution, and alternative policies, controls, hypotheses, or complete systems that might resolve or ameliorate issues
- Analysis of impacts of alternative policies, courses of action, or complete systems
- Interpretation or evaluation of the utility of alternatives and their impacts upon the affected stakeholder group, and selection of a set of action alternatives for implementation

We could also associate feedback and learning steps to interconnect these steps one to another. The systems process is typically very iterative. We shall not explicitly show feedback and learning in our conceptual models of the systems process, although it is ideally always there.

Here we have described a three-dimensional morphology of systems engineering. There are a number of systems engineering morphologies or frameworks. In many of these, the logic dimension is divided into a larger number of steps that are iterative in nature. A particular seven-step framework involves

1. *Problem definition*, in which a descriptive and normative scenario of needs, constraints, and alterables associated with an issue is developed. Problem definition clarifies the issues under consideration to allow other steps of a systems engineering effort to be carried out.

2. *Value system design*, in which objectives and objectives measures or attributes with which to determine success in achieving objectives are determined. Also, the interrelationship between objectives and objectives measures, and the interaction between objectives and the elements in the problem-definition step, are determined. This establishes a measurement framework, which is needed to establish the extent to which the impacts of proposed policies or decisions will achieve objectives.

3. *System synthesis*, in which candidate or alternative decisions, hypotheses, options, policies, or systems that might result in needs satisfaction and objective attainment are postulated.

4. *Systems analysis and modeling*, in which models are constructed to allow determination of the consequences of pursuing policies. Systems analysis and modeling determines the behavior or subsequent conditions resulting from alternative policies and systems. Forecasting and impact analysis are, therefore, the most important objectives of systems analysis and modeling.

5. *Optimization or refinement* of each alternative, in which the individual policies and/or systems are tuned, often by means of parameter-adjustment methods, so that each individual

---

*As noted in Refs. 1 and 2, there are life cycles for systems engineering efforts in research, development, test, and evaluation (RDT&E); systems acquisition, production, or manufacturing; and systems planning and marketing. Here, we restrict ourselves to discussions of the life cycle associated with acquisition, production, or manufacturing.

policy or system is refined in some "best" fashion in accordance with the value system that has been identified earlier.

6. *Evaluation and decision-making*, in which systems and/or policies and/or alternatives are evaluated in terms of the extent to which the impacts of the alternatives achieve objectives and satisfy needs. Needed to accomplish evaluation are the attributes of the impacts of proposed policies and associated objective and/or subjective measurement of attribute satisfaction for each proposed alternative. Often this results in a prioritization of alternatives, with one or more being selected for further planning and resource allocation.

7. *Planning for action*, in which implementation efforts, resource and management allocations, or plans for the next phase of a systems engineering effort are delineated.

More often than not, the information required to accomplish these seven steps is not perfect due to uncertainty, imprecision, or incompleteness effects. This presents a major challenge to the design of processes and to systems engineering practice.

Figure 26.4 illustrates a not-untypical 49-element morphological box for systems engineering. This is obtained by expanding our initial three systems engineering steps of formulation, analysis, and interpretation to the seven just discussed. The three basic phases of definition, development, and deployment are expanded to a total of seven phases. These seven steps, and the seven phases that we associate with them, are essentially those identified by Hall in his pioneering efforts in systems engineering.[5,6] The specific methods we need to use in each of these seven steps are clearly dependent upon the phase of activity that is being completed, and there are a plethora of systems engineering methods available.[3,4] Using a seven-phase, seven-step framework raises the number of activity cells to 49 for a single life cycle. A very large number of systems engineering methods may be needed to fill in this matrix, especially since more than one method will almost invariably be associated with many of the entries.

The requirements and specification phase of the systems engineering life cycle has as its goal the identification of client or stakeholder needs, activities, and objectives for the functionally operational system. This phase should result in the identification and description of preliminary conceptual design considerations for the next phase. It is necessary to translate operational deployment needs into requirements specifications so that these needs may be addressed by the system design efforts. As a result of the requirements specifications phase, there should exist a clear definition of development issues such that it becomes possible to make a decision concerning whether to undertake preliminary conceptual design. If the requirements specifications effort indicates that client needs can be satisfied in a functionally satisfactory manner, then documentation is typically prepared concerning system-level specifications for the preliminary conceptual design phase. Initial specifications for the following three phases of effort are typically also prepared, and a concept design team is selected to implement the next phase of the life-cycle effort. This effort is sometimes called *system-level architecting*.[7,8] Many[9,10] have discussed technical level architectures. It is only recently that the need for major attention to architectures at the systems level has also been identified.

Phases of Systems Engineering			Problem Definition	Value System Design	System Synthesis	Systems Analysis	Alternative Refinement	Decision Making	Planning for Action
			Formulation			**Analysis**		**Interpretation**	
Definition		Program Planning	1	2	3	4	5	6	7
Definition		Project Planning	8						
Development		System Development		16					
Development		Production							
Deployment		Distribution							
Deployment		Operations			38				
Deployment		Reengineering or Retirement						48	49

Fig. 26.4 The phases and steps in one 49-element two-dimensional systems engineering framework with activities shown sequentially for waterfall implementation of effort.

Preliminary conceptual system design typically includes, or results in, an effort to specify the content and associated architecture and general algorithms for the system product in question. The desired product of this phase of activity is a set of detailed design and architectural specifications that should result in a useful system product. There should exist a high degree of user confidence that a useful product will result from detailed design, or the entire design effort should be redone or possibly abandoned. Another product of this phase is a refined set of specifications for the evaluation and operational deployment phases of the life cycle. In the third phase, these are translated into detailed representations in logical form so that system development may occur. A product, process, or system is produced in the fourth phase of the life cycle. This is not the final system design, but rather the result of implementation of the design that resulted from the conceptual design effort.

Evaluation of the detailed design and the resulting product, process, or system is achieved in the sixth phase of the systems engineering life cycle. Depending upon the specific application being considered, an entire systems engineering life-cycle process could be called *design*, or *manufacturing*, or some other appropriate designator. *System acquisition* is an often-used term to describe the entire systems engineering process that results in an operational systems engineering product. Generally, an acquisition life cycle primarily involves knowledge practices or standard procedures to produce or manufacture a product based on established practices. An RDT&E life cycle is generally associated with an emerging technology and involves knowledge principles. A marketing life cycle is concerned with product planning and other efforts to determine market potential for a product or service, and generally involves knowledge perspectives.

The intensity of effort needed for the steps of systems engineering varies greatly with the type of problem being considered. Problems of large scale and scope will generally involve a number of perspectives. These interact and the intensity of their interaction and involvement with the issue under consideration determines the scope and type of effort needed in the various steps of the systems process. Selection of appropriate algorithms or approaches to enable completion of these steps and satisfactory transition to the next step, and ultimately to completion of each phase of the systems engineering effort, are major systems engineering tasks.

Each of these phases of a systems engineering life cycle is very important for sound development of physical systems or products and such service systems as information systems. Relatively less attention appears to have been paid to the requirement-specification phase than to the other phases of the systems engineering life-cycle process. In many ways, the requirement-specification phase of a systems engineering design effort is the most important. It is this phase that has as its goal the detailed definition of the needs, activities, and objectives to be fulfilled or achieved by the process to be ultimately developed. Thus, this phase strongly influences all the phases that follow. It is this phase that describes preliminary design considerations that are needed to achieve successfully the fundamental goals underlying a systems engineering study. It is in this phase that the information requirements and the method of judgment and choice used for selection of alternatives are determined. Effective systems engineering, which inherently involves design efforts, must also include an operational evaluation component that will consider the extent to which the product or service is useful in fulfilling the requirements that it is intended to satisfy.

26.3 SYSTEMS ENGINEERING OBJECTIVES

Ten performance objectives appear to be of primary importance to those who desire to evolve quality plans, forecasts, decisions, or alternatives for action implementation:

1. Identify needs, constraints, and alterables associated with the problem, issue, or requirement to be resolved (problem definition).

2. Identify a planning horizon or time interval for alternative action implementation, information flow, and objective satisfaction (planning horizon, identification).

3. Identify all significant objectives to be fulfilled, values implied by the choice of objectives, and objectives measures or attributes associated with various outcome states, with which to measure objective attainment (value system design).

4. Identify decisions, events, and event outcomes and the relations among them, such that a structure of the possible paths among options, alternatives, or decisions, and the possible outcomes of these, emerges (impact assessment).

5. Identify uncertainties and risks associated with the environmental influences affecting alternative decision outcomes (probability identification).

6. Identify measures associated with the costs and benefits or attributes of the various outcomes or impacts that result from judgment and choice (worth, value, or utility measurement).

7. Search for and evaluate new information, and the cost-effectiveness of obtaining this information, relevant to improved knowledge of the time-varying nature of event outcomes that follow decisions or choice of alternatives (information acquisition and evaluation).

8. Enable selection of a best course of action in accordance with a rational procedure (decision-assessment and choice-making).

9. Reexamine the expected effectiveness of all feasible alternative courses of action, including those initially regarded as unacceptable, prior to making a final alternative selection (sensitivity analysis).

10. Make detailed and explicit provisions for implementation of the selected action alternative, including contingency plans, as needed (planning for implementation of action).

These objectives are, of course, very closely related to the aforementioned steps of the framework for systems engineering. To accomplish them requires attention to and knowledge of the methods of systems engineering, such that we are able to design product systems and service systems. We also need to select an appropriate process, or product line, to use for management of the many activities associated with fielding a system. Also required is much effort at the level of systems management so that the resulting process is efficient, effective, equitable, and explicable. To ensure this, it is necessary to ensure that those involved in systems engineering efforts be concerned with technical knowledge of the issue under consideration, able to cope effectively with administrative concerns relative to the human elements of the issue, interested in and able to communicate across those actors involved in the issue, and capable of innovation and outscoping of relevant elements of the issue under consideration. These attributes (technical knowledge, human understanding and administrative ability, communicability, and innovativeness) are, of course, primary attributes of effective management.

26.4 SYSTEMS ENGINEERING METHODOLOGY AND METHODS

A variety of methods are suitable to accomplish the various steps of systems engineering. We shall briefly describe some of them here.

26.4.1 Issue Formulation

As indicated above, issue formulation is the step in the systems engineering effort in which the problem or issue is defined (problem definition) in terms of the objectives of a client group (value system design) and where potential alternatives that might resolve needs are identified (system synthesis). Many studies have shown that the way in which an issue is resolved is critically dependent on the way in which the issue is formulated or framed. The issue-formulation effort is concerned primarily with identification and description of the elements of the issue under consideration, with, perhaps, some initial effort at structuring these in order to enhance understanding of the relations among these elements. Structural concerns are also of importance in the analysis effort. The systems process is iterative and interactive, and the results of preliminary analysis are used to refine the issue-formulation effort. Thus, the primary intent of issue formulation is to identify relevant elements that represent and are associated with issue definition, the objectives that should be achieved in order to satisfy needs, and potential action alternatives.

There are at least four ways to accomplish issue formulation, or to identify requirements for a system, or to accomplish the initial part of the definition phase of systems engineering:

1. Asking stakeholders in the issue under consideration for the requirements
2. Descriptive identification of the requirements from a study of presently existing systems
3. Normative synthesis of the requirements from a study of documents describing what "should be," such as planning documents
4. Experimental discovery of requirements, based on experimentation with an evolving system

These approaches are neither mutually exclusive nor exhaustive. Generally, the most appropriate efforts will use a combination of these approaches.

There are conflicting concerns with respect to which blend of these requirements identification approaches is most appropriate for a specific task. The asking approach seems very appropriate when there is little uncertainty and imprecision associated with the issue under consideration, so that the issue is relatively well understood and may be easily structured, and where members of the client group possess much relevant expertise concerning the issue and the environment in which the issue is embedded. When these characteristics of the issue—lack of imprecision and presence of expert experiential knowledge—are present, then a direct declarative approach based on direct "asking" of "experts" is a simple and efficient approach. When there is considerable imprecision or a lack of experiential familiarity with the issue under concern, then the other approaches take on greater significance. The asking approach is also prone to a number of human information-processing biases, as will be discussed in Section 26.4.5. This is not as much of a problem in the other approaches.

Unfortunately, however, there are other difficulties with each of the other three approaches. Descriptive identification, from a study of existing systems of issue-formulation elements, will very likely result in a new system that is based or anchored on an existing system and tuned, adjusted, or perturbed from this existing system to yield incremental improvements. Thus, it is likely to result in incremental improvements to existing systems but not to result in major innovations or totally new systems and concepts.

Normative synthesis from a study of planning documents will result in an issue-formulation or requirements-identification effort that is based on what have been identified as desirable objectives and needs of a client group. A plan at any given phase may well not exist, or it may be flawed in any of several ways. Thus, the information base may well not be present, or may be flawed. When these circumstances exist, it will not be a simple task to accomplish effective normative synthesis of issue-formulation elements for the next phase of activity from a study of planning documents relative to the previous phase.

Often it is not easily possible to determine an appropriate set of issue-formulation elements or requirements. Often it will not be possible to define an appropriate set of issue-formulation efforts prior to actual implementation of a preliminary system design. There are many important issues where there is an insufficient experiential basis to judge the effectiveness and completeness of a set of issue-formulation efforts or requirements. Often, for example, clients will have difficulty in coping with very abstract formulation requirements and in visualizing the system that may ultimately evolve. Thus, it may be useful to identify an initial set of issue-formulation elements and accomplish subsequent analysis and interpretation based on these, without extraordinary concern for completeness of the issue-formulation efforts. A system designed with ease of adaptation and change as a primary requirement is implemented on a trial basis. As users become familiar with this new system or process, additions and modifications to the initially identified issue-formulation elements result. Such a system is generally known as a *prototype*. One very useful support for the identification of requirements is to build a prototype and allow the users of the system to be fielded to experiment with the prototype and, through this experimentation, to identify system requirements.[11] This heuristic approach allows users to identify the requirements for a system by experimenting with an easily changeable set of system-design requirements and to improve their identification of these issue-formulation elements as their experiential familiarity with the evolving prototype system grows.

The key parts of the problem-definition step of issue formulation involve identification of needs, constraints, and alterables, and determination of the interactions among these elements and the group that they impact. Need is a condition requiring supply or relief, or is a lack of something required, desired, or useful. In order to define a problem satisfactorily, we must determine the alterables or those items pertaining to the needs that can be changed. Alterables can be separated into those over which control is or is not possible. The controllable alterables are of special concern in systems engineering since they can be changed or modified to assist in achieving particular outcomes. To define a problem adequately, we must also determine the limitations or constraints under which the needs can or must be satisfied and the range over which it is permissible to vary the controllable alterables. Finally, we must determine relevant groups of people who are affected by a given problem.

Value system design is concerned with defining objectives, determining their interactions, and ordering these into a hierarchical structure. Objectives and their attainment are, of course, related to the needs, alterables, and constraints associated with problem definition. Thus, the objectives can, and should be, related to these problem-definition elements. Finally, a set of measures is needed whereby to measure objective attainment. Generally, these are called *attributes of objectives* or *objectives measures*. It is necessary to ensure that all needs are satisfied by attainment of at least one objective.

The first step in system synthesis is to identify activities and alternatives for attaining each of the objectives, or the postulation of complete systems to this end. It is then desirable to determine interactions among the proposed activities and to illustrate relationships between the activities and the needs and objectives. Activities measures are needed to gauge the degree of accomplishment of proposed activities. Systemic methods useful for problem-definition are generally useful for value system design and system synthesis as well. This is another reason that suggests the efficacy of aggregating these three steps under a single heading: *issue formulation*.

Complex issues will have a structure associated with them. In some problem areas, structure is well understood and well articulated. In other areas, it is not possible to articulate structure in such a clear fashion. There exists considerable motivation to develop techniques with which to enhance structure determination, as a system structure must always be dealt with by individuals or groups, regardless of whether the structure is articulated or not. Furthermore, an individual or a group can deal much more effectively with systems and make better decisions when the structure of the underlying system is well defined and exposed and communicated clearly. One of the fundamental objectives of systems engineering is to structure knowledge elements such that they are capable of being better understood and communicated.

We now discuss several formal methods appropriate for "asking" as a method of issue formulation. Most of these, and other, approaches are described in Refs. 1, 3, and 4. Then we shall very briefly contrast and compare some of these approaches. The methods associated with the other three generic approaches to issue formulation also involve approaches to analysis that will be discussed in the next subsection.

Several of the formal methods that are particularly helpful in the identification, through asking, of issue-formulation elements are based on principles of collective inquiry, in which a group of interested and motivated people is brought together to stimulate each other's creativity in generating issue-formulation elements. We may distinguish two groups of collective inquiry methods:

1. *Brainwriting, Brainstorming, Synectics, Nominal group technique,* and *Charette.* These approaches typically require a few hours of time, a group of knowledgeable people gathered in one place, and a group leader or facilitator. Brainwriting is typically better than brainstorming in reducing the influence of dominant individuals. Both methods can be very productive: 50–150 ideas or elements might be generated in less than an hour. Synectics, based on problem analogies, might be appropriate if there is a need for truly unconventional, innovative ideas. Considerable experience with the method is a requirement, however, particularly for the group leader. The nominal group technique is based on a sequence of idea generation, discussion, and prioritization. It can be very useful when an initial screening of a large number of ideas or elements is needed. Charette offers a conference or workshop-type format for generation and discussion of ideas and/or elements.

2. *Questionnaires, Surveys,* and *Delphi.* These three methods of collective-inquiry modeling do not require the group of participants to gather at one place and time, but they typically take more time to achieve results than the first group of methods. In questionnaires and surveys, a usually large number of participants is asked, on an individual basis, for ideas or opinions, which are then processed to achieve an overall result. There is no interaction among participants. Delphi usually provides for written interaction among participants in several rounds. Results of previous rounds are fed back to participants, who are asked to comment, revise their views as desired, and so on. A Delphi exercise can be very instructive, but usually takes several weeks or months to complete.

Use of most structuring methods, in addition to leading to greater clarity of the problem-formulation elements, will also typically lead to identification of new elements and revision of element definitions. As we have indicated, most structuring methods contain an analytical component; they may, therefore, be more properly labeled analysis methods. The following element-structuring aids are among the many modeling aids available:

- *Interaction matrices* may be used to identify clusters of closely related elements in a large set, in which case we have a self-interaction matrix; or to structure and identify the couplings between elements of different sets, such as objectives and alternatives. In this case, we produce cross-interaction matrices, such as shown in Fig. 26.5. Interaction matrices are useful for initial, comprehensive exploration of sets of elements. Learning about problem interrelationships during the process of constructing an interaction matrix is a major result of use of these matrices.

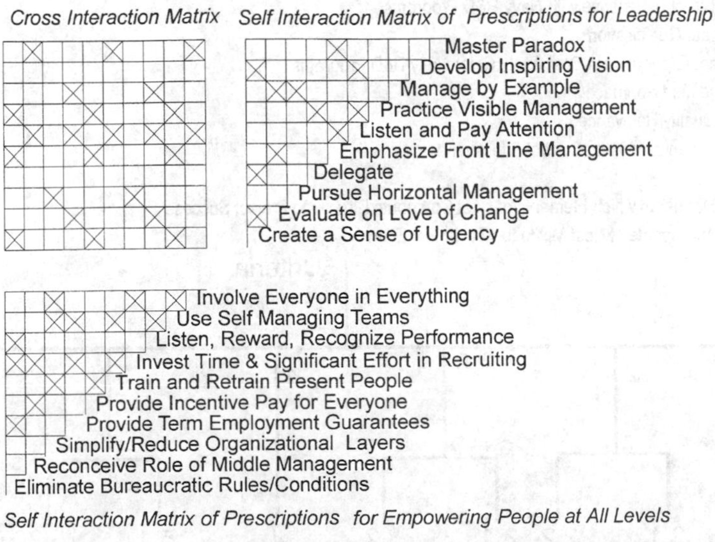

Cross Interaction Matrix   Self Interaction Matrix of Prescriptions for Leadership

Master Paradox
Develop Inspiring Vision
Manage by Example
Practice Visible Management
Listen and Pay Attention
Emphasize Front Line Management
Delegate
Pursue Horizontal Management
Evaluate on Love of Change
Create a Sense of Urgency

Involve Everyone in Everything
Use Self Managing Teams
Listen, Reward, Recognize Performance
Invest Time & Significant Effort in Recruiting
Train and Retrain Present People
Provide Incentive Pay for Everyone
Provide Term Employment Guarantees
Simplify/Reduce Organizational Layers
Reconceive Role of Middle Management
Eliminate Bureaucratic Rules/Conditions

Self Interaction Matrix of Prescriptions for Empowering People at All Levels

Fig. 26.5   Hypothetical self- and cross-interaction matrices for prescriptions for leadership and for empowering people at all levels.

- *Trees* are graphical aids particularly useful in portraying hierarchical or branching-type structures. They are excellent for communication, illustration, and clarification. Trees may be useful in all steps and phases of a systems effort. Figure 26.6 represents an attribute tree that represents those aspects of a proposal evaluation effort that will be formally considered in evaluation and prioritization of a set of proposals.
- *Causal loop diagrams*, or influence diagrams, represent graphical pictures of causal interactions between sets of variables. They are particularly helpful in making explicit one's perception of the causes of change in a system, and can serve very well as communication aids. A causal loop diagram is also useful as the initial part of a detailed simulation model. Figure 26.7 represents a causal loop diagram of a belief structure.

Two other descriptive methods are potentially useful for issue formulation:

- The *system definition matrix*, options profile, decision balance sheet, or checklist provides a framework for specification of the essential aspects, options, or characteristics of an issue, a plan, a policy, or a proposed or existing system. It can be helpful for the design and specification of alternative policies, designs, or other options or alternatives. The system definition matrix is just a table that shows important aspects of the options that are important for judgment relative to selection of approaches to issue formulation or requirements determination.
- *Scenario-writing* is based on narrative and creative descriptions of existing or possible situations or developments. Scenario descriptions can be helpful for clarification and communication of ideas and obtaining feedback on those ideas. Scenarios may also be helpful in conjunction with various analysis and forecasting methods, where they may represent alternative or opposing views.

1. Understanding of Problem
   1.1   Navy Cost Credentialing Process
   1.2   NAVELEX Cost Analysis Methodology
   1.3   DoD Procurement Procedures
2. Technical Approach
   2.1   Establishment of a Standard Methodology
   2.2   Compatibility with Navy Acquisition Process
3. Staff Experience
   3.1   Directly Related Experience in Cost Credentialing, Cost Analysis, and Procurement Procedures
   3.2   Direct Experience with Navy R&D Programs
4. Corporate Qualification
       Relevant Experience in Cost Analysis for Navy R&D Programs
5. Management Approach
   5.1   Quality/Relevance
   5.2   Organization and Control Effectiveness
6. Cost
   6.1   Manner in which Elements of Cost Contribute Directly to Project Success
   6.2   Appropriate of Cost Mix to the Technical Effort

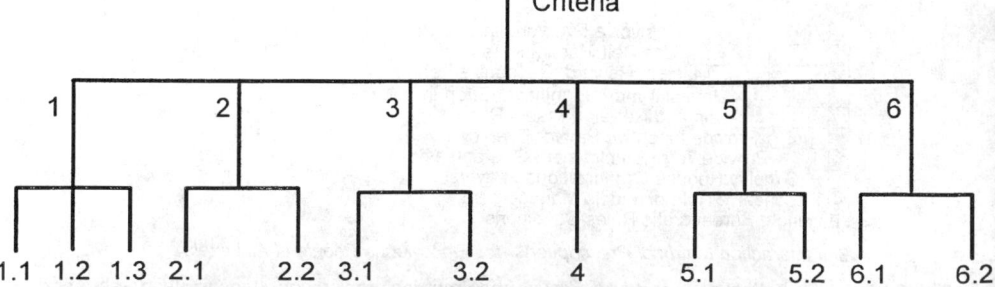

Fig. 26.6 Possible attribute tree for evaluation of proposals concerning cost credentialing.

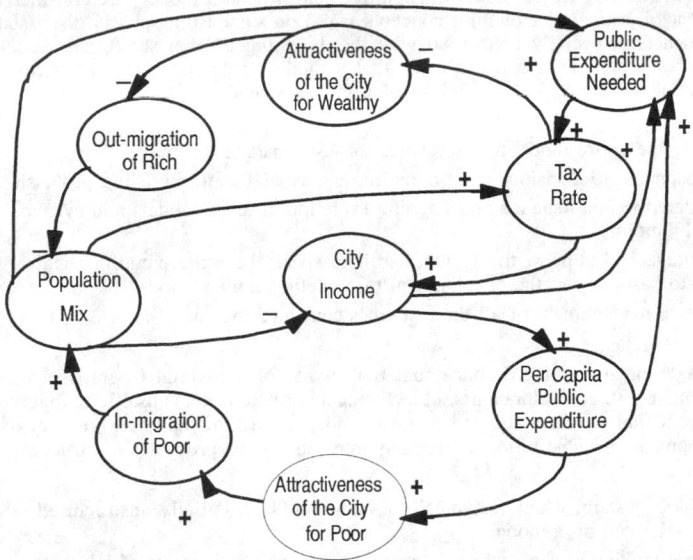

Fig. 26.7 Causal loop diagram of belief structure in a simple model of urban dynamics.

Clearly, successful formulation of issues through "asking" requires creativity. Creativity may be much enhanced through use of a structured systems engineering framework. For example, group meetings for issue formulation involve idea formulation, idea analysis, and idea interpretation. The structure of a group meeting may be conceptualized within a systems engineering framework. This framework is especially useful for visualizing the tradeoffs that must be made among allocation of resources for formulation, analysis, and interpretation of ideas in the issue-formulation step itself. If there is an emphasis on idea formulation, we shall likely generate too many ideas to cope with easily. This will lead to a lack of attention to detail. On the other hand, if idea formulation is de-emphasized, we shall typically encourage defensive avoidance through undue efforts to support the present situation, or a rapid unconflicted change to a new situation. An overemphasis on analysis of ideas is usually time-consuming and results in a meeting that seems to drown in details. There is inherent merit in encouraging a group to reach consensus, but the effort may also be inappropriate, since it may encourage arguments over concerns that are ineffective in influencing judgments.

De-emphasizing the analysis of identified ideas will usually result in disorganized meetings in which hasty, poorly thought-out ideas are accepted. Postmeeting disagreements concerning the results of the meeting are another common disadvantage. An emphasis on interpretation of ideas will produce a meeting that is emotional and people-centered. Misunderstandings will be frequent as issues become entrenched in an adversarial, personality-centered process. On the other hand, de-emphasizing the interpretation of ideas results in meetings in which important information is not elicited. Consequently, the meeting is awkward and empty, and routine acceptance of ideas is a likely outcome.

26.4.2 Issue Analysis

In systems engineering, issue analysis involves forecasting and assessing of the impacts of proposed alternative courses of action. In turn, this suggests construction, testing, and validation of models. Impact assessment in systems engineering includes system analysis and modeling, and optimization and ranking or refinement of alternatives. First, the options or alternatives defined in issue formulation are structured, often as part of the issue formulation effort, and then analyzed in order to assess the anticipated impacts that may result from their implementation. Second, a refinement or optimization effort is often desirable. This is directed toward refinement or fine-tuning a viable alternative, and parameters within an alternative, so as to obtain maximum needs satisfaction, within given constraints, from a proposed policy.

To determine the structure of systems in the most effective manner requires the use of quantitative analysis to direct the structuring effort along the most important and productive paths. This is especially needed when time available to construct structural models is limited. Formally, there are at least four types of self-interaction matrices: nondirected graphs, directed graphs (or digraphs), signed digraphs, and weighted digraphs. The theory of digraphs and structural modeling is authoritatively presented in Ref. 12 and a number of applications to what is called *interpretative structural modeling* are described in Refs. 3, 13, 14, and 15. Cognitive map structural models are considered in Ref. 16. A development of structural modeling concepts based on signed digraphs is discussed in Ref. 17. Geoffrion has been especially concerned with the development of a structured modeling methodology[18,19] and environment. He has noted[20] that a modeling environment needs five quality- and productivity-related properties. A modeling environment should

1. Nurture the entire modeling life cycle, not just a part of it
2. Be hospitable to decision- and policy-makers, as well as to modeling professionals
3. Facilitate the maintenance and ongoing evolution of those models and systems that are contained therein
4. Encourage and support those who use it to speak the same paradigm-neutral language, in order to best support the development of modeling applications
5. Facilitate management of all the resources contained therein

Structured modeling is a general conceptual framework for modeling. Cognitive maps, interaction matrices, intent structures, Delta charts, objective and attribute trees, causal loop diagrams, decision-outcome trees, signal flow graphs, and so on are all structural models that are very useful graphic aids to communications. The following are requirements for the processes of structural modeling:

1. An object system, which is typically a poorly defined, initially unstructured set of elements to be described by a model
2. A representation system, which is a presumably well-defined set of relations
3. An embedding of perceptions of some relevant features of the object system into the representation system

Structural modeling, which has been of fundamental concern for some time, refers to the systemic iterative application of typically graph-theoretic notions such that an easily communicable directed graph representation of complex patterns of a particular contextual relationship among a set of elements results. There are a number of computer software realizations of various structural modeling constructs, such as cognitive policy evaluation (COPE), interpretive structural modeling (ISM), and various multi-attribute utility theory-based representations, as typically found in most decision-aiding software.

Transformation of a number of identified issue-formulation elements, which typically represent unclear, poorly articulated mental models of a system, into visible, well-defined models useful for many purposes is the object of systems analysis and modeling. The principal objective of systems analysis and modeling is to create a process with which to produce information concerning consequences of proposed actions or policies. From the issue-formulation steps of problem definition, value system design, and system synthesis, we have various descriptive and normative scenarios available for use. Ultimately, as a part of the issue interpretation step, we wish to evaluate and compare alternative courses of action with respect to the value system through use of a systems model. A model is always a substitute for reality, but is, one hopes, descriptive enough of the system elements under consideration to be useful. By posing a variety of questions using the model, we can, from the results obtained, learn how to cope with that subset of the real world being modeled.

A model must depend on much more than the particular problem-definition elements being modeled; it must also depend strongly on the value system and the purpose behind construction and utilization of the model. These influence, generally strongly, the structure of the situation and the elements that comprise this structure. Which elements a client believes important enough to include in a model depend on the client's value system.

We wish to be able to determine correctness of predictions and forecasts that are based on model usage. Given the definition of a problem, a value system, and a set of proposed policies, we wish to be able to design a simulation model consisting of relevant elements of these three steps and to determine the results or impacts of implementing proposed policies. Following this, we wish to be able to validate a simulation model to determine the extent to which it represents reality sufficiently to be useful. Validation must, if we are to have confidence in what we are doing with a model, precede actual use of model-based results.

There are three essential steps in constructing a simulation model:

1. Determination of those problem definitions, value systems, and system-synthesis elements that are most relevant to a particular problem
2. Determination of the structural relationships among these elements
3. Determination of parametric coefficients within the structure

There are three uses to which models may normally be put. Model categories corresponding to these three uses are descriptive models, predictive or forecasting models, and policy or planning models. Representation and replication of important features of a given problem are the objects of a descriptive model. Good descriptive models are of considerable value in that they reveal much about the substance of complex issues and how, typically in a retrospective sense, change over time has occurred. One of the primary purposes behind constructing a descriptive model is to learn about the past. Often the past will be a good guide to the future.

In building a predictive or forecasting model, we must be especially concerned with determination of proper cause-and-effect relationships. If the future is to be predicted with integrity, we must have a method with which to determine exogenous variables, or input variables that result from external causes, accurately. Also, the model structure and parameters within the structure must be valid for the model to be valid. Often, it will not be possible to predict accurately all exogenous variables; in that case, conditional predictions can be made from particular assumed values of unknown exogenous variables.

The future is inherently uncertain. Consequently, predictive or forecasting models are often used to generate a variety of future scenarios, each a conditional prediction of the future based on some conditioning assumptions. In other words, we develop an "if-then" model.

Policy or planning models are much more than predictive or forecasting models, although any policy or planning model is also a predictive or forecasting model. The outcome from a policy or planning model must be evaluated in terms of a value system. Policy or planning efforts must not only predict outcomes from implementing alternative policies, but also present these outcomes in terms of the value system that is in a form useful and suitable for alternative ranking, evaluation, and decision making. Thus, a policy model must contain some provision for impact interpretation.

Model usefulness cannot be determined by objective truth criteria alone. Well-defined and well-stated functions and purposes for the simulation model are needed to determine simulation-model usefulness. Fully objective criteria for model validity do not typically exist. Development of a general-purpose, context-free simulation model appears unlikely; the task is simply far too complicated. We must build models for specific purposes, and thus the question of model validity is context-dependent.

Model credibility depends on the interaction between the model and model user. One of the major potential difficulties is that of building a model that reflects the outlook of the modeler. This activity is proscribed in effective systems engineering practice, since the purpose of a model is to describe systematically the "view" of a situation held by the client, not that held by the analyst.

A great variety of approaches have been designed and used for the forecasting and assessment that are the primary goals of systems analysis. There are basically two classes of methods that we describe here: expert-opinion methods and modeling and/or simulation methods.

Expert-opinion methods are based on the assumption that knowledgeable people will be capable of saying sensible things about the impacts of alternative policies on the system, as a result of their experience with, or insight into, the issue or problem area. These methods are generally useful, particularly when there are no established theories or data concerning system operation, precluding the use of more precise analytical tools. Among the most prominent expert-opinion-based forecasting methods are surveys and Delphi. There are, of course, many other methods of asking experts for their opinion—for example, hearings, meetings, and conferences. A particular problem with such methods is that cognitive bias and value incoherence are widespread, often resulting in inconsistent and self-contradictory results. There exists a strong need in the forecasting and assessment community to recognize and ameliorate, by appropriate procedures, the effects of cognitive bias and value incoherence in expert-opinion-modeling efforts. Expert-opinion methods are often appropriate for the "asking" approach to issue formulation. They may be of considerably less value, especially when used as stand-alone approaches, for impact assessment and forecasting.

Simulation and modeling methods are based on the conceptualization and use of an abstraction or model of the real world intended to behave in a similar way to the real system. Impacts of policy alternatives are studied in the model, which will, it is hoped, lead to increased insight with respect to the actual situation.

Most simulation and modeling methods use the power of mathematical formulations and computers to keep track of many pieces of information at the same time. Two methods in which the power of the computer is combined with subjective expert judgments are cross-impact analysis and workshop dynamic models. Typically, experts provide subjective estimates of event probabilities and event interactions. These are processed by a computer to explore their consequences and fed back to the analysts and thereafter to the experts for further study. The computer derives the resulting behavior of various model elements over time, giving rise to renewed discussion and revision of assumptions.

Expert judgment is virtually always included in all modeling methods. Scenario writing can be an expert-opinion-modeling method, but typically this is done in a less direct and explicit way than in Delphi, survey, ISM, cross-impact, or workshop dynamic models. As a result, internal inconsistency problems are reduced with those methods based on mathematical modeling. The following are additional forecasting methods based on mathematical modeling and simulation. In these methods, a structural model is generally formed on the basis of expert opinion and physical or social laws.

Available data are then processed to determine parameters within the structure. Unfortunately, these methods are sometimes very data-intensive and, therefore, expensive and time-consuming to implement.

Trend-extrapolation/time-series forecasting is particularly useful when sufficient data about past and present developments are available, but there is little theory about underlying mechanisms causing change. The method is based on the identification of a mathematical description or structure that will be capable of reproducing the data into the future, typically over the short to medium term.

Continuous-time dynamic simulation is based on postulation and qualification of a causal structure underlying change over time. A computer is used to explore long-range behavior as it follows from the postulated causal structure. The method can be very useful as a learning and qualitative forecasting device, but its application may be rather costly and time-consuming.

Discrete-event digital-simulation models are based on applications of queuing theory to determine the conditions under which system outputs or states will switch from one condition to another.

Input–output analysis has been specially designed for study of equilibrium situations and requirements in economic systems in which many industries are interdependent. Many economic data fit in directly to the method, which mathematically is relatively simple, and can handle many details.

Econometrics is another method mainly applied to economic description and forecasting problems. It is based on both theory and data, with, usually, the main emphasis on specification of structural relations based on macro-economic theory and the derivation of unknown parameters in behavioral equations from available economic data.

Micro-economic models represent an application of economic theories of firms and consumers who desire to maximize the profit and utility of their production and consumption alternatives.

Parameter estimation is a very important subject with respect to model construction and validation. Observation of basic data and estimation or identification of parameters within an assumed structure, often denoted as system identification, are essential steps in the construction and validation of system models. The simplest estimation procedure, in both concept and implementation, appears to be the least squares error estimator. Many estimation algorithms to accomplish this are available and are in actual use. The subjects of parameter estimation and system identification are being actively explored in both economics and systems engineering. There are numerous contemporary results, including algorithms for system identification and parameter estimation in very-large-scale systems representative of actual physical processes and organizations.

Verification of a model is necessary to ensure that the model behaves in a fashion intended by the model builder. If we can determine that the structure of the model corresponds to the structure of the elements obtained in the problem definition, value system design, and system synthesis steps, then the model is verified with respect to behaving in a gross, or structural, fashion, as the model builder intends.

Even if a model is verified in a structural as well as parametric sense, there is still no assurance that the model is valid in the sense that predictions made from the model will occur. We can determine validity only with respect to the past. That is all that we can possibly have available at the present. Forecasts and predictions inherently involve the future. Since there may be structural and parametric changes as the future evolves, and since knowledge concerning results of policies not implemented may never be available, there is usually no way to validate a model completely. Nevertheless, there are several steps that can be used to validate a model. These include a reasonableness test in which we determine that the overall model, as well as model subsystems, respond to inputs in a reasonable way, as determined by "knowledgeable" people. The model should also be valid according to statistical time series used to determine parameters within the model. Finally, the model should be epistemologically valid, in that the policy interpretations of the various model parameters, structure, and recommendations are consistent with ethical, professional, and moral standards of the group affected by the model.

Once a model has been constructed, it is often desirable to determine, in some best fashion, various policy parameters or controls that are subject to negotiation. The optimization or refinement-of-alternatives step is concerned with choosing parameters or controls to maximize or minimize a given performance index or criterion. Invariably, there are constraints that must be respected in seeking this extremum. As previously noted, the analysis step of systems engineering consists of systems analysis and modeling, and optimization or refinement of alternatives and related methods that are appropriate in aiding effective judgment and choice.

There exist a number of methods for fine-tuning, refinement, or optimization of individual specific alternative policies or systems. These are useful in determining the best (in terms of needs satisfaction) control settings or rules of operation in a well-defined, quantitatively describable system. A single scalar indicator of performance or desirability is typically needed. There are, however, approaches to multiple objective optimization that are based on welfare-type optimization concepts. It is these individually optimized policies or systems that are an input to the evaluation and decision-making effort in the interpretation step of systems engineering.

Among the many methods for optimization and refinement of alternatives are

- *Mathematical programming*, which is used extensively in operations research and analysis practice, for resource allocation under constraints, resolution of planning or scheduling problems, and similar applications. It is particularly useful when the best equilibrium or one-time setting has to be determined for a given policy or system.
- *Optimum systems control*, which addresses the problem of determining the best controls or actions when the system, the controls or actions, the constraints, and the performance index may change over time. A mathematical description of system change is necessary to use this approach. Optimum systems control is particularly suitable for refining controls or parameters in systems in which changes over time play an important part.

Application of the various refinement or optimization methods, like those described here, typically requires significant training and experience on the part of the systems analyst. Some of the many characteristics of analysis that are of importance for systemic efforts include the following:

1. Analysis methods are invaluable for understanding the impacts of proposed policy.
2. Analysis methods lead to consistent results if cognitive bias issues associated with expert forecasting and assessment methods are resolved.
3. Analysis methods may not necessarily lead to correct results since "formulation" may be flawed, perhaps by cognitive bias and value incoherence.

Unfortunately, however, large models and large optimization efforts are often expensive and difficult to understand and interpret. There are a number of possibilities for "paralysis through analysis" in the unwise use of systems analysis. On the other hand, models and associated analysis can help provide a framework for debate. It is important to note that small "back-of-the-envelope" models can be very useful. They have advantages that large models often lack, such as cost, simplicity, and ease of understanding and, therefore, explicability.

It is important to distinguish between analysis and interpretation in systems engineering efforts. Analysis cannot substitute, or will generally be a foolish substitute for, judgment, evaluation, and interpretation as exercised by a well-informed decision-maker. In some cases, refinement of individual alternative policies is not needed in the analysis step. But evaluation of alternatives is always needed, since, if there is but a single policy alternative, there really is no alternative at all. The option to do nothing at all must always be considered as a policy alternative. It is especially important to avoid a large number of cognitive biases, poor judgment heuristics, and value incoherence in the activities of evaluation and decision-making. The efforts involved in evaluation and choice-making interact strongly with the efforts in the other steps of the systems process, and these are also influenced by cognitive bias, judgment heuristics, and value incoherence. One of the fundamental tenets of the systems process is that making the complete issue-resolution process as explicit as possible makes it easier to detect and connect these deficiencies than it is in holistic intuitive processes.

26.4.3 Information Processing by Humans and Organizations

After completion of the analysis step, we begin the evaluation and decision-making effort of interpretation. Decisions must typically be made and policies formulated, evaluated, and applied in an atmosphere of uncertainty. The outcome of any proposed policy is seldom known with certainty. One of the purposes of analysis is to reduce, to the extent possible, uncertainties associated with the outcomes of proposed policies. Most planning, design, and resource-allocation issues will involve a large number of decision-makers who act according to their varied preferences. Often, these decision-makers will have diverse and conflicting data available to them and the decision situation will be quite fragmented. Furthermore, outcomes resulting from actions can often only be adequately characterized by a large number of incommensurable attributes. Explicit informed comparison of alternatives across these attributes by many stakeholders in an evaluation and choice-making process is typically most difficult.

As a consequence of this, people will often search for and use some form of a dominance structure to enable rejection of alternatives that are perceived to be dominated by one or more other alternatives. An alternative is said to be "dominated" by another alternative when the other alternative has attribute scores at least as large as those associated with the dominated alternative, and at least one attribute score that is larger. However, biases have been shown to be systematic and prevalent in most unaided cognitive activities. Decisions and judgments are influenced by differential weights of information and by a variety of human information-processing deficiencies, such as base rates, representativeness, availability, adjustment, and anchoring. Often it is very difficult to disaggregate values of policy outcomes from causal relations determining these outcomes. Often correlation is used to infer causality. Wishful thinking and other forms of selective perception encourage us not to obtain potentially disconfirming information. The resulting confounding of values with facts can lead to great difficulties in discourse and related decision-making.

It is especially important to avoid the large number of potential cognitive biases and flaws in the process of formulation, analysis, and interpretation for judgment and choice. These may well occur due to flaws in human information processing associated with the identification of problem elements, structuring of decision situations, and the probabilistic and utility assessment portions of the judgmental tasks of evaluation and decision-making.

Among the cognitive biases and information-processing flaws that have been identified are several that affect information formulation or acquisition, information analysis, and interpretation. These and related material are described in Ref. 21 and the references contained therein. Among these biases, which are not independent, are the following.

1. *Adjustment and anchoring.* Often a person finds that difficulty in problem-solving is due not to the lack of data and information, but rather to an excess of data and information. In such situations, the person often resorts to heuristics, which may reduce the mental efforts required to arrive at a solution. In using the anchoring and adjustment heuristic when confronted with a large number of data, the person selects a particular datum, such as the mean, as an initial or starting point or anchor, and then adjusts that value improperly in order to incorporate the rest of these data, resulting in flawed information analysis.

2. *Availability.* The decision-maker uses only easily available information and ignores sources of significant but not easily available information. An event is believed to occur frequently, that is, with high probability, if it is easy to recall similar events.

3. *Base rate.* The likelihood of occurrence of two events is often compared by contrasting the number of times the two events occur and ignoring the rate of occurrence of each event. This bias often arises when the decision-maker has concrete experience with one event but only statistical or abstract information on the other. Generally, abstract information will be ignored at the expense of concrete information. A base rate determined primarily from concrete information may be called a *causal base rate*, whereas that determined from abstract information is an *incidental base rate*. When information updates occur, this individuating information is often given much more weight than it deserves. It is much easier for the impact of individuating information to override incidental base rates than causal base rates.

4. *Conservatism.* The failure to revise estimates as much as they should be revised, based on receipt of new significant information, is known as *conservatism*. This is related to data-saturation and regression-effects biases.

5. *Data presentation context.* The impact of summarized data, for example, may be much greater than that of the same data presented in detailed, nonsummarized form. Also, different scales may be used to change the impact of the same data considerably.

6. *Data saturation.* People often reach premature conclusions on the basis of too small a sample of information while ignoring the rest of the data, which is received later, or stopping acquisition of data prematurely.

7. *Desire for self-fulfilling prophecies.* The decision-maker values a certain outcome, interpretation, or conclusion and acquires and analyzes only information that supports this conclusion. This is another form of selective perception.

8. *Ease of recall.* Data that can easily be recalled or assessed will affect perception of the likelihood of similar events reoccurring. People typically weigh easily recalled data more in decision-making than those data that cannot easily be recalled.

9. *Expectations.* People often remember and attach higher validity to information that confirms their previously held beliefs and expectations than they do to disconfirming information. Thus, the presence of large amounts of information makes it easier for one to selectively ignore disconfirming information such as to reach any conclusion and thereby prove anything that one desires to prove.

10. *Fact–value confusion.* Strongly held values may often be regarded and presented as facts. That type of information is sought that confirms or lends credibility to one's views and values. Information that contradicts one's views or values is ignored. This is related to wishful thinking in that both are forms of selective perception.

11. *Fundamental attribution error* (success/failure error). The decision-maker associates success with personal inherent ability and associates failure with poor luck in chance events. This is related to availability and representativeness.

12. *Habit.* Familiarity with a particular rule for solving a problem may result in reuse of the same procedure and selection of the same alternative when confronted with a similar type of problem and similar information. We choose an alternative because it has previously been acceptable for a perceived similar purpose or because of superstition.

13. *Hindsight.* People are often unable to think objectively if they receive information that an outcome has occurred and they are told to ignore this information. With hindsight, outcomes

that have occurred seem to have been inevitable. We see relationships much more easily in hindsight than in foresight and find it easy to change our predictions after the fact to correspond to what we know has occurred.

14. *Illusion of control.* A good outcome in a chance situation may well have resulted from a poor decision. The decision-maker may assume an unreasonable feeling of control over events.

15. *Illusion of correlation.* This is a mistaken belief that two events covary when they do not covary.

16. *Law of small numbers.* People are insufficiently sensitive to quality of evidence. They often express greater confidence in predictions based on small samples of data with nondisconfirming evidence than in much larger samples with minor disconfirming evidence. Sample size and reliability often have little influence on confidence.

17. *Order effects.* The order in which information is presented affects information retention in memory. Typically, the first piece of information presented (primacy effect) and the last presented (recency effect) assume undue importance in the mind of the decision-maker.

18. *Outcome-irrelevant learning system.* Use of an inferior processing or decision rule can lead to poor results that the decision-maker can believe are good because of inability to evaluate the impacts of the choices not selected and the hypotheses not tested.

19. *Representativeness.* When making inference from data, too much weight is given to results of small samples. As sample size is increased, the results of small samples are taken to be representative of the larger population. The "laws" of representativeness differ considerably from the laws of probability and violations of the conjunction rule $P(A \cap B) < P(A)$ are often observed.

20. *Selective perceptions.* People often seek only information that confirms their views and values and disregard or ignore disconfirming evidence. Issues are structured on the basis of personal experience and wishful thinking. There are many illustrations of selective perception. One is "reading between the lines"—for example, to deny antecedent statements and, as a consequence, accept "if you don't promote me, I won't perform well" as following inferentially from "I will perform well if you promote me."

Of particular interest are circumstances under which these biases occur and their effects on activities such as the identification of requirements for a system or for planning and design. Through this, it may be possible to develop approaches that might result in debiasing or amelioration of the effects of cognitive bias. A number of studies have compared unaided expert performance with simple quantitative models for judgment and decision-making. While there is controversy, most studies have shown that simple quantitative models perform better in human judgment and decision-making tasks, including information processing, than holistic expert performance in similar tasks. There are a number of prescriptions that might be given to encourage avoidance of possible cognitive biases and to debias those that do occur:

1. Sample information from a broad data base and be especially careful to include data bases that might contain disconfirming information.

2. Include sample size, confidence intervals, and other measures of information validity in addition to mean values.

3. Encourage use of models and quantitative aids to improve upon information analysis through proper aggregation of acquired information.

4. Avoid the hindsight bias by providing access to information at critical past times.

5. Encourage people to distinguish good and bad decisions from good and bad outcomes.

6. Encourage effective learning from experience. Encourage understanding of the decision situation and methods and rules used in practice to process information and make decisions so as to avoid outcome irrelevant learning systems.

A definitive discussion of debiasing methods for hindsight and overconfidence is presented by Fischhoff.[22] He suggests identifying faulty judges, faulty tasks, and mismatches between judges and tasks. Strategies for each of these situations are given.

Not everyone agrees with the conclusions just reached about cognitive human information processing and inferential behavior. Several arguments have been advanced for a decidedly less pessimistic view of human inference and decision. Jonathan Cohen,[23,24] for example, argues that all of this research is based upon a conventional model for probabilistic reasoning, which Cohen calls the "Pascalian" probability calculus. He expresses the view that human behavior does not appear "biased" at all when it is viewed in terms of other equally appropriate schemes for probabilistic reasoning, such as his own "inductive probability" system. Cohen states that human irrationality can never be demonstrated in laboratory experiments, especially experiments based upon the use of what he calls "probabilistic conundrums."

There are a number of other contrasting viewpoints as well. In their definitive study of behavioral and normative decision analysis, von Winterfelt and Edwards[25] refer to these information processing biases as "cognitive illusions." They indicate that there are four fundamental elements to every cognitive illusion:

1. A *formal operational* rule that determines *the* correct solution to an intellectual question
2. An intellectual question that almost invariably includes all of the information required to obtain the correct answer through use of the formal rule
3. A human judgment, generally made without the use of these analytical tools, that is intended to answer the posed question
4. A systematic and generally large and unforgivable discrepancy between the correct answer and the human judgment

They also, as does Phillips,[26] describe some of the ways in which subjects might have been put at a disadvantage in this research on cognitive heuristics and information-processing biases. Much of this centers around the fact that the subjects have little experiential familiarity with the tasks that they are asked to perform. It is suggested that as inference tasks are decomposed and better structured, it is very likely that a large number of information-processing biases will disappear. Thus, concern should be expressed about the structuring of inference and decision problems and the learning that is reflected by revisions of problem structure in the light of new knowledge. In any case, there is strong evidence that humans are very strongly motivated to understand, to cope with, and to improve themselves and the environment in which they function. One of the purposes of systems engineering is to aid in this effort.

26.4.4 Interpretation

While there are a number of fundamental limitations to systems engineering efforts to assist in bettering the quality of human judgment, choice, decisions, and designs, there are also a number of desirable activities. These have resulted in several important holistic approaches that provide formal assistance in the evaluation and interpretation of the impacts of alternatives, including the following.

- *Decision analysis*, which is a very general approach to option evaluation and selection, involves identification of action alternatives and possible consequence identification of the probabilities of these consequences, identification of the valuation placed by the decision-maker on these consequences, computation of the expected utilities of the consequences, and aggregating or summarizing these values for all consequences of each action. In doing this, we obtain an expected utility evaluation of each alternative act. The one with the highest value is the most preferred action or option. Figure 26.7 presents some of the salient features involved in the decision analysis of a simplified problem.

- *Multi-attribute utility theory* (MAUT) has been designed to facilitate comparison and ranking of alternatives with many attributes or characteristics. The relevant attributes are identified and structured and a weight or relative utility is assigned by the decision-maker to each basic attribute. The attribute measurements for each alternative are used to compute an overall worth or utility for each attribute. Multi-attribute utility theory allows for explicit recognition and incorporation of the decision-maker's attitude toward risk in the utility computations. There are a number of variants of MAUT; many of them are simpler, more straightforward processes in which risk and uncertainty considerations are not taken into account. The method is very helpful to the decision-maker in making values and preferences explicit and in making decisions consistent with those values. The tree structure of Fig. 26.6 also indicates some salient features of the MAUT approach for the particular case where there are no risks or uncertainties involved in the decision situation. We simply need to associate importance weights with the attributes and then provide scores for each alternative on each of the lowest-level attributes.

- *Policy capture* (or social judgment theory) has also been designed to assist decision-makers in making their values explicit and their decisions consistent with their values. In policy capture, the decision-maker is asked to rank order a set of alternatives in a gestalt or holistic fashion. Alternative attributes and associated attribute measures are then determined by elicitation from the decision-maker. A mathematical procedure involving regression analysis is used to determine that relative importance weight of each attribute that will lead to a ranking as specified by the decision-maker. The result is fed back to the decision-maker, who, typically, will express the view that his or her values are different. In an iterative learning process, preference weights and/or overall rankings are modified until the decision-maker is satisfied with both the weights and the overall alternative ranking.

There are many advantages to formal interpretation efforts in systems engineering, including the following:

1. Developing decision situation models to aid in making the choice-making effort explicit helps one both to identify and to overcome the inadequacies of implicit mental models.
2. The decision situation model elements, especially the attributes of the outcomes of alternative actions, remind us of information we need to obtain about alternatives and their outcomes.
3. We avoid such poor information processing heuristics as evaluating one alternative on attribute A and another on attribute B and then comparing them without any basis for compensatory tradeoffs across the different attributes.
4. We improve our ability to process information and, consequently, reduce the possibilities for cognitive bias.
5. We can aggregate facts and values in a prescribed systemic fashion rather than by adopting an agenda-dependent or intellect-limited approach.
6. We enhance brokerage, facilitation, and communication abilities among stakeholders to complex technological and social issues.

There is a plethora of literature describing the decision-assessment or decision-making part of the interpretation step of systems engineering. In addition to the discussions in Refs. 1, 2, and 3, excellent discussions are to be found in Refs. 27, 28, and 29.

26.4.5 The Central Role of Information in Systems Engineering

Information is certainly a key ingredient supporting quality decisions; all of systems engineering efforts are based on appropriate acquisition and use of information. There are three basic types of information, which are fundamentally related to the three-step framework of systems engineering:

1. Formulation information
   a. Information concerning the problem and associated needs, constraints, and alterables
   b. Information concerning the value system
   c. Information concerning possible option alternatives
   d. Information concerning possible future alternative outcomes, states and scenarios
2. Analysis information
   a. Information concerning probabilities of future scenarios
   b. Information concerning impacts of alternative options
   c. Information concerning the importance of various criteria or attributes
3. Interpretation information
   a. Information concerning evaluation and aggregation of facts and values
   b. Information concerning implementation

We see that useful and appropriate formulation, analysis, and interpretation of information is one of the most important and vital tasks in systems engineering efforts, since it is the efficient processing of information by the decision-maker that produces effective decisions. A useful definition of information for our purposes is that it is data of value for decision-making. The decision-making process is influenced by many contingency and environmental influences. A purpose of the management technology that is systems engineering is to provide systemic support processes to further enhance efficient decision-making activities.

After completion of evaluation and decision-making efforts, it is generally necessary to become involved in planning for action to implement the chosen alternative option or the next phase of a systems engineering effort. More often than not, it will be necessary to iterate the steps of systems engineering several times to obtain satisfactory closure upon one or more appropriate action alternatives. Planning for action also leads to questions concerning resource allocation, schedules, and management plans. There are, of course, a number of methods from systems science and operations research that support determination of schedules and implementation plans. Each of the steps is needed, with different focus and emphasis, at each phase of a systems effort. These phases depend on the particular effort under consideration, but will typically include such phases as policy and program planning, project planning, and system development.

There are a number of complexities affecting "rational" planning, design, and decision-making. We must cope with these in the design of effective systemic processes. The majority of these complexities involve systems management considerations. Many have indicated that the capacity of the human mind for formulating, analysis, and interpretation of complex large-scale issues is very small compared with the size and scope of the issues whose resolution is required for objective, substantive, and procedurally rational behavior. Among the limits to rationality are the fact that we can formulate, analyze, and interpret only a restricted amount of information; can devote only a limited amount of time to decision-making; and can become involved in many more activities than we can effectively consider and cope with simultaneously. We must therefore necessarily focus attention only on a portion of the major competing concerns. The direct effect of these is the presence of cognitive bias in information acquisition and processing and the use of cognitive heuristics for evaluation of alternatives.

Although in many cases these cognitive heuristics will be flawed, this is not necessarily so. One of the hoped-for results of the use of systems engineering approaches is the development of effective and efficient heuristics for enhanced judgment and choice through effective decision support systems.[30]

There are many cognitive biases prevalent in most information-acquisition activities. The use of cognitive heuristics and decision rules is also prevalent and necessary to enable us to cope with the many demands on our time. One such heuristic is satisfying or searching for a solution that is "good enough." This may be quite appropriate if the stakes are small. In general, the quality of cognitive heuristics will be task-dependent, and often the use of heuristics for evaluation will be both reasonable and appropriate. Rational decision-making requires time, skill, wisdom, and other resources. It must, therefore, be reserved for the more important decisions. A goal of systems engineering is to enhance information acquisition, processing, and evaluation so that efficient and effective use of information is made in a process that is appropriate to the cognitive styles and time constraints of management.

26.5 SYSTEM DESIGN

This section discusses several topics relevant to the design and evaluation of systems. In order to develop our design methodology, we first discuss the purpose and objectives of systems engineering and systems design. Development of performance objectives for quality systems is important, since evaluation of the logical soundness and performance of a system can be determined by measuring achievement of these objectives with and without the system. A discussion of general objectives for quality system design is followed by a presentation of a five-phase design methodology for system design. The section continues with leadership and training requirements for use of the resulting system and the impact of these requirements upon design considerations. While it is doubtless true that not every design process should, could, or would precisely follow each component in the detailed phases outlined here, we feel that this approach to systems design is sufficiently robust and generic that it can be used as a normative model of the design process and as a guide to the structuring and implementation of appropriate systems evaluation practices.

26.5.1 The Purposes of Systems Design

Contemporary issues that may result in the need for systems design are invariably complex. They typically involve a number of competing concerns, contain much uncertainty, and require expertise from a number of disparate disciplines for resolution. Thus, it is not surprising that intuitive and affective judgments, often based on incomplete data, form the usual basis used for contemporary design and associated choice-making. At the other extreme of the cognitive inquiry scale are the highly analytical, theoretical, and experimental approaches of the mathematical, physical, and engineering sciences. When intuitive judgment is skill-based, it is generally effective and appropriate. One of the major challenges in system design engineering is to develop processes that are appropriate for a variety of process users, some of whom may approach the design issue from a skill-based perspective, some from a rule-based perspective, and some from a knowledge-based perspective.

A central purpose of systems engineering and management is to incorporate appropriate methods and metrics into a methodology for problem solving, or a systems engineering process or life cycle, such that, when it is associated with human judgment through systems management, it results in a high-quality systems design procedure. By high-quality design, we mean one that will, with high probability, produce a system that is effective and efficient.

A systems design procedure must be specifically related to the operational environment for which the final system is intended. Control group testing and evaluation may serve many useful purposes with respect to determination of many aspects of algorithmic and behavioral efficacy of a system. Ultimate effectiveness involves user acceptability of the resulting system, and evaluation of this process effectiveness will often involve testing and evaluation in the environment, or at least a closely simulated model of the environment, in which the system would be potentially deployed.

The potential benefits of systems engineering approaches to design can be interpreted as attributes or criteria for evaluation of the design approach itself. Achievement of many of these attributes may often not be experimentally measured except by inference, anecdotal, or testimonial and case study evidence taken in the operational environment for which the system is designed. Explicit evaluation of attribute achievement is a very important part of the overall systemic design process. This section describes the following:

1. A methodological framework for the design of systems, such as planning and decision support systems

2. An evaluation methodology that may be incorporated with or used independently of the design framework

A number of characteristics of effective systems efforts can be identified. These form the basis for determining the attributes of systems and systemic design procedures. Some of these attributes will be more important for a given environment than others. Effective design must typically include an operational evaluation component that will consider the strong interaction between the system and the situational issues that led to the systems design requirement. This operational evaluation is needed in order to determine whether a product system or a service consisting of humans and machines

1. Is logically sound
2. Is matched to the operational and organizational situation and environment extant
3. Supports a variety of cognitive skills, styles, and knowledge of the humans who must use the system
4. Assists users of the system to develop and use their own cognitive skills, styles, and knowledge
5. Is sufficiently flexible to allow use and adaptation by users with differing cognitive skills, styles, and knowledge
6. Encourages more effective solution of unstructured and unfamiliar issues, allowing the application of job-specific experiences in a way compatible with various acceptability constraints
7. Promotes effective long-term management

It is certainly possible that the product, or system, that results from a systems engineering effort may be used as a process or life cycle in some other application. Thus, what we have to say here refers both to the design of products and to the design of processes.

26.5.2 Operational Environments and Decision Situation Models

In order to develop robust scenarios of planning and design situations in various operational environments, and specific instruments for evaluation, we first identify a mathematical and situational taxonomy of

- Algorithmic constructs used in systemic design
- Performance objectives for quality design
- Operational environments for design

One of the initial goals in systems design engineering is to obtain the conceptual specifications for a product such that development of the system will be based on customer or client information, objectives, and existing situations and needs. An aid to the process of design should assist in or support the evaluation of alternatives relative to some criteria. It is generally necessary that design information be described in ways that lead to effective structuring of the design problem. Of equal importance is the need to be aware of the role of the affective in design tasks such as to support different cognitive styles and needs, which vary from formal knowledge-based to rule-based to skill-based behavior.[31] We desire to design efficient and effective physical systems, problem-solving service systems, and interfaces between the two. This section is concerned with each of these.

Not all of the performance objectives for quality systems engineering will be, or need be, fully attained in all design instances, but it is generally true that the quality of a system or of a systems design process necessarily improves as more and more of these objectives are attained. Measures of quality of the resulting system, and therefore systems design process quality, may be obtained by assessing the degree of achievement of these performance criteria by the resulting system, generally in an operational environment. In this way, an evaluation of the effectiveness of a design decision support system may be conducted.

A taxonomy based on operational environments is necessary to describe particular situation models through which design decision support may be achieved. We are able to describe a large number of situations using elements or features of the three-component taxonomy described earlier. With these, we are able to evolve test instruments to establish quantitative and qualitative evaluations of a design support system within an operational environment. The structural and functional properties of such a system, or of the design process itself, must be described in order that a purposeful evaluation can be accomplished. This purposeful evaluation of a systemic process is obtained by embedding the process into specific operational planning, design, or decision situations. Thus, an evaluation effort also allows iteration and feedback to ultimately improve the overall systems design process. The evaluation methodology to be described is useful, therefore, as a part or phase of the design process. Also, it is useful, in and of itself, to evaluate and prioritize a set of systemic aids for planning, design, and decision support. It is also useful for evaluation of resulting system designs and operational systems providing a methodological framework both for the design and evaluation of physical systems and for systems that assist in the planning and design of systems.

26.5.3 The Development of Aids for the Systems Design Process

This section describes five important phases in the development of systems and systemic aids for the design process. These phases serve as a guide not only for the sound design and development of systems and systemic aids for design decision support, but for their evaluation and ultimate operational deployment as well.

- Requirements specification
- Preliminary conceptual design
- Detailed design, testing, and implementation
- Evaluation
- Operational deployment

These five phases are applicable to design in general. Although the five phases will be described as if they are to be sequenced in a chronological fashion, sound design practice will generally necessitate iteration and feedback from a given phase to earlier phases.

Requirements Specification Phase

The requirements specification phase has as its goal the detailed definition of those needs, activities, and objectives to be fulfilled or achieved by the system or process that is to result from the system design effort. Furthermore, the effort in this phase should result in a description of preliminary conceptual design considerations appropriate for the next phase. This must be accomplished in order to translate operational deployment needs, activities, and objectives into requirements specifications if, for example, that is the phase of the systems engineering design effort under consideration.

Among the many objectives of the requirements specifications phase of systems engineering are the following:

1. To define the problem to be solved, or range of problems to be solved, or issue to be resolved or ameliorated; including identification of needs, constraints, alterables, and stakeholder groups associated with operational deployment of the system or the systemic process
2. To determine objectives for operational system or the operational aids for planning, design, and decision support
3. To obtain commitment for prototype design of a system or systemic process aid from user group and management
4. To search the literature and seek other expert opinions concerning the approach that is most appropriate for the particular situation extant
5. To determine the estimated frequency and extent of need for the system or the systemic process
6. To determine the possible need to modify the system or the systemic process to meet changed requirements
7. To determine the degree and type of accuracy expected from the system or systemic process
8. To estimate expected effectiveness improvement or benefits due to the use of the system or systemic process
9. To estimate the expected costs of using the system or systemic process, including design and development costs, operational costs, and maintenance costs
10. To determine typical planning horizons and periods to which the system or systemic process must be responsive
11. To determine the extent of tolerable operational environment alteration due to use of the system or systemic process
12. To determine what particular planning, design, or decision process appears best
13. To determine the most appropriate roles for the system or systemic process to perform within the context of the planning, design, or decision situation and operational environment under consideration
14. To estimate potential leadership requirements for use of the final system itself
15. To estimate user group training requirements
16. To estimate the qualifications required of the design team
17. To determine preliminary operational evaluation plans and criteria
18. To determine political acceptability and institutional constraints affecting use of an aided support process, and those of the system itself

19. To document analytical and behavioral specifications to be satisfied by the support process and the system itself

20. To determine the extent to which the user group can require changes during and after system development

21. To determine potential requirements for contractor availability after completion of development and operational tests for additional needs determined by the user group, perhaps as a result of the evaluation effort

22. To develop requirements specifications for prototype design of a support process and the operational system itself

As a result of this phase, to which the four issue requirements identification approaches of Section 26.4.1 are fully applicable, there should exist a clear definition of typical planning, design, and decision issues, or problems requiring support, and other requirements specifications, so that it is possible to make a decision whether to undertake preliminary conceptual design. If the result of this phase indicates that the user-group or client needs can potentially be satisfied in a cost-effective manner, by a systemic process aid, for example, then documentation should be prepared concerning detailed specifications for the next phase, preliminary conceptual design, and initial specifications for the last three phases of effort. A design team is then selected to implement the next phase of the system life cycle. This discussion emphasizes the inherently coupled nature of these phases of the system life cycle and illustrates why it is not reasonable to consider the phases as if they are uncoupled.

Preliminary Conceptual Design Phase

The preliminary conceptual design phase includes specification of the mathematical and behavioral content and associated algorithms for the system or process that should ultimately result from the effort, as well as the possible need for computer support to implement these. The primary goal of this phase is to develop conceptualization of a prototype system or process in response to the requirements specifications developed in the previous phase. Preliminary design according to the requirements specifications should be achieved. Objectives for preliminary conceptual design include the following:

1. To search the literature and seek other expert opinion concerning the particular approach to design and implementation that is likely to be most responsive to requirements specifications

2. To determine the specific analytic algorithms to be implemented by the system or process

3. To determine the specific behavioral situation and operational environment in which the system or process is to operate

4. To determine the specific leadership requirements for use of the system in the operational environment extant

5. To determine specific hardware- and software-implementation requirements, including type of computer programming language and input devices

6. To determine specific information-input requirements for the system or process

7. To determine the specific type of output and interpretation of the output to be obtained from the system or process that will result from the design procedure

8. To reevaluate objectives obtained in the previous phase, to provide documentation of minor changes, and to conduct an extensive re-examination of the effort if major changes are detected that could result in major modification and iteration through requirements specification or even termination of effort

9. To develop a preliminary conceptual design of a prototype aid that is responsive to the requirements specification

The expected product of this phase is a set of detailed design and testing specifications that, if followed, should result in a usable prototype system or process. User-group confidence that an ultimately useful product should result from detailed design should be above some threshold, or the entire design effort should be redone. Another product of this phase is a refined set of specifications for the evaluation and operational deployment phases.

If the result of this phase is successful, the detailed design, testing, and implementation phase is begun. This phase is based on the products of the preliminary conceptual design phase, which should result in a common understanding among all interested parties about the planning and decision support design effort concerning the following:

1. Who the user group or responsive stakeholder is

2. The structure of the operational environment in which plans, designs, and decisions are made

3. What constitutes a plan, a design, or a decision

4. How plans, designs, and decisions are made without the process or system and how they will be made with it
5. What implementation, political acceptability, and institutional constraints affect the use of the system or process
6. What specific analysis algorithms will be used in the system or process and how these algorithms will be interconnected to form the methodological construction of the system or process

Detailed Design, Testing, and Implementation Phase

In the third phase of design, a system or process that is presumably useful in the operational environment is produced. Among the objectives to be attained in this phase are the following:

1. To obtain and design appropriate physical facilities (physical hardware, computer hardware, output device, room, etc.)
2. To prepare computer software
3. To document computer software
4. To prepare a user's guide to the system and the process in which the system is embedded
5. To prepare a leader's guide for the system and the associated process
6. To conduct control group or operational (simulated operational) tests of the system and make minor changes in the aid as a result of the tests
7. To complete detailed design and associated testing of a prototype system based on the results of the previous phase
8. To implement the prototype system in the operational environment as a process

The products of this phase are detailed guides to use of the system as well as, of course, the prototype system itself. It is very important that the user's guide and the leader's guide address, at levels appropriate for the parties interested in the effort, the way in which the performance objectives identified in Section 26.5.3 are satisfied. The description of system usage and leadership topics should be addressed in terms of the analytic and behavioral constructs of the system and the resulting process, as well as in terms of operational environment situation concerns. These concerns include

1. Frequency of occurrence of need for the system or process
2. Time available from recognition of need for a plan, design, or decision to identification of an appropriate plan, design, or decision
3. Time available from determination of an appropriate plan, design, or decision to implementation of the plan, design, or decision
4. Value of time
5. Possible interactions with the plans, designs, or decisions of others
6. Information base characteristics
7. Organizational structure
8. Top management support for the resulting system or process

It is especially important that the portion of this phase that concerns implementation of the prototype system specifically address important questions concerning cognitive style and organizational differences among parties at interest and institutions associated with the design effort. Stakeholder understanding of environmental changes and side effects that will result from use of the system is critical for ultimate success. This need must be addressed. Evaluation specification and operational deployment specifications will be further refined as a result of this phase.

Evaluation Phase

Evaluation of the system in accordance with evaluation criteria, determined in the requirements specification phase and modified in the subsequent two design phases, is accomplished in the fourth phase of systems development. This evaluation should always be assisted to the extent possible by all parties at interest to the systems design effort and the resultant systemic process. The evaluation effort must be adapted to other phases of the design effort so that it becomes an integral functional part of the overall design process. As noted, evaluation may well be an effort distinct from design that is used to determine usefulness or appropriateness for specified purposes of one or more previously designed systems. Among the objectives of system or process evaluation are the following:

1. To identify a methodology for evaluation
2. To identify criteria on which the success of the system or process may be judged

3. To determine effectiveness of the system in terms of success criteria
4. To determine an appropriate balance between the operational environment evaluation and the control group evaluation
5. To determine performance-objective achievement of the system
6. To determine behavioral or human-factor effectiveness of the system
7. To determine the most useful strategy for employment of the existing system
8. To determine user-group acceptance of the system
9. To suggest refinements in existing systems for greater effectiveness of the process in which the new system has been embedded
10. To evaluate the effectiveness of the system or process

These objectives are obtained from a critical evaluation issue specification or evaluation need specification, which is the first or problem-definition step of the evaluation methodology. Generally, the critical issues for evaluation are minor adaptations of the elements that are present in the requirements specifications step of the design process outlined in the previous section. A set of specific evaluation test requirements and tests is evolved from these objectives and needs. These must be such that each objective measure and critical evaluation issue component can be determined from at least one evaluation test instrument.

If it is determined that the system and the resulting process support cannot meet user needs, the systems design process iterates to an earlier phase and development continues. An important by-product of evaluation is the determination of ultimate performance limitations and the establishment of a protocol and procedure for use of the system that results in maximum user-group satisfaction. A report is written concerning results of the evaluation process, especially those factors relating to user group satisfaction with the designed system. The evaluation process should result in suggestions for improvement in design and in better methodologies for future evaluations.

Section 26.5.6 will present additional details of the methodologies framework for evaluation. These have applicability to cases where evaluation is a separate and independent effort as well as cases where it is one of the phases of the design process.

Operational Deployment Phase

The last phase of design concerns operational deployment and final implementation. This must be accomplished in such a way that all user groups obtain adequate instructions in use of the system and complete operating and maintenance documentation and instructions. Specific objectives for the operational deployment phase of the system design effort are

1. To enhance operational deployment
2. To accomplish final design of the system
3. To provide for continuous monitoring of post-implementation effectiveness of the system and the process into which the system is embedded
4. To provide for redesign of the system as indicated by effectiveness monitoring
5. To provide proper training and leadership for successful continued operational use of the system
6. To identify barriers to successful implementation of the final design product
7. To provide for "maintenance" of the system

26.5.4   Leadership Requirements for Design

The actual use, as contrasted with potential usefulness, of a system is directly dependent on the value that the user group of stakeholders associates with use of the system and the resulting process in an operational environment. This in turn is dependent, in part, on how well the system satisfies performance objectives and on how well it is able to cope with one or more of the pathologies or pitfalls of planning, design, and/or decision-making under potentially stressful operational environment conditions.

Quality planning, design, and decision support are dependent on the ability to obtain relatively complete identification of pertinent factors that influence plans, designs, and decisions. The careful, comprehensive formulation of issues and associated requirements for issue resolution will lead to identification of pertinent critical factors for system design. These factors are ideally illuminated in a relatively easy to understand fashion that facilitates the interpretation necessary to evaluate and subsequently select plans, designs, and decisions for implementation. Success in this is, however, strongly dependent on adroitness in use of the system. It is generally not fully meaningful to talk only of an algorithm or even a complete system—which is, typically, a piece of hardware and software, but which may well be a carefully written set of protocols and procedures—as useful by itself. It is meaningful to talk of a particular systemic process as being useful. This process involves the interaction of a methodology with systems management at the cognitive process or human judgment level. A systemic process depends on the system, the operational environment, and leadership associated with use of the system. A process involves design integration of a methodology with the behavioral concerns of human cognitive judgment in an operational environment.

Operational evaluation of a systemic process that involves human interaction, such as an integrated manufacturing complex, appears the only realistic way to extract truly meaningful information concerning process effectiveness of a given system design. This must necessarily include leadership and training requirements to use the system. There are necessary tradeoffs associated with leadership and training for using a system and these are addressed in operational evaluation.

26.5.5 System Evaluation

Previous subsections have described a framework for a general system design procedure. They have indicated the role of evaluation in this process. Successful evaluation, especially operational evaluation, is strongly dependent on explicit development of a plan for evaluation developed prior to, and perhaps modified and improved during the course of, an actual evaluation. This section will concern itself with development of a methodological framework for system evaluation, especially for operational evaluation of systemic processes for planning, design, and decision support.

Evaluation Methodology and Evaluation Criteria

Objectives for evaluation of a system concern the following:

1. Identification of a methodology for operational evaluation
2. Establishing criteria on which the success of the system may be judged
3. Determining the effectiveness of the support in terms of these criteria
4. Determining the most useful strategy for employment of an existing system and potential improvements such that effectiveness of the newly implemented system and the overall process might be improved

Figure 26.8 illustrates a partial intent structure or objectives tree, which contributes to system evaluation. The lowest-level objectives contribute to satisfaction of the 10 performance objectives for systems engineering and systems design outlined in Section 26.3. These lowest-level elements form pertinent criteria for the operational system evaluation. They concern the algorithmic effectiveness or performance objective achievement of the system, the behavioral or human factor effectiveness of the system in the operational environment, and the system efficacy. Each of these three elements become top level criteria or attributes and each should be evaluated to determine evaluation of the system itself.

Subcriteria that support the three lowest-level criteria of Fig. 26.8 may be identified. These are dependent on the requirements identified for the specific system that has been designed. Attainment of each of these criteria by the system may be measured by observation of the system within the operational environment and by test instruments and surveys of user groups involved with the operational system and process.

Algorithmic Effectiveness of Performance Objectives Achievement Evaluation

A number of performance objectives can be cited that, if achieved, should lead to a quality system. Achievement of these objectives is measured by logical soundness of the operational system and

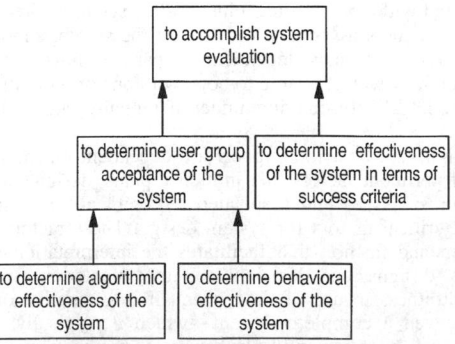

Fig. 26.8 Objectives tree for evaluation of decision support system.

process; improved system quality as a result of using the system; and improvements in the way an overall process functions, compared to the way it typically functions without the system or with an alternative system.

Behavioral or Human Factors Evaluation

A system may be well structured algorithmically in the sense of achieving a high degree of satisfaction of the performance objectives, yet the process incorporating the system may seriously violate behavioral and human factor sensibilities. This will typically result in misuse or underuse. There are many cases where technically innovative systems have failed to achieve broad scope objectives because of human factor failures. Strongly influencing the acceptability of system implementation in operational settings are such factors as organizational slack; natural human resistance to change; and the present sophistication, attitude, and past experience of the user group and its management with similar systems and processes. Behavioral or human factor evaluation criteria used to evaluate performance include political acceptability, institutional constraint satisfaction, implementability evaluation, human workload evaluation, management procedural change evaluation, and side-effect evaluation.

Efficacy Evaluation

Two of the three first-level evaluation criteria concern algorithmic effectiveness or performance-objective achievement and behavioral or human-factors effectiveness. It is necessary for a system to be effective in each of these for it to be potentially capable of truly aiding in terms of improving process quality and being acceptable for implementation in the operational environment for which it was designed. There are a number of criteria or attributes related to usefulness, service support, or efficacy to which a system must be responsive. Thus, evaluation of the efficacy of a system and the associated process is important in determining the service support value of the process. There are seven attributes of efficacy:

1. *Time requirements.* The time requirements to use a system form an important service-support criterion. If a system is potentially capable of excellent results, but the results can only be obtained after critical deadlines have passed, the overall process must be given a low rating with respect to a time-responsiveness criterion.

2. *Leadership and training.* Leadership and training requirements for use of a system are important design considerations. It is important that there be an evaluation component directed at assessing leadership and training needs and tradeoffs associated with the use of a system.

3. *Communication accomplishments.* Effective communication is important for two reasons. (1) Implementation action is often accomplished at a different hierarchical level, and therefore by a different set of actors, than the hierarchical level at which selection of alternative plans, designs, or decisions was made. Implementation-action agents often behave poorly when an action alternative is selected that they regard as threatening or arbitrary, either personally or professionally, on an individual or a group basis. Widened perspectives of a situation are made possible by effective communication. Enhanced understanding will often lead to commitment to successful action implementation as contrasted with unconscious or conscious efforts to subvert implementation action. (2) Recordkeeping and retrospective improvements to systems and processes are enhanced by the availability of well-documented constructions of planning and decision situations and communicable explanations of the rationale for the results of using the system.

4. *Educational accomplishments.* There may exist values to a system other than those directly associated with improvement in process quality. The participating group may, for example, learn a considerable amount about the issues for which a system was constructed. The possibility of enhanced ability and learning with respect to the issues for which the system was constructed should be evaluated.

5. *Documentation.* The value of the service support provided by a system will be dependent on the quality of the user's guide and its usefulness to potential users of the system.

6. *Reliability and maintainability.* To be operationally useful, a planning-and-decision-support system must be, and be perceived by potential users to be, reliable and maintainable.

7. *Convenience of access.* A system should be readily available and convenient to access, or usage will potentially suffer. While these last three service support measures are not of special significance with respect to justification of the need for a system, they may be important in determining operational usage and, therefore, operational effectiveness of a system and the associated process.

26.5.6 Evaluation Test Instruments

Several special evaluation test instruments to satisfy test requirements and measure achievement of the evaluation criteria will generally need to be developed. These include investigations of effectiveness in terms of performance objective attainment; selection of appropriate scenarios that affect use of the system, use of the system subject to these scenarios by a test group, and completion of evaluation questionnaires; and questionnaires and interviews with operational users of the system.

Every effort must be made to ensure, to the extent possible, that evaluation test results will be credible and valid. Intentional redundancy should be provided to allow correlation of results obtained from the test instruments to ensure maximum supportability and reliability of the facts and opinions to be obtained from test procedures.

The evaluation team should take advantage of every opportunity to observe use of the system within the operational environment. Evaluation of personnel reactions to the aid should be based on observations, designed to be responsive to critical evaluation issues, and the response of operational environment personnel to test questionnaires. When any of a number of constraints make it difficult to obtain real-time operational environment observation, experiential and anecdotal information becomes of increased value. Also, retrospective evaluation of use of a system is definitely possible and desirable if sufficiently documented records of past usage of an aided process are available.

Many other effectiveness questions will likely arise as an evaluation proceeds. Questions specific to a given evaluation are determined after study of the particular situation and the system being evaluated. It is, however, important to have an initial set of questions to guide the evaluation investigation and a purpose of this subsection to provide a framework for accomplishing this.

One of the important concerns in evaluation is that of those parts of the efficacy evaluation that deal with various "abilities" of a system. These include producibility, reliability, maintainability, and marketability. Figure 26.9 presents a listing of attributes that may be used to "score" the performance of systems on relevant effectiveness criteria.

26.6 CONCLUSIONS

In this chapter, we have discussed salient aspects concerning the systems engineering of large and complex systems. We have been concerned especially with systems design engineering and associated information processing and analysis efforts. To this end, we suggested a process for the design and evaluation of systems and how we might go about fielding a design decision support system. There are a number of effectiveness attributes or aspects of effective systems. Design of an effective large-scale system necessarily involves integration of operational environment concerns involving human behavior and judgment with mechanistic and physical science concerns. An effective systemic design process should

1. Allow a thorough and carefully conducted requirements specification effort to determine and specify needs of stakeholders prior to conceptual design of a system process to accomplish the desired task
2. Be capable of dealing with both quantitative and qualitative criteria representing costs and effectiveness from their economic, social, environmental, and other perspectives

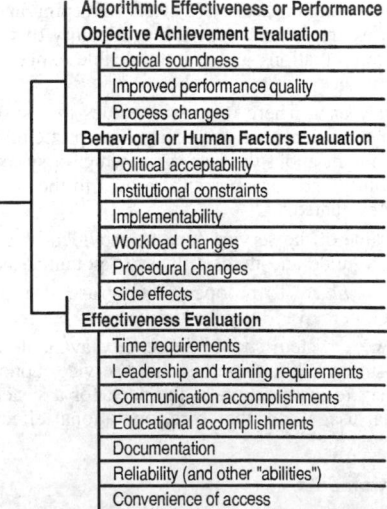

Fig. 26.9 Attribute tree and criteria for evaluation of decision support system for design and other end uses.

3. Be capable of minimizing opportunities for cognitive bias and provide debiasing procedures for those biases that occur

4. Allow separation of opinions and facts from values, and separation of ends from means, or values from alternative acts

5. Provide an objective communicable framework that allows identification, formulation, and display of the structure of the issue under consideration, as well as the rationale of the choice process

6. Allow for considerations of tradeoffs among conflicting and incommensurate criteria

7. Provide flexibility and monitoring support to allow design process evaluation rule selection with due consideration to the task structure and operational environment constraints on the decision-maker

8. Provide an open process to allow consideration of new criteria and alternatives as values change and broad-scope awareness of issues grows

There are a number of potential benefits of the systems approach that should follow high achievement of each of the criteria for effective systems design processes. An appropriate systems design process will

1. Provide structure to relatively unstructured issues

2. Facilitate conceptual formulation of issues

3. Provide cognitive cues to search and discovery

4. Encourage parsimonious collection, organization, and utilization of relevant data

5. Extend and debias information-processing abilities

6. Encourage vigilant cognitive style

7. Provide brokerage between parties at interest

There are many imperfections and limits to processes designed using the methodologies from what we know as systems engineering and systems analysis. Some of these have been documented in this chapter. Others are documented in the references provided here and in a recent handbook of systems engineering and management.[32] But what are the alternatives to appropriate systemic processes for the resolution of issues associated with the design of complex large-scale systems; and are not the fundamental limitations to these alternatives even greater?

REFERENCES

1. A. P. Sage, *Systems Engineering*, Wiley, New York, 1992.

2. A. P. Sage, *Systems Management for Information Technology and Software Engineering*, Wiley, New York, 1995.

3. A. P. Sage, *Methodology for Large Scale Systems*, McGraw-Hill, New York, 1977.

4. J. E. Armstrong and A. P. Sage, *Introduction to Systems Engineering*, Wiley, New York, 1998 (in press).

5. A. D. Hall, *A Methodology for Systems Engineering*, Van Nostrand, New York, 1962.

6. A. D. Hall, "A Three Dimensional Morphology of Systems Engineering," *IEEE Transactions on System Science and Cybernetics* **5** (2), 156–160 (April 1969).

7. E. Rechtin, *Systems Architecting: Creating and Building Complex Systems*, Prentice-Hall, Englewood Cliffs, NJ, 1991.

8. E. Rechtin, "Foundations of Systems Architecting," *Systems Engineering: The Journal of the National Council on Systems Engineering* **1** (1), 35–42 (July/September 1994).

9. W. R. Beam, *Systems Engineering: Architecture and Design*, McGraw-Hill, New York, 1990.

10. D. N. Chorfas, *Systems Architecture and Systems Design*, McGraw-Hill, New York, 1989.

11. A. P. Sage and J. D. Palmer, *Software Systems Engineering*, Wiley, New York, 1990.

12. F. Harary, R. Z. Norman, and D. Cartwright, *Structural Models: An Introduction to the Theory of Directed Graphs*, Wiley, New York, 1965.

13. J. N. Warfield, *Societal Systems: Planning, Policy, and Complexity*, Wiley, New York, 1976.

14. D. V. Steward, *Systems Analysis and Management: Structure, Strategy, and Design*, Petrocelli, New York, 1981.

15. G. G. Lendaris, "Structural Modeling: A Tutorial Guide," *IEEE Transactions on Systems, Man, and Cybernetics* **SMC 10**, 807–840 (1980).

16. C. Eden, S. Jones, and D. Sims, *Messing About in Problems*, Pergamon Press, Oxford, 1983.

17. F. M. Roberts, *Discrete Mathematical Models*, Prentice-Hall, Englewood Cliffs, NJ, 1976.

18. A. M. Geoffrion, "An Introduction to Structured Modeling," *Management Science* **33**, 547–588 (1987).

19. A. M. Geoffrion, "The Formal Aspects of Structured Modeling," *Operations Research* **37**(1), 30–51 (January 1989).

20. A. M. Geoffrion, "Computer Based Modeling Environments," *European Journal of Operations Research* **41**(1), 33–43 (July 1989).

21. A. P. Sage (ed.), *Concise Encyclopedia of Information Processing in Systems and Organizations*, Pergamon Press, Oxford, 1990.

22. D. Kahneman, P. Slovic, and A. Tversky (eds.), *Judgment Under Uncertainty: Heuristics and Biases*, Cambridge University Press, New York, 1981.

23. L. J. Cohen, "On the Psychology of Prediction: Whose Is the Fallacy," *Cognition* **7**, 385–407 (1979).

24. L. J. Cohen, "Can Human Irrationality Be Experimentally Demonstrated?," *The Behavioral and Brain Sciences* **4**, 317–370 (1981).

25. D. von Winterfeldt and W. Edwards, *Decision Analysis and Behavioral Research*, Cambridge University Press, Cambridge, 1986.

26. L. Phillips, "Theoretical Perspectives on Heuristics and Biases in Probabilistic Thinking," in *Analyzing and Aiding Decision Problems*, P. C. Humphries, O. Svenson, and O. Vari (eds.), North Holland, Amsterdam, 1984.

27. R. T. Clemen, *Making Hard Decisions: An Introduction to Decision Analysis*, Duxbury Press, Belmont, CA, 1986.

28. R. L. Keeney, *Value Focused Thinking: A Path to Creative Decision Making*, Harvard University Press, Cambridge, MA, 1992.

29. C. W. Kirkwood, *Strategic Decision Making: Multiobjective Decision Analysis with Spreadsheets*, Duxbury Press, Belmont, CA, 1997.

30. A. P. Sage, *Decision Support Systems Engineering*, Wiley, New York, 1991.

31. J. Rasmussen, *Information Processing and Human–Machine Interaction*, North Holland, Amsterdam, 1986.

32. A. P. Sage and W. B. Rouse (eds.), *Handbook of Systems Engineering and Management*, Wiley, New York, 1997.

CHAPTER 27

MATHEMATICAL MODELS OF DYNAMIC PHYSICAL SYSTEMS

K. Preston White, Jr.
Department of Systems Engineering
University of Virginia
Charlottesville, Virginia

27.1	**RATIONALE**	**795**	
27.2	**IDEAL ELEMENTS**	**796**	
	27.2.1 Physical Variables	796	
	27.2.2 Power and Energy	797	
	27.2.3 One-Port Element Laws	798	
	27.2.4 Multiport Elements	799	
27.3	**SYSTEM STRUCTURE AND INTERCONNECTION LAWS**	**802**	
	27.3.1 A Simple Example	802	
	27.3.2 Structure and Graphs	804	
	27.3.3 System Relations	806	
	27.3.4 Analogs and Duals	807	
27.4	**STANDARD FORMS FOR LINEAR MODELS**	**807**	
	27.4.1 I/O Form	808	
	27.4.2 Deriving the I/O Form—An Example	808	
	27.4.3 State-Variable Form	810	
	27.4.4 Deriving the "Natural" State Variables—A Procedure	811	
	27.4.5 Deriving the "Natural" State Variables—An Example	812	
	27.4.6 Converting from I/O to "Phase-Variable" Form	812	

27.5	**APPROACHES TO LINEAR SYSTEMS ANALYSIS**	**813**	
	27.5.1 Transform Methods	813	
	27.5.2 Transient Analysis Using Transform Methods	818	
	27.5.3 Response to Periodic Inputs Using Transform Methods	827	
27.6	**STATE-VARIABLE METHODS**	**829**	
	27.6.1 Solution of the State Equation	829	
	27.6.2 Eigenstructure	831	
27.7	**SIMULATION**	**840**	
	27.7.1 Simulation—Experimental Analysis of Model Behavior	840	
	27.7.2 Digital Simulation	841	
27.8	**MODEL CLASSIFICATIONS**	**846**	
	27.8.1 Stochastic Systems	846	
	27.8.2 Distributed-Parameter Models	850	
	27.8.3 Time-Varying Systems	851	
	27.8.4 Nonlinear Systems	852	
	27.8.5 Discrete and Hybrid Systems	861	

27.1 RATIONALE

The design of modern control systems relies on the formulation and analysis of mathematical models of dynamic physical systems. This is simply because a model is more accessible to study than the physical system the model represents. Models typically are less costly and less time consuming to construct and test. Changes in the structure of a model are easier to implement, and changes in the behavior of a model are easier to isolate and understand. A model often can be used to achieve insight when the corresponding physical system cannot, because experimentation with the actual system is too dangerous or too demanding. Indeed, a model can be used to answer "what if" questions about a system that has not yet been realized or actually cannot be realized with current technologies.

Mechanical Engineers' Handbook, 2nd ed., Edited by Myer Kutz.
ISBN 0-471-13007-9 © 1998 John Wiley & Sons, Inc.

The type of model used by the control engineer depends upon the nature of the system the model represents, the objectives of the engineer in developing the model, and the tools which the engineer has at his or her disposal for developing and analyzing the model. A mathematical model is a description of a system in terms of equations. Because the physical systems of primary interest to the control engineer are dynamic in nature, the mathematical models used to represent these systems most often incorporate difference or differential equations. Such equations, based on physical laws and observations, are statements of the fundamental relationships among the important variables that describe the system. Difference and differential equation models are expressions of the way in which the current values assumed by the variables combine to determine the future values of these variables.

Mathematical models are particularly useful because of the large body of mathematical and computational theory that exists for the study and solution of equations. Based on this theory, a wide range of techniques has been developed specifically for the study of control systems. In recent years, computer programs have been written that implement virtually all of these techniques. Computer software packages are now widely available for both simulation and computational assistance in the analysis and design of control systems.

It is important to understand that a variety of models can be realized for any given physical system. The choice of a particular model always represents a tradeoff between the fidelity of the model and the effort required in model formulation and analysis. This tradeoff is reflected in the nature and extent of simplifying assumptions used to derive the model. In general, the more faithful the model is as a description of the physical system modeled, the more difficult it is to obtain general solutions. In the final analysis, the best engineering model is not necessarily the most accurate or precise. It is, instead, the simplest model that yields the information needed to support a decision. A classification of various types of models commonly encountered by control engineers is given in Section 27.8.

A large and complicated model is justified if the underlying physical system is itself complex, if the individual relationships among the system variables are well understood, if it is important to understand the system with a great deal of accuracy and precision, and if time and budget exist to support an extensive study. In this case, the assumptions necessary to formulate the model can be minimized. Such complex models cannot be solved analytically, however. The model itself must be studied experimentally, using the techniques of computer simulation. This approach to model analysis is treated in Section 27.7.

Simpler models frequently can be justified, particularly during the initial stages of a control system study. In particular, systems that can be described by linear difference or differential equations permit the use of powerful analysis and design techniques. These include the transform methods of classical control theory and the state-variable methods of modern control theory. Descriptions of these standard forms for linear systems analysis are presented in Sections 27.4, 27.5, and 27.6.

During the past several decades, a unified approach for developing lumped-parameter models of physical systems has emerged. This approach is based on the idea of idealized system elements, which store, dissipate, or transform energy. Ideal elements apply equally well to the many kinds of physical systems encountered by control engineers. Indeed, because control engineers most frequently deal with systems that are part mechanical, part electrical, part fluid, and/or part thermal, a unified approach to these various physical systems is especially useful and economic. The modeling of physical systems using ideal elements is discussed further in Sections 27.2, 27.3, and 27.4.

Frequently, more than one model is used in the course of a control system study. Simple models that can be solved analytically are used to gain insight into the behavior of the system and to suggest candidate designs for controllers. These designs are then verified and refined in more complex models, using computer simulation. If physical components are developed during the course of a study, it is often practical to incorporate these components directly into the simulation, replacing the corresponding model components. An iterative, evolutionary approach to control systems analysis and design is depicted in Fig. 27.1.

27.2 IDEAL ELEMENTS

Differential equations describing the dynamic behavior of a physical system are derived by applying the appropriate physical laws. These laws reflect the ways in which energy can be stored and transferred within the system. Because of the common physical basis provided by the concept of energy, a general approach to deriving differential equation models is possible. This approach applies equally well to mechanical, electrical, fluid, and thermal systems and is particularly useful for systems that are combinations of these physical types.

27.2.1 Physical Variables

An idealized *two-terminal* or *one-port* element is shown in Fig. 27.2. Two *primary physical variables* are associated with the element: a through variable $f(t)$ and an across variable $v(t)$. *Through variables* represent quantities that are transmitted through the element, such as the force transmitted through a spring, the current transmitted through a resistor, or the flow of fluid through a pipe. Through variables have the same value at both ends or terminals of the element. *Across variables* represent the difference

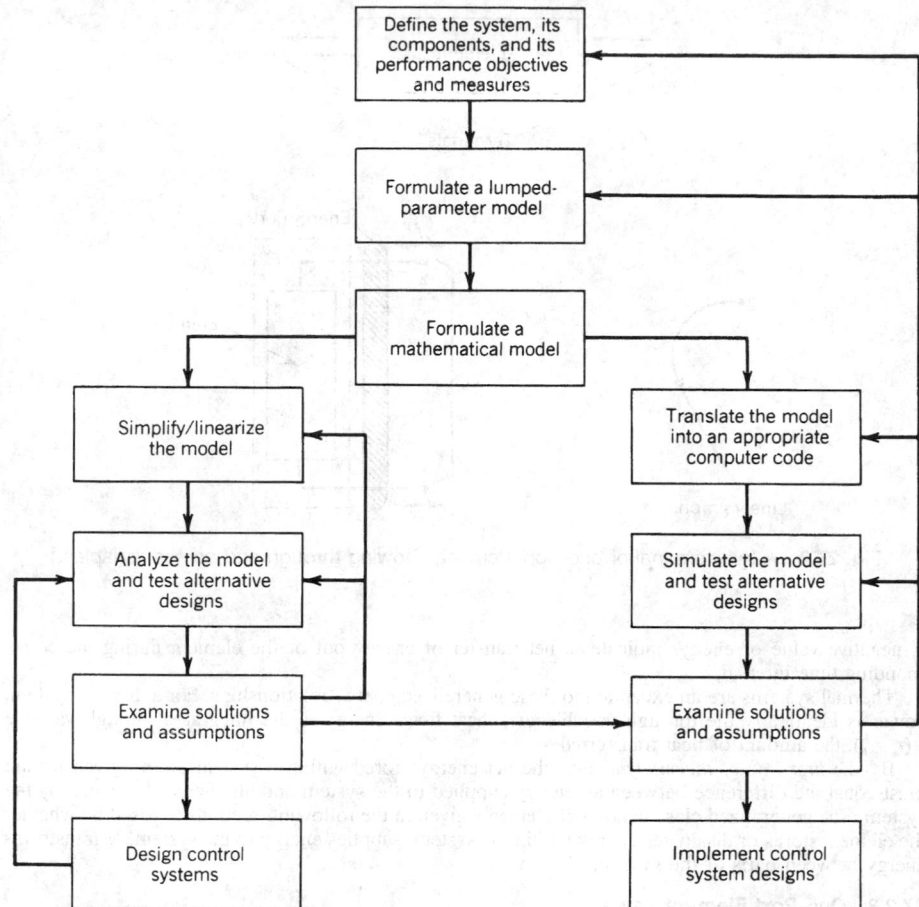

Fig. 27.1 An iterative approach to control system design, showing the use of mathematical analysis and computer simulation.

in state between the terminals of the element, such as the velocity difference across the ends of a spring, the voltage drop across a resistor, or the pressure drop across the ends of a pipe. *Secondary physical variables* are the integrated through variable $h(t)$ and the integrated across variable $x(t)$. These represent the accumulation of quantities within an element as a result of the integration of the associated through and across variables. For example, the momentum of a mass is an integrated through variable, representing the effect of forces on the mass integrated or accumulated over time. Table 27.1 defines the primary and secondary physical variables for various physical systems.

27.2.2 Power and Energy

The flow of *power* $P(t)$ into an element through the terminals 1 and 2 is the product of the through variable $f(t)$ and the difference between the across variables $v_2(t)$ and $v_1(t)$. Suppressing the notation for time dependence, this may be written as

$$P = f(v_2 - v_1) = fv_{21}$$

A negative value of power indicates that power flows out of the element. The *energy* $E(t_a, t_b)$ transferred to the element during the time interval from t_a to t_b is the integral of power, that is,

$$E = \int_{t_a}^{t_b} P \, dt = \int_{t_a}^{t_b} fv_{21} \, dt$$

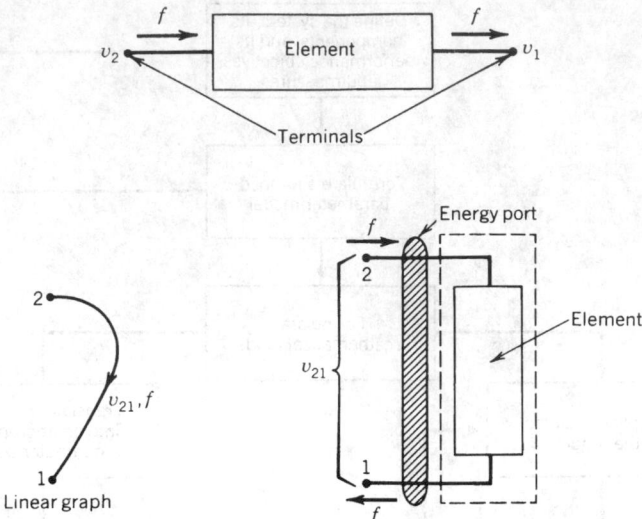

Fig. 27.2 A two-terminal or one-port element, showing through and across variables.[1]

A negative value of energy indicates a net transfer of energy out of the element during the corresponding time interval.

Thermal systems are an exception to these generalized energy relationships. For a thermal system, power is identically the through variable $q(t)$, heat flow. Energy is the integrated through variable $\mathcal{H}(t_a, t_b)$, the amount of heat transferred.

By the *first law of thermodynamics*, the net energy stored within a system at any given instant must equal the difference between all energy supplied to the system and all energy dissipated by the system. The generalized classification of elements given in the following sections is based on whether the element stores or dissipates energy within the system, supplies energy to the system, or transforms energy between parts of the system.

27.2.3 One-Port Element Laws

Physical devices are represented by idealized system elements, or by combinations of these elements. A physical device that exchanges energy with its environment through one pair of across and through variables is called a *one-port* or *two-terminal* element. The behavior of a one-port element expresses the relationship between the physical variables for that element. This behavior is defined mathematically by a *constitutive relationship*. Constitutive relationships are derived empirically, by experimentation, rather than from any more fundamental principles. The *element law,* derived from the corresponding constitutive relationship, describes the behavior of an element in terms of across and through variables and is the form most commonly used to derive mathematical models.

Table 27.1   Primary and Secondary Physical Variables for Various Systems[1]

System	Through Variable f	Integrated Through Variable h	Across Variable v	Integrated Across Variable x
Mechanical–translational	Force F	Translational momentum p	Velocity difference v_{21}	Displacement difference x_{21}
Mechanical–rotational	Torque T	Angular momentum h	Angular velocity difference Ω_{21}	Angular displacement difference Θ_{21}
Electrical	Current i	Charge q	Voltage difference v_{21}	Flux linkage λ_{21}
Fluid	Fluid flow Q	Volume V	Pressure difference P_{21}	Pressure–momentum Γ_{21}
Thermal	Heat flow q	Heat energy $\mathcal{H}$	Temperature difference θ_{21}	Not used in general

Table 27.2 summarizes the element laws and constitutive relationships for the one-port elements. Passive elements are classified into three types. *T-type* or *inductive storage* elements are defined by a single-valued constitutive relationship between the through variable $f(t)$ and the integrated across-variable difference $x_{21}(t)$. Differentiating the constitutive relationship yields the element law. For a linear (or ideal) *T*-type element, the element law states that the across-variable difference is proportional to the rate of change of the through variable. Pure translational and rotational compliance (springs), pure electrical inductance, and pure fluid inertance are examples of *T*-type storage elements. There is no corresponding thermal element.

A-type or *capacitive storage elements* are defined by a single-valued constitutive relationship between the across-variable difference $v_{21}(t)$ and the integrated through variable $h(t)$. These elements store energy by virtue of the across variable. Differentiating the constitutive relationship yields the element law. For a linear *A*-type element, the element law states that the through variable is proportional to the derivative of the across-variable difference. Pure translational and rotational inertia (masses), and pure electrical, fluid, and thermal capacitance are examples.

It is important to note that when a nonelectrical capacitance is represented by an *A*-type element, one terminal of the element must have a constant (reference) across variable, usually assumed to be zero. In a mechanical system, for example, this requirement expresses the fact that the velocity of a mass must be measured relative to a noninertial (nonaccelerating) reference frame. The constant velocity terminal of a pure mass may be thought of as being attached in this sense to the reference frame.

D-type or *resistive elements* are defined by a single-valued constitutive relationship between the across and the through variables. These elements dissipate energy, generally by converting energy into heat. For this reason, power always flows into a *D*-type element. The element law for a *D*-type energy dissipator is the same as the constitutive relationship. For a linear dissipator, the through variable is proportional to the across-variable difference. Pure translational and rotational friction (dampers or dashpots), and pure electrical, fluid, and thermal resistance are examples.

Energy-storage and energy-dissipating elements are called *passive* elements, because such elements do not supply outside energy to the system. The fourth set of one-port elements are *source elements,* which are examples of *active* or power-supplying elements. Ideal sources describe interactions between the system and its environment. A pure *A-type source* imposes an across-variable difference between its terminals, which is a prescribed function of time, regardless of the values assumed by the through variable. Similarly, a pure *T-type source* imposes a through-variable flow through the source element, which is a prescribed function of time, regardless of the corresponding across variable.

Pure system elements are used to represent physical devices. Such models are called *lumped-element models.* The derivation of lumped-element models typically requires some degree of approximation, since (1) there rarely is a one-to-one correspondence between a physical device and a set of pure elements and (2) there always is a desire to express an element law as simply as possible. For example, a coil spring has both mass and compliance. Depending on the context, the physical spring might be represented by a pure translational mass, or by a pure translational spring, or by some combination of pure springs and masses. In addition, the physical spring undoubtedly will have a nonlinear constitutive relationship over its full range of extension and compression. The compliance of the coil spring may well be represented by an ideal translational spring, however, if the physical spring is approximately linear over the range of extension and compression of concern.

27.2.4 Multiport Elements

A physical device that exchanges energy with its environment through two or more pairs of through and across variables is called a *multiport element.* The simplest of these, the idealized *four-terminal* or *two-port* element, is shown in Fig. 27.3. Two-port elements provide for transformations between the physical variables at different energy ports, while maintaining instantaneous continuity of power. In other words, net power flow into a two-port element is always identically zero:

$$P = f_a v_a + f_b v_b = 0$$

The particulars of the transformation between the variables define different categories of two-port elements.

A *pure transformer* is defined by a single-valued constitutive relationship between the integrated across variables or between the integrated through variables at each port:

$$x_b = f(x_a) \quad \text{or} \quad h_b = f(h_a)$$

For a linear (or ideal) transformer, the relationship is proportional, implying the following relationships between the primary variables:

$$v_b = n v_a, \qquad f_b = -\frac{1}{n} f_a$$

Table 27.2 Element Laws and Constitutive Relationships for Various One-Port Elements[1]

Type of element	Physical element	Linear graph	Diagram	Constitutive relationship	Energy or power function	Ideal elemental equation	Ideal energy or power
T-type energy storage $\varepsilon \geq 0$ Pure: $x_{21}=f(f)$ $\varepsilon=\int_0^f f\,dx_{21}$ Ideal: $x_{21}=Lf$ $\varepsilon=\frac{1}{2}Lf^2$	Translational spring	$2\;k\;\frac{v_{21}}{F}\;1$		$x_{21}=f(F)$	$\varepsilon=\int_0^F F\,dx_{21}$	$v_{21}=\frac{1}{k}\frac{dF}{dt}$	$\varepsilon=\frac{1}{2}\frac{F^2}{k}$
	Rotational spring	$2\;K\;\frac{\Omega_{21}}{T}\;1$		$\Theta_{21}=f(T)$	$\varepsilon=\int_0^T T\,d\Theta_{21}$	$\Omega_{21}=\frac{1}{K}\frac{dT}{dt}$	$\varepsilon=\frac{1}{2}\frac{T^2}{K}$
	Inductance	$2\;L\;\frac{v_{21}}{i}\;1$		$\lambda_{21}=f(i)$	$\varepsilon=\int_0^i i\,d\lambda_{21}$	$v_{21}=L\frac{di}{dt}$	$\varepsilon=\frac{1}{2}Li^2$
	Fluid inertance	$2\;I\;\frac{P_{21}}{Q}\;1$		$\Gamma_{21}=f(Q)$	$\varepsilon=\int_0^Q Q\,d\Gamma_{21}$	$P_{21}=I\frac{dQ}{dt}$	$\varepsilon=\frac{1}{2}IQ^2$
A-type energy storage $\varepsilon \geq 0$ Pure: $h=f(v_{21})$ $\varepsilon=\int_0^{v_{21}} v_{21}\,dh$ Ideal: $h=Cv_{21}$ $\varepsilon=\frac{1}{2}Cv_{21}^2$	Translational mass	$2\;m\;\frac{v_{21}}{F}\;1$		$p=f(v_2)$	$\varepsilon=\int_0^{v_2} v_2\,dp$	$F=m\frac{dv_2}{dt}$	$\varepsilon=\frac{1}{2}mv_2^2$
	Inertia	$2\;J\;\frac{\Omega_{21}}{T}\;1$		$h=f(\Omega_2)$	$\varepsilon=\int_0^{\Omega_2} \Omega_2\,dh$	$T=J\frac{d\Omega_2}{dt}$	$\varepsilon=\frac{1}{2}J\Omega_2^2$
	Electrical capacitance	$2\;C\;\frac{v_{21}}{i}\;1$		$q=f(v_{21})$	$\varepsilon=\int_0^{v_{21}} v_{21}\,dq$	$i=C\frac{dv_{21}}{dt}$	$\varepsilon=\frac{1}{2}Cv_{21}^2$
	Fluid capacitance	$2\;C_f\;\frac{P_{21}}{Q}\;1$		$V=f(P_2)$	$\varepsilon=\int_0^{P_2} P_2\,dV$	$Q=C_f\frac{dP_2}{dt}$	$\varepsilon=\frac{1}{2}C_fP_2^2$
	Thermal capacitance	$2\;C_t\;\frac{\theta_{21}}{q}\;1$		$\mathcal{H}=f(\theta_2)$	$\varepsilon=\int_0^{\theta_2} q\,dt=\mathcal{H}$	$q=C_t\frac{d\theta_2}{dt}$	$\varepsilon=C_t\theta_2$

D-type energy dissipators		General	Power	Ideal	Power
(diagram: resistor, $\mathcal{P} \geqslant 0$, f, v_2, v_1)	Pure: $f = f(v_{21})$, $\mathcal{P} = v_{21}f(v_{21})$ — Ideal: $f = \frac{1}{R}v_{21}$, $\mathcal{P} = \frac{1}{R}v_{21}^2 = Rf^2$				
Translational damper	(diagram: $2\ b\ 1$, v_{21}, F / F, v_1, v_2)	$F = f(v_{21})$	$\mathcal{P} = Fv_{21}$	$F = bv_{21}$	$\mathcal{P} = bv_{21}^2$
Rotational damper	(diagram: $2\ B\ 1$, Ω_{21}, T / T, Ω_1, Ω_2)	$T = f(\Omega_{21})$	$\mathcal{P} = T\Omega_{21}$	$T = B\Omega_{21}$	$\mathcal{P} = B\Omega_{21}^2$
Electrical resistance	(diagram: $2\ R\ 1$, v_{21}, i / i, v_1, v_2)	$i = f(v_{21})$	$\mathcal{P} = iv_{21}$	$i = \frac{1}{R}v_{21}$	$\mathcal{P} = \frac{1}{R}v_{21}^2$
Fluid resistance	(diagram: $2\ R_f\ 1$, P_{21}, Q / Q, P_1, P_2)	$Q = f(P_{21})$	$\mathcal{P} = QP_{21}$	$Q = \frac{1}{R_f}P_{21}$	$\mathcal{P} = \frac{1}{R_f}P_{21}^2$
Thermal resistance	(diagram: $2\ R_t\ 1$, θ_{21}, q / q, θ_1, θ_2)	$q = f(\theta_{21})$	$\mathcal{P} = q$	$q = \frac{1}{R_t}\theta_{21}$	$\mathcal{P} = \frac{1}{R_t}\theta_{21}$
Energy sources					
A-type across-variable source	(diagram: $2\ (v)\ 1$, $+\,v\,-$, v_1, v_2) $\mathcal{P} \gtrless 0$	$v_{21} = f(t)$	$\mathcal{P} = fv_{21}$		
T-type through-variable source	(diagram: $2\ (f)\ 1$, f, v_1, v_2) $\varepsilon \gtrless 0$	$f = f(t)$	$\mathcal{P} = fv_{21}$		

Nomenclature

λ = energy, $\mathcal{P}$ = power

f = generalized through-variable, F = force, T = torque, i = current, Q = fluid flow rate, q = heat flow rate

h = generalized integrated through-variable, p = translational momentum, h = angular momentum,
q = charge, Γ' = fluid volume displaced, $\mathcal{H}$ = heat

v = generalized across-variable, v = translational velocity, Ω = angular velocity, v = voltage, P = pressure, θ = temperature

x = generalized integrated across-variable, x = translational displacement, Θ = angular displacement,
λ = flux linkage, Γ = pressure-momentum

L = generalized ideal inductance, $1/k$ = reciprocal translational stiffness, $1/K$ = reciprocal rotational stiffness,
L = inductance, I = fluid inertance

C = generalized ideal capacitance, m = mass, J = moment of inertia, C = capacitance, C_f = fluid capacitance,
C_t = thermal capacitance

R = generalized ideal resistance, $1/b$ = reciprocal translational damping, $1/B$ = reciprocal rotational damping,
R = electrical resistance, R_f = fluid resistance, R_t = thermal resistance

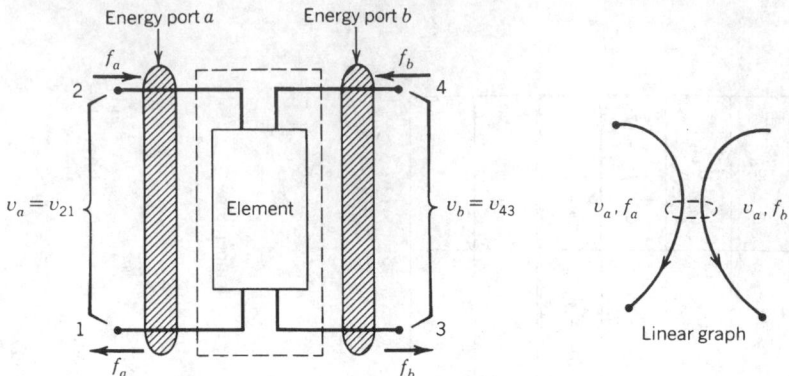

Fig. 27.3   A four-terminal or two-port element, showing through and across variables.

where the constant of proportionality n is called the *transformation ratio*. Levers, mechanical linkages, pulleys, gear trains, electrical transformers, and differential-area fluid pistons are examples of physical devices that typically can be approximated by pure or ideal transformers. Figure 27.4 depicts some examples. *Pure transmitters*, which serve to transmit energy over a distance, frequently can be thought of as transformers with $n = 1$.

A *pure gyrator* is defined by a single-valued constitutive relationship between the across variable at one energy port and the through variable at the other energy port. For a linear gyrator, the following relations apply:

$$v_b = rf_a, \qquad f_b = \frac{-1}{r} v_a$$

where the constant of proportionality is called the *gyration ratio* or *gyrational resistance*. Physical devices that perform pure gyration are not as common as those performing pure transformation. A mechanical gyroscope is one example of a system that might be modeled as a gyrator.

In the preceding discussion of two-port elements, it has been assumed that the type of energy is the same at both energy ports. A *pure transducer,* on the other hand, changes energy from one physical medium to another. This change may be accomplished either as a transformation or a gyration. Examples of *transforming transducers* are gears with racks (mechanical rotation to mechanical translation), and electric motors and electric generators (electrical to mechanical rotation and vice versa). Examples of *gyrating transducers* are the piston-and-cylinder (fluid to mechanical) and piezoelectric crystals (mechanical to electrical).

More complex systems may have a large number of energy ports. A common *six-terminal* or *three-port element* called a *modulator* is depicted in Fig. 27.5. The flow of energy between ports a and b is controlled by the energy input at the modulating port c. Such devices inherently dissipate energy, since

$$P_a + P_c \geq P_b$$

although most often the modulating power P_c is much smaller than the power input P_a or the power output P_b. When port a is connected to a pure source element, the combination of source and modulator is called a *pure dependent source*. When the modulating power P_c is considered the input and the modulated power P_b is considered the output, the modulator is called an *amplifier.* Physical devices that often can be modeled as modulators include clutches, fluid valves and couplings, switches, relays, transistors, and variable resistors.

27.3   SYSTEM STRUCTURE AND INTERCONNECTION LAWS

27.3.1   A Simple Example

Physical systems are represented by connecting the terminals of pure elements in patterns that approximate the relationships among the properties of component devices. As an example, consider the mechanical-translational system depicted in Fig. 27.6a, which might represent an idealized automobile suspension system. The inertial properties associated with the masses of the chassis, passenger compartment, engine, and so on, all have been lumped together as the pure mass m_1. The inertial prop-

System	Symbol	Pure transformer	Ideal transformer	Transformation ratio
Mechanical translation (lever)	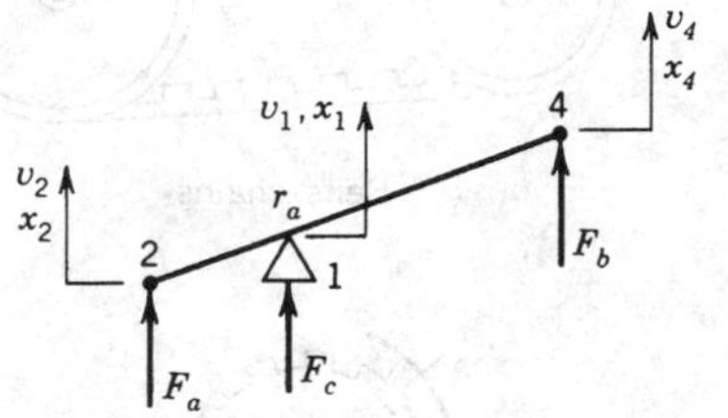	$x_{41} = f(x_{21})$	$v_{41} = n v_{21}$ $F_b = -\dfrac{1}{n} F_a$	$n = -\dfrac{r_b}{r_a}$ Lever ratio
Mechanical rotational (gears)	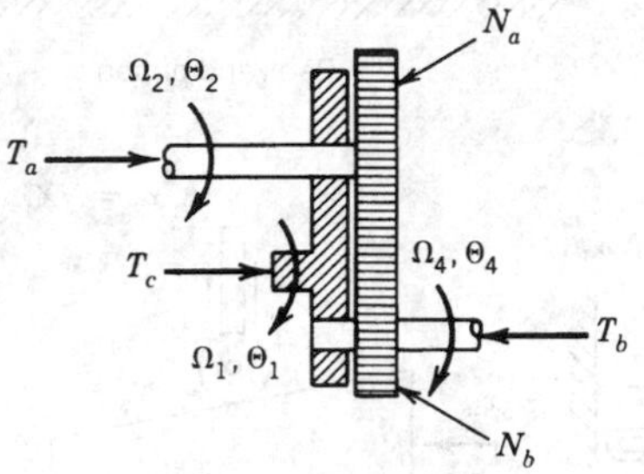	$\Theta_{41} = f(\Theta_2)$	$\Omega_{41} = n\Omega_{21}$ $T_b = -\dfrac{1}{n} T_a$	$n = -\dfrac{N_a}{N_b}$ Gear ratio
Electrical (magnetic)	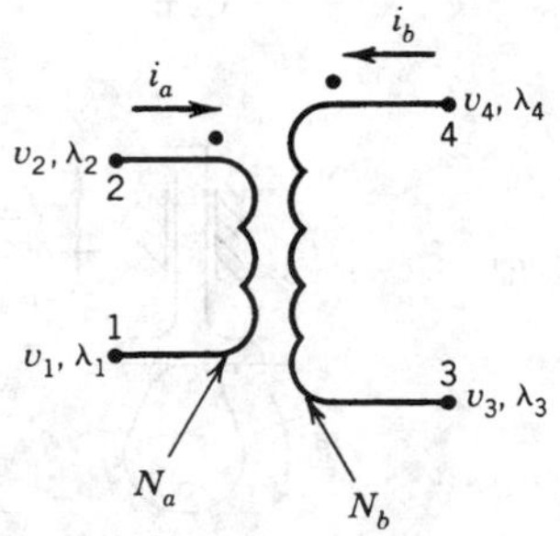	$\lambda_{43} = f(\lambda_{21})$	$v_{43} = n v_{21}$ $i_b = -\dfrac{1}{n} i_a$	$n = \dfrac{N_b}{N_a}$ Turns ratio
Fluid (differential piston)	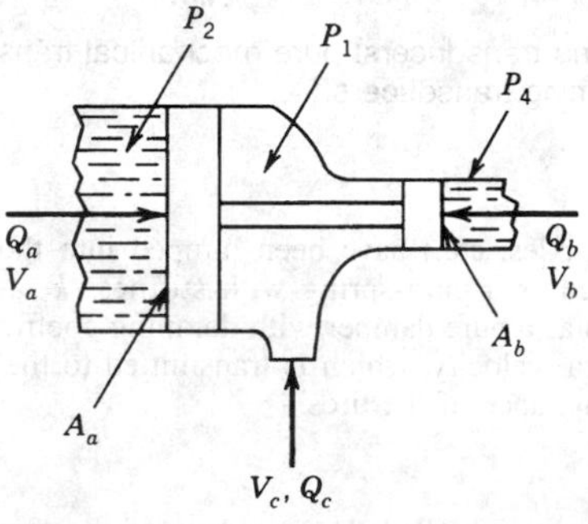	$V_b = f(V_a)$	$P_{41} = n P_{21}$ $Q_b = -\dfrac{1}{n} Q_a$	$n = \dfrac{A_a}{A_b}$ Area ratio

Fig. 27.4a Examples of transforms and transducers: pure transformers.[1]

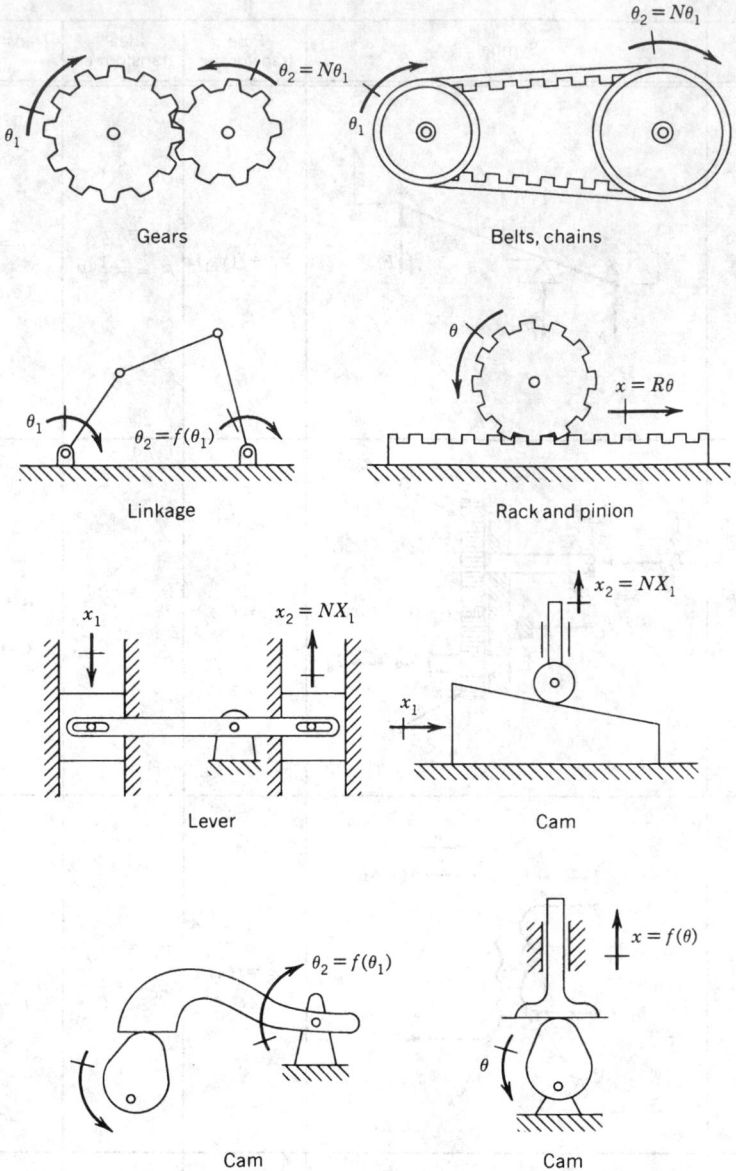

Fig. 27.4b Examples of transformers and transducers: pure mechanical transformers and transforming transducers.[2]

erties of the unsprung components (wheels, axles, etc.) have been lumped into the pure mass m_2. The compliance of the suspension is modeled as a pure spring with stiffness k_1 and the frictional effects (principally from the shock absorbers) as a pure damper with damping coefficient b. The road is represented as an input or source of vertical velocity, which is transmitted to the system through a spring of stiffness k_2, representing the compliance of the tires.

27.3.2 Structure and Graphs

The *pattern of interconnections* among elements is called the *structure* of the system. For a one-dimensional system, structure is conveniently represented by a *system graph*. The system graph for the idealized automobile suspension system of Fig. 27.6a is shown in Fig. 27.6b. Note that each distinct across variable (velocity) becomes a distinct *node* in the graph. Each distinct through variable

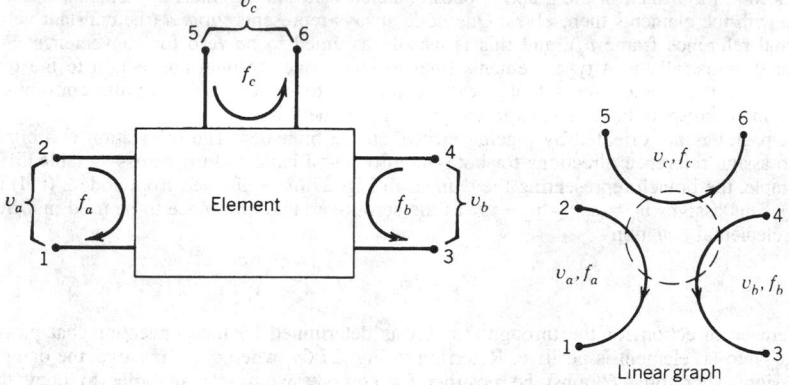

Fig. 27.5 A six-terminal or three-port element, showing through and across variables.

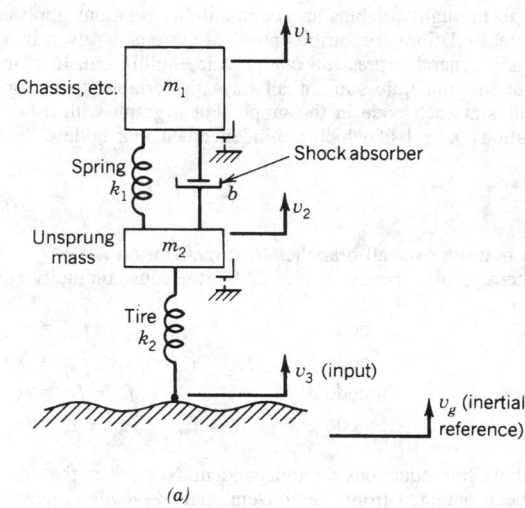

(a)

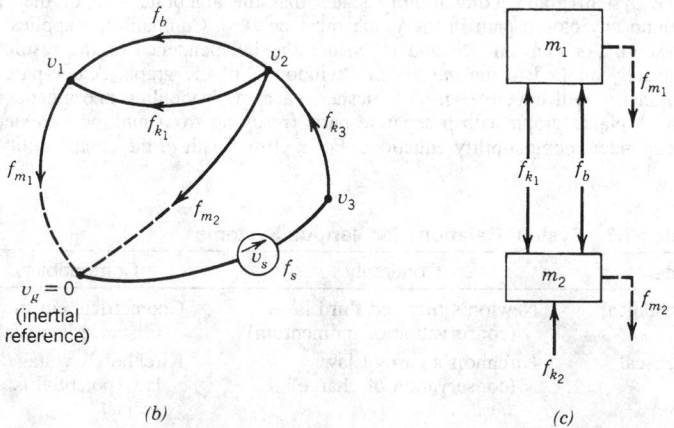

(b)   (c)

Fig. 27.6 An idealized model of an automobile suspension system: (a) lumped-element model, (b) system graph, (c) free-body diagram.

(force) becomes a *branch* in the graph. Nodes coincide with the terminals of elements and branches coincide with the elements themselves. One node always represents *ground* (the constant velocity of the inertial reference frame v_g), and this is usually assumed to be zero for convenience. For non-electrical systems, all the *A*-type elements (masses) have one terminal connection to the reference node. Because the masses are not physically connected to ground, however, the convention is to represent the corresponding branches in the graph by dashed lines.

System graphs are oriented by placing arrows on the branches. The orientation is arbitrary and serves to assign reference directions for both the through-variable and the across-variable difference. For example, the branch representing the damper in Fig. 27.6*b* is directed from node 2 (tail) to node 1 (head). This assigns $v_b = v_{21} = v_2 - v_1$ as the across-variable difference to be used in writing the damper elemental equation

$$f_b = bv_b = bv_{21}$$

The reference direction for the through variable is determined by the convention that power flow $P_b = f_b v_b$ into an element is positive. Referring to Fig. 27.6*a*, when v_{21} is positive, the damper is in compression. Therefore, f_b must be positive for compressive forces in order to obey the sign convention for power. By similar reasoning, tensile forces will be negative.

27.3.3 System Relations

The structure of a system gives rise to two sets of *interconnection laws* or *system relations*. Continuity relations apply to through variables and compatibility relations apply to across variables. The inter-pretation of system relations for various physical systems is given in Table 27.3.

Continuity is a general expression of dynamic equilibrium. In terms of the system graph, conti-nuity states that the algebraic sum of all through variables entering a given node must be zero. Continuity applies at each node in the graph. For a graph with *n* nodes, continuity gives rise to *n* continuity equations, $n - 1$ of which are independent. For node *i*, the continuity equation is

$$\sum_j f_{ij} = 0$$

where the sum is taken over all branches (i, j) incident on *i*.

For the system graph depicted in Fig. 27.6*b*, the four continuity equations are

$$\begin{aligned}
\text{node 1:} && f_{k_1} + f_b - f_{m_1} &= 0 \\
\text{node 2:} && f_{k_2} - f_{k_1} - f_b - f_{m_2} &= 0 \\
\text{node 3:} && f_s - f_{k_2} &= 0 \\
\text{node g:} && f_{m_1} + f_{m_2} - f_s &= 0
\end{aligned}$$

Only three of these four equations are independent. Note, also, that the equations for nodes 1 through 3 could have been obtained from the conventional *free-body diagrams* shown in Fig. 27.6*c*, where f_{m_1} and f_{m_2} are the *D'Alembert forces* associated with the pure masses. Continuity relations are also known as *vertex, node, flow,* and *equilibrium relations.*

Compatibility expresses the fact that the magnitudes of all across variables are scalar quantities. In terms of the system graph, compatibility states that the algebraic sum of the across-variable differences around any closed path in the graph must be zero. Compatibility applies to any closed path in the system. For convenience and to ensure the independence of the resulting equations, continuity is usually applied to the *meshes* or "windows" of the graph. A one-part graph with *n* nodes and *b* branches will have $b - n + 1$ meshes, each mesh yielding one independent compati-bility equation. A planar graph with *p* separate parts (resulting from multiport elements) will have $b - n + p$ independent compatibility equations. For a closed path *q*, the compatibility equation is

Table 27.3 System Relations for Various Systems

System	Continuity	Compatibility
Mechanical	Newton's first and third laws (conservation of momentum)	Geometrical constraints (distance is a scalar)
Electrical	Kirchhoff's current law (conservation of charge)	Kirchhoff's voltage law (potential is a scalar)
Fluid	Conservation of matter	Pressure is a scalar
Thermal	Conservation of energy	Temperature is a scalar

$$\sum_q v_{ij} = 0$$

where the summation is taken over all branches (i, j) on the path.

For the system graph depicted in Fig. 27.6b, the three compatibility equations based on the meshes are

$$\text{path } 1 \to 2 \to g \to 1: \qquad -v_b + v_{m_2} - v_{m_1} = 0$$
$$\text{path } 1 \to 2 \to 1: \qquad -v_{k_1} + v_b = 0$$
$$\text{path } 2 \to 3 \to g \to 2: \qquad -v_{k_2} - v_s - v_{m_2} = 0$$

These equations are all mutually independent and express apparent geometric identities. The first equation, for example, states that the velocity difference between the ends of the damper is identically the difference between the velocities of the masses it connects. Compatibility relations are also known as *path, loop,* and *connectedness* relations.

27.3.4 Analogs and Duals

Taken together, the element laws and system relations are a complete mathematical model of a system. When expressed in terms of generalized through and across variables, the model applies not only to the physical system for which it was derived, but to any physical system with the same generalized system graph. Different physical systems with the same generalized model are called *analogs*. The mechanical rotational, electrical, and fluid analogs of the mechanical translational system of Fig. 27.6a are shown in Fig. 27.7. Note that because the original system contains an inductive storage element, there is no thermal analog.

Systems of the same physical type, but in which the roles of the through variables and the across variables have been interchanged, are called *duals*. The analog of a dual—or, equivalently, the dual of an analog—is sometimes called a *dualog*. The concepts of analogy and duality can be exploited in many different ways.

27.4 STANDARD FORMS FOR LINEAR MODELS

The element laws and system relations together constitute a complete mathematical description of a physical system. For a system graph with n nodes, b branches, and s sources, there will be $b - s$

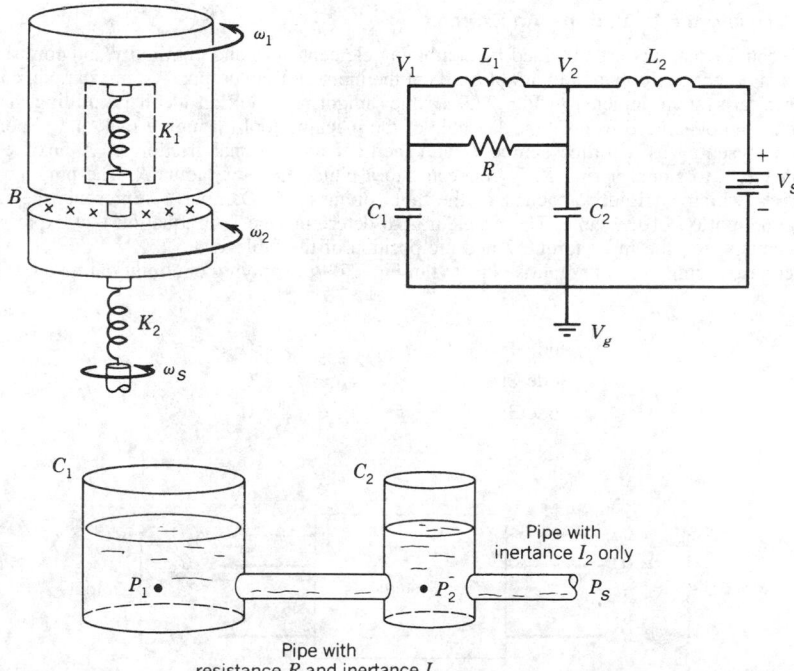

Fig. 27.7 Analogs of the idealized automobile suspension system depicted in Fig. 27.6.

element laws, $n - 1$ continuity equations, and $b - n + 1$ compatibility equations. This is a total of $2b - s$ differential and algebraic equations. For systems composed entirely of linear elements, it is always possible to reduce these $2b - s$ equations to either of two standard forms. The *input/output* or *I/O form* is the basis for *transform* or so-called *classical linear systems analysis*. The *state-variable form* is the basis for *state-variable* or so-called *modern linear systems analysis*.

27.4.1 I/O Form

The classical representation of a system is the "black box," depicted in Fig. 27.8. The system has a set of p inputs (also called *excitations* or *forcing functions*), $u_j(t)$, $j = 1, 2, \ldots, p$. The system also has a set of q outputs (also called *response variables*), $y_k(t)$, $k = 1, 2, \ldots, q$. Inputs correspond to sources and are assumed to be known functions of time. Outputs correspond to physical variables that are to be measured or calculated.

Linear systems represented in I/O form can be modeled mathematically by *I/O differential equations*. Denoting as $y_{kj}(t)$ that part of the kth output $y_k(t)$ that is attributable to the jth input $u_j(t)$, there are $(p \times q)$ I/O equations of the form

$$\frac{d^n y_{kj}}{dt^n} + a_{n-1}\frac{d^{n-1} y_{kj}}{dt^{n-1}} + \cdots + a_1 \frac{dy_{kj}}{dt} + a_0 y_{kj}(t) = b_m \frac{d^m u_j}{dt^m} + b_{m-1}\frac{d^{m-1} u_j}{dt^{m-1}} + \cdots + b_1 \frac{du_j}{dt} + b_0 u_j(t)$$

where $j = 1, 2, \ldots, p$ and $k = 1, 2, \ldots, q$. Each equation represents the dependence of one output and its derivatives on one input and its derivatives. By the *principle of superposition*, the kth output in response to all of the inputs acting simultaneously is

$$y_k(t) = \sum_{j=1}^{p} y_{kj}(t)$$

A system represented by nth-order I/O equations is called an n-*order system*. In general, the order of a system is determined by the number of *independent* energy-storage elements within the system, that is, by the combined number of T-type and A-type elements for which the initial energy stored can be independently specified.

The coefficients $a_0, a_1, \ldots, a_{n-1}$ and $b_0, b_1, \ldots, b_m$ are parameter groups made up of algebraic combinations of the system physical parameters. For a system with constant parameters, therefore, these coefficients are also constant. Systems with constant parameters are called *time-invariant* systems and are the basis for classical analysis.

27.4.2 Deriving the I/O Form—An Example

I/O differential equations are obtained by combining element laws and continuity and compatibility equations in order to eliminate all variables except the input and the output. As an example, consider the mechanical system depicted in Fig. 27.9a, which might represent an idealized milling machine. A rotational motor is used to position the table of the machine tool through a rack and pinion. The motor is represented as a torque source T with inertia J and internal friction B. A flexible shaft, represented as a torsional spring K, is connected to a pinion gear of radius R. The pinion meshes with a rack, which is rigidly attached to the table of mass m. Damper b represents the friction opposing the motion of the table. The problem is to determine the I/O equation that expresses the relationship between the input torque T and the position of the table x.

The corresponding system graph is depicted in Fig. 27.9b. Applying continuity at nodes 1, 2, and 3 yields

$$\begin{aligned}
\text{node 1:} \quad & T - T_J - T_B - T_K = 0 \\
\text{node 2:} \quad & T_K - T_p = 0 \\
\text{node 3:} \quad & -f_r - f_m - f_b = 0
\end{aligned}$$

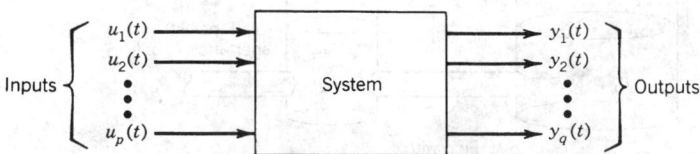

Fig. 27.8 Input/output (I/O) or "black box" representation of a dynamic system.

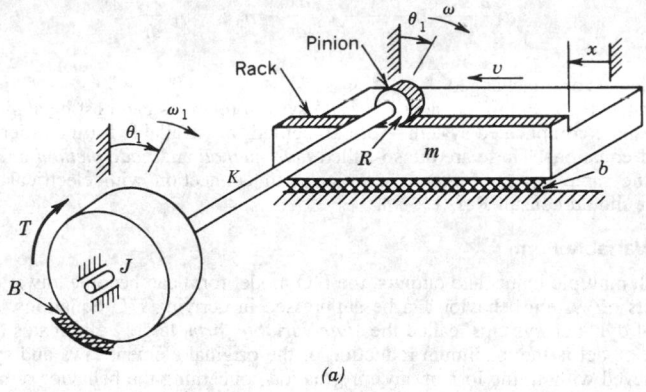

(a)

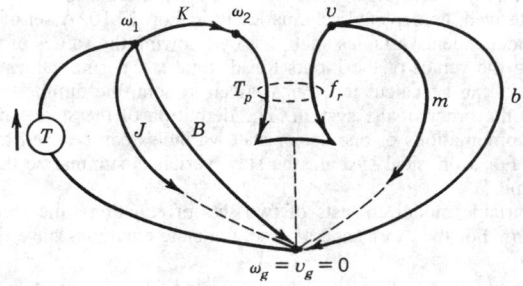

$$\omega_g = v_g = 0$$

Fig. 27.9 An idealized model of a milling machine: (a) lumped-element model,[3]
(b) system graph.

Substituting the elemental equation for each of the one-port elements into the continuity equations and assuming zero ground velocities yields

node 1: $T - J\dot\omega_1 - B\omega_1 - K \int (\omega_1 - \omega_2)dt = 0$

node 2: $K \int (\omega_1 - \omega_2)dt - T_p = 0$

node 3: $-f_r - m\dot v - bv = 0$

Note that the definition of the across variables for each element in terms of the node variables, as above, guarantees that the compatibility equations are satisfied. With the addition of the constitutive relationships for the rack and pinion

$$\omega_2 = \frac{1}{R} v \qquad \text{and} \qquad T_p = -Rf_r$$

there are now five equations in the five unknowns ω_1, ω_2, v, T_p, and f_r. Combining these equations to eliminate all of the unknowns except v yields, after some manipulation,

$$a_3 \frac{d^3v}{dt^3} + a_2 \frac{d^2v}{dt^2} + a_1 \frac{dv}{dt} + a_0v = b_1T$$

where

$$a_3 = Jm, \qquad a_1 = \frac{JK}{R^2} + Bb + mK, \qquad b_1 = \frac{K}{R}$$

$$a_2 = Jb + mB, \qquad a_0 = \frac{BK}{R^2} + Kb$$

Differentiating yields the desired I/O equation

$$a_3 \frac{d^3x}{dt^3} + a_2 \frac{d^2x}{dt^2} + a_1 \frac{dx}{dt} + a_0 x = b_1 \frac{dT}{dt}$$

where the coefficients are unchanged.

For many systems, combining element laws and system relations can best be achieved by *ad hoc* procedures. For more complicated systems, formal methods are available for the orderly combination and reduction of equations. These are the so-called *loop method* and *node method* and correspond to procedures of the same names originally developed in connection with electrical networks. The interested reader should consult Ref. 1.

27.4.3 State-Variable Form

For systems with multiple inputs and outputs, the I/O model form can become unwieldy. In addition, important aspects of system behavior can be suppressed in deriving I/O equations. The "modern" representation of dynamic systems, called the *state-variable form,* largely eliminates these problems. A state-variable model is the maximum reduction of the original element laws and system relations that can be achieved without the loss of any information concerning the behavior of a system. State-variable models also provide a convenient representation for systems with multiple inputs and outputs and for systems analysis using computer simulation.

State variables are a set of variables $x_1(t), x_2(t), \ldots, x_n(t)$ internal to the system from which any set of outputs can be derived, as depicted schematically in Fig. 27.10. A set of state variables is the minimum number of independent variables such that by knowing the values of these variables at any time t_0 and by knowing the values of the inputs for all time $t \geq t_0$, the values of the state variables for all future time $t \geq t_0$ can be calculated. For a given system, the number n of state variables is unique and is equal to the order of the system. The definition of the state variables is not unique, however, and various combinations of one set of state variables can be used to generate alternative sets of state variables. For a physical system, the state variables summarize the *energy state* of the system at any given time.

A complete state-variable model consists of two sets of equations, the *state* or *plant equations* and the *output equations.* For the most general case, the state equations have the form

$$\dot{x}_1(t) = f_1[x_1(t), x_2(t), \ldots, x_n(t), u_1(t), u_2(t), \ldots, u_p(t)]$$
$$\dot{x}_2(t) = f_2[x_1(t), x_2(t), \ldots, x_n(t), u_1(t), u_2(t), \ldots, u_p(t)]$$
$$\vdots$$
$$\dot{x}_n(t) = f_n[x_1(t), x_2(t), \ldots, x_n(t), u_1(t), u_2(t), \ldots, u_p(t)]$$

and the output equations have the form

$$y_1(t) = g_1[x_1(t), x_2(t), \ldots, x_n(t), u_1(t), u_2(t), \ldots, u_p(t)]$$
$$y_2(t) = g_2[x_1(t), x_2(t), \ldots, x_n(t), u_1(t), u_2(t), \ldots, u_p(t)]$$
$$\vdots$$
$$y_q(t) = g_q[x_1(t), x_2(t), \ldots, x_n(t), u_1(t), u_2(t), \ldots, u_p(t)]$$

These equations are expressed more compactly as the two vector equations

$$\dot{x}(t) = f[x(t), u(t)]$$
$$y(t) = g[x(t), u(t)]$$

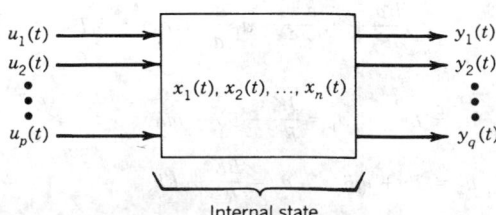

Internal state

Fig. 27.10 State-variable representation of a dynamic system.

where

$$\dot{x}(t) = \text{the } (n \times 1) \text{ state vector}$$
$$u(t) = \text{the } (p \times 1) \text{ input or control vector}$$
$$y(t) = \text{the } (q \times 1) \text{ output or response vector}$$

and f and g are vector-valued functions.

For linear systems, the state equations have the form

$$\dot{x}_1(t) = a_{11}(t)x_1(t) + \cdots + a_{1n}(t)x_n(t) + b_{11}(t)u_1(t) + \cdots + b_{1p}(t)u_p(t)$$
$$\dot{x}_2(t) = a_{21}(t)x_1(t) + \cdots + a_{2n}(t)x_n(t) + b_{21}(t)u_1(t) + \cdots + b_{2p}(t)u_p(t)$$
$$\vdots$$
$$\dot{x}_n(t) = a_{n1}(t)x_1(t) + \cdots + a_{nn}(t)x_n(t) + b_{n1}(t)u_1(t) + \cdots + b_{np}(t)u_p(t)$$

and the output equations have the form

$$y_1(t) = c_{11}(t)x_1(t) + \cdots + c_{1n}(t)x_n(t) + d_{11}(t)u_1(t) + \cdots + d_{1p}(t)u_p(t)$$
$$y_2(t) = c_{21}(t)x_1(t) + \cdots + c_{2n}(t)x_n(t) + d_{21}(t)u_1(t) + \cdots + d_{2p}(t)u_p(t)$$
$$\vdots$$
$$y_n(t) = c_{q1}(t)x_1(t) + \cdots + c_{qn}(t)x_n(t) + d_{q1}(t)u_1(t) + \cdots + d_{qp}(t)u_p(t)$$

where the coefficients are groups of parameters. The linear model is expressed more compactly as the two linear vector equations.

$$\dot{x}(t) = A(t)x(t) + B(t)u(t)$$
$$y(t) = C(t)x(t) + D(t)u(t)$$

where the vectors x, u, and y are the same as the general case and the matrices are defined as

$$A = [a_{ij}] \text{ is the } (n \times n) \text{ system matrix}$$
$$B = [b_{jk}] \text{ is the } (n \times p) \text{ control, input, or}$$
$$\quad \text{distribution matrix}$$
$$C = [c_{ij}] \text{ is the } (q \times n) \text{ output matrix}$$
$$D = [d_{ik}] \text{ is the } (q \times p) \text{ output distribution matrix}$$

For a time-invariant linear system, all of these matrices are constant.

27.4.4 Deriving the "Natural" State Variables—A Procedure

Because the state variables for a system are not unique, there are an unlimited number of alternative (but equivalent) state-variable models for the system. Since energy is stored only in generalized system storage elements, however, a natural choice for the state variables is the set of through and across variables corresponding to the independent T-type and A-type elements, respectively. This definition is sometimes called the set of *natural state variables* for the system.

For linear systems, the following procedure can be used to reduce the set of element laws and system relations to the natural state-variable model.

Step 1. For each independent T-type storage, write the element law with the derivative of the through variable isolated on the left-hand side, that is, $\dot{f} = L^{-1}v$.

Step 2. For each independent A-type storage, write the element law with the derivative of the across variable isolated on the left-hand side, that is, $\dot{v} = C^{-1}f$.

Step 3. Solve the compatibility equations, together with the element laws for the appropriate D-type and multiport elements, to obtain each of the across variables of the independent T-type elements in terms of the natural state variables and specified sources.

Step 4. Solve the continuity equations, together with the element laws for the appropriate D-type and multiport elements, to obtain the through variables of the A-type elements in terms of the natural state variables and specified sources.

Step 5. Substitute the results of step 3 into the results of step 1; substitute the results of step 4 into the results of step 2.

Step 6. Collect terms on the right-hand side and write in vector form.

27.4.5 Deriving the "Natural" State Variables—An Example

The six-step process for deriving a natural state-variable representation, outlined in the preceding section, is demonstrated for the idealized automobile suspension depicted in Fig. 27.6:

Step 1

$$f_{k_1} = k_1 v_{k_1}, \qquad f_{k_2} = k_2 v_{k_2}$$

Step 2

$$\dot{v}_{m_1} = m_1^{-1} f_{m_1}, \qquad \dot{v}_{m_2} = m_2^{-1} f_{m_2}$$

Step 3

$$v_{k_1} = v_b = v_{m_2} - v_{m_1}, \qquad v_{k_2} = -v_{m_2} - v_s$$

Step 4

$$f_{m_1} = f_{k_1} + f_b = f_{k_1} + b^{-1}(v_{m_2} - v_{m_1})$$
$$f_{m_2} = f_{k_2} - f_{k_1} - f_b = f_{k_2} - f_{k_1} - b^{-1}(v_{m_2} - v_{m_1})$$

Step 5

$$\dot{f}_{k_1} = k_1(v_{m_2} - v_{m_1}), \qquad \dot{v}_{m_1} = m_1^{-1}[f_{k_1} + b^{-1}(v_{m_2} - v_{m_1})]$$
$$\dot{f}_{k_2} = k_2(-v_{m_2} - v_s), \qquad \dot{v}_{m_2} = m_2^{-1}[f_{k_2} - f_{k_1} - b^{-1}(v_{m_2} - v_{m_1})]$$

Step 6

$$\frac{d}{dt}\begin{bmatrix} f_{k_1} \\ f_{k_2} \\ v_{m_1} \\ v_{m_2} \end{bmatrix} = \begin{bmatrix} 0 & 0 & -k_1 & k_1 \\ 0 & 0 & 0 & -k_2 \\ 1/m_1 & 0 & -1/m_1 b & 1/m_1 b \\ -1/m_2 & 1/m_2 & 1/m_2 b & -1/m_2 b \end{bmatrix}\begin{bmatrix} f_{k_1} \\ f_{k_2} \\ v_{m_1} \\ v_{m_2} \end{bmatrix} + \begin{bmatrix} 0 \\ -k_2 \\ 0 \\ 0 \end{bmatrix} v_s$$

27.4.6 Converting from I/O to "Phase-Variable" Form

Frequently, it is desired to determine a state-variable model for a dynamic system for which the I/O equation is already known. Although an unlimited number of such models is possible, the easiest to determine uses a special set of state variables called the *phase variables*. The phase variables are defined in terms of the output and its derivatives as follows:

$$x_1(t) = y(t)$$
$$x_2(t) = \dot{x}_1(t) = \frac{d}{dt} y(t)$$
$$x_3(t) = \dot{x}_2(t) = \frac{d^2}{dt^2} y(t)$$
$$\vdots$$
$$x_n(t) = \dot{x}_{n-1}(t) = \frac{d^{n-1}}{dt^{n-1}} y(t)$$

This definition of the phase variables, together with the I/O equation of Section 27.4.1, can be shown to result in a state equation of the form

$$\frac{d}{dt}\begin{bmatrix} x_1(t) \\ x_2(t) \\ \vdots \\ x_{n-1}(t) \\ x_n(t) \end{bmatrix} = \begin{bmatrix} 0 & 1 & 0 & \cdots & 0 \\ 0 & 0 & 1 & \cdots & 0 \\ \vdots & \vdots & \vdots & \ddots & \vdots \\ 0 & 0 & 0 & \cdots & 1 \\ -a_0 & -a_1 & -a_2 & \cdots & -a_{n-1} \end{bmatrix}\begin{bmatrix} x_1(t) \\ x_2(t) \\ \vdots \\ x_{n-1}(t) \\ x_n(t) \end{bmatrix} + \begin{bmatrix} 0 \\ 0 \\ \vdots \\ 0 \\ 1 \end{bmatrix} u(t)$$

and an output equation of the form

$$y(t) = [b_0 \quad b_1 \cdots b_m] \begin{bmatrix} x_1(t) \\ x_2(t) \\ \vdots \\ x_n(t) \end{bmatrix}$$

This special form of the system matrix, with ones along the upper off-diagonal and zeros elsewhere except for the bottom row, is called a *companion matrix*.

27.5 APPROACHES TO LINEAR SYSTEMS ANALYSIS

There are two fundamental approaches to the analysis of linear, time-invariant systems. *Transform methods* use rational functions obtained from the Laplace transformation of the system I/O equations. Transform methods provide a particularly convenient algebra for combining the component sub-models of a system and form the basis of so-called *classical control theory*. State-variable methods use the vector state and output equations directly. State-variable methods permit the adaptation of important ideas from linear algebra and form the basis for so-called *modern control theory*. Despite the deceiving names of "classical" and "modern," the two approaches are complementary. Both approaches are widely used in current practice and the control engineer must be conversant with both.

27.5.1 Transform Methods

A *transformation* converts a given mathematical problem into an equivalent problem, according to some well-defined rule called a *transform*. Prudent selection of a transform frequently results in an equivalent problem that is easier to solve than the original. If the solution to the original problem can be recovered by an inverse transformation, the three-step process of (1) transformation, (2) solution in the *transform domain*, and (3) inverse transformation, may prove more attractive than direct solution of the problem in the original problem domain. This is true for fixed linear dynamic systems under the *Laplace transform*, which converts differential equations into equivalent algebraic equations.

Laplace Transforms: Definition

The one-sided Laplace transform is defined as

$$F(s) = \mathscr{L}[f(t)] = \int_0^\infty f(t) e^{-st} \, dt$$

and the inverse transform as

$$f(t) = \mathscr{L}^{-1}[F(s)] = \frac{1}{2\pi j} \int_{\sigma - j\omega}^{\sigma + j\omega} F(s) e^{-st} \, ds$$

The Laplace transform converts the function $f(t)$ into the transformed function $F(s)$; the inverse transform recovers $f(t)$ from $F(s)$. The symbol $\mathscr{L}$ stands for the "Laplace transform of"; the symbol $\mathscr{L}^{-1}$ stands for "the inverse Laplace transform of."

The Laplace transform takes a problem given in the *time domain*, where all physical variables are functions of the *real variable* t, into the *complex-frequency domain*, where all physical variables are functions of the complex frequency $s = \sigma + j\omega$, where $j = \sqrt{-1}$ is the imaginary operator. Laplace transform pairs consist of the function $f(t)$ and its transform $F(s)$. Transform pairs can be calculated by substituting $f(t)$ into the defining equation and then evaluating the integral with s held constant. For a transform pair to exist, the corresponding integral must converge, that is,

$$\int_0^\infty |f(t)| e^{-\sigma^* t} \, dt < \infty$$

for some real $\sigma* > 0$. Signals that are physically realizable always have a Laplace transform.

Tables of Transform Pairs and Transform Properties

Transform pairs for functions commonly encountered in the analysis of dynamic systems rarely need to be calculated. Instead, pairs are determined by reference to a *table of transforms* such as that given in Table 27.4. In addition, the Laplace transform has a number of properties that are useful in determining the transforms and inverse transforms of functions in terms of the tabulated pairs. The most important of these are given in a *table of transform properties* such as that given in Table 27.5.

Table 27.4   Laplace Transform Pairs

$F(s)$	$f(t), t \geq 0$
1. 1	$\delta(t)$, the unit impulse at $t = 0$
2. $\dfrac{1}{s}$	1, the unit step
3. $\dfrac{n!}{s^{n+1}}$	t^n
4. $\dfrac{1}{s+a}$	e^{-at}
5. $\dfrac{1}{(s+a)^n}$	$\dfrac{1}{(n-1)!} t^{n-1} e^{-at}$
6. $\dfrac{a}{s(s+a)}$	$1 - e^{-at}$
7. $\dfrac{1}{(s+a)(s+b)}$	$\dfrac{1}{(b-a)} (e^{-at} - e^{-bt})$
8. $\dfrac{s+p}{(s+a)(s+b)}$	$\dfrac{1}{(b-a)} [(p-a)e^{-at} - (p-b)e^{-bt}]$
9. $\dfrac{1}{(s+a)(s+b)(s+c)}$	$\dfrac{e^{-at}}{(b-a)(c-a)} + \dfrac{e^{-bt}}{(c-b)(a-b)} + \dfrac{e^{-ct}}{(a-c)(b-c)}$
10. $\dfrac{s+p}{(s+a)(s+b)(s+c)}$	$\dfrac{(p-a)e^{-at}}{(b-a)(c-a)} + \dfrac{(p-b)e^{-bt}}{(c-b)(a-b)} + \dfrac{(p-c)e^{-ct}}{(a-c)(b-c)}$
11. $\dfrac{b}{s^2 + b^2}$	$\sin bt$
12. $\dfrac{s}{s^2 + b^2}$	$\cos bt$
13. $\dfrac{b}{(s+a)^2 + b^2}$	$e^{-at} \sin bt$
14. $\dfrac{s+a}{(s+a)^2 + b^2}$	$e^{-at} \cos bt$
15. $\dfrac{\omega_n^2}{s^2 + 2\zeta\omega_n s + \omega_n^2}$	$\dfrac{\omega_n}{\sqrt{1-\zeta^2}} e^{-\zeta\omega_n t} \sin \omega_n \sqrt{1-\zeta^2}\, t, \quad \zeta < 1$
16. $\dfrac{\omega_n^2}{s(s^2 + 2\zeta\omega_n s + \omega_n^2)}$	$1 + \dfrac{1}{\sqrt{1-\zeta^2}} e^{-\zeta\omega_n t} \sin(\omega_n \sqrt{1-\zeta^2}\, t + \phi)$ $$\phi = \tan^{-1} \dfrac{\sqrt{1-\zeta^2}}{\zeta} + \pi$$ (third quadrant)

Poles and Zeros

The response of a dynamic system most often assumes the following form in the complex-frequency domain

$$F(s) = \frac{N(s)}{D(s)} = \frac{b_m s^m + b_{m-1} s^{m-1} + \cdots + b_1 s + b_0}{s^n + a_{n-1} s^{n-1} + \cdots + a_1 s + a_0} \tag{27.1}$$

Functions of this form are called *rational functions,* because these are the ratio of two polynomials $N(s)$ and $D(s)$. If $n \geq m$, then $F(s)$ is a *proper rational function;* if $n > m$, then $F(s)$ is a *strictly proper rational function.*

In factored form, the rational function $F(s)$ can be written as

$$F(s) = \frac{N(s)}{D(s)} = \frac{b_m (s - z_1)(s - z_2) \cdots (s - z_m)}{(s - p_1)(s - p_2) \cdots (s - p_n)} \tag{27.2}$$

Table 27.5 Laplace Transform Properties

$f(t)$	$F(s) = \int_0^\infty f(t)e^{-st}\, dt$
1. $af_1(t) + bf_2(t)$	$aF_1(s) + bF_2(s)$
2. $\dfrac{df}{dt}$	$sF(s) - f(0)$
3. $\dfrac{d^2f}{dt^2}$	$s^2F(s) - sf(0) - \left.\dfrac{df}{dt}\right|_{t=0}$
4. $\dfrac{d^nf}{dt^n}$	$s^nF(s) - \displaystyle\sum_{k=1}^{n} s^{n-k}g_{k-1}$ $g_{k-1} = \left.\dfrac{d^{k-1}f}{dt^{k-1}}\right|_{t=0}$
5. $\displaystyle\int_0^t f(t)\, dt$	$\dfrac{F(s)}{s} + \dfrac{h(0)}{s}$ $h(0) = \left.\displaystyle\int f(t)\, dt\right|_{t=0}$
6. $\begin{cases}0, & t < D \\ f(t - D), & t \geqslant D\end{cases}$	$e^{-sD}F(s)$
7. $e^{-at}f(t)$	$F(s + a)$
8. $f\!\left(\dfrac{t}{a}\right)$	$aF(as)$
9. $f(t) = \displaystyle\int_0^t x(t - \tau)y(\tau)\, d\tau$ $ = \displaystyle\int_0^t y(t - \tau)x(\tau)\, d\tau$	$F(s) = X(s)Y(s)$
10.	$f(\infty) = \lim\limits_{s\to 0} sF(s)$
11.	$f(0+) = \lim\limits_{s\to\infty} sF(s)$

The roots of the numerator polynomial $N(s)$ are denoted by z_j, $j = 1, 2, \ldots, m$. These numbers are called the *zeros* of $F(s)$, since $F(z_j) = 0$. The roots of the denominator polynomial are denoted by p_i, $1, 2, \ldots, n$. These numbers are called the *poles* of $F(s)$, since $\lim_{s\to p_i} F(s) = \pm\infty$.

Inversion by Partial-Fraction Expansion

The *partial-fraction expansion theorem* states that a strictly proper rational function $F(s)$ with *distinct* (*nonrepeated*) poles p_i, $i = 1, 2, \ldots, n$, can be written as the sum

$$F(s) = \frac{A_1}{s - p_1} + \frac{A_2}{s - p_2} + \cdots + \frac{A_n}{s - p_n} = \sum_{i=1}^{n} A_i\left(\frac{1}{s - p_i}\right) \tag{27.3}$$

where the A_i, $i = 1, 2, \ldots, n$, are constants called *residues*. The inverse transform of $F(s)$ has the simple form

$$f(t) = A_1 e^{p_1 t} + A_2 e^{p_2 t} + \cdots + A_n e^{p_n t} = \sum_{i=1}^{n} A_i e^{p_i t}$$

The *Heaviside expansion theorem* gives the following expression for calculating the residue at the pole p_i,

$$A_i = (s - p_i)F(s)\big|_{s=p_i} \quad \text{for } i = 1, 2, \ldots, n$$

These values can be checked by substituting into Eq. (27.3), combining the terms on the right-hand side of Eq. (27.3), and showing the result yields the values for all the coefficients $b_j, j = 1, 2, \ldots, m$, originally specified in the form of Eq. (27.3).

Repeated Poles

When two or more poles of a strictly proper rational function are identical, the poles are said to be *repeated* or *nondistinct*. If a pole is repeated q times, that is, if $p_i = p_{i+1} = \cdots = p_{i+q-1}$, then the pole is said to be of *multiplicity q*. A strictly proper rational function with a pole of multiplicity q will contain q terms of the following form

$$\frac{A_{i1}}{(s - p_i)^q} + \frac{A_{i2}}{(s - p_i)^{q-1}} + \cdots + \frac{A_{iq}}{(s - p_i)}$$

in addition to the terms associated with the distinct poles. The corresponding terms in the inverse transform are

$$\left(\frac{1}{(q - 1)!} A_{i1} t^{(q-1)} + \frac{1}{(q - 2)!} A_{i2} t^{(q-2)} + \cdots + A_{iq} \right) e^{p_i t}$$

The corresponding residues are

$$A_{i1} = (s - p_i)^q F(s)|_{s=p_i}$$

$$A_{i2} = \left(\frac{d}{ds} [(s - p_i)^q F(s)] \right) \Big|_{s=p_i}$$

$$\vdots$$

$$A_{iq} = \frac{1}{(q - 1)!} \left(\frac{d^{(q-1)}}{ds^{(q-1)}} [(s - p_i)^q F(s)] \right) \Big|_{s=p_i}$$

Complex Poles

A strictly proper rational function with complex conjugate poles can be inverted using partial-fraction expansion. Using a method called *completing the square*, however, is almost always easier. Consider the function

$$F(s) = \frac{B_1 s + B_2}{(s + \sigma - j\omega)(s + \sigma + j\omega)}$$

$$= \frac{B_1 s + B_2}{s^2 + 2\sigma s + \sigma^2 + \omega^2}$$

$$= \frac{B_1 s + B_2}{(s + \sigma)^2 + \omega_2}$$

From the transform tables the Laplace inverse is

$$f(t) = e^{-\sigma t}[B_1 \cos \omega t + B_3 \sin \omega t]$$

$$= K e^{-\sigma t} \cos(\omega t + \phi)$$

where $B_3 = (1/\omega)(B_2 - aB_1)$
$K = \sqrt{B_1^2 + B_3^2}$
$\phi = -\tan^{-1}(B_3/B_1)$

Proper and Improper Rational Functions

If $F(s)$ is not a strictly proper rational function, then $N(s)$ must be divided by $D(s)$ using *synthetic division*. The result is

$$F(s) = \frac{N(s)}{D(s)} = P(s) + \frac{N^*(s)}{D(s)}$$

where $P(s)$ is a polynomial of degree $m - n$ and $N^*(s)$ is a polynomial of degree $n - 1$. Each term of $P(s)$ may be inverted directly using the transform tables. $N^*(s)/D(s)$ is a strictly proper rational function and may be inverted using partial-fraction expansion.

Initial-Value and Final-Value Theorems

The limits of $f(t)$ as time approaches zero or infinity frequently can be determined directly from the transform $F(s)$ without inverting. The *initial-value theorem* states that

$$f(0_+) = \lim_{s \to \infty} sF(s)$$

where the limit exists. If the limit does not exist (i.e., is infinite), the value of $f(0_+)$ is undefined. The *final-value theorem* states that

$$f(\infty) = \lim_{s \to 0} sF(s)$$

provided that (with the possible exception of a single pole at $s = 0$) $F(s)$ has no poles with nonnegative real parts.

Transfer Functions

The Laplace transform of system I/O equation may be written in terms of the transform $Y(s)$ of the system response $y(t)$ as

$$Y(s) = \frac{G(s)N(s) + F(s)D(s)}{P(s)D(s)}$$

$$= \left(\frac{G(s)}{P(s)}\right)\left(\frac{N(s)}{D(s)}\right) + \frac{F(s)}{P(s)}$$

where (a) $P(s) = a_n s^n + a_{n-1} + \cdots + a_1 s + a_0$ is the *characteristic polynomial* of the system,

(b) $G(s) = b_m s^m + b_{m-1} s^{m-1} + \cdots + b_1 s + b_0$ represents the *numerator dynamics* of the system,

(c) $U(s) = N(s)/D(s)$ is the transform of the input to the system, $u(t)$, assumed to be a rational function, and

(d) $F(s) = a_n y(0)s^{n-1} + \left(a_n \dfrac{dy}{dt}(0) + a_{n-1}y(0)\right)s^{n-2} + \cdots$
$\qquad + \left(a_n \dfrac{d^{n-1}y}{dt^{n-1}}(0) + a_{n-1}\dfrac{d^{n-2}y}{dt}(0) + \cdots + a_1 y(0)\right)$

reflects the initial system state [i.e., the initial conditions on $y(t)$ and its first $n - 1$ derivatives]. The transformed response can be thought of as the sum of two components,

$$Y(s) = Y_{zs}(s) + Y_{zi}(s)$$

where (e) $Y_{zs}(s) = [G(s)/P(s)][N(s)/D(s)] = H(s)U(s)$ is the transform of the *zero-state response,* that is, the response of the system to the input alone, and

(f) $Y_{zi}(s) = F(s)/P(s)$ is the transform of the *zero-input response,* that is, the response of the system to the initial state alone.

The rational function

(g) $H(s) = Y_{zs}(s)/U(s) = G(s)/P(s)$ is the *transfer function* of the system, defined as the Laplace transform of the ratio of the system response to the system input, assuming zero initial conditions.

The transfer function plays a crucial role in the analysis of fixed linear systems using transforms and can be written directly from knowledge of the system I/O equation as

$$H(s) = \frac{b_m s^m + \cdots + b_0}{a_n s^n + a_{n-1}s^{n-1} + \cdots + a_1 s + a_0}$$

Impulse Response

Since $U(s) = 1$ for a unit impulse function, the transform of the zero-state response to a unit impulse input is given by the relation (g) as

$$Y_{zs}(s) = H(s)$$

that is, the system transfer function. In the time domain, therefore, the unit *impulse response* is

$$h(t) = \begin{cases} 0 & \text{for } t \le 0 \\ \mathscr{L}^{-1}[H(s)] & \text{for } t > 0 \end{cases}$$

This simple relationship is profound for several reasons. First, this provides for a direct characterization of time-domain response $h(t)$ in terms of the properties (poles and zeros) of the rational function $H(s)$ in the complex-frequency domain. Second, applying the convolution transform pair (Table 27.5) to relation (e) above yields

$$Y_{zs}(t) = \int_0^t h(\tau)u(t - \tau)\, d\tau$$

In words, the zero-state output corresponding to an arbitrary input $u(t)$ can be determined by convolution with the impulse response $h(t)$. In other words, the impulse response completely characterizes the system. The impulse response is also called the system *weighing function*.

Block Diagrams

Block diagrams are an important conceptual tool for the analysis and design of dynamic systems, because block diagrams provide a graphic means for depicting the relationships among system variables and components. A block diagram consists of unidirectional blocks representing specified system components or subsystems, interconnected by arrows representing system variables. Causality follows in the direction of the arrows, as in Fig. 27.11, indicating that the output is caused by the input acting on the system defined in the block.

   Combining transform variables, transfer functions, and block diagrams provides a powerful graphical means for determining the overall transfer function of a system, when the transfer functions of its component subsystems are known. The basic blocks in such diagrams are given in Fig. 27.12. A block diagram comprising many blocks and summers can be reduced to a single transfer function block by using the diagram transformations given in Fig. 27.13.

27.5.2 Transient Analysis Using Transform Methods

Basic to the study of dynamic systems are the concepts and terminology used to characterize system behavior or performance. These ideas are aids in *defining* behavior, in order to consider for a given context those features of behavior which are desirable and undesirable; in *describing* behavior, in order to communicate concisely and unambiguously various behavioral attributes of a given system; and in *specifying* behavior, in order to formulate desired behavioral norms for system design. Characterization of dynamic behavior in terms of standard concepts also leads in many cases to analytical shortcuts, since key features of the system response frequently can be determined without actually solving the system model.

Parts of the Complete Response

A variety of names is used to identify terms in the response of a fixed linear system. The complete response of a system may be thought of alternatively as the sum of:

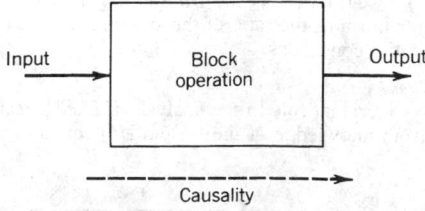

Fig. 27.11   Basic block diagram, showing assumed direction of causality or loading.

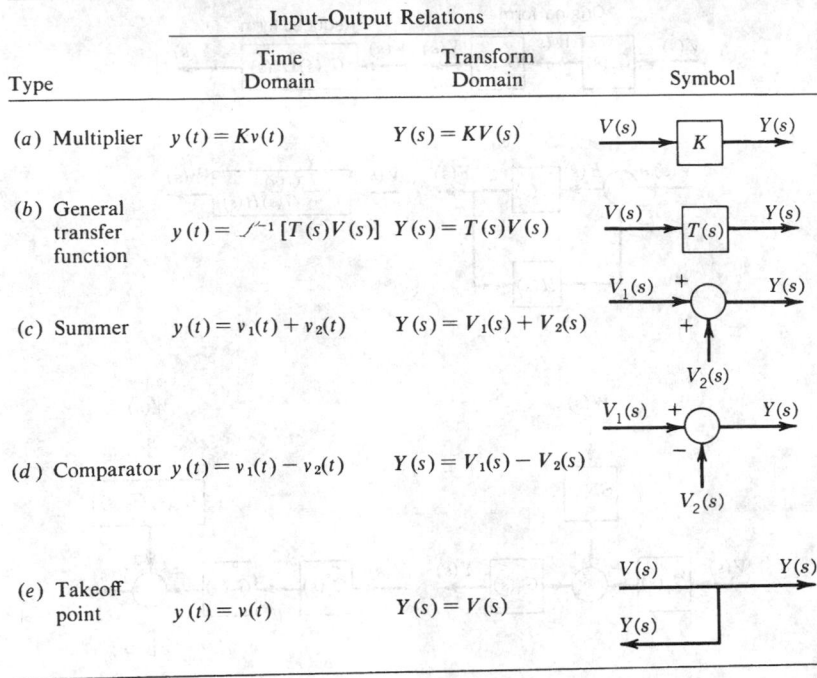

| | Input–Output Relations | | |
Type	Time Domain	Transform Domain	Symbol
(a) Multiplier	$y(t) = Kv(t)$	$Y(s) = KV(s)$	
(b) General transfer function	$y(t) = \mathcal{L}^{-1}[T(s)V(s)]$	$Y(s) = T(s)V(s)$	
(c) Summer	$y(t) = v_1(t) + v_2(t)$	$Y(s) = V_1(s) + V_2(s)$	
(d) Comparator	$y(t) = v_1(t) - v_2(t)$	$Y(s) = V_1(s) - V_2(s)$	
(e) Takeoff point	$y(t) = v(t)$	$Y(s) = V(s)$	

Fig. 27.12 Basic block diagram elements.[4]

1. The *free response* (or complementary or homogeneous solution) and the *forced response* (or particular solution). The free response represents the natural response of a system when inputs are removed and the system responds to some initial stored energy. The forced response of the system depends on the form of the input only.

2. The *transient response* and the *steady-state response*. The transient response is that part of the output that decays to zero as time progresses. The steady-state response is that part of the output that remains after all the transients disappear.

3. The *zero-state response* and the *zero-input response*. The zero-state response is the complete response (both free and forced responses) to the input when the initial state is zero. The zero-input response is the complete response of the system to the initial state when the input is zero.

Test Inputs or Singularity Functions

For a stable system, the response to a specific input signal will provide several measures of system performance. Since the actual inputs to a system are not usually known *a priori*, characterization of the system behavior is generally given in terms of the response to one of a standard set of *test input signals*. This approach provides a common basis for the comparison of different systems. In addition, many inputs actually encountered can be approximated by some combination of standard inputs. The most commonly used test inputs are members of the family of *singularity functions*, depicted in Fig. 27.14.

First-Order Transient Response

The standard form of the I/O equation for a first-order system is

$$\frac{dy}{dt} + \frac{1}{\tau} y(t) = \frac{1}{\tau} u(t)$$

where the parameter τ is called the system *time constant*. The response of this standard first-order system to three test inputs is depicted in Fig. 27.15, assuming zero initial conditions on the output $y(t)$. For all inputs, it is clear that the response approaches its steady state monotonically (i.e., without

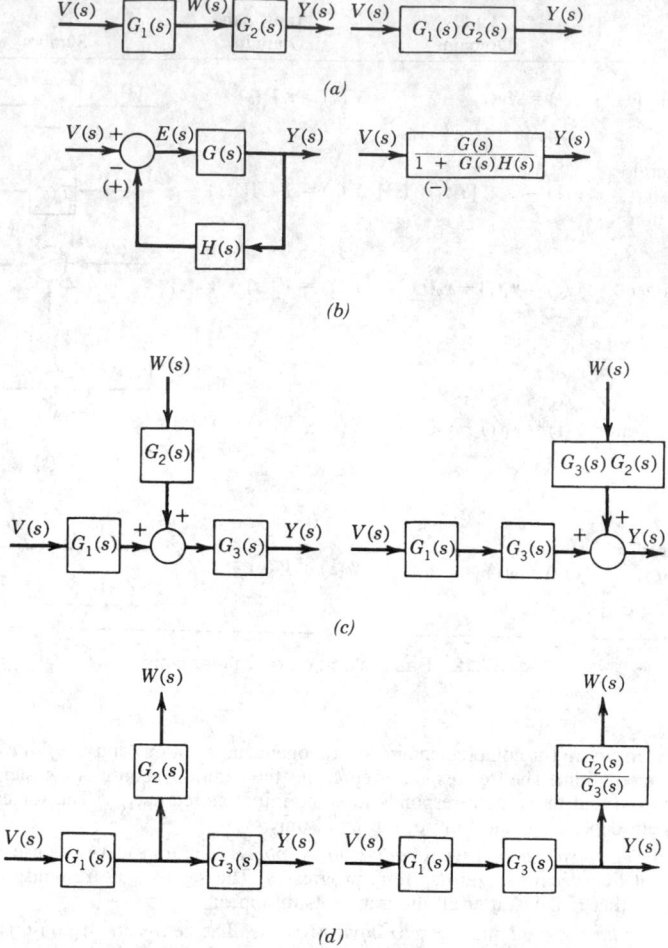

Fig. 27.13 Representative block diagram transformations: (a) series or cascaded elements, (b) feedback loop, (c) relocated summer, (d) relocated takeoff point.[4]

oscillations) and that the *speed of response* is completely characterized by the time constant τ. The transfer function of the system is

$$H(s) = \frac{Y(s)}{U(s)} = \frac{1/\tau}{s + 1/\tau}$$

and therefore $\tau = -p^{-1}$, where p is the system pole. As the absolute value of p increases, τ decreases and the response becomes faster.

The response of the standard first-order system to a step input of magnitude u for arbitrary initial condition $y(0) = y_0$ is

$$y(t) = y_{ss} - [y_{ss} - y_0]e^{-t/\tau}$$

where $y_{ss} = u$ is the steady-state response. Table 27.6 and Fig. 27.16 record the values of $y(t)$ and $\dot{y}(t)$ for $t = k\tau$, $k = 0, 1, \ldots, 6$. Note that over any time interval of duration τ, the response increases approximately 63% of the difference between the steady-state value and the value at the beginning of the time interval, that is,

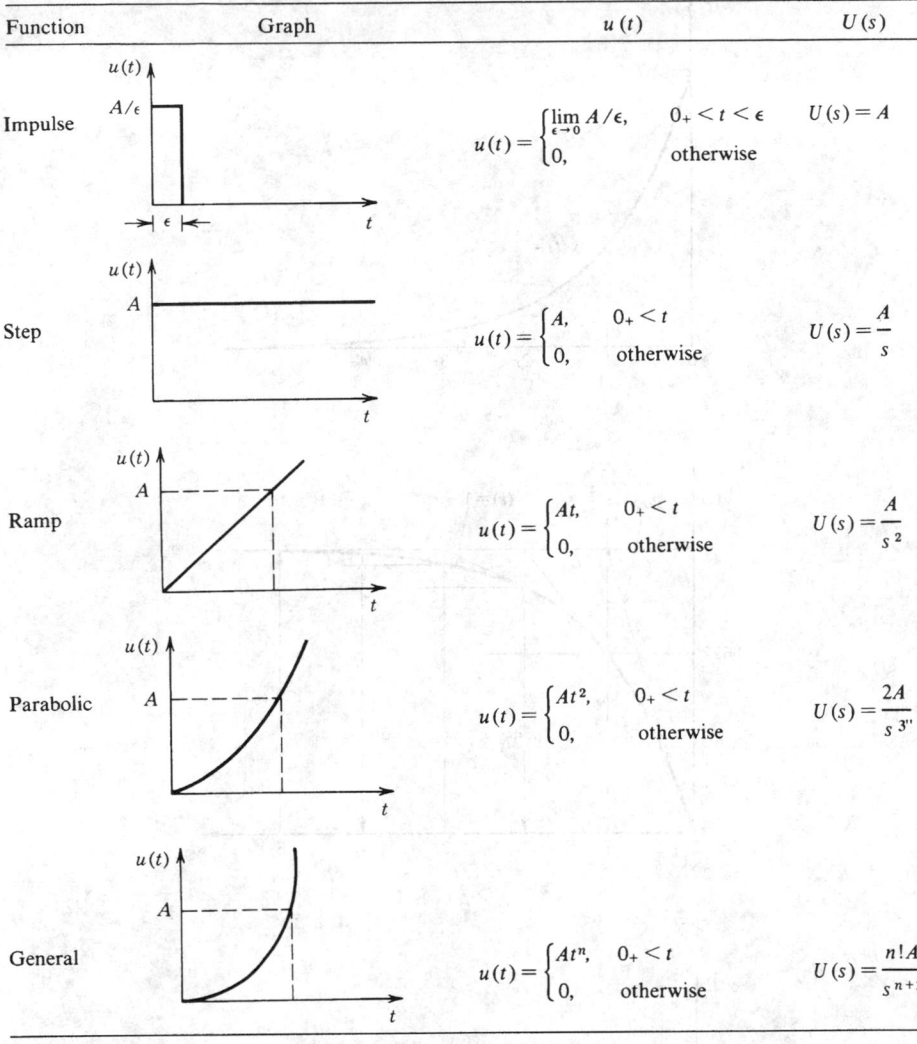

Fig. 27.14 Family of singularity functions commonly used as test inputs.

$$y(t + \tau) - y(t) \approx 0.63212[y_{ss} - y(t)]$$

Note also that the slope of the response at the beginning of any time interval of duration τ intersects the steady-state value y_{ss} at the end of the interval, that is,

$$\frac{dy}{dt}(t) = \frac{y_{ss} - y(t)}{\tau}$$

Finally, note that after an interval of four time constants, the response is within 98% of the steady-state value, that is,

$$y(4\tau) \approx 0.98168(y_{ss} - y_0)$$

For this reason, $T_s = 4\tau$ is called the (2%) *setting time*.

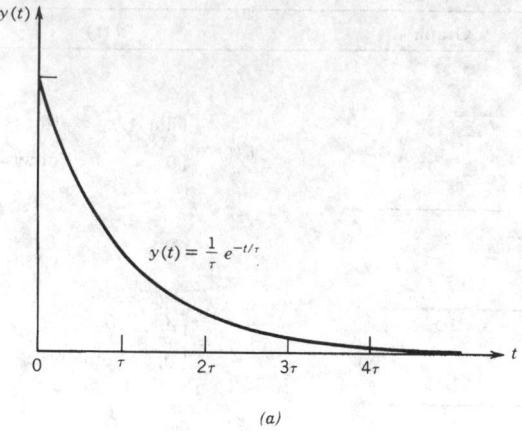

(a)

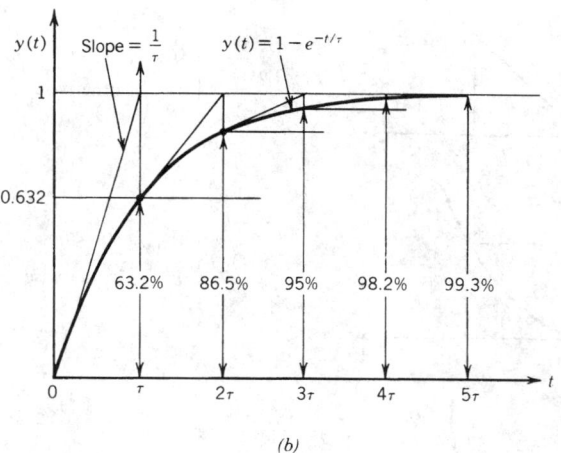

(b)

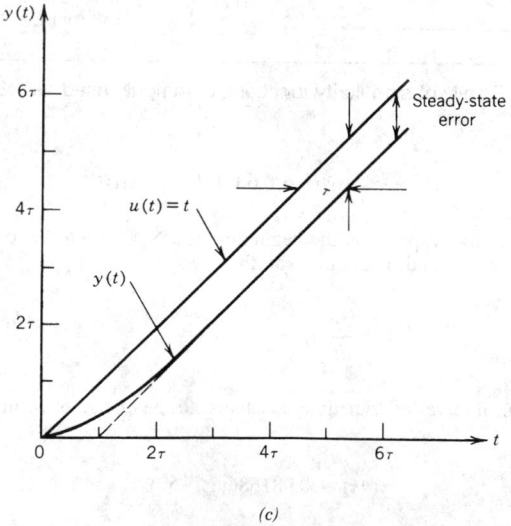

(c)

Fig. 27.15 Response of a first-order system to (a) unit impulse, (b) unit step, and (c) unit ramp inputs.

Table 27.6 Tabulated Values of the Response of a First-Order System to a Unit Step Input

t	$y(t)$	$\dot{y}(t)$
0	0	τ^{-1}
τ	0.632	$0.368\tau^{-1}$
2τ	0.865	$0.135\tau^{-1}$
3τ	0.950	$0.050\tau^{-1}$
4τ	0.982	$0.018\tau^{-1}$
5τ	0.993	$0.007\tau^{-1}$
6τ	0.998	$0.002\tau^{-1}$

Second-Order Transient Response

The standard form of the I/O equation for a second-order system is

$$\frac{d^2y}{dt^2} + 2\zeta\omega_n\frac{dy}{dt} + \omega_n^2 y(t) = \omega_n^2 u(t)$$

with transfer function

$$H(s) = \frac{Y(s)}{U(s)} = \frac{\omega_n^2}{s^2 + 2\zeta\omega_n s + \omega_n^2}$$

The system poles are obtained by applying the quadratic formula to the characteristic equation as

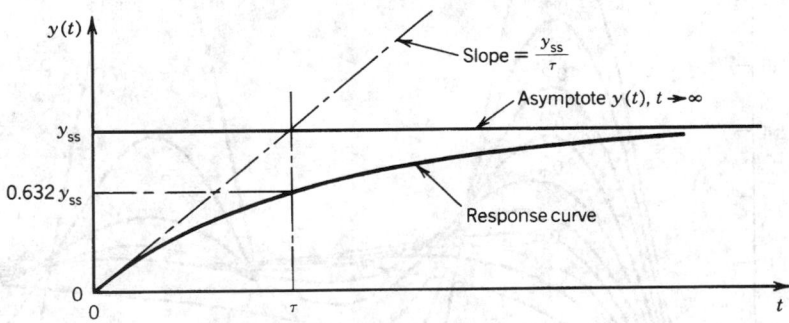

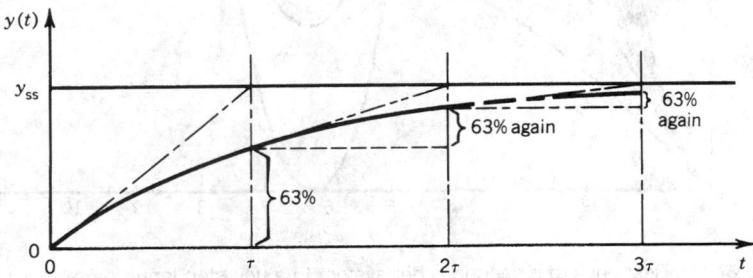

Fig. 27.16 Response of a first-order system to a unit step input, showing the relationship to the time constant.

$$p_{1,2} = -\zeta\omega_n \pm j\omega_n\sqrt{1 - \zeta^2}$$

where the following parameters are defined: ζ is the *damping ratio*, ω_n is the *natural frequency*, and $\omega_d = \omega_n\sqrt{1 - \zeta^2}$ is the *damped natural frequency*.

The nature of the response of the standard second-order system to a step input depends on the value of the damping ratio, as depicted in Fig. 27.17. For a stable system, four classes of response are defined.

1. *Overdamped Response* ($\zeta > 1$). The system poles are real and distinct. The response of the second-order system can be decomposed into the response of two cascaded first-order systems, as shown in Fig. 27.18.

2. *Critically Damped Response* ($\zeta = 1$). The system poles are real and repeated. This is the limiting case of overdamped response, where the response is as fast as possible without overshoot.

3. *Underdamped Response* ($1 > \zeta > 0$). The system poles are complex conjugates. The response oscillates at the damped frequency ω_d. The magnitude of the oscillations and the speed with which the oscillations decay depend on the damping ratio ζ.

4. *Harmonic Oscillation* ($\zeta = 0$). The system poles are pure imaginary numbers. The response oscillates at the natural frequency ω_n and the oscillations are undamped (i.e., the oscillations are sustained and do not decay).

The Complex s-Plane

The location of the system poles (roots of the characteristic equation) in the *complex s-plane* reveals the nature of the system response to test inputs. Figure 27.19 shows the relationship between the location of the poles in the complex plane and the parameters of the standard second-order system. Figure 27.20 shows the unit impulse response of a second-order system corresponding to various pole locations in the complex plane.

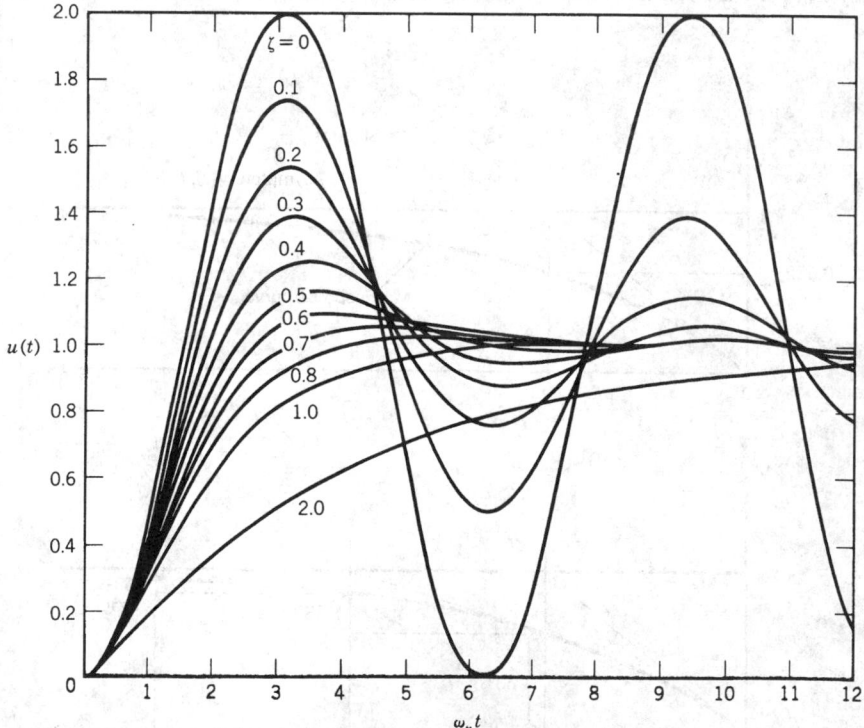

Fig. 27.17 Response of a second-order system to a unit step input for selected values of the damping ratio.

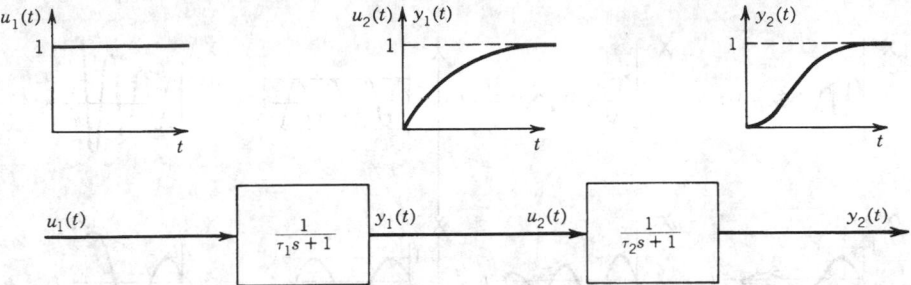

Fig. 27.18 Overdamped response of a second-order system decomposed into the responses of two first-order systems.

Transient Response of Higher-Order Systems

The response of third- and higher-order systems to test inputs is simply the sum of terms representing component first- and second-order responses. This is because the system poles must either be real, resulting in first-order terms, or complex, resulting in second-order underdamped terms. Furthermore, because the transients associated with those system poles having the largest real part decay the most slowly, these transients tend to dominate the output. The response of higher-order systems therefore tends to have the same form as the response to the *dominant poles*, with the response to the *subdominant poles* superimposed over it. Note that the larger the relative difference between the real parts of the dominant and subdominant poles, the more the output tends to resemble the dominant mode of response.

For example, consider a fixed linear third-order system. The system has three poles. The poles may either all be real, or one may be real while the other pair is complex conjugates. This leads to the three forms of step-response shown in Fig. 27.21, depending on the relative locations of the poles in the complex plane.

Transient Performance Measures

The transient response of a system is commonly described in terms of the measures defined in Table 27.7 and shown in Fig. 27.22. While these measures apply to any output, for a second-order system these can be calculated exactly in terms of the damping ratio and natural frequency, as shown in column three of the table. A common practice in control system design is to determine an initial design with dominant second-order poles that satisfy the performance specifications. Such a design can easily be calculated and then modified as necessary to achieve the desired performance.

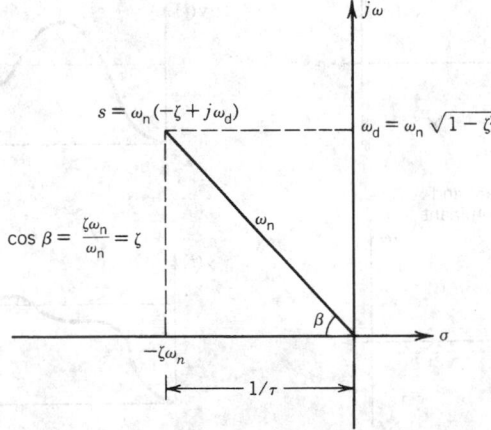

Fig. 27.19 Location of the upper complex pole in the s-plane in terms of the parameters of the standard second-order system.

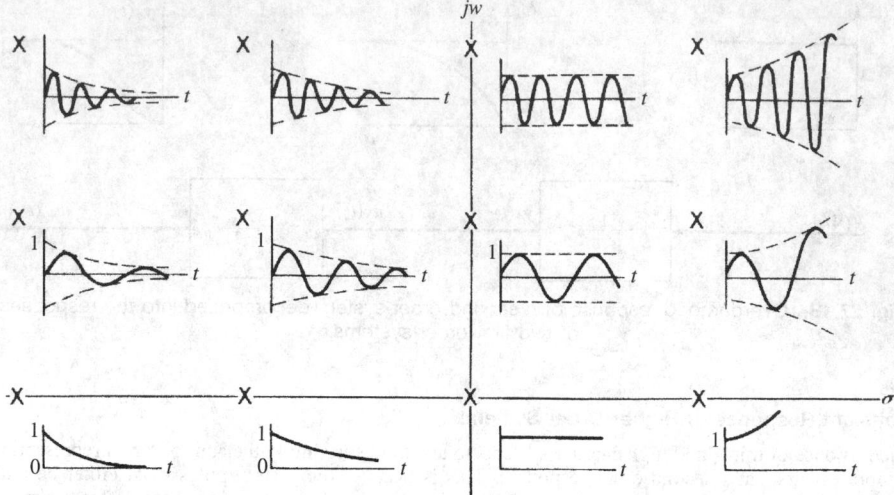

Fig. 27.20   Unit impulse response for selected upper complex pole locations in the s-plane.

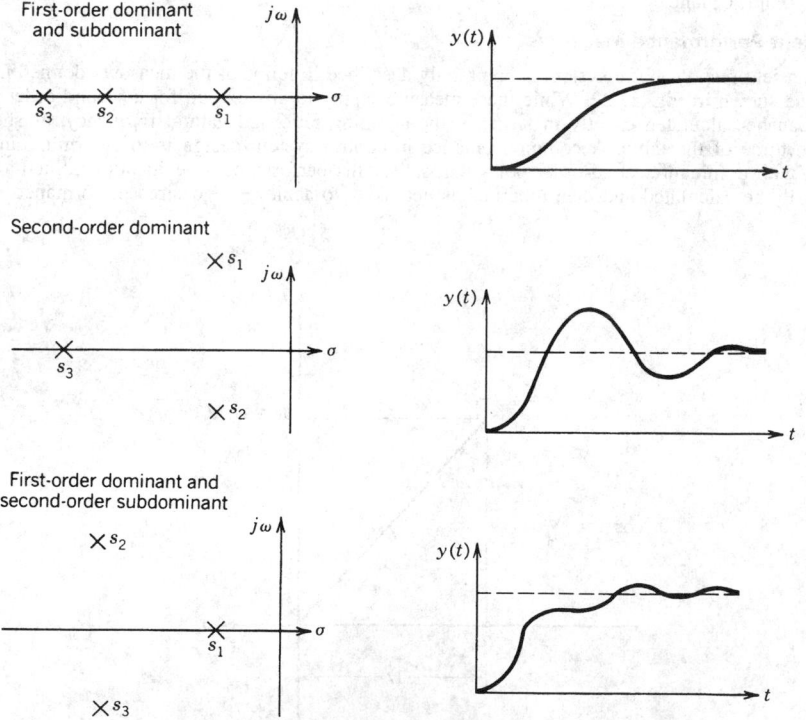

Fig. 27.21   Step response of a third-order system for alternative upper-complex pole locations in the s-plane.

Table 27.7 Transient Performance Measures Based Upon Step Response

Performance Measure	Definition	Formula for a Second-Order System
Delay time, t_d	Time required for the response to reach half the final value for the first time	
10–90% rise time, t_r	Time required for the response to rise from 10 to 90% of the final response (used for overdamped responses)	
0–100% rise time, t_r	Time required for the response to rise from 0 to 100% of the final response (used for underdamped responses)	$t_r = \dfrac{\pi - \beta}{\omega_d}$ where $\beta = \cos^{-1}\zeta$
Peak time, t_p	Time required for the response to reach the first peak of the overshoot	$t_p = \dfrac{\pi}{\omega_d}$
Maximum overshoot, M_p	The difference in the response between the first peak of the overshoot and the final response	$M_p = e^{-\zeta\pi/\sqrt{1 - \zeta^2}}$
Percent overshoot, PO	The ratio of maximum overshoot to the final response expressed as a percentage	$PO = 100e^{-\zeta\pi/\sqrt{1 - \zeta^2}}$
Setting time, t_s	The time required for the response to reach and stay within a specified band centered on the final response (usually a 2% or 5% of final response band)	$t_s = \dfrac{4}{\zeta\omega_n}$ (2% band) $t_s = \dfrac{3}{\zeta\omega_n}$ (5% band)

The Effect of Zeros on the Transient Response

Zeros arise in a system transfer function through the inclusion of one or more derivatives of $u(t)$ among the inputs to the system. By sensing the rate(s) of change of $u(t)$, the system in effect *anticipates* the future values of $u(t)$. This tends to increase the speed of response of the system relative to the input $u(t)$.

The effect of a zero is greatest on the modes of response associated with neighboring poles. For example, consider the second-order system represented by the transfer function

$$H(s) = K\frac{s - z}{(s - p_1)(s - p_2)}$$

If $z = p_1$, then the system responds as a first-order system with $\tau = -p_2^{-1}$; whereas if $z = p_2$, then the system responds as a first-order system with $\tau = -p_1^{-1}$. Such *pole-zero cancellation* can only be achieved mathematically, but it can be approximated in physical systems. Note that by diminishing the residue associated with the response mode having the larger time constant, the system responds more quickly to changes in the input, confirming our earlier observation.

27.5.3 Response to Periodic Inputs Using Transform Methods

The response of a dynamic system to periodic inputs can be a critical concern to the control engineer. An input $u(t)$ is *periodic* if $u(t + T) = u(t)$ for all time t, where T is a constant called the period. Periodic inputs are important because these are ubiquitous: rotating unbalanced machinery, reciprocating pumps and engines, ac electrical power, and a legion of noise and disturbance inputs can be approximated by periodic inputs. Sinusoids are the most important category of periodic inputs, because these are frequently occurring, easily analyzed, and form the basis for analysis of general periodic inputs.

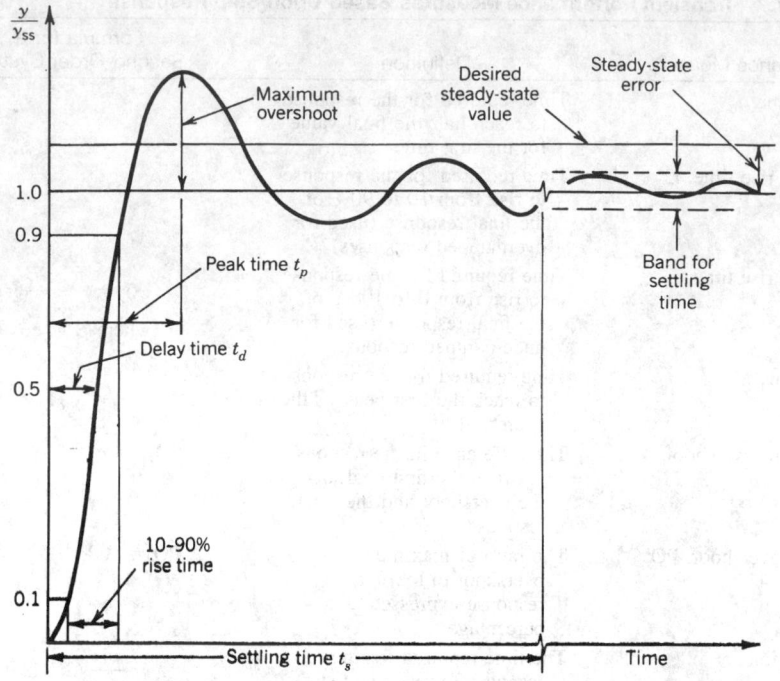

Fig. 27.22 Transient performance measures based on step response.

Frequency Response

The *frequency response* of a system is the steady-state response of the system to a sinusoidal input. For a linear system, the frequency response has the unique property that the response is a sinusoid of the same frequency as the input sinusoid, differing only in amplitude and phase. In addition, it is easy to show that the amplitude and phase of the response are functions of the input frequency, which are readily obtained from the system transfer function.

Consider a system defined by the transfer function $H(s)$. For an input

$$u(t) = A \sin \omega t$$

the corresponding steady-state output is

$$y_{ss}(t) = AM(\omega) \sin[\omega t + \phi(\omega)]$$

where $M(\omega) = |H(j\omega)|$ is called the *magnitude ratio*
$\phi(\omega) = \angle H(j\omega)$ is called the *phase angle*
$H(j\omega) = H(s)|_{s=j\omega}$ is called the *frequency transfer function*

The frequency transfer function is obtained by substituting $j\omega$ for s in the transfer function $H(s)$. If the complex quantity $H(j\omega)$ is written in terms of its real and imaginary parts as $H(j\omega) = \mathrm{Re}(\omega) + j\mathrm{Im}(\omega)$, then

$$M(\omega) = [\mathrm{Re}(\omega)^2 + \mathrm{Im}(\omega)^2]^{1/2}$$
$$\phi(\omega) = \tan^{-1}[\mathrm{Im}(\omega)/\mathrm{Re}(\omega)]$$

and in polar form

$$H(j\omega) = M(\omega)e^{j\phi(\omega)}$$

Frequency Response Plots

The frequency response of a fixed linear system is typically represented graphically, using one of three types of frequency response plots. A *polar plot* is simply a plot of the vector $H(j\omega)$ in the complex plane, where $\text{Re}(\omega)$ is the abscissa and $\text{Im}(\omega)$ is the ordinate. A *logarithmic plot* or *Bode diagram* consists of two displays: (1) the magnitude ratio in decibels $M_{db}(\omega)$ [where $M_{db}(\omega) = 20 \log M(\omega)$] versus $\log \omega$, and (2) the phase angle in degrees $\phi(\omega)$ versus $\log \omega$. Bode diagrams for normalized first- and second-order systems are given in Fig. 27.23. Bode diagrams for higher-order systems are obtained by adding these first- and second-order terms, appropriately scaled. A *Nichols diagram* can be obtained by cross plotting the Bode magnitude and phase diagrams, eliminating $\log \omega$. Polar plots and Bode and Nichols diagrams for common transfer functions are given in Table 27.8.

Frequency Response Performance Measures

Frequency response plots show that dynamic systems tend to behave like *filters*, "passing" or even amplifying certain ranges of input frequencies, while blocking or attenuating other frequency ranges. The range of frequencies for which the amplitude ratio is no less than 3 db of its maximum value is called the *bandwidth* of the system. The bandwidth is defined by upper and lower *cutoff frequencies* ω_c, or by $\omega = 0$ and an upper cutoff frequency if $M(0)$ is the maximum amplitude ratio. Although the choice of "down 3 db" used to define the cutoff frequencies is somewhat arbitrary, the bandwidth is usually taken to be a measure of the range of frequencies for which a significant portion of the input is felt in the system output. The bandwidth is also taken to be a measure of the system speed of response, since attenuation of inputs in the higher-frequency ranges generally results from the inability of the system to "follow" rapid changes in amplitude. Thus, a narrow bandwidth generally indicates a sluggish system response.

Response to General Periodic Inputs

The *Fourier series* provides a means for representing a general periodic input as the sum of a constant and terms containing sine and cosine. For this reason the *Fourier series,* together with the superposition principle for linear systems, extends the results of frequency response analysis to the general case of arbitrary periodic inputs. The Fourier series representation of a periodic function $f(t)$ with period $2T$ on the interval $t^* + 2T \ge t \ge t^*$ is

$$f(t) = \frac{a_0}{2} + \sum_{n=1}^{\infty} \left(a_n \cos \frac{n\pi t}{T} + b_n \sin \frac{n\pi t}{T} \right)$$

where

$$a_n = \frac{1}{T} \int_{t^*}^{t^*+2T} f(t) \cos \frac{n\pi t}{T} \, dt$$

$$b_n = \frac{1}{T} \int_{t^*}^{t^*+2T} f(t) \sin \frac{n\pi t}{T} \, dt$$

If $f(t)$ is defined outside the specified interval by a periodic extension of period $2T$, and if $f(t)$ and its first derivative are piecewise continuous, then the series converges to $f(t)$ if t is a point of continuity, or to $\frac{1}{2} [f(t_+) + f(t_-)]$ if t is a point of discontinuity. Note that while the Fourier series in general is infinite, the notion of bandwidth can be used to reduce the number of terms required for a reasonable approximation.

27.6 STATE-VARIABLE METHODS

State-variable methods use the vector state and output equations introduced in Section 27.4 for analysis of dynamic systems directly in the time domain. These methods have several advantages over transform methods. First, state-variable methods are particularly advantageous for the study of multivariable (multiple input/multiple output) systems. Second, state-variable methods are more naturally extended for the study of linear time-varying and nonlinear systems. Finally, state-variable methods are readily adapted to computer simulation studies.

27.6.1 Solution of the State Equation

Consider the vector equation of state for a fixed linear system:

$$\dot{x}(t) = Ax(t) + Bu(t)$$

The solution to this system is

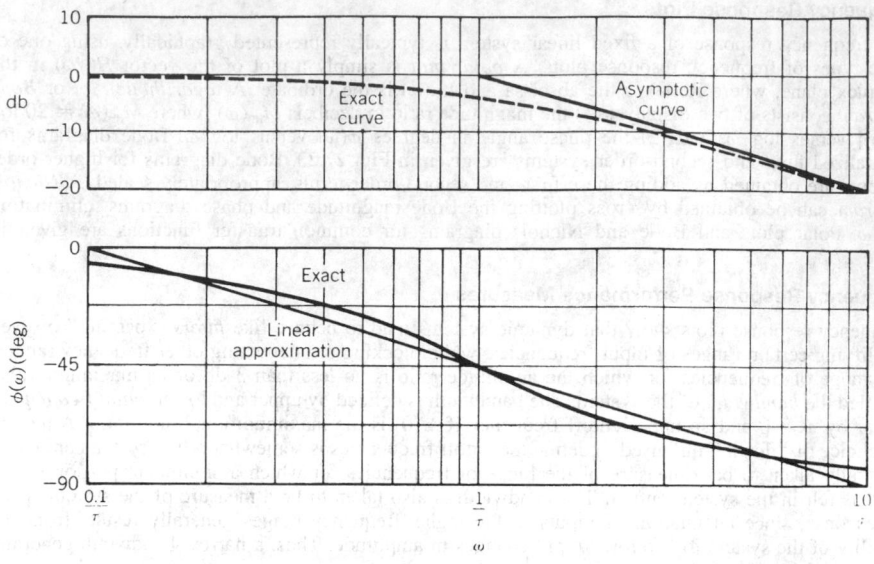

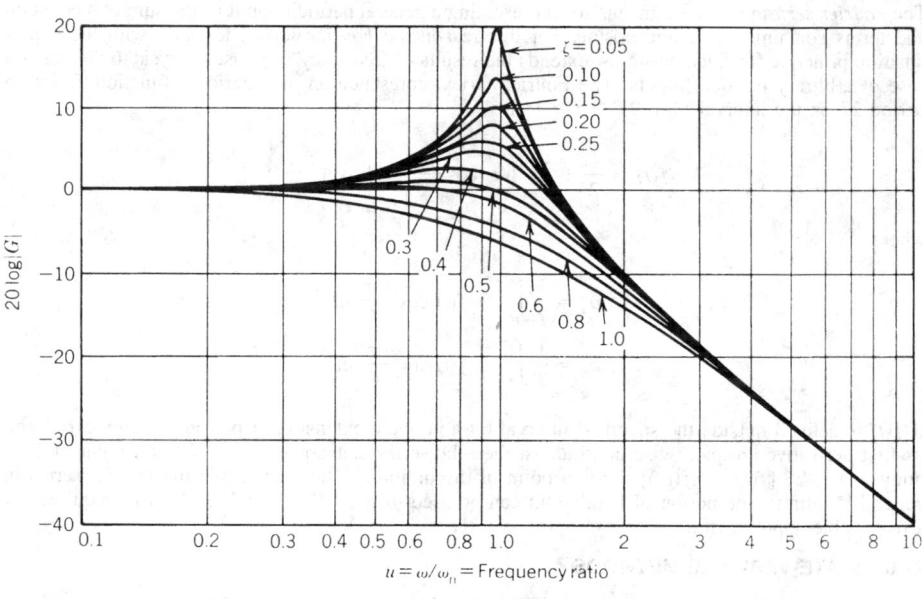

$$u = \omega/\omega_n = \text{Frequency ratio}$$

(a)

Fig. 27.23 Bode diagrams for normalized (a) first-order and (b) second-order systems.

$$x(t) = \Phi(t)x(0) + \int_0^t \Phi(t - \tau)Bu(\tau)\, d\tau$$

where the matrix $\Phi(t)$ is called the *state-transition matrix*. The state-transition matrix represents the free response of the system and is defined by the matrix exponential series

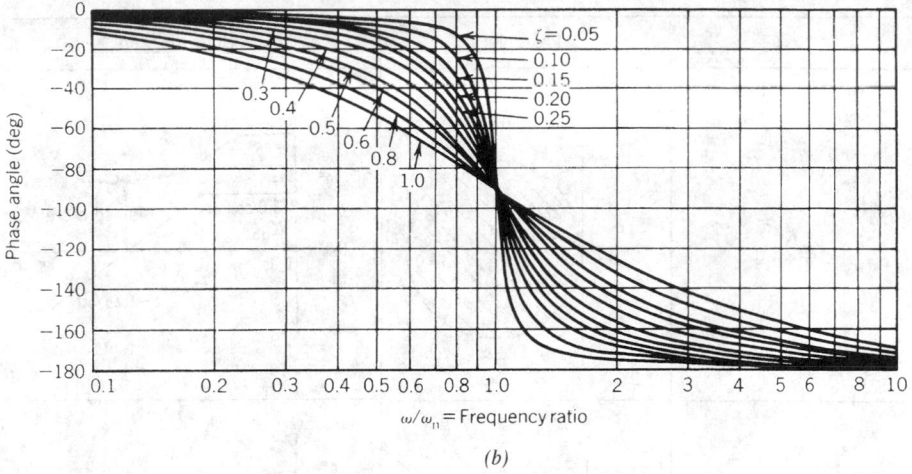

(b)

Fig. 27.23   (Continued)

$$\Phi(t) = e^{At} = I + At + \frac{1}{2!} A^2 t^2 + \cdots = \sum_{k=0}^{\infty} \frac{1}{k!} A^k t^k$$

where I is the identity matrix. The state transition matrix has the following useful properties:

$$\Phi(0) = I$$
$$\Phi^{-1}(t) = \Phi(-t)$$
$$\Phi^k(t) = \Phi(kt)$$
$$\Phi(t_1 + t_2) = \Phi(t_1)\Phi(t_2)$$
$$\Phi(t_2 - t_1)\Phi(t_1 - t_0) = \Phi(t_2 - t_0)$$
$$\dot{\Phi}(t) = A\Phi(t)$$

The Laplace transform of the state equation is

$$sX(s) - x(0) = AX(s) + BU(s)$$

The solution to the fixed linear system therefore can be written as

$$x(t) = \mathcal{L}^{-1}[X(s)]$$
$$= \mathcal{L}^{-1}[\Phi(s)]x(0) + \mathcal{L}^{-1}[\Phi(s)BU(s)]$$

where $\Phi(s)$ is called the *resolvent matrix* and

$$\Phi(t) = \mathcal{L}^{-1}[\Phi(s)] = \mathcal{L}^{-1}[sI - A]^{-1}$$

27.6.2   Eigenstructure

The internal structure of a system (and therefore its free response) is defined entirely by the system matrix A. The concept of matrix *eigenstructure,* as defined by the eigenvalues and eigenvectors of the system matrix, can provide a great deal of insight into the fundamental behavior of a system. In particular, the system eigenvectors can be shown to define a special set of first-order subsystems embedded within the system. These subsystems behave independently of one another, a fact that greatly simplifies analysis.

System Eigenvalues and Eigenvectors

For a system with system matrix A, the system *eigenvectors* v_i and associated *eigenvalues* λ_i are defined by the equation

Table 27.8   Transfer Function Plots for Representative Transfer Functions[5]

$G(s)$	Polar plot	Bode diagram
1. $\dfrac{K}{s\tau_1 + 1}$	-1 $\omega = \infty$ $\omega = 0$ $-\omega$ $+\omega$	$0°$ $-45°$ $-90°$ ϕ M 0 db/oct K_{db} $-180°$ 0 db $\frac{1}{\tau_1}$ Phase margin -6 db/oct log ω
2. $\dfrac{K}{(s\tau_1 + 1)(s\tau_2 + 1)}$	-1 $\omega = \infty$ $\omega = 0$ $-\omega$ $+\omega$	$0°$ ϕ M 0 $-180°$ 0 db $\frac{1}{\tau_1}$ $\frac{1}{\tau_2}$ -6 Phase margin -12 db/oct log ω
3. $\dfrac{K}{(s\tau_1 + 1)(s\tau_2 + 1)(s\tau_3 + 1)}$	-1 $\omega = \infty$ $\omega = 0$ $-\omega$ $+\omega$	$0°$ ϕ 0 M -6 Gain margin $-180°$ 0 db $\frac{1}{\tau_1}$ $\frac{1}{\tau_2}$ $\frac{1}{\tau_3}$ -12 db/oct $-270°$ Phase margin -18 db/oct log ω
4. $\dfrac{K}{s}$	$\omega = 0$ $-\omega$ -1 $\omega = \infty$ $+\omega$ $\omega \to 0$	$-90°$ ϕ M Phase margin $-180°$ 0 db log ω -6 db/oct

Table 27.8 (*Continued*)

Nichols diagram	Root locus	Comments
		Stable; gain margin $= \infty$
		Elementary regulator; stable; gain margin $= \infty$
		Regulator with additional energy-storage component; unstable, but can be made stable by reducing gain
		Ideal integrator; stable

Table 27.8 *(Continued)*

$G(s)$	Polar plot	Bode diagram
5. $\dfrac{K}{s(s\tau_1 + 1)}$		
6. $\dfrac{K}{s(s\tau_1 + 1)(s\tau_2 + 1)}$		
7. $\dfrac{K(s\tau_a + 1)}{s(s\tau_1 + 1)(s\tau_2 + 1)}$		
8. $\dfrac{K}{s^2}$		

Table 27.8 *(Continued)*

Nichols diagram	Root locus	Comments
M, ω, Phase margin, 0 db, $-180°$, $-90°$, ϕ, $\omega \to \infty$	ω, r_1, $-\frac{1}{T_1}$, σ, r_2	Elementary instrument servo; inherently stable; gain margin $= \infty$
Phase margin, M, ω, 0 db, $-180°$, $-90°$, ϕ, Gain margin, $\omega \to \infty$	ω, r_1, r_3, $-\frac{1}{T_2}$, $-\frac{1}{T_1}$, σ, r_2	Instrument servo with field-control motor or power servo with elementary Ward-Leonard drive; stable as shown, but may become unstable with increased gain
M, ω, Phase margin, 0 db, $-180°$, $-90°$, ϕ, $\omega \to \infty$	ω, r_1, r_3, $-\frac{1}{T_2}$, $-\frac{1}{T_a}$, $-\frac{1}{T_1}$, σ, r_2	Elementary instrument servo with phase-lead (derivative) compensator; stable
M, ω, Phase margin $=0$, 0 db, ϕ, $-270°$, $-180°$, $-90°$, $\omega \to \infty$	ω, r_1, Double pole, σ, r_2	Inherently unstable; must be compensated

Table 27.8   (Continued)

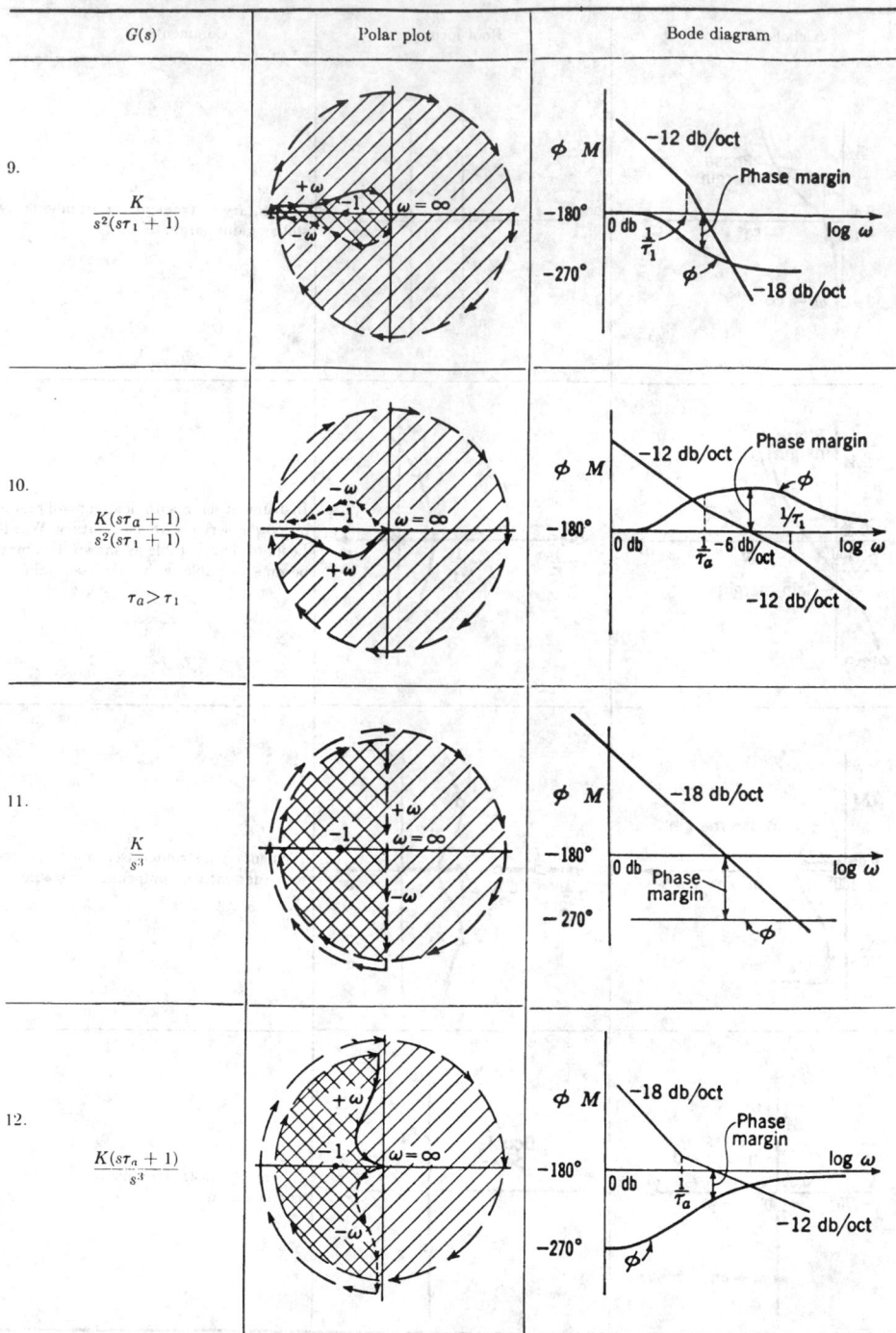

$G(s)$	Polar plot	Bode diagram
9. $\dfrac{K}{s^2(s\tau_1 + 1)}$		
10. $\dfrac{K(s\tau_a + 1)}{s^2(s\tau_1 + 1)}$ $\tau_a > \tau_1$		
11. $\dfrac{K}{s^3}$		
12. $\dfrac{K(s\tau_a + 1)}{s^3}$		

Table 27.8 *(Continued)*

Nichols diagram	Root locus	Comments
		Inherently unstable; must be compensated
		Stable for all gains
		Inherently unstable
		Inherently unstable

Table 27.8   *(Continued)*

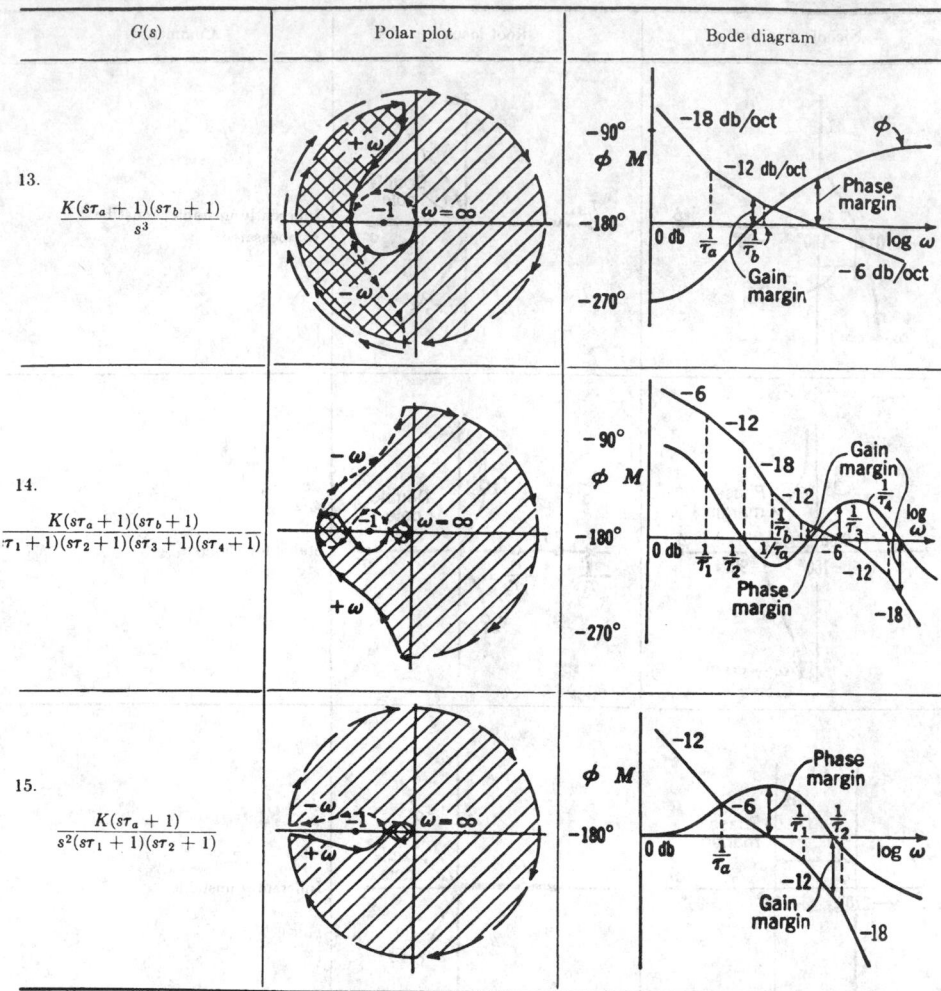

	$G(s)$	Polar plot	Bode diagram
13.	$\dfrac{K(s\tau_a + 1)(s\tau_b + 1)}{s^3}$		
14.	$\dfrac{K(s\tau_a + 1)(s\tau_b + 1)}{(s\tau_1 + 1)(s\tau_2 + 1)(s\tau_3 + 1)(s\tau_4 + 1)}$		
15.	$\dfrac{K(s\tau_a + 1)}{s^2(s\tau_1 + 1)(s\tau_2 + 1)}$		

$$Av_i = \lambda_i v_i$$

Note that the eigenvectors represent a set of special directions in the state space. If the state vector is aligned in one of these directions, then the homogeneous state equation becomes $\dot{v}_i = Av_i = \lambda v_i$, implying that each of the state variables changes at the *same* rate determined by the eigenvalue λ_i. This further implies that, in the absence of inputs to the system, a state vector that becomes aligned with a eigenvector will remain aligned with that eigenvector.

The system eigenvalues are calculated by solving the nth-order polynomial equation

$$|\lambda I - A| = \lambda^n + a_{n-1}\lambda^{n-1} + \cdots + a_1\lambda + a_0 = 0$$

This equation is called the *characteristic equation*. Thus the system eigenvalues are the roots of the characteristic equation, that is, the system eigenvalues are identically the system poles defined in transform analysis.

Each system eigenvector is determined by substituting the corresponding eigenvalue into the defining equation and then solving the resulting set of simultaneous linear equations. Only $n - 1$ of the n components of any eigenvector are independently defined, however. In other words, the magnitude of an eigenvector is arbitrary, and the eigenvector describes a direction in the state space.

Table 27.8 *(Continued)*

Nichols diagram	Root locus	Comments
		Conditionally stable; becomes unstable if gain is too low
		Conditionally stable; stable at low gain, becomes unstable as gain is raised, again becomes stable as gain is further increased, and becomes unstable for very high gains
		Conditionally stable; becomes unstable at high gain

Diagonalized Canonical Form

There will be one linearly independent eigenvector for each distinct (nonrepeated) eigenvalue. If all of the eigenvalues of an nth-order system are distinct, then the n independent eigenvectors form a new basis for the state space. This basis represents new coordinate axes defining a set of state variables $z_i(t)$, $i = 1, 2, \ldots, n$, called the *diagonalized canonical variables*. In terms of the diagonalized variables, the homogeneous state equation is

$$\dot{z}(t) = \Lambda z$$

where Λ is a diagonal system matrix of the eigenvectors, that is,

$$\Lambda = \begin{bmatrix} \lambda_1 & 0 & \cdots & 0 \\ 0 & \lambda_2 & \cdots & 0 \\ \vdots & \vdots & \ddots & \vdots \\ 0 & 0 & \vdots & \lambda_n \end{bmatrix}$$

The solution to the diagonalized homogeneous system is

$$z(t) = e^{\Lambda t}z(0)$$

where $e^{\Lambda t}$ is the diagonal state-transition matrix

$$e^{\Lambda t} = \begin{bmatrix} e^{\lambda_1 t} & 0 & \cdots & 0 \\ 0 & e^{\lambda_2 t} & \cdots & 0 \\ \vdots & \vdots & \ddots & \vdots \\ 0 & 0 & \cdots & e^{\lambda_n t} \end{bmatrix}$$

Modal Matrix

Consider the state equation of the nth-order system

$$\dot{x}(t) = Ax(t) + Bu(t)$$

which has real, distinct eigenvalues. Since the system has a full set of eigenvectors, the state vector $x(t)$ can be expressed in terms of the canonical state variables as

$$x(t) = v_1 z_1(t) + v_2 z_2(t) + \cdots + v_n z_n(t) = Mz(t)$$

where M is the $n \times n$ matrix whose columns are the eigenvectors of A, called the *modal matrix*. Using the modal matrix, the state-transition matrix for the original system can be written as

$$\Phi(t) = e^{\Lambda t} = Me^{\Lambda t}M^{-1}$$

where $e^{\Lambda t}$ is the diagonal state-transition matrix. This frequently proves to be an attractive method for determining the state-transition matrix of a system with real, distinct eigenvalues.

Jordan Canonical Form

For a system with one or more repeated eigenvalues, there is not in general a full set of eigenvectors. In this case, it is not possible to determine a diagonal representation for the system. Instead, the simplest representation that can be achieved is block diagonal. Let $L_k(\lambda)$ be the $k \times k$ matrix

$$L_k(\lambda) = \begin{bmatrix} \lambda & 1 & 0 & \cdots & 0 \\ 0 & \lambda & 1 & \cdots & 0 \\ \vdots & \vdots & \lambda & \ddots & 0 \\ \vdots & \vdots & \vdots & \ddots & 1 \\ 0 & 0 & 0 & 0 & \lambda \end{bmatrix}$$

Then for any $n \times n$ system matrix A there is certain to exist a nonsingular matrix T such that

$$T^{-1}AT = \begin{bmatrix} L_{k_1}(\lambda_1) & & & \\ & L_{k_2}(\lambda_2) & & \\ & & \ddots & \\ & & & L_{k_r}(\lambda_r) \end{bmatrix}$$

where $k_1 + k_2 + \cdots + k_r = n$ and λ_i, $i = 1, 2, \ldots, r$, are the (not necessarily distinct) eigenvalues of A. The matrix $T^{-1}AT$ is called the *Jordan canonical form*.

27.7 SIMULATION

27.7.1 Simulation—Experimental Analysis of Model Behavior

Closed-form solutions for nonlinear or time-varying systems are rarely available. In addition, while explicit solutions for time-invariant linear systems can always be found, for high-order systems this is often impractical. In such cases it may be convenient to study the dynamic behavior of the system using *simulation*.

Simulation is the *experimental* analysis of model behavior. A *simulation run* is a controlled experiment in which a specific realization of the model is manipulated in order to determine the response associated with that realization. A *simulation study* comprises *multiple runs*, each run for a different combination of model parameter values and/or initial conditions. The generalized solution of the model must then be inferred from a finite number of simulated data points.

Simulation is almost always carried out with the assistance of computing equipment. *Digital simulation* involves the *numerical solution* of model equations using a digital computer. *Analog simulation* involves solving model equations by analogy with the behavior of a physical system using an analog computer. *Hybrid simulation* employs digital and analog simulation together using a hybrid (part digital and part analog) computer.

27.7.2 Digital Simulation

Digital continuous-system simulation involves the approximate solution of a state-variable model over successive time steps. Consider the general state-variable equation

$$\dot{x}(t) = f[x(t), u(t)]$$

to be simulated over the time interval $t_0 \le t \le t_K$. The solution to this problem is based on the repeated solution of the single-variable, single-step subproblem depicted in Fig. 27.24. The subproblem may be stated formally as follows:

Given:

1. $\Delta t(k) = t_k - t_{k-1}$, the length of the kth *time step*.
2. $x_i(t) = f_i[x(t), u(t)]$ for $t_{k-1} \le t \le t_k$, the ith equation of state defined for the state variable $x_i(t)$ over the kth time step.
3. $u(t)$ for $t_{k-1} \le t \le t_k$, the input vector defined for the kth time step.
4. $\tilde{x}(k - 1) \simeq x(t_{k-1})$, an initial approximation for the state vector at the beginning of the time step.

Find:

5. $\tilde{x}_i(k) \simeq x_i(t_k)$, a final approximation for the state variable $x_i(t)$ at the end of the kth time step.

Solving this single-variable, single-step subproblem for each of the state variables $x_i(t)$, $i = 1, 2, \ldots, n$, yields a final approximation for the state vector $\tilde{x}(k) \simeq x(t_k)$ at the end of the kth time step. Solving the complete single-step problem K times over K time steps, beginning with the initial condition $\tilde{x}(0) = x(t_0)$ and using the final value of $\tilde{x}(t_k)$ from the kth time step as the initial value of the state for the $(k + 1)$st time step, yields a discrete succession of approximations $\tilde{x}(1) \simeq x(t_1)$, $\tilde{x}(2) \simeq x(t_2), \ldots, \tilde{x}(K) \simeq x(t_k)$ spanning the solution time interval.

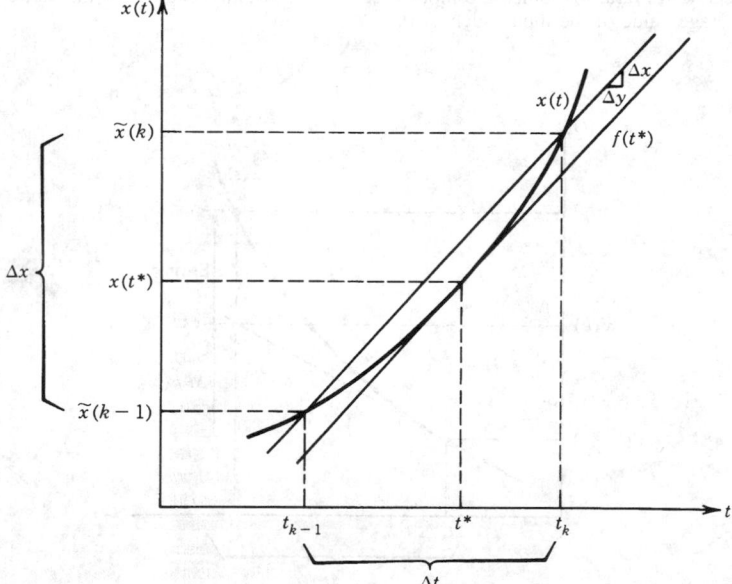

Fig. 27.24 Numerical approximation of a single variable over a single time step.

The basic procedure for completing the single-variable, single-step problem is the same regardless of the particular integration method chosen. It consists of two parts: (1) calculation of the average value of the ith derivative over the time step as

$$\dot{x}_i(t^*) = f_i[x(t^*), u(t^*)] = \frac{\Delta x_i(k)}{\Delta t(k)} \simeq \tilde{f}_i(k)$$

and (2) calculation of the final value of the simulated variable at the end of the time step as

$$\tilde{x}_i(k) = \tilde{x}_i(k - 1) + \Delta x_i(k)$$
$$\simeq \tilde{x}_i(k - 1) + \Delta t(k)\tilde{f}_i(k)$$

If the function $f_i[x(t), u(t)]$ is continuous, then t^* is guaranteed to be on the time step, that is, $t_{k-1} \leq t^* \leq t_k$. Since the value of t^* is otherwise unknown, however, the value of $x(t^*)$ can only be approximated as $\tilde{f}(k)$.

Different *numerical integration* methods are distinguished by the means used to calculate the approximation $f_i(k)$. A wide variety of such methods is available for digital simulation of dynamic systems. The choice of a particular method depends on the nature of the model being simulated, the accuracy required in the simulated data, and the computing effort available for the simulation study. Several popular classes of integration methods are outlined in the following subsections.

Euler Method

The simplest procedure for numerical integration is the Euler method. The standard Euler method approximates the average value of the ith derivative over the kth time step using the derivative evaluated at the beginning of the time step, that is,

$$\tilde{f}_i(k) = f_i[\tilde{x}(k - 1), u(t_{k-1})] \simeq f_i(t_{k-1})$$

$i = 1, 2, \ldots, n$ and $k = 1, 2, \ldots, K$. This is shown geometrically in Fig. 27.25 for the scalar single-step case. A modification of this method uses the newly calculated state variables in the derivative calculation as these new values become available. Assuming the state variables are computed in numerical order according to the subscripts, this implies

$$\tilde{f}_i(k) = f_i[\tilde{x}_1(k), \ldots, \tilde{x}_{i-1}(k), \tilde{x}_i(k - 1), \ldots, \tilde{x}_n(k - 1), u(t_{k-1})]$$

The modified Euler method is modestly more efficient than the standard procedure and, frequently, is more accurate. In addition, since the input vector $u(t)$ is usually known for the entire time step, using an average value of the input, such as

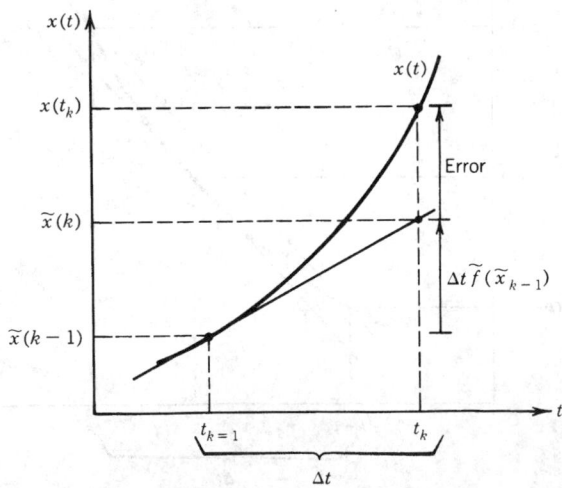

Fig. 27.25 Geometric interpretation of the Euler method for numerical integration.

$$u(k) = \frac{1}{\Delta t(k)} \int_{t_{k-1}}^{t_k} u(\tau) \, d\tau$$

frequently leads to a superior approximation of $\tilde{f}_i(k)$.

The Euler method requires the least amount of computational effort per time step of any numerical integration scheme. Local truncation error is proportional to Δt^2, however, which means that the error within each time step is highly sensitive to step size. Because the accuracy of the method demands very small time steps, the number of time steps required to implement the method successfully can be large relative to other methods. This can imply a large computational overhead and can lead to inaccuracies through the accumulation of roundoff error at each step.

Runge–Kutta Methods

Runge–Kutta methods precompute two or more values of $f_i[x(t), u(t)]$ in the time step $t_{k-1} \le t \le t_k$ and use some weighted average of these values to calculate $\tilde{f}_i(k)$. The *order* of a Runge–Kutta method refers to the number of derivative terms (or *derivative calls*) used in the scalar single-step calculation. A Runge–Kutta routine of order N therefore uses the approximation

$$\tilde{f}_i(k) = \sum_{j=1}^{N} w_j f_{ij}(k)$$

where the N approximations to the derivative are

$$f_{i1}(k) = f_i[\tilde{x}(k-1), u(t_{k-1})]$$

(the Euler approximation) and

$$f_{ij} = f_i\left[\tilde{x}(k-1) + \Delta t \sum_{t=1}^{j-1} Ib_{jt}f_{it}, u\left(t_{k-1} + \Delta t \sum_{t=1}^{j-1} b_{jt}\right)\right]$$

where I is the identity matrix. The weighting coefficients w_j and b_{jt} are not unique, but are selected such that the error in the approximation is zero when $x_i(t)$ is some specified Nth-degree polynomial in t. Coefficients commonly used for Runge–Kutta integration are given in Table 27.9.

Among the most popular of the Runge–Kutta methods is fourth-order Runge–Kutta. Using the defining equations for $N = 4$ and the weighting coefficients from Table 27.9 yields the derivative approximation

$$\tilde{f}_i(k) = \frac{1}{6}[f_{i1}(k) + 2f_{i2}(k) + 2f_{i3}(k) + f_{i4}(k)]$$

based on the four derivative calls

Table 27.9 Coefficients Commonly Used for Runge–Kutta Numerical Integration[6]

Common Name	N	b_{jl}	w_j
Open or explicit Euler	1	All zero	$w_1 = 1$
Improved polygon	2	$b_{21} = \frac{1}{2}$	$w_1 = 0$
			$w_2 = 1$
Modified Euler or Heun's method	2	$b_{21} = 1$	$w_1 = \frac{1}{2}$
			$w_2 = \frac{1}{2}$
Third-order Runge–Kutta	3	$b_{21} = \frac{1}{2}$	$w_1 = \frac{1}{6}$
		$b_{31} = -1$	$w_2 = \frac{2}{3}$
		$b_{32} = 2$	$w_3 = \frac{1}{6}$
Fourth-order Runge–Kutta	4	$b_{21} = \frac{1}{2}$	$w_1 = \frac{1}{6}$
		$b_{31} = 0$	$w_2 = \frac{1}{3}$
		$b_{32} = \frac{1}{2}$	$w_3 = \frac{1}{3}$
		$b_{43} = 1$	$w_4 = \frac{1}{6}$

$$f_{i1}(k) = f_i[\tilde{x}(k - 1), u(t_{k-1})]$$

$$f_{i2}(k) = f_i\left[\tilde{x}(k - 1) + \frac{\Delta t}{2} If_{i1}, u\left(t_{k-1} + \frac{\Delta t}{2}\right)\right]$$

$$f_{i3}(k) = f_i\left[\tilde{x}(k - 1) + \frac{\Delta t}{2} If_{i2}, u\left(t_{k-1} + \frac{\Delta t}{2}\right)\right]$$

$$f_{i4}(k) = f_i[\tilde{x}(k - 1) + \Delta t\, If_{i3}, u(t_k)]$$

where I is the identity matrix.

Because Runge–Kutta formulas are designed to be exact for a polynomial of order N, local truncation error is of the order Δt^{N+1}. This considerable improvement over the Euler method means that comparable accuracy can be achieved for larger step sizes. The penalty is that N derivative calls are required for each scalar evaluation within each time step.

Euler and Runge–Kutta methods are examples of *single-step methods* for numerical integration, so-called because the state $x(k)$ is calculated from knowledge of the state $x(k - 1)$, without requiring knowledge of the state at any time prior to the beginning of the current time step. These methods are also referred to as *self-starting methods*, since calculations may proceed from any known state.

Multistep Methods

Multistep methods differ from the single-step methods previously described in that multistep methods use the stored values of two or more previously computed states and/or derivatives in order to compute the derivative approximation $\tilde{f}_i(k)$ for the current time step. The advantage of multistep methods over Runge–Kutta methods is that these require only one derivative call for each state variable at each time step for comparable accuracy. The disadvantage is that multistep methods are not self-starting, since calculations cannot proceed from the initial state alone. Multistep methods must be started, or restarted in the case of discontinuous derivatives, using a single-step method to calculate the first several steps.

The most popular of the multistep methods are the *Adams–Bashforth predictor methods* and the *Adams–Moulton corrector methods.* These methods use the derivative approximation

$$\tilde{f}_i(k) = \sum_{j=0}^{N} b_j f_i[\tilde{x}(k - j), u(k - j)]$$

where the b_j are weighting coefficients. These coefficients are selected such that the error in the approximation is zero when $x_i(t)$ is a specified polynomial. Table 27.10 gives the values of the weighting coefficients for several Adams–Bashforth–Moulton rules. Note that the predictor methods employ an *open* or *explicit rule,* since for these methods $b_0 = 0$ and a prior estimate of $x_i(k)$ is not required. The corrector methods use a *closed* or *implicit rule,* since for these methods $b_i \neq 0$ and a prior estimate of $x_i(k)$ is required. Note also that for all of these methods $\sum_{j=0}^{N} b_j = 1$, ensuring unity gain for the integration of a constant.

Predictor–Corrector Methods

Predictor–corrector methods use one of the multistep predictor equations to provide an initial estimate (or "prediction") of $x(k)$. This initial estimate is then used with one of the multistep corrector equations to provide a second and improved (or "corrected") estimate of $x(k)$, before proceeding to

Table 27.10 Coefficients Commonly Used for Adams–Bashforth–Moulton Numerical Integration[6]

Common Name	Predictor or Corrector	Points	b_{-1}	b_0	b_1	b_2	b_3
Open or explicit Euler	Predictor	1	0	1	0	0	0
Open trapezoidal	Predictor	2	0	$3/2$	$-1/2$	0	0
Adams three-point predictor	Predictor	3	0	$23/12$	$-16/12$	$5/12$	0
Adams four-point predictor	Predictor	4	0	$55/24$	$-59/24$	$37/24$	$-9/24$
Closed or implicit Euler	Corrector	1	1	0	0	0	0
Closed trapezoidal	Corrector	2	$1/2$	$1/2$	0	0	0
Adams three-point corrector	Corrector	3	$5/12$	$8/12$	$-1/12$	0	0
Adams four-point corrector	Corrector	4	$9/24$	$19/24$	$-5/24$	$1/24$	0

the next step. A popular choice is the four-point Adams–Bashforth predictor together with the four-point Adams–Moulton corrector, resulting in a prediction of

$$\tilde{x}_i(k) = \tilde{x}_i(k - 1) + \frac{\Delta t}{24}\,[55\tilde{f}_i(k - 1) - 59\tilde{f}_i(k - 2) + 37\tilde{f}_i(k - 3) - 9\tilde{f}_i(k - 4)]$$

for $i = 1, 2, \ldots, n$, and a correction of

$$\tilde{x}_i(k) = \tilde{x}_i(k - 1) + \frac{\Delta t}{24}\,\{9f_i[\tilde{x}(k), u(k)] + 19\tilde{f}_i(k - 1) - 5\tilde{f}_i(k - 2) + \tilde{f}_i(k - 3)\}$$

Predictor–corrector methods generally incorporate a strategy for increasing or decreasing the size of the time step depending on the difference between the predicted and corrected $x(k)$ values. Such *variable time-step methods* are particularly useful if the simulated system possesses local time constants that differ by several orders of magnitude, or if there is little *a priori* knowledge about the system response.

Numerical Integration Errors

An inherent characteristic of digital simulation is that the discrete data points generated by the simulation $x(k)$ are only approximations to the exact solution $x(t_k)$ at the corresponding point in time. This results from two types of errors that are unavoidable in the numerical solutions. *Round-off errors* occur because numbers stored in a digital computer have finite word length (i.e., a finite number of bits per word) and therefore limited precision. Because the results of calculations cannot be stored exactly, round-off error tends to increase with the number of calculations performed. For a given total solution interval $t_0 \le t \le t_K$, therefore, round-off error tends to increase (1) with increasing integration-rule order (since more calculations must be performed at each time step) and (2) with decreasing step size Δt (since more time steps are required).

Truncation errors or numerical approximation errors occur because of the inherent limitations in the numerical integration methods themselves. Such errors would arise even if the digital computer had infinite precision. *Local or per-step truncation error* is defined as

$$e(k) = x(k) - x(t_k)$$

given that $x(k - 1) = x(t_{k-1})$ and that the calculation at the kth time step is infinitely precise. For many integration methods, local truncation errors can be approximated at each step. *Global or total truncation error* is defined as

$$e(K) = x(K) - x(t_K)$$

given that $x(0) = x(t_0)$ and the calculations for all K time steps are infinitely precise. Global truncation error usually cannot be estimated, neither can efforts to reduce local truncation errors be guaranteed to yield acceptable global errors. In general, however, truncation errors can be decreased by using more sophisticated integration methods and by decreasing the step size Δt.

Time Constants and Time Steps

As a general rule, the step size Δt for simulation must be less than the smallest local time constant of the model simulated. This can be illustrated by considering the simple first-order system

$$\dot{x}(t) = \lambda x(t)$$

and the difference equation defining the corresponding Euler integration

$$x(k) = x(k - 1) + \Delta t \lambda\, x(k - 1)$$

The continuous system is stable for $\lambda < 0$, while the discrete approximation is stable for $|1 + \lambda \Delta t| < 1$. If the original system is stable, therefore, the simulated response will be stable for

$$\Delta t \le 2\,|1/\lambda|$$

where the equality defines the *critical step size*. For larger step sizes, the simulation will exhibit *numerical instability*. In general, while higher-order integration methods will provide greater per-step accuracy, the critical step size itself will not be greatly reduced.

A major problem arises when the simulated model has one or more time constants $|1/\lambda_i|$ that are small when compared to the total solution time interval $t_0 \leq t \leq t_K$. Numerical stability will then require very small Δt, even though the transient response associated with the higher-frequency (larger λ_i) subsystems may contribute little to the particular solution. Such problems can be addressed either by neglecting the higher-frequency components where appropriate, or by adopting special numerical integration methods for *stiff systems*.

Selecting an Integration Method

The best numerical integration method for a specific simulation is the method that yields an acceptable global approximation error with the minimum amount of round-off error and computing effort. No single method is best for all applications. The selection of an integration method depends on the model simulated, the purpose of the simulation study, and the availability of computing hardware and software.

In general, for well-behaved problems with continuous derivatives and no stiffness, a lower-order Adams predictor is often a good choice. Multistep methods also facilitate estimating local truncation error. Multistep methods should be avoided for systems with discontinuities, however, because of the need for frequent restarts. Runge–Kutta methods have the advantage that these are self-starting and provide fair stability. For stiff systems where high-frequency modes have little influence on the global response, special stiff-system methods enable the use of economically large step sizes. Variable-step rules are useful when little is known *a priori* about solutions. Variable-step rules often make a good choice as general-purpose integration methods.

Round-off error usually is not a major concern in the selection of an integration method, since the goal of minimizing computing effort typically obviates such problems. Double-precision simulation can be used where round off is a potential concern. An upper bound on step size often exists because of discontinuities in derivative functions or because of the need for response output at closely spaced time intervals.

Continuous System Simulation Languages

Digital simulation can be implemented for a specific model in any high-level language such as FORTRAN or C. The general process for implementing a simulation is shown in Fig. 27.26. In addition, many special-purpose continuous system simulation languages are commonly available across a wide range of platforms. Such languages greatly simplify programming tasks and typically provide for good graphical output.

27.8 MODEL CLASSIFICATIONS

Mathematical models of dynamic systems are distinguished by several criteria which describe fundamental properties of model variables and equations. These criteria in turn prescribe the theory and mathematical techniques that can be used to study different models. Table 27.11 summarizes these distinguishing criteria. In the following sections, the approaches adopted for the analysis of important classes of systems are briefly outlined.

27.8.1 Stochastic Systems

Systems in which some of the dependent variables (input, state, output) contain random components are called *stochastic systems*. Randomness may result from environmental factors, such as wind gusts or electrical noise, or simply from a lack of precise knowledge of the system model, such as when a human operator is included within a control system. If the randomness in the system can be described by some rule, then it is often possible to derive a model in terms of probability distributions involving, for example, the means and variances of model variables or parameters.

State-Variable Formulation

A common formulation is the fixed, linear model with additive noise

$$\dot{x}(t) = Ax(t) + Bu(t) + w(t)$$
$$y(t) = Cx(t) + v(t)$$

where $w(t)$ is a zero-mean Gaussian disturbance and $v(t)$ is a zero-mean Gaussian measurement noise. This formulation is the basis for many *estimation problems,* including the problem of *optimal filtering.* Estimation essentially involves the development of a rule or algorithm for determining the best estimate of the past, current, or future values of measured variables in the presence of disturbances or noise.

Random Variables

In the following, important concepts for characterizing random signals are developed. A *random variable x* is a variable that assumes values that cannot be precisely predicted *a priori.* The likelihood

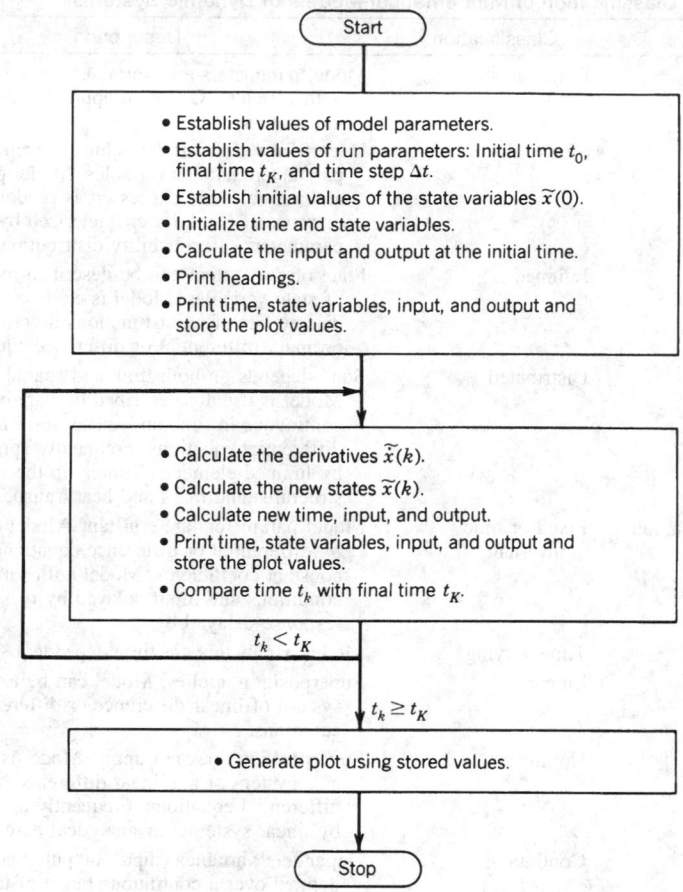

Fig. 27.26 General process for implementing digital simulation (adapted from Close and Frederick[3]).

that a random variable will assume a particular value is measured as the *probability* of that value. The probability *distribution function* $F(x)$ of a continuous random variable x is defined as the probability that x assumes a value no greater than x, that is,

$$F(x) = \Pr(X \le x) = \int_{-\infty}^{x} f(x)\, dx$$

The probability *density function* $f(x)$ is defined as the derivative of $F(x)$.

The *mean* or *expected value* of a probability distribution is defined as

$$E(X) = \int_{-\infty}^{\infty} x f(x)\, dx = \overline{X}$$

The mean is the first moment of the distribution. The *n-th moment* of the distribution is defined as

$$E(X^n) = \int_{-\infty}^{\infty} x^n f(x)\, dx$$

The mean square of the difference between the random variable and its mean is the *variance* or *second central moment* of the distribution,

Table 27.11 Classification of Mathematical Models of Dynamic Systems

Criterion	Classification	Description
Certainty	Deterministic	Model parameters and variables can be known with certainty. Common approximation when uncertainties are small.
	Stochastic	Uncertainty exists in the values of some parameters and/or variables. Model parameters and variables are expressed as random numbers or processes and are characterized by the parameters of probability distributions.
Spatial characteristics	Lumped	State of the system can be described by a finite set of state variables. Model is expressed as a discrete set of point functions described by ordinary differential or difference equations.
	Distributed	State depends on both time and spatial location. Model is usually described by variables that are continuous in time and space, resulting in partial differential equations. Frequently approximated by lumped elements. Typical in the study of structures and mass and heat transport.
Parameter variation	Fixed or time invariant	Model parameters are constant. Model described by differential or difference equations with constant coefficients. Model with same initial conditions and input delayed by t_d has the same response delayed by t_d.
	Time varying	Model parameters are time dependent.
Superposition property	Linear	Superposition applies. Model can be expressed as a system of linear difference or differential equations.
	Nonlinear	Superposition does not apply. Model is expressed as a system of nonlinear difference or differential equations. Frequently approximated by linear systems for analytical ease.
Continuity of independent variable (time)	Continuous	Dependent variables (input, output, state) are defined over a continuous range of the independent variable (time), even though the dependence is not necessarily described by a mathematically continuous function. Model is expressed as differential equations. Typical of physical systems.
	Discrete	Dependent variables are defined only at distinct instants of time. Model is expressed as difference equations. Typical of digital and nonphysical systems.
	Hybrid	System with continuous and discrete subsystems, most common in computer control and communication systems. Sampling and quantization typical in A/D (analog-to-digital) conversion; signal reconstruction for D/A conversion. Model frequently approximated as entirely continuous or entirely discrete.
Quantization of dependent variables	Nonquantized	Dependent variables are continuously variable over a range of values. Typical of physical systems at macroscopic resolution.
	Quantized	Dependent variables assume only a countable number of different values. Typical of computer control and communication systems (sample data systems).

$$\sigma^2(X) = E(X - \overline{X})^2 = \int_{-\infty}^{\infty} (x - \overline{X})^2 f(x) \, dx = E(X^2) - [E(X)]^2$$

The square root of the variance is the *standard deviation* of the distribution.

$$\sigma(X) = \sqrt{E(X^2) - [E(X)]^2}$$

The mean of the distribution therefore is a measure of the average magnitude of the random variable, while the variance and standard deviation are measures of the variability or dispersion of this magnitude.

The concepts of probability can be extended to more than one random variable. The *joint distribution* function of two random variables x and y is defined as

$$F(x, y) = \Pr(X < x \text{ and } Y < y) = \int_{-\infty}^{x} \int_{-\infty}^{y} f(x, y) \, dy \, dx$$

where $f(x, y)$ is the joint distribution. The ijth moment of the joint distribution is

$$E(X^i Y^j) = \int_{-\infty}^{\infty} x^i \int_{-\infty}^{\infty} y^j f(x, y) \, dy \, dx$$

The *covariance* of x and y is defined to be

$$E[(X - \overline{X})(Y - \overline{Y})]$$

and the normalized covariance or *correlation coefficient* as

$$\rho = \frac{E[(X - \overline{X})(Y - \overline{Y})]}{\sqrt{\sigma^2(X)\sigma^2(Y)}}$$

Although many distribution functions have proven useful in control engineering, far and away the most useful is the *Gaussian* or *normal distribution*

$$F(x) = \frac{1}{\sigma\sqrt{2\pi}} \exp[(-x - \mu)^2/2\sigma^2]$$

where μ is the mean of the distribution and σ is the standard deviation. The Gaussian distribution has a number of important properties. First, if the input to a linear system is Gaussian, the output also will be Gaussian. Second, if the input to a linear system is only approximately Gaussian, the output will tend to approximate a Gaussian distribution even more closely. Finally, a Gaussian distribution can be completely specified by two parameters, μ and σ, and therefore a zero-mean Gaussian variable is completely specified by its variance.

Random Processes

A *random process* is a set of random variables with time-dependent elements. If the statistical parameters of the process (such as σ for the zero-mean Gaussian process) do not vary with time, the process is *stationary*. The *autocorrelation function* of a stationary random variable $x(t)$ is defined by

$$\phi_{xx}(\tau) = \lim_{T \to \infty} \frac{1}{2T} \int_{-T}^{T} x(t)x(t + \tau) \, dt$$

a function of the fixed time interval τ. The autocorrelation function is a quantitative measure of the sequential dependence or time correlation of the random variable, that is, the relative effect of prior values of the variable on the present or future values of the variable. The autocorrelation function also gives information regarding how rapidly the variable is changing and about whether the signal is in part deterministic (specifically, periodic). The autocorrelation function of a zero-mean variable has the properties

$$\sigma^2 = \phi_{xx}(0) \geq \phi_{xx}(\tau), \qquad \phi_{xx}(\tau) = \phi_{xx}(-\tau)$$

In other words, the autocorrelation function for $\tau = 0$ is identically the variance and the variance is the maximum value of the autocorrelation function. From the definition of the function, it is clear that (1) for a purely random variable with zero mean, $\phi_{xx}(\tau) = 0$ for $\tau \neq 0$, and (2) for a deterministic variable, which is periodic with period T, $\phi_{xx}(k2\pi T) = \sigma^2$ for k integer. The concept of time correlation is readily extended to more than one random variable. The *cross-correlation function* between the random variables $x(t)$ and $y(t)$ is

$$\phi_{xy}(\tau) = \lim_{T\to\infty} \int_{-\infty}^{\infty} x(t)y(t + \tau)\, dt$$

For $\tau = 0$, the cross-correlation between two zero-mean variables is identically the covariance. A final characterization of a random variable is its *power spectrum,* defined as

$$G(\omega, x) = \lim_{T\to\infty} \frac{1}{2\pi T}\left|\int_{-T}^{T} x(t)e^{-j\omega t}\, dt\right|^2$$

For a stationary random process, the power spectrum function is identically the Fourier transform of the autocorrelation function

$$G(\omega, x) = \frac{1}{\pi}\int_{-\infty}^{\infty} \phi_{xx}(\tau)e^{-j\omega t}\, dt$$

with

$$\phi_{xx}(0) = \int_{-\infty}^{\infty} G(\omega, x)\, d\omega$$

27.8.2 Distributed-Parameter Models

There are many important applications in which the state of a system cannot be defined at a finite number of points in space. Instead, the system state is a continuously varying function of both time and location. When continuous spatial dependence is explicitly accounted for in a model, the independent variables must include spatial coordinates as well as time. The resulting *distributed-parameter model* is described in terms of *partial differential equations,* containing partial derivatives with respect to each of the independent variables.

Distributed-parameter models commonly arise in the study of mass and heat transport, the mechanics of structures and structural components, and electrical transmission. Consider as a simple example the unidirectional flow of heat through a wall, as depicted in Fig. 27.27. The temperature of the wall is not in general uniform, but depends on both the time t and position within the wall x, that is, $\theta = \theta(x, t)$. A distributed-parameter model for this case might be the first-order partial differential equation

$$\frac{d}{dt}\,\theta(x, t) = \frac{1}{C_t}\frac{\partial}{\partial x}\left[\frac{1}{R_t}\frac{\partial}{\partial x}\,\theta(x, t)\right]$$

where C_t is the thermal capacitance and R_t is the thermal resistance of the wall (assumed uniform).

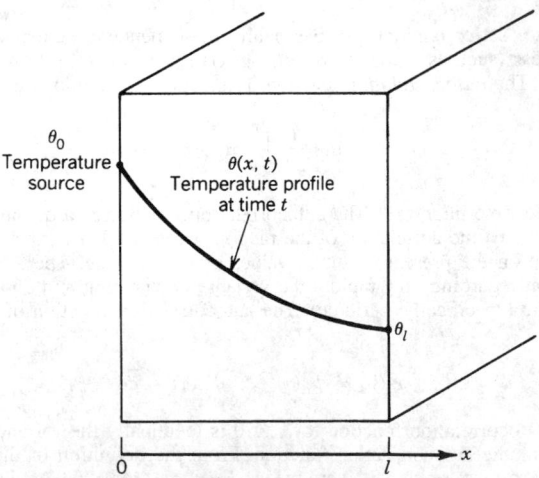

Fig. 27.27 Uniform heat transfer through a wall.

The complexity of distributed parameter models is typically such that these models are avoided in the analysis and design of control systems. Instead, distributed parameter systems are approximated by a finite number of spatial "lumps," each lump being characterized by some average value of the state. By eliminating the independent spatial variables, the result is a *lumped-parameter (or lumped-element) model* described by coupled ordinary differential equations. If a sufficiently fine-grained representation of the lumped microstructure can be achieved, a lumped model can be derived that will approximate the distributed model to any desired degree of accuracy. Consider, for example, the three temperature lumps shown in Fig. 27.28, used to approximate the wall of Fig. 27.27. The corresponding third-order lumped approximation is

$$
\frac{d}{dt}
\begin{bmatrix}
\theta_1(t) \\
\theta_2(t) \\
\theta_3(t)
\end{bmatrix}
=
\begin{bmatrix}
-\dfrac{9}{C_tR_t} & \dfrac{3}{C_tR_t} & 0 \\[2mm]
\dfrac{3}{C_tR_t} & -\dfrac{6}{C_tR_t} & \dfrac{3}{C_tR_t} \\[2mm]
0 & \dfrac{3}{C_tR_t} & -\dfrac{6}{C_tR_t}
\end{bmatrix}
\begin{bmatrix}
\theta_1(t) \\
\theta_2(t) \\
\theta_3(t)
\end{bmatrix}
+
\begin{bmatrix}
\dfrac{6}{C_tR_t} \\[2mm]
0 \\[2mm]
0
\end{bmatrix}
\theta_0(t)
$$

If a more detailed approximation is required, this can always be achieved at the expense of adding additional, smaller lumps.

27.8.3   Time-Varying Systems

Time-varying systems are those with characteristics that change as a function of time. Such variation may result from environmental factors, such as temperature or radiation, or from factors related to the operation of the system, such as fuel consumption. While in general a model with variable parameters can be either linear or nonlinear, the name time–varying is most frequently associated with linear systems described by the following state equation:

$$
\dot{x}(t) = A(t)x(t) + B(t)u(t)
$$

For this linear time-varying model, the superposition principle still applies. Superposition is a great aid in model formulation, but unfortunately does not prove to be much help in determining the model solution.

Paradoxically, the form of the solution to the linear time-varying equation is well known[7]:

$$
x(t) = \Phi(t, t_0)x(t_0) + \int_{t_0}^{t} \Phi(t, \tau)B(\tau)u(\tau)\, dt
$$

where $\Phi(t, t_0)$ is the time-varying state-transition matrix. This knowledge is typically of little value,

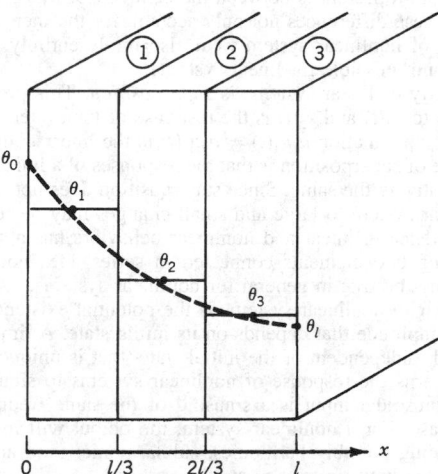

Fig. 27.28 Lumped-parameter model for uniform heat transfer through a wall.

however, since it is not usually possible to determine the state-transition matrix by any straightforward method. By analogy with the first-order case, the relationship

$$\Phi(t, t_0) = \exp\left(\int_{t_0}^{t} A(\tau)\, d\tau\right)$$

can be proven valid *if and only if*

$$A(t) \int_{t_0}^{t} A(\tau)\, d\tau = \int_{t_0}^{t} A(\tau)\, d\tau A(t)$$

that is, if and only if $A(t)$ and its integral commute. This is a very stringent condition for all but a first-order system and, as a rule, it is usually easiest to obtain the solution using simulation.

Most of the properties of the fixed transition matrix extend to the time-varying case:

$$\Phi(t, t_0) = I$$
$$\Phi^{-1}(t, t_0) = \Phi(t_0, t)$$
$$\Phi(t_2, t_1)\Phi(t_1, t_0) = \Phi(t_2, t_0)$$
$$\dot{\Phi}(t, t_0) = A(t)\Phi(t, t_0)$$

27.8.4 Nonlinear Systems

The theory of fixed, linear, lumped-parameter systems is highly developed and provides a powerful set of techniques for control system analysis and design. In practice, however, all physical systems are nonlinear to some greater or lesser degree. The linearity of a physical system is usually only a convenient approximation, restricted to a certain range of operation. In addition, nonlinearities such as dead zones, saturation, or on–off action are sometimes introduced into control systems intentionally, either to obtain some advantageous performance characteristic or to compensate for the effects of other (undesirable) nonlinearities.

Unfortunately, while nonlinear systems are important, ubiquitous, and potentially useful, the theory of nonlinear differential equations is comparatively meager. Except for specific cases, closed-form solutions to nonlinear systems are generally unavailable. The only universally applicable method for the study of nonlinear systems is *simulation*. As described in Section 27.7, however, simulation is an experimental approach, embodying all of the attending limitations of experimentation.

A number of special techniques are available for the analysis of nonlinear systems. All of these techniques are in some sense approximate, assuming, for example, either a restricted range of operation over which nonlinearities are mild or the relative isolation of lower-order subsystems. When used in conjunction with more complex simulation models, however, these techniques often provide insights and design concepts that would be difficult to discover through the use of simulation alone.[8]

Linear versus Nonlinear Behaviors

There are several fundamental differences between the behavior of linear and nonlinear systems that are especially important. These differences not only account for the increased difficulty encountered in the analysis and design of nonlinear systems, but also imply entirely new types of behavior for nonlinear systems that are not possible for linear systems.

The fundamental property of linear systems is *superposition*. This property states that if $y_1(t)$ is the response of the system to $u_1(t)$ and $y_2(t)$ is the response of the system to $u_2(t)$, then the response of the system to the linear combination $a_1u_1(t) + a_2u_2(t)$ is the linear combination $a_1y_1(t) + a_2y_2(t)$. An immediate consequence of superposition is that the responses of a linear system to inputs differing only in amplitude is qualitatively the same. Since superposition does not apply to nonlinear systems, the responses of a nonlinear system to large and small changes may be fundamentally different.

This fundamental difference in linear and nonlinear behaviors has a second consequence. For a linear system, interchanging two elements connected in series does not affect the overall system behavior. Clearly, this cannot be true in general for nonlinear systems.

A third property peculiar to nonlinear systems is the potential existence of *limit cycles*. A linear oscillator oscillates at an amplitude that depends on its initial state. A limit cycle is an oscillation of fixed amplitude and period, independent of the initial state, that is unique to the nonlinear system.

A fourth property concerns the response of nonlinear systems to sinusoidal inputs. For a linear system, the response to sinusoidal input is a sinusoid of the same frequency, potentially differing only in magnitude and phase. For a nonlinear system, the output will in general contain other frequency components, including possibly harmonics, subharmonics, and aperiodic terms. Indeed, the response need not contain the input frequency at all.

Linearizing Approximations

Perhaps the most useful technique for analyzing nonlinear systems is to approximate these with linear systems. While many linearizing approximations are possible, linearization can frequently be achieved by considering small excursions of the system state about a reference trajectory. Consider the nonlinear state equation

$$\dot{x}(t) = f[x(t), u(t)]$$

together with a reference trajectory $x^0(t)$ and reference input $u^0(t)$ that together satisfy the state equation

$$\dot{x}^0(t) = f[x^0(t), u^0(t)]$$

Note that the simplest case is to choose a static equilibrium or *operating point* $\bar{x}$ as the reference "trajectory," such that $0 = t(\bar{x}, 0)$. The actual trajectory is then related to the reference trajectory by the relationships

$$x(t) = x^0(t) + \delta x(t)$$
$$u(t) = u^0(t) + \delta u(t)$$

where $\delta x(t)$ is some small perturbation about the reference state and $\delta u(t)$ is some small perturbation about the reference input. If these perturbations are indeed small, then applying the Taylor's series expansion about the reference trajectory yields the linearized approximation

$$\delta \dot{x}(t) = A(t)\delta x(t) + B(t)\delta u(t)$$

where the state and distribution matrices are the *Jacobian matrices*

$$A(t) = \begin{bmatrix} \dfrac{\partial f_i}{\partial x_1} & \dfrac{\partial f_1}{\partial x_2} & \cdots & \dfrac{\partial f_1}{\partial x_n} \\[2ex] \dfrac{\partial f_2}{\partial x_1} & \dfrac{\partial f_2}{\partial x_2} & \cdots & \dfrac{\partial f_2}{\partial x_n} \\[2ex] \vdots & \vdots & & \vdots \\[2ex] \dfrac{\partial f_n}{\partial x_1} & \dfrac{\partial f_n}{\partial x_2} & \cdots & \dfrac{\partial f_n}{\partial x_n} \end{bmatrix}_{x(t)=x^0(t);\ u(t)=u^0(t)}$$

$$B(t) = \begin{bmatrix} \dfrac{\partial f_1}{\partial u_1} & \dfrac{\partial f_1}{\partial u_2} & \cdots & \dfrac{\partial f_1}{\partial u_m} \\[2ex] \dfrac{\partial f_2}{\partial u_1} & \dfrac{\partial f_2}{\partial u_2} & \cdots & \dfrac{\partial f_2}{\partial u_m} \\[2ex] \vdots & \vdots & & \vdots \\[2ex] \dfrac{\partial f_n}{\partial u_1} & \dfrac{\partial f_n}{\partial u_2} & \cdots & \dfrac{\partial f_n}{\partial u_m} \end{bmatrix}_{x(t)=x^0(t);\ u(t)=u^0(t)}$$

If the reference trajectory is a fixed operating point $\bar{x}$, then the resulting linearized system is time invariant and can be solved analytically. If the reference trajectory is a function of time, however, then the resulting system is linear, but time varying.

Describing Functions

The describing function method is an extension of the frequency transfer function approach of linear systems, most often used to determine the stability of limit cycles of systems containing nonlinearities. The approach is approximate and its usefulness depends on two major assumptions:

1. All the nonlinearities within the system can be aggregated mathematically into a single block, denoted as $N(M)$ in Fig. 27.29, such that the equivalent gain and phase associated with this block

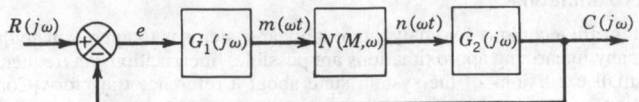

Fig. 27.29 General nonlinear system for describing function analysis.

depend only on the amplitude M_d of the sinusoidal input $m(\omega t) = M \sin (\omega t)$ and are independent of the input frequency ω.

2. All the harmonics, subharmonics, and any dc component of the output of the nonlinear block are filtered out by the linear portion of the system, such that the effective output of the nonlinear block is well approximated by a periodic response having the same fundamental period as the input.

Although these assumptions appear to be rather limiting, the technique gives reasonable results for a large class of control systems. In particular, the second assumption is generally satisfied by higher-order control systems with symmetric nonlinearities, since (a) symmetric nonlinearities do not generate dc terms, (b) the amplitudes of harmonics are generally small when compared with the fundamental term and subharmonics are uncommon, and (c) feedback within a control system typically provides low-pass filtering to further attenuate harmonics, especially for higher-order systems. Because the method is relatively simple and can be used for systems of any order, describing functions have enjoyed wide practical application.

The describing function of a nonlinear block is defined as the ratio of the fundamental component of the output to the amplitude of a sinusoidal input. In general, the response of the nonlinearity to the input

$$m(\omega t) = M \sin \omega t$$

is the output

$$n(\omega t) = N_1 \sin(\omega t + \phi_1) + N_2 \sin(2\omega t + \phi_2) + N_3 \sin(3\omega t + \phi_3) + \cdots$$

and, hence, the describing function for the nonlinearity is defined as the complex quantity

$$N(M) = \frac{N_1}{M} e^{j\phi_1}$$

Derivation of the approximating function typically proceeds by representing the fundamental frequency by the Fourier series coefficients

$$A_1(M) = \frac{2}{T} \int_{-T/2}^{T/2} n(\omega t) \cos \omega t \, d(\omega t)$$

$$B_1(M) = \frac{2}{T} \int_{-T/2}^{T/2} n(\omega t) \sin \omega t \, d(\omega t)$$

The describing function is then written in terms of these coefficients as

$$N(M) = \frac{B_1(M)}{M} + j \frac{A_1(M)}{M} = \left[\left(\frac{B_1(M)}{M} \right)^2 + \left(\frac{A_1(M)}{M} \right)^2 \right]^{1/2} \exp \left[j \tan^{-1} \left(\frac{A_1(M)}{B_1(M)} \right) \right]$$

Note that if $n(\omega t) = -n(-\omega t)$, then the describing function is odd, $A_1(M) = 0$, and there is no phase shift between the input and output. If $n(\omega t) = n(-\omega t)$, then the function is even, $B_1(M) = 0$, and the phase shift is $\pi/2$.

The describing functions for a number of typical nonlinearities are given in Fig. 27.30. Reference 9 contains an extensive catalog. The following derivation for a dead zone nonlinearity demonstrates the general procedure for deriving a describing function. For the saturation element depicted in Fig. 27.30a, the relationship between the input $m(\omega t)$ and output $n(\omega t)$ can be written as

$$n(\omega t) = \begin{cases} 0, & \text{for } -D < m < D \\ K_1 M(\sin \omega t - \sin \omega_1 t), & \text{for } m > D \\ K_1 M(\sin \omega t + \sin \omega_1 t), & \text{for } m < -D \end{cases}$$

Since the function is odd, $A_1 = 0$. By the symmetry over the four quarters of the response period,

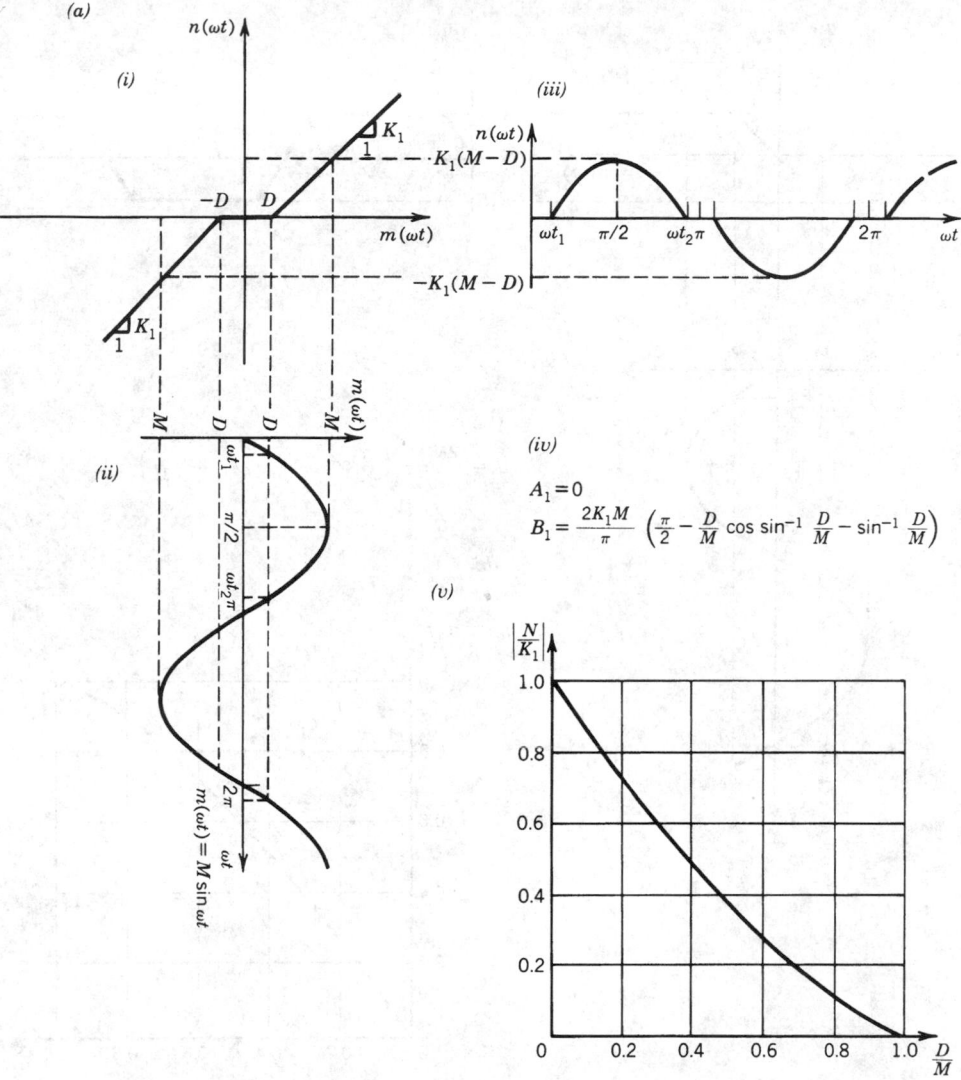

Fig. 27.30a Describing functions for typical nonlinearities (after Refs. 9 and 10). Dead zone nonlinearity: (*i*) nonlinear characteristic; (*ii*) sinusoidal input wave shape; (*iii*) output wave shape; (*iv*) describing-function coefficients; (*v*) normalized describing function.

$$B_1 = 4 \left[\frac{2}{\pi/2} \int_0^{\pi/2} n(\omega t) \sin \omega t \, d(\omega t) \right]$$

$$= \frac{4}{\pi} \left[\int_0^{\omega t_1} (0) \sin \omega t \, d(\omega t) + \int_{\omega t_1}^{\pi/2} K_1 M(\sin \omega t - \sin \omega_1 t) \sin \omega t \, d(\omega t) \right]$$

where $\omega t_1 = \sin^{-1}(D/M)$. Evaluating the integrals and dividing by M yields the describing function listed in Fig. 27.30.

Phase-Plane Method

The *phase-plane method* is a graphical application of the state-space approach used to characterize the free-response of second-order nonlinear systems. While any convenient pair of state variables can be used, the *phase variables* originally were taken to be the displacement and velocity of the mass

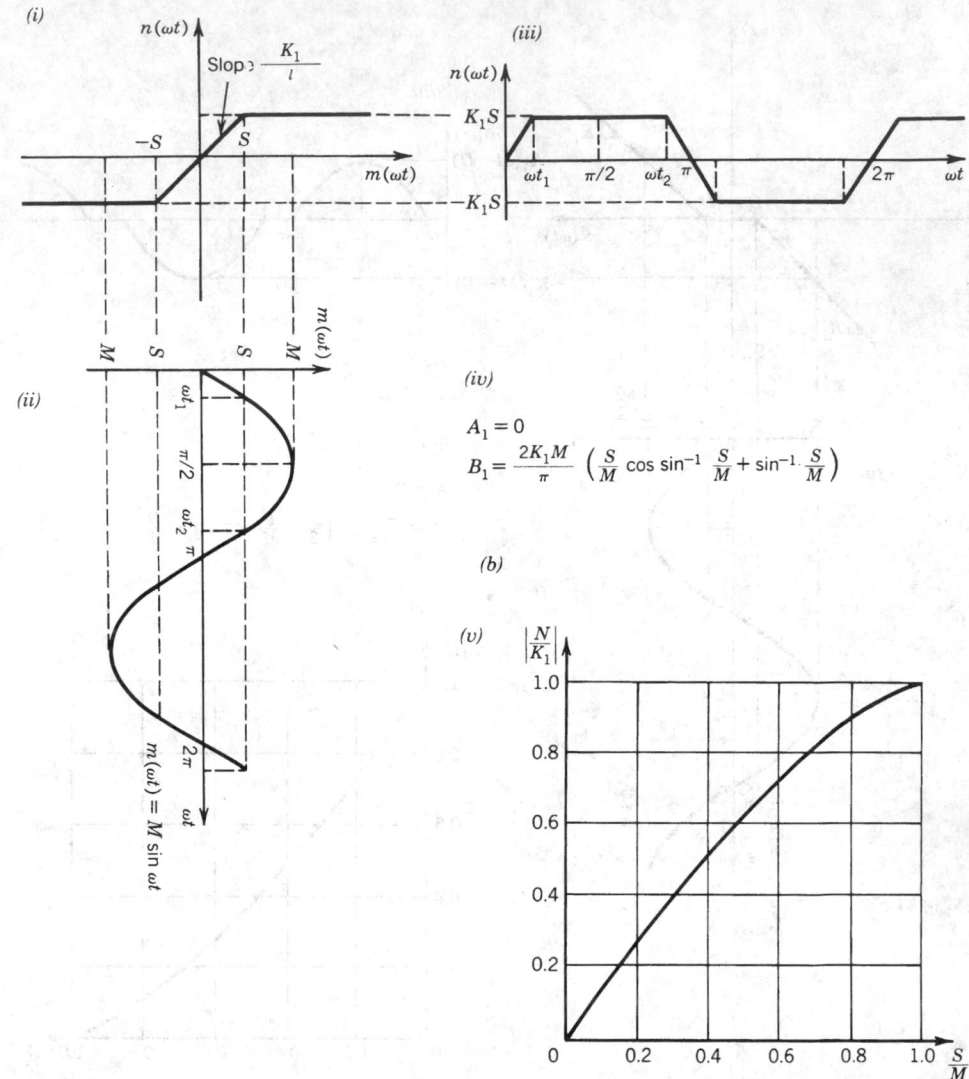

Fig. 27.30b Saturation nonlinearity: (*i*) nonlinear characteristic; (*ii*) sinusoidal input wave shape; (*iii*) output wave shape; (*iv*) describing-function coefficients; (*v*) normalized describing function.

of a second-order mechanical system. Using the two state variables as the coordinate axis, the transient response of a system is captured on the *phase plane* as the plot of one variable against the other, with time implicit on the resulting curve. The curve for a specific initial condition is called a *trajectory* in the phase plane; a representative sample of trajectories is called the *phase portrait* of the system. The phase portrait is a compact and readily interpreted summary of the system response. Phase portraits for a sample of typical nonlinearities are shown in Fig. 27.31.

Four methods can be used to construct a phase portrait: (1) direct solution of the differential equation, (2) the graphical *method of isoclines,* (3) transformation of the second-order system (with time as the independent variable) into an equivalent first-order system (with one of the phase variables as the independent variable), and (4) numerical solution using simulation. The first and second methods are usually impractical; the third and fourth methods are frequently used in combination. For example, consider the second-order model

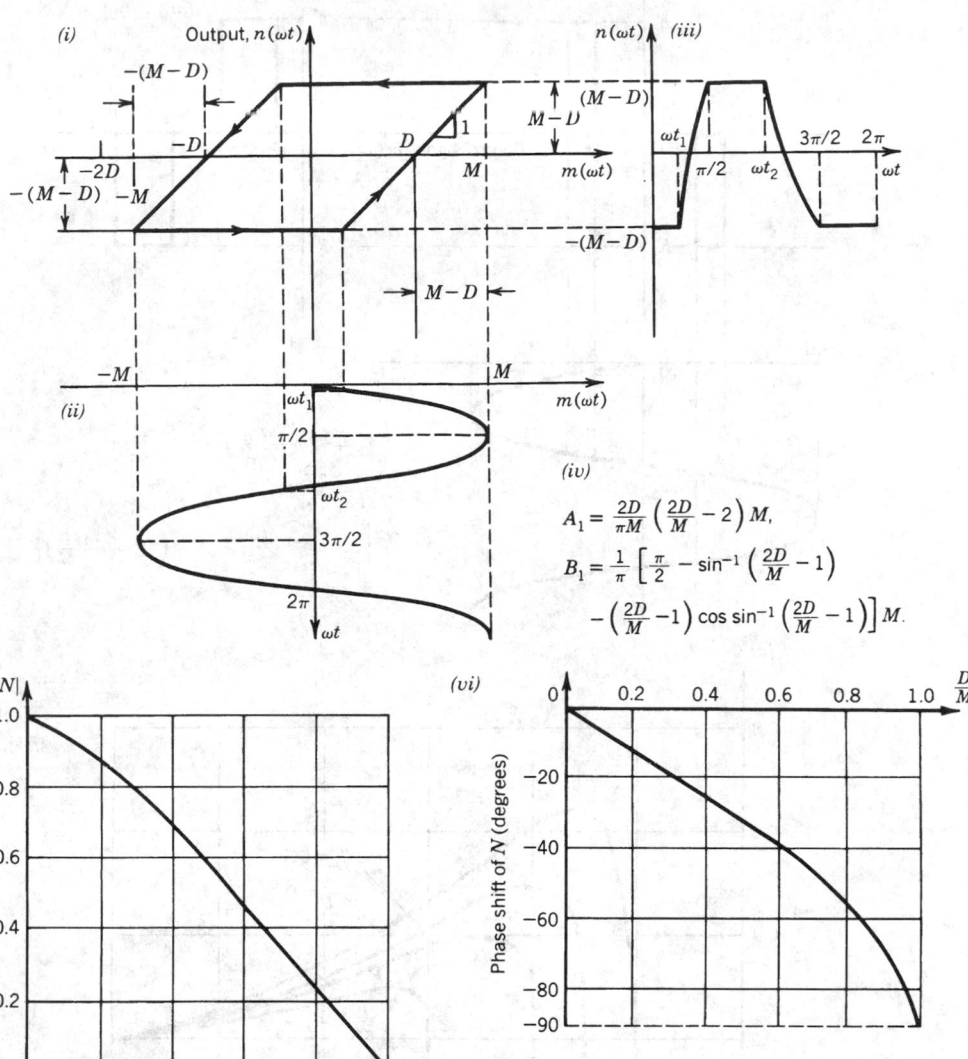

Fig. 27.30c Backlash nonlinearity: (*i*) nonlinear characteristic; (*ii*) sinusoidal input wave shape; (*iii*) output wave shape; (*iv*) describing-function coefficients; (*v*) normalized amplitude characteristics for the describing function; (*vi*) normalized phase characteristics for the describing function.

$$\frac{dx_1}{dt} = f_1(x_1, x_2), \qquad \frac{dx_2}{dt} = f_2(x_1, x_2)$$

Dividing the second equation by the first and eliminating the *dt* terms yields

$$\frac{dx_2}{dx_1} = \frac{f_2(x_1, x_2)}{f_1(x_1, x_2)}$$

This first-order equation describes the phase-plane trajectories. In many cases it can be solved analytically. If not, it always can be simulated.

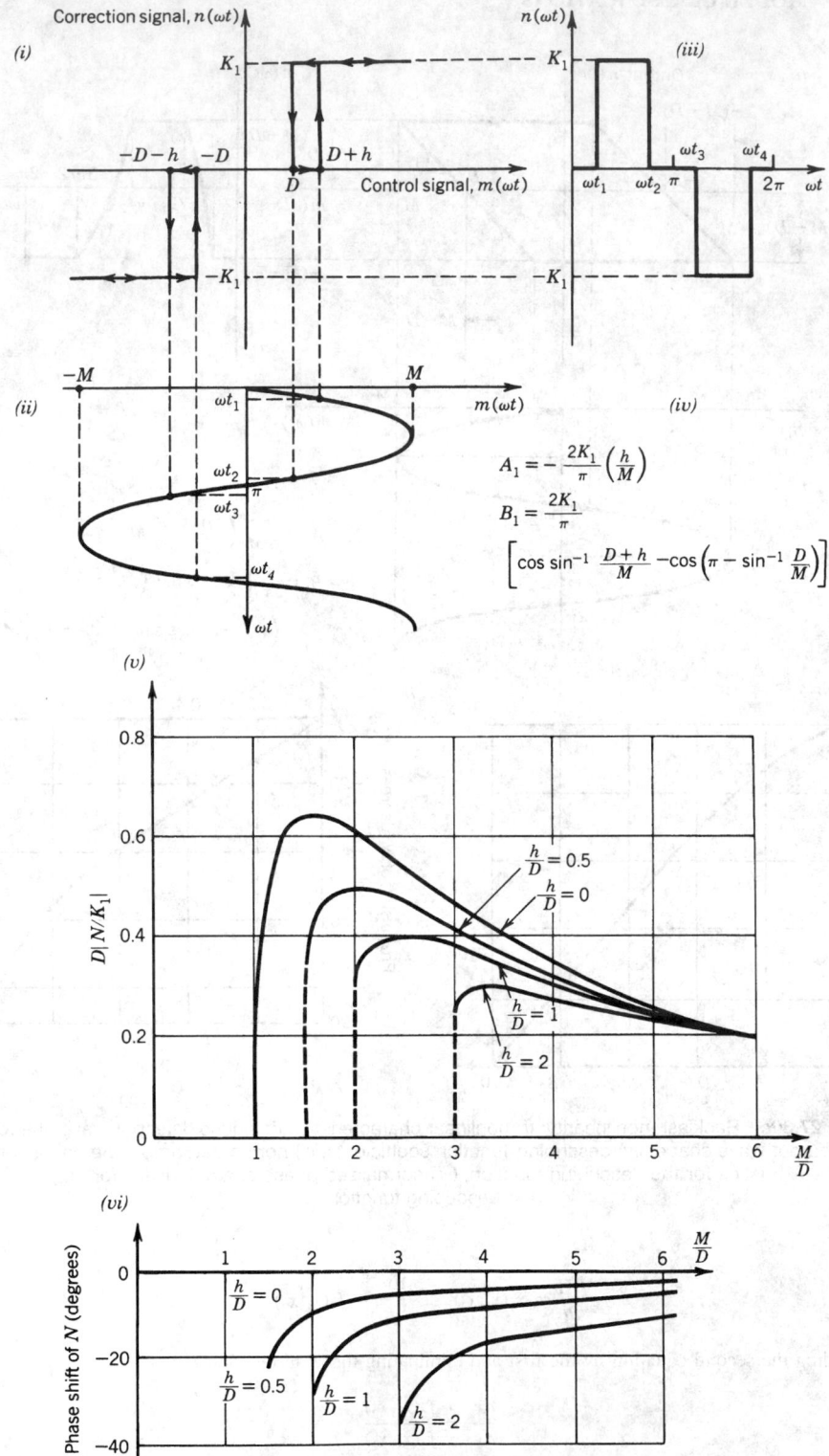

Fig. 27.30d Three-position on–off device with hysteresis: (*i*) nonlinear characteristic; (*ii*) sinusoidal input wave shape; (*ii*) output wave shape; (*iv*) describing-function coefficients; (*v*) normalized amplitude characteristics for the describing function; (*vi*) normalized phase characteristics for the describing function.

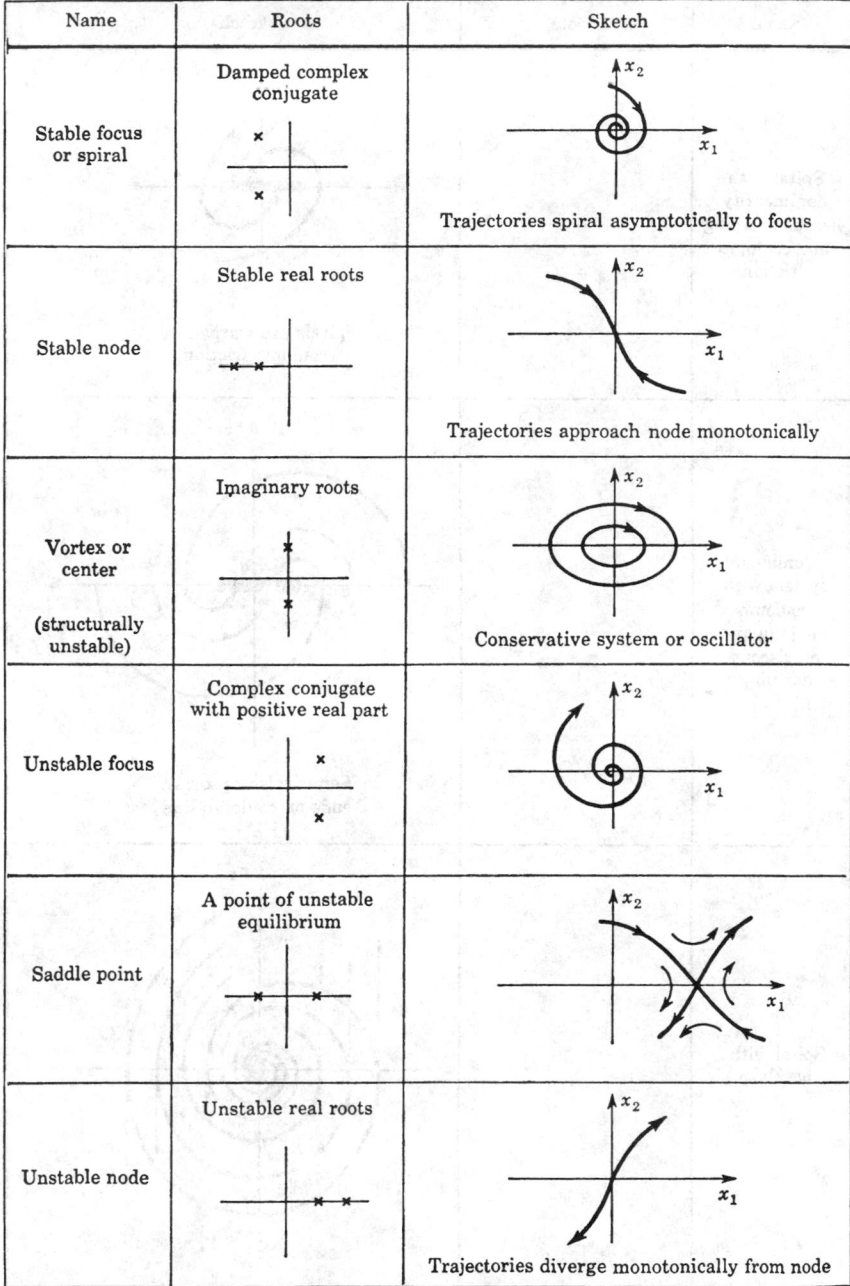

Name	Roots	Sketch
Stable focus or spiral	Damped complex conjugate	Trajectories spiral asymptotically to focus
Stable node	Stable real roots	Trajectories approach node monotonically
Vortex or center (structurally unstable)	Imaginary roots	Conservative system or oscillator
Unstable focus	Complex conjugate with positive real part	
Saddle point	A point of unstable equilibrium	
Unstable node	Unstable real roots	Trajectories diverge monotonically from node

Fig. 27.31 Typical phase-plane plots for second-order systems.[9]

The phase-plane method complements the describing-function approach. A describing function is an approximate representation of the sinusoidal response for systems of any order, while the phase plane is an exact representation of the (free) transient response for first- and second-order systems. Of course, the phase-plane method theoretically can be extended for higher-order systems, but the difficulty of visualizing the nth order state space typically makes such a direct extension impractical.

Name	Roots	Sketch
Spiral with nonlinearity viscous damping and coulomb friction		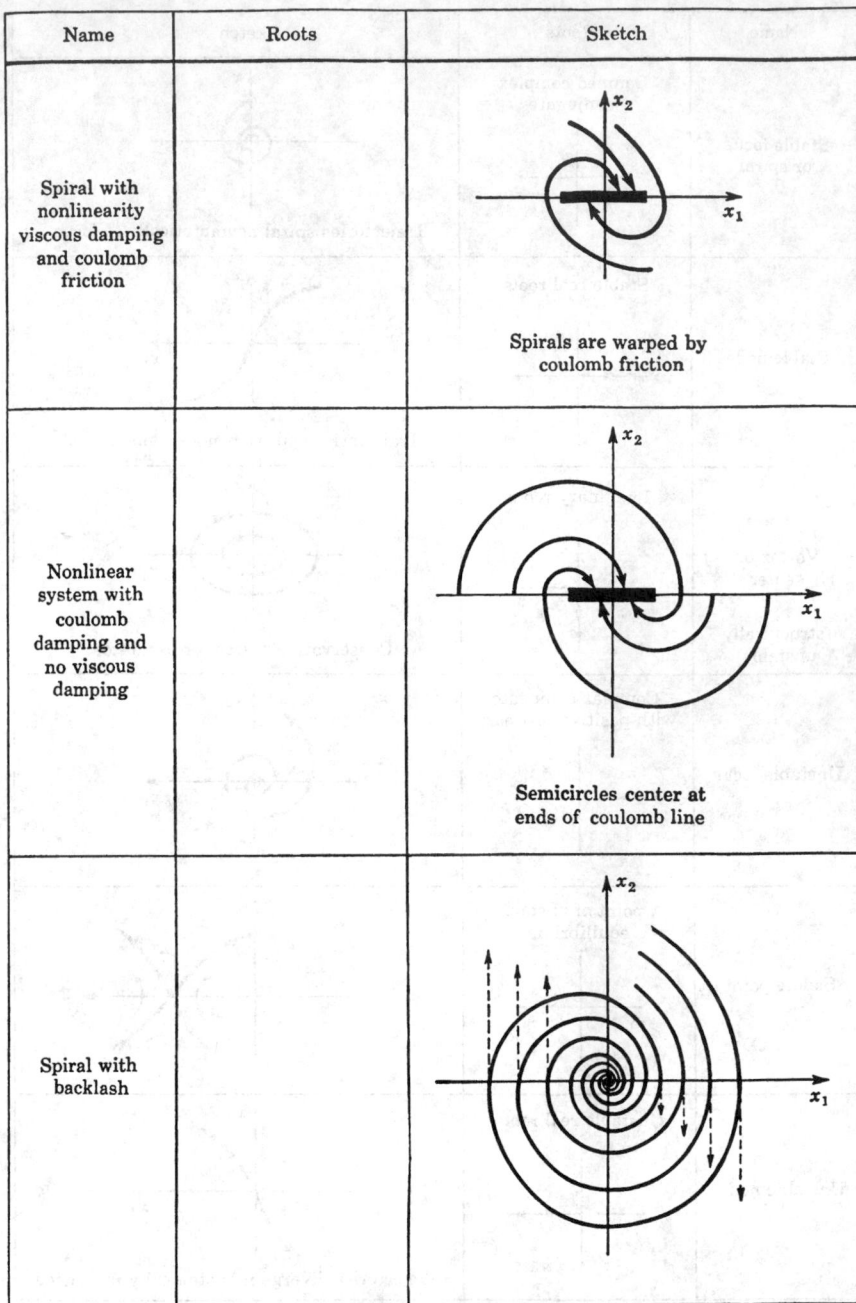
Nonlinear system with coulomb damping and no viscous damping		
Spiral with backlash		

Fig. 27.31 *(Continued)*

An approximate extension of the method has been used with some considerable success,[8] however, in order to explore and validate the relationships among pairs of variables in complex simulation models. The approximation is based on the assumptions that the paired variables define a second-order subsystem which, for the purposes of analysis, is weakly coupled to the remainder of the system.

27.8.5 Discrete and Hybrid Systems

A *discrete-time system* is one for which the dependent variables are defined only at distinct instants of time. Discrete-time models occur in the representation of systems that are inherently discrete, in the analysis and design of digital measurement and control systems, and in the numerical solution of differential equations (see Section 27.7). Because most control systems are now implemented using digital computers (especially microprocessors), discrete-time models are extremely important in dynamic systems analysis. The discrete-time nature of a computer's sampling of continuous physical signals also leads to the occurrence of *hybrid systems*, that is, systems that are in part discrete and in part continuous. Discrete-time models of hybrid systems are called *sampled-data systems*.

Difference Equations

Dynamic models of discrete-time systems most naturally take the form of *difference equations*. The input–output (I/O) form of an nth order difference equation model is

$$f[y(k + n), y(k + n - 1), \ldots, y(k), u(k + n - 1), \ldots, u(k)] = 0$$

which expresses the dependence of the $(k + n)$th value of the output, $y(k + n)$, on the n preceding values of the output y and input u. For a linear system, the I/O form can be written as

$$y(k + n) + a_{n-1}(k)y(k + n - 1) + \cdots + a_1(k)y(k + 1) + a_0(k)y(k)$$
$$= b_{n-1}(k)u(k + n - 1) + \cdots + b_0(k)u(k)$$

In state-variable form, the discrete-time model is the vector difference equation

$$x(k + 1) = f[x(k), u(k)]$$
$$y(k) = g[x(k), u(k)]$$

where x is the state-vector, u is the vector of inputs, and y is the vector of outputs. For a linear system, the discrete state-variable form can be written as

$$x(k + 1) = A(k)x(k) + B(k)u(k)$$
$$y(k) = C(k)x(k) + D(k)u(k)$$

The mathematics of difference equations parallels that of differential equations in many important respects. In general, the concepts applied to differential equations have direct analogies for difference equations, although the mechanics of their implementation may vary (see Ref. 11 for a development of dynamic modeling based on difference equations). One important difference is that the general solution of nonlinear and time-varying difference equations can usually be obtained through *recursion*. For example, consider the discrete nonlinear model

$$y(k + 1) = \frac{y(k)}{1 + y(k)}$$

Recursive evaluation of the equation beginning with the initial condition $y(0)$ yields

$$y(1) = \frac{y(0)}{1 + y(0)}$$

$$y(2) = \frac{y(1)}{1 + y(1)} = \left[\frac{y(0)}{1 + y(0)}\right] \bigg/ \left[1 + \frac{y(0)}{1 + y(0)}\right] = \frac{y(0)}{1 + 2y(0)}$$

$$y(3) = \frac{y(2)}{1 + y(2)} = \frac{y(0)}{1 + 3y(0)}$$

$$\vdots$$

the pattern of which reveals, by induction,

$$y(k) = \frac{y(0)}{1 + ky(0)}$$

as the general solution.

Uniform Sampling

Uniform sampling is the most common mathematical approach to *analog-to-digital (A/D) conversion*, that is, to extracting the discrete time approximation $y*(k)$ of the form

$$y*(k) = y(t = kT)$$

from the continuous-time signal $y(t)$, where T is a constant interval of time called the *sampling period*. If the sampling period is too large, however, it may not be possible to represent the continuous signal accurately. The *sampling theorem* guarantees that $y(t)$ can be reconstructed from the uniformly sampled values $y*(k)$ if the sampling period satisfies the inequality

$$T \leq \frac{\pi}{\omega_u}$$

where ω_u is the highest frequency contained in the Fourier transform $Y(\omega)$ of $y(t)$, that is, if

$$Y(\omega) = 0 \qquad \text{for all} \quad \omega > \omega_u$$

The Fourier transform of a signal is defined to be

$$\mathcal{F}[y(t)] = Y(\omega) = \int_{-\infty}^{\infty} y(t)e^{-j\omega t} \, dt$$

Note that if $y(t) = 0$ for $t \geq 0$, and if the region of convergence for the Laplace transform includes the imaginary axis, then the Fourier transform can be obtained from the Laplace transform as

Table 27.12   z-Transform Pairs

	$X(s)$	$x(t)$ or $x(k)$	$X(z)$
1	1	$\delta(t)$	1
2	e^{-kTs}	$\delta(t - kT)$	z^{-k}
3	$\dfrac{1}{s}$	$1(t)$	$\dfrac{z}{z - 1}$
4	$\dfrac{1}{s^2}$	t	$\dfrac{Tz}{(z - 1)^2}$
5	$\dfrac{1}{s + a}$	e^{-at}	$\dfrac{z}{z - e^{-aT}}$
6	$\dfrac{a}{s(s + a)}$	$1 - e^{-at}$	$\dfrac{(1 - e^{-aT})z}{(z - 1)(z - e^{-aT})}$
7	$\dfrac{\omega}{s^2 + \omega^2}$	$\sin \omega t$	$\dfrac{z \sin \omega T}{z^2 - 2z \cos \omega T + 1}$
8	$\dfrac{s}{s^2 + \omega^2}$	$\cos \omega t$	$\dfrac{z(z - \cos \omega T)}{z^2 - 2z \cos \omega T + 1}$
9	$\dfrac{1}{(s + a)^2}$	te^{-at}	$\dfrac{Tze^{-aT}}{(z - e^{-aT})^2}$
10	$\dfrac{\omega}{(s + a)^2 + \omega^2}$	$e^{-at} \sin \omega t$	$\dfrac{ze^{-aT} \sin \omega T}{z^2 - 2ze^{-aT} \cos \omega T + e^{-2aT}}$
11	$\dfrac{s + a}{(s + a)^2 + \omega^2}$	$e^{-at} \cos \omega t$	$\dfrac{z^2 - ze^{-aT} \cos \omega T}{z^2 - 2ze^{-aT} \cos \omega T + e^{-2aT}}$
12	$\dfrac{2}{s^3}$	t^2	$\dfrac{T^2z(z + 1)}{(z - 1)^3}$
13		a	$\dfrac{z}{z - a}$
14		$a^k \cos k\pi$	$\dfrac{z}{z + a}$

Table 27.13 z-Transform Properties

	$x(t)$ or $x(k)$	$\mathfrak{Z}\,[x(t)]$ or $\mathfrak{Z}\,[x(k)]$
1	$ax(t)$	$aX(z)$
2	$x_1(t) + x_2(t)$	$X_1(z) + X_2(z)$
3	$x(t + T)$ or $x(k + 1)$	$zX(z) - zx(0)$
4	$x(t + 2T)$	$z^2X(z) - z^2x(0) - zx(T)$
5	$x(k + 2)$	$z^2X(z) - z^2x(0) - zx(1)$
6	$x(t + kT)$	$z^kX(z) - z^kx(0) - z^{k-1}x(T) - \cdots - zx(kT - T)$
7	$x(k + m)$	$z^mX(z) - z^mx(0) - z^{m-1}x(1) - \cdots - zx(m - 1)$
8	$tx(t)$	$-Tz\dfrac{d}{dz}[X(z)]$
9	$kx(k)$	$-z\dfrac{d}{dz}[X(z)]$
10	$e^{-at}x(t)$	$X(ze^{aT})$
11	$e^{-ak}x(k)$	$X(ze^a)$
12	$a^kx(k)$	$X\left(\dfrac{z}{a}\right)$
13	$ka^kx(k)$	$-z\dfrac{d}{dz}\left[X\left(\dfrac{z}{a}\right)\right]$
14	$x(0)$	$\lim\limits_{z\to\infty} X(z)$ if the limit exists
15	$x(\infty)$	$\lim\limits_{z\to1}\,[(z - 1)X(z)]$ if $\dfrac{z - 1}{z}X(z)$ is analytic on and outside the unit circle
16	$\sum\limits_{k=0}^{\infty} x(k)$	$X(1)$
17	$\sum\limits_{k=0}^{n} x(kT)y(nT - kT)$	$X(z)Y(z)$

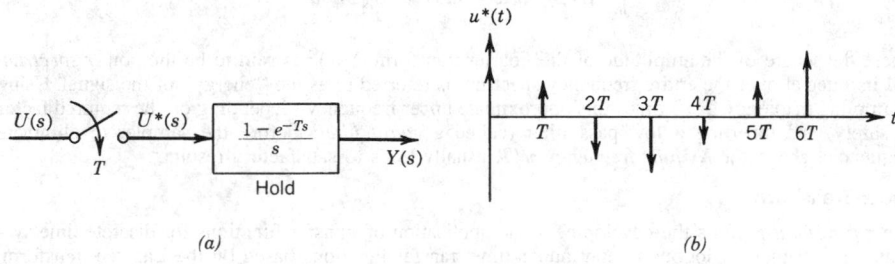

(a)   *(b)*

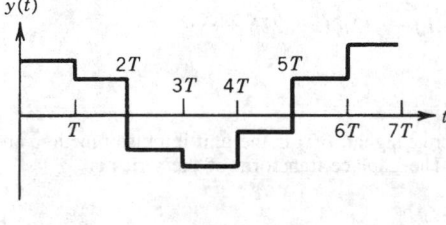

(c)

Fig. 27.32 Zero-order hold: (a) block diagram of hold with a sampler, (b) sampled input sequence, (c) analog output for the corresponding input sequence.[4]

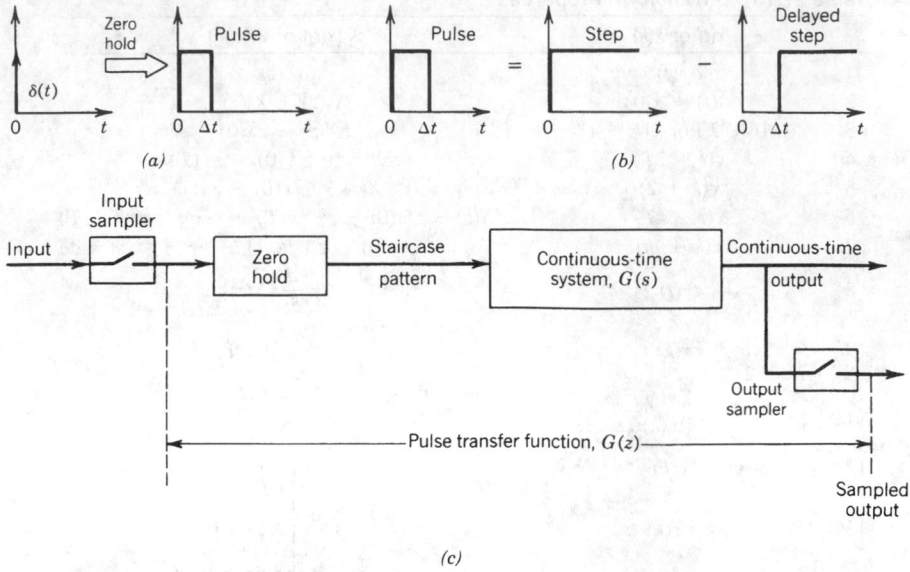

Fig. 27.33 Pulse transfer function of a continuous system with sampler and zero hold.[12]

$$Y(\omega) = [Y(s)]_{s=j\omega}$$

For cases where it is impossible to determine the Fourier transform analytically, such as when the signal is described graphically or by a table, numerical solution based on the *fast Fourier transform (FFT) algorithm* is usually satisfactory.

In general, the condition $T \leq \pi/\omega_u$ cannot be satisfied exactly, since most physical signals have no finite upper frequency ω_u. A useful approximation is to define the upper frequency as the frequency for which 99% of the signal "energy" lies in the frequency spectrum $0 \leq \omega \leq \omega_u$. This approximation is found from the relation

$$\int_0^{\omega_u} |Y(\omega)|^2 \, d\omega = 0.99 \int_0^{\infty} |Y(\omega)|^2 \, d\omega$$

where the square of the amplitude of the Fourier transform $|Y(\omega)|^2$ is said to be the *power spectrum* and its integral over the entire frequency spectrum is referred to as the "energy" of the signal. Using a sampling frequency 2–10 times this approximate upper frequency (depending on the required factor of safety) and inserting a low pass filter (called a *guard filter*) before the sampler to eliminate frequencies above the *Nyquist frequency* π/T, usually leads to satisfactory results.[4]

The z-Transform

The *z-transform* permits the development and application of transfer functions for discrete-time systems, in a manner analogous to continuous-time transfer functions based on the Laplace transform. A discrete signal may be represented as a series of impulses

$$y^*(t) = y(0)\delta(t) + y(1)\delta(t - T) + y(2)\delta(t - 2T) + \cdots$$

$$= \sum_{k=0}^{N} y(k)\delta(t - kT)$$

where $y(k) = y^*(t = kT)$ are the values of the discrete signal, $\delta(t)$ is the unit impulse function, and N is the number of samples of the discrete signal. The Laplace transform of the series is

$$Y^*(s) = \sum_{k=0}^{N} y(k)e^{-ksT}$$

where the shifting property of the Laplace transform has been applied to the pulses. Defining the *shift* or *advance operator* as $z = e^{st}$, $Y^*(s)$ may now be written as a function of z

$$Y^*(z) = \sum_{k=0}^{N} \frac{y(k)}{z^k} = \mathscr{Z}[y(t)]$$

where the transformed variable $Y^*(z)$ is called the z-transform of the function $y^*(t)$. The inverse of the shift operator $1/z$ is called the *delay operator* and corresponds to a time delay of T.

The z-transforms for many sampled functions can be expressed in closed form. A listing of the transforms of several commonly encountered functions is given in Table 27.12. Properties of the z-transform are listed in Table 27.13.

Pulse Transfer Functions

The transfer function concept developed for continuous systems has a direct analog for sampled-data systems. For a continuous system with sampled output $u(t)$ and sampled input $y(t)$, the *pulse* or *discrete transfer function* $G(z)$ is defined as the ratio of the z-transformed output $Y(z)$ to the z-transformed input $U(z)$, assuming zero initial conditions. In general, the pulse transfer function has the form

$$G(z) = \frac{Y(z)}{U(z)} = \frac{b_0 + b_1 z^{-1} + b_2 z^{-2} + \cdots + b_m z^{-m}}{1 + a_1 z^{-1} + a_2 z^{-1} + \cdots + a_n z^{-n}}$$

Zero-Order Hold

The *zero-order data hold* is the most common mathematical approach to *digital-to-analog (D/A) conversion,* that is, to creating a piecewise continuous approximation $u(t)$ of the form

$$u(t) = u^*(k) \qquad \text{for} \quad kT \leq t < (k+1)T$$

from the discrete time signal $u^*(k)$, where T is the period of the hold. The effect of the zero-order hold is to convert a sequence of discrete impulses into a staircase pattern, as shown in Fig. 27.32. The transfer function of the zero-order hold is

$$G(s) = \frac{1}{s}(1 - e^{-Ts}) = \frac{1 - z^{-1}}{s}$$

Using this relationship, the pulse transfer function of the sampled-data system shown in Fig. 27.33 can be derived as

$$G(z) = (1 - z^{-1})\mathscr{Z}\left[\mathscr{L}^{-1} \frac{G(s)}{s}\right]$$

The continuous system with transfer function $G(s)$ has a sampler and a zero-order hold at its input and a sampler at its output. This is a common configuration in many computer-control applications.

REFERENCES

1. J. L. Schearer, A. T. Murphy, and H. H. Richardson, *Introduction to System Dynamics,* Addison-Wesley, Reading, MA, 1971.
2. E. O. Doebelin, *System Dynamics: Modeling and Response,* Merrill, Columbus, OH, 1972.
3. C. M. Close and D. K. Frederick, *Modeling and Analysis of Dynamic Systems,* 2nd ed., Houghton Mifflin, Boston, 1993.
4. W. J. Palm III, *Modeling, Analysis, and Control of Dynamic Systems,* Wiley, New York, 1983.
5. G. J. Thaler and R. G. Brown, *Analysis and Design of Feedback Control Systems,* 2nd ed., McGraw-Hill, New York, 1960.
6. G. A. Korn and J. V. Wait, *Digital Continuous System Simulation,* Prentice-Hall, Englewood Cliffs, NJ, 1975.
7. B. C. Kuo, *Automatic Control Systems,* 7th ed., Prentice-Hall, Englewood Cliffs, NJ, 1995.
8. W. Thissen, "Investigation into the World3 Model: Lessons for Understanding Complicated Models," *IEEE Transactions on Systems, Man, and Cybernetics,* **SMC-8** (3) (1978).
9. J. E. Gibson, *Nonlinear Automatic Control,* McGraw-Hill, New York, 1963.
10. S. M. Shinners, *Modern Control System Theory and Design,* Wiley, New York, 1992.
11. D. G. Luenberger, *Introduction to Dynamic Systems: Theory, Models, and Applications,* Wiley, New York, 1979.
12. D. M. Auslander, Y. Takahashi, and M. J. Rabins, *Introducing Systems and Control,* McGraw-Hill, New York, 1974.

BIBLIOGRAPHY

Bateson, R. N., *Introduction to Control System Technology,* Macmillan, New York, 1993.

Brogan, W. L., *Modern Control Theory,* 3rd ed., Prentice-Hall, Englewood Cliffs, NJ, 1991.

Cannon, Jr., R. H., *Dynamics of Physical Systems,* McGraw-Hill, New York, 1967.

DeSilva, C. W., *Control Sensors and Actuators,* Prentice-Hall, Englewood Cliffs, NJ, 1989.

Doebelin, E. O., *Measurement Systems,* 4th ed., McGraw-Hill, New York, 1990.

Dorf, R. C., and R. H. Bishop, *Modern Control Systems,* 7th ed., Addison-Wesley, Reading, MA, 1995.

Franklin, G. F., J. D. Powell, and A. Emami-Naeini, *Feedback Control of Dynamic Systems,* 3rd ed., Addison-Wesley, Reading, MA, 1994.

Grace, A., A. J. Laub, J. N. Little, and C. Thompson, *Control System Toolbox Users Guide,* The Mathworks, Natick, MA, 1990.

Hartley, T. T., G. O. Beale, and S. P. Chicatelli, *Digital Simulation of Dynamic Systems: A Control Theory Approach,* Prentice-Hall, Englewood Cliffs, NJ, 1994.

Kheir, N. A., *Systems Modeling and Computer Simulation,* 2nd ed., Marcel Dekker, New York, 1996.

Lay, D. C., *Linear Algebra and Its Applications,* Addison-Wesley, Reading, MA, 1994.

Ljung, L., and T. Glad, *Modeling Simulation of Dynamic Systems,* Prentice-Hall, Englewood Cliffs, NJ, 1994.

Mathworks, The, *The Student Edition of MATLAB 4 User's Guide,* Prentice-Hall, Englewood Cliffs, NJ, 1995.

Mathworks, The, *The Student Edition of SIMULINK User's Guide,* Prentice-Hall, Englewood Cliffs, NJ, 1996.

Pegden, C. D., R. E. Shannon, and R. P. Sadowski, *Introduction to Simulation Using SIMAN,* 2nd ed., McGraw-Hill, New York, 1995.

Phillips, C. L., and R. D. Harbor, *Feedback Control Systems,* 3rd ed., Prentice-Hall, Englewood Cliffs, NJ, 1996.

Phillips, C. L., and H. T. Nagle, *Digital Control System Analysis and Design,* 3rd ed., Prentice-Hall, Englewood Cliffs, NJ, 1995.

Van Loan, C., *Computational Frameworks for the Fast Fourier Transform,* SIAM, Philadelphia, PA, 1992.

Wolfram, S., *Mathematica: A System for Doing Mathematics by Computer,* 2nd ed., Addison-Wesley, Reading, MA, 1991.

CHAPTER 28

BASIC CONTROL SYSTEMS DESIGN

William J. Palm III
Mechanical Engineering Department
University of Rhode Island
Kingston, Rhode Island

28.1 INTRODUCTION   868

28.2 CONTROL SYSTEM
     STRUCTURE   869
     28.2.1 A Standard Diagram   870
     28.2.2 Transfer Functions   870
     28.2.3 System-Type Number and
             Error Coefficients   871

28.3 TRANSDUCERS AND ERROR
     DETECTORS   872
     28.3.1 Displacement and Velocity
             Transducers   872
     28.3.2 Temperature Transducers   874
     28.3.3 Flow Transducers   874
     28.3.4 Error Detectors   874
     28.3.5 Dynamic Response of
             Sensors   875

28.4 ACTUATORS   875
     28.4.1 Electromechanical
             Actuators   875
     28.4.2 Hydraulic Actuators   876
     28.4.3 Pneumatic Actuators   878

28.5 CONTROL LAWS   880
     28.5.1 Proportional Control   881
     28.5.2 Integral Control   883
     28.5.3 Proportional-Plus-Integral
             Control   884
     28.5.4 Derivative Control   884
     28.5.5 PID Control   885

28.6 CONTROLLER HARDWARE   886
     28.6.1 Feedback Compensation
             and Controller Design   886
     28.6.2 Electronic Controllers   886
     28.6.3 Pneumatic Controllers   887
     28.6.4 Hydraulic Controllers   887

28.7 FURTHER CRITERIA FOR
     GAIN SELECTION   887
     28.7.1 Performance Indices   889
     28.7.2 Optimal Control Methods   891

28.7.3 The Ziegler–Nichols
        Rules   891
28.7.4 Nonlinearities and
        Controller Performance   892
28.7.5 Reset Windup   893

28.8 COMPENSATION AND
     ALTERNATIVE CONTROL
     STRUCTURES   893
     28.8.1 Series Compensation   893
     28.8.2 Feedback
             Compensation
             and Cascade Control   893
     28.8.3 Feedforward
             Compensation   894
     28.8.4 State-Variable Feedback   895
     28.8.5 Pseudoderivative
             Feedback   896

28.9 GRAPHICAL DESIGN
     METHODS   896
     28.9.1 The Nyquist Stability
             Theorem   896
     28.9.2 Systems with Dead-Time
             Elements   898
     28.9.3 Open-Loop Design for
             PID Control   898
     28.9.4 Design with the Root
             Locus   899

28.10 PRINCIPLES OF DIGITAL
      CONTROL   901
      28.10.1 Digital Controller
               Structure   902
      28.10.2 Digital Forms of PID
               Control   902

28.11 UNIQUELY DIGITAL
      ALGORITHMS   903
      28.11.1 Digital Feedforward
               Compensation   904
      28.11.2 Control Design in the
               z-Plane   904
      28.11.3 Direct Design of Digital
               Algorithms   908

Revised from William J. Palm III, *Modeling, Analysis and Control of Dynamic Systems,* Wiley, 1983,
by permission of the publisher.

Mechanical Engineers' Handbook, 2nd ed., Edited by Myer Kutz.
ISBN 0-471-13007-9 © 1998 John Wiley & Sons, Inc.

28.12   **HARDWARE AND**   28.13   **FUTURE TRENDS IN**
        SOFTWARE FOR DIGITAL   **CONTROL SYSTEMS**   **912**
        CONTROL   **909**   28.13.1   Fuzzy Logic Control   913
        28.12.1   Digital Control   28.13.2   Neural Networks   914
                  Hardware   909   28.13.3   Nonlinear Control   914
        28.12.2   Software for Digital   28.13.4   Adaptive Control   914
                  Control   911   28.13.5   Optimal Control   914

28.1 INTRODUCTION

The purpose of a *control system* is to produce a desired *output.* This output is usually specified by the command *input,* and is often a function of time. For simple applications in well-structured situations, *sequencing* devices like timers can be used as the control system. But most systems are not that easy to control, and the controller must have the capability of reacting to disturbances, changes in its environment, and new input commands. The key element that allows a control system to do this is *feedback,* which is the process by which a system's output is used to influence its behavior. Feedback in the form of the room-temperature measurement is used to control the furnace in a thermostatically controlled heating system. Figure 28.1 shows the *feedback loop* in the system's *block diagram,* which is a graphical representation of the system's control structure and logic. Another commonly found control system is the pressure regulator shown in Fig. 28.2.

Feedback has several useful properties. A system whose individual elements are nonlinear can often be modeled as a linear one over a wider range of its variables with the proper use of feedback. This is because feedback tends to keep the system near its reference operation condition. Systems that can maintain the output near its desired value despite changes in the environment are said to have good *disturbance rejection.* Often we do not have accurate values for some system parameter, or these values might change with age. Feedback can be used to minimize the effects of parameter changes and uncertainties. A system that has both good disturbance rejection and low sensitivity to parameter variation is *robust.* The application that resulted in the general understanding of the properties of feedback is shown in Fig. 28.3. The electronic amplifier gain A is large, but we are uncertain of its exact value. We use the resistors R_1 and R_2 to create a feedback loop around the amplifier, and pick R_1 and R_2 to create a feedback loop around the amplifier, and pick R_1 and R_2 so that $AR_2/R_1 \gg 1$. Then the input–output relation becomes $e_o \approx R_1 e_i/R_2$, which is independent of A as long as A remains large. If R_1 and R_2 are known accurately, then the system gain is now reliable.

Figure 28.4 shows the block diagram of a *closed-loop* system, which is a system with feedback. An *open-loop* system, such as a timer, has no feedback. Figure 28.4 serves as a focus for outlining the prerequisites for this chapter. The reader should be familiar with the *transfer-function* concept based on the Laplace transform, the *pulse-transfer* function based on the *z*-transform, for digital control, and the differential equation modeling techniques needed to obtain them. It is also necessary to understand block-diagram algebra, characteristic roots, the final-value theorem, and their use in evaluating system response for common inputs like the step function. Also required are stability analysis techniques such as the Routh criterion, and transient performance specifications, such as the damping ratio ζ, natural frequency ω_n, dominant time constant τ, maximum overshoot, settling time, and bandwidth. The above material is reviewed in the previous chapter. Treatment in depth is given in Refs. 1, 2, and 3.

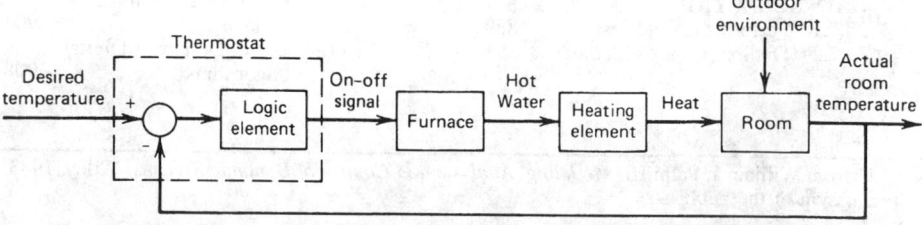

Fig. 28.1   Block diagram of the thermostat system for temperature control.[1]

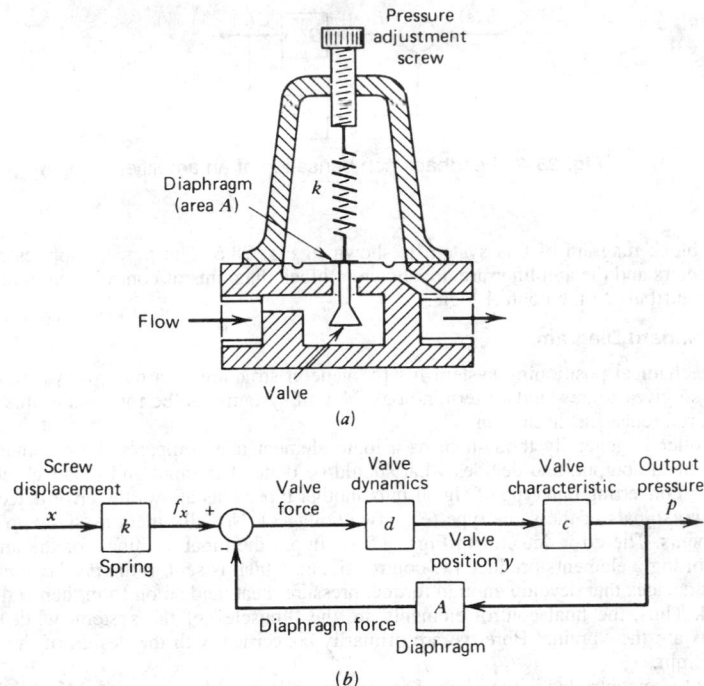

Fig. 28.2   Pressure regulator: (a) cutaway view; (b) block diagram.[1]

28.2   CONTROL SYSTEM STRUCTURE

The electromechanical position control system shown in Fig. 28.5 illustrates the structure of a typical control system. A load with an inertia I is to be positioned at some desired angle θ_r. A dc motor is provided for this purpose. The system contains viscous damping, and a disturbance torque T_d acts on the load, in addition to the motor torque T. Because of the disturbance, the angular position θ of the load will not necessarily equal the desired value θ_r. For this reason, a potentiometer, or some other sensor such as an encoder, is used to measure the displacement θ. The potentiometer voltage representing the controlled position θ is compared to the voltage generated by the command potentiometer. This device enables the operator to dial in the desired angle θ_r. The amplifier sees the difference e between the two potentiometer voltages. The basic function of the amplifier is to increase the small error voltage e up to the voltage level required by the motor and to supply enough current required by the motor to drive the load. In addition, the amplifier may shape the voltage signal in certain ways to improve the performance of the system.

The control system is seen to provide two basic functions: (1) to respond to a command input that specifies a new desired value for the controlled variable, and (2) to keep the controlled variable near the desired value in spite of disturbances. The presence of the feedback loop is vital to both

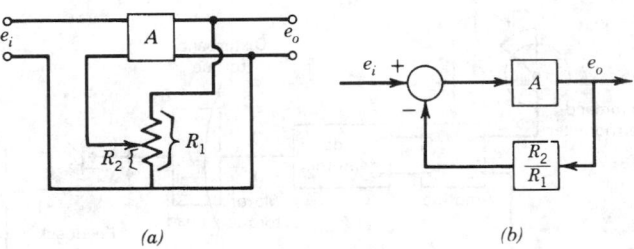

Fig. 28.3   A closed-loop system.

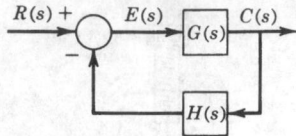

Fig. 28.4 Feedback compensation of an amplifier.

functions. A block diagram of this system is shown in Fig. 28.6. The power supplies required for the potentiometers and the amplifier are not shown in block diagrams of control system logic because they do not contribute to the control logic.

28.2.1 A Standard Diagram

The electromechanical positioning system fits the general structure of a control system (Fig. 28.7). This figure also gives some standard terminology. Not all systems can be forced into this format, but it serves as a reference for discussion.

The controller is generally thought of as a logic element that compares the command with the measurement of the output, and decides what should be done. The input and feedback elements are transducers for converting one type of signal into another type. This allows the error detector directly to compare two signals of the same type (e.g., two voltages). Not all functions show up as separate physical elements. The error detector in Fig. 28.5 is simply the input terminals of the amplifier.

The control logic elements produce the control signal, which is sent to the *final control elements.* These are the devices that develop enough torque, pressure, heat, and so on to influence the elements under control. Thus, the final control elements are the "muscle" of the system, while the control logic elements are the "brain." Here we are primarily concerned with the design of the logic to be used by this brain.

The object to be controlled is the *plant.* The *manipulated* variable is generated by the final control elements for this purpose. The disturbance input also acts on the plant. This is an input over which the designer has no influence, and perhaps for which little information is available as to the magnitude, functional form, or time of occurrence. The disturbance can be a random input, such as wind gust on a radar antenna, or deterministic, such as Coulomb friction effects. In the latter case, we can include the friction force in the system model by using a nominal value for the coefficient of friction. The disturbance input would then be the deviation of the friction force from this estimated value and would represent the uncertainty in our estimate.

Several control system classifications can be made with reference to Fig. 28.7. A *regulator* is a control system in which the controlled variable is to be kept constant in spite of disturbances. The command input for a regulator is its *set point.* A *follow-up system* is supposed to keep the control variable near a command value that is changing with time. An example of a follow-up system is a machine tool in which a cutting head must trace a specific path in order to shape the product properly. This is also an example of a *servomechanism,* which is a control system whose controlled variable is a mechanical position, velocity, or acceleration. A thermostat system is not a servomechanism, but a *process-control system,* where the controlled variable describes a thermodynamic process. Typically, such variables are temperature, pressure, flow rate, liquid level, chemical concentration, and so on.

28.2.2 Transfer Functions

A transfer function is defined for each input–output pair of the system. A specific transfer function is found by setting all other inputs to zero and reducing the block diagram. The *primary* or *command* transfer function for Fig. 28.7 is

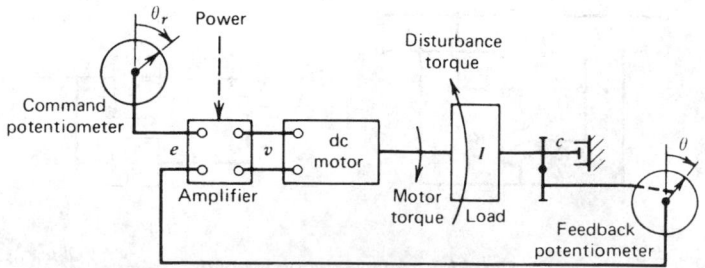

Fig. 28.5 Position-control system using a dc motor.[1]

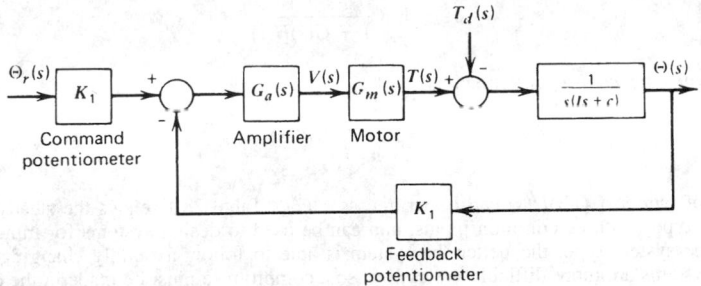

Fig. 28.6 Block diagram of the position-control system shown in Fig. 28.5.[1]

$$\frac{C(s)}{V(s)} = \frac{A(s)G_a(s)G_m(s)G_p(s)}{1 + G_a(s)G_m(s)G_p(s)H(s)} \tag{28.1}$$

The *disturbance* transfer function is

$$\frac{C(s)}{D(s)} = \frac{-Q(s)G_p(s)}{1 + G_a(s)G_m(s)G_p(s)H(s)} \tag{28.2}$$

The transfer functions of a given system all have the same denominator.

28.2.3 System-Type Number and Error Coefficients

The error signal in Fig. 28.4 is related to the input as

$$E(s) = \frac{1}{1 + G(s)H(s)} R(s) \tag{28.3}$$

If the final value theorem can be applied, the steady-state error is

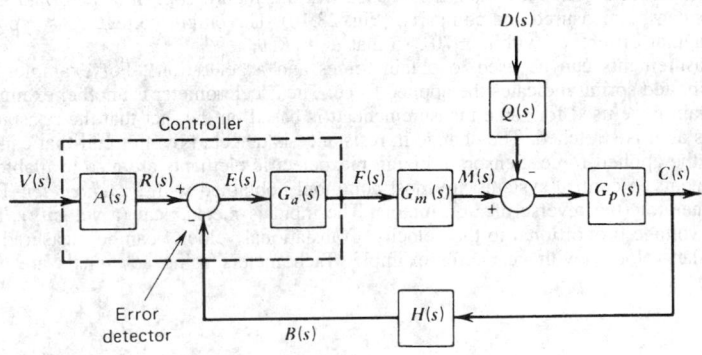

Elements		Signals	
$A(s)$	Input elements	$B(s)$	Feedback signal
$G_a(s)$	Control logic elements	$C(s)$	Controlled variable or output
$G_m(s)$	Final control elements	$D(s)$	Disturbance input
$G_p(s)$	Plant elements	$E(s)$	Error or actuating signal
$H(s)$	Feedback elements	$F(s)$	Control signal
$Q(s)$	Disturbance elements	$M(s)$	Manipulated variable
		$R(s)$	Reference input
		$V(s)$	Command input

Fig. 28.7 Terminology and basic structure of a feedback-control system.[1]

$$e_{ss} = \lim_{s \to 0} \frac{sR(s)}{1 + G(s)H(s)} \tag{28.4}$$

The *static error coefficient* c_i is defined as

$$c_i = \lim_{s \to 0} s^i G(s)H(s) \tag{28.5}$$

A system is of *type n* if $G(s)H(s)$ can be written as $s^n F(s)$. Table 28.1 relates the steady-state error to the system type for three common inputs, and can be used to design systems for minimum error. The higher the system type, the better the system is able to follow a rapidly changing input. But higher-type systems are more difficult to stabilize, so a compromise must be made in the design. The coefficients c_0, c_1, and c_2 are called the *position, velocity,* and *acceleration error coefficients.*

28.3 TRANSDUCERS AND ERROR DETECTORS

The control system structure shown in Fig. 28.7 indicates a need for physical devices to perform several types of functions. Here we present a brief overview of some available transducers and error detectors. Actuators and devices used to implement the control logic are discussed in Sections 28.4 and 28.5.

28.3.1 Displacement and Velocity Transducers

A *transducer* is a device that converts one type of signal into another type. An example is the potentiometer, which converts displacement into voltage, as in Fig. 28.8. In addition to this conversion, the transducer can be used to make measurements. In such applications, the term *sensor* is more appropriate. Displacement can also be measured electrically with a *linear variable differential transformer* (LVDT) or a *synchro*. An LVDT measures the linear displacement of a movable magnetic core through a primary winding and two secondary windings (Fig. 28.9). An ac voltage is applied to the primary. The secondaries are connected together and also to a detector that measures the voltage and phase difference. A phase difference of 0° corresponds to a positive core displacement, while 180° indicates a negative displacement. The amount of displacement is indicated by the amplitude of the ac voltage in the secondary. The detector converts this information into a dc voltage e_o, such that $e_o = Kx$. The LVDT is sensitive to small displacements. Two of them can be wired together to form an error detector.

A synchro is a rotary differential transformer, with angular displacement as either the input or output. They are often used in *pairs* (a *transmitter* and a *receiver*) where a remote indication of angular displacement is needed. When a transmitter is used with a synchro *control transformer,* two angular displacements can be measured and compared (Fig. 28.10). The output voltage e_0 is approximately linear with angular difference within $\pm 70°$, so that $e_0 = K(\theta_1 - \theta_2)$.

Displacement measurements can be used to obtain forces and accelerations. For example, the displacement of a calibrated spring indicates the applied force. The accelerometer is another example. Still another is the strain gage used for force measurement. It is based on the fact that the resistance of a fine wire changes as it is stretched. The change in resistance is detected by a circuit that can be calibrated to indicate the applied force. Sensors utilizing piezoelectric elements are also available.

Velocity measurements in control systems are most commonly obtained with a *tachometer.* This is essentially a dc generator (the reverse of a dc motor). The input is mechanical (a velocity). The output is a generated voltage proportional to the velocity. Translational velocity can be measured by converting it to angular velocity with gears, for example. Tachometers using ac signals are also available.

Table 28.1 Steady-State Error e_{ss} for Different System-Type Numbers

R(s)	System Type Number n			
	0	1	2	3
Step $1/s$	$\dfrac{1}{1 + C_0}$	0	0	0
Ramp $1/s^2$	∞	$\dfrac{1}{C_1}$	0	0
Parabola $1/s^3$	∞	∞	$\dfrac{1}{C_2}$	0

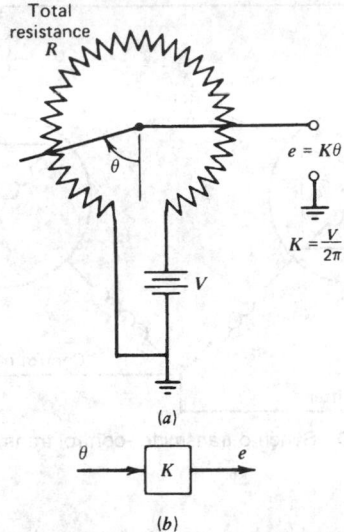

Fig. 28.8   Rotary potentiometer.[1]

Other velocity transducers include a magnetic pickup that generates a pulse every time a gear tooth passes. If the number of gear teeth is known, a pulse counter and timer can be used to compute the angular velocity. This principle is also employed in turbine flowmeters.

A similar principle is employed by *optical encoders*, which are especially suitable for digital control purposes. These devices use a rotating disk with alternating transparent and opaque elements whose passage is sensed by light beams and a photo-sensor array, which generates a binary (on–off) train of pulses. There are two basic types: the absolute encoder and the incremental encoder. By counting the number of pulses in a given time interval, the incremental encoder can measure the rotational speed of the disk. By using multiple tracks of elements, the absolute encoder can produce a binary digit that indicates the amount of rotation. Hence, it can be used as a position sensor.

Most encoders generate a train of TTL voltage level pulses for each channel. The incremental encoder output contains two channels that each produce N pulses every revolution. The encoder is mechanically constructed so that pulses from one channel are shifted relative to the other channel by a quarter of a pulse width. Thus, each pulse pair can be divided into four segments called *quadratures*. The encoder output consists of 4N *quadrature counts per revolution*. The pulse shift also allows the

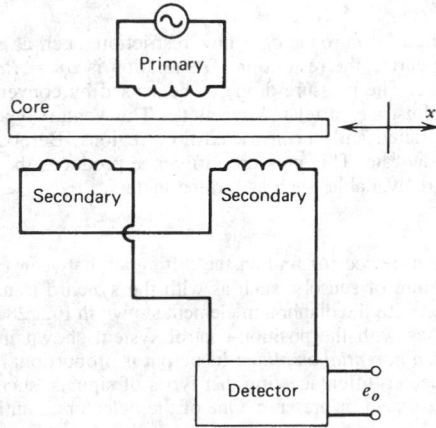

Fig. 28.9   Linear variable differential transformer (LVDT).[1]

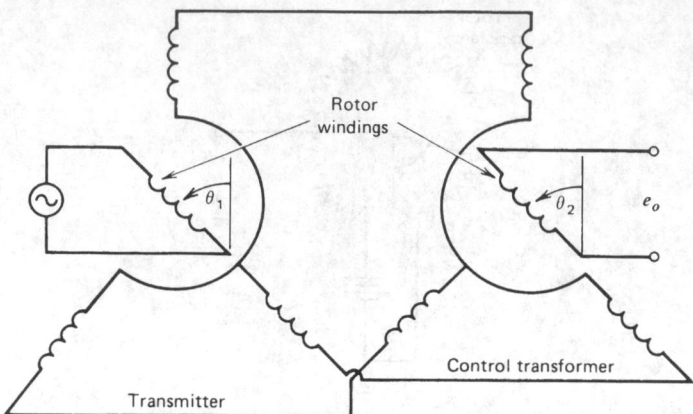

Fig. 28.10 Synchro transmitter-control transformer.[1]

direction of rotation to be determined by detecting which channel leads the other. The encoder might contain a third channel, known as the zero, index, or marker channel, that produces a pulse once per revolution. This is used for initialization.

The gain of such an incremental encoder is $4N/2\pi$. Thus, an encoder with 1000 pulses per channel per revolution has a gain of 636 counts per radian. If an absolute encoder produces a binary signal with n bits, the maximum number of positions it can represent is $2n$, and its gain is $2^n/2\pi$. Thus, a 16-bit absolute encoder has a gain of $2^{16}/2\pi = 10,435$ counts per radian.

28.3.2 Temperature Transducers

When two wires of dissimilar metals are joined together, a voltage is generated if the junctions are at different temperatures. If the reference junction is kept at a fixed, known temperature, the thermocouple can be calibrated to indicate the temperature at the other junction in terms of the voltage v. Electrical resistance changes with temperature. Platinum gives a linear relation between resistance and temperature, while nickel is less expensive and gives a large resistance change for a given temperature change. Seminconductors designed with this property are called *thermistors*. Different metals expand at different rates when the temperature is increased. This fact is used in the bimetallic strip transducer found in most home thermostats. Two dissimilar metals are bonded together to form the strip. As the temperature rises, the strip curls, breaking contact and shutting off the furnace. The temperature gap can be adjusted by changing the distance between the contacts. The motion also moves a pointer on the temperature scale of the thermostat. Finally, the pressure of a fluid inside a bulb will change as its temperature changes. If the bulb fluid is air, the device is suitable for use in pneumatic temperature controllers.

28.3.3 Flow Transducers

A flow rate q can be measured by introducing a flow restriction, such as an orifice plate, and measuring the pressure drop Δp across the restriction. The relation is $\Delta p = Rq^2$, where R can be found from calibration of the device. The pressure drop can be sensed by converting it into the motion of a diaphragm. Figure 28.11 illustrates a related technique. The Venturi-type flowmeter measures the static pressures in the constricted and unconstricted flow regions. Bernoulli's principle relates the pressure difference to the flow rate. This pressure difference produces the diaphragm displacement. Other types of flowmeters are available, such as turbine meters.

28.3.4 Error Detectors

The error detector is simply a device for finding the difference between two signals. This function is sometimes an integral feature of sensors, such as with the synchro transmitter–transformer combination. This concept is used with the diaphragm element shown in Fig. 28.11. A detector for voltage difference can be obtained, as with the position-control system shown in Fig. 28.5. An amplifier intended for this purpose is a *differential amplifier.* Its output is proportional to the difference between the two inputs. In order to detect differences in other types of signals, such as temperature, they are usually converted to a displacement or pressure. One of the detectors mentioned previously can then be used.

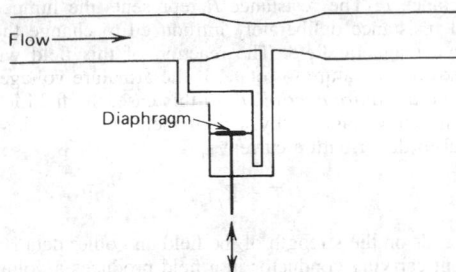

Fig. 28.11 Venturi-type flowmeter. The diaphragm displacement indicates the flow rate.[1]

28.3.5 Dynamic Response of Sensors

The usual transducer and detector models are static models, and as such imply that the components respond instantaneously to the variable being sensed. Of course, any real component has a dynamic response of some sort, and this response time must be considered in relation to the controlled process when a sensor is selected. If the controlled process has a time constant at least 10 times greater than that of the sensor, we often would be justified in using a static sensor model.

28.4 ACTUATORS

An *actuator* is the final control element that operates on the low-level control signal to produce a signal containing enough power to drive the plant for the intended purpose. The armature-controlled dc motor, the hydraulic servomotor, and the pneumatic diaphragm and piston are common examples of actuators.

28.4.1 Electromechanical Actuators

Figure 28.12 shows an electromechanical system consisting of an armature-controlled dc motor driving a load inertia. The rotating armature consists of a wire conductor wrapped around an iron core.

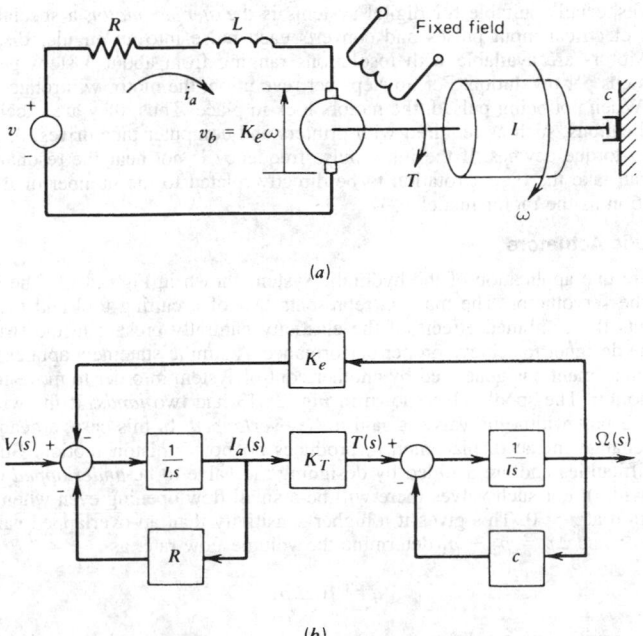

Fig. 28.12 Armature-controlled dc motor with a load, and the system's block diagram.[1]

This winding has an inductance L. The resistance R represents the lumped value of the armature resistance and any external resistance deliberately introduced to change the motor's behavior. The armature is surrounded by a magnetic field. The reaction of this field with the armature current produces a torque that causes the armature to rotate. If the armature voltage v is used to control the motor, the motor is said to be *armature-controlled*. In this case, the field is produced by an electromagnet supplied with a constant voltage or by a permanent magnet. This motor type produces a torque T that is proportional to the armature current i_a:

$$T = K_T i_a \qquad (28.6)$$

The torque constant K_T depends on the strength of the field and other details of the motor's construction. The motion of a current-carrying conductor in a field produces a voltage in the conductor that opposes the current. This voltage is called the *back emf* (electromotive force). Its magnitude is proportional to the speed and is given by

$$e_b = K_e \omega \qquad (28.7)$$

The transfer function for the armature-controlled dc motor is

$$\frac{\Omega(s)}{V(s)} = \frac{K_T}{LIs^2 + (RI + cL)s + cR + K_e K_T} \qquad (28.8)$$

Another motor configuration is the *field-controlled* dc motor. In this case, the armature current is kept constant and the field voltage v is used to control the motor. The transfer function is

$$\frac{\Omega(s)}{V(s)} = \frac{K_T}{(Ls + R)(Is + c)} \qquad (28.9)$$

where R and L are the resistance and inductance of the field circuit, and K_T is the torque constant. No back emf exists in this motor to act as a self-braking mechanism.

Two-phase ac motors can be used to provide a low-power, variable-speed actuator. This motor type can accept the ac signals directly from LVDTs and synchros without demodulation. However, it is difficult to design ac amplifier circuitry to do other than proportional action. For this reason, the ac motor is not found in control systems as often as dc motors. The transfer function for this type is of the form of Eq. (28.9).

An actuator especially suitable for digital systems is the *stepper motor*, a special dc motor that takes a train of electrical input pulses and converts each pulse into an angular displacement of a fixed amount. Motors are available with resolutions ranging from about 4 steps per revolution to more than 800 steps per revolution. For 36 steps per revolution, the motor will rotate by 10° for each pulse received. When not being pulsed, the motors lock in place. Thus, they are excellent for precise positioning applications, such as required with printers and computer tape drives. A disadvantage is that they are low-torque devices. If the input pulse frequency is not near the resonant frequency of the motor, we can take the output rotation to be directly related to the number of input pulses and use that description as the motor model.

28.4.2 Hydraulic Actuators

Machine tools are one application of the hydraulic system shown in Fig. 28.13. The applied force f is supplied by the servomotor. The mass m represents that of a cutting tool and the power piston, while k represents the combined effects of the elasticity naturally present in the structure and that introduced by the designer to achieve proper performance. A similar statement applies to the damping c. The valve displacement z is generated by another control system in order to move the tool through its prescribed motion. The spool valve shown in Fig. 28.13 had two *lands*. If the width of the land is greater than the port width, the valve is said to be *overlapped*. In this case, a dead zone exists in which a slight change in the displacement z produces no power piston motion. Such dead zones create control difficulties and are avoided by designing the valve to be *underlapped* (the land width is less the port width). For such valves there will be a small flow opening even when the valve is in the neutral position at $z = 0$. This gives it a higher sensitivity than an overlapped valve.

The variables z and $\Delta p = p_2 - p_1$ determine the volume flow rate, as

$$q = f(z, \Delta p)$$

For the reference equilibrium condition ($z = 0$, $\Delta p = 0$, $q = 0$), a linearization gives

$$q = C_1 z - C_2 \Delta p \qquad (28.10)$$

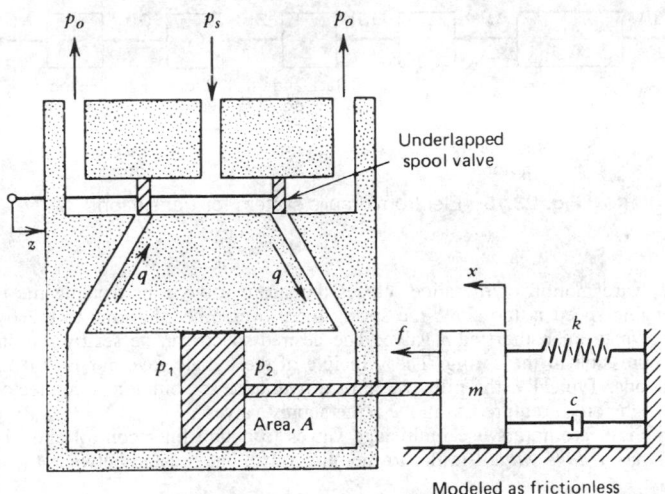

Fig. 28.13 Hydraulic servomotor with a load.[1]

The linearization constants are available from theoretical and experimental results.[4] The transfer function for the system is[1,2]

$$T(s) = \frac{X(s)}{Z(s)} = \frac{C_1}{\dfrac{C_2 m}{A} s^2 + \left(\dfrac{cC_2}{A} + A\right) s + \dfrac{C_2 k}{A}} \tag{28.11}$$

The development of the steam engine led to the requirement for a speed-control device to maintain constant speed in the presence of changes in load torque or steam pressure. In 1788, James Watt of Glasgow developed his now-famous flyball governor for this purpose (Fig. 28.14). Watt took the principle of sensing speed with the centrifugal pendulum of Thomas Mead and used it in a feedback loop on a steam engine. As the motor speed increases, the flyballs move outward and pull the slider

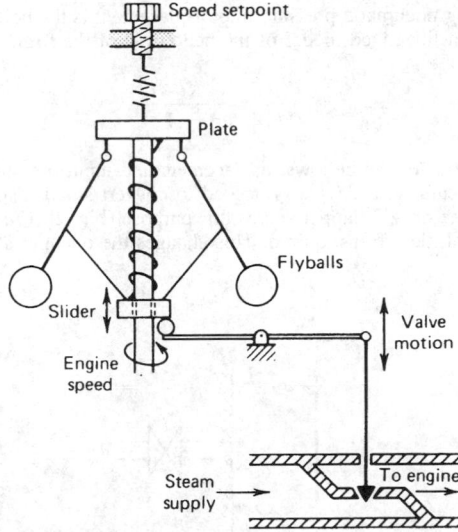

Fig. 28.14 James Watt's flyball governor for speed control of a steam engine.[1]

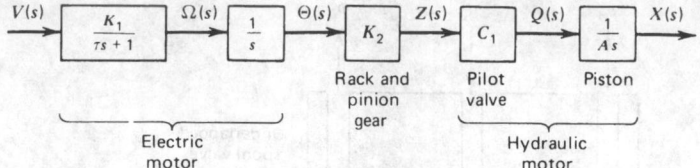

Fig. 28.15 Electrohydraulic system for translation.[1]

upward. The upward motion of the slider closes the steam valve, thus causing the engine to slow down. If the engine speed is too slow, the spring force overcomes that due to the flyballs, and the slider moves down to open the steam valve. The desired speed can be set by moving the plate to change the compression in the spring. The principle of the flyball governor is still used for speed-control applications. Typically, the pilot valve of a hydraulic servomotor is connected to the slider to provide the high forces required to move large supply valves.

Many hydraulic servomotors use multistage valves to obtain finer control and higher forces. A *two-stage valve* has a *slave value,* similar to the pilot valve, but situated between the pilot valve and the power piston.

Rotational motion can be obtained with a *hydraulic motor,* which is, in principle, a pump acting in reverse (fluid input and mechanical rotation output). Such motors can achieve higher torque levels than electric motors. A hydraulic pump driving a hydraulic motor constitutes a *hydraulic transmission.*

A popular actuator choice is the *electrohydraulic* system, which uses an electric actuator to control a hydraulic servomotor or transmission by moving the pilot valve or the swash-plate angle of the pump. Such systems combine the power of hydraulics with the advantages of electrical systems. Figure 28.15 shows a hydraulic motor whose pilot valve motion is caused by an armature-controlled dc motor. The transfer function between the motor voltage and the piston displacement is

$$\frac{X(s)}{V(s)} = \frac{K_1 K_2 C_1}{As^2(\tau s + 1)}$$

(28.12)

If the rotational inertia of the electric motor is small, then $\tau \approx 0$.

28.4.3 Pneumatic Actuators

Pneumatic actuators are commonly used because they are simple to maintain and use a readily available working medium. Compressed air supplies with the pressures required are commonly available in factories and laboratories. No flammable fluids or electrical sparks are present, so these devices are considered the safest to use with chemical processes. Their power output is less than that of hydraulic systems, but greater than that of electric motors.

A device for converting pneumatic pressure into displacement is the bellows shown in Fig. 28.16. The transfer function for a linearized model of the bellows is of the form

$$\frac{X(s)}{P(s)} = \frac{K}{\tau s + 1}$$

(28.13)

where x and p are deviations of the bellows displacement and input pressure from nominal values.

In many control applications, a device is needed to convert small displacements into relatively large pressure changes. The nozzle-flapper serves this purpose (Fig. 28.17a). The input displacement y moves the flapper, with little effort required. This changes the opening at the nozzle orifice. For a

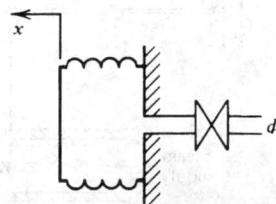

Fig. 28.16 Pneumatic bellows.[1]

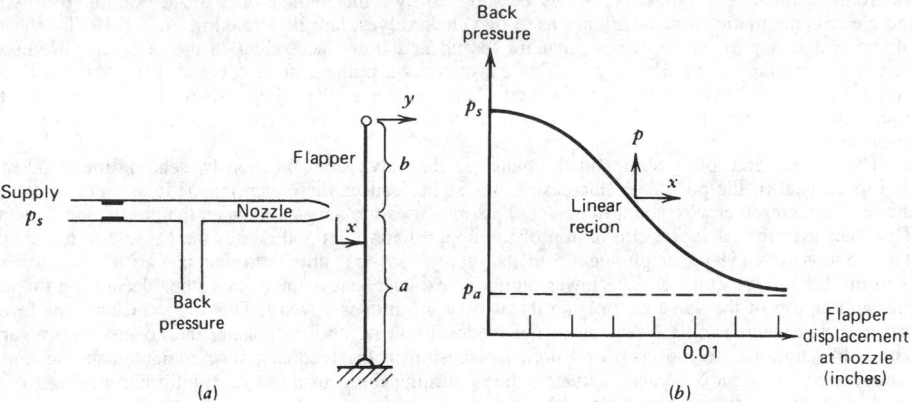

Fig. 28.17 Pneumatic nozzle-flapper amplifier and its characteristic curve.[1]

large enough opening, the nozzle back pressure is approximately the same as atmospheric pressure p_a. At the other extreme position with the flapper completely blocking the orifice, the back pressure equals the supply pressure p_s. This variation is shown in Fig. 28.17b. Typical supply pressures are between 30 and 100 psia. The orifice diameter is approximately 0.01 in. Flapper displacement is usually less than one orifice diameter.

The nozzle-flapper is operated in the linear portion of the back pressure curve. The linearized back pressure relation is

$$p = -K_f x \qquad (28.14)$$

where $-K_f$ is the slope of the curve and is a very large number. From the geometry of similar triangles, we have

$$p = -\frac{aK_f}{a + b} y \qquad (28.15)$$

In its operating region, the nozzle-flapper's back pressure is well below the supply pressure.

The output pressure from a pneumatic device can be used to drive a final control element like the pneumatic actuating valve shown in Fig. 28.18. The pneumatic pressure acts on the upper side of the diaphragm and is opposed by the return spring.

Formerly, many control systems utilized pneumatic devices to implement the control law in analog form. Although the overall, or higher-level, control algorithm is now usually implemented in digital form, pneumatic devices are still frequently used for final control corrections at the actuator level,

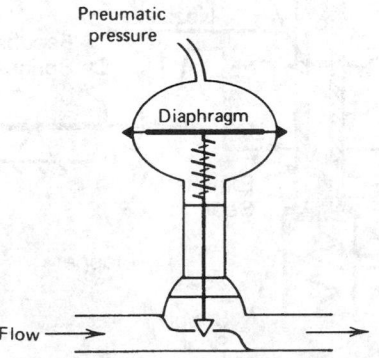

Fig. 28.18 Pneumatic flow-control valve.[1]

where the control action must eventually be supplied by a mechanical device. An example of this is the electro-pneumatic valve positioner used in Valtek valves, and illustrated in Fig. 28.19. The heart of the unit is a pilot valve capsule that moves up and down according to the pressure difference across its two supporting diaphragms. The capsule has a plunger at its top and at its bottom. Each plunger has an exhaust seat at one end and a supply seat at the other. When the capsule is in its equilibrium position, no air is supplied to or exhausted from the valve cylinder, so the valve does not move.

The process controller commands a change in the valve stem position by sending the 4–20 ma dc input signal to the positioner. Increasing this signal causes the electromagnetic actuator to rotate the lever counterclockwise about the pivot. This increases the air gap between the nozzle and flapper. This decreases the back pressure on top of the upper diaphragm and causes the capsule to move up. This motion lifts the upper plunger from its supply seat and allows the supply air to flow to the bottom of the valve cylinder. The lower plunger's exhaust seat is uncovered, thus decreasing the air pressure on top of the valve piston, and the valve stem moves upward. This motion causes the lever arm to rotate, increasing the tension in the feedback spring and decreasing the nozzle-flapper gap. The valve continues to move upward until the tension in the feedback spring counteracts the force produced by the electromagnetic actuator, thus returning the capsule to its equilibrium position.

A decrease in the dc input signal causes the opposite actions to occur, and the valve moves downward.

28.5 CONTROL LAWS

The control logic elements are designed to act on the error signal to produce the control signal. The algorithm that is used for this purpose is called the *control law*, the *control action,* or the *control algorithm*. A nonzero error signal results from either a change in command or a disturbance. The general function of the controller is to keep the controlled variable near its desired value when these occur. More specifically, the control objectives might be stated as follows:

1. Minimize the steady-state error.
2. Minimize the settling time.
3. Achieve other transient specifications, such as minimizing the overshoot.

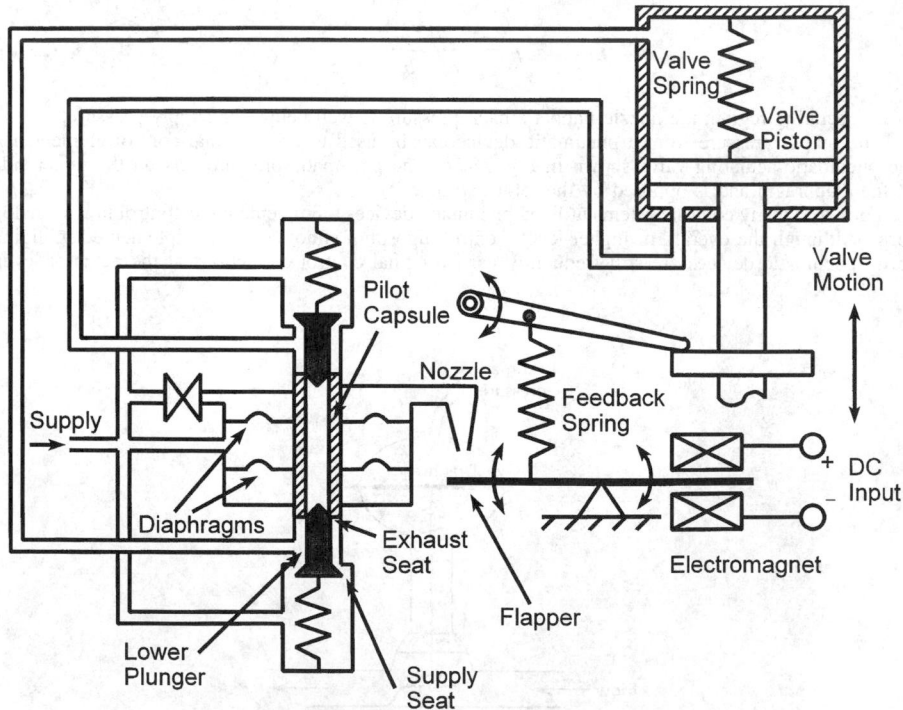

Fig. 28.19 An electro-pneumatic valve positioner.

In practice, the design specifications for a controller are more detailed. For example, the bandwidth might also be specified along with a safety margin for stability. We never know the numerical values of the system's parameters with true certainty, and some controller designs can be more sensitive to such parameter uncertainties than other designs. So a parameter sensitivity specification might also be included.

The following control laws form the basis of most control systems.

28.5.1 Proportional Control

Two-position control is the most familiar type, perhaps because of its use in home thermostats. The control output takes on one of two values. With the *on–off controller,* the controller output is either on or off (e.g., fully open or fully closed). Two-position control is acceptable for many applications in which the requirements are not too severe. However, many situations require finer control.

Consider a liquid-level system in which the input flowrate is controlled by a valve. We might try setting the control valve manually to achieve a flow rate that balances the system at the desired level. We might then added a controller that adjusts this setting in proportion to the deviation of the level from the desired value. This is *proportional control,* the algorithm in which the change in the control signal is proportional to the error. Block diagrams for controllers are often drawn in terms of the deviations from a zero-error equilibrium condition. Applying this convention to the general terminology of Fig. 28.6, we see that proportional control is described by

$$F(s) = K_P E(s)$$

where $F(s)$ is the deviation in the control signal and K_P is the *proportional gain.* If the total valve displacement is $y(t)$ and the manually created displacement is x, then

$$y(t) = K_P e(t) + x$$

The percent change in error needed to move the valve full scale is the *proportional band.* It is related to the gain as

$$K_P = \frac{100}{\text{band}\%}$$

The zero-error valve displacement x is the *manual reset.*

Proportional Control of a First-Order System

To investigate the behavior of proportional control, consider the speed-control system shown in Fig. 28.20; it is identical to the position controller shown in Fig. 28.6, except that a tachometer replaces the feedback potentiometer. We can combine the amplifier gains into one, denoted K_P. The system is thus seen to have proportional control. We assume the motor is field-controlled and has a negligible electrical time constant. The disturbance is a torque T_d, for example, resulting from friction. Choose the reference equilibrium condition to be $T_d = T = 0$ and $\omega_r = w = 0$. The block diagram is shown in Fig. 28.21. For a meaningful error signal to be generated, K_1 and K_2 should be chosen to be equal. With this simplification the diagram becomes that shown in Fig. 28.22, where $G(s) = K = K_1 K_P K_T/R$. A change in desired speed can be simulated by a unit step input for ω_r. For $\Omega_r(s) = 1/s$, the velocity approaches the steady-state value $\omega_{ss} = K/(c + K) < 1$. Thus, the final value is less than the desired value of 1, but it might be close enough if the damping c is small. The time required to

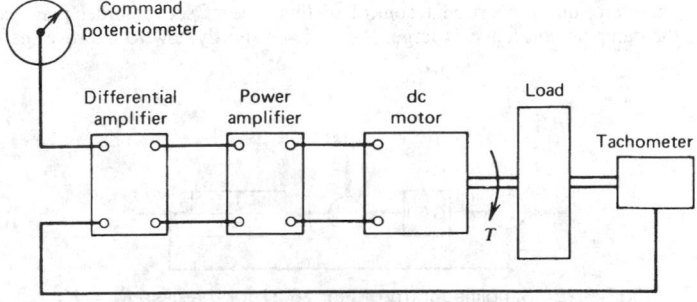

Fig. 28.20 Velocity-control system using a dc motor.[1]

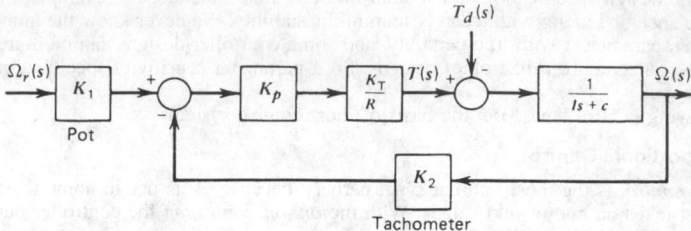

Fig. 28.21 Block diagram of the velocity-control system of Fig. 28.20.[1]

reach this value is approximately four time constants, or $4\tau = 4I/(c + K)$. A sudden change in load torque can also be modeled by a unit step function $T_d(s) = 1/s$. The steady-state response due solely to the disturbance is $-1/(c + K)$. If $(c + K)$ is large, this error will be small.

The performance of the proportional control law thus far can be summarized as follows. For a first-order plant with step function inputs:

1. The output never reaches its desired value if damping is present $(c \neq 0)$, although it can be made arbitrarily close by choosing the gain K large enough. This is called *offset error*.
2. The output approaches its final value without oscillation. The time to reach this value is inversely proportional to K.
3. The output deviation due to the disturbance at steady state is inversely proportional to the gain K. This error is present even in the absence of damping $(c = 0)$.

As the gain K is increased, the time constant becomes smaller and the response faster. Thus, the chief disadvantage of proportional control is that it results in steady-state errors and can only be used when the gain can be selected large enough to reduce the effect of the largest expected disturbance. Since proportional control gives zero error only for one load condition (the reference equilibrium), the operator must change the manual reset by hand (hence the name). An advantage to proportional control is that the control signal responds to the error instantaneously (in theory at least). It is used in applications requiring rapid action. Processes with time constants too small for the use of two-position control are likely candidates for proportional control. The results of this analysis can be applied to any type of first-order system (e.g., liquid-level, thermal, etc.) having the form in Fig. 28.22.

Proportional Control of a Second-Order System

Proportional control of a neutrally stable second-order plant is represented by the position controller of Fig. 28.6 if the amplifier transfer function is a constant $G_a(s) = K_a$. Let the motor transfer function be $G_m(s) = K_T/R$, as before. The modified block diagram is given in Fig. 28.23 with $G(s) = K = K_1 K_a K_T/R$. The closed-loop system is stable if I, c, and K are positive. For no damping $(c = 0)$, the closed-loop system is neutrally stable. With no disturbance and a unit step command, $\Theta_r(s) = 1/s$, the steady-state output is $\omega_{ss} = 1$. The offset error is thus zero if the system is stable $(c > 0, K > 0)$. The steady-state output deviation due to a unit step disturbance is $-1/K$. This deviation can be reduced by choosing K large. The transient behavior is indicated by the damping ratio, $\zeta = c/2\sqrt{IK}$.

For slight damping, the response to a step input will be very oscillatory and the overshoot large. The situation is aggravated if the gain K is made large to reduce the deviation due to the disturbance. We conclude, therefore, that proportional control of this type of second-order plant is not a good choice unless the damping constant c is large. We will see shortly how to improve the design.

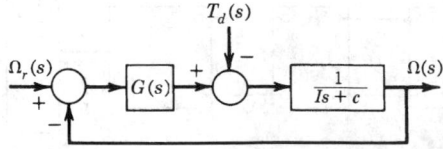

Fig. 28.22 Simplified form of Fig. 28.21 for the case $K_1 = K_2$.

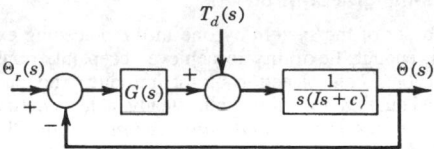

Fig. 28.23 Position servo.

28.5.2 Integral Control

The offset error that occurs with proportional control is a result of the system reaching an equilibrium in which the control signal no longer changes. This allows a constant error to exist. If the controller is modified to produce an increasing signal as long as the error is nonzero, the offset might be eliminated. This is the principle of *integral control*. In this mode the change in the control signal is proportional to the *integral* of the error. In the terminology of Fig. 28.7, this gives

$$F(s) = \frac{K_I}{s} E(s) \tag{28.16}$$

where $F(s)$ is the deviation in the control signal and K_I is the *integral gain*. In the time domain, the relation is

$$f(t) = K_I \int_0^t e(t)\, dt \tag{28.17}$$

if $f(0) = 0$. In this form, it can be seen that the integration cannot continue indefinitely because it would theoretically produce an infinite value of $f(t)$ if $e(t)$ does not change sign. This implies that special care must be taken to reinitialize a controller that uses integral action.

Integral Control of a First-Order System

Integral control of the velocity in the system of Fig. 28.20 has the block diagram shown in Fig. 28.22, where $G(s) = K/s$, $K = K_1 K_I K_T/R$. The integrating action of the amplifier is physically obtained by the techniques to be presented in Section 28.6, or by the digital methods presented in Section 28.10. The control system is stable if I, c, and K are positive. For a unit step command input, $\omega_{ss} = 1$; so the offset error is zero. For a unit step disturbance, the steady-state deviation is zero if the system is stable. Thus, the steady-state performance using integral control is excellent for this plant with step inputs. The damping ratio is $\zeta = c/2\sqrt{IK}$. For slight damping, the response will be oscillatory rather than exponential as with proportional control. Improved steady-state performance has thus been obtained at the expense of degraded transient performance. The conflict between steady-state and transient specifications is a common theme in control system design. As long as the system is underdamped, the time constant is $\tau = 2I/c$ and is not affected by the gain K, which only influences the oscillation frequency in this case. It might be physically possible to make K small enough so that $\zeta >> 1$, and the nonoscillatory feature of proportional control recovered, but the response would tend to be sluggish. Transient specifications for fast response generally require that $\zeta < 1$. The difficulty with using $\zeta < 1$ is that τ is fixed by c and I. If c and I are such that $\zeta < 1$, then τ is large if $I >> c$.

Integral Control of a Second-Order System

Proportional control of the position servomechanism in Fig. 28.23 gives a nonzero steady-state deviation due to the disturbance. Integral control [$G(s) = K/s$] applied to this system results in the command transfer function

$$\frac{\Theta(s)}{\Theta_r(s)} = \frac{K}{Is^3 + cs^2 + K} \tag{28.18}$$

With the Routh criterion, we immediately see that the system is not stable because of the missing s term. Integral control is useful in improving steady-state performance, but in general it does not improve and may even degrade transient performance. Improperly applied, it can produce an unstable control system. It is best used in conjunction with other control modes.

28.5.3 Proportional-Plus-Integral Control

Integral control raised the order of the system by one in the preceding examples, but did not give a characteristic equation with enough flexibility to achieve acceptable transient behavior. The instantaneous response of proportional control action might introduce enough variability into the coefficients of the characteristic equation to allow both steady-state and transient specifications to be satisfied. This is the basis for using *proportional-plus-integral control* (PI control). The algorithm for this two-mode control is

$$F(s) = K_P E(s) + \frac{K_I}{s} E(s) \tag{28.19}$$

The integral action provides an automatic, not manual, reset of the controller in the presence of a disturbance. For this reason, it is often called *reset action*.

The algorithm is sometimes expressed as

$$F(s) = K_P \left(1 + \frac{1}{T_I s} \right) E(s) \tag{28.20}$$

where T_I is the *reset time*. The reset time is the time required for the integral action signal to equal that of the proportional term, if a constant error exists (a hypothetical situation). The reciprocal of reset time is expressed as repeats per minute and is the frequency with which the integral action repeats the proportional correction signal.

The proportional control gain must be reduced when used with integral action. The integral term does not react instantaneously to a zero-error signal but continues to correct, which tends to cause oscillations if the designer does not take this effect into account.

PI Control of a First-Order System

PI action applied to the speed controller of Fig. 28.20 gives the diagram shown in Fig. 28.21 with $G(s) = K_P + K_I/s$. The gains K_P and K_I are related to the component gains, as before. The system is stable for positive values of K_P and K_I. For $\Omega_r(s) = 1/s$, $\omega_{ss} = 1$, and the offset error is zero, as with integral action only. Similarly, the deviation due to a unit step disturbance is zero at steady state. The damping ratio is $\zeta = (c + K_P)/2\sqrt{IK_I}$. The presence of K_P allows the damping ratio to be selected without fixing the value of the dominant time constant. For example, if the system is underdamped ($\zeta < 1$), the time constant is $\tau = 2I/(c + K_P)$. The gain K_P can be picked to obtain the desired time constant, while K_I used to set, the damping ratio. A similar flexibility exists if $\zeta = 1$. Complete description of the transient response requires that the numerator dynamics present in the transfer functions be accounted for.[1,2]

PI Control of a Second-Order System

Integral control for the position servomechanism of Fig. 28.23 resulted in a third-order system that is unstable. With proportional action, the diagram becomes that of Fig. 28.22, with $G(s) = K_P + K_I/s$. The steady-state performance is acceptable, as before, if the system is assumed to be stable. This is true if the Routh criterion is satisfied; that is, if I, c, K_P, and K_I are positive and $cK_P - IK_I > 0$. The difficulty here occurs when the damping is slight. For small c, the gain K_P must be large in order to satisfy the last condition, and this can be difficult to implement physically. Such a condition can also result in an unsatisfactory time constant. The root-locus method of Section 28.9 provides the tools for analyzing this design further.

28.5.4 Derivative Control

Integral action tends to produce a control signal even after the error has vanished, which suggests that the controller be made aware that the error is approaching zero. One way to accomplish this is to design the controller to react to the derivative of the error with *derivative control* action, which is

$$F(s) = K_D s E(s) \tag{28.21}$$

where K_D is the *derivative gain*. This algorithm is also called *rate action*. It is used to damp out oscillations. Since it depends only on the error rate, derivative control should never be used alone. When used with proportional action, the following PD-control algorithm results:

$$F(s) = (K_P + K_D s)E(s) = K_P(1 + T_D s)E(s) \tag{28.22}$$

where T_D is the *rate time* or *derivative time*. With integral action included, the *proportional-plus-integral-plus-derivative* (PID) control law is obtained.

$$F(s) = \left(K_P + \frac{K_I}{s} + K_D s \right) E(s) \tag{28.23}$$

This is called a *three-mode controller.*

PD Control of a Second-Order System

The presence of integral action reduces steady-state error, but tends to make the system less stable. There are applications of the position servomechanism in which a nonzero derivation resulting from the disturbance can be tolerated, but an improvement in transient response over the proportional control result is desired. Integral action would not be required, but rate action can be added to improve the transient response. Application of PD control to this system gives the block diagram of Fig. 28.23 with $G(s) = K_P + K_D s$.

The system is stable for positive values of K_D and K_P. The presence of rate action does not affect the steady-state response, and the steady-state results are identical to those with P control; namely, zero offset error and a deviation of $-1/K_P$, due to the disturbance. The damping ratio is $\zeta = (c + K_D)/2\sqrt{IK_P}$. For P control, $\zeta = c/2\sqrt{IK_P}$. Introduction of rate action allows the proportional gain K_P to be selected large to reduce the steady-state deviation, while K_D can be used to achieve an acceptable damping ratio. The rate action also helps to stabilize the system by adding damping (if $c = 0$ the system with P control is not stable).

The equivalent of derivative action can be obtained by using a tachometer to measure the angular velocity of the load. The block diagram is shown in Fig. 28.24. The gain of the amplifier–motor–potentiometer combination is K_1, and K_2 is the tachometer gain. The advantage of this system is that it does not require signal differentiation, which is difficult to implement if signal noise is present. The gains K_1 and K_2 can be chosen to yield the desired damping ratio and steady-state deviation, as was done with K_P and K_I.

28.5.5 PID Control

The position servomechanism design with PI control is not completely satisfactory because of the difficulties encountered when the damping c is small. This problem can be solved by the use of the full PID-control law, as shown in Fig. 28.23 with $G(s) = K_P + K_D s + K_I/s$.

A stable system results if all gains are positive and if $(c + K_D)K_P - IK_I > 0$. The presence of K_D relaxes somewhat the requirement that K_P be large to achieve stability. The steady-state errors are zero, and the transient response can be improved because three of the coefficients of the characteristic equation can be selected. To make further statements requires the root locus technique presented in Section 28.9.

Proportional, integral, and derivative actions and their various combinations are not the only control laws possible, but they are the most common. PID controllers will remain for some time the standard against which any new designs must compete.

The conclusions reached concerning the performance of the various control laws are strictly true only for the plant model forms considered. These are the first-order model without numerator dynamics and the second-order model with a root at $s = 0$ and no numerator zeros. The analysis of a control law for any other linear system follows the preceding pattern. The overall system transfer functions are obtained, and all of the linear system analysis techniques can be applied to predict the system's performance. If the performance is unsatisfactory, a new control law is tried and the process repeated. When this process fails to achieve an acceptable design, more systematic methods of altering the system's structure are needed; they are discussed in later sections. We have used step functions as the test signals because they are the most common and perhaps represent the severest test of system performance. Impulse, ramp, and sinusoidal test signals are also employed. The type to use should be made clear in the design specifications.

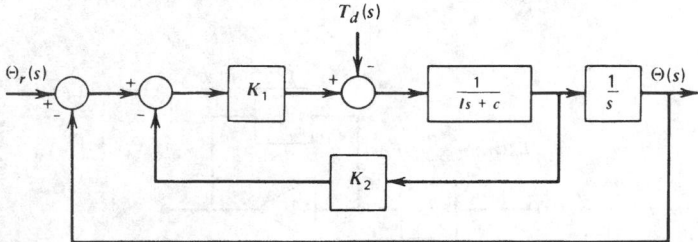

Fig. 28.24 Tachometer feedback arrangement to replace PD control for the position servo.[1]

28.6 CONTROLLER HARDWARE

The control law must be implemented by a physical device before the control engineer's task is complete. The earliest devices were purely kinematic and were mechanical elements such as gears, levers, and diaphragms that usually obtained their power from the controlled variable. Most controllers now are analog electronic, hydraulic, pneumatic, or digital electronic devices. We now consider the analog type. Digital controllers are covered starting in Section 28.10.

28.6.1 Feedback Compensation and Controller Design

Most controllers that implement versions of the PID algorithm are based on the following feedback principle. Consider the single-loop system shown in Fig. 28.1. If the open-loop transfer function is large enough that $|G(s)H(s)| \gg 1$, the closed-loop transfer function is approximately given by

$$T(s) = \frac{G(s)}{1 + G(s)H(s)} \approx \frac{G(s)}{G(s)H(s)} = \frac{1}{H(s)} \tag{28.24}$$

The principle states that a power unit $G(s)$ can be used with a feedback element $H(s)$ to create a desired transfer function $T(s)$. The power unit must have a gain high enough that $|G(s)H(s)| \gg 1$, and the feedback elements must be selected so that $H(s) = 1/T(s)$. This principle was used in Section 28.1 to explain the design of a feedback amplifier.

28.6.2 Electronic Controllers

The *operational amplifier* (*op amp*) is a high-gain amplifier with a high input impedance. A diagram of an op amp with feedback and input elements with impedances $T_f(s)$ and $T_i(s)$ is shown in Fig. 28.25. An approximate relation is

$$\frac{E_o(s)}{E_i(s)} = -\frac{T_f(s)}{T_i(s)}$$

The various control modes can be obtained by proper selection of the impedances. A proportional controller can be constructed with a *multiplier,* which uses two resistors, as shown in Fig. 28.26. An *inverter* is a multiplier circuit with $R_f = R_i$. It is sometimes needed because of the sign reversal property of the op amp. The multiplier circuit can be modified to act as an adder (Fig. 28.27).

PI control can be implemented with the circuit of Fig. 28.28. Figure. 28.29 shows a complete system using op amps for PI control. The inverter is needed to create an error detector. Many industrial controllers provide the operator with a choice of control modes, and the operator can switch from one mode to another when the process characteristics or control objectives change. When a switch occurs, it is necessary to provide any integrators with the proper initial voltages, or else undesirable transients will occur when the integrator is switched into the system. Commercially available controllers usually have built-in circuits for this purpose.

In theory, a differentiator can be created by interchanging the resistance and capacitance in the integrating op amp. The difficulty with this design is that no electrical signal is "pure." Contamination always exists as a result of voltage spikes, ripple, and other transients generally categorized as "noise." These high-frequency signals have large slopes compared with the more slowly varying primary signal, and thus they will dominate the output of the differentiator. In practice, this problem is solved by filtering out high-frequency signals, either with a low-pass filter inserted in cascade with the differentiator, or by using a redesigned differentiator such as the one shown in Fig. 28.30. For the ideal PD controller, $R_1 = 0$. The attenuation curve for the ideal controller breaks upward at $\omega = 1/R_2 C$ with a slope of 20 db/decade. The curve for the practical controller does the same but then becomes flat for $\omega > (R_1 + R_2)/R_1 R_2 C$. This provides the required limiting effect at high frequencies.

PID control can be implemented by joining the PI and PD controllers in parallel, but this is expensive because of the number of op amps and power supplies required. Instead, the usual implementation is that shown in Fig. 28.31. The circuit limits the effect of frequencies above $\omega = 1/$

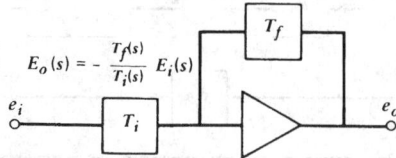

Fig. 28.25 Operational amplifier (op amp).[1]

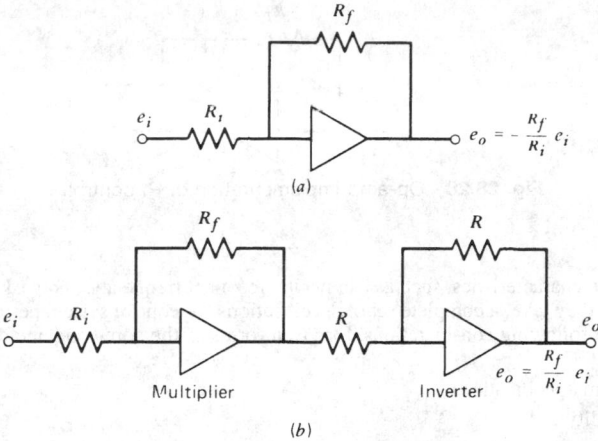

Fig. 28.26 Op-amp implementation of proportional control.[1]

$\beta R_1 C_1$. When $R_1 = 0$, ideal PID control results. This is sometimes called the *noninteractive* algorithm because the effect of each of the three modes is additive, and they do not interfere with one another. The form given for $R_1 \neq 0$ is the *real* or *interactive* algorithm. This name results from the fact that historically it was difficult to implement noninteractive PID control with mechanical or pneumatic devices.

28.6.3 Pneumatic Controllers

The nozzle-flapper introduced in Section 28.4 is a high-gain device that is difficult to use without modification. The gain K_f is known only imprecisely and is sensitive to changes induced by temperature and other environmental factors. Also, the linear region over which Eq. (28.14) applies is very small. However, the device can be made useful by compensating it with feedback elements, as was illustrated with the electropneumatic valve positioner shown in Fig. 28.19.

28.6.4 Hydraulic Controllers

The basic unit for synthesis of hydraulic controllers is the hydraulic servomotor. The nozzle-flapper concept is also used in hydraulic controllers.[4] A PI controller is shown in Fig. 28.32. It can be modified for P-action. Derivative action has not seen much use in hydraulic controllers. This action supplies damping to the system, but hydraulic systems are usually highly damped intrinsically because of the viscous working fluid. PI control is the algorithm most commonly implemented with hydraulics.

28.7 FURTHER CRITERIA FOR GAIN SELECTION

Once the form of the control law has been selected, the gains must be computed in light of the performance specifications. In the examples of the PID family of control laws in Section 28.5, the damping ratio, dominant time constant, and steady-state error were taken to be the primary indicators of system performance in the interest of simplicity. In practice, the criteria are usually more detailed. For example, the rise time and maximum overshoot, as well as the other transient response specifications of the previous chapter, may be encountered. Requirements can also be stated in terms of

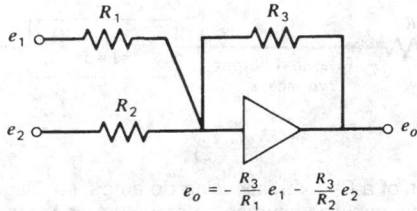

Fig. 28.27 Op-amp adder circuit.[1]

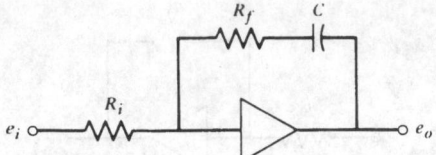

Fig. 28.28 Op-amp implementation of PI control.[1]

frequency response characteristics, such as bandwidth, resonant frequency, and peak amplitude. Whatever specific form they take, a complete set of specifications for control system performance generally should include the following considerations, for given forms of the command and disturbance inputs:

1. Equilibrium specifications
   (a) Stability
   (b) Steady-state error
2. Transient specifications
   (a) Speed of response
   (b) Form of response
3. Sensitivity specifications
   (a) Sensitivity to parameter variations
   (b) Sensitivity to model inaccuracies
   (c) Noise rejection (bandwidth, etc.)

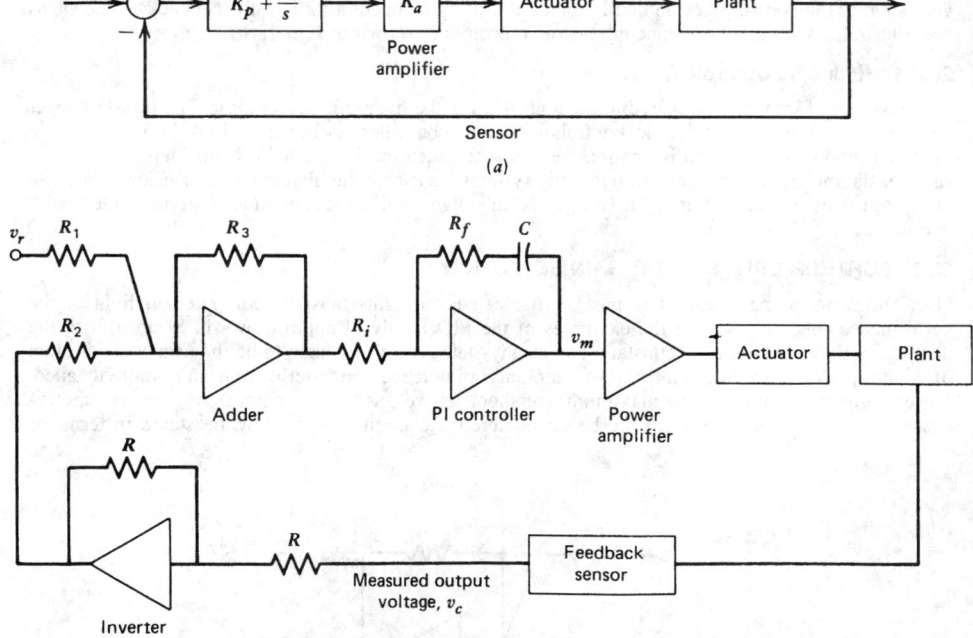

Fig. 28.29 Implementation of a PI-controller using op amps. (a) Diagram of the system. (b) Diagram showing how the op amps are connected.[2]

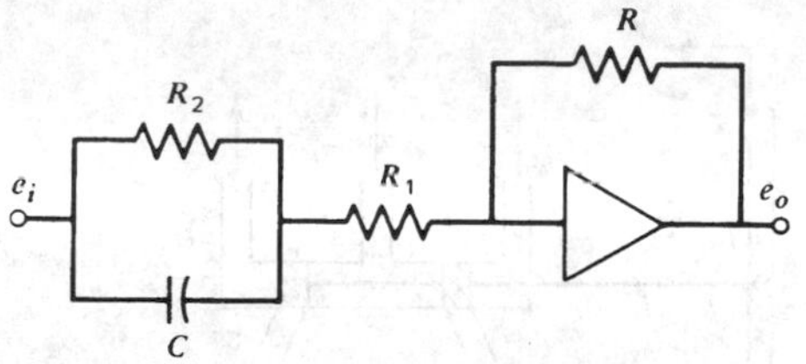

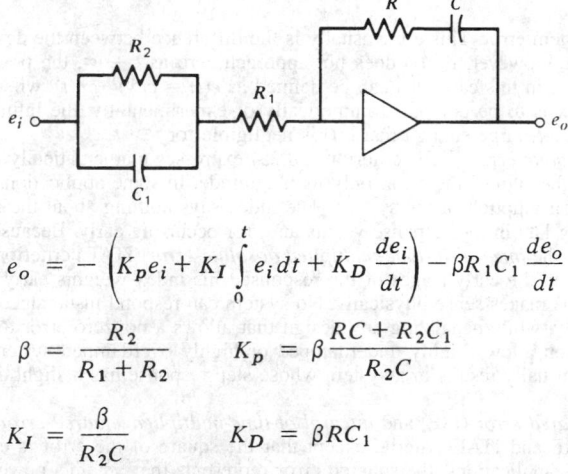

$$e_o = -K_p\left(e_i + T_D\,\frac{de_i}{dt}\right) - \alpha T_D\,\frac{de_o}{dt}$$

$$K_p = \frac{R}{R_1 + R_2} \qquad T_D = R_2 C \qquad \alpha = \frac{R_1}{R_1 + R_2}$$

Fig. 28.30 Practical op-amp implementation of PD control.[1]

In addition to these performance stipulations, the usual engineering considerations of initial cost, weight, maintainability, and so on must be taken into account. The considerations are highly specific to the chosen hardware, and it is difficult to deal with such issues in a general way.

Two approaches exist for designing the controller. The proper one depends on the quality of the analytical description of the plant to be controlled. If an accurate model of the plant is easily developed, we can design a specialized controller for the particular application. The range of adjustment of controller gains in this case can usually be made small because the accurate plant model allows the gains to be precomputed with confidence. This technique reduces the cost of the controller and can often be applied to electromechanical systems.

The second approach is used when the plant is relatively difficult to model, which is often the case in process control. A standard controller with several control modes and wide ranges of gains is used, and the proper mode and gain settings are obtained by testing the controller on the process in the field. This approach should be considered when the cost of developing an accurate plant model might exceed the cost of controller tuning in the field. Of course, the plant must be available for testing for this approach to be feasible.

28.7.1 Performance Indices

The performance criteria encountered thus far require a set of conditions to be specified—for example, one for steady-state error, one for damping ratio, and one for the dominant time constant. If there

$$e_o = -\left(K_p e_i + K_I \int_0^t e_i\,dt + K_D\,\frac{de_i}{dt}\right) - \beta R_1 C_1\,\frac{de_o}{dt}$$

$$\beta = \frac{R_2}{R_1 + R_2} \qquad K_p = \beta\,\frac{RC + R_2 C_1}{R_2 C}$$

$$K_I = \frac{\beta}{R_2 C} \qquad K_D = \beta R C_1$$

Fig. 28.31 Practical op-amp implementation of PID control.[1]

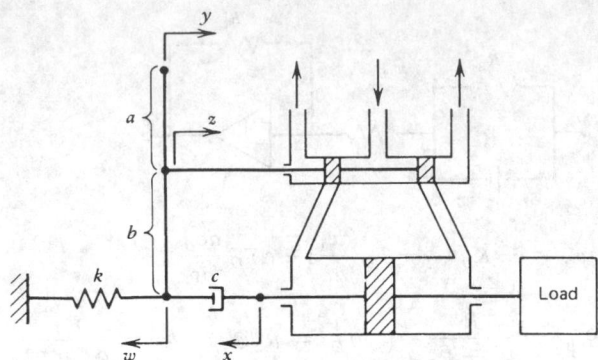

Fig. 28.32 Hydraulic implementation of PI control.[1]

are many such conditions, and if the system is of high order with several gains to be selected, the design process can get quite complicated because transient and steady-state criteria tend to drive the design in different directions. An alternative approach is to specify the system's desired performance by means of one analytical expression called a *performance index*. Powerful analytical and numerical methods are available that allow the gains to be systematically computed by minimizing (or maximizing) this index.

To be useful, a performance index must be selective. The index must have a sharply defined extremum in the vicinity of the gain values that give the desired performance. If the numerical value of the index does not change very much for large changes in the gains from their optimal values, the index will not be selective.

Any practical choice of a performance index must be easily computed, either analytically, numerically, or experimentally. Four common choices for an index are the following:

$$J = \int_0^\infty |e(t)| \, dt \quad \text{(IAE Index)} \tag{28.25}$$

$$J = \int_0^\infty t|e(t)| \, dt \quad \text{(ITAE Index)} \tag{28.26}$$

$$J = \int_0^\infty [e(t)]^2 \, dt \quad \text{(ISE Index)} \tag{28.27}$$

$$J = \int_0^\infty t[e(t)]^2 \, dt \quad \text{(ITSE Index)} \tag{28.28}$$

where $e(t)$ is the system error. This error usually is the difference between the desired and the actual values of the output. However, if $e(t)$ does not approach zero as $t \to \infty$, the preceding indices will not have finite values. In this case, $e(t)$ can be defined as $e(t) = c(\infty) - c(t)$, where $c(t)$ is the output variable. If the index is to be computed numerically or experimentally, the infinite upper limit can be replaced by a time t_f large enough that $e(t)$ is negligible for $t > t_f$.

The *integral absolute-error* (IAE) criterion (28.25) expresses mathematically that the designer is not concerned with the sign of the error, only its magnitude. In some applications, the IAE criterion describes the fuel consumption of the system. The index says nothing about the relative importance of an error occurring late in the response versus an error occurring early. Because of this, the index is not as selective as the *integral-of-time-multiplied absolute-error* (ITAE) criterion (28.26). Since the multiplier t is small in the early stages of the response, this index weights early errors less heavily than later errors. This makes sense physically. No system can respond instantaneously, and the index is lenient accordingly, while penalizing any design that allows a nonzero error to remain for a long time. Neither criterion allows highly underdamped or highly overdamped systems to be optimum. The ITAE criterion usually results in a system whose step response has a slight overshoot and well-damped oscillations.

The *integral squared-error* (ISE) and *integral-of-time-multiplied squared-error* (ITSE) criteria are analogous to the IAE and ITAE criteria, except that the square of the error is employed, for three reasons: (1) in some applications, the squared error represents the system's power consumption; (2) squaring the error weights large errors much more heavily than small errors; (3) the squared error is much easier to handle analytically. The derivative of a squared term is easier to compute than that of an absolute value and does not have a discontinuity at $e = 0$. These differences are important when the system is of high order with multiple error terms.

The closed-form solution for the response is not required to evaluate a performance index. For a given set of parameter values, the response and the resulting index value can be computed numerically. The optimum solution can be obtained using systematic computer search procedures; this makes this approach suitable for use with nonlinear systems.

28.7.2 Optimal Control Methods

Optimal control theory includes a number of algorithms for systematic design of a control law to minimize a performance index, such as the following generalization of the ISE index, called the *quadratic* index:

$$J = \int_0^\infty (\mathbf{x}^T \mathbf{Q} \mathbf{x} + \mathbf{u}^T \mathbf{R} \mathbf{u}) \, dt \tag{28.29}$$

where **x** and **u** are the deviations of the state and control vectors from the desired reference values. For example, in a servomechanism, the state vector might consist of the position and velocity, and the control vector might be a scalar—the force or torque produced by the actuator. The matrices **Q** and **R** are chosen by the designer to provide relative weighting for the elements of **x** and **u**. If the plant can be described by the linear state-variable model

$$\dot{\mathbf{x}} = \mathbf{A}\mathbf{x} + \mathbf{B}\mathbf{u} \tag{28.30}$$
$$\mathbf{y} = \mathbf{C}\mathbf{x} + \mathbf{D}\mathbf{u} \tag{28.31}$$

where **y** is the vector of outputs—for example, position and velocity—then the solution of this *linear-quadratic* control problem is the linear control law:

$$\mathbf{u} = \mathbf{K}\mathbf{y} \tag{28.32}$$

where **K** is a matrix of gains that can be found by several algorithms.[1,5,6] A valid solution is guaranteed to yield a stable closed-loop system, a major benefit of this method.

Even if it is possible to formulate the control problem in this way, several practical difficulties arise. Some of the terms in (28.29) might be beyond the influence of the control vector **u**; the system is then *uncontrollable*. Also, there might not be enough information in the output equation (28.31) to achieve control, and the system is then *unobservable*. Several tests are available to check controllability and observability. Not all of the necessary state variables might be available for feedback, or the feedback measurements might be noisy or biased. Algorithms known as *observers, state reconstructors, estimators,* and *digital filters* are available to compensate for the missing information. Another source of error is the uncertainty in the values of the coefficient matrices **A**, **B**, **C**, and **D**. Identification schemes can be used to compare the predicted and the actual system performance, and to adjust the coefficient values "on-line."

28.7.3 The Ziegler–Nichols Rules

The difficulty of obtaining accurate transfer function models for some processes has led to the development of empirically based rules of thumb for computing the optimum gain values for a controller. Commonly used guidelines are the *Ziegler–Nichols rules,* which have proved so helpful that they are still in use 50 years after their development. The rules actually consist of two separate methods. The first method requires the open-loop step response of the plant, while the second uses the results of experiments performed with the controller already installed. While primarily intended for use with systems for which no analytical model is available, the rules are also helpful even when a model can be developed.

Ziegler and Nichols developed their rules from experiments and analysis of various industrial processes. Using the IAE criterion with a unit step response, they found that controllers adjusted according to the following rules usually had a step response that was oscillatory but with enough damping so that the second overshoot was less than 25% of the first (peak) overshoot. This is the *quarter-decay* criterion and is sometimes used as a specification.

The first method is the *process-reaction* method and relies on the fact that many processes have an open-loop step response like that shown in Fig. 28.33. This is the *process signature* and is characterized by two parameters, R and L. R is the slope of a line tangent to the steepest part of the response curve, and L is the time at which this line intersects the time axis. First- and second-order linear systems do not yield positive values for L, and so the method cannot be applied to such

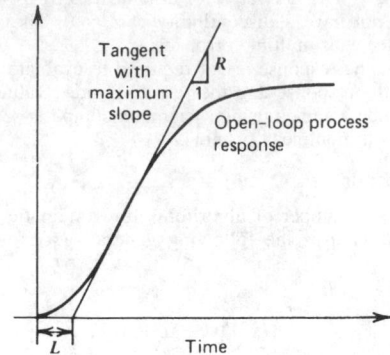

Fig. 28.33   Process signature for a unit step input.[1]

systems. However, third- and higher-order linear systems with sufficient damping do yield such a response. If so, the Zielger–Nichols rules recommend the controller settings given in Table 28.2.

The *ultimate-cycle* method uses experiments with the controller in place. All control modes except proportional are turned off, and the process is started with the proportional gain K_P set at a low value. The gain is slowly increased until the process begins to exhibit sustained oscillations. Denote the period of this oscillation by P_u and the corresponding *ultimate gain* by K_{P_u}. The Ziegler–Nichols recommendations are given in Table 28.2 in terms of these parameters. The proportional gain is lower for PI control than for P control, and is higher for PID control because I action increases the order of the system and thus tends to destabilize it; thus, a lower gain is needed. On the other hand, D action tends to stabilize the system; hence, the proportional gain can be increased without degrading the stability characteristics. Because the rules were developed for a typical case out of many types of processes, final tuning of the gains in the field is usually necessary.

28.7.4   Nonlinearities and Controller Performance

All physical systems have nonlinear characteristics of some sort, although they can often be modeled as linear systems provided the deviations from the linearization reference condition are not too great. Under certain conditions, however, the nonlinearities have significant effects on the system's performance. One such situation can occur during the start-up of a controller if the initial conditions are much different from the reference condition for linearization. The linearized model is then not accurate, and nonlinearities govern the behavior. If the nonlinearities are mild, there might not be much of a problem. Where the nonlinearities are severe, such as in process control, special consideration must be given to start-up. Usually, in such cases, the control signal sent to the final control elements is manually adjusted until the system variables are within the linear range of the controller.

Table 28.2   The Ziegler–Nichols Rules

Controller transfer function $G(s) = K_p \left(1 + \dfrac{1}{T_I s} + T_D s \right)$

Control Mode	Process-Reaction Method	Ultimate-Cycle Method
P control	$K_p = \dfrac{1}{RL}$	$K_p = 0.5 K_{pu}$
PI control	$K_p = \dfrac{0.9}{RL}$	$K_p = 0.45 K_{pu}$
	$T_I = 3.3L$	$T_I = 0.83 P_u$
PID control	$K_p = \dfrac{1.2}{RL}$	$K_p = 0.6 K_{pu}$
	$T_I = 2L$	$T_I = 0.5 P_u$
	$T_D = 0.5L$	$T_D = 0.125 P_u$

Then the system is switched into automatic mode. Digital computers are often used to replace the manual adjustment process because they can be readily coded to produce complicated functions for the start-up signals. Care must also be taken when switching from manual to automatic. For example, the integrators in electronic controllers must be provided with the proper initial conditions.

28.7.5   Reset Windup

In practice, all actuators and final control elements have a limited operating range. For example, a motor–amplifier combination can produce a torque proportional to the input voltage over only a limited range. No amplifier can supply an infinite current; there is a maximum current and thus a maximum torque that the system can produce. The final control elements are said to be *overdriven* when they are commanded by the controller to do something they cannot do. Since the limitations of the final control elements are ultimately due to the limited rate at which they can supply energy, it is important that all system performance specifications and controller designs be consistent with the energy-delivery capabilities of the elements to be used.

Controllers using integral action can exhibit the phenomenon called *reset windup* or *integrator buildup* when overdriven, if they are not properly designed. For a step change in set point, the proportional term responds instantly and saturates immediately if the set-point change is large enough. On the other hand, the integral term does not respond as fast, It integrates the error signal and saturates some time later if the error remains large for a long enough time. As the error decreases, the proportional term no longer causes saturation. However, the integral term continues to increase as long as the error has not changed sign, and thus the manipulated variable remains saturated. Even though the output is very near its desired value, the manipulated variable remains saturated until after the error has reversed sign. The result can be an undesirable overshoot in the response of the controlled variable.

Limits on the controller prevent the voltages from exceeding the value required to saturate the actuator, and thus protect the actuator, but they do not prevent the integral build-up that causes the overshoot. One way to prevent integrator build-up is to select the gains so that saturation will never occur. This requires knowledge of the maximum input magnitude that the system will encounter. General algorithms for doing this are not available; some methods for low-order systems are presented in Ref. 1, Chap. 7, and Ref. 2, Chap. 7. Integrator build-up is easier to prevent when using digital control; this is discussed in Section 28.10.

28.8   COMPENSATION AND ALTERNATIVE CONTROL STRUCTURES

A common design technique is to insert a *compensator* into the system when the PID control algorithm can be made to satisfy most but not all of the design specifications. A compensator is a device that alters the response of the controller so that the overall system will have satisfactory performance. The three categories of compensation techniques generally recognized are *series compensation, parallel* (or *feedback*) *compensation,* and *feedforward compensation*. The three structures are loosely illustrated in Fig. 28.34, where we assume the final control elements have a unity transfer function. The transfer function of the controller is $G_1(s)$. The feedback elements are represented by $H(s)$, and the compensator by $G_c(s)$. We assume that the plant is unalterable, as is usually the case in control system design. The choice of compensation structure depends on what type of specifications must be satisfied. The physical devices used as compensators are similar to the pneumatic, hydraulic, and electrical devices treated previously. Compensators can be implemented in software for digital control applications.

28.8.1   Series Compensation

The most commonly used series compensators are the *lead,* the *lag,* and the *lead-lag* compensators. Electrical implementations of these are shown in Fig. 28.35. Other physical implementations are available. Generally, the lead compensator improves the speed of response; the lag compensator decreases the steady-state error; and the lead-lag affects both. Graphical aids, such as the root locus and frequency response plots, are usually needed to design these compensators (Ref. 1, Chap. 8; Ref. 2, Chap. 9).

28.8.2   Feedback Compensation and Cascade Control

The use of a tachometer to obtain velocity feedback, as in Fig. 28.24, is a case of feedback compensation. The feedback-compensation principle of Fig. 28.3 is another. Another form is *cascade control,* in which another controller is inserted within the loop of the original control system (Fig. 28.36). The new controller can be used to achieve better control of variables within the forward path of the system. Its set point is manipulated by the first controller.

Cascade control is frequently used when the plant cannot be satisfactorily approximated with a model of second order or lower. This is because the difficulty of analysis and control increases rapidly with system order. The characteristic roots of a second-order system can easily be expressed in analytical form. This is not so for third order or higher, and few general design rules are available.

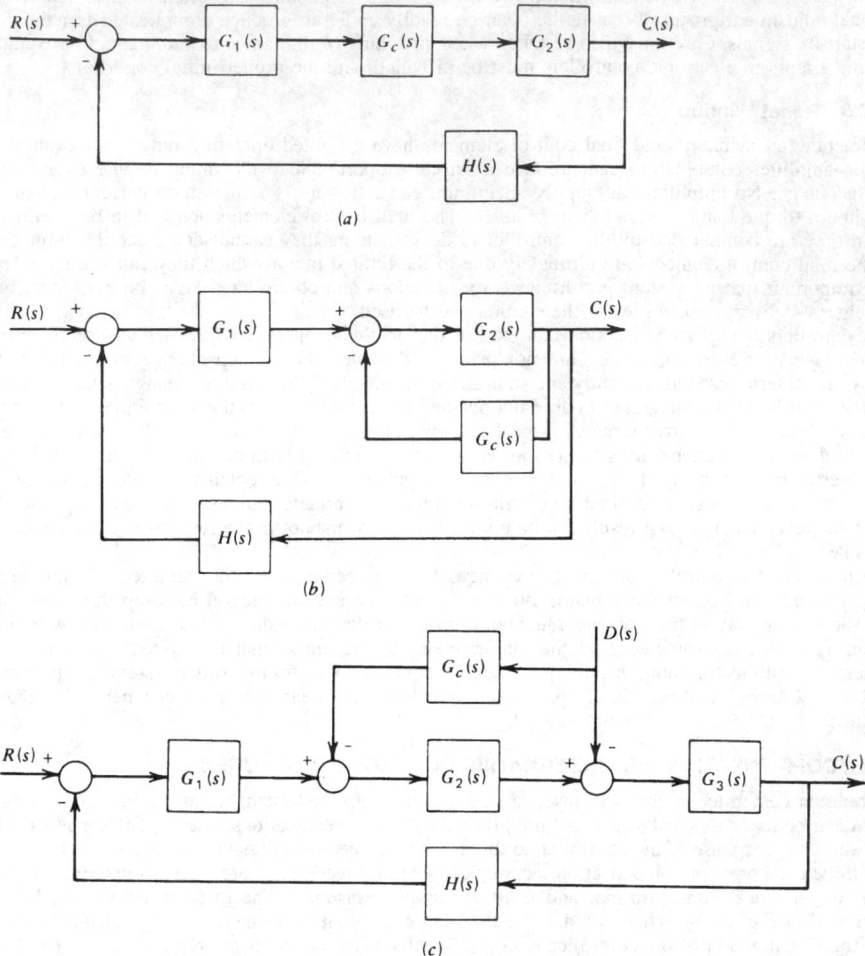

Fig. 28.34 General structures of the three compensation types: (a) series; (b) parallel (or feedback); (c) feed-forward. The compensator transfer function is $G_c(s)$.[1]

When faced with the problem of controlling a high-order system, the designer should first see if the performance requirements can be relaxed so that the system can be approximated with a low-order model. If this is not possible, the designer should attempt to divide the plant into subsystems, each of which is second order or lower. A controller is then designed for each subsystem. An application using cascade control is given in Section 28.11.

28.8.3 Feedforward Compensation

The control algorithms considered thus far have counteracted disturbances by using measurements of the output. One difficulty with this approach is that the effects of the disturbance must show up in the output of the plant before the controller can begin to take action. On the other hand, if we can measure the disturbance, the response of the controller can be improved by using the measurement to augment the control signal sent from the controller to the final control elements. This is the essence of feedforward compensation of the disturbance, as shown in Fig. 28.34c.

Feedforward compensation modified the output of the main controller. Instead of doing this by measuring the disturbance, another form of feedforward compensation utilizes the command input. Figure 28.37 is an example of this approach. The closed-loop transfer function is

$$\frac{\Omega(s)}{\Omega_r(s)} = \frac{K_f + K}{Is + c + K} \tag{28.33}$$

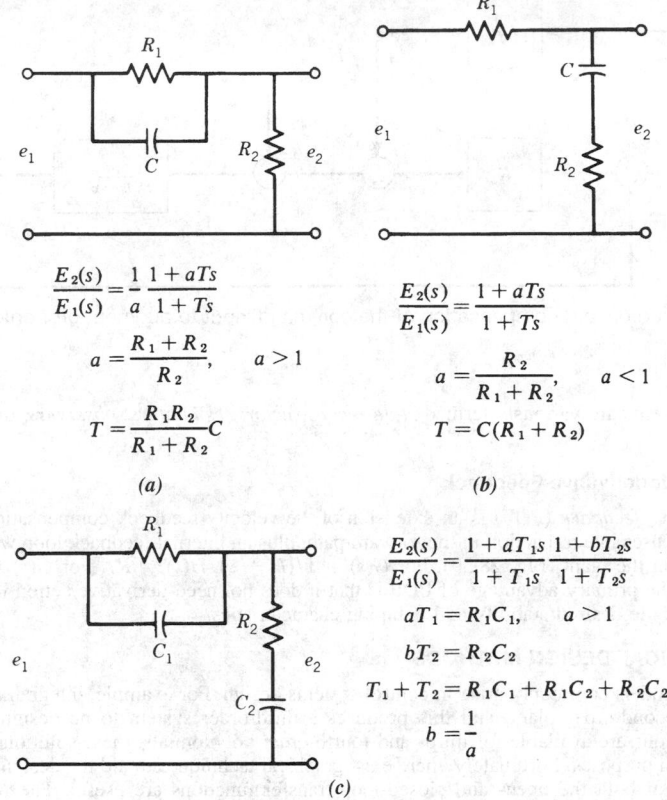

$$\frac{E_2(s)}{E_1(s)} = \frac{1}{a}\frac{1+aTs}{1+Ts}$$

$$a = \frac{R_1 + R_2}{R_2}, \qquad a > 1$$

$$T = \frac{R_1 R_2}{R_1 + R_2}C$$

(a)

$$\frac{E_2(s)}{E_1(s)} = \frac{1+aTs}{1+Ts}$$

$$a = \frac{R_2}{R_1 + R_2}, \qquad a < 1$$

$$T = C(R_1 + R_2)$$

(b)

$$\frac{E_2(s)}{E_1(s)} = \frac{1+aT_1s}{1+T_1s}\frac{1+bT_2s}{1+T_2s}$$

$$aT_1 = R_1C_1, \qquad a > 1$$

$$bT_2 = R_2C_2$$

$$T_1 + T_2 = R_1C_1 + R_1C_2 + R_2C_2$$

$$b = \frac{1}{a}$$

(c)

Fig. 28.35 Passive electrical compensators: (a) lead; (b) lag; (c) lead–lag.

For a unit-step input, the steady-state output is $\omega_{ss} = (K_f + K)/(c + K)$. Thus, if we choose the feedforward gain K_f to be $K_f = c$, then $\omega_{ss} = 1$ as desired, and the error is zero. Note that this form of feed forward compensation does not affect the disturbance response. Its effectiveness depends on how accurately we know the value of c. A digital application of feedforward compensation is presented in Section 28.11.

28.8.4   State-Variable Feedback

There are techniques for improving system performance that do not fall entirely into one of the three compensation categories considered previously. In some forms these techniques can be viewed as a type of feedback compensation, while in other forms they constitute a modification of the control law. *State-variable feedback* (SVFB) is a technique that uses information about all the system's state variables to modify either the control signal or the actuating signal. These two forms are illustrated in Fig. 28.38. Both forms require that the state vector x be measurable or at least derivable from other information. Devices or algorithms used to obtain state variable information other than directly

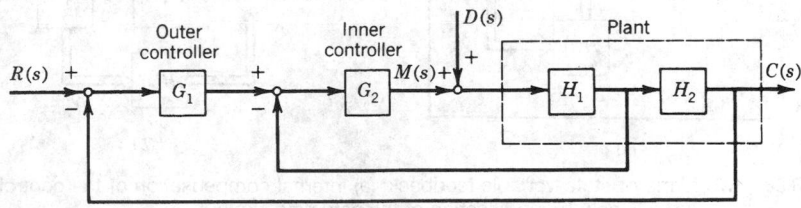

Fig. 28.36 Cascade control structure.

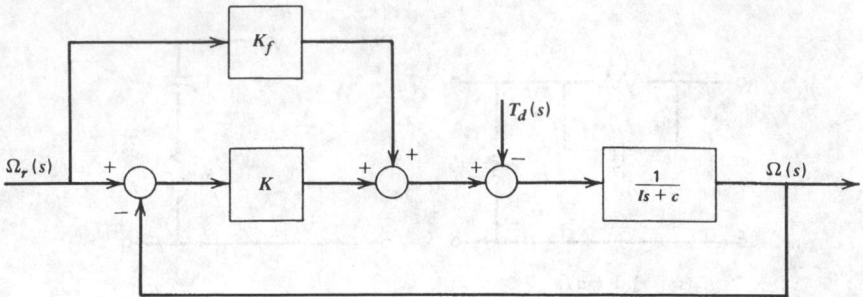

Fig. 28.37 Feedforward compensation of the command input to augment proportional control.[2]

from measurements are variously termed *state reconstructors, estimators, observers,* or *filters* in the literature.

28.8.5 Pseudoderivative Feedback

Pseudoderivative feedback (PDF) is an extension of the velocity feedback compensation concept of Fig. 28.24.[1,2] It uses integral action in the forward path, plus an internal feedback loop whose operator $H(s)$ depends on the plant (Fig. 28.39). For $G(s) = 1/(Is + c)$, $H(s) = K_1$. For $G(s) = 1/Is^2$, $H(s) = K_1 + K_2s$. The primary advantage of PDF is that it does not need derivative action in the forward path to achieve the desired stability and damping characteristics.

28.9 GRAPHICAL DESIGN METHODS

Higher-order models commonly arise in control systems design. For example, integral action is often used with a second-order plant, and this produces a third-order system to be designed. Although algebraic solutions are available for third- and fourth-order polynomials, these solutions are cumbersome for design purposes. Fortunately, there exist graphical techniques to aid the designer. Frequency response plots of both the open- and closed-loop transfer functions are useful. The *Bode plot* and the *Nyquist plot* all present the frequency response information in different forms. Each form has its own advantages. The root locus plot shows the location of the characteristic roots for a range of values of some parameters, such as a controller gain. A tabulation of these plots for typical transfer functions is given in the previous chapter (Fig. 27.8). The design of two-position and other nonlinear control systems is facilitated by the *describing function,* which is a linearized approximation based on the frequency response of the controller (see Section 27.8.4). Graphical design methods are discussed in more detail in Refs. 1, 2, and 3.

28.9.1 The Nyquist Stability Theorem

The Nyquist stability theorem is a powerful tool for linear system analysis. If the open-loop system has no poles with positive real parts, we can concentrate our attention on the region around the point $-1 + i0$ on the polar plot of the open-loop transfer function. Figure 28.40 shows the polar plot of the open-loop transfer function of an arbitrary system that is assumed to be open-loop stable. The Nyquist stability theorem is stated as follows:

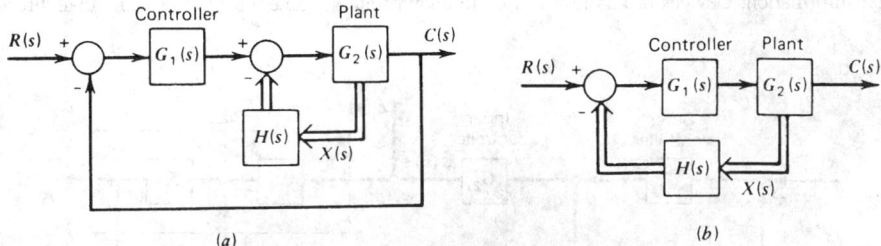

Fig. 28.38 Two forms of state-variable feedback: (*a*) internal compensation of the control signal; (*b*) modification of the actuating signal.[1]

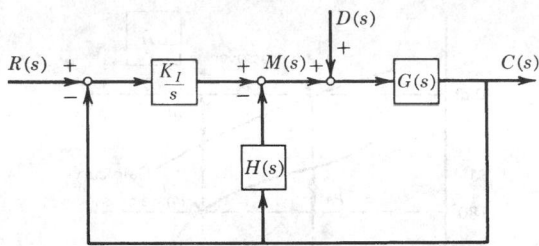

Fig. 28.39 Structure of pseudoderivative feedback (PDF).

A system is closed-loop stable if and only if the point $-1 + i0$ lies to the left of the open-loop Nyquist plot relative to an observer traveling along the plot in the direction of increasing frequency ω.

Therefore, the system described by Fig. 28.39 is closed-loop stable.

The Nyquist theorem provides a convenient measure of the relative stability of a system. A measure of the proximity of the plot to the $-1 + i0$ point is given by the angle between the negative real axis and a line from the origin to the point where the plot crosses the unit circle (see Fig. 28.39). The frequency corresponding to this intersection is denoted ω_g. This angle is the *phase margin* (PM) and is positive when measured down from the negative real axis. The phase margin is the phase at the frequency ω_g where the magnitude ratio or "gain" of $G(i\omega)H(i\omega)$ is unity, or 0 decibels (db). The frequency ω_p, the *phase crossover frequency,* is the frequency at which the phase angle is $-180°$. The *gain margin* (GM) is the difference in decibels between the unity gain condition (0 db) and the value of $|G(\omega_p)H(\omega_p)|$ db at the phase crossover frequency ω_p. Thus,

$$\text{gain margin} = -|G(\omega_p)H(\omega_p)| \quad (\text{db}) \tag{28.34}$$

A system is stable only if the phase and gain margins are both positive.

The phase and gain margins can be illustrated on the Bode plots shown in Fig. 28.41. The phase and gain margins can be stated as safety margins in the design specifications. A typical set of such specifications is as follows:

$$\text{gain margin} \geq 8 \text{ db} \quad \text{and} \quad \text{phase margin} \geq 30° \tag{28.35}$$

In common design situations, only one of these equalities can be met, and the other margin is allowed to be greater than its minimum value. It is not desirable to make the margins too large, because this results in a low gain, which might produce sluggish response and a large steady-state error. Another commonly used set of specifications is

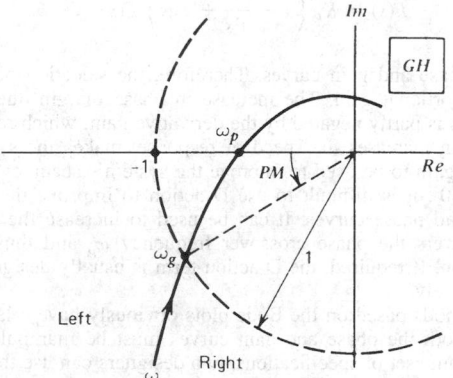

Fig. 28.40 Nyquist plot for a stable system.[1]

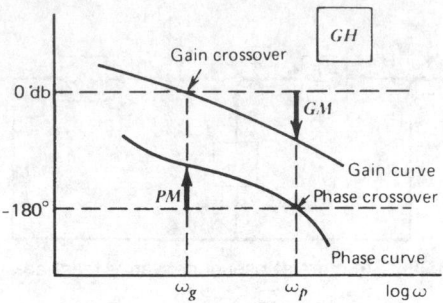

Fig. 28.41 Bode plot showing definitions of phase and gain margin.[1]

$$\text{gain margin} \geq 6 \text{ db} \quad \text{and} \quad \text{phase margin} \geq 40° \tag{28.36}$$

The 6-db limit corresponds to the quarter amplitude decay response obtained with the gain settings given by the Ziegler–Nichols ultimate-cycle method (Table 28.2).

28.9.2 Systems with Dead-Time Elements

The Nyquist theorem is particularly useful for systems with dead-time elements, especially when the plant is of an order high enough to make the root-locus method cumbersome. A delay D in either the manipulated variable or the measurement will result in an open-loop transfer function of the form

$$G(s)H(s) = e^{-Ds}P(s) \tag{28.37}$$

Its magnitude and phase angle are

$$|G(i\omega)H(i\omega)| = |P(i\omega)||e^{-i\omega D}| = |P(i\omega)| \tag{28.38}$$

$$\angle G(i\omega)H(i\omega)) = \angle P(i\omega) + \angle e^{-i\omega D} = \angle P(i\omega) - \omega D \tag{28.39}$$

Thus, the dead time decreases the phase angle proportionally to the frequency ω, but it does not change the gain curve. This makes the analysis of its effects easier to accomplish with the open-loop frequency response plot.

28.9.3 Open-Loop Design for PID Control

Some general comments can be made about the effects of proportional, integral, and derivative control actions on the phase and gain margins. P action does not affect the phase curve at all and thus can be used to raise or lower the open-loop gain curve until the specifications for the gain and phase margins are satisfied. If I action or D action is included, the proportional gain is selected last. Therefore, when using this approach to the design, it is best to write the PID algorithm with the proportional gain factored out, as

$$F(s) = K_P \left(1 + \frac{1}{T_I s} + T_D s \right) E(s) \tag{28.40}$$

D action affects both the phase and gain curves. Therefore, the selection of the derivative gain is more difficult than the proportional gain. The increase in phase margin due to the positive phase angle introduced by D action is partly negated by the derivative gain, which reduces the gain margin. Increasing the derivative gain increases the speed of response, makes the system more stable, and allows a larger proportional gain to be used to improve the system's accuracy. However, if the phase curve is too steep near $-180°$, it is difficult to use D action to improve the performance. I action also affects both the gain and phase curves. It can be used to increase the open-loop gain at low frequencies. However, it lowers the phase crossover frequency ω_p and thus reduces some of the benefits provided by D action. If required, the D-action term is usually designed first, followed by I action and P action, respectively.

The classical design methods based on the Bode plots obviously have a large component of trial and error because usually both the phase and gain curves must be manipulated to achieve an acceptable design. Given the same set of specifications, two designers can use these methods and arrive at substantially different designs. Many rules of thumb and ad hoc procedures have been developed, but a general foolproof procedure does not exist. However, an experienced designer can often obtain a good design quickly with these techniques. The use of a computer plotting routine greatly speeds up the design process.

28.9.4 Design with the Root Locus

The effect of D action as a series compensator can be seen with the root locus. The term $(1 + T_D s)$ in Fig. 28.32 can be considered as a series compensator to the proportional controller. The D action adds an open-loop zero at $s = -1/T_D$. For example, a plant with the transfer function $1/s(s + 1)(s + 2)$, when subjected to proportional control, has the root locus shown in Fig. 28.42a. If the proportional gain is too high, the system will be unstable. The smallest achievable time constant corresponds to the root $s = -0.42$, and is $\tau = 1/0.42 = 2.4$. If D action is used to put an open-loop zero at $s = -1.5$, the resulting root locus is given by Fig. 28.42b. The D action prevents the system from becoming unstable, and allows a smaller time constant to be achieved (τ can be made close to $1/0.75 = 1.3$ by using a high proportional gain).

The integral action in PI control can be considered to add an open-loop pole at $s = 0$, and a zero at $s = -1/T_I$. Proportional control of the plant $1/(s + 1)(s + 2)$ gives a root locus like that shown in Fig. 28.43, with $a = 1$ and $b = 2$. A steady-state error will exist for a step input. With the PI compensator applied to this plant, the root locus is given by Fig. 28.42b, with $T_I = 2/3$. The steady-state error is eliminated, but the response of the system has been slowed because the dominant paths of the root locus of the compensated system lie closer to the imaginary axis than those of the uncompensated system.

As another example, let the plant-transfer function be

$$G_P(s) = \frac{1}{s^2 + a_2 s + a_1} \qquad (28.41)$$

where $a_1 > 0$ and $a_2 > 0$. PI control applied to this plant gives the closed-loop command transfer function

$$T_1(s) = \frac{K_P s + K_I}{s^3 + a_2 s^2 + (a_1 + K_P)s + K_I} \qquad (28.42)$$

Note that the Ziegler–Nichols rules cannot be used to set the gains K_P and K_I. The second-order plant, Eq. (28.41), does not have the S-shaped signature of Fig. 28.33, so the process-reaction method does not apply. The ultimate-cycle method requires K_I to be set to zero and the ultimate gain K_{P_u}

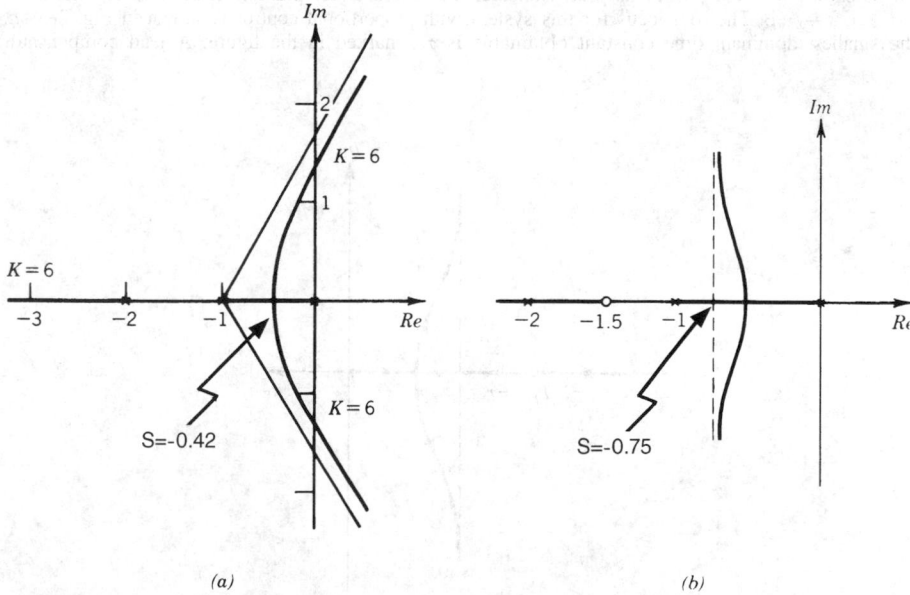

(a)   (b)

Fig. 28.42 (a) Root locus plot for $s(s + 1)(s + 2) + K = 0$, for $K \geq 0$. (b) The effect of PD control with $T_D = 2/3$.

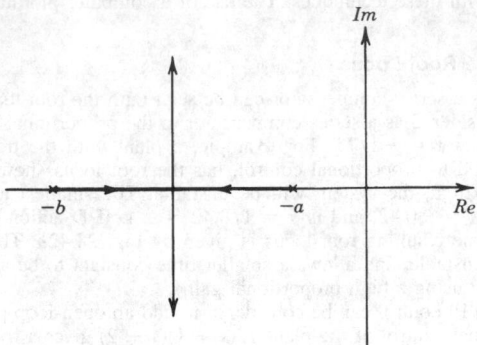

Fig. 28.43 Root-locus plot for $(s + a)(s + b) + K = 0$.

determined. With $K_I = 0$ in Eq. (28.42) the resulting system is stable for all $K_P > 0$, and thus a positive ultimate gain does not exist.

Take the form of the PI-control law given by Eq. (28.42) with $T_D = 0$, and assume that the characteristic roots of the plant (Fig. 28.44) are real values $-r_1$ and $-r_2$ such that $-r_2 < -r_1$. In this case the open-loop transfer function of the control system is

$$G(s)H(s) = \frac{K_P(s + 1/T_I)}{s(s + r_1)(s + r_2)} \qquad (28.43)$$

One design approach is to select T_I, and plot the locus with K_P as the parameter. If the zero at $s = -1/T_I$ is located to the right of $s = -r_1$, the dominant time constant cannot be made as small as is possible with the zero located between the poles at $s = -r_1$ and $s = -r_2$ (Fig. 28.44). A large integral gain (small T_I and/or large K_P) is desirable for reducing the overshoot due to a disturbance, but the zero should not be placed to the left of $s = -r_2$ because the dominant time constant will be larger than that obtainable with the placement shown in Fig. 28.44 for large values of K_P. Sketch the root-locus plots to see this. A similar situation exists if the poles of the plant are complex.

The effects of the lead compensator in terms of time-domain specifications (characteristic roots) can be shown with the root-locus plot. Consider the second-order plant with the real distinct roots $s = -\alpha$, $s = -\beta$. The root locus for this system with proportional control is shown in Fig. 28.45a. The smallest dominant time constant obtainable is τ_1, marked in the figure. A lead compensator

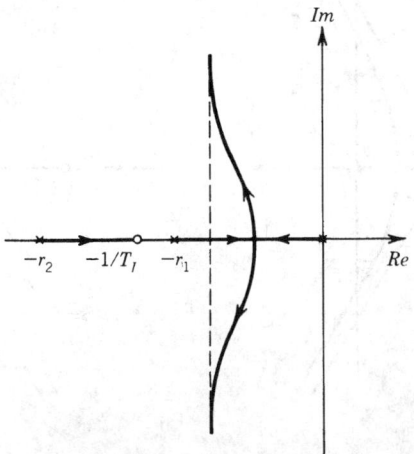

Fig. 28.44 Root-locus plot for PI control of a second-order plant.

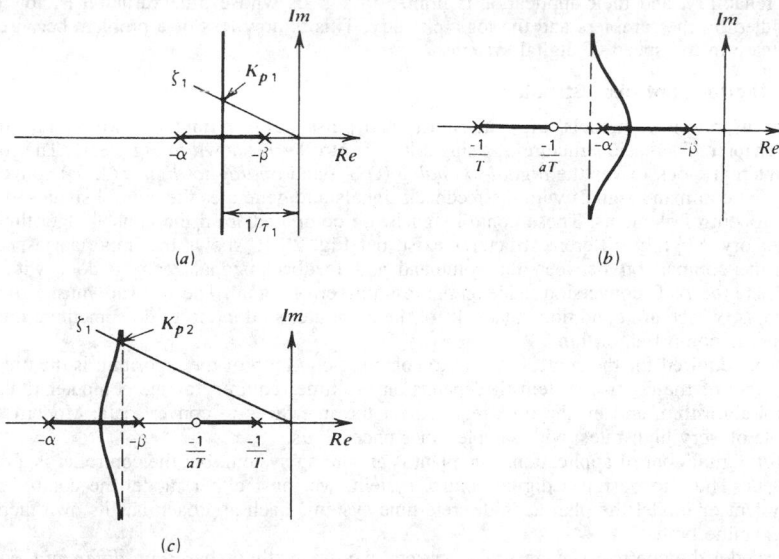

Fig. 28.45 Effects of series lead and lag compensators: (a) uncompensated system's root locus; (b) root locus with lead compensation; (c) root locus with lag compensation.

introduces a pole at $s = -1/T$ and a zero at $s = -1/aT$, and the root locus becomes that shown in Fig. 28.45b. The pole and zero introduced by the compensator reshape the locus so that a smaller dominant time constant can be obtained. This is done by choosing the proportional gain high enough to place the roots close to the asymptotes.

With reference to the proportional control system whose root locus is shown in Fig. 28.45a, suppose that the desired damping ratio ζ_1 and desired time constant τ_1 are obtainable with a proportional gain of K_{P1}, but the resulting steady-state error $\alpha\beta/(\alpha\beta + K_{P1})$ due to a step input is too large. We need to increase the gain while preserving the desired damping ratio and time constant. With the lag compensator, the root locus is as shown in Fig. 28.45c. By considering specific numerical values, one can show that for the compensated system, roots with a damping ratio ζ_1 correspond to a high value of the proportional gain. Call this value K_{P2}. Thus $K_{P2} > K_{P1}$, and the steady-state error will be reduced. If the value of T is chosen large enough, the pole at $s = -1/T$ is approximately canceled by the zero at $s = -1/aT$, and the open-loop transfer function is given approximately by

$$G(s)H(s) = \frac{aK_P}{(s + \alpha)(s + \beta)} \tag{28.44}$$

Thus, the system's response is governed approximately by the complex roots corresponding to the gain value K_{P2}. By comparing Fig. 28.45a with 28.45c, we see that the compensation leaves the time constant relatively unchanged. From Eq. (28.44) it can be seen that since $a < 1$, K_P can be selected as the larger value K_{P2}. The ratio of K_{P1} to K_{P2} is approximately given by the parameter a.

Design by pole–zero cancellation can be difficult to accomplish because a response pattern of the system is essentially ignored. The pattern corresponds to the behavior generated by the canceled pole and zero, and this response can be shown to be beyond the influence of the controller. In this example, the canceled pole gives a stable response because it lies in the left-hand plane. However, another input not modeled here, such as a disturbance, might excite the response and cause unexpected behavior. The designer should therefore proceed with caution. None of the physical parameters of the system are known exactly, so exact pole–zero cancellation is not possible. A root-locus study of the effects of parameter uncertainty and a simulation study of the response are often advised before the design is accepted as final.

28.10   PRINCIPLES OF DIGITAL CONTROL

Digital control has several advantages over analog devices. A greater variety of control algorithms is possible, including nonlinear algorithms and ones with time-varying coefficients. Also, greater accuracy is possible with digital systems. However, their additional hardware complexity can result in lower reliability, and their application is limited to signals whose time variation is slow enough to be handled by the samplers and the logic circuitry. This is now less of a problem because of the large increase in the speed of digital systems.

28.10.1 Digital Controller Structure

Sampling, discrete-time models, the z-transform, and pulse transfer functions were outlined in the previous chapter. The basic structure of a single-loop controller is shown in Fig. 28.46. The computer with its internal clock drives the *digital-to-analog* (D/A) and *analog-to-digital* (A/D) converters. It compares the command signals with the feedback signals and generates the control signals to be sent to the final control elements. These control signals are computed from the control algorithm stored in the memory. Slightly different structures exist, but Fig. 28.46 shows the important aspects. For example, the comparison between the command and feedback signals can be done with analog elements, and the A/D conversion made on the resulting error signal. The software must also provide for *interrupts,* which are conditions that call for the computer's attention to do something other than computing the control algorithm.

The time required for the control system to complete one loop of the algorithm is the time T, the *sampling time* of the control system. It depends on the time required for the computer to calculate the control algorithm, and on the time required for the interfaces to convert data. Modern systems are capable of very high rates, with sample times under 1 μs.

In most digital control applications, the plant is an analog system, but the controller is a discrete-time system. Thus, to design a digital control system, we must either model the controller as an analog system or model the plant as a discrete-time system. Each approach has its own merits, and we will examine both.

If we model the controller as an analog system, we use methods based on *differential* equations to compute the gains. However, a digital control system requires *difference* equations to describe its behavior. Thus, from a strictly mathematical point of view, the gain values we will compute will not give the predicted response exactly. However, if the sampling time is small compared to the smallest time constant in the system, then the digital system will act like an analog system, and our designs will work properly. Because most physical systems of interest have time constants greater than 1 ms, and controllers can now achieve sampling times less than 1 μs, controllers designed with analog methods will often be adequate.

28.10.2 Digital Forms of PID Control

There are a number of ways that PID control can be implemented in software in a digital control system, because the integral and derivative terms must be approximated with formulas chosen from a variety of available algorithms. The simplest integral approximation is to replace the integral with a sum of rectangular areas. With this rectangular approximation, the error integral is calculated as

$$\int_0^{(k+1)T} e(t)\, dt \approx Te(0) + Te(t_1) + Te(t_2) + \cdots + Te(t_k) = T\sum_{i=0}^{k} e(t_i) \qquad (28.45)$$

where $t_k = kT$ and the width of each rectangle is the sampling time $T = t_{i+1} - t_i$. The times t_i are the times at which the computer updates its calculation of the control algorithm after receiving an updated command signal and an updated measurement from the sensor through the A/D interfaces. If the time T is small, then the value of the sum in (28.45) is close to the value of the integral. After the control algorithm calculation is made, the calculated value of the control signal $f(t_k)$ is sent to the actuator via the output interface. This interface includes a D/A converter and a *hold* circuit that "holds" or keeps the analog voltage corresponding to the control signal applied to the actuator until

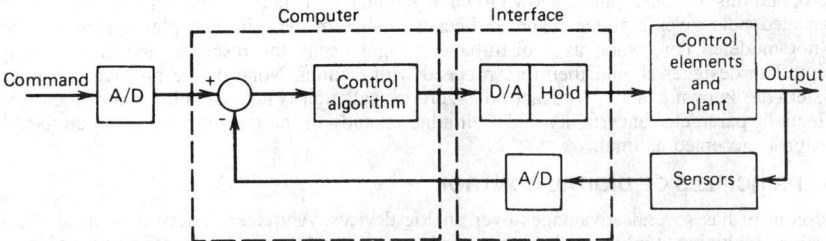

Fig. 28.46 Structure of a digital control system.[1]

the next updated value is passed along from the computer. The simplest digital form of PI control uses (28.45) for the integral term. It is

$$f(t_k) = K_P e(t_k) + K_I T \sum_{i=0}^{k} e(t_i) \tag{28.46}$$

This can be written in a more efficient form by noting that

$$f(t_{k-1}) = K_P e(t_{k-1}) + K_I T \sum_{i=0}^{k-1} e(t_i)$$

and subtracting this from (28.46) to obtain

$$f(t_k) = f(t_{k-1}) + K_P[e(t_k) - e(t_{k-1})] + K_I T e(t_k) \tag{28.47}$$

This form—called the *incremental* or *velocity* algorithm—is well suited for incremental output devices such as stepper motors. Its use also avoids the problem of integrator buildup, the condition in which the actuator saturates but the control algorithm continues to integrate the error.

The simplest approximation to the derivative is the first-order difference approximation

$$\frac{de}{dt} \approx \frac{e(t_k) - e(t_{k-1})}{T} \tag{28.48}$$

The corresponding PID approximation using the rectangular integral approximation is

$$f(t_k) = K_P e(t_k) + K_I T \sum_{i=0}^{k} e(t_i) + \frac{K_D}{T}[e(t_k) - e(t_{k-1})] \tag{28.49}$$

The accuracy of the integral approximation can be improved by substituting a more sophisticated algorithm, such as the following trapezoidal rule.

$$\int_{0}^{(k+1)T} e(t) \, dt \approx T \sum_{i=0}^{k} \frac{1}{2}[e(t_{i+1}) + e(t_i)] \tag{28.50}$$

The accuracy of the derivative approximation can be improved by using values of the sampled error signal at more instants. Using the four-point central difference method (Refs. 1 and 2), the derivative term is approximated by

$$\frac{de}{dt} \approx \frac{1}{6T}[e(t_k) + 3e(t_{k-1}) - 3e(t_{k-2}) - e(t_{k-3})]$$

The derivative action is sensitive to the resulting rapid change in the error samples that follows a step input. This effect can be eliminated by reformulating the control algorithm as follows (Refs. 1 and 2):

$$f(t_k) = f(t_{k-1}) + K_P[c(t_{k-1}) - c(t_k)]$$
$$+ K_I T[r(t_k) - c(t_k)]$$
$$+ \frac{K_D}{T}[-c(t_k) + 2c(t_{k-1}) - c(t_{k-2})] \tag{28.51}$$

where $r(t_k)$ is the command input and $c(t_k)$ is the variable being controlled. Because the command input $r(t_k)$ appears in this algorithm only in the integral term, we cannot apply this algorithm to PD control; that is, the integral gain K_I must be nonzero.

28.11 UNIQUELY DIGITAL ALGORITHMS

Development of analog control algorithms was constrained by the need to design physical devices that could implement the algorithm. However, digital control algorithms simply need to be programmable, and are thus less constrained than analog algorithms.

28.11.1 Digital Feedforward Compensation

Classical control system design methods depend on linear models of the plant. With linearization we can obtain an approximately linear model, which is valid only over a limited operating range. Digital control now allows us to deal with nonlinear models more directly, using the concepts of feedforward compensation discussed in Section 28.8.

Computed Torque Method

Figure 28.47 illustrates a variation of feedforward compensation of the disturbance called the *computed torque method*. It is used to control the motion of robots. A simple model of a robot arm is the following nonlinear equation.

$$I\ddot{\theta} = T - mgL \sin \theta \tag{28.52}$$

where θ is the arm angle, I is its inertia, mg is its weight, L is the distance from its mass center to the arm joint where the motor acts. The motor supplies the torque T. To position the arm at some desired angle θ_r, we can use PID control on the angle error $\theta_r - \theta$. This works well if the arm angle θ is never far from the desired angle θ_r so that we can linearize the plant model about θ_r. However, the controller will work for large-angle excursions if we compute the nonlinear gravity torque term $mgL \sin \theta$ and add it to the PID output. That is, part of the motor torque will be computed specifically to cancel the gravity torque, in effect producing a linear system for the PID algorithm to handle. The nonlinear torque calculations required to control multidegree-of-freedom robots are very complicated, and can be done only with a digital controller.

Feedforward Command Compensation

Computers can store lookup tables, which can be used to control systems that are difficult to model entirely with differential equations and analytical functions. Figure 28.48 shows a speed-control system for an internal combustion engine. The fuel flow rate required to achieve a desired speed depends in a complicated way on many variables not shown in the figure, such as temperature, humidity, and so on. This dependence can be summarized in tables stored in the control computer and can be used to estimate the required fuel flow rate. A PID algorithm can be used to adjust the estimate based on the speed error. This application is an example of feedforward compensation of the command input, and it requires a digital computer.

28.11.2 Control Design in the z-Plane

There are two common approaches to designing a digital controller:

1. The performance is specified in terms of the desired continuous-time response, and the controller design is done entirely in the s-plane, as with an analog controller. The resulting control law is then converted to discrete-time form, using approximations for the integral and derivative terms. This method can be successfully applied if the sampling time is small. The technique is widely used for two reasons. When existing analog controllers are converted to

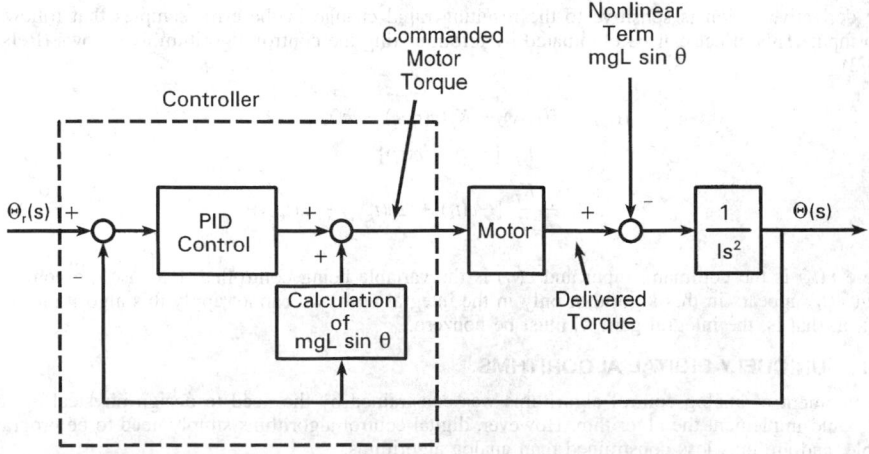

Fig. 28.47 The computed torque method applied to robot arm control.

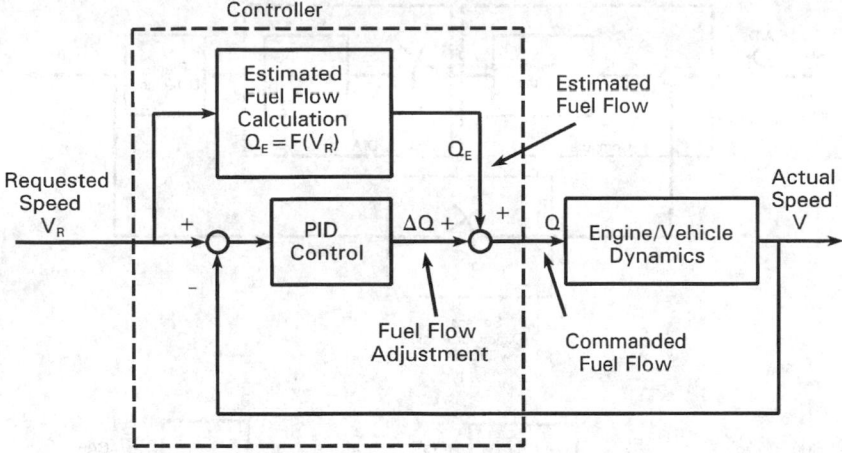

Fig. 28.48 Feedforward compensation applied to engine control.

digital control, the form of the control law and the values of its associated gains are known to have been satisfactory. Therefore, the digital version can use the same control law and gain values. Second, because analog design methods are well established, many engineers prefer to take this route and then convert the design into a discrete-time equivalent.

2. The performance specifications are given in terms of the desired continuous-time response and/or desired root locations in the s-plane. From these the corresponding root locations in the z-plane are found and a discrete control law is designed. This method avoids the derivative and integral approximation errors that are inherent in the first method, and is the preferred method when the sampling time T is large. However, the algebraic manipulations are more cumbersome.

The second approach uses the z-transform and pulse transfer functions, which were outlined in the previous chapter. If we have an analog model of the plant, with its transfer function $G(s)$, we can obtain its pulse transfer function $G(z)$ by finding the z-transform of the impulse response $g(t) = \mathcal{L}^{-1}[G(s)]$; that is, $G(z) = \mathcal{Z}[g(t)]$. Table 27.12 in the previous chapter facilitates this process; see also Refs. 1 and 2. Figure 28.49a shows the basic elements of a digital control system. Part (b) of the figure is an equivalent diagram with the analog transfer functions inserted. Part (c) represents the same system in terms of pulse transfer functions. From the diagram we can find the closed-loop pulse transfer function. It is

$$\frac{C(z)}{R(z)} = \frac{G(z)P(z)}{1 + G(z)P(z)} \tag{28.53}$$

The variable z is related to the Laplace variable s by

$$z = e^{sT} \tag{28.54}$$

If we know the desired root locations and the sampling time T, we can compute the z roots from this equation.

Digital PI Control Design

For example, the first-order plant $1/(2s + 1)$ with a zero-order hold has the following pulse transfer function (Refs. 1 and 2).

$$P(z) = \frac{1 - e^{-0.5T}}{z - e^{-0.5T}} \tag{28.55}$$

Suppose we use a control algorithm described by the following pulse transfer function:

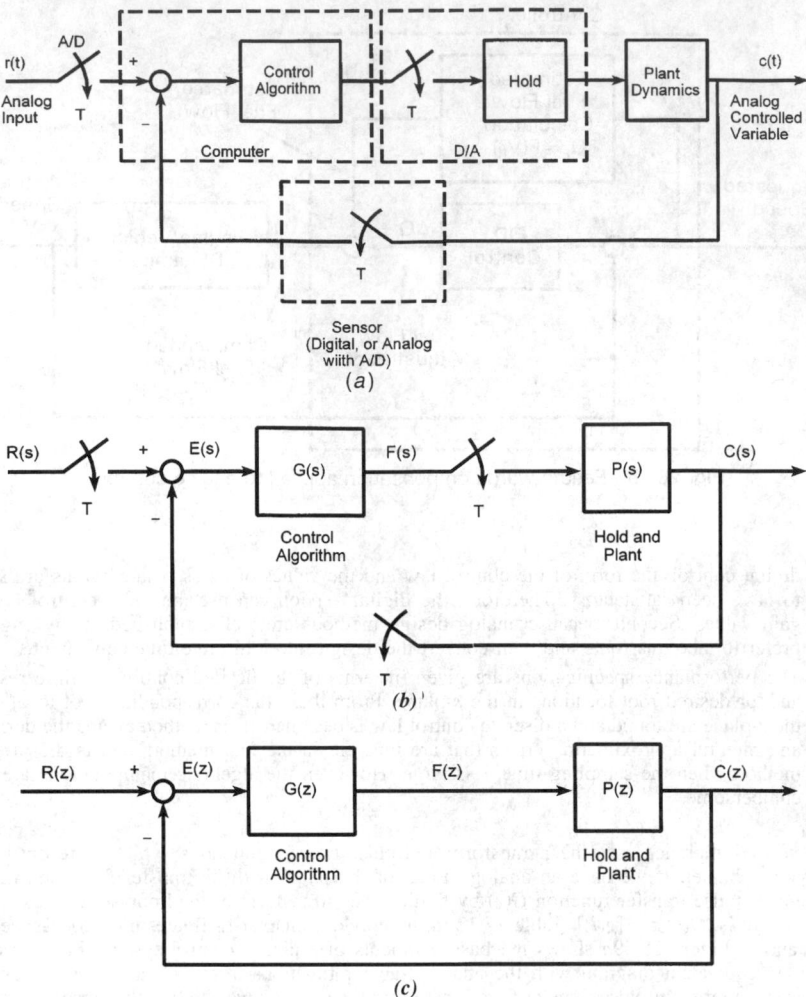

Fig. 28.49 Block diagrams of a typical digital controller. (*a*) Diagram showing the components. (*b*) Diagram of the *s*-plane relations. (*c*) Diagram of the *z*-plane relations.

$$G(z) = \frac{F(z)}{E(z)} = \frac{K_1 z + K_2}{z - 1} = \frac{K_1 + K_2 z^{-1}}{1 - z^{-1}} \tag{28.56}$$

The corresponding difference equation that the control computer must implement is

$$f(t_k) = f(t_{k-1}) + K_1 e(t_k) + K_2 e(t_{k-1}) \tag{28.57}$$

where $e(t_k) = r(t_k) - c(t_k)$. By comparing (28.57) with (28.47), it can be seen that this is the digital equivalent of PI control, where $K_P = -K_2$ and $K_I = (K_1 + K_2)/T$. Using the form of $G(z)$ given by (28.56), the closed loop transfer function is

$$\frac{C(z)}{R(z)} = \frac{(1 - b)(K_1 z + K_2)}{z^2 + (K_1 - 1 - b - bK_1)z + b + K_2 - bK_2} \tag{28.58}$$

where $b = e^{-05T}$.

If the design specifications call for $\tau = 1$ and $\zeta = 1$, then the desired s roots are $s = -1, -1$, and the analog PI gains required to achieve these roots are $K_P = 3$ and $K_I = 2$. Using a sampling time of $T = 0.1$, the z roots must be $z = e^{-0.1}, e^{-0.1}$. To achieve these roots, the denominator of the transfer function (28.58) must be $z^2 - 2e^{-0.1}z + e^{-0.2}$. Thus the control gains must be $K_1 = 2.903$ and $K_2 = -2.717$. These values of K_1 and K_2 correspond to $K_P = 2.72$ and $K_I = 1.86$, which are close to the PI gains computed for an analog controller. If we had used a sampling time smaller than 0.1, say $T = 0.01$, the values of K_P and K_I computed from K_1 and K_2 would be $K_P = 2.97$ and $K_I = 1.98$, which are even closer to the analog gain values. This illustrates the earlier claim that analog design methods can be used when the sampling time is small enough.

Digital Series Compensation

Series compensation can be implemented digitally by applying suitable discrete-time approximations for the derivative and integral to the model represented by the compensator's transfer function $G_c(s)$. For example, the form of a lead or a lag compensator's transfer function is

$$G_c(s) = \frac{M(s)}{F(s)} = K\frac{s + c}{s + d} \tag{28.59}$$

where $m(t)$ is the actuator command and $f(t)$ is the control signal produced by the main (PID) controller. The differential equation corresponding to (28.59) is

$$\dot{m} + dm = K(\dot{f} + cf) \tag{28.60}$$

Using the simplest approximation for the derivative, Eq. (28.48), we obtain the following difference equation that the digital compensator must implement:

$$\frac{m(t_k) - m(t_{k-1})}{T} + dm(t_k) = K\left[\frac{f(t_k) - f(t_{k-1})}{T} + cf(t_k)\right]$$

In the z-plane, the equation becomes

$$\frac{1 - z^{-1}}{T}M(z) + dM(z) = K\left[\frac{1 - z^{-1}}{T}F(z) + cF(z)\right] \tag{28.61}$$

The compensator's pulse transfer function is thus seen to be

$$G_c(z) = \frac{M(z)}{F(z)} = \frac{K(1 - z^{-1}) + cT}{1 - z^{-1} + dT}$$

which has the form

$$G_c(z) = K_c\frac{z + a}{z + b} \tag{28.62}$$

where K_c, a, and b can be expressed in terms of K, c, d, and T if we wish to use analog design methods to design the compensator. When using commercial controllers, the user might be required to enter the values of the gain, the pole, and the zero of the compensator. The user must ascertain whether these values should be entered as s-plane values (i.e., K, c, and d) or as z-plane values (K_c, a, and b).

Note that the digital compensator has the same number of poles and zeros as the analog compensator. This is a result of the simple approximation used for the derivative. Note that Eq. (28.61) shows that when we use this approximation, we can simply replace s in the analog transfer function with $1 - z^{-1}$. Because the integration operation is the inverse of differentiation, we can replace $1/s$ with $1/(1 - z^{-1})$ when integration is used. (This is equivalent to using the rectangular approximation for the integral, and can be verified by finding the pulse transfer function of the incremental algorithm (28.47) with $K_P = 0$.)

Some commercial controllers treat the PID algorithm as a series compensator, and the user is expected to enter the controller's values not as PID gains, but as pole and zero locations in the z-plane. The PID transfer function is

$$\frac{F(s)}{E(s)} = K_P + \frac{K_I}{s} + K_D s \tag{28.63}$$

Making the indicated replacements for the s terms, we obtain

$$\frac{F(z)}{E(z)} = K_P + \frac{K_I}{1 - z^{-1}} + K_D(1 - z^{-1})$$

which has the form

$$\frac{F(z)}{E(z)} = K_c \frac{z^2 - az + b}{z - 1} \tag{28.64}$$

where K_c, a, and b can be expressed in terms of K_P, K_I, K_D, and T. Note that the algorithm has two zeros and one pole, which is fixed at $z = 1$. Sometimes the algorithm is expressed in the more general form

$$\frac{F(z)}{E(z)} = K_c \frac{z^2 - az + b}{z - c} \tag{28.65}$$

to allow the user to select the pole as well.

Digital compensator design can be done with frequency response methods, or with the root-locus plot applied to the z-plane rather than the s-plane. However, when better approximations are used for the derivative and integral, the digital series compensator will have more poles and zeros than its analog counterpart. This means that the root-locus plot will have more root paths, and the analysis will be more difficult. This topic is discussed in more detail in Refs. 1, 2, 3, and 7.

28.11.3 Direct Design of Digital Algorithms

Because almost any algorithm can be implemented digitally, we can specify the desired response and work backward to find the required control algorithm. This is the *direct design* method. If we let $D(z)$ be the desired form of the closed loop transfer function $C(z)/R(z)$, and solve (28.53) for the controller transfer function $G(z)$, we obtain

$$G(z) = \frac{D(z)}{P(z)[1 - D(z)]} \tag{28.66}$$

We can pick $D(z)$ directly or obtain it from the specified input transform $R(z)$ and the desired output transform $C(z)$, because $D(z) = C(z)/R(z)$.

Finite Settling Time Algorithm

This method can be used to design a controller to compensate for the effects of process dead time. A plant having such a response can often be approximately described by a first-order model with a dead-time element; that is,

$$G_P(s) = K \frac{e^{-Ds}}{\tau s + 1} \tag{28.67}$$

where D is the dead time. This model also approximately describes the S-shaped response curve used with the Ziegler–Nichols method (Figure 28.33). When combined with a zero-order hold, this plant has the following pulse transfer function:

$$P(z) = Kz^{-n} \frac{1 - a}{z - a} \tag{28.68}$$

where $a = \exp(-T/\tau)$ and $n = D/T$. If we choose $D(z) = z^{-(n+1)}$, then with a step command input, the output $c(k)$ will reach its desired value in $n + 1$ sample times, one more than is in the dead time D. This is the fastest response possible. From (28.66) the required controller transfer function is

$$G(z) = \frac{1}{K(1 - a)} \frac{1 - az^{-1}}{1 - z^{-(n+1)}} \tag{28.69}$$

The corresponding difference equation that the control computer must implement is

$$f(t_k) = f(t_{k-n-1}) + \frac{1}{K(1 - a)} [e(t_k) - ae(t_{k-1})] \tag{28.70}$$

This algorithm is called a *finite settling time* algorithm, because the response reaches its desired value in a finite, prescribed time. The maximum value of the manipulated variable required by this algorithm occurs at $t = 0$ and is $1/K(1 - a)$. If this value saturates the actuator, this method will not work as predicted. Its success depends also on the accuracy of the plant model.

Dahlin's Algorithm

This sensitivity to plant modeling errors can be reduced by relaxing the minimum response time requirement. For example, choosing $D(z)$ to have the same form as $P(z)$, namely,

$$D(z) = K_d z^{-n} \frac{1 - a_d}{z - a_d} \tag{28.71}$$

we obtain from (28.66) the following controller transfer function:

$$G(z) = \frac{K_d(1 - a_d)}{K(1 - a)} \frac{1 - az^{-1}}{1 - a_d z^{-1} - K_d(1 - a_d)z^{-(n+1)}} \tag{28.72}$$

This is *Dahlin's algorithm.*[3] The corresponding difference equation that the control computer must implement is

$$f(t_k) = a_d f(t_{k-1}) + K_d(1 - a_d)f(t_{k-n-1})$$

$$+ \frac{K_d(1 - a_d)}{K(1 - a)} [e(t_k) - ae(t_{k-1})] \tag{28.73}$$

Normally we would first try setting $K_d = K$ and $a_d = a$, but since we might not have good estimates of K and a, we can use K_d and a_d as tuning parameters to adjust the controller's performance. The constant a_d is related to the time constant τ_d of the desired response: $a_d = \exp(-T/\tau_d)$. Choosing τ_d smaller gives faster response.

   Algorithms such as these are often used for system startup, after which the control mode is switched to PID, which is more capable of handling disturbances.

28.12 HARDWARE AND SOFTWARE FOR DIGITAL CONTROL

This section provides an overview of the general categories of digital controllers that are commercially available. This is followed by a summary of the software currently available for digital control and for control system design.

28.12.1 Digital Control Hardware

Commercially available controllers have different capabilities, such as different speeds and operator interfaces, depending on their targeted application.

Programmable Logic Controllers (PLCs)

These are controllers that are programmed with relay ladder logic, which is based on Boolean algebra. Now designed around microprocessors, they are the successors to the large relay panels, mechanical counters, and drum programmers used up to the 1960s for sequencing control and control applications requiring only a finite set of output values (for example, opening and closing of valves). Some models now have the ability to perform advanced mathematical calculations required for PID control, thus allowing them to be used for modulated control as well as finite state control. There are numerous manufacturers of PLCs.

Digital Signal Processors (DSPs)

A modern development is the *Digital Signal Processor* (DSP), which has proved useful for feedback control as well as signal processing.[8] This special type of processor chip has separate buses for moving data and instructions, and is constructed to perform rapidly the kind of mathematical operations required for digital filtering and signal processing. The separate buses allow the data and the instructions to move in parallel rather than sequentially. Because the PID control algorithm can be written in the form of a digital filter, DSPs can also be used as controllers.

   The DSP architecture was developed to handle the types of calculations required for digital filters and discrete Fourier transforms, which form the basis of most signal processing operations. DSPs usually lack the extensive memory-management capabilities of general-purpose computers, because they need not store large programs or large amounts of data. Some DSPs contain A/D and D/A converters, serial ports, timers, and other features. They are programmed with specialized software that runs on popular personal computers. Low-cost DSPs are now widely used in consumer electronics and automotive applications, with Texas Instruments being a major supplier.

Motion Controllers

Motion controllers are specialized control systems that provide feedback control for one or more motors. They also provide a convenient operator interface for generating the commanded trajectories. Motion controllers are particularly well suited for applications requiring coordinated motion of two or more axes, and for applications where the commanded trajectory is complicated. A higher-level host computer might transmit required distance, speed, and acceleration rates to the motion controller, which then constructs and implements the continuous position profile required for each motor. For example, the host computer would supply the required total displacement, the acceleration and deceleration times, and the desired slew speed (the speed during the zero acceleration phase). The motion controller would generate the commanded position versus time for each motor. The motion controller also has the task of providing feedback control for each motor, to ensure that the system follows the required position profile.

Figure 28.50 shows the functional elements of a typical motion controller, such as those built by Galil Motion Control, Inc. Provision for both analog and digital input signals allows these controllers to perform other control tasks besides motion control. Compared to DSPs, such controllers generally have greater capabilities for motion control and have operator interfaces that are better suited for such applications. Motion controllers are available as plug-in cards for most computer bus types. Some are available as stand-alone units.

Motion controllers use a PID control algorithm to provide feedback control for each motor (some manufacturers call this algorithm a "filter"). The user enters the values of the PID gains (some manufacturers provide preset gain values, which can be changed; others provide tuning software that assists in selecting the proper gain values). Such controllers also have their own language for programming a variety of motion profiles and other applications.[15] For example, they provide for linear and circular interpolation for 2D coordinated motion, motion smoothing (to eliminate jerk), contouring, helical motion, and electronic gearing. The latter is a control mode that emulates mechanical gearing in software, in which one motor (the slave) is driven in proportion to the position of another motor (the master) or an encoder.

Process Controllers

Process controllers are designed to handle inputs from sensors, such as thermocouples, and outputs to actuators, such as valve positioners, that are commonly found in process control applications. Figure 28.51 illustrates the input–output capabilities of a typical process controller such as those manufactured by Honeywell, which is a major supplier of such devices. This device is a stand-alone

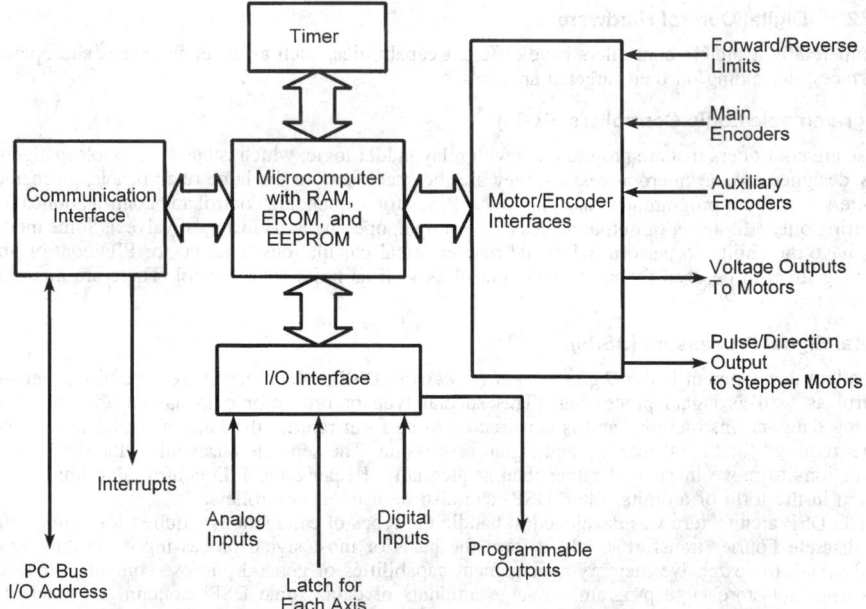

Fig. 28.50 Functional diagram of a motion controller.

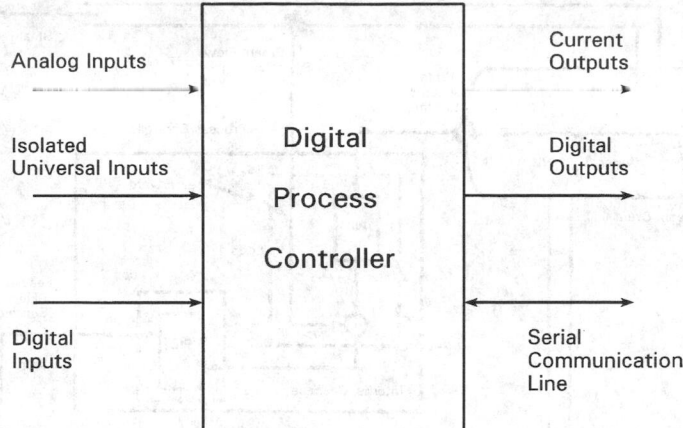

Fig. 28.51   Functional diagram of a digital process controller.

unit designed to be mounted in an instrumentation panel. The voltage and current ranges of the analog inputs are those normally found with thermocouple-based temperature sensors. The current outputs are designed for devices like valve positioners, which usually require 4–20 ma signals.

The controller contains a microcomputer with built-in math functions normally required for process control, such as thermocouple linearization, weighted averaging, square roots, ratio/bias calculations, and the PID control algorithm. These controllers do not have the same software and memory capabilities as desktop computers, but they are less expensive. Their operator interface consists of a small keypad with typically fewer than 10 keys, a small graphical display for displaying bargraphs of the setpoints and the process variables, indicator lights, and an alphanumeric display for programming the controller.

The PID gains are entered by the user. Some units allow multiple sets of gains to be stored; the unit can be programmed to switch between gain settings when certain conditions occur. Some controllers have an adaptive tuning feature that is supposed to adjust the gains to prevent overshoot in startup mode, to adapt to changing process dynamics, and to adapt to disturbances. However, at this time, adaptive tuning cannot claim a 100% success rate, and further research and development in adaptive control is needed.

Some process controllers have more than one PID control loop for controlling several variables. Figure 28.52 illustrates a boiler feedwater control application for a controller with two PID loops arranged in a cascade control structure. Loop 1 is the main or outer-loop controller for maintaining the desired water volume in the boiler. It uses sensing of the steam flow rate to implement feedforward compensation. Loop 2 is the inner-loop controller that directly controls the feedwater control valve.

28.12.2   Software for Digital Control

The software available to the modern control engineer is quite varied and powerful, and can be categorized according to the following tasks:

1. Control algorithm design, gain selection, and simulation
2. Tuning
3. Motion programming
4. Instrumentation configuration
5. Real-time control functions

Many analysis and simulation packages now contain algorithms of specific interest to control system designers. *Matlab* is one such package that is widely used. It contains built-in functions for generating root-locus and frequency-response plots, system simulation, digital filtering, calculation of control gains, and data analysis. It can accept model descriptions in the form of transfer functions or as state variable equations.

Other popular control system design and simulation packages include *Program CC*, *Matrix$_x$*, and *ACSL*. Some manufacturers provide software to assist the engineer in sizing and selecting components. An example is the *Motion Component Selector* (MCS) sold by Galil Motion Control, Inc. It

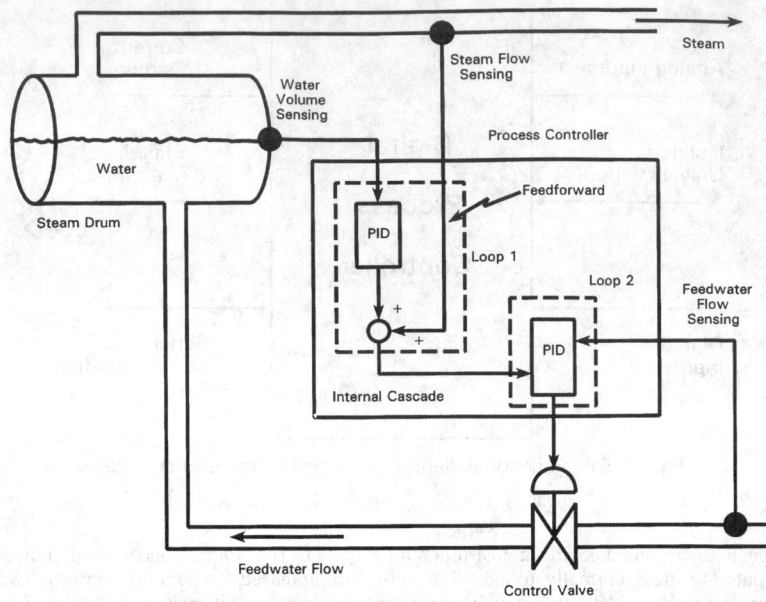

Fig. 28.52 Application of a two-loop process controller for feedwater control.

assists the engineer in computing the load inertia, including the effects of the mechanical drive, and then selects the proper motor and amplifier based on the user's description of the desired motion profile.

Some hardware manufacturers supply software to assist the engineer in selecting control gains and modifying (*tuning*) them to achieve good response. This might require that the system to be controlled be available for experiments prior to installation. Some controllers, such as some Honeywell process controllers, have an autotuning feature that adjusts the gains in real time to improve performance.

Motion programming software supplied with motion controllers was mentioned previously. Some packages, such as Galil's, allow the user to simulate a multi-axis system having more than one motor and to display the resulting trajectory.

Instrumentation configuration software, such as *LabView* and *Dadisp,* provides specialized programming languages for interacting with instruments and for creating graphical real-time displays of instrument outputs.

Until recently, development of real-time digital control software involved tedious programming, often in assembly language. Even when implemented in a higher-level language, such as Fortran or C, programming real-time control algorithms can be very challenging, partly because of the need to provide adequately for interrupts. Software packages are now available that provide real-time control capability, usually a form of the PID algorithm, that can be programmed through user-friendly graphical interfaces. Examples include the Galil motion controllers and the add-on modules for Labview and Dadisp.

28.13 FUTURE TRENDS IN CONTROL SYSTEMS

Microprocessors have rejuvenated the development of controllers for mechanical systems. Currently, there are several applications areas in which new control systems are indispensable to the product's success:

1. Active vibration control
2. Noise cancellation
3. Adaptive optics
4. Robotics
5. Micromachines
6. Precision engineering

Most of the design techniques presented here comprise "classical" control methods. These methods are widely used because when they are combined with some testing and computer simulation, an experienced engineer can rapidly achieve an acceptable design. Modern control algorithms, such as state variable feedback and the linear–quadratic optimal controller, have had some significant mechanical engineering applications—for example, in the control of aerospace vehicles. The current approach to multivariable systems like the one shown in Fig. 28.53 is to use classical methods to design a controller for each subsystem, because they can often be modeled with low-order linearized models. The coordination of the various low-level controllers is a nonlinear problem. High-order, nonlinear, multivariable systems that cannot be controlled with classical methods cannot yet be handled by modern control theory in a general way, and further research is needed.

In addition to the improvements, such as lower cost, brought on by digital hardware, microprocessors have allowed designers to incorporate algorithms of much greater complexity into control systems. The following is a summary of the areas currently receiving much attention in the control systems community.

28.13.1 Fuzzy Logic Control

In classical set theory, an object's membership in a set is clearly defined and unambiguous. *Fuzzy logic control* is based on a generalization of classical set theory to allow objects to belong to several sets with various degrees of membership. Fuzzy logic can be used to describe processes that defy precise definition or precise measurement, and thus it can be used to model the inexact and subjective aspects of human reasoning. For example, room temperature can be described as cold, cool, just right, warm, or hot. Development of a fuzzy logic temperature controller would require the designer to specify the membership functions that describe "warm" as a function of temperature, and so on. The control logic would then be developed as a linguistic algorithm that models a human operator's decision process (for example, if the room temperature is "cold," then "greatly" increase the heater output; if the temperature is "cool," then increase the heater output "slightly").

Fuzzy logic controllers have been implemented in a number of applications. Proponents of fuzzy logic control point to its ability to convert a human operator's reasoning process into computer code. Its critics argue that because all the controller's fuzzy calculations must eventually reduce to a specific output that must be given to the actuator (e.g., a specific voltage value or a specific valve position), why not be unambiguous from the start, and define a "cool" temperature to be the range between 65° and 68°, for example? Perhaps the proper role of fuzzy logic is at the human operator interface. Research is active in this area, and the issue is not yet settled.[10,11]

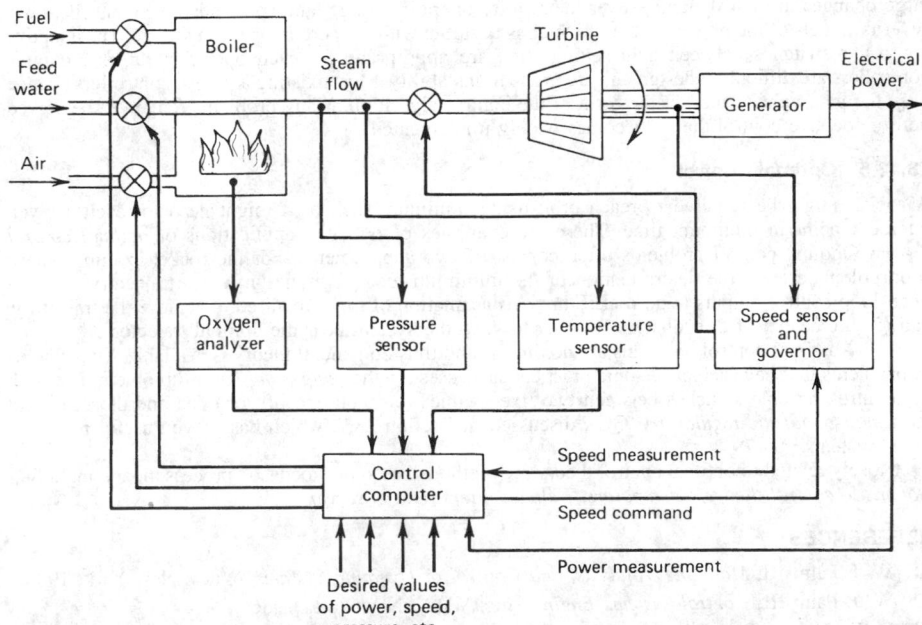

Fig. 28.53 Computer control system for a boiler–generator. Each important variable requires its own controller. The interaction between variables calls for coordinated control of all loops.[1]

28.13.2 Neural Networks

Most digital computers contain a small number of sophisticated processors capable of executing large sets of instructions. In contrast, *neural networks* are devices that contain a large number of simple processors, called *nodes* or *artificial neurons*.[11,12] In conventional computers, processing and memory functions are performed by distinct elements that communicate with each other via a common bus. In neural networks, processing, memory, and communication are all handled by the nodes. This allows for faster computation, while the nonlinear behavior of the nodes provides more versatile computational capabilities.

Few neural networks have yet been implemented in hardware; most are simulations using conventional digital computers. They show promise for control systems applications requiring pattern recognition, such as image processing, tactile sensing, and nonlinear control of robot arms.

28.13.3 Nonlinear Control

Most real systems are nonlinear, which means that they must be described by nonlinear differential equations. Control systems designed with the linear control theory described in this chapter depend on a linearized approximation to the original nonlinear model. This linearization can be explicitly performed, or implicitly made, as when we use the small-angle approximation: $\sin \theta \approx \theta$. This approach has been enormously successful because a well-designed controller will keep the system in the operating range where the linearization was done, thus preserving the accuracy of the linear model. However, it is difficult to control some systems accurately in this way, because their operating range is too large. Robot arms are a good example.[13,14] Their equations of motion are very nonlinear, due primarily to the fact that their inertia varies greatly as their configuration changes.

Nonlinear systems encompass everything that is "not linear," and thus there is no general theory for nonlinear systems. There have been many nonlinear control methods proposed; too many to summarize here.[9] *Liapunov's stability theory* and Popov's method play a central role in many such schemes. Adaptive control is a subcase of nonlinear control (see below).

The high speeds of modern digital computers now allow us to implement nonlinear control algorithms not possible with earlier hardware. An example is the computed-torque method for controlling robot arms, which was discussed in Section 28.11 (see Fig. 28.47).

28.13.4 Adaptive Control

The term *adaptive control*, which unfortunately has been loosely used, describes control systems that can change the form of the control algorithm or the values of the control gains in real time, as the controller improves its internal model of the process dynamics or in response to unmodeled disturbances.[16] Constant control gains do not provide adequate response for some systems that exhibit large changes in their dynamics over their entire operating range, and some adaptive controllers use several models of the process, each of which is accurate within a certain operating range. The adaptive controller switches between gain settings that are appropriate for each operating range. Adaptive controllers are difficult to design and are prone to instability. Most existing adaptive controllers change only the gain values, and not the form of the control algorithm. Many problems remain to be solved before adaptive control theory becomes widely implemented.

28.13.5 Optimal Control

A rocket might be required to reach orbit using minimum fuel, or it might need to reach a given intercept point in minimum time. These are examples of potential applications of *optimal control theory*. Optimal control problems often consist of two subproblems. For the rocket example, these subproblems are (1) the determination of the minimum-fuel (or minimum-time) trajectory, and the open-loop control outputs (e.g., rocket thrust as a function of time) required to achieve the trajectory, and (2) the design of a feedback controller to keep the system near the optimal trajectory.

Many optimal control problems are nonlinear, and thus no general theory is available. Two classes of problems that have achieved some practical successes are the *bang-bang control* problem, in which the control variable switches between two fixed values (e.g., on and off, or open and closed),[5] and the *linear-quadratic-regulator* (LQG), discussed in Section 28.7, which has proven useful for high-order systems.[1,5]

Closely related to optimal control theory are methods based on stochastic process theory, including *stochastic control theory*,[17] *estimators*, *Kalman filters*, and *observers*.[1,5,17]

REFERENCES

1. W. J. Palm III, *Modeling, Analysis, and Control of Dynamic Systems*, Wiley, New York, 1983.
2. W. J. Palm III, *Control Systems Engineering*, Wiley, New York, 1986.
3. D. E. Seborg, T. F. Edgar, and D. A. Mellichamp, *Process Dynamics and Control*, Wiley, New York, 1989.
4. D. McCloy and H. Martin, *The Control of Fluid Power*, 2nd ed., Halsted Press, London, 1980.

5. A. E. Bryson and Y. C. Ho, *Applied Optimal Control,* Blaisdell, Waltham, MA, 1969.

6. F. Lewis, *Optimal Control,* Wiley, New York, 1986.

7. K. J. Astrom and B. Wittenmark, *Computer Controlled Systems,* Prentice-Hall, Englewood Cliffs, NJ, 1984.

8. Y. Dote, *Servo Motor and Motion Control Using Digital Signal Processors,* Prentice-Hall, Englewood Cliffs, NJ, 1990.

9. J. Slotine and W. Li, *Applied Nonlinear Control,* Prentice-Hall, Englewood Cliffs, NJ, 1991.

10. G. Klir and B. Yuan, *Fuzzy Sets and Fuzzy Logic,* Prentice-Hall, Englewood Cliffs, NJ, 1995.

11. B. Kosko, *Neural Networks and Fuzzy Systems,* Prentice-Hall, Englewood Cliffs, NJ, 1992.

12. A. Cichocki, and R. Unbehauen, *Neural Networks for Optimization and Signal Processing,* Wiley, New York, 1994.

13. J. Craig, *Introduction to Robotics,* 2nd ed., Addison-Wesley, Reading, MA, 1989.

14. M. W. Spong and M. Vidyasagar, *Robot Dynamics and Control,* Wiley, New York, 1989.

15. J. Tal, *Step by Step Design of Motion Control Systems,* Galil Motion Control, Sunnyvale, CA, 1994.

16. K. J. Astrom, *Adaptive Control,* Addison-Wesley, Reading, MA, 1989.

17. R. Stengel, *Stochastic Optimal Control,* Wiley, New York, 1986.

CHAPTER 29

MEASUREMENTS

E. L. Hixson
E. A. Ripperger
University of Texas
Austin, Texas

29.1	**STANDARDS AND ACCURACY**	**917**		29.3.3	Use of Normal Distribution to Calculate the Probable Error in X	924
	29.1.1 Standards	917		29.3.4	External Estimates	925
	29.1.2 Accuracy and Precision	918				
	29.1.3 Sensitivity or Resolution	918	**29.4**	**APPENDIX**		**928**
	29.1.4 Linearity	919		29.4.1	Vibration Measurement	928
				29.4.2	Acceleration Measurement	928
29.2	**IMPEDANCE CONCEPTS**	**919**		29.4.3	Shock Measurement	928
				29.4.4	Sound Measurement	928
29.3	**ERROR ANALYSIS**	**923**				
	29.3.1 Introduction	923				
	29.3.2 Internal Estimates	923				

29.1 STANDARDS AND ACCURACY

29.1.1 Standards

Measurement is the process by which a quantitative comparison is made between a standard and a measurand. The measurand is the particular quantity of interest—the thing that is to be quantified. The standard of comparison is of the same character as the measurand and, so far as mechanical engineering is concerned, the standards are defined by law and maintained by the National Institute of Science and Technology (NIST).* The four independent standards that have been defined are *length, time, mass* and *temperature*.[1] All other standards are derived from these four. Before 1960, the standard for length was the *international prototype meter*, kept at Sevres, France. In 1960, the meter was redefined as 1,650,763.73 wavelengths of krypton light. Then, in 1983, the 17th General Conference on Weights and Measures, adopted and immediately put into effect a new standard: "meter is the distance traveled in a vacuum by light in 1/299,792,458 seconds."[2] However, there is a copy of the international prototype meter, known as the *National Prototype Meter*, kept at the National Institute of Science and Technology. Below that level there are several bars known as National Reference Standards and below that there are the working standards. Interlaboratory standards in factories and laboratories are sent to the National Institute of Science and Technology for comparison with the working standards. These interlaboratory standards are the ones usually available to engineers.

Standards for the other three basic quantities have also been adopted by the National Institute of Science and Technology and accurate measuring devices for those quantities should be calibrated against those standards.

The *standard mass* is a cylinder of platinum–iridium, the international kilogram, also kept at Sevres, France. It is the only one of the basic standards that is still established by a prototype. In the United States, the basic unit of mass is the U.S. basic prototype kilogram No. 20. There are working copies of this standard that are used to determine the accuracy of interlaboratory standards. Force is not one of the fundamental quantities, but in the United States the standard unit of force is

---

*Formerly known as the "National Bureau of Standards."

Mechanical Engineers' Handbook, 2nd ed., Edited by Myer Kutz.
ISBN 0-471-13007-9 © 1998 John Wiley & Sons, Inc.

the pound, defined as the gravitational attraction for a certain platinum mass at sea level and 45° latitude.

Absolute time, or the time when some event occurred in history, is not of much interest to engineers. They are more likely to need to measure time intervals, that is, the time between two events. At one time the *second,* the basic unit for time measurements, was defined as 1/86400 of the average period of rotation of the earth on its axis, but that is not a practical standard. The period varies and the earth is slowing down. Consequently, a new standard based on the oscillations associated with a certain transition within the cesium atom has been defined and adopted. The second is now "the duration of 9,192,631,770 periods of the radiation corresponding to the transition between two hyperfine levels of the fundamental state of cesium 133." [3] Thus, the cesium "clock" is the basic frequency standard, but tuning forks, crystals, electronic oscillators, and so on may be used as secondary standards. For the convenience of anyone who requires a time signal of a high order of accuracy, the National Institute of Science and Technology broadcasts continuously time signals of different frequencies from stations WWV, WWVB, and WWVL, located in Fort Collins, Colorado, and WWVH, located in Hawaii. Other nations also broadcast timing signals. For details on the time signal broadcasts, potential users should consult the National Institute of Science and Technology.[4]

Temperature is one of four fundamental quantities in the international measuring system. Temperature is fundamentally different in nature from length, time, and mass. It is an intensive quantity, whereas the others are extensive. Join two bodies that have the same temperature together and you will have a larger body at that same temperature. Join two bodies that have a certain mass and you will have one body of twice the mass of the original body. Two bodies are said to be at the same temperature if they are in thermal equilibrium. The *International Practical Temperature Scale* (IPTS-68), adopted in 1968 by the International Committee on Weights and Measurement,[5] is the one now in effect and the one with which engineers are primarily concerned. In this system, the kelvin (K) is the basic unit of temperature. It is 1/273.16 of the temperature at the triple point of water, which is the temperature at which the solid, liquid, and vapor phases of water exist in equilibrium. Degrees celsius (°C) are related to degrees kelvin by the equation

$$t = T - 273.15$$

where t = degrees celsius
$\quad\ T$ = degrees kelvin

Zero celsius is the temperature established between pure ice and air-saturated pure water at normal atmospheric pressure. The IPTS-68 established six primary fixed reference temperatures and procedures for interpolating between them. These are the temperatures and procedures used for calibrating precise temperature-measuring devices.

29.1.2 Accuracy and Precision

In measurement practice, four terms are frequently used to describe an instrument: *accuracy, precision, sensitivity,* and *linearity.* Accuracy, as applied to an instrument, is the closeness with which a reading approaches the true value. Since there is some error in every reading, the "true value" is never known. In the discussion of error analysis that follows later, methods of estimating the "closeness" with which the determination of a measured value approaches the true value will be presented. Precision is the degree to which readings agree among themselves. If the same value is measured many times and all the measurements agree very closely, the instrument is said to have a high degree of precision. It may not, however, be a very accurate instrument. Accurate calibration is necessary for accurate measurement. Measuring instruments must, for accuracy, be compared to a standard from time to time. These will usually be laboratory or company standards, which are in turn compared from time to time with a working standard at the National Institute of Science and Technology. This chain can be thought of as the pedigree of the instrument, and the calibration of the instrument is said to be traceable to NIST.

29.1.3 Sensitivity or Resolution

These two terms, as applied to a measuring instrument, refer to the smallest change in the measured quantity to which the instrument responds. Obviously, the accuracy of an instrument will depend to some extent on the sensitivity. If, for example, the sensitivity of a pressure transducer is one kilopascal, any particular reading of the transducer has a potential error of at least one kilopascal. If the readings expected are in the range of 100 kilopascals and a possible error of 1% is acceptable, then the transducer with a sensitivity of one kilopascal may be acceptable, depending upon what other sources of error may be present in the measurement. A highly sensitive instrument is difficult to use. Therefore, an instrument with a sensitivity significantly greater than that necessary to obtain the desired accuracy is no more desirable than one with insufficient sensitivity.

Many instruments in use today have digital readouts. For such instruments the concepts of sensitivity and resolution are defined somewhat differently than they are for analog-type instruments.

For example, the resolution of a digital voltmeter depends on the "bit" specification and the voltage range. The relationship between the two is expressed by the equation[6]

$$\epsilon = V/2^n$$

where V = voltage range
n = number of bits

Thus, an 8-bit instrument on a one-volt scale would have a resolution of $1/256$, or 0.004 volts. On a ten-volt scale that would increase to 0.04 volts. As in analog instruments, the higher the resolution, the more difficult it is to use the instrument, so if the choice is available, one should take the instrument which just gives the desired resolution and no more.

29.1.4 Linearity

The calibration curve for an instrument does not have to be a straight line. However, conversion from a scale reading to the corresponding measured value is most convenient if it can be done by multiplying by a constant rather than by referring to a nonlinear calibration curve, or by computing from an equation. Consequently, instrument manufacturers generally try to produce instruments with a linear readout, and the degree to which an instrument approaches this ideal is indicated by its "linearity." Several definitions of "linearity" are used in instrument-specification practice.[7] So-called "independent linearity" is probably the most commonly used in specifications. For this definition, the data for the instrument readout versus the input are plotted and then a "best straight line" fit is made using the method of least squares. Linearity is then a measure of the maximum deviation of any of the calibration points from this straight line. This deviation can be expressed as a percentage of the actual reading or a percentage of the full scale reading. The latter is probably the most commonly used, but it may make an instrument appear to be much more linear than it actually is. A better specification is a combination of the two. Thus, linearity $= \pm A\%$ of reading or $\pm B\%$ of full scale, whichever is greater.

Sometimes the term *independent linearity* is used to describe linearity limits based on actual readings. Since both are given in terms of a fixed percentage, an instrument with $A\%$ proportional linearity is much more accurate at low reading values than an instrument with $A\%$ independent linearity.

It should be noted that although specifications may refer to an instrument as having $A\%$ linearity, what is really meant is $A\%$ nonlinearity. If the linearity is specified as independent linearity, the user of the instrument should try to minimize the error in readings by selecting a scale, if that option is available, such that the actual reading is close to full scale. Never take a reading near the low end of a scale if it can possibly be avoided.

For instruments that use digital processing, linearity is still an issue since the analog to digital converter used can be nonlinear. Thus linearity specifications are still essential.

29.2 IMPEDANCE CONCEPTS[7]

A basic question that must be considered when any measurement is made is how the measured quantity has been affected by the instrument used to measure it. Is the quantity the same as it would have been had the instrument not been there? If the answer to the question is no, the effect of the instrument is called "loading." To characterize the loading, the concepts of "stiffness" and "input impedance" are used. At the input of each component in a measuring system there exists a variable q_{i1}, which is the one we are primarily concerned with in the transmission of information. At the same point, however, there is associated with q_{i1} another variable q_{i2} such that the product $q_{i1} q_{i2}$ has the dimensions of power and represents the rate at which energy is being withdrawn from the system. When these two quantities are identified, the generalized input impedance Z_{gi} can be defined by

$$Z_{gi} = q_{i1}/q_{i2} \qquad (29.1)$$

if q_{i1} is an "effort variable." The effort variable is also sometimes called the "across variable." The quantity q_{i2} is called the "flow variable" or "through variable."

The application of these concepts is illustrated by the example in Fig. 29.1. The output of the linear network in blackbox (a) is the open circuit voltage E_0 until the load Z_L is attached across the terminals $A–B$. If Thevenin's theorem is applied after the load Z_L is attached, the system in Fig. 29.1b is obtained. For that system the current is given by

$$i_m = E_0/[Z_{AB} + Z_L] \qquad (29.2)$$

and the voltage E_L across Z_L is

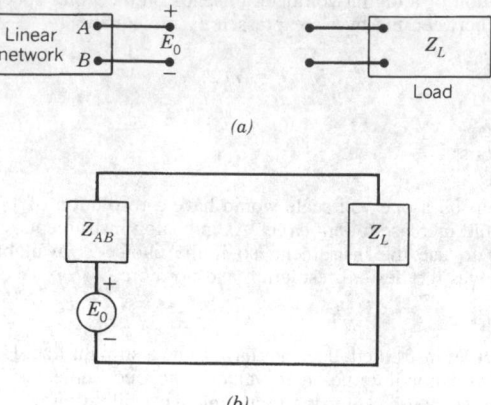

(a)

(b)

Fig. 29.1 Application of Thevenin's theorem.

$$E_L = i_m Z_L = E_0 Z_L / [Z_{AB} + Z_L]$$

or

$$E_L = E_0 / [1 + Z_{AB}/Z_L] \qquad (29.3)$$

In a measurement situation, E_L would be voltage indicated by the voltmeter, Z_L would be the input impedance of the voltmeter, and Z_{AB} would be the output impedance of the linear network. The true output voltage, E_0, has been reduced by the voltmeter, but it can be computed from the voltmeter reading if Z_{AB} and Z_L are known. From Eq. (29.3) it is seen that the effect of the voltmeter on the reading is minimized by making Z_L as large as possible.

If the generalized input and output impedances Z_{gi} and Z_{go} are defined for nonelectrical systems as well as electrical systems, Eq. (29.3) can be generalized to

$$q_{im} = q_{iu} / [1 + Z_{go}/Z_{gi}] \qquad (29.4)$$

where q_{im} is the measured value of the effort variable and q_{iu} is the undisturbed value of the effort variable. The output impedance Z_{go} is not always defined or easy to determine; consequently, Z_{gi} should be large. If it is large enough, knowing Z_{go} is unimportant. However, Z_{go} and Z_{gi} can be measured[8] and Eq. 29.4 can be applied.

If q_{i1} is a flow variable rather than an effort variable (current is a flow variable, voltage an effort variable), it is better to define an input admittance

$$Y_{gi} = q_{i1}/q_{i2} \qquad (29.5)$$

rather than the generalized input impedance

$$Z_{gi} = \text{effort variable/flow variable}$$

The power drain of the instrument is

$$P = q_{i1}q_{i2} = q_{i2}^2 / Y_{gi} \qquad (29.6)$$

Hence, to minimize power drain, Y_{gi} must be large. For an electrical circuit

$$I_m = I_u / [1 + Y_o/Y_i] \qquad (29.7)$$

where I_m = measured current
I_u = actual current
Y_o = output admittance of the circuit
Y_i = input admittance of the meter

When the power drain is zero, as in structures in equilibrium—as, for example, when deflection is to be measured—the concepts of impedance and admittance are replaced with the concepts of "static stiffness" and "static compliance." Consider the idealized structure in Fig. 29.2.

To measure the force in member K_2, an elastic link with a spring constant K_m is inserted in series with K_2. This link would undergo a deformation proportional to the force in K_2. If the link is very soft in comparison with K_1, no force can be transmitted to K_2. On the other hand, if the link is very stiff, it does not affect the force in K_2 but it will not provide a very good measure of the force. The measured variable is an effort variable and in general when it is measured it is altered somewhat. To apply the impedance concept a flow variable whose product with the effort variable gives power is selected. Thus,

$$\text{flow variable} = \text{power/effort variable}$$

Mechanical impedance is then defined as force divided by velocity, or

$$Z = \text{force/velocity}$$

This is the equivalent of electrical impedance. However, if the static mechanical impedance is calculated for the application of a constant force, the impossible result

$$Z = \text{force}/0 = \infty$$

is obtained.

This difficulty is overcome if energy rather than power is used in defining the variable associated with the measured variable. In that case, the static mechanical impedance becomes the "stiffness" and

$$\text{stiffness} = S_g = \text{effort}/\int \text{flow } dt$$

In structures,

$$S_g = \text{effort variable/displacement}$$

When these changes are made the same formulas used for calculating the error caused by the loading of an instrument in terms of impedances can be used for structures by inserting S for Z. Thus

$$q_{im} = q_{iu}/(1 + S_{go}/S_{gi}) \tag{29.8}$$

where q_{im} = measured value of the effort variable
$\quad\quad q_{iu}$ = undisturbed value of the effort variable
$\quad\quad S_{go}$ = static output stiffness of the measured system
$\quad\quad S_{gi}$ = static stiffness of the measuring system

For an elastic-force-measuring device such as a load cell, S_{gi} is the spring constant K_m. As an example, consider the problem of measuring the reactive force at the end of a propped cantilever beam, as in Fig. 29.3.

According to Eq. 29.8, the force indicated by the load cell will be

$$F_m = F_u/(1 + S_{go}/S_{gi})$$
$$S_{gi} = K_m \quad \text{and} \quad S_{go} = 3EI/L^3$$

The latter is obtained by noting that the deflection at the tip of a tip-loaded cantilever is given by

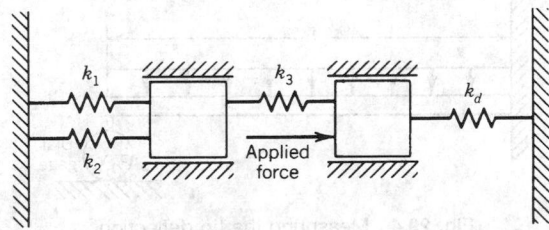

Fig. 29.2 Idealized elastic structure.

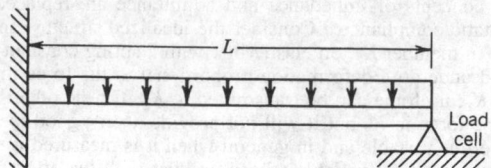

Fig. 29.3 Measuring the reactive force at the tip.

$$\delta = PL^3/3EI$$

where P = tip load
E = modulus of elasticity of the beam material
I = moment of inertia of the beam cross section

The stiffness is the quantity by which the deflection must be multiplied to obtain the force producing the deflection.
For the cantilever beam, then,

$$F_m = F_u/(1 + 3EI/K_mL^3) \qquad (29.9)$$

or

$$F_u = F_m(1 + 3EI/K_mL^3) \qquad (29.10)$$

Clearly, if $K_m \gg 3EI/L^3$, the effect of the load cell on the measurement will be negligible.
To measure displacement rather than force, introduce the concept of compliance and define it as

$$C_g = \text{flow variable}/\int \text{effort variable } dt$$

then

$$q_m = q_u/(1 + C_{go}/C_{gi}) \qquad (29.11)$$

If displacements in an elastic structure are considered, the compliance becomes the reciprocal of stiffness, or the quantity by which the force must be multiplied to obtain the displacement caused by the force. The cantilever beam in Fig. 29.4 again provides a simple illustrative example.
If the deflection at the tip of this cantilever is to be measured using a dial gage with a spring constant K_m,

$$C_{gi} = 1/K_m \qquad \text{and} \qquad C_{go} = L^3/3EI$$

Thus,

$$\delta_m = \delta_u(1 + K_mL^3/3EI) \qquad (29.12)$$

Not all interactions between a system and a measuring device lend themselves to this type of analysis. A pitot tube, for example, inserted into a flow field distorts the flow field but does not

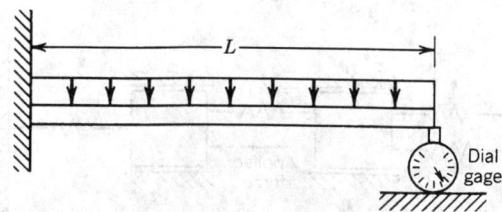

Fig. 29.4 Measuring the tip deflection.

extract energy from the field. Impedance concepts cannot be used to determine how the flow field will be affected.

There are also applications in which it is not desirable for a force-measuring system to have the highest possible stiffness. A subsoil pressure gage, for example, if it is much stiffer than the surrounding soil, will take a disproportionate share of the total load and will consequently indicate a higher pressure than would have existed in the soil if the gage had not been there.

29.3 ERROR ANALYSIS

29.3.1 Introduction

It may be accepted as axiomatic that there will always be errors in measured values. Thus, if a quantity X is measured, the correct value q and X will differ by some amount e. Hence,

$$\pm(q - X) = e$$

or

$$q = X \pm e \tag{29.13}$$

It is essential, therefore, in all measurement work that a realistic estimate of e be made. Without such an estimate, the measurement of X is of no value. There are two ways of estimating the error in a measurement. The first is the external estimate or ϵ_E, where $\epsilon = e/q$. This estimate is based on knowledge of the experiment and measuring equipment, and to some extent on the internal estimate ϵ_I.

The internal estimate is based on an analysis of the data using statistical concepts.

29.3.2 Internal Estimates

If a measurement is repeated many times, the repeat values will not, in general, be the same. Engineers, it may be noted, do not usually have the luxury of repeating measurements many times. Nevertheless, the standardized means for treating results of repeated measurements are useful, even in the error analysis for a single measurement.[9]

If some quantity is measured many times and it is assumed that the errors occur in a completely random manner, that small errors are more likely to occur than large errors, and that errors are just as likely to be positive as negative, the distribution of errors can be represented by the well-known bell-shaped error curve. The equation of the curve is

$$F(X) = Y_0 e^{-(X-\bar{X})/2\sigma^2} \tag{29.14}$$

where $F(X)$ = number of measurements for a given value of $(X - \bar{X})$
Y_0 = maximum height of the curve or the number of measurements for which $X = \bar{X}$
$\bar{X}$ = value of X at the point where maximum height of the curve occurs
σ determines the lateral spread of the curve

This curve is the normal, or Gaussian, frequency distribution. The area under the curve between X and δX represents the number of data points which fall between these limits and the total area under the curve denotes the total number of measurements made. If the normal distribution is defined so that the area between X and $X + \delta X$ is the probability that a data point will fall between those limits, the total area under the curve will be unity and

$$F(X) = \frac{\exp - (X - \bar{X})^2/2\sigma^2}{\sigma\sqrt{2\pi}} \tag{29.15}$$

and

$$P_x = \int_{-x}^{x} \frac{\exp - (X - \bar{X})^2/2\sigma^2}{\sigma\sqrt{2\pi}} \, dx \tag{29.16}$$

Now, if $\bar{X}$ is defined as the average of all the measurements and σ as the standard deviation, then

$$\sigma = [\Sigma (X - \bar{X})^2/N]^{1/2} \tag{29.17}$$

where N is the total number of measurements. Actually, this definition is used as the best estimate for a universe standard deviation, that is, for a very large number of measurements. For smaller subsets of measurements, the best estimate of σ is given by

$$\sigma = [\Sigma\ (X - \bar{X})^2/(n - 1)]^{1/2} \tag{29.18}$$

where n is the number of measurements in the subset. Obviously the difference between the two values of σ becomes negligible as n becomes very large (or as $n \to N$).

The probability curve based on these definitions is shown in Fig. 29.5.

The area under this curve between $-\sigma$ and $+\sigma$ is 0.68. Hence, 68% of the measurements can be expected to have errors that fall in the range of $\pm\sigma$. Thus, the chances are 68:32, or better than 2:1, that the error in a measurement will fall in this range. For the range $\pm 2\sigma$ the area is 0.95. Hence, 95% of all the measurement errors will fall in this range and the odds are about 20:1 that a reading will be within this range. The odds are about 384:1 that any given error will be in the range of $\pm 3\sigma$.

Some other definitions related to the normal distribution curve are:

1. *Probable error.* The error likely to be exceeded in half of all the measurements and not reached in the other half of the measurements. This error in Fig. 29.5 is about 0.67σ.
2. *Mean error.* The arithmetic mean of all the errors regardless of sign. This is about 0.8σ.
3. *Limit of error.* The error that is so large it is most unlikely ever to occur. It is usually taken as 4σ.

29.3.3   Use of Normal Distribution to Calculate the Probable Error in X

The foregoing statements apply strictly only if the number of measurements is very large. Suppose that n measurements have been made. That is, a sample of n data points out of an infinite number. From that sample, $\bar{X}$ and σ are calculated as above. How good are these numbers? To determine that, proceed as follows:

Let

$$\bar{X} = F\ (X_1, X_2, X_3, \ldots, X_n) = (\Sigma X_i)/n \tag{29.19}$$

$$e_{\bar{X}} = \sum_{i=1}^{n} \frac{\partial F}{\partial X_i}\ e_{x_i} \tag{29.20}$$

where $e_{\bar{x}}$ = the error in $\bar{X}$
e_{x_i} = the error in X_i

$$(e_{\bar{x}})^2 = \sum_{i=1}^{n} (\partial F/\partial X_i e_{x_i})^2 + \sum_{i=1, j=1}^{n} (\partial F/\partial X_i e_{x_i})\ (\partial F/\partial X_j e_{x_j}) \tag{29.21}$$

where $i \neq j$

If the errors e_i to e_n are independent and symmetrical, the cross-product terms will tend to disappear and

$$(e_{\bar{x}})^2 = \sum_{i=1}^{n} (\partial F/\partial X_i e_{x_i})^2 \tag{29.22}$$

Since $\partial F/\partial X_i = 1/n$

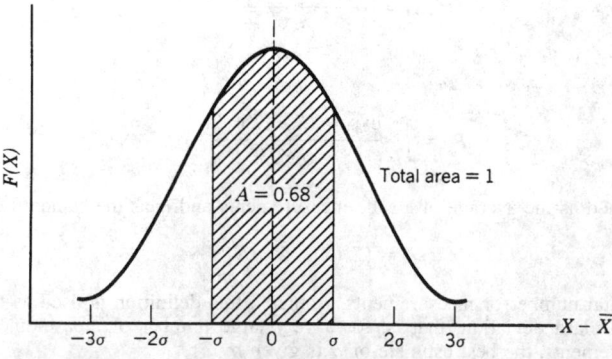

Fig. 29.5 Probability curve.

$$e_{\bar{x}} = \left[\sum_{i=1}^{n} (1/n)^2 e_{x_i}^2 \right]^{1/2} \qquad (29.23)$$

or

$$e_{\bar{x}} = \left[(1/n)^2 \sum_{i=1}^{n} (e_{x_i})^2 \right]^{1/2} \qquad (29.24)$$

from the definition of σ

$$\sum (e_{x_i})^2 = n\sigma^2 \qquad (29.25)$$

and

$$e_{\bar{x}} = \sigma/\sqrt{n}$$

This equation must be corrected because the real errors in X are not known. If the number n were to approach infinity, the equation would be correct. Since n is a finite number, the corrected equation is written as

$$e_{\bar{x}} = \sigma/(n - 1)^{1/2} \qquad (29.26)$$

and

$$q = \bar{X} \pm \sigma/(n - 1)^{1/2} \qquad (29.27)$$

This says that if one reading is likely to differ from the true value by an amount σ, then the average of 10 readings will be in error by only $\sigma/3$ and the average of 100 readings will be in error by $\sigma/10$. To reduce the error by a factor of 2, the number of readings must be increased by a factor of 4.

29.3.4 External Estimates

In almost all experiments, several steps are involved in making a measurement. It may be assumed that in each measurement there will be some error, and if the measuring devices are adequately calibrated, errors are as likely to be positive as negative. The worst condition insofar as accuracy of the experiment is concerned would be for all errors to have the same sign. In that case, assuming the errors are all much less than one, the resultant error will be the sum of the individual errors, that is,

$$\epsilon_E = \epsilon_1 + \epsilon_2 + \epsilon_3 + \cdots \qquad (29.28)$$

It would be very unusual for all errors to have the same sign. Likewise, it would be very unusual for the errors to be distributed in such a way that

$$\epsilon_E = 0$$

Following is a general method for treating problems that involve a combination of errors to determine what error is to be expected as a result of the combination:
Suppose that

$$V = F\,(a, b, c, d, e, \ldots, x, y, z) \qquad (29.29)$$

where $a, b, c, \ldots, x, y, z$ represent quantities that must be individually measured to determine V.

$$\delta V = \sum_{n=a}^{z} (\partial F/\partial n)\delta n$$

$$\epsilon_E = \sum_{n=a}^{z} (\partial F/\partial n)e_n \qquad (29.30)$$

The sum of the squares of the error contributions is given by

$$e_E^2 = \left(\sum_{n=a}^{z} (\partial F/\partial n)e_n \right)^2 \qquad (29.31)$$

Now, as in the discussion of internal errors, assume that errors e_n are independent and symmetrical. This justifies taking the sum of the cross products as zero.

$$\sum_{n=a,m=a}^{z} (\partial F/\partial n)(\partial F/\partial m)\, e_n e_m = 0 \tag{29.32}$$

$$n \neq m$$

Hence,

$$(e_E)^2 = \sum_{n=a}^{z} (\partial F/\partial N)^2 e_n^2$$

or

$$e_E = \left[\sum_{n=a}^{z} (\partial F/\partial n)^2 e_n^2 \right]^{1/2} \tag{29.33}$$

This is the "most probable value" of e_E. It is much less than the worst case

$$\epsilon_e = [|\epsilon_a| + |\epsilon_b| + |\epsilon_c| + \ldots + |\epsilon_g|] \tag{29.34}$$

As an application, the determination of g, the local acceleration of gravity, by use of a simple pendulum will be considered

$$g = 4\pi^2 L/T^2 \tag{29.35}$$

where L = the length of the pendulum
T = the period of the pendulum

If an experiment is performed to determine g, the length L and the period T would be measured. To determine how the accuracy of g will be influenced by errors in measuring L and T, write

$$\partial g/\partial L = 4\pi^2/T^2 \quad \text{and} \quad \partial g/\partial T = -8\pi^2 L/T^3 \tag{29.36}$$

The error in g is the variation in g, written as follows:

$$\delta g = (\partial g/\partial L)\,\Delta L + (\partial g/\partial T)\,\Delta T \tag{29.37}$$

or

$$\delta g = (4\pi^2/T^2)\,\Delta L - (8\pi^2 L/T^3)\,\Delta T \tag{29.38}$$

It is always better to write the errors in terms of percentages. Consequently, Eq. (29.38) is rewritten

$$\delta g = (4\pi^2 L/T^2)\,\Delta L/L - 2(4\pi^2 L/T^2)\,\Delta T/T \tag{29.39}$$

or

$$\delta g/g = \Delta L/L - 2\Delta T/T \tag{29.40}$$

then

$$e_g = [e_L^2 + (2e_T)^2]^{1/2} \tag{29.41}$$

where e_g is the "most probable error" in the measured value of g. That is to say,

$$g = 4\pi^2 L/T^2 \pm e_g \tag{29.42}$$

where L and T are the measured values. Note that even though a positive error in T causes a negative error in the calculated value of g, the contribution of the error in T to the "most probable error" is taken as positive. Note also that an error in T contributes four times as much to the "most probable error" as an error in L contributes. It is fundamental in measurements of this type that those quantities which appear in the functional relationship raised to some power greater than unity contribute more heavily to the "most probable error" than other quantities and must, therefore, be measured with greater care.

The determination of the "most probable error" is simple and straightforward. The question is, how are the errors, such as $\Delta L/L$ and $\Delta T/T$, determined. If the measurements could be repeated often enough, the statistical methods discussed in the internal error evaluation could be used to arrive at a value. Even in that case it would be necessary to choose some representative error, such as the standard deviation or the mean error. Unfortunately, as was noted previously, in engineering experiments it usually is not possible to repeat measurements enough times to make statistical treatments meaningful. Engineers engaged in making measurements will have to use what knowledge they have of the measuring instruments and the conditions under which the measurements are made to make a reasonable estimate of the accuracy of each measurement. When all of this has been done and a "most probable error" has been calculated, it should be remembered that the result is not the actual error in the quantity being determined, but is, rather, the engineer's best estimate of the magnitude of the uncertainty in the final result.[10,11]

Consider again the problem of determining g. Suppose that the length L of the pendulum has been determined by means of a meter stick with 1-mm calibration marks and the error in the calibration is considered negligible in comparison with other errors. Suppose the value of L is determined to be 91.7 cm. Since the calibration marks are 1 mm apart, it can be assumed that ΔL is no greater than 0.5 mm. Hence, maximum $\Delta L/L = 5.5 \times 10^{-4}$. Suppose T is determined with the pendulum swinging in a vacuum with an arc of $\pm 5°$ using a stopwatch that has an inherent accuracy of 1 part in 10,000. (If the arc is greater than $\pm 5°$, a nonisochronous swing error enters the picture.) This means that the error in the watch reading will be no more than 10^{-4} sec. However, errors are introduced in the period determination by human error in starting and stopping the watch as the pendulum passes a selected point in the arc. This error can be minimized by selecting the highest point in the arc because the pendulum has zero velocity at that point, and timing a large number of swings so as to spread the error out over that number of swings. Human reaction time may vary from as low as 0.2 sec to as high as 0.7 sec. A value of 0.5 sec will be assumed. Thus, the estimated maximum error in starting and stopping the watch will be 1 sec (± 0.5 sec at the start and ± 0.5 sec at the stop). A total of 100 swings will be timed. Thus, the estimated maximum error in the period will be 1/100 sec. If the period is determined to be 1.92 sec, the estimated maximum error will be $0.01/1.92 = .005$. Compared to this, the error in the period due to the inherent inaccuracy of the watch is negligible. The nominal value of g calculated from the measured values of L and T is 982.03 cm/sec². The "most probable error" is

$$[4(0.005)^2 + (5.5 \times 10^{-4})^2]^{1/2} = 0.01 \qquad (29.43)$$

The uncertainty in the value of g is then ± 9.82 cm/sec², or, in other words, the value of g will be somewhere between 972.21 and 991.85 cm/sec².

Often it is necessary for the engineer to determine in advance how accurately the measurements must be made in order to achieve a given accuracy in the final calculated result. For example, in the pendulum problem it may be noted that the contribution of the error in T to the "most probable error" is more than 300 times the contribution of the error in the length measurement. This suggests, of course, that the uncertainty in the value of g could be greatly reduced if the error in T could be reduced. Two possibilities for doing this might be (1) to find a way to do the timing that does not involve human reaction time, or (2) if that is not possible, to increase the number of cycles timed. If the latter alternative is selected and the other factors remain the same, the error in T, timed over 200 swings, is 1/200 or 0.005 sec. As a percentage the error is $0.005/1.92 = 0.0026$. The "most probable error" in g then becomes

$$e_g = [4 \times (2.6 \times 10^{-3})^2 + (5.5 \times 10^{-4})^2]^{1/2} = 0.005 \qquad (29.44)$$

This is approximately one-half of the "most probable error" in the result obtained by timing just 100 swings. With this new value of e_g, the uncertainty in the value of g becomes ± 4.91 cm/sec² and g then can be said to be somewhere between 977.12 and 986.94 cm/sec². The procedure for reducing this uncertainty still further is now self-evident.

Clearly the value of this type of error analysis depends upon the skill and objectivity of the engineer in estimating the errors in the individual measurements. Such skills are acquired only by practice and careful attention to all the details of the measurements.

29.4 APPENDIX

29.4.1 Vibration Measurement

See Section 23.3.

29.4.2 Acceleration Measurement

For acceleration measurements see Section 23.4. For the measurement of mechanical shock and vibration and acceleration see Chapter 23 on vibration and shock of this handbook.

29.4.3 Shock Measurement

See Section 23.5.

29.4.4 Sound Measurement

Introduction

Sound[12] is an oscillation about the mean of pressure or stress in a fluid or solid medium. Sound is also an auditory sensation produced by those oscillations. Acoustics[12] is the science of sound, which includes its generation, transmission, and effects. Most appliances, machines, and vehicles generate unwanted sound, called acoustic noise.[12] It is the purpose here to present methods for the accurate measurement of acoustic noise as a means for determining noise exposure of people in work spaces, as an aid in reducing noise, and for its use in diagnosing vibration problems.

Sound pressure is commonly determined by a microphone that converts pressure to an electrical signal, which is processed and presented on a visual display. Such devices are known as *sound level meters*. Three types of handheld meters, specified in an American National Standard,[13] are commercially available. The type required depends on the intended use. For example, the type 1 meter is required for precision laboratory measurements, while the type 3 is intended for sound surveys where high accuracy is unwarranted. Sound level meters are calibrated to give accurate measures of sound pressure and means are provided in some cases for a calibration check before making measurements.

Because of the extremely large range of pressures that can be sensed by the human ear, sound pressures are usually measured and expressed in logarithmic units. The *decibel notation* is used and the measured quantity is compared to a reference pressure. Logarithmic qualities are referred to as *levels*. *Sound pressure level* (SPL) is defined as follows and denoted by the symbol Lp

$$Lp = 20 \log_{10} \frac{P_x}{P_{ref}} \tag{29.45}$$

where P_x is the measured pressure and P_{ref} is 20×10^{-6} Pa (N/m). The units of Lp are dB referenced to 20 μPa.

Since acoustic noise has a major impact on people, measurements are usually made in the frequency range of hearing. This is normally taken as 20 Hz to 20 kHz. The knowledge of the distribution of sound energy with frequency is important when dealing with noise impact on people and using radiated sound to identify machine vibrations. Thus, sound level meters have some means of determining the frequency distribution of the spectral components of sound pressure.

The basic configuration of a sound level meter is that shown in Fig. 29.6. The weighting networks are used to shape the noise spectrum so that a measure that relates to the hearing characteristic can be made. The display may be an analog meter, numeric display, or digital read out. The frequency-versus-amplitude characteristics of the weighting networks are shown in Fig. 29.7. The A weighting that produced a level referred to as L_A in dBA is shaped as approximately the inverse of the loudness contour of the ear at the threshold of hearing. The B and C weightings are matched to successively

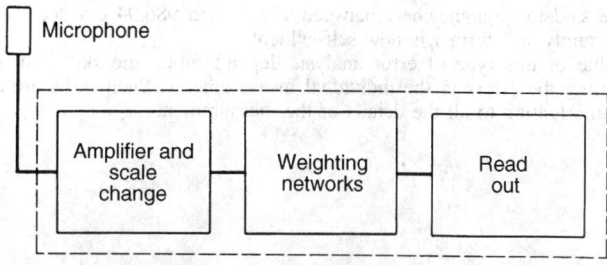

Fig. 29.6 Sound level meter.

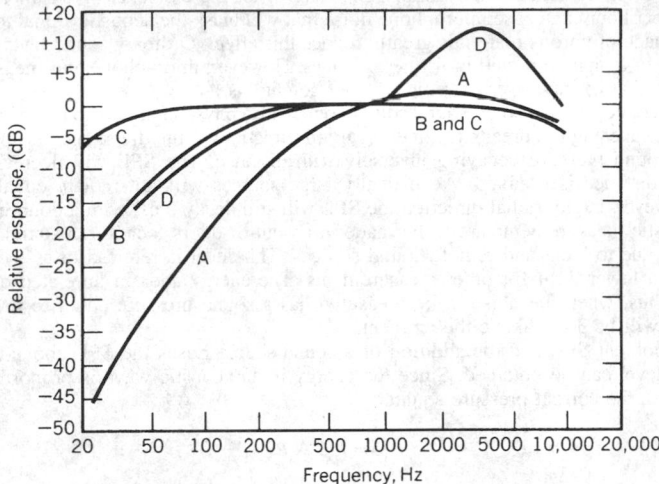

Fig. 29.7 Frequency characteristics for weighting functions of sound level meters.

higher loudness levels; however, most noise regulations require data to be presented in dBA. Some meters may have the D weighting, which is used as a measure of annoyance due to noise from aircraft flyovers. Some meters may have a "flat" weighting position, which eliminates the weighting networks. Then the frequency response is limited only by the characteristics of the microphone and amplifiers. In addition, some meters provide for an external filter or spectrum analyzer for a more complete frequency analysis.

Several other types of sound level meters are available that process the acoustic signal for special purposes. There are meters that will respond to instantaneous peak pressures and "hold" for convenient readout. Some have a response function that gives increased accuracy for impulsive repetitive noise. Others will integrate over time to give an equivalent sound level (L_{eq}) for noise with large and random variations with time. A final type worn by a worker gives a continuous readout of the accumulated dose of excessive noise.

Measurements in Open Spaces[14]

In this situation, the sound source is usually not inside a building and the acoustic waves propagate freely over appreciable distances. The measuring microphone is not in an enclosure. The noise impact may concern

1. Community noise
2. Transportation noise
3. Industrial noise

In open spaces, the microphone will sense the pressure produced by a passing acoustic wave. At high frequencies, when the microphone size is an appreciable fraction of the acoustic wavelength, waves are diffracted by the microphone and it becomes directional. For a 1-in. diameter microphone, these effects occur at frequencies about 2 kHz and higher. Under these conditions, one should use a microphone designed for *free-field* conditions[15] and correct for angle of wave arrival.

An accurate measure of the sound pressure level of an acoustic wave cannot be made when there are reflecting surfaces nearby. An operator holding a sound level meter can cause reflections. Holding the microphone at arm's length to the side of the body minimizes this problem; however, a microphone on a remote cable placed on the ground is recommended.

Large plane surfaces near the propagation path or behind the operator can cause serious interference effects. At a particular frequency, a direct and reflected path difference of one half-wavelength can cause almost complete cancellation. Thus, errors will depend on the frequency content of the sound wave.

The ground is a large reflecting surface and is usually present. However, if the microphone is placed directly on the ground, only the direct wave will be measured. A relatively open space without diffracting objects with the microphone placed a very few inches above the surface is recommended for accurate measurements.

Open-space measurements usually imply exposure to all the elements of weather. Wind speeds of a few miles per hour can cause microphone noise that will mask the acoustic signal to be measured. *Wind screens* made of porous materials greatly reduce this effect. Ordinary sound level meters should not be exposed to temperature and moisture extremes. However, microphone and measuring systems can be obtained that operate over extreme weather conditions.

The sound radiated from a small sound source, one whose physical dimensions are small compared to the acoustic wavelength, spreads or diverges in all directions. This divergence is spherical, which results in the sound pressure decaying inversely with distance. The SPL will decrease 6 dB when the measuring distance is doubled. Acoustically large sources will not radiate equally well in all directions. However, in any radial direction the SPL will still decay 6 dB with a doubling of distance.

When acoustic measurements are to be made in an out-of-doors location, a sound pressure level will be present due to local and remote sound sources. This level is referred to as the *ambient*, and it will provide a lower limit for other measurements. The energy adds in unrelated or uncorrelated sound fields. Thus, when the noise to be measured has a sound pressure equal to the ambient, the measured SPL will be 3 dB above the ambient.

When the ambient SPL and the addition of a sound source raises the SPL above the ambient, a correct source level can be obtained. Since the energy in an acoustic wave is proportional to sound pressure squared, the correct pressure squared is

$$P_s^2 = P_m^2 - P_a^2 \tag{29.46}$$

where P_s is the correct source pressure, P_m is the measured pressure, and P_a is the ambient. P can be obtained from Lp as follows:

$$P = P_{ref}10^{(Lp/20)} \tag{29.47}$$

where P_{ref} is 20 μPa. The corrected Lp is

$$(Lp)_s = 20 \log_{10} \frac{P_s}{P_{ref}} \tag{29.48}$$

When it is necessary to do a significant analysis on a sound source, a simple level correction as above is not adequate. When the frequency distribution of energy and statistical amplitude distributions are desired, the sound pressure level should be a least 20 dB above the ambient noise. Thus, a measuring distance must be chosen to ensure this signal-to-noise ratio.

Measurement in Enclosed Spaces[14]

In many noisy situations, a sound source radiates into a completely enclosed space, such as a room with closed doors and windows. Since sound waves travel at about 330 m/sec, it only takes a few milliseconds for a large number of acoustic rays to be traversing the paths between reflecting surfaces and filling the room with a *reverberant sound field*. The sound pressure level continues to increase until the absorbing surfaces in the room absorb the acoustic energy at the same rate as the source is supplying it. Thus, there will be a region near the source where the directly radiated sound will dominate over the reverberant sound, and the pressure will decay inversely with distance. At some distance, determined by the acoustic absorption in the room, the direct sound will equal the reverberant sound. In the rest of the room the reverberant sound will be relatively uniform and independent of the distance from the source. An exception to this uniformity occurs when large single-frequency components produce dominant standing waves in the room. More details on acoustic characteristics of rooms can be found in the acoustic literature.[16,17]

The sound pressure levels measured in enclosed spaces depend on many factors. The sound sources and the acoustical absorption in the room will determine the levels in the direct field close to the sources and in the reverberant field. All the sources in the room will contribute to the reverberant field. If the sound pressure level near a noisy machine is greater than the reverberant sound pressure level, the measurement is characteristic of that machine. If it is lower, no information is obtained. If other noise sources can be turned off, a single machine can be studied.

When the noise exposure of workers in a noisy environment is to be determined, it is essential to measure the sound pressure level at the work station at ear level. If the worker stays at one place, a single measurement may suffice. However, for a mobile worker, an acoustic dosimeter worn by the individual is essential.

Acoustic Intensity—Diagnosis of Vibrations

The acoustic power radiated by a noisy machine and the frequency spectrum of that power are very useful quantities. When the directive nature of radiated power is known, the source pressure level in the vicinity of such a machine placed in an open space can be determined. When such a machine is in an enclosed space, the reverberant sound field can be predicted. In this case, the total absorption in the room controls the reverberant sound field and the directive nature is not needed.

When the spectral content of the sound power of a noisy machine is measured, sources of vibrations may be located. Rotating unbalanced parts, oscillating structures, and repetitive impacts will cause vibration of machine surfaces that radiate sound. The period of a dominant frequency noise component can be related to the rotational, oscillating, or repetition frequency or a harmonic of that frequency. Such correlation can be used to locate problems due to excess wear in an old machine and locating the trouble areas in the design of quiet machines. For example, worn bearings and gears can be located by the characteristic noise they generate.

Vibrational diagnosis can be done with a single microphone used to measure sound pressure. The "ac" signal out of a sound level meter may be directed to a spectrum analyzer or filter set for this purpose. The determination of sound power, however, requires the quantitative measure of acoustic intensity. Intensity is the vector product of the real part of acoustic pressure times acoustic particle velocity:

$$I = \text{Re } (pv)\text{w}/\text{m}^2 \tag{29.49}$$

Intensity is thus the power per unit area flowing through a surface surrounding a noisy machine. Since intensity is a directed quantity, the component in the direction away from the machine will be radiated as acoustic power. When the machine surface is subdivided into small areas as in Fig. 29.8, the average intensity times that area plus the products for the rest of the small areas give total power radiated:

$$w = \sum_{i}^{n} I_i A_i \text{ w} \tag{29.50}$$

Small areas should be selected that have small intensity variations; then the average value can be used to calculate power flow through the area.

Since measurements can be made near to a noise source and the product of pressure and velocity tend to discriminate against uncorrelated signals, sound power determination can be made in enclosed spaces. This is particularly advantageous for large machines that cannot be placed in anechoic rooms or reverberation chambers.

The acoustic variables are measured by two microphones, as shown in Fig. 29.9. The average of the two measured pressures is the pressure between them. The acoustic partial velocity in the direction of the microphone axes is related to the pressure gradient. Then

$$v_x = \frac{1}{\rho_0 d} \int (p_1 - p_2) \, dt \text{ m}/\text{sec} \tag{29.51}$$

where ρ_0 is air density and d is the microphone spacing. Then the average intensity in the x direction becomes

$$I_x = \frac{1}{T} \int_0^T p_{av} \, v_x \, dt \tag{29.52}$$

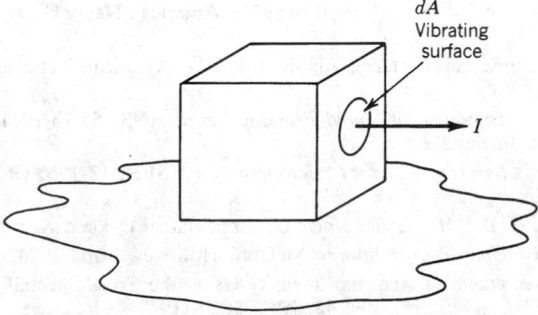

Fig. 29.8 Intensity and power measurements.

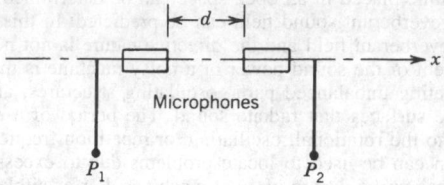

Fig. 29.9 Intensity measuring probe.

where averaging time T is long compared to the time scale of p and v. Some commercial instruments measure intensity in this way.

When a two-channel spectrum analyzer is available, intensity can be determined from the two microphone signals in another way. The frequency-domain representation of intensity can be obtained by the Fourier transform of the two pressure signals.[18,19]

$$I_x(f) = \frac{\mathrm{Im}(S_{12})}{2\rho_0 d\omega} \tag{29.53}$$

where $\mathrm{Im}(S_{12})$ is the imaginary part of the cross-spectral density $S_{12} = P_1(f)\, P_2^*(f)$ (* designates the complex conjugate). The angular frequency is ω. The frequency-domain representation is very informative, since the frequency of predominant energy components is displayed. Vibration diagnosis is then readily accomplished.

REFERENCES

1. W. A. Wildhack, "NBS Source of American Standards," *ISA Journal* **8**(2) (February 1961).

2. P. Giacomo, "News from the BIPM," *Metrologia* **20**(1) (April 1984).

3. *NBS Tech. News Bull.* **52**(1) (January 1968).

4. G. Kamas and S. L. Howe (eds.), *Time and Frequency User's Manual,* NBS Special Publication 559, U.S. Government Printing Office, Washington, DC, 1979.

5. "The International Temperature Scale of 1968: A Committee Report," *Metrologia* **5**(2) (April 1969).

6. T. G. Beckwith, R. D. Marangoni, and J. H. Lienhard, *Mechanical Measurements,* 5th ed., Addison-Wesley Publishing, Reading, MA, 1993, p. 365.

7. E. O. Doebelin, *Measurement Systems—Applications and Design,* McGraw-Hill, New York, 1966, pp. 38–101.

8. C. M. Harris and C. E. Crede, "Mechanical Impedance," in *Shock and Vibration Handbook,* McGraw-Hill, New York, 1985, Chap. 10.

9. N. H. Cook and E. Rabinowicz, *Physical Measurement and Analysis,* Addison-Wesley, Reading, MA, 1963, pp. 29–68.

10. S. J. Kline and F. A. McClintock, "Describing Uncertainties in Single Sample Experiments," *Mech. Engr.* **75**(3) (January 1953).

11. *ASME Performance Test Codes,* ANSI/ASME Pt. Supplement on Instruments and Apparatus, Part 1, Measurement Uncertainty.

12. *Acoustical Terminology,* ANSI S1.1-1960 (R1971), American National Standards Institute, New York.

13. *Specification for Sound Level Meters,* ANSI S1.4-1971, American National Standards Institute, New York.

14. *Methods for the Measurement of Sound Pressure Level,* ANSI S1.13-1971 (R1986), American National Standards Institute, New York.

15. *Specifications for Laboratory Standard Microphones,* ANSI S1.12-1967 (R1986), American National Standards Institute, New York.

16. L. H. Bell and D. H. Bell, *Industrial Noise Control,* Marcel Dekker, New York, 1994.

17. C. M. Harris, *Noise Control in Buildings,* McGraw-Hill, New York, 1994.

18. F. J. Foley, "Measurement of Acoustic Intensity Using the Cross-Spectral Density of Two Microphone Signals," *J. Acous. Soc. Am.* **62**, 1057–1059 (1977).

19. G. Krishnoppa, "Cross-Spectral Method of Measuring Acoustic Intensity by Correcting Phase and Gain Mismatch Errors by Microphone Calibration," *J. Acous. Soc. Am.* **69**, 307–310 (1981).

PART 3

MANUFACTURING ENGINEERING

CHAPTER 30

PRODUCT DESIGN FOR MANUFACTURING AND ASSEMBLY (DFM&A)

Gordon Lewis
Digital Equipment Corporation
Maynard, Massachusetts

30.1	**INTRODUCTION**	**935**	30.2.2	Getting the DFM&A Process Started	942
30.2	**DESIGN FOR MANUFACTURING AND ASSEMBLY**	**936**	30.2.3	The DFM&A Road Map	946
	30.2.1 What is DFM&A?	937	**30.3**	**WHY IS DFM&A IMPORTANT?**	**950**

30.1 INTRODUCTION

Major changes in product design practices are occurring in all phases of the new product development process. These changes will have a significant impact on how all products are designed and the development of the related manufacturing processes over the next decade. The high rate of technology changes has created a dynamic situation that has been difficult to control for most organizations. There are some experts who openly say that if we have no new technology for the next five years, corporate America might just start to catch up. The key to achieving benchmark time to market, cost, and quality is in up-front technology, engineering, and design practices that encourage and support a wide latitude of new product development processes. These processes must capture modern manufacturing technologies, piece parts that are designed for ease of assembly, and parts that can be fabricated using low-cost manufacturing processes. Optimal new product design occurs when the designs of machines and of the manufacturing processes that produce those machines are congruent.

The obvious goal of any new product development process is to turn a profit by converting raw material into finished products. This sounds simple, but it has to be done efficiently and economically. Many companies do not know how much it costs to manufacture a new product until well after the production introduction. Rule #1: the product development team must be given a cost target at the start of the project. We will call this cost the *unit manufacturing cost* (UMC) target. Rule #3: the product development team must be held accountable for this target cost. What happened to rule #2? We'll discuss that shortly. In the meantime, we should understand what UMC is.

$$UMC = BL + MC + TA$$

where BL = burdened assembly labor rate per hour; this is the direct labor cost of labor, benefits, and all appropriate overhead cost

MC = material cost; this is the cost of all materials used in the product

TA = tooling amortization; this is the cost of fabrication tools, molds and Assembly Tooling, divided by the forecast volume build of the product

UMC is the direct burdened assembly labor (direct wages, benefits, and overhead) plus the material cost. Material cost must include the cost of the transformed material plus piece part packaging plus duty, freight, and insurance (DIF). Tooling amortization should be included in the UMC target cost calculation, based on the forecast product life volume.

Mechanical Engineers' Handbook, 2nd ed., Edited by Myer Kutz.
ISBN 0-471-13007-9 © 1998 John Wiley & Sons, Inc.

Example UMC Calculation *BL + MC + TA*

Burdened assembly labor cost calculation (BL)

$$BL = (\$18.75 + \overset{\text{Labor}}{138\%}) = \$44.06/hr$$
$$\text{Wages+Benefits overhead}$$

Burdened assembly labor is made up of the direct wages and benefits paid to the hourly workers, plus a percentage added for direct overhead and indirect overhead. The overhead added percentage will change from month to month based on plant expenses.

Material cost calculation (MC)

$$\begin{array}{lllll}
& \text{(Part Cost + Packaging)} + & \text{DIF} & + \text{Mat. Acq. Cost} = \\
MC = & (\$2.45 + \$.16) & + 12\% + & 6\% & = \\
MC = & \$2.61 & + \$.31 + \$.15 & & = \$3.07 \\
& & & & \text{Material FOB Assm. Plant}
\end{array}$$

Material cost should include the cost of the parts and all necessary packaging. This calculation should also include a percent adder for duty, freight, and insurance (DFI) and an adder for the acquisition of the materials (Mat. Acq.). DFI typically is between 4% and 12% and Mat. Acq. typically is in the range of 6% to 16%. It is important to understand the MC because material is the largest expense in the UMC target.

Tooling amortization cost calculations (TA)

$$\begin{array}{ll}
& \text{(Tool Cost)} \quad \text{\# of parts} \\
TA = & TC \; / \quad PL \\
TA = & \$56,000/10,000 = \$5.60 \text{ per assembly}
\end{array}$$

TC is the cost of tooling and PL is the estimated number of parts expected to be produced on this tooling. Tooling cost is the total cost of dies and mold used to fabricate the component parts of the new product. This also should include the cost of plant assembly fixtures and test and quality inspection fixtures.

The question is, "How can the product development team quickly and accurately measure UMC during the many phases of the project?" What is needed is a tool that provides insight into the product structure and at the same time exposes high-cost areas of the design.

30.2 DESIGN FOR MANUFACTURING AND ASSEMBLY

Designing for Manufacturing and Assembly (DFM&A) is a technique for reducing the cost of a product by breaking the product down into its simplest components. All members of the design team can understand the product's assembly sequence and material flow early in the design process.

DFM&A tools lead the development team in reducing the number of individual parts that make up the product and ensure that any additional or remaining parts are easy to handle and insert during the assembly process. DFM&A encourages the integration of parts and processes, which helps reduce the amount of assembly labor and cost. DFM&A efforts include programs to minimize the time it takes for the total product development cycle, manufacturing cycle, and product life-cycle costs. Additionally, DFM&A design programs promote team cooperation and supplier strategy and business considerations at an early stage in the product development process.

The DFM&A process is composed of two major components: *design for assembly* (DFA) and *design for manufacturing* (DFM). DFA is the labor side of the product cost. This is the labor needed to transform the new design into a customer-ready product. DFM is the material and tooling side of the new product. DFM breaks the parts fabrication process down into its simplest steps, such as the type of equipment used to produce the part and fabrication cycle time to produce the part, and calculates a cost for each functional step in the process. The program team should use the DFM tools to establish the material target cost before the new product design effort starts.

Manufacturing costs are born in the early design phase of the project. Many different studies have found that as much as 80% of a new product's cost is set in concrete at the first drawing release phase of the product. Many organizations find it difficult to implement changes to their new product development process. The old saying applies: "only wet babies want to change, and they do it screaming and crying." Figure 30.1 is a memo that was actually circulated in a company trying to implement a DFM&A process. Only the names have been changed.

It is clear from this memo that neither the engineering program manager nor the manufacturing program manager understood what DFM&A was or how it should be implemented in the new product development process. It seems that their definition of concurrent engineering is, "Engineering creates the design and manufacturing is forced to concur with it with little or no input." This is not what DFM&A is.

Memorandum: *Ajax Bowl Corporation*

DATE: January 26, 1997
TO: Manufacturing Program Manager, Auto Valve Project
FROM: Engineering Program Manager, Auto Valve Project
RE: Design for Manufacturing & Assembly support for Auto Valve Project
CC: Director, Flush Valve Division

Due to the intricate design constraints placed on the Auto Valve project engineering feels they will not have the resources to apply the Design for Manufacturing and Assembly process. Additionally, this program is strongly schedule driven. The budget for the project is already approved as are other aspects of the program that require it to be on-time in order to achieve the financial goals of upper management.

In the meeting on Tuesday, engineering set down the guidelines for manufacturing involvement on the Auto Valve project. This was agreed to by several parties (not manufacturing) at this meeting.

The manufacturing folks wish to be tied early into the Auto Valve design effort:

1. This will allow manufacturing to be familiar with what is coming.
2. Add any ideas or changes that would reduce overall cost or help schedule.
3. Work vendor interface early, manufacturing owns the vendor issues when the product comes to the plant, anyways.

Engineering folks like the concept of new ideas, but fear:

1. Inputs that get pushed without understanding of all properly weighted constraints.
2. Drag on schedule due to too many people asking to change things.
3. Spending time defending and arguing the design.

PROPOSAL—Turns out this is the way we will do it.

Engineering shall on a few planned occasions address manufacturing inputs through one manufacturing person. Most correspondence will be written and meeting time will be minimal. It is understood that this program is strongly driven by schedule, and many cost reduction efforts are already built into the design so that the published budget can be met.

The plan for Engineering:

- When drawings are ready, Engineering Program Manager (EPM) will submit them to Manufacturing Program Manager (MPM).
- MPM gathers inputs from manufacturing people and submits them back in writting to EPM. MPM works questions through EPM to minimize any attention units that Engineering would have to spend.
- EPM submits suggestions to Engineering, for one quick hour of discussion/acceptance/veto.
- EPM submits written response back to MPM and works any Design continues under ENG direction.
- When a prototype parts arrives, the EPM will allow the MPM to use it in manufacturing discussions.
- MPM will submit written document back to EPM to describe issues and recommendations.
- Engineering will incorporate any changes that they can handle within the schedule that they see fit.

Fig. 30.1

30.2.1 What is DFM&A?

DFM&A is not a magic pill. It is a tool that, when used properly, will have a profound effect on the design philosophy of any product. The main goal of DFM&A is to lower product cost by examining the product design and structure at the early concept stages of a new product. DFM&A also leads to improvements in serviceability, reliability, and quality of the end product. It minimizes the total product cost by targeting assembly time, part cost, and the assembly process in the early stages of the product development cycle.

The life of a product begins with defining a set of product needs, which are then translated into a set of product concepts. Design engineering takes these product concepts and refines them into a detailed product design. Considering that from this point the product will most likely be in production for a number of years, it makes sense to take time out during the design phase to ask, "How should this design be put together?" Doing so will make the rest of the product life, when the design is complete and handed off to production and service, much smoother. To be truly successful, the DFM&A process should start at the early concept development phase of the project. True, it will take time during the hectic design phase to apply DFM&A, but the benefits easily justify additional time.

DFM&A is used as a tool by the development team to drive specific assembly benefits and identify drawbacks of various design alternatives, as measured by characteristics such as total number of parts, handling and insertion difficulty, and assembly time. DFM&A converts time into money, which should be the common metric used to compare alternative designs, or redesigns of an existing concept. The early DFM&A analysis provides the product development team with a baseline to which comparisons can be made. This early analysis will help the designer to understand the specific parts or concepts in the product that require further improvement, by keeping an itemized tally of each part's effect on the whole assembly. Once a user becomes proficient with a DFM&A tool and the concepts become second nature, the tool is still an excellent means of solidifying what is by now second nature to DFA veterans, and helps them present their ideas to the rest of the team in a common language: cost.

DFM&A is an interactive learning process. It evolves from applying a specific method to a change in attitude. Analysis is tedious at first, but as the ideas become more familiar and eventually ingrained, the tool becomes easier to use and leads to questions: questions about the assembly process and about established methods that have been accepted or existing design solutions that have been adopted. In the team's quest for optimal design solutions, the DFM&A process will lead to uncharted ways of doing things. Naturally, then, the environment in which DFA is implemented must be ripe for challenging pat solutions and making suggestions for new approaches. This environment must evolve from the top down, from upper management to the engineer. Unfortunately, this is where the process too often fails.

Figure 30.2 illustrates the ideal process for applying DFM&A. The development of any new product must go through four major phases before it reaches the marketplace: concept, design, development, and production. In the concept phase, product specifications are created and the design team creates a design layout of the new product. At this point, the first design for assembly analysis should be completed. This analysis will provide the design team with a theoretical minimum parts count and pinpoint high-assembly areas in the design.

At this point, the design team needs to review the DFA results and adjust the design layout to reflect the feedback of this preliminary analysis. The next step is to complete a design for manufacturing analysis on each unique part in the product. This will consist of developing a part cost and tooling cost for each part. It should also include doing a producibility study of each part. Based on the DFM analysis, the design team needs to make some additional adjustments in the design layout. At this point, the design team is now ready to start the design phase of the project. The DFM&A input at this point has developed a preliminary bill of material (BOM) and established a target cost for all the unique new parts in the design. It has also influenced the product architecture to improve the sequence of assembly as it flows through the manufacturing process.

The following case study illustrates the key elements in applying DFM&A. Figure 30.3 shows a product called the *motor drive assembly*. This design consists of 17 parts and assemblies. Outwardly it looks as if it can be assembled with little difficulty. The product is made up of two sheet metal parts and one aluminum machined part. It also has a motor assembly and a sensor, both bought from an outside supplier. In addition, the *motor drive assembly* has nine hardware items that provide other functions—or do they?

At this point, the design looks simple enough. It should take minimal engineering effort to design and detail the unique parts and develop an assembly drawing. Has a UMC been developed yet? Has a DFM&A analysis been performed? The DFA analysis will look at each process step, part, and subassembly used to build the product. It will analyze the time it takes to "get" and "handle" each part and the time it takes to insert each part in the assembly (see Table 30.1). It will point out areas where there are difficulties handling, aligning, and securing each and every part and subassembly. The DFM analysis will establish a cost for each part and estimate the cost of fabrication tooling. The analysis will also point out high-cost areas in the fabrication process so that changes can be made.

At this point, the DFA analysis suggested that this design could be built with fewer parts. A review of Table 30.2, column 5, shows that the design team feels it can eliminate the bushings, standoffs, end-plate screws, grommet, cover, and cover screws. Also by replacing the end plate with a new snap-on plastic cover, they can eliminate the need to turn the (reorientation) assembly over to install the end plate and two screws. Taking the time to eliminate parts and operations is the most powerful part of performing a DFA analysis. This is rule #2, which was left out above: DFM&A is a team sport. Bringing all members of the new product development team together and understanding the sequence of assembly, handling, and insertion time for each part will allow each team member to better understand the function of every part.

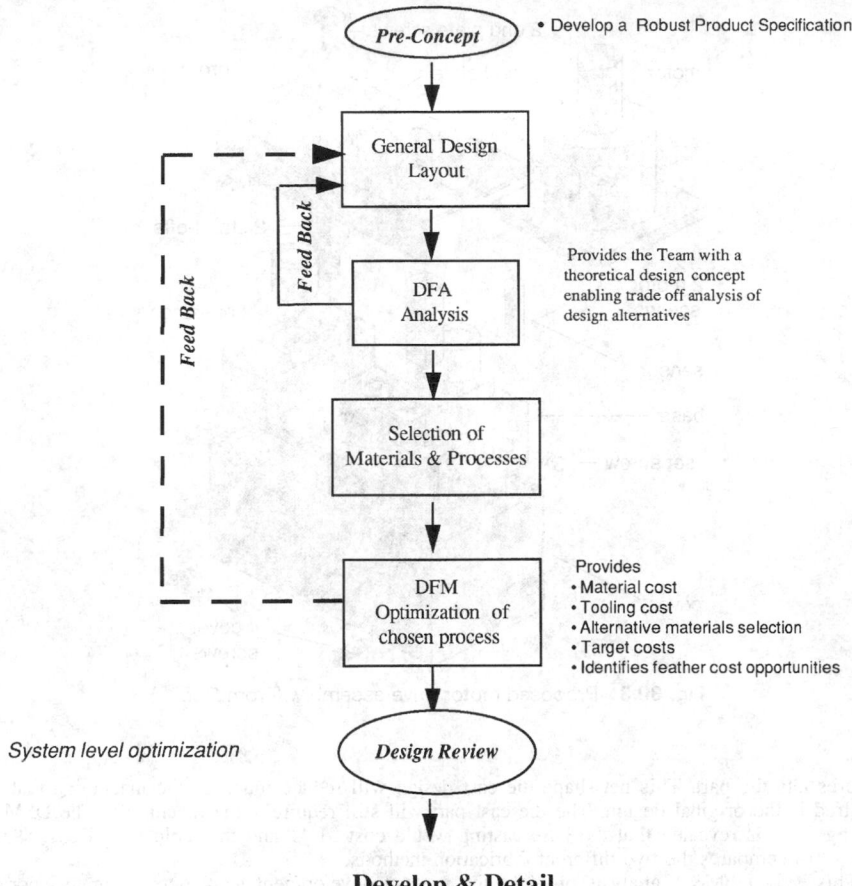

Develop & Detail

Fig. 30.2 Key components of the DFM&A process.

DFM Analysis

The DFM analysis provided the input for the fabricated part cost. As an example, the base is machined from a piece of solid aluminum bar stock. As designed, the base has 11 different holes drilled in it and 8 of them require taping. The DFM analysis (see Table 30.3) shows that it takes 17.84 minutes to machine this part from the solid bar stock. The finished machined base costs $10.89 in lots of 1,000 parts. The ideal process for completing a DFM analysis might be as follows.

In the case of the base, the design engineer created the solid geometry in Matra Data's Euliked CAD system (see Fig. 30.4). The design engineer then sent the solid database as an STL file to the manufacturing engineer, who then brought the STL file into a viewing tool called *Solid View* (see Fig. 30.5). SolidView allowed the ME to get all the dimensioning and geometry inputs needed to complete the Boothroyd Dewhurst design for manufacturing machining analysis of the base part. SolidView also allowed the ME to take cut sections of the part and then step through it to insure that no producibility rules had been violated.

Today all of the major CAD supplies provide the STL file output format. There are many new CAD viewing tools like SolidView available, costing about $500 to $1,000. These viewing tools will take STL or IGS files. The goal is to link all of the early product development data together so each member can have fast, accurate inputs to influence the design in its earliest stage.

In this example, it took the ME a total of 20 minutes to pull the STL files into SolidView and perform the DFM analysis. Engineering in the past has complained that DFM&A takes too much time and slows the design team down. The ME then analyzes the base as a die casting part, following the producibility rule. By designing the base as a die casting, it is possible to mold many of the part

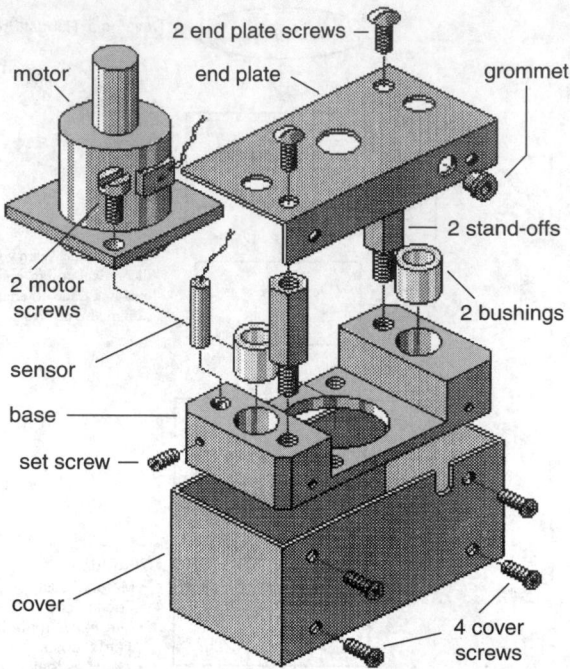

Fig. 30.3 Proposed motor drive assembly. (From Ref. 1.)

features into the part. This net shape die cast design will reduce much of the machining that was required in the original design. The die cast part will still require some machining. The DFM die casting analysis revealed that the base casting would cost $1.41 and the mold would cost $9,050. Table 30.4 compares the two different fabrication methods.

This early DFM&A analysis provides the product development team with accurate labor and material estimates at the start of the project. It removes much of the complexity of the assembly and allows each member of the design team to visualize every component's function. By applying the basic principles of DFA, such as

- Combining or eliminating parts
- Eliminating assembly adjustments
- Designing part with self-locating features
- Designing parts with self-fastening features

Table 30.1

Motor Drive Assembly	
Number of parts and assemblies	19
Number of reorientation or adjustment	1
Number of special operations	2
Total assembly time in seconds	213.4
Total cost of fabrication and assembly tooling	$3,590
Tool amortization at 10K assemblies	$ 0.36
Total cost of labor at $74.50/hr	$ 4.42
Total cost of materials	$ 42.44
Total cost of labor and materials	$ 46.86
Total UMC	$ 47.22

Table 30.2 Motor Drive Assembly: Design for Assembly Analysis

1	2	3	4	5	6	7	8	9	10	11	12	13	14	15	16	17
Name	Sub No. Entry No.	Type	Repeat Count	Minimum Items	Tool Fetching Time, sec	Handling Time, sec	Insertion or Op'n Time, sec	Total Time, sec	Labor Cost $	Ass'y Tool or Fixture Cost, $	Item Cost	Total Item Cost, $	Manuf. Tool Cost, $	Target Cost	Part Number	Description
Base	1.1	Part	1	1	0	1.95	1.5	3.45	$ .07	$500	$10.89	$10.89	$950	$7.00	1P033-01	Add base to fixture
Bushing	1.2	Part	2	0	0	1.13	6.5	15.26	$ .32	$0	$1.53	$3.06	$0	$0.23	16P024-01	Add & press fit
Motor	1.3	Sub	1	1	0	7	6	13	$ .27	$0	$18.56	$18.56	$0	$12.00	121S021-02	Add & hold down
Motor screw	1.4	Part	2	2	2.9	1.5	9.6	25.1	$ .52	$0	$0.08	$0.16	$0	$0.08	112W0223-06	Add & thread
Sensor	1.5	Sub	1	1	0	5.6	6	11.6	$ .24	$0	$2.79	$2.79	$0	$2.79	124S223-01	Add & hold down
Set screw	1.6	Part	1	1	2.9	3	9.2	15.1	$ .31	$0	$0.05	$0.05	$0	$0.05	111W0256-02	Add & thread
Stand-off	1.7	Part	2	0	2.9	1.5	9.6	25.1	$ .52	$0	$0.28	$0.56	$0	$0.18	110W0334-07	Add & thread
End plate	1.8	Part	1	1	0	1.95	5.2	7.15	$ .15	$0	$2.26	$2.26	$0	$0.56	15P067-01	Add & hold down
End plate screw	1.9	Part	2	0	2.9	1.8	5.7	17.9	$ .37	$0	$0.03	$0.06	$0	$0.03	110W0777-04	Add & thread
Grommet	1.1	Part	1	0	0	1.95	11	12.95	$ .27	$0	$0.12	$0.03	$0	$0.12	116W022-08	Add & push fit
Dress wires—grommet	1.11	Oper	2	2	—	—	—	18.79	$ .39	$350	$0.00	$0.00	$0	$0.00		Library operation
Reorientation	1.12	Oper	1	0	—	—	4.5	4.5	$ .09	$0	$0.00	$0.00	$0	$0.00		Reorient & adjust
Cover	1.13	Part	1	0	0	2.3	8.3	10.6	$ .22	$0	$3.73	$3.73	$1,230	$1.20	2P033-01	Add
Cover screw	1.14	Part	4	0	2.9	1.8	5.7	32.9	$ .68	$0	$0.05	$0.18	$0	$0.05	112W128-03	Add & thread
Totals =			22	9				213.4	$4.42	$850		$42.33	$2,740	$24.28		

UMC = $47.22

Production life volume = 10,000
Annual build volume = 3,000
Assm. labor rate $/hr = $74.50

Note: The information presented in this table was developed from the Boothroyd Dewhurst DFA software program, version 8.0.[2]

Table 30.3 Machining Analysis Summary Report

Set-Ups	Time Minutes	Cost $
Machine Tool Set-Ups		
Set-up	0.22	0.10
Nonproductive	10.63	4.87
Machining	6.77	3.10
Tool wear	—	0.31
Additional cost/part	—	0.00
Special tool or fixture	—	0.00
Library Operation Set-Ups		
Set-up	0.03	0.02
Process	0.20	0.13
Additional cost/part	—	0.03
Special tool or fixture	—	0.00
Material	—	2.34
Totals	17.84	10.89
Material		Gen aluminum alloy
Part number		5678
Initial hardness		55
Form of workpiece		Rectangular bar
Material cost, $/lb		2.75
Cut length, in.		4.000
Section height, in.		1.000
Section width, in.		2.200
Product life volume		10000
Number of machine tool set-ups		3
Number of library operation set-ups		1
Workpiece weight, lb		0.85
Workpiece volume, cu in.		8.80
Material density, lb/cu in.		0.097

- Facilitating handling of each part
- Eliminating reorientation of the parts during assembly
- Specifying standard parts,

the design team is able to rationalize the motor drive assembly with fewer parts and assembly steps. Figure 30.6 show a possible redesign of the original motor drive assembly. The DFM&A analysis (Table 30.5) provided the means for the design team to question the need and function of every part. As a result, the design team now has a new focus and an incentive to change the original design.

Table 30.6 shows the before-and-after DFM&A results.

If the motor drive product meets its expected production life volume of 10,000 units, the company will save $170,100. By applying principles of DFM&A to both the labor and material on the motor drive, the design team is able to achieve about a 35% cost avoidance on this program.

30.2.2 Getting the DFM&A Process Started

Management from All of the Major Disciplines Must Be on Your Side

In order for the DFM&A process to succeed, upper management must understand, accept, and encourage the DFM&A way of thinking. They must *want* it. It is difficult, if not impossible, for an individual or group of individuals to perform this task without management support, since the process requires the cooperation of so many groups working together. The biggest challenge of implementing DFM&A is the cooperation of so many individuals towards a common goal. This does not come naturally, especially if it is not perceived by the leaders as an integral part of the business's success. In many companies, management does not understand what DFM&A is. They believe it is a manufacturing process. It is not; it is a new product development process, which *must* include all disciplines (engineering, service, program managers, and manufacturing) to yield significant results. The simplest

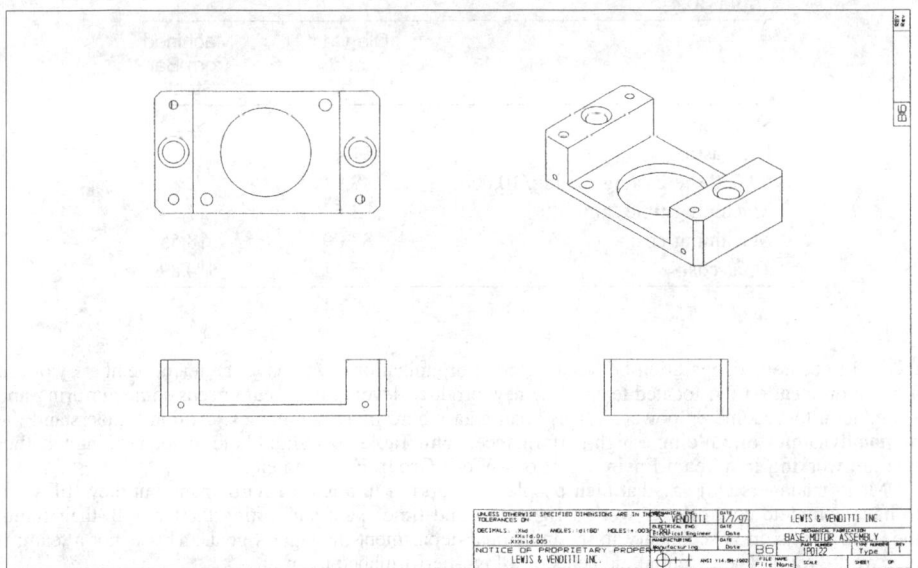

Fig. 30.4

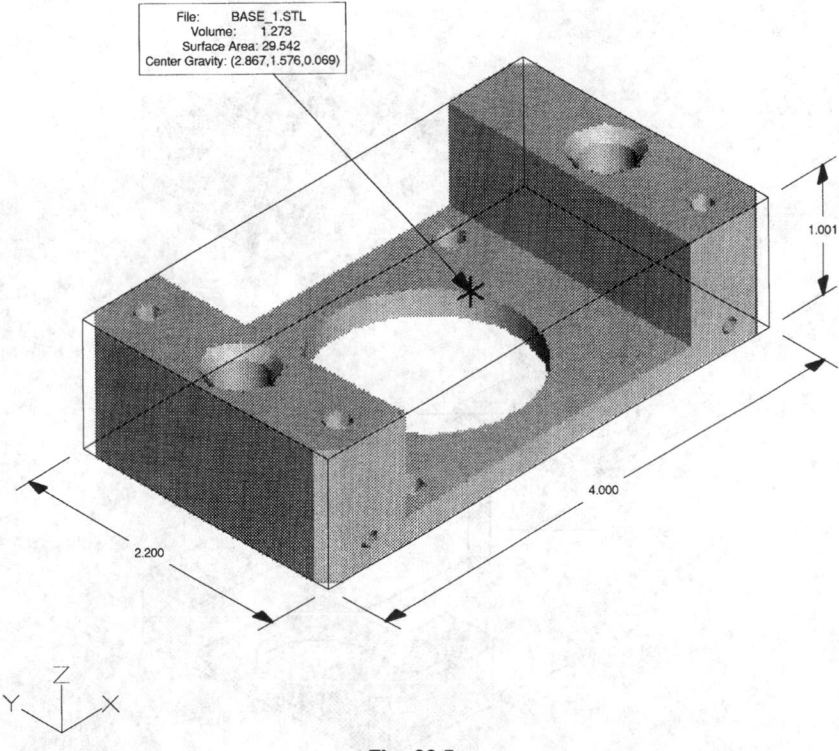

File: BASE_1.STL
Volume: 1.273
Surface Area: 29.542
Center Gravity: (2.867,1.576,0.069)

1.001

4.000

2.200

Fig. 30.5

Table 30.4

	Die Cast and Machined	Machined from Bar Stock
Stock cost		$2.34
Die casting	$1.41	
$9,050 die casting tooling/10,000	$.91	
Machining time, min	3.6	17.84
Machining cost	$3.09	$8.55
Total cost	5.41	$10.89

method to achieve cooperation between different organizations is to have the team members work in a common location (co-located team). The new product development team needs some nurturing and stimulation to become empowered. This is an area where most companies just don't understand the human dynamics of building a high-performance team. Table 30.7 should aid in determining whether you are working in a Team Environment or a Work Group Environment.

Many managers will say that their people are working in a team environment, but they still want to have complete control over work assignments and time spent supporting the team. In their mind, the team's mission is secondary to the individual department manager's goals. This is not a team; it is a work group. The essential elements of a high-performance team are

- A clear understanding of the team's goals (a defined set of goals and tasks assigned to each individual team member)
- A feeling of openness, trust, and communication

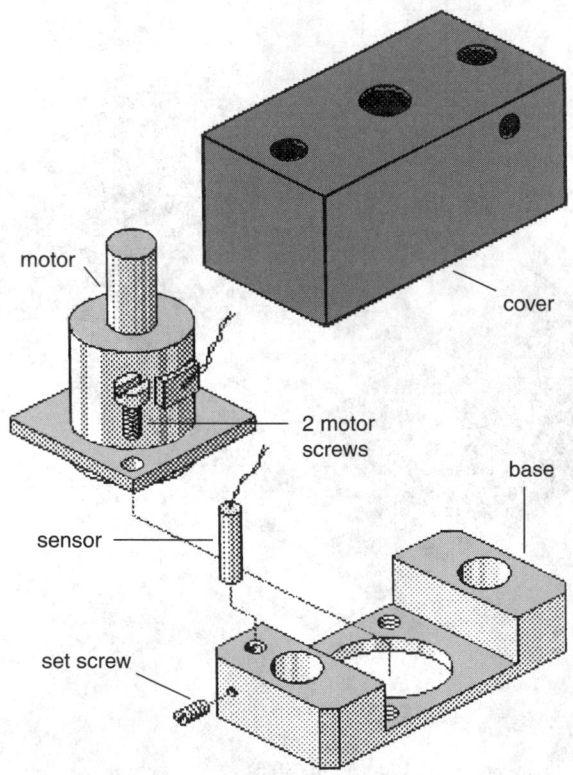

Fig. 30.6 Redesign of motor assembly.

Table 30.5 Redesign of Motor Drive Assembly: Design for Assembly Analysis

1	2	3	4	5	6	7	8	9	10	11	12	13	14	15	16	17
Name	Sub No. Entry No.	Type	Repeat Count	Minimum Items	Tool Fetching Time, sec	Handling Time, sec	Insertion or Op'n Time, sec	Total Time, sec	Labor Cost $	Ass'y Tool or Fixture Cost, $	Item Cost $	Total Item Cost, $	Manuf. Tool Cost, $	Target Cost	Part Number	Description
Base, casting	1.1	Part	1	1	0	1.95	1.5	3.45	$0.07	$850	$3.09	$3.09	$9,050	$3.09	1P033-02	Add base to fixture
Motor	1.2	Sub	1	1	0	3	6	9	$0.19		$18.56	$18.56		$15.00	121S021-02	Add & hold down
Motor screw	1.3	Part	2	2	2.9	1.5	9.6	25.1	$0.52		$0.08	$0.16		$0.08	112W0223-06	Add & thread
Sensor	1.4	Sub	1	1	0	5.6	6	11.6	$0.24		$2.79	$2.79		$2.79	124S223-01	Add & hold down
Set screw	1.5	Part	1	1	2.9	2.55	9.2	14.65	$0.30		$0.05	$0.05		$0.05	111W0256-02	Add & thread
Push/pull wire—easy	1.6	Oper	2	2	—	—	—	18.79	$0.39		$0.00	$0.00		$0.00		Library operation
Cover	1.7	Part	1	1	0	1.95	1.8	3.75	$0.08		$1.98	$1.98	$7,988	$1.70	2P033-02	Add & snap fit
Totals =			9	9				86.34	$1.79	$850		$26.63	$17,038	$22.71		
								UMC = $30.21								

Production life volume = 10,000

Annual build volume = 3,000

Assm. labor rate = $74.50

Note: The information presented in this table was developed from the Boothroyd Dewhurst DFA software program, version 8.0.[2]

Table 30.6   Comparison of DFM&A Results

	Motor Drive Assembly	Redesign of Motor Drive Assembly
Number of parts and assemblies	19	7
Number of reorientation or adjustment	1	0
Number of special operations	2	2
Total assembly time in seconds	213.4	86.34
Total cost of labor at $74.50/hr	$4.42	$1.79
Total cost of materials	$42.44	$26.63
Total cost of labor and material	$46.86	$28.42
Total cost of fabrication tooling	$3,590	$17,888
Tool amortization at 10K assemblies	$0.36	$1.79
Total UMC	$47.22	$30.21
Savings =	$17.01	

- Shared decision-making (consensus)
- A well-understood problem-solving process
- A leader who legitimizes the team-building process

Management must recognize that to implement DFM&A in their organization, they must be prepared to change the way they do things. Management's reluctance to accept the need for change is one reason DFM&A has been so slow to succeed in many companies. Training is one way of bringing DFM&A knowledge to an organization, but training alone cannot be expected to effect the change.

30.2.3   The DFM&A Road Map

The DFM&A Methodology (A Product Development Philosophy)

- Form a multifunctional team
- Establish the product goals through competitive benchmarking
- Perform a design for assembly analysis
- Segment the product into manageable subassemblies or levels of assembly
- As a team, apply the design for assembly principles
- Use creativity techniques to enhance the emerging design
- As a team, evaluate and select the best ideas
- Ensure economical production of every piece part
- Establish a target cost for every part in the new design

Table 30.7   Human Factors Test (check the box where your team fits)

Yes	Team Environment	Yes	Work Group Environment
	Are the team members committed to group's common goals?	✓	Are members loyal to outside groups with conflicting interests (functional managers)?
	Is there open communication with all members of the team?	✓	Is information unshared?
	Is there flexible, creative leadership?	✓	Is there a dominating leadership?
	Is the team rewarded as a group?	✓	Is there individual recognition?
	Is there a high degree of confidence and trust between members?	✓	Are you unsure of the group's authority?

- Start the detailed design of the emerging product
- Apply design for producibility guidelines
- Reapply the process at the next logical point in the design
- Provide the team with a time for reflection and sharing results

This DFM&A methodology incorporates all of the critical steps needed to insure a successful implementation.

Develop a multifunctional team of all key players before the new product architecture is defined. This team must foster a creative climate that will encourage ownership of the new product's design. The first member of this team should be the project leader, the person who has the authority for the project. This individual must control the resources of the organization, should hand-pick the people who will work on the team, and should have the authority to resolve problems within the team.

The team leader should encourage and develop a creative climate. It is of utmost importance to assemble a product development team that has the talent to make the right decisions, the ability to carry them out, and the persistence and dedication to bring the product to a successful finish. Although these qualities are invaluable, it is of equal importance that these individuals be allowed as much freedom as possible to germinate creative solutions to the design problem as early as possible in the product design cycle.

The product development team owns the product design and the development process. The DFM&A process is most successful when implemented by a multifunctional team, where each person brings to the product design process his or her specific area of expertise. The team should embrace group dynamics and the group decision-making process for DFM&A to be most effective.

Emphasis has traditionally been placed on the design team as the people who drive and own the product. Designers need to be receptive to team input and share the burden of the design process with other team members.

The team structure depends on the nature and complexity of the product. Disciplines that might be part of a product team include

- Engineering
- Manufacturing
- Field service and support
- Quality
- Human factors or ergonomics
- Purchasing
- Industrial design and packaging
- Distribution
- Sales
- Marketing

Although it is not necessary for all of these disciplines to be present all of the time, they should have an idea of how things are progressing during the design process.

Clearly, there can be drawbacks to multidisciplinary teams, such as managing too many opinions, difficulty in making decisions, and factors in general that could lengthen the product development cycle. However, once a team has worked together and has an understanding of individual responsibilities, there is much to gain from adopting the team approach. Groups working together can pool their individual talents, skills, and insight so that more resources are brought to bear on a problem. Group discussion leads to a more thorough understanding of problems, ideas, and potential solutions from a variety of standpoints. Group decision-making results in a greater commitment to decisions, since people are more motivated to support and carry out a decision that they helped make. Groups allow individuals to improve existing skills and learn new ones.

Having the team located together in one facility makes the process work even better. This co-location improves the team's morale and also makes communication easier. Remembering to call someone with a question, or adding it to a meeting agenda, is more difficult than mentioning it when passing in the hallway. Seeing someone reminds one of an issue that may have otherwise been forgotten. These benefits may seem trivial, but the difference that co-location makes is significant.

As a team, establish product goals through a competitive benchmarking process: Concept development. Competitive benchmarking is the continuous process of measuring your own products, services, and practices against the toughest competition, or the toughest competition in a particular area. The benchmarking process will help the team learn who the "best" are and what they do. It gives the team a means to understand how this new product measures up to other products in the marketplace. It identifies areas of opportunities that need changing in the current process. It allows the team to set targets and provides an incentive for change. Using a DFM&A analysis process for the competitive evaluation provides a means for relative comparison between your products and your competitors'. You determine exactly where the competition is better.

Before performing a competitive teardown, decide on the characteristics that are most important to review, what the group wants to learn from the teardown, and the metrics that will be noted. Also keep the teardown group small. It's great to have many people walk through and view the results, but a small group can better manage the initial task of disassembly and analysis. Ideally, set aside a conference room for several days so the product can be left out unassembled, with a data sheet and metrics available.

Perform a Design For Assembly analysis of the proposed product that identifies possible candidate parts for elimination or redesign and pinpoints high-cost assembly operations. Early in this chapter, the motor drive assembly DFM&A analysis was developed. This example illustrates the importance of using a DFA tool to identify, size, and track the cost-savings opportunities. This leads to an important question: Do you need a formal DFA analysis software tool? Some DFM&A consultants will tell you that it is not necessary to use a formal DFA analysis tool. It is my supposition that these consultants want to sell you a lot of their consulting services rather than teach the process. It just makes no sense *not* to use a formal DFA tool for evaluating and tracking the progress of the new product design through its evolution. The use of DFA software provides the team with a focus that is easily updated as design improvements are captured. The use of DFA software does not exclude the need for a good consultant to get the new team off to a good start. The selection of a DFA tool is a very important decision. The cost of buying a quality DFA software tool is easily justified by the savings from applying the DFA process on just one project.

At this point, the selection of the manufacturing site and type of assembly process should be completed. Every product must be designed with a thorough understanding of the capabilities of the manufacturing site. It is thus of paramount importance to choose the manufacturing site at the start of product design. This is a subtle point that is frequently overlooked at the start of a program, but it is of tremendous importance, since to build a partnership with the manufacturing site, the site needs to have been chosen! Also, manufacturing facilities have vastly different processes, capabilities, strengths, and weaknesses, that affect, if not dictate, design decisions. When selecting a manufacturing site, the process by which the product will be built is also being decided.

As a team, apply the design for assembly principles to every part and operation to generate a list of possible cost opportunities. The generic list of DFA principles includes the following:

- Designing parts with self-locating features
- Designing parts with self-fastening features
- Increasing the use of multifunctional parts
- Eliminating assembly adjustments
- Driving standardization of fasteners, components, materials, finishes, and processes

It is important that the team develop its own set of DFA principles that relate to the specific product it is working on. Ideally, the design team decides on the product characteristics it needs to meet based on input from product management and marketing. The product definition process involves gathering information from competitive benchmarking and teardowns, customer surveys, and market research. Competitive benchmarking illustrates which product characteristics are necessary.

Principles should be set forth early in the process as a contract that the team draws up together. It is up to the team to adopt many principles or only a few, and how lenient to be in granting waivers.

Use brainstorming or other creativity techniques to enhance the emerging design and identify further design improvements. The team must avoid the temptation to start engineering the product before developing the DFM&A analysis and strategy. As a team, evaluate and select the best ideas from the brainstorming, thus narrowing and focusing the product goals.

With the aid of DFM software, cost models and competitive benchmarking, establish a target cost for every part in the new design. Make material and manufacturing process selections. Start the early supplier involvement process to ensure economical production of every piece part. Start the detailed design of the emerging product. Model, test, and evaluate the new design for fit, form, and function. Apply design for producibility guidelines to the emerging parts design to ensure that cost and performance targets are met.

Table 30.8   DFM&A Metrics

	Old Design	Competitive	New Design
Number of Parts & Assemblies			
Number of Separate Assm. Operations			
Total Assembly Time			
Total Material Cost			
Totals			

Table 30.9   DFM&A New Products Checklist

Design for Manufacturing and Assembly Consideration	Yes	No
Design for assembly analysis completed:		
Has this design been analyzed for minimal part count?	☐	☐
Have all adjustments been eliminated?	☐	☐
Are more than 85% common parts and assemblies used in this design?	☐	☐
Has assembly sequence been provided?	☐	☐
Have assembly and part reorientations been minimized?	☐	☐
Have more than 96% preferred screws been used in this design?	☐	☐
Have all parts been analyzed for ease of insertion during assembly?	☐	☐
Have all assembly interferences been eliminated?	☐	☐
Have location features been provided?	☐	☐
Have all parts been analyzed for ease of handling?	☐	☐
Have part weight problems been identified?	☐	☐
Have special packaging requirements been addressed for problem parts?	☐	☐
Are special tools needed for any assembly steps?	☐	☐

Ergonomics Considerations	Yes	No
Does design capitalize on self-alignment features of mating parts?	☐	☐
Have limited physical and visual access conditions been avoided?	☐	☐
Does design allow for access of hands and tools to perform necessary assembly steps?	☐	☐
Has adequate access been provided for all threaded fasteners and drive tooling?	☐	☐
Have all operator hazards been eliminated (sharp edges)?	☐	☐

Wire Management	Yes	No
Has adequate panel pass-through been provided to allow for easy harness/cable routing?	☐	☐
Have harness/cable supports been provided?	☐	☐
Have keyed connectors been provided at all electrical interconnections?	☐	☐
Are all harnesses/cables long enough for ease of routing, tie down, plug in, and to eliminate strain relief on interconnects?	☐	☐
Does design allow for access of hands and tools to perform necessary wiring operations?	☐	☐
Does position of cable/harness impede air flow?	☐	☐

Design for Manufacturing and Considerations	Yes	No
Have all unique design parts been analyzed for producibility?	☐	☐
Have all unique design parts been analyzed for cost?	☐	☐
Have all unique design parts been analyzed for their impact of tooling/mold cost?	☐	☐

Assembly Process Consideration	Yes	No
Has assembly tryout been performed prior to scheduled prototype build?	☐	☐
Have assembly views and pictorial been provided to support assembly documentation?	☐	☐
Has opportunity defects analysis been performed on process build?	☐	☐
Has products cosmetics been considered (paint match, scratches)?	☐	☐

Provide the team with a time for reflection and sharing results.   Each team member needs to understand that there will be a final review of the program, at which time members will be able to make constructive criticism. This time helps the team determine what worked and what needs to be changed in the process.

Use DFM&A Metrics

The development of some DFM&A metrics is important. The team needs a method to measure the before-and-after results of applying the DFM&A process, thus justifying the time spent on the project. Table 30.8 shows the typical DFM&A metrics that should be used to compare your old product design against a competitive product and a proposed new redesign.

The total number of parts in an assembly is an excellent and widely used metric. If the reader remembers only one thing from this chapter, let it be to strive to reduce the quantity of parts in every product designed. The reason limiting parts count is so rewarding is that when parts are reduced, considerable overhead costs and activities that burden that part also disappear. When parts are reduced, quality of the end product is increased, since each part that is added to an assembly is an opportunity to introduce a defect into the product. Total assembly time will almost always be lowered by reducing the quantity of parts.

A simple method to test for potentially unnecessary parts is to ask the following three questions for each part in the assembly. The theory is that if the answers to all of the three questions are "no," then theoretically the part in question is not needed and is therefore a candidate for combination with other parts in the assembly or elimination from the design. The questions are:

1. During the products operation, does the part move relative to all other parts already assembled?   (*answer yes or no*)
2. Does the part need to be made from a different material or be isolated from all other parts already assembled?   (*answer yes or no*)
3. Must the part be separate from all other parts already assembled because of necessary assembly or disassembly of other parts?   (*answer yes or no*)

You must answer the questions above for each part in the assembly. If your answer is "no" for all three questions, then that part is a candidate for elimination.

The total time it takes to assemble a product is an important DFM&A metric. Time is money, and the lower the time needed to assemble the product, the better. Since some of the most time-consuming assembly operations are fastening operations, discrete fasteners are always candidates for elimination from a product. By examining the assembly time of each and every part in the assembly, the designer can target specific areas for improvement. Total material cost is self-explanatory.

The new product DFM&A checklist (Table 30.9) is a good review of how well your team did with applying the DFM&A methodology. Use this check sheet during all phases of the product development process; it is a good reminder. At the end of the project you should have checked most of the *yes* boxes.

30.3   WHY IS DFM&A IMPORTANT?

DFM&A is a powerful tool in the design team's repertoire. If used effectively, it can yield tremendous results, the least of which is that the product will be easy to assemble! The most beneficial outcome of DFM&A is to reduce part count in the assembly, which in turn will simplify the assembly process, lower manufacturing overhead, reduce assembly time, and increase quality by lessening the opportunities for introducing a defect. Labor content is also reduced because with fewer parts, there are fewer and simpler assembly operations. Another benefit to reducing parts count is a shortened product development cycle because there are fewer parts to design. The philosophy encourages simplifying the design and using standard, off-the-shelf parts whenever possible. In using DFM&A, renewed emphasis is placed on designing each part so it can be economically produced by the selected manufacturing process.

REFERENCES

1. G. Boothroyd, P. Dewhurst, and W. Knight, *Product Design for Manufacturing and Assembly,* Marcel Dekker, New York, 1994.
2. Boothroyd Dewhurst Inc., *Design for Assembly Software,* Version 8.0, Wakefield, RI, 1996.

CHAPTER 31

CLASSIFICATION SYSTEMS

Dell K. Allen
Manufacturing Engineering Department
Retired from Brigham Young University
Provo, Utah

31.1	**PART FAMILY CLASSIFICATION AND CODING**	**951**
	31.1.1 Introduction	951
	31.1.2 Application	952
	31.1.3 Classification Theory	954
	31.1.4 Part Family Code	955
	31.1.5 Tailoring the System	962
31.2	**ENGINEERING MATERIALS TAXONOMY**	**962**
	31.2.1 Introduction	962
	31.2.2 Material Classification	962
	31.2.3 Material Code	964
	31.2.4 Material Properties	965
	31.2.5 Material Availability	966
	31.2.6 Material Processability	966
31.3	**FABRICATION PROCESS TAXONOMY**	**967**
	31.3.1 Introduction	967
	31.3.2 Process Divisions	969
	31.3.3 Process Taxonomy	970
	31.3.4 Process Code	973
	31.3.5 Process Capabilities	973
31.4	**FABRICATION EQUIPMENT CLASSIFICATION**	**974**
	31.4.1 Introduction	974
	31.4.2 Standard and Special Equipment	976
	31.4.3 Equipment Classification	976
	31.4.4 Equipment Code	977
	31.4.5 Equipment Specification Sheets	978
31.5	**FABRICATION TOOL CLASSIFICATION AND CODING**	**981**
	31.5.1 Introduction	981
	31.5.2 Standard and Special Tooling	982
	31.5.3 Tooling Taxonomy	982
	31.5.4 Tool Coding	982
	31.5.5 Tool Specification Sheets	984

31.1 PART FAMILY CLASSIFICATION AND CODING

31.1.1 Introduction

History

Classification and coding practices are as old as the human race. They were used by Adam, as recorded in the Bible, to classify and name plants and animals, by Aristotle to identify basic elements of the earth, and in more modern times to classify concepts, books, and documents. But the classification and coding of manufactured pieceparts is relatively new. Early pioneers associated with workpiece classification are Mitrafanov of the USSR, Gombinski and Brisch, both of the United Kingdom, and Opitz of Germany. In addition, there are many who have espoused the principles developed by these men, adapted them and enlarged upon them, and created comprehensive workpiece classification systems. It is reported that over 100 such classification systems have been created specifically for machined parts, others for castings or forgings, and still others for sheet metal parts, and so on. In the United States there have been several workpiece classification systems commercially developed and used, and a large number of proprietary systems created for specific companies.

Why are there so many different part-classification systems? In attempting to answer this question, it should be pointed out that different workpiece classification systems were initially developed for

Mechanical Engineers' Handbook, 2nd ed., Edited by Myer Kutz.
ISBN 0-471-13007-9 © 1998 John Wiley & Sons, Inc.

different purposes. For example, Mitrafanov apparently developed his system to aid in formulating group production cells and in facilitating the design of standard tooling packages; Opitz developed his system for ascertaining the workpiece shape/size distribution to aid in designing suitable production equipment. The Brisch system was developed to assist in design retrieval. More recent systems are production-oriented.

Thus, the intended application perceived by those who have developed workpiece classification systems has been a major factor in their proliferation. Another significant factor has been personal preferences in identification of attributes and relationships. Few system developers totally agree as to what should or should not be the basis of classification. For example: Is it better to classify a workpiece by function as "standard" or "special" or by geometry as "rotational" or "non-rotational"? Either of these choices makes a significant impact on how a classification system will be developed.

Most classification systems are hierarchal, proceeding from the general to the specific. The hierarchal classification has been referred to by the Brisch developers as a monocode system. In an attempt to derive a workpiece code that addressed the question of how to include several related, but non-hierarchal, workpiece features, the feature code or polycode concept was developed. Some classification systems now include both polycode and monocode concepts.

A few classification systems are quite simple and yield a short code of five or six digits. Other part-classification systems are very comprehensive and yield codes of up to 32 digits. Some part codes are numeric and some are alphanumeric. The combination of such factors as application, identified attributes and relationships, hierarchal versus feature coding, comprehensiveness, and code format and length have resulted in a proliferation of classification systems.

31.1.2 Application

Identification of intended applications for a workpiece classification system are critical to the selection, development, or tailoring of a system.

It is not likely that any given system can readily satisfy both known present applications and unknown future applications. Nevertheless, a classification system can be developed in such a way as to minimize problems of adaptation. To do this, present and anticipated applications must be identified. It should be pointed out that development of a classification system for a narrow, specific application is relatively straightforward. Creation of a classification system for multiple applications, on the other hand, can become very complex and costly.

Figure 31.1 is a matrix illustrating this principle. As the applications increase, the number of required attributes also generally increases. Consequently, system complexity also increases, but often at a geometric or exponential rate, owing to the increased number of combinations possible. Therefore, it is important to establish reasonable application requirements first while avoiding unnecessary requirements and, at the same time, to make provision for adaptation to future needs.

In general, a classification system can be used to aid (1) design, (2) process planning, (3) materials control, and (4) management planning. A brief description of selected applications follows.

Design Retrieval

Before new workpieces are introduced into the production system, it is important to retrieve similar designs to see if a suitable one already exists or if an existing design may be slightly altered to accommodate new requirements. Potential savings from avoiding redundant designs range in the thousands of dollars.

Design retrieval also provides an excellent starting point for standardization and modularization. It has been stated that "only 10–20% of the geometry of most workpieces relates to the product function." The other 80–90% of the geometric features are often a matter of individual designer taste or preference. It is usually in this area that standardization could greatly reduce production costs, improve product reliability, increase ease of maintenance, and provide a host of other benefits.

One potential benefit of classification is in meeting the product liability challenge. If standard analytic tools are developed for each part family, and if product performance records are kept for those families, then the chances of negligent or inaccurate design are greatly reduced.

The most significant production savings in manufacturing enterprise begin with the design function. The function must be carefully integrated with the other functions of the company, including materials requisition, production, marketing, and quality assurance. Otherwise, suboptimization will likely occur, with its attendant frequent redesign, rework, scrap, excess inventory, employee frustration, low productivity, and high costs.

Generative Process Planning

One of the most challenging and yet potentially beneficial applications of workpiece classification is that of process planning. The workpiece class code can provide the information required for logical, consistent process selection and operation planning.

The various segments of the part family code may be used as keywords on a comprehensive process-classification taxonomy. Candidate processes are those that satisfy the conditions of the given

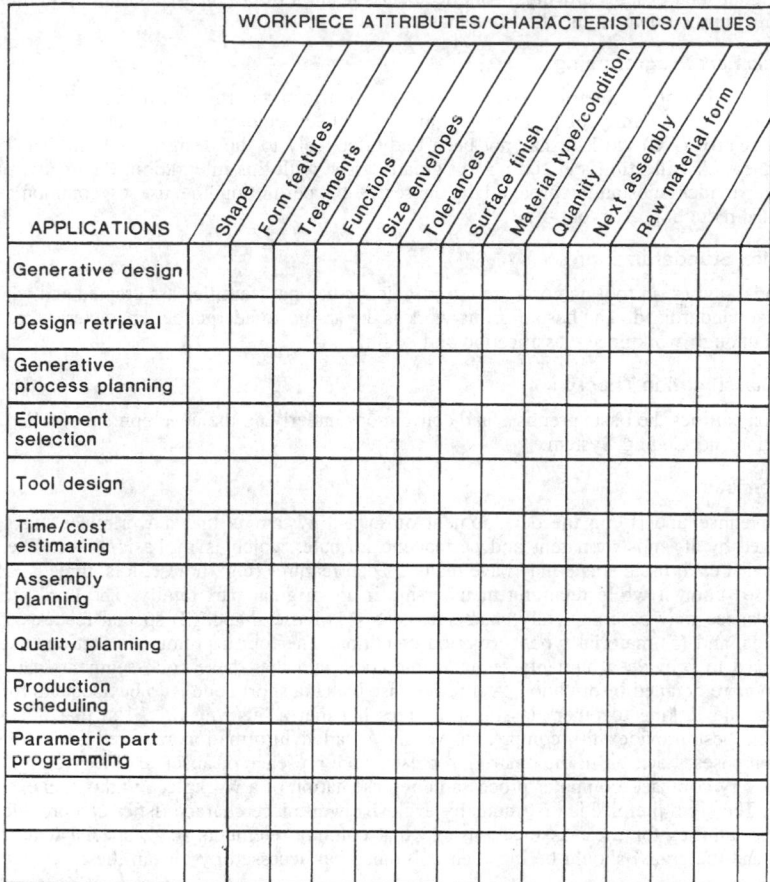

Fig. 31.1 Attribute selection matrix.

basic shape *and* the special features *and* the size *and* the precision *and* the material type *and* the form *and* the quality/time requirements.

After outputting the suitable processes, economic or other considerations may govern final process selection. When the suitable process has been selected, the codes for form features, heat treatments, coatings, surface finish, and part tolerance govern computerized selection of fabrication and inspection operations. The result is a generated process plan.

Production Estimating

Estimating of production time and cost is usually an involved and laborious task. Often the results are questionable because of unknown conditions, unwarranted assumptions, or shop deviations from the operation plan. The part family code can provide an index to actual production times and costs for each part family. A simple regression analysis can then be used to provide an accurate predictor of costs for new parts falling in a given part family. Feedback of these data to the design group could provide valuable information for evaluating alternative designs prior to their release to production.

Parametric and Generative Design

Once the product mix of a particular manufacturing enterprise has been established, high-cost, low-profit items can be singled out. During this sorting and characterization process, it is also possible to establish tabular or parametric designs for each basic family. Inputting of dimensional values and other data to a computer graphics system can result in the automatic production of a drawing for a given part. Taking this concept back one more step, it is conceivable that merely inputting a product name, specifications, functional requirements, and some dimensional data would result in the generation of a finished design drawing. Workpiece classification offers many exciting opportunities for productivity improvement in the design arena.

Parametric Part Programming

A logical extension of parametric design is that of parametric part programming. Although parametric part programming or family of parts programming has been employed for some time in advanced numerical control (NC) work, it has not been tied effectively to the design database. It is believed that workpiece classification and coding can greatly assist with this integration. Parametric part programming provides substantial productivity increases by permitting the use of common program modules and reduction of tryout time.

Tool Design Standardization

The potential savings in tooling costs are astronomical when part families are created and when form features are standardized. The basis for this work is the ability to adequately characterize component pieceparts through workpiece classification and coding.

31.1.3   Classification Theory

This section outlines the basic premises and conventions underlying the development of a Part Family Classification and Coding System.

Basic Premises

The first premise underlying the development of such a system is that a workpiece may be best characterized by its most apparent and permanent attribute, which is its basic shape. The second premise is that each basic shape may have many special features (e.g., holes, slots, threads, coatings) superimposed upon it while retaining membership in its original part family. The third premise is that a workpiece may be completely characterized by (1) basic shape, (2) special features, (3) size, (4) precision, and (5) material type, form, and condition. The fourth premise is that code segments can be linked to provide a humanly recognizable code, and that these code segments can provide pointers to more detailed information. A fifth premise is that a short code is to be adequate for human monitoring, and linking to other classification trees but that a bitstring (0's, 1's) that is computer-recognizable best provides the comprehensive and detailed information required for retrieval and planning purposes. Each bit in the bitstring represents the presence or absence of a given feature and provides a very compact, computer-processable representation of a workpiece without an excessively long code. The sixth premise is that mutually exclusive workpiece characteristics can provide unique basic shape families for the classification, and that common elements (e.g., special features, size, precision, and materials) should be included only once but accessed by all families.

E-Tree Concept

Hierarchal classification trees with mutually exclusive data (E-trees) provide the foundation for establishing the basic part shape (Fig. 31.2). Although a binary-type hierarchal tree is preferred because it is easy to use, it is not uncommon to find three or more branches.

It should be pointed out, however, that because the user must select only one branch, more than two branches require a greater degree of discrimination. With two branches, the user may say, "Is it this or that?" With five branches, the user must consider, "Is it this or this or this or this or this?" The reading time and error rate likely increase with the number of branches at each node. The E-tree is very useful for dividing a large collection of items into mainly exclusive families or sets.

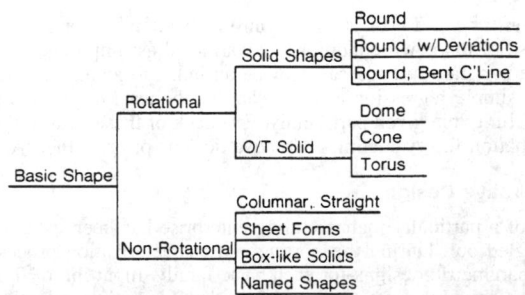

Fig. 31.2   E-tree concept applied to basic shape classification.

N-Tree Concept

The N-tree concept is based on a hierarchal tree with nonmutually exclusive paths (i.e., all paths may be selected concurrently). This type of tree (Fig. 31.3) is particularly useful for representing the common attributes mentioned earlier (e.g., form features, heat treatments, surface finish, size, precision, and material type, form, and condition).

In the example shown in Fig. 31.3, the keyword is Part Number (P/N) 101. The attributes selected are shown by means of an asterisk (*). In this example the workpiece is characterized as having a "bevel," a "notch," and a "tab."

Bitstring Representation

During the traversal of either an E-tree or an N-tree, a series of 1's and 0's are generated, depending on the presence or absence of particular characteristics or attributes. The keyword (part number) and its associated bitstring might look something like this:

$$P/N-101 = 100101 \cdots 010$$

The significance of the bitstring is twofold. First, one 16-bit computer word can contain as many as 16 different workpiece attributes. This represents a significant reduction in computer storage space compared with conventional representation. Second, the bitstring is in the proper format for rapid computer processing and information retrieval. The conventional approach is to use lists and pointers. This requires relatively large amounts of computation and a fast computer is necessary to achieve a reasonable response time.

Keywords

A keyword is an alphanumeric label with its associated bitstring. The label may be descriptive of a concept (e.g., stress, speed, feed, chip-thickness ratio), or it may be descriptive of an entity (e.g., cutting tool, vertical mill, 4340 steel, P/N-101). In conjunction with the Part Family Classification and Coding System, a number of standard keywords are provided. To conserve space and facilitate data entry, some of these keywords consist of one- to three-character alphanumeric codes. For example, the keyword code for a workpiece that is rotational and concentric, with two outside diameters and one bore diameter, is "B11." The keyword code for a family of low-alloy, low-carbon steels is A1. These codes are easy to use and greatly facilitate concise communication. They may be used as output keys or input keys to provide the very powerful capability of linking to other types of hierarchal information trees, such as those used for process selection, equipment selection, or automated time standard setting.

31.1.4 Part Family Code

Purpose

Part classification and coding is considered by many to be a prerequisite to the introduction of group technology, computer-aided process planning, design retrieval, and many other manufacturing activ-

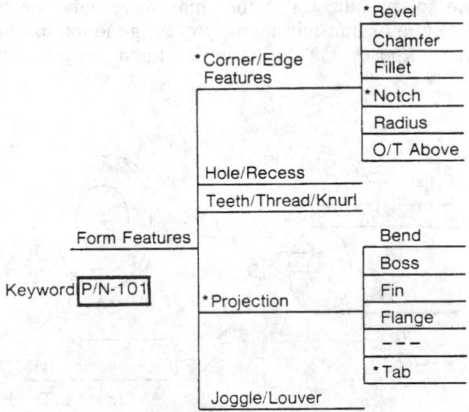

Fig. 31.3 N-tree concept applied to form features.

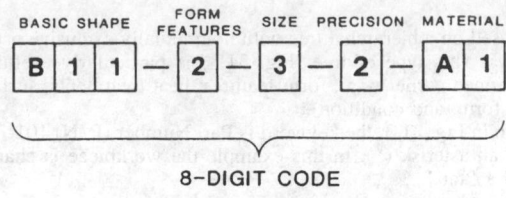

8-DIGIT CODE

Fig. 31.4 Part family code.

ities. Part classification and coding is aimed at improving productivity, reducing unnecessary variety, improving product quality, and reducing direct and indirect cost.

Code Format and Length

The part family code shown in Fig. 31.4 is composed of a five-section alphanumeric code. The first section of the code gives the basic shape. Other sections provide for form features, size, precision, and material. Each section of the code may be used as a pointer to more detailed information or as an output key for subsequent linking with related decision trees. The code length is eight digits. Each digit place has been carefully analyzed so that a compact code would result that is suitable for human communication and yet sufficiently comprehensive for generative process planning. The three-digit basic shape code provides for 240 standard families, 1160 custom families, and 1000 functional or named families. In addition, the combination of 50 form features, 9 size ranges, 5 precision classes, and 79 material types makes possible 2.5×10^{71} unique combinations! This capability should satisfy even the most sophisticated user.

Basic Shape

The basic shapes may be defined as those created from primitive solids and their derivatives (Fig. 31.5) by means of a basic founding process (cast, mold, machine). Primitives have been divided into rotational and nonrotational shapes. Rotational primitives include the cylinder, sphere, cone, ellipsoid, hyperboloid, and toroid. The nonrotational primitives include the cube (parallelepiped), polyhedron, warped (contoured) surfaces, free forms, and named shapes. The basic shape families are subdivided on the basis of predominant geometric characteristics, including external and internal characteristics.

The derivative concentric cylinder shown in Fig. 31.5 may have several permutations. Each permutation is created by merely changing dimensional ratios as illustrated or by adding form features. The rotational cylindrical shape shown may be thought of as being created from the intersection of a negative cylinder with a positive cylinder.

Figure 31.5a, with a length/diameter (L/D) ratio of 1:1, could be a spacer; Fig. 31.5b, with an L/D ratio of 0.1:1, would be a washer; and Fig. 31.5c, with an L/D ratio of 5:1, could be a thin-walled tube. If these could be made using similar processes, equipment, and tooling, they could be said to constitute a family of parts.

Name or Function Code

Some geometric shapes are so specialized that they may serve only one function. For example, a crankshaft has the major function of transmitting reciprocating motion to rotary motion. It is difficult to use a crankshaft for other purposes. For design retrieval and process planning purposes, it would

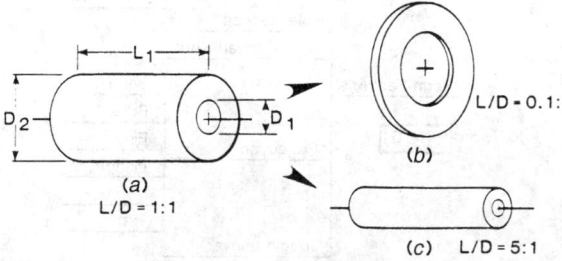

Fig. 31.5 Permutations of concentric cylinders.

probably be well to classify all crankshafts under the code name "crankshaft." Of course, it may still have a geometric code such as "P75," but the descriptive code will aid in classification and retrieval. A controlled glossary of function codes with cross references, synonyms, and preferred labels would aid in using name and function codes and avoid unnecessary proliferation.

Special Features

To satisfy product design requirements, the designer creates the basic shape of a workpiece and selects the engineering material of which it is to be made. The designer may also require special processing treatments to enhance properties of a given material. In other words, the designer adds special features. Special features of a workpiece include form features heat treatments, and special surface finishes.

Form features may include holes, notches, splines, threads, and so on. The addition of a form feature does not change the basic part shape (family), but does enable it to satisfy desired functional requirements. Form features are normally imparted to the workpiece subsequent to the basic founding process.

Heat treatments are often given to improve strength, hardness, and wear resistance of a material. Heat treatments, such as stress relieving or normalizing, may also be given to aid in processing the workpiece.

Surface finishing treatments, such as plating, painting, and anodizing, are given to enhance corrosion resistance, improve appearance, or meet some other design requirement.

The special features are contained in an N-tree format with an associated complexity-evaluation and classification feature. This permits the user to select many special features while still maintaining a relatively simple code. Basically, nine values (1–9) have been established as the special feature complexity codes. As the user classifies the workpiece and identifies the special features required, the number of features is tallied and an appropriate complexity code is stored. Figure 31.6 shows the number count for special features and the associated feature code.

The special feature complexity code is useful in conveying to the user some idea of the complexity of the workpiece. The associated bitstring contains detailed computer-interpretable information on all features. (Output keys may be generated for each individual feature.) This information is valuable for generative process planning and for estimating purposes.

Size Code

The size code is contained in the third section of the part family code. This code consists of one numeric digit. Values range from 1 to 9, with 9 representing very large parts (Fig. 31.7). The main purpose of the size code is to give the code user a feeling for the overall size envelope for the coded part. The size code is also useful in selecting production equipment of the appropriate size.

Precision Class Code

The precision class code is contained in the fourth segment of the part family code. It consists of a single numeric digit with values ranging from 1 to 5. Precision in this instance represents a composite

FEATURE COMPLEXITY CODE	NO. SPECIAL FEATURES
1	1
2	• 2
3	3
4	5
5	8
6	13
7	21
8	34
9	GT 34

Fig. 31.6 Complexity code for special features.

PART FAMILY SIZE CLASSIFICATION

SIZE CODE	MAXIMUM DIMENSION		DESCRIPTION	EXAMPLES
	ENGLISH (in.)	METRIC (mm)		
1	.5	10	Sub-miniature	Capsules
2	2	50	Miniature	Paper clip box
3	4	100	Small	Large match box
4	10	250	Medium-small	Shoe box
5	20	500	Medium	Bread box
6	40	1000	Medium-large	Washing machine
7	100	2500	Large	Pickup truck
8	400	10000	Extra-large	Moving van
9	1000	25000	Giant	Railroad box-car

Fig. 31.7 Part family size classification.

of tolerance and surface finish. Class 1 precision represents very close tolerances and a precision-ground or lapped-surface finish. Class 5, on the other hand, represents a rough cast or flame-cut surface with a tolerance of greater than 1/32 in. High precision is accompanied by multiple processing operations and careful inspection operations. Production costs increase rapidly as closer tolerances and finer surface finishes are specified. Care is needed by the designer to ensure that high precision is warranted. The precision class code is shown in Fig. 31.8.

Material Code

The final two digits of the part family code represent the material type. The material form and condition codes are captured in the associated bitstring.

Seventy-nine distinct material families have been coded (Fig. 31.9). Each material family or type is identified by a two-digit code consisting of a single alphabetic character and a single numeric digit.

The stainless-steel family, for example, is coded "A6." The tool steel family is "A7." This code provides a pointer to specification sheets containing comprehensive data on material properties, availability, and processability.

The material code provides a set of standard interface codes to which may be appended a given industry class code when appropriate. For example, the stainless-steel code may have appended to it a specific material code to uniquely identify it as follows: "A6-430" represents a chromium-type, ferritic, non-hardenable stainless steel.

PRECISION CLASS CODE

CLASS CODE	TOLERANCE	SURFACE FINISH
1	LE .0005"	LE 4 RMS
2	.0005"-.002"	4-32 RMS
3	.002"-.010"	32-125 RMS
4	.010"-.030"	125-500 RMS
5	GT .030"	GT 500 RMS

Fig. 31.8 Precision class code.

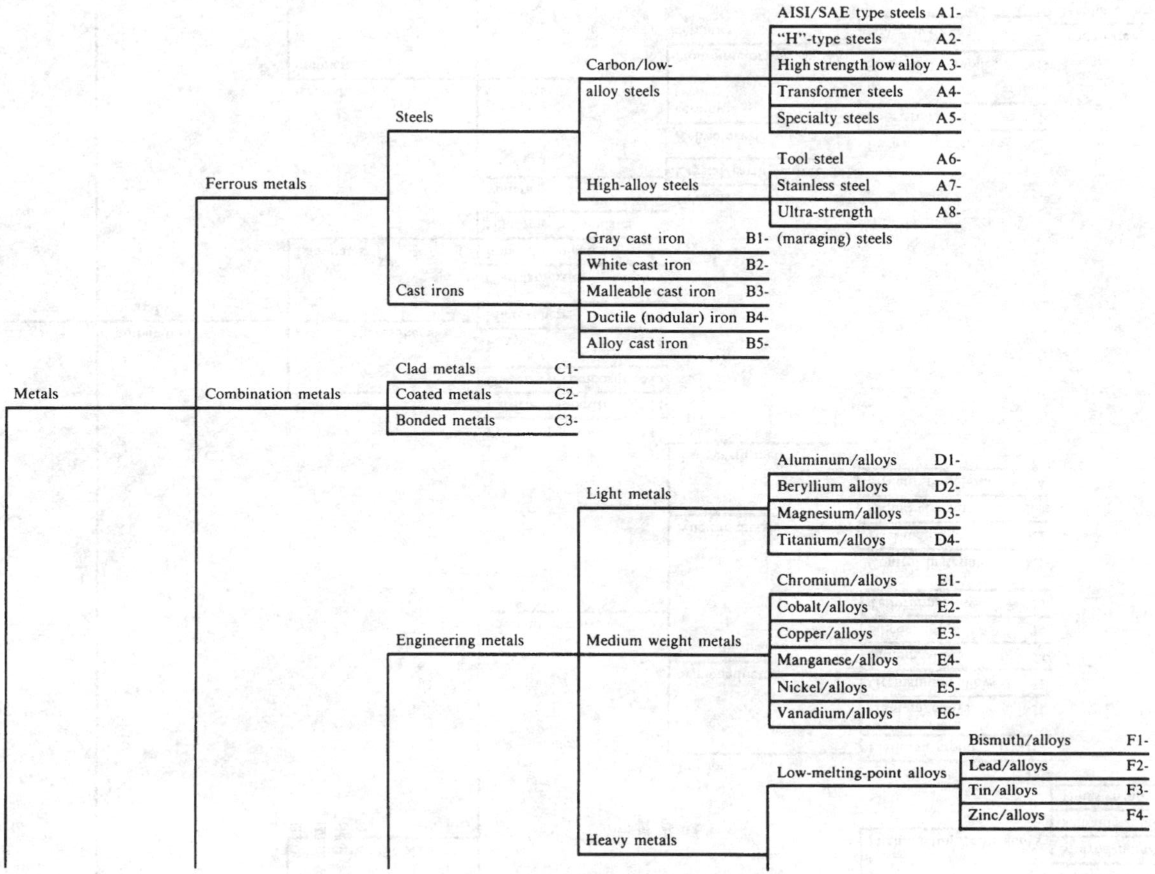

Ferrous metals

Steels

Carbon/low-alloy steels
- AISI/SAE type steels A1-
- "H"-type steels A2-
- High strength low alloy A3-
- Transformer steels A4-
- Specialty steels A5-

High-alloy steels
- Tool steel A6-
- Stainless steel A7-
- Ultra-strength A8-
- (maraging) steels

Cast irons
- Gray cast iron B1-
- White cast iron B2-
- Malleable cast iron B3-
- Ductile (nodular) iron B4-
- Alloy cast iron B5-

Combination metals
- Clad metals C1-
- Coated metals C2-
- Bonded metals C3-

Engineering metals

Light metals
- Aluminum/alloys D1-
- Beryllium alloys D2-
- Magnesium/alloys D3-
- Titanium/alloys D4-

Medium weight metals
- Chromium/alloys E1-
- Cobalt/alloys E2-
- Copper/alloys E3-
- Manganese/alloys E4-
- Nickel/alloys E5-
- Vanadium/alloys E6-

Heavy metals

Low-melting-point alloys
- Bismuth/alloys F1-
- Lead/alloys F2-
- Tin/alloys F3-
- Zinc/alloys F4-

Metals

Fig. 31.9 Engineering materials.

959

Engineering materials — Combination materials

- **Nonferrous metals** — Specialty metals
  - High-melting-point alloys H1-
    - Niobium (columbium) G1-
    - Molybdenum/alloys G2-
    - Tantalum/alloys G3-
    - Tungsten/alloys G4-
  - Precious metals
    - Noble metals H2-
    - Platinum group J1-
  - Semiconductor/specialty metals
    - Gallium/alloys J2-
    - Germanium/alloys J3-
    - Indium/alloys J4-
    - Silicon/alloys J5-
    - Tellurium/alloys
  - Nuclear metals
    - Control materials K1-
    - Fuel material K2-
    - Liquid coolants K3-
    - Structural materials K4-
  - Rare-earth metals L1-

- **Composites**
  - Fiber composite M1-
  - Particle composite M2-
  - Dispersion composite M3-
  - Foams, microspheres
    - Foams M4-
    - Microspheres M5-
  - Laminates
    - Clad laminates
    - Bonded laminates M6-
    - Honeycomb laminates

- **Crystalline nonmetals**
  - Minerals
    - Crystals N1-
    - Crystal/earth mixture N2-
  - Ceramics
    - Refractory ceramics
      - Furnace refractories N3-
      - Super-refractories N4-
    - Nonrefractory ceramics
      - Structural ceramics N5-
      - Nonstructural
        - Whiteware ceramics N6-
        - Technical ceramics N7-

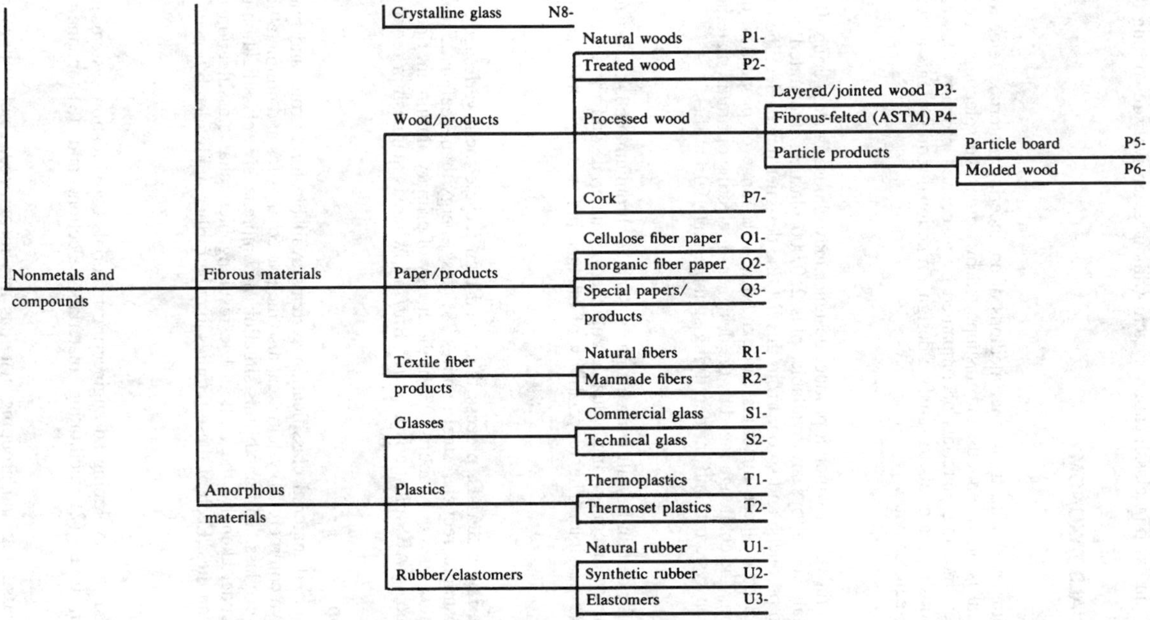

Fig. 31.9 (*Continued*)

Nonmetals and compounds	Crystalline glass			N8-
	Fibrous materials	Wood/products	Natural woods	P1-
			Treated wood	P2-
			Processed wood	Layered/jointed wood P3-
				Fibrous-felted (ASTM) P4-
				Particle products — Particle board P5-
				Molded wood P6-
			Cork	P7-
		Paper/products	Cellulose fiber paper	Q1-
			Inorganic fiber paper	Q2-
			Special papers/products	Q3-
		Textile fiber products	Natural fibers	R1-
			Manmade fibers	R2-
	Amorphous materials	Glasses	Commercial glass	S1-
			Technical glass	S2-
		Plastics	Thermoplastics	T1-
			Thermoset plastics	T2-
		Rubber/elastomers	Natural rubber	U1-
			Synthetic rubber	U2-
			Elastomers	U3-

31.1.5 Tailoring the System

It has been found that nearly all classification systems must be customized to meet the needs of each individual company or user. This effort can be greatly minimized by starting with a general system and then tailoring it to satisfy unique user needs. The Part Family Classification and Coding System permits this customizing. It is easy to add new geometric configurations to the existing family of basic shapes. It is likewise simple to add additional special features or to modify the size or precision class ranges. New material codes may be readily added if necessary.

The ability to modify easily an existing classification system without extensively reworking the system is one test of its design.

31.2 ENGINEERING MATERIALS TAXONOMY

31.2.1 Introduction

Serious and far-reaching problems exist with traditional methods of engineering materials selection. The basis for selecting a material is often tenuous and unsupported by defensible selection criteria and methods. A taxonomy of engineering materials accompanied by associated property files can greatly assist the designer in choosing materials to satisfy a design's functional requirements as well as procurement and processing requirements.

Material Varieties

The number of engineering materials from which a product designer may choose is staggering. It is estimated that over 40,000 metals and alloys are available, plus 250,000 plastics, uncounted composites, ceramics, rubbers, wood products, and so on. From this list, the designer must select the one for use with the new product. Each of these materials can exhibit a wide range of properties, depending on its form and condition. The challenge faced by the designer in selecting optimum materials can be reduced by a classification system to aid in identifying suitable material families.

Material Shortages

Dependency on foreign nations for certain key alloying elements, such as chromium, cobalt, tungsten and tin, points up the critical need for conserving valuable engineering materials and for selecting less strategic materials wherever possible. The recyclability of engineering materials has become another selection criterion.

Energy Requirements

The energy required to produce raw materials, process them, and then recycle them varies greatly from material to material. For example, recycled steel requires 75% less energy than steel made from iron ore, and recycled aluminum requires only about 10% of the energy of primary aluminum. Energy on a per-volume basis for producing ABS plastic is 2×10^6 Btu/in.^{3}, whereas magnesium requires 8×10^6 Btu/in.^{3}

31.2.2 Material Classification

Although there are many specialized material classification systems available for ferrous and nonferrous metals, there are no known commercial systems that also include composites and nonmetallics such as ceramic, wood, plastic, or glass. To remedy this situation, a comprehensive classification of all engineering materials was undertaken by the author. The resulting hierarchal classification or taxonomy provides 79 material families. Each of these families may be further subdivided by specific types as desired.

Objectives

Three objectives were established for developing an engineering materials classification system, including (1) minimizing search time, (2) facilitating materials selection, and (3) enhancing communication.

Minimize Search Time. Classifying and grouping materials into recognized, small subgroups having similar characteristic properties (broadly speaking) minimizes the time required to identify and locate other materials having similar properties. The classification tree provides the structure and codes to which important procedures, standards, and critical information may be attached or referenced. The information explosion has brought a superabundance of printed materials. Significant documents and information may be identified and referenced to the classification tree to aid in bringing new or old reference information to the attention of users.

Facilitate Materials Selection. One of the significant problems confronting the design engineer is that of selecting materials. The material chosen should ideally meet several selection criteria, including satisfying the design functional requirements, producibility, availability, and the more recent constraints for life-cycle costing, including energy and ecological considerations.

Materials selection is greatly enhanced by providing materials property tables in a format that can be used manually or that can be readily converted to computer usage. A secondary goal is to reduce material proliferation and provide for standard materials within an organization, thus reducing unnecessary materials inventory.

Enhance Communication. The classification scheme is intended to provide the logical grouping of materials for coding purposes. The material code associated with family of materials provides a pointer to the specific material desired and to its condition, form, and properties.

Basis of Classification

Although it is possible to use a fairly consistent basis of classification within small subgroups (e.g., stainless steels), it is difficult to maintain the same basis with divergent groups of materials (e.g., nonmetals). Recognizing this difficulty, several bases for classification were identified, and the one that seemed most logical (or that was used industrially) was chosen. This subgroup base was then cross-examined relative to its usefulness in meeting objectives cited in the preceding subsection.

The various bases for classification considered for the materials taxonomy are shown in Fig. 31.10. The particular basis selected for a given subgroup depends on the viewpoint chosen. The overriding viewpoint for each selection was (1) Will it facilitate material selection for design purposes? and (2) Does it provide a logical division that will minimize search time in locating materials with a predominant characteristic or property?

Taxonomy of Engineering Materials

An intensive effort to produce a taxonomy of engineering materials has resulted in the classification shown in Fig. 31.11. The first two levels of this taxonomy classify all engineering materials into the broad categories of metals, nonmetals and compounds, and combination materials. Metals are further subdivided into ferrous "nonferrous" and combination metals. Nonmetals are classified as crystalline, fibrous, and amorphous.

Combination materials are categorized as composites, foams, microspheres, and laminates. Each of these groups is further subdivided until a relatively homogeneous materials family is identified. At this final level a family code is assigned.

Customizing

The Engineering Materials Taxonomy may be easily modified to fit a unique user's needs. For example, if it were desirable to further subdivide "fiber-reinforced composites," it could easily be done

Base		Example
A.	State	Solid–liquid–gas
B.	Structure	Fibrous–crystalline–amorphous
C.	Origin	Natural–synthetic
D.	Application	Adhesive–paint–fuel–lubricant
E.	Composition	Organic–inorganic
F.	Structure	Metal–nonmetal
G.	Structure	Ferrous–nonferrous
H.	Processing	Cast–wrought
I.	Processing response	Water-hardening–oil-hardening–air-hardening, etc.
J.	Composition	Low alloy–high alloy
K.	Application	Nuclear–semiconducting–precious
L.	Property	Light weight–heavy
M.	Property	Low melting point–high melting point
N.	Operating environment	Low-temperature–high-temperature
O.	Operating environment	Corrosive–noncorrosive

Fig. 31.10 Basis for classifying engineering materials.

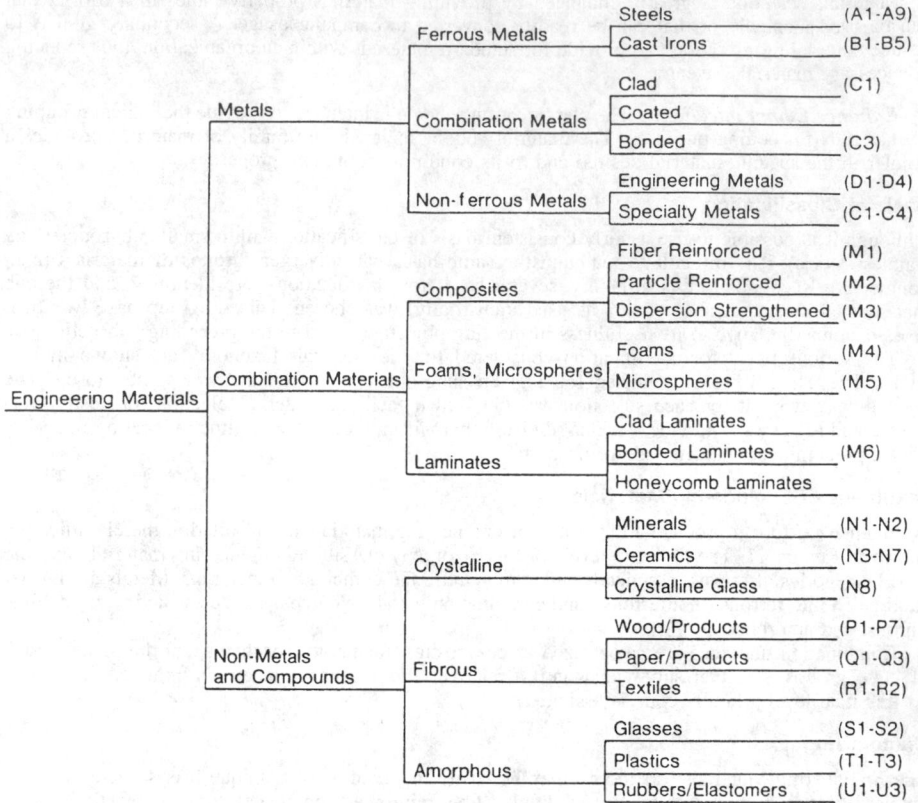

Fig. 31.11 Engineering materials taxonomy—three levels.

on the basis of type of filament used (e.g., boron, graphite, glass) and further by matrix employed (polymer, ceramic, metal). The code "M1," representing fiber-reinforced composites, could have appended to it a dash number uniquely identifying the specific material desired. Many additional material families may also be added if desired.

31.2.3 Material Code

As was mentioned earlier, there are many material classification systems, each of which covers only a limited segment of the spectrum of engineering materials available. The purpose of the Engineering Materials Taxonomy is to overcome this limitation. Furthermore, each of the various materials systems has its own codes. This creates additional problems. To solve this coding compatibility problem, a two-character alphanumeric code is provided as a standard interface code to which any industry or user code may be appended. This provides a very compact standard code so that any user will recognize the basic material family even though perhaps not recognizing a given industry code.

Material Code Format

The format used for the material code is shown in Fig. 31.12. The code consists of four basic fields of information. The first field contains a two-character interface code signifying the material family. The second field is to contain the specific material type based on composition or property. This code may be any five-character alphanumeric code. The third field contains a two-digit code containing the material condition (e.g., hot-worked, as-cast, ¾-hard). The fourth and final field of the code contains a one-digit alphabetic code signifying the material form (e.g., bar, sheet, structural shape).

Material Families

Of the 79 material families identified, 13 are ferrous metals, 30 are nonferrous metals, 6 are combination materials (composites, foams, laminates), and 26 are nonmetals and compounds.

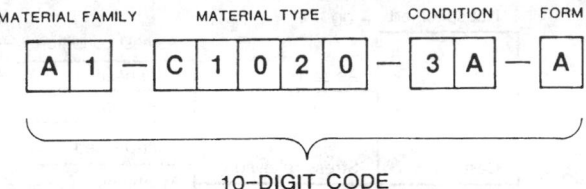

10-DIGIT CODE

Fig. 31.12 Format for engineering materials code.

The five-digit code space reserved for material type is sufficient to accommodate the UNS (Unified Numbering System) recently developed by ASTM, SAE, and others for metals and alloys. It will also accommodate industry or user-developed codes for nonmetals or combination materials. An example of the code (Fig. 31.10) for an open-hearth, low-carbon steel would be "A1-C1020," with the first two digits representing the steel family and the last five digits the specific steel alloy.

Material Condition

The material condition code consists of a two-digit code derived for each material family. The intent of this code is to reflect processes to which the material has been subjected and its resultant structure. Because of the wide variety of conditions that do exist for each family of materials, the creation of a D-tree for each of the 79 families seems to be the best approach. The D-tree can contain processing treatments along with resulting grain size, microstructure, or surface condition if desired. Typical material condition codes for steel family "Al" are given in Fig. 31.13.

Material form code consists of a single alphabetic character to represent this raw material form (e.g., rod, bar, tubing, sheet, structural shape). Typical forms are shown in Fig. 31.14.

31.2.4 Material Properties

Material properties have been divided into three broad classes: (1) mechanical properties, (2) physical properties, and (3) chemical properties. Each of these will be discussed briefly.

Mechanical Properties

The mechanical properties of an engineering material describe its behavior or quality when subjected to externally applied forces. Mechanical properties include strength, hardness, fatigue, elasticity, and plasticity. Figure 31.15 shows representative mechanical properties. Note that each property has been identified with a unique code number to reduce confusion in communicating precisely which property is intended. Confusion often arises because of the multiplicity of testing procedures that have been devised to assess the value of a desired property. For example, there are at least 15 different penetration hardness tests in common usage, each of which yields different numerical results from the others. The code uniquely identifies the property and the testing method used to ascertain it.

Each property of a material is intimately related to its composition, surface condition, internal condition, and material form. These factors are all included in the material code. A modification of any of these factors, either by itself or in combination, can result in quite different mechanical properties.

Thus, each material code combination is treated as a unique material. As an example of this, consider the tensile strength of a heat-treated 6061 aluminum alloy: in the wrought condition, the ultimate tensile strength is 19,000 psi; with the T4-temper, the ultimate tensile strength is 35,000 psi; and in the T913 condition, the ultimate tensile strength is 68,000 psi.

Physical Properties

The physical properties of an engineering material have to do with the intrinsic or structure-insensitive properties. These include melting point, expansion characteristics, dielectric strength, and density. Figure 31.16 shows representative physical properties.

Again, each property has been coded to aid in communication. Magnetic properties and electrical properties are included in this section for the sake of simplicity.

Chemical Properties

The chemical properties of an engineering material deal with its reactance to other materials or substances, including its operating environment. These properties include chemical reactivity, corrosion characteristics, and chemical compatibility. Atomic structure factors, chemical valence, and related factors useful in predicting chemical properties may also be included in the broad category of chemical properties. Figure 31.17 shows representative chemical properties.

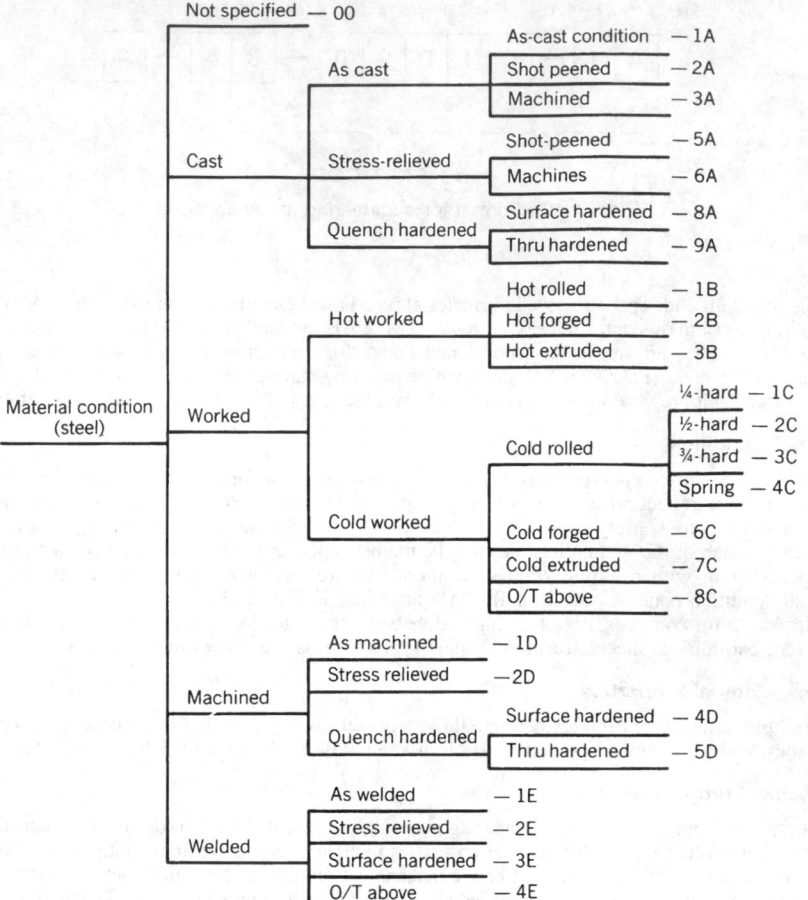

Fig. 31.13   Material condition for steel family "A1."

31.2.5   Material Availability

The availability of an engineering material is a prime concern in materials selection and use. Material availability includes such factors as stock shapes, sizes, and tolerances; material condition and finish; delivery; and price.

Other factors of increasing significance are energy requirements for winning the material from nature and recyclability. Figure 31.18 shows representative factors for assessing material availability.

31.2.6   Material Processability

Relative processability ratings for engineering materials in conjunction with material properties and availability can greatly assist the engineering designer in selecting materials that will meet essential design criteria. All too often, the processability of a selected engineering material is unknown to the designer. As likely as not, the materials may warp during welding or heat treatment and be difficult to machine, which may result in undesirable surface stresses because of tearing or cracking during drawing operations. Many of these problems could be easily avoided if processability ratings of various materials were ascertained, recorded, and used by the designer during the material selection process. Figure 31.19 shows relative processability ratings. These ratings include machinability, weldability, castability, moldability, formability, and heat-treatability. Relative ratings are established through experience for each family. Ratings must not be compared between families. For example, the machinability rating of two steels may be compared, but they should not be evaluated against brass or aluminum.

O—Unspecified

Rotational Solids

A—Rod/wire
B—Tubing/pipe

Flat Solids

C—Bar, flats
D—Hexagon/octagon
E—Sheet/plate

Structural Shapes

F—Angle
G—T section
H—Channel
I—H, I sections
J—Z sections
K—Special sections (extruded, rolled, etc.)

Fabricated Solid Shapes

L—Forging
M—Casting/ingot
N—Weldment
P—Powder metal
Q—Laminate
R—Honeycomb
S—Foam

Special Forms

T—Resin, liquid, granules
U—Fabric, roving, filament
V—Putty, clay
W—Other
Y—Reserved
Z—Reserved

Fig. 31.14 Raw material forms.

31.3 FABRICATION PROCESS TAXONOMY

31.3.1 Introduction

Purpose

The purpose of classifying manufacturing processes is to form families of related processes to aid in process selection, documentation of process capabilities, and information retrieval. A taxonomy or classification of manufacturing processes can aid in process selection by providing a display of potential manufacturing options available to the process planner.

Documentation of process capabilities can be improved by providing files containing the critical attributes and parameters for each classified process. Information retrieval and communication relative to various processes can be enhanced by providing a unique code number for each process. Process information can be indexed, stored, and retrieved by this code.

Classification and coding is an art and, as such, it is difficult to describe the steps involved, and even more difficult to maintain consistency in the results. The anticipated benefits to users of a well-

| Mechanical Properties | D | 1 | - | 0 | 6 | 0 | 6 | 1 | - | 1 | B | - | C |

Material Family/Type: Aluminum 6061-T6

| Prepared by: | Date: | Approved by: | Date: |

Revision No./Date:

Code	Description	Value	Units
11.02	Brinell hardness number	95	HB
12.06	Yield strength, 0.2% offset	40,000	psi
12.11	Ultimate tensile strength	45,000	psi
12.20	Ultimate shear (bearing) strength	30,000	psi
12.30	Impact energy (Charpy V-notch)		ft-lb
12.60	Fatigue (endurance limit)	14,000	psi
12.70	Creep strength		psi
13.01	Modulus of elasticity (tensile)	10.0×10^6	psi
13.02	Modulus of elasticity (compressive)	10.2×10^6	psi
13.20	Poisson's ratio	—	—
14.02	Elongation	15	%
14.10	Reduction of area	—	%
14.30	Strain hardening coefficient	—	%
14.40	Springback	—	%

Fig. 31.15   Representative mechanical properties.

planned process classification outweigh the anticipated difficulties, and thus the following plan is being formulated to aid in uniform and consistent classification and coding of manufacturing processes.

Primary Objectives

There are three primary objectives for classifying and coding manufacturing processes: (1) facilitating process planning, (2) improving process capability assessment, and (3) aiding in information retrieval.

Facilitate Process Selection.   One of the significant problems confronting the new process planner is process selection. The planner must choose, from many alternatives, the basic process, equipment, and tooling required to produce a given product of the desired quality and quantity in the specified time.

Although there are many alternative processes and subprocesses from which to choose, the process planner may be well acquainted with only a small number of them. The planner may thus continue to select these few rather than become acquainted with many of the newer and more competitive processes. The proposed classification will aid in bringing to the attention of the process planner all the processes suitable for modifying the shape of a material or for modifying its properties.

Improve Process Capability Assessment.   One of the serious problems facing manufacturing managers is that they can rarely describe their process capabilities. As a consequence, there is commonly a mismatch between process capability and process needs. This may result in precision parts being produced on unsuitable equipment, with consequent high scrap rates, or parts with no critical tolerances being produced on highly accurate and expensive machines, resulting in high manufacturing costs.

Process capability files may be prepared for each family of processes to aid in balancing capacity with need.

Aid Information Retrieval.   The classification and grouping of manufacturing processes into subgroups having similar attributes will minimize the time required to identify and retrieve similar processes. The classification tree will provide a structure and branches to which important information may be attached or referenced regarding process attributes, methods, equipment, and tooling.

The classification tree provides a logical arrangement for coding existing processes as well as a place for new processes to be added.

Physical Properties	D	1	-	0	6	0	6	1		

Material Family/Type: Aluminum 6061-T6

Prepared by:		Date:	Approved by:		Date:

Revision No./Date:

Code	Description	Value	Units
21.01	Coefficient of linear expansion	13×10^{-6}	in./in./°F
21.05	Thermal conductivity	1070	Btu/in./ft²/°F/hr
21.40	Minimum service temperature	−320	°F
21.50	Maximum service temperature	700	°F
21.66	Melting range	1080–1200	°F
21.80	Recrystallization temperature	650	°F
21.90	Annealing temperature	775	°F, 2–3 hr
21.92	Stress-relieving temperature	450	°F, 1 hr
21.95	Solution heat treatment	970	°F
21.96	Precipitation heat treatment	350	°F, 6–10 hr
22.01	Electrical conductivity (weight)	40	%
22.02	Electrical conductivity (volume)	135	%
22.10	Electrical resistivity (volume)	26	ohms mil, ft
26.01	Specific weight	0.098	lb/in.³
26.03	Specific gravity	270	gm/cm³
26.35	Crystal (lattice) system	f.c.c.	—
26.70	Damping index	0.03	Very low
26.71	Strength-to-weight ratio		
26.72	Basic refining energy	100,000	Btu/lb
26.73	Recycling energy	10,000	Btu/lb

Fig. 31.16 Representative physical properties.

31.3.2 Process Divisions

Manufacturing processes can be broadly grouped into two categories: (1) shaping processes and (2) nonshaping processes. Shaping processes are concerned primarily with modifying the shape of the plan material into the desired geometry of the finished part. Nonshaping processes are primarily concerned with modifying material properties.

Shaping Processes

Processes available for shaping the raw material to produce a desired geometry may be classified into three subdivisions: (1) mass-reducing processes, (2) mass-conserving processes, and (3) mass-increasing or joining processes. These processes may then be further subdivided into mechanical, thermal, and chemical processes.

Mass-reducing processes include cutting, shearing, melting or vaporizing, and dissolving or ionizing processes. Mass-conserving processes include casting, molding, compacting, deposition, and laminating processes. Mass-increasing or, more commonly, joining, processes include pressure and thermal welding, brazing, soldering, and bonding. The joining processes are those that produce a megalithic structure not normally disassembled.

Nonshaping Processes

Nonshaping processes that are available for modifying material properties or appearance may be classified into two broad subdivisions: (1) heat-treating processes and (2) surface-finishing processes.

Heat-treating processes are designed primarily to modify mechanical properties, or the processability ratings, of engineering materials. Heat-treating processes may be subdivided into (1) annealing (softening) processes, (2) hardening processes, and (3) other processes. The "other" category includes sintering, firing/glazing, curing/bonding, and cold treatments. Annealing processes are designed to

Chemical Properties	D	1	-	0	6	0	6	1	

Material Family/Type: Aluminum 6061-T6

Prepared by:	Date:	Approved by:	Date:

Revision No./Date:

Code	Description	Value[a]	Units
32.01	Resistance to high-temperature corrosion	C	
32.02	Resistance to stress corrosion cracking	C	
32.03	Resistance to corrosion pitting	B	
32.04	Resistance to intergranular corrosion	B	
32.10	Resistance to fresh water	A	
32.11	Resistance to salt water	A	
32.15	Resistance to acids	A	
32.20	Resistance to alkalies	C	
32.25	Resistance to petrochemicals	A	
32.30	Resistance to organic solvents	A	
32.35	Resistance to detergents	B	
33.01	Resistance to weathering	A	

[a] *Key:* A = fully resistant; B = slightly attacked; C = unsatisfactory.

Fig. 31.17 Representative chemical properties.

soften the work material, relieve internal stresses, or change the grain size. Hardening treatments, on the other hand, are often designed to increase strength and resistance to surface wear or penetration. Hardening treatments may be applied to the surface of a material or the treatments may be designed to change material properties throughout the section.

Surface-finishing processes are those used to prepare the workpiece surface for subsequent operations, to coat the surface, or to modify the surface. Surface-preparation processes include descaling, deburring, and degreasing. Surface coatings include organic and inorganic; metallic coatings applied by spraying, electrostatic methods, vacuum deposition, and electroplating; and coatings applied through chemical-conversion methods.

Surface-modification processes include burnishing, brushing, peening, and texturing. These processes are most often used for esthetic purposes, although some peening processes are used to create warped surfaces or to modify surface stresses.

31.3.3 Process Taxonomy

There are many methods for classifying production processes. Each may serve unique purposes. The Fabrication Process Taxonomy is the first known comprehensive classification of all processes used for the fabrication of discrete parts for the durable goods manufacturing industries.

Basis of Classification

The basis for process classification may be the source of *energy* (i.e., mechanical, electrical, or chemical); the *temperature* at which the processing is carried out (i.e, hot-working, cold-working); the type of material to be processed (i.e., plastic, steel, wood, zinc, or powdered metal); or another basis of classification.

The main purpose of the hierarchy is to provide functional groupings without drastically upsetting recognized and accepted families of processes within a given industry. For several reasons, it is difficult to select only one basis for classification and apply it to all processes and achieve usable results. Thus, it will be noted that the fabrication process hierarchy has several bases for classification, each depending on the level of classification and on the particular family of processes under consideration.

Classification Rules and Procedures

Rule 1. Processes are classified as either shaping or nonshaping, with appropriate mutually exclusive subdivisions.

Availability	D	1	-	0	6	0	6	1	

Material Family/Type: Aluminum 6061-T6

Prepared by:	Date:	Approved by:	Date:

Revision No./Date:

Surface Condition
  Cold worked
  Hot worked
  Cast
  Clad
  Peened
  Chromate
  Anodized
  Machined

Internal Condition
  Annealed
  Solution treated—naturally aged
  Solution treated—artificially aged
  Stress relieved
  Cold worked

Forms Available
  Sheet
  Plate
  Bar
  Tubing
  Wire
  Rod
  Extrusions
  Ingot

Fig. 31.18 Factors relating to material availability.

Rule 2. Processes are classified as independent of materials and temperature as possible.

Rule 3. Critical attributes of various processes are identified early to aid in forming process families.

Rule 4. Processes are subdivided at each level to show the next options available.

Rule 5. Each process definition is in terms of relevant critical attributes.

Rule 6. Shaping process attributes include

    6.1   Geometric shapes produced

    6.2   Form features or treatments imparted to the workpiece

    6.3   Size, weight, volume, or perimeter of parts

    6.4   Part precision class

    6.5   Production rates

    6.6   Set-up time

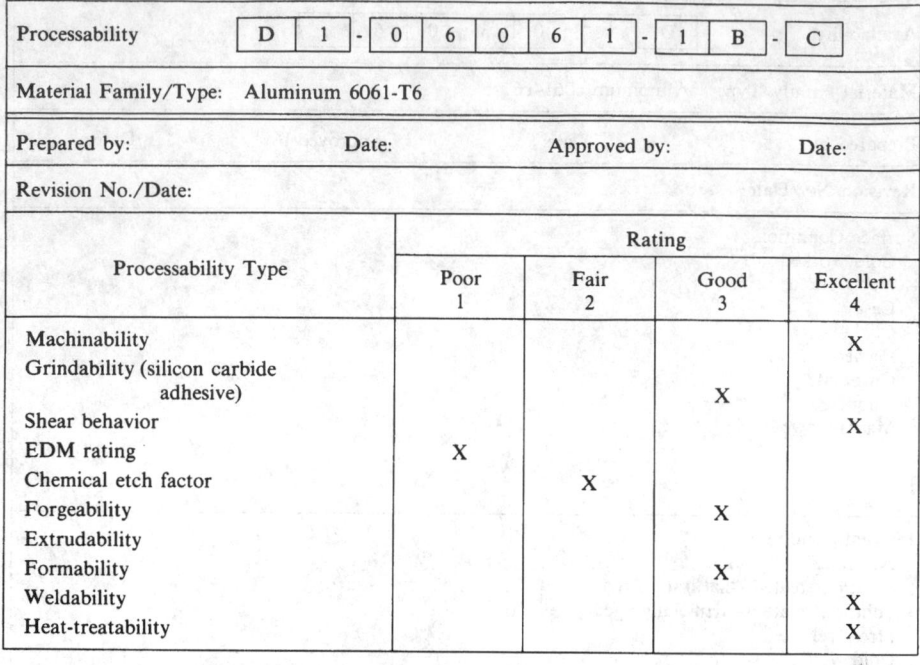

Fig. 31.19   Relative processability ratings.

	D	1	-	0	6	0	6	1	-	1	B	-	C

Processability

Material Family/Type: Aluminum 6061-T6

Prepared by: Date: **Approved by:**   Date:

Revision No./Date:

Processability Type	Rating			
	Poor 1	Fair 2	Good 3	Excellent 4
Machinability				X
Grindability (silicon carbide adhesive)			X	
Shear behavior				X
EDM rating	X			
Chemical etch factor		X		
Forgeability			X	
Extrudability				
Formability			X	
Weldability				X
Heat-treatability				X

    6.7   Tooling costs
    6.8   Relative labor costs
    6.9   Scrap and waste material costs
    6.10 Unit costs versus quantities of 10, 100, 1 K, 10 K, 100 K
Rule 7. All processes are characterized by
    7.1   Prerequisite processes
    7.2   Materials that can be processed, including initial form
    7.3   Basic energy source: mechanical, thermal, or chemical
    7.4   Influence of process on mechanical properties such as strength, hardness, or toughness
    7.5   Influence of process on physical properties such as conductivity, resistance, change in density, or color
    7.6   Influence of process on chemical properties such as corrosion resistance
Rule 8. At the operational level, the process may be fully described by the operation description and sequence, equipment, tooling, processing parameters, operating instructions, and standard time.

The procedure followed in creating the taxonomy was first to identify all the processes that were used in fabrication processes. These processes were then grouped on the basis of relevant attributes. Next, most prominent attributes were selected as the parent node label. Through a process of selection, grouping, and classification, the taxonomy was developed. The taxonomy was evaluated and found to readily accommodate new processes that subsequently were identified. This aided in verifying the generic design of the system. The process taxonomy was further cross-checked with the equipment and tooling taxonomies to see if related categories existed. In several instances, small modifications were required to ensure that the various categories were compatible. Following this method for cross-checking among processes, equipment, and tooling, the final taxonomy was prepared. As the taxonomy was subsequently typed and checked, and large charts were developed for printing, small remaining discrepancies were noted and corrected. Thus, the process taxonomy is presented as the best that is currently available. Occasionally, a process is identified that could be classified in one or more categories. In this case, the practice is to classify it with the preferred group and cross-reference it in the second group.

31.3.4 Process Code

The process taxonomy is used as a generic framework for creating a unique and unambiguous numeric code to aid in communication and information retrieval.

The process code consists of a three-digit numeric code. The first digit indicates the basic process division and the next two digits indicate the specific process group. The basic process divisions are as follows:

000	Material identification and handling
100	Material removal processes
200	Consolidation processes
300	Deformation processes
400	Joining processes
500	Heat-treating processes
600	Surface-finishing processes
700	Inspection
800	Assembly
900	Testing

The basic process code may be extended with the addition of an optional decimal digit similar to the Dewey Decimal System. The process code is organized as shown in Fig. 31.20.

The numeric process code provides a unique, easy-to-use shorthand communication symbol that may be used for manual or computer-assisted information retrieval. Furthermore, the numeric code can be used on routing sheets, in computer databases, for labeling of printed reports for filing and retrieval purposes, and for accessing instructional materials, process algorithms, appropriate mathematical and graphical models, and the like.

31.3.5 Process Capabilities

Fundamental to process planning is an understanding of the capabilities of various fabrication processes. This understanding is normally achieved through study, observation, and industrial experience. Because each planner has different experiences and observes processes through different eyes, there is considerable variability in derived process plans.

Fabrication processes have been grouped into families having certain common attributes. A study of these common attributes will enable the prospective planner to learn quickly the significant characteristics of the process without becoming confused by the large amount of factual data that may be available about the given process.

Also, knowledge about other processes in a given family will help the prospective planner learn about a specific process by inference. For example, if the planner understands that "turning" and "boring" are part of the family of single-point cutting operations and has learned about cutting-speed calculations for turning processes, the planner may correctly infer that cutting speeds for boring operations would be calculated in a similar manner, taking into account the rigidity of each setup. It is important at this point to let the prospective planner know the boundaries or exceptions for such generalizations.

A study of the common attributes and processing clues associated with each of these various processes will aid the planner. For example, an understanding of the attributes of a given process and recognition of process clues such as "feed marks," "ejector-pin marks," or "parting lines" can help the prospective planner to identify quickly how a given part was produced.

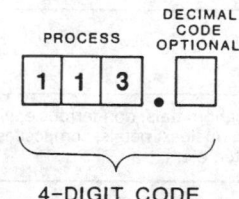

Fig. 31.20 Basic process code.

Figures 31.21 and 31.22 show a process capability sheet that has been designed for capturing information relative to each production process.

31.4 FABRICATION EQUIPMENT CLASSIFICATION

31.4.1 Introduction

Utilization of Capital Resources

One of the primary purposes for equipment classification systems is to better utilize capital resources. The amount of capital equipment and tooling per manufacturing employee has been reported to range from $30,000 to $50,000. An equipment classification system can be a valuable aid in capacity planning, equipment selection, equipment maintenance scheduling, equipment replacement, elimination of unnecessary equipment, tax depreciation, and amortization.

PROCESS CAPABILITY SHEET

Process: Turning/Facing	Code: 101

Prepared by:	Date:	Approved by:	Date:

Revision No. & Date:

Schematic:	Attributes:
	• Single point cutting tool • Chips removed from external surface • Helical or annular (tree-ring) feed marks are present.

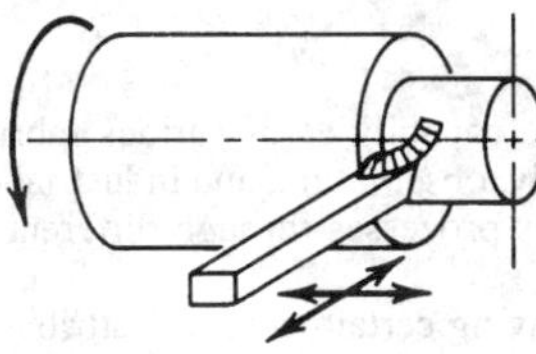

Basic Shapes Produced:
  Surfaces of revolution (cylindrical, tapered, spherical) or flat shoulders or ends.
  May have discontinuities in surfaces (interrupted cut).

Form Features or Treatments:
  Bead, boss, chf'r, groove , lip, radius, thread

Size Range:
  1-6

Precision Class:
  1-4

Raw Material Type: Steel, cast iron, light metals, non-ferrous engineering metals, low-m.p. metals, refractory metals, nuclear metals, composites, refractories, wood, polymers, rubbers and elastometers

Fig. 31.21 Process capability sheet.

Process: Turning/Facing		Code: 101			

Raw Material Condition:
Hot rolled, cold-rolled, forged, cast

Raw Material Form:
rod, tubing, forgings, castings

Production Rate		1 A	10 B	100 C	1,000 D	10,000 E	100,000 F
Tooling Costs	High-3						
	Med-2						
	Low-1						
Set-up Time	High-3						
	Med-2						
	Low-1						
Labor Costs	High-3						
	Med-2						
	Low-1						
Scrap & Waste Material Costs	High-3						
	Med-2						
	Low-1						
Unit Costs	High-3						
	Med-2						
	Low-1						

Prerequisite Processes:
Hot-rolling, cold rolling, forging, casting, p/m compacting

Influence on Mechanical Properties:
Creates very thin layer of stressed work material. Grains may be slightly deformed, and built-up edge may be present on work surface.

Influence on Physical Properties:
N/A

Influence on Chemical Properties:
Highly stressed work surface may promote corrosion.

Fig. 31.22 Process capability sheet.

Equipment Selection

A key factor in equipment selection is a knowledge of the various types of equipment and their capabilities. This knowledge may be readily transmitted through the use of an equipment classification tree showing the various types of equipment and through equipment specification sheets that capture significant information regarding production capabilities.

Equipment selection may be regarded as matching—the matching of production needs with equipment capabilities. Properly defined needs based on current and anticipated requirements, when coupled with an equipment classification system, provide a logical, consistent strategy for equipment selection.

Manufacturing Engineering Services

Some of the manufacturing engineering services that can be greatly benefitted by the availability of an equipment classification system include process planning, tool design, manufacturing development, industrial engineering, and plant maintenance. The equipment classification code can provide an index pointer to performance records, tooling information, equipment specification sheets (mentioned previously), and other types of needed records.

Quality Assurance Activities

Acceptance testing, machine tool capability assessment, and quality control are three important functions that can be enhanced by means of an equipment classification and coding system. As before, the derived code can provide a pointer to testing and acceptance procedures appropriate for the given family of machines.

31.4.2 Standard and Special Equipment

The classification system described below can readily accommodate both standard and special equipment. *Standard fabrication equipment* includes catalog items such as lathes, milling machines, drills, grinders, presses, furnaces, and welders. Furthermore, they can be used for making a variety of products. Although these machines often have many options and accessories, they are still classified as standard machines. *Special fabrication machines,* on the other hand, are custom designed for a special installation or application. These machines are usually justified for high-volume production or special products that are difficult or costly to produce on standard equipment. Examples of special machines include transfer machines, special multistation machines, and the like.

31.4.3 Equipment Classification

The relationship among the fabrication process, equipment, and tooling is shown graphically in Fig. 31.23 The term *process* is basically a concept and requires equipment and tooling for its physical embodiment. For example, the grinding process cannot take place without equipment and tooling. In some instances, the process can be implemented without equipment, as in "hand-deburring." The hierarchal relationships shown in Fig. 31.23 between the process, equipment, and tooling provide a natural linkage for generative process planning. Once the required processes have been identified for reproducing a given geometric shape and its associated form features and special treatments, the selection of equipment and tooling is quite straightforward.

Rationale

Two major functions of an equipment classification system are for process planning and tool design. These functions are performed each time a new product or piecepart is manufactured. Consequently, the relationships between processes, equipment, and tooling have been selected as primary in development of the equipment classification system.

The equipment taxonomy parallels the process taxonomy as far as possible. Primary levels of classification include those processes whose intent is to change the form of the material—for example, shaping processes and those processes whose intent is to modify or enhance the material properties. These nonshaping processes include heat treatments and coating processes, along with attendant cleaning and deburring.

As each branch of a process tree is traversed, it soon becomes apparent that there is a point at which an equipment branch must be grafted in. It is at this juncture that the basis for equipment classification must be carefully considered. There are a number of possible bases for classification of equipment, including

1. Form change (shaping, nonshaping)
2. Mass change (reduction, consolidation, joining)
3. Basic process (machine, cast, forge)
4. Basic subprocess (deep hole drill, precision drill)
5. Machine type (gang drill, radial drill)
6. Energy source (chemical, electrical, mechanical)

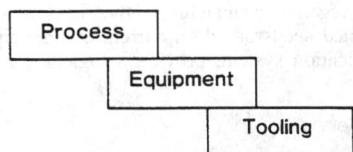

Fig. 31.23 Relationships between process, equipment, and tooling.

7. Energy transfer mechanism (mechanical, hydraulic, pneumatic)
8. Raw material form (sheet metal, forging, casting)
9. Shape produced (gear shaper, crankshaft lathe)
10. Speed of operation (high speed, low speed)
11. Machine orientation (vertical, horizontal)
12. Machine structure (open-side, two-column)
13. General purpose/special purpose (universal mill, spar mill)
14. Kinematics/motions (moving head, moving bed)
15. Control type (automatic, manual NC)
16. Feature machined (surface, internal)
17. Operating temperature (cold rolling, hot rolling)
18. Material composition (plastic molding, aluminum die casting)
19. Machine size (8-in. chucker, 12-in. chucker)
20. Machine power (600 ton, 100 ton)
21. Manufacturer (Landis, Le Blond, Gisholt)

In reviewing these bases of classification, it is apparent that some describe fundamental characteristics for dividing the equipment population into families, and others are simply attributes of a given family. For example, the features of "shaping," "consolidation," and "die casting" are useful for subdivision of the population into families (E-tree), whereas attributes such as "automatic," "cold-chamber," "aluminum," "100-ton," and "Reed-Prentice" are useful for characterizing equipment within a given family (N-tree). "Automatic" is an attribute of many machines; likewise, "100-ton" could apply to general-purpose presses, forging presses, powder-metal compacting presses, and so on. Similarly, the label "Reed-Prentice" could be applied equally well to lathes, die casting machines, or injection molders. In other words, these terms are not very useful for development of a taxonomy but are useful for characterizing a family.

Equipment Taxonomy

The first major division, paralleling the processes, is shaping or nonshaping. The second level for shaping is (1) mass-reducing, (2) mass-conserving, and (3) mass-increasing. The intent of this subdivision, as was mentioned earlier, is to classify equipment whose intent is to change the form or shape of the workpiece. The second level for nonshaping equipment includes (1) heat treating and (2) surface finishing. The intent of this subdivision is to classify equipment designed to modify or enhance material properties or appearance. The existing taxonomy identifies 257 unique families of fabrication equipment.

Customizing

As with other taxonomies described herein, the equipment taxonomy is designed to readily accommodate new classes of machines. This may be accomplished by traversing the tree until a node point is reached where the new class or equipment must appropriately fit. The new equipment with its various subclasses may be grafted in at this point and an appropriate code number assigned. It should be noted that code numbers have been intentionally reserved for this purpose.

31.4.4 Equipment Code

The code number for fabrication equipment consists of a nine-character code. The first three digits identify the basic process, leaving the remaining six characters to identify uniquely any given piece of equipment. As can be seen in Fig. 31.24, the code consists of four fields. Each of these fields will be briefly described in the following paragraphs.

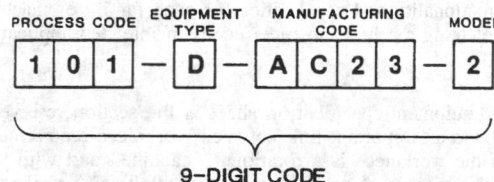

PROCESS CODE | EQUIPMENT TYPE | MANUFACTURING CODE | MODEL

1 0 1 — D — A C 2 3 — 2

9-DIGIT CODE

Fig. 31.24 Fabrication equipment code.

Process Code

The process code is a three-digit code that refers to one of the 222 fabrication processes currently classified. For example, code 111 = drilling and 121 = grinding. Appended to this code is a code for the specific type of equipment required to implement the given process.

Equipment Family Code

The code for equipment type consists of a one-character alphabetic code. For example, this provides for up to 26 types of turning machines, 101-A through 101-Z, 26 types of drilling machines, and so on. Immediately following the code for equipment type is a code that uniquely identifies the manufacturer.

Manufacturer Code

The manufacturer code consists of a four-digit alphanumeric code. The first character in the code is an alphabetic character (A–Z) representing the first letter in the name of the manufacturing company. The next three digits are used to identify uniquely a given manufacturing company. In-house-developed equipment would receive a code for your own company.

Model Number

The final character in the code is used to identify a particular manufacturer model number. Thus the nine-digit code is designed to provide a shorthand designation as to the basic process, type of equipment, manufacturer, and model number. The code can serve as a pointer to more detailed information, as might be contained on specification sheets, installation instructions, maintenance procedures, and so on.

31.4.5 Equipment Specification Sheets

In the preceding subsections the rationale for equipment classification was discussed, along with the type of information useful in characterizing a given piece of equipment. It has been found that this characterization information is best captured by means of a series of equipment specification sheets.

The philosophy has been that each family must have a tailored list of features or specifications to characterize it adequately. For example, the terms *swing, center-distance, rpm,* and *feed per revolution* are appropriate for a lathe family but not for forming presses, welding machines, electrical discharge machines, or vibratory deburring machines. Both common attributes and selected ones are described in the following paragraphs and shown in Figs. 31.25 and 31.26.

Equipment Family/Code

The equipment family label consists of the generic family name, subgroup name (e.g., "drill," "radial"), and the nine-digit equipment code.

Equipment Identification

Equipment name, make, model, serial number, and location provide a unique identification description for a given piece of equipment. The equipment name may be the familiar name given to the piece of equipment and may differ from the generic equipment family name.

Acquisition Data

The acquisition data include capital cost, date acquired, estimated life, and year of manufacture. This information can be used for amortization and depreciation purposes.

Facilities/Utilities

Facilities and utilities required for installing and operating the equipment include power (voltage, current, phases, frequency), and other connections, floor space, height, and weight.

Specifications

The specifications for functionality and capabilities for each family of machines must be carefully defined to be useful in selecting machines to meet needs of intended applications.

Operation Codes

A special feature of the equipment specification sheet is the section reserved for operation codes. The operation codes provide an important link between workpiece requirements and equipment capability. For example, if the workpiece is a rotational, machined part with threads, grooves, and a milled slot, and if a lathe is capable of operations for threading and grooving but not a milling slot, then it follows that either two machines are required to produce the part or a milling attachment must

Equipment Family: Mill, NC, vertical	Code: 113-K

Identification:

Name: Cintimatic, Single Spindle	Serial No.: 364
Make/Model: Cincinnati Milling Machine Co.	Location: 115 SNLB

Acquisition:

Capital Cost: $25,000	Estimated Life: 15 yr
Date Acquired: 15 Aug. 1963	Year of Manufacture: 1963

Maintenance:

Condition: U2	Reevaluation Date: 15 January 1981
Date: 15 Aug. 1963	

Facilities:

Voltage: 230 volts, 3 ph, 60 Hz	Floor Space: 63 in. × 74 in.
Current: 3 hp	Height: 101 in.
Other Connections: Air, 40 psi	Weight:

Specifications:

Working surface	22 in. × 36 in.
Throat	16¼ in.
Table top to spindle	14–24 in. (8 in. travel)
Weight capacity, max	1000 lb
Spindle:	
Axes	One axis
Range	85–3800 rpm
Rate	1–40 ipm
Table:	
Axes	Two axes
Range	15 in. × 25 in.
Rate	Feed—0–40 ipm, rapid travel 200 ipm
Motor hp:	
Drive motor	3.0 hp
Feed	Hydraulic servo motors
Coolant	Air mist spray
Spindle taper	#40 NMTB
T-slots	3 in X axis, $^{11}/_{16}$ in. wide
Accuracy	±0.001 in. in 24 in.
Control type	Accramatic Series 200 control

Fig. 31.25 Equipment specifications.

be installed on the lathe. A significant benefit of the operation code is that it can aid process planners in selecting the minimum number of machines required to produce a given workpiece. This fact must, of course, be balanced with production requirements and production rates. The main objective is to reduce transportation and waiting time and minimize cost. (See Fig. 31.26.)

Photograph or Sketch

A photograph or line drawing provides considerable data to aid in plant layout, tool design, process planning, and other production planning functions. Line drawings often provide information regarding T-slot size, spindle arbor size, limits of machine motion, and other information useful in interfacing the machine with tooling, fixtures, and the workpiece.

Equipment Family: Mill, N/C, vertical	Code: 113-K

Operation Codes:
  102 (boring), 104 (grooving), 111 (drilling), 112 (reaming), 113 (milling), 115 (tapping)

Photo or Sketch

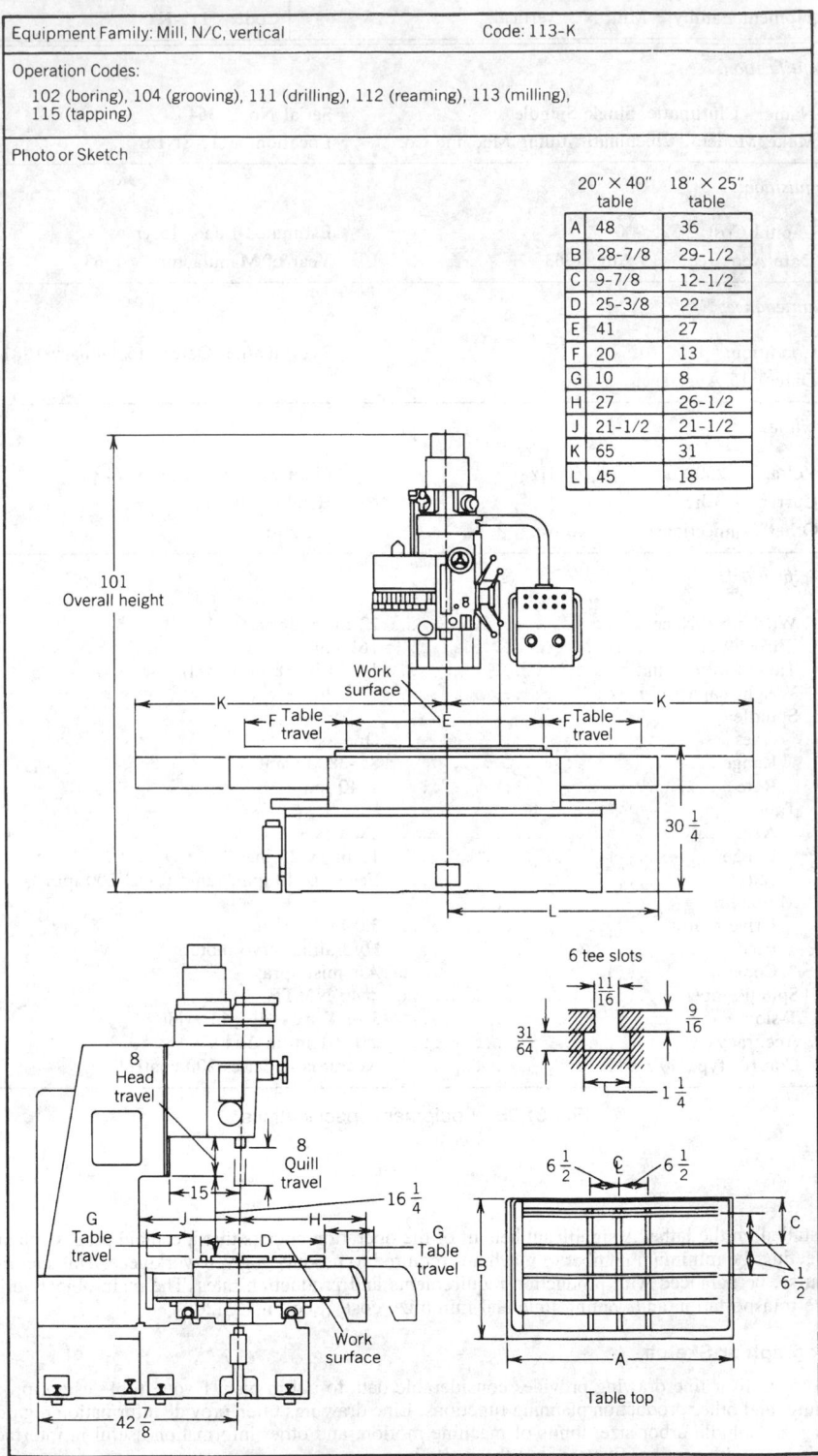

	20" × 40" table	18" × 25" table
A	48	36
B	28-7/8	29-1/2
C	9-7/8	12-1/2
D	25-3/8	22
E	41	27
F	20	13
G	10	8
H	27	26-1/2
J	21-1/2	21-1/2
K	65	31
L	45	18

Fig. 31.26 Operation codes and machine illustration.

31.5 FABRICATION TOOL CLASSIFICATION AND CODING

31.5.1 Introduction

Because standard and special tooling represent a sizable investment, it is prudent to minimize redundant tooling, to evaluate performance of perishable tooling, and to provide good storage and retrieval practices to avoid excessive tool-float and loss of valuable tooling.

The use of a standard tool-classification system could provide many benefits for both the supplier and the user. The problem is to derive a comprehensive tool-classification system that is suitable for the extremely wide variety of tools available to industry and that is agreeable to all suppliers and users. Although no general system exists, most companies have devised their own proprietary tool-classification schemes. This has resulted in much duplicate effort. Because of the difficulty of developing general systems that are expandable to accommodate new tooling categories, many of these existing schemes for tool classification are found to be inadequate.

This section describes a new classification and coding system for fabrication tooling. Assembly, inspection, and testing tools are not included. This new system for classifying fabrication tools is a derivative of the work on classifying fabrication processes and fabrication equipment. Furthermore, special tooling categories are directly related through a unique coding system to the basic shape of the workpieces they are used to fabricate.

Investment

The investment a manufacturing company must make for standard and special tooling is usually substantial. Various manufacturing companies may carry in stock from 5,000 to 10,000 different tools and may purchase several thousand special tools, as required. As a rule of thumb, the investment in standard tooling for a new machine tool is often 20–30% of the basic cost of the machine. Special tooling costs may approach or even exceed the cost of certain machines. For instance, complex die-casting molds costing from $50,000 to $250,000 are quite commonplace.

The use of a tool classification system can aid a manufacturing enterprise by helping to get actual cost data for various tooling categories and thus begin to monitor and control tooling expenditures. The availability of good tooling is essential for economical and productive manufacturing. The intent of monitoring tooling costs should be to ensure that funds are available for such needed tooling and that these funds are wisely used. The intent should not be the miserly allocation of tooling money.

Tool Control

Tool control is a serious challenge in almost all manufacturing enterprises. Six important aspects that must be addressed in any good tool control system include*

    Tool procurement
    Tool storage
    Tool identification and marking
    Tool dispensing
    Tool performance measurement
    Tool maintenance

The availability of a standard, comprehensive tool classification and coding system can greatly aid each of these elements of tool control. For example, tool-procurement data may be easily cross-indexed with a standard tool number, thus reducing problems in communication between the user and the purchasing department.

Tool storage and retrieval may be enhanced by means of standard meaningful codes to identify tools placed in a given bin or at a given location. This problem is especially acute with molds, patterns, fixtures, and other special tooling.

Meaningful tool identification markings aid in preventing loss or misplacement of tools. Misplaced tools can quickly be identified and returned to their proper storage locations. Standard tool codes may also be incorporated into bar codes or other machine-readable coding systems if desirable.

The development of an illustrated tooling manual can be a great asset to both the user and to the tool crib personnel in identifying and dispensing tools. The use of a cross-referenced standard tool code can provide an ideal index to such a manual.

Tool-performance measures require the use of some sort of coding system for each type of tool to be evaluated. Comparison of tools within a given tool family may be facilitated by means of

*"Small Tools Planning and Control," in *Tool Engineers Handbook,* McGraw-Hill, New York, 1959, Section 3.

expanded codes describing the specific application and the various types of failures. Such extended codes may be easily tied to the standard tool family code.

Tool-maintenance and repair costs can be best summarized when they are referenced to a standard tool family code. Maintenance and acquisition costs could be easily reduced to obtain realistic life-cycle costs for tools of a given family or type.

In summary, tool control in general may be enhanced with a comprehensive, meaningful tool classification and coding system.

31.5.2 Standard and Special Tooling

Although tooling may be classified in many ways, such as "durable or perishable tooling," "fabrication or assembly tooling," and "company-owned or customer-owned tooling," the fabrication tooling system described below basically classifies tooling as "standard tooling" or "special tooling."

Definition of Terms

STANDARD TOOLING. Standard tooling is defined as that which is basically off the shelf and may be used by different users or a variety of products. Standard tooling is usually produced in quantity, and the cost is relatively low.

SPECIAL TOOLING. Special tooling is that which is designed and built for a specific application, such as a specific product or family of products. Delivery on such tooling may be several weeks, and tooling costs are relatively high.

Examples

Examples of standard tooling are shown in Fig. 31.27. Standard tooling usually includes cutting tools, die components, nozzles, certain types of electrodes, rollers, brushes, tool holders, laps, chucks, mandrels, collets, centers, adapters, arbors, vises, step-blocks, parallels, angle-plates, and the like.

Examples of special tooling are shown in Fig. 31.28. This usually includes dies, molds, patterns, jigs, fixtures, cams, templates, N/C programs, and the like. Some standard tooling may be modified to perform a special function. When this modification is performed in accordance with a specified design, then the tooling is classified as special tooling.

31.5.3 Tooling Taxonomy

The tooling taxonomy is based on the same general classification system used for fabrication processes and for fabrication equipment. The first-level divisions are "shaping" and "nonshaping." The second-level divisions for shaping are "mass reduction," "mass conserving," and "mass increasing" (joining, laminating, etc.). Second-level divisions for nonshaping are "heat treatment'" and "surface finishing." Third-level subdivisions are more variable but include "mechanical," "chemical," and "thermal," among other criteria for subdivisions.

Rationale

The basic philosophy has been to create a tooling classification system that is related to fabrication processes, to fabrication equipment, and to fabrication products insofar as possible. The statement "Without the process, there is no product" has aided in clarifying the importance of a process classification to all phases of manufacturing. It was recognized early that "process" is really a concept and that only through the application of "equipment and tooling" could a process ever be implemented. Thus, this process taxonomy was used as the basis of both equipment classification and this tooling classification. Most tooling is used in conjunction with given families of equipment and in that way is related to the equipment taxonomy. Special tooling is also related to the workpiece geometry through a special coding system that will be explained later. Standard tooling may be applied to a number of product families and may be used on a variety of different machine tools.

31.5.4 Tool Coding

The tool code is a shorthand notation used for identification and communication purposes. It has been designed to provide the maximum amount of information in a short, flexible code. Complete tool information may be held in a computer database or charts and tables. The code provides a pointer to this information.

Tool Code Format

The format used for the tool code is shown in Fig. 31.29. The code consists of three basic fields of information. The first field contains a three-digit process code that identifies the process for which the tooling is to be used. The second field consists of a one-digit code that indicates the tool type. Tool types are explained in the next subsection. The last field consists of either a three-digit numeral code for standard tooling or a three-digit alphanumeric code for special tooling. Standard tool codes have been designed to accommodate further subdivision of tool families if so desired. For example, the tool code for single-point turning inserts is 101-1-020. The last three digits could be amplified

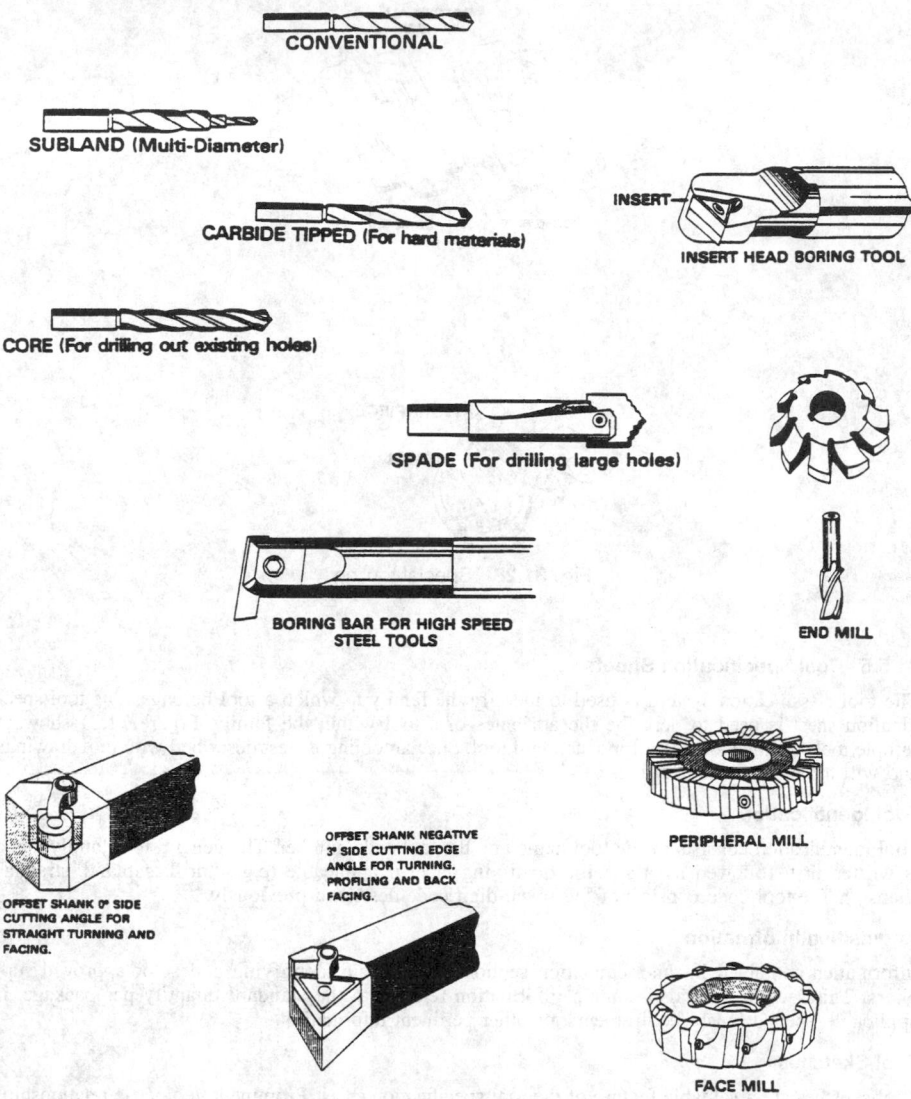

Fig. 31.27 Standard tooling.

for given insert geometry (e.g., triangular, −021; square, −022; round, −023). A dash number may further be appended to these codes to uniquely identify a given tool, as shown on the tool specification sheets.

Special tool codes are identified in the charts by a box containing three small squares. It is intended that the first three digits of this part family code will be inserted in this box, thus indicating the basic shape family for which the tool is designed. This way it will be possible to identify tool families and benefit from the application of group technology principles.

Tool Types

A single-digit alphanumeric code is used to represent the tool type. A code type of −1, for example, indicates that the tool actually contacts the workpiece, while a −2 indicates that the tool is used indirectly in shaping the workpiece. A foundry patten is an example of this: the pattern creates this mold cavity into which molten metal is introduced. The various tool type codes are shown in Fig. 31.30.

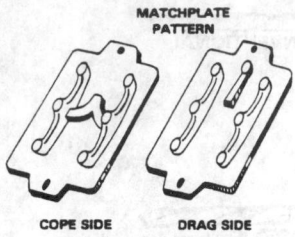

MATCHPLATE
PATTERN

COPE SIDE   DRAG SIDE

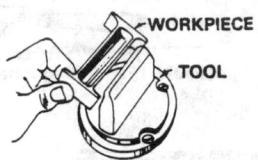

WORKPIECE

TOOL

Fig. 31.28   Special tooling.

31.5.5   Tool Specification Sheets

The tool classification system is used to identify the family to which a tool belongs. The tool specification sheet is used to describe the attributes of a tool within the family. Figure 31.31 shows a sample tool specification sheet for a standard tool. Special tooling is best described with tool drawings and will not be discussed further.

Tool Identification

Tool identification consists of the tool name and the tool code number. The general tool family name is written first, followed by a specific qualifying label if applicable (e.g., "drill, subland, straight-shank"). The tool code consists of the seven-digit code described previously.

Acquisition Information

Information contained in this acquisition section may contain identifying codes for approved suppliers. This section may also contain information relating to the standard quantity per package, if applicable, special finish requirements, or other pertinent information.

Tool Sketches

Tool sketches are a valuable feature of the tool specification sheet. Prominent geometric relationships and parameters are shown in the sketch. Information relative to interfacing this tool to other devices or adapters should also be shown. This may include type and size or shape, key slot size, and tool capacity.

Tool Parameters

Tool parameters are included on the tool specification sheet to aid selection of the most appropriate tools. Because it is expensive to stock all possible tools, the usual practice is to identify preferred

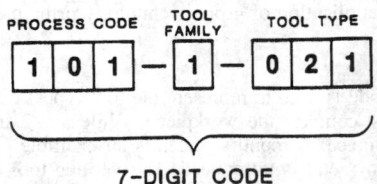

PROCESS CODE   TOOL   TOOL TYPE
                  FAMILY

1 0 1 — 1 — 0 2 1

7-DIGIT CODE

Fig. 31.29   Fabrication tool code format.

TOOL FAMILY	
-1-	TOOL/DIE (CONTACTS WORKPIECE)
-2-	MOLDS, PATTERNS, NEGATIVES
-3-	TIPS, NOZZLES
-4-	ELECTRODES
-5-	OTHER TOOLS
-6-	RESERVED
-7-	TOOL HOLDERS
-8-	WORK HOLDERS
-9-	RESERVED
-A-	N/C PROGRAMS
-B-	CAMS
-C-	TRACING TEMPLATES (2-D)
-D-	TRACING PATTERNS (3-D)

Fig. 31.30 Tool types.

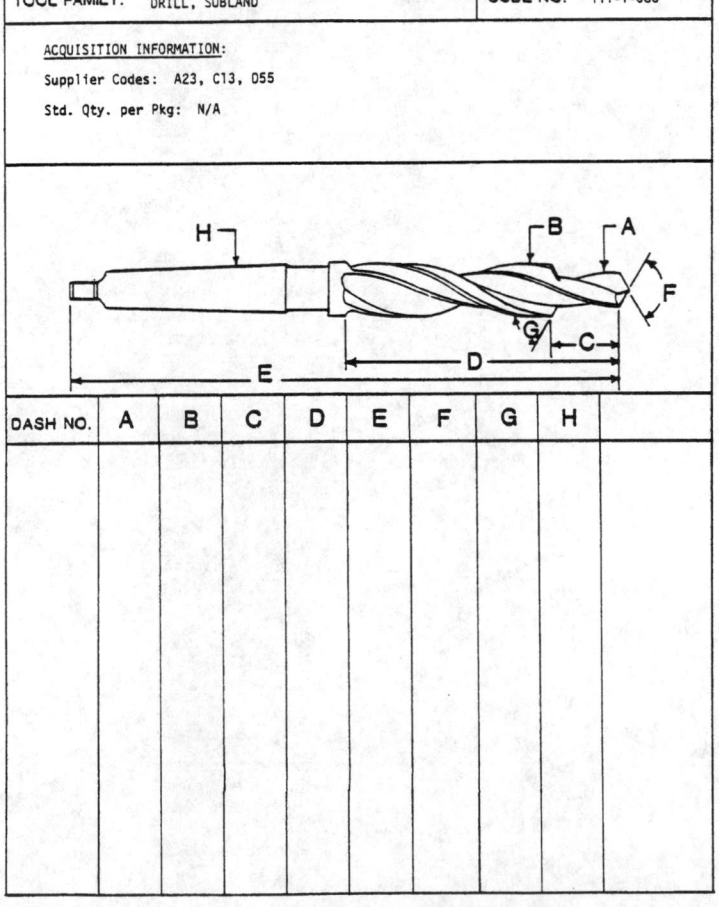

Fig. 31.31 Sample tool specification sheet.

tools and to store these. Preferred tools may be so indicated by a special symbol such as an asterisk (*) in the dash-number column.

Tool parameters must be selected that are appropriate for each tool family and that match product requirements. Tool parameters must also be identified that will aid in interfacing tools with fabrication equipment, as was explained earlier. Typical tool parameters for a subland drill are shown in Fig. 31.31.

CHAPTER 32

PRODUCTION PLANNING

Dennis B. Webster
Thomas G. Ray
Department of Industrial & Manufacturing Systems Engineering
Louisiana State University
Baton Rouge, Louisiana

32.1	**INTRODUCTION**	**987**		32.4.5	Aggregate Planning Dilemma	1005
32.2	**FORECASTING**	**988**	**32.5**	**MATERIALS REQUIREMENTS**		
	32.2.1 General Concepts	988		**PLANNING**		**1006**
	32.2.2 Qualitative Forecasting	988		32.5.1 Procedures and Required		
	32.2.3 Quantitative Forecasting	988		Inputs		1006
	32.2.4 Forecasting Error Analysis	993		32.5.2 Calculations		1010
	32.2.5 Conclusions on Forecasting	994		32.5.3 Conclusions on MRP		1010
				32.5.4 Lot-Sizing Techniques		1010
32.3	**INVENTORY MODELS**	**994**				
	32.3.1 General Discussion	994	**32.6**	**JOB SEQUENCING AND**		
	32.3.2 Types of Inventory Models	995		**SCHEDULING**		**1014**
	32.3.3 The Modeling Approach	996		32.6.1 Structure of the General		
				Sequencing Problem		1014
32.4	**AGGREGATE**			32.6.2 Single-Machine Problem		1015
	PLANNING—MASTER			32.6.3 Flow Shops		1017
	SCHEDULING	**1003**		32.6.4 Job Shops		1018
	32.4.1 Alternative Strategies to			32.6.5 Heuristics/Priority		
	Meet Demand Fluctuations	1003		Dispatching Rules		1020
	32.4.2 Aggregate Planning Costs	1004		32.6.6 Assembly Line Balancing		1023
	32.4.3 Approaches to Aggregate					
	Planning	1004	**32.7**	**OTHER RELATED TOPICS**		**1029**
	32.4.4 Levels of Aggregation and			32.7.1 Japanese Manufacturing		
	Disaggregation	1005		Philosophy		1029
				32.7.2 Time-Based Competition		1030

32.1 INTRODUCTION

Changes that were unforeseen prior to the 1970s are now sweeping the field of manufacturing. Competition from outside the United States is driving the forces of change. There is a relentless push for improvement in total quality, which includes the quality of service to the customer. Service to the customer is related to two concepts: the delivered product quality and the timeliness of customer service.

The topics discussed in this section relate primarily to the second concept of customer service, timeliness. These topics relate either directly or indirectly to accomplishing the job in a timely manner. Forecasting, for example, provides the manufacturer with a basis for anticipating consumer demand so as to have adequate product on hand when it is demanded. Of course, the preferred approach would be to wait for a customer order and then produce and ship immediately when the order arrives. This approach is, for practical purposes, impossible for products with any significant lead time. What the manufacturer must do is to perform as well or better than his competition for the business area.

Job sequencing is an approach to reduce the completion times of the jobs to be performed. Materials requirements planning (MRP) is a technique for assuring that adequate inventory is available to complete the work required on products needed to meet a customer schedule. Inventory models are used in an effort to provide components for a manufacturing process in a timely manner at minimum cost when the demand for the item is constant.

Mechanical Engineers' Handbook, 2nd ed., Edited by Myer Kutz.
ISBN 0-471-13007-9   © 1998 John Wiley & Sons, Inc.

In a similar manner, each of the topics in this section relates to the subject of meeting customer demand, on time and at the lowest cost possible.

32.2 FORECASTING

32.2.1 General Concepts

The function of production planning and control is based upon establishing a plan, revising it as required, and adhering to it to accomplish desired objectives. Plans are based upon a forecast of future demand for the related products or services. Good forecasts are a requirement for a plan to be valid and functionally useful. Managers, when faced with a forecast, plan what actions must be taken to meet the requirements of the forecast. These actions prepare the organization to cope with the anticipated future state of nature that is predicated upon the forecast.

Forecasting methods are traditionally grouped into one of three categories: *qualitative techniques, time-series analysis,* or *causal methods.* Qualitative techniques are normally based on opinions or surveys. Time-series analysis is based on historical data and the study of its trends, cycles, and seasons. Causal methods try to find relationships between independent and dependent variables, determining which variables are predictive of the dependent variable of concern. The method selected for forecasting must relate to the type of information available for analysis.

Definitions

DESEASONALIZATION. The removal of seasonal effects from the data for the purpose of further study of the residual data.

ERROR ANALYSIS. The evaluation of errors in the historic forecasts, done as a part of forecasting model evaluation.

EXPONENTIAL SMOOTHING. An iterative procedure for the fitting of polynomials to data for use in forecasting.

FORECAST. Estimation of a future outcome.

HORIZON. A future time period or periods for which a forecast is required.

INDEX NUMBER. A statistical measure used to compare an outcome which is measured by a cardinal number with the same outcome in another period of time, geographic area, profession, etc.

MOVING AVERAGE. A forecasting method in which the forecast is an average of the data for the most recent n periods.

QUALITATIVE FORECAST. A forecast made without using a quantitative model.

QUANTITATIVE FORECAST. A forecast prepared by the use of a mathematical model.

REGRESSION ANALYSIS. A method of fitting a mathematical model to data by minimizing the sums of the squares of the data from a theoretical line.

SEASONAL DATA. Data that cycle over a known seasonal period, such as a year.

SMOOTHING. A process for eliminating unwanted fluctuations in data; normally accomplished by calculating a moving average or a weighted moving average.

TIME-SERIES ANALYSIS. A procedure for determining a mathematical model *for* data correlated with time.

TIME-SERIES FORECAST. Forecast prepared with a mathematical model *from* data correlated with time.

TREND. Underlying patterns of movement of historic data that become the basis for prediction of future forecasts.

32.2.2 Qualitative Forecasting

These forecasts are normally used for purposes other than production planning. Their validity is more in the area of policy-making or in dealing with generalities to be made from qualitative data. Among these techniques are the Delphi method, market research, consensus methods, and other techniques based upon opinion or historic relationships other than quantitative data.

The Delphi method is one of a number of nominal group techniques. It involves prediction with feedback to the group that gives the predictor's reasoning. Upon each prediction, the group is again polled to see if a consensus has been reached. If no common ground for agreement has occurred, the process continues moving from member to member until agreement is reached.

Surveys may be conducted of relevant groups and their results analyzed to develop the basis for a forecast. One group appropriate for analysis is customers. If a company has relatively few customers, this select number can be an effective basis for forecasting. Customers are surveyed and their responses combined to form a forecast.

Many other techniques are available for nonquantitative forecasting. An appropriate area to search if these methods seem relevant to a subjective problem at hand is the area of *nominal group techniques.*

32.2.3 Quantitative Forecasting

Quantitative forecasting involves working with numerical data to prepare a forecast. This area is further divided into two subgroups of techniques, according to the data type involved. If historic data are available and it is believed that the dependent variable to be forecast relates only to time, time-series analysis is used. If the data available suggest relationships of the dependent variable to be forecast to one or more independent variables, then the techniques used fall into the category of causal analysis. The most commonly used method in this group is regression analysis.

Methods of Analysis of Time Series

The following material will discuss in general several methods for analysis of time series. These methods provide ways of removing the various components of the series, isolating them, and providing information for their consideration should it be desired to reconstruct the time series from its components.

The movements of a time series are classified into four types: long-term or *trend* movements, *cyclical* movements, *seasonal* movements, and *irregular* movements. Each of these components can be isolated or analyzed separately. Various methods exist for the analysis of the time series. These methods decompose the time series into its components by assuming that the components are either multiplicative or additive. Assuming that the components are multiplicative, the following relationship holds:

$$Y = T \times C \times S \times I$$

Where Y is the outcome of the time series, T is the trend value of the time series, and C, S, and I are indices respectively for cyclical, seasonal, and irregular variations.

To process data for this type of analysis, it is best first to plot the raw data in order to observe their form. If the data are yearly, they need no deseasonalization. If the data are monthly or quarterly, they can be converted into yearly data by summing the data points that would add to a year before plotting. (Seasonal index numbers can be calculated to seasonalize the data later if required.) By plotting yearly data, the period of apparent data cycles can be determined or approximated. A centered moving average of appropriate order can be used to remove the cyclical effect in the data. Further, cyclical indices can be calculated when the order of the cycle has been determined. At this point, the data contain only the trend and irregular components of variation. Regression analysis can be used to estimate the trend component of the data, leaving only the irregular, which is essentially forecasting error.

Index Numbers. Index numbers are calculated by grouping data of the same season together, calculating the average over the season for which the index is to be prepared, and then calculating the overall average of the data over each of the seasons. Once the seasonal and overall averages are obtained, the seasonal index is determined by dividing the seasonal average by the overall average.

Example Problem 32.1

A business has been operational for 24 months. The sales data in thousands of dollars for each of the monthly periods are as shown in Table 32.1.

Table 32.1

	Year 1	Year 2
Jan.	20	24
Feb.	23	27
Mar.	28	30
Apr.	32	35
May	35	36
Jun.	26	28
Jul.	25	27
Aug.	23	23
Sep.	19	17
Oct.	21	22
Nov.	18	19
Dec.	12	14

The overall average is 584 divided by 24, or 24.333. The index for January would be

$$I_{Jan} = \frac{(20 + 24)}{24.333}$$
$$= .904$$

The index for March would be

$$I_{Mar} = \frac{(28 + 30)}{24.333}$$
$$= 1.192$$

To use the index, a trend value for the year's sales would be calculated, the average monthly sales would be obtained, and then this figure would be multiplied by the index for the appropriate month to give the month's forecast.

It should be noted that a season can be defined as any period for which the data is available for appropriate analysis. If there are seasons within a month, i.e., four weeks in which the sales vary considerably according to a pattern, a forecast could be indexed within the monthly pattern also. This would be a second indexing within the overall forecast. Further, seasons could be chosen as quarters rather than months or weeks. This choice of the period for the analysis is dependent upon the requirements for the forecast.

Data given on a seasonal basis can be deseasonalized by dividing them by the appropriate seasonal index. Once this has been done, they are labeled *deseasonalized data.* They still contain the trend, cyclical, and irregular components after this adjustment.

Moving Average. A moving average can normally be used to remove the seasonal or cyclical components of variation. This removal is dependent upon the choice of a moving average that contains sufficient data points to bridge the season or cycle. For example, a seven-period-centered moving average should be sufficient to remove seasonal variation from monthly data. A disadvantage to the use of moving averages is the loss of data points due to the inclusion of multiple points into the calculation of a single point. For the monthly data related to the calculation of index numbers given in the previous section, only the months of April of Year 1 through September of Year 2 would be available for analysis when a seven-month-centered moving average is used. Six data points are not available for calculation due to the requirements of the method.

Example Problem 32.2

See Table 32.2. Note that in this case the five-year moving average lost four data points, two on each end of the data series. Observation of the moving average indicated a steady downward trend in the data. The raw data had fluctuations that might tend to confuse an observer, initially due to the apparent positive changes from time to time.

Weighted Moving Average. A major disadvantage of the moving average method, the effect of extreme data points, can be overcome by using a weighted moving average. In this average, the effect of the extreme data points may be decreased by weighing them less than the data points at the center

Table 32.2

Year	Data	5-Year Moving Total	5-Year Moving Average
1	60		
2	56.5		
3	53.0	275.3	55.1
4	54.6	269.2	53.8
5	51.2	261.1	52.2
6	53.9	257.2	51.4
7	48.4	250.9	50.2
8	49.1	242.1	48.4
9	48.3	232.8	46.6
10	42.4		
11	44.6		

Table 32.3

Year	Data	5-Year Moving Total	5-Year Total Less Center Value	Weighted Average (.5 Col 2/4 + .5 Col 4)
1	60			
2	56.5			
3	53.0	275.3	222.3	54.3
4	54.6	269.2	214.6	54.1
5	51.2	261.1	209.9	51.8
6	53.9	257.2	203.3	52.4
7	48.4	250.9	202.5	49.5
8	49.1	242.1	193.0	48.7
9	48.3	232.8	184.5	47.2
10	42.4			
11	44.6			

of the group. There are many ways to do this. One method would be to weight the center point of a five-period average as 50% of the total, with the remaining points weighted for the remaining 50%. For the example in the previous section, the yield would be as shown in Table 32.3.

Example Problem 32.3

See Table 32.3.

Table 32.4 displays the two forecasts. The results are very comparable, with the weighted average forecast distinguishing a slight upswing from period 5 to 6 that was ignored by the moving average method.

Exponential Smoothing. This method determines the forecast (F) for the next period as the weighted average of the last forecast and the current demand (D). The current demand is weighted by a constant α and the last forecast is weighted by the quantity $1 - \alpha$ $(0 \le \alpha \le 1.0)$.

new forecast = α (demand for current period) + $(1 - \alpha)$ (forecast for current period)

This can be expressed symbolically as

$$F_t = \alpha D_{t-1} + (1 - \alpha) F_{t-1}$$

Normally the forecast for the first period is taken to be the actual demand for that period (i.e., forecast and demand are the same for the initial data point). The smoothing constant is chosen as a result of analysis of error by a method such as mean absolute deviation coupled with the judgment of the analyst. A high value of α makes the forecast very responsive to the occurrence in the last period. Similarly, a small value would lead to a lack of significant response to the current demand. Evaluations must be made in light of the cost effects of the errors to determine what value of α is best for a given situation. The following example problem shows the relationship between actual data and forecasts for various values of α.

Table 32.4

Period	Moving Average Forecast	Weighted Average Forecast
3	55.1	54.3
4	53.8	54.1
5	52.2	51.8
6	51.4	52.4
7	50.2	49.5
8	48.4	48.7
9	46.6	47.2

Example Problem 32.4

See Table 32.5.

Causal Methods

These methods assume that there are certain factors that have a cause–effect relationship with the outcome of the quantity to be forecast and that a knowledge of these factors will allow a more accurate prediction of the dependent quantity. The statistical models of regression analysis fall within this category of forecasting.

Basic Regression Analysis. The simplest model for regression analysis is the linear model. The basic approach involves the determination of a theoretical line that passes through a group of data points that appear to follow a linear relationship. The desire of the modeler is to determine the equation for the line that would minimize the sums of the squares of the deviations of the actual points from the corresponding theoretical points. The values for the theoretical points are obtained by substituting the values of the independent variable x_i into the functional relationship

$$\hat{Y}_i = a + bx_i$$

The difference between the data and the forecasted value of point i is

$$Y_i - \hat{Y}_i$$

Squaring this value and summing the relationship over the N related points yields

$$L = \sum_{i=1}^{N} (Y_i - \hat{Y}_i)^2$$

Substituting the functional relationship for the forecasted value of Y gives

$$L = \sum_{i=1}^{N} (Y_i - \hat{a} + \hat{b}x_i)^2$$

By using this relationship, taking the partial derivatives of L with respect to a and b and solving the resulting equations simultaneously, the normal equations for least squares for the linear regression case are obtained. These are

$$\Sigma Y = aN + b \Sigma X$$
$$\Sigma XY = a \Sigma X + b \Sigma X^2$$

Solving these equations yields values for a and b. These values are given by

Table 32.5

Period	Demand	Forecasts for Various α Values		
		$\alpha = .1$	$\alpha = .2$	$\alpha = .3$
1	85	85	85	85
2	102	85	85	85
3	110	86.7	88.4	90.1
4	90	89.0	92.7	96.1
5	105	89.1	92.2	94.3
6	95	90.7	94.8	97.5
7	115	91.1	94.8	96.8
8	120	93.5	98.8	102.3
9	80	96.2	103.0	107.6
10	95	94.6	98.4	99.3

$$a = \frac{\Sigma X^2 \Sigma Y - \Sigma X \Sigma XY}{n \Sigma X^2 - (\Sigma X)^2}$$

and

$$b = \frac{n \Sigma XY - \Sigma X \Sigma Y}{N \Sigma X^2 - (\Sigma X)^2}$$

The regression equation is then $Y_i = a + bx_i$ and the correlation coefficient r, which gives the relative importance of the relationship between x and y as

$$r = \frac{n \Sigma XY - \Sigma X \Sigma Y}{\sqrt{[N \Sigma X^2 - (\Sigma X)^2][N \Sigma Y^2 - (\Sigma Y)^2]}}$$

This value of r can range from $+1$ to -1. The plus sign would indicate a positive correlation (i.e., large values of x are associated with large values of y; a negative correlation implies that large values of x are associated with small values of y) and the negative sign negative correlation.

Quadratic Regression. This regression model is used when the data appear to follow a simple curvilinear trend and the fit of a linear model is not adequate. The procedure for deriving the normal equations for quadratic regression is very similar to that for linear regression. The quadratic model has three parameters that must be estimated, however. These are the constant term a, the coefficient of the linear term b, and the coefficient of the square term c. The model is

$$Y_i = a + bx_i + cx_i^2$$

Its normal equations are

$$\Sigma Y = Na + b \Sigma X + c \Sigma X^2$$
$$\Sigma XY = a \Sigma X + b \Sigma X^2 + c \Sigma X^3$$
$$\Sigma X^2Y = a \Sigma X^2 + b \Sigma X^3 + c \Sigma X^4$$

The normal equations for least squares for a cubic curve, quartic curve, and so on can be generalized from the expressions for the linear and quadratic models.

32.2.4 Forecasting Error Analysis

One common method of evaluation of forecast accuracy is termed *mean absolute deviation* (MAD) from the procedure used in its calculation. For each available data point, a comparison of the forecasted value is made to the actual value. The absolute value of the differences is calculated. This absolute difference is then summed over all values and its average calculated to give the evaluation.

$$MAD = \frac{\text{Sum of the absolute deviations}}{\text{number of deviations}}$$

$$= \Sigma \frac{|(Y_i - \hat{Y}_i)|}{N}$$

Alternative forecasts can be analyzed to determine the value of MAD and a comparison can be made using this quantity as an evaluation criterion. Other criteria can also be calculated. Among these are the mean square of error (MSE) and the standard error of the forecast (s_{yx}). These evaluation criteria are calculated as shown below:

$$MSE = \sum_{i=1}^{N} \frac{(Y_i - \hat{Y}_i)^2}{N}$$

and

$$s_{yx} = \frac{\sqrt{\Sigma Y^2 - a \Sigma Y - b \Sigma XY}}{N - 2}$$

In general, these techniques are used to evaluate the forecast and then the results of the various evaluations, together with the data and forecasts, are studied. Conclusions may then be drawn as to which method is preferred or the results of the various methods compared to determine what they in effect distinguish.

32.2.5 Conclusions on Forecasting

A number of factors should be considered in choosing a method of forecasting. One of the most important factors is *cost*. The problem of valuing an accurate forecast is presented. If the question "How will the forecast help and how will it save money?" can be answered, a decision can be made regarding the allocation of a percent of the savings to the cost of the forecasting process. Further, concern must be directed to the required *accuracy* of a forecast in order to achieve desired cost reductions. Analysis of past data and the testing of the proposed model using this historic data provide a possible scenario for hypothetical testing of the effects of cost of variations of actual occurrences from the plan value (forecast).

In many cases, an inadequate data base will prohibit significant analysis. In others the data base may not be sufficient for the desired projection into the future.

The answers to each of these questions are affected by the type of product or service for which the forecast is to be made, as well as the value of the forecast to the planning process.

32.3 INVENTORY MODELS

32.3.1 General Discussion

Normally items waiting to be purchased or sold are considered to be in inventory. One of the most pressing problems in the manufacturing and sale of goods is the control of this inventory. Many companies experience financial difficulties each year due to a lack of an adequate control in this area. Whether the items in question be raw material used to manufacture a product or products waiting to be sold, problems arise when too many or too few items are available. The greatest number of problems arise when too many items are held in inventory.

The primary factor in the reduction of inventory costs is deciding when to order, how much to order, and if back ordering is permissible. Inventory control involves the making of decisions by management as to the source from which the inventory is to be procured and the quantity to be procured at the time. This source could be from another division of the company handled as an intrafirm transfer, outside purchase from any of a number of possible vendors, or manufacture of the product in-house.

The basic decisions to be made once a source has been determined are how much to order and when to order. Inherent in this previous analysis is the concept of demand. Demand can be known or unknown, probabilistic or deterministic, constant or lumpy. Each of these characteristics affects the method of approaching the inventory problem.

For the *unknown demand* case, a decision must be made as to how much the firm is willing to risk. Normally, the decision would be to produce some "k" units for sale and then determine, after some period of time, to produce more or to discontinue production due to insufficient demand. This amounts to the reduction of the unknown demand situation to one of a lumpy demand case after the decision has been made to produce a batch of finite size. Similarly, if a decision is made to begin production at a rate of n per day until further notice, the unknown demand situation has been changed to a constant known demand case.

Lumpy demand, or demand that occurs periodically with quantities varying, is frequently encountered in manufacturing and distribution operations. It is distinguished from the known demand case. This second case is that of a product which has historic data from which forecasts of demand can be prepared. Factor of concern in these situations are the lead time and the unit requirement on a periodic basis. The following are the major factors to be considered in the modeling of the inventory situation.

Demand

Demand is the primary stimulus on the procurement and inventory system; it is, in fact, the justification for its existence. Specifically, the system may exist to meet the demand of customers, the spare parts demand of an operational weapons system, the demand of the next step in a manufacturing process, and so on. The characteristic of demand, although independent of the source chosen to replenish inventories, will depend upon the nature of the environment giving rise to the demand.

The simplest demand pattern may be classified as deterministic. In this special case, the future demand for an item may be predicted with certainty. Demand considered in this restricted sense is only an approximation of reality. In the general case, demand may be described as a random variable that takes on values in accordance with a specific probability distribution.

Procurement Quantity

Procurement quantity is the order quantity, which in effect determines the frequency of ordering and is related directly to the maximum inventory level.

Maximum Shortage

The maximum shortage quantity is also related to the inventory level.

Item Cost

Item cost is the basic purchase cost of a unit delivered to the location of use. In some cases, delivery cost will not be included if that cost is insignificant in relation to the unit cost. In these cases, the delivery cost will be added to overhead and not treated as a part of direct material costs.

Holding Cost

Inventory holding costs are incurred as a function of the quantity on hand and the time duration involved. Included in these costs are the real out-of-pocket costs, such as insurance, taxes, obsolescence, and warehouse rental and other space charges, and operating costs, such as light, heat, maintenance, and security. In addition, capital investment in inventories is unavailable for investment elsewhere. The rate of return forgone represents a cost of carrying inventory.

The inventory holding cost per unit of time may be thought of as the sum of several cost components. Some of these may depend upon the maximum inventory level incurred. Others may depend upon the average inventory level. Still others, like the cost of capital invested, will depend on the value of the inventory during the time period. The determination of holding cost per unit for a specified time period depends on a detailed analysis of each cost component.

Ordering Cost

Ordering cost is the cost incurred when an order is placed. It is composed of the cost of time and materials, and any expense of communication in placing an order. In the case of a manufacturing model it is replaced by *setup cost,* which is the cost incurred when a machine's tooling or jigs and fixtures must be changed to accommodate the production of a different part or product.

Shortage Cost

Shortage cost is the penalty incurred for being unable to meet a demand when it occurs. This cost does not depend upon the source chosen to replenish the stock, but is a function of the number of units short and the time duration involved.

The specific dollar penalty incurred when a shortage exists depends on the nature of the demand. For instance, if the demand is that of customers of a retail establishment, the shortage cost will include the loss of goodwill. In this case, the shortage cost will be small relative to the cost of the item. If, however, the demand is that of the next step of a manufacturing process, the cost of the shortage may be high relative to the cost of the item. Being unable to meet the requirements for a raw material or a component part may result in lost production or even closing of the plant. Therefore, in establishing shortage cost, the seriousness of the shortage condition and its time duration must be considered.

32.3.2 Types of Inventory Models

Deterministic

Deterministic models assume that quantities used in the determination of relationships for the model are all known. These quantities include demand per unit of time, lead time for product arrival, and costs associated with such occurrences as a product shortage, the cost of holding the product in inventory, and the cost associated with placing an order for a product.

Constant Demand

Constant demand is one case that can be analyzed within the category of deterministic models. It represents very effectively the case for some components or parts in an inventory that are used in multiple parents, these multiple parent components having a composite demand that is fairly constant over time.

Lumpy Demand

Lumpy demand is varying demand that occurs at irregular points in time. This type of demand is normally a dependent demand that is driven by an irregular production schedule affected by actual customer requirements. Although the same assumptions are made regarding the knowledge of related quantities as in the constant demand case, this type of situation is analyzed separately under the topic of materials requirements planning (MRP). This separation of methodology is due to the different inputs to the modeling process in that the knowledge about demand is approached by different methods in the two cases.

Probabilistic

Probabilistic models consider the same quantities as do the deterministic models, but treat the quantities that are not cost-related as random variables. Hence, demand and lead time have their associated probability distributions. The added complexity of the probabilistic values requires that these models be analyzed by radically different methods.

Definitions

The following terms are defined in order to clarify their usage in sections of the material related to inventory that follow. Where appropriate, a literal symbol is assigned to represent the term.

INVENTORY (I). Stock held for the purpose of meeting a demand either internal or external to the organization.

LEAD TIME (T). The time required to replenish an item of inventory by either purchasing from a vendor or manufacturing the item in-house.

DEMAND (D). The number of units of an inventory item required per unit of time.

RE-ORDER POINT. The point at which an order must be placed in order for the procured quantity to arrive at the proper time or, for the manufacturing case, the finished product to begin flowing into inventory at the proper time.

RE-ORDER QUANTITY (Q). The quantity for which an order is placed when the re-order point is reached.

DEMAND DURING LEAD TIME. This quantity is the product of lead time and demand. It represents the number of units that will be required to fulfill demand during the time that it takes to receive an order that has been placed with a vendor.

REPLENISHMENT RATE (R). This quantity is the rate at which replenishment occurs when an order has been placed. For a purchase situation it is infinite (when an order arrives, in an instant the stock level rises from 0 to Q). For the manufacturing situation it is finite.

SHORTAGE. The units of unsatisfied demand that occur when there is an out-of-stock situation.

BACK ORDER. One method of treating demand in a shortage situation when it is acceptable to the customer. (A notice is sent to the customer saying that the item is out of stock and will be shipped as soon as it becomes available.)

LUMPY DEMAND. Demand that occurs in an aperiodic manner for quantities whose volume may or may not be known in advance. Constant demand models should normally never be used in a lumpy demand situation. The exception would be a component that is used for products that experience lumpy demand, but that itself experiences constant demand. The area of MRP (materials requirements planning) was developed to deal with the lumpy demand situations.

32.3.3 The Modeling Approach

Modeling in operations research involves the representation of reality by the construction of a model in one of several alternative ways. These models may be iconic, symbolic, or mathematical. For inventory models, the mathematical model is normally the selection of choice. The model is developed to represent a concept whose relationships are to be studied. As much detail can be included in a particular model as is required to represent the situation effectively. The detail omitted must be of little significance as to its effect on the model. The model's fidelity is the extent to which it accurately represents the situation for which it is constructed.

Inventory modeling involves building mathematical models to represent the interactions of the variables of the inventory situation to give results adequate for the application at hand. In this section, treatment will be limited to deterministic models for inventory control. Probabilistic or stochastic models may be required for some analysis. References 1–3 may be consulted if more sophisticated models are required.

General

Using the terminology defined above, a basic logic model of the general case inventory situation will be developed. The objective of inventory management will normally be to determine an operating policy that will provide a means to reduce inventory costs. To reduce costs, a determination must first be made as to what costs are present. The general model is as follows:

$$\text{total cost} = \text{cost of items} + \text{cost of ordering} + \text{cost of holding items in stock} + \text{cost of shortage}$$

This cost is stated without a base period specified. Normally it will be stated as a per-period cost, with the period being the same period as the demand rate (D) period.

Models of Inventory Situations

Purchase Model with Shortage Prohibited. This model is also known as an infinite replenishment rate model with infinite shortage costs because of the slope of the replenishment rate line (it is vertical) when the order arrives. The quantity on hand instantaneously changes from zero to "Q". The shortage condition is preempted by the assignment of an infinite value to shortage cost. See Fig. 32.1. For this case, the item cost per period is symbolically

$$C_iD$$

The ordering cost is

$$(C_pD)/Q$$

The shortage cost is zero since shortage is prohibited and the inventory holding cost is

$$(C_hQ)/2$$

The equation for total cost is then

$$TC(Q) = C_iD + (C_pD)/Q + (C_hQ)/2$$

Analysis of this model reveals that the first component of cost, the cost of items, does not vary with Q. (Here we are assuming a constant unit cost; purchase discounts models will be covered later.) The second component of cost, the cost of ordering, will vary on a per-period basis with the size of the order (Q). For larger values of Q, the cost will be smaller since fewer orders will be required to receive the fixed demand for the period. The third component of cost, cost of holding items in stock, will increase with increasing order size Q and conversely decrease with smaller order sizes. The fourth component of cost, cost of shortage, is affected by the re-order point. It is not affected by the order size and for this case shortage is not permitted.

It should be noted that this equation is essentially obtained by determining the cost of each of the component costs on a per cycle basis and then dividing that expression by the number of periods per cycle (Q/D). To obtain the extreme point(s) of the function, it is necessary to take the derivative of $TC(Q)$ with respect to Q, equate this quantity to zero, and solve for the corresponding value(s) of Q. This yields

$$0 = 0 + (C_pD)/Q^2 + C_h/2$$

or

$$\hat{Q} = (2C_pD)/C_h$$
$$L = DT$$

Inspection of the sign of the second derivative of this function reveals that the extreme point is a

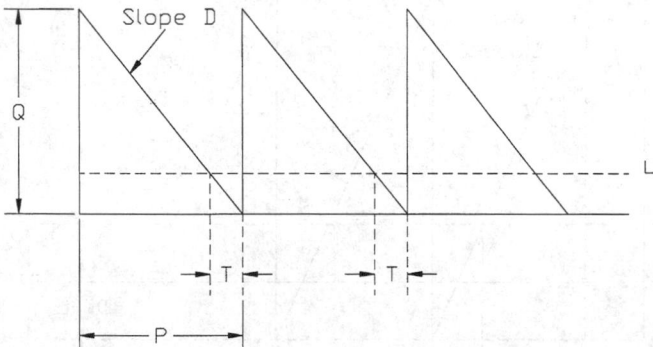

Fig. 32.1 Purchase model with shortage prohibited.

minimum. This fits the objective of the model formulation. The quantity to be ordered at any point in time is then $\hat{Q}$ and the time to place the order will be when the inventory level drops to L (the units consumed during the lead time for receiving the order).

Purchase Model with Shortage Permitted.   This model is also known as an infinite replenishment rate model with finite shortage costs. See Fig. 32.2. For this model, the product cost and the ordering cost are the same as for the previous model

$$C_iD + (C_pD)/Q$$

The holding cost is different, however. It is given by

$$C_h[Q - (DT - L)]^2/2Q$$

This represents the unit periods of holding per cycle times the holding cost per unit period. The unit periods of holding is obtained from the area of the triangle whose altitude is $Q - (DT - L)$ and whose base is the same quantity divided by the slope of the hypotenuse. The unit periods of shortage is calculated in the same manner. For that case, the altitude is $(DT - L)$ and the base is $(DT - L)$ divided by D. The shortage cost component is then

$$C_s(DT - L)^2/2Q$$

The total cost per period is given by

$$TC(Q,DT - L) = C_iD + \frac{(C_pD)}{Q} + \frac{C_h[Q - (DT - L)]^2}{2Q} + \frac{C_s[Q - (DT - L)]^2}{2Q}$$

Note that the quantity $(DT - L)$ is used as a variable. This is done for the purposes of simplifying the equations that result when the partial derivatives are taken for the function. Taking these derivatives and solving the resulting equations simultaneously for the values of Q and $(DT - L)$ yields the following relationships:

$$\hat{Q} = \sqrt{\frac{(2C_pD)}{C_h} + \frac{C_pD}{C_s}}$$

$$\hat{L} = DT - \sqrt{\frac{2C_nC_pD}{C_s(C_h + C_s)}}$$

Manufacturing Model with Shortage Prohibited.   This model is also known as a finite replenishment rate model with infinite shortage costs. Figure 32.3 illustrates the situation.

$$\hat{Q} = \sqrt{\frac{2C_pD}{C_h(1 - D/R)}}$$

$$L = DT$$

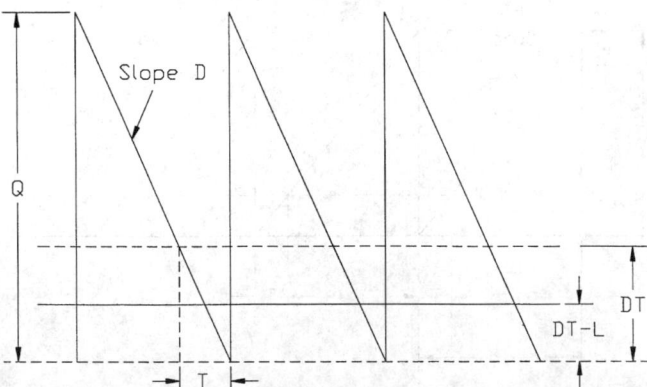

Fig. 32.2 Purchase model with shortage permitted.

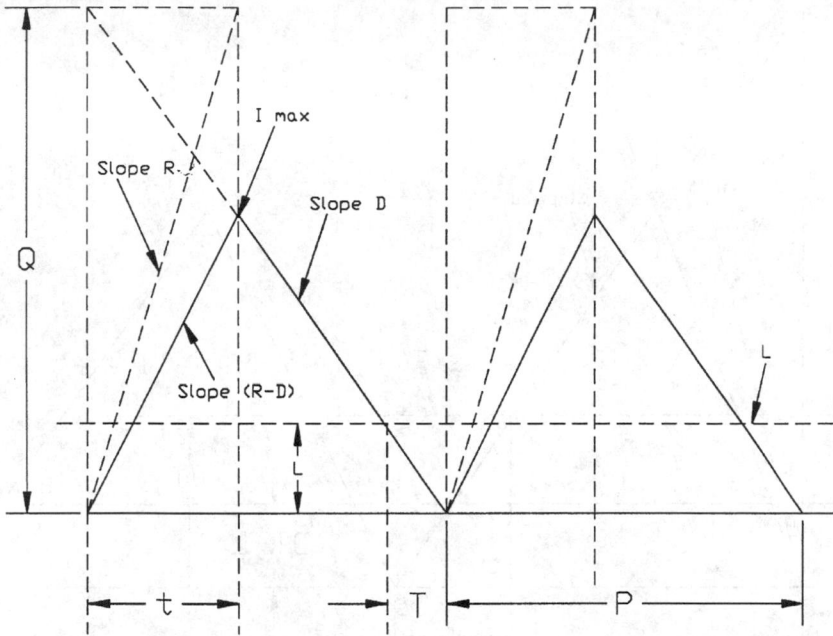

Fig. 32.3 Manufacturing model with shortage prohibited.

Manufacturing Model with Shortage Permitted. This model is also known as a finite replenishment rate model with finite shortage costs. It is the most complex of the models treated here, as it is the general case model. All of the other models can be obtained from it by properly defining the replenishment rate and shortage cost. For example, the purchase model with shortage prohibited is obtained by defining the manufacturing rate and the shortage cost as infinite. When this is done, the equations reduce to those appropriate for the stated situation. See Fig. 32.4.

For this model, the expressions for Q and L are as shown below:

$$\hat{Q} = \sqrt{\frac{1}{1 - D/R}} \quad \sqrt{\frac{2C_p D}{C_h} + \frac{2C_p D}{C_s}}$$

$$\hat{L} = DT \sqrt{1 - D/R} \quad \sqrt{\frac{2C_p D}{C_s\left(1 + \frac{C_s}{C_h}\right)}}$$

Models for Purchase Discounts

Purchase Discount Model with Fixed Holding Cost. In this situation, the hold cost (C_w) is assumed to be fixed, not a function of units costs.

A supplier offers a discount for ordering a larger quantity. The normal situation is as shown in Table 32.6.

The decision-maker must apply the appropriate EOQ purchase model, either finite or infinite shortage cost. Upon choice of the appropriate model, the following procedure will apply.

1. Evaluate $\hat{Q}$ and calculate $TC(\hat{Q})$.
2. Evaluate $TC(q_{k+1})_1$ where q_{k+1} is the smallest quantity in the price break interval above that interval where $\hat{q}$ lies.
3. If $TC(\hat{q}) < TC(qk + 1)$, the ordering quantity will be $\hat{q}$. If not, go to step 4.
4. Since the total cost of the minimum quantity in the next interval above that interval containing $\hat{q}$ is a basic amount, an evaluation must be made successively of total costs of the minimum quantities in the succeeding procurement intervals until one reflects an increase in cost or the last choice is found to be the minimum. For example, in a situation where shortage is not

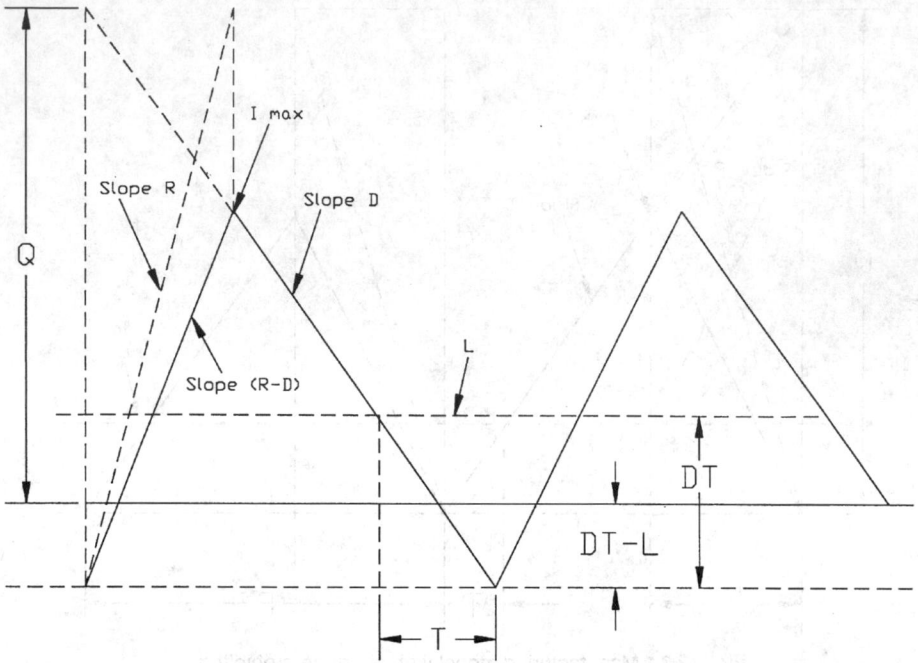

Fig. 32.4 Manufacturing model with shortage permitted.

permitted, the ordering cost is $50, the holding cost is $1 per unit year, and the demand is 10,000 units per year.

$$\hat{Q} = \sqrt{\frac{2C_p D}{C_w}}$$

$$= \sqrt{\frac{2(\$50)10,000}{1}}$$

$$= 1000$$

$$TC(Q) = C_i D + \frac{C_n Q}{2} + \frac{D}{Q} C_p$$

$$TC(\hat{Q}) = \$20(10,000) + \$1 \left(\frac{1000}{2}\right) + \$50 \left(\frac{\$810,000}{1,000}\right)$$

$$= \$201,000$$

Table 32.6

Range of Quantity Purchased	Price
$1-q_1$	P_1
$q_1 + 1-q_2$	P_2
$q_2 + 1-q_3$	P_3
.	.
.	.
.	.
$q_{m-1} + 1-q_m$	P_m

Table 32.7

Q	P
0–500	22.00
501–1199	20.00
1200–1799	18.00
1800 +	16.50

The question is then whether the smallest quantity in the next discount interval (1200–1799) would give a lower total cost. See Table 32.7.

$$TC(1200) = \$18(10,000) + \$1\left(\frac{1200}{2}\right) + \$50\left(\frac{10,000}{1200}\right)$$
$$= 181,033$$

Since this a lower cost, an evaluation must be made of the smallest quantity in the next interval, 1800.

$$TC(1800) = 16.50(10,000) + \$1\left(\frac{1800}{2}\right) + \$50\left(\frac{10,000}{1800}\right)$$
$$TC(1800) = 166,175$$

Since there are no further intervals for analysis, this is the lowest total cost and its associated q, 1800, should be chosen as the optimal $\hat{Q}$.

The total cost function for this model is shown in Fig. 32.5.

Quantity Discount Model with Variable Holding Cost. In this case, the holding cost is variable with unit cost, i.e., $C_w = KC_i$. Again, the appropriate model must be chosen for shortage conditions. For the infinite shortage case,

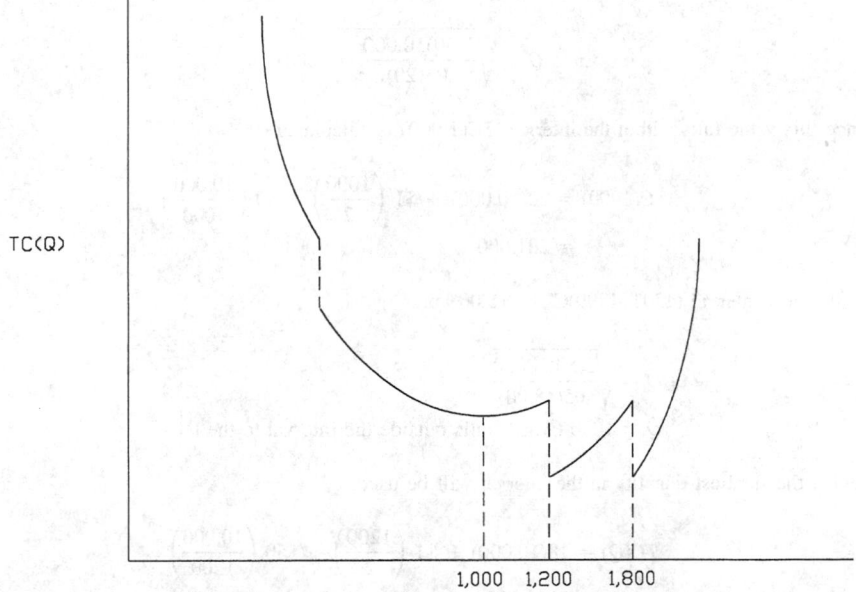

Fig. 32.5 Total cost function for a quantity discount model with fixed percentage holding cost.

$$Q = \sqrt{\frac{2C_pD}{KC_i}}$$

To obtain the optimal value of Q in this situation, the following procedure must be followed:

1. Evaluate $\hat{Q}$ using the expression above and the item cost for the first interval.
   a. If the value of $\hat{Q}$ falls within the interval for C_i, use this $TC(\hat{Q})$ for the smallest cost in the interval.
   b. If $\hat{Q}$ is greater than the maximum quantity in the interval, use Q_{max} where Q_{max} is the greatest quantity in the interval and evaluate $TC(Q_{max})$ as the lowest cost point in the interval.
   c. If $\hat{Q}$ is less than the smallest Q in the interval, use Q_{min} where Q_{min} is the smallest quantity in the interval as the best quantity and evaluate $TC(Q_{min})$.
2. For each cost interval, follow the steps of part 1 to evaluate best values in the interval.
3. Choose the minimum total cost from the applications of steps 1 and 2.

Example Problem 32.5

Using the same data as in the previous example, assume $C_h = .05C_I$

$$\hat{Q} = \sqrt{\frac{2(50)10,000}{(.05)(22)}}$$
$$\cong 990$$

Since $990 > 500$, the smallest cost in the interval will be at

$$TC(500) = 22.00(10,000) + \$1\left(\frac{500}{2}\right) + \$50\left(\frac{10,000}{500}\right)$$
$$= 220,000 + 250 + 1000$$
$$= \$221,250$$

Using the second interval,

$$\hat{Q} = \sqrt{\frac{2(50)10,000}{.05(20)}} = 1000$$

Since this value falls within the interval, $TC(1,000)$ is calculated

$$TC(1000) = 20(10,000) + \$1\left(\frac{1000}{2}\right) + 50\left(\frac{10,000}{1000}\right)$$
$$= 201,000$$

for the next interval $(1201-1799)C_{\pm} = 18.00$ and

$$\hat{Q} = \sqrt{\frac{2(50)10000}{.05(18.00)}}$$
$$\hat{Q} \cong 1050 \text{ (which falls outside the interval to the left)}$$

Hence, the smallest quantity in the interval will be used.

$$TC(Q) = 18(10,000) + \$1\left(\frac{1200}{2}\right) + \$50\left(\frac{10,000}{1200}\right)$$
$$\cong \$181,016$$

This is lower than either interval previously evaluated. For the next interval

$$\hat{Q} = \sqrt{\frac{2(50)10,000}{.05(16.50)}}$$

$\hat{Q} \cong 1101$ (evaluate on calculator)

Hence, the smallest quantity in the interval must be used.

$$TQ(1800) = 16.50(10,000) + \$1\left(\frac{1800}{2}\right) + \$50\left(\frac{10,000}{1800}\right)$$

$$= 165,000 + 900 + 272$$

$$\cong 166,172$$

This is chosen because it is the last interval for evaluation and yields the lowest total cost. Where the previous total cost function for fixed holding cost was a segmented curve with offsets, this is a combination of different curves, each valid over a specific range. In the case of the fixed holding cost, it had only one minimum point, yet the offsets in the total cost due to changes in applicable unit cost in an interval could change the overall minimum. In this situation, there are different values of $\hat{Q}$ for each unit price. The question becomes whether the value of $\hat{Q}$ falls within the price domain. If it does, the total cost function is evaluated at that point; if not, a determination must be made as to whether the value of $\hat{Q}$ lies to the left or right of the range. If it is to the left, the smallest value in the range is used to determine the minimum cost in the range. If it is to the right, the maximum value in the range is used.

Conclusions Regarding Inventory Models

It should be noted that the material discussed here has covered only a very small percentage of the class of deterministic inventory models, albeit these models represent a large percentage of applications. Should the models discussed here not adequately represent the situation under study, further research should be directed at finding a model with improved fidelity for the situation. Other models are covered in Refs. 1, 4, and 5.

32.4 AGGREGATE PLANNING—MASTER SCHEDULING

Aggregate planning is the process of determining overall production, inventory, and workforce levels required to meet forecasted demand over a finite time horizon while trying to minimize the associated production, inventory, and workforce costs. Inputs to the aggregate planning process are forecasted demand for the products (either aggregated or individual); outputs from aggregate planning after disaggregation into the individual products are the scheduled end products for the master production schedule. The time horizon for aggregate planning normally ranges from 6–18 months, with a 12-month average.

The difficulties associated with aggregate planning are numerous. Product demand forecasts vary widely in their accuracy; the process of developing a suitable aggregate measure to use for measuring the value or quantity of production in a multiple product environment is not always possible; actual production does not always meet scheduled production; and unexpected events occur, including material shortages, equipment breakdowns, and employee illness.

Nevertheless, some form of aggregate planning is often required because seldom is there a match between the timing and quantity for product demand versus product manufacture. How the organization should staff and produce to meet this imbalance between production and fluctuating demand is what aggregate planning is about.

32.4.1 Alternative Strategies to Meet Demand Fluctuations

Manufacturing managers use numerous approaches to meet changes in demand patterns for both short and intermediate time horizons. Among the more common are the following.

1. Produce at a constant production rate with a constant workforce, allowing inventories to build during periods of low demand and supplying demand from inventories during periods of high demand. This approach is used by firms with tight labor markets, but customer service may be adversely affected and levels of inventories may widely fluctuate between being excessively high to being out of stock.

2. Maintain a constant workforce, but vary production within defined limits by using overtime, scheduled idle time, and, potentially, subcontracting of production requirements. This strategy allows for rapid reaction to small or modest changes in production when faced with similar demand changes. It is the approach generally favored by many firms, if overall costs can be kept within reasonable limits.

3. Produce to demand, letting the workforce fluctuate by hiring and firing, while trying to minimize inventory carrying costs. This approach is used by firms that typically use low skilled labor where the availability of labor is not an issue. Employee morale and loyalty, however, will always be degraded if this strategy is followed.

32.4.2 Aggregate Planning Costs

Aggregate planning costs can be grouped into one or more of the following categories:

1. *Production costs* include all of those items that are directly related to or necessary in the production of the product, such as labor and material costs. Supplies, equipment, tooling, utilities, and other indirect costs are also included, generally through the addition of an overhead term. Production costs are usually divided into fixed and variable costs, depending upon whether the cost is directly related to production volume.

2. *Inventory costs* include the same ordering, carrying, and shortage costs discussed in Section 32.3.

3. *Costs associated with workforce and production rate changes* are in addition to the regular production costs and include the additional costs incurred when new employees are hired and exiting employees are fired or paid overtime premiums. They may also include costs when employees are temporarily laid off or given alternative work that underutilizes the employees' skills, or when production is subcontracted to an outside vendor.

32.4.3 Approaches to Aggregate Planning

Researchers and practitioners alike have been intrigued by aggregate planning problems, and numerous approaches have been developed over the decades. Although difficult to categorize, most approaches can be grouped as in Table 32.8.

Optimal formulations take many forms. Linear programming models are popular formulations and range from the very basic,[7,8] which assume deterministic demand, a fixed workforce, and no shortages, to complicated models, which use piecewise linear approximations to quadratic cost functions, variable demand, and shortages.[9,10] The linear decision rule (LDR) technique, developed in an extensive project,[11-13] is one of the few instances where the approach was implemented. Nevertheless, due to the very extensive data-collection, updating, and processing requirements to develop and maintain the rules, no other implementation has been reported. Lot-size models usually are either of the capacitated (fixed capacity)[14] or uncapacitated (variable capacity)[15] variety. Although a number of lot-size models have been developed and refined, including some limited implementation,[16,17] computational complexity constrains consideration to relatively small problems. Goal-programming models are attempts at developing more realistic formulations by including multiple goals and objectives. Essentially these models possess the same advantages and disadvantages of LP models, with the additional benefit of allowing tradeoffs among multiple objectives.[18,19] Other optional approaches have modeled the aggregate planning problem using queueing,[20] dynamic programming,[21-23] and Lagrangian techniques.[24,25]

Nonoptional approaches have included the use of search techniques (ST), simulation models (SM), production switching heuristics (PSH), and management coefficient models (MCM). STs involve first the development of a simulation model that describes the system under study to develop the system's response under various operating conditions. A standard search technique is then used to find the parameter settings that maximize or minimize the desired response.[26,27] SMs also develop a model describing the firm and are usually run using a restricted set of schedules to see which performs best. SMs allow the development of very complex systems, but computationally may be so large as to disallow exhaustive testing.[28] PSHs were developed to avoid frequent rescheduling of workforce sizes and production rates.[29] For example, the production rate P_t in period t is determined by[30]

Table 32.8 Classification of Aggregate Planning Approaches[a]

Optimal	Nonoptimal
A. Linear programming	A. Search techniques
B. Linear decision rules	B. Simulation models
C. Lot-size models	C. Production switching heuristics
D. Goal programming	D. Management coefficient models
E. Other analytical approaches	

[a]Modified from Ref. 6.

$$P_t = \begin{cases} L \text{ if } F_t - I_{t-1} < N - C \\ H \text{ if } F_t - I_{t-1} > N - A \\ N \text{ otherwise} \end{cases}$$

where F_t = demand forecast for period t
    I_{t-1} = net inventory level (inventory on-hand minus backorder) at the beginning of period t
    L = low-level production rate
    N = normal-level production rate
    H = high-level production rate
    A = minimum acceptable target inventory level
    C = maximum acceptable target inventory level

Although this example shows three levels of production, fewer or more levels could be specified. The fewer the levels, the less rescheduling and vice versa. However, with more levels, the technique should perform better, because of its ability to better track fluctuations in demand and inventory levels. MCMs were developed by attempting to model and duplicate management's past behavior.[31] However, consistency in past performance is required before valid models can be developed, and it has been argued that if consistency is present, the model is not required.[32]

32.4.4 Levels of Aggregation and Disaggregation

It should be obvious from the previous discussion that different levels of aggregation and disaggregation can be derived from use of the various models. For example, many of the linear programming formulations assume aggregate measures for multiple production and demand units such as production-hours, and provide output in terms of the number of production-hours that must be generated per planning period. For the multiple-product situation, therefore, a scheduler at the plant level would have to disaggregate this output into the various products by planning periods to generate the master production schedule. However, if data were available to support it, a similar, albeit more complex, model could be developed that considered the individual products in the original formulation, doing away with the necessity of disaggregation. This is not often done because of the increased complexity of the resultant model, the increased data requirements, and the increased time and difficulty in solving the formulation. Also, it should be noted that aggregate forecasts that are used as input to the planning process are generally more accurate than forecasts for individual products.

A major task facing the planner, therefore, is determining the level of aggregation and disaggregation required. Normally this is determined by the following:

1. The decision requirements and the level of detail required. Aggregate planning at a corporate level is usually more gross than that done at a division level.

2. The amount, form, and quality of data available to support the aggregate planning process. The better the data, the better the likelihood that more complex models can be supported. Complex aggregate models may also require less disaggregation.

3. The timing, frequency, and resources available to the planner. Generally, the more repetitive the planning, the simpler the approach becomes. Data and analysis requirements as well as analyst's capabilities significantly increase as the complexity of the approach increases.

32.4.5 Aggregate Planning Dilemma

Although aggregate planning models have been available since 1955 and many variations have been developed in the ensuing decades, few implementations of these models have been reported. Aggregate planning is still an important production-planning process, but many managers are unimpressed by the modeling approach. Why? One answer is that aggregate planning does occur throughout the organizational structure, but is done by different individuals at different levels in the organization for different purposes. For example, a major aggregate planning decision is that of plant capacity, which is a constraint on all lower-level aggregate planning decisions. Determining if and when new plant facilities are to be added is generally a corporate decision. However, input for the decision comes from both division and plant levels. Division-level decision-makers may then choose between competing plant facilities in their aggregate planning process in determining which plants will produce which quantity of which products within certain time frames, with input from the individual plant facilities. Plant level managers may aggregate plan their production facilities for capacity decisions, but then must disaggregate these into a master production schedule for their facility. This schedule is constrained by corporate and division decisions.

Most models developed to date do not explicitly recognize that aggregate planning is a hierarchical decision-making process performed on different levels by different people. Therefore, AP is not performed by one individual in the organization, as implicitly assumed by many modeling approaches, but by many people with different objectives in mind.

Other reasons that have been given for the lack of general adoption for AP models include:

1. The AP modeling approach is viewed as a top-down process, whereas many organizations operate AP as a bottom-up process.
2. The assumption used in many of the models, such as linear cost structures, the aggregation of all production into a common measure, or that all workers are equal, are too simplistic or unrealistic.
3. Data requirements are too extensive or costly to obtain and maintain.
4. Decision-makers are intimidated or unwilling to deal with the complexity of the models' formulations and required analyses.

Given this, therefore, it is not surprising that few modeling approaches have been adopted in industrial settings. Although research continues on AP, there is little to indicate any significant modeling breakthrough in the near future that will dramatically change this situation.

One direction, however, is to recognize the hierarchical decision-making structure of AP and to design modeling approaches that utilize it. These systems may be different for different organizations and will be difficult to design, but currently appear to be one approach for dealing with the complexity necessary in the aggregate planning process if a modeling approach is to be followed. For a comprehensive discussion of hierarchical planning systems, see Ref. 33.

32.5   MATERIALS REQUIREMENTS PLANNING

Materials requirements planning (MRP) is a procedure for converting the output of the aggregate planning process, the master production schedule, into a meaningful schedule for releasing orders for component inventory items to vendors or to the production department as required to meet the delivery requirements of the master production schedule.

Materials requirements planning is used in situations where the demand for a product is irregular and highly varying as to the quantity required at a given time. In these situations, the normal inventory models for quantities manufactured or purchased do not apply. Recall that those models assume a constant demand and are inappropriate for the situation where demand is unknown and highly variable. The basic difference between the independent and dependent demand systems is the manner in which the product demand is assumed to occur. For the constant demand case, it is assumed that the daily demand is the same. For dependent demand, a forecast of required units over a planning horizon is used. Treating the dependent demand situation differently allows the business to maintain a much lower inventory level in general than would be required for the same situation under an assumed constant demand. This is so because the average inventory level will be much less in the case where MRP is applied. With MRP, the business will procure inventory to meet high demand just in advance of the requirement and at other times maintain a much lower level of average inventory.

Definitions

AVAILABLE UNITS.   Units of stock that are in inventory and are not in the category of buffer or safety stock and are not otherwise committed.
GROSS REQUIREMENTS.   The quantity of material required at a particular time that does not consider any available units.
INVENTORY UNIT.   A unit of any product that is maintained in inventory.
LEAD TIME.   The time requirement for the conversion of inventory units into required subassemblies or the time required to order and receive an inventory unit.
MRP.   Materials Requirements Planning: a method for converting the end item schedule for a finished product into schedules for the components that make up the final product.
MRP-II.   Manufacturing Resources Planning: a procedural approach to the planning of all resource requirements for the manufacturing firm.
NET REQUIREMENTS.   The units of a requirement that must be satisfied by either purchasing or manufacturing.
PRODUCT STRUCTURE TREE.   A diagram representing the hierarchical structure of the product. The trunk of the tree would represent the final product as assembled from the subassemblies and inventory units that are represented by level one, which come from sub-subassemblies, and inventory units that come from the second level, and so on ad infinitum.
SCHEDULED RECEIPTS.   Material that is scheduled to be delivered in a given time bucket of the planning horizon.
TIME BUCKET.   The smallest distinguishable time period of the planning horizon for which activities are coordinated.

32.5.1   Procedures and Required Inputs

The *master production schedule* is devised to meet the production requirements for a product during a given planning horizon. It is normally prepared from fixed orders in the short run and product requirements forecasts for the time past that for which firm product orders are available. This master production schedule, together with information regarding inventory status and the product structure tree and/or the bill of materials, are used to produce a planned order schedule. An example of a master production schedule is shown in Table 32.9.

The MRP schedule is the basic document used to plan the scheduling of requirements for meeting the MPS. An example is shown in Table 32.10. Each horizontal section of this schedule is related to a single product, part, or subassembly from the product structure tree. The first section of the first form would be used for the parent product. The following sections of the form and required additional forms would be used for the children of this parent. This process is repeated until all parts and assemblies are listed.

To use the MRP schedule, it is necessary to complete a schedule first for the parent part. Upon completion of this level zero schedule, the "bottom line" becomes the input into the schedule for each child of the parent. This procedure is followed until each component, assembly, or purchased part has been scheduled for ordering or production in accordance with the time requirements and other limitations that are imposed by the problem parameters. It should be noted that if a part is used at more than one place in the assembly or manufacture of the final product, it has only one MRP schedule, which is the sum of the requirements at the various levels. The headings of the MRP schedule are as follows:

Item code. The company-assigned designation of the part or subassembly as shown on the product structure tree or the bill of materials.

Level code. The level of the product structure tree at which the item is introduced into the process. The completed product is designated level 0, subassemblies or parts that go together to make up the completed product are level 1, sub-subassemblies and parts that make up level 1 subassemblies are level 2, etc.

Lot size. The size of the lot that is purchased when an order is placed. This quantity may be an economic order quantity or a lot-for-lot purchase. (This later expression is used for a purchase quantity equal to the number required and no more.)

Lead time. The time required to receive an order from the time the order is placed. This order may be placed internally for manufacturing or externally for purchase.

On hand. The total of all units of stock in inventory.

Safety stock. Stock on hand that is set aside to meet emergency requirements.

Allocated (stock). Stock on hand that has been previously allocated for use, such as for repair parts for customer parts orders.

The rows related to a specific item code are designated as follows:

Gross requirements. The unit requirements for the specific item code in the specific time bucket, which are obtained from the MPS for the level 0 items. For item codes at levels other than level 0, the gross requirements are obtained from the planned order releases for the parent item. Where an item is used at more than one level in the product, its gross requirements would be the summation of the planned order releases of the items containing the required part.

Scheduled receipts. This quantity is defined at the beginning of the planning process for products that are on order at that time. Subsequently it is not used.

Available. Those units of a given item code that are not safety stock and are not dedicated for other uses.

Table 32.9 Example of a Master Production Schedule for a Given Product

Part Number	Quantity Needed	Due Date
A000	25	3
A000	30	5
A000	30	8
A000	30	10
A000	40	12
A000	40	15

Table 32.10 Example MRP Schedule Format

Item Code	Level Code	Lot Size	Lead Time (weeks)	On Hand	Safety Stock	Allocated		1	2	3	4	5	6	7	8	9	10	11	12
							Gross requirements												
							Scheduled receipts												
							Available												
							Net requirements												
							Planned order receipts												
							Planned order releases												
							Gross requirements												
							Scheduled receipts												
							Available												
							Net requirements												
							Planned order receipts												
							Planned order releases												
							Gross requirements												
							Scheduled receipts												
							Available												
							Net requirements												
							Planned order receipts												
							Planned order releases												
							Gross requirements												
							Scheduled receipts												
							Available												
							Net requirements												
							Planned order receipts												
							Planned order releases												

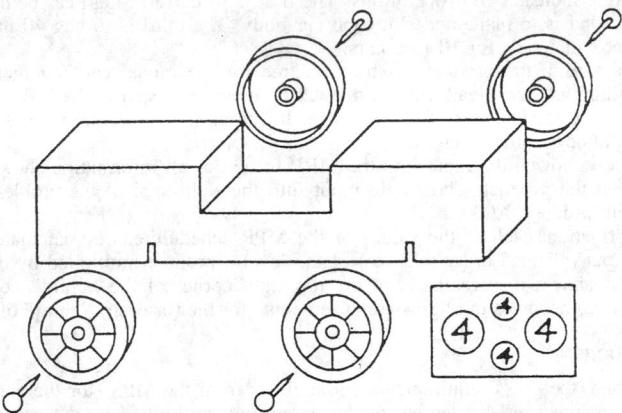

Fig. 32.6 Diagram of model car indicating all parts. (From Ref. 34.)

Net requirements. For a given item code, this is the difference between gross requirements and the quantity available.

Planned order receipts. An order quantity sufficient to meet the net requirements, determined by comparing the net requirements to the lot size (ordering quantity) for the specific item code. If the net requirements are less than the ordering quantity, an order of the size as shown as the lot size will be placed; if the lot size is LFL (lot-for-lot), a quantity equal to the net requirements will be placed.

Planned order releases. This row provides for the release of the order discussed in planned order receipts, to be released in the proper time bucket such that it will arrive appropriately to meet the need of its associated planned order receipt. Note also that this planned order release provides the input information for the requirements of those item codes that are the children of this unit in subsequent generations if such generations exist in the product structure.

Example Problem 32.6 (From Ref. 34, pp. 239–240)

If you were a Cub Scout, you may remember building and racing a little wooden race car. Such cars come 10 in a box. Each box has 10 preformed wood blocks, 40 wheels, 40 nails for axles, and a sheet of 10 vehicle number stickers. The problem is the manufacture and boxing of these race-car kits. An assembly explosion and manufacturing tree are given in Figs. 32.6 and 32.7.

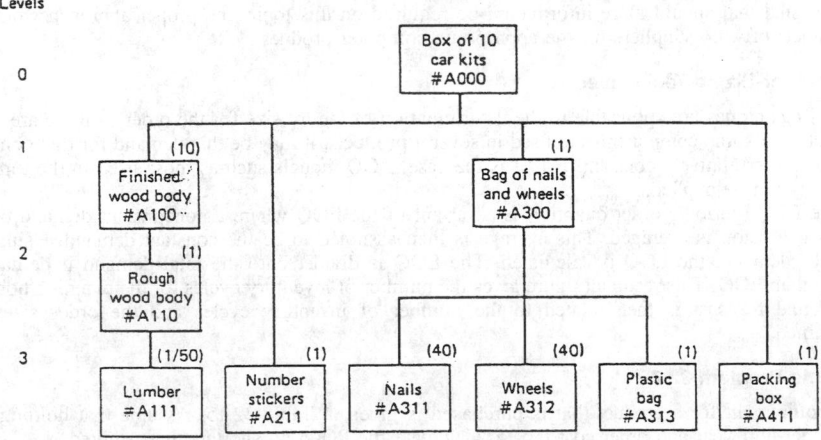

Fig. 32.7 Product structure tree. (From Ref. 34.)

Studying the tree indicates four operations. The first is to cut 50 rough car bodies from a piece of lumber. The second is to plane and slot each car body. The third is to bag 40 nails and wheels. The fourth is to box materials for 10 race cars.

The information from the production structure tree for the model car, together with available information regarding lot sizes, lead time, and stock on hand, is posted to the MRP schedule format to provide information for analysis of the problem. In the problem, no safety stock was prescribed and no stock was allocated for other use.

This information allowed the input into the MRP format of all information shown below for the eight item codes of the product. The single input into the right side of the problem format is the MPS for the parent product, A000.

With this information, each of the values of the MPR schedule can be calculated. It should be noted that the output (planned order releases) of the level 0 product multiplied by the requirements per parent unit (as shown in parenthesis at the top right corner of the "child" component in the product structure tree) becomes the "gross requirements" for the (or each) "child" of the parent part.

32.5.2   Calculations

As previously stated, the gross requirements come either from the MPS (for the parent part) or the calculation of the planned order releases for the parent part multiplied by the per-unit requirement of the current child, per parent part. The scheduled receipts are receipts scheduled from a previous MRP plan. The available units are those on hand from a previous period plus the scheduled receipts from previous MRP. The net requirements are gross requirements less the available units. If this quantity is negative, indicating that there is more than enough, it is set to zero. If it is positive, it is necessary to include an order in a previous period of quantity equal to or greater than the lot size, sufficient to meet the current need. This is accomplished by backing up a number of periods equal to the lead time for the component and placing an order in the planned order releases now that it is equal to or greater than the lot size for the given component.

It should be noted that scheduled receipts and planned order receipts are essentially the arrival of product. The distinction between the two is that scheduled receipts are orders that were made on a previous MRP plan. The planned order receipts are those that are scheduled on the current plan.

Further, in order to keep the system operating smoothly, the MRP plan must be reworked as soon as new information becomes available regarding demand for the product for which the MPS is prepared. This essentially, provides an ability to respond and to keep materials in the "pipeline" for delivery.

Without updating, the system becomes cumbersome and unresponsive. For example, most of the component parts are exhausted at the end of the 15-week period; hence, to respond in the 16th week would require considerable delay if the schedule were not updated.

The results of this process are shown in Tables 32.11, 32.12, and 32.13.

The *planned order release schedule* (Table 32.14) is the result of the MRP procedure. It is essentially the summation of the bottom lines for the individual components from the MRP schedules. It displays an overall requirement for meeting the original master production schedule.

32.5.3   Conclusions on MRP

It should be noted that this process is highly detailed and requires a large time commitment for even a simple product. It becomes intractable for doing by hand in realistic situations. Computerized MRP applications are available that are specifically designed for certain industries and product groups. It is suggested that should more information be required on this topic, the proper approach would be to contact software suppliers for the appropriate computer product.

32.5.4   Lot-Sizing Techniques

Several techniques are applicable to the determination of the lot size for the order. If there are many products and some components are used in several products, it may be that demand for that common component is relatively constant. If that is the case, EOQ models such as those used in the topic on inventory can be applied.

The POQ (periodic order quantity) is a variant of the EOQ where a nonconstant demand over a planning horizon is averaged. This average is then assumed to be the constant demand. Using this value of demand, the EOQ is calculated. The EOQ is divided into the total demand if demand is greater than EOQ. This resultant figure gives the number of inventory cycles for the planning horizon. The actual forecast is then related to the number of inventory cycles and the order sizes are determined.

Example Problem 32.7

The requirement for a product that is purchased is given in Table 32.15. Assume that holding cost is $10 per unit year and order cost is $25. Calculate the POQ; no shortage is permitted.

Using the basic EOQ formula:

Table 32.11

Item Code	Level Code	Lot Size	Lead Time (weeks)	On Hand	Safety Stock	Allocated		1	2	3	4	5	6	7	8	9	10	11	12	13	14	15
A000	0	50	1	20	0	0	Gross requirements			25			15	15	30		30		40			40
							Schedule receipts															
							Available	20	20	45	45	45	30	15	35	35	5	5	15	15	15	25
							Net requirements			5					15				35			25
							Planned order receipts			50					50				50			50
							Planned order releases		50					50				50			50	
A100	1	50	1	100	0	0	Gross requirements		500					500				500			500	
							Scheduled receipts															
							Available	100	0	0	0	0	0	0	0	0	0	0	0	0	0	0
							Net requirements		400					500				500			500	
							Planned order receipts		400					500				500			500	
							Planned order releases	400					500				500			500		
A300	1	50	1	150	0	0	Gross requirements		50					50				50			50	
							Scheduled receipts															
							Available	150	100	100	100	100	100	50	50	50	50	0	0	0	0	0
							Net requirements		0					0				0			50	
							Planned order receipts														50	
							Planned order releases													50		
A110	2	100	1	200	0	0	Gross requirements	400					500				500			500		
							Scheduled receipts	200*														
							Available	400	0	0	0	0	0	0	0	0	0	0	0	0	0	0
							Net requirements	0					500				500			500		
							Planned order receipts	0					500				500			500		
							Planned order releases	0				500				500			500			

*Order on a previous schedule

Table 32.12

Item Code	Level Code	Lot Size	Lead Time (weeks)	On Hand	Safety Stock	Allocated		1	2	3	4	5	6	7	8	9	10	11	12	13	14	15
A111	3	10	3	5	0	0	Gross requirements					10				10			10			
							Scheduled receipts															
							Available	5	5	5	5	5	5	5	5	5	5	5	5	5	5	5
							Net requirements					5				5			5			
							Planned order receipts					10				10			10			
							Planned order releases		10				10			10						
A211	3	500	10	500	0	0	Gross requirements		50					50				50			50	
							Scheduled receipts															
							Available	500	500	450	450	450	450	450	400	400	400	400	350	350	350	300
							Net requirements															
							Planned order receipts															
							Planned order releases															
A311	3	500	2	300	0	0	Gross requirements													2000		
							Scheduled receipts															
							Available	300	300	300	300	300	300	300	300	300	300	300	300	300	300	300
							Net requirements													1700		
							Planned order receipts													2000		
							Planned order releases											2000				
A312	3	500	2	200	0	0	Gross requirements													2000		
							Scheduled receipts															
							Available	200	200	200	200	200	200	200	200	200	200	200	200	200	200	200
							Net requirements													1800		
							Planned order receipts													2000		
							Planned order releases											2000				

Table 32.13

Item Code	Level Code	Lot Size	Lead Time (weeks)	On Hand	Safety Stock	Allocated		1	2	3	4	5	6	7	8	9	10	11	12	13	14	15
A313	3	500	3	30	0	0	Gross requirements														50	
							Scheduled receipts															
							Available	30	30	30	30	30	30	30	30	30	30	30	30	30	30	480
							Net requirements														20	
							Planned order receipts														500	
							Planned order releases											500				
A411	3	500	5	40	0	0	Gross requirements		50					50				50			50	
							Scheduled receipts															
							Available	40	40	490	490	490	490	490	440	440	440	440	390	390	390	340
							Net requirements		10													
							Planned order receipts		500*													
							Planned order releases															
							Gross requirements															
							Scheduled receipts															
							Available															
							Net requirements															
							Planned order receipts															
							Planned order releases															
							Gross requirements															
							Scheduled receipts															
							Available															
							Net requirements															
							Planned order receipts															
							Planned order releases															

*Ordered on a previous schedule

Table 32.14 Planned Ordered Release Schedule

	Week														
	1	2	3	4	5	6	7	8	9	10	11	12	13	14	15
A000		50						50			50			50	
A100	400					500				500			500		
A300														50	
A110					500				500				500		
A111		10					10		10						
A211															
A311											2000				
A312											2000				
A313											500				
A411															

Note: An advance order of 200 units of item 110 would have to have been made on a previous MRP schedule.

$$\hat{Q} = \sqrt{\frac{2C_p D}{C_n}}$$

$$= \sqrt{\frac{2(\$25)29(52)}{\$10}}$$

$$= 86.8 \approx 87$$

$$\frac{348 \quad units}{87 \quad units/order} = 4 \ orders$$

Lot for Lot (LFL) is the approach to the variable demand situation that merely requires that an order size equal to the required number of products be placed.

The first order would be 25 + 29 + 34 = 88 units. The second would be 26 + 24 + 32 = 82 units. The third and fourth orders would be 81 and 97, respectively.

It is coincidental that the number of orders turned out to be an integer. Had a non-integer occurred, it could have been rounded to the nearest integer. An economic evaluation can be made, if costs are significant, of which rounding (up or down) would yield the lower cost option. Other methods exist in the area of lot-sizing.

32.6 JOB SEQUENCING AND SCHEDULING

Sequencing and scheduling problems are among the most common situations found in service and manufacturing facilities. Determining the order and deciding when activities or tasks should be done are part of the normal functions and responsibilities of management and increasingly of the employees themselves. These terms are often used interchangeably, but it is important to note the difference. *Sequencing* is determining the order of a set of activities to be performed, whereas *scheduling* also includes determining the specific times when each activity will be done. Thus, scheduling includes sequencing; that is, to be able to develop a schedule for a set of activities, you must also know the sequence in which those activities are to be completed.

32.6.1 Structure of the General Sequencing Problem

The job sequencing problem is usually stated as follows: Given *n* jobs to be processed on *m* machines, each job having a *setup time, processing time, due date* for the completion of the job, and requiring

Table 32.15

Period (week)	Demand	Price	Demand	Period	Demand
1	25	5	24	9	8
2	29	6	32	10	35
3	34	7	28	12	32
4	26	8	25	12	30

processing on one or more of the machines, determine the sequence for processing the jobs on the machines to optimize the *performance criterion.*

The factors, therefore, used to describe a sequencing problem are

1. The number of machines in the shop, *m*
2. The number of jobs, *n*
3. The type of shop or facility, i.e., job shop or flow shop
4. The manner in which jobs arrive at the shop, i.e., static or dynamic
5. The performance criterion used to measure the performance of the shop

Usual assumptions for the sequencing problem include the following:

1. Setup times for the jobs on each machine are independent of sequence and can be included in the processing times.
2. All jobs are available at time zero to begin processing.
3. All setup times, processing times, and due dates are known and are deterministic.
4. Once a job begins processing on a machine, it will not be preempted by another job on that machine.
5. Machines are continuously available for processing; i.e., no breakdowns occur.

Commonly used performance criteria include the following:

1. *Mean flow time* $(\overline{F})$ is the average time a set of jobs spends in the shop, which includes processing and waiting times.
2. *Mean idle time of machines* $(\overline{I})$ is the average idle time for the set of machines in the shop.
3. *Mean lateness of jobs* $(\overline{L})$ is the difference between the actual completion time (C_j) for a job and its due date (d_j), i.e., $L_j = C_j - d_j$. A negative value means that the job is completed early. Therefore,

$$\overline{L} = \sum_{j=1}^{n} \frac{(C_j - d_j)}{n}$$

4. *Mean tardiness of jobs* $(\overline{T})$ is the maximum of 0 or its value of lateness, i.e., $T_j = \max\{O, L_j\}$. Therefore,

$$\overline{T} = \sum_{j=1}^{n} \max \frac{\{0, L_j\}}{n}$$

5. *Mean number of jobs late.*
6. *Percentage of jobs late.*
7. *Mean number of jobs in the system.*
8. *Variance of lateness* (s^2_L), for a set of jobs and a given sequence, is the variance calculated for the corresponding L_j's, i.e.,

$$\sum_{j=1}^{n} \frac{(L_j - \overline{L})^2}{(n-1)}$$

The following material will cover the broad range of sequencing problems, from the simple to the complex. The discussion will begin with the single-machine problem and progress through multiple machines. It will include quantitative and heuristic results for both flow shop and job shop environments.

32.6.2 Single-Machine Problem

In many instances, the single-machine sequencing problem is still a viable problem. For example, if one were trying to maximize production through a bottleneck operation, consideration of the bottleneck as a single machine might be a reasonable assumption. For the single-machine problem, i.e., *n* jobs one machine, results include the following.

Mean Flow Time

To minimize the mean flow time, jobs should be sequenced so that they are in increasing shortest processing time (SPT) order. For example, see the jobs and processing times (t_j's) for the jobs in Table 32.16.

Table 32.16

Job	t_j (days)
1	7
2	6
3	8
4	5

In Table 32.17, the jobs are processed in shortest processing-time order, i.e., 4, 2, 1, 3.

From Table 32.17 we conclude that $\overline{F} = 60/4 = 15$ days. Any other sequence will only increase $\overline{F}$. Proof of this is available in Ref. 35.

Mean Lateness

Note that as a result of the definition of lateness, SPT sequencing will minimize mean lateness $(\overline{L})$ in the single-machine shop.

Weighted Mean Flow Time

The above results assumed all jobs were of equal importance. What if, however, jobs should be weighted according to some measure of importance? Some jobs may be more important because of customer priority or profitability. If this importance can be measured by a weight assigned to each job, a weighted mean flow time measure, $\overline{F}_w$, can be defined as

$$\overline{F}_w = \frac{\sum_{j=1}^{n} w_j F_j}{\sum_{j=1}^{n} w_j}$$

To minimize weighted mean flow time $(\overline{F}_w)$, jobs should be sequenced in increasing order of weighted shortest processing time, i.e.,

$$\frac{t_{[1]}}{w_{[1]}} \leq \frac{t_{[2]}}{w_{[2]}} \leq \ldots \leq \frac{t_{[n]}}{w_{[n]}}$$

where the brackets indicate the first, second, etc., jobs in sequence.

As an example, consider the problem given in Table 32.18.

If jobs 2 and 6 are considered three times as important as the rest of the job, what sequence should be selected? The solution is given in Table 31.19.

Maximum Lateness/Maximum Tardiness

Other elementary results given without proof or example include the following. To minimize the maximum job lateness (L_{max}) or the maximum job tardiness (T_{max}) for a set of jobs, the jobs should be sequenced in order of non-decreasing due dates, i.e.,

$$d_{[1]} \leq d_{[2]} \leq \ldots \leq d_{[n]}$$

Minimize the Number of Tardy Jobs

If the sequence above, known as the earliest due date sequence, results in zero or one tardy job, then it is also an optional sequence for the number of tardy jobs, N_T. In general, however, to find an optional sequence minimizing N_T, an algorithm attributed to Moore and Hodgson[36] can be used. The

Table 32.17

Job	t_j (days)	C_j
4	5	5
2	6	11
1	7	18
3	8	26
		$\Sigma C_j = 60$

Table 32.18

Job	1	2	3	4	5	6
t_j (days)	20	27	16	6	15	24

algorithm divides all jobs into two sets: Set E, where all the jobs are either early or on time, and Set T, where all the jobs are tardy. The optional sequence then consists of Set E jobs followed by Set T jobs. The algorithm is as follows.

Step 1. Begin by placing all jobs in Set E in non-decreasing due date order, i.e., earliest due date order. Note that Step T is empty.

Step 2. If no jobs in Set E are tardy, stop: the sequence in Set E is optional. Otherwise, identify the first tardy job in Set E, labeling this job k.

Therefore, the job processing sequence should be 4, 6, 2, 5, 3 and 1.

Step 3. Find the job with the longest processing time among the first k jobs in sequence in Set E. Remove this job from Set E and place it in Set T. Revise the job completion times of the jobs remaining in Set E and go back to step 2 above. As an example in the use of this algorithm, consider the information given in Table 32.20.

The solution to the problem in Table 32.20 is:

Step 1. E = {3, 1, 4, 2}; T = {φ}
Step 2. Job 4 is first late job
Step 3. Job 1 is removed from E
            E = {3, 4, 2}; T = {1}
Step 2. Job 2 is first late job
Step 3. Job 2 is removed from E
            E = {3, 4}; T = {1, 2}
Step 2. No jobs in E are now late

Therefore, optional sequences are either (3, 4, 1, 2) or (3, 4, 2, 1)

32.6.3 Flow Shops

General flow shops can be depicted as in Fig. 32.8. All products being produced through these systems flow in the same direction without backtracking. For example, in a four-machine general flow shop, product 1 may require processing on machines 1, 2, 3, and 4; product 2 requires machines 1, 3, and 4; product 3 requires machines 1 and 2 only. Thus, a flow shop processes jobs much as a production line does, but, because it often processes jobs in batches, may look more like a job shop.

Two Machines/*n* Jobs

The most famous result in sequencing literature is concerned with two-machine flow shops and is known as *Johnson's Sequencing Algorithm*.[37] This algorithm will develop an optional sequence using makespan as the performance criterion. Makespan is defined as the time required to complete the set of jobs through all machines.

Steps for the algorithm are as follows:

1. List all processing times for the job set for machines 1 and 2.
2. Find the minimum processing time for all jobs.
3. If the minimum processing time is on machine 1, place the job first or as early as possible in the sequence. If it is on machine 2, place that job last or as late as possible in the sequence. Remove that job for further consideration.

Table 32.19

Job	1	2	3	4	5	6
w_j	1	3	1	1	1	3
t_j/w_j	20	9	16	6	15	8

Table 32.20

Job	t_j (days)	d_j (days)
1	10	14
2	18	27
3	2	4
4	6	16

4. Continue, by going back to step 2, until all jobs have been sequenced.

As an example, consider the five-job problem shown in Table 32.21. Applying the algorithm will give an optional sequence of 2, 4, 5, 3, 1 through the two machines, with a makespan of 26 time units.

Three Machines/n Jobs

Johnson's Sequencing Algorithm can be extended to a three-machine flow shop and may generate an optional solution with makespan as the criterion.

The extension consists of creating a two-machine flow shop from the three machines by summing the processing times for all jobs for the first two machines for artificial machine 1 and, likewise, summing the processing times for all jobs for the last two machines for artificial machine 2. Johnson's Sequencing Algorithm is then used on the two-artificial-machine flow shop problem.

For example, consider the following three-machine flow shop problem given in Table 32.22. The results of forming the five-job, two-artificial-machine problem are shown in Table 32.23. Therefore, the sequence using Johnson's Sequencing Algorithm is 3, 1, 4, 5, 2. It has been shown that the sequence obtained using this extension is optimal with respect to makespan if one of the following conditions holds:

1. min $t_{1j} \geq$ max t_{2j}, or
2. min $t_{3j} \geq$ max t_{2j}, or
3. If the sequence using $\{t_{1j}, t_{2j}\}$, i.e., only the first two machines, is the same sequence as using only $\{t_{2j}, t_{3j}\}$, i.e., only the last two machines, as two, two-machine flow shops using Johnson's Sequencing Algorithm.

The reader should check to see that the sequence obtained above is optimal using these conditions.

More Than Three Machines

Once the number of machines exceeds three, there are few ways to find optimal sequences in a flow shop environment. Enumeration procedures, such as branch and bound, are generally the only practical approach that has been successfully used, and then only in problems with five or fewer machines. The more usual approach is to develop heuristic procedures or using assignment rules such as priority dispatching rules. See Section 32.6.5 for more details.

32.6.4 Job Shops

General job shops can be represented as in Fig. 32.9. Products being produced in these systems may begin with any machine or process, followed by a succession of processing operations on any other sequence of machines. It is the most flexible form of production, but experience has shown that it is also the most difficult to control and to operate efficiently.

Two Machines/n Jobs

Johnson's Sequencing Algorithm can also be extended to a two-machine job shop to generate optimal schedules when makespan is the criterion. The steps to do this are as follows:

Step 1. Divide the job set into four sets, i.e.,
Set {A}—jobs that require only one processing operation and that on machine 1.
Set {B}—jobs that require only 1 processing operation and that on machine 2.
Set {AB}—jobs that require two processing operations, the first on machine 1, the second on machine 2.
Set {BA}—jobs that require two processing operations, the first on machine 2, the second on machine 1.

Step 2. Sequence jobs in Set {AB} and Set {BA} using Johnson's Sequencing Algorithm (note that in Set {BA}, machine 2 is the first machine in the process).

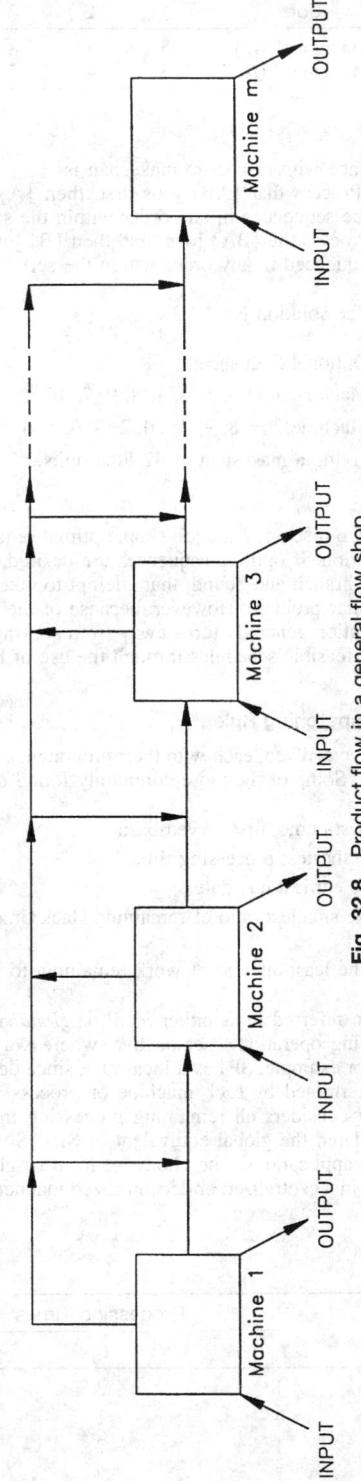

Fig. 32.8 Product flow in a general flow shop.

Table 32.21

Job		1	2	3	4	5
Processing	Machine 1 (t_{1j})	5	1	8	2	7
Times	Machine 2 (t_{2j})	3	4	5	6	6

Step 3. The optional sequence with respect to makespan is
On machine 1 ⇒ Process that {AB} jobs first, then {A} jobs, then {BA} jobs (note that {A} jobs can be sequenced in any order within the set).
On machine 2 ⇒ Process the {BA} jobs first, then {B} jobs, then {AB} jobs (note that {B} jobs can be sequenced in any order within the set).

For example, see Table 32.24. The solution is:

> Optional Sequence:
> Machine 1 ⇒ 6, 5, 4, 1, 8, 9, 7, 10
> Machine 2 ⇒ 8, 9, 7, 10, 2, 3, 6, 5, 4
> giving a makespan of 42 time units.

m Machines/*n* Jobs

Once the problem size exceeds two machines in a job shop, optimal sequences are difficult to develop even with makespan as the criterion. If optimal sequences are desired, the only options are usually enumeration techniques, such as branch and bound, that attempt to take account of the special structure that may exist in the particular problem. However, because of the complexity involved in these larger problems, sequencing attention generally turns away from seeking the development of optimal schedules to the development of feasible schedules through the use of heuristic decision rules called *priority dispatching rules.*

32.6.5 Heuristics/Priority Dispatching Rules

A large number of these rules have evolved, each with their proponents, given certain shop conditions and desired performance criteria. Some of the more commonly found ones are

> *FCFS*—select the job on a first-come, first-served basis
> *SPT*—select the job with the shortest processing time
> *EDD*—select the job with the earliest due date
> *STOP*—select the job with the smallest ratio of remaining slack time to the number of remaining operations
> *LWKR*—select the job with the least amount of work remaining to be done

Rules such as these are often referred to as either *local* or *global* rules. A local rule is applied from each machine's or processing operation's perspective, whereas a global view is applied from the overall shop's perspective. For example, SPT is a local rule, since deciding which of the available jobs to next process will be determined by each machine or process operator. On the other hand, LWKR is a global rule, since it considers all remaining processing that must be done on the job. Therefore, LWKR can be considered the global equivalent of SPT. Some rules, such as FCFS, can be used in either local or global applications. The choice of local or global use is often a matter of whether shop scheduling is done in a centralized or decentralized manner and whether the information

Table 32.22

Job	Processing Times		
	t_{1j}	t_{2j}	t_{3j}
1	1	3	8
2	4	1	3
3	1	2	3
4	7	2	7
5	6	1	5

Table 32.23

Job	Processing Times	
	t^1_{Aj}	t^1_{Bj}
1	4	11
2	5	4
3	3	5
4	9	9
5	7	6

system will support centralized scheduling. Implicit within these concepts is the fact that centralized scheduling requires more information to be distributed to individual workstations and is inherently a more complex scheduling environment requiring more supervisory oversight. Global scheduling intuitively should produce better system's schedules, but empirical evidence seems to indicate that local rules are generally more effective.

Whichever rule may be selected, the use of priority assignments is to *resolve conflicts*. As an example, consider the three-machine, four-job sequencing problem given in Table 32.25. Assuming all jobs are available at time zero, the initial job loading is shown in Fig. 32.10. As shown, there is no conflict on machines 1 and 2, so the first operation for jobs 1 and 2 would be assigned to these machines. However, there is a conflict on machine 3. If SPT were being used, the first operation for job 3 would be assigned on machine 3, and the earliest that the first operation of job 4 could be assigned to machine 3 is at time equal to 2 days.

Continuing this example following these assignments, the situation shown in Fig. 32.11 would then exist. If we continue to use SPT, we would assign the second operation of job 2 to machine 1 to resolve the conflict. (Note: if FCFS were being used, the second operation of job 3 would have been assigned.) Even though there is no conflict at this stage on machine 2, no job would normally

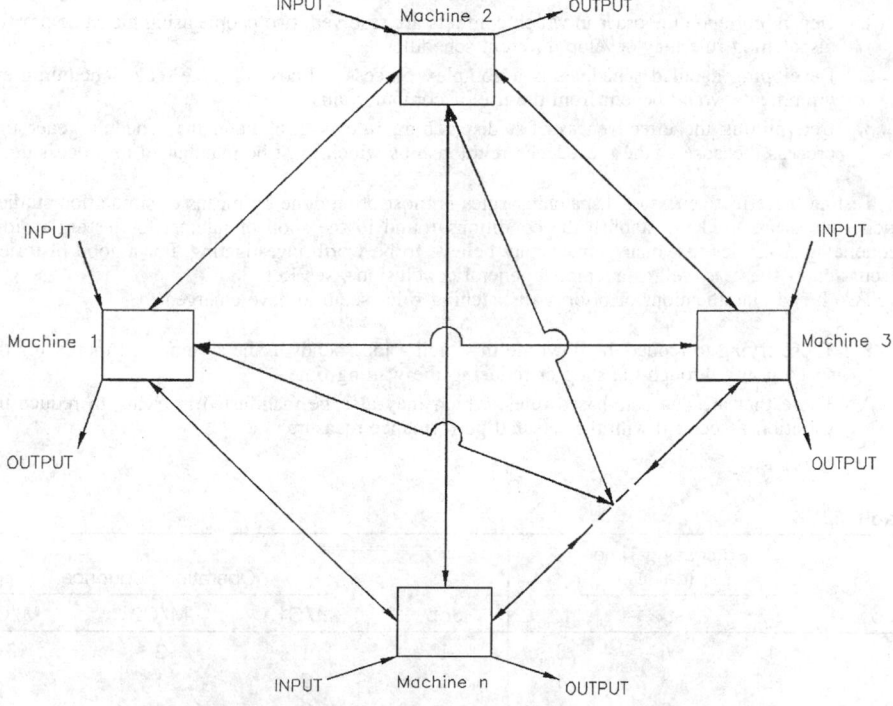

Fig. 32.9 Product flow in a general-job shop.

Table 32.24

Job Set	Job	Processing Times t_{1j}	t_{2j}
{A}	1	3	
{B}	2		2
	3		4
{AB}	4	4	2
	5	6	5
	6	3	7
{BA}	7	3	8
	8	4	1
	9	7	9
	10	2	4

be assigned at this time because it would introduce idle time unnecessarily within the schedule. The first operation for job 4, however, would be assigned to machine 3.

With these assignments, the schedule now appears in Fig. 32.12. The assignments made at this stage would include

1. Second operation, job 4 to machine 2
2. Third operation, job 2 to machine 3, since no other job could be processed on machine 3 during the idle time from 3–6 days

Note that the second operation for job 3 may or may not be scheduled at this time because the third operation for job 4 would also be available to begin processing on machine 1 at the 6th day. Because of this, there would be a conflict at the beginning of the sixth day, and if SPT is being used, the third operation for job 4 would be selected over the second operation for job 3.

Several observations for this partial example can be made:

1. Depending upon the order in which conflicts are resolved, two people using the same priority dispatching rule may develop different schedules.
2. Developing detailed schedules is a complex process. Almost all large-scale scheduling environments would benefit from the use of computer aids.
3. Determining the effectiveness of a dispatching rule is difficult in the schedule-generating process, because of the precedence relationships which must be maintained in processing.

Testing the effectiveness of dispatching rules is most often done by means of simulation studies. Such studies are used to establish the conditions found in the shop of interest for testing various sequencing strategies that management may believe to be worth investigating. For a good historical discussion of these as well as attempted general conclusions, see Ref. 38.

Two broad classifications of priority dispatching rules seem to have emerged:

1. Those trying to reduce the flowtime in which a job spends in the system, i.e., increasing the speed going through the shop or reducing the waiting time
2. Those that are due date-based rules, which may also be manifested in trying to reduce the variation associated with the selected performance measure

Table 32.25

Job	Processing Times (days) t_{1j}	t_{2j}	t_{3j}	Job	Operation Sequence M/C1	M/C2	M/C3
1	4	6	8	1	1	2	3
2	2	3	4	2	2	1	3
3	4	2	1	3	2	3	1
4	3	3	2	4	3	2	1

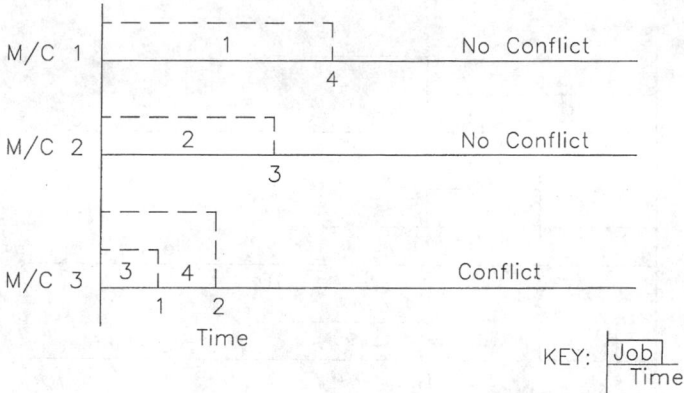

Fig. 32.10 Example Problem: Stage 1.

Although simulation has proven to be effective in evaluating the effectiveness of dispatching rules in a particular environment, few general conclusions have been drawn. When maximum throughput or speed is the primary criterion, SPT is often a good rule to use, even in situations when the quality of information is poor concerning due dates and processing times. When due date rules are of interest, selection is much more difficult. Results have been developed showing that when shop loads are heavy, SPT still may do well; when shop loads are moderate, STOP was preferable. Other research has shown that the manner in which due dates are set, as well the tightness of the due dates, can greatly affect the performance of the rule. Overall, the conclusions that can be drawn are:

1. It is generally more difficult to select an effective due date-based rule than a flowtime-based rule.
2. If time and resources are available, the best course of action is to develop a valid model of the particular shop of interest, and experiment with the various candidate rules to determine which are most effective, given that situation.

32.6.6 Assembly Line Balancing

Assembly lines are viewed as one of the purest forms of production lines. A usual form is visualized as shown in Fig. 32.14, where work moves continuously by means of a powered conveyor through a series of workstations where the assigned work is performed.

Definitions

CYCLE TIME (C). The time available for a workstation to perform its assigned work, assumed to be the same for each workstation. The cycle time must be greater than or equal to the longest

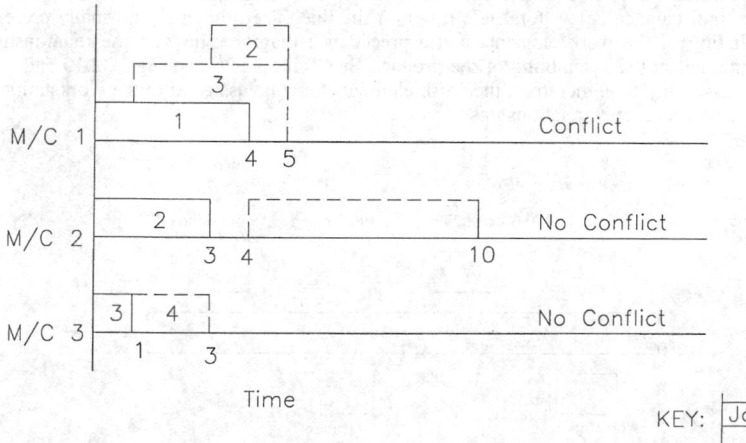

Fig. 32.11 Example Problem: Stage 2.

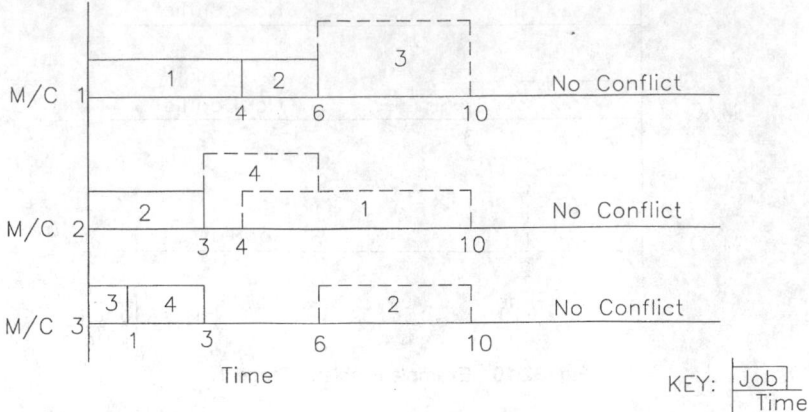

Fig. 32.12 Example Problem: Stage 3.

work element for the product. Note that it is also the time between successive completions of units of product on the line.

BALANCE DELAY OF A WORKSTATION. The difference between the cycle time (C) and the station time (S_j) for a workstation, i.e., the idle time for the station $= (C - S_j)$.

STATION TIME (S_j). The total amount of work assigned to station j, which consists of one or more of the work elements necessary for completion of the product. Note that each S_j must be less than or equal to C.

WORK ELEMENT (i). An amount of work necessary in the completion of a unit of product. It is usually considered indivisible. I is the total number of work elements necessary to complete one unit of product.

WORK ELEMENT TIME (t_i). The amount of time required to complete work element i. Therefore, the sum of all of the work elements, i.e., the total work content,

$$T = \sum_{i=1}^{I} t_i$$

is the time necessary to complete one unit of product.

WORKSTATION (j). A location on the line where assigned work elements on the product are performed $(1 \le j \le J)$.

Structure of the Assembly Line Balancing Problem

The objective of assembly line balancing is to assign work elements to the workstations so as to minimize the total balance delay (total idle time) on the line. The problem is normally presented by means of a listing of the work elements and a precedence diagram showing the relationships that must be maintained in the assembling of the product. See Table 32.26 and Figs. 32.15 and 32.16. In designing the assembly line, therefore, the work elements must be assigned to the workstations while adhering to these precedence relationships.

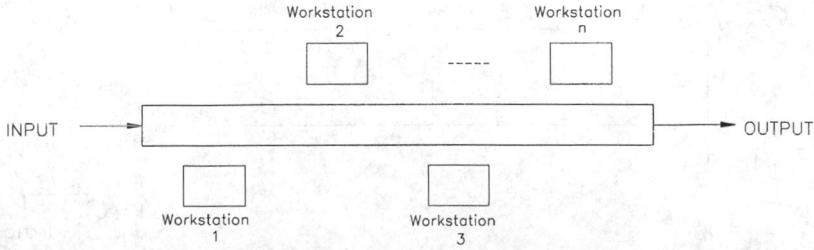

Fig. 32.13 Typical assembly line configuration.

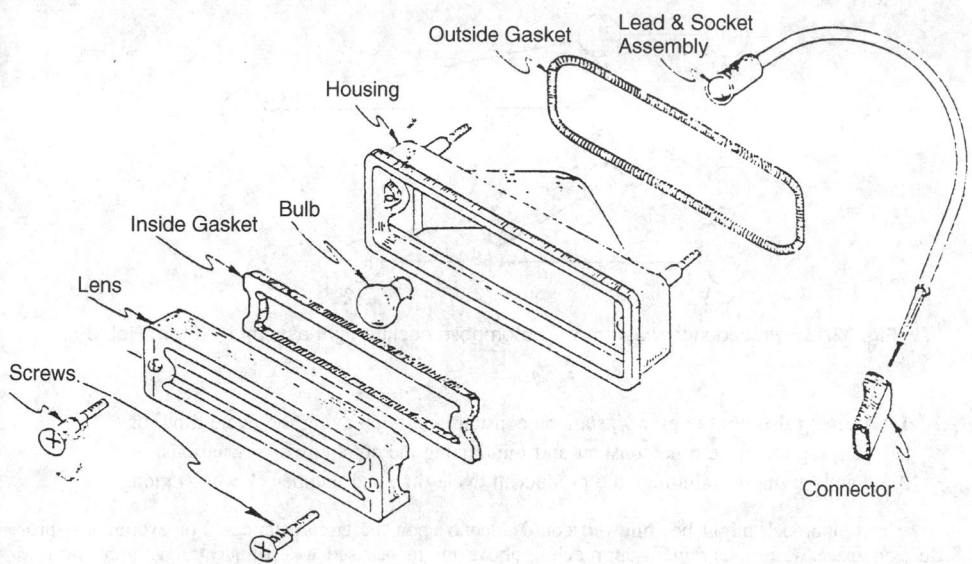

Fig. 32.14 Automobile back-up light assembly. (From Ref. 39.)

Note that if the balance delay is summed up over the entire production line, the total balance delay (total idle time) is equal to

$$\sum_{j=1}^{J} (C - S_j)$$

So, to minimize the total balance delay is the same as

$$Min. \sum_{j=1}^{J} (C - S_j) = JC - \sum_{j=1}^{J} S_j$$

$$= JC - \text{total work content for one unit of product}$$

$$= JC - \text{a constant}$$

Therefore, minimizing the total balance delay is equivalent to

Table 32.26

Workstation	Work Elements Assigned	Station Time	Balance Delay Time
1	10	.104	
	20	.105	
		.209	0.016
2	30	.102	
	70	.081	
		.183	0.042
3	40	.100	
	50	.053	
	60	.048	
		.201	0.024
4	80	.112	
	90	.097	
		.209	0.016

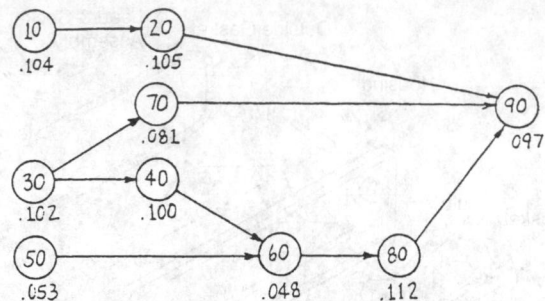

Fig. 32.15 Precedence diagram for automobile backup-light assembly. (From Ref. 39.)

1. Keeping the number of workstations constant and minimizing the cycle time, or
2. Keeping the cycle time constant and minimizing the number of workstations, or
3. Jointly trying to minimize the product of cycle time and number of workstations

Which approach might be followed could depend upon the circumstances. For example, if production space were constrained, approach 1 above might be used to estimate the volume the line would be capable of producing. Approach 2 might be used if the primary concern were ensuring a certain volume of product could be produced in a certain quantity of time. Approach 3 could be used in developing alternative assignments by trading off faster line speed (shorter cycle times, more workstations and greater production) for slower line speeds (fewer workstations, longer cycle times, and less production).

Designing the Assembly Line

Given the above structure and definitions, the following must hold.

1. $\{\max t_j\} \le C \le T$.
2. Minimum number of work stations $= [T/C]$, where the brackets indicate the value is rounded to the next largest integer.
3. $C_{max} = \dfrac{\text{production time available}}{\text{production volume required}}$ (C_{max} is the maximum value the cycle time can be, if the line is to generate the specified quantity in the specified time.)

As an example, consider the data provided in Table 32.25 and Figs. 32.14 and 32.15. Designing a line to produce 2000 units in seven-and-a-half-hour shift would give the following:

From condition 3:

$$C_{max} = \frac{(7\frac{1}{2}\ \text{hr/shift})\ (60\ \text{min/hr})}{2000\ \text{units (shift)}} = .225\ \text{min/unit}$$

From condition 2:

$$\text{minimum no. of workstations} = \frac{.802}{.225} = [3.56] = 4$$

Also note condition 1 is satisfied, i.e.,

$$.112 \le .225 \le .802$$

Line Balancing Techniques

Efforts have been made to optimally model variations of these problems, but currently no procedures exist that guarantee optimal solutions to these types of problems. Practitioners, therefore, have developed a variety of heuristic procedures (for examples, see Ref. 40). A general approach in making the assignment of work elements to workstations is to select a cycle time and to start assigning work elements where precedence restrictions are satisfied to the first workstation. Combinations of work elements may be explored in order to reduce the idle time present to the lowest level possible before going to the next workstation and repeating the procedure. This process is continued until all work elements have been assigned.

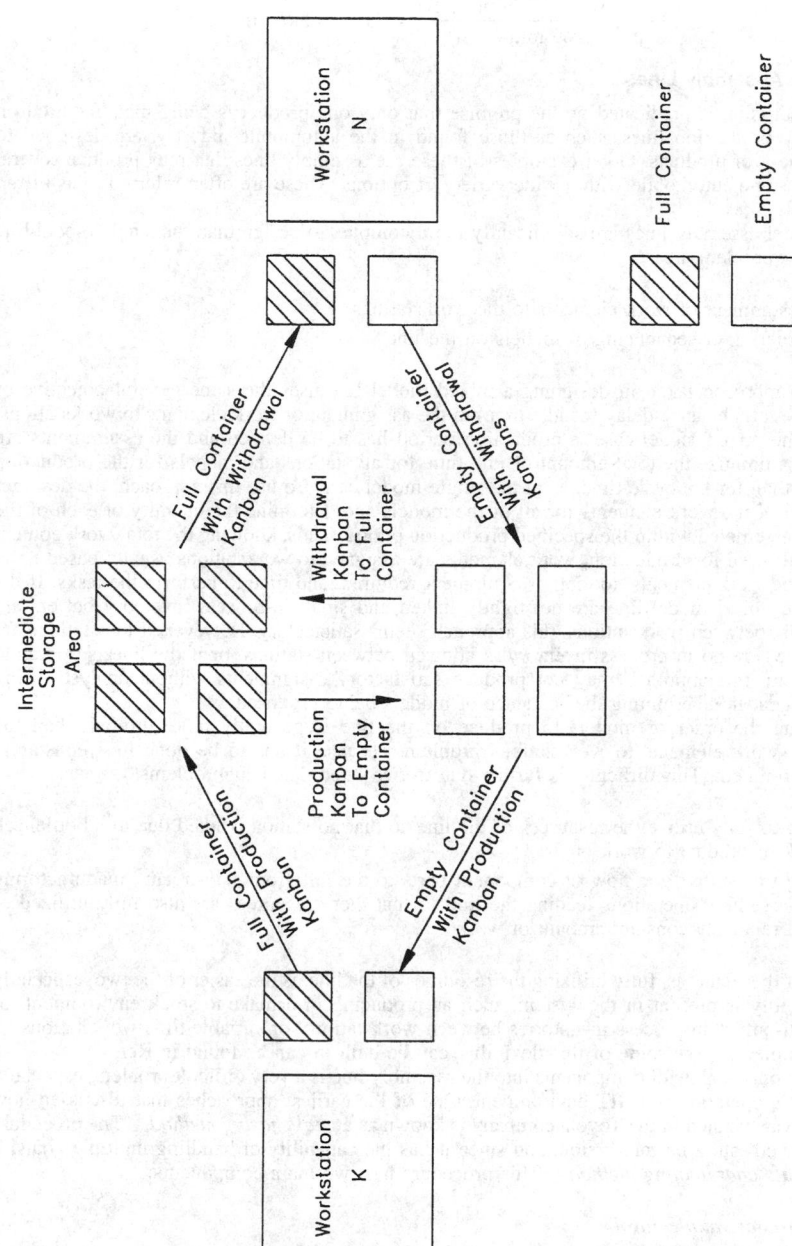

Fig. 32.16 Example of a two-card Kanban system.

Workstation
N

Full Container

Empty Container

Full Container
With Withdrawal
Kanban

Empty Container
With Withdrawal
Kanban

Withdrawal
Kanban
To Full
Container

Intermediate
Storage
Area

Production
Kanban
To Empty
Container

Full Container
With Production
Kanban

Empty Container
With Production
Kanban

Workstation
K

Applying this procedure to the data given in Table 32.25 and Fig. 32.14 would give the following solution, for a cycle time of 0.225 min. Also note that the cycle time could be reduced to 0.209 min with this assignment and would theoretically reduce the total balance delay by .016 min × 4 workstations = .064 min, resulting in a production increase to a total of

$$\frac{(7\frac{1}{2} \text{ hr/shift) (60 min/shift)}}{.209 \text{ min}} = 2153 \text{ units/shift}$$

Mixed Model Assembly Lines

The above discussion is predicated on the premise that only one product is being manufactured on the line. Many production lines, such as those found in the automobile industry, are designed to produce a variety of products. Good examples of these are assembly lines that may produce several models of the same automobile with a wide variety of options. These are often referred to as *mixed model assembly lines.*

Mixed model assembly lines are significantly more complex to design than the single model line because of two problem areas:

1. The assignment of work elements to the workstations
2. The ordering or sequencing of models on the line

One usual approach taken in designing a mixed model line uses the same general objective of minimizing the total balance delay (or idle time) in the assignment of work elements to workstations. However, in the mixed model case, a production period has to be defined and the assignments are made so as to minimize the total amount of idle time for all stations and models for the production period, rather than for the cycle time, as in the single model case. To use this approach, the designer must define all of the work elements for all of the models and determine the quantity of each of the models being assembled within the specified production period. Thus, knowing the total work content and the time allowed for production, work elements are assigned to workstations usually based upon similarity of the work elements, tooling or equipment required, and time to perform the tasks. If the stations on the mixed model line are not tightly linked and small in-process inventory buffers are allowed to exist between workstations, this approach seems satisfactory. However, if the stations are tightly linked where no in-process inventory is allowed between stations, or if the line is operating as a JIT system, this approach may not produce satisfactory assignments without analysts being especially diligent in determining the sequence of models being produced.

Determining the order of models to produce on the line is generally more difficult than the assignment of work elements to workstations problem, because it has to be done in a constantly changing environment. This difficulty is further due to two interrelated subproblems:

1. Trying to fully utilize the resources of the line so that no station is idled due to a bottleneck or lack of product to work on
2. Trying to even out the flow of component parts to the line from "upstream" manufacturing or subassembly operations feeding the line so that these resources are also fully utilized or have a relatively constant amount of work

The first of these, that is, fully utilizing the resources of the line, is the easier of the two, especially if some flexibility is present in the system, such as producing in a make-to-stock environment, or allowing small-buffer, in-process inventories between workstations, or variable-time workstations on the line. Examples of how some of this flexibility can be built in can be found in Ref. 41.

Smoothing out the flow of components into the assembly line is a very difficult problem, especially if the facility is operating in a JIT environment. One of the earliest approaches that discussed how this problem was handled in the Toyota company is known as a *goal chasing method.*[42] The procedure has since evolved into a newer version, and since it has the capability of handling multiple goals, it is called a *goals-coordinating method.*[43] This procedure has two main components:

1. *Appearance ratio control*
2. *Continuation and interval controls*

Appearance ratio control is a heuristic that determines the sequence of models on the line by attempting to minimize the variances of the components used for those products, that is, minimize the actual variation in component usage around a calculated average usage. A production schedule of end products is built by starting with the first end product to be scheduled, then working toward the last end product. For each step in determining the sequence, the following is calculated for each product, with the minimum D determining the next product to be produced:

$$D_{Ki} = \sqrt{\sum_{j=1}^{\beta} \left(\frac{KN_j}{Q} - X_{j,K-1} - b_{ij} \right)^2}$$

where D_{Ki} = the distance to be minimized for sequence number K and for end product i
β = the number of different components required
K = the sequence number of the current end product in the schedule
N_j = the total number of component j required for all products in the final schedule
Q = the total production quantity of all end products in the final schedule
$X_{j,K}$ = the cumulative number of component j actually used through assembly sequence K
b_{ij} = the number of component j required to make one unit of end product i

However, while this approach results in a smoothed production for the majority of the schedule, it will potentially cause very uneven use of components during the final phases of the day's schedule. To prevent this, continuation and interval controls are applied as constraints that may override the appearance ratio control and introduce other type models on the line. Continuation controls ensure that no more than a designated number of consecutive end products that use a particular component are scheduled (a maximum sequencing number condition), whereas interval controls ensure that at least a designated minimum number of certain end products are scheduled between other end products that require a particular component (a minimum sequencing condition).

The overall sequencing selection process then works as follows.

Step 1. Appearance ratio control is used to determine the first (or next) end product in sequence.

Step 2. If the selected end product also satisfies the continuation and interval controls, the end product is assigned that position in the sequence. Unless all end products have been scheduled, go to step 1. Otherwise, stop; the schedule is complete.

Step 3. If the selected end product does not satisfy both the continuation and interval controls, the appearance ratio control is applied to the remaining end products, while ignoring the component that violated the continuation and/or interval controls. Out of the end products that do not require the component in question, the end product that minimizes the amount of total deviation in the following formula would be selected as the next (Kth) in sequence (j = component number).

$$\sum_{j=1}^{n} \left| \frac{\text{total number of end product of the specified component } i}{\text{total number of end product}} \times K - \frac{\text{accumulated number of component } j}{\text{up to } (k-1)\text{th}} + \frac{\text{number of component } j \text{ of } K\text{th additional end product}}{} \right|$$

Unless all end products have been scheduled, go to step 1. Otherwise, stop; the schedule is complete.

As the number of models and components increase, the difficulties of developing satisfactory solutions for leveling production for mixed model lines also increase. As this occurs, the response is often to shorten the scheduled time period from, say, a day to every hour, to reduce the number of alternatives being investigated. While this may seem desirable, particularly if the facility is operating in a JIT environment, there is a danger that the resulting schedules may become so inefficient that they degrade the overall performance of the line. The leveling of production on mixed model lines remains an active research topic, with much of the research focusing on developing better or more efficient heuristic scheduling procedures.[44,45]

32.7 OTHER RELATED TOPICS

Within the arena of manufacturing, a number of new approaches have revolutionized thinking toward designing and controlling manufacturing organizations. Foremost have been the Japanese, who have developed and perfected a whole new philosophy. Some of the more important concepts related to production planning and control are presented below.

32.7.1 Japanese Manufacturing Philosophy

The central concepts are to design manufacturing systems as simply as possible and then to design simple control procedures to control them. This does not mean that Japanese manufacturing systems are simple, but that the design is well engineered to perform the required functions and the system is neither overdesigned nor underdesigned.

Central to this philosophy is the *Just in Time* (JIT) concept. JIT is a group of beliefs and management practices that attempt to eliminate all forms of "waste" in a manufacturing enterprise, where waste is defined as anything not necessary in the manufacturing organization. Waste in practice may include inventories, waiting times, equipment breakdowns, scrap, defective products, and excess equipment changeover times. The elimination of waste and the resulting simplification of the manufacturing organization are the results of implementing the following related concepts usually considered as defining or making up JIT.

1. *Kanban* (the word means "card") is used to control the movement and quantity of inventory through the shop, since a kanban card must be attached to each container of parts. The amount of production and in-process inventory, therefore, is controlled by the number of cards that are issued to the plant floor. An additional, major benefit of using Kanban is the very significant reduction in the information system that has to be used to control production.

   Various forms of Kanban exist, but the most frequently encountered are variations of the single-card or two-card system. One example of a two-card Kanban system is presented in Fig. 32.16. This example consists of two workstations, K and N. For simplicity, it is assumed that the production from Workstation K is used at Workstation N. The containers that move between these workstations have been sized to hold only a certain quantity of product. The two different types of Kanbans used are a *withdrawal* and a *production* Kanban. To control the amount of production for a given period of time, say one day, Workstation N is issued a predetermined number of withdrawal kanbans. The system operates as follows:

   a. When Workstation N needs parts, the operator takes an empty container, places a withdrawal kanban on it, and takes it to the storage area.

   b. The full containers in the storage area each have a production kanban on them. He removes the production kanban from a full container and places it on the empty container, and removes the withdrawal kanban and places it on the full container.

   c. He then transports the full container (now with the withdrawal kanban) back to Workstation N.

   d. Workstation K checks the production kanbans (on the empty containers) when checking for work to do. If a production kanban is present, this is his signal to begin production. If no production kanbans are present, Workstation K does not continue to produce parts.

   For this system to work, certain rules have to be adhered to:

   i. Each workstation works as long as there are *parts to work on* and a *container in which to put them*. If one or the other is missing, production stops.

   ii. There must be the same number of kanban cards as there are containers.

   iii. Containers are conveyed either full, with only their standard quantities, or empty.

2. *Lot size reduction* is used to reduce the amount of in-process inventory in concert with Kanban, by selecting the proper-sized containers to use, and increase the flexibility of the shop to change from one product to another. Overall benefits from using reduced lot sizes include shorter throughput times for product and thus smaller leadtimes required in satisfying customer orders.

3. *Scheduling* is used to schedule small lot production to increase the flexibility of the shop to be able to react to changes in demand and to produce the quantity of goods at just the time they are needed.

4. *Setup time reduction* is used to reduce the times required for machines to change from one product to another so as to allow lot size reduction and JIT scheduling. Reducing changeover times between products is critical to operating the production facility more like a flow shop and less like a job shop.

5. *Total quality management and maintenance* is used to reduce the disturbances to the manufacturing system by attempting to eliminate the making of defective products and breakdown of equipment. Central to the Japanese manufacturing philosophy is an obsession with maintenance and quality issues. For such a tightly controlled system to work, it is imperative that equipment function when it is supposed to and that components and products be produced that meet or exceed customers' requirements. Unexpected breakdowns or the production of bad parts is considered waste, and causes of such happenings are always high on the list for elimination in the quest for continuous improvement of the manufacturing processes.

6. *Employee cross training* is used to provide flexibility in the workforce to allow the organization to be able to react to changes in product demand and its resultant effect on the type and quantity of employee skills required. Multiskilled workers are necessary prerequisites in any form of JIT implementation.

32.7.2 Time-Based Competition

Following on the heels of JIT and the Japanese manufacturing philosophy is a business strategy called *time-based competition* (TBC). The successes of these earlier approaches were primarily grounded in providing the customer with better, more consistent-quality products that might also be less expensive in certain cases. Quality and cost were the major attributes of competitiveness for the organization that successfully employed these techniques. Although being competitive in quality and cost will always be important, some industries are finding that they alone are not enough to maintain an edge over their competitors, since many of their competitors also have gained these benefits by implementing JIT and related concepts. A third element is being introduced—that of time. TBC seeks a competitive advantage by the reduction of lead times associated with getting product to customers. TBC attempts to achieve reductions in the times required to design, manufacture, sell, and deliver products for its customers by analyzing and redesigning the processes that perform these functions.

TBC is seen as a natural evolution of JIT in that the implementation of JIT was most often found in production. Realizing that time spent on the shop floor represents less than one-half of the time it takes to get a product to the customer for most industries, TBC is a form of extension of JIT to the rest of the manufacturing organization, including such areas as design, sales, and distribution. Wherever in the organization lead times exist that lengthen the time it takes to get the desired product to the customer, the TBC approach seeks to reduce them.

Two forms of TBC exist: *first to market for new products* (FM) and *first to customer for existing products* (FC). Companies that seek to gain a competitive advantage through FM tend to be in dynamic industries such as automobile manufacturers and consumer products. For these industries, new innovations, developments, and improvements are important for their product's image, and are necessary to maintain and increase product sales. Companies employing FC as a competitive advantage tend to be in more stable industries, where innovations and new product developments are less frequent and dramatic. Thus, the products that competitors sell are very similar and competitive in terms of features, price, and quality. Here the emphasis is on speed—reducing the time it takes to get the product in the customer's hands from the time it was ordered is key. There is nothing, of course, preventing a company from employing both FM and FC approaches, and in the continuous improvement context, both approaches will be necessary if the full benefits of TBC are to be realized.

REFERENCES

1. W. A. Silver and R. Peterson, *Decision Systems for Inventory Management and Production Planning,* Wiley, New York, 1985.
2. H. A. Taha, *Operations Research,* 4th ed., Macmillan, New York, 1987.
3. D. P. Gover and G. L. Thompson, *Programming and Probability Models for Operations Research,* Wadsworth, Belmont, CA, 1973.
4. A. M. Wagner, *Principles of Management Science,* Prentice-Hall, Englewood Cliffs, NJ, 1970.
5. W. J. Fabrycky, P. M. Ghare, and P. E. Torgersen, *Applied Operations Research and Management Science,* Prentice-Hall, Englewood Cliffs, NJ, 1984.
6. S. Nam and R. Logendran, "Aggregate Production Planning—A Survey of Models and Methodologies," *European Journal of Operations Research* **61**, 255–272 (1992).
7. E. H. Bowman, "Production Scheduling by the Transportation Method of Linear Programming," *Operations Research* **4**, 100–103 (1956).
8. A. Charnes, W. W. Cooper, and B. Mellon, "A Model for Optimizing Production by Reference to Cost Surrogates," *Econometrics* **23**, 307–323 (1955).
9. M. E. Posner and W. Szwarc, "A Transportation Type Aggregate Production Model with Backordering," *Management Science* **29**, 188–199 (1983).
10. K. Sighal and V. Adlakha, "Cost and Shortage Trade-offs in Aggregate Production Planning," *Decision Science* **20**, 158–164 (1989).
11. C. C. Holt, F. Modigliani, and H. A. Simon, "A Linear Decision Rule for Production and Employment Scheduling," *Management Science* **2**, 1–30 (1955).
12. C. C. Holt, F. Modigliani, and J. F. Muth, "Derivation of a Linear Decision Rule for Production and Employment," *Management Science* **2**, 159–177 (1956).
13. C. C. Holt, F. Modigliani, J. F. Muth, and H. A. Simon. *Planning Production Inventories and Work Force,* Prentice-Hall, Englewood Cliffs, NJ, 1960.
14. A. S. Manne, "Programming of Economic Lot Sizes," *Management Science* **4**, 115–135 (1958).
15. H. M. Wagner and T. M. Whitin, "Dynamic Version of the Economic Lot Size Model," *Management Science* **5**, 89–96 (1958).
16. S. Gorenstein, "Planning Tire Production," *Management Science* **17**, B72–B82 (1970).
17. L. S. Lasdon and R. C. Terjung, "An Efficient Algorithm for Multi-item Scheduling," *Operations Research* **19**, 946–966 (1971).
18. S. M. Lee and L. J. Moore, "A Practical Approach to Production Scheduling," *Journal of Production and Inventory Management* **15**, 79–92 (1974).
19. R. F. Deckro and J. E. Hebert, "Goal Programming Approaches to Solving Linear Decision Rule Based Aggregate Production Planning Models," *IIE Transactions* **16**, 308–315 (1984).

20. D. P. Gaver, "Operating Characteristics of a Simple Production, Inventory-Control Model," *Operations Research* **9**, 635–649 (1961).
21. W. I. Zangwill, "A Deterministic Multiproduct, Multifacility Production and Inventory Model," *Operations Research* **14**, 486–507 (1966).
22. W. I. Zangwill, "A Deterministic Multiperiod Production Scheduling Model with Backlogging," *Management Science* **13**, 105–119 (1966).
23. W. I. Zangwill, "Production Smoothing of Economic Lot Sizes with Non-Decreasing Requirements," *Management Science* **13**, 191–209 (1966).
24. G. D. Eppen and F. J. Gould, "A Lagrangian Application to Production Models," *Operations Research* **16**, 819–829 (1968).
25. D. R. Lee and D. Orr, "Further Results on Planning Horizons in the Production Smoothing Problem," *Management Science* **23**, 490–498 (1977).
26. C. H. Jones, "Parametric Production Planning," *Management Science* **13**, 843–866 (1967).
27. A. D. Flowers and S. E. Preston, "Work Force Scheduling with the Search Decision Rule," *OMEGA* **5**, 473–479 (1977).
28. W. B. Lee and B. M. Khumawala, "Simulation Testing of Aggregate Production Planning Models in an Implementation Methodology," *Management Science* **20**, 903–911 (1974).
29. J. M. Mellichamp and R. M. Love, "Production Switching Heuristics for the Aggregate Planning Problem," *Management Science* **24**, 1242–1251 (1978).
30. H. Hwang and C. N. Cha, "An Improved Version of the Production Switching Heuristic for the Aggregate Production Planning Problems," *International Journal of Production Research* **33**, 2567–2577 (1995).
31. E. H. Bowman, "Consistency and Optimality in Management Decision Making," *Management Science* **9**, 310–321 (1963).
32. S. Eilon, "Five Approaches to Aggregate Production Planning," *AIIE Transactions* **7**, 118–131 (1975).
33. A. C. Hax and D. Candea, *Production and Inventory Management,* Prentice-Hall, Englewood Cliffs, NJ, 1984.
34. D. D. Bedworth and J. E. Bailey, *Integrated Production Control Systems,* Wiley, New York, 1982.
35. K. Baker, *Introduction to Sequencing and Scheduling,* Wiley, New York, 1974.
36. J. M. Moore, "Sequencing in Jobs on One Machine to Minimize the Number of Tardy Jobs," *Management Science* **17**(1), 102–109 (1968).
37. S. M. Johnson, "Optional Two-and-Three Stage Production Schedules with Setup Times Included," *Naval Research Logistics Quarterly* **1**, (1) (1954).
38. S. S. Panwalker and W. Iskander, "A Survey of Scheduling Rules," *Operations Research* **25**, 45–61 (1977).
39. J. Lorenz and D. Poock, "Assembly Line Balancing," in *Production Handbook,* 4th ed., J. White (ed.), Wiley, New York, 1987, pp. 3.176–3.189.
40. E. Elsayed and T. Boucher, *Analysis and Control of Production Systems,* Prentice-Hall, Englewood Cliffs, NJ, 1994.
41. N. Thomopoulos, "Mixed Model Line Balancing with Smoothed Station Assignments," *Management Science* **16**, 593–603 (1970).
42. Y. Moden, *Toyota Production System: Practical Approach to Production Management,* Industrial Engineering and Management Press, Atlanta, 1983.
43. Y. Moden, *Toyota Production System: An Integrated Approach to Just-In-Time,* 2nd ed., Industrial Engineering and Management Press, Atlanta, 1993.
44. R. T. Sumichrast and R. S. Russell, "Evaluating Mixed-Model Assembly Line Sequencing Heuristics for Just-in-Time Production Systems," *Journal of Operation Management* **9**, 371–386 (1990).
45. J. F. Bard, A. Shtub, and S. B. Joshi, "Sequencing Mixed-Model Assembly Lines to Level Parts Usage and Minimize Line Length," *International Journal of Production Research* **32**, 2431–2454 (1994).

BIBLIOGRAPHY

Adam, E. E., Jr., and R. J. Ebert, *Production and Operations Management,* 4th ed., Prentice-Hall, Englewood Cliffs, NJ, 1989.
Carter, P., S. Melnyk, and R. Handfield, "Identifying the Basic Process Strategies for Time Based Competition," *Production and Inventory Management Journal* **36**(1), 1st Qtr., 65–70 (1995).
Eilon, S., "Aggregate Production Scheduling," in *Handbook of Industrial Engineering,* G. Salvendy (ed.), Wiley Interscience, New York, 11.3.1–11.3.23.

French, S., *Sequencing and Scheduling: An Introduction to the Mathematics of the Job Shop,* Halsted Press, New York, 1982.

Gaither, N., *Production and Operations Management,* 6th ed., Dryden Press, Fort Worth, 1992.

Hax, A. C., "Aggregate Production Planning," in *Production Handbook,* 4th ed., J. White (ed.), Wiley, New York, 1987, pp. 3.116–3.127.

Johnson, L. A., and D. C. Montgomery, *Operations Research in Production Planning, Scheduling and Inventory Control,* Wiley, New York, 1974.

Nahmias, S., *Production and Operations Analysis,* 2nd ed., Irwin, Hinsdale, IL, 1993.

Schonberger, R. J., *Japanese Manufacturing Techniques,* Free Press, New York, 1983, pp. 219–245.

Vollmann, T. E., W. L. Berry, and D. C. Whybark, *Manufacturing Planning and Control Systems,* 3rd ed., Irwin, Homewood, IL, 1992.

CHAPTER 33

PRODUCTION PROCESSES AND EQUIPMENT

Magd E. Zohdi
Industrial Engineering Department
Louisiana State University
Baton Rouge, Louisiana

William E. Biles
Industrial Engineering Department
University of Louisville
Louisville, Kentucky

Dennis B. Webster
Industrial Engineering Department
Louisiana State University
Baton Rouge, Louisiana

33.1	METAL-CUTTING PRINCIPLES	1036
33.2	MACHINING POWER AND CUTTING FORCES	1039
33.3	TOOL LIFE	1041
33.4	METAL-CUTTING ECONOMICS	1043
	33.4.1 Cutting Speed for Minimum Cost (V_{min})	1043
	33.4.2 Tool Life Minimum Cost (T_m)	1043
	33.4.3 Cutting Speed for Maximum Production (V_{max})	1044
	33.4.4 Tool Life for Maximum Production (T_{max})	1046
33.5	CUTTING-TOOL MATERIALS	1046
	33.5.1 Cutting-Tool Geometry	1046
	33.5.2 Cutting Fluids	1047
	33.5.3 Machinability	1048
	33.5.4 Cutting Speeds and Feeds	1048
33.6	TURNING MACHINES	1048
	33.6.1 Lathe Size	1051
	33.6.2 Break-Even (BE) Conditions	1051
33.7	DRILLING MACHINES	1051
	33.7.1 Accuracy of Drills	1057
33.8	MILLING PROCESSES	1060
33.9	GEAR MANUFACTURING	1063
	33.9.1 Machining Methods	1063
	33.9.2 Gear Finishing	1067
33.10	THREAD CUTTING AND FORMING	1067
	33.10.1 Internal Threads	1067
	33.10.2 Thread Rolling	1068
33.11	BROACHING	1068
33.12	SHAPING, PLANING, AND SLOTTING	1070
33.13	SAWING, SHEARING, AND CUTTING OFF	1073
33.14	MACHINING PLASTICS	1074
33.15	GRINDING, ABRASIVE MACHINING, AND FINISHING	1074
	33.15.1 Abrasives	1074
	33.15.2 Temperature	1078
33.16	NONTRADITIONAL MACHINING	1079
	33.16.1 Abrasive Flow Machining	1079
	33.16.2 Abrasive Jet Machining	1079
	33.16.3 Hydrodynamic Machining	1079

Mechanical Engineers' Handbook, 2nd ed., Edited by Myer Kutz.
ISBN 0-471-13007-9   © 1998 John Wiley & Sons, Inc.

33.16.4	Low-Stress Grinding	1079
33.16.5	Thermally Assisted Machining	1084
33.16.6	Electromechanical Machining	1084
33.16.7	Total Form Machining	1085
33.16.8	Ultrasonic Machining	1086
33.16.9	Water-Jet Machining	1086
33.16.10	Electrochemical Deburring	1087
33.16.11	Electrochemical Discharge Grinding	1088
33.16.12	Electrochemical Grinding	1088
33.16.13	Electrochemical Honing	1089
33.16.14	Electrochemical Machining	1089
33.16.15	Electrochemical Polishing	1090
33.16.16	Electrochemical Sharpening	1090
33.16.17	Electrochemical Turning	1091
33.16.18	Electro-Stream	1091
33.16.19	Shaped-Tube Electrolytic Machining	1091
33.16.20	Electron-Beam Machining	1092
33.16.21	Electrical Discharge Grinding	1093
33.16.22	Electrical Discharge Machining	1093
33.16.23	Electrical Discharge Sawing	1094
33.16.24	Electrical Discharge Wire Cutting (Traveling Wire)	1094
33.16.25	Laser-Beam Machining	1095
33.16.26	Laser-Beam Torch	1096
33.16.27	Plasma-Beam Machining	1096
33.16.28	Chemical Machining: Chemical Milling, Chemical Blanking	1096
33.16.29	Electropolishing	1098
33.16.30	Photochemical Machining	1098
33.16.31	Thermochemical Machining	1099

33.1 METAL-CUTTING PRINCIPLES

Material removal by chipping process began as early as 4000 BC, when the Egyptians used a rotating bowstring device to drill holes in stones. Scientific work developed starting about the mid-19th century. The basic chip-type machining operations are shown in Fig. 33.1.

Figure 33.2 shows a two-dimensional type of cutting in which the cutting edge is perpendicular to the cut. This is known as *orthogonal* cutting, as contrasted with the three-dimensional *oblique* cutting shown in Fig. 33.3. The main three cutting velocities are shown in Fig. 33.4. The metal-cutting factors are defined as follows:

α	rake angle
β	friction angle
γ	strain
λ	chip compression ratio, t_2/t_1
μ	coefficient of friction
ψ	tool angle
τ	shear stress
ϕ	shear angle
Ω	relief angle
A_o	cross section, wt_1
e_m	machine efficiency factor
f	feed rate ipr (in./revolution), ips (in./stroke), mm/rev (mm/revolution), or mm/stroke
f_t	feed rate (in./tooth, mm/tooth) for milling and broaching
F	feed rate, in./min (mm/sec)
F_c	cutting force
F_f	friction force
F_n	normal force on shear plane
F_s	shear force
F_t	thrust force
HP_c	cutting horsepower
Hp_g	gross horsepower

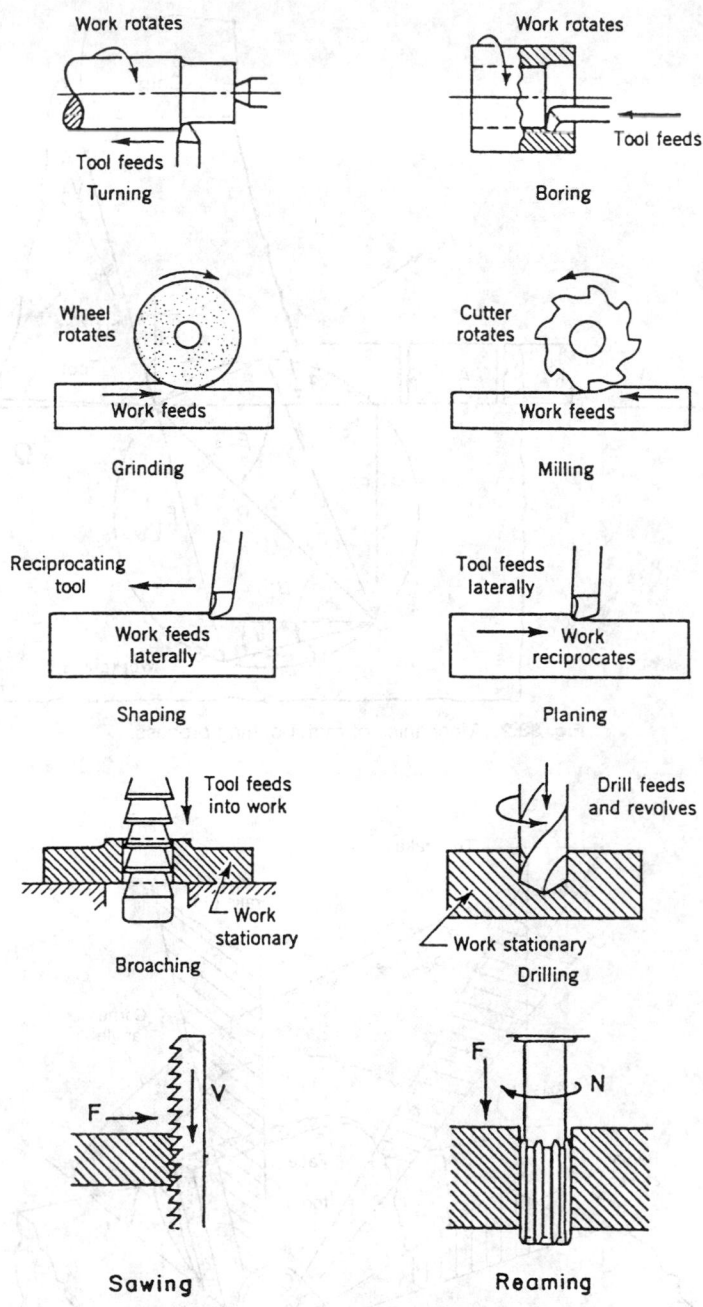

Fig. 33.1 Conventional machining processes.

Hp_{μ}	unit horsepower
N	revolutions per minute
Q	rate of metal removal, in.^{3}/min
R	resultant force
T	tool life in minutes
t_1	depth of cut

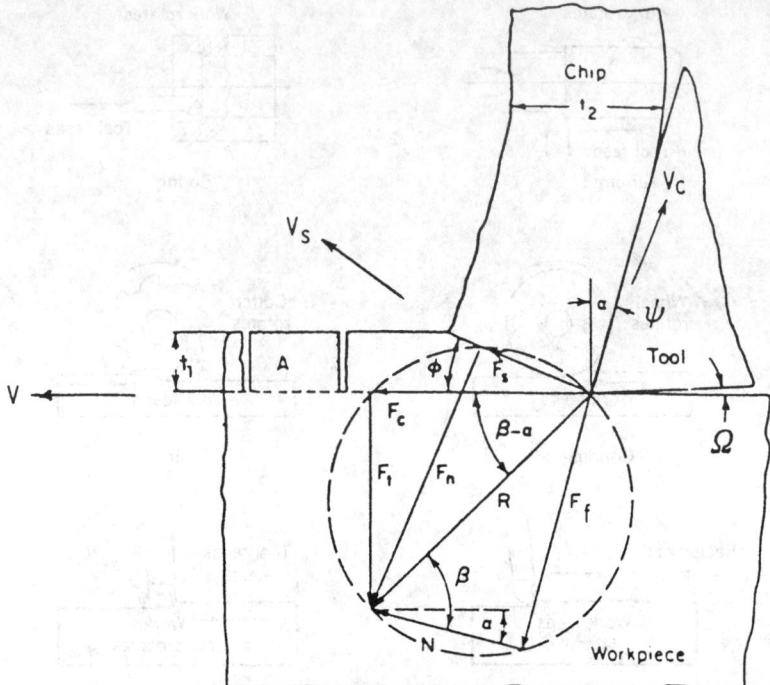

Fig. 33.2 Mechanics of metal-cutting process.

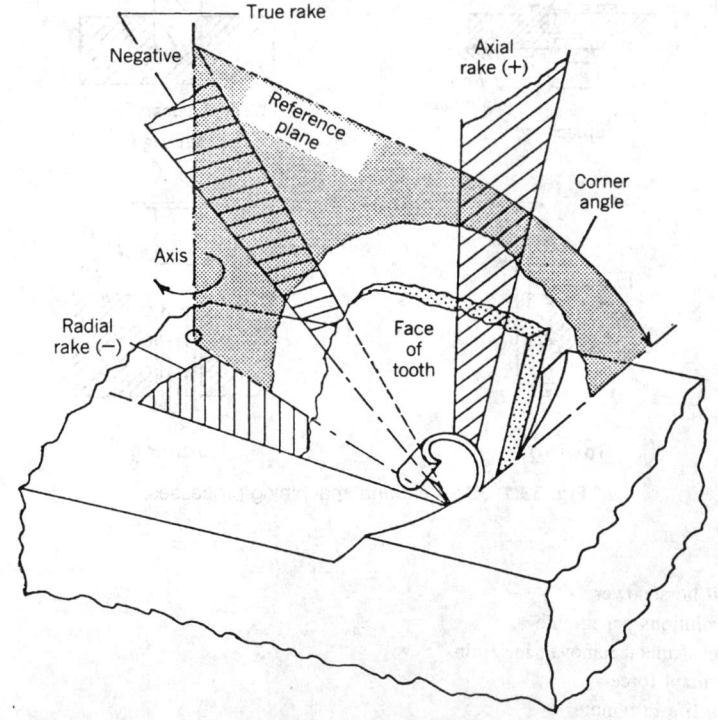

Fig. 33.3 Oblique cutting.

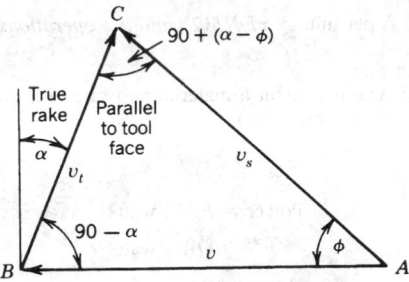

Fig. 33.4 Cutting velocities.

t_2	chip thickness
V	cutting speed, ft/min
V_c	chip velocity
V_s	shear velocity

The *shear angle* ϕ controls the thickness of the chip and is given by

$$\tan \phi = \frac{\cos \alpha}{\lambda - \sin \alpha} \tag{33.1}$$

The *strain* γ that the material undergoes in shearing is given by

$$\gamma = \cot \phi + \tan(\phi - \alpha)$$

The *coefficient of friction* μ on the face of the tool is

$$\mu = \frac{F_t + F_c \tan \alpha}{F_c - F_t \tan \alpha} \tag{33.2}$$

The *friction force* F_t along the tool is given by

$$F_t = F_t \cos \alpha + F_c \sin \alpha$$

Cutting forces are usually measured with dynamometers and/or wattmeters. The shear stress τ in the shear plane is

$$\tau = \frac{F_c \sin \phi \cos \phi - F_t \sin^2 \phi}{A}$$

The speed relationships are

$$\frac{V_c}{V} = \frac{\sin \phi}{\cos(\phi - \alpha)}$$

$$V_c = V/\lambda \tag{33.3}$$

33.2 MACHINING POWER AND CUTTING FORCES

Estimating the power required is useful when planning machining operations, optimizing existing ones, and specifying new machines. The power consumed in cutting is given by

$$\text{power} = F_c V \tag{33.4}$$

$$HP_c = \frac{F_c V}{33,000} \tag{33.5}$$

$$= Q \, HP_\mu \tag{33.6}$$

where F_c = cutting force, lb
V = cutting speed, ft per min = $\pi DN/12$ (*rotating operations*)
D = diameter, in.
N = revolutions per min
HP_μ = specific power required to cut a material at a rate of 1 cu in. per min
Q = material removal rate, cu in./min

For SI units,

$$Power = F_c V \quad watts \tag{33.7}$$

$$= QW \quad watts \tag{33.8}$$

where F_c = cutting force, newtons
V = m per sec = $2\pi RN$
W = specific power required to cut a material at a rate of 1 cu mm per sec
Q = material removal rate, cu mm per sec

The specific energies for different materials, using sharp tools, are given in Table 33.1.

$$power = F_c V = F_c 2\pi RN$$

$$= F_c R 2\pi N$$

$$= M 2\pi N \tag{33.9}$$

$$= \frac{MN}{63,025} \quad HP \tag{33.10}$$

where M = torque, in.–lbf
N = revolutions per min

In SI units,

$$= \frac{MN}{9549} \quad KW \tag{33.11}$$

Table 33.1 Average Values of Energy per Unit Material Removal Rate

Material	Bhn	HP_c/in.^{3} per min	W/mm^3 per sec
Aluminum alloys	50–100	0.3	0.8
	100–150	0.4	1.1
Cast iron	125–190	0.5	1.6
	190–250	1.6	4.4
Carbon steels	150–200	1.1	3.0
	200–250	1.4	3.8
	250–350	1.6	4.4
Leaded steels	150–175	0.7	1.9
Alloy steels	180–250	1.6	4.4
	250–400	2.4	6.6
Stainless steels	135–275	1.5	4.1
Copper	125–140	1.0	2.7
Copper alloys	100–150	0.8	2.2
Leaded brass	60–120	0.7	1.9
Unleaded brass	50	1.0	2.7
Magnesium alloys	40–70	0.2	0.55
	70–160	0.4	1.1
Nickel alloys	100–350	2.0	5.5
Refractory alloys (Tantalum, Columbium, Molybdenum)	210–230	2.0	5.5
Tungsten	320	3.0	8.0
Titanium alloys	250–375	1.3	3.5

where M = newton–meter

HP/cu in./min 2.73 = ? W/(cu mm/sec)

$M = F_cR$ = power/$2\pi N$

$F_c = M/R$

$$\text{gross power} = \text{cutting power}/e_m \tag{33.13}$$

The cutting horsepowers for different machining operations are given below.
    For turning, planing, and shaping,

$$HP_c = (HP_\mu)12CWVfd \tag{33.14}$$

For milling,

$$HP_c = (HP_\mu)CWFwd \tag{33.15}$$

For drilling,

$$HP_c = (HP_\mu)CW(N)f\left(\frac{\pi D^2}{4}\right) \tag{33.16}$$

For broaching,

$$HP_c = (HP_\mu)12CWVn_cwd_t \tag{33.17}$$

where V = cutting speed, fpm
    C = feed correction factor
    f = feed, ipr (turning and drilling), ips (planing and shaping)
    F = feed, ipm = $f \times N$
    d = depth of cut, in.
    d_t = maximum depth of cut per tooth, in.
    n_c = number of teeth engaged in work
    w = width of cut, in.
    W = tool wear factor

Specific energy is affected by changes in feed rate. Table 33.2 gives feed correction factor (C). Cutting speed and depth of cut have no significant effect on power. Tool wear effect factor (W) is given in Table 33.3.
    The gross power is calculated by applying the overall efficiency factor (e_m).

33.3 TOOL LIFE

Tool life is a measure of the length of time a tool will cut satisfactorily, and may be measured in different ways. Tool wear, as in Fig. 33.5, is a measure of tool failure if it reaches a certain limit. These limits are usually 0.062 in. (1.58 mm) for high-speed tools and 0.030 in. (0.76 mm) for carbide tools. In some cases, the life is determined by surface finish deterioration and an increase in cutting

Table 33.2 Feed Correction (C)

Factors for Turning, Milling, Drilling, Planing, and Shaping		
Feed (ipr or ips)	mm/rev or mm/stroke	Factor
0.002	0.05	1.4
0.005	0.12	1.2
0.008	0.20	1.05
0.012	0.30	1.0
0.020	0.50	0.9
0.030	0.75	0.80
0.040	1.00	0.80
0.050	1.25	0.75

Table 33.3 Tool Wear Factors (W)

Type of Operations[a]	W
Turning	
Finish turning (light cuts)	1.10
Normal rough and semifinish turning	1.30
Extra-heavy-duty rough turning	1.60–2.00
Milling	
Slab milling	1.10
End milling	1.10
Light and medium face milling	1.10–1.25
Extra-heavy-duty face milling	1.30–1.60
Drilling	
Normal drilling	1.30
Drilling hard-to-machine materials and drilling with a very dull drill	1.50
Broaching	
Normal broaching	1.05–1.10
Heavy-duty surface broaching	1.20–1.30

[a]For all operations with sharp cutting tools.

forces. The cutting speed is the variable that has the greatest effect on tool life. The relationship between tool life and cutting speed is given by the Taylor equation.

$$VT^n = C \tag{33.18}$$

where V = cutting speed, fpm (m/sec)
$\quad T$ = tool life, min (sec)
$\quad n$ = exponent depending on cutting condition
$\quad C$ = constant, the cutting speed for a tool life of 1 min

Table 33.4 gives the approximate ranges for the exponent n. Taylor's equation is equivalent to

$$\log V = C - n \log T \tag{33.19}$$

which when plotted on log–log paper gives a straight line, as shown in Fig. 33.6.
Equation (33.20) incorporates the size of cut:

$$K = VT^n f^{n1} d^{n2} \tag{33.20}$$

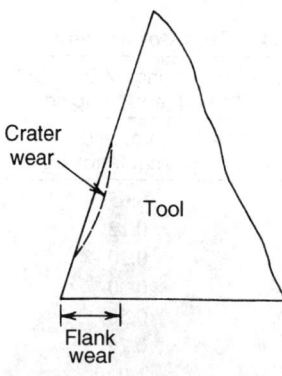

Fig. 33.5 Types of tool wear.

Table 33.4 Average Values of n

Tool Material	Work Material	n
HSS (18-4-1)	Steel	0.15
	C.I.	0.25
	Light metals	0.40
Cemented carbide	Steel	0.30
	C.I.	0.25
Sintered carbide	Steel	0.50
Ceramics	Steel	0.70

Average values for $n_1 = .5-.8$
$\quad\quad\quad\quad\quad\quad n_2 = .2-.4$

Equation (33.21) incorporates the hardness of the workpiece:

$$K = VT^n f^{n1} d^{n2} (BHN)^{1.25} \tag{33.21}$$

33.4 METAL-CUTTING ECONOMICS

The efficiency of machine tools increases as cutting speeds increase, but tool life is reduced. The main objective of metal-cutting economics is to achieve the optimum conditions, that is, the minimum cost while considering the principal individual costs: machining cost, tool cost, tool-changing cost, and handling cost. Figure 33.7 shows the relationships among these four factors.

$$\text{machining cost} = C_o t_m \tag{33.22}$$

where C_o = operating cost per minute, which is equal to the machine operator's rate plus appropriate overhead
$\quad\quad t_m$ = machine time in minutes, which is equal to $L/(fN)$, where L is the axial length of cut

$$\text{tool cost per operation} = C_t \frac{t_m}{T} \tag{33.23}$$

where C_t = tool cost per cutting edge
$\quad\quad T$ = tool life, which is equal to $(C/V)^{1/n}$

$$\text{tool changing cost} = C_o t_c (t_m/T) \tag{33.24}$$

where t_c = tool changing time, min

$$\text{handling cost} = C_o t_h$$

where t_h = handling time, min

The average unit cost C_u will be equal to

$$C_u = C_o t_m + \frac{t_m}{T}(C_t + C_o t_c) + C_o t_h \tag{33.25}$$

33.4.1 Cutting Speed for Minimum Cost (V_{min})

Differentiating the costs with respect to cutting speed and setting the results equal to zero will result in V_{min}:

$$V_{min} = \frac{C}{\left(\dfrac{1}{n} - 1\right)\left(\dfrac{C_o t + C_t}{C_o}\right)^n} \tag{33.26}$$

33.4.2 Tool Life Minimum Cost (T_m)

Since the constant C is the same in Taylor's equation and Eq. (33.23), and if V corresponds to V_{min}, then the tool life that corresponds to the cutting speed for minimum cost is

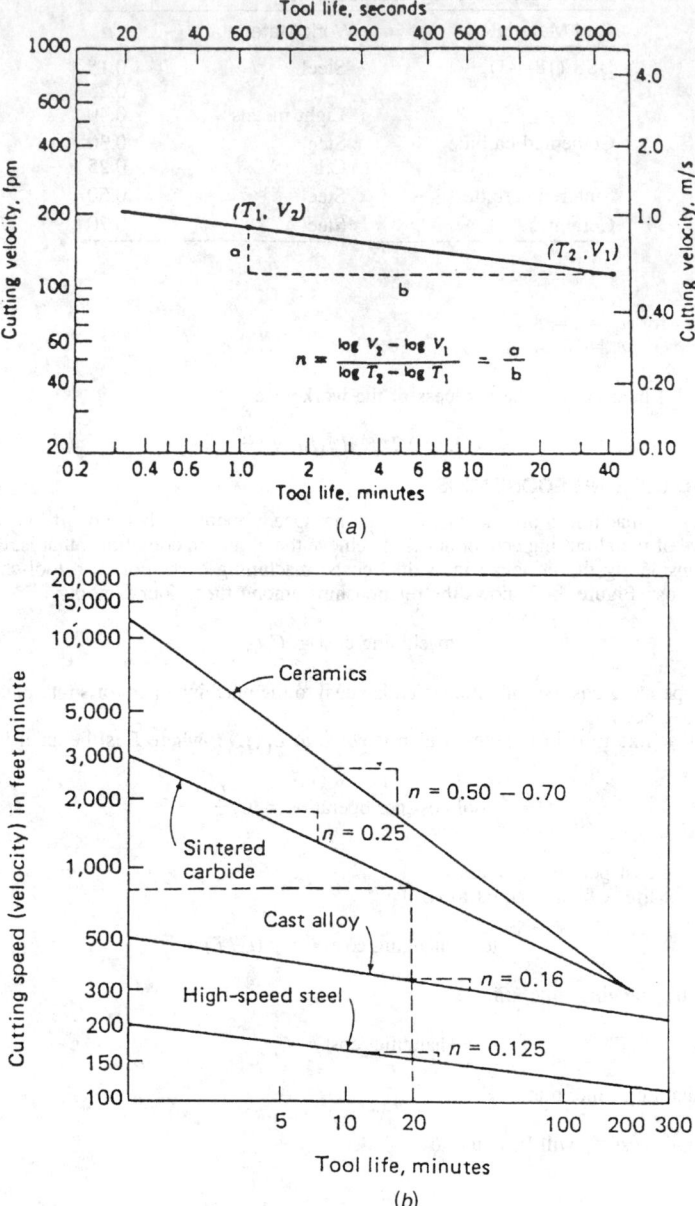

Fig. 33.6 Cutting speed/tool life relationship.

$$T_{\min} = \left(\frac{1}{n} - 1\right)\left(\frac{C_o t_c + C_t}{C_o}\right)$$ (33.27)

33.4.3 Cutting Speed for Maximum Production ($V_{\max}$)

This speed can be determined from Eq. (33.26) for the cutting speed for minimum cost by assuming that the tool cost is negligible, that is, by setting $C_1 = 0$:

$$V_{\max} = \frac{C}{\left[\left(\dfrac{1}{n} - 1\right) t_c\right]^n}$$ (33.28)

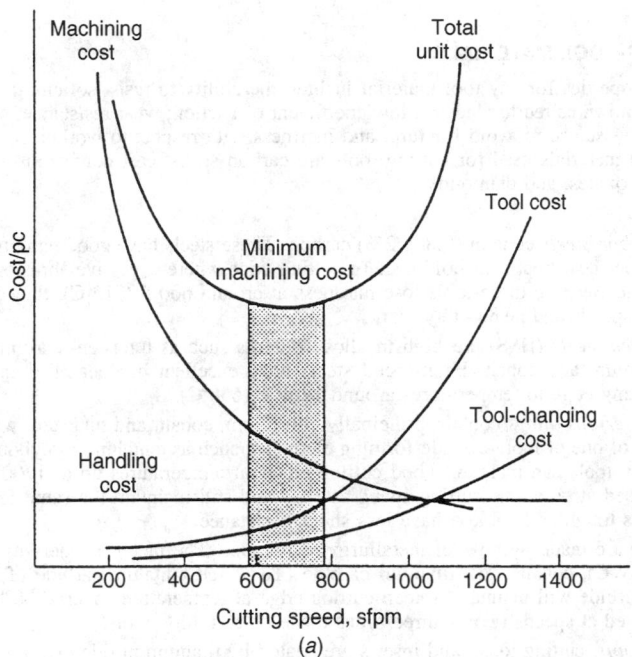

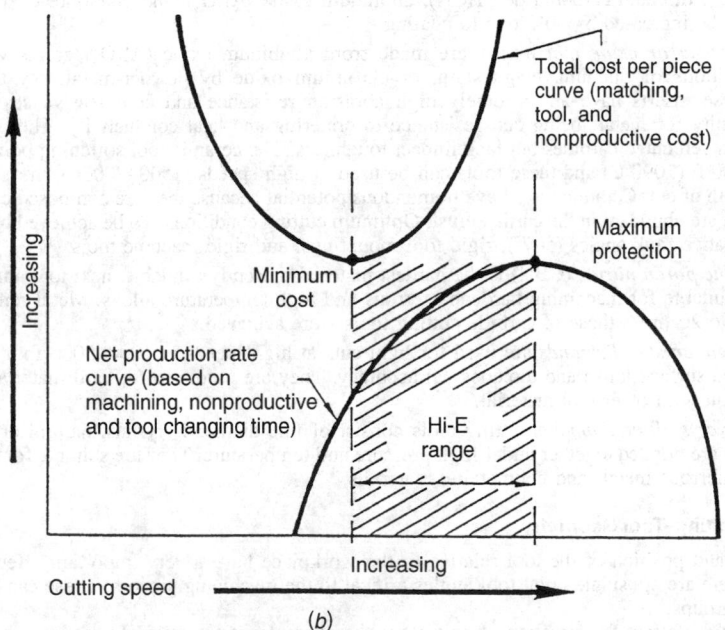

Fig. 33.7 Cost factors.

33.4.4 Tool Life for Maximum Production (T_{max})

By analogy to Taylor's equation, the tool life that corresponds to the maximum production rate is given by

$$T_{max} = \left(\frac{1}{n} - 1\right) t_c \tag{33.29}$$

33.5 CUTTING-TOOL MATERIALS

The desirable properties for any tool material include the ability to resist softening at high temperature, which is known as red hardness; a low coefficient of friction; wear resistance; sufficient toughness and shock resistance to avoid fracture; and inertness with respect to workpiece material.

The principal materials used for cutting tools are carbon steels, cast nonferrous alloys, carbides, ceramic tools or oxides, and diamonds.

1. *High-carbon steels* contain (0.8–1.2%) carbon. These steels have good hardening ability, and with proper heat treatment hold a sharp cutting edge where excessive abrasion and high heat are absent. Because these tools lose hardness at around 600°F (315°C), they are not suitable for high speeds and heavy-duty work.

2. *High-speed steels* (HSS) are high in alloy contents such as tungsten, chromium, vanadium, molybdenum, and cobalt. High-speed steels have excellent hardenability and will retain a keen cutting edge to temperatures around 1200°F (650°C).

3. *Cast nonferrous alloys* contain principally chromium, cobalt, and tungsten, with smaller percentages of one or more carbide-forming elements, such as tantalum, molybdenum, or boron. Cast-alloy tools can maintain good cutting edges at temperatures up to 1700°F (935°C) and can be used at twice the cutting speed as HSS and still maintain the same feed. Cast alloys are not as tough as HSS and have less shock resistance.

4. *Carbides* are made by powder-metallurgy techniques. The metal powders used are tungsten carbide (WC), cobalt (Co), titanium carbide (TiC), and tantalum carbide (TaC) in different ratios. Carbide will maintain a keen cutting edge at temperatures over 2200°F (1210°C) and can be used at speeds two or three times those of cast alloy tools.

5. *Coated tools,* cutting tools, and inserts are coated by titanium nitride (TiN), titanium carbide (TiC), titanium carbonitride (TiCN), aluminum oxide (Al_2O_3), and diamond. Cutting speeds can be increased by 50% due to coating.

6. *Ceramic or oxide tool* inserts are made from aluminum oxide (Al_2O_3) grains with minor additions of titanium, magnesium, or chromium oxide by powder-metallurgy techniques. These inserts have an extremely high abrasion resistance and compressive strength, lack affinity for metals being cut, resistance to cratering and heat conductivity. They are harder than cemented carbides but lack impact toughness. The ceramic tool softening point is above 2000°F (1090°C) and these tools can be used at high speeds (1500–2000 ft/min) with large depth of cut. Ceramic tools have tremendous potential because they are composed of materials that are abundant in the earth's crust. Optimum cutting conditions can be achieved by applying negative rank angles (5–7°), rigid tool mountings, and rigid machine tools.

7. *Cubic boron nitride* (CBN) is the hardest material presently available, next to diamond. CBN is suitable for machining hardened ferrous and high-temperature alloys. Metal removal rates up to 20 times those of carbide cutting tools were achieved.

8. *Single-crystal diamonds* are used for light cuts at high speeds of 1000–5000 fpm to achieve good surface finish and dimensional accuracy. They are used also for hard materials difficult to cut with other tool material.

9. *Polycrystalline diamond* cutting tools consist of fine diamond crystals, natural or synthetic, that are bonded together under high pressure and temperature. They are suitable for machining nonferrous metals and nonmetallic materials.

33.5.1 Cutting-Tool Geometry

The shape and position of the tool relative to the workpiece have a very important effect in metal cutting. There are six single-point tool angles critical to the machining process. These can be divided into three groups.

Rake angles affect the direction of chip flow, the characteristics of chip formation, and tool life. Positive rake angles reduce the cutting forces and direct the chip flow away from the material. Negative rake angles increase cutting forces but provide greater strength, as is recommended for hard materials.

Relief angles avoid excessive friction between the tool and workpiece and allow better access of coolant to tool–work interface.

The *side cutting-edge angle* allows the full load of the cut to be built up gradually. The *end cutting-edge angle* allows sufficient clearance so that the surface of the tool behind the cutting point will not rub over the work surface.

The purpose of the *nose radiuses* is to give a smooth surface finish and to increase the tool life by increasing the strength of the cutting edge. The elements of the single-point tool are written in the following order: back rake angle, side rake angle, end relief angle, side relief angle, end cutting-edge angle, side cutting-edge angle, and nose radius. Figure 33.8 shows the basic tool geometry.

Cutting tools used in various machining operations often appear to be very different from the single-point tool in Figure 33.8. Often they have several cutting edges, as in the case of drills, broaches, saws, and milling cutters. Simple analysis will show that such tools are comprised of a number of single-point cutting edges arranged so as to cut simultaneously or sequentially.

33.5.2 Cutting Fluids

The major roles of the cutting fluids—liquids or gases—are

1. Removal of the heat friction and deformation
2. Reduction of friction among chip, tool, and workpiece
3. Washing away chips
4. Reduction of possible corrosion on both workpiece and machine
5. Prevention of built-up edges

Cutting fluids work as coolants and lubricants. Cutting fluids applied depend primarily on the kind of material being used and the type of operation. The four major types of cutting fluids are

1. Soluble oil emulsions with water-to-oil ratios of 20:1 to 80:1
2. Oils
3. Chemicals and synthetics
4. Air

At low cutting speeds (40 ft/min and below), oils are highly recommended, especially in tapping, reaming, and gear and thread machining. Cutting fluids with the maximum specific heat, such as soluble oil emulsions, are recommended at high speeds.

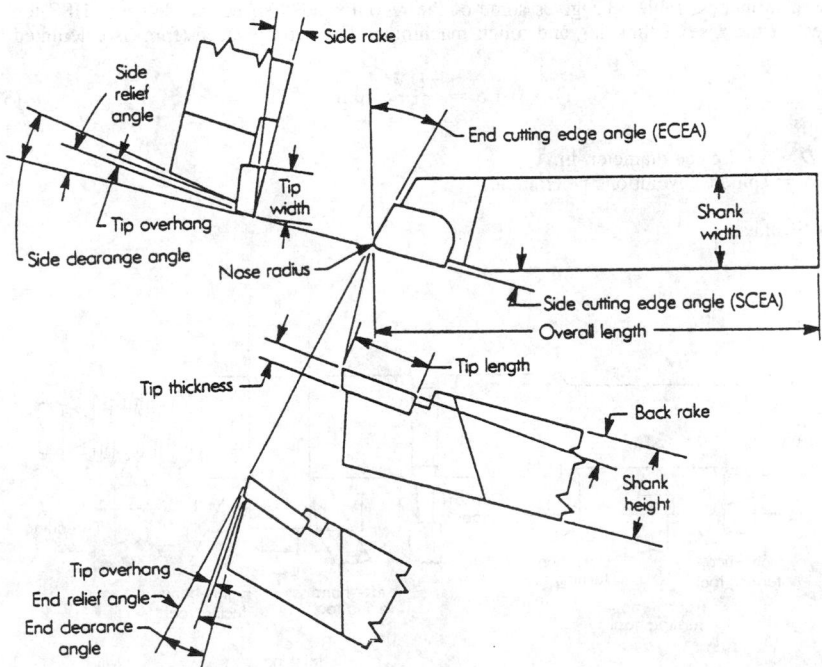

Fig. 33.8 Basic tool geometry.

33.5.3 Machinability

Machinability refers to a system for rating materials on the basis of their relative ability to be machined easily, long tool life, low cutting forces, and acceptable surface finish. Additives such as lead, manganese sulfide, or sodium sulfide with percentages less than 3% can improve the machinability of steel and copper-based alloys, such as brass and bronze. In aluminum alloys, additions up to 1–3% of zinc and magnesium improve their machinability.

33.5.4 Cutting Speeds and Feeds

Cutting speed is expressed in feet per minute (m/sec) and is the relative surface speed between the cutting tool and the workpiece. It may be expressed by the simple formula $CS = \pi DN/12$ fpm in., where D is the diameter of the workpiece in inches in case of turning or the diameter of the cutting tool in case of drilling, reaming, boring, and milling, and N is the revolutions per minute. If D is given in millimeters, the cutting speed is $CS = \pi DN/60,000$ m/sec.

Feed refers to the rate at which a cutting tool advances along or into the surface of the workpiece. For machines in which either the workpiece or the tool turns, feed is expressed in inches per revolution (ipr) (mm/rev). For reciprocating tools or workpieces, feed is expressed in inches per stroke (ips) (mm/stroke).

The recommended cutting speeds, and depth of cut that resulted from extensive research, for different combinations of tools and materials under different cutting conditions can be found in many references, including Society of Manufacturing Engineers (SME) publications such as *Tool and Manufacturing Engineers Handbook*;[1] *Machining Data Handbook*;[2] Metcut Research Associates, Inc.; *Journal of Manufacturing Engineers*; *Manufacturing Engineering Transactions*; *American Society for Metals (ASM) Handbook*;[3] *American Machinist's Handbook*;[4] *Machinery's Handbook*;[5] American Society of Mechanical Engineering (ASME) publications; Society of Automotive Engineers (SAE) Publications; and *International Journal of Machine Tool Design and Research*.

33.6 TURNING MACHINES

Turning is a machining process for generating external surfaces of revolution by the action of a cutting tool on a rotating workpiece, usually held in a lathe. Figure 33.9 shows some of the external operations that can be done on a lathe. When the same action is applied to internal surfaces of revolution, the process is termed *boring*. Operations that can be performed on a lathe are turning, facing, drilling, reaming, boring, chamfering, taping, grinding, threading, tapping, and knurling.

The primary factors involved in turning are speed, feed, depth of cut, and tool geometry. Figure 33.10 shows the tool geometry along with the feed (f) and depth of cut (d). The cutting speed (CS) is the surface speed in feet per minute (sfm) or meters per sec (m/s). The feed (f) is expressed in inches of tool advance per revolution of the spindle (ipr) or (mm/rev). The depth of cut (d) is expressed in inches. Table 33.5 gives some of the recommended speeds while using HSS tools and carbides for the case of finishing and rough machining. The cutting speed (fpm) is calculated by

$$CS = \frac{\pi DN}{12} \quad \text{fpm} \tag{33.30}$$

where D = workpiece diameter, in.
$\quad\quad N$ = spindle revolutions per minute

For SI units,

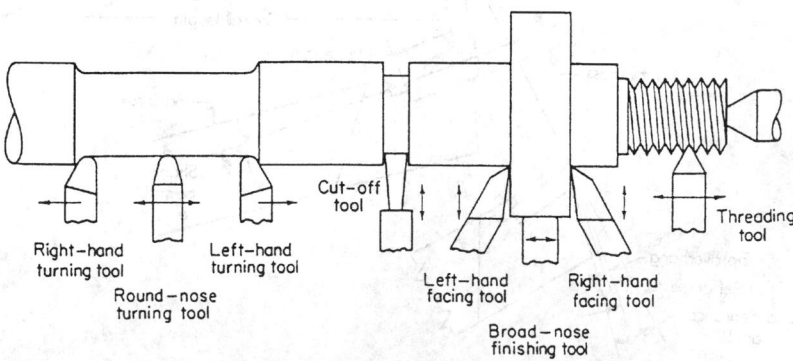

Fig. 33.9 Common lathe operations.

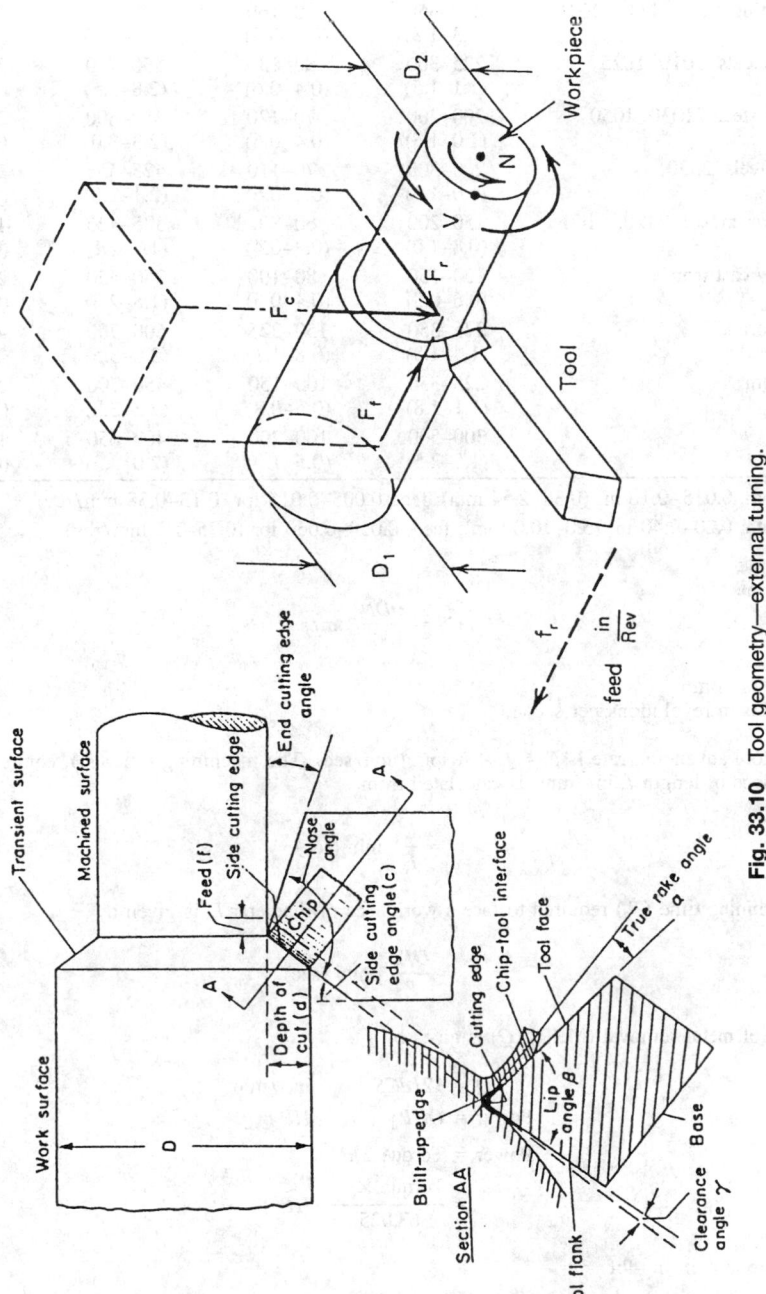

Fig. 33.10 Tool geometry—external turning.

Table 33.5 Typical Cutting Speeds ft/min (m/sec)

Material	High-Speed Steel		Carbide	
	Finish[a]	Rough[b]	Finish[a]	Rough[b]
Free cutting steels, 1112, 1315	250–350 (1.3–1.8)	80–160 (0.4–0.8)	600–750 (3.0–3.8)	350–500 (1.8–2.5)
Carbon steels, 1010, 1025	225–300 (1.1–1.5)	80–130 (0.4–0.6)	550–700 (2.8–3.5)	300–450 (1.5–2.3)
Medium steels, 1030, 1050	200–300 (1.0–1.5)	70–120 (0.4–0.6)	450–600 (2.3–3.0)	250–400 (1.3–2.0)
Nickel steels, 2330	200–300 (1.0–1.5)	70–110 (0.4–0.6)	425–550 (2.1–2.8)	225–350 (1.1–1.8)
Chromium nickel, 3120, 5140	150–200 (0.8–1.0)	60–80 (0.3–0.4)	325–425 (1.7–2.1)	175–300 (0.9–1.5)
Soft gray cast iron	120–150 (0.6–0.8)	80–100 (0.4–0.5)	350–450 (1.8–2.3)	200–300 (1.0–1.5)
Brass, normal	275–350 (1.4–1.8)	150–225 (0.8–1.1)	600–700 (3.0–3.5)	400–600 (2.0–3.0)
Aluminum	225–350 (1.1–1.8)	100–150 (0.5–0.8)	450–700 (2.3–3.5)	200–350 (1.0–1.8)
Plastics	300–500 (1.5–2.5)	100–200 (0.5–1.0)	400–650 (2.0–3.3)	150–300 (0.8–1.5)

[a]Cut depth, 0.015–0.10 in. (0.38–2.54 mm); feed 0.005–0.015 ipr (0.13–0.38 mm/rev).
[b]Cut depth, 0.20–0.40 in. (5.0–10.0 mm); feed, 0.030–0.060 ipr (0.75–1.5 mm/rev).

$$CS = \frac{\pi DN}{1000} \quad \text{m/s} \tag{33.31}$$

where D is in mm
$\quad N$ is in revolutions per second

The tool advancing rate is $F = f \times N$ ipm (mm/sec). The machining time (T_1) required to turn a workpiece of length L in. (mm) is calculated from

$$T_1 = \frac{L}{F} \quad \text{min (sec)} \tag{33.32}$$

The machining time (T_2) required to face a workpiece of diameter D is given by

$$T_2 = \frac{D/2}{F} \quad \text{min (sec)} \tag{33.33}$$

The rate of metal removal (MRR) (Q) is given by

$$Q = 12 f d CS \quad \text{in.}^3/\text{min} \tag{33.34}$$
$$\text{Power} = Q H P_\mu \quad HP \tag{33.35}$$
$$\text{Power} = \text{Torque } 2\pi N$$
$$= \frac{\text{Torque} \times N}{63{,}025} \quad HP \tag{33.36}$$

where torque is in in.–lbf

For SI units,

$$\text{Power} = \frac{\text{Torque} \times N}{9549} \quad KW \tag{33.37}$$

where torque is in newton–meter and N in rev/min
$\quad$ torque $= F_c \times R$

$$F_c = \frac{\text{Torque}}{R} \qquad (33.38)$$

where R = radius of workpiece

To convert to SI units,

$$HP \times 746 = ? \text{ Watt (W)}$$
$$f \text{ (lb)} \times 4.448 = ? \text{ newtons}$$
$$\text{torque (in.--lb)} \times 0.11298 = ? \text{ newton--meter (Nm)}$$
$$HP/(\text{cu in./min}) \times 2.73 = ? \text{ W/(cu mm/sec)}$$
$$\text{ft/min} \times .00508 = ? \text{ m/sec}$$
$$\text{in.}^3 \times 16{,}390 = ? \text{ mm}^3$$

Alignment charts were developed for determining metal removal rate and motor power in turning. Figures 33.11 and 33.12 show the method of using these charts either for English or metric units. The unit power (P) is the adjusted unit power with respect to turning conditions and machine efficiency.

33.6.1 Lathe Size

The size of a lathe is specified in terms of the diameter of the work it will swing and the workpiece length it can accommodate. The main types of lathes are engine, turret, single-spindle automatic, automatic screw machine, multispindle automatic, multistation machines, boring, vertical, and tracer. The level of automation can range from semiautomatic to tape-controlled machining centers.

33.6.2 Break-Even (BE) Conditions

The selection of a specific machine for the production of a required quantity q must be done in a way to achieve minimum cost per unit produced. The incremental setup cost is given by ΔC_t, C_1 is the machining cost per unit on the first machine, and C_2 is the machining cost for the second machine, the break-even point will be calculated as follows:

$$BE = \Delta C \sqrt{(C_1 - C_2)}$$

33.7 DRILLING MACHINES

Drills are used as the basic method of producing holes in a wide variety of materials. Figure 33.13 indicates the nomenclature of a standard twist drill and its comparison with a single-point tool. Knowledge of the thrust force and torque developed in the drilling process is important for design consideration. Figure 33.14 shows the forces developed during the drilling process. From the force diagram, the thrust force must be greater than $2P_y + P_y^1$ to include the friction on the sides and to be able to penetrate in the metal. The torque required is equal to P_2X. It is reported in the *Tool and Manufacturing Engineers Handbook*[1] that the following relations reasonably estimate the torque and thrust requirements of sharp twist drills of various sizes and designs.

Torque:

$$M = KF^{0.8}d^{1.8}A \quad \text{in.--lbf} \qquad (33.39)$$

Thrust:

$$T = 2Kf^{0.8}d^{0.8}B + kd^2E \quad \text{lb} \qquad (33.40)$$

The thrust force has a large effect upon the required strength, rigidity, and accuracy, but the power required to feed the tool axially is very small.

Cutting power:

$$HP = \frac{MN}{63{,}025} \qquad (33.41)$$

where K = work--material constant
$\quad f$ = drill feed, ipr
$\quad d$ = drill diameter, in.
A, B, E = design constants
$\quad N$ = drill speed, rpm

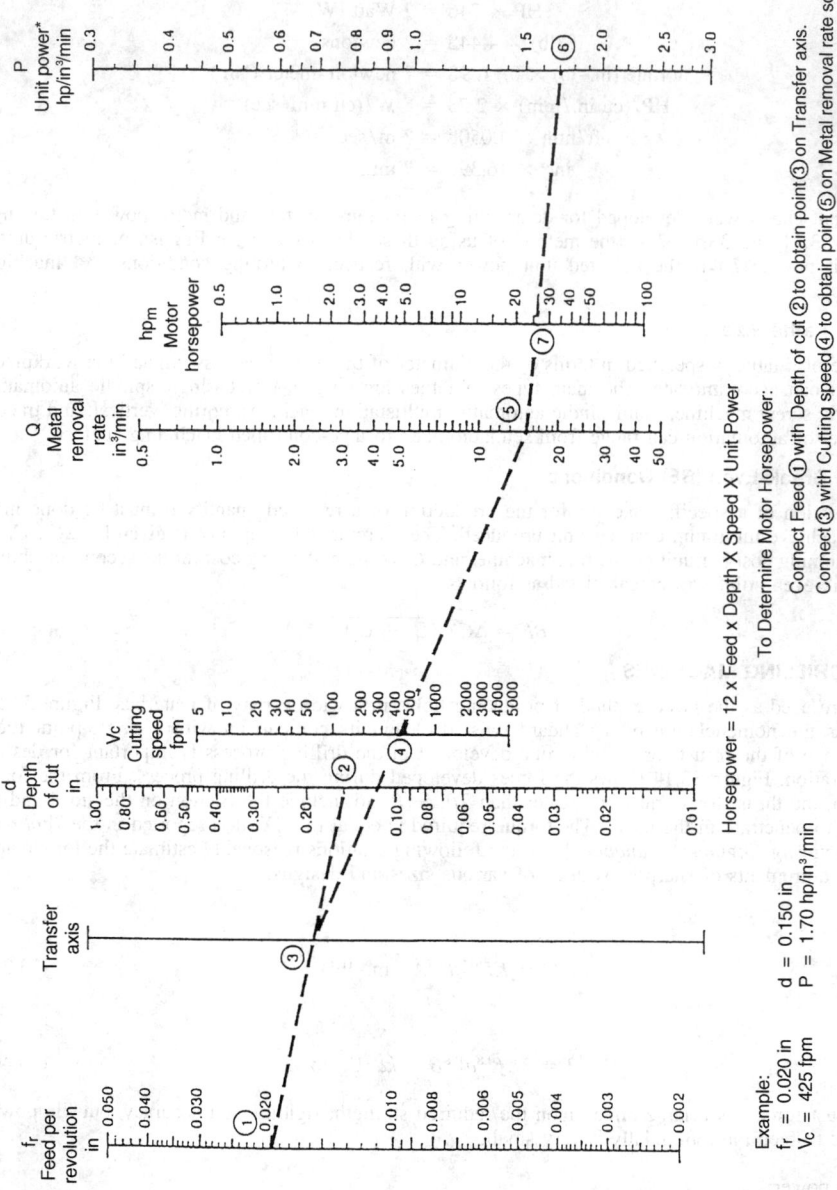

Horsepower = 12 x Feed x Depth X Speed X Unit Power

To Determine Motor Horsepower:

Connect Feed ① with Depth of cut ② to obtain point ③ on Transfer axis.
Connect ③ with Cutting speed ④ to obtain point ⑤ on Metal removal rate scale.
Connect point ⑤ with Unit power ⑥ to obtain 26 Motor horsepower at point ⑦.

Example:
d = 0.150 in
f_r = 0.020 in   P = 1.70 hp/in³/min
V_c = 425 fpm

Fig. 33.11 Alignment chart for determining metal removal rate and motor horsepower in turning—English units.

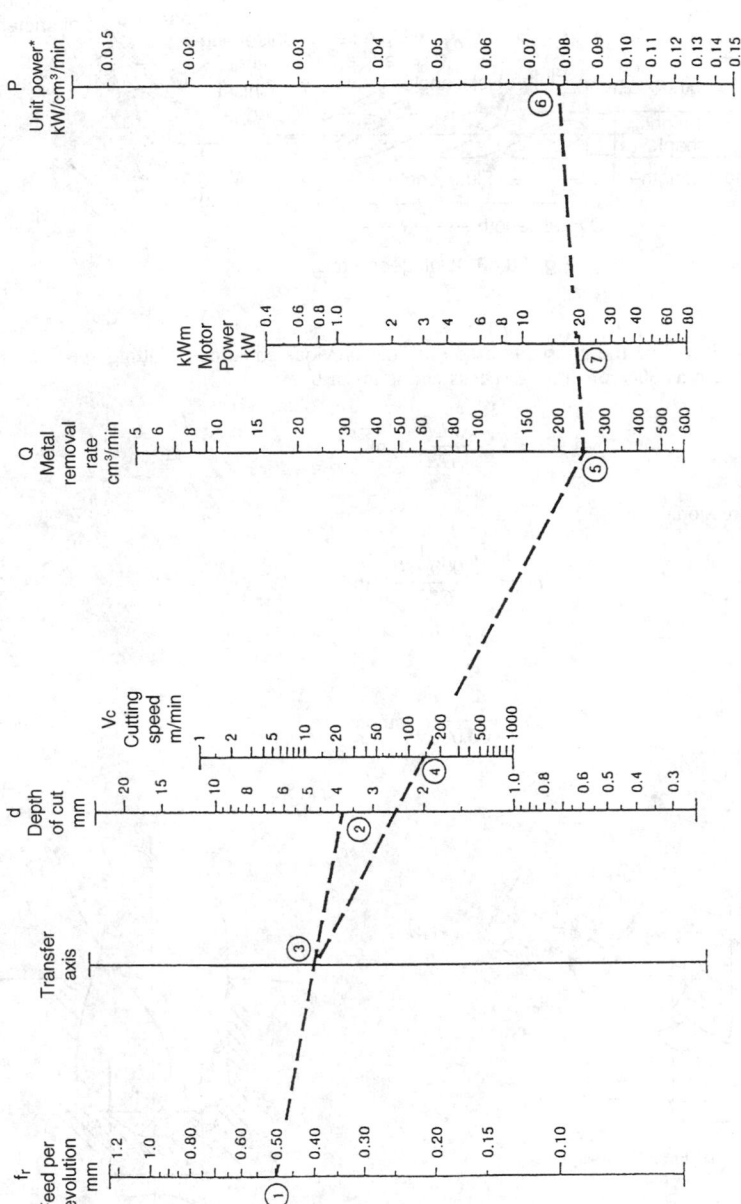

Power = Feed × Depth × Speed × Unit power

To Determine Motor Horsepower:

Connect Feed ① with Depth of cut ② to obtain point ③ on Transfer axis.
Connect ③ with Cutting speed ④ to obtain point ⑤ on Metal removal rate scale.
Connect point ⑤ with Unit power ⑥ to obtain 19.02 kW at Motor, point ⑦.

Example:
f_r = 0.5 mm   d = 3.8 mm
V_c = 130 m/min   P = 0.077 kW/cm³/min

Fig. 33.12 Alignment chart for determining metal removal rate and motor power in turning—metric units.

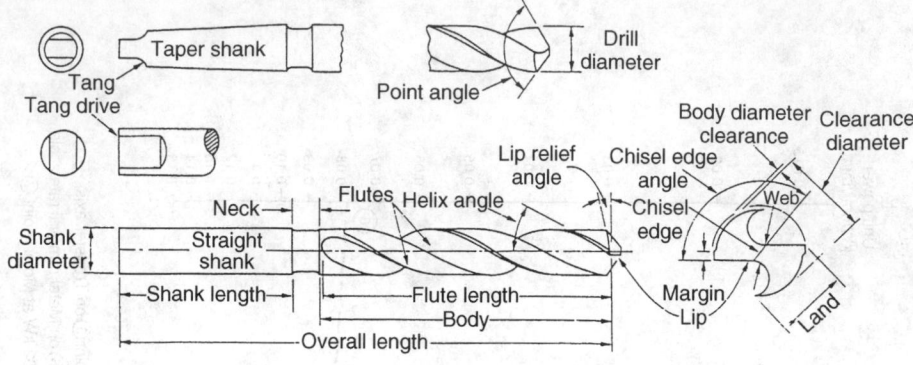

Fig. 33.13 Drill geometry.

Tables 33.6 and 33.7 give the constants used with the previous equations. Cutting speed at the surface is usually taken as 80% of turning speeds and is given by

$$CS = \frac{\pi dN}{12} \quad \text{fpm}$$

Force in cutting direction:

$$F_c = \frac{33,000 \, HP}{CS} \quad \text{lb} \tag{33.42}$$

For SI units,

$$CS = \frac{\pi d_1 N}{60,000} \quad \text{m/sec} \tag{33.43}$$

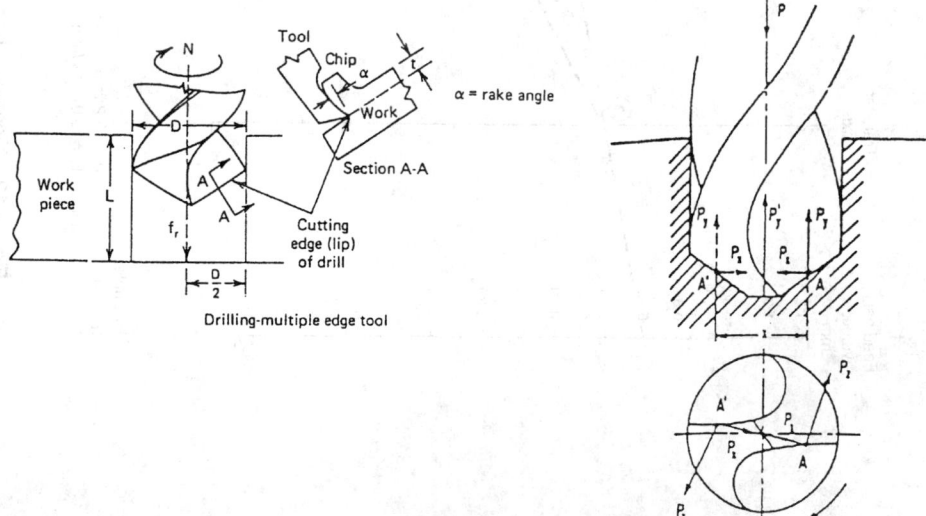

Fig. 33.14 Thrust forces and torque in drilling operation.

Table 33.6 Work-Material Constants for Calculating Torque and Thrust (National Twist Drill)

Work Material	K
Steel, 200 Bhn	24,000
Steel, 300 Bhn	31,000
Steel, 400 Bhn	34,000
Most aluminum alloys	7,000
Most magnesium alloys	4,000
Most brasses	14,000
Leaded brass	7,000
Cast iron, 65 Bhn	15,000
Free-machining mild steel, resulfurized	18,000
Austenitic stainless steel (type 316)	34,000

c = chisel-edge length, in.

d = drill diameter, in.

w = web thickness, in.

d_1 = drill diameter, in mm

Unit HP (hp/in.^{3}/min) $\times$ 2.73 = ? unit power (kW/cm^3/s)

$$kW = \frac{MN}{9549} \tag{33.44}$$

M = torque   Nm

For drills of regular proportion the ratio c/d is = 0.18 and c = 1.15 w, approximately.

It is a common practice to feed drills at a rate that is proportional to the drill diameter in accordance with

$$f = \frac{d}{65} \tag{33.45}$$

For holes that are longer than 3d, feed should be reduced. Also feeds and speeds should be adjusted due to differences in relative chip volume, material structure, cutting fluid effectiveness, depth of hole, and conditions of drill and machine. The advancing rate is

Table 33.7 Torque and Thrust Constants Based on Ratios c/d or w/d (National Twist Drill)

c/d	w/d	Torque Constant A	Thrust Constant B	Thrust Constant E
0.03	0.025	1.000	1.100	0.001
0.05	0.045	1.005	1.140	0.003
0.08	0.070	1.015	1.200	0.006
0.10	0.085	1.020	1.235	0.010
0.13	0.110	1.040	1.270	0.017
0.15	0.130	1.080	1.310	0.022
0.18	0.155	1.085	1.355	0.030
0.20	0.175	1.105	1.380	0.040
0.25	0.220	1.155	1.445	0.065
0.30	0.260	1.235	1.500	0.090
0.35	0.300	1.310	1.575	0.120
0.40	0.350	1.395	1.620	0.160

$$F = f \times N \quad \text{ipm} \tag{33.46}$$

The recommended feeds are given in Table 33.8.

The time T required to drill a hole of depth h is given by

$$T = \frac{h + 0.3d}{F} \quad \text{min} \tag{33.47}$$

The extra distance of $0.3d$ is approximately equal to the distance from the tip to the effective diameter of the tool. The rate of metal removal in case of blind holes is given by

$$Q = \left(\frac{\pi d^2}{4}\right) F \quad \text{in.}^3/\text{min} \tag{33.48}$$

When torque is unknown, the horsepower requirement can be calculated by

$$HP_c = Q \times C \times W \times (HP_\mu) \quad \text{hp}$$

C, W, HP_μ are given in previous sections.

$$\text{Power} = HP_c \times 396,000 \quad \text{in.–lb/min} \tag{33.49}$$

$$\text{Torque} = \frac{\text{Power}}{2\pi N} \quad \text{in–lbf}$$

$$F_c = \frac{\text{Torque}}{R} \quad \text{lb}$$

Along the cutting edge of the drill, the cutting speed is reduced toward the center as the diameter is reduced. The cutting speed is actually zero at the center. To avoid the region of very low speed and to reduce high thrust forces that might affect the alignment of the finished hole, a pilot hole is usually drilled before drilling holes of medium and large sizes. For the case of drilling with a pilot hole

$$Q = \frac{\pi}{4} (d^2 - d_p^2)F$$

$$= \frac{\pi}{4} (d + d_p)(d - d_p)F \quad \text{in.}^3/\text{min} \tag{33.50}$$

Due to the elimination of the effects of the chisel-edge region, the equations for torque and thrust can be estimated as follows:

Table 33.8 Recommended Feeds for Drills

Diameter		Feed	
(in.)	(mm)	(ipr)	(mm/rev)
Under ⅛	3.2	0.001–0.002	0.03–0.05
⅓–¼	3.2–6.4	0.002–0.004	0.05–0.10
¼–½	6.4–12.7	0.004–0.007	0.10–0.18
½–1	12.7–25.4	0.007–0.015	0.18–0.38
Over 1	25.4	0.015–0.025	0.38–0.64

$$M_p = M \left[\frac{1 - \left(\frac{d_p}{d}\right)^2}{\left(1 + \frac{d_1}{d}\right)^{0.2}} \right] \tag{33.51}$$

$$T_p = T \left[\frac{1 - \frac{d_1}{d}}{\left(1 + \frac{d_1}{d}\right)^{0.2}} \right] \tag{33.52}$$

where d_p = pilot hole diameter

Alignment charts were developed for determining motor power in drilling. Figures 33.15 and 33.16 show the use of these charts either for English or metric units. The unit power* (P) is the adjusted unit power with respect to drilling conditions and machine efficiency.

For English units,

$$HP_m = \frac{\pi D^2}{4} \times f \times N \times P*$$

$$N = \frac{12V}{\pi D}$$

As
$$HP_m = \frac{\pi D^2}{4} \times f \times \frac{12V}{\pi D} \times P*$$

$$= 3\, D \times f \times V \times P*$$

For metric units,

$$HP_m = \frac{\pi D^2}{4 \times 100} \times \frac{f}{10} \times N \times P$$

$$P* \text{ in kW/cm}^3/\text{min}$$

$$N = \frac{1000V}{\pi D}$$

$$HP_m = \frac{\pi D^2}{4 \times 100} \times \frac{f}{10} \times \frac{100V}{\pi D} \times P*$$

$$= 0.25\, D \times f \times V \times P*$$

33.7.1 Accuracy of Drills

The accuracy of holes drilled with a two-fluted twist drill is influenced by many factors, including the accuracy of the drill point; the size of the drill, the chisel edge, and the jigs used; the workpiece material; the cutting fluid used; the rigidity and accuracy of the machine used; and the cutting speed. Usually, when drilling most materials, the diameter of the drilled holes will be oversize. Table 33.9 provided the results of tests reported by The Metal Cutting Tool Institute for holes drilled in steel and cast iron.

Gun drills differ from conventional drills in that they are usually made with a single flute. A hole provides a passageway for pressurized coolant, which serves as a means of both keeping the cutting edge cool and flushing out the chips, especially in deep cuts.

Spade drills (Fig. 33.17) are made by inserting a spade-shaped blade into a shank. Some advantages of spade drills are (1) efficiency in making holes up to 15 in. in diameter; (2) low cost, since only the insert is replaced; (3) deep hole drilling; and (4) easiness of chip breaking on removal.

Trepanning is a machining process for producing a circular hole, groove, disk, cylinder, or tube from solid stock. The process is accomplished by a tool containing one or more cutters, usually single-point, revolving around a center. The advantages of trepanning are (1) the central core left is solid material, not chips, which can be used in later work; and (2) the power required to produce a given hole diameter is highly reduced because only the annulus is actually cut.

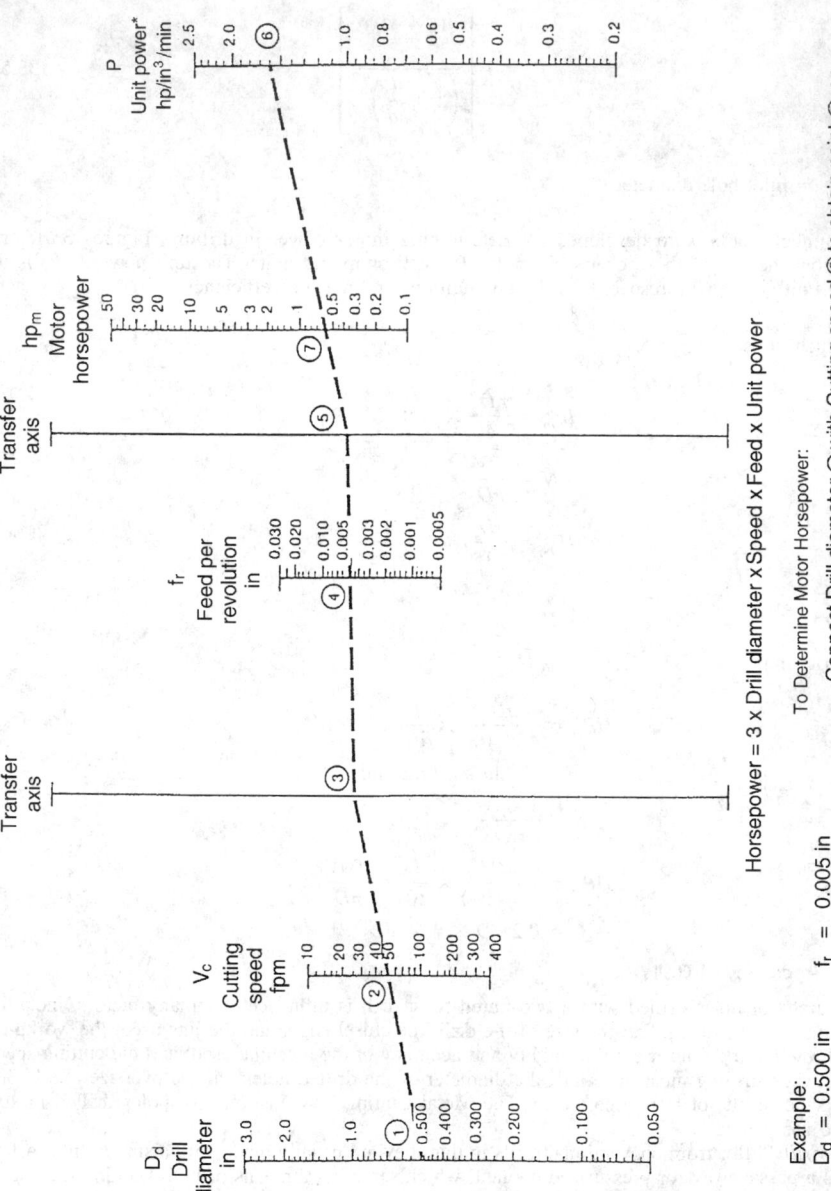

Horsepower = 3 × Drill diameter × Speed × Feed × Unit power

To Determine Motor Horsepower:

Connect Drill diameter ① with Cutting speed ② to obtain point ③ on Transfer axis. Connect ③ with Feed ④ to obtain point ⑤ on Transfer axis. Connect ⑤ with Unit power ⑥ to obtain 0.60 Motor horsepower, point ⑦.

Example:
D_d = 0.500 in f_r = 0.005 in
V_c = 50 fpm P = 1.6 hp/in³/min

Fig. 33.15 Alignment chart for determining motor horsepower in drilling—English units.

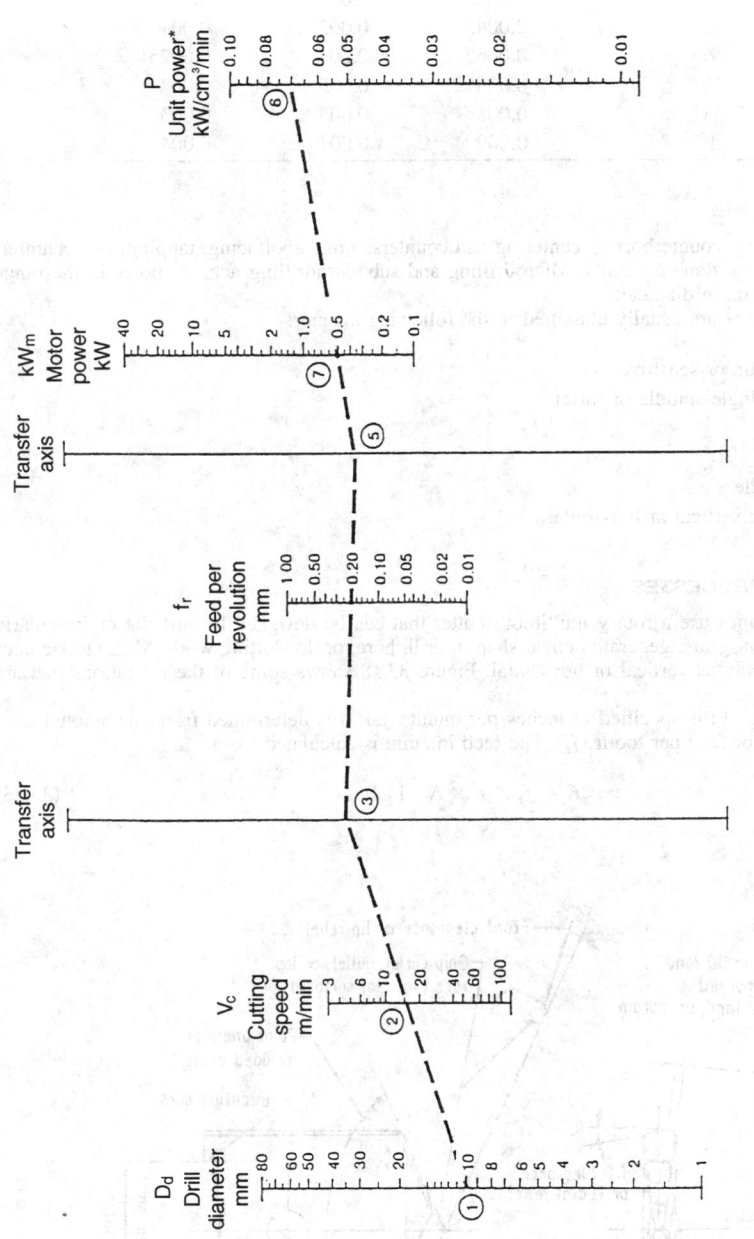

Fig. 33.16 Alignment chart for determining motor power in drilling—metric units.

Motor Power = 0.25 × Drill diameter × Speed × Feed × Unit Power

To Determine Motor Horsepower:

Connect Drill diameter ① with Cutting speed ② to obtain point ③ on Transfer axis. Connect ③ with Feed ④ to obtain point ⑤ on Transfer axis. Connect ⑤ with Unit power ⑥ to obtain 0.53 kW at motor power, point ⑦.

Example:

D_d = 10 mm f_r = 0.2 mm
V_c = 15 m/min   P = 0.07 kW/cm³/min

Table 33.9 Oversize Diameters in Drilling

Drill Diameter (in.)	Amount Oversize (in.)		
	Average Max.	Mean	Average Min.
$\frac{1}{16}$	0.002	0.0015	0.001
$\frac{1}{8}$	0.0045	0.003	0.001
$\frac{1}{4}$	0.0065	0.004	0.0025
$\frac{1}{2}$	0.008	0.005	0.003
$\frac{3}{4}$	0.008	0.005	0.003
1	0.009	0.007	0.004

Reaming, boring, counterboring, centering and countersinking, spotfacing, tapping, and chamfering processes can be done on drills. Microdrilling and submicrodrilling achieve holes in the range of 0.000025–0.20 in. in diameter.

Drilling machines are usually classified in the following manner:

1. Bench: plain or sensitive
2. Upright: single-spindle or turret
3. Radial
4. Gang
5. Multispindle
6. Deep-hole: vertical or horizontal
7. Transfer

33.8 MILLING PROCESSES

The milling machines use a rotary multitooth cutter that can be designed to mill flat or irregularly shaped surfaces, cut gears, generate helical shapes, drill, bore, or do slotting work. Milling machines are classified broadly as vertical or horizontal. Figure 33.18 shows some of the operations that are done on both types.

Feed in milling (F) is specified in inches per minute, but it is determined from the amount each tooth can remove or feed per tooth (f_t). The feed in./min is calculated from

$$F = f_t \times n \times N \quad \text{in./min} \tag{33.53}$$

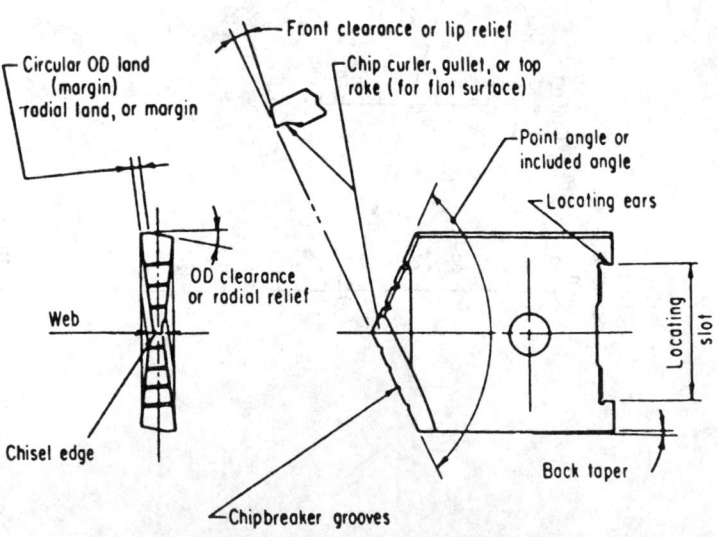

Fig. 33.17 Spade-drill blade elements.

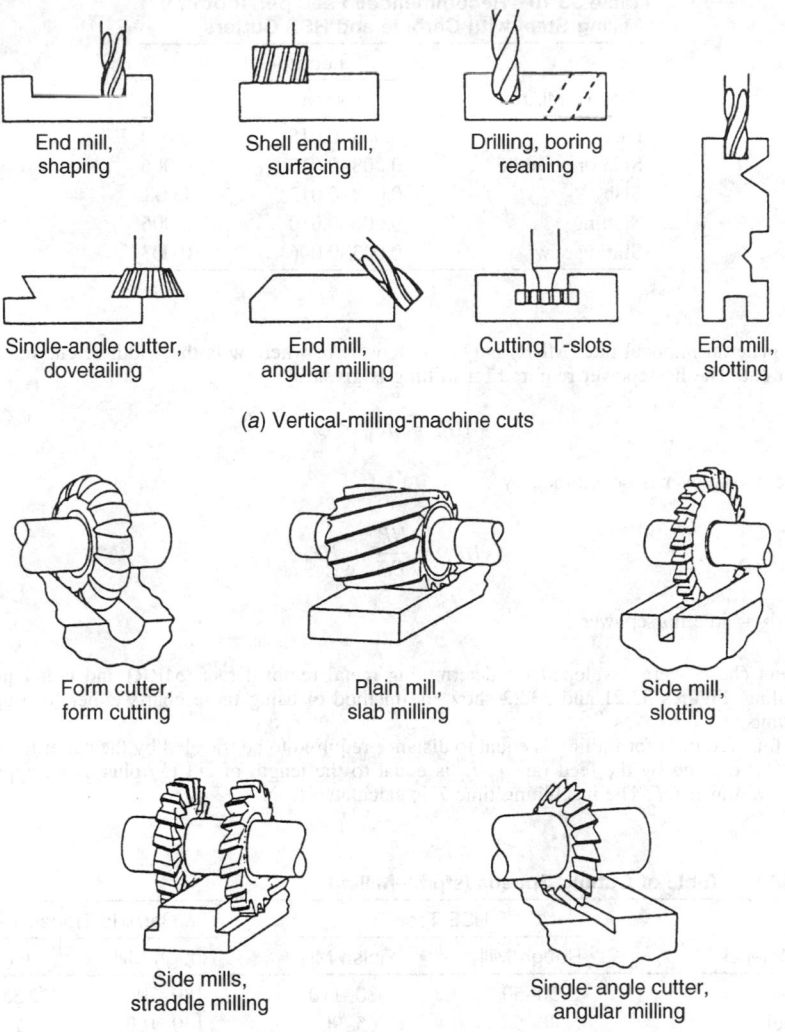

(a) Vertical-milling-machine cuts

(b) Horizontal-milling-machine cuts

Fig. 33.18 Applications of (a) vertical; (b) horizontal milling machines.

where n = number of teeth in cutter
N = rpm

Table 33.10 gives the recommended f_t for carbides and HSS tools. The cutting speed CS is calculated as follows:

$$CS = \frac{\pi DN}{12} \quad \text{fpm}$$

where D = tool diameter, in.

Table 33.11 gives the recommended cutting speeds while using HSS and carbide-tipped tools. The relationship between cutter rotation and feed direction is shown in Fig. 33.19. In climb milling or down milling, the chips are cut to maximum thickness at initial engagement and decrease to zero thickness at the end of engagement. In conventional or up milling, the reverse occurs. Because of the initial impact, climb milling requires rigid machines with backlash eliminators.

Table 33.10 Recommended Feed per Tooth for Milling Steel with Carbide and HSS Cutters

Type of Milling	Feed per Tooth	
	Carbides	HSS
Face	0.008–0.015	0.010
Side or straddle	0.008–0.012	0.006
Slab	0.008–0.012	0008
Slotting	0.006–0.010	0.006
Slitting saw	0.003–0.006	0.003

The material removal rate (MRR) is $Q = F \times w \times d$, where w is the width of cut and d is the depth of cut. The horsepower required for milling is given by

$$HP_c = HP_\mu \times Q$$

Machine horsepower is determined by

$$HP_m = \frac{HP_c}{Eff.} + HP_i \qquad (33.54)$$

where Hp_i = idle horsepower

Alignment charts were developed for determining metal removal rate (MRR) and motor power in face milling. Figures 33.21 and 33.22 show the method of using these charts either for English or metric units.

The time required for milling is equal to distance required to be traveled by the cutter to complete the cut (L_1) divided by the feed rate F. L_1 is equal to the length of cut (L) plus cutter approach A and the overtravel OT. The machining time T is calculated from

Table 33.11 Table of Cutting Speeds (sfpm)–Milling

Work Material	HSS Tools		Carbide-Tipped Tools	
	Rough Mill	Finish Mill	Rough Mill	Finish Mill
Cast iron	50–60	80–110	180–200	350–400
Semisteel	40–50	65–90	140–160	250–300
Malleable iron	80–100	110–130	250–300	400–500
Cast steel	45–60	70–90	150–180	200–250
Copper	100–150	150–200	600	1000
Brass	200–300	200–300	600–1000	600–1000
Bronze	100–150	150–180	600	1000
Aluminum	400	700	800	1000
Magnesium	600–800	1000–1500	1000–1500	1000–5000
SAE steels				
1020 (coarse feed)	60–80	60–80	300	300
1020 (fine feed)	100–120	100–120	450	450
1035	75–90	90–120	250	250
X-1315	175–200	175–200	400–500	400–500
1050	60–80	100	200	200
2315	90–110	90–110	300	300
3150	50–60	70–90	200	200
4150	40–50	70–90	200	200
4340	40–50	60–70	200	200
Stainless steel	60–80	100–120	240–300	240–300
Titanium	30–70		200–350	

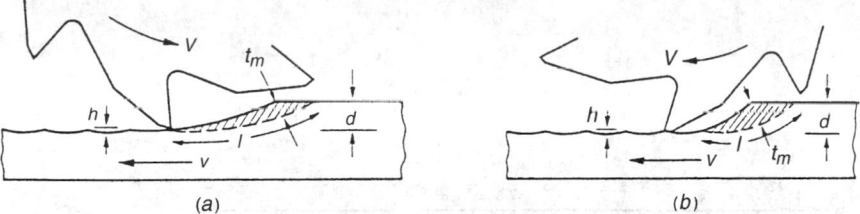

Fig. 33.19 Cutting action in up-and-down milling.

$$T = \frac{L + A + OT}{F} \quad \text{min} \tag{33.55}$$

OT depends on the specific milling operation.

The milling machines are designed according to the longitudinal table travel. Milling machines are built in different types, including:

1. Column-and-knee: vertical, horizontal, universal, and ram
2. Bed-type, multispindle
3. Planer
4. Special, turret, profilers, and duplicators
5. Numerically controlled

33.9 GEAR MANUFACTURING

Gears are made by various methods, such as machining, rolling, extrusion, blanking, powder metallurgy, casting, or forging. Machining still is the unsurpassed method of producing gears of all types and sizes with high accuracy. Roll forming can be used only on ductile materials; however, it has been highly developed and widely adopted in recent years. Casting, powder metallurgy, extruding, rolling, grinding, molding, and stamping techniques are used commercially in gear production.

33.9.1 Machining Methods

There are three basic methods for machining gears.

Form cutting uses the principle illustrated in Fig. 33.23. The equipment and cutters required are relatively simple, and standard machines, usually milling, are often used. Theoretically, there should be different-shaped cutters for each size of gear for a given pitch, as there is a slight change in the curvature of the involute. However, one cutter can be used for several gears having different numbers of teeth without much sacrifice in their operating action. The eight standard involute cutters are listed in Table 33.12. On the milling machine, the index or dividing head is used to rotate the gear blank through a certain number of degrees after each cut. The rule to use is: turns of index handle = $40/N$, where N is the number of teeth. Form cutting is usually slow.

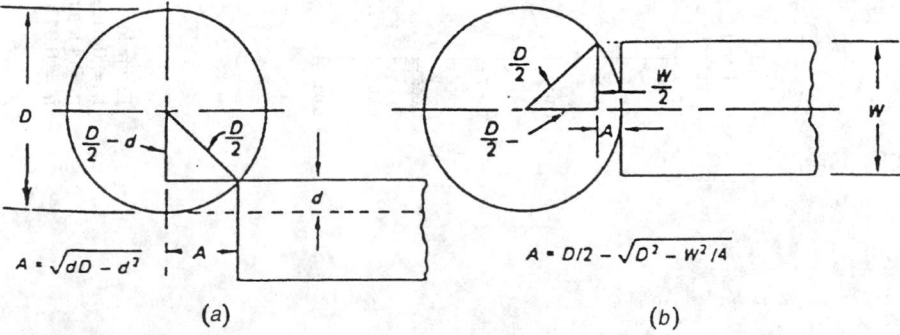

Fig. 33.20 Allowance for approach in (a) plain or slot milling; (b) face milling.

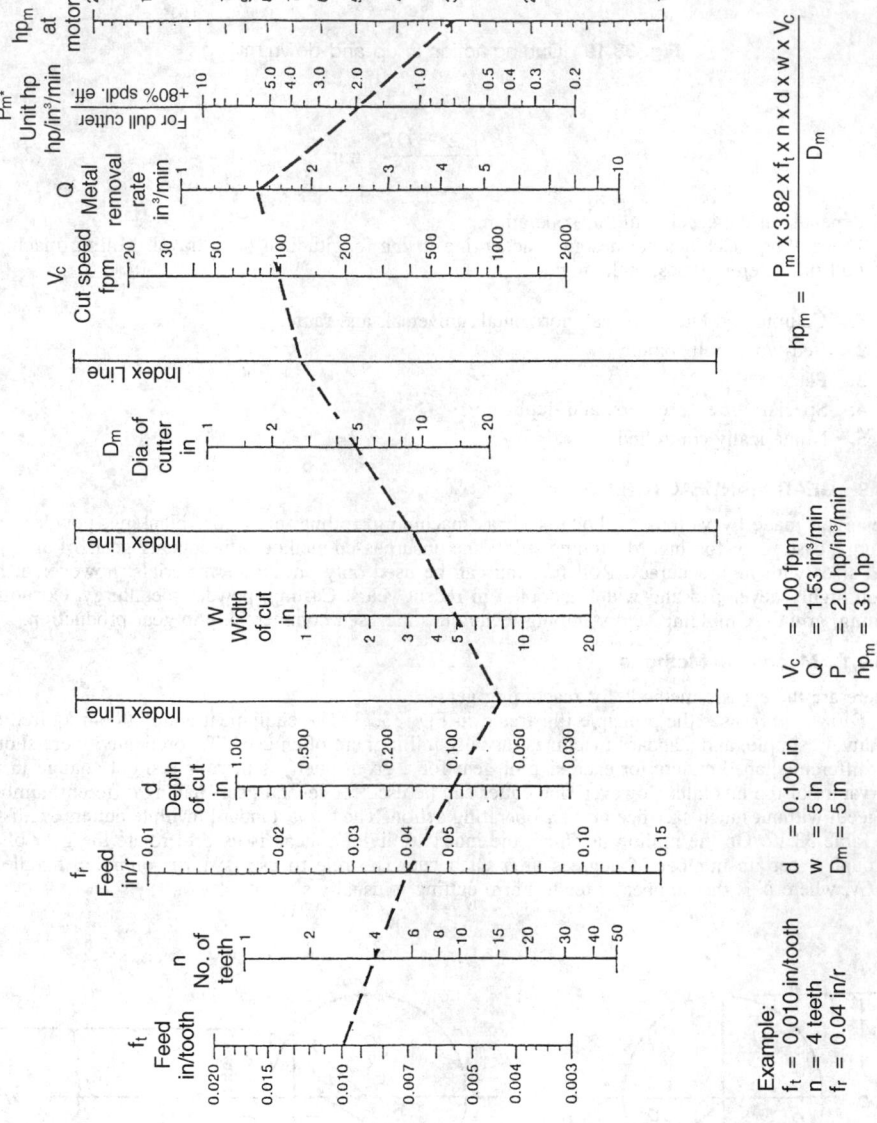

Fig. 33.21 Alignment chart for determining metal removal rate and motor horsepower in face milling—English units.

Example:

f_t = 0.010 in/tooth	d = 0.100 in	V_c = 100 fpm
n = 4 teeth	w = 5 in	Q = 1.53 in³/min
f_r = 0.04 in/r	D_m = 5 in	P = 2.0 hp/in³/min
		hp_m = 3.0 hp

$$hp_m = \frac{P_m \times 3.82 \times f_t \times n \times d \times w \times V_c}{D_m}$$

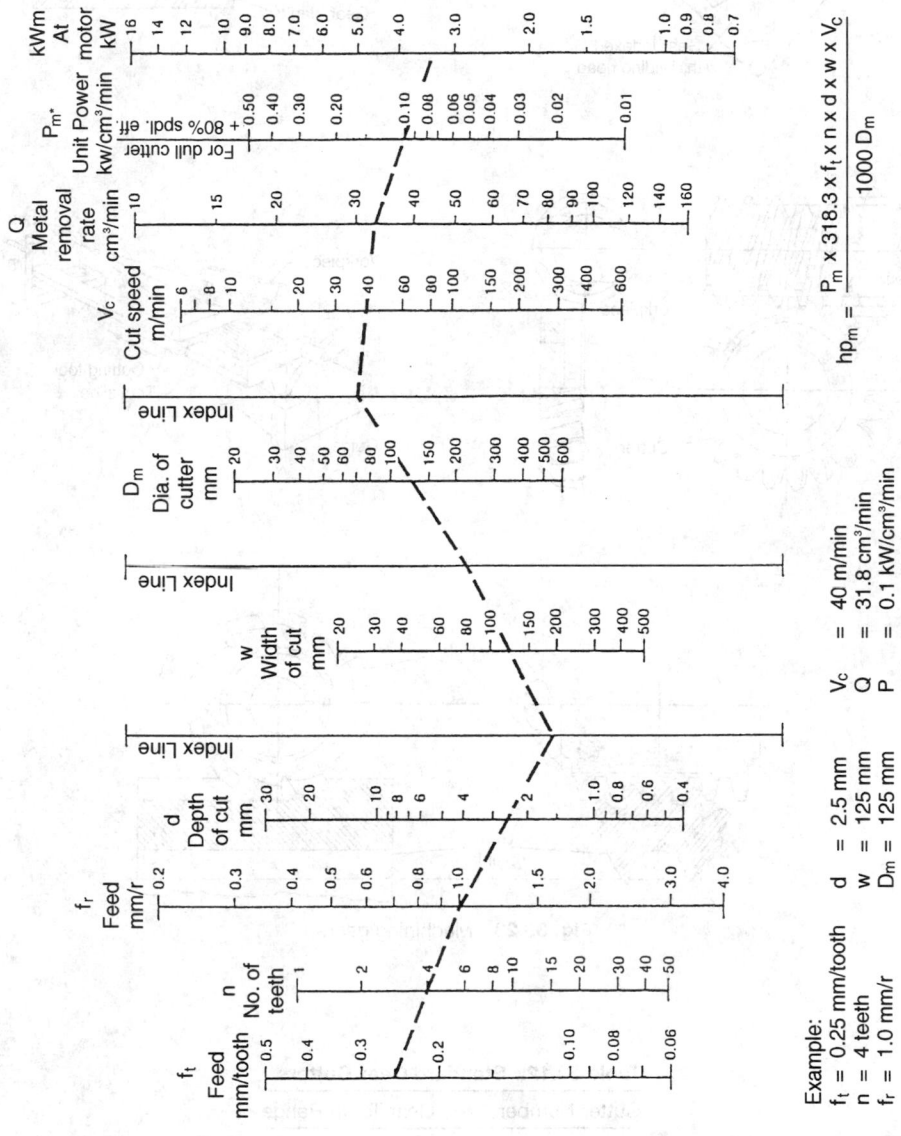

$$hp_m = \frac{P_m \times 318.3 \times f_t \times n \times d \times w \times V_c}{1000\ D_m}$$

Example:

f_t = 0.25 mm/tooth   d = 2.5 mm V_c = 40 m/min

n = 4 teeth   w = 125 mm   Q = 31.8 cm³/min

f_r = 1.0 mm/r D_m = 125 mm   P = 0.1 kW/cm³/min

                                   kW_m = 3.18 kW

Fig. 33.22 Alignment chart for determining metal removal rate and motor power in face milling—metric units.

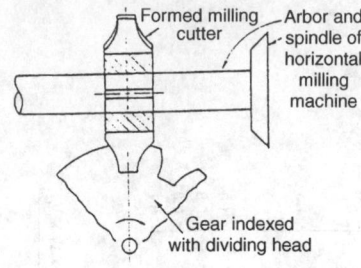

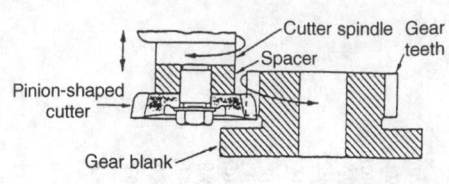

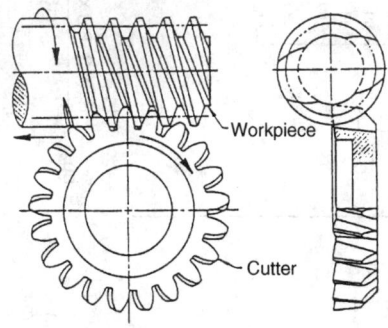

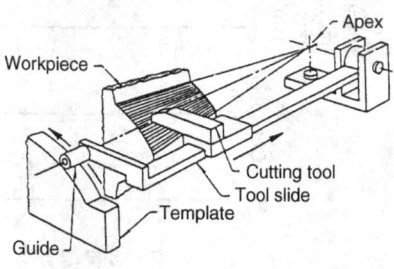

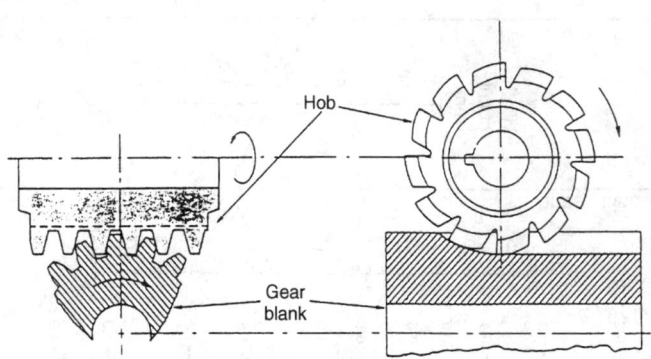

Fig. 33.23 Machining gears.

Table 33.12   Standard Gear Cutters

Cutter Number	Gear Tooth Range
1	135 teeth to rack
2	55–34
3	35–54
4	26–34
5	21–25
6	17–20
7	14–16
8	12–13

Template machining utilizes a simple, single-point cutting tool that is guided by a template. However, the equipment is specialized, and the method is seldom used except for making large-bevel gears.

The *generating process* is used to produce most high-quality gears. This process is based on the principle that any two involute gears, or any gear and a rack, of the same diametral pitch will mesh together. Applying this principle, one of the gears (or the rack) is made into a cutter by proper sharpening and is used to cut into a mating gear blank and thus generate teeth on the blank. Gear shapers (pinion or rack), gear-hobbing machines, and bevel-gear generating machines are good examples of the gear generating machines.

33.9.2 Gear Finishing

To operate efficiently and have satisfactory life, gears must have accurate tooth profile and smooth and hard faces. Gears are usually produced from relatively soft blanks and are subsequently heat-treated to obtain greater hardness, if it is required. Such heat treatment usually results in some slight distortion and surface roughness. *Grinding and lapping* are used to obtain very accurate teeth on hardened gears. Gear-shaving and burnishing methods are used in gear finishing. Burnishing is limited to unhardened gears.

33.10 THREAD CUTTING AND FORMING

Three basic methods are used for the manufacturing of threads; *cutting, rolling,* and *casting.* Die casting and molding of plastics are good examples of casting. The largest number of threads are made by rolling, even though it is restricted to standardized and simple parts, and ductile materials. Large numbers of threads are cut by the following methods:

1. Turning
2. Dies: manual or automatic (external)
3. Milling
4. Grinding (external)
5. Threading machines (external)
6. Taps (internal)

33.10.1 Internal Threads

In most cases, the hole that must be made before an internal thread is tapped is produced by drilling. The hole size determines the depth of the thread, the forces required for tapping, and the tap life. In most applications, a drill size is selected that will result in a thread having about 75% of full thread depth. This practice makes tapping much easier, increases the tap's life, and only slightly reduces the resulting strength. Table 33.13 gives the drill sizes used to produce 75% thread depth for several sizes of UNC threads. The feed of a tap depends on the lead of the screw and is equal to 1/lead ipr.

Cutting speeds depend on many factors, such as

1. Material hardness
2. Depth of cut
3. Thread profile

Table 33.13 Recommended Tap-Drill Sizes for Standard Screw-Thread Pitches (American National Coarse-Thread Series)

Number or Diameter	Threads per Inch	Outside Diameter of Screw	Tap Drill Sizes	Decimal Equivalent of Drill
6	32	0.138	36	0.1065
8	32	0.164	29	0.1360
10	24	0.190	25	0.1495
12	24	0.216	16	0.1770
¼	20	0.250	7	0.2010
⅜	16	0.375	5/16	0.3125
½	13	0.500	27/64	0.4219
¾	10	0.750	21/32	0.6562
1	8	1.000	⅞	0.875

4. Tooth depth
5. Hole depth
6. Fineness of pitch
7. Cutting fluid

Cutting speeds can range from lead 3 ft/min (1 m/min) for high-strength steels to 150 ft/min (45 m/min) for aluminum alloys. Long-lead screws with different configurations can be cut successfully on milling machines, as in Fig. 33.24. The feed per tooth is given by the following equation:

$$f_t = \frac{\pi d S}{nN} \tag{33.56}$$

where d = diameter of thread
n = number of teeth in cutter
N = rpm of cutter
S = rpm of work

33.10.2 Thread Rolling

In thread rolling, the metal on the cylindrical blank is cold-forged under considerable pressure by either rotating cylindrical dies or reciprocating flat dies. The advantages of thread rolling include improved strength, smooth surface finish, less material used (~19%), and high production rate. The limitations are that blank tolerance must be close, it is economical only for large quantities, it is limited to external threads, and it is applicable only for ductile materials, less than Rockwell C37.

33.11 BROACHING

Broaching is unique in that it is the only one of the basic machining processes in which the feed of the cutting edges is built into the tool. The machined surface is always the inverse of the profile of the broach. The process is usually completed in a single, linear stroke. A broach is composed of a series of single-point cutting edges projecting from a rigid bar, with successive edges protruding farther from the axis of the bar. Figure 33.25 illustrates the parts and nomenclature of the broach. Most broaching machines are driven hydraulically and are of the pull or push type.

The maximum force an internal pull broach can withstand without damage is given by

$$P = \frac{A_y F_y}{s} \quad \text{lb} \tag{33.57}$$

where A_y = minimum tool selection, in.^{2}
F_y = tensile yield strength of tool steel, psi
s = safety factor

The maximum push force is determined by the minimum tool diameter (D_y), the length of the broach (L), and the minimum compressive yield strength (F_y). The ratio L/D_y should be less than 25 so that the tool will not bend under load. The maximum allowable pushing force is given by

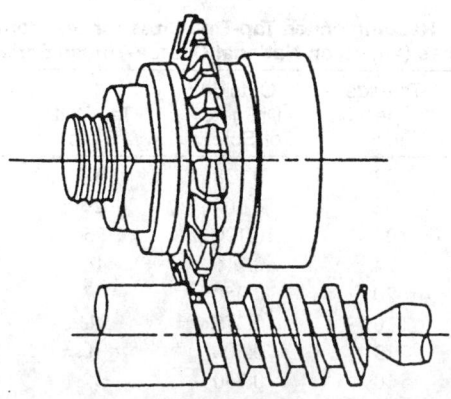

Fig. 33.24 Single-thread milling cutter.

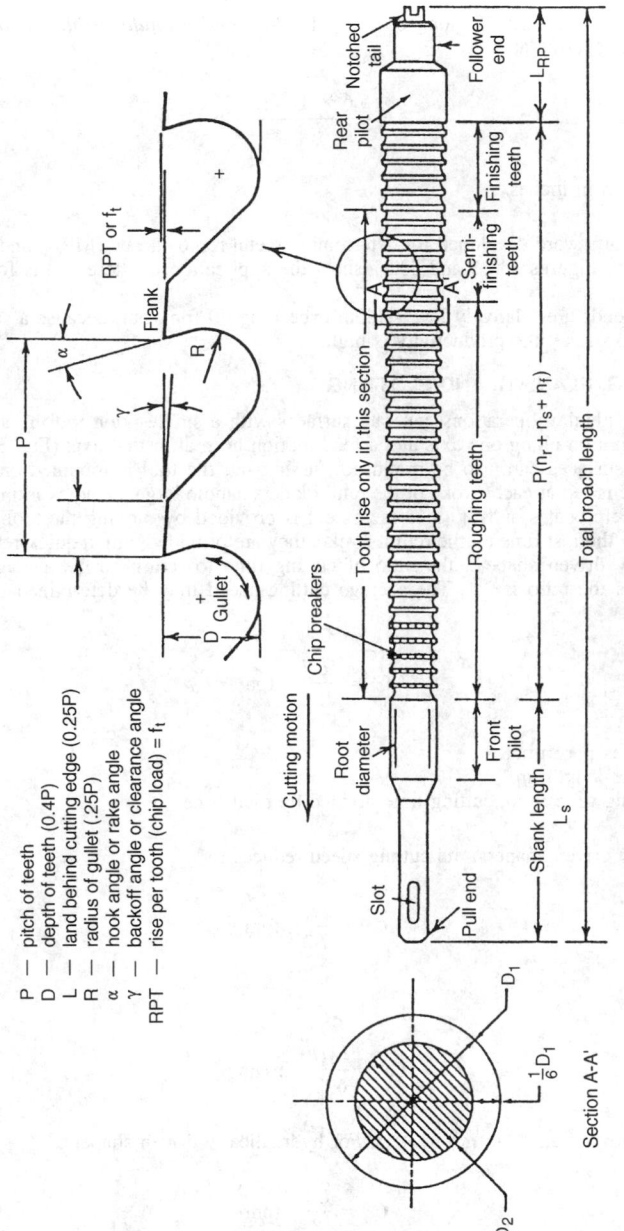

P — pitch of teeth
D — depth of teeth (0.4P)
L — land behind cutting edge (0.25P)
R — radius of gullet (.25P)
α — hook angle or rake angle
γ — backoff angle or clearance angle
RPT — rise per tooth (chip load) = f_t

Fig. 33.25 Standard broach part and nomenclature.

1069

$$P = \frac{A_y F_y}{s} \quad \text{lb} \tag{33.58}$$

where F_y is minimum compressive yield strength.

If L/D_y ratio is greater than 25 (long broach), the *Tool and Manufacturing Engineers Handbook* gives the following formula:

$$P = \frac{5.6 \times 10^7 D_r^4}{sL^2} \quad \text{lb} \tag{33.59}$$

D_r and L are given in inches.

Alignment charts were developed for determining metal removal rate (MRR) and motor power in surface broaching. Figures 33.26 and 33.27 show the application of these charts for either English or metric units.

Broaching speeds are relatively low, seldom exceeding 50 fpm, but, because a surface is usually completed in one stroke, the productivity is high.

33.12 SHAPING, PLANING, AND SLOTTING

The shaping and planing operations generate surfaces with a single-point tool by a combination of a reciprocating motion along one axis and a feed motion normal to that axis (Fig. 33.28). Slots and limited inclined surfaces can also be produced. In shaping, the tool is mounted on a reciprocating ram and the table is fed at each stroke of the ram. Planers handle large, heavy workpieces. In planing, the workpiece reciprocates and the feed increment is provided by moving the tool at each reciprocation. To reduce the lost time on the return stroke, they are provided with a quick-return mechanism. For mechanically driven shapers, the ratio of cutting time to return stroke averages 3:2, and for hydraulic shapers the ratio is 2:1. The average cutting speed may be determined by the following formula:

$$CS = \frac{LN}{12C} \quad \text{fpm} \tag{33.60}$$

where N = strokes per minute
L = stroke length, in.
C = cutting time ratio, cutting time divided by total time

For mechanically driven shapers, the cutting speed reduces to

$$CS = \frac{LN}{7.2} \quad \text{fpm} \tag{33.61}$$

or

$$CS = \frac{L_1 N}{600} \quad \text{m/min} \tag{33.62}$$

where L_1 is the stroke length in millimeters. For hydraulically driven shapers,

$$CS = \frac{LN}{8} \quad \text{fpm} \tag{33.63}$$

or

$$CS = \frac{L_1 N}{666.7} \quad \text{m/min} \tag{33.64}$$

The time T required to machine a workpiece of width W (in.) is calculated by

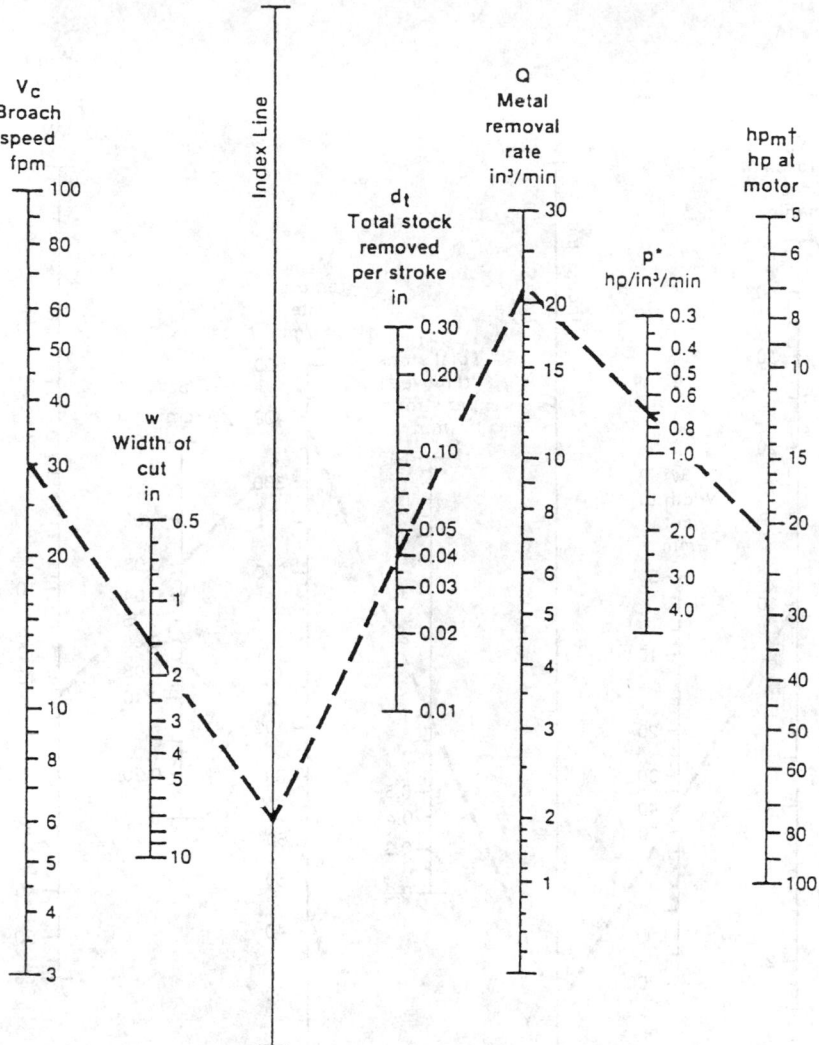

Example:

Material: Cast iron — HSS tools

Chipload 0.005 in/tooth

V_c = 30 fpm   w   = 1.5 in

d_t = 0.040 in   Q   = 22 in³/min

P = 0.7 hp/in³/min hp_m = 22 hp

$$Q = 12 V_c \times w \times d_t \text{ in}^3/\text{min}$$

$$hp_m = \frac{Q \times P}{E} = \frac{Q \times P}{0.7}$$

Fig. 33.26 Alignment chart for determining metal removal rate and motor horsepower in surface broaching with high-speed steel broaching tools—English units.

$$T = \frac{W}{N \times f} \quad \text{min} \tag{33.65}$$

where f = feed, in. per stroke

The number of strokes (S) required to complete a job is then

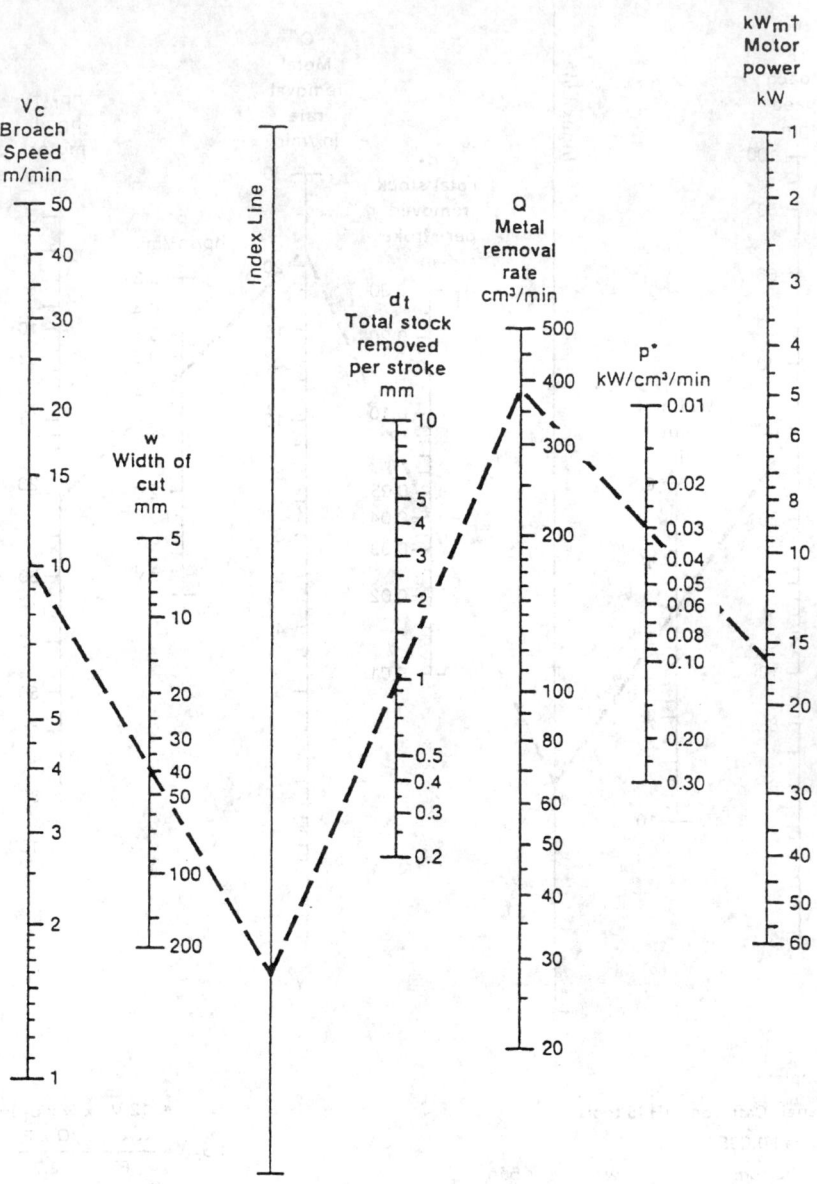

Example:

Material: Cast iron — HSS tools

Chipload 0.13 mm/tooth

V_c = 10 m/min   w = 38 mm

d_t = 1 mm   Q = 380 cm³/min

P = 0.03 kW/cm³/min P_m = 16.3 kW

$$Q = V_c \times w \times d_t \ \text{cm}^3/\text{min}$$

$$P_m = \frac{Q \times P}{E} = \frac{Q \times P}{0.7}$$

Fig. 33.27 Alignment chart for determining metal removal rate and motor power in surface broaching with high-speed steel broaching tools—metric units.

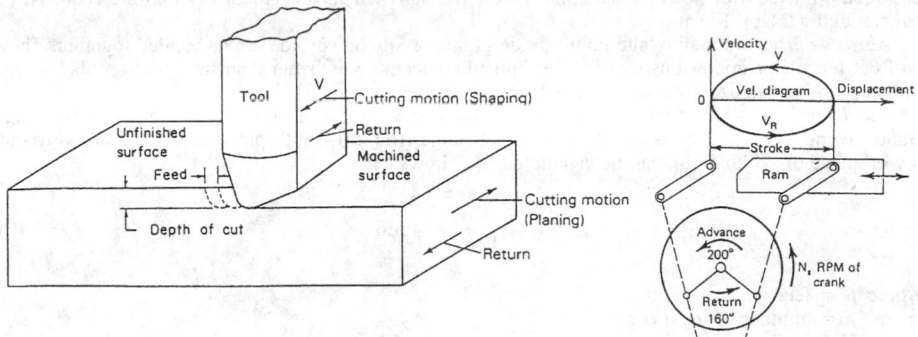

Fig. 33.28 Basic relationships of tool motion, feed, and depth of cut in shaping and planing.

$$S = \frac{W}{f} \qquad (33.66)$$

The power required can be approximated by

$$HP_c = Kdf(CS) \qquad (33.67)$$

where d = depth of cut, in.
CS = cutting speed, fpm
K = cutting constant, for medium cast iron, 3; free-cutting steel, 6; and bronze, 1.5 or

$$HP_c = 12f \times d \times CS \times HP_\mu$$

$$F_c = \frac{33,000 \, HP_c}{CS}$$

33.13 SAWING, SHEARING, AND CUTTING OFF

Saws are among the most common of machine tools, even though the surfaces they produce often require further finishing operations. Saws have two general areas of applications: contouring and cutting off. There are three basic types of saws: hacksaw, circular, and band saw.

The *reciprocating power hacksaw* machines can be classified as either positive or uniform-pressure feeds. Most of the new machines are equipped with a quick-return action to reduce idle time.

The machining time required to cut a workpiece of width W in. is calculated as follows:

$$T = \frac{W}{fN} \quad \text{min} \qquad (33.68)$$

where F = feed, in./stroke
N = number of strokes per min

Circular saws are made of three types: metal saws, steel friction disks, and abrasive disks. Solid metal saws are limited in size, not exceeding 16 in. in diameter. Large circular saws have either replaceable inserted teeth or segmented-type blades. The machining time required to cut a workpiece of width W in. is calculated as follows:

$$T = \frac{W}{f_t nN} \quad \text{min} \qquad (33.69)$$

where f_t = feed per tooth
n = number of teeth
N = rpm

Steel friction disks operate at high peripheral speeds ranging from 18,000–25,000 fpm (90–125 m/sec). The heat of friction quickly softens a path through the part. The disk, which is sometimes provided with teeth or notches, pulls and ejects the softened metal. About 0.5 min are required to cut through a 24-in. I-beam.

Abrasive disks are mainly aluminum oxide grains or silicon carbide grains bonded together. They will cut ferrous or nonferrous metals. The finish and accuracy is better than steel friction blades, but they are limited in size compared to steel friction blades.

Band saw blades are of the continuous type. Band sawing can be used for cutting and contouring. Band-sawing machines operate with speeds that range from 50–1500 fpm. The time required to cut a workpiece of width W in. can be calculated as follows:

$$T = \frac{W}{12f_t nV} \quad \text{min} \tag{33.70}$$

where f_t = feed, in. per tooth
 n = number of teeth per in.
 V = cutting speed, fpm

Cutting can also be achieved by band-friction cutting blades with a surface speed up to 15,000 fpm. Other band tools include band filing, diamond bands, abrasive bands, spiral bands, and special-purpose bands.

33.14 MACHINING PLASTICS

Most plastics are readily formed, but some machining may be required. Plastic's properties vary widely. The general characteristics that affect their machinability are discussed below.

First, all plastics are poor heat conductors. Consequently, little of the heat that results from chip formation will be conducted away through the material or carried away in the chips. As a result, cutting tools run very hot and may fail more rapidly than when cutting metal. Carbide tools frequently are more economical to use than HSS tools if cuts are of moderately long duration or if high-speed cutting is to be done.

Second, because considerable heat and high temperatures do develop at the point of cutting, thermoplastics tend to soften, swell, and bind or clog the cutting tool. Thermosetting plastics give less trouble in this regard.

Third, cutting tools should be kept very sharp at all times. Drilling is best done by means of straight-flute drills or by "dubbing" the cutting edge of a regular twist drill to produce a zero rake angle. Rotary files and burrs, saws, and milling cutters should be run at high speeds in order to improve cooling, but with feed carefully adjusted to avoid jamming the gullets. In some cases, coolants can be used advantageously if they do not discolor the plastic or cause gumming. Water, soluble oil and water, and weak solutions of sodium silicate in water are used. In turning and milling plastics, diamond tools provide the best accuracy, surface finish, and uniformity of finish. Surface speeds of 500–600 fpm with feeds of 0.002–0.005 in. are typical.

Fourth, filled and laminated plastics usually are quite abrasive and may produce a fine dust that may be a health hazard.

33.15 GRINDING, ABRASIVE MACHINING, AND FINISHING

Abrasive machining is the basic process in which chips are removed by very small edges of abrasive particles, usually synthetic. In many cases, the abrasive particles are bonded into wheels of different shapes and sizes. When wheels are used mainly to produce accurate dimensions and smooth surfaces, the process is called *grinding*. When the primary objective is rapid metal removal to obtain a desired shape or approximate dimensions, it is termed *abrasive machining*. When fine abrasive particles are used to produce very smooth surfaces and to improve the metallurgical structure of the surface, the process is called *finishing*.

33.15.1 Abrasives

Aluminum oxide (Al_2O_3), usually synthetic, performs best on carbon and alloy steels, annealed malleable iron, hard bronze, and similar metals. Al_2O_3 wheels are not used in grinding very hard materials, such as tungsten carbide, because the grains will get dull prior to fracture. Common trade names for aluminum oxide abrasives are *Alundum* and *Aloxite*.

Silicon carbide (SiC), usually synthetic, crystals are very hard, being about 9.5 on the Moh's scale, where diamond hardness is 10. SiC crystals are brittle, which limits their use. Silicon carbide wheels are recommended for materials of low tensile strength, such as cast iron, brass, stone, rubber, leather, and cemented carbides.

Cubic boron nitride (CBN) is the second-hardest natural or manmade substance. It is good for grinding hard and tough-hardened tool-and-die steels.

Diamonds may be classified as natural or synthetic. Commercial diamonds are now manufactured in high, medium, and low impact strength.

Grain Size

To have uniform cutting action, abrasive grains are graded into various sizes, indicated by the numbers 4–600. The number indicates the number of openings per linear inch in a standard screen through which most of the particles of a particular size would pass. Grain sizes from 4–24 are termed coarse; 30–60, medium; and 70–600, fine. Fine grains produce smoother surfaces than coarse ones but cannot remove as much metal.

Bonding materials have the following effects on the grinding process: (1) they determine the strength of the wheel and its maximum speed; (2) they determine whether the wheel is rigid or flexible; and (3) they determine the force available to pry the particles loose. If only a small force is needed to release the grains, the wheel is said to be soft. Hard wheels are recommended for soft materials and soft wheels for hard materials. The bonding materials used are vitrified, silicate, rubber, resinoid, shellac, and oxychloride.

Structure or Grain Spacing

Structure relates to the spacing of the abrasive grain. Soft, ductile materials require a wide spacing to accommodate the relatively large chips. A fine finish requires a wheel with a close spacing. Figure 33.29 shows the standard system of grinding wheels as adopted by the American National Standards Institute.

Speeds

Wheel speed depends on the wheel type, bonding material, and operating conditions. Wheel speeds range between 4500 and 18,000 sfpm (22.86 and 27.9 m/s). 5500 sfpm (27.9 m/s) is generally recommended as best for all disk-grinding operations. Work speeds depend on type of material, grinding operation, and machine rigidity. Work speeds range between 15 and 200 fpm.

Feeds

Cross feed depends on the width of grinding wheel. For rough grinding, the range is one-half to three-quarters of the width of the wheel. Finer feed is required for finishing, and it ranges between one-tenth and one-third of the width of the wheel. A cross feed between 0.125 and 0.250 in. is generally recommended.

Depth of Cut

Rough-grinding conditions will dictate the maximum depth of cut. In the finishing operation, the depth of cut is usually small, 0.0002–0.001 in. (0.005–0.025 mm). Good surface finish and close tolerance can be achieved by "sparking out" or letting the wheel run over the workpiece without increasing the depth of cut till sparks die out. The *grinding ratio* (*G*-ratio) refers to the ratio of the cubic inches of stock removed to the cubic inches of grinding wheel worn away. *G*-ratio is important in calculating grinding and abrasive machining cost, which may be calculated by the following formula:

$$C = \frac{C_a}{G} + \frac{L}{tq}$$
(33.71)

where C = specific cost of removing a cu in. of material
C_a = cost of abrasive, \$/in.^{3}
G = grinding ratio
L = labor and overhead charge, \$/hr
q = machining rate, in.^{3}/hr
t = fraction of time the wheel is in contact with workpiece

Power Requirement

$$\text{Power} = (u)(\text{MRR}) = F_c \times R \times 2\pi N$$

$$\text{MRR} = \text{material removal rate} = d \times w \times v$$

where d = depth of cut
w = width of cut
v = work speed
u = specific energy for surface grinding. Table 33.14 gives the approximate specific energy requirement for certain metals.
R = radius of wheel
N = rev/unit time

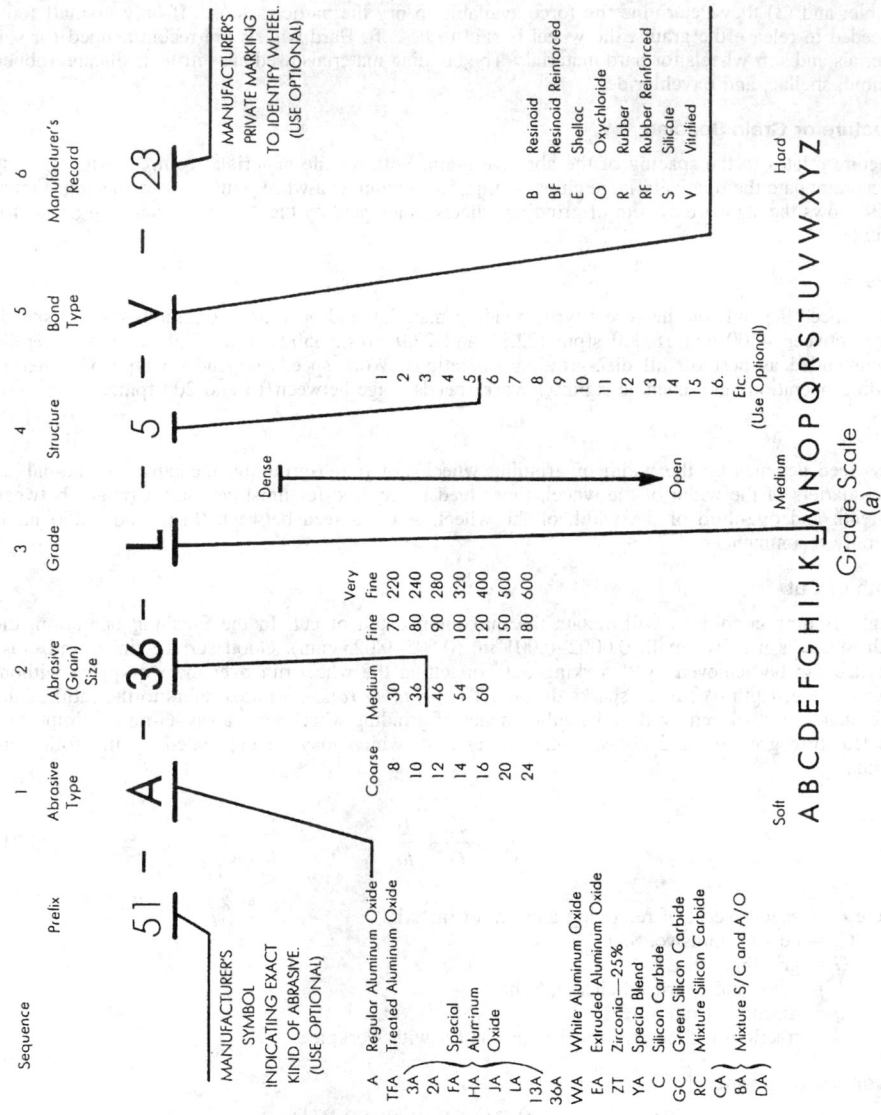

Grade Scale
(a)

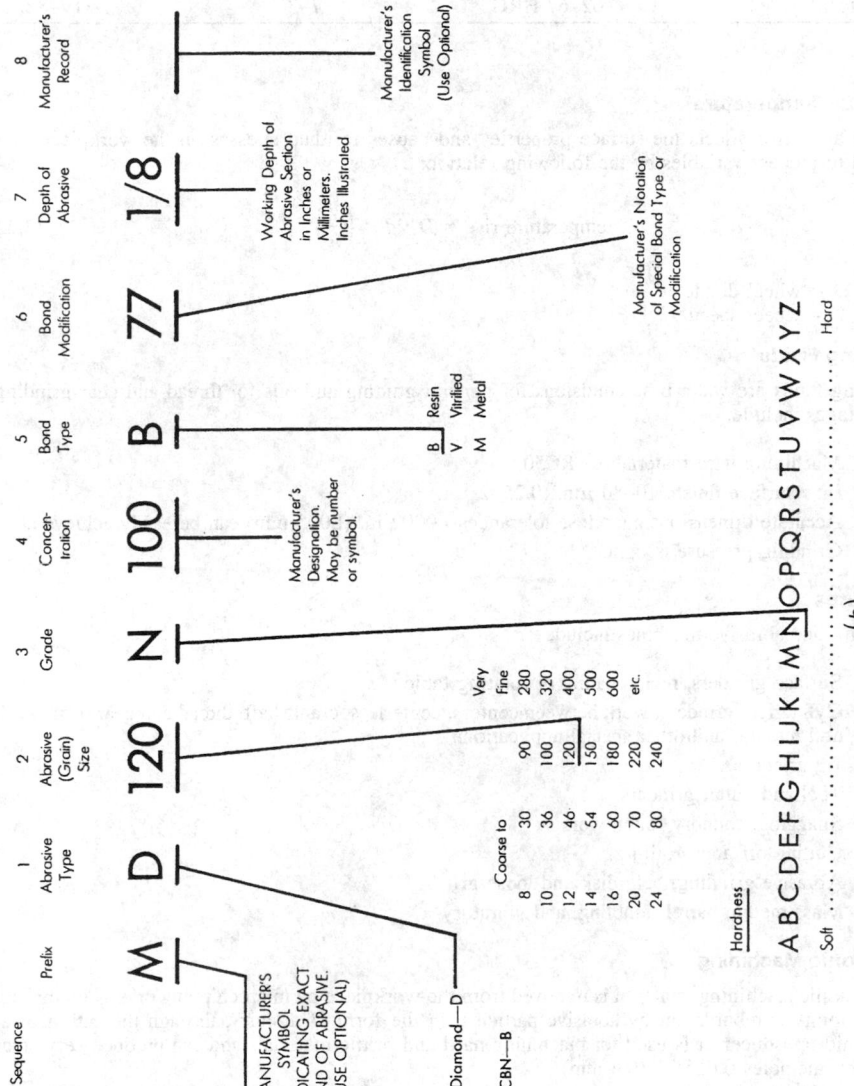

Fig. 33.29 Standard systems for grinding wheels. (a) aluminum oxide, silicon carbide; (b) diamond, CBN.

Sequence	Prefix	1 Abrasive Type	2 Abrasive (Grain) Size	3 Grade	4 Concen- tration	5 Bond Type	6 Bond Modification	7 Depth of Abrasive	8 Manufacturer's Record
	M	D	120	N	100	B	77	1/8	

MANUFACTURER'S SYMBOL INDICATING EXACT KIND OF ABRASIVE. (USE OPTIONAL)

Diamond—D
CBN—B

Coarse to		Very fine	
8	30	90	280
10	36	100	320
12	46	120	400
14	54	150	500
16	60	180	600
20	70	220	etc.
24	80	240	

Hardness

A B C D E F G H I J K L M N O P Q R S T U V W X Y Z

Soft .... Hard

Manufacturer's Designation. May be number or symbol.

B Resin
V Vitrified
M Metal

Manufacturer's Notation of Special Bond Type or Modification

Working Depth of Abrasive Section in Inches or Millimeters. Inches Illustrated

Manufacturer's Identification Symbol (Use Optional)

(b)

Table 33.14 Approximate Specific Energy Required for Surface Grinding

Workpiece Material	Hardness	hp (in.³/min)	W/(mm³/sec)
Aluminum	150 HB	3–10	8–27
Steel	(110–220) HB	6–24	16–66
Cast iron	(140–250) HB	5–22	14–60
Titanium alloy	300 HB	6–20	16–55
Tool steel	62–67 HRC	7–30	19–82

33.15.2 Temperature

Temperature rise affects the surface properties and causes residual stresses on the workpiece. It is related to process variables by the following relation:

$$\text{temperature rise} \propto D^{1/4}d^{3/4}\left(\frac{V}{v}\right)^{1/2} \tag{33.72}$$

where D = wheel diameter
V = wheel speed

Grinding Fluids

Grinding fluids are water-base emulsions for general guiding and oils for thread and gear grinding. Advantages include:

1. Machining hard materials > RC50.
2. Fine surface finish, 10–80 μin. (0.25–2 μm).
3. Accurate dimensions and close tolerances I.0002 in. (I.005 mm) can be easily achieved.
4. Grinding pressure is light.

Machines

Grinding and abrasive machines include

1. Surface grinders, reciprocating or rotating table
2. Cylindrical grinders, work between centers, centerless, crankshaft, thread and gear form work, and internal and other special applications
3. Jig grinders
4. Tool and cutter grinders
5. Snagging, foundry rough work
6. Cutting off and profiling
7. Abrasive grinding, belt, disk and loose grit
8. Mass media, barrel tumbling, and vibratory

Ultrasonic Machining

In ultrasonic machining, material is removed from the workpiece by microchipping or erosion through high-velocity bombardment by abrasive particles, in the form of a slurry, through the action of an ultrasonic transducer. It is used for machining hard and brittle materials and can produce very small and accurate holes 0.015 in. (0.4 mm).

Surface Finishing

Finishing processes produce an extra-fine surface finish; in addition, tool marks are removed and very close tolerances are achieved. Some of these processes follow.

Honing is a low-velocity abrading process. It uses fine abrasive stones to remove very small amounts of metals usually left from previous grinding processes. The amount of metal removed is usually less than 0.005 in. (0.13 mm). Because of low cutting speeds, heat and pressure are minimized, resulting in excellent sizing and metallurgical control.

Lapping is an abrasive surface-finishing process wherein fine abrasive particles are charged in some sort of a vehicle, such as grease, oil, or water, and are embedded into a soft material, called a *lap*. Metal laps must be softer than the work and are usually made of close-grained gray cast iron. Other materials, such as steel, copper, and wood, are used where cast iron is not suitable. As the charged lap is rubbed against a surface, small amounts of material are removed from the harder surface. The amount of material removed is usually less than 0.001 in. (0.03 mm).

Superfinishing is a surface-improving process that removes undesirable fragmentation, leaving a base of solid crystalline metal. It uses fine abrasive stones, like honing, but differs in the type of motion. Very rapid, short strokes, very light pressure, and low-viscosity lubricant–coolant are used in superfinishing. It is essentially a finishing process and not a dimensional one, and can be superimposed on other finishing operations.

Buffing

Buffing wheels are made from a variety of soft materials. The most widely used is muslin, but flannel, canvas, sisal, and heavy paper are used for special applications. Buffing is usually divided into two operations: cutting down and coloring. The first is used to smooth the surface and the second to produce a high luster. The abrasives used are extremely fine powders of aluminum oxide, tripoli (an amorphous silicon), crushed flint or quartz, silicon carbide, and red rouge (iron oxide). Buffing speeds range between 6,000 and 12,000 fpm.

Electropolishing is the reverse of electroplating; that is, the work is the anode instead of the cathode and metal is removed rather than added. The electrolyte attacks projections on the workpiece surface at a higher rate, thus producing a smooth surface.

33.16 NONTRADITIONAL MACHINING

Nontraditional, or nonconventional, machining processes are material-removal processes that have recently emerged or are new to the user. They have been grouped for discussion here according to their primary energy mode; that is, mechanical, electrical, thermal, or chemical, as shown in Table 33.15.

Nontraditional processes provide manufacturing engineers with additional choices or alternatives to be applied where conventional processes are not satisfactory, such as when

- Shapes and dimensions are complex or very small
- Hardness of material is very high (>400 HB)
- Tolerances are tight and very fine surface finish is desired
- Temperature rise and residual stresses must be avoided
- Cost and production time must be reduced

Figure 33.30 and Table 33.16 demonstrate the relationships among the conventional and the nontraditional machining processes with respect to surface roughness, dimensional tolerance, and metal-removal rate.

The *Machinery Handbook*[6] is an excellent reference for nontraditional machining processes, values, ranges, and limitations.

33.16.1 Abrasive Flow Machining

Abrasive flow machining (AFM) is the removal of material by a viscous, abrasive medium flowing, under pressure, through or across a workpiece. Figure 33.31 contains a schematic presentation of the AFM process. Generally, the putty-like medium is extruded through or over the workpiece with motion usually in both directions. Aluminum oxide, silicon carbide, boron carbide, or diamond abrasives are used. The movement of the abrasive matrix erodes away burrs and sharp corners and polishes the part.

33.16.2 Abrasive Jet Machining

Abrasive jet machining (AJM) is the removal of material through the action of a focused, high-velocity stream of fine grit or powder-loaded gas. The gas should be dry, clean, and under modest pressure. Figure 33.32 shows a schematic of the AJM process. The mixing chamber sometimes uses a vibrator to promote a uniform flow of grit. The hard nozzle is directed close to the workpiece at a slight angle.

33.16.3 Hydrodynamic Machining

Hydrodynamic machining (HDM) removes material by the stroking of high-velocity fluid against the workpiece. The jet of fluid is propelled at speeds up to Mach 3. Figure 33.33 shows a schematic of the HDM operation.

33.16.4 Low-Stress Grinding

Low-stress grinding (LSG) is an abrasive material-removal process that leaves a low-magnitude, generally compressive residual stress in the surface of the workpiece. Figure 33.34 shows a schematic of the LSG process. The thermal effects from conventional grinding can produce high tensile stress

Table 33.15 Current Commercially Available Nontraditional Material Removal Processes

	Mechanical		Electrical		Thermal		Chemical
AFM	Abrasive flow machining	ECD	Electrochemical deburring	EBM	Electron-beam machining	CHM	Chemical machining: chemical milling, chemical blanking
AJM	Abrasive jet machining	ECDG	Electrochemical discharge grinding	EDG	Electrical discharge grinding		
HDM	Hydrodynamic machining						
LSG	Low-stress grinding	ECG	Electrochemical grinding	EDM	Electrical discharge machining	ELP	Electropolish
RUM	Rotary ultrasonic machining	ECH	Electrochemical honing			PCM	Photochemical machining
		ECM	Electrochemical machining	EDS	Electrical discharge sawing	TCM	Thermochemical machining (or TEM, thermal energy method)
TAM	Thermally assisted machining	ECP	Electrochemical polishing	EDWC	Electrical discharge wire cutting		
		ECS	Electrochemical sharpening				
TFM	Total form machining	ECT	Electrochemical turning	LBM	Laser-beam machining		
USM	Ultrasonic machining	ES	Electro-stream™	LBT	Laser-beam torch		
WJM	Water-jet machining	STEM™	Shaped tube electrolytic machining	PBM	Plasma-beam machining		

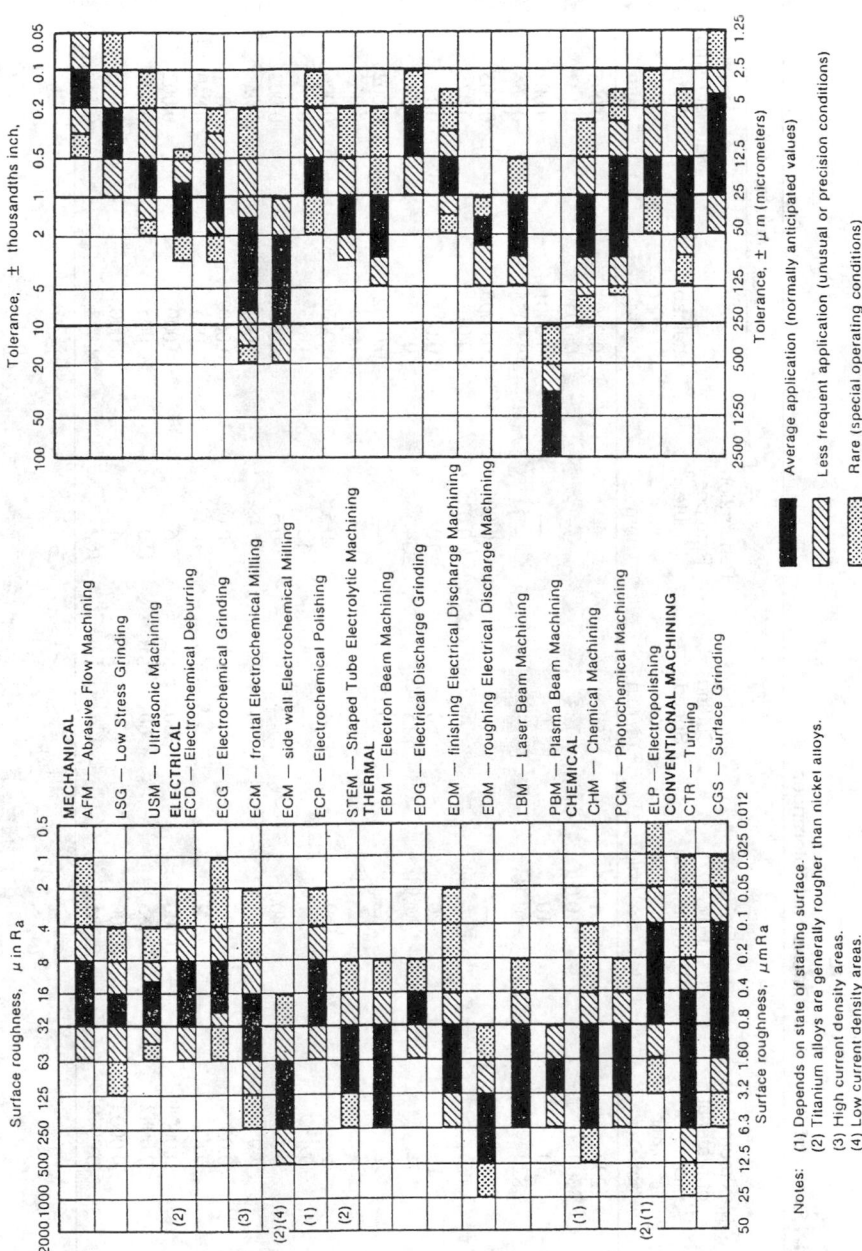

Fig. 33.30 Typical surface roughness and tolerances produced by nontraditional machining.

1081

Table 33.16 Material Removal Rates and Dimensional Tolerances

Process	Maximum Rate of Material Removal		Typical Power Consumption		Cutting Speed		Penetration Rate per Minute		Accuracy ±				Typical Machine Input	
									Attainable		At Maximum Material Removal Rate			
	in.³/min	cm³/min	hp/in.³/min	kW/cm³/min	fpm	m/min	in.	mm	in.	mm	in.	mm	hp	kW
Conventional turning	200	3300	1	0.046	250	76	—	—	0.0002	0.005	0.005	0.13	30	22
Conventional grinding	50	820	10	0.46	10	3	—	—	0.0001	0.0025	0.002	0.05	25	20
CHM	30	490	—	—	—	—	0.001	0.025	0.0005	0.013	0.003	0.075	—	—
PBM	10	164	20	0.91	50	15	10	254	0.02	0.5	0.1	2.54	200	150
ECG	2	33	2	0.019	0.25	0.08	—	—	0.0002	0.005	0.0025	0.063	4	3
ECM	1	16.4	160	7.28	—	—	0.5	12.7	0.0005	0.013	0.006	0.15	200	150
EDM	0.3	4.9	40	1.82	—	—	0.5	12.7	0.00015	0.004	0.002	0.05	15	11
USM	0.05	0.82	200	9.10	—	—	0.02	0.50	0.0002	0.005	0.0015	0.040	15	11
EBM	0.0005	0.0082	10,000	455	200	60	6	150	0.0002	0.005	0.002	0.050	10	7.5
LBM	0.0003	0.0049	60,000	2,731	—	—	4	102	0.0005	0.013	0.005	0.13	20	15

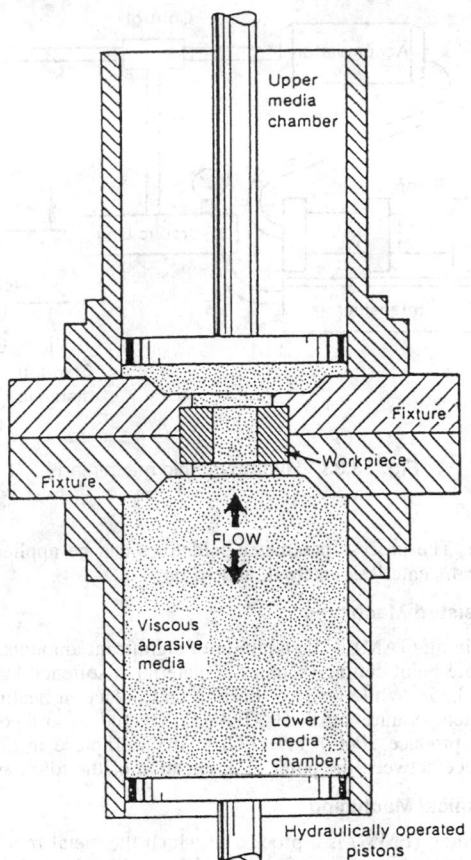

Fig. 33.31 Abrasive flow machining.

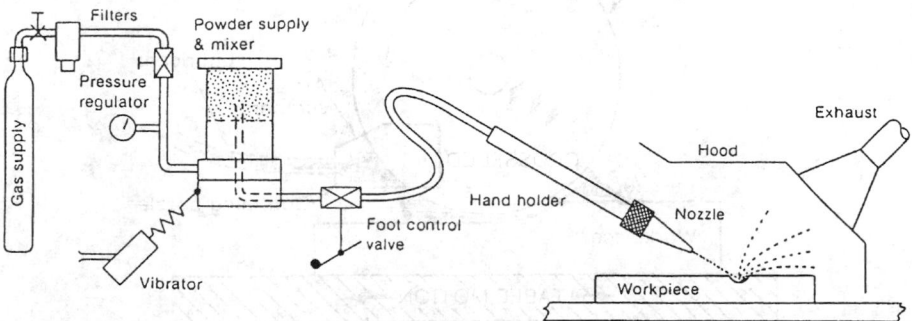

Fig. 33.32 Abrasive jet machining.

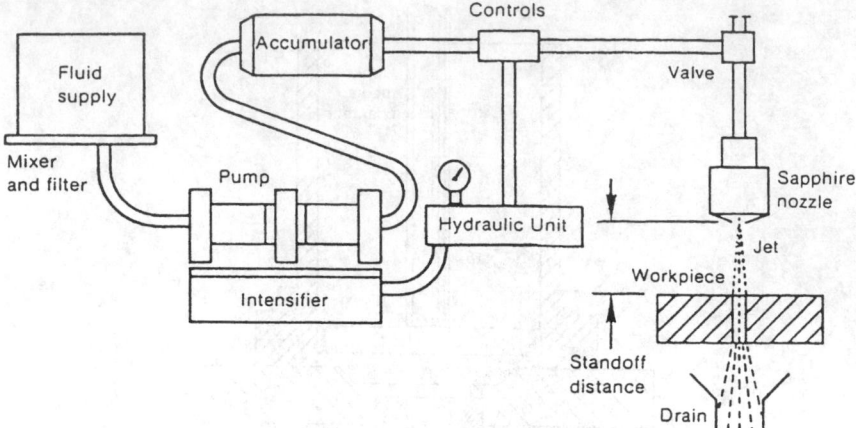

Fig. 33.33 Hydrodynamic machining.

in the workpiece surface. The process parameter guidelines can be applied to any of the grinding modes: surface, cylindrical, centerless, internal, and so on.

33.16.5 Thermally Assisted Machining

Thermally assisted machining (TAM) is the addition of significant amounts of heat to the workpiece immediately prior to single-point cutting so that the material is softened but the strength of the tool bit is unimpaired (Fig. 33.35). While resistive heating and induction heating offer possibilities, the plasma arc has a core temperature of 14,500°F (8000°C) and a surface temperature of 6500°F (3600°C). The torch can produce 2000°F (1100°C) in the workpiece in approximately one-quarter revolution of the workpiece between the point of application of the torch and the cutting tool.

33.16.6 Electromechanical Machining

Electromechanical machining (EMM) is a process in which the metal removal is effected in a conventional manner except that the workpiece is electrochemically polarized. When the applied voltage

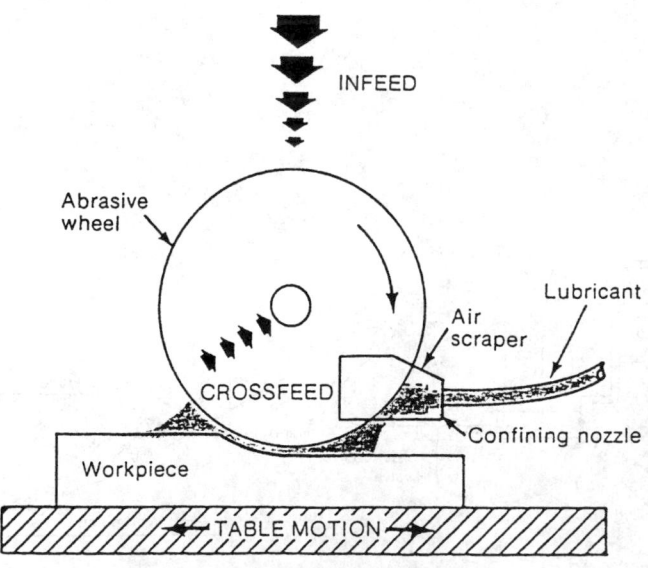

Fig. 33.34 Low-stress grinding.

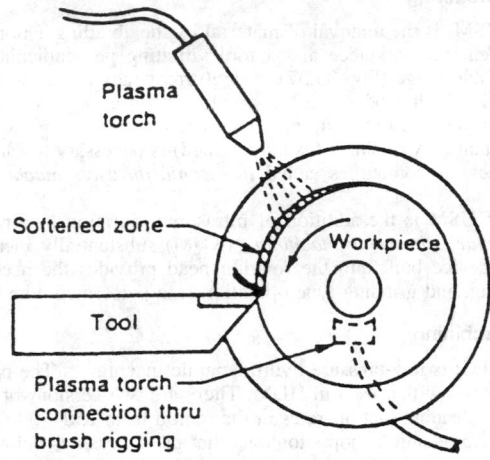

Fig. 33.35 Thermally assisted machining.

and the electrolytic solution are controlled, the surface of the workpiece can be changed to achieve the characteristics suitable for the machining operation.

33.16.7 Total Form Machining

Total form machining (TFM) is a process in which an abrasive master abrades its full three-dimensional shape into the workpiece by the application of force while a full-circle, orbiting motion is applied to the workpiece via the worktable (Fig. 33.36). The cutting master is advanced into the work until the desired depth of cut is achieved. Uniformity of cutting is promoted by the fluid that continuously transports the abraded particles out of the working gap. Adjustment of the orbiting cam drive controls the precision of the overcut from the cutting master. Cutting action takes place simultaneously over the full surface of abrasive contact.

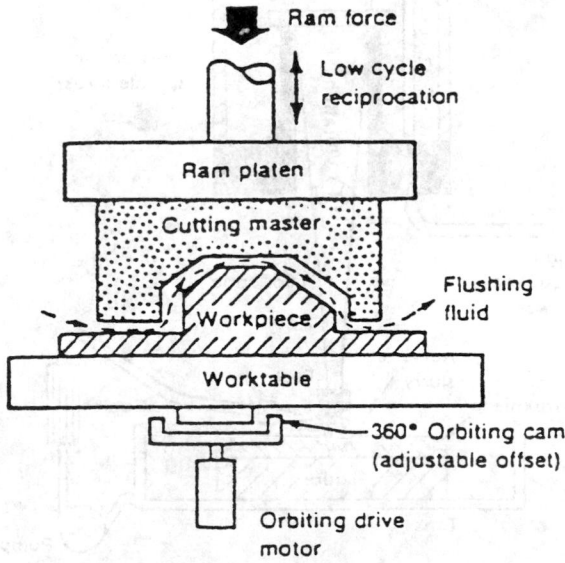

Fig. 33.36 Total form machining.

33.16.8 Ultrasonic Machining

Ultrasonic machining (USM) is the removal of material by the abrading action of a grit-loaded liquid slurry circulating between the workpiece and a tool vibrating perpendicular to the workface at a frequency above the audible range (Fig. 33.37). A high-frequency power source activates a stack of magnetostrictive material, which produces a low-amplitude vibration of the toolholder. This motion is transmitted under light pressure to the slurry, which abrades the workpiece into a conjugate image of the tool form. A constant flow of slurry (usually cooled) is necessary to carry away the chips from the workface. The process is sometimes called *ultrasonic abrasive machining* (UAM) or *impact machining*.

A prime variation of USM is the addition of ultrasonic vibration to a rotating tool—usually a diamond-plated drill. *Rotary ultrasonic machining* (RUM) substantially increases the drilling efficiency. A piezoelectric device built into the rotating head provides the needed vibration. Milling, drilling, turning, threading, and grinding-type operations are performed with RUM.

33.16.9 Water-Jet Machining

Water-jet machining (WJM) is low-pressure hydrodynamic machining. The pressure range for WJM is an order of magnitude below that used in HDM. There are two versions of WJM: one for mining, tunneling, and large-pipe cleaning that operates in the region from 250–1000 psi (1.7–6.9 Mpa); and one for smaller parts and production shop situations that uses pressures below 250 psi (1.7 Mpa).

The first version, or high-pressure range, is characterized by use of a pumped water supply with hoses and nozzles that generally are hand-directed. In the second version, more production-oriented and controlled equipment, such as that shown in Fig. 33.38, is involved. In some instances, abrasives are added to the fluid flow to promote rapid cutting. Single or multiple-nozzle approaches to the workpiece depend on the size and number of parts per load. The principle is that WJM is high-volume, not high-pressure.

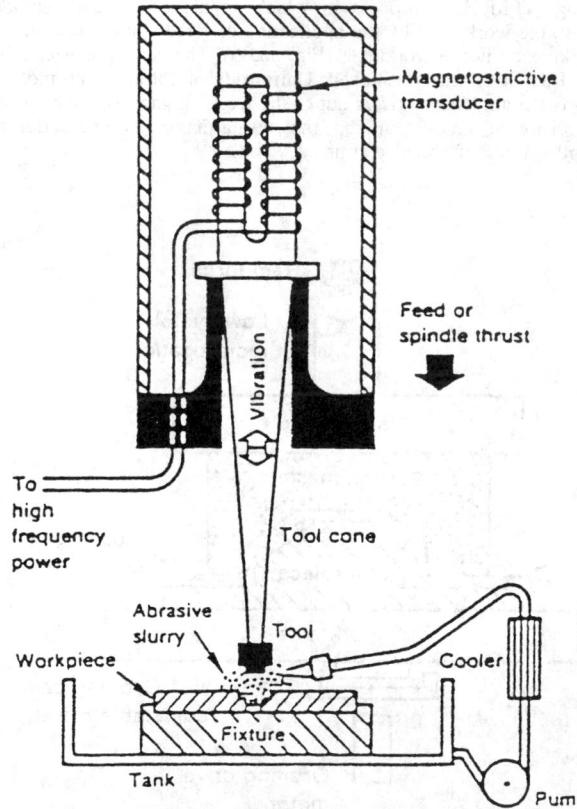

Fig. 33.37 Ultrasonic machining.

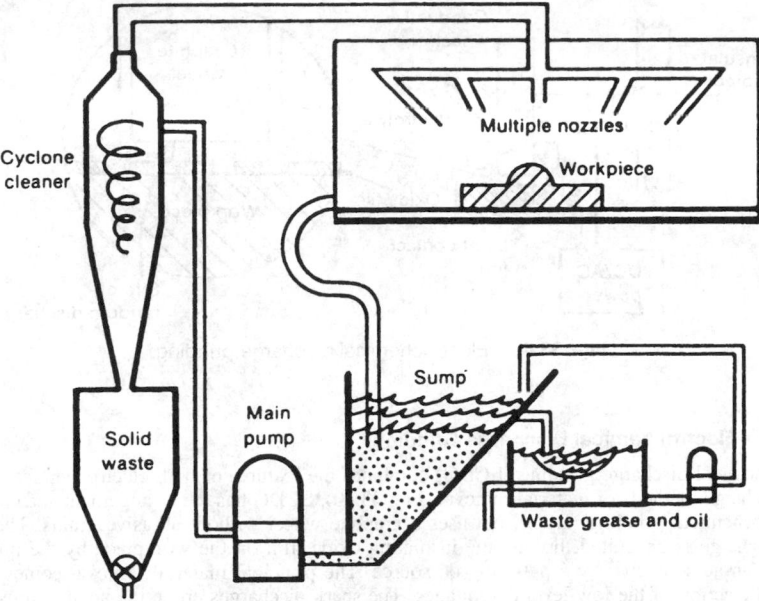

Fig. 33.38   Water-jet machining.

33.16.10   Electrochemical Deburring

Electrochemical deburring (ECD) is a special version of ECM (Fig. 33.39). ECD was developed to remove burrs and fins or to round sharp corners. Anodic dissolution occurs on the workpiece burrs in the presence of a closely placed cathodic tool whose configuration matches the burred edge. Normally, only a small portion of the cathode is electrically exposed, so a maximum concentration of the electrolytic action is attained. The electrolyte flow usually is arranged to carry away any burrs that may break loose from the workpiece during the cycle. Voltages are low, current densities are high, electrolyte flow rate is modest, and electrolyte types are similar to those used for ECM. The electrode (tool) is stationary, so equipment is simpler than that used for ECM. Cycle time is short for deburring. Longer cycle time produces a natural radiusing action.

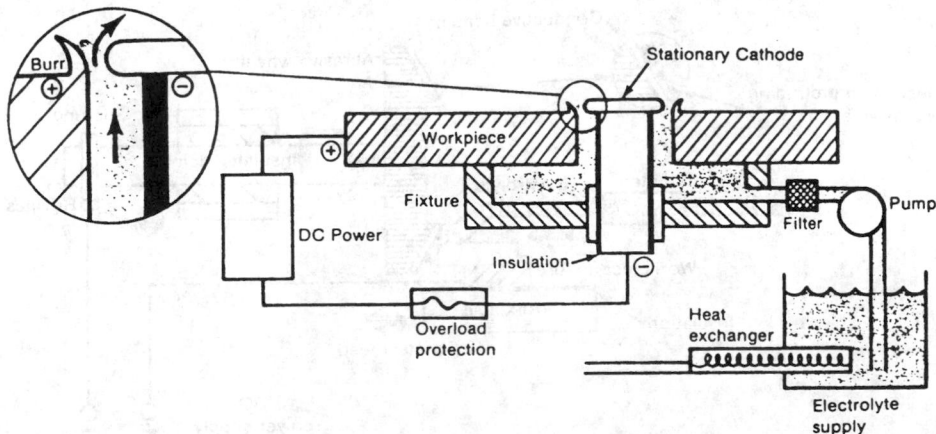

Fig. 33.39   Electrochemical deburring.

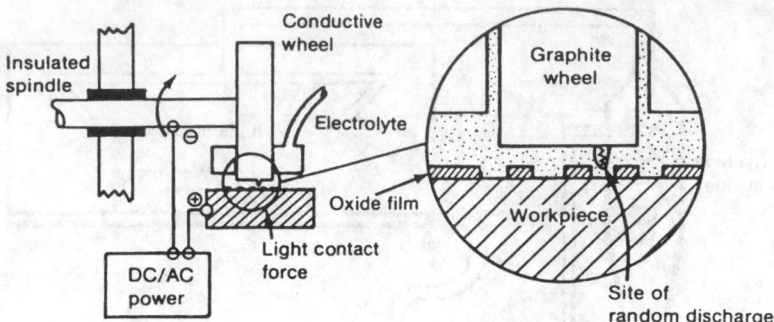

Fig. 33.40 Electrochemical discharge grinding.

33.16.11 Electrochemical Discharge Grinding

Electrochemical discharge grinding (ECDG) combines the features of both electrochemical and electrical discharge methods of material removal (Fig. 33.40). ECDG has the arrangement and electrolytes of electrochemical grinding (ECG), but uses a graphite wheel without abrasive grains. The random spark discharge is generated through the insulating oxide film on the workpiece by the power generated in an ac source or by a pulsating dc source. The principal material removal comes from the electrolytic action of the low-level dc voltages. The spark discharges erode the anodic films to allow the electrolytic action to continue.

33.16.12 Electrochemical Grinding

Electrochemical grinding (ECG) is a special form of electrochemical machining in which the conductive workpiece material is dissolved by anodic action, and any resulting films are removed by a rotating, conductive, abrasive wheel (Fig. 33.41). The abrasive grains protruding from the wheel form the insulating electrical gap between the wheel and the workpiece. This gap must be filled with electrolyte at all times. The conductive wheel uses conventional abrasives—aluminum oxide (because it is nonconductive) or diamond (for intricate shapes)—but lasts substantially longer than wheels used in conventional grinding. The reason for this is that the bulk of material removal (95–98%) occurs by deplating, while only a small amount (2–5%) occurs by abrasive mechanical action. Maximum wheel contact arc lengths are about ¾–1 in. (19–25 mm) to prevent overheating the electrolyte. The fastest material removal is obtained by using the highest attainable current densities without boiling the electrolyte. The corrosive salts used as electrolytes should be filtered and flow rate should be controlled for the best process control.

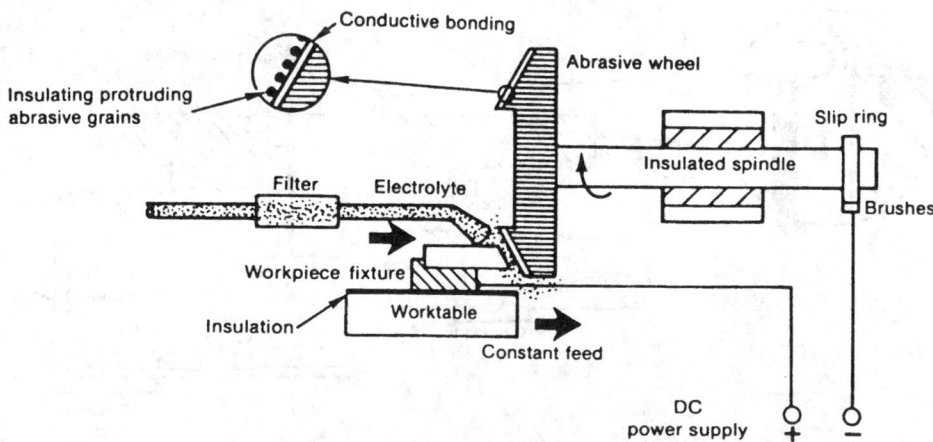

Fig. 33.41 Electrochemical grinding.

33.16.13 Electrochemical Honing

Electrochemical honing (ECH) is the removal of material by anodic dissolution combined with mechanical abrasion from a rotating and reciprocating abrasive stone (carried on a spindle, which is the cathode) separated from the workpiece by a rapidly flowing electrolyte (Fig. 33.42). The principal material removal action comes from electrolytic dissolution. The abrasive stones are used to maintain size and to clean the surfaces to expose fresh metal to the electrolyte action. The small electrical gap is maintained by the nonconducting stones that are bonded to the expandable arbor with cement. The cement must be compatible with the electrolyte and the low dc voltage. The mechanical honing action uses materials, speeds, and pressures typical of conventional honing.

33.16.14 Electrochemical Machining

Electrochemical machining (ECM) is the removal of electrically conductive material by anodic dissolution in a rapidly flowing electrolyte, which separates the workpiece from a shaped electrode (Fig. 33.43). The filtered electrolyte is pumped under pressure and at controlled temperature to bring a controlled-conductivity fluid into the narrow gap of the cutting area. The shape imposed on the workpiece is nearly a mirror or conjugate image of the shape of the cathodic electrode. The electrode is advanced into the workpiece at a constant feed rate that exactly matches the rate of dissolution of the work material. Electrochemical machining is basically the reverse of electroplating.

Calculation of Metal Removal and Feed Rates in ECM

$$\text{current } I = \frac{V}{R} \text{ amp}$$

$$\text{resistance } R = \frac{g \times r}{A}$$

where g = length of gap (cm)
r = electrolyte resistivity
A = area of current path (cm^2)
V = voltage
R = resistance

$$\text{current density } S = \frac{I}{A} = \frac{V}{r \times g} \text{ amp/cm}^2$$

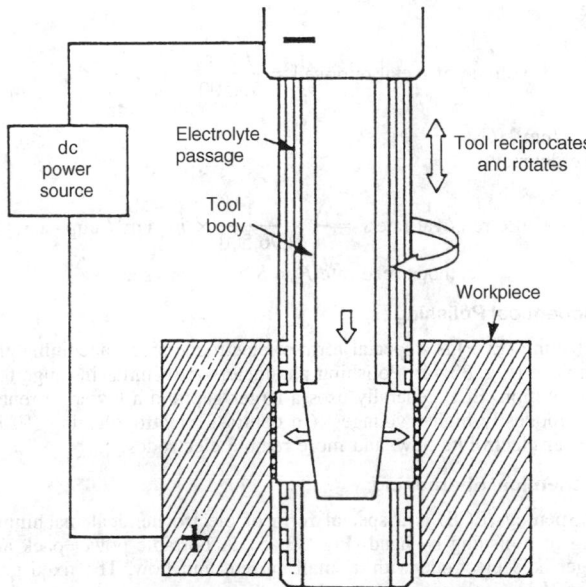

Fig. 33.42 Electrochemical honing.

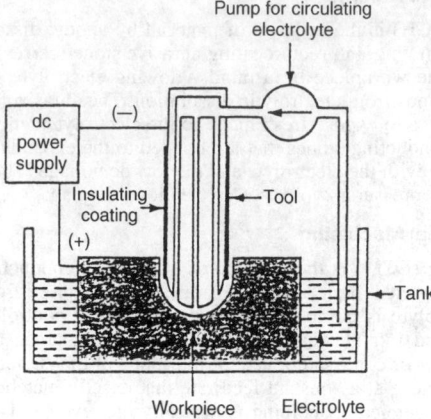

Fig. 33.43 Electrochemical machining.

The amount of material deposited or dissolved is proportional to the quantity of electricity passed (current × time).

$$\text{amount of material} = C \times I \times t$$

where C = constant
t = time, sec

The amount removed or deposited by one faraday (96,500 coulombs = 96,500 amp–sec) is 1 gram-equivalent weight (G)

$$G = \frac{N}{n} \quad \text{(for 1 faraday)}$$

where N = atomic weight
n = valence

$$\text{volume of metal removed} = \frac{I \times t}{96,500} \times \frac{N}{n} \times \frac{1}{d} \times h$$

where d = density, g/cm³
h = current efficiency

$$\text{specific removal rate } s = \frac{N}{n} \times \frac{1}{96,500} \times h \quad \text{cm}^3/\text{amp–sec}$$

$$\text{cathode feed rate } F = S \times s \quad \text{cm/sec.}$$

33.16.15 Electrochemical Polishing

Electrochemical polishing (ECP) is a special form of electrochemical machining arranged for cutting or polishing a workpiece (Fig. 33.44). Polishing parameters are similar in range to those for cutting, but without the feed motion. ECP generally uses a larger gap and a lower current density than does ECM. This requires modestly higher voltages. (In contrast, electropolishing (ELP) uses still lower current densities, lower electrolyte flow, and more remote electrodes.)

33.16.16 Electrochemical Sharpening

Electrochemical sharpening (ECS) is a special form of electrochemical machining arranged to accomplish sharpening or polishing by hand (Fig. 33.45). A portable power pack and electrolyte reservoir supply a finger-held electrode with a small current and flow. The fixed gap incorporated on the several styles of shaped electrodes controls the flow rate. A suction tube picks up the used electrolyte for recirculation after filtration.

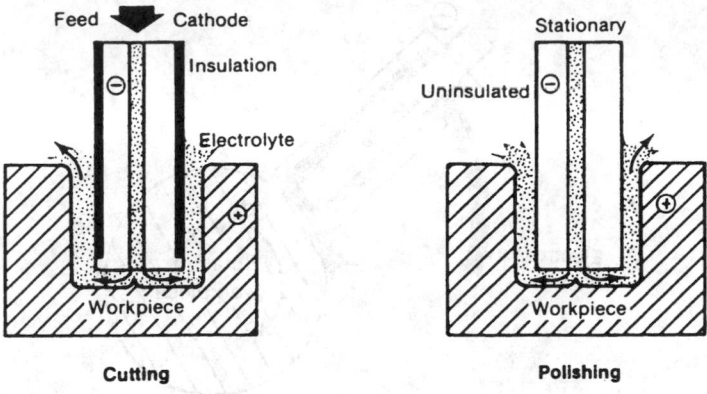

Fig. 33.44 Electrochemical polishing.

33.16.17 Electrochemical Turning

Electrochemical turning (ECT) is a special form of electrochemical machining designed to accommodate rotating workpieces (Fig. 33.46). The rotation provides additional accuracy but complicates the equipment with the method of introducing the high currents to the rotating part. Electrolyte control may also be complicated because rotating seals are needed to direct the flow properly. Otherwise, the parameters and considerations of electrochemical machining apply equally to the turning mode.

33.16.18 Electro-stream

Electro-stream (ES) is a special version of electrochemical machining adapted for drilling very small holes using high voltages and acid electrolytes (see Fig. 33.47). The voltages are more than 10 times those employed in ECM or STEM, so special provisions for containment and protection are required. The tool is a drawn-glass nozzle, 0.001–0.002 in. smaller than the desired hole size. An electrode inside the nozzle or the manifold ensures electrical contact with the acid. Multiple-hole drilling is achieved successfully by ES.

33.16.19 Shaped-Tube Electrolytic Machining

Shaped-tube electrolytic machining (STEM™) is a specialized ECM technique for "drilling" small, deep holes by using acid electrolytes (Fig. 33.48). Acid is used so that the dissolved metal will go into the solution rather than form a sludge, as is the case with the salt-type electrolytes of ECM. The electrode is a carefully straightened acid-resistant metal tube. The tube is coated with a film of enamel-type insulation. The acid is pressure-fed through the tube and returns via a narrow gap between the tube insulation and the hole wall. The electrode is fed into the workpiece at a rate exactly equal to the rate at which the workpiece material is dissolved. Multiple electrodes, even of varying

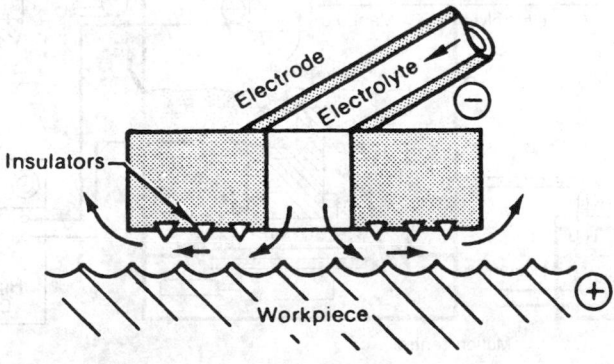

Fig. 33.45 Electrochemical sharpening.

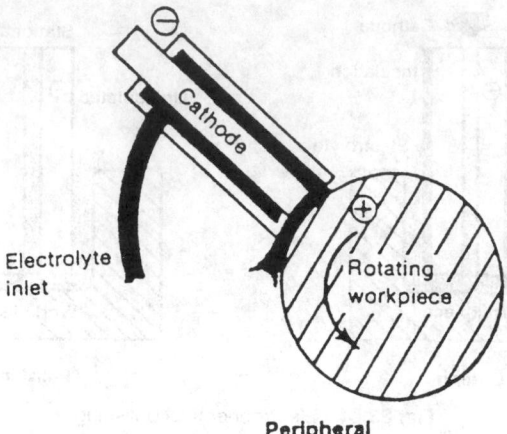

Peripheral

Fig. 33.46   Electrochemical turning.

diameters or shapes, may be used simultaneously. A solution of sulfuric acid is frequently used as the electrolyte when machining nickel alloys. The electrolyte is heated and filtered, and flow monitors control the pressure. Tooling is frequently made of plastics, ceramics, or titanium alloys to withstand the electrified hot acid.

33.16.20   Electron-Beam Machining

Electron-beam machining (EBM) removes material by melting and vaporizing the workpiece at the point of impingement of a focused stream of high-velocity electrons (Fig. 33.49). To eliminate scattering of the beam of electrons by contact with gas molecules, the work is done in a high-vacuum chamber. Electrons emanate from a triode electron-beam gun and are accelerated to three-fourths the speed of light at the anode. The collision of the electrons with the workpiece immediately translates their kinetic energy into thermal energy. The low-inertia beam can be simply controlled by electromagnetic fields. Magnetic lenses focus the electron beam on the workpiece, where a 0.001-in. (0.025-mm) diameter spot can attain an energy density of up to 10^9 W/in.^{2} (1.55×10^8 W/cm^2) to melt and vaporize any material. The extremely fast response time of the beam is an excellent companion for three-dimensional computer control of beam deflection, beam focus, beam intensity, and workpiece motion.

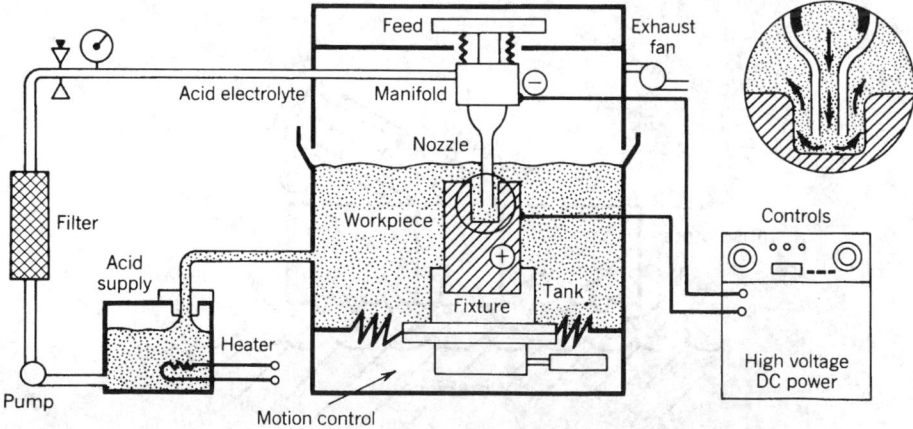

Fig. 33.47   Electro-stream.

Metal tube
electrode
(cathode)
Dielectric
coating
Electrolyte
flow
through
stem
electrode
Flow of metal
ions from
hole
Flow of metal
ions from
hole
Workpiece
(anode)

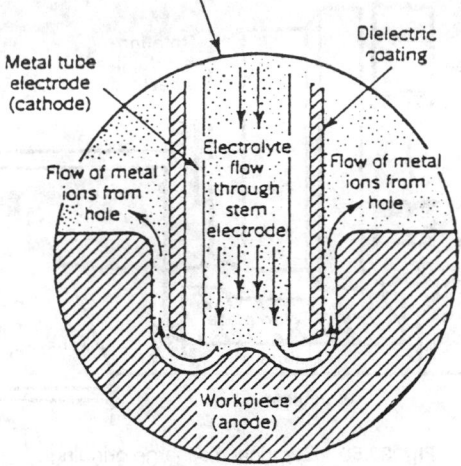

Fig. 33.48 Shaped-tube electrolytic machining.

33.16.21 Electrical Discharge Grinding

Electrical discharge grinding (EDG) is the removal of a conductive material by rapid, repetitive spark discharges between a rotating tool and the workpiece, which are separated by a flowing dielectric fluid (Fig. 33.50). (EDG is similar to EDM except that the electrode is in the form of a grinding wheel and the current is usually lower.) The spark gap is servocontrolled. The insulated wheel and the worktable are connected to the dc pulse generator. Higher currents produce faster cutting, rougher finishes, and deeper heat-affected zones in the workpiece.

33.16.22 Electrical Discharge Machining

Electrical discharge machining (EDM) removes electrically conductive material by means of rapid, repetitive spark discharges from a pulsating dc power supply with dielectric flowing between the workpiece and the tool (Fig. 33.51). The cutting tool (electrode) is made of electrically conductive material, usually carbon. The shaped tool is fed into the workpiece under servocontrol. A spark discharge then breaks down the dielectric fluid. The frequency and energy per spark are set and

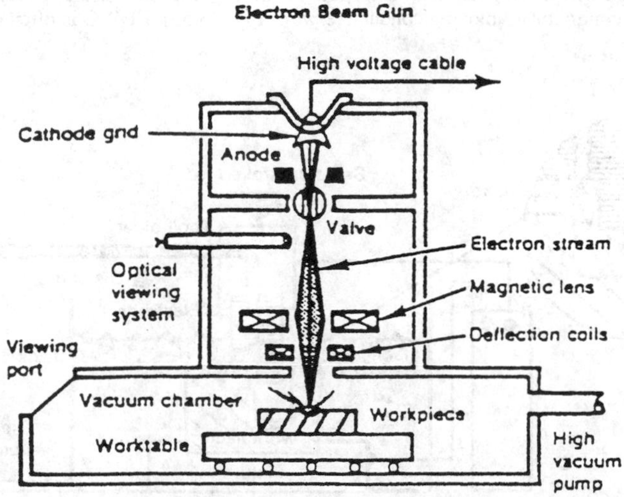

Electron Beam Gun
High voltage cable
Cathode grid
Anode
Valve
Electron stream
Optical
viewing
system
Magnetic lens
Deflection coils
Viewing
port
Vacuum chamber
Workpiece
Worktable
High
vacuum
pump

Fig. 33.49 Electron-beam machining.

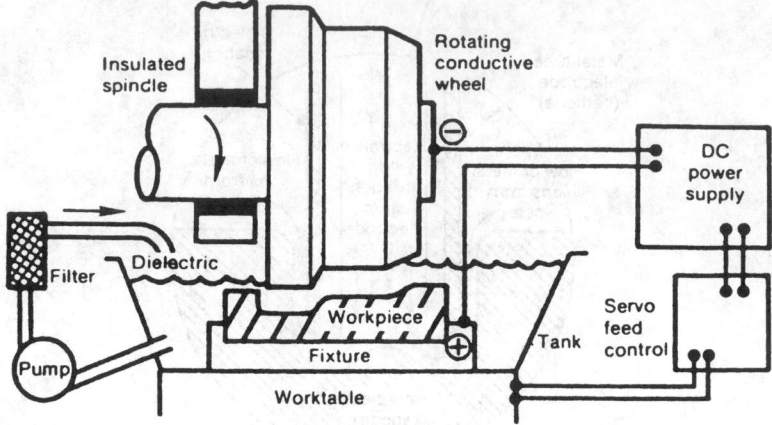

Fig. 33.50 Electrical discharge grinding.

controlled with a dc power source. The servocontrol maintains a constant gap between the tool and the workpiece while advancing the electrode. The dielectric oil cools and flushes out the vaporized and condensed material while reestablishing insulation in the gap. Material removal rate ranges from 16–245 cm³/h. EDM is suitable for cutting materials regardless of their hardness or toughness. Round or irregular-shaped holes 0.002 in. (0.05 mm) diameter can be produced with L/D ratio of 20:1. Narrow slots as small as 0.002–0.010 in. (0.05–0.25 mm) wide are cut by EDM.

33.16.23   Electrical Discharge Sawing

Electrical discharge sawing (EDS) is a variation of electrical discharge machining (EDM) that combines the motion of either a band saw or a circular disk saw with electrical erosion of the workpiece (Fig. 33.52). The rapid-moving, untoothed, thin, special steel band or disk is guided into the workpiece by carbide-faced inserts. A kerf only 0.002–0.005 in. (0.050–0.13 mm) wider than the blade or disk is formed as they are fed into the workpiece. Water is used as a cooling quenchant for the tool, swarf, and workpiece. Circular cutting is usually performed under water, thereby reducing noise and fumes. While the work is power-fed into the band (or the disk into the work), it is not subjected to appreciable forces because the arc does the cutting, so fixturing can be minimal.

33.16.24   Electrical Discharge Wire Cutting (Traveling Wire)

Electrical discharge wire cutting (EDWC) is a special form of electrical discharge machining wherein the electrode is a continuously moving conductive wire (Fig. 33.53). EDWC is often called *traveling*

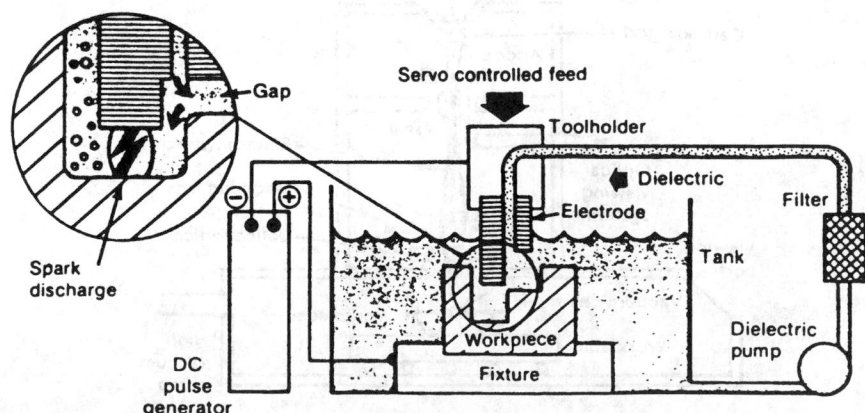

Fig. 33.51 Electrical discharge machining.

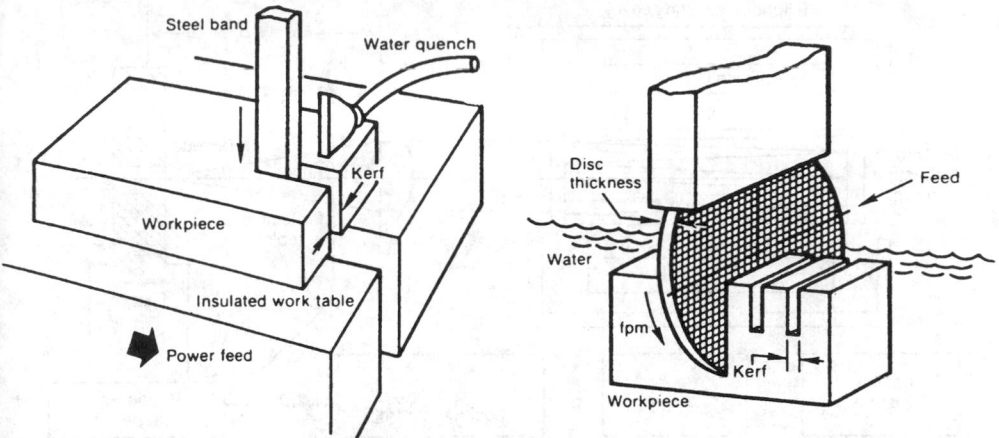

Fig. 33.52 Electrical discharge sawing.

wire EDM. A small-diameter tension wire, 0.001–0.012 in. (0.03–0.30 mm), is guided to produce a straight, narrow-kerf size 0.003–0.015 in. (0.075–0.375 mm). Usually, a programmed or numerically controlled motion guides the cutting, while the width of the kerf is maintained by the wire size and discharge controls. The dielectric is oil or deionized water carried into the gap by motion of the wire. Wire EDM is able to cut plates as thick as 12 in. (300 mm) and issued for making dies from hard metals. The wire travels with speed in the range of 6–300 in./min (0.15–8 mm/min). A typical cutting rate is 1 in.² (645 mm²) of cross-sectional area per hour.

33.16.25 Laser-Beam Machining

Laser-beam machining (LBM) removes material by melting, ablating, and vaporizing the workpiece at the point of impingement of a highly focused beam of coherent monochromatic light (Fig. 33.54). Laser is an acronym for "light amplification by stimulated emission of radiation." The electromagnetic radiation operates at wavelengths from the visible to the infrared. The principal lasers used for material removal are the ND:glass (neodymium–glass), the Nd:YAG (neodymium:yttrium-aluminum-garnet), the ruby and the carbon dioxide (CO_2). The last is a gas laser (most frequently used as a torch with an assisting gas—see LBT, laser-beam torch), while others are solid-state lasing materials.

For pulsed operation, the power supply produces short, intense bursts of electricity into the flash lamps, which concentrate their light flux on the lasing material. The resulting energy from the excited

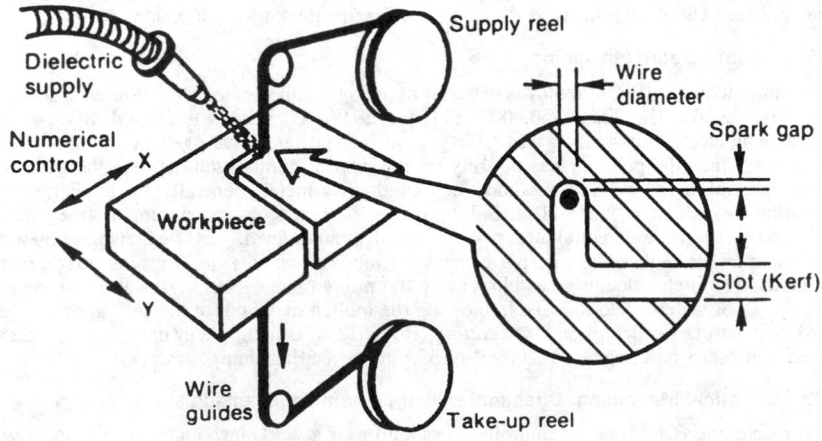

Fig. 33.53 Electrical discharge wire cutting.

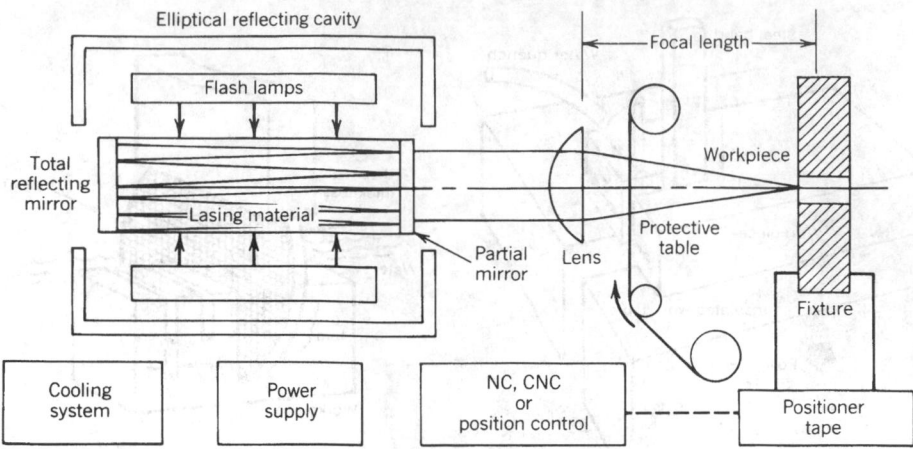

Fig. 33.54 Laser-beam machining.

atoms is released at a characteristic, constant frequency. The monochromatic light is amplified during successive reflections from the mirrors. The thoroughly collimated light exits through the partially reflecting mirror to the lens, which focuses it on or just below the surface of the workpiece. The small beam divergence, high peak power, and single frequency provide excellent, small-diameter spots of light with energy densities up to 3×10^{10} W/in.^{2} (4.6×10^9 W/cm^2), which can sublime almost any material. Cutting requires energy densities of 10^7–10^9 W/in.^{2} (1.55×10^6–1.55×10^8 W/cm^2), at which rate the thermal capacity of most materials cannot conduct energy into the body of the workpiece fast enough to prevent melting and vaporization. Some lasers can instantaneously produce 41,000°C (74,000°F). Holes of 0.001 in. (0.025 mm), with depth-to-diameter 50 to 1 are typically produced in various materials by LBM.

33.16.26   Laser-Beam Torch

Laser-beam torch (LBT) is a process in which material is removed by the simultaneous focusing of a laser beam and a gas stream on the workpiece (see Fig. 33.55). A continuous-wave (CW) laser or a pulsed laser with more than 100 pulses per second is focused on or slightly below the surface of the workpiece, and the absorbed energy causes localized melting. An oxygen gas stream promotes an exothermic reaction and purges the molten material from the cut. Argon or nitrogen gas is sometimes used to purge the molten material while also protecting the workpiece.

Argon or nitrogen gas is often used when organic or ceramic materials are being cut. Close control of the spot size and the focus on the workpiece surface is required for uniform cutting. The type of gas used has only a modest effect on laser penetrating ability. Typically, short laser pulses with high peak power are used for cutting and welding. The CO_2 laser is the laser most often used for cutting. Thin materials are cut at high rates, ⅛–⅜ in. (3.2–9.5 mm) thickness is a practical limit.

33.16.27   Plasma-Beam Machining

Plasma-beam machining (PBM) removes material by using a superheated stream of electrically ionized gas (Fig. 33.56). The 20,000–50,000°F (11,000–28,000°C) plasma is created inside a water-cooled nozzle by electrically ionizing a suitable gas, such as nitrogen, hydrogen, or argon, or mixtures of these gases. Since the process does not rely on the heat of combustion between the gas and the workpiece material, it can be used on almost any conductive metal. Generally, the arc is transferred to the workpiece, which is made electrically positive. The plasma—a mixture of free electrons, positively charged ions, and neutral atoms—is initiated in a confined, gas-filled chamber by a high-frequency spark. The high-voltage dc power sustains the arc, which exits from the nozzle at near-sonic velocity. The high-velocity gases blow away the molten metal "chips." Dual-flow torches use a secondary gas or water shield to assist in blowing the molten metal out of the kerf, giving a cleaner cut. PBM is sometimes called *plasma-arc cutting* (PAC). PBM can cut plates up to 6.0 in. (152 mm) thick. Kerf width can be as small as 0.06 in. (1.52 mm) in cutting thin plates.

33.16.28   Chemical Machining: Chemical Milling, Chemical Blanking

Chemical machining (CHM) is the controlled dissolution of a workpiece material by contact with a strong chemical reagent (Fig. 33.57). The thoroughly cleaned workpiece is covered with a strippable,

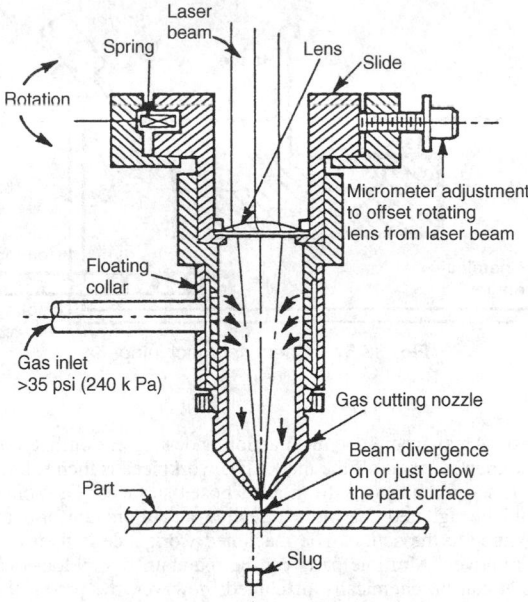

Fig. 33.55 Laser-beam torch.

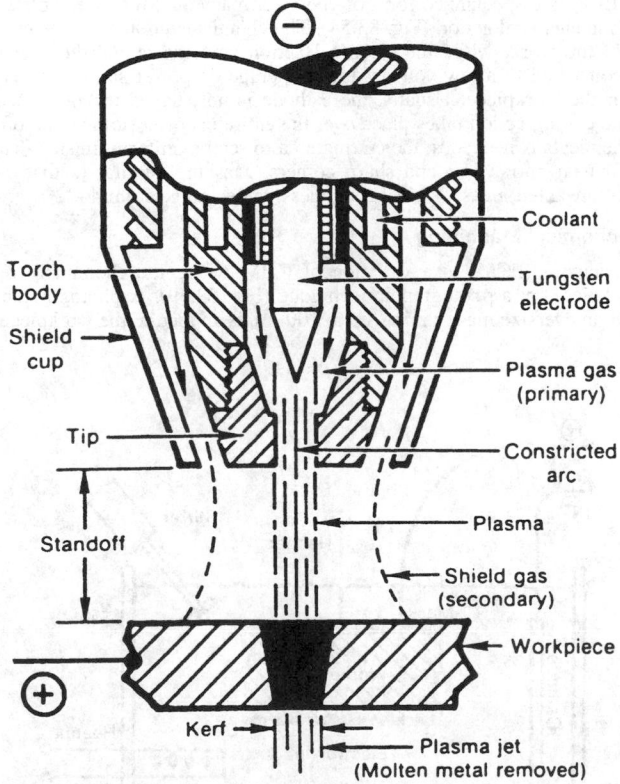

Fig. 33.56 Plasma-beam machining.

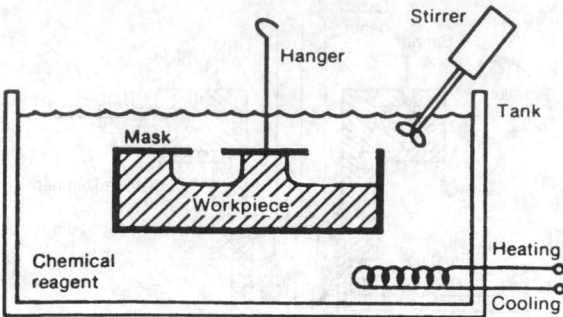

Fig. 33.57 Chemical machining.

chemically resistant mask. Areas where chemical action is desired are outlined on the workpiece with the use of a template and then stripped off the mask. The workpiece is then submerged in the chemical reagent to remove material simultaneously from all exposed surfaces. The solution should be stirred or the workpiece should be agitated for more effective and more uniform action. Increasing the temperatures will also expedite the action. The machined workpiece is then washed and rinsed, and the remaining mask is removed. Multiple parts can be maintained simultaneously in the same tank. A wide variety of metals can be chemically machined; however, the practical limitations for depth of cut are 0.25–0.5 in. (6.0–12.0 mm) and typical etching rate is 0.001 in./min (0.025 mm/min).

In chemical blanking, the material is removed by chemical dissolution instead of shearing. The operation is applicable to production of complex shapes in thin sheets of metal.

33.16.29 Electropolishing

Electropolishing (ELP) is a specialized form of chemical machining that uses an electrical deplating action to enhance the chemical action (Fig. 33.58). The chemical action from the concentrated heavy acids does most of the work, while the electrical action smooths or polishes the irregularities. A metal cathode is connected to a low-voltage, low-amperage dc power source and is installed in the chemical bath near the workpiece. Usually, the cathode is not shaped or conformed to the surface being polished. The cutting action takes place over the entire exposed surface; therefore, a good flow of heated, fresh chemicals is needed in the cutting area to secure uniform finishes. The cutting action will concentrate first on burrs, fins, and sharp corners. Masking, similar to that used with CHM, prevents cutting in unwanted areas. Typical roughness values range from 4–32 μin. (0.1–0.8 μm).

33.16.30 Photochemical Machining

Photochemical machining (PCM) is a variation of CHM where the chemically resistant mask is applied to the workpiece by a photographic technique (Fig. 33.59). A photographic negative, often a reduced image of an oversize master print (up to 100 ×), is applied to the workpiece and developed.

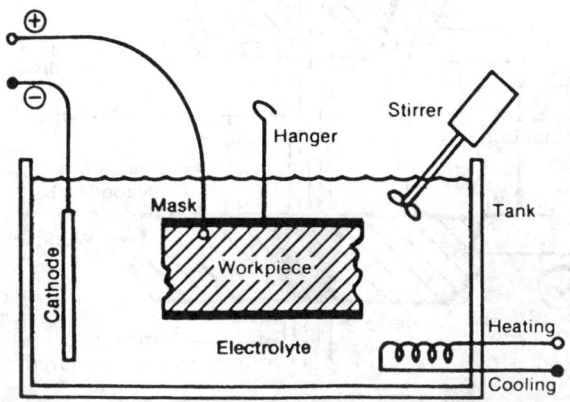

Fig. 33.58 Electropolishing.

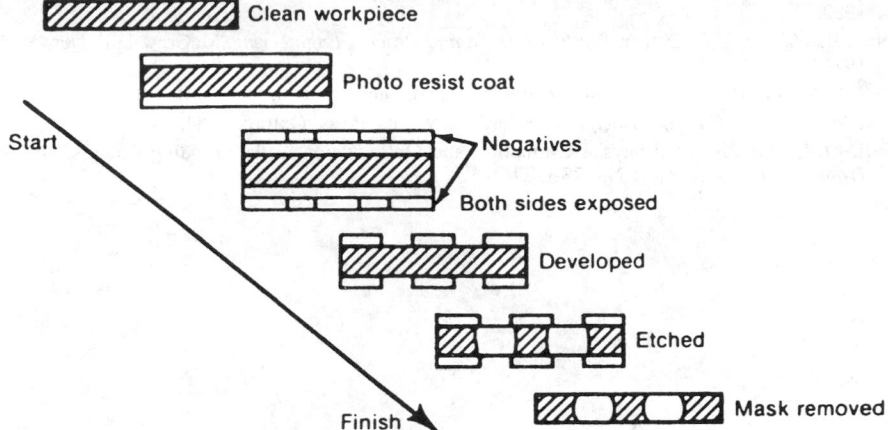

Fig. 33.59 Photochemical machining.

Precise registry of duplicate negatives on each side of the sheet is essential for accurately blanked parts. Immersion or spray etching is used to remove the exposed material. The chemicals used must be active on the workpiece, but inactive against the photoresistant mask. The use of PCM is limited to thin materials—up to $\frac{1}{16}$ in. (1.5 mm).

33.16.31 Thermochemical Machining

Thermochemical machining (TCM) removes the workpiece material—usually only burrs and fins—by exposure of the workpiece to hot, corrosive gases. The process is sometimes called *combustion machining, thermal deburring,* or *thermal energy method* (TEM). The workpiece is exposed for a very short time to extremely hot gases, which are formed by detonating an explosive mixture. The ignition of the explosive—usually hydrogen or natural gas and oxygen—creates a transient thermal wave that vaporizes the burrs and fins. The main body of the workpiece remains unaffected and relatively cool because of its low surface-to-mass ratio and the shortness of the exposure to high temperatures.

REFERENCES

1. Society of Manufacturing Engineers, *Tool and Manufacturing Engineers Handbook,* Vol. 1, *Machining,* McGraw-Hill, New York, 1985.
2. *Machining Data Handbook,* 3rd ed., Machinability Data Center, Cincinnati, OH, 1980.
3. *Metals Handbook,* 8th ed., Vol. 3, Machining American Society for Metals, Metals Park, OH, 1985.
4. R. LeGrand (ed.), *American Machinist's Handbook,* 3rd ed., McGraw-Hill, New York, 1973.
5. *Machinery's Handbook,* 21st ed., Industrial Press, New York, 1979.
6. *Machinery Handbook,* Vol. 2, Machinability Data Center, Cincinnati, Department of Defense, 1983.

BIBLIOGRAPHY

Alting, L., *Manufacturing Engineering Processes,* Marcel Dekker, New York, 1982.

Amstead, B. H., P. F. Ostwald, and M. L. Begeman, *Manufacturing Processes,* 8th ed., Wiley, New York, 1988.

DeGarmo, E. P., J. T. Black, and R. A. Kohser, *Material and Processes in Manufacturing,* 7th ed., Macmillan, New York, 1988.

Doyle, L. E., G. F. Schrader, and M. B. Singer, *Manufacturing Processes and Materials for Engineers,* 3rd ed., Prentice-Hill, Englewood Cliffs, NJ, 1985.

Kalpakjian, S., *Manufacturing Processes for Engineering Materials,* Addison-Wesley, Reading, MA, 1994.

Kronenberg, M., *Machining Science and Application,* Pergamon, London, 1966.

Lindberg, R. A., *Processes and Materials of Manufacture,* 2nd ed., Allyn and Bacon, Boston, MA, 1977.

Moore, H. D., and D. R. Kibbey, *Manufacturing Materials and Processes,* 3rd ed., Wiley, New York, 1982.

Niebel, B. W., and A. B. Draper, *Product Design and Process Engineering,* McGraw-Hill, New York, 1974.

Schey, J. A., *Introduction to Manufacturing Processes,* McGraw-Hill, New York, 1977.

Shaw, M. C., *Metal Cutting Principles,* Oxford University Press, Oxford, 1984.

Zohdi, M. E., "Statistical Analysis, Estimation and Optimization in the Grinding Process," *ASME Transactions,* 1973, Paper No. 73-DET-3.

CHAPTER 34

METAL FORMING, SHAPING, AND CASTING

Magd E. Zohdi
Dennis B. Webster
Industrial Engineering Department
Louisiana State University
Baton Rouge, Louisiana

William E. Biles
Industrial Engineering Department
University of Louisville
Louisville, Kentucky

34.1	**INTRODUCTION**	**1101**		34.4.3	Permanent-Mold Casting	1123
				34.4.4	Plaster-Mold Casting	1125
34.2	**HOT-WORKING PROCESSES**	**1102**		34.4.5	Investment Casting	1125
	34.2.1 Classification of Hot-Working Processes	1103	**34.5**	**PLASTIC-MOLDING PROCESSES**		**1126**
	34.2.2 Rolling	1103		34.5.1	Injection Molding	1126
	34.2.3 Forging	1105		34.5.2	Coinjection Molding	1126
	34.2.4 Extrusion	1107		34.5.3	Rotomolding	1126
	34.2.5 Drawing	1107		34.5.4	Expandable-Bead Molding	1126
	34.2.6 Spinning	1110		34.5.5	Extruding	1126
	34.2.7 Pipe Welding	1111		34.5.6	Blow Molding	1126
	34.2.8 Piercing	1111		34.5.7	Thermoforming	1127
				34.5.8	Reinforced-Plastic Molding	1127
34.3	**COLD-WORKING PROCESSES**	**1112**		34.5.9	Forged-Plastic Parts	1127
	34.3.1 Classification of Cold-Working Operations	1112				
	34.3.2 Squeezing Processes	1113	**34.6**	**POWDER METALLURGY**		**1127**
	34.3.3 Bending	1114		34.6.1	Properties of P/M Products	1127
	34.3.4 Shearing	1116				
	34.3.5 Drawing	1118	**34.7**	**SURFACE TREATMENT**		**1128**
				34.7.1	Cleaning	1128
34.4	**METAL CASTING AND MOLDING PROCESSES**	**1120**		34.7.2	Coatings	1130
	34.4.1 Sand Casting	1120		34.7.3	Chemical Conversions	1132
	34.4.2 Centrifugal Casting	1121				

34.1 INTRODUCTION

Metal-forming processes use a remarkable property of metals—their ability to flow plastically in the solid state without concurrent deterioration of properties. Moreover, by simply moving the metal to the desired shape, there is little or no waste. Figure 34.1 shows some of the metal-forming processes. Metal-forming processes are classified into two categories: hot-working processes and cold-working processes.

Mechanical Engineers' Handbook, 2nd ed., Edited by Myer Kutz.
ISBN 0-471-13007-9   © 1998 John Wiley & Sons, Inc.

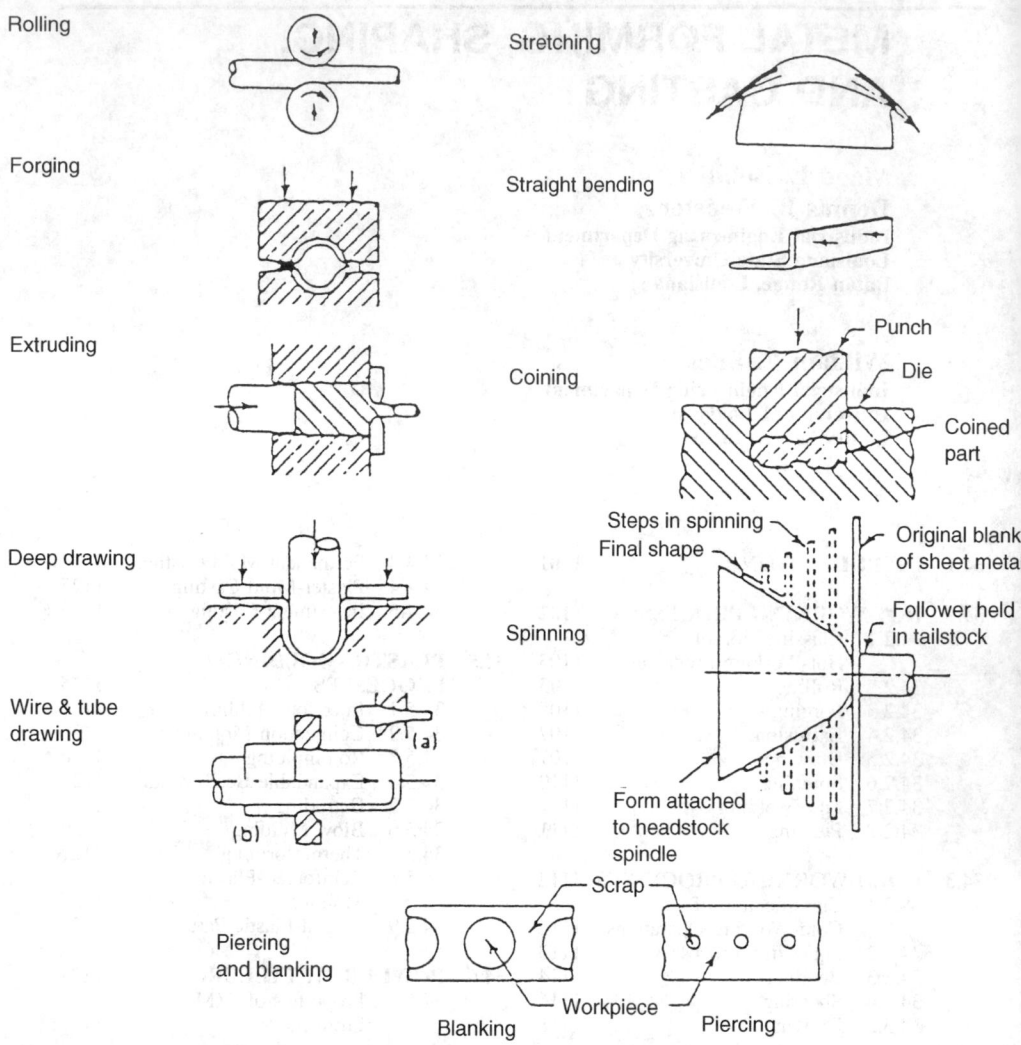

Fig. 34.1 Metal-forming processes.

34.2 HOT-WORKING PROCESSES

Hot working is defined as the plastic deformation of metals above their recrystallization temperature. Here it is important to note that the crystallization temperature varies greatly with different materials. Lead and tin are hot worked at room temperature, while steels require temperatures of 2000°F (1100°C). Hot working does not necessarily imply high absolute temperatures.

Hot working can produce the following improvements:

1. Production of randomly oriented, spherical-shaped grain structure, which results in a net increase not only in the strength but also in ductility and toughness.
2. The reorientation of inclusions or impurity material in metal. The impurity material often distorts and flows along with the metal.

This material, however, does not recrystallize with the base metal and often produces a fiber structure. Such a structure clearly has directional properties, being stronger in one direction than in another. Moreover, an impurity originally oriented so as to aid crack movement through the metal is often reoriented into a "crack-arrestor" configuration perpendicular to crack propagation.

34.2.1 Classification of Hot-Working Processes

The most obvious reason for the popularity of hot working is that it provides an attractive means of forming a desired shape. Some of the hot-working processes that are of major importance in modern manufacturing are

1. Rolling
2. Forging
3. Extrusion and upsetting
4. Drawing
5. Spinning
6. Pipe welding
7. Piercing

34.2.2 Rolling

Hot rolling (Fig. 34.2) consists of passing heated metal between two rolls that revolve in opposite directions, the space between the rolls being somewhat less than the thickness of the entering metal. Many finished parts, such as hot-rolled structural shapes, are completed entirely by hot rolling. More often, however, hot-rolled products, such as sheets, plates, bars, and strips, serve as input material for other processes, such as cold forming or machining.

In hot rolling, as in all hot working, it is very important that the metal be heated uniformly throughout to the proper temperature, a procedure known as *soaking*. If the temperature is not uniform, the subsequent deformation will also be nonuniform, the hotter exterior flowing in preference to the cooler and, therefore, stronger, interior. Cracking, tearing, and associated problems may result.

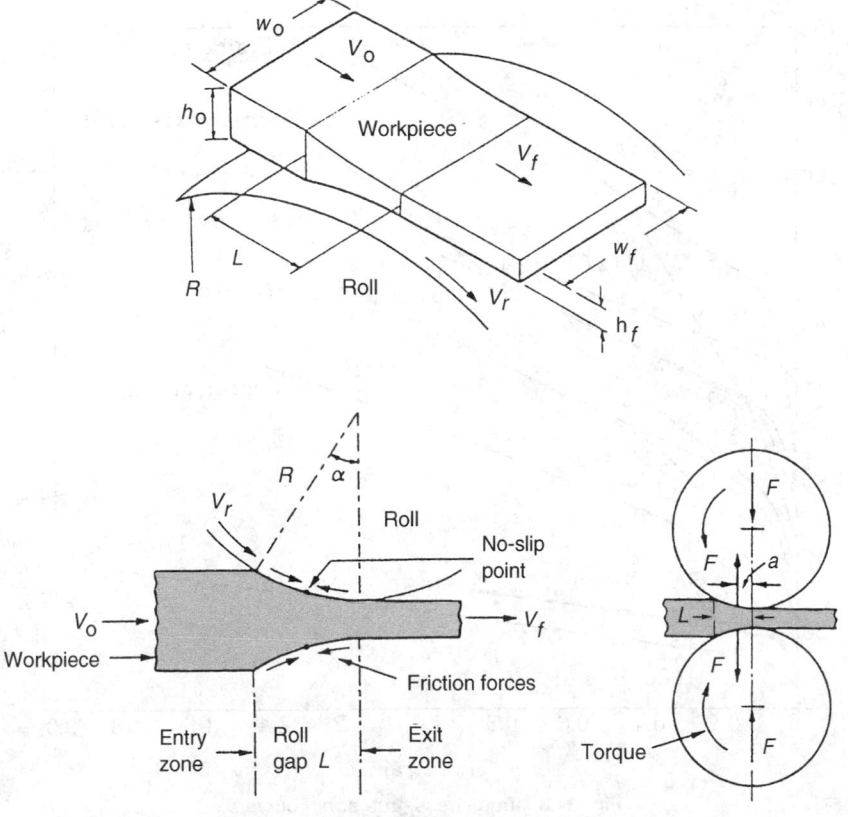

Fig. 34.2 Hot rolling.

Isothermal Rolling

The ordinary rolling of some high-strength metals, such as titanium and stainless steels, particularly in thicknesses below about 0.150 in. (3.8 mm), is difficult because the heat in the sheet is transferred rapidly to the cold and much more massive rolls. This has been overcome by isothermal rolling. Localized heating is accomplished in the area of deformation by the passage of a large electrical current between the rolls, through the sheet. Reductions up to 90% per roll have been achieved. The process usually is restricted to widths below 2 in. (50 mm).

The rolling strip contact length is given by

$$L \simeq \sqrt{R(h_0 - h)}$$

where R = roll radius
h_0 = original strip thickness
h = reduced thickness

The roll-force F is calculated by

$$F = LwY_{\text{avg}} \tag{34.1}$$

where w = width
Y_{avg} = average true stress

Figure 34.3 gives the true stress for different material at the true stress ϵ. The true stress ϵ is given by

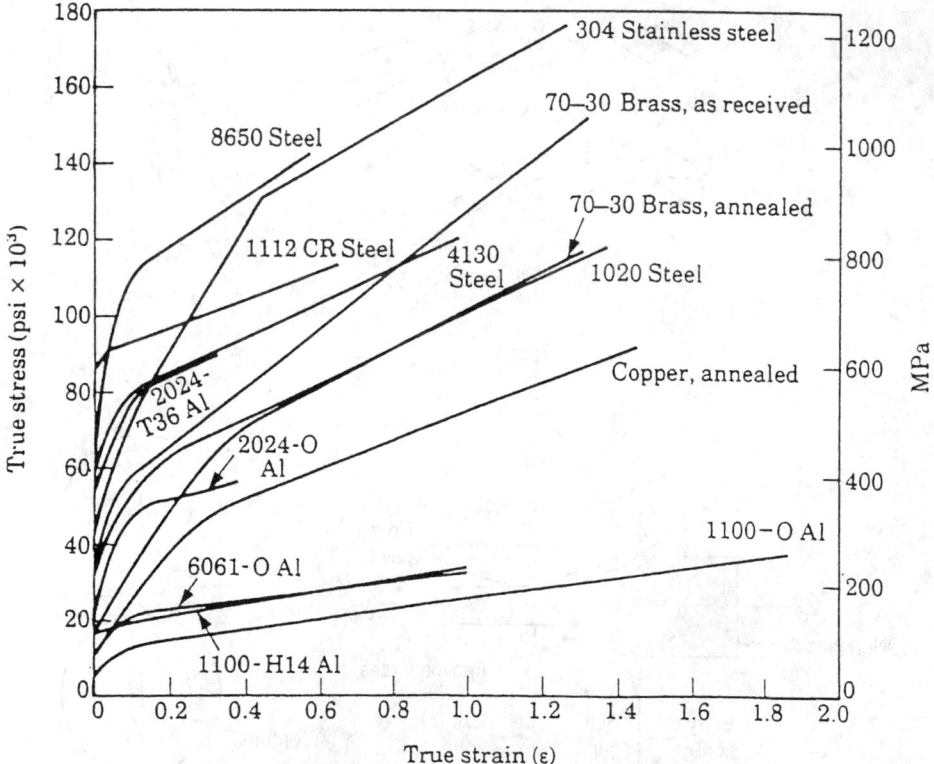

Fig. 34.3 True stress–true strain curves.

$$\epsilon = \ln\left(\frac{h_o}{h}\right)$$

$$\text{power/roll} = \frac{2\pi FLN}{60,000} \quad \text{kW} \tag{34.2}$$

where F = newtons
$\quad L$ = meters
$\quad N$ = rev per min

or

$$\text{power} = \frac{2\pi FLN}{33,000} \quad \text{hp} \tag{34.3}$$

where F = lb
$\quad L$ = ft

34.2.3 Forging

Forging is the plastic working of metal by means of localized compressive forces exerted by manual or power hammers, presses, or special forging machines.

Various types of forging have been developed to provide great flexibility, making it economically possible to forge a single piece or to mass produce thousands of identical parts. The metal may be

1. Drawn out, increasing its length and decreasing its cross section
2. Upset, increasing the cross section and decreasing the length, or
3. Squeezed in closed impression dies to produce multidirectional flow

The state of stress in the work is primarily uniaxial or multiaxial compression.

The common forging processes are

1. Open-die hammer
2. Impression-die drop forging
3. Press forging
4. Upset forging
5. Roll forging
6. Swaging

Open-Die Hammer Forging

Open-die forging, (Fig. 34.4) does not confine the flow of metal, the hammer and anvil often being completely flat. The desired shape is obtained by manipulating the workpiece between blows. Specially shaped tools or a slightly shaped die between the workpiece and the hammer or anvil are used to aid in shaping sections (round, concave, or convex), making holes, or performing cutoff operations.

The force F required for an open-die forging operation on a solid cylindrical piece can be calculated by

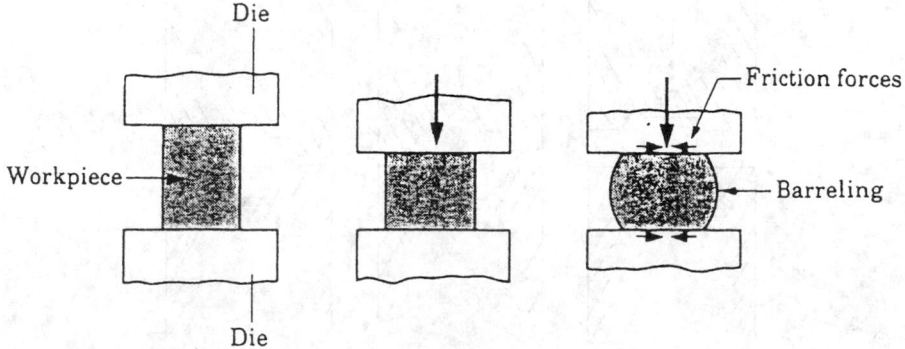

Fig. 34.4 Open-die hammer forging.

$$F = Y_f \pi r^2 \left(1 + \frac{2\mu r}{3h} \right) \qquad (34.4)$$

where Y_f = flow stress at the specific ϵ [$\epsilon = \ln(h_0/h)$]
$\quad \mu$ = coefficient of friction
r and h = radius and height of workpiece

Impression-Die Drop Forging

In impression-die or closed-die drop forging (Fig. 34.5), the heated metal is placed in the lower cavity of the die and struck one or more blows with the upper die. This hammering causes the metal to flow so as to fill the die cavity. Excess metal is squeezed out between the die faces along the periphery of the cavity to form a flash. When forging is completed, the flash is trimmed off by means of a trimming die.

The forging force F required for impression-die forging can be estimated by

$$F = KY_f A \qquad (34.5)$$

where K = multiplying factor (4–12) depending on the complexity of the shape
$\quad Y_f$ = flow stress at forging temperature
$\quad A$ = projected area, including flash

Press Forging

Press forging employs a slow-squeezing action that penetrates throughout the metal and produces a uniform metal flow. In hammer or impact forging, metal flow is a response to the energy in the hammer–workpiece collision. If all the energy can be dissipated through flow of the surface layers of metal and absorption by the press foundation, the interior regions of the workpiece can go undeformed. Therefore, when the forging of large sections is required, press forging must be employed.

Upset Forging

Upset forging involves increasing the diameter of the end or central portion of a bar of metal by compressing its length. Upset-forging machines are used to forge heads on bolts and other fasteners, valves, couplings, and many other small components.

Roll Forging

Roll forging, in which round or flat bar stock is reduced in thickness and increased in length, is used to produce such components as axles, tapered levers, and leaf springs.

Swaging

Swaging involves hammering or forcing a tube or rod into a confining die to reduce its diameter, the die often playing the role of the hammer. Repeated blows cause the metal to flow inward and take the internal form of the die.

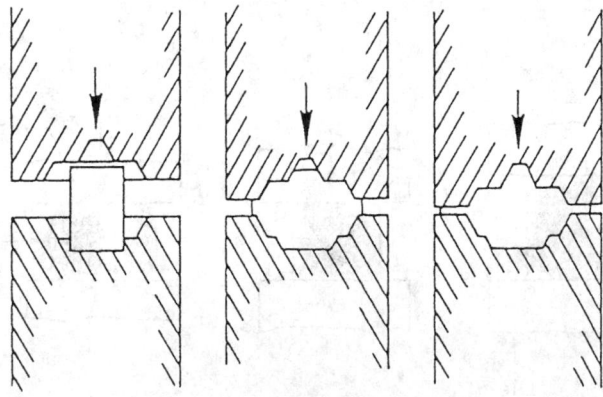

Fig. 34.5 Impression-die drop forging.

34.2.4 Extrusion

In the extrusion process (Fig. 34.6), metal is compressively forced to flow through a suitably shaped die to form a product with reduced cross section. Although it may be performed either hot or cold, hot extrusion is employed for many metals to reduce the forces required, to eliminate cold-working effects, and to reduce directional properties. The stress state within the material is triaxial compression.

Lead, copper, aluminum, and magnesium, and alloys of these metals, are commonly extruded, taking advantage of the relatively low yield strengths and extrusion temperatures. Steel is more difficult to extrude. Yield strengths are high and the metal has a tendency to weld to the walls of the die and confining chamber under the conditions of high temperature and pressures. With the development and use of phosphate-based and molten glass lubricants, substantial quantities of hot steel extrusions are now produced. These lubricants adhere to the billet and prevent metal-to-metal contact throughout the process.

Almost any cross-section shape can be extruded from the nonferrous metals. Hollow shapes can be extruded by several methods. For tubular products, the stationary or moving mandrel process is often employed. For more complex internal cavities, a spider mandrel or torpedo die is used. Obviously, the cost for hollow extrusions is considerably greater than for solid ones, but a wide variety of shapes can be produced that cannot be made by any other process.

The extrusion force F can be estimated from the formula

$$F = A_0 k \ln \left(\frac{A_0}{A}\right) \tag{34.6}$$

where k = extrusion constant depends on material and temperature (see Fig. 34.7)
A_0 = billet area
A_f = finished extruded area

34.2.5 Drawing

Drawing (Fig. 34.8) is a process for forming sheet metal between an edge-opposing punch and a die (draw ring) to produce a cup, cone, box, or shell-like part. The work metal is bent over and wrapped around the punch nose. At the same time, the outer portions of the blank move rapidly toward the center of the blank until they flow over the die radius as the blank is drawn into the die cavity by the punch. The radial movement of the metal increases the blank thickness as the metal moves toward the die radius; as the metal flows over the die radius, this thickness decreases because of the tension in the shell wall between the punch nose and the die radius and (in some instances) because of the clearance between the punch and the die.

The force (load) required for drawing a round cup is expressed by the following empirical equation:

$$L = \pi dtS \left(\frac{D}{d} - k\right) \tag{34.7}$$

where L = press load, lbs
d = cup diameter, in.

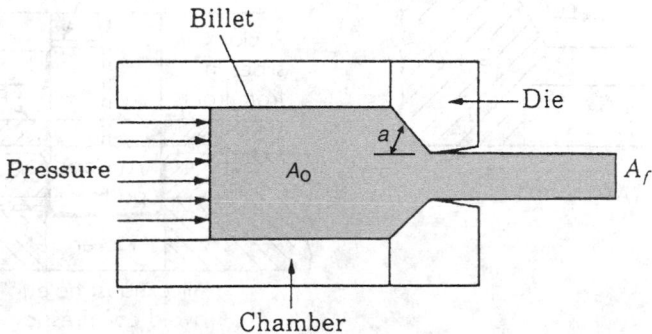

Fig. 34.6 Extrusion process.

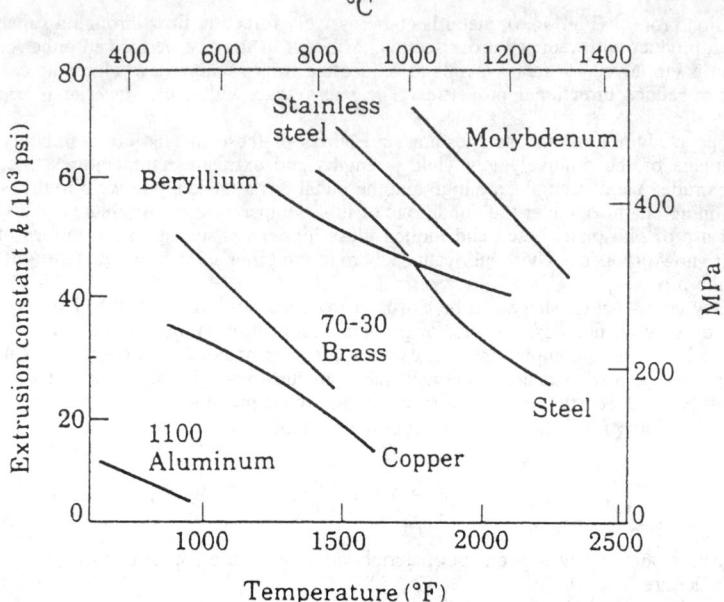

Fig. 34.7 Extrusion constant *k*.

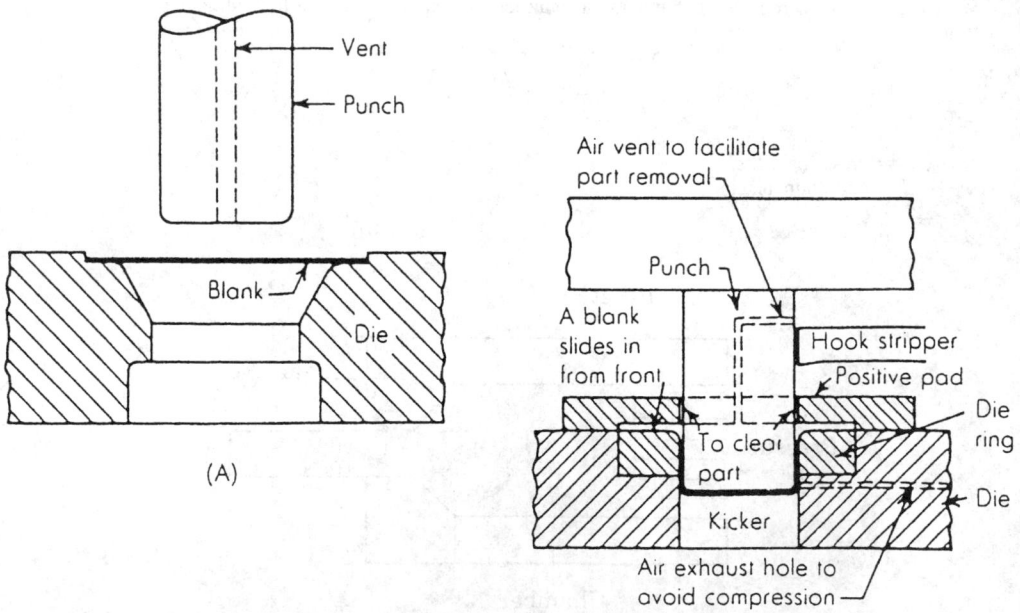

Fig. 34.8 Drawing process.

D = blank diameter, in.
t = work-metal thickness, in.
S = tensile strength, lbs/in.^{2}
k = a constant that takes into account frictional and bending forces, usually 0.6–0.7

The force (load) required for drawing a rectangular cup can be calculated from the following equation:

$$L = tS(2\pi Rk_A + lk_B)$$
(34.8)

where L = press load, lbs
t = work-metal thickness, in.
S = tensile strength, lbs/in.^{2}
R = corner radius of the cup, in.
l = the sum of the lengths of straight sections of the sides, in.
k_A and k_B = constants

Values for k_A range from 0.5 (for a shallow cup) to 2.0 (for a cup of depth five to six times the corner radius). Values for k_B range from 0.2 (for easy draw radius, ample clearance, and no blank-holding force) and 0.3 (for similar free flow and normal blankholding force of about $L/3$) to a maximum of 1.0 (for metal clamped too tightly to flow).

Figure 34.9 can be used as a general guide for computing maximum drawing load for a round shell. These relations are based on a free draw with sufficient clearance so that there is no ironing, using a maximum reduction of 50%. The nomograph gives the load required to fracture the cup (1 ton = 8.9 KN).

Blank Diameters

The following equations may be used to calculate the blank size for cylindrical shells of relatively thin metal. The ratio of the shell diameter to the corner radius (d/r) can affect the blank diameter and should be taken into consideration. When d/r is 20 or more,

$$D = \sqrt{d^2 + 4dh}$$
(34.9)

When d/r is between 15 and 20,

$$D = \sqrt{d^2 + 4dh - 0.5r}$$
(34.10)

When d/r is between 10 and 15,

$$D = \sqrt{d^2 + 4dh - r}$$
(34.11)

When d/r is below 10,

$$D = \sqrt{(d - 2r)^2 + 4d(h - r) + 2\pi r(d - 0.7r)}$$
(34.12)

where D = blank diameter
d = shell diameter
h = shell height
r = corner radius

The above equations are based on the assumption that the surface area of the blank is equal to the surface area of the finished shell.

In cases where the shell wall is to be ironed thinner than the shell bottom, the volume of metal in the blank must equal the volume of the metal in the finished shell. Where the wall-thickness reduction is considerable, as in brass shell cases, the final blank size is developed by trial. A tentative blank size for an ironed shell can be obtained from the equation

$$D = \sqrt{d^2 + 4dh\,\frac{t}{T}}$$
(34.13)

where t = wall thickness
T = bottom thickness

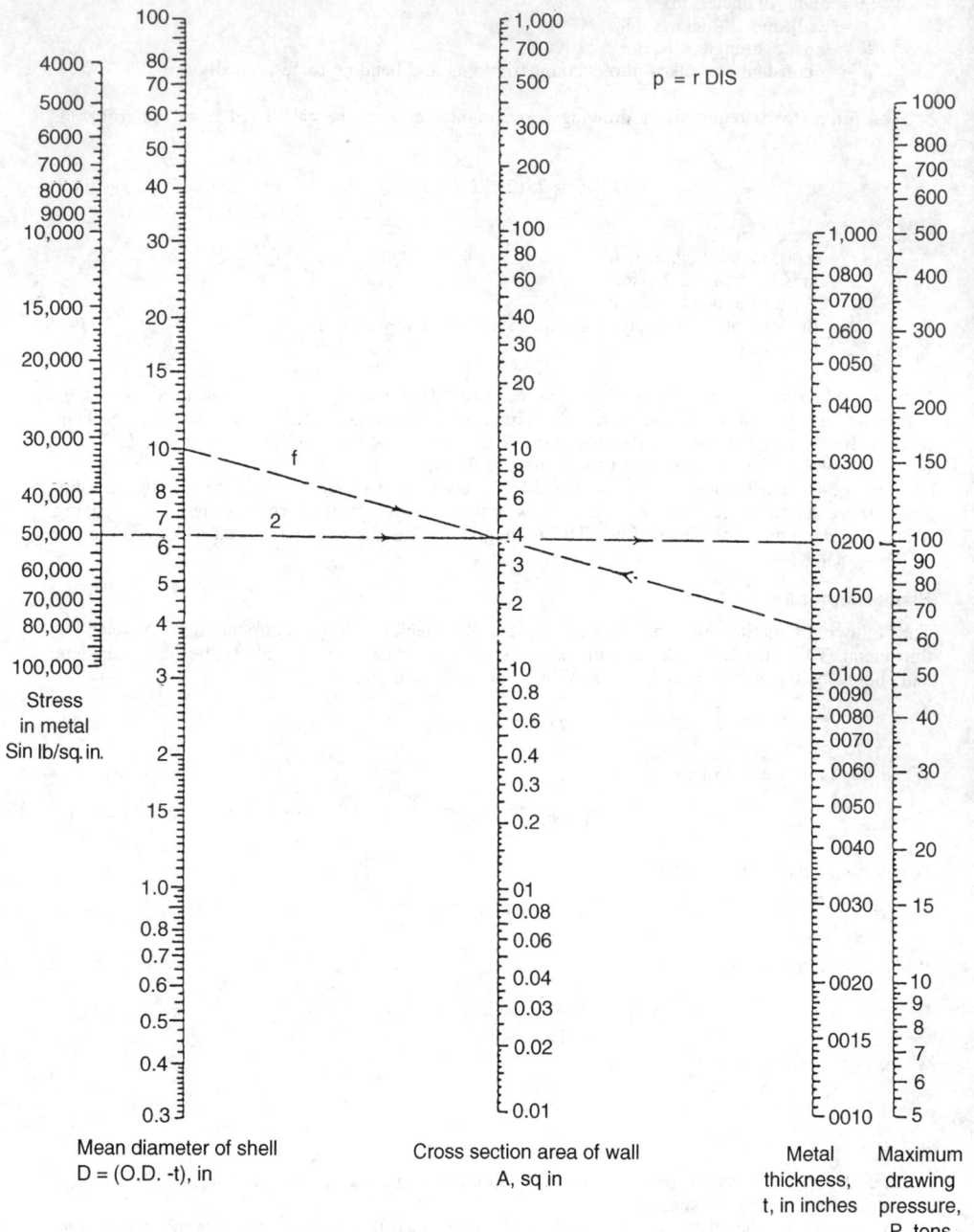

Fig. 34.9 Nomograph for estimating drawing pressures.

34.2.6 Spinning

Spinning is a method of forming sheet metal or tubing into seamless hollow cylinders, cones, hemispheres, or other circular shapes by a combination of rotation and force. On the basis of techniques used, applications, and results obtainable, the method may be divided into two categories: *manual spinning* (with or without mechanical assistance to increase the force) and *power spinning*.

Manual spinning entails no appreciable thinning of metal. The operation ordinarily done in a lathe consists of pressing a tool against a circular metal blank that is rotated by the headstock.

Power spinning is also known as *shear spinning* because in this method metal is intentionally thinned, by shear forces. In power spinning, forces as great as 400 tons are used.

The application of shear spinning to conical shapes is shown schematically in Fig. 34.10. The metal deformation is such that forming is in accordance with the sine law, which states that the wall thickness of the starting blank and that of the finished workpiece are related as

$$t_2 = t_1 (\sin \alpha) \tag{34.14}$$

where t_1 = the thickness of the starting blank
$\quad\quad\;\; t_2$ = the thickness of the spun workpiece
$\quad\quad\;\; \alpha$ = one-half the apex angle of the cone

Tube Spinning

Tube spinning is a rotary-point method of extruding metal, much like cone spinning, except that the sine law does not apply. Because the half-angle of a cylinder is zero, tube spinning follows a purely volumetric rule, depending on the practical limits of deformation that the metal can stand without intermediate annealing.

34.2.7 Pipe Welding

Large quantities of small-diameter steel pipe are produced by two processes that involve hot forming of metal strip and welding of its edges through utilization of the heat contained in the metal. Both of these processes, *butt welding* and *lap welding* of pipe, utilize steel in the form of skelp—long and narrow strips of the desired thickness. Because the skelp has been previously hot rolled and the welding process produces further compressive working and recrystallization, pipe welding by these processes is uniform in quality.

In the butt-welded pipe process, the skelp is unwound from a continuous coil and is heated to forging temperatures as it passes through a furnace. Upon leaving the furnace, it is pulled through forming rolls that shape it into a cylinder. The pressure exerted between the edges of the skelp as it passes through the rolls is sufficient to upset the metal and weld the edges together. Additional sets of rollers size and shape the pipe. Normal pipe diameters range from ⅛–3 in. (3–75 mm).

The lap-welding process for making pipe differs from butt welding in that the skelp has beveled edges and a mandrel is used in conjunction with a set of rollers to make the weld. The process is used primarily for larger sizes of pipe, from about 2–14 in. (50–400 mm) in diameter.

34.2.8 Piercing

Thick-walled and seamless tubing is made by the piercing process. A heated, round billet, with its leading end center-punched, is pushed longitudinally between two large, convex-tapered rolls that

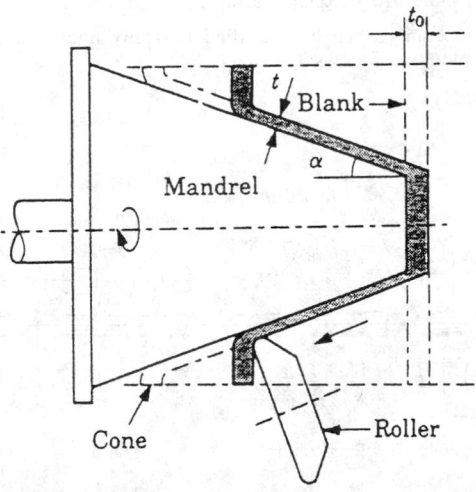

Fig. 34.10 Setup and dimensional relations for one-operation power spinning of a cone.

revolve in the same direction, their axes being inclined at opposite angles of about 6° from the axis of the billet. The clearance between the rolls is somewhat less than the diameter of the billet. As the billet is caught by the rolls and rotated, their inclination causes the billet to be drawn forward into them. The reduced clearance between the rolls forces the rotating billet to deform into an elliptical shape. To rotate with an elliptical cross section, the metal must undergo shear about the major axis, which causes a crack to open. As the crack opens, the billet is forced over a pointed mandrel that enlarges and shapes the opening, forming a seamless tube (Fig. 34.11).

This procedure applies to seamless tubes up to 6 in. (150 mm) in diameter. Larger tubes up to 14 in. (355 mm) in diameter are given a second operation on piercing rolls. To produce sizes up to 24 in. (610 mm) in diameter, reheated, double-pierced tubes are processed on a rotary rolling mill, and are finally completed by reelers and sizing rolls, as described in the single-piercing process.

34.3 COLD-WORKING PROCESSES

Cold working is the plastic deformation of metals below the recrystallization temperature. In most cases of manufacturing, such cold forming is done at room temperature. In some cases, however, the working may be done at elevated temperatures that will provide increased ductility and reduced strength, but will be below the recrystallization temperature.

When compared to hot working, cold-working processes have certain distinct advantages:

1. No heating required
2. Better surface finish obtained
3. Superior dimension control
4. Better reproducibility and interchangeability of parts
5. Improved strength properties
6. Directional properties can be imparted
7. Contamination problems minimized

Some disadvantages associated with cold-working processes include:

1. Higher forces required for deformation
2. Heavier and more powerful equipment required
3. Less ductility available
4. Metal surfaces must be clean and scale-free
5. Strain hardening occurs (may require intermediate anneals)
6. Imparted directional properties may be detrimental
7. May produce undesirable residual stresses

34.3.1 Classification of Cold-Working Operations

The major cold-working operations can be classified basically under the headings of squeezing, bending, shearing, and drawing, as follows:

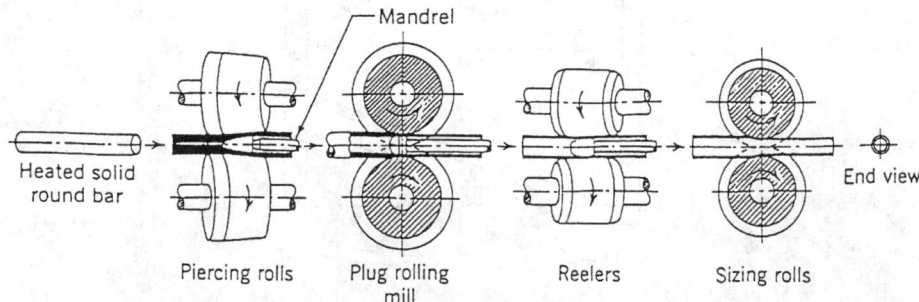

Piercing rolls   Plug rolling mill   Reelers   Sizing rolls

Fig. 34.11 Principal steps in the manufacture of seamless tubing.

Squeezing	Bending	Shearing	Drawing
1. Rolling	1. Angle	1. Shearing	1. Bar and tube drawing
2. Swaging	2. Roll	Slitting	2. Wire drawing
3. Cold forging	3. Roll forming	2. Blanking	3. Spinning
4. Sizing	4. Drawing	3. Piercing	4. Embossing
5. Extrusion	5. Seaming	Lancing	5. Stretch forming
6. Riveting	6. Flanging	Perforating	6. Shell drawing
7. Staking	7. Straightening	4. Notching	7. Ironing
8. Coining		Nibbling	8. High energy rate forming
9. Peening		5. Shaving	
10. Burnishing		6. Trimming	
11. Die hobbing		7. Cutoff	
12. Thread rolling		8. Dinking	

34.3.2 Squeezing Processes

Most of the cold-working squeezing processes have identical hot-working counterparts or are extensions of them. The primary reasons for deforming cold rather than hot are to obtain better dimensional accuracy and surface finish. In many cases, the equipment is basically the same, except that it must be more powerful.

Cold Rolling

Cold rolling accounts for by far the greatest tonnage of cold-worked products. Sheets, strip, bars, and rods are cold-rolled to obtain products that have smooth surfaces and accurate dimensions.

Swaging

Swaging basically is a process for reducing the diameter, tapering, or pointing round bars or tubes by external hammering. A useful extension of the process involves the formation of internal cavities. A shaped mandrel is inserted inside a tube and the tube is then collapsed around it by swaging (Fig. 34.12).

Cold Forging

Extremely large quantities of products are made by cold forging, in which the metal is squeezed into a die cavity that imparts the desired shape. Cold heading is used for making enlarged sections on the ends of rod or wire, such as the heads on bolts, nails, rivets, and other fasteners.

Sizing

Sizing involves squeezing areas of forgings or ductile castings to a desired thickness. It is used principally on basses and flats, with only enough deformation to bring the region to a desired dimension.

Extrusion

This process is often called *impact extrusion* and was first used only with the low-strength ductile metals, such as lead, tin, and aluminum, for producing such items as collapsible tubes for toothpaste, medications, and so forth; small "cans" such as are used for shielding in electronics and electrical apparatus; and larger cans for food and beverages. In recent years, cold extrusion has been used for forming mild steel parts, often being combined with cold heading.

Another type of cold extrusion, known as *hydrostatic extrusion,* used high fluid pressure to extrude a billet through a die, either into atmospheric pressure or into a lower-pressure chamber. The pressure-

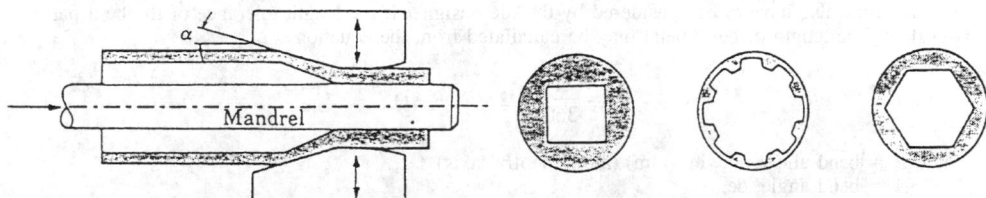

Fig. 34.12 Cross sections of tubes produced by swaging on shaped mandrels. Rifling (spiral grooves) in small gun barrels can be made by this process.

to-pressure process makes possible the extrusion of relatively brittle materials, such as molybdenum, beryllium, and tungsten. Billet-chamber friction is eliminated, billet-die lubrication is enhanced by the pressure, and the surrounding pressurized atmosphere suppresses crack initiation and growth.

Riveting

In riveting, a head is formed on the shank end of a fastener to provide a permanent method of joining sheets or plates of metal together. Although riveting usually is done hot in structural work, in manufacturing it almost always is done cold.

Staking

Staking is a commonly used cold-working method for permanently fastening two parts together where one protrudes through a hole in the other. A shaped punch is driven into one of the pieces, deforming the metal sufficiently to squeeze it outward.

Coining

Coining involves cold working by means of positive displacement punch while the metal is completely confined within a set of dies.

Peening

Peening involves striking the surface repeated blows by impelled shot or a round-nose tool. The highly localized blows deform and tend to stretch the metal surface. Because the surface deformation is resisted by the metal underneath, the result is a surface layer under residual compression. This condition is highly favorable to resist cracking under fatigue conditions, such as repeated bending, because the compressive stresses are subtractive from the applied tensile loads. For this reason, shafting, crankshafts, gear teeth, and other cyclic-loaded components are frequently peened.

Burnishing

Burnishing involves rubbing a smooth, hard object under considerable pressure over the minute surface protrusions that are formed on a metal surface during machining or shearing, thereby reducing their depth and sharpness through plastic flow.

Hobbing

Hobbing is a cold-working process that is used to form cavities in various types of dies, such as those used for molding plastics. A male hob is made with the contour of the part that ultimately will be formed by the die. After the hob is hardened, it is slowly pressed into an annealed die block by means of hydraulic press until the desired impression is produced.

Thread Rolling

Threads can be rolled in any material sufficiently plastic to withstand the forces of cold working without disintegration. Threads can be rolled by flat or roller dies.

34.3.3   Bending

Bending is the uniform straining of material, usually flat sheet or strip metal, around a straight axis that lies in the neutral plane and normal to the lengthwise direction of the sheet or strip. Metal flow takes place within the plastic range of the metal, so that the bend retains a permanent set after removal of the applied stress. The inner surface of the bend is in compression; the outer surface is in tension.

Terms used in bending are defined and illustrated in Fig. 34.13. The neutral axis is the plane area in bent metal where all strains are zero.

Bend Allowances

Since bent metal is longer after bending, its increased length, generally of concern to the product designer, may also have to be considered by the die designer if the length tolerance of the bent part is critical. The length of bent metal may be calculated from the equation

$$B = \frac{A}{360} \times 2\pi(R_i + Kt) \tag{34.14}$$

where B = bend allowance, in. (mm) (along neutral axis)
$\quad A$ = bend angle, deg
$\quad R_i$ = inside radius of bend, in. (mm)
$\quad t$ = metal thickness, in. (mm)
$\quad K$ = 0.33 when R_i is less than $2t$ and is 0.50 when R_i is more than $2t$

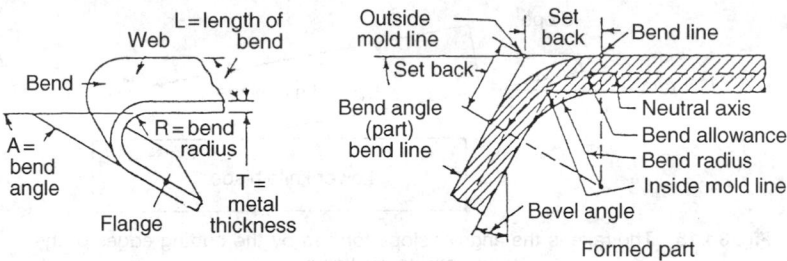

Fig. 34.13 Bend terms.

Bending Methods

Two bending methods are commonly made use of in press tools. Metal sheet or strip, supported by a V block (Fig. 34.14), is forced by a wedge-shaped punch into the block.

Edge bending (Fig. 34.14) is cantilever loading of a beam. The bending punch (1) forces the metal against the supporting die (2).

Bending Force

The force required for V bending is as follows:

$$P = \frac{KLSt^2}{W} \tag{34.15}$$

where P = bending force, tons (for metric usage, multiply number of tons by 8.896 to obtain kilonewtons)
 K = die opening factor: 1.20 for a die opening of 16 times metal thickness, 1.33 for an opening of eight times metal thickness
 L = length of part, in.
 S = ultimate tensile strength, tons/in.²
 W = width of V or U die, in.
 t = metal thickness, in.

For U bending (channel bending), pressures will be approximately twice those required. For U bending, edge bending is required about one-half those needed for V bending. Table 34.1 gives the ultimate strength = S for various materials.

Several factors must be considered when designing parts that are to be made by bending. Of primary importance is the minimum radius that can be bent successfully without metal cracking. This, of course, is related to the ductility of the metal.

Angle Bending

Angle bends up to 150° in the sheet metal under about 1/16 in. (1.5 mm) in thickness may be made in a bar folder. Heavier sheet metal and more complex bends in thinner sheets are made on a press brake.

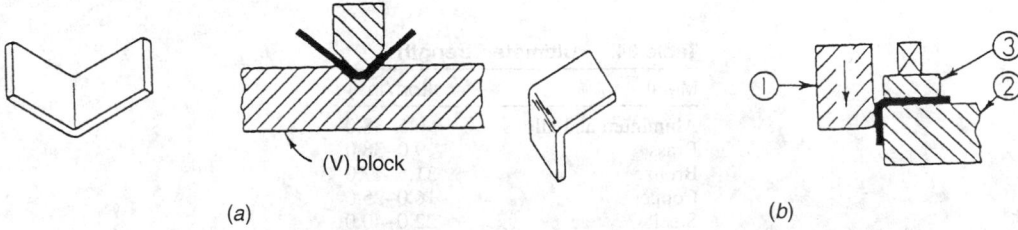

(a)   (b)

Fig. 34.14 Bending methods. (a) V bending; (b) edge bending.

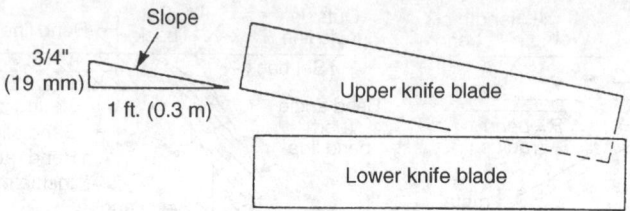

Fig. 34.15 The rake is the angular slope formed by the cutting edges of the upper and lower knives.

Roll Bending

Plates, heavy sheets, and rolled shapes can be bent to a desired curvature on forming rolls. These usually have three rolls in the form of a pyramid, with the two lower rolls being driven and the upper roll adjustable to control the degree of curvature. Supports can be swung clear to permit removal of a closed shape from the rolls. Bending rolls are available in a wide range of sizes, some being capable of bending plate up to 6 in. (150 mm) thick.

Cold-Roll Forming

This process involves the progressive bending of metal strip as it passes through a series of forming rolls. A wide variety of moldings, channeling, and other shapes can be formed on machines that produce up to 10,000 ft (3000 m) of product per day.

Seaming

Seaming is used to join ends of sheet metal to form containers such as cans, pails, and drums. The seams are formed by a series of small rollers on seaming machines that range from small hand-operated types to large automatic units capable of producing hundreds of seams per minute in the mass production of cans.

Flanging

Flanges can be rolled on sheet metal in essentially the same manner as seaming is done. In many cases, however, the forming of flanges and seams involves drawing, since localized bending occurs on a curved axis.

Straightening

Straightening or flattening has as its objective the opposite of bending and often is done before other cold-forming operations to ensure that flat or straight material is available. Two different techniques are quite common. *Roll straightening* or *roller leveling* involves a series of reverse bends. The rod, sheet, or wire is passed through a series of rolls having decreased offsets from a straight line. These bend the metal back and forth in all directions, stressing it slightly beyond its previous elastic limit and thereby removing all previous permanent set.

Sheet may also be straightened by a process called *stretcher leveling*. The sheets are grabbed mechanically at each end and stretched slightly beyond the elastic limit to remove previous stresses and thus produce the desired flatness.

34.3.4 Shearing

Shearing is the mechanical cutting of materials in sheet or plate form without the formation of chips or use of burning or melting. When the two cutting blades are straight, the process is called *shearing*.

Table 34.1 Ultimate Strength

Metal	(ton/in.²)
Aluminum and alloys	6.5–38.0
Brass	19.0–38.0
Bronze	31.5–47.0
Copper	16.0–25.0
Steel	22.0–40.0
Tin	1.1–1.4
Zinc	9.7–13.5

Other processes, in which the shearing blades are in the form of curved edges or punches and dies, are called by other names, such as *blanking, piercing, notching, shaving,* and *trimming.* These all are basically shearing operations, however.

The required shear force can be calculated as

$$F = \left(\frac{S \times P \times t^2 \times 12}{R} \right) \left(1 - \frac{P}{2} \right)$$ (34.16)

where F = shear force, lb
S = shear strength (stress), psi
P = penetration of knife into material, %
t = thickness of material, in.
R = rake of the knife blade, in./ft (Fig. 34.13)

For SI units, the force is multiplied by 4.448 to obtain newtons (N). Table 34.2 gives the values of P and S for various materials.

Blanking

A blank is a shape cut from flat or preformed stock. Ordinarily, a blank serves as a starting workpiece for a formed part; less often, it is a desired end product.

Calculation of the forces and the work involved in blanking gives average figures that are applicable only when (a) the correct shear strength for the material is used, and (b) the die is sharp and the punch is in good condition, has correct clearance, and is functioning properly.

The total load on the press, or the press capacity required to do a particular job, is the sum of the cutting force and other forces acting at the same time, such as the blankholding force exerted by a die cushion.

Cutting Force: Square-End Punches and Dies

When punch and die surfaces are flat and at right angles to the motion of the punch, the cutting force can be found by multiplying the area of the cut section by the shear strength of the work material:

Table 34.2 Values of Percent Penetration and Shear Strength for Various Materials

Material	Percent Penetration	Shear Strength, psi (MPa)
Lead alloys	50	3500 (24.1)–6000 (41.3)
Tin alloys	40	5000 (34.5)–10,000 (69)
Aluminum alloys	60	8000 (55.2)–45,000 (310)
Titanium alloys	10	60,000 (413)–70,000 (482)
Zinc	50	14,000 (96.5)
Cold worked	25	19,000 (131)
Magnesium alloys	50	17,000 (117)–30,000 (207)
Copper	55	22,000 (151.7)
Cold worked	30	28,000 (193)
Brass	50	32,000 (220.6)
Cold worked	30	52,000 (358.5)
Tobin bronze	25	36,000 (248.2)
Cold worked	17	42,000 (289.6)
Steel, 0.10C	50	35,000 (241.3)
Cold worked	38	43,000 (296.5)
Steel, 0.40C	27	62,000 (427.5)
Cold worked	17	78,000 (537.8)
Steel, 0.80C	15	97,000 (668.8)
Cold worked	5	127,000 (875.6)
Steel, 1.00C	10	115,000 (792.9)
Cold worked	2	150,000 (1034.2)
Silicon steel	30	65,000 (448.2)
Stainless steel	30	57,000 (363)–128,000 (882)
Nickel	55	35,000 (241.3)

$$L = Stl \qquad (34.17)$$

where L = load on the press, lb (cutting force)
$\quad S$ = shear strength of the stock, psi
$\quad t$ = stock thickness, in.
$\quad l$ = the length or perimeter of cut, in.

Piercing

Piercing is a shearing operation wherein the shearing blades take the form of closed, curved lines on the edges of a punch and die. Piercing is basically the same as blanking except that the piece punched out is the scrap and the remainder of the strip becomes the desired workpiece.

Lancing

Lancing is a piercing operation that may take the form of a slit in the metal or an actual hole. The purpose of lancing is to permit adjacent metal to flow more readily in subsequent forming operations.

Perforating

Perforating consists of piercing a large number of closely spaced holes.

Notching

Notching is essentially the same as piercing except that the edge of the sheet of metal forms a portion of the periphery of the piece that is punched out. It is used to form notches of any desired shape along the edge of a sheet.

Nibbling

Nibbling is a variation of notching in which a special machine makes a series of overlapping notches, each farther into the sheet of metal.

Shaving

Shaving is a finished operation in which a very small amount of metal is sheared away around the edge of a blanked part. Its primary use is to obtain greater dimensional accuracy, but it also may be used to obtain a square of smoother edge.

Trimming

Trimming is used to remove the excess metal that remains after a drawing, forging, or casting operation. It is essentially the same as blanking.

Cutoff

A cutoff operation is one in which a stamping is removed from a strip of stock by means of a punch and die. The cutoff punch and die cut across the entire width of the strip. Frequently, an irregularly shaped cutoff operation may simultaneously give the workpiece all or part of the desired shape.

Dinking

Dinking is a modified shearing operation that is used to blank shapes from low-strength materials, primarily rubber, fiber, and cloth.

34.3.5 Drawing

Cold Drawing

Cold drawing is a term that can refer to two somewhat different operations. If the stock is in the form of sheet metal, cold drawing is the forming of parts wherein plastic flow occurs over a curved axis. This is one of the most important of all cold-working operations because a wide range of parts, from small caps to large automobile body tops and fenders, can be drawn in a few seconds each. Cold drawing is similar to hot drawing, but the higher deformation forces, thinner metal, limited ductility, and closer dimensional tolerance create some distinctive problems.

If the stock is wire, rod, or tubing, *cold drawing* refers to the process of reducing the cross section of the material by pulling it through a die, a sort of tensile equivalent to extrusion.

Cold Spinning

Cold spinning is similar to hot spinning, discussed above.

Stretch Forming

In stretch forming, only a single male form block is required. The sheet of metal is gripped by two or more sets of jaws that stretch it and wrap it around the form block as the latter raises upward.

Various combinations of stretching, wrapping, and upward motion of the blocks are used, depending on the shape of the part.

Shell or Deep Drawing

The drawing of closed cylindrical or rectangular containers, or a variation of these shapes, with a depth frequently greater than the narrower dimension of their opening, is one of the most important and widely used manufacturing processes. Because the process had its earliest uses in manufacturing artillery shells and cartridge cases, it is sometimes called *shell drawing*. When the depth of the drawn part is less than the diameter, or minimum surface dimension, of the blank, the process is considered to be *shallow drawing*. If the depth is greater than the diameter, it is considered to be *deep drawing*.

The design of complex parts that are to be drawn has been aided considerably by computer techniques, but is far from being completely and successfully solved. Consequently, such design still involves a mix of science, experience, empirical data, and actual experimentation. The body of known information is quite substantial, however, and is being used with outstanding results.

Forming with Rubber or Fluid Pressure

Several methods of forming use rubber or fluid pressure (Fig. 34.16) to obtain the desired information and thereby eliminate either the male or female member of the die set. Blanks of sheet metal are placed on top of form blocks, which usually are made of wood. The upper ram, which contains a pad of rubber 8–10 in. (200–250 mm) thick in a steel container, then descends. The rubber pad is confined and transmits force to the metal, causing it to bend to the desired shape. Since no female die is used and form blocks replace the male die, die cost is quite low.

The hydroform process or "rubber bag forming" replaces the rubber pad with a flexible diaphragm backed by controlled hydraulic pressure. Deeper parts can be formed with truly uniform fluid pressure.

The bulging oil or rubber is used for applying an internal bulging force to expand a metal blank or tube outward against a female mold or die, thereby eliminating the necessity for a complicated, multiple-piece male die member.

Ironing

Ironing is the name given to the process of thinning the walls of a drawn cylinder by passing it between a punch and a die where the separation is less than the original wall thickness. The walls are elongated and thinned while the base remains unchanged. The most common example of an ironed product is the thin-walled all-aluminum beverage can.

Embossing

Embossing is a method for producing lettering or other designs in thin sheet metal. Basically, it is a very shallow drawing operation, usually in open dies, with the depth of the draw being from one to three times the thickness of the metal.

High-Energy-Rate Forming

A number of methods have been developed for forming metals through the release and application of large amounts of energy in a very short interval (Fig. 34.17). These processes are called *high-energy-rate forming processes* (HERF). Many metals tend to deform more readily under the ultrarapid rates of load application used in these processes, a phenomenon apparently related to the relative rates of load application and the movement of dislocations through the metal. As a consequence, HERF makes it possible to form large workpieces and difficult-to-form metals with less expensive equipment and tooling than would otherwise be required.

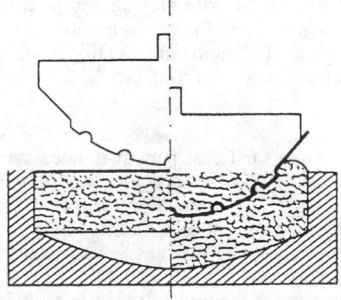

Fig. 34.16 Form with rubber.

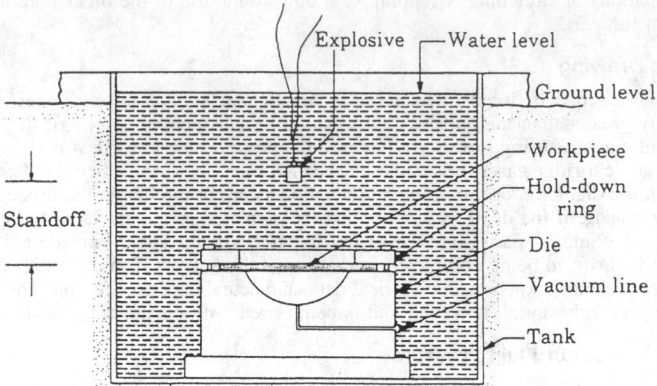

Fig. 34.17 High-energy-rate forming.

The high energy-release rates are obtained by five methods:

1. Underwater explosions
2. Underwater spark discharge (electrohydraulic techniques)
3. Pneumatic–mechanical means
4. Internal combustion of gaseous mixtures
5. Rapidly formed magnetic fields (electromagnetic techniques)

34.4 METAL CASTING AND MOLDING PROCESSES

Casting provides a versatility and flexibility that have maintained casting position as a primary production method for machine elements. Casting processes are divided according to the specific type of molding method used in casting, as follows:

1. Sand
2. Centrifugal
3. Permanent
4. Die
5. Plaster-mold
6. Investment

34.4.1 Sand Casting

Sand casting consists basically of pouring molten metal into appropriate cavities formed in a sand mold (Fig. 34.18). The sand may be natural, synthetic, or an artificially blended material.

Molds

The two common types of sand molds are the *dry sand mold* and the *green sand mold*. In the dry sand mold, the mold is dried thoroughly prior to closing and pouring, while the green sand mold is used without any preliminary drying. Because the dry sand mold is more firm and resistant to collapse than the green sand mold, core pieces for molds are usually made in this way. Cores are placed in mold cavities to form the interior surfaces of castings.

Patterns

To produce a mold for a conventional sand cast part, it is necessary to make a pattern of the part. Patterns are made from wood or metal to suit a particular design, with allowances to compensate for such factors as natural metal shrinkage and contraction characteristics. These and other effects, such as mold resistance, distortion, casting design, and mold design, which are not entirely within the range of accurate prediction, generally make it necessary to adjust the pattern in order to produce castings of the required dimensions.

Access to the mold cavity for entry of the molten metal is provided by sprues, runners, and gates.

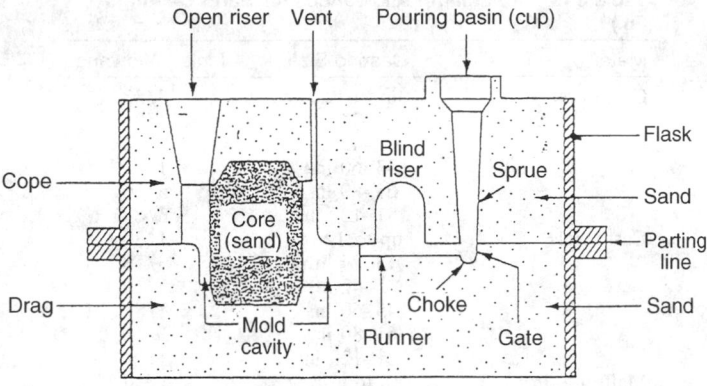

Fig. 34.18 Sectional view of casting mold.

Shrinkage

Allowances must be made on patterns to counteract the contraction in size as the metal cools. The amount of shrinkage is dependent on the design of the coating, type of metal used, solidification temperature, and mold resistance. Table 34.3 gives average shrinkage allowance values used in sand casting. Smaller values apply generally to large or cored castings of intricate design. Larger values apply to small to medium simple castings designed with unrestrained shrinkage.

Machining

Allowances are required in many cases because of unavoidable surface impurities, warpage, and surface variations. Average machining allowances are given in Table 34.4. Good practice dictates use of minimum section thickness compatible with the design. The normal minimum section recommended for various metals is shown in Table 34.5.

34.4.2 Centrifugal Casting

Centrifugal casting consists of having a sand, metal, or ceramic mold that is rotated at high speeds. When the molten metal is poured into the mold, it is thrown against the mold wall, where it remains until it cools and solidifies. The process is increasingly being used for such products as cast-iron pipes, cylinder liners, gun barrels, pressure vessels, brake drums, gears, and flywheels. The metals used include almost all castable alloys. Most dental tooth caps are made by a combined lost-wax process and centrifugal casting.

Advantages and Limitations

Because of the relatively fast cooling time, centrifugal castings have a fine grain size. There is a tendency for the lighter nonmetallic inclusion, slag particles, and dross to segregate toward the inner

Table 34.3 Pattern Shrinkage Allowance (in./ft)

Metal	Shrinkage
Aluminum alloys	1/10–5/32
Beryllium copper	1/8–5/32
Copper alloys	3/16–7/32
Everdur	3/16
Gray irons	1/8
Hastelloy alloys	1/4
Magnesium alloys	1/8–11/64
Malleable irons	1/16–3/16
Meehanite	1/10–5/32
Nickel and nickel alloys	1/4
Steel	1/8–1/4
White irons	3/16–1/4

Table 34.4 Machining Allowances for Sand Castings (in.)

Metal	Casting Size	Finish Allowance
Cast irons	up to 12 in.	3/32
	13–24 in.	1/8
	25–42 in.	3/16
	43–60 in.	1/4
	61–80 in.	5/16
	81–120 in.	3/8
Cast steels	up to 12 in.	1/8
	13–24 in.	3/16
	25–42 in.	5/16
	43–60 in.	3/8
	61–80 in.	7/16
	81–120 in.	1/2
Malleable irons	up to 8 in.	1/16
	9–12 in.	3/32
	13–24 in.	1/8
	25–36 in.	3/16
Nonferrous metals	up to 12 in.	1/16
	13–24 in.	1/8
	25–36 in.	5/32

radius of the castings (Fig. 34.19), where it can be easily removed by machining. Owing to the high purity of the outer skin, centrifugally cast pipes have a high resistance to atmospheric corrosion. Figure 34.19 shows a schematic sketch of how a pipe would be centrifugally cast in a horizontal mold.

Parts that have diameters exceeding their length are produced by vertical-axis casting (see Fig. 34.20).

If the centrifugal force is too low or too great, abnormalities will develop. Most horizontal castings are spun so that the force developed is about 65 g's. Vertically cast parts force is about 90–100 g's.

The centrifugal force (*CF*) is calculated from

$$CF = \frac{mv^2}{r}\ lb$$

$$m = \text{Mass} = \frac{W}{g} = \frac{\text{Weight, lb}}{\text{Acceleration of gravity (ft/s)}^2} = \frac{W}{32.2}$$

where v = velocity, ft/s = r × w
r = radius, ft = ½ D
w = angular velocity, rad/s
w = $2\pi/60$ × rpm
D = inside diameter, ft

The number of g's is

$$g's = CF/W$$

Table 34.5 Minimum Sections for Sand Castings (in.)

Metal	Section
Aluminum alloys	3/16
Copper alloys	3/32
Gray irons	1/8
Magnesium alloys	5/32
Malleable irons	1/8
Steels	1/4
White irons	1/8

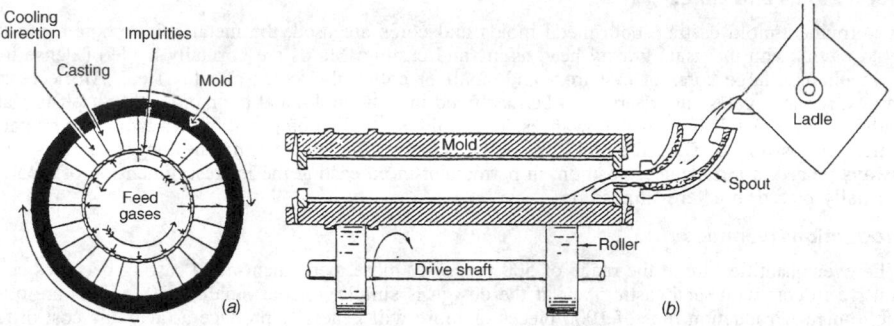

Fig. 34.19 The principle of centrifugal casting is to produce the high-grade metal by throwing the heavier metal outward and forcing the impurities to congregate inward (a). Shown at (b) is a schematic of how a horizontal-bond centrifugal casting is made.

Hence,

$$g's = \frac{1}{W} \times \left[\frac{W}{32.2 \times r} \left(\frac{r \times 2\pi}{60} \right)^2 \right]$$
$$= r \times 3.41 \times 10^{-4} \text{ rpm}^2$$
$$= 1.7 \times 10^{-4} \times D \times (\text{rpm})^2$$

The spinning speed for horizontal-axis molds may be found in English units from the equation

$$N = \sqrt{(\text{Number of g's}) \times \frac{70,500}{D}}$$

where N = rpm
D = inside diameter of mold, ft

34.4.3 Permanent-Mold Casting

As demand for quality castings in production quantities increased, the attractive possibilities of metal molds brought about the development of the permanent-mold process. Although not as flexible regarding design as sand casting, metal-mold casting made possible the continuous production of quantities of casting from a single mold as compared to batch production of individual sand molds.

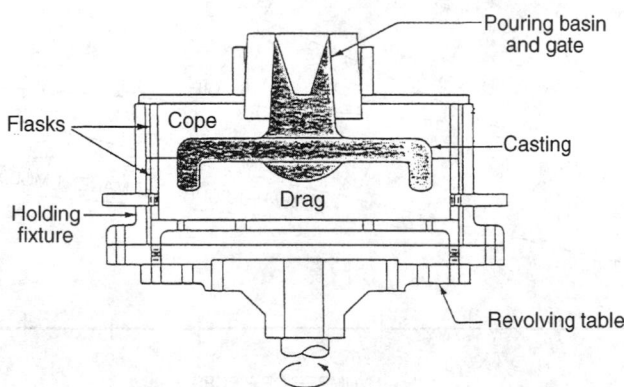

Fig. 34.20 Floor-type vertical centrifugal casting machine for large-diameter parts.

Metal Molds and Cores

In permanent-mold casting, both metal molds and cores are used, the metal being poured into the mold cavity with the usual gravity head as in sand casting. Molds are normally made of dense iron or meehanite, large cores of cast iron, and small or collapsible cores of alloy steel. All necessary sprues, runners, gates, and risers must be machined into the mold, and the mold cavity itself is made with the usual metal-shrinkage allowances. The mold is usually composed of one, two, or more parts, which may swing or slide for rapid operation. Whereas in sand casting the longest dimension is always placed in a horizontal position, in permanent-mold casting the longest dimension of a part is normally placed in a vertical position.

Production Quantities

Wherever quantities are in the range of 500 pieces or more, permanent-mold casting becomes competitive in cost with sand casting, and if the design is simple, runs as small as 200 pieces are often economical. Production runs of 1000 pieces or more will generally produce a favorable cost difference. High rates of production are possible, and multiple-cavity dies with as many as 16 cavities can be used. In casting gray iron in multiple molds, as many as 50,000 castings per cavity are common with small parts. With larger parts of gray iron, weighing from 12–15 lb, single-cavity molds normally yield 2000–3000 pieces per mold on an average. Up to 100,000 parts per cavity or more are not uncommon with nonferrous metals, magnesium providing the longest die life. Low-pressure permanent mold casting is economical for quantities up to 40,000 pieces (Fig. 34.21).

Die Casting

Die casting may be classified as a permanent-mold casting system; however, it differs from the process just described in that the molten metal is forced into the mold or die under high pressure [1000–30,000 psi (6.89–206.8 MPa)]. The metal solidifies rapidly (within a fraction of a second) because the die is water-cooled. Upon solidification, the die is opened and ejector pins automatically knock the casting out of the die. If the parts are small, several of them may be made at one time in what is termed a *multicavity die.*

There are two main types of machines used: the hot-chamber and the cold-chamber types.

Hot-Chamber Die Casting. In the hot-chamber machine, the metal is kept in a heated holding pot. As the plunger descends, the required amount of alloy is automatically forced into the die. As the piston retracts, the cylinder is again filled with the right amount of molten metal. Metals such as aluminum, magnesium, and copper tend to alloy with the steel plunger and cannot be used in the hot chamber.

Cold-Chamber Die Casting. This process gets its name from the fact that the metal is ladled into the cold chamber for each shot. This procedure is necessary to keep the molten-metal contact time with the steel cylinder to a minimum. Iron pickup is prevented, as is freezing of the plunger in the cylinder.

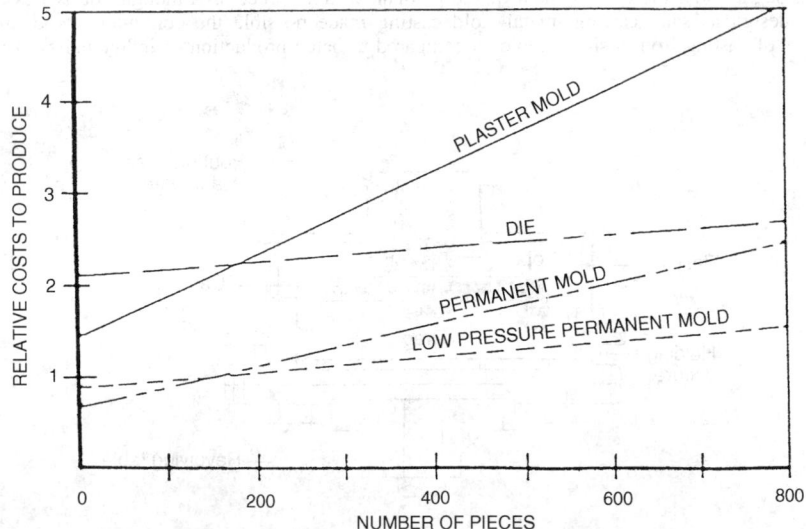

Fig. 34.21 Cost comparison of various casting systems.

Advantages and Limitations

Die-casting machines can produce large quantities of parts with close tolerances and smooth surfaces. The size is limited only by the capacity of the machine. Most die castings are limited to about 75 lb (34 kg) of zinc; 65 lb (30 kg) of aluminum; and 44 lb (20 kg) of magnesium. Die castings can provide thinner sections than any other casting process. Wall thickness as thin as 0.015 in. (0.38 mm) can be achieved with aluminum in small items. However, a more common range on larger sizes will be 0.105–0.180 in. (2.67–4.57 mm).

Some difficulty is experienced in getting sound castings in the larger capacities. Gases tend to be entrapped, which results in low strength and annoying leaks. Of course, one way to reduce metal sections without sacrificing strength is to design in ribs and bosses. Another approach to the porosity problem has been to operate the machine under vacuum. This process is now being developed.

The surface quality is dependent on that of the mold. Parts made from new or repolished dies may have a surface roughness of 24 μin. (0.61 μm). The high surface finish available means that, in most cases, coatings such as chromeplating, anodizing, and painting may be applied directly. More recently, decorative finishes of texture, as obtained by photoetching, have been applied. The technique has been used to simulate woodgrain finishes, as well as textile and leather finishes, and to obtain checkering and crosshatching.

34.4.4 Plaster-Mold Casting

In general, the various methods of plaster-mold casting are similar. The plaster, also known as *gypsum* or *calcium sulfate,* is mixed dry with other elements, such as talc, sand, asbestos, and sodium silicate. To this mix is added a controlled amount of water to provide the desired permeability in the mold. The slurry that results is heated and delivered through a hose to the flasks, all surfaces of which have been sprayed with a parting compound. The plaster slurry readily fills in and around the most minute details in the highly polished brass patterns. Following filling, the molds are subjected to a short period of vibration and the slurry sets in 5–10 min.

Molds

Molds are extracted from the flask with a vacuum head, following which drying is completed in a continuous oven. Copes and drags are then assembled, with cores when required, and the castings are poured. Upon solidification, the plaster is broken away and any cores used are washed out with a high-pressure jet of water.

34.4.5 Investment Casting

Casting processes in which the pattern is used only once are variously referred to as *lost-wax* or *precision-casting* processes. They involve making a pattern of the desired form out of wax or plastic (usually polystyrene). The expendable pattern may be made by pressing the wax into a split mold or by the use of an injection-molding machine. The patterns may be gated together so that several parts can be made at once. A metal flask is placed around the assembled patterns and a refractory mold slurry is poured in to support the patterns and form the cavities. A vibrating table equipped with a vacuum pump is used to eliminate all the air from the mold. Formerly, the standard procedure was to dip the patterns in the slurry several times until a coat was built up. This is called the *investment process.* After the mold material has set and dried, the pattern material is melted and allowed to run out of the mold.

The completed flasks are heated slowly to dry the mold and to melt out the wax, plastic, or whatever pattern material was used. When the molds have reached a temperature of 100°F (37.8°C), they are ready for pouring. Vacuum may be applied to the flasks to ensure complete filling of the mold cavities.

When the metal has cooled, the investment material is removed by vibrating hammers or by tumbling. As with other castings, the gates and risers are cut off and ground down.

Ceramic Process

The ceramic process is somewhat similar to the investment-casting in that a creamy ceramic slurry is poured over a pattern. In this case, however, the pattern, made out of plastic, plaster, wood, metal, or rubber, is reusable. The slurry hardens on the pattern almost immediately and becomes a strong green ceramic of the consistency of vulcanized rubber. It is lifted off the pattern while it is still in the rubberlike phase. The mold is ignited with a torch to burn off the volatile portion of the mix. It is then put in a furnace and baked at 1800°F (982°C), resulting in a rigid refractory mold. The mold can be poured while still hot.

Full-Mold Casting

Full-mold casting may be considered a cross between conventional sand casting and the investment technique of using lost wax. In this case, instead of a conventional pattern of wood, metals, or plaster, a polystyrene foam or styrofoam is used. The pattern is left in the mold and is vaporized by the molten metal as it rises in the mold during pouring. Before molding, the pattern is usually coated with a zirconite wash in an alcohol vehicle. The wash produces a relatively tough skin separating the metal from the sand during pouring and cooling. Conventional foundry sand is used in backing up the mold.

34.5 PLASTIC-MOLDING PROCESSES

Plastic molding is similar in many ways to metal molding. For most molding operations, plastics are heated to a liquid or a semifluid state and are formed in a mold under pressure. Some of the most common molding processes are discussed below.

34.5.1 Injection Molding

The largest quantity of plastic parts is made by injection molding. Plastic compound is fed in powdered or granular form from a hopper through metering and melting stages and then injected into a mold. After a brief cooling period, the mold is opened and the solidified part is ejected.

34.5.2 Coinjection Molding

Coinjection molding makes it possible to mold articles with a solid skin of one thermoplastic and a core of another thermoplastic. The skin material is usually solid while the core material contains blowing agents.

The basic process may be one-, two-, or three-channel technology. In one-channel technology, the two melts are injected into the mold, one after the other. The skin material cools and adheres to the colder surface; a dense skin is formed under proper parameter settings. The thickness of the skin can be controlled by adjustment of injection speed, stock temperature, mold temperature, and flow compatibility of the two melts.

In two- and three-channel techniques, both plastic melts may be introduced simultaneously. This allows for better control of wall thickness of the skin, especially in gate areas on both sides of the part.

Injection-Molded Carbon-Fiber Composites

By mixing carbon or glass fibers in injection-molded plastic parts, they can be made lightweight yet stiffer than steel.

34.5.3 Rotomolding

In rotational molding, the product is formed inside a closed mold that is rotated about two axes as heat is applied. Liquid or powdered thermoplastic or thermosetting plastic is poured into the mold, either manually or automatically.

34.5.4 Expandable-Bead Molding

The expandable-bead process consists of placing small beads of polystyrene along with a small amount of blowing agent in a tumbling container. The polystyrene beads soften under heat, which allows a blowing agent to expand them. When the beads reach a given size, depending on the density required, they are quickly cooled. This solidifies the polystyrene in its larger foamed size. The expanded beads are then placed in a mold until it is completely filled. The entrance port is then closed and steam is injected, resoftening the beads and fusing them together. After cooling, the finished, expanded part is removed from the mold.

34.5.5 Extruding

Plastic extrusion is similar to metal extrusion in that a hot material (plastic melt) is forced through a die having an opening shaped to produce a desired cross section. Depending on the material used, the barrel is heated anywhere from 250–600°F (121–316°C) to transform the thermoplastic from a solid to a melt. At the end of the extruder barrel is a screen pack for filtering and building back pressure. A breaker plate serves to hold the screen pack in place and straighten the helical flow as it comes off the screen.

34.5.6 Blow Molding

Blow molding is used extensively to make bottles and other lightweight, hollow plastic parts. Two methods are used: injection blow molding and extrusion blow molding.

Injection blow molding is used primarily for small containers. The parison (molten-plastic pipe) or tube is formed by the injection of plasticized material around a hollow mandrel. While the material is still molten and still on the mandrel, it is transferred into the blowing mold where air is used to inflate it. Accurate threads may be formed at the neck.

In extrusion-type blow molding, parison is inflated under relatively low pressure inside a split-metal mold. The die closes, pinching the end and closing the top around the mandrel. Air enters through the mandrel and inflates the tube until the plastic contacts the cold wall, where it solidifies. The mold opens, the bottle is ejected, and the tailpiece falls off.

34.5.7 Thermoforming

Thermoforming refers to heating a sheet of plastic material until it becomes soft and pliable and then forming it either under vacuum, by air pressure, or between matching mold halves.

34.5.8 Reinforced-Plastic Molding

Reinforced plastics generally refers to polymers that have been reinforced with glass fibers. Other materials used are asbestos, sisal, synthetic fibers such as nylon and polyvinyl chloride, and cotton fibers. High-strength composites using graphite fibers are now commercially available with moduli of 50,000,000 psi (344,700,000 MPa) and tensile strengths of about 300,000 psi (2,068,000 MPa). They are as strong as or stronger than the best alloy steels and are lighter than aluminum.

34.5.9 Forged-Plastic Parts

The forging of plastic materials is a relatively new process. It was developed to shape materials that are difficult or impossible to mold and is used as a low-cost solution for small production runs.

The forging operation starts with a blank or billet of the required shape and volume for the finished part. The blank is heated to a preselected temperature and transferred to the forging dies, which are closed to deform the work material and fill the die cavity. The dies are kept in the closed position for a definite period of time, usually 15–60 sec. When the dies are opened, the finished forging is removed. Since forging involves deformation of the work material in a heated and softened condition, the process is applicable only to thermoplastics.

34.6 POWDER METALLURGY

In powder metallurgy (P/M), fine metal powders are pressed into a desired shape, usually in a metal die and under high pressure, and the compacted powder is then heated (sintered), with a protective atmosphere. The density of sintered compacts may be increased by repressing. Repressing is also performed to improve the dimensional accuracy, either concurrently or subsequently, for a period of time at a temperature below the melting point of the major constituent. P/M has a number of distinct advantages that account for its rapid growth in recent years, including (1) no material is wasted, (2) usually no machining is required, (3) only semiskilled labor is required, and (4) some unique properties can be obtained, such as controlled degrees of porosity and built-in lubrication.

A crude form of powder metallurgy appears to have existed in Egypt as early as 3000 BC, using particles of sponge iron. In the 19th century, P/M was used for producing platinum and tungsten wires. However, its first significant use related to general manufacturing was in Germany, following World War I, for making tungsten carbide cutting-tool tips. Since 1945 the process has been highly developed, and large quantities of a wide variety of P/M products are made annually, many of which could not be made by any other process. Most are under 2 in. (50.8 mm) in size, but many are larger, some weighing up to 50 lb (22.7 kg) and measuring up to 20 in. (508 mm).

Powder metallurgy normally consists of four basic steps:

1. Producing a fine metallic powder
2. Mixing and preparing the powder for use
3. Pressing the powder into the desired shape
4. Heating (sintering) the shape at an elevated temperature

Other operations can be added to obtain special results.

The pressing and sintering operations are of special importance. The pressing and repressing greatly affect the density of the product, which has a direct relationship to the strength properties. Sintering strips contaminants from the surface of the powder particles, permitting diffusion bonding to occur and resulting in a single piece of material. Sintering usually is done in a controlled, inert atmosphere, but sometimes it is done by the discharge of spark through the powder while it is under compaction in the mold.

34.6.1 Properties of P/M Products

Because the strength properties of powder metallurgy products depend on so many variables—type and size of powder, pressing pressure, sintering temperature, finishing treatments, and so on—it is difficult to give generalized information. In general, the strength properties of products that are made from pure metals (unalloyed) are about the same as those made from the same wrought metals. As alloying elements are added, the resulting strength properties of P/M products fall below those of wrought products by varying, but usually substantial, amounts. The ductility usually is markedly less, as might be expected because of the lower density. However, tensile strengths of 40,000–50,000 psi (275.8–344.8 MPa) are common, and strengths above 100,000 psi (689.5 MPa) can be obtained. As larger presses and forging combined with P/M preforms are used, to provide greater density, the strength properties of P/M materials will more nearly equal those of wrought materials. Coining can also be used to increase the strength properties of P/M products and to improve their dimensional accuracy.

34.7 SURFACE TREATMENT

Products that have been completed to their proper shape and size frequently require some type of surface finishing to enable them to satisfactorily fulfill their function. In some cases, it is necessary to improve the physical properties of the surface material for resistance to penetration or abrasion.

Surface finishing may sometimes become an intermediate step in processing. For instance, cleaning and polishing are usually essential before any kind of plating process. Another important need for surface finishing is for corrosion protection in a variety of environments. The type of protection provided will depend largely on the anticipated exposure, with due consideration to the material being protected and the economic factors involved.

Satisfying the above objectives necessitates the use of many surface-finishing methods that involve chemical change of the surface; mechanical work affecting surface properties, cleaning by a variety of methods; and the application of protective coatings organic and metallic.

34.7.1 Cleaning

Few, if any, shaping and sizing processes produce products that are usable without some type of cleaning unless special precautions are taken. Figure 34.22 indicates some of the cleaning methods available. Some cleaning methods provide multiple benefits. Cleaning and finish improvements are often combined. Probably of even greater importance is the combination of corrosion protection with finish improvement, although corrosion protection is more often a second step that involves coating an already cleaned surface with some other material or chemical conversion.

Liquid and Vapor Baths

Liquid and Vapor Solvents. The most widely used cleaning methods make use of a cleaning medium in liquid or vapor form. These methods depend on a solvent or chemical action between the surface contaminants and the cleaning material.

Petroleum Solvents. Among the more common cleaning jobs required is the removal of grease and oil deposited during manufacturing or intentionally coated on the work to provide protection. One of the most efficient ways to remove this material is by use of solvents that dissolve the grease and oil but have no effect on the base metal. Petroleum derivatives, such as Stoddard solvent and kerosene, are common for this purpose, but, since they introduce some danger of fire, chlorinated solvents, such as trichlorethylene, that are free of this fault are sometimes substituted.

Conditioned Water. One of the most economical cleaning materials is water. However, it is seldom used alone, even if the contaminant is fully water soluble, because the impurity of the water itself may contaminate the work surface. Depending on its use, water is treated with various acids and alkalies to suit the job being performed.

Pickling. Water containing sulfuric acid in a concentration from about 10–25% and at a temperature of approximately 149°F (65°C) is commonly used in a process called *pickling* for removal of surface oxides or scale or iron and steel.

Mechanical Work Frequently Combined with Chemical Action. Spraying, brushing, and dipping methods are also used with liquid cleaners. In nearly all cases, mechanical work to cause surface film breakdown and particle movement is combined with chemical and solvent action. The mechanical work may be agitation of the product, as in dipping, movement of the cleaning agent, as in spraying, or use of a third element, as in rubbing brushing. In some applications, sonic or ultrasonic vibrations are applied to either the solution or the workpieces to speed the cleaning action. Chemical activity is increased with higher temperatures and optimum concentration of the cleaning agent, both of which must in some cases be controlled closely for efficient action.

Blasting

The term *blasting* is used to refer to all those cleaning methods in which the cleaning medium is accelerated to high velocity and impinged against the surface to be cleaned. The high velocity may be provided by air or water directed through a nozzle or by mechanical means with a revolving slinger. The cleaning agent may be either dry or wet solid media, such as sand, abrasive, steel grit, or shot, or may be liquid or vapor solvents combined with abrasive material. In addition to cleaning, solid particles can improve finish and surface properties of the material on which they are used. Blasting tends to increase the surface area and thus set up compressive stresses that may cause a warping of thin sections, but in other cases, it may be very beneficial in reducing the likelihood of

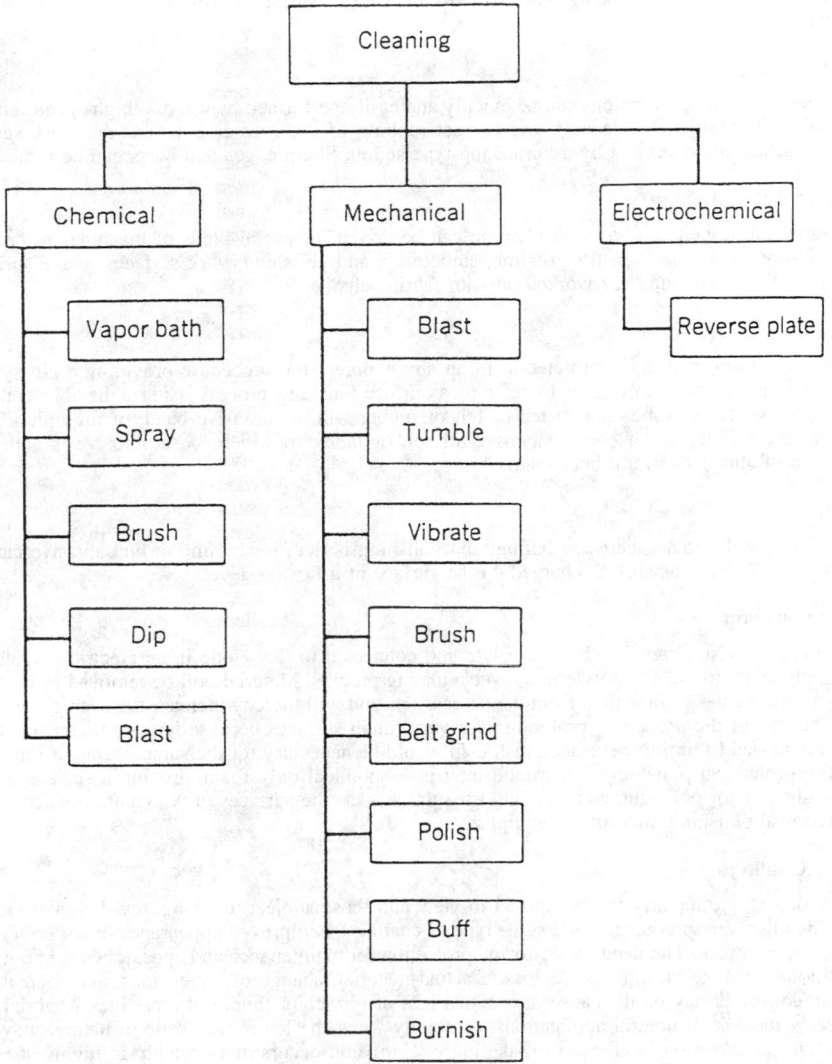

Fig. 34.22 Cleaning methods.

fatigue failure. When used for the latter purpose, the process is more commonly known as *shot peening*.

Water Slurries. Liquid or vaporized solvents may, by themselves, be blasted against a surface for high-speed cleaning of oil and grease films with both chemical and mechanical action. Water containing rust-inhibiting chemicals may carry, in suspension, fine abrasive particles that provide a grinding cutting-type action for finish improvement along with cleaning. The blasting method using this medium is commonly known as *liquid honing*.

Abrasive Barrel Finishing

Barrel finishing, rolling, tumbling, and *rattling* are terms used to describe similar operations that consist of packing parts together with some cleaning media in a cylinder or drum, which can be rotated to cause movement among them. The media may be abrasive (either fine or coarse); metal stars, slugs, or balls; stones; wood chips; sawdust; or cereals. The work may be done wet or dry, depending on the materials being worked with, the kind of surface finish desired, and the kind of equipment available.

Wire Brushing

A number of cleaning operations can be quickly and easily performed by use of a high-speed rotating wire brush. In addition to cleaning, the contact rubbing of the wire ends across the work surface produce surface improvement by a burnishing-type action. Sharp edges and burrs can be removed.

Abrasive Belt Finishing

Continuous fabric belts coated with abrasive can be driven in several kinds of machines to provide a straight-line cutting motion for grinding, smoothing, and polishing work surfaces. Plane surfaces are the most common surfaces worked on with fabric belts.

Polishing

The term *polishing* may be interpreted to mean any nonprecision procedure providing a glossy surface, but is most commonly used to refer to a surface-finishing process using a flexible abrasive wheel. The wheels may be constructed of felt or rubber with an abrasive band, of multiple coated abrasive discs, of leaves of coated abrasive, of felt or fabric to which loose abrasive is added as needed, or of abrasives in a rubber matrix.

Buffing

About the only difference between buffing and polishing is that, for buffing, a fine abrasive carried in wax or a similar substance is charged on the surface of a flexible level.

Electropolishing

If a workpiece is suspended in an electrolyte and connected to the anode in an electrical circuit, it will supply metal to the electrolyte in a reverse plating process. Material will be removed faster from the high spots of the surface than from the depressions and will thereby increase the average smoothness. The cost of the process is prohibitive for very rough surfaces because larger amounts of metal must be removed to improve surface finish than would be necessary for the same degree of improvement by mechanical polishing. Electropolishing is economical only for improving a surface that is already good or for polishing complex and irregular shapes, the surfaces of which are not accessible to mechanical polishing and buffing equipment.

34.7.2 Coatings

Many products, particularly those exposed to view and those subject to change by the environment with which they are in contact, need some type of coating for improved appearance or for protection from chemical attack. The need for corrosion protection for maintenance and appearance is important. In addition to change of appearance, loss of actual material, change of dimensions, and decrease of strength, corrosion may be the cause of eventual loss of service or failure of a product. Material that must carry loads in structural applications, especially when the loads are cyclic in nature, may fail with fatigue if corrosion is allowed to take place. Corrosion occurs more readily in highly stressed material, where it attacks grain boundaries in such a way as to form points of stress concentration that may be nuclei for fatigue failure.

Harness and wear resistance, however, can be provided on a surface by plating with hard metals. Chromium plating of gages and other parts subject to abrasion is frequently used to increase their wear life. Coatings of plastic material and asphaltic mixtures are sometimes placed on surfaces to provide sound-deadening. The additional benefit of protection from corrosion is usually acquired at the same time.

Plastics of many kinds, mostly of the thermoplastic type because they are easier to apply and also easier to remove later if necessary, are used for mechanical protection. Highly polished material may be coated with plastic, which may be stripped off later, to prevent abrasion and scratches during processing. It is common practice to coat newly sharpened cutting edges of tools by dipping them in thermoplastic material to provide mechanical protection during handling and storage.

Organic Coatings

Organic coatings are used to provide pleasing colors, to smooth surfaces, to provide uniformity in both color and texture, and to act as a protective film for control of corrosion. Organic resin coatings do not ordinarily supply any chemical-inhibiting qualities. Instead, they merely provide a separating film between the surface to be protected and the corrosive environment. The important properties, therefore, are continuity, permeability, and adhesion characteristics.

Paints, Varnishes, and Enamels

Paints. Painting is a generic term that has come to mean the application of almost any kind of organic coating by any method. Because of this interpretation, it is also used generally to describe a broad class of products. As originally defined and as used most at present, paint is a mixture of pigment in a drying oil. The oil serves as a carrier for the pigment and in addition creates a tough continuous film as it dries. Drying oils, one of the common ones of which is linseed oil, become solid when large surface areas are exposed to air. Drying starts with a chemical reaction of oxidation. Nonreversible polymerization accompanies oxidation to complete the change from liquid to solid.

Varnish. Varnish is a combination of natural or synthetic resins and drying oil, sometimes containing volatile solvents as well. The material dries by a chemical reaction in the drying oil to a clear or slightly amber-colored film.

Enamel. Enamel is a mixture of pigment in varnish. The resins in the varnish cause the material to dry to a smoother, harder, and glossier surface than is produced by ordinary paints. Some enamels are made with thermosetting resins that must be baked for complete dryness. These baking enamels provide a toughness and durability not usually available with ordinary paints and enamels.

Lacquers

The term *lacquer* is used to refer to finishes consisting of thermoplastic materials dissolved in fast-drying solvents. One common combination is cellulose nitrate dissolved in butyl acetate. Present-day lacquers are strictly air-drying and form films very quickly after being applied, usually by spraying. No chemical change occurs during the hardening of lacquers; consequently, the dry film can be redissolved in the thinner. Cellulose acetate is used in place of cellulose nitrate in some lacquers because it is nonflammable. Vinyls, chlorinated hydrocarbons, acrylics, and other synthetic thermoplastic resins are also used in the manufacture of lacquers.

Vitreous Enamels

Vitreous, or porcelain, enamel is actually a thin layer of glass fused onto the surface of a metal, usually steel or iron. Shattered glass, ball milled in a fine particle size, is called *frit*. Frit is mixed with clay, water, and metal oxides, which produce the desired color, to form a thin slurry called *slip*. This is applied to the prepared metal surface by dipping or spraying and, after drying, is fired at approximately 1470°F (800°C) to fuse the material to the metal surface.

Metallizing

Metal spraying, or metallizing, is a process in which metal wire or powder is fed into an oxyacetylene heating flame and then, after melting, is carried by high-velocity air to be impinged against the work surface. The small droplets adhere to the surface and bond together to build up a coating.

Vacuum Metallizing

Some metals can be deposited in very thin films, usually for reflective or decorative purposes, as a vapor deposit. The metal is vaporized in a high-vacuum chamber containing the parts to be coated. The metal vapor condenses on the exposed surfaces in a thin film that follows the surface pattern. The process is cheap for coating small parts, considering the time element only, but the cost of special equipment needed is relatively high.

Aluminum is the most used metal for deposit by this method and is used frequently for decorating or producing a mirror surface on plastics. The thin films usually require mechanical protection by covering with lacquer or some other coating material.

Hot-Dip Plating

Several metals, mainly zinc, tin, and lead, are applied to steel for corrosion protection by a hot-dip process. Steel in sheet, rod, pipe, or fabricated form, properly cleansed and fluxed, is immersed in molten plating metal. As the work is withdrawn, the molten metal that adheres solidifies to form a protective coat. In some of the large mills, the application is made continuously to coil stock that is fed through the necessary baths and even finally inspected before being recoiled or cut into sheets.

Electroplating

Coatings of many metals can be deposited on other metals, and on nonmetals when suitably prepared, by electroplating. The objectives of plating are to provide protection against corrosion, to improve appearance, to establish wear- and abrasion-resistant surfaces, to add material for dimensional increase, and to serve as an intermediate step of multiple coating. Some of the most common metals deposited in this way are copper, nickel, cadmium, zinc, tin, silver, and gold. The majority are used to provide some kind of corrosion protection but appearance also plays a strong part in their use.

Temporary Corrosion Protection

It is not uncommon in industry for periods of time, sometimes quite long periods, to elapse between manufacture, assembly, shipment, and use of parts. Unless a new processing schedule can be worked out, about the only cure for the problem is corrosion protection suitable for the storage time and exposure. The coatings used are usually nondrying organic materials, called *shushing compounds,* that can be removed easily. The two principal types of compounds used for this purpose are petroleum-based materials, varying from extremely light oils to semisolids, and thermoplastics. The most common method of application of shushing compounds for small parts is by dipping. Larger parts that cannot be handled easily may be sprayed, brushed, or flow coated with the compound.

34.7.3 Chemical Conversions

A relatively simple and often fully satisfactory method for protection from corrosion is by conversion of some of the surface material to a chemical composition that resists from the environment. These converted metal surfaces consist of relatively thin (seldom more than 0.001 in. (0.025 mm) thick) inorganic films that are formed by chemical reaction with the base material. One important feature of the conversion process is that the coatings have little effect on the product dimensions.

Anodizing

Aluminum, magnesium, and zinc can be treated electrically in a suitable electrolyte to produce a corrosion-resistant oxide coating. The metal being treated is connected to the anode in the circuit, which provides the name *anodizing* for the process. Aluminum is commonly treated by anodizing that produces an oxide film thicker than, but similar to, that formed naturally with exposure to air. Anodizing of zinc has very limited use. The coating produced on magnesium is not as protective as that formed on aluminum, but does provide some protective value and substantially increases protection when used in combination with paint coatings.

Chromate Coatings

Zinc is usually considered to have relatively good corrosion resistance. This is true when the exposure is to normal outdoor atmosphere where a relatively thin corrosion film forms. Contact with either highly aerated water films or immersion in stagnant water containing little oxygen causes uneven corrosion and pitting. The corrosion products of zinc are less dense than the base material, so that heavy corrosion not only destroys the product appearance, but also may cause malfunction by binding moving parts. Corrosion of zinc can be substantially slowed by the production of chromium salts on its surface. The corrosion resistance of magnesium alloys can be increased by immersion of anodic treatment in acid baths containing dichromates. Chromate treatment of both zinc and magnesium improves corrosion resistance, but is used also to improve adhesion of paint.

Phosphate Coatings

Phosphate coatings, used mostly on steel, result from a chemical reaction of phosphoric acid with the metal to form a nonmetallic coating that is essentially phosphoric salts. The coating is produced by immersing small items or spraying large items with the phosphating solution. Phosphate surfaces may be used alone for corrosion resistance, but their most common application is as a base for paint coatings. Two of the most common application methods are called *parkerizing* and *bonderizing.*

Chemical Oxide Coatings

A number of proprietary blacking processes, used mainly on steel, produce attractive black oxide coatings. Most of the processes involve the immersing of steel in a caustic soda solution, heated to about 300°F (150°C) and made strongly oxidizing by the addition of nitrites or nitrates. Corrosion resistance is rather poor unless improved by application of oil, lacquer, or wax. As in the case of most of the other chemical-conversion procedures, this procedure also finds use as a base for paint finishes.

BIBLIOGRAPHY

Alting, L., *Manufacturing Engineering Processes,* Marcel Dekker, New York, 1982.

Amstead, B. H., P. F. Ostwald, and M. L. Begeman, *Manufacturing Processes,* 8th ed., Wiley, New York, 1988.

DeGarmo, E. P., J. T. Black, and R. A. Kohser, *Material and Processes in Manufacturing,* 7th ed., Macmillan, New York, 1988.

Doyle, L. E., G. F. Schrader, and M. B. Singer, *Manufacturing Processes and Materials for Engineers,* Prentice-Hall, Englewood Cliffs, NJ, 1985.

LeGrand, R. (ed.), *American Machinist's Handbook,* McGraw-Hill, New York, 1973.

Kalpakjian, S., *Manufacturing Processes for Engineering Materials,* Addison-Wesley, Reading, MA, 1994.

Kronenberg, M., *Machining Science and Application,* Pergamon, London, 1966.

Lindberg, R. A., *Processes and Materials of Manufacture,* 2nd ed., Allyn and Bacon, Boston, MA, 1977.

Machining Data Handbook, 3rd ed., Machinability Data Center, Cincinnati, OH, 1980.

Metals Handbook, 8th ed., Vol. 4, *Forming,* Vol. 5, *Forging and Casting,* American Society for Metals, Metals Park, OH, 1985.

Moore, H. D., and D. R. Kibbey, *Manufacturing Materials and Processes,* 3rd ed., Wiley, New York, 1982.

Niebel, B. W., and A. B. Draper, *Product Design and Process Engineering,* McGraw-Hill, New York, 1974.

Schey, J. A., *Introduction to Manufacturing Processes,* McGraw-Hill, New York, 1977.

Shaw, M. C., *Metal Cutting Principles,* Oxford University Press, Oxford, 1984.

Society of Manufacturing Engineers, *Tool and Manufacturing Engineers Handbook,* Vol. 2, *Forming;* Vol. 3, *Materials, Finishing and Coating,* McGraw-Hill, New York, 1985.

Zohdi, M. E., "Statistical Analysis, Estimation and Optimization of Surface Finish," in *Proceedings of International Conference on Development of Production Systems,* Copenhagen, Denmark, 1974.

CHAPTER 35

MECHANICAL FASTENERS

Murray J. Roblin
Chemical and Materials Engineering Department
California State Polytechnic University
Pomona, California

35.1	INTRODUCTION	1136	
35.2	BOLTED AND RIVETED JOINT TYPES	1137	
35.3	EFFICIENCY	1138	
35.4	STRENGTH OF A SIMPLE LAP JOINT (BEARING-TYPE CONNECTION)	1138	
35.5	SAMPLE PROBLEM OF A COMPLEX BUTT JOINT (BEARING-TYPE CONNECTION)	1139	
	35.5.1 Preliminary Calculations	1140	
35.6	FRICTION-TYPE CONNECTIONS	1142	
35.7	UPPER LIMITS ON CLAMPING FORCE	1144	
	35.7.1 Yield Strength of the Bolt	1144	
	35.7.2 Thread Stripping Strength	1144	
	35.7.3 Design-Allowable Bolt Stress and Assembly Stress Limits	1144	
	35.7.4 Torsional Stress Factor	1144	
	35.7.5 Shear Stress Allowance	1145	
	35.7.6 Flange Rotation	1145	
	35.7.7 Gasket Crush	1145	
	35.7.8 Stress Cracking	1145	
	35.7.9 Combined Loads	1145	

35.8	THEORETICAL BEHAVIOR OF THE JOINT UNDER TENSILE LOADS	1146	
	35.8.1 Critical External Load	1148	
	35.8.2 Very Large External Loads	1149	
35.9	EVALUATION OF SLIP CHARACTERISTICS	1153	
35.10	INSTALLATION OF HIGH-STRENGTH BOLTS	1153	
35.11	TORQUE AND TURN TOGETHER	1155	
35.12	ULTRASONIC MEASUREMENT OF BOLT STRENGTH OR TENSION	1156	
35.13	FATIGUE FAILURE AND DESIGN FOR CYCLICAL TENSION LOADS	1158	
	35.13.1 Rolled Threads	1158	
	35.13.2 Fillets	1158	
	35.13.3 Perpindicularity	1158	
	35.13.4 Overlapping Stress Concentrations	1158	
	35.13.5 Thread Run-Out	1158	
	35.13.6 Thread Stress Distribution	1158	
	35.13.7 Bending	1159	
	35.13.8 Corrosion	1159	
	35.13.9 Surface Conditions	1159	
	35.13.10 Reduce Load Excursions	1159	

Mechanical Engineers' Handbook, 2nd ed., Edited by Myer Kutz.
ISBN 0-471-13007-9 © 1998 John Wiley & Sons, Inc.

35.14 WELDED JOINTS   **1159**   **35.15 COOLING RATES AND**
   35.14.1 Submerged Arc   **THE HEAT-AFFECTED ZONE**
          Welding (SAW)   1160   **(HAZ)**
   35.14.2 Gas Metal Arc Welding 1162   **IN WELDMENTS**   **1170**
   35.14.3 Flux-Cored Arc
          Welding: FCAW   1166
   35.14.4 Shielded Metal
          Arc Welding
          (SMAW)   1167

35.1 INTRODUCTION

Most of the information in this chapter is not original. I am merely passing along the information I have gained from many other people and from extensive reading in this subject.

For an in-depth understanding of this inexact field of study, I would recommend two excellent books that I used extensively in the preparation of this chapter.[1,2] For a full comprehension of this topic, it is necessary to read both volumes, as they approach the topic from distinctly different points of view.

Two or more components may need to be joined in such a way that they may be taken apart during the service life of the part. In these cases, the assembly must be fastened mechanically. Other reasons for choosing mechanical fastening over welding could be:

1. Ease of part replacement, repair, or maintenance
2. Ease or lower cost to manufacture
3. Designs requiring movable joints
4. Designs requiring adjustable joints

The most common mechanical joining methods are bolts (threaded fasteners), rivets, and welding (welding will be covered in a later section).

To join two members by bolting or riveting requires holes to be drilled in the parts to accommodate the rivets and bolts. These holes reduce the load-carrying cross-sectional area of the members to be joined. Because this reduction in area as a result of the holes is at least 10–15%, the load-carrying capacity of the bolted structure, is reduced, which must be accounted for in the design. Alternatively, when one inserts bolts into the holes, only the cross section of the bolt or rivet supports the load. In this case, the reduction in the strength of the joint is reduced even further than 15%.

Even more critical are the method and care taken in drilling the holes. When one drills a hole in metal, not only is the cross-sectional area reduced, but the hole itself introduces "stress risers" and/or flaws in/on the surface of the holes that may substantially endanger the structure. First, the hole places the newly created surface in tension, and if any defects are created as a result of drilling, they must be accounted for in a quantitative way. Unfortunately, it is very difficult to obtain definitive information on the inside of a hole that would allow characterization of the introduced defect.

The only current solution is to make certain that the hole is properly prepared which means not only drilling or subpunching to the proper size, but also *reaming* the surface of the hole. To be absolutely certain that the hole is not a problem, one needs to put the surface of the hole in residual compression by expanding it slightly with an expansion tool or by pressing the bolt, which is just slightly larger than the hole. This method causes the hole to expand during insertion, creating a hole whose surface is in residual compression. While there are fasteners designed to do this, it is not clear that all of the small surface cracks of the hole have been removed to prevent flaws/stress risers from existing in the finished product.

Using bolts and rivets in an assembly can also provide an ideal location for water to exist in the crevices between the two parts joined. This trapped water, under conditions where chlorides and sodium exist, can cause "crevice corrosion," which is a serious problem if encountered.

Obviously, in making the holes as perfect as possible, you increase the cost of a bolted and/or riveted joint significantly, which makes welding or adhesive joining a more attractive option. Of course, as will be shown below, welding and joining have their own set of problems that can degrade the joint strength.

The analysis of the strength of a bolted, riveted, or welded joint involves many indeterminate factors resulting in inexact solutions. However, by making certain simplifying assumptions, we can obtain solutions that are acceptable and practical. We discuss two types of solutions: *bearing-type connections,* which use ordinary or unfinished bolts or rivets, and *friction-type connections,* which

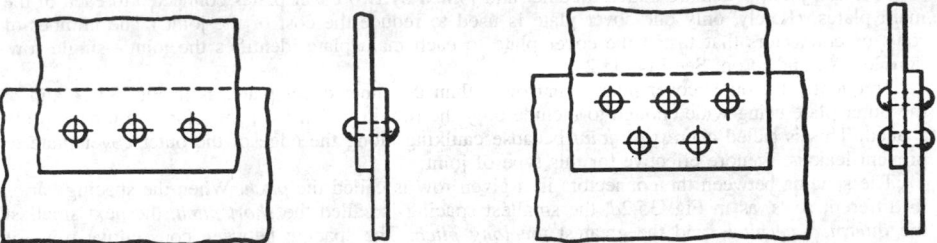

Fig. 35.1 Lap joints. Connectors are shown as rivets only for convenience.

use high-strength bolts. Today, economy and efficiency are obtained by using high-strength bolts for field connections together with welding in the shop. With the advent of lighter-weight welding power supplies, the use of field welding combined with shop welding is finding increasing favor.

While riveted joints do show residual clamping forces (even in cold-driven rivets), the clamping forces in the rivet is difficult to control, is not as great as that developed by high-strength bolts, and cannot be relied upon. Installation of hot-driven rivets involves many variables, such as the initial or driving temperature, driving time, finishing temperature, and driving method. Studies have shown that the holes are almost completely filled for short rivets. As the grip length is increased, the clearances between rivet and plate material tend to increase.

35.2 BOLTED AND RIVETED JOINT TYPES

There are two types of riveted and bolted joints: *lap joints* and *butt joints*. See Figs. 35.1 and 35.2 for lap and butt joints, respectively. Note that there can be one or more rows of connectors, as shown in Fig. 35.2a and b.

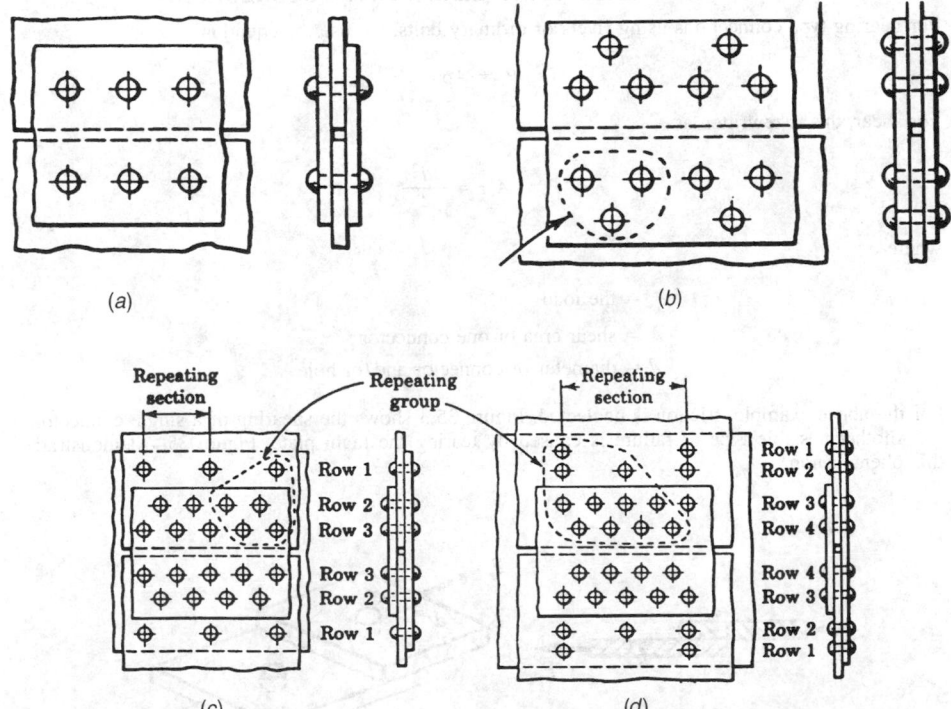

Fig. 35.2 Butt joints: (a) single-row; (b) double-row; (c) triple-row (pressure-type); (d) quadruple row (pressure-type).

In a butt joint, plates are butted together and joined by two cover plates connected to each of the main plates. (Rarely, only one cover plate is used to reduce the cost of the joint.) The number of rows of connectors that fasten the cover plate to each main plate identifies the joint—single row, double row, and so on. See Fig. 35.2.

Frequently the outer cover plate is narrower than the inner cover plate, as in Fig. 35.2c and d, the outer plate being wide enough to include only the row in which the connectors are most closely spaced. This is called a *pressure joint* because caulking along the edge of the outer cover plate to prevent leakage is more effective for this type of joint.

The spacing between the connectors in a given row is called the *pitch.* When the spacing varies in different rows, as in Fig. 35.2d, the smallest spacing is called the *short pitch,* the next smallest the *intermediate pitch,* and the greatest the *long pitch.* The spacing between consecutive rows of connectors is called the *back pitch.* When the connectors (rivets or bolts) in consecutive rows are staggered, the distance between their centers is the *diagonal pitch.*

In determining the strength of a joint, computations are usually made for the length of a joint corresponding to a repeating pattern of connectors. The length of the repeating pattern, called the *repeating section,* is equal to the long pitch.

To clarify how many connectors belong in a repeating section, see Fig. 35.2c, which shows that there are five connectors effective in each half of the triple row—that is, two half connectors in row 1, two whole connectors in row 2, and one whole and two half connectors in row 3. Similarly, there are 11 connectors effective in each half of the repeating section in Fig. 35.2d.

When rivets are used in joints, the holes are usually drilled or, punched, and reamed out to a diameter of $\frac{1}{16}$ in. (1.5 mm) larger than the nominal rivet size. The rivet is assumed to be driven so tightly that it fills the hole completely. Therefore, in calculations the diameter of the hole is used because the rivet fills the hole. This is not true for a bolt unless it is very highly torqued. In this case, a different approach needs to be taken, as delineated later in this chapter.

35.3 EFFICIENCY

Efficiency compares the strength of a joint to that of the solid plate as follows:

$$\text{Efficiency} = \frac{\text{strength of the joint}}{\text{strength of solid plate}}$$

35.4 STRENGTH OF A SIMPLE LAP JOINT (BEARING-TYPE CONNECTION)

For bearing-type connections using rivets or ordinary bolts, we use the equation

$$P_s = A\sigma$$

For shear, this is rewritten as

$$P_s = A_s\tau = \frac{\pi d^2\tau}{4}$$

where

$$P_s = \text{the load}$$

$$A = \text{shear area of one connector}$$

$$d = \text{diameter of connector and/or hole}$$

For the above example, friction is neglected. Figure 35.3 shows the shearing of a single connector.

Another possible type of failure is caused by tearing the main plate. Figure 35.4 demonstrates this phenomenon.

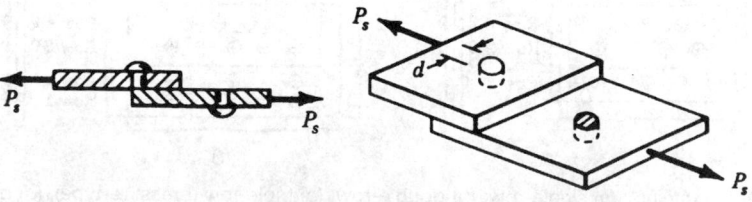

Fig. 35.3 Shear failure.

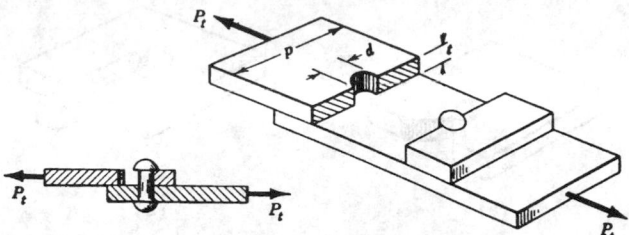

Fig. 35.4 Tear of plate at section through connector hole. $P_t = A_t\sigma_t = (p - d)t\sigma_t$.

The above failure occurs on a section through the connector hole because this region has the minimum tearing resistance. If p is the width of the plate or the length of a repeating section, the resisting area is the product of the net width of the plate $(p - d)$ times the thickness t. The failure load in tension therefore is

$$P_{\text{tension}} = A_t\sigma_t = (p - d)t(\sigma_t)$$

A third type of failure, called a *bearing failure,* is shown in Fig. 35.5. For this case, there is relative motion between the main plates or enlargement of the connector hole caused by an excessive tensile load. Actually, the stress that the connector bears against the edges of the hole varies from zero at the edges of the hole to the maximum value at the center of the bolt or rivet. However, common practice assumes the stress as uniformly distributed over the projected area of the hole. See Fig. 35.5.

The failure load in the bearing area can be expressed by

$$P_b = A_b\sigma_b = (td)\sigma_b$$

Other types of failure are possible but will not occur in a properly designed joint. These are tearing of the edge of the plate back of the connector hole (Fig. 35.6a) or a shear failure behind the connector hole (Fig. 35.6b) or a combination of both. Failures of this type occur when the distance from the edge of the plate is ~2 or less multiplied by the diameter of the connector or hole.

35.5 SAMPLE PROBLEM OF A COMPLEX BUTT JOINT (BEARING-TYPE CONNECTION)

The strength of a bearing-type connection is limited by the capacity of the rivets or ordinary bolts to transmit load between the plates or by the tearing resistance of the plates themselves, depending on which is smaller. The calculations are divided as follows:

1. Preliminary calculations to determine the load that can be transmitted by one rivet or bolt in shear or bearing *neglecting friction* between the plates
2. Calculations to determine which mode of failure is most likely

A repeating section 180 mm long of a riveted triple row butt joint of the pressure type is illustrated in Fig. 35.7. The rivet hole diameter $d = 20.5$ mm, the thickness of the main plate $t = 14$ mm, and the thickness of each cover plate $t = 10$ mm. The ultimate stresses in shear, bearing, and tension are respectively $\tau = 300$ MPa, $\sigma_b = 650$ MPa, and $\sigma_t = 400$ MPa. Using a factor of safety of 5, determine

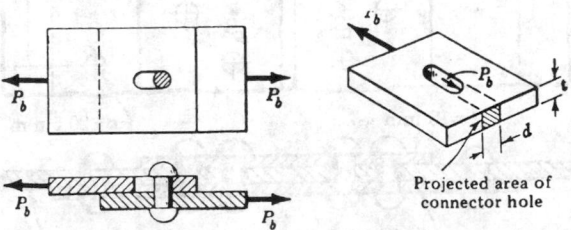

Fig. 35.5 Exaggerated bearing deformation of upper plate. $P_b = A_b\sigma_b = (td)\sigma_b$.

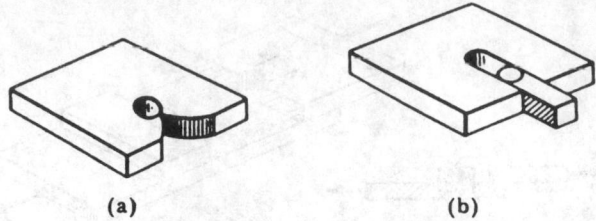

Fig. 35.6 Possible types of failure if connector hole is too close to edge of plate: (a) tear out; (b) shear behind connector.

the strength of a repeating section, the efficiency of the joint, and the maximum internal pressure that can be carried in a 1.5 m diameter boiler where this joint is the longitudinal seam.

Solution: The use of ultimate stresses will determine the ultimate load, which is then divided by the factor of safety (in this case 5) to determine the safe working load. An alternative but preferable procedure is to use allowable stresses to determine the safe working load directly, which involves smaller numbers. Thus, dividing the ultimate stressed by 5, we find that the allowable stresses in shear, bearing, and tension, respectively, are $\tau = 300/5 = 60$ MPa, $\sigma_b = 650/5 = 130$ MPa, and $\sigma_t = 400/5 = 80$ MPa. The ratio of the shear strength τ to the tensile strength σ of a rivet is about .75.

35.5.1 Preliminary Calculations

To single shear one rivet,

$$P_s = \frac{\pi d^2}{4} \tau = \frac{\pi}{4}(20.5 \times 10^{-3})^2(60 \times 10^6) = 19.8 \ kN$$

To double shear one rivet,

$$P_s = 2 \times 19.8 = 39.6 \ kN$$

To crush one rivet in the main plate,

$$P_B = (td)\sigma_b = (14.0 \times 10^{-3})(20.5 \times 10^{-3})(130 \times 10^6) = 37.3 \ kN$$

To crush one rivet in one cover plate,

$$P'_b = (t'd)\sigma_b = (10 \times 10^{-3})(20.5 \times 10^{-3})(130 \times 10^6) = 26.7 \ kN$$

Rivet capacity solution: The strength of a single rivet in row 1 in a repeating section is determined

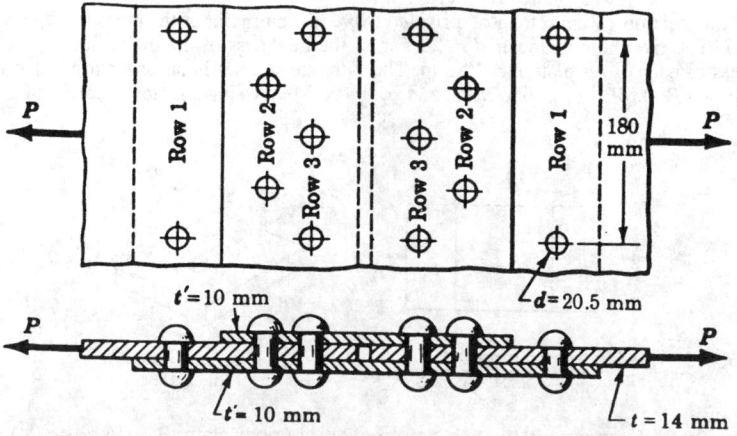

Fig. 35.7

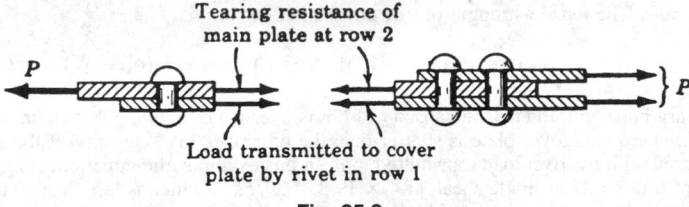

Fig. 35.8

by the lowest value of the load that will single shear the rivet, crush it in the main plate, or crush it in one of the cover plates. Based on the values in the preceding calculations, this value is 19.8 kN per rivet.

The strength of each of the two rivets in row 2 depends on the lowest value required to double shear the rivet, crush it the main plate, or crush it in both cover plates. From the above preliminary calculations, this value is 37.3 kN per rivet or $2 \times 37.3 + 74.6$ kN for both rivets in row 2.

Each of the two rivets in the repeating section in row 3 transmits the load between the main plate and the cover plate in the same manner as those in row 2; hence for row 3, the strength $= 74.6$ kN.

The total rivet capacity is the sum of the rivet strengths in all rows (rows 1, 2, 3), as follows:

$$P_{total} = 19.8 + 74.6 + 74.6 = 169.0 \; kN$$

Tearing capacity: The external load applied to the joint acts directly to tear the main plate at row 1, and the failure would be similar to Fig. 35.4. This is calculated as follows:

$$P_{tearing} = (p - d)\,\sigma_t = [(180 \times 10^{-3}) - (20.5 \times 10^{-3})](14 \times 10^{-3})(80 \times 10^6) = 178.6 \; kN$$

The external load applied does not act directly to tear the main plate at row 2 because part of the load is absorbed or transmitted by the rivet in row 1. Hence, if the main plate is to tear at row 2, the external load must be the sum of the tearing resistance of the main plate at row 2 plus the load transmitted by the rivet in row 1. See Figs. 35.8 and 35.9.

Thus,

$$P_{tearing2} = (p - 2d)t\sigma_t + \text{rivet strength in row 1}$$
$$= [(180 \times 10^{-3}) - 2(20.5 \times 10^{-3})](14 \times 10^{-3})(80 \times 10^6)$$
$$+ 19.8 \times 10^3 = 175.5 \; kN$$

Similarly, the external load required to tear the main plate at row 3 must include the rivet resistance in rows 1 and 2 or

$$P_3 = [(180 \times 10^{-3}) - 2(20.5 \times 10^{-3})](14 \times 10^{-3})(80 \times 10^6) + (19.8 \times 10^3) + (74.6 \times 10^3)$$
$$= 250.1 \; kN$$

It is obvious that this computation need not be made because the tearing resistance of the main plates at rows 2 and 3 is equal, thus giving a larger value.

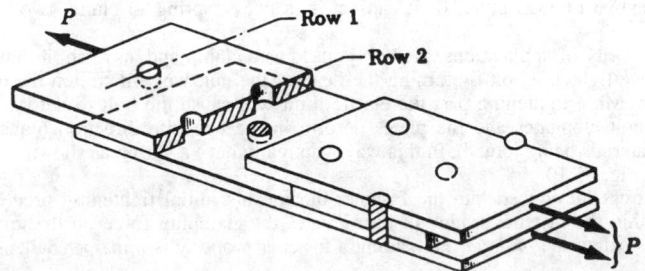

Fig. 35.9 Failure by shear of rivet in row 1 plus tear of main plate in row 2.

At row 3, the tearing resistance of the cover plates is resisted by the tensile strength of the reduced section of that row. The tensile strength of one cover plate is

$$P_c = [(180 \times 10^{-3}) - 2(20.5 \times 10^{-3})](10 \times 10^{-3})(80 \times 10^6) = 111.2 \; kN$$

In an ordinary butt joint, the tensile capacity of both cover plates is twice this value. In a pressure joint, however, where one cover plate is shorter than the other, the load capacity of the shorter plate must be compared with the rivet load transmitted to it. In this example, the upper cover plate transmits the rivet load of four rivets in single shear, or $4 \times 19.8 = 79.2 \; kN$, which is less than its tear capacity of 111.2 kN. Hence, the load capacity of both cover plates becomes

$$P_c = 79.2 + 111.2 = 190.4 \; kN$$

determined by rivet shear in the upper plate and by tension at row 3 in the lower plate.

Thus, the safe load is the lowest of these several values = 169.0 kN, which is the rivet strength in shear.

$$\text{Efficiency} = \frac{\text{safe load}}{\text{strength of solid plate}} = \frac{169 \times 10^3}{(180 \times 10^3)(14 \times 10^{-3})(80 \times 10^6)} = 83.8\%$$

In this discussion, we have neglected friction and assumed that the rivets or bolts only act as pins in the structure or joint—in essence like spot welds spaced in the same way as the rivets or bolts are spaced.

35.6 FRICTION-TYPE CONNECTIONS

In friction-type connections, high-strength bolts (generally high-strength medium carbon steel bolts plain, weathering, or galvanized finished, designated as A325 ASTM grade, or alloy steel bolts designated as A490 ASTM grade) are used and are tightened to high tensile stresses, thereby causing a large resultant normal force between the plates. Tightening of the bolts to a predetermined initial tension is usually done using a calibrated torque wrench or by turn-of-the nut methods.

If done properly (as will be discussed later), the load is now transferred by the friction between the plates and not by shear and the bearing of the bolt, as described in the previous sections. Heretofore, even though the bolts are not subject to shear, design codes, as a matter of convenience, specified an allowable shearing stress to be applied over the cross-sectional area of the bolt. Thus, friction-type joints were analyzed by the same procedures used for bearing-type joints and the frictional forces that existed, were taken as an extra factor of safety. In the ASME code, the "allowable stresses" listed in several places are not intended to limit assembly stresses in the bolts. These allowables are intended to force flange designers to overdesign the joint to use more and/or larger bolts and thicker flange members than they might otherwise be inclined to use.

Only in the non-mandatory Appendix S does section VIII of the code deal with assembly stresses, and then in relatively general terms.

The closest Appendix S comes to quantifying assembly stresses in the bolts is in Eq. (18.6) which suggests that the amount of stress you might expect to produce in the bolts at assembly is given by

$$S_A = \frac{45,000}{D^{1/2}}$$

where S_A = stress created in the bolts at assembly (psi)
D = the nominal diameter of the fastener (in.)

Structurally, a bolt serves one of two purposes: it can act as a pin to keep two or more members from slipping relative to each other, or it can act as a heavy spring to clamp two or more pieces together.

In the vast majority of applications, the bolt is used as a clamp and, as such, it must be tightened properly. When we tighten a bolt by turning the head or the nut, we will stretch the bolt initially in the elastic region. More tightening past the elastic limit will cause the bolt to deform plastically. In either case, the bolt elongates and the plates deform in the opposite direction (equal compressive stresses in the materials being joined). In this way, you really have a spring as shown (with substantial exaggeration) in Fig. 35.10.

The tensile stress introduced into the fastener during this initial tightening process results in a tension force within the fastener, which in turn creates the clamping force on the joint. This initial clamping force is called the *preload*. Preloading a fastener properly is a major challenge that will be discussed later.

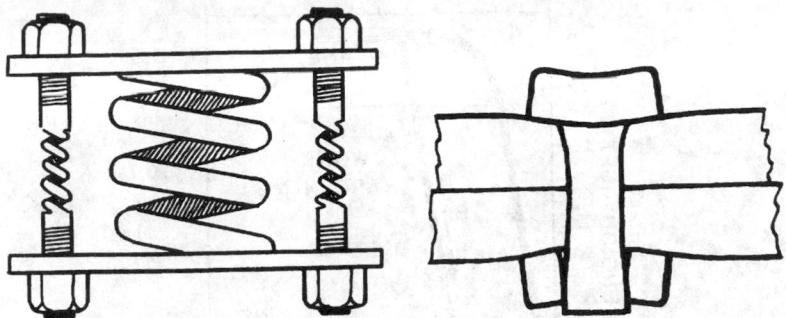

Fig. 35.10   When analyzing the behavior of a bolted joint, pretend the members are a large spring being compressed (clamped) by a group of smaller springs (bolts). When tightened, these springs distort somewhat as shown but grossly exaggerated on the right.

When a bolt is loaded in tension in a tensile testing machine, we generate a tension versus a change in length curve, as shown in Fig. 35.11. The initial straight line portion of the elastic curve is called the *elastic region.* Loading and unloading a bolt within this range of tension never results in a permanent deformation of the bolt because elastic deformation is recoverable. The upper limit of this curve ends at the *proportional limit* or *elastic limit.* Loading beyond or above this limit results in *plastic deformation* of the bolt, which is not recoverable; thus, the bolt has a permanent "set" (it is longer than it was originally even though the load is completely removed). At the *yield point,* the bolt has a specific amount of permanent plastic deformation, normally defined as 0.2 or 0.5% of the initial length. This permanent plastic deformation will increase up until the *ultimate tensile strength* (normally called the *ultimate strength* of the bolt), which is the maximum tension that can be created in the bolt. The UTS is always greater than the yield stress—sometimes as much as twice yield. The final point on the curve is the *failure* or *rupture stress,* where the bolt breaks under the applied load.

If we load the bolt well into the plastic region of its curve and then remove the load, it will behave as shown in Fig. 35.12, returning to the zero load point along a line parallel to the original elastic line but offset by the amount of plastic strain the bolt has set.

On reloading the bolt below the previous load but above the original yield point, the behavior of the bolt will follow this new offset stress strain line and the bolt will behave elastically well beyond the original load that caused plastic deformation in the first place. The difference between the original yield strength of the material and the new yield strength is a function of the "work hardening" that occurred by taking it past the original yield strength on the first cycle. By following the above procedure, we have made the bolt stronger, at least as far as static loads are concerned.

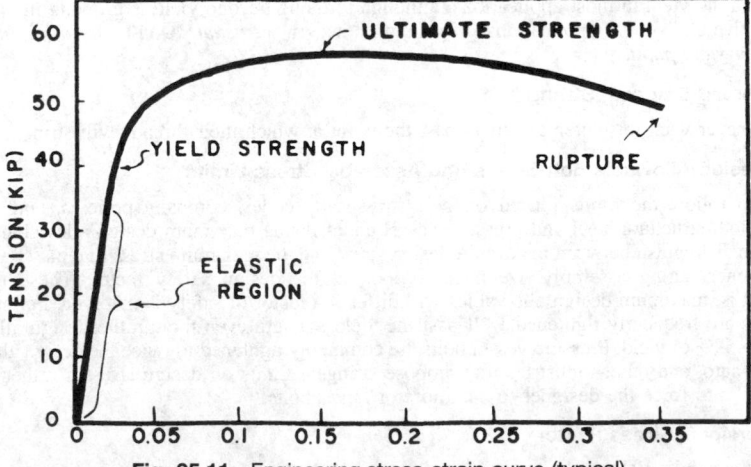

Fig. 35.11   Engineering stress-strain curve (typical).

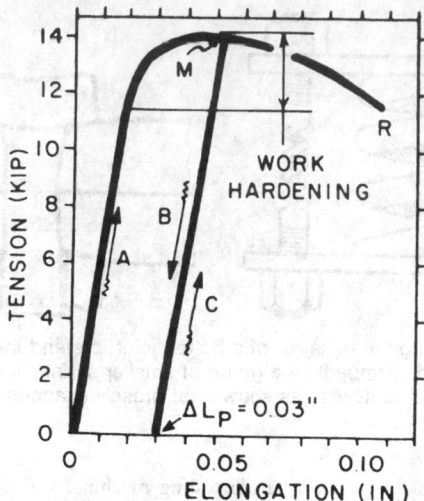

Fig. 35.12 Elastic curve for a ⅜–16 × 4 socket-head cap screw loaded (A) to point M well past the yield strength and then unloaded (B) to give permanent deformation L_p = 0.03 in. If reloaded, it will follow path (C).

This is not wise practice, however, for more brittle materials can suffer a loss of strength by such treatments. Loss of strength in ASTM A490 bolts, because of repeated cycling past the yield (under water and wind loads), has been publicly cited as a contributing factor in the 1979 collapse of the roof on the Kemper Auditorium in Kansas City.

The answer to the question "how much preload" we should place on the joint is currently impossible to answer other than in generalities, ranging from "We always want the maximum clamping force the parts can stand" to "The more the better, up to, but *probably not exceeding the yield stress.*"

35.7 UPPER LIMITS ON CLAMPING FORCE

When determining the amount of clamping force required to combat self-loosening or slip or a leak, we are establishing the essential minimum of force. In each of these situations, additional clamping force is usually desirable from an added safety point of view or is at least acceptable. Commonly used criteria for setting the upper limit follow.

35.7.1 Yield Strength of the Bolt

There is currently a lot of debate about this in the bolting world, as most feel it is unwise to tighten bolts beyond the yield in most applications, although torquing beyond yield is growing in popularity for automotive and similar applications. In general, however, *we usually don't want to tighten bolts beyond their yield point.*

35.7.2 Thread Stripping Strength

We would never want to tighten fasteners past the point at which their threads will strip.

35.7.3 Design-Allowable Bolt Stress and Assembly Stress Limits

We need to follow the limits placed on bolt stresses by codes, company policies, and standard practices. Both structural steel and pressure vessel codes define maximum design allowable stresses for bolts. To distinguish between maximum design stress and the maximum stress that may be allowed in the fastener during assembly, we need to look at the design safety factor. These two will differ—that is, maximum design allowables will differ if a factor of safety is involved. For structural steels, bolts are frequently tightened well past the yield strength even though the design allowables are only 35–58% of yield. Pressure vessel bolts are commonly tightened to twice the design allowable. Aerospace, auto, and other industries may impose stringent limits on design stresses rather than on actual stresses to force the designer to use more or larger bolts.

35.7.4 Torsional Stress Factor

If the bolts are to be tightened by turning the nut or the head, they will experience a torsion stress as well as a tensile stress during assembly. If tightened to the yield stress, they will yield under this combination. If we plan to tighten to or near the yield stress, we must reduce the maximum tensile stresses allowed by a "torquing factor." If using as received steel on steel bolts, then a reduction in the allowable tensile stress of 10% is reasonable. If the fasteners are to be lubricated, use 5%.

35.7.5 Shear Stress Allowance

Since we are designing a clamp, we are primarily interested in the clamping force that the bolts are going to exert on this joint, and therefore the tensile stress. If the bolts will also be exposed to a shear stress, we must take this into account in defining the maximum clamping force, since the shear stress will reduce the amount of strength capacity available for the tensile stress. The following equation can be used to determine how much shear stress the bolt can stand if subject to a given tensile stress or vice versa:

$$\frac{S_T^2}{G^2} + T_T^2 = 1.0$$

where

S_T = the ratio between the shear stress in the shear plane of the bolt and the UTS of the bolt
T_T = the ratio between the tensile stress in the bolt and the UTS of the bolt
G = the ratio between the shear strength and the tensile strength of the bolt (0.5—0.62 typically) if computed on the thread stress area. It is best to compute both S_T and T_T using the equivalent thread stress area rather than the shank area

35.7.6 Flange Rotation

Excessive bolt load can rotate raised face flanges so much that the ID of the gasket is unloaded, opening a leak path. The threat of rotation, therefore, can place an upper limit on planned or specified clamping forces.

35.7.7 Gasket Crush

Excessive preload can so compress a gasket that it will not be able to recover when the internal pressure or a thermal cycle partially unloads it. Contact the gasket manufacturer for upper limits. Note that these will be a function of the service temperature.

35.7.8 Stress Cracking

Stress cracking is encouraged by excessive tension in the bolts, particularly if service loads exceed 50% of the yield stress at least for low alloy quenched and tempered steels.

35.7.9 Combined Loads

These loads include weight, inertial affects, thermal effects, pressure, shock, earthquake loading, and so on. Both static and dynamic loads must be estimated. Load intensifiers such as prying and eccentricity should be acknowledged if present. Joint diagrams can be used (see later section) to add external loads and preloading. The parts designed must be able to withstand *worst case combinations of these pre- and service loads.*

Figure 35.13 shows the many factors considered above: each as a function of the UTS of the bolt. The chart describes a hypothetical situation in which a code limit, reduced by a safety factor, was used to pick a desired upper limit on bolt stress of 62% of the UTS of the bolt. Also shown is the minimum clamping required to prevent leaking, vibration, or fatigue problems, again modified by a factor of safety. This suggests a desired lower limit on clamp force corresponding to a bolt tension of 48% of the UTS of the bolt. Remember that these loads *are in-service bolt loads.* Know also that when we preload a joint, we can do so only approximately because (1) when tightening one joint at a time loading one joint influences the load on those joints already preloaded, and (2) after preloading, stress relaxation occurs, reducing the load from the initial torqued value to something less. Both of these factors have a large influence on the total preload achieved in the joint.

In general, fasteners relax rapidly following initial tightening, then relax at a slower rate, as shown in Fig. 35.13.

Figure 35.14 shows the residual stresses in a group of 90 2¼-12 × 29 4330 studs that were tightened by stretching them 79% of their yield stress with a hydraulic tensioner. The studs and nuts were not new but had been tightened several times before these data were taken. Relaxation varied from 5% to 43% of the initial tension applied in these apparently identical studs. In many cases, similar scatter in relaxation also occurs after torquing.

Charts of this sort can be constructed on the basis of individual bolts or multibolt joints. Limits can be defined in terms of force, stress, yield instead of UTS, or even assembly torque.

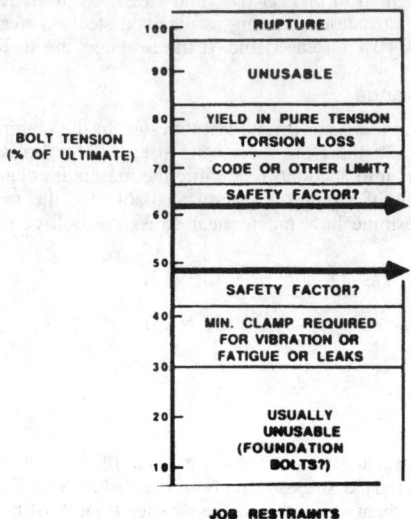

Fig. 35.13 Chart summarizing the design decisions made for a hypothetical joint.

35.8 THEORETICAL BEHAVIOR OF THE JOINT UNDER TENSILE LOADS

In this section, we will examine the way a joint responds when exposed to the external loads it has been designed to support. This will be done by examining the elastic behavior of the joint. When we tighten a bolt on a flange, the bolt is placed in tension; it gets longer. The joint compresses in the vicinity of the bolt.

We need to plot separate elastic curves for the bolt and joint members by plotting the absolute value of the force in each of the two vertical axes and the deformation of each (elongation in the bolt and compression in the joint) on the horizontal axes. See Fig. 35.16.

Three things should be noted.

1. Typically the slope (K_B) of the bolts elastic curve is only ⅓ to ⅕ of the slope (K_J) of the joints elastic curve; i.e., the stiffness of the bolt is only ⅓ to ⅕ that of the joint.

2. The clamping force exerted by the bolt on the joint is opposed by an equal and opposite force exerted by the joint members on the bolt. (The bolt wants to shrink back to its original length and the joint wants to expand to its original thickness.)

3. If we continue to tighten the bolt, it or the joint will ultimately start to yield plastically, as suggested by the dotted lines. In future diagrams, we will operate only in the elastic region of each curve.

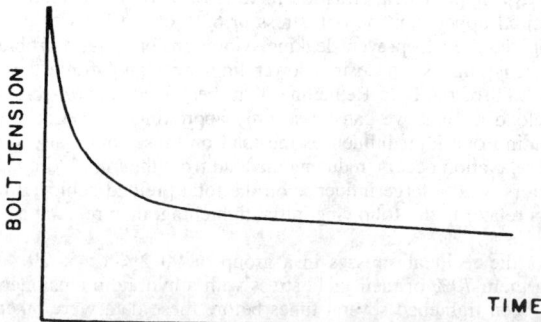

Fig. 35.14 Most short-term relaxation occurs in the first few seconds or minutes following initial tightening, but continues at a lesser rate for a long period of time.

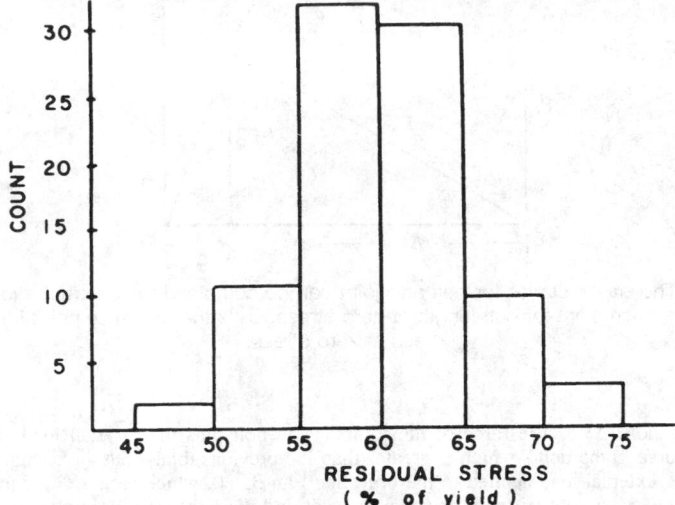

Fig. 35.15   Residual stress as a percentage of yield strength, following removal of tension. Studs were all tensioned to 79% of yield. Torqued to 500 lb–ft.

Rotscher first demonstrated what is called a *joint diagram* (Fig. 35.17).

In Fig. 35.17, the tensile force in the bolt is called the *preload* in the bolt and is equal and opposite to the compressive force in the joint.

If we apply an additional tension force to the bolt, this added load partially relieves the load on the joint, allowing it (if enough load is applied) to return to its original thickness while the bolt gets longer. Note that the increase in the length of the bolt is equal to the increase in thickness (reduction in compression) in the joint. In other words, the *joint expands to follow the nut as the bolt lengthens.*

Because the stiffness of the bolt is only $\frac{1}{3}$ to $\frac{1}{5}$ that of the joint, for an equal change in the strain, the change in load in the bolt must be only $\frac{1}{3}$ to $\frac{1}{5}$ of the change in the load in the joint. This is shown in Fig. 35.18.

The external tension load (L_X) required to produce this change of force and strain in the bolt and joint members is equal to the increase in the force on the bolt (ΔF_B) plus the reduction of force in the joint (ΔF_J):

$$L_X = \Delta F_B + \Delta F_J$$

The above relationship is demonstrated in Fig. 35.19.

Any external tension load, no matter how small, will be partially absorbed in replacing the force in the bolt (ΔF_B), and partially absorbed in replacing the reduction of force that the joint originally

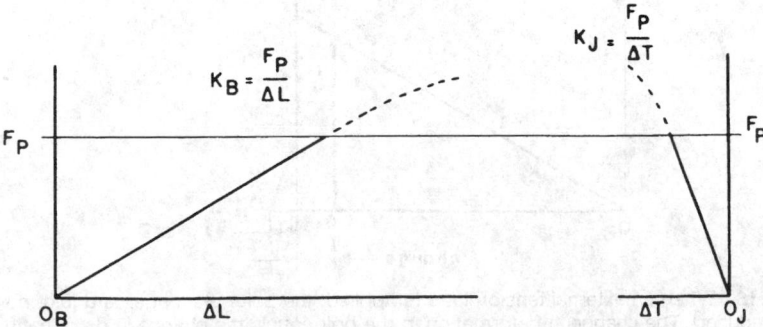

Fig. 35.16   Elastic curves for bolt and joint members.

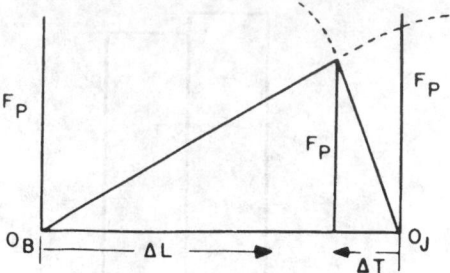

Fig. 35.17 The elastic curves for bolt and joint can be combined to construct a joint diagram. O_B is the reference point for bolt length at zero stress. O_J is the reference point for joint thickness at zero stress.

exerted on the bolt (ΔF_J). The force of the joint on the bolt plus the external load equal the new total tension force in the bolt—which is greater than the previous total—but the change in bolt force is less than the external load applied to the bolt. See Fig. 35.20, which recaps all of this.

We can change the joint stiffness between the bolt and the joint by making the bolt much stiffer (i.e., a bolt with a larger diameter). The new joint diagram resulting from this change is shown in Fig. 35.21.

Note that the bolt now absorbs a larger percentage of the same external load. For another example, see Fig. 35.22 for a softer bolt (less stiff). The joint will see a smaller percentage of a given load.

That the bolt sees only a part of the external load, and that the amount it sees is dependent on the stiffness ratio, between the bolt and the joint, have many implications for joint design, joint failure, measurement of residual preloads, and so on.

35.8.1 Critical External Load

If we keep adding external load to the original joint, we reach a point where the joint members are fully unloaded, as in Fig. 35.23. This is the critical external load, which is not equal to the original preload in the bolt but is often equal to the preload for several reasons.

1. In many joints, the bolt has a low spring rate compared to the joint members. Under these conditions, there is a very small difference between the preload in the bolt and the critical external load that frees the joint.

2. Joints almost always relax after first tightening with relaxations of 10–20% of the initial preload being not uncommon. If a bolt has $\frac{1}{5}$ the stiffness of the joint, then the critical external load to free the joint members is 20% greater than the residual preload in the bolt when the load is applied. Therefore, the difference between the critical external load and the present preload is just about equal and opposite to the loss in preload caused by bolt relax-

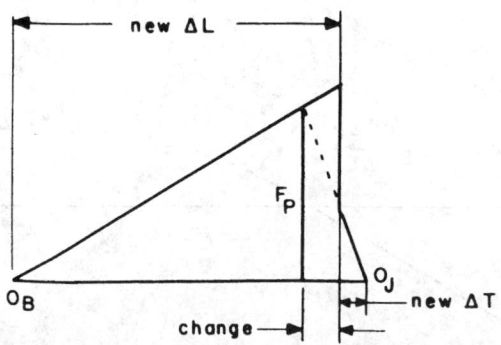

Fig. 35.18 When an external tension load is applied, the bolt gets longer and joint compression is reduced. The change in deformation in the bolt equals the change in deformation in the joint.

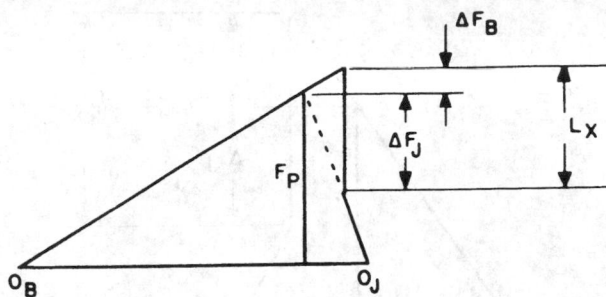

Fig. 35.19 Because the bolt and joint have different stiffness, equal changes in deformation mean an unequal change in force. ΔF_B is the increase in bolt force; ΔF_J is the decrease in clamping force in the joint. L_X is the external load.

ation. In other words, the critical external load equals the original preload before bolt relaxation.

35.8.2 Very Large External Loads

Any additional external load we add beyond the critical point will all be absorbed by the bolt. Although it is usually ignored in joint calculations, there is another curve we should be aware of. The compressive spring rate of many joint members is not a constant. A more accurate joint diagram would show this. See Fig. 35.24.

For joint diagrams, as shown in Fig. 35.19, we can make the following calculations where

$$\begin{aligned}
F_P &= \text{initial preload (lb, } N) \\
L_X &= \text{external tension load (lb, } N) \\
\Delta F_B &= \text{change in load in bolt (lb, } N) \\
\Delta F_J &= \text{change in load in joint (lb, } N)
\end{aligned}$$

$\Delta L, \Delta L' =$ elongation of the bolt before and after application of the external load (in., mm)

$\Delta T, \Delta T' =$ compression of joint members before and after application of the external load (in., mm)

$L_{\text{xcritical}} =$ external load required to completely unload the joint (lb, N) (not shown in diagram)

The stiffness (spring constants) of the bolt and joint are defined as follows:

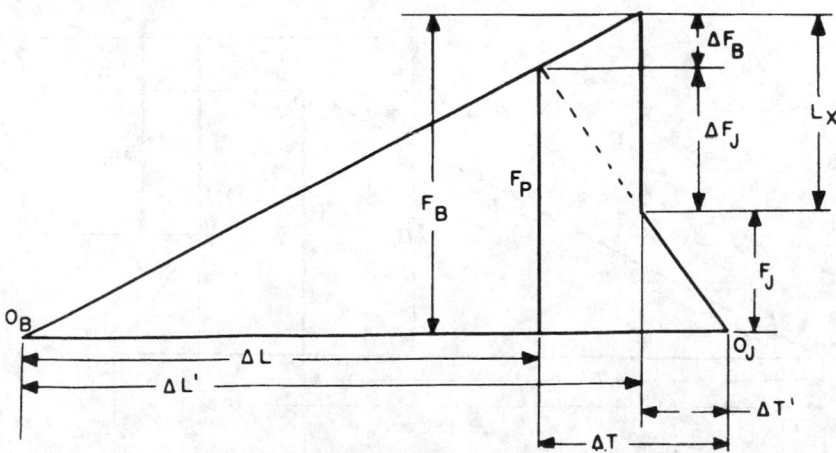

Fig. 35.20 Summary diagram. $F_P =$ initial preload; $F_B =$ present bolt load; $F_J =$ present joint load; $L_X =$ external tension load applied to the bolt.

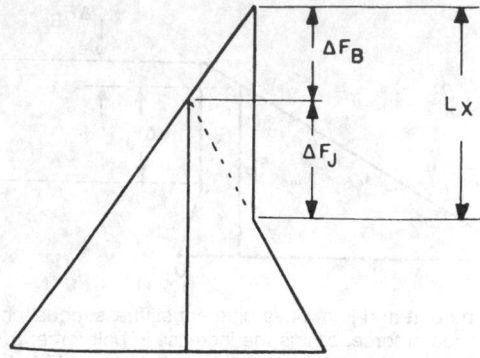

Fig. 35.21 Joint diagram when stiffness of the bolt nearly equals that of joint.

$$\text{for the bolt } K_B = \frac{F_P}{\Delta L} \text{ for the joint } K_J = \frac{F_P}{\Delta T}$$

$$\text{by manipulation } \Delta F_B = \frac{(K_B)}{(K_B + K_J)} \times L_X \text{ until joint separation, after which}$$

$$\Delta F_B = \Delta L_x \text{ and } L_{x \text{ critical}} = F_P \left\{ 1 + \frac{K_B}{K_J} \right\}$$

Different diagrams must be used if the external load is applied at different planes; but the bolt loads and critical load computed for these situations are only a little different than those calculated from the above diagrams and are often used in joint design and analysis even though they are based on external load applications points that will probably never be encountered in practice. (Remember that we have assumed that the external loads are applied to both ends of the bolt.)

An accurate description of loads created by pressure, weight, shock, inertia, and so on are transferred to the joint by the connected members. This description requires a detailed stress analysis (finite element). A far simpler way to place the load is to define "loading planes" parallel to the joint interface and located somewhere between the outer and contact surfaces of each joint member. Designers then assume that the tensile load on the joint is applied to those loading planes. Joint material between the loading planes will then be unloaded by a tensile load. Joint material outboard of the planes will be trapped between plane of application of the load and the head or nut of the

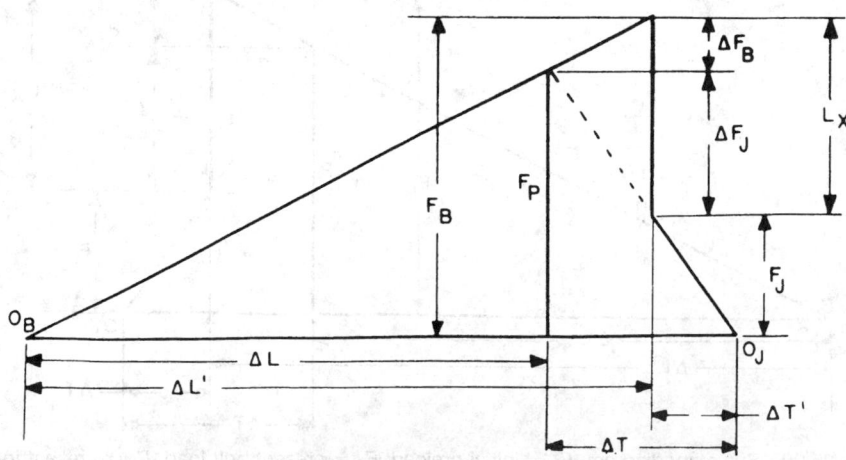

Fig. 35.22 Joint diagram with softer bolt and stiff joint.

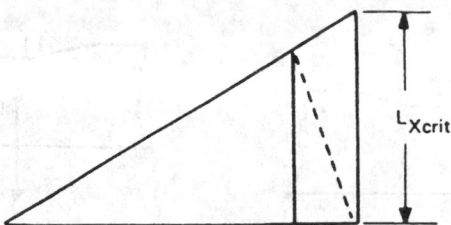

Fig. 35.23 A critical external load (L_{xcrit}) fully unloads the joint (but not the bolt).

fastener. Therefore, loading planes coincide with the interface between upper and lower joint members. To illustrate this, see Fig. 35.25, where (*a*) represents the original assumption and (*b*) the new assumption.

If the same tensile force, however, were applied *to the interface* between the upper and lower joint members, both the bolt and the joint (in compression) would be loaded by the external load. In this case, we are relieving these flange-on-flange forces rather than adding to them. In the joint, this means that the external load reduces the flange-on-flange force without increasing the total force in either of the flange members or the bolt. The joint diagram for the above situation is shown in Fig. 35.26.

In the above case, both elastic curves (bolt and joint) are drawn on the same side of the common vertical axis that represents the original preload or F_P. This is done because both springs are loaded by the external force.

When the external load equals the original preload in the bolt, it will have replaced all of the force that each joint member was exerting on the other. Note that neither bolt deformation nor joint deformation has changed to this point.

Increasing the external load beyond this point now adds to the original deformation of both the bolt and the joint members. That is, the bolt gets longer and the joint compresses more. The joint diagram merely gets larger. See Fig. 35.27 (dashed lines represent the original joint diagram; the solid lines represent the new joint diagram). Note that at all times, both the bolt and the joint see the same total load.

The mathematics of a tension load at the interface are very simple and can be determined by inspection of the joint diagram above. The change in the bolt force is

$$\Delta F_B = 0$$

until the external load exceeds the preload (F_P), after which

$$\Delta F_B = L_X - F_P$$

or further ΔF_B = further ΔL_X

The critical external load required to cause joint separation is

$$L_{x \text{ critical}} = F_P$$

and this is true regardless of the spring constants, or spring constant ratios of the bolt and joint members.

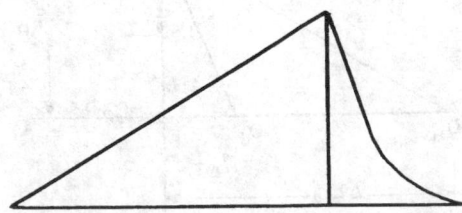

Fig. 35.24 The spring rate of the joint is frequently nonlinear for small deflections.

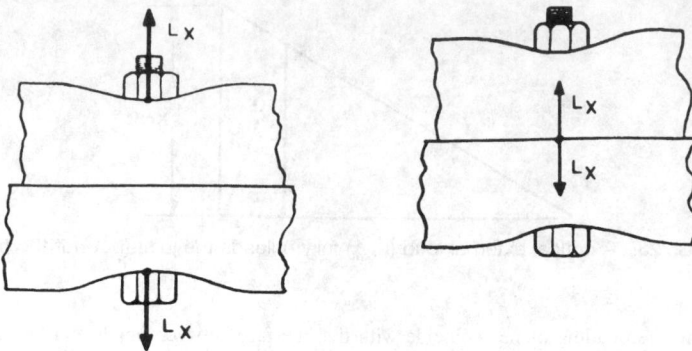

Fig. 35.25 The external tension load is applied at the joint interface.

For our first joint diagrams, the keys were that the change in elongation of the bolt under an external load equaled the change in compression of the joint, but the changes in force were unequal. In the present case, where the tension load is applied at the interface, the deflections are not equal, but the force in the bolt is always equal to the forces in the flange. This is the inverse of the first joint diagrams.

The mathematics above tells us that there will be no change in the force seen by the bolt when tension loads are applied at the interface *until the external load exceeds the original preload in the bolt*. On the other hand, interface loading gives us a critical external load (i.e., joint separation) that is equal to the preload. This is less than the load required for separation when the tension loads are applied at joint surface. The load capacity of the interface joint is *less* than the load capacity of the original joint.

Maximum bolt load, working change in bolt load, and the critical external load are important design factors. They are different for different loading planes; hence the importance of loading planes in our calculations. Loading planes that are in between the two situations that we have looked at will result in "in-between" values for all factors of interest. Designers can assume that the true values lie somewhere between the two conditions we have shown (depending on the position of the actual loading plane)—which, by the way, represent the limiting conditions; that is, head-to-nut and joint interface. Assuming the "worst case" scenario, which is given in Fig. 35.23 (or even a loading plane halfway to the interface) can be so conservative as to seriously affect the assumptions about the amount of change in bolt load actually created by an external load. Experimental data on "compact" and other pressure vessel and piping flanges report that the changes in bolt load in such rigid joints are often only about one-tenth of the change that would be predicted by Fig. 35.18 and the following equation:

$$\Delta F_B = \frac{(K_B)}{(K_B + K_J)} \times L_X$$

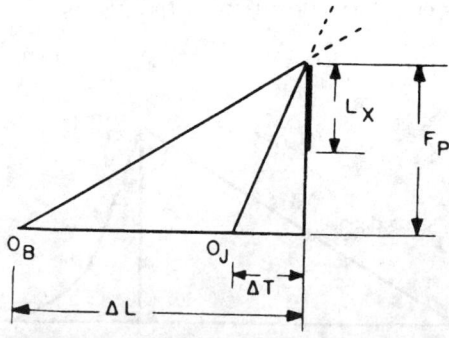

Fig. 35.26 Joint diagram when external tension load is applied at joint interface. ΔL elongation of bolt; ΔT compression of joint; F_P = original preload.

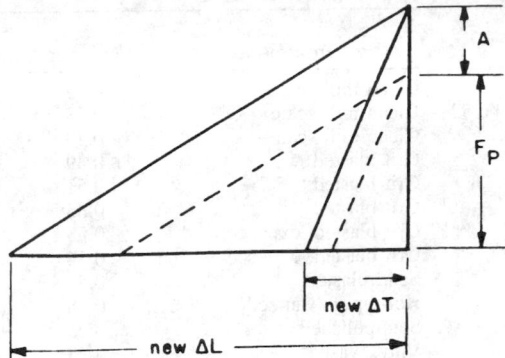

Fig. 35.27 The external load applied to the joint interface has exceeded the critical load by an amount = A.

This suggests that a joint designed to the above equation might have larger and/or more numerous bolts than necessary to support pressure loads the bolts will never see. The ASME Boiler and Pressure Vessel Code takes an even more conservative point of view than that described by the above equation to introduce a factor of safety. This code assumes that the bolts see 100% of external load L_X, not an amount reduced by the stiffness ratio.

35.9 EVALUATION OF SLIP CHARACTERISTICS

A slip-resistant joint is one that has a low probability of slip at any time during the life of the structure. In this type of joint, the external applied load usually acts in a plane perpendicular to the bolt axis. The load is completely transmitted by frictional forces acting on the contact area of the plates fastened by the bolts. This frictional resistance is dependent on (1) the bolt preload and (2) the slip resistance of the fraying surfaces.

Slip-resistant joints are often used in connections subjected to stress reversals, severe stress fluctuations, or in any situation wherein slippage of the structure into a "bearing mode" would produce intolerable geometric changes. A slip load of a simple tension splice is given by

$$P_{\text{slip}} = k_s m \sum_{i=1}^{n} T_i$$

where

$$k_s = \text{slip coefficient}$$

$$m = \text{number of slip planes}$$

$$\sum_{i=1}^{n} T_i = \text{the sum of the bolt tensions}$$

If the bolt tension is equal in all bolts, then

$$P_{\text{slip}} = k_s \, m \, n \, T_i$$

where

$$n = \text{the number of bolts in the joint}$$

The slip coefficient K_s varies from joint to joint, depending on the type of steel, different surface treatments, and different surface conditions, and along with the clamping force T_i shows considerable variation from its mean value. The slip coefficient K_s can only be determined experimentally, but some values are now available, as shown in Table 35.1.

35.10 INSTALLATION OF HIGH-STRENGTH BOLTS

Prior to 1985, North American practice had been to require that all high-strength bolts be installed and provide a high level of preload, regardless whether or not it was needed. The advantages in such an arrangement were that a standard bolt installation procedure was provided for all types of con-

Table 35.1 Summary of Slip Coefficients

Type of Steel	Treatment	Average	Standard Deviation	Number of Tests
A7, A36, A440	Clean mill scale	0.32	0.06	180
A7, A36, A440, Fe37, Fe.52	Clean mill scale	0.33	0.07	327
A 588	Clean mill scale	0.23	0.03	31
Fe 37	Grit blasted	0.49	0.07	167
A36, Fe37, Fe52	Grit blasted	0.51	0.09	186
A514	Grit blasted	0.33	0.04	17
A36, Fe37	Grit blasted, exposed	0.53	0.06	51
A36, Fe37, Fe52	Grit blasted, exposed	0.54	0.06	83
A7, A36, A514, A572	Sand blasted	0.52	0.09	106
A36, Fe37	Hot-dip galvanized	0.18	0.04	27
A7, A36	Semipolished	0.28	0.04	12
A36	Vinyl wash	0.28	0.02	15
	Cold zinc plated	0.30	–	3
	Metallized	0.48	–	2
	Galvanized and sand blasted	0.34	–	1
	Sand blasted treated with linseed oil (exposed)	0.26	0.01	3
	Red lead paint	0.06	–	6

nections and that a slightly stiffer structure probably resulted. Obviously, when a slip-resistant bolted structure was not needed, the disadvantages were the additional cost and inspection time for this type of installation. Since 1985, only fasteners that are to be used in slip-critical connections or in connections subject to direct tension loading have needed to be preloaded to the original preload, equal to 70% of the minimum specified tensile strength of the bolt. Bolts to be used in bearing-type connections only need to be tightened to the snug-tight condition.

When the high-strength bolt was first introduced, installation was primarily by methods of torque control. Approximate torque values were suggested, but tests performed and field experience confirmed the great variability of the torque–tension relationship, as much as ±30% from the mean tension desired. This variance is caused mainly by the variability of the thread conditions, surface conditions under the nut, lubrication, and other factors that cause energy dissipation without inducing tension in the bolt.

For a period of five years, the calibrated wrench method was banned in favor of turn-of-nut method or by use of direct tension indicators that depend on strain or displacement control versus torque control. However, in 1985, the RCSC (Research Council on Riveted and Bolted Structural Joints of the Engineering Foundation) specification again permitted the use of the calibrated wrench method, but with a clearer statement of the requirements of the method and its limitations.

The calibrated wrench method still has a number of drawbacks. Because the method is essentially one of torque control, factors such as friction between the nut and bolt threads and between the nut and washer are of major importance, as well as the type of lubricant used and the method of application, presence of dirt. These problems are not reflected in the calibration procedures.

To overcome the variability of torque control, efforts were made to develop a more reliable tightening procedure and testing began on the turn-of-nut method. (This is a strain-control method.) Initially it was believed that one turn from the snug position was the key, but because of out-of-flatness, thread imperfections, and dirt accumulation, it was difficult to determine the hand-tight position (the starting point—from the snug position). Many believe that turn control is better than torque control, but this is not true. In fact pure turn control is no more accurate than pure torque control. Current practice is as follows: run the nut up to a snug position using an impact wrench rather than the finger-tight condition (elongations are still within the elastic range). From the snug position, turn the nut in accordance with Table 35.2, provided by the RCSC specification.

Nut rotation is relative to bolt, regardless of the element (nut or bolt) being turned. For bolts installed by ⅔ turn and less, the tolerance should be ±30°; for bolts installed by ⅔ turn and more, the tolerance should be ±45°. All material within the grip of the bolt must be steel.

No research work has been performed by the council to establish the turn-of-nut procedure when bolt length exceeds 12 diameters. Therefore, the required rotation must be determined by actual tests in a suitable tension device simulating the actual conditions.

A325 bolts can be reused once or twice, providing that proper control on the number of reuses can be established. For A490 bolts, reuse is not recommended.

Washers are not required for A325 bolts because the galling in bolts that are tightened directly against the connected parts is not detrimental to the static or fatigue strength of the joint. If bolts are

Table 35.2 Nut Rotation from Snug-Tight Condition

Bolt Length (as measured from underside of head to extreme end of point)	Both Faces Normal to Bolt Axis	One Face Normal to Bolt Axis and Other Face Sloped Not More Than 1:20 (bevel washer not used)	Both Faces Sloped Not More Than 1:20 from Normal to Bolt Axis (bevel washers not used)
Up to and including 4 diameters	⅓ turn	½ turn	⅔ turn
Over 4 diameters but not exceeding 8 diameters	½ turn	⅔ turn	⅚ turn
Over 8 diameters but not exceeding 12 diameters	⅔ turn	⅚ turn	1 turn

tightened by the calibrated wrench method, a washer should be used under the turned element—that is, the nut or the bolt head. For A490 bolts, washers are required under both the head and nut when they are used to connect material with a yield point of less than 40 ksi. This prevents galling and brinelling of the connected parts. For higher strength steel assembled using high-strength bolts (higher than 40 ksi yield point), washers are only required to prevent galling of the turned element.

When bolts pass through a sloping interface greater than 1:20, a beveled washer is required to compensate for the lack of parallelism. As noted in Table 35.2, bolts require additional nut rotation to ensure that tightening will achieve the required minimum preload.

35.11 TORQUE AND TURN TOGETHER

Measuring of torque and turn at the same time can improve our control over preload. The final variation in preload in a large number of bolts is closer to ±5% than the 25–30% if we used torque or turn control alone. For this reason the torque–turn method is widely used today, especially in structural steel applications.

In this procedure, the nut is first snugged with a torque that is expected to stretch the fastener to a minimum of 75% of its ultimate strength. The nut is then turned (half a turn) or the like, which stretches the bolt well past its yield point. See Fig. 35.28.

This torque–turn method cannot be used on brittle bolts, but only on ductile bolts having long plastic regions. Therefore, it is limited to A325 fasteners used in structural steel work. Furthermore, it should never be used unless you can predict the working loads that the bolt will see in service. Anything that loads the bolts above the original tension will create additional plastic deformation in the bolt. If the overloads are high enough, the bolt will break.

A number of knowledgeable companies have developed manual torque–turn procedures that they call "turn of the nut" but that do not involve tightening the fasteners past the yield point. Experience shows that some of these systems provide additional accuracy over turn or torque alone.

Other methods have also been developed to control the amount of tension produced in bolts during assembly, namely *stretch* and *tension control*.[1] All of these methods have drawbacks and limitations, but each is good enough for many applications. However, in more and more applications,

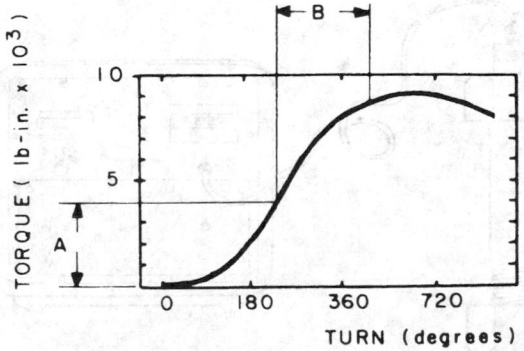

Fig. 35.28 In turn-of-nut techniques, the nut is first tightened with an approximate torque (A) and then further tightened with a measured turn (B).

we need to find a better way to control bolt tension and/or clamping forces. Fortunately, that better way is emerging, namely *ultrasonic measurement of bolt stretch or tension.*

35.12 ULTRASONIC MEASUREMENT OF BOLT STRETCH OR TENSION

Ultrasonic techniques, while not in common use, allow us to get past dozens of the variables that affect the results we achieve with torque and/or torque and turn control.

The basic concepts are simple. The two most common systems are *pulse-echo* and *transit time* instruments. In both, a small acoustic transducer is placed against one end of the bolt being tested. See Fig. 35.29. An electronic instrument delivers a voltage pulse to the transducer, which emits a very brief burst of ultrasound that passes down the bolt, echoes off the far end, and returns to the transducer. An electronic instrument measures precisely the amount of time required for the sound to make its round trip in the bolt.

As the bolt is tightened, the amount of time required for the round trip increases for two reasons:

1. The bolt stretches as it is tightened, so the path length increases.
2. The average velocity of sound within the bolt decreases because the average stress level in the bolt has increased.

Both of these changes are linear functions of the preload in the fastener, so that the total change in transit time is also a linear function of preload.

The instrument is designed to measure the change in transit time that occurs during tightening and to report the results as

1. A change in length of the fastener
2. A change in the stress level within the threaded region of the fastener
3. A change in tension within the fastener

Using such an instrument is relatively easy. A drop of coupling fluid is placed on one end of the fastener to reduce the acoustic impedance between the transducer and the bolt. The transducer is placed on the puddle of fluid and held against the bolt, mechanically or magnetically. The instrument is zeroed for this particular bolt (because each bolt will have a slightly different acoustic length). If you wish to measure residual preload, or relaxation, or external loads at some later date, you record the length of the fastener at zero load at this time. Next the bolt is tightened. If the transducer can remain in place during tightening, the instrument will show you the buildup of stretch or tension in the bolt. If it must be removed, it is placed on the bolt after tightening to show the results achieved by torque, turn, or tension.

If, at some later date, you wish to measure the present tension, you dial in the original length of *that* bolt into the instrument and place the transducer back on the bolt. The instrument will then show you the difference in length or stress that now exists in the bolt.

Because ultrasonic equipment is not in common use at this time, it is used primarily in applications involving relatively few bolts in critically important joints or quality control audits. Operator training in the use of this equipment is necessary and is a low-cost alternative to strain-gaged bolts in all sorts of studies.

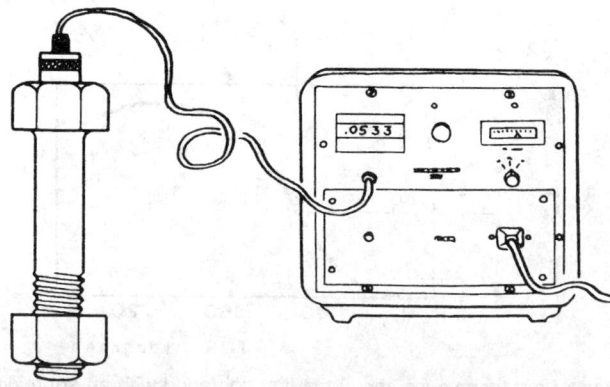

Fig. 35.29 An acoustic transducer is held against one end of the fastener to measure the fastener's change in length as it is tightened.

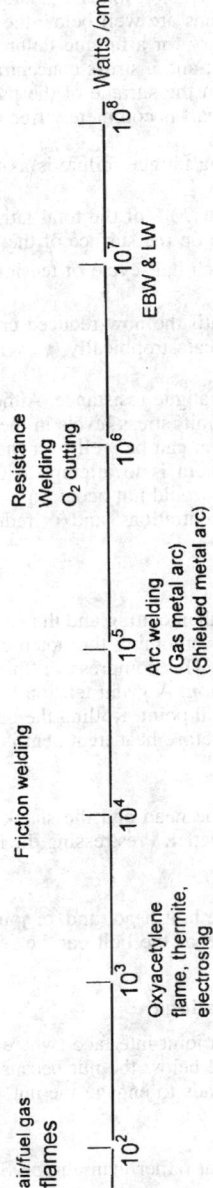

Fig. 35.30 Spectrum of practical heat intensities used for fusion welding.

These instruments are new to the field, so you must be certain to find out from the manufacturers exactly what the equipment will or will not do as well as precise information needed for use or equipment calibration. Training is essential not only for the person ordering the equipment, but for all who will use it in the field or laboratory. Proper calibration is essential. If the equipment can only measure transit time, you must tell it how to interpret transit time data *for your application.*

35.13   FATIGUE FAILURE AND DESIGN FOR CYCLICAL TENSION LOADS

A fastener subjected to repeated cyclical tension loads can suddenly break. These failures are generally catastrophic in kind, even if the loads are well below the yield strength of the material.

Three essential conditions are necessary for a fatigue failure: cyclical tensile loads; stress levels above the endurance limit of the material; and a stress concentration region (such as a sharp corner, a hole, a surface scratch or other mark on the surface of the part, corrosion pits, an inclusion and/ or a flaw in the material). Essentially no part is completely free of these types of defects unless great care has been taken to remove them.

The sequence of events leading up to a fatigue failure is as follows:

1. Crack inititation begins after about 90% of the total fatigue life (total number of cycles) has occurred. This crack always starts on the surface of the part.

2. The crack begins to grow with each half-cycle of tension stress, leaving beach marks on the part.

3. Growth of the crack continues until the now-reduced cross section is unable to support the load, at which time the part fails catastrophically (very rapidly).

A bolt is a very poor shape for good fatigue resistance. Although the average stress levels in the body may be well below the endurance limit, stress levels in the stress concentration points, such as thread roots, head to body fillets, and so on can be well over the endurance limit. One thing we can do to reduce or eliminate a fatigue problem is to attempt to overcome one or more of the three essential conditions without which failure would not occur. In general, most of the steps are intended to reduce stress levels, reduce stress concentrations, and/or reduce the load excursions seen by the bolt.

35.13.1   Rolled Threads

Rolling provides a smoother thread finish than cutting and thus lowers the stress concentrations found at the root of the thread. In addition to overcoming the notch effect of cut threads, rolling induces compressive stresses on the surface rolled. This compressive "preload" must be overcome by tension forces before the roots will be in net tension. A given tension load on the bolt, therefore, will result in a smaller tension excursion at this critical point. Rolling the threads is best done after heat treating the bolt, but it is more difficult. Rolling before heat treatment is possible on larger-diameter bolts.

35.13.2   Fillets

Use bolts with generous fillets between the head and the shank. An elliptical fillet is better than a circular one and the larger the radius the better. Prestressing the fillet is wise (akin to thread rolling).

35.13.3   Perpendicularity

If the face of the nut, the underside of the bolt head, and/or joint surfaces are not perpendicular to thread axes and bolt holes, the fatigue life of the bolt can be seriously affected. For example, a 2° error reduces the fatigue life by 79%.[3]

35.13.4   Overlapping Stress Concentrations

Thread run-out should not coincide with a joint interface (where shear loads exist) and there should be at least two full bolt threads above and below the nut because bolts normally see stress concentrations at (1) thread run-out; (2) first threads to engage the nut, and head-to-shank fillets.

35.13.5   Thread Run-Out

The run-out of the thread should be gradual rather than abrupt. Some people suggest a maximum of 15° to minimize stress concentrations.

35.13.6   Thread Stress Distribution

Most of the tension in a conventional bolt is supported by the first two or three nut threads. Anything that increases the number of active threads will reduce the stress concentration and increase the fatigue life. Some of the possibilities are

1. Using so-called "tension nuts," which create nearly uniform stress in all threads.

2. Modifying the nut pitch so that it is slightly different than the pitch of the bolt, i.e., thread of nut 11.85 threads/in. used with a bolt having 12 threads/in.

3. Using a nut slightly softer than the bolt (this is the usual case); however, select still softer nuts if you can stand the loss in proof load capability.

4. Using a jam nut, which improves thread stress distribution by preloading the threads in a direction opposite to that of the final load.

5. Tapering the threads slightly. This can distribute the stresses more uniformly and increase the fatigue life. The taper is 15°.

35.13.7 Bending

Reduce bending by using a spherical washer because nut angularity hurts fatigue life.

35.13.8 Corrosion

Anything that can be done to reduce corrosion will reduce the possibilities of crack initiation and/or crack growth and will extend fatigue life. Corrosion can be more rapid at points of high stress concentration, which is also the point where fatigue failure is most prevalent. Fatigue and corrosion aid each other and it is difficult to tell which mechanism initiated or resulted in a failure.

35.13.9 Surface Conditions

Any surface treatment that reduces the number and size of incipient cracks will improve fatigue life significantly, so that polishing of the surface will greatly improve the fatigue life of a part. This is particularly important for punched or drilled holes, which can be improved by reaming and expanding to put the surface in residual compression. Shot peening of bolts or any surface smooths out sharp discontinuites and puts the surface in residual compression. Handling of bolts in such a way as not to ding one against the other is also important.

35.13.10 Reduce Load Excursions

It is necessary to identify the maximum safe preload that your joint can stand by estimating fastener strength, joint strength, and external loads. Also do whatever is required to minimize the bolt-to-joint stiffness ratio so that most of the excursion and external load will be seen by the joint and not the bolt. Use long, thin bolts even if it means using more bolts. Eliminate gaskets and/or use stiffer gaskets.

While there are methods available for estimating the endurance limit of a bolt, it is best to base your calculations on actual fatigue tests of the products you are going to use or your own experience with those products.

For the design criteria for fatigue loading of slip resistant joints, see Refs. 1 and 2.

35.14 WELDED JOINTS

In industry, welding is the most widely used and cost-effective means for joining sections of metals to produce an assembly that will perform as if made from a single solid piece.

A perfect joint is indistinguishable from the material surrounding it, but a perfect joint is indeed a very rare case. Diffusion bonding can achieve results that are close to this ideal, but are either expensive or restricted to use on just a few materials. There is no universal process that performs adequately on all materials in all geometries. Nevertheless, any material can be joined in some way, although joint properties equal to those of the bulk material cannot always be achieved.

Generally, any two solids will bond if their surfaces are flat enough that atom-to-atom contact can be made. Two factors exist to make this currently impossible.

1. Even the most carefully machined, polished, and lapped surfaces have random hills and valleys differing in elevation by 100–1000 atomic diameters.

2. Any fresh surface is immediately contaminated by formation of a nonmetallic film a few atomic diameters thick, consisting of a brittle oxide layer, a water vapor layer, a layer of absorbed CO_2, and hydrocarbons, which forms in about 10^{-3} seconds after cleaning.

If large enough compressive forces were applied to the surfaces, the underlying aspirates (regions where two hills, one on each surface, meet) would flow plastically, fragmenting the intermediate, brittle oxide layer. On increasing the compressive force, isolated regions of metal-to-metal contact would occur, separated by volumes of accumulated debris from the oxide and absorbed-moisture films. Upon release of the compressive load, the isolated regions of coalescence would be ruptured by the action of the compressive residuals in unbonded areas. In diffusion bonding, the compressive forces are maintained while heating the material very near to its melting temperature, causing the aspirates to grow by means of recrystallization and grain growth. But this still leaves regions where the fragmented oxides remain, thus reducing the overall bonded joint length.

In order to produce a satisfactory metallic bond between two metal objects, it is first necessary to dissipate all nonmetallic films from the interface. In fusion welding, intimate interfacial contact is achieved by placing a liquid metal, of essentially the same composition as the base metal, between the two solid pieces. If the surface contamination is soluble, it is dissolved in the liquid; if not then it will float away from the liquid solid interface. While floating away the oxide is an attractive procedure, it does not preclude cleaning all surfaces to be welded as well as you possibly can before applying the heat source to the joint to be welded.

One distinguishing feature of all fusion welding processes is the intensity of the heat source used to melt the solid into a liquid. It is generally found that heat source power densities of approximately 1000 watts/cm^2 are necessary to melt most metals.

At the high end of the power densities, heat intensities of 10^6 or 10^7 watts/cm^2 will vaporize most materials within a few microseconds and all of the solid that interacts with the heat source will vaporize. Around the hole thus created, a molten pool is developed that will flow into the hole once the beam has moved ahead, allowing the weld to be made. This is the case for electron-beam and laser welding. Power densities of the order of 10^3 watts/cm^2, such as oxyacetylene or electroslag welding, require interaction times of 25 seconds with steel. This is why welders begin their training with the oxyacetylene process. It is inherently slow and does not require a rapid response from the new welder in order to control the molten puddle. Much greater skill is needed to control the arc in the faster arc processes.

The selection of materials for welded construction applications involves a number of considerations, including design codes and specifications, where they exist. In every design situation, economics—choosing the correct material for the life cycle of the part and its cost of fabrication—is of prime importance. Design codes or experience frequently offer an adequate basis for material selection. For new or specialized applications, the engineer encounters problems of an unusual nature and thus must rely on basic properties of the material, such as strength, corrosion or erosion resistance, ductility, and toughness. Welding processes may be significant in meeting the design goals.

The processes that are most frequently used in the welding of large structures are normally limited to four or five fusion welding methods. These methods will be discussed starting from the most automatic, cheapest method progressing to semiautomatic and finally to those methods that are manual only.

35.14.1 Submerged Arc Welding (SAW)

This method is the workhorse of heavy metal fabrication and used as a semi-automatic or fully automatic operation, although most installations are fully automatic. Its cost per unit length of weld is the lowest of all the processes, but it has the disadvantage of operating only in the downhand position. Thus, it requires manipulation of the parts into positions where welding can be accomplished in the horizontal position. It is suitable for shop welding, but not field welding.

Heat is provided by an arc between a bare solid metal consumable electrode and the workpiece. The arc is maintained in a cavity of molten flux or slag, which refines the weld deposit and protects it from atmospheric contamination. Alloy ingredients in the flux may be present to enhance the mechanical properties and crack resistance of the weld deposit. See Fig. 33.31.

A layer of granular flux, deep enough to prevent flash-through, is deposited in front of the arc. The electrode wire is fed through a contact tube. The current can be ac, dc reverse, or straight polarity. The figure shows the melting and solidification sequence. After welding, the unfused slag and flux may be collected, crushed, and blended back into the new flux. To increase the deposition and welding rate, more than one wire (one in front of the other) can be fed simultaneously into the same weld pool. Each electrode has its own power supply and contact tip. Two, three, or even four wire feeds are frequently used.

Advantages of the Process

1. The arc, which is under a blanket of flux, eliminates arc flash, spatter, and fumes. This is attractive from an environmental point of view.
2. High current densities increase penetration and decrease the need for edge preparation.
3. High deposition rates and welding speeds are possible.
4. Cost per unit length of weld is low.
5. The flux deoxidizes contaminates such as O_2, N_2, and sulfur.
6. Low hydrogen welds can be produced.
7. The shielding provided by the flux is substantial and not sensitive to wind, and UV light emissions are low.
8. The training requirements are lower than for other welding procedures.
9. The slag can be collected, reground, and sized back into new flux.

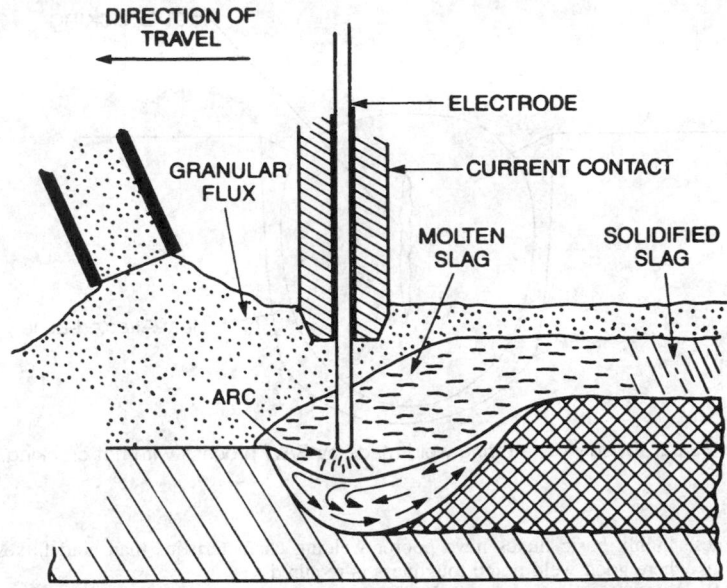

Fig. 35.31 Diagrammatic sketch of the submerged arc welding process (SAW). Sketch illustrates electrode deposition on a thick plate, Arrows drawn on weld pool show the usual hydrodynamic motion of the molten metal.

Disadvantages or Limitations of the Process

1. Initial cost of all equipment required is high.
2. Must be welded in the flat or horizontal position.
3. The slag must be removed between passes.
4. Most commonly used to join steels ¼ inch thick or greater.

This process is most commonly used to join plain carbon steels and low alloy steels, but alloy steels can be welded if care is taken to limit the heat input as required to prevent grain coarsening in the heat-affected zone (HAZ). It can also be used to weld stainless steels and nonferrous alloys or to provide overlays on the top of a base metal. To prevent porosity, the surface to be welded should be clean and free of all grease, oil, paints, moisture, and oxides.

Because SAW is used to join thick steel sections, it is primarily used for shipbuilding, pipe fabrication, pressure vessels, and structural components for bridges and buildings. It is also used to overlay, with stainless steel or wear-resistant steel, such things as rolls for continuous casting, pressure vessels, rail car wheels, and equipment for mining, mineral processing, construction, and agriculture.

Power sources consist of a dc constant voltage power supply that is self-regulating, so it can be used with a constant-speed wire feeder. No voltage or current sensing is necessary. The current is controlled by the wire diameter, the amount of stick-out, and the wire speed feed. Constant current ac machines can also be used, but require voltage-sensing variable wire speed controls. On newer solid state power supplies, the current and voltage outputs both approximate square waves, with instantaneous polarity reversal reducing arc initiation problems.

Fluxes interact with the molten steel in very similar ways to those in open-hearth refining of steel. These processes need to be understood for the best selection of the flux depending on the material being welded. For this chapter it suffices to say that acid fluxes are typically preferred for single-pass SAW welding because of their superior operating and bead wetting characteristics. In addition, these fluxes have more resistance to porosity caused by oil contamination of the material to be welded, rust, and mill scale.

Basic fluxes tend to give better impact properties, and this is evident on large multipass welds. Highly basic (see Boniszewski basicity index) fluxes produce weld metals with very good impact

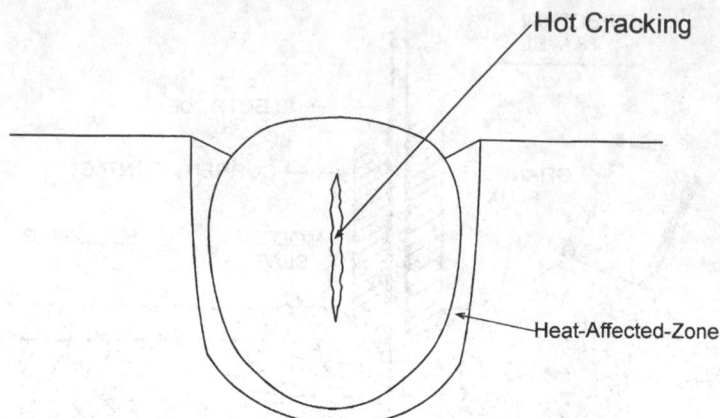

Fig. 35.32 Deeply penetrating weld made by SAW process with hot cracking.

properties. These highly basic fluxes have poorer welding characteristics than acid fluxes and are limited to cases where good weld notch toughness is required.

While SAW is the most inexpensive and efficient process for making large, long, and repetitive welds, much time is required to prepare the joint. Care must be taken to line up all joints for a consistent gap in groove welds and to provide backing plates and flux dams to prevent the spillage of molten metal and/or flux. Once all the pieces are clamped or tacked in place, welding procedures and specifications need consultation before welding begins.

The fact that SAW is a high heat input process, under a protective blanket of flux, greatly decreases the chance of weld defects. However, defects such as lack of fusion, slag entrapment, solidification cracking, and hydrogen cracking occasionally occur. See Figs. 35.32 and 35.33 for two examples of defects.

Welds with a high depth/width ratio may have unfavorable bulbous X-sectional shape that is susceptible to cracking at center from microshrinkage and segregation of low melting consituents. Note—crack does not extend to surface.

35.14.2 Gas Metal Arc Welding

See Fig. 35.34.

The GMAW process allows welds to be made with the continuous deposition of filler metal from a spool of consumable electrode wire that is pushed or pulled automatically through the torch. Thus, the process is semi-automatic and/or automatic and avoids the problem of removing the slag, which is required in the SAW process (and required in the two other processes to be mentioned in this

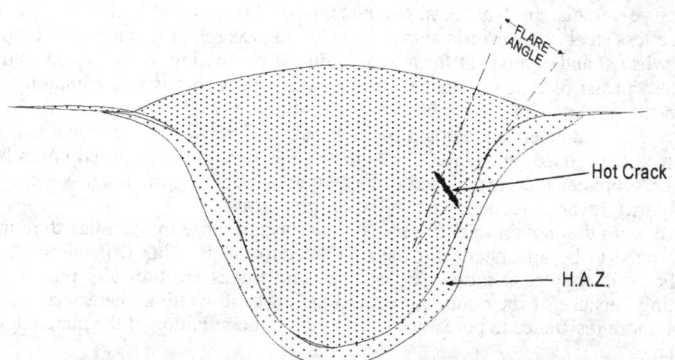

Fig. 35.33 Weld made by SAW with flare angle and hot cracking at juncture of 2 solidification fronts.

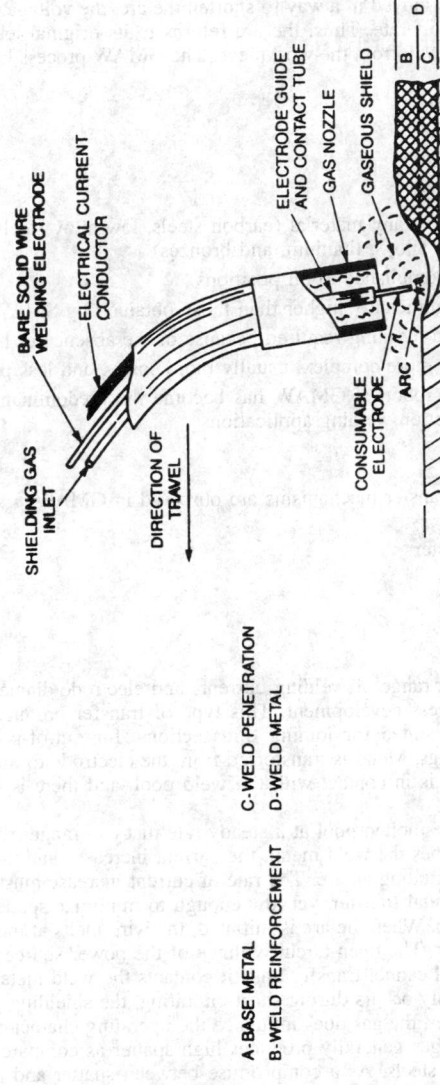

Fig. 35.34 Diagrammatic sketch of (GMA) welding process.

SHIELDING GAS INLET

BARE SOLID WIRE WELDING ELECTRODE

ELECTRICAL CURRENT CONDUCTOR

ELECTRODE GUIDE AND CONTACT TUBE

GAS NOZZLE

GASEOUS SHIELD

CONSUMABLE ELECTRODE

ARC

DIRECTION OF TRAVEL

A - BASE METAL
B - WELD REINFORCEMENT
C - WELD PENETRATION
D - WELD METAL

chapter). In addition, the process provides an inert shielding gas through nozzles in the torch or gun. The process is generally a dc method using reverse polarity. Initially the process used argon as the shielding gas, although helium was tried because the spatter was excessive using argon. Today pure argon is not used but is doped with oxygen (as little as .5% with a maximum of 2%) or carbon dioxide (no more than 10%). These additions assure confinement of the arc to a clear weld pool, which makes the arc steady and produces sound weld deposits. The reverse polarity of this process delivers enough heat input into the cathodic base metal for good penetration, and it provides a positive-ion cleaning effect. Alternating current is not used for GMAW because the arc extinguishes each ½ cycle. The preferred method is a dc constant voltage supply because the arc is self-regulating. This means that if the gun is moved in a way to shorten the arc, the voltage decreases and the current increases, as does the burn-off rate. Thus, the arc returns to its original length. The opposite effect occurs if the gun is moved away from the workpiece. The GMAW process has the following features:

1. High deposition rates
2. Good quality welds
3. Deep penetration
4. Self-regulating feature
5. Adaptability to almost any material (carbon steels, low alloy steels, stainless steels, aluminum, copper, nickel alloys, titanium, and bronzes)
6. Welding can be accomplished in all positions
7. Deposition rates significantly higher than those obtained by SMAW but less than SAW
8. Minimum post-weld cleaning required because of the absence of heavy slag
9. Welding equipment more complex, usually more costly, and less portable than in SMAW
10. With the advent of robotics, GMAW has become the predominant process choice for automatic high-production welding applications

Metal Transfer

Three types of filler metal transfer mechanisms are observed in GMAW:

1. Short-circuiting transfer
2. Globular transfer
3. Spray transfer

Short-Circuiting Transfer

This encompasses the lowest range of welding currents and electrode diameters associated with this process and is a recent process development. This type of transfer produces a small, fast-freezing weld pool that is generally suited for joining thin sections, for out-of-position welding, and for bridging of large root openings. Metal is transferred from the electrode to the workpiece only during the period that the electrode is in contact with the weld pool, and there is no metal transfer across the arc gap.

The electrode contacts the molten pool at a steady rate that can range from 20 to over 200 times per second. As the wire touches the weld metal, the current increases and the liquid metal at the top of the wire is pinched off, initiating an arc. The rate of current increase must be high enough to heat the electrode and promote metal transfer, yet low enough to minimize spatter caused by the violent separation of the molten drop. When the arc is initiated, the wire melts at the top as it is fed forward towards the next short circuit. The open-circuit voltage of the power source must be low enough so that the drop of molten metal cannot transfer until it contacts the weld metal.

Because metal transfer only occurs during short-circuiting, the shielding gas has very little effect on the transfer itself. However, the gas does influence the operating characteristics of the arc and the base metal penetration. CO_2 gas generally produces high spatter as compared to the inert gases, but allows deeper penetration in steels. As a compromise between spatter and penetration, mixtures of carbon dioxide and argon are often used. For nonferrous metals, argon–helium mixtures achieve the same compromise. See Fig. 35.35.

While less expensive CO_2 serves well for shielding low carbon and low alloy steels, argon–CO_2 mixtures improve conditions for welding thinner sections of base metal. The ease of welding in all positions is so advantageous that this variation of GMAW is applied to many different structures and the weld metal deposited displays normal strength and good toughness. One frequent defect has sharply dampened enthusiasm for this process, namely small areas of incomplete fusion between the interface of the base and weld metal. These areas occur when welding with relatively low heat input, as is the case when penetration is minimal or welding steel thicker than ¼ inch. The unfused areas are difficult to detect. Many code-writing bodies either forbade GMAW-S (short-circuit) welding on heavy sections of steel or called for stringent testing before permitting its use.

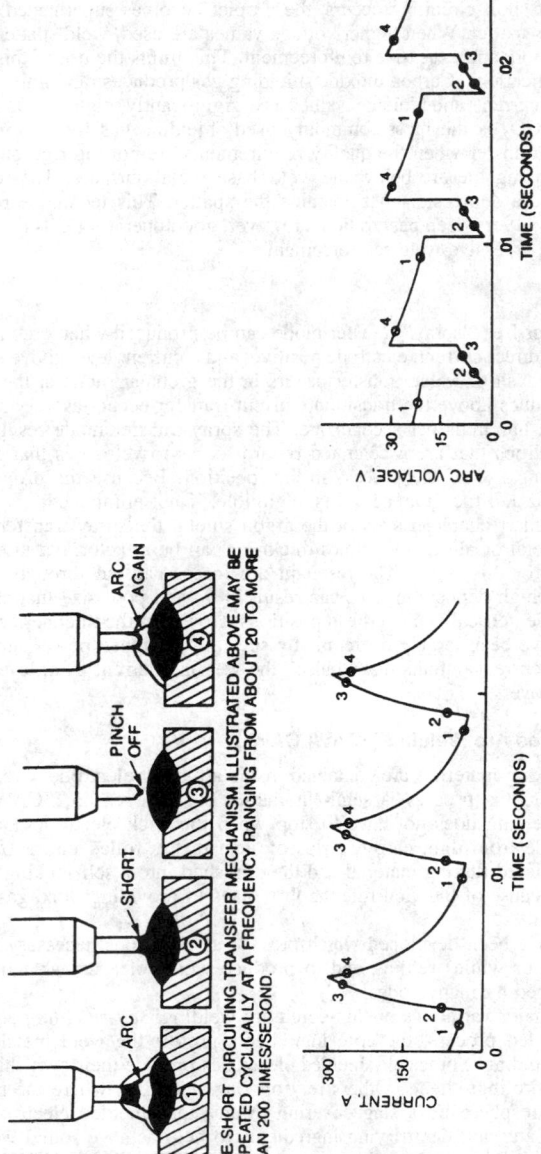

THE SHORT CIRCUITING TRANSFER MECHANISM ILLUSTRATED ABOVE MAY BE REPEATED CYCLICALLY AT A FREQUENCY RANGING FROM ABOUT 20 TO MORE THAN 200 TIMES/SECOND.

Fig. 35.35 GMAW process operating in the short-circuiting transfer mode.

Globular Transfer

With a positive electrode, globular transfer takes place when the current density is relatively low, regardless of the shielding gas used. However, the use of CO_2 or helium results in this type of transfer at all usable welding currents. Globular transfer is characterized by drop size with a diameter greater than that of the electrode. This large drop is easily acted upon by gravity, which means that transfer occurs successfully only in the flat position (downhand).

The arc length must be long enough to avoid the large droplet size from causing short-circuiting to the workpiece. If short-circuiting occurs, the droplet becomes superheated and disintegrates, producing considerable spatter. When higher voltage values are used, weld spatter causes lack of fusion, insufficient penetration, and excessive reinforcement. This limits the use of this transfer mode to very few production applications. Carbon dioxide shielding gas produces randomly directed globular transfer when welding current and voltage values are significantly higher than those used for short-circuiting transfer. CO_2 is the most commonly used shielding gas for welding carbon steels using this mode of metal transfer when the quality requirements are not too rigorous. The spatter problem is controlled by placing the arc below the weld/base metal surfaces. The resulting arc forces are adequate to produce a depression that contains the spatter. This technique requires relatively high currents and results in very deep penetration. However, good operator skills are required. Poor wetting action can result in excessive weld reinforcement.

Spray Transfer

A very stable, spatter-free "spray" transfer mode can be produced when argon-rich shielding is used. This mode requires direct current, electrode positive, and a current level above the critical "transition" value. Below this transition value transfer occurs in the globular mode at the rate of a few droplets every second. At values above the transition, current transfer occurs as very small drops are formed and detached at the hundreds per second rate. The spray-transfer mode results in a highly directed stream of discrete drops that are accelerated by arc forces to velocities that overcome the effect of gravity, thus allowing use of the process in any position. Because the drops are separated, short-circuits do not occur and the spatter level is negligible, if not eliminated.

Because of the inert characteristics of the argon shield, the spray transfer mode can be used to weld almost any metal or alloy. Sometimes thickness can be a factor, because of the relatively high currents necessary for this mode. The resultant arc forces can cut through, rather than weld, thin sheets. In addition, high deposition rates can result in a weld pool size that cannot be supported by surface tension in the vertical and overhead positions. However, the thickness and position limitations of spray transfer have been largely overcome by specially designed power supplies. These machines produce controlled current outputs that "pulse" the welding current from levels below the transition current to levels above.

35.14.3   Flux-Cored Arc Welding (FCAW) CAW

In this case, the fluxing materials are contained inside a tubular electrode wire. When combined with an automatic feeder, this process is semi-automatic. For this reason, FCAW is quickly replacing SMAW, because welding does not have to stop when the stick electrodes of SMAW are used up. Such tubes are made from thin, narrow strips of steel that is rolled into a U-shaped configuration, filled with the powdered fluxing material and then finished into a self-locking tube. This filling must be without gaps, because of the vital role the flux plays in providing: flux, gas coverage, deoxidants, and alloying elements.

Flux systems have been developed which provide all functions necessary for these electrodes to exhibit good behavior while welding and to produce welds of acceptable quality and mechanical properties, as required by many codes.

There are two major variations of flux-cored arc welding: *self-shielding* and *gas shielding*.

In the self-shielded process, the core ingredients protect the weld metal from the atmosphere without external shielding. Some self-shielded electrodes provide their own shielding gas through the decomposition of core ingredients. Others rely on slag shielding, where the metal droplets are protected from the atmosphere by a slag covering. Many self-shielded electrodes contain substantial amounts of deoxidizing and denitrifying ingredients to help achieve sound weld metal. These electrodes also contain stabilizers and alloying elements. Self-shielded FCAW-S electrodes are highly desirable because the cost and practical problems associated with supplying and manipulating the gas-shielded torch are eliminated. Although certain self-shielded electrodes contain gas-generating constituents in their core, primary protection from oxygen and nitrogen is provided by the addition of aluminum and possibly others, such as titanium and silicon. If added gas protection is used with these electrodes, the high aluminum content can be great enough to suppress austenite formation, producing abnormally large grains in the weld microstructure. This is not desirable because of the reduction in strength, ductility, and toughness.

The gas-shielded process uses an externally supplied gas to assist in shielding the arc from nitrogen and oxygen in the atmosphere. Generally, the core ingredients are slag formers, deoxidizers, arc stabilizers, and alloying elements. These electrodes must not be used without the gas covering because the core formulation has neither the gas-generating constituents nor the deoxidizers needed to cope with the exposure to air.

Advantages of FCAW

Because it combines the productivity of continuous welding with the benefits of having a flux present, the FCAW process has several advantages over other welding processes and has largely supplanted SMAW welding as the most popular method for welding low-carbon and low-alloy steels. These advantages are

1. High deposition rates, especially for out-of-position welding
2. Less operator skill required than for GMAW
3. Simpler and more adaptable than SAW
4. Deeper penetration than SMAW
5. More tolerant of rust and mill scale than GMAW

The disadvantages of the FCAW process are

1. Slag must be removed from each pass of the weld and disposed of.
2. More smoke and fume produced than in GMAW and SAW, requiring fume extraction.
3. Equipment is more complex and much less portable than SMAW equipment, particularly for the FCAW-G option.

Applications

This process enjoys widespread use. Both the gas-shielded and self-shielded processes are used to fabricate structures from carbon and low-alloy steels. Self-shielding is preferred for field use. Carbon steel, low-alloy steels, and stainless steels for the construction of pressure vessels and piping are welded by this process for the petroleum, refining, and power-generation industries. Some nickel-based alloys are also welded by FCAW, as well as in the heavy equipment industry for the fabrication of frame members, wheel rims, suspension components, and other parts. Automatic FCAW equipment is also common.

Figure 35.36 shows both the electrodes and the two different types of equipment.

35.14.4   Shielded Metal Arc Welding (SMAW)

This process is commonly called "stick" welding. An arc is struck between a flux-covered solid consumable electrode and the workpiece. The process uses the decomposition of the flux covering to generate a shielding gas and fluxing elements to protect the molten weld metal droplets and the weld pool with a slag covering. The flux coating on the electrode can be provided for gaseous shielding, arc stabilization, fluxing action, and slag formation.

SMAW was the most widely used welding process for steel fabrication for many years, but over the past 10 years, other, more efficient, cost-effective processes have been taking over, in particular the semi-automatic process FCAW and the GMAW process in welding robotics.

Advantages and Limitations: SMAW

The equipment investment is relatively small and welding electrodes for all but the most reactive metals, such as magnesium and titanium, are available for virtually all manufacturing, construction, or maintenance applications. The most important advantage of the process is that it can be used in all positions, with virtually all base metal thicknesses and in areas of limited accessibility.

Because this process is manual, the skill of the welder is of paramount importance in obtaining an acceptable weld. The skill level of the welder needs to be much higher than in most other processes. It is perhaps the most difficult in terms of welder training and skill level requirements of all the processes discussed in this chapter.

The SMAW process is diagrammed in Fig. 35.37.

Covered electrodes for manual welding are made in standard sizes, ranging from $3/32$ to $3/8$ in. in the United States, defined by the diameter of the metal core. The thickness of the flux covering is determined by the requirements of the specification to which the electrode is marketed and its handling characteristics. This range of electrode sizes can be used for welding base metals from thin sheet to very heavy plate. Electrodes as thick as $3/4$ in. are made for special applications, such as filling large cavities in castings or modifying the shape of metalworking dies. These electrodes are normally 14 in. in length, but smaller and longer sizes are available. One end of the electrode is stripped of coating for about 1 in. to allow electrical contact when gripped by the holder. The opposite end has the coating chamfered with the exposed end cleaned of flux to permit touch starting. When the arc is first struck, by touching it to the workpiece, the first small increment of filler metal deposited

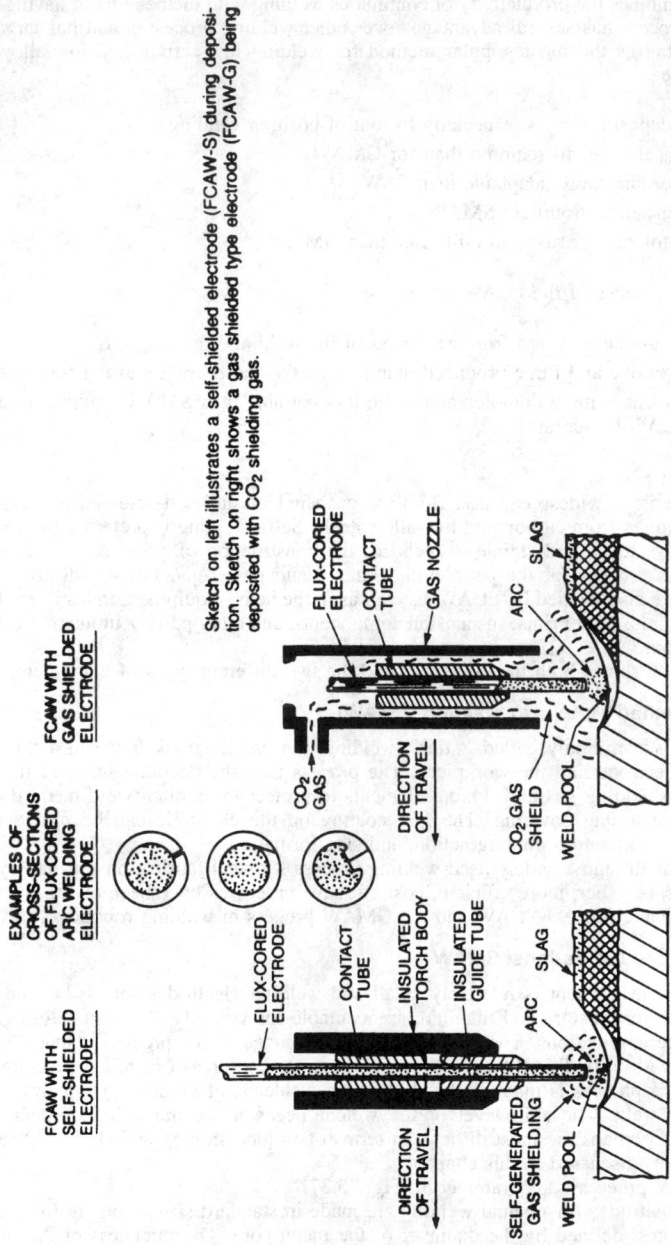

EXAMPLES OF
CROSS-SECTIONS
OF FLUX-CORED
ARC WELDING
ELECTRODE

FCAW WITH
SELF-SHIELDED
ELECTRODE

FLUX-CORED
ELECTRODE

CONTACT
TUBE

INSULATED
TORCH BODY

INSULATED
GUIDE TUBE

ARC

SLAG

DIRECTION
OF TRAVEL

SELF-GENERATED
GAS SHIELDING

WELD POOL

FCAW WITH
GAS SHIELDED
ELECTRODE

FLUX-CORED
ELECTRODE

CONTACT
TUBE

GAS NOZZLE

ARC

SLAG

CO₂
GAS

DIRECTION
OF TRAVEL

CO₂ GAS
SHIELD

WELD POOL

Sketch on left illustrates a self-shielded electrode (FCAW-S) during deposition. Sketch on right shows a gas shielded type electrode (FCAW-G) being deposited with CO₂ shielding gas.

Fig. 35.36   Diagrammatic sketches of the flux-cored arc welding process.

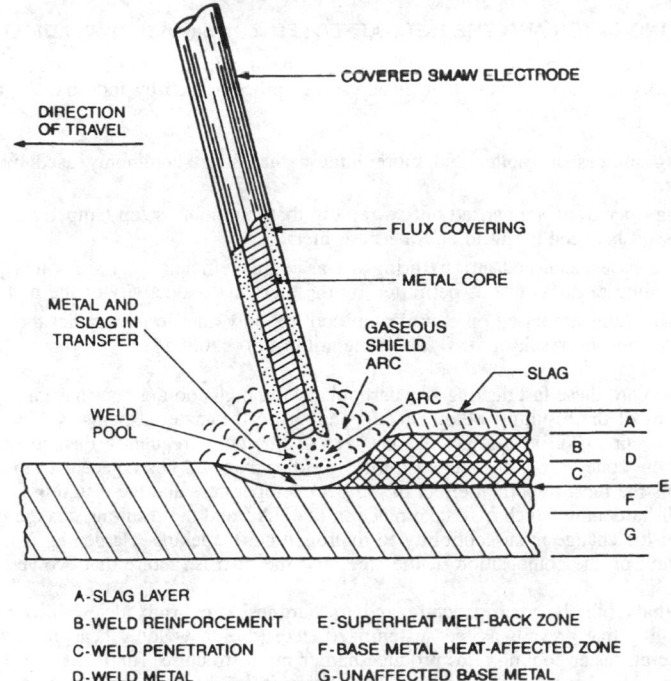

Fig. 35.37 Diagrammatic sketch of the shielded metal arc welding process (SMAW) using a consumable flux-covered electrode.

will not receive normal shielding, nor will it contain intended additions of deoxidizing and alloying elements. This can cause porosity and inadequate alloying at the very start of the weld pass. This problem can be aggravated on restarting a partially used electrode if covering fragmentation results in a bare core for a short length of the electrode. The coverings on most SMAW electrodes are formulated to resist melting just enough to form a short conical projection around and beyond the arcing core, thus accomplishing two objectives: providing greater protection of the emitting end of the metal core, which helps direct the flight of molten droplets toward the weld pool; and preventing short-circuiting of the electrode should it happen to touch the base metal surface.

Hydrogen can be troublesome in SMAW welds. Cellulose-type coverings, widely used for carbon steel electrodes, may contain as much as 30% organic materials and must be carefully baked to retain a small but important percentage of combined moisture necessary to achieve good deposition characteristics. Low-hydrogen rods are manufactured to avoid this problem, but unless very carefully packaged and handled properly after opening, they tend to pick up moisture.

Another problem with SMAW is slag entrapment in the solidifying metal. This problem is dependent on operator skill. Unless the operator can get the slag to float on the top of the molten weld pool and completely remove it by wire brushing, subsequent passes over the first pass may trap slag and become a weld defect.

Applications

Most manufacturing operations that require welding will strive to utilize mechanized processes that offer greater productivity, higher and more consistent quality, and therefore more cost-effective methods. For these reasons, the SMAW process has been replaced wherever possible. However, the simplicity and the ability of this process to achieve welds in restricted accessibility means that it still finds considerable use in certain situations and applications. Heavy construction, such as shipbuilding, and welding in the field, away from support services that would provide shielding gas, cooling water, and on-line electricity and other necessities all rely on SMAW process to a great extent. The process, while primarily designed to join steels, including low-carbon, high-strength steel, quench and tempered steels, high-alloy steels, stainless steels and many cast irons, is also used to join nickel, copper, and their alloys. While rarely used, the process will also weld aluminum.

Table 35.3 is a process selection guide for the processes just discussed above.

35.15 COOLING RATES AND THE HEAT-AFFECTED ZONE (HAZ) IN WELDMENTS[4]

Unusual combinations of time and temperature must be dealt with in welding because the temperature changes are wider and more drastic than in any other process used by industry, for the following reasons:

1. Welding sources are hotter and more intense than most commonly used by industry for heating.
2. Welding operations are carried out so rapidly that extremely steep temperature gradients are established between the weld and the base metal.
3. Both the base metal and any fixturing act as highly efficient heat sinks that promote very high cooling rates (as fast as permitted by the thermal conductivity of the metals involved).
4. Phase diagrams are based on equilibrium cooling only and do not predict the structures that will develop as a result of very fast (nonequilibrium) cooling.

For the most part, these fast heating and particularly the high cooling rates have a negative impact on the properties of the resulting weldment, depending on the material being welded. These undesirable effects occur in the heat-affected zone (HAZ) far more frequently than in the weld metal itself. (This is the zone that is not melted during welding but that lies adjacent to the molten weld zone.) For steels the most important effect of the peak temperature and the resulting fast cooling rate is the degree of hardening which may take place in the HAZ and the resulting change in the fracture toughness or in the change in susceptibility to hydrogen cold cracking. Hardening (forming martensite) is dependent on the composition of the steel and the microstructure that evolves from the fast rate of cooling.

We know that both the microstructure and its hardness are firmly determined by the rate of continuous cooling that prevails as the austenitized steel (due to welding heat) undergoes transformation. In general, faster cooling rates produce harder microstructures (up to RC 60-65 as the maximum hardness). This hardness cooling rate relationship is reliable and repeatable, and is dependent on the given steel's *hardenability*. Hardenability is a characteristic of carbon, low-alloy, and alloyed steels that is governed primarily by the composition of the steel and the austenite grain size. Classically, a guide to the hardenability of a steel can be obtained by calculating the carbon equivalent. Several carbon equivalent formulas have been developed for different classes of steel such as

$$CE = \%C + \frac{\%Mn}{6} + \frac{\% Cr + \% Mo + \% V}{5} + \frac{\%Si + \%Ni + \% Cu}{15}$$

The calculated carbon equivalent can be related to hydrogen-sensitive microstructures. That is, as the carbon equivalent increases, the microstructures that are evolved during cooling through the transformation temperature (from 800 to 500°C), become increasingly sensitive or susceptible to hydrogen induced cracking. At high levels of CE, martensitic structures can be expected. For this formula, when the carbon equivalent exceeds 0.35%, preheats are recommended to minimize susceptibility to hydrogen cracking. At higher levels of CE, both preheats and postheats may be required.

A special carbon equivalent has been developed by Yorioka to establish the critical Δt_{8-5} for a martensite-free HAZ in low-carbon alloy steels. This carbon equivalent is given as

$$CE^* = \% C^* + \frac{\%Mn}{3.6} + \frac{\% Cu}{20} + \frac{\% Ni}{9} + \frac{\% Cr}{5} + \frac{\% Mo}{4}$$

where

$$\% C^* = 5C \text{ for carbon} \leq 0.30\%$$

$$\% C^* = \% C/6 = .25 \text{ for } \% C \geq .3\%$$

The critical time length in seconds Δt_{8-5} for the formation of martensite is given as

$$\log \Delta t_{8-5} = 2.69 \, CE^* = 0.321$$

When CE* is known, welding parameters and preheat temperatures for a given thickness of material can be established to produce cooling rates that avoid formation of martensite in the HAZ. For more details, see Ref. 5.

The use of preheating and postheating to minimize the susceptibility to hydrogen-induced cracking is an accepted welding procedure for many steels. Preheating controls and lowers the cooling rate

Table 35.3

Parameters	SAW	GMAW	FCAW	SMAW
Usability	Limited to flat and horizontal positions. Semi-automatic version has some adaptability but is most often mechanized. Limited portability. Minimum thickness 1/16 in. Joint preparation required on material 1/2 in. and thicker. Process lends self to weld thicker materials.	All position process in the short-arc or pulsed mode. Moderately adaptable but use outside, where shielding can be lost. Usable for steels to 0.010 thick. Above 3/16 in. thick requires joint preparation. No upper limit on plate thickness.	All-position process. Equipment similar to GMAW but self-shielded version has better portability and is usable outdoors. Minimum plate thickness 18 gage. For self-shielded material above 1/4 in. requires joint prep. With CO_2 shielding metals above 1/2 in. requires joint prep. No upper limit on plate thickness.	Very adaptable, all-position. Can be used outdoors. Excellent joint accessibility. Very portable. Can be used on carbon steels to 18 gage. Joint preparation required on thickness over 1/8 in. Unlimited upper thickness, but other processes are usually more economical.
Cost Factors	Deposition rates are very high (over 100 LB/h) with multiwire systems. Deposition efficiency is 99% but doesn't include flux. Usually mechanized, high operator factor, cost moderate for single wire systems. High welding speeds. Higher housekeeping costs (slag and unused flux).	Deposition rates to 35 lb/h. Deposition efficiency 90–95%. Operator factor 50%. Equipment and spares cost are moderate to high. Pulsed arc power supplies are higher-cost. Welding speeds moderate to high. Cleanup minimal.	Deposition rates to 40 lb/h are higher than GMAW. Deposition efficiencies 80–90%. Operator factors 50%. Equipment cost moderate. Good out-of-position deposition rates. Welding speeds moderate to high. Slag and spatter removal and disposal required.	Low deposition rate 20 lb/h with low deposit efficiency 65%. Low operator factor. Equipment cost low. Spares minimal. Welding speeds are low. Housekeeping is required to deslag and dispose of flux and electrode stubs.
Weld metal quality	Very good with good toughness, possible. Handles rust and mill scale well with proper flux. High-dilution process.	Very good quality. Porosity or lack of fusion can be a problem. Less tolerant of rust and mill scale than flux-using processes. Very good toughness achievable.	Good quality. Weld metal toughness is fair to good (best with basic electrode). Slag inclusions are a potential problem.	Strongly dependent on skill of operator. Lack of fusion or slag inclusions are potential problems. Small beads result in high percentage of refining in multipass welding and very good toughness is achievable with some electrodes.
Effect on base metal	Higher heat inputs can result in large HAZ and possible deterioration of baseplate properties. Flux is a source of hydrogen.	Generally a low-hydrogen process.	Flux core can contribute hydrogen.	Low heat inputs can cause rapid HAZ cooling. Flux coatings are a potential source of hydrogen.
General comments	A high-deposition, high-penetration process, but thin material can be welded at high speeds. Easily mechanized. Natural for welding thick plates. Housekeeping and position limitation can be a problem.	Relatively versatile. Equipment more expensive, complex and less portable than SMAW. Easily mechanized. A clean process with high deposition rates and good efficiencies.	Relatively versatile. High deposition rates but high fumes also. Welds easily in out-of-position. Readily mechanized.	Very versatile, low-cost process. Especially strong on nonroutine jobs. Usually not economical on standard thick welds or repetitive jobs that can be mechanized.

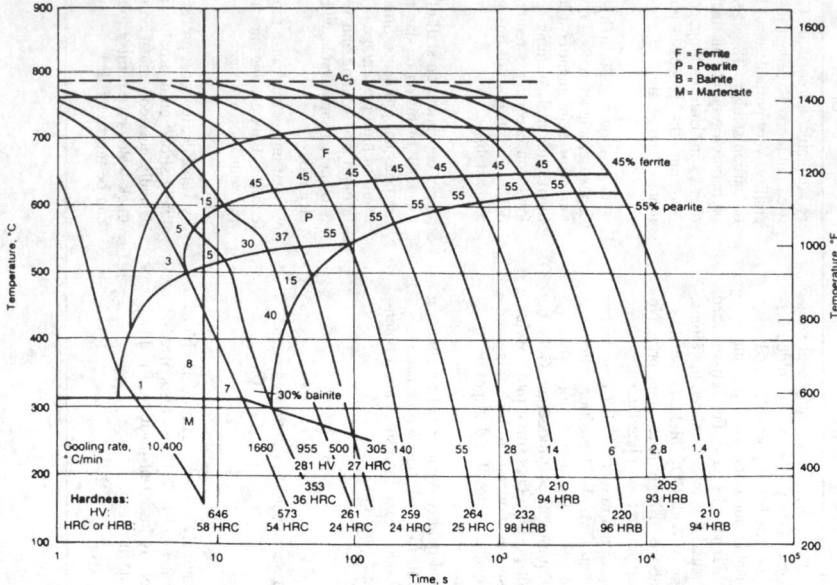

Fig. 35.38 Conventional CCT diagram for an AISI 1541 (.39 C), ASTM grain size 6, austenitized at 980°C. For each of the cooling curves the transformation start and end temperature are given as well as the hardness.

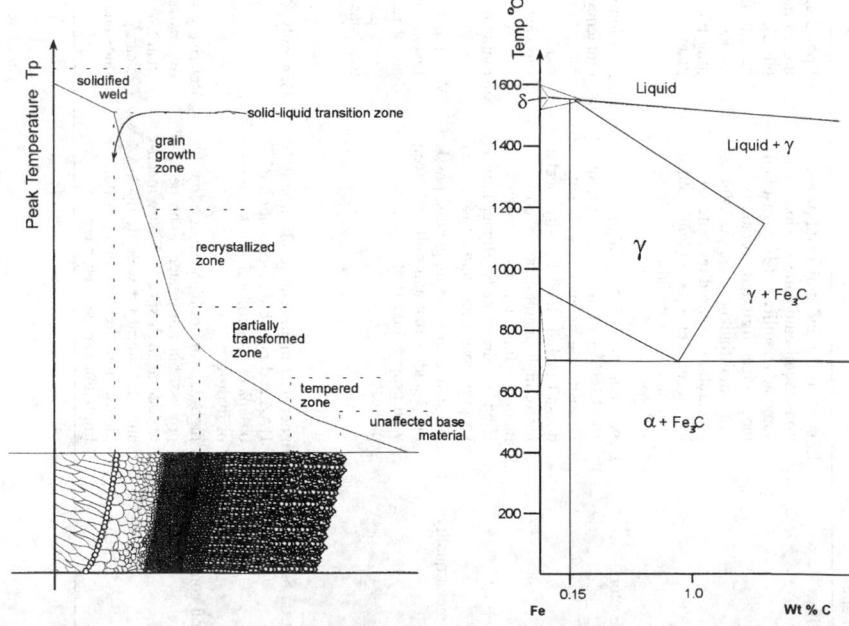

Fig. 35.39 A schematic of the various subzones of the HAZ for the alloy indicated on the Fe-Fe$_3$C equilibrium diagram.

through the transformation temperature range—in other words, 800 to 500°C. Thicker sections often require preheating because of the greater heat-sinking capacity of thicker sections. Preheating may also lower the level of residual stresses in the welded assembly. Preheat temperatures to 150°C (300°F) are not uncommon for low-alloy steels and may increase to 425°C (800°F) for higher-carbon equivalent steels. The practical limits to preheating are that high base metal temperatures can make welding difficult, particularly for manual welding. Furthermore, a high preheat temperature will result in a flattening of the temperature gradient, increasing the time spent in the austenite range, which will cause excessive grain growth in the HAZ and can lead to a loss of fracture toughness independent of hydrogen content.

Two graphs will give a better idea of the problem. A concise method of describing the transformation behavior of a steel is by a continuous cooling transformation diagram (CCT), as shown in Fig. 35.38. A warning needs to be issued: Fig. 35.38 cannot be used to describe the transformation behavior in a weldment of the same material because weld thermal cycles are very different from those used in generating conventional CCT diagrams.

In Fig. 35.39 we have a peak temperature–cooling time diagram. In this case, thermocouples have been embedded in the base metal at known distances from the molten pool interface. A weld is then made and the peak temperatures at each distance from the molten pool are plotted and correlated to the iron–carbon phase diagram. The HAZ is shown for the case of a single pass weld in a .15% carbon steel, showing each subzone. Each subzone refers to a different microstructure, and each is likely to possess different mechanical properties.

REFERENCES

1. J. H. Bickford, *An Introduction to the Design and Behavior of Bolted Joints,* 2nd ed., Marcel Dekker, New York, 1990.
2. G. L. Kulak, J. W. Fisher, and J. H. A. Struik, *Guide to Design Criteria for Bolted and Riveted Joints,* Wiley, New York, 1987.
3. *SPS Fastener Facts,* Standard Pressed Steel Co., Jenkintown, PA, Section IV-C-4.
4. G. Linnert, *Welding, Metallurgy, Carbon and Alloy Steels,* Vol. 4, American Welding Society, Miami, FL, 1994, Chap. 7.
5. N. Yurioka, "Weldability of Modern High Strength Steels," in *First US Japan Symposium on Advances in Welding Metallurgy,* American Welding Society, Miami, FL, 1990, pp. 79–100.

CHAPTER 36

STATISTICAL QUALITY CONTROL

Magd E. Zohdi
Department of Industrial Engineering and Manufacturing
Louisiana State University
Baton Rouge, Louisiana

36.1	MEASUREMENTS AND QUALITY CONTROL	1175	36.6	CONTROL CHARTS FOR ATTRIBUTES	1180
				36.6.1 The p and np Charts	1182
36.2	DIMENSION AND TOLERANCE	1175		36.6.2 The c and u Charts	1183
			36.7	ACCEPTANCE SAMPLING	1183
36.3	QUALITY CONTROL	1175		36.7.1 Double Sampling	1184
	36.3.1 $\bar{X}$, R, and σ Charts	1175		36.7.2 Multiple and Sequential Sampling	1184
36.4	INTERRELATIONSHIP OF TOLERANCES OF ASSEMBLED PRODUCTS	1179	36.8	DEFENSE DEPARTMENT ACCEPTANCE SAMPLING BY VARIABLES	1184
36.5	OPERATION CHARACTERISTIC CURVE (OC)	1180			

36.1 MEASUREMENTS AND QUALITY CONTROL

The metric and English measuring systems are the two measuring systems commonly used throughout the world. The metric system is universally used in most scientific applications, but, for manufacturing in the United States, has been limited to a few specialties, mostly items that are related in some way to products manufactured abroad.

36.2 DIMENSION AND TOLERANCE

In dimensioning a drawing, the numbers placed in the dimension lines are only approximate and do not represent any degree of accuracy unless so stated by the designer. To specify the degree of accuracy, it is necessary to add tolerance figures to the dimension. Tolerance is the amount of variation permitted in the part or the total variation allowed in a given dimension.

Dimensions given close tolerances mean that the part must fit properly with some other part. Both must be given tolerances in keeping with the allowance desired, the manufacturing processes available, and the minimum cost of production and assembly that will maximize profit. Generally speaking, the cost of a part goes up as the tolerance is decreased.

Allowance, which is sometimes confused with tolerance, has an altogether different meaning. It is the minimum clearance space intended between mating parts and represents the condition of tightest permissible fit.

36.3 QUALITY CONTROL

When parts must be inspected in large numbers, 100% inspection of each part is not only slow and costly, but does not eliminate all of the defective pieces. Mass inspection tends to be careless; operators become fatigued; and inspection gages become worn or out of adjustment more frequently. The risk of passing defective parts is variable and of unknown magnitude, whereas, in a planned sampling procedure, the risk can be calculated. Many products, such as bulbs, cannot be 100%

Mechanical Engineers' Handbook, 2nd ed., Edited by Myer Kutz.
ISBN 0-471-13007-9 © 1998 John Wiley & Sons, Inc.

inspected, since any final test made on one results in the destruction of the product. Inspection is costly and nothing is added to a product that has been produced to specifications.

Quality control enables an inspector to sample the parts being produced in a mathematical manner and to determine whether or not the entire stream of production is acceptable, provided that the company is willing to allow up to a certain known number of defective parts. This number of acceptable defectives is usually taken as 3 out of 1000 parts produced. Other values might be used.

36.3.1 $\overline{X}$, R, and σ Charts

To use quality techniques in inspection, the following steps must be taken (see Table 36.1).

1. Sample the stream of products by taking m samples, each of size n.
2. Measure the desired dimension in the sample, mainly the central tendency.
3. Calculate the deviations of the dimensions.
4. Construct a control chart.
5. Plot succeeding data on the control chart.

The arithmetic mean of the set of n units is the main measure of central tendency. The symbol $\overline{X}$ is used to designate the arithmetic mean of the sample and may be expressed in algebraic terms as

$$\overline{X}_i = (X_1 + X_2 + X_3 + \ldots + X_n)/n \tag{36.1}$$

where X_1, X_2, X_3, etc. represent the specific dimensions in question. The most useful measure of dispersion of a set of numbers is the standard deviation σ. It is defined as the root–mean–square deviation of the observed numbers from their arithmetic mean. The standard deviation σ is expressed in algebraic terms as

$$\sigma_i = \sqrt{\frac{(X_1 - \overline{X})^2 + (X_2 - \overline{X})^2 + \ldots + (X_n - \overline{X})^2}{n}} \tag{36.2}$$

Another important measure of dispersion, used particularly in control charts, is the range R. The range is the difference between the largest observed value and the smallest observed in a specific sample.

$$R = X_i(\text{max}) - X_i(\text{min}) \tag{36.3}$$

Even though the distribution of the X values in the universe can be of any shape, the distribution of the $\overline{X}$ values tends to be close to the normal distribution. The larger the sample size and the more nearly normal the universe, the closer will the frequency distribution of the average $\overline{X}$'s approach the normal curve, as in Fig. 36.1.

According to the statistical theory, (the Central Limit Theory) in the long run, the average of the $\overline{X}$ values will be the same as μ, the average of the universe. And in the long run, the standard deviation of the frequency distribution $\overline{X}$ values, $\sigma_{\overline{x}}$, will be given by

$$\sigma_{\overline{x}} = \frac{\sigma}{\sqrt{n}} \tag{36.4}$$

where σ is the standard deviation of the universe. To construct the control limits, the following steps are taken:

Table 36.1 Computational Format for Determining $\overline{X}$, R, and σ

Sample Number	Sample Values	Mean $\overline{X}$	Range R	Standard Deviation σ'
1	$X_{11}, X_{12}, \ldots, X_{1n}$	$\overline{X}_1$	R_1	σ'_1
2	$X_{21}, X_{22}, \ldots, X_{2n}$	$\overline{X}_2$	R_2	σ'_2
.	$\ldots$	.	.	.
.	$\ldots$	.	.	.
.	$\ldots$	.	.	.
m	$X_{m1}, X_{m2}, \ldots, X_{mn}$	$\overline{X}_m$	R_m	σ'_m

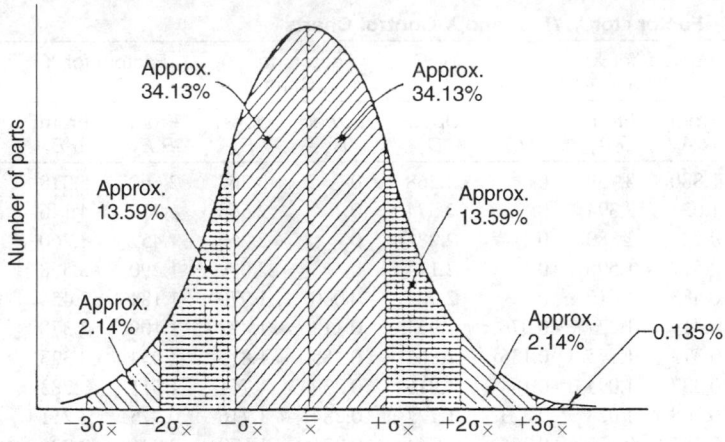

Fig. 36.1 Normal distribution and percentage of parts that will fall within σ limits.

1. Calculate the average of the average $\overline{\overline{X}}$ as follows:

$$\overline{\overline{X}} = \sum_{1}^{m} \overline{X}_i/m \qquad i = 1, 2, \ldots, m \tag{36.5}$$

2. Calculate the average deviation, $\overline{\sigma}$ where

$$\overline{\sigma} = \sum_{1}^{m} \sigma_i'/m \qquad i = 1, 2, \ldots, m \tag{36.6}$$

Statistical theory predicts the relationship between $\overline{\sigma}$ and $\sigma_{\overline{x}}$. The relationship for the $3\sigma_{\overline{x}}$ limits or the 99.73% limits is

$$A_1\overline{\sigma} = 3\sigma_{\overline{x}} \tag{36.7}$$

This means that control limits are set so that only 0.27% of the produced units will fall outside the limits. The value of $3\sigma_{\overline{x}}$ is an arbitrary limit that has found acceptance in industry.

The value of A_1 calculated by probability theory is dependent on the sample size and is given in Table 36.2. The formula for 3σ control limits using this factor is

$$CL(\overline{X}) = \overline{\overline{X}} \pm A_1\overline{\sigma} \tag{36.8}$$

Once the control chart (Fig. 36.2) has been established, data ($\overline{X}_i$'s) that result from samples of the same size n are recorded on it. It becomes a record of the variation of the inspected dimensions over a period of time. The data plotted should fall in random fashion between the control limits 99.73% of the time if a stable pattern of variation exists.

So long as the points fall between the control lines, no adjustments or changes in the process are necessary. If five to seven consecutive points fall on one side of the mean, the process should be checked. When points fall outside of the control lines, the cause must be located and corrected immediately.

Statistical theory also gives the expected relationship between $\overline{R}$ ($\Sigma R_i/m$) and $\sigma_{\overline{x}}$. The relationship for the $3\sigma_{\overline{x}}$ limits is

$$A_2\overline{R} = 3\sigma_{\overline{x}} \tag{36.9}$$

The values for A_2 calculated by probability theory, for different sample sizes, are given in Table 36.2. The formula for 3σ control limits using this factor is

$$CL(\overline{X}) = \overline{\overline{X}} \pm A_2\overline{R} \tag{36.10}$$

In control chart work, the ease of calculating R is usually much more important than any slight

Table 36.2   Factors for $\overline{X}$, R, σ, and X Control Charts

Sample Size n	Factors for $\overline{X}$ Chart		Factors for R Chart		Factors for σ' Chart		Factors for X Chart		$\sigma = \overline{R}/d_2$
	From $\overline{R}\,A_2$	From $\overline{\sigma}\,A_1$	Lower D_3	Upper D_4	Lower B_3	Upper B_4	From $\overline{R}\,E_2$	From $\overline{\sigma}\,E_1$	d_2
2	1.880	3.759	0	3.268	0	3.267	2.660	5.318	1.128
3	1.023	2.394	0	2.574	0	2.568	1.772	4.146	1.693
4	0.729	1.880	0	2.282	0	2.266	1.457	3.760	2.059
5	0.577	1.596	0	2.114	0	2.089	1.290	3.568	2.326
6	0.483	1.410	0	2.004	0.030	1.970	1.184	3.454	2.539
7	0.419	1.277	0.076	1.924	0.118	1.882	1.109	3.378	2.704
8	0.373	1.175	0.136	1.864	0.185	1.815	1.054	3.323	2.847
9	0.337	1.094	0.184	1.816	0.239	1.761	1.011	3.283	2.970
10	0.308	1.028	0.223	1.777	0.284	1.716	0.975	3.251	3.078
11	0.285	0.973	0.256	1.744	0.321	1.679	0.946	3.226	3.173
12	0.266	0.925	0.284	1.717	0.354	1.646	0.921	3.205	3.258
13	0.249	0.884	0.308	1.692	0.382	1.618	0.899	3.188	3.336
14	0.235	0.848	0.329	1.671	0.406	1.594	0.881	3.174	3.407
15	0.223	0.817	0.348	1.652	0.428	1.572	0.864	3.161	3.472
16	0.212	0.788	0.364	1.636	0.448	1.552	0.848	3.152	3.532
17	0.203	0.762	0.380	1.621	0.466	1.534	0.830	3.145	3.588
18	0.194	0.738	0.393	1.608	0.482	1.518	0.820	3.137	3.640
19	0.187	0.717	0.404	1.597	0.497	1.503	0.810	3.130	3.687
20	0.180	0.698	0.414	1.586	0.510	1.490	0.805	3.122	3.735
21	0.173	0.680	0.425	1.575	0.523	1.477	0.792	3.114	3.778
22	0.167	0.662	0.434	1.566	0.534	1.466	0.783	3.105	3.819
23	0.162	0.647	0.443	1.557	0.545	1.455	0.776	3.099	3.858
24	0.157	0.632	0.451	1.548	0.555	1.445	0.769	3.096	3.895
25	0.153	0.619	0.459	1.540	0.565	1.435	0.765	3.095	3.931

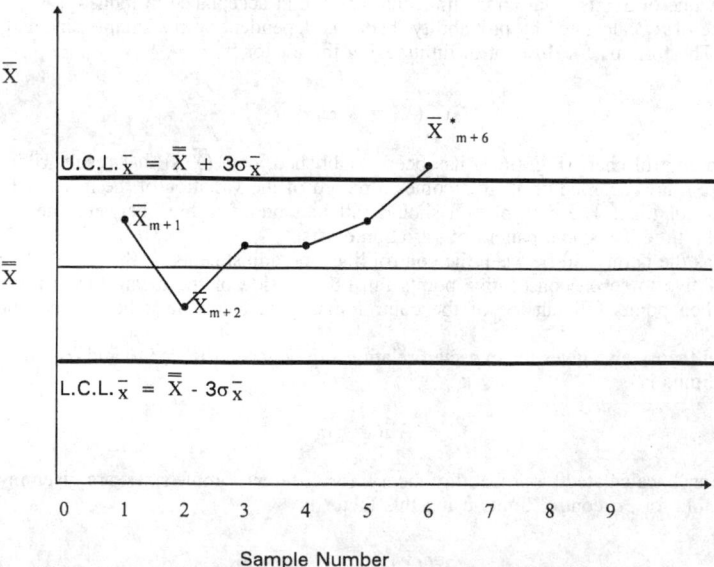

Fig. 36.2   Control chart $\overline{X}$.

theoretical advantage that might come from the use of σ. However, in some cases where the measurements are costly and it is necessary that the inferences from a limited number of tests be as reliable as possible, the extra cost of calculating σ is justified. It should be noted that, because Fig. 36.2 shows the averages rather than individual values, it would have been misleading to indicate the tolerance limits on this chart. It is the individual article that has to meet the tolerances, not the average of a sample. Tolerance limits should be compared to the machine capability limits. Capability limits are the limits on a single unit and can be calculated by

$$\text{capability limits} = \overline{\overline{X}} \pm 3\sigma \tag{36.11}$$

$$\sigma = \overline{R}/d_2$$

Since $\sigma' = \sqrt{n}\,\sigma_{\overline{x}}$, the capability limits can be given by

$$\text{capability limits } (X) = \overline{\overline{X}} \pm 3\sqrt{n}\,\sigma_{\overline{x}} \tag{36.12}$$

$$= \overline{\overline{X}} \pm E_1\overline{\sigma} \tag{36.13}$$

$$= \overline{\overline{X}} \pm E_2\overline{R} \tag{36.14}$$

The values for d_2, E_1, and E_2 calculated by probability theory, for different sample sizes, are given in Table 36.2.

Figure 36.3 shows the relationships among the control limits, the capability limits, and assumed tolerance limits for a machine that is capable of producing the product with this specified tolerance. Capability limits indicate that the production facility can produce 99.73% of its products within these limits. If the specified tolerance limits are greater than the capability limits, the production facility is capable of meeting the production requirement. If the specified tolerance limits are tighter than the capability limits, a certain percentage of the production will not be usable and 100% inspection will be required to detect the products outside the tolerance limits.

To detect changes in the dispersion of the process, the R and σ charts are often employed with $\overline{X}$ and X charts.

The upper and lower control limits for the R chart are specified as

$$U\,C\,L\,(R) = D_4\overline{R} \tag{36.15}$$

$$L\,C\,L\,(R) = D_3\overline{R} \tag{36.16}$$

Figure 36.4 shows the $\overline{R}$ chart for samples of size 5.

The upper and lower control for the T chart are specified as

$$U\,C\,L\,(\sigma) = B_4\overline{\sigma} \tag{36.17}$$

$$L\,C\,L\,(\sigma) = B_3\overline{\sigma} \tag{36.18}$$

The values for D_3, D_4, B_3, and B_4 calculated by probability theory, for different sample sizes, are given in Table 36.2.

36.4   INTERRELATIONSHIP OF TOLERANCES OF ASSEMBLED PRODUCTS

Mathematical statistics states that the dimension on an assembled product may be the sum of the dimensions of the several parts that make up the product. It states also that the standard deviation of the sum of any number of independent variables is the square root of the sum of the squares of the standard deviations of the independent variables. So if

$$X = X_1 \pm X_2 \pm \ldots X_n \tag{36.19}$$

$$\overline{X} = \overline{X}_1 \pm \overline{X}_2 \pm \ldots \overline{X}_n \tag{36.20}$$

$$\sigma(X) = \sqrt{(\sigma_1)^2 + (\sigma_2)^2 + \ldots + (\sigma_n)^2} \tag{36.21}$$

Whenever it is reasonable to assume that the tolerance ranges of the parts are proportional to their respective σ' values, such tolerance ranges may be combined by taking the square root of the sum of the squares:

$$T = \sqrt{T_1^2 + T_2^2 + T_3^2 + \ldots + T_n^2} \tag{36.22}$$

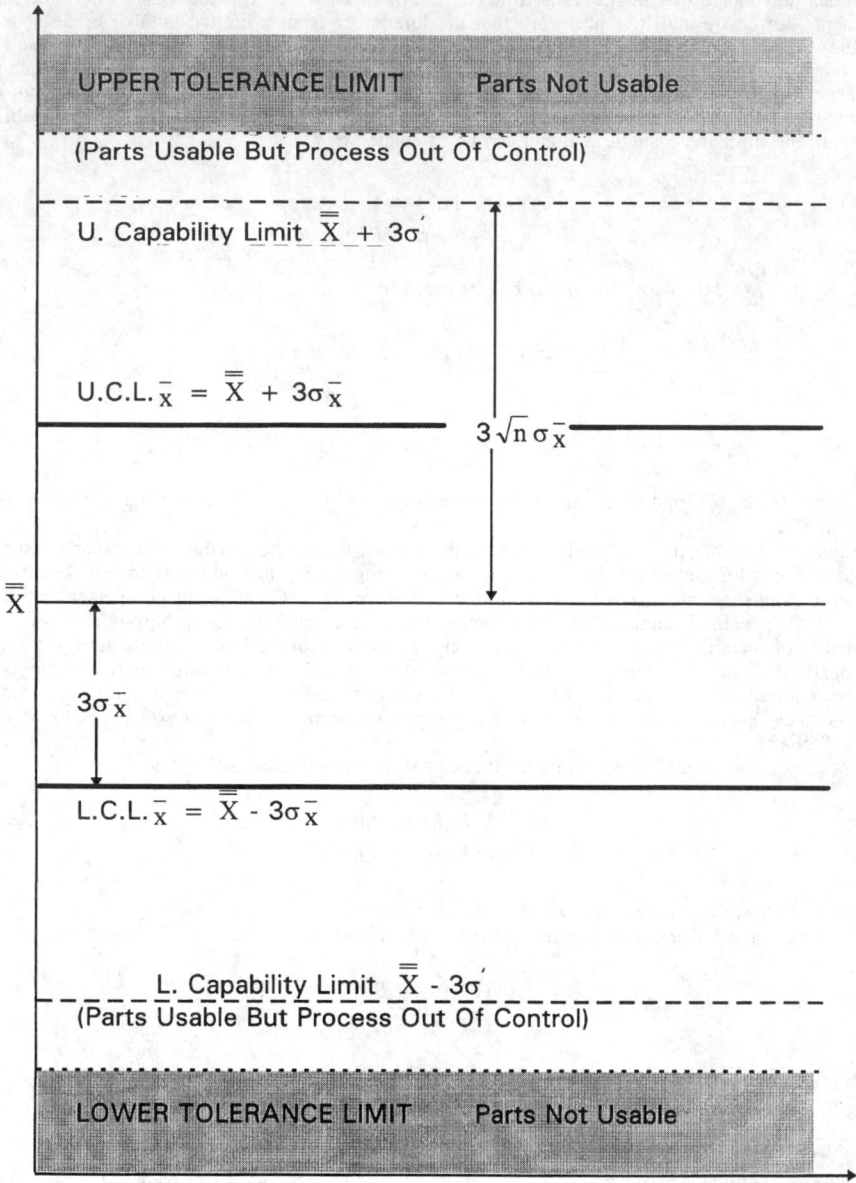

UPPER TOLERANCE LIMIT   Parts Not Usable

(Parts Usable But Process Out Of Control)

U. Capability Limit $\overline{\overline{X}} + 3\sigma'$

U.C.L.$\overline{X}$ = $\overline{\overline{X}}$ + $3\sigma_{\overline{X}}$

$3\sqrt{n}\,\sigma_{\overline{X}}$

$\overline{\overline{X}}$

$3\sigma_{\overline{X}}$

L.C.L.$\overline{X}$ = $\overline{\overline{X}}$ - $3\sigma_{\overline{X}}$

L. Capability Limit $\overline{\overline{X}}$ - $3\sigma'$
(Parts Usable But Process Out Of Control)

LOWER TOLERANCE LIMIT   Parts Not Usable

Fig. 36.3 Control, capability, and tolerance (specification limits).

36.5 OPERATION CHARACTERISTIC CURVE (OC)

Control charts detect changes in a pattern of variation. If the chart indicates that a change has occurred when it has not, Type I error occurs. If three-sigma limits are used, the probability of making a Type I error is approximately 0.0027.

The probability of the chart indicating no change, when in fact it has, is the probability of making a Type II error. The operation characteristic curves are designed to indicate the probability of making a Type II error. An OC curve for an $\overline{X}$ chart of three-sigma limits is illustrated in Fig. 36.5.

36.6 CONTROL CHARTS FOR ATTRIBUTES

Testing may yield only one of two defined classes: within or outside certain limits, acceptable or defective, working or idle. In such a classification system, the proportion of units falling in one class may be monitored with a p chart.

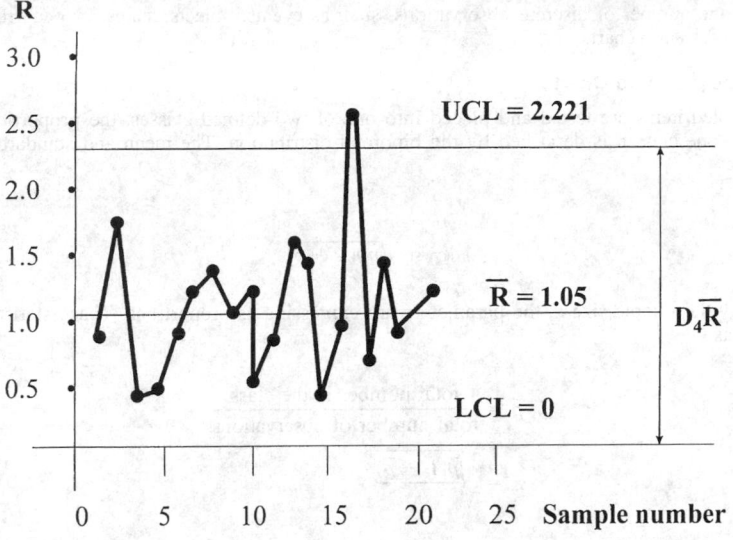

Fig. 36.4 *R* Chart for samples of 5 each.

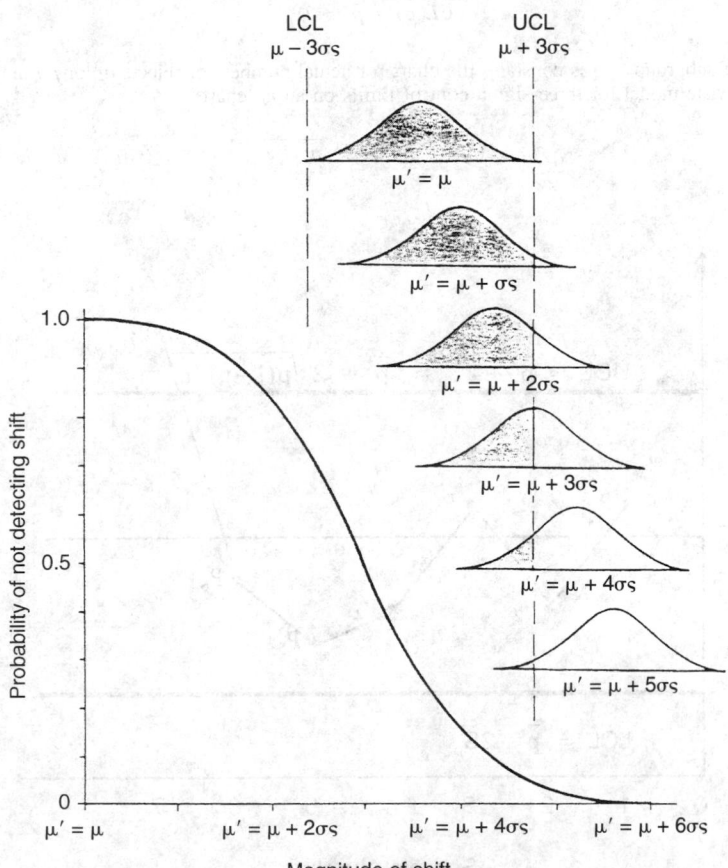

Fig. 36.5 Operating characteristic curve for 3σ limit.

In other cases, observation may yield a multivalued, but still discrete, classification system. In such case, the number of discrete observations, such as events, objects, states, or occurrences, may be monitored by a c chart.

36.6.1 The p and np Charts

When sampled items are tested and placed into one of two defined classes, the proportion of units falling into one class p is described by the binomial distribution. The mean and standard deviation are given as

$$\mu = np$$
$$\sigma = \sqrt{np(1 - p)}$$

Dividing by the sample size n, the parameters are expressing as proportions. These statistics can be expressed as

$$\bar{p} = \frac{\text{total number in the class}}{\text{total number of observations}} \tag{36.23}$$

$$s_p = \sqrt{\frac{\bar{p}(1 - \bar{p})}{n}} \tag{36.24}$$

The control limits are either set at two sigma limits with Type I error as 0.0456 or at three-sigma limits with Type I error as 0.0027. The control limits for the p chart with two-sigma limits (Fig. 36.6) are defined as

$$CL(p) = \bar{p} \pm 2S_p \tag{36.25}$$

However, if subgroup size is constant, the chart for actual numbers of rejects np or pn may be used. The appropriate model for three-sigma control limits on an np chart is

$$CL(np) = n\bar{p} \pm 3\sqrt{n\bar{p}(1 - \bar{p})} \tag{36.26}$$

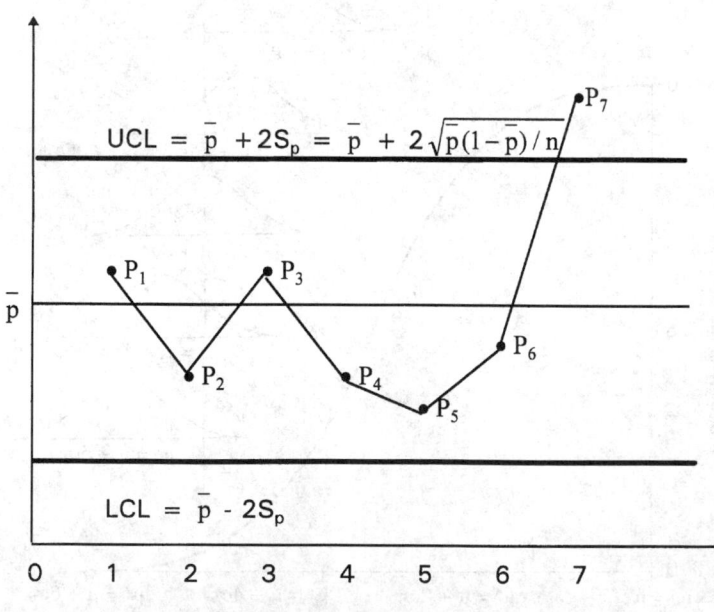

Fig. 36.6 P charts.

36.6.2 The c and u Charts

The random variable process that provides numerical data that are recorded as a number c rather than a proportion p is described by the Poisson distribution. The mean and the variance of the Poisson distribution are equal and expressed as $\mu = \sigma^2 = np$. The Poisson distribution is applicable in any situation when n and p cannot be determined separately, but their product np can be established. The mean and variable can be estimated as

$$\bar{c} = S_c^2 = \frac{\sum\limits_{1}^{m} C_i}{m} = \frac{\sum\limits_{1}^{m} (np)_i}{m} \tag{36.27}$$

The control limits (Fig. 36.7) are defined as

$$CL(c) = \bar{C} \pm 3S_c \tag{36.28}$$

If there is change in the area of opportunity for occurrence of a nonconformity from subgroup to subgroup, such as number of units inspected or the lengths of wires checked, the conventional c chart showing only the total number of nonconformities is not applicable. To create some standard measure of the area of opportunity, the nonconformities per unit (c/n) or u is used as the control statistic. The control limits are

$$CL(u)\bar{u} \pm 3\,\frac{\sqrt{\bar{u}}}{\sqrt{n_i}} \tag{36.29}$$

$$\text{where } \bar{u} = \frac{\sum C_i}{\sum n_i} = \frac{\text{total nonconformities found}}{\text{total units inspected}}$$

$$c = nu \text{ is Poisson-distributed, } u \text{ is not}$$

36.7 ACCEPTANCE SAMPLING

The objective of acceptance sampling is to determine whether a quantity of the output of a process is acceptable according to some criterion of quality. A sample from the lot is inspected and the lot is accepted or rejected in accordance with the findings of the sample.

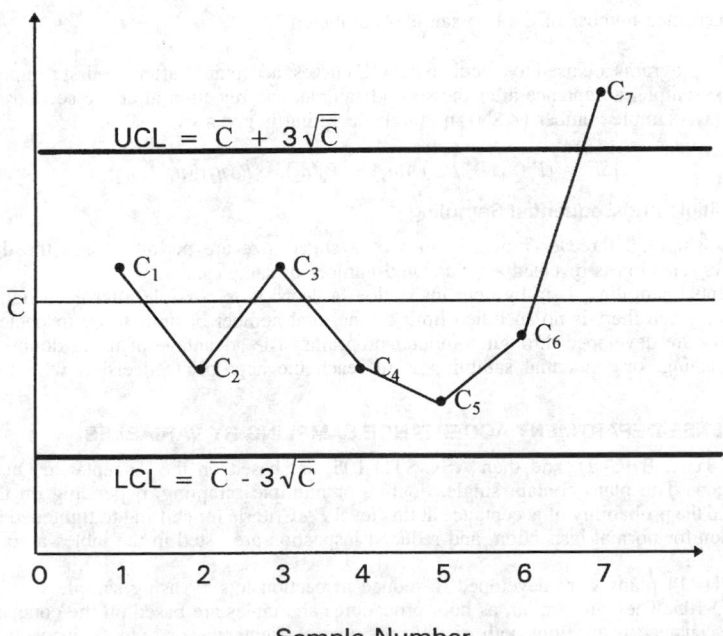

Sample Number

Fig. 36.7 C charts.

Acceptance sampling plans call for the random selection of sample of size n from a lot containing N items. The lot is accepted if the number of defectives found in the sample $\leq c$, the acceptance number. A rejected lot can either be returned to the producer, nonrectifying inspections, or it can be retained and subjected to a 100% screening process, rectifying inspection plan improves the outgoing quality. A second attribute-inspection plan might use two samples before requiring the acceptance or rejection of a lot. A third plan might use multiple samples or a sequential sampling process in evaluating a lot. Under rectifying inspection programs, the average outgoing quality level (AOQ), the average inspection lot (I), and the average outgoing quality limit (AOQL) can be predicted for varying levels of incoming fraction defective p.

Assuming that all lots arriving contain the same proportion of defectives p, and that rejected lots will be subjected to 100% inspection, AOQ and I are given below:

$$\text{AOQ} = \frac{P_a p(N - n)}{N - pn - (1 - P_a)p(Nn)} \tag{36.30}$$

$$I = n + (1 - P_a)(N - n) \tag{36.31}$$

The average outgoing quality (AOQ) increases as the proportion defective in incoming lots increases until it reaches a maximum value and then starts to decrease. This maximum value is referred to as the average outgoing quality limit (AOQL). The hypergeometric distribution is the appropriate distribution to calculate the probability of acceptance P_a; however, the Poisson distribution is used as an approximation.

Nonrectifying inspection program does not significantly improve the quality level of the lots inspected.

36.7.1 Double Sampling

Double sampling involves the possibility of putting off the decision on the lot until a second sample has been taken. A lot may be accepted at once if the first sample is good enough or rejected at once if the first sample is bad enough. If the first sample is neither, the decision is based on the evidence of the first and second samples combined.

The symbols used in double sampling are

N = lot size

n_1 = first sample

c_1 = acceptance number for first sample

n_2 = second sample

c_2 = acceptance number of the two samples combined

Computer programs are used to calculate the OC curves; acceptance after the first sample, rejection after the first sample, acceptance after the second sample, and rejection after the second sample.

The average sample number (ASN) in double sampling is given by

$$ASN = [P_a(n_1) + P_r(n_1)]n_1 + [P_a(n_2) + P_r(n_2)](n_1 + n_2) \tag{36.32}$$

36.7.2 Multiple and Sequential Sampling

In multiple sampling, three or more samples of a stated size are permitted and the decision on acceptance or rejection is revealed after a stated number of samples.

In sequential sampling, item-by-item inspection, a decision is possible after each item has been inspected and when there is no specified limit on the total number of units to be inspected.

OC curves are developed through computer programs. The advantage of using double sampling, multiple sampling, or sequential sampling is to reach the appropriate decision with fewer items inspected.

36.8 DEFENSE DEPARTMENT ACCEPTANCE SAMPLING BY VARIABLES

MIL-STD-105 A, B, C, D, and then ABC-STD-105, are based on the Acceptance Quality Level (AQL) concept. The plans contain single, double, or multiple sampling, depending on the lot size and AQL and the probability of acceptance at this level P_a. Criteria for shifting to tightened inspection, requalification for normal inspection, and reduced inspection are listed in the tables associated with plan.

MIL-STD-414 plans were developed to reduce inspection lots by using sample sizes compared to MIL-STD-105. They are similar, as both procedures and tables are based on the concept of AQL; lot-by-lot acceptance inspection; both provide for normal, tightened, or reduced inspection; sample sizes are greatly influenced by lot size; several inspection levels are available; and all plans are identified by sample size code letter. MIL-STD-414 could be applied either with a single specification limit, L or U, or with two specification limits. Known-sigma plans included in the standard were designated as having "variability known." Unknown-sigma plans were designated as having "variability unknown." In the latter-type plans, it was possible to use either the standard deviation method or the range method in estimating the lot variability.

BIBLIOGRAPHY

ASTM Manual on Presentation of Data and Control Chart Analysis, Special Technical Pub. 15D, American Society for Testing and Materials, Philadelphia, PA, 1976.

Besterfield, D. H., *Quality Control,* 4th ed., Prentice Hall, Englewood Cliffs, NJ, 1994.

Clements, R., *Quality ITQM/ISO 9000,* Prentice-Hall, Englewood Cliffs, NJ, 1995.

Control Chart Method of Controlling Quality During Production, ANSI Standard 21.3-1975, American National Standards Institute, New York, 1975.

Devore, J. L., *Probability and Statistics for Engineering and the Sciences,* Duxburg Press, New York, 1995 (software included).

Dodge, H. F., *A General Procedure for Sampling Inspection by Attributes—Based on the AQL Concept,* Technical Report No. 10, The Statistics Center, Rutgers—The State University, New Brunswick, NJ, 1959.

Romig, H. G., *Sampling Inspection Tables—Single and Double Sampling,* 2nd ed., Wiley, New York, 1959.

Duncan, A. J., *Quality Control and Industrial Statistics,* 5th ed., Richard D. Irwin, Homewood, IL, 1986.

Feigenbaum, A. V., *Total Quality Control—Engineering and Mangement,* 3rd ed., McGraw-Hill, New York, 1991.

Grant, E. L., and R. S. Leavenworth, *Statistical Quality Control,* 7th ed., McGraw Hill, New York, 1996 (software included).

Juran, J. M., and F. M. Gryna, Jr., *Quality Control Handbook,* McGraw-Hill, New York, 1988.

Lamprecht, J., *Implementing the ISO 9000 Series,* Marcel Dekker, New York, 1995.

Military Standard 105E, Sampling Procedures and Tables for Inspection by Attributes, Superintendent of Documents, Government Printing Office, Washington, DC, 1989 (nominally, the ABC standard).

Military Standard 414, Sampling Procedures and Tables for Inspection by Variables for Percent Defective, Superintendent of Documents, Government Printing Office, Washington, DC, 1957.

Military Standard 690-B, Failure Rate Sampling Plans and Procedures, Superintendent of Documents, Government Printing Office, Washington, DC, 1969.

Military Standard 781-C, Reliability Design Qualification and Production Acceptance Tests, Exponential Distribution, Superintendent of Documents, Government Printing Office, Washington, DC, 1977.

Military Standard 1235B, Single-and-Multi-Level Continuous Sampling Procedures and Tables for Inspection by Attributes, Superintendent of Documents, Government Printing Office, Washington, DC, 1981.

Montgomery, D. C., and G. C. Runger, *Applied Statistics and Probability for Engineers,* Wiley, New York, 1994.

Rabbit, J., and P. Bergh, *The ISO 9000 Book,* ME, Dearborn, MI, 1993.

Society of Manufacturing Engineers, *Quality Control and Assembly,* 4th ed., Dearborn, MI, 1994.

Supply and Logistic Handbook—Inspection H 105. Administration of Sampling Procedures for Acceptance Inspection, Superintendent of Documents, Government Printing Office, Washington, DC, 1954.

Zohdi, M. E., *Manufacturing Processes Quality Evaluation and Testing,* International Conference, Operations Research, January 1976.

CHAPTER 37

COMPUTER-INTEGRATED MANUFACTURING

William E. Biles
Department of Industrial Engineering
University of Louisville
Louisville, Kentucky

Magd E. Zohdi
Department of Industrial and Manufacturing Engineering
Louisiana State University
Baton Rouge, Louisiana

37.1	**INTRODUCTION**	**1187**		**37.4**	**INDUSTRIAL ROBOTS**	**1195**
					37.4.1 Definition	1195
37.2	**DEFINITIONS AND**				37.4.2 Robot Configurations	1196
	CLASSIFICATONS	**1188**			37.4.3 Robot Control and	
	37.2.1 Automation	1188			Programming	1197
	37.2.2 Production Operations	1189			37.4.4 Robot Applications	1197
	37.2.3 Production Plants	1190				
	37.2.4 Models for Production			**37.5**	**COMPUTERS IN**	
	Operations	1190			**MANUFACTURING**	**1197**
					37.5.1 Hierarchical Computer	
37.3	**NUMERICAL-CONTROL**				Control	1197
	MANUFACTURING SYSTEMS	**1192**			37.5.2 CNC and DNC Systems	1198
	37.3.1 Numerical Control	1192			37.5.3 The Manufacturing Cell	1198
	37.3.2 The Coordinate System	1192			37.5.4 Flexible Manufacturing	
	37.3.3 Selection of Parts for NC				Systems	1198
	Machining	1193				
	37.3.4 CAD/CAM Part			**37.6**	**GROUP TECHNOLOGY**	**1199**
	Programming	1193			37.6.1 Part Family Formation	1200
	37.3.5 Programming by Scanning				37.6.2 Parts Classification and	
	and Digitizing	1194			Coding	1200
	37.3.6 Adaptive Control	1194			37.6.3 Production Flow Analysis	1201
	37.3.7 Machinability Data				37.6.4 Types of Machine Cell	
	Prediction	1195			Designs	1201
					37.6.5 Computer-Aided Process	
					Planning	1203

37.1 INTRODUCTION

Modern manufacturing systems are advanced automation systems that use computers as an integral part of their control. Computers are a vital part of automated manufacturing. They control stand-alone manufacturing systems, such as various machine tools, welders, laser-beam cutters, robots, and automatic assembly machines. They control production lines and are beginning to take over control of the entire factory. The computer-integrated-manufacturing system (CIMS) is a reality in the modern industrial society. As illustrated in Fig. 37.1, CIMS combines computer-aided design (CAD), computer-aided manufacturing (CAM), computer-aided inspection (CAI), and computer-aided pro-

Mechanical Engineers' Handbook, 2nd ed., Edited by Myer Kutz.
ISBN 0-471-13007-9 © 1998 John Wiley & Sons, Inc.

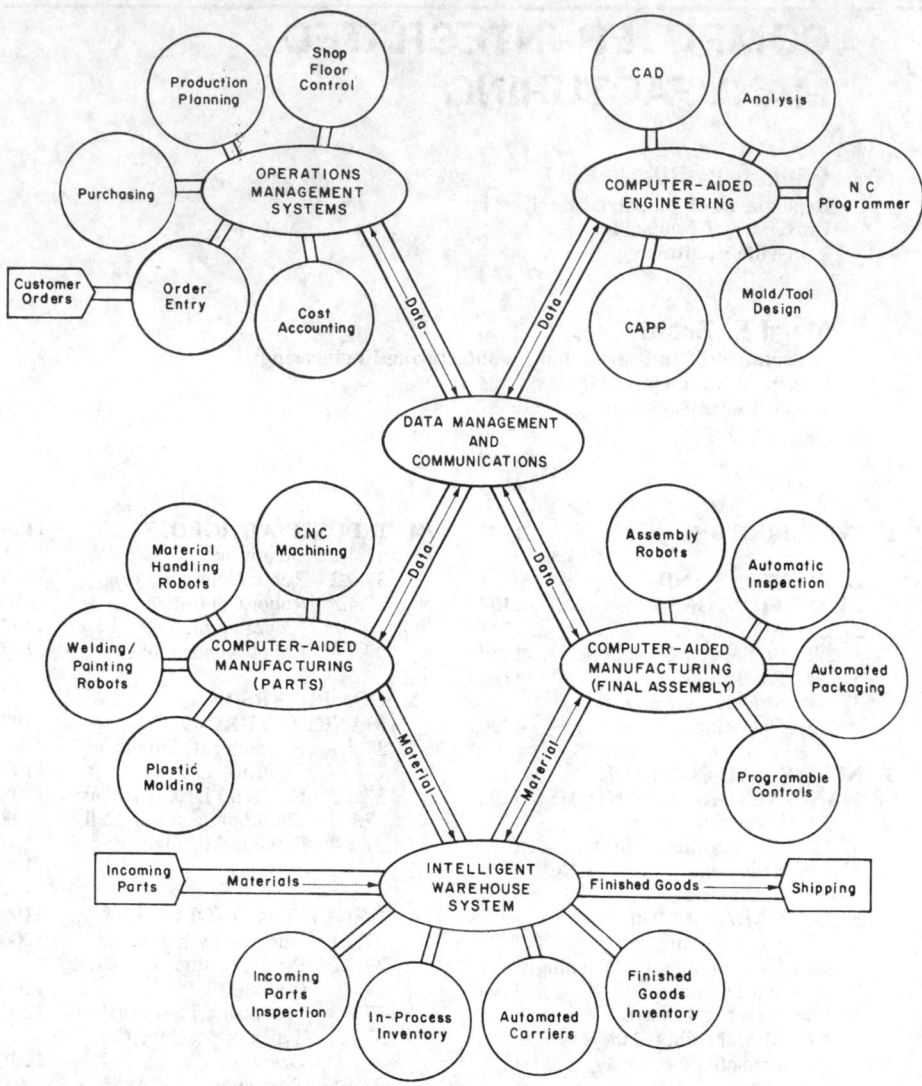

Fig. 37.1 Computer-integrated manufacturing system.

duction planning (CAPP), along with automated material handling. This chapter focuses on computer-aided manufacturing for both parts fabrication and assembly, as shown in Fig. 37.1. It treats numerical-control (NC) machining, robotics, and group technology. It shows how to integrate these functions with automated material storage and handling to form a CIM system.

37.2 DEFINITIONS AND CLASSIFICATIONS

37.2.1 Automation

Automation is a relatively new word, having been coined in the 1930s as a substitute for the word *automatization*, which referred to the introduction of automatic controls in manufacturing. Automation implies the performance of a task without human assistance. Manufacturing processes are classified as manual, semiautomatic, or automatic, depending on the extent of human involvement in the ongoing operation of the process.

The primary reasons for automating a manufacturing process are to

1. Reduce the cost of the manufactured product, through savings in both material and labor
2. Improve the quality of the manufactured product by eliminating errors and reducing the variability in product quality
3. Increase the rate of production
4. Reduce the lead time for the manufactured product, thus providing better service for customers
5. Make the workplace safer

The economic reality of the marketplace has provided the incentive for industry to automate its manufacturing processes. In Japan and in Europe, the shortage of skilled labor sparked the drive toward automation. In the United States, stern competition from Japanese and European manufacturers, in terms of both product cost and product quality, has necessitated automation. Whatever the reasons, a strong movement toward automated manufacturing processes is being witnessed throughout the industrial nations of the world.

37.2.2 Production Operations

Production is a transformation process in which raw materials are converted into the goods demanded in the marketplace. Labor, machines, tools, and energy are applied to materials at each of a sequence of steps that bring the materials closer to a marketable final state. These individual steps are called *production operations*.

There are three basic types of industries involved in transforming raw materials into marketable products:

1. *Basic producers.* These transform natural resources into raw materials for use in manufacturing industry—for example, iron ore to steel ingot in a steel mill.
2. *Converters.* These take the output of basic producers and transform the raw materials into various industrial products—for example, steel ingot is converted into sheet metal.
3. *Fabricators.* These fabricate and assemble final products—for example, sheet metal is fabricated into body panels and assembled with other components into an automobile.

The concept of a computer-integrated-manufacturing system as depicted in Fig. 37.1 applies specifically to a "fabricator" type of industry. It is the "fabricator" industry that we focus on in this chapter.

The steps involved in creating a product are known as the "manufacturing cycle." In general, the following functions will be performed within a firm engaged in manufacturing a product:

1. *Sales and marketing.* The order to produce an item stems either from customer orders or from production orders based on product demand forecasts.
2. *Product design and engineering.* For proprietary products, the manufacturer is responsible for development and design, including component drawings, specifications, and bill of materials.
3. *Manufacturing engineering.* Ensuring manufacturability of product designs, process planning, design of tools, jigs, and fixtures, and "troubleshooting" the manufacturing process.
4. *Industrial engineering.* Determining work methods and time standards for each production operation.
5. *Production planning and control.* Determining the master production schedule, engaging in material requirements planning, operations scheduling, dispatching job orders, and expediting work schedules.
6. *Manufacturing.* Performing the operations that transform raw materials into finished goods.
7. *Material handling.* Transporting raw materials, in-process components, and finished goods between operations.
8. *Quality control.* Ensuring the quality of raw materials, in-process components, and finished goods.
9. *Shipping and receiving.* Sending shipments of finished goods to customers, or accepting shipments of raw materials, parts, and components from suppliers.
10. *Inventory control.* Maintaining supplies of raw materials, in-process items, and finished goods so as to provide timely availability of these items when needed.

Thus, the task of organizing and coordinating the activities of a company engaged in the manufacturing enterprise is complex. The field of industrial engineering is devoted to such activities.

37.2.3 Production Plants

There are several ways to classify production facilities. One way is to refer to the volume or rate of production. Another is to refer to the type of plant layout. Actually, these two classification schemes are related, as will be pointed out.

In terms of the volume of production, there are three types of manufacturing plants:

1. *Job shop production.* Commonly used to meet specific customer orders; great variety of work; production equipment must be flexible and general purpose; high skill level among workforce—for example, aircraft manufacturing.

2. *Batch production.* Manufacture of product in medium lot sizes; lots produced only once at regular intervals; general-purpose equipment, with some specialty tooling—for example, household appliances, lawn mowers.

3. *Mass production.* Continuous specialized manufacture of identical products; high production rates; dedicated equipment; lower labor skills than in a job shop or batch manufacturing—for example, automotive engine blocks.

In terms of the arrangement of production resources, there are three types of plant layouts. These include

1. *Fixed-position layout.* The item is placed in a specific location and labor and equipment are brought to the site. Job shops often employ this type of plant layout.

2. *Process layout.* Production machines are arranged in groups according to the general type of manufacturing process; forklifts and hand trucks are used to move materials from one work center to the next. Batch production is most often performed in process layouts.

3. *Product-flow layout.* Machines are arranged along a line or in a U or S configuration, with conveyors transporting work parts from one station to the next; the product is progressively fabricated as it flows through the succession of workstations. Mass production is usually conducted in a product-flow layout.

37.2.4 Models for Production Operations

In this section, we will examine three types of models by which we can examine production operations, including graphical models, manufacturing process models, and mathematical models of production activity.

Process-flow charts depict the sequence of operations, storages, transportations, inspections, and delays encountered by a workpart of assembly during processing. As illustrated in Fig. 37.2, a process-flow chart gives no representation of the layout or physical dimensions of a process, but

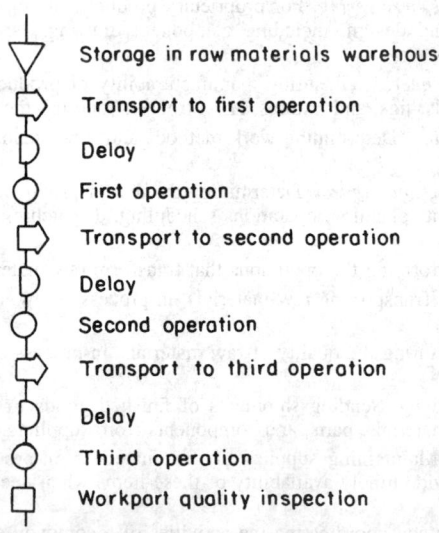

Fig. 37.2   Flow process chart for a sample workpart.

focuses on the succession of steps seen by the product. It is useful in analyzing the efficiency of the process, in terms of the proportion of time spent in transformation operations as opposed to transportations, storages, and delays.

The manufacturing-process model gives a graphical depiction of the relationship among the several entities that comprise the process. It is an input–output model. Its inputs are raw materials, equipment (machine tools), tooling and fixtures, energy, and labor. Its outputs are completed workpieces, scrap, and waste. These are shown in Fig. 37.3. Also shown in this figure are the controls that are applied to the process to optimize the utilization of the inputs in producing completed workpieces, or in maximizing the production of completed workpieces at a given set of values describing the inputs.

Mathematical models of production activity quantify the elements incorporated into the process-flow chart. We distinguish between operation elements, which are involved whenever the work part is on the machine and correspond to the circles in the process-flow chart, and nonoperation elements, which include storages, transportations, delays, and inspections. Letting T_o represent operation time per machine, T_{no} the nonoperation time associated with each operation, and n_m the number of machines or operations through which each part must be processed, then the total time required to process the part through the plant [called the manufacturing lead time (T_l)] is

$$T_l = n_m(T_o + T_{no})$$

If there is a batch of p parts,

$$T_l = n_m(pT_o + T_{no})$$

If a setup of duration T_{su} is required for each batch,

$$T_l = n_m(T_{su} + pT_o + T_{no})$$

The total batch time per machine, T_b, is given by

$$T_b = T_{su} + pT_o$$

The average production time T_a per part is therefore

$$T_a = \frac{T_{su} + pT}{p}$$

The average production rate for each machine is

$$R_a = 1/T_a$$

As an example, a part requires six operations (machines) through the machine shop. The part is produced in batches of 100. A setup of 2.5 hr is needed. Average operation time per machine is 4.0 min. Average nonoperation time is 3.0 hr. Thus,

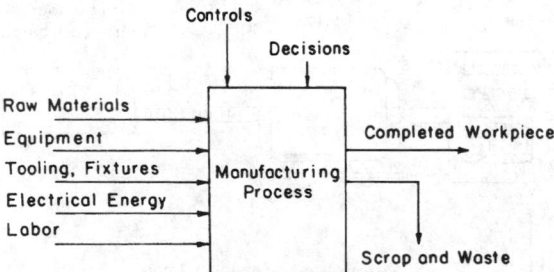

Fig. 37.3 General input-output model of the manufacturing process.

$$n_m = 6 \text{ machines}$$
$$p = 100 \text{ parts}$$
$$T_{su} = 2.5 \text{ hr}$$
$$T_o = 4/60 \text{ hr}$$
$$T_{no} = 3.0 \text{ hr}$$

Therefore, the total manufacturing lead time for this batch of parts is

$$T_l = 6[2.5 + 100(0.06667) + 3.0] = 73.0 \text{ hr}$$

If the shop operates on a 40-hr week, almost two weeks are needed to complete the order.

37.3 NUMERICAL-CONTROL MANUFACTURING SYSTEMS

37.3.1 Numerical Control

The most commonly accepted definition of numerical control (NC) is that given by the Electronic Industries Association (EIA): A system in which motions are controlled by the direct insertion of numerical data at some point. The system must automatically interpret at least some portion of these data.

The numerical control system consists of five basic, interrelated components, as follows:

1. Data input devices
2. Machine control unit
3. Machine tool or other controlled equipment
4. Servo-drives for each axis of motion
5. Feedback devices for each axis of motion

The major components of a typical NC machine tool system are shown in Fig. 37.4.

The programmed codes that the machine control unit (MCU) can read may be perforated tape or punched tape, magnetic tape, tabulating cards, or signals directly from computer logic or some computer peripherals, such as disk or drum storage. Direct computer control (DCC) is the most recent development, and one that affords the help of a computer in developing a part program.

37.3.2 The Coordinate System

The Cartesian coordinate system is the basic system in NC control. The three primary linear motions for an NC machine are given as X, Y, and Z. Letters A, B, and C indicate the three rotational axes, as in Fig. 37.5.

NC machine tools are commonly classified as being either point-to-point or continuous path. The simplest form of NC is the point-to-point machine tool used for operations such as drilling, tapping, boring, punching, spot welding, or other operations that can be completed at a fixed coordinate position with respect to the workpiece. The tool does not contact the workpiece until the desired coordinate position has been reached; consequently, the exact path by which this position is reached is not important.

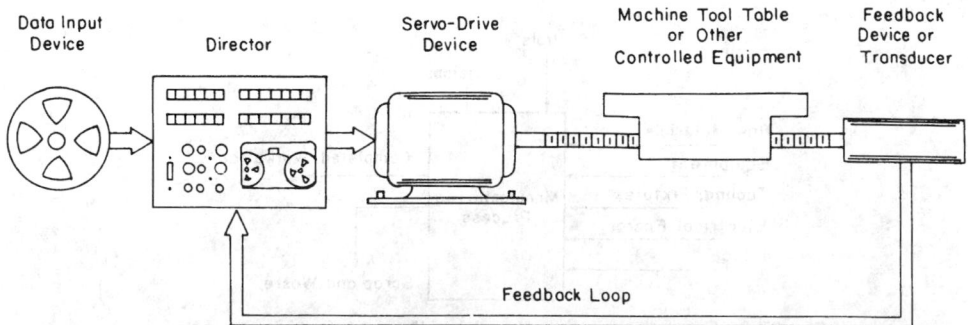

Fig. 37.4 Simplified numerical control system.

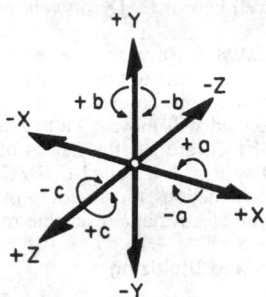

Fig. 37.5 An example of typical axis nomenclature for machine tools.

With continuous-path (contouring) NC systems, there is contact between the workpiece and the tool as the relative movements are made. Continuous-path NC systems are used primarily for milling and turning operations that can profile and sculpture workpieces. Other NC continuous-path operations include flame cutting, sawing, grinding, and welding, and even operations such as the application of adhesives. We should note that continuous-path systems can be programmed to perform point-to-point operations, although the reverse (while technically possible) is infrequently done.

37.3.3 Selection of Parts for NC Machining

Parts selection for NC should be based on an economic evaluation, including scheduling and machine availability. Economic considerations affecting NC part selection including alternative methods, tooling, machine loadings, manual versus computer-assisted part programming, and other applicable factors.

Thus, NC should be used only where it is more economical or does the work better or faster, or where it is more accurate than other methods. The selection of parts to be assigned to NC has a significant effect on its payoff. The following guidelines, which may be used for parts selection, describe those parts for which NC may be applicable.

1. Parts that require *substantial tooling costs* in relation to the total manufacturing costs by conventional methods
2. Parts that require *long setup times* compared to the machine run time in conventional machining
3. Parts that are machined in *small or variable lots*
4. A *wide diversity of parts* requiring frequent changes of machine setup and a large tooling inventory if conventionally machined
5. Parts that are *produced at intermittent times* because demand for them is cyclic
6. Parts that have *complex configurations* requiring close tolerances and intricate relationships
7. Parts that have *mathematically defined complex contours*
8. Parts that require *repeatability* from part to part and lot to lot
9. *Very expensive* parts where human error would be very costly and increasingly so as the part nears completion
10. *High-priority* parts where lead time and flow time are serious considerations
11. Parts with *anticipated design changes*
12. Parts that involve a *large number of operations* or *machine setups*
13. Parts where *non-uniform cutting conditions* are required
14. Parts that require 100% *inspection* or require measuring many checkpoints, resulting in high inspection costs
15. *Family of parts*
16. *Mirror-image parts*
17. *New parts* for which conventional tooling does not already exist
18. Parts that are suitable for *maximum machining* on NC machine tools

37.3.4 CAD/CAM Part Programming

Computer-Aided Design (CAD) consists of using computer software to produce drawings of parts or products. These drawings provide the dimensions and specifications needed by the machinist to produce the part or product. Some well-known CAD software products include *AutoCAD, Cadkey,* and *Mastercam.*

Computer-Aided Manufacturing (CAM) involves the use of software by NC programmers to create programs to be read by a CNC machine in order to manufacture a desired shape or surface. The end product of this effort is an NC program stored on disk, usually in the form of G codes, that when loaded into a CNC machine and executed will move a cutting tool along the programmed path to create the desired shape. If the CAM software has the means of creating geometry, as opposed to importing the geometry from a CAD system, it is called *CAD/CAM.* CAD/CAM software, such as Mastercam, is capable of producing instructions for a variety of machines, including lathes, mills, drilling and tapping machines, and wire electrostatic discharge machining (EDM) processes.

37.3.5 Programming by Scanning and Digitizing

Programming may be done directly from a drawing, model, pattern, or template by digitizing or scanning. An optical reticle or other suitable viewing device connected to an arm is placed over the drawing. Transducers will identify the location and translate it either to a tape puncher or other suitable programming equipment. Digitizing is used in operations such as sheet-metal punching and hole drilling. A scanner enables an operator to program complex free-form shapes by manually moving a tracer over the contour of a model or premachined part. Data obtained through the tracer movements are converted into tape by a minicomputer. Digitizing and scanning units have the capability of editing, modifying, or revising the basic data gathered.

37.3.6 Adaptive Control

Optimization processes have been developed to improve the operational characteristics of NC machine-tool systems. Two distinct methods of optimization are adaptive control and machinability data prediction. Although both techniques have been developed for metal-cutting operations, adaptive control finds application in other technological fields.

The adaptive control (AC) system is an evolutionary outgrowth of numerical control. AC optimizes an NC process by sensing and logically evaluating variables that are not controlled by position and velocity feedback loops. Essentially, an adaptive control system monitors process variables, such as cutting forces, tool temperatures, or motor torque, and alters the NC commands so that optimal metal removal or safety conditions are maintained.

A typical NC configuration (Fig. 37.6a) monitors position and velocity output of the servo system, using feedback data to compensate for errors between command response. The AC feedback loop (Fig. 37.6b) provides sensory information on other process variables, such as workpiece–tool air gaps, material property variations, wear, cutting depth variations, or tool deflection. This information is determined by techniques such as monitoring forces on the cutting tool, motor torque variations,

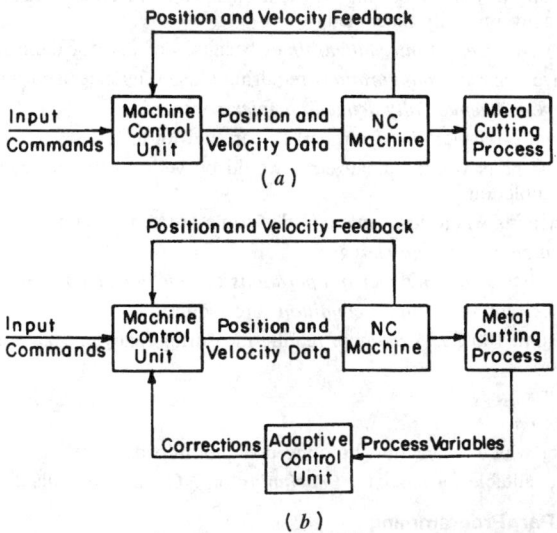

Fig. 37.6 Schematic diagrams for conventional and adaptive NC systems.

or tool–workpiece temperatures. The data are processed by an adaptive controller that converts the process information into feedback data to be incorporated into the Machine Control Unit output.

37.3.7 Machinability Data Prediction

The specification of suitable feeds and speeds is essentially in conventional and NC cutting operations. Machinability data are used to aid in the selection of metal-cutting parameters based on the machining operation, the tool and workpiece material, and one or more production criteria. Techniques used to select machinability data for conventional machines have two important drawbacks in relation to NC applications: data are generally presented in a tabular form that requires manual interpolation, checkout, and subsequent revisions; and tests on the machine tool are required to find optimum conditions.

Specialized machinability data systems have been developed for NC application to reduce the need for machinability data testing and to decrease expensive NC machining time. Part programming time is also reduced when machinability information is readily available.

A typical process schematic showing the relationship between machinability data and NC process flow is illustrated in Fig. 37.7.

37.4 INDUSTRIAL ROBOTS

37.4.1 Definition

As defined by the Robot Institute of America, "a robot is a reprogrammable, multifunctional manipulator designed to handle material, parts, tools or specialized devices through variable programmed motions for the performance of a variety of tasks."

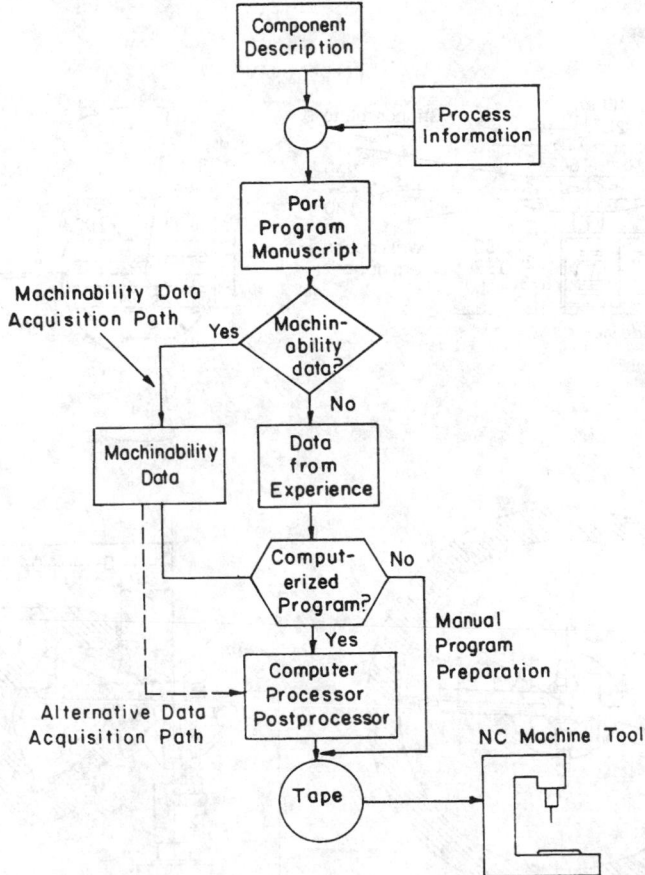

Fig. 37.7 Acquisition of machinability data in the NC process flow.

Robots have the following components:

1. *Manipulator.* The mechanical unit or "arm" that performs the actual work of the robot, consisting of mechanical linkages and joints with actuators to drive the mechanism directly through gears, chains, or ball screws.
2. *Feedback Devices.* Transducers that sense the positions of various linkages or joints and transmit this information to the controller.
3. *Controller.* Computer used to initiate and terminate motion, store data for position and sequence, and interface with the system in which the robot operates.
4. *Power Supply.* Electric, pneumatic, and hydraulic power systems used to provide and regulate the energy needed for the manipulator's actuators.

37.4.2 Robot Configurations

Industrial robots have one of three mechanical configurations, as illustrated in Fig. 37.8. Cylindrical coordinate robots have a work envelope that is composed of a portion of a cylinder. Spherical coordinate robots have a work envelope that is a portion of a sphere. Jointed-arm robots have a work envelope that approximates a portion of a sphere. There are six motions or degrees of freedom in the design of a robot—three arm and body motions and three wrist movements.

Arm and body motions:

1. *Vertical traverse*—an up-and-down motion of the arm
2. *Radial traverse*—an in-and-out motion of the arm
3. *Rotational traverse*—rotation about the vertical axis (right or left swivel of the robot body)

Wrist motions:

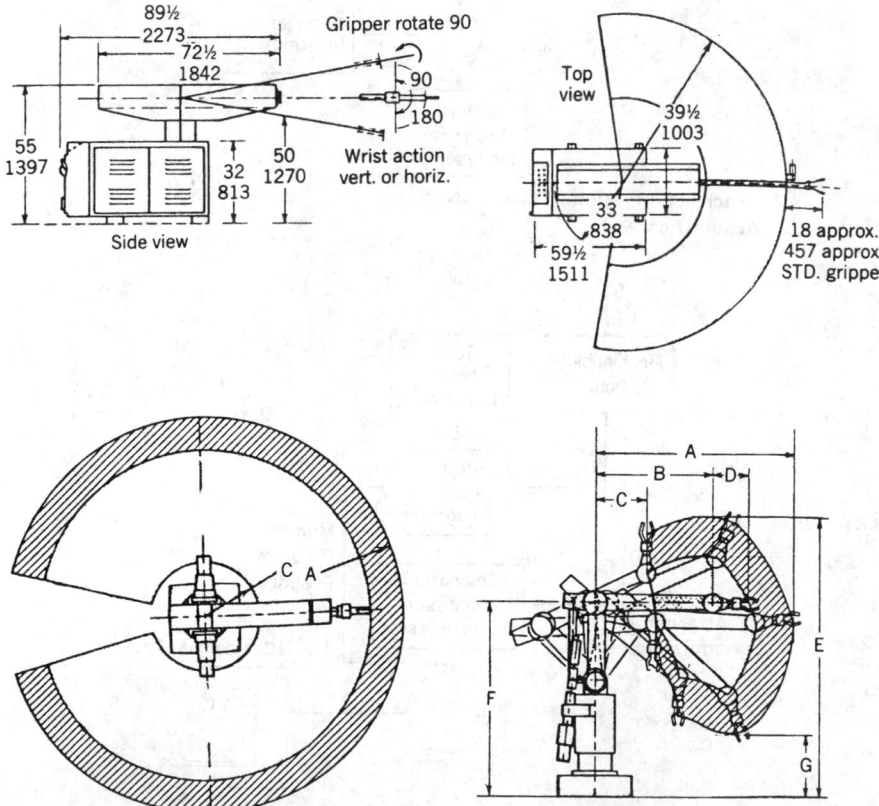

Fig. 37.8 Mechanical configurations of industrial robots.

4. *Wrist swivel*—rotation of the wrist
5. *Wrist bend*—up-and-down movement of the wrist
6. *Wrist yaw*—right or left swivel of the wrist

The mechanical hand movement, usually opening and closing, is not considered one of the basic degrees of freedom of the robot.

37.4.3 Robot Control and Programming

Robots can also be classified according to type of control. Point-to-point robot systems are controlled from one programmed point in the robot's control to the next point. These robots are characterized by high load capacity, large working range, and relative ease of programming. They are suitable for pick-and-place, material handling, and machine loading tasks.

Contouring robots, on the other hand, possess the capacity to follow a closely spaced locus of points that describe a smooth, continuous path. The control of the path requires a large memory to store the locus of points. Continuous-path robots are therefore more expensive than point-to-point robots, but they can be used in such applications as seam welding, flame cutting, and adhesive beading.

There are three principal systems for programming robots:

1. *Manual method.* Used in older, simpler robots, the program is set up by fixing stops, setting switches, and so on.
2. *Walk-through.* The programmer "teaches" the robot by actually moving the hand through a sequence of motions or positions, which are recorded in the memory of the computer.
3. *Lead-through.* The programmer drives the robot through a sequence of motions or positions using a console or teach pendant. Each move is recorded in the robot's memory.

37.4.4 Robot Applications

A current directory of robot applications in manufacturing includes the following:

1. Material handling
2. Machine loading and unloading
3. Die casting
4. Investment casting
5. Forging and heat treating
6. Plastic molding
7. Spray painting and electroplating
8. Welding (spot welding and seam welding)
9. Inspection
10. Assembly

Research and development efforts are under way to provide robots with sensory perception, including voice programming, vision and "feel." These capabilities will no doubt greatly expand the inventory of robot applications in manufacturing.

37.5 COMPUTERS IN MANUFACTURING

Flexible manufacturing systems combined with automatic assembly and product inspection, on the one hand, and integrated CAD/CAM systems, on the other hand, are the basic components of the computer-integrated manufacturing system. The overall control of such systems is predicated on hierarchical computer control, such as illustrated in Fig. 37.9.

37.5.1 Hierarchical Computer Control

The lowest level of the hierarchical computer control structure illustrated in Fig. 37.9 contains stand-alone computer control systems of manufacturing processes and industrial robots. The computer control of processes includes all types of CNC machine tools, welders, electrochemical machining (ECM), electrical discharge machining (EDM), and laser-cutting machines.

When a set of NC or CNC machine tools is placed under the direct control of a single computer, the resulting system is known as a *direct-numerical-control* (DNC) system. DNC systems can produce several different categories of parts or products, perhaps unrelated to one another. When several CNC machines and one or more robots are organized into a system for the production of a single part or family of parts, the resulting system is called a *manufacturing cell*. The distinction between DNC systems and a manufacturing cell is that in DNC systems the same computer receives data from and issues instructions to several separate machines, whereas in manufacturing cells the computer coor-

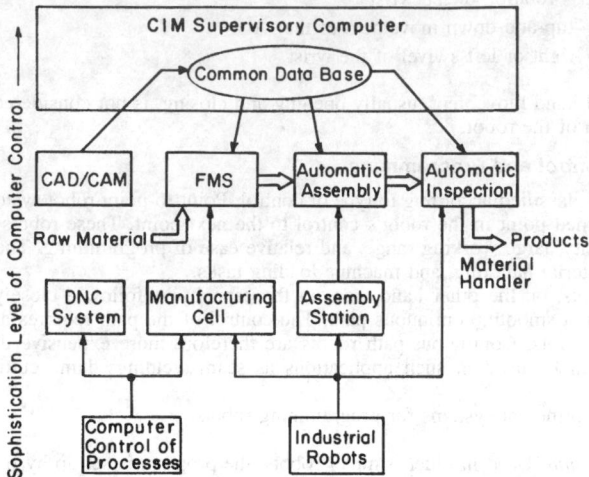

Fig. 37.9 Hierarchical computer control in manufacturing.

dinates the movements of several machines and robots working in concert. The computer receives "completion of job" signals from the machines and issues instructions to the robot to unload the machines and change their tools. The software includes strategies for handling machine breakdowns, tool wear, and other special situations.

The operation of several manufacturing cells can be coordinated by a central computer in conjunction with an automated material-handling system. This is the next level of control in the hierarchical structure and is known as a *flexible manufacturing system* (FMS). The FMS receives incoming workpieces and processes them into finished parts, completely under computer control.

The parts fabricated in the FMS are then routed on a transfer system to automatic assembly stations, where they are assembled into subassemblies or final product. These assembly stations can also incorporate robots for performing assembly operations. The subassemblies and final product may also be tested at automatic inspection stations.

As shown in Fig. 37.9, FMS, automatic assembly, and automatic inspection are integrated with CAD/CAM systems to minimize production lead time. These four functions are coordinated by means of the highest level of control in the hierarchical structure-computer-integrated-manufacturing (CIM) systems. The level of control is often called *supervisory computer control*.

The increase in productivity associated with CIM systems will not come from a speedup of machining operations, but rather from minimizing the direct labor employed in the plant. Substantial savings will also be realized from reduced inventories, with reductions in the range of 80–90%.

37.5.2 CNC and DNC Systems

The distinguishing feature of a CNC system is a dedicated computer, usually a microcomputer, associated with a single machine tool, such as a milling machine or a lathe. Programming the machine tools is managed through punched or magnetic tape, or directly from a keyboard.

DNC is another step beyond CNC, in that a number of CNC machines, ranging from a few to as many as 100, are connected directly to a remote computer. NC programs are downloaded directly to the CNC machine, which then processes a prescribed number of parts.

37.5.3 The Manufacturing Cell

The concept of a manufacturing cell is based on the notion of cellular manufacturing, wherein a group of machines served by one or more robots manufactures one part or one part family. Figure 37.10 depicts a typical manufacturing cell consisting of a CNC lathe, a CNC milling machine, a CNC drill, open conveyor to bring workparts into the cell, another to remove completed parts from the cell, and a robot to serve all these components. Each manufacturing cell is self-contained and self-regulating. The cell is usually made up of 10 or fewer machines. Those cells that are not completely automated are usually staffed with fewer personnel than machines, with each operator trained to handle several machines or processes.

37.5.4 Flexible Manufacturing Systems

Flexible manufacturing systems (FMS) combine many different automation technologies into a single production system. These include NC and CNC machine tools, automatic material handling between

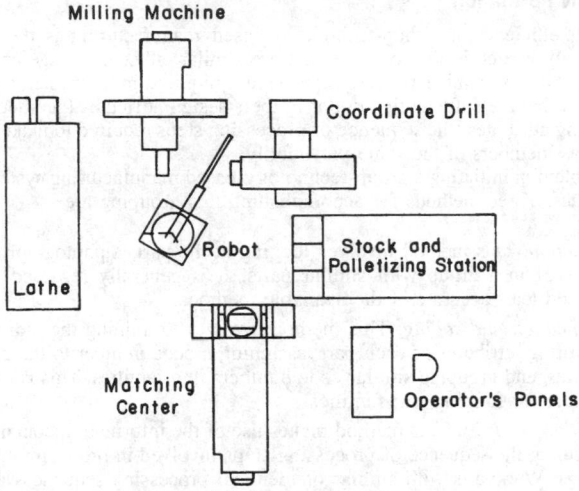

Fig. 37.10 A typical manufacturing cell.

machines, computer control over the operation of the material handling system and machine tools, and group technology principles. Unlike the manufacturing cell, which is typically dedicated to the production of a single parts family, the FMS is capable of processing a variety of part types simultaneously under NC control at the various workstations.

Human labor is used to perform the following functions to support the operation of the FMS:

- Load raw workparts into the system
- Unload finished workparts from the system
- Change tools and tool settings
- Equipment maintenance and repair

Robots can be used to replace human labor in certain areas of these functions, particularly those involving material or tool handling. Figure 37.11 illustrates a sample FMS layout.

37.6 GROUP TECHNOLOGY

Group technology is a manufacturing philosophy in which similar parts are identified and grouped together to take advantage of similarities in design and/or manufacture. Similar parts are grouped into part families. For example, a factory that produces as many as 10,000 different part numbers can group most of these parts into as few as 50 distinct part families. Since the processing of each family would be similar, the production of part families in dedicated manufacturing cells facilitates workflow. Thus, group technology results in efficiencies in both product design and process design.

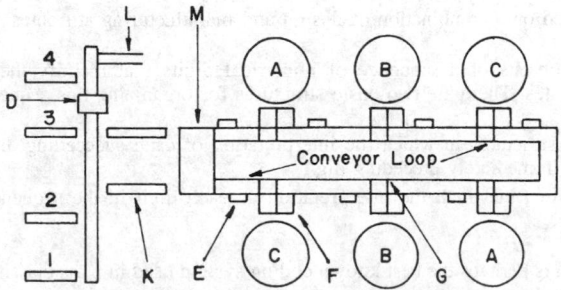

Fig. 37.11 A flexible manufacturing system.

37.6.1 Part Family Formation

The key to gaining efficiency in group-technology-based manufacturing is the formation of part families. A part family is a collection of parts that are similar either due to geometric features such as size and shape or because similar processing steps are required in their manufacture. Parts within a family are different, but are sufficiently similar in their design attributes (geometric size and shape) and/or manufacturing attributes (the sequence of processing steps required to make the part) to justify their identification as members of the same part family.

The biggest problem in initiating a group-technology-based manufacturing system is that of grouping parts into families. Three methods for accomplishing this grouping are

1. *Visual inspection.* This method involves looking at the part, a photograph, or a drawing and placing the part in a group with similar parts. It is generally regarded as the most time-consuming and least accurate of the available methods.
2. *Parts classification and coding.* This method involves examining the individual design and/or manufacturing attributes of each part, assigning a code number to the part on the basis of these attributes, and grouping similar code numbers into families. This is the most commonly used procedure for forming part families.
3. *Production flow analysis.* This method makes use of the information contained on the routing sheets describing the sequence of processing steps involved in producing the part, rather than part drawings. Workparts with similar or identical processing sequences are grouped into a part family.

37.6.2 Parts Classification and Coding

As previously stated, parts classification and coding is the most frequently applied method for forming part families. Such a system is useful in both design and manufacture. In particular, parts coding and classification, and the resulting coding system, provide a basis for interfacing CAD and CAM in CIM systems. Parts classification systems fall into one of three categories:

1. Systems based on part design attributes:
    Basic external shape
    Basic internal shape
    Length/diameter ratio
    Material type
    Part function
    Major dimensions
    Minor dimensions
    Tolerances
    Surface finish
2. Systems based on part manufacturing attributes:
    Primary process
    Minor processes
    Major dimensions
    Length/diameter ratio
    Surface finish
    Machine tool
    Operation sequence
    Production time
    Batch size
    Annual production requirement
    Fixtures needed
    Cutting tools
3. Systems based on a combination of design and manufacturing attributes.

The part code consists of a sequence of numerical digits that identify the part's design and manufacturing attributes. There are two basic structures for organizing this sequence of digits:

1. Hierarchical structures in which the interpretation of each succeeding digit depends on the value of the immediately preceding digit
2. Chain structures in which the interpretation of each digit in the sequence is position-wise fixed

The Opitz system is perhaps the best known coding system used in parts classification and coding. The code structure is

12345 6789 ABCD

The first nine digits constitute the basic code that conveys both design and manufacturing data. The first five digits, 12345, are called the *form code* and give the primary design attributes of the part. The next four digits, 6789, constitute the *supplementary code* and indicate some of the manufacturing attributes of the part. The next four digits, ABCD, are called the *secondary code* and are used to indicate the production operations of type and sequence. Figure 37.12 gives the basic structure for the Opitz coding system. Note that digit 1 establishes two primary categories of parts, rotational and non-rotational, among nine separate part classes.

The MICLASS (Metal Institute Classification System) was developed by the Netherlands Organization for Applied Scientific Research to help automate and standardize a number of design, manufacturing, and management functions. MICLASS codes range from 12 to 30 digits, with the first 12 constituting a universal code that can be applied to any part. The remaining 18 digits can be made specific to any company or industry. The organization of the first 12 digits is as follows:

1st digit	main shape
2nd and 3rd digits	shape elements
4th digit	position of shape elements
5th and 6th digits	main dimensions
7th digit	dimension ratio
8th digit	auxiliary dimension
9th and 10th digits	tolerance codes
11th and 12th digits	material codes

MICLASS allows computer-interactive parts coding, in which the user responds to a series of questions asked by the computer. The number of questions asked depends on the complexity of the part and ranges from as few as 7 to more than 30, with an average of about 15.

37.6.3   Production Flow Analysis

Production flow analysis (PFA) is a method for identifying part families and associated grouping of machine tools. PFA is used to analyze the operations sequence of machine routing for the parts produced in a shop. It groups parts that have similar sequences and routings into a part family. PFA then establishes machine cells for the producing part families. The PFA procedure consists of the following steps:

1. Data collection is the gathering of part numbers and machine routings for each part produced in the shop
2. Sorting process routings into "packs" according to similarity
3. Constructing a PFA chart, such as depicted in Fig. 37.13, that shows the process sequence (in terms of machine code numbers) for each pack (denoted by a letter)
4. Analysis of the PFA chart in an attempt to identify similar packs. This is done by rearranging the data on the original PFA chart into a new pattern that groups packs having similar sequences. Figure 37.14 shows the rearranged PFA chart. The machines grouped together within the blocks in this figure form logical machine cells for producing the resulting part family

37.6.4   Types of Machine Cell Designs

The organization of machines into cells, whether based on parts classification and coding or PFA, follows one of three general patterns:

1. Single-machine cell
2. Group-machine layout
3. Flow-line cell layout

The single-machine pattern can be used for workparts whose attributes allow them to be produced using a single process. For example, a family composed of 40 different machine bolts can be produced on a single turret lathe.

The group-machine layout was illustrated in Fig. 37.13. The cell contains the necessary grouping of machine tools and fixtures for processing all parts in a given family, but material handling between machines is not fixed. The flow-line cell design likewise contains all machine tools and fixtures needed to produce a family of parts, but these are arranged in a fixed sequence with conveyors providing the flow of parts through the cell.

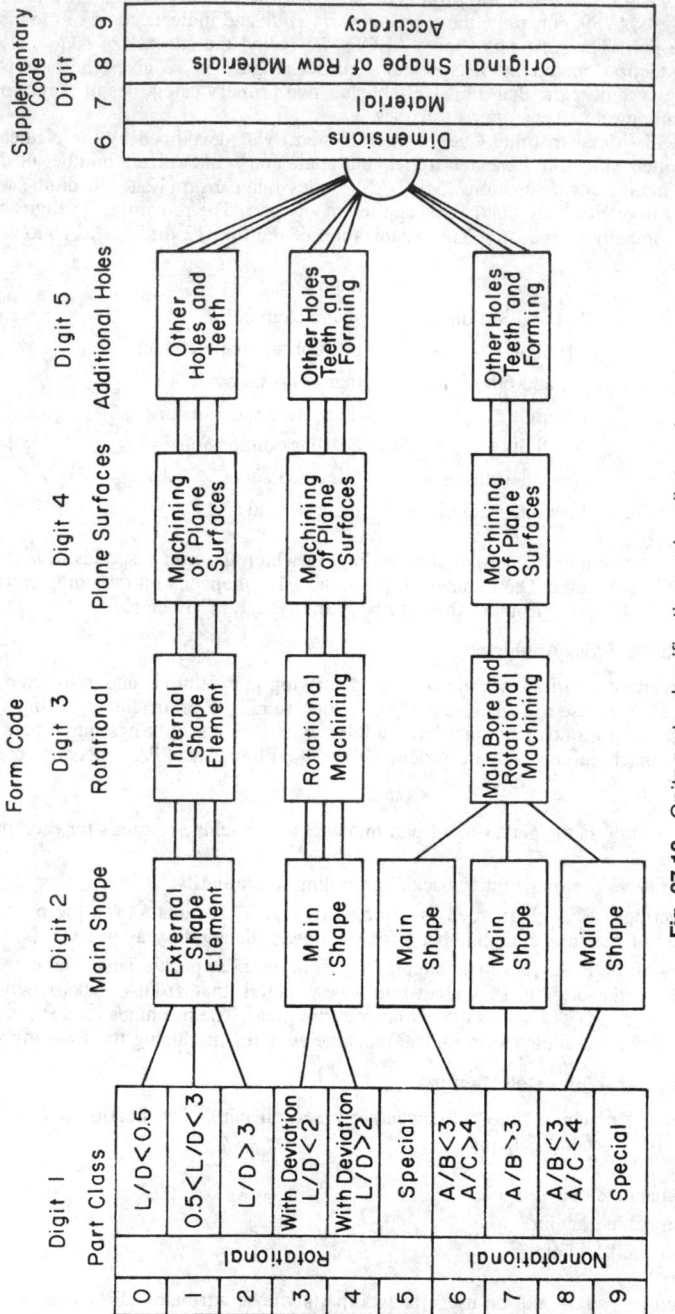

Fig. 37.12 Opitz parts classification and coding system.

Machine \ Part No.	1	2	3	4	5	6	7	8	9	10	11	12	13	14	15	16	17	18	19	20
Lathe	x	x		x	x		x	x	x		x	x		x	x		x	x	x	x
Milling Mach. I	x	x	x		x	x	x		x		x		x	x		x				x
Milling Mach. II			x	x				x		x		x	x		x		x	x	x	
Drilling Mach.	x	x	x	x		x	x	x		x	x	x	x	x		x	x	x		x
Grinding Mach.	x	x	x	x		x			x			x	x		x			x		x

Fig. 37.13 PFA chart.

Machine \ Part No.	1	2	20	7	11	14	9	5	4	18	12	8	17	15	19	3	13	6	16	10
Lathe	x	x	x	x	x	x	x	x	x											
Milling Mach. I	x	x	x	x	x	x	x	x	x											
Drilling Mach.	x	x	x	x	x	x														
Grinding Mach.	x	x	x				x													
Lathe								x	x	x	x	x	x	x						
Milling Mach. II								x	x	x	x	x	x	x						
Drilling Mach.								x	x	x	x	x								
Grinding Mach.								x	x	x			x							
Milling Mach. I														x	x	x	x			
Milling Mach. II														x	x				x	
Drilling Mach.														x	x	x	x	x	x	
Grilling Mach.														x	x	x				

Fig. 37.14 Rearranged PFA chart.

37.6.5 Computer-Aided Process Planning

Computer-aided process planning (CAPP) involves the use of a computer to automatically generate the operation sequence (routing sheet) based on information about the workpart. CAPP systems require some form of classification and coding system, together with standard process plans for specific part families. The flow of information in a CAPP system is initiated by having the user enter the part code for the workpart to be processed. The CAPP program then searches the part family matrix file to determine if a match exists. If so, the standard machine routing and the standard operation sequence are extracted from the computer file. If no such match exists, the user must then search the file for similar code numbers and manually prepare machine routings and operation sequences for dissimilar segments. Once this process has been completed, the new information becomes part of the master file so that the CAPP system generates an ever-growing data file.

BIBLIOGRAPHY

Asfahl, C. R., *Robots and Manufacturing Automation,* Wiley, New York, 1985.

CAD/CAM: Meeting Today's Productivity Challenge, Society of Manufacturing Engineers, Dearborn, MI, 1980.

Childs, J. J., *Principles of Numerical Control,* Industrial Press, New York, 1969.

Compact II, Programming Manual, Manufacturing Data Systems, Inc., Ann Arbor, MI, 1984.

Computer Integrated Manufacturing Systems: Selected 1985 Readings, Industrial Engineering and Management Press, Atlanta, GA, 1985.

Groover, M. P., *Automation, Production Systems, and Computer-Aided Manufacturing,* Prentice-Hall, Englewood Cliffs, NJ, 1980.

Groover, M. P., and E. W. Zimmers, Jr., *CAD/CAM: Computer-Aided Design and Manufacturing.* Tech Tran Corporation, Naperville, IL, 1984.

Hunt, V. D., *Smart Robots,* Tech Tran Corporation, Naperville, IL, 1984.

Industrial Robots: A Summary and Forecast, Tech Tran Corporation, Naperville, IL, 1983.

Kafrissen, E., and M. Stephens, *Industrial Robots and Robotics,* Tech Tran Corporation, Naperville, IL, 1984.

Koren, Y., *Computer Control of Manufacturing Systems,* McGraw-Hill, New York, 1983.

Lindberg, R. A., *Processes and Materials of Manufacturing,* 2nd ed., Allyn and Bacon, Boston, MA, 1977.

Machining Data Handbook, 3rd ed., Machinability Data Center, Cincinnati, OH, 1980.

Morgan, C., *Robots: Planning and Implementation,* Tech Tran Corporation, Naperville, IL, 1984.

Numerical Control, Vol. 1, *Fundamentals,* Society of Manufacturing Engineers, Dearborn, MI, 1981.

Numerical Control, Vol. 2, *Applications,* Society of Manufacturing Engineers, Dearborn, MI, 1981.

Pao, Y. C., *Elements of Computer-Aided Design and Manufacturing,* Wiley, New York, 1984.

Pressman, R. S., and J. E. Williams, *Numerical Control and Computer-Aided Manufacturing,* Wiley, New York, 1977.

Roberts and Prentice, *Programming for Numerical Control Machines,* McGraw-Hill, New York, 1978.

Tool and Manufacturing Handbook, Society of Manufacturing Engineers, Dearborn, MI, 1984.

CHAPTER 38

MATERIAL HANDLING

William E. Biles
Mickey R. Wilhelm
Department of Industrial Engineering
University of Louisville
Louisville, Kentucky

Magd E. Zohdi
Department of Industrial and Manufacturing Engineering
Louisiana State University
Baton Rouge, Louisiana

38.1	**INTRODUCTION**	**1205**
38.2	**BULK MATERIAL HANDLING**	**1206**
38.2.1	Conveying of Bulk Solids	1206
38.2.2	Screw Conveyors	1207
38.2.3	Belt Conveyors	1207
38.2.4	Bucket Elevators	1208
38.2.5	Vibrating or Oscillating Conveyors	1208
38.2.6	Continuous-Flow Conveyors	1208
38.2.7	Pneumatic Conveyors	1208
38.3	**BULK MATERIALS STORAGE**	**1212**
38.3.1	Storage Piles	1212
38.3.2	Storage Bins, Silos, and Hoppers	1212
38.3.3	Flow-Assisting Devices and Feeders	1214
38.3.4	Packaging of Bulk Materials	1214
38.3.5	Transportation of Bulk Materials	1218
38.4	**UNIT MATERIAL HANDLING**	**1219**
38.4.1	Introduction	1219
38.4.2	Analysis of Systems for Material Handling	1220
38.4.3	Identifying and Defining the Problem	1220
38.4.4	Collecting Data	1220
38.4.5	Unitizing Loads	1223
38.5	**MATERIAL-HANDLING EQUIPMENT CONSIDERATIONS AND EXAMPLES**	**1225**
38.5.1	Developing the Plan	1225
38.5.2	Conveyors	1226
38.5.3	Hoists, Cranes and Monorails	1233
38.5.4	Industrial Trucks	1234
38.5.5	Automated Guided Vehicle Systems	1234
38.5.6	Automated Storage and Retrieval Systems	1234
38.5.7	Carousel Systems	1236
38.5.8	Shelving, Bin, Drawer, and Rack Storage	1238
38.6	**IMPLEMENTING THE SOLUTION**	**1239**

38.1 INTRODUCTION

Material handling is defined by the Materials Handling Institute (MHI) as the movement, storage, control, and protection of materials and products throughout the process of their manufacture, distribution, consumption, and disposal. The five commonly recognized aspects of material handling are:

Mechanical Engineers' Handbook, 2nd ed., Edited by Myer Kutz.
ISBN 0-471-13007-9 © 1998 John Wiley & Sons, Inc.

1. Motion. Parts, materials, and finished products that must be moved from one location to another should be moved in an efficient manner and at minimum cost.

2. Time. Materials must be where they are needed at the moment they are needed.

3. Place. Materials must be in the proper location and positioned for use.

4. Quantity. The rate of demand varies between the steps of processing operations. Materials must be continually delivered to, or removed from, operations in the correct weights, volumes, or numbers of items required.

5. Space. Storage space, and its efficient utilization, is a key factor in the overall cost of an operation or process.

The science and engineering of material handling is generally classified into two categories, depending upon the form of the material handled. *Bulk solids handling* involves the movement and storage of solids that are flowable, such as fine, free-flowing materials (e.g., wheat flour or sand), pelletized materials (e.g., soybeans or soap flakes), or lumpy materials (e.g., coal or wood bark). *Unit handling* refers to the movement and storage of items that have been formed into unit loads. A *unit load* is a single item, a number of items, or bulk material that is arranged or restrained so that the load can be stored, picked up, and moved between two locations as a single mass. The handling of liquids and gases is usually considered to be in the domain of fluid mechanics, whereas the movement and storage of containers of liquid or gaseous material properly comes within the domain of unit material handling.

38.2 BULK MATERIAL HANDLING

The handling of bulk solids involves four main areas: (1) conveying, (2) storage, (3) packaging, and (4) transportation.

38.2.1 Conveying of Bulk Solids

The selection of the proper equipment for conveying bulk solids depends on a number of interrelated factors. First, alternative types of conveyors must be evaluated and the correct model and size must be chosen. Because standardized equipment designs and complete engineering data are available for many types of conveyors, their performance can be accurately predicted when they are used with materials having well-known conveying characteristics. Some of the primary factors involved in conveyor equipment selection are as follows:

1. *Capacity requirement.* The rate at which material must be transported (e.g., tons per hour). For instance, belt conveyors can be manufactured in relatively large sizes, operate at high speeds, and deliver large weights and volumes of material economically. On the other hand, screw conveyors can become very cumbersome in large sizes, and cannot be operated at high speeds without severe abrasion problems.

2. *Length of travel.* The distance material must be moved from origin to destination. For instance, belt conveyors can span miles, whereas pneumatic and vibrating conveyors are limited to hundreds of feet.

3. *Lift.* The vertical distance material must be transported. Vertical bucket elevators are commonly applied in those cases in which the angle of inclination exceeds 30°.

4. *Material characteristics.* The chemical and physical properties of the bulk solids to be transported, particularly flowability.

5. *Processing requirements.* The treatment material incurs during transport, such as heating, mixing, and drying.

6. *Life expectancy.* The period of performance before equipment must be replaced; typically, the economic life of the equipment.

7. *Comparative costs.* The installed first cost and annual operating costs of competing conveyor systems must be evaluated in order to select the most cost-effective configuration.

Table 38.1 lists various types of conveyor equipment for certain common industrial functions. Table 38.2 provides information on the various types of conveyor equipment used with materials having certain characteristics.

The choice of the conveyor itself is not the only task involved in selecting a conveyor system. Conveyor drives, motors, and auxiliary equipment must also be chosen. Conveyor drives comprise from 10%–30% of the total cost of the conveyor system. Fixed-speed drives and adjustable speed drives are available, depending on whether changes in conveyor speed are needed during the course of normal operation. Motors for conveyor drives are generally three-phase, 60-cycle, 220-V units; 220/440-V units; 550-V units; or four-wire, 208-V units. Also available are 240-V and 480-V ratings.

Table 38.1 Types of Conveyor Equipment and Their Functions

Function	Conveyor Type
Conveying materials horizontally	Apron, belt, continuous flow, drag flight, screw, vibrating, bucket, pivoted bucket, air
Conveying materials up or down an incline	Apron, belt, continuous flow, flight, screw, skip hoist, air
Elevating materials	Bucket elevator, continuous flow, skip hoist, air
Handling materials over a combination horizontal and vertical path	Continuous flow, gravity-discharge bucket, pivoted bucket, air
Distributing materials to or collecting materials from bins, bunkers, etc.	Belt, flight, screw, continuous flow, gravity-discharge bucket, pivoted bucket, air
Removing materials from railcars, trucks, etc.	Car dumper, grain-car unloader, car shaker, power shovel, air

Auxiliary equipment includes such items as braking or arresting devices on vertical elevators to prevent reversal of travel, torque-limiting devices or electrical controls to limit power to the drive motor, and cleaners on belt conveyors.

38.2.2 Screw Conveyors

A screw conveyor consists of a helical shaft mount within a pipe or trough. Power may be transmitted through the helix, or in the case of a fully enclosed pipe conveyor through the pipe itself. Material is forced through the channel formed between the helix and the pipe or trough. Screw conveyors are generally limited to rates of flow of about 10,000 ft^3/hr. Figure 38.1 shows a chute-fed screw conveyor, one of several types in common use. Table 38.3 gives capacities and loading conditions for screw conveyors on the basis of material classifications.

38.2.3 Belt Conveyors

Belt conveyors are widely used in industry. They can traverse distances up to several miles at speeds up to 1000 ft/min and can handle thousands of tons of material per hour. Belt conveyors are generally placed horizontally or at slopes ranging from 10°–20°, with a maximum incline of 30°. Direction changes can occur readily in the vertical plane of the belt path, but horizontal direction changes must be managed through such devices as connecting chutes and slides between different sections of belt conveyor.

Belt-conveyor design depends largely on the nature of the material to be handled. Particle-size distribution and chemical composition of the material dictate selection of the width of the belt and the type of belt. For instance, oily substances generally rule out the use of natural rubber belts. Conveyor-belt capacity requirements are based on peak load rather than average load. Operating conditions that affect belt-conveyor design include climate, surroundings, and period of continuous service. For instance, continuous service operation will require higher-quality components than will intermittent service, which allows more frequent maintenance. Belt width and speed depend on the bulk density of the material and lump size. The horsepower to drive the belt is a function of the following factors:

1. Power to drive an empty belt

Table 38.2 Material Characteristics and Feeder Type

Material Characteristics	Feeder Type
Fine, free-flowing materials	Bar flight, belt, oscillating or vibrating, rotary vane, screw
Nonabrasive and granular materials, materials with some lumps	Apron, bar flight, belt, oscillating or vibrating, reciprocating, rotary plate, screw
Materials difficult to handle because of being hot, abrasive, lumpy, or stringy	Apron, bar flight, belt, oscillating or vibrating, reciprocating
Heavy, lumpy, or abrasive materials similar to pit-run stone and ore	Apron, oscillating or vibrating, reciprocating

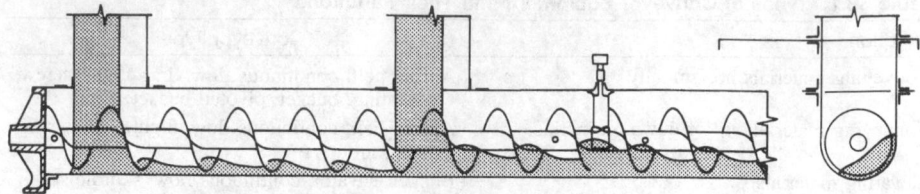

Fig. 38.1 Chute-fed screw conveyor.

2. Power to move the load against the friction of the rotating parts
3. Power to elevate and lower the load
4. Power to overcome inertia in placing material in motion
5. Power to operate a belt-driven tripper

Table 38.4 provides typical data for estimating belt-conveyor and design requirements. Figure 38.2 illustrates a typical belt-conveyor loading arrangement.

38.2.4 Bucket Elevators

Bucket elevators are used for vertical transport of bulk solid materials. They are available in a wide range of capacities and may operate in the open or totally enclosed. They tend to be acquired in highly standardized units, although specifically engineered equipment can be obtained for use with special materials, unusual operating conditions, or high capacities. Figure 38.3 shows a common type of bucket elevator, the spaced-bucket centrifugal-discharge elevator. Other types include spaced-bucket positive-discharge elevators, V-bucket elevators, continuous-bucket elevators, and super-capacity continuous-bucket elevators. The latter handle high tonnages and are usually operated at an incline to improve loading and discharge conditions.

Bucket elevator horsepower requirements can be calculated for space-bucket elevators by multiplying the desired capacity (tons per hour) by the lift and dividing by 500. Table 38.5 gives bucket elevator specifications for spaced-bucket, centrifugal-discharge elevators.

38.2.5 Vibrating or Oscillating Conveyors

Vibrating conveyors are usually directional-throw devices that consist of a spring-supported horizontal pan or trough vibrated by an attached arm or rotating weight. The motion imparted to the material particles abruptly tosses them upward and forward so that the material travels in the desired direction. The conveyor returns to a reference position, which gives rise to the term *oscillating conveyor.* The capacity of the vibrating conveyor is determined by the magnitude and frequency of trough displacement, angle of throw, and slope of the trough, and the ability of the material to receive and transmit through its mass the directional "throw" of the trough. Classifications of vibrating conveyors include (1) mechanical, (2) electrical, and (3) pneumatic and hydraulic vibrating conveyors. Capacities of vibrating conveyors are very broad, ranging from a few ounces or grams for laboratory-scale equipment to thousands of tons for heavy industrial applications. Figure 38.4 depicts a leaf-spring mechanical vibrating conveyor, and provides a selection chart for this conveyor.

38.2.6 Continuous-Flow Conveyors

The continuous-flow conveyor is a totally enclosed unit that operates on the principle of pulling a surface transversely through a mass of bulk solids material, such that it pulls along with it a cross section of material that is greater than the surface of the material itself. Figure 38.5 illustrates a typical configuration for a continuous-flow conveyor. Three common types of continuous flow conveyors are (1) closed-belt conveyors, (2) flight conveyors, and (3) apron conveyors. These conveyors employ a chain-supported transport device, which drags through a totally enclosed boxlike tunnel.

38.2.7 Pneumatic Conveyors

Pneumatic conveyors operate on the principle of transporting bulk solids suspended in a stream of air over vertical and horizontal distances ranging from a few inches or centimeters to hundreds of feet or meters. Materials in the form of fine powders are especially suited to this means of conveyance, although particle sizes up to a centimeter in diameter can be effectively transported pneumatically. Materials with bulk densities from one to more than 100 lb/ft^3 can be transported through pneumatic conveyors.

The capacity of a pneumatic conveying system depends on such factors as the bulk density of the product, energy within the conveying system, and the length and diameter of the conveyor.

Table 38.3 Capacity and Loading Conditions for Screw Conveyors

Capacity tons/hr	ft³/hr	Diam. of Flights (in.)	Diam. of Pipe (in.)	Diam. of Shafts (in.)	Hanger Centers (ft)	Max. Size Lumps			Speed (rpm)	Max. Torque Capacity (in.-lb)	Feed Section Diam. (in.)	Hp at Motor					Max. Hp Capacity at Speed Listed
						All Lumps	Lumps 20–25%	Lumps 10% or Less				15 ft Max. Length	30 ft Max. Length	45 ft Max. Length	60 ft Max. Length	75 ft Max. Length	
5	200	9	2½	2	10	¾	1½	2¼	40	7,600	6	0.43	0.85	1.27	1.69	2.11	4.8
10	400	10	2½	2	10	¾	1½	2½	55	7,600	9	0.85	1.69	2.25	3.00	3.75	6.6
15	600	10	2½	2	10	¾	1½	2½	80	7,600	9	1.27	2.25	3.38	3.94	4.93	9.6
		12	2½	2	12	1	2	3	45	7,600	10	1.27	2.25	3.38	3.94	4.93	5.4
		12	3½	3						16,400		1.27	2.25	3.38	3.94	4.93	11.7
20	800	12	2½	2	12	1	2	3	60	7,600	10	1.69	3.00	3.94	4.87	5.63	7.2
		12	3½	3						16,400		1.69	3.00	3.94	4.87	5.63	15.6
25	1000	12	2½	2	12	1	2	3	75	7,600	10	2.12	3.75	4.93	5.63	6.55	9.0
			3½	3						16,400		2.12	3.75	4.93	5.63	6.55	19.5
30	1200	14	3½	3		1¼	2½	3½	45	16,400	12	2.12	3.75	4.93	5.63	6.55	11.7
35	1400	14	3½	3	12	1¼	2½	3½	55	16,400	12	2.25	3.94	5.05	6.75	7.50	14.3
		14	3½	3	12	1¼	2½	3½	65	16,400	12	2.62	4.58	5.90	7.00	8.75	16.9
40	1600	16	3½	3	12	1½	3	4	50	16,400	14	3.00	4.50	6.75	8.00	10.00	13.0

Table 38.4 Data for Estimating Belt Conveyor Design Requirements

Belt Width (in.)	Cross-Sectional Area of Load (ft²)	Belt Speed Normal Operating Speed (ft/min)	Belt Speed Max. Advisable Speed (ft/min)	Belt Plies Min.	Belt Plies Max.	Max. Size Lump (in.) Sized Material 80% Under	Max. Size Lump (in.) Unsized Material Not Over 20%	Belt Speed (ft/min)	50 lb/ft³ Material Capacity (tons/hr)	50 lb/ft³ hp 10-ft Lift	50 lb/ft³ hp 100-ft Centers	100 lb/ft³ Material Capacity (tons/hr)	100 lb/ft³ hp 10-ft Lift	100 lb/ft³ hp 100-ft Centers	Add hp for Tripper
14	0.11	200	300	3	5	2	3	100	16	0.17	0.22	32	0.34	0.44	1.00
								200	32	0.34	0.44	64	0.68	0.88	
								300	48	0.52	0.66	96	1.04	1.32	
16	0.14	200	300	3	5	2½	4	100	22	0.23	0.28	44	0.46	0.56	1.25
								200	44	0.45	0.56	88	0.90	1.12	
								300	66	0.68	0.84	132	1.36	1.68	
18	0.18	250	350	4	6	3	5	100	27	0.29	0.35	54	0.58	0.7	1.50
								250	67	0.71	0.88	134	1.42	1.76	
								350	95	1.00	1.21	190	2.00	2.42	
20	0.22	250	350	4	6	3½	6	100	33	0.35	0.42	66	0.70	0.84	1.60
								250	82	0.86	1.03	164	1.72	2.06	
								350	115	1.22	1.45	230	2.44	2.9	
24	0.33	300	400	4	7	4½	8	100	49	0.51	0.51	98	1.02	1.02	1.75
								300	147	1.53	1.52	294	3.06	3.04	
								400	196	2.04	2.02	392	4.08	4.04	
30	0.53	300	450	4	8	7	12	100	79	0.80	0.75	158	1.60	1.5	2.50
								300	237	2.40	2.25	474	4.80	4.5	
								450	355	3.60	3.37	710	7.20	6.74	
36	0.78	400	600	4	9	8	15	100	115	1.22	0.80	230	2.44	1.59	3.53
								400	460	4.87	3.18	920	9.74	6.36	
								600	690	7.30	4.76	1380	14.6	9.52	
42	1.09	400	600	4	10	10	18	100	165	1.75	1.14	330	3.50	2.28	4.79
								400	660	7.00	4.56	1320	14.0	9.12	
								600	990	11.6	6.84	1980	23.2	13.68	
48	1.46	400	600	4	12	12	21	100	220	2.33	1.52	440	4.66	3.04	6.42
								400	880	9.35	6.07	1760	18.7	12.14	
								600	1320	14.0	9.10	2640	28.0	18.2	
54	1.90	450	600	6	12	14	24	100	285	3.02	1.97	570	6.04	3.94	10.56
								450	1282	13.6	8.85	2564	27.2	17.7	
								600	1710	18.1	11.82	3420	36.2	23.6	
60	2.40	450	600	6	13	16	28	100	360	3.82	2.49	720	7.64	4.98	
								450	1620	17.2	11.20	3240	34.4	22.4	
								600	2160	22.9	14.95	4320	45.8	29.9	

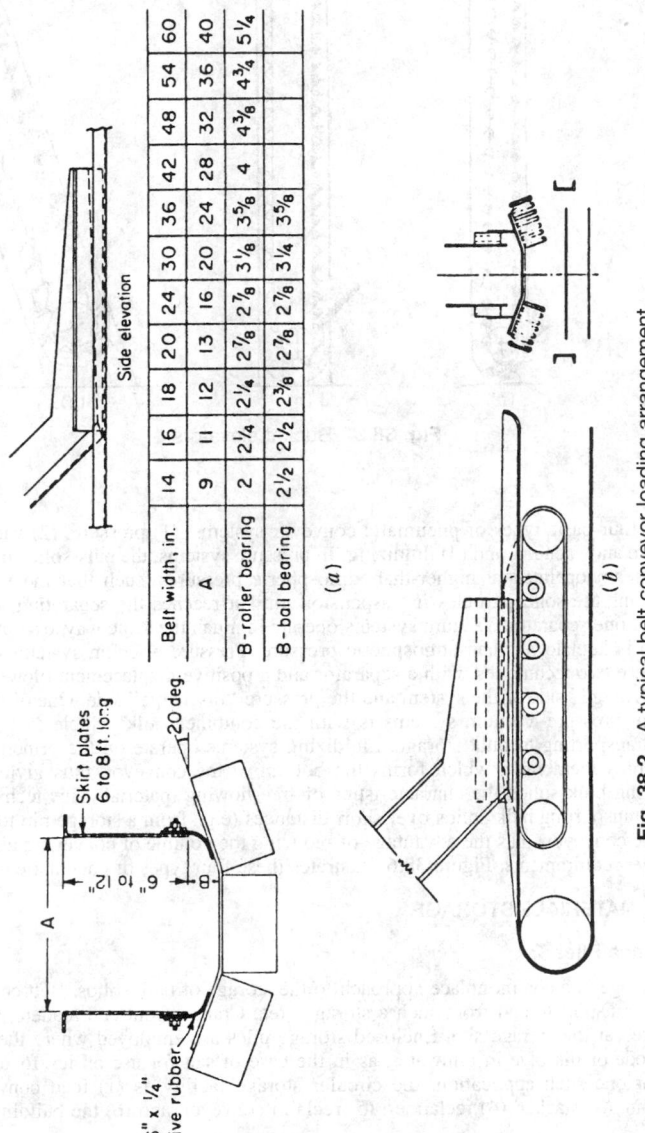

Belt width, in.	14	16	18	20	24	30	36	42	48	54	60
A	9	11	12	13	16	20	24	28	32	36	40
B - roller bearing	2	2¼	2¼	2⅞	2⅞	3⅛	3⅝	4	4⅜	4¾	5¼
B - ball bearing	2½	2½	2⅜	2⅞	2⅞	3¼	3⅝				

Side elevation

(*a*)

(*b*)

Fig. 38.2 A typical belt conveyor loading arrangement.

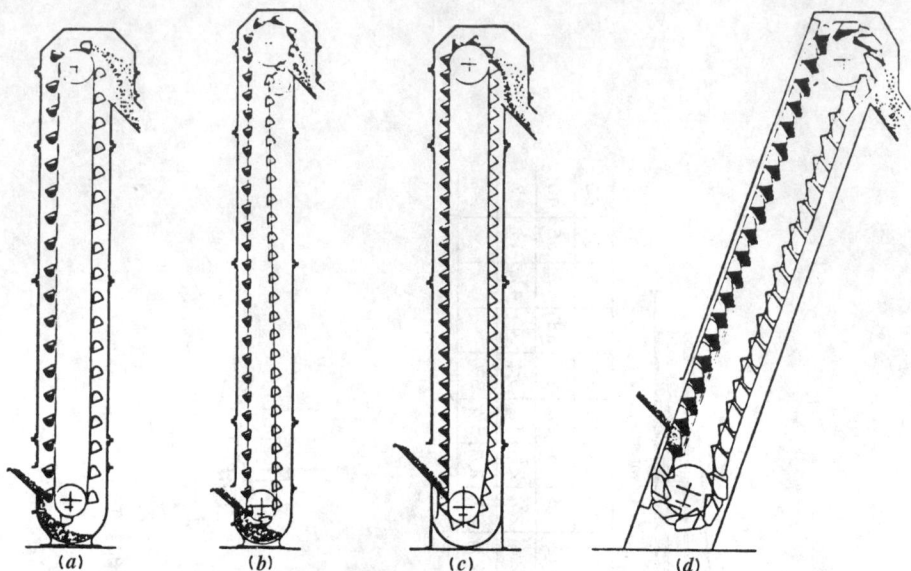

Fig. 38.3 Bucket elevators.

There are four basic types of pneumatic conveyor systems: (1) pressure, (2) vacuum, (3) combination pressure and vacuum, and (4) fluidizing. In pressure systems, the bulk solids material is charged into an air stream operated at higher-than-atmospheric pressures, such that the velocity of the air stream maintains the solid particles in suspension until it reaches the separating vessel, usually an air filter or cyclone separator. Vacuum systems operate in much the same way, except that the pressure of the system is kept lower than atmospheric pressure. Pressure–vacuum systems combine the best features of these two techniques, with a separator and a positive-displacement blower placed between the vacuum "charge" side of the system and the pressure "discharge" side. One of the most common applications of pressure–vacuum systems is with the combined bulk vehicle (e.g., hopper car) unloading and transporting to bulk storage. Fluidizing systems operate on the principle of passing air through a porous membrane, which forms the bottom of the conveyor, thus giving finely divided, non-free-flowing bulk solids the characteristics of free-flowing material. This technique, commonly employed in transporting bulk solids over short distances (e.g., from a storage bin to the charge point to a pneumatic conveyor), has the advantage of reducing the volume of conveying air needed, thereby reducing power requirements. Figure 38.6 illustrates these four types of pneumatic conveyor systems.

38.3 BULK MATERIALS STORAGE

38.3.1 Storage Piles

Open-yard storage is a commonplace approach to the storage of bulk solids. Belt conveyors are most often used to transport to and from such a storage area. Cranes, front-end loaders, and draglines are commonly used at the storage site. Enclosed storage piles are employed where the bulk solids materials can erode or dissolve in rainwater, as in the case of salt for use on icy roads. The necessary equipment for one such application, the circular storage facility, is (1) feed conveyor, (2) central support column, (3) stacker, (4) reclaimer, (5) reclaim conveyor, and (6) the building or dome cover.

38.3.2 Storage Bins, Silos, and Hoppers

A typical storage vessel for bulk solids materials consists of two components—a bin and a hopper. The bin is the upper section of the vessel and has vertical sides. The hopper is the lower part of the vessel, connecting the bin and the outlet, and must have at least one sloping side. The hopper serves as the means by which the stored material flows to the outlet channel. Flow is induced by opening the outlet port and using a feeder device to move the material, which drops through the outlet port.

If all material stored in the bin moves whenever material is removed from the outlet port, *mass flow* is said to prevail. However, if only a portion of the material moves, the condition is called *funnel flow*. Figure 38.7 illustrates these two conditions.

Many flow problems in storage bins can be reduced by taking the physical characteristics of the bulk material into account. Particle size, moisture content, temperature, age, and oil content of the

Table 38.5 Bucket Elevator Specifications

Size of Bucket (in.)[a]	Elevator Centers (ft)	Capacity (tons/hr) Material Weighing 100 lb/ft[b]	Size Lumps Handled (in.)[c]	Bucket Speed (ft/min)	rpm Head Shaft	Horsepower[b] Required at Head Shaft	Additional Horsepower[b] per Foot for Intermediate Lengths	Bucket Spacing (in.)	Shaft Diameter (in.)		Diameter of Pulleys (in.)		Belt Width (in.)
									Head	Tail	Head	Tail	
	25	14	$\frac{3}{4}$	225	43	1.0	0.02	12	$1\frac{15}{16}$	$1\frac{11}{16}$	20	14	7
	50	14	$\frac{3}{4}$	225	43	1.6	0.02	12	$1\frac{15}{16}$	$1\frac{11}{16}$	20	14	7
$6 \times 4 \times 4\frac{1}{4}$	75	14	$\frac{3}{4}$	225	43	2.1	0.02	12	$1\frac{15}{16}$	$1\frac{11}{16}$	20	14	7
	25	27	1	225	43	1.6	0.04	14	$1\frac{15}{16}$	$1\frac{11}{16}$	20	14	9
	50	30	1	260	41	3.5	0.05	14	$1\frac{15}{16}$	$1\frac{11}{16}$	24	14	9
$8 \times 5 \times 5\frac{1}{2}$	75	30	1	260	41	4.8	0.05	14	$2\frac{7}{16}$	$1\frac{11}{16}$	24	14	9
	25	45	$1\frac{1}{4}$	225	43	3.0	0.063	16	$1\frac{15}{16}$	$1\frac{15}{16}$	20	16	11
	50	52	$1\frac{1}{4}$	260	41	5.2	0.07	16	$2\frac{7}{16}$	$1\frac{15}{16}$	24	16	11
$10 \times 6 \times 6\frac{1}{4}$	75	52	$1\frac{1}{4}$	260	41	7.2	0.07	16	$2\frac{15}{16}$	$1\frac{15}{16}$	24	16	11
	25	75	$1\frac{1}{2}$	260	41	4.7	0.1	18	$2\frac{7}{16}$	$1\frac{15}{16}$	24	18	13
	50	84	$1\frac{1}{2}$	300	38	8.9	0.115	18	$2\frac{15}{16}$	$1\frac{15}{16}$	30	18	13
$12 \times 7 \times 7\frac{1}{4}$	75	84	$1\frac{1}{2}$	300	38	11.7	0.115	18	$3\frac{7}{16}$	$2\frac{7}{16}$	30	18	13
	25	100	$1\frac{3}{4}$	300	38	7.3	0.14	18	$2\frac{15}{16}$	$2\frac{7}{16}$	30	18	15
	50	100	$1\frac{3}{4}$	300	38	11.0	0.14	18	$3\frac{7}{16}$	$2\frac{7}{16}$	30	18	15
$14 \times 7 \times 7\frac{1}{4}$	75	100	$1\frac{3}{4}$	300	38	14.3	0.14	18	$3\frac{7}{16}$	$2\frac{7}{16}$	30	18	15
	25	150	2	300	38	8.5	0.165	18	$2\frac{15}{16}$	$2\frac{7}{16}$	30	20	18
	50	150	2	300	38	12.6	0.165	18	$3\frac{7}{16}$	$2\frac{7}{16}$	30	20	18
$16 \times 8 \times 8\frac{1}{2}$	75	150	2	300	38	16.7	0.165	18	$3\frac{15}{16}$	$2\frac{7}{16}$	30	20	18

[a] Size of buckets given: width × projection × depth.

[b] Capacities and horsepowers given for materials weighing 100 lb/ft^3. For materials of other weights, capacity and horsepower will vary in direct proportion. For example, an elevator handling coal weighing 50 lb/ft^3 will have half the capacity and will require approximately half the horsepower listed above.

[c] If volume of lumps averages less than 15% of total volume, lumps of twice size listed may be handled.

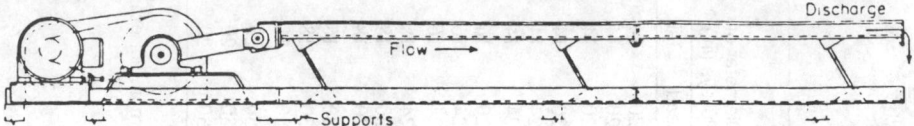

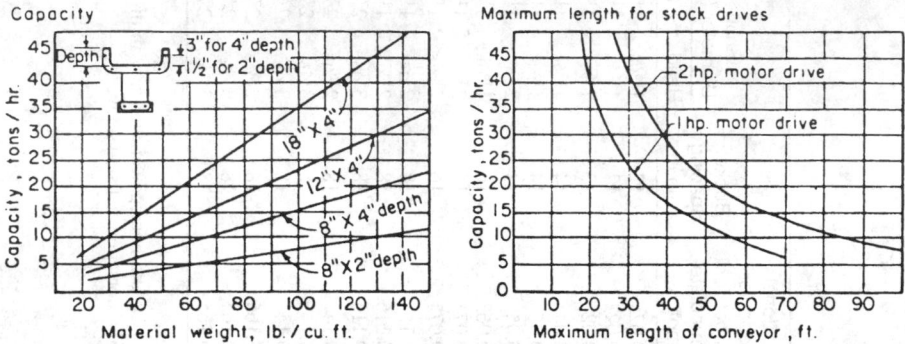

Fig. 38.4   Leaf-spring mechanical vibrating conveyor.

stored material affect flowability. Flow-assisting devices and feeders are usually needed to overcome flow problems in storage bins.

38.3.3   Flow-Assisting Devices and Feeders

To handle those situations in which bin design alone does not produce the desired flow characteristics, flow-assisting devices are available. Vibrating hoppers are one of the most important types of flow-assisting devices. These devices fall into two categories: *gyrating devices,* in which vibration is applied perpendicular to the flow channel; and *whirlpool devices,* which apply a twisting motion and a lifting motion to the material, thereby disrupting any bridges that might tend to form. Screw feeders are used to assist in bin unloading by removing material from the hopper opening.

38.3.4   Packaging of Bulk Materials

Bulk materials are often transported and marketed in containers, such as bags, boxes, and drums. Packaged solids lend themselves to material handling by means of unit material handling.

Bags

Paper, plastic, and cloth bags are common types of containers for bulk solids materials. Multiwall paper bags are made from several plies of kraft paper. Bag designs include valve and open-mouth designs. Valve-type bags are stitched or glued at both ends prior to filling, and are filled through a

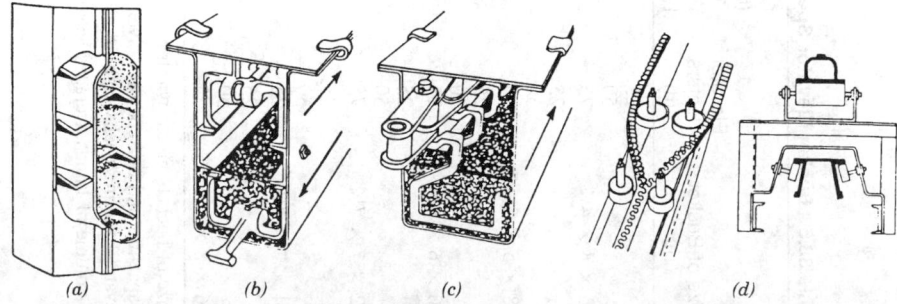

(a)   *(b)*   *(c)*   *(d)*

Fig. 38.5   Continuous-flow conveyor.

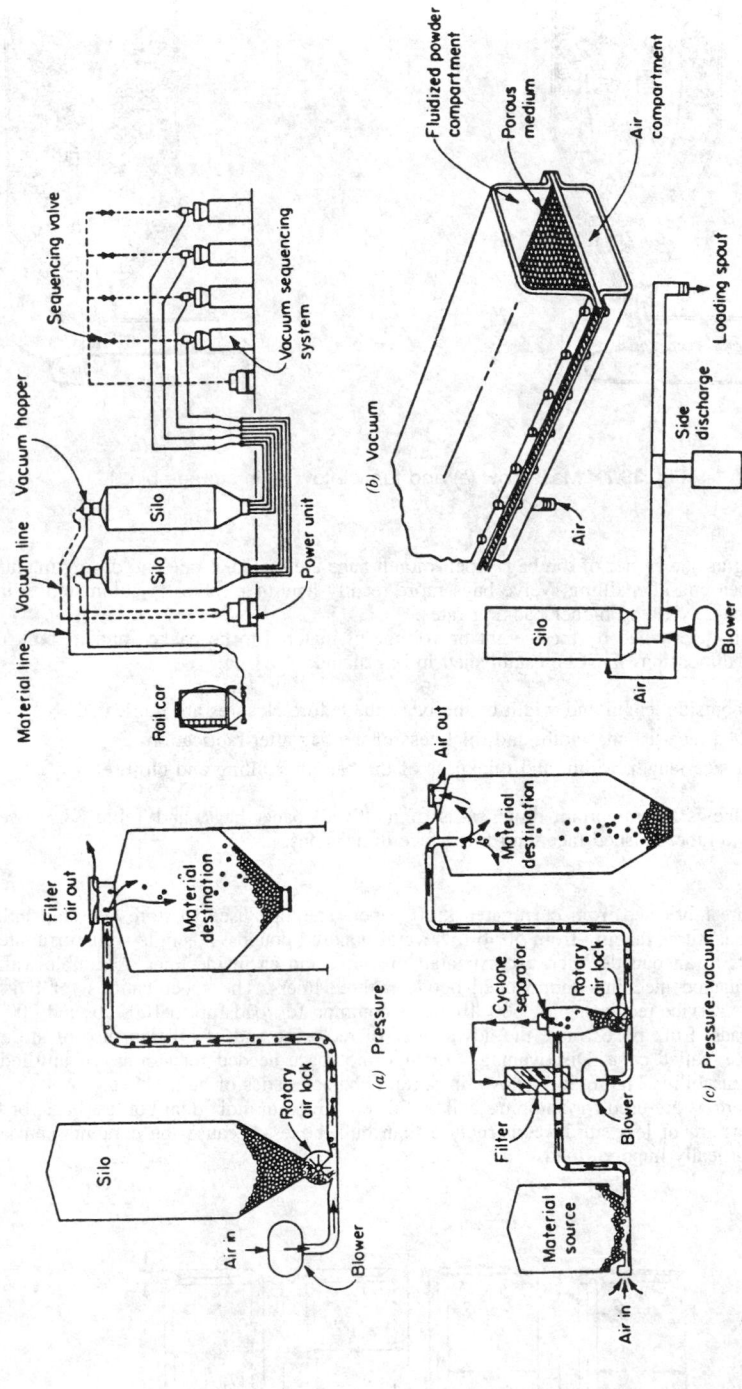

Fig. 38.6 Four types of pneumatic conveyor systems.

(a) Pressure

(b) Vacuum

(c) Pressure-vacuum

(d) Fluidizing system

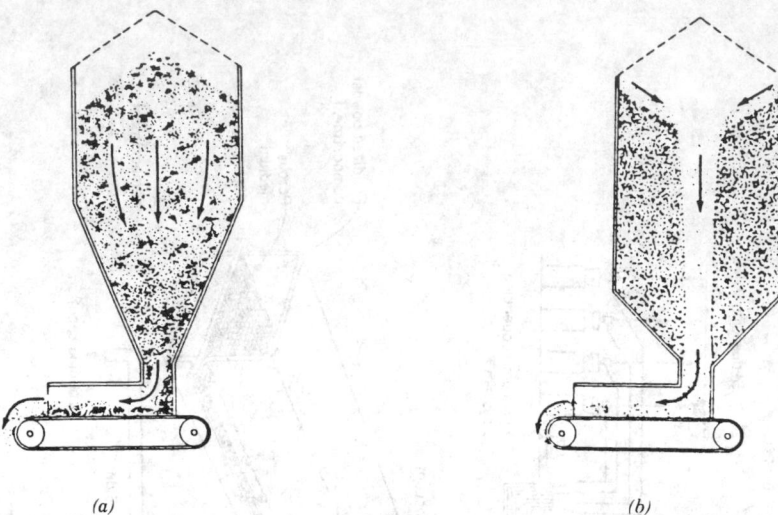

Fig. 38.7 Mass-flow (a) and funnel-flow (b) in storage bins.

valve opening at one corner of the bag. Open-mouth bags are sealed at one end during manufacture, and at the open end after filling. Valve bags more readily lend themselves to automated filling than open-mouth bags, yielding higher packing rates.

Bag size is determined by the weight or volume of material to be packed and its bulk density. Three sets of dimensions must be established in bag sizing:

1. Tube–outside length and width of the bag tube before closures are fabricated
2. Finished face-length, width, and thickness of the bag after fabrication
3. Filled face-length, width, and thickness of the bag after filling and closure

Figure 38.8 shows the important dimensions of multiwall paper bags, and Table 38.6 gives their relationships to tube, finished face, and filled face dimensions.

Boxes

Bulk boxes are fabricated from corrugated kraft paper. They are used to store and ship bulk solid materials in quantities ranging from 50 lb to several hundred pounds. A single-wall corrugated kraft board consists of an outside liner, a corrugated medium, and an inside liner. A double-wall board has two corrugated mediums sandwiched between three liners. The specifications for bulk boxes depend on the service requirements; 600 lb/in.^{2} is common for loads up to 1000 lb, and 200 lb/in.^{2} for 100-lb loads. Bulk boxes have the advantages of reclosing and of efficient use of storage and shipping space, called *cube*. Disadvantages include the space needed for storage of unfilled boxes and limited reusability. Figure 38.9 shows important characteristics of bulk boxes.

Folding cartons are used for shipping bulk solids contained in individual bottles, bags, or folding boxes. Cartons are of less sturdy construction than bulk boxes, because the contents can assist in supporting vertically imposed loads.

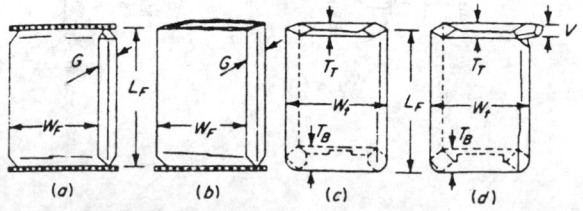

Fig. 38.8 Dimensions of multiwall paper bags.

Table 38.6 Dimensions of Multiwall Paper Bags

Bag Type	Tube Dimensions	Finished-Face Dimensions	Filled-Face Dimensions	Valve Dimensions
Sewn open-mouth	Width = $W_t = W_f + G_f$ Length = $L_t = L_f$	Width = $W_f = W_t - G_f$ Length = $L_f = L_t$ Gusset = G_f	Width = $W_F = W_f + \frac{1}{2}$ in. Length = $L_F = L_f - 0.67 G_f$ Thickness = $G_F = G_f + \frac{1}{2}$ in.	
Sewn valve	Width = $W_t = W_f + G_f$ Length = $L_t = L_f$	Width = $W_f = W_t - G_f$ Length = $L_f = L_t$ Gusset = G_f	Width = $W_F = W_f + 1$ in. Length = $L_F = L_f - 0.67 G_f$ Thickness = $G_F = G_f + 1$ in.	Width = $V = G_f \pm \frac{1}{2}$ in.
Pasted valve	Width = $W_t = W_f$ Length = $L_t = L_f$	Width = $W_f = W_f$ Length = $L_f = L_t - (T_T + T_B)/2 - 1$ Thickness at top = T_T Thickness at bottom = T_B	Width = $W_F = W_f - T_T + 1$ in. Length = $L_F = L_f - T_T + 1$ in. Thickness = $T_F = T_T + \frac{1}{2}$ in.	Width = $V = T_T \begin{cases} +0 \text{ in.} \\ -1 \text{ in.} \end{cases}$

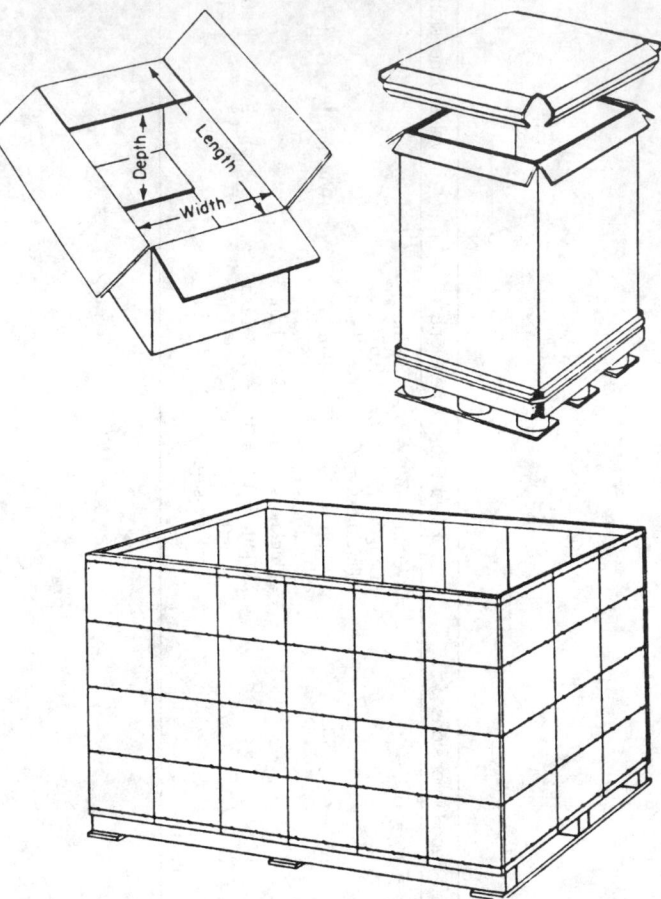

Fig. 38.9 Bulk boxes and cartons.

38.3.5 Transportation of Bulk Materials

The term *transportation of bulk materials* refers to the movement of raw materials, fuels, and bulk products by land, sea, and air. A useful definition of a bulk shipment is any unit greater than 4000 lb or 40 ft³. The most common bulk carriers are railroad hopper cars, highway hopper trucks, portable bulk bins, barges, and ships. Factors affecting the choice of transportation include the characteristics of material size of shipment, available transportation routes from source to destination (e.g., highway, rail, water), and the time available for shipment.

Railroad Hopper Cars

Railroad hopper cars are of three basic designs:

1. Covered, with bottom-unloading ports
2. Open, with bottom-unloading ports
3. Open, without unloading ports

Gravity, pressure differential, and fluidizing unloading systems are available with railroad hopper cars. Loading of hopper cars can be done with most types of conveyors: belt, screw, pneumatic, and so on. Unloading of bottom-unloading hopper cars can be managed by constructing a special dumping pit beneath the tracks with screw or belt takeaway conveyors.

Hopper Trucks

Hopper trucks are used for highway transportation of bulk solids materials. The most common types include (1) closed type with a pneumatic conveyor unloading system and (2) the open dump truck.

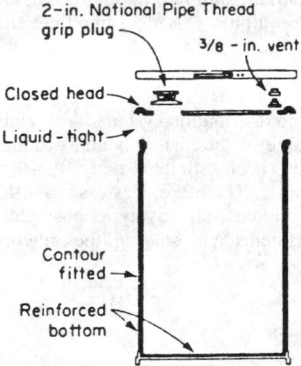

Drum type	Outside dimensions	
	Dia., in.	Height, in.
55 – gal. lever top	21	40 3/4
55 – gal. lever top	23 1/2	30 3/4
55 – gal. lever top	22	34 3/4
41 – gal. lever top	20 1/2	30 1/4
30 – gal. lever top	19	26 1/4
6.28 – cu. ft. rectangular	17 5/8*	37 1/2
55 – gal. liquid	22	37 1/2
30 – gal. liquid	19	28
55 – gal. fiber	20 3/8	40 3/4
30 – gal. fiber	17 3/8	30 3/4

* Side dimension, square

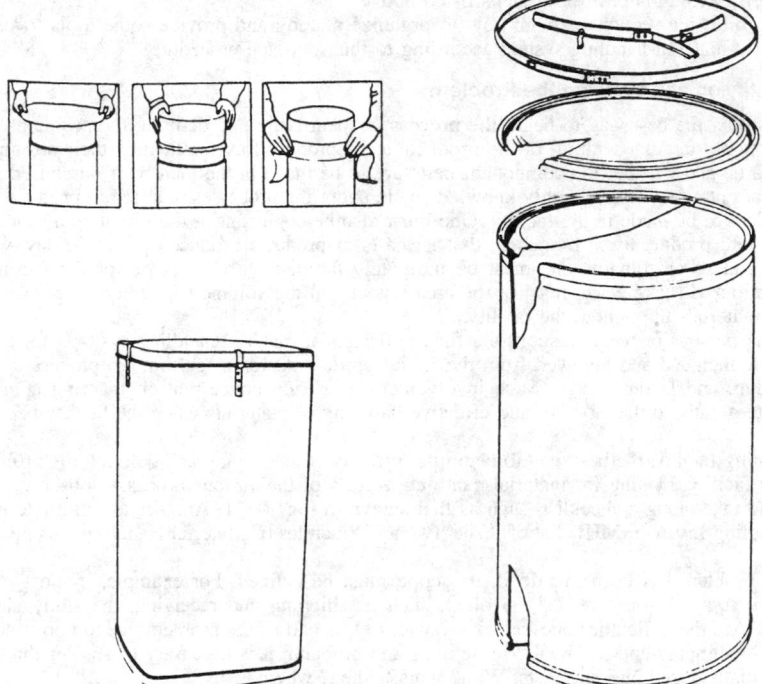

Fig. 38.10 Storage drums.

With the first type, a truck can discharge its cargo directly into a storage silo. The shipment weights carried by trucks depend on state highway load limits, usually from 75,000–125,000 lb.

38.4 UNIT MATERIAL HANDLING

38.4.1 Introduction

Unit material handling involves the movement and storage of unit loads, as defined in Section 38.1. Examples include automobile body components, engine blocks, bottles, cans, bags, pallets of boxes, bins of loose parts, and so on. As the previous definition implies, the word *unit* refers to the single entity that is handled. That entity can consist of a single item or numerous items that have been unitized for purposes of movement and storage.

This section discusses some of the procedures employed in material-handling system design, and describes various categories, with examples, of material-handling equipment used in handling unit loads.

38.4.2 Analysis of Systems for Material Handling

Material handling is an indispensable element in most production and distribution systems. Yet, while material handling is generally considered to add nothing to the value of the materials and products that flow through the system, it does add to their cost. In fact, it has been estimated that 30%–60% of the end-price of a product is related to the cost of material handling. Therefore, it is essential that material handling systems be designed and operated as efficiently and cost-effectively as possible.

The following steps can be used in analyzing production systems and solving the inherent material-handling problems:

1. Identify and define the problem(s).
2. Collect relevant data.
3. Develop a plan.
4. Implement the solution.

Unfortunately, when most engineers perceive that a material-handling problem exists, they skip directly to step 4; that is, they begin looking for material-handling equipment that will address the symptoms of the problem without looking for the underlying root causes of the problem, which may be uncovered by execution of all four steps listed above.

Thus, the following sections explain how to organize a study and provide some tools to use in an analysis of a material-handling system according to this four-step procedure.

38.4.3 Identifying and Defining the Problem

For a new facility, the best way to begin the process of identifying and defining the problems is to become thoroughly familiar with all of the products to be produced by the facility, their design and component parts, and whether the component parts are to be made in the facility or purchased from vendors. Then, one must be thoroughly knowledgeable about the processes required to produce each part and product to be made in the facility. One must also be cognizant of the production schedules for each part and product to be produced; that is, parts or products produced per shift, day, week, month, year, and so on. Finally, one must be intimately familiar with the layout of the facility in which production will take place; not just the area layout, but the volume (or cubic space) available for handling materials throughout the facility.

Ideally, the persons or teams responsible for the design of material-handling systems for a new facility will be included and involved from the initial product design stage through process design, schedule design, and layout design. Such involvement in a truly concurrent engineering approach will contribute greatly to the efficient and effective handling of materials when the facility becomes operational.

In an existing facility, the best way to begin the process of identifying and defining the problems is to tour the facility, looking for material-handling aspects of the various processes observed. It is a good idea to take along a checklist, such as that shown in Fig. 38.11. Another useful guide is the Material Handling Institute (MHI) list of "The Twenty Principles of Material Handling," as given in Fig. 38.12.

Once the problem has been identified, its scope must be defined. For example, if most of the difficulties are found in one area of the plant, such as shipping and receiving, the study can be focused there. Are the difficulties due to lack of space? Or is part of the problem due to poor training of personnel in shipping and receiving? In defining the problem, it is necessary to answer the basic questions normally asked by journalists: Who? what? when? where? why?

38.4.4 Collecting Data

In attempting to answer the journalistic questions above, all relevant data must be collected and analyzed. At a minimum, the data collection and analysis must be concerned with the products to be produced in the facility, the processes (fabrication, assembly, and so on) used to produce each product, the schedule to be met in producing the products, and the facility layout (three-dimensional space allocation) supporting the production processes.

Some useful data can be obtained by interviewing management, supervisors, operators, vendors, and competitors, by consulting available technical and sales literature, and through personal observation. However, most useful data are acquired by systematically charting the flows of materials and the movements that take place within the plant. Various graphical techniques are used to record and analyze this information.

An assembly chart, shown in Fig. 38.13, is used to illustrate the composition of the product, the relationship among its component parts, and the sequence in which components are assembled.

Material Handling Checklist

☐ Is the material handling equipment more than 10 years old?

☐ Do you use a wide variety of makes and models which require a high spare parts inventory?

☐ Are equipment breakdowns the result of poor preventive maintenance?

☐ Do the lift trucks go too far for servicing?

☐ Are there excessive employee accidents due to manual handling of materials?

☐ Are materials weighing more than 50 pounds handled manually?

☐ Are there many handling tasks that require 2 or more employees?

☐ Are skilled employees wasting time handling materials?

☐ Does material become congested at any point?

☐ Is production work delayed due to poorly scheduled delivery and removal of materials?

☐ Is high storage space being wasted?

☐ Are high demurrage charges experienced?

☐ Is material being damaged during handling?

☐ Do shop trucks operate empty more than 20% of the time?

☐ Does the plant have an excessive number of rehandling points?

☐ Is power equipment used on jobs that could be handled by gravity?

☐ Are too many pieces of equipment being used, because their scope of activity is confined?

☐ Are many handling operations unnecessary?

☐ Are single pieces being handled where unit loads could be used?

☐ Are floors and ramps dirty and in need of repair?

☐ Is handling equipment being overloaded?

☐ Is there unnecessary transfer of material from one container to another?

☐ Are inadequate storage areas hampering efficient scheduling of movement?

☐ Is it difficult to analyze the system because there is no detailed flow chart?

☐ Are indirect labor costs too high?

Fig. 38.11   Material-handling checklist.

The operations process chart, shown in Fig. 38.14, provides an even more detailed depiction of material flow patterns, including sequences of production and assembly operations. It begins to afford an idea of the relative space requirements for the process.

The flow process chart, illustrated in Fig. 38.15, tabulates the steps involved in a process, using a set of standard symbols adopted by the American Society of Mechanical Engineers (ASME). Shown at the top of the chart, these five symbols allow one to ascribe a specific status to an item at each step in processing. The leftmost column in the flow process chart lists the identifiable activities comprising the process, in sequential order. In the next column, one of the five standard symbols is selected to identify the activity as an operation, transportation, inspection, delay, or storage. The remaining columns permit the recording of more detailed information.

Note that in the flow process chart in Fig. 38.16, for each step recorded as a "transport," a distance (in feet) is recorded. Also, in some of the leftmost columns associated with a transport activity, the type of material handling equipment used to make the move is recorded—for example, "fork lift." However, material-handling equipment could be used for any of the activities shown in this chart. For example, automated storage and retrieval systems (AS/RSs) can be used to store materials, accumulating conveyors can be used to queue materials during a delay in processing, or conveyors can be configured as a moving assembly line so that operations can be performed on the product while it is being transported through the facility.

In the columns grouped under the heading *possibilities,* opportunities for improvement or simplification of each activity can be noted.

The flow diagram, depicted in Fig. 38.16, provides a graphical record of the sequence of activities required in the production process, superimposed upon an area layout of a facility. This graphical technique uses the ASME standard symbol set and augments the flow process chart.

The "from–to" chart, illustrated in Fig. 38.17, provides a matrix representation of the required number of material moves (unit loads) in the production process. A separate from–to chart can also be constructed that contains the distances materials must be moved between activities in the production process. Of course, such a chart will be tied to a specific facility layout and usually contains assumptions about the material-handling equipment to be used in making the required moves.

The activity relationship chart, shown in Fig. 38.18, can be used to record qualitative information regarding the flow of materials between activities or departments in a facility. Read like a highway mileage table in a typical road atlas, which indicates the distances between pairs of cities, the activity relationship chart allows the analyst to record a qualitative relationship that should exist between each pair of activities or departments in a facility layout. The relationships recorded in this chart show the importance that each pair of activities be located at varying degrees of closeness to each

The 20 Principles of Material Handling

1. Planning Principle. Plan all material handling and storage activities to obtain maximum overall operating efficiency.

2. Systems Principle. Integrate as many handling activities as is practical into a coordinated system of operations, covering vendor, receiving, storage, production, inspection, packaging, warehousing, shipping, transportation, and customer.

3. Material Flow Principle. Provide an operation sequence and equipment layout optimizing material flow.

4. Simplification Principle. Simplify handling by reducing, eliminating, or combining unnecessary movements and/or equipment.

5. Gravity Principle. Utilize gravity to move material wherever practical.

6. Space Utilization Principle. Make optimum utilization of building cube.

7. Unit Size Principle. Increase the quantity, size, or weight of unit loads or flow rate.

8. Mechanization Principle. Mechanize handling operations.

9. Automation Principle. Provide automation to include production, handling, and storage functions.

10. Equipment Selection Principle. In selecting handling equipment consider all aspects of the material handled — the movement and the method to be used.

11. Standardization Principle. Standardize handling methods as well as types and sizes of handling equipment.

12. Adaptability Principle. Use methods and equipment that can best perform a variety of tasks and applications where special purpose equipment is not justified.

13. Dead Weight Principle. Reduce ratio of dead weight of mobile handling equipment to load carried.

14. Utilization Principle. Plan for optimum utilization of handling equipment and manpower.

15. Maintenance Principle. Plan for preventive maintenance and scheduled repairs of all handling equipment.

16. Obsolescence Principle. Replace obsolete handling methods and equipment when more efficient methods or equipment will improve operations.

17. Control Principle. Use material handling activities to improve control of production, inventory and order handling.

18. Capacity Principle. Use handling equipment to help achieve desired production capacity.

19. Performance Principle. Determine effectiveness of handling performance in terms of expense per unit handled.

20. Safety Principle. Provide suitable methods and equipment for safe handling.

Fig. 38.12   Twenty principles of material handling.

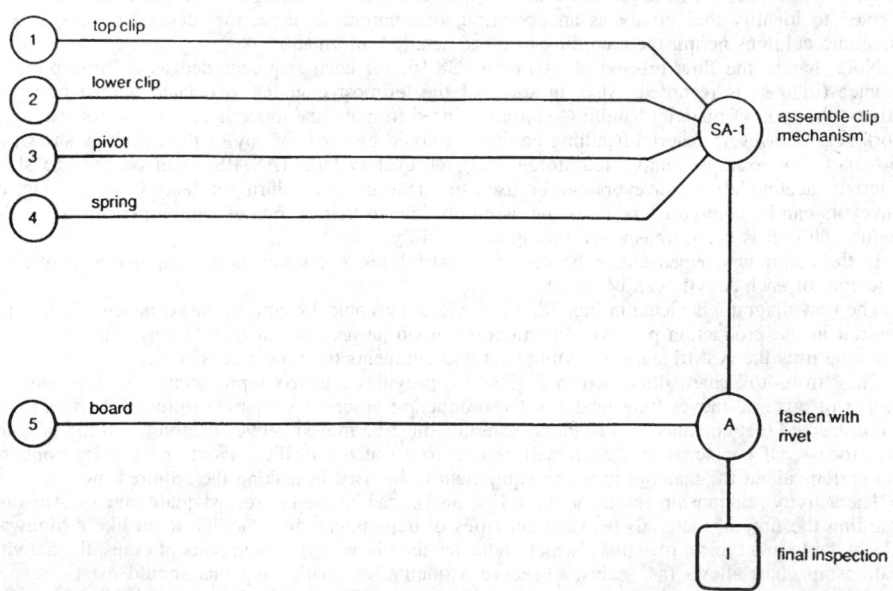

Fig. 38.13   Assembly chart.

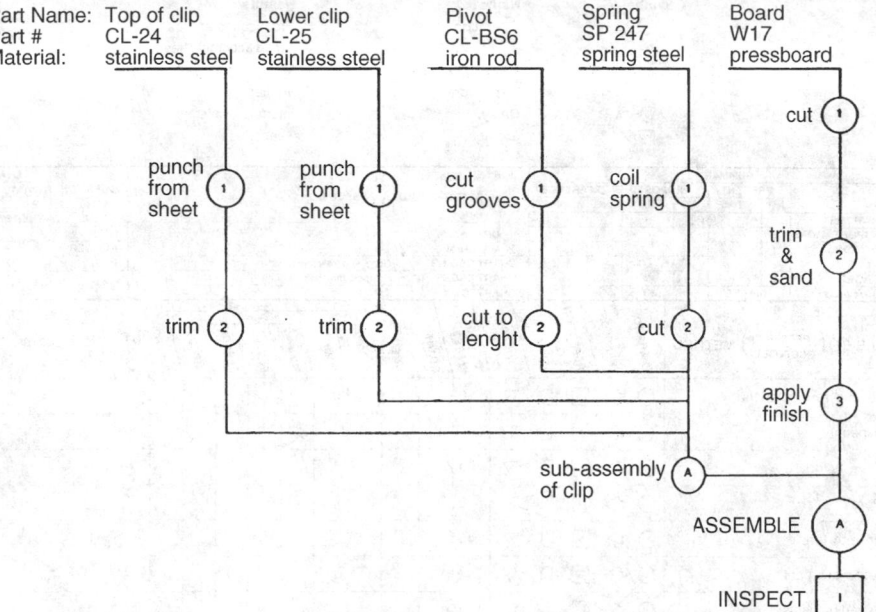

Fig. 38.14 Operations process chart.

other (using an alphabetic symbol) and the reason for the assignment of that rating (using a numeric symbol). Together these charting techniques provide the analyst extensive, qualitative data about the layout to support a production process. This is very useful from the standpoint of designing a material handling system.

38.4.5 Unitizing Loads

Principle number 7 of the MHI Twenty Principles of Material Handling (Fig. 38.12) is the *unit size principle,* also known as the *unit load principle,* which states, " Increase the quantity, size, or weight of unit loads or flow rate." The idea behind this principle is that if materials are consolidated into large quantities or sizes, fewer moves of this material will have to be made to meet needs of the production processes. Therefore, less time will be required to move the unitized material than that required to move the same quantity of non-unitized material. So, unitizing materials usually results in low-cost, efficient material-handling practices.

The decision to unitize is really a design decision in itself, as illustrated in Fig. 38.19. Unitization can consist of individual pieces through unit packs, inner packs, shipping cartons, tiers on pallets, pallet loads, containers of pallets, truckloads, and so on. The material-handling system must then be designed to accommodate the level of unitized parts at each step of the production process.

As shown in Fig. 38.19, once products or components have been unitized into shipping cartons, further consolidation may easily be achieved by placing the cartons on a pallet, slip sheet, or some other load-support medium for layers (or tiers) of cartons comprising the unit load. Since the unit load principle requires the maximum utilization of the area on the pallet surface, another design problem is to devise a carton stacking pattern that achieves this objective. Examples of pallet loading patterns that can achieve optimal surface utilization are illustrated in Fig. 38.20. Charts of such patterns are available from the U.S. Government (General Services Administration). There are also a number of providers of computer software programs for personal computers that generate pallet-loading patterns.

Highly automated palletizer machines as well as palletizing robots are available that can be programmed to form unit loads in any desired configuration. Depending upon the dimensions of the cartons to be palletized, and the resulting optimal loading pattern selected, the palletized load may be inherently stable due to overlapping of cartons in successive tiers; for example, the various pin-wheel patterns shown in Fig. 38.20.

However, other pallet-loading patterns may be unstable, such as the block pattern in Fig. 38.21, particularly when cartons are stacked several tiers high. In such instances, the loads may be stabilized by stretch-wrapping the entire pallet load with plastic film, or by placing bands around the individual

Symbol	Name	Results
○	Operation	Produces, prepares, and accomplishes
◇	Transportation	Moves
□	Inspection	Verifies
D	Delay	Interfere, waits
▽	Storage	Keeps, retains

SUMMARY							JOB				ANALYSIS	
	PRESENT		PROPOSED		DIFERENCE		Manufacture of a tissue box		QUESTION EACH DETAIL	WHAT? WHERE?	WHEN? WHO?
	NO.	TIME	NO.	TIME	NO.	TIME				WHY?	HOW?
○ OPERATIONS	5								DATE		
◇ TRANSPORTATIONS	9						☐ OPERATOR		NUMBER		
☐ INSPECTIONS	1						☒ MATERIAL		PAGE 1	OF 1	
D DELAYS	2						CHART BEGINS　Receiving (raw materials)				
▽ STORAGES	3						CHART ENDS　Shipping (finished product)				
Distance Traveled	1485 FT.		FT.		FT.		CHARTED BY　T.P.C.				

DETAILS OF (PRESENT / PROPOSED) METHOD	OPERATION	TRANSPORT	INSPECTION	DELAY	STORAGE	DISTANCE IN FEET	QUANTITY	TIME	ELIMINATE	COMBINE	SEQUENCE	PLACE	PERSON	IMPROVE	SAFER?	$ SAVED?	NOTES
1. Receive raw materials	○	◆	□	D	▽	50											
2. Inspect	○	◇	◈	D	▽												
3. Move by fork lift	○	◆	□	D	▽	40											
4. Store	○	◇	□	D	▼												
5. Move by fork lift	○	◆	□	D	▽	45											
6. Set up and print	◉	◇	□	D	▽												
7. Moved by printer	○	◆	□	D	▽	120											
8. Stack at end of printer	◉	◇	□	D	▽												
9. Move to stripping	○	◆	□	D	▽	165											
10. Delay	○	◇	□	◗	▽												
11. Being stripped	◉	◇	□	D	▽												
12. Move to temp. storage	○	◆	□	D	▽	150											
13. Storage	○	◇	□	D	▼												
14. Move to folders	○	◆	□	D	▽	200											
15. Delay	○	◇	□	◗	▽												
16. Set up, fold, glue	◉	◇	□	D	▽												
17. Mechanically moved	○	◆	□	D	▽	90											
18. Stack, count, crate	◉	◇	□	D	▽												
19. Move by fork lift	○	◆	□	D	▽	525											
20. Storage	○	◇	□	D	▽												

Fig. 38.15　Flow process chart.

tiers. The wrapping or banding operations themselves can be automated by use of equipment that exists in the market today.

Once the unit load has been formed, there are only four basic ways it can be handled while being moved. These are illustrated in Fig. 38.22 and consist of the following:

1. Support the load from below.
2. Support or grasp the load from above.
3. Squeeze opposing sides of the load.
4. Pierce the load.

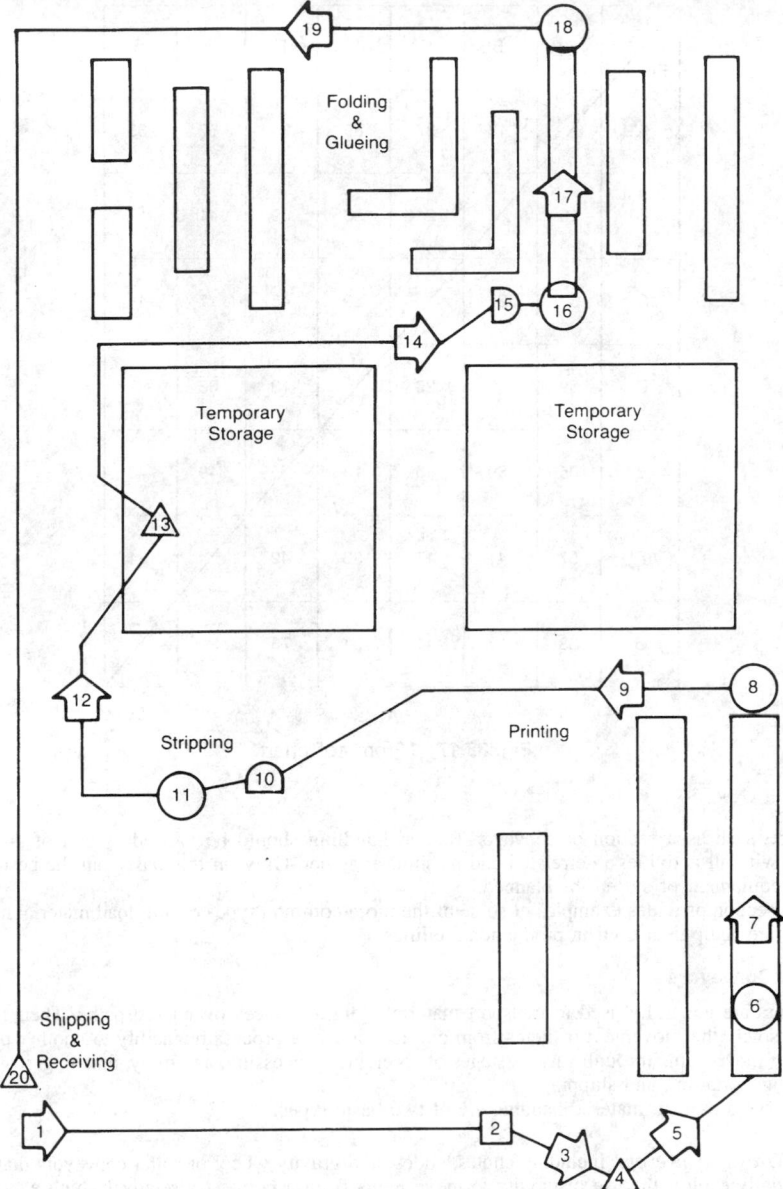

Fig. 38.16 Flow diagram.

These handling methods are implemented individually, or in combination, by commercially available material-handling equipment types.

38.5 MATERIAL-HANDLING EQUIPMENT CONSIDERATIONS AND EXAMPLES

38.5.1 Developing the Plan

Once the material-handling problem has been identified and the relevant data have been collected and analyzed, the next step in the design process is to develop a plan for solving the problem. This usually involves the design and/or selection of appropriate types, sizes, and capacities of material-handling equipment. In order to properly select material handling equipment, it must be realized that in most cases, the solution to the problem does not consist merely of selecting a particular piece of

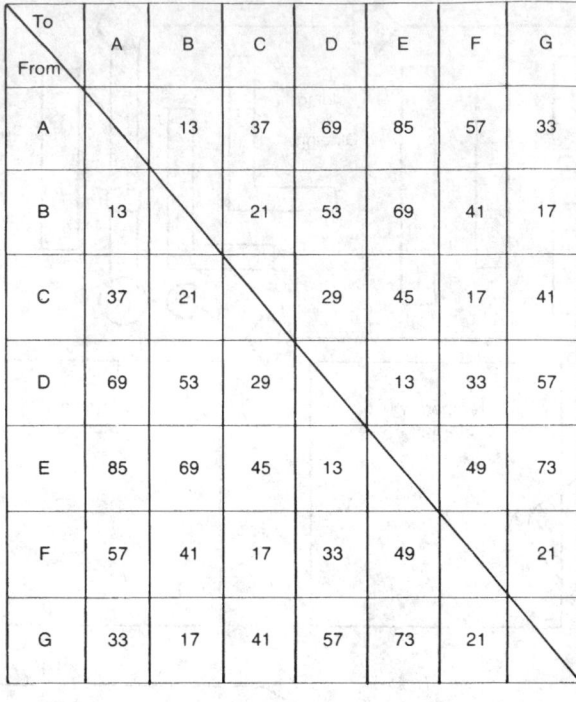

To From	A	B	C	D	E	F	G
A		13	37	69	85	57	33
B	13		21	53	69	41	17
C	37	21		29	45	17	41
D	69	53	29		13	33	57
E	85	69	45	13		49	73
F	57	41	17	33	49		21
G	33	17	41	57	73	21	

A

Fig. 38.17 "From–to" chart.

hardware, such as a section of conveyor. Rather, handling should be viewed as part of an overall system, with all activities interrelated and meshing together. Only on this basis can the best overall type of equipment or system be planned.

This section provides examples of some of the more common types of unit load material handling and storage equipment used in production facilities.

38.5.2 Conveyors

Conveyors are generally used to transport materials long distances over fixed paths. Their function may be solely the movement of items from one location in a process or facility to another point, or they may move items through various stages of receiving, processing, assembly, finishing, inspection, packaging, sortation, and shipping.

Conveyors used in material handing are of two basic types:

1. *Gravity conveyors,* including chutes, slides, and gravity wheel or roller conveyors that essentially exploit the use of gravity to move items from a point at a relatively high elevation to another point at a lower elevation. As listed in Fig. 38.12, MHI Principle 5 indicates that one should maximize the use of gravity in designing material-handling systems.

2. *Powered conveyors,* which generally use electric motors to drive belts, chains, or rollers in a variety of in-floor, floor-mounted, or overhead configurations.

In general, conveyors are employed in unit material handling when

1. Loads are uniform.
2. Materials move continuously.
3. Routes do not vary.
4. Load is constant.
5. Movement rate is relatively fixed.
6. Cross traffic can be bypassed.

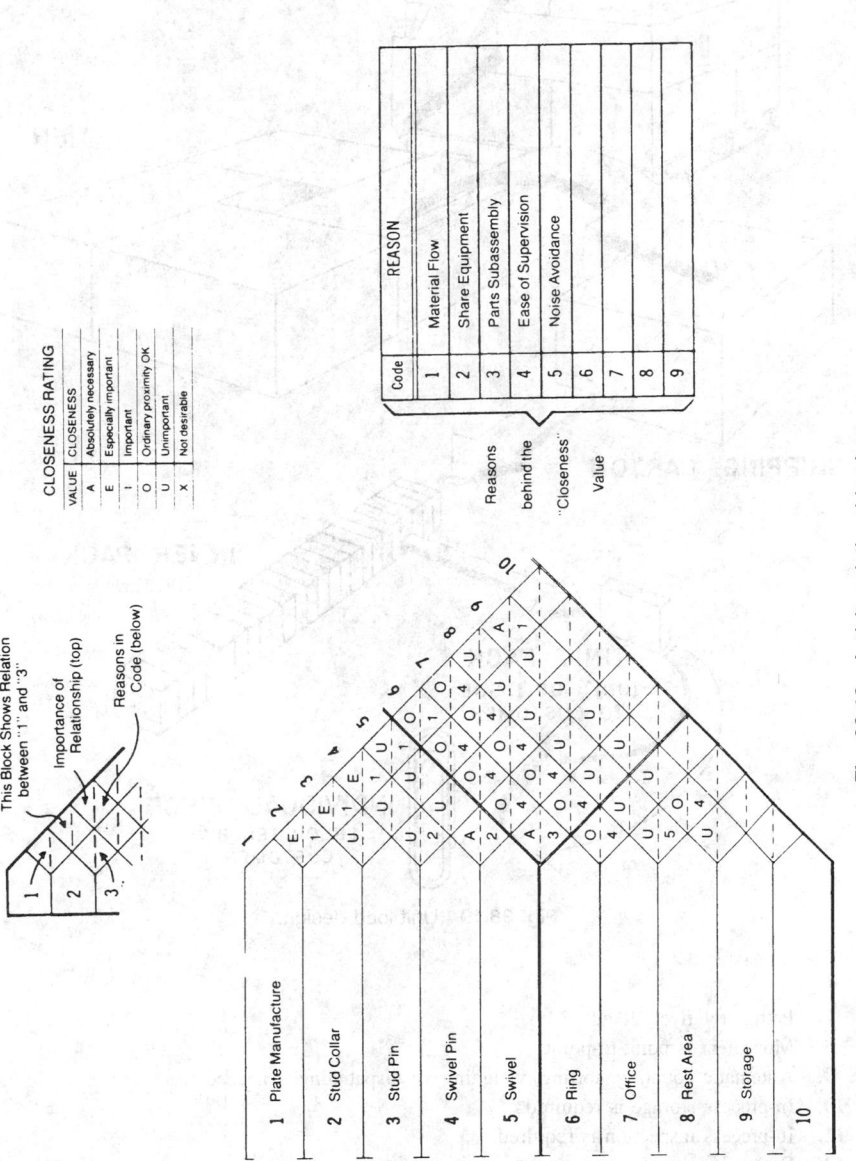

CLOSENESS RATING

VALUE	CLOSENESS
A	Absolutely necessary
E	Especially important
I	Important
O	Ordinary proximity OK
U	Unimportant
X	Not desirable

Code	REASON
1	Material Flow
2	Share Equipment
3	Parts Subassembly
4	Ease of Supervision
5	Noise Avoidance
6	
7	
8	
9	

Reasons behind the "Closeness" Value

This Block Shows Relation between "1" and "3"

Importance of Relationship (top)

Reasons in Code (below)

1 Plate Manufacture
2 Stud Collar
3 Stud Pin
4 Swivel Pin
5 Swivel
6 Ring
7 Office
8 Rest Area
9 Storage
10

Fig. 38.18 Activity relationship chart.

1227

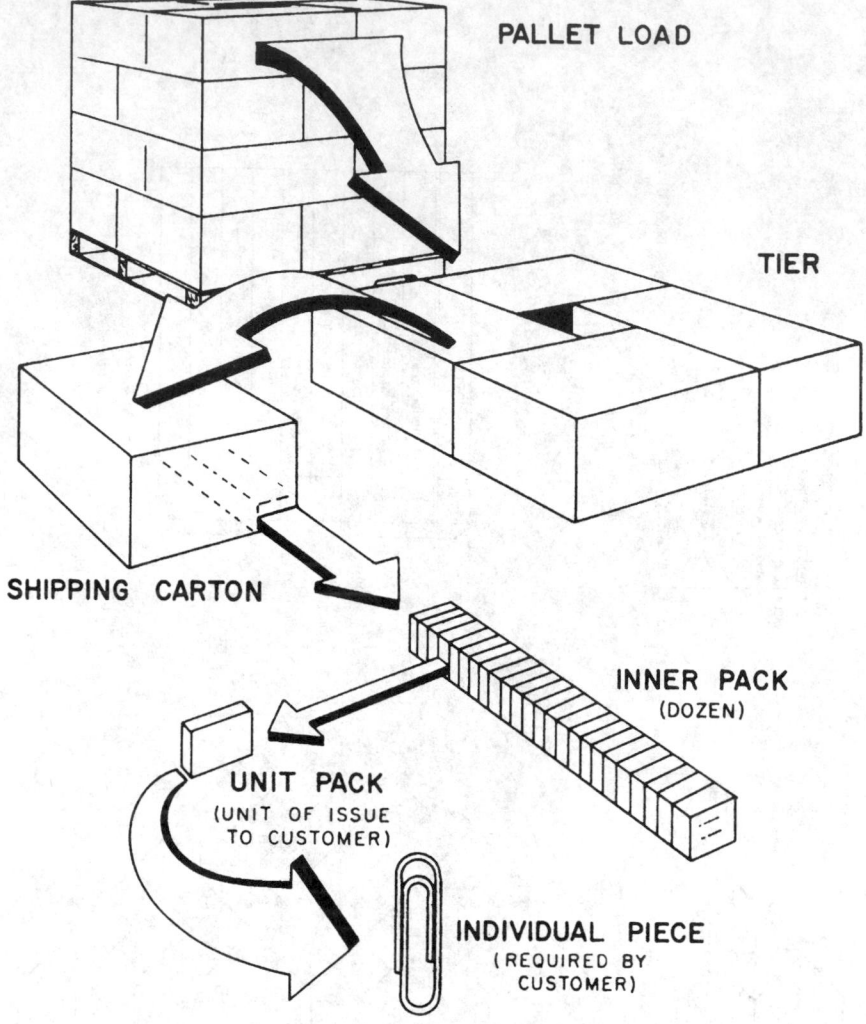

PALLET LOAD

TIER

SHIPPING CARTON

INNER PACK
(DOZEN)

UNIT PACK
(UNIT OF ISSUE
TO CUSTOMER)

INDIVIDUAL PIECE
(REQUIRED BY
CUSTOMER)

Fig. 38.19 Unit load design.

7. Path is relatively fixed.
8. Movement is point-to-point.
9. Automatic counting, sorting, weighing, or dispatching is needed.
10. In-process storage is required.
11. In-process inspection is required.
12. Production pacing is necessary.
13. Process control is required.
14. Controlled flow is needed.
15. Materials are handled at extreme temperatures, or other adverse conditions.
16. Handling is required in a hazardous area.
17. Hazardous materials are handled.
18. Machines are integrated into a system.
19. Robots are integrated into a system.
20. Materials are moved between workplaces.
21. Manual handling and/or lifting is undesirable.

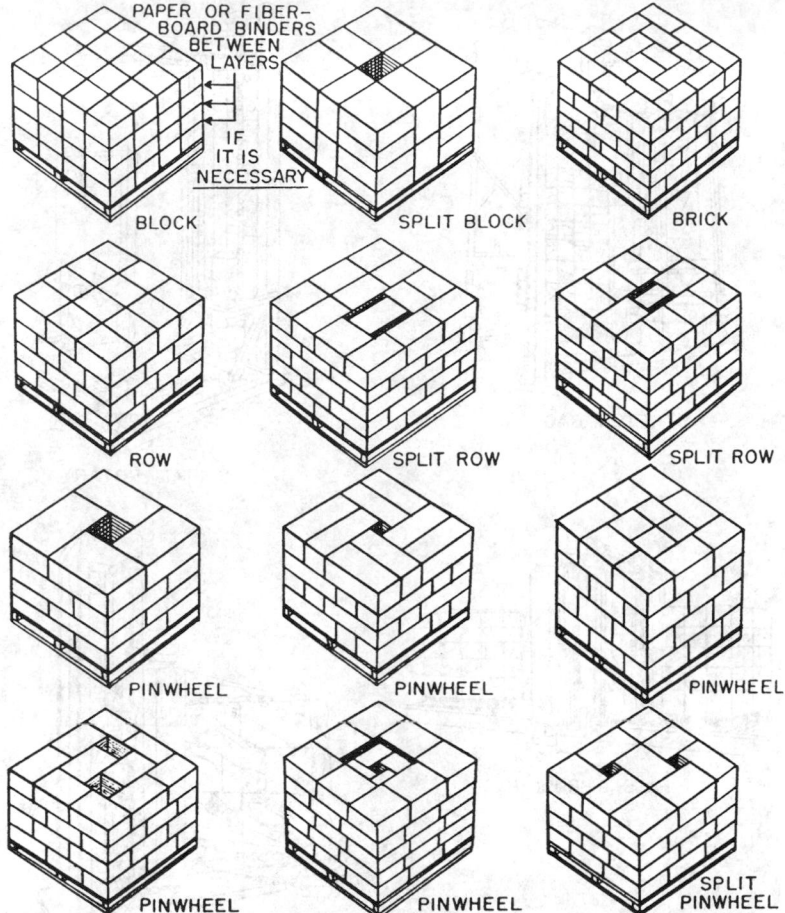

Fig. 38.20 Example pallet-loading patterns.

22. Changes in production volume or pace are needed.
23. Visual surveillance of a production process is required.
24. Floor space can be saved by utilizing overhead space.
25. Flexibility is required to meet changes in production processes.
26. Integration between computer-aided design and computer-aided manufacturing is required.

This section further details essential information on four main classes of conveyors used in unit material handling:

1. Gravity conveyors
2. Powered conveyors
3. Chain-driven conveyors
4. Power-and-free conveyors

Gravity Conveyors

Gravity conveyors exploit gravity to move material without the use of other forms of energy. Chutes, skate wheel conveyors, and roller conveyors are the most common forms of gravity conveyors. Figure 38.23 illustrates wheel and roller conveyors. Advantages of gravity conveyors are low cost, relatively low maintenance, and negligible breakdown rate. The main requirement for using gravity conveyors is the ability to provide the necessary gradient in the system configuration at the point at which gravity units are placed.

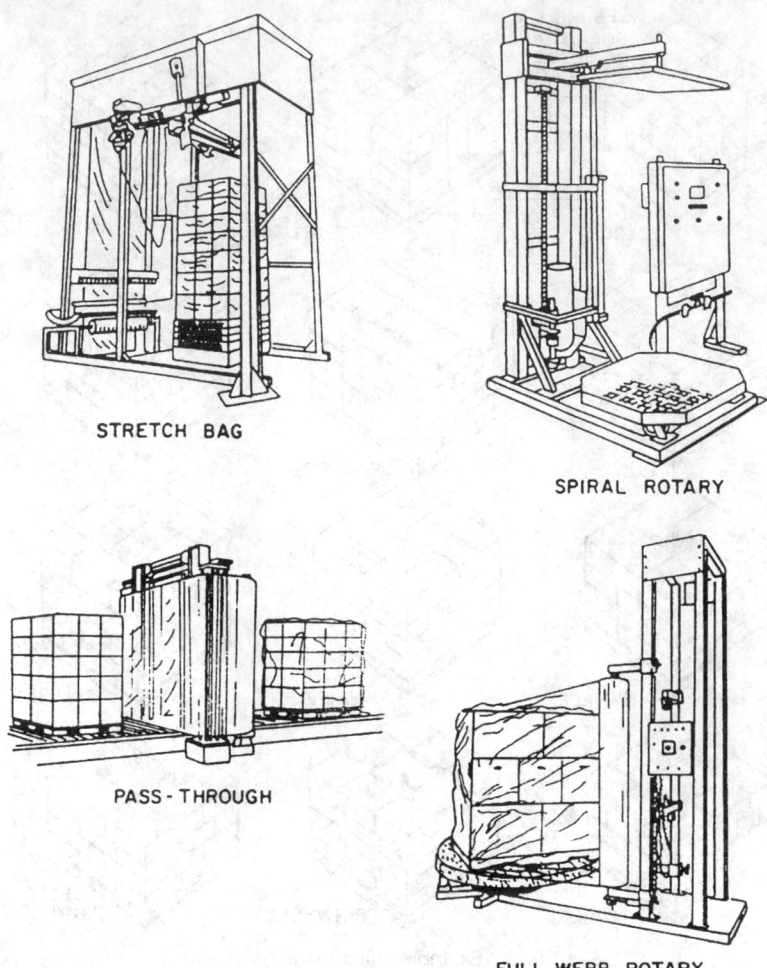

STRETCH BAG

SPIRAL ROTARY

PASS-THROUGH

FULL WEBB ROTARY

Fig. 38.21   Stretch wrap equipment.

Powered Conveyors

The two principal types of powered conveyors are belt conveyors and roller conveyors, as shown in Fig. 38.24. Electric motors provide the energy to drive the belt or rollers on these conveyors.

Belt conveyors are used either in the horizontal plane or with inclines up to 30°. They can range in length from a few feet to hundreds of feet, and are usually unidirectional. Changes in direction must be managed through the use of connecting chutes or diverters to conveyors running in another direction.

Roller conveyors are used for heavier loads than can be moved with belt conveyors, and are generally of sturdier construction. When used as accumulating conveyors, roller conveyors can also be used to provide spacing between items. Inclines are possible to about 10°; declines of about 15° are possible.

Powered conveyors should be operated at about 65 ft/min (about 1 mile/hr).

Chain-Driven Conveyors

Chain conveyors are those in which closed-loop systems of chain, usually driven by electric motors, are used to pull items or carts along a specified path. The three principal types of chain-driven conveyors used in unit material handling are flight conveyors, overhead towlines and monorails, and in-floor towlines. Figure 38.25 illustrates an overhead towline type of chain-driven conveyor.

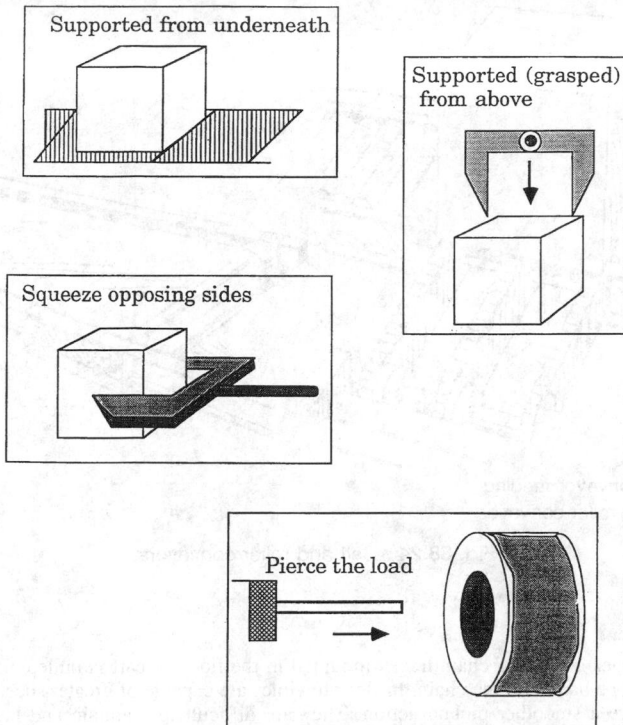

Fig. 38.22 Ways to handle a load.

Flight conveyors consist of one or more endless strands of chain with spaced transverse flights or scrapers attached, which push the material along through a trough. Used primarily in bulk material handling, its primary function in unit material handling includes movement of cans or bottles in food canning and bottling. A flight conveyor is generally limited to speeds of up to 120 ft/min.

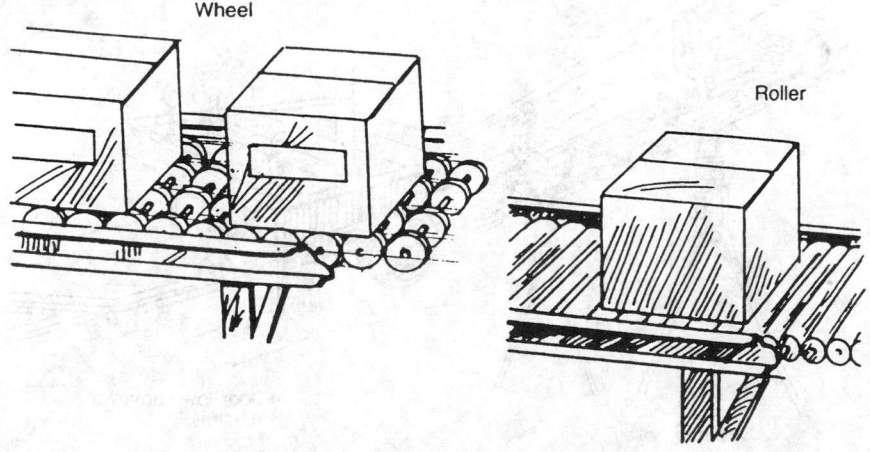

Fig. 38.23 Gravity conveyors.

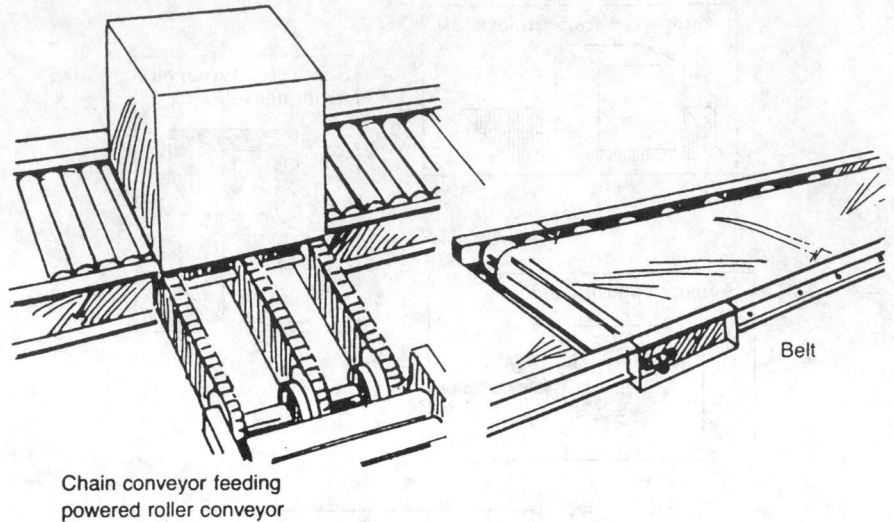

Chain conveyor feeding
powered roller conveyor

Fig. 38.24   Belt and roller conveyors.

In-floor towlines consist of chain tracks mounted in the floor. A cart is pulled along the track by attaching a pin-type handle to the chain. In-floor towlines are capable of greater speeds than overhead towlines and have a smoother pickup action. They are difficult to maintain and lack flexibility for rerouting.

Overhead towlines consists of a track mounted 8–9 ft above the floor. Carts on the floor are attached to the chain, which moves through the overhead track. Overhead towlines free the floor for other uses, and are less expensive and more flexible than in-floor towlines.

Power-and-Free Conveyors

Power-and-free conveyors are a combination of powered trolley conveyors and unpowered monorail-type conveyors. Two sets of tracks are used, one positioned above the other. The upper track carries the powered trolley or monorail-type conveyor, which is chain-driven. The lower track is the free,

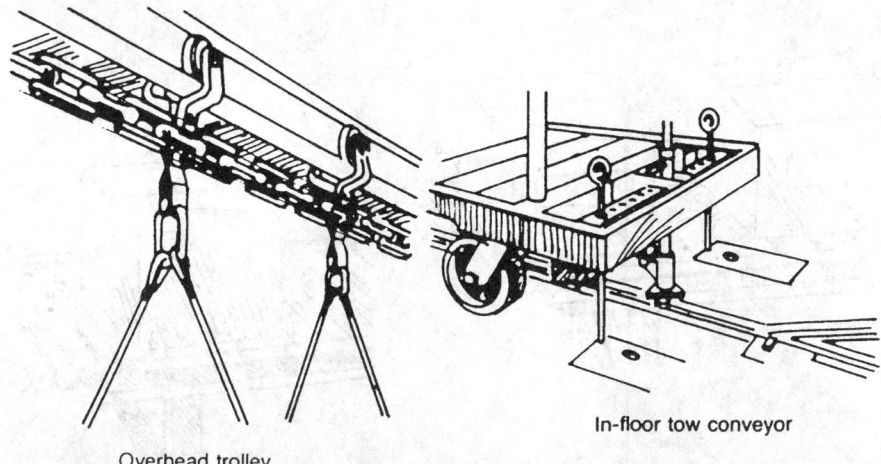

Overhead trolley

In-floor tow conveyor

Fig. 38.25   Chain-driven conveyors.

unpowered monorail. Load-carrying free trolleys are engaged by pushers attached to the powered trolley conveyors. Load trolleys can be switched to and from adjacent unpowered free tracks.

Interconnections on power-and-free conveyors may be manually or automatically controlled. Track switches may divert trolleys from "power" to "free" tracks. Speeds may vary from one "power" section to another, and programmable logic controllers (PLC) or computers can be used to control power-and-free conveyors.

Power-and-free conveyors, shown in Fig. 38.26, are relatively expensive and are costly to relocate.

38.5.3 Hoists, Cranes, and Monorails

Overhead material handling system components (e.g., tracks, carriers/trolleys, hoists, monorails, and cranes), can make effective use of otherwise unused overhead space to move materials in a facility. This can free up valuable floor space for other uses than material handling, reduce floor-based traffic, and reduce handling time by employing "crow-fly" paths between activities or departments.

Hoists, cranes, and monorails are used for a variety of overhead handling tasks. A hoist is a device for lifting and/or lowering a load and typically consists of an electric or pneumatically powered motor; a lifting medium, such as a chain, cable, or rope; a drum for reeling the chain, cable, or rope; and a handling device at the end of the lifting medium, such as a hook, scissor clamp mechanism, grapple, and so on. Hoists may be manually operated or automatically controlled by PLC or computer.

A monorail is a single-beam overhead track that provides a horizontal single path or route for a load as it is moved through a facility. The lower flange of the rail serves as a runway for a trolley-mounted hoist. A monorail system is used to move, store, and queue material overhead.

Through the use of switches, turntables, and other path-changing devices, an overhead monorail can be made to follow multiple predetermined paths, carrying a series of trolleys through various

Fig. 38.26 Power-and-free conveyor.

stations in processing or assembly. A chain-driver overhead monorail is very similar to the overhead towline in its configuration, except that it generally carries uniformly spaced trolleys overhead instead of pulling carts along the floor.

However, newer monorail technology has led to the development of individually powered and controllable trolleys that travel on the monorail (see Fig. 38.27). These devices are termed *automated electrified monorails* (AEM). The speed of the individually powered AEM vehicles can be changed en route and can function in nonsynchronous or flexible production environments.

Monorails can be made to dip down at specific points to deliver items to machines or other processing stations.

A crane also involves a hoist mounted on a trolley. Frequently, the trolley may be transported, as in the case of the bridge cranes shown in Fig. 38.28. Cranes may be manually, electrically, or pneumatically powered.

A jib crane has a horizontal beam on which a hoist trolley rides. The beam is cantilevered from a vertical support mast about which the beam can rotate or pivot (see the wall bracket-type jib crane in Fig. 38.28). This rotation permits the jib crane a broad range of coverage within the cylindrical work envelope described by the degrees of freedom of the beam, hoist, and mast.

38.5.4 Industrial Trucks

Industrial trucks provide flexible handling of materials along variable (or random) flow paths. The two main categories of industrial trucks are hand trucks and powered trucks, illustrated by the examples in Fig. 38.29.

Four-wheeled and multiple-wheeled carts and trucks include dollies, platform trucks, and skid platforms equipped with jacks. Hand-operated lift trucks include types equipped with hand-actuated hydraulic cylinders, and others having mechanical-lever systems.

Perhaps the most familiar type of powered truck is the forklift, which uses a pair of forks—capable of variable spacing—riding on a vertical mast to engage, lift, lower, and move loads. Lift trucks may be manually propelled or powered by electric motors, gasoline, liquified propane, or diesel-fueled engines. With some models, the operator walks behind the truck. On others, he or she rides on the truck, in either a standing or sitting position. Figure 38.30 depicts several types of forklift trucks.

Lift trucks are very effective in lifting, stacking, and unloading materials from storage racks, highway vehicles, railroad cars, and other equipment. Some lift trucks are designed for general-purpose use, while others are designed for specific tasks, such as narrow-aisle or high-rack handling.

38.5.5 Automated Guided Vehicle Systems

An automated guided vehicle system (AGVS) has similar uses as an industrial truck-based material-handling system. However, as implied by their name, the vehicles in an AGVS are under automatic control and do not require operators to guide them. In general, the vehicles in an AGVS are battery-powered, driverless, and capable of being automatically routed between, and positioned at, selected pickup or dropoff stations strategically located within a facility. Most of the vehicles in industrial use today are transporters of unit loads. However, when properly equipped, AGVs can provide a number of other functions, such as serving as automated storage devices or assembly platforms.

The four commonly recognized operating environments for AGVSs are distribution warehouses, manufacturing storerooms and delivery systems, flexible manufacturing systems, and assembly systems. Vehicles are guided by inductive-loop wires embedded in the floor of a facility, a chemical stripe painted on the floor, or laser-based navigation systems. All vehicular motion, as well as load pickup and delivery interfaces, are under computer control.

Examples of typical unit load AGVs are shown in Fig. 38.31 equipped with various types of load-handling decks that can be used.

38.5.6 Automated Storage and Retrieval Systems

An automated storage and retrieval system (AS/RS) consists of a set of racks or shelves arrayed along either side of an aisle through which a machine travels that is equipped with devices for storing or retrieving unit loads from the rack or shelf locations. As illustrated in Fig. 38.32, the AS/RS machine resembles a vertically oriented bridge crane (mast) with one end riding on a rail mounted on the floor and the other end physically connected to a rail or channel at the top of the rack structure. The shuttle mechanism travels vertically along the mast as it, in turn, travels horizontally through the aisle. In this manner, it carries a unit load from an input station to the storage location in the rack structure, then extends into the rack to place the load. The procedure is reversed for a retrieval operation; that is, the empty shuttle is positioned at the correct rack location by the mast, then it is extended to withdraw the load from storage and transport it to the output station, usually located at the end of the aisle.

The AS/RS machines can have people on board to control the storage/retrieval operations, or they can be completely controlled by a computer. The objective in using AS/RSs is to achieve very dense storage of unit loads while simultaneously exercising very tight control of the inventory stored in these systems.

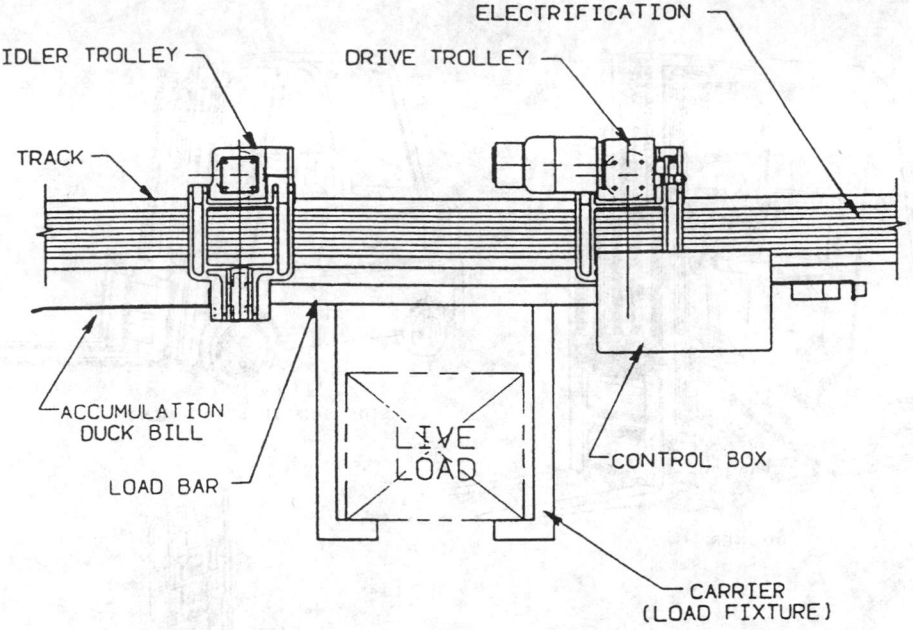

Fig. 38.27 Automated electrified monorail.

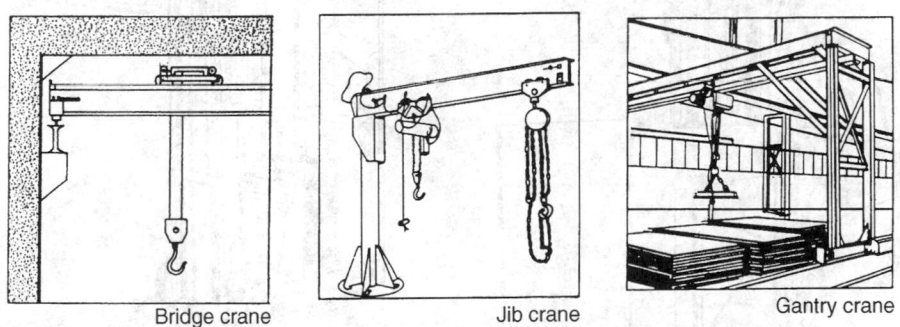

Bridge crane   Jib crane   Gantry crane

Fig. 38.28 Cranes.

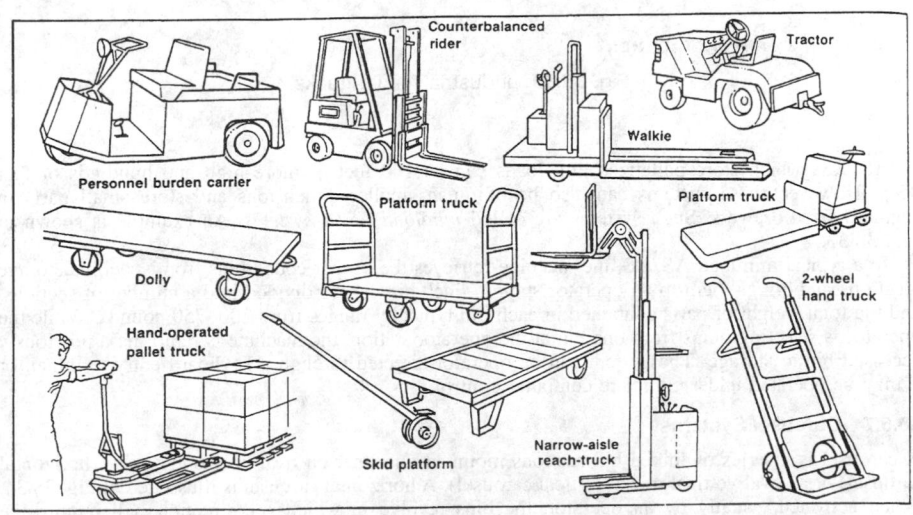

Fig. 38.29 Industrial truck equipment.

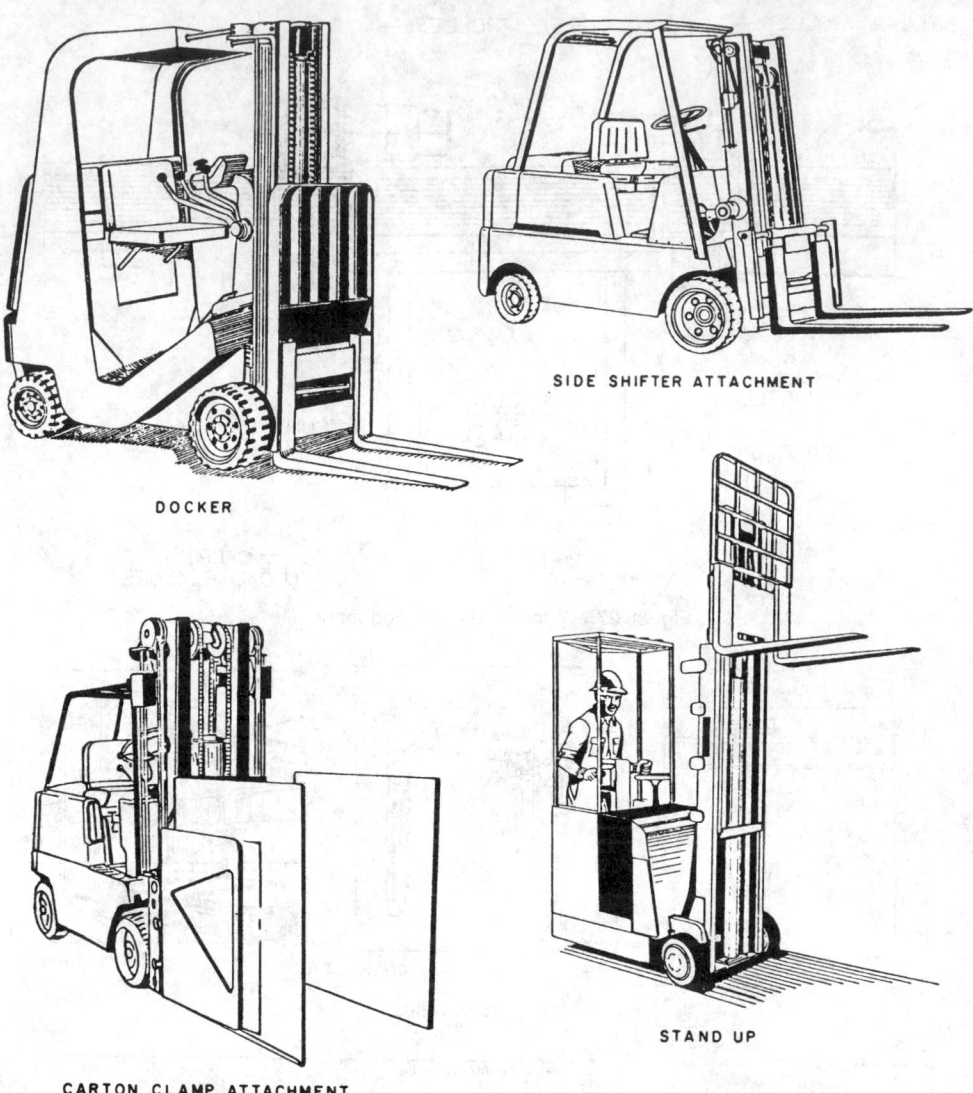

DOCKER

SIDE SHIFTER ATTACHMENT

CARTON CLAMP ATTACHMENT

STAND UP

Fig. 38.30 Industrial forklift trucks.

AS/RSs which store palletized unit loads can be 100 feet or more high and hundreds of feet deep. However, these systems can also be of much smaller dimensions and store small parts in standard-sized drawers. Such systems are called *miniload AS/R systems.* An example is shown in Fig. 38.33.

In a typical miniload AS/RS, the machine retrieves the proper coded bins from specified storage locations and brings them to an operator station. Each bin can be divided into a number of sections, and the total weight of parts contained in each bin typically ranges from 200–750 pounds. While the operator is selecting items from one bin at the operator station, the machine is returning a previously accessed bin to storage. The system can be operator-directed through a keyboard entry terminal, or it may be operated under complete computer control.

38.5.7 Carousel Systems

A carousel is a series of linked bin sections mounted on either an oval horizontal track (horizontal carousel) or an oval vertical track (vertical carousel). A horizontal carousel is illustrated in Fig. 38.34. When activated, usually by an operator, the bins revolve in whichever direction will require the

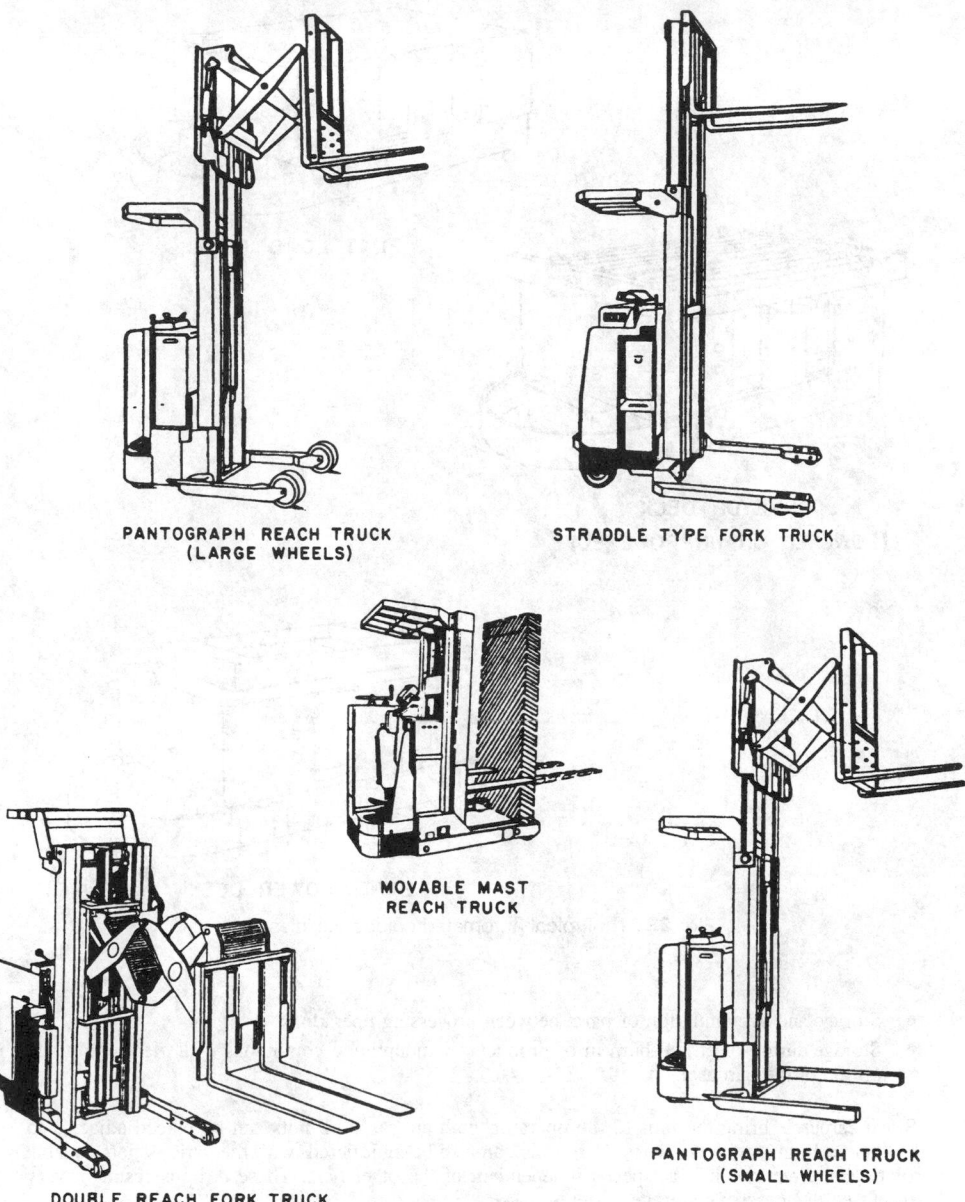

PANTOGRAPH REACH TRUCK
(LARGE WHEELS)

STRADDLE TYPE FORK TRUCK

MOVABLE MAST
REACH TRUCK

DOUBLE REACH FORK TRUCK

PANTOGRAPH REACH TRUCK
(SMALL WHEELS)

Fig. 38.30 *(Continued)*

minimum travel distance to bring the desired bin to the operator. The operator then either picks or puts away stock into the bin selected.

Some of the standard applications for carousels are

- Picking less than full case lots of small items for customer or dealer orders
- Storing small parts or subassemblies on the shop floor or in stockrooms
- Storing tools, maintenance parts, or other items that require limited access or security
- Storing work-in-process kits for assembly operations
- Storing documents, tapes, films, manuals, blueprints, etc.

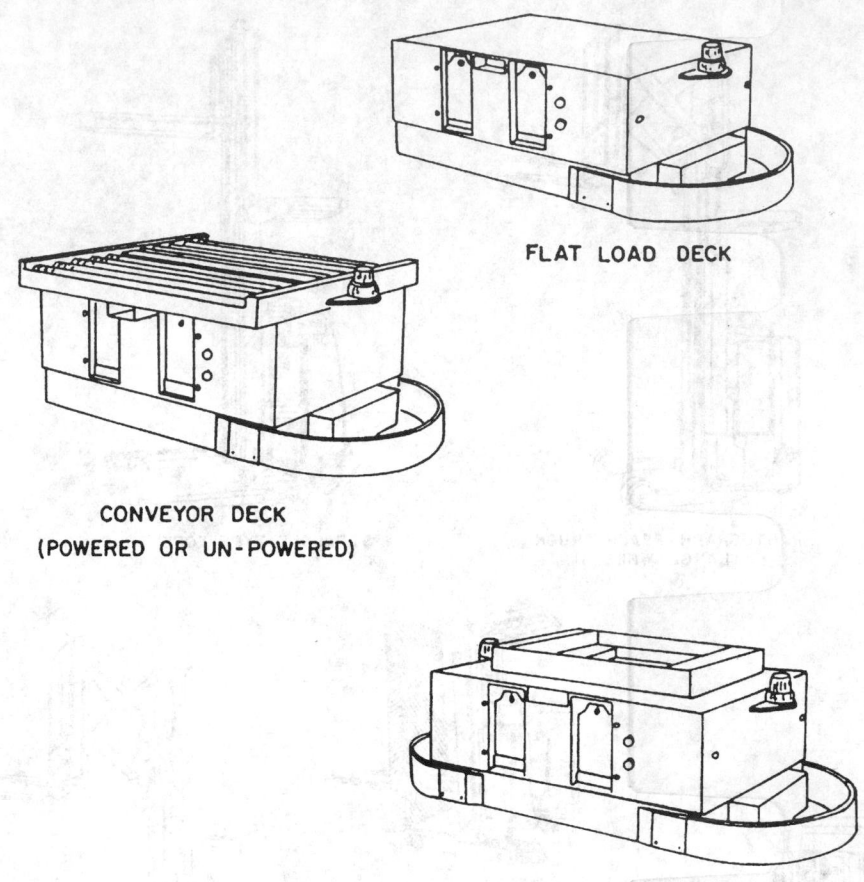

FLAT LOAD DECK

CONVEYOR DECK
(POWERED OR UN-POWERED)

LIFT / LOWER DECK

Fig. 38.31 Typical automated guided vehicles.

- Storage and accumulation of parts between processing operations
- Storage during electrical burn-in of products by mounting a continuous oval electrified track on top of the carousel

Since carousels bring the bins to the operator, multiple carousel units can be placed adjacent to each other (≤18 in.), with no aisles. Carousels can also be multitiered, with the various tiers capable of rotating in directions and at speeds independent of the other tiers. These designs result in very dense but easily accessible storage systems.

Carousels can be of almost any length, from a minimum of 10 feet to over 100 feet. However, most are in the 30- to 50-foot range so as to minimize bin access times. Carousels may be arranged so that one operator can pick or put away items in bins located on different systems or tiers while the other systems or tiers are positioning the next bin for access.

38.5.8 Shelving, Bin, Drawer, and Rack Storage

Shelving is used to economically store small, hand-stackable items that are generally not suited to mechanized handling and storage due to their handling characteristics, activity, or quantity. Standard shelving units are limited to about seven feet in height, but mezzanines can be used to achieve multiple storage levels and density.

Bin storage is, in most instances, identical in application to shelf storage, but is generally used for small items that do not require the width of a conventional shelf module. Bin storage usually represents a small part of the total storage system in terms of physical space, but it may represent a

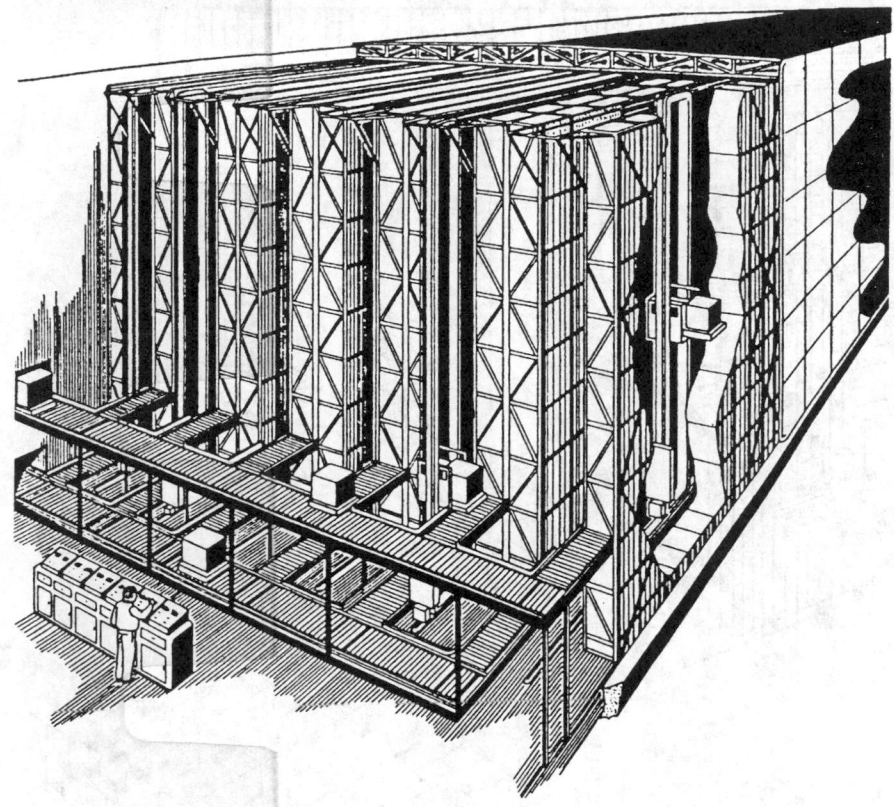

Fig. 38.32 Automated storage and retrieval system.

significant portion of the total storage in terms of the item positions used or stock keeping units (SKUs).

Modular drawer cabinets provide the advantages of increased security and density of storage over shelving and bin storage. As illustrated in Fig. 38.35, drawers provide the operator a clear view of all parts stored in them when pulled out. They can be configured to hold a large number of different SKU parts by partitioning the drawer volume into separate storage cells with dividers. This provides high storage density, good organization, and efficient utilization of storage space for small parts in applications such as tool cribs, maintenance shops, and parts supply rooms.

A pallet rack, as illustrated in Fig. 38.36, is a framework designed primarily to facilitate the storage of unit loads. This framework consists of upright columns and horizontal members for supporting the loads, and diagonal bracing for stability. The structural members may be bolted into place or lock-fitted.

Standard pallet racks can also be equipped with shelf panel inserts to facilitate their use for storage of binnable or shelvable materials. They may be loaded or unloaded by forklift trucks, by AS/R machines, or by hand. They may be fixed into position or made to slide along a track for denser storage. The primary purpose in using rack storage is to provide a highly organized unit material storage system that facilitates highly efficient operations in either manufacturing or distribution.

38.6 IMPLEMENTING THE SOLUTION

After the best system for solving the material handling problem has been designed, it is recommended that computer simulation be used to test the design before implementation. Although somewhat expensive to build and time-consuming to use, a valid simulation model can effectively test the overall operation of the material handling system as designed. It can identify potential bottleneck flows or choke points, isolate other costly design errors, determine efficient labor distribution, and evaluate various operating conditions that can be encountered. In other words, simulation enables the material-handling system designer to look into the future and get a realistic idea of how the system will

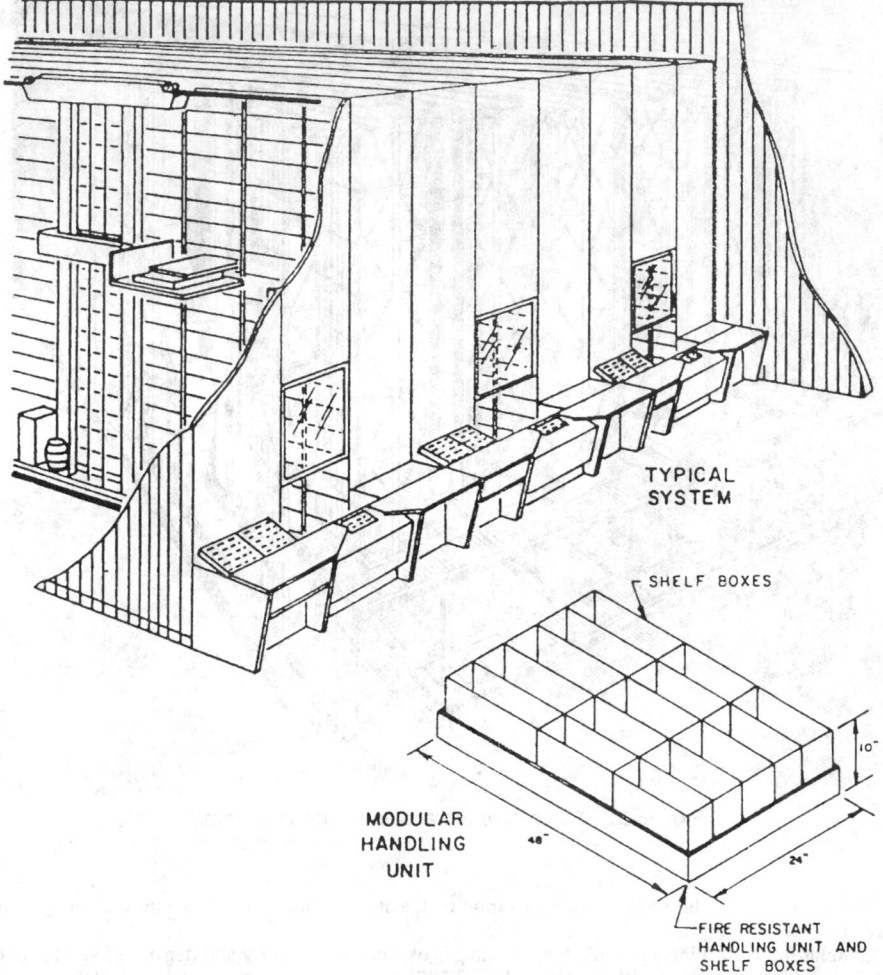

TYPICAL
SYSTEM

SHELF BOXES

MODULAR
HANDLING
UNIT

FIRE RESISTANT
HANDLING UNIT AND
SHELF BOXES

Fig. 38.33 Miniload AS/RS.

operate before proceeding to cost justification—or before the millwrights start bolting the wrong equipment together.

The final step is to implement the solution. Once total system costs—initial costs, recurring costs and salvage costs—have been calculated, an engineering economic analysis should be done to justify the investments required. Then the justification must be presented to, and approved by, appropriate managers. Once approval is obtained, a carefully prepared, written bid specification called a *request for quotation* (RFQ) is typically sent to several qualified vendors or contractors. Competing bids or proposals submitted by the vendors or contractors must then be evaluated carefully to ascertain whether they all are quoting on the same type and grade of equipment and components.

Each step of the equipment-acquisition process must be closely monitored to ensure that any construction is accomplished in a correct and timely manner and that equipment-installation procedures are faithfully followed. Once the completed facility is operational, it should be fully tested before final acceptance from the vendors and contractors. Operating personnel must be fully trained to use systems installed in the new facility.

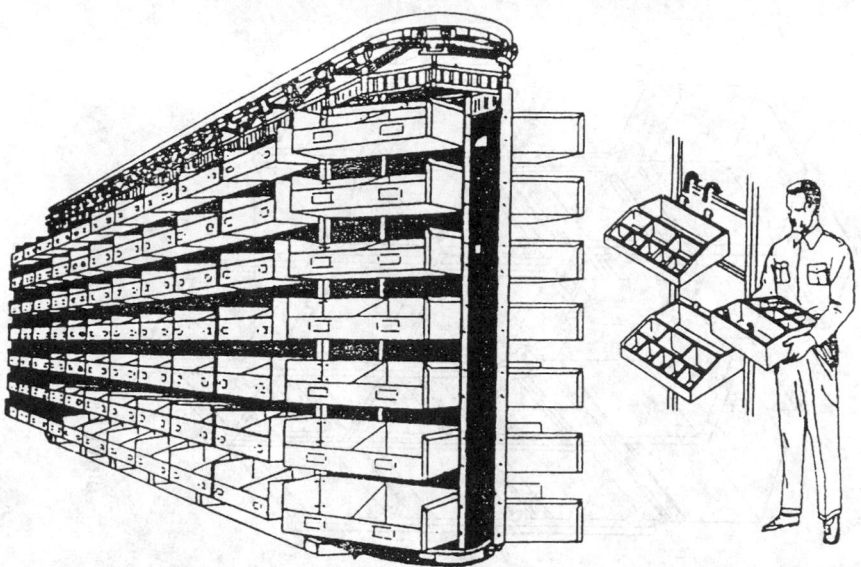

Fig. 38.34   Horizontal carousel.

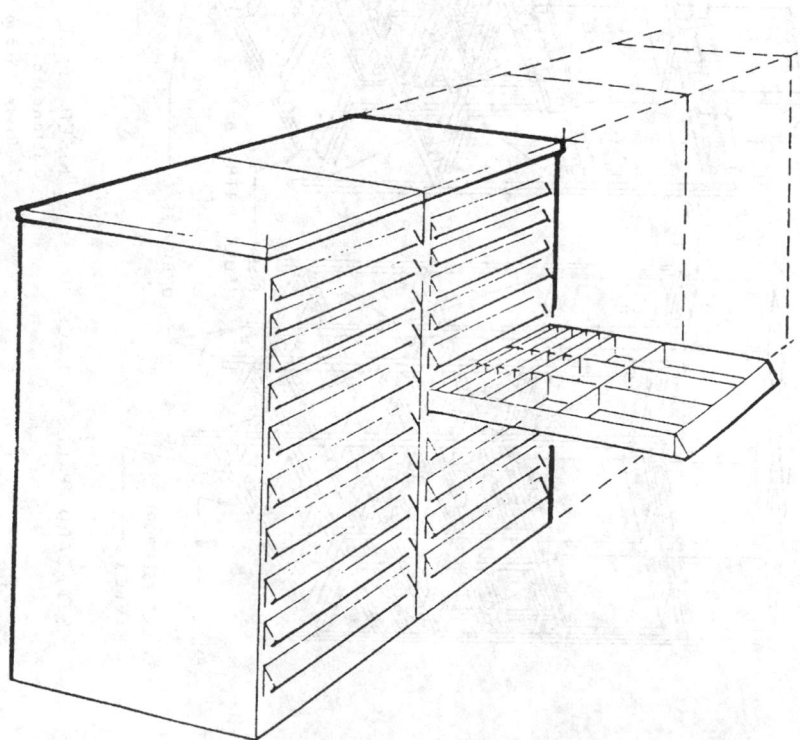

Fig. 38.35   Modular cabinet drawer storage.

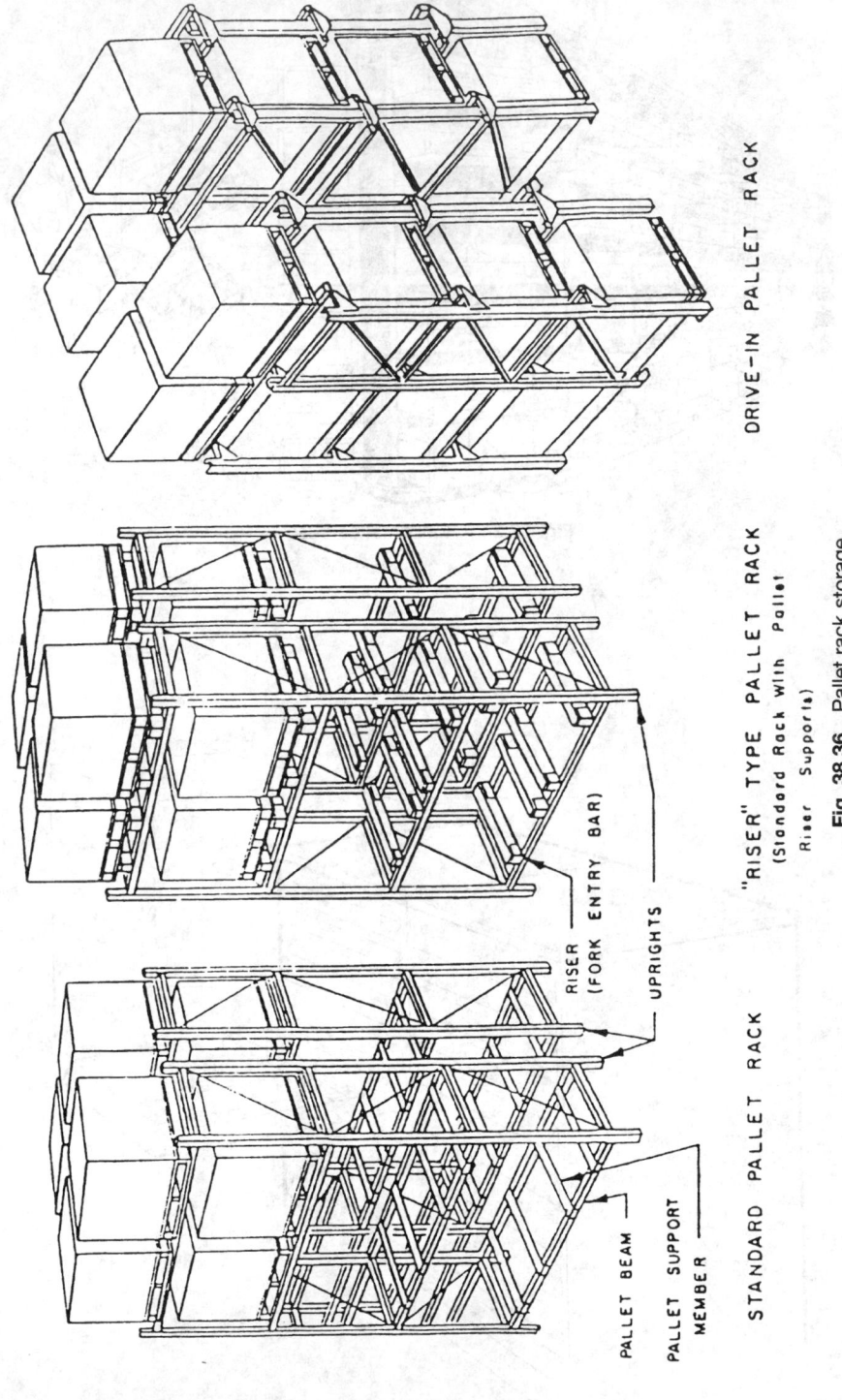

STANDARD PALLET RACK

"RISER" TYPE PALLET RACK
(Standard Rack With Pallet
Riser Supports)

DRIVE-IN PALLET RACK

PALLET BEAM

PALLET SUPPORT
MEMBER

RISER
(FORK ENTRY BAR)

UPRIGHTS

Fig. 38.36 Pallet rack storage.

PART 4
ENERGY, POWER, AND POLLUTION CONTROL TECHNOLOGY

CHAPTER 39

THERMOPHYSICAL PROPERTIES OF FLUIDS

Peter E. Liley
School of Mechanical Engineering
Purdue University
West Lafayette, Indiana

TABLE 39.1	CONVERSION FACTORS	1247	TABLE 39.9	THERMODYNAMIC PROPERTIES OF SATURATED MERCURY	1259
TABLE 39.2	PHASE TRANSITION DATA FOR THE ELEMENTS	1247	FIGURE 39.2	ENTHALPY–LOG PRESSURE DIAGRAM FOR MERCURY	1260
TABLE 39.3	PHASE TRANSITION DATA FOR COMPOUNDS	1249	TABLE 39.10	THERMODYNAMIC PROPERTIES OF SATURATED METHANE	1261
TABLE 39.4	THERMODYNAMIC PROPERTIES OF LIQUID AND SATURATED VAPOR AIR	1251	TABLE 39.11	THERMOPHYSICAL PROPERTIES OF METHANE AT ATMOSPHERIC PRESSURE	1262
TABLE 39.5	IDEAL GAS THERMOPHYSICAL PROPERTIES OF AIR	1252			
TABLE 39.6	THERMOPHYSICAL PROPERTIES OF THE U.S. STANDARD ATMOSPHERE	1255	TABLE 39.12	THERMOPHYSICAL PROPERTIES OF SATURATED REFRIGERANT 22	1263
TABLE 39.7	THERMOPHYSICAL PROPERTIES OF CONDENSED AND SATURATED VAPOR CARBON DIOXIDE FROM 200 K TO THE CRITICAL POINT	1256	TABLE 39.13	THERMOPHYSICAL PROPERTIES OF REFRIGERANT 22 AT ATMOSPHERIC PRESSURE	1264
TABLE 39.8	THERMOPHYSICAL PROPERTIES OF GASEOUS CARBON DIOXIDE AT 1 BAR PRESSURE	1257	FIGURE 39.3	ENTHALPY–LOG PRESSURE DIAGRAM FOR REFRIGERANT 22	1264
FIGURE 39.1	ENTHALPY–LOG PRESSURE DIAGRAM FOR CARBON DIOXIDE	1258	TABLE 39.14	THERMODYNAMIC PROPERTIES OF SATURATED REFRIGERANT 134a	1265

Mechanical Engineers' Handbook, 2nd ed., Edited by Myer Kutz.
ISBN 0-471-13007-9 © 1998 John Wiley & Sons, Inc.

TABLE 39.15 INTERIM THERMOPHYSICAL PROPERTIES OF REFRIGERANT 134a 1266

FIGURE 39.4 COMPRESSIBILITY FACTOR OF REFRIGERANT 134a 1268

FIGURE 39.5 ENTHALPY–LOG PRESSURE DIAGRAM FOR REFRIGERANT 134a 1269

TABLE 39.16 THERMODYNAMIC PROPERTIES OF SATURATED SODIUM 1270

TABLE 39.17 THERMODYNAMIC PROPERTIES OF ICE/WATER 1271

TABLE 39.18 THERMODYNAMIC PROPERTIES OF SATURATED STEAM/WATER 1272

TABLE 39.19 THERMOPHYSICAL PROPERTIES OF MISCELLANEOUS SUBSTANCES AT ATMOSPHERIC PRESSURE 1274

TABLE 39.20 PHYSICAL PROPERTIES OF NUMBERED REFRIGERANTS 1276

TABLE 39.21 SPECIFIC HEAT (kJ/kg · K) AT CONSTANT PRESSURE OF SATURATED LIQUIDS 1279

TABLE 39.22 RATIO OF PRINCIPAL SPECIFIC HEATS, c_p/c_v, FOR LIQUIDS AND GASES AT ATMOSPHERIC PRESSURE 1280

TABLE 39.23 SURFACE TENSION (N/m) OF LIQUIDS 1281

TABLE 39.24 THERMAL CONDUCTIVITY (W/m · K) OF SATURATED LIQUIDS 1282

TABLE 39.25 VISCOSITY (10^{-4} Pa · sec) OF SATURATED LIQUIDS 1283

TABLE 39.26 THERMOCHEMICAL PROPERTIES AT 1.013 bar, 298.15 K 1284

TABLE 39.27 IDEAL GAS SENSIBLE ENTHALPIES (kJ/kg · mol) OF COMMON PRODUCTS OF COMBUSTION 1285

FIGURE 39.6 PSYCHROMETRIC CHART 1287

In this chapter, information is usually presented in the System International des Unités, called in English the International System of Units and abbreviated SI. Various tables of conversion factors from other unit systems into the SI system and vice versa are available. The following table is only intended to enable rapid conversion to be made with moderate, that is, five significant figure, accuracy, usually acceptable in most engineering calculations. The references listed should be consulted for more exact conversions and defintions.

Table 39.1   Conversion Factors[a]

Density: 1 kg/m^3 = 0.06243 lb$_m$/ft^3 = 0.01002 lb$_m$/U.K. gallon = 8.3454 × 10^{-3} lb$_m$/U.S. gallon = 1.9403 × 10^{-3} slug/ft^3 = 10^{-3} g/cm^3

Energy: 1 kJ = 737.56 ft · lb$_f$ = 239.01 cal$_{th}$ = 0.94783 Btu = 3.7251 × 10^{-4} hp hr = 2.7778 × 10^{-4} kWhr

Specific energy: 1 kJ/kg = 334.54 ft · lb$_f$/lb$_m$ = 0.4299 Btu/lb$_m$ = 0.2388 cal/g

Specific energy per degree: 1 kJ/kg · K = 0.23901 Btu$_{th}$/lb · °F = 0.23901 cal$_{th}$/g · °C

Mass: 1 kg = 2.20462 lb$_m$ = 0.06852 slug = 1.1023 × 10^{-3} U.S. ton = 10^{-3} tonne = 9.8421 × 10^{-4} U.K. ton

Pressure: 1 bar = 10^5 N/m^2 = 10^5 Pa = 750.06 mm Hg at 0°C = 401.47 in. H$_2$O at 32°F = 29.530 in. Hg at 0°C = 14.504 lb/in.^{2} = 14.504 psia = 1.01972 kg/cm^2 = 0.98692 atm = 0.1 MPa

Temperature: T(K) = T(°C) + 273.15 = [T(°F) + 459.69]/1.8 = T(°R)/1.8

Temperature difference: ΔT(K) = ΔT(°C) = ΔT(°F)/1.8 = ΔT(°R)/1.8

Thermal conductivity: 1 W/m · K = 0.8604 kcal/m · hr · °C = 0.5782 Btu/ft · hr · °F = 0.01 W/cm · K = 2.390 × 10^{-3} cal/cm · sec · °C

Thermal diffusivity: 1 m^2/sec = 38750 ft^2/hr = 3600 m^2/hr = 10.764 ft^2/sec

Viscosity, dynamic: 1 N · sec/m^2 = 1 Pa · sec = 10^7 μP = 2419.1 lb$_m$/ft · hr = 10^3 cP = 75.188 slug/ft · hr = 10 P = 0.6720 lb$_m$/ft · sec = 0.02089 lb$_f$ · sec/ft^2

Viscosity, kinematic (*see* thermal diffusivity)

[a]E. Lange, L. F. Sokol, and V. Antoine, *Information on the Metric System and Related Fields*, 6th ed., G. C. Marshall Space Flight Center, AL (exhaustive bibliography); C. H. Page and P. Vigoureux, *The International System of Units*, NBS S.P. 330, Wahsington, D.C., 1974; E. A. Mechtly, *The International System of Units. Physical Constants and Conversion Factors*, NASA S.P. 9012, 1973. Numerous revisions periodically appear: see, for example, *Pure Appl. Chem.*, **51**, 1–41 (1979) and later issues.

Table 39.2   Phase Transition Data for the Elements[a]

Name	Symbol	Formula Weight	T_m (K)	Δh_{fus} (kJ/kg)	T_b (K)	T_c (K)
Actinium	Ac	227.028	1323	63	3475	
Aluminum	Al	26.9815	933.5	398	2750	7850
Antimony	Sb	121.75	903.9	163	1905	5700
Argon	Ar	39.948	83	30	87.2	151
Arsenic	As	74.9216	885			2100
Barium	Ba	137.33	1002	55.8		4450
Beryllium	Be	9.01218	1560	1355	2750	6200
Bismuth	Bi	208.980	544.6	54.0	1838	4450
Boron	B	10.81	2320	1933	4000	3300
Bromine	Br	159.808	266	66.0	332	584
Cadmium	Cd	112.41	594	55.1	1040	2690
Calcium	Ca	40.08	1112	213.1	1763	4300
Carbon	C	12.011	3810		4275	7200
Cerium	Ce	140.12	1072	390		9750
Cesium	Cs	132.905	301.8	16.4	951	2015
Chlorine	Cl$_2$	70.906	172	180.7	239	417
Chromium	Cr	51.996	2133	325.6	2950	5500
Cobalt	Co	58.9332	1766	274.7	3185	6300

Table 39.2 *(Continued)*

Name	Symbol	Formula Weight	T_m (K)	Δh_{fus} (kJ/kg)	T_b (K)	T_c (K)
Copper	Cu	63.546	1357	206.8	2845	8280
Dysprosium	Dy	162.50	1670	68.1	2855	6925
Erbium	Er	167.26	1795	119.1	3135	7250
Europium	Eu	151.96	1092	60.6	1850	4350
Fluorine	F_2	37.997	53.5	13.4	85.0	144
Gadolinum	Gd	157.25	1585	63.8	3540	8670
Gallium	Ga	69.72	303	80.1	2500	7125
Germanium	Ge	72.59	1211	508.9	3110	8900
Gold	Au	196.967	1337	62.8	3130	7250
Hafnium	Hf	178.49	2485	134.8	4885	10400
Helium	He	4.00260	3.5	2.1	4.22	5.2
Holmium	Ho	164.930	1744	73.8	2968	7575
Hydrogen	H_2	2.0159	14.0		20.4	
Indium	In	114.82	430	28.5	2346	6150
Iodine	I_2	253.809	387	125.0	457	785
Iridium	Ir	192.22	2718	13.7	4740	7800
Iron	Fe	55.847	1811	247.3	3136	8500
Krypton	Kr	83.80	115.8	19.6	119.8	209.4
Lanthanum	La	138.906	1194	44.6	3715	10500
Lead	Pb	207.2	601	23.2	2025	5500
Lithium	Li	6.941	454	432.2	1607	3700
Lutetium	Lu	174.967	1937	106.6	3668	
Magnesium	Mg	24.305	922	368.4	1364	3850
Manganese	Mn	54.9380	1518	219.3	2334	4325
Mercury	Hg	200.59	234.6	11.4	630	1720
Molybdenum	Mo	95.94	2892	290.0	4900	1450
Neodymium	Nd	144.24	1290	49.6	3341	7900
Neon	Ne	20.179	24.5	16.4	27.1	44.5
Neptunium	Np	237.048	910		4160	12000
Nickel	Ni	58.70	1728	297.6	3190	8000
Niobium	Nb	92.9064	2740	283.7	5020	12500
Nitrogen	N_2	28.013	63.2	25.7	77.3	126.2
Osmium	Os	190.2	3310	150.0	5300	12700
Oxygen	O_2	31.9988	54.4	13.8	90.2	154.8
Palladium	Pd	106.4	1826	165.0	3240	7700
Phosphorus	P	30.9738	317		553	995
Platinum	Pt	195.09	2045	101	4100	10700
Plutonium	Pu	244	913	11.7	3505	10500
Potassium	K	39.0983	336.4	60.1	1032	2210
Praseodymium	Pr	140.908	1205	49	3785	8900
Promethium	Pm	145	1353		2730	
Protactinium	Pa	231	1500	64.8	4300	
Radium	Ra	226.025	973		1900	
Radon	Rn	222	202	12.3	211	377
Rhenium	Re	186.207	3453	177.8	5920	18900
Rhodium	Rh	102.906	2236	209.4	3980	7000
Rubidium	Rb	85.4678	312.6	26.4	964	2070
Ruthenium	Ru	101.07	2525	256.3	4430	9600
Samarium	Sm	150.4	1345	57.3	2064	5050
Scandium	Sc	44.9559	1813	313.6	3550	6410

Table 39.2 (Continued)

Name	Symbol	Formula Weight	T_m (K)	Δh_{fus} (kJ/kg)	T_b (K)	T_c (K)
Selenium	Se	78.96	494	66.2	958	1810
Silicon	Si	28.0855	1684	1802	3540	5160
Silver	Ag	107.868	1234	104.8	2435	6400
Sodium	Na	22.9898	371	113.1	1155	2500
Strontium	Sr	87.62	1043	1042	1650	4275
Sulfur	S	32.06	388	53.4	718	1210
Tantalum	Ta	180.948	3252	173.5	5640	16500
Technetium	Tc	98	2447	232	4550	11500
Tellurium	Te	127.60	723	137.1	1261	2330
Terbium	Tb	158.925	1631	67.9	3500	8470
Thallium	Tl	204.37	577	20.1	1745	4550
Thorium	Th	232.038	2028	69.4	5067	14400
Thulium	Tm	168.934	1819	99.6	2220	6450
Tin	Sn	118.69	505	58.9	2890	7700
Titanium	Ti	47.90	1943	323.6	3565	5850
Tungsten	W	183.85	3660	192.5	5890	15500
Uranium	U	238.029	1406	35.8	4422	12500
Vanadium	V	50.9415	2191	410.7	3680	11300
Xenon	Xe	131.30	161.3	17.5	164.9	290
Ytterbium	Yb	173.04	1098	44.2	1467	4080
Yttrium	Y	88.9059	1775	128.2	3610	8950
Zinc	Zn	65.38	692.7	113.0	1182	
Zirconium	Zr	91.22	2125	185.3	4681	10500

$^a T_m$ = normal melting point; Δh_{fus} = enthalpy of fusion; T_b = normal boiling point; T_c = critical temperature.

Table 39.3 Phase Transition Data for Compounds[a]

Substance	T_m (K)	Δh_m (kJ/kg)	T_b (K)	Δh_v (kJ/kg)	T_c (K)	P_c (bar)
Acetaldehyde	149.7	73.2	293.4	584	461	55.4
Acetic acid	289.9	195.3	391.7	405	594	57.9
Acetone	178.6	98	329.5	501	508	47
Acetylene		96.4	189.2	687	309	61.3
Air	60				133	37.7
Ammonia	195.4	331.9	239.7	1368	405.6	112.8
Aniline	267.2	113.3	457.6	485	699	53.1
Benzene	267.7	125.9	353.3	394	562	49
n-Butane	134.8	80.2	261.5	366	425.2	38
Butanol	188	125.2	391.2	593	563	44.1
Carbon dioxide	216.6	184	194.7	573	304.2	73.8
Carbon disulfide	161	57.7	319.6	352	552	79
Carbon monoxide	68.1	29.8	81.6	215	133	35
Carbon tetrachloride	250.3	173.9	349.9	195	556	45.6
Carbon tetrafluoride	89.5		145.2	138	227.9	37.4
Chlorobenzene	228		405	325	632.4	45.2
Chloroform	210	77.1	334.4	249	536.6	54.7
m-Cresol	285.1		475.9	421	705	45.5
Cyclohexane	279.6	31.7	356	357	554.2	40.7
Cyclopropane	145.5	129.4	240.3	477	397.8	54.9

Table 39.3 (*Continued*)

Substance	T_m (K)	Δh_m (kJ/kg)	T_b (K)	Δh_v (kJ/kg)	T_c (K)	P_c (bar)
n-Decane	243.2	202.1	447.3	276	617	21
Ethane	89.9	94.3	184.6	488	305.4	48.8
Ethanol	158.6	109	351.5	840	516	63.8
Ethyl acetate	190.8	119	350.2	366	523.3	38.3
Ethylene	104	119.5	169.5	480	283.1	51.2
Ethylene oxide	161.5	117.6	283.9	580	469	71.9
Formic acid	281.4	246.4	373.9	502	576	34.6
Heptane	182.6	140.2	371.6	316	540	27.4
Hexane	177.8	151.2	341.9	335	507	29.7
Hydrazine	274.7	395	386.7	1207	653	147
Hydrogen peroxide	271.2	310	431	1263		
iso-Butane	113.6	78.1	272.7	386	408.1	36.5
Methane	90.7	58.7	111.5	512	191.1	46.4
Methanol	175.4	99.2	337.7	1104	513.2	79.5
Methyl acetate	174.8		330.2	410	507	46.9
Methyl bromide	180	62.9	277.7	252	464	71.2
Methyl chloride	178.5	127.4	249.3	429	416.3	66.8
Methyl formate	173.4	125.5	305	481	487.2	60
Methylene chloride	176.4	54.4	312.7	328	510.2	60.8
Naphthalene	353.2	148.1	491	341	747	39.5
Nitric oxide	111	76.6	121.4	460	180.3	65.5
Octane	216.4	180.6	398.9	303	569.4	25
Pentane	143.7	116.6	309.2	357	469.8	33.7
Propane	86	80	231.1	426	370	42.5
Propanol	147	86.5	370.4	696	537	51.7
Propylene	87.9	71.4	225.5	438	365	46.2
Refrigerant 12	115	34.3	243.4	165	385	41.2
Refrigerant 13	92		191.8	148	302.1	38.7
Refrigerant 13B1	105.4		215.4	119	340	39.6
Refrigerant 21	138		282.1	242	451.7	51.7
Refrigerant 22	113	47.6	232.4	234	369	49.8
Steam/water	273.2	334	373.2	2257	647.3	221.2
Sulfuric acid	283.7	100.7	v	v	v	v
Sulfur dioxide	197.7	115.5	268.4	386	430.7	78.8
Toluene	178.2		383.8	339	594	41

[a]v = variable; T_m = normal melting point; Δh_m = enthalpy of fusion; T_b = normal boiling point; Δh_v = enthalpy of vaporization; T_c = critical temperature; P_c = critical pressure.

Table 39.4 Thermodynamic Properties of Liquid and Saturated Vapor Air[a]

T (K)	P_f (MPa)	P_g (MPa)	v_f (m³/kg)	v_g (m³/kg)	h_f (kJ/kg)	h_g (kJ/kg)	s_f (kJ/kg·K)	s_g (kJ/kg·K)
60	0.0066	0.0025	0.001027	6.876	−144.9	59.7	2.726	6.315
65	0.0159	0.0077	0.001078	2.415	−144.8	64.5	2.727	6.070
70	0.0340	0.0195	0.001103	1.021	−144.0	69.1	2.798	5.875
75	0.0658	0.0424	0.001130	0.4966	−132.8	73.5	2.896	5.714
80	0.1172	0.0826	0.001160	0.2685	−124.4	77.5	3.004	5.580
85	0.1954	0.1469	0.001193	0.1574	−115.2	81.0	3.114	5.464
90	0.3079	0.2434	0.001229	0.0983	−105.5	84.1	3.223	5.363
95	0.4629	0.3805	0.001269	0.0644	−95.4	86.5	3.331	5.272
100	0.6687	0.5675	0.001315	0.0439	−84.8	88.2	3.436	5.189
105	0.9334	0.8139	0.001368	0.0308	−73.8	89.1	3.540	5.110
110	1.2651	1.1293	0.001432	0.0220	−62.2	89.1	3.644	5.033
115	1.6714	1.5238	0.001512	0.0160	−49.7	87.7	3.750	4.955
120	2.1596	2.0081	0.001619	0.0116	−35.9	84.5	3.861	4.872
125	2.743	2.614	0.001760	0.0081	−19.7	78.0	3.985	4.770
132.5[b]	3.770	3.770	0.00309	0.0031	33.6	33.6	4.38	4.38

[a]v = specific volume; h = specific enthalpy; s = specific entropy; f = saturated liquid; g = saturated vapor. 1 MPa = 10 bar.
[b]Approximate critical point. Air is a multicomponent mixture.

Table 39.5 Ideal Gas Thermophysical Properties of Air[a]

T (K)	v (m³/kg)	h (kJ/kg)	s (kJ/kg·K)	c_p (kJ/kg·K)	γ	$\bar{v}_s$ (m/sec)	η (N·sec/m²)	λ (W/m·K)	Pr
200	0.5666	−103.0	6.4591	1.008	1.398	283.3	1.33.−5[b]	0.0183	0.734
210	0.5949	−92.9	6.5082	1.007	1.399	290.4	1.39.−5	0.0191	0.732
220	0.6232	−82.8	6.5550	1.006	1.399	297.3	1.44.−5	0.0199	0.730
230	0.6516	−72.8	6.5998	1.006	1.400	304.0	1.50.−5	0.0207	0.728
240	0.6799	−62.7	6.6425	1.005	1.400	310.5	1.55.−5	0.0215	0.726
250	0.7082	−52.7	6.6836	1.005	1.400	317.0	1.60.−5	0.0222	0.725
260	0.7366	−42.6	6.7230	1.005	1.400	323.3	1.65.−5	0.0230	0.723
270	0.7649	−32.6	6.7609	1.004	1.400	329.4	1.70.−5	0.0237	0.722
280	0.7932	−22.5	6.7974	1.004	1.400	335.5	1.75.−5	0.0245	0.721
290	0.8216	−12.5	6.8326	1.005	1.400	341.4	1.80.−5	0.0252	0.720
300	0.8499	−2.4	6.8667	1.005	1.400	347.2	1.85.−5	0.0259	0.719
310	0.8782	7.6	6.8997	1.005	1.400	352.9	1.90.−5	0.0265	0.719
320	0.9065	17.7	6.9316	1.006	1.399	358.5	1.94.−5	0.0272	0.719
330	0.9348	27.7	6.9625	1.006	1.399	364.0	1.99.−5	0.0279	0.719
340	0.9632	37.8	6.9926	1.007	1.399	369.5	2.04.−5	0.0285	0.719
350	0.9916	47.9	7.0218	1.008	1.398	374.8	2.08.−5	0.0292	0.719
360	1.0199	57.9	7.0502	1.009	1.398	380.0	2.12.−5	0.0298	0.719
370	1.0482	68.0	7.0778	1.010	1.397	385.2	2.17.−5	0.0304	0.719
380	1.0765	78.1	7.1048	1.011	1.397	390.3	2.21.−5	0.0311	0.719
390	1.1049	88.3	7.1311	1.012	1.396	395.3	2.25.−5	0.0317	0.719
400	1.1332	98.4	7.1567	1.013	1.395	400.3	2.29.−5	0.0323	0.719
410	1.1615	108.5	7.1817	1.015	1.395	405.1	2.34.−5	0.0330	0.719
420	1.1898	118.7	7.2062	1.016	1.394	409.9	2.38.−5	0.0336	0.719
430	1.2181	128.8	7.2301	1.018	1.393	414.6	2.42.−5	0.0342	0.718
440	1.2465	139.0	7.2535	1.019	1.392	419.3	2.46.−5	0.0348	0.718
450	1.2748	149.2	7.2765	1.021	1.391	423.9	2.50.−5	0.0355	0.718
460	1.3032	159.4	7.2989	1.022	1.390	428.3	2.53.−5	0.0361	0.718
470	1.3315	169.7	7.3209	1.024	1.389	433.0	2.57.−5	0.0367	0.718
480	1.3598	179.9	7.3425	1.026	1.389	437.4	2.61.−5	0.0373	0.718

490	1.3882	190.2	7.3637	1.028	1.388	441.8	2.65.−5	0.0379	0.718
500	1.4165	200.5	7.3845	1.030	1.387	446.1	2.69.−5	0.0385	0.718
520	1.473	221.1	7.4249	1.034	1.385	454.6	2.76.−5	0.0398	0.718
540	1.530	241.8	7.4640	1.038	1.382	462.9	2.83.−5	0.0410	0.718
560	1.586	262.6	7.5018	1.042	1.380	471.0	2.91.−5	0.0422	0.718
580	1.643	283.5	7.5385	1.047	1.378	479.0	2.98.−5	0.0434	0.718
600	1.700	304.5	7.5740	1.051	1.376	486.8	3.04.−5	0.0446	0.718
620	1.756	325.6	7.6086	1.056	1.374	494.4	3.11.−5	0.0458	0.718
640	1.813	346.7	7.6422	1.060	1.371	501.9	3.18.−5	0.0470	0.718
660	1.870	368.0	7.6749	1.065	1.369	509.3	3.25.−5	0.0482	0.717
680	1.926	389.3	7.7067	1.070	1.367	516.5	3.32.−5	0.0495	0.717
700	1.983	410.8	7.7378	1.075	1.364	523.6	3.38.−5	0.0507	0.717
720	2.040	432.3	7.7682	1.080	1.362	530.6	3.45.−5	0.0519	0.716
740	2.096	453.9	7.7978	1.084	1.360	537.5	3.51.−5	0.0531	0.716
760	2.153	475.7	7.8268	1.089	1.358	544.3	3.57.−5	0.0544	0.716
780	2.210	497.5	7.8551	1.094	1.356	551.0	3.64.−5	0.0556	0.716
800	2.266	519.4	7.8829	1.099	1.354	557.6	3.70.−5	0.0568	0.716
820	2.323	541.5	7.9101	1.103	1.352	564.1	3.76.−5	0.0580	0.715
840	2.380	563.6	7.9367	1.108	1.350	570.6	3.82.−5	0.0592	0.715
860	2.436	585.8	7.9628	1.112	1.348	576.8	3.88.−5	0.0603	0.715
880	2.493	608.1	7.9885	1.117	1.346	583.1	3.94.−5	0.0615	0.715
900	2.550	630.4	8.0136	1.121	1.344	589.3	4.00.−5	0.0627	0.715
920	2.606	652.9	8.0383	1.125	1.342	595.4	4.05.−5	0.0639	0.715
940	2.663	675.5	8.0625	1.129	1.341	601.5	4.11.−5	0.0650	0.714
960	2.720	698.1	8.0864	1.133	1.339	607.5	4.17.−5	0.0662	0.714
980	2.776	720.8	8.1098	1.137	1.338	613.4	4.23.−5	0.0673	0.714
1000	2.833	743.6	8.1328	1.141	1.336	619.3	4.28.−5	0.0684	0.714
1050	2.975	800.8	8.1887	1.150	1.333	633.8	4.42.−5	0.0711	0.714
1100	3.116	858.5	8.2423	1.158	1.330	648.0	4.55.−5	0.0738	0.715
1150	3.258	916.6	8.2939	1.165	1.327	661.6	4.68.−5	0.0764	0.715
1200	3.400	975.0	8.3437	1.173	1.324	675.4	4.81.−5	0.0789	0.715
1250	3.541	1033.8	8.3917	1.180	1.322	688.6	4.94.−5	0.0814	0.716
1300	3.683	1093.0	8.4381	1.186	1.319	701.6	5.06.−5	0.0839	0.716

Table 39.5 (Continued)

T (K)	v (m³/kg)	h (kJ/kg)	s (kJ/kg·K)	c_p (kJ/kg·K)	γ	$\bar{v}_s$ (m/sec)	η (N·sec/m²)	λ (W/m·K)	Pr
1350	3.825	1152.3	8.4830	1.193	1.317	714.4	5.19.−5	0.0863	0.717
1400	3.966	1212.2	8.5265	1.199	1.315	726.9	5.31.−5	0.0887	0.717
1450	4.108	1272.3	8.5686	1.204	1.313	739.2	5.42.−5	0.0911	0.717
1500	4.249	1332.7	8.6096	1.210	1.311	751.3	5.54.−5	0.0934	0.718
1550	4.391	1393.3	8.6493	1.215	1.309	763.2	5.66.−5	0.0958	0.718
1600	4.533	1454.2	8.6880	1.220	1.308	775.0	5.77.−5	0.0981	0.717
1650	4.674	1515.3	8.7256	1.225	1.306	786.5	5.88.−5	0.1004	0.717
1700	4.816	1576.7	8.7622	1.229	1.305	797.9	5.99.−5	0.1027	0.717
1750	4.958	1638.2	8.7979	1.233	1.303	809.1	6.10.−5	0.1050	0.717
1800	5.099	1700.0	8.8327	1.237	1.302	820.2	6.21.−5	0.1072	0.717
1850	5.241	1762.0	8.8667	1.241	1.301	831.1	6.32.−5	0.1094	0.717
1900	5.383	1824.1	8.8998	1.245	1.300	841.9	6.43.−5	0.1116	0.717
1950	5.524	1886.4	8.9322	1.248	1.299	852.6	6.53.−5	0.1138	0.717
2000	5.666	1948.9	8.9638	1.252	1.298	863.1	6.64.−5	0.1159	0.717
2050	5.808	2011.6	8.9948	1.255	1.297	873.5	6.74.−5	0.1180	0.717
2100	5.949	2074.4	9.0251	1.258	1.296	883.8	6.84.−5	0.1200	0.717
2150	6.091	2137.3	9.0547	1.260	1.295	894.0	6.95.−5	0.1220	0.717
2200	6.232	2200.4	9.0837	1.263	1.294	904.0	7.05.−5	0.1240	0.718
2250	6.374	2263.6	9.1121	1.265	1.293	914.0	7.15.−5	0.1260	0.718
2300	6.516	2327.0	9.1399	1.268	1.293	923.8	7.25.−5	0.1279	0.718
2350	6.657	2390.5	9.1672	1.270	1.292	933.5	7.35.−5	0.1298	0.719
2400	6.800	2454.0	9.1940	1.273	1.291	943.2	7.44.−5	0.1317	0.719
2450	6.940	2517.7	9.2203	1.275	1.291	952.7	7.54.−5	0.1336	0.720
2500	7.082	2581.5	9.2460	1.277	1.290	962.2	7.64.−5	0.1354	0.720

[a] v = specific volume; h = specific enthalpy; s = specific entropy; c_p = specific heat at constant pressure; γ = specific heat ratio, c_p/c_v (dimensionless); $\bar{v}_s$ = velocity of sound; η = dynamic viscosity; λ = thermal conductivity; Pr = Prandtl number (dimensionless). Condensed from S. Gordon, *Thermodynamic and Transport Properties of Hydrocarbons with Air*, NASA Technical Paper 1906, 1982, Vol. 1. These properties are based on constant gaseous composition. The reader is reminded that, at the higher temperatures, the influence of pressure can affect the composition and the thermodynamic properties.

[b] The notation 1.33.−5 signifies 1.33×10^{-5}.

Table 39.6 Thermophysical Properties of the U.S. Standard Atmosphere[a]

Z (m)	H (m)	T (K)	P (bar)	ρ (kg/m³)	g (m/sec²)	$\bar{v}_s$ (m/sec)
0	0	288.15	1.0133	1.2250	9.8067	340.3
1000	1000	281.65	0.8988	1.1117	9.8036	336.4
2000	1999	275.15	0.7950	1.0066	9.8005	332.5
3000	2999	268.66	0.7012	0.9093	9.7974	328.6
4000	3997	262.17	0.6166	0.8194	9.7943	324.6
5000	4996	255.68	0.5405	0.7364	9.7912	320.6
6000	5994	249.19	0.4722	0.6601	9.7882	316.5
7000	6992	242.70	0.4111	0.5900	9.7851	312.3
8000	7990	236.22	0.3565	0.5258	9.7820	308.1
9000	8987	229.73	0.3080	0.4671	9.7789	303.9
10000	9984	223.25	0.2650	0.4135	9.7759	299.5
11000	10981	216.77	0.2270	0.3648	9.7728	295.2
12000	11977	216.65	0.1940	0.3119	9.7697	295.1
13000	12973	216.65	0.1658	0.2667	9.7667	295.1
14000	13969	216.65	0.1417	0.2279	9.7636	295.1
15000	14965	216.65	0.1211	0.1948	9.7605	295.1
16000	15960	216.65	0.1035	0.1665	9.7575	295.1
17000	16954	216.65	0.0885	0.1423	9.7544	295.1
18000	17949	216.65	0.0756	0.1217	9.7513	295.1
19000	18943	216.65	0.0647	0.1040	9.7483	295.1
20000	19937	216.65	0.0553	0.0889	9.7452	295.1
22000	21924	218.57	0.0405	0.0645	9.7391	296.4
24000	23910	220.56	0.0297	0.0469	9.7330	297.7
26000	25894	222.54	0.0219	0.0343	9.7269	299.1
28000	27877	224.53	0.0162	0.0251	9.7208	300.4
30000	29859	226.51	0.0120	0.0184	9.7147	301.7
32000	31840	228.49	0.00889	0.01356	9.7087	303.0
34000	33819	233.74	0.00663	0.00989	9.7026	306.5
36000	35797	239.28	0.00499	0.00726	9.6965	310.1
38000	37774	244.82	0.00377	0.00537	9.6904	313.7
40000	39750	250.35	0.00287	0.00400	9.6844	317.2
42000	41724	255.88	0.00220	0.00299	9.6783	320.7
44000	43698	261.40	0.00169	0.00259	9.6723	324.1
46000	45669	266.93	0.00131	0.00171	9.6662	327.5
48000	47640	270.65	0.00102	0.00132	9.6602	329.8
50000	49610	270.65	0.00080	0.00103	9.6542	329.8

[a]Z = geometric attitude; H = geopotential attitude; ρ = density; g = acceleration of gravity; $\bar{v}_s$ = velocity of sound. Condensed and in some cases converted from *U.S. Standard Atmosphere 1976*, National Oceanic and Atmospheric Administration and National Aeronautics and Space Administration, Washington, DC. Also available as NOAA-S/T 76-1562 and Government Printing Office Stock No. 003-017-00323-0.

Table 39.7 Thermophysical Properties of Condensed and Saturated Vapor Carbon Dioxide from 200 K to the Critical Point[a]

T (K)	P (bar)	Specific Volume Condensed[b]	Vapor	Specific Enthalpy Condensed[b]	Vapor	Specific Entropy Condensed[b]	Vapor	Specific Heat (c_p) Condensed[b]	Vapor	Thermal Conductivity Liquid	Vapor	Viscosity Liquid	Vapor	Prandtl Number Liquid	Vapor
200	1.544	0.000644	0.2362	164.8	728.3	1.620	4.439								
205	2.277	0.000649	0.1622	171.5	730.0	1.652	4.379								
210	3.280	0.000654	0.1135	178.2	730.9	1.682	4.319								
215	4.658	0.000659	0.0804	185.0	731.3	1.721	4.264								
216.6	5.180	0.000661	0.0718	187.2	731.5	1.736	4.250								
216.6	5.180	0.000848	0.0718	386.3	731.5	2.656	4.250	1.707	0.958	0.182	0.011	2.10	0.116	1.96	0.96
220	5.996	0.000857	0.0624	392.6	733.1	2.684	4.232	1.761	0.985	0.178	0.012	1.86	0.118	1.93	0.97
225	7.357	0.000871	0.0515	401.8	735.1	2.723	4.204	1.820	1.02	0.171	0.012	1.75	0.120	1.87	0.98
230	8.935	0.000886	0.0428	411.1	736.7	2.763	4.178	1.879	1.06	0.164	0.013	1.64	0.122	1.84	0.99
235	10.75	0.000901	0.0357	420.5	737.9	2.802	4.152	1.906	1.10	0.160	0.013	1.54	0.125	1.82	1.01
240	12.83	0.000918	0.0300	430.2	738.9	2.842	4.128	1.933	1.15	0.156	0.014	1.45	0.128	1.80	1.02
245	15.19	0.000936	0.0253	440.1	739.4	2.882	4.103	1.959	1.20	0.148	0.015	1.36	0.131	1.80	1.04
250	17.86	0.000955	0.0214	450.3	739.6	2.923	4.079	1.992	1.26	0.140	0.016	1.28	0.134	1.82	1.06
255	20.85	0.000977	0.0182	460.8	739.4	2.964	4.056	2.038	1.34	0.134	0.017	1.21	0.137	1.84	1.08
260	24.19	0.001000	0.0155	471.6	738.7	3.005	4.032	2.125	1.43	0.128	0.018	1.14	0.140	1.89	1.12
265	27.89	0.001026	0.0132	482.8	737.4	3.047	4.007	2.237	1.54	0.122	0.019	1.08	0.144	1.98	1.17
270	32.03	0.001056	0.0113	494.4	735.6	3.089	3.981	2.410	1.66	0.116	0.020	1.02	0.150	2.12	1.23
275	36.59	0.001091	0.0097	506.5	732.8	3.132	3.954	2.634	1.81	0.109	0.022	0.96	0.157	2.32	1.32
280	41.60	0.001130	0.0082	519.2	729.1	3.176	3.925	2.887	2.06	0.102	0.024	0.91	0.167	2.57	1.44
285	47.10	0.001176	0.0070	532.7	723.5	3.220	3.891	3.203	2.40	0.095	0.028	0.86	0.178	2.90	1.56
290	53.15	0.001241	0.0058	547.6	716.9	3.271	3.854	3.724	2.90	0.088	0.033	0.79	0.191	3.35	1.68
295	59.83	0.001322	0.0047	562.9	706.3	3.317	3.803	4.68		0.081	0.042	0.71	0.207	4.1	1.8
300	67.10	0.001470	0.0037	585.4	690.2	3.393	3.742			0.074	0.065	0.60	0.226		
304.2[c]	73.83	0.002145	0.0021	636.6	636.6	3.558	3.558								

[a]Specific volume, m³/kg; specific enthalpy, kJ/kg; specific entropy, kJ/kg · K; specific heat at constant pressure, kJ/kg · K; thermal conductivity, W/m · K; viscosity, 10⁻⁴ Pa · sec. Thus, at 250 K the viscosity of the saturated liquid is 1.28×10^{-4} N · sec/m² = 0.000128 N · sec/m² = 0.000128 Pa · sec. The Prandtl number is dimensionless.

[b]Above the solid line the condensed phase is solid; below the line, it is liquid.

[c]Critical point.

Table 39.8 Thermophysical Properties of Gaseous Carbon Dioxide at 1 Bar Pressure[a]

	\(T\) (K)														
	300	350	400	450	500	550	600	650	700	750	800	850	900	950	1000
v (m³/kg)	0.5639	0.6595	0.7543	0.8494	0.9439	1.039	1.133	1.228	1.332	1.417	1.512	1.606	1.701	1.795	1.889
h (kJ/kg)	809.3	853.1	899.1	947.1	997.0	1049	1102	1156	1212	1269	1327	1386	1445	1506	1567
s (kJ/kg·K)	4.860	4.996	5.118	5.231	5.337	5.435	5.527	5.615	5.697	5.775	5.850	5.922	5.990	6.055	6.120
c_p (kJ/kg·K)	0.852	0.898	0.941	0.980	1.014	1.046	1.075	1.102	1.126	1.148	1.168	1.187	1.205	1.220	1.234
λ (W/m·K)	0.0166	0.0204	0.0243	0.0283	0.0325	0.0364	0.0407	0.0445	0.0481	0.0517	0.0551	0.0585	0.0618	0.0650	0.0682
μ (10^{-4} Pa·sec)	0.151	0.175	0.198	0.220	0.242	0.261	0.281	0.299	0.317	0.334	0.350	0.366	0.381	0.396	0.410
Pr	0.778	0.770	0.767	0.762	0.755	0.750	0.742	0.742	0.742	0.742	0.742	0.742	0.742	0.743	0.743

[a] v = specific volume; h = enthalpy; s = entropy; c_p = specific heat at constant pressure; λ = thermal conductivity; η = viscosity (at 300 K the gas viscosity is 0.0000151 N·sec/m² = 0.0000151 Pa·sec); Pr = Prandtl number.

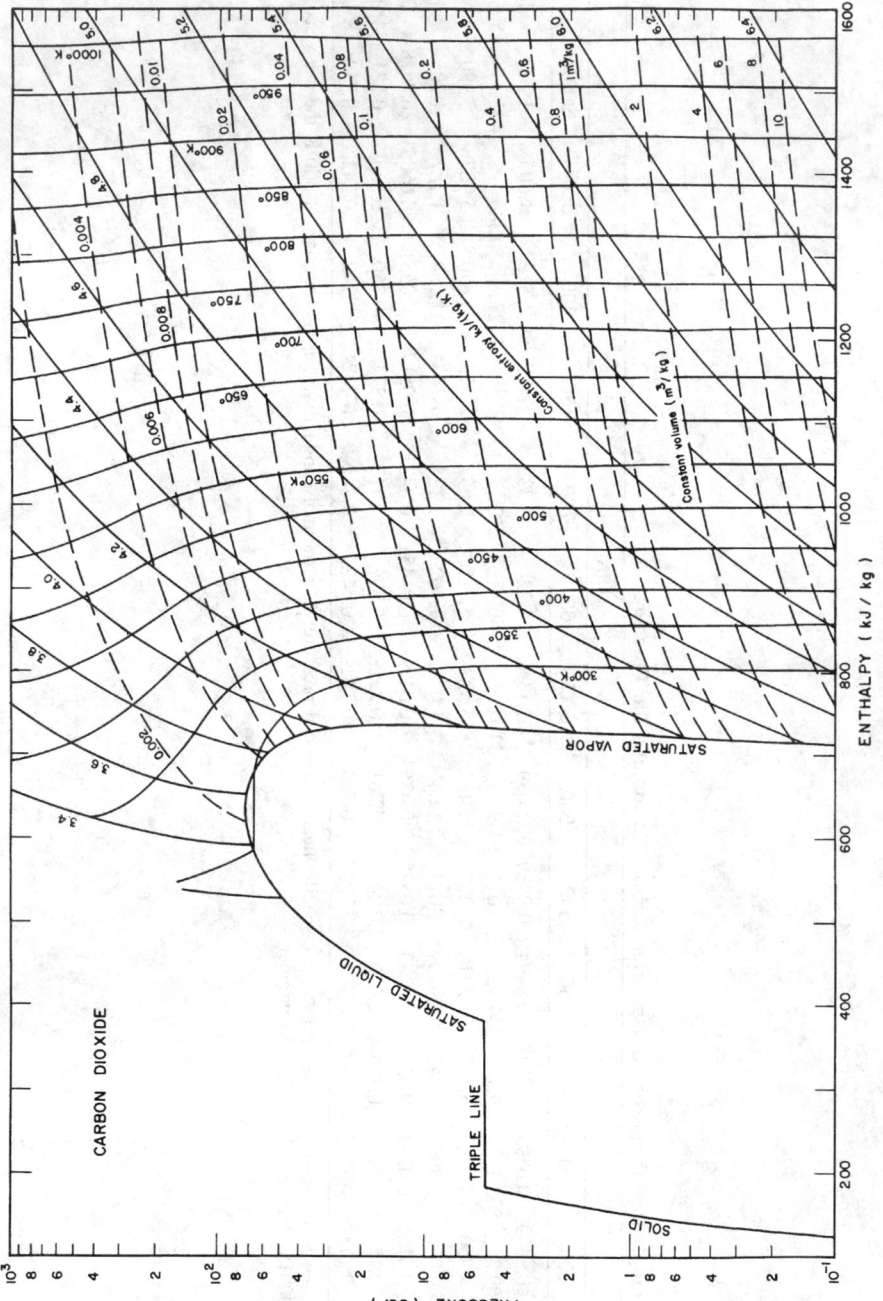

Fig. 39.1 Enthalpy–log pressure diagram for carbon dioxide.

Table 39.9 Thermodynamic Properties of Saturated Mercury[a]

T (K)	P (bar)	v (m³/kg)	h (kJ/kg)	s (kJ/kg · K)	c_p (kJ/kg · K)
0		6.873.−5[b]	0	0	0
20		6.875.−5	0.466	0.0380	0.0513
40		6.884.−5	1.918	0.0868	0.0894
60		6.897.−5	3.897	0.1267	0.1067
80		6.911.−5	6.129	0.1588	0.1156
100		6.926.−5	8.497	0.1852	0.1209
120		6.942.−5	10.956	0.2076	0.1248
140		6.958.−5	13.482	0.2270	0.1278
160		6.975.−5	16.063	0.2443	0.1304
180		6.993.−5	18.697	0.2598	0.1330
200		7.013.−5	21.386	0.2739	0.1360
220		7.034.−5	24.139	0.2870	0.1394
234.3	7.330.−10	7.050.−5	26.148	0.2959	0.1420
234.3	7.330.−10	7.304.−5	37.585	0.3447	0.1422
240	1.668.−9	7.311.−5	38.395	0.3481	0.1420
260	6.925.−8	7.339.−5	41.224	0.3595	0.1409
280	5.296.−7	7.365.−5	44.034	0.3699	0.1401
300	3.075.−6	7.387.−5	46.829	0.3795	0.1393
320	1.428.−5	7.413.−5	49.609	0.3885	0.1386
340	5.516.−5	7.439.−5	52.375	0.3969	0.1380
360	1.829.−4	7.472.−5	55.130	0.4048	0.1375
380	5.289.−4	7.499.−5	57.874	0.4122	0.1370
400	1.394.−3	7.526.−5	60.609	0.4192	0.1366
450	0.01053	7.595.−5	67.414	0.4352	0.1357
500	0.05261	7.664.−5	74.188	0.4495	0.1353
550	0.1949	7.735.−5	80.949	0.4624	0.1352
600	0.5776	7.807.−5	87.716	0.4742	0.1356
650	1.4425	7.881.−5	94.508	0.4850	0.1360
700	3.153	7.957.−5	101.343	0.4951	0.1372
750	6.197	8.036.−5	108.242	0.5046	0.1382
800	11.181	8.118.−5	115.23	0.5136	0.1398
850	18.816	8.203.−5	122.31	0.5221	0.1416
900	29.88	8.292.−5	129.53	0.5302	0.1439
950	45.23	8.385.−5	137.16	0.5381	0.1464
1000	65.74	8.482.−5	144.41	0.5456	0.1492

[a]v = specific volume; h = specific enthalpy; s = specific entropy, c_p = specific heat at constant pressure. Properties above the solid line are for the solid; below they are for the liquid. Condensed, converted, and interpolated from the tables of M. P. Vukalovich, A. I. Ivanov, L. R. Fokin, and A. T. Yakovlev, *Thermophysical Properties of Mercury,* Standartov, Moscow, USSR, 1971.

[b]The notation 6.873.−5 signifies 6.873×10^{-5}.

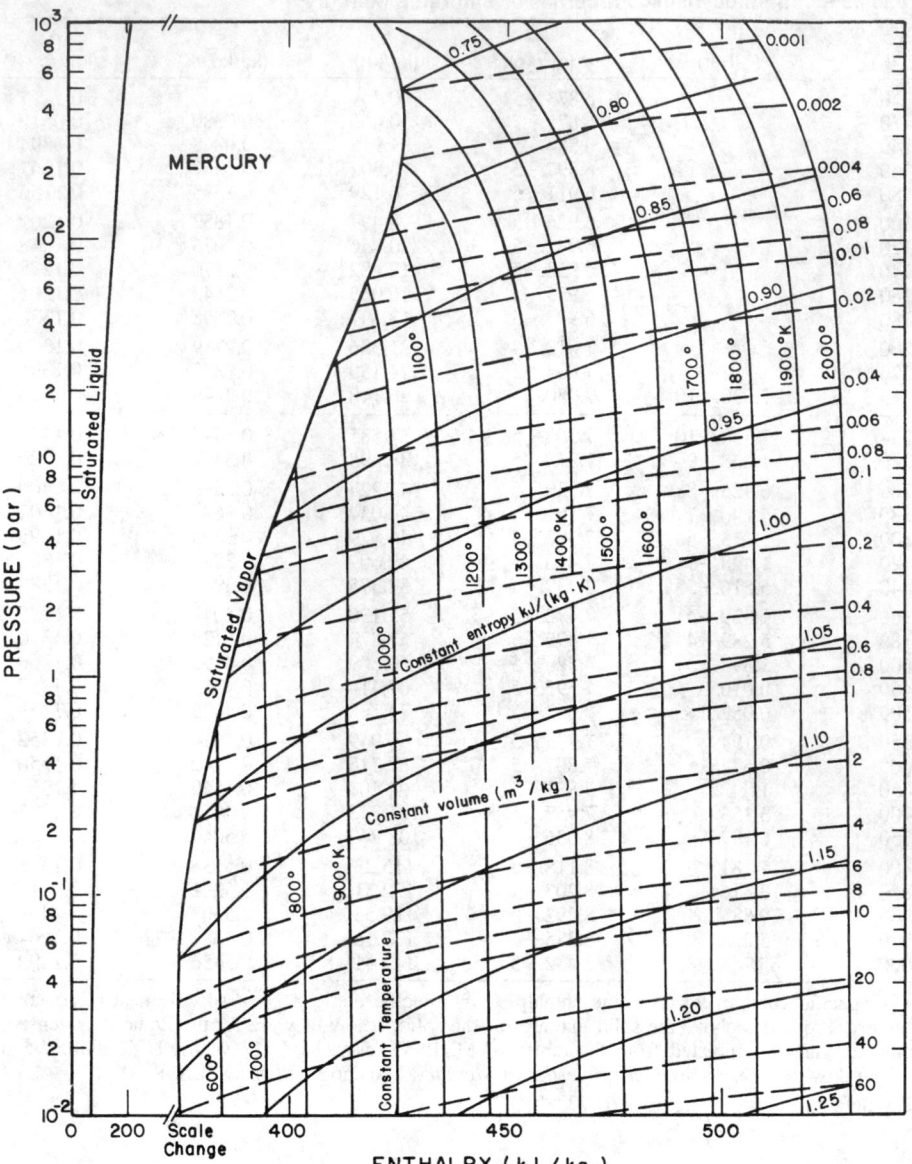

Fig. 39.2 Enthalpy–log pressure diagram for mercury.

Table 39.10 Thermodynamic Properties of Saturated Methane[a]

T (K)	P (bar)	v_f (m³/kg)	v_g (m³/kg)	h_f (kJ/kg)	h_g (kJ/kg)	s_f (kJ/kg·K)	s_g (kJ/kg·K)	c_{pr} (kJ/kg·K)	$\bar{v}_s$ (m/sec)
90.68	0.117	2.215.−3[b]	3.976	216.4	759.9	4.231	10.225	3.288	1576
92	0.139	2.226.−3	3.410	220.6	762.4	4.279	10.168	3.294	1564
96	0.223	2.250.−3	2.203	233.2	769.5	4.419	10.006	3.326	1523
100	0.345	2.278.−3	1.479	246.3	776.9	4.556	9.862	3.369	1480
104	0.515	2.307.−3	1.026	259.6	784.0	4.689	9.731	3.415	1437
108	0.743	2.337.−3	0.732	273.2	791.0	4.818	9.612	3.458	1393
112	1.044	2.369.−3	0.536	287.0	797.7	4.944	9.504	3.497	1351
116	1.431	2.403.−3	0.401	301.1	804.2	5.068	9.405	3.534	1308
120	1.919	2.438.−3	0.306	315.3	810.8	5.187	9.313	3.570	1266
124	2.523	2.475.−3	0.238	329.7	816.2	5.305	9.228	3.609	1224
128	3.258	2.515.−3	0.187	344.3	821.6	5.419	9.148	3.654	1181
132	4.142	2.558.−3	0.150	359.1	826.5	5.531	9.072	3.708	1138
136	5.191	2.603.−3	0.121	374.2	831.0	5.642	9.001	3.772	1093
140	6.422	2.652.−3	0.0984	389.5	834.8	5.751	8.931	3.849	1047
144	7.853	2.704.−3	0.0809	405.2	838.0	5.858	8.864	3.939	999
148	9.502	2.761.−3	0.0670	421.3	840.6	5.965	8.798	4.044	951
152	11.387	2.824.−3	0.0558	437.7	842.2	6.072	8.733	4.164	902
156	13.526	2.893.−3	0.0467	454.7	843.2	6.177	8.667	4.303	852
160	15.939	2.971.−3	0.0392	472.1	843.0	6.283	8.601	4.470	802
164	18.647	3.059.−3	0.0326	490.1	841.6	6.390	8.533	4.684	749
168	21.671	3.160.−3	0.0278	508.9	839.0	6.497	8.462	4.968	695
172	25.034	3.281.−3	0.0234	528.6	834.6	6.606	8.385	5.390	637
176	28.761	3.428.−3	0.0196	549.7	827.9	6.720	8.301	6.091	570
180	32.863	3.619.−3	0.0162	572.9	818.1	6.843	8.205	7.275	500
184	37.435	3.890.−3	0.0131	599.7	802.9	6.980	8.084	9.831	421
188	42.471	4.361.−3	0.0101	634.0	776.4	7.154	7.912	19.66	327
190.56	45.988	6.233.−3	0.0062	704.4	704.4	7.516	7.516		

[a] v = specific volume; h = specific enthalpy; s = specific entropy; c_p = specific heat at constant pressure; $\bar{v}_s$ = velocity of sound; f = saturated liquid; g = saturated vapor. Condensed and converted from R. D. Goodwin, N.B.S. Technical Note 653, 1974.

[b] The notation 2.215.−3 signifies 2.215×10^{-3}.

Table 39.11 Thermophysical Properties of Methane at Atmospheric Pressure[a]

	Temperature (K)									
	250	300	350	400	450	500	550	600	650	700
v	1.275	1.532	1.789	2.045	2.301	2.557	2.813	3.068	3.324	3.580
h	1090	1200	1315	1437	1569	1709	1857	2016	2183	2359
s	11.22	11.62	11.98	12.30	12.61	12.91	13.19	13.46	13.73	14.00
c_p	2.04	2.13	2.26	2.43	2.60	2.78	2.96	3.16	3.35	3.51
Z	0.997	0.998	0.999	1.000	1.000	1.000	1.000	1.000	1.000	1.000
$\bar{v}_s$	413	450	482	511	537	562	585	607	629	650
λ	0.0276	0.0342	0.0417	0.0486	0.0571	0.0675	0.0768	0.0863	0.0956	0.1052
η	0.095	0.112	0.126	0.141	0.154	0.168	0.180	0.192	0.202	0.214
Pr	0.701	0.696	0.683	0.687	0.690	0.693	0.696	0.700	0.706	0.714

[a] v = specific volume (m³/kg); h = specific enthalpy (kJ/kg); s = specific entropy (kJ/kg·K); c_p = specific heat at constant pressure (kJ/kg·K); Z = compressibility factor = Pv/RT; $\bar{v}_s$ = velocity of sound (m/sec); λ = thermal conductivity (W/m·K); η = viscosity 10^{-4} N·sec/m² (thus, at 250 K the viscosity is 0.095×10^{-4} N·sec/ m² = 0.0000095 Pa·sec); Pr = Prandtl number.

Table 39.12 Thermophysical Properties of Saturated Refrigerant 22[a]

T (K)	P (bar)	v (m³/kg) Liquid	v (m³/kg) Vapor	h (kJ/kg) Liquid	h (kJ/kg) Vapor	s (kJ/kg·K) Liquid	s (kJ/kg·K) Vapor	c_p (Liquid)	η (Liquid)	λ (Liquid)	τ (Liquid)
150	0.0017	6.209.−4[b]	83.40	268.2	547.3	3.355	5.215	1.059		0.161	
160	0.0054	6.293.−4	28.20	278.2	552.1	3.430	5.141	1.058		0.156	
170	0.0150	6.381.−4	10.85	288.3	557.0	3.494	5.075	1.057	0.770	0.151	
180	0.0369	6.474.−4	4.673	298.7	561.9	3.551	5.013	1.058	0.647	0.146	
190	0.0821	6.573.−4	2.225	308.6	566.8	3.605	4.963	1.060	0.554	0.141	
200	0.1662	6.680.−4	1.145	318.8	571.6	3.657	4.921	1.065	0.481	0.136	0.024
210	0.3116	6.794.−4	0.6370	329.1	576.5	3.707	4.885	1.071	0.424	0.131	0.022
220	0.5470	6.917.−4	0.3772	339.7	581.2	3.756	4.854	1.080	0.378	0.126	0.021
230	0.9076	7.050.−4	0.2352	350.6	585.9	3.804	4.828	1.091	0.340	0.121	0.019
240	1.4346	7.195.−4	0.1532	361.7	590.5	3.852	4.805	1.105	0.309	0.117	0.0172
250	2.174	7.351.−4	0.1037	373.0	594.9	3.898	4.785	1.122	0.282	0.112	0.0155
260	3.177	7.523.−4	0.07237	384.5	599.0	3.942	4.768	1.143	0.260	0.107	0.0138
270	4.497	7.733.−4	0.05187	396.3	603.0	3.986	4.752	1.169	0.241	0.102	0.0121
280	6.192	7.923.−4	0.03803	408.2	606.6	4.029	4.738	1.193	0.225	0.097	0.0104
290	8.324	8.158.−4	0.02838	420.4	610.0	4.071	4.725	1.220	0.211	0.092	0.0087
300	10.956	8.426.−4	0.02148	432.7	612.8	4.113	4.713	1.257	0.198	0.087	0.0071
310	14.17	8.734.−4	0.01643	445.5	615.1	4.153	4.701	1.305	0.186	0.082	0.0055
320	18.02	9.096.−4	0.01265	458.6	616.7	4.194	4.688	1.372	0.176	0.077	0.0040
330	22.61	9.535.−4	9.753.−3	472.4	617.3	4.235	4.674	1.460	0.167	0.072	0.0026
340	28.03	1.010.−3	7.479.−3	487.2	616.5	4.278	4.658	1.573	0.151	0.067	0.0014
350	34.41	1.086.−3	5.613.−3	503.7	613.3	4.324	4.637	1.718	0.130	0.062	0.0008
360	41.86	1.212.−3	4.036.−3	523.7	605.5	4.378	4.605	1.897	0.106		
369.3	49.89	2.015.−3	2.015.−3	570.0	570.0	4.501	4.501	∞	—	—	0

[a] c_p in units of kJ/kg·K; η = viscosity (10^{-4} Pa·sec); λ = thermal conductivity (W/m·K); τ = surface tension (N/m). Sources: P, v, T, h, s interpolated and extrapolated from I. I. Perelshteyn, *Tables and Diagrams of the Thermodynamic Properties of Freons 12, 13, 22*, Moscow, USSR, 1971. c_p, η, λ interpolated and converted from *Thermophysical Properties of Refrigerants*, ASHRAE, New York, 1976. τ calculated from V. A. Gruzdev et al., *Fluid Mech. Sov. Res.*, **3**, 172 (1974).

[b] The notation 6.209.−4 signifies 6.209×10^{-4}.

Table 39.13 Thermophysical Properties of Refrigerant 22 at Atmospheric Pressure[a]

	Temperature (K)					
	250	300	350	400	450	500
v	0.2315	0.2802	0.3289	0.3773	0.4252	0.4723
h	597.8	630.0	664.5	702.5	740.8	782.3
s	4.8671	4.9840	5.0905	5.1892	5.2782	5.3562
c_p	0.587	0.647	0.704	0.757	0.806	0.848
Z	0.976	0.984	0.990	0.994	0.995	0.996
$\bar{v}_s$	166.4	182.2	196.2	209.4	220.0	233.6
λ	0.0080	0.0110	0.0140	0.0170	0.0200	0.0230
η	0.109	0.130	0.151	0.171	0.190	0.209
Pr	0.800	0.765	0.759	0.761	0.766	0.771

[a]v = specific volume (m³/kg); h = specific enthalpy (kJ/kg); s = specific entropy (kJ/kg·K); c_p = specific heat at constant pressure (kJ/kg·K); Z = compressibility factor = Pv/RT; $\bar{v}_s$ = velocity of sound (m/sec); λ = thermal conductivity (W/m·K); η = viscosity 10^{-4} N·sec/m² (thus, at 250 K the viscosity is 0.109×10^{-4} N sec/m² = 0.0000109 Pa·sec); Pr = Prandtl number.

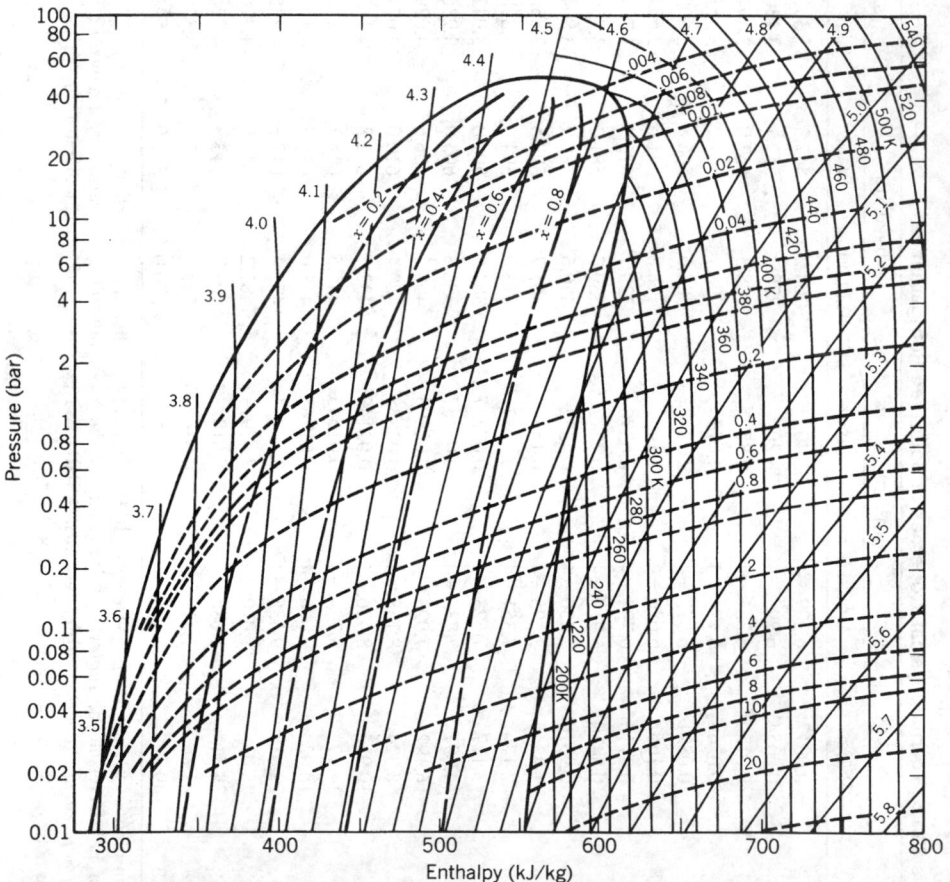

Fig. 39.3 Enthalpy–log pressure diagram for Refrigerant 22.

Table 39.14 Thermodynamic Properties of Saturated Refrigerant 134a[a]

P (bar)	t (°C)	v_f (m³/kg)	v_g (m³/kg)	h_f (kJ/kg)	h_g (kJ/kg)	s_f (kJ/kg·K)	s_g (kJ/kg·K)	c_{pf} (kJ/kg·K)	c_{pg} (kJ/kg·K)
0.5	−40.69	0.000 707	0.3690	−0.9	225.5	−0.0036	0.9698	1.2538	0.7476
0.6	−36.94	0.000 712	0.3109	3.3	227.5	0.0165	0.9645	1.2600	0.7584
0.7	−33.93	0.000 716	0.2692	7.7	229.4	0.0324	0.9607	1.2654	0.7680
0.8	−31.12	0.000 720	0.2375	11.2	231.2	0.0472	0.9572	1.2707	0.7771
0.9	−28.65	0.000 724	0.2126	14.4	233.0	0.0601	0.9545	1.2755	0.7854
1	−26.52	0.000 729	0.1926	16.8	234.2	0.0720	0.9519	1.2799	0.7931
1.5	−17.26	0.000 742	0.1314	28.6	240.0	0.1194	0.9455	1.2992	0.8264
2	−10.18	0.000 754	0.1001	37.8	243.6	0.1547	0.9379	1.3153	0.8540
2.5	−4.38	0.000 764	0.0809	45.5	247.3	0.1834	0.9340	1.3297	0.8782
3	0.59	0.000 774	0.0679	52.1	250.1	0.2077	0.9312	1.3428	0.9002
4	8.86	0.000 791	0.0514	63.3	254.9	0.2478	0.9270	1.3670	0.9400
5	15.68	0.000 806	0.0413	72.8	258.8	0.2804	0.9241	1.3894	0.9761
6	21.54	0.000 820	0.0344	81.0	262.1	0.3082	0.9219	1.4108	0.9914
8	31.35	0.000 846	0.0257	95.0	267.1	0.3542	0.9185	1.4526	1.0750
10	39.41	0.000 871	0.0204	106.9	270.9	0.3921	0.9157	1.4948	1.1391
12	46.36	0.000 894	0.01672	117.5	273.9	0.4246	0.9132	1.539	1.205
14	52.48	0.000 918	0.01411	127.0	276.2	0.4533	0.9107	1.586	1.276
16	57.96	0.000 941	0.01258	135.7	277.9	0.4794	0.9081	1.637	1.353
18	62.94	0.000 965	0.01056	143.9	279.3	0.5031	0.9052	1.695	1.439
20	67.52	0.000 990	0.00929	151.4	280.1	0.5254	0.9021	1.761	1.539
22.5	72.74	0.001 023	0.00800	160.9	280.8	0.5512	0.8976	1.859	1.800
25	77.63	0.001 058	0.00694	169.8	280.8	0.5756	0.8925	1.983	1.883
27.5	82.04	0.001 096	0.00605	178.2	280.3	0.5989	0.8865	2.151	2.149
30	86.20	0.001 141	0.00528	186.5	279.4	0.6216	0.8812	2.388	2.527
35	93.72	0.001 263	0.00397	203.7	274.1	0.6671	0.8589	3.484	4.292
40	100.34	0.001 580	0.00256	227.4	257.2	0.7292	0.8090	26.33	37.63
40.59	101.06	0.001 953	0.00195	241.5	241.5	0.7665	0.7665		

[a]Converted and reproduced from R. Tillner-Roth and H. D. Baehr, *J. Phys. Chem. Ref. Data*, **23** (5), 657–730 (1994). $h_f = s_f = 0$ at 233.15K $= -40°C$.

Table 39.15 Interim Thermophysical Properties of Refrigerant 134a[a]

t (°C)	Property	P (bar)							
		1	2.5	5	7.5	10	12.5	15	Sat. Vapor
0	c_p (kJ/kg·K)	0.8197	0.8740						0.8975
	μ (10^{-6} Pas)	11.00	10.95						10.94
	k (W/m·K)	0.0119	0.0120						0.0120
	Pr	0.763	0.798						0.809
10	c_p (kJ/kg·K)	0.8324	0.8815						0.9408
	μ (10^{-6} Pas)	11.38	11.42						11.42
	k (W/m·K)	0.0126	0.0127						0.0129
	Pr	0.753	0.786						0.821
20	c_p (kJ/kg·K)	0.8458	0.8726	0.9642					0.9864
	μ (10^{-6} Pas)	11.78	11.83	11.91					11.93
	k (W/m·K)	0.0134	0.0135	0.0138					0.0139
	Pr	0.747	0.774	0.830					0.838
30	c_p (kJ/kg·K)	0.8602	0.8900	0.9587	1.044				1.048
	μ (10^{-6} Pas)	12.27	12.28	12.29	12.36				12.37
	k (W/m·K)	0.0141	0.0143	0.0145	0.0150				0.0150
	Pr	0.743	0.764	0.805	0.857				0.859
40	c_p (kJ/kg·K)	0.8747	0.8998	0.9547	1.027	1.134			1.145
	μ (10^{-6} Pas)	12.57	12.61	12.66	12.75	12.88			12.89
	k (W/m·K)	0.0148	0.0150	0.0153	0.0156	0.0161			0.0161
	Pr	0.740	0.757	0.789	0.839	0.907			0.916

50							
c_p (kJ/kg·K)	0.8891	0.9112	0.9555	1.017	1.120	1.213	1.246
μ (10^{-6} Pas)	12.96	13.00	13.05	13.14	13.23	13.33	13.47
k (W/m·K)	0.0156	0.0158	0.0160	0.0163	0.0167	0.0171	0.0173
Pr	0.739	0.752	0.778	0.820	0.887	0.946	0.960
60							
c_p (kJ/kg·K)	0.9045	0.9230	0.9589	1.003	1.060	1.151	1.387
μ (10^{-6} Pas)	13.35	13.39	13.44	13.51	13.60	13.75	14.16
k (W/m·K)	0.0164	0.0165	0.0167	0.0170	0.0173	0.0178	0.0185
Pr	0.739	0.750	0.772	0.801	0.829	0.889	1.059
70							
c_p (kJ/kg·K)	0.9201	0.9359	0.9652	0.9972	1.046	1.100	1.606
μ (10^{-6} Pas)	13.74	13.77	13.82	13.89	13.97	14.10	15.04
k (W/m·K)	0.0171	0.0172	0.0174	0.0176	0.0179	0.0183	0.0197
Pr	0.739	0.750	0.759	0.787	0.813	0.848	1.226
80							
c_p (kJ/kg·K)	0.9359	0.9520	0.9715	0.9992	1.038	1.222	2.026
μ (10^{-6} Pas)	14.11	14.14	14.19	14.25	14.33	14.43	16.31
k (W/m·K)	0.0178	0.0179	0.0180	0.0182	0.0185	0.0188	0.0205
Pr	0.741	0.752	0.757	0.782	0.804	0.938	1.612
sat. vapor							
c_p (kJ/kg·K)	0.7931	0.8782	0.9761	1.059	1.139	1.223	1.314
μ (10^{-6} Pas)			11.72	12.34	12.86	13.34	13.81
k (W/m·K)			0.0136	0.0149	0.0161	0.0173	0.0184
Pr			0.841	0.877	0.910	0.943	0.986
						—	—
						—	—
						—	—

^{a}Some significant differences presently exist between different sets of property measurements. Definitive values were still awaited in 1997.
At 0°C, 1 bar the viscosity is 11×10^{-6} Pas.; Pr = Prandtl number.

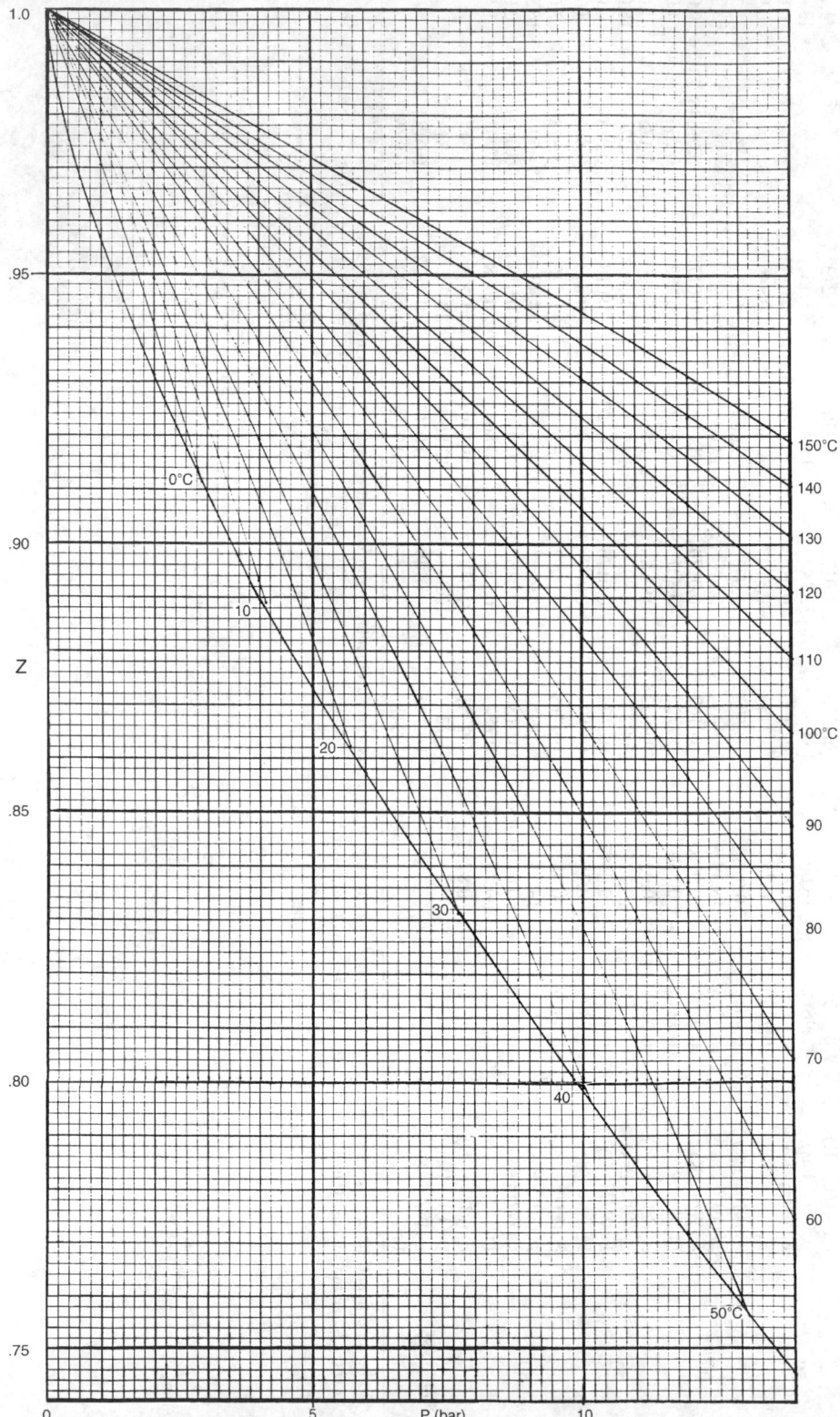

Fig. 39.4 Compressibility factor of Refrigerant 134a.

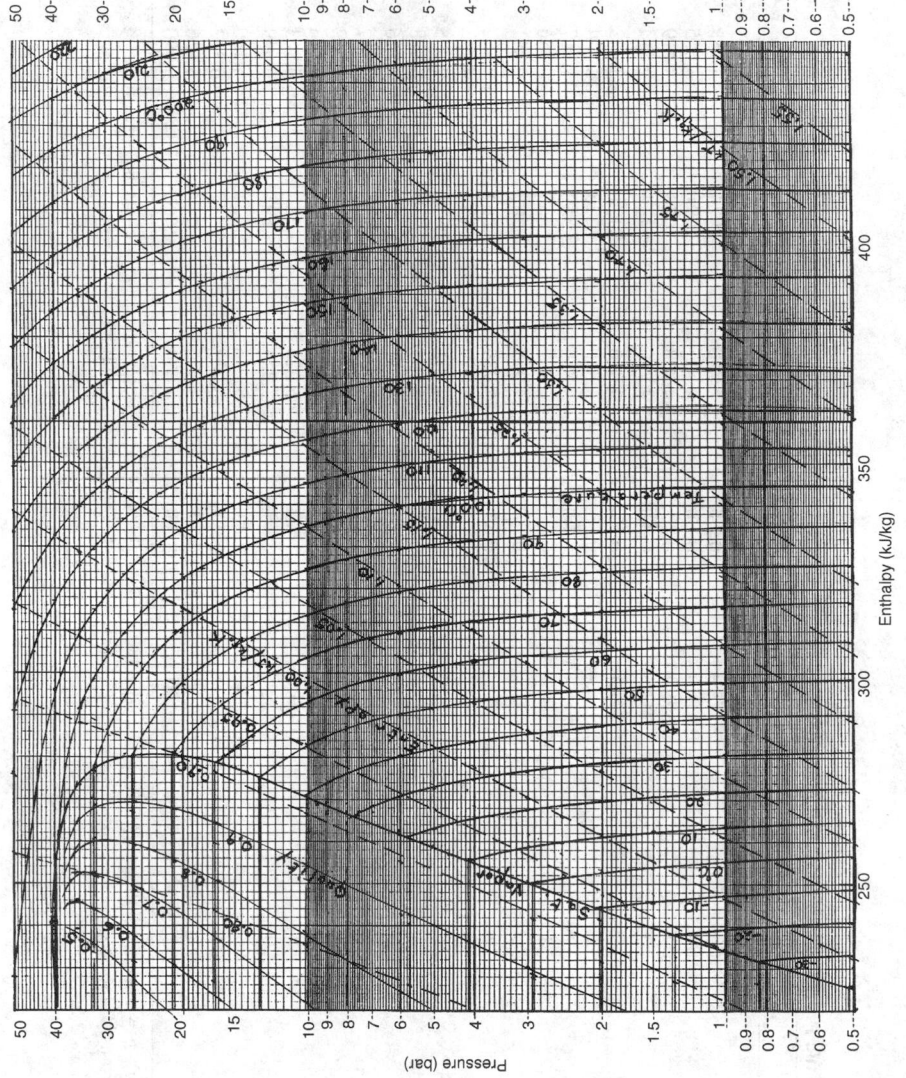

Fig. 39.5 Enthalpy–log pressure diagram for Refrigerant 134a.

Enthalpy (kJ/kg)

Pressure (bar)

Table 39.16 Thermodynamic Properties of Saturated Sodium[a]

T (K)	P (bar)	v_f	v_g	h_f	h_g	s_f	s_g	c_{pf}	c_{pg}
380	2.55.−10[b]	1.081.−3	5.277.+9	0.219	4.723	2.853	14.705	1.384	0.988
400	1.36.−9	1.086.−3	2.173.+9	0.247	4.740	2.924	14.158	1.374	1.023
420	6.16.−9	1.092.−3	2.410.+8	0.274	4.757	2.991	13.665	1.364	1.066
440	2.43.−8	1.097.−3	6.398.+7	0.301	4.773	3.054	13.219	1.355	1.117
460	8.49.−8	1.103.−3	1.912.+7	0.328	4.790	3.114	12.814	1.346	1.176
480	2.67.−7	1.109.−3	6.341.+6	0.355	4.806	3.171	12.443	1.338	1.243
500	7.64.−7	1.114.−3	2.304.+6	0.382	4.820	3.226	12.104	1.330	1.317
550	7.54.−6	1.129.−3	2.558.+5	0.448	4.856	3.352	11.367	1.313	1.523
600	5.05.−5	1.145.−3	41511	0.513	4.887	3.465	10.756	1.299	1.745
650	2.51.−4	1.160.−3	9001	0.578	4.915	3.569	10.241	1.287	1.963
700	9.87.−4	1.177.−3	2449	0.642	4.939	3.664	9.802	1.278	2.160
750	0.00322	1.194.−3	794	0.705	4.959	3.752	9.424	1.270	2.325
800	0.00904	1.211.−3	301	0.769	4.978	3.834	9.095	1.264	2.452
850	0.02241	1.229.−3	128.1	0.832	4.995	3.910	8.808	1.260	2.542
900	0.05010	1.247.−3	60.17	0.895	5.011	3.982	8.556	1.258	2.597
1000	0.1955	1.289.−3	16.84	1.021	5.043	4.115	8.137	1.259	2.624
1200	1.482	1.372.−3	2.571	1.274	5.109	4.346	7.542	1.281	2.515
1400	6.203	1.469.−3	0.688	1.535	5.175	4.546	7.146	1.330	2.391
1600	17.98	1.581.−3	0.258	1.809	5.225	4.728	6.863	1.406	2.301
1800	40.87	1.709.−3	0.120	2.102	5.255	4.898	6.649	1.516	2.261
2000	78.51	1.864.−3	0.0634	2.422	5.256	5.064	6.480	1.702	2.482
2200	133.5	2.076.−3	0.0362	2.794	5.207	5.235	6.332	2.101	3.307
2400	207.6	2.480.−3	0.0196	3.299	5.025	5.447	6.166	3.686	8.476
2500	251.9	3.323.−3	0.0100	3.850	4.633	5.666	5.980		

[a] v = specific volume (m^3/kg); h = specific enthalpy (MJ/kg); s = specific entropy (kJ/kg·K); c_p = specific heat at constant pressure (kJ/kg·K); f = saturated liquid; g = saturated vapor. Converted from the tables of J. K. Fink, Argonne Nat. Lab. rept. ANL-CEN-RSD-82-4, 1982.

[b] The notation 2.55.−10 signifies 2.55×10^{-10}.

Table 39.17 Thermodynamic Properties of Ice/Water[a]

T (K)	P (bar)	v (m³/kg)	h (kJ/kg)	s (kJ/kg · K)	c_p (kJ/kg · K)
150	6.30.−11[b]	0.001073	94.7	1.328	1.224
160	7.72.−10	0.001074	107.3	1.409	1.291
170	7.29.−9	0.001076	120.6	1.489	1.357
180	5.38.−8	0.001077	134.5	1.569	1.426
190	3.23.−7	0.001078	149.1	1.648	1.495
200	1.62.−6	0.001079	164.4	1.726	1.566
210	7.01.−6	0.001081	180.4	1.805	1.638
220	2.65.−5	0.001082	197.1	1.882	1.711
230	8.91.−5	0.001084	214.6	1.960	1.785
240	2.73.−4	0.001085	232.8	2.038	1.860
250	7.59.−4	0.001087	251.8	2.115	1.936
260	0.00196	0.001088	271.5	2.192	2.013
270	0.00469	0.001090	292.0	2.270	2.091
273.15	0.00611	0.001091	298.7	2.294	2.116
273.15	0.00611	0.001000	632.2	3.515	4.217
280	0.00990	0.001000	661.0	3.619	4.198
290	0.01917	0.001001	702.9	3.766	4.184
300	0.03531	0.001003	744.7	3.908	4.179

[a]v = specific volume; h = specific enthalpy; s = specific entropy; c_p = specific heat at constant pressure. Ice values ($T \leq 273.15$ K) converted and rounded off from S. Gordon, NASA Tech. Paper 1906, 1982.

[b]The notation 6.30.−11 signifies 6.30×10^{-11}.

Table 39.18 Thermophysical Properties of Saturated Steam/Water[a]

P (bar)	T (K)	v_f	v_g	h_f	h_g	η_g	λ_f	λ_g	Pr_f	Pr_g
1.0	372.78	1.0434.-3[b]	1.6937	417.5	2675.4	0.1202	0.6805	0.0244	1.735	1.009
1.5	384.52	1.0530.-3	1.1590	467.1	2693.4	0.1247	0.6847	0.0259	1.538	1.000
2.0	393.38	1.0608.-3	0.8854	504.7	2706.3	0.1280	0.6866	0.0268	1.419	1.013
2.5	400.58	1.0676.-3	0.7184	535.3	2716.4	0.1307	0.6876	0.0275	1.335	1.027
3.0	406.69	1.0735.-3	0.6056	561.4	2724.7	0.1329	0.6879	0.0281	1.273	1.040
3.5	412.02	1.0789.-3	0.5240	584.3	2731.6	0.1349	0.6878	0.0287	1.224	1.050
4.0	416.77	1.0839.-3	0.4622	604.7	2737.6	0.1367	0.6875	0.0293	1.185	1.057
4.5	421.07	1.0885.-3	0.4138	623.2	2742.9	0.1382	0.6869	0.0298	1.152	1.066
5	424.99	1.0928.-3	0.3747	640.1	2747.5	0.1396	0.6863	0.0303	1.124	1.073
6	432.00	1.1009.-3	0.3155	670.4	2755.5	0.1421	0.6847	0.0311	1.079	1.091
7	438.11	1.1082.-3	0.2727	697.1	2762.0	0.1443	0.6828	0.0319	1.044	1.105
8	445.57	1.1150.-3	0.2403	720.9	2767.5	0.1462	0.6809	0.0327	1.016	1.115
9	448.51	1.1214.-3	0.2148	742.6	2772.1	0.1479	0.6788	0.0334	0.992	1.127
10	453.03	1.1274.-3	0.1943	762.6	2776.1	0.1495	0.6767	0.0341	0.973	1.137
12	461.11	1.1386.-3	0.1632	798.4	2782.7	0.1523	0.6723	0.0354	0.943	1.156
14	468.19	1.1489.-3	0.1407	830.1	2787.8	0.1548	0.6680	0.0366	0.920	1.175
16	474.52	1.1586.-3	0.1237	858.6	2791.8	0.1569	0.6636	0.0377	0.902	1.191
18	480.26	1.1678.-3	0.1103	884.6	2794.8	0.1589	0.6593	0.0388	0.889	1.206
20	485.53	1.1766.-3	0.0995	908.6	2797.2	0.1608	0.6550	0.0399	0.877	1.229
25	497.09	1.1972.-3	0.0799	962.0	2800.9	0.1648	0.6447	0.0424	0.859	1.251
30	506.99	1.2163.-3	0.0666	1008.4	2802.3	0.1684	0.6347	0.0449	0.849	1.278
35	515.69	1.2345.-3	0.0570	1049.8	2802.0	0.1716	0.6250	0.0472	0.845	1.306
40	523.48	1.2521.-3	0.0497	1087.4	2800.3	0.1746	0.6158	0.0496	0.845	1.331
45	530.56	1.2691.-3	0.0440	1122.1	2797.7	0.1775	0.6068	0.0519	0.849	1.358
50	537.06	1.2858.-3	0.0394	1154.5	2794.2	0.1802	0.5981	0.0542	0.855	1.386
60	548.70	1.3187.-3	0.0324	1213.7	2785.0	0.1854	0.5813	0.0589	0.874	1.442
70	558.94	1.3515.-3	0.0274	1267.4	2773.5	0.1904	0.5653	0.0638	0.901	1.503
80	568.12	1.3843.-3	0.0235	1317.1	2759.9	0.1954	0.5499	0.0688	0.936	1.573
90	576.46	1.4179.-3	0.0205	1363.7	2744.6	0.2005	0.5352	0.0741	0.978	1.651
100	584.11	1.4526.-3	0.0180	1408.0	2727.7	0.2057	0.5209	0.0798	1.029	1.737
110	591.20	1.4887.-3	0.0160	1450.6	2709.3	0.2110	0.5071	0.0859	1.090	1.837
120	597.80	1.5268.-3	0.0143	1491.8	2689.2	0.2166	0.4936	0.0925	1.163	1.963
130	603.98	1.5672.-3	0.0128	1532.0	2667.0	0.2224	0.4806	0.0998	1.252	2.126
140	609.79	1.6106.-3	0.0115	1571.6	2642.4	0.2286	0.4678	0.1080	1.362	2.343
150	615.28	1.6579.-3	0.0103	1611.0	2615.0	0.2373	0.4554	0.1172	1.502	2.571
160	620.48	1.7103.-3	0.0093	1650.5	2584.9	0.2497	0.4433	0.1280	1.688	3.041
170	625.41	1.7696.-3	0.0084	1691.7	2551.6	0.2627	0.4315	0.1404	2.098	3.344
180	630.11	1.8399.-3	0.0075	1734.8	2513.9	0.2766	0.4200	0.1557	2.360	3.807
190	634.58	1.9260.-3	0.0067	1778.7	2470.6	0.2920	0.4087	0.1749	2.951	8.021
200	638.85	2.0370.-3	0.0059	1826.5	2410.4	0.3094	0.3976	0.2007	4.202	12.16

Table 39.18 (*Continued*)

P (bar)	s_f	s_g	c_{pf}	c_{pg}	η_f	γ_f'	γ_g	$\bar{v}_{sf}'$	$\bar{v}_{sg}$	τ
1.0	1.3027	7.3598	4.222	2.048	2.801	1.136	1.321	438.74	472.98	0.0589
1.5	1.4336	7.2234	4.231	2.077	2.490	1.139	1.318	445.05	478.73	0.0566
2.0	1.5301	7.1268	4.245	2.121	2.295	1.141	1.316	449.51	482.78	0.0548
2.5	1.6071	7.0520	4.258	2.161	2.156	1.142	1.314	452.92	485.88	0.0534
3.0	1.6716	6.9909	4.271	2.198	2.051	1.143	1.313	455.65	488.36	0.0521
3.5	1.7273	6.9392	4.282	2.233	1.966	1.143	1.311	457.91	490.43	0.0510
4.0	1.7764	6.8943	4.294	2.266	1.897	1.144	1.310	459.82	492.18	0.0500
4.5	1.8204	6.8547	4.305	2.298	1.838	1.144	1.309	461.46	493.69	0.0491
5	1.8604	6.8192	4.315	2.329	1.787	1.144	1.308	462.88	495.01	0.0483
6	1.9308	6.7575	4.335	2.387	1.704	1.144	1.306	465.23	497.22	0.0468
7	1.9918	6.7052	4.354	2.442	1.637	1.143	1.304	467.08	498.99	0.0455
8	2.0457	6.6596	4.372	2.495	1.581	1.142	1.303	468.57	500.55	0.0444
9	2.0941	6.6192	4.390	2.546	1.534	1.142	1.302	469.78	501.64	0.0433
10	2.1382	6.5821	4.407	2.594	1.494	1.141	1.300	470.76	502.64	0.0423
12	2.2161	6.5194	4.440	2.688	1.427	1.139	1.298	472.23	504.21	0.0405
14	2.2837	6.4651	4.472	2.777	1.373	1.137	1.296	473.18	505.33	0.0389
16	2.3436	6.4175	4.504	2.862	1.329	1.134	1.294	473.78	506.12	0.0375
18	2.3976	6.3751	4.534	2.944	1.291	1.132	1.293	474.09	506.65	0.0362
20	2.4469	6.3367	4.564	3.025	1.259	1.129	1.291	474.18	506.98	0.0350
25	2.5543	6.2536	4.640	3.219	1.193	1.123	1.288	473.71	507.16	0.0323
30	2.6455	6.1837	4.716	3.407	1.143	1.117	1.284	472.51	506.65	0.0300
35	2.7253	6.1229	4.792	3.593	1.102	1.111	1.281	470.80	505.66	0.0280
40	2.7965	6.0685	4.870	3.781	1.069	1.104	1.278	468.72	504.29	0.0261
45	2.8612	6.0191	4.951	3.972	1.040	1.097	1.275	466.31	502.68	0.0244
50	2.9206	5.9735	5.034	4.168	1.016	1.091	1.272	463.67	500.73	0.0229
60	3.0273	5.8908	5.211	4.582	0.975	1.077	1.266	457.77	496.33	0.0201
70	3.1219	5.8162	5.405	5.035	0.942	1.063	1.260	451.21	491.31	0.0177
80	3.2076	5.7471	5.621	5.588	0.915	1.048	1.254	444.12	485.80	0.0156
90	3.2867	5.6820	5.865	6.100	0.892	1.033	1.249	436.50	479.90	0.0136
100	3.3606	5.6198	6.142	6.738	0.872	1.016	1.244	428.24	473.67	0.0119
110	3.4304	5.5595	6.463	7.480	0.855	0.998	1.239	419.20	467.13	0.0103
120	3.4972	5.5002	6.838	8.384	0.840	0.978	1.236	409.38	460.25	0.0089
130	3.5616	5.4408	7.286	9.539	0.826	0.956	1.234	398.90	453.00	0.0076
140	3.6243	5.3803	7.834	11.07	0.813	0.935	1.232	388.00	445.34	0.0064
150	3.6859	5.3178	8.529	13.06	0.802	0.916	1.233	377.00	437.29	0.0053
160	3.7471	5.2531	9.456	15.59	0.792	0.901	1.235	366.24	428.89	0.0043
170	3.8197	5.1855	11.30	17.87	0.782	0.867	1.240	351.19	420.07	0.0034
180	3.8765	5.1128	12.82	21.43	0.773	0.838	1.248	336.35	410.39	0.0026
190	3.9429	5.0332	15.76	27.47	0.765	0.808	1.260	320.20	399.87	0.0018
200	4.0149	4.9412	22.05	39.31	0.758	0.756	1.280	298.10	387.81	0.0011

$^a v$ = specific volume (m³/kg); h = specific enthalpy (kJ/kg); s = specific entropy (kJ/kg · K); c_p = specific heat at constant pressure (kJ/kg · K); η = viscosity (10^{-4} Pa · sec); λ = thermal conductivity (w/m · K); Pr = Prandtl number; $\gamma = c_p/c_v$ ratio; $\bar{v}_s$ = velocity of sound (m/sec); τ = surface tension (N/m); f' = wet saturated vapor; g = saturated vapor. Rounded off from values of C. M. Tseng, T. A. Hamp, and E. O. Moeck, Atomic Energy of Canada Report AECL-5910, 1977.

^{b} The notation 1.0434.-3 signifies 1.0434×10^{-3}.

Table 39.19 Thermophysical Properties of Miscellaneous Substances at Atmospheric Pressure[a]

	Temperature (K)					
	250	300	350	400	450	500
n-Butane						
v	0.0016	0.411	0.485	0.558	0.630	0.701
h	236.6	718.9	810.7	913.1	1026	1149
s	3.564	5.334	5.616	5.889	6.155	6.414
c_p	2.21	1.73	1.94	2.15	2.36	2.56
Z	0.005	0.969	0.982	0.988	0.992	0.993
$\bar{v}_s$	1161	211	229	245	259	273
λ	0.0979	0.0161	0.0220	0.0270	0.0327	0.0390
η	2.545	0.076	0.088	0.101	0.111	0.124
Pr	5.75	0.84	0.82	0.81	0.80	0.80
Ethane						
v	0.672	0.812	0.950	1.088	1.225	1.362
h	948.7	1068	1162	1265	1380	1505
s	7.330	7.634	7.854	8.198	8.467	8.730
c_p	1.58	1.76	1.97	2.18	2.39	2.60
Z	0.986	0.992	0.995	0.997	0.998	0.998
$\bar{v}_s$	287	312	334	355	374	392
λ	0.0103	0.0157	0.0219	0.0288	0.0361	0.0438
η	0.079	0.094	0.109	0.123	0.135	0.148
Pr	1.214	1.056	0.978	0.932	0.900	0.878
Ethylene						
v	0.734	0.884	1.034	1.183	1.332	1.482
h	966.8	1039	1122	1215	1316	1415
s	7.556	7.832	8.085	8.331	8.568	8.800
c_p	1.40	1.57	1.75	1.93	2.10	2.26
Z	0.991	0.994	0.997	0.998	0.999	1.000
$\bar{v}_s$	306	330	353	374	394	403
λ	0.0149	0.0206	0.0271	0.0344	0.0425	0.0506
η	0.087	0.103	0.119	0.134	0.148	0.162
Pr	0.816	0.785	0.767	0.751	0.735	0.721
n-Hydrogen						
v	10.183	12.218	14.253	16.289	18.324	20.359
h	3517	4227	4945	5669	6393	7118
s	67.98	70.58	72.79	74.72	76.43	77.96
c_p	14.04	14.31	14.43	14.48	14.50	14.51
Z	1.000	1.000	1.000	1.000	1.000	1.000
$\bar{v}_s$	1209	1319	1423	1520	1611	1698
λ	0.162	0.187	0.210	0.230	0.250	0.269
η	0.079	0.089	0.099	0.109	0.118	0.127
Pr	0.685	0.685	0.685	0.684	0.684	0.684
Nitrogen						
v	0.7317	0.8786	1.025	1.171	1.319	1.465
h	259.1	311.2	363.3	415.4	467.8	520.4
s	6.650	6.840	7.001	7.140	7.263	7.374
c_p	1.042	1.041	1.042	1.045	1.050	1.056
Z	0.9992	0.9998	0.9998	0.9999	1.0000	1.0002
$\bar{v}_s$	322	353	382	407	432	455
λ	0.0223	0.0259	0.0292	0.0324	0.0366	0.0386
η	0.155	0.178	0.200	0.220	0.240	0.258
Pr	0.724	0.715	0.713	0.710	0.708	0.706

Table 39.19 *(Continued)*

	Temperature (K)					
	250	300	350	400	450	500
Oxygen						
v	0.6402	0.7688	0.9790	1.025	1.154	1.282
h	226.9	272.7	318.9	365.7	413.1	461.3
s	6.247	6.414	6.557	6.682	6.793	6.895
c_p	0.915	0.920	0.929	0.942	0.956	0.972
Z	0.9987	0.9994	0.9996	0.9998	1.0000	1.0000
λ	0.0226	0.0266	0.0305	0.0343	0.0380	0.0416
η	0.179	0.207	0.234	0.258	0.281	0.303
Pr	0.725	0.716	0.713	0.710	0.708	0.707
Propane						
v	0.451	0.548	0.644	0.738	0.832	0.926
h	877.2	957.0	1048	1149	1261	1384
s	5.840	6.131	6.409	6.680	6.944	7.202
c_p	1.50	1.70	1.96	2.14	2.35	2.55
Z	0.970	0.982	0.988	0.992	0.994	0.996
$\bar{v}_s$	227	248	268	285	302	317
λ	0.0128	0.0182	0.0247	0.0296	0.0362	0.0423
η	0.070	0.082	0.096	0.108	0.119	0.131
Pr	0.820	0.772	0.761	0.765	0.773	0.793
Propylene						
v	0.482	0.585	0.686	0.786	0.884	0.972
h	891.8	957.6	1040	1131	1235	1338
s	6.074	6.354	6.606	6.851	7.095	7.338
c_p	1.44	1.55	1.73	1.91	2.09	2.27
Z	0.976	0.987	0.992	0.994	0.995	0.996
$\bar{v}_s$	247	257	278	298	315	333
λ	0.0127	0.0177	0.0233	0.0296	0.0363	0.0438
η	0.072	0.087	0.101	0.115	0.128	0.141
Pr	0.814	0.769	0.754	0.742	0.731	0.728

$^a v$ = specific volume (m³/kg); h = specific enthalpy (kJ/kg); s = specific entropy (kJ/kg · K); c_p = specific heat at constant pressure (kJ/kg · K); Z = compressibility factor = Pv/RT; $\bar{v}_s$ = velocity of sound (m/sec); λ = thermal conductivity (W/m · K); η = viscosity (10^{-4})(N · sec/m²) (thus, at 250 K for *n*-butane the viscosity is 2.545×10^{-4} N · sec/m² = 0.0002545 Pa · sec); Pr = Prandtl number.

Table 39.20 Physical Properties of Numbered Refrigerants[a]

Number	Formula, Composition, Synonym	Molecular Weight	n.b.p. (°C)	crit. P (bar)	crit. T (°C)
4	R-32/125/134a/143a (10/33/21/36)	94.50	−49.4	40.1	77.5
10	CCl_4 (carbon tetrachloride)	153.8	76.8	45.6	283.2
CFC 11	CCl_3F	137.37	23.8	44.1	198.0
11B1	$CBrCl_2F$	181.82	52		
11B2	CBr_2ClF	226.27	80		
11B3	CBr_3F	270.72	107		
CFC 12	CCl_2F_2	120.91	−29.8	41.1	112.0
12B1	$CBrClF_2$	165.36	−2.5	42.5	153.0
12B2	CF_2Br_2	209.81	24.5	40.7	204.9
CFC 13	$CClF_3$	104.46	−81.4	38.7	28.8
BFC 13B1	$CBrF_3$ (Halon 1301)	148.91	−57.8	39.6	67.0
FC 14	CF_4 (carbon tetrafluoride)	88.00	−127.9	37.5	−45.7
20	$CHCl_3$ (chloroform)	119.38	61.2	54.5	263.4
21	$CHCl_2F$	102.92	8.9	51.7	178.5
HCFC 22	$CHClF_2$	86.47	−40.8	49.9	96.2
HFC 23	CHF_3	70.01	−82.1	48.7	26.3
HCC 30	CH_2Cl_2 (methylene chloride)	84.93	40.2	60.8	237.0
31	CH_2ClF	68.47	−9.1	56.2	153.8
HFC 32	CH_2F_2	52.02	−51.7	58.0	78.2
33	R-22/124/152a (40/43/17)	96.62	−28.8		
40	CH_3Cl (methyl chloride)	50.49	−12.4	66.7	143.1
FX 40	R-32/125/143a (10/45/45)	90.70	−48.4	40.5	72.0
HFC 41	CH_3F (methyl fluoride)	34.03	−78.4	58.8	44.3
50	CH_4 (methane)	16.04	−161.5	46.4	−82.5
FX 57	R-22/124/142b (65/25/10)	96.70	−35.2	47.0	105.0
110	CCl_3CCl_3	236.8	185	33.4	401.8
111	CCl_3CCl_2F	220.2	137		
112	CCl_2FCCl_2F	203.8	92.8	33.4	278
CFC 113	$CClF_2CCl_2F$	187.38	47.6	34.4	214.1
113a	CCl_3CF_3	187.36	47.5		
CFC 114	$CClF_2CClF_2$	170.92	3.8	32.5	145.7
114a	CF_3CCl_2F	170.92	3.0	33.0	145.5
CFC 115	$CClF_2CF_3$	154.47	−39.1	31.5	79.9
FC 116	CF_3CF_3 (perfluoroethane)	138.01	−78.2	30.4	19.9
120	$CHCl_2CCl_3$	202.3	162	34.8	373
121	$CHCl_2CCl_2F$	185.84	116.6		
122	$CClF_2CHCl_2$	131.39	72.0		
HCFC 123	$CHCl_2CF_3$	152.93	27.9	36.7	183.7
HCFC 123a	$CHClFCClF_2$	152.93	28.0	44.7	188.5
HCFC 124	$CHClFCF_3$	136.48	−12.0	36.4	122.5
E 125	CHF_2OCF_2	136.02	−41.9	33.3	80.4
HFC 125	CHF_2CF_3	120.02	−48.1	36.3	66.3
131	$CHCl_2CHClF$	151.4	102.5		
132	$CHClFCHClF$	134.93	58.5		
133	$CHClFCHF_2$	118.5	17.0		
E 134	CHF_2OCHF_2	118.03	6.2	42.3	153.5
HFC 134	CHF_2CHF_2	102.03	−23.0	46.2	118.7
HFC 134a	CH_2FCF_3	102.03	−26.1	40.6	101.1
141	$CH_2ClCHClF$	116.95	76		

Table 39.20 *(Continued)*

Number	Formula, Composition, Synonym	Molecular Weight	*n.b.p.* (°C)	crit. *P* (bar)	crit. *T* (°C)
141a	$CHCl_2CH_2F$	116.95			
HCFC 141b	CH_3CCl_2F	116.95	32.2	42.5	204.4
142	CHF_2CH_2Cl	100.49	35.1		
142a	$CHClFCH_2F$	100.49			
HCFC 142b	CH_3CClF_2	100.50	−9.8	41.2	137.2
143	CHF_2CH_2F	84.04	5.		
E 143a	CH_3OCF_3	100.04	−24.1	35.9	104.9
HFC 143a	CH_3CF_3	84.04	−47.4	38.3	73.6
151	CH_2FCH_2Cl	82.50	53.2		
152	CH_2FCH_2F	66.05	10.5	43.4	171.8
HFC 152a	CH_3CHF_2	66.05	−24.0	45.2	113.3
HCC 160	CH_3CH_2Cl	64.51	12.4	52.4	186.6
HFC 161	CH_3CH_2F	48.06	−37.1	47.0	102.2
E 170	CH_3OCH_3 (dimethyl ether)	46.07	−24.8	53.2	128.8
HC 170	CH_3CH_3 (ethane)	30.07	−88.8	48.9	32.2
216	$C_3Cl_2F_6$	220.93	35.7	27.5	180.0
FC 218	$CF_3CF_2CF_3$	188.02	−36.7	26.8	71.9
HFC 227ca	$CHF_2CF_2CF_3$	170.03	−17.0	28.7	106.3
HFC 227ea	CF_3CHFCF_3	170.03	−18.3	29.5	103.5
234da	$CF_3CHClCHClF$	114.03	70.1		
235ca	$CF_3CF_2CH_2Cl$	156.46	28.1		
HFC 236ca	$CHF_2CF_2CHF_2$	152.04	5.1	34.1	153.2
HFC 236cb	$CH_2FCF_2CF_3$	152.04	−1.4	31.2	130.2
HFC 236ea	CHF_2CHFCF_3	152.04	6.6	35.3	141.2
HFC 235fa	$CF_3CH_2CF_3$	152.04	−1.1	31.8	130.7
HFC 245ca	$CH_2FCF_2CHF_2$	134.05	25.5	38.6	178.5
E 245cb	$CHF_2OCH_2CF_3$	150.05	34.0		185.2
HFC 245cb	$CH_3CF_2CF_3$	134.05	−18.3	32.6	108.5
E 245fa	$CHF_2OCH_2CF_3$	150.05	29.2	37.3	170.9
HFC 245fa	$CHF_2CH_2CF_3$	134.05	15.3	36.4	157.6
HFC 254cb	$CH_3CF_2CHF_2$	116.06	−0.8	37.5	146.2
HC 290	$CH_3CH_2CH_3$ (propane)	44.10	−42.1	42.5	96.8
RC 318	cyclo-$CF_2CF_2CF_2CF_2$	200.04	−5.8	27.8	115.4
400	R-12/114				
R-401a	R-22/124/152a (53/34/13)	94.44	−33.1	46.0	108.0
R-401b	R-22/124/152a (61/28/11)	92.84	−34.7	46.8	106.1
R 401c	R 22/124/152a (33/52/15)	101.04	−28.4	43.7	112.7
R 402a	R 22/125/290 (38/60/2)	101.55	−49.2	41.3	75.5
R 402b	R 22/125/290 (60/38/2)	94.71	−47.4	44.5	82.6
R 403a	R 22/218/290 (75/20/5)	91.06	−50.0	50.8	93.3
R 403b	R 22/218/290 (56/39/5)	102.06	−49.5	50.9	90.0
R 404a	R 125/134a/143a (44/4/52)	97.60	−46.5	37.3	72.1
R 405a	R 22/142b/152a/C318 (45/5/7/43)	116.00	−27.3	42.6	106.1
R 406a	R 22/142b/600a (55/41/4)	89.85	−30.0	47.4	123.0
R 407a	R 32/125/134a (20/40/40)	90.10	−45.5	45.4	82.8
R 407b	R 32/125/134a (10/70/20)	102.94	−47.3	41.6	75.8
R 407c	R 32/125/134a (23/25/52)	86.20	−43.6	46.2	86.7
R 408a	R 22/125/143a (47/7/46)	87.02	−43.5	43.4	83.5
R 409a	R 22/124/142b (60/25/15)	97.45	−34.2	45.0	107.0

Table 39.20 *(Continued)*

Number	Formula, Composition, Synonym	Molecular Weight	n.b.p. (°C)	crit. P (bar)	crit. T (°C)
R 410a	R 32/125 (50/50)	72.56	−50.5	49.6	72.5
R 410b	R 32/125 (45/55)				
R 411a	R 22/152a/1270 (88/11/2)				
R 411b	R 22/152a/1270 (94/3/3)				
R 412a	R 22/142b/218 (70/25/5)				
R 500	R 12/152a (74/26)	99.31	−33.5	44.2	105.5
R 501	R 12/22 (25/75)				
R 502	R 22/115 (49/51)	111.64	−45.4	40.8	82.2
R 503	R 13/23 (60/40)	87.28	−87.8	43.6	19.5
R 504	R 32/115 (48/52)	79.2	−57.2	47.6	66.4
R 505	R 12/31 (78/22)	103.5	−30		
R 506	R 31/114 (55/45)	93.7	−12		
R 507	R 125/143a (50/50)	98.90	−46.7	37.9	70.9
R 508	R 23/116 (39/61)	100.10	−85.7		23.1
R 509	R 22/218 (44/56)	124.0	−47		
R 600	$CH_3CH_2CH_2CH_3$ (butane)	58.13	−0.4	38.0	152.0
R 600a	$CH_3CH_3CH_3CH$ (isobutane)	58.13	−11.7	36.5	135.0
R 610	$C_4H_{10}O$ (ethyl ether)	74.12	−116.3	36.0	194.0
R 611	$C_2H_4O_2$ (methyl formate)	60.05	31.8	59.9	204
630	CH_3NH (methyl amine)	31.06	−6.7	74.6	156.9
631	$C_2H_5NH_2$ (ethyl amine)	45.08	16.6	56.2	183.0
702	H_2 (hydrogen)	2.016	−252.8	13.2	−239.9
702p	para-hydrogen	2.016	−252.9	12.9	−240.2
704	He (helium)	4.003	−268.9	2.3	−267.9
717	NH_3 (ammonia)	17.03	−33.3	114.2	133.0
718	H_2O (water)	18.02	100.0	221.0	374.2
720	Ne (neon)	20.18	−246.1	34.0	−228.7
728	N_2 (nitrogen)	28.01	−198.8	34.0	−146.9
728a	CO (carbon monoxide)	28.01	−191.6	35.0	−140.3
729	- (air)	28.97	−194.3	37.6	−140.6
732	O_2 (oxygen)	32.00	−182.9	50.8	−118.4
740	A (argon)	39.95	−185.9	49.0	−122.3
744	CO_2 (carbon dioxide)	44.01	−78.4	73.7	31.1
744a	N_2O (nitrous oxide)	44.02	−89.5	72.2	36.5
R 764	SO_2 (sulfur dioxide)	64.07	−10.0	78.8	157.5
1113	C_2ClF_3	116.47	−27.9	40.5	106
1114	C_2F_4	100.02	−76.0	39.4	33.3
1120	$CHClCCl_2$	131.39	87.2	50.2	271.1
1130	CHClCHCl	96.95	47.8	54.8	243.3
1132a	$C_2H_2F_2$	64.03	−85.7	44.6	29.7
1141	C_2H_3F (vinyl fluoride)	46.04	−72.2	52.4	54.7
1150	C_2H_4 (ethylene)	28.05	−103.7	51.1	9.3
1270	C_3H_6 (propylene)	42.09	−185	46.2	91.8
R 7146	SF_6 (sulfur hexafluoride)	146.05	−63.8	37.6	45.6

[a]Refrigerant numbers in some cases are tentative and subject to revision. Compositions rounded to nearest weight percent.

Based upon data supplied by M.O. McLinden, N.I.S.T., Boulder, CO, PCR Chemicals, Gainesville, FL, G.H. Thomson, DIPPR, Bartlesville, OK, and literature sources.

Table 39.21 Specific Heat (kJ/kg · K) at Constant Pressure of Saturated Liquids

Substance	Temperature (K)															
	250	260	270	280	290	300	310	320	330	340	350	360	370	380	390	400
Acetic acid	—[a]	—	—	—	2.03	2.06	2.09	2.12	2.16	2.19	2.23	2.26	2.29	2.33	2.36	2.39
Acetone	2.05	2.07	2.10	2.13	2.16	2.19	2.22	2.26	2.30	2.35	2.40					
Ammonia	4.48	4.54	4.60	4.66	4.73	4.82	4.91	5.02	5.17	5.37	5.64	6.04	6.68	7.80	10.3	21
Aniline	—	—	2.03	2.04	2.05	2.07	2.10	2.13	2.16	2.19	2.22	2.26	2.31	2.38	2.47	2.58
Benzene	—	—	—	1.69	1.71	1.73	1.75	1.78	1.81	1.84	1.87	1.91	1.94	1.98	2.03	2.09
n-Butane	2.19	2.23	2.27	2.32	2.37	2.43	2.50	2.58	2.67	2.76	2.86	2.97	3.08			
Butanol	2.13	2.17	2.22	2.27	2.33	2.38	2.44	2.51	2.58	2.65	2.73	2.82	2.93	3.06	3.20	3.36
Carbon tetrachloride	0.833	0.838	0.843	0.848	0.853	0.858	0.864	0.870	0.879	0.891	0.912	0.941	0.975			
Chlorobenzene	1.29	1.31	1.32	1.32	1.33	1.33	1.34	1.36	1.38	1.40	1.42	1.44	1.46	1.47	1.49	1.51
m-Cresol	—	—	—	—	2.04	2.07	2.11	2.14	2.18	2.21	2.24	2.27	2.30	2.32	2.35	2.38
Ethane	2.97	3.20	3.50	4.00	5.09	9.92										
Ethanol	—	2.24	2.28	2.33	2.38	2.45	2.54	2.64	2.75	2.86	2.99	3.12	3.26	3.41	3.56	3.72
Ethyl acetate				1.89	1.92	1.94	1.97	2.00	2.03	2.06	2.09	2.13	2.16	2.20	2.24	2.28
Ethyl sulfide	1.96	1.97	1.97	1.98	2.00	2.01	2.02	2.03								
Ethylene	3.25	3.78	5.0													
Formic acid	—	—	—	—	2.15	2.16	2.17	2.18	2.20	2.22	2.24	2.26	2.28	2.30	2.33	2.36
Heptane	2.08	2.10	2.13	2.17	2.20	2.24	2.28	2.32	2.36	2.41	2.45	2.49	2.54	2.59	2.64	2.70
Hexane	2.09	2.12	2.15	2.19	2.22	2.26	2.31	2.36	2.41	2.46	2.51	2.56	2.62	2.69	2.76	2.83
Methanol	2.31	2.34	2.37	2.41	2.46	2.52	2.58	2.65	2.73	2.82	2.91	3.01	3.12	3.24	3.36	3.49
Methyl formate					2.16	2.16										
Octane	2.07	2.10	2.13	2.16	2.19	2.22	2.26	2.31	2.35	2.39	2.43	2.47	2.52	2.57	2.62	2.69
Oil, linseed	1.58	1.61	1.65	1.69	1.73	1.78	1.82	1.87	1.91	1.95	1.99	2.03	2.08	2.13	2.17	2.21
Oil, olive	1.90	1.92	1.95	1.98	2.01	2.05	2.09	2.13	2.16	2.20	2.24	2.28	2.32	2.37	2.41	2.46
Pentane	1.96	2.02	2.08	2.14	2.21	2.28	2.35	2.42	2.49	2.56	2.63	2.70	2.77	2.84	2.91	2.98
Propane	2.35	2.41	2.48	2.56	2.65	2.76	2.89	3.06	3.28	3.62	4.23	5.98				
Propanol	2.04	2.10	2.16	2.24	2.32	2.41	2.51	2.62	2.74	2.86	2.99	3.12	3.26	3.40	3.55	3.71
Propylene	2.22	2.27	2.34	2.43	2.55	2.69	2.87	3.12	3.44	3.92	4.75	6.75				
Sulfuric acid	—	—	—	—	1.39	1.41	1.43	1.45	1.46	1.48	1.50	1.51	1.53	1.54	1.56	1.57
Sulfur dioxide	1.31	1.32	1.33	1.34	1.36	1.39	1.42	1.46	1.50	1.55	1.61	1.68	1.76	1.85	1.99	2.14
Turpentine	1.60	1.65	1.70	1.75	1.80	1.85	1.90	1.95	2.00	2.05	2.10	2.15	2.20	2.25	2.30	2.35

[a]Dashes indicate inaccessible states.

Table 39.22 Ratio of Principal Specific Heats, c_p/c_v, for Liquids and Gases at Atmospheric Pressure

Substance	Temperature (K)															
	200	220	240	260	280	300	320	340	360	380	400	420	440	460	480	500
Acetylene	1.313	1.294	1.277	1.261	1.247	1.234	1.222	1.211	1.200							
Air	1.399	1.399	1.399	1.399	1.399	1.399	1.399	1.398	1.397	1.396	1.395	1.394	1.392	1.390	1.388	1.386
Ammonia							1.307	1.299	1.291	1.284	1.278	1.271	1.265	1.260	1.255	1.249
Argon	1.663	1.665	1.666	1.666	1.666	1.666	1.666	1.666	1.666	1.666	1.666	1.666	1.666	1.666	1.666	1.666
i-Butane	1.357	1.356	1.357	1.362	1.113	1.103	1.096	1.089	1.084	1.079	1.075	1.071	1.068	1.065	1.063	1.060
n-Butane	1.418	1.413	1.409	1.406	1.112	1.103	1.096	1.089	1.084	1.079	1.075	1.072	1.069	1.066	1.063	1.061
Carbon dioxide		1.350	1.332	1.317	1.303	1.290	1.282	1.273	1.265	1.258	1.253	1.247	1.242	1.238	1.233	1.229
Carbon monoxide	1.405	1.404	1.403	1.402	1.402	1.401	1.401	1.400	1.399	1.398	1.396	1.395	1.393	1.391	1.389	1.387
Ethane		1.250	1.233	1.219	1.205	1.193	1.182	1.172	1.163	1.155	1.148	1.141	1.135	1.130	1.125	1.120
Ethylene				1.268	1.251	1.236	1.223	1.212	1.201	1.192	1.183	1.175	1.168	1.162	1.156	1.151
Fluorine	1.393	1.387	1.380	1.374	1.368	1.362										
Helium	1.667	1.667	1.667	1.667	1.667	1.667	1.667	1.667	1.667	1.667	1.667	1.667	1.667	1.667	1.667	1.667
n-Hydrogen	1.439	1.428	1.418	1.413	1.409	1.406	1.403	1.402	1.401	1.400	1.399	1.398	1.398	1.398	1.397	1.397
Krypton	1.649	1.654	1.657	1.659	1.661	1.662	1.662	1.662	1.662	1.662	1.662	1.663	1.663	1.664	1.666	1.667
Methane	1.337	1.333	1.328	1.322	1.314	1.306	1.296	1.287	1.278	1.268	1.258	1.249	1.241	1.233	1.226	1.219
Neon	1.667	1.667	1.667	1.667	1.667	1.667	1.667	1.667	1.667	1.667	1.667	1.667	1.667	1.667	1.667	1.667
Nitrogen	1.399	1.399	1.399	1.399	1.399	1.399	1.399	1.399	1.398	1.398	1.397	1.396	1.395	1.393	1.392	1.391
Oxygen	1.398	1.397	1.397	1.396	1.395	1.394	1.392	1.389	1.387	1.384	1.381	1.378	1.375	1.371	1.368	1.365
Propane	1.513	1.506	1.173	1.158	1.145	1.135	1.126	1.118	1.111	1.105	1.100	1.095	1.091	1.087	1.084	1.081
Propylene			1.133	1.122	1.111	1.101	1.091	1.082	1.072	1.063	1.055	1.046	1.038	1.030	1.023	1.017
R12			1.171	1.159	1.148	1.139	1.132	1.127	1.122	1.118	1.115	1.112	1.110	1.108	1.106	1.104
R21						1.179	1.165	1.156	1.148	1.142	1.137	1.133	1.129	1.126	1.123	1.120
R22					1.204	1.190	1.174	1.167	1.160	1.153	1.148	1.144	1.140	1.136	1.132	1.129
Steam									1.323	1.321	1.319	1.317	1.315	1.313	1.311	1.309
Xenon	1.623	1.632	1.639	1.643	1.651	1.655	1.658	1.661	1.662	1.662	1.662	1.662	1.662	1.662	1.662	1.662

Table 39.23 Surface Tension (N/m) of Liquids

Substance	Temperature (K)															
	250	260	270	280	290	300	310	320	330	340	350	360	370	380	390	400
Acetone	0.0291	0.0279	0.0266	0.0253	0.0240	0.0228	0.0214	0.0201	0.0187	0.0174	0.0162	0.0150	0.0139	0.0128	0.0117	0.0106
Ammonia	0.0317	0.0294	0.0271	0.0248	0.0226	0.0203	0.0181	0.0159	0.0138	0.0117	0.0099	0.0080	0.0059	0.0040	0.0021	0.0003
Benzene			0.0320	0.0306	0.0292	0.0278	0.0265	0.0252	0.0239	0.0227	0.0215	0.0203	0.0191	0.0179	0.0167	0.0155
Butane	0.0177	0.0165	0.0153	0.0141	0.0129	0.0116	0.0104	0.0092	0.0080	0.0069	0.0059	0.0049	0.0040	0.0031	0.0023	0.0016
CO$_2$	0.0092	0.0071	0.0051	0.0032	0.0016	0.0003	—	—	—	—	—	—	—	—	—	—
Chlorine	0.0244	0.0228	0.0213	0.0198	0.0183	0.0168	0.0153	0.0138	0.0123	0.0108	0.0094	0.0080	0.0066	0.0052	0.0044	0.0037
Ethane	0.0059	0.0047	0.0035	0.0024	0.0013	0.0005	—	—	—	—	—	—	—	—	—	—
Ethanol	0.0271	0.0261	0.0251	0.0242	0.0232	0.0223	0.0214	0.0205	0.0196	0.0186	0.0177	0.0167	0.0158	0.0148	0.0137	0.0126
Ethylene	0.0032	0.0019	0.0009	0.0002	—	—	—	—	—	—	—	—	—	—	—	—
Heptane	0.0244	0.0234	0.0224	0.0214	0.0205	0.0195	0.0185	0.0176	0.0166	0.0156	0.0147	0.0137	0.0127	0.0118	0.0109	0.0099
Hexane	0.0229	0.0218	0.0207	0.0197	0.0186	0.0175	0.0165	0.0154	0.0145	0.0135	0.0125	0.0115	0.0106	0.0096	0.0086	0.0076
Mercury	0.474	0.472	0.470	0.468	0.466	0.464	0.462	0.460	0.458	0.456	0.454	0.452	0.450	0.448	0.446	0.444
Methanol					0.0223	0.0223	0.0214	0.0205	0.0196	0.0187	0.0178	0.0168	0.0159	0.0149	0.0139	0.0128
Octane	0.0251	0.0243	0.0234	0.0225	0.0216	0.0207	0.0197	0.0188	0.0179	0.0170	0.0161	0.0152	0.0143	0.0135	0.0127	0.0120
Propane	0.0132	0.0118	0.0104	0.0091	0.0079	0.0067	0.0056	0.0046	0.0037	0.0028	0.0018	0.0009	—	—	—	—
Propylene	0.0133	0.0120	0.0106	0.0091	0.0078	0.0065	0.0053	0.0042	0.0032	0.0023	0.0014	0.0006	—	—	—	—
R12	0.0148	0.0135	0.0122	0.0109	0.0096	0.0083	0.0070	0.0058	0.0047	0.0036	0.0027	0.0018	0.0010	0.0003	—	—
R13	0.0057	0.0042	0.0029	0.0018	0.0009	0.0003	—	—	—	—	—	—	—	—	—	—
Toluene	0.0342	0.0327	0.0312	0.0298	0.0285	0.0272	0.0260	0.0249	0.0236	0.0225	0.0214	0.0203	0.0193	0.0183	0.0173	0.0163
Water			—	0.0746	0.0732	0.0716	0.0699	0.0684	0.0666	0.0650	0.0636	0.0614	0.0595	0.0575	0.0555	0.0535

Table 39.24 Thermal Conductivity (W/m · K) of Saturated Liquids

Substance	Temperature (K)															
	250	260	270	280	290	300	310	320	330	340	350	360	370	380	390	400
Acetic acid	—	—	—	—	0.165	0.164	0.162	0.161	0.160	0.158	0.157	0.156	0.154	0.153	0.152	0.150
Acetone	0.179	0.175	0.171	0.167	0.163	0.160	0.156	0.152	0.149	0.145	0.141	0.137	0.133	0.130	0.126	0.122
Ammonia	0.562	0.541	0.522	0.503	0.484	0.466	0.445	0.424	0.403	0.382	0.359	0.335	0.309	0.284	0.252	0.227
Aniline	—	—	0.175	0.175	0.174	0.173	0.172	0.172	0.171	0.170	0.169	0.168	0.167	0.166	0.165	0.165
Benzene	—	—	—	0.151	0.149	0.145	0.141	0.138	0.135	0.132	0.129	0.126	0.123	0.120	0.117	0.114
Butane	0.130	0.126	0.122	0.118	0.115	0.111	0.108	0.104	0.101	0.097	0.094	0.090	0.086	0.082	0.078	0.075
Butanol	0.161	0.159	0.157	0.155	0.153	0.151	0.149	0.147	0.145	0.143	0.141	0.139	0.138	0.136	0.135	0.133
Carbon tetrachloride	0.114	0.112	0.110	0.107	0.105	0.103	0.100	0.098	0.096	0.094	0.092	0.089	0.087	0.085	0.083	0.080
Chlorobenzene	0.137	0.135	0.133	0.131	0.129	0.127	0.125	0.123	0.121	0.119	0.117	0.115	0.113	0.111	0.109	0.107
m-Cresol	—	—	—	—	0.150	0.149	0.148	0.148	0.147	0.146	0.146	—	—	—	—	—
Ethane	0.105	0.098	0.091	0.084	0.074	0.065	—	—	—	—	—	—	—	—	—	—
Ethanol	—	0.178	0.176	0.173	0.169	0.166	0.163	0.160	0.158	0.155	0.153	0.151	0.149	0.147	0.145	0.143
Ethyl acetate	0.158	0.155	0.152	0.149	0.146	0.143	0.140	0.137	0.134	0.131	0.128	0.124	0.121	0.118	0.115	0.112
Ethyl sulfide	0.144	0.142	0.140	0.138	0.136	0.134	—	—	—	—	—	—	—	—	—	—
Ethylene	0.100	0.090	0.078	0.064	—	—	—	—	—	—	—	—	—	—	—	—
Formic acid	—	—	—	—	0.293	0.279	0.255	0.241	—	—	—	—	—	—	—	—
Heptane	0.138	0.136	0.133	0.130	0.127	0.125	0.122	0.119	0.117	0.114	0.111	0.108	0.106	0.104	0.102	0.100
Hexane	0.138	0.135	0.132	0.129	0.125	0.122	0.119	0.116	0.112	0.110	0.107	0.104	0.101	0.099	0.096	0.093
Methanol	0.218	0.215	0.211	0.208	0.205	0.202	0.199	0.196	0.194	0.191	0.188	0.186	0.183	0.180	0.178	0.176
Methyl formate	0.202	0.199	0.196	0.193	0.190	0.187	0.184	0.181	0.178	0.175	0.172	—	—	—	—	—
Octane	0.141	0.138	0.136	0.133	0.130	0.128	0.125	0.123	0.120	0.118	0.115	0.112	0.110	0.107	0.105	0.103
Oil, linseed	0.175	0.173	0.171	0.169	0.167	0.165	0.163	0.161	0.159	0.157	0.154	0.152	0.150	0.148	0.146	0.144
Oil, olive	0.170	0.170	0.169	0.169	0.168	0.168	0.167	0.167	0.166	0.165	0.165	0.164	0.163	0.163	0.162	0.161
Pentane	0.133	0.128	0.124	0.120	0.116	0.112	0.108	0.106	0.102	0.098	0.094	0.090	0.086	0.082	0.078	0.074
Propane	0.120	0.116	0.112	0.107	0.102	0.097	0.093	0.088	0.084	0.079	0.075	0.070	—	—	—	—
Propanol	0.166	0.163	0.161	0.158	0.156	0.154	0.152	0.150	0.148	0.146	0.144	0.142	0.141	0.139	0.138	0.136
Propylene	0.137	0.131	0.127	0.122	0.118	0.111	0.107	0.101	0.096	0.090	0.083	0.077	—	—	—	—
Sulfuric acid	—	—	—	—	0.324	0.329	0.332	0.336	0.339	0.342	0.345	0.349	0.352	0.355	0.358	0.361
Sulfur dioxide	0.227	0.221	0.214	0.208	0.202	0.196	0.190	0.184	0.172	0.158						
Turpentine				0.128	0.127	0.127	0.126									

Table 39.25 Viscosity (10^{-4} Pa · sec) of Saturated Liquids

Substance	Temperature (K)															
	250	260	270	280	290	300	310	320	330	340	350	360	370	380	390	400
Acetic acid	5.27	—	—	—	13.1	11.3	9.77	8.49	7.40	6.48	5.70	5.03	4.45	3.95	3.52	3.15
Acetone	—	4.63	4.11	3.68	3.32	3.01	2.73	2.49	2.28	2.10	1.94	—	—	—	—	—
Ammonia	2.20	1.94	1.74	1.58	1.44	1.31	1.20	1.08	0.98	0.88	0.78	0.70	0.64	0.58	0.53	0.48
Aniline	—	—	96.9	71.6	53.5	40.3	30.7	23.6	18.2	14.2	11.2	8.84	7.03	5.63	4.54	3.68
Benzene	—	—	—	8.08	6.80	5.84	5.13	4.52	4.05	3.60	3.29	2.98	2.70	2.45	2.24	2.04
Butane	2.63	2.37	2.15	1.95	1.77	1.61	1.56	1.33	1.21	1.10	1.00	0.90	0.81	0.73	0.66	0.60
Butanol	94.2	70.9	53.9	41.4	32.1	25.1	19.8	15.7	12.6	10.1	8.22	6.71	5.50	4.53	3.75	3.13
Carbon tetrachloride	—	—	14.0	11.9	10.2	8.78	7.70	6.81	6.06	5.43	4.89	4.41	4.00	3.65	3.34	3.07
Chlorobenzene	13.9	12.2	10.7	9.50	8.44	7.53	6.74	6.05	5.46	4.93	4.47	4.07	3.70	3.39	3.10	2.85
m-Cresol	—	—	—	—	261	136	72.0	39.0	21.5	12.1	6.90	—	—	—	—	—
Ethane	0.81	0.70	0.60	0.51	0.42	0.35	—	—	—	—	—	—	—	—	—	—
Ethanol	—	23.3	18.9	15.5	12.8	10.6	8.81	7.39	6.24	5.29	4.51	3.86	3.31	2.86	2.48	2.15
Ethyl acetate	7.21	6.53	5.86	5.28	4.78	4.33	3.95	3.60	3.30	3.03	2.79	2.57	2.38	2.20	2.04	1.90
Ethyl sulfide	7.69	6.68	5.86	5.19	4.64	4.18	3.78	3.45	3.16	2.92	2.70	2.51	2.34	2.20	2.06	1.95
Ethylene	0.582	0.505	0.418	0.31	—	—	—	—	—	—	—	—	—	—	—	—
Formic acid	48.3	37.4	29.5	23.6	19.2	15.9	13.3	11.2	9.56	8.24	7.16	6.27	5.53	4.91	4.39	3.94
Heptane	7.25	6.25	5.46	4.82	4.28	3.83	3.45	3.14	2.87	2.62	2.39	2.20	2.03	1.87	1.72	1.59
Hexane	5.03	4.46	3.97	3.57	3.22	2.93	2.68	2.45	2.24	2.07	1.92	1.78	1.63	1.50	1.38	1.26
Methanol	12.3	10.1	8.43	7.15	6.14	5.32	4.60	4.09	3.60	3.19	2.84	2.53	2.28	2.06	1.85	1.65
Methyl formate	5.71	5.04	4.50	4.04	3.66	3.34	3.06	2.82	2.61	2.43	2.27	2.13	2.01	1.89	1.79	1.70
Octane	10.5	8.80	7.49	6.45	5.66	5.01	4.48	4.03	3.65	3.33	3.05	2.80	2.58	2.38	2.18	2.02
Oil, linseed	2300	1500	1000	700	490	356	263	198	152	118	93.5	74.8	60.6	49.6	41.1	34.3
Oil, olive	8350	4600	2600	1600	1000	630	410	278	193	136	98.3	72.2	54.0	40.9	31.0	24.6
Pentane	3.50	3.16	2.87	2.62	2.39	2.20	2.03	1.87	1.73	1.59	1.46	1.34	1.22	1.11	1.01	0.92
Propane	1.71	1.53	1.37	1.23	1.10	0.98	0.87	0.76	0.66	0.58	0.51	0.45	—	—	—	—
Propanol	67.8	50.0	37.7	29.0	22.7	18.0	14.6	11.9	9.87	8.27	6.99	5.97	5.14	4.46	3.90	3.44
Propylene	1.41	1.27	1.16	1.05	0.94	0.84	0.75	0.67	0.60	0.53	0.47	0.42	—	—	—	—
Sulfuric acid	—	—	—	363	259	189	141	107	82.6	64.7	51.4	41.4	33.7	27.7	23.1	19.4
Sulfur dioxide	5.32	4.53	3.90	3.39	2.85	2.42	2.09	1.83	1.62	1.45	1.30	1.16	1.02	0.89	0.77	0.66
Turpentine	36.3	28.8	23.3	19.1	15.9	13.4	11.4	9.80	8.50	7.44	6.56	5.83	5.21	4.68	4.23	3.85

Table 39.26 Thermochemical Properties at 1.013 bar, 298.15 K

Substance	Formula	ΔH_f° (kJ/kg · mol)	ΔG_f° (kJ/kg · mol)	S° (kJ/kg · mol · K)
Acetaldehyde	$C_2H_4O(g)$	−166,000	−132,900	265.2
Acetic acid	$C_2H_4O_2(g)$	−436,200	+315,500	282.5
Acetone	$C_3H_6O(l)$	−248,000	−155,300	200.2
Acetylene	$C_2H_2(g)$	+266,740	+209,190	200.8
Ammonia	$NH_3(g)$	−46,190	−16,590	192.6
Aniline	$C_6H_7N(l)$	+35,300	+153,200	191.6
Benzene	$C_6H_6(l)$	+49,030	+117,000	172.8
Butanol	$C_4H_{10}O(l)$	−332,400	−168,300	227.6
n-Butane	$C_4H_{10}(l)$	−105,900		
n-Butane	$C_4H_{10}(g)$	−126,200	−17,100	310.1
i-Butane	$C_4H_{10}(g)$	−134,500	−20,900	294.6
Carbon dioxide	$CO_2(g)$	−393,510	−394,390	213.7
Carbon disulfide	$CS_2(g)$	−109,200		237.8
Carbon monoxide	$CO(g)$	−110,520	−137,160	197.6
Carbon tetrachloride	$CCl_4(g)$	−103,000	−66,100	311.3
Carbon tetrafluoride	$CF_4(g)$	−921,300	−878,200	261.5
Chloroform	$CHCl_3(g)$	−104,000	−70,500	295.6
Cyclohexane	$C_6H_{12}(g)$	−123,100	+31,800	298.2
Cyclopropane	$C_3H_6(g)$	+53,300	+104,300	237.7
n-Decane	$C_{10}H_{22}(l)$	−332,600	−17,500	425.5
Diphenyl	$C_{12}H_{10}(g)$	−172,800	−283,900	348.5
Ethane	$C_2H_6(g)$	−84,670	−32,900	229.5
Ethanol	$C_2H_6O(g)$	−235,200	−168,700	282.7
Ethanol	$C_2H_6O(l)$	−277,600	−174,600	160.7
Ethyl acetate	$C_4H_8O_2(g)$	−432,700	−325,800	376.8
Ethyl chloride	$C_2H_5Cl(g)$	−107,600	−55,500	274.8
Ethyl ether	$C_4H_{10}O(g)$	−250,800	−118,400	352.5
Ethylene	$C_2H_4(g)$	+52,280	+68,130	219.4
Ethylene oxide	$C_2H_4O(g)$	−38,500	−11,600	242.9
Heptane	$C_7H_{16}(g)$	−187,800	−427,800	166.0
Hexane	$C_6H_{18}(g)$	−208,400	16,500	270.7
Hydrazine	$N_2H_4(l)$	−50,600	−149,200	121.2
Hydrazine	$N_2H_4(g)$	+95,400	+159,300	238.4
Hydrogen peroxide	$H_2O_2(l)$	−187,500	−120,400	109.6
Methane	$CH_4(g)$	−74,840	−50,790	186.2
Methanol	$CH_4O(l)$	−238,600	−126,800	81.8
Methanol	$CH_4O(g)$	−200,900	−162,100	247.9
Methyl acetate	$C_3H_6O_2(l)$	−444,300		
Methyl bromide	$CH_3Br(g)$	−36,200	−25,900	246.1
Methyl chloride	$CH_3Cl(g)$	−86,300	−63,000	234.2
Methyl formate	$C_2H_4O_2(g)$	−335,100	−301,000	292.8
Methylene chloride	$CH_2Cl_2(g)$	−94,000	−67,000	270.2
Naphthalene	$C_{10}H_8(g)$	−151,500	−224,200	336.5
Nitric oxide	$NO(g)$	+90,300	86,600	210.6
Nitrogen peroxide	$NO_2(g)$	+33,300		240.0
Nitrous oxide	$N_2O(g)$	+82,000	+104,000	219.9
Octane	$C_8H_{18}(l)$	−250,000	6,610	360.8

Table 39.26 (*Continued*)

Substance	Formula	ΔH_f^o (kJ/kg · mol)	ΔG_f^o (kJ/kg · mol)	S^o (kJ/kg · mol · K)
Octane	$C_8H_{18}(g)$	−208,400	16,500	466.7
i-Pentane	$C_5H_{12}(g)$	−154,500	−14,600	343.6
n-Pentane	$C_5H_{12}(g)$	−146,400	−8,370	348.9
Propane	$C_3H_8(g)$	−103,800	−107,200	269.9
Propanol	$C_3H_8(g)$	−258,800	−164,100	322.6
Propylene	$C_3H_6(g)$	+20,400	+62,700	267.0
R11	$CFCl_3(g)$	−284,500	−238,000	309.8
R12	$CCl_2F_2(g)$	−468,600	−439,300	300.9
R13	$CClF_3(g)$	−715,500	−674,900	285.6
R13B1	$CF_3Br(g)$	−642,700	−616,300	297.6
R23	$CHF_3(g)$	−682,000	−654,900	259.6
Sulfur dioxide	$SO_2(g)$	−296,900	−300,200	248.1
Sulfur hexafluoride	$SF_6(g)$	−1,207,900	−1,105,000	291.7
Toluene	$C_7H_8(g)$	−50,000	−122,300	320.2
Water	$H_2O(l)$	−285,830	−237,210	70.0
Water	$H_2O(g)$	−241,820	−228,600	188.7

Table 39.27 Ideal Gas Sensible Enthalpies (kJ/kg · mol) of Common Products of Combustion[a,b]

T (K)	\multicolumn{8}{c}{Substance}							
	CO	CO_2	$H_2O(g)$	NO	NO_2	N_2	O_2	SO_2
200	−2858	−3414	−3280	−2950	−3494	−2855	−2860	−3736
220	−2276	−2757	−2613	−2345	−2803	−2270	−2279	−3008
240	−1692	−2080	−1945	−1749	−2102	−1685	−1698	−2266
260	−1110	−1384	−1277	−1148	−1390	−1105	−1113	−1508
280	−529	−667	−608	−547	−668	−525	−529	−719
300	54	67	63	54	67	55	60	75
320	638	822	736	652	813	636	650	882
340	1221	1594	1411	1248	1573	1217	1238	1702
360	1805	2383	2088	1847	2344	1800	1832	2541
380	2389	3187	2768	2444	3130	2383	2429	3387
400	2975	4008	3452	3042	3929	2967	3029	4251
420	3562	4841	4138	3641	4739	3554	3633	5123
440	4152	5689	4828	4247	5560	4147	4241	6014
460	4642	6550	5522	4849	6395	4737	4853	6923
480	5334	7424	6221	5054	7243	5329	5468	7827
500	5929	8314	6920	6058	8100	5921	6088	8749
550	7427	10580	8695	7589	10295	7401	7655	11180
600	8941	12915	10500	9146	12555	8902	9247	13545
650	10475	15310	12325	10720	14875	10415	10865	16015
700	12020	17760	14185	12310	17250	11945	12500	18550
750	13590	20270	16075	13890	19675	13490	14160	21150
800	15175	22820	17990	15550	22140	15050	15840	23720
850	16780	25410	19945	17200	24640	16630	17535	26390
900	18400	28040	21920	18860	27180	18220	19245	29020
950	20030	30700	23940	20540	29740	19835	20970	31700
1000	21690	33410	25980	22230	32340	21460	22710	34430

Table 39.27　(*Continued*)

T (K)	CO	CO_2	$H_2O(g)$	NO	NO_2	N_2	O_2	SO_2
							Substance	
1050	23350	36140	28060	23940	34950	23100	24460	37180
1100	25030	38890	30170	25650	37610	24750	26180	39920
1150	26720	41680	32310	27380	40260	26430	27990	42690
1200	28430	44480	34480	29120	42950	28110	29770	45460
1250	30140	47130	36680	30870	45650	29810	31560	48270
1300	31870	50160	38900	32630	48350	31510	33350	51070
1350	33600	53030	41130	34400	51090	33220	35160	53900
1400	35340	55910	43450	36170	53810	34940	36970	56720
1450	37090	58810	45770	37950	56560	36670	38790	59560
1500	38850	61710	48100	39730	59310	38410	40610	62400
1550	40610	64800	50460	41520	62070	40160	42440	65260
1600	42380	67580	52840	43320	64850	41910	44280	68120
1650	44156	70530	55240	45120	67640	43670	46120	71000
1700	45940	73490	57680	46930	70420	45440	47970	73870
1750	47727	76460	60130	48740	73220	47210	49830	76760
1800	49520	79440	62610	50560	76010	49000	51690	79640
1850	51320	82430	65100	52380	78810	50780	53560	82540
1900	53120	85430	67610	54200	81630	52570	55440	85440
1950	54930	88440	70140	56020	84450	54370	57310	88350
2000	56740	91450	72690	57860	87260	56160	59200	91250
2100	60380	97500	77830	61530	92910	59760	62990	97080
2200	64020	103570	83040	65220	98580	63380	66800	102930
2300	67680	109670	88290	68910	104260	67010	70630	108790
2400	71350	115790	93600	72610	109950	70660	74490	114670
2500	75020	121930	98960	76320	115650	74320	78370	120560
2600	78710	128080	104370	80040	121360	77990	82270	126460
2700	82410	134260	109810	83760	127080	81660	86200	132380
2800	86120	140440	115290	87490	132800	85360	90140	138300
2900	89830	146650	120810	91230	138540	89050	94110	144240
3000	93540	152860	126360	94980	144270	92750	98100	150180

[a]Converted and usually rounded off from *JANAF Thermochemical Tables,* NSRDS-NBS-37, 1971.

[b]To illustrate the term *sensible enthalpy,* which is the difference between the actual enthalpy and the enthalpy at the reference temperature, 298.15 K (= 25°C = 77°F = 537°R), the magnitude of the heat transfer, in kJ/kg · mol fuel and in kJ/kg fuel, will be calculated for the steady-state combustion of acetylene in excess oxygen, the reactants entering at 298.15 K and the products leaving at 2000 K. All substances are in the gaseous phase.

The basic equation is

$$Q + W = \sum_P n_i \, (\Delta h_f^\circ + \Delta h_s)_i - \sum_R n_i (\Delta h_f^\circ + \Delta h_s)_i$$

where P signifies products and R reactants, s signifies sensible enthalpy, and the Δh_s are looked up in the table for the appropriate temperatures.

If the actual reaction was

$$C_2H_2 + 1\frac{1}{2}O_2 \rightarrow 2CO_2 + H_2O + 3O_2$$

then $W = 0$ and $Q = 2\,(-393{,}510 + 91{,}450) + 1(-241{,}810 + 72{,}690) + 3(0 + 59{,}200) - (226{,}740 + 0) - 1\frac{1}{2}(0 + 0) = -604{,}120 + (-169{,}120) + 177{,}600 - 226{,}740 = -822{,}380$ kJ/mg mol. C_2H_2 $= -31{,}584$ kJ/kg C_2H_2. Had the fuel been burnt in air one would write the equation with an additional 3.76(5.5) N_2 on each side of the equation. In the above, the enthalpy of formation of the stable elements at 298.15 K has been set equal to zero. For further information, most undergraduate engineering thermodynamics texts may be consulted.

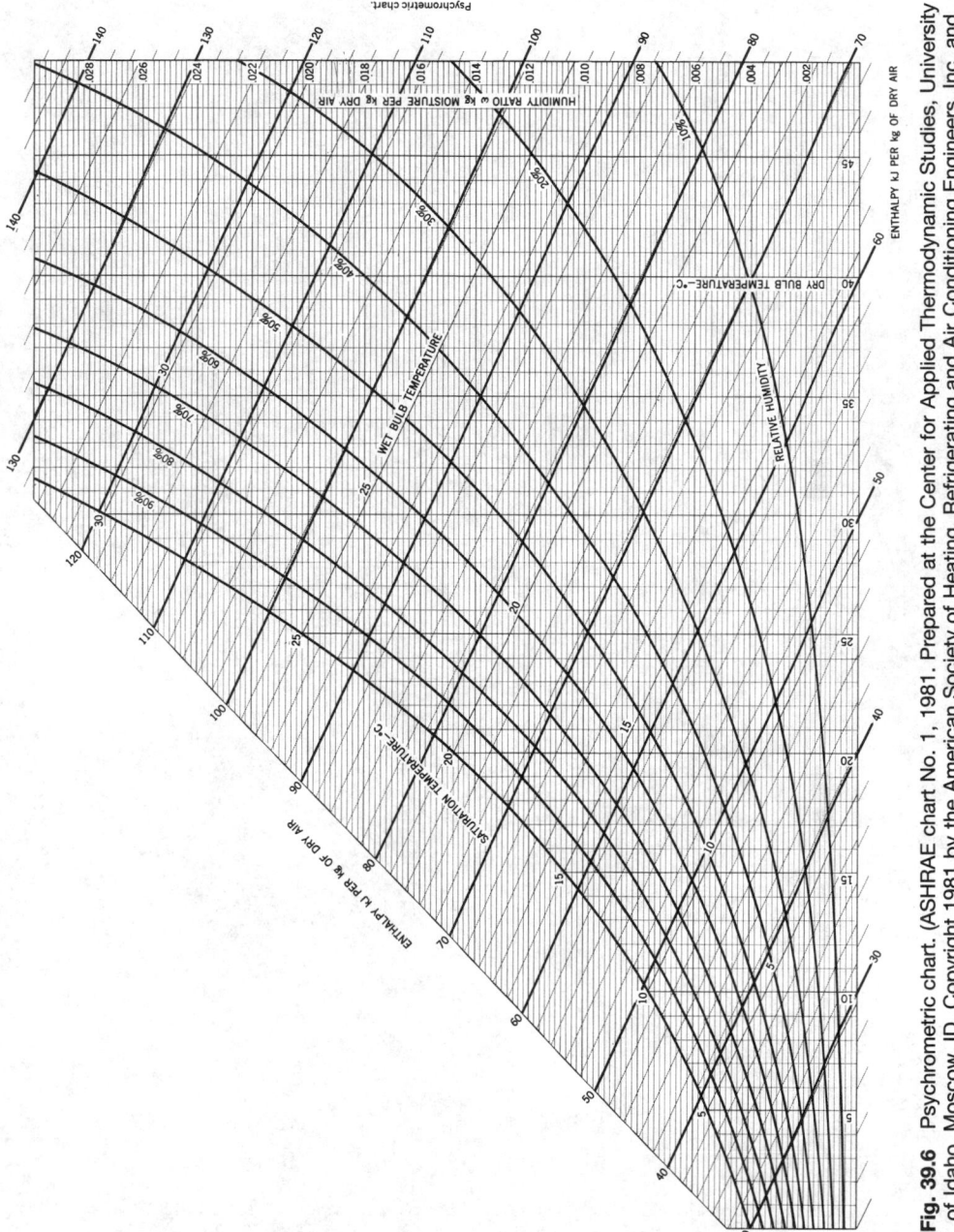

Fig. 39.6 Psychrometric chart. (ASHRAE chart No. 1, 1981. Prepared at the Center for Applied Thermodynamic Studies, University of Idaho, Moscow, ID. Copyright 1981 by the American Society of Heating, Refrigerating and Air Conditioning Engineers, Inc. and reproduced by permission of the copyright owner.)

CHAPTER 40

FLUID MECHANICS

Reuben M. Olson
College of Engineering and Technology
Ohio University
Athens, Ohio

40.1	**DEFINITION OF A FLUID**	**1290**
40.2	**IMPORTANT FLUID PROPERTIES**	**1290**
40.3	**FLUID STATICS**	**1290**
	40.3.1 Manometers	1291
	40.3.2 Liquid Forces on Submerged Surfaces	1291
	40.3.3 Aerostatics	1293
	40.3.4 Static Stability	1293
40.4	**FLUID KINEMATICS**	**1294**
	40.4.1 Velocity and Acceleration	1295
	40.4.2 Streamlines	1295
	40.4.3 Deformation of a Fluid Element	1295
	40.4.4 Vorticity and Circulation	1297
	40.4.5 Continuity Equations	1298
40.5	**FLUID MOMENTUM**	**1298**
	40.5.1 The Momentum Theorem	1299
	40.5.2 Equations of Motion	1300
40.6	**FLUID ENERGY**	**1301**
	40.6.1 Energy Equations	1301
	40.6.2 Work and Power	1302
	40.6.3 Viscous Dissipation	1302
40.7	**CONTRACTION COEFFICIENTS FROM POTENTIAL FLOW THEORY**	**1303**
40.8	**DIMENSIONLESS NUMBERS AND DYNAMIC SIMILARITY**	**1304**
	40.8.1 Dimensionless Numbers	1304
	40.8.2 Dynamic Similitude	1305
40.9	**VISCOUS FLOW AND INCOMPRESSIBLE BOUNDARY LAYERS**	**1307**
	40.9.1 Laminar and Turbulent Flow	1307
	40.9.2 Boundary Layers	1307
40.10	**GAS DYNAMICS**	**1310**
	40.10.1 Adiabatic and Isentropic Flow	1310
	40.10.2 Duct Flow	1311
	40.10.3 Normal Shocks	1311
	40.10.4 Oblique Shocks	1313
40.11	**VISCOUS FLUID FLOW IN DUCTS**	**1313**
	40.11.1 Fully Developed Incompressible Flow	1315
	40.11.2 Fully Developed Laminar Flow in Ducts	1315
	40.11.3 Fully Developed Turbulent Flow in Ducts	1316
	40.11.4 Steady Incompressible Flow in Entrances of Ducts	1319
	40.11.5 Local Losses in Contractions, Expansions, and Pipe Fittings; Turbulent Flow	1319
	40.11.6 Flow of Compressible Gases in Pipes with Friction	1320
40.12	**DYNAMIC DRAG AND LIFT**	**1323**
	40.12.1 Drag	1323
	40.12.2 Lift	1323
40.13	**FLOW MEASUREMENTS**	**1324**
	40.13.1 Pressure Measurements	1324
	40.13.2 Velocity Measurements	1325
	40.13.3 Volumetric and Mass Flow Fluid Measurements	1326

All figures and tables produced, with permission, from *Essentials of Engineering Fluid Mechanics,* Fourth Edition, by Reuben M. Olson, copyright 1980, Harper & Row, Publishers.

Mechanical Engineers' Handbook, 2nd ed., Edited by Myer Kutz.
ISBN 0-471-13007-9 © 1998 John Wiley & Sons, Inc.

40.1 DEFINITION OF A FLUID

A solid generally has a definite shape; a fluid has a shape determined by its container. Fluids include liquids, gases, and vapors, or mixtures of these. A fluid continuously deforms when shear stresses are present; it cannot sustain shear stresses at rest. This is characteristic of all real fluids, which are viscous. Ideal fluids are nonviscous (and nonexistent), but have been studied in great detail because in many instances viscous effects in real fluids are very small and the fluid acts essentially as a nonviscous fluid. Shear stresses are set up as a result of relative motion between a fluid and its boundaries or between adjacent layers of fluid.

40.2 IMPORTANT FLUID PROPERTIES

Density ρ and surface tension σ are the most important fluid properties for liquids at rest. Density and viscosity μ are significant for all fluids in motion; surface tension and vapor pressure are significant for cavitating liquids; and bulk elastic modulus K is significant for compressible gases at high subsonic, sonic, and supersonic speeds.

Sonic speed in fluids is $c = \sqrt{K/\rho}$. Thus, for water at 15°C, $c = \sqrt{2.18 \times 10^9/999} = 1480$ m/sec. For a mixture of a liquid and gas bubbles at nonresonant frequencies, $c_m = \sqrt{K_m/\rho_m}$, where m refers to the mixture. This becomes

$$c_m = \sqrt{\frac{p_g K_l}{[xK_l + (1-x)p_g][x\rho_g + (1-x)\rho_l]}}$$

where the subscript l is for the liquid phase and g is for the gas phase. Thus, for water at 20°C containing 0.1% gas nuclei by volume at atmospheric pressure, $c_m = 312$ m/sec. For a gas or a mixture of gases (such as air), $c = \sqrt{kRT}$, where $k = c_p/c_v$, R is the gas constant, and T is the absolute temperature. For air at 15°C, $c = \sqrt{(1.4)(287.1)(288)} = 340$ m/sec. This sonic property is thus a combination of two properties, density and elastic modulus.

Kinematic viscosity is the ratio of dynamic viscosity and density. In a Newtonian fluid, simple laminar flow in a direction x at a speed of u, the shearing stress parallel to x is $\tau_L = \mu(du/dy) = \rho\nu(du/dy)$, the product of dynamic viscosity and velocity gradient. In the more general case, $\tau_L = \mu(\partial u/\partial y + \partial v/\partial x)$ when there is also a y component of velocity v. In turbulent flows the shear stress resulting from lateral mixing is $\tau_T = -\rho\overline{u'v'}$, a Reynolds stress, where u' and v' are instantaneous and simultaneous departures from mean values $\overline{u}$ and $\overline{v}$. This is also written as $\tau_T = \rho\epsilon(du/dy)$, where ϵ is called the turbulent eddy viscosity or diffusivity, an indirectly measurable flow parameter and not a fluid property. The eddy viscosity may be orders of magnitude larger than the kinematic viscosity. The total shear stress in a turbulent flow is the sum of that from laminar and from turbulent motion: $\tau = \tau_L + \tau_T = \rho(\nu + \epsilon)du/dy$ after Boussinesq.

40.3 FLUID STATICS

The differential equation relating pressure changes dp with elevation changes dz (positive upward parallel to gravity) is $dp = -\rho g\,dz$. For a constant-density liquid, this integrates to $p_2 - p_1 = -\rho g (z_2 - z_1)$ or $\Delta p = \gamma h$, where γ is in N/m³ and h is in m. Also $(p_1/\gamma) + z_1 = (p_2/\gamma) + z_2$; a constant piezometric head exists in a homogeneous liquid at rest, and since $p_1/\gamma - p_2/\gamma = z_2 - z_1$, a change in pressure head equals the change in potential head. Thus, horizontal planes are at constant pressure when body forces due to gravity act. If body forces are due to uniform linear accelerations or to centrifugal effects in rigid-body rotations, points equidistant below the free liquid surface are all at the same pressure. Dashed lines in Figs. 40.1 and 40.2 are lines of constant pressure.

Pressure differences are the same whether all pressures are expressed as gage pressure or as absolute pressure.

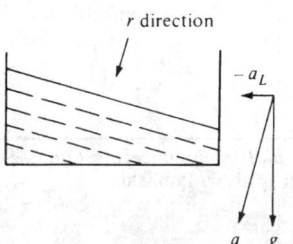

Fig. 40.1 Constant linear acceleration.

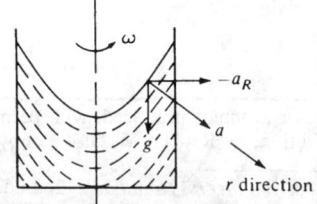

Fig. 40.2 Constant centrifugal acceleration.

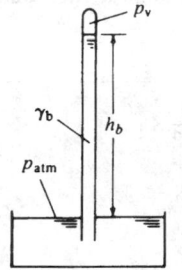

Fig. 40.3 Barometer.

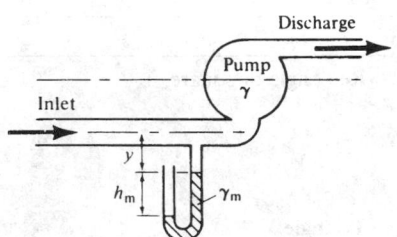

Fig. 40.4 Open manometer.

40.3.1 Manometers

Pressure differences measured by barometers and manometers may be determined from the relation $\Delta p = \gamma h$. In a barometer, Fig. 40.3, $h_b = (p_a - p_v)/\gamma_b$ m.

An open manometer, Fig. 40.4, indicates the inlet pressure for a pump by $p_{inlet} = -\gamma_m h_m - \gamma y$ Pa gage. A differential manometer, Fig. 40.5, indicates the pressure drop across an orifice, for example, by $p_1 - p_2 = h_m(\gamma_m - \gamma_0)$ Pa.

Manometers shown in Figs. 40.3 and 40.4 are a type used to measure medium or large pressure differences with relatively small manometer deflections. Micromanometers can be designed to produce relatively large manometer deflections for very small pressure differences. The relation $\Delta p = \gamma \Delta h$ may be applied to the many commercial instruments available to obtain pressure differences from the manometer deflections.

40.3.2 Liquid Forces on Submerged Surfaces

The liquid force on any flat surface submerged in the liquid equals the product of the gage pressure at the centroid of the surface and the surface area, or $F = \bar{p}A$. The force F is not applied at the centroid for an inclined surface, but is always below it by an amount that diminishes with depth. Measured parallel to the inclined surface, $\bar{y}$ is the distance from 0 in Fig. 40.6 to the centroid and $y_F = \bar{y} + I_{CG}/A\bar{y}$, where I_{CG} is the moment of inertia of the flat surface with respect to its centroid. Values for some surfaces are listed in Table 40.1.

For curved surfaces, the horizontal component of the force is equal in magnitude and point of application to the force on a projection of the curved surface on a vertical plane, determined as above. The vertical component of force equals the weight of liquid above the curved surface and is applied at the centroid of this liquid, as in Fig. 40.7. The liquid forces on opposite sides of a submerged surface are equal in magnitude but opposite in direction. These statements for curved surfaces are also valid for flat surfaces.

Buoyancy is the resultant of the surface forces on a submerged body and equals the weight of fluid (liquid or gas) displaced.

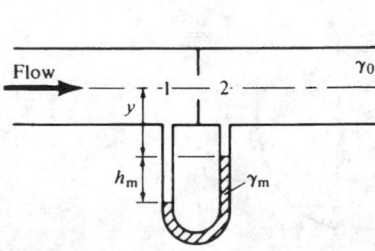

Fig. 40.5 Differential manometer.

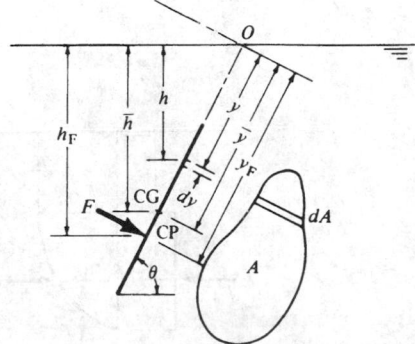

Fig. 40.6 Flat inclined surface submerged in a liquid.

Table 40.1   Moments of Inertia for Various Plane Surfaces about Their Center of Gravity

Surface		I_{CG}
Rectangle or square		$\dfrac{1}{12}Ah^2$
Triangle		$\dfrac{1}{18}Ah^2$
Quadrant of circle (or semicircle)		$\left(\dfrac{1}{4} - \dfrac{16}{9\pi^2}\right)Ar^2 = 0.0699\,Ar^2$
Quadrant of ellipse (or semiellipse)		$\left(\dfrac{1}{4} - \dfrac{16}{9\pi^2}\right)Aa^2 = 0.0699\,Aa^2$
Parabola		$\left(\dfrac{3}{7} - \dfrac{9}{25}\right)Ah^2 = 0.0686\,Ah^2$
Circle		$\dfrac{1}{16}Ad^2$
Ellipse		$\dfrac{1}{16}Ah^2$

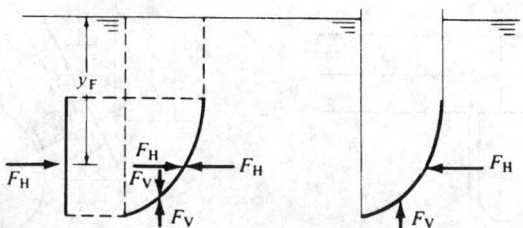

Fig. 40.7 Curved surfaces submerged in a liquid.

40.3.3 Aerostatics

The U.S. standard atmosphere is considered to be dry air and to be a perfect gas. It is defined in terms of the temperature variation with altitude (Fig. 40.8), and consists of isothermal regions and polytropic regions in which the polytropic exponent n depends on the lapse rate (temperature gradient).

Conditions at an upper altitude z_2 and at a lower one z_1 in an isothermal atmosphere are obtained by integrating the expression $dp = -\rho g \, dz$ to get

$$\frac{p_2}{p_1} = \exp \frac{-g(z_2 - z_1)}{RT}$$

In a polytropic atmosphere where $p/p_1 = (\rho/\rho_1)^n$,

$$\frac{p_2}{p_1} = \left[1 - g \frac{(n-1)}{n} \frac{(z_2 - z_1)}{RT_1} \right]^{n/(n-1)}$$

from which the lapse rate is $(T_2 - T_1)/(z_2 - z_1) = -g(n-1)/nR$ and thus n is obtained from $1/n = 1 + (R/g)(dt/dz)$. Defining properties of the U.S. standard atmosphere are listed in Table 40.2.

The U.S. standard atmosphere is used in measuring altitudes with altimeters (pressure gages) and, because the altimeters themselves do not account for variations in the air temperature beneath an aircraft, they read too high in cold weather and too low in warm weather.

40.3.4 Static Stability

For the *atmosphere* at rest, if an air mass moves very slowly vertically and remains there, the atmosphere is neutral. If vertical motion continues, it is unstable; if the air mass moves to return to its initial position, it is stable. It can be shown that atmospheric stability may be defined in terms of the polytropic exponent. If $n < k$, the atmosphere is stable (see Table 40.2); if $n = k$, it is neutral (adiabatic); and if $n > k$, it is unstable.

The stability of a body *submerged* in a fluid at rest depends on its response to forces which tend to tip it. If it returns to its original position, it is stable; if it continues to tip, it is unstable; and if it remains at rest in its tipped position, it is neutral. In Fig. 40.9 G is the center of gravity and B is the center of buoyancy. If the body in (a) is tipped to the position in (b), a couple Wd restores the body toward position (a) and thus the body is stable. If B were below G and the body displaced, it would move until B becomes above G. Thus stability requires that G is below B.

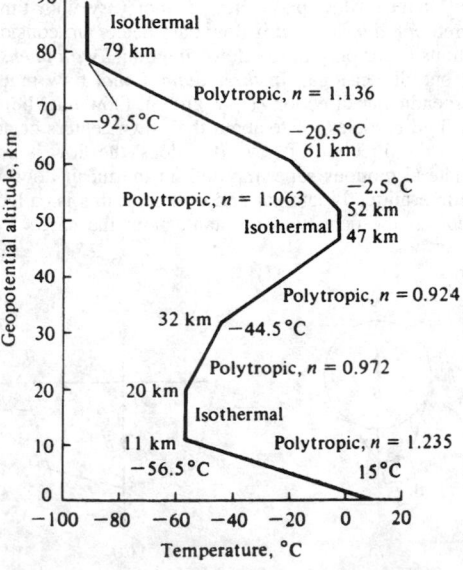

Fig. 40.8 U.S. standard atmosphere.

Table 40.2 Defining Properties of the U.S. Standard Atmosphere

Altitude (m)	Temperature (°C)	Type of Atmosphere	Lapse Rate (°C/km)	$\bar{g}$ (m/s²)	n	Pressure p(Pa)	Density ρ(kg/m³)
0	15.0					1.013×10^5	1.225
		Polytropic	−6.5	9.790	1.235		
11,000	−56.5					2.263×10^4	3.639×10^{-1}
		Isothermal	0.0	9.759			
20,000	−56.5					5.475×10^3	8.804×10^{-2}
		Polytropic	+1.0	9.727	0.972		
32,000	−44.5					8.680×10^2	1.323×10^{-2}
		Polytropic	+2.8	9.685	0.924		
47,000	−2.5					1.109×10^2	1.427×10^{-3}
		Isothermal	0.0	9.654			
52,000	−2.5					5.900×10^1	7.594×10^{-4}
		Polytropic	−2.0	9.633	1.063		
61,000	−20.5					1.821×10^1	2.511×10^{-4}
		Polytropic	−4.0	9.592	1.136		
79,000	−92.5					1.038	2.001×10^{-5}
		Isothermal	0.0	9.549			
88,743	−92.5					1.644×10^{-1}	3.170×10^{-6}

Floating bodies may be stable even though the center of buoyancy B is below the center of gravity G. The center of buoyancy generally changes position when a floating body tips because of the changing shape of the displaced liquid. The floating body is in equilibrium in Fig. 40.10a. In Fig. 40.10b the center of buoyancy is at B_1, and the restoring couple rotates the body toward its initial position in Fig. 40.10a. The intersection of BG is extended and a vertical line through B_1 is at M, the metacenter, and GM is the metacentric height. The body is stable if M is above G. Thus, the position of B relative to G determines stability of a submerged body, and the position of M relative to G determines the stability of floating bodies.

40.4 FLUID KINEMATICS

Fluid flows are classified in many ways. Flow is *steady* if conditions at a point do not vary with time, or for turbulent flow, if mean flow parameters do not vary with time. Otherwise the flow is *unsteady.* Flow is considered *one dimensional* if flow parameters are considered constant throughout a cross section, and variations occur only in the flow direction. *Two-dimensional* flow is the same in parallel planes and is not one dimensional. In *three-dimensional* flow gradients of flow parameters exist in three mutually perpendicular directions (x, y, and z). Flow may be *rotational* or *irrotational,* depending on whether the fluid particles rotate about their own centers or not. Flow is *uniform* if the velocity does not change in the direction of flow. If it does, the flow is *nonuniform. Laminar* flow exists when there are no lateral motions superimposed on the mean flow. When there are, the flow is *turbulent.* Flow may be intermittently laminar and turbulent; this is called flow in *transition.* Flow is considered *incompressible* if the density is constant, or in the case of gas flows, if the density

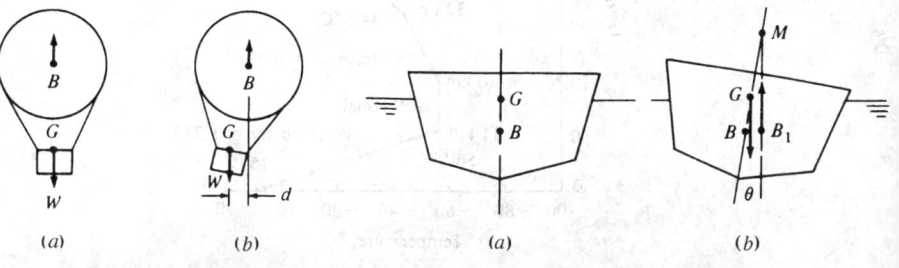

(a)   (b)   (a)   (b)

Fig. 40.9 Stability of a submerged body.   **Fig. 40.10** Floating body.

variation is below a specified amount throughout the flow, 2–3%, for example. Low-speed gas flows may be considered essentially incompressible. Gas flows may be considered as *subsonic, transonic, sonic, supersonic,* or *hypersonic* depending on the gas speed compared with the speed of sound in the gas. Open-channel water flows may be designated as *subcritical, critical,* or *supercritical* depending on whether the flow is less than, equal to, or greater than the speed of an elementary surface wave.

40.4.1   Velocity and Acceleration

In Cartesian coordinates, velocity components are u, v, and w in the x, y, and z directions, respectively. These may vary with position and time, such that, for example, $u = dx/dt = u(x, y, z, t)$. Then

$$du = \frac{\partial u}{\partial x} dx + \frac{\partial u}{\partial y} dy + \frac{\partial u}{\partial z} dz + \frac{\partial u}{\partial t} dt$$

and

$$a_x = \frac{du}{dt} = \frac{\partial u}{\partial x}\frac{dx}{dt} + \frac{\partial u}{\partial y}\frac{dy}{dt} + \frac{\partial u}{\partial z}\frac{dz}{dt} + \frac{\partial u}{\partial t}$$

$$= \frac{Du}{Dt} = u\frac{\partial u}{\partial x} + v\frac{\partial u}{\partial y} + w\frac{\partial u}{\partial z} + \frac{\partial u}{\partial t}$$

The first three terms on the right hand side are the *convective* acceleration, which is zero for uniform flow, and the last term is the *local* acceleration, which is zero for steady flow.

In natural coordinates (streamline direction s, normal direction n, and meridional direction m normal to the plane of s and n), the velocity V is always in the streamline direction. Thus, $V = V(s,t)$ and

$$dV = \frac{\partial V}{\partial s} ds + \frac{\partial V}{\partial t} dt$$

$$a_s = \frac{dV}{dt} = V\frac{\partial V}{\partial s} + \frac{\partial V}{\partial t}$$

where the first term on the right-hand side is the *convective* acceleration and the last is the *local* acceleration. Thus, if the fluid velocity changes as the fluid moves throughout space, there is a convective acceleration, and if the velocity at a point changes with time, there is a local acceleration.

40.4.2   Streamlines

A *streamline* is a line to which, at each instant, velocity vectors are tangent. A *pathline* is the path of a particle as it moves in the fluid, and for steady flow it coincides with a streamline.

The equations of streamlines are described by stream functions ψ, from which the velocity components in two-dimensional flow are $u = -\partial\psi/\partial y$ and $v = +\partial\psi/\partial x$. Streamlines are lines of constant stream function. In polar coordinates

$$v_r = -\frac{1}{r}\frac{\partial\psi}{\partial\theta} \quad \text{and} \quad v_\theta = +\frac{\partial\psi}{\partial r}$$

Some streamline patterns are shown in Figs. 40.11, 40.12, and 40.13. The lines at right angles to the streamlines are potential lines.

40.4.3   Deformation of a Fluid Element

Four types of deformation or movement may occur as a result of spatial variations of velocity: translation, linear deformation, angular deformation, and rotation. These may occur singly or in combination. Motion of the face (in the x-y plane) of an elemental cube of sides δx, δy, and δz in a time dt is shown in Fig. 40.14. Both translation and rotation involve motion or deformation without a change in shape of the fluid element. Linear and angular deformations, however, do involve a change in shape of the fluid element. Only through these linear and angular deformations are heat generated and mechanical energy dissipated as a result of viscous action in a fluid.

For linear deformation the relative change in volume is at a rate of

$$(\Psi_{dt} - \Psi_0)/\Psi_0 = \frac{\partial u}{\partial x} + \frac{\partial v}{\partial y} + \frac{\partial w}{\partial z} = \text{div } V$$

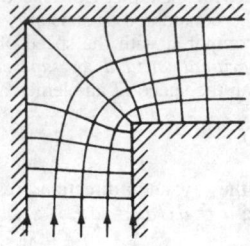

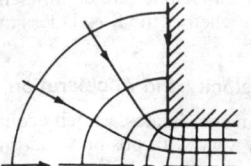

Fig. 40.11 Flow around a corner in a duct.   **Fig. 40.12** Flow around a corner into a duct.

which is zero for an incompressible fluid, and thus is an expression for the continuity equation. Rotation of the face of the cube shown in Fig. 40.14d is the average of the rotations of the bottom and left edges, which is

$$\frac{1}{2} \left(\frac{\partial v}{\partial x} - \frac{\partial u}{\partial y} \right) dt$$

The rate of rotation is the angular velocity and is

$$\omega_z = \frac{1}{2} \left(\frac{\partial v}{\partial x} - \frac{\partial u}{\partial y} \right) \quad \text{about the } z \text{ axis in the } x\text{-}y \text{ plane}$$

$$\omega_x = \frac{1}{2} \left(\frac{\partial w}{\partial y} - \frac{\partial v}{\partial z} \right) \quad \text{about the } x \text{ axis in the } y\text{-}z \text{ plane}$$

and

$$\omega_y = \frac{1}{2} \left(\frac{\partial u}{\partial z} - \frac{\partial w}{\partial x} \right) \quad \text{about the } y \text{ axis in the } x\text{-}z \text{ plane}$$

These are the components of the angular velocity vector Ω,

$$\Omega = \tfrac{1}{2} \text{ curl } V = \frac{1}{2} \begin{vmatrix} \mathbf{i} & \mathbf{j} & \mathbf{k} \\ \dfrac{\partial}{\partial x} & \dfrac{\partial}{\partial y} & \dfrac{\partial}{\partial z} \\ u & v & w \end{vmatrix} = \omega_x \mathbf{i} + \omega_y \mathbf{j} + \omega_z \mathbf{k}$$

If the flow is irrotational, these quantities are zero.

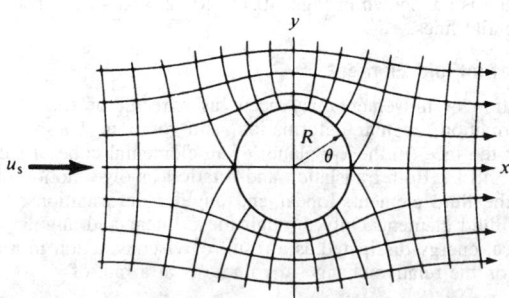

Fig. 40.13 Inviscid flow past a cylinder.

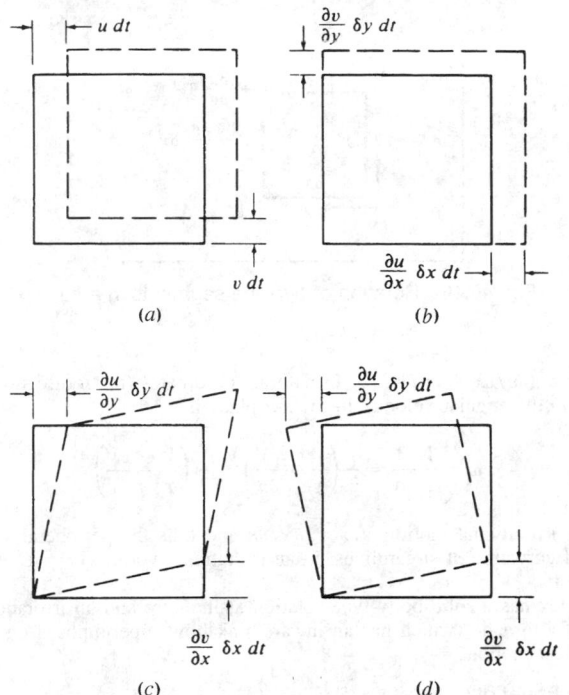

Fig. 40.14 Movements of the face of an elemental cube in the *x-y* plane: (*a*) translation; (*b*) linear deformation; (*c*) angular deformation; (*d*) rotation.

40.4.4 Vorticity and Circulation

Vorticity is defined as twice the angular velocity, and thus is also zero for irrotational flow. Circulation is defined as the line integral of the velocity component along a closed curve and equals the total strength of all vertex filaments that pass through the curve. Thus, the vorticity at a point within the curve is the circulation per unit area enclosed by the curve. These statements are expressed by

$$\Gamma = \oint \mathbf{V} \cdot d\mathbf{l} = \oint (u\,dx + v\,dy + w\,dz) \qquad \text{and} \qquad \zeta_A = \lim_{A \to 0} \frac{\Gamma}{A}$$

Circulation—the product of vorticity and area—is the counterpart of volumetric flow rate as the product of velocity and area. These are shown in Fig. 40.15.

Physically, fluid rotation at a point in a fluid is the instantaneous average rotation of two mutually perpendicular infinitesimal line segments. In Fig. 40.16 the line δx rotates positively and δy rotates

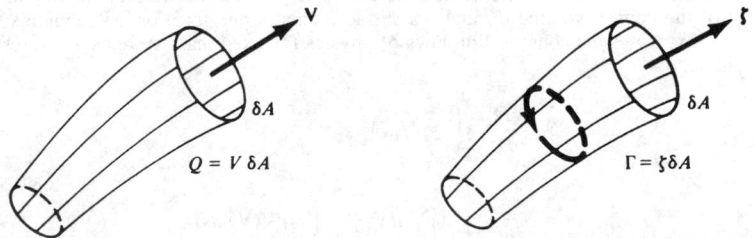

Fig. 40.15 Similarity between a stream filament and a vortex filament.

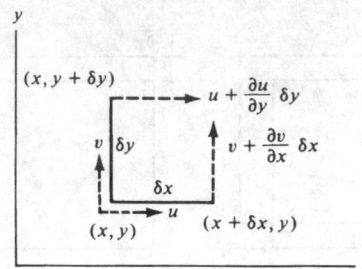

Fig. 40.16   Rotation of two line segments in a fluid.

negatively. Then $\omega_x = (\partial v/\partial x - \partial u/\partial y)/2$. In natural coordinates (the n direction is opposite to the radius of curvature r) the angular velocity in the s-n plane is

$$\omega = \frac{1}{2}\frac{\Gamma}{\delta A} = \frac{1}{2}\left(\frac{V}{r} - \frac{\partial V}{\partial n}\right) = \frac{1}{2}\left(\frac{V}{r} + \frac{\partial V}{\partial r}\right)$$

This shows that for irrotational motion $V/r = \partial V/\partial n$ and thus the peripheral velocity V increases toward the center of curvature of streamlines. In an irrotational vortex, $Vr = C$ and in a solid-body-type or rotational vortex, $V = \omega r$.

A combined vortex has a solid-body-type rotation at the core and an irrotational vortex beyond it. This is typical of a tornado (which has an inward sink flow superimposed on the vortex motion) and eddies in turbulent motion.

40.4.5   Continuity Equations

Conservation of mass for a fluid requires that in a *material* volume, the mass remains constant. In a *control* volume the net rate of influx of mass into the control volume is equal to the rate of change of mass in the control volume. Fluid may flow into a control volume either through the control surface or from internal sources. Likewise, fluid may flow out through the control surface or into internal sinks. The various forms of the continuity equations listed in Table 40.3 do not include sources and sinks; if they exist, they must be included.

The most commonly used forms for duct flow are $\dot{m} = VA\rho$ in kg/sec where V is the average flow velocity in m/sec, A is the duct area in m³, and ρ is the fluid density in kg/m³. In differential form this is $dV/V + dA/A + d\rho/\rho = 0$, which indicates that all three quantities may not increase nor all decrease in the direction of flow. For incompressible duct flow $Q = VA$ m³/sec where V and A are as above. When the velocity varies throughout a cross section, the average velocity is

$$V = \frac{1}{A}\int u\,dA = \frac{1}{n}\sum_{i=1}^{n} u_i$$

where u is a velocity at a point, and u_i are point velocities measured at the centroid of n equal areas. For example, if the velocity is u at a distance y from the wall of a pipe of radius R and the centerline velocity is u_m, $u = u_m(y/R)^{1/7}$ and the average velocity is $V = {}^{49}\!/_{60}\, u_m$.

40.5   FLUID MOMENTUM

The momentum theorem states that the net external force acting on the fluid within a control volume equals the time rate of change of momentum of the fluid plus the net rate of momentum flux or transport out of the control volume through its surface. This is one form of the Reynolds transport theorem, which expresses the conservation laws of physics for fixed mass systems to expressions for a control volume:

$$\Sigma\mathbf{F} = \frac{D}{Dt}\int\limits_{\substack{\text{material}\\\text{volume}}} \rho\mathbf{V}\,d\Psi$$

$$= \frac{\partial}{\partial t}\int\limits_{\substack{\text{control}\\\text{volume}}} \rho\mathbf{V}\,d\Psi + \int\limits_{\substack{\text{control}\\\text{surface}}} \rho\mathbf{V}(\mathbf{V}\cdot d\mathbf{s})$$

Table 40.3 Continuity Equations

General	$\dfrac{\partial \rho}{\partial t} + \nabla \cdot \rho \mathbf{V} = 0$ or $\dfrac{D\rho}{Dt} + \rho \nabla \cdot \mathbf{V} = 0$	Vector
Unsteady, compressible	$\dfrac{\partial \rho}{\partial t} + \dfrac{\partial(\rho u)}{\partial x} + \dfrac{\partial(\rho v)}{\partial y} + \dfrac{\partial(\rho w)}{\partial z} = 0$	Cartesian
	$\dfrac{\partial \rho}{\partial t} + \dfrac{\partial(\rho v_r)}{\partial r} + \dfrac{1}{r}\dfrac{\partial(\rho v_\theta)}{\partial \theta} + \dfrac{\partial(\rho v_z)}{\partial z} + \dfrac{\rho v_r}{r} = 0$	Cylindrical
	$\dfrac{\partial(\rho A)}{\partial t} + \dfrac{\partial}{\partial s}(\rho \mathbf{V} \cdot \mathbf{A}) = 0$	Duct
Steady, compressible	$\nabla \cdot \rho \mathbf{V} = 0$	Vector
	$\dfrac{\partial(\rho u)}{\partial x} + \dfrac{\partial(\rho v)}{\partial y} + \dfrac{\partial(\rho w)}{\partial z} = 0$	Cartesian
	$\dfrac{\partial(\rho v_r)}{\partial r} + \dfrac{1}{r}\dfrac{\partial(\rho v_\theta)}{\partial \theta} + \dfrac{\partial(\rho v_z)}{\partial z} + \dfrac{\rho v_r}{r} = 0$	Cylindrical
	$\rho \mathbf{V} \cdot \mathbf{A} = \dot{m}$	
Incompressible	$\nabla \cdot \mathbf{V} = 0$	Vector
Steady or unsteady	$\dfrac{\partial u}{\partial x} + \dfrac{\partial v}{\partial y} + \dfrac{\partial w}{\partial z} = 0$	Cartesian
	$\dfrac{\partial v_r}{\partial r} + \dfrac{1}{r}\dfrac{\partial v_\theta}{\partial \theta} + \dfrac{\partial v_z}{\partial z} + \dfrac{v_r}{r} = 0$	Cylindrical
	$\mathbf{V} \cdot \mathbf{A} = Q$	Duct

40.5.1 The Momentum Theorem

For steady flow the first term on the right-hand side of the preceding equation is zero. Forces include normal forces due to pressure and tangential forces due to viscous shear over the surface S of the control volume, and body forces due to gravity and centrifugal effects, for example. In scalar form the net force equals the total momentum flux leaving the control volume minus the total momentum flux entering the control volume. In the x direction

$$\Sigma F_x = (\dot{m}V_x)_{\text{leaving } S} - (\dot{m}V_x)_{\text{entering } S}$$

or when the same fluid enters and leaves,

$$\Sigma F_x = \dot{m}(V_{x\,\text{leaving } S} - V_{x\,\text{entering } S})$$

with similar expressions for the y and z directions.

For one-dimensional flow $\dot{m}V_x$ represents momentum flux passing a section and V_x is the average velocity. If the velocity varies across a duct section, the true momentum flux is $\int_A (u\rho dA)u$, and the ratio of this value to that based upon average velocity is the momentum correction factor β,

$$\beta = \frac{\displaystyle\int_A u^2\,dA}{V^2 A} \geq 1$$

$$\approx \frac{1}{V^2 n}\sum_{i=1}^{n} u_i^2$$

For laminar flow in a circular tube, $\beta = \frac{1}{3}$; for laminar flow between parallel plates, $\beta = 1.20$; and for turbulent flow in a circular tube, β is about $1.02 - 1.03$.

40.5.2 Equations of Motion

For steady irrotational flow of an incompressible nonviscous fluid, Newton's second law gives the Euler equation of motion. Along a streamline it is

$$V\frac{\partial V}{\partial s} + \frac{1}{\rho}\frac{\partial p}{\partial s} + g\frac{\partial z}{\partial s} = 0$$

and normal to a streamline it is

$$\frac{V^2}{r} + \frac{1}{\rho}\frac{\partial p}{\partial n} + g\frac{\partial z}{\partial n} = 0$$

When integrated, these show that the sum of the kinetic, displacement, and potential energies is a constant along streamlines as well as across streamlines. The result is known as the Bernoulli equation:

$$\frac{V^2}{2} + \frac{p}{\rho} + gz = \text{constant energy per unit mass}$$

$$\frac{\rho V_1^2}{2} + p_1 + \rho g z_1 = \frac{\rho V_2^2}{2} + p_2 + \rho g z_2 = \text{constant total pressure}$$

and

$$\frac{V_1^2}{2g} + \frac{p_1}{g\rho} + z_1 = \frac{V_2^2}{2g} + \frac{p_2}{g\rho} + z_2 = \text{constant total head}$$

For a reversible adiabatic compressible gas flow with no external work, the Euler equation integrates to

$$\frac{V_1^2}{2} + \frac{k}{k-1}\left(\frac{p_1}{\rho_1}\right) + gz_1 = \frac{V_2^2}{2} + \frac{k}{k-1}\left(\frac{p_2}{\rho_2}\right) + gz_2$$

which is valid whether the flow is reversible or not, and corresponds to the steady-flow energy equation for adiabatic no-work gas flow.

Newton's second law written normal to streamlines shows that in horizontal planes $dp/dr = \rho V^2/r$, and thus dp/dr is positive for both rotational and irrotational flow. The pressure increases away from the center of curvature and decreases toward the center of curvature of curvilinear streamlines. The radius of curvature r of straight lines is infinite, and thus no pressure gradient occurs across these.

For a liquid rotating as a solid body

$$-\frac{V_1^2}{2g} + \frac{p_1}{\rho g} + z_1 = -\frac{V_2^2}{2g} + \frac{p_2}{\rho g} + z_2$$

The negative sign balances the increase in velocity and pressure with radius.

The differential equations of motion for a viscous fluid are known as the Navier–Stokes equations. For incompressible flow the x-component equation is

$$\frac{\partial u}{\partial t} + u\frac{\partial u}{\partial x} + v\frac{\partial u}{\partial y} + w\frac{\partial u}{\partial z} = X - \frac{1}{\rho}\frac{\partial p}{\partial x} + v\left(\frac{\partial^2 u}{\partial x^2} + \frac{\partial^2 u}{\partial y^2} + \frac{\partial^2 u}{\partial z^2}\right)$$

with similar expressions for the y and z directions. X is the body force per unit mass. Reynolds developed a modified form of these equations for turbulent flow by expressing each velocity as an average value plus a fluctuating component ($u = \bar{u} + u'$ and so on). These modified equations indicate shear stresses from turbulence ($\tau_T = -\rho u'v'$, for example) known as the Reynolds stresses, which have been useful in the study of turbulent flow.

40.6 FLUID ENERGY

The Reynolds transport theorem for fluid passing through a control volume states that the heat added to the fluid less any work done by the fluid increases the energy content of the fluid in the control volume or changes the energy content of the fluid as it passes through the control surface. This is

$$Q - Wk_{done} = \frac{\partial}{\partial t} \int_{\substack{control \\ volume}} (e\rho) \, dV + \int_{\substack{control \\ surface}} e\rho(V \cdot dS)$$

and represents the first law of thermodynamics for control volume. The energy content includes kinetic, internal, potential, and displacement energies. Thus, mechanical and thermal energies are included, and there are no restrictions on the direction of interchange from one form to the other implied in the first law. The second law of thermodynamics governs this.

40.6.1 Energy Equations

With reference to Fig. 40.17, the steady flow energy equation is

$$\alpha_1 \frac{V_1^2}{2} + p_1 v_1 + gz_1 + u_1 + q - w = \alpha_2 \frac{V_2^2}{2} + p_2 v_2 + gz_2 + u_2$$

in terms of energy per unit mass, and where α is the kinetic energy correction factor:

$$\alpha = \frac{\int_A u^3 \, dA}{V^3 A} \approx \frac{1}{V^3 n} \sum_{i=1}^{n} u_i^3 \geq 1$$

For laminar flow in a pipe, $\alpha = 2$; for turbulent flow in a pipe, $\alpha = 1.05 - 1.06$; and if one-dimensional flow is assumed, $\alpha = 1$.

For one-dimensional flow of compressible gases, the general expression is

$$\frac{V_1^2}{2} + h_1 + gz_1 + q - w = \frac{V_2^2}{2} + h_2 + gz_2$$

For adiabatic flow, $q = 0$; for no external work, $w = 0$; and in most instances changes in elevation z are very small compared with changes in other parameters and can be neglected. Then the equation becomes

$$\frac{V_1^2}{2} + h_1 = \frac{V_2^2}{2} + h_2 = h_0$$

where h_0 is the stagnation enthalpy. The stagnation temperature is then $T_0 = T_1 + V_1^2/2c_p$, in terms of the temperature and velocity at some point 1. The gas velocity in terms of the stagnation and static temperatures, respectively, is $V_1 = \sqrt{2c_p(T_0 - T_1)}$. An increase in velocity is accompanied by a decrease in temperature, and vice versa.

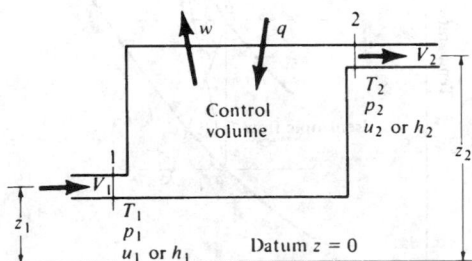

Fig. 40.17 Control volume for steady-flow energy equation.

For one-dimensional flow of liquids and constant-density (low-velocity) gases, the energy equation generally is written in terms of energy per unit weight as

$$\frac{V_1^2}{2g} + \frac{p_1}{\gamma} + z_1 - w = \frac{V_2^2}{2g} + \frac{p_2}{\gamma} + z_2 + h_L$$

where the first three terms are velocity, pressure, and potential heads, respectively. The head loss $h_L = (u_2 - u_1 - q)/g$ and represents the mechanical energy dissipated into thermal energy irreversibly (the heat transfer q is assumed zero here). It is a positive quantity and increases in the direction of flow.

Irreversibility in compressible gas flows results in an entropy increase. In Fig. 40.18 reversible flow between pressures p' and p is from a to b or from b to a. Irreversible flow from p' to p is from b to d, and from p to p' it is from a to c. Thus, frictional duct flow from one pressure to another results in a higher final temperature, and a lower final velocity, in both instances. For frictional flow between given temperatures (T_a and T_b, for example), the resulting pressures are lower than for frictionless flow (p_c is lower than p_a and p_f is lower than p_b).

40.6.2 Work and Power

Power is the rate at which work is done, and is the work done per unit mass times the mass flow rate, or the work done per unit weight times the weight flow rate.

Power represented by the work term in the energy equation is $P = w(VA\gamma) = w(VA\rho)$W.

Power in a jet at a velocity V is $P = (V^2/2)(VA\rho) = (V^2/2g)(VA\gamma)$W.

Power loss resulting from head loss is $P = h_L(VA\gamma)$W.

Power to overcome a drag force is $P = FV$W.

Power available in a hydroelectric power plant when water flows from a headwater elevation z_1 to a tailwater elevation z_2 is $P = (z_1 - z_2)(Q\gamma)$W, where Q is the volumetric flow rate.

40.6.3 Viscous Dissipation

Dissipation effects resulting from viscosity account for entropy increases in adiabatic gas flows and the heat loss term for flows of liquids. They can be expressed in terms of the rate at which work is done—the product of the viscous shear force on the surface of an elemental fluid volume and the corresponding component of velocity parallel to the force. Results for a cube of sides dx, dy, and dz give the dissipation function Φ:

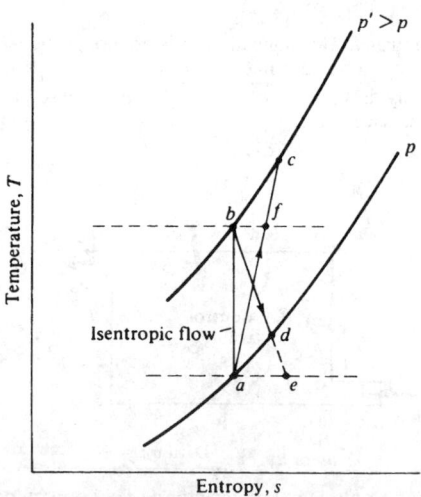

Fig. 40.18 Reversible and irreversible adiabatic flows.

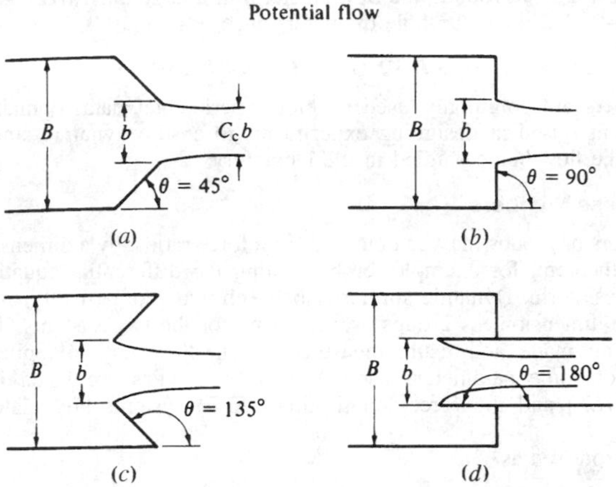

Potential flow

Fig. 40.19 Geometry of two-dimensional jets.

$$\Phi = 2\mu\left[\left(\frac{\partial u}{\partial x}\right)^2 + \left(\frac{\partial v}{\partial y}\right)^2 + \left(\frac{\partial w}{\partial z}\right)^2\right]$$
$$+ \mu\left[\left(\frac{\partial v}{\partial x} + \frac{\partial u}{\partial y}\right)^2 + \left(\frac{\partial w}{\partial y} + \frac{\partial v}{\partial z}\right)^2 + \left(\frac{\partial u}{\partial z} + \frac{\partial w}{\partial x}\right)^2\right]$$
$$- \frac{2}{3}\mu\left(\frac{\partial u}{\partial x} + \frac{\partial v}{\partial y} + \frac{\partial w}{\partial z}\right)^2$$

The last term is zero for an incompressible fluid. The first term in brackets is the linear deformation, and the second term in brackets is the angular deformation and in only these two forms of deformation is there heat generated as a result of viscous shear within the fluid. The second law of thermodynamics precludes the recovery of this heat to increase the mechanical energy of the fluid.

40.7 CONTRACTION COEFFICIENTS FROM POTENTIAL FLOW THEORY

Useful engineering results of a conformal mapping technique were obtained by von Mises for the contraction coefficients of two-dimensional jets for nonviscous incompressible fluids in the absence of gravity. The ratio of the resulting cross-sectional area of the jet to the area of the boundary opening is called the *coefficient of contraction, C_c*. For flow geometries shown in Fig. 40.19, von Mises calculated the values of C_c listed in Table 40.4. The values agree well with measurements for low-viscosity liquids. The results tabulated for two-dimensional flow may be used for axisymmetric jets if C_c is defined by $C_c = b_{jet}/b = (d_{jet}/d)^2$ and if d and D are diameters equivalent to widths b and

Table 40.4 Coefficients of Contraction for Two-Dimensional Jets

b/B	C_c $\theta = 45°$	C_c $\theta = 90°$	C_c $\theta = 135°$	C_c $\theta = 180°$
0.0	0.746	0.611	0.537	0.500
0.1	0.747	0.612	0.546	0.513
0.2	0.747	0.616	0.555	0.528
0.3	0.748	0.622	0.566	0.544
0.4	0.749	0.631	0.580	0.564
0.5	0.752	0.644	0.599	0.586
0.6	0.758	0.662	0.620	0.613
0.7	0.768	0.687	0.652	0.646
0.8	0.789	0.722	0.698	0.691
0.9	0.829	0.781	0.761	0.760
1.0	1.000	1.000	1.000	1.000

B, respectively. Thus, if a small round hole of diameter d in a large tank ($d/D \approx 0$), the jet diameter would be $(0.611)^{1/2} = 0.782$ times the hole diameter, since $\theta = 90°$.

40.8 DIMENSIONLESS NUMBERS AND DYNAMIC SIMILARITY

Dimensionless numbers are commonly used to plot experimental data to make the results more universal. Some are also used in designing experiments to ensure dynamic similarity between the flow of interest and the flow being studied in the laboratory.

40.8.1 Dimensionless Numbers

Dimensionless numbers or groups may be obtained from force ratios, by a dimensional analysis using the Buckingham Pi theorem, for example, or by writing the differential equations of motion and energy in dimensionless form. Dynamic similarity between two geometrically similar systems exists when the appropriate dimensionless groups are the same for the two systems. This is the basis on which model studies are made, and results measured for one flow may be applied to similar flows.

The dimensions of some parameters used in fluid mechanics are listed in Table 40.5. The mass–length–time (MLT) and the force–length–time (FLT) systems are related by $F = Ma = ML/T^2$ and $M = FT^2/L$.

Force ratios are expressed as

$$\frac{\text{inertia force}}{\text{viscous force}} = \frac{\rho L^2 V^2}{\mu VL} = \frac{\rho LV}{\mu}, \quad \text{the Reynolds number Re}$$

$$\frac{\text{inertia force}}{\text{gravity force}} = \frac{\rho L^2 V^2}{\rho L^3 g} = \frac{V^2}{Lg} \quad \text{or} \quad \frac{V}{\sqrt{Lg}}, \quad \text{the Froude number Fr}$$

$$\frac{\text{pressure force}}{\text{inertia force}} = \frac{\Delta p L^2}{\rho L^2 V^2} = \frac{\Delta p}{\rho V^2} \quad \text{or} \quad \frac{\Delta p}{\rho V^2/2}, \quad \text{the pressure coefficient } C_p$$

$$\frac{\text{inertia force}}{\text{surface tension force}} = \frac{\rho L^2 V^2}{\sigma L} = \frac{V^2}{\sigma/\rho L} \quad \text{or} \quad \frac{V}{\sqrt{\sigma/\rho L}}, \quad \text{the Weber number We}$$

$$\frac{\text{inertia force}}{\text{compressibility force}} = \frac{\rho L^2 V^2}{KL^2} = \frac{V^2}{K/\rho} \quad \text{or} \quad \frac{V}{\sqrt{K/\rho}}, \quad \text{the Mach number M}$$

If a system includes n quantities with m dimensions, there will be at least $n - m$ independent dimensionless groups, each containing m repeating variables. Repeating variables (1) must include all the m dimensions, (2) should include a geometrical characteristic, a fluid property, and a flow characteristic and (3) should not include the dependent variable.

Thus, if the pressure gradient $\Delta p/L$ for flow in a pipe is judged to depend on the pipe diameter D and roughness k, the average flow velocity V, and the fluid density ρ, the fluid viscosity μ, and compressibility K (for gas flow), then $\Delta p/L = f(D, k, V, \rho, \mu, K)$ or in dimensions, $F/L^3 = f(L, L, L/T, FT^2/L^4, FT/L^2, F/L^2)$, where $n = 7$ and $m = 3$. Then there are $n - m = 4$ independent groups to be sought. If D, ρ, and V are the repeating variables, the results are

$$\frac{\Delta p}{\rho V^2/2} = f\left(\frac{DV\rho}{\mu}, \frac{k}{D}, \frac{V}{\sqrt{K/\rho}}\right)$$

or that the friction factor will depend on the Reynolds number of the flow, the relative roughness, and the Mach number. The actual relationship between them is determined experimentally. Results may be determined analytically for laminar flow. The seven original variables are thus expressed as four dimensionless variables, and the Moody diagram of Fig. 40.32 shows the result of analysis and experiment. Experiments show that the pressure gradient does depend on the Mach number, but the friction factor does not.

The Navier–Stokes equations are made dimensionless by dividing each length by a characteristic length L and each velocity by a characteristic velocity U. For a body force X due to gravity, $X = g_x = g(\partial z/\partial x)$. Then $x' = x/L$, etc., $t' = t(L/U)$, $u' = u/U$, etc., and $p' = p/\rho U^2$. Then the Navier–Stokes equation (x component) is

$$u'\frac{\partial u'}{\partial x'} + v'\frac{\partial u'}{\partial y'} + w'\frac{\partial u'}{\partial z'} + \frac{\partial u'}{\partial t'}$$

$$= \frac{gL}{U^2} - \frac{\partial p'}{\partial x'} + \frac{\mu}{\rho UL}\left(\frac{\partial^2 u'}{\partial x'^2} + \frac{\partial^2 u'}{\partial y'^2} + \frac{\partial^2 u'}{\partial z'^2}\right)$$

$$= \frac{1}{\text{Fr}^2} - \frac{\partial p'}{\partial x'} + \frac{1}{\text{Re}}\left(\frac{\partial^2 u'}{\partial x'^2} + \frac{\partial^2 u'}{\partial y'^2} + \frac{\partial^2 u'}{\partial z'^2}\right)$$

Table 40.5   Dimensions of Fluid and Flow Parameters

	FLT	MLT
Geometrical characteristics		
Length (diameter, height, breadth, chord, span, etc.)	L	L
Angle	None	None
Area	L^2	L^2
Volume	L^3	L^3
Fluid properties[a]		
Mass	FT^2/L	M
Density (ρ)	FT^2/L^4	M/L^3
Specific weight (γ)	F/L^3	M/L^2T^2
Kinematic viscosity (v)	L^2/T	L^2/T
Dynamic viscosity (μ)	FT/L^2	M/LT
Elastic modulus (K)	F/L^2	M/LT^2
Surface tension (σ)	F/L	M/T^2
Flow characteristics		
Velocity (V)	L/T	L/T
Angular velocity (ω)	$1/T$	$1/T$
Acceleration (a)	L/T^2	L/T^2
Pressure (Δp)	F/L^2	M/LT^2
Force (drag, lift, shear)	F	ML/T^2
Shear stress (τ)	F/L^2	M/LT^2
Pressure gradient ($\Delta p/L$)	F/L^3	M/L^2T^2
Flow rate (Q)	L^3/T	L^3/T
Mass flow rate ($\dot{m}$)	FT/L	M/T
Work or energy	FL	ML^2/T^2
Work or energy per unit weight	L	L
Torque and moment	FL	ML^2/T^2
Work or energy per unit mass	L^2/T^2	L^2/T^2

[a]Density, viscosity, elastic modulus, and surface tension depend upon temperature, and therefore temperature will not be considered a property in the sense used here.

Thus for incompressible flow, similarity of flow in similar situations exists when the Reynolds and the Froude numbers are the same.

For compressible flow, normalizing the differential energy equation in terms of temperatures, pressure, and velocities gives the Reynolds, Mach, and Prandtl numbers as the governing parameters.

40.8.2   Dynamic Similitude

Flow systems are considered to be dynamically similar if the appropriate dimensionless numbers are the same. Model tests of aircraft, missiles, rivers, harbors, breakwaters, pumps, turbines, and so forth are made on this basis. Many practical problems exist, however, and it is not always possible to achieve complete dynamic similarity. When viscous forces govern the flow, the Reynolds number should be the same for model and prototype, the length in the Reynolds number being some characteristic length. When gravity forces govern the flow, the Froude number should be the same. When surface tension forces are significant, the Weber number is used. For compressible gas flow, the Mach number is used; different gases may be used for the model and prototype. The pressure coefficient $C_p = \Delta p/(\rho V^2/2)$, the drag coefficient $C_D = drag/(\rho V^2/2)A$, and the lift coefficient $C_L = lift/(\rho V^2/2)A$ will be the same for model and prototype when the appropriate Reynolds, Froude, or Mach number is the same. A cavitation number is used in cavitation studies, $\sigma_v = (p - p_v)/(\rho V^2/2)$ if vapor pressure p_v is the reference pressure or $\sigma_c = (p - p_c)/(\rho V^2/2)$ if a cavity pressure is the reference pressure.

Modeling ratios for conducting tests are listed in Table 40.6. Distorted models are often used for rivers in which the vertical scale ratio might be 1/40 and the horizontal scale ratio 1/100, for example, to avoid surface tension effects and laminar flow in models too shallow.

Incomplete similarity often exists in Froude–Reynolds models since both contain a length parameter. Ship models are tested with the Froude number parameter, and viscous effects are calculated for both model and prototype.

The specific speed of pumps and turbines results from combining groups in a dimensional analysis of rotary systems. That for pumps is $N_{s\,(pump)} = N\sqrt{Q}/e^{3/4}$ and for turbines it is $N_{s\,(turbines)} = N\sqrt{power}/\rho^{1/2}e^{5/4}$, where N is the rotational speed in rad/sec, Q is the volumetric flow rate in m³/

Table 40.6 Modeling Ratios[a]

		Modeling Parameter			
Ratio	Reynolds Number	Froude Number, Undistorted Model[b]	Froude Number, Distorted Model[b]	Mach Number, Same Gas[d]	Mach Number, Different Gas[d]
Velocity $\dfrac{V_m}{V_p}$	$\dfrac{L_p}{L_m}\dfrac{\rho_p}{\rho_m}\dfrac{\mu_m}{\mu_p}$	$\left(\dfrac{L_m}{L_p}\right)^{1/2}$	$\left(\dfrac{L_m}{L_p}\right)^{1/2}_V$	$\left(\dfrac{\theta_m}{\theta_p}\right)^{1/2}$	$\left(\dfrac{k_m R_m \theta_m}{k_p R_p \theta_p}\right)^{1/2}$
Angular velocity $\dfrac{\omega_m}{\omega_p}$	$\left(\dfrac{L_p}{L_m}\right)^2\dfrac{\rho_p}{\rho_m}\dfrac{\mu_m}{\mu_p}$	$\left(\dfrac{L_p}{L_m}\right)^{1/2}$	—[c]	$\left(\dfrac{\theta_m}{\theta_p}\right)^{1/2}\dfrac{L_p}{L_m}$	$\left(\dfrac{k_m R_m \theta_m}{k_p R_p \theta_p}\right)^{1/2}\dfrac{L_p}{L_m}$
Volumetric flow rate $\dfrac{Q_m}{Q_p}$	$\dfrac{L_m}{L_p}\dfrac{\rho_p}{\rho_m}\dfrac{\mu_m}{\mu_p}$	$\left(\dfrac{L_m}{L_p}\right)^{5/2}$	$\left(\dfrac{L_m}{L_p}\right)^{3/2}_V\left(\dfrac{L_m}{L_p}\right)_H$	—[c]	—[c]
Time $\dfrac{t_m}{t_p}$	$\left(\dfrac{L_m}{L_p}\right)^2\dfrac{\rho_m}{\rho_p}\dfrac{\mu_p}{\mu_m}$	$\left(\dfrac{L_m}{L_p}\right)^{1/2}\left(\dfrac{g_p}{g_m}\right)^{1/2}$	$\left(\dfrac{L_m}{L_p}\right)_H\left(\dfrac{L_p}{L_m}\right)^{1/2}_V\left(\dfrac{g_p}{g_m}\right)^{1/2}$	$\left(\dfrac{\theta_p}{\theta_m}\right)^{1/2}\dfrac{L_m}{L_p}$	$\left(\dfrac{k_p R_p \theta_p}{k_m R_m \theta_m}\right)^{1/2}\dfrac{L_m}{L_p}$
Force $\dfrac{F_m}{F_p}$	$\left(\dfrac{\mu_m}{\mu_p}\right)^2\dfrac{\rho_p}{\rho_m}$	$\left(\dfrac{L_m}{L_p}\right)^3\dfrac{\rho_m}{\rho_p}$	$\dfrac{\rho_m}{\rho_p}\left(\dfrac{L_m}{L_p}\right)_H\left(\dfrac{L_m}{L_p}\right)^2_V$	$\dfrac{\rho_m}{\rho_p}\dfrac{\theta_m}{\theta_p}\left(\dfrac{L_m}{L_p}\right)^2$	$\dfrac{K_m}{K_p}\left(\dfrac{L_m}{L_p}\right)^2$

[a]Subscript m indicates model, subscript p indicates prototype.

[b]For the same value of gravitational acceleration for model and prototype.

[c]Of little importance.

[d]Here θ refers to temperature.

sec, and e is the energy in J/kg. North American practice uses N in rpm, Q in gal/min, e as energy per unit weight (head in ft), power as brake horsepower rather than watts, and omits the density term in the specific speed for turbines. The numerical value of specific speed indicates the type of pump or turbine for a given installation. These are shown for pumps in North America in Fig. 40.20. Typical values for North American turbines are about 5 for impulse turbines, about 20–100 for Francis turbines, and 100–200 for propeller turbines. Slight corrections in performance for higher efficiency of large pumps and turbines are made when testing small laboratory units.

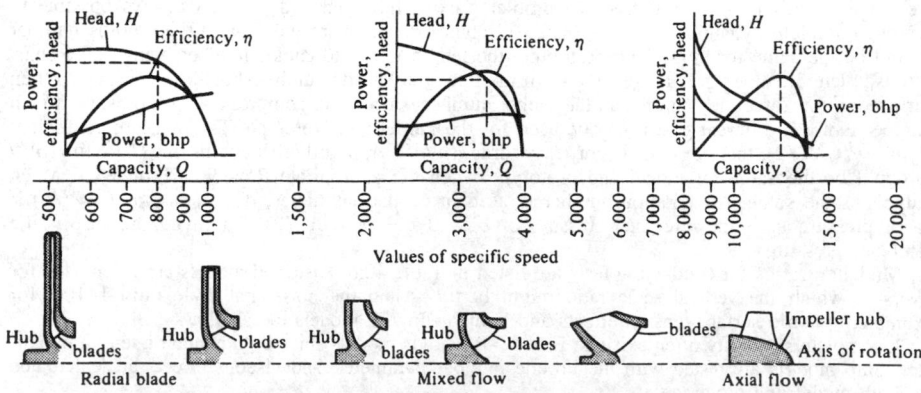

Fig. 40.20 Pump characteristics and specific speed for pump impellers. (Courtesy Worthington Corporation)

40.9   VISCOUS FLOW AND INCOMPRESSIBLE BOUNDARY LAYERS

In viscous flows, adjacent layers of fluid transmit both normal forces and tangential shear forces, as a result of relative motion between the layers. There is no relative motion, however, between the fluid and a solid boundary along which it flows. The fluid velocity varies from zero at the boundary to a maximum or free stream value some distance away from it. This region of retarded flow is called the boundary layer.

40.9.1   Laminar and Turbulent Flow

Viscous fluids flow in a laminar or in a turbulent state. There are, however, transition regimes between them where the flow is intermittently laminar and turbulent. Laminar flow is smooth, quiet flow without lateral motions. Turbulent flow has lateral motions as a result of eddies superimposed on the main flow, which results in random or irregular fluctuations of velocity, pressure, and, possibly, temperature. Smoke rising from a cigarette held at rest in still air has a straight threadlike appearance for a few centimeters; this indicates a laminar flow. Above that the smoke is wavy and finally irregular lateral motions indicate a turbulent flow. Low velocities and high viscous forces are associated with laminar flow and low Reynolds numbers. High speeds and low viscous forces are associated with turbulent flow and high Reynolds numbers. Turbulence is a characteristic of flows, not of fluids. Typical fluctuations of velocity in a turbulent flow are shown in Fig. 40.21.

The axes of eddies in turbulent flow are generally distributed in all directions. In *isotropic* turbulence they are distributed equally. In flows of low turbulence, the fluctuations are small; in highly turbulent flows, they are large. The turbulence level may be defined as (as a percentage)

$$T = \frac{\sqrt{(\overline{u'^2} + \overline{v'^2} + \overline{w'^2})/3}}{\overline{u}} \times 100$$

where u', v', and w' are instantaneous fluctuations from mean values and $\overline{u}$ is the average velocity in the main flow direction (x, in this instance).

Shear stresses in turbulent flows are much greater than in laminar flows for the same velocity gradient and fluid.

40.9.2   Boundary Layers

The growth of a boundary layer along a flat plate in a uniform external flow is shown in Fig. 40.22. The region of retarded flow, δ, thickens in the direction of flow, and thus the velocity changes from zero at the plate surface to the free stream value u_s in an increasingly larger distance δ normal to the plate. Thus, the velocity gradient at the boundary, and hence the shear stress as well, decreases as the flow progresses downstream, as shown. As the laminar boundary thickens, instabilities set in and the boundary layer becomes turbulent. The transition from the laminar boundary layer to a turbulent boundary layer does not occur at a well-defined location; the flow is intermittently laminar and turbulent with a larger portion of the flow being turbulent as the flow passes downstream. Finally, the flow is completely turbulent, and the boundary layer is much thicker and the boundary shear greater in the turbulent region than if the flow were to continue laminar. A viscous sublayer exists within the turbulent boundary layer along the boundary surface. The shape of the velocity profile also changes when the boundary layer becomes turbulent, as shown in Fig. 40.22. Boundary surface roughness, high turbulence level in the outer flow, or a decelerating free stream causes transition to occur nearer the leading edge of the plate. A surface is considered rough if the roughness elements have an effect outside the viscous sublayer, and smooth if they do not. Whether a surface is rough or smooth depends not only on the surface itself but also on the character of the flow passing it.

A boundary layer will separate from a continuous boundary if the fluid within it is caused to slow down such that the velocity gradient du/dy becomes zero at the boundary. An adverse pressure gradient will cause this.

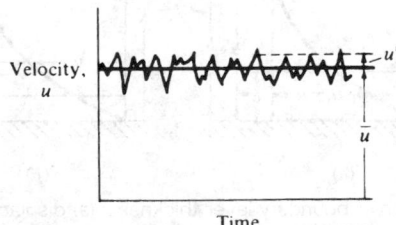

Fig. 40.21   Velocity at a point in steady turbulent flow.

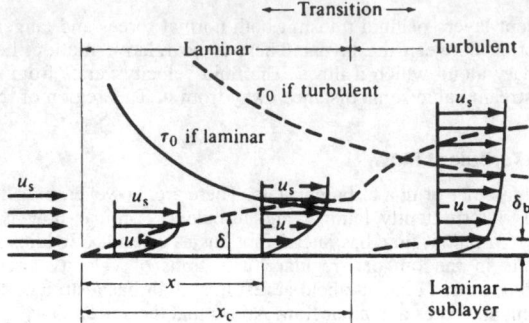

Fig. 40.22 Boundary layer development along a flat plate.

One parameter of interest is the boundary layer thickness δ, the distance from the boundary in which the flow is retarded, or the distance to the point where the velocity is 99% of the free stream velocity (Fig. 40.23). The displacement thickness is the distance the boundary is displaced such that the boundary layer flow is the same as one-dimensional flow past the displaced boundary. It is given by (see Fig. 40.23)

$$\delta_1 = \frac{1}{u_s} \int_0^\delta (u_s - u)\, dy = \int_0^\delta \left(1 - \frac{u}{u_s}\right) dy$$

A momentum thickness is the distance from the boundary such that the momentum flux of the free stream within this distance is the deficit of momentum of the boundary layer flow. It is given by (see Fig. 40.23)

$$\delta_2 = \int_0^\delta \left(1 - \frac{u}{u_s}\right) \frac{u}{u_s}\, dy$$

Also of interest is the viscous shear drag $D = C_f(\rho u_s^2/2)A$, where C_f is the average skin friction drag coefficient and A is the area sheared.

These parameters are listed in Table 40.7 as functions of the Reynolds number $Re_x = u_s \rho x/\mu$, where x is based on the distance from the leading edge. For Reynolds numbers between 1.8×10^5 and 4.5×10^7, $C_f = 0.045/Re_x^{1/6}$, and for Re_x between 2.9×10^7 and 5×10^8, $C_f = 0.0305/Re_x^{1/7}$. These results for turbulent boundary layers are obtained from pipe flow friction measurements for smooth pipes, by assuming the pipe radius equivalent to the boundary layer thickness, the centerline pipe velocity equivalent to the free stream boundary layer flow, and appropriate velocity profiles. Results agree with measurements.

When a turbulent boundary layer is preceded by a laminar boundary layer, the drag coefficient is given by the Prandtl–Schlichting equation:

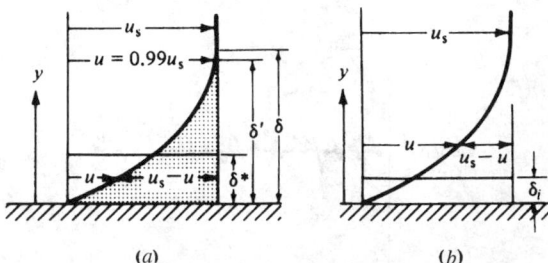

Fig. 40.23 Definition of boundary layer thickness: (*a*) displacement thickness; (*b*) momentum thickness.

Table 40.7 Boundary Layer Parameters

Parameter	Laminar Boundury Layer	Turbulent Boundary Layer
$\dfrac{\delta}{x}$	$\dfrac{4.91}{Re_x^{1/2}}$	$\dfrac{0.382}{Re_x^{1/5}}$
$\dfrac{\delta_1}{x}$	$\dfrac{1.73}{Re_x^{1/2}}$	$\dfrac{0.048}{Re_x^{1/5}}$
$\dfrac{\delta_2}{x}$	$\dfrac{0.664}{Re_x^{1/2}}$	$\dfrac{0.037}{Re_x^{1/5}}$
C_f	$\dfrac{1.328}{Re_x^{1/2}}$	$\dfrac{0.074}{Re_x^{1/5}}$
Re_x range	Generally not over 10^6	Less than 10^7

$$C_f = \frac{0.455}{(\log_{10} Re_x)^{2.58}} - \frac{A}{Re_x}$$

where A depends on the Reynolds number Re_c at which transition occurs. Values of A for various values of $Re_c = u_s x_c/v$ are

Re_c	3×10^5	5×10^5	9×10^5	1.5×10^6
A	1035	1700	3000	4880

Some results are shown in Fig. 40.24 for transition at these Reynolds numbers for completely laminar boundary layers, for completely turbulent boundary layers, and for a typical ship hull. (The other curves are applicable for smooth model ship hulls.) Drag coefficients for flat plates may be used for other shapes that approximate flat plates.

The thickness of the viscous sublayer δ_b in terms of the boundary layer thickness is approximately

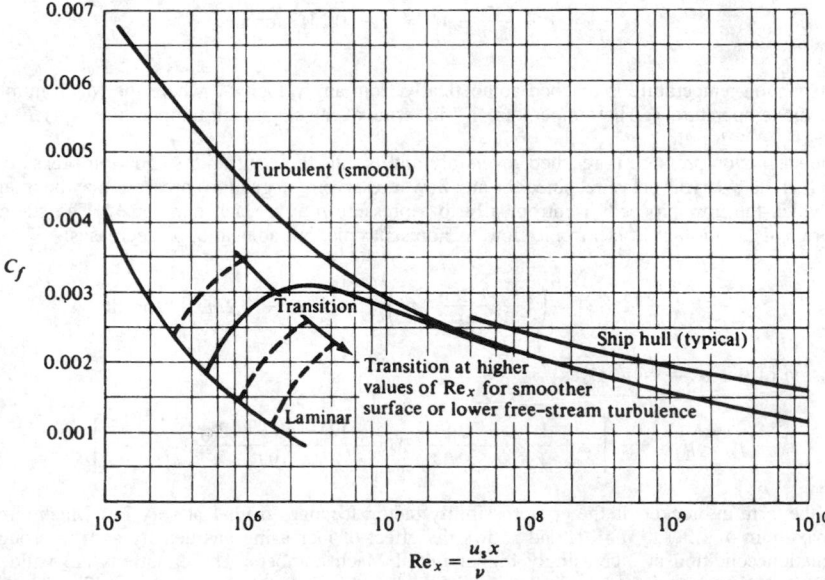

Fig. 40.24 Drag coefficients for smooth plane surfaces parallel to flow.

$$\frac{\delta_b}{\delta} = \frac{80}{(\text{Re}_x)^{7/10}}$$

At $\text{Re}_x = 10^6$, $\delta_b/\delta = 0.0050$ and when $\text{Re}_x = 10^7$, $\delta_b/\delta = 0.001$, and thus the viscous sublayer is very thin.

Experiments show that the boundary layer thickness and local drag coefficient for a turbulent boundary layer preceded by a laminar boundary layer at a given location are the same as though the boundary layer were turbulent from the beginning of the plate or surface along which the boundary layer grows.

40.10 GAS DYNAMICS

In gas flows where density variations are appreciable, large variations in velocity and temperature may also occur and then thermodynamic effects are important.

40.10.1 Adiabatic and Isentropic Flow

In adiabatic flow of a gas with no external work and with changes in elevation negligible, the steady-flow energy equation is

$$\frac{V_1^2}{2} + h_1 = \frac{V_2^2}{2} + h_2 = h_0 = \text{constant}$$

for flow from point 1 to point 2, where V is velocity and h is enthalpy. Subscript 0 refers to a stagnation condition where the velocity is zero.

The speed of sound is $c = \sqrt{(\partial p/\partial s)_{\text{isentropic}}} \sqrt{K/\rho} = \sqrt{kp/\rho} = \sqrt{kRT}$. For air, $c = 20.04\sqrt{T}$ m/sec, where T is in degrees kelvin. A local Mach number is then $M = V/c = V/\sqrt{kRT}$.

A gas at rest may be accelerated adiabatically to any speed, including sonic ($M = 1$) and theoretically to its maximum speed when the temperature reduces to absolute zero. Then,

$$c_p T_0 = c_p T + \frac{V^2}{2} = c_p T* + \frac{V*^2}{2} = \frac{V_{\text{max}}^2}{2}$$

where the asterisk (*) refers to a sonic state where the Mach number is unity.

The stagnation temperature T_0 is $T_0 = T + V^2/2c_p$, or in terms of the Mach number $[c_p = Rk/(k-1)]$

$$\frac{T_0}{T} = 1 + \frac{k-1}{2} M^2 = 1 + 0.2M^2 \text{ for air}$$

The stagnation temperature is reached adiabatically from any velocity V where the Mach number is M and the temperature T. The temperature $T*$ in terms of the stagnation temperature T_0 is $T*/T_0 = 2/(k+1) = \frac{5}{6}$ for air.

The stagnation pressure is reached reversibly and is thus the isentropic stagnation pressure. It is also called the reservoir pressure, since for any flow a reservoir (stagnation) pressure may be imagined from which the flow proceeds isentropically to a pressure p at a Mach number M. The stagnation pressure p_0 is a constant in isentropic flow; if nonisentropic, but adiabatic, p_0 decreases:

$$\frac{p_0}{p} = \left(\frac{T_0}{T}\right)^{k/(k-1)} = \left(1 + \frac{k-1}{2} M^2\right)^{k/(k-1)} = (1 + 0.2M^2)^{3.5} \text{ for air}$$

Expansion of this expression gives

$$p_0 = p + \frac{\rho V^2}{2}\left[1 + \frac{1}{4} M^2 + \frac{2-k}{24} M^4 + \frac{(2-k)(3-2k)}{192} M^6 + \cdots\right]$$

where the term in brackets is the compressibility factor. It ranges from 1 at very low Mach numbers to a maximum of 1.27 at $M = 1$, and shows the effect of increasing gas density as it is brought to a stagnation condition at increasingly higher initial Mach numbers. The equations are valid to or from a stagnation state for subsonic flow, and from a stagnation state for supersonic flow at M^2 less than $2/(k-1)$, or M less than $\sqrt{5}$ for air.

40.10.2 Duct Flow

Adiabatic flow in short ducts may be considered reversible, and thus the relation between velocity and area changes is $dA/dV = (A/V)(M^2 - 1)$. For subsonic flow, dA/dV is negative and velocity changes relate to area changes in the same way as for incompressible flow. At supersonic speed, dA/dV is positive and an expanding area is accompanied by an increasing velocity; a contracting area is accompanied by a decreasing velocity, the opposite of incompressible flow behavior. Sonic flow in a duct (at $M = 1$) can exist only when the duct area is constant $(dA/dV = 0)$, in the throat of a nozzle or in a pipe. It can also be shown that velocity and Mach numbers always increase or decrease together, that temperature and Mach numbers change in opposite directions, and that pressure and Mach numbers also change in opposite directions.

Isentropic gas flow tables give pressure ratios p/p_0, temperature ratios T/T_0, density ratios ρ/ρ_0, area ratios A/A^*, and velocity ratios V/V^* as functions of the upstream Mach number M_x and the specific heat ratio k for gases.

The mass flow rate through a converging nozzle from a reservoir with the gas at a pressure p_0 and temperature T_0 is calculated in terms of the pressure at the nozzle exit from the equation $\dot{m} = (VA\rho)_{\text{exit}}$, where $\rho_e = p_e/RT_e$ and the exit temperature is $T_e = T_0(p_e/p_0)^{(k-1)/k}$ and the exit velocity is

$$V_e = \sqrt{2c_p T_0 \left[1 - \left(\frac{p_e}{p_0}\right)^{(k-1)/k} \right]}$$

The mass flow rate is maximum when the exit velocity is sonic. This requires the exit pressure to be critical, and the receiver pressure to be critical or below. Supersonic flow in the nozzle is impossible. If the receiver pressure is below critical, flow is not affected in the nozzle, and the exit flow remains sonic. For air at this condition, the maximum flow rate is $\dot{m} = 0.0404A_1 p_0/\sqrt{T_0}$ kg/sec.

Flow through a converging–diverging nozzle (Fig. 40.25) is subsonic throughout if the throat pressure is above critical (dashed lines in Fig. 40.25). When the receiver pressure is at A, the exit pressure is also, and sonic flow exists at the throat, but is subsonic elsewhere. Only at B is there sonic flow in the throat with isentropic expansion in the diverging part of the nozzle. The flow rate is the same whether the exit pressure is at A or B. Receiver pressures below B do not affect the flow in the nozzle. Below A (at C, for example) a shock forms as shown and then the flow is isentropic to the shock, and beyond it, but not through it. When the throat flow is sonic, the mass flow rate is given by the same equation as for a converging nozzle with sonic exit flow. The pressures at A and B in terms of the reservoir pressure p_0 are given in isentropic flow tables as a function of the ratio of exit area to throat area, A_e/A^*.

40.10.3 Normal Shocks

The plane of a normal shock is at right angles to the flow streamlines. These shocks may occur in the diverging part of a nozzle, the diffuser of a supersonic wind tunnel, in pipes and forward of blunt-nosed bodies. In all instances the flow is supersonic upstream and subsonic downstream of the shock. Flow through a shock is not isentropic, although nearly so for very weak shocks. The abrupt changes in gas density across a shock allow for optical detection. The interferometer responds to density changes, the Schlieren method to density gradients, and the spark shadowgraph to the rate of change of density gradient. Density ratios across normal shocks in air are 2 at $M = 1.58$, 3 at $M = 2.24$, and 4 at $M = 3.16$ to a maximum value of 6.

Changes in fluid and flow parameters across normal shocks are obtained from the continuity, energy, and momentum equations for adiabatic flow. They are expressed in terms of upstream Mach numbers with upstream conditions designated with subscript x and downstream with subscript y. Mach numbers M_x and M_y are related by

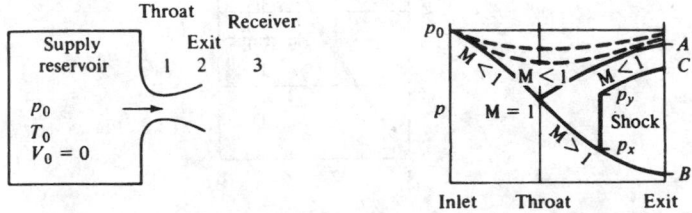

Fig. 40.25 Gas flow through converging–diverging nozzle.

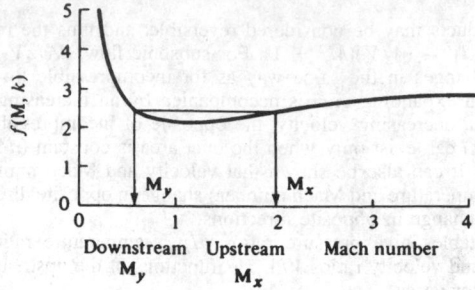

Fig. 40.26 Mach numbers across a normal shock, $k = 1.4$.

$$\frac{1 + kM_x^2}{M_x \left(1 + \dfrac{k-1}{2} M_x^2\right)^{1/2}} = \frac{1 + kM_y^2}{M_y \left(1 + \dfrac{k-1}{2} M_y^2\right)^{1/2}} = f(M,k)$$

which is plotted in Fig. 40.26. The requirement for an entropy increase through the shock indicates M_x to be greater than M_y. Thus, the higher the upstream Mach number, the lower the downstream Mach number, and vice versa. For normal shocks, values of downstream Mach number M_y; temperature ratios T_y/T_x; pressure ratios p_y/p_x, p_{0y}/p_x, and p_{0y}/p_{0x}; and density ratios ρ_y/ρ_x depend only on the upstream Mach number M_x and the specific heat ratio k of the gas. These values are tabulated in books on gas dynamics and in books of gas tables.

The density ratio across the shock is given by the Rankine–Hugoniot equation

$$\frac{\rho_y}{\rho_x} = \left[\left(\frac{k+1}{k-1}\right) \frac{p_y}{p_x} + 1 \right] \Big/ \left[\frac{p_y}{p_x} + \left(\frac{k+1}{k-1}\right) \right]$$

and is plotted in Fig. 40.27, which shows that weak shocks are nearly isentropic, and that the density ratio approaches a limit of 6 for gases with $k = 1.4$.

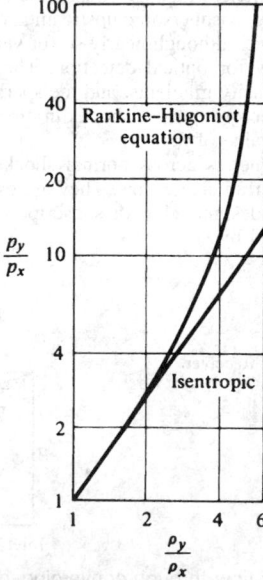

Fig. 40.27 Rankine–Hugoniot curve, $k = 1.4$.

Gas tables show that at an upstream Mach number of 2 for air, $M_y = 0.577$, the pressure ratio is $p_y/p_x = 4.50$, the density ratio is $\rho_y/\rho_x = 2.66$, the temperature ratio is $T_y/T_x = 1.68$, and the stagnation pressure ratio is $p_{0y}/p_{0x} = 0.72$, which indicates an entropy increase of $s_y - s_x = -R \ln(p_{0y}/p_{0x}) = 94$ J/kg.

40.10.4 Oblique Shocks

Oblique shocks are inclined from a direction normal to the approaching streamlines. Figure 40.28 shows that the normal velocity components are related by the normal shock relations. From a momentum analysis, the tangential velocity components are unchanged through oblique shocks. The upstream Mach number M_1 is given in terms of the deflection angle θ, the shock angle β, and the specific heat ratio k for the gas as

$$\frac{1}{M_1^2} = \sin^2 \beta - \frac{(k+1)}{2} \frac{\sin \beta \sin \theta}{\cos(\beta - \theta)}$$

The geometry is shown in Fig. 40.29, and the variables in this equation are illustrated in Fig. 40.30. For each M_1 there is the possibility of two wave angles β for a given deflection angle θ. The larger wave angle is for strong shocks, with subsonic downstream flow. The smaller wave angle is for weak shocks, generally with supersonic downstream flow at a Mach number less than M_1.

Normal shock tables are used for oblique shocks if M_x is used for $M_1 \sin \beta$. Then $M_y = M_2 \sin(\beta - \theta)$ and other ratios of property values (pressure, temperature, and density) are the same as for normal shocks.

40.11 VISCOUS FLUID FLOW IN DUCTS

The development of flow in the entrance of a pipe with the development of the boundary layer is shown in Fig. 40.31. Wall shear stress is very large at the entrance, and generally decreases in the flow direction to a constant value, as does the pressure gradient dp/dx. The velocity profile also

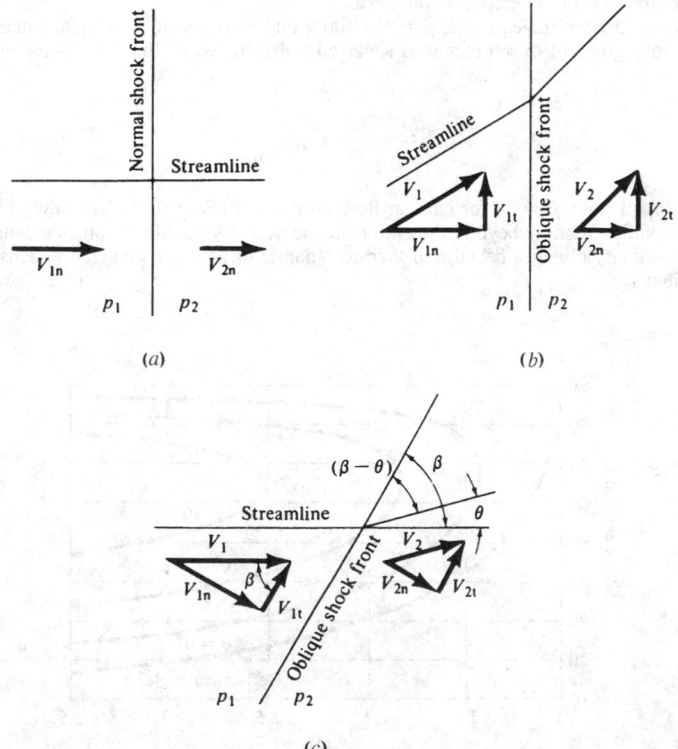

Fig. 40.28 Oblique shock relations from normal shock; (a) normal shock; (b) oblique shock; (c) oblique shock angles.

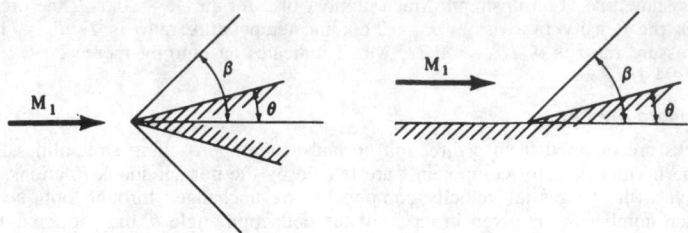

Fig. 40.29 Supersonic flow past a wedge and an inside corner.

changes and becomes adjusted to a fixed shape. When these have reached constant conditions, the flow is called *fully developed* flow.

The momentum equation for a pipe of diameter D gives the pressure gradient as

$$-\frac{dp}{dx} = \frac{4}{D}\,\tau_0 + \rho V^2 \frac{d\beta}{dx} + \beta \rho V \frac{dV}{dx}$$

which shows that a pressure gradient overcomes wall shear and increases momentum of the fluid either as a result of changing the shape of the velocity profile ($d\beta/dx$) or by changing the mean velocity along the pipe (dV/dx is not zero for gas flows).

For fully developed incompressible flow

$$-\frac{dp}{dx} = \frac{\Delta p}{L} = \frac{4\tau_0}{D}$$

and a pressure drop simply overcomes wall shear.

For developing flow in the entrance, $\beta = 1$ initially and increases to a constant value downstream. Thus, the pressure gradient overcomes wall shear and also increases the flow momentum according to

$$-\frac{dp}{dx} = \frac{4\tau_0}{D} + \rho V^2 \frac{d\beta}{dx}$$

For fully developed flow, $\beta = \frac{4}{3}$ for laminar flow and $\beta \approx 1.03$ for turbulent flow in round pipes.

For compressible gas flow beyond the entrance, the velocity profile becomes essentially fixed in shape, but the velocity changes because of thermodynamic effects that change the density. Thus, the pressure gradient is

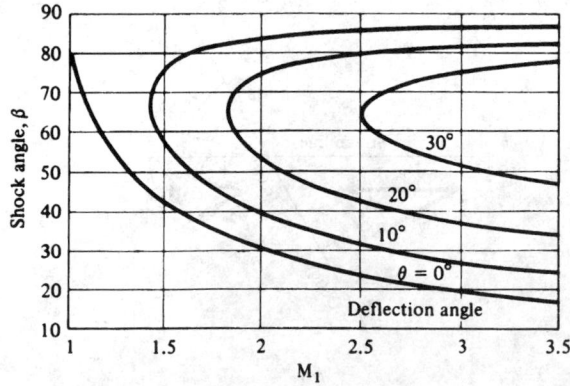

Fig. 40.30 Oblique shock relations, $k = 1.4$.

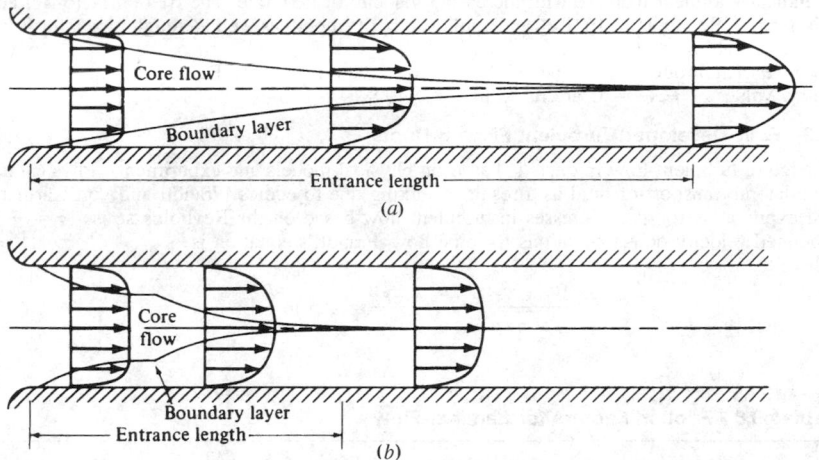

Fig. 40.31 Growth of boundary layers in a pipe: (a) laminar flow; (b) turbulent flow.

$$-\frac{dp}{dx} = \frac{4\tau_0}{D} + \beta\rho V\frac{dV}{dx}$$

Here β is essentially constant but dV/dx may be significant.

40.11.1 Fully Developed Incompressible Flow

The pressure drop is $\Delta p = (fL/D)(\rho V^2/2)$ Pa, where f is the Darcy friction factor. The Fanning friction factor $f' = f/4$ and then $\Delta p = (4f'/D)(\rho V^2/2)$, and the head loss from pipe friction is

$$h_f = \frac{\Delta p}{\gamma} = f\left(\frac{L}{D}\right)\frac{V^2}{2g} = (4f')\left(\frac{L}{D}\right)\frac{V^2}{2g}\quad \text{m}$$

The shear stress varies linearly with radial position, $\tau = (\Delta p/L)(r/2)$, so that the wall shear is $\tau_0 = (\Delta p/L)(D/4)$, which may then be written $\tau_0 = f\rho V^2/8 = f'\rho V^2/2$.

A shear velocity is defined as $v \cdot = \sqrt{\tau_0/\rho} = V\sqrt{f/8} = V\sqrt{f'/2}$ and is used as a normalizing parameter.

For noncircular ducts the diameter D is replaced by the hydraulic or equivalent diameter $D_h = 4.4/P$, where A is the flow cross section and P is the wetted perimeter. Thus, an annulus between pipes of diameter D_1 and D_2, D_1 being larger, the hydraulic diameter is $D_2 - D_1$.

40.11.2 Fully Developed Laminar Flow in Ducts

The velocity profile in circular tubes is that of a parabola, and the centerline velocity is

$$u_{\max} = \frac{\Delta p}{L}\left(\frac{R^2}{4\mu}\right)$$

and the velocity profile is

$$\frac{u}{u_{\max}} = 1 - \left(\frac{r}{R}\right)^2$$

where r is the radial location in a pipe of radius R. The average velocity is one-half the maximum velocity, $V = u_{\max}/2$.

The pressure gradient is

$$\frac{\Delta p}{L} = \frac{128\mu Q}{\pi D^4}$$

which indicates a linear increase with increasing velocity or flow rate. The friction factor for circular ducts is $f = 64/\text{Re}_D$ or $f' = 16/\text{Re}_D$ and applies to both smooth as well as rough pipes, for Reynolds numbers up to about 2000.

For *noncircular* ducts the value of the friction factor is $f = C/\text{Re}$ and depends on the duct geometry. Values of $f\,\text{Re} = C$ are listed in Table 40.8.

40.11.3 Fully Developed Turbulent Flow in Ducts

Knowledge of turbulent flow in ducts is based on physical models and experiments. Physical models describe lateral transport of fluid as a result of mixing due to eddies. Prandtl and von Kármán both derived expressions for shear stresses in turbulent flow based on the Reynolds stress ($\tau = -\rho\overline{u'v'}$) and obtained velocity defect equations for pipe flow. Prandtl's equation is

$$\frac{u_{max} - u}{\sqrt{\tau_0/\rho}} = \frac{u_{max} - u}{v\cdot} = 2.5 \ln \frac{R}{y}$$

Table 40.8 Friction Factors for Laminar Flow

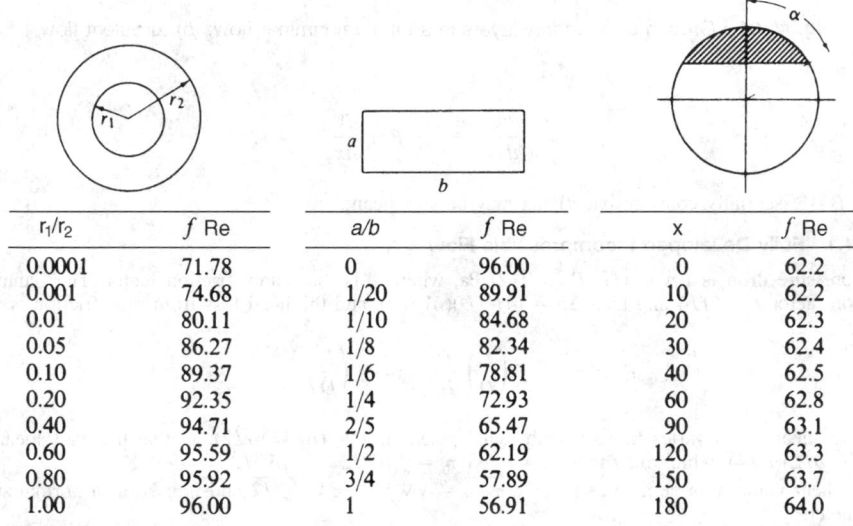

r_1/r_2	f Re	a/b	f Re	x	f Re
0.0001	71.78	0	96.00	0	62.2
0.001	74.68	1/20	89.91	10	62.2
0.01	80.11	1/10	84.68	20	62.3
0.05	86.27	1/8	82.34	30	62.4
0.10	89.37	1/6	78.81	40	62.5
0.20	92.35	1/4	72.93	60	62.8
0.40	94.71	2/5	65.47	90	63.1
0.60	95.59	1/2	62.19	120	63.3
0.80	95.92	3/4	57.89	150	63.7
1.00	96.00	1	56.91	180	64.0

	Circular Sector	Isosceles Triangle	Right Triangle
α	f Re	f Re	f Re
0	48.0	48.0	48.0
10	51.8	51.6	49.9
20	54.5	52.9	51.2
30	56.7	53.3	52.0
40	58.4	52.9	52.4
50	59.7	52.0	52.4
60	60.8	51.1	52.0
70	61.7	49.5	51.2
80	62.5	48.3	49.9
90	63.1	48.0	48.0

where $u_{\max}$ is the centerline velocity and u is the velocity a distance y from the pipe wall. von Kármán's equation is

$$\frac{u_{\max} - u}{\sqrt{\tau_0/\rho}} = \frac{u_{\max} - u}{v\cdot}$$

$$= -\frac{1}{\kappa}\left[\ln\left(1 - \sqrt{1 - \frac{y}{R}}\right) + \sqrt{1 - \frac{y}{R}}\right]$$

In both, κ is an experimentally determined constant equal to 0.4 (some experiments show better agreement when $\kappa = 0.36$). Similar expressions apply to external boundary layer flow when the pipe radius R is replaced by the boundary layer thickness δ. Friction factors for smooth pipes have been developed from these results. One is the Blasius equation for $\text{Re}_D = 10^5$ and is $f = 0.316/\text{Re}_D^{1/4}$ obtained by using a power-law velocity profile $u/u_{\max} = (y/R)^{1/7}$. The value 7 here increases to 10 at higher Reynolds numbers. The use of a logarithmic form of velocity profile gives the Prandtl law of pipe friction for smooth pipes:

$$\frac{1}{\sqrt{f}} = 2\log(\text{Re}_D\sqrt{f}) - 0.8$$

which agrees well with experimental values. A more explicit formula by Colebrook is $1/\sqrt{f} = 1.8$ $\log(\text{Re}_D/6.9)$, which is within 1% of the Prandtl equation over the entire range of turbulent Reynolds numbers.

The logarithmic velocity defect profiles apply for *rough* pipes as well as for smooth pipes, since the velocity defect $(u_{\max} - u)$ decreases linearly with the shear velocity $v\cdot$, keeping the ratio of the two constant.

A relation between the centerline velocity and the average velocity is $u_{\max}/V = 1 + 133\sqrt{f}$, which may be used to estimate the average velocity from a single centerline measurement.

The Colebrook–White equation encompasses all turbulent flow regimes, for both smooth and rough pipes:

$$\frac{1}{\sqrt{f}} = 1.74 - 2\log\left(\frac{2k}{D} + \frac{18.7}{\text{Re}_D\sqrt{f}}\right)$$

and this is plotted in Fig. 40.32, where k is the equivalent sand-grain roughness. A simpler equation by Haaland is

$$\frac{1}{\sqrt{f}} = -1.8\log\left[\frac{6.9}{\text{Re}_D} + \left(\frac{k}{3.7D}\right)^{1.11}\right]$$

which is explicit in f and is within 1.5% of the Colebrook–White equation in the range $4000 \leq \text{Re}_D \leq 10^8$ and $0 \leq k/D \leq 0.05$.

Three types of problems may be solved:

1. *The Pressure Drop or Head Loss.* The Reynolds number and relative roughness are determined and calculations are made directly.
2. *The Flow Rate for Given Fluid and Pressure Drops or Head Loss.* Assume a friction factor, based on a high Re_D for a rough pipe, and determine the velocity from the Darcy equation. Calculate a Re_D, get a better f, and repeat until successive velocities are the same. A second method is to assume a flow rate and calculate the pressure drop or head loss. Repeat until results agree with the given pressure drop or head loss. A plot of Q versus h_L, for example, for a few trials may be used.
3. *A Pipe Size.* Assume a pipe size and calculate the pressure drop or head loss. Compare with given values: Repeat until agreement is reached. A plot of D versus h_L, for example, for a few trials may be used. A second method is to assume a reasonable friction factor and get a first estimate of the diameter from

$$D = \left[\frac{8fLQ^2}{\pi^2 g h_f}\right]^{1/5}$$

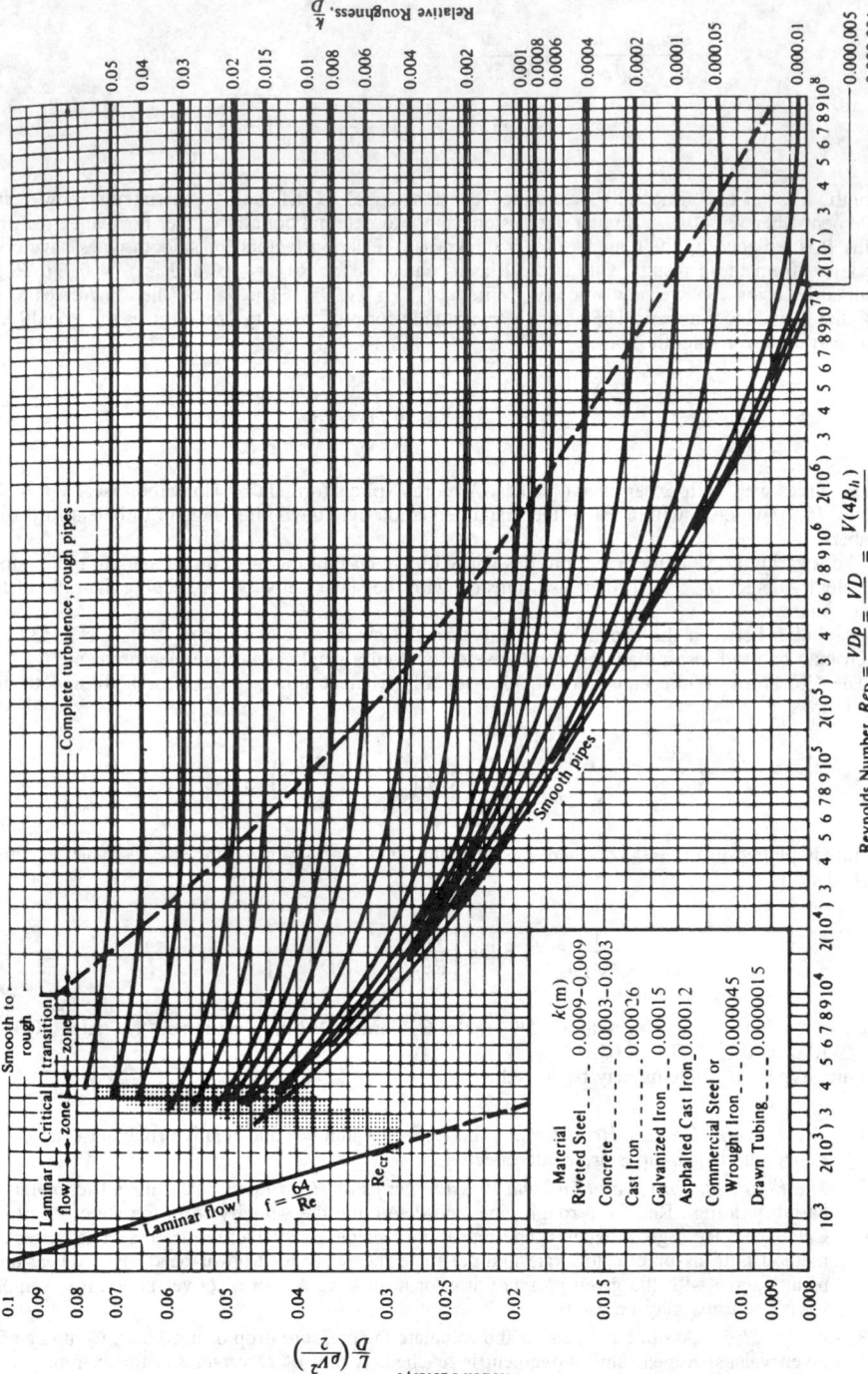

Fig. 40.32 Friction factors for commercial pipe. [From L. F. Moody, "Friction Factors for Pipe Flow," *Trans. ASME*, 66 (1944). Courtesy of The American Society of Mechanical Engineers.]

Material	k(m)
Riveted Steel	0.0009–0.009
Concrete	0.0003–0.003
Cast Iron	0.00026
Galvanized Iron	0.00015
Asphalted Cast Iron	0.00012
Commercial Steel or Wrought Iron	0.000045
Drawn Tubing	0.0000015

$$R_{TD} = \frac{VD\rho}{\mu} = \frac{VD}{\nu} = \frac{V(4R_h)}{\nu}$$

From the first estimate of D, calculate the Re_D and k/D to get a better value of f. Repeat until successive values of D agree. This is a rapid method.

Results for circular pipes may be applied to noncircular ducts if the hydraulic diameter is used in place of the diameter of a circular pipe. Then the relative roughness is k/D_h and the Reynolds number is $\mathrm{Re} = VD_h/v$. Results are reasonably good for square ducts, rectangular ducts of aspect ratio up to about 8, equilateral ducts, hexagonal ducts, and concentric annular ducts of diameter ratio to about 0.75. In eccentric annular ducts where the pipes touch or nearly touch, and in tall narrow triangular ducts, both laminar and turbulent flow may exist at a section. Analyses mentioned here do not apply to these geometries.

40.11.4 Steady Incompressible Flow in Entrances of Ducts

The increased pressure drop in the entrance region of ducts as compared with that for the same length of fully developed flow is generally included in a correction term called a loss coefficient, k_L. Then,

$$\frac{p_1 - p}{\rho V^2/2} = \frac{fL}{D_h} + k_L$$

where p_1 is the pressure at the duct inlet and p is the pressure a distance L from the inlet. The value of k_L depends on L but becomes a constant in the fully developed region, and this constant value is of greatest interest.

For *laminar* flow the pressure drop in the entrance length L_e is obtained from the Bernoulli equation written along the duct axis where there is no shear in the core flow. This is

$$p_1 - p_e = \frac{\rho u_{\max}^2}{2} - \frac{\rho V^2}{2} = \left[\left(\frac{u_{\max}}{V}\right)^2 - 1\right]\frac{\rho V^2}{2}$$

for any duct for which $u_{\max}/V$ is known. When both friction factor and k_L are known, the entrance length is

$$\frac{L_e}{D_h} = \frac{1}{f}\left[\left(\frac{u_{\max}}{V}\right)^2 - 1 - k_L\right]$$

For a circular duct, experiments and analyses indicate that $k_L \approx 1.30$. Thus, for a circular duct, $L_e/D = (\mathrm{Re}_D/64)(2^2 - 1 - 1.30) = 0.027\mathrm{Re}_D$. The pressure drop for fully developed flow in a length L_e is $\Delta p = 1.70\rho V^2/2$ and thus the pressure drop in the entrance is $3/1.70 = 1.76$ times that in an equal length for fully developed flow. Entrance effects are important for short ducts.

Some values of k_L and $(L_e/D_h)\mathrm{Re}$ for laminar flow in various ducts are listed in Table 40.9.

For turbulent flow, loss coefficients are determined experimentally. Results are shown in Fig. 40.33. Flow separation accounts for the high loss coefficients for the square and reentrant shapes for circular tubes and concentric annuli. For a rounded entrance, a radius of curvature of $D/7$ or more precludes separation. The boundary layer starts laminar then changes to turbulent, and the pressure drop does not significantly exceed the corresponding value for fully developed flow in the same length. (It may even be less with the laminar boundary layer—a trip or slight roughness may force a turbulent boundary layer to exist at the entrance.)

Entrance lengths for circular ducts and concentric annuli are defined as the distance required for the pressure gradient to become within a specified percentage of the fully developed value (5%, for example). On this basis L_e/D_h is about 30 or less.

40.11.5 Local Losses in Contractions, Expansions, and Pipe Fittings; Turbulent Flow

Calculations of local head losses generally are approximate at best unless experimental data for given fittings are provided by the manufacturer.

Losses in *contractions* are given by $h_L = k_L V^2/2g$. Loss coefficients for a sudden contraction are shown in Fig. 40.34. For gradually contracting sections k_L may be as low as 0.03 for D_2/D_1 of 0.5 or less.

Losses in expansions are given by $h_L = k_L(V_1 - V_2)^2/2g$, section 1 being upstream. For a sudden expansion, $k_L = 1$, and for gradually expanding sections with divergence angles of 7° or 8°, k_L may be as low as 0.14 or even 0.06 for diffusers for low-speed wind tunnels or cavitation-testing water tunnels with curved inlets to avoid separation.

Losses in pipe fittings are given in the form $h_L = k_L V^2/2g$ or in terms of an equivalent pipe length by pipe-fitting manufacturers. Typical values for various fittings are given in Table 40.10.

Table 40.9 Entrance Effects, Laminar Flow (See Table 40.8 for Symbols)

r_1/r_2	k_L		a/b	k_L	$L_c D_h$ Re		x	k_L
0.0001	1.13		0	0.69	0.0059		0	1.74
0.001	1.07		1/8	0.88	0.0094		10	1.73
0.01	0.97		1/5	1.00	0.0123		20	1.72
0.05	0.86		1/4	1.08	0.0146		30	1.69
0.10	0.81		1/2	1.38	0.0254		40	1.65
0.20	0.75		3/4	1.52	0.0311		60	1.57
0.40	0.71		1	1.55	0.0324		90	1.46
0.60	0.69						120	1.39
0.80	0.69						150	1.34
1.00	0.69						180	1.33

α	Circular Sector k_L	Isosceles Triangle k_L	Right Triangle k_L
0	2.97	2.97	2.97
10	2.06	2.14	2.40
20	1.71	1.85	2.09
30	1.58	1.79	1.94
40	1.53	1.83	1.88
50	1.50	1.95	1.88
60	1.49	2.14	1.94
70	1.48	2.38	2.09
80	1.47	2.72	2.40
90	1.46	2.97	2.97

40.11.6 Flow of Compressible Gases in Pipes with Friction

Subsonic gas flow in pipes involves a decrease in gas density and an increase in gas velocity in the direction of flow. The momentum equation for this flow may be written as

$$\frac{dp}{\rho V^2/2} + f\,\frac{dx}{D} + 2\,\frac{dV}{V} = 0$$

For *isothermal* flow the first term is $(2/\rho_1 V_1^2 p_1)p\,dp$, where the subscript 1 refers to an upstream section where all conditions are known. For $L = x_2 - x_1$, integration gives

$$p_1^2 - p_2^2 = \rho_1 V_1^2 p_1 \left(f\,\frac{L}{D} - 2\ln\frac{p_2}{p_1} \right)$$

or, in terms of the initial Mach number,

$$p_1^2 - p_2^2 = kM_1^2 p_1^2 \left(f\,\frac{L}{D} - 2\ln\frac{p_2}{p_1} \right)$$

The downstream pressure p_2 at a distance L from section 1 may be obtained by trial by neglecting

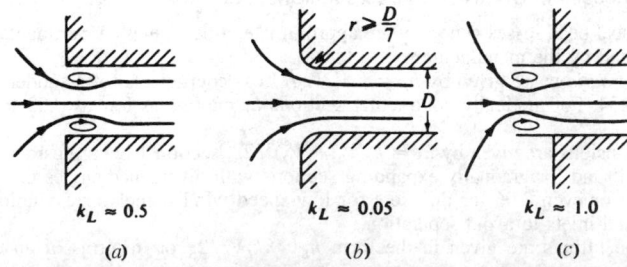

$k_L \approx 0.5$ $k_L \approx 0.05$ $k_L \approx 1.0$

(a)   (b)   (c)

Fig. 40.33 Pipe entrance flows: (a) square entrance; (b) round entrance; (c) reentrant inlet.

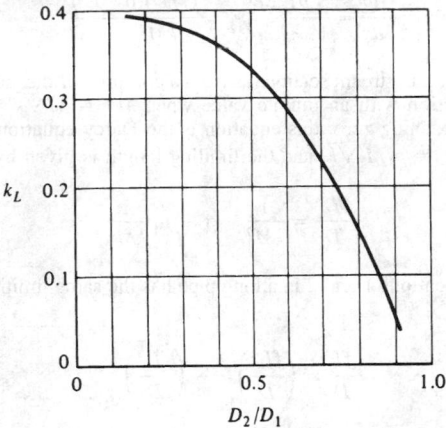

Fig. 40.34 Loss coefficients for abrupt contract in pipes.

the term $2 \ln(p_2/p_1)$ initially to get a p_2, then including it for an improved value. The distance L is a section where the pressure is p_2 is obtained from

$$f \frac{L}{D} = \frac{1}{kM_1^2} \left[1 - \left(\frac{p_2}{p_1} \right)^2 \right] - 2 \ln \frac{p_1}{p_2}$$

A limiting condition (designated by an asterisk) at a length L^* is obtained from an expression dp/dx to get

Table 40.10 Typical Loss Coefficients for Valves and Fittings

Valve or Fitting	Nominal Diameter, CM					
	2.5	5	10	15	20	25
Globe valve, wide open:						
Screwed	9	7	5.5			
Flanged	12	9	6	6	5.5	5.5
Gate valve, wide open:						
Screwed	0.24	0.18	0.13			
Flanged		0.35	0.16	0.11	0.08	0.06
Foot valve, wide open			0.80 for all sizes			
Swing check valve, wide open						
Screwed	3.0	2.3	2.1			
Flanged			2.0 for all sizes			
Angle valve, wide open:						
Screwed	4.5	2.1	1.0			
Flanged		2.4	2.1	2.1	2.1	2.1
Regular elbow, 90°						
Screwed	1.5	1.0	0.65			
Flanged	0.42	0.37	0.31	0.28	0.26	0.25
Long-radius elbow, 90°						
Screwed	0.75	0.4	0.25			
Flanged		0.3	0.22	0.18	0.15	0.14

Note: The k_L values listed may be expressed in terms of an equivalent pipe length for a given installation and flow by equating $k_L = fL_c/D$ so that $L_e = k_L D/f$.

SOURCE: Reproduced, with permission, from Engineering Data Book: Pipe Friction Manual (Cleveland: Hydraulic Institute, 1979).

$$\frac{dp}{dx} = \frac{pf/2D}{1 - p/\rho V^2} = \frac{(f/D)(\rho V^2/2)}{kM^2 - 1}$$

For a low subsonic flow at an upstream section (as from a compressor discharge) the pressure gradient increases in the flow direction with an infinite value when $M^* = 1/\sqrt{k} = 0.845$ for $k = 1.4$ (air, for example). For M approaching zero, this equation is the Darcy equation for incompressible flow. The limiting pressure is $p^* = p_1 M_1 \sqrt{k}$, and the limiting length is given by

$$\frac{fL^*}{D} = \frac{1}{kM_1^2} - 1 - \ln \frac{1}{kM_1^2}$$

Since the gas at any two locations 1 and 2 in a long pipe has the same limiting condition, the distance L between them is

$$\frac{fL}{D} = \left(\frac{fL^*}{D}\right)_{M_1} - \left(\frac{fL^*}{D}\right)_{M_2}$$

Conditions along a pipe for various initial Mach numbers are shown in Fig. 40.35.

For *adiabatic* flow the limiting Mach number is $M^* = 1$. This is from an expression for dp/dx for adiabatic flow:

$$\frac{dp}{dx} = -\frac{fkp}{2D} M^2 \left[\frac{1 + (k - 1)M^2}{1 - M^2}\right] = -\frac{f}{D} \frac{\rho V^2}{2} \left[\frac{1 + (k - 1)M^2}{1 - M^2}\right]$$

The limiting pressure is

$$\frac{p^*}{p_1} = M_1 \sqrt{\frac{2[1 + \frac{1}{2}(k - 1)M_1^2]}{k + 1}}$$

and the limiting length is

$$\frac{\bar{f}L^*}{D} = \frac{1 - M_1^2}{kM_1^2} + \frac{k + 1}{2k} \ln \frac{(k + 1)M_1^2}{2[1 + \frac{1}{2}(k - 1)M_1^2]}$$

Except for subsonic flow at high Mach numbers, isothermal and adiabatic flow do not differ appreciably. Thus, since flow near the limiting condition is not recommended in gas transmission

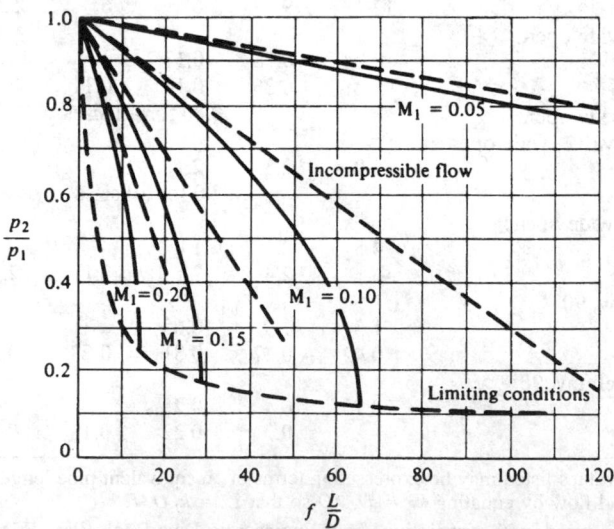

Fig. 40.35 Isothermal gas flow in a pipe for various initial Mach numbers, $k = 1.4$.

pipelines because of the excessive pressure drop, and since purely isothermal or purely adiabatic flow is unlikely, either adiabatic or isothermal flow may be assumed in making engineering calculations. For example, for methane from a compressor at 2000 kPa absolute pressure, 60°C temperature and 15 m/sec velocity ($M_1 = 0.032$) in a 30-cm commercial steel pipe, the limiting pressure is 72 kPa absolute at $L^* = 16.9$ km for isothermal flow, and 59 kPa at $L^* = 17.0$ km for adiabatic flow. A pressure of 500 kPa absolute would exist at 16.0 km for either type of flow.

40.12 DYNAMIC DRAG AND LIFT

Two types of forces act on a body past which a fluid flows: a pressure force normal to any infinitesimal area of the body and a shear force tangential to this area. The components of these two forces integrate over the entire body in a direction parallel to the approach flow is the *drag* force, and in a direction normal to it is the *lift* force. *Induced* drag is associated with a lift force on finite airfoils or blank elements as a result of downwash from tip vortices. Surface waves set up by ships or hydrofoils, and compression waves in gases such as Mach cones are the source of *wave* drag.

40.12.1 Drag

A drag force is $D = C(\rho u_s^2/2)A$, where C is the drag coefficient, $\rho u_s^2/2$ is the dynamic pressure of the free stream, and A is an appropriate area. For pure viscous shear drag C is C_f, the skin friction drag coefficient of Section 40.9.2 and A is the area sheared. In general, C is designated C_D, the drag coefficient for drag other than that from viscous shear only, and A is the chord area for lifting vanes or the projected frontal area for other shapes.

The drag coefficient for incompressible flow with pure pressure drag (a flat plate normal to a flow, for example) or for combined skin friction and pressure drag, which is called *profile* drag, depends on the body shape, the Reynolds number, and, usually, the location of boundary layer transition.

Drag coefficients for spheres and for flow normal to infinite circular cylinders are shown in Fig. 40.36. For spheres at $Re_D < 0.1$, $C_D = 24/Re_D$ and for $Re_D < 100$, $C_D = (24/Re_D)(1 + 3 Re_D/16)^{1/2}$. The boundary layer for both shapes up to and including the flat portion of the curves before the rather abrupt drop in the neighborhood of $Re_D = 10^5$ is laminar. This is called the *subcritical* region; beyond that is the *supercritical* region. Table 40.11 lists typical drag coefficients for two-dimensional shapes, and Table 40.12 lists them for three-dimensional shapes.

The drag of spheres, circular cylinders, and streamlined shapes is affected by boundary layer separation, which, in turn, depends on surface roughness, the Reynolds number, and free stream turbulence. These factors contribute to uncertainties in the value of the drag coefficient.

40.12.2 Lift

Lift in a nonviscous fluid may be produced by prescribing a circulation around a cylinder or lifting vane. In a viscous fluid this may be produced by spinning a ping-pong ball, a golf ball, or a baseball, for example, Circulation around a lifting vane in a viscous fluid results from the bound vortex or countercirculation that is equal and opposite to the starting vortex, which peels off the trailing edge of the vane. The lift is calculated from $L = C_L(\rho u_s^2/2)A$, where C_L is the lift coefficient, $\rho u_s^2/2$ is the

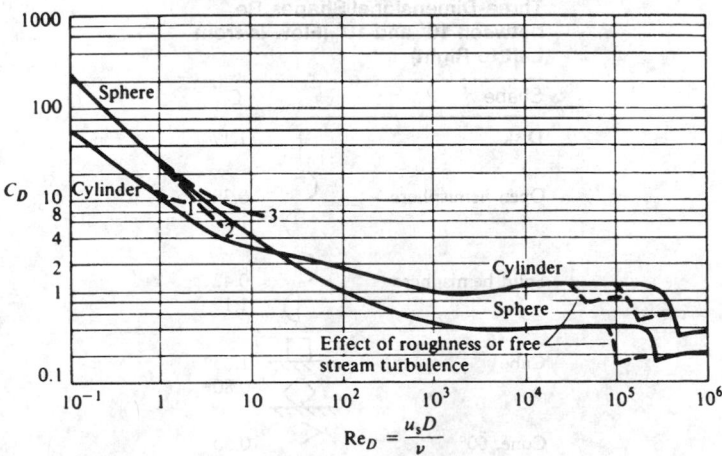

Fig. 40.36 Drag coefficients for infinite circular cylinders and spheres: (1) Lamb's solution for cylinder; (2) Stokes' solution for sphere; (3) Oseen's solution for sphere.

Table 40.11 Drag Coefficients for Two-Dimensional Shapes at Re = 10^5 Based on Frontal Projected Area (Flow is from Left to Right)

Shape		C_D	Shape	C_D	
Plate	|	2.0	Rectangle		
			1:1	1.18	
Open tube	(	1.2	5:1	1.2	
	)	2.3	10:1	1.3	
			20:1	1.5	
Half cylinder	(|	1.16	Elliptical	Below	Above
	|)	1.7	Cylinder	Re_c	Re_c
			2:1	0.6	0.20
Square cylinder	▢	2.05	4:1	0.36	0.10
	◇	1.55	8:1	0.26	0.10
Equilateral	▷	2.0			
triangle	◁	1.6			

dynamic pressure of the free stream, and A is the chord area of the lifting vane. Typical values of C_L as well as C_D are shown in Fig. 40.37. The induced drag and the profile drag are shown. The profile drag is the difference between the dashed and solid curves. The induced drag is zero at zero lift.

40.13 FLOW MEASUREMENTS

Fluid flow measurements generally involve determining static pressures, local and average velocities, and volumetric or mass flow rates.

40.13.1 Pressure Measurements

Static pressures are measured by means of a small hole in a boundary surface connected to a sensor—a manometer, a mechanical pressure gage, or an electrical transducer. The surface may be a duct wall or the outer surface of a tube, such as those shown in Fig. 40.38. In any case, the surface past which the fluid flows must be smooth, and the tapped holes must be at right angles to the surface.

Table 40.12 Drag Coefficients for Three-Dimensional Shapes Re between 10^4 and 10^6 (Flow is from Left to Right)

Shape		C_D
Disk	|	1.17
Open hemisphere	(	0.38
	)	1.42
Solid hemisphere	(|	0.42
	|)	1.17
Cube	▢	1.05[a]
	◇	0.80[a]
Cone, 60°	<	0.50

[a]Mounted on a boundary wall.

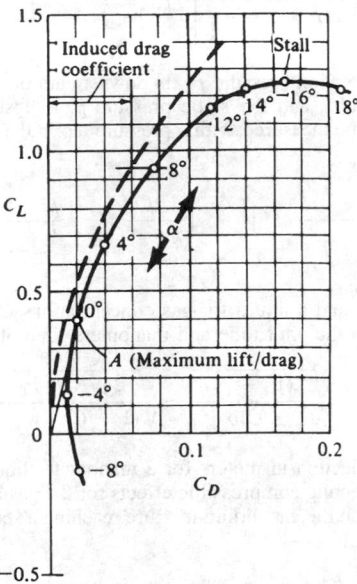

Fig. 40.37 Typical polar diagram showing lift–drag characteristics for an airfoil of finite span.

Total or stagnation pressures are easily measured accurately with an open-ended tube facing into the flow, as shown in Fig. 40.38.

40.13.2 Velocity Measurements

A combined pitot tube (Fig. 40.38) measures or detects the difference between the total or stagnation pressure p_0 and the static pressure p. For an incompressible fluid the velocity being measured is $V = \sqrt{2(p_0 - p)/\rho}$. For subsonic gas flow the velocity of a stream at a temperature T and pressure p in

$$V = \sqrt{\frac{2kRT}{k-1} \left[\left(\frac{p_0}{p} \right)^{(k-1)/k} - 1 \right]}$$

and the corresponding Mach number is

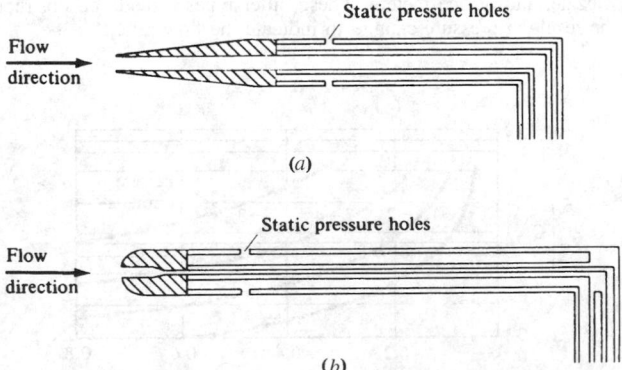

Fig. 40.38 Combined pitot tubes: (a) Brabbee's design; (b) Prandtl's design—accurate over a greater range of yaw angles.

$$M = \sqrt{\frac{2}{k-1}\left[\left(\frac{p_0}{p}\right)^{(k-1)/k} - 1\right]}$$

For supersonic flow the stagnation pressure p_{0y} is downstream of a shock, which is detached and ahead of the open stagnation tube, and the static pressure p_x is upstream of the shock. In a wind tunnel the static pressure could be measured with a pressure tap in the tunnel wall. The Mach number M of the flow is

$$\frac{p_{0y}}{p} = \left(\frac{k+1}{2}M^2\right)^{k/(k-1)}\left(\frac{2k}{k+1}M^2 - \frac{k-1}{k+1}\right)^{1/(1-k)}$$

which is tabulated in gas tables.

In a mixture of gas bubbles and a liquid for gas concentrations C no more than 0.6 by volume, the velocity of the mixture with the pitot tube and manometer free of bubbles is

$$V_{\text{mixture}} = \sqrt{\frac{2(p_0 - p_1)}{(1-C)\rho_{\text{liquid}}}} = \sqrt{\frac{2gh_m}{(1-C)}\left(\frac{\gamma_m}{\gamma_{\text{liquid}}} - 1\right)}$$

where h_m is the manometer deflection in meters for a manometer liquid of specific weight γ_m. The error in this equation from neglecting compressible effects for the gas bubbles is shown in Fig. 40.39. A more correct equation based on the gas–liquid mixture reaching a stagnation pressure isentropically is

$$\frac{V_1^2}{2} = \frac{p_0 - p_1}{\rho_u(1-C)} + \frac{C}{1-C}\left(\frac{p_1}{\rho_u}\right)\left[\frac{k}{k-1}\left(\frac{p_0}{p_1}\right)^{(k-1)/k} - \frac{1}{k-1} - \left(\frac{p_0}{p_1}\right)\right]$$

but is cumbersome to use. As indicated in Fig. 40.39 the error in using the first equation is very small for high concentrations of gas bubbles at low speeds and for low concentrations at high speeds.

If n velocity readings are taken at the centroid of n subareas in a duct, the average velocity V from the point velocity readings u_i is

$$V = \frac{1}{n}\sum_{i=1}^{n} u_i$$

In a circular duct, readings should be taken at $(r/R)^2 = 0.055, 0.15, 0.25, \ldots, 0.95$. Velocities measured at other radial positions may be plotted versus $(r/R)^2$, and the area under the curve may be integrated numerically to obtain the average velocity.

Other methods of measuring fluid velocities include length–time measurements with floats or neutral-buoyancy particles, rotating instruments such as anemometers and current meters, hot-wire and hot-film anemometers, and laser-doppler anemometers.

40.13.3 Volumetric and Mass Flow Fluid Measurements

Liquid flow rates in pipes are commonly measured with commercial water meters; with rotameters; and with venturi, nozzle, and orifice meters. These latter types provide an obstruction in the flow and make use of the resulting pressure change to indicate the flow rate.

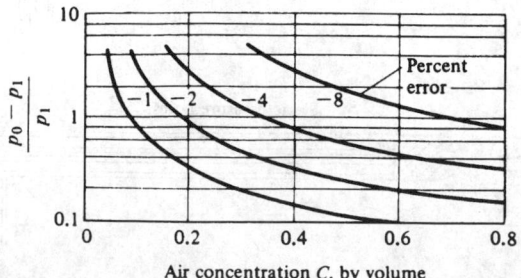

Fig. 40.39 Error in neglecting compressibility of air in measuring velocity of air–water mixture with a combined pitot tube.

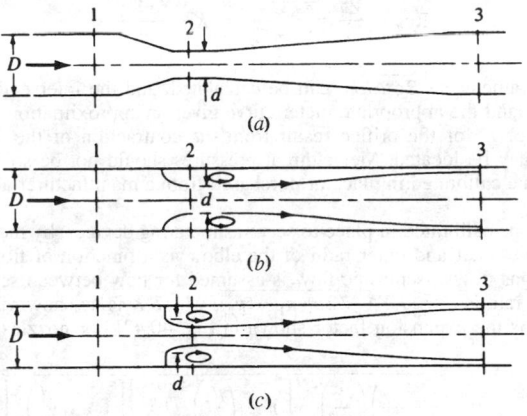

Fig. 40.40 Pipe flow meters: (a) venturi; (b) nozzle; (c) concentric orifice.

The continuity and Bernoulli equations for liquid flow applied between sections 1 and 2 in Fig. 40.40 give the ideal volumetric flow rate as

$$Q_{ideal} = \frac{A_2\sqrt{2g\,\Delta h}}{\sqrt{1 - (A_2/A_1)^2}}$$

where Δh is the change in piezometric head. A form of this equation generally used is

$$Q = K\left(\frac{\pi d^2}{4}\right)\sqrt{2g\,\Delta h}$$

where K is the flow coefficient, which depends on the type of meter, the diameter ratio d/D, and the viscous effects given in terms of the Reynolds number. This is based on the length parameter d and the velocity V through the hole of diameter d. Approximate flow coefficients are given in Fig. 40.41. The relation between the flow coefficient K and this Reynolds number is

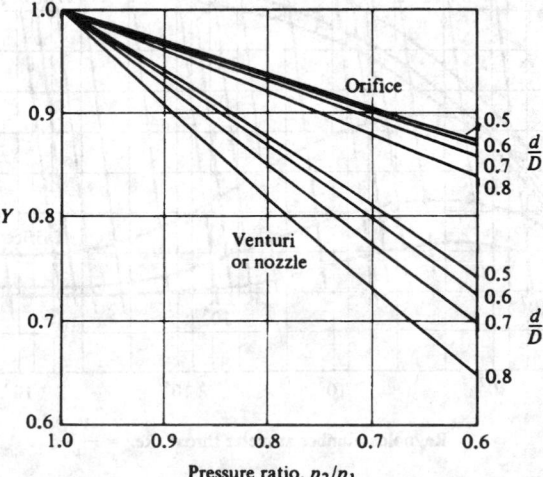

Fig. 40.41 Approximate flow coefficients for pipe meters.

$$\mathrm{Re}_d = \frac{Vd}{v} = \frac{Qd}{\frac{1}{4}\pi d^2 v} = K\frac{d\sqrt{2g\,\Delta h}}{v}$$

The dimensionless parameter $d\sqrt{2g\,\Delta h}/v$ can be calculated, and the intersection of the appropriate line for this parameter and the appropriate meter curve gives an approximation to the flow coefficient K. The lower values of K for the orifice result from the contraction of the jet beyond the orifice where pressure taps may be located. Meter throat pressures should not be so low as to create cavitation. Meters should be calibrated in place or purchased from a manufacturer and installed according to instructions.

Elbow meters may be calibrated in place to serve as metering devices, by measuring the difference in pressure between the inner and outer radii of the elbow as a function of flow rate.

For compressible gas flows, isentropic flow is assumed for flow between sections 1 and 2 in Fig. 40.40. The mass flow rate is $\dot{m} = KYA_2\sqrt{2\rho_1(p_1 - p_2)}$, where K is as shown in Fig. 40.41 and $Y = Y(k, p_2/p_1, d/D)$ and is the expansion factor shown in Fig. 40.42. For nozzles and venturi tubes

$$Y = \sqrt{\frac{\left(\dfrac{k}{k-1}\right)\left(\dfrac{p_2}{p_1}\right)^{2/k}\left[1 - \left(\dfrac{p_2}{p_1}\right)^{(k-1)/k}\right]\left[1 - \left(\dfrac{d}{D}\right)^4\right]}{\left[1 - \left(\dfrac{p_2}{p_1}\right)\right]\left[1 - \left(\dfrac{d}{D}\right)^4\left(\dfrac{p_2}{p_1}\right)^{2/k}\right]}}$$

and for orifice meters

$$Y = 1 - \frac{1}{k}\left[0.41 + 0.35\left(\frac{d}{D}\right)^4\right]\left(1 - \frac{p_2}{p_1}\right)$$

These are the basic principles of fluid flow measurements. Utmost care must be taken when accurate measurements are necessary, and reference to meter manufacturers' pamphlets or measurements handbooks should be made.

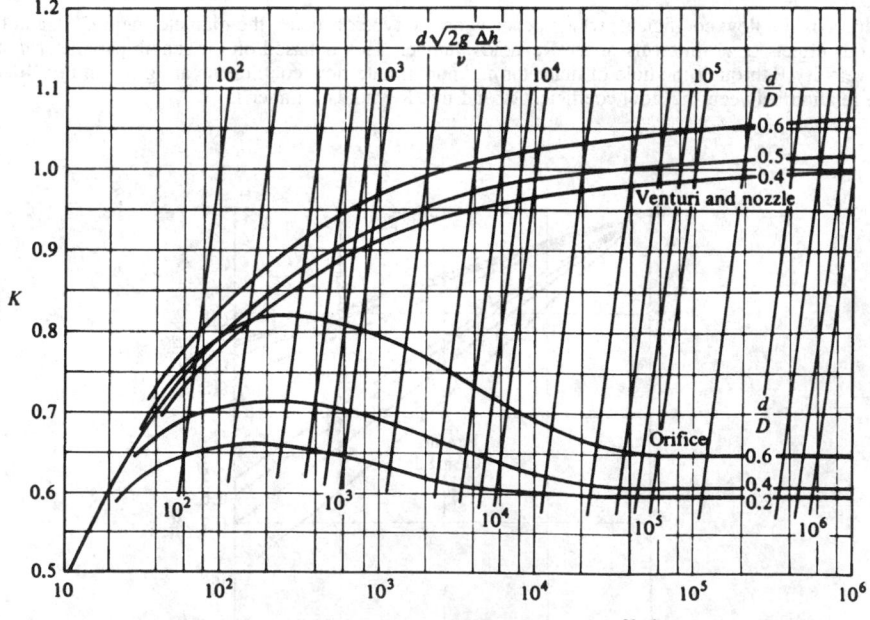

Fig. 40.42 Expansion factors for pipe meters, $k = 1.4$.

BIBLIOGRAPHY

General

Olson, R. M., *Essentials of Engineering Fluid Mechanics,* 4th ed., Harper and Row, New York, 1980.

Streeter, V. L. (ed.), *Handbook of Fluid Dynamics,* McGraw-Hill, New York, 1961.

Streeter, V. L., and E. B. Wylie, *Fluid Mechanics,* McGraw-Hill, New York, 1979.

Section 40.9

Schlichting, H., *Boundary Layer Theory* (translated by J. Kestin), 7th ed., McGraw-Hill, New York, 1979.

Section 40.10

Shapiro, A. H., *The Dynamics and Thermodynamics of Compressible Fluid Flow,* Ronald Press, New York, 1953, Vol. I.

Section 40.12

Hoerner, S. F., *Fluid-Dynamic Drag,* S. F. Hoerner, Midland Park, NJ, 1958.

Section 40.13

Miller, R. W., *Flow Measurement Engineering Handbook,* McGraw-Hill, New York, 1983.

Ower, E., and R. C. Pankhurst, *Measurement of Air Flow,* Pergamon Press, Elmsford, NY, 1977.

CHAPTER 41

THERMODYNAMICS FUNDAMENTALS

Adrian Bejan
Department of Mechanical Engineering and Materials Science
Duke University
Durham, North Carolina

41.1	INTRODUCTION	1331	41.5	RELATIONS AMONG THERMODYNAMIC PROPERTIES	1339
41.2	THE FIRST LAW OF THERMODYNAMICS FOR CLOSED SYSTEMS	1333	41.6	IDEAL GASES	1341
41.3	THE SECOND LAW OF THERMODYNAMICS FOR CLOSED SYSTEMS	1335	41.7	INCOMPRESSIBLE SUBSTANCES	1344
			41.8	TWO-PHASE STATES	1344
41.4	THE LAWS OF THERMODYNAMICS FOR OPEN SYSTEMS	1338	41.9	ANALYSIS OF ENGINEERING SYSTEM COMPONENTS	1347

41.1 INTRODUCTION

Thermodynamics has historically grown out of man's determination—as Sadi Carnot put it—to capture "the motive power of fire." Relative to mechanical engineering, thermodynamics describes the relationship between mechanical work and other forms of energy. There are two facets of contemporary thermodynamics that must be stressed in a review such as this. The first is the equivalence of *work* and *heat* as two possible forms of energy exchange. This facet is encapsulated in the first law of thermodynamics. The second aspect is the irreversibility of all processes (changes) that occur in nature. As summarized by the second law of thermodynamics, irreversibility or entropy generation is what prevents us from extracting the most possible work from various sources; it is also what prevents us from doing the most with the work that is already at our disposal. The objective of this chapter is to review the first and second laws of thermodynamics and their implications in mechanical engineering, particularly with respect to such issues as energy conversion and conservation. The analytical aspects (the formulas) of engineering thermodynamics are reviewed primarily in terms of the behavior of a pure substance, as would be the case of the working fluid in a heat engine or in a refrigeration machine. In the next chapter we review in greater detail the newer field of entropy generation minimization (thermodynamic optimization).

SYMBOLS AND UNITS

c	specific heat of incompressible substance, $J/(kg \cdot K)$
c_P	specific heat at constant pressure, $J/(kg \cdot K)$
c_T	constant temperature coefficient, m^3/kg
c_v	specific heat at constant volume, $J/(kg \cdot K)$
COP	coefficient of performance
E	energy, J
f	specific Helmholtz free energy $(u - Ts)$, J/kg
$\overline{F}$	force vector, N

Mechanical Engineers' Handbook, 2nd ed., Edited by Myer Kutz.
ISBN 0-471-13007-9 © 1998 John Wiley & Sons, Inc.

g	gravitational acceleration, m/sec^2
g	specific Gibbs free energy ($h - Ts$), J/kg
h	specific enthalpy ($u + Pv$), J/kg
K	isothermal compressibility, m^2/N
m	mass of closed system, kg
$\dot{m}$	mass flow rate, kg/sec
m_i	mass of component in a mixture, kg
M	mass inventory of control volume, kg
M	molar mass, g/mol or kg/kmol
n	number of moles, mol
N_0	Avogadro's constant
P	pressure
δQ	infinitesimal heat interaction, J
$\dot{Q}$	heat transfer rate, W
$\bar{r}$	position vector, m
R	ideal gas constant, J/(kg $\cdot$ K)
s	specific entropy, J/(kg $\cdot$ K)
S	entropy, J/K
S_{gen}	entropy generation, J/K
$\dot{S}_{\text{gen}}$	entropy generation rate, W/K
T	absolute temperature, K
u	specific internal energy, J/kg
U	internal energy, J
v	specific volume m^3/kg
$\bar{v}$	specific volume of incompressible substance, m^3/kg
V	volume, m^3
V	velocity, m/sec
δW	infinitesimal work interaction, J
$\dot{W}_{\text{lost}}$	rate of lost available work, W
$\dot{W}_{\text{sh}}$	rate of shaft (shear) work transfer, W
x	linear coordinate, m
x	quality of liquid and vapor mixture
Z	vertical coordinate, m
β	coefficient of thermal expansion, 1/K
γ	ratio of specific heats, c_p/c_v
η	"efficiency" ratio
η_{I}	first-law efficiency
η_{II}	second-law efficiency
θ	relative temperature, °C

SUBSCRIPTS

$()_{\text{in}}$	inlet port
$()_{\text{out}}$	outlet port
$()_{\text{rev}}$	reversible path
$()_H$	high-temperature reservoir
$()_L$	low-temperature reservoir
$()_{\text{max}}$	maximum
$()_T$	turbine
$()_C$	compressor
$()_N$	nozzle
$()_D$	diffuser
$()_1$	initial state
$()_2$	final state
$()_0$	reference state

$()_f$ saturated liquid state (f = "fluid")
$()_g$ saturated vapor state (g = "gas")
$()_s$ saturated solid state (s = "solid")
$()_*$ moderately compressed liquid state
$()_+$ slightly superheated vapor state

Definitions

THERMODYNAMIC SYSTEM is the region or the collection of matter in space selected for analysis.
ENVIRONMENT is the thermodynamic system external to the system of interest, that is, external to the region selected for analysis or for discussion.
BOUNDARY is the real or imaginary surface delineating the system of interest. The boundary separates the system from its environment. The boundary is an unambiguously defined surface. The boundary has zero thickness.
CLOSED SYSTEM is the thermodynamic system whose boundary is not penetrated (crossed) by the flow of mass.
OPEN SYSTEM, or flow system, is the thermodynamic system whose boundary is permeable to mass flow. Open systems have their own nomenclature, so that the thermodynamic system is usually referred to as the *control volume*, the boundary of the open system is the *control surface*, and the particular regions of the boundary that are crossed by mass flows are the *inlet* or *outlet ports*.
STATE is the condition (the being) of a thermodynamic system at a particular point in time, as described by an ensemble of quantities called *thermodynamic properties* (e.g., pressure, volume, temperature, energy, enthalpy, entropy). Thermodynamic properties are only those quantities that depend solely on the instantaneous state of the system. Thermodynamic properties do not depend on the "history" of the system between two different states. Quantities that depend on the system evolution (path) between states are not thermodynamic properties (examples of nonproperties are the work, heat, and mass transfer interactions; the entropy transfer interactions; the entropy generation; and the lost available work—see also the definition of *process*).
EXTENSIVE PROPERTIES are properties whose values depend on the size of the system (e.g., mass, volume, energy, enthalpy, entropy).
INTENSIVE PROPERTIES are properties whose values do not depend on the system size (e.g., pressure, temperature). The collection of all intensive properties (or the properties of an infinitesimally small element of the system, including the per-unit-mass properties, such as specific energy and specific entropy) constitutes the *intensive state*.
PHASE is the collection of all system elements that have the same intensive state (e.g., the liquid droplets dispersed in a liquid–vapor mixture have the same intensive state, that is, the same pressure, temperature, specific volume, specific entropy, etc.).
PROCESS is the change of state from one initial state to a final state. In addition to the end states, knowledge of the process implies knowledge of the *interactions* experienced by the system while in communication with its environment (e.g., work transfer, heat transfer, mass transfer, and entropy transfer). To know the process also means to know the *path* (the history, or the succession of states) followed by the system from the initial to the final state.
CYCLE is a special process in which the final state coincides with the initial state.

41.2 THE FIRST LAW OF THERMODYNAMICS FOR CLOSED SYSTEMS

The first law of thermodynamics is a statement that brings together three concepts in thermodynamics: work transfer, heat transfer, and energy change. Of these concepts, only energy change or, simply, energy, is, in general, a thermodynamic property. Before stating the first law and before writing down the *equation* that accounts for this statement, it is necessary to review[1] the concepts of work transfer, heat transfer, and energy change.

Consider the force F_x experienced by a certain system at a point on its boundary. Relative to this system, the infinitesimal *work transfer* interaction between system and environment is

$$\delta W = -F_x dx$$

where the boundary displacement dx is defined as positive in the direction of the force F_x. When the force $\overline{F}$ and the displacement of its point of application $d\overline{r}$ are not collinear, the general definition of infinitesimal work transfer is

$$\delta W = -\overline{F} \cdot d\overline{r}$$

The work transfer interaction is considered positive when the system does work on its environment—in other words, when $\overline{F}$ and $d\overline{r}$ point in opposite directions. This sign convention has its origin in heat engine engineering, since the purpose of heat engines as thermodynamic systems is to deliver work while receiving heat.

In order for a system to experience work transfer, two things must occur: (1) a force must be present on the boundary, and (2) the point of application of this force (hence, the boundary) must move. The mere presence of forces on the boundary, without the displacement or the deformation of the boundary, does not mean work transfer. Likewise, the mere presence of boundary displacement without a force opposing or driving this motion does not mean work transfer. For example, in the free expansion of a gas into an evacuated space, the gas system does not experience work transfer because throughout the expansion the pressure at the imaginary system–environment interface is zero.

If a closed system can interact with its environment only via work transfer (i.e., in the absence of heat transfer δQ discussed later), then it is observed that the work transfer during a change of state from state 1 to state 2 is the same for all processes linking states 1 and 2,

$$-\left(\int_1^2 \delta W\right)_{\delta Q = 0} = E_2 - E_1$$

In this special case the work transfer interaction $(W_{1-2})_{\delta Q=0}$ is a property of the system, since its value depends solely on the end states. This thermodynamic property is the *energy change* of the system, $E_2 - E_1$. The statement that preceded the last equation is the first law of thermodynamics for closed systems that do not experience heat transfer.

Heat transfer is, like work transfer, an energy interaction that can take place between a system and its environment. The distinction between δQ and δW is made by the second law of thermodynamics discussed in the next section: Heat transfer is the energy interaction accompanied by entropy transfer, whereas work transfer is the energy interaction taking place in the absence of entropy transfer. The transfer of heat is driven by the *temperature difference* established between the system and its environment.[2] The system temperature is measured by placing the system in thermal communication with a test system called *thermometer.* The result of this measurement is the *relative temperature* θ expressed in degrees Celsius, $\theta(°C)$, or Fahrenheit, $\theta(°F)$; these alternative temperature readings are related through the conversion formulas

$$\theta(°C) = \frac{5}{9}[\theta(°F) - 32]$$
$$\theta(°F) = \frac{9}{5}\theta(°C) + 32$$
$$1°F = \frac{5}{9}°C$$

The boundary that prevents the transfer of heat, regardless of the magnitude of the system–environment temperature difference, is termed *adiabatic.* Conversely, the boundary that is the locus of heat transfer even in the limit of vanishingly small system–environment temperature difference is termed *diathermal.*

It is observed that a closed system undergoing a change of state $1 \rightarrow 2$ in the absence of work transfer experiences a heat-transfer interaction whose magnitude depends solely on the end states:

$$\left(\int_1^2 \delta Q\right)_{\delta W = 0} = E_2 - E_1$$

In the special case of zero work transfer, the heat-transfer interaction is a thermodynamic property of the system, which is by definition equal to the energy change experienced by the system in going from state 1 to state 2. The last equation is the first law of thermodynamics for closed systems incapable of experiencing work transfer. Note that, unlike work transfer, the heat transfer is considered positive when it increases the energy of the system.

Most thermodynamic systems do not manifest the purely mechanical ($\delta Q = 0$) or purely thermal ($\delta W = 0$) behavior discussed to this point. Most systems manifest a *coupled* mechanical and thermal behavior. The preceding first-law statements can be used to show that the first law of thermodynamics for a process executed by a closed system experiencing both work transfer and heat transfer is

$$\underbrace{\int_1^2 \delta Q}_{\substack{\text{heat} \\ \text{transfer}}} - \underbrace{\int_1^2 \delta W}_{\substack{\text{work} \\ \text{transfer}}} = \underbrace{E_2 - E_1}_{\substack{\text{energy} \\ \text{change}}}$$

$$\underbrace{}_{\substack{\text{energy interactions} \\ \text{(nonproperties)}}} \quad \text{(property)}$$

The first law means that the net heat transfer into the system equals the work done by the system on the environment, plus the increase in the energy of the system. The first law of thermodynamics for a cycle or for an integral number of cycles executed by a closed system is

$$\oint \delta Q = \oint \delta W = 0$$

Note that the net change in the thermodynamic property energy is zero during a cycle or an integral number of cycles.

The energy change term $E_2 - E_1$ appearing on the right-hand side of the first law can be replaced by a more general notation that distinguishes between macroscopically identifiable forms of energy storage (kinetic, gravitational) and energy stored internally,

$$E_2 - E_1 = \underbrace{U_2 - U_1}_{} + \underbrace{\frac{mV_2^2}{2} - \frac{mV_1^2}{2}}_{} + \underbrace{mgZ_2 - mgZ_1}_{}$$

| energy change | internal energy change | kinetic energy change | gravitational energy change |

If the closed system expands or contracts *quasi-statically* (i.e., slowly enough, in mechanical equilibrium internally and with the environment) so that at every point in time the pressure P is uniform throughout the system, then the work transfer term can be calculated as being equal to the work done by all the boundary pressure forces as they move with their respective points of application,

$$\int_1^2 \delta W = \int_1^2 P \, dV$$

The work-transfer integral can be evaluated provided the path of the quasi-static process, $P(V)$, is known; this is another illustration that the work transfer is path-dependent (i.e., not a thermodynamic property).

41.3   THE SECOND LAW OF THERMODYNAMICS FOR CLOSED SYSTEMS

A *temperature reservoir* is a thermodynamic system that experiences only heat-transfer interactions and whose temperature remains constant during such interactions. Consider first a closed system executing a cycle or an integral number of cycles *while in thermal communication with no more than one temperature reservoir*. To state the second law for this case is to observe that the net work transfer during each cycle cannot be positive,

$$\oint \delta W \leq 0$$

In other words, a closed system cannot deliver work during one cycle, while in communication with one temperature reservoir or with no temperature reservoir at all. Examples of such cyclic operation are the vibration of a spring–mass system, or the bouncing of a ball on the pavement: in order for these systems to return to their respective initial heights, that is, in order for them to execute cycles, the environment (e.g., humans) must perform work on them. The limiting case of frictionless cyclic operation is termed *reversible*, because in this limit the system returns to its initial state without intervention (work transfer) from the environment. Therefore, the distinction between reversible and irreversible cycles executed by closed systems in communication with no more than one temperature reservoir is

$$\oint \delta W = 0 \qquad \text{(reversible)}$$
$$\oint \delta W < 0 \qquad \text{(irreversible)}$$

To summarize, the first and second laws for closed systems operating cyclically in contact with no more than one temperature reservoir are (Fig. 41.1)

$$\oint \delta W = \oint \delta Q \leq 0$$

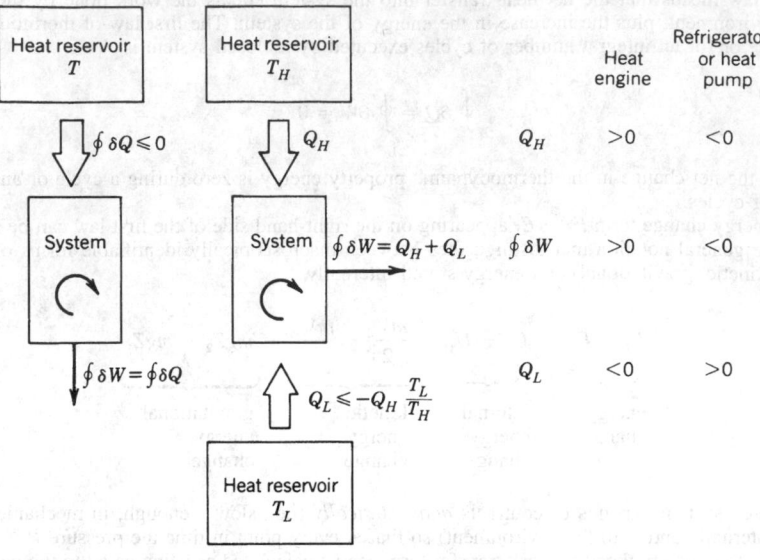

Fig. 41.1 The first and second laws of thermodynamics for a closed system operating cyclically while in communication with one or two heat reservoirs.

This statement of the second law can be used to show that in the case of a closed system executing one or an integral number of cycles *while in communication with two temperature reservoirs*, the following inequality holds (Fig. 41.1)

$$\frac{Q_H}{T_H} + \frac{Q_L}{T_L} \le 0$$

where H and L denote the high-temperature and the low-temperature reservoirs, respectively. Symbols Q_H and Q_L stand for the value of the cyclic integral $\oint \delta Q$, where δQ is in one case exchanged only with the H reservoir, and in the other with the L reservoir. In the reversible limit, the second law reduces to $T_H/T_L = -Q_H/Q_L$, which serves as definition for the absolute *thermodynamic temperature* scale denoted by symbol T. Absolute temperatures are expressed either in degrees Kelvin, $T(K)$, or in degrees Rankine, $T(R)$; the relationships between absolute and relative temperatures are

$$T(K) = \theta(°C) + 273.15 \text{ K} \qquad T(R) = \theta(°F) + 459.67 \text{ R}$$

$$1 \text{ K} = 1°C \qquad\qquad 1 \text{ R} = 1°F$$

A *heat engine* is a special case of a closed system operating cyclically while in thermal communication with two temperature reservoirs, a system that during each cycle intakes heat and delivers work:

$$\oint \delta W = \oint \delta Q = Q_H + Q_L > 0$$

The goodness of the heat engine can be described in terms of the heat engine efficiency or the first-law efficiency

$$\eta_1 = \frac{\oint \delta W}{Q_H} \le 1 - \frac{T_L}{T_H}$$

Alternatively, the second-law efficiency of the heat engine is defined as[1,3,4]

$$\eta_{\mathrm{II}} = \frac{\oint \delta W}{\left(\oint \delta W\right)_{\text{maximum (reversible case)}}} = \frac{\eta_h}{1 - T_L/T_H}$$

A *refrigerating machine* or a *heat pump* operates cyclically between two temperature reservoirs in such a way that during each cycle it intakes work and delivers net heat to the environment,

$$\oint \delta W = \oint \delta Q = Q_H + Q_L < 0$$

The goodness of such machines can be expressed in terms of a coefficient of performance (COP)

$$\text{COP}_{\text{refrigerator}} = \frac{Q_L}{-\oint \delta W} \leq \frac{1}{T_H/T_L - 1}$$

$$\text{COP}_{\text{heat pump}} = \frac{-Q_H}{-\oint \delta W} \leq \frac{1}{1 - T_L/T_H}$$

Generalizing the second law for closed systems operating cyclically, one can show that if during each cycle the system experiences any number of heat transfer interactions Q_i with any number of temperature reservoirs whose respective absolute temperatures are T_i, then

$$\sum_i \frac{Q_i}{T_i} \leq 0$$

Note that T_i is the absolute temperature of the boundary region crossed by Q_i. Another way to write the second law in this case is

$$\oint \frac{\delta Q}{T} \leq 0$$

where, again, T is the temperature of the boundary pierced by δQ. Of special interest is the reversible cycle limit, in which the second law states ($\oint \delta Q/T)_{\text{rev}} = 0$). According to the definition of thermodynamic property, the second law implies that during a reversible process the quantity $\delta Q/T$ is the infinitesimal change in a property of the system: by definition, that property is the *entropy change*

$$dS = \left(\frac{\delta Q}{T}\right)_{\text{rev}} \quad \text{or} \quad S_2 - S_1 = \left(\int_1^2 \frac{\delta Q}{T}\right)_{\text{rev}}$$

Combining this definition with the second-law statement for a cycle, $\oint \delta Q/T \leq 0$, yields the second law of thermodynamics for *any process* executed by a closed system,

$$\underbrace{S_2 - S_1}_{\substack{\text{entropy} \\ \text{change} \\ \text{(property)}}} - \underbrace{\int_1^2 \frac{\delta Q}{T}}_{\substack{\text{entropy} \\ \text{transfer} \\ \text{(nonproperty)}}} \geq 0$$

The left-hand side in this inequality is by definition as the *entropy generation* associated with the process,

$$S_{\text{gen}} = S_2 - S_1 - \int_1^2 \frac{\delta Q}{T}$$

The entropy generation is a measure of the inequality sign in the second law and hence a measure of the irreversibility of the process. As shown in the next section and chapter 42, the entropy generation is proportional to the useful work destroyed during the process.[1,3,4] Note again that any heat-transfer interaction (δQ) is accompanied by entropy transfer ($\delta Q/T$), whereas the work transfer δW is not.

41.4 THE LAWS OF THERMODYNAMICS FOR OPEN SYSTEMS

If $\dot{m}$ represents the mass flow rate of working fluid through a port in the control surface, then the principle of *mass conservation* in the control volume reads

$$\underbrace{\sum_{in} \dot{m} - \sum_{out} \dot{m}}_{\text{mass transfer}} = \underbrace{\frac{\partial M}{\partial t}}_{\text{mass change}}$$

Subscripts in and out refer to summation over all the inlet and outlet ports, respectively, while M stands for the instantaneous mass inventory of the control volume.

The first law of thermodynamics is more general than the statement encountered earlier for closed systems, because this time we must account for the flow of energy associated with the $\dot{m}$ streams.

$$\underbrace{\sum_{in} \dot{m} \left(h + \frac{V^2}{2} + gZ \right) - \sum_{out} \dot{m} \left(h + \frac{V^2}{2} + gZ \right) + \sum_{i} \dot{Q}_i - \dot{W}}_{\text{energy transfer}} = \underbrace{\frac{\partial E}{\partial t}}_{\substack{\text{energy} \\ \text{change}}}$$

On the left-hand side we have the energy transfer interactions: heat, work, and the energy transfer associated with mass flow across the control surface. The specific enthalpy h, fluid velocity V, and height Z are evaluated right at the boundary. On the right-hand side, E is the instantaneous system energy integrated over the control volume.

The second law of thermodynamics for an open system assumes the form

$$\underbrace{\sum_{in} \dot{m}s - \sum_{out} \dot{m}s + \sum_{i} \frac{\dot{Q}_i}{T_i}}_{\text{entropy transfer}} \leq \underbrace{\frac{\partial S}{\partial t}}_{\text{entropy change}}$$

The specific entropy s is representative of the thermodynamic state of each stream right at the system boundary. The *entropy generation rate* defined as

$$\dot{S}_{gen} = \frac{\partial S}{\partial t} + \sum_{out} \dot{m}s - \sum_{in} \dot{m}s - \sum_{i} \frac{\dot{Q}_i}{T_i}$$

is a measure of the irreversibility of open system operation. The engineering importance of $\dot{S}_{gen}$ stems from its proportionality to the rate of one-way destruction of available work. If the following parameters are fixed—all the mass flows ($\dot{m}$), the peripheral conditions (h, s, V, Z), and the heat-transfer interactions (Q_i, T_i) except (Q_0, T_0)— then one can use the first law and the second law to show that the work transfer rate cannot exceed a theoretical maximum:

$$\dot{W} \leq \sum_{in} \dot{m} \left(h + \frac{V^2}{2} + gZ - T_0 s \right) - \sum_{out} \dot{m} \left(h + \frac{V^2}{2} + gZ - T_0 s \right) - \frac{\partial}{\partial t} (E - T_0 s)$$

The right-hand side in this inequality is the maximum work transfer rate $\dot{W}_{sh,max}$, which exists only in the ideal limit of reversible operation. The rate of *lost work*, or exergy (availability) destruction, is defined as

$$\dot{W}_{lost} = \dot{W}_{max} - \dot{W}$$

Again, using both laws, one can show that lost work is directly proportional to entropy generation,

$$\dot{W}_{lost} = T_0 \dot{S}_{gen}$$

This result is known as the Gouy-Stodola theorem.[1,3] Conservation of useful (available) work in thermodynamic systems can only be achieved based on the systematic minimization of entropy generation in all the components of the system. Engineering applications of entropy generation minimization as a thermodynamic optimization philosophy may be found in Refs. 1, 3, and 4, and in the next chapter.

41.5 RELATIONS AMONG THERMODYNAMIC PROPERTIES

The analytical forms of the first and second laws of thermodynamics contain properties such as internal energy, enthalpy, and entropy, which cannot be measured directly. The values of these properties are derived from measurements that can be carried out in the laboratory (e.g., pressure, volume, temperature, specific heat); the formulas connecting the derived properties to the measurable properties are reviewed in this section. Consider an infinitesimal change of state experienced by a closed system. If kinetic and gravitational energy changes can be neglected, the first law reads

$$\delta Q_{\text{any path}} - \delta W_{\text{any path}} = dU$$

which emphasizes that dU is path-independent. In particular, for a reversible path (rev), the same dU is given by

$$\delta Q_{\text{rev}} - \delta W_{rev} = dU$$

Note that from the second law for closed systems we have $\delta Q_{rev} = T \, dS$. Reversibility (or zero entropy generation) also requires internal mechanical equilibrium at each stage during the process; hence, $\delta W_{\text{rev}} = PdV$, as for a quasi-static change in volume. The infinitesimal change in U is therefore

$$T \, dS - P \, dV = dU$$

Note that this formula holds for an infinitesimal change of state along any path (because dU is path-independent); however, $T \, dS$ matches δQ and $P \, dV$ matches δW only if the path is reversible. In general, $\delta Q < T \, dS$ and $\delta W < P \, dV$, as shown in Fig. 41.2. The formula derived above for dU can be written for a unit mass: $Tds - P \, dv = du$. Additional identities implied by this relation are

$$T = \left(\frac{\partial u}{\partial s}\right)_v \qquad -P = \left(\frac{\partial u}{\partial v}\right)_s$$
$$\frac{\partial^2 u}{\partial s \, \partial v} = \left(\frac{\partial T}{\partial v}\right)_s = -\left(\frac{\partial P}{\partial s}\right)_v$$

where the subscript indicates which variable is held constant during partial differentiation. Similar relations and partial derivative identities exist in conjunction with other derived functions such as enthalpy, Gibbs free energy, and Helmholtz free energy:

- Enthalpy (defined as $h = u + Pv$)

$$dh = T \, ds + v \, dP$$

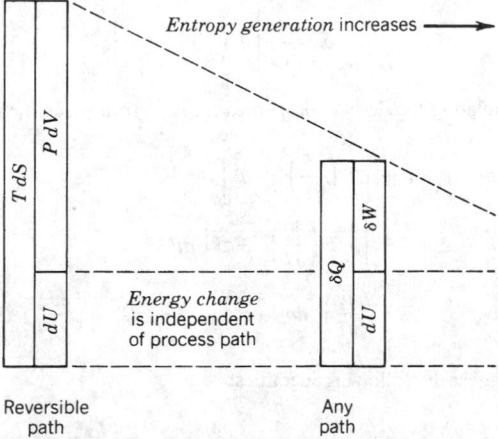

Fig. 41.2 The applicability of the relation $dU = T \, dS - P \, dV$ to any infinitesimal process. (In this drawing, all the quantities are assumed positive.)

$$T = \left(\frac{\partial h}{\partial s}\right)_P \qquad v = \left(\frac{\partial h}{\partial P}\right)_s$$

$$\frac{\partial^2 h}{\partial s \partial P} = \left(\frac{\partial T}{\partial P}\right)_s = \left(\frac{\partial v}{\partial s}\right)_P$$

- Gibbs free energy (defined as $g = h - Ts$)

$$dg = -s\,dT + v\,dP$$

$$-s = \left(\frac{\partial g}{\partial T}\right)_P \qquad v = \left(\frac{\partial g}{\partial P}\right)_T$$

$$\frac{\partial^2 g}{\partial T \partial P} = -\left(\frac{\partial s}{\partial P}\right)_T = \left(\frac{\partial v}{\partial T}\right)_P$$

- Helmholtz free energy (defined as $f = u - Ts$)

$$df = -s\,dT - P\,dv$$

$$-s = \left(\frac{\partial f}{\partial T}\right)_v \qquad -P = \left(\frac{\partial f}{\partial v}\right)_T$$

$$\frac{\partial^2 f}{\partial T \partial v} = -\left(\frac{\partial s}{\partial v}\right)_T = -\left(\frac{\partial P}{\partial T}\right)_v$$

In addition to the (P, v, T) surface, which can be determined based on measurements, (Fig. 41.3), the following partial derivatives are furnished by special experiments:[1]

- The specific heat at constant volume, $c_v = (\partial u / \partial T)_v$, follows directly from the constant volume ($\delta W = 0$) heating of a unit mass of pure substance.
- The specific heat at constant pressure, $c_P = (\partial h / \partial T)_P$, is determined during the constant-pressure heating of a unit mass of pure substance.
- The Joule-Thomson coefficient, $\mu = (\partial T / \partial P)_h$, is measured during a throttling process, that is, during the flow of a stream through an adiabatic duct with friction (see the first law for an open system in the steady state).
- The coefficient of thermal expansion, $\beta = (1/v)(\partial v / \partial T)_P$.
- The isothermal compressibility, $K = (-1/v)(\partial v / \partial P)_T$.
- The constant temperature coefficient, $c_T = (\partial h / \partial P)_T$.

Two noteworthy relationships between some of the partial-derivative measurements are

$$c_P - c_v = \frac{Tv\beta^2}{K}$$

$$\mu = \frac{1}{c_P}\left[T\left(\frac{\partial v}{\partial T}\right)_P - v\right]$$

The general equations relating the derived properties (u, h, s) to measurable quantities are

$$du = c_v\,dT + \left[T\left(\frac{\partial P}{\partial T}\right)_v - P\right]dv$$

$$dh = c_P\,dT + \left[-T\left(\frac{\partial v}{\partial T}\right)_P + v\right]dP$$

$$ds = \frac{c_v}{T}\,dT + \left(\frac{\partial v}{\partial T}\right)_v dv \quad \text{or} \quad ds = \frac{c_P}{T}\,dT - \left(\frac{\partial v}{\partial T}\right)_P dP$$

These relations also suggest the following identities:

$$\left(\frac{\partial u}{\partial T}\right)_v = T\left(\frac{\partial s}{\partial T}\right) = c_v \qquad \left(\frac{\partial h}{\partial T}\right)_P = T\left(\frac{\partial s}{\partial T}\right)_P = c_P$$

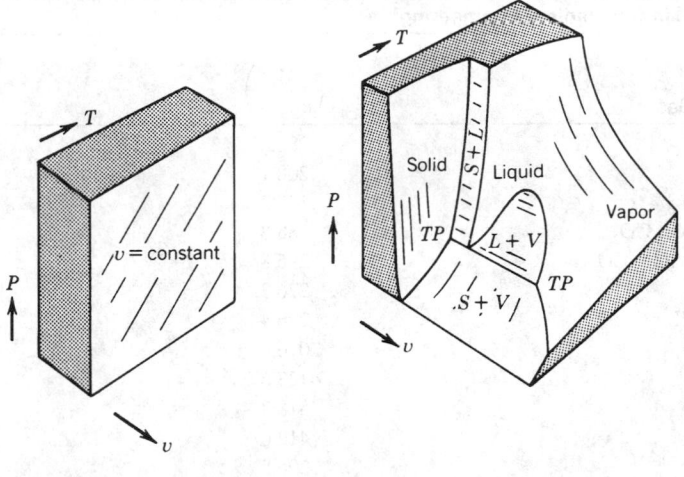

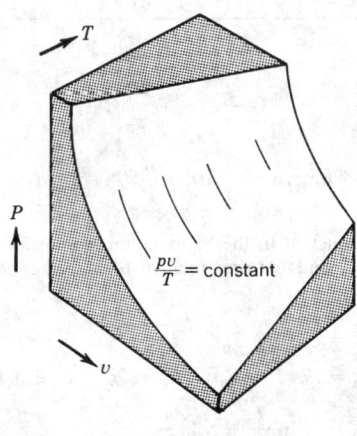

Fig. 41.3 The (P, v, T) surface for a pure substance that contracts upon freezing, showing regions of ideal gas and incompressible fluid behavior. (In this figure, S = solid, V = vapor, L = liquid, TP = triple point.)

41.6 IDEAL GASES

The relationships between thermodynamic properties and the analysis associated with applying the laws of thermodynamics are simplified considerably in cases where the pure substance exhibits ideal gas behavior. As shown in Fig. 41.3, this behavior sets in at sufficiently high temperatures and low pressures; in this limit, the (P, v, T) surface is fitted closely by the simple expression

$$\frac{Pv}{T} = R \quad \text{(constant)}$$

where R is the ideal gas constant of the substance of interest (Table 41.1). The formulas for internal energy, enthalpy, and entropy, which concluded the preceding section, assume the following form in the ideal-gas limit:

Table 41.1   Values of Ideal-Gas Constant and Specific Heat at Constant Volume for Gases Encountered in Mechanical Engineering[1]

Ideal Gas	R $\left(\dfrac{J}{kg \cdot K}\right)$	c_v $\left(\dfrac{J}{kg \cdot K}\right)$
Air	286.8	715.9
Argon, Ar	208.1	316.5
Butane, $C_4 H_{10}$	143.2	1595.2
Carbon dioxide, CO_2	188.8	661.5
Carbon monoxide, CO	296.8	745.3
Ethane, C_2H_6	276.3	1511.4
Ethylene, C_2H_4	296.4	1423.5
Helium, He_2	2076.7	3152.7
Hydrogen, H	4123.6	10216.0
Methane, CH_4	518.3	1687.3
Neon, Ne	412.0	618.4
Nitrogen, N_2	296.8	741.1
Octane, C_8H_{18}	72.85	1641.2
Oxygen, O_2	259.6	657.3
Propane, C_3H_8	188.4	1515.6
Steam, H_2O	461.4	1402.6

$$du = c_v \, dT; \quad c_v = c_v(T)$$
$$dh = c_P \, dT; \quad c_P = c_P(T) = c_v + R$$
$$ds = \frac{c_v}{T} \, dT + \frac{R}{v} \, dv \quad \text{or} \quad ds = \frac{c_P}{T} \, dT - \frac{R}{P} \, dP \quad \text{or} \quad ds = \frac{c_v}{P} \, dP + \frac{c_P}{v} \, dv$$

If the coefficients c_v and c_P are constant in the temperature domain of interest, then the *changes* in specific internal energy, enthalpy, and entropy relative to a reference state $( )_0$ are given by the formulas

$$u - u_0 = c_v \, (T - T_0)$$
$$h - h_0 = c_P \, (T - T_0) \quad \text{(where } h_0 = u_0 + RT_0 \text{)}$$
$$s - s_0 = c_v \ln \frac{T}{T_0} + R \ln \frac{v}{v_0}$$
$$s - s_0 = c_P \ln \frac{T}{T_0} - R \ln \frac{P}{P_0}$$
$$s - s_0 = c_v \ln \frac{P}{P_0} + c_P \ln \frac{v}{v_0}$$

The ideal-gas model rests on two empirical constants, c_v and c_P, or c_v and R, or c_P and R. The ideal-gas limit is also characterized by

$$\mu = 0, \quad \beta = \frac{1}{T}, \quad K = \frac{1}{P}, \quad c_T = 0$$

The extent to which a thermodynamic system destroys available work is intimately tied to the system's entropy generation, that is, to the system's departure from the theoretical limit of reversible operation. Idealized processes that can be modeled as reversible occupy a central role in engineering thermodynamics, because they can serve as standard in assessing the goodness of real processes. Two benchmark reversible processes executed by closed ideal-gas systems are particularly simple and useful. A *quasi-static adiabatic process* $1 \rightarrow 2$ executed by a closed ideal-gas system has the following characteristics:

- Energy interactions

$$\int_1^2 \delta Q = 0$$

$$\int_1^2 \delta W = \frac{P_2 V_2}{\gamma - 1} \left[\left(\frac{V_2}{V_1} \right)^{\gamma - 1} - 1 \right]$$

where $\gamma = c_P / c_v$
- Path

$$PV^\gamma = P_1 V_1^\gamma = P_2 V_2^\gamma \quad \text{(constant)}$$

- Entropy change

$$S_1 - S_1 = 0$$

hence the name *isoentropic* or *isentropic* for this process
- Entropy generation

$$S_{\text{gen}_{1 \to 2}} = S_2 - S_1 - \int_1^2 \frac{\delta Q}{T} = 0 \quad \text{(reversible)}$$

A *quasi-static isothermal process* $1 \to 2$ executed by a closed ideal-gas system in communication with a single temperature reservoir T is characterized by

- Energy interactions

$$\int_1^2 \delta Q = \int_1^2 \delta W = m\, RT \ln \frac{V_2}{V_1}$$

- Path

$$T = T_1 = T_2 \text{ (constant) or } PV = P_1 V_1 = P_2 V_2 \text{ (constant)}$$

- Entropy change

$$S_2 - S_1 = m\, R \ln \frac{V_2}{V_1}$$

- Entropy generation

$$S_{\text{gen}_{1 \to 2}} = S_2 - S_1 - \int_1^2 \frac{\delta Q}{T} = 0 \quad \text{(reversible)}$$

Mixtures of ideal gases also behave as ideal gases in the high-temperature, low-pressure limit. If a certain mixture of mass m contains ideal gases mixed in mass proportions m_i, and if the ideal-gas constants of each component are (c_{v_i}, c_{P_i}, R_i), then the equivalent ideal gas constants of the mixture are

$$c_v = \frac{1}{m} \sum_i m_i\, c_{v_i}$$

$$c_P = \frac{1}{m} \sum_i m_i\, c_{P_i}$$

$$R = \frac{1}{m} \sum_i m_i\, R_i$$

where $m = \Sigma_i\, m_i$.

One mole is the amount of substance of a system that contains as many elementary entities (e.g., molecules) as there are in 12 g of carbon 12; the number of such entities is Avogadro's constant, $N_0 \cong 6.022 \times 10^{23}$. The mole is not a mass unit, since the mass of 1 mole is not the same for all substances. The *molar mass M* of a given molecular species is the mass of 1 mole of that species, so that the total mass m is equal to M times the number of moles n,

$$m = nM$$

Thus, the ideal-gas equation of state can be written as

$$PV = nMRT$$

where the product MR is the *universal gas constant*

$$\bar{R} = MR = 8.314 \frac{J}{mol \cdot K}$$

The equivalent molar mass of a mixture of ideal gases with individual molar masses M_i is

$$M = \frac{1}{n} \Sigma n_i M_i$$

where $n = \Sigma n_i$. The molar mass of air, as a mixture of nitrogen, oxygen, and traces of other gases, is 28.966 g/mol (or 28.966 kg/kmol). A more useful model of the air gas mixture relies on only nitrogen and oxygen as constituents, in the proportion 3.76 moles of nitrogen to every mole of oxygen; this simple model is used frequently in the field of combustion.

41.7 INCOMPRESSIBLE SUBSTANCES

At the opposite end of the spectrum, that is, at sufficiently high pressures and low temperatures in Fig. 41.3, solids and liquids behave so that their density or specific volume is practically constant. In this limit the (P, v, T) surface is adequately represented by the equation

$$v = \bar{v} \quad (\text{constant})$$

The formulas for calculating changes in internal energy, enthalpy, and entropy become (see the end of the section on relations among thermodynamic properties)

$$du = c \, dT$$
$$dh = c \, dT + \bar{v} \, dP$$
$$ds = \frac{c}{T} \, dT$$

where c is the sole specific heat of the incompressible substance,

$$c = c_v = c_P$$

The specific heat c is a function of temperature only. In a sufficiently narrow temperature range where c can be regarded as constant, the finite changes in internal energy, enthalpy, and entropy relative to a reference state denoted by $(\ )_0$ are

$$u - u_0 = c \, (T - T_0)$$
$$h - h_0 = c \, (T - T_0) + \bar{v}(P - P_0) \quad (\text{where } h_0 = u_0 + P_0\bar{v})$$
$$s - s_0 = c \ln \frac{T}{T_0}$$

The incompressible substance model rests on two empirical constants, c and $\bar{v}$.

41.8 TWO-PHASE STATES

As shown in Fig. 41.3, the domains in which the pure substance behaves either as an ideal gas or as an incompressible substance are bounded by regions where the substance exists as a mixture of two phases, liquid and vapor, solid and liquid, or solid and vapor. The two-phase regions themselves intersect along the *triple point* line labeled *TP–TP* on the middle sketch of Fig. 41.3. In engineering cycle calculations, more useful are the projections of the (P, v, T) surface on the *P–v* plane or, through the relations reviewed earlier, on the *T–s* plane. The terminology associated with two-phase equilibrium states is easier to understand by focusing on the *P–v* diagram of Fig. 41.4a and by

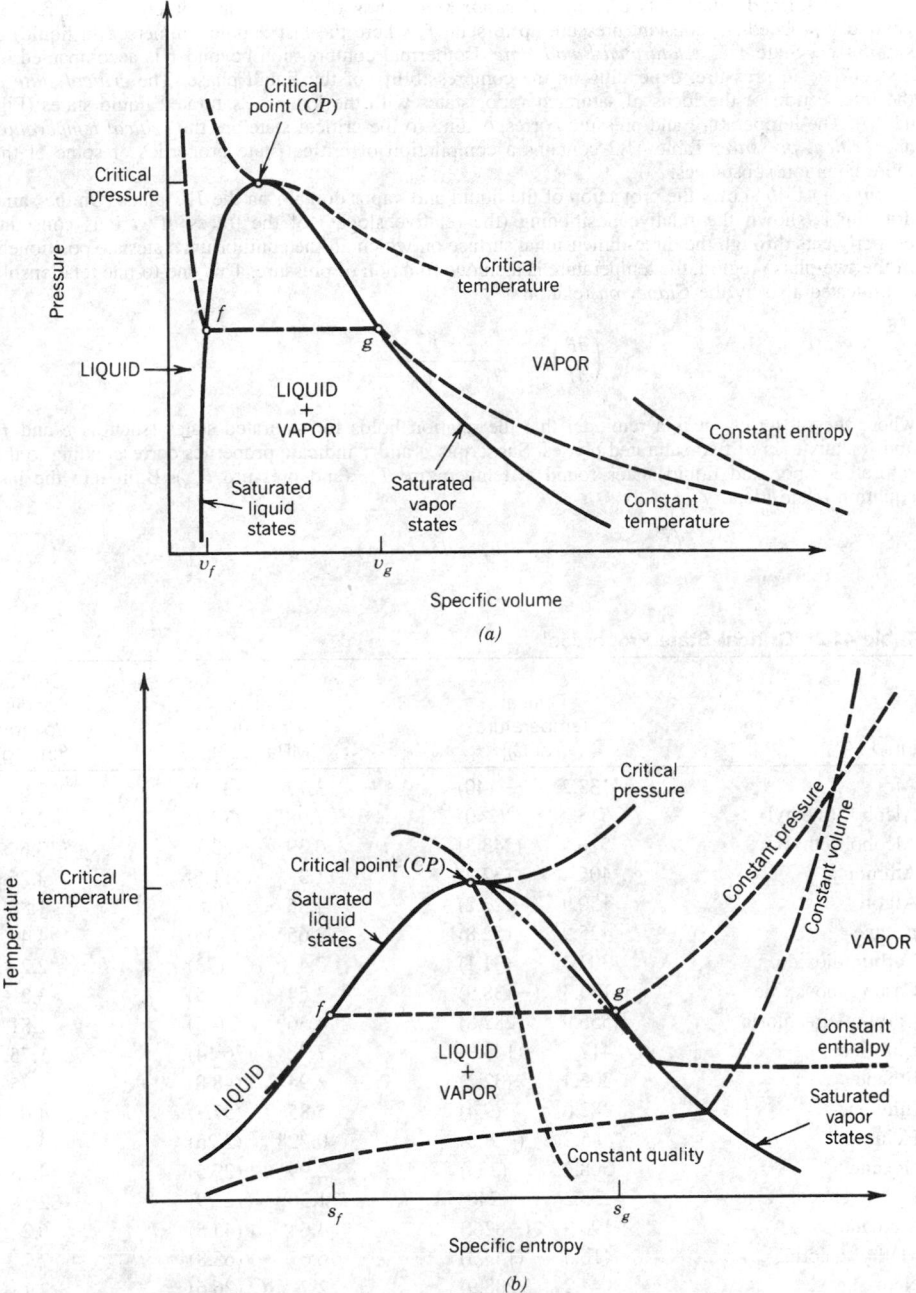

Fig. 41.4 The locus of two-phase (liquid and vapor) states, projected on (a) the P–v plane, and (b) the T–s plane.

imagining the isothermal compression of a unit mass of substance (a closed system). As the specific volume v decreases, the substance ceases to be a pure vapor at state g, where the first droplets of liquid are formed. State g is a *saturated vapor state*. It is observed that isothermal compression beyond g proceeds at constant pressure up to state f, where the last bubble (immersed in liquid) is suppressed. State f is a *saturated liquid state*. Isothermal compression beyond f is accompanied by a steep rise in pressure, depending on the compressibility of the liquid phase. The *critical state* is the intersection of the locus of saturated vapor states with the locus of saturated liquid states (Fig. 41.4a). The temperature and pressure corresponding to the critical state are the *critical temperature* and *critical pressure*. Table 41.2 contains a compilation of critical-state properties of some of the more common substances.

Figure 41.4b shows the projection of the liquid and vapor domain on the T–s plane. On the same drawing is shown the relative positioning (the relative slopes) of the traces of various constant-property cuts through the three-dimensional surface on which all the equilibrium states are positioned. In the two-phase region, the temperature is a unique function of pressure. This one-to-one relationship is indicated also by the *Clapeyron* relation

$$\left(\frac{dP}{dT}\right)_{sat} = \frac{h_g - h_f}{T(v_g - v_f)} = \frac{s_g - s_f}{v_g - v_f}$$

where the subscript sat is a reminder that the relation holds for saturated states (such as g and f) and for mixtures of two saturated phases. Subscripts g and f indicate properties corresponding to the saturated vapor and liquid states found at temperature T_{sat} (and pressure P_{sat}). Built into the last equation is the identity

$$h_g - h_f = T(s_g - s_f)$$

Table 41.2   Critical-State Properties[1]

Fluid	Critical Temperature [K (°C)]		Critical Pressure [MPa (atm)]		Critical Specific Volume (cm³/g)
Air	133.2	(−140)	3.77	(37.2)	2.9
Alcohol (methyl)	513.2	(240)	7.98	(78.7)	3.7
Alcohol (ethyl)	516.5	(243.3)	6.39	(63.1)	3.6
Ammonia	405.4	(132.2)	11.3	(111.6)	4.25
Argon	150.9	(−122.2)	4.86	(48)	1.88
Butane	425.9	(152.8)	3.65	(36)	4.4
Carbon dioxide	304.3	(31.1)	7.4	(73)	2.2
Carbon monoxide	134.3	(−138.9)	3.54	(35)	3.2
Carbon tetrachloride	555.9	(282.8)	4.56	(45)	1.81
Chlorine	417	(143.9)	7.72	(76.14)	1.75
Ethane	305.4	(32.2)	4.94	(48.8)	4.75
Ethylene	282.6	(9.4)	5.85	(57.7)	4.6
Helium	5.2	(−268)	0.228	(2.25)	14.4
Hexane	508.2	(235)	2.99	(29.5)	4.25
Hydrogen	33.2	(−240)	1.30	(12.79)	32.3
Methane	190.9	(−82.2)	4.64	(45.8)	6.2
Methyl chloride	416.5	(143.3)	6.67	(65.8)	2.7
Neon	44.2	(−288.9)	2.7	(26.6)	2.1
Nitric oxide	179.2	(−93.9)	6.58	(65)	1.94
Nitrogen	125.9	(−147.2)	3.39	(33.5)	3.25
Octane	569.3	(296.1)	2.5	(24.63)	4.25
Oxygen	154.3	(−118.9)	5.03	(49.7)	2.3
Propane	368.7	(95.6)	4.36	(43)	4.4
Sulfur dioxide	430.4	(157.2)	7.87	(77.7)	1.94
Water	647	(373.9)	22.1	(218.2)	3.1

which is equivalent to the statement that the Gibbs free energy is the same for the saturated states and their mixtures found at the same temperature, $g_g = g_f$.

The properties of a two-phase mixture depend on the proportion in which saturated vapor, m_g, and saturated liquid, m_f, enter the mixture. The composition of the mixture is described by the property *quality*

$$x = \frac{m_g}{m_f + m_g}$$

whose value varies between 0 at state f and 1 at state g. Other properties of the mixture can be calculated in terms of the properties of the saturated states found at the same temperature.

$$u = u_f + xu_{fg} \qquad s = s_f + xs_{fg}$$
$$h = h_f + xh_{fg} \qquad v = v_f + xv_{fg}$$

with the notation $()_{fg} = ()_g - ()_f$.

Similar relations can be used to calculate the properties of two-phase states other than liquid and vapor, namely, solid and vapor or solid and liquid. For example, the enthalpy of a solid and liquid mixture is given by $h = h_s + xh_{sf}$, where subscript s stands for the *saturated solid state* found at the same temperature as for the two-phase state, and where h_{sf} is the latent heat of melting or solidification.

In general, the states situated immediately outside the two-phase dome sketched in Figs. 41.3 and 41.4 do not follow very well the limiting models discussed already (ideal gas, incompressible substance). Since the properties of closely neighboring states are usually not available in tabular form, the following approximate calculation proves useful. For a *moderately compressed liquid state*, which is indicated by the subscript $()_*$, that is, for a state situated close to the left of the dome in Fig. 41.4, the properties may be calculated as slight deviations from those of the saturated liquid state found at the same temperature as the compressed liquid state of interest,

$$h_* \cong (h_f)_{T*} + (v_f)_{T*}[P_* - (P_f)_{T*}]$$
$$s \cong (s_f)_{T*}$$

For a *slightly superheated vapor state*, that is, a state situated close to the right of the dome in Fig. 41.4, the properties may be estimated in terms of those of the saturated vapor state found at the same temperature,

$$h_+ \cong (h_g)_{T+}$$
$$s_+ \cong (s_g)_{T+} + \left(\frac{P_g v_g}{T_g}\right)_{T+} \ln \frac{(P_g)_{T+}}{P_+}$$

In these expressions, subscript $()_+$ indicates the properties of the slightly superheated vapor state.

41.9 ANALYSIS OF ENGINEERING SYSTEM COMPONENTS

This section contains a summary of the equations obtained by applying the first and second laws of thermodynamics to the components encountered in most engineering systems, such as power cycles and refrigeration cycles. It is assumed that each component operates in *steady flow*.

- *Valve* (throttle) or adiabatic duct with friction (Fig. 41.5*a*):

    First law $\qquad h_1 = h_2$

    Second law $\quad \dot{S}_{gen} = \dot{m}(s_2 - s_1) > 0$

- *Expander* or *turbine* with negligible heat transfer to the ambient (Fig. 41.5*b*):

    First law $\qquad \dot{W}_T = \dot{m}(h_1 - h_2)$

    Second law $\quad \dot{S}_{gen} = \dot{m}(s_2 - s_1) \geq 0$

    Efficiency $\qquad \eta_T = \dfrac{h_1 - h_2}{h_1 - h_{2,\text{rev}}} \leq 1$

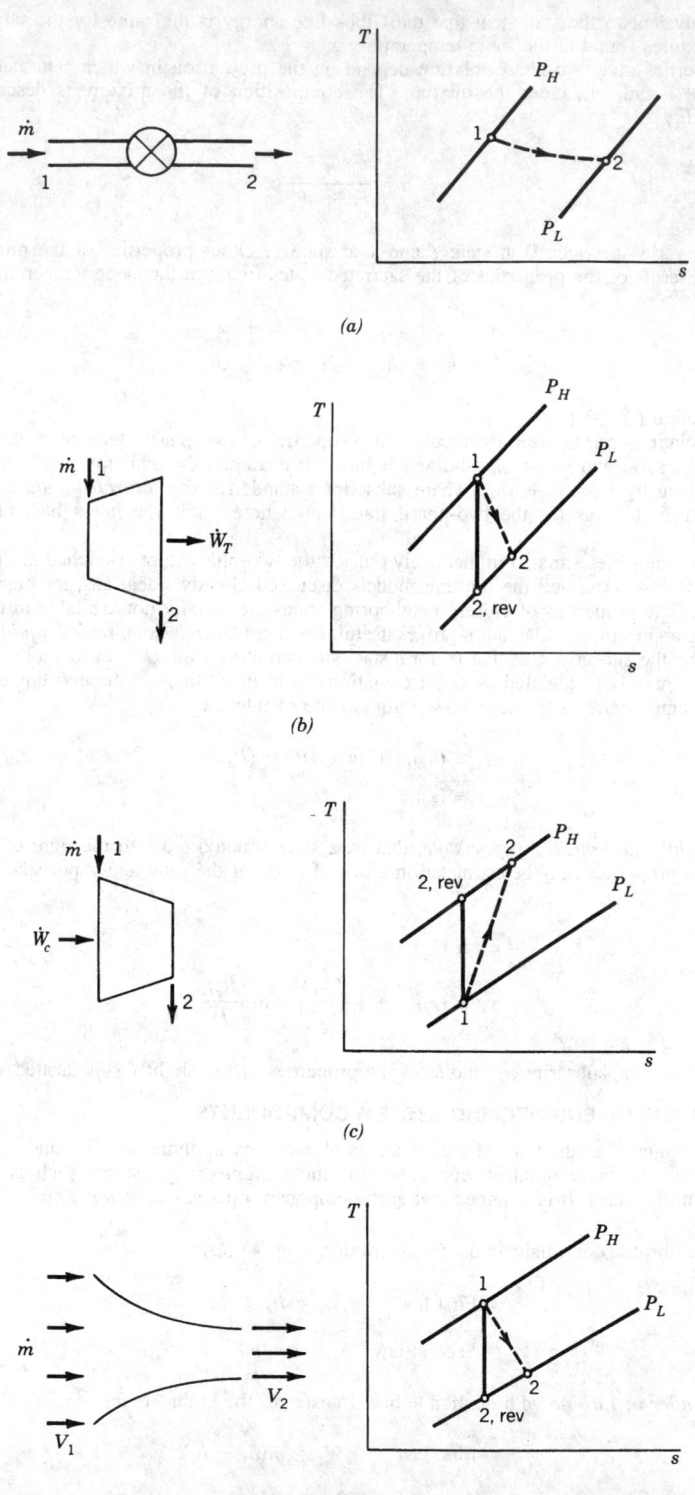

Fig. 41.5 Engineering system components, and their inlet and outlet states on the T–s plane, (P_H = high pressure; P_L = low pressure.)

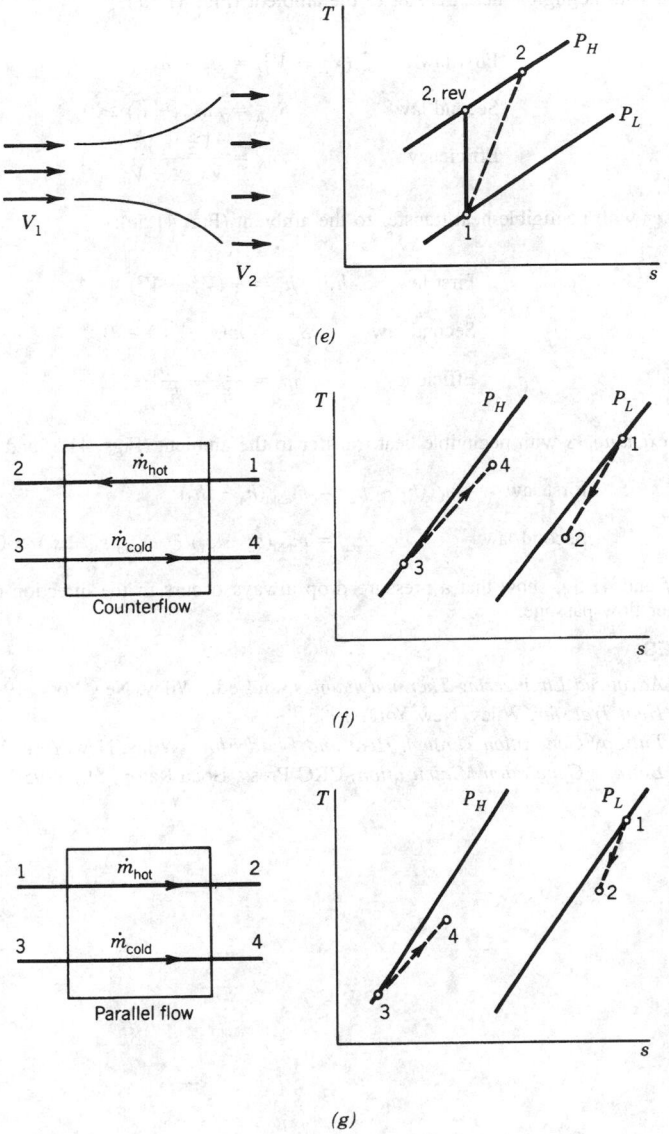

(e)

(f)

(g)

Fig. 41.5 (Continued)

- *Compressor* or *pump* with negligible heat transfer to the ambient (Fig. 41.5c):

$$\text{First law} \qquad \dot{W}_C = \dot{m}(h_2 - h_1)$$

$$\text{Second law} \qquad \dot{S}_{gen} = \dot{m}(s_2 - s_1) \geq 0$$

$$\text{Efficiency} \qquad \eta_c = \frac{h_{2,rev} - h_1}{h_2 - h_1} \leq 1$$

- *Nozzle* with negligible heat transfer to the ambient (Fig. 41.5*d*):

  First law $\dfrac{1}{2}(V_2^2 - V_1^2) = h_1 - h_2$

  Second law $\dot{S}_{gen} = \dot{m}(s_2 - s_1) \geq 0$

  Efficiency $\eta_N = \dfrac{V_2^2 - V_1^2}{V_{2,rev}^2 - V_1^2} \leq 1$

- *Diffuser* with negligible heat transfer to the ambient (Fig. 41.5*e*):

  First law $h_2 - h_1 = \dfrac{1}{2}(V_1^2 - V_2^2)$

  Second law $\dot{S}_{gen} = \dot{m}(s_2 - s_1) \geq 0$

  Efficiency $\eta_D = \dfrac{h_{2,rev} - h_1}{h_2 - h_1} \leq 1$

- *Heat exchangers* with negligible heat transfer to the ambient (Figs. 41.5*f* and 41.5*g*)

  First law $\dot{m}_{hot}(h_1 - h_2) = \dot{m}_{cold}(h_4 - h_3)$

  Second law $\dot{S}_{gen} = \dot{m}_{hot}(s_2 - s_1) + \dot{m}_{cold}(s_4 - s_3) \geq 0$

Figures 41.5*f* and 41.5*g*, show that a pressure drop always occurs in the direction of flow, in any heat exchanger flow passage.

REFERENCES

1. A. Bejan, *Advanced Engineering Thermodynamics*, 2nd ed., Wiley, New York, 1997.
2. A. Bejan, *Heat Transfer*, Wiley, New York, 1993.
3. A. Bejan, *Entropy Generation Through Heat and Fluid Flow*, Wiley, New York, 1982.
4. A. Bejan, *Entropy Generation Minimization*, CRC Press, Boca Raton, FL, 1995.

CHAPTER 42

EXERGY ANALYSIS AND ENTROPY GENERATION MINIMIZATION

Adrian Bejan
Department of Mechanical Engineering and Materials Science
Duke University
Durham, North Carolina

42.1	INTRODUCTION	1351	42.6 HEAT TRANSFER	1359
42.2	PHYSICAL EXERGY	1353	42.7 STORAGE SYSTEMS	1361
42.3	CHEMICAL EXERGY	1355	42.8 SOLAR ENERGY CONVERSION	1362
42.4	ENTROPY GENERATION MINIMIZATION	1357	42.9 POWER PLANTS	1362
42.5	CRYOGENICS	1358		

42.1 INTRODUCTION

In this chapter, we review two important methods that account for much of the newer work in engineering thermodynamics and thermal design and optimization. The method of *exergy analysis* rests on thermodynamics alone. The first law, the second law, and the environment are used simultaneously in order to determine (i) the theoretical operating conditions of the system in the reversible limit and (ii) the entropy generated (or exergy destroyed) by the actual system, that is, the departure from the reversible limit. The focus is on *analysis*. Applied to the system components individually, exergy analysis shows us quantitatively how much each component contributes to the overall irreversibility of the system.[1-3]

Entropy generation minimization (EGM) is a method of *modeling* and *optimization*. The entropy generated by the system is first developed as a function of the physical characteristics of the system (dimensions, materials, shapes, constraints). An important preliminary step is the construction of a system model that incorporates not only the traditional building blocks of engineering thermodynamics (systems, laws, cycles, processes, interactions), but also the fundamental principles of fluid mechanics, heat transfer, mass transfer and other transport phenomena. This combination makes the model "realistic" by accounting for the inherent irreversibility of the actual device. Finally, the minimum entropy generation design ($S_{gen,min}$) is determined for the model, and the approach of any other design (S_{gen}) to the limit of realistic ideality represented by $S_{gen,min}$ is monitored in terms of the entropy generation number $N_S = S_{gen} / S_{gen,min} > 1$.

To calculate S_{gen} and minimize it, the analyst does not need to rely on the concept of exergy. The EGM method represents an important step beyond thermodynamics. It is a new method[4] that combines thermodynamics, heat transfer, and fluid mechanics into a powerful technique for modeling and optimizing real systems and processes. The use of the EGM method has expanded greatly during the last two decades.[5]

SYMBOLS AND UNITS

a	specific nonflow availability, J/kg
A	nonflow availability, J

Mechanical Engineers' Handbook, 2nd ed., Edited by Myer Kutz.
ISBN 0-471-13007-9 © 1998 John Wiley & Sons, Inc.

A	area, m^2
b	specific flow availability, J/kg
B	flow availability, J
B	duty parameter for plate and cylinder
B_s	duty parameter for sphere
B_0	duty parameter for tube
Be	dimensionless group, $\dot{S}'''_{gen,\Delta T}/(\dot{S}'''_{gen,\Delta T} + \dot{S}'''_{gen,\Delta P})$
c_p	specific heat at constant pressure, J/(kg · K)
C	specific heat of incompressible substance, J/(kg · K)
C	heat leak thermal conductance, W/K
C^*	time constraint constant, sec/kg
D	diameter, m
e	specific energy, J/kg
E	energy, J
$\bar{e}_{ch}$	specific flow chemical exergy, J/kmol
$\bar{e}_t$	specific total flow exergy, J/kmol
e_x	specific flow exergy, J/kg
$\bar{e}_x$	specific flow exergy, J/kmol
E_Q	exergy transfer via heat transfer, J
$\dot{E}_W$	exergy transfer rate, W
E_x	flow exergy, J
EGM	the method of entropy generation minimization
f	friction factor
F_D	drag force, N
g	gravitational acceleration, m/sec^2
G	mass velocity, kg/(sec · m^2)
h	specific enthalpy, J/kg
h	heat transfer coefficient, W/(m^{2}K)
$h°$	total specific enthalpy, J/kg
$H°$	total enthalpy, J
k	thermal conductivity, W/(m K)
L	length, m
m	mass, kg
$\dot{m}$	mass flow rate, kg/sec
M	mass, kg
N	mole number, kmol
$\dot{N}$	molal flow rate, kmol/sec
N_S	entropy generation number, $S_{gen}/S_{gen,min}$
Nu	Nusselt number
N_{tu}	number of heat transfer units
P	pressure, N/m^2
Pr	Prandtl number
q'	heat transfer rate per unit length, W/m
Q	heat transfer, J
$\dot{Q}$	heat transfer rate, W
r	dimensionless insulation resistance
R	ratio of thermal conductances
Re$_D$	Reynolds number
s	specific entropy, J/(kg · K)
S	entropy, J/K
S_{gen}	entropy generation, J/K
$\dot{S}_{gen}$	entropy generation rate, W/K
$\dot{S}'_{gen}$	entropy generation rate per unit length, W/(m · K)

$\dot{S}'''_{\text{gen}}$	entropy generation rate per unit volume, $W/(m^3\ K)$
t	time, sec
t_c	time constraint, sec
T	temperature, K
U	overall heat transfer coefficient, $W/(m^2\ K)$
U_∞	free stream velocity, m/sec
v	specific volume, m^3/kg
V	volume, m^3
V	velocity, m/sec
$\dot{W}$	power, W
x	longitudinal coordinate, m
z	elevation, m
ΔP	pressure drop, N/m^2
ΔT	temperature difference, K
η	first law efficiency
η_{II}	second law efficiency
θ	dimensionless time
μ	viscosity, $kg/(sec \cdot m)$
μ_i^*	chemical potentials at the restricted dead state, J/kmol
$\mu_{0,i}$	chemical potentials at the dead state, J/kmol
ν	kinematic viscosity, m^2/sec
ξ	specific nonflow exergy, J/kg
Ξ	nonflow exergy, J
Ξ_{ch}	nonflow chemical exergy, J
Ξ_t	nonflow total exergy, J
ρ	density, kg/m^3

Subscripts

$(\ )_B$	base
$(\ )_C$	collector
$(\ )_C$	Carnot
$(\ )_H$	high
$(\ )_L$	low
$(\ )_m$	melting
$(\ )_{\text{max}}$	maximum
$(\ )_{\text{min}}$	minimum
$(\ )_{\text{opt}}$	optimal
$(\ )_p$	pump
$(\ )_{\text{rev}}$	reversible
$(\ )_t$	turbine
$(\ )_0$	environment
$(\ )_\infty$	free stream

42.2 PHYSICAL EXERGY

Figure 42.1 shows the general features of an open thermodynamic system that can interact thermally ($\dot{Q}_0$) and mechanically ($P_0\ dV/dt$) with the atmospheric temperature and pressure reservoir (T_0, P_0). The system may have any number of inlet and outlet ports, even though only two such ports are illustrated. At a certain point in time, the system may be in communication with any number of additional temperature reservoirs ($T_1, \ldots, T_n$), experiencing the instantaneous heat transfer interactions, $\dot{Q}_1, \ldots, \dot{Q}_n$. The work transfer rate $\dot{W}$ represents all the possible modes of work transfer, specifically, the work done on the atmosphere ($P_0\ dV/dt$) and the remaining (useful, deliverable) portions such as $P\ dV/dt$, shaft work, shear work, electrical work, and magnetic work. The useful part is known as available work (or simply exergy) or, on a unit time basis,

$$\dot{E}_W = \dot{W} - P_0 \frac{dV}{dt}$$

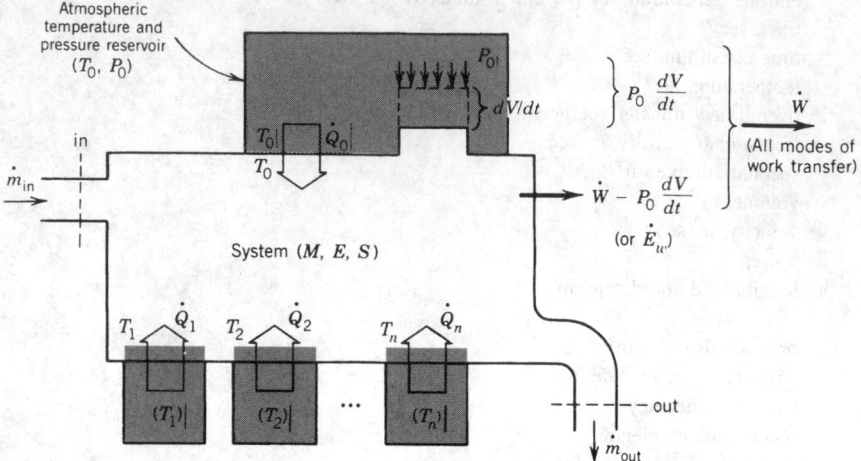

Fig. 42.1 Open system in thermal and mechanical communication with the ambient. (From A. Bejan, *Advanced Engineering Thermodynamics.* © 1997 John Wiley & Sons, Inc. Reprinted by permission.)

The first law and the second law of thermodynamics can be combined to show that the available work transfer rate from the system of Fig. 42.1 is given by the $\dot{E}_W$ equation:[1-3]

$$\dot{E}_W = \underbrace{-\frac{d}{dt}(E - T_0 S + P_0 V)}_{\substack{\text{Accumulation} \\ \text{of nonflow exergy}}} + \underbrace{\sum_{i=1}^{n}\left(1 - \frac{T_0}{T_i}\right)\dot{Q}_i}_{\substack{\text{Exergy transfer} \\ \text{via heat transfer}}}$$

$$+ \underbrace{\sum_{\text{in}}\dot{m}(h^\circ - T_0 s)}_{\substack{\text{Intake of} \\ \text{flow exergy via} \\ \text{mass flow}}} - \underbrace{\sum_{\text{out}}\dot{m}(h^\circ - T_0 s)}_{\substack{\text{Release of} \\ \text{flow exergy via} \\ \text{mass flow}}} - \underbrace{T_0 \dot{S}_{\text{gen}}}_{\substack{\text{Destruction} \\ \text{of exergy}}}$$

where E, V, and S are the instantaneous energy, volume, and entropy of the system, and h° is shorthand for the specific enthalpy plus the kinetic and potential energies of each stream, $h^\circ = h + \frac{1}{2}V^2 + gz$. The first four terms on the right-hand side of the $\dot{E}_W$ equation represent the energy rate delivered as useful power (to an external user) in the limit of reversible operation ($\dot{E}_{W,\text{rev}}$, $\dot{S}_{\text{gen}} = 0$). It is worth noting that the $\dot{E}_W$ equation is a restatement of the Gouy–Stodola theorem (see Section 41.4), or the proportionality between the rate of exergy (work) destruction and the rate of entropy generation

$$\dot{E}_{W,\text{rev}} - \dot{E}_W = T_0 \dot{S}_{\text{gen}}$$

A special exergy nomenclature has been devised for the terms formed on the right side of the $\dot{E}_W$ equation. The exergy content associated with a heat transfer interaction (Q_i, T_i) and the environment (T_0) is the *exergy of heat transfer*,

$$E_{Q_i} = Q_i\left(1 - \frac{T_0}{T_i}\right)$$

This means that the heat transfer with the environment (Q_0, T_0) carries zero exergy relative to the environment T_0.

Associated with the system extensive properties (E, S, V) and the two specified intensive properties of the environment (T_0, P_0) is a new extensive property: the thermomechanical or physical *nonflow availability*,

$$A = E - T_0 S + P_0 V$$
$$a = e - T_0 s + P_0 v$$

Let A_0 represent the nonflow availability when the system is at the *restricted dead state* (T_0, P_0), that is, in thermal and mechanical equilibrium with the environment, $A_0 = E_0 - T_0 S_0 + P_0 V_0$. The difference between the nonflow availability of the system in a given state and its nonflow availability in the restricted dead state is the thermomechanical or physical *nonflow exergy*,

$$\Xi = A - A_0 = E - E_0 - T_0(S - S_0) + P_0(V - V_0)$$
$$\xi = a - a_0 = e - e_0 - T_0(s - s_0) + P_0(v - v_0)$$

The nonflow exergy represents the most work that would become available if the system were to reach its restricted dead state reversibly, while communicating thermally only with the environment. In other words, the nonflow exergy represents the exergy content of a given closed system relative to the environment.

Associated with each of the streams entering or exiting an open system is the thermomechanical or physical *flow availability*,

$$B = H^\circ - T_0 S$$
$$b = h^\circ - T_0 s$$

At the restricted dead state, the nonflow availability of the stream is $B_0 = H_0^\circ - T_0 S_0$. The difference $B - B_0$ is known as the thermomechanical or physical *flow exergy* of the stream,

$$E_x = B - B_0 = H^\circ - H_0^\circ - T_0(S - S_0)$$
$$e_x = b - b_0 = h^\circ - h_0^\circ - T_0(s - s_0)$$

Physically, the flow exergy represents the available work content of the stream relative to the restricted dead state (T_0, P_0). This work could be extracted in principle from a system that operates reversibly in thermal communication only with the environment (T_0), while receiving the given stream $(\dot{m}, h^\circ, s)$ and discharging the same stream at the environmental pressure and temperature $(\dot{m}, h_0^\circ, s_0)$.

In summary, the $\dot{E}_W$ equation can be rewritten more simply as

$$\dot{E}_W = -\frac{d\Xi}{dt} + \sum_{i=1}^{n} \dot{E}_{Q_i} + \sum_{in} \dot{m} e_x - \sum_{out} \dot{m} e_x - T_0 \dot{S}_{gen}$$

Examples of how these exergy concepts are used in the course of analyzing component by component the performance of complex systems can be found in Refs. 1–3. Figure 42.2 shows one such example.[1] The upper part of the drawing shows the traditional description of the four components of a simple Rankine cycle. The lower part shows the exergy streams that enter and exit each component, with the important feature that the heater, the turbine and the cooler destroy significant portions (shaded, fading away) of the entering exergy streams. The numerical application of the $\dot{E}_W$ equation to each component tells the analyst the exact widths of the exergy streams to be drawn in Fig. 42.2. In graphical or numerical terms, the "exergy wheel" diagram[1] shows not only *how much* exergy is being destroyed but also *where*. It tells the designer how to rank order the components as candidates for optimization according to the method of *entropy generation minimization* (Sections 42.4–42.9).

To complement the traditional (first law) energy conversion efficiency, $\eta = (\dot{W}_t - \dot{W}_p)/\dot{Q}_H$ in Fig. 42.2, exergy analysis recommends as figure of merit the *second law efficiency*,

$$\eta_{II} = \frac{\dot{W}_t - \dot{W}_p}{\dot{E}_{Q_H}}$$

where $\dot{W}_t - \dot{W}_p$ is the net power output (i.e., $\dot{E}_W$ earlier in this section). The second law efficiency can have values between 0 and 1, where 1 corresponds to the reversible limit. Because of this limit, η_{II} describes very well the fundamental difference between the method of exergy analysis and the method of entropy generation minimization (EGM), because in EGM the system always operates *irreversibly*. The question in EGM is how to change the system such that its $\dot{S}_{gen}$ value (always finite) approaches the minimum $\dot{S}_{gen}$ allowed by the system constraints.

42.3 CHEMICAL EXERGY

Consider now a nonflow system that can experience heat, work, and mass transfer in communication with the environment. The environment is represented by T_0, P_0, and the n chemical potentials $\mu_{0,i}$

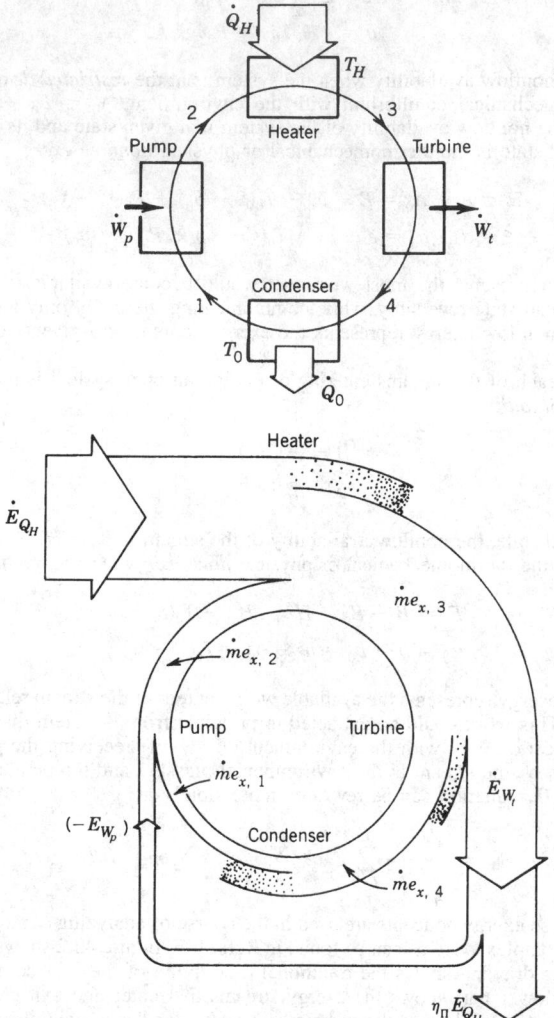

Fig. 42.2 The exergy wheel diagram of a simple Rankine cycle. Top: the traditional notation and energy interactions. Bottom: the exergy flows and the definition of the second law efficiency. (From A. Bejan, *Advanced Engineering Thermodynamics.* © 1997 John Wiley & Sons, Inc. Reprinted by permission.)

of the environmental constituents that are also present in the system. Taken together, the $n + 2$ intensive properties of the environment (T_0, P_0, $\mu_{0,i}$) are known as the *dead state*.

Reading Fig. 42.3 from left to right, we see the system in its initial state represented by E, S, V and its composition (mole numbers $N_1, \ldots, N_n$), and its $n + 2$ intensities (T, P, μ_i). The system can reach its dead state in two steps. In the first, it reaches only thermal and mechanical equilibrium with the environment (T_0, P_0), and delivers the nonflow exergy Ξ defined in the preceding section. At the end of this first step, the chemical potentials of the constituents have changed to μ_i^* ($i = 1, \ldots, n$). During the second step, mass transfer occurs (in addition to heat and work transfer) and, in the end, the system reaches chemical equilibrium with the environment, in addition to thermal and mechanical equilibrium. The work made available during this second step is known as *chemical exergy,*[1-3]

$$\Xi_{\text{ch}} = \sum_{i=1}^{n} (\mu_i^* - \mu_{0,i}) N_i$$

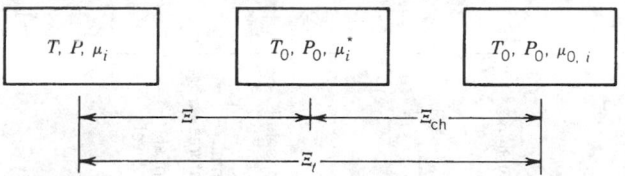

Fig. 42.3   The relationship between the nonflow total (Ξ_t), physical (Ξ), and chemical (Ξ_{ch}) exergies. (From A. Bejan, *Advanced Engineering Thermodynamics*. © 1997 John Wiley & Sons, Inc. Reprinted by permission.)

The total exergy content of the original nonflow system (E, S, V, N_i) relative to the environmental dead state (T_0, P_0, $\mu_{0,i}$) represents the *total nonflow exergy*,

$$\Xi_t = \Xi + \Xi_{ch}$$

Similarly, *the total flow exergy of* a mixture stream of total molal flow rate $\dot{N}$ (composed of n species, with flow rates $\dot{N}_i$) and intensities T, P and μ_i ($i = 1, \ldots, n$) is, on a mole of mixture basis,

$$\bar{e}_t = \bar{e}_x + \bar{e}_{ch}$$

where the physical flow exergy $\bar{e}_x$ was defined in the preceding section, and $\bar{e}_{ch}$ is the *chemical exergy* per mole of mixture,

$$\bar{e}_{ch} = \sum_{i=1}^{n} (\mu_i^* - \mu_{0,i}) \frac{\dot{N}_i}{\dot{N}}$$

In the $\bar{e}_{ch}$ expression μ_i^* ($i = 1, \ldots, n$) are the chemical potentials of the stream constituents at the restricted dead state (T_0, P_0). The chemical exergy is the additional work that could be extracted (reversibly) as the stream evolves from the restricted dead state to the dead state (T_0, P_0, $\mu_{0,i}$) while in thermal, mechanical, and chemical communication with the environment. Applications of the concepts of chemical exergy and total exergy can be found in Refs. 1–3.

42.4   ENTROPY GENERATION MINIMIZATION

The EGM method[4,5] is distinct from exergy analysis, because in exergy analysis the analyst needs only the first law, the second law, and a convention regarding the values of the intensive properties of the environment. The critically new aspects of the EGM method are system modeling, the development of S_{gen} as a function of the physical parameters of the model, and the *minimization* of the calculated entropy generation rate. To minimize the irreversibility of a proposed design, the engineer must use the relations between temperature differences and heat transfer rates, and between pressure differences and mass flow rates. The engineer must relate the degree of thermodynamic nonideality of the design to the physical characteristics of the system, namely, to finite dimensions, shapes, materials, finite speeds, and finite-time intervals of operation. For this, the engineer must rely on heat transfer and fluid mechanics principles, in addition to thermodynamics. Only by varying one or more of the physical characteristics of the system can the engineer bring the design closer to the operation characterized by minimum entropy generation subject to finite-size and finite-time constraints.

The modeling and optimization progress made in EGM is illustrated by some of the simplest and most fundamental results of the method, which are reviewed in the following sections. The structure of the EGM field is summarized in Fig. 42.4 by showing on the vertical the expanding list of applications. On the horizontal, we see the two modeling approaches that are being used. One approach is to focus from the start on the total system, to "divide" the system into compartments that account for one or more of the irreversibility mechanisms, and to declare the "rest" of the system irreversibility-free. In this approach, success depends fully on the modeler's intuition, as there are not one-to-one relationships between the assumed compartments and the pieces of hardware of the real system.

In the alternative approach (from the right in Fig. 42.4), modeling begins with dividing the system into its real components, and recognizing that each component may contain large numbers of one or more elemental features. The approach is to minimize S_{gen} in a fundamental way at each level, starting from the simple and proceeding toward the complex. Important to note is that when a component or elemental feature is imagined separately from the larger system, the quantities assumed specified at the points of separation act as constraints on the optimization of the smaller system. The principle

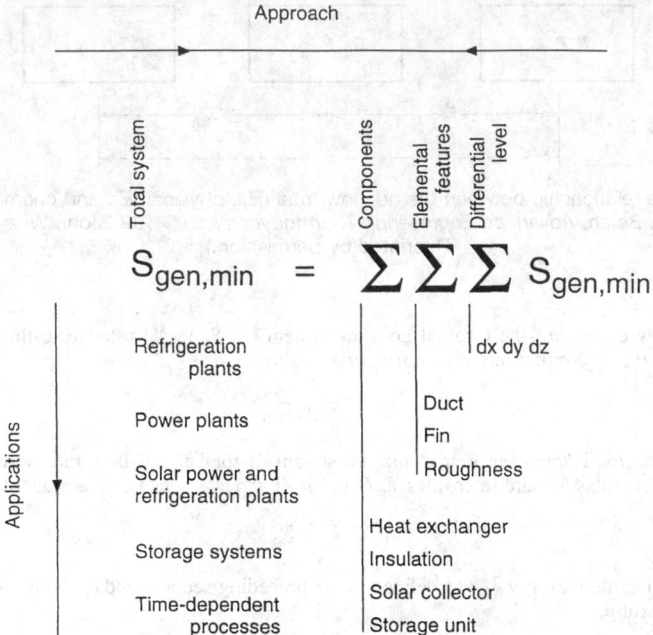

Fig. 42.4 Approaches and applications of the method of entropy generation minimization (EGM). (Reprinted by permission from A. Bejan, *Entropy Generation Minimization*. Copyright CRC Press, Boca Raton, Florida. © 1996.)

of thermodynamic isolation (Ref. 5, p. 125) must be kept in mind during the later stages of the optimization procedure, when the optimized elements and components are integrated into the total system, which itself is optimized for *minimum cost* in the final stage.[3]

42.5 CRYOGENICS

The field of low-temperature refrigeration was the first where EGM became an established method of modeling and optimization. Consider a path for heat leak ($\dot{Q}$) from room temperature (T_H) to the cold end (T_L) of a low-temperature refrigerator or liquefier. Examples of such paths are mechanical supports, insulation layers without or with radiation shields, counterflow heat exchangers, and electrical cables. The total rate of entropy generation associated with the heat leak path is

$$\dot{S}_{gen} = \int_{T_L}^{T_H} \frac{\dot{Q}}{T^2}\, dT$$

where $\dot{Q}$ is in general a function of the local temperature T. The proportionality between the heat leak and the local temperature gradient along its path, $\dot{Q} = kA\, (dT/dx)$, and the finite size of the path [length L, cross section A, material thermal conductivity $k(T)$] are accounted for by the integral constraint

$$\int_{T_L}^{T_H} \frac{k(T)}{\dot{Q}(T)}\, dT = \frac{L}{A} \quad \text{(constant)}$$

The optimal heat leak distribution that minimizes $\dot{S}_{gen}$ subject to the finite-size constraint is[4,5]

$$\dot{Q}_{opt}(T) = \left(\frac{A}{L} \int_{T_L}^{T_H} \frac{k^{1/2}}{T}\, dT \right) k^{1/2} T$$

$$\dot{S}_{gen,min} = \frac{A}{L} \left(\int_{T_L}^{T_H} \frac{k^{1/2}}{T}\, dT \right)^2$$

The technological applications of the variable heat leak optimization principle are numerous and important. In the case of a mechanical support, the optimal design is approximated in practice by placing a stream of cold helium gas in counterflow (and in thermal contact) with the conduction path. The heat leak varies as $d\dot{Q}/dT = \dot{m}c_p$, where $\dot{m}c_p$ is the capacity flow rate of the stream. The practical value of the EGM theory is that it guides the designer to an optimal flow rate for minimum entropy generation. To illustrate, if the support conductivity is temperature-independent, then the optimal flow rate is $\dot{m}_{opt} = (Ak/Lc_p) \ln (T_H/T_L)$. In reality, the conductivity of cryogenic structural materials varies strongly with the temperature, and the single-stream intermediate cooling technique can approach $\dot{S}_{gen,min}$ only approximately.[4,5]

Other applications include the optimal cooling (e.g., optimal flow rate of boil-off helium) for cryogenic current leads, and the optimal temperatures of cryogenic radiation shields. The main counterflow heat exchanger of a low-temperature refrigeration machine is another important path for heat leak in the end-to-end direction ($T_H \rightarrow T_L$). In this case, the optimal variable heat leak principle translates into[4,5]

$$\left(\frac{\Delta T}{T}\right)_{opt} = \frac{\dot{m}c_p}{UA} \ln \frac{T_H}{T_L}$$

where ΔT is the local stream-to-stream temperature difference of the counterflow, $\dot{m}c_p$ is the capacity flow rate through one branch of the counterflow, and UA is the fixed size (total thermal conductance) of the heat exchanger. Other EGM applications in the field of cryogenics are reviewed in Refs. 4 and 5.

42.6 HEAT TRANSFER

The field of heat transfer adopted the techniques developed in cryogenic engineering and applied them to a vast selection of devices for promoting heat transfer. The EGM method was applied to complete components (e.g., heat exchangers) and elemental features (e.g., ducts, fins). For example, consider the flow of a single-phase stream ($\dot{m}$) through a heat exchanger tube of internal diameter D. The heat transfer rate per unit of tube length q' is given. The entropy generation rate per unit of tube length is

$$\dot{S}'_{gen} = \frac{q'^2}{\pi k T^2 \text{Nu}} + \frac{32 \dot{m}^3 f}{\pi^2 \rho^2 T D^5}$$

where Nu and f are the Nusselt number and the friction factor, $\text{Nu} = hD/k$ and $f = (-dP/dx) \rho D/(2G^2)$ with $G = \dot{m}/(\pi D^2/4)$. The $\dot{S}'_{gen}$ expression has two terms, in order, the irreversibility contributions made by heat transfer and fluid friction. These terms compete against one another such that there is an optimal tube diameter for minimum entropy generation rate,[4,5]

$$\text{Re}_{D,opt} \cong 2B_0^{0.36} \, \text{Pr}^{-0.07}$$

$$B_0 = \frac{q' \dot{m} \rho}{(kT)^{1/2} \, \mu^{5/2}}$$

where $\text{Re}_D = VD/\nu$ and $V = \dot{m}/(\rho \pi D^2/4)$. This result is valid in the range $2500 < \text{Re}_D < 10^6$ and $\text{Pr} > 0.5$. The corresponding entropy generation number is

$$N_S = \frac{\dot{S}'_{gen}}{\dot{S}'_{gen,min}} = 0.856 \left(\frac{\text{Re}_D}{\text{Re}_{D,opt}}\right)^{-0.8} + 0.144 \left(\frac{\text{Re}_D}{\text{Re}_{D,opt}}\right)^{4.8}$$

where $\text{Re}_D/\text{Re}_{D,opt} = D_{opt}/D$ because the mass flow rate is fixed. The N_S criterion was used extensively in the literature to monitor the approach of actual designs to the optimal irreversible designs conceived subject to the same constraints.[4,5]

The EGM of elemental features was extended to the optimization of augmentation techniques such as extended surfaces (fins), roughened walls, spiral tubes, twisted tape inserts, and full-size heat exchangers that have such features. For example, the entropy generation rate of a body with heat transfer and drag in an external stream (U_∞, T_∞) is

$$\dot{S}_{gen} = \frac{\dot{Q}_B(T_B - T_\infty)}{T_B T_\infty} + \frac{F_D U_\infty}{T_\infty}$$

where $\dot{Q}_B$, T_B and F_D are the heat transfer rate, body temperature, and drag force. The relation between $\dot{Q}_B$ and temperature difference ($T_B - T_\infty$) depends on body shape and external fluid and flow, and is provided by the field of convective heat transfer.[6] The relation between F_D, U_∞, geometry and fluid type comes from fluid mechanics.[6] The $\dot{S}_{gen}$ expression has the expected two-term structure, which leads to an optimal body size for minimum entropy generation rate.

The simplest example is the selection of the swept length L of a plate immersed in a parallel stream (Fig. 42.5 inset). The results for $\text{Re}_{L,\text{opt}} = U_\infty L_{\text{opt}}/\nu$ are shown in Fig. 42.5 where B is the constraint (duty parameter)

$$B = \frac{\dot{Q}_B/W}{U_\infty (k\mu T_\infty \text{Pr}^{1/3})^{1/2}}$$

and W is the plate dimension perpendicular to the figure. The same figure shows the corresponding results for the optimal diameter of a cylinder in cross flow, where $\text{Re}_{D,\text{opt}} = U_\infty D_{\text{opt}}/\nu$ and B is given by the same equation as for the plate. The optimal diameter of the sphere is referenced to the sphere duty parameter defined by

$$B_s = \frac{\dot{Q}_B}{\nu (k\mu T_\infty \text{Pr}^{1/3})^{1/2}}$$

The fins built on the surfaces of heat exchanges act as bodies with heat transfer in external flow. The size of a fin of given shape can be optimized by accounting for the internal heat transfer characteristics (longitudinal conduction) of the fin, in addition to the two terms (convective heat and fluid flow) shown in the last $\dot{S}_{\text{gen}}$ formula. The EGM method has also been applied to complete heat exchangers and heat exchanger networks. This vast literature is reviewed in Ref. 5. One technological benefit of EGM is that it shows how to select certain dimensions of a device such that the device destroys minimum power while performing its assigned heat and fluid flow duty.

Several computational heat and fluid flow studies recommended that future commercial CFD packages have the capability of displaying entropy generation rate fields (maps) for both laminar and turbulent flows. For example, Paoletti et al.[7] recommend the plotting of contour lines for constant values of the dimensionless group $\text{Be} = \dot{S}'''_{\text{gen},\Delta T}/(\dot{S}'''_{\text{gen},\Delta T} + \dot{S}'''_{\text{gen},\Delta P})$ where $\dot{S}'''_{\text{gen}}$ means local (volumetric) entropy generation rate, and ΔT and ΔP refer to the heat transfer and fluid flow irreversibilities, respectively.

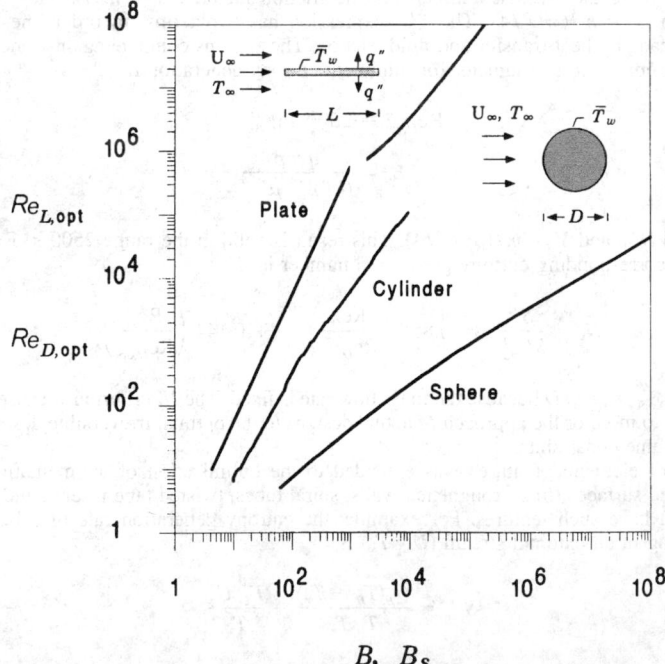

Fig. 42.5   The optimal size of a plate, cylinder and sphere for minimum entropy generation. (From A. Bejan, G. Tsatsaronis, and M. Moran, *Thermal Design and Optimization*. © 1996 John Wiley & Sons, Inc. Reprinted by permission.)

42.7 STORAGE SYSTEMS

In the optimization of time-dependent heating or cooling processes the search is for optimal historics, that is, optimal ways of executing the processes. Consider as a first example the sensible heating of an amount of incompressible substance (mass M, specific heat C), by circulating through it a stream of hot ideal gas ($\dot{m}$, c_p, T_∞) (Fig. 42.6). Initially, the storage material is at the ambient temperature T_0. The total thermal conductance of the heat exchanger placed between the storage material and the gas stream is UA and the pressure drop is negligible. After flowing through the heat exchanger, the gas stream is discharged into the atmosphere. The entropy generated from $t = 0$ until a time t reaches a minimum when t is of the order of $MC/(\dot{m}c_p)$. Charts for calculating the optimal heating (storage) time interval are available in Refs. 4 and 5. For example, when $(T_\infty - T_0) << T_0$, the optimal heating time is given by

$$\theta_{opt} = \frac{1.256}{1 - \exp(-N_{tu})}$$

where $\theta_{opt} = t_{opt}\,\dot{m}c_p/(MC)$ and $N_{tu} = UA/(\dot{m}c_p)$.

Another example is the optimization of a sensible-heat cooling process subject to an overall time constraint. Consider the cooling of an amount of incompressible substance (M, C) from a given initial temperature to a given final temperature, during a prescribed time interval t_c. The coolant is a stream of cold ideal gas with flow rate $\dot{m}$ and specific heat $c_p(T)$. The thermal conductance of the heat exchanger is UA; however, the overall heat transfer coefficient generally depends on the instantaneous temperature, $U(T)$. The cooling process requires a minimum amount of coolant m (or minimum refrigerator work for producing the cryogen m),

$$m = \int_0^{t_c} \dot{m}(t)\, dt$$

when the gas flow rate has the optimal history[4,5]

$$\dot{m}_{opt}(t) = \left[\frac{U(T)A}{C^* c_p(T)}\right]^{1/2}$$

In this expression, $T(t)$ is the corresponding optimal temperature history of the object that is being cooled and C^* is a constant that can be evaluated based on the time constraint, as shown in Refs. 4 and 5. The optimal flow rate history result ($\dot{m}_{opt}$) tells the operator that at temperatures where U is small the flow rate should be decreased. Furthermore, since during cooldown the gas c_p increases, the flow rate should decrease as the end of the process nears.

In the case of energy storage by melting there is an optimal melting temperature (i.e., optimal type of storage material) for minimum entropy generation during storage. If T_∞ and T_0 are the temperatures of the heat source and the ambient, the optimal melting temperature of the storage material has the value $T_{m,opt} = (T_\infty T_0)^{1/2}$

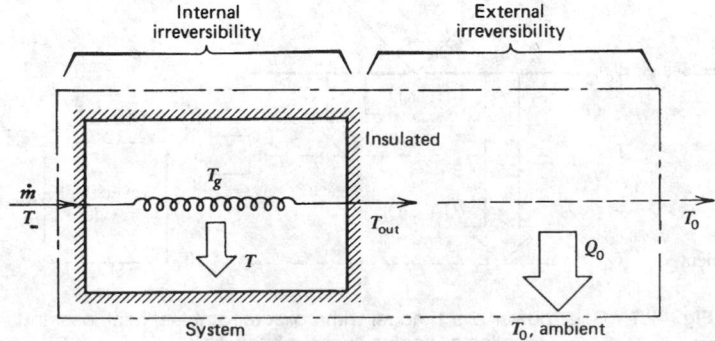

Fig. 42.6 Entropy generation during sensible-heat storage.[4]

42.8 SOLAR ENERGY CONVERSION

The generation of power and refrigeration based on energy from the sun has been the subject of some of the oldest EGM studies, which cover a vast territory. A characteristic of these EGM models is that they account for the irreversibility due to heat transfer in the two temperature gaps (sun-collector and collector-ambient) and that they reveal an optimal *coupling* between the collector and the rest of the plant.

Consider, for example, the steady operation of a power plant driven by a solar collector with convective heat leak to the ambient, $\dot{Q}_0 = (UA)_c(T_c - T_0)$, where $(UA)_c$ is the collector-ambient thermal conductance and T_c is the collector temperature (Fig. 42.7). Similarly, there is a finite size heat exchanger $(UA)_i$ between the collector and the hot end of the power cycle (T), such that the heat input provided by the collector is $\dot{Q} = (UA)_i(T_c - T)$. The power cycle is assumed reversible. The power output $\dot{W} = \dot{Q}\,(1 - T_0/T)$ is maximum, or the total entropy generation rate is minimum, when the collector has the optimal temperature[4,5]

$$\frac{T_{c,\text{opt}}}{T_0} = \frac{\theta_{\max}^{1/2} + R\theta_{\max}}{1 + R}$$

where $R = (UA)_c/(UA)_i$, $\theta_{\max} = T_{c,\max}/T_0$ and $T_{c,\max}$ is the maximum (stagnation) temperature of the collector.

Another type of optimum is discovered when the overall size of the installation is fixed. For example, in an extraterrestrial power plant with collector area A_H and radiator area A_L, if the total area is constrained[1]

$$A_H + A_L = A \quad (\text{constant})$$

the optimal way to allocate the area is $A_{H,\text{opt}} = 0.35A$ and $A_{L,\text{opt}} = 0.65A$. Other examples of optimal allocation of hardware between various components subject to overall size constraints are given in Ref. 5. The progress on the thermodynamic optimization of solar energy (thermal and photovoltaic) is reviewed in Refs. 1 and 5.

42.9 POWER PLANTS

There are several EGM models and optima of power plants that have fundamental implications. The loss of heat from the hot end of a power plant can be modeled by using a thermal resistance in parallel with an irreversibility-free compartment that accounts for the power output $\dot{W}$ of the actual power plant (Fig. 42.8). The hot-end temperature of the working fluid cycle T_H can vary. The heat input $\dot{Q}_H$ is fixed. The bypass heat leak is proportional to the temperature difference, $\dot{Q}_C = C(T_H - T_L)$, where C is the thermal conductance of the power plant insulation. The power output is maximum (and $\dot{S}_{\text{gen}}$ is minimum) when the hot-end temperature reaches the optimal level[4]

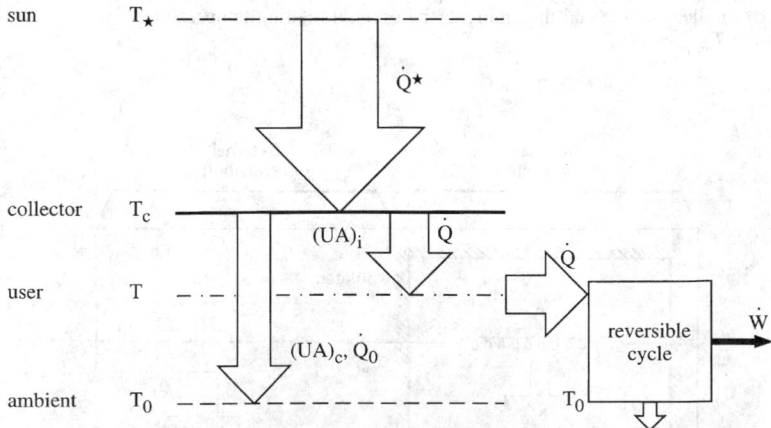

Fig. 42.7 Solar power plant model with collector-ambient heat loss and collector-engine heat exchanger.[4]

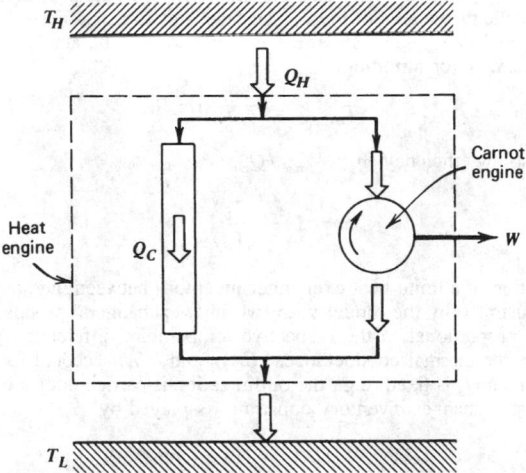

Fig. 42.8 Power plant model with bypass heat leak.[4]

$$T_{H,opt} = T_L \left(1 + \frac{\dot{Q}_H}{CT_L} \right)^{1/2}$$

The corresponding efficiency ($\dot{W}_{max}/\dot{Q}_H$) is

$$\eta = \frac{(1 + r)^{1/2} - 1}{(1 + r)^{1/2} + 1}$$

where $r = \dot{Q}_H/(CT_L)$ is a dimensionless way of expressing the size (thermal resistance) of the power plant. An optimal T_H value exists because when $T_H < T_{H,opt}$, the Carnot efficiency of the power producing compartment is too low, while when $T_H > T_{H,opt}$, too much of the unit heat input $\dot{Q}_H$ bypasses the power compartment.

Another optimal hot-end temperature is revealed by the power plant model shown in Fig. 42.9 (e.g., Ref. 1, p. 357). The power plant is driven by a stream of hot single-phase fluid of inlet temperature T_H and constant specific heat c_p. The model has two compartments. The one sandwiched

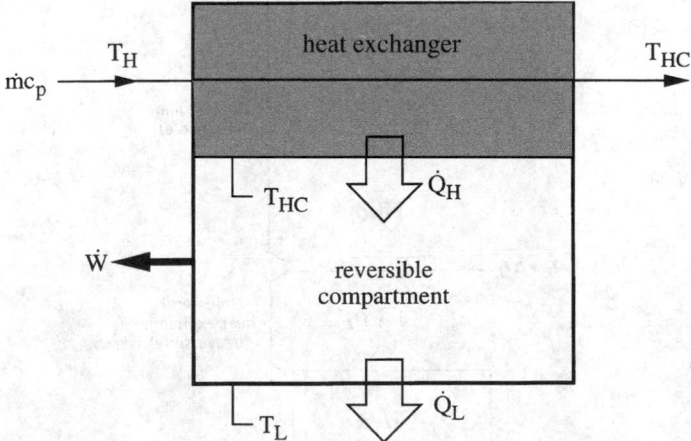

Fig. 42.9 Power plant driven by a stream of hot single-phase fluid.[1,5]

between the heat exchanger surface (T_{HC}) and the ambient (T_L) operates reversibly. The other is a heat exchanger: for simplicity, the area of the T_{HC} surface is assumed sufficiently large that the stream outlet temperature is equal to T_{HC}. The stream is discharged into the ambient. The optimal hot-end temperature for maximum $\dot{W}$ (or minimum $\dot{S}_{gen}$) is[1,5]

$$T_{HC,opt} = (T_H T_L)^{1/2}$$

The corresponding first-law efficiency, $\eta = \dot{W}_{max}/\dot{Q}_H$, is[5]

$$\eta = 1 - \left(\frac{T_L}{T_H}\right)^{1/2}$$

The optimal allocation of a finite heat exchanger inventory between the hot end and the cold end of a power plant is illustrated by the model with two heat exchangers[4] proposed in Fig. 42.10. The heat transfer rates are proportional to the respective temperature differences, $\dot{Q}_H = (UA)_H \Delta T_H$ and $\dot{Q}_L = (UA)_L \Delta T_L$, where the thermal conductances $(UA)_H$ and $(UA)_L$ account for the sizes of the heat exchangers. The heat input $\dot{Q}_L$ is fixed (e.g., the optimization is carried out for one unit of fuel burnt). The role of overall heat exchanger inventory constraint is played by[1]

$$(UA)_H + (UA)_L = UA \quad \text{(constant)}$$

where UA is the total thermal conductance available. The power output is maximized, and the entropy generation rate is minimized, when UA is allocated according to the rule[1,5]

$$(UA)_{H,opt} = (UA)_{L,opt} = \tfrac{1}{2} UA$$

The corresponding maximum efficiency is, as expected, lower than the Carnot efficiency,

$$\eta = 1 - \frac{T_L}{T_H}\left(1 - \frac{4\dot{Q}_H}{T_H UA}\right)^{-1}$$

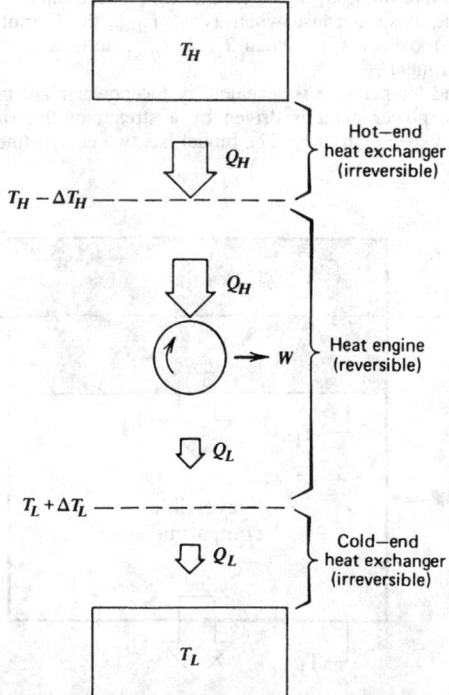

Fig. 42.10 Power plant with two finite-size heat exchangers.[4]

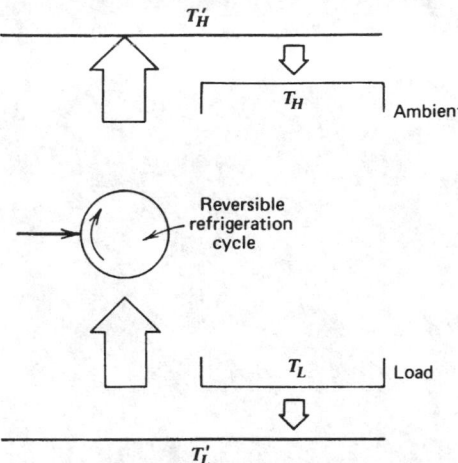

Fig. 42.11 Refrigerator model with two finite-size heat exchangers.[4]

The EGM modeling and optimization progress on power plants is extensive, and is reviewed in Ref. 5. Similar models have also been used in the field of refrigeration, as we saw already in Section 42.5. For example, in a steady-state refrigeration plant with two heat exchangers (Fig. 42.11) subjected to the total UA constraint listed above, the refrigerator power input is minimum when UA is divided equally among the two heat exchangers, $(UA)_{H,\text{opt}} = \frac{1}{2}UA = (UA)_{L,\text{opt}}$.

REFERENCES

1. A. Bejan, *Advanced Engineering Thermodynamics*, 2nd ed., Wiley, New York, 1997.
2. M. J. Moran, *Availability Analysis: A Guide to Efficient Energy Use*, ASME Press, New York, 1989.
3. A. Bejan, G. Tsatsaronis, and M. Moran, *Thermal Design and Optimization*, Wiley, New York, 1996.
4. A. Bejan, *Entropy Generation through Heat and Fluid Flow*, Wiley, New York, 1982.
5. A. Bejan, *Entropy Generation Minimization*, CRC Press, Boca Raton, FL, 1996.
6. A. Bejan, *Convection Heat Transfer*, 2nd ed., Wiley, New York, 1995.
7. S. Paoletti, F. Rispoli, and E. Sciubba, "Calculation of Exergetic Losses in Compact Heat Exchanger Passages," *ASME AES* **10**(2), 21–29 (1989).

CHAPTER 43

HEAT TRANSFER FUNDAMENTALS

G. P. "Bud" Peterson
Executive Associate Dean and Associate Vice Chancellor of Engineering
Texas A&M University
College Station, Texas

43.1	**SYMBOLS AND UNITS**	**1367**	
43.2	**CONDUCTION HEAT TRANSFER**	**1369**	
	43.2.1 Thermal Conductivity	1370	
	43.2.2 One-Dimensional Steady-State Heat Conduction	1375	
	43.2.3 Two-Dimensional Steady-State Heat Conduction	1377	
	43.2.4 Heat Conduction with Convection Heat Transfer on the Boundaries	1381	
	43.2.5 Transient Heat Conduction	1383	
43.3	**CONVECTION HEAT TRANSFER**	**1385**	
	43.3.1 Forced Convection—Internal Flow	1385	
	43.3.2 Forced Convection—External Flow	1393	
	43.3.3 Free Convection	1397	

	43.3.4 The Log Mean Temperature Difference	1400	
43.4	**RADIATION HEAT TRANSFER**	**1400**	
	43.4.1 Black-Body Radiation	1400	
	43.4.2 Radiation Properties	1404	
	43.4.3 Configuration Factor	1407	
	43.4.4 Radiative Exchange among Diffuse-Gray Surfaces in an Enclosure	1410	
	43.4.5 Thermal Radiation Properties of Gases	1415	
43.5	**BOILING AND CONDENSATION HEAT TRANSFER**	**1417**	
	43.5.1 Boiling	1420	
	43.5.2 Condensation	1423	
	43.5.3 Heat Pipes	1424	

43.1 SYMBOLS AND UNITS

A	area of heat transfer
Bi	Biot number, hL/k, dimensionless
C	circumference, m, constant defined in text
C_p	specific heat under constant pressure, J/kg · K
D	diameter, m
e	emissive power, W/m^2
f	drag coefficient, dimensionless
F	cross flow correction factor, dimensionless
F_{i-j}	configuration factor from surface i to surface j, dimensionless
Fo	Fourier number, $\alpha t A^2/V^2$, dimensionless
$F_{o-\lambda T}$	radiation function, dimensionless
G	irradiation, W/m^2; mass velocity, kg/m^2 · sec
g	local gravitational acceleration, 9.8 m/sec^2
g_c	proportionality constant, 1 kg · m/N · sec^2
Gr	Grashof number, $gL^3\beta\Delta t/\nu^2$, dimensionless
h	convection heat transfer coefficient, equals $q/A\Delta T$, W/m^2 · K
h_{fg}	heat of vaporization, J/kg
J	radiocity, W/m^2
k	thermal conductivity, W/m · K

Mechanical Engineers' Handbook, 2nd ed., Edited by Myer Kutz.
ISBN 0-471-13007-9 © 1998 John Wiley & Sons, Inc.

K   wick permeability, m^2
L length, m
Ma   Mach number, dimensionless
N   screen mesh number, m^{-1}
Nu   Nusselt number, $Nu_L = hL/k$, $Nu_D = hD/k$, dimensionless
$\overline{Nu}$ Nusselt number averaged over length, dimensionless
P   pressure, N/m^2, perimeter, m
Pe   Peclet number, RePr, dimensionless
Pr   Prandtl number, $C_p\mu/k$, dimensionless
q rate of heat transfer, W
q'' rate of heat transfer per unit area, W/m^2
R distance, m; thermal resistance, K/W
r radial coordinate, m; recovery factor, dimensionless
Ra   Rayleigh number, GrPr; $Ra_L = Gr_L Pr$, dimensionless
Re   Reynolds number, $Re_L = \rho VL/\mu$, $Re_D = \rho VD/\mu$, dimensionless
S conduction shape factor, m
T temperature, K or °C
t time, sec
T_{as} adiabatic surface temperature, K
T_{sat} saturation temperature, K
T_b fluid bulk temperature or base temperature of fins, K
T_e excessive temperature, $T_s - T_{sat}$, K or °C
T_f film temperature, $(T_\infty + T_s)/2$, K
T_i initial temperature; at $t = 0$, K
T_o stagnation temperature, K
T_s surface temperature, K
T_∞ free stream fluid temperature, K
U overall heat transfer coefficient, $W/m^2 \cdot K$
V fluid velocity, m/sec; volume, m^3
w   groove width, m; or wire spacing, m
We   Weber number, dimensionless
x one of the axes of Cartesian reference frame, m

Greek Symbols

α thermal diffusivity, $k/\rho C_p$, m^2/sec; absorptivity, dimensionless
β coefficient of volume expansion, 1/K
Γ mass flow rate of condensate per unit width, $kg/m \cdot sec$
γ specific heat ratio, dimensionless
ΔT temperature difference, K
δ thickness of cavity space, groove depth, m
ϵ emissivity, dimensionless
ε wick porosity, dimensionless
λ wavelength, μm
η_f fin efficiency, dimensionless
μ viscosity, $kg/m \cdot sec$
ν kinematic viscosity, m^2/sec
ρ reflectivity, dimensionless; density, kg/m^3
σ surface tension, N/m; Stefan–Boltzmann constant, 5.729×10^{-8} $W/m^2 \cdot K^4$
τ transmissivity, dimensionless, shear stress, N/m^2
Ψ angle of inclination, degrees or radians

Subscripts

a adiabatic section, air
b boiling, black body
c convection, capillary, capillary limitation, condenser
e entrainment, evaporator section
eff   effective
f fin
i inner
l liquid
m mean, maximum
n nucleation
o outer
0   stagnation condition
p pipe
r radiation

s	surface, sonic or sphere
w	wire spacing, wick
v	vapor
λ	spectral
∞	free stream
$-$	axial hydrostatic pressure
$+$	normal hydrostatic pressure

The science or study of heat transfer is that subset of the larger field of transport phenomena that focuses on the energy transfer occurring as a result of a temperature gradient. This energy transfer can manifest itself in several forms, including *conduction,* which focuses on the transfer of energy through the direct impact of molecules; *convection,* which results from the energy transferred through the motion of a fluid; and *radiation,* which focuses on the transmission of energy through electromagnetic waves. In the following review, as is the case with most texts on heat transfer, *phase change heat transfer,* that is, *boiling* and *condensation,* will be treated as a subset of convection heat transfer.

43.2 CONDUCTION HEAT TRANSFER

The exchange of energy or heat resulting from the kinetic energy transferred through the direct impact of molecules is referred to as *conduction* and takes place from a region of high energy (or temperature) to a region of lower energy (or temperature). The fundamental relationship that governs this form of heat transfer is *Fourier's law of heat conduction,* which states that in a one-dimensional system with no fluid motion, the rate of heat flow in a given direction is proportional to the product of the temperature gradient in that direction and the area normal to the direction of heat flow. For conduction heat transfer in the *x*-direction this expression takes the form

$$q_x = -kA\frac{\partial T}{\partial x}$$

where q_x is the heat transfer in the *x*-direction, A is the area normal to the heat flow, $\partial T/\partial x$ is the temperature gradient, and k is the thermal conductivity of the substance.

Writing an energy balance for a three-dimensional body, and utilizing Fourier's law of heat conduction, yields an expression for the transient diffusion occurring within a body or substance.

$$\frac{\partial}{\partial x}\left(k\frac{\partial T}{\partial x}\right) + \frac{\partial}{\partial y}\left(k\frac{\partial T}{\partial y}\right) + \frac{\partial}{\partial z}\left(k\frac{\partial T}{\partial z}\right) + \dot{q} = \rho c_p \frac{\partial}{\partial x}\frac{\partial T}{\partial t}$$

This expression, usually referred to as the *heat diffusion equation* or heat equation, provides a basis for most types of heat conduction analysis. Specialized cases of this equation, such as the case where the thermal conductivity is a constant

$$\frac{\partial^2 T}{\partial x^2} + \frac{\partial^2 T}{\partial y^2} + \frac{\partial^2 T}{\partial z^2} + \frac{\dot{q}}{k} = \frac{\rho c_p}{k}\frac{\partial T}{\partial t}$$

steady-state with heat generation

$$\frac{\partial}{\partial x}\left(k\frac{\partial T}{\partial x}\right) + \frac{\partial}{\partial y}\left(k\frac{\partial T}{\partial y}\right) + \frac{\partial}{\partial z}\left(k\frac{\partial T}{\partial z}\right) + \dot{q} = 0$$

steady-state, one-dimensional heat transfer with heat transfer to a heat sink (i.e., a fin)

$$\frac{\partial}{\partial x}\left(\frac{\partial T}{\partial x}\right) + \frac{\dot{q}}{k} = 0$$

or one-dimensional heat transfer with no internal heat generation

$$\frac{\partial}{\partial x}\left(\frac{\partial T}{\partial x}\right) = \frac{\rho c_p}{k}\frac{\partial T}{\partial t}$$

can be utilized to solve many steady-state or transient problems. In the following sections, this equation will be utilized for several specific cases. However, in general, for a three-dimensional body of constant thermal properties without heat generation under steady-state heat conduction, the temperature field satisfies the expression

$$\nabla^2 T = 0$$

43.2.1 Thermal Conductivity

The ability of a substance to transfer heat through conduction can be represented by the constant of proportionality k, referred to as the thermal conductivity. Figures 43.1a, b, and c illustrate the characteristics of the thermal conductivity as a function of temperature for several solids, liquids and gases, respectively. As shown, the thermal conductivity of solids is higher than liquids, and liquids higher than gases. Metals typically have higher thermal conductivities than nonmetals, with pure metals having thermal conductivities that decrease with increasing temperature, while the thermal conductivities of nonmetallic solids generally increase with increasing temperature and density. The addition of other metals to create alloys, or the presence of impurities, usually decreases the thermal conductivity of a pure metal.

In general, the thermal conductivities of liquids decrease with increasing temperature. Alternatively, the thermal conductivities of gases and vapors, while lower, increase with increasing temperature and decrease with increasing molecular weight. The thermal conductivities of a number of commonly used metals and nonmetals are tabulated in Tables 43.1 and 43.2, respectively. Insulating materials, which are used to prevent or reduce the transfer of heat between two substances or a substance and the surroundings are listed in Tables 43.3 and 43.4, along with the thermal properties. The thermal conductivities for liquids, molten metals, and gases are given in Tables 43.5, 43.6 and 43.7, respectively.

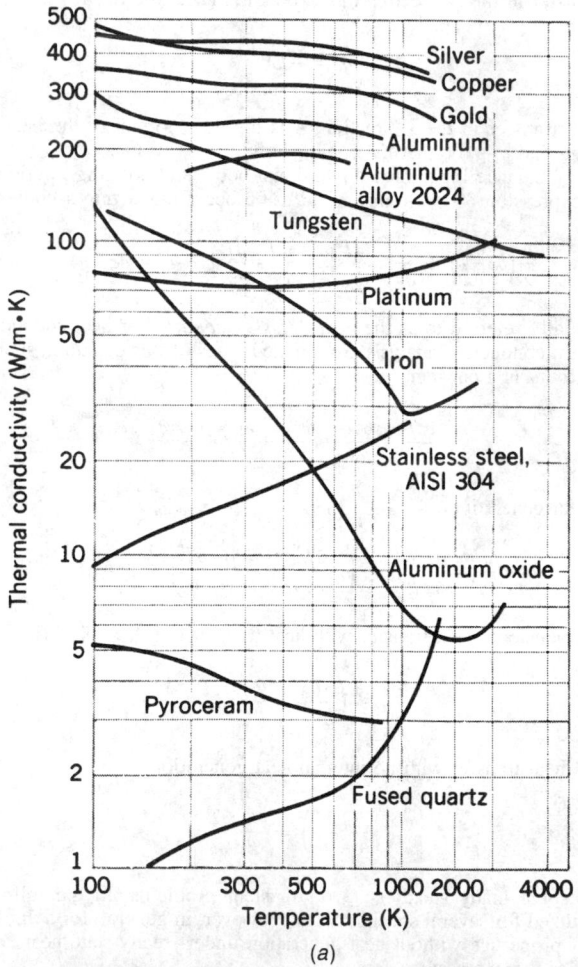

(a)

Fig. 43.1a Temperature dependence of the thermal conductivity of selected solids.

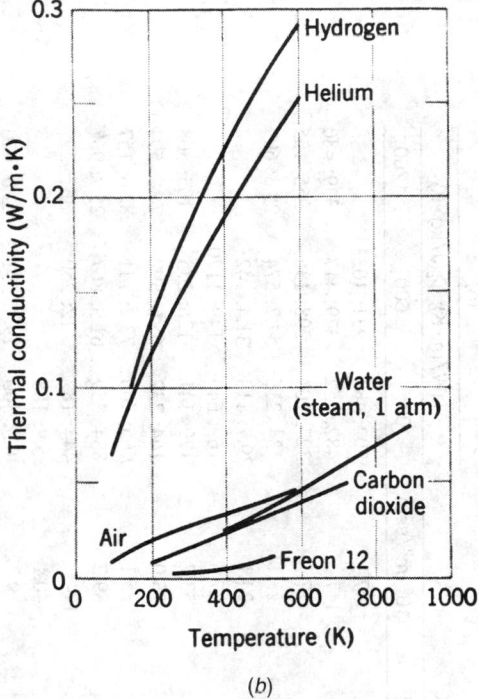

(b)

Fig. 43.1b Selected nonmetallic liquids under saturated conditions.

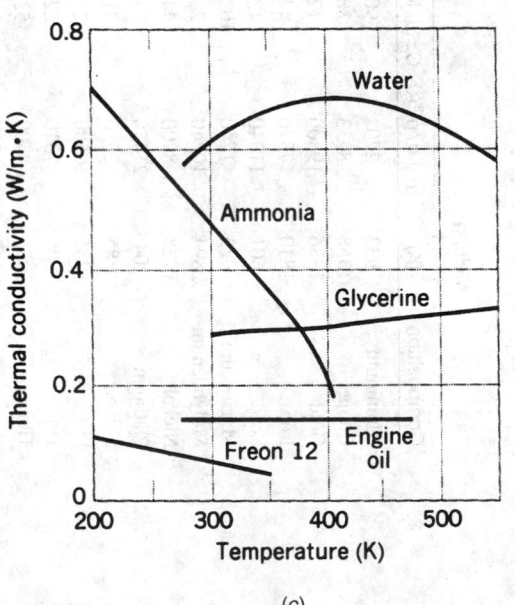

(c)

Fig. 43.1c Selected gases at normal pressures.[1]

Table 43.1 Thermal Properties of Metallic Solids[a]

Composition	Melting Point (K)	Properties at 300 K ρ (kg/m³)	Cp (J/kg·K)	k (W/m·K)	α × 10⁶ (m²/sec)	Properties at Various Temperatures (K) k(W/m·K); Cp(J/kg·K) 100	600	1200
Aluminum	933	2702	903	237	97.1	302; 482	231; 1033	
Copper	1358	8933	385	401	117	482; 252	379; 417	339; 480
Gold	1336	19300	129	317	127	327; 109	298; 135	255; 155
Iron	1810	7870	447	80.2	23.1	134; 216	54.7; 574	28.3; 609
Lead	601	11340	129	35.3	24.1	39.7; 118	31.4; 142	
Magnesium	923	1740	1024	156	87.6	169; 649	149; 1170	
Molybdenum	2894	10240	251	138	53.7	179; 141	126; 275	105; 308
Nickel	1728	8900	444	90.7	23.0	164; 232	65.6; 592	76.2; 594
Platinum	2045	21450	133	71.6	25.1	77.5; 100	73.2; 141	82.6; 157
Silicon	1685	2330	712	148	89.2	884; 259	61.9; 867	25.7; 967
Silver	1235	10500	235	429	174	444; 187	412; 250	361; 292
Tin	505	7310	227	66.6	40.1	85.2; 188		
Titanium	1953	4500	522	21.9	9.32	30.5; 300	19.4; 591	22.0; 620
Tungsten	3660	19300	132	174	68.3	208; 87	137; 142	113; 152
Zinc	693	7140	389	116	41.8	117; 297	103; 436	

[a]Adapted from F. P. Incropera and D. P. Dewitt, *Fundamentals of Heat Transfer.* © 1981 John Wiley & Sons, Inc. Reprinted by permission.

Table 43.2 Thermal Properties of Nonmetals

Description/Composition	Temperature (K)	Density ρ (kg/m³)	Thermal Conductivity k (W/m·K)	Specific Heat C_p (J/kg·K)	$\alpha \times 10^6$ (m²/sec)
Bakelite	300	1300	0.232	1465	0.122
Brick, refractory					
Carborundum	872	—	18.5	—	—
Chrome-brick	473	3010	2.32	835	0.915
Fire clay brick	478	2645	1.0	960	0.394
Clay	300	1460	1.3	880	1.01
Coal, anthracite	300	1350	0.26	1260	0.153
Concrete (stone mix)	300	2300	1.4	880	0.692
Cotton	300	80	0.059	1300	0.567
Glass, window	300	2700	0.78	840	0.344
Rock, limestone	300	2320	2.15	810	1.14
Rubber, hard	300	1190	0.160	—	—
Soil, dry	300	2050	0.52	1840	0.138
Teflon	300	2200	0.35	—	—
	400	—	0.45	—	—

Table 43.3 Thermal Properties of Building and Insulating Materials (at 300K)[a]

Description/Composition	Density ρ (kg/m³)	Thermal Conductivity k (W/m·K)	Specific Heat C_p (J/kg·K)	$\alpha \times 10^6$ (m²/sec)
Building boards				
Plywood	545	0.12	1215	0.181
Acoustic tile	290	0.058	1340	0.149
Hardboard, siding	640	0.094	1170	0.126
Woods				
Hardwoods (oak, maple)	720	0.16	1255	0.177
Softwoods (fir, pine)	510	0.12	1380	0.171
Masonry materials				
Cement mortar	1860	0.72	780	0.496
Brick, common	1920	0.72	835	0.449
Plastering materials				
Cement plaster, sand aggregate	1860	0.72	—	—
Gypsum plaster, sand aggregate	1680	0.22	1085	0.121
Blanket and batt				
Glass fiber, paper faced	16	0.046	—	—
Glass fiber, coated; duct liner	32	0.038	835	1.422
Board and slab				
Cellular glass	145	0.058	1000	0.400
Wood, shredded/cemented	350	0.087	1590	0.156
Cork	120	0.039	1800	0.181
Loose fill				
Glass fiber, poured or blown	16	0.043	835	3.219
Vermiculite, flakes	80	0.068	835	1.018

[a]Adapted from F. P. Incropera and D. P. Dewitt, *Fundamentals of Heat Transfer*. © 1981 John Wiley & Sons, Inc. Reprinted by permission.

Table 43.4 Thermal Conductivities for Some Industrial Insulating Materials[a]

Description/Composition	Maximum Service Temperature (K)	Typical Density (kg/m³)	Typical Thermal Conductivity, k (W/m · K), at Various Temperatures (K)			
			200	300	420	645
Blankets						
Blanket, mineral fiber, glass; fine fiber organic bonded	450	10		0.048		
		48		0.033		
Blanket, alumina–silica fiber	1530	48				0.105
Felt, semirigid; organic bonded	480	50–125		0.038	0.063	
Felt, laminated; no binder	920	120			0.051	0.087
Blocks, boards, and pipe insulations						
Asbestos paper, laminated and corrugated, 4-ply	420	190		0.078		
Calcium silicate	920	190			0.063	0.089
Polystyrene, rigid						
Extruded (R-12)	350	56	0.023	0.027		
Molded beads	350	16	0.026	0.040		
Rubber, rigid foamed	340	70		0.032		
Insulating cement						
Mineral fiber (rock, slag, or glass)						
With clay binder	1255	430			0.088	0.123
With hydraulic setting binder	922	560			0.123	
Loose fill						
Cellulose, wood or paper pulp	—	45		0.039		
Perlite, expanded	—	105	0.036	0.053		
Vermiculite, expanded	—	122		0.068		

[a]Adapted from F. P. Incropera and D. P. Dewitt, *Fundamentals of Heat Transfer*. © 1981 John Wiley & Sons, Inc. Reprinted by permission.

Table 43.5 Thermal Properties of Saturated Liquids[a]

T (K)	ρ (kg/m³)	C_p (kJ/kg·K)	$v \times 10^6$ (m²/sec)	$k \times 10^3$ (W/m·K)	$\alpha \times 10^7$ (m²/sec)	Pr	$\beta \times 10^3$ (K⁻¹)
Ammonia, Nh₃							
223	703.7	4.463	0.435	547	1.742	2.60	2.45
323	564.3	5.116	0.330	476	1.654	1.99	2.45
Carbon Dioxide, CO₂							
223	1,156.3	1.84	0.119	85.5	0.402	2.96	14.0
303	597.8	36.4	0.080	70.3	0.028	28.7	14.0
Engine Oil (Unused)							
273	899.1	1.796	4,280	147	0.910	47,000	0.70
430	806.5	2.471	5.83	132	0.662	88	0.70
Ethylene Glycol, C₂H₄(OH)₂							
273	1,130.8	2.294	57.6	242	0.933	617.0	0.65
373	1,058.5	2.742	2.03	263	0.906	22.4	0.65
Glycerin, C₃H₅(OH)₃							
273	1,276.0	2.261	8,310	282	0.977	85,000	0.47
320	1,247.2	2.564	168	287	0.897	1,870	0.50
Freon (Refrigerant-12), CCl₂F₂							
230	1,528.4	0.8816	0.299	68	0.505	5.9	1.85
320	1,228.6	1.0155	0.190	68	0.545	3.5	3.50

[a]Adapted from Ref. 2. See Table 43.23 for H_2O.

43.2.2 One-Dimensional Steady-State Heat Conduction

The rate of heat transfer for steady-state heat conduction through a homogeneous material can be expressed as $q = \Delta T/R$, where ΔT is the temperature difference and R is the *thermal resistance*. This thermal resistance, is the reciprocal of the *thermal conductance* ($C = 1/R$) and is related to the thermal conductivity by the cross-sectional area. Expressions for the thermal resistance, the temperature distribution, and the rate of heat transfer are given in Table 43.8 for a plane wall, a cylinder, and a sphere. For the plane wall, the heat transfer is assumed to be one-dimensional (i.e., conducted only in the x-direction) and for the cylinder and sphere, only in the radial direction.

In addition to the heat transfer in these simple geometric configurations, another common problem encountered in practice is the heat transfer through a layered or composite wall consisting of N layers where the thickness of each layer is represented by Δx_n and the thermal conductivity by k_n for $n = 1, 2, \ldots, N$. Assuming that the interfacial resistance is negligible (i.e., there is no thermal resistance at the contacting surfaces), the overall thermal resistance can be expressed as

$$R = \sum_{n=1}^{N} \frac{\Delta x_n}{k_n A}$$

Similarly, for conduction heat transfer in the radial direction through N *concentric cylinders* with negligible interfacial resistance, the overall thermal resistance can be expressed as

$$R = \sum_{n=1}^{N} \frac{\ln(r_{n+1}/r_n)}{2\pi k_n L}$$

where r_1 = inner radius
$\quad r_{N+1}$ = outer radius

For N *concentric spheres* with negligible interfacial resistance, the thermal resistance can be expressed as

$$R = \sum_{n=1}^{N} \left(\frac{1}{r_n} - \frac{1}{r_{n+1}} \right) \Big/ 4\pi k$$

where r_1 = inner radius
$\quad r_{N+1}$ = outer radius

Table 43.6 Thermal Properties of Liquid Metals[a]

Composition	Melting Point (K)	T (K)	ρ (kg/m³)	C_p (kJ/kg·K)	$v \times 10^7$ (m²/sec)	k (W/m·K)	$\alpha \times 10^5$ (m²/sec)	Pr
Bismuth	544	589	10,011	0.1444	1.617	16.4	0.138	0.0142
		1033	9,467	0.1645	0.8343	15.6	1.001	0.0083
Lead	600	644	10,540	0.159	2.276	16.1	1.084	0.024
		755	10,412	0.155	1.849	15.6	1.223	0.017
Mercury	234	273	13,595	0.140	1.240	8.180	0.429	0.0290
		600	12,809	0.136	0.711	11.95	0.688	0.0103
Potassium	337	422	807.3	0.80	4.608	45.0	6.99	0.0066
		977	674.4	0.75	1.905	33.1	6.55	0.0029
Sodium	371	366	929.1	1.38	7.516	86.2	6.71	0.011
		977	778.5	1.26	2.285	59.7	6.12	0.0037
NaK (56%/44%)	292	366	887.4	1.130	6.522	25.6	2.55	0.026
		977	740.1	1.043	2.174	28.9	3.74	0.0058
PbBi (44.5%/55.5%)	398	422	10,524	0.147	—	9.05	0.586	—
		644	10,236	0.147	1.496	11.86	0.790	0.189

[a]Adapted from *Liquid Metals Handbook*, The Atomic Energy Commission, Department of the Navy, Washington, DC, 1952.

Table 43.7 Thermal Properties of Gases at Atmospheric Pressure [a]

T (K)	ρ (kg/m³)	C_p (kJ/kg·K)	$v \times 10^6$ (m²/sec)	k (W/m·K)	$\alpha \times 10^4$ (m²/sec)	Pr
Air						
100	3.6010	1.0266	1.923	0.009246	0.0250	0.768
300	1.1774	1.0057	16.84	0.02624	0.2216	0.708
2500	0.1394	1.688	543.0	0.175	7.437	0.730
Ammonia, Nh_3						
220	0.3828	2.198	19.0	0.0171	0.2054	0.93
473	0.4405	2.395	37.4	0.0467	0.4421	0.84
Carbon Dioxide						
220	2.4733	0.783	4.490	0.01081	0.0592	0.818
600	0.8938	1.076	30.02	0.04311	0.4483	0.668
Carbon Monoxide						
220	1.5536	1.0429	8.903	0.01906	0.1176	0.758
600	0.5685	1.0877	52.06	0.04446	0.7190	0.724
Helium						
33	1.4657	5.200	3.42	0.0353	0.04625	0.74
900	0.05286	5.200	781.3	0.298	10.834	0.72
Hydrogen						
30	0.8472	10.840	1.895	0.0228	0.02493	0.759
300	0.0819	14.314	109.5	0.182	1.554	0.706
1000	0.0819	14.314	109.5	0.182	1.554	0.706
Nitrogen						
100	3.4808	1.0722	1.971	0.009450	0.02531	0.786
300	1.1421	1.0408	15.63	0.0262	0.204	0.713
1200	0.2851	1.2037	156.1	0.07184	2.0932	0.748
Oxygen						
100	3.9918	0.9479	1.946	0.00903	0.02388	0.815
300	1.3007	0.9203	15.86	0.02676	0.2235	0.709
600	0.6504	1.0044	52.15	0.04832	0.7399	0.704
Steam (H_2O Vapor)						
380	0.5863	2.060	21.6	0.0246	0.2036	1.060
850	0.2579	2.186	115.2	0.0637	1.130	1.019

[a]Adapted from Ref. 2.

43.2.3 Two-Dimensional Steady-State Heat Conduction

Two-dimensional heat transfer in an isotropic, homogeneous material with no internal heat generation requires solution of the heat diffusion equation of the form $\partial^2 T/\partial X^2 + \partial T/\partial y^2 = 0$, referred to as the *Laplace equation*. For certain geometries and a limited number of fairly simple combinations of boundary conditions, exact solutions can be obtained analytically. However, for anything but simple geometries or for simple geometries with complicated boundary conditions, development of an appropriate analytical solution can be difficult and other methods are usually employed. Among these are solution procedures involving the use of *graphical* or *numerical* approaches. In the first of these, the rate of heat transfer between two isotherms T_1 and T_2 is expressed in terms of the conduction shape factor, defined by

$$q = kS(T_1 - T_2)$$

Table 43.9 illustrates the shape factor for a number of common geometric configurations. By combining these shape factors, the heat transfer characteristics for a wide variety of geometric configurations can be obtained.

Prior to the development of high-speed digital computers, shape factor and analytical methods were the most prevalent methods utilized for evaluating steady-state and transient conduction problems. However, more recently, solution procedures for problems involving complicated geometries

Table 43.8 One-Dimensional Heat Conduction

Geometry	Heat-Transfer Rate and Temperature Distribution	Heat-Transfer Rate and Overall Heat-Transfer Coefficient with Convection at the Boundaries
Plane wall	$q = \dfrac{T_1 - T_2}{(x_2 - x_1)/kA}$ $T = T_1 + \dfrac{T_2 - T_1}{x_x - x_1}(x - x_1)$ $R = (x_x - x_1)/kA$	$q = UA(T_{\infty,1} - T_{\infty,2})$ $U = \dfrac{1}{\dfrac{1}{h_1} + \dfrac{x_2 - x_2}{k} + \dfrac{1}{h_2}}$
Hollow cylinder	$q = \dfrac{T_1 - T_2}{[\ln(r_2/r_1)]/2\pi kL}$ $T = \dfrac{T_2 - T_1}{\ln(r_2/r_1)}\ln\dfrac{r}{r_1}$ $R = \dfrac{\ln(r_2/r_1)}{2\pi kL}$	$q = 2\pi r_1 L U_1(T_{\infty,1} - T_{\infty,2})$ $\;\; = 2\pi r_1 L U_2(T_{\infty,1} - T_{\infty,2})$ $U_1 = \dfrac{1}{\dfrac{1}{h_1} + \dfrac{r_1\ln(r_2/r_1)}{k} + \dfrac{r_1}{r_2}\dfrac{1}{h_2}}$ $U_2 = \dfrac{1}{\left(\dfrac{r_2}{r_1}\right)\dfrac{1}{h_1} + \dfrac{r_2\ln(r_2/r_1)}{k} + \dfrac{1}{h_2}}$
Hollow sphere	$q = \dfrac{T_2 - T_2}{\left(\dfrac{1}{r_1} - \dfrac{1}{r_2}\right)\Big/ 4\pi k}$ $T = \dfrac{1}{\left(\dfrac{1}{r_1} - \dfrac{1}{r_2}\right)}\left[\dfrac{r_1}{r}(T_1 - T_2) + \left(T_2 - T_1\dfrac{r_1}{r_2}\right)\right]$ $R = \left(\dfrac{1}{r_1} - \dfrac{1}{r_2}\right)\Big/ 4\pi k$	$q = 4\pi r_1^2 U_1(T_{\infty,1} - T_{\infty,2})$ $\;\; = 4\pi r_2^2 U_2(T_{\infty,1} - T_{\infty,2})$ $U_1 = \dfrac{1}{\dfrac{1}{h_1} + r_1^2\left(\dfrac{1}{r_1} - \dfrac{1}{r_2}\right)\Big/k + \left(\dfrac{r_1}{r_2}\right)^2\dfrac{1}{h_2}}$ $U_2 = \dfrac{1}{\left(\dfrac{r_2}{r_1}\right)^2\dfrac{1}{h_1} + r_2^2\left(\dfrac{1}{r_1} - \dfrac{1}{r_2}\right)\Big/k + \dfrac{1}{h_2}}$

Table 43.9 Conduction Shape Factors

System	Schematic	Restrictions	Shape Factor
Isothermal sphere buried in a semi-infinite medium having isothermal surface		$z > D/2$	$\dfrac{2\pi D}{1 - D/4z}$
Horizontal isothermal cylinder of length L buried in a semi-infinite medium having isothermal surface		$\left.\begin{array}{l} L \gg D \\ L \gg D \\ z > 3D/2 \end{array}\right\}$	$\dfrac{2\pi L}{\cosh^{-1}(2z/D)}$ $\dfrac{2\pi L}{\ln(4z/D)}$
The cylinder of length L with eccentric bore		$L \gg D_1,\, D_2$	$\dfrac{2\pi L}{\cosh^{-1}\!\left(\dfrac{D_1^2 + D_2^2 - 4\varepsilon^2}{2D_1 D_2}\right)}$

Table 43.9 *(Continued)*

System	Schematic	Restrictions	Shape Factor
Conduction between two cylinders of length L in infinite medium		$L \gg D_1, D_2$ $L \gg W$	$\dfrac{2\pi L}{\cosh^{-1}\left(\dfrac{4W^2 - D_1^2 - D_2^2}{2D_1 D_2}\right)}$
Circular cylinder of length L in a square solid		$w > D$	$\dfrac{2\pi L}{\ln(1.08\, w/D)}$
Conduction through the edge of adjoining walls		$D > L/5$	$0.54D$
Conduction through corner of three walls with inside and outside temperature, respectively, at T_1 and T_2		$L \ll$ length and width of wall	$0.15L$

or boundary conditions have utilized the finite difference method (FDM). In this method, the solid object is divided into a number of distinct or discrete regions, referred to as *nodes*, each with a specified boundary condition. An energy balance is then written for each nodal region and these equations are solved simultaneously. For interior nodes in a two-dimensional system with no internal heat generation, the energy equation takes the form of the Laplace equation, discussed earlier. However, because the system is characterized in terms of a nodal network, a finite difference approximation must be used. This approximation is derived by substituting the following equation for the x-direction rate of change expression

$$\frac{\partial^2 T}{\partial x^2}\bigg|_{m,n} \approx \frac{T_{m+1,n} + T_{m-1,n} - 2T_{m,n}}{(\Delta x)^2}$$

and for the y-direction rate of change expression

$$\frac{\partial^2 T}{\partial y^2}\bigg|_{m,n} \quad \frac{T_{m,n+1} + T_{m,n-1} + T_{m,n}}{(\Delta y)^2}$$

Assuming $\Delta x = \Delta y$ and substituting into the Laplace equation results in the following expression:

$$T_{m,n+1} + T_{m,n-1} + T_{m+1,n} + T_{m-1,n} - 4T_{m,n} = 0$$

which reduces the exact difference equation to an approximate algebraic expression.

Combining this temperature difference with Fourier's law yields an expression for each internal node:

$$T_{m,n+1} + T_{m,n+1} + T_{m-1,n} + T_{m-1,n} + \frac{\dot{q}\Delta x \cdot \Delta y \cdot 1}{k} - 4T_{m,n} = 0$$

Similar equations for other geometries (i.e., corners) and boundary conditions (i.e., convection) and combinations of the two are listed in Table 43.10. These equations must then be solved using some form of matrix inversion technique, Gauss–Seidel iteration method, or other method for solving large numbers of simultaneous equations.

43.2.4 Heat Conduction with Convection Heat Transfer on the Boundaries

In physical situations where a solid is immersed in a fluid, or a portion of the surface is exposed to a liquid or gas, heat transfer will occur by convection (or when there is a large temperature difference, through some combination of convection and/or radiation). In these situations, the heat transfer is governed by *Newton's law of cooling,* which is expressed as

$$q = hA\Delta T$$

where h is the *convection heat transfer coefficient* (Section 43.2), ΔT is the temperature difference between the solid surface and the fluid, and A is the surface area in contact with the fluid. The resistance occurring at the surface abounding the solid and fluid is referred to as the *thermal resistance* and is given by $1/hA$, i.e., the *convection resistance*. Combining this resistance term with the appropriate conduction resistance yields an *overall heat transfer coefficient* U. Usage of this term allows the overall heat transfer to be defined as $q = UA\Delta T$.

Table 43.8 shows the overall heat transfer coefficients for some simple geometries. Note that U may be based either on the inner surface (U_1) or on the outer surface (U_2) for the cylinders and spheres.

Critical Radius of Insulation for Cylinders

A large number of practical applications involve the use of insulation materials to reduce the transfer of heat to or from cylindrical surfaces. This is particularly true of steam or hot water pipes, where concentric cylinders of insulation are typically added to the outside of the pipes to reduce the heat loss. Beyond a certain thickness, however, the continued addition of insulation may not result in continued reductions in the heat loss. To optimize the thickness of insulation required for these types of applications, a value typically referred to as the *critical radius,* defined as $r_{cr} = k/h$, is used. If the outer radius of the object to be insulated is less than r_{cr}, then the addition of insulation will increase the heat loss, while for cases where the outer radii is greater than r_{cr}, any additional increases in insulation thickness will result in a decrease in heat loss.

Table 43.10 Summary of Nodal Finite-Difference Equations

Configuration	Finite-Difference Equation for $\Delta x = \Delta y$
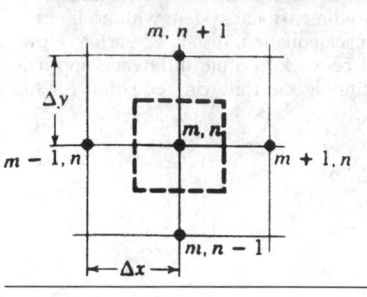	$T_{m,n+1} + T_{m,n-1} + T_{m-1,n}$ $-4T_{m,n} = 0$ **Case 1.** Interior node.
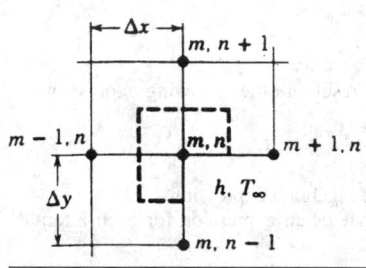	$2(T_{m-1,n} + T_{m,n+1}) + (T_{m+1,n} + T_{m,n-1})$ $+ 2\dfrac{h\Delta x}{k} T_\infty - 2\left(3 + \dfrac{h\Delta x}{k}\right) T_{m,n} = 0$ **Case 2.** Node at an internal corner with convection.
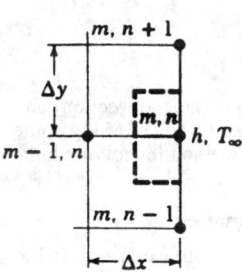	$(2T_{m-1,n} + T_{m,n+1} + T_{m,n-1}) + \dfrac{2h\Delta x}{k} T_\infty$ $- 2\left(\dfrac{h\Delta x}{k} + 2\right) T_{m,n} = 0$ **Case 3.** Node at a plane surface with convection.
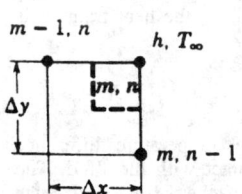	$(T_{m,n-1} \text{ pl } T_{m-1,n}) + 2\dfrac{h\Delta x}{k} T_\infty$ $-2\left(\dfrac{h\Delta x}{k} + 1\right) T_{m,n} = 0$ **Case 4.** Node at an external corner with convection.
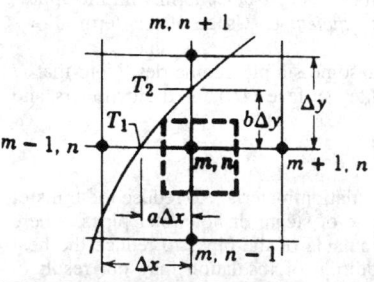	$\dfrac{2}{a+1} T_{m+1,n} + \dfrac{2}{b+1} T_{m,n-1}$ $+ \dfrac{2}{a(a+1)} T_1 + \dfrac{2}{b(b+1)} T_2$ $-\left(\dfrac{2}{a} + \dfrac{2}{b}\right) T_{m,n} = 0$ **Case 5.** Node near a curved surface maintained at a nonuniform temperature.

Extended Surfaces

In examining Newton's law of cooling, it is clear that the rate of heat transfer between a solid and the surrounding ambient fluid may be increased by increasing the surface area of the solid that is exposed to the fluid. This is typically done through the addition of extended surfaces or fins to the primary surface. Numerous examples exist, including the cooling fins on air-cooled engines, such as motorcycles or lawn mowers, or the fins attached to automobile radiators.

Figure 43.2 illustrates a common uniform cross-section extended surface, fin, with a constant base temperature T_b, a constant cross-sectional area A, a circumference of $C = 2W + 2t$, and a length L that is much larger than the thickness t. For these conditions, the temperature distribution in the fin must satisfy the following expression:

$$\frac{d^2T}{dx^2} - \frac{hC}{kA}(T - T_\infty) = 0$$

The solution of this equation depends upon the boundary conditions existing at the tip, that is, at $x = L$. Table 43.11 shows the temperature distribution and heat transfer rate for fins of uniform cross section subjected to a number of different tip conditions, assuming a constant value for the heat transfer coefficient h.

Two terms are used to evaluate fins and their usefulness. *Fin effectiveness* is defined as the ratio of heat transfer rate with the fin to the heat transfer rate that would exist if the fin were not used. For most practical applications, the use of a fin is justified only when the fin effectiveness is significantly greater than 2. *Fin efficiency* η_f represents the ratio of the actual heat transfer rate from a fin to the heat transfer rate that would occur if the entire fin surface could be maintained at a uniform temperature equal to the temperature of the base of the fin. For this case, Newton's law of cooling can be written as

$$q = \eta_f h A_f (T_b - T_\infty)$$

where A_f is the total surface area of the fin and T_b is the temperature of the fin at the base. The application of fins for heat removal can be applied to either forced or natural convection of gases, and while some advantages can be gained in terms of increasing the liquid–solid or solid–vapor surface area, fins as such are not normally utilized for situations involving phase change heat transfer, such as boiling or condensation.

43.2.5 Transient Heat Conduction

If a solid body, all at some uniform temperature $T_{\infty i}$, is immersed in a fluid of different temperature T_∞, the surface of the solid body may be subject to heat losses (or gains) through convection from

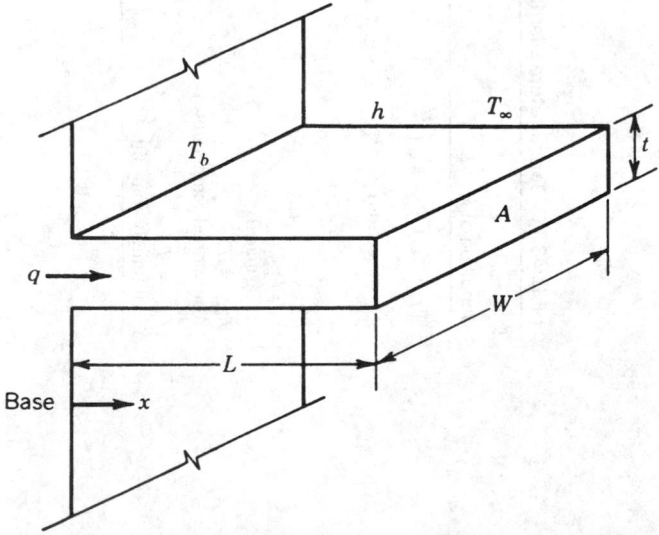

Fig. 43.2 Heat transfer by extended surfaces.

Table 43.11 Temperature Distribution and Heat Transfer Rate at the Fin Base ($m = \sqrt{hc/kA}$)

Condition at x = L	$\dfrac{T - T_\infty}{T_b - T_\infty}$	Heat Transfer Rate $q/mkA\,(T_b - T_\infty)$
$h(T_{x=L} - T_\infty) = -k\left(\dfrac{dT}{dx}\right)_{x=L}$ (convection)	$\dfrac{\cosh m(L-x) + \dfrac{h}{mk}\sinh m(L-x)}{\cosh mL + \dfrac{h}{mk}\sinh mL}$	$\dfrac{\sinh mL + \dfrac{h}{mk}\cosh mL}{\cosh mL + \dfrac{h}{mk}\sinh mL}$
$\left(\dfrac{dT}{dx}\right)_{x=L} = 0$ (insulated)	$\dfrac{\cosh m(L-x)}{\cosh mL}$	$\tanh mL$
$T_{x=L} = T_L$ (prescribed temperature)	$\dfrac{(T_L - T_\infty)/(T_b - T_\infty)\sinh mx + \sinh m(L - x)}{\sinh mL}$	$\dfrac{\cosh mL - (T_L - T_\infty)/(T_b - T_\infty)}{\sinh mL}$
$T_{x=L} = T_\infty$ (infinitely long fin, $L \to \infty$)	e^{-mx}	1

the surface. In this situation, the heat lost (or gained) at the surface results from the conduction of heat from inside the body. To determine the significance of these two heat transfer modes, a dimensionless parameter referred to as the *Biot number* is used. This dimensionless number, defined as $\text{Bi} = hL/k$ where $L = V/A$ or the ratio of the volume of the solid to the surface area of the solid, really represents a comparative relationship of the importance of convection from the outer surface to the conduction occurring inside. When this value is less than 0.1, the temperature of the solid may be assumed uniform and dependent on time alone. When this value is greater than 0.1, there is some spatial temperature variation that will affect the solution procedure.

For the first case, that is, $\text{Bi} < 0.1$, an approximation referred to as the *lumped heat-capacity* method may be used. In this method, the temperature of the solid is given by the expression

$$\frac{T - T_\infty}{T_i - T_\infty} = \exp\left(\frac{-t}{\tau_t}\right) = \exp\left(-\text{BiFo}\right)$$

where τ_t is the *time constant* and is equal to $\rho C_p V/hA$. Increasing the value of the time constant, τ_t, will result in a decrease in the thermal response of the solid to the environment and hence will increase the time required to reach thermal equilibrium (i.e., $T = T_\infty$). In this expression, Fo represents the dimensionless time and is called the *Fourier number,* the value of which is equal to $\alpha t A^2/V^2$. The Fourier number, along with the Biot number, can be used to characterize transient heat conduction problems. The total heat flow through the surface of the solid over the time interval from $t = 0$ to time t can be expressed as

$$Q = \rho V C_p (T_i - T\infty)[1 - \exp(-t/\tau_t]$$

Transient Heat Transfer for Infinite Plate, Infinite Cylinder, and Sphere Subjected to Surface Convection

Generalized analytical solutions to transient heat transfer problems involving infinite plates, cylinders, and finite diameter spheres subjected to surface convection have been developed. These solutions can be presented in graphical form through the use of the *Heisler charts,*[3] illustrated in Figs. 43.3–43.11 for plane walls, cylinders, and spheres, respectively. In this procedure, the solid is assumed to be at a uniform temperature T_i at time $t = 0$ and then is suddenly subjected or immersed in a fluid at a uniform temperature T_∞. The convection heat-transfer coefficient h is assumed to be constant, as is the temperature of the fluid. Combining Figs. 43.3 and 43.4 for plane walls; Figs. 43.6 and 43.7 for cylinders; Figs. 43.9 and 43.10 for spheres, allows the resulting time-dependent temperature of any point within the solid to be found. The total amount of energy Q transferred to or from the solid surface from time $t = 0$ to time t can be found from Figs. 43.5, 43.8, and 43.11.

43.3 CONVECTION HEAT TRANSFER

As discussed earlier, convection heat transfer is the mode of energy transport in which the energy is transferred by means of fluid motion. This transfer can be the result of the random molecular motion or bulk motion of the fluid. If the fluid motion is caused by external forces, the energy transfer is called *forced convection*. If the fluid motion arises from a buoyancy effect caused by density differences, the energy transfer is called *free convection* or *natural convection*. For either case, the heat-transfer rate, q, can be expressed in terms of the surface area, A, and the temperature difference, ΔT, by Newton's law of cooling:

$$q = hA\Delta T$$

In this expression, h is referred to as the convection heat-transfer coefficient or film coefficient, which is a function of the velocity and physical properties of the fluid and the shape and nature of the surface. The nondimensional heat-transfer coefficient $\text{Nu} = hL/k$ is called the *Nusselt number,* where L is a characteristic length and k is the thermal conductivity of the fluid.

43.3.1 Forced Convection—Internal Flow

For internal flow in a tube or pipe, the convection heat-transfer coefficient is typically defined as a function of the temperature difference existing between the temperature at the surface of the tube and the *bulk* or *mixing-cup temperature* T_b, that is, $\Delta T = T_s - T_b$, which can be defined as

$$T_b = \frac{\int C_p T dm}{\int C_p dm}$$

where m is the axial flow rate. Using this value, the heat transfer between the tube and the fluid can be written as $q = hA(T_s - T_b)$.

In the entrance region of a tube or pipe, the flow is quite different from that occurring downstream from the entrance. The rate of heat transfer differs significantly depending on whether the flow is

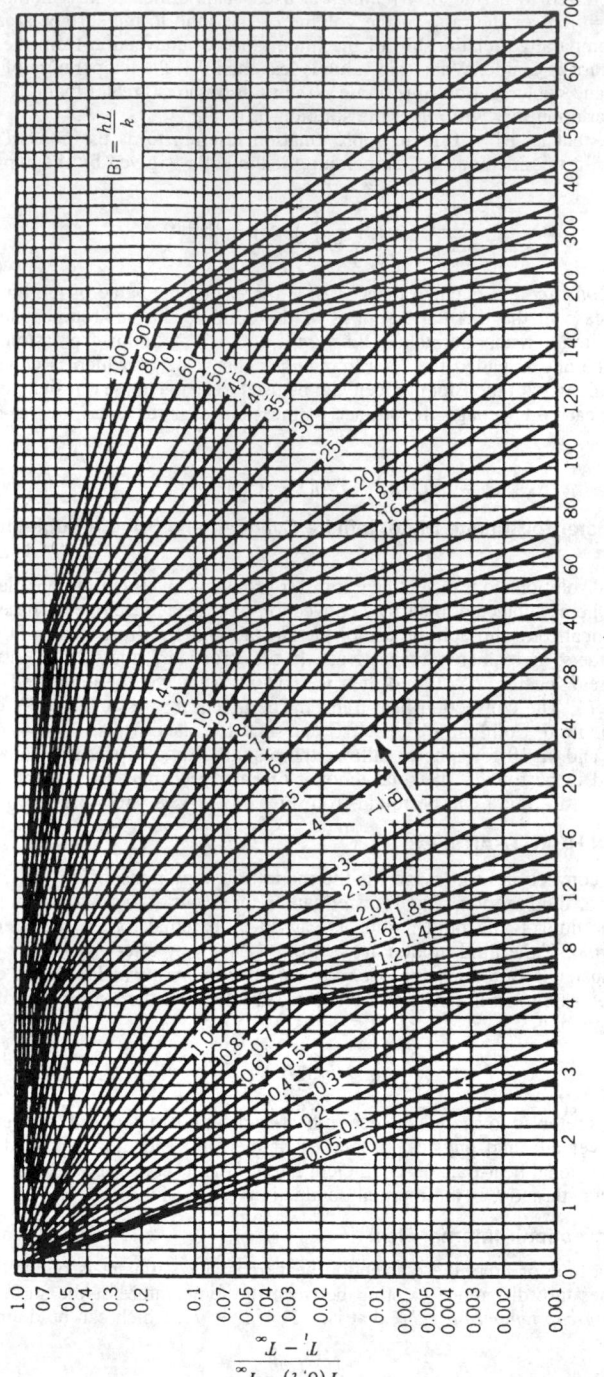

Fig. 43.3 Midplane temperature as a function of time for a plane wall of thickness 2L. (Adapted from Heisler.[3])

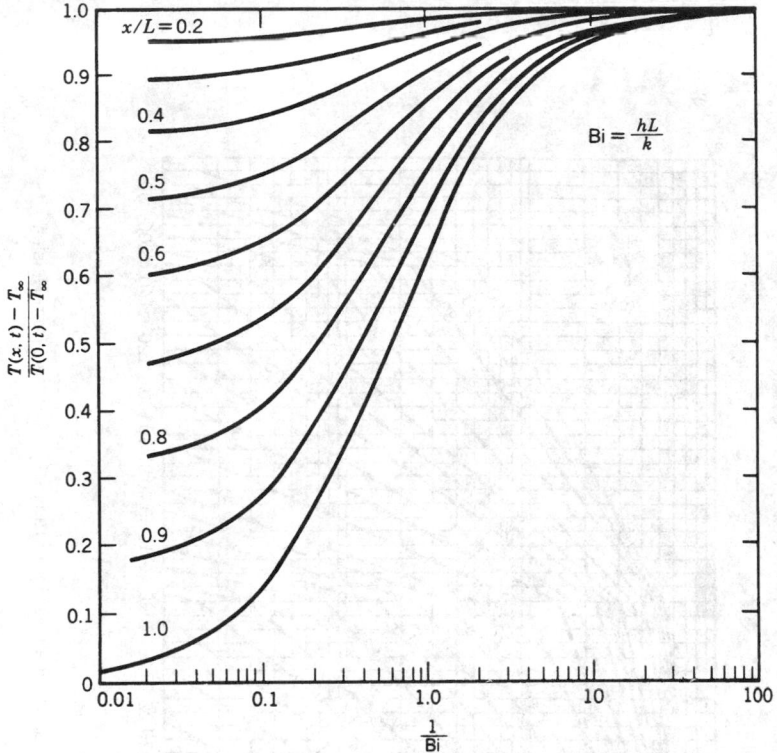

Fig. 43.4 Temperature distribution in a plane wall of thickness 2L. (Adapted from Heisler.[3])

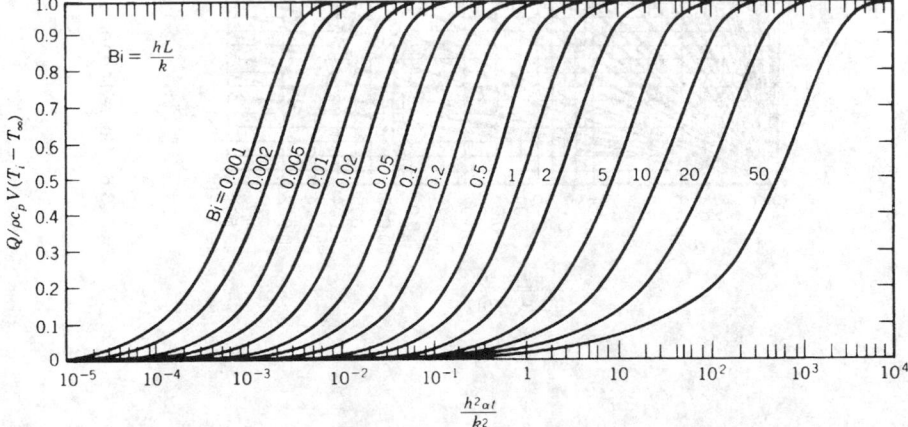

Fig. 43.5 Internal energy change as a function of time for a plane wall of thickness 2L.[4] (Used with the permission of McGraw-Hill Book Company.)

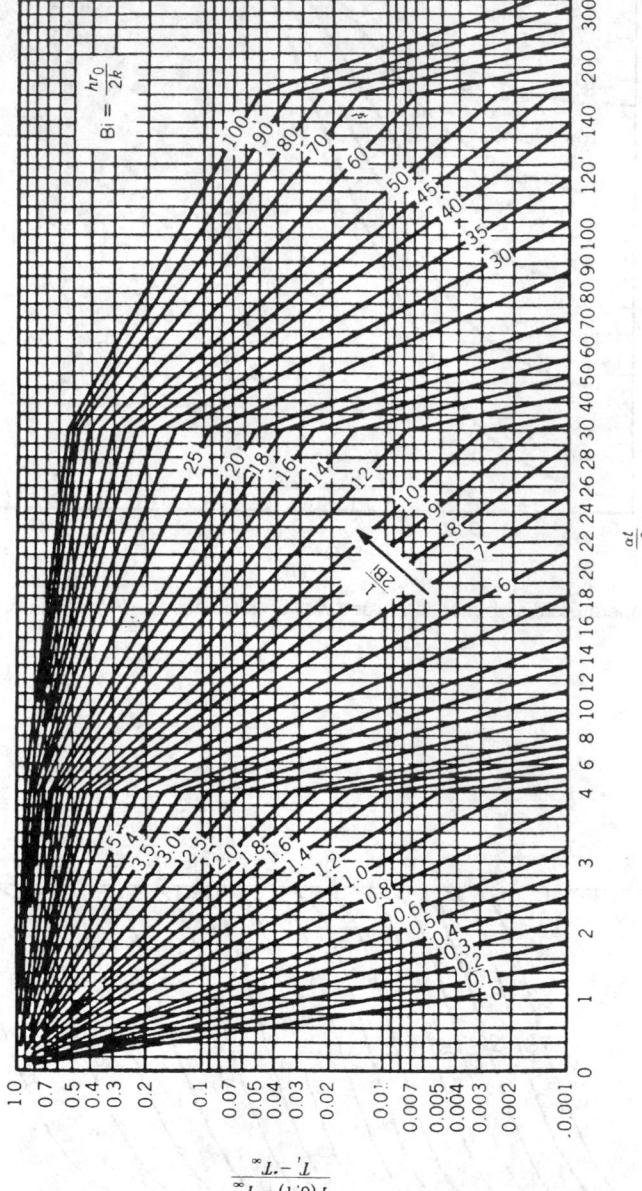

Fig. 43.6 Centerline temperature as a function of time for an infinite cylinder of radius r_o. (Adapted from Heisler.[3])

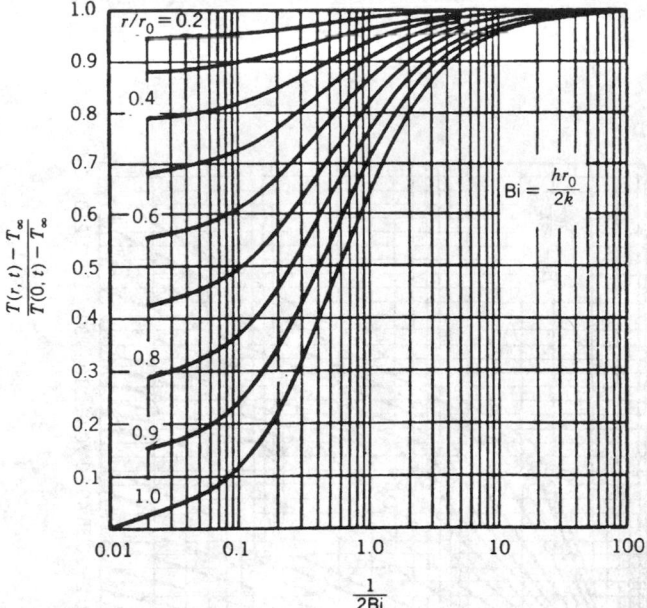

Fig. 43.7 Temperature distribution in an infinite cylinder of radius r_o. (Adapted from Heisler.[3])

laminar or *turbulent*. From fluid mechanics, the flow is considered to be turbulent when $Re_D = V_m D/v > 2300$ for a smooth tube. This transition from laminar to turbulent, however, also depends on the roughness of tube wall and other factors. The generally accepted range for transition is $2000 < Re_D < 4000$.

Laminar Fully Developed Flow

For situations where both the thermal and velocity profiles are fully developed, the Nusselt number is constant and depends only on the thermal boundary conditions. For *circular tubes* with $Pr \geq 0.6$ and $x/DRe_D \, Pr > 0.05$, the Nusselt numbers have been shown to be $Nu_D = 3.66$ and 4.36 for constant temperature and constant heat flux conditions, respectively. Here, the fluid properties are based on the mean bulk temperature.

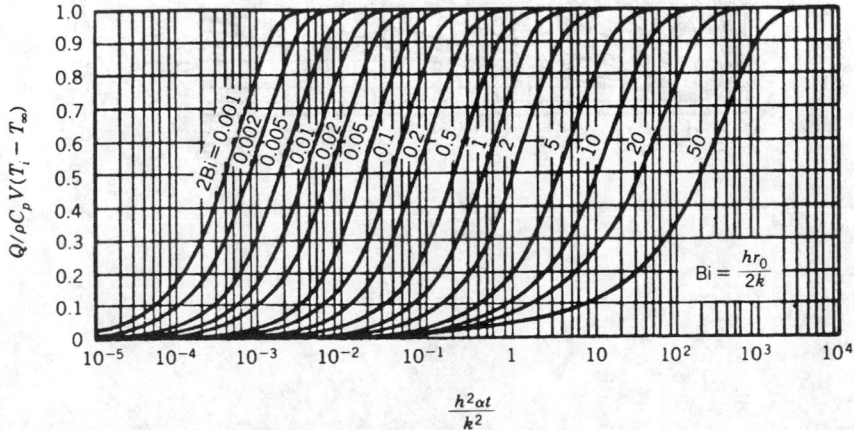

Fig. 43.8 Internal energy change as a function of time for an infinite cylinder of radius r_o.[4] (Used with the permission of McGraw-Hill Book Company.)

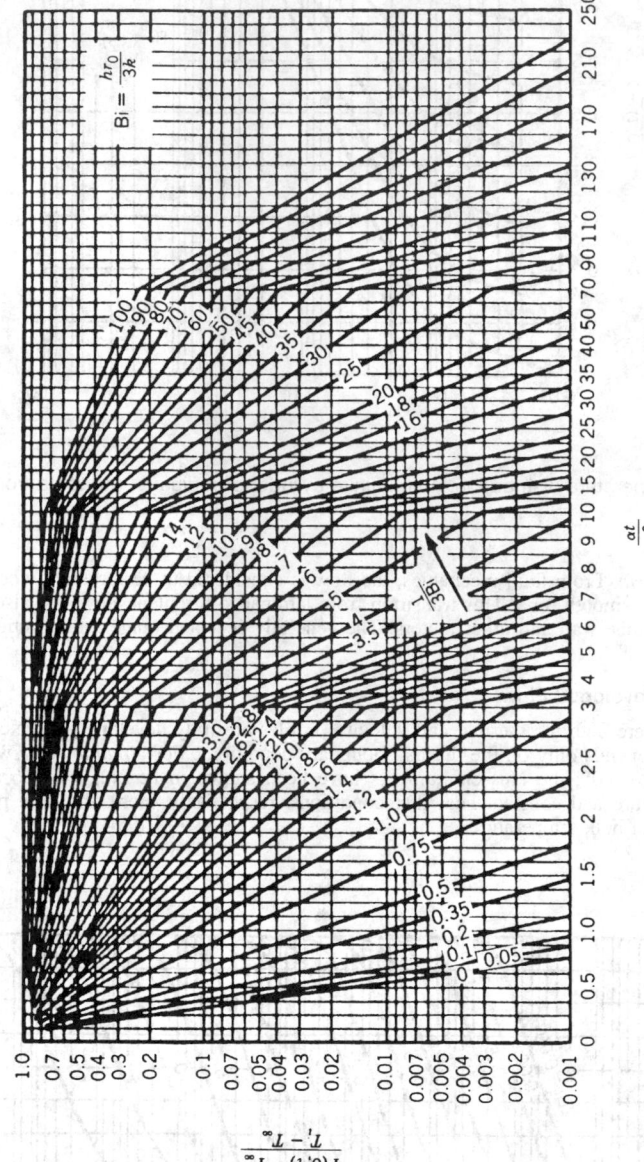

Fig. 43.9 Center temperature as a function of time in a sphere of radius r_o. (Adapted from Heisler.[3])

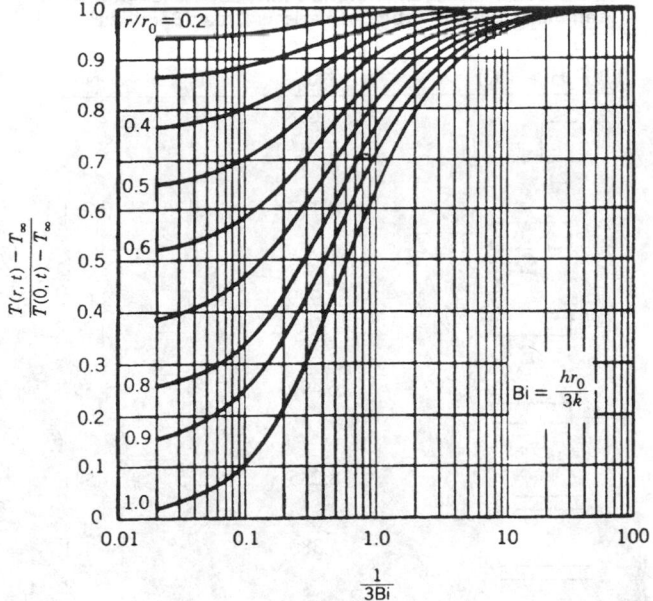

Fig. 43.10 Temperature distribution in a sphere of radius r_o. (Adapted from Heisler.[3])

For *noncircular tubes,* the hydraulic diameter, $D_h = 4 \times$ the flow cross-sectional area / wetted perimeter, is used to define the Nusselt number Nu_D and the Reynolds number Re_D. Table 43.12 shows the Nusselt numbers based on hydraulic diameter for various cross-sectional shapes.

Laminar Flow for Short Tubes

At the entrance of a tube, the Nusselt number is infinite, and decreases asymptotically to the value for fully developed flow as the flow progresses down the tube. The Sieder–Tate equation[5] gives good correlation for the combined entry length, that is, that region where the thermal and velocity profiles are both developing or for short tubes:

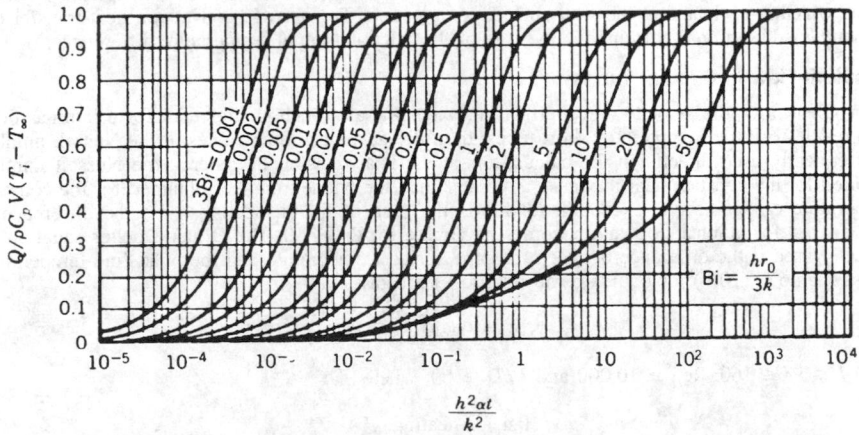

Fig. 43.11 Internal energy change as a function of time for a sphere of radius r_o.[4] (Used with the permission of McGraw-Hill Book Company.)

Table 43.12   Nusselt Numbers for Fully Developed Laminar Flow for Tubes of Various Cross Sections[a]

Geometry (L/DH > 100)	Nu_{H1}	Nu_{H2}	Nu_r
$2b\ \square\ \dfrac{2b}{2a}=1$	3.608	3.091	2.976
$2b\ \square\ \dfrac{2b}{2a}=\dfrac{1}{2}$	4.123	3.017	3.391
$2b\ \square\ \dfrac{2b}{2a}=\dfrac{1}{4}$	5.099	4.35	3.66
$2b\ \square\ \dfrac{2b}{2a}=\dfrac{1}{8}$	6.490	2.904	5.597
$\dfrac{2b}{2a}=0$	8.235	8.235	7.541
$\dfrac{b}{a}=0$	5.385	—	4.861
$\bigcirc$	4.364	4.364	3.657

[a]Nu_{H1} = average Nusselt number for uniform heat flux in flow direction and uniform wall temperature at paticular flow cross section.

Nu_{H2} = average Nusselt number for uniform heat flux both in flow direction and around periphery.

Nu_{Hrr} = average Nusselt number for uniform wall temperature.

$$\overline{Nu_D} = \frac{\overline{h}D}{k} = 1.86(ReDPr)^{1/3}\left(\frac{D}{L}\right)^{1/3}\left(\frac{\mu}{\mu_s}\right)^{0.14}$$

for T_s = constant, $0.48 < Pr < 16{,}700$, $0.0044 < \mu/\mu_s < 9.75$, and $(Re_D\ Pr\ D/L)^{1/3}\ (\mu/\mu_s)^{0.14} > 2$.

In this expression, all of the fluid properties are evaluated at the mean bulk temperature except for μ_s which is evaluated at the wall surface temperature. The average convection heat-transfer coefficient $\overline{h}$ is based on the arithmetic average of the inlet and outlet temperature differences.

Turbulent Flow in Circular Tubes

In turbulent flow, the velocity and thermal entry lengths are much shorter than for a laminar flow. As a result, with the exception of short tubes, the fully developed flow values of the Nusselt number are frequently used directly in the calculation of the heat transfer. In general, the Nusselt number obtained for the constant heat flux case is greater than the Nusselt number obtained for the constant temperature case. The one exception to this is the case of liquid metals, where the difference is smaller than for laminar flow and becomes negligible for $Pr > 1.0$. The Dittus–Boelter equation[6] is typically used if the difference between the pipe surface temperature and the bulk fluid temperature is less than 6°C (10°F) for liquids or 56°C (100°F) for gases:

$$Nu_D = 0.023\ Re_D^{0.8}\ Pr^n$$

for $0.7 \le Pr \le 160$, $Re_D \ge 10{,}000$ and $L/D \ge 60$, where

$$n = 0.4 \text{ for heating, } T_s > T_b$$
$$= 0.3 \text{ for cooling, } T_s < T_b$$

For temperature differences greater than those specified above, use[5]

$$\mathrm{Nu}_D = 0.027 \mathrm{Re}_D^{0.8} \mathrm{Pr}^{1/3} \left(\frac{\mu}{\mu_s}\right)^{0.14}$$

for $0.7 \le \mathrm{Pr} \le 16{,}700$, $\mathrm{Re}_D \ge 10{,}000$ and $L/D \ge 60$.

In this expression, the properties are all evaluated at the mean bulk fluid temperature, with the exception of μ_s, which is again evaluated at the tube surface temperature.

For *concentric tube annuli*, the hydraulic diameter $D_h = D_o - D_i$ (outer diameter–inner diameter) must be used for Nu_D and Re_D, and the coefficient h at either surface of the annulus must be evaluated from the Dittus–Boelter equation. Here, it should be noted that the foregoing equations apply for smooth surfaces and that the heat-transfer rate will be larger for rough surfaces and are not applicable to liquid metals.

Fully Developed Turbulent Flow of Liquid Metals in Circular Tubes

Because the Prandtl number for liquid metals is on the order of 0.01, the Nusselt number is primarily dependent upon a dimensionless parameter number referred to as the *Peclet number*, which in general is defined as $\mathrm{Pe} = \mathrm{RePr}$:

$$\mathrm{Nu}_D = 5.0 + 0.025 Pe_D^{0.8}$$

which is valid for situations where $T_s = $ a constant and $\mathrm{Pe}_D > 100$ and $L/D > 60$.

For $q'' = $ constant and $3.6 \times 10^3 < \mathrm{Re}_D < 9.05 \times 10^5$, $10^2 < \mathrm{Pe}_D < 10^4$, and $L/D > 60$, the Nusselt number can be expressed as

$$\mathrm{Nu}_D = 4.8 + 0.0185 Pe_D^{0.827}$$

43.3.2 Forced Convection—External Flow

In forced convection heat transfer, the heat transfer coefficient h is based on the temperature difference between the wall surface temperature and the fluid temperature in the free stream outside the thermal boundary layer. The total heat-transfer rate from the wall to the fluid is given by $q = hA\,(T_s - T_\infty)$. The Reynolds numbers are based on the free stream velocity. The fluid properties are evaluated either at the free stream temperature T_∞ or at the film temperature $T_f = (T_s + T_\infty)/2$.

Laminar Flow on a Flat Plate

When the flow velocity along a constant temperature semi-infinite plate is uniform, the boundary layer originates from the leading edge and is laminar and the flow remains laminar until the local Reynolds number, $\mathrm{Re}_x = U_\infty x / v$, reaches the *critical Reynolds number* Re_c. When the surface is smooth, the Reynolds number is generally assumed to be $\mathrm{Re}_c = 5 \times 10^5$; however, the value will depend on several parameters, including the surface roughness.

For a given distance x from the leading edge, the *local Nusselt number* and the *average Nusselt number* between $x = 0$ and $x = L$ are given below (Re_x and $\mathrm{Re}_L \le 5 \times 10^5$):

$$\left.\begin{array}{l} \mathrm{Nu}_x = hx/k = 0.332\mathrm{Re}_x^{0.5}\,\mathrm{Pr}^{1/3} \\ \overline{\mathrm{Nu}}_L = \bar{h}L/k = 0.664\mathrm{Re}_L^{0.5}\,\mathrm{Pr}^{1/3} \end{array}\right\} \quad \text{for } \mathrm{Pr} \ge 0.6$$

$$\left.\begin{array}{l} \mathrm{Nu}_x = 0.565(\mathrm{Re}_x\,\mathrm{Pr})^{0.5} \\ \overline{\mathrm{Nu}}_L = 1.13(\mathrm{Re}_L\,\mathrm{Pr})^{0.5} \end{array}\right\} \quad \text{for } \mathrm{Pr} \le 0.6$$

Here, all of the fluid properties are evaluated at the mean or average film temperature.

Turbulent Flow on a Flat Plate

When the flow over a flat plate is turbulent from the leading edge, expressions for the local Nusselt number can be written as

$$\mathrm{Nu}_x = 0.0292\mathrm{Re}_x^{0.8}\,\mathrm{Pr}^{1/3}$$

$$\overline{\mathrm{Nu}}_L = 0.036\mathrm{Re}_L^{0.8}\,\mathrm{Pr}^{1/3}$$

where the fluid properties are all based on the mean film temperature and $5 \times 10^5 \le \mathrm{Re}_x$ and $\mathrm{Re}_L \le 10^8$ and $0.6 \le \mathrm{Pr} \le 60$.

The Average Nusselt Number Between $x = 0$ and $x = L$ with Transition

For situations where transition occurs immediately once the critical Reynolds number Re_c has been reached,[7]

$$\overline{\mathrm{Nu}}_L = 0.036\mathrm{Pr}^{1/3}[\mathrm{Re}_L^{0.8} - \mathrm{Re}_c^{0.8} + 18.44\mathrm{Re}_c^{0.5}]$$

provided that $5 \times 10^5 \leq Re_L \leq 10^8$ and $0.6 \leq Pr \leq 60$. Specialized cases exist for this situation, such as

$$\overline{Nu}_L = 0.036 Pr^{1/3}(Re_L^{0.8} - 18,700)$$

for $Re_c = 4 \times 10^5$ or

$$\overline{Nu}_L = 0.036 Pr^{1/3}(Re_L^{0.8} - 23,000)$$

for $Re_c = 5 \times 10^5$. Again, all fluid properties are evaluated at the mean film temperature.

Circular Cylinders in Cross Flow

For circular cylinders in cross flow, the Nusselt number is based upon the diameter and can be expressed as

$$\overline{Nu}_D = (0.4 Re_D^{0.5} + 0.06 Re^{2/3}) Pr^{0.4}(\mu_\infty/\mu_s)^{0.25}$$

for $0.67 < Pr < 300$, $10 < Re_D < 10^5$, and $0.25 < 5.2$. Here, the fluid properties are evaluated at the free stream temperature, except μ_s, which is evaluated at the surface temperature.[8]

Cylinders of Noncircular Cross Section in Cross Flow of Gases

For noncircular cylinders in cross flow, the Nusselt number is again based upon the diameter, but is expressed as

$$\overline{Nu}_D = C(Re_D)^m Pr^{1/3}$$

where C and m are listed in Table 43.13, and the fluid properties are evaluated at the mean film temperature.[9]

Flow Past a Sphere

For flow over a sphere, the Nusselt number is based upon the sphere diameter and can be expressed as

$$\overline{Nu}_D = 2 + (0.4 Re_D^{0.5} + 0.006 Re_D^{2/3}) Pr^{0.4}(\mu_\infty/\mu_s)^{0.25}$$

for the case of $3.5 < Re_D < 8 \times 10^4$, $0.7 < Pr < 380$, and $1.0 < \mu_\infty/\mu_s < 3.2$. The fluid properties are calculated at the free stream temperature, except μ_s, which is evaluated at the surface temperature.[8]

Table 43.13 Constants and m for Noncircular Cylinders in Cross Flow

Geometry	Re_D	C	m
Square			
$V \rightarrow \diamond\, \updownarrow D$	$5 \times 10^3 - 10^5$	0.246	0.588
	$5 \times 10^3 - 10^5$	0.102	0.675
$V \rightarrow \square\, \dfrac{D}{\uparrow}$			
Hexagon			
$V \rightarrow\ \updownarrow D$	$5 \times 10^3 - 1.95 \times 10^4$	0.160	0.638
	$1.95 \times 10^4 - 10^5$	0.0385	0.782
$V \rightarrow\ \dfrac{D}{\uparrow}$			
Vertical Plate			
$V \rightarrow \square\ \updownarrow D$	$5 \times 10^3 - 10^5$	0.153	0.638
	$4 \times 10^3 - 1.5 \times 10^4$	0.228	0.721

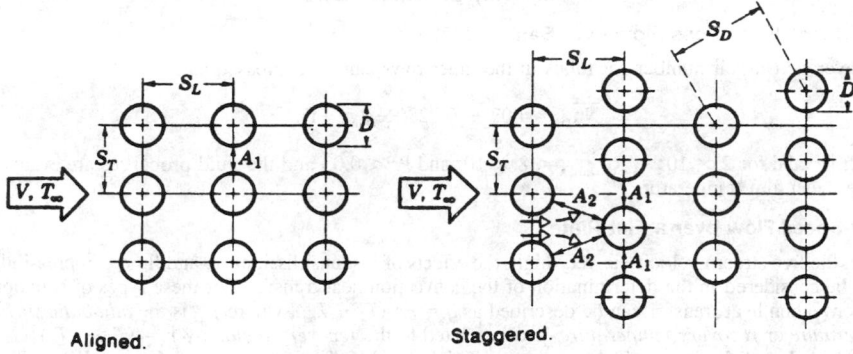

Fig. 43.12 Tube arrangement.

Flow across Banks of Tubes

For banks of tubes, the tube arrangement may be either *staggered* or *aligned* (Fig. 43.12), and the heat transfer coefficient for the first row is approximately equal to that for a single tube. In turbulent flow, the heat transfer coefficient for tubes in the first row is smaller than that of the subsequent rows. However, beyond the fourth or fifth row, the heat transfer coefficient becomes approximately constant. For tube banks with more than twenty rows, $0.7 < \text{Pr} < 500$, and $1000 < \text{Re}_{D,\text{max}} < 2 \times 10^6$, the average Nusselt number for the entire tube bundle can be expressed as[10]

$$\overline{\text{Nu}}_D = C(\text{Re}_{D,\text{max}})^m \, \text{Pr}^{0.36}(\text{Pr}_\infty/\text{Pr}_s)^{0.25}$$

where all fluid properties are evaluated at T_∞ except Pr_s, which is evaluated at the surface temperature. The constants C and m used in this expression are listed in Table 43.14, and the Reynolds number is based on the maximum fluid velocity occurring at the minimum free flow area available for the fluid. Using the nomenclature shown in Fig. 43.12, the maximum fluid velocity can be determined by

$$V_{\text{max}} = \frac{S_T}{S_T - D} V$$

for the aligned or staggered configuration provided

$$\sqrt{S_L^2 + (S_T/2)^2} > (S_T + D)/2$$

or as

$$V_{\text{max}} = \frac{S_T}{\sqrt[2]{S_L^2 + (S_T/2)^2}} V$$

for staggered if

Table 43.14 Constants C and m of Heat-Transfer Coefficient for the Banks in Cross Flow

Configuration	$\text{Re}_{D,\text{max}}$	C	m
Aligned	10^3–2×10^5	0.27	0.63
Staggered ($S_T/S_L < 2$)	10^3–2×10^5	$0.35(S_T/S_L)^{1/5}$	0.60
Staggered ($S_G/S_L > 2$)	10^3–2×10^5	0.40	0.60
Aligned	2×10^5–2×10^6	0.21	0.84
Staggered	2×10^5–2×10^6	0.022	0.84

$$\sqrt{S_L^2 + (S_T/2)^2} < (S_T + D)/2$$

Liquid Metals in Cross Flow over Banks of Tubes

The average Nusselt number for tubes in the inner rows can be expressed as

$$\overline{Nu}_D = 4.03 + 0.228(Re_{D,max} \, Pr)^{0.67}$$

which is valid for $2 \times 10^4 < Re_{D,max} < 8 \times 10^4$ and $Pr < 0.03$ and the fluid properties are evaluated at the mean film temperature.[11]

High-Speed Flow over a Flat Plate

When the free stream velocity is very high, the effects of viscous dissipation and fluid compressibility must be considered in the determination of the convection heat transfer. For these types of situations, the convection heat transfer can be described as $q = hA \, (T_s - T_{as})$, where T_{as} is the *adiabatic surface temperature* or *recovery temperature,* and is related to the *recovery factor* by $r = (T_{as} - T_\infty)/(T_0 - T_\infty)$. The value of the *stagnation temperature* T_0 is related to the free stream static temperature T_∞ by the expression

$$\frac{T_0}{T_\infty} = 1 + \frac{\gamma - 1}{2} M_\infty^2$$

where γ is the specific heat ratio of the fluid and M_∞ is the ratio of the free stream velocity and the acoustic velocity. For the case where $0.6 < Pr < 15$,

$$r = Pr^{1/2} \quad \text{for laminar flow } (Re_x < 5 \times 10^5)$$
$$r = Pr^{1/3} \quad \text{for turbulent flow } (Re_x > 5 \times 10^5)$$

Here, all of the fluid properties are evaluated at the reference temperature $T_{ref} = T_\infty + 0.5(T_s - T_\infty) + 0.22(T_{as} - T_\infty)$. Expressions for the local heat-transfer coefficients at a given distance x from the leading edge are given as:[2]

$$Nu_x = 0.332 Re_x^{0.5} \, Pr^{1/3} \qquad \text{for } Re_x < 5 \times 10^5$$
$$Nu_x = 0.0292 Re_x^{0.8} \, Pr^{1/3} \qquad \text{for } 5 \times 10^5 < Re_x < 10^7$$
$$Nu_x = 0.185 Re_x (\log_{10} Re_x)^{-2.584} \qquad \text{for } 10^7 < Re_x < 10^9$$

In the case of gaseous fluids flowing at very high free stream velocities, dissociation of the gas may occur and will cause large variations in the properties within the boundary layer. For these cases, the heat-transfer coefficient must be defined in terms of the enthalpy difference, namely, $q = hA \, (i_s - i_{as})$, and the recovery factor will be given by $r = (i_s - i_{as})/(i_0 - i_\infty)$, where i_{as} represents the enthalpy at the adiabatic wall conditions. Similar expressions to those shown above for Nu_x can be used by substituting the properties evaluated at a reference enthalpy defined as $i_{ref} = i_\infty + 0.5(i_s - i_\infty) + 0.22(i_{g}a_s - i_\infty)$.

High-Speed Gas Flow Past Cones

For the case of high-speed gaseous flows over conical-shaped objects, the following expressions can be used:

$$Nu_x = 0.575 Re_x^{0.5} \, Pr^{1/3} \qquad \text{for } Re_x < 10^5$$
$$Nu_x = 0.0292 Re_x^{0.8} \, Pr^{1/3} \qquad \text{for } Re_x > 10^5$$

where the fluid properties are evaluated at T_{ref}, as in the plate.[12]

Stagnation Point Heating for Gases

When the conditions are such that the flow can be assumed to behave as *incompressible*, the Reynolds number is based on the free stream velocity and $\bar{h}$ is defined as $q = \bar{h}A(T_s - T_\infty)$.[13] Estimations of the Nusselt number can be made using the following relationship:

$$Nu_D = C Re_D^{0.5} \, Pr^{0.4}$$

where $C = 1.14$ for cylinders and 1.32 for spheres, and the fluid properties are evaluated at the mean film temperature. When the flow becomes *supersonic*, a bow shock wave will occur just off the front of the body. In this situation, the fluid properties must be evaluated at the stagnation state occurring behind the bow shock and the Nusselt number can be written as

$$\overline{Nu}_D = C Re_D^{0.5} \, Pr^{0.5}(\rho_\infty/\rho_0)^{0.25}$$

where $C = 0.95$ for cylinders and 1.28 for spheres, ρ_∞ is the free stream gas density, and ρ_o is the stagnation density of the stream behind the bow shock. The heat-transfer rate for this case is given by $q = \overline{h}A(T_s - T_0)$.

43.3.3 Free Convection

In free convection, the fluid motion is caused by the buoyant force resulting from the density difference near the body surface, which is at a temperature different from that of the free fluid far removed from the surface, where the velocity is zero. In all free convection correlations, except for the enclosed cavities, the fluid properties are usually evaluated at the mean film temperature $T_f = (T_1 + T_\infty)/2$. The thermal expansion coefficient β, however, is evaluated at the free fluid temperature T_∞. The convection heat transfer coefficient h is based on the temperature difference between the surface and the free fluid.

Free Convection from Flat Plates and Cylinders

For free convection from flat plates and cylinders, the average Nusselt number $\overline{Nu}_L$ can be expressed as[4]

$$\overline{Nu}_L = C(Gr_L \, Pr)^m$$

where the constants C and m are given as shown in Table 43.15. The *Grashof Prandtal number product*, $(Gr_L Pr)$ is called the *Rayleigh number* (Ra_L) and for certain ranges of this value, Figs. 43.13 and 43.14 are used instead of the above equation. Reasonable approximations for other types of *three-dimensional shapes,* such as short cylinders and blocks, can be made for $10^4 < Ra_L < 10^9$, by using this expression and $C = 0.6$, $m = \frac{1}{4}$, provided that the characteristic length, L, is determined from $1/L = 1/L_{hor} + 1/L_{ver}$, where L_{ver} is the height and L_{hor} is the horizontal dimension of the object in question.

For *unsymmetrical horizontal* square, rectangular, or circular surfaces, the characteristic length L can be calculated from the expression $L = A/P$, where A is the area and P is the wetted perimeter of the surface.

Free Convection from Spheres

For free convection from spheres, the following correlation has been developed:

$$\overline{Nu}_D = 2 + 0.43(Gr_D \, Pr)^{0.25} \qquad \text{for } 1 < Gr_D < 10^5$$

Although this expression was designed primarily for gases, $Pr \approx 1$, it may be used to approximate the values for liquids as well.[15]

Free Convection in Enclosed Spaces

Heat transfer in an enclosure occurs in a number of different situations and with a variety of configurations. When a temperature difference is imposed on two opposing walls that enclose a space filled with a fluid, convective heat transfer will occur. For small values of the Rayleigh number, the heat transfer may be dominated by conduction, but as the Rayleigh number increases, the contribution made by free convection will increase. Following are a number of correlations, each designed for a specific geometry. For all of these, the fluid properties are evaluated at the average temperature of the two walls.

Cavities between Two Horizontal Walls at Temperatures T_1 and T_2 Separated by Distance $\delta(T_1$ for Lower Wall, $T_1 > T_2$)

$$q'' = \overline{h}(T_1 - T_2)$$
$$\overline{Nu}_\delta = 0.069 Ra_\delta^{1/3} \, Pr^{0.074} \qquad \text{for } 3 \times 10^5 < Ra_\delta < 7 \times 10^9$$
$$= 1.0 \qquad\qquad\quad \text{for } Ra_\delta < 1700$$

where $Ra_\delta = g\beta(T_1 - T_2)\delta^3/\alpha\nu$ and δ is the thickness of the space.[16]

Table 43.15 Constants for Free Convection from Flat Plates and Cylinders

Geometry	$Gr_K Pr$	C	m	L
Vertical flat plates and cylinders	10^{-1}–10^4	Use Fig. 43.12	Use Fig. 43.12	Height of plates and cylinders; restricted to $D/L \geq 35/Gr_L^{1/4}$ for cylinders
	10^4–10^9	0.59	¼	
	10^9–10^{13}	0.10	⅓	
Horizontal cylinders	0–10^{-5}	0.4	0	Diameter D
	10^{-5}–10^4	Use Fig. 43.13	Use Fig. 43.13	
	10^4–10^9	0.53	¼	
	10^9–10^{13}	0.13	⅓	
Upper surface of heated plates or lower surface of cooled plates	2×10^4–8×10^6	0.54	¼	Length of a side for square plates, the average length of the two sides for rectangular plates
	8×10^6–10^{11}	0.15	⅓	
Lower surface of heated plates or upper surface of cooled plates	10^5–10^{11}	0.58	⅕	$0.9D$ for circular disks

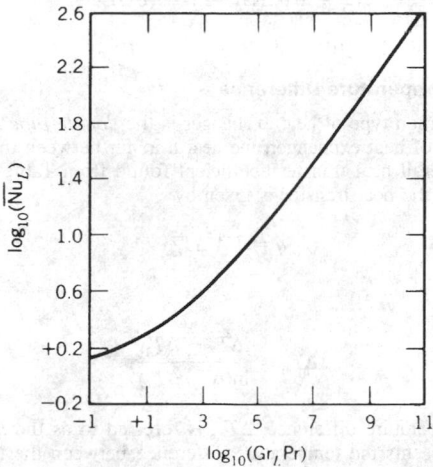

Fig. 43.13 Free convection heat-transfer correlation for heated vertical plates and cylinders. Adapted from Ref. 14. (Used with permission of McGraw-Hill Book Company.)

Cavities between Two Vertical Walls of Height H at Temperatures by Distance T_1 and T_2 Separated by Distance δ[17,18]

$$q'' = soh(T_1 - T_2)$$

$$\overline{Nu}_\delta = 0.22 \left(\frac{Pr}{0.2 + Pr} Ra_\delta \right)^{0.28} \left(\frac{\delta}{H} \right)^{0.25}$$

for $2 < H/\delta < 10$, $Pr < 10^5$ $Ra_\delta < 10^{10}$;

$$\overline{Nu}_\delta = 0.18 \left(\frac{Pr}{0.2 + Pr} Ra_\delta \right)^{0.29}$$

for $1 < H/\delta < 2$, $10^3 < Pr < 10^5$, and $10^3 < Ra_\delta Pr/(0.2 + Pr)$;

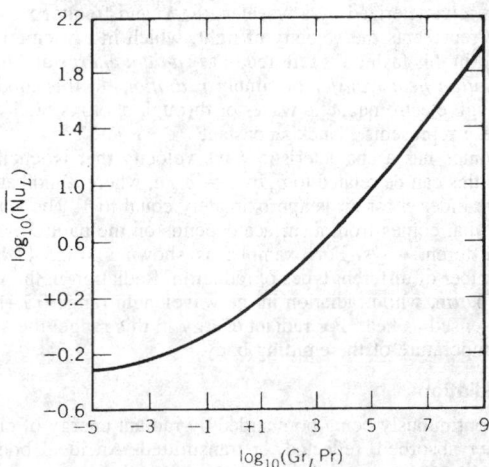

Fig. 43.14 Free convection heat-transfer correlation from heated horizontal cylinders. (Adapted from Ref. 14. Used with permission of McGraw-Hill Book Company.)

$$\overline{\text{Nu}}_\delta = 0.42 \text{Ra}_\delta^{0.25} \, \text{Pr}^{0.012} (\delta/H)^{0.3}$$

for $10 < H/\delta < 40$, $1 < \text{Pr} < 2 \times 10^4$, and $10^4 < \text{Ra}_\delta < 10^7$.

43.3.4   The Log Mean Temperature Difference

The simplest and most common type of heat exchanger is the *double-pipe heat exchanger*, illustrated in Fig. 43.15. For this type of heat exchanger, the heat transfer between the two fluids can be found by assuming a constant overall heat transfer coefficient found from Table 43.8 and a constant fluid specific heat. For this type, the heat transfer is given by

$$q = UA \, \Delta T_m$$

where

$$\Delta T_m = \frac{\Delta T_2 - \Delta T_1}{\ln(\Delta T_2 / \Delta T_1)}$$

In this expression, the temperature difference, ΔT_m, is referred to as the *log-mean temperature difference* (LMTD); ΔT_1 represents the temperature difference between the two fluids at one end and ΔT_2 at the other end. For the case where the ratio $\Delta T_2/\Delta T_1$ is less than two, the *arithmetic mean temperature difference* $(\Delta T_2 + \Delta T_1)/2$ may be used to calculate the heat-transfer rate without introducing any significant error. As shown in Fig. 43.15,

$$\Delta T_1 = T_{h,i} - T_{c,i} \qquad \Delta T_2 = T_{h,o} - T_{c,o} \qquad \text{for parallel flow}$$
$$\Delta T_1 = T_{h,i} - T_{c,o} \qquad \Delta T_2 = T_{h,o} - T_{c,i} \qquad \text{for counterflow}$$

Cross-Flow Coefficient

In other types of heat exchangers, where the values of the overall heat transfer coefficient, U, may vary over the area of the surface, the LMTD may not be representative of the actual average temperature difference. In these cases, it is necessary to utilize a correction factor such that the heat transfer, q, can be determined by

$$q = UAF \, \Delta T_m$$

Here the value of ΔT_m is computed assuming counterflow conditions, $\Delta T_1 = T_{h,i} - T_{c,i}$ and $\Delta T_2 = T_{h,o} - T_{c,o}$. Figures 43.16 and 43.17 illustrate some examples of the *correction factor*, F, for various multiple-pass heat exchangers.

43.4   RADIATION HEAT TRANSFER

Heat transfer can occur in the absence of a participating medium through the transmission of energy by electromagnetic waves, characterized by a wavelength, λ, and frequency, ν, which are related by $c = \lambda\nu$. The parameter c represents the velocity of light, which in a vacuum is $c_o = 2.9979 \times 10^8$ m/sec. Energy transmitted in this fashion is referred to as *radiant energy* and the heat transfer process that occurs is called *radiation heat transfer* or simply *radiation*. In this mode of heat transfer, the energy is transferred through electromagnetic waves or through photons, with the energy of a photon being given by $h\nu$, where h represents Planck's constant.

In nature, every substance has a characteristic wave velocity that is smaller than that occurring in a vacuum. These velocities can be related to c_o by $c = c_o/n$, where n indicates the refractive index. The value of the refractive index n for air is approximately equal to 1. The wavelength of the energy given or for the radiation that comes from a surface depends on the nature of the source and various wavelengths sensed in different ways. For example, as shown in Fig. 43.18 the electromagnetic spectrum consists of a number of different types of radiation. Radiation in the visible spectrum occurs in the range $\lambda = 0.4$–0.74 μm, while radiation in the wavelength range 0.1–100 μm is classified as *thermal radiation* and is sensed as heat. For radiant energy in this range, the amount of energy given off is governed by the temperature of the emitting body.

43.4.1   Black-Body Radiation

All objects in space are continuously being bombarded by radiant energy of one form or another and all of this energy is either absorbed, reflected, or transmitted. An ideal body that absorbs all the radiant energy falling upon it, regardless of the wavelength and direction, is referred to as a *black body*. Such a body emits the maximum energy for a prescribed temperature and wavelength. Radiation from a black body is independent of direction and is referred to as a *diffuse emitter*.

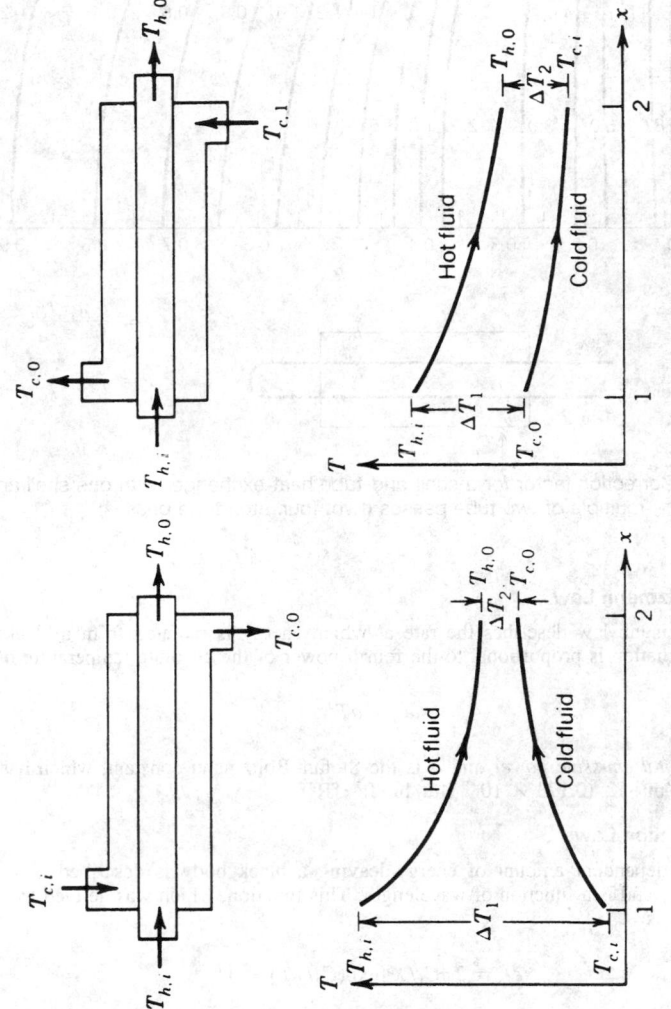

Fig. 43.15 Temperature profiles for parallel flow and counterflow in double-pipe heat exchanger.

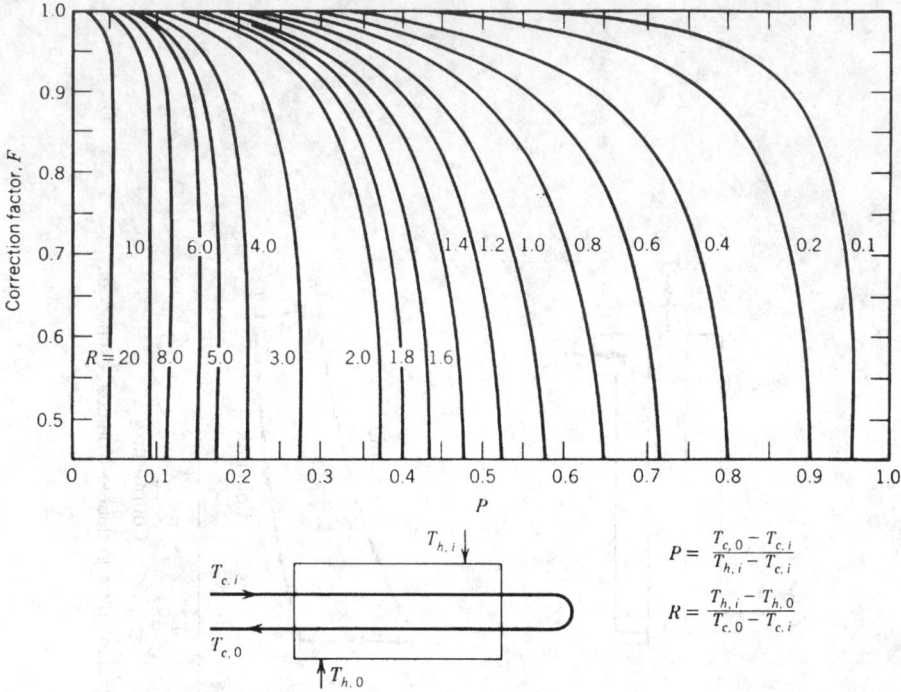

Fig. 43.16 Correction factor for a shell-and-tube heat exchanger with one shell and any multiple of two tube passes (two, four, etc., tube passes).

The Stefan–Boltzmann Law

The Stefan–Boltzmann law describes the rate at which energy is radiated from a black body and states that this radiation is proportional to the fourth power of the absolute temperature of the body

$$e_b = \sigma T^4$$

where e_b is the *total emissive power* and σ is the Stefan–Boltzmann constant, which has the value 5.729×10^{-8} W/m$^2 \cdot$ K^4 (0.173×10^{-8} Btu/hr $\cdot$ ft$^2 \cdot$ °R^4).

Planck's Distribution Law

The temperature dependent amount of energy leaving a black body is described as the *spectral emissive power* $e_{\delta b}$ and is a function of wavelength. This function, which was derived from quantum theory by Planck, is

$$e_{\lambda b} = 2\pi C_1 / \lambda^5 [\exp(C_2 / \lambda T) - 1]$$

where $e_{\delta b}$ has a unit W/m$^2 \cdot \mu$m (Btu/hr $\cdot$ ft$^2 \cdot \mu$m).

Values of the constants C_1 and C_2 are 0.59544×10^{-16} W $\cdot$ m^2 (0.18892×10^8 Btu $\cdot \mu$m^4/hr ft^2) and 14,388 μm $\cdot$ K (25,898 μm $\cdot$ °R), respectively. The distribution of the spectral emissive power from a black body at various temperatures is shown in Fig. 43.19, where, as shown, the energy emitted at all wavelengths increases as the temperature increases. The maximum or peak values of the constant temperature curves illustrated in Fig. 43.20 shift to the left for shorter wavelengths as the temperatures increase.

The fraction of the emissive power of a black body at a given temperature and in the wavelength interval between λ_1 and λ_2 can be described by

$$F_{\lambda_1 T - \lambda_2 T} = \frac{1}{\sigma T^4} \left(\int_o^{\delta_1} e_{\lambda b} \, d\lambda - \int_o^{\lambda_2} e_{\lambda b} \, d\lambda \right) = F_{o-\lambda_1 T} - F_{o-\lambda_2 T}$$

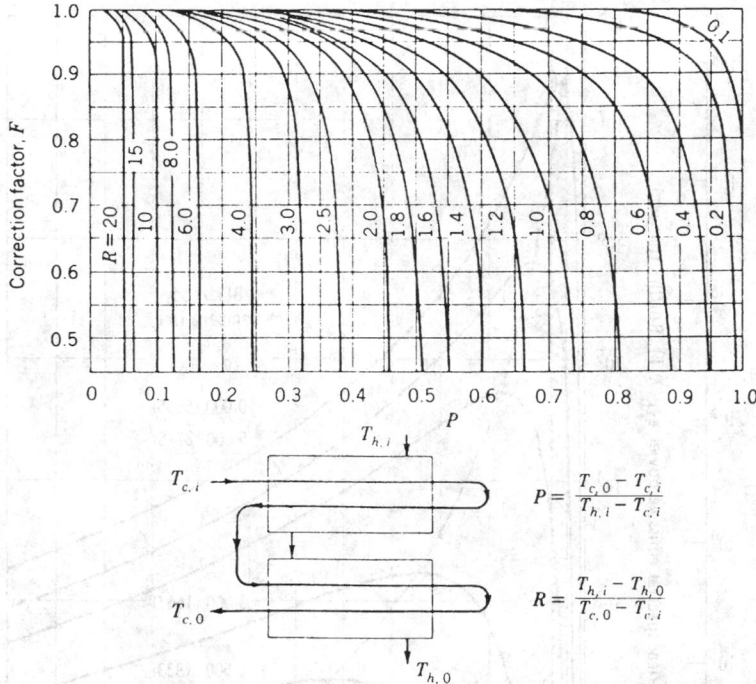

Fig. 43.17 Correction factor for a shell-and-tube heat exchanger with two shell passes and any multiple of four tubes passes (four, eight, etc., tube passes).

where the function $F_{o-\lambda T} = (1/\sigma T^4) \int_o^\lambda e_{\lambda b}\, d\lambda$ is given in Table 43.16. This function is useful for the evaluation of total properties involving integration on the wavelength in which the spectral properties are piecewise constant.

Wien's Displacement Law

The relationship between these peak or maximum temperatures can be described by *Wien's displacement law,*

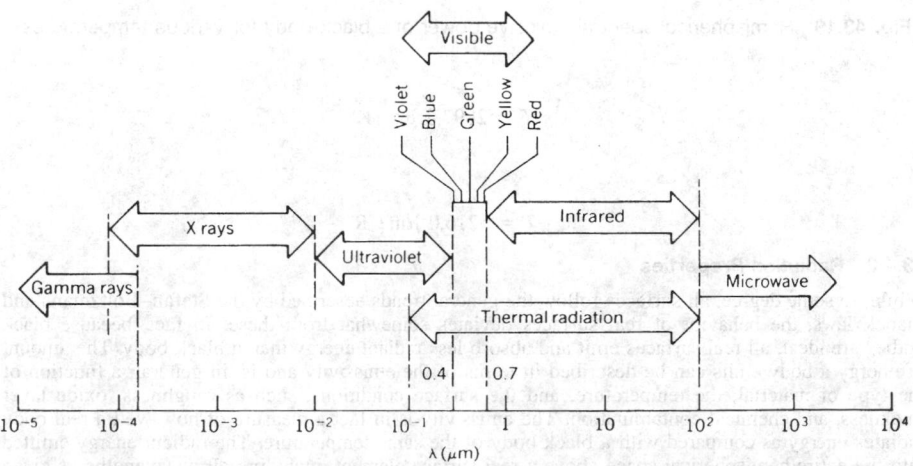

Fig. 43.18 Electromagnetic radiation spectrum.

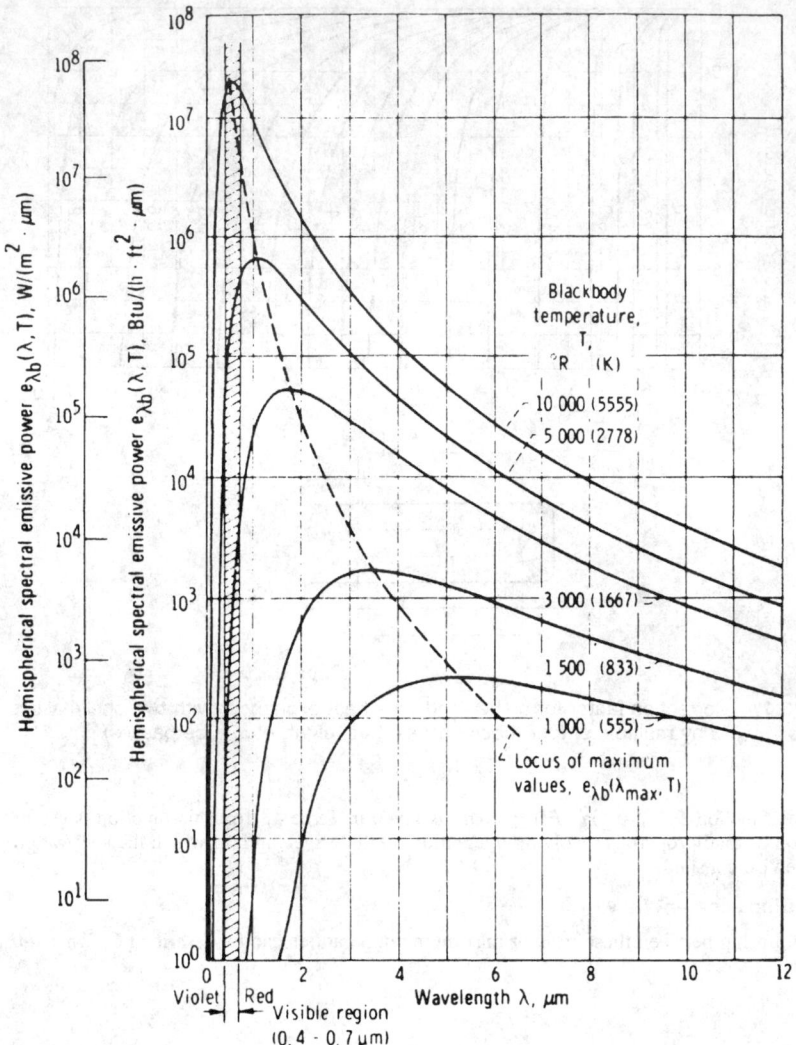

Fig. 43.19 Hemispherical spectral emissive power of a black-body for various temperatures.

$$\lambda_{max} T = 2897.8 \ \mu m \cdot K$$

or

$$\lambda_{max} T = 5216.0 \ \mu m \cdot °R$$

43.4.2 Radiation Properties

While, to some degree, all surfaces follow the general trends described by the Stefan–Boltzmann and Planck laws, the behavior of real surfaces deviates somewhat from these. In fact, because black bodies are ideal, all real surfaces emit and absorb less radiant energy than a black body. The amount of energy a body emits can be described in terms of the emissivity and is, in general, a function of the type of material, the temperature, and the surface conditions, such as roughness, oxide layer thickness, and chemical contamination. The emissivity is in fact a measure of how well a real body radiates energy as compared with a black body of the same temperature. The radiant energy emitted into the entire hemispherical space above a real surface element, including all wavelengths, is given

$$dA_i \, dF_{di \, -dj} = dA_j \, dF_{dj \, -di}$$

$$dA_i F_{di \, -j} = A_j \, dF_{j \, -di}$$

$$A_i F_{i \, -j} = A_j F_{j \, -i}$$

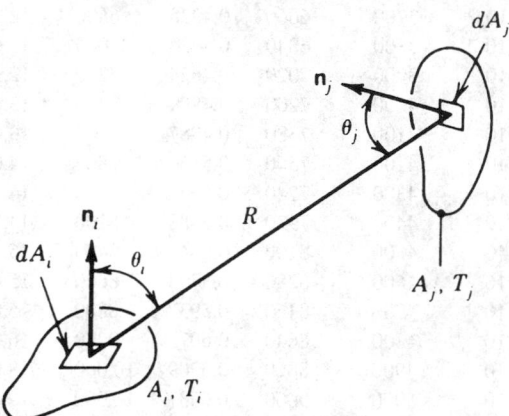

Fig. 43.20 Configuration factor for radiation exchange between surfaces of area dA_i and dA_j.

by $e = \varepsilon \sigma T^4$, where ε is less than 1.0, and is called the *hemispherical emissivity* (or *total hemispherical emissivity* to indicate integration over the total wavelength spectrum). For a given wavelength, the *spectral hemispherical emissivity* ε_λ of a real surface is defined as

$$\varepsilon_\lambda = e_l / e_{\lambda b}$$

where e_λ is the hemispherical emissive power of the real surface and $e_{\lambda b}$ is that of a black body at the same temperature.

Spectral irradiation G_λ (W/m$^2 \cdot \mu$m) is defined as the rate at which radiation is incident upon a surface per unit area of the surface, per unit wavelength about the wavelength λ, and encompasses the incident radiation from all directions.

Spectral hemispherical reflectivity ρ_l is defined as the radiant energy reflected per unit time, per unit area of the surface, per unit wavelength/G_λ.

Spectral hemispherical absorptivity α_λ, is defined as the radiant energy absorbed per unit area of the surface, per unit wavelength about the wavelength/G_λ.

Spectral hemispherical transmissivity is defined as the radiant energy transmitted per unit area of the surface, per unit wavelength about the wavelength/G_λ.

For any surface, the sum of the reflectivity, absorptivity and transmissivity must equal unity, that is,

$$\alpha_\lambda + \rho_\lambda \tau_\lambda = 1$$

When these values are integrated over the entire wavelength from $\lambda = 0$ to ∞ they are referred to as *total* values. Hence, the *total hemispherical reflectivity, total hemispherical absorptivity,* and *total hemispherical transmissivity* can be written as

$$\rho = \int_o^\infty \rho_\lambda G_\lambda d\lambda / G$$

$$\alpha = \int_o^\infty \alpha_\lambda G_\lambda d\lambda / G$$

and

Table 43.16   Radiation Function $F_{o-\lambda T}$

λT			λT			λT		
$\mu m \cdot K$	$\mu m \cdot °R$	$F_{o-\lambda T}$	$\mu m \cdot K$	$\mu m \cdot °R$	$F_{o-\lambda T}$	$\mu m \cdot K$	$\mu m \cdot °R$	$F_{o-\lambda T}$
400	720	0.1864×10^{-11}	3400	6120	0.3617	6400	11,520	0.7692
500	900	0.1298×10^{-8}	3500	6300	0.3829	6500	11,700	0.7763
600	1080	0.9290×10^{-7}	3600	6480	0.4036	6600	11,880	0.7832
700	1260	0.1838×10^{-5}	3700	6660	0.4238	6800	12,240	0.7961
800	1440	0.1643×10^{-4}	3800	6840	0.4434	7000	12,600	0.8081
900	1620	0.8701×10^{-4}	3900	7020	0.4624	7200	12,960	0.8192
1000	1800	0.3207×10^{-3}	4000	7200	0.4809	7400	13,320	0.8295
1100	1980	0.9111×10^{-3}	4100	7380	0.4987	7600	13,680	0.8391
1200	2160	0.2134×10^{-2}	4200	7560	0.5160	7800	14,040	0.8480
1300	2340	0.4316×10^{-2}	4300	7740	0.5327	8000	14,400	0.8562
1400	2520	0.7789×10^{-2}	4400	7920	0.5488	8200	14,760	0.8640
1500	2700	0.1285×10^{-1}	4500	8100	0.5643	8400	15,120	0.8712
1600	2880	0.1972×10^{-1}	4600	8280	0.5793	8600	15,480	0.8779
1700	3060	0.2853×10^{-1}	4700	8460	0.5937	8800	15,840	0.8841
1800	3240	0.3934×10^{-1}	4800	8640	0.6075	9000	16,200	0.8900
1900	3420	0.5210×10^{-1}	4900	8820	0.6209	10,000	18,000	0.9142
2000	3600	0.6673×10^{-1}	5000	9000	0.6337	11,000	19,800	0.9318
2100	3780	0.8305×10^{-1}	5100	9180	0.6461	12,000	21,600	0.9451
2200	3960	0.1009	5200	9360	0.6579	13,000	23,400	0.9551
2300	4140	0.1200	5300	9540	0.6694	14,000	25,200	0.9628
2400	4320	0.1402	5400	9720	0.6803	15,000	27,000	0.9689
2500	4500	0.1613	5500	9900	0.6909	20,000	36,000	0.9856
2600	4680	0.1831	5600	10,080	0.7010	25,000	45,000	0.9922
2700	4860	0.2053	5700	10,260	0.7108	30,000	54,000	0.9953
2800	5040	0.2279	5800	10,440	0.7201	35,000	63,000	0.9970
2900	5220	0.2505	5900	10,620	0.7291	40,000	72,000	0.9979
3000	5400	0.2732	6000	10,800	0.7378	45,000	81,000	0.9985
3100	5580	0.2058	6100	10,980	0.7461	50,000	90,000	0.9989
3200	5760	0.3181	6200	11,160	0.7541	55,000	99,000	0.9992
3300	5940	0.3401	6300	11,340	0.7618	60,000	108,000	0.9994

$$\tau = \int_o^\infty \tau_\lambda G_\lambda \, d\lambda / G$$

respectively, where

$$G = \int_o^\infty G_\lambda \, d\lambda$$

As was the case for the wavelength-dependent parameters, the sum of the total reflectivity, total absorptivity, and total transmissivity must be equal to unity, that is,

$$\alpha + \rho + \tau = 1$$

It is important to note that while the emissivity is a function of the material, temperature, and surface conditions, the absorptivity and reflectivity depend on both the surface characteristics and the nature of the incident radiation.

The terms *reflectance, absorptance,* and *transmittance* are used by some authors for the real surfaces and the terms *reflectivity, absorptivity,* and *transmissivity* are reserved for the properties of the ideal surfaces (i.e., those optically smooth and pure substances perfectly uncontaminated). Surfaces that allow no radiation to pass through are referred to as *opaque,* that is, $\tau_\lambda = 0$, and all of the incident energy will be either reflected or absorbed. For such a surface,

$$\alpha_\lambda + \rho_\lambda = 1$$

and

$$\alpha + \rho = 1$$

Light rays reflected from a surface can be reflected in such a manner that the incident and reflected rays are symmetric with respect to the surface normal at the point of incidence. This type of radiation is referred to as *specular.* The radiation is referred to as *diffuse* if the intensity of the reflected radiation is uniform over all angles of reflection and is independent of the incident direction, and the surface is called a *diffuse surface* if the radiation properties are independent of the direction. If they are independent of the wavelength, the surface is called a *gray surface,* and a *diffuse-gray surface* absorbs a fixed fraction of incident radiation from any direction and at any wavelength, and $\alpha_\lambda = \varepsilon_\lambda = \alpha = \varepsilon$.

Kirchhoff's Law of Radiation

The directional characteristics can be specified by the addition of a ′ to the value. For example the spectral emissivity for radiation in a particular direction would be denoted by α'_λ. For radiation in a particular direction, the spectral emissivity is equal to the directional spectral absorptivity for the surface irradiated by a black body at the same temperature. The most general form of this expression states that $\alpha'_\lambda = \varepsilon'_\lambda$. If the incident radiation is independent of angle or if the surface is diffuse, then $\alpha_\lambda = \varepsilon_\lambda$ for the hemispherical properties. This relationship can have various conditions imposed, depending on whether the spectral, total, directional, or hemispherical quantities are being considered.[19]

Emissivity of Metallic Surfaces

The properties of pure smooth metallic surfaces are often characterized by low emissivity and absorptivity values and high values of reflectivity. The spectral emissivity of metals tends to increase with decreasing wavelength and exhibits a peak near the visible region. At wavelengths $\lambda > \sim 5\ \mu m$, the spectral emissivity increases with increasing temperature; however, this trend reverses at shorter wavelengths ($\lambda < \sim 1.27\ \mu m$). Surface roughness has a pronounced effect on both the hemispherical emissivity and absorptivity, and large *optical roughnesses,* defined as the mean square roughness of the surface divided by the wavelength, will increase the hemispherical emissivity. For cases where the optical roughness is small, the directional properties will approach the values obtained for smooth surfaces. The presence of impurities, such as oxides or other nonmetallic contaminants, will change the properties significantly and increase the emissivity of an otherwise pure metallic body. A summary of the normal total emissivities for metals is given in Table 43.17. It should be noted that the hemispherical emissivity for metals is typically 10–30% higher than the values typically encountered for normal emissivity.

Emissivity of Nonmetallic Materials

Large values of total hemispherical emissivity and absorptivity are typical for nonmetallic surfaces at moderate temperatures and, as shown in Table 43.18, which lists the normal total emissivity of some nonmetals, the temperature dependence is small.

Absorptivity for Solar Incident Radiation

The spectral distribution of solar radiation can be approximated by black-body radiation at a temperature of approximately 5800 K (10,000°R) and yields an average solar irradiation at the outer limit of the atmosphere of approximately 1353 W/m² (429 Btu/ft²·hr). This solar irradiation is called the *solar constant* and is greater than the solar irradiation received at the surface of the earth, due to the radiation scattering by air molecules, water vapor, and dust, and the absorption by O_3, H_2O, and CO_2 in the atmosphere. The absorptivity of a substance depends not only on the surface properties but also on the sources of incident radiation. Since solar radiation is concentrated at a shorter wavelength, due to the high source temperature, the absorptivity for certain materials when exposed to solar radiation may be quite different from that for low-temperature radiation, where the radiation is concentrated in the longer-wavelength range. A comparison of absorptivities for a number of different materials is given in Table 43.19 for both solar and low-temperature radiation.

43.4.3 Configuration Factor

The magnitude of the radiant energy exchanged between any two given surfaces is a function of the emissivity, absorptivity, and transmissivity. In addition, the energy exchange is a strong function of

Table 43.17 Normal Total Emissivity of Metals[a]

Materials	Surface Temperature (K)	Normal Total Emissivity
Aluminum		
Highly polished plate	480–870	0.038–0.06
Polished plate	373	0.095
Heavily oxidized	370–810	0.20–0.33
Bismuth, bright	350	0.34
Chromium, polished	310–1370	0.08–0.40
Copper		
Highly polished	310	0.02
Slightly polished	310	0.15
Black oxidized	310	0.78
Gold, highly polished	370–870	0.018–0.035
Iron		
Highly polished, electrolytic	310–530	0.05–0.07
Polished	700–760	0.14–0.38
Wrought iron, polished	310–530	0.28
Cast iron, rough, strongly oxidized	310–530	0.95
Lead		
Polished	310–530	0.06–0.08
Rough unoxidized	310	0.43
Mercury, unoxidized	280–370	0.09–0.12
Molybdenum, polished	310–3030	0.05–0.29
Nickel		
Electrolytic	310–530	0.04–0.06
Electroplated on iron, not polished	293	0.11
Nickel oxide	920–1530	0.59–0.86
Platinum, electrolytic	530–810	0.06–0.10
Silver, polished	310–810	0.01–0.03
Steel		
Polished sheet	90–420	0.07–0.14
Mild steel, polished	530–920	0.27–0.31
Sheet with rough oxide layer	295	0.81
Tin, polished sheet	310	0.05
Tungsten, clean	310–810	0.03–0.08
Zinc		
Polished	310–810	0.02–0.05
Gray oxidized	295	0.23–0.28

[a]Adapted from Ref. 19.

how one surface is viewed from the other. This aspect can be defined in terms of the *configuration factor* (sometimes called the *radiation shape factor, view factor, angle factor,* or *interception factor*). As shown in Fig. 43.20, the configuration factor F_{i-j} is defined as that fraction of the radiation leaving a black surface i that is intercepted by a black or gray surface j, and is based upon the relative geometry, position, and shape of the two surfaces. The configuration factor can also be expressed in terms of the differential fraction of the energy or dF_{i-dj}, which indicates the differential fraction of energy from a finite area A_i that is intercepted by an infinitesimal area dA_j. Expressions for a number of different cases are given below for several common geometries.

Infinitesimal area dA_i to infinitesimal area dA_j

$$dF_{di-dj} = \frac{\cos\theta_i \cos\theta_j}{\pi R^2} \, dA_j$$

Infinitesimal area dA_i to finite area A_j

Table 43.18 Normal Total Emissivity of Nonmetals[a]

Materials	Surface Temperature (K)	Normal Total Emissivity
Asbestos, board	310	0.96
Brick		
White refractory	1370	0.29
Rough red	310	0.93
Carbon, lampsoot	310	0.95
Concrete, rough	310	0.94
Ice, smooth	273	0.966
Magnesium oxide, refractory	420–760	0.69–0.55
Paint		
Oil, all colors	373	0.92–0.96
Lacquer, flat black	310–370	0.96–0.98
Paper, white	310	0.95
Plaster	310	0.91
Porcelain, glazed	295	0.92
Rubber, hard	293	0.92
Sandstone	310–530	0.83–0.90
Silicon carbide	420–920	0.83–0.96
Snow	270	0.82
Water, deep	273–373	0.96
Wood, sawdust	310	0.75

[a]Adapted from Ref. 19.

Table 43.19 Comparison of Absorptivities of Various Surfaces to Solar and Low-Temperature Thermal Radiation[a]

Surface	Absorptivity	
	For Solar Radiation	For Low-Temperature Radiation (~300 K)
Aluminum, highly polished	0.15	0.04
Copper, highly polished	0.18	0.03
Tarnished	0.65	0.75
Cast iron	0.94	0.21
Stainless steel, No. 301, polished	0.37	0.60
White marble	0.46	0.95
Asphalt	0.90	0.90
Brick, red	0.75	0.93
Gravel	0.29	0.85
Flat black lacquer	0.96	0.95
White paints, various types of pigments	0.12–0.16	0.90–0.95

[a]Adapted from Ref. 20 after J. P. Holman, *Heat Transfer*, McGraw-Hill, New York, 1981.

$$F_{di-j} = \int_{A_j} \frac{\cos\theta_i \cos\theta_j}{\pi R^2} dA_j$$

Finite area A_i to finite area A_j

$$F_{i-j} = \frac{1}{A_i} \int_{A_i} \int_{A_j} \frac{\cos\theta_i \cos\theta_j}{\pi R^2} dA_i dA_j$$

Analytical expressions of other configuration factors have been found for a wide variety of simple geometries. A number of these are presented in Figs. 43.21–43.24 for surfaces that emit and reflect diffusely.

Reciprocity Relations

The configuration factors can be combined and manipulated using algebraic rules referred to as *configuration factor geometry*. These expressions take several forms, one of which is the reciprocal properties between different configuration factors that allow one configuration factor to be determined from knowledge of the others:

$$dA_i dF_{di-dj} = dA_j dF_{dj-di}$$
$$dA_i dF_{di-j} = A_j dF_{j-di}$$
$$A_i F_{i-j} = A_j F_{j-i}$$

These relationships can be combined with other basic rules to allow the determination of the configuration of an infinite number of complex shapes and geometries form a few select, known geometries. These are summarized in the following sections.

The Additive Property

For a surface A_i subdivided into N parts $(A_{i_1}, A_{i_2}, \ldots, A_{i_N})$ and a surface A_j subdivided into M parts $(A_{j_1}, A_{j_2}, \ldots, A_{j_M})$,

$$A_i F_{i-j} = \sum_{n=1}^{N} \sum_{m=1}^{M} A_{i_n} F_{i_n-j_m}$$

Relation in an Enclosure

When a surface is completely enclosed, the surface can be subdivided into N parts having areas A_1, $A_2, \ldots, A_N$, respectively, and

$$\sum_{j=1}^{N} F_{i-j} = 1$$

Black-Body Radiation Exchange

For black surfaces A_i and A_j at temperatures T_i and T_j, respectively, the net radiative exchange q_{ij} can be expressed as

$$q_{ij} = A_i F_{i-j} \sigma(T_i^4 - T_j^4)$$

and for a surface completely enclosed and subdivided into N surfaces maintained at temperatures $T_1, T_2, \ldots, T_N$, the net radiative heat transfer q_i to surface area A_i is

$$q_i = \sum_{j=1}^{N} A_i F_{i-j} \sigma(T_i^4 - T_j^4) = \sum_{j=1}^{N} q_{ij}$$

43.4.4   Radiative Exchange among Diffuse-Gray Surfaces in an Enclosure

One method for solving for the radiation exchange between a number of surfaces or bodies is through the use of the *radiosity J*, defined as the total radiation that leaves a surface per unit time and per unit area. For an opaque surface, this term is defined as

$$J = \varepsilon \sigma T^4 + (1 - \varepsilon)G$$

For an enclosure consisting of N surfaces, the irradiation on a given surface i can be expressed as

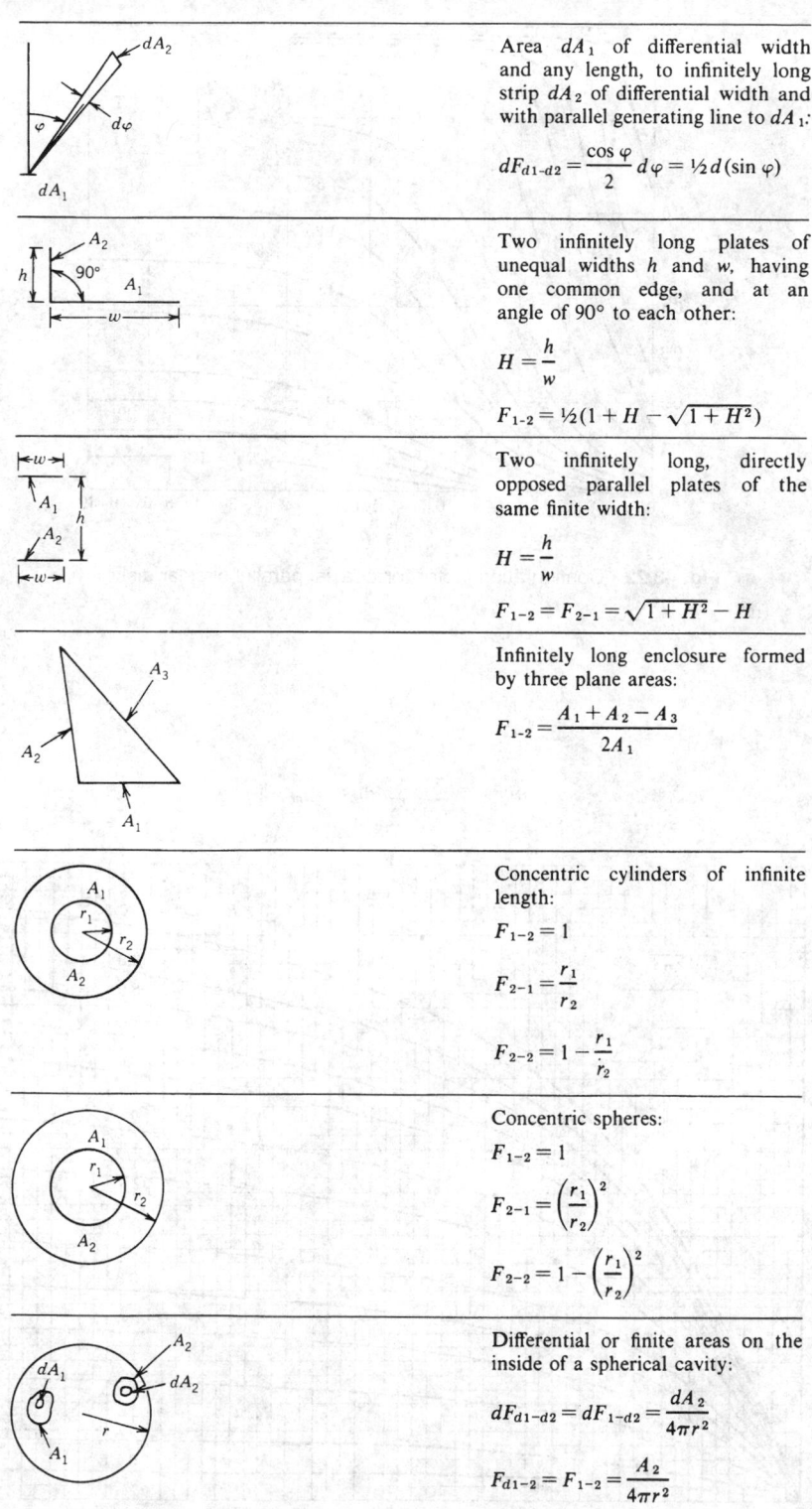

Area dA_1 of differential width and any length, to infinitely long strip dA_2 of differential width and with parallel generating line to dA_1:

$$dF_{d1-d2} = \frac{\cos \varphi}{2} d\varphi = \tfrac{1}{2} d(\sin \varphi)$$

Two infinitely long plates of unequal widths h and w, having one common edge, and at an angle of 90° to each other:

$$H = \frac{h}{w}$$

$$F_{1-2} = \tfrac{1}{2}(1 + H - \sqrt{1 + H^2})$$

Two infinitely long, directly opposed parallel plates of the same finite width:

$$H = \frac{h}{w}$$

$$F_{1-2} = F_{2-1} = \sqrt{1 + H^2} - H$$

Infinitely long enclosure formed by three plane areas:

$$F_{1-2} = \frac{A_1 + A_2 - A_3}{2A_1}$$

Concentric cylinders of infinite length:

$$F_{1-2} = 1$$

$$F_{2-1} = \frac{r_1}{r_2}$$

$$F_{2-2} = 1 - \frac{r_1}{r_2}$$

Concentric spheres:

$$F_{1-2} = 1$$

$$F_{2-1} = \left(\frac{r_1}{r_2}\right)^2$$

$$F_{2-2} = 1 - \left(\frac{r_1}{r_2}\right)^2$$

Differential or finite areas on the inside of a spherical cavity:

$$dF_{d1-d2} = dF_{1-d2} = \frac{dA_2}{4\pi r^2}$$

$$F_{d1-2} = F_{1-2} = \frac{A_2}{4\pi r^2}$$

Fig. 43.21 Configuration factors for some simple geometries.[19]

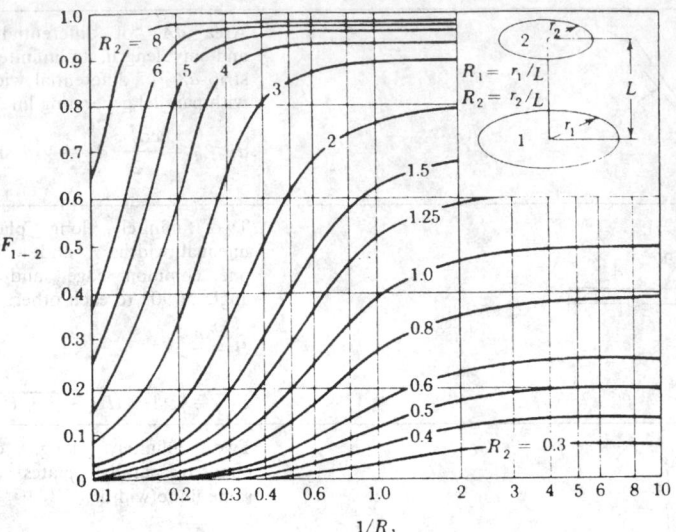

Fig. 43.22 Configuration factor for coaxial parallel circular disks.

$$A_i F_{i-j} = \sum_{n=1}^{N} \sum_{m=1}^{M} A_{i_n} F_{i_n-j_m}$$

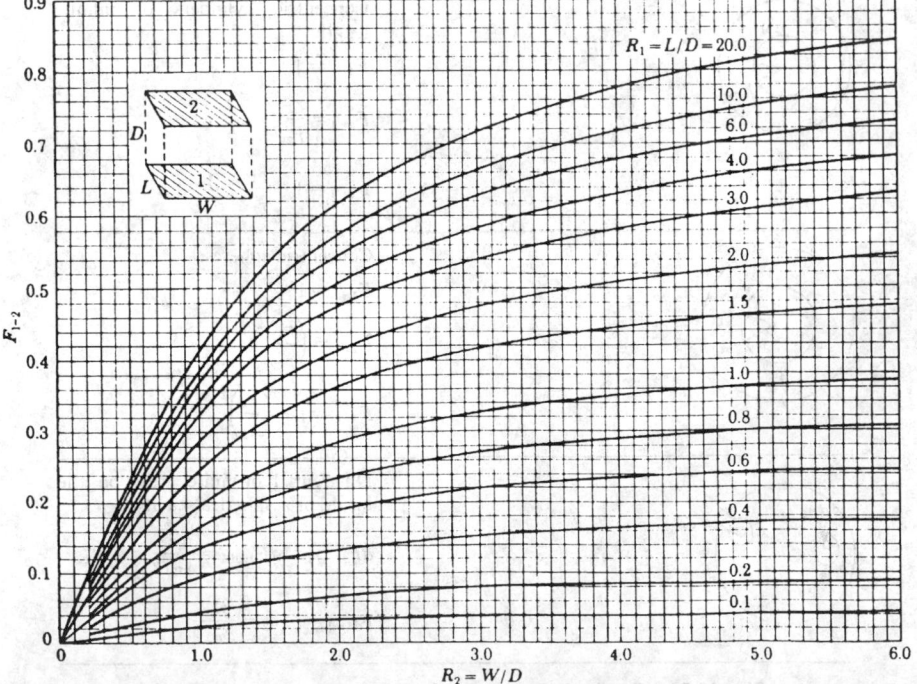

Fig. 43.23 Configuration factor for aligned parallel rectangles.

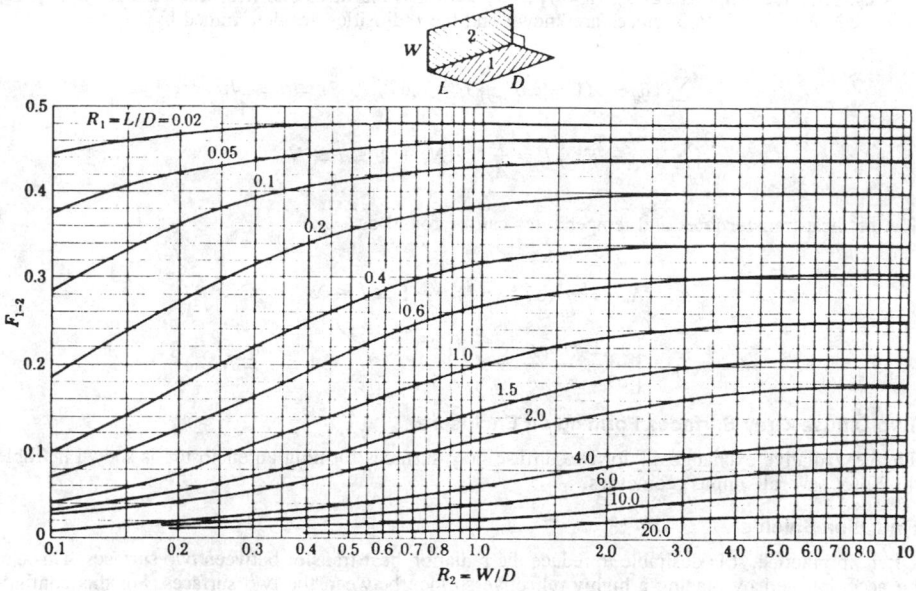

Fig. 43.24 Configuration factor for rectangles with common edge.

$$G_i = \sum_{j=1}^{N} J_j F_{i-j}$$

and the net radiative heat-transfer rate at given surface i is

$$q_i = A_i(J_i - G_i) = \frac{\varepsilon_i A_i}{1 - \varepsilon_i}(\sigma T_i^4 - J_i)$$

For every surface in the enclosure, a uniform temperature or a constant heat transfer rate can be specified. If the surface temperature is given, the heat transfer rate can be determined for that surface and vice versa. Shown below are several specific cases that are commonly encountered

Case I: Temperatures T_i ($i = 1,2, \ldots, N$) are known for each of the N surfaces and the values of the radiosity J_i are solved from the expression

$$\sum_{j=1}^{N} \{\delta_{ij} - (1 - \varepsilon_i)F_{i-j}\} J_i = \varepsilon_i \sigma T_i^4 \qquad 1 \le i \le N$$

The net heat-transfer rate to surface i can then be determined from the fundamental relationship

$$q_i = A_i \frac{\varepsilon_i}{1 - \varepsilon_i}(\sigma T_i^4 - J_i) \qquad 1 \le i \le N$$

where $\delta_{ij} = 0$ for $i \pm j$ and $\delta_{ij} = 1$ for $i = j$.

Case II: The heat transfer rates q_i ($i = 1, 2, \ldots, N$) are known for each of the N surfaces and the values of the radiosity J_i are determined from

$$\sum_{j=1}^{N} \{\delta_{ij} - F_{i-j}\} J_j = q_i/A_i \qquad 1 \le i \le N$$

The surface temperature can then be determined from

$$T_i = \left[\frac{1}{\sigma}\left(\frac{1 - \varepsilon_i}{\varepsilon_i}\frac{q_i}{A_i} + J_i\right)\right]^{1/4} \qquad 1 \le i \le N$$

Case III: The temperatures T_i. $(i = 1, \ldots, N_1)$ for N_i surfaces and heat-transfer rates q_i $(i = N_1 + 1, \ldots, N)$ for $(N-N_i)$ surfaces are known and the radiosities are determined by

$$\sum_{j=1}^{N} \{\delta_{ij} - (1 - \varepsilon_i)F_{i-j}\} J_j = \varepsilon_i \alpha T_i^4 \qquad 1 \le i \le N_1$$

$$\sum_{j=1}^{N} \{\delta_{ij} - F_{i-j}\} J_j = \frac{q_i}{A_i} \qquad N_1 + 1 \le i \le N$$

The net heat-transfer rates and temperatures can be found as

$$q_i = A_i \frac{\varepsilon_i}{1 - \varepsilon_i} (\sigma T_i^4 - J_i) \qquad 1 \le i \le N_1$$

$$T_i = \left[\frac{1}{\sigma}\left(\frac{1 - \varepsilon_i}{\varepsilon_i}\frac{q_i}{A_i} + J_i\right)\right]^{1/4} \qquad N_1 + 1 \le i \le N$$

Two Diffuse-Gray Surfaces Forming an Enclosure

The net radiative exchange q_{12} for two diffuse-gray surfaces forming an enclosure is shown in Table 43.20 for several simple geometries.

Radiation Shields

Often, in practice, it is desirable to reduce the radiation heat transfer between two surfaces. This can be accomplished by placing a highly reflective surface between the two surfaces. For this configuration, the ratio of the net radiative exchange with the shield to that without the shield can be expressed by the relationship

$$\frac{q_{12 \text{ with shield}}}{q_{12 \text{ without shield}}} = \frac{1}{1 + \chi}$$

Values for this ratio χ for shields between parallel plates, concentric cylinders, and concentric spheres

---

Table 43.20   Net Radiative Exchange between Two Surfaces Forming an Enclosure

Large (Infinite) Parallel Planes

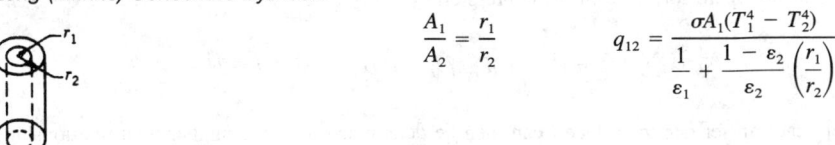

$A_1 = A_2 = A$

$$q_{12} = \frac{A\sigma(T_1^4 - T_2^4)}{\dfrac{1}{\varepsilon_1} + \dfrac{1}{\varepsilon_2} - 1}$$

Long (Infinite) Concentric Cylinders

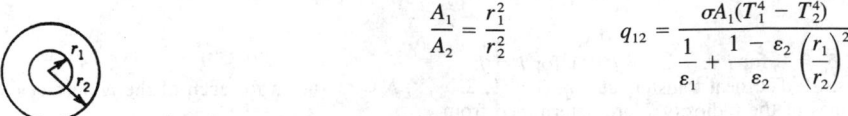

$\dfrac{A_1}{A_2} = \dfrac{r_1}{r_2}$

$$q_{12} = \frac{\sigma A_1(T_1^4 - T_2^4)}{\dfrac{1}{\varepsilon_1} + \dfrac{1 - \varepsilon_2}{\varepsilon_2}\left(\dfrac{r_1}{r_2}\right)}$$

Concentric Sphere

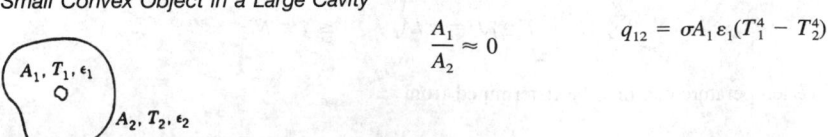

$\dfrac{A_1}{A_2} = \dfrac{r_1^2}{r_2^2}$

$$q_{12} = \frac{\sigma A_1(T_1^4 - T_2^4)}{\dfrac{1}{\varepsilon_1} + \dfrac{1 - \varepsilon_2}{\varepsilon_2}\left(\dfrac{r_1}{r_2}\right)^2}$$

Small Convex Object in a Large Cavity

$\dfrac{A_1}{A_2} \approx 0$

$$q_{12} = \sigma A_1 \varepsilon_1 (T_1^4 - T_2^4)$$

are summarized in Table 43.21. For the special case of parallel plates involving more than one or N shields, where all of the emissivities are equal, the value of χ equals N.

Radiation Heat-Transfer Coefficient

The rate at which radiation heat transfer occurs can be expressed in a form similar to Fourier's law or Newton's law of cooling by expressing it in terms of the temperature difference $T_1 - T_2$ or as

$$q = h_r A(T_1 - T_2)$$

where h_r is the radiation heat-transfer coefficient or *radiation film coefficient*. For the case of radiation between two large parallel plates with emissivities, respectively, of ε_1 and ε_2,

$$h_r = \frac{(T_1^4 - T_2^4)}{(T_1 - T_2)\left(\dfrac{1}{\varepsilon_1} + \dfrac{1}{\varepsilon_2} - 1\right)}$$

43.4.5 Thermal Radiation Properties of Gases

All of the previous expressions assumed that the medium present between the surfaces did not affect the radiation exchange. In reality, gases such as air, oxygen (O_2), hydrogen (H_2), and nitrogen (N_2) have a symmetrical molecular structure and neither emit nor absorb radiation at low to moderate temperatures. Hence, for most engineering applications, such *non-participating gases* can be ignored. However, polyatomic gases, such as water vapor (H_2O), carbon dioxide (CO_2), carbon monoxide (CO), sulfur dioxide (SO_2), and various hydrocarbons, emit and absorb significant amounts of radiation. These *participating gases* absorb and emit radiation in limited spectral ranges, referred to as spectral *bands*. In calculating the emitted or absorbed radiation for a gas layer, its thickness, shape, surface area, pressure, and temperature distribution must be considered. Although a precise method for calculating the effect of these participating media is quite complex, an approximate method developed by Hottel[21] will yield results that are reasonably accurate.

The effective total emissivities of carbon dioxide and water vapor are a function of the temperature and the product of the partial pressure and the mean beam length of the substance, as indicated in Figs. 43.24 and 43.25, respectively. The *mean beam length L_e* is the characteristic length that corresponds to the radius of a hemisphere of gas, such that the energy flux radiated to the center of the base is equal to the average flux radiated to the area of interest by the actual gas volume. Table 43.22 lists the mean beam lengths of several simple shapes. For a geometry for which L_e has not been determined, it is generally approximated by $L_e = 3.6V/A$ for an entire gas volume V radiating to its entire boundary surface A. The data in Figs. 43.25 and 43.26 were obtained for a total pressure of 1

Table 43.21 Values of X for Radiative Shields

Geometry	X	
Shield	$\dfrac{\dfrac{1}{\varepsilon_{s1}} + \dfrac{1}{\varepsilon_{s2}} - 1}{\dfrac{1}{\varepsilon_1} + \dfrac{1}{\varepsilon_2} - 1}$	Infinitely long parallel plates
	$\dfrac{\left(\dfrac{r_1}{r_2}\right)^2\left(\dfrac{1}{\varepsilon_{s1}} + \dfrac{1}{\varepsilon_{s2}} - 1\right)}{\dfrac{1}{\varepsilon_1} + \left(\dfrac{1}{\varepsilon_2} - 1\right)\left(\dfrac{r_1}{r_2}\right)^2}$	$n = 1$ for infinitely long concentric cylinders $n = 2$ for concentric spheres

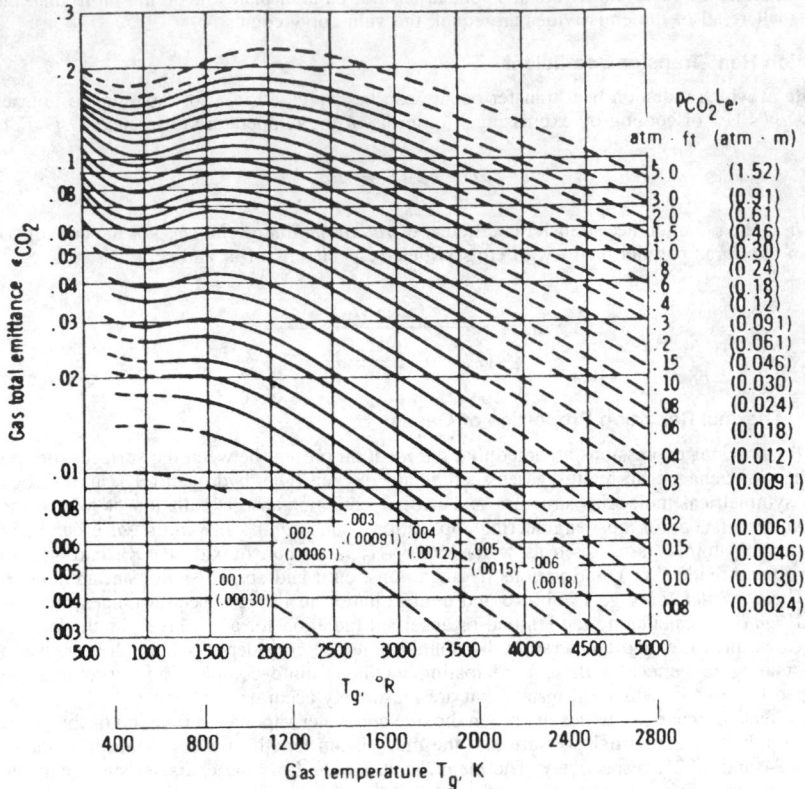

Fig. 43.25 Total emissivity of CO_2 in a mixture having a total pressure of 1 atm. From Ref. 21. (Used with the permission of McGraw-Hill Book Company.)

atm and zero partial pressure of the water vapor. For other total and partial pressures, the emissivities are corrected by multiplying C_{CO_2} (Fig. 43.27) and C_{H_2O} (Fig. 43.28), respectively, to ε_{CO_2} and ε_{H_2O}, which are found from Figs. 43.25 and 43.26.

These results can be applied when water vapor or carbon dioxide appear separately or in a mixture with other non-participating gases. For mixtures of CO_2 and water vapor in a non-participating gas, the total emissivity of the mixture ε_g can be estimated from the expression

$$\varepsilon_g = C_{CO_2}\varepsilon_{CO_2} + C_{H_2O}\varepsilon_{H_2O} - \Delta\varepsilon$$

where $\Delta\varepsilon$ is a correction factor given in Fig. 43.29.

Radiative Exchange between Gas Volume and Black Enclosure of Uniform Temperature

When radiative energy is exchanged between a gas volume and a black enclosure, the exchange per unit area q'' for a gas volume at uniform temperature T_g and a uniform wall temperature T_w is given by

$$q'' = \varepsilon_g(T_g)\sigma T_g^4 - \alpha_g(T_w)\sigma T_w^4$$

where $\varepsilon_g(T_g)$ is the gas emissivity at a temperature T_g and $\alpha_g(T_w)$ is the absorptivity of gas for the radiation from the black enclosure at T_w. As a result of the nature of the band structure of the gas, the absorptivity α_g for black radiation at a temperature T_w is different from the emissivity ε_g at a gas temperature of T_g. When a mixture of carbon dioxide and water vapor is present, the empirical expression for α_g is

$$\alpha_g = \alpha_{CO_2} + \alpha_{H_2O} - \Delta\alpha$$

where

Table 43.22 Mean Beam Length[a]

Geometry of Gas Volume	Characteristic Length	L_e
Hemisphere radiating to element at center of base	Radius R	R
Sphere radiating to its surface	Diameter D	$0.65D$
Circular cylinder of infinite height radiating to concave bounding surface	Diameter D	$0.95D$
Circular cylinder of semi-infinite height radiating to:		
Element at center of base	Diameter D	$0.90D$
Entire base	Diameter D	$0.65D$
Circular cylinder of height equal to diameter radiating to:		
Element at center of base	Diameter D	$0.71D$
Entire surface	Diameter D	$0.60D$
Circular cylinder of height equal to two diameters radiating to:		
Plane end	Diameter D	$0.60D$
Concave surface	Diameter D	$0.76D$
Entire surface	Diameter D	$0.73D$
Infinite slab of gas radiating to:		
Element on one face	Slab thickness D	$1.8D$
Both bounding planes	Slab thickness D	$1.8D$
Cube radiating to a face	Edge X	$0.6X$
Gas volume surrounding an infinite tube bundle and radiating to a single tube:		
Equilateral triangular array:	Tube diameter D and spacing between tube centers, S	
$S = 2D$		$3.0(S - D)$
$S = 3D$		$3.8(S - D)$
Square array:		$3.5(S - D)$
$S = 2D$		

[a]Adapted from Ref. 19.

$$\alpha_{CO_2} = C_{CO_2} \varepsilon'_{CO_2} \left(\frac{T_g}{T_w}\right)^{0.65}$$

$$\alpha_{H_2O} = C_{H_2O} \varepsilon'_{H_2O} \left(\frac{T_g}{T_w}\right)^{0.45}$$

where $\Delta\alpha = \Delta\varepsilon$ and all properties are evaluated at T_w. In this expression, the values of ε'_{CO_2} and ε'_{H_2O} can be found from Figs. 43.25 and 43.26 using an abscissa of T_w, but substituting the parameters $p_{CO_2}L_e T_w/T_g$ and $p_{H_2O}L_e T_w/T_g$ for $p_{CO_2}L_e$ and $p_{H_2O}L_e$, respectively.

Radiative Exchange between a Gray Enclosure and a Gas Volume

When the emissivity of the enclosure ε_w is larger than 0.8, the rate of heat transfer may be approximated by

$$q_{gray} = \left(\frac{\varepsilon_w + 1}{2}\right) q_{black}$$

where q_{gray} is the heat-transfer rate for gray enclosure and q_{black} is that for black enclosure. For values of $\varepsilon_w < 0.8$, the band structures of the participating gas must be taken into account for heat-transfer calculations.

43.5 BOILING AND CONDENSATION HEAT TRANSFER

Boiling and condensation are both forms of convection in which the fluid medium is undergoing a change of phase. When a liquid comes into contact with a solid surface maintained at a temperature above the saturation temperature of the liquid, the liquid may vaporize, resulting in boiling. This process is always accompanied by a change of phase from the liquid to the vapor state and results

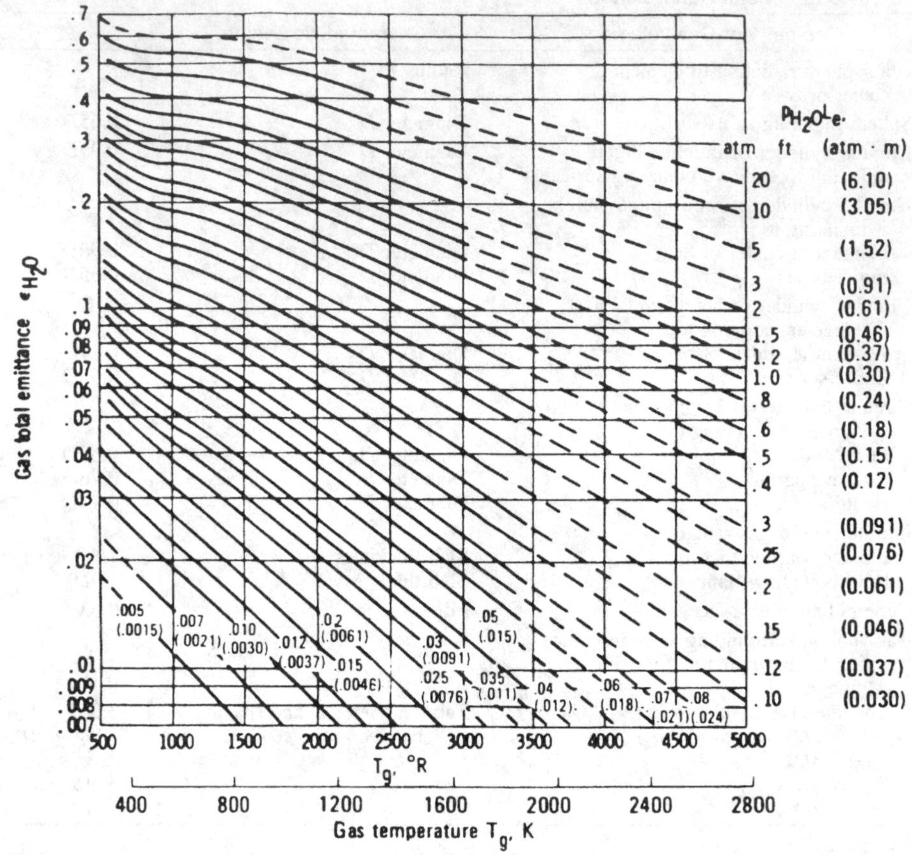

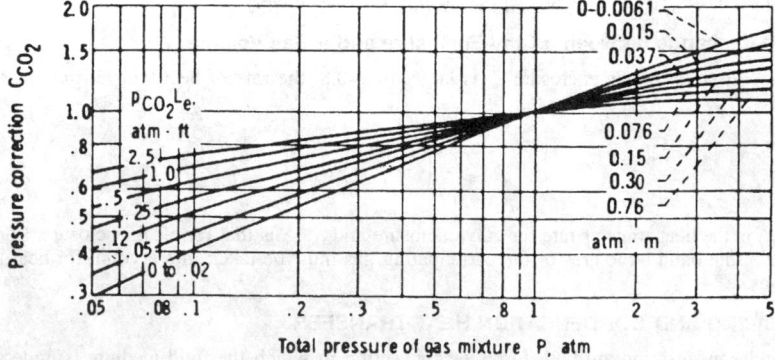

Fig. 43.26 Total emissivity of H_2O at 1 atm total pressure and zero partial pressure. (From Ref. 21. Used with the permission of McGraw-Hill Book Company.)

Fig. 43.27 Pressure correction for CO_2 total emissivity for values of P other than 1 atm. Adapted from Ref. 21. (Used with the permission of McGraw-Hill Book Company.)

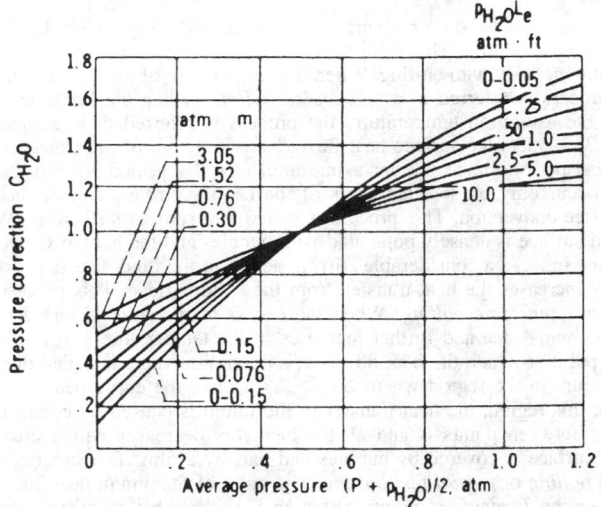

Fig. 43.28 Pressure correction for water vapor total emissivity for values of P_{H2O} and P other than 0 and 1 atm. Adapted from Ref. 21. (Used with the permission of McGraw-Hill Book Company.)

in large rates of heat transfer from the solid surface, due to the latent heat of vaporization of the liquid. The process of condensation is usually accomplished by allowing the vapor to come into contact with a surface at a temperature below the saturation temperature of the vapor, in which case the liquid undergoes a change in state from the vapor state to the liquid state, giving up the latent heat of vaporization.

The heat-transfer coefficients for condensation and boiling are generally larger than that for convection without phase change, sometimes by as much as several orders of magnitude. Application of boiling and condensation heat transfer may be seen in a closed-loop power cycle or in a device referred to as a *heat pipe*, which will be discussed in the following section. In power cycles, the liquid is vaporized in a boiler at high pressure and temperature. After producing work by means of expansion through a turbine, the vapor is condensed to the liquid state in a condenser and then returned to the boiler, where the cycle is repeated.

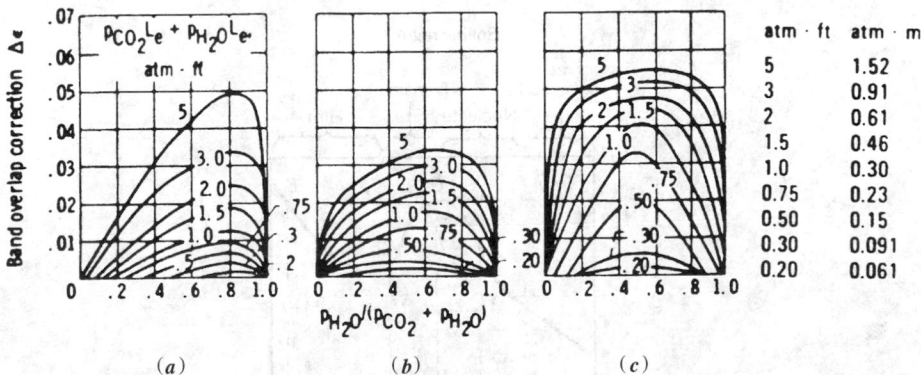

Fig. 43.29 Correction on total emissivity for band overlap when both CO_2 and water vapor are present: (a) gas temperature $T_g = 400$ K (720°R); (b) gas temperature $T_g = 810$ K (1460°R); (c) gas temperature $T_g = 1200$ K (2160°R). Adapted from Ref. 21. (Used with the permission of McGraw-Hill Book Company.)

43.5.1 Boiling

The formation of vapor bubbles on a hot surface in contact with a quiescent liquid without external agitation it is called *pool boiling*. This differs from *forced-convection boiling,* in which forced convection occurs simultaneously with boiling. When the temperature of the liquid is below the saturation temperature, the process is referred to as *subcooled boiling.* When the liquid temperature is maintained or exceeds the saturation temperature, the process is referred to as *saturated or saturation boiling.* Figure 43.30 depicts the surface heat flux q'' as a function of the excess temperature $\delta T_e = T_s - T_{sat}$, for typical pool boiling of water using an electrically heated wire. In the region $0 < \Delta T_e < \Delta T_{e,A}$, bubbles occur only on selected spots of the heating surface and the heat transfer occurs primarily through free convection. This process is called *free convection boiling.* When $\Delta T_{e,A} < \Delta T_e < \Delta T_{e,C}$, the heated surface is densely populated with bubbles and the bubble separation and eventual rise due to buoyancy induces a considerable stirring action in the fluid near the surface. This stirring action substantially increases the heat transfer from the solid surface. This process or region of the curve is referred to as *nucleate boiling.* When the excess temperature is raised to $\Delta T_{e,C}$, the heat flux reaches a maximum value and further increases in the temperature will result in a decrease in the heat flux. The point at which the heat flux is at a maximum value is called the *critical heat flux.*

Film boiling occurs in the region where $\Delta T_e > \Delta T_{e,D}$, and the entire heating surface is covered by a vapor film. In this region, the heat transfer to the liquid is caused by conduction and radiation through the vapor. Between points C and D, the heat flux decreases with increasing ΔT_e. In this region, part of the surface is covered by bubbles and part by a film. The vaporization in this region is called *transition boiling* or *partial film boiling.* The point of maximum heat flux, point C, is called the *burnout point* or the *Liedenfrost point.* Although it is desirable to operate vapor generators at heat fluxes close to q_c'', to permit the maximum use of the surface area, in most engineering applications, it is necessary to control the heat flux and great care is taken to avoid reaching this point. The primary reason for this is that, as illustrated, when the heat flux is increased gradually, the temperature rises steadily until point C is reached. Any increase of heat flux beyond the value of q_c'', however, will dramatically change the surface temperature to $T_s = T_{sat} + T_{e,E}$, typically exceeding the solid melting point and leading to failure of the material in which the liquid is held or from which the heater is fabricated.

Nucleate Pool Boiling

The heat flux data are best correlated by[25]

$$q'' = \mu_l h_g \left(\frac{g(\rho_l - \rho_v)}{g_c \sigma} \right)^{1/2} \left(\frac{c_{p,l} \Delta T_e}{C h_{fg} \, Pr_l^{1.7}} \right)^3$$

where the subscripts l and v denote saturated liquid and vapor, respectively. The surface tension of the liquid is σ (N/m). The quantity g_c is the proportionality constant equal to 1 kg · m/N · sec². The quantity g is the local gravitational acceleration in m/sec². The values of C are given in Table 43.24. The above equation may be applied to different geometries such as plates, wire, or cylinders.

The *critical heat flux* (point C of Fig. 43.30) is given by[27]

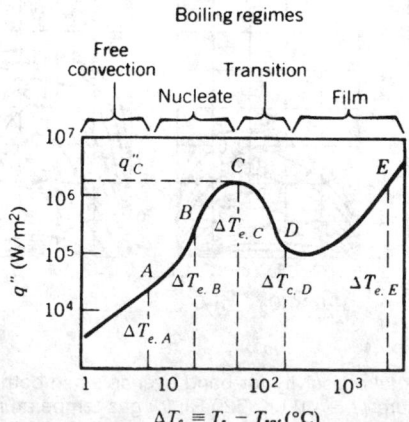

Fig. 43.30 Typical boiling curve for a wire in a pool of water at atmospheric pressure.

Table 43.23 Thermophysical Properties of Saturated Water

Temperature T (K)	Pressure P (bar)[a]	Specific Volume (m³/kg)		Heat of Vaporization h_{jf} (kJ/kg)	Specific Heat (kJ/kg·K)		Viscosity (N·sec/m²)		Thermal Conductivity (W/m·K)		Prandtl Number		Surface Tension $\sigma_l \times 10^3$ (N/m)	Expansion Coefficient $\beta1 \times 10^6$ (K⁻¹)
		$vf \times 10^3$	v_u		$C_{p,l}$	$C_{p,u}$	$\mu_l \times 10^6$	$\mu_v \times 10^3$	$k_l \times 10^3$	$k_v \times 10^3$	Pr_l	Pr_v		
273.15	0.00611	1.000	206.3	2502	4.217	1.854	1750	8.02	659	18.2	12.99	0.815	75.5	−68.05
300	0.03531	1.003	39.13	2438	4.179	1.872	855	9.09	613	19.6	5.83	0.857	71.7	276.1
320	0.1053	1.011	13.98	2390	4.180	1.895	577	9.89	640	21.0	3.77	0.894	68.3	436.7
340	0.2713	1.021	5.74	2342	4.188	1.930	420	10.69	660	22.3	2.66	0.925	64.9	566.0
360	0.6209	1.034	2.645	2291	4.203	1.983	324	11.49	674	23.7	2.02	0.960	61.4	697.9
380	1.2869	1.049	1.337	2239	4.226	2.057	260	12.29	683	25.4	1.61	0.999	57.6	788
400	2.455	1.067	0.731	2183	4.256	2.158	217	13.05	688	27.2	1.34	1.033	63.6	896
450	9.319	1.123	0.208	2024	4.40	2.56	152	14.85	678	33.1	0.99	1.14	42.9	
500	26.40	1.203	0.0766	1825	4.66	3.27	118	16.59	642	42.3	0.86	1.28	31.6	
550	61.19	1.323	0.0317	1564	5.24	4.64	97	18.6	580	58.3	0.87	1.47	19.7	
600	123.5	1.541	0.0137	1176	7.00	8.75	81	22.7	497	92.9	1.14	2.15	8.4	
647.3	221.2	3.170	0.0032	0	∞	∞	45	45	238	238	∞	∞	0.0	

Table 43.24 Values of the Constant C for Various Liquid–Surface Combinations[a]

Fluid-Heating Surface Combinations	C
Water with polished copper, platinum, or mechanically polished stainless steel	0.0130
Water with brass or nickel	0.006
Water with ground and polished stainless steel	0.008
Water with Teflon-plated stainless steel	0.008

[a]Adapted from Ref. 26.

$$q_c'' = \frac{\pi}{24} h_{fg} \rho_v \left(\frac{\sigma g g_c (\rho_l - \rho_v)}{\rho_v^2} \right)^{0.25} \left(1 + \frac{\rho_v}{\rho_l} \right)^{0.5}$$

For a water–steel combination, $q_c'' \approx 1290$ KW/m^2 and $\Delta T_{e,c} \approx 30°C$. For water–chrome-plated copper, $q_c'' \approx 940–1260$ KW/m^2 and $\Delta T_{e,c} \approx 23–28°C$.

Film Pool Boiling

The heat transfer from a surface to a liquid is due to both convection and radiation. A total heat-transfer coefficient is defined by the combination of convection and radiation heat-transfer coefficients of the following form[28] for the outside surfaces of horizontal tubes:

$$h^{4/3} = h_c^{4/3} + h_r h^{1/3}$$

where

$$h_c = 0.62 \left(\frac{k_v^3 \rho_v (\rho_l - \rho_v) g (h_{fg} + 0.4 c_{p,v} \Delta T_e)}{\mu_v D \Delta T_e} \right)^{1/4}$$

and

$$h_r = \frac{5.73 \times 10^{-8} \varepsilon (T_s^4 - T_{sat}^r)}{T_s - T_{sat}}$$

The vapor properties are evaluated at the film temperature $T_f = (T_s + T_{sat})/2$. The temperatures T_s and T_{sat} are in Kelvins for the evaluation of h_r. The emissivity of the metallic solids can be found from Table 43.17. Note that $q = hA(T_s - T_{sat})$.

Nucleate Boiling in Forced Convection

The total heat-transfer rate can be obtained by simply superimposing the heat transfer due to nucleate boiling and forced convection:

$$q'' = q''_{\text{boiling}} + q''_{\text{forced convection}}$$

For forced convection, it is recommended that the coefficient 0.023 be replaced by 0.014 in the Dittus–Boelter equation (Section 43.2.1). The above equation is generally applicable to forced convection where the bulk liquid temperature is subcooled (*local forced convection boiling*).

Simplified Relations for Boiling in Water

For *nucleate boiling*,[29]

$$h = C(\Delta T_e)^n \left(\frac{p}{p_a} \right)^{0.4}$$

where p and p_a are, respectively, the system pressure and standard atmospheric pressure. The constants C and n are listed in Table 43.25.

For *local forced convection boiling inside vertical tubes*, valid over a pressure range of 5–170 atm (Ref. 29, Vol. 2, p. 584),

Table 43.25 Values of C and n for Simplified Relations for Boiling in Water[a]

Surface	q'' (KW/m²)	C	n
Horizontal	$q'' < 16$	1042	⅓
	$16 < q'' < 240$	5.56	3
Vertical	$q'' < 3$	5.7	½
	$3 < q'' < 63$	7.96	3

[a]Adapted from Ref. 29.

$$h = 2.54(\Delta T_e)^3 e^{p/1.551}$$

where h has the unit W/m² · °C, ΔT_e is in °C, and p is the pressure in 10^6 N/m³.

43.5.2 Condensation

Depending on the surface conditions, the condensation may be a *film condensation* or a *dropwise condensation*. Film condensation usually occurs when a vapor, relatively free of impurities, is allowed to condense on a clean, uncontaminated surface. Dropwise condensation occurs on highly polished surfaces or on surfaces coated with substances that inhibit wetting. The condensate provides a resistance to heat transfer between the vapor and the surface. Therefore, it is desirable to use short vertical surfaces or horizontal cylinders to prevent the condensate from growing too thick. The heat-transfer rate for dropwise condensation is usually an order of magnitude larger than that for film condensation under similar conditions. Silicones, Teflon, and certain fatty acids can be used to coat the surfaces to promote dropwise condensation. However, such coatings may lose their effectiveness owing to oxidation or outright removal. Thus, except under carefully controlled conditions, film condensation may be expected to occur in most instances, and the condenser design calculations are often based on the assumption of film condensation.

For condensation on a surface at temperature T_s, the total heat-transfer rate to the surface is given by $q = \bar{h}_L A (T_{sat} - T_s)$, where T_{sat} is the saturation temperature of the vapor. The mass flow rate is determined by $\dot{m} = q/h_{fg}$; h_{fg} is the latent heat of vaporization of the fluid (see Table 43.23 for saturated water). Correlations are based on the evaluation of liquid properties at $T_f = (T_s + T_{sat})/2$ except h_{fg}, which is to be taken at T_{sat}.

Film Condensation on a Vertical Plate

The Reynolds number for *condensate flow* is defined by $\text{Re}_\Phi = \rho_l V_m D_h / \mu_l$, where ρ_l and μ_l are the density and viscosity of the liquid, V_m is the average velocity of the condensate, and D_h is the hydraulic diameter defined by $D_h = 4 \times$ condensate film cross-sectional area/wetted perimeter. For the condensation on a vertical plate, $\text{Re}_\Gamma = 4\Gamma/\mu_l$, where Γ is the mass flow rate of condensate per unit width evaluated at the lowest point on the condensing surface. The condensate flow is generally considered to be laminar for $\text{Re}_\Gamma < 1800$ and turbulent for $\text{Re}_\Gamma > 1800$. The average Nusselt number is given by[14]

$$\overline{Nu}_L = 1.13 \left[\frac{g\rho_l(\rho_l - \rho_v)h_{fg}L^3}{\mu_l k_l(T_{sat} - T_s)} \right]^{0.25} \qquad \text{for } \text{Re}_\Gamma < 1800$$

$$\overline{Nu}_L = 0.0077 \left[\frac{g\rho_l(\rho_l - \rho_v)L^3}{\mu_l^2} \right]^{1/3} \text{Re}_\Gamma^{0.4} \qquad \text{for } \text{Re}_\Gamma > 1800$$

Film Condensation on the Outside of Horizontal Tubes and Tube Banks

$$\overline{Nu}_D = 0.725 \left[\frac{g\rho_l(\rho_l - \rho_v)h_{fg}D^3}{N\mu_l k_l(T_{sat} - T_s)} \right]^{0.25}$$

where N is the number of horizontal tubes placed one above the other; $N = 1$ for a single tube.[23]

Film Condensation Inside Horizontal Tubes

For low vapor velocities such that Re_D based on the vapor velocities at the pipe inlet is less than 3500,[23]

$$\overline{Nu}_D = 0.555 \left[\frac{g\rho_l(\rho_l - \rho_v)h'_{fg}}{\mu_l k_l(T_{sat} - T_s)} D^3 \right]^{0.25}$$

where $h'_{fg} + \tfrac{3}{8}C_{p,l}(T_{sat} - T_s)$.

For higher flow rates,[24] $Re_G > 5 \times 10^4$,

$$\overline{Nu}_D = 0.0265\ Re_G^{0.8}\ Pr^{1/3}$$

where the Reynolds number $Re_G = GD/\mu_l$ is based on the equivalent mass velocity $G = G_l + G_v(\rho_l/\rho_v)^{0.5}$. The mass velocity for the liquid G_l and for vapor G_v are calculated as if each occupied the entire flow area alone.

The Effect of Noncondensable Gases

If noncondensable gas such as air is present in a vapor, even in a small amount, the heat transfer coefficient for condensation may be greatly reduced. It has been found that the presence of a few percent of air by volume in steam reduces the coefficient by 50% or more. Therefore, it is desirable in the condenser design to vent the noncondensable gases as much as possible.

43.5.3 Heat Pipes

Heat pipes are a two-phase heat transfer device that operate on a closed two-phase cycle[31] and come in a wide variety of sizes and shapes.[31,32] As shown in Fig. 43.31, they typically consist of three distinct regions, the evaporator or heat addition region, the condenser or heat rejection region, and the adiabatic or isothermal region. Heat added to the evaporator region of the container causes the working fluid in the evaporator wicking structure to be vaporized. The high temperature and corresponding high pressure in this region result in flow of the vapor to the other, cooler end of the container, where the vapor condenses, giving up its latent heat of vaporization. The capillary forces existing in the wicking structure then pump the liquid back to the evaporator section. Other similar devices, referred to as *two-phase thermosyphons,* have no wick, and utilize gravitational forces to provide the liquid return. Thus the heat pipe functions as a nearly isothermal device, adjusting the evaporation rate to accommodate a wide range of power inputs, while maintaining a relatively constant source temperature.

Transport Limitations

The transport capacity of a heat pipe is limited by several important mechanisms, including the capillary wicking, viscous, sonic, entrainment, and boiling limits. The capillary wicking limit and viscous limits deal with the pressure drops occurring in the liquid and vapor phases, respectively. The sonic limit results from the occurrence of choked flow in the vapor passage, while the entrainment limit is due to the high liquid vapor shear forces developed when the vapor passes in counter-flow over the liquid saturated wick. The boiling limit is reached when the heat flux applied in the evap-

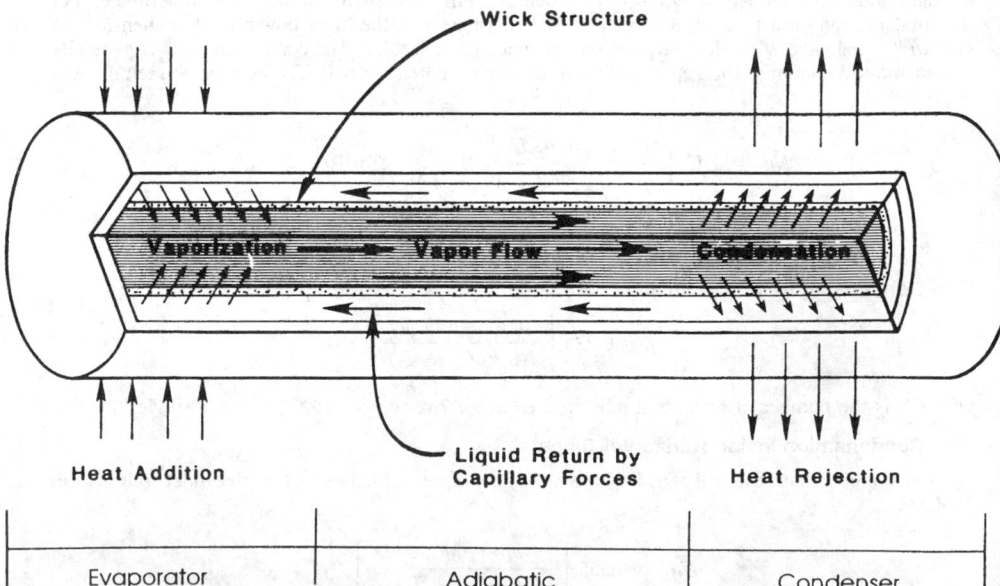

Fig. 43.31 Typical heat pipe construction and operation.[33]

orator portion is high enough that nucleate boiling occurs in the evaporator wick, creating vapor bubbles that partially block the return of fluid.

In order to function properly, the net capillary pressure difference between the condenser and the evaporator in a heat pipe must be greater than the pressure losses throughout the liquid and vapor flow paths. This relationship can be expressed as

$$\Delta P_c \geq \Delta P_+ + \Delta P_- + \Delta P_l + \Delta P_v$$

where ΔP_c = net capillary pressure difference
  ΔP_+ = normal hydrostatic pressure drop
  ΔP_- = axial hydrostatic pressure drop
  ΔP_l = viscous pressure drop occurring in the liquid phase
  ΔP_v = viscous pressure drop occurring in the vapor phase.

If these conditions are not met, the heat pipe is said to have reached the *capillary limitation*.

Expressions for each of these terms have been developed for steady-state operation, and are summarized below.

Capillary Pressure

$$\Delta P_{c,m} = \left(\frac{2\sigma}{r_{c,e}}\right)$$

Values for the effective capillary radius r_c can be found theoretically for simple geometries or experimentally for pores or structures of more complex geometry. Table 43.26 gives values for some common wicking structures.

Normal and Axial Hydrostatic Pressure Drop

$$\Delta P_+ + \rho_l g d_v \cos \psi$$
$$\Delta P_- = \rho_l g L \sin \psi$$

In a gravitational environment, the axial hydrostatic pressure term may either assist or hinder the capillary pumping process, depending upon whether the tilt of the heat pipe promotes or hinders the flow of liquid back to the evaporator (i.e., the evaporator lies either below or above the condenser). In a zero-g environment, both this term and the normal hydrostatic pressure drop term can be neglected because of the absence of body forces.

Liquid Pressure Drop

$$\Delta P_l = \left(\frac{\mu_l}{K A_w h_{fg} \rho_l}\right) L_{\text{eff}} q$$

where L_{eff} = the effective heat pipe length, defined as

Table 43.26 Expressions for the Effective Capillary Radius for Several Wick Structures

Structure	r_c	Data
Circular cylinder (artery or tunnel wick)	r	r = radius of liquid flow passage
Rectangular groove	ω	ω = groove width
Triangular groove	$\omega/\cos \beta$	ω = groove width β = half-included angle
Parallel wires	ω	ω = wire spacing
Wire screens	$(\omega + d_\omega)/2 = 1/2N$	d = wire diameter N = screen mesh number ω = wire spacing
Packed spheres	$0.41 \, r_s$	r_s = sphere radius

$$L_{\text{eff}} = 0.5L_e + L_a + 0.5L_c$$

and K is the liquid permeability as shown in Table 43.26.

Vapor Pressure Drop

$$\Delta P_v = \left(\frac{C(f_v Re_v)\mu_v}{2(r_{h,v})^2 A v \rho_v h_{fg}}\right) L_{\text{eff}}\, q$$

Although during steady-state operation the liquid flow regime is always laminar, the vapor flow may be either laminar or turbulent. It is therefore necessary to determine the vapor flow regime as a function of the heat flux. This can be accomplished by evaluating the local axial Reynolds and Mach numbers and substituting the values as shown below:

$$Re_v < 2300,\ Ma_v < 0.2$$
$$(f_v Re_v) = 16$$
$$C = 1.00$$

$$Re_v < 2300,\ Ma_v > 0.2$$
$$(f_v Re_v) = 16$$
$$C = \left[1 + \left(\frac{\gamma_v - 1}{2}\right) Ma_v^2\right]^{-1/2}$$

$$Re_v > 2300,\ M_v < 0.2$$
$$(f_v Re_v) = 0.038 \left(\frac{2(r_{h,v})q}{A_v \mu_v h_{fg}}\right)^{3/4}$$
$$C = 1.00$$

$$Re_v > 2300,\ Ma_v > 0.2$$
$$(f_v Re_v) = 0.038 \left(\frac{2(r_{h,v})q}{A_v \mu_v h_{fg}}\right)^{3/4}$$
$$C = \left[1 + \left(\frac{\gamma_v - 1}{2}\right) Ma_v^2\right]^{-1/2}$$

Since the equations used to evaluate both the Reynolds number and the Mach number are functions of the heat transport capacity, it is necessary to first assume the conditions of the vapor flow. Using these assumptions, the maximum heat transport capacity $q_{c,m}$ can be determined by substituting the values of the individual pressure drops into equation (1) and solving for $q_{c,m}$. Once the value of $q_{c,m}$ is known, it can then be substituted into the expressions for the vapor Reynolds number and Mach number to determine the accuracy of the original assumption. Using this iterative approach, accurate values for the capillary limitation as a function of the operating temperature can be determined in units of watt-m or watts for $(qL)_{c,m}$ and $q_{c,m}$ respectively.

The *viscous limitation* in heat pipes occurs when the viscous forces within the vapor region are dominant and limit the heat pipe operation. The expression

$$\frac{\Delta P_v}{Pv} < 0.1$$

can be used to determine when this limit might be of a concern. Due to the operating temperature range, this limitation will normally be of little consequence in the design of heat pipes for use in the thermal control of electronic components and devices, but may be important in liquid metal heat pipes.

The *sonic limitation* in heat pipes is analogous to the sonic limitation in a converging–diverging nozzle and can be determined from

$$q_{s,m} = A_v \rho_v h_{fg} \left(\frac{\gamma_v R_v T_v}{2(\gamma_v + 1)}\right)^{1/2}$$

where T_v is the mean vapor temperature within the heat pipe.

Since the liquid and vapor flow in opposite directions in a heat pipe, at high enough vapor velocities, liquid droplets may be picked up or entrained in the vapor flow. This entrainment results in excess liquid accumulation in the condenser and, hence, dryout of the evaporator wick. Using the Weber number, We, defined as the ratio of the viscous shear force to the force resulting from the liquid surface tension, an expression for the *entrainment limit* can be found as

$$q_{e,m} = A_v h_{fg} \left(\frac{\sigma \rho_v}{2(r_{h,w})} \right)^{1/2}$$

where $(r_{h,w})$ is the hydraulic radius of the wick structure, defined as twice the area of the wick pore at the wick–vapor interface divided by the wetted perimeter at the wick–vapor interface.

The *boiling limit* occurs when the input heat flux is so high that nucleate boiling occurs in the wicking structure and bubbles may become trapped in the wick, blocking the liquid return and resulting in evaporator dryout. This phenomenon differs from the other limitations previously discussed in that it depends on the evaporator heat flux as opposed to the axial heat flux. This expression, which is a function of the fluid properties, can be written as

$$q_{b,m} = \left(\frac{2\pi L_{eff} k_{eff} T_v}{h_{fg} \rho_v \ln(r_i/r_v)} \right) \left(\frac{2\sigma}{r_n} - \Delta P_{c,m} \right)$$

where k_{eff} is the effective thermal conductivity of the liquid–wick combination, given in Table 43.27, r_i is the inner radius of the heat pipe wall, and r_n is the nucleation site radius.

After the power level associated with each of the four limitations is established, determination of the maximum heat transport capacity is only a matter of selecting the lowest limitation for any given operating temperature.

Heat Pipe Thermal Resistance

The *heat pipe thermal resistance* can be found using an analogous electrothermal network. Figure 43.32 illustrates the electrothermal analog for the heat pipe illustrated in Fig. 43.31. As shown, the overall thermal resistance is comprised of nine different resistances arranged in a series/parallel combination, which can be summarized as follows:

R_{pe}—the radial resistance of the pipe wall at the evaporator

R_{we}—the resistance of the liquid–wick combination at the evaporator

R_{ie}—the resistance of the liquid–vapor interface at the evaporator

R_{ya}—the resistance of the adiabatic vapor section

R_{pa}—the axial resistance of the pipe wall

R_{wa}—the axial resistance of the liquid–wick combination

R_{ic}—the resistance of the liquid–vapor interface at the condenser

R_{wc}—the resistance of the liquid–wick combination at the condenser

R_{pc}—the radial resistance of the pipe wall at the condenser

Table 43.27 Wick Permeability for Several Wick Structures

Structure	K	Data
Circular cylinder (artery or tunnel wick)	$r^2/8$	r = radius of liquid flow passage
Open rectangular grooves	$2\varepsilon(r_{h,1})^2/(f_1 Re_1) = \omega/s$	ε = wick porosity ω = groove width s = groove pitch δ = groove depth $(r_{h,1}) = 2\omega\delta/(\omega + 2\delta)$
Circular annular wick	$2(r_{h,1})^2/(f_1 Re_1)$	$(r_{h,1}) = r_1 - r_2$
Wrapped screen wick	$\frac{1}{122} d_\omega^2 \varepsilon^3/(1 - \varepsilon)^2$	d_ω = wire diameter $\varepsilon = 1-(1.05\pi N d\omega/4)$ N = mesh number
Packed sphere	$\frac{1}{37.5} r_s^2 \varepsilon^3/(1 - \varepsilon)^2$	r_s = sphere radius ε = porosity (dependent on packing mode)

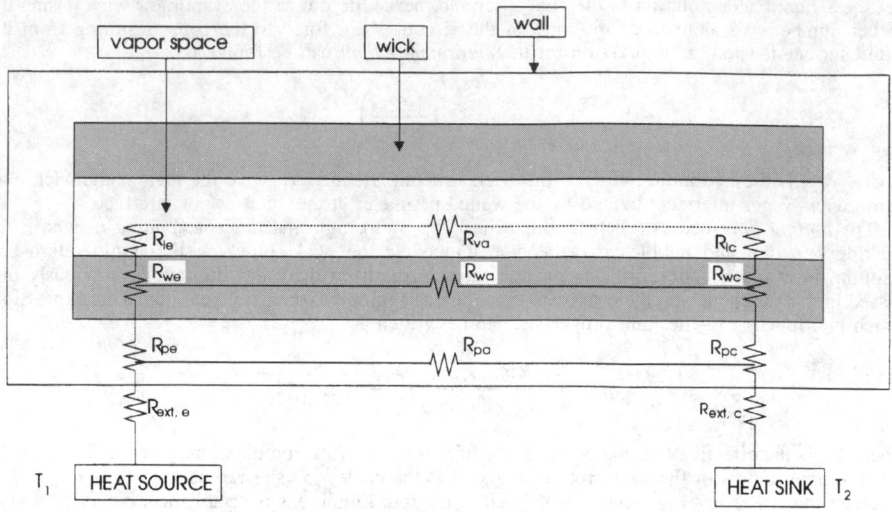

Fig. 43.32 Equivalent thermal resistance of a heat pipe.

Because of the comparative magnitudes of the resistance of the vapor space and the axial resistances of the pipe wall and liquid–wick combinations, the axial resistance of both the pipe wall and the liquid–wick combination may be treated as open circuits and neglected. Also, because of the comparative resistances, the liquid–vapor interface resistances and the axial vapor resistance can, in most situations, be assumed to be negligible. This leaves only the pipe wall radial resistances and the liquid–wick resistances at both the evaporator and condenser. The radial resistances at the pipe wall can be computed from Fourier's law as

$$R_{pe} = \frac{\delta}{k_p A_e}$$

for flat plates, where δ is the plate thickness and A_e is the evaporator area, or

$$R_{pe} = \frac{\ln(D_o/D_i)}{2\pi L_e k_p}$$

for cylindrical pipes, where L_e is the evaporator length. An expression for the equivalent thermal resistance of the liquid–wick combination in circular pipes is

$$R_{we} = \frac{\ln(D_o/D_i)}{2\pi L_e k_{\text{eff}}}$$

where values for the effective conductivity k_{eff} can be found in Table 43.27. The adiabatic vapor resistance, although usually negligible, can be found as

$$R_{va} = \frac{T_v(P_{v,e} - P_{v,c})}{\rho_v h_{fg} q}$$

where $P_{v,e}$ and $P_{v,c}$ are the vapor pressures at the evaporator and condenser. Combining these individual resistances provides a mechanism by which the overall thermal resistance can be computed and hence the temperature drop associated with various axial heat fluxes can be determined.

Table 43.28 Effective Thermal Conductivity for Liquid-Saturated Wick Structures

Wick Structures	k_{eff}
Wick and liquid in series	$\dfrac{k_l k_w}{\varepsilon k_w + k_l(1 - \varepsilon)}$
Wick and liquid in parallel	$\varepsilon k_l + k_w(1 - \varepsilon)$
Wrapped screen	$\dfrac{k_l[(k_l + k_w) - (1 - \varepsilon)(k_l - k_w)]}{(k_l + k_w) + (1 - \varepsilon)(k_l - k_w)]}$
Packed spheres[34]	$\dfrac{k_l[(2k_l + k_w) - 2(1 - \varepsilon)(k_l - k_w)]}{(2k_l + k_w) + (1 - \varepsilon)(k_l - k_w)}$
Rectangular grooves	$\dfrac{w_f k_l k_w \delta) + w k_l(0.185 \, w_f k_w + \delta k_l)}{(w + w_f)(0.185 \, w_f k_f + \delta k_l)}$

REFERENCES

1. F. P. Incropera and D. P. Dewitt, *Fundamentals of Heat Transfer*, Wiley, New York, 1981.
2. E. R. G. Eckert and R. M. Drake, Jr., *Analysis of Heat and Mass Transfer*, McGraw-Hill, NY, 1972.
3. M. P. Heisler, "Temperature Charts for Induction and Constant Temperature Heating," *Trans. ASME* **69**, 227 (1947).
4. H. Grober and S. Erk, *Fundamentals of Heat Transfer*, McGraw-Hill, New York, 1961.
5. E. N. Sieder and C. E. Tate, "Heat Transfer and Pressure Drop of Liquids in Tubes," *Ind. Eng. Chem.* **28**, 1429 (1936).
6. F. W. Dittus and L. M. K. Baelter, *Univ. Calif, Berkeley, Pub. Eng.* **2**, 443 (1930).
7. A. J. Chapman, *Heat Transfer*, Macmillan, New York, 1974.
8. S. Whitaker, *AICHE J.* **18**, 361 (1972).
9. M. Jakob, *Heat Transfer*, Vol. 1, Wiley, New York, 1949.
10. A. Zhukauska, "Heat Transfer from Tubes in Cross Flow," in *Advances in Heat Transfer*, Vol. 8, J. P. Hartnett and T. F. Irvine, Jr. (eds.), Academic, New York, 1972.
11. F. Kreith, *Principles of Heat Transfer*, Harper and Row, New York, 1973.
12. H. A. Johnson and M. W. Rubesin, "Aerodynamic Heating and Convective Heat Transfer," *Trans. ASME*, **71**, 447 (1949).
13. C. C. Lin (ed.), *Turbulent Flows and Heat Transfer, High Speed Aerodynamics and Jet Propulsion*, Vol. 5, Princeton University Press, Princeton, NJ, 1959.
14. W. H. McAdams (ed.), *Heat Transmission*, 3rd ed., McGraw-Hill, New York, 1954.
15. T. Yuge, "Experiments on Heat Transfer from Spheres Including Combined Natural and Forced Convection," *J. Heat Transfer* **82**, 214 (1960).
16. S. Globe and D. Dropkin, "Natural Convection Heat Transfer in Liquids Confined between Two Horizontal Plates," *J. Heat Transfer* **81C**, 24 (1959).
17. I. Catton, "Natural Convection in Enclosures," in *Proc. 6th International Heat Transfer Conference*, 6, Toronto, Canada, 1978.
18. R. K. MacGregor and A. P. Emery, "Free Convection through Vertical Plane Layers: Moderate and High Prandtl Number Fluids," *J. Heat Transfer* **91**, 391(1969).
19. R. Siegel and J. R. Howell, *Thermal Radiation Heat Transfer*, McGraw-Hill, New York, 1981.
20. G. G. Gubareff, J. E. Janssen, and R. H. Torborg, *Thermal Radiation Properties Survey*, 2nd ed., Minneapolis Honeywell Regulator Co., Minneapolis, MN, 1960.
21. H. C. Hottel, in *Heat Transmission*, W. C. McAdams (ed.), McGraw-Hill, New York, 1954, Chap. 2.
22. W. M. Rohsenow, "Film Condensation," in *Handbook of Heat Transfer*, W. M. Rohsenow and J. P. Hartnett (eds.), McGraw-Hill, New York, 1973.
23. J. C. Chato, "Laminar Condensation Inside Horizontal and Inclined Tubes," *J. Am. Soc. Heating Refrig. Aircond. Engrs.* **4**, 52 (1962).

24. W. W. Akers, H. A. Deans, and O. K. Crosser, "Condensing Heat Transfer Within Horizontal Tubes," *Chem. Eng. Prog., Sym. Ser.* **55** (29), 171 (1958).
25. W. M. Rohsenow, "A Method of Correlating Heat Transfer Data for Surface Boiling Liquids," *Trans. ASME* **74**, 969 (1952).
26. J. P. Holman, *Heat Transfer*, McGraw-Hill, New York, 1981.
27. N. Zuber, "On the Stability of Boiling Heat Transfer," *Trans. ASME* **80**, 711 (1958).
28. L. A. Bromley, "Heat Transfer in Stable Film Boiling," *Chem. Eng. Prog.* **46**, 221 (1950).
29. M. Jacob and G. A. Hawkins, *Elements of Heat Transfer,* Wiley, New York, 1957.
30. G. P. Peterson, *An Introduction to Heat Pipes: Modeling, Testing and Applications*, Wiley, New York, 1994.
31. G. P. Peterson, A. B. Duncan and M. H. Weichold, "Experimental Investigation of Micro Heat Pipes Fabricated in Silicon Wafers," *ASME J. Heat Transfer* **115**, 751 (1993).
32. G. P. Peterson, "Capillary Priming Characteristics of a High Capacity Dual Passage Heat Pipe," *Chemical Engineering Communications* **27**(1), 119 (1984).
33. G. P. Peterson and L. S. Fletcher, "Effective Thermal Conductivity of Sintered Heat Pipe Wicks," *AIAA J. of Thermophysics and Heat Transfer* **1**, 36 (1987).

BIBLIOGRAPHY

American Society of Heating, Refrigerating and Air Conditioning Engineering, *ASHRAE Handbook of Fundamentals,* 1972.

Arpaci, V. S., *Conduction Heat Transfer,* Addison-Wesley, Reading, MA, 1966.

Carslaw, H. S., and J. C. Jager, *Conduction of Heat in Solid,* Oxford University Press, Oxford, 1959.

Chi, S. W., *Heat Pipe Theory and Practice,* McGraw-Hill, New York, 1976.

Duffie, J. A., and W. A. Beckman, *Solar Engineering of Thermal Process*, Wiley, New York, 1980.

Dunn, P. D., and D. A. Reay, *Heat Pipes*, 3rd ed., Pergamon Press, New York, 1983.

Gebhart, B., *Heat Transfer*, McGraw-Hill, New York, 1971.

Hottel, H. C., and A. F. Saroffin, *Radiative Transfer*, McGraw-Hill, New York, 1967.

Kays, W. M., *Convective Heat and Mass Transfer,* McGraw-Hill, New York, 1966.

Knudsen, J. G., and D. L. Katz, *Fluid Dynamics and Heat Transfer*, McGraw-Hill, New York, 1958.

Ozisik, M. N., *Radiative Transfer and Interaction with Conduction and Convection,* Wiley, New York, 1973.

————, *Heat Conduction,* Wiley, New York, 1980.

Peterson, G. P., *An Introduction to Heat Pipes: Modeling, Testing and Applications*, Wiley, New York, 1994.

Planck, M., *The Theory of Heat Radiation,* Dover, New York, 1959.

Rohsenow, W. M., and H. Y. Choi, *Heat, Mass, and Momentum Transfer*, Prentice-Hall, Englewood Cliffs, NJ, 1961.

Rohsenow, W. M., and J. P. Hartnett, *Handbook of Heat Transfer,* McGraw-Hill, New York, 1973.

Schlichting, H., *Boundary-Layer Theory,* McGraw-Hill, New York, 1979.

Schneider, P. J., *Conduction Heat Transfer,* Addison-Wesley, Reading, MA, 1955.

Sparrow, E. M., and R. D. Cess, *Radiation Heat Transfer,* Wadsworth, Belmont, CA, 1966.

Tien, C. L., "Fluid Mechanics of Heat Pipes," *Ann. Rev. Fluid Mechanics*, 167 (1975).

Turner, W. C., and J. F. Malloy, *Thermal Insulation Handbook*, McGraw-Hill, New York, 1981.

Vargafik, N. B., *Table of Thermophysical Properties of Liquids and Gases*, Hemisphere, Washington, DC, 1975.

Wiebelt, J. A., *Engineering Radiation Heat Transfer*, Holt, Rinehart and Winston, New York, 1966.

CHAPTER 44

COMBUSTION

Richard J. Reed
North American Manufacturing Company
Cleveland, Ohio

44.1 FUNDAMENTALS OF
COMBUSTION   1431
    44.1.1 Air–Fuel Ratios   1431
    44.1.2 Fuels   1433

44.2 PURPOSES OF COMBUSTION 1435

44.3 BURNERS   1439
    44.3.1 Burners for Gaseous Fuels 1439
    44.3.2 Burners for Liquid Fuels   1441

44.4 SAFETY CONSIDERATIONS   1442

44.5 OXY-FUEL FIRING   1447

44.1 FUNDAMENTALS OF COMBUSTION

44.1.1 Air–Fuel Ratios

Combustion is rapid oxidation, usually for the purpose of changing chemical energy into thermal energy—heat. This energy usually comes from oxidation of carbon, hydrogen, sulfur, or compounds containing C, H, and/or S. The oxidant is usually O_2—molecular oxygen from the air.

The stoichiometry of basic chemical equation balancing permits determination of the air required to burn a fuel. For example,

$$1CH_4 + 2O_2 \rightarrow 1CO_2 + 2H_2O$$

where the units are moles or volumes; therefore, 1 ft³ of methane (CH_4) produces 1 ft³ of CO_2; or 1000 m³ CH_4 requires 2000 m³ O_2 and produces 2000 m³ H_2O. Knowing that the atomic weight of C is 12, H is 1, N is 14, O is 16, and S is 32, it is possible to use the balanced chemical equation to predict weight flow rates: 16 lb/hr CH_4 requires 64 lb/hr O_2 to burn to 44 lb/hr CO_2 and 36 lb/hr H_2O.

If the oxygen for combustion comes from air, it is necessary to know that air is 20.99% O_2 by volume and 23.20% O_2 by weight, most of the remainder being nitrogen.

It is convenient to remember the following ratios:

$$air/O_2 = 100/20.99 = 4.76 \text{ by volume}$$
$$N_2/O_2 = 3.76 \text{ by volume}$$
$$air/O_2 = 100/23.20 = 4.31 \text{ by weight}$$
$$N_2/O_2 = 3.31 \text{ by weight}$$

Rewriting the previous formula for combustion of methane,

$$1CH_4 + 2O_2 + 2(3.76)N_2 \rightarrow 1CO_2 + 2H_2O + 2(3.76)N_2$$

or

$$1CH_4 + 2(4.76)air \rightarrow 1CO_2 + 2H_2O + 2(3.76)N_2$$

Table 44.1 lists the amounts of air required for stoichiometric (quantitatively and chemically

Mechanical Engineers' Handbook, 2nd ed., Edited by Myer Kutz.
ISBN 0-471-13007-9 © 1998 John Wiley & Sons, Inc.

Table 44.1 Proper Combining Proportions for Perfect Combustion[a]

Fuel	$\dfrac{\text{vol } O_2}{\text{vol fuel}}$	$\dfrac{\text{vol air}}{\text{vol fuel}}$	$\dfrac{\text{wt } O_2}{\text{wt fuel}}$	$\dfrac{\text{wt air}}{\text{wt fuel}}$	$\dfrac{\text{ft}^3 O_2}{\text{lb fuel}}$	$\dfrac{\text{ft}^3 \text{ air}}{\text{lb fuel}}$	$\dfrac{\text{m}^2 O_2}{\text{kg fuel}}$	$\dfrac{\text{m}^3 \text{ air}}{\text{kg fuel}}$
Acetylene, C_2H_2	2.50	11.9	3.08	13.3	36.5	174	2.28	10.8
Benzene, C_6H_6	7.50	35.7	3.08	13.3	36.5	174	2.28	10.8
Butane, C_4H_{10}	6.50	31.0	3.59	15.5	42.5	203	2.65	12.6
Carbon, C	—	—	2.67	11.5	31.6	150	1.97	9.39
Carbon monoxide, CO	0.50	2.38	0.571	2.46	6.76	32.2	0.422	2.01
Ethane, C_2H_6	3.50	16.7	3.73	16.1	44.2	210	2.76	13.1
Hydrogen, H_2	0.50	2.38	8.00	34.5	94.7	451	5.92	28.2
Hydrogen sulfide, H_2S	1.50	7.15	1.41	6.08	16.7	79.5	1.04	4.97
Methane, CH_4	2.00	9.53	4.00	17.2	47.4	226	2.96	14.1
Naphthalene, $C_{10}H_8$	—	—	3.00	12.9	35.5	169	2.22	10.6
Octane, C_8H_{18}	—	—	3.51	15.1	41.6	198	2.60	12.4
Propane, C_3H_8	5.00	23.8	3.64	15.7	43.1	205	2.69	12.8
Propylene, C_3H_6	4.50	21.4	3.43	14.8	40.6	193	2.54	12.1
Sulfur, S	—	—	1.00	4.31	11.8	56.4	0.74	3.52

[a]Reproduced with permission from *Combustion Handbook*.[1] (See Ref. 1)

correct) combustion of a number of pure fuels, calculated by the above method. (Table 46.1c lists similar information for typical fuels that are mixtures of compounds, calculated by the above method, but weighted for the percentages of the various compounds in the fuels.)

The stoichiometrically correct (perfect, ideal) air/fuel ratio from the above formula is therefore $2 + 2(3.76) = 9.52$ volumes of air per volume of the fuel gas. More than that is called a "lean" ratio, and includes excess air and produces an oxidizing atmosphere. For example, if the actual air/fuel ratio were $10:1$, the %excess air would be

$$\frac{10 - 9.52}{9.52} \times 100 = 5.04\%$$

Communications problems sometimes occur because some people think in terms of air/fuel ratios, others in fuel/air ratios; some in weight ratios, others in volume ratios; and some in mixed metric units (such as normal cubic meters of air per metric tonne of coal), others in mixed American units (such as ft³ air/gal of oil). To avoid such confusions, the following method from Ref. 1 is recommended.

It is more convenient to specify air/fuel ratio in unitless terms such as %air (%aeration), %excess air, %deficiency of air, or equivalence ratio. Those experienced in this field prefer to converse in terms of %excess air. The scientific community favors equivalence ratio. The %air is easiest to use and explain to newcomers to the field: "*100% air*" is the correct (stoichiometric) amount; 200% air is twice as much as necessary, or 100% excess air. *Equivalence ratio,* widely used in combustion research, is the actual amount of fuel expressed as a fraction percent of the stoichiometrically correct amount of fuel. The Greek letter phi, ϕ, is usually used: $\phi = 0.9$ is lean; $\phi = 1.1$ is rich; and $\phi = 1.0$ is "on-ratio."

Formulas relating %air, ϕ, %excess air (%XS), and %deficiency of air (%def) are

$$\%air = 100/\phi = \%XS + 100 = 100 - \%def$$

$$\phi = \frac{100}{\%XS + 100} = \frac{1}{1 - (\%def/100)}$$

$$\%XS = \%air - 100 = \frac{1 - \phi}{\phi} \times 100$$

$$\%def = 100 - \%air = \frac{\phi - 1}{\phi} \times 100$$

Table 44.2 lists a number of equivalent terms for convenience in converting values from one "language" to another.

Excess air is undesirable, because, like N_2, it passes through the combustion process without chemical reaction; yet it absorbs heat, which it carries out the flue. The percent available heat (best possible fuel efficiency) is highest with zero excess air. (See Fig. 44.1.)

Excess fuel is even more undesirable because it means there is a deficiency of air and some of the fuel cannot be burned. This results in formation of soot and smoke. The accumulation of unburned fuel or partially burned fuel can represent an explosion hazard.

Enriching the oxygen content of the combustion "air" above the normal 20.9% reduces the nitrogen and thereby reduces the loss due to heat carried up the stack. This also raises the flame temperature, improving heat transfer, especially that by radiation.

Vitiated air (containing less than the normal 20.9% oxygen) results in less fuel efficiency, and may result in flame instability. Vitiated air is sometimes encountered in incineration of fume streams or in staged combustion, or with flue gas recirculation.

44.1.2 Fuels

Fuels used in practical industrial combustion processes have such a major effect on the combustion that they must be studied simultaneously with combustion. Fuels are covered in detail in later chapters, so the treatment here is brief, relating only to the aspects having direct bearing on the combustion process.

Gaseous fuels are generally easier to burn, handle, and control than are liquid or solid fuels. Molecular mixing of a gaseous fuel with oxygen need not wait for vaporization nor mass transport within a solid. Burning rates are limited only by mixing rates and the kinetics of the combustion reactions; therefore, combustion can be compact and intense. Reaction times as short as 0.001 sec and combustion volumes from 10^4 to 10^7 Btu/hr · ft³ are possible at atmospheric pressure.[2] Gases of low calorific value may require such large volumes of air that their combustion rates will be limited by the mixing time.

Combustion stability means that a flame lights easily and then burns steadily and reliably after the pilot (or direct spark) is programmed off. Combustion stability depends on burner geometry, plus

Table 44.2 Equivalent Ways to Express Fuel-to-Air or Air-to-Fuel Ratios[1]

	φ	%air	%def	%XS
Fuel rich	2.50	40	60	
(air lean)	1.67	60	40	
	1.25	80	20	
	1.11	90	10	
	1.05	95	5	
Stoichiometric	1.00	100	0	0
Fuel lean	0.95	105		5
(air rich)	0.91	110		10
	0.83	120		20
	0.78	130		30
	0.71	140		40
	0.62	160		60
	0.56	180		80
	0.50	200		100
	0.40	250		150
	0.33	300		200
	0.25	400		300
	0.20	500		400
	0.167	600		500
	0.091	1100		1000
	0.048	2100		2000

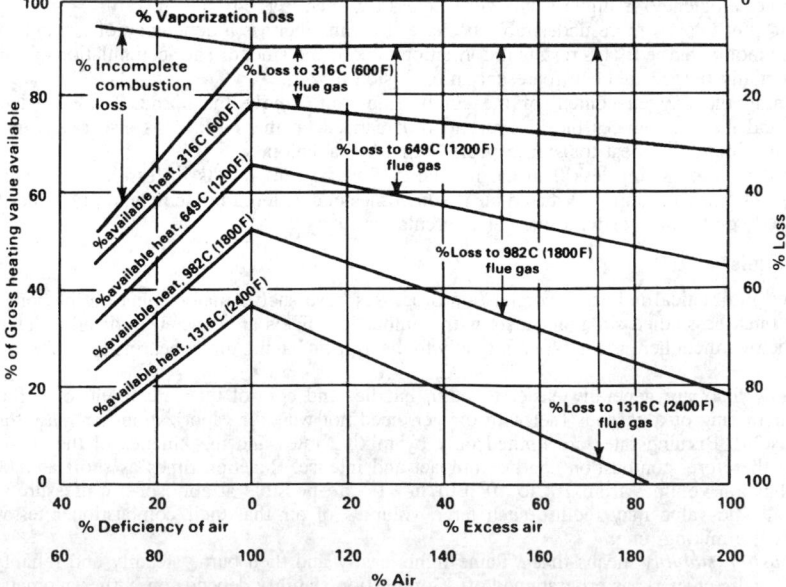

Fig. 44.1 Percent available heat (best possible efficiency) peaks at stoichiometric air/fuel ratio.[1]

air and fuel flow controls that maintain the point(s) of flame initiation (a) above the fuel's minimum ignition temperature, (b) within the fuel's flammability limits, and (c) with feed speed equal to flame speed—throughout the burner's full range of firing rates and conditions. (Fuel properties are discussed and tabulated in Chapters 46 and 47.)

Liquid fuels are usually not as easily burned, handled, or controlled as are gaseous fuels. Mixing with oxygen can occur only after the liquid fuel is evaporated; therefore, burning rates are limited by vaporization rates. In practice, combustion intensities are usually less with liquid fuels than with high calorific gaseous fuels such as natural gas.

Because vaporization is such an integral part of most liquid fuel burning processes, much of the emphasis in evaluating liquid fuel properties is on factors that relate to vaporization, including viscosity, which hinders good atomization, the primary method for enhancing vaporization. Much concern is also devoted to properties that affect storage and handling because, unlike gaseous fuels that usually come through a public utility's mains, liquid fuels must be stored and distributed by the user.

The stability properties (ignition temperature, flammability limits, and flame velocity) are not readily available for liquid fuels, but flame stability is often less critical with liquid fuels.

Solid fuels are frequently more difficult to burn, handle, and control than liquid or gaseous fuels. After initial volatilization, the combustion reaction rate depends on diffusion of oxygen into the remaining char particle, and the diffusion of carbon monoxide back to its surface, where it burns as a gas. Reaction rates are usually low and required combustion volumes high, even with pulverized solid fuels burned in suspension. Some fluidized bed and cyclone combustors have been reported to reach the intensities of gas and oil flames.[2]

Most commonly measured solid fuel properties apply to handling in stokers or pulverizers. See Chapter 48.

Wastes, by-product fuels, and gasified solids are being used more as fuel costs rise. Operations that produce such materials should attempt to consume them as energy sources. Handling problems, the lack of a steady supply, and pollution problems often complicate such fuel usage.

For the precise temperature control and uniformity required in many industrial heating processes, the burning of solids, especially the variable quality solids found in wastes, presents a critical problem. Such fuels are better left to very large combustion chambers, particularly boilers. When solids and wastes must be used as heat sources in small and accurate heating processes, a better approach is to convert them to low-Btu (producer) gas, which can be cleaned and then controlled more precisely.

44.2 PURPOSES OF COMBUSTION

The purposes of combustion, for the most part, center around elevating the temperature of something. This includes the first step in all successive combustion processes—the pilot flame—and, similarly, the initiation of incineration. Elevating the temperature of something can also make it capable of transmitting light or thermal energy (radiation and convection heat transfer), or it can cause chemical dissociation of molecules in the products of combustion to generate a special atmosphere gas for protection of materials in industrial heat processing.

All of the above functions of combustion are minor in comparison to the heating of air, water and steam, metals, nonmetallic minerals, and organics for industrial processing, and for space comfort conditioning. For all of these, it is necessary to have a workable method for evaluating the heat available from a combustion process.

Available heat is the heat accessible for the load (useful output) and to balance all losses other than stack losses. (See Fig. 44.2.) The available heat per unit of fuel is

$$AH = HHV - \text{total stack gas loss} = LHV - \text{dry stack gas loss}$$
$$\% \text{ available heat} = 100(AH/HHV)$$

where AH = available heat, HHV = higher heating value, and LHV = lower heating value, as defined in Chapter 47. Figure 44.3 shows values of % available heat for a typical natural gas; Fig. 44.4 for a typical residual oil; and Fig. 53.2 in Chapter 53, for a typical distillate oil.

Example 44.1

A process furnace is to raise the heat content of 10,000 lb/hr of a load from 0 to 470 Btu/lb in a continuous furnace (no wall storage) with a flue gas exit temperature of 1400°F. The sum of wall loss and opening loss is 70,000 Btu/hr. There is no conveyor loss. Estimate the fuel consumption using 1000 Btu/ft³ natural gas with 10% excess air.

Solution: From Fig. 44.3, % available heat = 58.5%. In other words, the flue losses are 100% − 58.5% = 41.5%. The sum of other losses and useful output = 70,000 + (10,000)(470) = 4,770,000 Btu/hr. This constitutes the "available heat" required. The required gross input is therefore 4,770,000/0.585 = 8,154,000 Btu/hr, of 8154 ft³/hr of natural gas (and about 81.540 ft³/hr of air).

The use of the above precalculated % available heats has proved to be a practical way to avoid long iterative methods for evaluating stack losses and what is therefore left for useful heat output

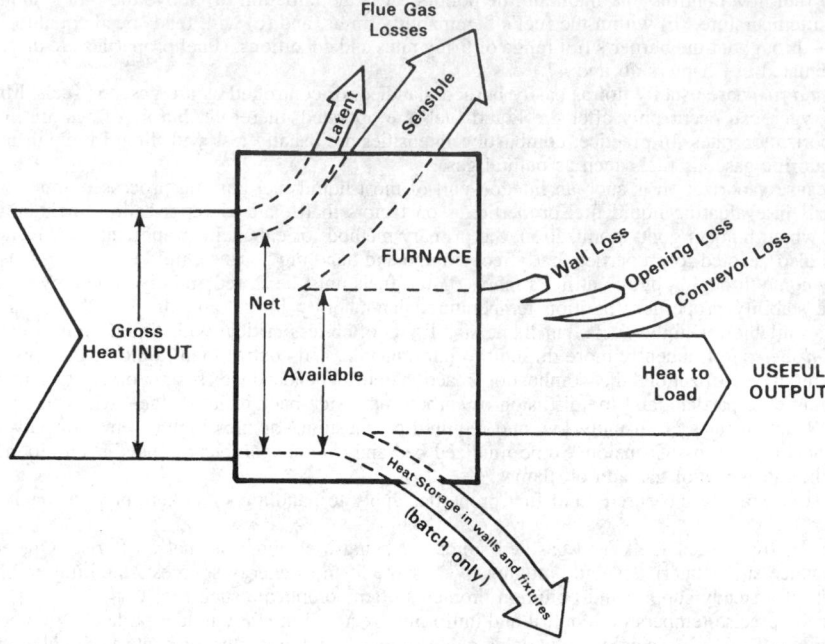

Fig. 44.2 Sankey diagram for a furnace, oven, kiln, incinerator, boiler, or heater—a qualitative and roughly quantitative method for analyzing efficiency of fuel-fired heating equipment.

and to balance other losses. For low exit gas temperatures such as encountered in boilers, ovens, and dryers, the dry stack gas loss can be estimated by assuming the total exit gas stream has the specific heat of nitrogen, which is usually a major component of the poc (products of combustion).

$$\frac{\text{dry stack loss}}{\text{unit of fuel}} = \left(\frac{\text{lb dry poc}}{\text{unit fuel}}\right)\left(\frac{0.253 \text{ Btu}}{\text{lb poc (°F)}}\right)(T_{\text{exit}} - T_{\text{in}})$$

or

$$\left(\frac{\text{scf dry proc}}{\text{unit fuel}}\right)\left(\frac{0.0187 \text{ Btu}}{\text{scf poc (°F)}}\right)(T_{\text{exit}} - T_{\text{in}})$$

For a gaseous fuel, the "unit fuel" is usually scf (standard cubic foot), where "standard" is at 29.92 in. Hg and 60°F or nm³ (normal cubic meter), where "normal" is at 1.013 bar and 15°C.

Heat transferred from combustion takes two forms: radiation and convection. Both phenomena involve transfer to a surface.

Flame radiation comes from particle radiation and gas radiation. The visible yellow-orange light normally associated with a flame is actually from solid soot or char particles in the flame, and the "working" portion of this form of heat transfer is in the infrared wavelength range. Because oils have higher C/H ratios than gaseous fuels, oil flames are usually more yellow than gas flames (although oil flames can be made blue). Gas flames can be made yellow, by a delayed-mixing burner design, for the purpose of increasing their radiating capability.

Particulate radiation follows the Stefan-Boltzmann law for solids, but depends on the concentration of particles within the flame. Estimating or measuring the particle temperature and concentration is difficult.

Gas radiation and blue flame radiation contain more ultraviolet radiation and tend to be less intense. Triatomic gases (CO_2, H_2O, and SO_2) emit radiation that is largely invisible. Gases beyond the tips of both luminous and nonluminous flames continue to emit this gas radiation. As a very broad generalization, blue or nonluminous flames tend to be hotter, smaller, and less intense radiators than luminous flames. Gas radiation depends on the concentrations (or partial pressures) of the triatomic molecules and the beam thickness of their "cloud." Their temperatures are very transient.

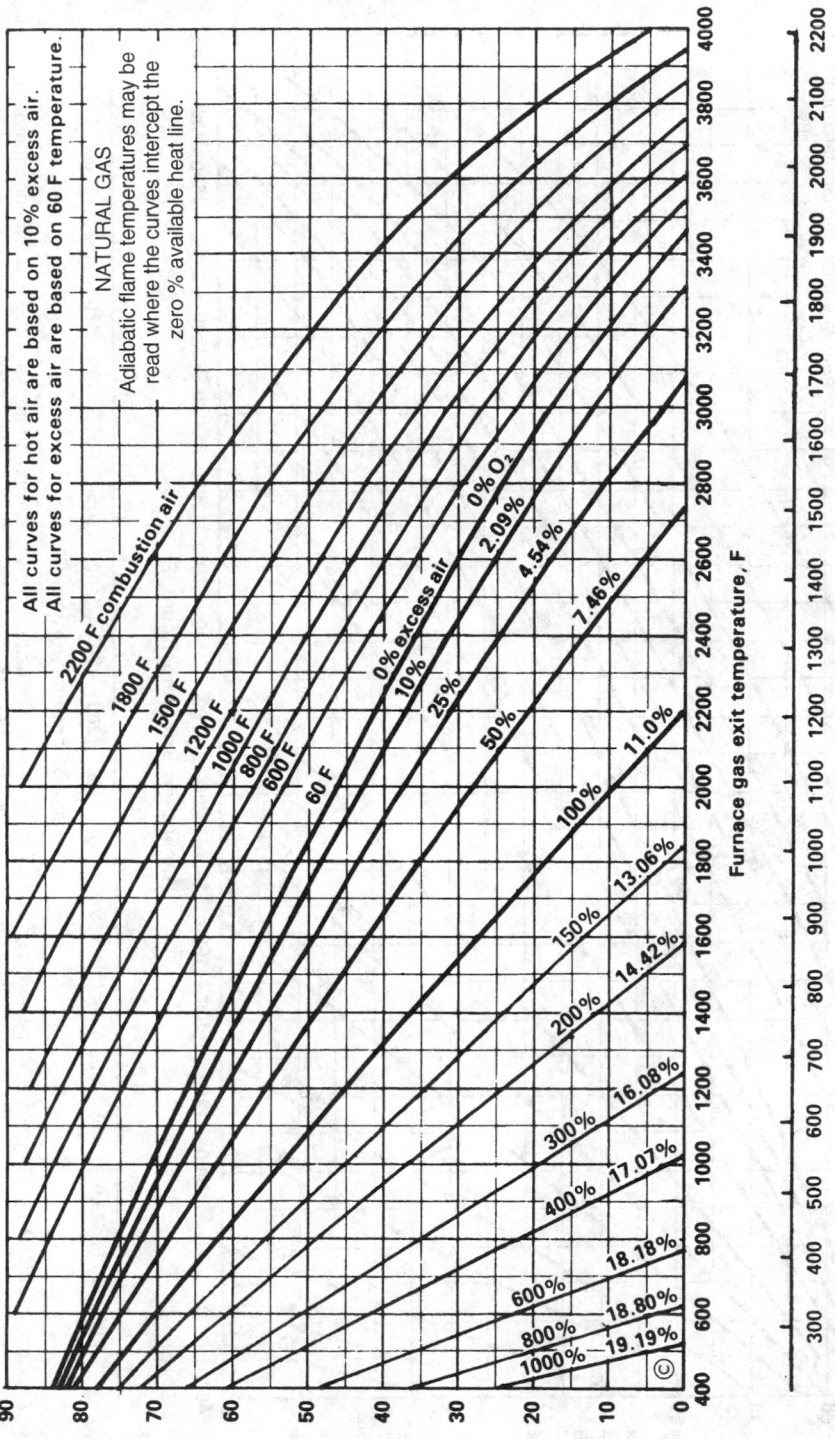

Fig. 44.3 Available heat for 1000 Btu/ft³ natural gas. Examples: In a furnace with 1600°F flue temperature, 60°F air, and 10% excess air, read that 54% of the gross heat input is available for heating the load and balancing the losses other than stack losses; and, at the x-intercept, read that the adiabatic flame temperature will be 3310°F. If the combustion air were 1200°F instead of 60°F, read that the available heat would be 77% and that the adiabatic flame temperature would be 3750°F. It is enlightening to compare this graph with Fig. 44.16 for oxy-fuel firing and oxygen enrichment.

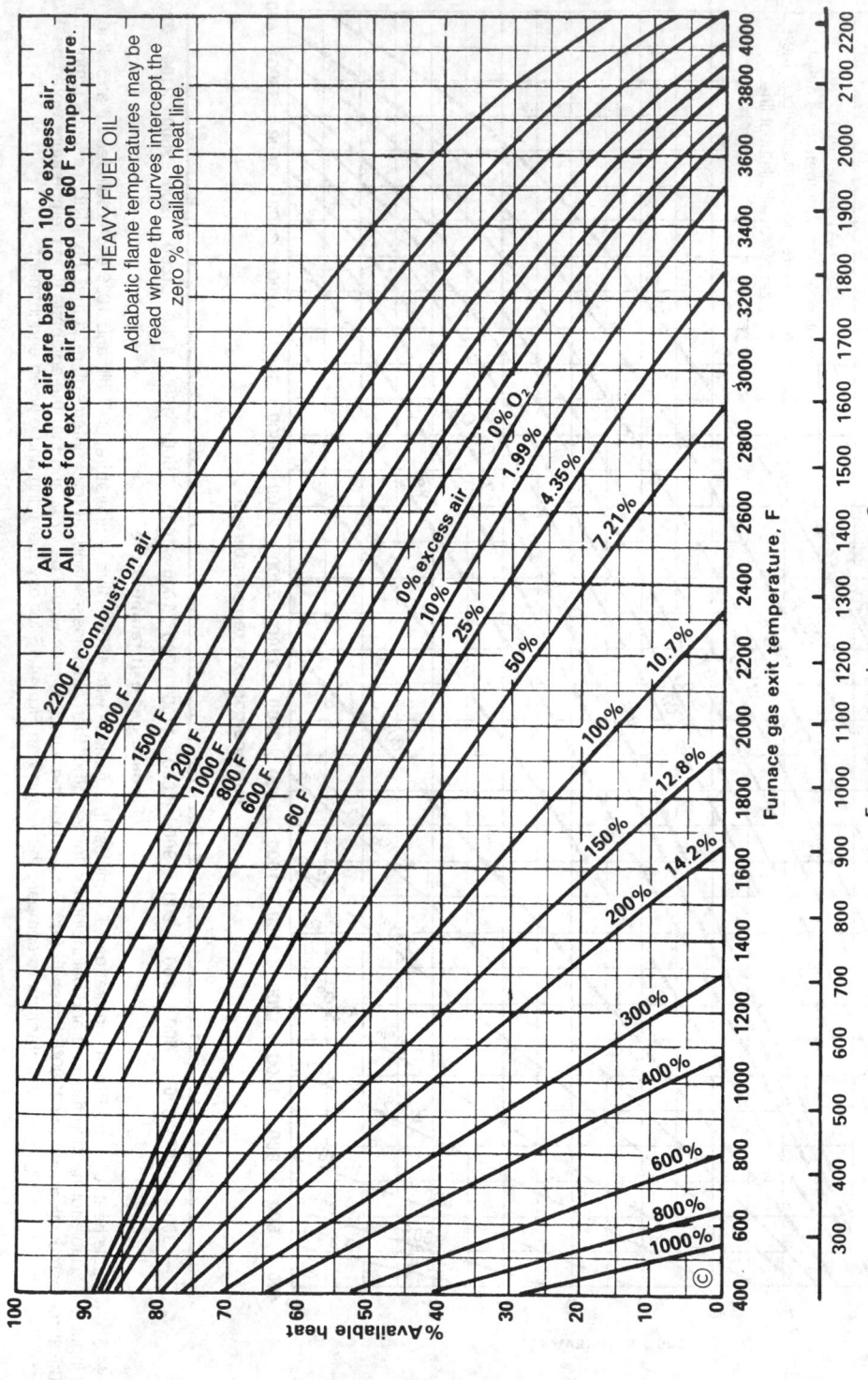

Fig. 44.4 Available heat for 153,120 gross Btu/gal residual fuel oil (heavy, No. 6). With 2200°F gases leaving a furnace, 1000°F air entering the burners, and 10% excess air, 62% of the 153,120 is available; 100% − 62% = 38% is stack loss.

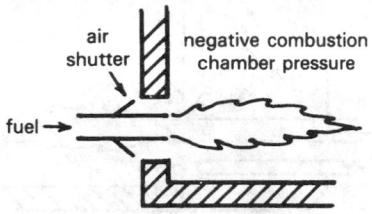

Fig. 44.5 Open, natural draft-type burner.

Convection from combustion produces beyond the flame tip follows conventional convection formulas—largely a function of velocity. This is the reason for recent emphasis on high-velocity burners. Flame convection by actual flame impingement is more difficult to evaluate because (a) flame temperatures change so rapidly and are so difficult to measure or predict, and (b) this involves extrapolating many convection formulas into ranges where good data are lacking.

Refractory radiation is a second stage of heat transfer. The refractory must first be heated by flame radiation and/or convection. A gas mantle, so-called "infrared" burners, and "radiation burners" use flame convection to heat some solid (refractory or metal) to incandescence so that they become good radiators.

44.3 BURNERS

In some cases, a burner may be nothing more than a nozzle. Some would say it includes a mixing device, windbox, fan, and controls. In some configurations, it is difficult to say where the burner ends and the combustion chamber or furnace begins. In this section, the broadest sense of the terms will generally be used.

A combustion system provides (1) fuel, (2) air, (3) mixing, (4) proportioning, (5) ignition, and (6) flame holding. In the strictest sense, a burner does only function 6; in the broadest sense, it may do any or all of these functions.

44.3.1 Burners for Gaseous Fuels

Open and natural draft-type burners rely on a negative pressure in the combustion chamber to pull in the air required for combustion, usually through adjustable shutters around the fuel nozzles. The suction in the chamber may be natural draft (chimney effect) or induced draft fans. A crude "burner" may be nothing more than a gas gun and/or atomizer inserted through a hole in the furnace wall. Fuel–air mixing may be poor, and fuel–air ratio control may be nonexistent. Retrofitting for addition of preheated combustion air is difficult. (See Fig. 44.5.)

Sealed-in and power burners have no intentional "free" air inlets around the burner, nor are there air inlets in the form of louvers in the combustion chamber wall. All air in-flow is controlled, usually by a forced draft blower or fan pushing the air through pipes or a windbox.

These burners usually have a higher air pressure drop at the burner, so air velocities are higher, enabling more through mixing and better control of flame geometry. Air flow can be measured, so automatic air–fuel ratio control is easy. (See Fig. 44.6.)

Windbox burners often consist of little more than a long atomizer and a gas gun or gas ring. These are popular for boilers and air heaters where economic reasons have dictated that the required large volumes of air be supplied at very low pressure (2–10 in. wc) (in. wc = inches of water column). Precautions are necessary to avoid fuel flowback into the windbox. (See Fig. 44.7.)

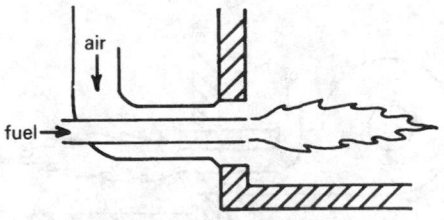

Fig. 44.6 Sealed-in, power burner.

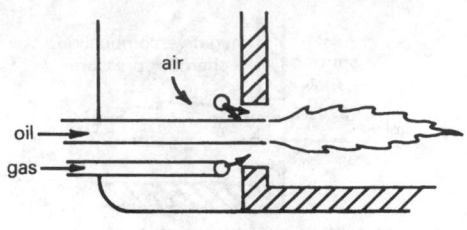

Fig. 44.7 Windbox burner.

Packaged burners usually consist of bolt-on arrangements with an integral fan and perhaps integral controls. These are widely used for new and retrofit installations from very small up to about 50×10^6 Btu/hr. (See Fig. 44.8.)

Premix burner systems may be found in any of the above configurations. Gas and air are thoroughly mixed upstream of the flame-holding nozzle. Most domestic appliances incorporate premixing, using some form of gas injector or inspirator (gas pressure inducing air through a venturi). Small industrial multiport burners of this type facilitate spreading a small amount of heat over a large area, as for heating kettles, vats, rolls, small boilers, moving webs, and low-temperature processing of conveyorized products. (See Fig. 44.9.) Large single port premix burners have been replaced by nozzle-mix burners. Better fuel–air ratio control is possible by use of aspirator mixers. (Air injection provides the energy to draw in the proper proportion of gas.) (See Fig. 44.10.) Many small units have undersized blowers, relying on furnace draft to provide secondary air. As fuel costs rise, the unwarranted excess air involved in such arrangements makes them uneconomical.

Larger than a 4-in. (100-mm) inside-diameter mixture manifold is usually considered too great an explosion risk. For this reason, mixing in a fan inlet is rarely used.

Nozzle-mix burner systems constitute the most common industrial gas burner arrangement today. Gas and air are mixed as they enter the combustion chamber through the flame holder. (See Fig. 44.11.) They permit a broad range of fuel–air ratios, a wide variety of flame shapes, and multifuel firing. A very wide range of operating conditions are now possible with stable flames, using nozzle-mix burners. For processes requiring special atmospheres, they can even operate with very rich (50% excess fuel) or lean (1500% excess air). They can be built to allow very high velocities (420,000 scfh/in.² of refractory nozzle opening) for emphasizing convection heat transfer. (See Fig. 44.12.) Others use centrifugal and coanda effects to cause the flame to scrub an adjacent refractory wall contour, thus enhancing wall radiation. (See Fig. 44.13.) By engineering the mixing configuration, nozzle-mix burner designers are able to provide a wide range of mixing rates, from a fast, intense ball of flame ($L/D = 1$) to conventional feather-shaped flame ($L/D = 5$–10) to long flames ($L/D = 20$–50). Changeable flame patterns are also possible.

Delayed-mix burners are a special form of nozzle mix, in which mixing is intentionally slow. (A raw gas torch is an unintentional form of delayed mixing.) Ignition of a fuel with a shortage of air results in polymerization or thermal cracking that forms soot particles only a few microns in diameter. These solids in the flame absorb heat and glow immediately, causing a delayed mix flame to be yellow or orange. The added luminosity enhances flame radiation heat transfer, which is one of the reasons for using delayed-mix flames. The other reason is that delayed mixing permits stretching the heat release over a great distance for uniform heating down the length of a radiant tube or a long kiln or furnace that can only be fired from one end.

Fuel-Directed Burners

Most industrial process burners have traditionally used energy from the air stream to maintain flame stability and flame shape. Now that most everyone has access to higher-pressure fuel supplies, it

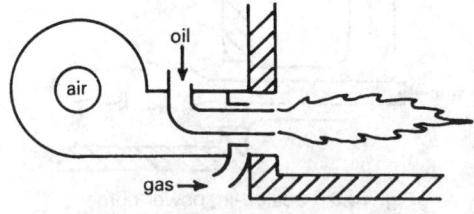

Fig. 44.8 Integral fan burner.

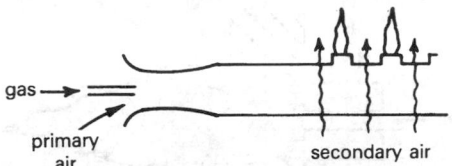

Fig. 44.9 Premix burners with inspirator mixer.

makes sense to use the energy in the fuel stream for controlling flame stability and shape, thereby permitting use of lower pressure air sources.

Figure 44.14 shows a fuel-directed burner for gas and preheated air. Multiple supply passages and outlet port positions permit changing the flame pattern during operation for optimum heat transfer during the course of a furnace cycle. Oil burners or dual-fuel combination burners can be constructed in a similar manner using two-fluid atomizers with compressed air or steam as the atomizing medium.

44.3.2 Burners for Liquid Fuels

Much of what has been said above for gas burners applies as well for oil burning. Liquids do not burn; therefore, they must be vaporized first. Kettle boiling or hot air can be used to produce a hot vapor stream that is directly substitutable for gas in premix burners. Unless there are many burners or they are very small, it is generally more practical (less maintenance) to convert to combination (dual-fuel) burners of the nozzle-mix type.

Vaporization by Atomization

Almost all industrial liquid fuel burners use atomization to aid vaporization by exposing the large surface area (relative to volume) of millions of droplets in the size range of 100–400 μm. Mass transfer then occurs at a rapid rate even if the droplets are not exposed to furnace radiation or hot air.

Pressure atomization (as with a garden hose) uses the pressure energy in the liquid steam to cause the kinetic energy to overcome viscous and surface tension forces. If input is turned down by reducing fuel pressure, however, atomizing quality suffers; therefore, this method of atomization is limited to on–off units or cases where more than 250 psi fuel pressure is available.

Two-fluid atomization is the method most commonly used in industrial burners. Viscous friction by a high-velocity second fluid surrounding the liquid fuel stream literally tears it into droplets. The second fluid may be low-pressure air (<2 psi, or <13.8 kPa), compressed air, gaseous fuel, or steam. Many patented atomizer designs exist—for a variety of spray angles, sizes, turndown ranges, droplet sizes. Emulsion mixing usually gives superior atomization (uniformly small drops with relatively small consumption of atomizing medium) but control is complicated by interaction of the pressures and flows of the two streams. External mixing is just the opposite. A compromise called tip-emulsion atomization is the current state of the art.

Rotary-cup atomization delivers the liquid fuel to the center of a fast spinning cup surrounded by an air stream. Rotational speed and air pressure determine the spray angle. This is still used in some large boilers, but the moving parts near the furnace heat have proved to be too much of a maintenance problem in higher-temperature process furnaces and on smaller installations where a strict preventive maintenance program could not be effected.

Sonic and ultrasonic atomization systems create very fine drops, but impart very little motion to them. For this reason they do not work well with conventional burner configurations, but require an all new design.

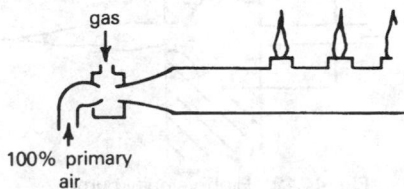

Fig. 44.10 Premix burners with aspirator mixer.

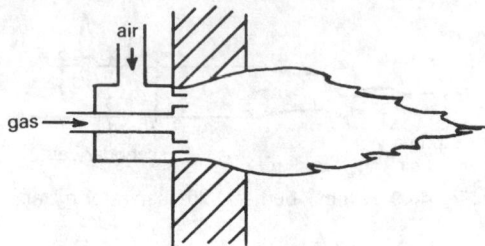

Fig. 44.11   Air-directed nozzle-mix burner.

Liquid Fuel Conditioning

A variety of additives can be used to reduce fuel degeneration in storage, minimize slagging, lessen surface tension, reduce pollution, and lower the dew point. Regular tank draining and cleaning and the use of filters are recommended.

Residual oils must be heated to reduce their viscosity for pumping, usually to 500 SSU (100 cSt). For effective atomization, burner manufacturers specify viscosities in the range of 100–150 SSU (22–32 cSt). In all but tropic climates, blended oils (Nos. 4 and 5) also require heating. In Arctic situations, distillate oils need heating. Figure 44.15 enables one to predict the oil temperature necessary for a specified viscosity. It is best, however, to install extra heating capacity because delivered oil quality may change.

Oil heaters can be steam or electric. If oil flow stops, the oil may vaporize or char. Either reduces heat transfer from the heater surfaces, which can lead to catastrophic failure in electric heaters. Oil must be circulated through heaters, and the system must be fitted with protective limit controls.

Hot oil lines must be insulated and traced with steam, induction, or resistance heating. The purpose of tracing is to balance heat loss to the environment. Rarely will a tracing system have enough capacity to heat up an oil line from cold. When systems are shut down, arrangements must be made to purge the heavy oil from the lines with steam, air, or (preferably) distillate oil. Oil standby systems should be operated regularly, whether needed or not. In cold climates, they should be started before the onset of cold weather and kept circulating all winter.

44.4 SAFETY CONSIDERATIONS

Operations involving combustion must be concerned about all the usual safety hazards of industrial machinery, plus explosions, fires, burns from hot surfaces, and asphyxiation. Less immediately severe, but long-range health problems related to combustion result from overexposure to noise and pollutants.

Preventing explosions should be *the* primary operating and design concern of *every* person in any way associated with combustion operations, because an explosion can be so devastating as to eliminate *all* other goals of anyone involved. The requirements for an explosion include the first five requirements for combustion (Table 44.3); therefore, striving to do a good job of combustion may set you up for an explosion. The statistical probability of having all seven explosion requirements at the same time and place is so small that people become careless, and therein lies the problem. Continuing training and retraining is the only answer.

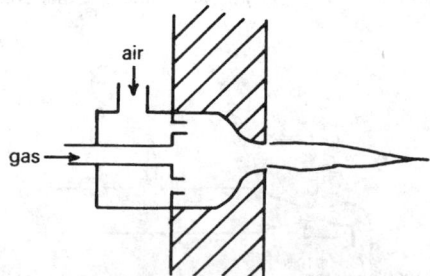

Fig. 44.12   High-velocity burner.

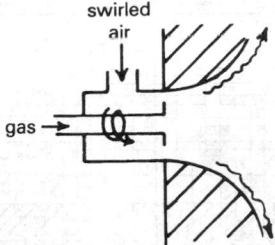

Fig. 44.13 Wall radiating burner (flat flame or coanda type).

The lower and upper limits of flammability are the same as the lower and upper explosive limits for any combustible gas or vapor. Table 46.3 lists these values for gases. Table 44.4 lists similar information for some common liquids. References 3, 4, and 5 list explosion-related data for many industrial solvents and off-gases.

Electronic safety control programs for most industrial combustion systems are generally designed (a) to prevent accumulation of unburned fuel when any source of ignition is present or (b) to immediately remove any source of ignition when something goes wrong, causing fuel accumulation. Of course, this is impossible in a furnace operating above 1400°F. If a burner in such a furnace should snuff out because it happened to go too rich, requirement number 3 is negated and there can be no explosion until someone (untrained) opens a port or shuts off the fuel. The only safe procedure is to gradually flood the chamber with steam or inert gas (gradually, so as not to change furnace pressure and thereby cause more air in-flow).

(a) The best way to prevent unburned fuel accumulation is to have a reliable automatic fuel–air ratio control system coordinated with automatic furnace pressure control and with input control so that input cannot range beyond the capabilities of either automatic system. The emergency back-up system consists of a trip valve that stops fuel flow in the event of flame failure or any of many other interlocks such as low air flow, or high or low fuel flow.

(b) Removal of ignition sources is implemented by automatic shutoff of other burner flames, pilot flames, spark igniters, and glow plugs. In systems where a single flame sensor monitors either main flame or pilot flame, the pilot flame must be programmed out when the main flame is proven. If this is not done, such a "constant" or "standing" pilot can "fool" the flame sensor and cause an explosion.

Most codes and insuring authorities insist on use of flame monitoring devices for combustion chambers that operate at temperatures below 1400°F. Some of these authorities point out that even high-temperature furnaces must go through this low-temperature range on their way to and from their

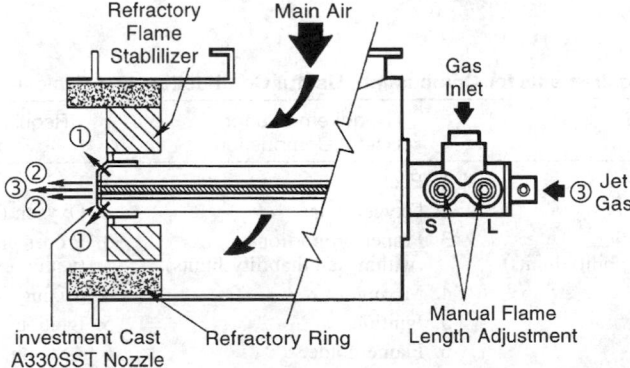

Fig. 44.14 Low NO$_x$ fuel-directed gas burner for use with preheated air. ① Increasing tangential gas flow (adjustment screw S) shortens flame. ② Increasing forward gas flow (adjustment screw L) lengthens flame. ③ Jet gas—to maintain flame definition as input is reduced. (Courtesy of North American Mfg. Company.)

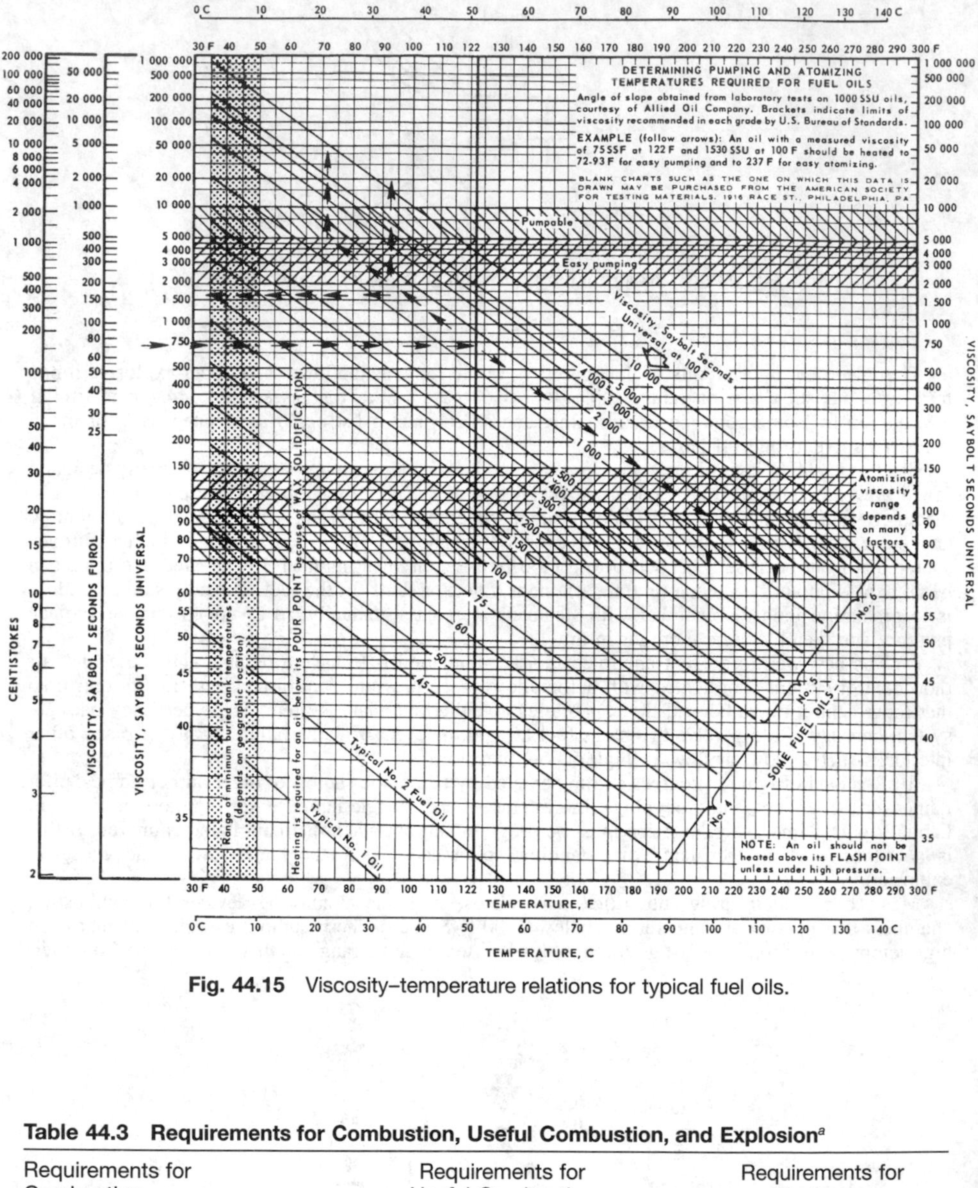

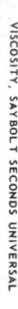

Fig. 44.15 Viscosity–temperature relations for typical fuel oils.

Table 44.3   Requirements for Combustion, Useful Combustion, and Explosion[a]

Requirements for Combustion	Requirements for Useful Combustion	Requirements for Explosion
1. Fuel	1. Fuel	1. Fuel
2. Oxygen (air)	2. Oxygen (air)	2. Oxygen (air)
3. Proper proportion (within flammability limits)	3. Proper proportion (within flammability limits)	3. Proper proportion (within explosive limits)
4. Mixing	4. Mixing	4. Mixing
5. Ignition	5. Ignition	5. Ignition
	6. Flame holder	6. Accumulation
		7. Confinement

[a]There have been incidents of disastrous explosions of unconfined fast-burning gases, but most of the damage from industrial explosions comes from the fragments of the containing furnace that are propelled like shrapnel. Lightup explosions are often only "puffs" if large doors are kept open during startup.

Table 44.4 Flammability Data for Liquid Fuels

Liquid Fuel	Flash Point, °F (°C) (Closed Cup Method)	Flammability Limits (%) Volume in Air		Autoignition Temperature, °F (°C)	Vapor density, G(air = 1)	Boiling Temperature, °F (°C)
		Lower	Upper			
Butane, -n	−76 (−60)	1.9	8.5	761 (405)	2.06	31 (−1)
Butane, -iso	−117 (−83)	1.8	8.4	864 (462)	2.06	11 (−12)
Ethyl alcohol (ethanol)	55 (13)	3.5	19	737 (392)	1.59	173 (78)
Ethyl alcohol, 30% in water	85 (29)	3.6	10	—	—	203 (95)
Fuel oil, #1	114–185 (46–85)	0.6	5.6	445–560 (229–293)	—	340–555 (171–291)
Fuel oil (diesel), #1-D	>100 (>38)	1.3	6.0	350–625 (177–329)	—	<590 (<310)
Fuel oil, #2	126–230 (52–110)	—	—	500–705 (260–374)	—	340–640 (171–338)
Fuel oil (diesel), #2-D	>100 (>38)	1.3	6.0	490–545 (254–285)	—	380–650 (193–343)
Fuel oil, #4	154–240 (68–116)	1	5	505 (263)	—	425–760 (218–404)
Fuel oil, #5	130–310 (54–154)	1	5	—	—	—
Fuel oil, #6	150–430 (66–221)	1	5	765 (407)	—	—
Gasoline, automotive	−50± (−46±)	1.3–1.4	6.0–7.6	700 (371)	3–4	91–403 (33–206)
Gasoline, aviation	−50± (−46±)	1	6.0–7.6	800–880 (427–471)	3–4	107–319 (42–159)
Jet fuel, JP-4	−2 (−19)	0.8	6.2	468 (242)	—	140–490 (60–254)
Jet fuel, JP-5	105 (41)	0.6	4.6	400 (204)	—	370–530 (188–277)
Jet fuel, JP-6	127 (53)	—	—	500 (260)	—	250–500 (121–260)
Kerosene	110–130 (43–54)	0.6	5.6	440–560 (227–293)	4.5	350–550 (177–288)
Methyl alcohol (methanol)	54 (12)	5.5	36.5	878 (470)	1.11	147 (64)
Methyl alcohol, 30% in water	75 (24)	—	—	—	—	167 (75)
Naphtha, dryclean	100–110 (38–43)	0.8	5.0	440–500 (227–260)	—	300–400 (149–204)
Naphtha, 76% vm&p	20–45 (−7–+7)	0.9	6.0	450–500 (232–260)	3.75	200–300 (93–149)
Nonane, -n	88 (31)	0.74	2.9	403 (206)	4.41	303 (151)
Octane, -iso	10 (−12)	1.0	6.0	784 (418)	3.93	190–250 (88–121)
Propane	−156 (−104)	2.2	9.6	871 (466)	1.56	−44 (−42)
Propylene	−162 (−108)	2.0	11.1	927–952 (497–511)	1.49	−54 (−48)

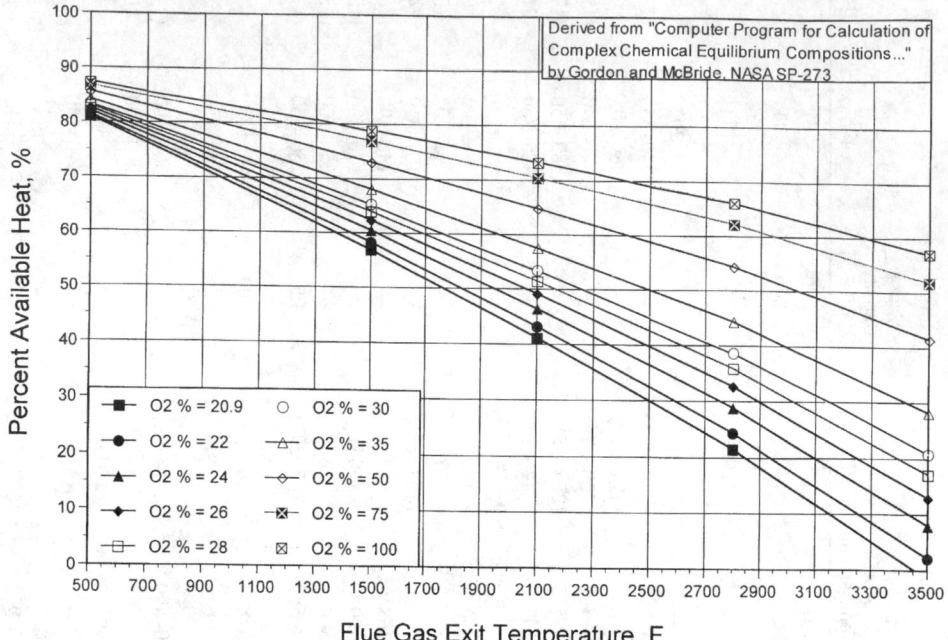

Fig. 44.16 Percents available heat for Phi = .95 or 5% excess air with average natural gas, HHV = 1025 Btu/cu ft, with various degrees of oxygen enrichment and for oxy-fuel firing (top curve).

normal operating temperature. Another situation where safety regulations and economic reality have not yet come to agreement involves combustion chambers with dozens or even hundreds of burners, such as refinery heaters, ceramic kilns, and heat-transfer furnaces.

Avoiding fuel-fed fires first requires preventing explosions, which often start such fires. (See previous discussion.) Every building containing a fuel-fired boiler, oven, kiln, furnace, heater, or incinerator should have a spring-operated manual reset fuel shutoff valve outside the building with panic buttons at the guardhouse or at exits to allow shutting off fuel as one leaves the burning building. Gas-fuel lines should be overhead where crane and truck operators cannot rupture them, or underground. If underground, keep and use records of their locations to avoid digging accidents. Overhead fuel lines must have well-marked manual shutoff valves where they can be reached without a ladder. Liquid-fuel lines should be underground; otherwise, a rupture will pour or spray fuel on a fire.

The greatest contributor to fuel-fed fires is the fuel shutoff valve that will not work. All such valves, manual or automatic, must be tested on a regular maintenance schedule. Such testing may cause nuisance shutdowns of related equipment; therefore, a practical procedure is to have the maintenance crew do the day's end (or week's end) shutdown about once each month. The same test can check for leaking. If it is a fully automatic or manual-reset automatic valve, shutdown should be accomplished by simulating flame failure and, in succession, each of the interlocks.

Because so much depends on automatic fuel shutoff valves, it makes sense (a) to have a backup valve (blocking valve), and (b) to replace it before it hangs up—at least every 5 years, more often in adverse environments.

Maintenance and management people must be ever alert for open side panels and covers on safety switches and fuel shutoff valves. Remove wedges, wires, or blocks that are holding circuits closed or valves open. Remove jumper wires unless the manufacturer's wiring diagram specifies that they be left in. Eliminate valve by-passes unless they contain a similar valve. Then, get to the root of the problem by finding the cause of the nuisance that caused someone to try to bypass the safety system.

Storage of LP gas, oils, or solid fuels requires careful attention to applicable codes. If the point of use, an open line, or a large leak is below the oil storage elevation, large quantities may siphon out and flow into an area where there is a water heater or other source of ignition. Steam heaters in heavy oil tanks need regular inspections, or leaks can emulsify the oil, causing an overflow. It is advisable to make provision for withdrawing the heater for repair without having to drain the whole tank.

LP gas is heavier than air. Workers have been suffocated by this invisible gas when it leaked into access pits below equipment.

Codes and regulations are proliferating, many by local authorities or insuring groups. Most refer, as a base, to publications of the NFPA, the National Fire Protection Association, Batterymarch Park, Quincy, MA 02269. Their publications usually represent the consensus of technically competent volunteer committees from industries involved with the topic.

44.5 OXY-FUEL FIRING

Commercially "pure" oxygen (90–99% oxygen) is sometimes substituted for air (20.9% oxygen), (a) to achieve higher flame temperature, (b) to get higher % available heat (best possible efficiency), or (c) to try to eliminate nitrogen from the furnace atmosphere, and thereby reduce the probability of NO_x pollution.

Figure 44.16 shows percents available heat and adiabatic flame temperatures (x-intercepts) for various amounts of oxygen enrichment of an existing air supply, and for "oxy-fuel firing". The latter is with 100% oxygen, and is the only way to use oxygen to reduce NO_x; oxygen enrichment usually causes more NO_x.

Oxy-fuel burners have been water-cooled in the past, but their propensity to spring leaks and do terrible damage has led to use of better materials to avoid water cooling. Oxygen burner nozzles and tiles are subject to much higher temperatures and more oxidizing atmospheres than are air burner nozzles and tiles. Control valves, regulators, and piping for oxygen require special cleaning and material selection.

REFERENCES

1. R. J. Reed (ed.), *Combustion Handbook,* Vol. I, 3rd Edition. North American Mfg. Co., Cleveland, OH, 1986, pp. 5, 68, 255.
2. R. H. Essenhigh, "An Introduction to Stirred Reactor Theory Applied to Design of Combustion Chambers," in *Combustion Technology,* H. B. Palmer and J. M. Beer (eds.), Academic, New York, 1974, pp. 389–391.
3. National Fire Protection Association, NFPA No. 86, *Standard for Ovens and Furnaces,* National Fire Protection Association, Quincy, MA, 1995.
4. National Fire Protection Association, *Flash Point Index of Trade Name Ligands,* National Fire Protection Association, Quincy, MA, 1978.
5. National Fire Protection Association, NFPA No. 325M, *Fire Hazard Properties of Flammable Liquids, Gases, Volatile Solids,* National Fire Protection Association, Quincy, MA, 1994.
6. Factory Mutual Engineering Corp., *Handbook of Industrial Loss Prevention,* McGraw-Hill, New York, 1967.

CHAPTER 45

FURNACES

Carroll Cone
Toledo, Ohio

45.1	**SCOPE AND INTENT**	**1450**		**45.9**	**FLUID FLOW**	**1485**
					45.9.1 Preferred Velocities	1485
45.2	**STANDARD CONDITIONS**	**1450**			45.9.2 Centrifugal Fan	
	45.2.1 Probable Errors	1450			Characteristics	1486
					45.9.3 Laminar and Turbulent	
45.3	**FURNACE TYPES**	**1450**			Flows	1487
45.4	**FURNACE CONSTRUCTION**	**1453**		**45.10**	**BURNER AND CONTROL EQUIPMENT**	**1488**
45.5	**FUELS AND COMBUSTION**	**1454**			45.10.1 Burner Types	1489
					45.10.2 Burner Ports	1494
45.6	**OXYGEN ENRICHMENT OF COMBUSTION AIR**	**1459**			45.10.3 Combustion Control Equipment	1494
					45.10.4 Air Pollution Control	1496
45.7	**THERMAL PROPERTIES OF MATERIALS**	**1460**		**45.11**	**WASTE HEAT RECOVERY SYSTEMS**	**1496**
45.8	**HEAT TRANSFER**	**1462**			45.11.1 Regenerative Air Preheating	1496
	45.8.1 Solid-State Radiation	1464			45.11.2 Recuperator Systems	1497
	45.8.2 Emissivity–Absorptivity	1465			45.11.3 Recuperator Combinations	1498
	45.8.3 Radiation Charts	1465				
	45.8.4 View Factors for Solid-State Radiation	1465		**45.12**	**FURNACE COMPONENTS IN COMPLEX THERMAL PROCESSES**	**1499**
	45.8.5 Gas Radiation	1466				
	45.8.6 Evaluation of Mean Emissivity–Absorptivity	1471		**45.13**	**FURNACE CAPACITY**	**1501**
	45.8.7 Combined Radiation Factors	1472		**45.14**	**FURNACE TEMPERATURE PROFILES**	**1501**
	45.8.8 Steady-State Conduction	1472				
	45.8.9 Non-Steady-State Conduction	1474		**45.15**	**REPRESENTATIVE HEATING RATES**	**1501**
	45.8.10 Heat Transfer with Negligible Load Thermal Resistance	1477		**45.16**	**SELECTING NUMBER OF FURNACE MODULES**	**1502**
	45.8.11 Newman Method	1477				
	45.8.12 Furnace Temperature Profiles	1479		**45.17**	**FURNACE ECONOMICS**	**1502**
	45.8.13 Equivalent Furnace Temperature Profiles	1480			45.17.1 Operating Schedule	1503
	45.8.14 Convection Heat Transfer	1481			45.17.2 Investment in Fuel-Saving Improvements	1503
	45.8.15 Fluidized-Bed Heat Transfer	1483				
	45.8.16 Combined Heat-Transfer Coefficients	1483				

Mechanical Engineers' Handbook, 2nd ed., Edited by Myer Kutz.
ISBN 0-471-13007-9 © 1998 John Wiley & Sons, Inc.

45.1 SCOPE AND INTENT

This chapter has been prepared for the use of engineers with access to an electronic calculator and to standard engineering reference books, but not necessarily to a computer terminal. The intent is to provide information needed for the solution of furnace engineering problems in areas of design, performance analysis, construction and operating cost estimates, and improvement programs.

In selecting charts and formulas for problem solutions, some allowance has been made for probable error, where errors in calculations will be minor compared with errors in the assumptions on which calculations are based. Conscientious engineers are inclined to carry calculations to a far greater degree of accuracy than can be justified by probable errors in data assumed. Approximations have accordingly been allowed to save time and effort without adding to probable margins for error. The symbols and abbreviations used in this chapter are given in Table 45.1.

45.2 STANDARD CONDITIONS

Assuming that the user will be using English rather than metric units, calculations have been based on pounds, feet, Btu's, and degrees Fahrenheit, with conversion to metric units provided in the following text (see Table 45.2).

Assumed standard conditions include: ambient temperature for initial temperature of loads, for heat losses from furnace walls or open cooling of furnace loads—70°F.

Condition of air entering system for combustion or convection cooling: temperature, 70°F; absolute pressure, 14.7 psia; relative humidity, 60% at 70°F, for a water vapor content of about 1.4% by volume.

45.2.1 Probable Errors

Conscientious furnace engineers are inclined to carry calculations to a far greater degree of accuracy than can be justified by uncertainties in basic assumptions such as thermal properties of materials, system temperatures and pressures, radiation view factors and convection coefficients. Calculation procedures recommended in this chapter will, accordingly, include some approximations, identified in the text, that will result in probable errors much smaller than those introduced by basic assumptions, where such approximations will expedite problem solutions.

45.3 FURNACE TYPES

Furnaces may be grouped into two general types:

1. As a source of energy to be used elsewhere, as in firing steam boilers to supply process steam, or steam for electric power generation, or for space heating of buildings or open space
2. As a source of energy for industrial processes, other than for electric power

The primary concern of this chapter will be the design, operation, and economics of industrial furnaces, which may be classified in several ways:

By function:

Heating for forming in solid state (rolling, forging)

Melting metals or glass

Heat treatment to improve physical properties

Preheating for high-temperature coating processes, galvanizing, vitreous enameling, other coatings

Smelting for reduction of metallic ores

Firing of ceramic materials

Incineration

By method of load handling:

Batch furnaces for cyclic heating, including forge furnaces arranged to heat one end of a bar or billet inserted through a wall opening, side door, stationary-hearth-type car bottom designs

Continuous furnaces with loads pushed through or carried by a conveyor

Tilting-type furnace

To avoid the problem of door warpage or leakage in large batch-type furnaces, the furnace can be a refractory-lined box with an associated firing system, mounted above a stationary hearth, and arranged to be tilted around one edge of the hearth for loading and unloading by manual handling, forklift trucks, or overhead crane manipulators.

Table 45.1 Symbols and Abbreviations

A	area in ft^2
a	absorptivity for radiation, as fraction of black body factor for receiver temperature:
	a_g combustion gases
	a_w furnace walls
	a_s load surface
	a_m combined emissivity–absorptivity factor for source and receiver
C	specific heat in Btu/lb · °F or cal/g · °C
cfm	cubic feet per minute
D	diameter in ft or thermal diffusivity (k/dC)
d	density in lb/ft^3
e	emissivity for radiation as fraction of black-body factor for source temperature, with subscripts as for a above
F	factor in equations as defined in text
fpm	velocity in ft/min
G	mass velocity in lb/ft^2 · hr
g	acceleration by gravity (32.16 ft/sec^2)
H	heat-transfer coefficient (Btu/hr · ft^2 · °F)
	H_r for radiation
	H_c for convection
	H_t for combined $H_r + H_c$
HHV	higher heating value of fuel
h	pressure head in units as defined
k	thermal conductivity (Btu/hr · ft · °F)
L	length in ft, as in effective beam length for radiation, decimal rather than feet and inches
LHV	lower heating value of fuel
ln	logarithm to base e
MTD	log mean temperature difference
N	a constant as defined in text
psi	pressure in lb/in^2
	psig, pressure above atmospheric
	psia, absolute pressure
Pr	Prandtl number ($\mu C/k$)
Q	heat flux in Btu/hr
R	thermal resistance (r/k) or ratio of external to internal thermal resistance (k/rH)
Re	Reynolds number (DG/μ)
r	radius or depth of heat penetration in ft
T	temperature in °F, except for radiation calculations where °S = (°F + 460)/100
	T_g, combustion gas temperature
	T_w, furnace wall temperature
	T_s, heated load surface
	T_c, core or unheated surface of load
t	time in hr
μ	viscosity in lb/hr · ft
wc	inches of water column as a measure of pressure
V	volume in ft^3
v	velocity in ft/sec
W	weight in lb
X	time factor for nonsteady heat transfer (tD/r^2)
x	horizontal coordinate
y	vertical coordinate
z	coordinate perpendicular to plane xy

For handling heavy loads by overhead crane, without door problems, the furnace can be a portable cover unit with integral firing and temperature control. Consider a cover-type furnace for annealing steel strip coils in a controlled atmosphere. The load is a stack of coils with a common vertical axis, surrounded by a protective inner cover and an external heating cover. To improve heat transfer parallel to coil laminations, they are loaded with open coil separators between them, with heat transferred from the inner cover to coil ends by a recirculating fan. To start the cooling cycle, the heating cover

Table 45.2 Conversion of Metric to English Units

Length	1 m = 3.281 ft
	1 cm = 0.394 in
Area	1 m^2 = 10.765 ft^2
Volume	1 m^3 = 35.32 ft^3
Weight	1 kg = 2.205 lb
Density	1 g/cm^3 = 62.43 lb/ft^2
Pressure	1 g/cm^2 = 2.048 lb/ft^2 = 0.0142 psi
Heat	1 kcal = 3.968 Btu
	1 kwh = 3413 Btu
Heat content	1 cal/g = 1.8 Btu/lb
	1 kcal/m^2 = 0.1123 Btu/ft^3
Heat flux	1 W/cm^2 = 3170 Btu/hr · ft^2
Thermal conductivity	$\dfrac{1 \text{ cal}}{\text{sec cm °C}} = \dfrac{242 \text{ Btu}}{\text{hr ft °F}}$
Heat transfer	$\dfrac{1 \text{ cal}}{\text{sec cm}^2 \text{ °C}} = \dfrac{7373 \text{ Btu}}{\text{hr ft}^2 \text{ °F}}$
Thermal diffusivity	$\dfrac{1 \text{ cal/sec} \cdot \text{cm} \cdot \text{°C}}{C \cdot \text{g/cm}^3} = \dfrac{3.874 \text{ Btu/hr} \cdot \text{ft} \cdot \text{°F}}{C \cdot \text{lb/ft}^3}$

is removed by an overhead crane, while atmosphere circulation by the base fan continues. Cooling may be enhanced by air-blast cooling of the inner cover surface.

For heating heavy loads of other types, such as weldments, castings, or forgings, car bottom furnaces may be used with some associated door maintenance problems. The furnace hearth is a movable car, to allow load handling by an overhead traveling crane. In one type of furnace, the door is suspended from a lifting mechanism. To avoid interference with an overhead crane, and to achieve some economy in construction, the door may be mounted on one end of the car and opened as the car is withdrawn. This arrangement may impose some handicaps in access for loading and unloading.

Loads such as steel ingots can be heated in pit-type furnaces, preferably with units of load separated to allow radiating heating from all sides except the bottom. Such a furnace would have a cover displaced by a mechanical carriage and would have a compound metal and refractory recuperator arrangement. Loads are handled by overhead crane equipped with suitable gripping tongs.

Continuous-Type Furnaces

The simplest type of continuous furnace is the hearth-type pusher furnace. Pieces of rectangular cross section are loaded side by side on a charge table and pushed through the furnace by an external mechanism. In the design shown, the furnace is fired from one end, counterflow to load travel, and is discharged through a side door by an auxiliary pusher lined up by the operator.

Furnace length is limited by thickness of the load and alignment of abutting edges, to avoid buckling up from the hearth.

A more complex design would provide multiple zone firing above and below the hearth, with recuperative air preheating.

Long loads can be conveyed in the direction of their length in a roller-hearth-type furnace. Loads can be bars, tubes, or plates of limited width, heated by direct firing, by radiant tubes, or by electric-resistor-controlled atmosphere, and conveyed at uniform speed or at alternating high and low speeds for quenching in line.

Sequential heat treatment can be accomplished with a series of chain or belt conveyors. Small parts can be loaded through an atmosphere seal, heated in a controlled atmosphere on a chain belt conveyor, discharged into an oil quench, and conveyed through a washer and tempering furnace by a series of mesh belts without intermediate handling.

Except for pusher-type furnaces, continuous furnaces can be self-emptying. To secure the same advantage in heating slabs or billets for rolling and to avoid scale loss during interrupted operation, loads can be conveyed by a walking-beam mechanism. Such a walking-beam-type slab heating furnace would have loads supported on water-cooled rails for over- and underfiring, and would have an overhead recuperator.

Thin strip materials, joined in continuous strand form, can be conveyed horizontally or the strands can be conveyed in a series of vertical passes by driven support rolls. Furnaces of this type can be incorporated in continuous galvanizing lines.

Unit loads can be individually suspended from an overhead conveyor, through a slot in the furnace roof, and can be quenched in line by lowering a section of the conveyor.

Small parts or bulk materials can be conveyed by a moving hearth, as in the rotary-hearth-type or tunnel kiln furnace. For roasting or incineration of bulk materials, the shaft-type furnace provides a simple and efficient system. Loads are charged through the open top of the shaft and descend by gravity to a discharge feeder at the bottom. Combustion air can be introduced at the bottom of the furnace and preheated by contact with the descending load before entering the combustion zone, where fuel is introduced through sidewalls. Combustion gases are then cooled by contact with the descending load, above the combustion zone, to preheat the charge and reduce flue gas temperature.

With loads that tend to agglomerate under heat and pressure, as in some ore-roasting operations, the rotary kiln may be preferable to the shaft-type furnace. The load is advanced by rolling inside an inclined cylinder. Rotary kilns are in general use for sintering ceramic materials.

Classification by Source of Heat

The classification of furnaces by source of heat is as follows:

Direct-firing with gas or oil fuels

Combustion of material in process, as by incineration with or without supplemental fuel

Internal heating by electrical resistance or induction in conductors, or dielectric heating of nonconductors

Radiation from electric resistors or radiant tubes, in controlled atmospheres or under vacuum

45.4 FURNACE CONSTRUCTION

The modern industrial furnace design has evolved from a rectangular or cylindrical enclosure, built up of refractory shapes and held together by a structural steel binding. Combustion air was drawn in through wall openings by furnace draft, and fuel was introduced through the same openings without control of fuel/air ratios except by the judgment of the furnace operator. Flue gases were exhausted through an adjacent stack to provide the required furnace draft.

To reduce air infiltration or outward leakage of combustion gases, steel plate casings have been added. Fuel economy has been improved by burner designs providing some control of fuel/air ratios, and automatic controls have been added for furnace temperature and furnace pressure. Completely sealed furnace enclosures may be required for controlled atmosphere operation, or where outward leakage of carbon monoxide could be an operating hazard.

With the steadily increasing costs of heat energy, wall structures are being improved to reduce heat losses or heat demands for cyclic heating. The selection of furnace designs and materials should be aimed at a minimum overall cost of construction, maintenance, and fuel or power over a projected service life. Heat losses in existing furnaces can be reduced by adding external insulation or rebuilding walls with materials of lower thermal conductivity. To reduce losses from intermittent operation, the existing wall structure can be lined with a material of low heat storage and low conductivity, to substantially reduce mean wall temperatures for steady operation and cooling rates after interrupted firing.

Thermal expansion of furnace structures must be considered in design. Furnace walls have been traditionally built up of prefired refractory shapes with bonded mortar joints. Except for small furnaces, expansion joints will be required to accommodate thermal expansion. In sprung arches, lateral expansion can be accommodated by vertical displacement, with longitudinal expansion taken care of by lateral slots at intervals in the length of the furnace. Where expansion slots in furnace floors could be filled by scale, slag, or other debris, they can be packed with a ceramic fiber that will remain resilient after repeated heating.

Differential expansion of hotter and colder wall surfaces can cause an inward-bulging effect. For stability in self-supporting walls, thickness must not be less than a critical fraction of height.

Because of these and economic factors, cast or rammed refractories are replacing prefired shapes for lining many types of large, high-temperature furnaces. Walls can be retained by spaced refractory shapes anchored to the furnace casing, permitting reduced thickness as compared to brick construction. Furnace roofs can be suspended by hanger tile at closer spacing, allowing unlimited widths.

Cast or rammed refractories, fired in place, will develop discontinuities during initial shrinkage that can provide for expansion from subsequent heating, to eliminate the need for expansion joints.

As an alternate to cast or rammed construction, insulating refractory linings can be gunned in place by jets of compressed air and retained by spaced metal anchors, a construction increasingly popular for stacks and flues.

Thermal expansion of steel furnace casings and bindings must also be considered. Where the furnace casing is constructed in sections, with overlapping expansion joints, individual sections can be separately anchored to building floors or foundations. For gas-tight casings, as required for controlled atmosphere heating, the steel structure can be anchored at one point and left free to expand elsewhere. In a continuous galvanizing line, for example, the atmosphere furnace and cooling zone can be anchored to the foundation near the casting pot, and allowed to expand toward the charge end.

45.5 FUELS AND COMBUSTION

Heat is supplied to industrial furnaces by combustion of fuels or by electrical power. Fuels now used are principally fuel oil and fuel gas. Because possible savings through improved design and operation are much greater for these fuels than for electric heating or solid fuel firing, they will be given primary consideration in this section.

Heat supply and demand may be expressed in units of *Btu* or *kcal* or as gallons or barrels of fuel oil, tons of coal or *kwh* of electric power. For the large quantities considered for national or world energy loads, a preferred unit is the "quad," one quadrillion or 10^{15} Btu. Conversion factors are:

$$
\begin{aligned}
1 \text{ quad} &= 10^{15} \text{ Btu} \\
&= 172 \times 10^6 \text{ barrels of fuel oil} \\
&= 44.34 \times 10^6 \text{ tons of coal} \\
&= 10^{12} \text{ cubic feet of natural gas} \\
&= 2.93 \times 10^{11} \text{ kwh electric power}
\end{aligned}
$$

At 30% generating efficiency, the fuel required to produce 1 quad of electrical energy is 3.33 quads. One quad fuel is accordingly equivalent to 0.879×10^{11} kwh net power.

Fuel demand, in the United States during recent years, has been about 75 quads per year from the following sources:

Coal	15 quads
Fuel oil	
Domestic	18 quads
Imported	16 quads
Natural gas	23 quads
Other, including nuclear	3 quads

Hydroelectric power contributes about 1 quad net additional. Combustion of waste products has not been included, but will be an increasing fraction of the total in the future.

Distribution of fuel demand by use is estimated at:

Power generation	20 quads
Space heating	11 quads
Transportation	16 quads
Industrial, other than power	25 quads
Other	4 quads

Net demand for industrial furnace heating has been about 6%, or 4.56 quads, primarily from gas and oil fuels.

The rate at which we are consuming our fossil fuel assets may be calculated as (annual demand)/(estimated reserves). This rate is presently highest for natural gas, because, besides being available at wellhead for immediate use, it can be transported readily by pipeline and burned with the simplest type of combustion system and without air pollution problems. It has also been delivered at bargain prices, under federal rate controls.

As reserves of natural gas and fuel oil decrease, with a corresponding increase in market prices, there will be an increasing demand for alternative fuels such as synthetic fuel gas and fuel oil, waste materials, lignite, and coal.

Synthetic fuel gas and fuel oil are now available from operating pilot plants, but at costs not yet competitive.

As an industrial fuel, coal is primarily used for electric power generation. In the form of metallurgical coke, it is the source of heat and the reductant in the blast furnace process for iron ore reduction, and as fuel for cupola furnaces used to melt foundry iron. Powdered coal is also being used as fuel and reductant in some new processes for solid-state reduction of iron ore pellets to make synthetic scrap for steel production.

Since the estimated life of coal reserves, particularly in North America, is so much greater than for other fossil fuels, processes for conversion of coal to fuel gas and fuel oil have been developed almost to the commercial cost level, and will be available whenever they become economical. Processes for coal gasification, now being tried in pilot plants, include:

1. *Producer Gas.* Bituminous coal has been commercially converted to fuel gas of low heating value, around 110 Btu/scf LHV, by reacting with insufficient air for combustion and steam as a source of hydrogen. Old producers delivered a gas containing sulfur, tar volatiles, and suspended ash, and have been replaced by cheap natural gas. By reacting coal with a mixture of oxygen and steam, and removing excess carbon dioxide, sulfur gases, and tar, a clean fuel gas of about 300 Btu/scf LHV can be supplied. Burned with air preheated to 1000°F and with a flue gas temperature of 2000°F, the available heat is about 0.69 HHV, about the same as for natural gas.

2. *Synthetic Natural Gas.* As a supplement to dwindling natural gas supplies, a synthetic fuel gas of similar burning characteristics can be manufactured by adding a fraction of hydrogen to the product of the steam–oxygen gas producer and reacting with carbon monoxide at high temperature and pressure to produce methane. Several processes are operating successfully on a pilot plant scale, but with a product costing much more than market prices for natural gas. The process may yet be practical for extending available natural gas supplies by a fraction, to maintain present market demands. For gas mixtures or synthetic gas supplies to be interchangeable with present gas fuels, without readjustment of fuel/air ratio controls, they must fit the Wobbe Index:

$$\frac{\text{HHV Btu/scf}}{(\text{specific gravity})^{0.5}}$$

The fuel gas industry was originally developed to supply fuel gas for municipal and commercial lighting systems. Steam was passed through incandescent coal or coke, and fuel oil vapors were added to provide a luminous flame. The product had a heating value of around 500 HHV, and a high carbon monoxide content, and was replaced as natural gas or coke oven gas became available. Coke oven gas is a by-product of the manufacture of metallurgical coke that can be treated to remove sulfur compounds and volatile tar compounds to provide a fuel suitable for pipeline distribution. Blast furnace gas can be used as an industrial or steam-generating fuel, usually after enrichment with coke oven gas. Gas will be made from replaceable sources such as agricultural and municipal wastes, cereal grains, and wood, as market economics for such products improve.

Heating values for fuels containing hydrogen can be calculated in two ways:

1. Higher heating value (HHV) is the total heat developed by burning with standard air in a ratio to supply 110% of net combustion air, cooling products to ambient temperature, and condensing all water vapor from the combustion of hydrogen.

2. Lower heating value (LHV) is equal to HHV less heat from the condensation of water vapor. It provides a more realistic comparison between different fuels, since flue gases leave most industrial processes well above condensation temperatures.

HHV factors are in more general use in the United States, while LHV values are more popular in most foreign countries.

For example, the HHV value for hydrogen as fuel is 319.4 Btu/scf, compared to a LHV of 270.2.

The combustion characteristics for common fuels are tabulated in Table 45.3, for combustion with 110% standard air. Weights in pounds per 10^6 Btu HHV are shown, rather than corresponding volumes, to expedite calculations based on mass flow. Corrections for flue gas and air temperatures other than ambient are given in charts to follow.

The heat released in a combustion reaction is:

total heats of formation of combustion products − total heats of formation of reactants

Heats of formation can be conveniently expressed in terms of Btu per pound mol, with the pound mol for any substance equal to a weight in pounds equal to its molecular weight. The heat of formation for elemental materials is zero. For compounds involved in common combustion reactions, values are shown in Table 45.4.

Data in Table 45.4 can be used to calculate the higher and lower heating values of fuels. For methane:

$$CH_4 + 2O_2 = CO_2 + 2H_2O$$

Table 45.3   Combustion Characteristics of Common Fuels

Fuel	Btu/scf	Weight in lb/10⁶ Btu		
		Fuel	Air	Flue Gas
Natural gas (SW U.S.)	1073	42	795	837
Coke oven gas	539	57	740	707
Blast furnace gas	92	821	625	1446
Mixed blast furnace and coke oven gas:				
Ratio CO/BF 1/1	316	439	683	1122
1/3	204	630	654	1284
1/10	133	752	635	1387
Hydrogen	319	16	626	642
	Btu/lb			
No. 2 fuel oil	19,500	51	810	861
No. 6 fuel oil	18,300	55	814	
With air atomization				869
With steam atomization at 3 lb/gal				889
Carbon	14,107	71	910	981

HHV

$$169{,}290 + (2 \times 122{,}976) - 32{,}200 = 383{,}042 \text{ Btu/lb} \cdot \text{mol}$$
$$383{,}042/385 = 995 \text{ Btu/scf}$$

LHV

$$169{,}290 + (2 \times 104{,}040) - 32{,}200 = 345{,}170 \text{ Btu/lb} \cdot \text{mol}$$
$$345{,}170/385 = 897 \text{ Btu/scf}$$

Available heats from combustion of fuels, as a function of flue gas and preheated air temperatures, can be calculated as a fraction of the HHV. The net ratio is one plus the fraction added by preheated air less the fraction lost as sensible heat and latent heat of water vapor, from combustion of hydrogen, in flue gas leaving the system.

Available heats can be shown in chart form, as in the following figures for common fuels. On each chart, the curve on the right is the fraction of HHV available for combustion with 110% cold air, while the curve on the left is the fraction added by preheated air, as functions of air or flue gas temperatures. For example, the available heat fraction for methane burned with 110% air preheated to 1000°F, and with flue gas out at 2000°F, is shown in Fig. 45.1: 0.41 + 0.18 − 0.59 HHV.

Values for other fuels are shown in charts that follow:

Fig. 45.2, fuel oils with air or steam atomization

Fig. 45.3, by-product coke oven gas

Fig. 45.4, blast furnace gas

Fig. 45.5, methane

Table 45.4   Heats of Formation

Material	Formula	Molecular Weight	Heats of Formation (Btu/lb · mol[a])
Methane	CH_4	16	32,200
Ethane	C_2H_6	30	36,425
Propane	C_3H_8	44	44,676
Butane	C_4H_{10}	58	53,662
Carbon monoxide	CO	28	47,556
Carbon dioxide	CO_2	44	169,290
Water vapor	H_2O	18	104,040
Liquid water			122,976

[a]The volume of 1 lb mol, for any gas, is 385 scf.

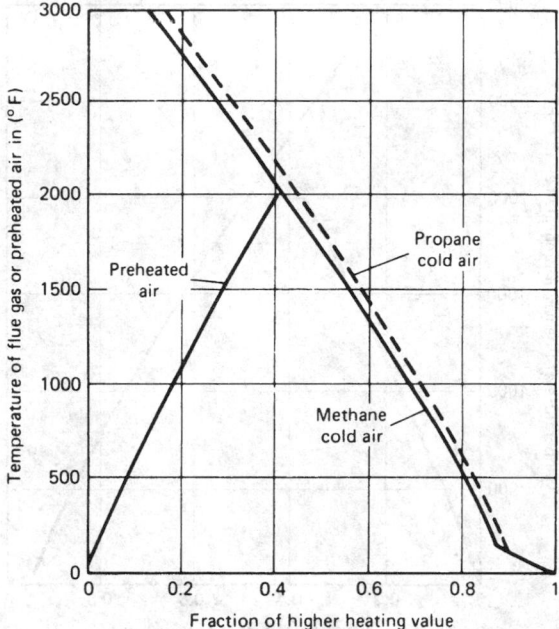

Fig. 45.1 Available heat for methane and propane combustion. Approximate high and low limits for commercial natural gas.[1]

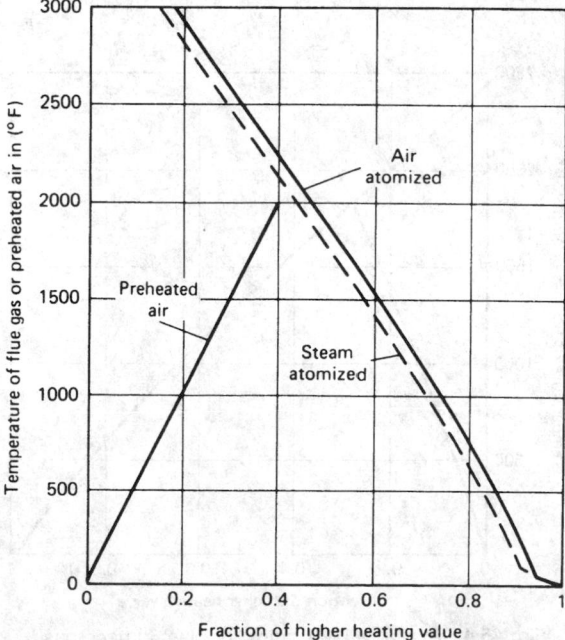

Fig. 45.2 Available heat ratios for fuel oils with air or steam atomization.[1]

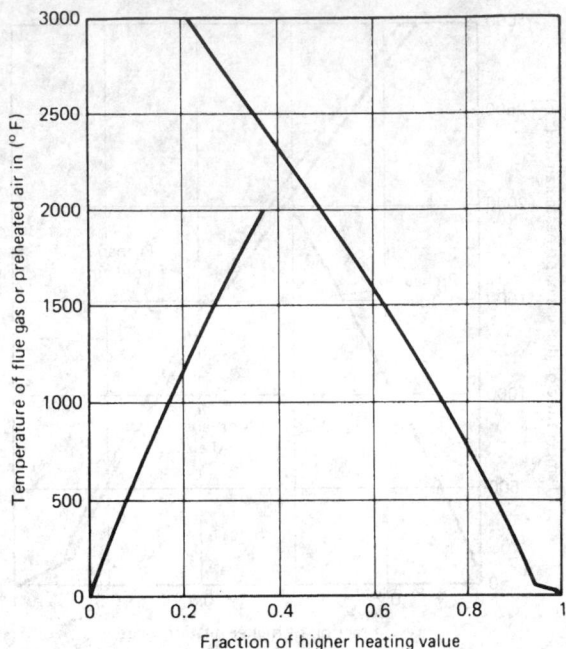

Fig. 45.3   Available heat ratios for by-product coke oven gas.[1]

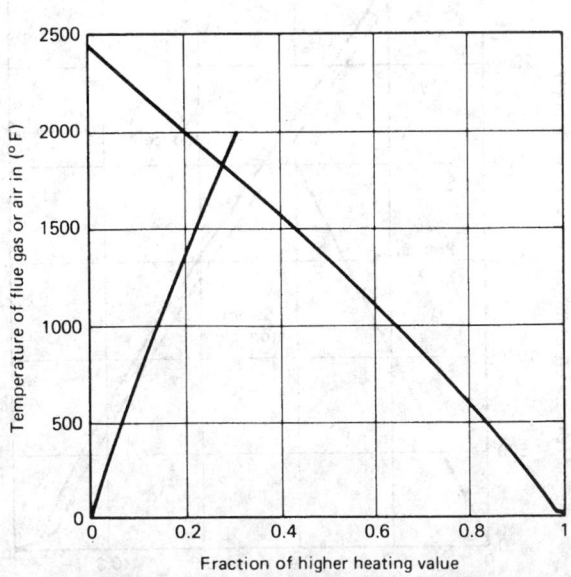

Fig. 45.4   Available heat ratios for blast furnace gas.[1]

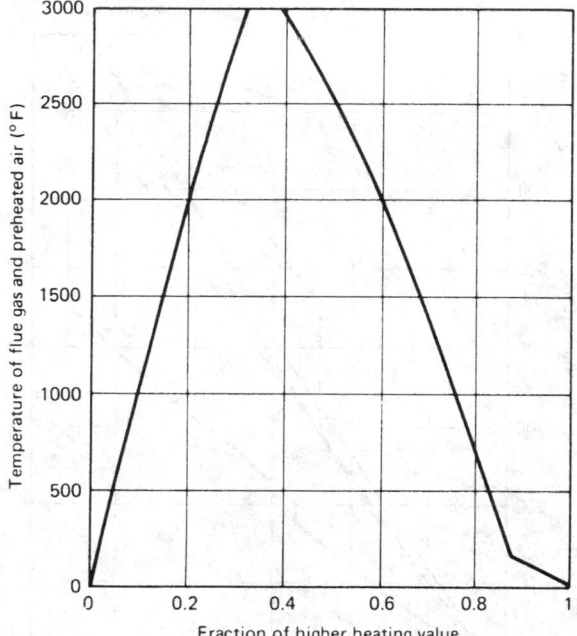

Fig. 45.5 Available heat ratios for combustion of methane with 110% air containing 35% O_2.[1]

For combustion with other than 110% of net air demand, the corrected available heat can be calculated as follows. For methane with preheated air at 1000°F and flue gas out at 2000°F and 150% net air supply:

$$
\begin{array}{lr}
\text{Available heat from Fig. 58.1} & 0.59 \\
\text{Add excess air} + 0.18\,(1.5 - 1.1) = & 0.072 \\
- 0.41\,(1.5 - 1.1) = & \underline{-0.164} \\
\text{Net total at 150\%} & 0.498
\end{array}
$$

Available heats for fuel gas mixtures can be calculated by adding the fractions for either fuel and dividing by the combined volume. For example, a mixture of one-quarter coke oven gas and three-quarters blast furnace gas is burned with 110% combustion air preheated to 1000°F, and with flue gas out at 2000°F. Using data from Table 45.3 and Figs. 45.3 and 45.4:

$$
\begin{array}{rl}
\text{CO } (539 \times 0.25 = 134.75)\,(0.49 + 0.17) = & 88.93 \\
\text{BF } (92 \times 0.75 = \underline{69.00})\,(0.21 + 0.144) = & \underline{24.43} \\
\text{HHV } 203.75 \qquad \text{Available} = & 113.36
\end{array}
$$

Net: $113.36/203.75 = 0.556$ combined HHV

45.6 OXYGEN ENRICHMENT OF COMBUSTION AIR

The available heats of furnace fuels can be improved by adding oxygen to combustion air. Some studies have been based on a total oxygen content of 35%, which can be obtained by adding 21.5 scf pure oxygen or 25.45 scf of 90% oxygen per 100 scf of dry air. The available heat ratios are shown in the chart in Fig. 45.5.

At present market prices, the power needed to concentrate pure oxygen for enrichment to 35% will cost more than the fuel saved, even with metallurgical oxygen from an in-plant source. As plants are developed for economical concentration of oxygen to around 90%, the cost balance may become favorable for very-high-temperature furnaces.

In addition to fuel savings by improvement of available heat ratios, there will be additional savings in recuperative furnaces by increasing preheated air temperature at the same net heat demand, de-

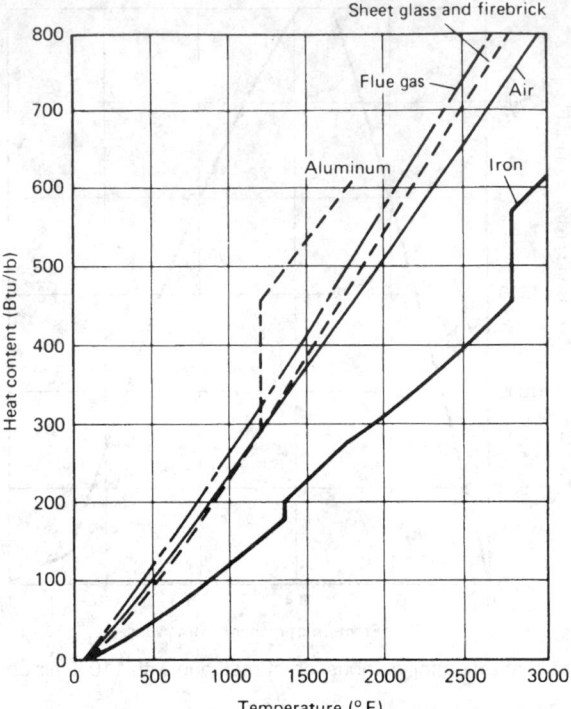

Fig. 45.6 Heat content of materials at temperature.[1]

pending on the ratio of heat transfer by convection to that by gas radiation in the furnace and recuperator.

45.7 THERMAL PROPERTIES OF MATERIALS

The heat content of some materials heated in furnaces or used in furnace construction is shown in the chart in Fig. 45.6, in units of Btu/lb. Vertical lines in curves represent latent heats of melting or other phase transformations. The latent heat of evaporation for water in flue gas has been omitted from the chart. The specific heat of liquid water is, of course, about 1.

Thermal conductivities in English units are given in reference publications as: $(Btu/(ft^2 \cdot hr))/$ ($°F/in.$) or as $(Btu/(ft^2 \cdot hr))/(°F/ft)$. To keep dimensions consistent, the latter term, abbreviated to $k = Btu/ft \cdot hr \cdot °F$ will be used here. Values will be $\frac{1}{12}$th of those in terms of $°F/in$.

Thermal conductivities vary with temperature, usually inversely for iron, steel, and some alloys, and conversely for common refractories. At usual temperatures of use, average values of k in Btu/ ($ft \cdot hr \cdot °F$) are in Table 45.5.

Table 45.5 Average Values of k (Btu/ft · hr · °F)

	Mean Temperature (°F)				
	100	1000	1500	2000	2500
Steel, SAE 1010	33	23	17	17	
Type HH HRA	8	11	14	16	
Aluminum	127	133			
Copper	220	207	200		
Brass, 70/30	61	70			
Firebrick	0.81	0.82	0.85	0.89	0.93
Silicon carbide	11	10	9	8	6
Insulating firebrick	0.12	0.17	0.20	0.24	

To expedite calculations for nonsteady conduction of heat, it is convenient to use the factor for "thermal diffusivity," defined as

$$D = \frac{k}{dC} = \frac{\text{thermal conductivity}}{\text{density} \times \text{specific heat}}$$

in consistent units. Values for common furnace loads over the usual range of temperatures for heating are:

Carbon steels, 70–1650°F	0.32
70–2300°F	0.25
Low-alloy steels, 70–2000°F	0.23
Stainless steels, 70–2000°F	
300 type	0.15
400 type	0.20
Aluminum, 70–1000°F	3.00
Brass, 70/30, 70–1500°F	1.20

In calculating heat losses through furnace walls with multiple layers of materials with different thermal conductivities, it is convenient to add thermal resistance $R = r/k$, where r is thickness in ft. For example,

	r	k	r/k
9-in. firebrick	0.75	0.9	0.833
4½-in. insulating firebrick	0.375	0.20	1.875
2¼-in. block insulation	0.208	0.15	1.387
Total R for wall materials			4.095

Overall thermal resistance will include the factor for combined radiation and convection from the outside of the furnace wall to ambient temperature. Wall losses as a function of wall surface temperature, for vertical surfaces in still air, are shown in Fig. 45.7, and are included in the overall heat loss data for furnace walls shown in the chart in Fig. 45.8.

The chart in Fig. 45.9 shows the thermodynamic properties of air and flue gas, over the usual range of temperatures, for use in heat-transfer and fluid flow problems. Data for other gases, in formula form, are available in standard references.

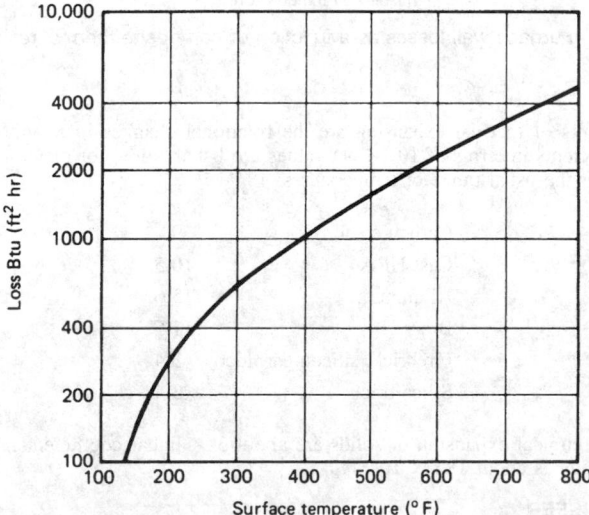

Fig. 45.7 Furnace wall losses as a function of surface temperature.[1]

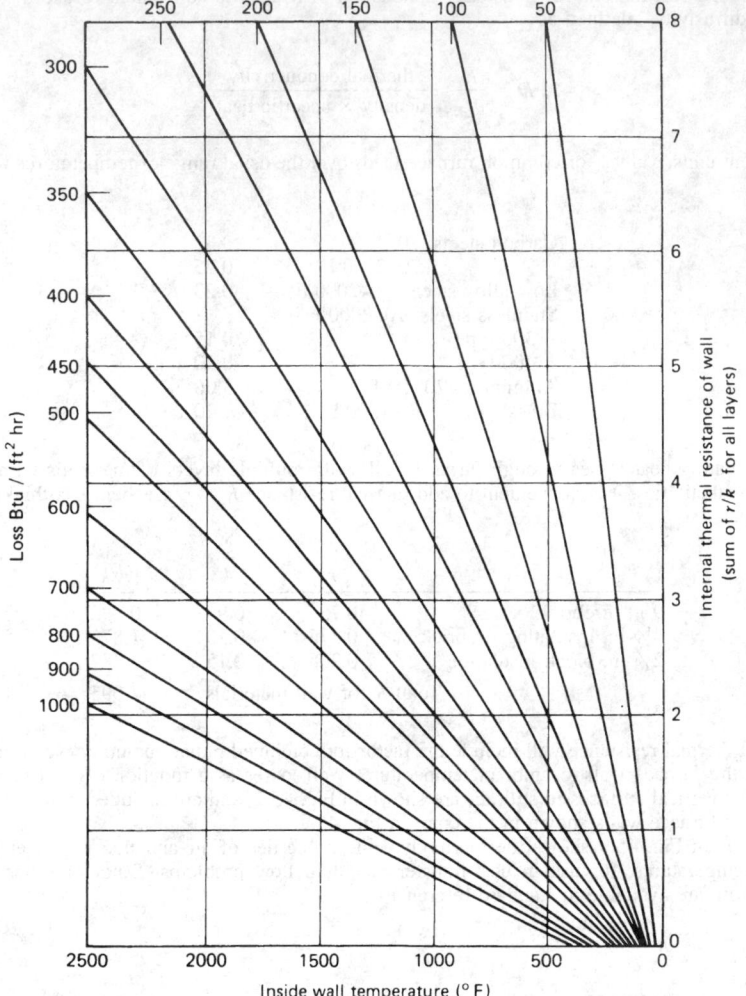

Fig. 45.8   Furnace wall losses as a function of composite thermal resistance.[1]

Linear coefficients of thermal expansion are the fractional changes in length per °F change in temperature. Coefficients in terms of $10^6 \times$ net values are listed below for materials used in furnace construction and for the usual range of temperatures:

Carbon steel	9
Cast HRA	10.5
Aluminum	15.6
Brass	11.5
Firebrick, silicon carbide	3.4
Silica brick	3.4

Coefficients for cubical expansion of solids are about $3 \times$ linear coefficients. The cubical coefficient for liquid water is about 185×10^{-6}.

45.8   HEAT TRANSFER

Heat may be transmitted in industrial furnaces by radiation—gas radiation from combustion gases to furnace walls or direct to load, and solid-state radiation from walls, radiant tubes, or electric heating

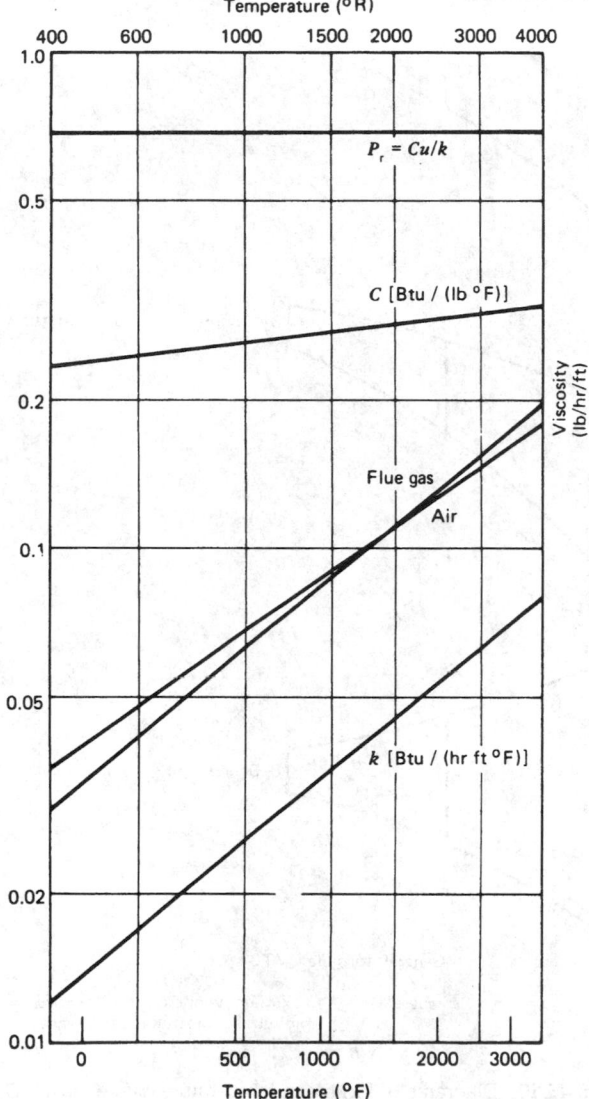

Fig. 45.9 Thermodynamic properties of air and flue gas.[1]

elements to load—or by convection—from combustion gases to walls or load. Heat may be generated inside the load by electrical resistance to an externally applied voltage or by induction, with the load serving as the secondary circuit in an alternating current transformer. Nonconducting materials may be heated by dielectric heating from a high-frequency source.

Heat transfer in the furnace structure or in solid furnace loads will be by conduction. If the temperature profile is constant with time, the process is defined as "steady-state conduction." If temperatures change during a heating cycle, it is termed "non-steady-state conduction."

Heat flow is a function of temperature differentials, usually expressed as the "log mean temperature difference" with the symbol MTD. MTD is a function of maximum and minimum temperature differences that can vary with position or time. Three cases encountered in furnace design are illustrated in Fig. 45.10. If the maximum differential, in any system of units, is designated as A and the minimum is designated by B:

$$\text{MTD} = \frac{A - B}{\ln (A/B)}$$

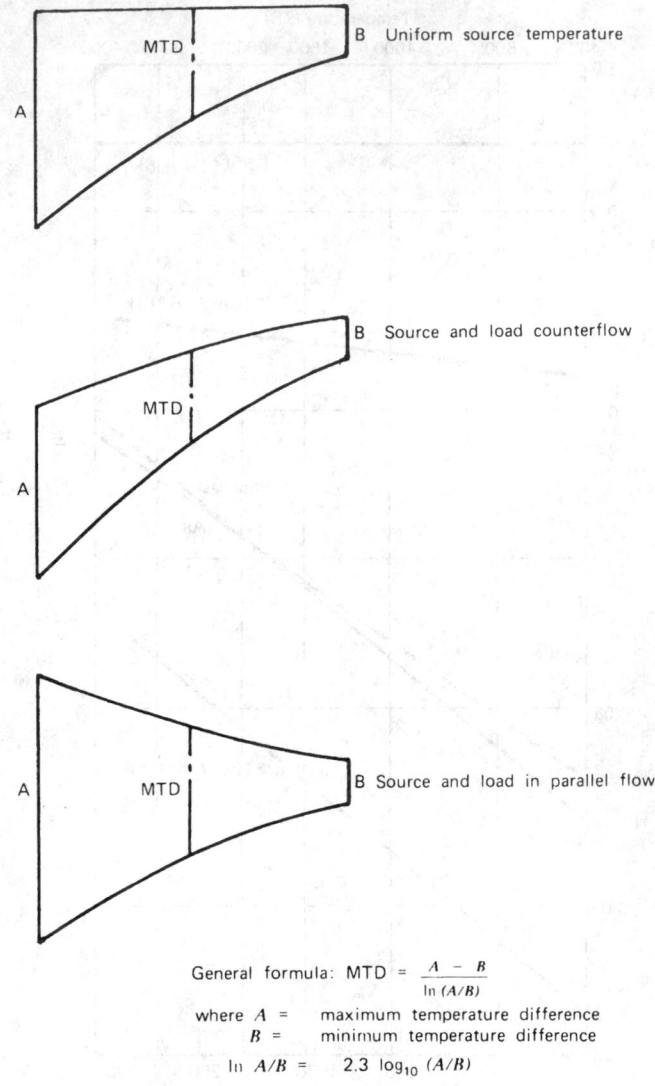

General formula: MTD $= \dfrac{A - B}{\ln (A/B)}$

where $A =$ maximum temperature difference
$B =$ minimum temperature difference

$\ln A/B =$ 2.3 $\log_{10} (A/B)$

Fig. 45.10 Diagrams of log mean temperature difference (MTD).[1]

45.8.1 Solid-State Radiation

"Black-body" surfaces are those that absorb all radiation received, with zero reflection, and exist only as limits approached by actual sources or receivers of solid radiation. Radiation between black bodies is expressed by the Stefan–Boltzmann equation:

$$Q/A = N(T^4 - T_0^4) \text{Btu/hr} \cdot \text{ft}^2$$

where N is the Stefan–Boltzmann constant, now set at about 0.1713×10^{-8} for T and T_0, source and receiver temperatures, in °R. Because the fourth powers of numbers representing temperatures in °R are large and unwieldy, it is more convenient to express temperatures in °S, equivalent to (°F + 460)/100. The constant N is then reduced to 0.1713.

With source and receiver temperatures identified as T_s and T_r in °S, and with allowance for emissivity and view factors, the complete equation becomes:

$$Q/A = 0.1713 \times em \times Fr(T_s^4 - T_r^4) \quad \text{Btu/hr} \cdot \text{ft}^2$$

at the receiving surface,

where em = combined emissivity and absorptivity factors for source and receiving surfaces
Fr = net radiation view factor for receiving surface
T_s and T_r = source and receiving temperature in °S

The factor em will be somewhat less than e for the source or a for the receiving surface, and can be calculated:

$$em = 1 \bigg/ \frac{1}{a} + \frac{A_r}{A_s}\left(\frac{1}{e} - 1\right)$$

where a = receiver absorptivity at T_r
A_r/A_s = area ratio, receiver/source
e = source emissivity at T_s

45.8.2 Emissivity—Absorptivity

While emissivity and absorptivity values for solid materials vary with temperatures, values for materials commonly used as furnace walls or loads, in the usual range of temperatures, are:

Refractory walls	0.80–0.90
Heavily oxidized steel	0.85–0.95
Bright steel strip	0.25–0.35
Brass cake	0.55–0.60
Bright aluminum strip	0.05–0.10
Hot-rolled aluminum plate	0.10–0.20
Cast heat-resisting alloy	0.75–0.85

For materials such as sheet glass, transparent in the visible light range, radiation is reflected at both surfaces at about 4% of incident value, with the balance absorbed or transmitted. Absorptivity decreases with temperature, as shown in Fig. 45.11.
The absorptivity of liquid water is about 0.96.

45.8.3 Radiation Charts

For convenience in preliminary calculations, black-body radiation, as a function of temperature in °F, is given in chart form in Fig. 45.12. The value for the receiver surface is subtracted from that of the source to find net interchange for black-body conditions, and the result is corrected for emissivity and view factors. Where heat is transmitted by a combination of solid-state radiation and convection, a black-body coefficient, in Btu/hr · °F, is shown in the chart in Fig. 45.13. This can be added to the convection coefficient for the same temperature interval, after correcting for emissivity and view factor, to provide an overall coefficient (H) for use in the formula

$$Q/A = H(T - T_r)$$

45.8.4 View Factors for Solid-State Radiation

For a receiving surface completely enclosed by the source of radiation, or for a flat surface under a hemispherical radiating surface, the view factor is unity. Factors for a wide range of geometrical configurations are given in available references. For cases commonly involved in furnace heat-transfer calculations, factors are shown by the following charts.
For two parallel planes, with edges in alignment as shown in Fig. 45.14a, view factors are given in Fig. 45.15 in terms of ratios of x, y, and z. For two surfaces intersecting at angle of 90° at a common edge, the view factor is shown in Fig. 45.16. If surfaces do not extend to a common intersection, the view factor for the missing areas can be calculated and deducted from that with surfaces extended as in the figure, to find the net value for the remaining areas.
For spaced cylinders parallel to a furnace wall, as shown in Fig. 45.17, the view factor is shown in terms of diameter and spacing, including wall reradiation. For tubes exposed on both sides to source or receiver radiation, as in some vertical strip furnaces, the following factors apply if sidewall reradiation is neglected:

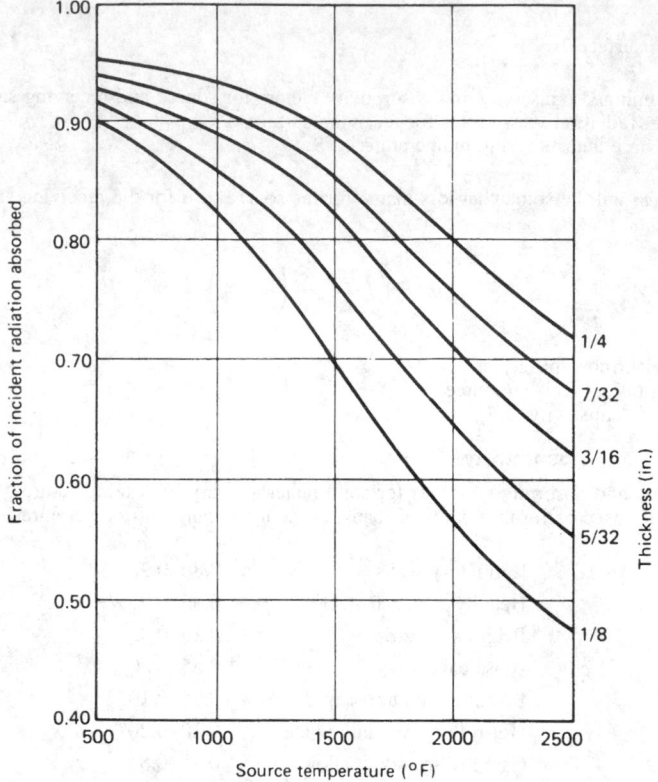

Fig. 45.11 Radiation absorptivity of sheet glass with surface reflection deducted.[1]

Ratio C/D	1.0	1.5	2.0	2.5	3.0
Factor	0.67	0.793	0.839	0.872	0.894

For ribbon-type electric heating elements, mounted on a back-up wall as shown in Fig. 45.18, exposure factors for projected wall area and for total element surface area are shown as a function of the (element spacing)/(element width) ratio. Wall reradiation is included, but heat loss through the back-up wall is not considered. The emission rate from resistor surface will be W/in.^{2} = $Q/$ 491A, where

$$\frac{Q}{A} = \frac{\text{Btu/hr}}{\text{ft}^2}$$

For parallel planes of equal area, as shown in Fig. 45.14, connected by reradiating walls on four sides, the exposure factor is increased as shown in Fig. 45.19. Only two curves, for $z/x = 1$ and $z/x = 10$ have been plotted for comparison with Fig. 45.13.

45.8.5 Gas Radiation

Radiation from combustion gases to walls and load can be from luminous flames or from nonluminous products of combustion. Flame luminosity results from suspended solids in combustion gases, either incandescent carbon particles or ash residues, and the resulting radiation is in a continuous spectrum corresponding to that from solid-state radiation at the same source temperature. Radiation from non-luminous gases is in characteristic bands of wavelengths, with intensity depending on depth and density of the radiating gas layer, its chemical composition, and its temperature.

For combustion of hydrocarbon gases, flame luminosity is from carbon particles formed by cracking of unburned fuel during partial combustion, and is increased by delayed mixing of fuel and air in the combustion chamber. With fuel and air thoroughly premixed before ignition, products of combustion will be nonluminous in the range of visible light, but can radiate strongly in other

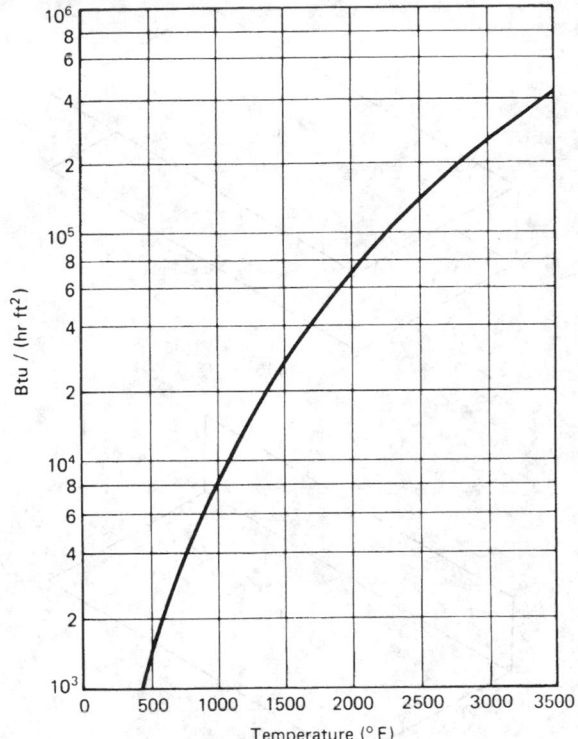

Fig. 45.12 Black-body radiation as function of load surface temperature.

wavelength bands for some products of combustion including carbon dioxide and water vapor. Published data on emissivities of these gases show intensity of radiation as a function of temperature, partial pressure, and beam length. The combined emissivity for mixtures of carbon dioxide and water vapor requires a correction factor for mutual absorption. To expedite calculations, a chart has been prepared for the overall emissivity of some typical flue gases, including these correction factors. The chart in Fig. 45.20 has been calculated for products of combustion of methane with 110% of net air demand, and is approximately correct for other hydrocarbon fuels of high heating value, including

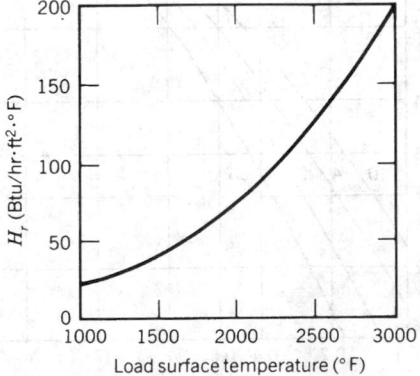

Fig. 45.13 Black-body radiation coefficient for source temperature uniform at 50–105° above final load surface temperature.

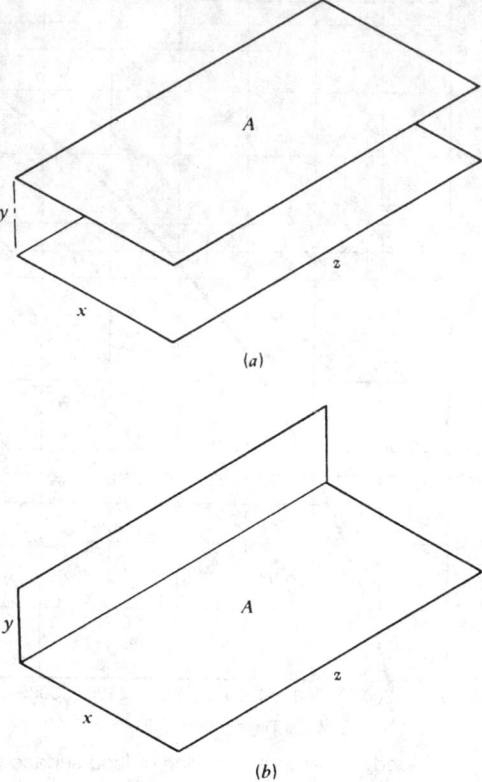

Fig. 45.14 Diagram of radiation view factors for parallel and perpendicular planes.[1]

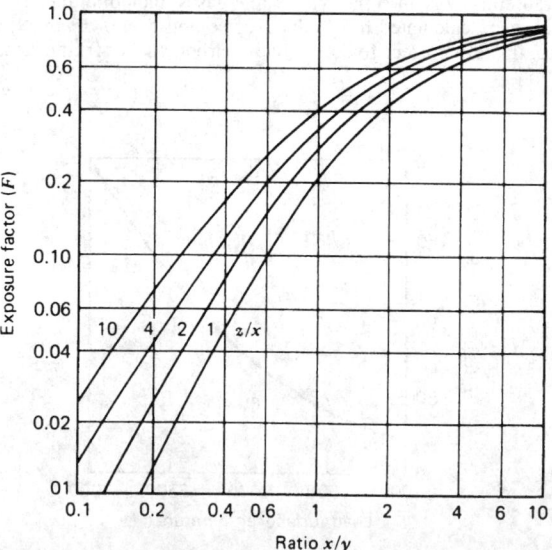

Fig. 45.15 Radiation view factors for parallel planes.[1]

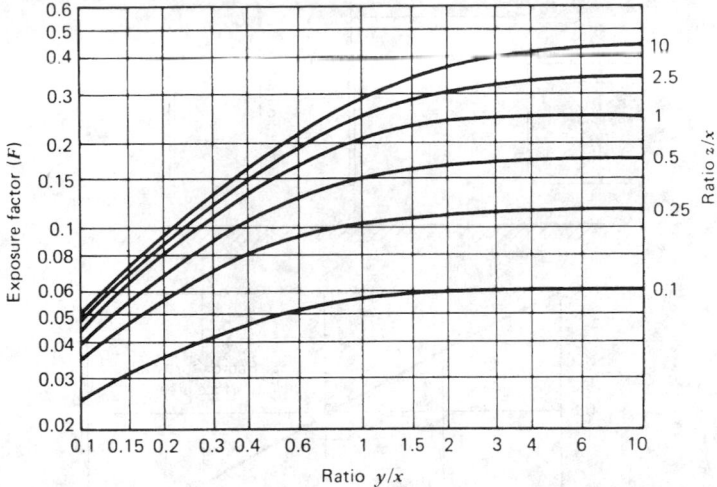

Fig. 45.16   Radiation view factors for perpendicular planes.[1]

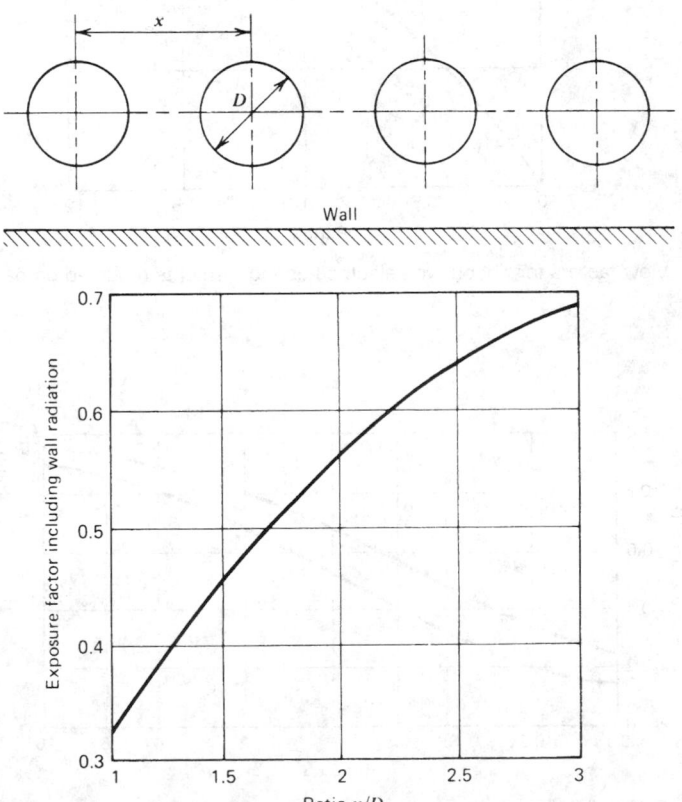

Fig. 45.17   View factors for spaced cylinders with back-up wall.[1]

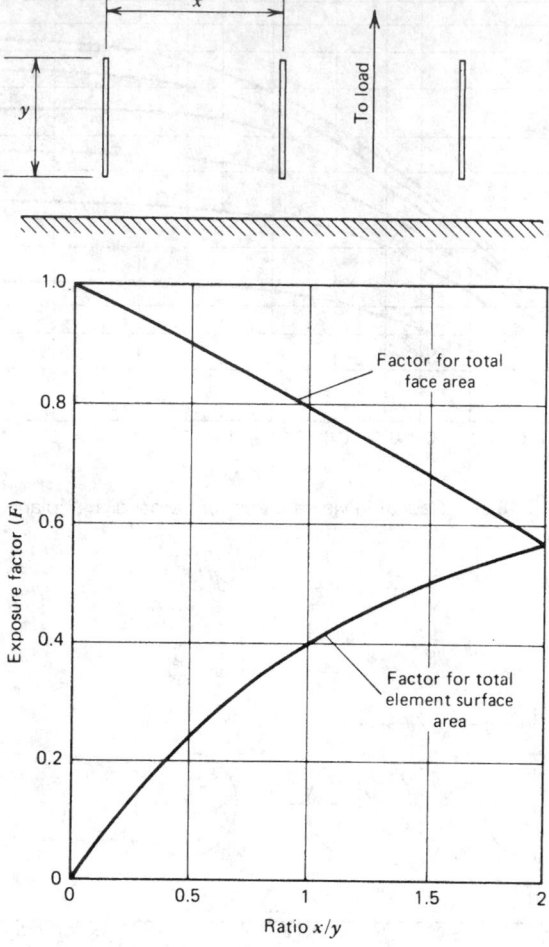

Fig. 45.18 View factors for ribbon-type electric heating elements mounted on back-up wall.[1]

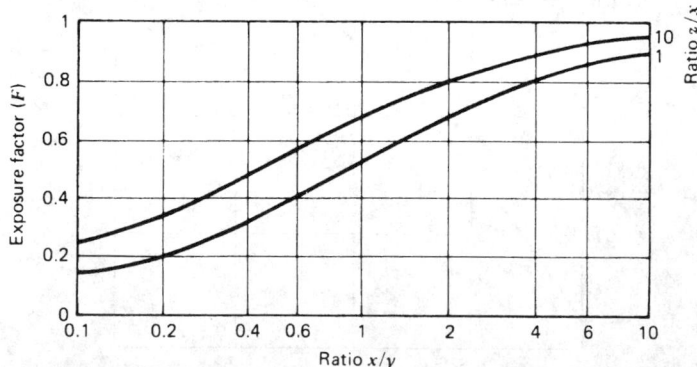

Fig. 45.19 View factors for parallel planes connected by reradiating sidewalls.[1]

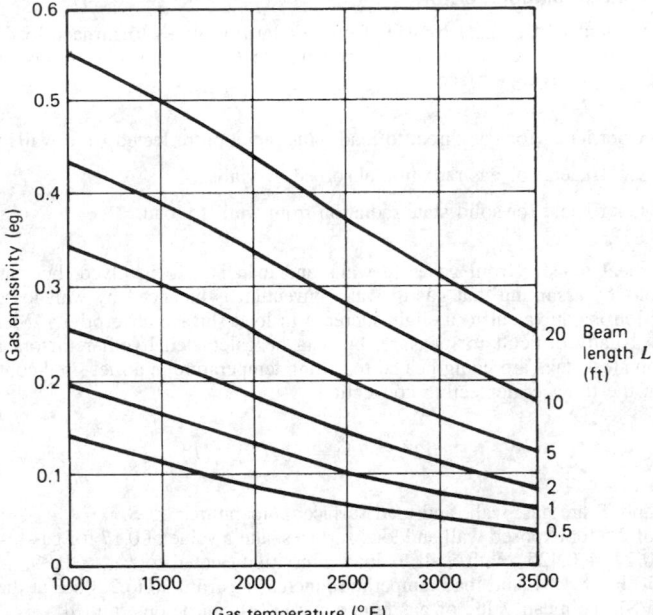

Fig. 45.20   Gas-emissivity for products of combustion of methane burned with 110% air. Approximate for fuel oils and coke oven gas.[1]

coke oven gas and fuel oils. Emissivities for producer gas and blast furnace gas will be lower, because of dilution of radiating gases by nitrogen.

The emissivity of a layer of combustion gases does not increase directly with thickness or density, because of partial absorption during transmission through the depth of the layer. The chart provides several curves for a range of values of L, the effective beam length in feet, at a total pressure of 1 atm. For other pressures, the effective beam length will vary directly with gas density.

Beam lengths for average gas densities will be somewhat less than for very low density because of partial absorption. For some geometrical configurations, average beam lengths are:

Between two large parallel planes, 1.8 × spacing

Inside long cylinder, about 0.85 × diameter in feet

For rectangular combustion chambers, $3.4V/A$ where V is volume in cubic feet and A is total wall area in square feet

Transverse radiation to tube banks, with tubes of D outside diameter spaced at x centers: L/D ranges from 1.48 for staggered tubes at $x/D = 1.5$ to 10.46 for tubes in line and $x/D = 3$ in both directions

45.8.6   Evaluation of Mean Emissivity–Absorptivity

For a gas with emissivity e_g radiating to a solid surface at a temperature of T_s °F, the absorptivity a_g will be less than e_g at T_s because the density of the gas is still determined by T_g. The effective PL becomes $T_s/T_g \times PL$ at T_s. Accurate calculation of the combined absorptivity for carbon dioxide and water vapor requires a determination of a_g for either gas and a correction factor for the total. For the range of temperatures and PL factors encountered in industrial heat transfer, the net heat transfer can be approximated by using a factor e_{gm} somewhat less than e_g at T_g in the formula:

$$Q/A = 0.1713e_{gm} F(T_g^4 - T_s^4)$$

where T_g is an average of gas temperatures in various parts of the combustion chamber; the effective emissivity will be about $e_{gm} = 0.9e_g$ at T_g and can be used with the chart in Fig. 45.20 to approximate net values.

45.8.7 Combined Radiation Factors

For a complete calculation of heat transfer from combustion gases to furnace loads, the following factors will need to be evaluated in terms of the equivalent fraction of black-body radiation per unit area of the exposed receiving surface:

F_{gs} = Coefficient for gas direct to load, plus radiation reflected from walls to load.

F_{gw} = Coefficient for gas radiation absorbed by walls.

F_{ws} = Coefficient for solid-state radiation from walls to load.

Convection heat transfer from gases to walls and load is also involved, but can be eliminated from calculations by assuming that gas to wall convection is balanced by wall losses, and that gas to load convection is equivalent to a slight increase in load surface absorptivity. Mean effective gas temperature is usually difficult to measure, but can be calculated if other factors are known. For example, carbon steel slabs are being heated to rolling temperature in a fuel-fired continuous furnace. At any point in the furnace, neglecting convection,

$$F_{gw} (T_g^4 - T_w^4) = F_{ws} (T_w^4 - T_s^4)$$

where T_g, T_w, and T_s are gas, wall, and load surface temperatures in °S.

For a ratio of 2.5 for exposed wall and load surfaces, and a value of 0.17 for gas-to-wall emissivity, F_{gw} = 2.5 × 0.17 = 0.425. With wall to load emissitivity equal to F_{ws} = 0.89, wall temperature constant at 2350°F (28.1°S), and load temperature increasing from 70 to 2300°F at the heated surface (T_s = 5.3–27.6°S), the mean value of gas temperature (T_g) can be determined:

$$\text{MTD, walls to load} = \frac{2280 - 50}{\ln(2280/50)} = 584°F$$

Mean load surface temperature T_{sm} = 2350 − 584 = 1766°F (22.26°S)

Q/A per unit of load surface, for reradiation:

0.425 × 0.1713(T_g^4 − 28.1²) = 0.89 × 0.1713(28.1⁴ − 22.26⁴) = 57,622 Btu/hr · ft²

T_g = 34.49°S (2989°F)

With a net wall emissivity of 0.85, 15% of gas radiation will be reflected to the load, with the balance being absorbed and reradiated. Direct radiation from gas to load is then

1.15 × 0.17 × 0.1713(34.49⁴ − 22.26⁴) = 47,389 Btu/hr · ft²

Total radiation: 57,622 + 47,389 = 105,011 Btu/hr · ft²

For comparison, black-body radiation from walls to load, without gas radiation, would be 64,743 Btu/hr · ft² or 62% of the combined total.

With practical furnace temperature profiles, in a counterflow, direct-fired continuous furnace, gas and wall temperatures will be depressed at the load entry end to reduce flue gas temperature and stack loss. The resulting net heating rates will be considered in Section 45.8.12.

Overall heat-transfer coefficients have been calculated for constant wall temperature, in the upper chart in Fig. 45.21, or for constant gas temperature in the lower chart. Coefficients vary with mean gas emissivity and with A_w/A_s, the ratio of exposed surface for walls and load, and are always less than one for overall radiation from gas to load, or greater than one for wall to load radiation. Curves can be used to find gas, wall, or mean load temperatures when the other two are known.

45.8.8 Steady-State Conduction

Heat transfer through opaque solids and motionless layers of liquids or gases is by conduction. For constant temperature conditions, heat flow is by "steady-state" conduction and does not vary with time. For objects being heated or cooled, with a continuous change in internal temperature gradients, conduction is termed "non-steady-state."

Thermal conduction in some solid materials is a combination of heat flow through the material, radiation across internal space resulting from porosity, and convection within individual pores or through the thickness of porous layers.

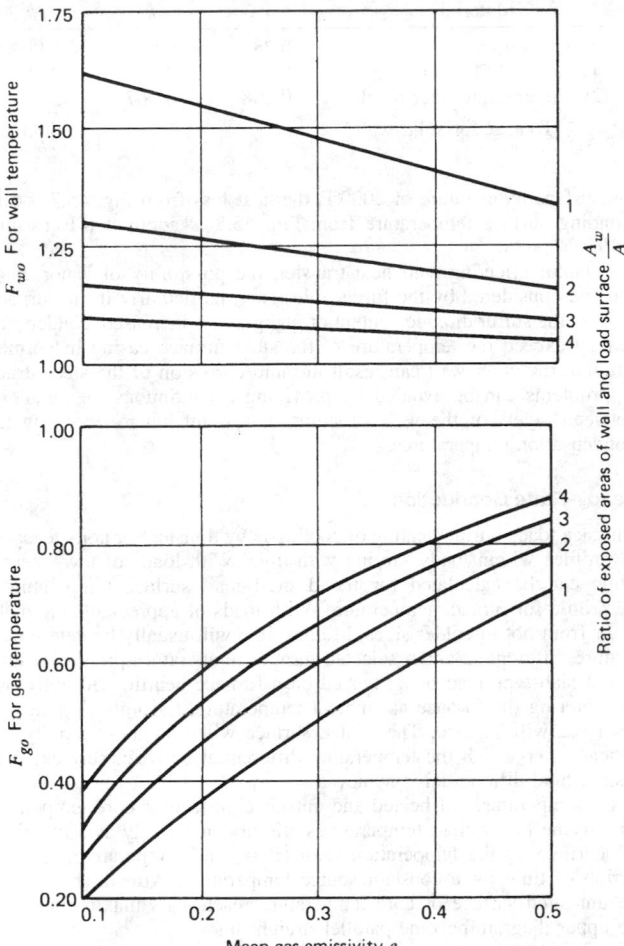

Fig. 45.21 Overall heat-transfer coefficients for gas and solid radiation, as function of gas emissivity and wall-to-load area ratio, for uniform gas or wall temperature, compared to black-body radiation.[1]

Conductivities of refractory and insulating materials tend to increase with temperature, because of porosity effects. Values for most metals decrease with temperature, partly because of reduced density. Conductivity coefficients for some materials used in furnace construction or heated in furnaces are listed in Table 45.5.

A familiar problem in steady-state conduction is the calculation of heat losses through furnace walls made up of multiple layers of materials of different thermal conductivities. A convenient method of finding overall conductance is to find the thermal resistance (r/k = thickness/conductivity in consistent units) and add the total for all layers. Because conductivities vary with temperature, mean temperatures for each layer can be estimated from a preliminary temperature profile for the composite wall. Overall resistance will include the effects of radiation and conduction between the outer wall surface and its surroundings.

A chart showing heat loss from walls to ambient surrounding at 70°F, combining radiation and convection for vertical walls, is shown in Fig. 45.7. The corresponding thermal resistance is included in the overall heat-transfer coefficient shown in Fig. 45.8 as a function of net thermal resistance of the wall structure and inside face temperature.

As an example of application, assume a furnace wall constructed as follows:

Material	r	k	r/k
9 in. firebrick	0.75	0.83	0.90
4½ in. 2000°F insulation	0.375	0.13	2.88
2½ in. ceramic fiber block	0.208	0.067	3.10
Total R for solid wall			6.88

With an inside surface temperature of 2000°F, the heat loss from Fig. 45.7 is about 265 Btu/ft · hr². The corresponding surface temperature from Fig. 45.8 is about 200°F, assuming an ambient temperature of 70°F.

Although not a factor affecting wall heat transfer, the possibility of vapor condensation in the wall structure must be considered by the furnace designer, particularly if the furnace is fired with a sulfur-bearing fuel. As the sulfur dioxide content of fuel gases is increased, condensation temperatures increase to what may exceed the temperature of the steel furnace casing in normal operation. Resulting condensation at the outer wall can result in rapid corrosion of the steel structure.

Condensation problems can be avoided by providing a continuous membrane of aluminum or stainless steel between layers of the wall structure, at a point where operating temperatures will always exceed condensation temperatures.

45.8.9 Non-Steady-State Conduction

Heat transfer in furnace loads during heating or cooling is by transient or non-steady-state conduction, with temperature profiles within loads varying with time. With loads of low internal thermal resistance, heating time can be calculated for the desired load surface temperature and a selected time–temperature profile for furnace temperature. With loads of appreciable thermal resistance from surface to center, or from hot to colder sides, heating time will usually be determined by a specified final load temperature differential, and a selected furnace temperature profile for the heating cycle.

For the case of a slab-type load being heated on a furnace hearth, with only one side exposed, and with the load entering the furnace at ambient temperature, the initial gradient from the heated to the unheated surface will be zero. The heated surface will heat more rapidly until the opposite surface starts to heat, after which the temperature differential between surfaces will taper off with time until the desired final differential is achieved.

In Fig. 45.22 the temperatures of heated and unheated surface or core temperature are shown as a function of time. In the lower chart temperatures are plotted directly as a function of time. In the upper chart the logarithm of the temperature ratio (Y = load temperature/source temperature) is plotted as a function of time for a constant source temperature. After a short initial heating time, during which the unheated surface or core temperature reaches its maximum rate of increase, the two curves in the upper diagram become parallel straight lines.

Factors considered in non-steady-state conduction and their identifying symbols are listed in Table 45.6.

Charts have been prepared by Gurney-Lurie, Heisler, Hottel, and others showing values for Y_s and Y_c for various R factors as a function of X. Separate charts are provided for Y_s and Y_c, with a series of curves representing a series of values of R. These curves are straight lines for most of their length, curving to intersect at $Y = 1$ and $X = 0$. If straight lines are extended to $Y = 1$, the curves for Y_c at all values of R converge at a point near $X = 0.1$ on the line for $Y_c = 1$. It is accordingly possible to prepare a single line chart for $-\ln Y_c/(X - 0.1)$ to fit selected geometrical shapes. This has been done in Fig. 45.23 for slabs, long cylinders, and spheres. Values of Y_c determined with this chart correspond closely with those from conventional charts for $X - 0.1$ greater than 0.2.

Because the ratio Y_s/Y_c remains constant as a function of R after initial heating, it can be shown in chart form, as in Fig. 45.24, to allow Y_s to be determined after Y_c has been found.

By way of illustration, a carbon steel slab 8 in. thick is being heated from cold to $T_s = 2350$°F in a furnace with a constant wall temperature of 2400°F, with a view factor of 1 and a mean emissivity–absorptivity factor of 0.80. The desired final temperature of the unheated surface is 2300°F, making the Y_c factor

$$Y_c = \frac{2400 - 2300}{2400 - 70} = 0.0429$$

From Fig. 45.23 $H_r = 114 \times 0.80 = 91$; $r = \frac{8}{12} = 0.67$; R is assumed at 17. The required heating time is determined from Fig. 45.24:

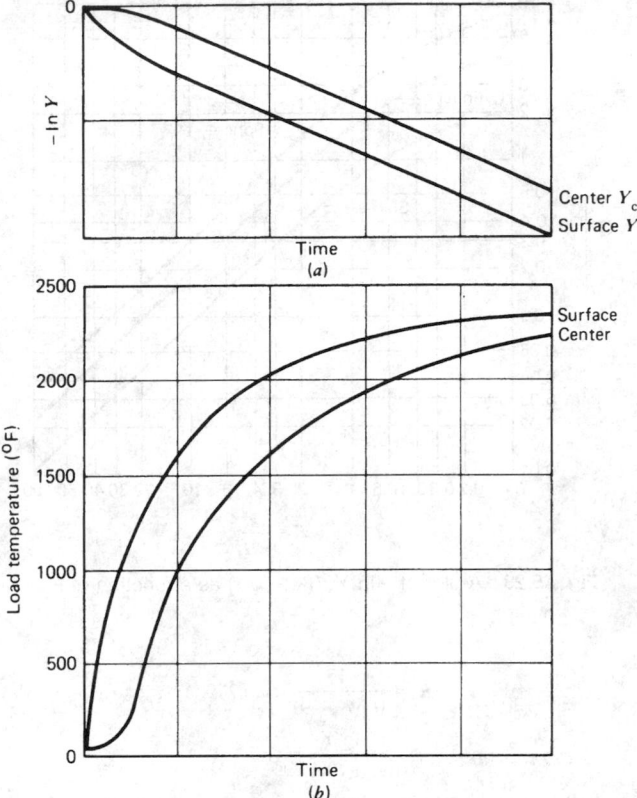

Fig. 45.22 Maximum and minimum load temperatures, and $-\ln Y_s$ or $-\ln Y_c$ as a function of heating time with constant source temperature.[1]

Table 45.6 Non-Steady-State Conduction Factors and Symbols

T_f = Furnace temperature, gas or wall as defined

T_s = Load surface temperature

T_c = Temperature at core or unheated side of load

T_0 = Initial load temperature with all temperatures in units of (°F − 460)/100 or °S

$$Y_s = \frac{T_f - T_s}{T_f - T_0}$$

$$Y_c = \frac{T_f - T_c}{T_f - T_0}$$

R = External/internal thermal resistance ratio = k/rH

X = Time factor = tD/r^2

D = Diffusivity as defined in Section 45.7

r = Depth of heat penetration in feet

k = Thermal conductivity of load (Btu/ft · hr · °F)

H = External heat transfer coefficient (Btu/ft² · hr · °F)

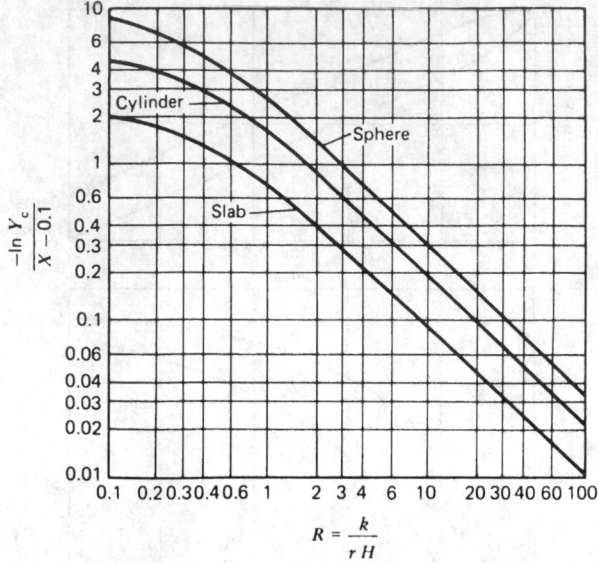

Fig. 45.23 A plot of $-\ln Y_c/(X - 0.1)$ as a function of R.[1]

$$R = \frac{17}{0.67 \times 91} = 0.279$$

$$\frac{-\ln Y_c}{X - 0.1} = 1.7$$

and

$$X = \frac{-\ln 0.0429}{1.7} + 0.1 = 1.95 = tD/r^2$$

With $D = 0.25$, from Section 45.7,

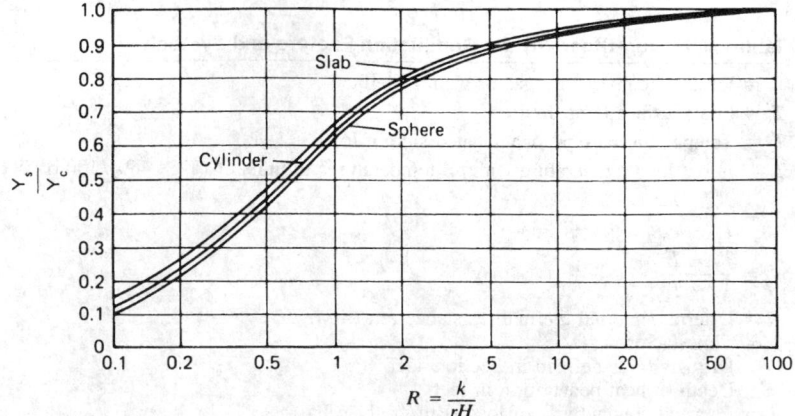

Fig. 45.24 The ratio Y_s/Y_c plotted as a function of R.[1]

$$t = \frac{Xr^2}{D} = \frac{1.95 \times 0.67^2}{0.25} = 3.50 \text{ hr}$$

Slabs or plates heated from two sides are usually supported in the furnace in a horizontal position on spaced conveyor rolls or rails. Support members may be uncooled, in which case radiation to the bottom surface will be reduced by the net view factor. If supports are water cooled, the additional heat input needed to balance heat loss from load to supports can be balanced by a higher furnace temperature on the bottom side. In either case, heating times will be greater than for a uniform input from both sides.

Furnace temperatures are normally limited to a fraction above final load temperatures, to avoid local overheating during operating delays. Without losses to water cooling, top and bottom furnace temperature will accordingly be about equal.

45.8.10 Heat Transfer with Negligible Load Thermal Resistance

When heating thin plates or small-diameter rods, with internal thermal resistance low enough to allow heating rates unlimited by specified final temperature differential, the non-steady-state-conduction limits on heating rates can be neglected. Heating time then becomes

$$t = \frac{W \times C \times (T_s - T_0)}{A \times H \times \text{MTD}}$$

The heat-transfer coefficient for radiation heating can be approximated from the chart in Fig. 45.13 or calculated as follows:

$$H_r = \frac{0.1713 e_m F_s [T_f^4 - (T_f - \text{MTD})^4]}{\text{MTD} \times A_s}$$

As an illustration, find the time required to heat a steel plate to 2350°F in a furnace at a uniform temperature of 2400°F. The plate is 0.25 in. thick with a unit weight of 10.2 lb/ft² and is to be heated from one side. Overall emissivity–absorptivity is $e_m = 0.80$. Specific heat is 0.165. The view factor is $F_s = 1$. MTD is

$$\frac{(2400 - 70) - (2400 - 2350)}{\ln(2400 - 70)/(2400 - 2350)} = 588°F$$

$$H_r = \frac{0.1713 \times 0.80 \times 1[28.6^4 - (28.6 - 5.88)^4]}{588} = 93.8$$

$$t = \frac{10.2 \times 0.165(2350 - 70)}{1 \times 93.8 \times 588} = 0.069 \text{ hr}$$

45.8.11 Newman Method

For loads heated from two or more perpendicular sides, final maximum temperatures will be at exposed corners, with minimum temperatures at the center of mass for heating from all sides, or at the center of the face in contact with the hearth for hearth-supported loads heated equally from the remaining sides. For surfaces not fully exposed to radiation, the corrected H factor must be used.

The Newman method can be used to determine final load temperatures with a given heating time t. To find time required to reach specified maximum and minimum final load temperatures, trial calculations with several values of t will be needed.

For a selected heating time t, the factors Y_s and Y_c can be found from charts in Figs. 45.23 and 45.24 for the appropriate values of the other variables—T_s, T_c, H, k, and r—for each of the heat flow paths involved—r_x, r_y, and r_z. If one of these paths is much longer than the others, it can be omitted from calculations:

$$Y_c = Y_{cx} \times Y_{cy} \times Y_{cz}$$
$$Y_s = Y_{sx} \times Y_{sy} \times Y_{sz}$$

For two opposite sides with equal exposure only one is considered. With T_c known, T_s and T_f (furnace temperature, T_g or T_w) can be calculated.

As an example, consider a carbon steel ingot, with dimensions 2 ft × 4 ft × 6 ft, being heated in a direct-fired furnace. The load is supported with one 2 ft × 4 ft face in contact with the refractory hearth and other faces fully exposed to gas and wall radiation. Maximum final temperature will be at an upper corner, with minimum temperature at the center of the 2 ft × 4 ft bottom surface.

Assuming that the load is a somewhat brittle steel alloy, the initial heating rate should be suppressed and heating with a constant gas temperature will be assumed. Heat-transfer factors are then

Flow paths r_s = 1 ft and r_y = 2 ft, the contribution of vertical heat flow, on axis r_z, will be small enough to be neglected.

Desired final temperatures: T_c = 2250°F and T_s (to be found) about 2300°F, with trial factor t = 9 hr.

H from gas to load = 50

k mean value for load = 20 and D = 0.25

Radial heat flow path	rx	ry
r	1	2
$X = tD/r^2$	2.25	0.5625
$R = k/H_r$	0.4	0.2
$-\ln Y_c/(X - 0.1)$ from Fig. 45.23	1.3	1.7
Y_s/Y_c from Fig. 45.24	0.41	0.26
Y_c	0.0611	0.455
Y_s	0.025	0.119

Combined factors:

$$Y_c = 0.0611 \times 0.455 = 0.0278 = \frac{T_g - T_c}{T_g - 70}$$

$$Y_s = 0.025 \times 0.119 = 0.003 = \frac{T_g - T_s}{T_g - 70}$$

For T_c = 2250°F, T_g = 2316°F
T_s = 2309°F

This is close enough to the desired T_s = 2300°F.

The time required to heat steel slabs to rolling temperature, as a function of the thickness heated from one side and the final load temperature differential, is shown in Fig. 45.25. Relative heating times for various hearth loading arrangements, for square billets, are shown in Fig. 45.26. These have

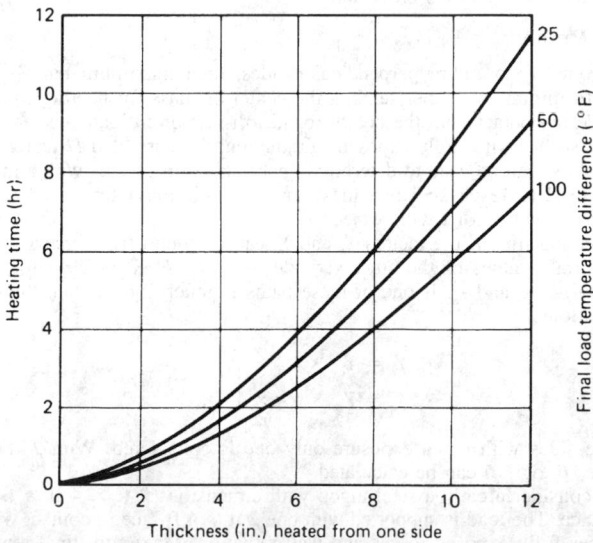

Fig. 45.25 Relative heating time for square billets as a function of loading pattern.[1]

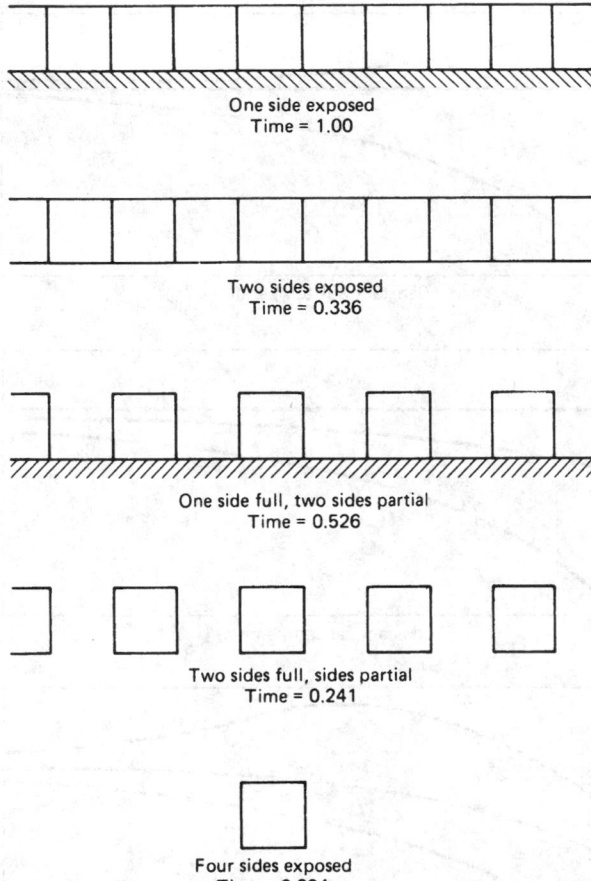

One side exposed
Time = 1.00

Two sides exposed
Time = 0.336

One side full, two sides partial
Time = 0.526

Two sides full, sides partial
Time = 0.241

Four sides exposed
Time = 0.204

Fig. 45.26 Heating time for carbon steel slabs to final surface temperature of 2300°F, as a function of thickness and final load temperature differential.[1]

been calculated by the Newman method, which can also be used to evaluate other loading patterns and cross sections.

45.8.12 Furnace Temperature Profiles

To predict heating rates and final load temperatures in either batch or continuous furnaces, it is convenient to assume that source temperatures, gas (T_g) or furnace wall (T_w), will be constant in time. Neither condition is achieved with contemporary furnace and control system designs. With constant gas temperature, effective heating rates are unnecessarily limited, and the furnace temperature control system is dependent on measurement and control of gas temperatures, a difficult requirement. With uniform wall temperatures, the discharge temperature of flue gases at the beginning of the heating cycle will be higher than desirable. Three types of furnace temperature profiles, constant T_g, constant T_w, and an arbitrary pattern with both variables, are shown in Fig. 45.27.

Contemporary designs of continuous furnaces provide for furnace temperature profiles of the third type illustrated, to secure improved capacity without sacrificing fuel efficiency. The firing system comprises three zones of length: a preheat zone that can be operated to maintain minimum flue gas temperatures in a counterflow firing arrangement, a firing zone with a maximum temperature and firing rate consistent with furnace maintenance requirements and limits imposed by the need to avoid overheating of the load during operating delays, and a final or soak zone to balance furnace temperature with maximum and minimum load temperature specifications. In some designs, the preheat zone is unheated except by flue gases from the firing zone, with the resulting loss of furnace capacity offset by operating the firing zone at the maximum practical limit.

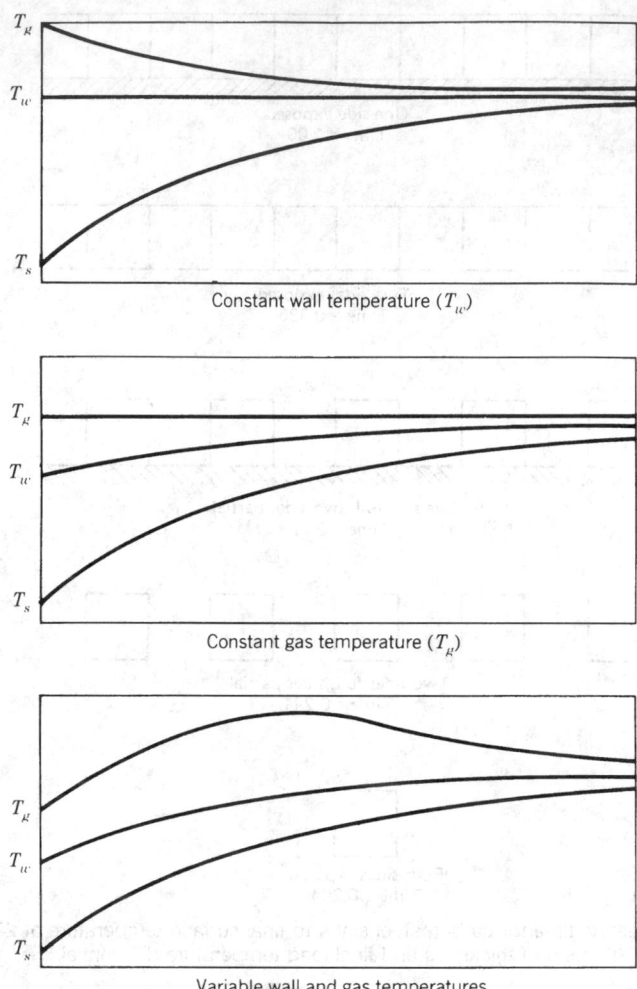

Fig. 45.27   Furnace temperature profiles.

45.8.13   Equivalent Furnace Temperature Profiles

Furnace heating capacities are readily calculated on the assumption that furnace temperature, either combustion gases or radiating walls, is constant as a function of position or time. Neither condition is realized in practice; and to secure improved capacity with reduced fuel demand in a continuous furnace, contemporary designs are based on operation with a variable temperature profile from end to end, with furnace wall temperature reduced at the load charge and flue gas discharge end, to improve available heat of fuel, and at the load discharge end, to balance the desired maximum and minimum load temperatures. Any loss in capacity can be recovered by operating the intermediate firing zones at a somewhat elevated temperature.

Consider a furnace designed to heat carbon steel slabs, 6 in. thick, from the top only to final temperatures of 2300°F at top and 2250°F at the bottom. To hold exit flue gas temperature to about 2000°F, wall temperature at the charge end will be about 1400°F. The furnace will be fired in four zones of length, each 25 ft long for an effective total length of 100 ft. The preheat zone will be unfired, with a wall temperature tapering up to 2400°F at the load discharge end. That temperature will be held through the next two firing zones and dropped to 2333°F to balance final load temperatures in the fourth or soak zone. With overall heating capacity equal to the integral of units of length times their absolute temperatures, effective heat input will be about 87% of that for a uniform temperature of 2400°F for the entire length.

Heat transfer from combustion gases to load will be by direct radiation from gas to load, including reflection of incident radiation from walls, and by radiation from gas to walls, absorbed and reradiated from walls to load. Assuming that wall losses will be balanced by convection heat transfer from gases, gas radiation to walls will equal solid-state radiation from walls to load:

$$A_w/A_s \times 0.1713 \times e_{gm}(T_g^4 - T_w^4) = e_{ws} \times 0.1713(T_w^4 - T_s^4)$$

where A_w/A_s = exposed area ratio for walls and load
$\quad\quad e_{gm}$ = emissivity–absorptivity, gas to walls
$\quad\quad e_{ws}$ = emissivity–absorptivity, walls to load

At the midpoint in the heating cycle, MTD = 708°F and mean load surface temperature = T_{sm} = 1698°F.

With a_s = 0.85 for refractory walls, 15% of gas radiation will be reflected to load, and total gas to load radiation will be:

$$1.15 \times e_{gm} \times 0.1713(T_g^4 - T_s^4)$$

For A_w/A_s = 2.5, e_{gm} = 0.17, and e_{ws} = 0.89 from walls to load, the mean gas temperature = T_g = 3108°F, net radiation, gas to load = 47,042 Btu/hr · ft^2 and gas to walls = walls to load = 69,305 Btu/hr · ft^2 for a total of 116,347 Btu/hr · ft^2. This illustrates the relation shown in Fig. 45.21, since black-body radiation from walls to load, without gas radiation, would be 77,871 Btu/hr · ft^2. Assuming black-body radiation with a uniform wall temperature from end to end, compared to combined radiation with the assumed wall temperature, overall heat transfer ratio will be

$$(0.87 \times 116,347)/77,871 = 1.30$$

As shown in Fig. 45.26, this ratio will vary with gas emissivity and wall to load areas exposed. For the range of possible values for these factors, and for preliminary estimates of heating times, the chart in Fig. 45.26 can be used to indicate a conservative heating time as a function of final load temperature differential and depth of heat penetration, for a furnace temperature profile depressed at either end.

Radiation factors will determine the mean coefficient of wall to load radiation, and the corresponding non-steady-state conduction values. For black-body radiation alone, H_r is about 77,871/708 = 110. For combined gas and solid-state radiation, in the above example, it becomes 0.87 × 116,347/708 = 143. Values of R for use with Figs. 45.23 and 45.24, will vary correspondingly ($R = k/4H$).

45.8.14 Convection Heat Transfer

Heat transferred between a moving layer of gas and a solid surface is identified by "convection." Natural convection occurs when movement of the gas layer results from differentials in gas density of the boundary layer resulting from temperature differences and will vary with the position of the boundary surface: horizontal upward, horizontal downward, or vertical. A commonly used formula is:

$$H_c = 0.27(T_g - T_s)^{0.25}$$

where H_c = Btu/hr · ft^2 · °F
$\quad T_g - T_s$ = temperature difference between gas and surface, in °F

Natural convection is a significant factor in estimating heat loss from the outer surface of furnace walls or from uninsulated pipe surfaces.

"Forced convection" is heat transfer between gas and a solid surface, with gas velocity resulting from energy input from some external source, such as a recirculating fan.

Natural convection can be increased by ambient conditions such as building drafts and gas density. Forced convection coefficients will depend on surface geometry, thermal properties of the gas, and Reynolds number for gas flow. For flow inside tubes, the following formula is useful:

$$H_c = 0.023 \frac{k}{D} \text{Re}^{0.8}\text{Pr}^{0.4} \text{Btu/hr} \cdot \text{ft}^2 \cdot °F$$

where k = thermal conductivity of gas
   D = inside diameter of tube in ft
   Re = Reynolds number
   Pr = Prandtl number

Forced convection coefficients are given in chart form in Fig. 45.28 for a Prandtl number assumed at 0.70.

For forced convection over plane surfaces, it can be assumed that the preceding formula will apply for a rectangular duct of infinitely large cross section, but only for a length sufficient to establish uniform velocity over the cross section and a velocity high enough to reach the Re value needed to promote turbulent flow.

In most industrial applications, the rate of heat transfer by forced convection as a function of power demand will be better for perpendicular jet impingement from spaced nozzles than for parallel flow. For a range of dimensions common in furnace design, the heat-transfer coefficient for jet impingement of air or flue gas is shown in Fig. 45.29, calculated for impingement from slots 0.375 in. wide spaced at 18–24 in. centers and with a gap of 8 in. from nozzle to load.

Forced convection factors for gas flow through banks of circular tubes are shown in the chart in Fig. 45.30 and for tubes spaced as follows:

A: staggered tubes with lateral spacing equal to diagonal spacing.
B: tubes in line, with equal spacing across and parallel to direction of flow.
C: tubes in line with lateral spacing less than half longitudinal spacing.
D: tubes in line with lateral spacing over twice longitudinal spacing.

With F the configuration factor from Fig. 45.30, heat-transfer coefficients are

$$H_c = Fk\,\mathrm{Re}^{0.6}/D$$

Convection coefficients from this formula are approximately valid for 10 rows of tubes or more, but are progressively reduced to a factor of 0.65 for a single row.

For gas to gas convection in a cross-flow tubular heat exchanger, overall resistance will be the sum of factors for gas to the outer diameter of tubes, tube wall conduction, and inside diameter of

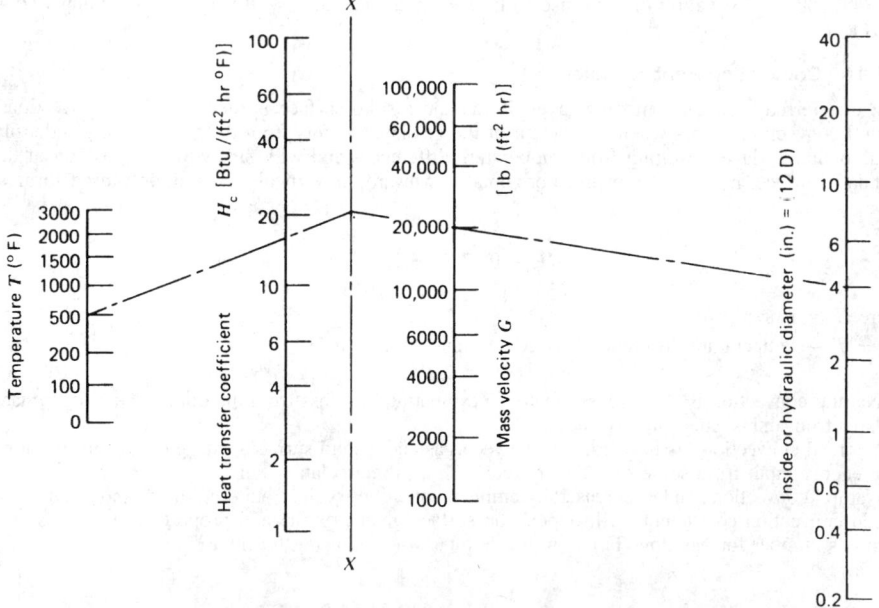

Fig. 45.28   Convection coefficient (H_c) for forced convection inside tubes to air or flue gas.[1]

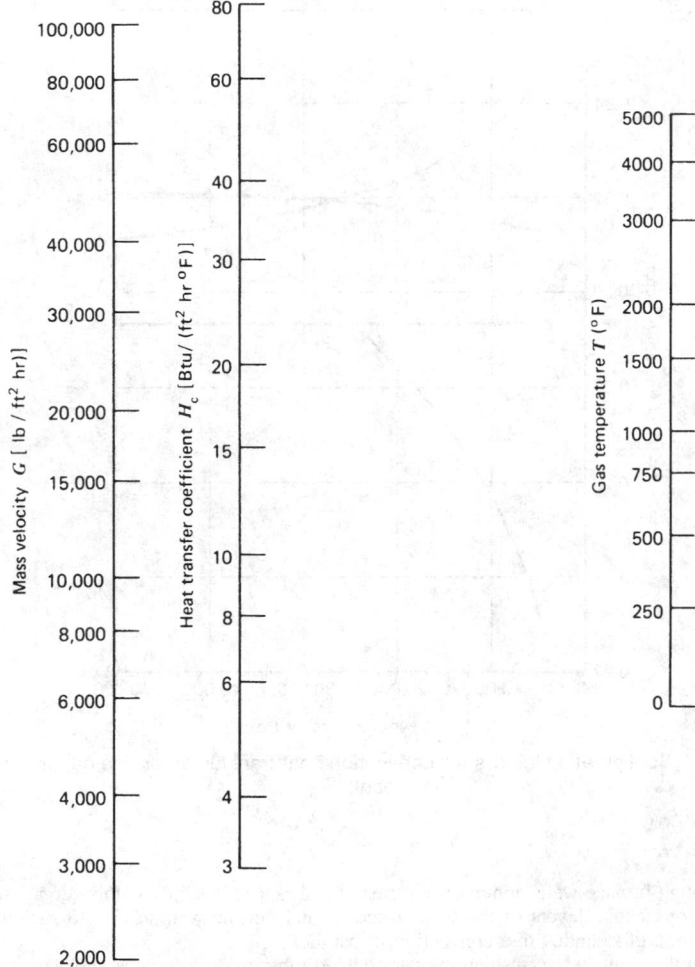

Fig. 45.29 Convection coefficient (H_c) for jet impingement of air or flue gas on plane surfaces, for spaced slots, 0.375 in. wide at 18–24 in. centers, 8 in. from load.[1]

tubes to gas. Factors for the outer diameter of tubes may include gas radiation as calculated in Section 45.7.5.

45.8.15 Fluidized-Bed Heat Transfer

For gas flowing upward through a particular bed, there is a critical velocity when pressure drop equals the weight of bed material per unit area. Above that velocity, bed material will be suspended in the gas stream in a turbulent flow condition. With the total surface area of suspended particles on the order of a million times the inside surface area of the container, convection heat transfer from gas to bed material is correspondingly large. Heat transfer from suspended particles to load is by conduction during repeated impact. The combination can provide overall coefficients upward of 10 times those available with open convection, permitting the heating of thick and thin load sections to nearly uniform temperatures by allowing a low gas to load thermal head.

45.8.16 Combined Heat-Transfer Coefficients

Many furnace heat-transfer problems will combine two or more methods of heat transfer, with thermal resistances in series or in parallel. In a combustion chamber, the resistance to radiation from gas to load will be parallel to the resistance from gas to walls to load, which is two resistances in series.

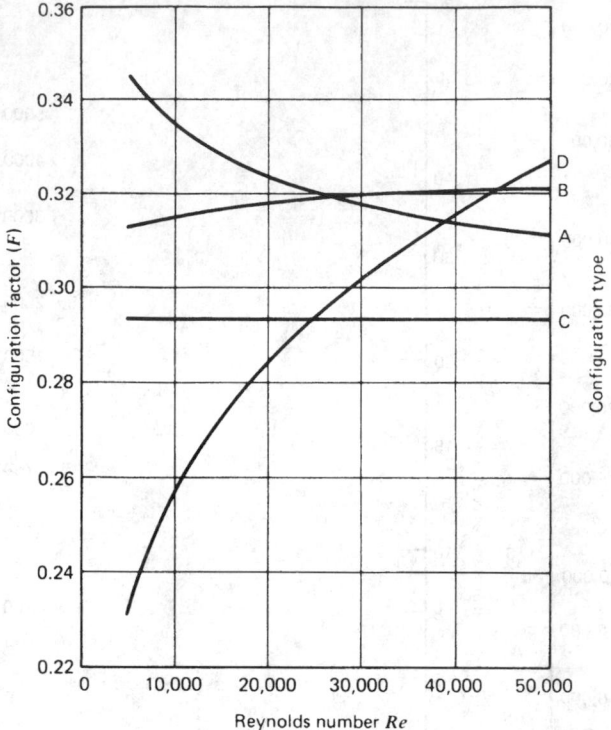

Fig. 45.30 Configuration factors for convection heat transfer, air or flue gas through tube banks.[1]

Heat flow through furnace walls combines a series of resistances in series, combustion gases to inside wall surface, consecutive layers of the wall structure, and outside wall surface to surroundings, the last a combination of radiation and convection in parallel.

As an example, consider an insulated, water-cooled tube inside a furnace enclosure. With a tube outside diameter of 0.5 ft and a cylindrical insulation enclosure with an outside diameter of 0.75 ft, the net thickness will be 0.125 ft. The mean area at midthickness is $\pi(0.5 + 0.75)/2$, or 1.964 ft² per ft of length. Outer surface area of insulation is 0.75π, or 2.36 ft² per linear foot. Conductivity of insulation is $k = 0.20$. The effective radiation factor from gas to surface is assumed at 0.5 including reradiation from walls. For the two resistances in series,

$$0.1713 \times 0.5 \times 2.36(29.6^4 - T_s^4) = 1.964(T_s - 150) \times \frac{0.20}{0.125}$$

By trial, the receiver surface temperature is found to be about 2465°F. Heat transfer is about 7250 Btu/hr · linear ft or 9063 Btu/hr · ft² water-cooled tube surface.

If the insulated tube in the preceding example is heated primarily by convection, a similar treatment can be used to find receiver surface temperature and overall heat transfer.

For radiation through furnace wall openings, heat transfer in Btu/hr · ft² · °F is reduced by wall thickness, and the result can be calculated similarly to the problem of two parallel planes of equal size connected by reradiating walls, as shown in Fig. 45.19.

Heat transfer in internally fired combustion tubes ("radiant tubes") is a combination of convection and gas radiation from combustion gases to tube wall. External heat transfer from tubes to load will be direct radiation and reradiation from furnace walls, as illustrated in Fig. 45.19. The overall factor for internal heat transfer can be estimated from Fig. 45.31, calculated for 6 in. and 8 in. inside diameter tubes. The convection coefficient increases with firing rate and to some extent with tem-

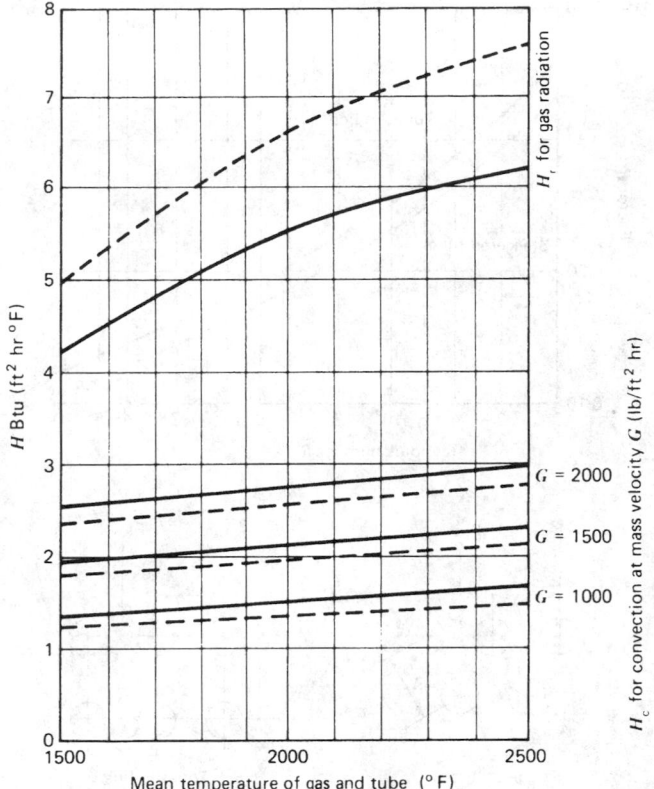

Fig. 45.31 Gas radiation (H_r) and convection (H_c) coefficients for flue gas inside radiant tubes.[1]

perature. The gas radiation factor depends on temperature and inside diameter. The effect of flame luminosity has not been considered.

45.9 FLUID FLOW

Fluid flow problems of interest to the furnace engineer include the resistance to flow of air or flue gas, over a range of temperatures and densities through furnace ductwork, stacks and flues, or recuperators and regenerators. Flow of combustion air and fuel gas through distribution piping and burners will also be considered. Liquid flow, of water and fuel oil, must also be evaluated in some furnace designs but will not be treated in this chapter.

To avoid errors resulting from gas density at temperature, velocities will be expressed as mass velocities in units of $G = \text{lb/hr} \cdot \text{ft}^2$. Because the low pressure differentials in systems for flow of air or flue gas are usually measured with a manometer, in units of inches of water column (in. H_2O), that will be the unit used in the following discussion.

The relation of velocity head h_v in in. H_2O to mass velocity G is shown for a range of temperatures in Fig. 45.32. Pressure drops as multiples of h_v are shown, for some configurations used in furnace design, in Figs. 45.33 and 45.34. The loss for flow across tube banks, in multiples of the velocity head, is shown in Fig. 45.35 as a function of the Reynolds number.

The Reynolds number Re is a dimensionless factor in fluid flow defined as $\text{Re} = DG/\mu$, where D is inside diameter or equivalent dimension in feet, G is mass velocity as defined above, and μ is viscosity as shown in Fig. 45.9. Values for Re for air or flue gas, in the range of interest, are shown in Fig. 45.36. Pressure drop for flow through long tubes is shown in Fig. 45.37 for a range of Reynolds numbers and equivalent diameters.

45.9.1 Preferred Velocities

Mass velocities used in contemporary furnace design are intended to provide an optimum balance between construction costs and operating costs for power and fuel; some values are listed on the next page:

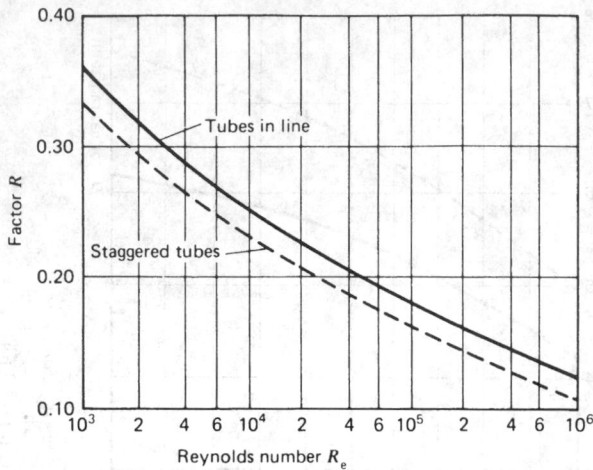

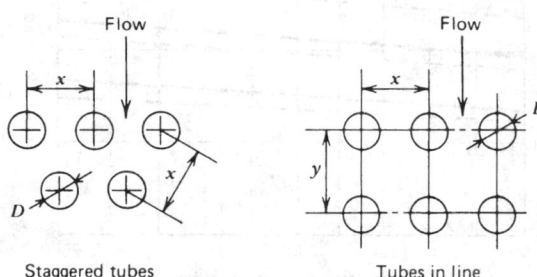

Fig. 45.32 Heat loss for flow of air or flue gas across tube banks at atmospheric pressure (velocity head) $\times F \times R$.

Medium	Mass Velocity G	Velocity Head (in. H_2O)
Cold air	15,000	0.7
800°F air	10,000	0.3
2200°F flue gas	1,750	0.05
1500°F flue gas	2,000	0.05

The use of these factors will not necessarily provide an optimum cost balance. Consider a furnace stack of self-supporting steel construction, lined with 6 in. of gunned insulation. For $G = 2000$ and $h_v = 0.05$ at 1500°F, an inside diameter of 12 ft will provide a flow of 226,195 lb/hr. To provide a net draft of 1 in. H_2O with stack losses of about 1.75 h_v or 0.0875 in., the effective height from Fig. 45.38 is about 102 ft. By doubling the velocity head to 0.10 in. H_2O, G at 1500°F becomes 3000. For the same mass flow, the inside diameter is reduced to 9.8 ft. The pressure drop through the stack increases to about 0.175 in., and the height required to provide a net draft of 1 in. increases to about 110 ft. The outside diameter area of the stack is reduced from 4166 ft² to $11 \times 3.1416 \times 110 = 3801$ ft². If the cost per square foot of outside surface is the same for both cases, the use of a higher stack velocity will save construction costs. It is accordingly recommended that specific furnace designs receive a more careful analysis before selecting optimum mass velocities.

Stack draft, at ambient atmospheric temperature of 70°F, is shown in Fig. 45.38 as a function of flue gas temperature. Where greater drafts are desirable with a limited height of stack, a jet-type stack can be used to convert the momentum of a cold air jet into stack draft. Performance data are available from manufacturers.

45.9.2 Centrifugal Fan Characteristics

Performance characteristics for three types of centrifugal fans are shown in Fig. 45.39. More exact data are available from fan manufacturers. Note that the backward curved blade has the advantage

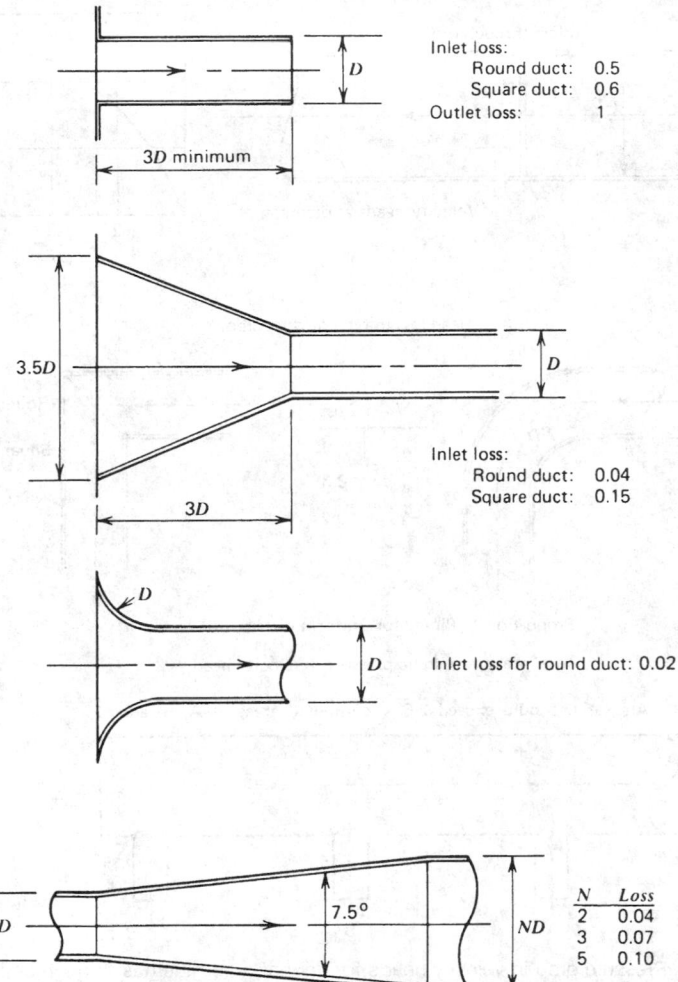

Fig. 45.33 Pressure drop in velocity heads for flow of air or flue gas through entrance configurations or expansion sections.[1]

of limited horsepower demand with reduced back pressure and increasing volume, and can be used where system resistance is unpredictable. The operating point on the pressure–volume curve is determined by the increase of duct resistance with flow, matched against the reduced outlet pressure, as shown in the upper curve.

45.9.3 Laminar and Turbulent Flows

The laminar flow of a fluid over a boundary surface is a shearing process, with velocity varying from zero at the wall to a maximum at the center of cross section or the center of the top surface for liquids in an open channel. Above a critical Reynolds number, between 2000 and 3000 in most cases, flow becomes a rolling action with a uniform velocity extending almost to the walls of the duct, and is identified as turbulent flow.

With turbulent flow the pressure drop is proportional to D; the flow in a large duct can be converted from turbulent to laminar by dividing the cross-sectional area into a number of parallel channels. If flow extends beyond the termination of these channels, the conversion from laminar to turbulent flow will occur over some distance in the direction of flow.

Radial mixing with laminar flow is by the process of diffusion, which is the mixing effect that occurs in a chamber filled with two different gases separated by a partition after the partition is removed. Delayed mixing and high luminosity in the combustion of hydrocarbon gases can be ac-

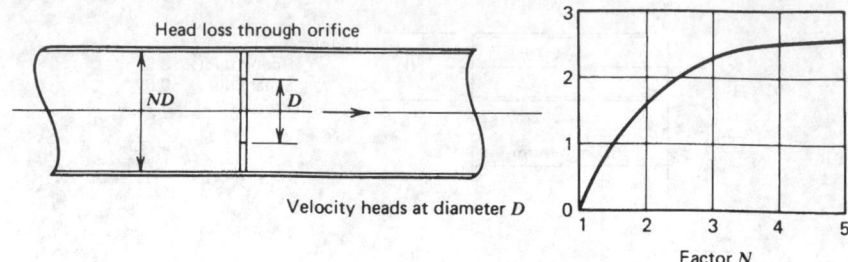

Head loss through orifice

Velocity heads at diameter D

Head loss in pipe or duct elbows

N	$Loss$
0.5	1.0
1	0.3
2	0.2

Round: 1.0

Square: 1.2

Proportioning Piping for uniform distribution

Total pressure = static pressure + velocity head

Area at D should exceed 2.5 × combined areas of A, B, and C

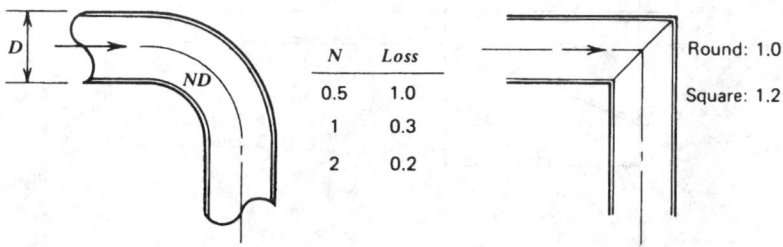

Fig. 45.34 Pressure drop in velocity heads for flow of air or flue gas through orifices, elbows, and lateral outlets.[1]

Staggered Tubes		Tubes in Line		Factor F for x/D		
x/D	Factor F	y/D	1.5	2	3	4
1.5	2.00	1.25	1.184	0.576	0.334	0.268
2	1.47	1.5	1.266	0.656	0.387	0.307
3	1.22	2	1.452	0.816	0.497	0.390
4	1.14	3	1.855	1.136	0.725	0.572
		4	2.273	1.456	0.957	0.761

complished by "diffusion combustion," in which air and fuel enter the combustion chamber in parallel streams at equal and low velocity.

45.10 BURNER AND CONTROL EQUIPMENT

With increasing costs of fuel and power, the fraction of furnace construction and maintenance costs represented by burner and control equipment can be correspondingly increased. Burner designs should be selected for better control of flame pattern over a wider range of turndown and for complete combustion with a minimum excess air ratio over that range.

Furnace functions to be controlled, manually or automatically, include temperature, internal pressure, fuel/air ratio, and adjustment of firing rate to anticipated load changes. For intermittent operation, or for a wide variation in required heating capacity, computer control may be justified to

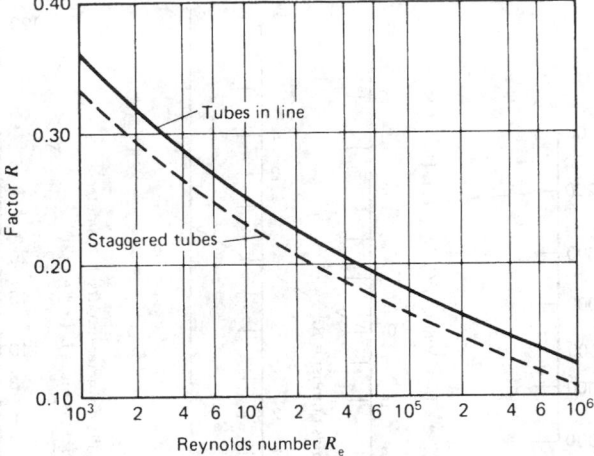

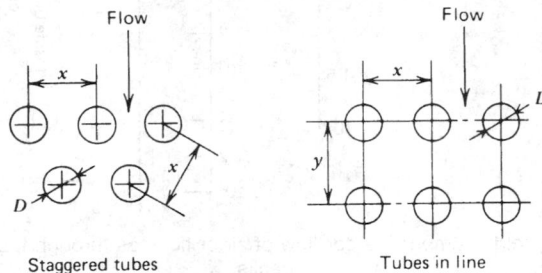

Fig. 45.35 Pressure drop factors for flow of air or flue gas through tube banks.[1]

| Staggered Tubes | | Tubes in Line | | Factor F for x/D | | |
x/D	Factor F	y/D	1.5	2	3	4
1.5	2.00	1.25	1.184	0.576	0.334	0.268
2	1.47	1.5	1.266	0.656	0.387	0.307
3	1.22	2	1.452	0.816	0.497	0.390
4	1.14	3	1.855	1.136	0.725	0.572
		4	2.273	1.456	0.957	0.761

anticipate required changes in temperature setting and firing rates, particularly in consecutive zones of continuous furnaces.

45.10.1 Burner Types

Burners for gas fuels will be selected for the desired degree of premixing of air and fuel, to control flame pattern, and for the type of flame pattern, compact and directional, diffuse or flat flame coverage of adjacent wall area. Burners for oil fuels, in addition, will need provision for atomization of fuel oil over the desired range of firing rates.

The simplest type of gas burner comprises an opening in a furnace wall, through which combustion air is drawn by furnace draft, and a pipe nozzle to introduce fuel gas through that opening. Flame pattern will be controlled by gas velocity at the nozzle and by excess air ratio. Fuel/air ratio will be manually controlled for flame appearance by the judgment of the operator, possibly supplemented by continuous or periodic flue gas analysis. In regenerative furnaces, with firing ports serving alternately as exhaust flues, the open pipe burner may be the only practical arrangement.

For one-way fired furnaces, with burner port areas and combustion air velocities subject to control, fuel/air ratio control can be made automatic over a limited range of turndown with several systems, including:

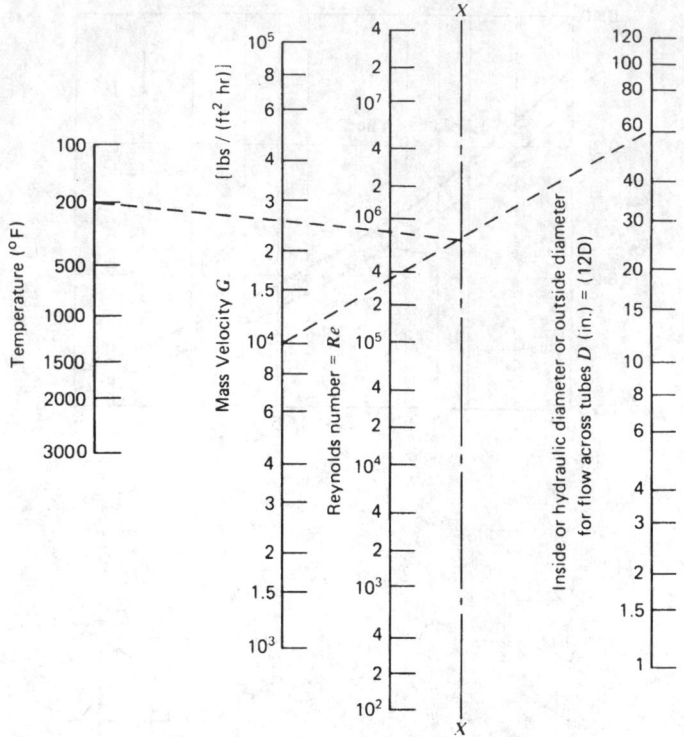

Fig. 45.36 Reynolds number (Re) for flow of air or flue gas through tubes or across tube banks.[1]

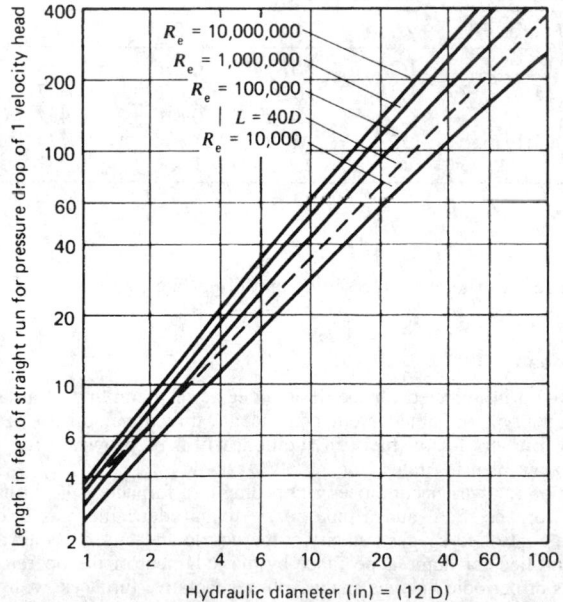

Fig. 45.37 Length in feet for pressure drop of one velocity head, for flow of air or flue gas, as a function of Re and D.[1]

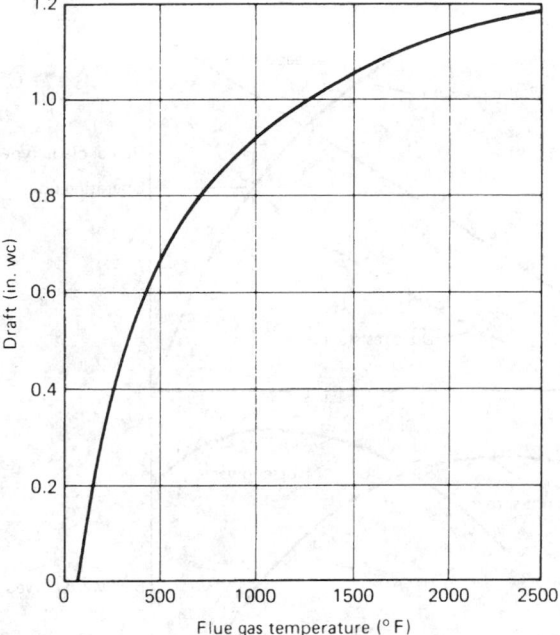

Fig. 45.38 Stack draft for ambient $T_g = 70°F$ and psia $= 14.7$ lb/in.^{2}.[1]

Mixing in venturi tube, with energy supplied by gas supply inducing atmospheric air. Allows simplest piping system with gas available at high pressure, as from some natural gas supplies.

Venturi mixer with energy from combustion air at intermediate pressure. Requires air supply piping and distribution piping from mixing to burners.

With both combustion air and fuel gas available at intermediate pressures, pressure drops through adjustable orifices can be matched or proportioned to hold desired flow ratios. For more accurate control, operation of flow control valves can be by an external source of energy.

Proportioning in venturi mixers depends on the conservation of momentum—the product of flow rate and velocity or of orifice area and pressure drop. With increased back pressure in the combustion chamber, fuel/air ratio will be increased for the high pressure gas inspirator, or decreased with air pressure as the source of energy, unless the pressure of the induced fluid is adjusted to the pressure in the combustion chamber.

The arrangement of a high-pressure gas inspirator system is illustrated in Fig. 45.40. Gas enters the throat of the venturi mixer through a jet on the axis of the opening. Air is induced through the surrounding area of the opening, and ratio control can be adjusted by varying the air inlet opening by a movable shutter disk. A single inspirator can supply a number of burners in one firing zone, or a single burner.

For the air primary mixing system, a representative arrangement is shown in Fig. 45.41. The gas supply is regulated to atmospheric, or to furnace gas pressure, by a diaphragm-controlled valve. Ratio control is by adjustment of an orifice in the gas supply line. With air flow the only source of energy, errors in proportioning can be introduced by friction in the gas-pressure control valve. Each mixer can supply one or more burners, representing a control zone.

With more than one burner per zone, the supply manifold will contain a combustible mixture that can be ignited below a critical port velocity to produce a backfire that can extinguish burners and possibly damage the combustion system. This hazard has made the single burner per mixer combination desirable, and many contemporary designs combine mixer and burner in a single structure.

With complete premixing of fuel and air, the flame will be of minimum luminosity, with combustion complete near the burner port. With delayed mixing, secured by introducing fuel and air in separate streams, through adjacent openings in the burner, or by providing a partial premix of fuel with a fraction of combustion air, flame luminosity can be controlled to increase flame radiation.

In a burner providing no premix ahead of the combustion chamber, flame pattern is determined by velocity differentials between air and fuel streams, and by the subdivision of air flow into several

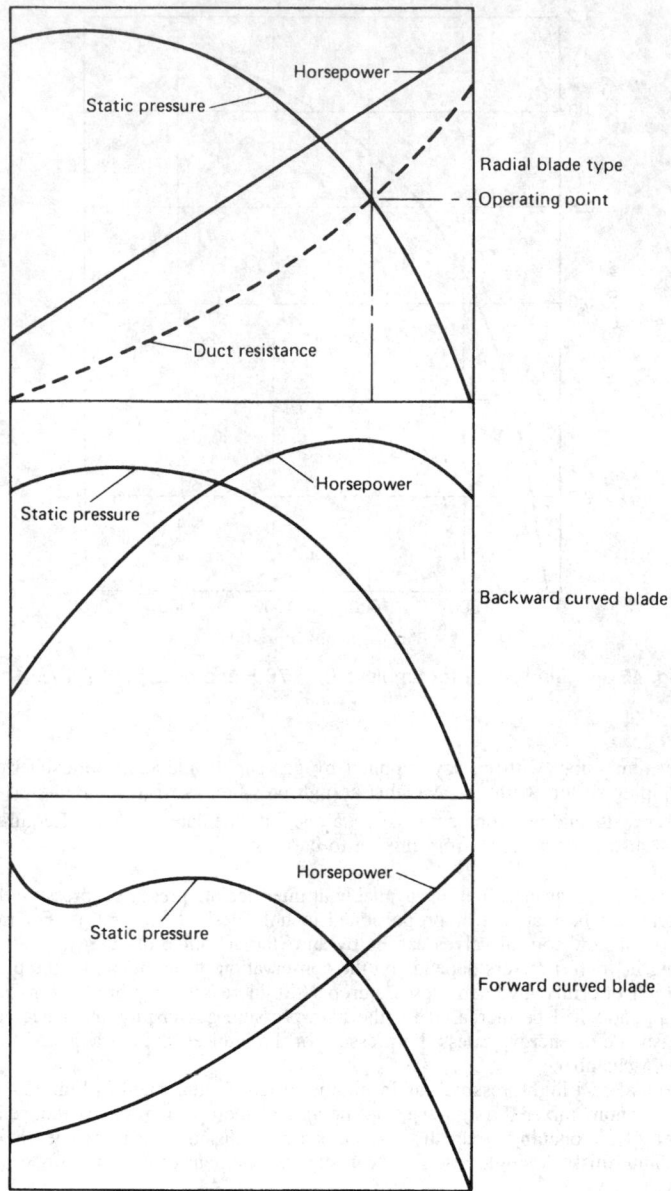

Radial blade type
Operating point

Backward curved blade

Forward curved blade

Fig. 45.39 Centrifugal fan characteristics.[1]

parallel streams. This type of burner is popular for firing with preheated combustion air, and can be insulated for that application.

Partial premix can be secured by dividing the air flow between a mixing venturi tube and a parallel open passage.

With the uncertainty of availability of contemporary fuel supplies, dual fuel burners, optionally fired with fuel gas or fuel oil, can be used. Figure 45.42 illustrates the design of a large burner for firing gas or oil fuel with preheated air. For oil firing, an oil-atomizing nozzle is inserted through the gas tube. To avoid carbon buildup in the oil tube from cracking of residual oil during gas firing, the oil tube assembly is removable.

Oil should be atomized before combustion in order to provide a compact flame pattern. Flame length will depend on burner port velocity and degree of atomization. Atomization can be accom-

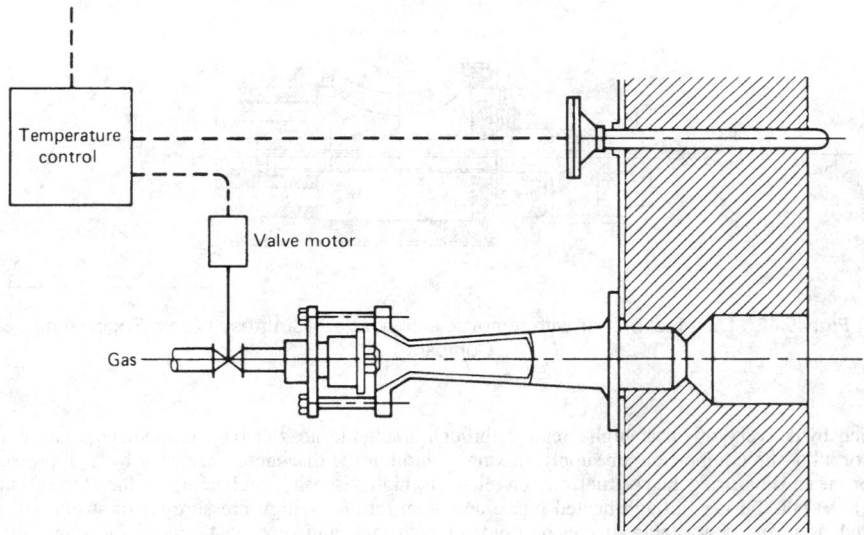

Fig. 45.40 Air/gas ratio control by high-pressure gas inspirator.[1]

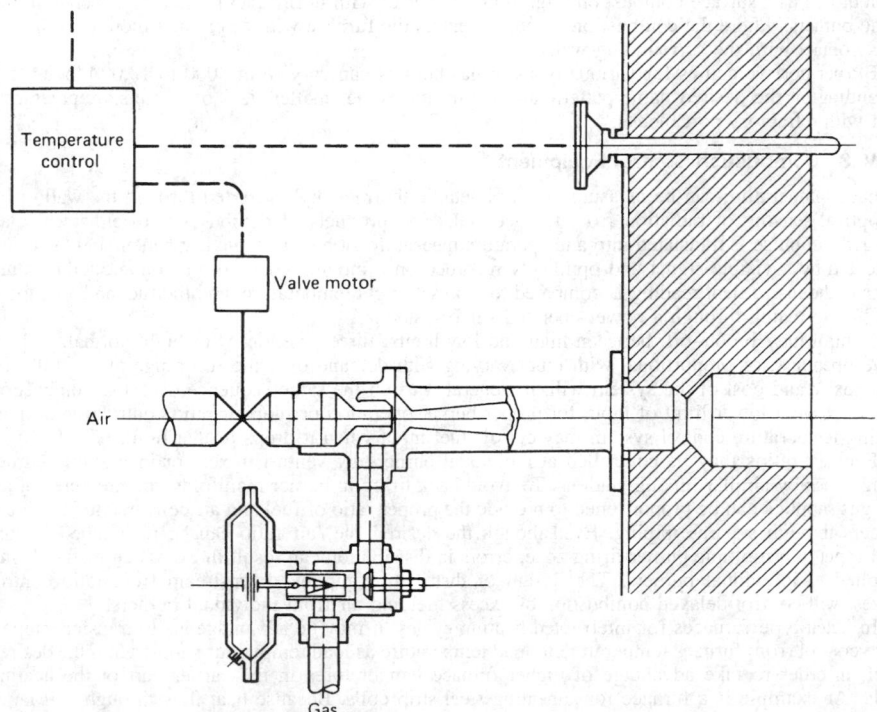

Fig. 45.41 Air/gas ratio control by air inspirator.[1]

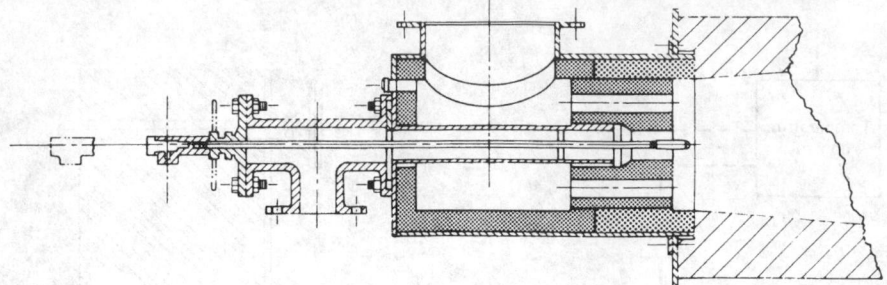

Fig. 45.42 Dual fuel burner with removable oil nozzle.[1] (Courtesy Bloom Engineering Company.)

plished by delivery of oil at high pressure through a suitable nozzle; by intermediate pressure air, part or all of the combustion air supply, mixing with oil at the discharge nozzle; or by high-pressure air or steam. For firing heavy fuel oils of relatively high viscosity, preheating in the storage tank, delivery to the burner through heated pipes, and atomization by high-pressure air or steam will be needed. If steam is available, it can be used for both tank and pipe heating and for atomization. Otherwise, the tank and supply line can be electrically heated, with atomization by high-pressure air.

45.10.2 Burner Ports

A major function of fuel burners is to maintain ignition over a wide range of demand and in spite of lateral drafts at the burner opening. Ignition can be maintained at low velocities by recirculation of hot products of combustion at the burner nozzle, as in the bunsen burner, but stability of ignition is limited to low port velocities for both the entering fuel/air mixture and for lateral drafts at the point of ignition. Combustion of a fuel/air mixture can be catalyzed by contact with a hot refractory surface. A primary function of burner ports is to supply that source of ignition. Where combustion of a completely mixed source of fuel and air is substantially completed in the burner port, the process is identified as "surface combustion." Ignition by contact with hot refractory is also effective in flat flame burners, where the combustion air supply enters the furnace with a spinning motion and maintains contact with the surrounding wall.

Burner port velocities for various types of gas burners can vary from 3000 to 13,000 lb/hr · ft², depending on the desired flame pattern and luminosity. Some smaller sizes of burners are preassembled with refractory port blocks.

45.10.3 Combustion Control Equipment

Furnace temperature can be measured by a bimetallic thermocouple inserted through the wall or by an optical sensing of radiation from furnace walls and products of combustion. In either case, an electrical impulse is translated into a temperature measurement by a suitable instrument and the result indicated by a visible signal and optionally recorded on a moving chart. For automatic temperature control, the instrument reading is compared to a preset target temperature, and the fuel and air supply adjusted to match through a power-operated valve system.

Control may be on–off, between high and low limits; three position, with high, normal, and off valve openings; or proportional with input varying with demand over the full range of control. The complexity and cost of the system will, in general, vary in the same sequence. Because combustion systems have a lower limit of input for proper burner operation or fuel/air ratio control, the proportioning temperature control system may cut off fuel input when it drops to that limit.

Fuel/air ratios may be controlled at individual burners by venturi mixers or in multiple burner firing zones by similar mixing stations. To avoid back firing in burner manifolds, the pressures of air and gas supplies can be proportioned to provide the proper ratio of fuel and air delivered to individual burners through separate piping. Even though the desired fuel/air ratio can be maintained for the total input to a multiple burner firing zone, errors in distribution can result in excess air or fuel being supplied to individual burners. The design of distribution piping, downstream from ratio control valves, will control delayed combustion of excess fuel and air from individual burners.

In batch-type furnaces for interrupted heating cycles, it may be advantageous to transfer temperature control from furnace temperature to load temperature as load temperature approaches the desired level, in order to take advantage of higher furnace temperatures in the earlier part of the heating cycle. An example is a furnace for annealing steel strip coils. Because heat flow through coil laminations is a fraction of that parallel to the axis of the coil, coils may be stacked vertically with open coil separators between them, to provide for heat transfer from recirculated furnace atmosphere to the end surfaces of coils. For bright annealing, the furnace atmosphere will be nonoxidizing, and the load will be enclosed in an inner cover during heating and cooling, with the atmosphere recirculated by a centrifugal fan in the load support base, to transfer heat from the inner cover to end faces of coils. There will also be some radiation heat transfer from the inner cover to the cylindrical surface of the coil stack.

Inner covers are usually constructed of heat-resisting alloy, with permissible operating temperatures well above the desired final load temperature. A preferred design provides for initial control of furnace inside wall temperature from a thermocouple inserted through the furnace wall, with control switched to a couple in the support base, in control with the bottom of the coil stack, after load temperature reaches a present level below the desired final temperature.

To avoid leakage of combustion gases outward through furnace walls, with possible overheating of the steel enclosure, or infiltration of cold air that could cause nonuniform wall temperatures, control of internal furnace pressure to slightly above ambient is desirable. This can be accomplished by an automatic damper in the outlet flue, adjusted to hold the desired pressure at the selected point in the furnace enclosure. In furnaces with door openings at either end, the point of measurement should be close to hearth level near the discharge end. A practical furnace pressure will be 0.01–0.05 in. H_2O.

With recuperative or regenerative firing systems, the preferred location of the control damper will be between the waste-heat recovery system and the stack, to operate at minimum temperature. In high-temperature furnaces without waste-heat recovery, a water-cooled damper may be needed.

With combustion air preheated before distribution to several firing zones, the ratio control system for each zone will need adjustment to entering air temperature. However, if each firing zone has a separate waste-heat recovery system, the zone air supply can be measured before preheating to maintain the balance with fuel input.

The diagram of a combustion control system in Fig. 45.43 shows how these control functions can be interlocked with the required instrumentation.

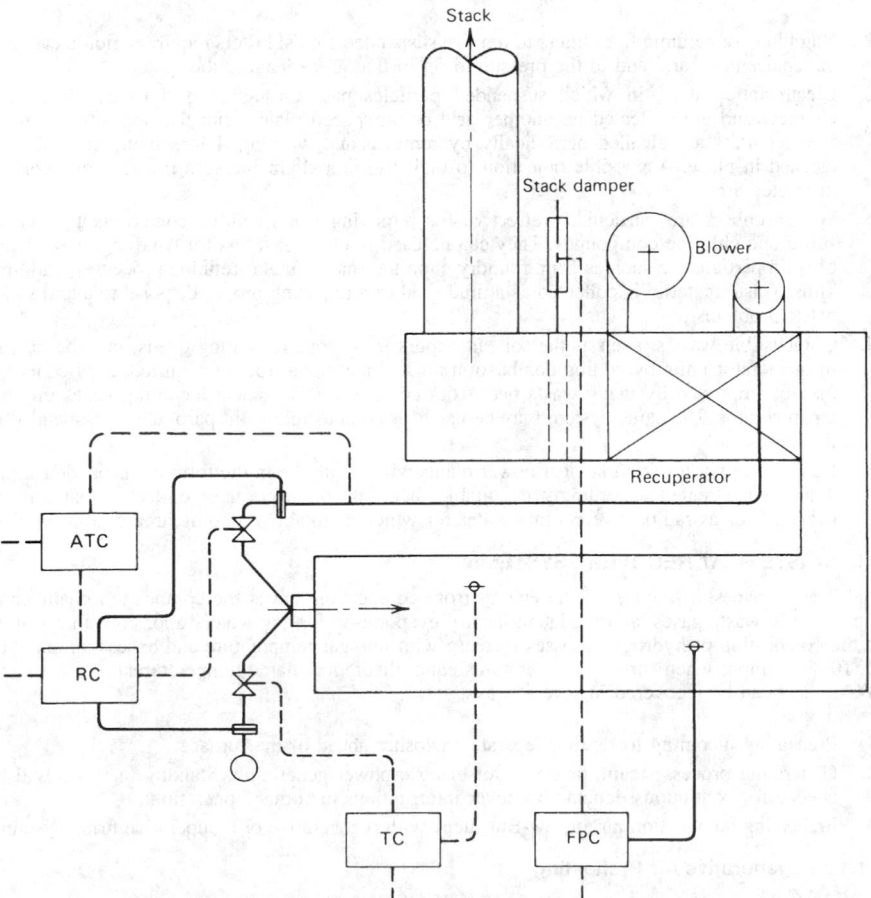

Fig. 45.43 Combustion control diagram for recuperative furnace.[1]

For automatic furnace pressure control to be effective, it should be used in combination with proportioning-type temperature control. With on–off control, for example, the control of furnace pressure at zero firing rate cannot be accomplished by damper adjustment, and with a continuous variation in firing rate between maximum and minimum limits, or between maximum and off, the adjustment of damper position to sudden changes in firing rate will involve a time-lag factor that can make control ineffective.

An important function of a furnace control system is to guard against safety hazards, such as explosions, fires, and personal injury. Requirements have been well defined in codes issued by industrial insurers, and include provision for continuous ignition of burners in low-temperature furnaces, purging of atmosphere furnaces and combustion of hydrogen or carbon monoxide in effluent atmospheres, and protection of operating personnel from injury by burning, mechanical contact, electrical shock, poisoning by inhalation of toxic gases, or asphyxiation. Plants with extensive furnace operation should have a safety engineering staff to supervise selection, installation, and maintenance of safety hazard controls and to coordinate the instruction of operating personnel in their use.

45.10.4　Air Pollution Control

A new and increasing responsibility of furnace designers and operators is to provide controls for toxic, combustible, or particulate materials in furnace flue gases, to meet federal or local standards for air quality. Designs for furnaces to be built in the immediate future should anticipate a probable increase in restrictions of air pollution in coming years.

Toxic contaminants include sulfur and chlorine compounds, nitrogen oxides, carbon monoxide, and radioactive wastes. The epidermic of "acid rain" in areas downwind from large coal-burning facilities is an example.

Combustible contaminants include unburned fuel, soot, and organic particulates from incinerators, and the visible constituents of smoke, except for steam. Other particulates include suspended ash and suspended solids from calcination processes.

Types of control equipment include:

1. Bag filters or ceramic fiber filters to remove suspended solids. Filters require periodic cleaning or replacement, and add to the pressure drop in flue gases leaving the system.
2. Electrostatic filters, in which suspended particles pass through a grid to be electrically charged, and are collected on another grid or on spaced plates with the opposite potential. Smaller units are cleaned periodically by removal and washing. Large industrial units are cleaned in place. A possible objection to their use is a slight increase in the ozone content of treated air.
3. Wet scrubbers are particularly effective for removing water-soluble contaminants such as sulfur and chlorine compounds. They can be used in place of filters for handling heavy loads of solid particulates such as from foundry cupola furnaces, metal-refining processes, and lime kilns. Waste material is collected as a mud or slurry, requiring proper disposal to avoid solid-waste problems.
4. Combustible wastes, such as the solvent vapors from organic coating ovens, may be burned in incinerator units by adding combustion air and additional fuel as required. Fuel economy may be improved by using waste heat from combustion to preheat incoming gases through a recuperator. The same system may be used for combustible solid particulates suspended in flue gases.
5. Radioactive wastes from nuclear power plants will usually be in the form of suspended solids that can be treated accordingly if suitable facilities for disposal of collected material are available, or as radioactive cooling water for which a suitable dumping area will be needed.

45.11　WASTE HEAT RECOVERY SYSTEMS

In fuel-fired furnaces, a fraction of the energy from combustion leaves the combustion chamber as sensible heat in waste gases, and the latent heat of evaporation for any water vapor content resulting from the combustion of hydrogen. Losses increase with flue gas temperature and excess air, and can reach 100% of input when furnace temperatures equal theoretical flame temperatures.

Waste heat can be recovered in several ways:

1. Preheating incoming loads in a separate enclosure ahead of the furnace.
2. Generating process steam, or steam for electric power generation. Standby facilities will be needed for continuous demand, to cover interruptions of furnace operation.
3. Preheating combustion air, or low-Btu fuels, with regenerative or recuperative firing systems.

45.11.1　Regenerative Air Preheating

For the high flue gas temperatures associated with glass- and metal-melting processes, for which metallic recuperators are impractical, air may be preheated by periodical reversal of the direction of firing, with air passing consecutively through a hot refractory bed or checker chamber, the furnace combustion chamber, and another heat-storage chamber in the waste-gas flue. The necessary use of the furnace firing port as an exhaust port after reversal limits the degree of control of flame patterns and the accuracy of fuel/air control in multiple port furnaces. Regenerative firing is still preferred, however, for open hearth furnaces used to convert blast furnace iron to steel, for large glass-melting furnaces, and for some forging operations.

A functional diagram of a regenerative furnace is shown in Fig. 45.44. The direction of flow of combustion air and flue gas is reversed by a valve arrangement, connecting the low-temperature end of the regenerator chamber to either the combustion air supply or the exhaust stack. Fuel input is reversed simultaneously, usually by an interlocked control. Reversal can be in cycles of from 10 to 30 min duration, depending primarily on furnace size.

45.11.2 Recuperator Systems

Recuperative furnaces are equipped with a heat exchanger arranged to transfer heat continuously from outgoing flue gas to incoming combustion air. Ceramic heat exchangers, built up of refractory tubes or refractory block units arranged for cross flow of air and flue gas, have the advantage of higher temperature limits for incoming flue gas, and the disadvantage of leakage of air or flue gas between passages, with leakage usually increasing with service life and pressure differentials. With the improvement in heat-resistant alloys to provide useful life at higher temperatures, and with better control of incoming flue gas temperatures, metallic recuperators are steadily replacing ceramic types.

Metal recuperators can be successfully used with very high flue gas temperatures if entering temperatures are reduced by air dilution or by passing through a high-temperature waste-heat boiler.

Familiar types of recuperators are shown in the accompanying figures:

Figure 45.45: radiation or stack type. Flue gases pass through an open cylinder, usually upward, with heat transfer primarily by gas radiation to the surrounding wall. An annular passage is provided between inner and outer cylinders, in which heat is transferred to air at high velocity by gas radiation and convection, or by solid-state radiation from inner to outer cylinders and convection. The radiation recuperator has the advantage of acting as a portion of an exhaust stack, usually with flue gas and air counterflow. Disadvantages are distortion and resulting uneven distribution of air flow, resulting from differential thermal expansion of the inner tube, and the liability of damage from secondary combustion in the inner chamber.

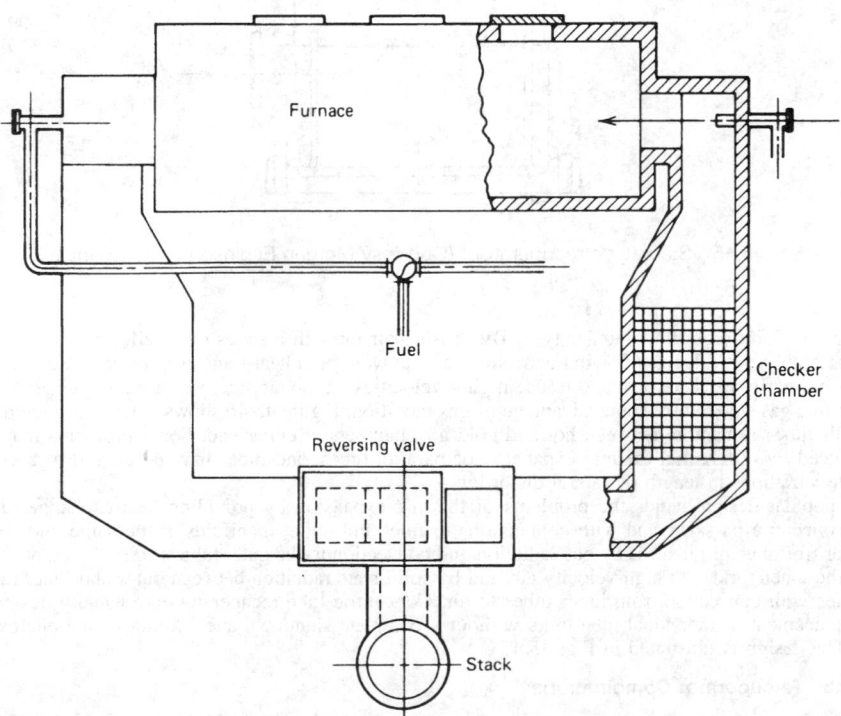

Fig. 45.44 Regenerative furnace diagram.[1]

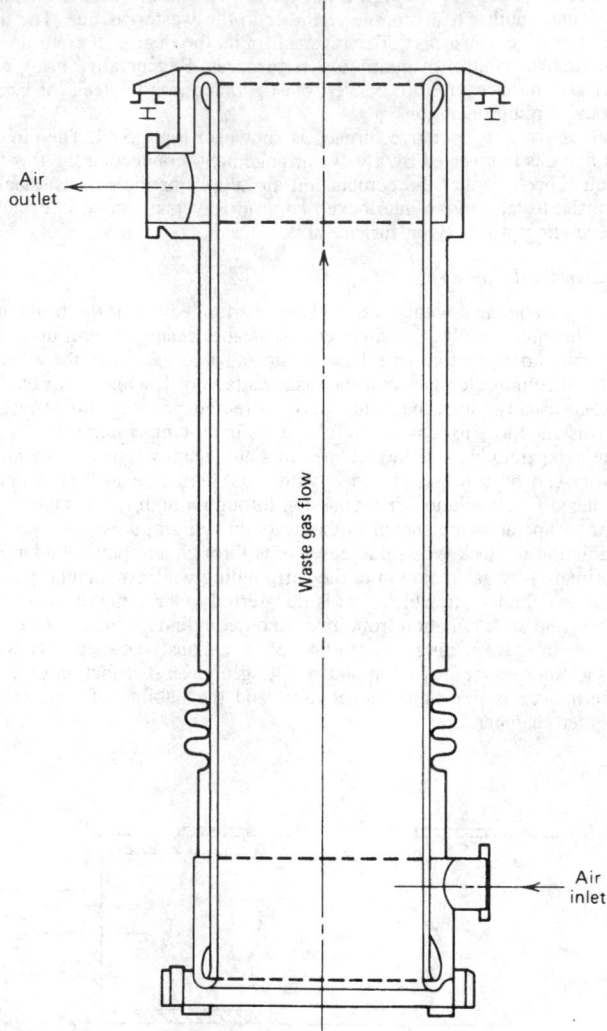

Fig. 45.45 Stack-type recuperator.[1] (Courtesy Morgan Engineering Company.)

Figure 45.46: cross-flow tubular type. By passing air through a series of parallel passes, as in a tube assembly, with flue gas flowing across tubes, relatively high heat-transfer rates can be achieved. It will ordinarily be more practical to use higher velocities on the air side, and use an open structure on the flue gas side to take some advantage of gas radiation. Figure 45.46 shows a basic arrangement, with air tubes in parallel between hot and cold air chambers at either end. Some problems may be introduced by differential thermal expansion of parallel tubes, and tubes may be curved to accommodate variations in length by lateral distortion.

A popular design avoids the problems of thermal expansion by providing heat-exchange tubes with concentric passages and with connections to inlet and outlet manifolds at the same end. Heat transfer from flue gas to air is by gas radiation and convection to the outer tube surface, by convection from the inner surface to high-velocity air, and by solid-state radiation between outer and inner tubes in series with convection from inner tubes to air. Concentric tube recuperators are usually designed for replacement of individual tube units without a complete shutdown and cooling of the enclosing flue. The design is illustrated in Fig. 45.47.

45.11.3 Recuperator Combinations

To provide preheated air at pressure required for efficient combustion, without excessive air leakage from the air to the flue gas side in refractory recuperators, the air pressure can be increased between

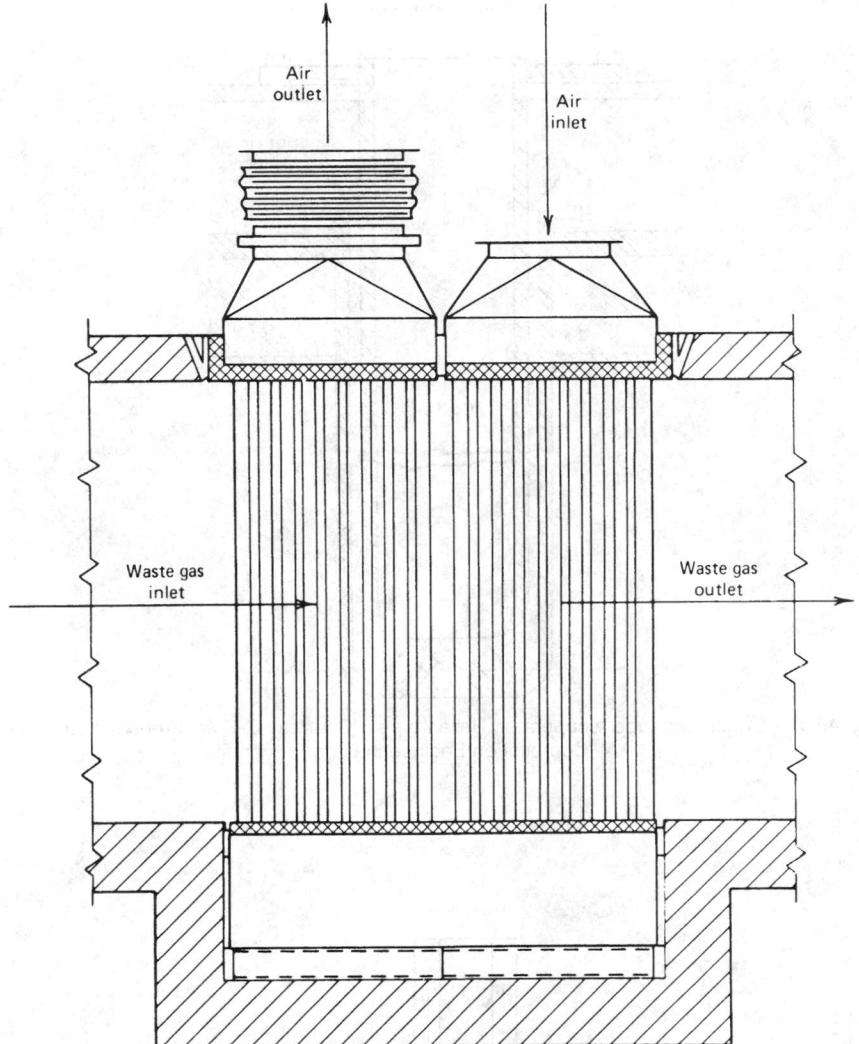

Fig. 45.46 Cross-flow-type recuperator in waste-gas flue.[1] (Courtesy Thermal Transfer Corporation.)

the recuperator and burner by a booster fan. Top air temperatures will be limited by fan materials. As an alternative, air temperatures can be boosted by a jet pump with tolerance for much higher temperatures.

In a popular design for recuperator firing of soaking pits, flue gases pass through the refractory recuperator at low pressure, with air flowing counterflow at almost the same pressure. Air flow is induced by a jet pump, and, to increase the jet pump efficiency, the jet air can be preheated in a metal recuperator between the refractory recuperator and the stack. Because the metal recuperator can handle air preheated to the limit of the metal structure, power demand can be lowered substantially below that for a cold air jet.

Radiant tubes can be equipped with individual recuperators, as shown in Fig. 45.48. Some direct-firing burners are available with integral recuperators.

45.12 FURNACE COMPONENTS IN COMPLEX THERMAL PROCESSES

An industrial furnace, with its auxiliaries, may be the principal component in a thermal process with functions other than heating and cooling. For example, special atmosphere treatment of load surfaces, to increase or decrease carbon content of ferrous alloys, can be accomplished in a furnace heated by

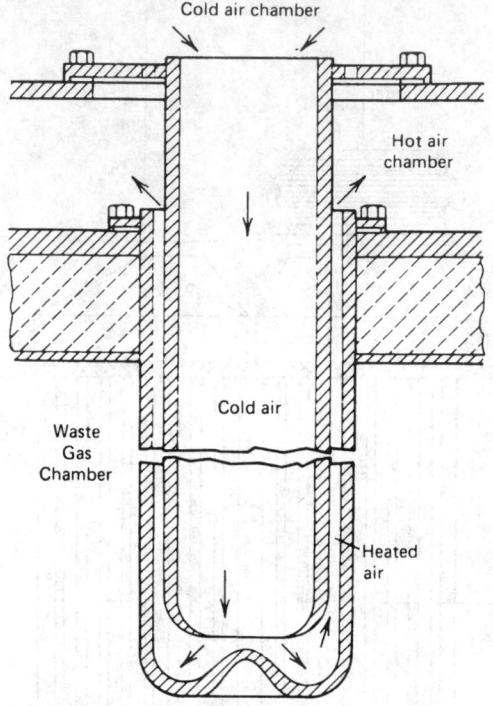

Fig. 45.47 Concentric tube recuperator, Hazen type.[1] (Courtesy C-E Air Preheater Division, Combustion Engineering, Inc.)

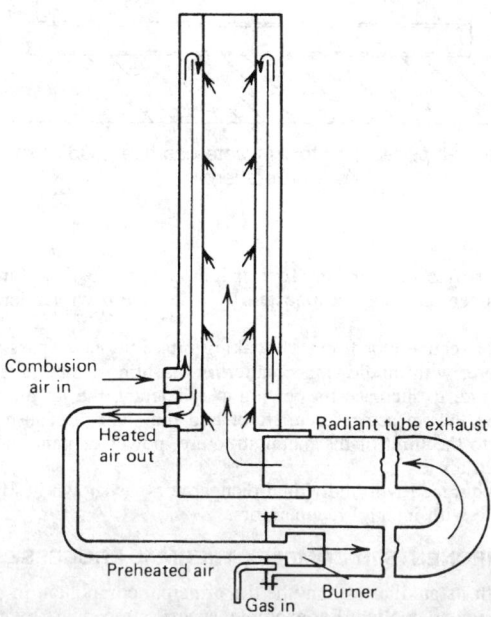

Fig. 45.48 Radiant tube recuperator.[1] (Courtesy Holcroft Division, Thermo-Electron Corp.)

radiant tubes or electrical heating elements or by electric induction. A source of the required controlled atmosphere is usually part of the furnace process equipment, designed and supplied by the furnace manufacturer.

Continuous heat treatment of strip or wire, to normalize or anneal ferrous materials, followed by coating in molten metal, such as zinc or aluminum, or electroplating can be accomplished by one of two arrangements for furnace coating lines. One arrangement has a sequence of horizontal passes, with a final cooling zone to regulate strip temperature to the approximate temperature of the coating bath, and an integral molten-metal container. Strip is heat treated in a controlled atmosphere to avoid oxidation, with the same atmosphere maintained to the point of immersion in molten metal. The second arrangement is for higher velocities and longer strands in heating and cooling passes. In this arrangement, strip may be processed in a series of vertical strands, supported by conveyor rolls.

Furnace lines designed for either galvanizing or aluminum coating may be designed with two molten-metal pots, with the entry strand arranged to be diverted to either one, and with the cooling zone adjustable to discharge the strand to either pot at the required temperature.

Thermal processing lines may include furnace equipment for heating the load to the temperature required for annealing, normalizing, or hardening, a quench tank for oil or water cooling to develop hardness, a cleaning station to remove quench oil residues, and a separate tempering furnace to develop the desired combination of hardness and toughness. Loads may be in continuous strand form, or in units carried by trays or fixtures that go through the entire process or carried on a series of conveyors. The required atmosphere generator will be part of the system.

Where exposure to hydrogen or nitrogen in furnace atmospheres may be undesirable, as in heat treatment of some ferrous alloys, heating and cooling can be done in a partial vacuum, usually with heat supplied by electrical resistors. Quenching can be done in a separate chamber with a controlled atmosphere suitable for brief exposure.

Systems for collecting operating data from one or more furnaces, and transmitting the data to a central recording or controlling station, may also be part of the responsibility of the furnace supplier.

45.13   FURNACE CAPACITY

Factors limiting the heating capacity of industrial furnaces include building space limitations, available fuel supplies, limited temperature of heat sources such as electric resistors or metal radiant tubes, and limits on final load temperature differentials. Other factors under more direct control by furnace designers are the choice between batch and continuous heating cycles; time–temperature cycles to reach specified final load temperatures; fuel firing arrangements; and control systems for furnace temperature, furnace pressure, and fuel/air ratios. In addition, the skills and motivation of furnace operating personnel, as the result of training, experience, and incentive policies, will directly affect furnace efficiency.

45.14   FURNACE TEMPERATURE PROFILES

Time–temperature patterns can be classified as uniform wall temperature (T_w), uniform combustion gas temperature (T_g), or variable T_w and T_g designed to secure the best combination of heating capacity and fuel efficiency.

In a batch-type furnace with fairly massive loads, the temperature control system can be arranged to allow firing at the maximum burner capacity until a preset wall temperature limit is reached, adjusting firing rate to hold that wall temperature, until load temperature approaches the limit for the heated surface, and reducing the wall temperature setting to hold maximum load temperature T_s while the minimum T_c reaches the desired level.

In continuous furnaces, control systems have evolved from a single firing zone, usually fired from the discharge end with flue gas vented from the load charge end, to two or three zone firing arranged for counterflow relation between furnace loads and heating gases.

Progress from single to multiple zone firing has improved heating rates, by raising furnace temperatures near the charge end, while increasing fuel demand by allowing higher temperatures in flue gas leaving the preheat zone. Load temperature control has been improved by allowing lower control temperatures in the final zone at the discharge end.

With multiple zone firing, the control system can be adjusted to approach the constant-gas-temperature model, constant wall temperature, or a modified system in which both T_g and T_w vary with time and position. Because gas temperatures are difficult to measure directly, the constant-gas-temperature pattern can be simulated by an equivalent wall temperature profile. With increasing fuel costs, temperature settings in a three-zone furnace can be arranged to discharge flue gases at or below the final load temperature, increasing the temperature setting in the main firing zone to a level to provide an equilibrium wall and load temperature, close to the desired final load temperature, during operating delays, and setting a temperature in the final or soak zone slightly above the desired final load surface temperature.

45.15   REPRESENTATIVE HEATING RATES

Heating times for various furnace loads, loading patterns, and time–temperature cycles can be calculated from data on radiation and non-steady-state conduction. For preliminary estimates, heating times for steel slabs to rolling temperatures, with a furnace temperature profile depressed at the entry end, have been estimated on a conservative basis as a function of thickness heated from one side and final load temperature differential and are shown in Fig. 45.26. The ratios for heating time required for square steel billets, in various loading patterns, are shown in Fig. 45.25. For other rectangular cross sections and loading patterns, heating times can be calculated by the Newman method.

Examples of heating times required to reach final load temperatures of T_s = 2300°F and T_c = 2350°F, with constant furnace wall temperatures, are:

1. 12-in.-thick carbon steel slab on refractory hearth with open firing: 9 hr at 54.4 lb/hr · ft².
2. 4-in.-thick slab, same conditions as 1: 1.5 hr at 109 lb/hr · ft².
3. 4 in. square carbon steel billets loaded at 8 in. centers on a refractory hearth: 0.79 hr at 103 lb/hr · ft².
4. 4 in. square billets loaded as in 3, but heated to T_s = 1650°F and T_c = 1600°F for normalizing: 0.875 hr at 93 lb/hr · ft².
5. Thin steel strip, heated from both sides to 1350°F by radiant tubes with a wall temperature of 1700°F, total heating rate for both sides: 70.4 lb/hr · ft².
6. Long aluminum billets, 6 in. diameter, are to be heated to 1050°F. Billets will be loaded in multiple layers separated by spacer bars, with wind flow parallel to their length. With billets in lateral contact and with wind at a mean temperature of 1500°F, estimated heating time is 0.55 hr.
7. Small aluminum castings are to be heated to 1000°F on a conveyor belt, by jet impingement of heated air. Assuming that the load will have thick and thin sections, wind temperature will be limited to 1100°F to avoid overheating thinner sections. With suitable nozzle spacing and wind velocity, the convection heat-transfer coefficient can be H_c = 15 Btu/hr · ft² and the heating rate 27 lb/hr · ft².

45.16 SELECTING NUMBER OF FURNACE MODULES

For a given heating capacity and with no limits on furnace size, one large furnace will cost less to build and operate than a number of smaller units with the same total hearth area. However, furnace economy may be better with multiple units. For example, where reheating furnaces are an integral part of a continuous hot strip mill, the time required for furnace repairs can reduce mill capacity unless normal heating loads can be handled with one of several furnaces down for repairs. For contemporary hot strip mills, the minimum number of furnaces is usually three, with any two capable of supplying normal mill demand.

Rolling mills designed for operation 24 hr per day may be supplied by batch-type furnaces. For example, soaking-pit-type furnaces are used to heat steel ingots for rolling into slabs. The mill rolling rate is 10 slabs/hr. Heating time for ingots with residual heat from casting averages 4 hr, and the time allowed for reloading an empty pit is 2 hr, requiring an average turnover time of 6 hr. The required number of ingots in pits and spaces for loading is accordingly 60, requiring six holes loaded 10 ingots per hole.

If ingots are poured after a continuous steelmaking process, such as open hearth furnaces or oxygen retorts, and are rolled on a schedule of 18 turns per week, it may be economical at present fuel costs to provide pit capacity for hot storage of ingots cast over weekends, rather than reheating them from cold during the following week.

With over- and underfired slab reheating furnaces, with slabs carried on insulated, water-cooled supports, normal practice has been to repair pipe insulation during the annual shutdown for furnace maintenance, by which time some 50% of insulation may have been lost. By more frequent repair, for example, after 10% loss of insulation, the added cost of lost furnace time, material, and labor may be more than offset by fuel savings, even though total furnace capacity may be increased to offset idle time.

45.17 FURNACE ECONOMICS

The furnace engineer may be called on to make decisions, or submit recommendations for the design of new furnace equipment or the improvement of existing furnaces. New furnaces may be required for new plant capacity or addition to existing capacity, in which case the return on investment will not determine the decision to proceed. Projected furnace efficiency will, however, influence the choice of design.

If new furnace equipment is being considered to replace obsolete facilities, or if the improvement of existing furnaces is being considered to save fuel or power, or to reduce maintenance costs, return on investment will be the determining factor. Estimating that return will require evaluation of these factors:

Projected service life of equipment to be improved

Future costs of fuel, power, labor for maintenance, or operating supervision and repairs, for the period assumed

Cost of production lost during operating interruptions for furnace improvement or strikes by construction trades

Cost of money during the improvement program and interest available from alternative investments

Cost of retraining operating personnel to take full advantage of furnace improvements

45.17.1 Operating Schedule

For a planned annual capacity, furnace size will depend on the planned hours per year of operation, and fuel demand will increase with the ratio of idle time to operating time, particularly in furnaces with water-cooled load supports. If furnace operation will require only a two- or three-man crew, and if furnace operation need not be coordinated with other manufacturing functions, operating costs may be reduced by operating a smaller furnace two or three turns per day, with the cost of overtime labor offset by fuel savings.

On the other hand, where furnace treatment is an integral part of a continuous manufacturing process, the provision of standby furnace capacity to avoid plant shutdown for furnace maintenance or repairs may be indicated.

If furnace efficiency deteriorates rapidly between repairs, as with loss of insulation from water-cooled load supports, the provision of enough standby capacity to allow more frequent repairs may reduce overall costs.

45.17.2 Investment in Fuel-Saving Improvements

At present and projected future costs of gas and oil fuels, the added cost of building more efficient furnaces or modifying existing furnaces to improve efficiency can usually be justified. Possible improvements include better insulation of the furnace structure, modified firing arrangements to reduce flue gas temperatures or provide better control of fuel/air ratios, programmed temperature control to anticipate load changes, more durable insulation of water-cooled load supports and better maintenance of insulation, proportioning temperature control rather than the two position type, and higher preheated air temperatures. For intermittent furnace operation, the use of a low-density insulation to line furnace walls and roofs can result in substantial savings in fuel demand for reheating to operating temperature after idle periods.

The relative costs and availability of gas and oil fuels may make a switch from one fuel to another desirable at any future time, preferably without interrupting operations. Burner equipment and control systems are available, at some additional cost, to allow such changeovers.

The replacement of existing furnaces with more fuel-efficient designs, or the improvement of existing furnaces to save fuel, need not be justified in all cases by direct return on investment. Where present plant capacity may be reduced by future fuel shortages, or where provision should be made for increasing capacity with fuel supplies limited to present levels, cost savings by better fuel efficiency may be incidental.

Government policies on investment tax credits or other incentives to invest in fuel-saving improvements can influence the return on investment for future operation.

REFERENCE

1. C. Cone, *Energy Management for Industrial Furnaces,* Wiley, New York, 1980.

CHAPTER 46

GASEOUS FUELS

Richard J. Reed
North American Manufacturing Company
Cleveland, Ohio

46.1	**INTRODUCTION**	**1505**		46.2.7	Net Heating Value	1507
				46.2.8	Flame Stability	1509
46.2	**NATURAL GAS**	**1505**		46.2.9	Gas Gravity	1509
	46.2.1 Uses and Distribution	1505		46.2.10	Wobbe Index	1512
	46.2.2 Environmental Impact	1505		46.2.11	Flame Temperature	1512
	46.2.3 Sources, Supply, and			46.2.12	Minimum Ignition	
	Storage	1507			Temperature	1512
	46.2.4 Types and Composition	1507		46.2.13	Flammability Limits	1512
	46.2.5 Properties	1507				
	46.2.6 Calorific Value or		46.3	**LIQUEFIED PETROLEUM**	
	Heating Value	1507		**GASES**		**1514**

46.1 INTRODUCTION

Gaseous fuels are generally easier to handle and burn than are liquid or solid fuels. Gaseous fossil fuels include natural gas (primarily methane and ethane) and liquefied petroleum gases (LPG; primarily propane and butane). Gaseous man-made or artificial fuels are mostly derived from liquid or solid fossil fuels. Liquid fossil fuels have evolved from animal remains through eons of deep underground reaction under temperature and pressure, while solid fuel evolved from vegetable remains. Figure 46.1, adapted from Ref. 1, shows the ranges of hydrogen/carbon ratios for most fuels.

46.2 NATURAL GAS

46.2.1 Uses and Distribution

Although primarily used for heating, natural gas is also frequently used for power generation (via steam turbines, gas turbines, diesel engines, and Otto cycle engines) and as feedstock for making chemicals, fertilizers, carbon-black, and plastics. It is distributed through intra- and intercontinental pipe lines in a high-pressure gaseous state and via special cryogenic cargo ships in a low-temperature, high-pressure liquid phase (LNG).

Final street-main distribution for domestic space heating, cooking, water heating, and steam generation is at regulated pressures on the order of a few inches of water column to a few pounds per square inch, gage, depending on local facilities and codes. Delivery to commercial establishments and institutions for the same purposes, plus industrial process heating, power generation, and feedstock, may be at pressures as high as 100 or 200 psig (800 or 1500 kPa absolute). A mercaptan odorant is usually added so that people will be aware of leaks.

Before the construction of cross-country natural gas pipe lines, artificial gases were distributed through city pipe networks, but gas generators are now usually located adjacent to the point of use.

46.2.2 Environmental Impact

The environmental impact of natural gas combustion is generally less than that of liquid or solid fuels. Pollutants from natural gas may be (a) particulates, if burners are poorly adjusted or controlled (too rich, poor mixing, quenching), or (b) nitrogen oxides, in some cases with intense combustion, preheated air, or oxygen enrichment.

Mechanical Engineers' Handbook, 2nd ed., Edited by Myer Kutz.
ISBN 0-471-13007-9 © 1998 John Wiley & Sons, Inc.

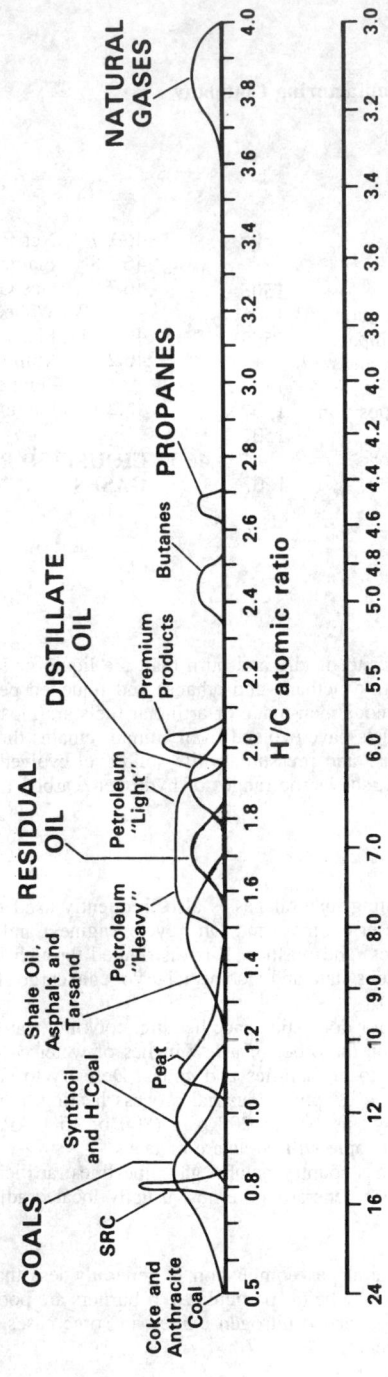

Fig. 46.1 Hydrogen/carbon ratios of fossil and synthetic fuels. (Adapted from Ref. 1.)

46.2.3 Sources, Supply, and Storage

Natural gas is found with oil deposits (animal fossils) and coal deposits (plant fossils). As-yet untapped supplies are known to exist (1) near the coast of the Gulf of Mexico in very deep geopressured/geothermal aquifers and (2) in difficult-to-separate Appalachian shale formations.

Except for these hard-to-extract supplies, U.S. natural gas supplies have been variously predicted to last 10–20 years, but such predictions are questionable because of the effects of economic and regulatory variations on consumption, production, and exploration. Except for transoceanic LNG vessels, distribution is by pipe line, using a small fraction of the fuel in compressors to provide pumping power.

Storage facilities are maintained by many local gas utilities as a cushion for changing demand. These may be low-pressure gas holders with floating bell-covers, old wells or mines (medium pressure), or cryogenic vessels for high-pressure liquefied gas.

46.2.4 Types and Composition

Natural gases are classified as "sweet" or "sour," depending on their content of sulfur compounds. Most such compounds are removed before distribution. Odorants added (so that leaks can be detected) are usually sulfur compounds, but the amount is so minute that it has no effect on performance or pollution.

Various geographic sources yield natural gases that may be described as "high methane," "high Btu," or "high inert."

46.2.5 Properties

Properties that concern most users of natural gases relate to the heat available from their combustion, flow characteristics, and burnability in a variety of burner types. Strangely, few people pay attention to the properties of their gas until they are forced to substitute another fuel for it. Some properties are listed in Table 46.1.[2]

46.2.6 Calorific Value or Heating Value

The gross or higher heating value (HHV) is usually measured in a steady-state calorimeter, which is a small fire-tube heat exchanger with a water-cooled surface area so large that it cools the products of combustion to the temperature at which they entered as fuel and air (usually 60°F). HHV can be calculated from a volumetric analysis and the calorific values of the pure compounds in the gas (Table 46.2). For example, for a natural gas having the analysis shown in column 2 below, the tabulation shows how a weighted average method can be used to determine the calorific value of the mixture:

Col. 1, Constituent	Col. 2, % Volume	Col. 3, HHV from Table 46.2 (Btu/ft³)	Col. 4 = (Col. 3 × Col. 2)/100
Methane, CH_4	90	1013	912
Ethane, C_2H_6	6	1763	106
Nitrogen, N_2	4	0	0
Total	100%		1018 Btu/ft³

It is a convenient coincidence that most solid fossil fuels release about 96–99 gross Btu/ft³ of standard *air;* liquid fossil fuels release about 101–104 Btu/ft³; gaseous fossil fuels about 104–108 Btu/ft³.

This would say that the natural gas in the example above should require about 1017 Btu/ft³ gas divided by 106 Btu/ft³ air = 9.6 ft³ air/ft³ gas. Precise stoichiometric calculations would say 0.909(9.53) + 0.06(16.7) = 9.58 ft³ air/ft³ gas.

46.2.7 Net Heating Value

Because a calorimeter cools the exit gases below their dew point, it retrieves the latent heat of condensation of any water vapor therein. But that latent heat is not recapturable in most practical heating equipment because of concern about corrosion; therefore, it is more realistic to subtract the latent heat from HHV, yielding a net or lower heating value, LHV. This is approximately

$$\frac{LHV}{\text{unit of fuel}} = \frac{HHV}{\text{unit of fuel}} - \left(\frac{970 \text{ Btu}}{\text{lb } H_2O} \times \frac{\text{lb } H_2O}{\text{unit of fuel}}\right)$$

Values for the latter term are listed in Table 46.2. (Note that available heat was discussed in Chapter 44.)

Table 46.1a Analyses of Typical Gaseous Fuels[2]

Type of Gas	Analysis in % by Volume								
	CH_4	C_2H_6	C_3H_8	C_4H_{10}	CO	H_2	CO_2	O_2	N_2
Acetylene, commercial	(97.1% C_2H_2, 2.5% C_3H_6O)							0.084	0.28
Blast furnace	—	—	—	—	27.5	1.0	11.5	—	60.0
Blue (water), bituminous	4.6	—	—	0.7	28.2	32.5	5.5	0.9	27.6
Butane, commercial, natural gas	—	—	6.0	70.7 n-, 23.3 iso-	—	—	—	—	—
Butane, commercial, refinery gas	—	—	5.0	50.1 n-, 16.5 iso-	(28.3% C_4H_6)				
Carbureted blue, low gravity	10.9	2.5	—	6.1	21.9	49.6	3.6	0.4	5.0
Carbureted blue, heavy oil	13.5	—	—	8.2	26.8	32.2	6.0	0.9	12.4
Coke oven, by-product	32.3	—	—	3.2	5.5	51.9	2.0	0.3	4.8
Mapp	—	—	15.0	10.0	(66.0% C_3H_4, 9.0% C_3H_6)				
Natural, Alaska	99.6	—	—	—	—	—	—	—	0.4
Natural, Algerian LNG, Canvey	87.20	8.61	2.74	1.07	—	—	—	—	0.36
Natural, Gaz de Lacq	97.38	2.17	0.10	0.05	—	—	—	—	0.30
Natural, Groningen, Netherlands	81.20	2.90	0.36	0.14	—	—	0.87	—	14.40
Natural, Libyan LNG	70.0	15.0	10.0	3.5	—	—	—	—	0.90
Natural, North Sea, Bacton	93.63	3.25	0.69	0.27	—	—	0.13	—	1.78
Natural, Birmingham, AL	90.0	5.0	—	—	—	—	—	—	5.0
Natural, Cleveland, OH	82.9	11.9		0.3	—	—	0.2	0.3	4.4
Natural, Kansas City, MO	84.1	6.7	—	—	—	—	0.8	—	8.4
Natural, Pittsburgh, PA	83.4	15.8	—	—	—	—	—	—	0.8
Producer, Koppers–Totzek[a]	0.09	—	—	—	55.1	33.7	9.8	—	1.3
Producer, Lurgi[b]	5.0	—	—	—	16.0	25.0	14.0	—	40.0
Producer, W-G, bituminous[b]	2.7	—	—	—	28.6	15.0	3.4	0	50.3
Producer, Winkler[b]	1	—	—	—	10	12	22	—	55
Propane, commercial, natural gas	—	2.2	97.3	0.5	—	—	—	—	—
Propane, commercial, refinery gas	—	2.0	72.9	0.8	(24.3% C_3H_6)				
Sasol, South Africa	28.0	—	—	—	22.0	48.9	1.0		
Sewage, Decatur	68.0	—	—	—	—	2.0	22.0	—	6.0
SNG, no methanation	79.9	—	—	—	1.2	19.0	0.5	—	—

[a]O_2-blown.
[b]Air-blown.

Table 46.1b Properties of Typical Gaseous Fuels[2]

Type of Gas	Gas Gravity	Calorific Value				Gross Btu/ft³ of Standard Air	Gross kcal/m³ of Standard Air
		Btu/ft³		kcal/m³			
		Gross	Net	Gross	Net		
Acetylene, commercial	0.94	1410	1360	12548	12105	115.4	1027
Blast furnace	1.02	92	91	819	819	135.3	1204
Blue (water), bituminous	0.70	260	239	2314	2127	126.2	1121
Butane, commercial, natural gas	2.04	3210	2961	28566	26350	104.9	932.6
Butane, commercial, refinery gas	2.00	3184	2935	28334	26119	106.1	944.2
Carbureted blue, low gravity	0.54	536	461	4770	4102	106.1	944.2
Carbureted blue, heavy oil	0.66	530	451	4716	4013	101.7	905.0
Coke oven, by-product	0.40	569	509	5064	4530	105.0	934
Mapp	1.48	2406	2282	21411	20308	113.7	1011.86
Natural, Alaska	0.55	998	906	8879	8063	104.8	932.6
Natural, Algerian LNG, Canvey	0.64	1122	1014	9985	9024	104.3	928.2
Natural, Gaz de Lacq	0.57	1011	911	8997	8107	104.1	927.3
Natural, Groningen, Netherlands	0.64	875	789	7787	7021	104.4	927.3
Natural, Libyan LNG	0.79	1345	1223	11969	10883	106.1	928.2
Natural, North Sea, Bacton	0.59	1023	922	9104	8205	105.0	934.4
Natural, Birmingham, AL	0.60	1002	904	8917	8045	106.1	945.1
Natural, Cleveland, OH	0.635	1059	959	9424	8534	106.2	942.4
Natural, Kansas City, MO	0.63	974	879	8668	7822	106.3	946.0
Natural, Pittsburgh, PA	0.61	1129	1021	10047	9086	106.3	945.1
Producer, Koppers–Totzek[a]	0.78	288	271	2563	2412	135.2	1203
Producer, Lurgi[b]	0.80	183	167	1629	1486	125.3	1115
Producer, W-G, bituminous[b]	0.84	168	158	1495	1406	129.2	1150
Producer, Winkler[b]	0.98	117	111	1041	988	188.7	1679
Propane, commercial, natural gas	1.55	2558	2358	22764	20984	107.5	956.6
Propane, commercial, refinery gas	1.77	2504	2316	22283	20610	108.0	961.1
Sasol, South Africa	0.55	500	448	4450	3986	114.9	1022
Sewage, Decatur	0.79	690	621	6140	5526	105.3	936.2
SNG, no methanation	0.47	853	765	7591	6808	105.8	943.3

[a]O_2-blown.
[b]Air-blown.

46.2.8 Flame Stability

Flame stability is influenced by burner and combustion chamber configuration (aerodynamic and heat transfer characteristics) and by the fuel properties tabulated in Table 46.3.

46.2.9 Gas Gravity

Gas gravity, G (Table 46.1), is the ratio of the actual gas density relative to the density of dry air at standard temperature and pressure (0.0765 lb/ft³). This should not be confused with "specific gravity," which is the ratio of actual density relative to that of water. Gas gravity for natural gases typically ranges from 0.58 to 0.64, and is used in determination of flow rates and pressure drops through pipe lines, orifices, burners, and regulators:

$$\text{flow} = \text{flow coefficient} \times \text{area (ft}^2) \times \sqrt{2g(\text{psf pressure drop})/\rho}$$

where $g = 32.2$ ft/sec² and $\rho = $ gas gravity $\times 0.0765$. Unless otherwise emphasized, gas gravity is measured and specified at standard temperature and pressure (60°F and 29.92 in Hg).

Table 46.1c Combustion Characteristics of Typical Gaseous Fuels[2]

Type of Gas	Wobbe Index	Vol. Air. Req'd per Vol. Fuel	Stoichiometric Products of Combustion				Total Vol. Vol. Fuel	Flame Temperature (°F)[b]
			% CO_2 Dry[a]	% H_2O Wet	% N_2 Wet			
Acetylene, commercial	1559	12.14	17.4	8.3	75.8		12.66	3966
Blast furnace	91.0	0.68	25.5	0.7	74.0		1.54	2559
Blue (water), bituminous	310.8	2.06	17.7	16.3	68.9		2.77	3399
Butane, commercial, natural gas	2287	30.6	14.0	14.9	73.2		33.10	3543
Butane, commercial, refinery gas	2261	30.0	14.3	14.4	73.4		32.34	3565
Carbureted blue, low gravity	729.4	5.05	14.0	18.9	69.8		5.79	3258
Carbureted blue, heavy oil	430.6	5.21	15.7	16.6	70.3		6.03	3116
Coke oven, by-product	961.2	5.44	10.8	21.4	70.1		6.20	3525
Mapp	1947	21.25	15.6	11.9	74.4		22.59	3722
Natural, Alaska	1352	9.52	11.7	18.9	71.6		10.52	3472
Natural, Algeria LNG, Canvey	1423	10.76	12.1	18.3	71.9		11.85	3483
Natural, Gaz de Lacq	1365	9.71	11.7	18.8	71.6		10.72	3474
Natural, Groningen, Netherlands	1107	8.38	11.7	18.4	72.0		9.40	3446
Natural, Kuwait, Burgan	1364	10.33	12.2	18.3	71.7		10.40	3476
Natural, Libya LNG	1520	12.68	12.5	17.4	72.2		13.90	3497
Natural, North Sea, Bacton	1345	9.74	11.8	18.7	71.7		10.77	3473

Natural, Birmingham, AL	1291	9.44	11.7	18.6	71.8	10.47	3468
Natural, East Ohio	1336	9.70	11.9	18.7	71.7	10.72	3472
Natural, Kansas City, MO	1222	9.16	11.8	18.5	71.9	10.19	3461
Natural, Pittsburgh, PA	1446	10.62	12.0	18.3	71.9	11.70	3474
Producer, BCR, W. Kentucky	444	3.23	23.3	14.7	66.0	3.88	3514
Producer, IGT, Lignite	562	4.43	18.7	17.5	67.0	5.24	3406
Producer, Koppers–Totzek	326.1	2.13	27.7	12.6	63.2	2.69	3615
Producer, Lurgi	204.6	1.46	18.4	15.5	68.9	2.25	3074
Producer, Lurgi, subbituminous	465	2.49	23.4	19.6	61.5	3.20	3347
Producer, W-G, bituminous	183.6	1.30	18.5	9.8	73.5	2.08	3167
Producer, Winkler	118.2	0.62	24.1	9.3	68.9	1.51	3016
Propane, commercial, natural gas	2029	23.8	13.7	15.5	73.0	25.77	3532
Propane, commercial, refinery gas	2008	23.2	14.0	14.9	73.2	25.10	3560
Sasol, South Africa	794.4	4.30	12.8	21.0	68.8	4.94	3584
Sewage, Decatur	791.5	6.55	14.7	18.4	69.7	7.52	3368
SNG, no methanation	1264	8.06	11.3	19.8	71.1	8.96	3485

[a]Ultimate.

[b]Theoretical (calculated) flame temperatures, dissociation considered, with stoichiometrically correct air/fuel ratio. Although these temperatures are lower than those reported in the literature, they are all computed on the same basis; so they offer a comparison of the relative flame temperatures of various fuels.

Table 46.2   Calorific Properties of Some Compounds Found in Gaesous Fuels

Compound	Wobbe Index	Gross Heating Value[d] (Btu/ft³)	Net Heating Value (Btu/ft³)	Pounds, Dry poc[a] per std ft³ of Fuel	Pounds H_2O per std ft³ of Fuel	Air Volume per Fuel Volume
Methane, CH_4	1360	1013	921	0.672	0.0950	9.56
Ethane, C_2H_6	1729	1763	1625	1.204	0.1425	16.7
Propane, C_3H_8	2034	2512	2328	1.437	0.1900	23.9
Butane, C_4H_{10}	2302	3264	3034	2.267	0.2375	31.1
Carbon Monoxide, CO	328	323	323	0.255	0	2.39
Hydrogen, H_2	1228	325	279	0	0.0474	2.39
Hydrogen Sulfide, H_2S	588	640	594	0.5855	0.0474	7.17
N_2, O_2, H_2O, CO_2, SO_2	0	0	0	[b]	[c]	0

[a]poc = products of combustion.
[b]Weight of N_2, O_2, CO_2, and SO_2 in fuel.
[c]Weight of H_2O in fuel.
[d]Higher heating value (HHV).

46.2.10   Wobbe Index

Wobbe index or Wobbe number (Table 46.2) is a convenient indicator of heat input considering the flow resistance of a gas-handling system. Wobbe index is equal to gross heating value divided by the square root of gas gravity; $W = HHV/\sqrt{G}$.

If air can be mixed with a substitute gas to give it the same Wobbe index as the previous gas, the existing burner system will pass the same gross Btu/hr input. This is often invoked when propane–air mixtures are used as standby fuels during natural gas curtailments. To be precise, the amount of air mixed with the propane should then be subtracted from the air supplied through the burner.

The Wobbe index is also used to maintain a steady input despite changing calorific value and gas gravity. Because most process-heating systems have automatic input control (temperature control), maintaining steady input may not be as much of a problem as maintaining a constant furnace atmosphere (oxygen or combustibles).

46.2.11   Flame Temperature

Flame temperature depends on burner mixing aerodynamics, fuel-air ratio, and heat loss to surroundings. It is very difficult to measure with repeatability. Calculated adiabatic flame temperatures, corrected for dissociation of CO_2 and H_2O, are listed in Tables 46.1 and 46.3 for 60°F air; in Chapter 53 it is listed for elevated air temperatures. Obviously, higher flame temperatures produce better heat-transfer rates from flame to load.

46.2.12   Minimum Ignition Temperature

Minimum ignition temperature, Table 46.3, relates to safety in handling, ease of light-up, and ease of continuous self-sustained ignition (without pilot or igniter, which is preferred). In mixtures of gaseous compounds, such as natural gas, the minimum ignition temperature of the mixture is that of the compound with the lowest ignition temperature.

46.2.13   Flammability Limits

Flammability limits (Table 46.3, formerly termed "limits of inflammability") spell out the range of air-to-fuel proportions that will burn with continuous self-sustained ignition. "Lower" and "upper" flammability limits [also termed lower explosive limit (LEL) and upper explosive limit (UEL)] are designated in % gas in a gas–air mixture. For example, the flammability limits of a natural gas are 4.3% and 15%. The 4.3% gas in a gas–air mixture means 95.7% must be air; therefore, the "lean limit" or "lower limit" air/fuel ratio is 95.7/4.3 = 22.3:1, which means that more than 22.3:1 (volume ratio) will be too lean to burn. Similarly, less than (100 − 15)/15 = 5.67:1 is too rich to burn.

Table 46.3 Fuel Properties That Influence Flame Stability[2,a]

Fuel	Minimum Ignition Temperature, °F(°C)	Calculated Flame Temperature, °F(°C)[b]		Flammability Limits, % Fuel Gas by Volume[c]		Laminar Flame Velocity, fps(m/sec)		Percent Theoretical Air for Maximum Flame Velocity
		In Air	In O_2	Lower	Upper	In Air	In O_2	
Acetylene, C_2H_2	581(305)	4770(2632)	5630(3110)	2.5	81.0	8.75(2.67)	—	83
Blast furnace gas	—	2650(1454)	—	35.0	73.5	—	—	—
Butane, commercial	896(480)	3583(1973)	—	1.86	8.41	2.85(0.87)	—	97
Butane, n-C_4H_{10}	761(405)	3583(1973)	—	1.86	8.41	1.3(0.40)	—	55
Carbon monoxide, CO	1128(609)	3542(1950)	—	12.5	74.2	1.7(0.52)	—	55
Carbureted water gas	—	3700(2038)	5050(2788)	6.4	37.7	2.15(0.66)	—	90
Coke oven gas	—	3610(1988)	—	4.4	34.0	2.30(0.70)	—	90
Ethane, C_2H_6	882(472)	3540(1949)	—	3.0	12.5	1.56(0.48)	—	98
Gasoline	536(280)	—	—	1.4	7.6	—	—	—
Hydrogen, H_2	1062(572)	4010(2045)	5385(2974)	4.0	74.2	9.3(2.83)	—	57
Hydrogen sulfide, H_2S	558(292)	—	—	4.3	45.5	—	—	—
Mapp gas, C_3H_4	850(455)	—	5301(2927)	3.4	10.8	—	15.4(4.69)	90
Methane, CH_4	1170(632)	3484(1918)	—	5.0	15.0	1.48(0.45)	14.76(4.50)	90
Methanol, CH_3OH	725(385)	3460(1904)	—	6.7	36.0	—	1.6(0.49)	—
Natural gas	—	3525(1941)	4790(2643)	4.3	15.0	1.00(0.30)	15.2(4.63)	100
Producer gas	—	3010(1654)	—	17.0	73.7	0.85(0.26)	—	90
Propane, C_3H_8	871(466)	3573(1967)	5130(2832)	2.1	10.1	1.52(0.46)	12.2(3.72)	94
Propane, commercial	932(500)	3573(1967)	—	2.37	9.50	2.78(0.85)	—	—
Propylene, C_3H_6	—	—	5240(2893)	—	—	—	—	—
Town gas (Br. coal)	700(370)	3710(2045)	—	4.8	31.0	—	—	—

[a]For combustion with air at standard temperature and pressure.

[b]Flame temperatures are theoretical—calculated for stoichiometric ratio, dissociation considered.

[c]In a fuel-air mix. Example for methane: the lower flammability limit or lower explosive limit, LEL = 5% or 95 volumes air/5 volumes gas = 19.1 air/gas ratio. From Table 46.2, stoichiometric ratio is 9.56:1. Therefore excess air is 19 − 9.56 = 9.44 ft³ air/ft³ gas or 9.44/9.56 × 100 = 99.4% excess air.

Table 46.4a   Physical Properties[a] of LP Gases[b,5]

	Propane	iso-Butane	Butane
Molecular weight	44.09	58.12	58.12
Boiling point, °F	−43.7	+10.9	+31.1
Boiling point, °C	−42.1	−11.7	−0.5
Freezing point, °F	−305.8	−255.0	−216.9
Density of liquid			
Specific gravity, 60°F/60°F	0.508	0.563	0.584
Degrees, API	147.2	119.8	110.6
Lb/gal	4.23	4.69	4.87
Density of vapor (ideal gas)			
Specific gravity (air = 1)	1.522	2.006	2.006
Ft³ gas/lb	8.607	6.53	6.53
Ft³ gas/gal of liquid	36.45	30.65	31.8
Lb gas/1000 ft³	116.2	153.1	153.1
Total heating value (after vaporization)			
Btu/ft³	2,563	3,369	3,390
Btu/lb	21,663	21,258	21,308
Btu/gal of liquid	91,740	99,790	103,830
Critical constants			
Pressure, psia	617.4	537.0	550.1
Temperature, °F	206.2	272.7	306.0
Specific heat, Btu/lb, °F			
c_p, vapor	0.388	0.387	0.397
c_v, vapor	0.343	0.348	0.361
c_p/c_v	1.13	1.11	1.10
c_p, liquid 60°F	0.58	0.56	0.55
Latent heat of vaporization at boiling point, Btu/lb	183.3	157.5	165.6
Vapor pressure, psia			
0°F	37.8	11.5	7.3
70°F	124.3	45.0	31.3
100°F	188.7	71.8	51.6
100°F (ASTM), psig max	210		70
130°F	274.5	109.5	80.8

[a]Properties are for commercial products and vary with composition.
[b]All values at 60°F and 14.696 psia unless otherwise stated.

For the flammability limits of fuel mixtures other than those listed in Table 46.3, the Le Chatelier equation[3] and U.S. Bureau of Mines data[4] can be used.

46.3 LIQUEFIED PETROLEUM GASES

LP gases (LPG) are by-products of natural gas production and of refineries. They consist mainly of propane (C_3H_8), with some butane, propylene, and butylene. They are stored and shipped in liquefied form under high pressure; therefore, their flow rates are usually measured in gallons per hour or pounds per hour. When expanded and evaporated, LPG are heavier than air. Workmen have been asphixiated by LPG in pits beneath leaking LPG equipment.

The rate of LPG consumption is much less than that of natural gas or fuel oils. Practical economics usually limit use to (a) small installations inaccessible to pipe lines, (b) transportation, or (c) standby for industrial processes where oil burning is difficult or impossible.

LPG can usually be burned in existing natural gas burners, provided the air/gas ratio is properly readjusted. On large multiple burner installations an automatic propane–air mixing station is usually installed to facilitate quick changeover without changing air–gas ratios. (See the discussion of Wobbe index, Section 46.2.10.) Some fuel must be consumed to produce steam or hot water to operate a vaporizer for most industrial installations.

Table 46.4 lists some properties of commercial LPG, but it is suggested that more specific information be obtained from the local supplier.

Table 46.4b   Physical Properties[a] of LP Gases[b,5]

	Propane	iso-Butane	Butane
Flash temperature, °F (calculated)	−156	−117	−101
Ignition temperature, °F	932	950	896
Maximum flame temperature in air, °F			
Observed	3497	3452	3443
Calculated	3573	3583	3583
Flammability limits, % gas in air			
Lower	2.37	1.80	1.86
Higher	9.50	8.44	8.41
Maximum rate flame propagation in 1 in. tube			
Inches per second	32	33	33
Percentage gas in air	4.6–4.8	3.6–3.8	3.6–3.8
Required for complete combustion (ideal gas)			
Air, ft^3 per ft^3 gas	23.9	31.1	31.1
lb per lb gas	15.7	15.5	15.5
Oxygen, ft^3 per ft^3 gas	5.0	6.5	6.5
lb per lb gas	3.63	3.58	3.58
Products of combustion (ideal gas)			
Carbon dioxide, ft^3 per ft^3 gas	3.0	4.0	4.0
lb per lb gas	2.99	3.03	3.03
Water vapor, ft^3 per ft^3 gas	4.0	5.0	5.0
lb per lb gas	1.63	1.55	1.55
Nitrogen, ft^3 per ft^3 gas	18.9	24.6	24.6
lb per lb gas	12.0	11.8	11.8

[a]Properties are for commercial products and vary with composition.
[b]All values at 60°F and 14.696 psia unless otherwise stated.

REFERENCES

1. M. G. Fryback, "Synthetic Fuels—Promises and Problems," *Chemical Engineering Progress* (May 1981).
2. R. J. Reed (ed.), *Combustion Handbook,* Vol. I, North American Mfg. Co., Cleveland, OH, 1986, pp. 12, 36–38.
3. F. E. Vandeveer and C. G. Segeler, "Combustion," in *Gas Engineers Handbook,* C. G. Segeler (ed.), Industrial Press, New York, 1965, pp. 2/75–2/76.
4. H. F. Coward and G. W. Jones, *Limits of Flammability of Gases and Vapors* (U.S. Bureau of Mines Bulletin 503), U.S. Government Printing Office, Washington, DC, 1952, pp. 20–81.
5. E. W. Evans and R. W. Miller, "Testing and Properties of LP-Gases," in *Gas Engineers Handbook,* C. G. Segeler (ed.), Industrial Press, New York, 1965, p. 5/11.

CHAPTER 47

LIQUID FOSSIL FUELS FROM PETROLEUM

Richard J. Reed
North American Manufacturing Company
Cleveland, Ohio

47.1	**INTRODUCTION**	**1517**	**47.3 SHALE OILS**	**1528**
47.2	**FUEL OILS**	**1517**	**47.4 OILS FROM TAR SANDS**	**1528**
	47.2.1 Kerosene	1519		
	47.2.2 Aviation Turbine Fuels	1525	**47.5 OIL–WATER EMULSIONS**	**1528**
	47.2.3 Diesel Fuels	1526		
	47.2.4 Summary	1528		

47.1 INTRODUCTION

The major source of liquid fuels is crude petroleum; other sources are shale and tar sands. Synthetic hydrocarbon fuels—gasoline and methanol—can be made from coal and natural gas. Ethanol, some of which is used as an automotive fuel, is derived from vegetable matter.

Crude petroleum and refined products are a mix of a wide variety of hydrocarbons—aliphatics (straight- or branched-chained paraffins and olefins), aromatics (closed rings, six carbons per ring with alternate double bonds joining the ring carbons, with or without aliphatic side chains), and naphthenic or cycloparaffins (closed single-bonded carbon rings, five to six carbons).

Very little crude petroleum is used in its natural state. Refining is required to yield marketable products that are separated by distillation into fractions including a specific boiling range. Further processing (such as cracking, reforming, and alkylation) alters molecular structure of some of the hydrocarbons and enhances the yield and properties of the refined products.

Crude petroleum is the major source of liquid fuels in the United States now and for the immediate future. Although the oil embargo of 1973–1974 intensified development of facilities for extraction of oil from shale and of hydrocarbon liquids from coal, the economics do not favor early commercialization of these processes. Their development has been slowed by an apparently adequate supply of crude oil. Tar sands are being processed in small amounts in Canada, but no commercial facility exists in the United States. (See Table 47.1.)

Except for commercial propane and butane, fuels for heating and power generation are generally heavier and less volatile than fuels used in transportation. The higher the "flash point," the less hazardous is handling of the fuel. (Flash point is the minimum temperature at which the fuel oil will catch fire if exposed to naked flame. Minimum flash points are stipulated by law for safe storage and handling of various grades of oils.) See Table 44.4, Flammability Data for Liquid Fuels.

Properties of fuels reflect the characteristics of the crude. Paraffinic crudes have a high concentration of straight-chain hydrocarbons, which may leave a wax residue with distillation. Aromatic and naphthenic crudes have concentrations of ring hydrocarbons. Asphaltic crudes have a preponderance of heavier ring hydrocarbons and leave a residue after distillation. (See Table 47.2.)

47.2 FUEL OILS

Liquid fuels in common use are broadly classified as follows:

1. Distillate fuel oils derived directly or indirectly from crude petroleum

---

For most of the information in this chapter, the author is deeply indebted to John W. Thomas, retired Chief Mechanical Engineer of the Standard Oil Company (Ohio).

Mechanical Engineers' Handbook, 2nd ed., Edited by Myer Kutz.
ISBN 0-471-13007-9 © 1998 John Wiley & Sons, Inc.

Table 47.1   Principal Uses of Liquid Fuels

Heat and Power

Fuel oil	Space heating (residential, commercial, industrial)
	Steam generation for electric power
	Industrial process heating
	Refinery and chemical feedstock
Kerosene	Supplemental space heating
Turbine fuel	Stationary power generation
Diesel fuel	Stationary power generation
Liquid propane[a]	Isolated residential space heating
	Standby industrial process heating

Transportation

Jet fuel	Aviation turbines
Diesel fuel	Automotive engines
	Marine engines
	Truck engines
Gasoline	Automotive
	Aviation
Liquid propane and butane[a]	Limited automotive use

[a]See Chapter 46 on gaseous fossil fuels.

2. Residual fuel oils that result after crude petroleum is topped; or viscous residuums from refining operations
3. Blended fuel oils, mixtures of the above

The distillate fuels have lower specific gravity and are less viscous than residual fuel oils. Petroleum refiners burn a varying mix of crude residue and distilled oils in their process heaters. The changing gravity and viscosity require maximum oil preheat for atomization good enough to assure complete combustion. Tables 47.5–47.8 describe oils in current use. Some terms used in those tables are defined below.

Aniline point is the lowest Fahrenheit temperature at which an oil is completely miscible with an equal volume of freshly distilled aniline.

API gravity is a scale of specific gravity for hydrocarbon mixtures referred to in "degrees API" (for American Petroleum Institute). The relationships between API gravity, specific gravity, and density are:

Table 47.2   Ultimate Chemical Analyses of Various Crudes[a,b]

Crude Petroleum Source	% wt of					Specific Gravity (at temperature, °F)	Base
	C	H	N	O	S		
Baku, USSR	86.5	12.0		1.5		0.897	
California	86.4	11.7	1.14		0.60	0.951 (at 59°F)	Naphthene
Colombia, South America	85.62	11.91	0.54				
Kansas	85.6	12.4			0.37	0.912	Mixed
Mexico	83.0	11.0	___1.7___		4.30	0.97 (at 59°F)	Naphthene
Oklahoma	85.0	12.9			0.76		Mixed
Pennsylvania	85.5	14.2				0.862 (at 59°F)	Paraffin
Texas	85.7	11.0	___2.61___		0.70	0.91	Naphthene
West Virginia	83.6	12.9		3.6		0.897 (at 32°F)	Paraffin

[a]See, also, Table 47.7.

Table 47.3 Some Properties of Liquid Fuels[2]

Property	Gaso-line	Kero-sene	Diesel Fuel	Light Fuel Oil	Heavy Fuel Oil	Coal Tar Fuel	Bituminous Coal (for Comparison)
Analysis, % wt							
C	85.5	86.3	86.3	86.2	86.2	90.0	80.0
H	14.4	13.6	12.7	12.3	11.8	6.0	5.5
N						1.2	1.5
O						2.5	7
S	0.1	0.1	1.0	1.5	2.0	0.4	1
Boiling range, °F	104–365	284–536	356 up	392 up	482 up	392 up	
Flash point, °F	−40	102	167	176	230	149	
Gravity specific at 59°F	0.73	0.79	0.87	0.89	0.95	1.1	1.25
Heat value, net							
cal/g	10,450	10,400	10,300	10,100	9,900	9,000	7,750
Btu/lb	18,810	18,720	18,540	18,180	17,820	16,200	13,950
Btu/US gal	114,929	131,108	129,800	131,215	141,325		
Residue, % wt at 662°F			15	50	60	60	
Viscosity, kinematic							
Centistokes at 59°F	0.75	1.6	5.0	50	1,200	1,500	
Centistokes at 212°F		0.6	1.2	3.5	20	18	

$$\text{sp gr } 60/60°F = \frac{141.5}{°API + 131.5}$$

where °API is measured at 60°F (15.6°C).

$$\text{sp gr } 60/60°F = \frac{lb/ft^3}{62.3}$$

where lb/ft^3 is measured at 60°F (15.6°C).

SSU (or SUS) is seconds, Saybolt Universal, a measure of kinematic viscosity determined by measuring the time required for a specified quantity of the sample oil to flow by gravity through a specified orifice at a specified temperature. For heavier, more viscous oils, a larger (Furol) orifice is used, and the results are reported as SSF (seconds, Saybolt Furol).

$$\text{kin visc in centistokes} = 0.226 \times SSU - 195/SSU, \text{ for SSU } 32–100$$
$$\text{kin visc in centistokes} = 0.220 \times SSU - 135/SSU, \text{ for SSU} > 100$$
$$\text{kin visc in centistokes} = 2.24 \times SSF - 184/SSF, \text{ for SSF } 25–40$$
$$\text{kin visc in centistokes} = 2.16 \times SSF - 60/SSF, \text{ for SSF} > 40$$
$$1 \text{ centistoke (cSt)} = 0.000001 \text{ m}^2/\text{sec}$$

Unlike distillates, residual oils contain noticeable amounts of inorganic matter, ash content ranging from 0.01% to 0.1%. Ash often contains vanadium, which causes serious corrosion in boilers and heaters. (A common specification for refinery process heaters requires 50% nickel–50% chromium alloy for tube supports and hangers when the vanadium exceeds 150 ppm.) V_2O_5 also lowers the eutectic of many refractories, causing rapid disintegration. Crudes that often contain high vanadium are

Venezuela, Bachaqoro	350 ppm
Iran	350–440 ppm
Alaska, North Slope	80 ppm

47.2.1 Kerosene

Kerosene is a refined petroleum distillate consisting of a homogeneous mixture of hydrocarbons. It is used mainly in wick-fed illuminating lamps and kerosene burners. Oil for illumination and for

Table 47.4 Gravities and Related Properties of Liquid Petroleum Products

°API	Specific Gravity 60°F/60°F (15.6°C/15.6°C)	lb/gal	kg/m³	Gross Btu/gal[a]	Gross kcal/liter[a]	H, wt[a] %	Net Btu/gal[a]	Net kcal/liter[a]	Specific Heat @ 40°F	Specific Heat @ 300°F	Temperature Correction °API/°F[a]	ft³ 60°F air/gal	Ultimate % CO₂
0	1.076	8.969	1075	160,426	10,681	8.359	153,664	10,231	0.391	0.504	0.045	1581	—
2	1.060	8.834	1059	159,038	10,589	8.601	152,183	10,133	0.394	0.508	—	—	—
4	1.044	8.704	1043	157,692	10,499	8.836	150,752	10,037	0.397	0.512	—	—	18.0
6	1.029	8.577	1028	156,384	10,412	9.064	149,368	9,945	0.400	0.516	0.048	1529	17.6
8	1.014	8.454	1013	155,115	10,328	9.285	148,028	9,856	0.403	0.519	0.050	1513	17.1
10[b]	1.000[b]	8.335[b]	1000[b]	153,881	10,246	10.00	146,351	9,744	0.406	0.523	0.051	1509	16.7
12	0.986	8.219	985.0	152,681	10,166	10.21	145,100	9,661	0.409	0.527	0.052	1494	16.4
14	0.973	8.106	971.5	151,515	10,088	10.41	143,888	9,580	0.412	0.530	0.054	1478	16.1
16	0.959	7.996	958.3	150,380	10,013	10.61	142,712	9,502	0.415	0.534	0.056	1463	15.8
18	0.946	7.889	945.5	149,275	9,939	10.80	141,572	9,426	0.417	0.538	0.058	1448	15.5
20	0.934	7.785	933.0	148,200	9,867	10.99	140,466	9,353	0.420	0.541	0.060	1433	15.2
22	0.922	7.683	920.9	147,153	9,798	11.37	139,251	9,272	0.423	0.545	0.061	1423	14.9
24	0.910	7.585	909.0	146,132	9,730	11.55	138,210	9,202	0.426	0.548	0.063	1409	14.7
26	0.898	7.488	897.5	145,138	9,664	11.72	137,198	9,135	0.428	0.552	0.065	1395	14.5
28	0.887	7.394	886.2	144,168	9,599	11.89	136,214	9,069	0.431	0.555	0.067	1381	14.3
30	0.876	7.303	875.2	143,223	9,536	12.06	135,258	9,006	0.434	0.559	0.069	1368	14.0
32	0.865	7.213	864.5	142,300	9,475	12.47	134,163	8,933	0.436	0.562	0.072	1360	13.8
34	0.855	7.126	854.1	141,400	9,415	12.63	133,259	8,873	0.439	0.566	0.074	1347	13.6
36	0.845	7.041	843.9	140,521	9,356	12.78	132,380	8,814	0.442	0.569	0.076	1334	13.4
38	0.835	6.958	833.9	139,664	9,299	12.93	131,524	8,757	0.444	0.572	0.079	1321	13.3
40	0.825	6.887	824.2	138,826	9,243	13.07	130,689	8,702	0.447	0.576	0.082	1309	13.1
42	0.816	6.798	814.7	138,007	9,189	—	—	—	0.450	0.579	0.085	—	13.0
44	0.806	6.720	805.4	137,207	9,136	—	—	—	0.452	0.582	0.088	—	12.8

Typical Ranges for (Diesel Fuels, Aviation Turbine Fuels, Fuel Oils):

Fuel Oils: #6, #5, #4, #2, #1

Aviation Turbine Fuels: JET A (48) (47), JP5 (48), JP4 (48) (56)

Diesel Fuels: 1D (48), 2D

[a]For gravity measured at 60°F (15.6°C) only.
[b]Same as H₂O.

1520

Table 47.5 Heating Requirements for Products Derived from Petroleum[3]

Commercial Fuels	Specific Gravity at 60°F/60°F (15.6°C)	Distillation Range, °F(°C)	Vapor Pressure,[a] psia(mm Hg)	Latent Btu/gal[b] to Vaporize	Btu/gal[b] to Heat from 32°F (0°C) to		
					Pumping Temperature	Atomizing Temperature	Vapor
No. 6 oil	0.965	600–1000(300–500)	0.054 (2.8)	764	371	996	3619[c]
No. 5 oil	0.945	600–1000(300–500)	0.004 (0.2)	749	133	635	3559[c]
No. 4 oil	0.902	325–1000(150–500)	0.232 (12)	737	—	313	2725[c]
No. 2 oil	0.849	325–750(150–400)	0.019 (1)	743	—	—	2704[c]
Kerosene	0.780	256–481(160–285)	0.039 (2)	750	—	—	1303[c]
Gasoline	0.733	35–300(37–185)	0.135 (7)	772	—	—	1215[c]
Methanol	0.796	148 (64)	4.62 (239)	3140	—	—	3400[d]
Butane	0.582	31 (0)	31(1604)	808	—	—	976[d]
Propane	0.509	−44 (−42)	124(6415)	785	—	—	963[d]

[a]At the atomizing temperature or 60°F, whichever is lower. Based on a sample with the lowest boiling point from column 3.

[b]To convert Btu/US gallon to kcal/liter, multiply by 0.666. To convert Btu/US gallon to Btu/lb, divide by 8.335 × sp gr, from column 2. To convert Btu/US gallon to kcal/kg, divide by 15.00 × sp gr, from column 2.

[c]Calculated for boiling at midpoint of distillation range, from column 3.

[d]Includes latent heat plus sensible heat of the vapor heated from boiling point to 60°F (15.6°C).

Table 47.6 Analyses and Characteristics of Selected Fuel Oils[3]

Source	Ultimate Analysis (% Weight)						ppm if > 50	% wt Asphaltine	% wt C Residue	°API at 60°F	Flash Point, °F	HV, Btu/lb		Pour Point, °F	Viscosity, SSU	
	C	H	N	S	Ash	O[a]						Gross	Net		At 140°F	At 210°F
Alaska	86.99	12.07	0.007	0.31	<0.001	0.62	—	—	—	33.1	—	—	—	—	33.0	29.5
California	86.8	12.52	0.053	0.27	<0.001	0.36	—	—	—	32.6	—	19,330	—	—	30.8	29.5
West Texas	88.09	9.76	0.026	1.88	<0.001	0.24	—	—	—	18.3	—	—	—	—	32.0	28.8
Alaska	86.04	11.18	0.51	1.63	0.034	0.61	50 Ni 67 V	5.6	12.9	15.6	215	18,470	17,580	38	1071	194
California	86.66	10.44	0.86	0.99	0.20	0.85	b	8.62	15.2	12.6	180	18,230	17,280	42	720	200
DFM (shale)	86.18	13.00	0.24	0.51	0.003	1.07	—	0.036	4.1	33.1	182	19,430	18,240	40	36.1	30.7
Gulf of Mexico	84.62	10.77	0.36	2.44	0.027	1.78	—	7.02	14.8	13.2	155	18,240	17,260	40	835	181
Indo/Malaysia	86.53	11.93	0.24	0.22	0.036	1.04	101 V	0.74	3.98	21.8	210	19,070	17,980	61	199	65
Middle East[c]	86.78	11.95	0.18	0.67	0.012	0.41	—	3.24	6.0	19.8	350	19,070	17,980	48	490	131.8
Pennsylvania[d]	84.82	11.21	0.34	2.26	0.067	1.3	65 Na 82 V	4.04	12.4	15.4	275	18,520	17,500	66	1049	240
Venezuela	85.24	10.96	0.40	2.22	0.081	1.10	52 Ni 226 V	8.4	6.8	14.1	210	18,400	17,400	58	742	196.7
Venezuela desulfurized	85.92	12.05	0.24	0.93	0.033	0.83	101 V	2.59	5.1	23.3	176	18,400	17,300	48	113.2	50.5

[a]By difference.
[b]91 Ca, 77 Fe, 88 Ni, 66 V.
[c]Exxon.
[d]Amerada Hess.

Table 47.7 ASTM Fuel Oil Specifications[8]

Grade of Fuel Oil[a]	Flash Point, °C (°F) Min	Pour Point, °C (°F) Max	Water and Sediment, Vol % Max	Carbon Residue on 10% Bottoms, % Max	Ash, Weight % Max	Distillation Temperatures, °C (°F)			Saybolt Viscosity, s[d]				Kinematic Viscosity, cSt[d]						Specific Gravity, 60/60°F (deg API) Max	Copper Strip Corrosion Max	Sulfur, % Max
						10% Point Max	90% Point Min	90% Point Max	Universal at 38°C (100°F)		Furol at 50°C (122°F)		At 38°C (100°F)		At 40°C (104°F)		At 50°C (122°F)				
									Min	Max	Min	Max	Min	Max	Min	Max	Min	Max			
No. 1 A distillate oil intended for vaporizing pot-type burners and other burners requiring this grade of fuel	38 (100)	−18[c] (0)	0.05	0.15	—	215 (420)	—	288 (550)	—	—	—	—	1.4	2.2	1.3	2.1	—	—	0.8499 (35 min)	No. 3	0.5
No. 2 A distillate oil for general purpose heating for use in burners not requiring No. 1 fuel oil	38 (100)	−6[c] (20)	0.05	0.35	—	—	282[c] (540)	338 (640)	(32.6)	(37.9)	—	—	2.0[c]	3.6	1.9[c]	3.4	—	—	0.8762 (30 min)	No. 3	0.5[b]
No. 4 (Light) Preheating not usually required for handling or burning	38 (100)	−6[c] (20)	0.50	—	0.05	—	—	—	(32.6)	(45)	—	—	2.0	5.8	—	—	—	—	0.876[g] (30 max)	—	—
No. 4 Preheating not usually required for handling or burning	55 (130)	−6[c] (20)	0.50	—	0.10	—	—	—	(45)	(125)	—	—	5.8	26.4[h]	5.5	24.0[f]	—	—	—	—	—

Grade																	
No. 5 (Light) Preheating may be required depending on climate and equipment	55 (130)	—	—	1.00	0.10	—	—	(>125) (300)	—	—	>26.4 65^f	>24.0 58^f	—	—			
No. 5 (Heavy) Preheating may be required for burning and, in cold climates, may be required for handling	55 (130)	—	—	1.00	0.10	—	—	(>300) (900)	(23) (40)	>65 194^f	58 168^f	(42) (81)					
No. 6 Preheating required for burning and handling	60 (140)	g	2.00^e	—	—	—	—	(>900) (9000)	(>45) (300)	—	—	—	—	>92 638^f			

^{a}It is the intent of these classifications that failure to meet any requirement of a given grade does not automatically place an oil in the next lower grade unless in fact it meets all requirements of the lower grade.

^{b}In countries outside the United States other sulfur limits may apply.

^{c}Lower or higher pour points may be specified whenever required by conditions of storage or use. When pour point less than −18°C (0°F) is specified, the minimum viscosity for grade No. 2 shall be 1.7 cSt (31.5 SUS) and the minimum 90% point shall be waived.

^{d}Viscosity values in parentheses are for information only and not necessarily limiting.

^{e}The amount of water by distillation plus the sediment by extraction shall not exceed 2.00%. The amount of sediment by extraction shall not exceed 0.50%. A deduction in quantity shall be made for all water and sediment in excess of 1.0%.

^{f}Where low-sulfur fuel oil is required, fuel oil falling in the viscosity range of a lower numbered grade down to and including No. 4 may be supplied by agreement between purchaser and supplier. The viscosity range of the initial shipment shall be identified and advance notice shall be required when changing from one viscosity range to another. This notice shall be in sufficient time to permit the user to make the necessary adjustments.

^{g}This limit guarantees a minimum heating value and also prevents misrepresentation and misapplication of this product as Grade No. 2.

^{h}Where low-sulfur fuel oil is required, Grade 6 fuel oil will be classified as low pour +15°C (60°F) max or high pour (no max). Low-pour fuel oil should be used unless all tanks and lines are heated.

Table 47.8 Application of ASTM Fuel Oil Grades, as Described by One Burner Manufacturer

Fuel Oil	Description
No. 1	Distillate oil for vaporizing-type burners
No. 2	Distillate oil for general purpose use, and for burners not requiring No. 1 fuel oil
No. 4	Blended oil intended for use without preheating
No. 5	Blended residual oil for use with preheating; usual preheat temperature is 120–220°F
No. 6	Residual oil for use with preheaters permitting a high-viscosity fuel; usual preheat temperature is 180–260°F
Bunker C	Heavy residual oil, originally intended for oceangoing ships

domestic stoves must be high in paraffins to give low smoke. The presence of naphthenic and especially aromatic hydrocarbons increases the smoking tendency. A "smoke point" specification is a measure of flame height at which the tip becomes smoky. The "smoke point" is about 73 mm for paraffins, 34 mm for naphthalenes, and 7.5 mm for aromatics and mixtures.

Low sulfur content is necessary in kerosenes because:

1. Sulfur forms a bloom on glass lamp chimneys and promotes carbon formation on wicks.
2. Sulfur forms oxides in heating stoves. These swell, are corrosive and toxic, creating a health hazard, particularly in nonvented stoves.

Kerosene grades[9] (see Table 47.9) in the United States are:

No. 1 K: A special low-sulfur grade kerosene suitable for critical kerosene burner applications
No. 2 K: A regular-grade kerosene suitable for use in flue-connected burner applications and for use in wick-fed illuminating lamps

47.2.2 Aviation Turbine Fuels

The most important requirements of aircraft jet fuel relate to freezing point, distillation range, and level of aromatics. Fluidity at low temperature is important to ensure atomization. A typical upper viscosity limit is 7–10 cSt at 0°F, with the freezing point as low as −60°F.

Aromatics are objectionable because (1) coking deposits from the flame are most pronounced with aromatics of high C/H ratio and less pronounced with short-chain compounds, and (2) they must be controlled to keep the combustor liner at an acceptable temperature.

Jet fuels for civil aviation are identified as Jet A and A1 (high-flash-point, kerosene-type distillates), and Jet B (a relatively wide boiling range, volatile distillate).

Jet fuels for military aviation are identified as JP4 and JP5. The JP4 has a low flash point and a wide boiling range. The JP5 has a high flash point and a narrow boiling range. (See Table 47.10.)

Table 47.9 ASTM Chemical and Physical Requirements for Kerosene[9]

Property	Limit
Distillation temperature	
10% recovered	401°F (205°C)
Final boiling point	572°F (300°C)
Flash point	100°F (38°C)
Freezing point	−22°F (−30°C)
Sulfur, % weight	
No. 1 K	0.04 maximum
No. 2 K	0.30 maximum
Viscosity, kinematic at 104°F (40°C), centistokes	1.0 min/1.9 max

Table 47.10   ASTM Specifications[10] and Typical Properties[7] of Aviation Turbine Fuels

| Property | Specifications | | | Typical, 1979 | | |
	Jet A	Jet A1	Jet B	26 Samples JP4	7 Samples JP5	60 Samples Jet A
Aromatics, % vol	20	20	20	13.0	16.4	17.9
Boiling point, final, °F	572	572	—	—	—	—
Distillation, max temperature, °F						
For 10% recovered	400	400	—	208	387	375
For 20% recovered	—	—	290	—	—	—
For 50% recovered	—	—	370	293	423	416
For 90% recovered	—	—	470	388	470	473
Flash point, min, °F	100	100	—	—	—	—
Freezing point, max, °F	−40	−53	−58	−110	−71	−56
Gravity, API, max	51	51	57	53.5	41.2	42.7
Gravity, API, min	37	37	45	—	—	—
Gravity, specific 60°F min	0.7753	0.7753	0.7507	0.765	0.819	0.812
Gravity, specific 60°F max	0.8398	0.8398	0.8017	—	—	—
Heating value, gross Btu/lb	—	—	—	18,700	18,530	18,598
Heating value, gross Btu/lb min	18,400	18,400	18,400	—	—	—
Mercaptan, % wt	0.003	0.003	0.003	0.0004	0.0003	0.0008
Sulfur, max % wt	0.3	0.3	0.3	0.030	0.044	0.050
Vapor pressure, Reid, psi	—	—	3	2.5	—	0.2
Viscosity, max SSU						
At −4°F	52	—	—	—	—	—
At −30°F	—	—	—	34–37	60.5	54.8

Gas turbine fuel oils for other than use in aircraft must be free of inorganic acid and low in solid or fibrous materials. (See Tables 47.11 and 47.12.) All such oils must be homogeneous mixtures that do not separate by gravity into light and heavy components.

47.2.3   Diesel Fuels

Diesel engines, developed by Rudolf Diesel, rely on the heat of compression to achieve ignition of the fuel. Fuel is injected into the combustion chamber in an atomized spray at the end of the compression stroke, after air has been compressed to 450–650 psi and has reached a temperature, due to compression, of at least 932°F (500°C). This temperature ignites the fuel and initiates the piston's power stroke. The fuel is injected at about 2000 psi to ensure good mixing.

Diesels are extensively used in truck transport, rail trains, and marine engines. They are being used more in automobiles. In addition, they are employed in industrial and commercial stationary power plants.

Fuels for diesels vary from kerosene to medium residual oils. The choice is dictated by engine characteristics, namely, cylinder diameter, engine speed, and combustion wall temperature. High-

Table 47.11   Nonaviation Gas Turbine Fuel Grades per ASTM[11]

Grade	Description
No. 0-GT	A naphtha or low-flash-point hydrocarbon liquid
No. 1-GT	A distillate for gas turbines requiring cleaner burning than No. 2-GT
No. 2-GT	A distillate fuel of low ash suitable for gas turbines not requiring No. 1-GT
No. 3-GT	A low ash fuel that may contain residual components
No. 4-GT	A fuel containing residual components and having higher vanadium content than No. 3-GT

Table 47.12 ASTM Specifications[11] for Nonaviation Gas Turbine Fuels

Property	Specifications				
	0-GT	1-GT	2-GT	3-GT	4-GT
Ash, max % wt	0.01	0.01	0.01	0.03	—
Carbon residue, max % wt	0.15	0.15	0.35	—	—
Distillation, 90% point, max °F	—	(550)[a]	(640)	—	—
Distillation, 90% point, min °F	—	—	(540)	—	—
Flash point, min °F	—	(100)	(100)	(130)	(150)
Gravity, API min	—	(35)	(30)	—	—
Gravity, spec 60°F max	—	0.850	0.876	—	—
Pour point, max °F	—	(0)	(20)	—	—
Viscosity, kinematic					
Min SSU at 100°F	—	—	(32.6)	(45)	(45)
Max SSU at 100°F	—	(34.4)	(40.2)	—	—
Max SSF at 122°F	—	—	—	(300)	(300)
Water and sediment, max % vol	0.05	0.05	0.05	1.0	1.0

[a]Values in parentheses are approximate.

speed small engines require lighter fuels and are more sensitive to fuel quality variations. Slow-speed, larger industrial and marine engines use heavier grades of diesel fuel oil.

Ignition qualities and viscosity are important characteristics that determine performance. The ignition qualities of diesel fuels may be assessed in terms of their cetane numbers or diesel indices. Although the diesel index is a useful indication of ignition quality, it is not as reliable as the cetane number, which is based on an engine test:

$$\text{diesel index} = (\text{aniline point, °F}) \times (\text{API gravity}/100)$$

The diesel index is an arbitrary figure having a significance similar to cetane number, but having a value 1–5 numbers higher.

The cetane number is the percentage by volume of cetane in a mixture of cetane with an ethyl-naphthalene that has the same ignition characteristics as the fuel. The comparison is made in a diesel engine equipped either with means for measuring the delay period between injection and ignition or with a surge chamber, separated from the engine intake port by a throttle in which the critical measure below which ignition does not occur can be measured. Secondary reference fuels with specific cetane numbers are available. Cetane number is a measure of ignition quality and influences combustion roughness.

The use of a fuel with too low a cetane number results in accumulation of fuel in the cylinder before combustion, causing "diesel knock." Too high a cetane number will cause rapid ignition and high fuel consumption.

The higher the engine speed, the higher the required fuel cetane number. Suggested rpm values for various fuel cetane numbers are shown in Table 47.13.[5] Engine size and operating conditions are important factors in establishing approximate ignition qualities of a fuel.

Too viscous an oil will cause large spray droplets and incomplete combustion. Too low a viscosity may cause fuel leakage from high-pressure pumps and injection needle valves. Preheating permits use of higher viscosity oils.

Table 47.13 ASTM Fuel Cetane Numbers for Various Engine Speeds[5]

Engine Speed (rpm)	Cetane Number
Above 1500	50–60
500–1500	45–55
400–800	35–50
200–400	30–45
100–200	15–40
Below 200	15–30

To minimize injection system wear, fuels are filtered to remove grit. Fine gage filters are considered adequate for engines up to 8 Hz, but high-speed engines usually have fabric or felt filters. It is possible for wax to crystallize from diesel fuels in cold weather, therefore, preheating before filtering is essential.

To minimize engine corrosion from combustion products, control of fuel sulfur level is required. (See Tables 47.14 and 47.15.)

47.2.4 Summary

Aviation jet fuels, gas turbine fuels, kerosenes, and diesel fuels are very similar. The following note from Table 1 of Ref. 11 highlights this:

> *No. 0-GT includes naphtha, Jet B fuel, and other volatile hydrocarbon liquids. No. 1-GT corresponds in general to Spec D396 Grade No. 1 fuel and Classification D975 Grade No. 1-D Diesel fuel in physical properties. No. 2-GT corresponds in general to Spec D396 Grade No. 2 fuel and Classification D975 Grade No. 2 Diesel fuel in physical properties. No. 3-GT and No. 4-GT viscosity range brackets Spec D396 and Grade No. 4, No. 5 (light), No. 5 (heavy), No. 6, and Classification D975 Grade No. 4-D Diesel fuel in physical properties.*

47.3 SHALE OILS

As this is written, there is no commercial producing shale oil plant in the United States. Predictions are that the output products will be close in characteristics and performance to those made from petroleum crudes.

Table 47.16 lists properties of a residual fuel oil (DMF) from one shale pilot operation and of a shale crude oil.[13] Table 47.17 lists ultimate analyses of oils derived from shales from a number of locations.[14] Properties will vary with the process used for extraction from the shale. The objective of all such processes is only to provide feedstock for refineries. In turn, the refineries' subsequent processing will also affect the properties.

If petroleum shortages occur, they will probably provide the economic impetus for completion of developments already begun for the mining, processing, and refining of oils from shale.

47.4 OILS FROM TAR SANDS

At the time that this is written, the only commercially practical operation for extracting oil from tar sands is at Athabaska, Alberta, Canada, using surface mining techniques. When petroleum supplies become short, economic impetus therefrom will push completion of developments already well under way for mining, processing, and refining of oils from tar sands.

Table 47.18 lists chemical and physical properties of several tar sand bitumens.[15] Further refining will be necessary because of the high density, viscosity, and sulfur content of these oils.

Extensive deposits of tar sands are to be found around the globe, but most will have to be recovered by some *in situ* technique, fireflooding, or steam flooding. Yields tend to be small and properties vary with the recovery method, as illustrated in Table 47.19.[15]

47.5 OIL–WATER EMULSIONS

Emulsions of oil have offered some promise of low fuel cost and alternate fuel supply for some time. The following excerpts from Ref. 16 provide introductory information on a water emulsion with an

Table 47.14 ASTM Diesel Fuel Descriptions[12]

Grade	Description
No. 1D	A volatile distillate fuel oil for engines in service requiring frequent speed and load changes
No. 2D	Distillate fuel oil of lower volatility for engines in industrial and heavy mobile service
No. 4D	A fuel oil for low and medium speed diesel engines
Type CB	For buses, essentially 1D
Type TT	For trucks, essentially 2D
Type RR	For railroads, essentially 2D
Type SM	For stationary and marine use, essentially 2D or heavier

Table 47.15a ASTM Detailed Requirements for Diesel Fuel Oils[a,h,12]

Grade of Diesel Fuel Oil	Flash Point, °C (°F) Min	Cloud Point, °C (°F) Max	Water and Sediment, Vol% Max	Carbon Residue on 10% Residuum, % Max	Ash, Weight % Max	Distillation Temperatures, °C (°F), 90% Point		Viscosity				Sulfur,[d] Weight % Max	Copper Strip Corrosion Max	Cetane Number[e] Min
								Kinematic, cSt[g] at 40°C		Saybolt, SUS at 100°F				
						Min	Max	Min	Max	Min	Max			
No. 1-D A volatile distillate fuel oil for engines in service requiring frequent speed and load changes	38 (100)	b	0.05	0.15	0.01	—	288 (550)	1.3	2.4	—	34.4	0.50	No.3	40[f]
No. 2-D A distillate fuel oil of lower volatility for engines in industrial and heavy mobile service	52 (125)	b	0.05	0.35	0.01	282[c] (540)	338 (640)	1.9	4.1	32.6	40.1	0.50	No.3	40[f]
No. 4-D A fuel oil for low and medium speed engines	55 (130)	b	0.50	—	0.10	—	—	5.5	24.0	45.0	125.0	2.0	—	30[f]

[a]To meet special operating conditions, modifications of individual limiting requirements may be agreed upon between purchaser, seller, and manufacturer.

[b]It is unrealistic to specify low-temperature properties that will ensure satisfactory operation on a broad basis. Satisfactory operation should be achieved in most cases if the cloud point (or wax appearance point) is specified at 6°C above the tenth percentile minimum ambient temperature for the area in which the fuel will be used. This guidance is of a general nature; some equipment designs, using improved additives, fuel properties, and/or operations, may allow higher or require lower cloud point fuels. Appropriate low-temperature operability properties should be agreed on between the fuel supplier and purchaser for the intended use and expected ambient temperatures.

[c]When cloud point less than −12°C (10°F) is specified, the minimum viscosity shall be 1.7 cSt (or mm²/sec) and the 90% point shall be waived.

[d]In countries outside the United States, other sulfur limits may apply.

[e]Where cetane number by Method D613 is not available, ASTM Method D976, Calculated Cetane Index of Distillate Fuels may be used as an approximation. Where there is disagreement, Method D613 shall be the referee method.

[f]Low-atmospheric temperatures as well as engine operation at high altitudes may require use of fuels with higher cetane ratings.

[g]cSt = 1 mm²/sec.

[h]The values stated in SI units are to be regarded as the standard. The values in U.S. customary units are for information only.

Table 47.15b ASTM Typical Properties of Diesel Fuels[7]

	All United States, 1981						Eastern United States, 1981											
	48 Samples			112 Samples			24 Samples			44 Samples			13 Samples			4 Samples		
	No. 1D			No. 2D			Type CB			Type TT			Type RR			Type SM		
Property	Min	Avg	Max	Min	Avg	Max	Min	Avg	Max	Min	Avg	Max	Min	Avg	Max	Min	Avg	Max
Ash, % wt	0.000	0.001	0.005	0.000	0.002	0.020	—	0.001	0.005	—	0.002	0.015	—	0.000	0.001	—	0.001	0.001
Carbon residue, % wt	0.000	0.059	0.067	0.000	0.101	0.300	—	—	0.21	0.101	—	0.25	—	0.121	0.23	—	0.148	0.21
Cetane number	36	46.7	53.0	29.0	45.6	52.4	—	49.8	—	—	45.6	—	—	44.8	—	—	—	—
Distillation, 90% point, °F	445	448	560	493	587	640	451	512	640	451	571	640	540	590	640	482	577	640
Flash point, °F	104	138	176	132	166	240	120	140	240	120	162	240	156	164	192	136	162	180
Gravity, API	37.8	42.4	47.9	22.8	34.9	43.1	—	41.5	—	—	36.3	—	—	33.8	—	—	35.3	—
spec, 60/60°F	0.836	0.814	0.789	0.917	0.850	0.810	—	0.818	—	—	0.843	—	—	0.856	—	—	0.848	—
Sulfur, % wt	0.000	0.070	0.25	0.010	0.283	0.950	—	0.086	0.24	—	0.198	0.46	—	0.283	0.580	—	0.155	0.28
Viscosity, SSU at 100°F	32.6	33.3	35.7	33.8	36.0	40.3	32.9	34.3	40.2	32.9	35.7	40.2	34.2	36.0	37.8	36.0	—	37.8

Table 47.16   Properties of Shale Oils[13]

Property	DMF Residual	Crude
Ultimate analysis		
Carbon, % wt	86.18	84.6
Hydrogen, % wt	13.00	11.3
Nitrogen, % wt	0.24	2.08
Sulfur, % wt	0.51	0.63
Ash, % wt	0.003	0.026
Oxygen, % wt by difference	1.07	1.36
Conradson carbon residue, %	4.1	2.9
Asphaltene, %	0.036	1.33
Calcium, ppm	0.13	1.5
Iron, ppm	6.3	47.9
Manganese, ppm	0.06	0.17
Magnesium, ppm	—	5.40
Nickel, ppm	0.43	5.00
Sodium, ppm	0.09	11.71
Vanadium, ppm	0.1	0.3
Flash point, °F	182	250
Pour point, °F	40	80
API gravity at 60°F	33.1	20.3
Viscosity, SSU at 140°F	36.1	97
SSU at 210°F	30.7	44.1
Gross heating value, Btu/lb	19,430	18,290
Net heating value, Btu/lb	18,240	17,260

Table 47.17   Elemental Content of Shale Oils, % wt[14]

Source	Carbon, C			Hydrogen, H			Nitrogen, N			Sulfur, S			Oxygen, O		
	Min	Avg	Max	Min	Avg	Max	Min	Avg	Max	Min	Avg	Max	Min	Avg	Max
Colorado	83.5	84.2	84.9	10.9	11.3	11.7	1.6	1.8	1.9	0.7	1.2	1.7	1.3	1.7	2.1
Utah	84.1	84.7	85.2	10.9	11.5	12.0	1.6	1.8	2.0	0.5	0.7	0.8	1.2	1.6	2.0
Wyoming	81.3	83.1	84.4	11.2	11.7	12.2	1.4	1.8	2.2	0.4	1.0	1.5	1.7	2.0	2.3
Kentucky	83.6	84.4	85.2	9.6	10.2	10.7	1.0	1.3	1.6	1.4	1.9	2.4	1.8	2.3	2.7
Queensland Australia (four locations)	80.0	82.2	85.5	10.0	11.1	12.8	1.0	1.2	1.6	0.3	1.9	6.0	1.1	3.0	6.6
Brazil		85.3			11.2			0.9			1.1			1.5	
Karak, Jordan	77.6	78.3	79.0	9.4	9.7	9.9	0.5	0.7	0.8	9.3	10.0	10.6	0.9	1.4	1.9
Timahdit Morocco	79.5	80.0	80.4	9.7	9.8	9.9	1.2	1.4	1.6	6.7	7.1	7.4	1.8	2.0	2.2
Sweden	86.5	86.5	86.5	9.0	9.4	9.8	0.6	0.7	0.7	1.7	1.9	2.1	1.4	1.6	1.7

oil from the vicinity of the Orinoco River in Venezuela. It is being marketed as "Orimulsion" by Petrtoleos de Veneauels SA and Bitor America Corp of Boca Raton, Florida. It is a natural bitumen, like a liquid coal that has been emulsified with water to make it possible to extract it from the earth and to transport it.

Table 47.20 shows some of its properties and contents. Although its original sulfur content is high, the ash is low. A low C/H ratio promises less CO_2 emission. Because of handleability concerns, it will probably find use mostly in large steam generators.

Table 47.18   Chemical and Physical Properties of Several Tar Sand Bitumens[15]

	Uinta Basin, Utah	South-east Utah	Athabasca, Alberta	Trapper Canyon, WY[a]	South TX	Santa Rosa, NM[a]	Big Clifty, KY	Bellamy, MO
Carbon, % wt	85.3	84.3	82.5	82.4	—	85.6	82.4	86.7
Hydrogen, % wt	11.2	10.2	10.6	10.3	—	10.1	10.8	10.3
Nitrogen, % wt	0.96	0.51	0.44	0.54	0.36	0.22	0.64	0.10
Sulfur, % wt	0.49	4.46	4.86	5.52	~10	2.30	1.55	0.75
H/C ratio	1.56	1.44	1.53	1.49	1.34	1.41	1.56	1.42
Vanadium, ppm	23	151	196	91	85	25	198	—
Nickel, ppm	96	62	82	53	24	23	80	—
Carbon residue, % wt	10.9	19.6	13.7	14.8	24.5	22.1	16.7	—
Pour point, °F	125	95	75	125	180	—	85	—
API gravity	11.6	9.2	9.5	5.4	−2.0	5	8.7	10

Viscosities range from 50,000 to 600,000 SSF (100,000 to 1,300,000 cSt).

[a]Outcrop samples.

Table 47.19   Elemental Composition of Bitumen and Oils Recovered from Tar Sands by Methods C and S[a,15]

	Bitumen	Light Oil C[b]	Heavy Oil C 1–4 Mo.	Heavy Oil C 5–6 Mo.	Product Oil C	Product Oil S[c]
Carbon, % wt	86.0	86.7	86.1	86.7	86.6	85.9
Hydrogen, % wt	11.2	12.2	11.8	11.3	11.6	11.3
Nitrogen, % wt	0.93	0.16	0.82	0.66	0.82	1.17
Sulfur, % wt	0.45	0.30	0.39	0.33	0.43	0.42
Oxygen, % wt	1.42	0.64	0.89	1.01	0.55	1.21

[a]These percentages are site and project specific.
[b]C = reverse-forward combustion.
[c]S = steamflood.
[d]By difference.

Table 47.20   Orimulsion Fuel Characteristics

Density	63 lb/ft³	
Apparent Viscosity	41F/20sec-1—700 mPa	
	86F/20sec-1—450 mPa	
	158F/100sec-1—105 mPa	
Flash point	266°F	
Pour point	32°F	
Higher heating value	12,683 Btu/lb	
Lower heating value	11,694 Btu/lb	
Weight analysis	Carbon	60%
	Hydrogen	7.5%
	Sulfur	2.7%
	Nitrogen	0.5%
	Oxygen	0.2%
	Ash	0.25%
	Water	30%
	Vanadium	300 ppm
	Sodium	70 ppm
	Magnesium	350 ppm

REFERENCES

1. "Journal Forecast Supply & Demand," *Oil and Gas Journal,* 131 (Jan. 25, 1982).
2. J. D. Gilchrist, *Fuels and Refractories,* Macmillan, New York, 1963.
3. R. J. Reed, *Combustion Handbook,* 3rd ed., Vol. 1. North American Manufacturing Co., Cleveland, OH, 1986.
4. Braine and King, *Fuels—Solid, Liquid, Gaseous,* St. Martin's Press, New York, 1967.
5. *Kempe's, Engineering Yearbook,* Morgan Grompium, London.
6. W. L. Nelson, *Petroleum Refinery Engineering,* McGraw-Hill, New York, 1968.
7a. E. M. Shelton, *Diesel Oils, DOE/BETC/PPS—81/5,* U.S. Department of Energy, Washington, DC, 1981.
7b. E. M. Shelton, *Heating Oils, DOE/BETC/PPS—80/4,* U.S. Department of Energy, Washington, DC, 1980.
7c. E. M. Shelton, *Aviation Turbine Fuel, DOE/BETC/PPS—80/2,* Department of Energy, Washington, DC, 1979.
8. ANSI/ASTM D396, *Standard Specification for Fuel Oils,* American Society for Testing and Materials, Philadelphia, PA, 1996.
9. ANSI/ASTM D3699, *Standard Specification for Kerosene,* American Society for Testing and Materials, Philadelphia, PA, 1996.
10. ANSI/ASTM D1655, *Standard Specification for Aviation Turbine Fuels,* American Society for Testing and Materials, Philadelphia, PA, 1996.
11. ANSI/ASTM D2880, *Standard Specification for Gas Turbine Fuel Oils,* American Society for Testing and Materials, Philadelphia, PA, 1996.
12. ANSI/ASTM D975, *Standard Specification for Diesel Fuel Oils,* American Society for Testing and Materials, Philadelphia, PA, 1996.
13. M. Heap et al., *The Influence of Fuel Characteristics on Nitrogen Oxide Formation—Bench Scale Studies,* Energy and Environmental Research Corp., Irvine, CA, 1979.
14. H. Tokairin and S. Morita, "Properties and Characterizations of Fischer-Assay-Retorted Oils from Major World Deposits," in *Synthetic Fuels from Oil Shale and Tar Sands,* Institute of Gas Technology, Chicago, IL, 1983.
15. K. P. Thomas et al., "Chemical and Physical Properties of Tar Sand Bitumens and Thermally Recovered Oils," in *Synthetic Fuels from Oil Shale and Tar Sands,* Institute of Gas Technology, Chicago, IL, 1983.
16. J. Makansi, "New Fuel Could Find Niche between Oil, Coal," *POWER* (Dec. 1991).

CHAPTER 48

COALS, LIGNITE, PEAT

James G. Keppeler, P.E.
Progress Energy Corporation

48.1	**INTRODUCTION**	**1535**	**48.4**	**PHYSICAL AND CHEMICAL**
	48.1.1 Nature	1535		**PROPERTIES—DESCRIPTION**
	48.1.2 Reserves—Worldwide			**AND TABLES OF SELECTED**
	and United States	1535		**VALUES** 1540
	48.1.3 Classifications	1537		
			48.5	**BURNING**
48.2	**CURRENT USES—HEAT,**			**CHARACTERISTICS** 1541
	POWER, STEELMAKING,			
	OTHER	**1539**	**48.6**	**ASH CHARACTERISTICS** 1543
48.3	**TYPES**	**1539**	**48.7**	**SAMPLING** 1545
			48.8	**COAL CLEANING** 1546

48.1 INTRODUCTION

48.1.1 Nature

Coal is a dark brown to black sedimentary rock derived primarily from the unoxidized remains of carbon-bearing plant tissues. It is a complex, combustible mixture of organic, chemical, and mineral materials found in strata, or "seams," in the earth, consisting of a wide variety of physical and chemical properties.

The principal types of coal, in order of metamorphic development, are lignite, subbituminous, bituminous, and anthracite. While not generally considered a coal, peat is the first development stage in the "coalification" process, in which there is a gradual increase in the carbon content of the fossil organic material, and a concomitant reduction in oxygen.

Coal substance is composed primarily of carbon, hydrogen, and oxygen, with minor amounts of nitrogen and sulfur, and varying amounts of moisture and mineral impurities.

48.1.2 Reserves—Worldwide and United States

According to the World Coal Study (see Ref. 3), the total geological resources of the world in "millions of tons of coal equivalent" (mtce) is 10,750,212, of which 662,932, or 6%, is submitted as "Technically and Economically Recoverable Resources."

Millions of tons of coal equivalent is based on the metric ton (2205 lb) with a heat content of 12,600 Btu/lb (7000 kcal/kg).

A summary of the percentage of technically and economically recoverable reserves and the percentage of total recoverable by country is shown in Table 48.1.

As indicated in Table 48.1, the United States possesses over a quarter of the total recoverable reserves despite the low percentage of recovery compared to other countries.

It is noted that the interpretation of "technical and economic" recovery is subject to considerable variation and also to modification, as technical development and changing economic conditions dictate. It should also be noted that there are significant differences in density and heating values in various coals, and, therefore, the mtce definition should be kept in perspective.

In 1977, the world coal production was approximately 2450 mtce,[3] or about ¹⁄₂₇₀th of the recoverable reserves.

Mechanical Engineers' Handbook, 2nd ed., Edited by Myer Kutz.
ISBN 0-471-13007-9 © 1998 John Wiley & Sons, Inc.

Table 48.1

	Percentage of Recoverable[a] of Geological Resources	Percentage of Total Recoverable Reserves
Australia	5.5	4.9
Canada	1.3	0.6
Peoples Republic of China	6.8	14.9
Federal Republic of Germany	13.9	5.2
India	15.3	1.9
Poland	42.6	9.0
Republic of South Africa	59.7	6.5
United Kingdom	23.7	6.8
United States	6.5	25.2
Soviet Union	2.2	16.6
Other Countries	24.3	8.4
		100.0

[a]Technically and economically recoverable reserves. Percentage indicated is based on total geological resources reported by country.

Source: World Coal Study, *Coal—Bridge to the Future,* 1980.

Table 48.2 Demonstrated Reserve Base[a] of Coal in the United States on January, 1980, by Rank (Millions of Short Tons)

State[b]	Anthracite	Bituminous	Subbituminous	Lignite	Total[c]
Alabama[d]	—	3,916.8	—	1,083.0	4,999.8
Alaska	—	697.5	5,443.0	14.0	6,154.5
Arizona	—	410.0	—	—	410.0
Arkansas	96.4	288.7	—	25.7	410.7
Colorado[d]	25.5	9,086.1	3,979.9	4,189.9	17,281.3
Georgia	—	3.6	—	—	3.6
Idaho	—	4.4	—	—	4.4
Illinois[d]	—	67,606.0	—	—	67,606.0
Indiana	—	10,586.1	—	—	10,586.1
Iowa	—	2,197.1	—	—	2,197.7
Kansas	—	993.8	—	—	993.8
Kentucky					
Eastern[d]	—	12,927.5	—	—	12,927.5
Western	—	21,074.4	—	—	21,074.4
Maryland	—	822.4	—	—	822.4
Michigan[d]	—	127.7	—	—	127.7
Missouri	—	6,069.1	—	—	6,069.1
Montana	—	1,385.4	103,277.4	15,765.2	120,428.0
New Mexico[d]	2.3	1,835.7	2,683.4	—	4,521.4
North Carolina	—	10.7	—	—	10.7
North Dakota	—	—	—	9,952.3	9,952.3
Ohio[d]	—	19,056.1	—	—	19,056.1
Oklahoma	—	1,637.8	—	—	1,637.8
Oregon	—	—	17.5	—	17.5
Pennsylvania	7,092.0	23,188.8	—	—	30,280.8
South Dakota	—	—	—	366.1	366.1
Tennessee[d]	—	983.7	—	—	983.7
Texas[d]	—	—	—	12,659.7	12,659.7
Utah[d]	—	6,476.5	1.1	—	6,477.6
Virginia	125.5	3,345.9	—	—	3,471.4
Washington[d]	—	303.7	1,169.4	8.1	1,481.3
West Virginia	—	39,776.2	—	—	39,776.2
Wyoming[d]	—	4,460.5	65,463.5	—	69,924.0
Total[c]	7,341.7	239,272.9	182,035.0	44,063.9	472,713.6

[a]Includes measured and indicated resource categories defined by USBM and USGS and represents 100% of the coal in place.

[b]Some coal-bearing states where data are not sufficiently detailed or where reserves are not currently economically recoverable.

[c]Data may not add to totals due to rounding.

[d]Data not completely reconciled with demonstrated reserve base data.

According to the U.S. Geological Survey, the remaining U.S. Coal Reserves total almost 4000 billion tons,[4] with overburden to 6000 ft in seams of 14 in. or more for bituminous and anthracite and in seams of 2½ ft or more for subbituminous coal and lignite. The U.S. Bureau of Mines and U.S. Geological Survey have further defined "Reserve Base" to provide a better indication of the technically and economically minable reserves, where a higher degree of identification and engineering evaluation is available.

A summary of the reserve base of U.S. coal is provided in Table 48.2.[5]

48.1.3 Classifications

Coals are classified by "rank," according to their degree of metamorphism, or progressive alteration, in the natural series from lignite to anthracite. Perhaps the most widely accepted standard for classification of coals is ASTM D388, which ranks coals according to fixed carbon and calorific value (expressed in Btu/lb) calculated to the mineral-matter-free basis. Higher-rank coals are classified according to fixed carbon on the dry basis; the lower-rank coals are classed according to calorific value on the moist basis. Agglomerating character is used to differentiate between certain adjacent groups. Table 48.3 shows the classification requirements.

Agglomerating character is determined by examination of the residue left after the volatile determination. If the residue supports a 500-g weight without pulverizing or shows a swelling or cell structure, it is said to be "agglomerating."

The mineral-matter-free basis is used for ASTM rankings, and formulas to convert Btu, fixed carbon, and volatile matter from "as-received" bases are provided. Parr formulas—Eqs. (48.1)–(48.3) are appropriate in case of litigation. Approximation formulas—Eqs. (48.4)–(48.6) are otherwise acceptable.

Parr formulas

$$\text{Dry, MM-Free } FC = \frac{FC - 0.15S}{100 - (M + 1.08A + 0.55S)} \times 100 \qquad (48.1)$$

$$\text{Dry, MM-Free } VM = 100 - \text{Dry, MM-Free } FC \qquad (48.2)$$

$$\text{Moist, MM-Free Btu} = \frac{\text{Btu} - 50S}{100 - (1.08A + 0.55S)} \times 100 \qquad (48.3)$$

Approximation formulas

$$\text{Dry, MM-Free } FC = \frac{FC}{100 - (M + 1.1A + 0.1S)} \times 100 \qquad (48.4)$$

$$\text{Dry, MM-Free } VM = 100 - \text{Dry, MM-Free } FC \qquad (48.5)$$

$$\text{Moist, MM-Free Btu} = \frac{\text{Btu}}{100 - (1.1A + 0.1S)} \times 100 \qquad (48.6)$$

where MM = mineral matter
  Btu = British thermal unit
  FC = percentage of fixed carbon
  VM = percentage of volatile matter
  A = percentage of ash
  S = percentage of sulfur

Other classifications of coal include the International Classification of Hard Coals, the International Classification of Brown Coals, the "Lord" value based on heating value with ash, sulfur, and moisture removed, and the Perch and Russell Ratio, based on the ratio of Moist, MM-Free Btu to Dry, MM-Free VM.

Table 48.3 ASTM (D388) Classification of Coals by Rank[a]

Class	Group	Fixed Carbon Limits, Percent (Dry, Mineral-Matter-Free Basis) — Equal to or Greater Than	Fixed Carbon — Less Than	Volatile Matter Limits, Percent (Dry, Mineral-Matter-Free Basis) — Greater Than	Volatile Matter — Equal to or Less Than	Calorific Value Limits, Btu/lb (Moist,[b] Mineral-Matter-Free Basis) — Equal to or Greater Than	Calorific Value — Less Than	Agglomerating Character
I Anthracitic	1. Metaanthracite	98	—	—	2	—	—	
	2. Anthracite	92	98	2	8	—	—	Nonagglomerating
	3. Semianthracite[c]	86	92	8	14	—	—	
II Bituminous	1. Low-volatile bituminous	78	86	14	22	—	—	
	2. Medium-volatile bituminous	69	78	22	31	—	—	
	3. High-volatile A bituminous	—	69	31	—	14,000[d]	—	Commonly agglomerating[e]
	4. High-volatile B bituminous	—	—	—	—	13,000[d]	14,000	
	5. High-volatile C bituminous	—	—	—	—	11,500	13,000	
						10,500	11,500	Agglomerating
III Subbituminous	1. Subbituminous A	—	—	—	—	10,500	11,500	
	2. Subbituminous B	—	—	—	—	9,500	10,500	
	3. Subbituminous C	—	—	—	—	8,300	9,500	
IV Lignitic	1. Lignite A	—	—	—	—	6,300	8,300	
	2. Lignite B	—	—	—	—	—	6,300	

[a]This classification does not include a few coals, principally nonbanded varieties, that have unusual physical and chemical properties and that come within the limits of fixed carbon or calorific value of the high-volatile bituminous and subbituminous ranks. All of these coals either contain less than 48% dry, mineral-matter-free fixed carbon or have more than 15,500 moist, mineral-matter-free British thermal units per pound.

[b]Moist refers to coal containing its natural inherent moisture but not including visible water on the surface of the coal.

[c]If agglomerating, classify in the low-volatile group of the bituminous class.

[d]Coals having 69% or more fixed carbon on the dry, mineral-matter-free basis shall be classified according to fixed carbon, regardless of calorific value.

[e]It is recognized that there may be nonagglomerating varieties in these groups of the bituminous class and that there are notable exceptions in the high-volatile C bituminous group.

48.2 CURRENT USES—HEAT, POWER, STEELMAKING, OTHER

According to statistics compiled for the 1996 Keystone Coal Industry Manual, the primary use of coals produced in the United States in recent years has been for Electric Utilities; comprising almost 90% of the 926 million tons consumed in the U.S. in 1993. Industry accounted for about 8% of the consumption during that year in a variety of Standard Industrial Classification (SIC) Codes, replacing the manufacturing of coke (now about 3%) as the second largest coal market from the recent past. Industrial users typically consume coal for making process steam as well as in open-fired applications, such as in kilns and process heaters.

It should be noted that the demand for coal for coking purposes was greater than the demand for coal for utility use in the 1950's, and has steadily declined owing to more efficient steelmaking, greater use of scrap metal, increased use of substitute fuels in blast furnaces, and other factors. The production of coke from coal is accomplished by heating certain coals in the absence of air to drive off volatile matter and moisture. To provide a suitable by-product coke, the parent coal must possess quality parameters of low ash content, low sulfur content, low coking pressure, and high coke strength. By-product coking ovens, the most predominant, are so named for their ability to recapture otherwise wasted by-products driven off by heating the coal, such as coke oven gas, coal-tar, ammonia, oil, and useful chemicals. Beehive ovens, named for their shape and configuration, are also used, albeit much less extensively, in the production of coke.

48.3 TYPES

Anthracite is the least abundant of U.S. coal forms. Sometimes referred to as "hard" coal, it is shiny black or dark silver-gray and relatively compact. Inasmuch as it is the most advanced form in the coalification process, it is sometimes found deeper in the earth than bituminous. As indicated earlier, the ASTM definition puts upper and lower bounds of dry, mineral-matter-free fixed carbon percent at 98% and 86%, respectively, which limits volatile matter to not more than 14%. Combustion in turn is characterized by higher ignition temperatures and longer burnout times than bituminous coals.

Excepting some semianthracites that have a granular appearance, they have a consolidated appearance, unlike the layers seen in many bituminous coals. Typical Hardgrove Grindability Index ranges from 20 to 60 and specific gravity typically ranges 1.55 ± 0.10.

Anthracite coals can be found in Arkansas, Colorado, Pennsylvania, Massachusetts, New Mexico, Rhode Island, Virginia, and Washington, although by far the most abundant reserves are found in Pennsylvania.

Bituminous coal is by far the most plentiful and utilized coal form, and within the ASTM definitions includes low-, medium-, and high-volatile subgroups. Sometimes referred to as "soft" coal, it is named after the word bitumen, based on general tendency toward forming a sticky mass on heating.

At a lower stage of development in the coalification process, carbon content is less than the anthracites, from a maximum of 86% to less than 69% on a dry, mineral-matter-free basis. Volatile matter, at a minimum of 14% on this basis, is greater than the anthracites, and, as a result, combustion in pulverized form is somewhat easier for bituminous coals. Production of gas is also enhanced by their higher volatility.

The tendency of bituminous coals to produce a cohesive mass on heating lends them to coke applications. Dry, mineral-matter-free oxygen content generally ranges from 5% to 10%, compared to a value as low as 1% for anthracite. They are commonly banded with layers differing in luster.

The low-volatile bituminous coals are grainier and more subject to size reduction in handling.

The medium-volatile bituminous coals are sometimes distinctly layered, and sometimes only faintly layered and appearing homogeneous. Handling may or may not have a significant impact on size reduction.

The high-volatile coals (A, B, and C) are relatively hard and less sensitive to size reduction from handling than low- or medium-volatile bituminous.

Subbituminous coals, like anthracite and lignite, are generally noncaking. "Caking" refers to fusion of coal particles after heating in a furnace, as opposed to "coking," which refers to the ability of a coal to make a good coke, suitable for metallurgical purposes.

Oxygen content, on a dry, mineral-matter-free basis, is typically 10–20%.

Brownish black to black in color, this type coal is typically smooth in appearance with an absence of layers.

High in inherent moisture, it is ironic that these fuels are often dusty in handling and appear much like drying mud as they disintegrate on sufficiently long exposure to air.

The Healy coal bed in Wyoming has the thickest seam of coal in the United States at 220 ft. It is subbituminous, with an average heating value of 7884 Btu/lb, 28.5% moisture, 30% volatile matter, 33.9% fixed carbon, and 0.6% sulfur. Reported strippable reserves of this seam are approximately 11 billion tons.[4]

Lignites, often referred to as "brown coal," often retain a woodlike or laminar structure in which wood fiber remnants may be visible. Like subbituminous coals, they are high in seam moisture, up to 50% or more, and also disintegrate on sufficiently long exposure to air.

Both subbituminous coals and lignites are more susceptible than higher-rank coals to storage, shipping, and handling problems, owing to their tendency for slacking (disintegration) and spontaneous ignition. During the slacking, a higher rate of moisture loss at the surface than at the interior may cause higher rates and stresses at the outside of the particles, and cracks may occur with an audible noise.

Peat is decaying vegetable matter formed in wetlands; it is the first stage of metamorphosis in the coalification process. Development can be generally described as anaerobic, often in poorly drained flatlands or former lake beds. In the seam, peat moisture may be 90% or higher, and, therefore, the peat is typically "mined" and stacked for drainage or otherwise dewatered prior to consideration as a fuel. Because of its low bulk density at about 15 lb/ft^3 and low heating value at about 6000 Btu/lb (both values at 35% moisture), transportation distances must be short to make peat an attractive energy option.

In addition, it can be a very difficult material to handle, as it can arch in bins, forming internal friction angles in excess of 70°.

Chemically, peat is very reactive and ignites easily. It may be easily ground, and unconsolidated peat may create dusting problems.

48.4 PHYSICAL AND CHEMICAL PROPERTIES—DESCRIPTION AND TABLES OF SELECTED VALUES

There are a number of tests, qualitative and quantitative, used to provide information on coals; these tests will be of help to the user and/or equipment designer. Among the more common tests are the following, with reference to the applicable ASTM test procedure.

A. *"Proximate"* analysis (D3172) includes moisture, "volatile matter," "fixed carbon," and ash as its components.

Percent moisture (D3173) is determined by measuring the weight loss of a prepared sample (D2013) when heated to between 219°F (104°C) and 230°F (110°C) under rigidly controlled conditions. The results of this test can be used to calculate other analytical results to a dry basis. The moisture is referred to as "residual," and must be added to moisture losses incurred in sample preparation, called "air-dry losses" in order to calculate other analytical results to an "as-received" basis. The method which combines both residual and air dry moisture is D3302.

Percent volatile matter (D3175) is determined by establishing the weight loss of a prepared sample (D2013) resulting from heating to 1740°F (950°C) in the absence of air under controlled conditions. This weight loss is corrected for residual moisture, and is used for an indication of burning properties, coke yield, and classification by rank.

Percent ash (D3174) is determined by weighing the residue remaining after burning a prepared sample under rigidly controlled conditions. Combustion is in an oxidizing atmosphere and is completed for coal samples at 1290–1380°F (700–750°C).

Fixed carbon is a calculated value making up the fourth and final component of a proximate analysis. It is determined by subtracting the volatile, moisture, and ash percentages from 100.

Also generally included with a proximate analysis are calorific value and sulfur determinations.

B. *Calorific value,* Btu/lb (J/g, cal/g), is most commonly determined (D2015) in an "adiabatic bomb calorimeter," but is also covered by another method (D3286), which uses an "isothermal jacket bomb calorimeter." The values determined by this method are called gross or high heating values and include the latent heat of water vapor in the products of combustion.

C. *Sulfur* is determined by one of three methods provided by ASTM, all covered by D3177: the Eschka method, the bomb washing method, and a high-temperature combustion method.

The Eschka method requires that a sample be ignited with an "Eschka mixture" and sulfur be precipitated from the resulting solution as $BaSO_4$ and filtered, ashed, and weighed.

The bomb wash method requires use of the oxygen-bomb calorimeter residue, sulfur is precipitated as $BaSO_4$ and processed as in the Eschka method.

The high-temperature combustion method produces sulfur oxides from burning of a sample at 2460°F (1350°C), which are absorbed in a hydrogen peroxide solution for analysis. This is the most rapid of the three types of analysis.

D. *Sulfur forms* include sulfate, organic, and pyritic, and rarely, elemental sulfur. A method used to quantify sulfate, pyritic sulfur, and organic sulfur is D2492. The resulting data are sometimes used to provide a first indication of the maximum amount of sulfur potentially removable by mechanical cleaning.

E. *Ultimate analysis* (D3176) includes total carbon, hydrogen, nitrogen, oxygen, sulfur, and ash. These data are commonly used to perform combustion calculations to estimate combustion air requirements, products of combustion, and heat losses such as incurred by formation of water vapor by hydrogen in the coal.

Chlorine (D2361) and phosphorus (D2795) are sometimes requested with ultimate analyses, but are not technically a part of D3176.

F. *Ash mineral analysis* (D2795) includes the oxides of silica (SiO_2), alumina (Al_2O_3), iron (Fe_2O_3), titanium (TiO_2), phosphorus (P_2O_5), calcium (CaO), magnesium (MgO), sodium (Na_2O), and potassium (K_2O).

These data are used to provide several indications concerning ash slagging or fouling tendencies, abrasion potential, electrostatic precipitator operation, and sulfur absorption potention.

See Section 48.8 for further details.

G. *Grindability* (D409) is determined most commonly by the Hardgrove method to provide an indication of the relative ease of pulverization or grindability, compared to "standard" coals having grindability indexes of 40, 60, 80, and 110. As the index increases, pulverization becomes easier, that is, an index of 40 indicates a relatively hard coal; an index of 100 indicates a relatively soft coal.

Standard coals may be obtained from the U.S. Bureau of Mines.

A word of caution is given: grindability may change with ash content, moisture content, temperature, and other properties.

H. *Free swelling index* (D720) also referred to as a "coke-button" test, provides a relative index (1–9) of the swelling properties of a coal. A sample is burned in a covered crucible, and the resulting index increases as the swelling increases, determined by comparison of the button formed with standard profiles.

I. *Ash fusion temperatures* (D1857) are determined from triangular-core-shaped ash samples, in a reducing atmosphere and/or in an oxidizing atmosphere. Visual observations are recorded of temperatures at which the core begins to deform, called "initial deformation"; where height equals width, called "softening"; where height equals one-half width, called "hemispherical"; and where the ash is fluid.

The hemispherical temperature is often referred to as the "ash fusion temperature."

While not definitive, these tests provide a rough indication of the slagging tendency of coal ash.

Analysis of petrographic constituents in coals has been used to some extent in qualitative and semiquantitative analysis of some coals, most importantly in the coking coal industry. It is the application of macroscopic and microscopic techniques to identify maceral components related to the plant origins of the coal. The macerals of interest are vitrinite, exinite, resinite, micrinite, semifusinite, and fusinite. A technique to measure reflectance of a prepared sample of coal and calculate the volume percentages of macerals is included in ASTM Standard D2799.

Table 48.4 shows selected analyses of coal seams for reference.

48.5 BURNING CHARACTERISTICS

The ultimate analysis, described in the previous section, provides the data required to conduct fundamental studies of the air required for stoichiometric combustion, the volumetric and weight amounts of combustion gases produced, and the theoretical boiler efficiencies. These data assist the designer in such matters as furnace and auxiliary equipment sizing. Among the items of concern are draft equipment for supplying combustion air requirements, drying and transporting coal to the burners and exhausting the products of combustion, mass flow and velocity in convection passes for heat transfer and erosion considerations, and pollution control equipment sizing.

The addition of excess air must be considered for complete combustion and perhaps minimization of ash slagging in some cases. It is not uncommon to apply 25% excess air or more to allow operational flexibility.

As rank decreases, there is generally an increase in oxygen content in the fuel, which will provide a significant portion of the combustion air requirements.

The theoretical weight, in pounds, of combustion air required per pound of fuel for a stoichiometric condition is given by

$$11.53C + 34.34 [H_2 - \tfrac{1}{8}O_2] + 4.29S \tag{48.7}$$

where C, H_2, O_2, and S are percentage weight constituents in the ultimate analysis.

The resulting products of combustion, again at a stoichiometric condition and complete combustion, are

$$CO_2 = 3.66C \tag{48.8}$$

$$H_2O = 8.94H_2 + H_2O \text{ (wt\% } H_2O \text{ in fuel)} \tag{48.9}$$

$$SO_2 = 2.00S \tag{48.10}$$

$$N_2 = 8.86C + 26.41 (H_2 - \tfrac{1}{8}O_2) + 3.29S + N_2 \text{ (wt\% nitrogen in fuel)} \tag{48.11}$$

The combustion characteristics of various ranks of coal can be seen in Fig. 48.1, showing "burning profiles" obtained by thermal gravimetric analysis. As is apparent from this figure, ignition of lower rank coals occurs at a lower temperature and combustion proceeds at a more rapid rate than higher rank coals. This information is, of course, highly useful to the design engineer in determination of the size and configuration of combustion equipment.

The predominant firing technique for combustion of coal is in a pulverized form. To enhance ignition, promote complete combustion, and, in some cases, mitigate the effects of large particles on

Table 48.4   Selected Values—Coal and Peat Quality

Parameter	East Kentucky, Skyline Seam (Washed)	Pennsylvania, Pittsburgh #8 Seam (Washed)	Illinois, Harrisburg 5 (Washed)	Wyoming, Powder River Basin (Raw)	Florida Peat, Sumter County In situ	Florida Peat, Sumter County Dry
Moisture % (total)	8.00	6.5	13.2	25.92	86.70	—
Ash %	6.48	6.5	7.1	6.00	0.54	4.08
Sulfur %	0.82	1.62	1.28	0.25	0.10	0.77
Volatile %	36.69	34.40	30.6	31.27	8.74	65.73
Grindability (HGI)	45	55	54	57	36	69[a]
Calorific value (Btu/lb, as received)	12,500	13,100	11,700	8,500	1,503	11,297
Fixed carbon	48.83	52.60	49.1	37.23	4.02	30.19
Ash minerals						
SiO_2	50.87	50.10	48.90	32.02	58.29	
Al_2O_2	33.10	24.60	25.50	15.88	19.50	
TiO_2	2.56	1.20	1.10	1.13	1.05	
CaO	2.57	2.2	2.90	23.80	1.95	
K_2O	1.60	1.59	3.13	0.45	1.11	
MgO	0.80	0.70	1.60	5.73	0.94	
Na_2O	0.53	0.35	1.02	1.27	0.40	
P_2O_5	0.53	0.38	0.67	1.41	0.09	
Fe_2O_3	5.18	16.20	12.20	5.84	14.32	
SO_3	1.42	1.31	1.96	11.35	2.19	
Undetermined	0.84	1.37	1.02	1.12	0.16	
Ash	7.04	7.0	8.23	7.53	4.08	
Hydrogen	5.31	5.03	4.95	4.80	4.59	
Carbon	75.38	78.40	76.57	69.11	69.26	
Nitrogen	1.38	1.39	1.35	0.97	1.67	
Sulfur	0.89	1.73	1.47	0.34	0.77	
Oxygen (by difference)	9.95	6.35	7.03	17.24	19.33	
Chlorine	0.05	0.10	0.40	0.01	0.30	
Ash Fusion Temperatures (°F)						
Initial deformation (reducing)	2800+	2350	2240	2204	1950	
Softening ($H = W$) (reducing)	2800+	2460	2450	2226	2010	
Hemispherical ($H = \frac{1}{2}W$) (reducing)	2800+	2520	2500	2250	2060	
Fluid (reducing)	2800+	2580	2700+	2302	2100	

[a]At 9% H_2O.

Rate of weight loss,

Mg/minute

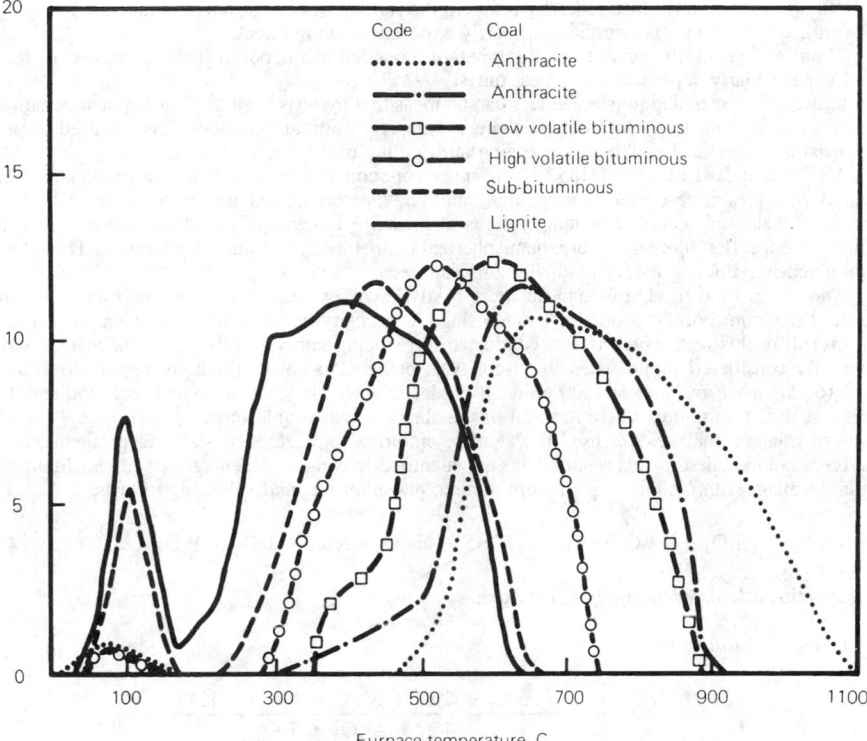

Fig. 48.1 Comparison of burning profiles for coals of different rank (courtesy of The Babcock and Wilcox Company).

slagging and particulate capture, guidelines are generally given by the boiler manufacturer for pulverizer output (burner input).

Typical guidelines are as follows:

Coal Class/Group	Percentage Passing a 200 Mesh Sieve	Percentage Retained on a 50 Mesh Sieve	Allowable Coal/Air Temperature (°F)
Anthracite	80	2.0	200
Low-volatile bituminous	70–75	2.0	180
High-volatile bituminous A	70–75	2.0	170
High-volatile bituminous C	65–72	2.0	150–160
Lignite	60–70	2.0	110–140

It is noted that these guidelines may vary for different manufacturers, ash contents, and equipment applications and, of course, the manufacturer should be consulted for fineness and temperature recommendations.

The sieve designations of 200 and 50 refer to U.S. Standard sieves. The 200 mesh sieve has 200 openings per linear inch, or 40,000 per square inch. The ASTM designations for these sieves are 75 and 300 micron, respectively.

Finally, agglomerating character may also have an influence on the fineness requirements, since this property might inhibit complete combustion.

48.6 ASH CHARACTERISTICS

Ash is an inert residue remaining after the combustion of coal and can result in significant challenges for designers and operators of the combustion, ash handling, and pollution control equipment. The quantity of ash in the coal varies widely from as little as 6% or less to more than 30% by weight. Additionally, diverse physical and chemical properties of ashes can pose substantial problems, with slagging, abrasion, and fouling of boilers. Electrostatic precipitators, used for pollution control, can experience material changes in collection efficiency depending on the mineral constituents of the ash.

"Slagging" is a term that generally refers to the formation of high-temperature fused ash deposits on furnace walls and other surfaces primarily exposed to radiant heat.

"Fouling" generally refers to high-temperature bonded ash deposits forming on convection tube banks, particularly superheat and reheat tubes.

Indication of ash-slagging tendencies can be measured by tests such as viscosity–temperature tests or by ash-softening tests. In addition, there are many empirical equations that are used to provide information as to the likelihood of slagging and fouling problems.

ASTM Standard number D1857 is the most common test used for slagging indication. In this test, ash samples are prepared as triangular cones and then are heated at a specified rate. Observations are then made and recorded of temperatures at prescribed stages of ash deformation, called initial deformation, softening temperature, hemispherical temperature, and fluid temperature. These tests are conducted in reducing and/or oxidizing atmospheres.

Another method used, although far more costly, involves measurement of the torque required to rotate a platinum bob suspended in molten slag. A viscosity–temperature relationship is established as a result of this test, which is also conducted in reducing and/or oxidizing atmospheres. A slag is generally considered liquid when its viscosity is below 250 poise, although tapping from a boiler may require a viscosity of 50–100 poise. It is plastic when its viscosity is between 250 and 10,000 poise. It is in this region where removal of the slag is most troublesome.

Ash mineral analyses are used to calculate empirical indicators of slagging problems. In these analyses are included metals reported as equivalent oxide weight percentages of silica, alumina, iron, calcium, magnesium, sodium, potassium, titania, phosphorous, and sulfur, as follows:

$$SiO_2 + Al_2O_3 + Fe_2O_3 + CaO + MgO + Na_2O + K_2O + TiO_2 + P_2O_5 + SO_3 = 100\%$$

Some ratios calculated using these data are:

Base: Acid Ratio, B/A

$$\frac{B}{A} = \frac{base}{acid} = \frac{Fe_2O_3 + CaO + MgO + Na_2O + K_2O}{SiO_2 + Al_2O_3 + TiO_2}$$

It has been reported[1] that a base/acid ratio in the range of 0.4 to 0.7 results typically in low ash fusibility temperatures and, hence, more slagging problems.

Slagging Factor, R_s

$$R_s = B/A \times \% \text{ sulfur, dry coal basis}$$

It has been reported[15] that coals with bituminous-type ashes exhibit a high slagging potential with a slagging factor above 2 and severe slagging potential with a slagging factor of more than 2.6. Bituminous-type ash refers to those ashes where iron oxide percentage is greater than calcium *plus* magnesium oxide.

Silica/Alumina Ratio

$$\frac{silica}{alumina} = \frac{SiO_2}{Al_2O_3}$$

It has been reported[1] that the silica in ash is more likely to form lower-melting-point compounds than is alumina and for two coals having the same base/acid ratio, the coal with a higher silica/alumina ratio should result in lower fusibility temperatures. However, it has also been reported[2] that for low base/acid ratios the opposite is true.

Iron/Calcium Ratio

$$\frac{iron}{dolomite} = \frac{Fe_2O_3}{CaO + MgO}$$

This ratio and its use are essentially the same as the iron/calcium ratio.

Silica Percentage (SP)

$$SP = \frac{SiO_2 \times 100}{SiO_2 + Fe_2O_3 + CaO + MgO}$$

This parameter has been correlated with ash viscosity. As silica ratio increases, the viscosity of slag increases. Graphical methods[2] are used in conjunction with this parameter to estimate the T_{250} temperature—the temperature where the ash would have a viscosity of 250 poise. Where the acidic content is less than 60% and the ash is lignitic, the *dolomite percentage (DP)* is used in preference to the silica percentage, along with graphs to estimate the T_{250}:

$$DP = \frac{(CaO + MgO) \times 100}{Fe_2O_3 + CaO + MgO + Na_2O + K_2O}$$

where the sum of the basic and acidic components are adjusted, if necessary, to equal 100%. For bituminous ash or lignitic-type ash having acidic content above 60%, the base/acid ratio is used in conjunction with yet another graph.

Fouling Factor (R_F)

$$R_F = acid\ base \times \%\ Na_2\ (bituminous\ ash)$$

or

$$R_F = \%\ Na_2O\ (lignitic\ ash)$$

For bituminous ash, the fouling factor[17] is "low" for values less than 0.1, "medium" for values between 0.1 and 0.25, "high" for values between 0.25 and 0.7, and "severe" values above 0.7. For lignitic-type ash, the percentage of sodium is used, and low, medium, and high values are <3.0, 3.0–6.0, and >6.0, respectively.

The basis for these factors is that sodium is the most important single factor in ash fouling, volatilizing in the furnace and subsequently condensing and sintering ash deposits in cooler sections.

Chlorine has also been used as an indicator of fouling tendency of eastern-type coals. If chlorine, from the ultimate analysis, is less than 0.15%, the fouling potential is low; if between 0.15 and 0.3, it has a medium fouling potential; and if above 0.3, its fouling potential is high.[1]

Ash resistivity can be predicted from ash mineral and coal ultimate analyses, according to a method described by Bickelhaupt.[24] Electrostatic precipitator sizing and/or performance can be estimated using the calculated resistivity. For further information, the reader is referred to Ref. 14.

48.7 SAMPLING

Coals are by nature heterogeneous and, as a result, obtaining a representative sample can be a formidable task. Its quality parameters such as ash, moisture, and calorific value can vary considerably from seam to seam and even within the same seam. Acquisition of accurate data to define the nature and ranges of these values adequately is further compounded by the effects of size gradation, sample preparation, and analysis accuracy.

Inasmuch as these data are used for such purposes as pricing, control of the operations in mines, and preparation plants, determination of power plant efficiency, estimation of material handling and storage requirements for the coal and its by-products, and in some cases for determination of compliance with environmental limitations, it is important that samples be taken, prepared, and analyzed in accordance with good practice.

To attempt to minimize significant errors in sampling, ASTM D2234 was developed as a standard method for the collection of a gross sample of coal and D2013 for preparation of the samples collected for analysis. It applies to lot sizes up to 10,000 tons (9080 mg) per gross sample and is intended to provide an accuracy of $\pm \frac{1}{10}$th of the average ash content in 95 of 100 determinations.

The number and weight of increments of sample comprising the gross sample to represent the lot, or consignment, is specified for nominal top sizes of $\frac{5}{8}$ in. (16 mm), 2 in. (50 mm), and 6 in. (150 mm), for raw coal or mechanically cleaned coal. Conditions of collection include samples taken from a stopped conveyor belt (the most desirable), full and partial stream cuts from moving coal consignments, and stationary samples.

One recommendation made in this procedure and worth special emphasis is that the samples be collected under the direct supervision of a person qualified by training and experience for this responsibility.

This method does not apply to the sampling of reserves in the ground, which is done by core drilling methods or channel sampling along outcrops as recommended by a geologist or mining engineer. It also does not apply to the sampling of coal slurries.

A special method (D197) was developed for the collection of samples of pulverized coals to measure size consist or fineness, which is controlled to maintain proper combustion efficiency.

Sieve analyses using this method may be conducted on No. 16 (1.18 mm) through No. 200 (750 mm) sieves, although the sieves most often referred to for pulverizer and classifier performance are the No. 50, the No. 100, and the No. 200 sieves. The number refers to the quantity of openings per linear inch.

Results of these tests should plot as a straight line on a sieve distribution chart. A typical fineness objective, depending primarily on the combustion characteristics might be for 70% of the fines to pass through a 200-mesh sieve and not more than 2% to be retained on the 50-mesh sieve.

D2013 covers the preparation of coal samples for analysis, or more specifically, the reduction and division of samples collected in accordance with D2234. A "referee" and "nonreferee" method are delineated, although the "nonreferee" method is the most commonly used. "Referee" method is used to evaluate equipment and the nonreferee method.

Depending on the amount of moisture and, therefore, the ability of the coal to pass freely through reduction equipment, samples are either predried in air drying ovens or on drying floors or are processed directly by reduction, air drying, and division. Weight losses are computed for each stage of air drying to provide data for determination of total moisture, in combination with D3173 or 3302 (see Section 48.6).

Samples must ultimately pass the No. 60 (250 mm) sieve (D3173) or the No. 8 (2.36 mm) sieve (D3302) prior to total moisture determination.

Care must be taken to adhere to this procedure, including the avoidance of moisture losses while awaiting preparation, excessive time in air drying, proper use of riffling or mechanical division equipment, and verification and maintenance of crushing equipment size consist.

48.8 COAL CLEANING

Partial removal of impurities in coal such as ash and pyritic sulfur has been conducted since before 1900, although application and development has intensified during recent years owing to a number of factors, including the tightening of emissions standards, increasing use of lower quality seams, and increasing use of continuous mining machinery. Blending of two or more fuels to meet tight emissions standards, or other reasons, often requires that each of the fuels is of a consistent grade, which in turn may indicate some degree of coal cleaning.

Coal cleaning may be accomplished by physical or chemical means, although physical coal cleaning is by far the most predominant.

Primarily, physical processes rely on differences between the specific gravity of the coal and its impurities. Ash, clay, and pyritic sulfur have a higher specific gravity than that of coal. For example, bituminous coal typically has a specific gravity in the range 1.12–1.35, while pyrite's specific gravity is between 4.8 and 5.2.

One physical process that does not benefit from specific gravity differences is froth flotation. It is only used for cleaning coal size fractions smaller than 28 mesh ($\frac{1}{2}$ mm). Basically the process requires coal fines to be agitated in a chamber with set amounts of air, water, and chemical reagents, which creates a froth. The coal particles are selectively attached to the froth bubbles, which rise to the surface and are skimmed off, dewatered, and added to other clean coal fractions.

A second stage of froth flotation has been tested successfully at the pilot scale for some U.S. coals. This process returns the froth concentrate from the first stage to a bank of cells where a depressant is used to sink the coal and a xanthate flotation collector is used to selectively float pyrite.

The predominant commercial methods of coal cleaning use gravity separation by static and/or dynamic means. The extent and cost of cleaning naturally depends on the degree of end product quality desired, the controlling factors of which are primarily sulfur, heating value, and ash content.

Although dry means may be used for gravity separation, wet means are by far the more accepted and used techniques.

The first step in designing a preparation plant involves a careful study of the washability of the coal. "Float and sink" tests are run in a laboratory to provide data to be used for judging application and performance of cleaning equipment. In these tests the weight percentages and composition of materials are determined after subjecting the test coal to liquid baths of different specific gravities.

Pyritic sulfur and/or total sulfur percent, ash percent, and heating value are typically determined for both the float (called "yield") and sink (called "reject") fractions.

Commonly, the tests are conducted on three or more size meshes, such as $1\frac{1}{2}$ in. $\times$ 0 mesh, $\frac{3}{8}$ in. $\times$ 100 mesh, and minus 14 mesh, and at three or more gravities of such as 1.30, 1.40, and 1.60. Percentage recovery of weight and heating value are reported along with other data on a cumulative basis for the float fractions. An example of this, taken from a Bureau of Mines study[11] of 455 coals is shown in Table 48.5.

Many coals have pyrite particles less than one micron in size (0.00004 in.), which cannot be removed practically by mechanical means. Moreover, the cost of coal cleaning increases as the particle size decreases, as a general rule, and drying and handling problems become more difficult. Generally, coal is cleaned using particle sizes as large as practical to meet quality requirements.

Table 48.5 Cumulative Washability Data[a]

Product	Recovery Weight	% Btu	Btu/lb	Ash %	Sulfur (%) Pyritic	Sulfur (%) Total	lb SO₂/ MBtu
Sample Crushed to Pass 1½ in.							
Float—1.30	55.8	58.8	14,447	3.3	0.26	1.09	1.5
Float—1.40	90.3	93.8	14,239	4.7	0.46	1.33	1.9
Float—1.60	94.3	97.3	14,134	5.4	0.53	1.41	2.0
Total	100.0	100.0	13,703	8.3	0.80	1.67	2.4
Sample Crushed to Pass ⅜ in.							
Float—1.30	58.9	63.0	14,492	3.0	0.20	1.06	1.5
Float—1.40	88.6	93.2	14,253	4.6	0.37	1.26	1.8
Float—1.60	92.8	96.8	14,134	5.4	0.47	1.36	1.9
Total	100.0	100.0	13,554	9.3	0.77	1.64	2.4
Sample Crushed to Pass 14 Mesh							
Float—1.30	60.9	65.1	14,566	2.5	0.16	0.99	1.4
Float—1.40	88.7	93.1	14,298	4.3	0.24	1.19	1.7
Float—1.60	93.7	97.1	14,134	5.4	0.40	1.29	1.8
Total	100.0	100.0	13,628	8.8	0.83	1.72	2.5

[a]State: Pennsylvania (bituminous); coal bed: Pittsburgh; county: Washington; raw coal moisture: 2.0%.

It is interesting to note that a 50-micron pyrite particle (0.002 in.) inside a 14-mesh coal particle (0.06 in.) does not materially affect the specific gravity of a pure coal particle of the same size. Washability data are usually organized and plotted as a series of curves, including:

1. Cumulative float–ash, sulfur %
2. Cumulative sink ash %
3. Elementary ash % (not cumulative)
4. % recovery (weight) versus specific gravity

Types of gravity separation equipment include jigs, concentrating tables, water-only cyclones, dense-media vessels, and dense-media cyclones.

In jigs, the coal enters the vessel in sizes to 8 in. and larger and stratification of the coal and heavier particles occurs in a pulsating fluid. The bottom layer, primarily rock, ash, and pyrite, is stripped from the mixture and rejected. Coal, the top layer, is saved. A middle layer may also be collected and saved or rejected depending on quality.

Concentrating tables typically handle coals in the ⅜ in. × 0 or ¼ in. × 0 range and use water cascading over a vibrating table tilted such that heavier particles travel to one end of the table while lighter particles, traveling more rapidly with the water, fall over the adjacent edge.

In "water-only" and dense-media cyclones, centrifugal force is used to separate the heavier particles from the lighter particles.

In heavy-media vessels, the specific gravity of the media is controlled typically in the range of 1.45 to 1.65. Particles floating are saved as clean coal, while those sinking are reject. Specific gravity of the media is generally maintained by the amount of finely ground magnetite suspended in the water, as in heavy-media cyclones. Magnetite is recaptured in the preparation plant circuitry by means of magnetic separators.

Drying of cleaned coals depends on size. The larger sizes, ¼ in. or ⅜ in. and larger, typically require little drying and might only be passed over vibrating screens prior to stockpiling. Smaller sizes, down to, say, 28 mesh, are commonly dried on stationary screens followed by centrifugal driers. Minus 28 mesh particles, the most difficult to dry, are processed in vibrating centrifuges, high-speed centrifugal driers, high-speed screens, or vacuum filters.

REFERENCES

1. J. G. Singer, *Combustion—Fossil Power Systems,* Combustion Engineering, Inc., Windsor, CT, 1981.

2. *Stream—Its Generation and Use,* Babcock and Wilcox, New York, 1992.

3. C. L. Wilson, *Coal—Bridge to the Future,* Vol. 1, Report of the World Coal Study, Ballinger, Cambridge, MA, 1980.

4. *Keystone Coal Industry Manual,* McGraw-Hill, New York, 1996.

5. *1979/1980 Coal Data,* National Coal Association and U.S. DOE, Washington, DC, 1982.

6. R. A. Meyers, *Coal Handbook,* Dekker, New York, 1981.

7. ASTM Standards, Part 26: *Gaseous Fuels; Coal and Coke, Atmospheric Analysis,* Philadelphia, PA 1982.

8. F. M. Kennedy, J. G. Patterson, and T. W. Tarkington, *Evaluation of Physical and Chemical Coal Cleaning and Flue Gas Desulfurization,* EPA 600/7-79-250, U.S. EPA, Washington, DC, 1979.

9. Gibbs and Hill, Inc., *Coal Preparation for Combustion and Conversion,* Electric Power Research Institute, Palo Alto, CA, 1978.

10. *Coal Data Book,* President's Commission on Coal, U.S. Government Printing Office, Washington, DC, 1980.

11. J. A. Cavallaro and A. W. Deubrouch, *Sulfur Reduction Potential of the Coals of the United States,* RI8118, U.S. Dept. of Interior, Washington, DC, 1976.

12. R. A. Schmidt, *Coal in America,* 1979.

13. Auth and Johnson, *Fuels and Combustion Handbook,* McGraw-Hill, New York, 1951.

14. R. E. Bickelhaupt, *A Technique for Predicting Fly Ash Resistivity,* EPA 600/7-79-204, Southern Research Institute, Birmingham, AL, August, 1979.

15. R. C. Attig and A. F. Duzy, *Coal Ash Deposition Studies and Application to Boiler Design,* Bobcock and Wilcox, Atlanta, GA, 1969.

16. A. F. Duzy and C. L. Wagoner, *Burning Profiles for Solid Fuels,* Babcock and Wilcox, Atlanta, GA, 1975.

17. E. C. Winegartner, *Coal Fouling and Slagging Parameters,* Exxon Research and Engineering, Baytown, TX, 1974.

18. J. B. McIlroy and W. L. Sage, Relationship of Coal-Ash Viscosity to Chemical Composition, *The Journal of Engineering for Power,* Babcock and Wilcox, New York, 1960.

19. R. P. Hensel, *The Effects of Agglomerating Characteristics of Coals on Combustion in Pulverized Fuel Boilers,* Combustion Engineering, Windsor, CT, 1975.

20. W. T. Reed, *External Corrosion and Deposits in Boilers and Gas Turbines,* Elsevier, New York, 1971.

21. O. W. Durrant, *Pulverized Coal—New Requirements and Challenges,* Babcock and Wilcox, Atlanta, GA, 1975.

CHAPTER 49

SOLAR ENERGY APPLICATIONS

Jan F. Kreider
Jan F. Kreider and Associates, Inc.
and Joint Center for Energy Management
University of Colorado
Boulder, Colorado

49.1 SOLAR ENERGY AVAILABILITY	**1549**	
49.1.1 Solar Geometry	1549	
49.1.2 Sunrise and Sunset	1552	
49.1.3 Quantitative Solar Flux Availability	1554	
49.2 SOLAR THERMAL COLLECTORS	**1560**	
49.2.1 Flat-Plate Collectors	1560	
49.2.2 Concentrating Collectors	1564	
49.2.3 Collector Testing	1568	
49.3 SOLAR THERMAL APPLICATIONS	**1569**	
49.3.1 Solar Water Heating	1569	

49.3.2 Mechanical Solar Space Heating Systems	1569	
49.3.3 Passive Solar Space Heating Systems	1571	
49.3.4 Solar Ponds	1571	
49.3.5 Industrial Process Applications	1575	
49.3.6 Solar Thermal Power Production	1575	
49.3.7 Other Thermal Applications	1576	
49.3.8 Performance Prediction for Solar Thermal Processes	1576	
49.4 NONTHERMAL SOLAR ENERGY APPLICATIONS	**1577**	

49.1 SOLAR ENERGY AVAILABILITY

Solar energy is defined as that radiant energy transmitted by the sun and intercepted by earth. It is transmitted through space to earth by electromagnetic radiation with wavelengths ranging between 0.20 and 15 microns. The availability of solar flux for terrestrial applications varies with season, time of day, location, and collecting surface orientation. In this chapter we shall treat these matters analytically.

49.1.1 Solar Geometry

Two motions of the earth relative to the sun are important in determining the intensity of solar flux at any time—the earth's rotation about its axis and the annual motion of the earth and its axis about the sun. The earth rotates about its axis once each day. A solar day is defined as the time that elapses between two successive crossings of the local meridian by the sun. The local meridian at any point is the plane formed by projecting a north–south longitude line through the point out into space from the center of the earth. The length of a solar day on the average is slightly less than 24 hr, owing to the forward motion of the earth in its solar orbit. Any given day will also differ from the average day owing to orbital eccentricity, axis precession, and other secondary effects embodied in the equation of time described below.

Declination and Hour Angle

The earth's orbit about the sun is elliptical with eccentricity of 0.0167. This results in variation of solar flux on the outer atmosphere of about 7% over the course of a year. Of more importance is the variation of solar intensity caused by the inclination of the earth's axis relative to the ecliptic plane of the earth's orbit. The angle between the ecliptic plane and the earth's equatorial plane is 23.45°. Figure 49.1 shows this inclination schematically.

Mechanical Engineers' Handbook, 2nd ed., Edited by Myer Kutz.
ISBN 0-471-13007-9   © 1998 John Wiley & Sons, Inc.

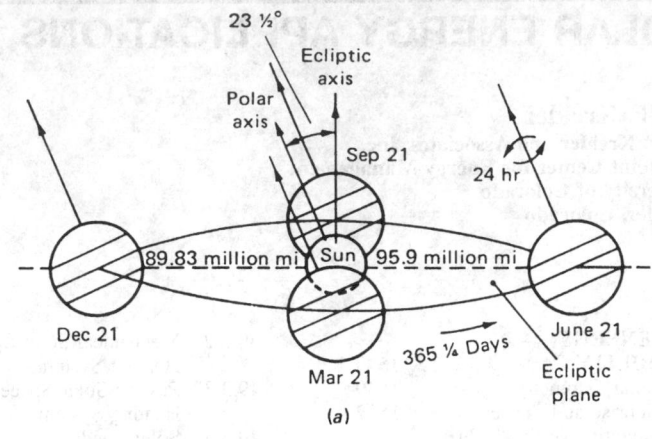

(a)

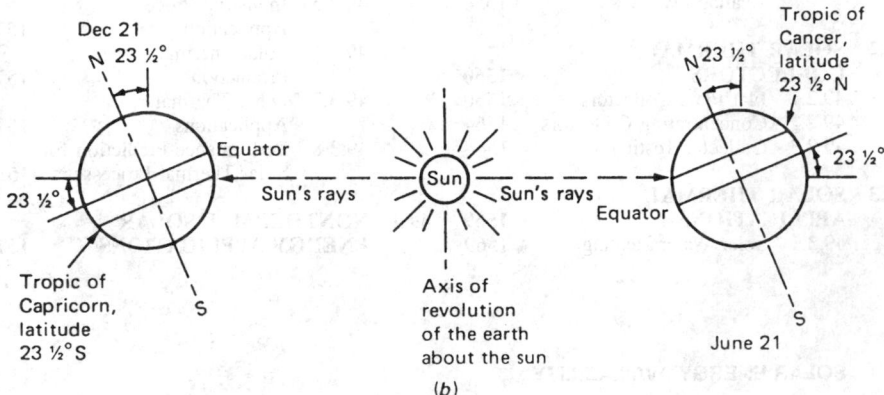

(b)

Fig. 49.1   (a) Motion of the earth about the sun. (b) Location of tropics. Note that the sun is so far from the earth that all the rays of the sun may be considered as parallel to one another when they reach the earth.

The earth's motion is quantified by two angles varying with season and time of day. The angle varying on a seasonal basis that is used to characterize the earth's location in its orbit is called the solar "declination." It is the angle between the earth–sun line and the equatorial plane as shown in Fig. 49.2. The declination δ_s is taken to be positive when the earth-sun line is north of the equator and negative otherwise. The declination varies between $+23.45°$ on the summer solstice (June 21 or 22) and $-23.45°$ on the winter solstice (December 21 or 22). The declination is given by

$$\sin \delta_s = 0.398 \cos [0.986(N - 173)] \tag{49.1}$$

in which N is the day number.

The second angle used to locate the sun is the solar-hour angle. Its value is based on the nominal $360°$ rotation of the earth occurring in 24 hr. Therefore, 1 hr is equivalent to an angle of $15°$. The hour angle is measured from zero at solar noon. It is denoted by h_s and is positive before solar noon and negative after noon in accordance with the right-hand rule. For example 2:00 PM corresponds to $h_s = -30°$ and 7:00 AM corresponds to $h_s = +75°$.

Solar time, as determined by the position of the sun, and clock time differ for two reasons. First, the length of a day varies because of the ellipticity of the earth's orbit; and second, standard time is determined by the standard meridian passing through the approximate center of each time zone. Any position away from the standard meridian has a difference between solar and clock time given by [(local longitude − standard meridian longitude)/15] in units of hours. Therefore, solar time and local standard time (LST) are related by

$$\text{solar time} = \text{LST} - \text{EoT} - (\text{local longitude} - \text{standard meridian longitude})/15 \tag{49.2}$$

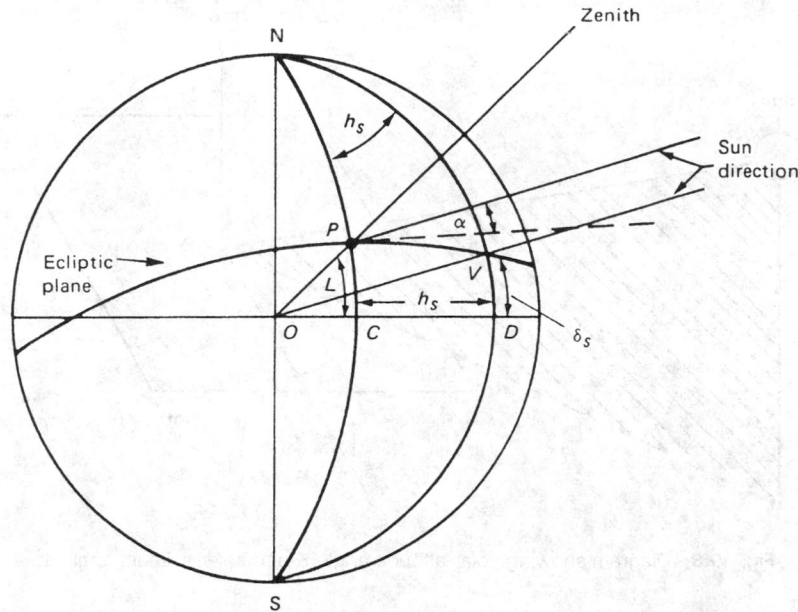

Fig. 49.2 Definition of solar-hour angle h_s (*CND*), solar declination δ_s (*VOD*), and latitude L (*POC*): P, site of interest. (Modified from J. F. Kreider and F. Kreith, *Solar Heating and Cooling*, revised 1st ed., Hemisphere, Washington, DC, 1977.)

in units of hours. EoT is the equation of time which accounts for difference in day length through a year and is given by

$$\text{EoT} = 12 + 0.1236 \sin x - 0.0043 \cos x + 0.1538 \sin 2x + 0.0608 \cos 2x \qquad (49.3)$$

in units of hours. The parameter x is

$$x = \frac{360(N - 1)}{365.24} \qquad (49.4)$$

where N is the day number counted from January 1 as $N = 1$.

Solar Position

The sun is imagined to move on the celestial sphere, an imaginary surface centered at the earth's center and having a large but unspecified radius. Of course, it is the earth that moves, not the sun, but the analysis is simplified if one uses this Ptolemaic approach. No error is introduced by the moving sun assumption, since the relative motion is the only motion of interest. Since the sun moves on a spherical surface, two angles are sufficient to locate the sun at any instant. The two most commonly used angles are the solar-altitude and azimuth angles (see Fig. 49.3) denoted by α and a_s, respectively. Occasionally, the solar-zenith angle, defined as the complement of the altitude angle, is used instead of the altitude angle.

The solar-altitude angle is related to the previously defined declination and hour angles by

$$\sin \alpha = \cos L \cos \delta_s \cos h_s + \sin L \sin \delta_s \qquad (49.5)$$

in which L is the latitude, taken positive for sites north of the equator and negative for sites south of the equator. The altitude angle is found by taking the inverse sine function of Eq. (49.5).

The solar-azimuth angle is given by[1]

$$\sin a_s = \frac{\cos \delta_s \sin h_s}{\cos \alpha} \qquad (49.6)$$

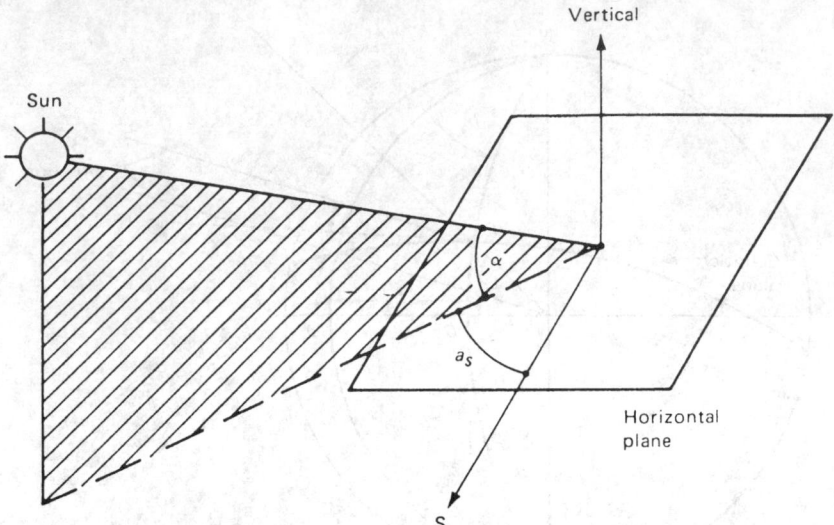

Fig. 49.3 Diagram showing solar-altitude angle α and solar-azimuth angle a_s.

To find the value of a_s, the location of the sun relative to the east-west line through the site must be known. This is accounted for by the following two expressions for the azimuth angle:

$$a_s = \sin^{-1}\left(\frac{\cos \delta_s \sin h_s}{\cos \alpha}\right), \qquad \cos h_s > \frac{\tan \delta_s}{\tan L} \qquad (49.7)$$

$$a_s = 180° - \sin^{-1}\left(\frac{\cos \delta_s \sin h_s}{\cos \alpha}\right), \qquad \cos h_s < \frac{\tan \delta_s}{\tan L} \qquad (49.8)$$

Table 49.1 lists typical values of altitude and azimuth angles for latitude $L = 40°$. Complete tables are contained in Refs. 1 and 2.

49.1.2 Sunrise and Sunset

Sunrise and sunset occur when the altitude angle $\alpha = 0$. As indicated in Fig. 49.4, this occurs when the center of the sun intersects the horizon plane. The hour angle for sunrise and sunset can be found from Eq. (49.5) by equating α to zero. If this is done, the hour angles for sunrise and sunset are found to be

$$h_{sr} = \cos^{-1}(-\tan L \tan \delta_s) = -h_{ss} \qquad (49.9)$$

in which h_{sr} is the sunrise hour angle and h_{ss} is the sunset hour angle.

Figure 49.4 shows the path of the sun for the solstices and the equinoxes (length of day and night are both 12 hr on the equinoxes). This drawing indicates the very different azimuth and altitude angles that occur at different times of year at identical clock times. The sunrise and sunset hour angles can be read from the figures where the sun paths intersect the horizon plane.

Solar Incidence Angle

For a number of reasons, many solar collection surfaces do not directly face the sun continuously. The angle between the sun–earth line and the normal to any surface is called the incidence angle. The intensity of off-normal solar radiation is proportional to the cosine of the incidence angle. For example, Fig. 49.5 shows a fixed planar surface with solar radiation intersecting the plane at the incidence angle i measured relative to the surface normal. The intensity of flux at the surface is $I_b \times \cos i$, where I_b is the beam radiation along the sun–earth line; I_b is called the direct, normal radiation. For a fixed surface such as that in Fig. 49.5 facing the equator, the incidence angle is given by

$$\cos i = \sin \delta_1(\sin L \cos \beta - \cos L \sin \beta \cos a_w)$$
$$+ \cos \delta_s \cos h_s(\cos L \cos \beta + \sin L \sin \beta \cos a_w) \qquad (49.10)$$
$$+ \cos \delta_s \sin \beta \sin a_w \sin h_s$$

Table 49.1 Solar Position for 40°N Latitude

Date	Solar Time AM	Solar Time PM	Solar Position Altitude	Solar Position Azimuth	Date	Solar Time AM	Solar Time PM	Solar Position Altitude	Solar Position Azimuth
January 21	8	4	8.1	55.3	July 21	5	7	2.3	115.2
	9	3	16.8	44.0		6	6	13.1	106.1
	10	2	23.8	30.9		7	5	24.3	97.2
	11	1	28.4	16.0		8	4	35.8	87.8
	12		30.0	0.0		9	3	47.2	76.7
February 21	7	5	4.8	72.7		10	2	57.9	61.7
	8	4	15.4	62.2		11	1	66.7	37.9
	9	3	25.0	50.2		12		70.6	0.0
	10	2	32.8	35.9	August 21	6	6	7.9	99.5
	11	1	38.1	18.9		7	5	19.3	90.9
	12		40.0	0.0		8	4	30.7	79.9
March 21	7	5	11.4	80.2		9	3	41.8	67.9
	8	4	22.5	69.6		10	2	51.7	52.1
	9	3	32.8	57.3		11	1	59.3	29.7
	10	2	41.6	41.9		12		62.3	0.0
	11	1	47.7	22.6	September 21	7	5	11.4	80.2
	12		50.0	0.0		8	4	22.5	69.6
April 21	6	6	7.4	98.9		9	3	32.8	57.3
	7	5	18.9	89.5		10	2	41.6	41.9
	8	4	30.3	79.3		11	1	47.7	22.6
	9	3	41.3	67.2		12		50.0	0.0
	10	2	51.2	51.4	October 21	7	5	4.5	72.3
	11	1	58.7	29.2		8	4	15.0	61.9
	12		61.6	0.0		9	3	24.5	49.8
May 21	5	7	1.9	114.7		10	2	32.4	35.6
	6	6	12.7	105.6		11	1	37.6	18.7
	7	5	24.0	96.6		12		39.5	0.0
	8	4	35.4	87.2	November 21	8	4	8.2	55.4
	9	3	46.8	76.0		9	3	17.0	44.1
	10	2	57.5	60.9		10	2	24.0	31.0
	11	1	66.2	37.1		11	1	28.6	16.1
	12		70.0	0.0		12		30.2	0.0
June 21	5	7	4.2	117.3	December 21	8	4	5.5	53.0
	6	6	14.8	108.4		9	3	14.0	41.9
	7	5	26.0	99.7		10	2	20.0	29.4
	8	4	37.4	90.7		11	1	25.0	15.2
	9	3	48.8	80.2		12		26.6	0.0
	10	2	59.8	65.8					
	11	1	69.2	41.9					
	12		73.5	0.0					

in which a_w is the "wall" azimuth angle and β is the surface tilt angle relative to the horizontal plane, both as shown in Fig. 49.5.

For fixed surfaces that face due south, the incidence angle expression simplifies to

$$\cos i = \sin(L - \beta)\sin \delta_s + \cos(L - \beta)\cos \delta_s \cos h_s \qquad (49.11)$$

A large class of solar collectors move in some fashion to track the sun's diurnal motion, thereby improving the capture of solar energy. This is accomplished by reduced incidence angles for properly tracking surfaces vis-á-vis a fixed surface for which large incidence angles occur in the early morning and late afternoon (for generally equator-facing surfaces). Table 49.2 lists incidence angle expressions for nine different types of tracking surfaces. The term "polar axis" in this table refers to an axis of

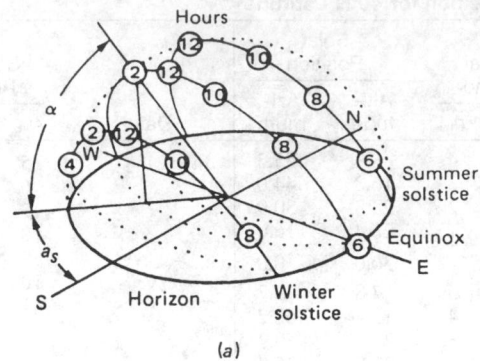

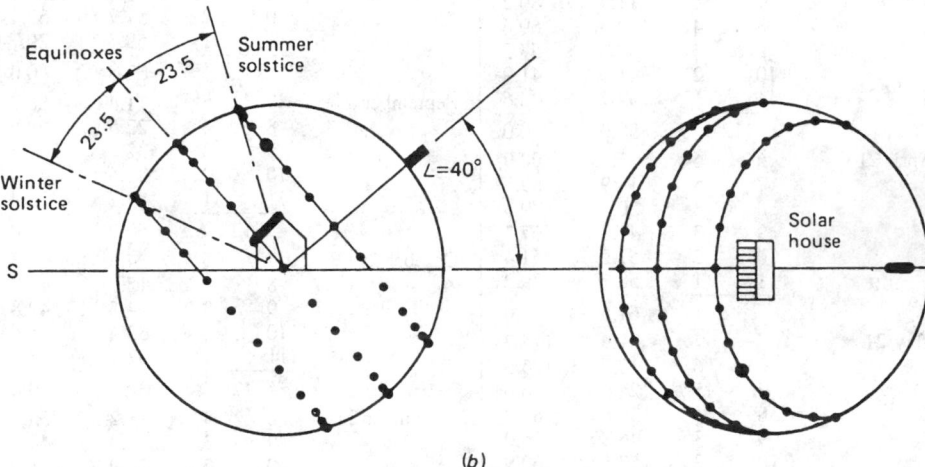

Fig. 49.4 Sun paths for the summer solstice (6/21), the equinoxes (3/21 and 9/21), and the winter solstice (12/21) for a site at 40°N; (a) isometric view; (b) elevation and plan views.

rotation directed at the north or south pole. This axis of rotation is tilted up from the horizontal at an angle equal to the local latitude. It is seen that normal incidence can be achieved (i.e., $\cos i = 1$) for any tracking scheme for which two axes of rotation are present. The polar case has relatively small incidence angles as well, limited by the declination to $\pm 23.45°$. The mean value of $\cos i$ for polar tracking is 0.95 over a year, nearly as good as the two-axis case for which the annual mean value is unity.

49.1.3 Quantitative Solar Flux Availability

The previous section has indicated how variations in solar flux produced by seasonal and diurnal effects can be quantified. However, the effect of weather on solar energy availability cannot be analyzed theoretically; it is necessary to rely on historical weather reports and empirical correlations for calculations of actual solar flux. In this section this subject is described along with the availability of solar energy at the edge of the atmosphere—a useful correlating parameter, as seen shortly.

Extraterrestrial Solar Flux

The flux intensity at the edge of the atmosphere can be calculated strictly from geometric considerations if the direct-normal intensity is known. Solar flux incident on a terrestrial surface, which has traveled from sun to earth with negligible change in direction, is called beam radiation and is denoted by I_b. The extraterrestrial value of I_b averaged over a year is called the solar constant, denoted by I_{sc}. Its value is 429 Btu/hr · ft² or 1353 W/m². Owing to the eccentricity of the earth's orbit, however, the extraterrestrial beam radiation intensity varies from this mean solar constant value. The variation of I_b over the year is given by

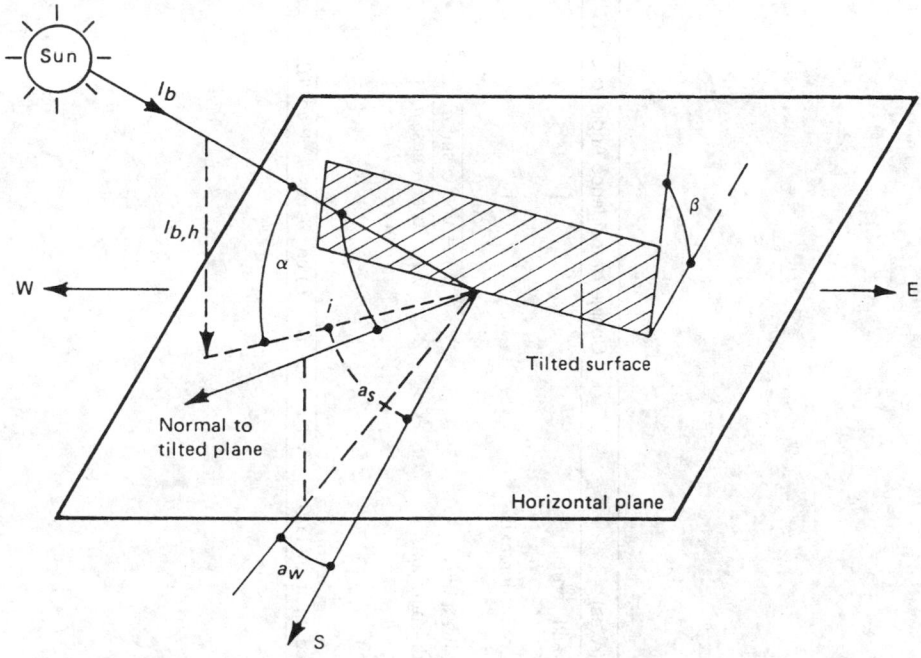

Fig. 49.5 Definition of incidence angle i, surface tilt angle β, solar-altitude angle α, wall-azimuth angle a_w, and solar-azimuth angle a_s for a non-south-facing tilted surface. Also shown is the beam component of solar radiation I_b and the component of beam radiation $I_{b,h}$ on a horizontal plane.

$$I_{b,0}(N) = \left[1 + 0.034 \cos\left(\frac{360\ N}{265}\right)\right] \times I_{sc} \qquad (49.12)$$

in which N is the day number as before.

In subsequent sections the total daily, extraterrestrial flux will be particularly useful as a nondimensionalizing parameter for terrestrial solar flux data. The instantaneous solar flux on a horizontal, extraterrestrial surface is given by

$$I_{b,h0} = I_{b,0}(N) \sin \alpha \qquad (49.13)$$

as shown in Fig. 49.5. The daily total, horizontal radiation is denoted by I_0 and is given by

$$I_0(N) = \int_{t_{sr}}^{t_{ss}} I_{b,0}(N) \sin \alpha \, dt \qquad (49.14)$$

$$I_0(N) = \frac{24}{\pi} I_{sc} \left[1 + 0.034 \cos\left(\frac{360\ N}{265}\right)\right] \times (\cos L \cos \delta_s \sin h_{sr} + h_{sr} \sin L \sin \delta_s) \qquad (49.15)$$

in which I_{sc} is the solar constant. The extraterrestrial flux varies with time of year via the variations of δ_s and h_{sr} with time of year. Table 49.3 lists the values of extraterrestrial, horizontal flux for various latitudes averaged over each month. The monthly averaged, horizontal, extraterrestrial solar flux is denoted by $\overline{H}_0$.

Terrestrial Solar Flux

Values of instantaneous or average terrestrial solar flux cannot be predicted accurately owing to the complexity of atmospheric processes that alter solar flux magnitudes and directions relative to their extraterrestrial values. Air pollution, clouds of many types, precipitation, and humidity all affect the values of solar flux incident on earth. Rather than attempting to predict solar availability accounting for these complex effects, one uses long-term historical records of terrestrial solar flux for design purposes.

Table 49.2 Solar Incidence Angle Equations for Tracking Collectors

Description	Axis (Axes)	Cosine of Incidence Angle ($\cos i$)
Movements in altitude and azimuth	Horizontal axis and vertical axis	1
Rotation about a polar axis and adjustment in declination	Polar axis and declination axis	1
Uniform rotation about a polar axis	Polar axis	$\cos \delta_s$
East–west horizontal	Horizontal, east–west axis	$\sqrt{1 - \cos^2 \alpha \sin^2 a_s}$
North–south horizontal	Horizontal, north–south axis	$\sqrt{1 - \cos^2 \alpha \cos^2 a_s}$
Rotation about a vertical axis of a surface tilted upward L (latitude) degrees	Vertical axis	$\sin (\alpha + L)$
Rotation of a horizontal collector about a vertical axis	Vertical axis	$\sin \alpha$
Rotation of a vertical surface about a vertical axis	Vertical axis	$\cos \alpha$
Fixed "tubular" collector	North–south tilted up at angle β	$\sqrt{1 - [\sin (\beta - L) \cos \delta_s \cos h_s + \cos (\beta - L) \sin \delta_s]^2}$

Table 49.3 Average Extraterrestrial Radiation on a Horizontal Surface $\overline{H}_0$ in SI Units and in English Units Based on a Solar Constant of 429 Btu/hr·ft² or 1.353 kW/m²

Latitude, Degrees	January	February	March	April	May	June	July	August	September	October	November	December
SI Units, W·hr/m²·Day												
20	7415	8397	9552	10,422	10,801	10,868	10,794	10,499	9791	8686	7598	7076
25	6656	7769	9153	10,312	10,936	11,119	10,988	10,484	9494	8129	6871	6284
30	5861	7087	8686	10,127	11,001	11,303	11,114	10,395	9125	7513	6103	5463
35	5039	6359	8153	9869	10,995	11,422	11,172	10,233	8687	6845	5304	4621
40	4200	5591	7559	9540	10,922	11,478	11,165	10,002	8184	6129	4483	3771
45	3355	4791	6909	9145	10,786	11,477	11,099	9705	7620	5373	3648	2925
50	2519	3967	6207	8686	10,594	11,430	10,981	9347	6998	4583	2815	2100
55	1711	3132	5460	8171	10,358	11,352	10,825	8935	6325	3770	1999	1320
60	963	2299	4673	7608	10,097	11,276	10,657	8480	5605	2942	1227	623
65	334	1491	3855	7008	9852	11,279	10,531	8001	4846	2116	544	97
English Units, Btu/ft²·Day												
20	2346	2656	3021	3297	3417	3438	3414	3321	3097	2748	2404	2238
25	2105	2458	2896	3262	3460	3517	3476	3316	3003	2571	2173	1988
30	1854	2242	2748	3204	3480	3576	3516	3288	2887	2377	1931	1728
35	1594	2012	2579	3122	3478	3613	3534	3237	2748	2165	1678	1462
40	1329	1769	2391	3018	3455	3631	3532	3164	2589	1939	1418	1193
45	1061	1515	2185	2893	3412	3631	3511	3070	2410	1700	1154	925
50	797	1255	1963	2748	3351	3616	3474	2957	2214	1450	890	664
55	541	991	1727	2585	3277	3591	3424	2826	2001	1192	632	417
60	305	727	1478	2407	3194	3567	3371	2683	1773	931	388	197
65	106	472	1219	2217	3116	3568	3331	2531	1533	670	172	31

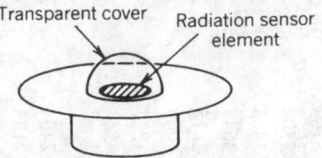

Fig. 49.6 Schematic drawing of a pyranometer used for measuring the intensity of total (direct plus diffuse) solar radiation.

The U.S. National Weather Service (NWS) records solar flux data at a network of stations in the United States. The pyranometer instrument, as shown in Fig. 49.6, is used to measure the intensity of horizontal flux. Various data sets are available from the National Climatic Center (NCC) of the NWS. Prior to 1975, the solar network was not well maintained; therefore, the pre-1975 data were rehabilitated in the late 1970s and are now available from the NCC on magnetic media. Also, for the period 1950–1975, synthetic solar data have been generated for approximately 250 U.S. sites where solar flux data were not recorded. The predictive scheme used is based on other widely available meteorological data. Finally, since 1977 the NWS has recorded hourly solar flux data at a 38-station network with improved instrument maintenance. In addition to horizontal flux, direct-normal data are recorded and archived at the NCC. Figure 49.7 is a contour map of annual, horizontal flux for the United States based on recent data. The appendix to this chapter contains tabulations of average, monthly solar flux data for approximately 250 U.S. sites.

The principal difficulty with using NWS solar data is that they are available for horizontal surfaces only. Solar-collecting surfaces normally face the general direction of the sun and are, therefore, rarely horizontal. It is necessary to convert measured horizontal radiation to radiation on arbitrarily oriented collection surfaces. This is done using empirical approaches to be described.

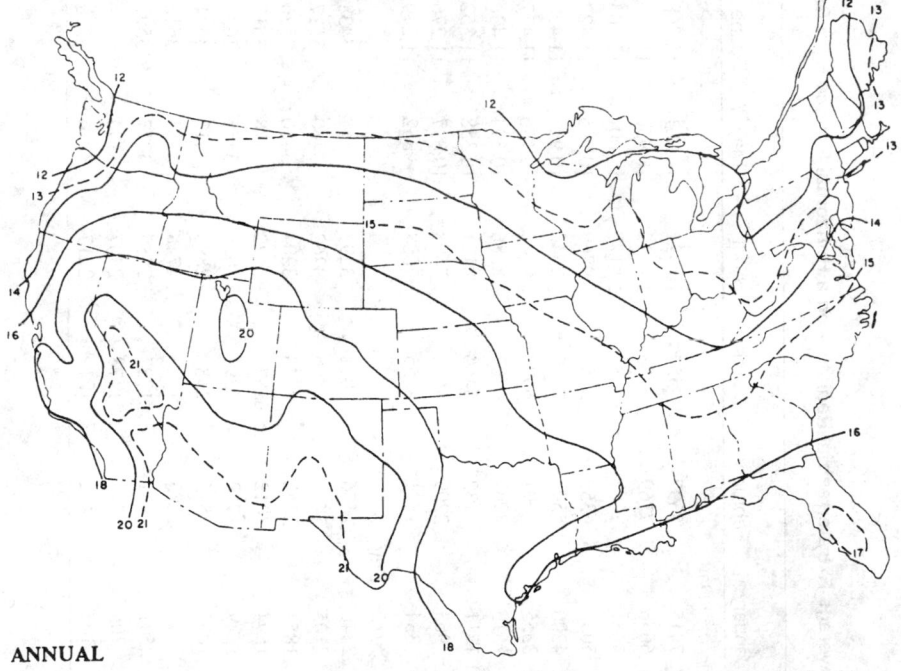

ANNUAL

*1 mJ/m² = 88.1 Btu/ft².

Fig. 49.7 Mean daily solar radiation on a horizontal surface in megajoules per square meter for the continental United States.

Hourly Solar Flux Conversions

Measured, horizontal solar flux consists of both beam and diffuse radiation components. Diffuse radiation is that scattered by atmospheric processes; it intersects surfaces from the entire sky dome, not just from the direction of the sun. Separating the beam and diffuse components of measured, horizontal radiation is the key difficulty in using NWS measurements.

The recommended method for finding the beam component of total (i.e., beam plus diffuse) radiation is described in Ref. 1. It makes use of the parameter k_T called the clearness index and defined as the ratio of terrestrial to extraterrestrial hourly flux on a horizontal surface. In equation form k_T is

$$k_T \equiv \frac{I_h}{I_{b,h0}} = \frac{I_h}{I_{b,0}(N) \sin \alpha} \tag{49.16}$$

in which I_h is the measured, total horizontal flux. The beam component of the terrestrial flux is then given by the empirical equation

$$I_b = (ak_T + b)I_{b,0}(N) \tag{49.17}$$

in which the empirical constants a and b are given in Table 49.4. Having found the beam radiation, the horizontal diffuse component $I_{d,h}$ is found by the simple difference

$$I_{d,h} = I_h - I_b \sin \alpha \tag{49.18}$$

The separate values of horizontal beam and diffuse radiation can be used to find radiation on any surface by applying appropriate geometric "tilt factors" to each component and forming the sum accounting for any radiation reflected from the foreground. The beam radiation incident on any surface is simply $I_b \cos i$. If one assumes that the diffuse component is isotropically distributed over the sky dome, the amount intercepted by any surface tilted at an angle β is $I_{d,h} \cos^2(\beta/2)$. The total beam and diffuse radiation intercepted by a surface I_c is then

$$I_c = I_b \cos i + I_{d,h} \cos^2(\beta/2) + \rho I_h \sin^2(\beta/2) \tag{49.19}$$

The third term in this expression accounts for flux reflected from the foreground with reflectance ρ.[1]

Monthly Averaged, Daily Solar Flux Conversions

Most performance prediction methods make use of monthly averaged solar flux values. Horizontal flux data are readily available (see the appendix), but monthly values on arbitrarily positioned surfaces must be calculated using a method similar to that previously described for hourly tilted surface calculations. The monthly averaged flux on a tilted surface $\bar{I}_c$ is given by

$$\bar{I}_c = \bar{R}\bar{H}_h \tag{49.20}$$

in which $\bar{H}_h$ is the monthly averaged, daily total of horizontal solar flux and $\bar{R}$ is the overall tilt factor given by Eq. (49.21) for a fixed, equator-facing surface:

Table 49.4 Empirical Coefficients for Eq. (49.17)

Interval for k_T	a	b
0.00, 0.05	0.04	0.00
0.05, 0.15	0.01	0.002
0.15, 0.25	0.06	-0.006
0.25, 0.35	0.32	-0.071
0.35, 0.45	0.82	-0.246
0.45, 0.55	1.56	-0.579
0.55, 0.65	1.69	-0.651
0.65, 0.75	1.49	-0.521
0.75, 0.85	0.27	0.395

$$\bar{R} = \left(1 - \frac{\bar{D}_h}{\bar{H}_h}\right)\bar{R}_b + \frac{\bar{D}_h}{\bar{H}_h}\cos^2\frac{\beta}{2} + \rho\sin^2\frac{\beta}{2} \tag{49.21}$$

The ratio of monthly averaged diffuse to total flux, $\bar{D}_h/\bar{H}_h$ is given by

$$\frac{\bar{D}_h}{\bar{H}_h} = 0.775 + 0.347\left[h_{sr} - \frac{\pi}{2}\right] - \left[0.505 + 0.261\left(h_{sr} - \frac{\pi}{2}\right)\right]\cos\left[(\bar{K}_T - 0.9)\frac{360}{\pi}\right] \tag{49.22}$$

in which $\bar{K}_T$ is the monthly averaged clearness index analogous to the hourly clearness index. $\bar{K}_T$ is given by

$$\bar{K}_T \equiv \bar{H}_h/\bar{H}_0$$

where H_0 is the monthly averaged, extraterrestrial radiation on a horizontal surface at the same latitude at which the terrestrial radiation $\bar{H}_h$ was recorded. The monthly averaged beam radiation tilt factor $\bar{R}_b$ is

$$\bar{R}_b = \frac{\cos(L - \beta)\cos\delta_s\sin h'_{sr} + h'_{sr}\sin(L - \beta)\sin\delta_s}{\cos L\cos\delta_s\sin h_{sr} + h_{sr}\sin L\sin\delta_s} \tag{49.23}$$

The sunrise hour angle is found from Eq. (49.9) and the value of h'_{sr} is the smaller of (1) the sunrise hour angle h_{sr} and (2) the collection surface sunrise hour angle found by setting $i = 90°$ in Eq. (49.11). That is, h'_{sr} is given by

$$h'_{sr} = \min\{\cos^{-1}[-\tan L\tan\delta_s], \cos^{-1}[-\tan(L - \beta)\tan\delta_s]\} \tag{49.24}$$

Expressions for solar flux on a tracking surface on a monthly averaged basis are of the form

$$\bar{I}_c = \left[r_T - r_d\left(\frac{\bar{D}_h}{\bar{H}_h}\right)\right]\bar{H}_h \tag{49.25}$$

in which the tilt factors r_T and r_d are given in Table 49.5. Equation (49.22) is to be used for the diffuse to total flux ratio $\bar{D}_h/\bar{H}_h$.

49.2 SOLAR THERMAL COLLECTORS

The principal use of solar energy is in the production of heat at a wide range of temperatures matched to a specific task to be performed. The temperature at which heat can be produced from solar radiation is limited to about 6000°F by thermodynamic, optical, and manufacturing constraints. Between temperatures near ambient and this upper limit very many thermal collector designs are employed to produce heat at a specified temperature. This section describes the common thermal collectors.

49.2.1 Flat-Plate Collectors

From a production volume standpoint, the majority of installed solar collectors are of the flate-plate design; these collectors are capable of producing heat at temperatures up to 100°C. Flat-plate collectors are so named since all components are planar. Figure 49.8a is a partial isometric sketch of a liquid-cooled flat-plate collector. From the top down it contains a glazing system—normally one pane of glass, a dark colored metal absorbing plate, insulation to the rear of the absorber and, finally, a metal or plastic weatherproof housing. The glazing system is sealed to the housing to prohibit the ingress of water, moisture, and dust. The piping shown is thermally bonded to the absorber plate and contains the working fluid by which the heat produced is transferred to its end use. The pipes shown are manifolded together so that one inlet and one outlet connection, only, are present. Figure 49.8b shows a number of other collector designs in common use.

The energy produced by flat-plate collectors is the difference between the solar flux absorbed by the absorber plate and that lost from it by convection and radiation from the upper (or "front") surface and that lost by conduction from the lower (or "back") surface. The solar flux absorbed is the incident flux I_c multiplied by the glazing system transmittance τ and by the absorber plate absorptance α. The heat lost from the absorber in steady state is given by an overall thermal conductance U_c multiplied by the difference in temperature between the collector absorber temperature T_c and the surrounding, ambient temperature T_a. In equation form the net heat produced q_u is then

$$q_u = (\tau\alpha)I_c - U_c(T_c - T_a) \tag{49.26}$$

The rate of heat production depends on two classes of parameters. The first—T_c, T_a, and I_c—having

Table 49.5 Concentrator Tilt Factors

Collector Type	$r_T^{a,b,c,d}$	r_d^e
Fixed aperture concentrators that do not view the foreground	$[\cos(L - \beta)/(d\cos L)]\{-ah_{coll}\cos h_{sr}(i = 90°) + [a - b\cos h_{sr}(i = 90°)]\sin h_{coll} + (b/2)(\sin h_{coll}\cos h_{coll} + h_{coll})\}$	$(\sin h_{coll}/d)\{[\cos(L + \beta)/\cos L] - [1/(CR)]\} + (h_{coll}/d)\{[\cos h_{sr}/(CR)] - [\cos(L - \beta)/\cos L]\cos h_{sr}(i = 90°)\}$
East–west axis tracking[f]	$(1/d)\int_0^{h_{coll}}\{[(a + b\cos x)/\cos L] \times \sqrt{\cos^2 x + \tan^2\delta_s}\}\,dx$	$(1/d)\int_0^{h_{coll}}\{(1/\cos L)\sqrt{\cos^2 x + \tan^2\delta_s} - [1/(CR)][\cos x - \cos h_{sr}]\}\,dx$
Polar tracking	$(ah_{coll} + b\sin h_{coll})/(d\cos L)$	$(h_{coll}/d)\{(1/\cos L) + [\cos h_{sr}/(CR)]\} - \sin h_{coll}/[d(CR)]$
Two-axis tracking	$(ah_{coll} + b\sin h_{coll})/(d\cos\delta_s\cos L)$	$(h_{coll}/d)[1/\cos\delta_s\cos L] + [\cos h_{sr}/(CR)] - h_{coll}/[d(CR)]$

[a] The collection hour angle value h_{coll} not used as the argument of trigonometric functions is expressed in radians; note that the total collection interval, $2h_{coll}$, is assumed to be centered about solar noon.

[b] $a = 0.409 + 0.5016\sin(h_{sr} - 60°)$.

[c] $c = 0.6609 - 0.4767\sin(h_{sr} - 60°)$.

[d] $d = \sin h_{sr} - h_{sr}\cos h_{sr}(i = 90°) = -\tan\delta_s\tan(L - \beta)$.

[e] CR is the collector concentration ratio.

[f] Use elliptic integral tables to evaluate terms of the form of $\int_0^h\sqrt{\cos^2 x + \tan^2\delta_s}\,dx$ contained in r_T and r_d.

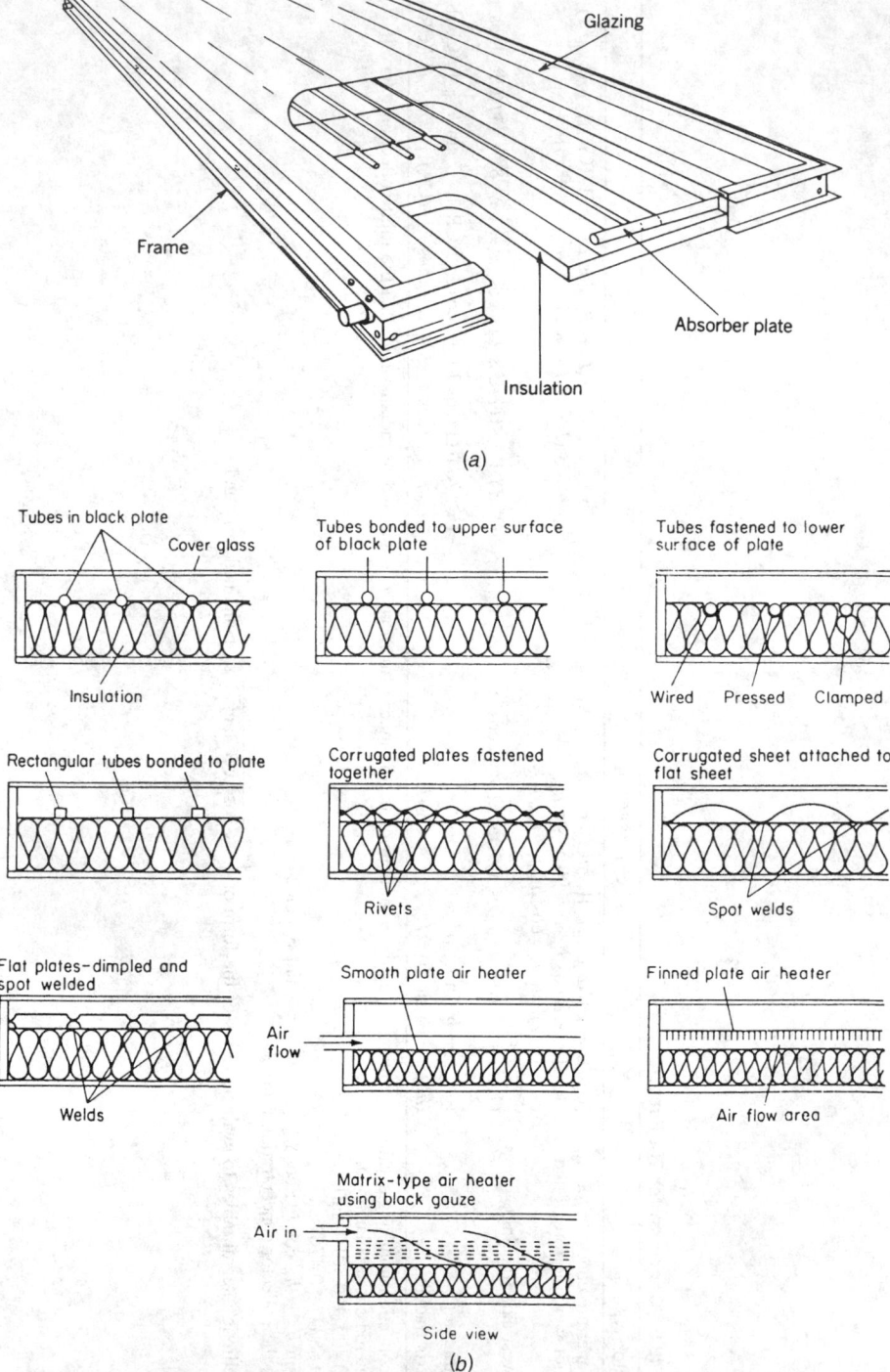

Fig. 49.8 (a) Schematic diagram of solar collector with one cover. (b) Cross sections of various liquid- and air-based flat-plate collectors in common use.

to do with the operational environment and the condition of the collector. The second—U_c and $\tau\alpha$—are characteristics of the collector independent of where or how it is used. The optical properties τ and α depend on the incidence angle, both dropping rapidly in value for $i > 50$–$55°$. The heat loss conductance can be calculated,[1,2] but formal tests, as subsequently described, are preferred for the determination of both $\tau\alpha$ and U_c.

Collector efficiency is defined as the ratio of heat produced q_u to incident flux I_c, that is,

$$\eta_c \equiv q_u/I_c \qquad (49.27)$$

Using this definition with Eq. (49.26) gives the efficiency as

$$\eta_c = \tau\alpha - U_c \left(\frac{T_c - T_a}{I_c} \right) \qquad (49.28)$$

The collector plate temperature is difficult to measure in practice, but the fluid inlet temperature $T_{f,i}$ is relatively easy to measure. Furthermore, $T_{f,i}$ is often known from characteristics of the process to which the collector is connected. It is common practice to express the efficiency in terms of $T_{f,i}$ instead of T_c for this reason. The efficiency is

$$\eta_c = F_R \left[\tau\alpha - U_c \left(\frac{T_{f,i} - T_a}{I_c} \right) \right] \qquad (49.29)$$

in which the heat removal factor F_R is introduced to account for the use of $T_{f,i}$ for the efficiency basis. F_R depends on the absorber plate thermal characteristics and heat loss conductance.[2]

Equation (49.29) can be plotted with the group of operational characteristics $(T_{f,i} - T_a)/I_c$ as the independent variable as shown in Fig. 49.9. The efficiency decreases linearly with the abscissa value. The intercept of the efficiency curve is the optical efficiency $\tau\alpha$ and the slope is $-F_R U_c$. Since the glazing transmittance and absorber absorptance decrease with solar incidence angle, the efficiency

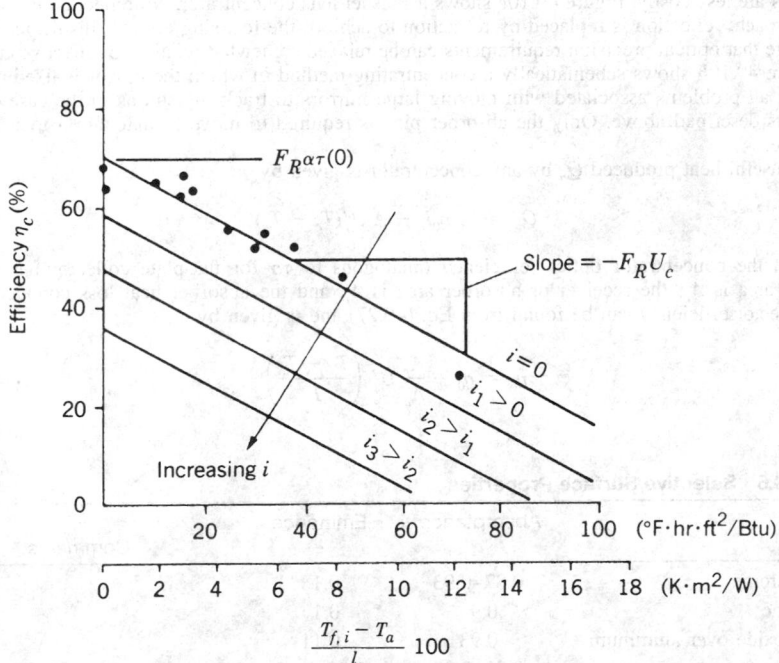

Fig. 49.9 Typical collector performance with 0° incident beam flux angle. Also shown qualitatively is the effect of incidence angle i, which may be quantified by $\overline{\tau\alpha}(i)/\overline{\tau\alpha}(0) = 1.0 + b_0(1/\cos i - 1.0)$, where b_0 is the incidence angle modifier determined experimentally (ASHRAE 93-77) or from the Stokes and Fresnel equations.

curve migrates toward the origin with increasing incidence angle, as shown in the figure. Data points from a collector test are also shown on the plot. The best-fit efficiency curve at normal incidence ($i = 0$) is determined numerically by a curve-fit method. The slope and intercept of the experimental curve, so determined, are the preferred values of the collector parameters as opposed to those calculated theoretically.

Selective Surfaces

One method of improving efficiency is to reduce radiative heat loss from the absorber surface. This is commonly done by using a low emittance (in the infrared region) surface having high absorptance for solar flux. Such surfaces are called (wavelength) selective surface and are used on very many flat-plate collectors to improve efficiency at elevated temperature. Table 49.6 lists emittance and absorptance values for a number of common selective surfaces. Black chrome is very reliable and cost effective.

49.2.2 Concentrating Collectors

Another method of improving the efficiency of solar collectors is to reduce the parasitic heat loss embodied in the second term of Eq. (49.29). This can be done by reducing the size of the absorber relative to the aperture area. Relatively speaking, the area from which heat is lost is smaller than the heat collection area and efficiency increases. Collectors that focus sunlight onto a relatively small absorber can achieve excellent efficiency at temperatures above which flat-plate collectors produce no net heat output. In this section a number of concentrators are described.

Trough Collectors

Figure 49.10 shows cross sections of five concentrators used for producing heat at temperatures up to 650°F at good efficiency. Figure 49.10a shows the parabolic "trough" collector representing the most common concentrator design available commercially. Sunlight is focused onto a circular pipe absorber located along the focal line. The trough rotates about the absorber centerline in order to maintain a sharp focus of incident beam radiation on the absorber. Selective surfaces and glass enclosures are used to minimize heat losses from the absorber tube.

Figures 49.10c and 49.10d show Fresnel-type concentrators in which the large reflector surface is subdivided into several smaller, more easily fabricated and shipped segments. The smaller reflector elements are easier to track and offer less wind resistance at windy sites; futhermore, the smaller reflectors are less costly. Figure 49.10e shows a Fresnel lens concentrator. No reflection is used with this approach; reflection is replaced by refraction to achieve the focusing effect. This device has the advantage that optical precision requirements can be relaxed somewhat relative to reflective methods.

Figure 49.10b shows schematically a concentrating method in which the mirror is fixed, thereby avoiding all problems associated with moving large mirrors to track the sun as in the case of concentrators described above. Only the absorber pipe is required to move to maintain a focus on the focal line.

The useful heat produced Q_u by any concentrator is given by

$$Q_u = A_a \, \eta_0 I_c - A_r U_c'(T_c - T_a) \tag{49.30}$$

in which the concentrator optical efficiency (analogous to $\tau\alpha$ for flat-plate collectors) is η_0, the aperture area is A_a, the receiver or absorber area is A_r, and the absorber heat loss conductance is U_c'. Collector efficiency can be found from Eq. (49.27) and is given by

$$\eta_c = \eta_0 - \frac{A_r}{A_a} U_c' \left(\frac{T_c - T_a}{I_c} \right) \tag{49.31a}$$

Table 49.6 Selective Surface Properties

Material	Absorptance[a] α	Emittance ϵ	Comments
Black chrome	0.87–0.93	0.1	
Black zinc	0.9	0.1	
Copper oxide over aluminum	0.93	0.11	
Black copper over copper	0.85–0.90	0.08–0.12	Patinates with moisture
Black chrome over nickel	0.92–0.94	0.07–0.12	Stable at high temperatures
Black nickel over nickel	0.93	0.06	May be influenced by moisture
Black iron over steel	0.90	0.10	

[a] Dependent on thickness.

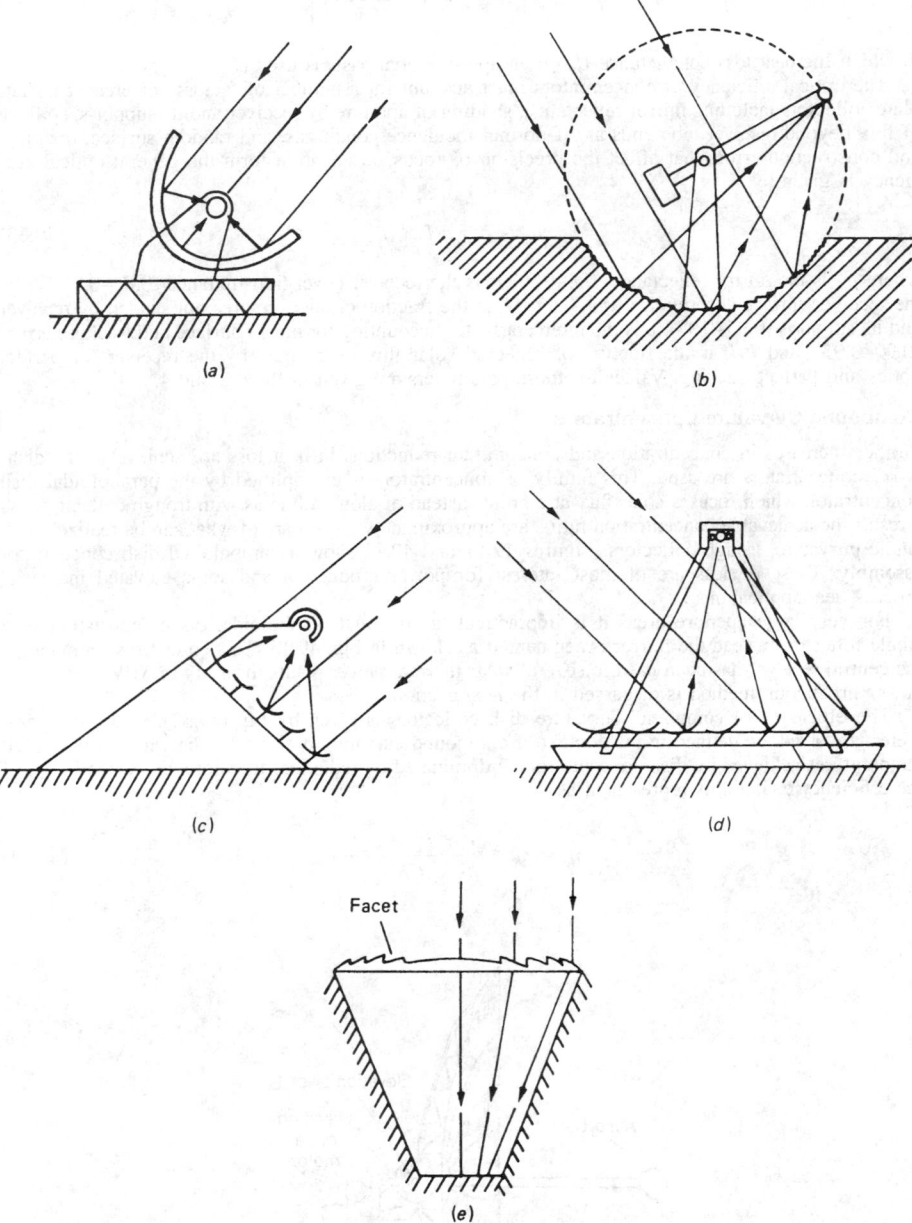

Fig. 49.10 Single-curvature solar concentrators: (*a*) parabolic trough; (*b*) fixed circular trough with tracking absorber; (*c*) and (*d*) Fresnel mirror designs; and (*e*) Fresnel lens.

The aperture area-receiver area ratio $A_a/A_r > 1$ is called the geometric concentration ratio CR. It is the factor by which absorber heat losses are reduced relative to the aperture area:

$$\eta_c = \eta_0 - \frac{U_c'}{CR}\left(\frac{T_c - T_a}{I_c}\right) \tag{49.31b}$$

As with flat-plate collectors, efficiency is most often based on collector fluid inlet temperature $T_{f,i}$. On this basis, efficiency is expressed as

$$\eta_c = F_R \left[\eta_0 - U_c \left(\frac{T_{f,i} - T_a}{I_c} \right) \right] \tag{49.32}$$

in which the heat loss conductance U_c on an aperture area basis is used ($U_c = U'_c / CR$).

The optical efficiency of concentrators must account for a number of factors not present in flat-plate collectors including mirror reflectance, shading of aperture by receiver and its supports, spillage of flux beyond receiver tube ends at off-normal incidence conditions, and random surface, tracking, and construction errors that affect the precision of focus. In equation form the general optical efficiency is given by

$$\eta_0 = \rho_m \tau_c a_r f_t \delta F(i) \tag{49.33}$$

where ρ_m is the mirror reflectance (0.8–0.9), τ_c is the receiver cover transmittance (0.85–0.92), α_r is the receiver surface absorptance (0.9–0.92), f_t is the fraction of aperture area not shaded by receiver and its supports (0.95–0.97), δ is the intercept factor accounting for mirror surface and tracking errors (0.90–0.95), and $F(i)$ is the fraction of reflected solar flux intercepted by the receiver for perfect optics and perfect tracking. Values for these parameters are given in Refs. 2 and 4.

Compound Curvature Concentrators

Further increases in concentration and concomitant reductions in heat loss are achievable if "dish-type" concentrators are used. This family of concentrators is exemplified by the paraboloidal dish concentrator, which focuses solar flux at a point instead of along a line as with trough collectors. As a result the achievable concentration ratios are approximately the square of what can be realized with single curvature, trough collectors. Figures 49.11 and 49.12 show a paraboloidal dish concentrator assembly. These devices are of most interest for power production and some elevated industrial process heat applications.

For very large aperture areas it is impractical to construct paraboloidal dishes consisting of a single reflector. Instead the mirror is segmented as shown in Fig. 49.13. This collector system called the central receiver has been used in several solar thermal power plants in the 1–15 MW range. This power production method is discussed in the next section.

The efficiency of compound curvature dish collectors is given by Eq. (49.32), where the parameters involved are defined in the context of compound curvature optics.[4] The heat loss term at high temperatures achieved by dish concentrators is dominated by radiation; therefore, the second term of the efficiency equation is represented as

$$\eta_c = \eta_0 - \frac{\epsilon_r \sigma (T_c^4 - T_a'^4)}{CR} \tag{49.34}$$

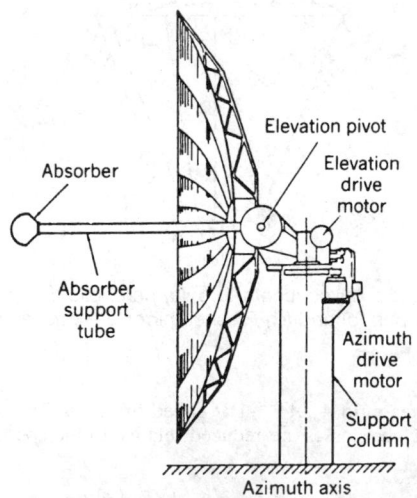

Fig. 49.11 Segmented mirror approximation to paraboloidal dish designed by Raytheon, Inc. Paraboloid is approximated by about 200 spherical mirrors. Average *CR* is 118, while maximum local *CR* is 350.

Fig. 49.12 Commercial paraboloidal solar concentrator. The receiver assembly has been removed from the focal zone for this photograph. (Courtesy of Omnium-G Corp., Anaheim, CA.)

where ϵ_r the infrared emittance of the receiver, σ is the Stefan-Boltzmann constant, and T_a' is the
equivalent ambient temperature for radiation depending on ambient humidity and cloud cover. For
clear, dry conditions T_a' is about 15–20°F below the ambient dry bulb temperature. As humidity
decreases, T_a approaches the dry bulb temperature.

The optical efficiency for the central receiver is expressed in somewhat different terms than those
used in Eq. (49.33). It is referenced to solar flux on a horizontal surface and therefore includes the
geometric tilt factor. For the central receiver, the optical efficiency is given by

$$\eta_0 = \phi \vartheta \rho_m \alpha_r f_t \delta \tag{49.35}$$

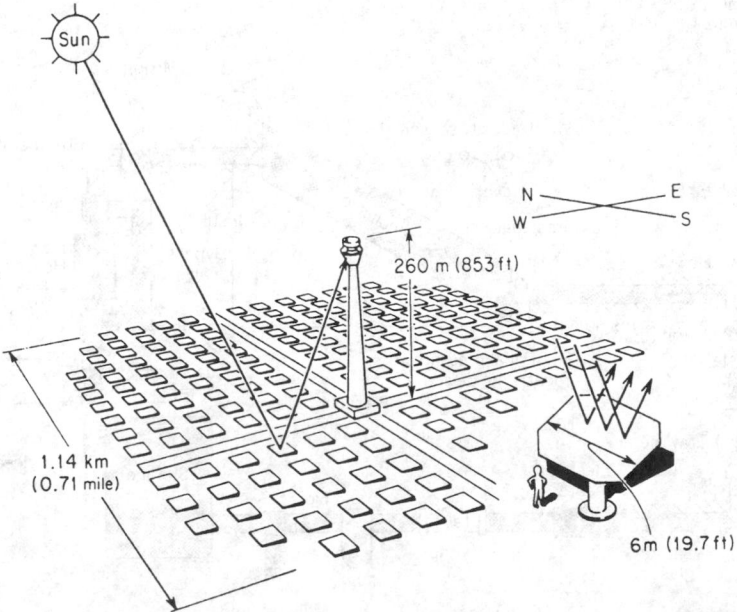

Fig. 49.13 Schematic diagram of a 50-MWe central receiver power plant. A single heliostat is
shown in the inset to indicate its human scale. (From Electric Power Research Institute (EPRI).)

in which the last four parameters are defined as in Eq. (49.33). The ratio of redirected flux to horizontal flux is $\mathscr{P}$ and is given approximately by

$$\mathscr{P} = 0.78 + 1.5(1 - \alpha/90)^2 \tag{49.36}$$

from Ref. 4. The ratio of mirror area to ground area ϕ depends on the size and economic factors applicable to a specific installation. Values for ϕ have been in the range 0.4–0.5 for installations made through 1985.

49.2.3 Collector Testing

In order to determine the optical efficiency and heat loss characteristics of flat-plate and concentrating collectors (other than the central receiver, which is difficult to test because of its size), testing under controlled conditions is preferred to theoretical calculations. Such test data are required if comparisons among collectors are to be made objectively. As of the mid-1980s very few consensus standards had been adopted by the U.S. solar industry. The ASHRAE Standard Number 93-77 applies to flat-plate collectors that contain either a liquid or a gaseous working fluid.[5] Collectors in which a phase change occurs are not included. In addition, the standards do not apply well to concentrators, since additional procedures are needed to find the optical efficiency and aging effects. Testing of concentrators uses sections of the above standard where applicable plus additional procedures as needed; however, no industry standard exists. (The ASTM has promulgated standard E905 as the first proposed standard for concentrator tests.) ASHRAE Standard Number 96-80 applies to very-low-temperature collectors manufactured without any glazing system.

Figure 49.14 shows the test loop used for liquid-cooled flat-plate collectors. Tests are conducted with solar flux at near-normal incidence to find the normal incidence optical efficiency $(\tau\alpha)_n$ along with the heat loss conductance U_c. Off-normal optical efficiency is determined in a separate test by orienting the collector such that several substantially off-normal values of $\tau\alpha$ or η_0 can be measured. The fluid used in the test is preferably that to be used in the installed application, although this is not always possible. If operational and test fluids differ, an analytical correction in the heat removal factor F_R is to be made.[2] An additional test is made after a period of time (nominal one month) to determine the effect of aging, if any, on the collector parameters listed above. A similar test loop and procedure apply to air-cooled collectors.[5]

The development of full system tests has only begun. Of course, it is the entire solar system (see next section) not just the collector that ultimately must be rated in order to compare solar and other

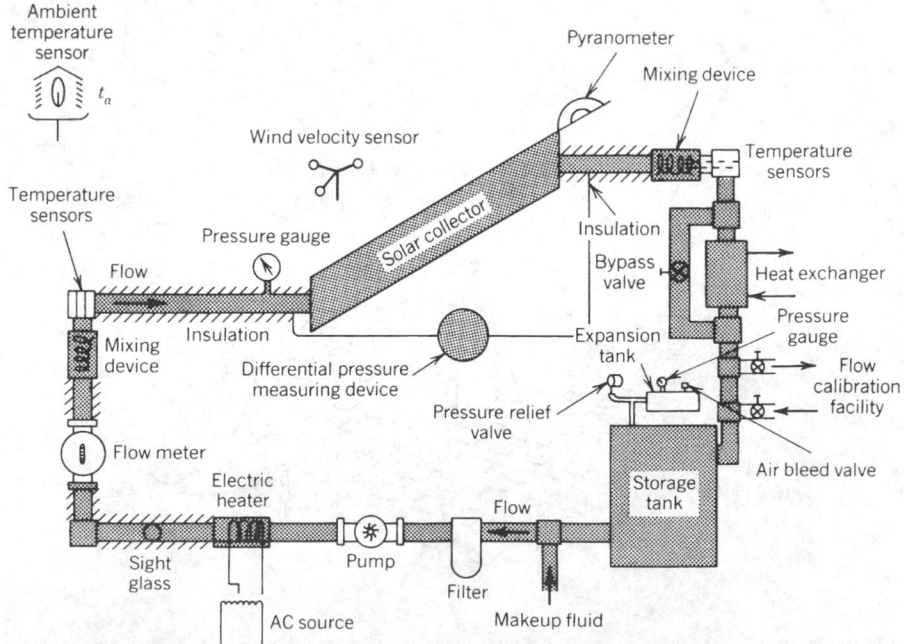

Fig. 49.14 Closed-loop testing configuration for the solar collector when the transfer fluid is a liquid.

energy-conversion systems. Testing of full-size solar systems is very difficult owing to their large size and cost. Hence, it is unlikely that full system tests will ever be practical except for the smallest systems such as residential water heating systems. For this one group of systems a standard test procedure (ASHRAE 95-81) exists. Larger-system performance is often predicted, based on component tests, rather than measured.

49.3 SOLAR THERMAL APPLICATIONS

One of the unique features of solar heat is that it can be produced over a very broad range of temperatures—the specific temperature being selected to match the thermal task to be performed. In this section the most common thermal applications will be described in summary form. These include low-temperature uses such as water and space heating (30–100°C), intermediate temperature industrial processes (100–300°C), and high-temperature thermal power applications (500–850°C and above). Methods for predicting performance, where available, will also be summarized. Nonthermal solar applications are described in the next section.

49.3.1 Solar Water Heating

The most often used solar thermal application is for the heating of water for either domestic or industrial purposes. Relatively simple systems are used, and the load exists relatively uniformly through a year resulting in a good system load factor. Figure 49.15a shows a single-tank water heater schematically. The key components are the collector (0.5–1.0 ft²/gal day load), the storage tank (1.0–2.0 gal/ft² of collector), a circulating pump, and controller. The check valve is essential to prevent backflow of collector fluid, which can occur at night when the pump is off if the collectors are located some distance above the storage tank. The controller actuates the pump whenever the collector is 15–30°F warmer than storage. Operation continues until the collector is only 1.5–5°F warmer than the tank, at which point it is no longer worthwhile to operate the pump to collect the relatively small amounts of solar heat available.

The water-heating system shown in Fig. 49.15a uses an electrical coil located near the top of the tank to ensure a hot water supply during periods of solar outage. This approach is only useful in small residential systems and where nonsolar energy resources other than electricity are not available. Most commercial systems are arranged as shown in Fig. 49.15b, where a separate preheat tank, heated only by solar heat, is connected upstream of the nonsolar, auxiliary water heater tank or plant steam-to-water heat exchanger. This approach is more versatile in that any source of backup energy whatever can be used when solar heat is not available. Additional parasitic heat loss is encountered, since total tank surface area is larger than for the single tank design.

The water-heating systems shown in Fig. 49.15 are of the indirect type, that is, a separate fluid is heated in the collector and heat thus collected is transferred to the end use via a heat exchanger. This approach is needed in locations where freezing occurs in winter and antifreeze solutions are required. The heat exchanger can be eliminated, thereby reducing cost and eliminating the unavoidable fluid temperature decrement between collector and storage fluid streams, if freezing will never occur at the application site. The exchanger can also be eliminated if the "drain-back" approach is used. In this system design the collectors are filled with water only when the circulating pump is on, that is, only when the collectors are warm. If the pump is not operating, the collectors and associated piping all drain back into the storage tank. This approach has the further advantage that heated water otherwise left to cool overnight in the collectors is returned to storage for useful purposes.

The earliest water heaters did not use circulating pumps, but used the density difference between cold collector inlet water and warmer collector outlet water to produce the flow. This approach is called a "thermosiphon" and is shown in Fig. 49.16. These systems are among the most efficient, since no parasitic use of electric pump power is required. The principal difficulty is the requirement that the large storage tank be located above the collector array, often resulting in structural and architectural difficulties. Few industrial solar water-heating systems have used this approach, owing to difficulties in balancing buoyancy-induced flows in large piping networks.

49.3.2 Mechanical Solar Space Heating Systems

Solar space heating is accomplished using systems similar to those for solar water heating. The collectors, storage tank, pumps, heat exchangers, and other components are larger in proportion to the larger space heat loads to be met by these systems in building applications. Figure 49.17 shows the arrangement of components in one common space heating system. All components except the solar collector and controller have been in use for many years in building systems and are not of special design for the solar application.

The control system is somewhat more complex than that used in nonsolar building heating systems, since two heat sources—solar and nonsolar auxiliary—are to be used under different conditions. Controls using simple microprocessors are available for precise and reliable control of solar space heating systems.

Air-based systems are also widely used for space heating. They are similar to the liquid system shown in Fig. 49.17 except that no heat exchanger is used and rock piles, not tanks of fluid, are the storage media. Rock storage is essential to efficient air-system operation since gravel (usually 1–2

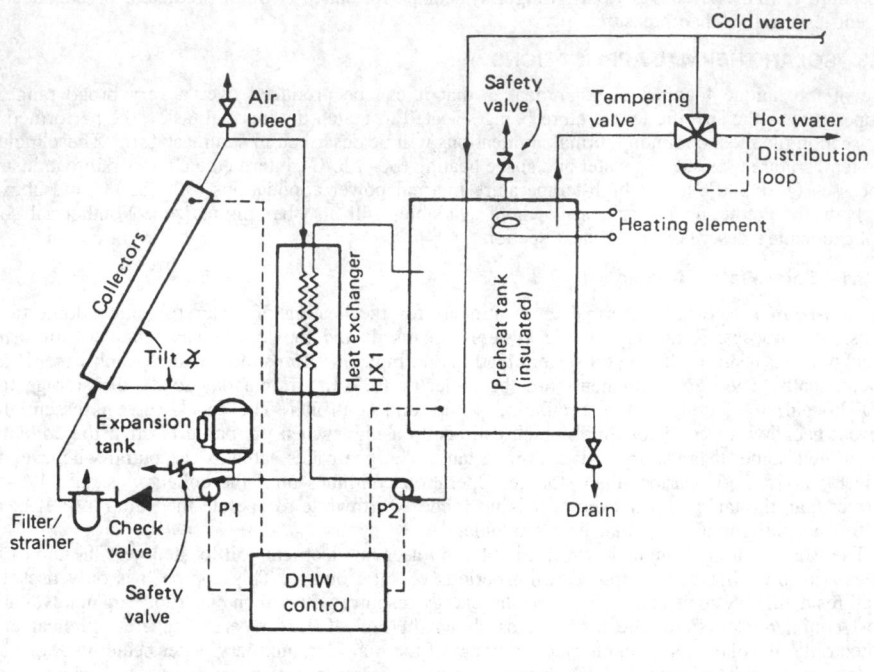

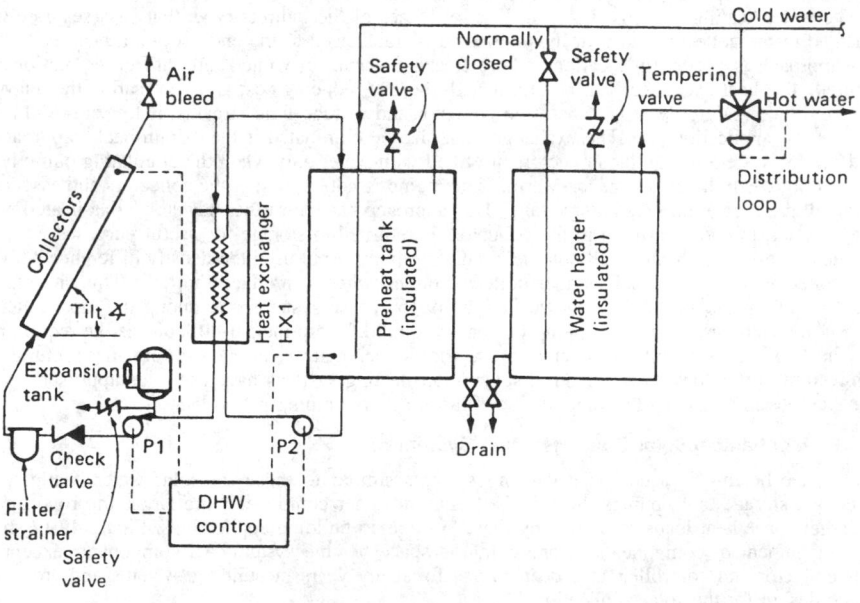

Fig. 49.15 (a) Single-tank indirect solar water-heating system. (b) Double-tank indirect solar water-heating system. Instrumentation and miscellaneous fittings are not shown.

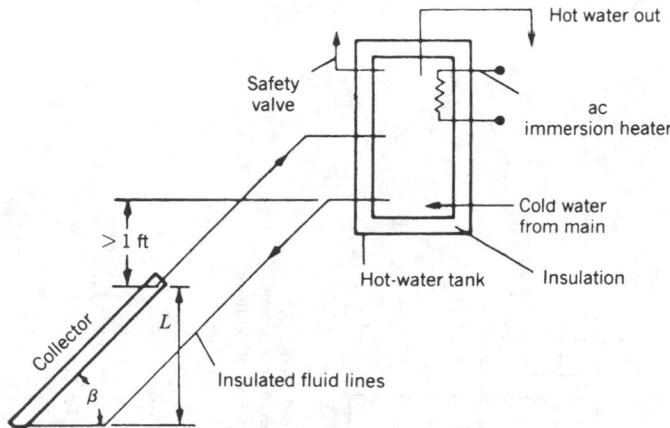

Fig. 49.16   Passive thermosiphon single-tank direct system for solar water heating. Collector is positioned below the tank to avoid reverse circulation.

in. in diameter) has a large surface-to-volume ratio necessary to offset the poor heat transfer characteristics of the air working fluid. Slightly different control systems are used for air-based solar heaters.

49.3.3   Passive Solar Space Heating Systems

A very effective way of heating residences and small commercial buildings with solar energy and without significant nonsolar operating energy is the "passive" heating approach. Solar flux is admitted into the space to be heated by large sun-facing apertures. In order that overheating not occur during sunny periods, large amounts of thermal storage are used, often also serving a structural purpose. A number of classes of passive heating systems have been identified and are described in this section.

Figure 49.18 shows the simplest type of passive system known as "direct gain." Solar flux enters a large aperture and is converted to heat by absorption on dark colored floors or walls. Heat produced at these wall surfaces is partly conducted into the wall or floor serving as stored heat for later periods without sun. The remaining heat produced at wall or floor surfaces is convected away from the surface thereby heating the space bounded by the surface. Direct-gain systems also admit significant daylight during the day; properly used, this can reduce artificial lighting energy use. In cold climates significant heat loss can occur through the solar aperture during long, cold winter nights. Hence, a necessary component of efficient direct-gain systems is some type of insulation system put in place at night over the passive aperture. This is indicated by the dashed lines in the figure.

The second type of passive system commonly used is variously called the thermal storage wall (TSW) or collector storage wall. This system, shown in Fig. 49.19, uses a storage mass interposed between the aperture and space to be heated. The reason for this positioning is to better illuminate storage for a significant part of the heating season and also to obviate the need for a separate insulation system; selective surfaces applied to the outer storage wall surface are able to control heat loss well in cold climates, while having little effect on solar absorption. As shown in the figure, a thermocirculation loop is used to transport heat from the warm, outer surface of the storage wall to the space interior to the wall. This air flow convects heat into the space during the day, while conduction through the wall heats the space after sunset. Typical storage media include masonry, water, and selected eutectic mixtures of organic and inorganic materials. The storage wall eliminates glare problems associated with direct-gain systems, also.

The third type of passive system in use is the attached greenhouse or "sunspace" as shown in Fig. 49.20. This system combines certain features of both direct-gain and storage wall systems. Night insulation may or may not be used, depending on the temperature control required during nighttime.

The key parameters determining the effectiveness of passive systems are the optical efficiency of the glazing system, the amount of directly illuminated storage and its thermal characteristics, the available solar flux in winter, and the thermal characteristics of the building of which the passive system is a part. In a later section, these parameters will be quantified and will be used to predict the energy saved by the system for a given building in a given location.

49.3.4   Solar Ponds

A "solar pond" is a body of water no deeper than a few meters configured in such a way that usual convection currents induced by solar absorption are suppressed. The oldest method for convection

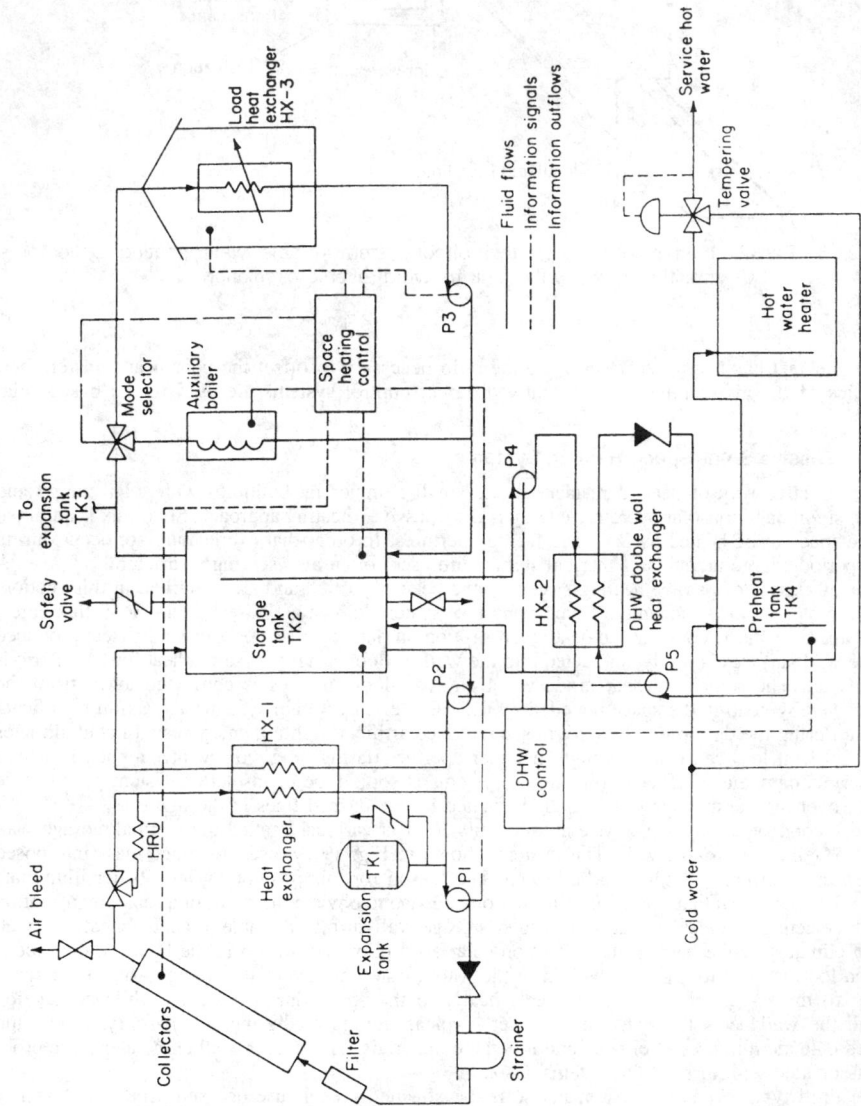

Fig. 49.17 Schematic diagram of a typical liquid-based space heating system with domestic water preheat.

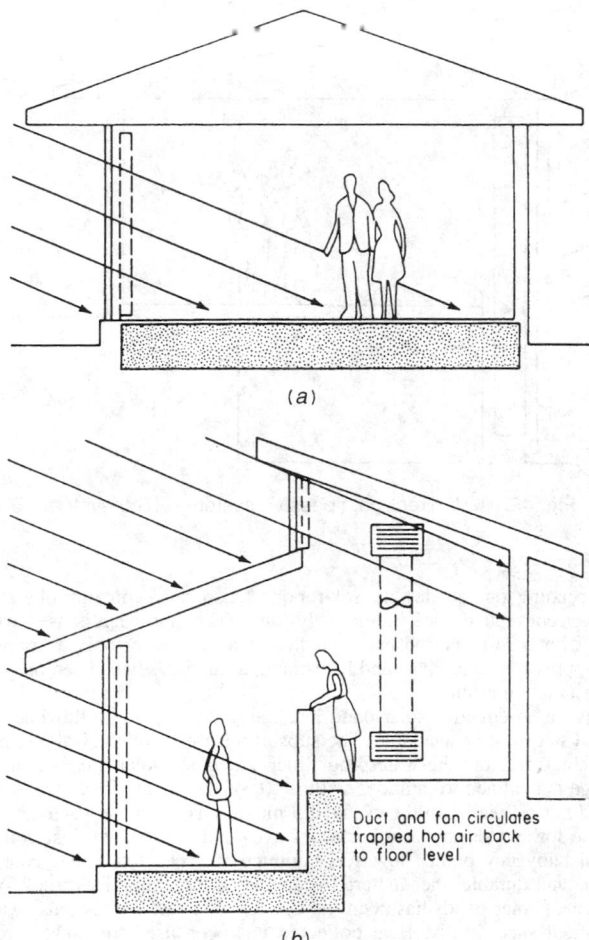

Fig. 49.18 Direct-gain passive heating systems: (a) adjacent space heating; (b) clerestory for north zone heating.

suppression is the use of high concentrations of soluble salts in layers near the bottom of the pond with progressively smaller concentrations near the surface. The surface layer itself is usually fresh water. Incident solar flux is absorbed by three mechanisms. Within a few millimeters of the surface the infrared component (about one-third of the total solar flux energy content) is completely absorbed. Another third is absorbed as the visible and ultraviolet components traverse a pond of nominal 2-m depth. The remaining one-third is absorbed at the bottom of the pond. It is this component that would induce convection currents in a freshwater pond thereby causing warm water to rise to the top where convection and evaporation would cause substantial heat loss. With proper concentration gradient, convection can be completely suppressed and significant heat collection at the bottom layer is possible. Salt gradient ponds are hydrodynamically stable if the following criterion is satisfied:

$$\frac{d\rho}{dz} = \frac{\partial \rho}{\partial s}\frac{ds}{dz} + \frac{\partial \rho}{\partial T}\frac{dT}{dz} > 0 \tag{49.37}$$

where s is the salt concentration, ρ is the density, T is the temperature, and z is the vertical coordinate measured positive downward from the pond surface. The inequality requires that the density must decrease upward.

Useful heat produced is stored in and removed from the lowest layer as shown in Fig. 49.21. This can be done by removing the bottom layer of fluid, passing it through a heat exchanger, and returning the cooled fluid to another point in the bottom layer. Alternatively, a network of heat-removal pipes can be placed on the bottom of the bond and the working fluid passed through for

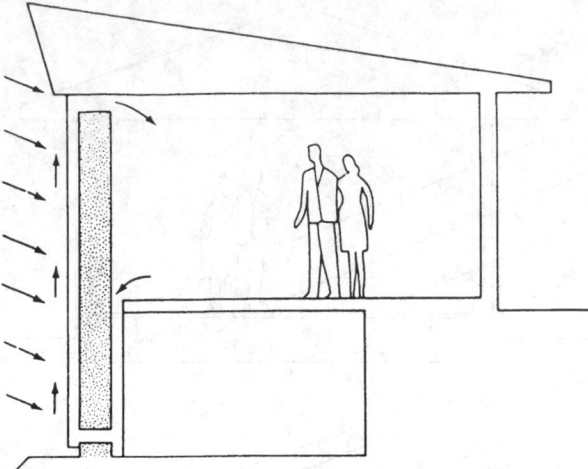

Fig. 49.19 Indirect-gain passive system—TSW system.

heat collection. Depending on the design, solar ponds also may contain substantial heat storage capability if the lower convective zone is relatively thick. This approach is used when uniform heat supply is necessary over a 24-hr period but solar flux is available for only a fraction of the period. Other convection-suppression techniques and heat-removal methods have been proposed but not used in more than one installation at most.

The requirements for an effective solar pond installation include the following. Large amounts of nearly free water and salt must be available. The subsoil must be stable in order to minimize changes in pond shape that could fracture the waterproof liner. Adequate solar flux is required year around; therefore, pond usage is confined to latitudes within 40° of the equator. Freshwater aquifers used for potable water should not be nearby in the event of a major leak of saline water into the groundwater. Other factors include low winds to avoid surface waves and windblown dust collection within the pond (at the neutral buoyancy point), low soil conductivity (i.e., low water content) to minimize conduction heat loss, and durable liner materials capable of remaining leakproof for many years.

The principal user of solar ponds has been the country of Israel. Ponds tens of acres in size have been built and operated successfully. Heat collected has been used for power production with an organic Rankine cycle, for space heating, and for industrial uses. A thorough review of solar pond technology is contained in Ref. 6. A method for predicting the performance of a solar pond is presented in the next section. The theory of pond optics, heat production, and heat loss is contained in Ref. 1.

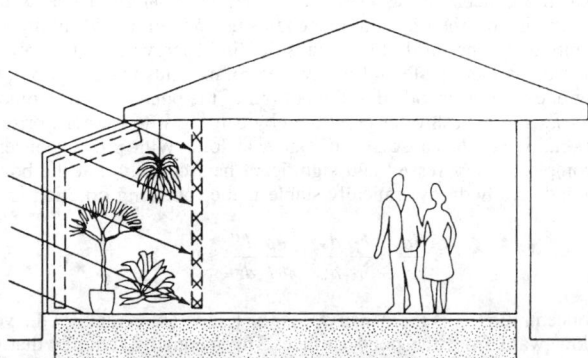

Fig. 49.20 Greenhouse or attached sun-space passive heating system using a combination of direct gain into the greenhouse and indirect gain through the thermal storage wall, shown by cross-hatching, between the greenhouse and the living space.

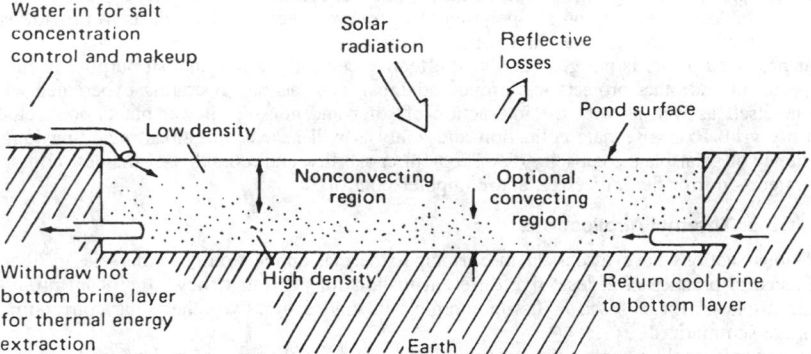

Fig. 49.21 Schematic diagram of a nonconvecting solar pond showing conduits for heat withdrawal, surface washing, and an optional convecting zone near the bottom.

49.3.5 Industrial Process Applications

Process heat up to 300°C for industry can be produced by commercially available trough-type concentrators. Most of the heat needed in the chemical, food processing, textile, pulp and paper, and petrochemical industries can, therefore, be provided by solar heat, in principle. In the United States about half of industrial heat is used at temperatures below 300°C. The viable applications below 300°C use collectors ranked in increasing temperature capability—flat-plates, solar ponds, evacuated tubes, and parabolic or other trough designs. Above 300°C, solar applications have been few in the major industries—primary metals, stone-lay-glass, electric power production.

Since many industrial processes operate around the clock, it may appear prudent at first glance to use substantial storage to permit high daily and annual load factors. This is appropriate from a thermal viewpoint but economic constraints have dictated that the most cost effective systems built to date have only sufficient storage to carry through solar transients of no longer than 30 min. The difficulty is the unavailability of an inexpensive, high-temperature, storage medium with proper heat-transport properties.

Solar industrial process heat was not yet a mature technology in the mid-1990s. Less than 100 systems existed worldwide and many of the earliest systems performed well below expectations. The key reasons were poor control function, lower than expected collector efficiency, parasitic heat losses in extensive piping networks, and poor durability of important components. However, the early experiments were very valuable in significantly improving this new technology. Later generation systems worked well, and the promise of solar heat applications is good under certain conditions of available land area for large arrays, adequate solar flux, and favorable economic conditions—advantageous tax consideration and expensive, nonsolar fuels. Significant reductions in system cost are needed for widespread application.

49.3.6 Solar Thermal Power Production

Solar energy has very high thermodynamic availability owing to the high effective temperature of the source. Therefore, production of shaft power and electric power therefrom is thermodynamically possible. Two fundamentally different types of systems can be used for power production: (1) a large array of concentrating collectors of several tens of meters in area connected by a fluid or electrical network and (2) a single, central receiver using mirrors distributed over a large area but producing heat and power only at one location. The determination of which approach is preferred depends on required plant capacity. For systems smaller than 10 MW the distributed approach appears more economical with existing steam turbines. For systems greater than 10 MW, the central receiver appears more economical.[7] However, if highly efficient Brayton or Stirling engines were available in the 10–20 kW range, the distributed approach would have lowest cost for any plant size. Such systems will be available by the year 2000.

The first U.S. central receiver began operating in the fall of 1982. Located in the Mojave Desert, this 10-MW plant (called "Solar One") is connected to the southern California electrical grid. The collection system consists of 1818 heliostats totaling 782,000 ft^2 in area. Each 430 ft^2 mirror is computer controlled to focus reflected solar flux onto the receiver located 300 ft above the desert floor. The receiver is a 23-ft diameter cyclinder whose outer surface is the solar absorber. The absorbing surface is coated with a special black paint selected for its reliability at the nominal 600°C operating temperature. Thermal storage consisting of a mixture of an industrial heat transfer oil for heat transport and of rock and sand has a nominal operating temperature of 300°C. Storage is used to extend the plant operating time beyond sunset (albeit at lower turbine efficiency) and to maintain the turbine, condenser, and piping at operating temperatures overnight as well as to provide startup steam the following morning. The plant was modernized in 1996.

Solar-produced power is not generally cost effective currently. The principal purpose of the Solar One experiment and other projects in Europe and Japan is to acquire operating experience with the solar plant itself as well as with the interaction of solar and nonsolar power plants connected in a large utility grid. Extensive data collection and analysis will answer questions regarding long-term net efficiency of solar plants, capacity displacement capability, and reliability of the new components of the system—mirror field, receiver, and computer controls.

49.3.7 Other Thermal Applications

The previous sections have discussed the principal thermal applications of solar energy that have been reduced to practice in at least five different installations and that show significant promise for economic displacement of fossil or fissile energies. In this section two other solar-conversion technologies are summarized.

Solar-powered cooling has been demonstrated in many installations in the United States, Europe, and Japan. Chemical absorption, organic Rankine cycle, and desiccant dehumidifaction processes have all been shown to be functional. Most systems have used flat-plate collectors, but higher coefficients of performance are achievable with mildly concentrating collectors. References 1, 2, and 8 describe solar-cooling technologies. To date, economic viability has not been generally demonstrated, but further research resulting in reduced cost and improved efficiency is expected to continue.

Thermal energy stored in the surface layers of the tropical oceans has been used to produce electrical power on a small scale. A heat engine is operated between the warmest layer at the surface and colder layers several thousand feet beneath. The available temperature difference is of the order of 20°C, therefore, the cycle efficiency is very low—only a few percent. However, this type of power plant does not require collectors or storage. Only a turbine capable of operating efficiently at low temperature is needed. Some cycle designs also require very large heat exchangers, but new cycle concepts without heat exchangers and their unavoidable thermodynamic penalties show promise.

49.3.8 Performance Prediction for Solar Thermal Processes

In a rational economy the single imperative for use of solar heat for any of the myriad applications outlined heretofore must be cost competitiveness with other energy sources—fossil and fissile. The amount of useful solar energy produced by a solar-conversion system must therefore be known along with the cost of the system. In this section the methods usable for predicting the performance of widely deployed solar systems are summarized. Special systems such as the central receiver, the ocean thermal power plant, and solar cooling are not included. The methods described here require a minimum of computational effort, yet embody all important parameters determining performance.

Solar systems are connected to end uses characterized by an energy requirement or "load" L and by operating temperature that must be achievable by the solar-heat-producing system. The amount of solar-produced heat delivered to the end use is the useful energy Q_u. This is the net heat delivery accounting for parasitic losses in the solar subsystem. The ratio of useful heat delivered to the requirement L is called the "solar fraction" denoted by f_s. In equation form the solar fraction is

$$f_s \equiv \frac{Q_u}{L} \tag{49.38}$$

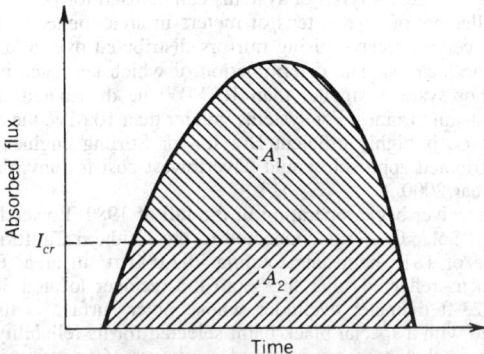

Fig. 49.22 Daily absorbed solar flux ($A_1 + A_2$) and useful solar flux (A_1) at intensities above I_{cr}.

Table 49.7 Empirical Solar Fraction Equations[a]

System Type	f_s Expression	Time Scale
Water heating and liquid-based space heating	$f_s = 1.029P_s - 0.065P_L - 0.245P_s^2 + 0.0018P_L^2 + 0.00215P_s^3$	Monthly
Space heating—air-based systems	$f_s = 1.040P_s - 0.065P_L - 0.159P_s^2 + 0.00187P_L^2 + 0.0095P_s^3$	Monthly
Passive direct gain	$f_s = PX + (1 - P)(3.082 - 3.142\bar{\phi})(1 - e^{-0.329x})$	Monthly
Passive storage wall	$f_s = Pf_\infty + 0.88(1 - P)(1 - e^{-1.26f_\infty})$	Monthly
Concentrating collector systems	$f_s = F_R\bar{\eta}_0\bar{I}_cA_cN\bar{\phi}'/L$	Monthly
Solar ponds (pond radius R to provide annual pond temperature T_p)	$R = \dfrac{2.2\,\overline{\Delta T} + [4.84(\Delta T)^2 + \bar{L}(0.3181\,\bar{I}_p - 0.1592\Delta T)]^{1/2}}{\bar{I}_p - 0.5\Delta T}$	Annual

[a] See Table 49.8 for symbol definitions.

Empirical equations have been developed relating the solar fraction to other dimensionless groups characterizing a given solar process. These are summarized shortly.

A fundamental concept used in many predictive methods is the solar "utilizability" defined as that portion of solar flux absorbed by a collector that is capable of providing heat to the specified end use. The key characteristic of the end use is its temperature. The collector must produce at least enough heat to offset losses when the collector is at the minimum temperature T_{min} usable by the given process. Figure 49.22 illustrates this idea schematically. The curve represents the flux absorbed over a day by a hypothetical collector. The horizontal line intersecting this curve represents the threshold flux that must be exceeded for a net energy collection to take place. In the context of the efficiency equation [Eq. (49.32)], this critical flux I_{cr} is that which results in a collector efficiency of exactly zero when the collector is at the minimum usable process temperature T_{min}. Any greater flux will result in net heat production. From Eq. (49.32) the critical intensity is

$$I_{cr} = \frac{U_c(T_{min} - T_a)}{\tau\alpha} \tag{49.39}$$

The solar utilizability is the ratio of the useful daily flux (area above I_{cr} line in Fig. 49.22) to the total absorbed flux (area $A_1 + A_2$) beneath the curve. The utilizability denoted by ϕ is

$$\phi = \frac{A_1}{A_1 + A_2} \tag{49.40}$$

This quantity is a solar radiation statistic depending on I_{cr}, characteristics of the incident solar flux and characteristics of the collection system. It is a very useful parameter in predicting the performance of solar thermal systems.

Table 49.7 summarizes empirical equations used for predicting the performance of the most common solar-thermal systems. These expressions are given in terms of the solar fraction defined above and dimensionless parameters containing all important system characteristics. The symbols used in this table are defined in Table 49.8. In the brief space available in this chapter, all details of these prediction methodologies cannot be included. The reader is referred to Refs. 1, 4, 9, 10, and 11 for details.

49.4 NONTHERMAL SOLAR ENERGY APPLICATIONS

In this section the principal nonthermal solar conversion technology is described. Photovoltaic cells are capable of converting solar flux directly into electric power. This process, first demonstrated in the 1950s, holds considerable promise for significant use in the future. Major cost reductions have been accomplished. In this section the important features of solar cells are described.

Photovoltaic conversion of sunlight to electricity occurs in a thin layer of semiconductor material exposed to solar flux. Photons free electric charges, which flow through an external circuit to produce useful work. The semiconductor materials used for solar cells are tailored to be able to convert the majority of terrestrial solar flux; however, low-energy photons in the infrared region are usually not usable. Figure 49.23 shows the maximum theoretical conversion efficiency of seven common mate-

Table 49.8 Definition of Symbols in Table 49.7

Parameters		Definition	Units[a]
P_L		$P_s = \dfrac{F_{hx}F_R U_c(T_r - \overline{T}_a)\Delta t}{L}$	None
	F_{hx}	$F_{hx} = \left\{\left[1 + \dfrac{F_R U_c A_c}{(\dot{m}C_p)_c}\right]\left[\dfrac{(\dot{m}C_p)_c}{(\dot{m}C_p)_{\min}\epsilon} - 1\right]\right\}^{-1}$, collector heat exchanger penalty factor	None
	$F_R U_c$	Collector heat-loss conductance	But/hr · ft² · °F
	A_c	Collector area	ft²
	$(\dot{m}C_p)_c$	Collector fluid capacitance rate	Btu/hr · °F
	$(\dot{m}C_p)_{\min}$	Minimum capacitance rate in collector heat exchanger	Btu/hr · °F
	ϵ	Collector heat-exchanger effectiveness	None
	T_r	Reference temperature, 212°F	°F
	$\overline{T}_a$	Monthly averaged ambient temperature	°F
	Δt	Number of hours per month	hr/month
	L	Monthly load	Btu/month
P_s		$P_s = \dfrac{F_{hx}F_R\overline{\tau\alpha}\overline{I}_c N}{L}$	None
	$F_R\overline{\tau\alpha}$	Monthly averaged collector optical efficiency	None
	$\overline{I}_c$	Monthly averaged, daily incident solar flux	Btu/day · ft²
	N	Number of days per month	day/month
$(P'_L$ — to be used for water heating only)		$P'_L = P_L\dfrac{(1.18T_{wo} + 3.86T_{wi} - 2.32\overline{T}_a - 66.2)}{212 - \overline{T}_a}$	None
	T_{wo}	Water output temperature	°F
	T_{wi}	Water supply temperature	°F
	P_L	(See above)	None
P		$P = (1 - e^{-0.294Y})^{0.652}$	None
	Y	Storage-vent ratio, $Y = \dfrac{C\Delta T}{\phi \overline{I}_c\overline{\tau\alpha}A_c}$	None
	C	Passive storage capacity	Btu/°F
	ΔT	Allowable diurnal temperature saving in heated space	°F
$\overline{\phi}$		Monthly averaged utilizability (see below)	Nonc
X		Solar-load ratio, $X = \dfrac{\overline{I}_c\overline{\tau\alpha}A_c N}{L}$	None
	L	Monthly space heat load	Btu/month
f_∞		Solar fraction with hypothetically infinite storage, $f_\infty = \dfrac{\overline{Q}_i + L_w}{L}$	None
	$\overline{Q}_i$	Net monthly heat flow through storage wall from outer surface to heated space	Btu/month
	L_w	Heat *loss* through storage wall	Btu/month
$F_R\overline{\eta}_0$		Monthly averaged concentrator optical efficiency	None
$\overline{\phi}'$		Monthly average utilizability for concentrators	None
R		Pond radius to provide diurnal average pond temperature $\overline{T}_p$	m
ΔT		$\overline{\Delta T} = \overline{T}_p - \overline{\overline{T}}_a$	°C
	$\overline{T}_p$	Annually averaged pond temperature	°C
	$\overline{\overline{T}}_a$	Annually averaged ambient temperature	°C
$\overline{L}$		Annual averaged load at $\overline{T}_p$	W
$\overline{I}_p$		Annual averaged insolation absorbd at pond bottom	W/m²

Table 49.8 (*Continued*)

Parameters	Definition	Units[a]
$\bar{\phi}$	Monthly flat-plate utilizability (equator facing collectors), $\bar{\phi} = \exp\{[A + B(\bar{R}_N/\bar{R})](\bar{X}_c + C\bar{X}_c^2)\}$	None
A	$A = 7.476 - 20.0\bar{K}_T + 11.188\bar{K}_T^2$	None
B	$B = -8.562 + 18.679\bar{K}_T - 9.948\bar{K}_T^2$	None
C	$C = -0.722 + 2.426\bar{K}_T + 0.439\bar{K}_T^2$	None
$\bar{R}$	Tilt factor, see Eq. (49.21)	None
$\bar{R}_N$	Monthly averaged tilt factor for hour centered about noon (see Ref. 9)	None
$\bar{X}_c$	Critical intensity ratio, $\bar{X}_c = \dfrac{I_{cr}}{r_{T,N}\bar{R}_N\bar{H}_h}$	None
$r_{T,N}$	Fraction of daily total radiation contained in hour about noon, $r_{T,N} = r_{d,n}[1.07 + 0.025 \sin(h_{sr} - 60)]$	day/hr
	$r_{d,n} = \dfrac{\pi}{24} \dfrac{1 - \cos h_{sr}}{h_{sr} - h_{sr}\cos h_{sr}}$	day/hr
I_{cr}	Critical intensity [see Eq. (49.39)]	Btu/hr · ft²
$\bar{\phi}'$	Monthly concentrator utilizability, $\bar{\phi}' = 1.0 - (0.049 + 1.49\bar{K}_T)\bar{X} + 0.341\bar{K}_T\bar{X}^2$ $0.0 < \bar{K}_T < 0.75, 0 < \bar{X} < 1.2$)	None
	$\bar{\phi}' = 1.0 - \bar{X}$ ($\bar{K}_T > 0.75, 0 < \bar{X} < 1.0$)	
$\bar{X}$	Concentrator critical intensity ratio, $\bar{X} = \dfrac{U_c(T_{f,i} - \bar{T}_a)\Delta t_c}{\bar{\eta}_0\bar{I}_c}$	None
$T_{f,i}$	Collector fluid inlet temperature—assumed constant	°F
Δt_c	Monthly averaged solar system operating time	hr/day

[a]USCS unit shown except for solar ponds; SI units may also be used for all parameters shown in USCS units.

rials used in the application. Each material has its own threshold band-gap energy, which is a weak function of temperature. The energy contained in a photon is $E = h\nu$. If E is greater than the band-gap energy shown in this figure, conversion can occur.

Figure 49.23 also shows the very strong effect of temperature on efficiency. For practical systems it is essential that the cell be maintained as near to ambient temperature as possible.

Solar cells produce current proportional to the solar flux intensity with wavelengths below the band-gap threshold. Figure 49.24 shows the equivalent circuit of a solar cell. Both internal shunt and series resistances must be included. These result in unavoidable parasitic loss of part of the power produced by the equivalent circuit current source of strength I_s. Solving the equivalent circuit for the power P produced and using an expression from Ref. 2 for the junction leakage I_J results in

$$P = [I_s - I_0(e^{e_0\nu/kT} - 1)]V \qquad (49.41)$$

in which e_0 is the electron charge, k is the Boltzmann constant, and T is the temperature. The current source I_s is given by

$$I_s = \eta_0(1 - \rho_c)\alpha e_0 n_p \qquad (49.42)$$

in which η_0 is the collector carrier efficiency, ρ_c is the cell surface reflectance, α is the absorptance of photons, and n_p is the flux density of sufficiently energetic photons.

In addition to the solar cell, complete photovoltaic systems also must contain electrical storage and a control system. The cost of storage presents another substantial cost problem in the widespread application of photovoltaic power production. The costs of the entire conversion system must be reduced by an order of magnitude in order to be competitive with other power sources. Vigorous research in the United States, Europe, and Japan has made significant gains in the past decade. The installed capacity of photovoltaic systems is expected to be at least 10,000 MW by the year 2000.

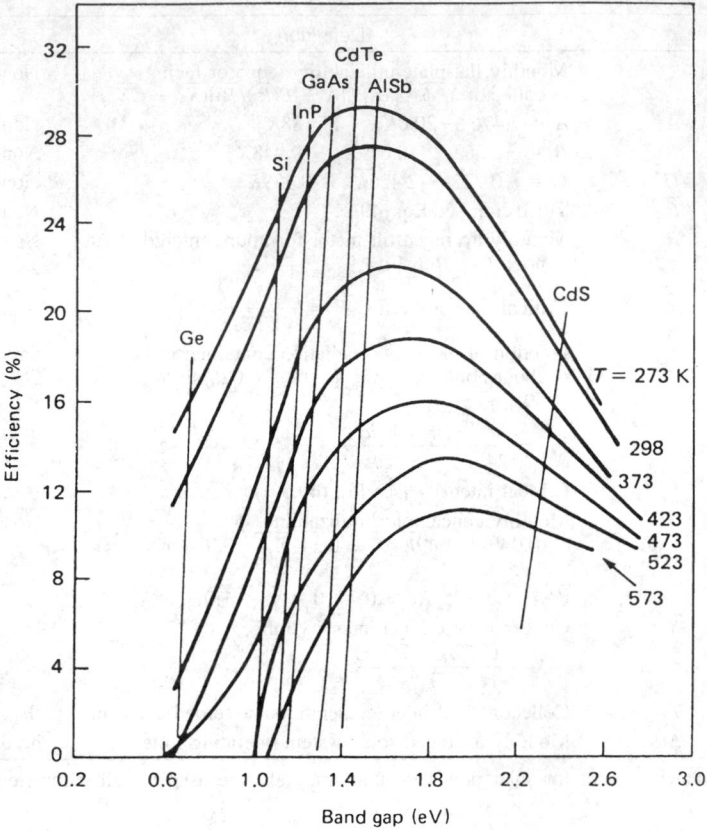

Fig. 49.23 Maximum theoretical efficiency of photovoltaic converters as a function of band-gap energy for several materials.

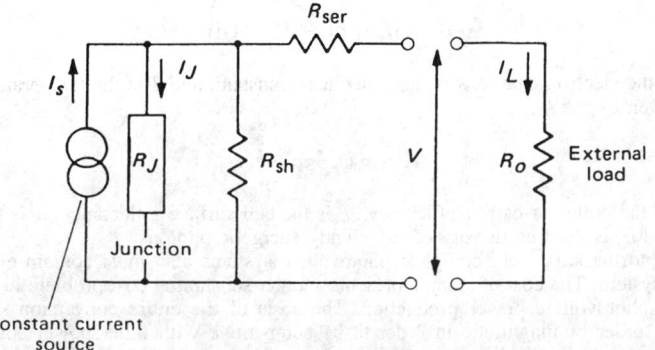

Fig. 49.24 Equivalent circuit of an illuminated *p-n* photocell with internal series and shunt resistances and nonlinear junction impedance R_J.

REFERENCES

1. J. F. Kreider and F. Kreith, *Solar Energy Handbook,* McGraw-Hill, New York, 1981.
2. F. Kreith and J. F. Kreider, *Principles of Solar Engineering,* Hemisphere/McGraw-Hill, New York, 1978.
3. M. Collares-Pereira and A. Rabl, "The Average Distribution of Solar Energy," *Solar Energy* **22,** 155–164 (1979).
4. J. F. Kreider, *Medium and High Temperature Solar Processes,* Academic Press, New York, 1979.
5. ASHRAE Standard 93-77, *Methods of Testing to Determine the Thermal Performance of Solar Collectors,* ASHRAE, Atlanta, GA, 1977.
6. H. Tabor, "Solar Ponds—Review Article," *Solar Energy* **27,** 181–194 (1981).
7. T. Fujita et al., *Projection of Distributed Collector Solar Thermal Electric Power Plant Economics to Years 1990–2000,* JPL Report No. DOE/JPL-1060-77/1, 1977.
8. J. F. Kreider and F. Kreith, *Solar Heating and Cooling,* Hemisphere/McGraw-Hill, New York, 1982.
9. W. A. Monsen, Master's Thesis, University of Wisconsin, 1980.
10. M. Collares-Pereira and A. Rabl, "Simple Procedure for Predicting Long Term Average Performance of Nonconcentrating and of Concentrating Solar Collectors," *Solar Energy* **23,** 235–253 (1979).
11. J. F. Kreider, *The Solar Heating Design Process,* McGraw-Hill, New York, 1982.

CHAPTER 50

GEOTHERMAL RESOURCES: AN INTRODUCTION

Peter D. Blair
Sigma Xi
The Scientific Research Society
Research Triangle Park, North Carolina

50.1	INTRODUCTON	1583	50.4 GEOPRESSURED RESOURCES	1585
50.2	HYDROTHERMAL RESOURCES	1584	50.5 GEOTHERMAL ENERGY CONVERSION	1587
	50.2.1 Vapor-Dominated Resources	1585	50.5.1 Direct Steam Conversion	1587
	50.2.2 Liquid-Dominated Resources	1585	50.5.2 Flashed Steam Conversion	1588
			50.5.3 Binary Cycle Conversion	1588
			50.5.4 Hybrid Fossil/Geothermal Plants	1590
50.3	HOT DRY ROCK RESOURCES	1585		

50.1 INTRODUCTON

Geothermal energy is the internal heat of the Earth. For centuries, geothermal energy was apparent only through anomalies in the Earth's crust that allow the heat from the Earth's molten core to venture close to the Earth's surface. Volcanoes, geysers, fumaroles, and hot springs are the most visible surface manifestations of these anomalies.

Geothermal energy has been used for centuries where it is available for aquaculture, greenhousing, industrial process heat, and space heating. It was first used for electricity production in 1904 in Lardarello, Italy.

Geothermal resources are traditionally divided into three basic classes:

1. Hydrothermal convection systems, including both vapor-dominated and liquid-dominated systems
2. Hot igneous resources, including hot dry rock and magma systems
3. Conduction-dominated resources, including geopressured and radiogenic resources

The three basic resource categories are distinguished by geologic characteristics and the manner in which heat is transferred to the Earth's surface (see Table 50.1). The following includes a discussion of the characteristics and location of these resource categories in the United States. Only the first of these resource types, hydrothermal resources, is commercially exploited today in the United States.

In 1975, the U.S. Geological Survey completed a national assessment of geothermal resources in the United States and published the results in USGS Circular 726 (subsequently updated in 1978 as Circular 790). This assessment defined a "geothermal resource base" for the United States based on geological estimates of all stored heat in the earth above 15°C and within six miles of the surface, ignoring recoverability. In addition, these resources were catalogued according to the classes given in Table 50.1. The end result is a set of 108 known geothermal resource areas (KGRAs) encompassing over three million acres in the 11 western states. Since the 1970s, many of these KGRAs have been explored extensively and some developed commercially for electric power production.

Mechanical Engineers' Handbook, 2nd ed., Edited by Myer Kutz.
ISBN 0-471-13007-9   © 1998 John Wiley & Sons, Inc.

Table 50.1 Geothermal Resource Classification

Resource Type	Temperature Characteristics
1. Hydrothermal convection resource (heat carried upward from depth by convection of water or steam)	
a. Vapor-dominated	About 240°C (464°F)
b. Hot-water dominated	
1. High temperature	150–350°C+ (300–660°F)
2. Intermediate temperature	90–150°C (290–300°F)
3. Low temperature	Less than 90°C (290°F)
2. Hot igneous resources (rock intruded in molten form from depth)	
a. Part still molten—"magma systems"	Higher than 650°C (1200°F)
b. Not molten—"hot dry rock" systems	90–650°F (190–1200°F)
3. Conduction-dominated resources (heat carried upward by conduction through rock)	
a. Radiogenic (heat generated by radioactive decay)	30–150°C (86–300°F)
b. Sedimentary basins (hot fluid in sedimentary rocks)	30–150°C (86–300°F)
c. Geopressured (hot fluid under high pressure)	150–200°C (300–390°F)

50.2 HYDROTHERMAL RESOURCES

Hydrothermal convection systems are formed when underground reservoirs carry the Earth's heat toward the surface by convective circulation of water (liquid-dominated resources) or steam (vapor-dominated resources). There are only seven known vapor-dominated resources in the world today, three of which are located in the United States: The Geysers and Mount Lassen in California and the Mud Volcano system in Yellowstone National Park. The remaining U.S. resources are liquid-dominated (see Fig. 50.1).

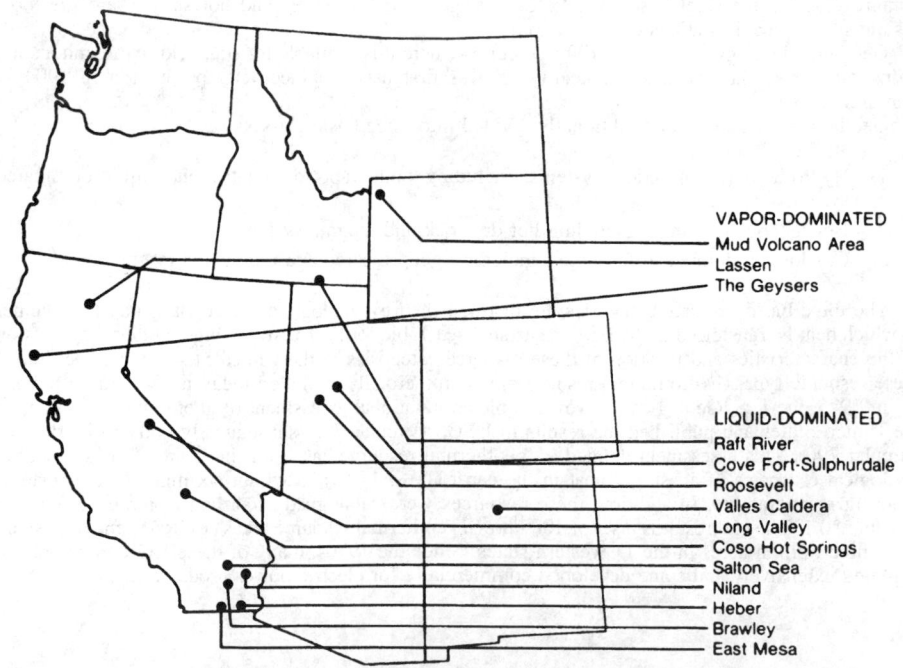

VAPOR-DOMINATED
— Mud Volcano Area
— Lassen
— The Geysers

LIQUID-DOMINATED
— Raft River
— Cove Fort-Sulphurdale
— Roosevelt
— Valles Caldera
— Long Valley
— Coso Hot Springs
— Salton Sea
— Niland
— Heber
— Brawley
— East Mesa

Fig. 50.1 Known major U.S. geothermal resource areas.

50.2.1 Vapor-Dominated Resources

In a vapor-dominated hydrothermal system, boiling of deep subsurface water produces water vapor, which is also often superheated by the hot surrounding rock. Geologists speculate that as the vapor moves toward the surface, a level of cooler near-surface rock may induce condensation, which, along with the cooler groundwater from the margins of the reservoir, serves to recharge the reservoir. Since fluid convection is taking place constantly, the temperature in the vapor-filled area of the reservoir is relatively uniform and a well drilled in this region will yield high-quality superheated steam.

The most developed geothermal resource in the world is The Geysers in northern California, which is a high-quality, vapor-dominated hydrothermal convection system. Currently, steam is produced from this resource at a depth of 5,000–10,000 feet and piped directly into turbine-generators to produce electricity. Geothermal power production capacity at The Geysers peaked in 1987 at over 2,000 mW, but since then has declined to about 1,800 mW.

Commercially produced vapor-dominated systems at The Geysers, Lardarello (Italy), and Matsukawa (Japan) are all characterized by reservoir temperatures in excess of 450°F. Accompanying the water vapor are small concentrations (i.e., less than 5%) of noncondensible gases (mostly carbon dioxide, hydrogen sulfide, and ammonia). The Mont Amiata field (Italy) is a different type of vapor-dominated resource, which is characterized by lower temperatures than The Geysers-type resource and by much higher gas content (hydrogen sulfide and carbon dioxide). The geology of this category of vapor-dominated resource is not yet well understood, but may turn out to be more common than The Geysers-type resource because its existence is much more difficult to detect.

50.2.2 Liquid-Dominated Resources

Hot-water or wet-steam hydrothermal resources are much more commonly found than dry-steam deposits. Hot-water systems are often associated with a hot spring that discharges at the surface. When wet steam deposits occur at considerable depths, the resource temperature is often well above the normal boiling point of water at atmospheric pressures. These temperatures are known to range from 100–700°F at pressures of 50–150 psig. When such resources penetrate to the surface, either through wells or through natural geologic anomalies, the water often flashes into steam.

The types of impurities found in wet-steam deposits vary dramatically. Commonly found dissolved salts and minerals include sodium, potassium, lithium, chlorides, sulfates, borates, bicarbonates, and silica. Salinity concentrations can vary from thousands to hundreds of thousands of parts per million. The Wairakei (New Zealand) and Cerro Prieto (Mexico) fields are examples of currently exploited liquid-dominated fields. In the United States, many of the liquid-dominated systems that have been identified are either being developed or are being considered for development (see Fig. 50.1).

50.3 HOT DRY ROCK RESOURCES

In some areas of the western United States, geologic anomalies such as tectonic plate movement and volcanic activity have created pockets of impermeable rocks covering a magma chamber within six miles of the surface. The temperature in these pockets increases with depth and proximity to the magma chamber, but, because of their impermeable nature, they lack a water aquifer. They are often referred to as *hot dry rock* (HDR) deposits. Several schemes for useful energy production from HDR resources have been proposed, but all basically involve creation of an artificial aquifer will be used to bring heat to the surface. The concept is being tested by the U.S. Department of Energy at Fenton Hill near Los Alamos, New Mexico, and is also being studied in England. The research so far indicates that it is technologically feasible to fracture a hot impermeable system though hydraulic fracturing from a deep well.

A typical two-well HDR system is shown in Fig. 50.2. Water is injected at high pressure through the first well to the reservoir and returns to the surface through the second well at approximately the temperature of the reservoir. The water (steam) is used to generate electric power and is then recirculated through the first well. The critical parameters affecting the ultimate commercial feasibility of HDR resources are the geothermal gradient and the achievable well flow rate.

50.4 GEOPRESSURED RESOURCES

Near the Gulf Coast of the United States are a number of deep sedimentary basins that are geologically very young, that is, less than 60 million years. In such regions, fluid located in subsurface rock formations carry a part of the overburden load, thereby increasing the pressure within the formation. Such formations are referred to as *geopressured* and are judged by some geologists to be promising sources of energy in the coming decades.

Geopressured basins exist in several areas within the United States, but those of current interest are located in the Texas–Louisiana coast. These are of particular interest because they are very large in terms of both areal extent and thickness, and the geopressured liquids appear to have a great deal of dissolved methane. In past investigations of the Gulf Coast, a number of "geopressured fairways" were identified; these are thick sandstone bodies expected to contain geopressured fluids of at least

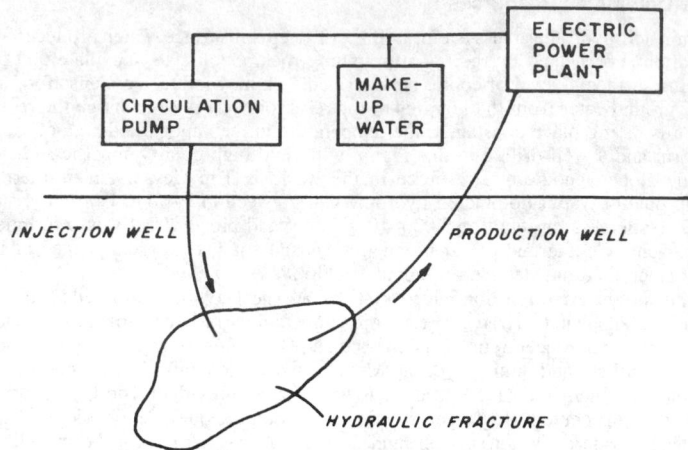

Fig. 50.2   Hot dry rock power plant configuration.

300°F. Detailed studies of the fairways of the Frio Formation in East Texas were carried out in 1979, although only one, Brazoria (see Fig. 50.3), met the requirements for further well testing. Within this fairway, a particularly promising site known as the Austin Bayou Prospect (ABP) was identified as an excellent candidate for a test well. This identification was based on the productive history of the neighboring oil and gas wells, and on the gradient of increasing temperature, permeability, and other resource characteristics.

If the water in these geopressured formations is also contained in insulating clay beds, as is the case in the Gulf Coast, the normal heat flow of the Earth can raise the temperature of this water to nearly 300°C. This water is typically of lower salinity than normal formations and, in many cases, is saturated with large amounts of recoverable natural methane gas. Hence, recoverable energy exists in geopressured formations in three forms: hydraulic pressure, heated water, and methane gas.

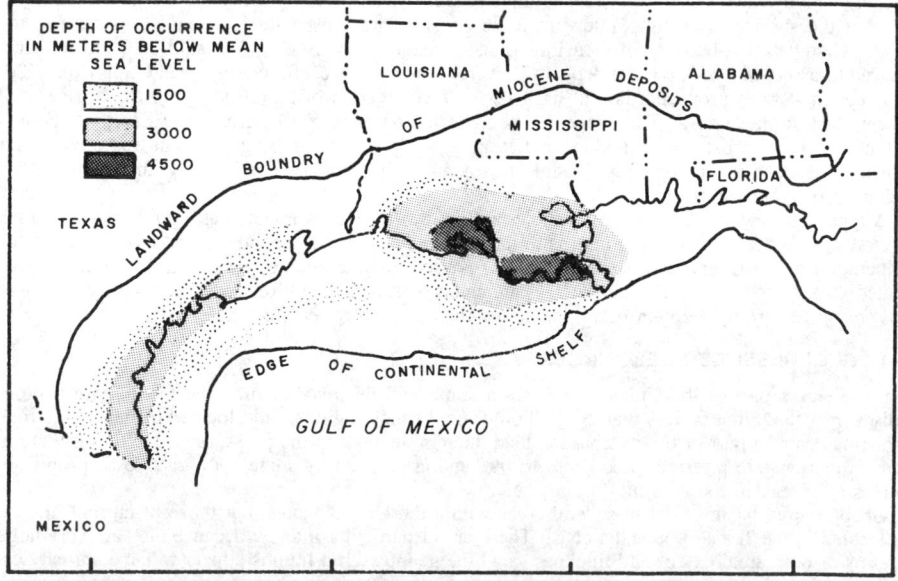

Fig. 50.3   Geopressured zones: Gulf of Mexico Basin.

The initial motivation for developing geopressured resources was to recover the entrained methane. Hence, a critical resource parameter affecting the commercial potential of a given prospect is the methane solubility, which is, in turn, a function of the geopressured reservoir's pressure, temperature, and brine salinity. The commercial potential of a prospect is also a function of the estimated volume of the reservoir, which dictates the amount of recoverable entrained methane; the "areal extent," which dictates how much methane can ultimately be recovered from the prospect site; and the "pay thickness," which dictates the initial production rate and the rate of production decline over time.

50.5 GEOTHERMAL ENERGY CONVERSION

The appropriate technology for converting geothermal energy to electricity depends on the nature of the resource. For vapor-dominated resources, it is possible to use direct steam conversion; for high-quality liquid-dominated resources, flashed steam or binary cycle technologies can be employed; and for lower quality liquid-dominated resources, a mixture of fossil and geothermal sources can be employed.

50.5.1 Direct Steam Conversion

The geothermal resources of central Italy and The Geysers are, as noted earlier, "vapor-dominated" resources, for which conversion of geothermal energy into electric energy is a straightforward process. The naturally pressurized steam is piped from wells to a power plant, where it expands through a turbine-generator to produce electric energy. The geothermal steam is supplied to the turbine directly, save for the relatively simple removal of entrained solids in gravity separators or the removal of noncondensible gases in degassing vessels. From the turbine, steam is exhausted to a condenser, condensed to its liquid state, and pumped from the plant. Usually this condensate is reinjected to the subterranean aquifer. Unfortunately, vapor-dominated geothermal resources occur infrequently in nature. To date, electric power from natural dry steam occurs at only one area, Matsukawa in Japan, other than central Italy and The Geysers.

A simplified flow diagram illustrating the direct steam conversion process is shown in Fig. 50.4. The major components of such systems are the steam turbine-generator, condenser, and cooling towers. Dry steam from the geothermal production well is expanded through the turbine, which drives an electric generator. The wet steam exhausting from the turbine is condensed and the condensate is piped from the plant for reinjection or other disposal. The cooling towers reject the waste heat released by condensation to the atmosphere. Additional plant systems not shown in Fig. 50.4 remove entrained solids from the steam prior to expansion and remove noncondensible gases from the condenser. The most recent power plants at The Geysers also include systems to control the release of hydrogen sulfide (a noncondensible gas contained in the steam) to the atmosphere.

Direct steam conversion is the most efficient type of geothermal electric power generation. One measure of plant efficiency is the level of electricity generated per unit of geothermal fluid used. The plants at The Geysers produce 50–55 Whr of electric energy per pound of 350° steam consumed. A second measure of efficiency used for geothermal power plants is the geothermal resource utilization efficiency, defined as the ratio of the net plant power output to the difference in thermodynamic availability of the geothermal fluid entering the plant and that of the fluid at ambient conditions. Plants at The Geysers operate at utilization efficiencies of 50–56%.

Release of hydrogen sulfide into the atmosphere is recognized as the most important environmental issue associated with direct steam conversion plants at The Geysers. Control measures are

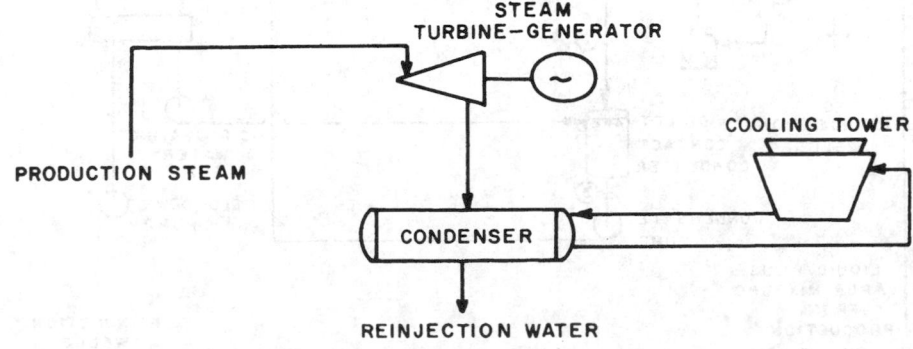

Fig. 50.4 Direct steam conversion.

required to meet California emission standards. Presently available control systems, which treat the steam after it has passed through the turbine, result in significant penalties in capital and operating cost. These systems include the iron/caustic/peroxide process, which has been installed on a number of Geysers units, and the Stretford process, which is used on several of the newer plants. Other, more economic, processes that treat the steam before it reaches the turbine are under development as well.

50.5.2 Flashed Steam Conversion

Most geothermal resources produce not dry steam, but a pressurized two-phase mixture of steam and water. The majority of plants currently operating at these liquid-dominated resources use a flashed steam energy conversion process. Figure 50.5 is a simplified schematic of a flashed steam plant. In addition to the turbine, condenser, and cooling towers found in the direct steam process, the flashed steam plant contains a separator or flash vessel. The geothermal fluid from the production wells first enters this vessel, where saturated steam is flashed from the liquid brine. This steam enters the turbine, while the unflashed brine is piped from the plant for reinjection or disposal. The remainder of the process is similar to the direct steam process.

Multiple stages of flash vessels are often used in the flashed steam systems to improve the plan efficiency and increase power output. Figure 50.6 shows a flow diagram of a two-stage flash plant. In this case, the unflashed brine leaving the initial flash vessel enters a second flash vessel that operates at a lower pressure, causing additional steam to be flashed. This lower-pressure steam is admitted to the low-pressure section of the turbine, recovering energy that would have been lost if a single-stage flash process had been used. In a design study for a geothermal plant to be located near Heber, California, the two-stage flash process resulted in a 37% improvement in plant performance over a single-stage flash process. Addition of a third flash stage showed an incremental improvement of 6% and was determined to be cost-effective.

50.5.3 Binary Cycle Conversion

Binary cycle conversion plants are an alternative approach to flashed steam plants for electric power generation at liquid-dominated geothermal resources. In this type of plant, a secondary fluid, usually a fluorocarbon or hydrocarbon, is used as a working fluid in a Rankine cycle, and the geothermal brine is used to heat this working fluid.

Figure 50.7 shows the main components and flow streams in a binary conversion process. Geothermal brine from production wells passes through a heat exchanger, where it transfers heat to the secondary working fluid. The cooled brine is then reinjected into the well field. The secondary working fluid is vaporized and superheated in the heat exchanger and expanded through a turbine, which drives an electric generator. The turbine exhaust is condensed in a surface condenser, and the condensate is pressurized and returned to the heat exchanger to complete the cycle. A cooling tower and circulating water system reject the heat of condensation to the atmosphere.

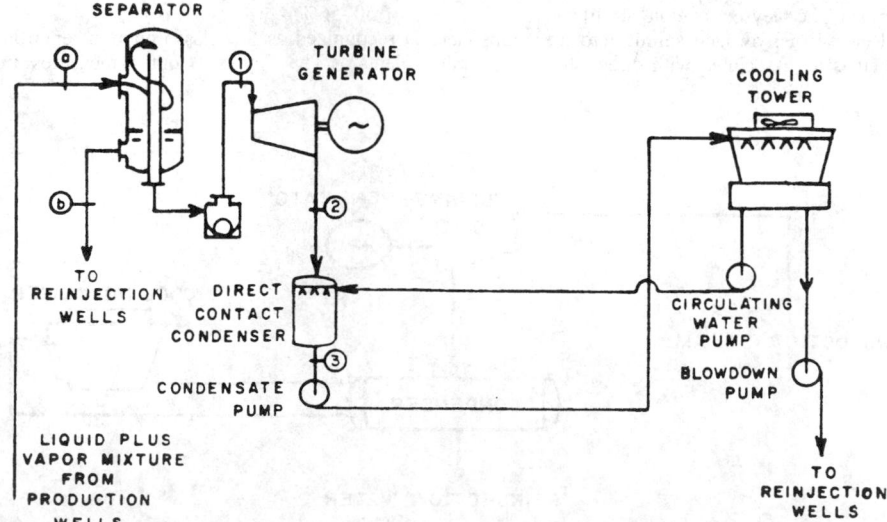

Fig. 50.5 Flashed steam conversion.

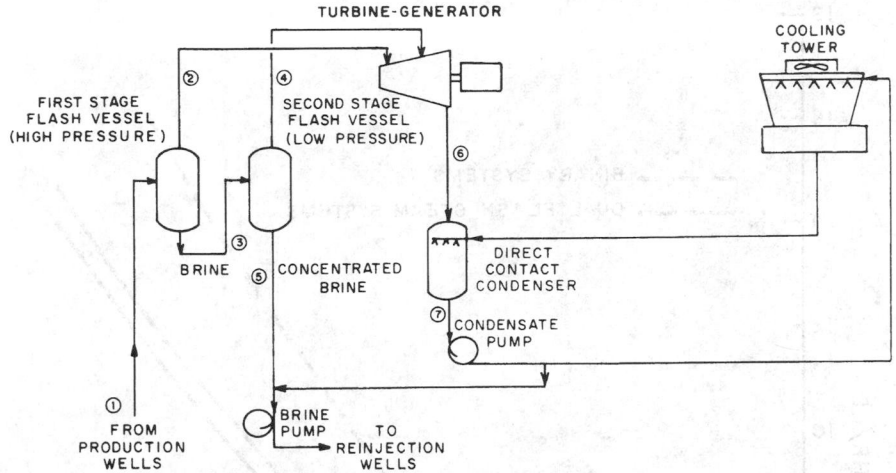

Fig. 50.6 Two-stage flash conversion.

Several variations of this cycle have been considered for geothermal power generation. A regenerator may be added between the turbine and condenser to recover energy from the turbine exhaust for condensate heating and to improve plant efficiency. The surface-type heat exchanger, which passes heat from the brine to the working fluid, may be replaced with a direct contact or fluidized-bed type exchanger to reduce plant cost. Hybrid plants combining the flashed steam and binary processes have also been evaluated.

The binary process may be an attractive alternative to the flashed steam process at geothermal resources producing high-salinity brine. Since the brine can remain in a pressurized liquid state throughout the process and it does not pass through the turbine, problems associated with salt precipitation and scaling as well as corrosion and erosion are greatly reduced. Binary cycles offer the additional advantage that a working fluid can be selected that has superior thermodynamic characteristics to steam, resulting in a more efficient cycle.

The binary cycle is not without disadvantages, however, as suitable secondary fluids are expensive and may be flammable or toxic. Plant complexity and cost are also increased by the requirement for two plant flow systems.

The efficiency of energy-conversion processes for liquid-dominated resources is dependent on resource temperature and to a lesser degree on brine salinity and noncondensible gas content. Additionally, conversion efficiency can be improved by system modifications at the penalty of additional plant complexity and cost. Figure 50.8 shows power production per unit of brine consumed for a two-stage flash system and for a binary system.

Emissions of hydrogen sulfide at liquid-dominated geothermal plants are lower than for direct steam processes. Flashed steam plants emit 30–50% less hydrogen sulfide than direct steam plants.

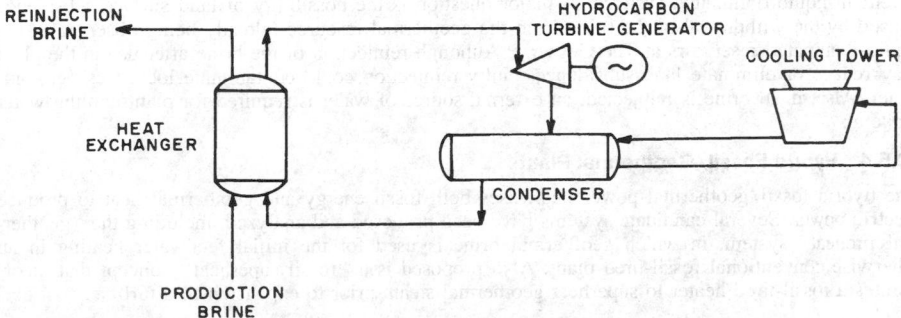

Fig. 50.7 Binary cycle conversion.

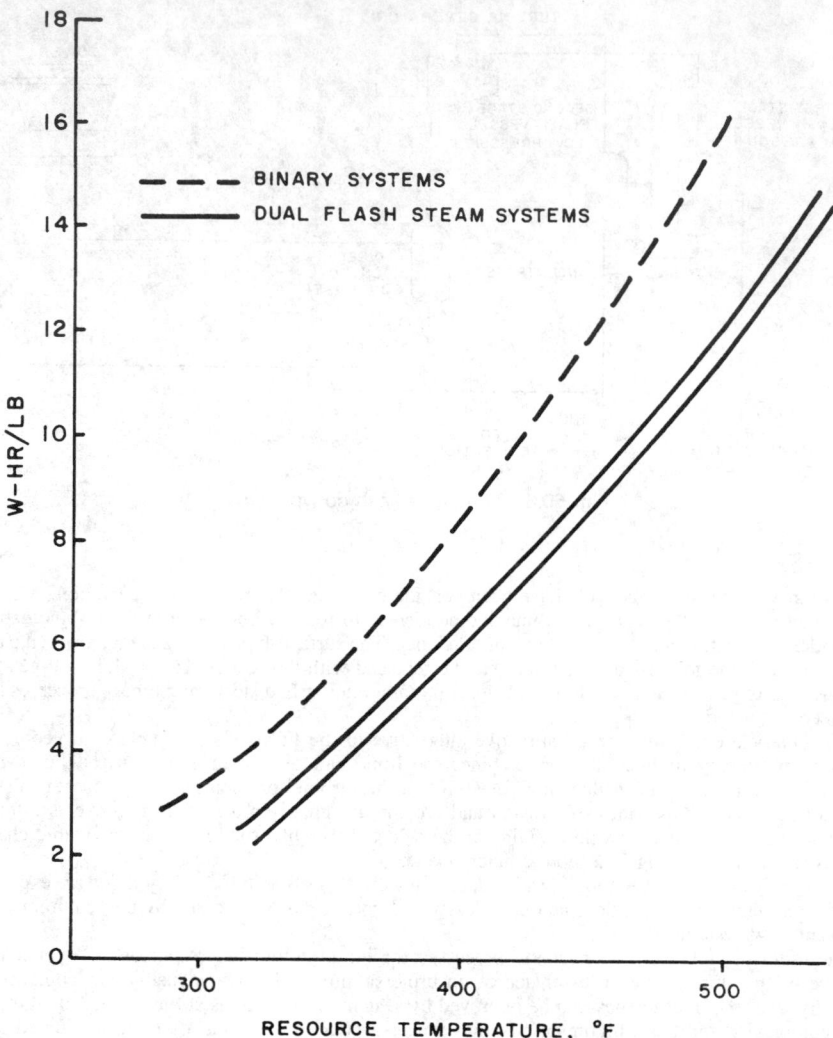

Fig. 50.8 Net geothermal brine effectiveness.

Binary plants would not routinely emit hydrogen sulfide because the brine would remain contained and pressurized throughout the process. However, there are other environmental considerations inherent in liquid-dominated systems. A major question is the possibility of land surface subsidence caused by the withdrawal of the brine from the geothermal resource (already being observed in the liquid-dominated reservoirs at The Geysers). Although reinjection of the brine after use in the plant may reduce or eliminate land subsidence, faulty reinjection could contaminate local fresh groundwater. Also, if all brine is reinjected, an external source of water is required for plant-cooling-water makeup.

50.5.4 Hybrid Fossil/Geothermal Plants

The hybrid fossil/geothermal power plant uses both fossil energy and geothermal heat to produce electric power. Several candidate systems have been proposed and analyzed, including the "geothermal preheat" system, in which geothermal brine is used for the initial feedwater heating in an otherwise conventional fossil-fired plant. Also proposed is a "fossil superheat" concept that incorporates a fossil-fired heater to superheat geothermal steam prior to expansion in a turbine.

CHAPTER 51
ENERGY AUDITING

Carl Blumstein
Universitywide Energy Research Group
University of California
Berkeley, California

Peter Kuhn
Kuhn and Kuhn,
Industrial Energy Consultants
Golden Gate Energy Center
Sausalito, California

51.1	**ENERGY MANAGEMENT AND THE ENERGY AUDIT**	**1591**
51.2	**PERFORMING AN ENERGY AUDIT—ANALYZING ENERGY USE**	**1592**
51.3	**PERFORMING AN ENERGY AUDIT—IDENTIFYING OPPORTUNITIES FOR SAVING ENERGY**	**1597**
51.3.1	Low-Cost Conservation	1598
51.3.2	Capital-Intensive Energy Conservation Measures	1600
51.4	**EVALUATING ENERGY CONSERVATION OPPORTUNITIES**	**1602**
51.5	**PRESENTING THE RESULTS OF AN ENERGY AUDIT**	**1604**

51.1 ENERGY MANAGEMENT AND THE ENERGY AUDIT

Energy auditing is the practice of surveying a facility to identify opportunities for increasing the efficiency of energy use. A facility may be a residence, a commercial building, an industrial plant, or other installation where energy is consumed for any purpose. Energy management is the practice of organizing financial and technical resources and personnel to increase the efficiency with which energy is used in a facility. Energy management typically involves the keeping of records on energy consumption and equipment performance, optimization of operating practices, regular adjustment of equipment, and replacement or modification of inefficient equipment and systems.

Energy auditing is a part of an energy management program. The auditor, usually someone not regularly associated with the facility, reviews operating practices and evaluates energy using equipment in the facility in order to develop recommendations for improvement. An energy audit can be, and often is, undertaken when no formal energy management program exists. In simple facilities, particularly residences, a formal program is impractical and informal procedures are sufficient to alter operating practices and make simple improvements such as the addition of insulation. In more complex facilities, the absence of a formal energy management program is usually a serious deficiency. In such cases a major recommendation of the energy audit will be to establish an energy management program.

There can be great variation in the degree of thoroughness with which an audit is conducted, but the basic procedure is universal. The first step is to collect data with which to determine the facility's major energy uses. These data always include utility bills, nameplate data from the largest energy-using equipment, and operating schedules. The auditor then makes a survey of the facility. Based on the results of this survey, he or she chooses a set of energy conservation measures that could be applied in the facility and estimates their installed cost and the net annual savings that they would

Mechanical Engineers' Handbook, 2nd ed., Edited by Myer Kutz.
ISBN 0-471-13007-9 © 1998 John Wiley & Sons, Inc.

provide. Finally, the auditor presents his or her results to the facility's management or operators. The audit process can be as simple as a walkthrough visit followed by a verbal report or as complex as a complete analysis of all of a facility's energy using equipment that is documented by a lengthy written report.

The success of an energy audit is ultimately judged by the resulting net financial return (value of energy saved less costs of energy saving measures). Since the auditor is rarely in a position to exercise direct control over operating and maintenance practices or investment decisions, his or her work can come to naught because of the actions or inaction of others. Often the auditor's skills in communication and interpersonal relations are as critical to obtaining a successful outcome from an energy audit as his or her engineering skills. The auditor should stress from the outset of his or her work that energy management requires a sustained effort and that in complex facilities a formal energy management program is usually needed to obtain the best results. Most of the auditor's visits to a facility will be spent in the company of maintenance personnel. These personnel are usually conscientious and can frequently provide much useful information about the workings of a facility. They will also be critical to the success of energy conservation measures that involve changes in operating and maintenance practices. The auditor should treat maintenance personnel with respect and consideration and should avoid the appearance of "knowing it all." The auditor must also often deal with nontechnical managers. These managers are frequently involved in the decision to establish a formal energy management program and in the allocation of capital for energy saving investments. The auditor should make an effort to provide clear explanations of his or her work and recommendations to nontechnical managers and should be careful to avoid the use of engineering jargon when communicating with them.

While the success of an energy audit may depend in some measure on factors outside the auditor's control, a good audit can lead to significant energy savings. Table 51.1 shows the percentage of energy saved as a result of implementing energy audit recommendations in 172 nonresidential buildings. The average savings is more than 20%. The results are especially impressive in light of the fact that most of the energy-saving measures undertaken in these buildings were relatively inexpensive. The median value for the payback on energy-saving investments was in the 1–2 year range (i.e., the value of the energy savings exceeded the costs in 1–2 years). An auditor can feel confident in stating that an energy saving of 20% or more is usually possible in facilities where systematic efforts to conserve energy have not been undertaken.

51.2 PERFORMING AN ENERGY AUDIT—ANALYZING ENERGY USE

A systematic approach to energy auditing requires that an analysis of existing energy-using systems and operating practices be undertaken before efforts are made to identify opportunities for saving energy. In practice, the auditor may shift back and forth from the analysis of existing energy-use patterns to the identification of energy-saving opportunities several times in the course of an audit—first doing the most simple analysis and identifying the most obvious energy-saving opportunities, then performing more complex analyses, and so on. This strategy may be particularly useful if the audit is to be conducted over a period of time that is long enough for some of the early audit recommendations to be implemented. The resultant savings can greatly increase the auditor's credi-

Table 51.1 The Percentage of Energy Saved as a Result of Implementing Energy Audit Recommendations in 172 Nonresidential Buildings[a,4]

Building Category	Site		Source	
	Savings (%)	Sample Size	Savings (%)	Sample Size
Elementary school	24	72	21	72
Secondary school	30	38	28	37
Large office	23	37	21	24
Hospital	21	13	17	10
Community center	56	3	23	18
Hotel	25	4	24	4
Corrections	7	4	5	4
Small office	33	1	30	1
Shopping center	11	1	11	1
Multifamily apartment	44	1	43	1

[a]Electricity is counted at 3413 Btu/kWhr for site energy and 11,500 Btu/kWhr for source energy (i.e., including generation and transmission losses).

bility with the facility's operators and management, so that he or she will receive more assistance in completing his or her work and his or her later recommendations will be attended to more carefully.

The amount of time devoted to analyzing energy use will vary, but, even in a walkthrough audit, the auditor will want to examine records of past energy consumption. These records can be used to compare the performance of a facility with the performance of similar facilities. Examination of the seasonal variation in energy consumption can give an indication of the fractions of a facility's use that are due to space heating and cooling. Records of energy consumption are also useful in determining the efficacy of past efforts to conserve energy.

In a surprising number of facilities the records of energy consumption are incomplete. Often records will be maintained on the costs of energy consumed but not on the quantities. In periods of rapidly escalating prices, it is difficult to evaluate energy performance with such records. Before visiting a facility to make an audit, the auditor should ask that complete records be assembled and, if the records are not on hand, suggest that they be obtained from the facility's suppliers. Good record keeping is an essential part of an energy management program. The records are especially important if changes in operation and maintenance are to be made, since these changes are easily reversed and often require careful monitoring to prevent backsliding.

In analyzing the energy use of a facility, the auditor will want to focus his or her attention on the systems that use the most energy. In industrial facilities these will typically involve production processes such as drying, distillation, or forging. Performing a good audit in an industrial facility requires considerable knowledge about the processes being used. Although some general principles apply across plant types, industrial energy auditing is generally quite specialized. Residential energy auditing is at the other extreme of specialization. Because a single residence uses relatively little energy, highly standardized auditing procedures must be used to keep the cost of performing an audit below the value of potential energy savings. Standardized procedures make it possible for audits to be performed quickly by technicians with relatively limited training.

Commercial buildings lie between these extremes of specialization. The term "commercial building" as used here refers to those nonresidential buildings that are not used for the production of goods and includes office buildings, schools, hospitals, and retail stores. The largest energy-using systems in commercial buildings are usually lighting and HVAC (heating, ventilating, and air conditioning). Refrigeration consumes a large share of the energy used in some facilities (e.g., food stores) and other loads may be important in particular cases (e.g., research equipment in laboratory buildings). Table 51.2 shows the results of a calculation of the amount of energy consumed in a relatively energy-efficient office building for lighting and HVAC in different climates. Office buildings (and other commercial buildings) are quite variable in their design and use. So, while the proportions of energy devoted to various uses shown in Table 51.2 are not unusual, it would be unwise to treat them (or any other proportions) as "typical." Because of the variety and complexity of energy-using systems in commercial buildings and because commercial buildings frequently use quite substantial amounts of energy in their operation, an energy audit in a commercial building often warrants the effort of a highly trained professional. In the remainder of this section commercial buildings will be used to illustrate energy auditing practice.

Lighting systems are often a good starting point for an analysis of energy in commercial buildings. They are the most obvious energy consumers, are usually easily accessible, and can provide good opportunities for energy saving. As a first step the auditor should determine the hours of operation of the lighting systems and the watts per square foot of floorspace that they use. These data, together with the building area, are sufficient to compute the energy consumption for lighting and can be used to compare the building's systems with efficient lighting practice. Next, lighting system maintenance practices should be examined. As shown in Fig. 51.1, the accumulation of dirt on lighting fixtures can significantly reduce light output. Fixtures should be examined for cleanliness and the auditor should determine whether or not a regular cleaning schedule is maintained. As lamps near the end of their rated life, they lose efficiency. Efficiency can be maintained by replacing lamps in groups

Table 51.2 Results of a Calculation of the Amount of Energy Consumed in a Relatively Energy-Efficient Office Building for Lighting and HVAC[5]

	Energy Use (kBtu/ft²/yr)			
	Miami	Los Angeles	Washington	Chicago
Lights	34.0	34.0	34.0	34.0
HVAC auxiliaries	8.5	7.7	8.8	8.8
Cooling	24.4	9.3	10.2	7.6
Heating	0.2	2.9	17.7	28.4
Total	67.1	53.9	70.7	78.8

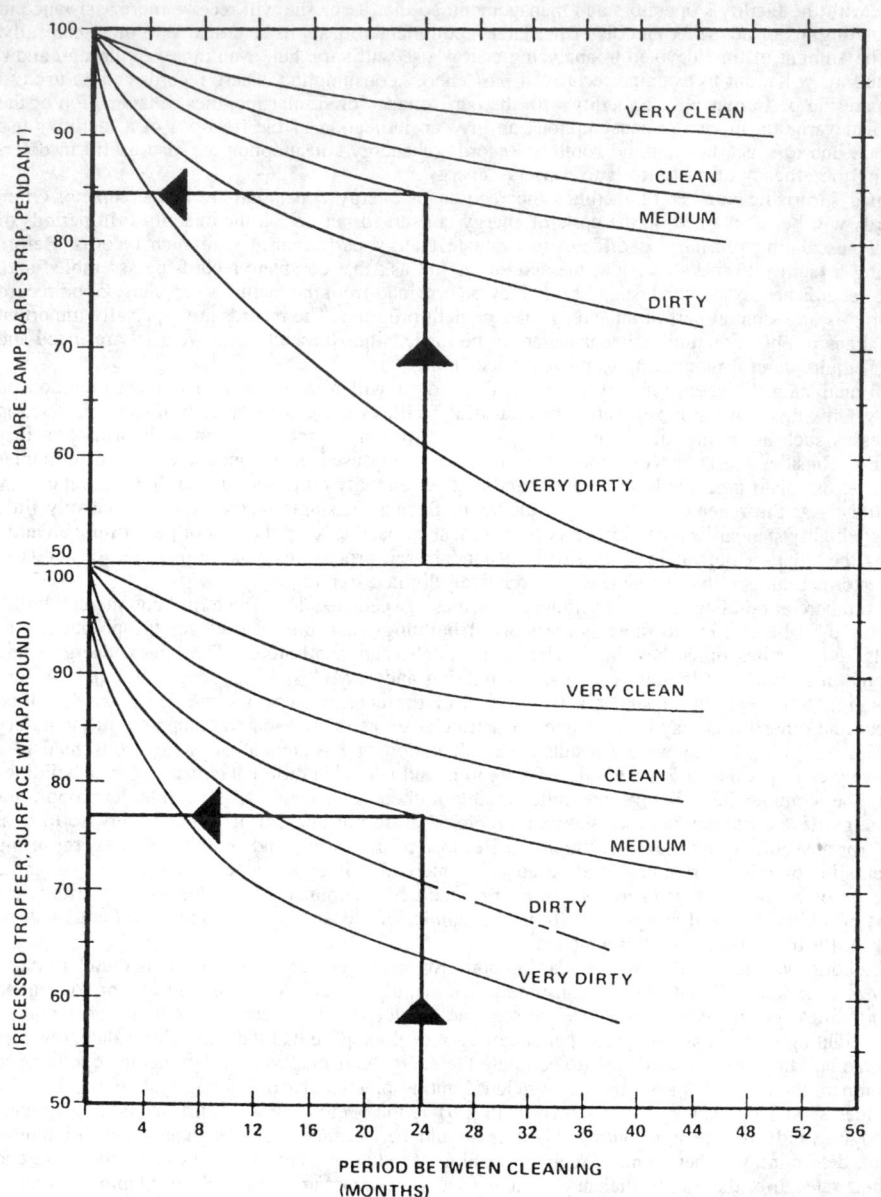

Fig. 51.1 Reduction in light output from fluorescent fixtures as a function of fixture cleaning frequency and the cleanliness of the fixture's surroundings.[3]

before they reach the end of their rated life. This practice also reduces the higher maintenance costs associated with spot relamping. Fixtures should be checked for lamps that are burned out or show signs of excessive wear, and the auditor should determine whether or not a group-relamping program is in effect.

After investigating lighting operation and maintenance practices, the auditor should measure the levels of illumination being provided by the lighting systems. These measurements can be made with a relatively inexpensive photometer. Table 51.3 gives recommended levels of illumination for a variety of activities. A level much in excess of these guidelines usually indicates an opportunity for saving energy. However, the auditor should recognize that good seeing also depends on other factors such as glare and contrast and that the esthetic aspects of lighting systems (i.e., their appearance and the

Table 51.3 Range of Illuminances Appropriate for Various Types of Activities and Weighting Factors for Choosing the Footcandle Level[a] within a Range of Illuminance[6]

Category	Range of Illuminances (Footcandles)	Type of Activity
A	2–3–5	Public areas with dark surroundings
B	5–7.5–10	Simple orientation for short temporary visits
C	10–15–20	Working spaces where visual tasks are only occasionally performed
D	20–30–50	Performance of visual tasks of high contrast or large size: for example, reading printed material, typed originals, handwriting in ink and good xerography; rough bench and machine work; ordinary inspection; rough assembly
E	50–75–100	Performance of visual tasks of medium contrast or small size: for example, reading medium-pencil handwriting, poorly printed or reproduced material; medium bench and machine work; difficult inspection; medium assembly
F	100–150–200	Performance of visual tasks of low contrast or very small size: for example, reading handwriting in hard pencil or very poorly reproduced material; very difficult inspection
G	200–300–500	Performance of visual tasks of low contrast and very small size over a prolonged period: for example, fine assembly; very difficult inspection; fine bench and machine work
H	500–750–1000	Performance of very prolonged and exacting visual tasks: for example, the most difficult inspection; extra-fine bench and machine work; extra-fine assembly
I	1000–1500–2000	Performance of very special visual tasks of extremely low contrast and small size: for example, surgical procedures

Weighting Factors

Worker or task charactristics	−1	0	+1
Workers' age	Under 40	40–65	Over 65
Speed and/or accuracy	Not important	Important	Critical
Reflectance of task background	Greater than 70%	30–70%	Less than 30%

[a]To determine a footcandle level within a range of illuminance, find the weighting factor for each worker or task characteristic and sum the weighting factors to obtain a score. If the score is −3 or −2, use the lowest footcandle level; if −1, 0, or 1, use the middle footcandle level; if 2 or 3, use the highest level.

effect they create) can also be important. More information about the design of lighting systems can be found in Ref. 1.

Analysis of HVAC systems in a commercial building is generally more complicated and requires more time and effort than lighting systems. However, the approach is similar in that the auditor will usually begin by examining operating and maintenance practices and then proceed to measure system performance.

Determining the fraction of a building's energy consumption that is devoted to the operation of its HVAC systems can be difficult. The approaches to this problem can be classified as either deterministic or statistical. In the deterministic approaches an effort is made to calculate HVAC energy consumption from engineering principles and data. First, the building's heating and cooling loads are calculated. These depend on the operating schedule and thermostat settings, the climate, heat gains and losses from radiation and conduction, the rate of air exchange, and heat gains from internal sources. Then energy use is calculated by taking account of the efficiency with which the HVAC systems meet these loads. The efficiency of the HVAC systems depends on the efficiency of equipment such as boilers and chillers and losses in distribution through pipes and ducts; equipment efficiency and distribution losses are usually dependent on load. In all but the simplest buildings, the calculation of HVAC energy consumption is sufficiently complex to require the use of computer programs; a number of such programs are available (see, for example, Ref. 2). The auditor will usually make some investigation of all of the factors necessary to calculate HVAC energy consumption. However, the effort involved in obtaining data that are sufficiently accurate and preparing them in suitable form for input to a computer program is quite considerable. For this reason, the deterministic approach is not recommended for energy auditing unless the calculation of savings from energy conservation measures requires detailed information on building heating and cooling loads.

Statistical approaches to the calculation of HVAC energy consumption involve the analysis of records of past energy consumption. In one common statistical method, energy consumption is analyzed as a function of climate. Regression analysis with energy consumption as the dependent variable and some function of outdoor temperature as the independent variable is used to separate "climate-dependent" energy consumption from "base" consumption. The climate-dependent fraction is considered to be the energy consumption for heating and cooling, and the remainder is assumed to be due to other uses. This method can work well in residences and in some small commercial buildings where heating and cooling loads are due primarily to the climate. It does not work as well in large commercial buildings because much of the cooling load in these buildings is due to internal heat gains and because a significant part of the heating load may be for reheat (i.e., air that is precooled to the temperature required for the warmest space in the building may have to be reheated in other spaces). The easiest statistical method to apply, and the one that should probably be attempted first, is to calculate the energy consumption for all other end uses (lighting, domestic hot water, office equipment, etc.) and subtract this from the total consumption; the remainder will be HVAC energy consumption. If different fuel types are used for heating and cooling, it will be easy to separate consumption for these uses; if not, some further analysis of the climate dependence of consumption will be required. Energy consumption for ventilation can be calculated easily if the operating hours and power requirements for the supply and exhaust fans are known.

Whatever approach is to be taken in determining the fraction of energy consumption that is used for HVAC systems, the auditor should begin his or her work on these systems by determining their operating hours and control settings. These can often be changed to save energy with no adverse effects on a building's occupants. Next, maintenance practices should be examined. This examination will usually be initiated by determining whether or not a preventive maintenance (PM) program is being conducted. If there is a PM program, much can be learned about the adequacy of maintenance practices by examining the PM records. Often only a few spot checks of the HVAC systems will be required to verify that the records are consistent with actual practice. If there is no PM program, the auditor will usually find that the HVAC systems are in poor condition and should be prepared to make extensive checks for energy-wasting maintenance problems. Establishment of a PM program as part of the energy management program is a frequent recommendation from an energy audit.

Areas for HVAC maintenance that are important to check include heat exchanger surfaces, fuel-air mixture controls in combustors, steam traps, and temperature controllers. Scale on the water side of boiler tubes and chiller condenser tubes reduces the efficiency of heat transfer. Losses of efficiency can also be caused by the buildup of dirt on finned-tube air-cooled condensers. Improper control of fuel-air mixtures can cause significant losses in combustors. Leaky steam traps are a common cause of energy losses. Figure 51.2 shows the annual rate of heat loss through a leaky trap as a function of the size of the trap orifice and steam pressure. Poorly maintained room thermostats and other controls such as temperature reset controllers can also cause energy waste. While major failures of thermostats can usually be detected as a result of occupant complaints or behavior (e.g., leaving windows open on cold days), drifts in these controls that are too small to cause complaints can still lead to substantial waste. Other controls, especially reset controls, can sometimes fail completely and cause an increase in energy consumption without affecting occupant comfort.

After investigating HVAC operation and maintenance practices, the auditor should make measurements of system performance. Typical measurements will include air temperature in rooms and ducts, water temperatures, air flow rates, pressure drops in air ducts, excess air in stack gases, and current drawn by electric motors operating fans and pumps. Instruments required include a thermometer, a pitot tube or anemometer, a manometer, a strobe light, a combustion test kit, and an ammeter. The importance of making measurements instead of relying on design data cannot be emphasized too strongly. Many, if not most, buildings operate far from their design points. Measurements may point to needed adjustments in temperature settings or air flow rates. Table 51.4 gives recommended air flow rates for various applications. Detailed analysis of the measured data requires a knowledge of HVAC system principles.

After measuring HVAC system performance, the auditor should make rough calculations of the relative importance of the different sources of HVAC system loads. These are primarily radiative and conductive heat gains and losses through the building's exterior surfaces, gains and losses from air exchange, and gains from internal heat sources. Rough calculations are usually sufficient to guide the auditor in selecting conservation measures for consideration. More detailed analyses can await the selection of specific measures.

While lighting and HVAC systems will usually occupy most of the auditor's time in a commercial building, other systems such as domestic hot water may warrant attention. The approach of first

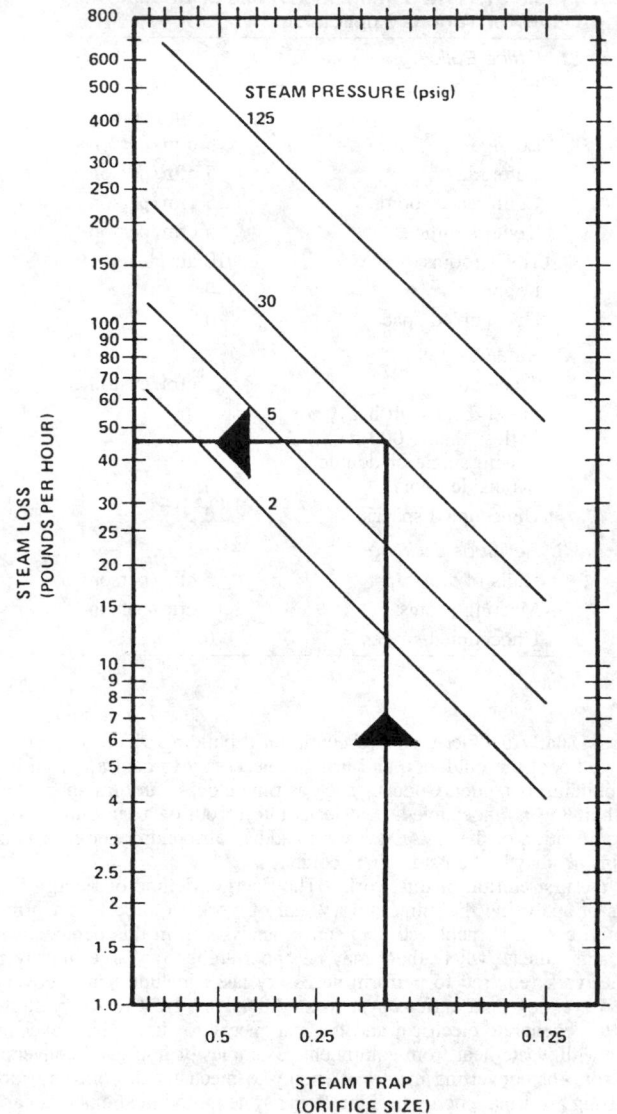

Fig. 51.2 Steam loss through leaking steam traps as a function of stem pressure and trap orifice size.[3]

investigating operation and maintenance practices and then measuring system performance is usually appropriate for these systems.

51.3 PERFORMING AN ENERGY AUDIT—IDENTIFYING OPPORTUNITIES FOR SAVING ENERGY

In almost every facility one can discover a surprisingly large number of opportunities to save energy. These opportunities range from the obvious such as use of light switches to exotic approaches involving advanced energy conversion technologies. Identification of ways to save energy requires imagination and resourcefulness as well as a sound knowledge of engineering principles.

The auditor's job is to find ways to *eliminate unnecessary energy-using tasks* and ways to *minimize the work required to perform necessary tasks.* Some strategies that can be used to eliminate unnecessary tasks are improved controls, "leak plugging," and various system modifications. Taking space conditioning as an example, it is necessary to provide a comfortable interior climate for building

Table 51.4   Recommended Rates of Outside-Air Flow for Various Applications[3]

1. Office Buildings	
Work space	5 cfm/person
Heavy smoking areas	15 cfm/person
Lounges	5 cfm/person
Cafeteria	5 cfm/person
Conference rooms	15 cfm/person
Doctors' offices	5 cfm/person
Toilet rooms	10 air changes/hr
Lobbies	0
Unoccupied spaces	0
2. Retail Stores	
Trade areas	6 cfm/customer
Street level with heavy use (less than 5,000 ft.² with single or double outside door)	0
Unoccupied spaces	0
3. Religious Buildings	
Halls of worship	5 cfm/person
Meeting rooms	10 cfm/person
Unoccupied spaces	0

occupants, but it is usually not necessary to condition a building when it is unoccupied, it is not necessary to heat and cool the outdoors, and it is not necessary to cool air from inside the building if air outside the building is colder. Controls such as time clocks can turn space-conditioning equipment off when a building is unoccupied, heat leaks into or out of a building can be plugged using insulation, and modification of the HVAC system to add an air-conditioner economizer can eliminate the need to cool inside air when outside air is colder.

Chapter 55 of the first edition of this work, "The Exergy Method of Energy Systems Analysis," discusses methods of analyzing the minimum amount of work required to perform tasks. While the theoretical minimum cannot be achieved in practice, analysis from this perspective can reveal inefficient operations and indicate where there may be opportunities for large improvements. Strategies for minimizing the work required to perform necessary tasks include heat recovery, improved efficiency of energy conversion, and various system modifications. Heat recovery strategies range from complex systems to cogenerate electrical and thermal energy to simple heat exchangers that can be used to heat water with waste heat from equipment. Examples of improved conversion efficiency are more efficient motors for converting electrical energy to mechanical work and more efficient light sources for converting electrical energy to light. Some system modifications that can reduce the work required to perform tasks are the replacement of resistance heaters with heat pumps and the replacement of dual duct HVAC systems with variable air volume systems.

There is no certain method for discovering all of the energy-saving opportunities in a facility. The most common approach is to review lists of energy conservation measures that have been applied elsewhere to see if they are applicable at the facility being audited. A number of such lists have been compiled (see, for example, Ref. 3). However, while lists of measures are useful, they cannot substitute for intelligent and creative engineering. The energy auditor's recommendations need to be tailored to the facility, and the best energy conservation measures often involve novel elements.

In the process of identifying energy saving opportunities, the auditor should concentrate first on low-cost conservation measures. The savings potential of these measures should be estimated before more expensive measures are evaluated. Estimates of the savings potential of the more expensive measures can then be made from the reduced level of energy consumption that would result from implementing the low-cost measures. While this seems obvious, there have been numerous occasions on which costly measures have been used but simpler, less expensive alternatives have been ignored.

51.3.1   Low-Cost Conservation

Low-cost conservation measures include turning off energy-using equipment when it is not needed, reducing lighting and HVAC services to recommended levels, rescheduling of electricity-intensive operations to off-peak hours, proper adjustment of equipment controls, and regular equipment maintenance. These measures can be initiated quickly, but their benefits usually depend on a sustained effort. An energy management program that assigns responsibility for maintaining these low-cost measures and monitors their performance is necessary to ensure good results.

In commercial buildings it is often possible to achieve very large energy savings simply by shutting down lighting and HVAC systems during nonworking hours. This can be done manually or, for HVAC systems, by inexpensive time clocks. If time clocks are already installed, they should be maintained in good working order and set properly. During working hours lights should be turned off in unoccupied areas. Frequent switching of lamps does cause some decrease in lamp life, but this decrease is generally not significant in comparison to energy savings. As a rule of thumb, lights should be turned out in a space that will be unoccupied for more than 5 min.

Measurements of light levels, temperatures, and air flow rates taken during the auditor's survey will indicate if lighting or HVAC services exceed recommended levels. Light levels can be decreased by relamping with lower-wattage lamps or by removing lamps from fixtures. In fluorescent fixtures, except for instant-start lamps, ballasts should also be disconnected because they use some energy when the power is on even when the lamps are removed.

If the supply of outside air is found to be excessive, reducing the supply can save heating and cooling energy (but see below on air-conditioner economizers). If possible, the reduction in air supply should be accomplished by reducing fan speed rather than by restricting air flow by the use of dampers, since the former procedure is more energy efficient. Also, too much air flow restriction can cause unstable operation in some fans.

Because most utilities charge more for electricity during their peak demand periods, rescheduling the operation of some equipment can save considerable amounts of money. It is not always easy to reschedule activities to suit the utility's peak demand schedule, since the peak demand occurs when most facilities are engaging in activities requiring electricity. However, a careful examination of major electrical equipment will frequently reveal some opportunities for rescheduling. Examples of activities that have been rescheduled to save electricity costs are firing of electric ceramic kilns, operation of swimming pool pumps, finish grinding at cement plants, and pumping of water from wells to storage tanks.

Proper adjustment of temperature and pressure controls in HVAC distribution systems can cut losses in these systems significantly. Correct temperature settings in air supply ducts can greatly reduce the energy required for reheat. Temperature settings in hot water distribution systems can usually be adjusted to reduce heat loss from the pipes. Temperatures are often set higher than necessary to provide enough heating during the coldest periods; during milder weather, the distribution temperature can be reduced to a lower setting. This can be done manually or automatically using a reset control. Reset controls are generally to be preferred, since they can adjust the temperature continuously. In steam distribution systems, lowering the distribution pressure will reduce heat loss from the flashing of condensate (unless the condensate return system is unvented) and also reduce losses from the surface of the pipes. Figure 51.3 shows the percentage of the heat in steam that is lost due to condensate flashing at various pressures. Raising temperatures in chilled-water distribution systems also saves energy in two ways. Heat gain through pipe surfaces is reduced, and the chiller's efficiency increases due to the higher suction head on the compressor (see Fig. 51.4).

A PM program is needed to ensure that energy-using systems are operating efficiently. Among the activities that should be conducted regularly in such a program are cleaning of heat exchange surfaces, surveillance of steam traps so that leaky traps can be found and repaired, combustion efficiency testing, and cleaning of light fixtures. Control equipment such as thermostats, time clocks, and reset controllers need special attention. This equipment should be checked and adjusted frequently.

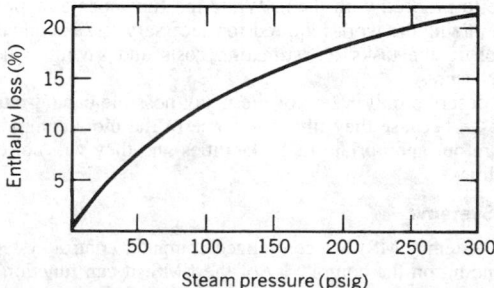

Fig. 51.3 Percentage of heat that is lost due to condensate flashing at various pressures.

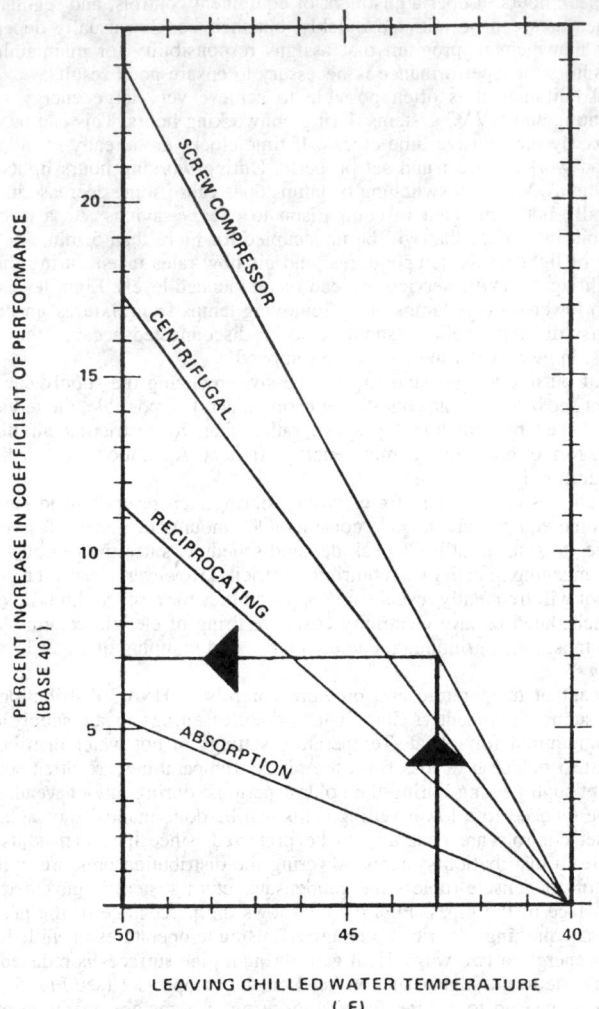

Fig. 51.4 Adjusting air-conditioner controls to provide higher chilled-water temperatures improves chiller efficiency.[3]

51.3.2 Capital-Intensive Energy Conservation Measures

Major additions, modifications, or replacement of energy-using equipment usually require significant amounts of capital. These measures consequently undergo a more detailed scrutiny before a facility's management will decide to proceed with them. While the fundamental approach of eliminating unnecessary tasks and minimizing the work required for necessary tasks is unchanged, the auditor must pay much more attention to the tasks of estimating costs and savings when considering capital-intensive conservation measures.

This subsection will describe only a few of the many possible capital-intensive measures. These measures have been chosen because they illustrate some of the more common approaches to energy saving. However, they are not appropriate in all facilities and they will not encompass the majority of savings in many facilities.

Energy Management Systems

An energy management system (EMS) is a centralized computer control system for building services, especially HVAC. Depending on the complexity of the EMS, it can function as a simple time clock to turn on equipment when necessary, it can automatically cycle the operation of large electrical equipment to reduce peak demand, and it can program HVAC system operation in response to outdoor and indoor temperature trends so that, for example, the "warm-up" heating time before a building is occupied in the morning is minimized. While such a system can be a valuable component of complex building energy service systems, the energy auditor should recognize that the functions of an EMS often duplicate the services of less costly equipment such as time clocks, temperature controls, and manual switches.

Air-Conditioner Economizers

In many areas, outdoor temperatures are lower than return air temperatures during a large part of the cooling season. An air-conditioner economizer uses outside air for cooling during these periods so that the load on the compressor is reduced or eliminated. The economizer is a system of automatic dampers on the return air duct that are controlled by return-air and outside-air temperature sensors. When the outside air is cooler than the return air, the dampers divert the return air to the outdoors and let in enough fresh outside air to supply all the air to the building. In humid climates, economizers must be fitted with "enthalpy" sensors that measure wet-bulb as well as dry-bulb temperature so that the economizer will not let in outside air when it is too humid for use in the building.

Building Exhaust-Air Heat Recovery Units

Exhaust-air heat recovery can be practical for facilities with large outside-air flow rates in relatively extreme climates. Hospitals and other facilities that are required to have once-through ventilation are especially good candidates. Exhaust-air heat recovery units reduce the energy loss in exhaust air by transferring heat between the exhaust air and the fresh air intake.

The common types of units available are heat wheels, surface heat exchangers, and heat-transfer-fluid loops. Heat wheels are revolving arrays of corrugated steel plates or other media. In the heating season, the plates absorb heat in the exhaust air duct, rotate to the intake air duct, and reject heat to the incoming fresh air. Surface heat exchangers are air-to-air heat exchangers. Some of these units are equipped with water sprays on the exhaust air side of the heat exchanger for indirect evaporative cooling. When a facility's exhaust- and fresh-air intakes are physically separated by large distances, heat-transfer-fluid loops (sometimes called run-around systems) are the only practical approach to exhaust-air heat recovery. With the fluid loop, heat exchangers are installed in both the exhaust and intake ducts and the fluid is circulated between the exchangers.

A key factor in estimating savings from exhaust air heat recovery is the unit's effectiveness, expressed as the percentage of the theoretically possible heat transfer that the unit actually achieves. With a 40°F temperature difference between the exhaust and intake air in the heating mode, a 60% effective unit will raise the intake air temperature by 24°F. In units with indirect evaporative cooling, the effectiveness indicates the extent to which the unit can reduce the difference between the intake air dry-bulb temperature and the exhaust air wet-bulb temperature. The effectiveness of commercially available exhaust air heat recovery units ranges from 50% to 80%; greater effectiveness is usually obtained at a higher price per unit of heat recovery capacity.

Refrigeration Heat Recovery

Heat recovery from refrigerators and air conditioners can replace fuel that would otherwise be consumed for low-temperature heating needs. Heat recovery units that generate hot water consist of water storage tanks with an integral refrigerant condenser that supplements or replaces the existing condenser on the refrigerator or air conditioner. These units reduce the facility's fuel or electricity consumption for water heating, and also increase the refrigeration or air conditioning system's efficiency due to the resulting cooler operating temperature of the condenser.

The most efficient condensing temperature will vary, depending on the compressor design and refrigerant, but in most cases it will be below 100°F. In facilities requiring water at higher temperatures, the refrigeration heat recovery unit can preheat water for the existing water heater, which will then heat the water to the final temperature.

Boiler Heat Recovery Devices

Part of the energy conversion losses in a boiler room can be reduced by installing a boiler economizer, air preheater, or a blowdown heat recovery unit. Both the economizer and the air preheater recover heat from the stack gases. The economizer preheats boiler feedwater and the air preheater heats combustion air. The energy savings from these devices are typically 5–10% of the boiler's fuel consumption. The savings depend primarily on the boiler's stack gas temperature. Blowdown heat recovery units are used with continuous blowdown systems and can either supply low-pressure steam to the deaerator or preheat makeup water for the boiler. Their energy savings are typically 1–2% of boiler fuel consumption. The actual savings will depend on the flow rate of the boiler blowdown and the boiler's steam pressure or hot-water temperature.

More Efficient Electric Motors

Replacement of integral-horsepower conventional electric motors with high-efficiency motors will typically yield an efficiency improvement of 2–5% at full load (see Table 51.5). While this saving is relatively small, replacement of fully loaded motors can still be economical for motors that operate

Table 51.5   Comparative Efficiencies and Power Factors (%) for U-Frame, T-Frame, and Energy-Efficient Motors[7]

For Smaller Motors

Horsepower Range: Speed: Type:	3–30 hp 3600 rpm			3–30 hp 1800 rpm			1.5–20 hp 1200 rpm		
	U	T	EEM	U	T	EEM	U	T	EEM
Efficiency									
4/4 load	84.0	84.7	86.9	86.0	86.2	89.2	84.1	82.9	86.1
3/4 load	82.6	84.0	87.4	85.3	85.8	91.1	83.5	82.3	86.1
1/2 load	79.5	81.4	85.9	82.8	83.3	83.3	81.0	79.6	83.7
Power factor									
4/4 load	90.8	90.3	86.6	85.3	83.5	85.8	78.1	77.0	73.7
3/4 load	88.7	87.8	84.1	81.5	79.2	81.9	72.9	70.6	67.3
1/2 load	83.5	81.8	77.3	72.8	70.1	73.7	60.7	59.6	56.7

For Larger Motors

Horsepower Range: Speed: Type:	40–100 hp 3600 rpm			40–100 hp 1800 rpm			25–75 hp 1200 rpm		
	U	T	EEM	U	T	EEM	U	T	EEM
Efficiency									
4/4 load	89.7	89.6	91.6	90.8	90.9	92.9	90.4	90.1	92.1
3/4 load	88.6	89.0	92.1	90.2	90.7	93.2	90.3	90.3	92.8
1/2 load	85.9	87.2	91.3	88.1	89.2	92.5	89.2	89.3	92.7
Power factor									
4/4 load	91.7	91.5	89.1	88.7	87.4	87.6	88.3	88.5	86.0
3/4 load	89.9	89.8	88.8	87.1	85.4	86.3	86.6	86.4	83.8
1/2 load	84.7	85.0	85.2	82.0	79.2	81.1	80.9	80.3	77.8

continuously in areas where electricity costs are high. Motors that are seriously underloaded are better candidates for replacement. The efficiency of conventional motors begins to fall sharply at less than 50% load, and replacement with a smaller high-efficiency motor can yield a quick return. Motors that must run at part load for a significant part of their operating cycle are also good candidates for replacement, since high-efficiency motors typically have better part-load performance than conventional motors.

High-efficiency motors typically run faster than conventional motors with the same speed rating because high-efficiency motors operate with less slip. The installation of a high-efficiency motor to drive a fan or pump may actually increase energy consumption due to the increase in speed, since power consumption for fans and pumps increases as the cube of the speed. The sheaves in the fan or pump drive should be adjusted or changed to avoid this problem.

More Efficient Lighting Systems

Conversion of lighting fixtures to more efficient light sources is often practical when the lights are used for a significant portion of the year. Table 51.6 lists some of the more common conversions and the difference in power consumption. Installation of energy-saving ballasts in fluorescent lights provides a small (5–12%) percentage reduction in fixture power consumption, but the cost can be justified by energy cost savings if the lights are on most of the time. Additional lighting controls such as automatic dimmers can reduce energy consumption by making better use of daylight. Attention should also be given to the efficiency of the luminaire and (for indoor lighting) interior wall surfaces in directing light to the areas where it is needed. Reference 1 provides data for estimating savings from more efficient luminaires and more reflective wall and ceiling surfaces.

51.4   EVALUATING ENERGY CONSERVATION OPPORTUNITIES

The auditor's evaluation of energy conservation opportunities should begin with a careful consideration of the possible effects of energy conservation measures on safety, health, comfort, and productivity within a facility. A conscientious effort should be made to solicit information from knowledgeable personnel and those who have experience with conservation measures in similar facilities. For energy conservation measures that do not interfere with the main business of a facility and the health and safety of its occupants, the determinant of action is the financial merit of a given measure.

Most decisions regarding the implementation of an energy conservation measure are based on the auditor's evaluation of the annual dollar savings and the initial capital cost, if any, associated with

Table 51.6 Some Common Lighting Conversions

	Present Fixture				Replacement Fixture		
Type	Power Consumption (W)	Light Output (lumens)	Lifetime (hr)	Type	Power Consumption (W)	Light Output (lumens)	Lifetime (hr)
100-W incandescent	100	1,740	750	8-in. 22-W circline adapter in same fixture	34	980	7,500
500-W incandescent	500	10,600	1,000	175-W metal halide fixture	205	12,000	7,500
Four 40-W rapid-start warm-white 48-in. fluorescent tubes in two-ballast fixture	184	12,800	18,000	Four 34-W energy saver rapid-start tubes in same fixture	160	11,600	18,000
Two 75-W slimline warm-white 96-in. fluorescent tubes in one-ballast fixture	172	12,800	12,000	Two 60-W energy saver slimline tubes in same fixture	142	10,680	12,000
150-W incandescent reflector flood-light (R-40)	150	1,200 (beam candlepower)	5,000	75-W incandescent projector flood-light (PAR-38)	75	1,430 (beam candlepower)	5,000
250-W mercury vapor streetlamp	285	10,400	24,000	150-W high-pressure sodium streetlamp	188	14,400	24,000

the measure. Estimation of the cost and savings from energy conservation measures is thus a critically important part of the analytical work involved in energy auditing.

When an energy conservation opportunity is first identified, the auditor should make a rough estimate of costs and savings in order to assess the value of further investigation. A rough estimate of the installed cost of a measure can often be obtained by consulting a local contractor or vendor who has experience with the type of equipment that the measure would involve. For commercial building energy conservation measures, a good guide to costs can be obtained from one of the annually published building construction cost estimating guides. The most valuable guides provide costs for individual mechanical, electrical, and structural components in a range of sizes or capacities. Rough estimates of the annual dollar savings from a measure can use simplified approaches to estimating energy savings such as assuming that a motor operates at its full nameplate rating for a specified percentage of the time.

If further analysis of a measure is warranted, a more accurate estimate of installed cost can be developed by preparing a clear and complete specification for the measure and obtaining quotations from experienced contractors or vendors. In estimating savings, one should be careful to calculate the measure's effect on energy use using accurate data for operating schedules, temperatures, flow rates, and other parameters. One should also give careful consideration to the measure's effect on maintenance requirements and equipment lifetimes, and include a dollar figure for the change in labor or depreciation costs in the savings estimate.

51.5 PRESENTING THE RESULTS OF AN ENERGY AUDIT

Effective presentation of the energy audit's results is crucial to achieving energy savings. The presentation may be an informal conversation with maintenance personnel, or it may be a formal presentation to management with a detailed financial analysis. In some cases the auditor may also need to make a written application to an outside funding source such as a government agency.

The basic topics that should be covered in most presentations are the following:

1. The facility's historical energy use, in physical and dollar amounts broken down by end use.
2. A review of the existing energy management program (if any) and recommendations for improvement.
3. A description of the energy conservation measures being proposed and the means by which they will save energy.
4. The cost of undertaking the measures and the net benefits the facility will receive each year.
5. Any other effects the measure will have on the facility's operation, such as changes in maintenance requirements or comfort levels.

The auditor should be prepared to address these topics with clear explanations geared to the interests and expertise of the audience. A financial officer, for example, may want considerable detail on cash flow analysis. A maintenance foreman, however, will want information on the equipment's record for reliability under conditions similar to those in his or her facility. Charts, graphs, and pictures may help to explain some topics, but they should be used sparingly to avoid inundating the audience with information that is of secondary importance.

The financial analysis will be the most important part of a presentation that involves recommendations of measures requiring capital expenditures. The complexity of the analysis will vary, depending on the type of presentation, from a simple estimate of the installed cost and annual savings to an internal rate of return or discounted cash flow calculation.

The more complex types of calculations involve assumptions regarding future fuel and electricity price increases, interest rates, and other factors. Because these assumptions are judgmental and may critically affect the results of the analysis, the more complex analyses should not be used in presentations to the exclusion of simpler indices such as simple payback time or after-tax return on investment. These methods do not involve numerous projections about the future.

REFERENCES

1. J. E. Kaufman (ed.), *IES Lighting Handbook,* Illuminating Engineers Society of North America, New York, 1981.
2. M. Lokmanhekim et al., *DOE-2: A New State-of-the-Art Computer Program for the Energy Utilization Analysis of Buildings,* Lawrence Berkeley Laboratory Report, LBL-8974, Berkeley, CA, 1979.
3. U.S. Department of Energy, *Architects and Engir·ers Guide to Energy Conservation in Existing Buildings,* Federal Energy Management Program Manual, U.S. Department of Energy, Federal Programs Office, Conservation and Solar Energy, NTIS Report DOE/CS-1302, February 1, 1980.
4. L. W. Wall and J. Flaherty, *A Summary Review of Building Energy Use Compilation and Analysis (BECA) Part C: Conservation Progress in Retrofitted Commercial Buildings,* Lawrence Berkeley Laboratory Report, LBL-15375, Berkeley, CA, 1982.

5. F. C. Winkelmann and M. Lokmanhekim, *Life-Cycle Cost and Energy-Use Analysis of Sun Control and Daylighting Options in a High-Rise Office Building,* Lawrence Berkeley Laboratory Report, LBL-12298, Berkeley, CA, 1981.

6. California Energy Commission, *Institutional Conservation Program Energy Audit Report: Minimum Energy Audit Guidelines,* California Energy Commission, Publication No. P400-82-022, Sacramento, CA, 1982.

7. W. C. Turner (ed.), *Energy Management Handbook,* Wiley-Interscience, New York, 1982.

CHAPTER 52

HEAT EXCHANGERS, VAPORIZERS, CONDENSERS

Joseph W. Palen
Heat Transfer Research, Inc.
College Station, Texas

52.1	**HEAT EXCHANGER TYPES AND CONSTRUCTION**	**1607**
52.1.1	Shell and Tube Heat Exchangers	1607
52.1.2	Plate-Type Heat Exchangers	1610
52.1.3	Spiral Plate Heat Exchangers	1610
52.1.4	Air-Cooled Heat Exchangers	1611
52.1.5	Compact Heat Exchangers	1611
52.1.6	Boiler Feedwater Heaters	1613
52.1.7	Recuperators and Regenerators	1613
52.2	**ESTIMATION OF SIZE AND COST**	**1613**
52.2.1	Basic Equations for Required Surface	1614
52.2.2	Mean Temperature Difference	1615
52.2.3	Overall Heat-Transfer Coefficient	1615
52.2.4	Pressure Drop	1616
52.3	**RATING METHODS**	**1616**
52.3.1	Shell and Tube Single-Phase Exchangers	1616
52.3.2	Shell and Tube Condensers	1619
52.3.3	Shell and Tube Reboilers and Vaporizers	1622
52.3.4	Air-Cooled Heat Exchangers	1625
52.3.5	Other Exchangers	1627
52.4	**COMMON OPERATIONAL PROBLEMS**	**1627**
52.4.1	Fouling	1627
52.4.2	Vibration	1628
52.4.3	Flow Maldistribution	1629
52.4.4	Temperature Pinch	1629
52.4.5	Critical Heat Flux in Vaporizers	1630
52.4.6	Instability	1630
52.4.7	Inadequate Venting, Drainage, or Blowdown	1630
52.5	**USE OF COMPUTERS IN THERMAL DESIGN OF PROCESS HEAT EXCHANGERS**	**1631**
52.5.1	Introduction	1631
52.5.2	Incrementation	1631
52.5.3	Main Convergence Loops	1631
52.5.4	Rating, Design, or Simulation	1632
52.5.5	Program Quality and Selection	1633
52.5.6	Determining and Organizing Input Data	1633

52.1 HEAT EXCHANGER TYPES AND CONSTRUCTION

Heat exchangers permit exchange of energy from one fluid to another, usually without permitting physical contact between the fluids. The following configurations are commonly used in the power and process industries.

52.1.1 Shell and Tube Heat Exchangers

Shell and tube heat exchangers normally consist of a bundle of tubes fastened into holes, drilled in metal plates called tubesheets. The tubes may be rolled into grooves in the tubesheet, welded to the tubesheet, or both to ensure against leakage. When possible, U-tubes are used, requiring only one

Mechanical Engineers' Handbook, 2nd ed., Edited by Myer Kutz.
ISBN 0-471-13007-9 © 1998 John Wiley & Sons, Inc.

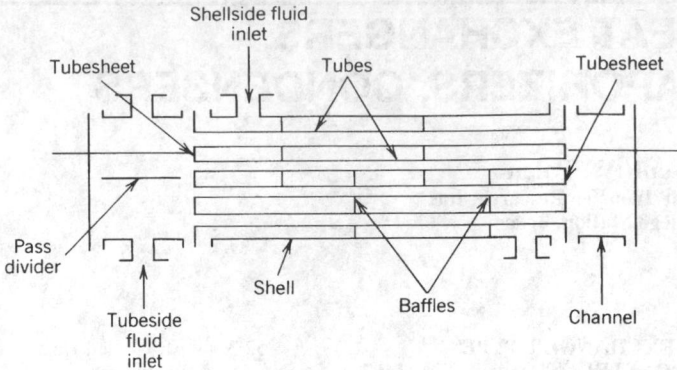

Fig. 52.1   Schematic illustration of shell and tube heat exchanger construction.

tubesheet. The tube bundle is placed inside a large pipe called a shell, see Fig. 52.1. Heat is exchanged between a fluid flowing inside the tubes and a fluid flowing outside the tubes in the shell.

When the tubeside heat-transfer coefficient is as high as three times the shellside heat-transfer coefficient, it may be advantageous to use low integral finned tubes. These tubes can have outside heat-transfer coefficients as high as plain tubes, or even higher, but increase the outside heat-transfer area by a factor of about 2.5–4. For design methods using finned tubes, see Ref. 11 for single-phase heat exchangers and Ref. 14 for condensers. Details of construction practices are described by Saunders.[58]

The Tubular Exchanger Manufacturers Association (TEMA) provides a manual of standards for construction of shell and tube heat exchangers,[1] which contains designations for various types of shell and tube heat exchanger configurations. The most common types are summarized below.

E-Type

The E-type shell and tube heat exchanger, illustrated in Figs. 52.2a and 52.2b, is the workhorse of the process industries, providing economical rugged construction and a wide range of capabilities.

Baffles support the tubes and increase shellside velocity to improve heat transfer. More than one pass is usually provided for tubeside flow to increase the velocity, Fig. 52.2a. However, for some cases, notably vertical thermosiphon vaporizers, a single tubepass is used, as shown in Fig. 52.2b.

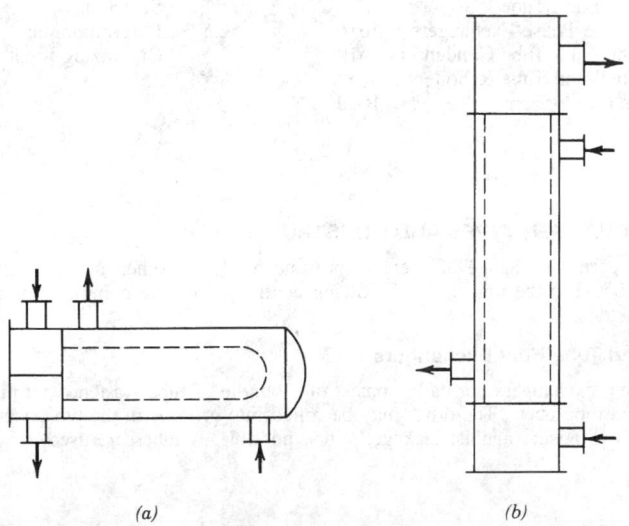

(a)   (b)

Fig. 52.2   TEMA E-type shell: (a) horizontal multitubepass; (b) vertical single tubepass.

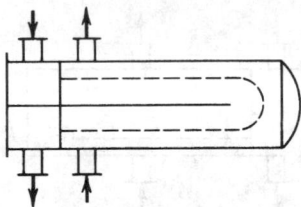

Fig. 52.3 TEMA F-type shell.

The E-type shell is usually the first choice of shell types because of lowest cost, but sometimes requires more than the allowable pressure drop, or produces a temperature "pinch" (see Section 52.4.4), so other, more complicated types are used.

F-Type Shell

If the exit temperature of the cold fluid is greater than the exit temperature of the hot fluid, a temperature cross is said to exist. A slight temperature cross can be tolerated in a multitubepass E-type shell (see below), but if the cross is appreciable, either units in series or complete countercurrent flow is required. A solution sometimes used is the F-type or two-pass shell, as shown in Fig. 52.3.

The F-type shell has a number of potential disadvantages, such as thermal and fluid leakage around the longitudinal baffle and high pressure drop, but it can be effective in some cases if well designed.

J-Type

When an E-type shell cannot be used because of high pressure drop, a J-type or divided flow exchanger, shown in Fig. 52.4, is considered. Since the flow is divided and the flow length is also cut in half, the shellside pressure drop is only about one-eighth to one-fifth that of an E-type shell of the same dimensions.

X-Type

When a J-type shell would still produce too high a pressure drop, an X-type shell, shown in Fig. 52.5, may be used. This type is especially applicable for vacuum condensers, and can be equipped with integral finned tubes to counteract the effect of low shellside velocity on heat transfer. It is usually necessary to provide a flow distribution device under the inlet nozzle.

G-Type

This shell type, shown in Fig. 52.6, is sometimes used for horizontal thermosiphon shellside vaporizers. The horizontal baffle is used especially for boiling range mixtures and provides better flow distribution than would be the case with the X-type shell. The G-type shell also permits a larger temperature cross than the E-type shell with about the same pressure drop.

H-Type

If a G-type is being considered but pressure drop would be too high, an H-type may be used. This configuration is essentially just two G-types in parallel, as shown in Fig. 52.7.

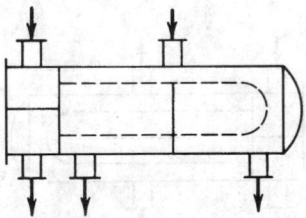

Fig. 52.4 TEMA J-type shell.

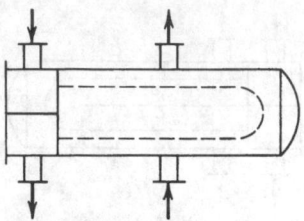

Fig. 52.5   TEMA X-type shell.

K-Type

This type is used exclusively for kettle reboilers and vaporizers, and is characterized by the oversized shell intended to separate vapor and liquid phases, Fig. 52.8. Shell-sizing relationships are given in Ref. 25. Usually, the shell diameter is about 1.6–2.0 times the bundle diameter. Design should consider amount of acceptable entrainment, height required for flow over the weir, and minimum clearance in case of foaming.

Baffle Types

Baffles are used to increase velocity of the fluid flowing outside the tubes ("shellside" fluid) and to support the tubes. Higher velocities have the advantage of increasing heat transfer and decreasing fouling (material deposit on the tubes), but have the disadvantage of increasing pressure drop (more energy consumption per unit of fluid flow). The amount of pressure drop on the shellside is a function of baffle spacing, baffle cut, and baffle type.

Baffle types commonly used are shown in Fig. 52.9, with pressure drop decreasing from Fig. 52.9a to Fig. 52.9c.

Baffle spacing is increased when it is necessary to decrease pressure drop. A limit must be imposed to prevent tube sagging or flow-induced tube vibration. Recommendations for maximum baffle spacing are given in Ref. 1. Tube vibration is discussed in more detail in Section 52.4.2. When the maximum spacing still produces too much pressure drop, a baffle type is considered that produces less cross flow and more longitudinal flow, for example, double segmental instead of segmental. Minimum pressure drop is obtained if baffles are replaced by rod-type tube supports.[52]

52.1.2 Plate-Type Heat Exchangers

Composed of a series of corrugated or embossed plates clamped between a stationary and a movable support plate, these exchangers were originally used in the food-processing industry. They have the advantages of low fouling rates, easy cleaning, and generally high heat-transfer coefficients, and are becoming more frequently used in the chemical process and power industries. They have the disadvantage that available gaskets for the plates are not compatible with all combinations of pressure, temperature, and chemical composition. Suitability for specific applications must be checked. The maximum operating pressure is usually considered to be about 1.5 MPa (220 psia).[3] However, welded plate versions are now available for much higher pressures. A typical plate heat exchanger is shown in Fig. 52.10.

52.1.3 Spiral Plate Heat Exchangers

These exchangers are also becoming more widely used, despite limitations on maximum size and maximum operating pressure. They are made by wrapping two parallel metal plates, separated by

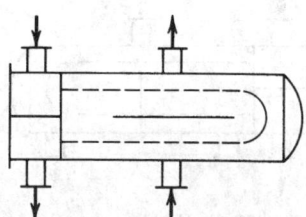

Fig. 52.6   TEMA G-type shell.

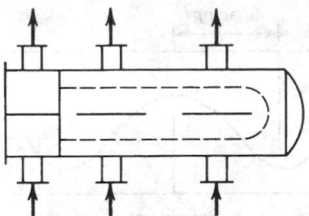

Fig. 52.7   TEMA H-type shell.

spacers, into a spiral to form two concentric spiral passages. A schematic example is shown in Fig. 52.11.

Spiral plate heat exchangers can provide completely countercurrent flow, permitting temperature crosses and close approaches, while maintaining high velocity and high heat-transfer coefficients. Since all flow for each fluid is in a single channel, the channel tends to be flushed of particles by the flow, and the exchanger can handle sludges and slurries more effectively than can shell and tube heat exchangers. The most common uses are for difficult-to-handle fluids with no phase change. However, the low-pressure-drop characteristics are beginning to promote some use in two-phase flow as condensers and reboilers. For this purpose the two-phase fluid normally flows axially in a single pass rather than spirally.

52.1.4 Air-Cooled Heat Exchangers

It is sometimes economical to condense or cool hot streams inside tubes by blowing air across the tubes rather than using water or other cooling liquid. They usually consist of a horizontal bank of finned tubes with a fan at the bottom (forced draft) or top (induced draft) of the bank, as illustrated schematically in Fig. 52.12.

Tubes in air-cooled heat exchangers (Fig. 52.12) are often 1 in. (25.4 mm) in outside diameter with ⅝ in. (15.9 mm) high annular fins, 0.4–0.5 mm thick. The fins are usually aluminum and may be attached in a number of ways, ranging from tension wrapped to integrally extruded (requiring a steel or alloy insert), depending on the severity of service. Tension wrapped fins have an upper temperature limit (~300°F) above which the fin may no longer be in good contact with the tube, greatly decreasing the heat-transfer effectiveness. Various types of fins and attachments are illustrated in Fig. 52.13.

A more detailed description of air-cooled heat exchanger geometries is given Refs. 2 and 3.

52.1.5 Compact Heat Exchangers

The term compact heat exchanger normally refers to one of the many types of plate fin exchangers used extensively in the aerospace and cryogenics industries. The fluids flow alternately between parallel plates separated by corrugated metal strips that act as fins and that may be perforated or interrupted to increase turbulence. Although relatively expensive to construct, these units pack a very large amount of heat-transfer surface into a small volume, and are therefore used when exchanger volume or weight must be minimized. A detailed description with design methods is given in Ref. 4.

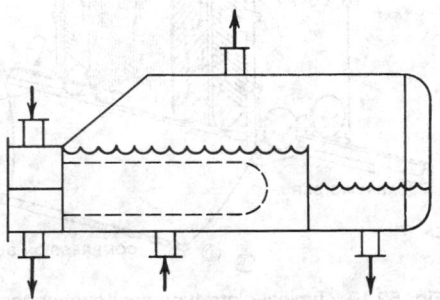

Fig. 52.8   TEMA K-type shell.

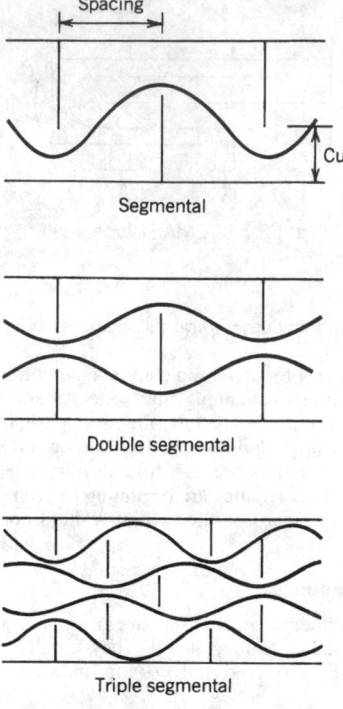

Fig. 52.9 Baffle types.

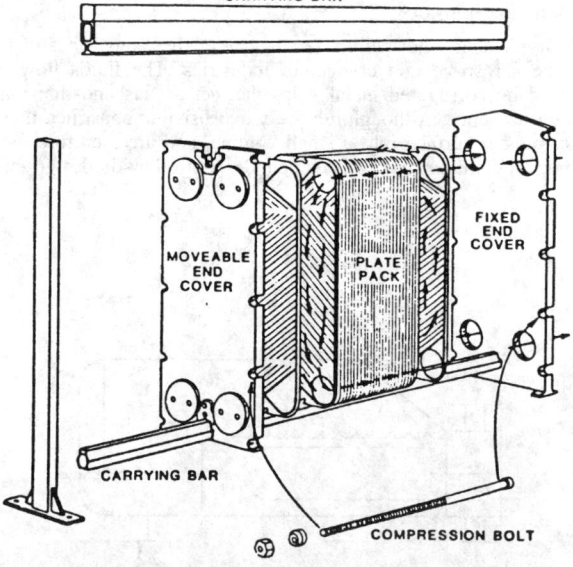

Fig. 52.10 Typical plate-type heat exchanger.

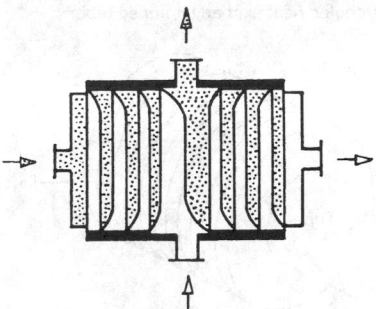

Fig. 52.11 Spiral plate heat exchanger.

52.1.6 Boiler Feedwater Heaters

Exchangers to preheat feedwater to power plant boilers are essentially of the shell and tube type but have some special features, as described in Ref. 5. The steam that is used for preheating the feedwater enters the exchanger superheated, is condensed, and leaves as subcooled condensate. More effective heat transfer is achieved by providing three zones on the shellside: desuperheating, condensing, and subcooling. A description of the design requirements of this type of exchanger is given in Ref. 5.

52.1.7 Recuperators and Regenerators

These heat exchangers are used typically to conserve heat from furnace off-gas by exchanging it against the inlet air to the furnace. A recuperator does this in the same manner as any other heat exchanger except the construction may be different to comply with requirements for low pressure drop and handling of the high-temperature, often dirty, off-gas stream.

The regenerator is a transient batch-type exchanger in which packed beds are alternately switched from the hot stream to the cold stream. A description of the operating characteristics and design of recuperators and regenerators is given in Refs. 6 and 59.

52.2 ESTIMATION OF SIZE AND COST

In determining the overall cost of a proposed process plant or power plant, the cost of heat exchangers is of significant importance. Since cost is roughly proportional to the amount of heat-transfer surface required, some method of obtaining an estimate of performance is necessary, which can then be translated into required surface. The term "surface" refers to the total area across which the heat is transferred. For example, with shell and tube heat exchangers "surface" is the tube outside circumference times the tube length times the total number of tubes. Well-known basic equations taken from Newton's law of cooling relate the required surface to the available temperature difference and the required heat duty.

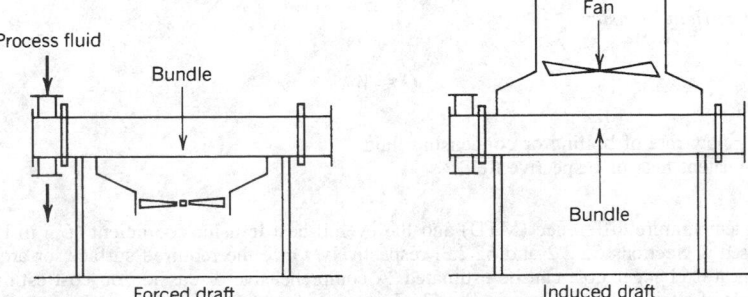

Fig. 52.12 Air-cooled heat exchangers.

Air-cooled heat exchanger finned tube

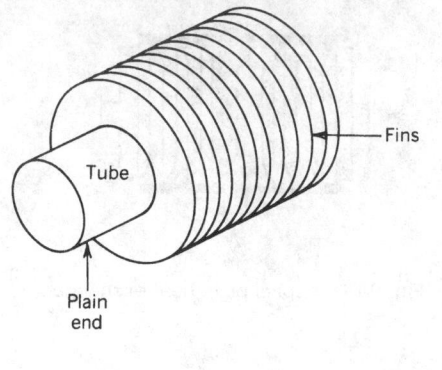

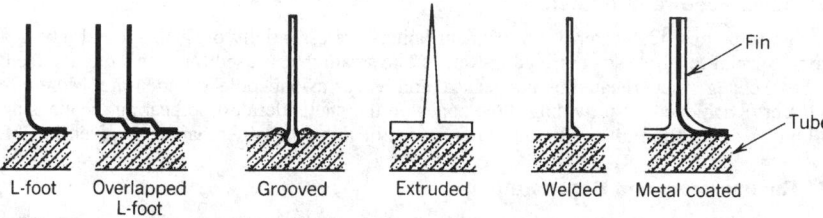

Fig. 52.13 Typical finned tube and attachments.

52.2.1 Basic Equations for Required Surface

The following well-known equation is used (equation terms are defined in the Nomenclature):

$$A_o = \frac{Q}{U_o \times \text{MTD}} \tag{52.1}$$

The required duty (Q) is related to the energy change of the fluids:

(a) *Sensible Heat Transfer*

$$Q = W_1 C_{p1}(T_2 - T_1) \tag{52.2a}$$

$$= W_2 C_{p2}(t_1 - t_2) \tag{52.2b}$$

(b) *Latent Heat Transfer*

$$Q = W\lambda \tag{52.3}$$

where W = flow rate of boiling or condensing fluid
λ = latent heat of respective fluid

The mean temperature difference (MTD) and the overall heat transfer coefficient (U_o) in Eq. (52.1) are discussed in Sections 52.2.2 and 52.2.3, respectively. Once the required surface, or area, (A_o) is obtained, heat exchanger cost can be estimated. A comprehensive discussion on cost estimation for several types of exchangers is given in Ref. 7. Cost charts for small- to medium-sized shell and tube exchangers, developed in 1982, are given in Ref. 8.

52.2.2 Mean Temperature Difference

The mean temperature difference (MTD) in Eq. (52.1) is given by the equation

$$\text{MTD} = \frac{F(T_A - T_B)}{\ln(T_A/T_B)} \tag{52.4}$$

where

$$T_A = T_1 - t_2 \tag{52.5}$$

$$T_B = T_2 - t_1 \tag{52.6}$$

The temperatures (T_1, T_2, t_1, t_2) are illustrated for the base case of countercurrent flow in Fig. 52.14.

The factor F in Eq. (52.4) is the multitubepass correction factor. It accounts for the fact that heat exchangers with more than one tubepass can have some portions in concurrent flow or cross flow, which produce less effective heat transfer than countercurrent flow. Therefore, the factor F is less than 1.0 for multitubepass exchangers, except for the special case of isothermal boiling or condensing streams for which F is always 1.0. Charts for calculating F are available in most heat-transfer textbooks. A comprehensive compilation for various types of exchangers is given by Taborek.[9]

In a properly designed heat exchanger, it is unusual for F to be less than 0.7, and if there is no temperature cross $(T_2 > t_2)$, F will be 0.8 or greater. As a first approximation for preliminary sizing and cost estimation, F may be taken as 0.85 for multitubepass exchangers with temperature change of both streams and 1.0 for other cases.

52.2.3 Overall Heat-Transfer Coefficient

The factor (U_o) in Eq. (52.1) is the overall heat-transfer coefficient. It may be calculated by procedures described in Section 52.3, and is the reciprocal of the sum of all heat-transfer resistances, as shown in the equation

$$U_o = 1/(R_{h_o} + R_{f_o} + R_w + R_{h_i} + R_{f_i}) \tag{52.7}$$

where

$$R_{h_o} = 1/h_o \tag{52.8}$$

$$R_{h_i} = (A_o/A_i h_i) \tag{52.9}$$

$$R_w = \frac{A_o x_w}{A_m k_w} \tag{52.10}$$

Calculation of the heat-transfer coefficients h_o and h_i can be time consuming, since they depend on the fluid velocities, which, in turn, depend on the exchanger geometry. This is usually done now by computer programs that guess correct exchanger size, calculate heat-transfer coefficients, check size, adjust, and reiterate until satisfactory agreement between guessed and calculated size is obtained.

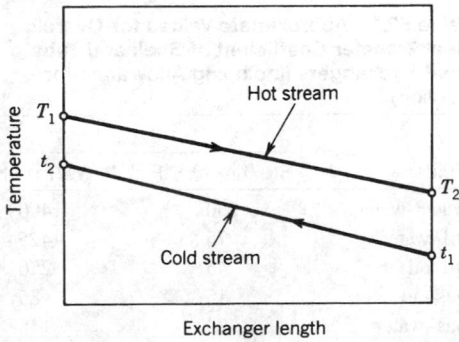

Fig. 52.14 Temperature profiles illustrated for countercurrent flow.

For first estimates by hand before size is known, values of h_o and h_i, as well as values of the fouling resistances, R_{f_o} and R_{f_i}, are recommended by Bell for shell and tube heat exchangers.[10]

Very rough, first approximation values for the overall heat-transfer coefficient are given in Table 52.1.

52.2.4 Pressure Drop

In addition to calculation of the heat-transfer surface required, it is usually necessary to consider the pressure drop consumed by the heat exchanger, since this enters into the overall cost picture. Pressure drop is roughly related to the individual heat-transfer coefficients by an equation of the form,

$$\Delta P = Ch^m + EX \tag{52.11}$$

where ΔP = shellside or tubeside pressure drop
$\quad\quad h$ = heat-transfer coefficient
$\quad\quad C$ = coefficient depending on geometry
$\quad\quad m$ = exponent depending on geometry—always greater than 1.0, and usually about 3.0
$\quad\quad EX$ = extra pressure drop from inlet, exit, and pass turnaround momentum losses

See Section 52.3 for actual pressure drop calculations.

Pressure drop is sensitive to the type of exchanger selected. In the final design it is attempted, where possible, to define the exchanger geometry so as to use all available pressure drop and thus maximize the heat-transfer coefficient. This procedure is subject to some constraints, however, as follows. The product of density times velocity squared ρv^2 is limited to minimize the possibility of erosion or tube vibration. A limit often used is $\rho v^2 < 4000$ lbm/ft · sec². This results in a velocity for liquids in the range of 7–10 ft/sec. For flow entering the shellside of an exchanger and impacting the tubes, an impingement plate is recommended to prevent erosion if $\rho v^2 > 1500$. Other useful design recommendations may be found in Ref. 1.

For condensing vapors, pressure drop should be limited to a fraction of the operating pressure for cases with close temperature approach to prevent severe decrease of the MTD owing to lowered equilibrium condensing temperature. As a safe "rule of thumb," the pressure drop for condensing is limited to about 10% of the operating pressure. For other cases, "reasonable" design pressure drops for heat exchangers roughly range from about 5 psi for gases and boiling liquids to as high as 20 psi for pumped nonboiling liquids.

52.3 RATING METHODS

After the size and basic geometry of a heat exchanger has been proposed, the individual heat-transfer coefficients h_o and h_i may be calculated based on actual velocities, and the required surface may be checked, based on these updated values. The pressure drops are also checked at this stage. Any inadequacies are adjusted and the exchanger is rechecked. This process is known as "rating." Different rating methods are used depending on exchanger geometry and process type, as covered in the following sections.

52.3.1 Shell and Tube Single-Phase Exchangers

Before the individual heat-transfer coefficients can be calculated, the heat exchanger tube geometry, shell diameter, shell type, baffle type, baffle spacing, baffle cut, and number of tubepasses must be

Table 52.1 Approximate Values for Overall Heat Transfer Coefficient of Shell and Tube Heat Exchangers (Including Allowance for Fouling)

Fluids	U_o	
	Btu/hr · ft² · °F	W/m² · K
Water–water	250	1400
Oil–water	75	425
Oil–oil	45	250
Gas–oil	15	85
Gas–water	20	115
Gas–gas	10	60

decided. As stated above, lacking other insight, the simplest exchanger—E-type with segmental baffles—is tried first.

Tube Length and Shell Diameter

For shell and tube exchangers the tube length is normally about 5–8 times the shell diameter. Tube lengths are usually 8–20 ft long in increments of 2 ft. However, very large size exchangers with tube lengths up to 40 ft are more frequently used as economics dictate smaller MTD and larger plants. A reasonable trial tube length is chosen and the number of tubes (NT) required for surface A_o, Section 52.2, is calculated as follows:

$$NT = \frac{A_o}{a_o L} \tag{52.12}$$

where a_o = the surface/unit length of tube.
    For plain tubes (as opposed to finned tubes),

$$a_o = \pi D_o \tag{52.13}$$

where D_o = the tube outside diameter
       L = the tube length

The tube bundle diameter (D_b) can be determined from the number of tubes, but also depends on the number of tubepasses, tube layout, and bundle construction. Tube count tables providing this information are available from several sources. Accurate estimation equations are given by Taborek.[11] A simple basic equation that gives reasonable first approximation results for typical geometries is the following:

$$D_b = P_t \left(\frac{NT}{\pi/4}\right)^{0.5} \tag{52.14}$$

where P_t = tube pitch (spacing between tube diameters). Normally, P_t/D_o = 1.25, 1.33, or 1.5.
    The shell diameter D_s is larger than the bundle diameter D_b by the amount of clearance necessary for the type of bundle construction. Roughly, this clearance ranges from about 0.5 in. for U-tube or fixed tubesheet construction to 3–4 in. for pull-through floating heads, depending on the design pressure and bundle diameter. (For large clearances, sealing strips are used to prevent flow bypassing the bundles.) After the bundle diameter is calculated, the ratio of length to diameter is checked to see if it is in an acceptable range, and the length is adjusted if necessary.

Baffle Spacing and Cut

Baffle spacing L_{bc} and cut B_c (see Fig. 52.9) cannot be decided exactly until pressure drop is evaluated. However, a reasonable first guess ratio of baffle spacing to shell diameter (L_{bc}/D_s) is about 0.45. The baffle cut (B_c, a percentage of D_s) required to give good shellside distribution may be estimated by the following equation:

$$B_c = 16.25 + 18.75 \left(\frac{L_{bc}}{D_s}\right) \tag{52.15}$$

For more detail, see the recommendations of Taborek.[11]

Cross-Sectional Flow Areas and Flow Velocities

The cross-sectional flow areas for tubeside flow S_t and for shellside flow S_s are calculated as follows:

$$S_t = \left(\frac{\pi}{4} D_i^2\right)\left(\frac{NT}{NP}\right) \tag{52.16}$$

$$S_s = 0.785(D_b)(L_{bc})(P_t - D_o)/P_t \tag{52.17}$$

where L_{bc} = baffle spacing.
    Equation (52.17) is approximate in that it neglects pass partition gaps in the tube field, it approximates the bundle average chord, and it assumes an equilateral triangular layout. For more accurate equations see Ref. 11.
    The tubeside velocity V_t and the shellside velocity V_s are calculated as follows:

$$V_t = \frac{W_t}{S_t \, \rho_t} \tag{52.18}$$

$$V_s = \frac{W_s}{S_s \, \rho_s} \tag{52.19}$$

Heat-Transfer Coefficients

The individual heat-transfer coefficients, h_o and h_i, in Eq. (52.1) can be calculated with reasonably good accuracy (± 20–30%) by semiempirical equations found in several design-oriented text-books.[11,12] Simplified approximate equations are the following:

(a) *Tubeside Flow*

$$\text{Re} = \frac{D_o V_t \, \rho_t}{\mu_t} \tag{52.20}$$

where μ_t = tubeside fluid viscosity.
  If Re < 2000, laminar flow,

$$h_i = 1.86 \left(\frac{k_f}{D_i}\right) \left(\text{Re Pr} \frac{D_i}{L}\right)^{0.33} \left(\frac{\mu_f}{\mu_w}\right)^{0.14} \tag{52.21}$$

If Re > 10,000, turbulent flow,

$$h_i = 0.024 \left(\frac{k_f}{D_i}\right) \text{Re}^{0.8} \, \text{Pr}^{0.4} \left(\frac{\mu_f}{\mu_w}\right)^{0.14} \tag{52.22}$$

If 2000 < Re < 10,000, prorate linearly.

(b) *Shellside Flow*

$$\text{Re} = \frac{D_o V_s \, \rho_s}{\mu_s} \tag{52.23}$$

If Re < 500, see Refs. 11 and 12.
If Re > 500,

$$h_o = 0.38 \, C_b^{0.6} \left(\frac{k_f}{D_o}\right) \text{Re}^{0.6} \, \text{Pr}^{0.33} \left(\frac{\mu_f}{\mu_w}\right)^{0.14} \tag{52.24}$$

The term *Pr* is the Prandtl number and is calculated as $C_p \, \mu/k$.
  The constant (C_b) in Eq. (52.24) depends on the amount of bypassing or leakage around the tube bundle.[13] As a first approximation, the values in Table 52.2 may be used.

Pressure Drop

Pressure drop is much more sensitive to exchanger geometry, and, therefore, more difficult to accurately estimate than heat transfer, especially for the shellside. The so-called Bell–Delaware method[11] is considered the most accurate method in open literature, which can be calculated by hand. The following very simplified equations are provided for a rough idea of the range of pressure drop, in order to minimize preliminary specification of unrealistic geometries.

(a) *Tubeside (contains about 30% excess for nozzles)*

Table 52.2 Approximate Bypass Coefficient for Heat Transfer, C_b

Bundle Type	C_b
Fixed tubesheet or U-tube	0.70
Split ring floating head, seal strips	0.65
Pull-through floating head, seal strips	0.55

$$\Delta P_t = \left[\frac{0.025(L)(NP)}{D_i} + 2(NP - 1) \right] \frac{\rho_t V_t^2}{g_c} \left(\frac{\mu_w}{\mu_f} \right)^{0.14} \tag{52.25}$$

where NP = number of tubepasses.

(b) Shellside (contains about 30% excess for nozzles)

$$\Delta P_s = \frac{0.24(L)(D_b)(\rho_s)(C_b V_s)^2}{g_c L_{bc} P_t} \left(\frac{\mu_w}{\mu_f} \right)^{0.14} \tag{52.26}$$

where g_c = gravitational constant (4.17×10^8 for velocity in ft/hr and density in lb/ft^3).

52.3.2 Shell and Tube Condensers

The condensing vapor can be on either the shellside or tubeside depending on process constraints. The "cold" fluid is often cooling tower water, but can also be another process fluid, which is sensibly heated or boiled. In this section, the condensing-side heat-transfer coefficient and pressure drop are discussed. Single-phase coolants are handled, as explained in the last section. Boiling fluids will be discussed in the next section.

Selection of Condenser Type

The first task in designing a condenser, before rating can proceed, is to select the condenser configuration. Mueller[14] presents detailed charts for selection based on the criteria of system pressure, pressure drop, temperature, fouling tendency of the coolant, fouling tendency of the vapor, corrosiveness of the vapor, and freezing potential of the vapor. Table 52.3 is an abstract of the recommendations of Mueller.

The suggestions in Table 52.3 may, of course, be ambiguous in case of more than one important criterion, for example, corrosive vapor together with a fouling coolant. In these cases, the most critical constraint must be respected, as determined by experience and engineering judgment. Corrosive vapors are usually put on the tubeside, and chemical cleaning used for the shellside coolant, if necessary. Since most process vapors are relatively clean (not always the case!), the coolant is usually the dirtier of the two fluids and the tendency is to put it on the tubeside for easier cleaning. Therefore, the most common shell and tube condenser is the shellside condenser using TEMA types E, J, or X, depending on allowable pressure drop; see Section 52.1. An F-type shell is sometimes specified if there is a large condensing range and a temperature cross (see below), but, owing to problems with the F-type, E-type units in series are often preferred in this case.

In addition to the above condenser types the vertical E-type tubeside condenser is sometimes used in a "reflux" configuration with vapor flowing up and condensate flowing back down inside the tubes. This configuration may be useful in special cases, such as when it is required to strip out condensable components from a vent gas that is to be rejected to the atmosphere. The disadvantage of this type of condenser is that the vapor velocity must be very low to prevent carryover of the condensate (flooding), so the heat-transfer coefficient is correspondingly low, and the condenser rather inefficient. Methods used to predict the limiting vapor velocity are given in Ref. 14.

Temperature Profiles

For a condensing pure component, if the pressure drop is less than about 10% of the operating pressure, the condensing temperature is essentially constant and the LMTD applied ($F = 1.0$) for the condensing section. If there are desuperheating and subcooling sections,[5] the MTD and surface for these sections must be calculated separately. For a condensing mixture, with or without noncon-

Table 52.3 Condenser Selection Chart

Process Condition	Suggested Condenser Type[a]
Potential coolant fouling	HS/E, J, X
High condensing pressure	VT/E
Low condensing pressure drop	HS/J, X
Corrosive or very-high- temperature vapors	VT/E
Potential condensate freezing	HS/E
Boiling coolant	VS/E or HT/K, G, H

[a]V, vertical; H, horizontal; S, shellside condensation; T, tubeside condensation; /E, J, H, K, X, TEMA shell styles.

densables, the temperature profile of the condensing fluid with respect to fraction condensed should be calculated according to vapor–liquid equilibrium (VLE) relationships.[15] A number of computer programs are available to solve VLE relationships; a version suitable for programmable calculator is given in Ref. 16.

Calculations of the condensing temperature profile may be performed either integrally, which assumes vapor and liquid phases are well mixed throughout the condenser, or differentially, which assumes separation of the liquid phase from the vapor phase. In most actual condensers the phases are mixed near the entrance where the vapor velocity is high and separated near the exit where the vapor velocity is lower. The "differential" curve produces a lower MTD than the "integral" curve and is safer to use where separation is expected.

For most accuracy, condensers are rated incrementally by stepwise procedures such as those explained by Mueller.[14] These calculations are usually performed by computers.[17] As a first approximation, to get an initial size, a straight-line temperature profile is often assumed for the condensing section (not including desuperheating or subcooling sections!). As illustrated in Fig. 52.15, the true condensing curve is usually more like curve I, which gives a larger MTD than the straight line, curve II, making the straight-line approximation conservative. However, a curve such as curve III is certainly possible, especially with immiscible condensates, for which the VLE should always be calculated. For the straight-line approximation, the condensing heat-transfer coefficient is calculated at average conditions, as shown below.

Heat-Transfer Coefficients, Pure Components

For condensers, it is particularly important to be able to estimate the two-phase flow regime in order to predict the heat-transfer coefficient accurately. This is because completely different types of correlations are required for the two major flow regimes.

Shear Controlled Flow.   The vapor shear force on the condensate is much greater than the gravity force. This condition can be estimated, according to Ref. 18, as,

$$J_g > 1.5 \tag{52.27}$$

where

$$J_g = \left[\frac{(Gy)^2}{gD_j \rho_v (\rho_l - \rho_v)} \right]^{0.5} \tag{52.28}$$

For shear-controlled flow, the condensate film heat-transfer coefficient (h_{cf}) is a function of the convective heat-transfer coefficient for liquid flowing alone and the two-phase pressure drop.[18]

$$h_{cf} = h_l (\phi_l^2)^{0.45} \tag{52.29}$$

$$h_l = h_i (1 - y)^{0.8} \tag{52.30}$$

or

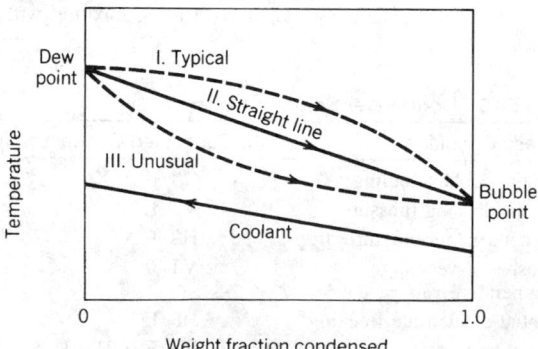

Fig. 52.15 Condensation profiles illustrated.

$$h_l = h_o(1 - y)^{0.6} \tag{52.31}$$

$$\phi_l^2 = 1 + \frac{C}{X_{tt}} + \frac{1}{X_{tt}^2} \tag{52.32}$$

$$C = 20 \text{ (tubeside flow)}, \qquad C = 9 \text{ (shellside flow)}$$

$$X_{tt} = \left[\frac{1-y}{y}\right]^{0.9} \left[\frac{\rho_v}{\rho_t}\right]^{0.5} \left[\frac{\mu_l}{\mu_v}\right]^{0.1} \tag{52.33}$$

$$\mu_l = \text{liquid viscosity}, \qquad \mu_v = \text{vapor viscosity}$$

Gravity Controlled Flow. The vapor shear force on the condensate is small compared to the gravity force, so condensate drains by gravity. This condition can be estimated, according to Ref. 18, when $J_g < 0.5$. Under gravity-controlled conditions, the condensate film heat-transfer coefficient is calculated as follows:

$$h_{cf} = F_g h_N \tag{52.34}$$

The term h_N is the heat-transfer coefficient from the well-known Nusselt derivation, given in Ref. 14 as

Horizontal Tubes

$$h_N = 0.725 \left[\frac{k_l^3 \rho_l (\rho_l - \rho_v) g \lambda}{\mu_l (T_s - T_w) D}\right]^{0.25} \tag{52.35}$$

where λ = latent heat.

Vertical Tubes

$$h_N = 1.1 k_l \left[\frac{\rho_l (\rho_l - \rho_v) g}{\mu_l^2 \, \text{Re}_c}\right]^{0.33} \tag{52.36}$$

$$\text{Re}_c = \frac{4 W_c}{\pi D \mu_l} \tag{52.37}$$

The term F_g in Eq. (52.34) is a correction for condensate loading, and depends on the exchanger geometry.[14]

On horizontal X-type tube bundles

$$F_g = N_{rv}^{-1/6} \tag{52.38}$$

(Ref. 12), where N_{rv} = number of tubes in a vertical row.
On baffled tube bundles (owing to turbulence)

$$F_g = 1.0 \qquad \text{(frequent practice)} \tag{52.39}$$

In horizontal tubes

$$F_g = \left[\frac{1}{1 + (1/(y-1))(\rho_v/\rho_l)^{0.667}}\right]^{0.75} \qquad \text{(from Ref. 14)} \tag{52.40}$$

or

$$F_g = 0.8 \qquad \text{(from Ref. 18)} \tag{52.41}$$

Inside or outside vertical tubes

$$F_g = 0.73 \, \text{Re}_c^{0.11} \qquad \text{(rippled film region)} \tag{52.42}$$

or

$$F_g = 0.021 \, \text{Re}_c^{0.58} \, \text{Pr}^{0.33} \qquad \text{(turbulent film region)} \qquad (52.43)$$

Use higher value of Eq. (52.42) or (52.43).

For quick hand calculations, the gravity-controlled flow equations may be used for h_{cf}, and will usually give conservative results.

Correction for Mixture Effects

The above heat-transfer coefficients apply only to the condensate film. For mixtures with a significant difference between the dew-point and bubble-point temperatures (condensing range), the vapor-phase heat-transfer coefficient must also be considered as follows:

$$h_c = \frac{1}{(1/h_{cf} + 1/h_v)} \qquad (52.44)$$

The vapor-phase heat-transfer rate depends on mass diffusion rates in the vapor. The well-known Colburn–Hougen method and other more recent approaches are summarized by Butterworth.[19] Methods for mixtures forming immiscible condensates are discussed in Ref. 20.

Diffusion-type methods require physical properties not usually available to the designer except for simple systems. Therefore, the vapor-phase heat-transfer coefficient is often estimated in practice by a "resistance-proration"-type method such as the Bell–Ghaly method.[21] In these methods the vapor-phase resistance is prorated with respect to the relative amount of duty required for sensible cooling of the vapor, resulting in the following expression:

$$h_v = (q_t/q_{sv})h_{sv} \qquad (52.44a)$$

For more detail in application of the resistance proration method for mixtures, see Refs. 14 or 21.

Pressure Drop

For the condensing vapor, pressure drop is composed of three components—friction, momentum, and static head—as covered in Ref. 14. An approximate estimate on the conservative side can be obtained in terms of the friction component, using the Martinelli separated flow approach:

$$\Delta P_f = \Delta P_l \, \phi_l^2 \qquad (52.45)$$

where ΔP_f = two-phase friction pressure drop
ΔP_l = friction loss for liquid phase alone

The Martinelli factor ϕ_l^2 may be calculated as shown in Eq. (52.32). Alternative methods for shellside pressure drop are presented by Diehl[22] and by Grant and Chisholm.[23] These methods were reviewed by Ishihara[24] and found reasonably representative of the available data. However, Eq. (52.32), also evaluated in Ref. 24 for shellside flow, should give about equivalent results.

52.3.3 Shell and Tube Reboilers and Vaporizers

Heat exchangers are used to boil liquids in both the process and power industries. In the process industry they are often used to supply vapors to distillation columns and are called reboilers. The same types of exchangers are used in many applications in the power industry, for example, to generate vapors for turbines. For simplicity these exchangers will all be called "reboilers" in this section. Often the heating medium is steam, but it can also be any hot process fluid from which heat is to be recovered, ranging from chemical reactor effluent to geothermal hot brine.

Selection of Reboiler Type

A number of different shell and tube configurations are in common use, and the first step in design of a reboiler is to select a configuration appropriate to the required job. Basically, the type of reboiler should depend on expected amount of fouling, operating pressure, mean temperature difference (MTD), and difference between temperatures of the bubble point and the dew point (boiling range).

The main considerations are as follows: (1) fouling fluids should be boiled on the tubeside at high velocity; (2) boiling either under deep vacuum or near the critical pressure should be in a kettle to minimize hydrodynamic problems unless means are available for very careful design; (3) at low MTD, especially at low pressure, the amount of static head must be minimized; (4) for wide boiling range mixtures, it is important to maximize both the amount of mixing and the amount of counter-current flow.

These and other criteria are discussed in more detail in Ref. 25, and summarized in a selection guide, which is abstracted in Table 52.4.

Table 52.4 Reboiler Selection Guide

Process Conditions	Suggested Reboiler Type[a]
Moderate pressure, MTD, and fouling	VT/E
Very high pressure, near critical	HS/K or (F)HT/E
Deep vacuum	HS/K
High or very low MTD	HS/K, G, H
Moderate to heavy fouling	VT/E
Very heavy fouling	(F)HT/E
Wide boiling range mixture	HS/G or /H
Very wide boiling range, viscous liquid	(F)HT/E

[a] V, vertical; H, horizontal; S, shellside boiling; T, tubeside boiling; (F), forced flow, else natural convection; /E, G, H, K, TEMA shell styles.

In addition to the above types covered in Ref. 25, falling film evaporators[26] may be preferred in cases with very low MTD, viscous liquids, or very deep vacuum for which even a kettle provides too much static head.

Temperature Profiles

For pure components or narrow boiling mixtures, the boiling temperature is nearly constant and the LMTD applies with $F = 1.0$. Temperature profiles for boiling range mixtures are very complicated, and although the LMTD is often used, it is not a recommended practice, and may result in under-designed reboilers unless compensated by excessive design fouling factors. Contrary to the case for condensers, using a straight-line profile approximation always tends to give too high MTD for reboilers, and can be tolerated only if the temperature rise across the reboiler is kept low through a high circulation rate.

Table 52.5 gives suggested procedures to determine an approximate MTD to use for initial size estimation, based on temperature profiles illustrated in Fig. 52.16. It should be noted that the MTD values in Table 52.5 are intended to be on the safe side and that excessive fouling factors are not necessary as additional safety factors if these values are used. See Section 52.4.1 for suggested fouling factor ranges.

Heat-Transfer Coefficients

The two basic types of boiling mechanisms that must be taken into account in determining boiling heat-transfer coefficients are nucleate boiling and convective boiling. A detailed description of both types is given by Collier.[27] For all reboilers, the nucleate and convective boiling contributions are additive, as follows:

$$h_b = \alpha h_{nb} + h_{cb} \tag{52.46a}$$

or

$$h_b = [h_{nb}^2 + h_{cb}^2]^{0.5} \tag{52.46b}$$

Equation (52.46a) includes a nucleate boiling suppression factor, α, that originally was correlated by Chen.[60]

Table 52.5 Reboiler MTD Estimation

Reboiler Type[a]	T_A	T_B	MTD
HS/K	$T_1 - t_2$	$T_2 - t_2$	Eq. (52.7), $F = 1$
HS/X, G, H	$T_1 - t_1$	$T_2 - t_2$	Eq. (52.7), $F = 0.9$
VT/E	$T_1 - t_2$	$T_2 - t_1$	Eq. (52.7), $F = 1$
(F)HT/E or (F)HS/E	$T_1 - t_2$	$T_2 - t_1$	Eq. (52.7), $F = 0.9$
All types	Isothermal	$T_A = T_B$	T_A

[a] V, vertical; H, horizontal; S, shellside boiling; T, tubeside boiling; (F), forced flow, else natural convection; /E, G, H, K, TEMA shell styles.

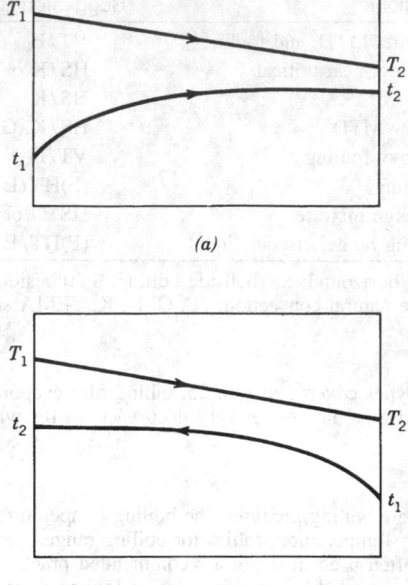

Fig. 52.16 Reboiler temperature profiles illustrated: (a) use for kettle and horizontal thermosiphon; (b) use for tubeside boiling vertical thermosiphon.

Equation (52.46b) is a simple asymptotic proration that was found to work well by Steiner and
Taborek.[61]

The convective boiling coefficient h_{cb} depends on the liquid-phase convective heat-transfer coefficient h_l, according to the same relationship, Eq. (52.29), given for shear-controlled condensation.
For all reboiler types, except forced flow, the flow velocities required to calculate h_l depend on
complex pressure balances for which computers are necessary for practical solution. Therefore, the
convective component is sometimes approximated as a multiplier to the nucleate boiling component
for quick estimations,[25] as in the following equation:

$$h_b = h_{nb} F_b \qquad (52.47)$$

$$F_b = \frac{h_{nb} + h_{cb}}{h_{nb}} \qquad (52.48)$$

where F_b is approximated as follows:

For tubeside reboilers (VT/E thermosiphon)

$$F_b = 1.5 \qquad (52.49)$$

For shellside reboilers (HS/X, G, H, K)

$$F_b = 2.0 \qquad (52.50)$$

Equations (52.49) and (52.50) are intended to give conservative results for first approximations. For
more detailed calculations see Refs. 28–30.

The nucleate boiling heat-transfer coefficient (h_{nb}) is dependent not only on physical properties,
but also on the temperature profile at the wall and the microscopic topography of the surface. For a
practical design, many simplifications must be made, and the approximate nature of the resulting
coefficients should be recognized. A reasonable design value is given by the following simple
equation[25]:

$$h_{nb} = 0.025F_c P_c^{0.69} q^{0.70}(P/P_c)^{0.17} \tag{52.51}$$

The term F_c is a correction for the effect of mixture composition on the boiling heat-transfer coefficient. The heat-transfer coefficient for boiling mixtures is lower than that of any of the pure components if boiled alone, as summarized in Ref. 27. This effect can be explained in terms of the change in temperature profile at the wall caused by the composition gradient at the wall, as illustrated in Ref. 31. Since the liquid-phase diffusional methods necessary to predict this effect theoretically are still under development and require data not usually available to the designer, an empirical relationship in terms of mixture boiling range (BR) is recommended in Ref. 25:

$$F_c = [1 + 0.018q^{0.15}\text{BR}^{0.75}]^{-1} \tag{52.52}$$

(BR = difference between dew-point and bubble-point temperatures, °F.)

Maximum Heat Flux

Above a certain heat flux, the boiling heat-transfer coefficient can decrease severely, owing to vapor blanketing, or the boiling process can become very unstable, as described in Refs. 27, 31, and 32. Therefore, the design heat flux must be limited to a practical maximum value. For many years the limit used by industry was in the range of 10,000–20,000 Btu/hr · ft² for hydrocarbons and about 30,000 Btu/hr · ft² for water. These rules of thumb are still considered reasonable at moderate pressures, although the limits, especially for water, are considerably conservative for good designs. However, at both very high and very low pressures the maximum heat fluxes can be severely decreased. Also, the maximum heat fluxes must be a function of geometry to be realistic. Empirical equations are presented in Ref. 25; the equations give much more accurate estimates over wide ranges of pressure and reboiler geometry.

(a) *For kettle (HS/K) and horizontal thermosiphon (HS/X, G, H)*

$$q_{max} = 803P_c \left(\frac{P}{P_c}\right)^{0.35} \left(1 - \frac{P}{P_c}\right)^{0.9} \phi_b \tag{52.53}$$

$$\phi_b = 3.1 \left[\frac{\pi D_b L}{A_o}\right] \tag{52.54}$$

In the limit, for $\phi_b > 1.0$, let $\phi_b = 1.0$. For $\phi_b < 0.1$, consider larger tube pitch or vapor relief channels.[25] Design heat flux should be limited to less than $0.7\ q_{max}$.

(b) *For vertical thermosiphon (VT/E)*

$$q_{max} = 16,080 \left(\frac{D_i^2}{L}\right)^{0.35} P_c^{0.61} \left(\frac{P}{P_c}\right)^{0.25} \left(1 - \frac{P}{P_c}\right) \tag{52.55}$$

In addition to the preceding check, the vertical tubeside thermosiphon should be checked to insure against mist flow (dryout). The method by Fair[28] was further confirmed in Ref. 33 for hydrocarbons. For water, extensive data and empirical correlations are available as described by Collier.[27] In order to determine the flow regime by these methods it is necessary to determine the flow rate, as described, for example, in Ref. 28. However, for preliminary specification, it may be assumed that the exit vapor weight fraction will be limited to less than 0.35 for hydrocarbons and less than 0.10 for aqueous solutions and that under these conditions dryout is unlikely.

52.3.4 Air-Cooled Heat Exchangers

Detailed rating of air-cooled heat exchangers requires selection of numerous geometrical parameters, such as tube type, number of tube rows, length, width, number and size of fans, etc., all of which involve economic and experience considerations beyond the scope of this chapter. Air-cooled heat exchangers are still designed primarily by the manufacturers using proprietary methods. However, recommendations for initial specifications and rating are given by Paikert[2] and by Mueller.[3] A preliminary rating method proposed by Brown[34] is also sometimes used for first estimates owing to its simplicity.

Heat-Transfer Coefficients

For a first approximation of the surface required, the bare-surface-based overall heat-transfer coefficients recommended by Smith[35] may be used. A list of these values from Ref. 3 is abstracted in Table 52.6. The values in Table 52.6 were based on performance of finned tubes, having a 1 in.

Table 52.6 Typical Overall Heat-Transfer Coefficients (U_o), Based on Bare Tube Surface, for Air-Cooled Heat Exchangers

	U_o	
Service	Btu/hr · ft^2 · °F	W/m^2 · K
Sensible Cooling		
Process water	105–120	600–680
Light hydrocarbons	75–95	425–540
Fuel oil	20–30	114–170
Flue gas, 10 psig	10	57
Condensation		
Steam, 0–20 psig	130–140	740–795
Ammonia	100–200	570–680
Light hydrocarbons	80–95	455–540
Refrigerant 12	60–80	340–455
Mixed hydrocarbons, steam, and noncondensables	60–70	340–397

outside diameter base tube on 2⅜ in. triangular pitch, ⅝ in. high aluminum fins (⅛ in. spacing between fin tips), with eight fins per inch. However, the values may be used as first approximations for other finned types.

As stated by Mueller, air-cooled heat exchanger tubes have had approximately the preceding dimensions in the past, but fin densities have tended to increase and now more typically range from 10 to 12 fins/in. For a more detailed estimate of the overall heat-transfer coefficient, the tubeside coefficients are calculated by methods given in the preceding sections and the airside coefficients are obtained as functions of fin geometry and air velocity from empirical relationships such as given by Gnielinski et al.[36] Rating at this level of sophistication is now done mostly by computer.

Temperature Difference

Air-cooled heat exchangers are normally "cross-flow" arrangements with respect to the type of temperature profile calculation. Charts for determination of the F-factor for such arrangements are presented by Taborek.[9] Charts for a number of arrangements are also given by Paikert[2] based on the "NTU method." According to Paikert, optimum design normally requires NTU to be in the range of 0.8–1.5, where,

$$\text{NTU} = \frac{t_2 - t_1}{\text{MTD}} \qquad (52.56)$$

For first approximations, a reasonable air-temperature rise $(t_2 - t_1)$ may be assumed, MTD calculated from Eq. (52.4) using $F = 0.9$–1.0, and NTU checked from Eq. (52.56). It is assumed that if the air-temperature rise is adjusted so that NTU is about 1, the resulting preliminary size estimation will be reasonable. Another design criterion often used is that the face velocity V_f should be in the range of 300–700 ft/min (1.5–3.5 m/sec):

$$V_f = \frac{W_a}{L \, W_d \rho_v} \qquad (52.57)$$

where W_a = air rate, lb/min
$\quad L$ = tube length, ft
$\quad W_d$ = bundle width, ft
$\quad \rho_v$ = air density, lb/ft^3

Fan Power Requirement

One or more fans may be used per bundle. Good practice requires that not less than 40–50% of the bundle face area be covered by the fan diameter. The bundle aspect ratio per fan should approach 1 for best performance. Fan diameters range from about 4 to 12 ft (1.2 to 3.7 m), with tip speeds usually limited to less than 12,000 ft/min (60 m/sec) to minimize noise. Pressure drops that can be handled are in the range of only 1–2 in. water (0.035–0.07 psi, 250–500 Pa). However, for typical bundle designs and typical air rates, actual bundle pressure drops may be in the range of only ¼–1 in. water.

Paikert[2] gives the expression for fan power as follows:

$$P_f = \frac{V(\Delta p_s + \Delta p_d)}{E_f} \qquad (52.58)$$

where V = volumetric air rate, m^3/sec
Δp_s = static pressure drop, Pa
Δp_d = dynamic pressure loss, often 40–60 Pa
E_f = fan efficiency, often 0.6–0.7
P_f = fan power, W

52.3.5 Other Exchangers

For spiral, plate, and compact heat exchangers the heat-transfer coefficients and friction factors are sensitive to specific proprietary designs and such units are best sized by the manufacturer. However, preliminary correlations have been published. For spiral heat exchangers, see Mueller[3] and Minton.[37] For plate-type heat exchangers, Figs. 52.9 and 52.10, recommendations are given by Cooper[38] and Marriott.[39] For plate-fin and other compact heat exchangers, a comprehensive treatment is given by Webb.[4] For recuperators and regenerators the methods of Hausen are recommended.[6] Heat pipes are extensively covered by Chisholm.[40] Design methods for furnaces and combustion chambers are presented by Truelove.[41] Heat transfer in agitated vessels is discussed by Penney.[42] Double-pipe heat exchangers are described by Guy.[43]

52.4 COMMON OPERATIONAL PROBLEMS

When heat exchangers fail to operate properly in practice, the entire process is often affected, and sometimes must be shut down. Usually, the losses incurred by an unplanned shutdown are many times more costly than the heat exchanger at fault. Poor heat-exchanger performance is usually due to factors having nothing to do with the heat-transfer coefficient. More often the designer has overlooked the seriousness of some peripheral condition not even addressed in most texts on heat-exchanger design. Although only long experience, and numerous "experiences," can come close to uncovering all possible problems waiting to plague the heat-exchanger designer, the following subsections relating the more obvious problems are included to help make the learning curve less eventful.

52.4.1 Fouling

The deposit of solid insulating material from process streams on the heat-transfer surface is known as fouling, and has been called "the major unresolved problem in heat transfer."[44] Although this problem is recognized to be important (see Ref. 45) and is even being seriously researched,[45,46] the nature of the fouling process makes it almost impossible to generalize. As discussed by Mueller,[3] fouling can be caused by (1) precipitation of dissolved substances, (2) deposit of particulate matter, (3) solidification of material through chemical reaction, (4) corrosion of the surface, (5) attachment and growth of biological organisms, and (6) solidification by freezing. The most important variables affecting fouling (besides concentration of the fouling material) are velocity, which affects types 1, 2, and 5, and surface temperature, which affects types 3–6. For boiling fluids, fouling is also affected by the fraction vaporized. As stated in Ref. 25, it is usually impossible to know ahead of time what fouling mechanism will be most important in a particular case. Fouling is sometimes catalyzed by trace elements unknown to the designer. However, most types of fouling are retarded if the flow velocity is as high as possible, the surface temperature is as low as possible (exception is biological fouling[48]), the amount of vaporization is as low as possible, and the flow distribution is as uniform as possible.

The expected occurrence of fouling is usually accounted for in practice by assignment of fouling factors, which are additional heat-transfer resistances, Eq. (52.7). The fouling factors are assigned for the purpose of oversizing the heat exchanger sufficiently to permit adequate on-stream time before cleaning is necessary. Often in the past the fouling factor has also served as a general purpose "safety factor" expected to make up for other uncertainties in the design. However, assignment of overly large fouling factors can produce poor operation caused by excessive overdesign.[49,50]

For shell and tube heat exchangers it has been common practice to rely on the fouling factors suggested by TEMA.[1] Fouling in plate heat exchangers is usually less, and is discussed in Ref. 38. The TEMA fouling factors have been used for over 30 years and, as Mueller states, must represent some practical validity or else complaints would have forced their revision. A joint committee of TEMA and HTRI members has reviewed the TEMA fouling recommendations and slightly updated for the latest edition. In addition to TEMA, fouling resistances are presented by Bell[10] and values recommended for reboiler design are given in Ref. 25. In general, the minimum value commonly used for design is 0.0005 °F · hr · ft^2/Btu for condensing steam or light hydrocarbons. Typical values for process streams or treated cooling water are around 0.001–0.002 °F · hr · ft²/Btu, and for heavily fouling streams values in the range of 0.003–0.01 °F · hr · ft²/Btu are used. For reboilers (which have been properly designed) a design value of 0.001 °F · hr · ft²/Btu is usually adequate, although for wide boiling mixtures other effects in addition to fouling tend to limit performance.

52.4.2 Vibration

A problem with shell and tube heat exchangers that is becoming more frequent as heat exchangers tend to become larger and design velocities tend to become higher is tube failure due to flow-induced tube vibration. Summaries including recommended methods of analysis are given by Chenoweth[51] and by Mueller.[3] In general, tube vibration problems tend to occur when the distance between baffles or tube-support plates is too great. Maximum baffle spacings recommended by TEMA were based on the maximum unsupported length of tube that will not sag significantly. Experience has shown that flow-induced vibration can still occur at TEMA maximum baffle spacing, but for less than about 0.7 times this spacing most vibration can be eliminated at normal design velocities (see Section 52.2.4). Taborek[11] gives the following equations for TEMA maximum unsupported tube lengths (L_{su}), inches.

Steel and Steel Alloy Tubes

$$\text{For } D_o = \text{¾–2 in.,}$$
$$L_{su} = 52D_o + 21 \tag{52.59}$$

$$\text{For } D_o = \text{¼–¾ in.,}$$
$$L_{su} = 68D_o + 9 \tag{52.60}$$

Aluminum and Copper Alloy Tubes

$$\text{For } D_o = \text{¾–2 in.,}$$
$$L_{su} = 46D_o + 17 \tag{52.61}$$

$$\text{For } D_o = \text{¼–¾ in.,}$$
$$L_{su} = 60D_o + 7 \tag{52.62}$$

For segmental baffles with tubes in the windows, Fig. 52.9, the maximum baffle spacing is one-half the maximum unsupported tube length.

For very large bundle diameters, segmental or even double segmental baffles may not be suitable, since the spacing required to prevent vibration may produce too high pressure drops. (In addition, flow distribution considerations require that the ratio of baffle spacing to shell diameter not be less than about 0.2.) In such cases, one commonly used solution is to eliminate tubes in the baffle windows so that intermediate support plates can be used and baffle spacing can be increased; see Fig. 52.17. Another solution, with many advantages is the rod-type tube support in which the flow is essentially longitudinal and the tubes are supported by a cage of rods. A proprietary design of this type exchanger (RODbaffle) is licensed by Phillips Petroleum Co. Calculation methods are published in Ref. 52.

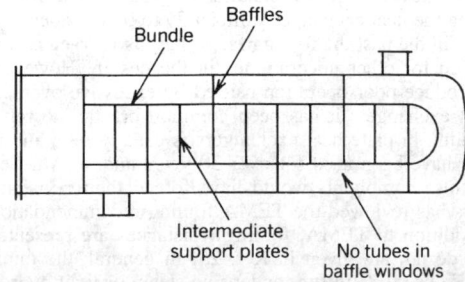

Fig. 52.17 Segmental baffles with no tubes in window.

52.4.3 Flow Maldistribution

Several types of problems can occur when the flow velocities or fluid phases become distributed in a way not anticipated by the designer. This occurs in all types of exchangers, but the following discussion is limited to shell and tube and air-cooled exchangers, in which maldistribution can occur on either shellside or tubeside.

Shellside Flow

Single-phase flow can be maldistributed on the shellside owing to bypassing around the tube bundle and leakage between tubes and baffle and between baffle and shell. Even for typical well-designed heat exchangers, these ineffective streams can comprise as much as 40% of the flow in the turbulent regime and as much as 60% of the flow in the laminar regime. It is especially important for laminar flow to minimize these bypass and leakage streams, which cause both lower heat-transfer coefficients and lower effective MTD.[13] This can, of course, be done by minimizing clearances, but economics dictate that more practical methods include use of bypass sealing strips, increasing tube pitch, increasing baffle spacing, and using an optimum baffle cut to provide more bundle penetration. Methods for calculating the effects of these parameters are described by Taborek.[11]

Another type of shellside maldistribution occurs in gas–liquid two-phase flow in horizontal shells when the flow velocity is low enough that the vapor and liquid phases separate, with the liquid flowing along the bottom of the shell. For condensers this is expected and taken into account. However, for some other types of exchangers, such as vapor–liquid contactors or two-phase reactor feed-effluent exchangers, separation may cause unacceptable performance. For such cases, if it is important to keep the phases mixed, a vertical heat exchanger is recommended. Improvement in mixing is obtained for horizontal exchangers if horizontal rather than vertical baffle cut is used.

Tubeside Flow

Several types of tubeside maldistribution have been experienced. For single-phase flow with axial nozzles into a single-tubepass exchanger, the dynamic head of the entering fluid can cause higher flow in the central tubes, sometimes even producing backflow in the peripheral tubes. This effect can be prevented by using an impingement plate on the centerline of the axial nozzle.

Another type of tubeside maldistribution occurs in cooling viscous liquids. Cooler tubes in parallel flow will tend to completely plug up in this situation, unless a certain minimum pressure drop is obtained, as explained by Mueller.[53]

For air-cooled single pass condensers, a backflow can occur owing to the difference in temperature driving force between bottom and top tube rows, as described by Berg and Berg.[54] This can cause an accumulation of noncondensables in air-cooled condensers, which can significantly affect performance, as described by Breber et al.[55] In fact, in severe cases, this effect can promote freezeup of tubes, or even destruction of tubes by water hammer. Backflow effects are eliminated if a small amount of excess vapor is taken through the main condenser to a backup condenser or if the number of fins per inch on bottom rows is less than on top rows to counteract the difference in temperature driving force.

For multipass tubeside condensers, or tubeside condensers in series, the vapor and liquid tend to separate in the headers with liquid running in the lower tubes. The fraction of tubes filled with liquid tends to be greater at higher pressures. In most cases the effect of this separation on the overall condenser heat-transfer coefficient is not serious. However, for multicomponent mixtures the effect on the temperature profile will be such as to decrease the MTD. For such cases, the temperature profile should be calculated by the differential flash procedure, Section 52.3.2. In general, because of unpredictable effects, entering a pass header with two phases should be avoided when possible.

52.4.4 Temperature Pinch

When the hot and cold streams reach approximately the same temperature in a heat exchanger, heat transfer stops. This condition is referred to as a temperature pinch. For shellside single-phase flow, unexpected temperature pinches can be the result of excessive bypassing and leakage combined with a low MTD and possibly a temperature cross. An additional factor, "temperature profile distortion factor," is needed as a correction to the normal F factor to account for this effect.[11,13] However, if good design practices are followed with respect to shellside geometry, this effect normally can be avoided.

In condensation of multicomponent mixtures, unexpected temperature pinches can occur in cases where the condensation curve is not properly calculated, especially when the true curve happens to be of type III in Fig. 52.15. This can happen when separation of liquid containing heavy components occurs, as mentioned above, and also when the condensing mixture has immiscible liquid phases with more than one dew point.[20] In addition, condensing mixtures with large desuperheating and subcooling zones can produce temperature pinches and must be carefully analyzed. In critical cases it is safer and may even be more effective to do desuperheating, condensing, and subcooling in separate heat exchangers. This is especially true of subcooling.[3]

Reboilers can also suffer from temperature-pinch problems in cases of wide boiling mixtures and inadequate liquid recirculation. Especially for thermosiphon reboilers, if poorly designed and the circulation rate is not as high as expected, the temperature rise across the reboiler will be greater than expected and a temperature pinch may result. This happens most often when the reboiler exit piping is too small and consumes an unexpectedly large amount of pressure drop. This problem normally can be avoided if the friction and momentum pressure drop in the exit piping is limited to less than 30% of the total driving head and the exit vapor fraction is limited to less than 0.25 for wide boiling range mixtures. For other recommendations, see Ref. 25.

52.4.5 Critical Heat Flux in Vaporizers

Owing to a general tendency to use lower temperature differences for energy conservation, critical heat flux problems are not now frequently seen in the process industries. However, for waste heat boilers, where the heating medium is usually a very hot fluid, surpassing the critical heat flux is a major cause of tube failure. The critical heat flux is that flux (Q/A_o) above which the boiling process departs from the nucleate or convective boiling regimes and a vapor film begins to blanket the surface, causing a severe rise in surface temperature, approaching the temperature of the heating medium. This effect can be caused by either of two mechanisms: (1) flow of liquid to the hot surface is impeded and is insufficient to supply the vaporization process or (2) the local temperature exceeds that for which a liquid phase can exist.[32] Methods of estimating the maximum design heat flux are given in Section 52.3.3, and the subject of critical heat flux is covered in great detail in Ref. 27. However, in most cases where failures have occurred, especially for shellside vaporizers, the problem has been caused by local liquid deficiency, owing to lack of attention to flow distribution considerations.

52.4.6 Instability

The instability referred to here is the massive large-scale type in which the fluid surging is of such violence as to at least disrupt operations, if not to cause actual physical damage. One version is the boiling instability seen in vertical tubeside thermosiphon reboilers at low operating pressure and high heat flux. This effect is discussed and analyzed by Blumenkrantz and Taborek.[56] It is caused when the vapor acceleration loss exceeds the driving head, producing temporary flow stoppage or backflow, followed by surging in a periodic cycle. This type of instability can always be eliminated by using more frictional resistance, a valve or orifice, in the reboiler feed line. As described in Ref. 32, instability normally only occurs at low reduced pressures, and normally will not occur if design heat flux is less than the maximum value calculated from Eq. (52.55).

Another type of massive instability is seen for oversized horizontal tubeside pure component condensers. When more surface is available than needed, condensate begins to subcool and accumulate in the downstream end of the tubes until so much heat-transfer surface has been blanketed by condensate that there is not enough remaining to condense the incoming vapor. At this point the condensate is blown out of the tube by the increasing pressure and the process is repeated. This effect does not occur in vertical condensers since the condensate can drain out of the tubes by gravity. This problem can sometimes be controlled by plugging tubes or injecting inert gas, and can always be eliminated by taking a small amount of excess vapor out of the main condenser to a small vertical backup condenser.

52.4.7 Inadequate Venting, Drainage, or Blowdown

For proper operation of condensers it is always necessary to provide for venting of noncondensables. Even so-called pure components will contain trace amounts of noncondensables that will eventually build up sufficiently to severely limit performance unless vented. Vents should always be in the vapor space near the condensate exit nozzle. If the noncondensable vent is on the accumulator after the condenser, it is important to ensure that the condensate nozzle and piping are large enough to provide unrestricted flow of noncondensables to the accumulator. In general, it is safer to provide vent nozzles directly on the condenser.

If condensate nozzles are too small, condensate can accumulate in the condenser. It is recommended that these nozzles be large enough to permit weir-type drainage (with a gas core in the center of the pipe) rather than to have a full pipe of liquid. Standard weir formulas[57] can be used to size the condensate nozzle. A rule of thumb used in industry is that the liquid velocity in the condensate piping, based on total pipe cross section, should not exceed 3 ft/sec (0.9 m/sec).

The problem of inadequate blowdown in vaporizers is similar to the problem of inadequate venting for condensers. Especially with kettle-type units, trace amounts of heavy, high-boiling, or nonboiling components can accumulate, not only promoting fouling but also increasing the effective boiling range of the mixture, thereby decreasing the MTD as well as the effective heat-transfer coefficient. Therefore, means of continuous or at least periodic removal of liquid from the reboiler (blowdown) should be provided to ensure good operation. Even for thermosiphon reboilers, if designed for low heat fluxes (below about 2000 BTU/hr/ft^2, 6300 W/m^2), the circulation through the reboiler may not be high enough to prevent heavy components from building up, and some provision for blowdown may be advisable in the bottom header.

52.5   USE OF COMPUTERS IN THERMAL DESIGN OF PROCESS HEAT EXCHANGERS

52.5.1   Introduction

The approximate methods for heat transfer coefficient and pressure drop given in the preceding sections will be used mostly for orientation. For an actual heat exchanger design, it only makes sense to use a computer. Standard programs can be obtained for most geometries in practical use. These allow reiterations and incrementation to an extent impossible by hand and also supply physical properties for a wide range of industrial fluids. However, computer programs by no means solve the whole problem of producing a workable efficient heat exchanger. Many experience-guided decisions must be made both in selection of the input data and in interpreting the output data before even the thermal design can be considered final. We will first review why a computer program is effective. This has to do with 1) incrementation and 2) convergence loops.

52.5.2   Incrementation

The method described in Section 52.2.1 for calculation of required surface can only be applied accurately to the entire exchanger if the overall heat transfer coefficient is constant and the temperature profiles for both streams are linear. This often is not a good approximation for typical process heat exchangers because of variation in physical properties and/or vapor fraction along the exchanger length. The rigorous expression for Eq. (52.1) is as follows:

$$A_o = \int \frac{dQ}{U_o \, \text{MTD}}$$

Practical solution of this integral equation requires dividing the heat transfer process into finite increments of ΔQ that are small enough so that U_o may be considered constant and the temperature profiles may be considered linear. The incremental area, Δa_o, is then calculated for each increment and summed to obtain the total required area. An analogous procedure is followed for the pressure drop. This procedure requires determining a full set of fluid physical properties for all phases of both fluids in each increment and the tedious calculations can be performed much more efficiently by computer. Furthermore, in each increment several trial and error convergence loops may be required, as discussed next.

52.5.3   Main Convergence Loops

Within each of the increments discussed above, a number of implicit equations must be solved, requiring convergence loops. The two main types of loops found in any heat exchanger calculation are as follows.

Intermediate Temperature Loops

These convergence loops normally are used to determine either wall temperature or, less commonly, interface temperature. The discussion here will be limited to the simpler case of wall temperature. Because of the variation of physical properties between the wall and the bulk of the fluid, heat transfer coefficients depend on the wall temperature. Likewise, the wall temperature depends on the relative values of the heat transfer coefficients of each fluid. Wall temperatures on each side of the surface can be estimated by the following equations:

$$T_{w, \, \text{hot}} = T_{\text{hot}} - \frac{U_o}{h_{\text{hot}}} (T_{\text{hot}} - T_{\text{cold}})$$

$$T_{w, \, \text{cold}} = T_{\text{cold}} + \frac{U_o}{h_{\text{cold}}} (T_{\text{hot}} - T_{\text{cold}})$$

It is assumed in the above equations that the heat transfer coefficient on the inside surface is corrected to the outside area. Convergence on the true wall temperature can be done in several ways. Figure 52.18 shows a possible convergence scheme.

Pressure Balance Loops

These convergence loops are needed whenever the equations to be solved are implicit with respect to velocity. The two most frequent cases encountered in heat exchanger design are 1) flow distribution and 2) natural circulation. The first case, flow distribution, is the heart of the shell and tube heat exchanger shellside flow calculations, and involves solution for the fraction of flow across the tube bundle, as opposed to the fraction of flow leaking around baffles and bypassing the bundle. Since the resistance coefficients of each stream are functions of the stream velocity, the calculation is reiterative. The second case, natural circulation, is encountered in thermosiphon and kettle reboilers where the flow rate past the heat transfer surface is a function of the pressure balance between the two-phase flow in the bundle, or tubes, and the liquid static head outside the bundle. In this case the

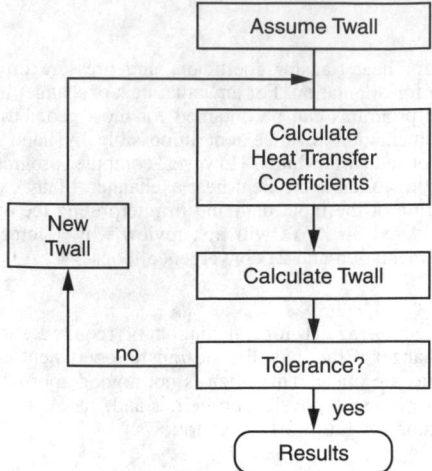

Fig. 52.18   Temperature convergence loop.

heat transfer coefficients that determine the vaporization rate are functions of the flow velocity, which is in turn a function of the amount of vaporization. Figure 52.19 shows a flow velocity convergence loop applicable to the flow distribution case.

52.5.4   Rating, Design, or Simulation

Several types of solutions are possible by computer. The better standard programs allow the user to choose. It is important to understand what the program is doing in order to properly interpret the results. The above three types of calculations are described as follows.

Rating

This is the normal mode for checking a vendor's bid. All geometry and all process conditions are specified. The program calculates the required heat transfer area and pressure drop and compares with the specified values. Normally this is done including the specified fouling factor. This means that on startup the amount of excess surface will be greater, sometimes excessively greater, causing severe operating adjustments. It is therefore advisable to review clean conditions also.

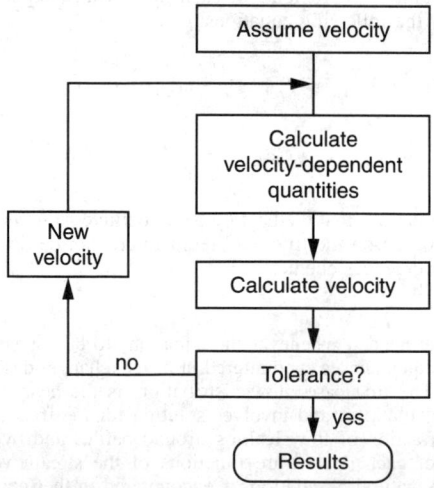

Fig. 52.19   Velocity convergence loop.

Design

This mode is used by the process engineer to obtain a size based on process heat transfer requirements. In this case, most of the geometry specifications still are not determined by the program and must be determined by the designer based on experience. Required, but unknown, specifications, in addition to the process requirements of temperatures, flow rates, and pressure drops, include

- Exchanger type (shell and tube, plate-and-frame, plate-fin, air-cooled, etc.)

If shell and tube

- TEMA shell type (E, F, J, G, H, X, K)
- TEMA front and rear head types (flat, dished, fixed tube sheet, split ring, pull-through)
- Baffle type (segmental, double segmental, triple segmental, rod, etc.)
- Tube type (plain, low-finned, enhanced surface, etc.)
- Tube length (usually standard lengths of 8, 12, 16, 20 ft)
- Tube diameter (usually ⅝, ¾, 1, 1¼ in. or 1.25 in.)
- Tube pitch (pitch ratios 1.25, 1.3, 1.5)
- Tube layout (30, 45, 60, 90°)
- Tube material (carbon steel, stainless steel, copper alloys, titanium, etc.)
- Exchanger orientation (horizontal, vertical)

As shown, even with a good computer program, an overwhelming number of combinations of geometry parameters is possible and presently the engineer is required to select the best combination based on mechanical considerations, process considerations, fouling tendencies, and allowable pressure drop. Some general guidelines are given in Section 52.5.6. Once the above parameters are specified to the computer program, it can proceed to calculate the number of tubes required and the baffle spacing and number of tube passes consistent with the required pressure drops for both streams.

Simulation

This mode of calculation is used most to predict the performance of a field heat exchanger under different operating conditions. Usually the engineer "zeros" the program first by adjusting fouling factors and friction factor multipliers to match existing field performance. Then the adjusted process conditions are imposed and the computer program predicts the heat transfer rates and pressure drops under the new conditions. This mode of calculation can also be used to monitor apparent fouling resistance increase on operating units in order to better schedule maintenance.

52.5.5 Program Quality and Selection

All heat exchanger programs are not created equal. Heat exchange is not yet an exact science and all of the heat transfer coefficients and friction factors used in calculation are from correlations with empirically determined constants. Therefore, the data base used for correlation development is important.

Methods Source

The methods used for the program should be available and documented in a readable form. Good methods will be based on theoretically derived equation forms that either are limited in range or automatically achieve theoretically justified limits. "Black box" methods, for which this may not be true, should be avoided.

Data Base

Good programs are also backed by a sizable data bank covering the range of conditions of interest as well as demonstrated successes in predicting field performance. No non-tested methods, including so-called rigorous incremental methods, should be accepted without some data-based support.

Suitability

Completely general programs that apply to all geometries and process conditions and fulfill the above data base requirements probably will not exist for sometime. The program manual should list recommended ranges of applicability. When in doubt, consult the supplier.

52.5.6 Determining and Organizing Input Data

As of this writing, available programs still require a large number of input data decisions to be made by the user. The quality of the answers obtained is crucially dependent on the quality of these input decisions.

Process Data

The basis for the calculation is the heat duty, which usually comes from the process flow sheet. There must, of course, be a heat balance between the hot and cold sides of the exchanger. The temperature profiles are much more significant to a good design than are the heat transfer coefficients. Only in rare cases are these straight lines. For multicomponent phase-change cases, the condensing or vaporization curves should be calculated by a good process simulator program containing state-of-the-art vapor–liquid equilibria methods. Most good heat exchanger programs will accept these curves as the basis for the heat-transfer calculations.

It is important to specify realistic pressure drop limitations, since the heat-transfer coefficient and the fouling rate are functions of the velocity, which is a function of the available pressure drop. For phase change, too much pressure drop can mean a significant loss in available temperature difference and one rule of thumb suggests a limit of 10% of the operating pressure. For liquid flow, erosion velocity often is the limiting factor, and this is usually taken to be in the range of 7–10 ft/sec tubeside or 3–5 ft/sec shellside. Velocities also are sometimes limited to a value corresponding to ρv^2 less than 4000, where ρ is in lb/ft^3 and v is in ft/sec.

Geometry Data

It is necessary for the program user to make a large number of geometry decisions, starting with the type of exchanger, which decides the type of program to be used. Only a brief list of suggestions can be accommodated in this chapter, so recommendations will be limited to some of the main shell-and-tube geometries mentioned in Section 52.5.4.

TEMA Shell Style.   The types E, J, and X are selected based on available pressure drop, highest E, lowest X, and intermediate J. Types G and H are used mostly for horizontal thermosiphon reboilers, although they also obtain a slightly better MTD correction factor than the E-type shell and are sometimes used even for single phase for that purpose. Pressure drop for G and E shells are about the same. For horizontal thermosiphon reboilers, the longitudinal baffle above the inlet nozzle prevents the light vaporizing component to shortcut directly to the exit nozzle. If pressure drop for the less expensive G-shell is too high, H-shell (two G's in parallel) is used. Type F is used when it is required to have a combination of countercurrent flow and two tube passes in a single shell. This type has the disadvantage of leakage around the longitudinal baffle, which severely decreases performance. A welded baffle prevents this but prevents bundle removal. Type K is used only for kettle reboilers.

TEMA Front and Rear Head Types.   These are selected based on pressure and/or maintenance considerations. TEMA Standards should be consulted. With respect to maintenance, rear heads permitting bundle removal should be specified for shellside fouling fluids. These are the split ring and pull-through types.

Baffle Types.   These are selected based on a combination of pressure drop and vibration considerations. In general, the less expensive, higher-velocity segmental baffle is tried first, going to the double segmental and possibly the triple segmental types if necessary to lower pressure drop. Allowable pressure drop is a very important design parameter and should not be allocated arbitrarily. In the absence of other process limits, the allowable pressure drop should be about 10% of the operating pressure or the ρv^2 should be less than about 4000 (lb/ft^3)(ft/sec)^2, whichever gives the lower velocity. However, vibration limits override these limits. Good thermal design programs also check for tube vibration and warn the user if vibration problems are likely due to high velocity or insufficient tube support. In case of potential vibration problems, it is necessary to decrease velocity or provide more tube support, the latter being preferable. The two best ways of eliminating vibration problems within allowable pressure drop limitations are 1) no-tube-in-window baffles, or 2) RoDbaffles, as discussed in Section 52.4.2.

Tube Types.   For low temperature differences and low heat-transfer coefficients, low-finned or enhanced tubes should be investigated. In proper applications these can decrease the size of the exchanger dramatically. Previously, enhanced tubes were considered only for very clean streams. However, recent research is beginning to indicate that finned tubes fare as well in fouling services as plain tubes, and sometimes much better, providing longer on-stream time and often even easier cleaning. In addition, the trend in the future will be to stop assigning arbitrary fouling factors, but rather to design for conditions minimizing fouling.

Tube Length.   This is usually limited by plant requirements. In general, longer exchangers are economically preferable within pressure drop restrictions, except possibly for vertical thermosiphon reboilers.

Tube Diameter.   Small diameters are more economical in the absence of restrictions. Cleaning restrictions normally limit outside diameters to not less than ⅝ or ¾ in. Pressure drop restrictions, especially in vacuum, may require larger sizes. Vacuum vertical thermosiphon reboilers often require

1¼-in. tubes, and vacuum falling film evaporators frequently use as large as 2-in. tubes. Excessive pressure drop can be quickly decreased by going to the next standard tube diameter, since pressure drop is inversely proportional to the fifth power of the inside diameter.

Tube Pitch. Tube pitch for shellside flow is analogous to tube diameter for tubeside flow. Small pitches are more economical and also can cause pressure drop or cleaning problems. In laminar flow, too-small tube pitch can prevent bundle penetration and force more bypassing and leakage. A pitch-to-tube diameter ratio of 1.25 or 1.33 is often used in absence of other restrictions depending on allowable pressure drop. For shellside reboilers operating at high heat flux, a ratio of as much as 1.5 is often required. Equation (52.54) shows that the maximum heat flux for kettle reboilers increases with increasing tube pitch.

Tube Layout. Performance is not critically affected by tube layout, although some minor differences in pressure drop and vibration characteristics are seen. In general, either 30 or 60° layouts are used for clean fluids, while 45 or 90° layouts are more frequently seen for fluids requiring shellside fouling maintenance.

Tube Material. The old standby for noncorrosive moderate-temperature hydrocarbons is the less expensive and sturdy carbon steel. Corrosive or very high-temperature fluids require stainless steel or other alloys. Titanium and hastelloy are becoming more frequently used for corrosion or high temperature despite the high cost, as a favorable economic balance is seen in comparison with severe problems of tube failure.

Exchanger Orientation. Exchangers normally are horizontal except for tubeside thermosiphons, falling film evaporators, and tubeside condensers requiring very low pressure drop or extensive subcooling. However, it is becoming more frequent practice to specify vertical orientation for two-phase feed-effluent exchangers to prevent phase separation, as mentioned in Section 52.4.3.

Fouling

All programs require the user to specify a fouling factor, which is the heat-transfer resistance across the deposit of solid material left on the inside and/or outside of the tube surface due to decomposition of the fluid being heated or cooled. Considerations involved in the determination of this resistance are discussed in Section 52.4.1. Since there are presently no thermal design programs available that can make this determination, the specification of a fouling resistance, or fouling factor, for each side is left up to the user. Unfortunately, this input is probably more responsible than any other for causing inefficient designs and poor operation. The major problem is that there is very little relationship between actual fouling and the fouling factor specified. Typically, the fouling factor contains a safety factor that has evolved from practice, lived a charmed life as it is passed from one handbook to another, and may no longer be necessary if modern accurate design programs are used. An example is the frequent use of a fouling factor of 0.001 hr ft² °F/Btu for clean overhead condenser vapors. This may have evolved as a safety or correction from the failure of early methods to account for mass transfer effects and is completely unnecessary with modern calculation methods. Presently, the practice is to use fouling factors from TEMA Standards. However, these often result in heat exchangers that are oversized by as much as 50% on startup, causing operating problems that actually tend to enhance fouling tendencies. Hopefully, with ongoing research on fouling threshold conditions, it will be possible to design exchangers to essentially clean conditions. In the meantime, the user of computer programs should use common sense in assigning fouling factors only to actual fouling conditions. Startup conditions should also be checked as an alternative case.

NOMENCLATURE

Note: Dimensional equations should use U.S. Units only.

	Description	U.S. Units	S.I. Units
A_i	Inside surface area	ft²	m²
A_m	Mean surface area	ft²	m²
A_o	Outside surface area	ft²	m²
a_o	Outside surface per unit length	ft	m
B_c	Baffle cut % of shell diameter	%	%
BR	Boiling range (dew–bubble points)	°F	(U.S. only)
C	Two-phase pressure drop constant	—	—
C_b	Bundle bypass constant	—	—
C_{p1}	Heat capacity, hot fluid	Btu/lb · °F	J/kg · K
C_{p2}	Heat capacity, cold fluid	Btu/lb · °F	J/kg · K
D	Tube diameter, general	ft	m
D_b	Bundle diameter	ft	m

D_i	Tube diameter, inside	ft	m
D_o	Tube diameter, outside	ft or in.	m or U.S. only
D_s	Shell diameter	ft	m
D_f	Effective length:	ft	m
	$= D_i$ for tubeside		
	$= P_t - D_o$ for shellside		
E_f	Fan efficiency (0.6–0.7, typical)	—	—
F	MTD correction factor	—	—
F_b	Bundle convection factor	—	—
F_c	Mixture correction factor	—	—
F_g	Gravity condensation factor	—	—
g	Acceleration of gravity	ft/hr^2	m/sec^2
G	Total mass velocity	lb/hr $\cdot$ ft^2	kg/sec $\cdot$ m^2
g_c	Gravitational constant	4.17×10^8 lb$_f \cdot$ ft/lb $\cdot$ hr^2	1.0
h_{hot}	Heat transfer coeff., hot fluid	Btu/hr $\cdot$ ft^2 $\cdot$ °F	W/m^2 $\cdot$ K
h_{cold}	Heat transfer coeff., cold fluid	Btu/hr $\cdot$ ft^2 $\cdot$ °F	W/m^2 $\cdot$ K
h_b	Heat transfer coeff., boiling	Btu/hr $\cdot$ ft^2 $\cdot$ °F	W/m^2 $\cdot$ K
h_c	Heat transfer coeff., condensing	Btu/hr $\cdot$ ft^2 $\cdot$ °F	W/m^2 $\cdot$ K
h_{cb}	Heat transfer coeff., conv. boiling	Btu/hr $\cdot$ ft^2 $\cdot$ °F	W/m^2 $\cdot$ K
h_{cf}	Heat transfer coeff., cond. film	Btu/hr $\cdot$ ft^2 $\cdot$ °F	W/m^2 $\cdot$ K
h_i	Heat transfer coeff., inside	Btu/hr $\cdot$ ft^2 $\cdot$ °F	W/m^2 $\cdot$ K
h_l	Heat transfer coeff., liq. film	Btu/hr $\cdot$ ft^2 $\cdot$ °F	W/m^2 $\cdot$ K
h_N	Heat transfer coeff., Nusselt	Btu/hr $\cdot$ ft^2 $\cdot$ °F	W/m^2 $\cdot$ K
h_{nb}	Heat transfer coeff., nucleate boiling	Btu/hr $\cdot$ ft^2 $\cdot$ °F	W/m^2 $\cdot$ K
h_o	Heat transfer coeff., outside	Btu/hr $\cdot$ ft^2 $\cdot$ °F	W/m^2 $\cdot$ K
h_{sv}	Heat transfer coeff., sens. vapor	Btu/hr $\cdot$ ft^2 $\cdot$ °F	W/m^2 $\cdot$ K
h_v	Heat transfer coeff., vapor phase	Btu/hr $\cdot$ ft^2 $\cdot$ °F	W/m^2 $\cdot$ K
J_g	Wallis dimensionless gas velocity	—	—
k_f	Thermal conductivity, fluid	Btu/hr $\cdot$ ft $\cdot$ °F	W/m $\cdot$ K
k_l	Thermal conductivity, liquid	Btu/hr $\cdot$ ft $\cdot$ °F	W/m $\cdot$ K
k_w	Thermal conductivity, wall	Btu/hr $\cdot$ ft $\cdot$ °F	W/m $\cdot$ K
L	Tube length	ft	m
L_{bc}	Baffle spacing	ft	m
L_{su}	Maximum unsupported length	in.	use U.S. only
MTD	Mean temperature difference	°F	K
NP	Number of tube passes	—	—
NT	Number of tubes	—	—
NTU	Number of transfer units	—	—
P	Pressure	psia	use U.S. only
P_c	Critical pressure	psia	use U.S. only
P_f	Fan power	use S.I. only	W
Pr	Prandtl number	—	—
P_t	Tube pitch	ft	m
q_{max}	Maximum allowable heat flux	Btu/hr $\cdot$ ft^2	use U.S. only
q	Heat flux	Btu/hr ft^2	use U.S. only
Q	Heat duty	Btu/hr	W
q_{sv}	Sensible vapor heat flux	Btu/hr ft^2	W/m^2
q_t	Total heat flux	Btu/hr ft^2	W/m^2
Re	Reynolds number	—	—
Re$_c$	Reynolds number, condensate	—	—
R_{fi}	Fouling resistance, inside	°F ft^2 hr/Btu	K m^2/W
R_{fo}	Fouling resistance, outside	°F ft^2 hr/Btu	K m^2/W
R_{hi}	Heat transfer resistance, inside	°F ft^2 hr/Btu	K m^2/W
R_{ho}	Heat transfer resistance, outside	°F ft^2 hr/Btu	K m^2/W
R_w	Heat transfer resistance, wall	°F ft^2 hr/Btu	K m^2/W
S_s	Crossflow area, shellside	ft^2	m^2
S_t	Crossflow area, tubeside	ft^2	m^2
t_1	Temperature, cold fluid inlet	°F	°C
T_1	Temperature, hot fluid inlet	°F	°C
t_2	Temperature, cold fluid outlet	°F	°C
T_2	Temperature, hot fluid outlet	°F	°C
T_A	Hot inlet—cold outlet temperature	°F	°C
T_B	Hot outlet—cold inlet temperature	°F	°C
T_{hot}	Temperature, hot fluid	°F	°C
T_{cold}	Temperature, cold fluid	°F	°C

T_s	Saturation temperature	°F	°C
T_w	Wall temperature	°F	°C
$T_{w,\,hot}$	Wall temperature, hot fluid side	°F	°C
$T_{w,\,cold}$	Wall temperature, cold fluid side	°F	°C
U_o	Overall heat transfer coefficient	Btu/hr · ft² · °F	W/m² · K
V	Volumetric flow rate	use S.I. only	m³/s
V_f	Face velocity	ft/min	use S.I. only
V_s	Shellside velocity	ft/hr	m/hr
V_t	Tubeside velocity	ft/hr	m/hr
W_a	Air flow rate	lb/min	use U.S. only
W_1	Flow rate, hot fluid	lb/hr	kg/hr
W_2	Flow rate, cold fluid	lb/hr	kg/hr
W_c	Flow rate, condensate	lb/hr	kg/hr
W_d	Air-cooled bundle width	ft	use U.S. only
W_s	Flow rate, shellside	lb/hr	kg/hr
W_t	Flow rate, tubeside	lb/hr	kg/hr
X_{tt}	Martinelli parameter	—	—
x_w	Wall thickness	ft	m
y	Weight fraction vapor	—	—
α	Nucleate boiling suppression factor	—	—
Δp_d	Dynamic pressure loss (typically 40–60 Pa)	use S.I.	Pa
ΔP_f	Two-phase friction pressure drop	psi	kPa
ΔP_l	Liquid phase friction pressure drop	psi	kPa
Δp_s	Static pressure drop, air cooler	use S.I. only	Pa
ΔP_s	Shellside pressure drop	lb/ft²	use U.S. only
ΔP_t	Tubeside pressure drop	lb/ft²	use U.S. only
λ	Latent heat	Btu/lb	J/kg
μ	Viscosity, general	lb/ft · hr	Pa
μ_f	Viscosity, bulk fluid	lb/ft · hr	Pa
μ_w	Viscosity, at wall	lb/ft · hr	Pa
ρ_l	Density, liquid	lb/ft³	kg/m³
ρ_s	Density, shellside fluid	lb/ft³	kg/m³
ρ_t	Density, tubeside fluid	lb/ft³	kg/m³
ρ_v	Density, vapor	lb/ft³	kg/m³
ϕ_b	Bundle vapor blanketing correction	—	—
ϕ_l	Two-phase pressure drop correction	—	—

REFERENCES

Note: Many of the following references are taken from the *Heat Exchanger Design Handbook* (HEDH), Hemisphere, Washington, DC, 1982, which will be referred to for simplicity as HEDH.

1. *Standards of Tubular Heat Exchanger Manufacturers Association,* 6th ed., TEMA, New York, 1978.
2. P. Paikert, "Air-Cooled Heat Exchangers," Section 3.8, HEDH.
3. A. C. Mueller, in *Handbook of Heat Transfer,* Rohsenow and Hartnet (eds.), McGraw-Hill, New York, 1983, Chap. 18.
4. R. L. Webb, "Compact Heat Exchangers," Section 3.9, HEDH.
5. F. L. Rubin, "Multizone Condensers, Desuperheating, Condensing, Subcooling," *Heat Transfer Eng.* **3**(1), 49–59 (1981).
6. H. Hausen, *Heat Transfer in Counterflow, Parallel Flow, and Crossflow,* McGraw-Hill, New York, 1983.
7. D. Chisholm et al., "Costing of Heat Exchangers," Section 4.8, HEDH.
8. R. S. Hall, J. Matley, and K. J. McNaughton, "Current Costs of Process Equipment," *Chem. Eng.* **89**(7), 80–116 (Apr. 5, 1982).
9. J. Taborek, "Charts for Mean Temperature Difference in Industrial Heat Exchanger Configurations," Section 1.5, HEDH.
10. K. J. Bell, "Approximate Sizing of Shell-and-Tube Heat Exchangers," Section 3.1.4, HEDH.
11. J. Taborek, "Shell and Tube Heat Exchangers, Single-Phase Flow," Section 3.3, HEDH.
12. D. Q. Kern, *Process Heat Transfer,* McGraw-Hill, New York, 1950.
13. J. W. Palen and J. Taborek, "Solution of Shellside Heat Transfer and Pressure Drop by Stream Analysis Method," *Chem. Eng. Prog. Symp. Series* **65**(92) (1969).

14. A. C. Mueller, "Condensers," Section 3.4, HEDH.
15. B. D. Smith, *Design of Equilibrium Stage Processes,* McGraw-Hill, New York, 1963.
16. V. L. Rice, "Program Performs Vapor-Liquid Equilibrium Calculations," *Chem. Eng.,* 77–86 (June 28, 1982).
17. R. S. Kistler and A. E. Kassem, "Stepwise Rating of Condensers," *Chem. Eng. Prog.* **77**(7), 55–59 (1981).
18. G. Breber, J. Palen, and J. Taborek, "Prediction of Horizontal Tubeside Condensation of Pure Components Using Flow Regime Criteria," *Heat Transfer Eng.* **1**(2), 72–79 (1979).
19. D. Butterworth, "Condensation of Vapor Mixtures," Section 2.6.3, HEDH.
20. R. G. Sardesai, "Condensation of Mixtures Forming Immiscible Liquids," Section 2.5.4, HEDH.
21. K. J. Bell and A. M. Ghaly, "An Approximate Generalized Design Method for Multi-component/Partial Condensers," *AIChE Symp. Ser.,* No. 131, 72–79 (1972).
22. J. E. Diehl, "Calculate Condenser Pressure Drop," *Pet. Refiner* **36**(10), 147–153 (1957).
23. I. D. R. Grant and D. Chisholm, "Two-Phase Flow on the Shell-side of a Segmentally Baffled Shell-and-Tube Heat Exchanger," *Trans. ASME J. Heat Transfer* **101**(1), 38–42 (1979).
24. K. Ishihara, J. W. Palen, and J. Taborek, "Critical Review of Correlations for Predicting Two-Phase Flow Pressure Drops Across Tube Banks," *Heat Transfer Eng.* **1**(3) (1979).
25. J. W. Palen, "Shell and Tube Reboilers," Section 3.6, HEDH.
26. R. A. Smith, "Evaporaters," Section 3.5, HEDH.
27. J. G. Collier, "Boiling and Evaporation," Section 2.7, HEDH.
28. J. R. Fair, "What You Need to Design Thermosiphon Reboilers," *Pet. Refiner* **39**(2), 105 (1960).
29. J. R. Fair and A. M. Klip, "Thermal Design of Horizontal Type Reboilers," *Chem. Eng. Prog.* **79**(3) (1983).
30. J. W. Palen and C. C. Yang, "Circulation Boiling Model of Kettle and Internal Reboiler Performance," Paper presented at the 21st National Heat Transfer Conference, Seattle, WA, 1983.
31. J. W. Palen, A. Yarden, and J. Taborek, "Characteristics of Boiling Outside Large Scale Multitube Bundles," *Chem. Eng. Prog. Symp. Ser.* **68**(118), 50–61 (1972).
32. J. W. Palen, C. C. Shih, and J. Taborek, "Performance Limitations in a Large Scale Thermosiphon Reboiler," *Proceedings of the 5th International Heat Transfer Conference,* Tokyo, 1974, Vol. 5, pp. 204–208.
33. J. W. Palen, C. C. Shih, and J. Taborek, "Mist Flow in Thermosiphon Reboilers," *Chem. Eng. Prog.* **78**(7), 59–61 (1982).
34. R. Brown, "A Procedure for Preliminary Estimate of Air-Cooled Heat Exchangers," *Chem. Eng.* **85**(8), 108–111 (Mar. 27, 1978).
35. E. C. Smith, "Air-Cooled Heat Exchangers," *Chem. Eng.* (Nov. 17, 1958).
36. V. Gnielinski, A. Zukauskas, and A. Skrinska, "Banks of Plain and Finned Tubes," Section 2.5.3, HEDH.
37. P. Minton, "Designing Spiral-Plate Heat Exchangers," *Chem. Eng.* **77**(9) (May 4, 1970).
38. A. Cooper and J. D. Usher, "Plate Heat Exchangers," Section 3.7, HEDH.
39. J. Marriott, "Performance of an Alfaflex Plate Heat Exchanger," *Chem. Eng. Prog.* **73**(2), 73–78 (1977).
40. D. Chisholm, "Heat Pipes," Section 3.10, HEDH.
41. J. S. Truelove, "Furnaces and Combustion Chambers," Section 3.11, HEDH.
42. W. R. Penney, "Agitated Vessels," Section 3.14, HEDH.
43. A. R. Guy, "Double-Pipe Heat Exchangers," Section 3.2, HEDH.
44. J. Taborek et al., "Fouling—The Major Unresolved Problem in Heat Transfer," *Chem. Eng. Prog.* **65**(92), 53–67 (1972).
45. *Proceedings of the Conference on Progress in the Prevention of Fouling in Process Plants,* sponsored by the Institute of Corrosion Science Technology and the Institute of Chemical Engineers, London, 1981.
46. J. W. Suitor, W. J. Marner, and R. B. Ritter, "The History and Status of Research in Fouling of Heat Exchangers in Cooling Water Service," *Canad. J. Chem. Eng.* **55** (Aug., 1977).
47. A. Cooper, J. W. Suitor, and J. D. Usher, "Cooling Water Fouling in Plate Exchangers," *Heat Transfer Eng.* **1**(3) (1979).
48. R. B. Ritter and J. W. Suitor, "Seawater Fouling of Heat Exchanger Tubes," in *Proceedings of the 2nd National Conference on Complete Water Reuse,* Chicago, 1975.
49. C. H. Gilmour, "No Fooling–No Fouling," *Chem. Eng. Prog.* **61**(7), 49–54 (1965).

50. J. V. Smith, "Improving the Performance of Vertical Thermosiphon Reboilers," *Chem. Eng. Prog.* **70**(7), 68–70 (1974).

51. J. C. Chenoweth, "Flow-Induced Vibration," Section 4.6, HEDH.

52. C. C. Gentry, R. K. Young, and W. M. Small, "RODbaffle Heat Exchanger Thermal-Hydraulic Predictive Methods," in *Proceedings of the 7th International Heat Transfer Conference,* Munich, 1982.

53. A. C. Mueller, "Criteria for Maldistribution in Viscous Flow Coolers," in *Proceedings of the 5th International Heat Transfer Conference,* HE 1.4, Tokyo, Vol. 5, pp. 170–174.

54. W. F. Berg and J. L. Berg, "Flow Patterns for Isothermal Condensation in One-Pass Air-Cooled Heat Exchangers," *Heat Transfer Eng.* **1**(4), 21–31 (1980).

55. G. Breber, J. W. Palen, and J. Taborek, "Study on Non-Condensable Vapor Accumulation in Air-Cooled Condensers," in *Proceedings of the 7th International Heat Transfer Conference,* Munich, 1982.

56. A. Blumenkrantz and J. Taborek, "Application of Stability Analysis for Design of Natural Circulation Boiling Systems and Comparison with Experimental Data," *AIChE Symp. Ser.* **68**(118) (1971).

57. V. L. Streeter, *Fluid Mechanics,* McGraw-Hill, New York, 1958.

58. E. A. D. Saunders, "Shell and Tube Heat Exchangers, Elements of Construction," Section 4.2, HEDH.

59. F. W. Schmidt, "Thermal Energy Storage and Regeneration," in *Heat Exchangers Theory and Practice,* J. Taborek et al. (eds.), Hemisphere, McGraw-Hill, New York.

60. J. C. Chen, "Correlation for Boiling Heat Transfer to Saturated Fluids in Convective Flow," *Ind. Eng. Chem. Proc. Design and Dev.* **5**(3), 322–339 (1966).

61. D. Steiner and J. Taborek, "Flow Boiling Heat Transfer in Vertical Tubes Correlated by an Asymptotic Method," *Heat Transfer Engineering* **13**(3), 43 (1992).

CHAPTER 53

AIR HEATING

Richard J. Reed
North American Manufacturing Company
Cleveland, Ohio

53.1	AIR-HEATING PROCESSES	1641	53.3 WARNINGS	1643
53.2	COSTS	1643	53.4 BENEFITS	1644

53.1 AIR-HEATING PROCESSES

Air can be heated by burning fuel or by recovering waste heat from another process. In either case, the heat can be transferred to air directly or indirectly. *Indirect air heaters* are heat exchangers wherein the products of combustion never contact or mix with the air to be heated. In waste heat recovery, the heat exchanger is termed a *recuperator.*

Direct air heaters or *direct-fired air heaters* heat the air by intentionally mixing the products or combustion of waste gas with the air to be heated. They are most commonly used for ovens and dryers. It may be impractical to use them for space heating or for preheating combustion air because of lack of oxygen in the resulting mixture ("vitiated air"). In some cases, direct-fired air heating may be limited by codes and/or by presence of harmful matter of undesirable odors from the heating stream. Direct-fired air heaters have lower first cost and lower operating (fuel) cost than indirect air heaters.

Heat requirements for direct-fired air heating. Table 53.1 lists the gross Btu of fuel input required to heat one standard cubic foot of air from a given inlet temperature to a given outlet temperature. It is based on natural gas at 60°F, having 1000 gross Btu/ft³, 910 net Btu/ft³, and stoichiometric air/gas ratio of 9.4:1. The oxygen for combustion is supplied by the air that is being heated. The hot outlet "air" includes combustion products obtained from burning sufficient natural gas to raise the air to the indicated outlet temperature.

Recovered waste heat from another nearby heating process can be used for process heating, space heating, or for preheating combustion air (Ref. 4). If the waste stream is largely nitrogen, and if the temperatures of both streams are between 0 and 800°F, where specific heats are about 0.24, a simplified heat balance can be used to evaluate the mixing conditions:

heat content of the waste stream + heat content of the fresh air = heat content of the mixture or

$$W_w T_w + W_f T_f = W_m T_m = (W_w + W_f) T_m$$

where W = weight and T = temperature of waste gas, fresh air, and mixture (subscripts w, f, and m).

Example 53.1

If a 600°F waste gas stream flowing at 100 lb/hr is available to mix with 10°F fresh air and fuel, how many pounds per hour of 110°F makeup air can be produced?

Solution:

$$(100 \times 600) + 10W_f = (100 + W_f) \times (110)$$

Mechanical Engineers' Handbook, 2nd ed., Edited by Myer Kutz.
ISBN 0-471-13007-9 © 1998 John Wiley & Sons, Inc.

Table 53.1 Heat Requirements for Direct-Fired Air Heating, Gross Btu of Fuel Input per scf of Outlet "Air."

Inlet Air Temperature, °F	Outlet Air Temperature, °F														
	100	200	300	400	500	600	700	800	900	1000	1100	1200	1300	1400	1500
−20	2.39	4.43	6.51	8.63	10.8	13.0	15.2	17.5	19.9	22.2	24.7	27.1	29.7	32.2	34.9
0	2.00	4.04	6.11	8.23	10.4	12.6	14.8	17.1	19.5	21.8	24.3	26.7	29.3	31.8	34.4
+20	1.60	3.64	5.71	7.83	9.99	12.2	14.4	16.7	19.0	21.4	23.8	26.3	28.8	31.4	34.0
40	1.20	3.24	5.31	7.43	9.58	11.8	14.0	16.3	18.6	21.0	23.4	25.9	28.4	31.0	33.6
60	0.802	2.84	4.91	7.02	9.18	11.4	13.6	15.9	18.2	20.6	23.0	25.5	28.0	30.6	33.2
80	0.402	2.43	4.51	6.62	8.77	11.0	13.2	15.5	17.8	20.2	22.6	25.1	27.6	30.1	32.7
100		2.03	4.10	6.21	8.36	10.6	12.8	15.1	17.4	19.8	22.2	24.6	27.2	29.7	32.3
200			2.06	4.17	6.31	8.50	10.7	13.0	15.3	17.7	20.1	22.5	25.0	27.6	30.2
300				2.10	4.23	6.41	8.63	10.9	13.2	15.5	17.9	20.4	22.9	25.4	28.0
400					2.13	4.30	6.51	8.76	11.1	13.4	15.8	18.2	20.7	23.2	25.8
500						2.16	4.36	6.61	8.90	11.2	13.6	16.0	18.5	21.0	23.6
600							2.19	4.43	6.71	9.03	11.4	13.8	16.3	18.8	21.3
700								2.23	4.50	6.81	9.16	11.6	14.0	16.5	19.0
800									2.26	4.56	6.91	9.30	11.7	14.2	16.7
900										2.29	4.63	7.01	9.43	11.9	14.4
1000											2.32	4.69	7.11	9.57	12.1

Example: Find the amount of natural gas required to heat 1000 scfm of air from 400°F to 1400°F.

Solution: From the table, read 23.2 gross Btu/scf air. Then $\left(\dfrac{23.2 \text{ gross Btu}}{\text{scf air}} \times \dfrac{1000 \text{ scf air}}{\text{min}} \times \dfrac{60 \text{ min}}{1 \text{ hr}} \right) \div \dfrac{1000 \text{ gross Btu}}{\text{ft}^3 \text{ gas}} = 1392 \text{ cfh gas.}$

The conventional formula derived from the specific heat equation is: $Q = wc\Delta T$; so Btu/hr = weight/hr × specific heat × temp rise = $\dfrac{\text{scf}}{\text{min}} \times \dfrac{60 \text{ min}}{\text{hr}} \times \dfrac{0.076 \text{ lb}}{\text{ft}^3} \times$

$\dfrac{0.24 \text{ Btu}}{\text{lb °F}} \times \text{°rise} = \text{scfm} \times 1.1 \times \text{°rise.}$

The table above incorporates many refinements not considered in the conventional formulas: (a) % available heat which corrects for heat loss to dry flue gases and the heat loss due to heat of vaporization in the water formed by combustion, (b) the specific heats of the products of combustion (N_2, CO_2, and H_2O) are not the same as that of air, and (c) the specific heats of the combustion products change at higher temperatures.

For the example above, the rule of thumb would give 1000 scfm × 1.1 × (1400 − 400) = 1 100 000 gross Btu/hr: whereas the example finds 1392 × 1000 = 1 392 000 gross Btu/hr required. *Reminder:* The fuel being burned adds volume and weight to the stream being heated.

Solving, we find W_f = 490 lb/hr of fresh air can be heated to 110°F, but the 100 lb/hr of waste gas will be mixed with it; so the delivered stream, W_m will be 100 + 490 = 590 lb/hr.

If "indirect" air heating is necessary, a heat exchanger (recuperator or regenerator) must be used. These may take many forms such as plate-type heat exchangers, shell and tube heat exchangers, double-pipe heat exchangers, heat-pipe exchangers, heat wheels, pebble heater recuperators, and refractory checkerworks. The supplier of the heat exchanger should be able to predict the air preheat temperature and the final waste gas temperature. The amount of heat recovered Q is then $Q = W c_p$ $(T_2 - T_1)$, where W is the weight of air heated, c_p is the specific heat of air (0.24 when below 800°F), T_2 is the delivered hot air temperature, and T_1 is the cold air temperature entering the heat exchanger. Tables and graphs later in this chapter permit estimation of fuel savings and efficiencies for cases involving preheating of combustion air.

If a waste gas stream is only a few hundred degrees Fahrenheit hotter than the air stream temperature required for heating space, an oven, or a dryer, such uses of recovered heat are highly desirable. For higher waste gas stream temperatures, however, the second law of thermodynamics would say that we can make better use of the energy by stepping it down in smaller temperature increments, and preheating combustion air usually makes more sense. This also simplifies accounting, since it returns the recovered heat to the process that generated the hot waste stream.

Preheating combustion air is a very logical method for recycling waste energy from flue gases in direct-fired industrial heating processes such as melting, forming, ceramic firing, heat treating, chemical and petroprocess heaters, and boilers. (It is always wise, however, to check the economics of using flue gases to preheat the load or to make steam in a waste heat boiler.)

53.2 COSTS

In addition to the cost of the heat exchanger for preheating the combustion air, there are many other costs that have to be weighed. Retrofit or add-on recuperators or regenerators may have to be installed overhead to keep the length of heat-losing duct and pipe to a minimum; therefore, extra foundations and structural work may be needed. If the waste gas or air is hotter than about 800°F, carbon steel pipe and duct should be insulated on the inside. For small pipes or ducts where this would be impractical, it is necessary to use an alloy with strength and oxidation resistance at the higher temperature, and to insulate on the outside.

High-temperature air is much less dense; therefore, the flow passages of burners, valves, and pipe must be greater for the same input rate and pressure drop. Burners, valves, and piping must be constructed of better materials to withstand the hot air stream. The front face of the burner is exposed to more intense radiation because of the higher flame temperature resulting from preheated combustion air.

If the system is to be operated at a variety of firing rates, the output air temperature will vary; so temperature-compensating fuel/air ratio controls are essential to avoid wasting fuel. Also, to protect the investment in the heat exchanger, it is only logical that it be protected with high-limit temperature controls.

53.3 WARNINGS

Changing temperatures from end to end of high-temperature heat exchangers and from time to time during high-temperature furnace cycles cause great thermal stress, often resulting in leaks and shortened heat-exchanger life. Heat-transfer surfaces fixed at both ends (welded or rolled in) can force something to be overstressed. Recent developments in the form of high-temperature slip seal methods, combined with sensible location of such seals in cool air entrance sections, are opening a whole new era in recuperator reliability.

Corrosion, fouling, and condensation problems continue to limit the applications of heat-recovery equipment of all kinds. Heat-transfer surfaces in air heaters are never as well cooled as those in water heaters and waste heat boilers; therefore, they must exist in a more hostile environment. However, they may experience fewer problems from acid-dew-point condensation. If corrosives, particulates, or condensables are emitted by the heating process at limited times, perhaps some temporary bypassing arrangement can be instituted. High waste gas design velocities may be used to keep particulates and condensed droplets in suspension until they reach an area where they can be safely dropped out.

Figure 53.1 shows recommended minimum temperatures to avoid "acid rain" in the heat exchanger.[2] Although a low final waste gas temperature is desirable from an efficiency standpoint, the shortened equipment life seldom warrants it. Acid forms from combination of water vapor with SO_3, SO_2, or CO_2 in the flue gases.

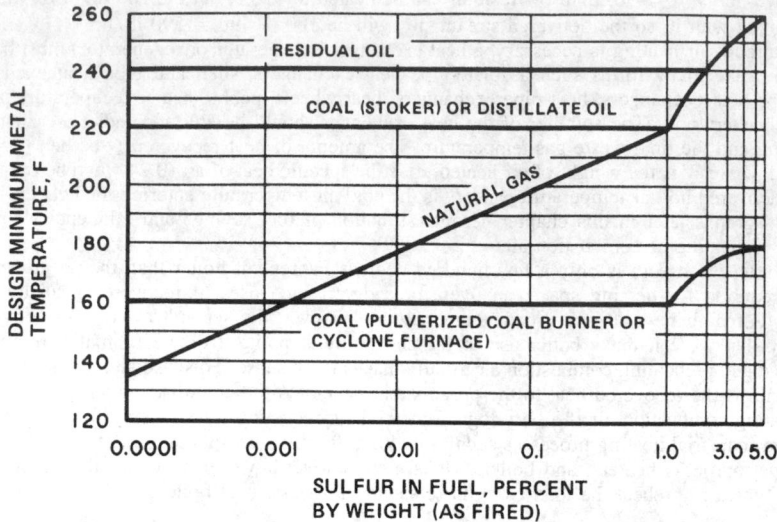

Fig. 53.1 Recommended minimum temperatures to avoid "acid rain" in heat exchangers.

53.4 BENEFITS

Despite all the costs and warnings listed above, combustion air preheating systems *do* pay. As fuel costs rise, the payback is more rewarding, even for small installations. Figure 53.2 shows percent available heat[3] (best possible efficiency) with various amounts of air preheat and a variety of furnace exit (flue) temperatures. All curves for hot air are based on 10% excess air.* The percentage of fuel saved by addition of combustion air preheating equipment can be calculated by the formula

$$\% \text{ fuel saved} = 100 \times \left(1 - \frac{\% \text{ available heat before}}{\% \text{ available heat after}}\right)$$

Table 53.2 lists fuel savings calculated by this method.[4]

Preheating combustion air raises the flame temperature and thereby enhances radiation heat transfer in the furnace, which should lower the exit gas temperature and further improve fuel efficiency. Table 53.3 and the x-intercepts of Fig. 53.2 show adiabatic flame temperatures when operating with 10% excess air,† but it is difficult to quantify the resultant saving from this effect.

Preheating combustion air has some lesser benefits. Flame stability is enhanced by the faster flame velocity and broader flammability limits. If downstream pollution control equipment is required (scrubber, baghouse), such equipment can be smaller and of less costly materials because the heat exchanger will have cooled the waste gas stream before it reaches such equipment.

*It is advisable to tune a combustion system for closer to stoichiometric air/fuel ratio *before* attempting to preheat combustion air. This is not only a quicker and less costly fuel conservation measure, but it then allows use of smaller heat-exchange equipment.

†Although 0% excess air (stoichiometric air/fuel ratio) is ideal, practical considerations usually dictate operation with 5–10% excess air. During changes in firing rate, time lag in valve operation may result in smoke formation if some excess air is not available prior to the change. Heat exchangers made of 300 series stainless steels may be damaged by alternate oxidation and reduction (particularly in the presence of sulfur). For these reasons, it is wise to have an accurate air/fuel ratio controller with very limited time-delay deviation from air/fuel ratio setpoint.

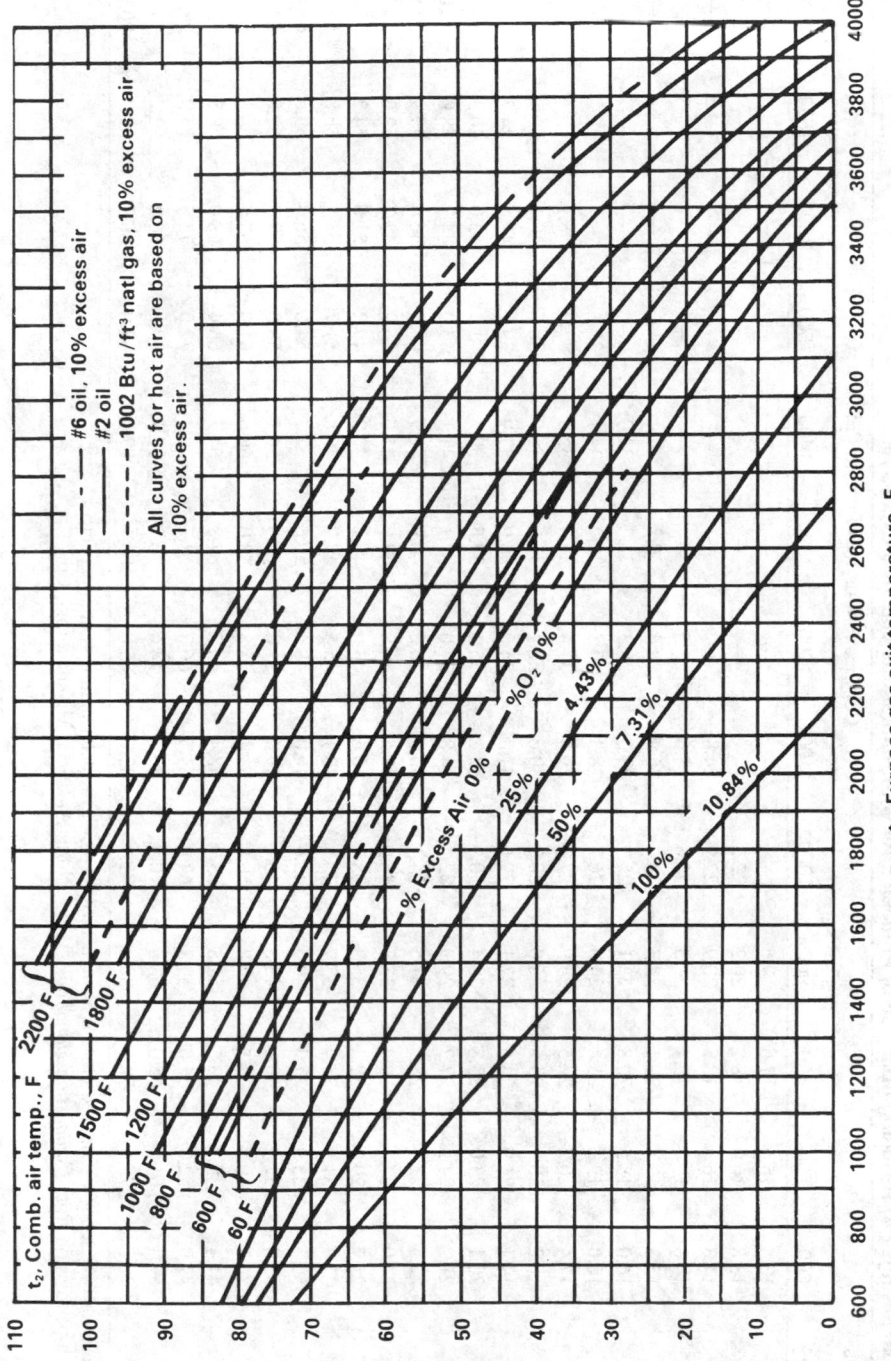

Fig. 53.2 Available heat with preheated combustion air at 10% excess air. Applicable only if there is no unburned fuel in the products of combustion. Corrected for dissociation. (Reproduced with permission from *Combustion Handbook*.[3]) See also Figs. 44.3 and 44.4.

Table 53.2 Fuel Savings (%) Resulting from Use of Preheated Air with Natural Gas and 10% Excess Air[a]

t_3, Furnace Gas Exit Temperature (°F)	t_2, Combustion Air Temperature (°F)													
	600	700	800	900	1000	1100	1200	1300	1400	1500	1600	1800	2000	2200
1000	13.4	15.5	17.6	19.6	—	—	—	—	—	—	—	—	—	—
1100	13.8	16.0	18.2	20.2	22.2	—	—	—	—	—	—	—	—	—
1200	14.3	16.6	18.7	20.9	22.9	24.8	—	—	—	—	—	—	—	—
1300	14.8	17.1	19.4	21.5	23.6	25.6	27.5	—	—	—	—	—	—	—
1400	15.3	17.8	20.1	22.3	24.4	26.4	28.4	30.2	—	—	—	—	—	—
1500	16.0	18.5	20.8	23.1	25.3	27.3	29.3	31.2	33.0	—	—	—	—	—
1600	16.6	19.2	21.6	24.0	26.2	28.3	30.3	32.2	34.1	35.8	—	—	—	—
1700	17.4	20.2	22.5	24.9	27.2	29.4	31.4	33.4	35.3	37.0	38.7	—	—	—
1800	18.2	20.9	23.5	26.0	28.3	30.6	32.7	34.6	36.5	38.3	40.1	—	—	—
1900	19.1	21.9	24.6	27.1	29.6	31.8	34.0	36.0	37.9	39.7	41.5	44.7	—	—
2000	20.1	23.0	25.8	28.4	30.9	33.2	35.4	37.5	39.4	41.3	43.0	46.3	—	—
2100	21.2	24.3	27.2	29.9	32.4	34.8	37.0	39.1	41.1	43.0	44.7	48.0	51.0	—
2200	22.5	25.7	28.7	31.5	34.1	36.5	38.8	40.9	42.9	44.8	46.6	49.9	52.8	—
2300	24.0	27.3	30.4	33.3	36.0	38.5	40.8	42.9	45.0	46.9	48.7	52.0	54.9	57.5
2400	25.7	29.2	32.4	35.3	38.1	40.6	43.0	45.2	47.2	49.2	51.0	54.2	57.1	59.7
2500	27.7	31.3	34.7	37.7	40.5	43.1	45.5	47.7	49.8	51.7	53.5	56.8	59.6	62.2
2600	30.1	33.9	37.3	40.5	43.4	46.0	48.4	50.6	52.7	54.6	56.4	59.6	62.4	64.9
2700	33.0	37.0	40.6	43.8	46.7	49.4	51.8	54.0	56.1	58.0	59.7	62.8	65.5	67.9
2800	36.7	40.8	44.5	47.8	50.8	53.4	55.8	58.0	60.0	61.9	63.5	66.5	69.1	71.3
2900	41.4	45.7	49.5	52.8	55.7	58.4	60.7	62.8	64.7	66.4	68.0	70.8	73.2	75.2
3000	47.9	52.3	56.0	59.3	62.1	64.6	66.7	68.7	70.4	72.0	73.5	75.9	78.0	79.8
3100	57.3	61.5	65.0	68.0	70.5	72.7	74.6	76.2	77.7	79.0	80.2	82.2	83.8	85.2
3200	72.2	75.6	78.3	80.4	82.2	83.7	85.0	86.1	87.1	87.9	88.7	89.9	90.9	91.8

[a]These figures are for evaluating a proposed change to preheated air—not for determining system capacity.
Reproduced with permission from *Combustion Handbook*, Vol. I, North American Manufacturing Co.

Table 53.3 Effect of Combustion Air Preheat on Flame Temperature

Excess Air (%)	Preheated Combustion Air Temperature (°F)	Adiabatic Flame Temperatured (°F)		
		With 1000 Btu/scf Natural Gas	With 137,010 Btu/gal Distillate Fuel Oil	With 153,120 Btu/gal Residual Fuel Oil
0	60	3468	3532	3627
10	60	3314	3374	3475
10	600	3542	3604	3690
10	700	3581	3643	3727
10	800	3619	3681	3763
10	900	3656	3718	3798
10	1000	3692	3754	3831
10	1100	3727	3789	3864
10	1200	3761	3823	3896
10	1300	3794	3855	3927
10	1400	3826	3887	3957
10	1500	3857	3918	3986
10	1600	3887	3948	4014
10	1700	3917	3978	4042
10	1800	3945	4006	4069
10	1900	3973	4034	4095
10	2000	4000	4060	4121
0	2000	4051	4112	4171

REFERENCES

1. *Heat Requirements for Direct-Fired Air Heating,* North American Mfg. Co., Cleveland, OH, 1981.
2. *Steam—Its Generation and Use,* Babcock and Wilcox, New York, 1978.
3. R. J. Reed, *Combustion Handbook,* 3rd ed., Vol. 1, North American Manufacturing Co., Cleveland, OH, 1986.
4. R. J. Reed, *Combustion Handbook,* 4th ed., Vol. 2, North American Manufacturing. Co., Cleveland, OH, 1997.

CHAPTER 54

COOLING ELECTRONIC EQUIPMENT

Allan Kraus
Allan D. Kraus Associates
Aurora, Ohio

54.1	**THERMAL MODELING**	**1649**		54.2.2	Forced Convection	1662
	54.1.1 Introduction	1649				
	54.1.2 Conduction Heat Transfer	1649	**54.3**	**THERMAL CONTROL TECHNIQUES**		**1667**
	54.1.3 Convective Heat Transfer	1652		54.3.1 Extended Surface and Heat Sinks		1672
	54.1.4 Radiative Heat Transfer	1655		54.3.2 The Cold Plate		1672
	54.1.5 Chip Module Thermal Resistances	1656		54.3.3 Thermoelectric Coolers		1674
54.2	**HEAT-TRANSFER CORRELATIONS FOR ELECTRONIC EQUIPMENT COOLING**	**1661**				
	54.2.1 Natural Convection in Confined Spaces	1661				

54.1 THERMAL MODELING

54.1.1 Introduction

To determine the temperature differences encountered in the flow of heat within electronic systems, it is necessary to recognize the relevant heat transfer mechanisms and their governing relations. In a typical system, heat removal from the active regions of the microcircuit(s) or chip(s) may require the use of several mechanisms, some operating in series and others in parallel, to transport the generated heat to the coolant or ultimate heat sink. Practitioners of the thermal arts and sciences generally deal with four basic thermal transport modes: conduction, convection, phase change, and radiation.

54.1.2 Conduction Heat Transfer

One-Dimensional Conduction

Steady thermal transport through solids is governed by the Fourier equation, which, in one-dimensional form, is expressible as

$$q = -kA \frac{dT}{dx} \quad (\text{W}) \quad (54.1)$$

where q is the heat flow, k is the thermal conductivity of the medium, A is the cross-sectional area for the heat flow, and dT/dx is the temperature gradient. Here, heat flow produced by a negative temperature gradient is considered positive. This convention requires the insertion of the minus sign in Eq. (54.1) to assure a positive heat flow, q. The temperature difference resulting from the steady state diffusion of heat is thus related to the thermal conductivity of the material, the cross-sectional area and the path length, L, according to

$$(T_1 - T_2)_{\text{cd}} = q \frac{L}{kA} \quad (\text{K}) \quad (54.2)$$

Mechanical Engineers' Handbook, 2nd ed., Edited by Myer Kutz.
ISBN 0-471-13007-9   © 1998 John Wiley & Sons, Inc.

The form of Eq. (54.2) suggests that, by analogy to Ohm's Law governing electrical current flow through a resistance, it is possible to define a thermal resistance for conduction, R_{cd}, as

$$R_{cd} \equiv \frac{(T_1 - T_2)}{q} = \frac{L}{kA} \tag{54.3}$$

One-Dimensional Conduction with Internal Heat Generation

Situations in which a solid experiences internal heat generation, such as that produced by the flow of an electric current, give rise to more complex governing equations and require greater care in obtaining the appropriate temperature differences. The axial temperature variation in a slim, internally heated conductor whose edges (ends) are held at a temperature T_o is found to equal

$$T = T_o + q_g \frac{L^2}{2k} \left[\left(\frac{x}{L} \right) - \left(\frac{x}{L} \right)^2 \right]$$

When the volumetic heat generation rate, q_g, in W/m^3 is uniform throughout, the peak temperature is developed at the center of the solid and is given by

$$T_{max} = T_o + q_g \frac{L^2}{8k} \quad (K) \tag{54.4}$$

Alternatively, because q_g is the volumetric heat generation, $q_g = q/LW\delta$, the center–edge temperature difference can be expressed as

$$T_{max} - T_o = q \frac{L^2}{8kLW\delta} = q \frac{L}{8kA} \tag{54.5}$$

where the cross-sectional area, A, is the product of the width, W, and the thickness, δ. An examination of Eq. (54.5) reveals that the thermal resistance of a conductor with a distributed heat input is only one quarter that of a structure in which all of the heat is generated at the center.

Spreading Resistance

In chip packages that provide for lateral spreading of the heat generated in the chip, the increasing cross-sectional area for heat flow at successive "layers" below the chip reduces the internal thermal resistance. Unfortunately, however, there is an additional resistance associated with this lateral flow of heat. This, of course, must be taken into account in the determination of the overall chip package temperature difference.

For the circular and square geometries common in microelectronic applications, an engineering approximation for the spreading resistance for a small heat source on a thick substrate or heat spreader (required to be 3 to 5 times thicker than the square root of the heat source area) can be expressed as[1]

$$R_{sp} = \frac{0.475 - 0.62\epsilon + 0.13\epsilon^2}{k\sqrt{A_c}} \quad (K/W) \tag{54.6}$$

where ϵ is the ratio of the heat source area to the substrate area, k is the thermal conductivity of the substrate, and A_c is the area of the heat source.

For relatively thin layers on thicker substrates, such as encountered in the use of thin lead-frames, or heat spreaders interposed between the chip and substrate, Eq. (54.6) cannot provide an acceptable prediction of R_{sp}. Instead, use can be made of the numerical results plotted in Fig 54.1 to obtain the requisite value of the spreading resistance.

Interface/Contact Resistance

Heat transfer across the interface between two solids is generally accompanied by a measurable temperature difference, which can be ascribed to a contact or interface thermal resistance. For perfectly adhering solids, geometrical differences in the crystal structure (lattice mismatch) can impede the flow of phonons and electrons across the interface, but this resistance is generally negligible in engineering design. However, when dealing with real interfaces, the asperities present on each of the surfaces, as shown in an artist's conception in Fig 54.2, limit actual contact between the two solids to a very small fraction of the apparent interface area. The flow of heat across the gap between two solids in nominal contact is thus seen to involve solid conduction in the areas of actual contact and fluid conduction across the "open" spaces. Radiation across the gap can be important in a vacuum environment or when the surface temperatures are high.

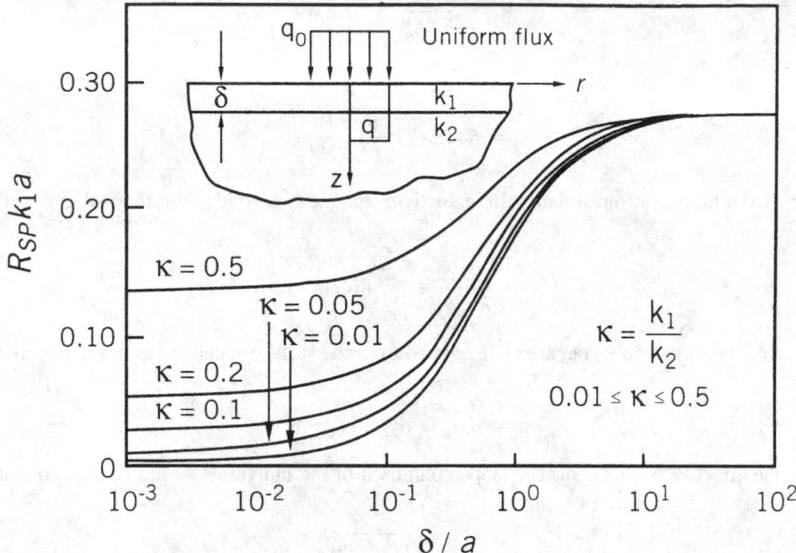

Fig. 54.1 The thermal resistance for a circular heat source on a two layer substrate (from Ref. 2).

The heat transferred across an interface can be found by adding the effects of the solid–to–solid conduction and the conduction through the fluid and recognizing that the solid–to–solid conduction, in the contact zones, involves heat flowing sequentially through the two solids. With the total contact conductance, h_{co}, taken as the sum of the solid–to–solid conductance, h_c, and the gap conductance, h_g

$$h_{co} = h_c + h_g \qquad (\text{W/m}^2 \cdot \text{K}) \qquad (54.7a)$$

the contact resistance based on the apparent contact area, A_a, may be defined as

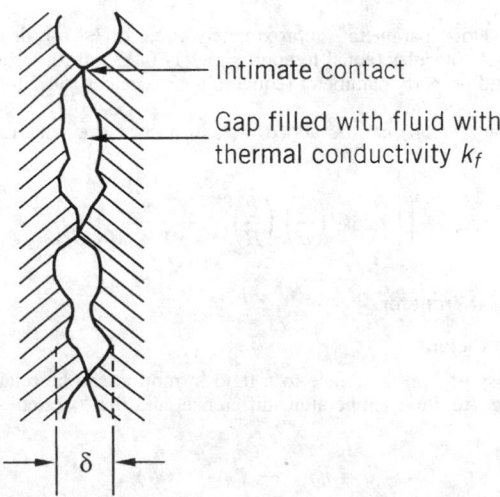

Fig. 54.2 Physical contact between two nonideal surfaces.

$$R_{co} \equiv \frac{1}{h_{co} A_a} \quad (K/W) \tag{54.7b}$$

In Eq. (54.7a), h_c is given by

$$h_c = 54.25 k_s \left(\frac{m}{\sigma}\right) \left(\frac{P}{H}\right)^{0.95} \tag{54.8a}$$

where k_s is the harmonic mean thermal conductivity for the two solids with thermal conductivities, k_1 and k_2,

$$k_s = \frac{2 k_1 k_2}{k_1 + k_2} \quad (W/m \cdot K)$$

σ is the effective rms surface roughness developed from the surface roughnesses of the two materials, σ_1 and σ_2,

$$\sigma = \sqrt{\sigma_1^2 + \sigma_2^2} \quad (\mu \cdot m)$$

and m is the effective absolute surface slope composed of the individual slopes of the two materials, m_1 and m_2,

$$m = \sqrt{m_1^2 + m_2^2}$$

where P is the contact pressure and H is the microhardness of the softer material, both in N/m^2. In the absence of detailed information, the σ/m ratio can be taken equal to 5–9 microns for relatively smooth surfaces.[1,2]

In Eq. (54.7a), h_g is given by

$$h_g = \frac{k_g}{Y + M} \tag{54.8b}$$

where k_g is the thermal conductivity of the gap fluid, Y is the distance between the mean planes (Fig. 54.2) given by

$$\frac{Y}{\sigma} = 54.185 \left[-\ln \left(3.132 \frac{P}{H} \right) \right]^{0.547}$$

and M is a gas parameter used to account for rarefied gas effects

$$M = \alpha \beta \Lambda$$

where α is an accommodation parameter (approximately equal to 2.4 for air and clean metals), Λ is the mean free path of the molecules (equal to approximately 0.06 μm for air at atmospheric pressure and 15°C), and β is a fluid property parameter (equal to approximately 54.7 for air and other diatomic gases).

Equations (54.8a) and (54.8b) can be added and, in accordance with Eq. (54.7b), the contact resistance becomes

$$R_{co} \equiv \left\{ \left[1.25 k_s \left(\frac{m}{\sigma}\right) \left(\frac{P}{H}\right)^{0.95} + \frac{k_g}{Y + M} \right] A_a \right\}^{-1} \tag{54.9}$$

54.1.3 Convective Heat Transfer

The Heat Transfer Coefficient

Convective thermal transport from a surface to a fluid in motion can be related to the heat transfer coefficient, h, the surface–to–fluid temperature difference, and the "wetted" surface area, S, in the form

$$q = hS(T_s - T_{fl}) \quad (W) \tag{54.10}$$

The differences between convection to a rapidly moving fluid, a slowly flowing or stagnant fluid, as well as variations in the convective heat transfer rate among various fluids, are reflected in the values of h. For a particular geometry and flow regime, h may be found from available empirical correlations and/or theoretical relations. Use of Eq. (54.10) makes it possible to define the convective thermal resistance as

$$R_{cv} \equiv \frac{1}{hS} \quad (K/W) \tag{54.11}$$

Dimensionless Parameters

Common dimensionless quantities that are used in the correlation of heat transfer data are the *Nusselt number*, Nu, which relates the convective heat transfer coefficient to the conduction in the fluid where the subscript, *fl*, pertains to a fluid property,

$$Nu \equiv \frac{h}{k_{fl}/L} = \frac{hL}{k_{fl}}$$

the *Prandtl number*, Pr, which is a fluid property parameter relating the diffusion of momentum to the conduction of heat,

$$Pr \equiv \frac{c_p \mu}{k_{fl}}$$

the *Grashof number*, Gr, which accounts for the bouyancy effect produced by the volumetric expansion of the fluid,

$$Gr \equiv \frac{\rho^2 \beta g L^3 \Delta T}{\mu^2}$$

and the *Reynolds number*, Re, which relates the momentum in the flow to the viscous dissipation,

$$Re \equiv \frac{\rho V L}{\mu}$$

Natural Convection

In natural convection, fluid motion is induced by density differences resulting from temperature gradients in the fluid. The heat transfer coefficient for this regime can be related to the buoyancy and the thermal properties of the fluid through the *Rayleigh number*, which is the product of the Grashof and Prandtl numbers,

$$Ra = \frac{\rho^2 \beta g c_p}{\mu k_{fl}} L^3 \Delta T$$

where the fluid properties, ρ, β, c_p, μ, and k, are evaluated at the fluid bulk temperature and ΔT is the temperature difference between the surface and the fluid.

Empirical correlations for the natural convection heat transfer coefficient generally take the form

$$h = C \left(\frac{k_{fl}}{L} \right) (Ra)^n \quad (W/m^2 \cdot K) \tag{54.12}$$

where n is found to be approximately 0.25 for $10^3 < Ra < 10^9$, representing laminar flow, 0.33 for $10^9 < Ra < 10^{12}$, the region associated with the transition to turbulent flow, and 0.4 for $Ra > 10^{12}$, when strong turbulent flow prevails. The precise value of the correlating coefficient, C, depends on fluid, the geometry of the surface, and the Rayleigh number range. Nevertheless, for common plate, cylinder, and sphere configurations, it has been found to vary in the relatively narrow range of 0.45–0.65 for laminar flow and 0.11–0.15 for turbulent flow past the heated surface.[42]

Natural convection in vertical channels such as those formed by arrays of longitudinal fins is of major significance in the analysis and design of heat sinks and experiments for this configuration have been conducted and confirmed.[4,5]

These studies have revealed that the value of the Nusselt number lies between two extremes associated with the separation between the plates or the channel width. For wide spacing, the plates appear to have little influence upon one another and the Nusselt number in this case achieves its *isolated plate limit*. On the other hand, for closely spaced plates or for relatively long channels, the fluid attains its *fully developed* value and the Nusselt number reaches its *fully developed limit*. Intermediate values of the Nusselt number can be obtained from a form of a correlating expression for smoothly varying processes and have been verified by detailed experimental and numerical studies.[19,20]

Thus, the correlation for the average value of h along isothermal vertical placed separated by a spacing, z

$$h = \frac{k_{fl}}{z} \left[\frac{576}{(\text{El})^2} + \frac{2.873}{(\text{El})^{1/2}} \right]^{1/2} \tag{54.13}$$

where El is the *Elenbaas number*

$$\text{El} \equiv \frac{\rho^2 \beta g c_p z^4 \Delta T}{\mu k_{fl} L}$$

and $\Delta T = T_s - T_{fl}$.

Several correlations for the coefficient of heat transfer in natural convection for various configurations are provided in Section 54.2.1.

Forced Convection

For forced flow in long, or very narrow, parallel-plate channels, the heat transfer coefficient attains an asymptotic value (a fully developed limit), which for symmetrically heated channel surfaces is equal approximately to

$$h = \frac{4k_{fl}}{d_e} \quad (\text{W/m}^2 \cdot \text{K}) \tag{54.14}$$

where d_e is the *hydraulic diameter* defined in terms of the flow area, A, and the wetted perimeter of the channel, P_w

$$d_e \equiv \frac{4A}{P_w}$$

Several correlations for the coefficient of heat transfer in forced convection for various configurations are provided in Section 54.2.2.

Phase Change Heat Transfer

Boiling heat transfer displays a complex dependence on the temperature difference between the heated surface and the saturation temperature (boiling point) of the liquid. In nucleate boiling, the primary region of interest, the ebullient heat transfer rate can be approximated by a relation of the form

$$q_\phi = C_{sf} A (T_s - T_{sat})^3 \quad (\text{W}) \tag{54.15}$$

where C_{sf} is a function of the surface/fluid combination and various fluid properties. For comparison purposes, it is possible to define a boiling heat transfer coefficient, h_ϕ,

$$h_\phi = C_{sf}(T_s - T_{sat})^2 \quad [\text{W/m}^2 \cdot \text{K}]$$

which, however, will vary strongly with surface temperature.

Finned Surfaces

A simplified discussion of finned surfaces is germane here and what now follows is not inconsistent with the subject matter contained Section 54.3.1. In the thermal design of electronic equipment, frequent use is made of finned or "extended" surfaces in the form of *heat sinks* or *coolers*. While such finning can substantially increase the surface area in contact with the coolant, resistance to heat flow in the fin reduces the average temperature of the exposed surface relative to the fin base. In the analysis of such finned surfaces, it is common to define a fin efficiency, η, equal to the ratio of the actual heat dissipated by the fin to the heat that would be dissipated if the fin possessed an infinite thermal conductivity. Using this approach, heat transferred from a fin or a fin structure can be expressed in the form

$$q_f = hS_f \eta (T_b - T_s) \quad (\text{W}) \tag{54.16}$$

where T_b is the temperature at the base of the fin and where T_s is the surrounding temperature and q_f is the heat entering the base of the fin, which, in the steady state, is equal to the heat dissipated by the fin.

The thermal resistance of a finned surface is given by

$$R_f \equiv \frac{1}{h S_f \eta} \qquad (54.17)$$

where η, the fin efficiency, is 0.627 for a thermally optimum rectangular cross section fin,[11]

Flow Resistance

The transfer of heat to a flowing gas or liquid that is not undergoing a phase change results in an increase in the coolant temperature from an inlet temperature of T_{in} to an outlet temperature of T_{out}, according to

$$q = m c_p (T_{out} - T_{in}) \qquad (W) \qquad (54.18)$$

Based on this relation, it is possible to define an effective flow resistance, R_{fl}, as

$$R_{fl} \equiv \frac{1}{\dot{m} c_p} \qquad (K/W) \qquad (54.19)$$

where $\dot{m}$ is in kg/sec.

54.1.4   Radiative Heat Transfer

Unlike conduction and convection, radiative heat transfer between two surfaces or between a surface and its surroundings is not linearly dependent on the temperature difference and is expressed instead as

$$q = \sigma S \mathcal{F}(T_1^4 - T_2^4) \qquad (W) \qquad (54.20)$$

where $\mathcal{F}$ includes the effects of surface properties and geometry and σ is the Stefan–Boltzman constant, $\sigma = 5.67 \times 10^{-8}$ W/m² · K⁴. For modest temperature differences, this equation can be linearized to the form

$$q = h_r S(T_1 - T_2) \qquad (W) \qquad (54.21)$$

where h_r is the effective "radiation" heat transfer coefficient

$$h_r = \sigma \mathcal{F}(T_1^2 + T_2^2)(T_1 + T_2) \qquad (W/m^2 \cdot K) \qquad (54.22a)$$

and, for small $\Delta T = T_1 - T_2$, h_r is approximately equal to

$$h_r = 4\sigma \mathcal{F}(T_1 T_2)^{3/2} \qquad (W/m^2 \cdot K) \qquad (54.22b)$$

It is of interest to note that for temperature differences of the order of 10 K, the radiative heat transfer coefficient, h_r, for an ideal (or "black") surface in an absorbing environment is approximately equal to the heat transfer coefficient in natural convection of air.

Noting the form of Eq. (54.21), the radiation thermal resistance, analogous to the convective resistance, is seen to equal

$$R_r \equiv \frac{1}{h_r S} \qquad (K/W) \qquad (54.23)$$

Thermal Resistance Network

The expression of the governing heat transfer relations in the form of thermal resistances greatly simplifies the first-order thermal analysis of electronic systems. Following the established rules for resistance networks, thermal resistances that occur sequentially along a thermal path can be simply summed to establish the overall thermal resistance for that path. In similar fashion, the reciprocal of the effective overall resistance of several parallel heat transfer paths can be found by summing the reciprocals of the individual resistances. In refining the thermal design of an electronic system, prime attention should be devoted to reducing the largest resistances along a specified thermal path and/or providing parallel paths for heat removal from a critical area.

While the thermal resistances associated with various paths and thermal transport mechanisms constitute the "building blocks" in performing a detailed thermal analysis, they have also found widespread application as "figures-of-merit" in evaluating and comparing the thermal efficacy of various packaging techniques and thermal management strategies.

54.1.5　Chip Module Thermal Resistances

Definition

The thermal performance of alternative chip and packaging techniques is commonly compared on the basis of the overall (junction-to-coolant) thermal resistance, R_T. This packaging figure-of-merit is generally defined in a purely empirical fashion,

$$R_T \equiv \frac{T_j - T_{fl}}{q_c} \quad (\text{K/W}) \tag{54.24}$$

where T_j and T_{fl} are the junction and coolant (fluid) temperatures, respectively, and q_c is the chip heat dissipation.

　　Unfortunately, however, most measurement techniques are incapable of detecting the actual junction temperature, that is, the temperature of the small volume at the interface of p-type and n-type semiconductors. Hence, this term generally refers to the average temperature or a representative temperature on the chip. To lower chip temperature at a specified power dissipation, it is clearly necessary to select and/or design a chip package with the lowest thermal resistance.

　　Examination of various packaging techniques reveals that the junction-to-coolant thermal resistance is, in fact, composed of an internal, largely conductive, resistance and an external, primarily convective, resistance. As shown in Fig. 54.3, the internal resistance, R_{jc}, is encountered in the flow of dissipated heat from the active chip surface through the materials used to support and bond the chip and on to the case of the integrated circuit package. The flow of heat from the case directly to the coolant, or indirectly through a fin structure and then to the coolant, must overcome the external resistance, R_{ex}.

　　The thermal design of single-chip packages, including the selection of die-bond, heat spreader, substrate, and encapsulant materials, as well as the quality of the bonding and encapsulating processes, can be characterized by the internal, or so-called junction–to–case, resistance. The convective heat removal techniques applied to the external surfaces of the package, including the effect of finned heat sinks and other thermal enhancements, can be compared on the basis of the external thermal resistance. The complexity of heat flow and coolant flow paths in a multichip module generally requires that the thermal capability of these packaging configurations be examined on the basis of overall, or chip-to-coolant, thermal resistance.

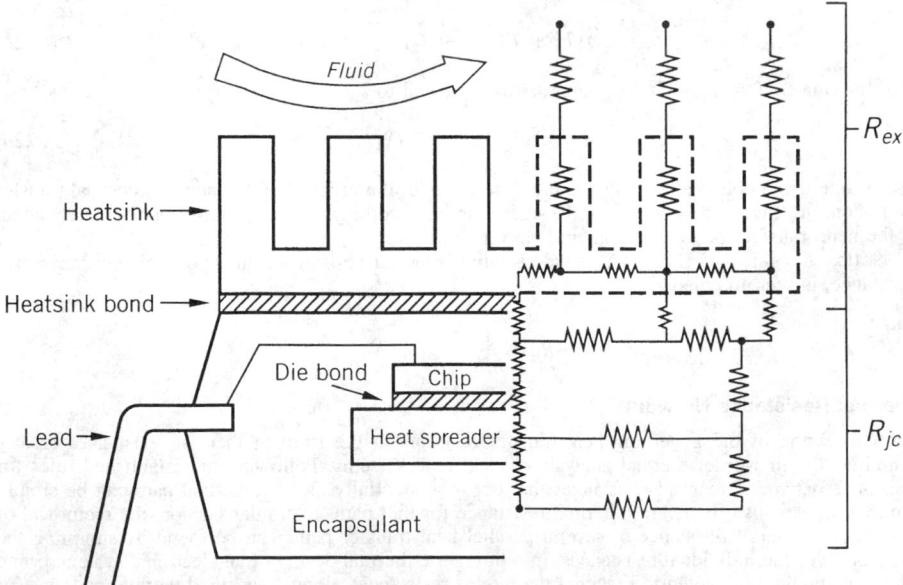

Fig. 54.3　Primary thermal resistances in a single chip package.

Internal Thermal Resistance

As discussed in Section 54.1.2, conductive thermal transport is governed by the Fourier equation, which can be used to define a conduction thermal resistance, as in Eq. (54.3). In flowing from the chip to the package surface or case, the heat encounters a series of resistances associated with individual layers of materials such as silicon, solder, copper, alumina, and epoxy, as well as the contact resistances that occur at the interfaces between pairs of materials. Although the actual heat flow paths within a chip package are rather complex and may shift to accommodate varying external cooling situations, it is possible to obtain a first-order estimate of the internal resistance by assuming that power is dissipated uniformly across the chip surface and that heat flow is largely one-dimensional. To the accuracy of these assumptions,

$$R_{jc} = \frac{T_j - T_c}{qc} = \sum \frac{x}{kA} \quad \text{(K/W)} \tag{54.25}$$

can be used to determine the internal chip module resistance where the summed terms represent the conduction thermal resistances posed by the individual layers, each with thickness x. As the thickness of each layer decreases and/or the thermal conductivity and cross-sectional area increase, the resistance of the individual layers decreases. Values of R_{cd} for packaging materials with typical dimensions can be found via Eq. (54.25) or Fig 54.4, to range from 2 K/W for a 1000 mm^2 by 1 mm thick layer of epoxy encapsulant to 0.0006 K/W for a 100 mm^2 by 25 micron (1 mil) thick layer of copper. Similarly, the values of conduction resistance for typical "soft" bonding materials are found to lie in the range of approximately 0.1 K/W for solders and 1–3 K/W for epoxies and thermal pastes for typical x/A ratios of 0.25 to 1.0.

Commercial fabrication practice in the late 1990s yields internal chip package thermal resistances varying from approximately 80 K/W for a plastic package with no heat spreader to 15–20 K/W for a plastic package with heat spreader, and to 5–10 K/W for a ceramic package or an especially designed plastic chip package. Large and/or carefully designed chip packages can attain even lower values of R_{jc}, down perhaps to 2 K/W.

Comparison of theoretical and experimental values of R_{jc} reveals that the resistances associated with compliant, low-thermal-conductivity bonding materials and the spreading resistances, as well as

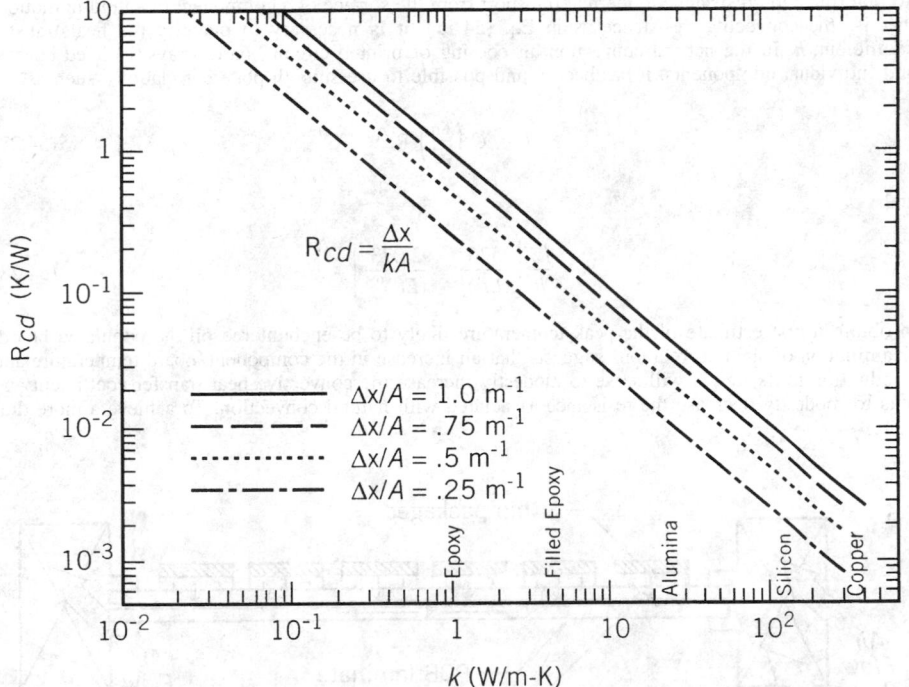

Fig. 54.4 Conductive thermal resistance for packaging materials.

the contact resistances at the lightly loaded interfaces within the package, often dominate the internal thermal resistance of the chip package. It is thus not only necessary to determine the bond resistance correctly but also to add the values of R_{sp}, obtained from Eq. (54.6) and/or Fig. 54.1, and R_{co} from Eq. (54.7b) or (54.9) to the junction–to–case resistance calculated from Eq. (54.25). Unfortunately, the absence of detailed information on the voidage in the die-bonding and heat-sink attach layers and the present inability to determine, with precision, the contact pressure at the relevant interfaces, conspire to limit the accuracy of this calculation.

Substrate or PCB Conduction

In the design of airborne electronic systems and equipment to be operated in a corrosive or damaging environment, it is often necessary to conduct the heat dissipated by the components down into the substrate or printed circuit board and, as shown in Fig. 54.5, across the substrate/PCB to a cold plate or sealed heat exchanger. For a symmetrically cooled substrate/PCB with approximately uniform heat dissipation on the surface, a first estimate of the peak temperature, at the center of the board, can be obtained by use of Eq. (54.5).

Setting the heat generation rate equal to the heat dissipated by all the components and using the volume of the board in the denominator, the temperature difference between the center at T_{ctr} and the edge of the substrate/PCB at T_o is given by

$$T_{ctr} - T_o = \left(\frac{Q}{LW\delta}\right)\left(\frac{L^2}{8k_e}\right) = \frac{QL}{8W\delta k_e} \tag{54.26}$$

where Q is the total heat dissipation, W, L, and δ are the width, length, and thickness, respectively, and k_e is the effective thermal conductivity of the board.

This relation can be used effectively in the determination of the temperatures experienced by conductively cooled substrates and conventional printed circuit boards, as well as PCBs with copper lattices on the surface, metal cores, or heat sink plates in the center. In each case it is necessary to evaluate or obtain the effective thermal conductivity of the conducting layer. As an example, consider an alumina substrate 0.20 m long, 0.15 m wide and 0.005 m thick with a thermal conductivity of 20 W/m · K, whose edges are cooled to 35°C by a cold-plate. Assuming that the substrate is populated by 30 components, each dissipating 1 W, use of Eq. (54.26) reveals that the substrate center temperature will equal 85°C.

External Resistance

To determine the resistance to thermal transport from the surface of a component to a fluid in motion, that is, the convective resistance as in Eq. (54.11), it is necessary to quantify the heat transfer coefficient, h. In the natural convection air cooling of printed circuit board arrays, isolated boards, and individual components, it has been found possible to use smooth-plate correlations, such as

$$h = C\left(\frac{k_{fl}}{L}\right)\text{Ra}^n \tag{54.27}$$

and

$$h = \frac{k_{fl}}{b}\left[\frac{576}{(El^1)^2} + \frac{2.073}{(El^1)^{0.5}}\right]^{-1/2} \tag{54.28}$$

to obtain a first estimate of the peak temperature likely to be encountered on the populated board. Examination of such correlations suggests that an increase in the component/board temperature and a reduction in its length will serve to modestly increase the convective heat transfer coefficient and thus to modestly decrease the resistance associated with natural convection. To achieve a more dra-

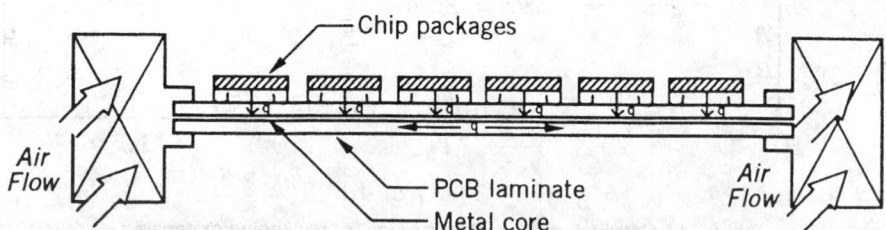

Fig. 54.5 Edge-cooled printed circuit board populated with components.

matic reduction in this resistance, it is necessary to select a high density coolant with a large thermal expansion coefficient—typically a pressurized gas or a liquid.

When components are cooled by forced convection, the laminar heat transfer coefficient, given by Eq. (54.17), is found to be directly proportional, to the square root of fluid velocity and inversely proportional to the square root of the characteristic dimension. Increases in the thermal conductivity of the fluid and in Pr, as are encountered in replacing air with a liquid coolant, will also result in higher heat transfer coefficients. In studies of low-velocity convective air cooling of simulated integrated circuit packages, the heat transfer coefficient, h, has been found to depend somewhat more strongly on Re (using channel height as the characteristic length) than suggested in Eq. (54.17), and to display a Reynolds number exponent of 0.54 to 0.72.[8-10] When the fluid velocity and the Reynolds number increase, turbulent flow results in higher heat transfer coefficients, which, following Eq. (54.19), vary directly with the velocity to the 0.8 power and inversely with the characteristic dimension to the 0.2 power. The dependence on fluid conductivity and Pr remains unchanged.

An application of Eq. (54.27) or (54.28) to the transfer of heat from the case of a chip module to the coolant shows that the external resistance, $R_{ex} = 1/hS$, is inversely proportional to the wetted surface area and to the coolant velocity to the 0.5 to 0.8 power and directly proportional to the length scale in the flow direction to the 0.5 to 0.2 power. It may thus be observed that the external resistance can be strongly influenced by the fluid velocity and package dimensions and that these factors must be addressed in any meaningful evaluation of the external thermal resistances offered by various packaging technologies.

Values of the external resistance, for a variety of coolants and heat transfer mechanisms are shown in Fig. 54.6 for a typical component wetted area of 10 cm² and a velocity range of 2–8 m/s. They are seen to vary from a nominal 100 K/W for natural convection in air, to 33 K/W for forced convection in air, to 1 K/W in fluorocarbon liquid forced convection, and to less than 0.5 K/W for boiling in fluorocarbon liquids. Clearly, larger chip packages will experience proportionately lower external resistances than the displayed values. Moreover, conduction of heat through the leads and package base into the printed circuit board or substrate will serve to further reduce the effective thermal resistance.

In the event that the direct cooling of the package surface is inadequate to maintain the desired chip temperature, it is common to attach finned heat sinks, or compact heat exchangers, to the chip package. These heat sinks can considerably increase the wetted surface area, but may act to reduce the convective heat transfer coefficient by obstructing the flow channel. Similarly, the attachment of a heat sink to the package can be expected to introduce additional conductive resistances, in the

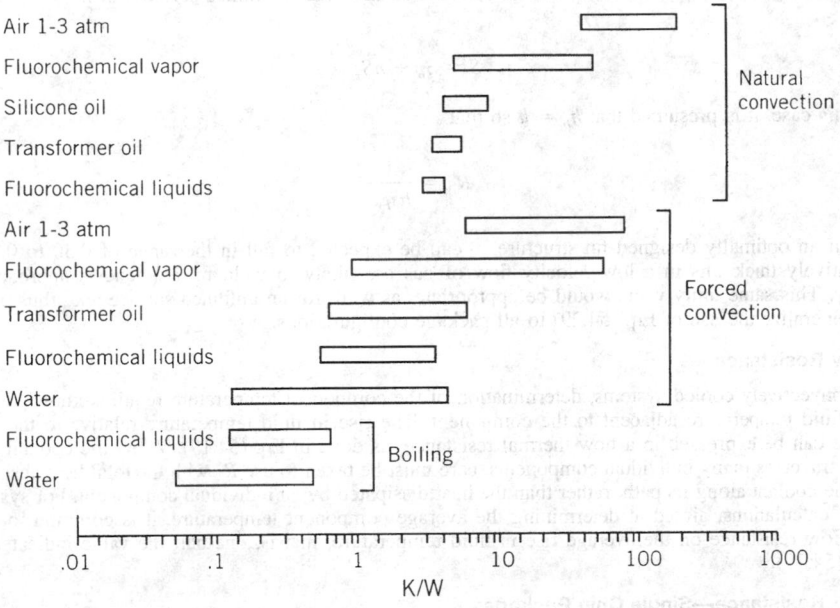

Note: For wetted area = 10 cm²

Fig. 54.6 Typical external (convective) thermal resistances for various coolants and cooling nodes.

adhesive used to bond the heat sink and in the body of the heat sink. Typical air-cooled heat sinks can reduce the external resistance to approximately 15 K/W in natural convection and to as low as 5 K/W for moderate forced convection velocities.

When a heat sink or compact heat exchanger is attached to the package, the external resistance accounting for the bond-layer conduction and the total resistance of the heat sink, R_{sk}, can be expressed as

$$R_{ex} = \frac{T_c - T_{fl}}{q_c} = \sum \left(\frac{x}{kA}\right)_b + R_{sk} \quad (K/W) \qquad (54.29)$$

where R_{sk}

$$R_{sk} = \left[\frac{1}{nhS_f\eta} + \frac{1}{h_bS_b}\right]^{-1}$$

is the the parallel combination of the resistance of the n fins

$$R_f = \frac{1}{nhS_f\eta}$$

and the *bare* or base surface not occupied by the fins

$$R_b = \frac{1}{h_bS_b}$$

Here, the base surface is $S_b = S - S_f$ and the heat transfer coefficient, h_b, is used because the heat transfer coefficient that is applied to the base surfaces is not necessarily equal to that applied to the fins.

An alternative expression for R_{sk} involves and *overall surface efficiency*, η_o, defined by

$$\eta_o = 1 - \frac{nS_f}{S}(1 - \eta)$$

where S is the total surface composed of the base surface and the finned surfaces of n fins

$$S = S_b + nS_f$$

In this case, it is presumed that $h_b = h$ so that

$$R_{sk} = \frac{1}{h\eta_o S}$$

In an optimally designed fin structure, η can be expected to fall in the range of 0.50 to 0.70.[11] Relatively thick fins in a low-velocity flow of gas are likely to yield fin efficiencies approaching unity. This same unity value would be appropriate, as well, for an unfinned surface and, thus, serve to generalize the use of Eq. (54.29) to all package configurations.

Flow Resistance

In convectively cooled systems, determination of the component temperature requires knowledge of the fluid temperature adjacent to the component. The rise in fluid temperature relative to the inlet value can be expressed in a flow thermal resistance, as done in Eq. (54.19). When the coolant flow path traverses many individual components, care must be taken to use R_{fl} with the total heat absorbed by the coolant along its path, rather than the heat dissipated by an individual component. For system-level calculations, aimed at determining the average component temperature, it is common to base the flow resistance on the average rise in fluid temperature, that is, one-half the value indicated by Eq. (54.19).

Total Resistance—Single Chip Packages

To the accuracy of the assumptions employed in the preceding development, the overall single-chip package resistance, relating the chip temperature to the inlet temperature of the coolant, can be found by summing the internal, external, and flow resistances to yield

$$R_T = R_{jc} + R_{ex} + R_{fl} = \sum \frac{x}{kA} + R_{int} + R_{sp}$$

$$+ R_{sk} + \left(\frac{Q}{q}\right)\left(\frac{1}{2\rho Q c_p}\right) \qquad (\text{K/W}) \qquad (54.30)$$

In evaluating the thermal resistance by this relationship, care must be taken to determine the effective cross-sectional area for heat flow at each layer in the module and to consider possible voidage in any solder and adhesive layers.

As previously noted in the development of the relationships for the external and internal resistances, Eq. (54.30) shows R_T to be a strong function of the convective heat transfer coefficient, the flowing heat capacity of the coolant, and geometric parameters (thickness and cross-sectional area of each layer). Thus, the introduction of a superior coolant, use of thermal enhancement techniques that increase the local heat transfer coefficient, or selection of a heat transfer mode with inherently high heat transfer coefficients (boiling, for example) will all be reflected in appropriately lower external and total thermal resistances. Similarly, improvements in the thermal conductivity and reduction in the thickness of the relatively low-conductivity bonding materials (such as soft solder, epoxy or silicone) would act to reduce the internal and total thermal resistances.

Frequently, however, even more dramatic reductions in the total resistance can be achieved simply by increasing the cross-sectional area for heat flow within the chip module (such as chip, substrate and heat spreader) as well as along the wetted, exterior surface. The implementation of this approach to reducing the internal resistance generally results in a larger package footprint or volume but is rewarded with a lower thermal resistance. The use of heat sinks is, of course, the embodiment of this approach to the reduction of the external resistance.

54.2 HEAT-TRANSFER CORRELATIONS FOR ELECTRONIC EQUIPMENT COOLING

The reader should use the material in this section which pertains to heat-transfer correlations in geometries peculiar to electronic equipment in conjunction with the correlations provided in Chapter 43.

54.2.1 Natural Convection in Confined Spaces

For natural convection in confined horizontal spaces the recommended correlations for air are[12]

$$\text{Nu} = 0.195(\text{Gr})^{1/4}, \quad 10^4 < \text{Gr} < 4 \times 10^5 \qquad (54.31)$$
$$\text{Nu} = 0.068(\text{Gr})^{1/3}, \qquad \text{Gr} > 10^5$$

where Gr is the Grashof number,

$$\text{Gr} = \frac{g\rho^2\beta L^2 \Delta T}{\mu^2} \qquad (54.32)$$

and where, in this case, the significant dimension L is the gap spacing in both the Nusselt and Grashof numbers.

For liquids[13]

$$\text{Nu} = 0.069(\text{Gr})^{1/3}\text{Pr}^{0.407}, \quad 3 \times 10^5 < \text{Ra} < 7 \times 10^9 \qquad (54.33a)$$

where Ra is the Rayleigh number,

$$\text{Ra} = \text{GrPr} \qquad (54.33b)$$

For horizontal gaps with Gr < 1700, the conduction mode predominates and

$$h = \frac{k}{b} \qquad (54.34)$$

where b is the gap spacing. For 1700 < Gr < 10,000, use may be made of the Nusselt-Grashof relationship given in Fig. 54.7.[14,15]

For natural convection in confined vertical spaces containing air, the heat-transfer coefficient depends on whether the plates forming the space are operating under isoflux or isothermal conditions.[16]

For the symmetric isoflux case, a case that closely approximates the heat transfer in an array of printed circuit boards, the correlation for Nu is formed by using the method of Churchhill and Usagi[17] by considering the isolated plate case[18-20] and the fully developed limit:[21]

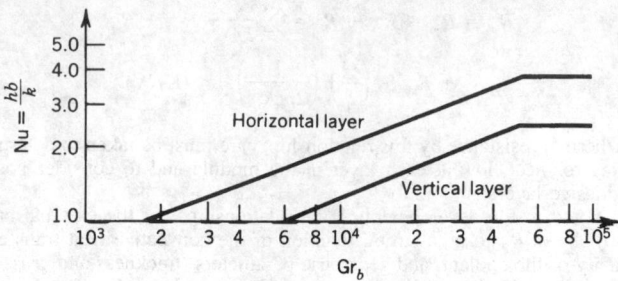

Fig. 54.7 Heat transfer through enclosed air layers.[14,15]

$$Nu = \left[\frac{12}{Ra''} + \frac{1.88}{(Ra'')^{2/5}} \right]^{-1/2} \tag{54.35}$$

where Ra'' is the modified channel Rayleigh number,

$$Ra'' = \frac{g\beta p^2 q'' c_p b^5}{\mu k^2 L} \tag{54.36}$$

The optimum spacing for the symmetrical isoflux case is

$$b_{opt} = 1.472 R^{-0.2} \tag{54.37}$$

where

$$R = \frac{g\beta p^2 c_p q''}{\mu k^2 L} \tag{54.38}$$

For the symmetric isothermal case, a case that closely approximates the heat transfer in a vertical array of extended surface or fins, the correlation is again formed using the Churchhill and Usagi[17] method by considering the isolated plate case[20] and the fully developed limit:[4,5,21]

$$Nu = \left[\frac{576}{(Ra')^2} + \frac{2.873}{(Ra')^{1/2}} \right]^{-1/2} \tag{54.39}$$

where Ra' is the channel Rayleigh number

$$Ra' = \frac{g\beta p^2 c_p b^4}{\mu k L} \tag{54.40}$$

The optimum spacing for the symmetrical isothermal case is

$$b_{opt} = \frac{2.714}{P^{1/4}} \tag{54.41}$$

where

$$P = \frac{g\beta p^2 c_p \Delta T}{\mu k L} \tag{54.42}$$

54.2.2 Forced Convection

External Flow on a Plane Surface

For an unheated starting length of the plane surface, x_0, in laminar flow, the local Nusselt number can be expressed by

Table 54.1 Constants for Eq. 54.11

Reynolds Number Range	B	n
1–4	0.891	0.330
4–40	0.821	0.385
40–4000	0.615	0.466
4000–40,000	0.174	0.618
40,000–400,000	0.0239	0.805

$$\mathrm{Nu}_x = \frac{0.332\mathrm{Re}^{1/2}\mathrm{Pr}^{1/3}}{[1 - (x_0/x)^{3/4}]^{1/3}} \tag{54.43}$$

Where Re is the Reynolds number, Pr is the Prandtl number, and Nu is the Nusselt number.

For flow in the inlet zones of parallel plate channels and along isolated plates, the heat transfer coefficient varies with L, the distance from the leading edge.[3] in the range Re $\leq 3 \times 10^5$,

$$h = 0.664 \left(\frac{k_{fl}}{L}\right) \mathrm{Re}^{0.5}\mathrm{Pr}^{0.33} \tag{54.44}$$

and for Re $> 3 \times 10^5$

$$h = 0.036 \left(\frac{k_{fl}}{L}\right) \mathrm{Re}^{0.8}\mathrm{Pr}^{0.33} \tag{54.45}$$

Cylinders in Crossflow

For airflow around single cylinders at all but very low Reynolds numbers, Hilpert[23] has proposed

$$\mathrm{Nu} = \frac{hd}{k_f} = B \left(\frac{\rho V_\infty d}{\mu_f}\right)^n \tag{54.46}$$

where V_∞ is the free stream velocity and where the constants B and n depend on the Reynolds number as indicated in Table 54.1.

It has been pointed out[12] that Eq. (54.46) assumes a natural turbulence level in the oncoming air stream and that the presence of augmentative devices can increase n by as much as 50%. The modifications to B and n due to some of these devices are displayed in Table 54.2.

Equation (54.46) can be extended to other fluids[24] spanning a range of $1 < \mathrm{Re} < 10^5$ and $0.67 < \mathrm{Pr} < 300$:

$$\mathrm{Nu} = \frac{hd}{k} = (0.4\mathrm{Re}^{0.5} + 0.06\mathrm{Re}^{0.67})\mathrm{Pr}^{0.4} \left(\frac{\mu}{\mu_w}\right)^{0.25} \tag{54.47}$$

where all fluid properties are evaluated at the free stream temperature except μ_w, which is the fluid viscosity at the wall temperature.

Noncircular Cylinders in Crossflow

It has been found[12] that Eq. (54.46) may be used for noncircular geometries in crossflow provided that the characteristic dimension in the Nusselt and Reynolds numbers is the diameter of a cylinder having the same wetted surface equal to that of the geometry of interest and that the values of B and n are taken from Table 54.3.

Table 54.2 Flow Disturbance Effects on B and n in Eq. (54.42)

Disturbance	Re Range	B	n
1. Longitudinal fin, 0.1d thick on front of tube	1000–4000	0.248	0.603
2. 12 longitudinal grooves, 0.7d wide	3500–7000	0.082	0.747
3. Same as 2 with burrs	3000–6000	0.368	0.86

Table 54.3   Values of B and n for Eq. (54.46)[a]

Flow Geometry	B	n	Range of Reynolds Number
	0.224	0.612	2,500–15,000
	0.085	0.804	3,000–15,000
◇	0.261	0.624	2,500–7,500
◇	0.222	0.588	5,000–100,000
□	0.160	0.699	2,500–8,000
□	0.092	0.675	5,000–100,000
○	0.138	0.638	5,000–100,000
○	0.144	0.638	5,000–19,500
○	0.035	0.782	19,500–100,000
|	0.205	0.731	4,000–15,000

[a]From Ref. 12.

Flow across Spheres

For airflow across a single sphere, it is recommended that the average Nusselt number when $17 < Re < 7 \times 10^4$ be determined from[22]

$$Nu = \frac{hd}{k_f} = 0.37 \left(\frac{\rho V_\infty d}{\mu_f}\right)^{0.6} \tag{54.48}$$

and for $1 < Re < 25^{25}$,

$$Nu = \frac{hd}{k} = 2.2Pr + 0.48Pr(Re)^{0.5} \tag{54.49}$$

For both gases and liquids in the range $3.5 < Re < 7.6 \times 10^4$ and $0.7 < Pr < 380^{24}$

$$Nu = \frac{hd}{k} = 2 + (4.0Re^{0.5} + 0.06Re^{0.67})Pr^{0.4} \left(\frac{\mu}{\mu_w}\right)^{0.25} \tag{54.50}$$

Flow across Tube Banks

For the flow of fluids flowing normal to banks of tubes,[26]

$$Nu = \frac{hd}{k_f} = C \left(\frac{\rho V_\infty d}{\mu_f}\right)^{0.6} \left(\frac{c_p \mu}{k}\right)_f^{0.33} \phi \tag{54.51}$$

which is valid in the range $2000 < Re < 32,000$.

For in-line tubes, $C = 0.26$, whereas for staggered tubes, $C = 0.33$. The factor ϕ is a correction factor for sparse tube banks, and values of ϕ are provided in Table 54.4.

For air in the range where Pr is nearly constant ($Pr \simeq 0.7$ over the range 25–200°C), Eq. (54.51) can be reduced to

$$Nu = \frac{hd}{k_f} = C' \left(\frac{\rho V_\infty d}{\mu_f}\right)^{n'} \tag{54.52}$$

where C' and n' may be determined from values listed in Table 54.5. This equation is valid in the range $2000 < Re < 40,000$ and the ratios x_L and x_T denote the ratio of centerline diameter to tube spacing in the longitudinal and transverse directions, respectively.

For fluids other than air, the curve shown in Fig. 54.8 should be used for staggered tubes.[22] For in-line tubes, the values of

$$j = \left(\frac{hd_0}{k}\right) \left(\frac{c_p \mu}{k}\right)^{-1/3} \left(\frac{\mu}{\mu_w}\right)^{-0.14}$$

should be reduced by 10%.

Table 54.4 Correlation Factor ϕ for Sparse Tube Banks

Number of Rows, N	In Line	Staggered
1	0.64	0.68
2	0.80	0.75
3	0.87	0.83
4	0.90	0.89
5	0.92	0.92
6	0.94	0.95
7	0.96	0.97
8	0.98	0.98
9	0.99	0.99
10	1.00	1.00

Flow across Arrays of Pin Fins

For air flowing normal to banks of staggered cylindrical pin fins or spines,[28]

$$Nu = \frac{hd}{k} = 1.40 \left(\frac{\rho v_\infty d}{\mu}\right)^{0.8} \left(\frac{c_p \mu}{k}\right)^{1/3} \tag{54.53}$$

Flow of Air over Electronic Components

For single prismatic electronic components, either normal or parallel to the sides of the component in a duct,[29] for $2.5 \times 10^3 < Re < 8 \times 10^3$,

$$Nu = 0.446 \left[\frac{Re}{(1/6) + (5A_n/6A_0)}\right]^{0.57} \tag{54.54}$$

where the Nusselt and Reynolds numbers are based on the prism side dimension and where A_0 and A_n are the gross and net flow areas, respectively.

For staggered prismatic components, Eq. (54.54) may be modified to[29]

$$Nu = 0.446 \left[\frac{Re}{(1/6) + (5A_n/A_0)}\right]^{0.57} \left[1 + 0.639 \left(\frac{S_T}{S_{T,\max}}\right)\left(\frac{d}{S_L}\right)^{0.172}\right] \tag{54.55}$$

Table 54.5 Values of the Constants C' and n' in Eq. (54.52)

$x_L = \dfrac{S_L}{d_0}$	$x_T = \dfrac{S_T}{d_0} = 1.25$		$x_T = \dfrac{S_T}{d_0} = 1.50$		$x_T = \dfrac{S_T}{d_0} = 2.00$		$x_T = \dfrac{S_T}{d_0} = 3.00$	
	C'	n'	C'	n'	C'	n'	C'	n'
Staggered								
0.600							0.213	0.636
0.900					0.446	0.571	0.401	0.581
1.000			0.497	0.558				
1.125					0.478	0.565	0.518	0.560
1.250	0.518	0.556	0.505	0.554	0.519	0.556	0.522	0.562
1.500	0.451	0.568	0.460	0.562	0.452	0.568	0.488	0.568
2.000	0.404	0.572	0.416	0.568	0.482	0.556	0.449	0.570
3.000	0.310	0.592	0.356	0.580	0.440	0.562	0.421	0.574
In Line								
1.250	0.348	0.592	0.275	0.608	0.100	0.704	0.0633	0.752
1.500	0.367	0.586	0.250	0.620	0.101	0.702	0.0678	0.744
2.000	0.418	0.570	0.299	0.602	0.229	0.632	0.198	0.648
3.000	0.290	0.601	0.357	0.584	0.374	0.581	0.286	0.608

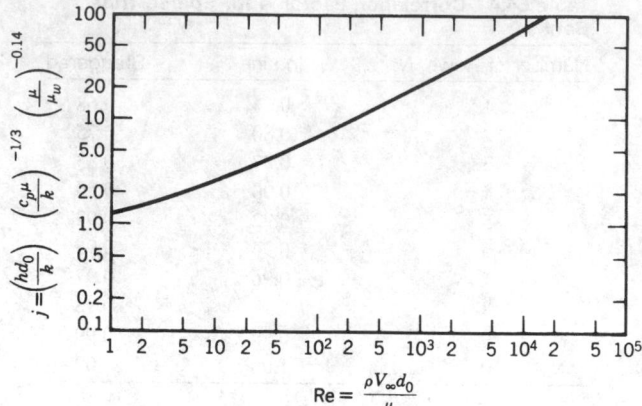

Fig. 54.8 Recommended curve for estimation of heat transfer coefficient for fluids flowing normal to staggered tubes 10 rows deep (from Ref. 22).

where d is the prism side dimension, S_L is the longitudinal separation, S_T is the transverse separation, and $S_{T,\text{max}}$ is the maximum transverse spacing if different spacings exist.

When cylindrical heat sources are encountered in electronic equipment, a modification of Eq. (54.46) has been proposed:[30]

$$\text{Nu} = \frac{hd}{k_f} = FB\left(\frac{\rho V_\infty d}{\mu}\right)^n \tag{54.56}$$

where F is an arrangement factor depending on the cylinder geometry (see Table 54.6) and where the constants B and n are given in Table 54.7.

Forced Convection in Tubes, Pipes, Ducts, and Annuli

For heat transfer in tubes, pipes, ducts, and annuli, use is made of the equivalent diameter

$$d_e = \frac{4A}{WP} \tag{54.57}$$

in the Reynolds and Nusselt numbers unless the cross section is circular, in which case d_e and $d_i = d$.

In the laminar regime[31] where $\text{Re} < 2100$,

Table 54.6 Values of F to Be Used in Eq. (54.56)[a]

Single cylinder in free stream: $F = 1.0$
Single cylinder in duct: $F = 1 + d/w$
In-line cylinders in duct:

$$F = \left(1 + \sqrt{\frac{1}{S_T}}\right)\left\{1 + \left(\frac{1}{S_L} - \frac{0.872}{S_L^2}\right)\left(\frac{1.81}{S_T^2} - \frac{1.46}{S_T} + 0.318\right)[\text{Re}^{0.526-(0.354/S_T)}]\right\}$$

Staggered cylinders in duct:

$$F = \left(1 + \sqrt{\frac{1}{S_T}}\right)\left\{1 + \left[\frac{1}{S_L}\left(\frac{15.50}{S_T^2} - \frac{16.80}{S_T} + 4.15\right) - \frac{1}{S_L}\left(\frac{14.15}{S_T^2} - \frac{15.33}{S_T} + 3.69\right)\right]\text{Re}^{0.13}\right\}$$

[a]Re to be evaluated at film temperature. S_L = ratio of longitudinal spacing to cylinder diameter. S_T = ratio of transverse spacing to cylinder diameter.

Table 54.7 Values of B and n for Use in Eq. (54.56)

Reynolds Number Range	B	n
1000–6000	0.409	0.531
6000–30,000	0.212	0.606
30,000–100,000	0.139	0.806

$$Nu = hd_e/k = 1.86[RePr(d_e/L)]^{1/3}(\mu/\mu_w)^{0.14} \qquad (54.58)$$

with all fluid properties except μ_w evaluated at the bulk temperature of the fluid.
For Reynolds numbers above transition, Re $>$ 2100,

$$Nu = 0.023(Re)^{0.8}(Pr)^{1/3}(\mu/\mu_w)^{0.14} \qquad (54.59)$$

and in the transition region, 2100 $<$ Re $<$ 10,000,[32]

$$Nu = 0.116[(Re)^{2/3} - 125](Pr)^{1/3}(\mu/\mu_w)^{0.14}[1 + (d_e/L)^{2/3}] \qquad (54.60)$$

London[33] has proposed a correlation for the flow of air in rectangular ducts. It is shown in Fig. 54.9. This correlation may be used for air flowing between longitudinal fins.

54.3 THERMAL CONTROL TECHNIQUES

54.3.1 Extended Surface and Heat Sinks

The heat flux from a surface, q/A, can be reduced if the surface area A is increased. The use of extended surface or fins in a common method of achieving this reduction. Another way of looking at this is through the use of Newton's law of cooling:

$$q = hA\Delta T \qquad (54.61)$$

and considering that ΔT can be reduced for a given heat flow q by increasing h, which is difficult for a specified coolant, or by increasing the surface area A.

The common extended surface shapes are the longitudinal fin of rectangular profile, the radial fin of rectangular profile, and the cylindrical spine shown, respectively, in Figs. 54.10a, e, and g.

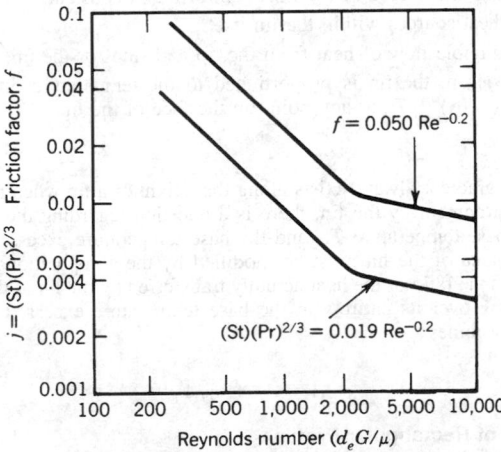

Fig. 54.9 Heat transfer and friction data for forced air through rectangular ducts. St is the stanton number, St $= hG/c_p$.

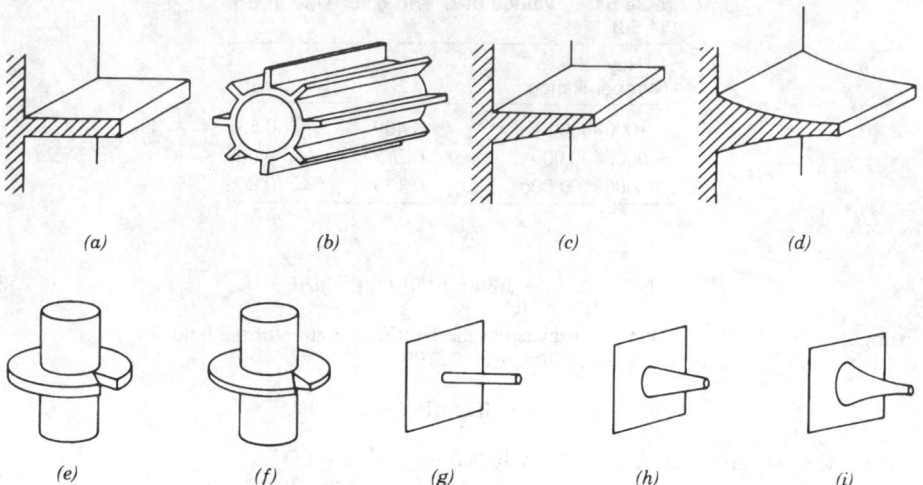

Fig. 54.10 Some typical examples of extended surfaces: (a) longitudinal fin of rectangular profile: (b) cylindrical tube equipped with longitudinal fins; (c) longitudinal fin of trapezoidal profile; (d) longitudinal fin of truncated concave parabolic profile; (e) cylindrical tube equipped with radial fin of rectangular profile; (f) cylindrical tube equipped with radial fin of truncated triangular profile; (g) cylindrical spine; (h) truncated conical spine; (i) truncated concave parabolic spine.

Assumptions in Extended Surface Analysis

The analysis of extended surface is subject to the following simplifying assumptions:[34,35]

1. The heat flow is steady; that is, the temperature at any point does not vary with time.
2. The fin material is homogeneous, and the thermal conductivity is constant and uniform.
3. The coefficient of heat transfer is constant and uniform over the entire face surface of the fin.
4. The temperature of the surrounding fluid is constant and uniform.
5. There are no temperature gradients within fin other than along the fin height.
6. There is no bond resistance to the flow of heat at the base of the fin.
7. The temperature at the base of the fin is uniform and constant.
8. There are no heat sources within the fin itself.
9. There is a negligible flow of heat from the tip and sides of the fin.
10. The heat flow from the fin is proportioned to the temperature difference or temperature excess, $\theta(x) = T(x) - T_s$, at any point on the face of the fin.

The Fin Efficiency

Because a temperature gradient always exists along the height of a fin when heat is being transferred to the surrounding environment by the fin, there is a question regarding the temperature to be used in Eq. (54.61). If the base temperature T_b (and the base temperature excess, $\theta_b = T_b - T_s$) is to be used, then the surface area of the fin must be modified by the computational artifice known as the fin efficiency, defined as the ratio of the heat actually transferred by the fin to the ideal heat transferred if the fin were operating over its entirety at the base temperature excess. In this case, the surface area A in Eq. (54.43) becomes

$$A = A_b + \eta_f A_f \tag{54.62}$$

The Longitudinal Fin of Rectangular Profile

With the origin of the height coordinate x taken at the fin tip, which is presumed to be adiabatic, the temperature excess at any point on the fin is

$$\theta(x) = \theta_b \frac{\cosh mx}{\cosh mb} \tag{54.63}$$

where

$$m = \left(\frac{2h}{k\delta}\right)^{1/2} \tag{54.64}$$

The heat dissipated by the fin is

$$q_b = Y_0 \theta_b \tanh mb \tag{54.65}$$

where Y_0 is called the characteristic admittance

$$Y_0 = (2hk\delta)^{1/2}L \tag{54.66}$$

and the fin efficiency is

$$\eta_f = \frac{\tanh mb}{mb} \tag{54.67}$$

The heat-transfer coefficient in natural convection may be determined from the symmetric isothermal case pertaining to vertical plates in Section 54.2.1. For forced convection, the London correlation described in Section 54.2.2 applies.

The Radial Fin of Rectangular Profile

With the origin of the radial height coordinate taken at the center of curvature and with the fin tip at $r = r_a$ presumed to be adiabatic, the temperature excess at any point on the fin is

$$\theta(r) = \theta_b \left[\frac{K_1(mr_a)I_0(mr) + I_1(mr_a)K_0(mr)}{I_0(mr_b)K_1(mr_a) + I_1(mr_a)K_0(mr_b)}\right] \tag{54.68}$$

where m is given by Eq. (54.64). The heat dissipated by the fin is

$$q_b = 2\pi r_b km\theta_b \left[\frac{I_1(mr_a)K_1(mr_b) - K_1(mr_a)I_1(mr_b)}{I_0(mr_b)K_1(mr_a) + I_1(mr_a)K_0(mr_b)}\right] \tag{54.69}$$

and the fin efficiency is

$$\eta f = \frac{2r_b}{m(r_a^2 - r_b^2)} \left[\frac{I_1(mr_a)K_1(mr_b) - K_1(mr_a)I_1(mr_b)}{I_0(mr_b)K_1(mr_a) + I_1(mr_a)K_0(mr_b)}\right] \tag{54.70}$$

Tables of the fin efficiency are available,[36] and they are organized in terms of two parameters, the radius ratio

$$\rho = \frac{r_b}{r_a} \tag{54.71a}$$

and a parameter ϕ

$$\phi = (r_a - r_b)\left(\frac{2h}{kA_p}\right)^{1/2} \tag{54.71b}$$

where A_p is the profile area of the fin:

$$A_p = \delta(r_a - r_b) \tag{54.71c}$$

For air under forced convection conditions, the correlation for the heat-transfer coefficient developed by Briggs and Young[37] is applicable:

$$\frac{h}{2r_b k} = \left(\frac{2\rho V r_b}{\mu}\right)^{0.681} \left(\frac{c_p \mu}{k}\right)^{1/3} \left(\frac{s}{r_a - r_b}\right)^{0.200} \left(\frac{s}{\delta}\right)^{0.1134} \quad (54.72)$$

where all thermal properties are evaluated at the bulk air temperature, s is the space between the fins, and r_a and r_b pertain to the fins.

The Cylindrical Spine

With the origin of the height coordinate x taken at the spine tip, which is presumed to be adiabatic, the temperature excess at any point on the spine is given by Eq. (54.61), but for the cylindrical spine

$$m = \left(\frac{4h}{kd}\right)^{1/2} \quad (54.73)$$

where d is the spine diameter. The heat dissipated by the spine is given by Eq. (54.65), but in this case

$$Y_0 = (\pi^2 h k d^3)^{1/2}/2 \quad (54.74)$$

and the spine efficiency is given by Eq. (54.67).

Algorithms for Combining Single Fins into Arrays

The differential equation for temperature excess that can be developed for any fin shape can be solved to yield a particular solution, based on prescribed initial conditions of fin base temperature excess and fin base heat flow, that can be written in matrix form[38,39] as

$$\begin{bmatrix} \theta_a \\ q_a \end{bmatrix} = [\Gamma] \begin{bmatrix} \theta_b \\ q_b \end{bmatrix} = \begin{bmatrix} \gamma_{11} & \gamma_{12} \\ \gamma_{21} & \gamma_{22} \end{bmatrix} \begin{bmatrix} \theta_b \\ q_b \end{bmatrix} \quad (54.75)$$

The matrix $[\Gamma]$ is called the thermal transmission matrix and provides a linear transformation from tip to base conditions. It has been cataloged for all of the common fin shapes.[38–40] For the longitudinal fin of rectangular profile

$$[\Gamma] = \begin{bmatrix} \cosh mb & -\dfrac{1}{Y_0} \sinh mb \\ -Y_0 \sinh mb & \cosh mb \end{bmatrix} \quad (54.76)$$

and this matrix possesses an inverse called the inverse thermal transmission matrix

$$[\Lambda] = [\Gamma]^{-1} = \begin{bmatrix} \cosh mb & \dfrac{1}{Y_0} \sinh mb \\ Y_0 \sinh mb & \cosh mb \end{bmatrix} \quad (54.77)$$

The assembly of fins into an array may require the use of any or all of three algorithms.[40–42] The objective is to determine the input admittance of the entire array

$$Y_{\text{in}} = \frac{q_b}{\theta_b} \bigg|_A \quad (54.78)$$

which can be related to the array (fin) efficiency by

$$\eta_f = \frac{Y_{\text{in}}}{h A_f} \quad (54.79)$$

The determination of Y_{in} can involve as many as three algorithms for the combination of individual fins into an array.

The Cascade Algorithm: For n fins in cascade as shown in Fig. 54.11a, an equivalent inverse thermal transmission matrix can be obtained by a simple matrix multiplication, with the individual fins closest to the base of the array acting as permultipliers:

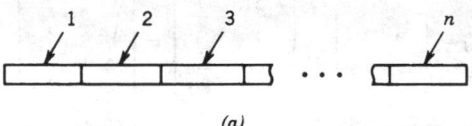

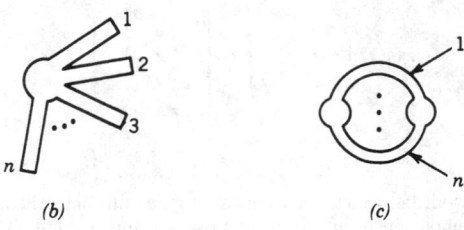

Fig. 54.11   (a) n fins in cascade, (b) n fins in cluster, and (c) n fins in parallel.

$$\{\Lambda\}_e = \{\Lambda\}_n\{\Lambda\}_{n-1}\{\Lambda\}_{n-2} \cdots \{\Lambda\}_2\{\Lambda\}_1 \tag{54.80}$$

For the case of the tip of the most remote fin adiabatic, the array input admittance will be

$$Y_{in} = \frac{\lambda_{21.e}}{\lambda_{11.e}} \tag{54.81}$$

If the tip of the most remote fin is not adiabatic, the heat flow to temperature excess ratio at the tip which is designated as μ

$$\mu = \frac{q_a}{\theta_a} \tag{54.82}$$

will be known. For example, for a fin dissipating to the environment through its tip designated by the subscript a:

$$\mu = hA_a \tag{54.83}$$

In this case, Y_{in} may be obtained through successive use of what is termed the reflection relationship (actually a bilinear transformation):

$$Y_{in,k-1} = \frac{\lambda_{21,k-1} + \lambda_{22,k-1}(q_a/\theta_a)}{\lambda_{11,k-1} + \lambda_{12,k-1}(q_a/\theta_a)} \tag{54.84}$$

The Cluster Algorithm.   For n fins in cluster, as shown in Fig. 54.11b, the equivalent thermal transmission ratio will be the sum of the individual fin input admittances:

$$\mu_e = \sum_{k=1}^{n} Y_{in,k} = \sum_{k=1}^{n} \frac{q_b}{\theta_b}\bigg|_k \tag{54.85}$$

Here, $Y_{in,k}$ can be determined for each individual fin via Eq. (54.82) if the fin has an adiabatic tip or via Eq. (54.84) if the tip is not adiabatic. It is obvious that this holds if subarrays containing more than one fin are in cluster.

The Parallel Algorithm.   For n fins in parallel, as shown in Fig. 54.11c, an equivalent thermal admittance matrix $[Y]_e$ can be obtained from the sum of the individual thermal admittance matrices:

$$[Y]_e = \sum_{k=1}^{n} [Y]_k \tag{54.86}$$

where the individual thermal admittance matrices can be obtained from

$$[Y] = \begin{bmatrix} y_{11} & y_{12} \\ y_{21} & y_{22} \end{bmatrix} = \begin{bmatrix} -\dfrac{\gamma_{11}}{\gamma_{12}} & \dfrac{1}{\gamma_{12}} \\ -\dfrac{1}{\gamma_{12}} & \dfrac{\gamma_{22}}{\gamma_{12}} \end{bmatrix} = \begin{bmatrix} \dfrac{\lambda_{22}}{\lambda_{12}} & -\dfrac{1}{\lambda_{12}} \\ \dfrac{1}{\lambda_{12}} & \dfrac{\lambda_{21}}{\lambda_{22}} \end{bmatrix} \tag{54.87}$$

If necessary, $[\Lambda]$ may be obtained from $[Y]$ using

$$[\Lambda] = \begin{bmatrix} \lambda_{11} & \lambda_{12} \\ \lambda_{21} & \lambda_{22} \end{bmatrix} = \begin{bmatrix} -\dfrac{y_{22}}{y_{21}} & \dfrac{1}{y_{21}} \\ -\dfrac{\Delta\gamma}{y_{21}} & \dfrac{y_{11}}{y_{21}} \end{bmatrix} \tag{54.88}$$

where $\Delta_Y = y_{11}y_{22} - y_{12}y_{21}$

Singular Fans. There will be occasions when a singular fin, one whose tip comes to a point, will be used as the most remote fin in an array. In this case the $[\Gamma]$ and $[\Lambda]$ matrices do not exist and the fin is characterized by its input admittance.[38-40] Such a fin is the longitudinal fin of triangular profile where

$$Y_{in} = \frac{q_b}{\theta_b} = \frac{2hI_1(2mb)}{mI_0(2mb)} \tag{54.89}$$

where

$$m = \left(\frac{2h}{k\delta_b}\right)^{1/2} \tag{54.90}$$

54.3.2 The Cold Plate

The cold plate heat exchanger or forced cooled electronic chassis is used to provide a "cold wall" to which individual components and, for that matter, entire packages of equipment may be mounted. Its design and performance evaluation follows a certain detailed procedure that depends on the type of heat loading and whether the heat loading is on one or two sides of the cold plate. These configurations are displayed in Fig. 54.12.

The design procedure is based on matching the available heat-transfer effectiveness ϵ to the required effectiveness ϵ determined from the design specifications. These effectivenesses are for the isothermal case in Fig. 54.12a

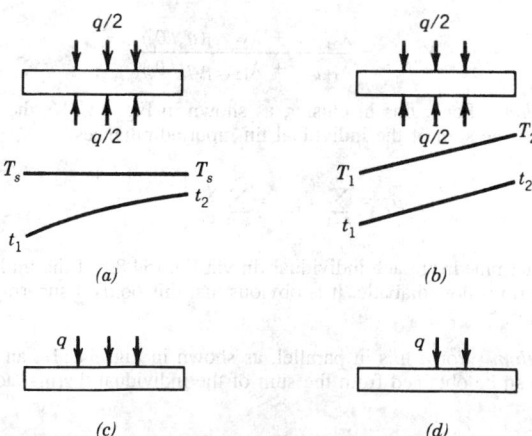

Fig. 54.12 (a) Double-sided, evenly loaded cold plate—isothermal case; (b) double-sided, evenly loaded cold plate—isoflux case; (c) single-sided, evenly loaded cold plate—isothermal case; and (d) single-sided, evenly loaded cold plate—isoflux case.

$$\epsilon = \frac{t_2 - t_1}{T_s - t_1} = e^{-NTU} \tag{54.91}$$

and for the isoflux case in Fig. 69.12*b*

$$\epsilon = \frac{t_2 - t_1}{T_2 - t_1} \tag{54.92}$$

where the "number of transfer units" is

$$NTU = \frac{h\eta_0 A}{W c_p} \tag{54.93}$$

and the overall passage efficiency is

$$\eta_0 = 1 - \frac{A_f}{A}(1 - \eta_f) \tag{54.94}$$

The surfaces to be used in the cold plate are those described by Kays and London[41] where physical, heat-transfer, and friction data are provided.

The detailed design procedure for the double-side-loaded isothermal case is as follows:

1. Design specification
   (a) Heat load, q, W
   (b) Inlet air temperature, t_1, °C
   (c) Airflow, W, kg/sec
   (d) Allowable pressure loss, cm H_2O
   (e) Overall envelope, H, W, D
   (f) Cold plate material thermal conductivity, k_m, W/m · °C
   (g) Allowable surface temperature, T_s, °C

2. Select surface[41]
   (a) Type
   (b) Plate spacing, b, m
   (c) Fins per meter, fpm
   (d) Hydraulic diameter, d_e, m
   (e) Fin thickness, δ, m
   (f) Heat transfer area/volume, β, m²/m³
   (g) Fin surface area/total surface area, A_f/A, m²/m³

3. Plot of j and f data[41]

$$j = (St)(Pr)^{2/3} = f_1(Re) = f_1\left(\frac{d_e G}{\mu}\right)$$

where St is the Stanton number

$$St = \frac{hG}{c_p} \tag{54.95}$$

and f is the friction factor

$$f = f_2(Re) = f_2\left(\frac{d_e G}{\mu}\right)$$

4. Establish physical data
   (a) $a = (b/H)\beta$, m²/m³
   (b) $r_h = d_e/4$, m
   (c) $\sigma = ar_h$
   (d) $A_{fr} = WH$, m²   (frontal area)
   (e) $A_c = \sigma A_{fr}$, m²   (flow areas)
   (f) $V = DWH$ (volume)
   (g) $A = aV$, m²   (total surface)

5. Heat balance
   (a) Assume average fluid specific heat, c_p, J/kg · °C
   (b) $\Delta t = t_2 - t_1 = q/Wc_p$, °C
   (c) $t_2 = t_1 + \Delta t$, °C
   (d) $t_{av} = \frac{1}{2}(t_1 + t_2)$
   (e) Check assumed value of c_p. Make another assumption if necessary

6. Fluid properties at t_{av}
   (a) c_p (already known), J/kg · °C
   (b) μ, N/sec · m²
   (c) k, W/m · °C
   (d) $(\text{Pr})^{2/3} = (c_p\mu/k)^{2/3}$

7. Heat-transfer coefficient
   (a) $G = W/A_c$, kg/sec · m²
   (b) $\text{Re} = d_eG/\mu$
   (c) Obtain j from curve (see item 3)
   (d) Obtain f from curve (see item 3)
   (e) $h = jGc_p/(\text{Pr})^{2/3}$, W/m² · °C

8. Fin efficiency
   (a) $m = (2h/k\delta)^{1/2}$, m⁻¹
   (b) $mb/2$ is a computation
   (c) $\eta_f = (\tanh mb/2)/mb/2$

9. Overall passage efficiency
   (a) Use Eq. (54.94)

10. Effectiveness
    (a) Required $\epsilon = (t_2 - t_1)/(T_s - T_1)$
    (b) Form NTU from Eq. (54.93)
    (c) Actual available $\epsilon = 1 - e^{-NTU}$
    (d) Compare required ϵ and actual ϵ and begin again with step 1 if the comparison fails. If comparison is satisfactory go on to pressure loss calculation.

11. Pressure loss
    (a) Establish v_1 (specific volume), m³/k
    (b) Establish v_2, m³/kg
    (c) $v_m = \frac{1}{2}(v_1 + v_2)$, m³/kg
    (d) Form v_m/v_1
    (e) Form v_2/v_1
    (f) Obtain K_c and K_e[41]
    (g) Determine ΔP, cm

$$\Delta P = 0.489 \frac{G^2v_1}{2g}\left[(1 + K_c - \sigma^2) + f\frac{A}{A_c}\frac{v_m}{v_1} + 2\left(\frac{v_2}{v_1} - 1\right) - (1 - \sigma^2 - K_c)\frac{v_2}{v_1}\right] \quad (54.96)$$

   (h) Compare ΔP with specified ΔP. If comparison fails select a different surface or adjust the dimensions and begin again with step 1

If the cold plate is loaded on one side only, an identical procedure is followed except in steps 8 and 9. For single-side loading and for double and triple stacks, use must be made of the cascade and cluster algorithms for the combination of fins described in Section 54.3.1. Detailed examples of both of the foregoing cases may be found in Kraus and Bar-Cohen.[11]

54.3.3 Thermoelectric Coolers

Two thermoelectric effects are traditionally considered in the design and performance evaluation of a thermoelectric cooler:

The Seebeck effect concerns the net conversion of thermal energy into electrical energy under zero current conditions when two dissimilar materials are brought into contact. When the junction temperature differs from a reference temperature, the effect is measured as a voltage called the Seebeck voltage E_s.

The Peltier effect concerns the reversible evolution or absorption of heat that occurs when an electric current traverses the junction between two dissimilar materials. The Peltier heat absorbed or rejected depends on and is proportional to the current flow. There is an additional thermoelectric effect known as the Thomson effect, which concerns the reversible evolution or absorption of heat that occurs when an electric current traverses a single homogeneous material in the presence of a temperature gradient. This effect, however, is a negligible one and is neglected in considerations of thermoelectric coolers operating over moderate temperature differentials.

Equations for the Thermoelectric Effects

Given a pair of thermoelectric materials, A and B, with each having a thermoelectric power α_A and α_B,[42] the Seebeck coefficient is

$$\alpha = |\alpha_A| + |\alpha_B| \tag{54.97}$$

The Seebeck coefficient is the proportionality constant between the Seebeck voltage and the junction temperature with respect to some reference temperature

$$dE_s = \pm \, \alpha dT$$

and it is seen that

$$\alpha = \frac{dE_s}{dt}$$

The Peltier heat is proportional to the current flow and the proportionality constant is Π, the Peltier-voltage

$$q_p = \pm \, \Pi I \tag{54.98}$$

The Thomson heat is proportional to a temperature difference dT and the proportionality constant is σ, the Thomson coefficient. With $dq_T = \pm \, \sigma I \, dt$, it is observed that $\sigma \, dT$ is a voltage and the Thomson voltage is defined by

$$E_T = \pm \int_{T_1}^{T_2} \sigma dT$$

Considerations of the second laws of thermodynamics and the Kirchhoff voltage laws show that the Peltier voltage is related to the Seebeck coefficient[42]

$$\Pi = \alpha T \tag{54.99}$$

and if the Seebeck coefficient is represented as a polynomial[42]

$$\alpha = a + bT + \cdots$$

then

$$\Pi = aT + bT^2 + \cdots$$

Design Equations

In Fig. 54.13, which shows a pair of materials arranged as a thermoelectric cooler, there is a cold junction at T_c and a hot junction at T_h. The materials possess a thermal conductivity k and an electrical resistivity ρ. A voltage is provided so that a current I flows through the cold junction from B to A and through the hot junction from A to B. This current direction is selected to guarantee that $T_c < T_h$.

The net heat absorbed at the cold junction is the Peltier heat

$$q_p = \Pi T_c = \alpha I T_c \tag{54.100a}$$

minus one-half of the I^2R loss (known as the Joule heat or Joule effect)

$$q_j = \tfrac{1}{2} I^2 R \tag{54.100b}$$

and minus the heat regained at the cold junction (known as the Fourier heat or Fourier effect) due to the temperature difference $\Delta T = T_h - T_c$

$$q_F = K \, \Delta T = K(T_h - T_c) \tag{54.100c}$$

Thus the net heat absorbed at the cold junction is

$$q = \alpha I T_c - \tfrac{1}{2} I^2 R - K \, \Delta T \tag{54.101}$$

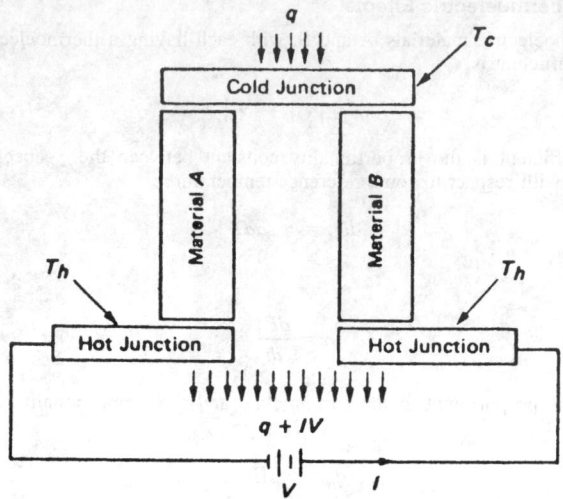

Fig. 54.13 Thermoelectric cooler.

where the total resistance of the couple is the series resistance of material A and material B having areas A_A and A_B, respectively (both have length L),

$$R = \left(\frac{\rho_A}{A_A} + \frac{\rho_B}{A_B}\right) L \tag{54.102}$$

and where the overall conductance K is the parallel conductance of the elements A and B:

$$K = \frac{1}{L} (k_A A_A + k_B A_B) \tag{54.103}$$

In order to power the device, a voltage equal to the sum of the Seebeck voltages at the hot and cold junctions plus the voltage necessary to overcome the resistance drop must be provided:

$$V = \alpha T_h - \alpha T_c + RI = \alpha \, \Delta T + RI$$

and the power is

$$P = VI = (\alpha \, \Delta T + RI)I \tag{54.104}$$

The coefficient of performance (COP) is the ratio of the net cooling effect to the power provided:

$$\text{COP} = \frac{q}{P} = \frac{\alpha T_c I - \frac{1}{2} I^2 R - K \, \Delta T}{\alpha \Delta T I + I^2 R} \tag{54.105}$$

Optimizations

The maximum possible temperature differential $\Delta T = T_h - T_c$ will occur when there is no net heat absorbed at the cold junction:

$$\Delta T_m = \frac{1}{2} z T_c^2 \tag{54.106}$$

where z is the figure of merit of the material

$$z = \frac{\alpha^2}{KR} \tag{54.107}$$

The current that yields the maximum amount of heat absorbed at the cold junction can be shown to be[42]

$$I = I_m = \frac{\alpha T_c}{R} \qquad (54.108)$$

and the coefficient of performance in this case will be

$$COP_m = \frac{1 - \Delta T / \Delta T_m}{2(1 + \Delta T / T_c)} \qquad (54.109)$$

The current that optimizes or maximizes the coefficient of performance can be shown to be

$$I_0 = \frac{\alpha \Delta T}{R[(1 + zT_a)^{1/2} - 1]} \qquad (54.110)$$

where $T_a = \frac{1}{2}(T_h + T_c)$. In this case, the optimum coefficient of performance will be

$$COP_0 = \frac{T_c}{\Delta T}\left[\frac{\gamma - (T_h/T_c)}{\gamma + 1}\right] \qquad (54.111)$$

where

$$\gamma = [1 + \frac{1}{2}z(T_h + T_c)]^{1/2} \qquad (54.112)$$

Analysis of Thermoelectric Coolers

In the event that a manufactured thermoelectric cooling module is being considered for a particular application, the designer will need to specify the number of junctions required. A detailed procedure for the selection of the number of junctions is as follows:

1. Design specifications
    (a) Total cooling load, q_T, W
    (b) cold-side temperature, T_c, °K
    (c) Hot-side temperature, T_h, °K
    (d) Cooler specifications
        i.   Materials A and B
        ii. α_A and α_B, V/°C
        iii. ρ_A and ρ_B, ohm · cm
        iv. k_A and k_B, W/cm · °C
        v. A_A and A_B, cm²
        vi. L, cm

2. Cooler calculations
    (a) Establish $\alpha = |\alpha_A| + |\alpha_B|$
    (b) Calculate R from Eq. (54.102)
    (c) Calculate K from Eq. (54.103)
    (d) Form $\Delta T = T_h - T_c$, K or °C
    (e) Obtain z from Eq. (54.107), 1/°C

3. For maximum heat pumping per couple
    (a) Calculate I_m from Eq. (54.108), A
    (b) Calculate the heat absorbed by each couple q, from Eq. (54.101), W
    (c) Calculate ΔT_m from Eq. (54.106), K or °C
    (d) Determine COP_m from Eq. (54.109)
    (e) The power required per couple will be $p = q/COP_m$, W
    (f) The heat rejected per couple will be $p + q$, W
    (g) The required number of couples will be $n = q_T/q$
    (h) The total power required will be $p_T = nP$, W
    (i) The total heat rejected will be $q_{RT} = nq_R$, W

3A. For optimum coefficient of performance
    (a) Determine $T_a = \frac{1}{2}(T_h + T_c)$, K
    (b) Calculate I_0 from Eq. (54.110), A
    (c) Calculate the heat absorbed by each couple, q, from Eq. (54.101), W
    (d) Determine γ from Eq. (54.112)
    (e) Determine COP_0 from Eq. (54.111)
    (f) The power required per couple will be $P = q/COP_0$, W
    (g) The heat rejected per couple will be $q_R = P + q$, W
    (h) The required number of couples will be $n = q_T/q$

(i) The total power required will be $P_T = nP$, W

(j) The total heat rejected will be $q_{RT} = nq_R$, W

REFERENCES

1. K. J. Negus, R. W. Franklin, and M. M. Yovanovich, "Thermal Modeling and Experimental Techniques for Microwave Bi-Polar Devices," *Proceedings of the Seventh Thermal and Temperature Symposium,*" San Diego, CA, 1989, pp. 63–72.

2. M. M. Yovanovich and V. W. Antonetti, "Application of Thermal Contact Resistance Theory to Electronic Packages," in *Advances in Thermal Modeling of Electronic Components and Systems,* A. Bar-Cohen and A. D. Kraus (eds.), Hemisphere, New York, 1988, pp. 79–128.

3. *Handbook of Chemistry and Physics (CRC),* Chemical Rubber Co., Cleveland, OH, 1954.

4. W. Elenbaas, "Heat Dissipation of Parallel Plates by Free Convection," *Physica* **9**(1), 665–671 (1942).

5. J. R. Bodoia and J. F. Osterle, "The Development of Free Convection Between Heated Vertical Plates," *J. Heat Transfer* **84**, 40–44 (1964).

6. A. Bar-Cohen, "Fin Thickness for an Optimized Natural Convection Array of Rectangular Fins," *J. Heat Transfer* **101**, 564–566.

7. A. Bar-Cohen and W. M. Rohsenow, "Thermally Optimum Arrays of Cards and Fins in Natural Convection," *Trans IEEE Chart, CHMT-6,* 154–158.

8. E. M. Sparrow, J. E. Niethhammer, and A. Chaboki, "Heat Transfer and Pressure Drop Characteristics of Arrays of Rectangular Modules Encountered in Electronic Equipment," *Int. J. of Heat and Mass Transfer* **25**(7), 961–973 (1982).

9. R. A. Wirtz and P. Dykshoorn, "Heat Transfer from Arrays of Flatpacks in Channel Flow," *Proceedings of the Fourth Int. Electronic Packaging Society Conference,* New York, 1984, pp. 318–326.

10. S. B. Godsell, R. J. Dischler, and S. M. Westbrook, "Implementing a Packaging Strategy for High Performance Computers," *High Performance Systems,* 28–31 (January 1990).

11. A. D. Kraus and A. Bar-Cohen, *Design and Analysis of Heat Sinks,* Wiley, New York, 1995.

12. M. Jakob, *Heat Transfer,* Wiley, New York, 1949.

13. S. Globe and D. Dropkin, "Natural Convection Heat Transfer in Liquids Confined by Two Horizontal Plates and Heated from Below," *J. Heat Transfer, Series C* **81**, 24–28 (1959).

14. W. Mull and H. Rieher, "Der Warmeschutz von Luftschichten," *Gesundh-Ing. Beihefte* **28** (1930).

15. J. G. A. DeGraaf and E. F. M. von der Held, "The Relation Between the Heat Transfer and the Convection Phenomena in Enclosed Plane Air Layers," *Appl. Sci. Res., Sec. A* **3**, 393–410 (1953).

16. A. Bar-Cohen and W. M. Rohsenow, "Thermally Optimum Spacing of Vertical, Natural Convection Cooled, Parallel Plates," *J. Heat Transfer* **106**, 116–123 (1984).

17. S. W. Churchhill and R. A. Usagi, "A General Expression for the Correlation of Rates of Heat Transfer and Other Phenomena,," *AIChE J* **18**(6), 1121–1138 (1972).

18. N. Sobel, F. Landis, and W. K. Mueller, "Natural Convection Heat Transfer in Short Vertical Channels Including the Effect of Stagger," *Proceedings of the Third International Heat Transfer Conference,* Vol. 2, Chicago, IL, 1966, pp. 121–125.

19. W. Aung, L. S. Fletcher, and V. Sernas, "Developing Laminar Free Convection Between Vertical Flat Plates with Asymmetric Heating," *Int. J. Heat Mass Transfer* **15**, 2293–2308 (1972).

20. O. Miyatake, T. Fujii, M. Fujii, and H. Tanaka, "Natural Convection Heat Transfer Between Vertical Parallel Plates—One Plate with a Uniform Heat Flux and the Other Thermally Insulated," *Heat Transfer Japan Research* **4**, 25–33 (1973).

21. W. Aung, "Fully Developed Laminar Free Convection Between Vertical Flat Plates Heated Asymetrically," *Int. J. Heat Mass Transfer* **15**, 1577–1580 (1972).

22. W. H. McAdams, *Heat Transmission,* 3rd ed., McGraw-Hill, New York, 1954.

23. R. Hilpert, Warmeabgue von Geheizten Drähten and Rohren in Lufstrom, *Forsch, Ing-Wes* **4**, 215–224 (1933).

24. S. Whitaker, "Forced Convection Heat Transfer Correlations for Flow in Pipes, Past Flat Plates, Single Cylinders, Single Spheres and for Flow in Packed Beds and Tube Bundles," *AIChE Journal* **18**, 361–371 (1972).

25. F. Kreith, *Principles of Heat Transfer,* International Textbook Co., Scranton, PA, 1959.

26. A. P. Colburn, "A Method of Correlating Forced Convection Heat Transfer Data and a Comparison of Fluid Friction," *Trans AIChE* **29**, 174–210 (1933).

27. "Standards of the Tubular Exchanger Manufacturer's Association," New York, 1949.

28. W. Drexel, "Convection Cooling," *Sperry Engineering Review* **14**, 25–30 (December 1961).

29. W. Robinson and C. D. Jones, *The Design of Arrangements of Prismatic Components for Cross-flow Forced Air Cooling,* Ohio State University Research Foundation Report No. 47, Columbus, OH, 1955.

30. W. Robinson, L. S. Han, R. H. Essig, and C. F. Heddleson, *Heat Transfer and Pressure Drop Data for Circular Cylinders in Ducts and Various Arrangements,* Ohio State University Research Foundation Report No. 41, Columbus, OH, 1951.

31. E. N. Sieder and G. E. Tate, "Heat Transfer and Pressure Drop of Liquids in Tubes," *Ind. Eng. Chem.* **28**, 1429–1436 (1936).

32. H. Hausen, *Z VDI, Beih. Verfahrenstech.* **4**, 91–98 (1943).

33. A. L. London, "Air Coolers for High Power Vacuum Tubes," *Trans. IRE* **ED-1**, 9–26 (April, 1954).

34. K. A. Gardner, "Efficiency of Extended Surfaces," *Trans. ASME* **67**, 621–631 (1945).

35. W. M. Murray, "Heat Transfer Through an Annular Disc or Fin of Uniform Thickness," *J. Appl. Mech.* **5**, A78–A80 (1938).

36. D. Q. Kern and A. D. Kraus, *Extended Surface Heat Transfer,* McGraw-Hill, New York, 1972.

37. D. E. Briggs and E. H. Young, "Convection Heat Transfer and Pressure Drop of Air Flowing across Triangular Pitch Banks of Finned Tubes," *Chem. Eng. Prog. Symp. Ser.* **41**(59), 1–10 (1963).

38. A. D. Kraus, A. D. Snider, and L. F. Doty, "An Efficient Algorithm for Evaluating Arrays of Extended Surface," *J. Heat Transfer* **100**, 288–293 (1978).

39. A. D. Kraus, *Analysis and Evaluation of Extended Surface Thermal Systems,* Hemisphere, New York, 1982.

40. A. D. Kraus and A. D. Snider, "New Parametrizations for Heat Transfer in Fins and Spines," *J. Heat Transfer* **102**, 415–419 (1980).

41. W. M. Kays and A. L. London, *Compact Heat Exchangers,* 3rd ed., McGraw-Hill, New York, 1984.

42. A. D. Kraus and A. Bar-Cohen, *Thermal Analysis and Control of Electronic Equipment,* Hemisphere, New York, 1983.

CHAPTER 55

PUMPS AND FANS

William A. Smith
College of Engineering
University of South Florida
Tampa, Florida

55.1	**PUMP AND FAN SIMILARITY**	**1681**
55.2	**SYSTEM DESIGN: THE FIRST STEP IN PUMP OR FAN SELECTION**	**1682**
	55.2.1 Fluid System Data Required	1682
	55.2.2 Determination of Fluid Head Required	1682
	55.2.3 Total Developed Head of a Fan	1684
	55.2.4 Engineering Data for Pressure Loss in Fluid Systems	1684
	55.2.5 Systems Head Curves	1684
55.3	**CHARACTERISTICS OF ROTATING FLUID MACHINES**	**1687**
	55.3.1 Energy Transfer in Rotating Fluid Machines	1687
	55.3.2 Nondimensional Performance Characteristics of Rotating Fluid Machines	1687
	55.3.3 Importance of the Blade Inlet Angle	1689
	55.3.4 Specific Speed	1690
	55.3.5 Modeling of Rotating Fluid Machines	1691
	55.3.6 Summary of Modeling Laws	1691
55.4	**PUMP SELECTION**	**1692**
	55.4.1 Basic Types: Positive Displacement and Centrifugal (Kinetic)	1692
	55.4.2 Characteristics of Positive Displacement Pumps	1692
	55.4.3 Characteristics of Centrifugal Pumps	1693
	55.4.4 Net Positive Suction Head (NPSH)	1693
	55.4.5 Selection of Centrifugal Pumps	1693
	55.4.6 Operating Performance of Pumps in a System	1694
	55.4.7 Throttling versus Variable Speed Drive	1695
55.5	**FAN SELECTION**	**1696**
	55.5.1 Types of Fans; Their Characteristics	1696
	55.5.2 Fan Selection	1696
	55.5.3 Control of Fans for Variable Volume Service	1698

55.1 PUMP AND FAN SIMILARITY

The performance characteristics of centrifugal pumps and fans (i.e., rotating fluid machines) are described by the same basic laws and derived equations and, therefore, should be treated together and not separately. Both fluid machines provide the input energy to create flow and a pressure rise in their respective fluid systems and both use the principle of fluid acceleration as the mechanism to add this energy. If the pressure rise across a fan is small (5000 Pa), then the gas can be considered as an incompressible fluid, and the equations developed to describe the process will be the same as for pumps.

Compressors are used to obtain large increases in a gaseous fluid system. With such devices the compressibility of the gas must be considered, and a new set of derived equations must be developed to describe the compressor's performance. Because of this, the subject of gas compressors will be included in a separate chapter.

Mechanical Engineers' Handbook, 2nd ed., Edited by Myer Kutz.
ISBN 0-471-13007-9 © 1998 John Wiley & Sons, Inc.

55.2 SYSTEM DESIGN: THE FIRST STEP IN PUMP OR FAN SELECTION

55.2.1 Fluid System Data Required

The first step in selecting a pump or fan is to finalize the design of the piping or duct system (i.e., the "fluid system") into which the fluid machine is to be placed. The fluid machine will be selected to meet the flow and developed head requirements of the fluid system. The developed head is the energy that must be added to the fluid by the fluid machine, expressed as the potential energy of a column of fluid having a height H_p (meters). H_p is the "developed head." Consequently, the following data must be collected before the pump or fan can be selected:

1. Maximum flow rate required and variations expected
2. Detailed design (including layout and sizing) of the pipe or duct system, including all elbows, valves, dampers, heat exchangers, filters, etc
3. Exact location of the pump or fan in the fluid system, including its elevation
4. Fluid pressure and temperature available at start of system (suction)
5. Fluid pressure and temperature required at end of system (discharge)
6. Fluid characteristics (density, viscosity, corrosiveness, and erosiveness)

55.2.2 Determination of Fluid Head Required

The fluid head required is calculated using both the Bernoulli and D'Arcy equations from fluid mechanics. The Bernoulli equation represents the total mechanical (nonthermal) energy content of the fluid at any location in the system:

$$E_{T(1)} = P_1 v_1 + Z_1 g + V_1^2/2 \qquad (55.1)$$

where $E_{T(1)}$ = total energy content of the fluid at location (1), J/kg
$\quad P_1$ = absolute pressure of fluid at (1), Pa
$\quad v_1$ = specific volume of fluid at (1), m³/kg
$\quad Z_1$ = elevation of fluid at (1), m
$\quad\ g$ = gravity constant, m/sec²
$\quad V_1$ = velocity of fluid at (1), m/sec

The D'Arcy equation expresses the loss of mechanical energy from a fluid through friction heating between any two locations in the system:

$$\bar{v}\Delta P_f(i,j) = f\, L_e(i - j)\, V^2/2D \qquad \text{J/kg·m} \qquad (55.2)$$

where $\bar{v}$ = average fluid specific volume between two locations (i and j) in the system, m³/kg
$\Delta P_f(i,j)$ = pressure loss due to friction between two locations (i and j) in the system, Pa
$\quad f$ = Moody's friction factor, an empirical function of the Reynolds number and the pipe roughness, nondimensional
$L_e(i - j)$ = equivalent length of pipe, valves, and fittings between two locations i and j in the system, m
$\quad D$ = pipe internal diameter (i.d.), m

An example best illustrates the method.

Example 55.1

A piping system is designed to provide 2.0 m³/sec of water (Q) to a discharge header at a pressure of 200 kPa. Water temperature is 20°C. Water viscosity is 0.0013 N·sec/m². Pipe roughness is 0.05 mm. The gravity constant (g) is 9.81 m/sec². Water suction is from a reservoir at atmospheric pressure (101.3 kPa). The level of the water in the reservoir is assumed to be at elevation 0.0 m. The pump will be located at elevation 1.0 m. The discharge header is at elevation 50.0 m. Piping from the reservoir to the pump suction flange consists of the following:

1 20 m length of 1.07 m i.d. steel pipe
3 90° elbows, standard radius
2 gate valves
1 check valve
1 strainer

Piping from the pump discharge flange to the discharge header inlet flange consists of the following:

1   100 m length of 1.07 m i.d. steel pipe
4   90° elbows, standard radius
1   gate valve
1   check valve

Determine the "total developed head," H_p (m), required of the pump.

Solution:

Let location (1) be the surface of the reservoir, the system "suction location."
Let location (2) be the inlet flange of the pump.
Let location (3) be the outlet flange of the pump.
Let location (4) be the inlet flange to the discharge header, the system "discharge location."

By energy balances

$$E_{T(1)} - \bar{v}\Delta P_f(1-2) = E_{T(2)}$$
$$E_{T(2)} + E_p = E_{T(3)}$$
$$E_{T(3)} - \bar{v}\Delta P_f(3-4) = E_{T(4)}$$

where E_p is the energy input required by the pump. When E_p is described as the potential energy equivalent of a height of liquid, this liquid height is the "total developed head" required of the pump.

$$H_p = E_p/g \text{ m}$$

where H_p = total developed head, m.
     For the data given, assuming incompressible flow:

$P_1 = 101.3$ kPa	$Z_2 = +1.0$ m
$v_1 = 0.001$ m^3/kg = constant	$Z_3 = +1.0$ m
$Z_1 = 0.0$ m	$Z_4 = +50.0$ m
$V_1 = 0.0$ m/sec	$P_4 = 200$ kPa

A_p = internal cross sectional area of the pipe, m^2
$V_2 = Q/A = (2.0)(4)/\pi(1.07)^2 = 2.22$ m/sec
Assume $V_3 = V_4 = V_2 = 2.22$ m/sec
Viscosity (μ) = 0.0013 N · sec/m^2
Reynolds number = $D V/v\mu$
                   = $(1.07)(2.22)/(0.001)(0.0013) = 1.82 \times 10$
Pipe roughness (ϵ) = 0.05 mm
$\epsilon/D = 0.05/(1000)(1.07) = 0.000047$
From Moody's chart, $f = 0.009$ (see references on fluid mechanics)

From tables of equivalent lengths (see references on fluid mechanics):

Fitting	Equivalent Length, L_e (m)
Elbow	1.6
Gate valve (open)	0.3
Check valve	0.3
Strainer	1.8

$$L_e(1\text{--}2) = 20 + (3)(1.6) + 2(0.3) + 0.3 + 1.8 = 27.5 \text{ m}$$
$$L_e(3\text{--}4) = 100 + (4)(1.6) + 0.3 + 0.3 = 107.0 \text{ m}$$
$$\bar{v}\Delta P(1\text{--}2) = (0.009)(27.5)(2.22)^2/(2)(1.07) = 0.57 \text{ J/kg}$$
$$\bar{v}\Delta P(3\text{--}4) = (0.009)(107.0)(2.22)^2/(2)(1.07) = 2.21 \text{ J/kg}$$
$$E_{T(1)} = P_1 v_1 + Z_1 g + V_1^2/2$$
$$= (101,300)(0.001) + 0 + 0 = 101.30 \text{ J/kg}$$
$$E_{T(2)} = E_{T(1)} - \bar{v}\Delta P_f(1\text{--}2)$$
$$= 101.3 - 0.57 = 100.7 \text{ J/kg}$$
$$E_{T(4)} = P_4 v_4 + Z_4 g + V_4^2/2$$
$$= (200,000)(0.001) + (50.0)(9.81) + (2.22)^2/2$$
$$= 692.9 \text{ J/kg}$$
$$E_{T(3)} = E_{T(4)} + \bar{v}\Delta P_f(3\text{--}4)$$
$$= 692.9 + 2.21 = 695.1 \text{ J/kg}$$
$$E_p = E_{T(3)} - E_{T(2)}$$
$$= 695.1 - 100.7 = 594.4 \text{ J/kg}$$
$$H_p = E_p/g = 594.4/9.81 = 60.6 \text{ m of water}$$

It is seen that a pump capable of providing 2.0 m³/sec flow with a developed head of 60.6 m of water is required to meet the demands of this fluid system.

55.2.3 Total Developed Head of a Fan

The procedure for finding the total developed head of a fan is identical to that described for a pump. However, the fan head is commonly expressed in terms of a height of water instead of a height of the gas being moved, since water manometers are used to measure gas pressures at the inlet and outlet of a fan. Consequently,

$$H_{fw} = (\rho_g/\rho_w)H_{fg}$$

where H_{fw} = developed head of the fan, expressed as a head of water, m

H_{fg} = developed head of the fan, expressed as a head of the gas being moved, m

ρ_g = density of gas, kg/m³

ρ_w = density of water in manometer, kg/m³

As an example, if the head required of a fan is found to be 100 m of air by the method described in Section 55.2.2, the air density is 1.21 kg/m³, and the water density in the manometer is 1000 kg/m³, then the developed head, in terms of the column of water, is

$$H_{fw} = (1.21/1000)(100) = 0.121 \text{ m of water}$$

In this example the air is assumed to be incompressible, since the pressure rise across the fan was small (only 0.12 m of water, or 1177 Pa).

55.2.4 Engineering Data for Pressure Loss in Fluid Systems

In practice, only rarely will an engineer have to apply the D'Arcy equation to determine pressure losses in fluid systems. Tables and figures for pressure losses of water, steam, and air in pipe and duct systems are readily available from a number of references. (See Figs. 55.1 and 55.2.)

55.2.5 Systems Head Curves

A systems head curve is a plot of the head required by the system for various flow rates through the system. This plot is necessary for analyzing system performance for variable flow application and is desirable for pump and fan selection and system analysis for constant flow applications.

The curve to be plotted is H versus Q, where

$$H = [E_{T(3)} - E_{T(2)}]/g \tag{55.3}$$

Assume that $V_1 = 0$ and $V = V_4$ in Eqs. (55.1) and (55.2), and letting $V = Q/A$, then Eq. (55.3) reduces to

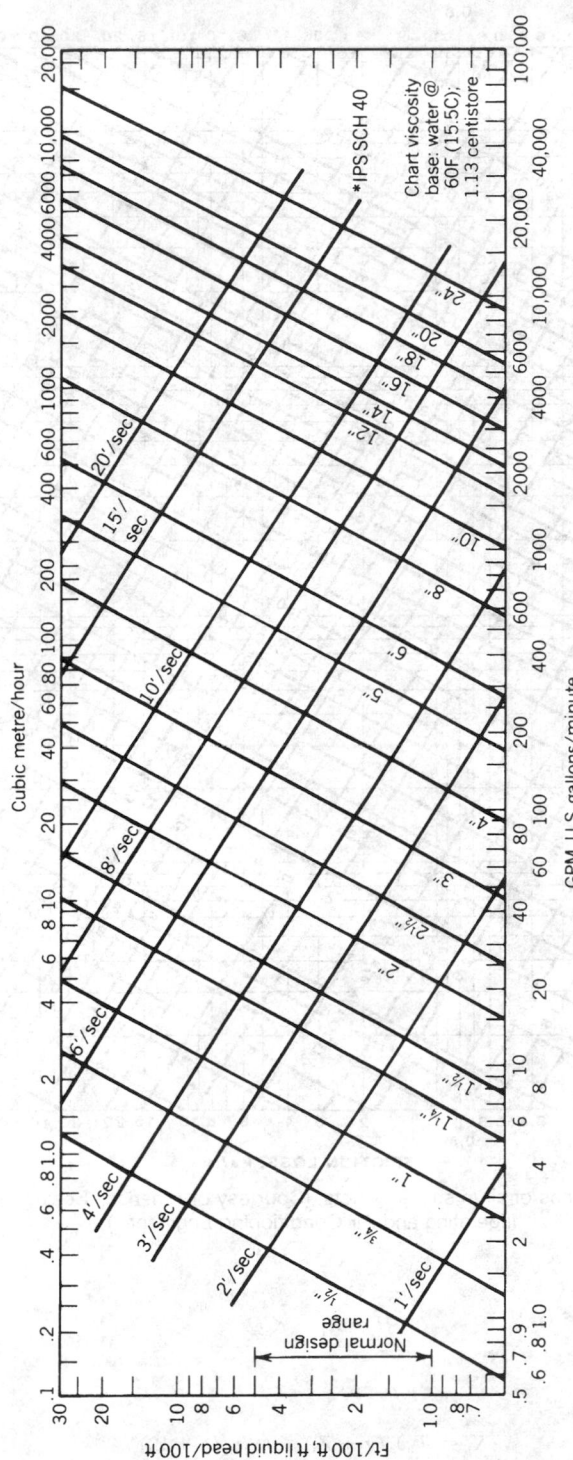

Fig. 55.1 Friction loss for water in commercial steel pipe (schedule 40). (Courtesy of American Society of Heating, Refrigerating and Air Conditioning Engineers.)

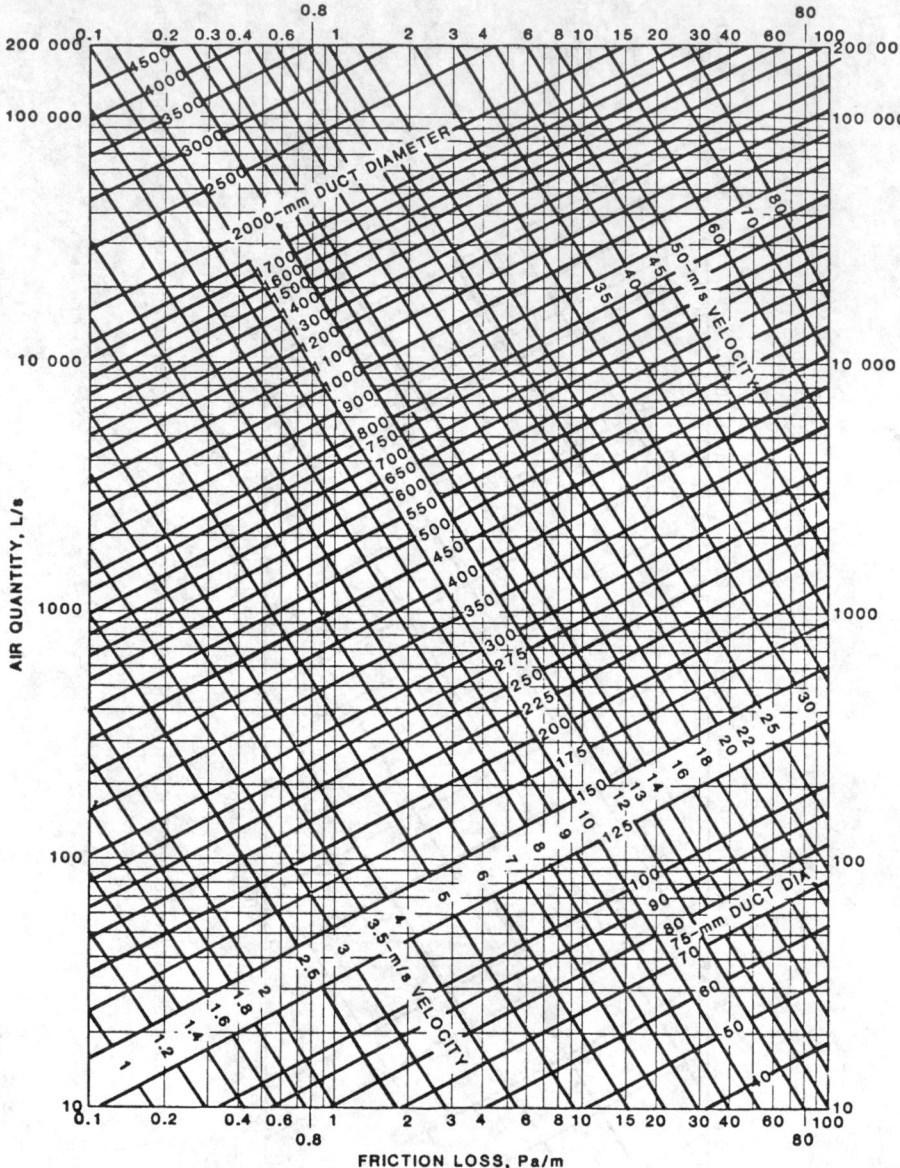

Fig. 55.2 Friction loss of air in straight ducts. (Courtesy of American Society of Heating, Refrigerating and Air Conditioning Engineers.)

$$H = K_1 + K_2 Q^2 \tag{55.4}$$

where

$$K_1 = (P_4 v_4 / g + Z_4) - (P_1 v_1 / g + Z_1)$$
$$K_2 = [f L_e (1\text{–}4) A^2 D g + 1/A^2 g](0.5)$$

However, K_2 is more easily calculated from

$$K_2 = (H - K_1)/Q^2$$

since both H and Q are known from previous calculations.
For example 55.1:

$$K_1 = (200,000)(0.001)/9.81 + 50 - (101,300)(0.001)/9.81 + 0$$

$$= 60.0 \text{ m}$$

$$K_2 = (60.6 - 60.0)/(7200)^2 = 0.012 \times 10^{-6} \text{ hr}^2/\text{m}$$

A plot of this curve [Eq. (55.4)] would show a shallow parabola displaced from the origin by 60.0 m. (This will be shown in Fig. 55.10. Its usefulness will be discussed in Sections 55.6 and 55.7.)

55.3   CHARACTERISTICS OF ROTATING FLUID MACHINES

55.3.1   Energy Transfer in Rotating Fluid Machines

Most pumps and fans are of the rotating type. In a centrifugal machine the fluid enters a rotor at its eye and is accelerated radially by centrifugal force until it leaves at high velocity. The high velocity is then reduced by an area increase (either a volute or diffuser ring of a pump, or scroll of a fan) in which, by Bernoulli's law, the pressure is increased. This pressure rise causes only negligible density changes, since liquids (in pumps) are nearly incompressible and gases (in fans) are not compressed significantly by the small pressure rise (up to 0.5 m of water, or 5000 Pa, or 0.05 bar) usually encountered. For fan pressure rises exceeding 0.5 m of water, compressibility effects should be considered, especially if the fan is a large one (above 50 kW).

The principle of increasing a fluid's velocity, and then slowing it down to get the pressure rise, is also used in mixed flow and axial flow machines. A mixed flow machine is one where the fluid acceleration is in both the radial and axial directions. In an axial machine, the fluid acceleration is intended to be axial but, in practice, is also partly radial, especially in those fans (or propellors) without any constraint (shroud) to prevent flow in the radial direction.

The classical equation for the developed head of a centrifugal machine is that given by Euler:

$$H = (C_{t_2}U_2 - C_{t_1}U_1)/g \quad \text{m} \tag{55.5}$$

where H is the developed head, m, of fluid in the machine; C_t is the tangential component of the fluid velocity C in the rotor; subscript 2 stands for the outer radius of the blade, r_2, and subscript 1 for the inner radius, r_1, m/sec; U is the tangential velocity of the blade, subscript 2 for outer tip and subscript 1 for the inner radius; and U_2 is the "tip speed," m/sec. The velocity vector relationships are shown in Fig. 55.3.

The assumptions made in the development of the theory are:

1. Fluid is incompressible
2. Angular velocity is constant
3. There is no rotational component of fluid velocity while the fluid is between the blades, that is, the velocity vector W exactly follows the curvature of the blade
4. No fluid friction

The weakness of the third assumption is such that the model is not good enough to be used for design purposes. However, it does provide a guidepost to designers on the direction to take to design rotors for various head requirements.

If it is assumed that C_{t_1} is negligible (and this is reasonable if there is no deliberate effort made to cause prerotation of the fluid entering the rotor eye), then Eq. (55.5) reduces to

$$gH = \pi^2 N^2 D^2 - NQ \cot(\beta/b) \tag{55.6}$$

where Q = the flow rate, m³/sec
       D = the outer diameter of the rotor, m
       b = the rotor width, m
       N = the rotational frequency, Hz

55.3.2   Nondimensional Performance Characteristics of Rotating Fluid Machines

Equation (55.6) can also be written as

$$(H/N^2D^2) = \pi^2/g - [D\cot(\beta/gb)](Q/ND^3) \tag{55.7}$$

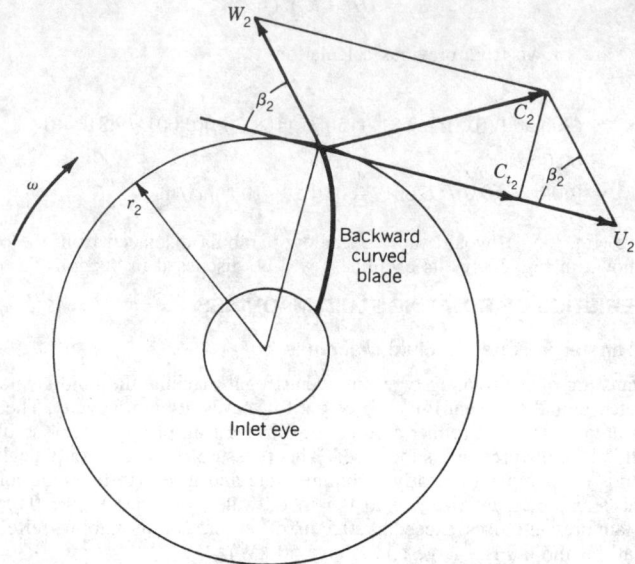

Fig. 55.3 Relationships of velocity vectors used in Euler's theory for the developed head in a centrifugal fluid machine; W is the fluid's velocity with respect to the blade; β is the blade angle, ω is the angular velocity, 1/sec.

In Eq. (55.7) H/N^2D^2 is called the "head coefficient" and Q/ND^3 is the "flow coefficient." The theoretical power, P (W), to drive the unit is given by $P = QgH$, and this reduces to

$$(P/\rho N^3D^5) = (\pi^2)\,(Q/ND^3) - [D\cot(\beta/b)](Q/ND^3)^2 \tag{55.8}$$

where $P/\rho N^3D^5$ is called the "power coefficient." Plots of Eqs. (55.7) and (55.8) for a given D/b ratio are shown in Fig. 55.4.

Analysis of Fig. 55.4 reveals that:

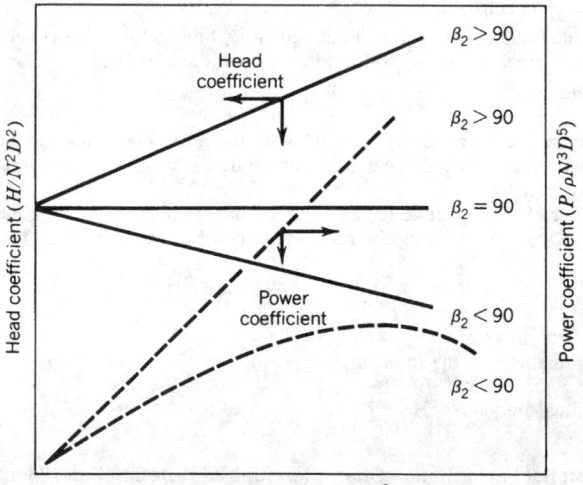

Fig. 55.4 Theoretical (Euler's) head and power coefficients plotted against the flow coefficient for constant D/b ratio and for values of $\beta < 90°$, equal to $90°$, and $>90°$.

1. For a given Q, N, and D, the developed head increases as β gets larger, that is, as the blade tips are curved more into the direction of rotation

2. For a given N and D, the head either rises, stays the same, or drops as Q increases, depending on the value of β

3. For a given N and D, the power required continuously increases as Q increases for β's of 90° or larger, but has a peak value if β is less than 90°

The practical applications of these guideposts appear in the designs offered by the fluid machine industry. Although there is a theoretical reason for using large values of β, there are practical reasons why β must be constrained. For liquids, β's cannot be too large or else there will be excessive turbulence, vibration, and erosion. Blades in pumps are always backward curved ($\beta < 90°$). For gases, however, β's can be quite large before severe turbulence sets in. Blade angles are constrained for fans not only by the turbulence but also by the decreasing efficiency of the fan and the negative economic effects of this decreasing efficiency. Many fan sizes utilize β's $> 90°$.

One important characteristic of fluid machines with blade angles less than 90° is that they are "limit load"; that is, there is a definite maximum power they will draw regardless of flow rate. This is an advantage when sizing a motor for them. For fans with radial (90°) or forward curved blades, the motor size selected for one flow rate will be undersized if the fan is operated at a higher flow rate. The result of undersizing a motor is overheating, deterioration of the insulation, and, if badly undersized, cutoff due to overcurrent.

55.3.3 Importance of the Blade Inlet Angle

While the outlet angle, β_2, sets the head characteristic the inlet angle, β_1, sets the flow characteristic, and by setting the flow characteristic, β_1 also sets the efficiency characteristic.

The inlet vector geometry is shown in Fig. 55.5.

If the rotor width is b at the inlet and there is no prerotation of the fluid prior to its entering the eye (i.e., $C_{t_1} = 0$), then the flow rate into the vector is given by $Q = D_1 b_1 c_1$ and β_1 is given by:

$$\beta_1 = \arctan(C_1/U_1) = \tan^{-1}(Q/ND_1^3)(D_1/b_1)(1/\pi^2) \qquad (55.9)$$

It is seen that β_1 is fixed by any choice of Q, N, D, and b_1. Also, a machine of fixed dimensions ($D_1 b_1$, β_1) and operated at one angular frequency (N) is properly designed for only one flow rate, Q. For flow rates other than its design value, the inlet geometry is incorrect, turbulence is created, and efficiency is reduced. A typical efficiency curve for a machine of fixed dimensions and constant angular velocity is shown in Fig. 55.6.

A truism of all fluid machines is that they operate at peak efficiency only in a narrow range of flow conditions (H and Q). It is the task of the system designer to select a fluid machine that operates at peak efficiency for the range of heads and flows expected in the operation of the fluid system.

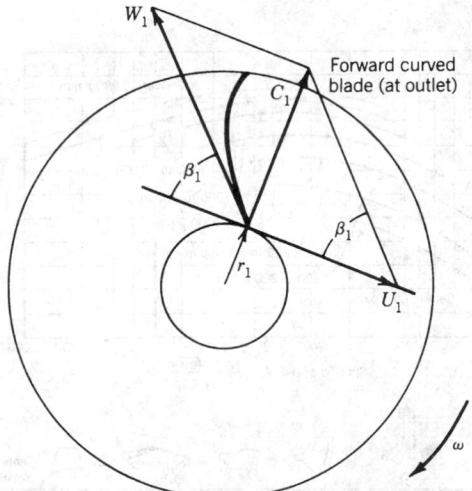

Fig. 55.5 Relationship of velocity vectors at the inlet to the rotor. Symbols are defined in Section 55.3.1.

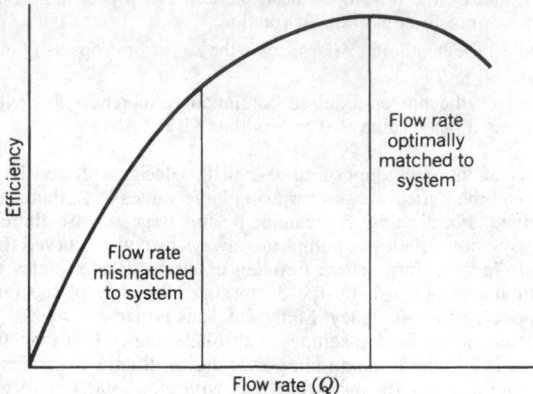

Fig. 55.6 Typical efficiency curve for fluid machines of fixed geometry and constant angular frequency.

55.3.4 Specific Speed

Besides the flow, head, and power coefficients, there is one other nondimensional coefficient that has been found particularly useful in describing the characteristics of rotating fluid machines, namely the specific speed N_s. Specific speed is defined as $NQ^{0.5}/H^{0.75}$ at peak efficiency. It is calculated by using the Q and H that a machine develops at its peak efficiency (i.e., when operated at a condition where its internal geometry is exactly right for the flow conditions required). The specific speed coefficient has usefulness when applying a fluid machine to a particular fluid system. Once the flow and head requirements of the system are known, the best selection of a fluid machine is that which has a specific speed equal to $NQ^{0.5}/H^{0.75}$, where the N, Q, and H are the actual operating parameters of the machine.

Since the specific speed of a machine is dependent on its structural geometry, the physical appearance of the machine as well as its application can be associated with the numerical value of its specific speed. Figure 55.7 illustrates this for a variety of pump geometries. The figure also gives approximate efficiencies to be expected from these designs for a variety of system flow rates (and pump sizes).

It is observed that centrifugal machines with large D/b ratios have low specific speeds and are suitable for high-head and low-flow applications. At the other extreme, the axial flow machines are suitable for low-head and large-flow applications. This statement holds for fans as well as pumps.

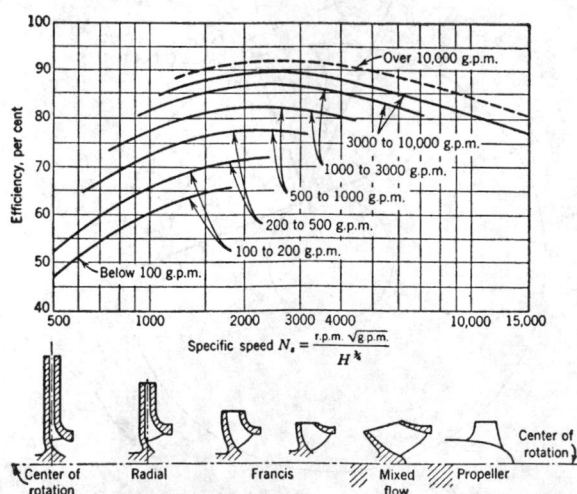

Fig. 55.7 Variation of physical appearance and expected efficiency with specific speed for a variety of pump designs and sizes. (Courtesy of Worthington Corporation.)

As an example of the use of specific speed, consider the pump application of Example 55.1. The head required was found to be 60.6 m. The flow was 7200 m^3/hr. If a pump selected for this service is to have a rotational frequency of 14.75 Hz, then it should have a (nondimensional) specific speed of

$$N_s = (14.75)(2\pi)(2.0)^{0.5}/(9.81)^{0.75}(60.6)^{0.75} = 1.089$$

Its dimensional equivalent in the English system of units (rpm, gpm, ft) is 2972. Looking at Fig. 55.7 it is seen that a pump for this service would be of the centrifugal type, with an impeller that is wide and not very large in diameter. It is a large pump (31,700 gpm) and its efficiency is expected to be high (90%).

Assuming an efficiency of 90%, then the power requirement (P) would be

$$P = \rho QgH/\text{eff} = (1000)(2.0)(9.81)(60.6/(0.9)(1000)$$
$$= 1321 \text{ kW (or 1770 hp)}$$

55.3.5 Modeling of Rotating Fluid Machines

A "family" of fluid machines is one in which each member has the same geometric proportions (and physical appearance) as every other member, except for overall size. The largest member is merely a blown-up version of the smallest member.

Since the geometric proportions of each are the same, all members of a family have the same specific speed. They also have (theoretically) the same performance characteristics (Q, H, P, N) when the performance characteristics are expressed nondimensionally. Practically, the performance characteristics between members of a family differ slightly owing to changes in clearance distances, relative roughness, and Reynolds number that occur between sizes. These differences are called "secondary effects."

Ignoring secondary effects (and structural effects such as vibrations) the performance of an as yet unbuilt, large prototype can be predicted from tests on a small-scale model. Assume that the test data on a pump model, expressed nondimensionally, are as given in Fig. 55.8. It can be assumed that these results will be identical to those obtained on the prototype. If the prototype is to have a diameter of 0.81 m and a rotational frequency of 14.75 Hz, then, at peak efficiency, it can be predicted that the prototype will have the following flow, head, and power characteristics:

$$(Q/ND^3)_{\text{prototype}} = (Q/ND^3)_{\text{model}}$$
$$Q_p = (Q/ND^3)_{\text{model}} (ND^3)_{\text{prototype}}$$
$$= (0.0406)(2\pi)(14.75)(0.81)^3 = 2.0 \text{ m}^3/\text{sec}$$
$$(H/N^2D^2)_{\text{prototype}} = (H/N^2D^2)_{\text{model}}$$
$$H_p = (H/N^2D^2)_{\text{model}} (N^2D^2)_{\text{prototype}}$$
$$= (0.1054)(2\pi)^2(14.75)^2(0.81)^2/9.81 = 60.6 \text{ m}$$
$$P = \rho QgH/\text{eff} = 1321 \text{ kW (from the previous section)}$$

If the model had a diameter of 0.1 m and a rotational frequency of 29 Hz, then, at peak efficiency, its flow, head, and power were:

$$Q_m = (Q/ND^3)(ND^3) = (2.0)(29/14.75)(0.1/0.81)^3$$
$$= 0.0074 \text{ m}^3/\text{sec}$$
$$H_m = (H/N^2D^2)p(N^2D^2)m = 60.6 (29/14.75)^2(0.1/0.81)^2$$
$$= 3.57 \text{ m}$$
$$P_m = (1000)(0.0074)(9.81)(3.57)/(0.9)(1000) = 0.29 \text{ kW}$$

Manufacturers of fluid machines often do not have facilities large enough (fluid quantities and power) to test their largest products. Consequently, the performance of such large machines is estimated from model tests.

55.3.6 Summary of Modeling Laws

Neglecting secondary effects (changes in Reynolds number, size, and clearance distances) the nondimensional performance relationships between a model and a prototype, for any single point of operation (i.e., one point on a common nondimensional curve of performance characteristics), can be summarized as follows:

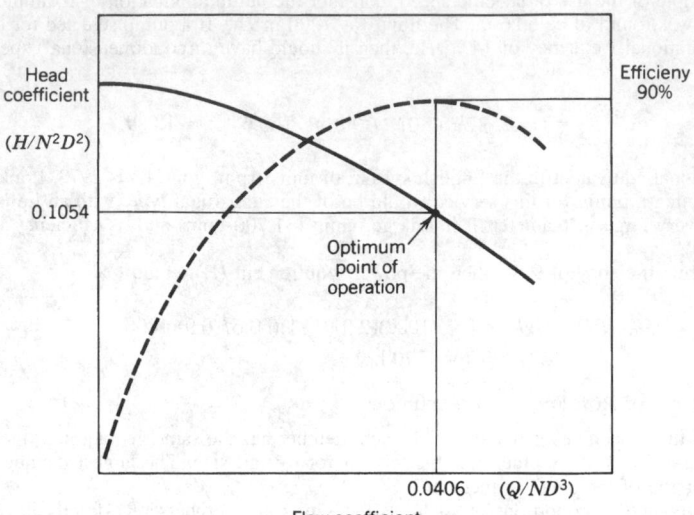

Fig. 55.8 Performance characteristics of a model pump, expressed nondimensionally.

Nondimensional Characteristic	Model	Prototype
Flow coefficient	Q/ND^3	$= Q/ND^3$
Head coefficient	H/N^2D^2	$= H/N^2/D^2$
Power coefficient	$P/\rho N^3 D^5$	$= P/\rho N^3 D^5$
Efficiency	eff	$=$ eff
Specific speed	$NQ^{0.5}/H^{0.75}$	$= NQ^{0.5}/H^{0.75}$

The same relationships can be used to determine the changes in flow, head, or power of a single-fluid machine whenever its diameter or angular frequency is changed in an unchanging fluid system. For a machine of constant diameter, the flow, head, and power will vary with angular frequency as follows:

$$Q \propto N$$
$$H \propto N^2$$
$$P \propto N^3$$

55.4 PUMP SELECTION

55.4.1 Basic Types: Positive Displacement and Centrifugal (Kinetic)

Positive displacement pumps are best suited for systems requiring high heads and low flow rates, or for use with very viscous fluids. The common types are reciprocating (piston and cylinder) and rotary (gears, lobes, vanes, screws). Centrifugal pumps are well suited for the majority of pumping services. The common types are radial, centrifugal, mixed flow, and propellor (axial flow).

55.4.2 Characteristics of Positive Displacement Pumps

Some advantages of positive displacement pumps, besides their inherent ability to provide high discharge pressures at low flow rates, are: ability to provide metered quantities of fluid at a wide range of viscosities; can handle non-Newtonian fluids (sludge, syrup, mash); can operate at slow speeds. Some disadvantages are: flow is pulsating; costs (initial and maintenance) are higher than for centrifugals; must have pressure relief valves in the discharge piping; tight seals and close tolerances are essential to prevent leak-back.

Overall efficiencies usually vary with pump size, being lowest (50%) for small pumps (2 kW) and highest (90%) for large pumps (250 kW). Efficiencies do not vary significantly with flow rate.

Pulsating flows of reciprocating pumps can be smoothed out somewhat by installing air chambers in the discharge. The volume of the air chamber should be recommended by the manufacturer, but is approximately three or four times the displacement volume of the piston. Pulsating flows can be further smoothed out by using double-acting reciprocating pumps that discharge fluid at both ends of the stroke. Rotary pumps have smooth flows.

55.4.3 Characteristics of Centrifugal Pumps

Centrifugal pumps are used in most pumping services. They can deliver small to large flow rates and operate against pressures up to 3000 psi when several impellers are staged in series. They do not work well on highly viscous or non-Newtonian fluids; they operate at high speeds; flow is smooth; clearances between impeller tip and casing are not critical; they do not develop dangerously high head pressures when the discharge valve is closed; and their initial and maintenance costs are lower than that for positive displacement pumps.

Efficiencies of centrifugal pumps are about the same as their corresponding-sized positive displacement pumps if they are carefully matched to their systems. However, their efficiencies vary significantly with flow rate when operated at constant speed, and their efficiencies can be very poor if mismatched to their system, as seen in Figs. 55.6 and 55.8.

55.4.4 Net Positive Suction Head (NPSH)

The liquid static pressure at the suction of both positive displacement and centrifugal pumps must be higher than the liquid's vapor pressure to prevent vaporization at the inlet. Vaporization at the inlet, called "cavitation," causes a drop in developed head, and, in severe cases, a complete loss of flow. Cavitation also causes pitting of the impeller that, in time and if severe enough, can destroy the impeller.

Net positive suction head (NPSH) is the difference between the static pressure and the vapor pressure of the liquid at the pump inlet flange, expressed in meters:

$$NPSH = (P_s - P_v)/\rho g \quad m \tag{55.10}$$

where P_s is the static pressure at the pump inlet flange, Pa; P_v is the liquid's vapor pressure, Pa; and ρ is the liquid density, kg/m^3.

There are two NPSHs that a system designer must consider. One is the NPSH available (NPSHA), which is dependent on the design of the piping system (most importantly the relative elevations of the pump and the source of liquid being pumped). The second is the NPSH required (NPSHR) by the pump selected for the service. There is a static pressure loss within the pump as the liquid passes through the inlet casing and enters the blades. The severity of this loss is dependent on the design of the casing and the amount of acceleration (and turbulence) that the liquid experiences as it enters the blading. Manufacturers test for the NPSHR for each model of pump and report these requirements on the engineering performance specification sheets for the model. The task of the system designer is to ensure that the NPSHA exceeds the NPSHR. Using the data in Example 55.1, the NPSHA is calculated as follows:

P_v at 20°C = 2237 Pa

P_s is found from the calculation for the total energy at the pump suction flange, $E_{t(2)}$

$E_{t(2)} = 100.7$ J/kg $= (P_s/\rho + Zg + V^2/2)$ at (2)

$P_s = (100.7)(1000) - (1)(9.81)(1000) - (2.21)^2(1000)/2 = 88,400$ Pa

NPSHA $= (P_s - P_v)/\rho g = (88,400 - 2,237)/(9.81)(1000) = 8.8$ m

This NPSHA (8.8 m) is considered large and quite adequate for most pump models. However, if, after a survey of available pumps, it is found that none can operate with this net positive suction head, then the design of the piping system will have to be changed: the pump will have to be placed at a lower elevation to ensure adequate suction static pressure.

55.4.5 Selection of Centrifugal Pumps

The pump selected for a fluid system must deliver the specified flow and required head at or near the pump's maximum efficiency, and have a NPSHR less than the NPSHA. However, only rarely will one find a pump model, even from a survey of several manufacturers, that exactly matches the system; that is, a pump whose flow and head at maximum efficiency exactly match the flow and head required.

The first step in pump selection is to contact several pump manufacturers and obtain the performance curves of the pumps they recommend for the specified service. A typical pump curve is shown in Fig. 55.9.

It is seen that on this one curve data are presented giving flow, head, efficiency, power, and NPSHR for a variety of impeller sizes (diameters). The curves for impeller sizes in between those shown can be estimated by extrapolation. Since clearance distances between the impeller tip and the

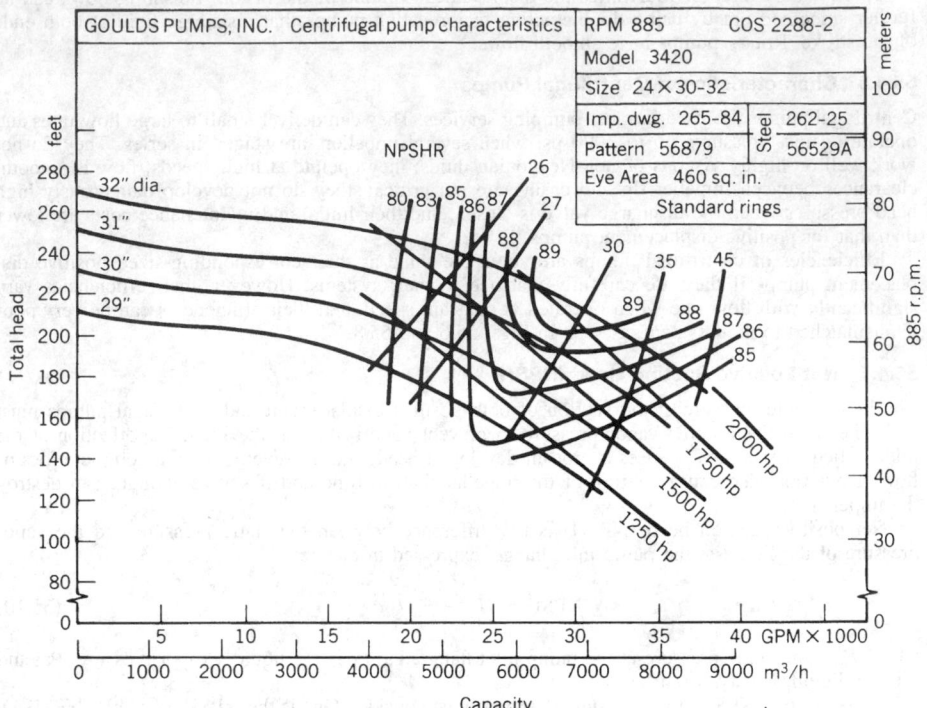

Fig. 55.9 Example of pump curve provided by manufacturer. (Courtesy Goulds Pumps, Inc., Seneca Falls, NY.)

pump casing are not critical, it is possible to install any of several different size impellers in one casing. It is also possible to cut down an existing impeller to a smaller size if it is found advantageous to do so after delivery of a pump and installation in its system.

An important selection parameter is the motor size. Note that there is a maximum power that the pump will require regardless of flow. It is advisable to specify a motor with a power rating at least equal to this maximum power required, since, in most applications, there will be times when the pump is called upon to deliver higher flow rates than originally expected.

For the pump in Example 55.1, the purchase specifications would be:

Flow	7200 m³/hr (2.0 m³/sec)
Head	60.6 m
NPSHA	8.8 m (28.9 ft)

It is seen, in Fig. 55.9, that the 32 in.-diameter impeller in the model 3420 would be adequate for the head and flow. However, the NPSHR is 10.29 m (33 ft), which is more than is available, and therefore not acceptable. The efficiency is 89%, which is close to what was expected. The usual procedure is to survey more manufacturers in hopes of finding a better match, a higher efficiency, and one requiring less NPSH. If the model 3420 were finally selected, it is recommended that the 2000 hp (1492 kW) motor be specified. Also, the piping system would have to be altered to lower the pump elevation and provide more NPSHA.

Referring again to Fig. 55.9, it is seen that the size is given by the numbers 24 × 30–32. It is standard practice in the industry to use a size designation number that gives, in order, the diameter of the discharge flange, the diameter of the suction flange, and then the impeller diameter, all in inches.

55.4.6 Operating Performance of Pumps in a System

The actual point of operation (head and flow) of a pump in a piping system is found from the intersection of the pump curve (Fig. 55.9) and the system head curve [Eq. (55.4)]. Both curves represent the energy required to cause a specified flow rate. By the law of conservation of energy the energy input to the fluid by the pump must equal the energy required by the piping system for a specified flow. Figure 55.10 shows both curves plotted on the same coordinate system and establishes the point of operation of the pump in the system.

The actual point of operation of the model 3420 pump and the system of Example 55.1 is 7450 m³/hr at 60.7 m. If this flow is too large, the system will have to be throttled (by closing in a valve) until the flow is reduced to the desired value. If the system is throttled to 7200 m³/hr, the actual developed head will be 63.1 m. This means the head loss created in the valve, at 7200 m³/hr, is 63.1–60.6 or 2.5 m. The power wasted in this throttling process is ρQHg/eff = (1000) (2)(2.5)(9.81)/(0.89)(1000) = 55.1 kW, which is converted into heat.

In large pumps (100 kW) even small differences in operating power (2%) can make large differences in operating economy. For this reason it is important for the purchaser of a pump to seek the best possible match of the pump to the system to optimize efficiency and avoid having to throttle the flow. For example, if the difference in operating power between two pumps capable of meeting a specified service (head and flow) is as small as 2 kW but the pump is operated continuously (8760 hours per year), then the energy difference is 17,520 kWhr which, at 50.05/kWhr, has a value of $876 per year. If the additional cost (if any) of the more economical pump can be amortized over its financial lifetime for less than $876 per year, then the better pump should be purchased.

55.4.7 Throttling versus Variable Speed Drive

If the pump is to be operated at reduced flow rates for extended periods of time, it may be economically justifiable to use a variable speed drive.

As an example, assume that the system in Example 55.1 is operated at 5000 m³/hr for 2500 hours per year. If throttled to 5000 m³/hr, the pump head (from Fig. 55.10) would be 73.5 m and the efficiency would be (about) 84%. The energy consumed at this point of operation would be $\rho QHgh$/eff = (1000)(5000/3600)(93.5)(9.81)(2500)/(0.84)(1000) = 2.98 × 10⁶ kWhr per year.

The operating points for the variable speed drive are determined by using the modeling laws (Section 55.3.5). If the diameter is constant, the H/N^2 and Q/N are constant and, for variable N, $H = KQ^2$, which is a parabola through the origin, as shown in Fig. 55.10. The operating points (1) and (2) on this parabola are related by the equations $H_1/N_1^2 = H_2/N_2^2$ and $Q_1/N_1 = Q_2/N_2$, where $H_2 = 60.0 + 0.012 \times 10^{-6} (5000)^2 = 60.3$ m. The K of the parabola in $H_2/Q_2^2 = 60.3/(5000)^2$. The intersection of this parabola with the original pump curve, point (1), is $H_1 = 72.0$ m and $Q_1 = 5400$ m³/hr. The reduced speed $N_1 = N_2Q_2/Q_1 = (14.75)(5000)/5400 = 13.66$ Hz. The efficiency at (2), 86%, equals the efficiency at (1) since all nondimensional parameters at (1) and (2) are the same.

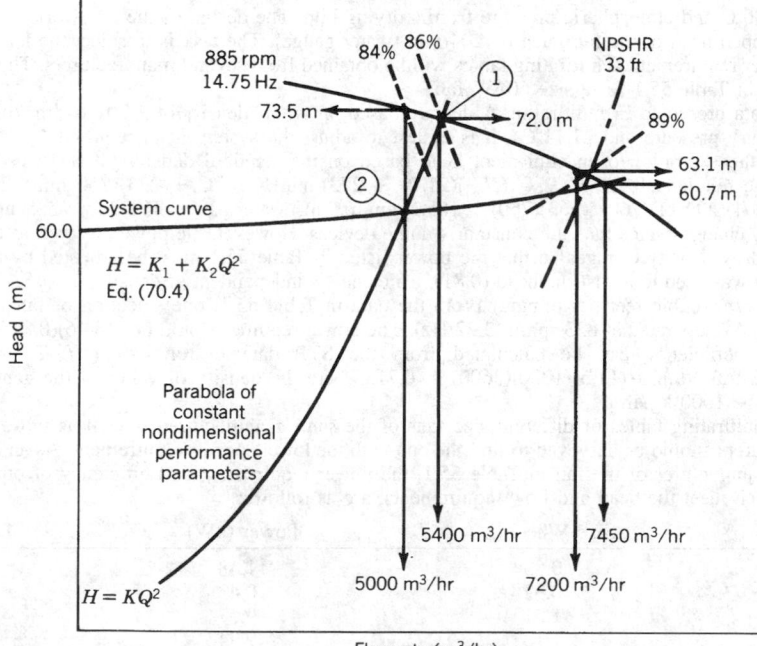

Fig. 55.10 Point of operation of a pump in a system.

The energy consumed at the reduced pump speed (13.66 Hz) to provide 5000 m^3/hr for 2500 hours per year is $(1000)(5000)/(3600)(60.3)(9.81)(2500)/(0.86)(1000) = 2.38 \times 10^6$ kWhr per year. The saving of 600,000 kWhr, at $0.05/kWhr, is worth $30,000 per year. If the cost of a variable speed drive in this example can be amortized over its financial lifetime for less than $30,000 per year, it should be purchased.

55.5 FAN SELECTION

55.5.1 Types of Fans; Their Characteristics

Fans, the same as pumps, are made in a large variety of types in order to serve a large variety of applications. There are also options in both cost and efficiency for applications requiring low power (5 kW). High-power applications require high efficiencies.

Fan types with low specific speeds (0.17) are suitable for high-head, low-flow application. These fans are usually centrifugal, with both forward and backward curved blades. Fan types with high specific speeds (16.75) are suitable for low-head, large-flow application. These fans are of the axial flow type (propellor blades). Higher heads can be achieved with axial flow fans if provision is made to recover, into head, the swirl (rotational) component of velocity imparted by the blades. Two methods of recovering this energy component are: (1) a set of fixed blades located either up- or downstream from the rotating blades (vane axial type); and (2) for maximum recovery, two sets of rotating blades, one turning in a reverse direction to the other (contrarotating propellors).

Characteristics of fans are similar to those of centrifugal pumps: they must be carefully matched to their system in order to achieve their best efficiencies; the basic modeling laws are used to predict their performance; clearances between the wheel tip and casing (cutoff) are not critical; their discharge ducts can be closed without causing high heads to develop; their flow is smooth; they can be used on gases and gas–particle mixtures (powders, dusts, lints); and their maintenance costs are low.

55.5.2 Fan Selection

The steps to follow in fan selection are the same as those for pump selection with two exceptions: (1) there is no net positive suction head to be concerned with; (2) a variety of speeds are usually available for each fan through the use of different size sheaves (using belt drives). This latter exception causes some inconvenience in determining optimum efficiency matches since the method of presenting performance data, called "multirating tables," does not include an efficiency parameter. However, the tables do list the power requirement so that a system designer can seek the best efficiency by seeking the lowest power requirement. If efficiencies are wanted, they can either be calculated or requested from the manufacturer in the form of performance curves (rather than tables).

An example best illustrates the method. Assume a fan is to be selected to exhaust 18,170 m^3/hr of air at 90°C and atmospheric pressure from a drying kiln. The design of the ductwork is such that the developed head of the fan must be 204 mm (water gauge). The task is to select the fan with the least power requirement. Multirating tables will be obtained from several manufacturers. They appear as shown in Table 55.1 for a size 60AW fan.

The data presented in multirating tables are based on an air density of 1.201 kg/m^3 (air at 760 mm mercury pressure and 21.11°C). It is easiest to adjust the system head required (at 1 atm and 90°C for the example) to an equivalent head based on the standard density (at STP) used in the tables. The relationship is: $H_{STP} = (H_{req})(21.11 + 273)(\text{mmHG})/(\text{°C} + 273)(760 \text{ mm})$. Therefore $H_{STP} = (204)(294.11)(760)/(363)(760) = 165.1$ mm (6.5 in.) water gauge. The flow rate is unaffected by density changes since fans are constant volume devices. However, the power, as well as the head, is affected by density changes so that the power listed in Table 55.1 must be adjusted by the same factor that was used to adjust the head (0.81). Efficiency is independent of density, and 18,170 m^3/hr is 10,000 cfm (cubic feet per minute) From the data in Table 55.1, one selection of fan would be the size 60AW operated at 823 rpm (13.72 Hz). The power required would be (15.46)(0.81) hp (9.34 kW). The efficiency can be calculated from the STP data by eff $= \rho QHg/P = (1000)$ $(18,170)(0.1651)(9.81)/(11.5)(1000)(3600) = 0.71$, where the density of water in the gauge is assumed to be 1000 kg/m^3.

The multirating tables of different size fans of the same manufacturer as well as those of other manufacturers should be surveyed to find the one with the lowest power requirement. As an example, for the manufacturer of the fan in Table 55.1, the power requirement and efficiency of other sizes, all of which meet the head and flow requirements, are as follows:

Model	Size Wheel (m)	Power (kW)	Efficiency
45AW	0.83	11.48	0.58
50AW	0.92	10.43	0.64
55AW	1.01	9.67	0.69
60AW	1.10	9.34	0.71
70AW	1.29	8.57	0.77
80AW	1.47	9.00	0.74
90AW	1.66	9.76	0.68

Table 55.1   Example of Multirating Table for Fans[a]

Capacity (cfm)	Outlet Velocity (fpm)	5½″ S.P.		6″ S.P.		6½″ S.P.	
		rpm	bhp	rpm	bhp	rpm	bhp
3000	921						
3700	1136	715	5.83				
4400	1351	715	6.37	746	7.10	777	7.84
5100	1566	715	6.98	747	7.73	777	8.48
5800	1781	717	7.63	748	8.41	778	9.24
6500	1996	722	8.40	752	9.21	781	10.04
7200	2211	730	9.26	759	10.11	787	10.98
7900	2426	738	10.17	766	11.07	794	12.00
8600	2641	748	11.12	776	12.12	803	13.08
9300	2856	759	12.18	786	13.21	812	14.21
10000	3071	771	13.29	798	14.37	823	15.46
10700	3286	783	14.45	809	15.60	835	16.76
11400	3501	796	15.69	822	16.87	847	18.11
12100	3716	810	16.99	836	18.28	860	19.52
12800	3931	825	18.40	850	19.71	874	21.07
13500	4146	841	19.91	865	21.27	888	22.64
14200	4361	857	21.45	881	22.92	904	24.35
14900	4576	873	23.03	896	24.60	919	26.16
15600	4791	890	24.75	913	26.34	935	27.98
16300	5006	907	26.48	930	28.22	952	29.88
17000	5221	924	28.22	947	30.10	969	31.92
17700	5436	941	30.18	963	32.00	985	33.96
18400	5651	960	32.30	981	34.13	1002	36.02
19100	5866	978	34.43	999	36.41	1020	38.31
19800	6081	997	36.87	1017	38.73	1038	40.77

[a]Courtesy of Buffalo Forge Co., Buffalo, NY.

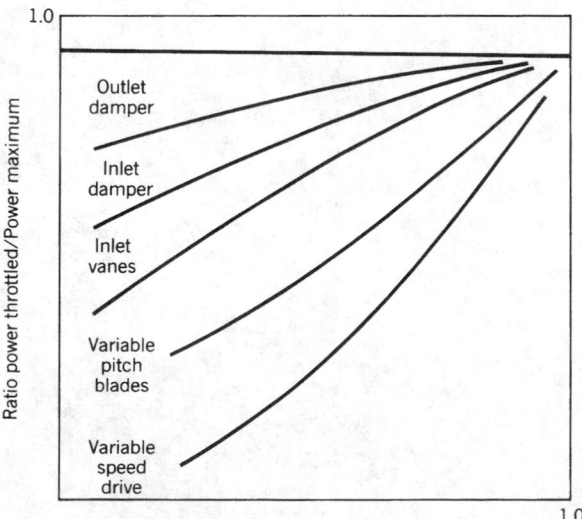

Fig. 55.11   Effectiveness of various methods of controlling fans in variable volume service.

The 70AW model is seen to be the best choice for this service.

55.5.3 Control of Fans for Variable Volume Service

Two common applications of fans requiring variable volume operation are combustion air to boilers and conditioned air to rooms in a building. Common methods of controlling air volume are outlet dampers; inlet dampers; inlet vanes (which impart a prerotation or swirl velocity to the air entering the wheel); variable pitch blades (on axial fans); and variable speed drives on the fan motor.

Figure 55.11 illustrates the effectiveness of these methods by comparing power ratios with flow ratios at reduced flows. The variable speed drive is the most effective method and, with the commercialization of solid-state motor controls (providing variable frequency and variable voltage electrical service to standard induction motors), is becoming the most popular method for fan speed control.

BIBLIOGRAPHY

ASHRAE Handbook of Fundamentals, American Society of Heating, Refrigerating and Air Conditioning Engineers, Atlanta, GA, 1980.

Cameron Hydraulic Data, Ingersoll Rand Co., Woodcliff Lake, NJ, 1977.

Csanady, G. T., *Theory of Turbo Machines,* McGraw-Hill, New York, 1964.

Fans and Systems, Publication 201; *Troubleshooting,* Publication 202; *Field Performance Measurements,* Publication 203; Air Moving and Conditioning Association, Arlington Heights, IL.

Fans in Air Conditioning, The Trane Co., La Crosse, WI.

Flow of Fluids Through Valves, Fittings and Pipe, Technical Paper No. 410, Crane Co., Chicago, IL, 1976.

Hicks, T. G., and T. W. Edwards, *Pump Application Engineering,* McGraw-Hill, New York, 1971.

Hydraulic Institute Standards, Hydraulic Institute, Cleveland, OH, 1975.

Karassick, I. J., *Centrifugal Pump Clinic,* Marcel Dekker, New York, 1981.

Laboratory Methods of Testing Fans for Ratings, Standard 210-74, Air Moving and Conditioning Association, Arlington Heights, IL, 1974.

CHAPTER 56

NUCLEAR POWER

William Kerr
Department of Nuclear Engineering
University of Michigan
Ann Arbor, Michigan

56.1	**HISTORICAL PERSPECTIVE**	**1699**
	56.1.1 The Birth of Nuclear Energy	1699
	56.1.2 Military Propulsion Units	1700
	56.1.3 Early Enthusiasm for Nuclear Power	1700
	56.1.4 U.S. Development of Nuclear Power	1700
56.2	**CURRENT POWER REACTORS, AND FUTURE PROJECTIONS**	**1701**
	56.2.1 Light-Water-Moderated Enriched-Uranium-Fueled Reactor	1701
	56.2.2 Gas-Cooled Reactor	1701
	56.2.3 Heavy-Water-Moderated Natural-Uranium-Fueled Reactor	1701
	56.2.4 Liquid-Metal-Cooled Fast Breeder Reactor	1701
	56.2.5 Fusion	1701
56.3	**CATALOG AND PERFORMANCE OF OPERATING REACTORS, WORLDWIDE**	**1701**
56.4	**U.S. COMMERCIAL REACTORS**	**1701**
	56.4.1 Pressurized-Water Reactors	1701
	56.4.2 Boiling-Water Reactors	1704
	56.4.3 High-Temperature Gas-Cooled Reactors	1705
	56.4.4 Constraints	1705
	56.4.5 Availability	1706
56.5	**POLICY**	**1707**
	56.5.1 Safety	1707
	56.5.2 Disposal of Radioactive Wastes	1708
	56.5.3 Economics	1709
	56.5.4 Environmental Considerations	1709
	56.5.5 Proliferation	1709
56.6	**BASIC ENERGY PRODUCTION PROCESSES**	**1710**
	56.6.1 Fission	1711
	56.6.2 Fusion	1712
56.7	**CHARACTERISTICS OF THE RADIATION PRODUCED BY NUCLEAR SYSTEMS**	**1712**
	56.7.1 Types of Radiation	1714
56.8	**BIOLOGICAL EFFECTS OF RADIATION**	**1714**
56.9	**THE CHAIN REACTION**	**1715**
	56.9.1 Reactor Behavior	1715
	56.9.2 Time Behavior of Reactor Power Level	1717
	56.9.3 Effect of Delayed Neutrons on Reactor Behavior	1717
56.10	**POWER PRODUCTION BY REACTORS**	**1718**
	56.10.1 The Pressurized-Water Reactor	1718
	56.10.2 The Boiling-Water Reactor	1720
56.11	**REACTOR SAFETY ANALYSIS**	**1720**

56.1 HISTORICAL PERSPECTIVE

56.1.1 The Birth of Nuclear Energy

The first large-scale application of nuclear energy was in a weapon. The second use was in submarine propulsion systems. Subsequent development of fission reactors for electric power production has

Mechanical Engineers' Handbook, 2nd ed., Edited by Myer Kutz.
ISBN 0-471-13007-9   © 1998 John Wiley & Sons, Inc.

been profoundly influenced by these early military associations, both technically and politically. It appears likely that the military connection, tenuous though it may be, will continue to have a strong political influence on applications of nuclear energy.

Fusion, looked on by many as a supplement to, or possibly as an alternative to fission for producing electric power, was also applied first as a weapon. Most of the fusion systems now being investigated for civilian applications are far removed from weapons technology. A very few are related closely enough that further civilian development could be inhibited by this association.

56.1.2 Military Propulsion Units

The possibilities inherent in an extremely compact source of fuel, the consumption of which requires no oxygen, and produces a small volume of waste products, was recognized almost immediately after World War II by those responsible for the improvement of submarine propulsion units. Significant resources were soon committed to the development of a compact, easily controlled, quiet, and highly reliable propulsion reactor. As a result, a unit was produced which revolutionized submarine capabilities.

The decisions that led to a compact, light-water-cooled and -moderated submarine reactor unit, using enriched uranium for fuel, were undoubtedly valid for this application. They have been adopted by other countries as well. However, the technological background and experience gained by U.S. manufacturers in submarine reactor development was a principal factor in the eventual decision to build commercial reactors that were cooled with light water and that used enriched uranium in oxide form as fuel. Whether this was the best approach for commercial reactors is still uncertain.

56.1.3 Early Enthusiasm for Nuclear Power

Until the passage, in 1954, of an amendment to the Atomic Energy Act of 1946, almost all of the technology that was to be used in developing commercial nuclear power was classified. The 1954 Amendment made it possible for U.S. industry to gain access to much of the available technology, and to own and operate nuclear power plants. Under the amendment the Atomic Energy Commission (AEC), originally set up for the purpose of placing nuclear weapons under civilian control, was given responsibility for licensing and for regulating the operation of these plants.

In December of 1953 President Eisenhower, in a speech before the General Assembly of the United Nations, extolled the virtues of peaceful uses of nuclear energy and promised the assistance of the United States in making this potential new source of energy available to the rest of the world. Enthusiasm over what was then viewed as a potentially inexpensive and almost inexhaustible new source of energy was a strong force which led, along with the hope that a system of international inspection and control could inhibit proliferation of nuclear weapons, to formation of the International Atomic Energy Agency (IAEA) as an arm of the United Nations. The IAEA, with headquarters in Vienna, continues to play a dual role of assisting in the development of peaceful uses of nuclear energy, and in the development of a system of inspections and controls aimed at making it possible to detect any diversion of special nuclear materials, being used in or produced by civilian power reactors, to military purposes.

56.1.4 U.S. Development of Nuclear Power

Beginning in the early 1950s the AEC, in its national laboratories, and with the participation of a number of industrial organizations, carried on an extensive program of reactor development. A variety of reactor systems and types were investigated analytically and several prototypes were built and operated.

In addition to the light water reactor (LWR), gas-cooled graphite-moderated reactors, liquid-fueled reactors with fuel incorporated in a molten salt, liquid-fueled reactors with fuel in the form of a uranium nitrate solution, liquid-sodium-cooled graphite-moderated reactors, solid-fueled reactors with organic coolant, and liquid-metal solid-fueled fast spectrum reactors have been developed and operated, at least in pilot plant form in the United States. All of these have had enthusiastic advocates. Most, for various reasons, have not gone beyond the pilot plant stage. Two of these, the high-temperature gas-cooled reactor (HTGR) and the liquid-metal-cooled fast breeder reactor (LMFBR), have been built and operated as prototype power plants.

Some of these have features associated either with normal operation, or with possible accident situations, which seem to make them attractive alternatives to the LWR. The HTGR, for example, operates at much higher outlet coolant temperature than the LWR and thus makes possible a significantly more efficient thermodynamic cycle as well as permitting use of a physically smaller steam turbine. The reactor core, primarily graphite, operates at a much lower power density than that of LWRs. This lower power density and the high-temperature capability of graphite make the HTGR's core much more tolerant of a loss-of-coolant accident than the LWR core.

The long, difficult, and expensive process needed to take a conceptual reactor system to reliable commercial operation has unquestionably inhibited the development of a number of alternative systems.

56.2 CURRENT POWER REACTORS, AND FUTURE PROJECTIONS

Although a large number of reactor types have been studied for possible use in power production, the number now receiving serious consideration is rather small.

56.2.1 Light-Water-Moderated Enriched-Uranium-Fueled Reactor

The only commercially viable power reactor systems operating in the United States today use LWRs. This is likely to be the case for the next decade or so. France has embarked on a construction program that will eventually lead to productions of about 90% of its electric power by LWR units. Great Britain has under consideration the construction of a number of LWRs. The Federal Republic of Germany has a number of LWRs in operation with additional units under construction. Russia and a number of other Eastern European countries are operating LWRs, and are constructing additional plants. Russia is also building a number of smaller, specially designed LWRs near several population centers. It is planned to use these units to generate steam for district heating. The first one of these reactors is scheduled to go into operation soon near Gorki.

56.2.2 Gas-Cooled Reactor

Several designs exist for gas-cooled reactors. In the United States the one that has been most seriously considered uses helium for cooling. Fuel elements are large graphite blocks containing a number of vertical channels. Some of the channels are filled with enriched uranium fuel. Some, left open, provide a passage for the cooling gas. One small power reactor of this type is in operation in the United States. Carbon dioxide is used for cooling in some European designs. Both metal fuels and graphite-coated fuels are used. A few gas-cooled reactors are being used for electric power production both in England and in France.

56.2.3 Heavy-Water-Moderated Natural-Uranium-Fueled Reactor

The goal of developing a reactor system that does not require enriched uranium led Canada to a natural-uranium-fueled, heavy-water-moderated, light-water-cooled reactor design dubbed Candu. A number of these are operating successfully in Canada. Argentina and India each uses a reactor power plant of this type, purchased from Canada, for electric power production.

56.2.4 Liquid-Metal-Cooled Fast Breeder Reactor

France, England, Russia, and the United States all have prototype liquid-metal-cooled fast breeder reactors (LMFBRs) in operation. Experience and analysis provide evidence that the plutonium-fueled LMFBR is the most likely, of the various breeding cycles investigated, to provide a commercially viable breeder. The breeder is attractive because it permits as much as 80% of the available energy in natural uranium to be converted to useful energy. The LWR system, by contrast, converts at most 3%–4%.

Because plutonium is an important constituent of nuclear weapons, there has been concern that development of breeder reactors will produce nuclear weapons proliferation. This is a legitimate concern, and must be dealt with in the design of the fuel cycle facilities that make up the breeder fuel cycle.

56.2.5 Fusion

It may be possible to use the fusion reaction, already successfully harnessed to produce a powerful explosive, for power production. Considerable effort in the United States and in a number of other countries is being devoted to development of a system that would use a controlled fusion reaction to produce useful energy. At the present stage of development the fusion of tritium and deuterium nuclei appears to be the most promising reaction of those that have been investigated. Problems in the design, construction, and operation of a reactor system that will produce useful amounts of economical power appear formidable. However, potential fuel resources are enormous, and are readily available to any country that can develop the technology.

56.3 CATALOG AND PERFORMANCE OF OPERATING REACTORS, WORLDWIDE

Worldwide, the operation of nuclear power plants in 1982 produced more than 10% of all the electrical energy used. Table 56.1 contains a listing of reactors in operation in the United States and in the rest of the world.

56.4 U.S. COMMERCIAL REACTORS

As indicated earlier, the approach to fuel type and core design used in LWRs in the United States comes from the reactors developed for marine propulsion by the military.

56.4.1 Pressurized-Water Reactors

Of the two types developed in the United States, the pressurized water reactor (PWR) and the boiling water reactor (BWR), the PWR is a more direct adaptation of marine propulsion reactors. PWRs are

Table 56.1   Operating Power Reactors (1995)

Country	Reactor Type[a]	Number in Operation	Net MWe
Argentina	PHWR	3	1627
Armenia	PWR	2	800
Belgium	PWR	7	5527
Brazil	PWR	1	626
Bulgaria	PWR	6	3420
Canada	PHWR	22	15439
China	PWR	3	2100
Czech Republic	PWR	4	1632
Finland	PWR	2	890
	BWR	2	1420
France	PWR	54	57140
Germany	PWR	14	15822
	BWR	7	6989
Hungary	PWR	4	1729
India	BWR	2	300
	PHWR	8	1395
Japan	PWR	22	17298
	BWR	26	22050
Korea	PWR	9	7541
	PHWR	1	629
Lithuania	LGR	2	2760
Mexico	BWR	2	1308
Netherlands	PWR	1	452
	BWR	1	55
Pakistan	PHWR	1	125
Russia	LGR	11	10175
	PWR	13	9064
	LMFBR	1	560
Slovenia	PWR	1	620
Slovokia	PWR	4	1632
South Africa	PWR	2	1840
Spain	BWR	2	1389
	PWR	7	5712
Sweden	BWR	9	7370
	PWR	3	2705
Switzerland	BWR	2	1385
	PWR	3	1665
Taiwan	BWR	4	3104
	PWR	2	1780
UK	GCR	20	3360
	AGR	14	8180
	PWR	1	1188
Ukraine	LGR	2	1850
	PWR	12	10245
United States	BWR	37	32215
	PWR	72	67458

[a]PWR = pressurized water reactor; BWR = boiling water reactor; AGR = advanced gas-cooled reactor; GCR = gas-cooled reactor; HTGR = high-temperature gas-cooled reactor; LMFBR = liquid-metal fast-breeder reactor; LGR = light-water-cooled graphite-moderated reactor; HWLWR = heavy-water-moderated light-water-cooled reactor; PHWR = pressurized heavy-water-moderated-and-cooled reactor; GCHWR = gas-cooled heavy-water-moderated reactor.

operated at pressures in the pressure vessel (typically about 2250 psi) and temperatures (primary inlet coolant temperature is about 564°F with an outlet temperature about 64°F higher) such that bulk boiling does not occur in the core during normal operation. Water in the primary system flows through the core as a liquid, and proceeds through one side of a heat exchanger. Steam is generated on the other side at a temperature slightly less than that of the water that emerges from the reactor vessel outlet. Figure 56.1 shows a typical PWR vessel and core arrangement. Figure 56.2 shows a steam generator.

The reactor pressure vessel is an especially crucial component. Current U.S. design and operational philosophy assumes that systems provided to ensure maintenance of the reactor core integrity

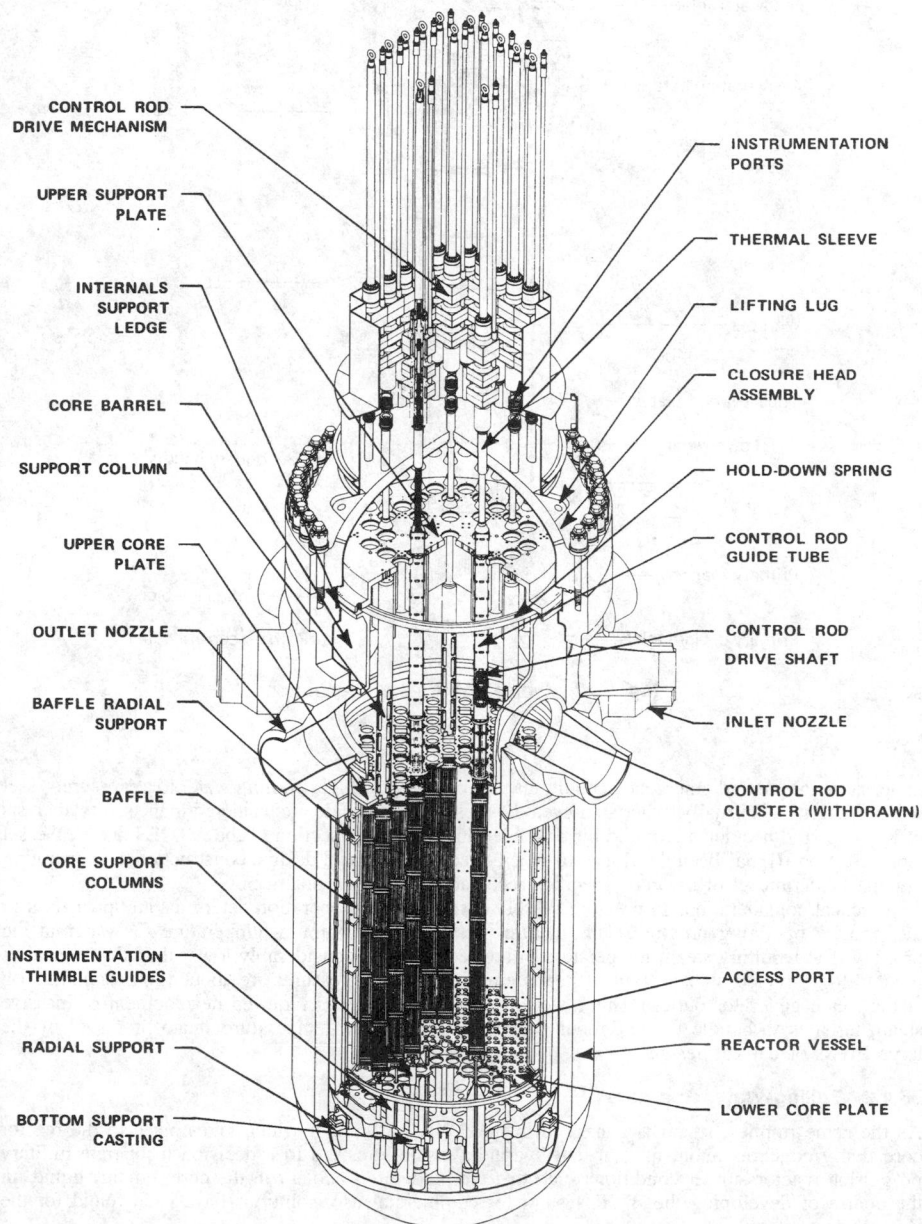

Fig. 56.1 Typical vessel and core configuration for PWR. (Courtesy Westinghouse.)

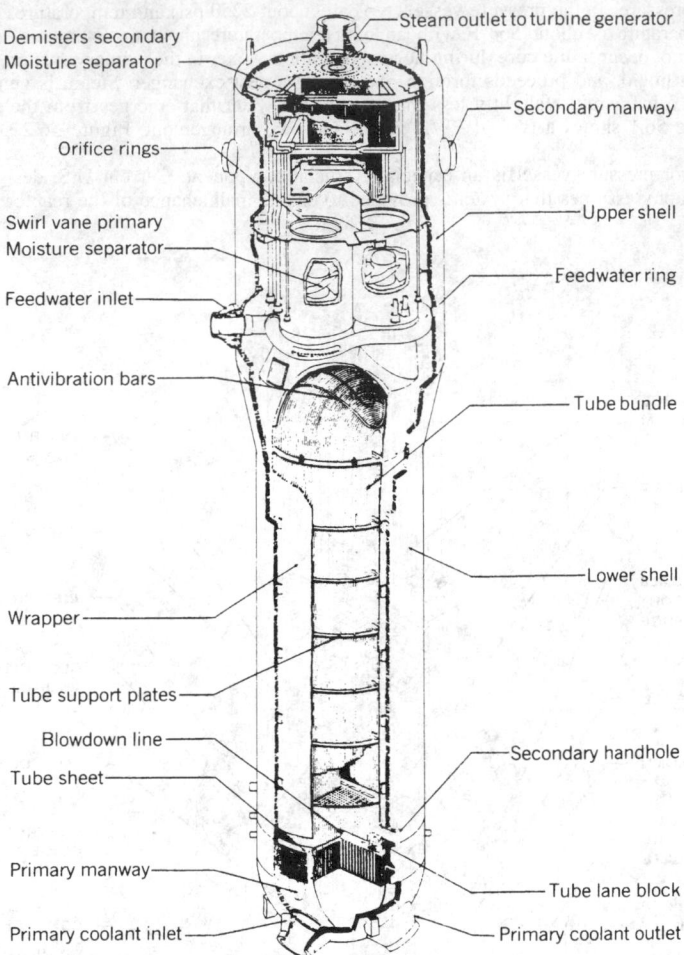

Fig. 56.2 Typical PWR steam generator.

under both normal and emergency conditions will be able to deliver cooling water to a pressure vessel whose integrity is virtually intact after even the most serious accident considered in the safety analysis of hypothesized accidents required by U.S. licensing. A special section of the ASME Pressure Vessel Code, Section III, has been developed to specify acceptable vessel design, construction, and operating practices. Section XI of the code specifies acceptable inspection practices.

Practical considerations in pressure vessel construction and operation determine an upper limit to the primary operating pressure. This in turn prescribes a maximum temperature for water in the primary. The resulting steam temperature in the secondary is considerably lower than that typical of modern fossil-fueled plants. (Typical steam temperatures and pressures are about 1100 psi and 556°F at the steam generator outlet.) This lower steam temperature has required development of massive steam turbines to handle the enormous steam flow of the low-temperature steam produced by the large PWRs of current design.

56.4.2 Boiling-Water Reactors

As the name implies, steam is generated in the BWR by boiling, which takes place in the reactor core. Early concerns about nuclear and hydraulic instabilities led to a decision to operate military propulsion reactors under conditions such that the moderator–coolant in the core remains liquid. In the course of developing the BWR system for commercial use, solutions have been found for the instability problems.

Although some early BWRs used a design that separates the core coolant from the steam which flows to the turbine, all modern BWRs send steam generated in the core directly to the turbine. This arrangement eliminates the need for a separate steam generator. It does, however, provide direct communication between the reactor core and the steam turbine and condenser, which are located outside the containment. This leads to some problems not found in PWRs. For example, the turbine–condenser system must be designed to deal with radioactive nitrogen-16 generated by an (n,p) reaction of fast neutrons in the reactor core with oxygen-16 in the cooling water. Decay of the short-lived nitrogen-16 (half-life 7.1 sec) produces high-energy (6.13-MeV) highly penetrating gamma rays. As a result, the radiation level around an operating BWR turbine requires special precautions not needed for the PWR turbine. The direct pathway from core to turbine provided by the steam pipes also affords a possible avenue of escape and direct release outside of containment for fission products that might be released from the fuel in a core-damaging accident. Rapid-closing valves in the steam lines are provided to block this path in case of such an accident.

The selection of pressure and temperature for the steam entering the turbine that are not markedly different from those typical of PWRs leads to an operating pressure for the BWR pressure vessel that is typically less than half that for PWRs. (Typical operating pressure at vessel outlet is about 1050 psi with a corresponding steam temperature of about 551°F.)

Because it is necessary to provide for two-phase flow through the core, the core volume is larger than that of a PWR of the same power. The core power density is correspondingly smaller. Figure 56.3 is a cutaway of a BWR vessel and core arrangement. The in-vessel steam separator for removing moisture from the steam is located above the core assembly. Figure 56.4 is a BWR fuel assembly. The assembly is contained in a channel box, which directs the two-phase flow. Fuel pins and fuel pellets are not very different in either size or shape from those for PWRs, although the cladding thickness for the BWR pin is somewhat larger than that of PWRs.

56.4.3 High-Temperature Gas-Cooled Reactors

Experience with the high-temperature gas-cooled reactor (HTGR) in the United States is limited. A 40-MWe plant was operated from 1967 to 1974. A 330-MWe plant has been in operation since 1976. A detailed design was developed for a 1000-MWe plant, but plans for its construction were abandoned.

Fuel elements for the plant in operation are hexagonal prisms of graphite about 31 in. tall and 5.5 in. across flats. Vertical holes in these blocks allow for passage of the helium coolant. Fuel elements for the larger proposed plant were similar. Figure 56.5 shows core and vessel arrangement. Typical helium-coolant outlet temperature for the reactor now in operation is about 1300°F. Typical steam temperature is 1000°F. The large plant was also designed to produce 1000°F steam.

The fuel cycle for the HTGR was originally designed to use fuel that combined highly enriched uranium with thorium. This cycle would convert thorium to uranium-233, which is also a fissile material, thereby extending fuel lifetime significantly. This mode of operation also produces uranium-233, which can be chemically separated from the spent fuel for further use. Recent work has resulted in the development of a fuel using low-enriched uranium in a once-through cycle similar to that used in LWRs.

The use of graphite as a moderator and helium as coolant allows operation at temperatures significantly higher than those typical of LWRs, resulting in higher thermal efficiencies. The large thermal capacity of the graphite core and the large negative temperature coefficient of reactivity make the HTGR insensitive to inadvertent reactivity insertions and to loss-of-coolant accidents. Operating experience to date gives some indication that the HTGR has advantages in increased safety and in lower radiation exposure to operating personnel. These possible advantages plus the higher thermal efficiency that can be achieved make further development attractive. However, the high cost of developing a large commercial unit, plus the uncertainties that exist because of the limited operating experience with this type reactor have so far outweighed the perceived advantages.

As the data in Table 56.1 indicate, there is significant successful operating experience with several types of gas-cooled reactors in a number of European countries.

56.4.4 Constraints

Reactors being put into operation today are based on designs that were originally conceived as much as 20 years earlier. The incredible time lag between the beginning of the design process and the operation of the plant is one of the unfortunate products of a system of industrial production and federal regulation that moves ponderously and uncertainly toward producing a power plant that may be technically obsolescent by the time it begins operation. The combination of the large capital investment required for plant construction, the long period during which this investment remains unproductive for a variety of reasons, and the high interest rates charged for borrowed money have recently led to plant capital costs some 5–10 times larger than those for plants that came on line in the early to mid 1970s. Added to the above constraints is a widespread concern about dangers of nuclear power. These concerns span a spectrum that encompasses fear of contribution to nuclear weapons proliferation, on the one hand, to a strong aversion to high technology, on the other hand.

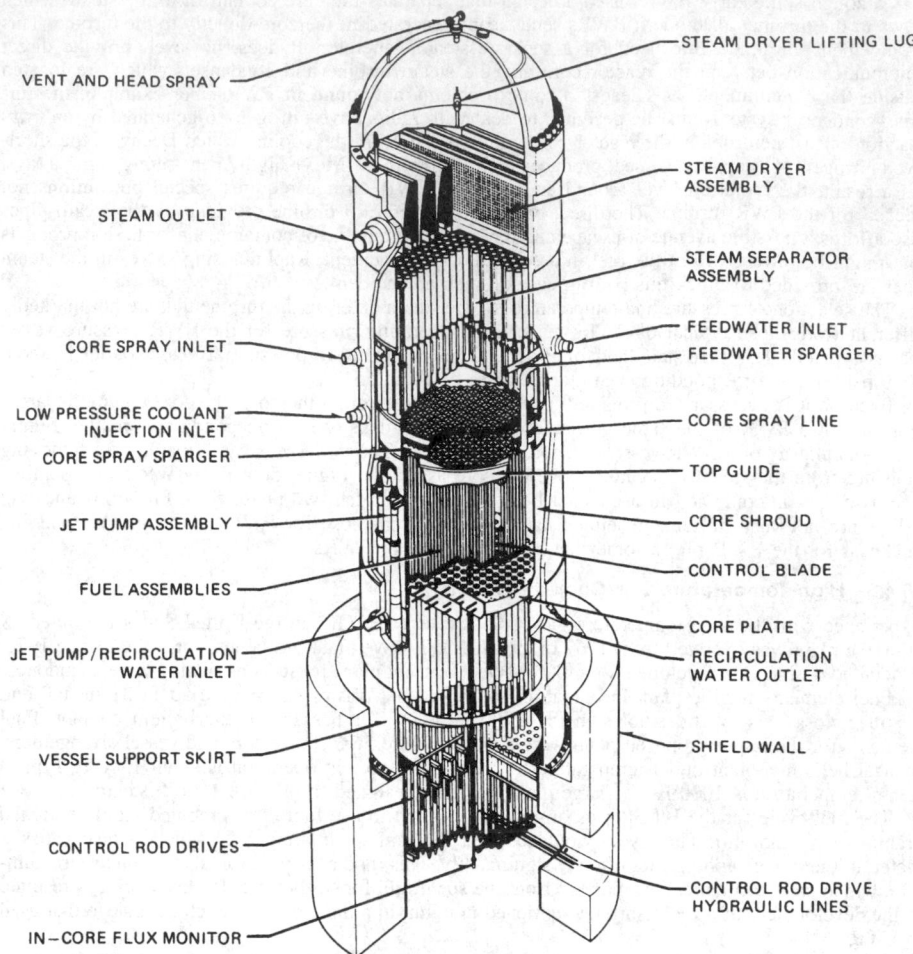

VENT AND HEAD SPRAY

STEAM OUTLET

CORE SPRAY INLET

LOW PRESSURE COOLANT
INJECTION INLET

CORE SPRAY SPARGER

JET PUMP ASSEMBLY

FUEL ASSEMBLIES

JET PUMP/RECIRCULATION
WATER INLET

VESSEL SUPPORT SKIRT

CONTROL ROD DRIVES

IN-CORE FLUX MONITOR

STEAM DRYER LIFTING LUG

STEAM DRYER
ASSEMBLY

STEAM SEPARATOR
ASSEMBLY

FEEDWATER INLET
FEEDWATER SPARGER

CORE SPRAY LINE

TOP GUIDE

CORE SHROUD

CONTROL BLADE

CORE PLATE

RECIRCULATION
WATER OUTLET

SHIELD WALL

CONTROL ROD DRIVE
HYDRAULIC LINES

Fig. 56.3 Typical BWR vessel and core configuration. (Courtesy General Electric.)

This combination of technical, economic, and political constraints places a severe burden on those working to develop this important alternative source of energy.

56.4.5 Availability

A significant determinant in the cost of electrical energy produced by nuclear power plants is the plant capacity factor. The capacity factor is defined as a fraction calculated by dividing actual energy production during some specified time period by the amount that would have been produced by continuous power production at 100% of plant capacity. Many of the early estimates of power cost for nuclear plants were made with the assumption of a capacity factor of 0.80. Experience indicates an average for U.S. power plants of about 0.60. The contribution of capital costs to energy production has thus been more than 30% higher than the early estimates. Since capital costs typically represent anywhere between about 40%–80% (depending on when the plant was constructed) of the total energy cost, this difference in goal and achievement is a significant factor in some of the recently observed cost increases for electricity produced by nuclear power. Examination of the experience of individual plants reveals a wide range of capacity factors. A few U.S. plants have achieved a cumulative capacity factor near 0.80. Some have capacity factors as low as 0.40. There is reason to believe that improvements can be made in many of those with low capacity factors. It should also be possible to go beyond 0.80. Capacity factor improvement is a fruitful area for better resource utilization and realization of lower energy costs.

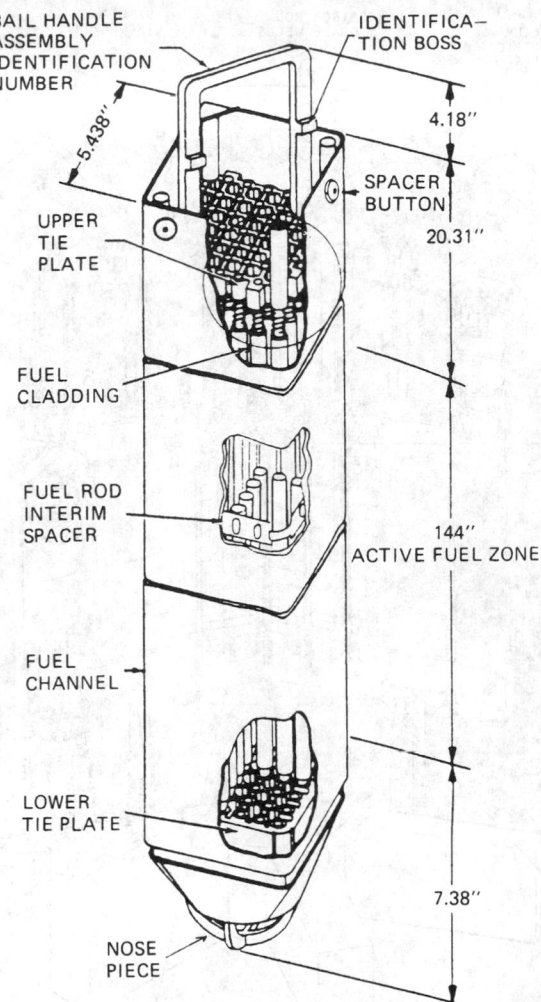

Fig. 56.4 BWR fuel assembly.

56.5 POLICY

The Congress, in the 1954 amendment to the Atomic Energy Act, made the development of nuclear power national policy. Responsibility for ensuring safe operation of nuclear power plants was originally given to the Atomic Energy Commission. In 1975 this responsibility was turned over to a Nuclear Regulatory Commission (NRC), set up for this purpose as an independent federal agency. Nuclear power is the most highly regulated of all the existing sources of energy. Much of the regulation is at the federal level. However, nuclear power plants and their operators are subject to a variety of state and local regulations as well. Under these circumstances nuclear power is of necessity highly responsive to any energy policy that is pursued by the federal government, or of local branches of government, including one of bewilderment and uncertainty.

56.5.1 Safety

The principal safety concern is the possibility of exposure of people to the radiation produced by the large (in terms of radioactivity) quantity of radioactive material produced by the fissioning of the reactor fuel. In normal operation of a nuclear power plant all but a minuscule fraction of this material is retained within the reactor fuel and the pressure vessel. Significant exposure of people outside the plant can occur only if a catastrophic and extremely unlikely accident should release a large fraction

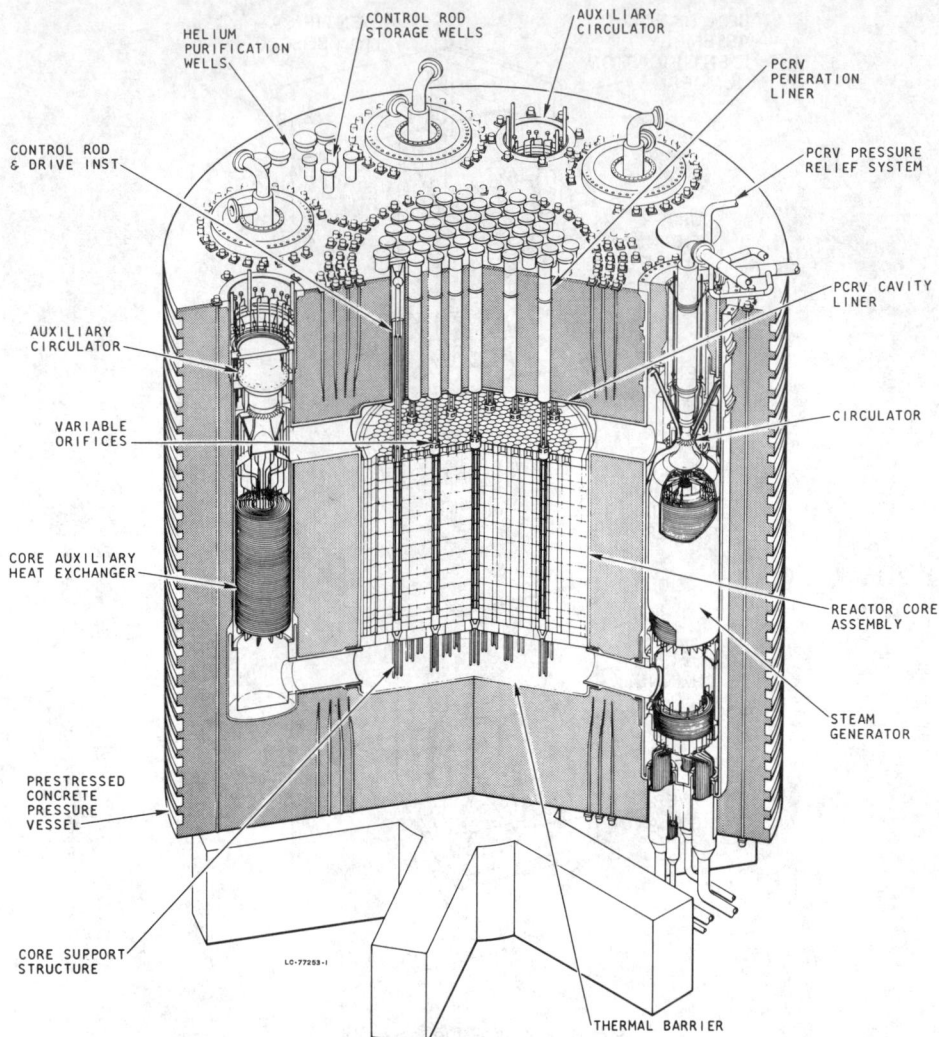

Fig. 56.5 HTGR pressure vessel and core arrangement. (Used by permission of Marcel Dekker, Inc., New York.)

of the radioactive fission products from the pressure vessel and from the surrounding containment system, and if these radioactive materials are then transported to locations where people are exposed to their radiation.

The uranium eventually used in reactor fuel is itself radioactive. The radioactive decay process, which begins with uranium, proceeds to produce several radioactive elements. One of these, radon-226, is a gas and can thus be inhaled by uranium miners. Hence, those who work in the mines are exposed to some hazard. Waste products of the mining and milling of uranium are also radioactive. When stored or discarded above ground, these wastes subject those in the vicinity to radon-226 exposure. These wastes or mill tailings must be dealt with to protect against this hazard. One method of control involves covering the wastes with a layer of some impermeable material such as asphalt.

The fresh fuel elements are also radioactive because of the contained uranium. However, the level of radioactivity is sufficiently low that the unused fuel assemblies can be handled safely without shielding.

56.5.2 Disposal of Radioactive Wastes

The used fuel from a power reactor is highly radioactive, although small in volume. The spent fuel produced by a year's operation of a 1000-MWe plant typically weighs about 40 tons and could be stored in a cube less than 5 ft on a side. It must be kept from coming in contact with people or other living organisms for long periods of time. (After 1000 years of storage the residual radioactivity of the spent fuel is about that of the original fresh fuel.) This spent fuel, or the radioactive residue that remains if most of the unused uranium and the plutonium generated during operation are chemically separated, is called high-level radioactive waste. Up to the present a variety of considerations, many of them political, have led to postponement of a decision on the choice of a permanent storage method for this material. The problem of safe storage has several solutions that are both technically and economically feasible. Technical solutions that currently exist include aboveground storage in air-cooled metal cannisters (for an indefinite period if desirable, with no decrease of safety over the period), as well as permanent disposal in deep strata of salt or of various impermeable rock formations. There have also been proposals to place the radioactive materials in deep ocean caverns. This method, although probably technically possible, is not yet developed. It would require international agreements not now in place. As indicated earlier, an operating plant also generates radioactive material in addition to fission products. Some of this becomes part of the various process streams that are part of the plant's auxiliary systems. These materials are typically removed by filters or ion-exchange systems, leaving filters or ion-exchange resins that contain radioactive materials. Tools, gloves, clothing, paper, and other materials may become slightly contaminated during plant operation. If the radioactive contamination has a half-life of more than a few weeks, these materials, described as low-level radioactive waste, must be stored or disposed of. The currently used disposal method involves burial in comparatively shallow trenches. Because of insufficient attention having been given to design and operation of some of the earlier burial sites, small releases of radioactive material have been observed. Several early burial sites are no longer in operation. Current federal legislation provides for compacts among several states that could lead to cooperative operation, by these states, of burial sites for low-level waste.

56.5.3 Economics

Nuclear power plants that began operation in the 1970s produce power at a cost considerably less than coal-burning plants of the same era. The current cost of power produced by oil-burning plants is two to three times as great as that produced by these nuclear plants. Nuclear power plants coming on line in the 1980s are much more expensive in capital cost (in some cases by a factor of 10!). The cost of the power they produce will be correspondingly greater. The two major contributors to the cost increase are high interest rates and the long construction period that has been required for most of these plants. Average construction time for plants now coming on line is about 11 years! It is likely that construction times can be decreased for new plants. The changes that were required as a result of the TMI accident have now been incorporated into regulations, into existing plants, and into new designs, eliminating the costly and time-consuming back fits that were required for plants under construction when the accident occurred. In Japan the average construction time for nuclear power plants is about 54 months. In Russia it is said to be about 77 months.

Standard plants are being designed and licensed that should make the licensing of an individual plant much faster and less involved. Concern over the pollution of the ecosphere caused by fossil-fueled plants (acid rain, CO_2) will call for additional pollution control, which will drive up costs of construction and operation of these plants. It is reasonable to expect nuclear power to be economically competitive with alternative methods of electric power generation in both the near and longer term.

56.5.4 Environmental Considerations

The environmental pollution produced by an operating nuclear power plant is far less than that caused by any other currently available method of producing electric power. The efficiency of the thermo-dynamic cycle for water reactors is lower than that of modern fossil-fuel plants because current design of reactor pressure vessels limits the steam temperature. Thus, the amount of waste heat rejected is greater for a nuclear plant than for a modern fossil-fuel plant of the same rated power. However, current methods of waste heat rejection (typically cooling towers) handle this with no particular environmental degradation. Nuclear power plants emit no carbon dioxide, no sulfur, no nitrous oxides. No large coal storage area is required. The tremendous volumes of sulfur compounds removed during coal combustion and the enormous quantities of ash produced by coal plants are problems with which those who operate nuclear plants do not have to deal.

Table 56.2 provides a comparison of emissions and wastes from a large coal-burning plant and from a nuclear power plant of the same rated power. Although there is a small release of radioactive material to the biosphere from the nuclear power plant, the resulting increase in exposure to a member of the population in the immediate vicinity of the plant is typically about 1% of that produced by naturally occurring background radiation.

56.5.5 Proliferation

Nuclear power plants are thought by some to increase the probability of nuclear weapons proliferation. It is true that a country with the trained engineers and scientists, the facilities, and the resources required to produce nuclear power can develop a weapons capability more rapidly than one without

Table 56.2   Waste Material from Different Types of 1000-MWe Power Plants (Capacity Factor = 0.8)

	Coal Fired	Water Reactor
Typical thermal efficiency, %	39	32
Thermal wastes (in thermal megawatts)		
To cooling water	1,170	1,970
To atmosphere	400	150
Total	1,570	2,120
Solid wastes		
Fly ash or slag, tons/year	330,000	0
cubic feet/year	7,350,000	0
railroad carloads/year	3,300	0
Radioactive wastes		
Fuel to reprocessing plant,		
assemblies/year	0	160
railroad carloads/year	0	5
Solid waste storage		
From reprocessing plant, cubic feet/year	0	100
From power plant, cubic feet/year	0	5,000
Gaseous and liquid wastes[a]		
(tons per day/10^6 cubic feet per day)		
Carbon monoxide	2/8	0
Carbon dioxide	21,000/53,200	0
Sulfur dioxide: 1% sulfur fuel	140/325	0
2.5% sulfur fuel	350/812	0
Nitrogen oxides	82/305	0
Particulates to atmosphere (tons/day)	0.4	0
Radioactive gases or liquids, equivalent dose mrem/year at plant boundary	Minor	5

[a]For 3,000,000 tons/year coal total ash content of 11%, fly ash precipitator efficiency of 99.5%, and 15% of sulfur remaining in ash.

this background. However, for a country starting from scratch, the development of nuclear power is a detour that would consume needless time and resources. None of the countries that now possess nuclear weapons capability has used the development of civil nuclear power as a route to weapons development.

Nevertheless, it must be recognized that plutonium, an important constituent of weapons, is produced in light-water nuclear power plants. Plutonium is the preferred fuel for breeder reactors. The development of any significant number of breeder reactors would thus involve the production and handling of large quantities of plutonium.

As will be discussed in a later section, plutonium-239 can be produced by the absorption of a neutron in uranium-238. Since most of the uranium in the core of an LWR is uranium-238, plutonium is produced during operation of the reactor. However, if the plutonium-239 is left in a power reactor core for the length of time typical of the fuel cycle used for LWRs or for breeders, neutrons are absorbed by some fraction of the plutonium to produce plutonium-240. This isotope also absorbs neutrons to produce plutonium-241. These heavier isotopes make the plutonium undesirable as weapons material. Thus, although the plutonium produced in power reactors can be separated chemically from the other materials in a used fuel element, it is not what would be considered weapons-grade material. A nation with the goal of developing weapons would almost certainly design and use a reactor and a fuel cycle designed specifically for producing weapons-grade material. On the other hand, if a drastic change in government produced a correspondingly drastic change in political objectives in a country that had a civil nuclear power program in operation, it would probably be possible to make use of power reactor plutonium to produce some sort of low-grade weapon.

56.6 BASIC ENERGY PRODUCTION PROCESSES

Energy can be produced by nuclear reactions that involve either fission (the splitting of a nucleus) or fusion (the fusing of two light nuclei to produce a heavier one). If energy is to result from fission, the resultant nuclei must have a smaller mass per nucleon (which means they are more tightly bound) than the original nucleus. If the fusion process is to produce energy, the fused nucleus must have a

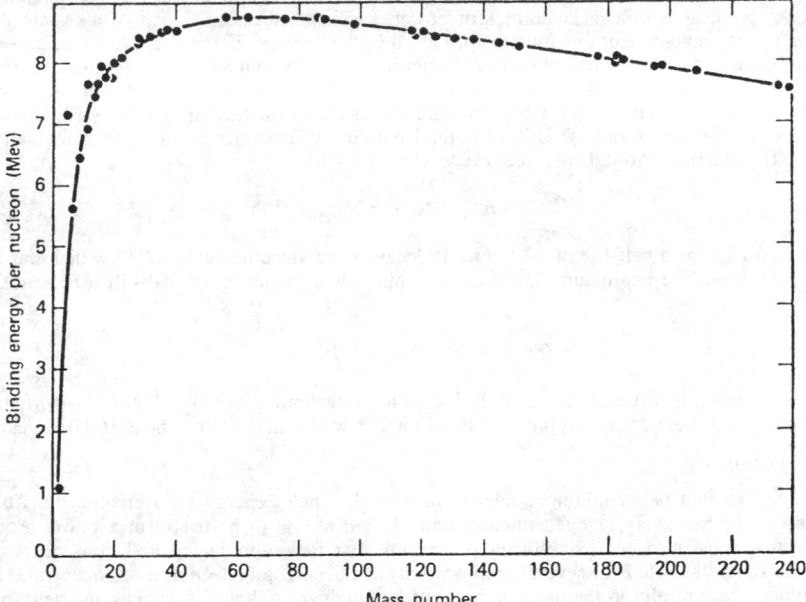

Fig. 56.6 Binding energy per nucleon versus mass number.

smaller mass per nucleon (i.e., be more tightly bound) than the original nuclei. Figure 56.6 is a curve of nuclear binding energies. Observe that only the heavy nuclei are expected to produce energy on fission, and that only the light nuclei yield energy in fusion. The differences in mass per nucleon before and after fission or fusion are available as energy.

56.6.1 Fission

In the fission process this energy is available primarily as kinetic energy of the fission fragments. Gamma rays are also produced as well as a few free neutrons, carrying a small amount of kinetic energy. The radioactive fission products decay (in most cases there is a succession of decays) to a stable nucleus. Gamma and beta rays are produced in the decay process. Most of the energy of these radiations is also recoverable as fission energy. Table 56.3 lists typical energy production due to fission of uranium by thermal neutrons, and indicates the form in which the energy appears. The quantity of energy available is of course related to the nuclear mass change by

$$\Delta E = \Delta mc^2$$

Table 56.3 Emitted and Recoverable Energies from Fission of ^{235}U

Form	Emitted Energy (MeV)	Recoverable Energy (MeV)
Fission fragments	168	168
Fission product decay		
β rays	8	8
γ rays	7	7
Neutrinos	12	—
Prompt γ rays	7	7
Fission neutrons (kinetic energy)	5	5
Capture γ rays	—	3–12
Total	207	198–207

Fission in reactors is produced by the absorption of a neutron in the nucleus of a fissionable atom. In order to produce significant quantities of power, fission must occur as part of a sustained chain reaction, that is, enough neutrons must be produced in the average fission event to cause at least one new fission event to occur when absorbed in fuel material. The number of nuclei that are available and that have the required characteristics to sustain a chain reaction is limited to uranium-235, plutonium-239, and uranium-233. Only uranium-235 occurs in nature in quantities sufficient to be useful. (And it occurs as only 0.71% of natural uranium.) The other two can be manufactured in reactors. The reactions are indicated below:

$$^{238}U + n \rightarrow {}^{239}U \rightarrow {}^{239}Np \rightarrow {}^{239}Pu$$

Uranium-239 has a half-life of 23.5 min. It decays to produce neptunium-239, which has a half-life of 2.35 days. The neptunium-239 decays to plutonium, which has a half-life of about 24,400 years.

$$^{232}Th + n \rightarrow {}^{233}Th \rightarrow {}^{233}Pa \rightarrow {}^{233}U$$

Thorium-233 has a half-life of 22.1 min. It decays to protactinium-233, which has a half-life of 27.4 days. The protactinium decays to produce uranium-233 with a half-life of about 160,000 years.

56.6.2 Fusion

Fusion requires that two colliding nuclei have enough kinetic energy to overcome the Coulomb repulsion of the positively charged nuclei. If the fusion rate is to be useful in a power-producing system, there must also be a significant probability that fusion-producing collisions occur. These conditions can be satisfied for several combinations of nuclei if a collection of atoms can be heated to a temperature typically in the neighborhood of hundreds of millions of degrees and held together for a time long enough for an appreciable number of fusions to occur. At the required temperature the atoms are completely ionized. This collection of hot, highly ionized particles is called a plasma. Since average collision rate can be related to the product of the density of nuclei, n, and the average containment time, τ, the $n \tau$ product for the contained plasma is an important parameter in describing the likelihood that a working system with these plasma characteristics will produce a useful quantity of energy.

Examination of the fusion probability, or the cross section for fusion, as a function of the temperature of the hot plasma shows that the fusion of deuterium (^{2}H) and tritium (^{3}H) is significant at temperatures lower than that for other candidates. Figure 56.7 shows fusion cross section as a function of plasma temperature (measured in electron volts) for several combinations of fusing nuclei. Table 56.4 lists several fusion reactions that might be used, together with the fusion products and the energy produced per fusion.

One of the problems with using the D–T reaction is the large quantity of fast neutrons that results, and the fact that a large fraction of the energy produced appears as kinetic energy of these neutrons. Some of the neutrons are absorbed in and activate the plasma-containment-system walls, making it highly radioactive. They also produce significant damage in most of the candidate materials for the containment walls. For these reasons there are some who advocate that work with the D–T reaction be abandoned in favor of the development of a system that depends on a set of reactions that is neutron-free.

Another problem with using the D–T reaction is that tritium does not occur in nature in sufficient quantity to be used for fuel. It must be manufactured. Typical systems propose to produce tritium by the absorption in lithium of neutrons resulting from the fusion process. Natural lithium consists of ^{6}Li (7.5%) and ^{7}Li (92.5%). The reactions are

$$^{6}Li + n \rightarrow {}^{4}He + {}^{3}H + 4.8 \text{ MeV (thermal neutrons)}$$

and

$$^{7}Li + n \rightarrow {}^{4}He + {}^{3}H + n + 2.47 \text{ MeV (threshold reaction)}$$

Considerations of neutron economy dictate that most of the neutrons produced in the fusion process be absorbed in lithium in order to breed the needed quantities of tritium. The reactions shown produce not only tritium, but also additional energy. The ($^{6}Li,n$) reaction, for example, produces 4.7 MeV per reaction. If this energy can be recovered, it effectively increases the average available energy per fusion by about 27%.

56.7 CHARACTERISTICS OF THE RADIATION PRODUCED BY NUCLEAR SYSTEMS

An important by-product of the processes used to generate nuclear power is a variety of radiations in the form of either particles or electromagnetic photons. These radiations can produce damage in

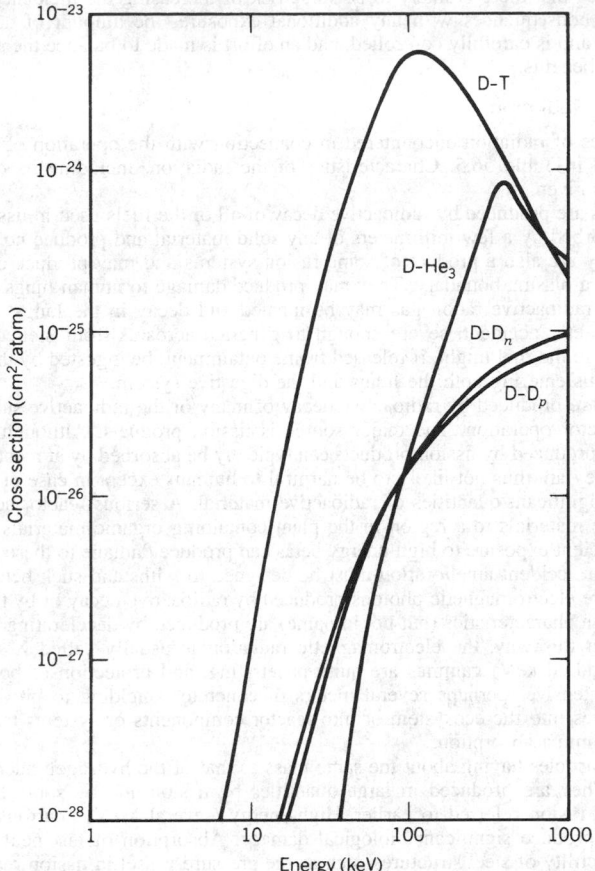

Fig. 56.7 Fusion cross section versus plasma temperature.

the materials that make up the systems and structures of the power reactors. High-energy neutrons, for example, absorbed in the vessel wall make the steel in the pressure vessel walls less ductile.

Radiation also causes damage to biological systems, including humans. Thus, most of the radiations must be contained within areas from which people are excluded. Since the ecosystem to which humans are normally exposed contains radiation as a usual constituent, it is assumed that some additional exposure can be permitted without producing undue risk. However, since the best scientific

Table 56.4 Fusion Reactions

$^3H + {}^2H \rightarrow n + {}^4He$	+17.6 MeV
$^3He + {}^2H \rightarrow p + {}^4He$	+18.4 MeV
$^2H + {}^2H \rightarrow p + {}^3H$	+4.0 MeV
$^2H + {}^2H \rightarrow n + {}^3He$	+3.3 MeV
$^6Li + p \rightarrow {}^3He + {}^4He$	+4.0 MeV
$^6Li + {}^3He \rightarrow {}^4He + p + {}^4He$	+16.9 MeV
$^6Li + {}^2H \rightarrow p + {}^7Li$	+5.0 MeV
$^6Li + {}^2H \rightarrow {}^3H + p + {}^4He$	+2.6 MeV
$^6Li + {}^2H \rightarrow {}^4He + {}^4He$	+22.4 MeV
$^6Li + {}^2H \rightarrow n + {}^7Be$	+3.4 MeV
$^6Li + {}^2H \rightarrow n + {}^3He + {}^4He$	+1.8 MeV

judgment concludes that there is likely to be some risk of increasing the incidence of cancer and of other undesirable consequences with any additional exposure, the amount of additional exposure permitted is small and is carefully controlled, and an effort is made to balance the permitted exposure against perceived benefits.

56.7.1 Types of Radiation

The principal types of radiation encountered in connection with the operation of fission and fusion systems are listed in Table 56.5. Characteristics of the radiation, including its charge and energy spectrum, are also given.

Alpha particles are produced by radioactive decay of all of the fuels used in fission reactors. They are, however, absorbed by a few millimeters of any solid material and produce no damage in typical fuel material. They are also a product of some fusion systems and may produce damage to the first wall that provides a plasma boundary. They may produce damage to human lungs during the mining of uranium when radioactive radon gas may be inhaled and decay in the lungs. In case of a catastrophic fission reactor accident, severe enough to generate aerosols from melted fuels, the alpha-emitting materials in the fuel might, if released from containment, be ingested by those in the vicinity of the accident, thus entering both the lungs and the digestive system.

Beta particles are produced by radioactive decay of many of the radioactive substances produced during fission reactor operation. The major source is fission products. Although more penetrating than alphas, betas produced by fission products can typically be absorbed by at most a few centimeters of most solids. They are thus not likely to be harmful to humans except in case of accidental release and ingestion of significant quantities of radioactive material. A serious reactor accident might also release radioactive materials to a region in the plant containing organic materials such as electrical insulation. A sufficient exposure to high-energy betas can produce damage to these materials. Reactor systems needed for accident amelioration must be designed to withstand such beta irradiations.

Gamma rays are electromagnetic photons produced by radioactive decay or by the fission process. Photons identical in characteristics (but not in name) are produced by decelerating electrons or betas. When produced in this way, the electromagnetic radiation is usually called X rays. High-energy (above several hundred keV) gammas are quite penetrating, and protection of both equipment and people requires extensive (perhaps several meters of concrete) shielding to prevent penetration of significant quantities into the ecosystem or into reactor components or systems that may be subject to damage from gamma absorption.

Neutrons are particles having about the same mass as that of the hydrogen nucleus or proton, but with no charge. They are produced in large quantities by fission and by some fusion interactions including the D-T fusion referred to earlier. High-energy (several MeV) neutrons are highly penetrating. They can produce significant biological damage. Absorption of fast neutrons can induce a decrease in the ductility of steel structures such as the pressure vessel in fission reactors or the inner wall of fusion reactors. Fast-neutron absorption also produces swelling in certain steel alloys.

56.8 BIOLOGICAL EFFECTS OF RADIATION

Observations have indicated that the radiations previously discussed can cause biological damage to a variety of living organisms, including humans. The damage that can be done to human organisms includes death within minutes or weeks if the exposure is sufficiently large, and if it occurs during an interval of minutes or at most a few hours.

Radiation exposure has also been found to increase the probability that cancer will develop. It is considered prudent to assume that the increase in probability is directly proportional to exposure. However, there is evidence to suggest that at very low levels of exposure, say an exposure comparable to that produced by natural background, the linear hypothesis is not a good representation. Radiation exposure has also been found to induce mutations in a number of biological organisms. Studies of the survivors of the two nuclear weapons exploded in Japan have provided the largest body of data

Table 56.5 Radiation Encountered in Nuclear Power Systems

Name	Description	Charge (in Units of Electron Charge)	Energy Spectrum (MeV)
Alpha	Helium nucleus	+2	0 to about 5
Beta	Electron	+1, −1	0 to several
Gamma	Electromagnetic radiation	0	0 to about 10
Neutron		0	0 to about 20

available for examining the question of whether harmful mutations are produced in humans by exposure of their forebears to radiation. Analyses of these data have led those responsible for the studies to conclude that the existence of an increase in harmful mutations has not been demonstrated unequivocally. However, current regulations of radiation exposure, in order to be conservative, assume that increased exposure will produce an increase in harmful mutations. There is also evidence to suggest that radiation exposure produces life shortening.

The Nuclear Regulatory Commission has the responsibility for regulating exposure due to radiation produced by reactors and by radioactive material produced by reactors. The standards used in the regulatory process are designed to restrict exposures to a level such that the added risk is not greater than that from other risks in the workplace or in the normal environment. In addition, effort is made to see that radiation exposure is maintained as "low as reasonably achievable."

56.9 THE CHAIN REACTION

Setting up and controlling a chain reaction is fundamental to achieving and controlling a significant energy release in a fission system. The chain reaction can be produced and controlled if a fission event, produced by the absorption of a neutron, produces more than one additional neutron. If the system is arranged such that one of these fission-produced neutrons produces, on the average, another fission, there exists a steady-state chain reaction. Competing with fission for the available neutrons are leakage out of the fuel region and absorptions that do not produce fission.

We observe that if only one of these fission-neutrons produces another fission, the average fission rate will be constant. If more than one produces fission, the average fission rate will increase at a rate that depends on the average number of new fissions produced for each preceding fission and the average time between fissions.

Suppose, for example, each fission produced two new fissions. One gram of uranium-235 contains 2.56×10^{21} nuclei. It would therefore require about 71 generations ($2^{71} \sim 2.4 \times 10^{21}$) to fission 1 g of uranium-235. Since fission of each nucleus produces about 200 MeV, this would result in an energy release of about 5.12×10^{23} MeV or 5.12×10^{10} J. The time interval during which this release takes place depends on the average generation time. Note, however, that in this hypothesized situation only about the last 10 generations contribute any significant fraction of the total energy. Thus, for example, if a generation could be made as short as 10^{-8} sec, the energy production rate could be nearly 5.12×10^{17} J/sec/g.

In power reactors the generation time is typically much larger than 10^{-8} sec by perhaps four or five orders of magnitude. Furthermore, the maximum number of new fissions produced per old fission is much less than two. Power reactors (in contrast to explosive devices) cannot achieve the rapid energy release hypothesized in the above example, for the very good reasons that the generation time and the multiplication inherent in these machines make it impossible.

56.9.1 Reactor Behavior

As indicated, it is neutron absorption in the nuclei of fissile material in the reactor core that produces fission. Furthermore, the fission process produces neutrons that can generate new fissions. This process sustains a chain reaction at a fixed level, if the relationship between neutrons produced by fission and neutrons absorbed in fission-producing material can be maintained at an appropriate level.

One can define neutron multiplication k as

$$k = \frac{\text{neutrons produced in a generation}}{\text{neutrons produced in the preceding generation}}$$

A reactor is said to be critical when k is 1. We examine the process in more detail by following neutron histories. The probability of interaction of neutrons with the nuclei of some designated material can be described in terms of a mean free path for interaction. The inverse, which is the interaction probability per unit path length, is also called macroscopic cross section. It has dimensions of inverse length.

We designate a cross section for absorption, Σ_a, a cross section for fission, Σ_f, and a cross section for scattering, Σ_s. If, then, we know the number of path lengths per unit time, per unit volume, traversed by neutrons in the reactor (for monoenergetic neutrons this will be nv, where n is neutron density and v is neutron speed), usually called the neutron flux, we can calculate the various interaction rates associated with these cross sections and with a prescribed neutron flux, as a product of the flux and the cross section.

A diagrammatic representation of neutron history, with the various possibilities that are open to the neutrons produced in the fission process, is shown below:

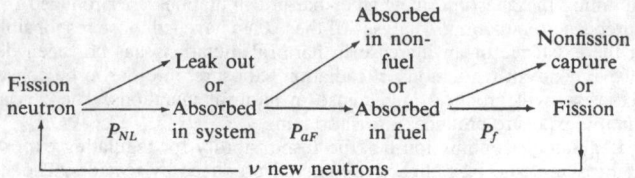

where P_{NL} = probability that neutron will not leak out of system before being absorbed
P_{aF} = probability that a neutron absorbed is absorbed in fuel
P_f = probability that a neutron absorbed in fuel produces a fission

In terms of the cross sections for absorption in fuel, Σ_a^F and for absorption, Σ_a

$$P_{aF} = f = \Sigma_a^F/\Sigma_a$$

where f is called the utilization factor. We can describe P_f as

$$P_f = \Sigma_f^F/\Sigma_a^F$$

Making use of the average number of neutrons produced per fission ν, we calculate the quantity η, the average number of neutrons produced per neutron absorbed in fuel, as

$$\eta = \nu \, \Sigma_f^F/\Sigma_a^F$$

With these definitions, and guided by the preceding diagram, we conclude that the number of offspring neutrons produced by a designated fission neutron can be calculated as

$$N = \eta f P_{NL}$$

We conclude that the multiplication factor k is thus equal to $N/1$ and write

$$k = \eta f P_{NL}$$

Alternatively, making use of the earlier definitions we write

$$k = (v\Sigma_f^F/\Sigma_a^F)(\Sigma_a^F/\Sigma_a)$$

and if we describe Σ_a as

$$\Sigma_a = \Sigma_a^F + \Sigma_a^{nF}$$

where Σ_a^{nF} is absorption in the nonfuel constituents of the core, we have

$$k = \nu\Sigma_f^F P_{NL}/(\Sigma_a^F + \Sigma_a^{nF})$$

Observe that from this discussion one can also define a neutron generation time l as

$$l = N(t)/L(t)$$

where $N(t)$ and $L(t)$ represent, respectively, the neutron population and the rate of neutron loss (through absorption and leakage) at a time t.

For large reactors, the size of those now in commercial power production, the nonleakage probability is high, typically about 97%. For many purposes it can be neglected. For example, small changes in multiplication, produced by small changes in concentration of fissile or nonfissile material in the core, can be assumed to have no significant effect on the nonleakage probability, P_{NL}. Under these circumstances, and assuming that appropriate cross-sectional averaging can be done, the following relationships can be shown to hold. If we rewrite an earlier equation for k as

$$k = \eta f \left(n_f\sigma_f \Big/ \sum_i n_i\sigma_i \right)$$

where η_f, η_i represent, respectively, the concentration of fissile and nonfissile materials, and

$$\eta_f \sigma_f = \Sigma_f$$
$$n_i \sigma_i = \Sigma_{a_i}$$

where the last equation in the macroscopic cross section of the ith nonfissile isotope.

Variation of k with the variation in concentration of the fissile material (i.e., n_f) is given by

$$\frac{\delta k}{k} = \frac{\delta n_f}{n_f}\left(\frac{\Sigma_i n_i \sigma_i}{\eta_f \sigma_f + \Sigma_i n_i \sigma_i}\right)$$

This says that the fractional change in multiplication is equal to the fractional change in concentration of the fissile isotope times the ratio of the neutrons absorbed in all of the nonfissile isotopes to the total neutrons absorbed in the core.

Variation of k with variation in concentration of the jth nonfissile isotope is given by

$$\frac{\delta k}{k} = -\frac{\delta m_j}{m_j}\left(\frac{n_j \sigma_j}{n_f \sigma_f + \Sigma_i n_i \sigma_i}\right), \quad i \neq j$$

This says that the fractional increase in multiplication is equal to the fractional decrease in concentration of the jth nonfissile isotope times the ratio of neutrons absorbed in that isotope to total neutrons absorbed. Although these are approximate expressions, they provide useful guidance in estimating effects of small changes in the material in the core on neutron multiplication.

56.9.2   Time Behavior of Reactor Power Level

We assume that the fission rate, and hence the reactor power, is proportional to neutron population. We express the rate of change of neutron population $N(t)$ as

$$l\left(\frac{dN}{dt}\right) = (k - 1)N$$

that is, in one generation the change in neutron population should be just the excess over the previous generation, times the multiplication k. The preceding equation has as a solution

$$N = N_0 \exp\left[\left(\frac{k - 1}{l}\right)t\right]$$

where N_0 is the neutron population at time zero. One observes an exponential increase or decrease, depending on whether k is larger or smaller than unity. The associated time constant or e-folding time is

$$\tau = l/(k - 1)$$

For a $k - 1$ of 0.001 and l of 10^{-4} sec, the e-folding time is 0.10 sec. Thus in 1 sec the power level increases by e^{10} or about 10^4.

56.9.3   Effect of Delayed Neutrons on Reactor Behavior

Dynamic behavior as rapid as that described by the previous equations would make a reactor almost impossible to control. Fortunately there is a mode of operation in which the time constant is significantly greater than that predicted by these oversimplified equations. A small fraction of neutrons produced by fission, typically about 0.7%, come from radioactive decay of fission products. Six such fission products are identified for uranium fission. The mean time for decay varies from about 0.3 to about 79 sec. For an approximate representation, it is reasonable to assume a weighted mean time to decay for the six of about 17 sec. Thus, about 99.3% of the neutrons (prompt neutrons) may have a generation time of, say, 10^{-4} sec, while 0.7% have an effective generation time of 17 sec plus that of the prompt neutrons. An effective mean lifetime can be estimated as

$$\bar{l} = (0.993)l + 0.007(l + \bar{\lambda}^{-1})$$

For an l of 10^{-4} sec and a $\bar{\lambda}^{-1}$ of 17 sec we calculate

$$l \approx 0.993 \times 10^{-4} + 0.007(10^{-4} + 17) \approx 0.12 \text{ sec}$$

This suggests, given a value for $k - 1$ of 10^{-3}, an e-folding time of 120 sec. Observe that with this model the delayed neutrons are a dominant factor in determining time behavior of the reactor power level. A more detailed examination of the situation reveals that for a reactor slightly subcritical on prompt neutrons alone, but supercritical when delayed neutrons are considered (such a reactor is said to be "delayed supercritical"), the delayed neutrons almost alone determine time behavior. If, however, the multiplication is increased to the point that the reactor is critical on prompt neutrons alone (i.e., "prompt critical"), the time behavior is determined by the prompt neutrons, and changes in power level may be too rapid to be controlled by any external control system. Reactors that are meant to be controlled are designed to be operated in a delayed critical mode. Fortunately, if the reactor should inadvertently be put in a prompt critical mode, there are inherent physical phenomena that decrease the multiplication to a controllable level when a power increase occurs.

56.10   POWER PRODUCTION BY REACTORS

Most of the nuclear-reactor-produced electric power in the United States, and in the rest of the world, comes from light-water-moderated reactors (LWRs). Nuclear power reactors produce heat that is converted, in a thermodynamic cycle, to electrical energy. The two types now in use, the pressurized water reactor (PWR) and the boiling water reactor (BWR), use fuel that is very similar, and produce steam having about the same temperature and pressure. In both types water serves both as a coolant and a moderator. We will examine some of the salient features of each system and identify some of the differences.

56.10.1   The Pressurized-Water Reactor

The arrangement of fuel in the reactor core, and of the core in the pressure vessel, are shown in Fig. 56.1. As indicated earlier, bulk boiling is avoided by operation at high pressures. Liquid water is circulated through the core by large electric-motor-driven pumps located outside the pressure vessel in the cold leg of the piping that connects the vessel to a steam generator. Current designs use from two to four separate loops, each containing a steam generator. Each loop contains at least one pump. One current design uses two pumps in the cold leg of each loop. A schematic of the arrangement is shown in Fig. 56.8. Reactor pressure vessel and primary coolant loops, including the steam generator, are located inside a large containment vessel. A typical containment structure is shown in Fig. 56.9.

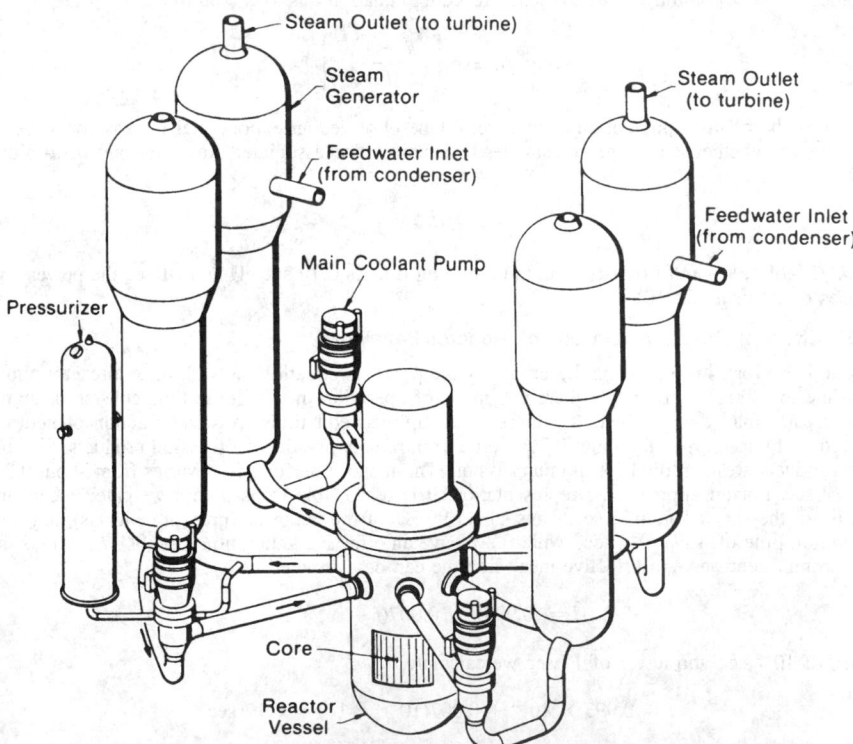

Fig. 56.8   Typical arrangement of PWR primary system.

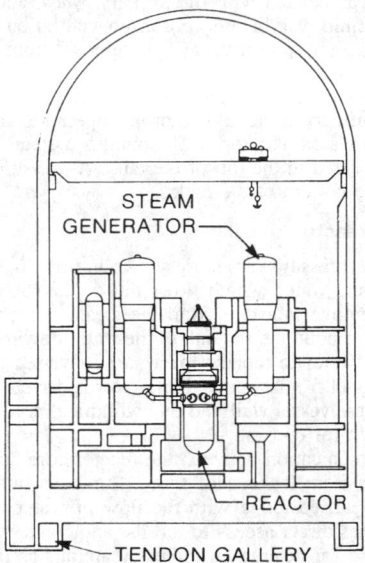

Fig. 56.9 Typical large dry PWR containment.

The containment, typically a massive 3–4-ft-thick structure of reinforced concrete, with a steel inner-liner, has two principal functions: protection of pressure vessel and primary loop from external damage (e.g., tornadoes, aircraft crashes) and containment of fission products that might be released outside the primary pressure boundary in case of serious damage to the reactor in an accident.

The steam generator is markedly different from the boiler in a fossil-fueled plant. It is essentially a heat exchanger containing several thousand metal tubes that carry the hot water coming from the reactor vessel outlet. Water surrounding the outside of these tubes is converted to steam. The rest of the energy-conversion cycle is similar in principle to that found in a fossil-fueled plant.

Experience with reactor operation has indicated that very careful control of water chemistry is necessary to preclude erosion and corrosion of the steam generator tubes (SGT). An important contributor to SGT damage has been leakage in the main condenser, which introduces impurities into the secondary water system. A number of early PWRs have retubed or otherwise modified their original condensers to reduce contamination caused by in-leakage of condenser cooling water.

The performance of steam generator tubes is of crucial importance because: (1) These tubes are part of the primary pressure boundary. SGT rupture can initiate a loss of coolant accident. (2) Leakage or rupture of SGTs usually leads to opening of the steam system safety valves because of the high primary system pressure. Since these valves are located outside containment, this accident sequence can provide an uncontrolled path for release of any radioactive material in the primary system directly to the atmosphere outside containment.

The reactor control system controls power level by a combination of solid control rods, containing neutron-absorbing materials that can be moved into and out of the core region, and by changing the concentration of a neutron-absorbing boron compound (typically boric acid) in the primary coolant. Control rod motion is typically used to achieve rapid changes in power. Slower changes, as well as compensation for burnup of uranium-235 in the core, are accomplished by boron-concentration changes. In the PWR the control rods are inserted from the top of the core. In operation enough of the absorber rods are held out of the core to produce rapid shutdown, or scram, when inserted. In an emergency, if rod drive power should be lost, the rods automatically drop into the core, driven by gravity.

The PWR is to some extent load following. Thus, for example, an increase in turbine steam flow caused by an increase in load produces a decrease in reactor coolant–moderator temperature. In the usual mode of operation a decrease in moderator temperature produces an increase in multiplication (the size of the effect depends on the boron concentration in the coolant-moderator), leading to an increase in power. The increase continues (accompanied by a corresponding increase in moderator-coolant temperature) until the resulting decrease in reactor multiplication leads to a return to criticality at an increased power level. Since the size of the effect changes significantly during the operating cycle (as fuel burnup increases the boron concentration is decreased), the inherent load-following characteristic must be supplemented by externally controlled changes in reactor multiplication.

A number of auxiliaries are associated with the primary. These include a water purification and makeup system, which also permits varying the boron concentration for control purposes, and an emergency cooling system to supply water for decay heat removal from the core in case of an accident that causes loss of the primary coolant.

Pressure in the primary is controlled by a pressurizer, which is a vertical cylindrical vessel connected to the hot leg of the primary system. In normal operation the bottom 60% or so of the pressurizer tank contains liquid water. The top 40% contains a steam bubble. System pressure can be decreased by water sprays located in the top of the tank. A pressure increase can be achieved by turning on electric heaters in the bottom of the tank.

56.10.2 The Boiling-Water Reactor

Fuel and core arrangement in the pressure vessel are shown in Fig. 56.3. Boiling in the core produces a two-phase mixture of steam and water, which flows out of the top of the core. Steam separators above the vessel water level remove moisture from the steam, which goes directly to the turbine outside of containment. Typically about one-seventh of the water flowing through the core is converted to steam during each pass. Feedwater to replace the water converted to steam is distributed around the inside near the top of the vessel from a spray ring. Water is driven through the core by jet pumps located in the annulus between the vessel wall and the cylindrical core barrel that surrounds the core and defines the upward flow path for coolant.

Because there is direct communication between the reactor core and the turbine, any radioactive material resulting, for example, from leakage of fission products out of damaged fuel pins, from neutron activation of materials carried along with the flow of water through the core, or from the nitrogen-16 referred to earlier, has direct access to turbine and condenser. Systems must be provided for removal from the coolant and for dealing with these materials as radioactive waste.

Unlike the PWR, the BWR is not load following. In fact, normal behavior in the reactor core produces an increase in the core void volume with an increase in steam flow to the turbine. This increased core voiding will increase neutron leakage, thereby decreasing reactor multiplication, and leading to a decrease in reactor power in case of increased demand. To counter this natural tendency of the reactor, a control system senses an increase in turbine steam flow and increases coolant flow through the core. The accompanying increase in core pressure decreases steam voiding, increasing multiplication and producing an increase in reactor power.

Pressure regulation in the BWR is achieved primarily by adjustment of turbine throttle setting to achieve constant pressure. An increase in load demand is sensed by the reactor control system and produces an increase in reactor power. Turbine valve position is adjusted to maintain constant steam pressure at the throttle. In rapid transients, which involve decreases in load demand, a bypass valve can be opened to send steam directly to the condenser, thus helping to maintain constant pressure.

BWRs in the United States make use of a pressure-suppression containment system. The hot water and steam released during a loss-of-coolant accident are forced to pass through a pool of water, condensing the steam. Pressure buildup is markedly less than if the two-phase mixture is released directly into containment. Figure 56.10 shows a Mark III containment structure of the type being used with the latest BWRs. Passing the fission products and the hot water and steam from the primary containment through water also results in significant removal of some of the fission products. The designers of this containment claim decontamination factors of 10,000 for some of the fission products that are usually considered important in producing radiation exposure following an accident.

As previously indicated, the control system handles normal load changes by adjusting coolant flow in the core. For rapid shutdown and for compensating for core burnup, the movable control rods are used. In a normal operating cycle several groups of control rods will be in the core at the beginning of core life, but will be completely out of the core at the end of the cycle. At any stage in core life some absorber rods are outside the core. These can be inserted to produce rapid shutdown.

Because of the steam-separator structure above the core, control rods in the BWR core must be inserted from the bottom. Insertion is thus not gravity assisted. Control rod drive is hydraulic. Compressed gas cylinders provide for emergency insertion if needed.

No neutron absorber is dissolved in the coolant, hence, control of absorber concentration is not required. However, cleanup of coolant containments, both solid and gaseous, is continuous. Maintaining a low oxygen concentration is especially important for the inhibition of stress-assisted corrosion cracking that has occurred in the primary system piping of a number of BWRs.

56.11 REACTOR SAFETY ANALYSIS

Under existing law the Nuclear Regulatory Commission has the responsibility for licensing power reactor construction and operation. (Those who operate the controls of the reactor and those who exercise immediate supervision of the operation must also be licensed by the Commission.) Commission policy provides for the granting of an operating license only after it has been formally determined by Commission review that the reactor power plant can be operated without undue risk to the health and safety of the public.

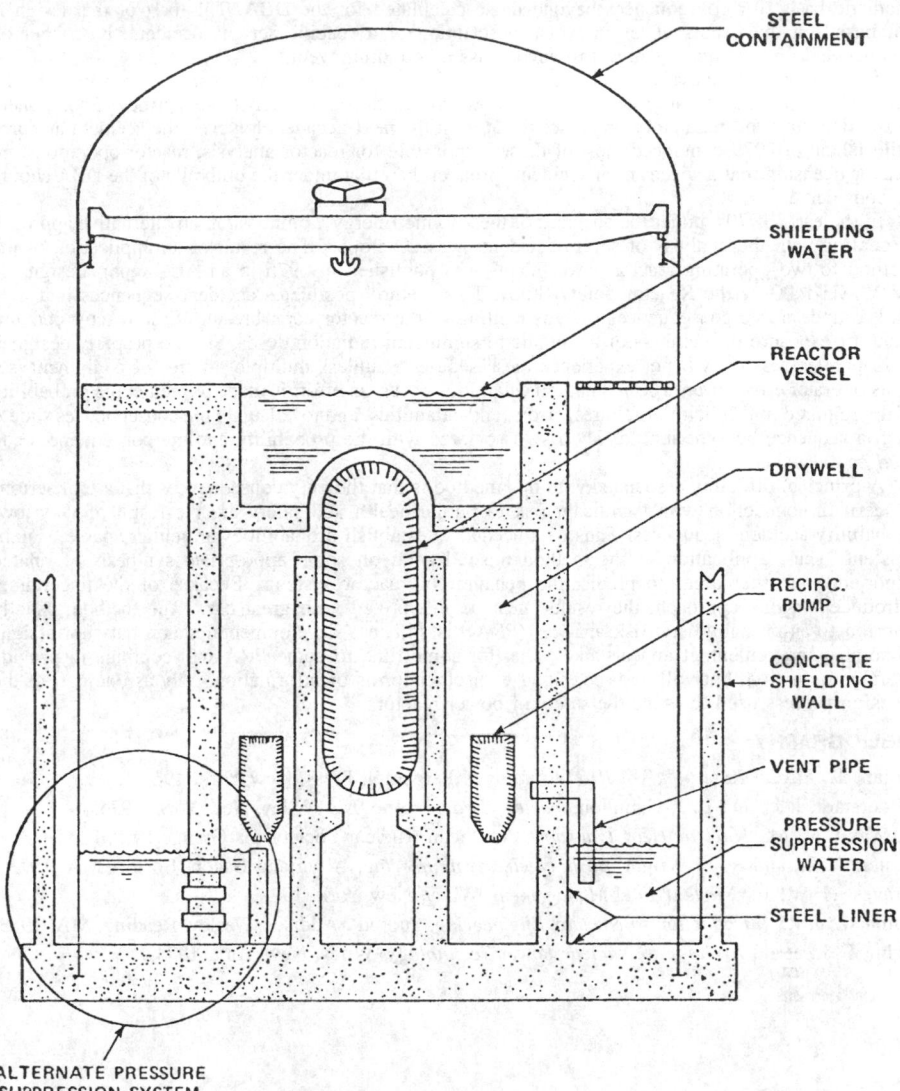

Fig. 56.10   Mark III containment for BWR. (Used by permission of Pergamon Press, Inc., New York.)

The current review process includes a detailed analysis of reactor system behavior under both normal and accident conditions. The existing approach involves the postulating of a set of design basis accidents (DBAs) and carrying out a deterministic analysis, which must demonstrate that the consequences of the hypothesized accidents are within a defined acceptable region. A number of the accident scenarios used for this purpose are of sufficiently low probability that they have not been observed in operating reactors. It is not practical to simulate the accidents using full-scale models or existing reactors. Analysis of reactor system behavior under the hypothesized situations must depend on analytical modeling. A number of large and complicated computer codes have been developed for this purpose.

Although the existing approach to licensing involves analysis of DBAs that can cause significant damage to the reactor power plant, none of the DBAs produces any calculable damage to personnel. Indeed core damage severe enough to involve melting of the core is not included in any of the sequences that are considered. However, in the design of the plant allowance is made, on a nonmechanistic basis, for consequences beyond those calculated for the DBA. This part of the design is not based on the results of an analytical description of a specific serious accident, but rather on nonmechanistic assumptions meant to encompass a bounding event.

This method of analysis, developed over a period of about two decades, has been used in the licensing and in the regulation of the reactors now in operation. It is likely to be a principal component of the licensing and regulatory processes for at least the next decade. However, the accident at Three Mile Island in 1979 convinced most of those responsible for reactor analysis, reactor operation, and reactor licensing that a spectrum of accidents broader than that under the umbrella of the DBA should be considered.

In the early 1970s, under the auspices of the Atomic Energy Commission, an alternative approach to dealing with the analysis of severe accidents was developed. The result of an application of the method to two operating reactor power plants was published in 1975 in an AEC report designated as WASH-1400 or the Reactor Safety Study. This method postulates accident sequences that may lead to undesirable consequences such as melting of the reactor core, breach of the reactor containment, or exposure of members of the public to significant radiation doses. Since a properly designed and operated reactor will not experience these sequences unless multiple failures of equipment, serious operator error, or unexpected natural calamities occur, an effort is made to predict the probability of the required multiplicity of failures, errors, and calamities, and to calculate the consequences should such a sequence be experienced. The risk associated with the probability and the consequences can then be calculated.

A principal difficulty associated with this method is that the only consequences that are of serious concern in connection with significant risk to public health and safety are the result of very-low-probability accident sequences. Thus data needed to establish probabilities are either sparse or nonexistent. Thus, application of the method must depend on some appropriate synthesis of related experience in other areas to predict the behavior of reactor systems. Because of the uncertainty introduced in this approach, the results must be interpreted with great care. The method, usually referred to as probabilitistic risk analysis (PRA), is still in a developmental stage, but shows signs of some improvement. It appears likely that for some time to come PRA will continue to provide useful information, but will be used, along with other forms of information, only as one part of the decision process used to judge the safety of power reactors.

BIBLIOGRAPHY

Dolan, T., *Fusion Research, Vol III (Technology),* Pergamon Press, New York, 1980.

Duderstadt, J. J., and L. J. Hamilton, *Nuclear Reactor Analysis,* Wiley, New York, 1975.

El-Wakil, M. M., *Nuclear Heat Transport,* American Nuclear Society, La Grange Park, IL, 1978.

Foster, A. L., and R. L. Wright, *Basic Nuclear Engineering,* Allyn and Bacon, Boston, MA, 1973.

Graves, H. W., Jr., *Nuclear Fuel Management,* Wiley, New York, 1979.

Lamarsh, J. R., *Introduction to Nuclear Engineering,* 2nd ed., Addison-Wesley, Reading, MA, 1983.

Rahn, F. J., et al., *A Guide to Nuclear Power Technology,* Wiley, New York, 1984.

CHAPTER 57

GAS TURBINES

Harold Miller
GE Power Systems
Schenectady, New York

57.1	**INTRODUCTION**	**1723**	**57.3**	**APPLICATIONS**	**1749**	
	57.1.1 Basic Operating Principles	1723		57.3.1 Use of Exhaust Heat in Industrial Gas Turbines	1749	
	57.1.2 A Brief History of Gas Turbine Development and Use	1727		57.3.2 Integrated Gasification Combined Cycle	1751	
	57.1.3 Components, Characteristics and Capabilities	1728		57.3.3 Applications in Electricity Generation	1753	
	57.1.4 Controls and Accessories	1737		57.3.4 Engines for Aircraft	1755	
	57.1.5 Gas Turbine Operation	1740		57.3.5 Engines for Surface Transportation	1757	
57.2	**GAS TURBINE PERFORMANCE**	**1740**	**57.4**	**EVALUATION AND SELECTION**	**1759**	
	57.2.1 Gas Turbine Configurations and Cycle Characteristics	1740		57.4.1 Maintenance Intervals, Availability, and Reliability	1759	
	57.2.2 Trends in Gas Turbine Design and Performance	1747		57.4.2 Selection of Engine and System	1761	

57.1 INTRODUCTION

57.1.1 Basic Operating Principles

Gas turbines are heat engines based on the Brayton thermodynamic cycle. This cycle is one of the four that account for most of the heat engines in use. Other cycles are the Otto, Diesel and Rankine. The Otto and Diesel cycles are cyclic in regard to energy content. Steady-flow, continuous energy transfer cycles are the Brayton (gas turbine) and Rankine (steam turbine) cycles. The Rankine cycle involves condensing and boiling of the working fluid, steam, and utilizes a boiler to transfer heat to the working fluid. The working fluid in the other cycles is generally air, or air plus combustion products. The Otto, Diesel and Brayton cycles are usually internal combustion cycles wherein the fuel is burned in the working fluid. In summary, the Brayton cycle is differentiated from the Otto and Diesel cycle in that it is continuous, and from the Rankine in that it relies on internal combustion, and does not involve a phase change in the working fluid.

In all cycles, the working fluid experiences induction, compression, heating, expansion, and exhaust. In a non-steady cycle, these processes are performed in sequence in the same closed space,

This chapter was written as an update to chapter 72 of the Handbook's previous edition. Much of the structure and significant portions of the text of the previous edition's chapter is retained. The new edition has increased emphasis on the most significant current and future projected gas turbine configurations and applications. Thermodynamic cycle variations are presented here in a consistent format, and the description of current cycles replaces the discussions of some interesting and historical, but less significant, cycles described in the earlier edition. In addition, there is a new discussion of economic and regulatory trends, of supporting technologies, and their interconnection with gas turbine development. The author of the previous version had captured the history of the gas turbine's development, and this history is repeated and supplemented here.

Mechanical Engineers' Handbook, 2nd ed., Edited by Myer Kutz.
ISBN 0-471-13007-9   © 1998 John Wiley & Sons, Inc.

formed by a piston and cylinder that operate on the working fluid one mass at a time. In contrast, the working fluid flows through a steam turbine power plant or gas turbine engine, without interruption, passing continuously from one single-purpose device to the next.

Gas turbines are used to power aircraft and land vehicles, to drive generators (alternators) to produce electric power, and to drive other devices, such as pumps and compressors. Gas turbines in production range in output from below 50 kW to over 200 MW. Design philosophies and engine configurations vary significantly across the industry. Aircraft engines are optimized for high power-to-weight ratios, while heavy-duty, industrial, and utility gas turbines are heavier, being designed for low cost and long life in severe environments.

The arrangement of a simple gas turbine engine is shown in Fig. 57.1a. The rotating compressor acts to raise the pressure of the working fluid and force it into the combustor. The turbine rotation is caused by the work produced by the fluid while expanding from the high pressure at the combustor discharge to ambient air pressure at the turbine exhaust. The resulting mechanical work drives the mechanically connected compressor and output load device.

The nomenclature of the gas turbine is not standardized. In this chapter, the term *blading* refers to all rotating and stationary airfoils in the gas path. Turbine (expander) section rotating blades are *buckets,* a term derived from steam turbine practice. Turbine section stationary blades are *nozzles.* The combustion components in contact with the working fluid are called *combustors;* major combustor components are fuel nozzles and combustion liners. Some combustors (Can-annular and silo-types) have transition pieces that conduct hot gas from the combustion liners to the first-stage nozzles. A stage of the compressor consists of a row of rotor blades, all at one axial position in the gas turbine, and the stationary blade row downstream of it. A turbine stage consists of a set of nozzles occupying one axial location and the set of buckets immediately downstream. Rotating blading is attached either to a monolithic rotor structure or to individual discs or wheels designed to support the blading against centrifugal force and the aerodynamic loads of the working fluid. The terms *discs* and *wheels* are used interchangeably.

Gas turbine performance is established by three basic parameters: mass flow, pressure ratio, and firing temperature. Compressor, combustor, and turbine efficiency have significant, but secondary, effects on performance, as do inlet and exhaust systems, turbine gas path and rotor cooling, and heat loss through turbine and combustor casings.

In gas turbine catalogues and other descriptive literature, mass flow is usually quoted as compressor inlet flow, although turbine exit flow is sometimes quoted. Output is proportional to mass flow.

Pressure ratio is quoted as the compressor pressure ratio. Aircraft engine practice is to define the ratio as the total pressure at the exit of the compressor blading divided by the total pressure at the inlet of the compressor blading. Industrial/utility turbine manufacturers generally refer to the static pressure in the plenum downstream of the compressor discharge diffuser (upstream of the combustor) divided by the total pressure downstream of the inlet filter and upstream of the inlet of the gas turbine. Similarly, there are various possibilities for defining turbine pressure ratio. All definitions yield values within 1 or 2% of one another. Pressure ratio is the primary determinant of simple cycle gas turbine efficiency. High pressure results in high simple cycle efficiency.

Firing temperature is defined differently by each manufacturer, and the differences are significant. Heavy-duty gas turbine manufacturers use three definitions. There is an ISO definition of firing temperature, which is a calculated temperature. The compressor discharge temperature is increased by a calculated enthalpy rise based on the compressor inlet air flow and the fuel flow. This definition is valuable in that it can be used to compare gas turbines or monitor changes in performance through calculations made on the basis of field measurements. To determine ISO firing temperature, one does not require knowledge of the secondary flows within the gas turbine. A widely used definition of

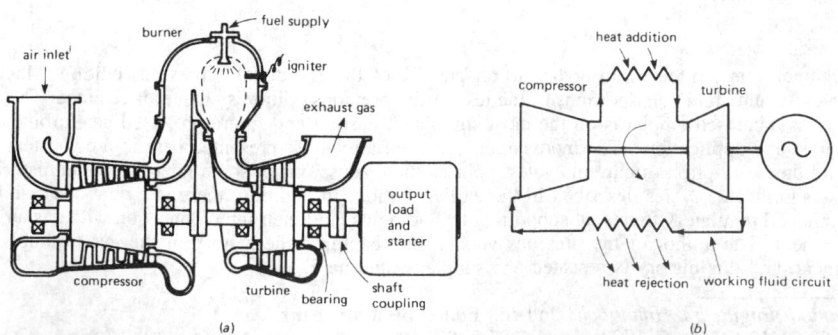

Fig. 57.1 Simple engine type: (*a*) open cycle; (*b*) closed cycle (diagrammatic).[1]

firing temperature is the average total temperature in the exit plane of the first stage nozzle. This definition is used by General Electric for its industrial engines. Westinghouse refers to "turbine inlet temperature," the temperature of the gas entering the first stage nozzle. Turbine inlet temperature is approximately 100°C above nozzle exit firing temperature, which is in turn approximately 100°C above ISO firing temperature. Since firing temperature is commonly used to compare the technology level of competing gas turbines, it is important to compare on one definition of this parameter.

Aircraft engines and aircraft-derivative industrial gas turbines have other definitions. One nomenclature establishes numerical stations—here, station 3.9 is combustor exit and station 4.0 is first-stage nozzle exit. Thus, $T_{3.9}$ is very close to "turbine inlet temperature" and $T_{4.0}$ is approximately equal to GE's "firing temperature." There are some subtle differences relating to the treatment of the leakage flows near the first-stage nozzle. This nomenclature is based on SAE ARP 755A, a recommended practice for turbine engine notation.

Firing temperature is a primary determiner of power density (specific work) and combined cycle (Brayton–Rankine) efficiency. High firing temperature increases the power produced by a gas turbine of a given physical size and mass flow. The pursuit of higher firing temperatures by all manufacturers of large, heavy-duty gas turbines used for electrical power generation is driven by the economics of high combined cycle efficiency.

Pressures and temperatures used in the following descriptions of gas turbine performance will be total pressures and temperatures. Absolute, stagnation, or total values are those measured by instruments that face into the approaching flow to give an indication of the energy in the fluid at any point. The work done in compression or expansion is proportional to the change of stagnation temperature in the working fluid, in the form of heating during a compression process or cooling during an expansion process. The temperature ratio, between the temperatures before and after the process, is related to the pressure ratio across the process by the expression $T_b/T_a = (P_b/P_a)^{(\gamma-1)/\gamma}$, where γ is the ratio of working fluid specific heats at constant pressure and volume. The temperature and pressure are stagnation values. It is the interaction between the temperature change and ratio, at different starting temperature levels, that permits the engine to generate a useful work output.

This relationship between temperature and pressure can be demonstrated by a simple numerical example using the Kelvin scale for temperature. For a starting temperature of 300°K (27°C), a temperature ratio of 1.5 yields a final temperature of 450°K and a change of 150°C. Starting instead at 400°K, the same ratio would yield a change of 200°C and a final temperature of 600°K. The equivalent pressure ratio would ideally be 4.13, as calculated from solving the preceding equation for P_b/P_a; $P_b/P_a = T_b/T_a^{1/\gamma-1} = 1.5^{1.4/0.4} = 4.13$. These numbers show that, working over the same temperature ratio, the temperature change and, therefore, the work involved in the process vary in proportion to the starting temperature level.[2]

This conclusion can be depicted graphically. If the temperature changes are drawn as vertical lines a–b and c–d, and are separated horizontally to avoid overlap, the resultant is Fig. 57.2a. Assuming the starting and finishing pressures to be the same for the two processes, the thin lines through a–d and b–c depict two of a family of lines of constant pressure, which diverge as shown. In this ideal case, expansion processes could be represented by the same diagram, simply by proceeding down the lines b–a and c–d. Alternatively, if a–b is taken as a compression process, b–c as heat addition, c–d as an expansion process, and d–a as a heat rejection process, then the figure a–b–c–d–a represents the ideal cycle to which the working fluid of the engine is subjected.

Over the small temperature range of this example, the assumption of constant gas properties is justified. In practice, the 327°C (600°K) level at point d is too low a temperature from which to start

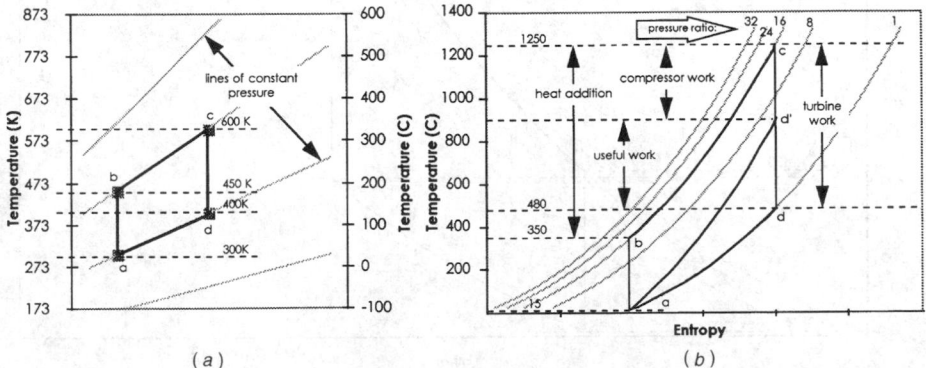

Fig. 57.2 Temperature changes and temperature–entropy diagram for ideal simple gas turbine cycles.

the expansion. Figure 57.2b is more realistic. Here, the lines of constant pressure have been constructed for ideal gas–air properties that are dependent upon temperature. Expansion begins from a temperature of 1250°C. With a pressure ratio of 16:1, the end point of the expansion is approximately 480°C. Now a–b represents the work input required by the compressor. Of the expansion work capacity c–d, only the fraction c–d' is required to drive the compressor. An optical illusion makes it appear otherwise, but line a–d' is displaced vertically from line b–c by the same distance everywhere. The remaining 435°C, line d'–d, is energy that can be used to perform useful external work, by further expansion through the turbine or by blowing through a nozzle to provide jet thrust.

Now consider line b–c. The length of its vertical projection is proportional to the heat added. The ability of the engine to generate a useful output arises from its use of the energy in the input fuel flow, but not all of the fuel energy can be recovered usefully. In this example, the heat added proportional to 1250–350 = 900°C compares with the excess output proportional to 435°C (line d'–d) to represent an efficiency of (435/900), or 48%. If more fuel could be used, raising the maximum temperature level at the same pressure, then more useful work could be obtained at nearly the same efficiency.

The line d–a represents heat rejection. This could involve passing the exhaust gas through a cooler before returning it to the compressor, and this would be a closed cycle. But, almost universally, d–a reflects discharge to the ambient conditions and intake of ambient air (Fig. 57.1b). Figure 57.1a shows an open-cycle engine, which takes air from the atmosphere and exhausts back to the atmosphere. In this case, line d–a still represents heat rejection, but the path from d to a involves the whole atmosphere and very little of the gas finds its way immediately from e to a. It is fundamental to this cycle that the remaining 465°C, the vertical projection of line d–a, is wasted heat because point d is at atmospheric pressure. The gas is therefore unable to expand further and so can do no more work.

Designers of simple cycle gas turbines—including aircraft engines—have pursued a course of reducing exhaust temperature through increasing cycle pressure ratio, which improves the overall efficiency. Figure 57.3 is identical to Fig. 57.2b except for the pressure ratio, which has been increased from 16:1 to 24:1. The efficiency is calculated in the same manner. The total turbine work is proportional to the temperature difference across the turbine, 1250–410 = 840°C. The compressor work, proportional to 430–15 = 415°C, is subtracted from the turbine temperature drop 840–415 = 425°C. The heat added to the cycle is proportional to 1250–430 = 820°C. The ratio of the net work to the heat added is 425/820 = 52%. The approximately 8% improvement in efficiency is accompanied by a 70°C drop in exhaust temperature. When no use is made of the exhaust heat, the 8% efficiency may justify the mechanical complexity associated with higher pressure ratios. Where there is value to the exhaust heat, as there is in combined Brayton–Rankine cycle power plants, the lower pressure ratio may be superior. Manufacturers forecast their customer requirements and understand

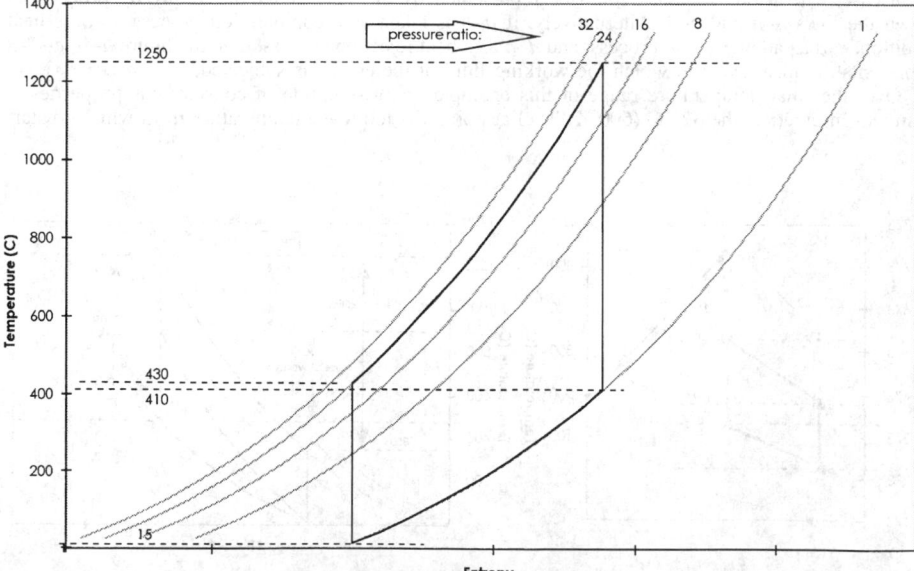

Fig. 57.3 Simple cycle gas turbine temperature–entropy diagram for high (24:1) pressure ratio and 1250°C firing temperature.

the costs associated with cycle changes and endeavor to produce gas turbines featuring the most economical thermodynamic designs.

The efficiency levels calculated in the preceding example are very high because many factors have been ignored for the sake of simplicity. Inefficiency of the compressor increases the compressor work demand, while turbine inefficiency reduces turbine work output, thereby reducing the useful work output and efficiency. The effect of inefficiency is that, for a given temperature change, the compressor generates less than the ideal pressure level while the turbine expands to a higher temperature for the same pressure ratio. There are also pressure losses in the heat addition and heat rejection processes. There may be variations in the fluid mass flow rate and its specific heat (energy input divided by consequent temperature rise) around the cycle. These factors can easily combine to reduce the overall efficiency.

57.1.2 A Brief History of Gas Turbine Development and Use

The use of a turbine driven by the rising flue gases above a fire dates back to Hero of Alexandria in 150 BC. It was not until AD 1791 that John Barber patented the forerunner of the gas turbine, proposing the use of a reciprocating compressor, a combustion system, and an impulse turbine. Even then, he foresaw the need to cool the turbine blades, for which he proposed water injection.

The year 1808 saw the introduction of the first explosion type of gas turbine, which in later forms used valves at entry and exit from the combustion chamber to provide intermittent combustion in a closed space. The pressure thus generated blew the gas through a nozzle to drive an impulse turbine. These operated successfully but inefficiently for Karavodine and Holzwarth from 1906 onward, and the type died out after a Brown, Boveri model was designed in 1939.[3]

Developments of the continuous flow machine suffered from lack of knowledge, as different configurations were tried. Stolze in 1872 designed an engine with a seven-stage axial flow compressor, heat addition through a heat exchanger by external combustion, and a 10-stage reaction turbine. It was tested from 1900 to 1904 but did not work because of its very inefficient compressor. Parsons was equally unsuccessful in 1884, when he tried to run a reaction turbine in reverse as a compressor. These failures resulted from the lack of understanding of aerodynamics prior to the advent of aircraft. As a comparison, in typical modern practice, a single-stage turbine drives about six or seven stages of axial compressor with the same mass flow.

The first successful dynamic compressor was Rateau's centrifugal type in 1905. Three assemblies of these, with a total of 25 impellers in series giving an overall pressure ratio of 4, were made by Brown, Boveri and used in the first working gas turbine engine, built by Armengaud and Lemale in the same year. The exhaust gas heated a boiler behind the turbine to generate low-pressure steam, which was directed through turbines to cool the blades and augment the power. Low component efficiencies and flame temperature (828°K) resulted in low work output and an overall efficiency of 3%. By 1939, the use of industrial gas turbines had become well established: experience with the Velox boiler led Brown, Boveri into diverging applications; a Hungarian engine (Jendrassik) with axial flow compressor and turbine used regeneration to achieve an efficiency of 0.21; and the Sun Oil Co. in the United States was using a gas turbine engine to improve a chemical process.[2]

The history of gas turbine engines for aircraft propulsion dates from 1930, when Frank Whittle saw that its exhaust gas conditions ideally matched the requirements for jet propulsion and took out a patent.[4] His first model was built by British Thomson-Houston and ran as the Power Jets Type U in 1937, with a double-sided centrifugal compressor, a long combustion chamber that was curled round the outside of the turbine and an exhaust nozzle just behind the turbine. Problems of low compressor and turbine efficiency were matched by hardware problems and the struggle to control the combustion in a very small space. Reverse-flow, can-annular combustors were introduced in 1938, the aim still being to keep the compressor and turbine as close together as possible to avoid shaft whirl problems (Fig. 57.4). Whittle's first flying engine was the W1, with 850 lb thrust, in 1941. It was made by Rover, whose gas turbine establishment was taken over by Rolls-Royce in 1943. A General Electric version of the W1 flew in 1941. A parallel effort at General Electric led to the development of a successful axial-flow compressor. This was incorporated in the first turboprop engine, the TG100, later designated the T31. This engine, first tested in May of 1943, produced 1200 horsepower from an engine weighing under 400 kg. Flight testing followed in 1949. An axial-compressor turbojet version was also constructed, designated the J35. It flew in 1946. The compressor of this engine evolved to the compressor of the GE MS3002 industrial engine, which was introduced in 1950 and is still in production.[5]

A Heinkel experimental engine flew in Germany in 1939. Several jet engines were operational by the end of the Second World War, but the first commercial engine did not enter service until 1953, the Rolls-Royce Dart turboprop in the Viscount, followed by the turbojet de Havilland Ghost in the Comet of 1954. The subsequent growth of the use of jet engines has been visible to most of the world, and has forced the growth of design and manufacturing technology.[6] By 1970, a range of standard configurations for different tasks had become established, and some aircraft engines were established in industrial applications and in ships.

Gas turbines entered the surface transportation fields also during their early stages of development. The first railway locomotive application was in Switzerland in 1941, with a 2200-hp Brown, Boveri

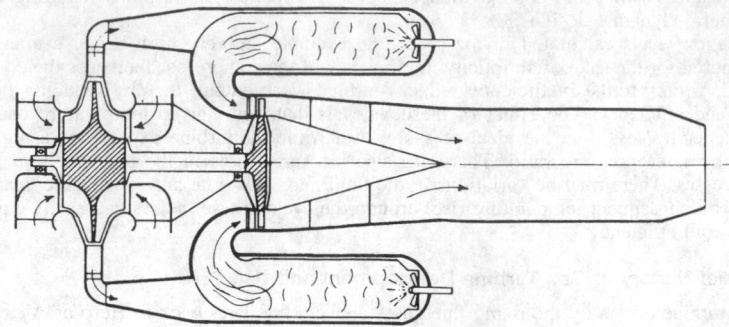

Fig. 57.4   Simplified arrangement of an early Whittle jet engine, with double-sided compressor and reverse-flow combustion chambers. (Redrawn from Ref. 4 by permission of the Council of the Institution of Mechanical Engineers.)

engine driving an electric generator and electric motors driving the wheels. The engine efficiency approached 19%, using regeneration. The next decade saw several similar applications of gas turbines by some 43 different manufacturers. A successful application of gas turbines to transportation was the 4500 draw-bar horsepower engine, based on the J35 compressor. Twenty-five locomotives so equipped were delivered to the Union Pacific railroad between 1952 and 1954. The most powerful locomotive gas turbine was the 8500-hp unit offered by General Electric to the Union Pacific railroad for long-distance freight service.[7] This became the basis of the MS5001 gas turbine, which is the most common heavy-duty gas turbine in use today. Railroad applications continue today, but relying on a significantly different system. Japan Railway uses large stationary gas turbines to generate power transmitted by overhead lines to their locomotives.

Automobile and road vehicle use started with a Rover car of 1950, followed by Chrysler and other companies, but commercial use has been limited to trucks, particularly by Ford. Automotive gas turbine development has been largely independent of other types, and has forced the pace of development of regenerators.

57.1.3   Component Characteristics and Capabilities

Compressors

Compressors used in gas turbines are of the dynamic type, wherein air is continuously ingested and raised to the required pressure level—usually, but not necessarily, between 8 and 40 atmospheres. Larger gas turbines use axial types; smaller ones use radial outflow centrifugal compressors. Some smaller gas turbines use both—an axial flow compressor upstream of a centrifugal stage.

Axial compressors feature an annular flowpath, larger in cross-section area at the inlet than at the discharge. Multiple stages of blades alternately accelerate the flow of air and allow it to expand, recovering the dynamic component and increasing pressure. Both rotating and stationary stages consist of cascades of airfoils, as can be seen in Fig. 57.5. Physical characteristics of the compressor determine many aspects of the gas turbine's performance. Inlet annulus area establishes the mass flow of the gas turbine. Rotor speed and mean blade diameter are interrelated, since optimum blade velocities exist. A wide range of pressure ratios can be provided, but today's machines feature compressions from 8:1 to as high as 40:1. The higher pressure ratios are achieved using two compressors operating in series at different rotational speeds. The number of stages required is partially dependent on the pressure ratio required, but also on the sophistication of the blade aerodynamic design that is applied. Generally, the length of the compressor is a function of pressure ratio, regardless of the number of stages. Older designs have stage pressure ratios of 1.15:1. Low-aspect ratio blading designed with three-dimensional analytical techniques have stage pressure ratios of 1.3:1. There is a trend toward fewer stages of blades of more complicated configuration. Modern manufacturing techniques make more complicated forms more practical to produce, and minimizing parts count usually reduces cost.

Centrifugal compressors are usually chosen for machines of below 2 or 3 MW in output, where their inherent simplicity and ruggedness can largely offset their lower compression efficiency. Such compressors feature a monolithic rotor with a shaped passage leading from the inlet circle or annulus to a volute at the outer radius, where the compressed air is collected and directed to the combustor. The stator contains no blades or passages and simply provides a boundary to the flow path, three sides of which are machined or cast into the rotor. Two or more rotors can be used in series to achieve the desired pressure ratio within the mechanical factors that limit rotor diameter at a given rotational speed.[8]

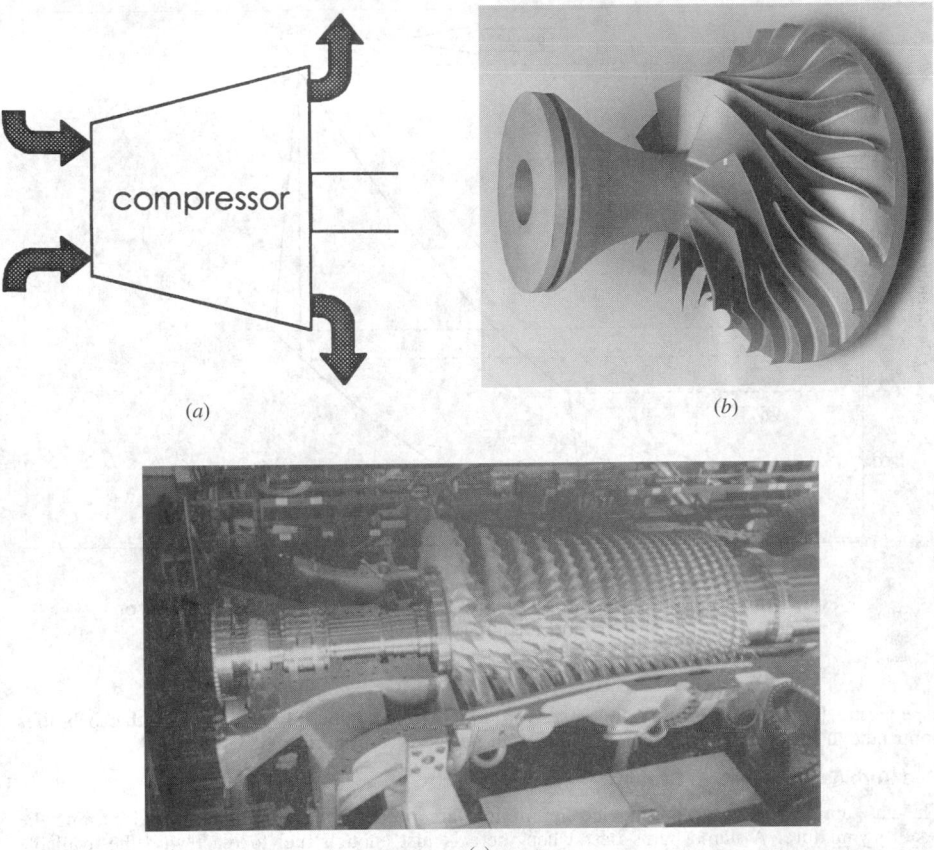

Fig. 57.5 Diagram, and photos of centrifugal compressor rotor (courtesy of Nuovo Pignone Company) and axial compressor during assembly (courtesy of General Electric Company).

Two efficiency definitions are used to describe compressor performance. Polytropic efficiency characterizes the aerodynamic efficiency of low-pressure-ratio individual stages of the compressor. Isentropic, or adiabatic, efficiency describes the efficiency of the first step of the thermodynamic process shown in Fig. 57.6 (the path from a to b). The isentropic efficiency can be calculated from the temperatures shown for the compression process on this figure. The isentropic temperature rise is for the line a–b: 335°C. The actual rise is shown by line a–b', and this rise is 372°C. The compressor efficiency η_c is the ratio $335/372 = 90\%$.

Successful compressor designs achieve high component efficiency while avoiding compressor surge or stall—the same phenomenon experienced when airplane wings are forced to operate at too high an angle of attack at too low a velocity. Furthermore, blade and rotor structures must be designed to avoid vibration problems. These problems occur when natural frequencies of components and assemblies are coincident with mechanical and aerodynamic stimuli, such as those encountered as blades pass through wakes of upstream blades. The stall phenomenon may occur locally in the compressor or even generally, whereupon normal flow through the machine is disrupted. A compressor must have good stall characteristics in order to operate at all ambient pressures and temperatures and to operate through the start, acceleration, load, load-change, unload, and shutdown phases of turbine operation. Compressors are designed with features and mechanisms for avoiding stall. These include air bleed at various points, variable-angle stator (as opposed to rotor) blades, and multiple spools.

Recent developments in the field of computational fluid dynamics (CFD) provide analytical tools that allow designers to substantially reduce aerodynamic losses due to shock waves in the supersonic flow regions. Using this technique, stages that have high tip Mach numbers can attain efficiencies comparable to those of completely subsonic designs. With these tools, compressors can be designed with higher tip diameters, hence higher flows. The same tools permit the design of low

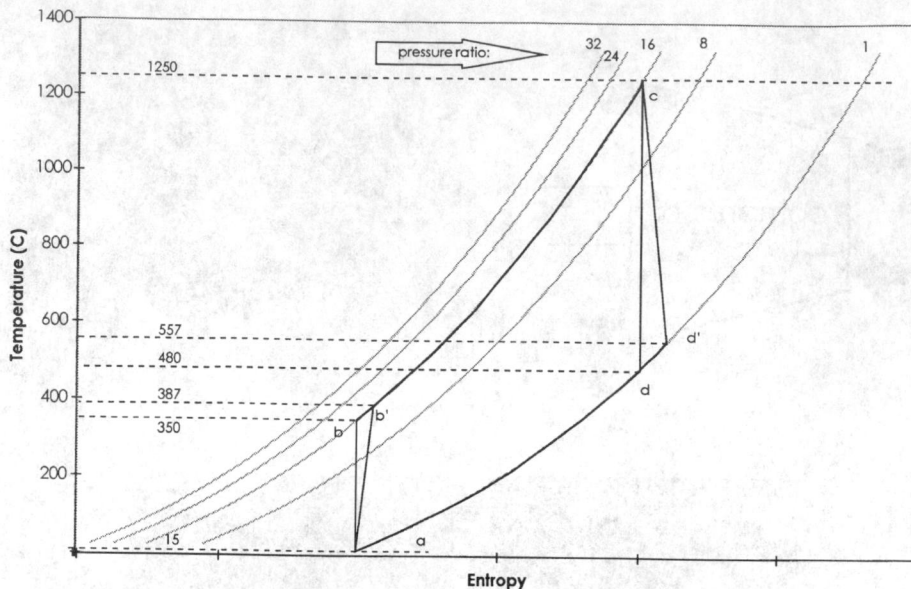

Fig. 57.6 Temperature–entropy diagram showing the effect of compressor and turbine efficiency.

aspect ratio, high stage pressure ratio blades for reducing the number of blade rows. Both capabilities contribute to lower cost gas turbine designs with no sacrifice in performance.

Gas Turbine Combustion System

The gas turbine combustor is a device for mixing large quantities of fuel and air and burning the resulting mixture. A flame burns best when there is just enough fuel to react with the available oxygen. This is called a *stoichiometric condition,* and combustion here produces the fastest chemical reaction and the highest flame temperatures, compared with excess air (fuel-lean) and excess fuel (fuel-rich) conditions, where reaction rates and temperatures are lower. The term *equivalence ratio* is used to describe the ratio of fuel to air relative to the stoichiometric condition. An equivalence ratio of 1.0 corresponds to the stoichiometric condition. Under fuel-lean conditions, the ratio is less than 1, and under fuel-rich conditions it is greater than 1. The European practice is to use the reciprocal, which is the Lambda value (λ).

In a gas turbine, since air is extracted from the compressor for cooling the combustor, buckets, nozzles, and other components and to dilute the flame—as well as support combustion—the overall equivalence ratio is far less than the value in the flame zone, ranging from 0.4–0.5 (λ = 2.5 to 2).[9]

Historically, the design of combustors required providing for the near-stoichiometric mixture of fuel and air locally. The combustion in this near-stoichiometric situation results in a diffusion flame of high temperature. Near-stoichiometric conditions produce a stable combustion front without requiring designers to provide significant flame-stabilizing features. Since the temperatures generated by the burning of a stoichiometric mixture greatly exceed those at which materials are structurally sound, combustors have to be cooled, and also the gas heated by the diffusion flame must be cooled by dilution before it becomes the working fluid of the turbine.

Gas turbine operation involves a startup cycle that features ignition of fuel at 20% of rated operating speed where air flow is proportionally lower. Loading, unloading, and part-load operation, however, require low fuel flow at full compressor speed, which means full air flow. Thermodynamic cycles are such that the lowest fuel flow per unit mass flow of air through the turbine exists at full speed and no-load. The fuel flow here is about 1/6 of the full-load fuel flow. Hence, the combustion system must be designed to operate over a 6:1 range of fuel flows with full rated air flow.

Manufacturers have differed on gas turbine combustor construction in significant ways. Three basic configurations have been used: annular, can-annular, and "silo" combustors. All have been used successfully in machines with firing temperatures up to 1100°C. Annular and can-annular combustors feature a combustion zone uniformly arranged about the centerline of the engine. All aircraft engines and most industrial gas turbine feature this type of design. A significant number of units equipped with silo combustors have been built as well. Here, one or two large combustion vessels are constructed on top of or beside the gas turbine. All manufacturers of large machines have now abandoned silo combustors in their state-of-the-art products. The can-annular, multiple combustion chamber assembly consists of an arrangement of cylindrical combustors, each with a fuel injection system, and a transition piece that provides a flow path for the hot gas from the combustor to the inlet of the turbine. Annular combustors have fuel nozzles at their upstream end and an inner and outer liner surface extending from the fuel nozzles to the entrance of the first-stage stationary blading. No transition piece is needed.

The current challenge to combustion designers is providing the cycle with a sufficiently high firing temperature while simultaneously limiting the production of oxides of nitrogen, NO_x, which refers to NO and NO_2. Very low levels of NO_x have been achieved in special low-emission combustors. NO_x is formed from the nitrogen and oxygen in the air when it is heated. The nitrogen and oxygen combine at a significant rate at temperatures above 1500°C, and the formation rate increases exponentially as temperature increases. Even with the high gas velocities in gas turbines, NO_x emissions will reach 200 parts per million by volume, dry (ppmvd), in gas turbines with conventional combustors and no NO_x abatement features. Emissions standards throughout the world vary, but many parts of the world require gas turbines to be equipped to control NO_x to below 25 parts per million by volume, dry (ppmvd) at base load.

Emissions

Combustion of common fuels necessarily results in the emission of water vapor and carbon dioxide. Combustion of near-stoichiometric mixtures results in very high temperatures. Oxides of nitrogen are formed as the oxygen and nitrogen in the air combine, and this happens at gas turbine combustion temperatures. Carbon monoxide forms when the combustion process is incomplete. Unburned hydrocarbons (UHC) are discharged as well when combustion is incomplete. Other pollutants are attributed to fuel; principal among these is sulfur. Gas turbines neither add nor remove sulfur; hence, what sulfur enters the gas turbine in the fuel exits as SO_2 in the exhaust.

Much of the gas turbine combustion research and development of the 1980s and 1990s focused on lowering NO_x production in mechanically reliable combustors while maintaining low CO and UHC emissions. Early methods of reducing NO_x emissions included removing it from the exhaust by selective catalytic reduction (SCR) and by diluent injection, that is, the injection of water or steam into the combustor. These methods continue to be employed. The lean-premix combustors now in general use are products of ongoing research.

Thermal NO_x is generally regarded as being generated by a chemical reaction sequence called the Zeldovich mechanism,[10] and the rate of NO_x formation is proportional to temperature, as shown in Fig. 57.7. In practical terms, a conventional gas turbine emits approximately 200 ppmvd when its combustors are not designed to control NO_x. This is because a significant portion of the combustion zone has stoichiometric or near-stoichiometric conditions, and temperatures are high. Additional oxygen, and of course nitrogen on the boundary of the flame, is heated to sufficiently high temperatures, and held at these temperatures for sufficient time, to produce NO_x.

Water- and steam-injected combustors achieve low flame temperatures by placing diluent in the neighborhood of the reacting fuel and air. Among low NO_x combustion systems operating today, water and steam injection is the most common means of flame temperature reduction. Several hundred large industrial turbines operating with steam or water injection have accumulated over 2-1/2 million hours of service. Water is not the only diluent used for NO_x control. In the case of integrated gasification combined cycle plants, nitrogen and CO_2 are available and can be introduced into the combustion region. The NO_x emissions measured at the Cool Water IGCC plant in the United States rival those of the cleanest natural gas plants in the world.[11]

Water or steam injection can achieve levels that satisfy all current standards, but water consumption is sometimes not acceptable to the operator because of cost, availability, or the impact on efficiency. Steam injection sufficient to reduce NO_x emissions to 25 ppmvd can increase fuel consumption in combined cycle power plants by over 3%. Water injection increases fuel use by over 4% for the same emissions level. In base-load power plants, fuel cost is so significant that it has caused the development of systems that do not require water.[12]

In all combustion processes, when a molecule of methane combines with two molecules of oxygen, a known and fixed amount of heat is released. When only these three molecules are present, a minimum amount of mass is present to absorb the energy not radiated and the maximum temperature is realized. Add to the neighborhood of the reaction the nitrogen as found in air (four times the volume of oxygen involved in the reaction) and the equilibrium temperature is lower. When even more air is added to the combustion region, more mass is available to absorb the energy and the resulting observable temperature is lower still. The same can be achieved through the use of excess fuel. Thus, moving away from the stoichiometric mixture means that observable flame temperature is lowered and the production of NO_x is also reduced. On a microscopic level, lean-burning low-NO_x combustors are designed to force the chemical reaction to take place in such a way that the energy released is in the neighborhood of as much mass not taking part in the reaction as possible. By transferring heat to neighboring material immediately, the time-at-temperature is reduced. On a larger

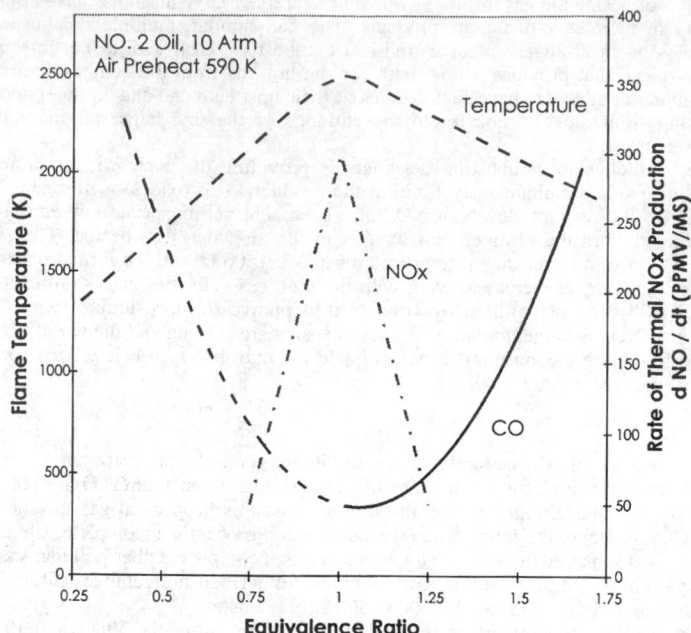

Fig. 57.7 NO$_x$ formation rate driven by temperature (drawn from figure in Ref. 9; courtesy of General Electric Company).

scale, a high measurable temperature will never be reached in a well-mixed lean system and thus NO$_x$ generation is minimized. Both rich-mixture and lean-mixture systems have led to low NO$_x$ schemes. Although those featuring rich flames followed by lean burning zones are sometimes suggested for situations where there is nitrogen in the fuel, most of today's systems are based on lean burning.

Early lean premix dry low-NO$_x$ combustors were operated in GE gas turbines at the Houston Light and Power Wharton Station in 1980 in the United States, in Mitsubishi units in Japan in 1983, and were introduced in Europe in 1986 by Siemens KWU. These combustors control the formation of NO$_x$ by premixing fuel with air prior to its ignition while conventional combustors mix essentially at the instant of ignition. Dry low-NO$_x$ combustors, as the name implies, achieve NO$_x$ control without consuming water and without imposing efficiency penalties on combined-cycle plants.

Figures 57.8 and 57.9 show dry low-NO$_x$ combustors developed for large gas turbines. In the GE system, several premixing chambers are located at the head end of the combustor. A fuel nozzle assembly is located in the center of each chamber. By the manipulation of valves external to the gas turbine, fuel can be directed to several combinations of chambers and to various parts of the fuel nozzles. This is to permit the initial ignition of the fuel and to maintain a relatively constant local fuel–air ratio at all load levels. There is one flame zone, immediately downstream of the premixing chambers. The Westinghouse combustor illustrated in Fig. 57.9 has three concentric premixing chambers. The two nearest the centerline of the combustor are designed to swirl the air passing through them in opposite directions and discharge into the primary combustion zone. The third, which has a longer passage, is directed to the secondary zone. Modulating fuel flow to the various mixing passages and combustion zones ensures low NO$_x$ production over a wide range of operating temperatures. Both the combustors shown are designed for state-of-the-art, high-firing-temperature gas turbines.

Low-NO$_x$ combustors feature multiple premixing features and a more complex control system than more conventional combustors, to achieve stable operation over the required range of operating conditions. The reason for this complexity is explained with the aid of Fig. 57.10. Conventional combustors operate with stability over a wide range of fuel–air mixtures—between the rich and lean flammability limits. A sufficiently wide range of fuel flows could be burned in a combustor with a fixed air flow, to match the range of load requirements from no-load to full-load. In a low-NO$_x$ combustor, the fuel–air mixture feeding the flame must be regulated between the point of flame loss and the point where the NO$_x$ limit is exceeded. When low gas turbine output is required, the air premixed with the fuel must be reduced to match the fuel flow corresponding to the low power output. The two combustors shown above hold nearly constant fuel–air ratios over the load range by

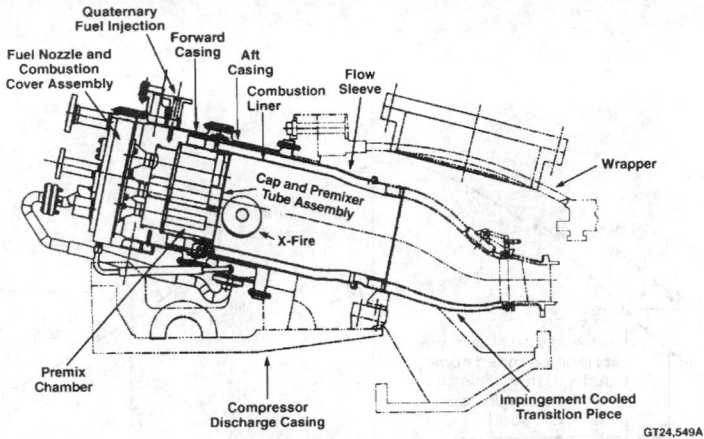

Fig. 57.8 GE DLN-2 lean-premix combustor designed for low emissions at high firing temperatures (courtesy of General Electric Company).

having multiple premixing chambers, each one flowing a constant fraction of the compressor discharge flow. By directing fuel to only some of these passages at low load, the design achieves both part load and optimum local fuel–air ratio. Three, four, or more sets of fuel passages are not uncommon, and premixed combustion is maintained to approximately 50% of the rated load of the machine.[9,13]

Catalytic combustion systems are under investigation for gas turbines. These systems have demonstrated stable combustion at lower fuel–air ratios than those using chamber, or nozzle, shapes to

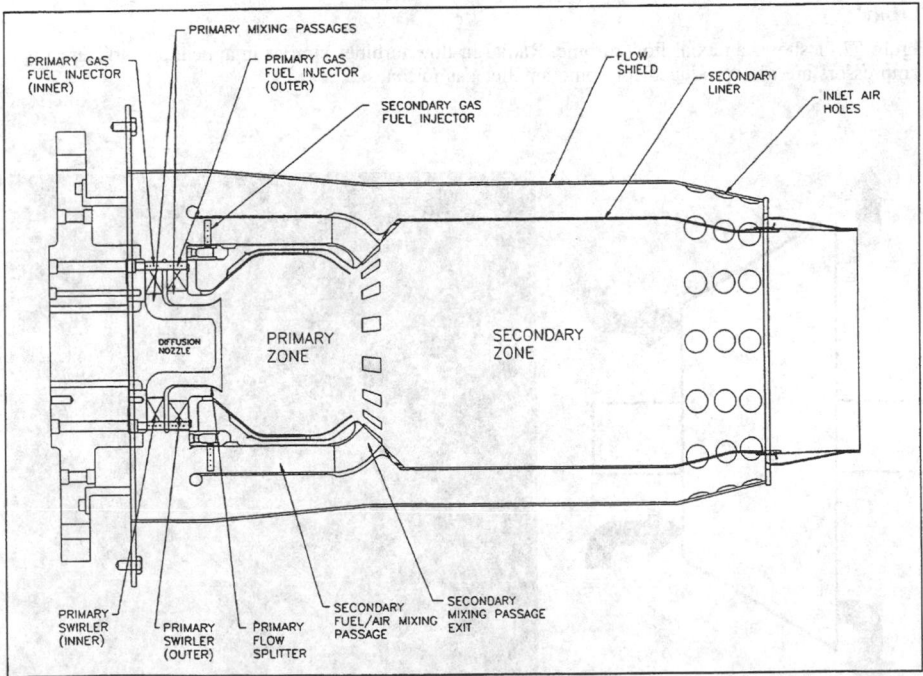

Fig. 57.9 Westinghouse dry low-NO_x combustor for advanced gas turbines (courtesy of Westinghouse Corporation).

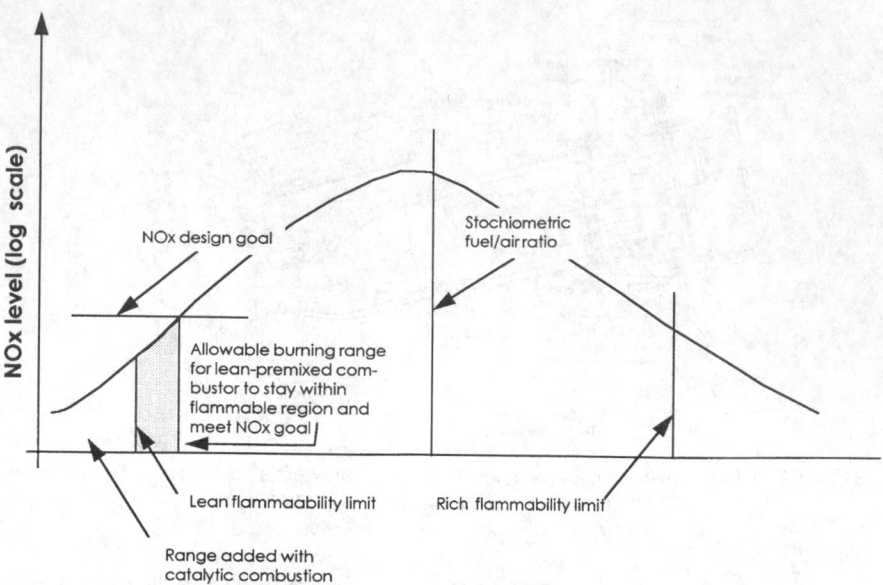

Fig. 57.10 Fuel–air mixture ranges for conventional and premixed combustors (courtesy of Westinghouse Corporation).

stabilize flames. They offer the promise of simpler fuel regulation and greater turn-down capability than low-NO$_x$ combustors now in use. In catalytic combustors, the fuel and air react in the presence of a catalytic material that is deposited on a structure having multiple parallel passages or mesh. Extremely low NO$_x$ levels have been observed in laboratories with catalytic combustion systems.

Turbine

Figure 57.11 shows an axial flow turbine. Radial in-flow turbines similar in appearance to centrifugal compressors are also produced for some smaller gas turbines.

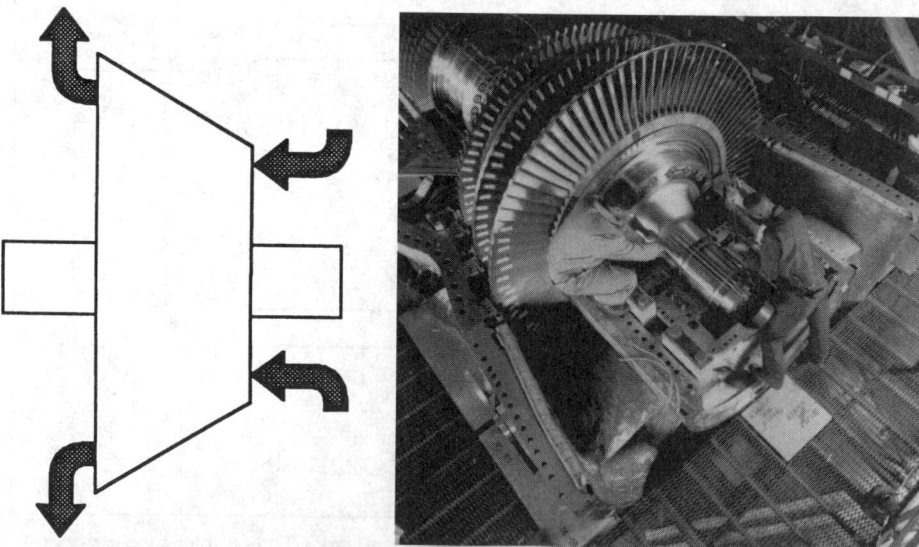

Fig. 57.11 Turbine diagram, and photo of an axial flow turbine during assembly (courtesy of General Electric Company).

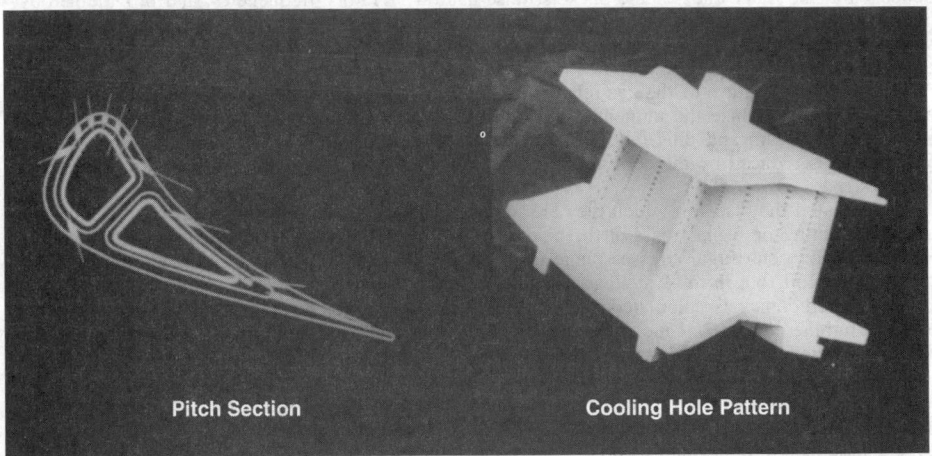

Pitch Section   Cooling Hole Pattern

Fig. 57.12 Gas turbine first stage nozzle. Sketch shows cooling system of one airfoil (courtesy of General Electric Company).

By the time the extremely hot gas leaves the combustor and enters the turbine, it has been mixed with compressor discharge air to cool it to temperatures that can be tolerated by the first-stage blading in the turbine: temperatures ranging from 950°C in first-generation gas turbines to over 1500°C in turbines currently being developed and in state-of-the-art aircraft engines. Less dilution flow is required as firing temperatures approach 1500°C.

The first-stage stationary blades, or nozzles, are located at the discharge of the combustor. Their function is to accelerate the hot working fluid and turn it so as to enter the following rotor stage at the proper angle. These first-stage nozzles are subjected to the highest gas velocity in the engine. The gas entering the first-stage nozzle can regularly be above the melting temperature of the structural metal. These conditions produce high heat transfer to the nozzles, so that cooling is necessary.

Nozzles (Fig. 57.12) are subjected to stresses imposed by aerodynamic flow of the working fluid, pressure loading of the cooling air, and thermal stresses caused by uneven temperatures over the nozzle structure. First-stage nozzles can be supported at both ends, by the inner and outer sidewalls. But later-stage nozzles, because of their location in the engine, can be supported only at the outer end, intensifying the effect of aerodynamic loading.

The rotating blades of the turbine, or buckets (Fig. 57.13), convert the kinetic energy of the hot gas exiting the nozzles to shaft power used to drive the compressor and load devices. The blade consists of an airfoil section in the gas path, a dovetail or other type of joint connecting the blade to the turbine disc, and often a shank between the airfoil and dovetail allowing the dovetail to run at lower temperature than the root of the airfoil. Some bucket designs employ tip shrouds to limit

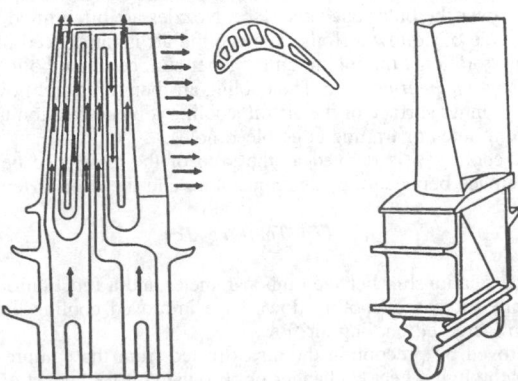

Fig. 57.13 Gas turbine first-stage air-cooled bucket. Cut-away view exposes serpentine cooling passages (courtesy of General Electric Company).

deflection at the outer ends of the buckets, raise natural vibratory frequencies, and provide aerodynamic benefits. Exceptions from this configuration are radial inflow turbines like those common to automotive turbochargers and axial turbines, wherein the buckets and wheels are made of one piece of metal or ceramic.

The total temperature of the gas relative to the bucket is lower than that relative to the preceding nozzles. This is because the tangential velocity of the rotor-mounted airfoil is in a direction away from the gas stream and thus reduces the dynamic component of total temperature. Also, the gas temperature is reduced by the cooling air provided to the upstream nozzle and the various upstream leakages.

Buckets and the discs on which they are mounted are subject to centrifugal stresses. The centrifugal force acting on a unit mass at the blades' midspan is 10,000 to 100,000 times that of gravity. Midspan airfoil centrifugal stresses range from 7 kg/mm^2 (10,000 psi) to over 28 kg/mm^2 (40,000 psi) at the airfoil root in the last stage (longest buckets).

Turbine efficiency is calculated similarly to compressor efficiency. Figure 57.6 also shows the effect of turbine efficiency. Line c–d represents the isentropic expansion process and c–d' the actual. Turbine efficiency η_t is the ratio of the vertical projections of the lines. Thus, (1250–557)/ (1250–480) = 90%. It is possible at this point to compute the effect of a 90% efficient compressor and a 90% efficient turbine upon the simple cycle efficiency of the gas turbine represented in the figure. The turbine work is proportional to 693°C and the compressor work to 372°C. The heat added by combustion is proportional to 887°C, the temperature rise from b' to c. The ratio of the useful work to the heat addition is thus 36.2%. It was shown previously that the efficiency with ideal components is approximately 48.3%.

The needs of gas turbine blading have been responsible for the rapid development of a special class of alloys. To tolerate higher metal temperatures without decrease in component life, materials scientists and engineers have developed, and continue to advance, families of temperature-resistant alloys, processes, and coatings. The "superalloys" were invented and continue to be developed primarily in response to turbine needs. These are usually based on Group VIIIA elements: cobalt, iron, and nickel. Bucket alloys are austenitic with gamma/gamma-prime, face-centered cubic structure (Ni$_3$Al). The elements titanium and columbium are present and partially take the place of aluminum, with beneficial hot corrosion effect. Carbides are present for grain boundary strength, along with some chromium to further enhance corrosion resistance. The turbine industry has also developed processes to produce single-crystal and directionally solidified components that have even better high-temperature performance. Coatings are now in universal use that enhance the corrosion and erosion performance of hot gas path components.[14]

Cooling

Metal temperature control is addressed primarily through airfoil cooling, with cooling air being extracted from the gas turbine flow ahead of the combustor. Since this air is not heated by the combustion process, and may even bypass some turbine stages, the cycle is less efficient than it would be without cooling. Further, as coolant re-enters the gas path, it produces quenching and mixing losses. Hence, for efficiency, the use of cooling air should be minimized. Turbine designers must make tradeoffs among cycle efficiency (firing temperature), parts lives (metal temperature), and component efficiency (cooling flow).

In early, first-generation gas turbines, buckets were solid metal, operating at the temperature of the combustion gases. In second-generation machines, cooling air was conducted through simple, radial passages to keep metal temperatures below those of the surrounding gas. In today's advanced-technology gas turbines, most manufacturers utilize serpentine air passages within the first-stage buckets, with cooling air flowing out the tip, leading, and trailing edges. Leading edge flow is used to provide a cooling film over the outer bucket surface. Nozzles are often fitted with perforated metal inserts attached to the inside of hollow airfoils. The cooling air is introduced inside of the inserts. It then flows through the perforations, impinging on the inner surface of the hollow airfoil. The cooling thus provided is called *impingement cooling*. The cooling air then turns and flows within the passage between the insert and the inner surface of the airfoil, cooling it by convection until it exits the airfoil in either leading edge film holes or trailing edge bleed holes.

The effectiveness of cooling η is defined as the ratio of the difference between gas and metal temperatures to the difference between the gas temperature and the coolant temperature:

$$\eta = (Tg - Tm)/(Tg - Tc)$$

Figure 57.14 portrays the relationship between this parameter and a function of the cooling air flow. It can be seen that, while increased cooling flows have improved cooling effectiveness, there are diminishing returns with increased cooling air flow.

Cooling can be improved by precooling the air extracted from the compressor. This is done by passing the extracted air through a heat exchanger prior to using it for bucket or nozzle cooling. This does increase cooling, but presents several challenges, such as increasing temperature gradients and

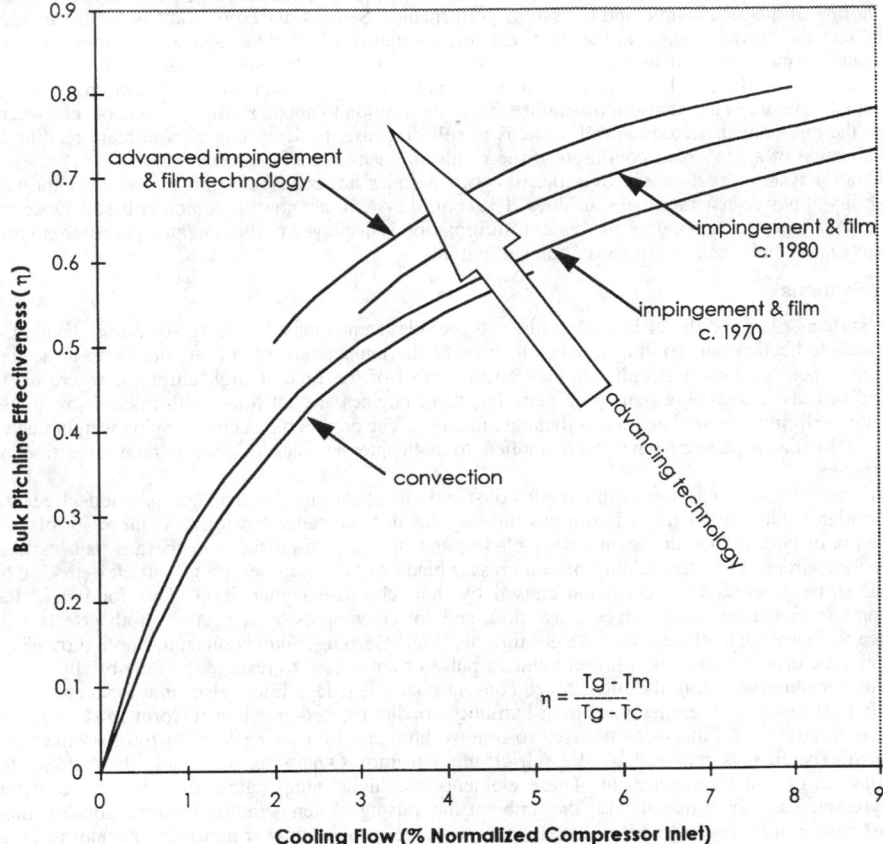

Fig. 57.14 Evolution of turbine airfoil cooling technology.

the cost and reliability of the cooling equipment. Recent advanced gas turbine products have been designed with both cooled and uncooled cooling air.

Other cooling media have been investigated. In the late 1970s, the U.S. Department of Energy sponsored the study and preliminary design of high-temperature turbines cooled by water and steam. Nozzles of the water-cooled turbine were cooled by water contained in closed passages and kept in the liquid state by pressurization; no water from the nozzle circuits entered the gas path. Buckets were cooled by two-phase flow; heat was absorbed as the coolant was vaporized and heated. Actual nozzles were successfully rig tested. Simulated buckets were tested in heated rotating rigs. Recent advanced land-based gas turbines have been configured with both buckets and nozzles cooled with a closed steam circuit. Steam, being a more effective cooling medium than air, permits high firing temperatures and, since it does not enter the gas path, eliminates the losses associated with cooling air mixing with the working fluid. The coolant, after being heated in the buckets and nozzles, returns to the steam cycle of a combined cycle plant. The heat carried away by the steam is recovered in a steam turbine.

57.1.4 Controls and Accessories

Controls

The control system is the interface between the operator and the gas turbine. More correctly, the control system of modern industrial and utility gas turbines interfaces between the operator and the entire power plant, including the gas turbine, generator, and all related accessories. In combined cycle power plants where a steam turbine, heat recovery steam generator, condensing system, and all related accessories are also present, the control system interfaces with these as well.

Functions provided are described below in Section 57.1.5 plus protection of the turbine from faults such as overspeed, overheating, combustion anomalies, cooling system failures, and high vibrations. Also, controls facilitate condition monitoring, problem identification and diagnosis, and monitoring of thermodynamic and emissions performance. Sensors placed on the gas turbine include speed pickups, thermocouples at the inlet, exhaust, compressor exit, wheelspaces, bearings, oil supplies and drains. Vibration monitors are placed on each bearing. Pressure is also monitored at the compressor exit. Multiple thermocouples in the exhaust can detect combustor malfunction by noting abnormal differences in exhaust temperature from one location to another. Multiple sensors elsewhere allow the more sophisticated control systems to self-diagnose, to determine if a problem reading is an indication of a dangerous condition or the result of a sensor malfunction.

Control system development over the past two decades has contributed greatly to the improved reliability of power-generation gas turbines. The control systems are now all computer based. Operator input is via keyboard and cursor movement. Information is displayed to the operator via color graphic displays and tabular and text data on color monitors.

Inlet Systems

Inlet systems filter and direct incoming air, and provide attenuation of compressor noise. They also can include heating and cooling devices to modify the temperature of the air drawn into the gas turbine. Since fixed-wing aircraft engines operate most of the time at high altitudes, where air is devoid of heavier and more damaging particles, these engines are not fitted with inlet air treatment systems performing more than an aerodynamic function. The premium placed on engine weight makes this so. Inertial separators have been applied to helicopter engines to reduce their ingestion of particulates.

Air near the surface of the earth contains dust and dirt of various chemical compositions. Because of the high volume of air taken into a gas turbine, this dirt can cause erosion of compressor blades, corrosion of both turbine and compressor blades, and plugging of passages in the gas path as well as cooling circuits. The roughening of compressor blade surfaces can be due to particles sticking to airfoil surfaces, erosion, or corrosion caused by their chemical composition. This fouling of the compressor can, over time, reduce mass flow and lower compressor efficiency. Both effects will reduce the output and efficiency of the gas turbine. "Self-cleaning" filters collect airborne dirt. When the pressure drop increases to a preset value, a pulse of air is used to reverse the flow briefly across the filter medium, cleaning the filter. More conventional, multistage filters also find application.

Under low-ambient-temperature, high-humidity conditions, it is possible to form frost or ice in the gas turbine inlet. Filters can be used to remove humidity by causing frost to form on the filter element. The frost is removed by the self-cleaning feature. Otherwise, a heating element can be installed in the inlet compartment. These elements use higher-temperature air extracted from the compressor. This air is mixed with the ambient air, raising its temperature. Compressors of most robust gas turbines are designed so that these systems are required only at part load or under unusual operating conditions.

Inlet chillers have been applied on gas turbines installed in high-ambient-temperature, low-humidity regions of the world. The incoming air is cooled by the evaporation of water. Cooling the inlet air increases its density and increases the output of the gas turbine.

Exhaust Systems

The exhaust systems of industrial gas turbines perform three basic functions. Personnel must be protected from the high-temperature gas and from the ducts that carry it. The exhaust gas must be conducted to an exhaust stack or to where the remaining heat from the gas turbine cycle can be effectively used. The exhaust system also contains baffles and other features employed to reduce the noise generated by the gas turbine.

Enclosures and Lagging

Gas turbines are enclosed for four reasons: noise, heat, fire protection, and visual aesthetics. Gas turbines are sometimes provided for outdoor installation, where the supplier includes a sheet metal enclosure that may be part of the factory-shipped package. Other times, gas turbines are installed in a building. Even in a building, the gas turbine is enclosed for the benefit of maintenance crews or other occupants. Some gas turbines are designed to accommodate an insulating wrapping that attaches to the casings of the gas turbine. This prevents maintenance crews from coming into contact with the hot casings when the turbine is operating and reduces some of the noise generated by the gas turbine. Proponents cite the benefit of lowering the heat transferred from the gas turbine to the environment. Theoretically, more heat is carried to the exhaust which can be used for other energy needs. Others contend that the larger internal clearances resulting from hotter casings would offset this gain by lower component efficiencies.

Where insulation is not attached to the casings, and sometimes when it is, a small building-like structure is provided. This structure is either attached to the turbine base or to the concrete foundation. Such a structure provides crew protection and noise control, and assists in fire protection. If a fire is detected on the turbine, within the enclosure, its relative small volume makes it possible to quickly flood the area with CO_2 or other firefighting chemical. The fire is thereby contained in a small volume and more quickly extinguished. Even in a building, the noise control provided by an enclosure is beneficial, especially in buildings containing additional gas turbines or other equipment. By lowering the noise 1 m from the enclosure to below 85 or 90 dba, it is possible to safely perform maintenance on this other equipment, yet continue to operate the gas turbine. Where no turbine enclosure is provided within a building, the building becomes part of the fire-protection and acoustic system.

Fuel Systems

The minimum functions required of a gas turbine fuel system are to deliver fuel from a tank or pipeline to the gas turbine combustor fuel nozzles at the required pressure and flow rate. The pressure required is somewhat above the compressor discharge pressure, and the flow rate is that called for by the controls. On annular and can-annular combustors, the same fuel flow must be distributed to each nozzle to ensure minimum variation in the temperature to which gas path components are exposed. Other fuel system requirements are related to the required chemistry and quality of the fuel.

Aircraft engine fuel quality and chemistry are closely regulated, so extensive on-board fuel conditioning systems are not required. Such is not the case in many industrial applications. Even the better grades of distillate oil may be delivered by oceangoing tanker and run the risk of sodium contamination from the salt water sometimes used for ballast. Natural gas now contains more of the heavier, LP gases. Gas turbines are also fueled with crude oil, heavy oils, and various blends. Some applications require the use of non-lubricating fuels such as naphtha. Most fuels today require some degree of on-site treatment.

Complete liquid fuel treatment includes washing to remove soluble trace metals, such as sodium, potassium, and certain calcium compounds. Filtering the fuel removes solid oxides and silicates. Inhibiting the vanadium in the fuel with magnesium compounds in a ratio of three parts of magnesium (by weight) to one part of vanadium limits the corrosive action of vanadium on the alloys used in high-temperature gas path parts.

Gas fuel is primarily methane, but it contains varying levels of propane, butane, and other heavier hydrocarbons. When levels of these heavier gases increase, the position of the flame in the combustor may change, resulting in local hot spots that could damage first-stage turbine stator blades. Also, sudden increases could cause problems for dry low-NO_x premixed combustors. These combustors depend on being able to mix fuel and air in a combustible mixture before the mixture is ignited. Under some conditions, heavier hydrocarbons can self-ignite in these mixtures at compressor exit temperatures, thus causing flame to exist in the premixing portion of the combustor. The flame in the premixing area would have to be extinguished and reestablished in the proper location. This process interferes with normal operation of the machine.

Lubricating Systems

Oil must be provided to the bearings of the gas turbine and its driven equipment. The lubricating system must maintain the oil at sufficiently low temperature to prevent deterioration of its properties. Contaminants must be filtered out. Sufficient volume of oil must be in the system so that any foam has time to settle out. Also, vapors must be dealt with; they are preferably recovered and the oil returned to the plenum. The oil tank for large industrial turbines is generally the base of the lubricating system package. Large utility machines are provided with tanks that hold over 12,000 liters of oil. The oil is generally replaced after approximately 20,000 hours of operation. More oil is required in applications where the load device is connected to the gas turbine by a gearbox.

The lubrication system package also contains filters and coolers. The turbine is fitted with mist-elimination devices connected to the bearing air vents. Bearings may be vented to the turbine exhaust, but this practice is disappearing for environmental reasons.

Cooling Water and Cooling Air Systems

Several industrial gas turbine applications require the cooling of some accessories. The accessories requiring cooling include the starting means, lubrication system, atomizing air, load equipment (generator/alternator), and turbine support structure. Water is circulated in the component requiring cooling, then conducted to where the heat can be removed from the coolant. The cooling system can be integrated into the industrial or powerplant hosting the gas turbine, or can be dedicated to the gas turbine. In this case, the system usually contains a water-to-air heat exchanger with fans to provide the flow of air past finned water tubes.

Water-Wash Systems

Compressor fouling related to deposition of particles that are not removed by the air filter can be dealt with by water-washing the compressor. A significant benefit in gas turbine efficiency over time can be realized by periodic cleaning of the compressor blades. This cleaning is most conveniently done when the gas turbine is fitted with an automatic water-wash system. Washing is initiated by the operator. The water is preheated and detergent is added. The gas turbine rotor is rotated at a low speed and the water is sprayed into the compressor. Drains are provided to remove waste water.

57.1.5　Gas Turbine Operation

Like other internal combustion engines, the gas turbine requires an outside source of starting power. This is provided by an electrical motor or diesel engine connected through a gear box to the shaft of the gas turbine (the high-pressure shaft in a multishaft configuration). Other devices can be used, including the generator of large electric utility gas turbines, by using a variable frequency power supply. Power is normally required to rotate the rotor past the gas turbine's ignition speed of 10–15% on to 40–80% of rated speed where the gas turbine is self-sustaining, meaning the turbine produces sufficient work to power the compressor and overcome bearing friction, drag, and so on. Below self-sustaining speed, the component efficiencies of the compressor and turbine are too low to reach or exceed this equilibrium.

When the operator initiates the starting sequence of a gas turbine, the control system acts by starting auxiliaries such as those that provide lubrication and the monitoring of sensors provided to ensure a successful start. The control system then calls for application of torque to the shaft by the starting means. In many industrial and utility applications, the rotor must be rotated for a period of time to purge the flow path of unburned fuel that may have collected there. This is a safety precaution. Thereafter, the light-off speed is achieved and ignition takes place and is confirmed by sensors. Ignition is provided by either a sparkplug type device or by an LP gas torch built into the combustor. Fuel flow is then increased to increase the rotor speed. In large gas turbines, a warmup period of one minute or so is required at approximately 20% speed. The starting means remains engaged, since the gas turbine has not reached its self-sustaining speed. This reduces the thermal gradients experienced by some of the turbine components and extends their low cycle fatigue life.

The fuel flow is again increased to bring the rotor to self-sustaining speed. For aircraft engines, this is approximately the idle speed. For power generation applications, the rotor continues to be accelerated to full speed. In the case of these alternator-driving gas turbines, this is set by the speed at which the alternator is synchronized with the power grid to which it is to be connected.

Aircraft engines' speed and thrust are interrelated. The fuel flow is increased and decreased to generate the required thrust. The rotor speed is principally a function of this fuel flow, but also depends on any variable compressor or exhaust nozzle geometry changes programmed into the control algorithms. Thrust is set by the pilot to match the current requirements of the aircraft, through takeoff, climb, cruise, maneuvering, landing, and braking.

At full speed, the power-generation gas turbine and its generator (alternator) must be synchronized with the power grid in both speed (frequency) and phase. This process is computer-controlled and involves making small changes in turbine speed until synchronization is achieved. At this point, the generator is connected with the power grid. The load of a power-generation gas turbine is set by a combination of generator (alternator) excitement and fuel flow. As the excitation is increased, the mechanical work absorbed by the generator increases. To maintain a constant speed (frequency), the fuel flow is increased to match that required by the generator. The operator normally sets the desired electrical output and the turbine's electronic control increases both excitation and fuel flow until the desired operating conditions are reached.

Normal shutdown of a power-generation gas turbine is initiated by the operator and begins with the reduction of load, reversing the loading process described immediately above. At a point near zero load, the breaker connecting the generator to the power grid is opened. Fuel flow is decreased and the turbine is allowed to decelerate to a point below 40% speed, whereupon the fuel is shut off and the rotor is allowed to stop. Large turbines' rotors should be turned periodically to prevent temporary bowing from uneven cool-down that will cause vibration on subsequent startups. Turning of the rotor for cool-down is accomplished by a ratcheting mechanism on smaller gas turbines, or by operation of a motor associated with shaft-driven accessories, or even the starting mechanism on others. Aircraft engine rotors do not tend to exhibit the bowing just described. Bowing is a phenomenon observed in massive rotors left stationary surrounded by cooling, still air that, due to free convection, is cooler at the 6:00 position than at the 12:00 position. The large rotor assumes a similar gradient and, because of proportional thermal expansion, assumes a bowed shape. Because of the massiveness of the rotor, this shape persists for several hours, and could remain present when the operator wishes to restart the turbine.

57.2　GAS TURBINE PERFORMANCE

57.2.1　Gas Turbine Configurations and Cycle Characteristics

There are several possible mechanical configurations for the basic simple cycle, or open cycle, gas turbine. There are also some important variants on the basic cycle: intercooled, regenerative, and reheat cycles.

The simplest configuration is shown in Fig. 57.15. Here the compressor and turbine rotors are connected directly to one another and to shafts by which turbine work in excess of that required to drive the compressor can be applied to other work-absorbing devices. Such devices are the propellers and gear boxes of turboprop engines, electrical generators, ships' propellers, pumps, gas compressors, vehicle gear boxes and driving wheels, and the like. A variation is shown in Fig. 57.16, where a jet

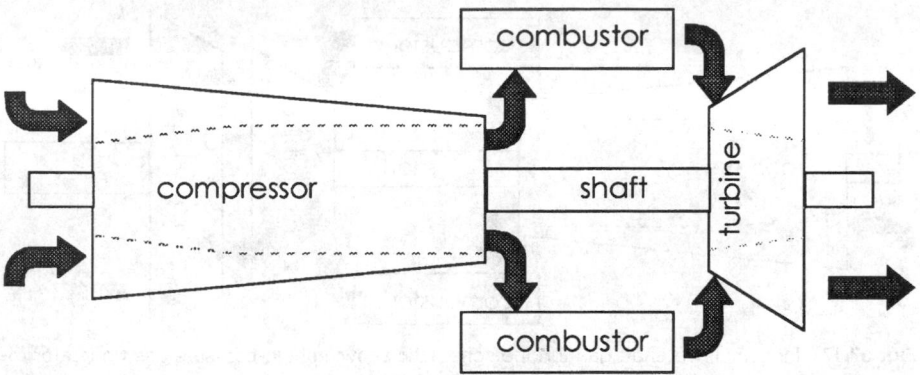

Fig. 57.15 Simple-cycle, single-shaft gas turbine schematic.

nozzle is added to generate thrust. Through aerodynamic design, the pressure drop between the turbine inlet and ambient air is divided so that part of the drop occurs across the turbine and the remainder across the jet nozzle. The pressure at the turbine exit is set so that there is only enough work extracted from the working fluid by the turbine to drive the compressor (and mechanical accessories). The remaining energy accelerates the exhaust flow through the nozzle to provide jet thrust.

The simplest of multishaft arrangements appears in Fig. 57.17. For decades, such arrangements have been used in heavy-duty turbines applied to various petrochemical and gas pipeline uses. Here, the turbine consists of a high-pressure and a low-pressure section. There is no mechanical connection between the rotors of the two turbines. The high-pressure (h.p.) turbine drives the compressor and the low-pressure (l.p.) turbine drives the load—usually a gas compressor for a process, gas well, or pipeline. Often, there is a variable nozzle between the two turbine rotors that can be used to vary the work split between the two turbines. This offers the user an advantage. When it is necessary to lower the load applied to the driven equipment—for example, when it is necessary to reduce the flow from a gas-pumping station—fuel flow would be reduced. With no variable geometry between the turbines, both would drop in speed until a new equilibrium between l.p. and h.p. speeds occurs. By changing the nozzle area between the rotors, the pressure drop split is changed and it is possible to keep the h.p. rotor at a high, constant speed and have all the speed drop occur in the l.p. rotor. By doing this, the compressor of the gas turbine continues to operate at or near its maximum efficiency, contributing to the overall efficiency of the gas turbine and providing high part-load efficiency. This two-shaft arrangement is one of those applied to aircraft engines in industrial applications. Here, the h.p. section is essentially identical to the aircraft turbojet engine or the core of a fan-jet engine. This h.p. section then becomes the *gas generator* and the free-turbine becomes what is referred to as the *power turbine.* The modern turbofan engine is somewhat similar in that a low-pressure turbine drives a fan that forces a concentric flow of air outboard of the gas generator aft, adding to the thrust provided by the engine. In the case of modern turbofans, the fan is upstream of the compressor and is driven by a concentric shaft inside the hollow shaft connecting the h.p. compressor and h.p. turbine.

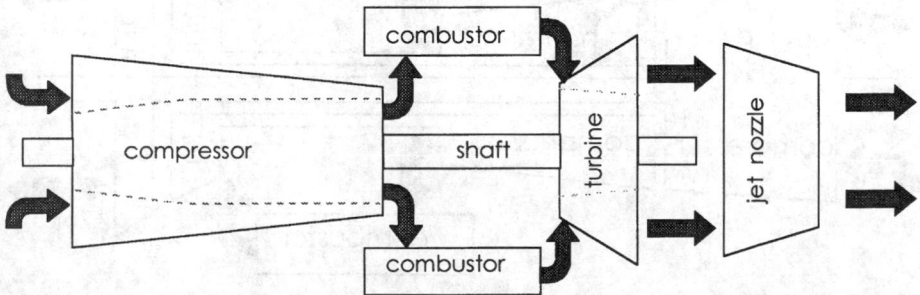

Fig. 57.16 Simple-cycle single-shaft, gas turbine with jet nozzle; simple
turbojet engine schematic.

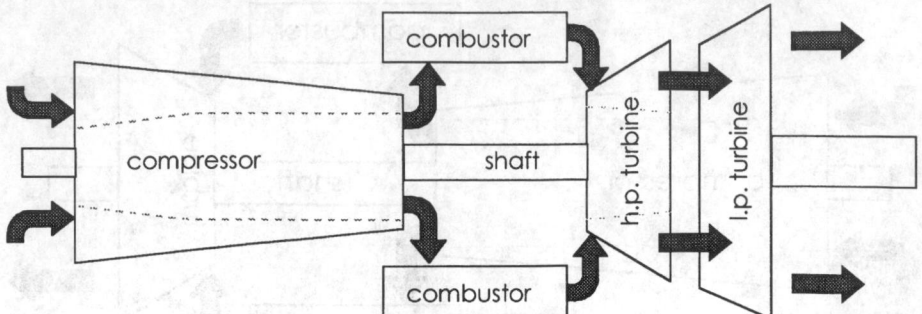

Fig. 57.17 Industrial two-shaft gas turbine schematic showing high-pressure gas generator rotor and separate free-turbine low-pressure rotor.

Figure 57.18 shows a multishaft arrangement common to today's high-pressure turbojet and turbofan engines. The h.p. compressor is connected to the h.p. turbine, and the l.p. compressor to the l.p. turbine, by concentric shafts. There is no mechanical connection between the two rotors (h.p. and l.p.) except via bearings and the associated supporting structure, and the shafts operate at speeds mechanically independent of one another. The need for this apparently complex structure arises from the aerodynamic design constraints encountered in very high-pressure-ratio compressors. By having the higher-pressure stages of a compressor rotating at a higher speed than the early stages, it is possible to avoid the low-annulus-height flow paths that contribute to poor compressor efficiency. The relationship between the speeds of the two shafts is determined by the aerodynamics of the turbines and compressors, the load on the loaded shaft and the fuel flow. The speed of the h.p. rotor is allowed to float, but is generally monitored. Fuel flow and adjustable compressor blade angles are used to control the l.p. rotor speed. Turbojet engines, and at least one industrial aero-derivative engine, have been configured just as shown in Fig. 57.18. Additional industrial aero-derivative engines have gas-generators configured as shown and have power turbines as shown in Fig. 57.17.

The next three configurations reflect deviations from the basic Brayton gas turbine cycle. To describe them, reference must be made back to the temperature–entropy diagram.

Intercooling is the cooling of the working fluid at one or more points during the compression process. Figure 57.19 shows a low-pressure compression, from points a to b. At point b, heat is removed at constant pressure. The result is moving to point c, where the remaining compression takes place (line c–d), after which heat is added by combustion (line d–e). Following combustion, expansion takes place (line e–f). Finally, the cycle is closed by discharge of air to the environment (line f–a), closing the cycle. Intercooling lowers the amount of work required for compression, because work is proportional to the sum of line a–b and line c–d, and this is less than that of line

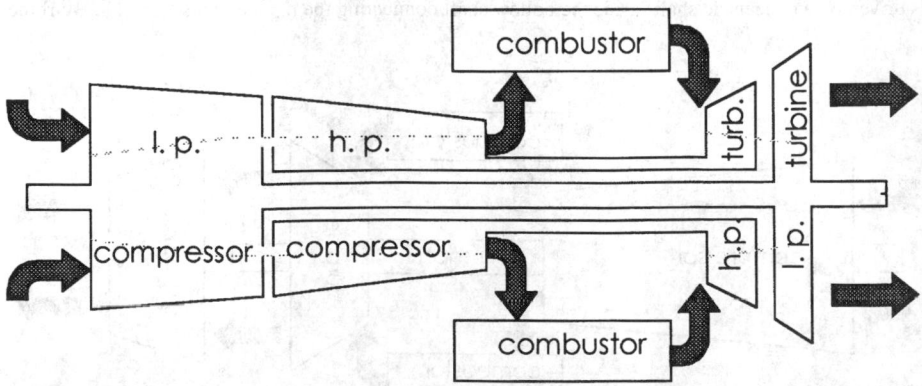

Fig. 57.18 Schematic of multishaft gas turbine arrangement typical of those used in modern high-pressure-ratio aircraft engines. Either a jet nozzle, for jet propulsion, or a free power turbine, for mechanical drive, can be added aft of the l.p. turbine.

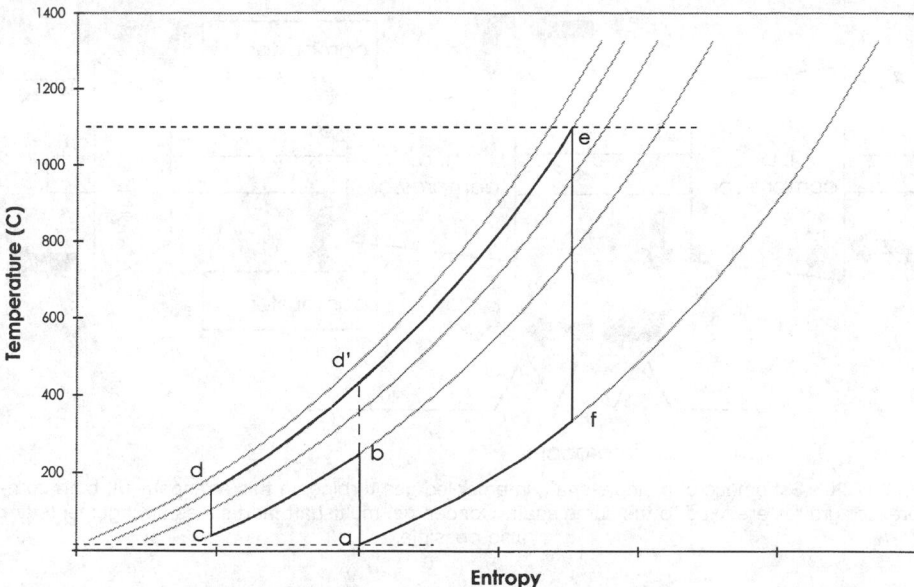

Fig. 57.19 Temperature–entropy diagram for intercooled gas turbine cycle. Firing temperature arbitrarily selected at 1100°C and pressure ratio at 24:1.

a–d', which would be the compression process without the intercooler. Lines of constant pressure are closer together at lower temperatures, due to the same phenomenon that explains higher turbine work than compressor work over the same pressure ratio. Although the compression process is more efficienct with intercooling, more fuel is required by this cycle. Note the line d–e as compared with the line d'–e. It is clear that the added vertical length of line d–e versus d'–e is greater than the reduced vertical distance achieved in the compression cycle. For this reason, when the heat in the partially compressed air is rejected, the efficiency of an intercooled cycle is lower than a simple cycle. Attempts to use the rejected, low-quality heat in a cost-effective manner are usually not successful.

The useful work, which is proportional to e–f less the sum of a–b and c–d, is greater than the useful work of the simple a–d'–e–f–a cycle. Hence for the same turbomachinery, more work is produced by the intercooled cycle—an increase in power density. This benefit is somewhat offset by the fact that relatively large heat-transfer devices are required to accomplish the intercooling. The intercoolers are roughly the size and volume of the turbomachinery and its accessories. Whether the intercooled cycle offers true economic advantage over simple-cycle applications depends on the details of the application, the design features of the equipment, and the existence of a use for the rejected heat.

An intercooled gas turbine is shown schematically in Fig. 57.20. A single-shaft arrangement is shown to demonstrate the principal, but a multishaft configuration could also be used. The compressor is divided at some point where air can be taken offboard, cooled, and brought back to the compressor for the remainder of the compression process. Combustion and turbine configurations are not affected.

The compressor-discharge temperature of the intercooled cycle (point d) is lower than that of the simple cycle (point d'). Often, cooling air, used to cool turbine and combustor components, is taken from, or from near, the compressor discharge. An advantage often cited for intercooled cycles is the lower volume of compressor air that has to be extracted. Critics of intercooling point out that the cooling of the cooling air only, rather than the full flow of the machine, would offer the same benefit with smaller heat exchangers. Only upon assessment of the details of the individual application can the point be settled.

The temperature–entropy diagram for a reheat, or refired, gas turbine is shown in Fig. 57.21. The cycle begins with the compression process shown by line a–b. The first combustion process is shown by line b–c. At point c, a turbine expands the fluid (line c–d) to a temperature associated with an intermediate pressure ratio. At point d, another combustion process takes place, returning the fluid to a high temperature (line d–e). At point e, the second expansion takes place, returning the fluid to ambient pressure (line e–f), whereafter the cycle is closed by discharge of the working fluid back to the atmosphere.

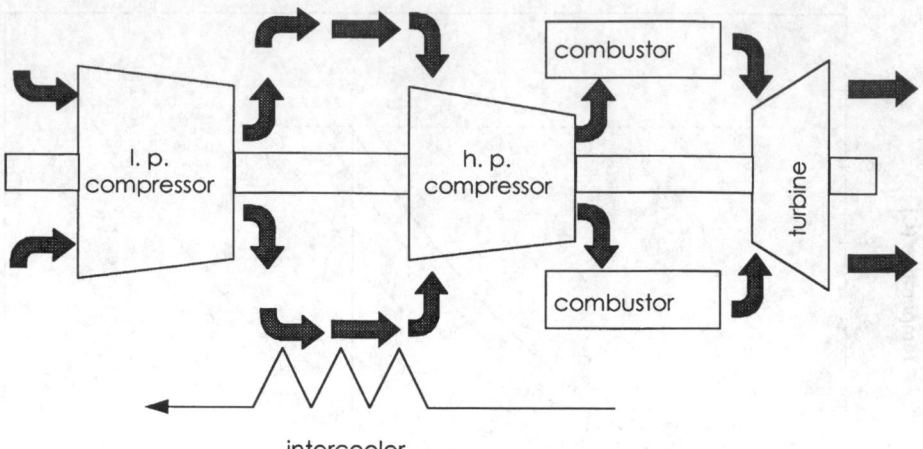

Fig. 57.20 Schematic of a single-shaft, intercooled gas turbine. In this arrangement, both compressor groups are fixed to the same shaft. Concentric, multishaft, and series arrangements are also possible.

An estimate of the cycle efficiency can be made from the temperatures corresponding to the process end points of the cycle in Fig. 57.21. Dividing the turbine temperature drops, less the compressor temperature rise, by the sum of the combustor temperature rises, one calculates an efficiency of approximately 48%. This, of course, reflects perfect compressor, combustor, and turbine efficiency and pure air as the working fluid. Actual efficiencies and properties, and consideration of turbine cooling produce less optimistic values.

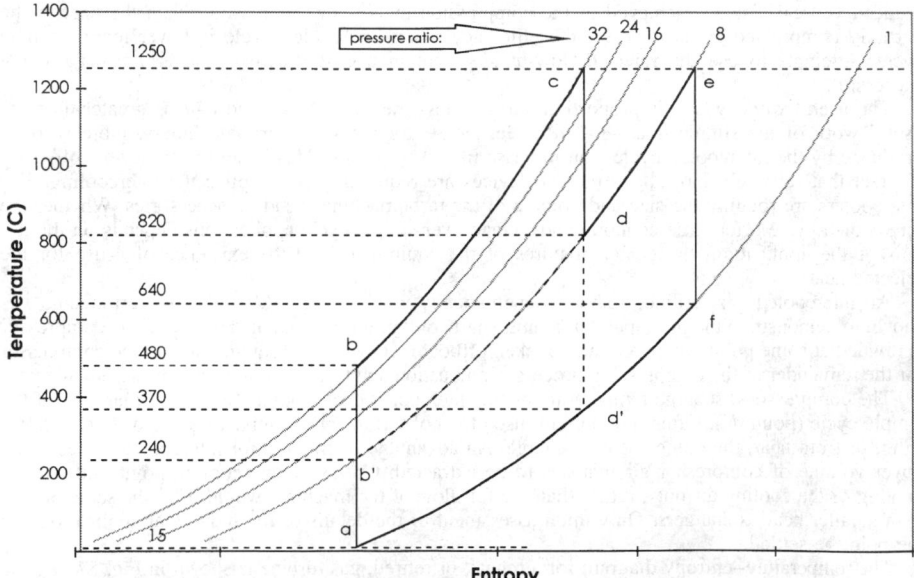

Fig. 57.21 Temperature–entropy diagram for a reheat, or refired, gas turbine. Firing temperatures were arbitrarily chosen to be equal, and to be 1250°C. The intermediate pressure ratio was chosen to be 8:1 and the overall pressure ratio to be 32:1. Dashed lines are used to illustrate comparable simple gas turbine cycles.

$$\text{Eff.} = \frac{(T_c - T_d) + (T_e - T_f) - (T_b - T_a)}{(T_c - T_b) + (T_e - T_d)}$$

A simple cycle with the same firing temperature and exhaust temperature would be described by the cycle $a–b'–e–f–a$. The efficiency calculated for this cycle is approximately 38%, significantly lower than for the reheat cycle. This is really not a fair comparison, since the simple cycle has a pressure of only 8:1, whereas the refired cycle operates at 32:1.

A simple-cycle gas turbine with the same pressure ratio and firing temperature would be described by the cycle $a–b–c–d'–a$. Computing the efficiency, one obtains a value of approximately 54%, more efficient than the comparable reheat cycle. However, there is another factor to be considered. The exhaust temperature of the reheat cycle is 270°C higher than for the simple cycle gas turbine. When applied in combined cycle power plants (these will be discussed later) this difference is sufficient to allow optimized reheat cycle-based plants more efficient than simple-cycle based plants of similar overall pressure ratio and firing temperature. Figure 57.22 shows the arrangement of a single-shaft reheat gas turbine.

Regenerators, or recuperators, are devices used to transfer the heat in a gas turbine exhaust to the working fluid, after it exits the compressor but before it is heated in the combustor. Figure 57.23 shows the schematic arrangement of a gas turbine with regenerator. Such gas turbines have been used extensively for compressor drives on natural gas pipelines and have been tested in road vehicle-propulsion applications. Regeneration offers the benefit of high efficiency from a simple, low-pressure gas turbine without resort to combining the gas turbine with a steam turbine and a boiler to make use of exhaust heat. Regenerative gas turbines with modest firing temperature and pressure ratio have comparable efficiency to advanced, aircraft-derived simple-cycle gas turbines.

The temperature–entropy diagram for an ideal, regenerative gas turbine appears in Fig. 57.24. Without regeneration, the 8:1 pressure ratio, 1000°C firing temperature gas turbine has an efficiency of $((1000-480)-(240-15))/(1000-240) = 38.8\%$ by the method used repeatedly above. Regeneration, if perfectly effective, would raise the compressor discharge temperature to the turbine exhaust temperature, 480°C. This would reduce the heat required from the combustor, reducing the denominator of this last equation from 760°C to 520°C and thereby increasing the efficiency to 56.7%. Such efficiency levels are not realized in practice because of real component efficiencies and heat transfer effectiveness in real regenerators. The relative increase in efficiency between simple and regenerative cycles is as indicated in this example.

Figure 57.24 has shown the benefit of regeneration in low-pressure ratio gas turbines. As the pressure ratio is increased, the exhaust temperature decreases and the compressor discharge temperature increases. The dashed line $a–b'–c'–d'–a$ shows the effect of increasing the pressure to 24:1. Note that the exhaust temperature d' is lower than the compressor discharge temperature b'. Here regeneration is impossible. As the pressure ratio (at constant firing temperature) is increased from 8:1 to nearly 24:1, the benefit of regeneration decreases and eventually vanishes. There is, of course, the possibility of intercooling the high-pressure ratio compressor, reducing its discharge temperature to where regeneration is again possible. Economic analysis and detailed analyses of the thermodynamic cycle with real component efficiencies are required to evaluate the benefits of the added costs of the heat transfer and air handling equipment.

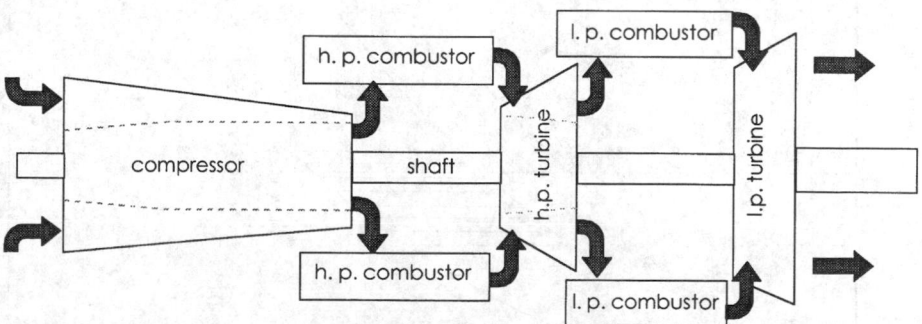

Fig. 57.22 Schematic of a reheat, or refired, gas turbine. This arrangement shows both turbines connected by a shaft. Variations include multiple shaft arrangements and independent components or component groups arranged in series.

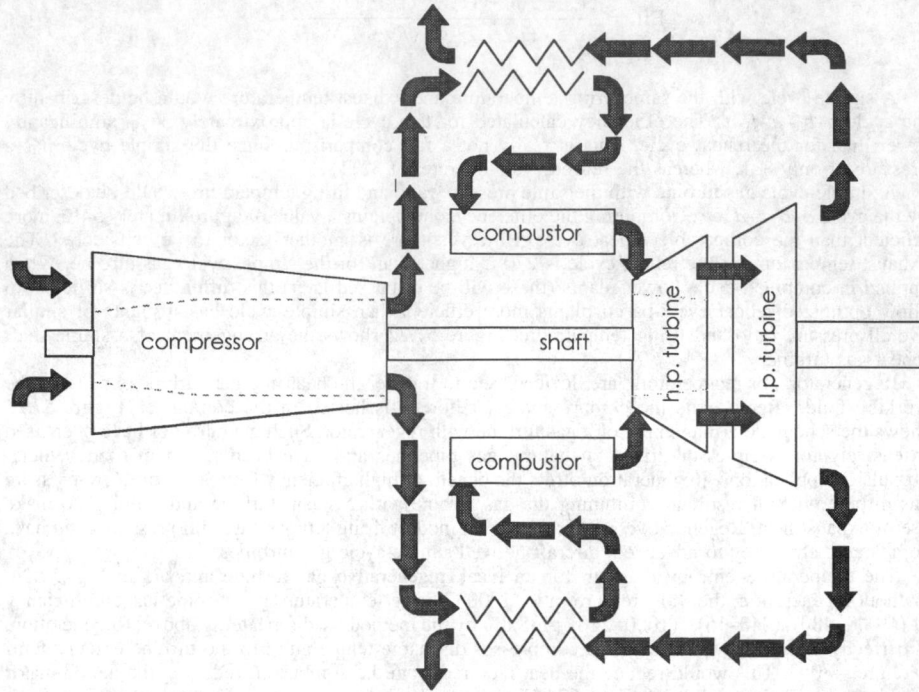

Fig. 57.23 Regenerative, multishaft gas turbine.

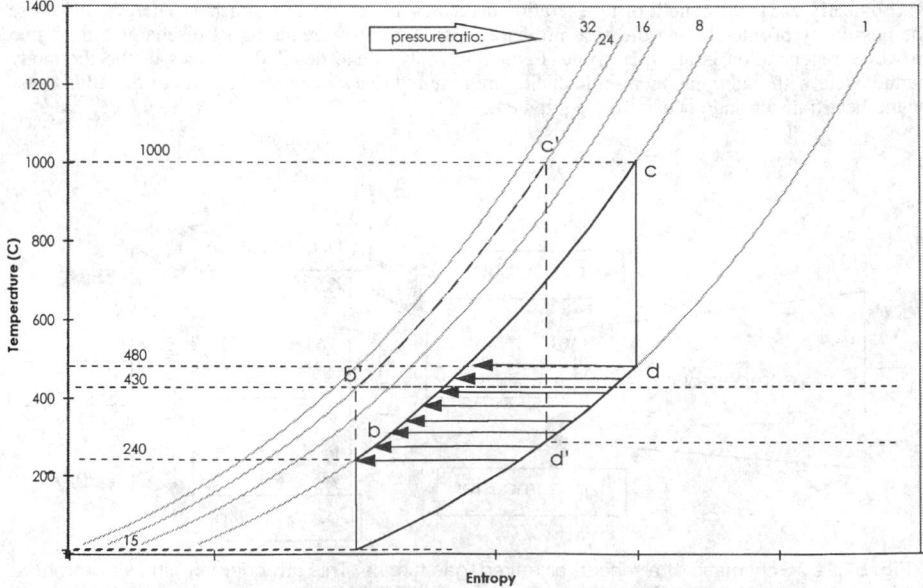

Fig. 57.24 Temperature–entropy diagram comparing an 8:1 pressure ratio, ideal, regenerative cycle with a 24:1 pressure ratio simple cycle, both at a firing temperature of 1000°C.

57.2.2   Trends in Gas Turbine Design and Performance

Output, or Size

The need for power in one location often exceeds the power produced by individual gas turbines. This is true in aircraft applications as well as power generation, and less true in gas pipelines. The specific cost (cost per unit power) of gas turbines decreases as size increases, as can be shown in Fig. 57.25. Note that the cost decreases, but at a decreasing rate; the slope remains negative at the maximum current output for a single gas turbine, around 240 MW. Output increases are accomplished by increased mass flow and increased firing temperature. Mass flow is limited by the inlet annulus area of the compressor. There are three ways of increasing annulus area:

1. Lowering rotor speed while scaling root and tip diameter proportionally. This results in geometric similarity and low risk, but is not possible in the case of synchronous gas turbines, where the shaft of the gas turbine must rotate at either 3600 rpm or 3000 rpm to generate 60 Hz or 50 Hz (respectively) alternating current.

2. Increasing tip diameter. Designers have been moving the tip velocity into the trans-sonic region. Modern airfoil design techniques have made this possible while maintaining good aerodynamic efficiency.

3. Decreasing hub diameter. This involves increasing the solidity near the root, since the cross section of blade roots must be large enough to support the outer portion of the blade against centrifugal force. The increased solidity interferes with aerodynamic efficiency. Also, where a drive shaft is designed into the front of the compressor (cold end drive) and where there is a large bearing at the outboard end of the compressor, there are mechanical limits to reducing the inlet inner diameter.

Firing Temperature

Firing temperature increases provide higher output per unit mass flow and higher combined cycle efficiency. Efficiency is improved by increased firing temperature wherever exhaust heat is put to

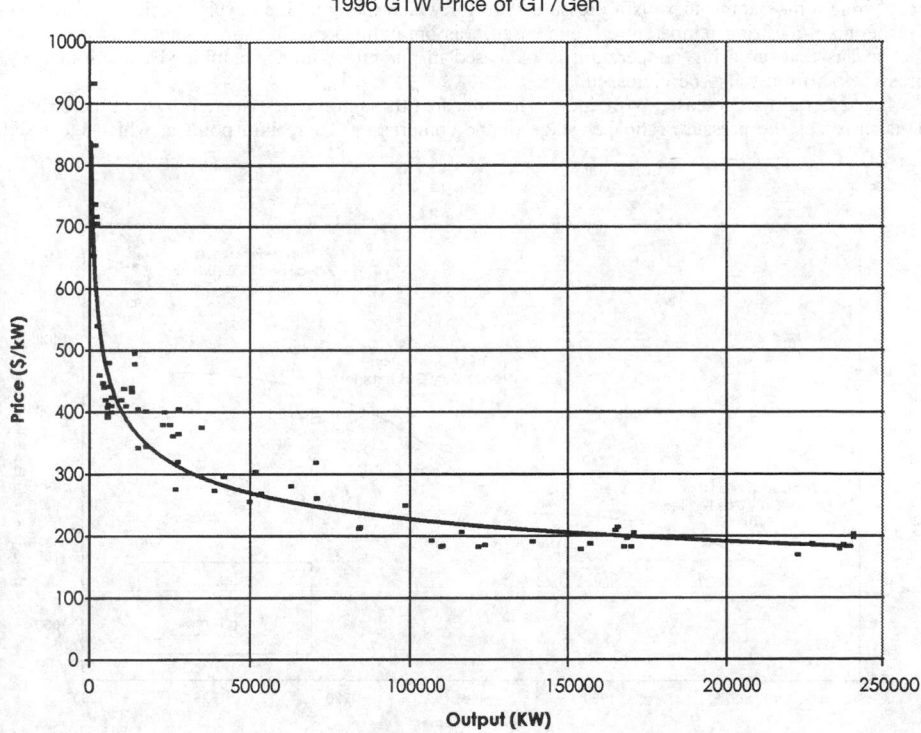

Fig. 57.25 Cost of simple cycle, generator-drive electric power generation equipment (plotted from data published by Gas Turbine World Magazine[15]).

use. Such uses include regeneration/recuperation, district heating, supplying heat to chemical and industrial processes, Rankine bottoming cycles, and adding a power turbine to drive a fan in an aircraft engine. The effect of firing temperature upon the evolution of combined Brayton–Rankine cycles for power generation is illustrated in Fig. 57.26.

Firing temperature increases when the fuel flow to the engine's combustion system is increased. The challenge faced by designers is to increase firing temperature without decreasing the reliability of the engine. A metal temperature increase of 15°C will reduce bucket creep life by 50%. Material advances and increasingly more aggressive cooling techniques must be employed to allow even small increases in firing temperature. These technologies have been discussed previously.

Maintenance practices represent a third means of keeping reliability high while increasing temperature. Sophisticated life-prediction methods and experience on identical or similar turbines are used to set inspection, repair, and replacement intervals. Coupled with design features that reduce the time required to perform maintenance, both planned and unplanned down time can be reduced to offset shorter parts lives, with no impact on reliability.

Increased firing temperature usually increases the cost of the buckets and nozzles (through exotic materials or complicated cooling configurations). Although these parts are expensive, they represent a small fraction of the cost of an entire power plant. The increased output permitted by the use of advanced buckets and nozzles is generally much higher, proportionally, than the increase in power-plant cost; hence, increased firing temperature tends to lower specific powerplant cost.

Pressure Ratio

Two factors drive the choice of pressure ratio. First is the primary dependence of simple-cycle efficiency on pressure ratio. Gas turbines intended for simple-cycle application, such as those used in aircraft propulsion, emergency power, and power where space or weight is a primary consideration, benefit from higher pressure ratios.

Combined-cycle power plants do not necessarily benefit from high pressure ratios. At a given firing temperature, an increase in pressure ratio lowers the exhaust temperature. Lower exhaust temperature means less power from the bottoming cycle and a lower efficiency bottoming cycle. So, as pressure ratio is increased, the gas turbine becomes more efficient and the bottoming cycle becomes less efficient. There is an optimum pressure ratio for each firing temperature, all other design rules held constant. Figure 57.27 shows how specific output and combined cycle efficiency are affected by gas turbine firing temperature and pressure ratio for a given type of gas turbine and steam cycle. At each firing temperature, there is a pressure ratio for which the combined cycle efficiency is highest. Furthermore, as firing temperature is increased, this optimum pressure ratio is higher as well. This fact means that, as firing temperature is increased in pursuit of higher combined cycle efficiency, pressure ratio must also be increased.

Pressure ratio is increased by reducing the flow area through the first-stage nozzle of the turbine. This increases the pressure ratio per stage of the compressor. There is a point at which increased

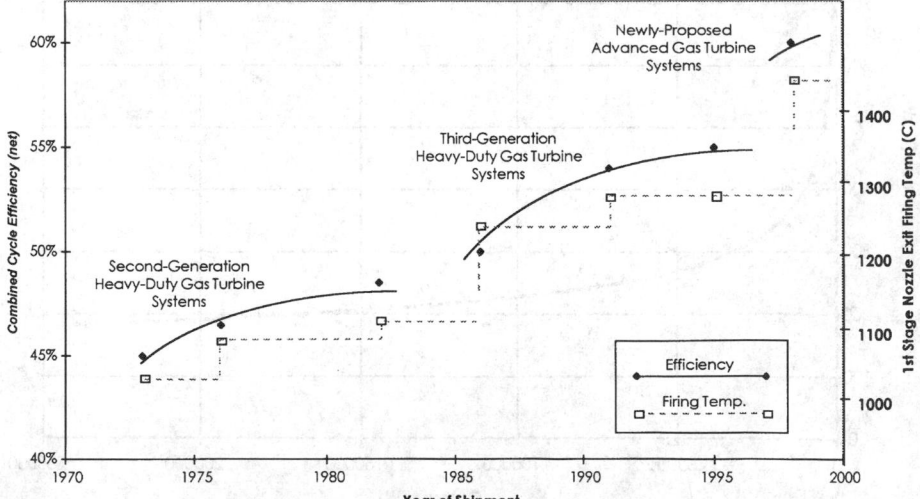

Fig. 57.26 History of power-generation, combined-cycle efficiency and firing temperature, illustrating the trend to higher firing temperature and its effect on efficiency.

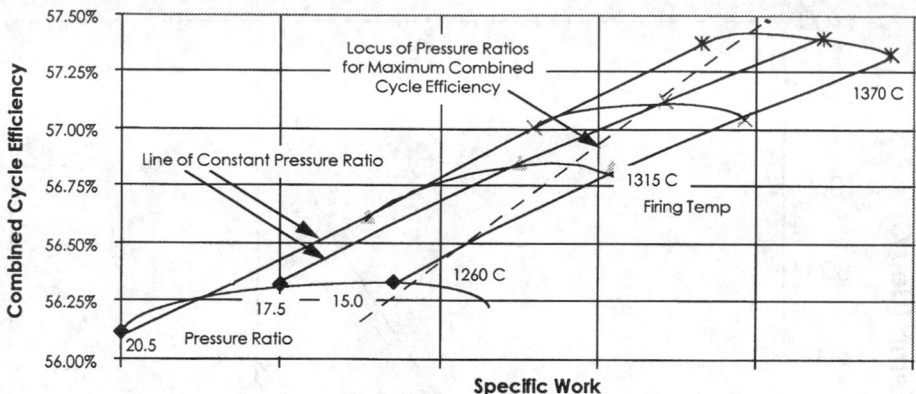

Fig. 57.27 Effect of pressure ratio and firing temperature on combined cycle efficiency and specific work.

pressure ratio causes the compressor airfoils to stall. Stall is avoided by either adding stages (reducing the pressure ratio per stage) or increasing the chord length, and applying advanced aerodynamic design techniques. For significant increases in pressure ratio, a simple, single-shaft rotor with fixed stationary airfoils cannot deliver the necessary combination of pressure ratio, stall margin, and operating flexibility. Features required to meet all design objectives simultaneously include variable-angle stationary blades in one or more stages; extraction features that can be used to bleed air from the compressor during low-speed operation; and multiple rotors that can be operated at different speeds.

Larger size, higher firing temperature, and higher pressure ratio are pursued by manufacturers to lower cost and increase efficiency. Materials and design features evolve to accomplish these advances with only positive impact on reliability.

57.3 APPLICATIONS

57.3.1 Use of Exhaust Heat in Industrial Gas Turbines

Adding equipment for converting exhaust energy to useful work can increase the thermal efficiency of a gas turbine-based power plant by 10 to over 30%. The schemes are numerous, but the most significant is the fitting of a heat-recovery steam generator (HRSG) to the exhaust of the gas turbine and delivering the steam produced to a steam turbine. Both the steam turbine and gas turbine drive electrical generators.

Figure 57.28 displays the combining of the Brayton and Rankine cycles. The Brayton cycle $a–b–c–d–a$ has been described already. It is important to point out that the line $d–a$ now represents heat transferred in the HRSG. In actual plants, the turbine work is reduced slightly by the backpressure associated with the HRSG. Point d would be above the 1:1 pressure curve, and the temperature drop proportionately reduced.

The Rankine cycle begins with the pumping of water into the HRSG, line $m–n$. This process is analogous to the compression in the gas turbine, but rather than absorbing 50% of the turbine work, consumes only about 5%, since the work required to pump a liquid is less than that required to compress a gas. The water is heated (line $n–o$) and evaporated ($o–p$). The energy for this is supplied in the HRSG by the exhaust gas of the gas turbine. More energy is extracted to superheat the steam, as indicated by line $p–r$. At this point, superheated steam is delivered to a steam turbine and expanded ($r–s$) to convert the energy therein to mechanical work.

The addition of the HRSG reduces the output of the gas turbine only slightly. The power required by the mechanical devices (like the feedwater pump) in the steam plant is also small. Therefore, most of the steam turbine work can be added to the net gas turbine work with almost no increase in fuel flow. For combined-cycle plants based on industrial gas turbines where exhaust temperature is in the 600°C class, the output of the steam turbine is about half that of the gas turbine. Their combined-cycle efficiency is approximately 50% higher than simple-cycle efficiency. For high-pressure ratio gas turbines with exhaust temperature near 450°C, the associated steam turbine output is close to 25% of the gas turbine output, and efficiency is increased by approximately 25%. The thermodynamic cycles of the more recent large industrial gas turbines have been optimized for high combined-cycle efficiency. They have moderate to high simple-cycle efficiency and relatively high exhaust temperatures. Figure 57.28 has shown that net combined-cycle efficiency (lower heating value) of

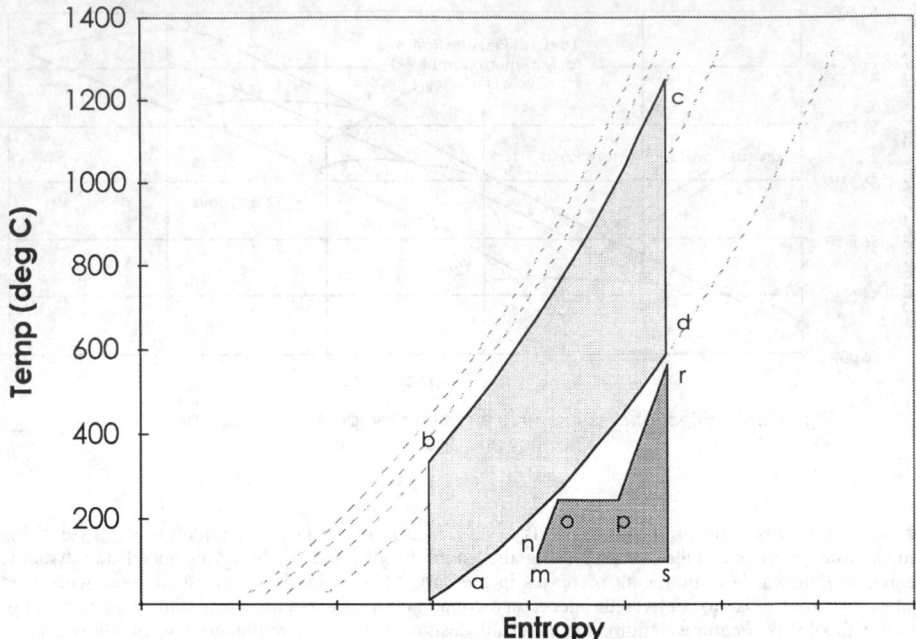

Fig. 57.28 Temperature–entropy diagram illustrating the combining of a gas turbine (a–b–c–d–a) and steam turbine cycle (m–n–o–p–r–s–m). The heat wasted in process d–a in simple-cycle turbines supplies the heat required by processes n–o, o–p, and p–r.

approximately 55% has been realized as of this writing, and levels of 60% and beyond are under development.

Figure 57.29 shows a simple combined-cycle arrangement, where the HRSG delivers steam at one pressure level. All the steam is supplied to a steam turbine. Here, there is neither steam reheat nor additional heat supplied to the HRSG. There are many alternatives.

Fired HRSGs, where heat is supplied both by the gas turbine and by a burner in the exhaust duct, have been constructed. This practice lowers overall efficiency, but accommodates the economics of some situations of variable load requirements and fuel availability. In other applications, steam from the HRSG is supplied to nearby industries or used for district heating, lowering the power generation efficiency but contributing to the overall economics in specific applications.

Efficiency of electric power generation benefits from more complicated steam cycles. Multiple pressure, non-reheat cycles improve efficiency as a result of additional heat transfer surface in the HRSG. Multiple pressure, reheat cycles, such as shown in Fig. 57.30 match the performance of higher exhaust temperature gas turbines (600°C). Such systems are the most efficient currently available, but are also the most costly. The relative performance for several combined-cycle arrangements is shown in Table 57.1.[16] The comparison was made for plants using a gas turbine in the 1250°C firing temperature class.

Plant costs for simple-cycle gas turbine generators is lower than that for steam turbines and most other types of powerplant. Since combined-cycle plants generate 2/3 of their power with the gas turbine, their cost is between that of simple-cycle gas turbine plants and steam turbine plants. Their efficiency is higher than either. The high efficiency and low cost combine to make combined-cycle plants extremely popular. Very large commitments to this type of plant have been made around the world. Table 57.2 shows some of the more recent to be put into service.

There are other uses for gas turbine exhaust energy. Regeneration, or recuperation, uses the exhaust heat to raise the temperature of the compressor discharge air before the combustion process. Various steam-injection arrangements have been used as well. Here, an HRSG is used as in the combined-cycle arrangements shown in Fig. 57.30, but instead of expanding the steam in a steam turbine, it is introduced into the gas turbine, as illustrated in Fig. 57.31. It may be injected into the combustor, where it lowers the generation of NO_x by cooling the combustion flame. This steam increases the mass flow of the turbine and its heat is converted to useful work as it expands through

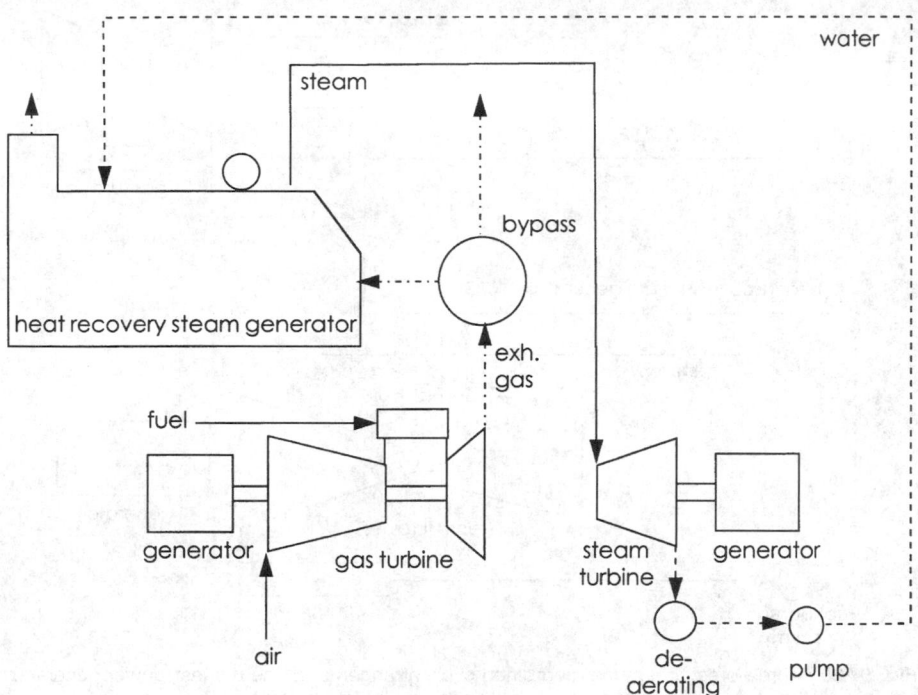

Fig. 57.29 Schematic of simple combined cycle power plant. A single-pressure, nonreheat cycle is shown.

the turbine section of the gas turbine. Steam can also be injected downstream of the combustor at various locations in the turbine, where it adds to the mass flow and heat of the working fluid. Many gas turbines can tolerate steam-injection levels of 5% of the mass flow of the air entering the compressor; others can accommodate 16% or more, if distributed appropriately along the gas path of the gas turbine. Gas turbines specifically designed for massive steam injection have been proposed and studied. These proposals arise from the fact that the injection of steam into gas turbines of existing designs has significant reliability implications. There is a limit to the level of steam injection into combustors without flame-stability problems and loss of flame. Adding steam to the gas flowing through the first-stage nozzle increases the pressure ratio of the machine and reduces the stall margin of the compressor. Addition of steam to the working fluid expanding in the turbine increases the heat-transfer coefficient on the outside surfaces of the blading, raising the temperature of these components. The higher work increases the aerodynamic loading on the blading, which may be an issue on latter-stage nozzles, and increases the torque applied to the shafts and rotor flanges. Design changes can be made to address the effects of steam in the gas path.[17]

Benefits of steam injection are an increase in both efficiency and output over those of the simple-cycle gas turbine. The improvements are less than those of the steam turbine and gas turbine combined-cycles, since the pressure ratio of the steam expansion is much higher in a steam turbine. Steam turbine pressures may be 100 atmospheres; gas turbines no higher than 40. Steam-injection cycles are less costly to produce since there is no steam turbine. There is, of course, higher water consumption with steam injection, since the expanded steam exits the plant in the gas turbine exhaust.

57.3.2 Integrated Gasification Combined Cycle

In many parts of the world, coal is the most abundant and lowest-cost fuel. Coal-fired boilers raising steam that is expanded in steam turbine generators are the conventional means of converting this fuel to electricity. Pulverized coal plants with flue gas desulfurization operate at over 40% overall efficiency and have demonstrated the ability to control sulfur emissions from conventional boiler systems. Gas turbine combined-cycle plants are operating with minimal environmental impact on natural gas at 55% efficiency, and 60% is expected with new technologies. A similar combined-cycle plant that could operate on solid fuel would be an attractive option.

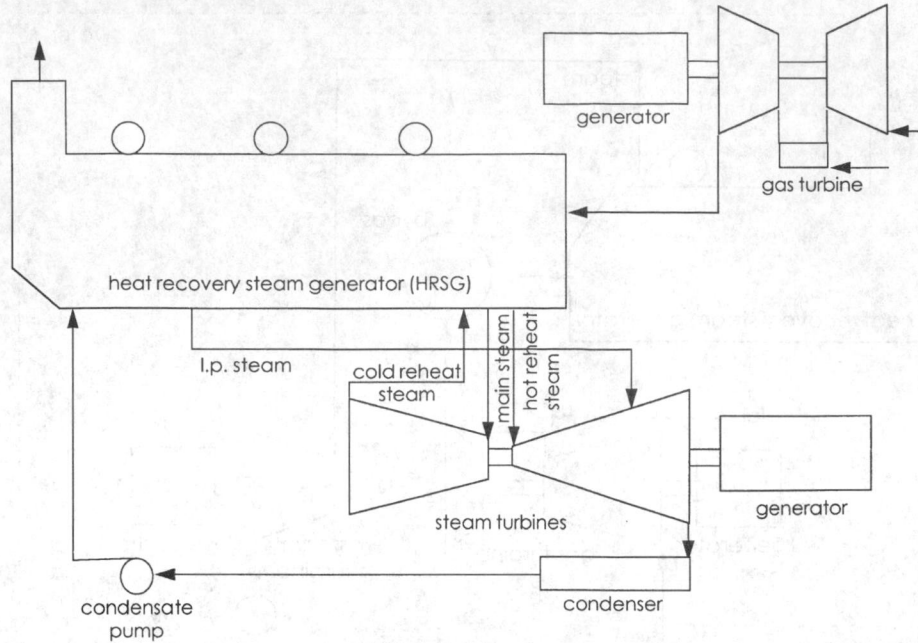

Fig. 57.30 Three-pressure, reheat combined cycle arrangement. The highest power generation efficiency is currently achieved by such plants.

Competing means of utilizing coal with gas turbines have included direct combustion, indirect firing, and gasification. Direct combustion in conventional, on-engine combustors has resulted in rapid, ash-caused erosion of bucket airfoils. Off-base combustion schemes, such as pressurized fluidized bed combustors, have not simultaneously demonstrated the high exit temperature needed for efficiency and low emissions. Indirect firing raises compressor discharge temperature by passing it through a heat exchanger. Metal heat exchangers are not compatible with the high turbine inlet temperature required for competitive efficiency. Ceramic heat exchangers have promise, but their use will necessitate the same types of emission controls required on conventional coal-fired plants. Power plants with the gasification process, desulfurization, and the combined-cycle machinery integrated have been successfully demonstrated, with high efficiency, low emissions, and competitive first cost. Significant numbers of integrated gasification combined-cycle (IGCC) plants are operating or under construction.

Fuels suitable for gasification include several types of coal and other low-cost fuels. Those studied include:

- Bituminous coal
- Sub-bituminous coal
- Lignite
- Petroleum coke
- Heavy oil
- Orimulsion
- Biomass

Fuel feed systems of several kinds have been used to supply fuel into the gasifier at the required pressure. Fuel type, moisture content size, and the particular gasification process need to be considered in selecting a feed system.

Several types of gasifiers have been designed to produce fuel with either air or oxygen provided. The system shown in Fig. 57.32 features a generic oxygen-blown gasifier and a system for extracting some of the air from the compressor discharge and dividing it into oxygen and nitrogen. An oxygen-blown gasifier produces a fuel about 1/3 of the heating value of natural gas. The fuel produced by the gasifier, after sulfur removal, is about 40% CO, 30% H$_2$. The remaining 30% is mostly H$_2$O and

Table 57.1 Comparison of Performance for Combined Cycle Arrangements Based on Third Generation (1250°C Firing Temperature) Industrial Gas Turbines

Steam Cycle	Relative Net Plant Output (%)	Relative Net Plant Efficiency (%)
Three pressure, reheat	Base	Base
Two pressure, reheat	−1.1	−1.1
Three pressure, non-reheat	−1.2	−1.2
Two pressure, non-reheat	−2.0	−2.0

CO_2, which are inert and act as diluents in the gas turbine combustor, reducing NO_x formation. A typical lower heating value is 1950 K-Cal/m^3. The fuel exits the gasifier at a temperature higher than that at which it can be cleaned. The gas is cooled by either quench or heat-exchange and cleaned. Cleaning is done by water spray scrubber or dry filtration to remove solids that are harmful to the turbine and potentially harmful to the environment. This is followed by a solvent process that absorbs H_2S.

Some gas turbine models can operate on coal-gas without modification. The implications for the gas turbine relate to the volume of fuel—three times higher, or more, than that of natural gas. When the volume flow through the first stage nozzle of the gas turbine increases, the backpressure on the compressor increases. This increases the pressure ratio of the cycle and decreases the stall margin of the compressor. Gas turbines with robust stall margins need no modification. Others can be adapted by reducing inlet flow by inlet heating or by closing off a portion of the inlet (variable inlet stator vanes can be rotated toward a closed position). The volume flow through the turbine increases as well. This increases the heat transfer to the buckets and nozzles. To preserve parts lives, depending on the robustness of the original design, the firing temperature may have to be reduced. The increased flow and decreased firing temperature, if required, result in higher gas turbine output than developed by the same gas turbine fired on natural gas.

57.3.3 Applications in Electricity Generation

Worldwide shipments of industrial gas turbines before 1965 were below 2 gW total output capacity per year. In 1992, more than 25 gW of capacity was shipped, and deliveries continue at this rate. Of the 1992 volume, nearly 90% was for electric power generation. Approximately 9% of all the world's current generating capacity is by gas turbines, either in simple-cycle or combined-cycle. Current electric power generation additions are increasingly provided by gas turbines. Over 10% of additions are by simple-cycle gas turbines, and over 25% by combined-cycle plants, which derive 2/3 of their capacity from gas turbines. Thus, between 25 and 30% of additions are gas turbine generators. This compares to 40 to 50% by steam turbine generators alone and in combined-cycle. The remaining additions are provided by hydroelectric plants, nuclear, and other means.

The tenfold increase in the volume of gas turbines shipments between 1965 and the present was due to several factors. First, in the late 1960s and early 1970s, there was a need for peaking power in the United States. Gas turbines, because of their low cost, low operating crew size, and fast installation time, were the engine of choice. Because of the seasonal and daily variations in the demand for electric power, generating companies could minimize their investment in plant and equipment by installing a mixture of expensive but efficient base load plants (steam and nuclear), run over 8000 hours per year, and far less expensive—but less efficient—plants that would operate only a few hundred hours per year.

Existing industrial gas turbines and newly designed larger units whose operating speed was chosen to match the requirements of a directly coupled alternator met the demand for peaking power. The experience on these early units resulted in improvements in efficiency, reliability, and cost-

Table 57.2 Recent Large Multiunit Gas Turbine-Based Combined-Cycle Power Plants

Plant	Country	Number of Gas Turbines	Plant Output (mW)
TEPCO—Yokohama	Japan	8	2800
TEPCO—Futtsu	Japan	14	2000
KEPCO—Ildo 1 & 2	Korea	8	1910
PLN—Gresik 1–3	Indonesia	9	1796
Chubu—Kawagoe 1–7	Japan	7	1695
Enron—Wilton on Teeside 1 & 2	U.K.	8	1644
Midland Cogen—Michigan	U.S.A.	12	1470

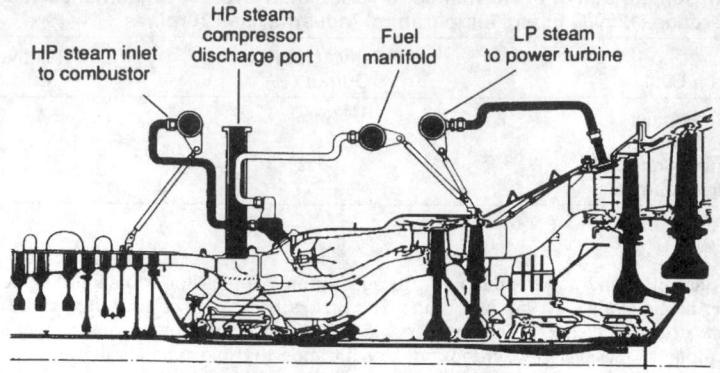

Fig. 57.31 LM500 aero-derivative gas turbine with steam injection (courtesy of General Electric Company).

effectiveness. Much of the technology needed to improve the value of industrial gas turbines came from aircraft engine developments, as it still does. Beginning in the 1970s, with the rapid rise in oil prices and associated natural gas prices, electric utilities focused on ways of improving the efficiency of generating plants. Combined-cycle plants are the most thermally efficient fuel-burning plants. Furthermore, their first-cost is lower than all other types of plants except simple-cycle gas turbine plants.

The only drawback to gas turbine plants was their requirement for more noble fuels; natural gas and light distillates are usually chosen to minimize maintenance requirements. Coal is abundant in many parts of the world and costs significantly less than oil or gas per unit energy. Experiments in the direct firing of gas turbines on coal have been conducted without favorable results. Other schemes

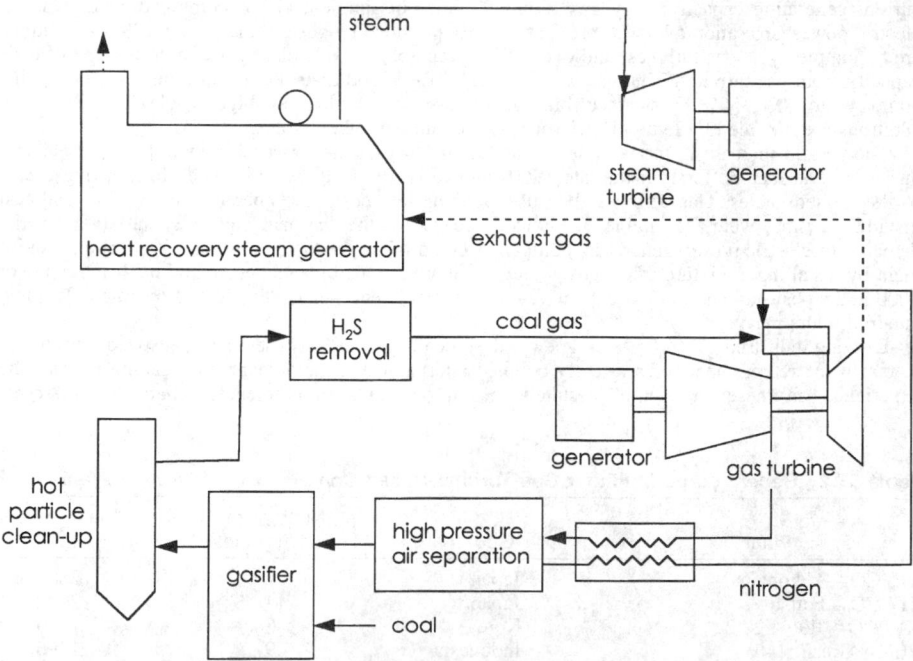

Fig. 57.32 Integrated gasification combined-cycle diagram. Air compressed in the gas turbine is cooled, oxygen separated and fed to the gasifier, nitrogen to gas turbine for NO_x control. Coal is partially burned in the gasifier. The gas produced is cleaned and flows to the gas turbine as fuel.

for using coal in gas turbines include indirect firing, integrating with a fluidized bed combustor, and integrated gasification. The last of these offers the highest efficiency due to its ability to deliver the highest-temperature gas to the turbine blading. Furthermore, integrated gasification is the most environmentally benign means of converting coal to electricity. The technology has been demonstrated in several plants, including early technology demonstration plants and commercial power-generating facilities.

57.3.4 Engines for Aircraft

Aircraft engines exert a forward thrust on the airframe in reaction to the rearward acceleration imparted to part of the passing airflow. This is achieved by means of a propeller or fan or by generating a high-pressure jet that emerges through a nozzle. Propellers can be driven by gas turbines (these are turboprop engines) as well as reciprocating engines. The term *fan* is used to describe a low-pressure compressor of one or more stages; part of its discharge flow bypasses the core compressor, combustor, and turbine of the engine. The acceleration of the air bypassing the core contributes to the engine's thrust. Variations include the duct-burner configurations. The choice of engine type is determined by the required flight speed, since propulsion efficiency of each engine type has different efficiency characteristics.

The appearance of aircraft engines and stationary gas turbines differ considerably. The premium placed on weight for all aircraft equipment results in different economic tradeoffs. This is particularly evident in the casings. Industrial engines, for the most part, are designed with thick cast casings of inexpensive material that are bolted together both horizontally and axially. The horizontal split line allows convenient field disassembly—a necessity in large equipment. Industrial engine casings are left as-cast on the outer surface, while aircraft engine casings are machined to close tolerances and have features that may be costly to machine, but are compensated for by the value placed on low weight. Aircraft engine casings are made of higher-strength material and can be thinner. Disks, especially in the compressor, are thinner than in industrial gas turbines. This is made possible by the use of higher strength materials, light alloy blades, and wheel-to-wheel attachment methods that may not be economical in larger sizes. Hot section (turbine and combustor) parts, however, are quite similar in appearance and material.

Aircraft engine performance is described in terms of thrust, fuel consumption rate, and engine weight.

- Net thrust:

$$F_n = W1(V_j - V_o) + Ae(P_j - P_o) \qquad \text{for turbojet}$$

$$F_n = W_c(V_j - V_o) + W_d(V_d - V_o) + Ae(P_j - P_o) \qquad \text{for turbofan}$$

- Specific thrust:

$$ST = F_n / W1$$

- Specific fuel consumption:

$$SFC = W_f / F_n \qquad \text{(general definition)}$$

$$SFC_{tp} = W_f / HP / hr \qquad \text{(sometimes applied)}$$

to turboprop and shaft engines

where $W1$ = total inlet mass flow rate
  W_c = portion of $W1$ passing through the core engine
  W_d = portion of $W1$ passing through bypass duct
  V_o = flight velocity
  V_j = exhaust jet velocity
  Ae = exhaust area
  P_j = exhaust jet static pressure
  P_o = ambient pressure
  W_f = fuel flow rate

Thrust multiplied by flight velocity is a measure of the work done by the propulsion system. Propulsion system efficiency is the ratio of this work to the fuel supplied per unit time. Hence,

$$\eta_{ps} = \text{const} \times (F_n V_o) / W_f$$

or

$$\eta_{ps} = \text{const} \times V_o / \text{SFC}$$

Efficiency is inversely proportional to specific fuel consumption and is a function of both SFC and aircraft velocity.[18,19]

Thrust characteristics and SFC characteristics vary as a function of aircraft velocity, and vary differently for each type of engine. This is because efficiency improves as the aircraft velocity approaches the engine exit velocity. The engine exhaust velocity must be above the aircraft velocity to generate thrust, but large differences in velocity relate to inefficient propulsion. Figures 57.33 and 57.34 show the specific thrust and specific fuel consumption for various engines as functions of aircraft velocity.

The figures show that at lower Mach numbers, the turbofan engines have relatively high propulsion efficiency (low SFC). By employing a large-diameter fan, a very large mass of air can be accelerated to relatively low discharge velocities, avoiding high mismatch between exhaust and aircraft velocity. The need for improved efficiency in the high-subsonic speed regime has produced a focus on turbofan engines rather than turbojets. At lower speeds, turboprop engines are preferred.

Figures 57.35a and b compare the engines selected for, or competing for, recent applications. All of the larger commercial transport applications and newer military applications are met by turbofan engines.[20]

The range of ratings for each engine designation is due to the practice of fine-tuning engine performance for particular applications, incremental performance gains over time, and optional features. This comparison is a snapshot of performance over a particular time. Relative ratings change often as manufacturers continue to apply new technologies and improve designs. One of the newest and most powerful turbofan engines is shown in Fig. 57.36. It is a two-rotor engine. The one-stage fan and three-stage low-pressure compressor are joined on one shaft connected to a six-stage low-pressure turbine. The 10-stage high-pressure compressor is driven by the two-stage high-pressure turbine, both joined on another shaft that can rotate at a higher speed. The ratio of the air mass flow through the duct to the air flowing through the compressor, combustor, and turbines is 9:1. The overall pressure ratio is 40:1 and the rated thrust is in the 85,000 pound class. The engine is over 3 m in diameter. A new feature for aircraft engines is the double-domed, lean-premixed, fuel-staged dry low-NO$_x$ combustor. The GE 90, PW4084, and 800-series RB.211 Trent high-bypass-ratio turbofan engines have been built for use on the Boeing 777 aircraft.

Supersonic flight requires a considerably increased jet velocity. The afterburner reheats the exhaust gas after the turbines, permitting it to accelerate to an appropriate level above the flight velocity and boosting the thrust to overcome the increased drag.

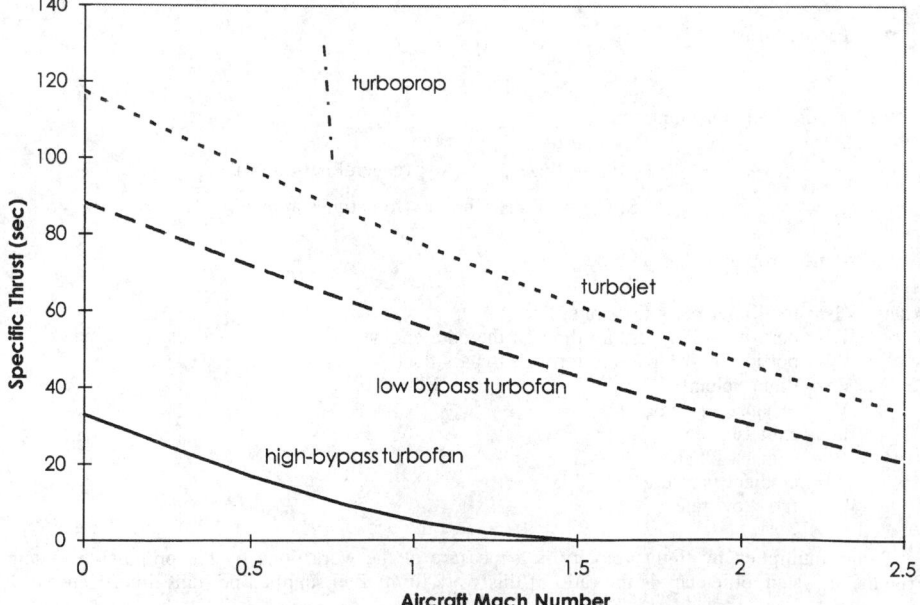

Fig. 57.33 Low values of specific thrust give higher propulsive efficiency at low Mach numbers (redrawn from a figure in Ref. 18).

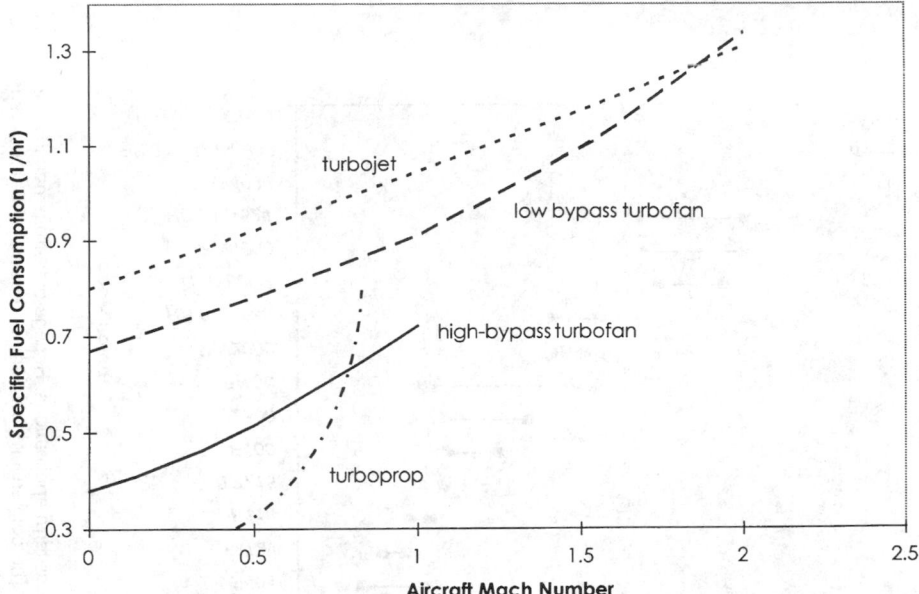

Fig. 57.34 As flight Mach number increases, higher specific thrusts are necessary to maintain high propulsive efficiency and reduce SFC (redrawn from a figure in Ref. 18).

Another aircraft engine type is the auxiliary power unit (APU). It is a small turboshaft engine that provides air-conditioning and electric and hydraulic power while the main engines are stopped, but its main function is to start the main engines without external assistance. The APU is usually started from batteries by an electric motor. When it reaches its operating speed, a substantial flow of air is bled from its compressor outlet and is ducted to drive the air turbine starters on the main engines. The fuel flow is increased when the turbine air supply is reduced by the air bleed, to provide the energy required for compression. These engines are also found on ground carts, which may be temporarily connected to an aircraft to service it. They may also have uses in industrial plants requiring air pressure at 3 or 4 bar.

57.3.5 Engines for Surface Transportation

This category includes engines for rail, road, offroad, and overwater transport. The low weight and high power density of gas turbines are assets in all cases, but direct substitution for Diesel or Otto cycle engines is unusual. When the economics of an application favor high power density, high driven-device speed, or when some heat recovery is possible, gas turbines become the engines of choice. Surface vehicle engines include the array of turboshaft and turboprop derivatives, free-turbine aero-derivative and industrial gas turbines, and purpose-built gas turbines. Applications exist for engines of from around 100 horsepower to nearly 40,000.

Truck, bus, and automobile gas turbine engines are, for the most part, in the development stage. Current U.S. Department of Energy initiatives are supporting development of gas turbine automobile engines of superior efficiency and low emissions. Production cost similar to current power plants is also a program goal. Additional requirements must be met, such as fast throttle response and low fuel consumption at idle. The balancing of efficiency, first-cost, size, and weight have led to different cycle and configuration choices than for aircraft or power-generation applications. Regenerative cycles with low pressure ratios have been selected. Parts count and component costs are addressed through the use of centrifugal compressors, integral blade-disk axial turbines, and radial inflow turbines. Low-pressure-ratio designs support the low stage count. It is possible to achieve the necessary pressure ratio with one centrifugal compression stage and in one turbine stage, or one each high pressure and power turbine. The small size of parts and the selection of radial inflow or integral blade-disk turbines make ceramic materials an option. Single-can combustors are also employed to control cost. Proto-types have been built and operated in the United States, Europe, and Japan.[21]

The most successful automotive application of gas turbines is the power plant for the M1 Abrams Main Battle Tank. The engine uses a two-spool, multistage, all-axial flow gas generator plus power turbine. The cycle is regenerative. Output and cost appear too high for highway vehicle application.

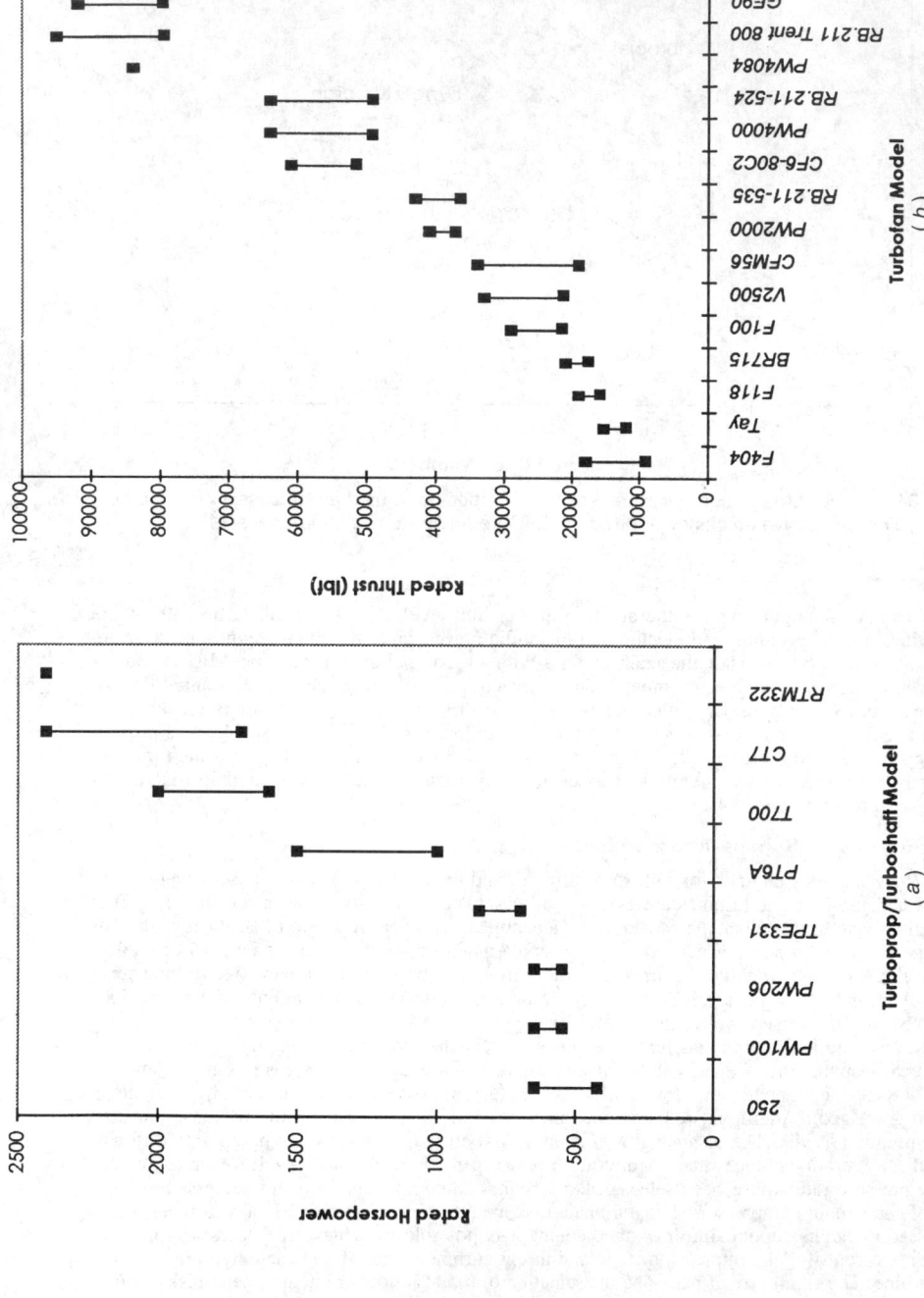

Fig. 57.35 Output performance ratings of some aircraft propulsion gas turbines in recent and near-term applications. Ratings data from industry publications. Note that the shaft engine ratings are in horsepower and the turbofans are rated by pounds of thrust. The conversion between horsepower and thrust varies, since it depends on propeller or rotor efficiency.

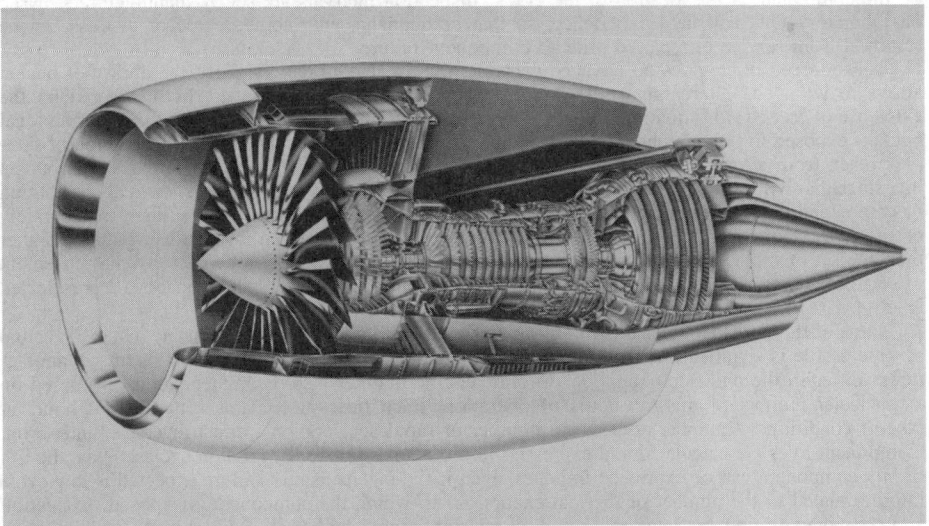

Fig. 57.36 Sectional view of the GE 90, a high-bypass-ratio two-shaft turbofan engine rated at over 80,000 lbf thrust (courtesy of General Electric Company).

Ship propulsion by gas turbine is more commonplace. One recent report summarizing orders and installations over an 18-month period listed 10 orders for a total of 64 gas turbines—775 MW in all. Military applications accounted for 55% of the total. The remaining 45% were applied to fast ferries and similar craft being built in Europe, Australia, and Hong Kong.[15] Gas turbine outputs in the 3–5 MW range, around 10 MW, and in the 20–30 MW range account for all the applications. Small industrial engines were selected in the 3–5 MW range and aero-derivative, free-turbine engines accounted for the remainder.

Successful application of gas turbines aboard ship requires protection from the effects of salt water and, in the case of military vessels, maneuver and sudden seismic loads. In addition to the common problems with salt water-induced corrosion of unprotected metal, airborne sodium, in combination with sulfur usually found in the fuel or air, presents a problem for buckets, nozzles, and combustors. Hot corrosion—also called *sulfidation*—has led to the development of alloys that combine the creep strength of typical aircraft engine bucket and nozzle alloys with superior corrosion resistance. Inconel 738 was the first such alloy. This set of alloys is used in marine propulsion engines. Special corrosion-resistant coatings are applied to further improve the corrosion resistance of nickel-based superalloy components. The level of sodium ingested by the engine can also be controlled with proper inlet design and filtration.[14]

Although there was a period when gas turbines were being applied as prime movers on railroad locomotives, the above report contained only one small railroad application.

57.4 EVALUATION AND SELECTION

57.4.1 Maintenance Intervals, Availability, and Reliability

Service requirements of aircraft and industrial gas turbines differ from other power plants principally in the fact that several key components, because they operate at very high temperatures, have limited lives and must be repaired or replaced periodically to avoid failures during operation. These components include combustion chambers, buckets, and nozzles. Other components, such as wheels or casings, may occasionally require inspection or retirement.

Wear-out mechanisms in hot gas path components include creep, low cycle fatigue, corrosion, and oxidation. All combustors, buckets, and nozzles, if operated for significantly longer than their design life, will eventually fail in one of these modes. Repair or replacement is required to avoid failure. Most of the failure mechanisms give some warning prior to loss of component integrity. Corrosion and oxidation are observable by visual inspection. The creep deflection of nozzles can be detected by measuring changes in clearances. Low-cycle fatigue cracks can occur in nozzles, buckets, and combustors without causing immediate failure of these components. These can be detected visually or by more sophisticated nondestructive inspection techniques. Tolerance of cracks depends on the particular component design, service conditions, and other forces or temperatures superimposed on the component at the location of the crack. Inspection intervals are set by manufacturers, based on laboratory data and field experience, so that components with some degree of distress can be removed from service or repaired prior to component failure.

Bucket creep often gives no advance warning, due to several factors. First, the ability of bucket alloys to withstand alternating stress and the rate of creep progression are both affected by the existence of creep void formation. Local creep void formation is difficult to observe even in individual buckets exposed to radiographic and other nondestructive inspections. Destructive inspection of samples taken from a turbine are not useful in predicting the conditions in the particular single bucket in a stage that will exhibit the most advanced creep conditions. This is due to the statistical distribution of creep conditions in a sample set. Such a large number of samples would be required to accurately predict the condition of the worst part in a set that the cost of such an inspection would be higher than the set of replacement components. Because of this, creep failure is avoided by the retirement of sets of buckets as the risk of the failure of the worst bucket in the set increases above a preselected level.

Some of the wear-out mechanisms are time-related, while others are start-related. Thus, the actual service profile is significant to determining when to inspect or retire gas path components. Manufacturers differ in the philosophy applied. Aircraft engine maintenance recommendations are based on a particular number of mission hours of operation. Each mission contains a number of hours at takeoff conditions, a number at cruise, a number of rapid accelerations, thrust-reversals, and so on. Component lives are calculated and expressed in terms of a number of mission cycles. Thus, the life of any component can be expressed in hours, even if the mechanism of failure expected is low-cycle fatigue related to the number of thermal excursions to which the component is exposed. Inspection and component retirement intervals, based on mission-hours, can be set to detect distress and remove or repair components before the actual failure is likely to occur.

Industrial gas turbine manufacturers have historically designed individual products to be suitable for both continuous duty and frequent starts and stops. A particular turbine model may be applied to missions ranging from twice-daily starts to continuous operation for over 8000 hours per year and virtually no start cycles. To deal with this, manufacturers of industrial gas turbines have developed two ways of expressing component life and inspection intervals. One is to set two criteria for inspection: one based on hours and one based on starts. The other is to develop a formula for "equivalent hours" that counts each start as a number of additional hours of operation. These two methods are illustrated in Fig. 57.37. The figure is a simplification in that it considers only normal starts and base-load hours. Both criteria evaluate hours of operation at elevated firing temperature, fast starts, and emergency shutdown events as more severe than normal operating hours and starts. Industry practice is to establish maintenance factors that can be used to account for effects that shorten the intervals between inspections. Table 57.3 gives typical values. The hours to inspection or starts to inspection in Fig. 57.37 would be divided by the factor in Table 57.3.

The values shown here are similar to those used by manufacturers but are only approximate, since recommendations are modified and updated periodically. Also, the number, extent, and types of inspections vary across the industry. To compare the frequency of inspection recommended for competing gas turbines, the evaluator must forecast the number of starts and hours expected during the evaluation period and, using the manufacturers' recommendation and other experience, determine the inspection frequency for the particular application.

Reliability and *availability* have specific definitions where applied to power generation equipment.[22]

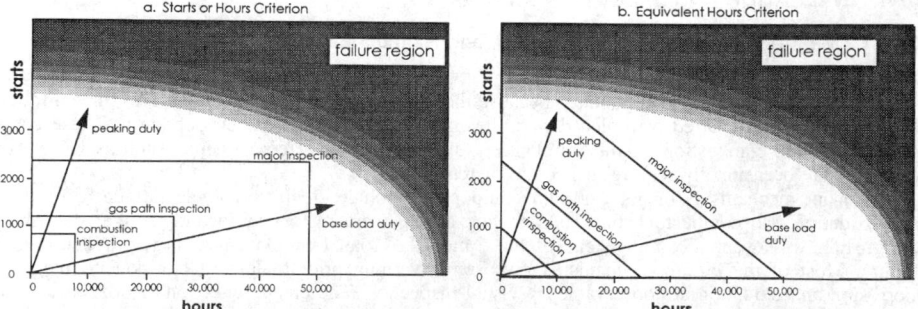

Fig. 57.37 Inspection interval criteria compared. Starts-or-hours criterion (*a*) requires major inspection after 48,000 hours or 2400 starts. Equivalent-hours criterion (*b*) reflects each start being equivalent to 10 hours, with major inspection required after 50,000 equivalent hours.

Table 57.3 Maintenance Factors—Industrial Gas Turbine Nozzles and Buckets

Factors Increasing Maintenance Frequency		
		Hours factors
Fuel	Natural gas	1
	Distillate	1–1.5
	Residual	3–4
Peak load	Elevated firing temp.	5–10
Diluent injection	Water or steam	1–2
		Starts Factors
Trip from full-load		6–10+
Fast load		2–4
Emergency start		10–20

Reliability = 1 − (FOH/PH) (expressed as a percentage)

FOH = total forced outage hours

PH = period hours (8760 hrs/year)

Reliability is used to reflect the percentage of time for which the equipment is not on forced outage. It is the probability of not being forced out of service when the unit is needed, and includes forced outage hours while in service, while on reserve shutdown, and while attempting to start.

Availability = 1 − (UH/PH) (expressed as a percentage)

UH = total unavailable hours (forced outage, failure to start,

unscheduled maintenance hours, maintenance hours)

PH = period hours

Availability reflects the probability of being available, independent of whether or not the unit is needed, and includes all unavailable hours normalized by period hours.

There are some minor differences in the definitions across the industry, reflecting the way different databases treat particular types of events, but the equations given above reasonably represent industry norms.

Availability and reliability figures used in power-generation industry literature reflect the performance not only of the turbomachinery, but of the generator, control system, and accessories. Historically, less than half of the unavailability and forced outage hours are due to the turbomachinery. Accessories, controls and driven devices account for the remainder.

Availability is affected by the frequency and duration of inspections as well as the duration of forced outages. Improvements in analytical capability, understanding of material behavior, operating practices, and design sophistication have led to improvements in both availability and reliability over the past decades. The availability of industrial gas turbines has grown from 80% in the early 1970s to better than 95% in the mid-1990s.

57.4.2 Selection of Engine and System

In the transportation field, gas turbines are the engine of choice in large and increasingly in small aircraft where the number of hours per year flown is sufficiently high that the higher speed and lower fuel and service costs attributable to gas turbines justify the higher first-cost. Private automobiles, which operate nominally 400 hours per year and where operating characteristics favor the Otto and Diesel cycles, are not likely to be candidates for gas turbine power. Exhaust-driven superchargers are a more acceptable application of turbomachinery technology to this market. Long-haul trucks, buses, and military applications may be served by gas turbines if the economics that made them commonplace on aircraft can be applied.

Gas turbine technology finds application in mechanical drive and electric power generation. In mechanical drive application, the turbine rotor shaft typically drives a pump, compressor, or drive system. Mechanical drive applications usually employ "two-shaft" gas turbines, in which the output shaft is controllable in speed to match the varying load/speed characteristic of the application. In electric power generation, the shaft drives an electrical generator at a constant synchronous speed. Mechanical drive applications typically find application for gas turbines in the 5–25 MW range. Over the last five years, this market has been approximately 1000 MW per year. Power-generation applications are typically in the larger size ranges, from 25–250 MW and have averaged over 20,000 MW per year.

Gas turbine technology competes with other technologies in both power generation and mechanical drive applications. In both applications, the process for selecting which thermodynamic cycle or engine type to apply is similar. Table 57.4 summarizes the four key choices in electric power generation.

Steam turbine technology utilizes an externally fired boiler to produce steam and drive a Rankine cycle. This technology has been used in power generation for nearly a century. Because the boiler is fired external to the working fluid, steam, any type of fuel may be used, including coal, distillate oil, residual oil, natural gas, refuse, and bio-mass. The thermal efficiencies are typically in the 30% range for small (20–40 MW) industrial and refuse plants to 35% for large (400 MW) power-generation units, to 40% for large, ultra-efficient, ultra-supercritical plants. These plants are largely assembled and erected at the plant site and have relatively high investment cost per kW of output. Local labor costs and labor productivity influence the plant cost. Thus, the investment cost can vary considerably throughout the world.

Diesel technology uses the Diesel cycle in a reciprocating engine. The diesels for power generation are typically medium speed (800 rpm). The diesel engine has efficiencies from 40–45% on distillate oil. If natural gas is the fuel, the ordinary Diesel cycle is not applicable, but a spark ignition system based on the Otto cycle's can be employed. The Otto cycle leads to three percentage points lower efficiency than the diesel. Diesel engines are available in smaller unit sizes than the gas turbines that account for most of the power generated for mechanical drive and power generation (1–10 MW). The investment cost of medium-speed diesels is relatively high per kW of output when compared with large gas turbines, but is lower than that of gas turbines in this size range. Maintenance cost of diesels per kW of output is typically higher than gas turbine technology.

The life-cycle cost of power-generation technology projects is the key factor in their application. The life-cycle cost includes the investment cost charges and the present worth of annual fuel and operating expenses. The investment cost charges are the present worth costs of financing, depreciation, and taxes. The fuel and operating expenses include fuel-consumption cost, maintenance expenses, operational material costs (lubricants, additives, etc.), and plant-operation and maintenance labor costs. For a combined-cycle technology plant, investment charges can contribute 20%, fuel 70%, and operation and maintenance costs 10%. The magnitude and composition of costs is very technology and geographic location dependent.

One way to evaluate the application of technology is to utilize a screening curve, as shown in Fig. 57.38. This chart represents one particular power output and set of economic conditions, and is used here to illustrate a principle, not to make a general statement on the relative merits of various power generation means. The screening curve plots the total $/kW/year annual life-cycle cost of a powerplant versus the number of hours per year of operation. At zero hours of operation (typical of a standby plant used only in the event of loss of power from other sources), the only life-cycle cost component is from investment financing charges and any operating expense associated with providing manpower to be at the site. As the operating hours increase toward 8000 hours per year, the costs of fuel, maintenance, labor, and direct materials are added into the annual life cycle cost.

Table 57.4   Fossil Fuel Technologies for Mechanical Drive and Electric Power Generation

Technology	Power Cycle	Performance Level	Primary Advantages	Primary Disadvantages
Steam Turbine	Rankine cycle	30–40%	Custom size Solid fuels Dirty fuels	Low efficiency Rel. high $/kW Slow load change
Gas Turbine	Brayton cycle	30–40%	Packaged power plant Low $/kW Med. fast starts Fast load delta	Clean fuels Ambient dependence
Combined Cycle	Brayton Topping/ Rankine Bottoming	45–60%	Highest efficiency Med. $/kW Limited fast load delta	Clean fuels Ambient dependence Med. start times
Diesel	Diesel cycle	40–50%	Rel. high efficiency Packaged power plant Fast start Fast construction	High maint. Small size (5 MW)

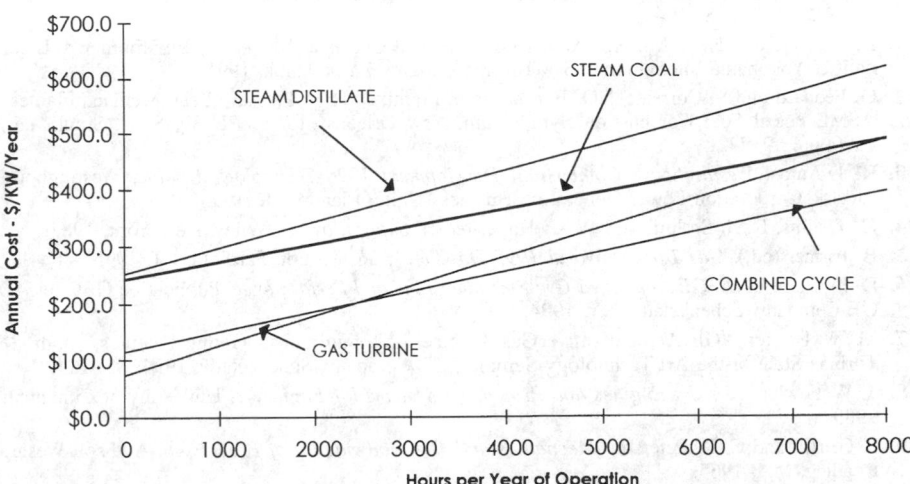

Typical Technology Screening Curve

Fig. 57.38 Hypothetical screening curve for selecting power-generation technology from among various thermodynamic cycles and fuel alternatives. This curve would indicate that the most economic choice for few operating hours per year is the simple-cycle gas turbine, and the combined cycle for base load applications.

If the application has only a few hours per year of operation (less than 2000) simple-cycle gas turbine technology has typically the lowest annual life-cycle cost and is therefore chosen. Simple-cycle gas turbine has the lowest annual life-cycle cost in this region in view of its low investment cost. If the application has more than 2000 hours per year of operation, then combined-cycle technology provides the lowest annual life-cycle cost and is selected for application.

Other technology choices are the higher investment cost alternatives of coal-fired steam turbine technology and IGCC technology. In the example of Fig. 57.38, these technologies do not have the lowest annual life-cycle cost in any region and consequently would not find application. However, the screening curve of Fig. 57.38 is based on a specific set of fuel prices and investment costs. In other regions of the world, coal prices may be lower or natural gas prices may be higher. In this case, the coal technologies may have the lowest annual life-cycle cost, in the 6000–8000-hour range. These technologies would then be selected for application.

In summary, there is a range of fuel prices and investment costs for power generation technology. This range influences the applicability of the power-generation technology. In some countries with large, low-priced coal resources, coal steam turbine technology is the most widely used. Where natural gas is available and modestly priced, gas turbine and combined-cycle technology is widely selected.

REFERENCES

1. R. T. C. Harman, *Gas Turbine Engineering,* Macmillan Press, Great Britain, 1981.
2. R. Harman, "Gas Turbines," in *Mechanical Engineers' Handbook,* M. Kutz (ed.), Wiley, New York, 1986, pp. 1984–2013.
3. A. Meyer, "The Combustion Gas Turbine, Its History and Development, *Proc. Instn. Mech. Engrs.* **141**, 197–222 (1939).
4. F. Whittle, "The Early History of the Whittle Jet Propulsion Gas Turbine," *Proc. Instn. Mech. Engrs.* **152**, 419–435 (1945).
5. D. E. Brandt, *The History of the Gas Turbine with an Emphasis on Schenectady General Electric Development,* GE Company, Schenectady, 1994.
6. *Gas Turbines—Status and Prospects,* Institute of Mechanical Engineers Conference Publications C.P. 1976–1.
7. A. N. Smith and J. D. Alrich, "Gas Turbine Electric Locomotives in Operation in the USA," *Combustion Engine Progress* (ca. 1957).
8. H. E. Miller and E. Benvenuti, "State of the Art and Trends in Gas-Turbine Design and Technology," European Powerplant Congress, Liège, 1993.

9. L. B. Davis, "Dry Low NO$_x$ Combustion Systems for GE Heavy-Duty Gas Turbines," 38th GE Turbine State-of-the-Art Technology Seminar, GE Company, Schenectady, 1994.

10. J. Zeldovich, "The Oxidation of Nitrogen in Combustion and Explosions," *Acta Physicochimica USSR* **21**, 577–628 (1946).

11. D. M. Todd and H. E. Miller, "Advanced Combined Cycles as Shaped by Environmental Externality," Yokohama International Gas Turbine Congress, Yokohama, 1991.

12. G. Leonard and S. Correa, "NO$_x$ Formation in Premixed High-Pressure Lean Methane Flames," ASME Fossil Fuel Combustion Symposium, New Orleans, *ASME/PD* **30**, S. N. Singh (ed.), 1990, pp. 69–74.

13. R. J. Antos, *Westinghouse Combustion Development 1996 Technology Update,* Westinghouse Electric Corporation Power Generation Business Unit, Orlando, FL, 1996.

14. C. T. Sims, N. S. Stoloff, and W. C. Hagel (eds.), *Superalloys II,* Wiley, New York, 1987.

15. B. Farmer (ed.), *Gas Turbine World 1995 Handbook,* **16**, Pequot, Fairfield, CT, 1995.

16. D. L. Chase et al., *GE Combined Cycle Product Line and Performance,* Publication GER-3574E, GE Company, Schenectady, NY, 1994.

17. M. W. Horner, "GE Aeroderivative Gas Turbines—Design and Operating Features," 38th GE Turbine State-of-the-Art Technology Seminar, GE Company, Schenectady, 1994. See Ref. 9.

18. T. W. Fowler (ed.), *Jet Engines and Propulsion Systems for Engineers,* University of Cincinnati, 1989.

19. P. G. Hill and C. R. Peterson, *Mechanics and Thermodynamics of Propulsion,* Addison-Wesley, Reading, MA, 1965.

20. D. M. North (ed.), *Aviation Week & Space Technology Aerospace Source Book* **144**(2), 95–106, (January 8, 1996).

21. D. G. Wilson, "Automotive Gas Turbines: Government Funding and the Way Ahead," *Global Gas Turbine News* **35**(4), (1995).

22. R. F. Hoeft, "Heavy-Duty Gas Turbine Operating and Maintenance Considerations," 38th GE Turbine State-of-the-Art Technology Seminar, GE Company, Schenectady, 1994. See Ref. 9.

BIBLIOGRAPHY

Cox, H. R., "British Aircraft Gas Turbines," *Journal of the Aeronautical Sciences* **13**(2) (February 1946).

Gusso, R., and H. E. Miller, "Dry Low NO$_x$ Combustion Systems for GE/Nuovo Pignone Heavy Duty Gas Turbines," Flowers '92 Gas Turbine Congress, Florence, Italy, 1992.

Houllier, C., *The Limitation of Pollutant Emissions into the Air from Gas Turbines—Draft Final Report,* CITEPA No. N 6611-90-007872, Paris, 1991.

CHAPTER 58

STEAM TURBINES

William G. Steltz
Turboflow International Inc.
Orlando, Florida

58.1	HISTORICAL BACKGROUND	1765
58.2	THE HEAT ENGINE AND ENERGY CONVERSION PROCESSES	1767
58.3	SELECTED STEAM THERMODYNAMIC PROPERTIES	1772
58.4	BLADE PATH DESIGN	1775
	58.4.1 Thermal to Mechanical Energy Conversion	1776

58.4.2	Turbine Stage Designs	1782
58.4.3	Stage Performance Characteristics	1784
58.4.4	Low-Pressure Turbine Design	1788
58.4.5	Flow Field Solution Techniques	1790
58.4.6	Field Test Verification of Flow Field Design	1791
58.4.7	Blade-to-Blade Flow Analysis	1796
58.4.8	Blade Aerodynamic Considerations	1796

58.1 HISTORICAL BACKGROUND

The process of generating power depends on several energy-conversion processes, starting with the chemical energy in fossil fuels or the nuclear energy within the atom. This energy is converted to thermal energy, which is then transferred to the working fluid, in our case, steam. This thermal energy is converted to mechanical energy with the help of a high-speed turbine rotor and a final conversion to electrical energy is made by means of an electrical generator in the electrical power-generation application. The presentation in this section focuses on the electrical power application, but is also relevant to other applications, such as ship propulsion.

Throughout the world, the power-generation industry relies primarily on the steam turbine for the production of electrical energy. In the United States, approximately 77% of installed power-generating capacity is steam turbine-driven. Of the remaining 23%, hydroelectric installations contribute 13%, gas turbines account for 9%, and the remaining 1% is split among geothermal, diesel, and solar power sources. In effect, over 99% of electric power generated in the United States is developed by turbomachinery of one design or another, with steam turbines carrying by far the greatest share of the burden.

Steam turbines have had a long and eventful life since their initial practical development in the late 19th century due primarily to efforts led by C. A. Parsons and G. deLaval. Significant developments came quite rapidly in those early days in the fields of ship propulsion and later in the power-generation industry. Steam conditions at the throttle progressively climbed, contributing to increases in power production and thermal efficiency. The recent advent of nuclear energy as a heat source for power production had an opposite effect in the late 1950s. Steam conditions tumbled to accommodate reactor designs, and unit heat rates underwent a step change increase. By this time, fossil unit throttle steam conditions had essentially settled out at 2400 psi and 1000°F with single reheat to 1000°F. Further advances in steam powerplants were achieved by the use of once-through boilers delivering supercritical pressure steam at 3500–4500 psi. A unique steam plant utilizing advanced steam con-

This chapter was previously published in J. A. Schetz and A. E. Fuhs (eds.), *Handbook of Fluid Dynamics and Fluid Machinery*, Vol. 3, *Applications of Fluid Dynamics*, New York, Wiley, 1996, Chapter 27.

Mechanical Engineers' Handbook, 2nd ed., Edited by Myer Kutz.
ISBN 0-471-13007-9 © 1998 John Wiley & Sons, Inc.

ditions is Eddystone No. 1, designed to deliver steam at 5000 psi and 1200°F to the throttle, with reheat to 1050°F and second reheat also to 1050°F.

Unit sizes increased rapidly in the period from 1950 to 1970; the maximum unit size increased from 200 mW to 1200 mW (a sixfold increase) in this span of 20 years. In the 1970s, unit sizes stabilized, with new units generally rated at substantially less than the maximum size. At the present time, however, the expected size of new units is considerably less, appearing to be in the range of 350–500 mW.

In terms of heat rate (or thermal efficiency), the changes have not been so dramatic. A general trend showing the reduction of power station heat rate over an 80-year period is presented in Fig. 58.1. The advent of regenerative feedwater heating in the 1920s brought about a step change reduction in heat rate. A further reduction was brought about by the introduction of steam reheating. Gradual improvements continued in steam systems and were recently supplemented by the technology of the combined cycle, the gas turbine/steam turbine system (see Fig. 58.2). In the same period of time that unit sizes changed by a factor of six (1950 to 1970), heat rate diminished by less than 20%, a change that includes the combined cycle. In reality, the improvement is even less, as environmental regulations and the energy required to satisfy them can consume up to 6% or so of a unit's generated power.

The rate of improvement of turbine cycle heat rate is obviously decreasing. Powerplant and machinery designers are working hard to achieve small improvements both in new designs and in retrofit and repowering programs tailored to existing units. Considering the worth of energy, what, then, are our options leading to thermal performance improvements and the management of our energy and financial resources? Exotic energy-conversion processes are a possibility: MHD, solar

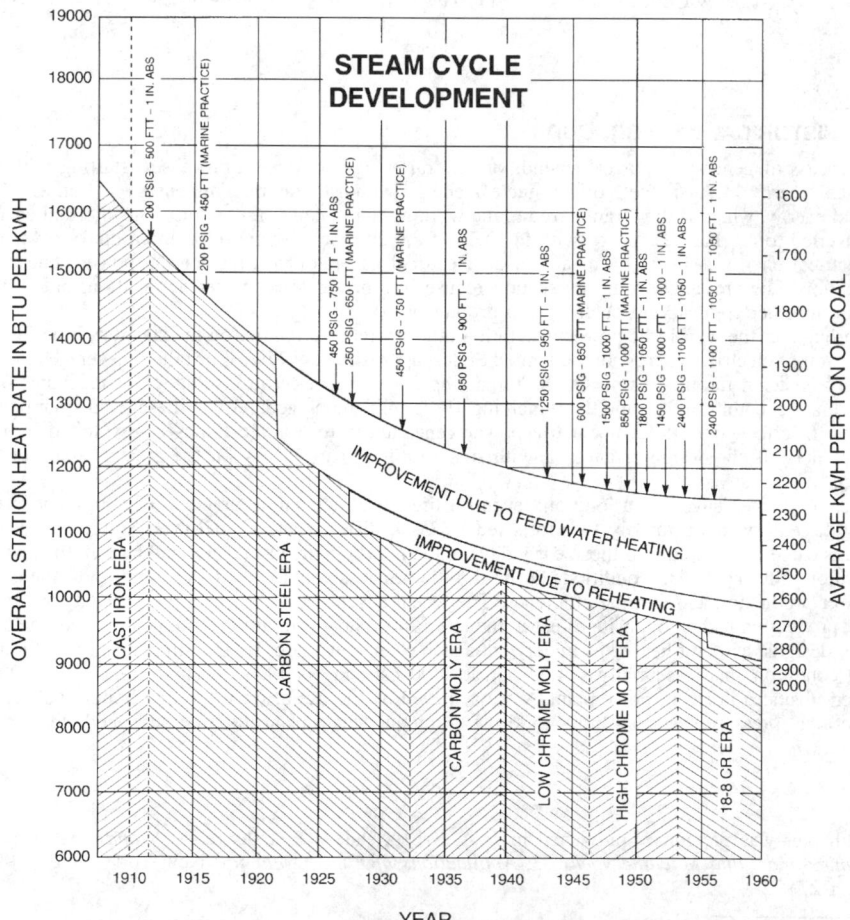

Fig. 58.1 Steam cycle development.

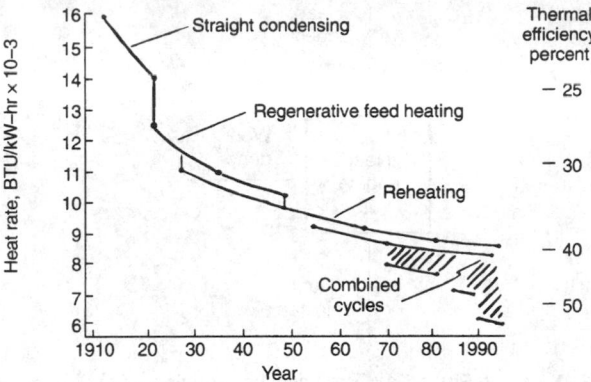

Fig. 58.2 Fossil-fueled unit heat rate as a function of time.

power, the breeder reactor, and fusion are some of the longer-range possibilities. A more near-term possibility is through the improvement (increase) of steam conditions. The effect of improved steam conditions on turbine cycle heat rate is shown in Fig. 58.3, where heat rate is plotted as a function of throttle pressure with parameters of steam temperature level. The plus mark indicates the placement of the Eddystone unit previously mentioned.

58.2 THE HEAT ENGINE AND ENERGY CONVERSION PROCESSES

The mechanism for conversion of thermal energy is the *heat engine,* a thermodynamic concept, defined and sketched out by Carnot and applied by many, the power generation industry in particular. The heat engine is a device that accepts thermal energy (heat) as input and converts this energy to useful work. In the process, it rejects a portion of this supplied heat as unusable by the work production process. The efficiency of the ideal conversion process is known as the *Carnot efficiency.* It serves as a guide to the practitioner and as a limit for which no practical process can exceed. The Carnot efficiency is defined in terms of the absolute temperatures of the heat source T_{hot} and the heat sink T_{cold} as follows:

$$\text{Carnot efficiency} = \frac{T_{hot} - T_{cold}}{T_{hot}} \tag{58.1}$$

Consider Fig. 58.4, which depicts a heat engine in fundamental terms consisting of a quantity of heat supplied, heat added, a quantity of heat rejected, heat rejected. and an amount of useful work done, work done. The *thermal efficiency* of this basic engine can be defined as

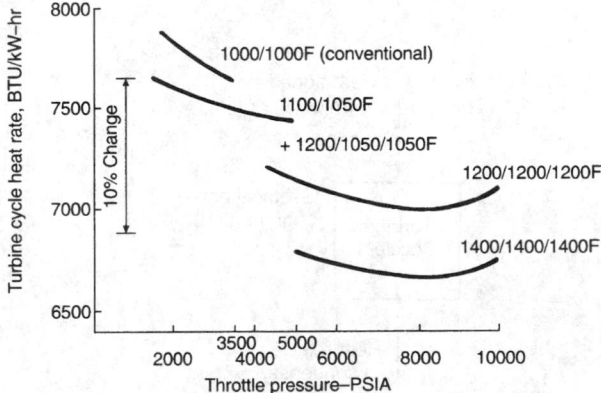

Fig. 58.3 Comparison of turbine cycle heat rate as a function of steam conditions.

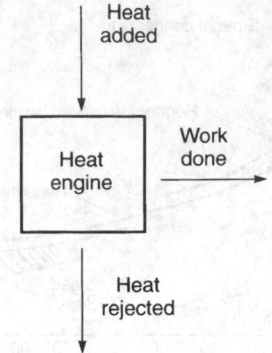

Fig. 58.4 The basic heat engine.

$$\text{efficiency} = \frac{\text{work done}}{\text{heat added}} \tag{58.2}$$

This thermal efficiency is fundamental to any heat engine and is, in effect, a measure of the heat rate of any turbine-generator unit of interest. Figure 58.5 is the same basic heat engine redefined in terms of turbine cycle terminology, that is, heat added is the heat input to the steam generator, heat rejected is the heat removed by the condenser, and the difference is the work done (power) produced by the turbine cycle. Figure 58.6 is a depiction of a simple turbine cycle showing the same parameters, but described in conventional terms. *Heat rate* is now defined as the quantity of heat input required to generate a unit of electrical power (kW).

$$\text{heat rate} = \frac{\text{heat added}}{\text{work done}} \tag{58.3}$$

The units of heat rate are usually in terms of Btu/kW·hr.

Further definition of the turbine cycle is presented in Fig. 58.7, which shows the simple turbine cycle with pumps and a feedwater heater included (of the open type). In this instance, two types of heat rate are identified: (1) a *gross heat rate,* in which the turbine-generator set's natural output (i.e., gross electrical power) is the denominator of the heat rate expression, and (2) a *net heat rate,* in which the gross power output has been debited by the power requirement of the boiler feed pump, resulting in a larger numeric value of heat rate. This procedure is conventional in the power-generation industry, as it accounts for the inner requirements of the cycle needed to make it operate. In other, more complex cycles, the boiler feed pump power might be supplied by a steam turbine-driven feed pump. These effects are then included in the heat balance describing the unit's performance.

The same accounting procedures are true for all cycles, regardless of their complexity. A typical 450-mW fossil unit turbine cycle heat balance is presented in Fig. 58.8. Steam conditions are 2415

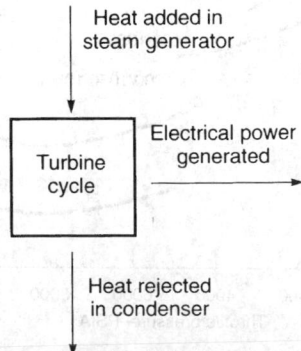

Fig. 58.5 The basic heat engine described in today's terms.

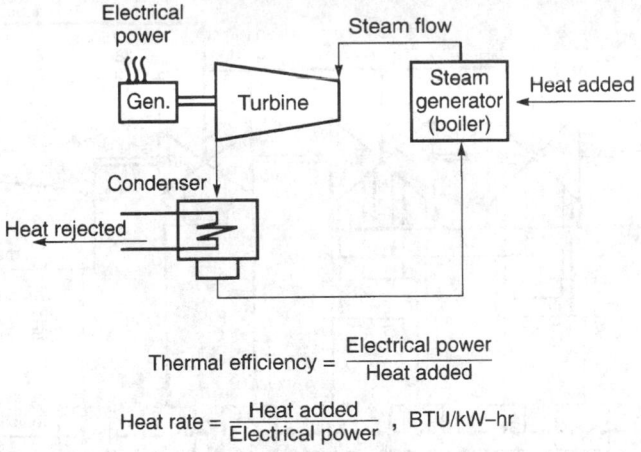

$$\text{Thermal efficiency} = \frac{\text{Electrical power}}{\text{Heat added}}$$

$$\text{Heat rate} = \frac{\text{Heat added}}{\text{Electrical power}} \text{ , BTU/kW–hr}$$

Fig. 58.6 A simple turbine cycle.

psia/1000°F/1000°F/2.5 inHga, and the cycle features seven feedwater heaters and a motor-driven boiler feed pump. Only pertinent flow and steam property parameters have been shown, in order to avoid confusion and to support the conceptual simplicity of heat rate. As shown in the two heat rate expressions, only two flow rates, four enthalpies, and two kW values are required to determine the gross and net heat rates of 8044 and 8272 Btu/kW·hr, respectively.

To supplement the fossil unit of Fig. 58.8, Fig. 58.9 presents a typical nuclear unit of 1000 mW capability. Again, only the pertinent parameters are included in this sketch for simplicity. Steam conditions at the throttle are 690 psi with ¼% moisture, and the condenser pressure is 3.0 inHga. The cycle features six feedwater heaters, a steam turbine-driven feed pump, and a moisture separator reheater (MSR). The reheater portion of the MSR takes throttle steam to heat the low-pressure (LP) flow to 473°F from 369°F (saturation at 164 psia). In this cycle, the feed pump is turbine-driven by steam taken from the MSR exit; hence, only one heat rate is shown, the net heat rate, 10,516 Btu/kW·hr. This heat rate comprises only four numbers, the throttle mass flow rate, the throttle enthalpy, the final feedwater enthalpy, and the net power output of the cycle.

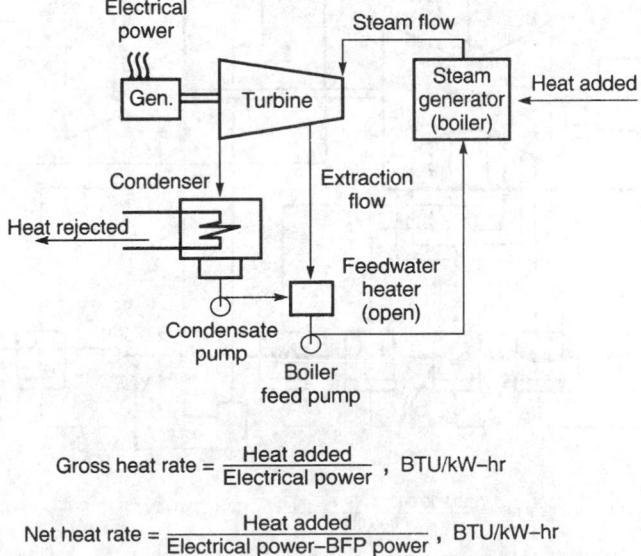

$$\text{Gross heat rate} = \frac{\text{Heat added}}{\text{Electrical power}} \text{ , BTU/kW–hr}$$

$$\text{Net heat rate} = \frac{\text{Heat added}}{\text{Electrical power–BFP power}} \text{ , BTU/kW–hr}$$

Fig. 58.7 A simple turbine cycle with an open heater and a boiler feed pump.

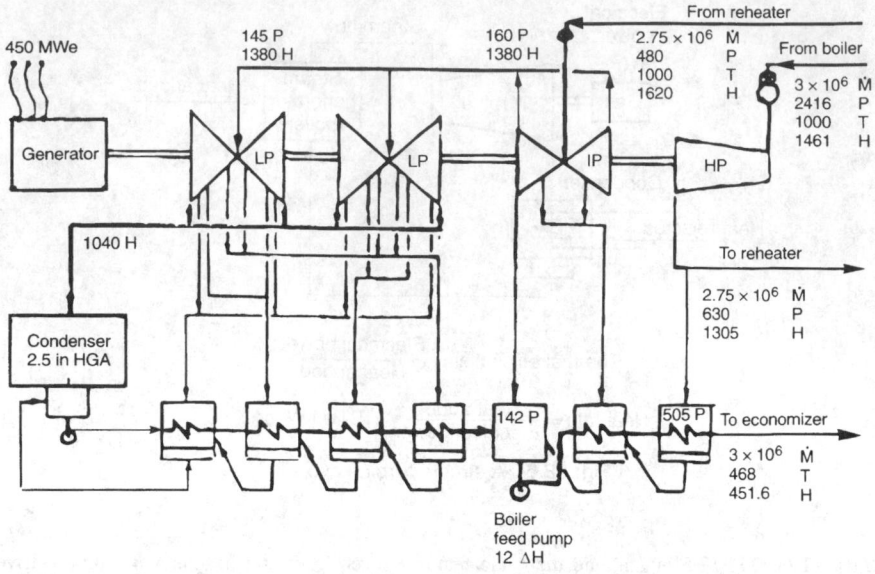

$$\text{Gross heat rate} = \frac{3,000,000\,(1461 - 451.6) + 2,760,000\,(1520 - 1305)}{450,000} = 8044 \text{ BTU/kW-hr}$$

$$\text{Net heat rate} = \frac{3,000,000\,(1461 - 451.6) + 2,760,000\,(1520 - 1305)}{450,000 - 12,400} = 8272 \text{ BTU/kW-hr}$$

Fig. 58.8 Typical fossil unit turbine cycle heat balance.

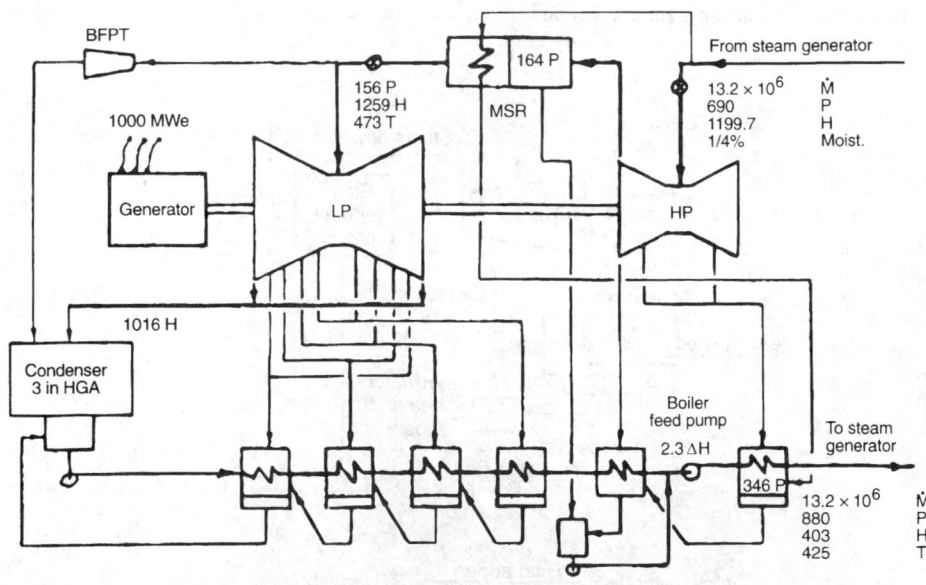

$$\text{Net heat rate} = \frac{13,200,000\,(1199.7 - 403)}{1,000,000} = 10516 \text{ BTU/kW-hr}$$

Fig. 58.9 Typical nuclear unit turbine cycle heat balance.

For comparative purposes, the expansion lines of the fossil and nuclear units of Figs. 58.8 and 58.9 have been superimposed on the *Mollier diagram* of Fig. 58.10. It is easy to see the great difference in steam conditions encompassed by the two designs and to relate the ratio of cold to hot temperatures to their Carnot efficiencies. In the terms of Carnot, the maximum fossil unit thermal efficiency would be 61% and the maximum nuclear unit thermal efficiency would be 40%. The ratio of these two Carnot efficiencies (1.53) compares somewhat favorably with the ratio of their net heat rates (1.27).

To this point, emphasis has been placed on the *conventional* steam turbine cycle, where conventional implies the central station power-generating unit whose energy source is either a fossil fuel (coal, oil, gas) or a fissionable nuclear fuel. Figure 58.2 has shown a significant improvement in heat rate attributable to *combined cycle* technology, that is, the marriage of the gas turbine used as a *topping unit* and the steam turbine used as a *bottoming unit*. The cycle efficiency benefits come from the high firing temperature level of the gas turbine, current units in service operating at 2300°F, and the utilization of its waste heat to generate steam in a heat-recovery steam generator (HRSG). Figure 58.11 is a heat balance diagram of a simplified combined cycle showing a two-pressure-level HRSG. The purpose of the two-pressure-level (or even three-pressure-level) HRSG is the minimization of the temperature differences existing between the gas turbine exhaust and the evaporating water/steam mixture. Second Law analyses (commonly termed *availability* or *exergy analyses*) result in improved cycle thermal efficiency when integrated average values of the various heat-exchanger temperature differences are small. The smaller, the better, from an efficiency viewpoint; however, the smaller the

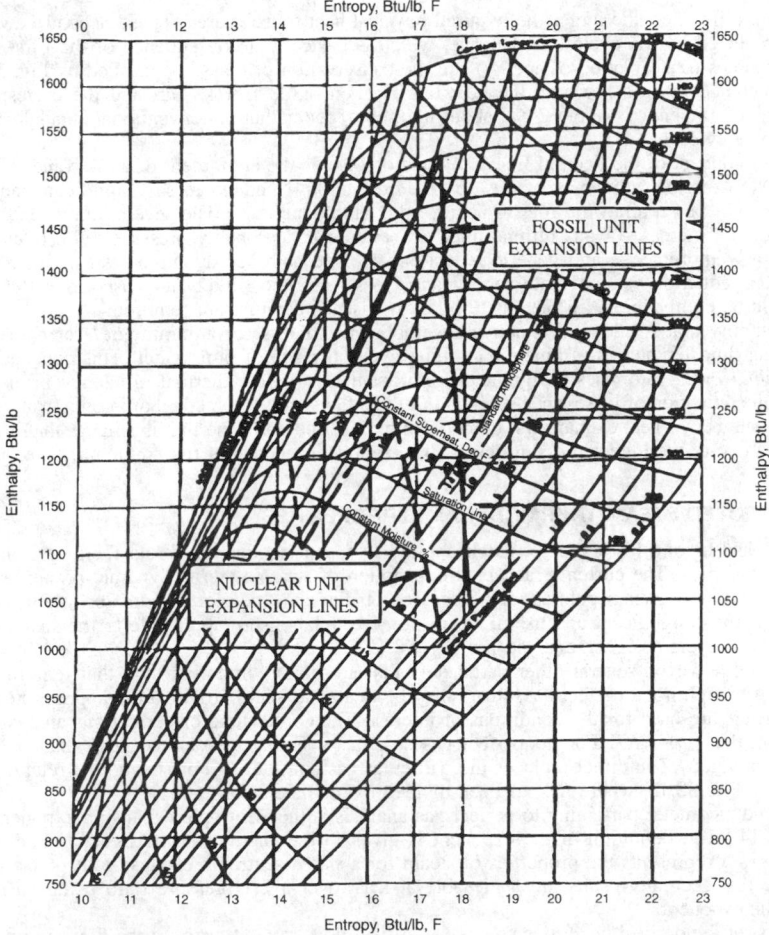

Fig. 58.10 Fossil and nuclear unit turbine expansion lines superimposed
on the Mollier diagram.

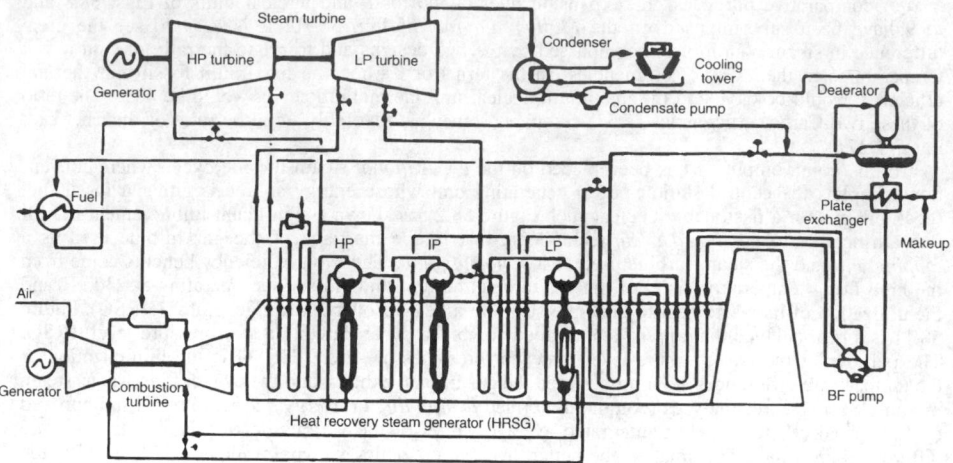

Fig. 58.11 A typical combined cycle plant schematic.

temperature difference, the larger the required physical heat transfer area. These Second Law results are then reflected by the cycle heat balance, which is basically a consequence of the First Law of thermodynamics (conservation of energy) and the conservation of mass. As implied by Fig. 58.11, a typical combined cycle schematic, the heat rate is about 6300 Btu/kW·hr, and the corresponding cycle thermal efficiency is about 54%, about ten points better than a conventional standalone fossil steam turbine cycle.

A major concept of the Federal Energy Policy of 1992 is the attainment of an Advanced Turbine System (ATS) thermal efficiency of 60% by the year 2000. Needless to say, significant innovative approaches will be required in order to achieve this ambitious level. The several approaches to this end include the increase of gas turbine inlet temperature and probably pressure ratio, reduction of cooling flow requirements, and generic reduction of blade path aerodynamic losses. On the steam turbine side, reduction of blade path aerodynamic losses and most likely increased inlet steam temperatures to be compatible with the gas turbine exhaust temperature are required.

A possibility that is undergoing active development is the use of an ammonia/water mixture as the working fluid of the gas turbine's bottoming cycle in place of pure water. This concept known as the *Kalina cycle*[1] promises a significant improvement to cycle thermal efficiency primarily by means of the reduction of losses in system *availability*. Physically, a practical ammonia/water system requires a number of heat exchangers, pumps and piping, and a turbine that is smaller than its steam counterpart due to the higher pressure levels that are a consequence of the ammonia/water working fluid.

58.3 SELECTED STEAM THERMODYNAMIC PROPERTIES

Steam has had a long history of research applied to the determination of its thermodynamic and transport properties. The currently accepted description of steam's thermodynamic properties is the ASME Properties of Steam publication.[2] The Mollier diagram, the plot of enthalpy versus entropy, is the single most significant and useful steam property relationship applicable to the steam turbine machinery and cycle designer/analyst (see Fig. 58.12).

There are, however, several other parameters that are just as important and that require special attention. Although not a perfect gas, steam may be treated as such, provided the appropriate perfect gas parameters are used for the conditions of interest. The cornerstone of perfect gas analysis is the requirement that $pv = RT$. For nonperfect gases, a factor Z may be defined such that $pv = RZT$ where the product RZ in effect replaces the particular gas constant R. For steam, this relationship is described in Fig. 58.13, where RZ has been divided by J, Joule's constant.

A second parameter pertaining to perfect gas analysis is the isentropic expansion exponent given in Fig. 58.14. (The definition of the exponent is given in the caption on the figure.) Note that the value of γ well represents the properties of steam for a *short* isentropic expansion. It is the author's experience that accurate results are achievable at least over a 2:1 pressure ratio using an average value of the exponent.

The first of the derived quantities relates the critical flow rate of steam[3] to the flow system's inlet pressure and enthalpy, as in Fig. 58.15. The critical (maximum) mass flow rate M, assuming an isentropic expansion process and equilibrium steam properties, is obtained by multiplying the ordinate value K by the inlet pressure p_1 in psia and the passage throat area A in square inches:

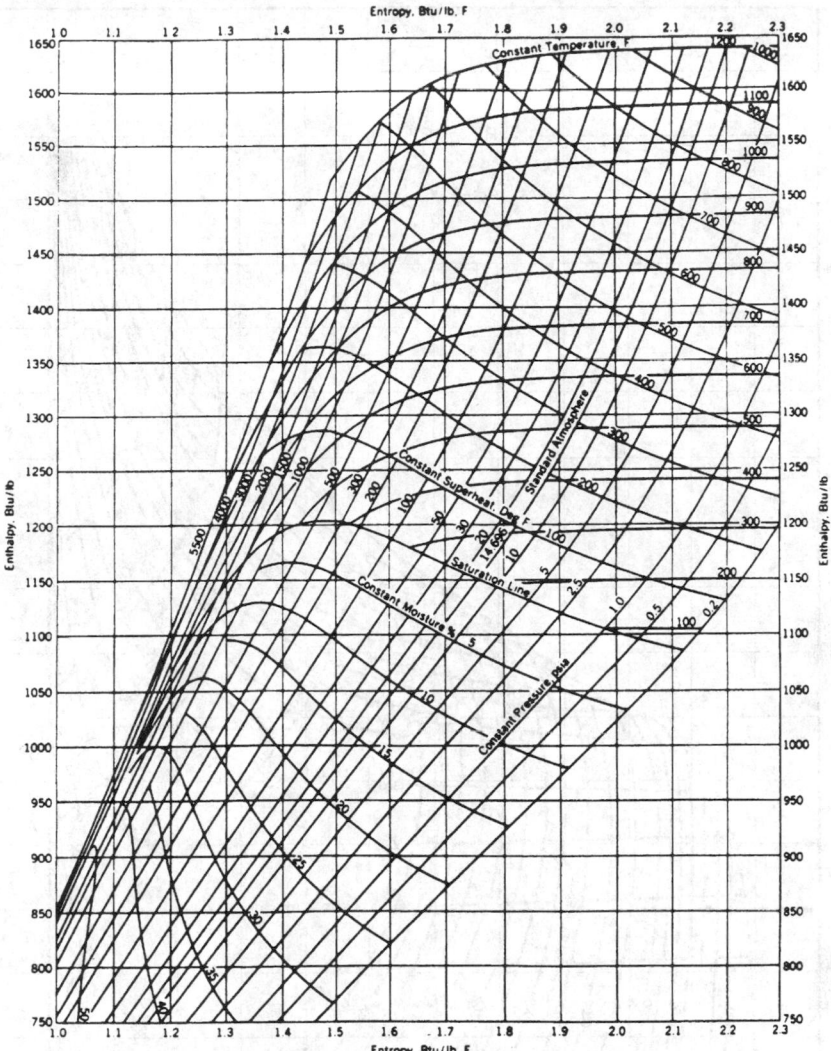

Fig. 58.12 Mollier diagram (h·s) for steam. (From Ref. 4.)

$$\dot{M}_{\text{critical}} = p_1 KA \tag{58.4}$$

The actual steam flow rate can then be determined as a function of actual operating conditions and geometry.

The corresponding choking velocity (acoustic velocity in the superheated steam region) is shown in Figs. 58.16 and 58.17 for superheated steam and wet steam, respectively. The range of Mach numbers experienced in steam turbines can be put in terms of the *wheel speed Mach number,* that is, the rotor tangential velocity divided by the local acoustic velocity. In the HP turbine, wheel speed is on the order of 600 ft/sec, while the acoustic velocity at 2000 psia and 975°F is about 2140 ft/sec; hence, the wheel speed Mach number is 0.28. For the last rotating blade of the LP turbine, its tip wheel speed could be as high as 2050 ft/sec. At a pressure level of 1.0 psia and an enthalpy of 1050 Btu/lb, the choking velocity is 1275 ft/sec; hence, the wheel speed Mach number is 1.60. As Mach numbers relative to either the stationary or rotating blading are approximately comparable, the steam turbine designer must negotiate flow regimes from incompressible flow, low subsonic Mach number of 0.3, to supersonic Mach numbers on the order of 1.6.

Another quite useful characteristic of steam is the product of pressure and specific volume plotted versus enthalpy in Figs. 58.18 and 58.19 for low-temperature/wet steam and superheated steam,

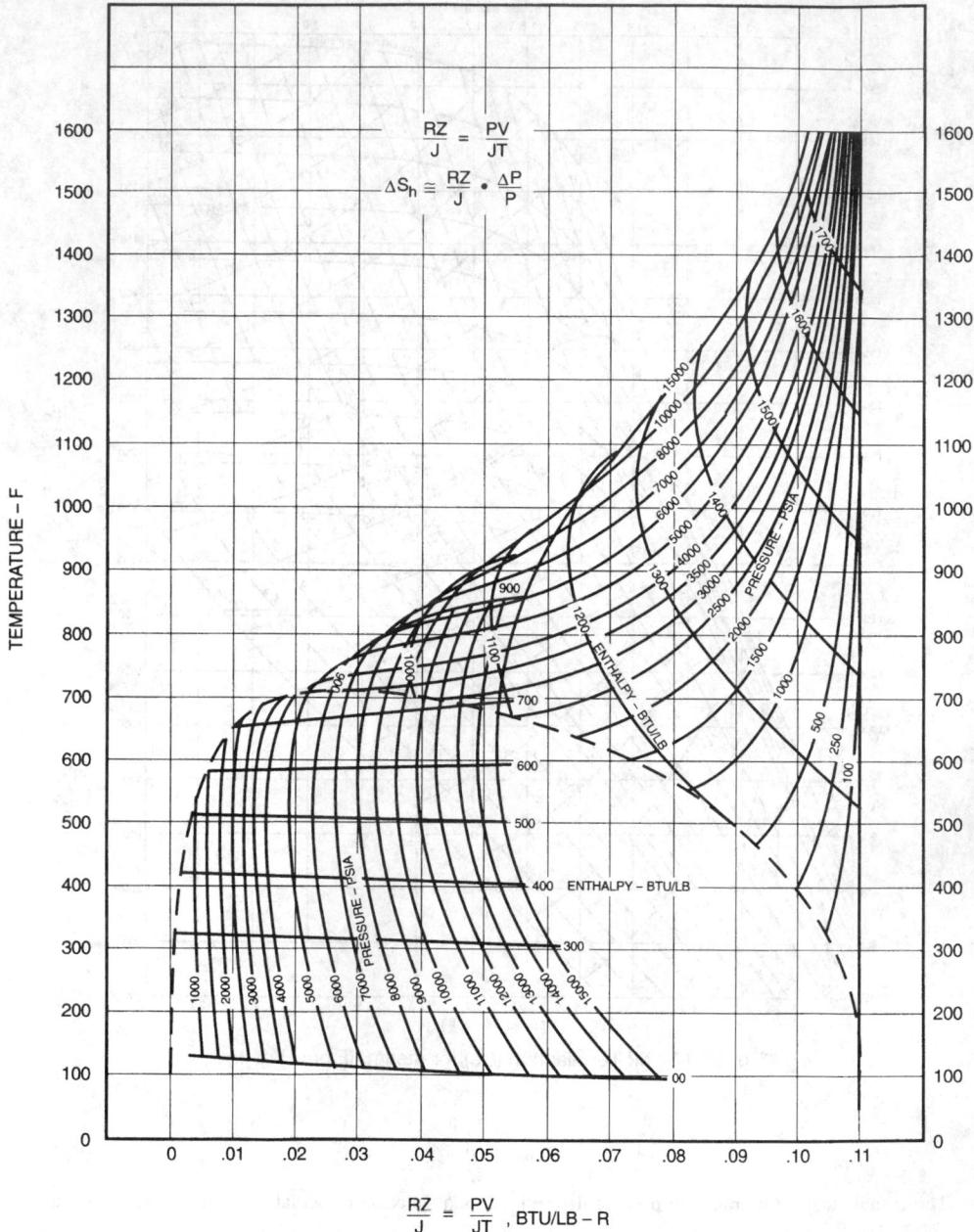

$$\frac{RZ}{J} = \frac{PV}{JT}$$

$$\Delta S_h \cong \frac{RZ}{J} \cdot \frac{\Delta P}{P}$$

$$\frac{RZ}{J} = \frac{PV}{JT} \text{ , BTU/LB – R}$$

Fig. 58.13 (RZ/J) for steam and water. (From Ref. 5.)

respectively. If the fluid were a perfect gas, this plot would be a straight line. In reality, it is a series of nearly straight lines, with pressure as a parameter. A significant change occurs in the wet steam region, where the pressure parameters spread out at a slope different from that of the superheated region. These plots are quite accurate for determining specific volume and for computing the often used *flow number*

$$\frac{M\sqrt{pv}}{p} \tag{58.5}$$

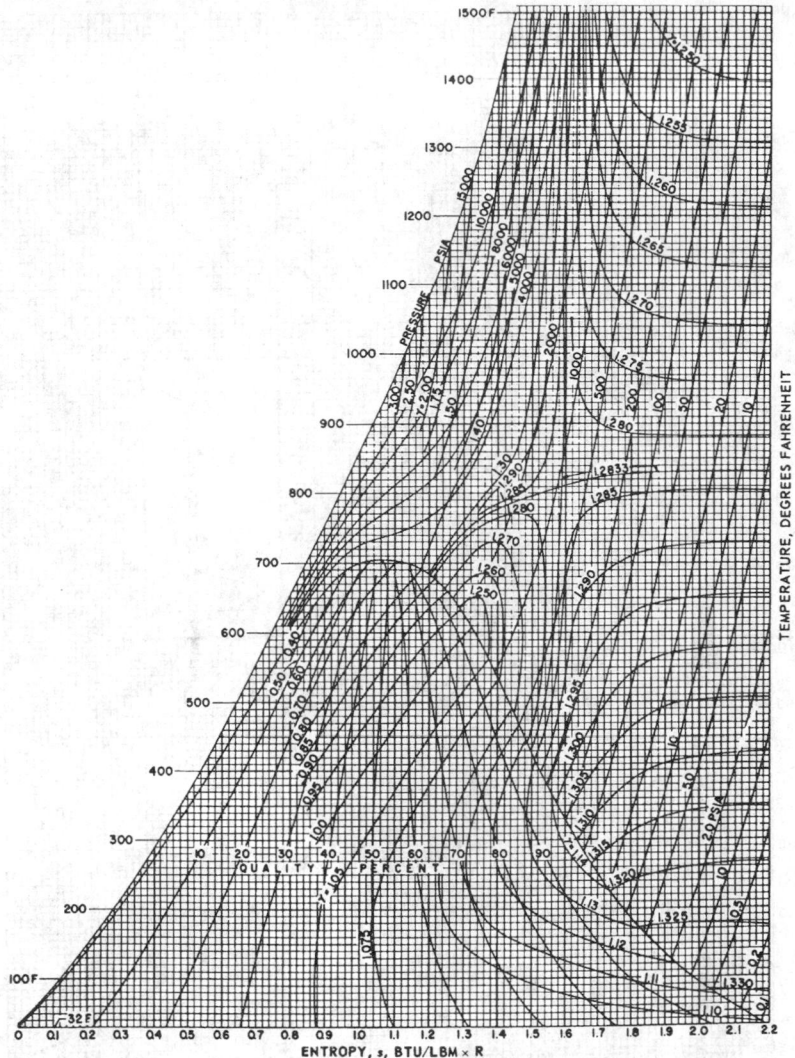

Fig. 58.14 Isentropic exponent, $\gamma = -\dfrac{v}{p}\left(\dfrac{\partial p}{\partial v}\right)_s$, $pv^\gamma = $ constant for a short expansion. (From Ref. 2.)

A direct application of the above-mentioned approximations is the treatment by perfect gas analysis techniques of applications where the working fluid is a mixture of air and a significant amount of steam (*significant* implies greater than 2–4%). Not to limit the application to air and steam, the working fluid could be the products of combustion and steam, or other arbitrary gases and steam.

58.4 BLADE PATH DESIGN

The accomplishment of the thermal to mechanical energy-conversion process in a steam turbine is, in general, achieved by successive expansion processes in alternate stationary and rotating blade rows. The turbine is a heat engine, working between the thermodynamic boundaries of maximum and minimum temperature levels, and as such is subject to the laws of thermodynamics prohibiting the achievement of engine efficiencies greater than that of a Carnot cycle. The turbine is also a dynamic machine in that the thermal to mechanical energy-conversion process depends on blading forces, traveling at rotor velocities, developed by the change of momentum of the fluid passing

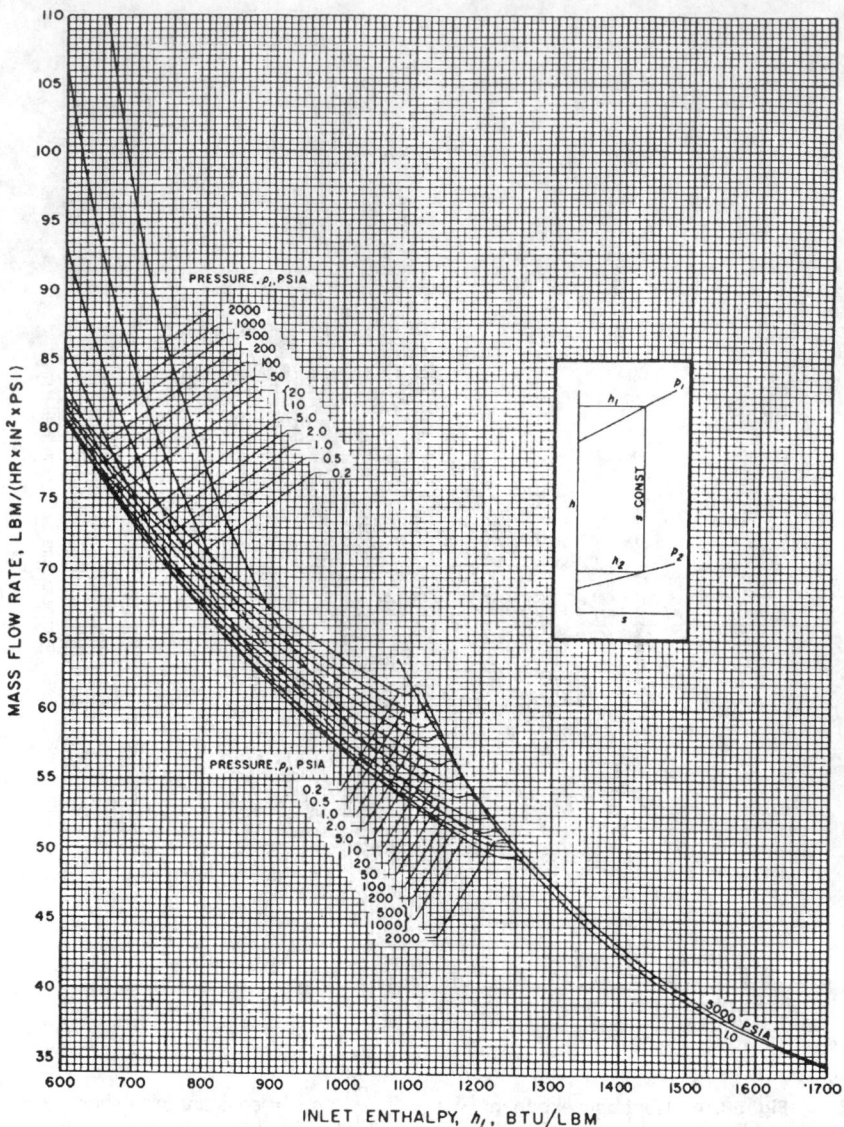

Fig. 58.15 Critical (choking) mass flow rate for isentropic process and equilibrium conditions. (From Ref. 2.)

through a blade passage. The laws of nature as expressed by Carnot and Newton govern the turbine designer's efforts and provide common boundaries for his achievements.

The purpose of this section is to present the considerations involved in the design of steam turbine blade paths and to indicate means by which these concerns have been resolved. The means of design resolution are not unique. An infinite number of possibilities exist to achieve the specific goals, such as efficiency, reliability, cost, and time. The real problem is the achievement of all these goals simultaneously in an optimum manner, and even then, there are many approaches and many very similar solutions.

58.4.1 Thermal to Mechanical Energy Conversion

The purpose of turbomachinery blading is the implementation of the conversion of thermal energy to mechanical energy. The process of conversion is by means of curved airfoil sections that accept

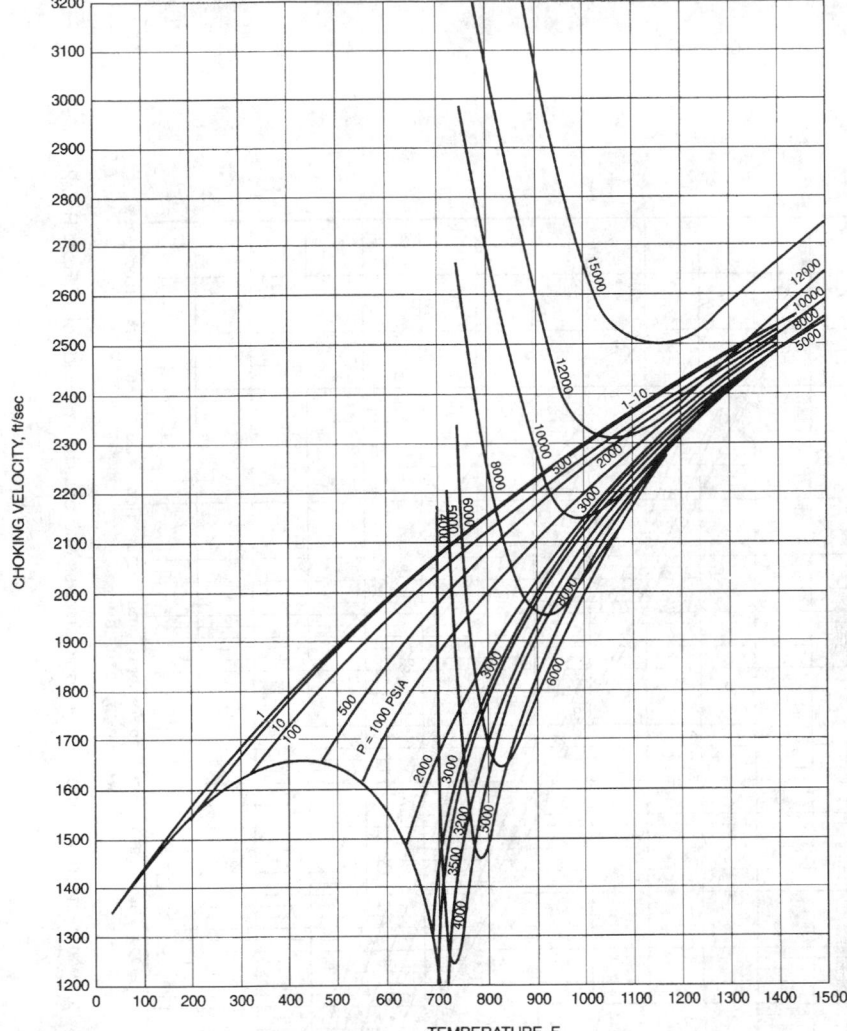

Fig. 58.16 Choking (sonic) velocity as a function of temperature and pressure.

incoming steam flow and redirect it in order to develop blading forces and resultant mechanical work. The force developed on the blading is equal and opposite to the force imposed on the steam flow, that is, the force is proportional to the change of fluid momentum passing through the turbine blade row. The magnitude of the force developed is determined by application of Newton's laws to the quantity of fluid occupying the flow passage between adjacent blades, a space that is in effect, a *streamtube*. The assumptions are made that the flow process is steady and that flow conditions are uniform upstream and downstream of the flow passage.

Figure 58.20 presents an isometric view of a turbine blade row defining the components of steam velocity at the inlet and exit of the blade passage streamtube. The tangential force impressed on the fluid is equal to the change of fluid momentum in the tangential direction by direct application of Newton's laws.

$$F_{\theta \text{on fluid}} = \int_{A_2} V_\theta d\dot{M} - \int_{A_1} V_\theta \, d\dot{M} \qquad (58.6)$$

$$F_{\theta \text{on fluid}} = \int_{A_2} \rho_2 V_{Z2} V_{\theta_2} \, dA - \int_{A_1} \rho_1 V_{Z1} V_{\theta_1} \, dA \qquad (58.6a)$$

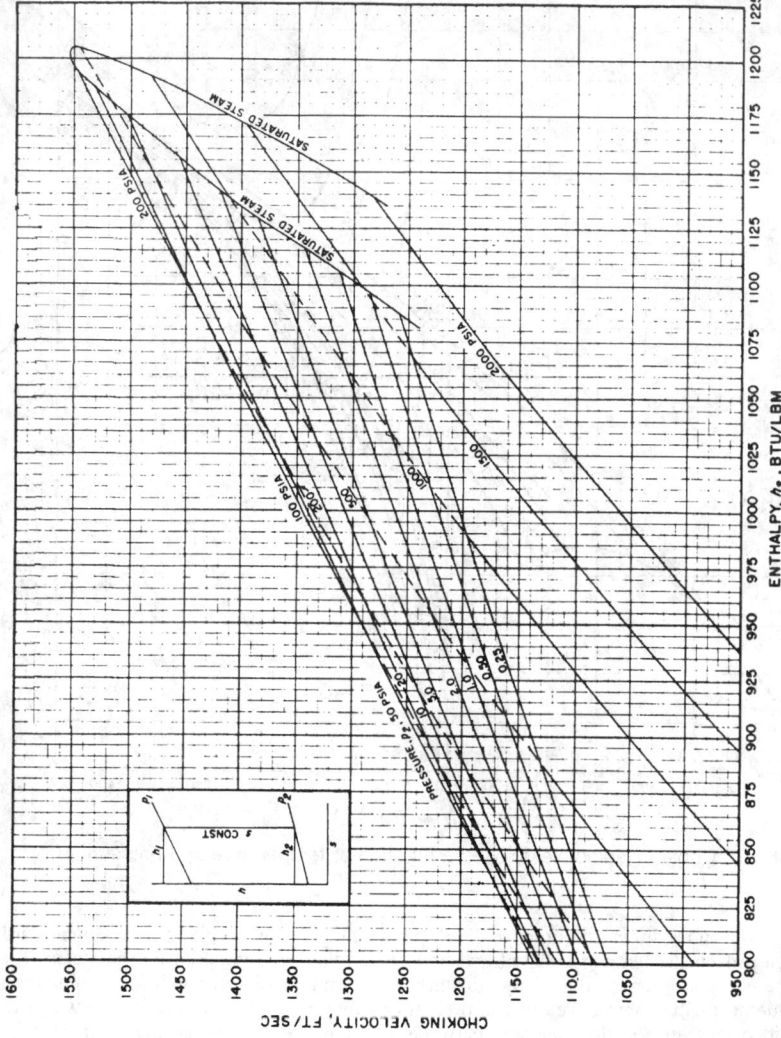

Fig. 58.17 Choking velocity for water–steam mixture for isentropic process and equilibrium conditions. (From Ref. 2.)

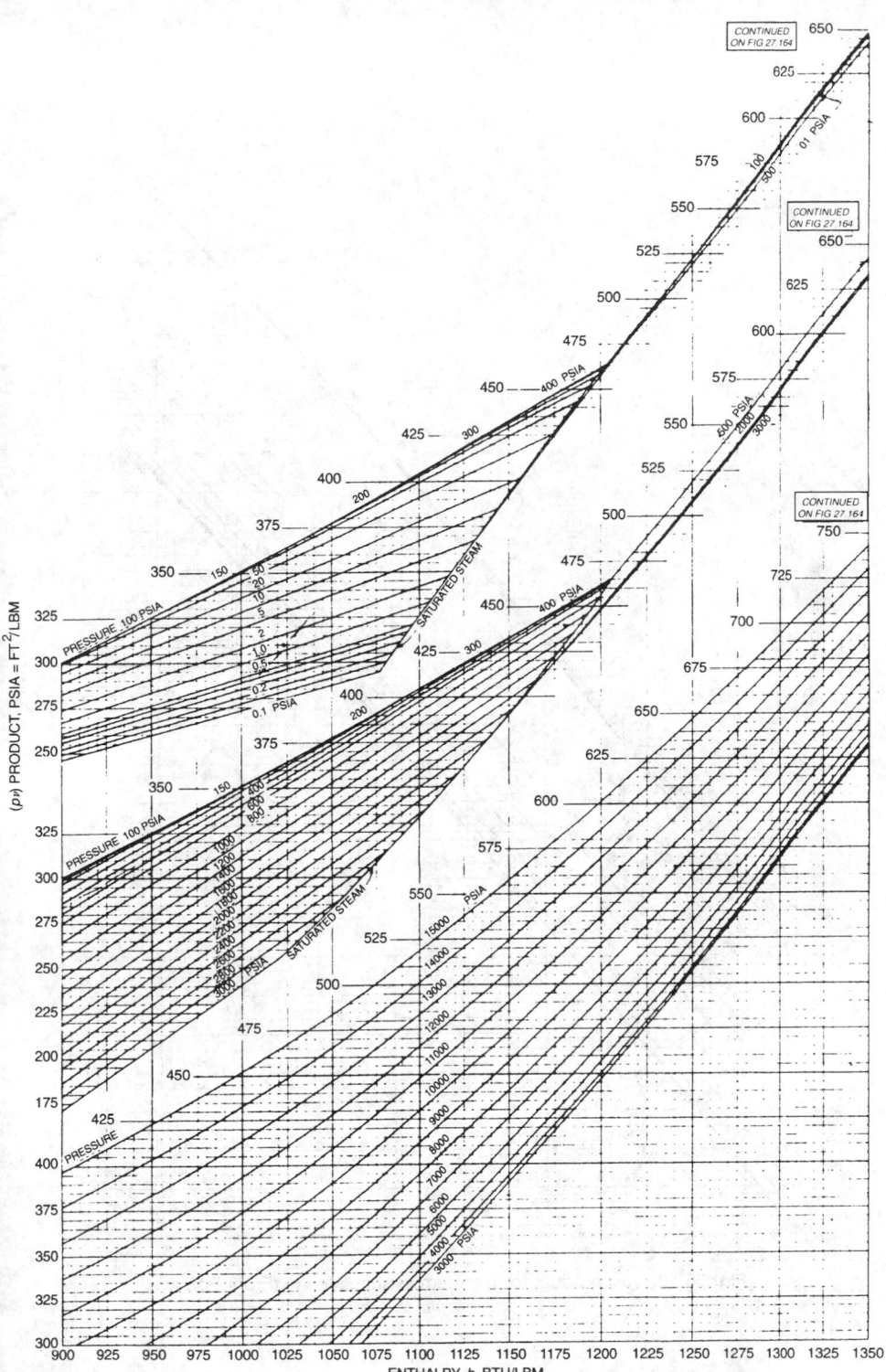

Fig. 58.18 (ρv) product for low-temperature steam. (From Ref. 2.)

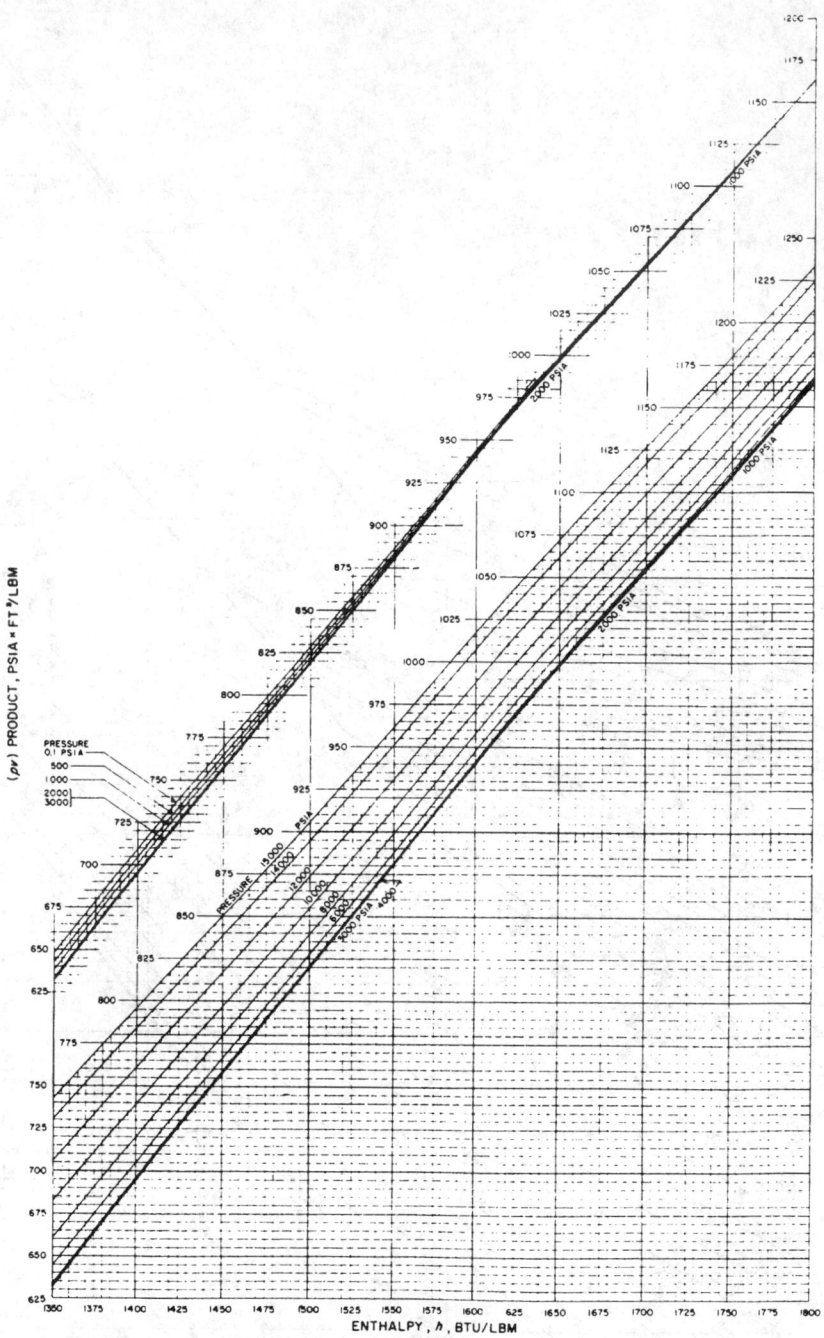

Fig. 58.19 (ρv) product for high-temperature steam. (From Ref. 2.)

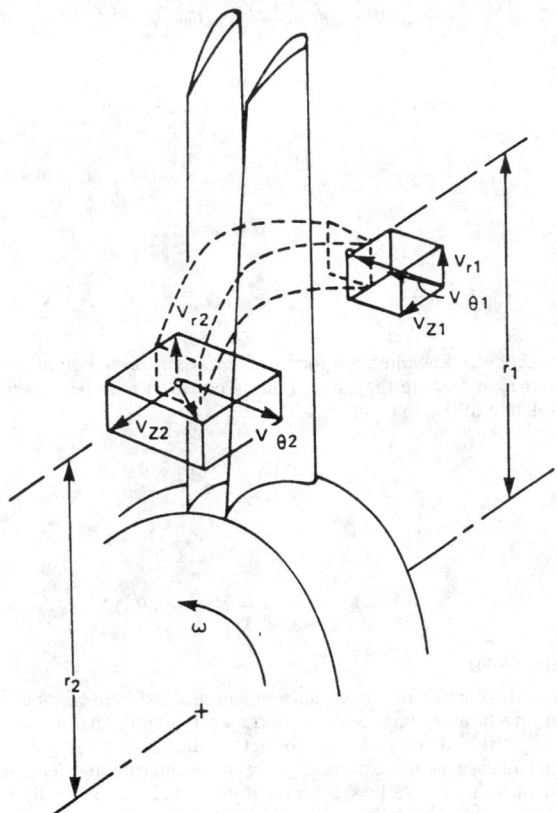

Fig. 58.20 Turbine blade row streamtube.

Since the flow is assumed steady and the velocity components are assumed constant within the streamtube upstream and downstream of the blade row, the force on the fluid is

$$F_{\theta\text{on fluid}} = \frac{\dot{M}}{g}(V_{\theta_2} - V_{\theta_1}) \tag{58.6b}$$

If the entering and leaving streamtube radii are identical, the power P developed by the moving blade row is

$$P = \frac{\dot{M}}{g}r\omega(V_{\theta 1} - V_{\theta 2}) \tag{58.7}$$

where the algebraic sign is changed to determine the fluid force on the blade. With unequal radii, implying unequal tangential blade velocities, the change in angular fluid momentum requires the power relation to be

$$P = \frac{\dot{M}}{g}\omega(r_1 V_{\theta 1} - r_2 V_{\theta 2}) \tag{58.7a}$$

Consideration of the general energy relationship (the First Law) expressed as

$$\delta Q - \delta W = du + d(pv) + \frac{VdV}{g} + dz \tag{58.8}$$

(Q is heat, W is work, and u is internal energy) and applied to the inlet and exit of the streamtube yields

$$\delta W = dh + \frac{VdV}{g} = dh_0 \tag{58.9}$$

where h is static enthalpy and h_0 is stagnation enthalpy. The change in total energy content of the fluid must equal that amount absorbed by the moving rotor blade in the form of mechanical energy

$$h_{01} - h_{02} = \frac{\omega}{g}(r_1 V_{\theta 1} - r_2 V_{\theta 2}) \tag{58.10}$$

or

$$h_1 + \frac{V_1^2}{2g} - h_2 - \frac{V_2^2}{2g} = \frac{\omega}{g}(r_1 V_{\theta 1} - r_2 V_{\theta 2}) \tag{58.10a}$$

Both of these may be recognized as alternate forms of *Euler's turbine equation.*

In the event the radii r_1 and r_2 are the same, Euler's equation may be expressed in terms of the velocity components relative to the rotor blade as

$$h_{01} - h_{02} = \frac{U}{g}(W_{\theta 1} - W_{\theta 2}) \tag{58.10b}$$

or

$$h_1 + \frac{V_1^2}{2g} - h_2 - \frac{V_2^2}{2g} = \frac{U}{g}(W_{\theta 1} - W_{\theta 2}) \tag{58.10c}$$

58.4.2 Turbine Stage Designs

The means of achieving the change in tangential momentum, and hence blade force and work, are many and result in varying turbine stage designs. Stage design and construction falls generally into two broad categories, *impulse* and *reaction*. The former implies that the stage pressure drop is taken entirely in the stationary blade passage, and that the relative entering and leaving rotor blade steam velocities are equal in magnitude. Work is achieved by the redirection of flow through the blade without incurring additional pressure decrease. At the other end of the scale, one could infer that for the reaction concept, all the stage pressure drop is taken across the rotor blade, while the stator blade merely redefines steam direction. The modern connotation of reaction is that of 50% *reaction,* wherein half the pressure drop is accommodated in both stator and rotor. Reaction can be defined in two ways, on an enthalpy-drop or a pressure-drop basis. Both definitions ratio the change occurring in the rotor blade to the stage change. The pressure definition is more commonly encountered in practice.

Figure 58.21 presents a schematic representation of impulse and reaction stage designs. The impulse design requires substantial stationary diaphragms to withstand the stage pressure drop and tight sealing at the inner diameter to minimize leakage. The reaction design (50%) can accept somewhat greater seal clearances under the stator and over the rotor blade and still achieve the same stage efficiency.

The work per stage potential of the impulse design is substantially greater than that of the reaction design. A comparison is presented in Fig. 58.22 of various types of impulse designs with the basic 50% reaction design. A typical impulse stage (*Rateau stage*) can develop 1.67 times the work output of the reaction stage; for a given energy availability, this results in 40% fewer stages.

Performance characteristics of these designs are fundamentally described by the variation of stage aerodynamic efficiency as a function of velocity ratio. The *velocity ratio* can be defined many ways dependent on the reference used to define the steam speed V. The velocity ratio in general is

$$\nu = \frac{U}{V} \tag{58.11}$$

where U is the blade tangential velocity.

The relative level of stage aerodynamic efficiency is indicated in Fig. 58.22 for these various geometries ranging from the symmetrical reaction design (50% reaction) to a three-rotating-row impulse design. As the work level per design increases, the aerodynamic efficiency decreases. Some qualifying practicalities prevent this from occurring precisely this way in practice. That is, the impulse concept can be utilized in the high-pressure region of the turbine, where densities are high and blade heights are short. When blade heights necessarily increase in order to pass the required mass flow, significant radial changes occur in the flow conditions, resulting in radial variations in pressure and, hence, reaction. All impulse designs have some amount of reaction dependent on the ratio of blade

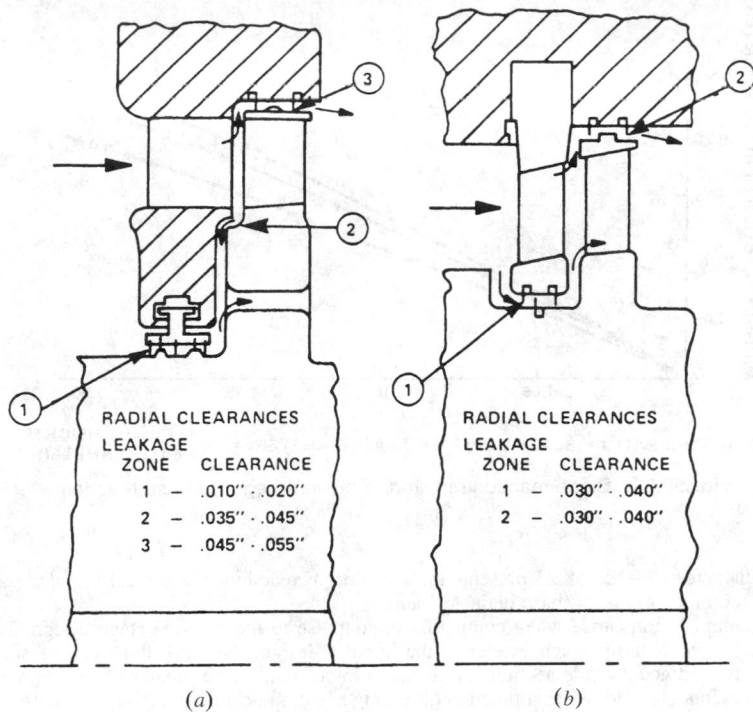

Fig. 58.21 Comparison of impulse and reaction stage geometries: (a) impulse construction; (b) reaction construction.

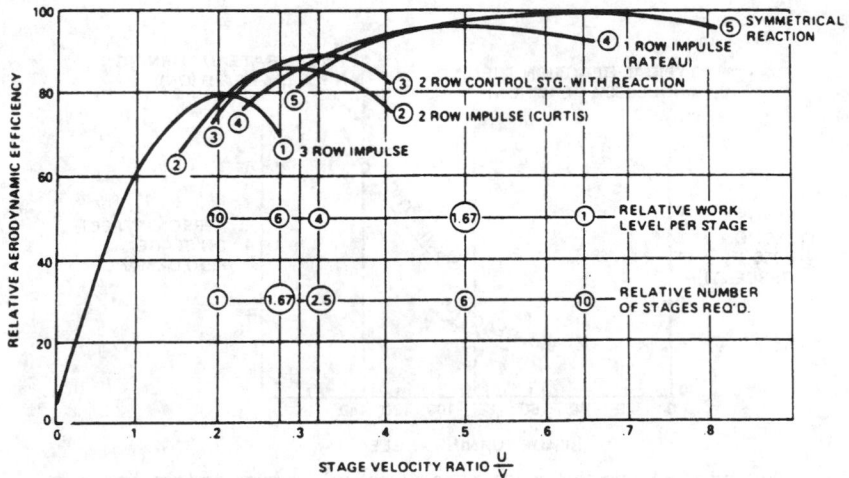

Fig. 58.22 The effect of stage design on aerodynamic efficiency.

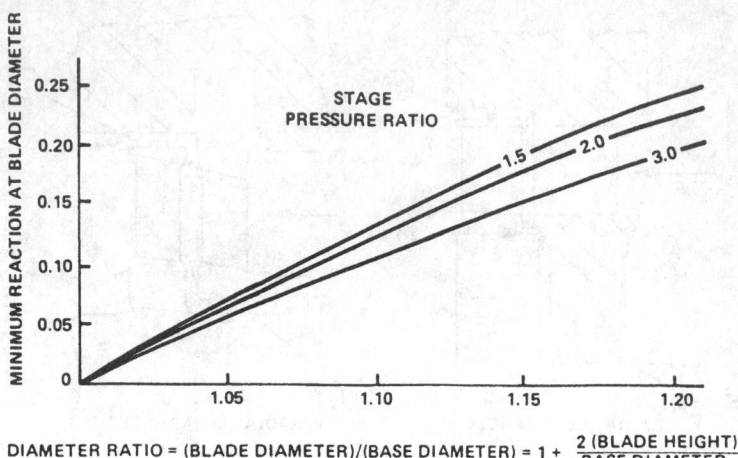

$$\text{DIAMETER RATIO} = \text{(BLADE DIAMETER)/(BASE DIAMETER)} = 1 + \frac{2 \text{ (BLADE HEIGHT)}}{\text{BASE DIAMETER}}$$

Fig. 58.23 Blade reaction required to prevent negative base reaction.

hub to tip diameters. Figure 58.23 presents the variation of reaction in a typical impulse stage as a function of the blade diameter/base diameter ratio.

Further complications arise when comparing the impulse to the reaction stage design. The rotor blade turning is necessarily much greater in the impulse design, the work done can be some 67% greater, and the attendant blade section aerodynamic losses tend to be greater. As an extreme case, Fig. 58.24 presents blade losses as a function of turning angle, indicating much greater losses incurred by the impulse rotor blade design.

Test data is, of course, the best source of information defining these performance relationships. Overall turbine test data provide the stage characteristic, which includes all losses incurred. The fundamental blade section losses are compounded by leakages around blade ends, three-dimensional losses due to finite length blades and their end walls, moisture losses if applicable, disk and shroud friction, and blading incidence losses induced by the variation in the velocity ratio itself. A detailed description of losses has been presented by Craig and Cox[6] and more recently by others including Kacker and Okapuu.[7]

58.4.3 Stage Performance Characteristics

Referring again to Fig. 58.20, the velocities relative to the rotor blade are related to the stator velocities and to the wheel speed by means of the appropriate velocity triangles, as in Fig. 58.25.

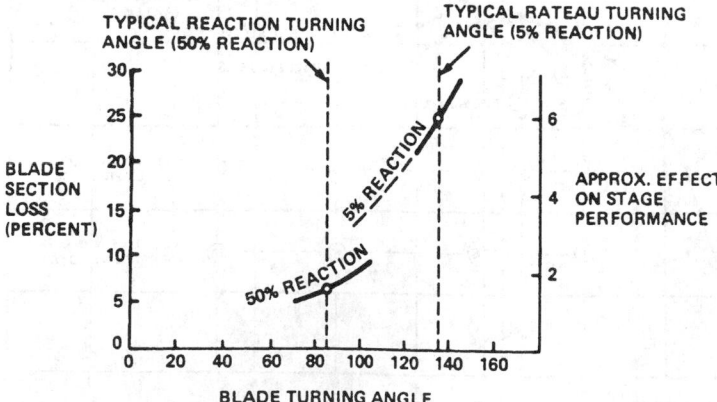

Fig. 58.24 Blade section losses as a function of turning and reaction.

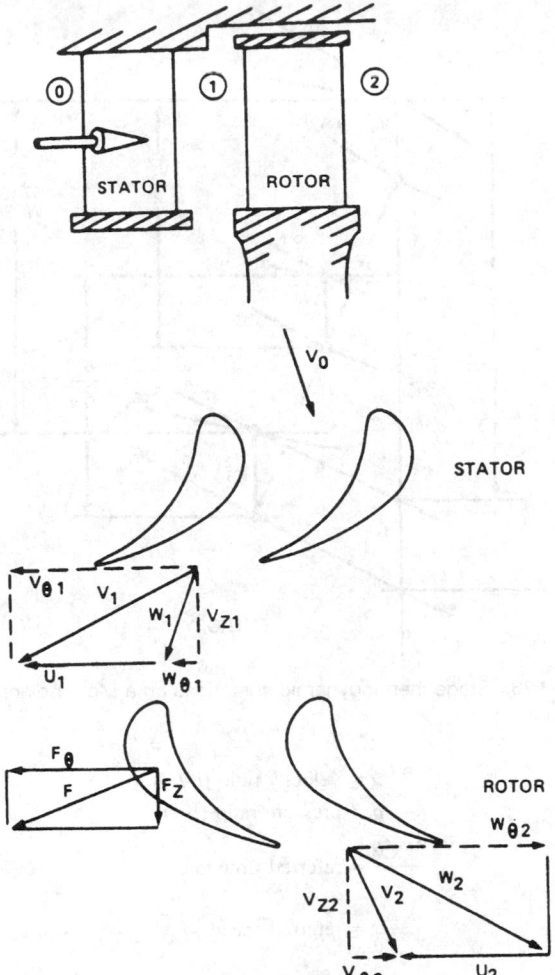

Fig. 58.25   Stage velocity triangles.

Absolute velocities are denoted by V and relative velocities by W. The blade sections schematically shown in this figure are representative of reaction blading. The concepts and relationships are representative of all blading.

The transposition of these velocities and accompanying steam conditions are presented on a Mollier diagram for superheated steam in Fig. 58.26. The stage work will be compatible with Euler's turbine equation shown in Eqs. (58.10). The *stage work* is the change in total enthalpy ($\Delta h_w = h_{00} - h_{02}$) of the fluid passing through the stage. The total pressures and temperatures before and after the stage, the local thermodynamic conditions, and total pressure and total temperature relative to the rotor blade (p_{0_r} and T_{0_r}) are all indicated in Fig. 58.26.

This ideal description of the steam expansion process defines the local conditions of a particular streamtube located at a certain blade height. For analysis calculations and the prediction of turbine performance, the mean diameter is usually chosen as representative of the stage performance. This procedure is, in effect, a one-dimensional analysis, and it represents the turbine performance well if appropriate corrections are made for three-dimensional effects, leakage, and moisture. In other words, the blade row, or stage efficiency, must be known or characterized as a function of operating parameters. The velocity ratio, for example, is one parameter of great significance. Dimensional analysis of the individual variables bearing on turbomachinery performance has resulted in several commonly used dimensionless quantities:

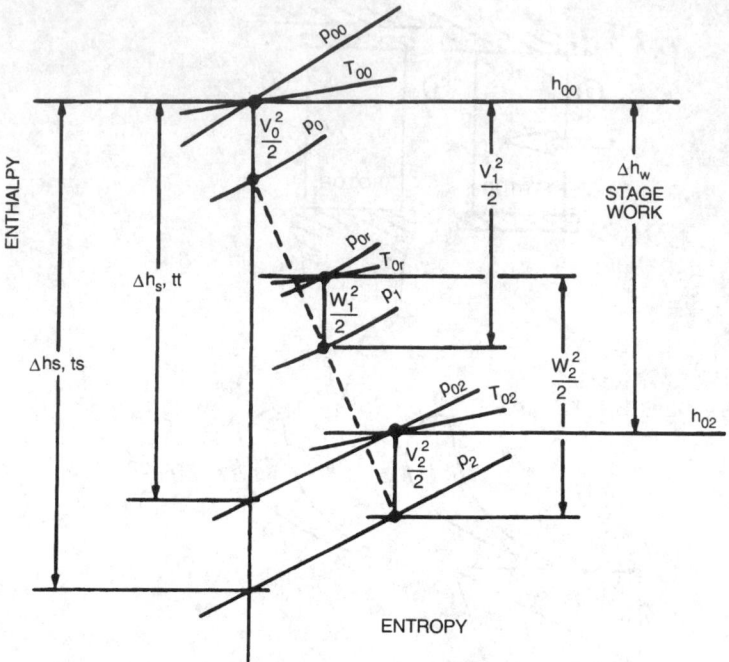

Fig. 58.26 Stage thermodynamic conditions on a Mollier diagram.

$$\nu = \text{velocity ratio } (U/V) \tag{58.12}$$

$$\rho = \text{pressure ratio } (p_0/p) \tag{58.13}$$

$$\frac{\dot{M}\sqrt{\theta}}{\delta} = \text{referred flow rate} \tag{58.14}$$

$$\frac{N}{\sqrt{\theta}} = \text{referred speed} \tag{58.15}$$

$$\eta = \text{efficiency} \tag{58.16}$$

where $\theta = T_0/T_{\text{ref}}$ and $\delta = p_0/p_{\text{ref}}$.

The *referred flow rate* is a function of pressure ratio. In gas dynamics, the referred flow rate through a converging nozzle is primarily a function of the pressure ratio across the nozzle (Fig. 58.27). When the pressure ratio p_0/p is approximately 2 (depending on the ratio of specific heats), the referred mass flow maximizes, that is,

$$\frac{\dot{M}\sqrt{T_0}}{p_0} = \text{constant} \tag{58.17}$$

This referred flow rate is also termed a *flow number* and takes the following form, which is more appropriate for steam turbine usage.

$$\frac{\dot{M}\sqrt{p_0 v_0}}{p_0} = \text{flow number} \tag{58.18}$$

The turbine behaves in a manner similar to a nozzle, but it is also influenced by the wheel-speed Mach number, which is represented by $N/\sqrt{T_0}$. In effect, a similar flow rate–pressure ratio relationship is experienced, as shown schematically in Fig. 58.28.

A combined plot describing turbine performance as a function of all these dimensionless quantities is shown schematically in Fig. 58.29. Power-producing steam turbines, however, run at constant speed and at essentially constant flow number, pressure ratio, and velocity ratio, and in effect operate at

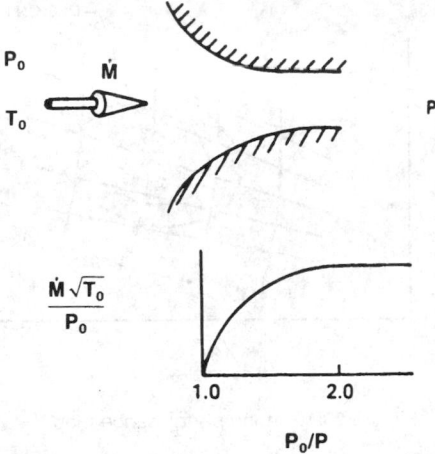

Fig. 58.27 Gas dynamic flow relationships.

nearly a fixed point on this performance map. Exceptions to this are the control stage, if the unit has one, and the last few stages of the machine. The control stage normally experiences a wide range of pressure ratios in the situation where throttle pressure is maintained at a constant value and a series of governing valves admits steam to the control stage blading through succeeding active nozzle arcs.

The last stage of the low pressure end exhausts to an essentially constant pressure zone maintained by the condenser. Some variation occurs as the condenser heat load changes with unit flow rate and load. At a point several stages upstream where the pressure ratio to the condenser is sufficiently high, the flow number is maintained at a constant value. As unit load reduces, the pressure level and mass flow rate decrease simultaneously and the pressure ratio across the last few stages reduces in value. This change in pressure ratio changes the stage velocity ratio, and the performance level of these last few stages change. Figure 58.30 indicates the trend of operation of these stages as a function of load change superimposed on the dimensionless turbine performance map. It is these last few stages, say two or three, that confront the steam turbine designer with the greatest challenge.

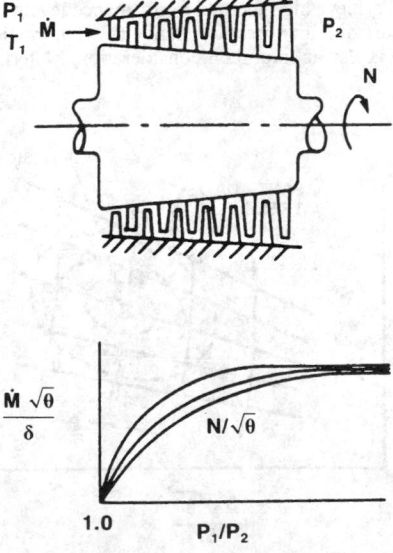

Fig. 58.28 Turbine flow relationships.

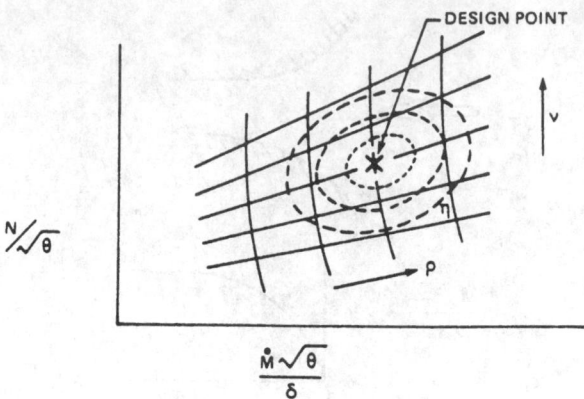

Fig. 58.29 Turbine performance map.

58.4.4 Low-Pressure Turbine Design

The low-pressure element of the steam turbine is normally a self-contained design, in the sense that it comprises a rotor carrying several stages, a cylinder with piping connections, and bearings supporting the rotor. Complete power-generating units may use from one to three low-pressure elements in combination with high-pressure and intermediate-pressure elements depending on the particular application.

Symmetric double-flow designs utilize from 5 to 10 stages on each end of the low-pressure element. The initial stage accepts flow from a centrally located plenum fed by large-diameter, low-pressure steam piping. In general, the upstream stages feature constant cross-section blading, while the latter stages require blades of varying cross section. Varying design requirements and criteria result in these latter twisted and tapered low pressure blades comprising from 40 to 60% of the stages in the elements.

The particular design criterion used significantly affects the end product. For example, the low-pressure end (one-half of the element) of a nuclear unit might contain 10 stages and rotate at 1800 rpm and generate 100 mW, while the fossil machine would spin at 3600 rpm, carry five or six stages, and develop 50 mW. The physical size could double and the weight increase by a factor of 8.

Further differences in design are immediately implied by the steam conditions to be accepted by the element. The nuclear unit inlet temperature would be approximately 550°F (288°C), as compared to the fossil condition of 750°F (399°C). The nuclear unit must also be designed to effectively remove blade path moisture in order to achieve high thermal performance levels (heat rate). Moisture-removal devices in both the high-pressure and low-pressure elements are normally incorporated for this purpose. Blading physical erosion is also an important consideration. Material removal by the high speed

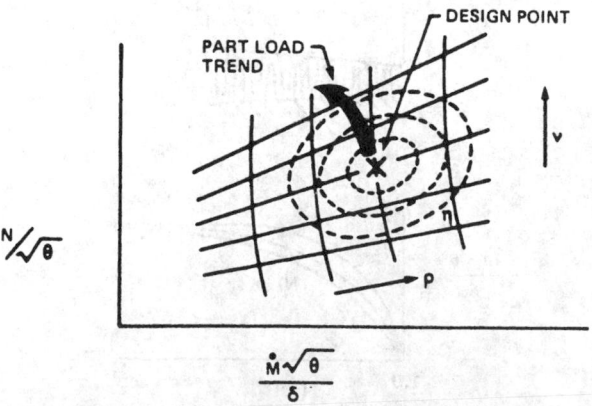

Fig. 58.30 Turbine performance map with part load trend.

impingement of water droplets can be minimized by proper design. Moore and Sieverding[8] describe the phenomena controlling wet steam flow in turbines.

Flow Field Considerations

Substantial advances have been achieved in the aerothermodynamic processes of low-pressure turbines. A significant factor is the availability of high-speed digital computers in the solution of the system of turbomachinery aerothermodynamic equations. The coupling of the solution of this complex equation system with the necessary verification as derived from experimental testing programs is the subject of this section.

The development of the system of aerothermodynamic equations is traced from basic considerations of the several conservation equations to working equations, which are then generally solved by numerical techniques. Pioneering work in the development of this type of equation systems was performed by Wu.[9] Subsequent refinement and adaptation of his fundamental approach has been made by several workers in recent years.

The general three-dimensional representation of the conservation equations of mass, momentum, and energy can be written for this adiabatic system

$$\frac{\partial \rho}{\partial t} + \nabla \cdot \rho \overline{V} = 0 \tag{58.19}$$

$$\frac{\partial \overline{V}}{\partial t} + \overline{V} \cdot \nabla \overline{V} = -\frac{1}{\rho} \nabla p \tag{58.20}$$

$$\nabla h = T\nabla s + \frac{1}{\rho} \nabla p \tag{58.21}$$

Assuming the flow to be steady and axisymmetric, the conservation equations can be expanded in cylindrical coordinates as

$$\frac{\partial}{\partial r}(\rho rbV_r) + \frac{\partial}{\partial z}(\rho rbV_z) = 0 \tag{58.22}$$

$$V_r \frac{\partial}{\partial r}(rV_\theta) + V_z \frac{\partial}{\partial z}(rV_\theta) = 0 \tag{58.23}$$

$$V_r \frac{\partial V_z}{\partial r} + V_z \frac{\partial V_z}{\partial z} = -\frac{1}{\rho} \frac{\partial p}{\partial z} \tag{58.24}$$

$$V_r \frac{\partial V_r}{\partial r} + V_z \frac{\partial V_r}{\partial z} - \frac{V_\theta^2}{r} = -\frac{1}{\rho} \frac{\partial p}{\partial r} \tag{58.25}$$

$$\frac{1}{\rho} dp = dh_0 - Tds - \frac{1}{2} d(V_\theta^2 + V_z^2 + V_r^2) \tag{58.26}$$

where

$$h_0 = h + \frac{1}{2}(V_\theta^2 + V_z^2 + V_r^2) \tag{58.27}$$

and b has been introduced to account for *blade blockage*. The properties of steam may be described by equations of state, such that $p = p(p, T)$ and $h = h(p, T)$. These functions are available in the form of the steam tables.[2]

As the increase in system entropy is a function of the several internal loss mechanisms inherent in the turbomachine, it is necessary that definitive known (or assumed) relationships be employed for its evaluation. In the design process, this need can usually be met, while the performance analysis of a given geometry usually introduces loss considerations more difficult to evaluate completely. The matter of loss relationship has been discussed with regard to stage design and will be further reviewed in a later section. In general, however, the entropy increase can be determined as a function of aerodynamic design parameters, that is,

$$\Delta s = f(V_i, W_i, \text{Mach number}) \tag{58.28}$$

Equations (58.22) through (58.25) form a system of nine equations in nine unknowns, V_r, V_θ, V_z, ρ, p, T, h, h_0, and s, the solution of which defines the low pressure turbine flow field.

The *meridional plane,* defined as that plane passing through the turbomachine axis and containing the radial and axial coordinates, can be used to describe an additional representation of the flow process. The velocity V_m (see Fig. 58.31) represents the streamtube meridional plane velocity with direction proportional to the velocity components V_r and V_z. If the changes in entropy and total enthalpy along the streamlines are known, or specified, it can be shown that Eqs. (58.24) and (58.25) are equivalent and that Eq. (58.23) is also satisfied. Application of Eq. (58.22) to the rotating blade in effect described Euler's turbine equation when equated to the blade force producing useful work. In this situation, it is most convenient to choose Eq. (58.25), the commonly known *radial equilibrium equation,* as the relationship for continued evaluation.

58.4.5   Flow Field Solution Techniques

Two commonly encountered techniques have been employed to solve this set of fundamental differential equations and relationships. They are usually referred to as the *streamline curvature* and *matrix solution* techniques. The streamline curvature technique is structured to evaluate meridional velocities and to trace streamlines in the flow field, allowing the calculation of the streamline curvature itself. The streamline curvature technique has been developed to a high degree of sophistication and has been applied to axial flow compressor design as well as to axial flow turbine problems.

The matrix approach utilizes a stream function satisfying the equation of continuity. The calculation procedure then determines the stream function throughout the flow field. This technique has also been developed satisfactorily with particular application to low-pressure steam turbines. The matrix approach asserts the existence of a stream function that is defined such that

$$V_r = -\frac{1}{\rho r b}\frac{\partial \psi}{\partial z} \tag{58.29}$$

and

$$V_z = \frac{1}{\rho r b}\frac{\partial \psi}{\partial r} \tag{58.30}$$

With this definition, ψ, V_r, and V_z identically satisfy Eq. (58.19). The combination of the equations of momentum, energy, and continuity, Eqs. (58.25), (58.26), (58.29), (58.30), and the definition of total enthalpy, Eq. (58.27), results in

$$\frac{\partial}{\partial r}\left(\frac{1}{\rho r b}\frac{\partial \psi}{\partial r}\right) + \frac{\partial}{\partial z}\left(\frac{1}{\rho r b}\frac{\partial \psi}{\partial z}\right) = \frac{1}{V_z}\left[\frac{\partial h_o}{\partial r} - \frac{V_\theta}{r}\frac{\partial(rV_\theta)}{\partial r} - T\frac{\partial s}{\partial r}\right] \tag{58.31}$$

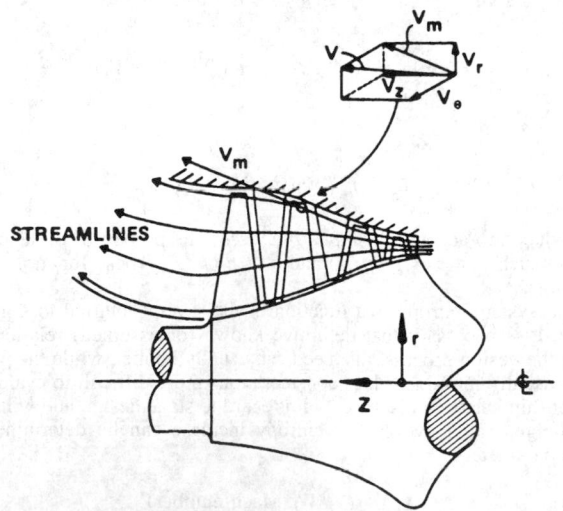

Fig. 58.31   Turbine blade path flow field.

This is the basic flow field equation, which can then be solved by numerical techniques. This equation is an elliptic differential equation, provided the meridional Mach number is less than unity. This is probably the case in all large steam turbines. The absolute Mach number can of course exceed unity and usually does in the last few blade rows of the low-pressure turbine.

The streamline curvature approach satisfies the same governing equations but solves for the meridional velocity V_m rather than the stream function ψ, as in the matrix approach. Equations (58.25) and (58.26) can be combined and expressed in terms of directions along the blade row leading or trailing edges. An equation used to describe the variation of meridional velocity is

$$\frac{dV_m^2}{dl} + A(l)V_m^2 = B(l) \tag{58.32}$$

where l is the coordinate along the blade edge and the coefficients A and B depend primarily on the slopes of the streamlines in the meridional plane. An iterative process is then employed to satisfy the governing equations.

58.4.6 Field Test Verification of Flow Field Design

The most conclusive verification of a design concept is the in-service evaluation of the product with respect to its design parameters. Application of a matrix-type flow field design program has been made to the design of the last three stages of a 3600-rpm low-pressure end. Figure 58.32 is indicative of the general layout of this high-speed fossil turbine.

The aerodynamic design process is to a great degree an iterative process, that is, the specification of the design parameters (for example, work per stage and radial work distribution) are continuously adjusted in order to optimize the flow field design. Just as important, feedback from mechanical analyses must be accommodated in order to achieve reliable long low-pressure blades.

A key design criterion is the requirement that low-pressure blades must be vibrationally tuned in that the lowest several natural modes of vibration must be sufficiently removed from harmonics of running speed during operation. This tuning process is best represented by the *Campbell diagram* of Fig. 58.33. The first four modes of the longest blade are represented as a function of running speed in rpm. At the design speed, the intersection of the mode lines and their band widths indicate nonresonant operation. A comparison is also shown between full-size laboratory test results and shop test data. The laboratory results[10] were obtained by means of strain gage signals transmitted by radio telemetry techniques. In the shop tests, the blades were excited once per revolution by a specially designed steam jet. The strain gage signals were delivered to recording equipment by means of slip rings.

If it turns out that the mode lines intersect a harmonic line at running speed, a resonant condition exists that could lead to a fatigue failure of the blade. The iterative process occurs prior to laboratory

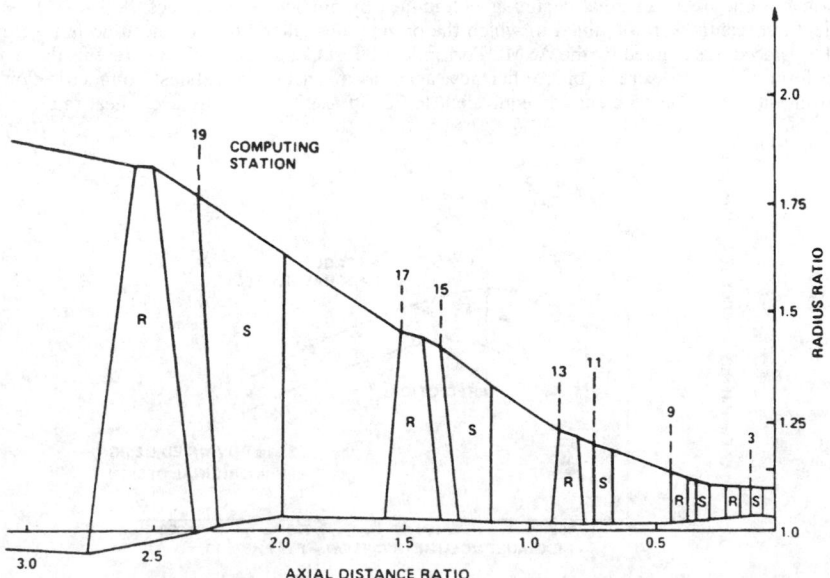

Fig. 58.32 Low-pressure turbine blade path.

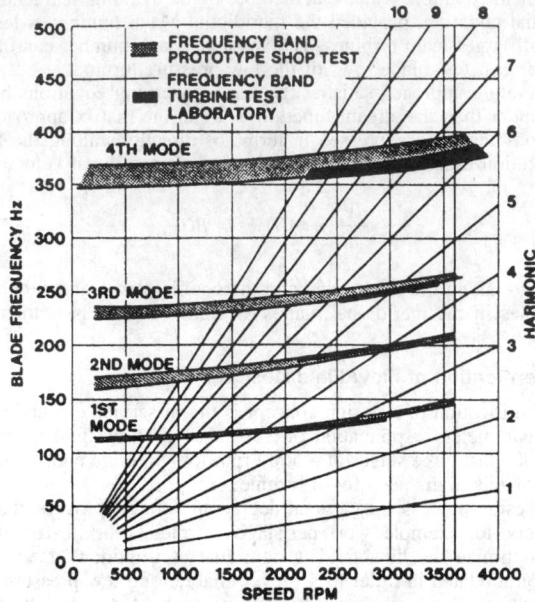

Fig. 58.33 Campbell diagram of low-pressure blade.

and shop verification. Analytic studies and guidance from experience gained from previous blade design programs enable the mechanical designer to determine blade shape changes that will eliminate the resonance problem. This information is then incorporated by the aerodynamicist into the flow field design. As this process continues, manufacturing considerations are incorporated to ensure the design can be produced satisfactorily.

Construction and testing then followed: test data was obtained from several sources, a field test and an in-house test program.[10] Figure 58.34 displays low-pressure turbine efficiency versus exhaust volumetric flow for the advanced design turbine, designed by means of the matrix-type flow field design process, and original design turbine. Although the same range of exhaust volumetric flow was not achievable in both test series, the improvement is apparent and is further indicated by the extended performance line of the original design as determined by predictive techniques.

Field test results were obtained in which the original design and the advanced design were evaluated by procedures defined by the ASME Performance Test Code. Figure 58.35 presents these results in the form of low-pressure turbine efficiency, again as a function of exhaust volumetric flow. The improvement in turbine efficiency is equivalent to 70 Btu/kWh in turbine cycle heat rate (18 kcal/kWh).

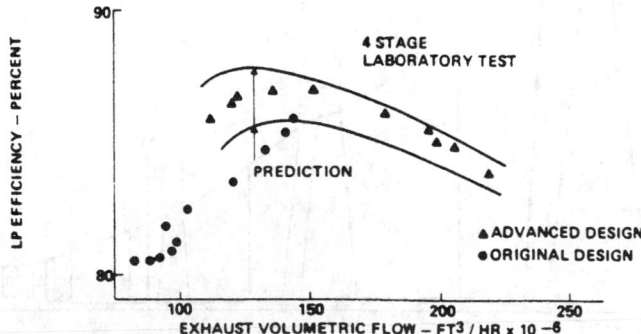

Fig. 58.34 Low-pressure turbine laboratory test results.

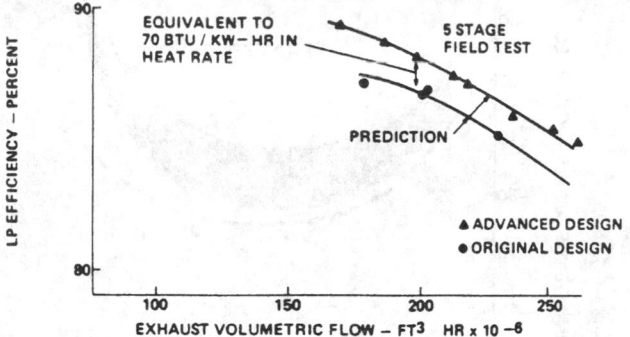

Fig. 58.35 Low-pressure turbine field test results.

Aerodynamic losses in the blade path are an integral part of performance determination, as they directly affect the overall efficiency of the turbine. The losses at part load, or off-design operation, are additive to those existing at the design point and may contribute significantly to a deterioration of performance. Many factors influence these incremental losses, but they are due primarily to blading flow conditions that are different from those at the optimum operating condition. For example, lower (or higher) mass flow rates through the machine have been shown to change the pressure levels, temperatures, velocities, flow angles, moisture content, and so on within the blading. These changes induce flow conditions relative to the blading, which create additional losses. As different radial locations on the same blade are affected to a different degree, the overall effect must be considered as a summation of all individual effects for that blade. In fact, different blade rows, either stationary or rotating, are affected in a like manner, and the complete effect would then be the summation of the individual effects of all the blade rows.

Detailed verification of the design process has been obtained from the analysis and comparison of these internal flow characteristics to expected conditions, as determined by off-design calculations employing the same principles as incorporated in the design procedure. From measurements of total and static pressure and a knowledge of the enthalpy level at the point of interest, it is possible to determine the steam velocity, Mach number, and local mass flow rate. The flow angle is measured simultaneously. Traverse data for the advanced design turbine are presented in Figs. 58.36 and 58.37 for the last stationary blade inlet. Flow incidence is presented as a function of blade height for high and low values of specific mass flow. Good agreement is indicated between prediction and the traverse data. As flow rate (and load) decrease, the correspondence between test and calculation becomes less convincing, indicating that loss mechanisms existing under part-load operation are not as well defined as at high load near the design point.

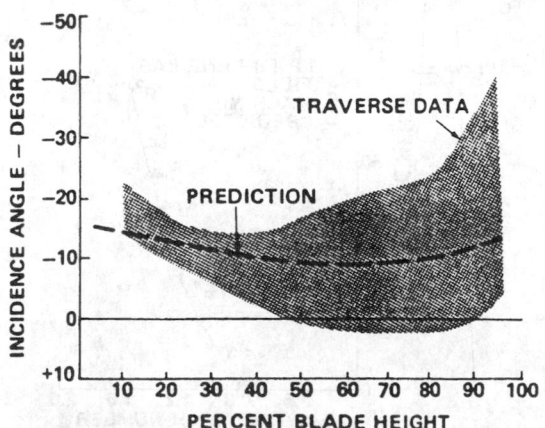

Fig. 58.36 Last stationary blade incidence angle at high-end loading—12,000 lb/hr-ft².

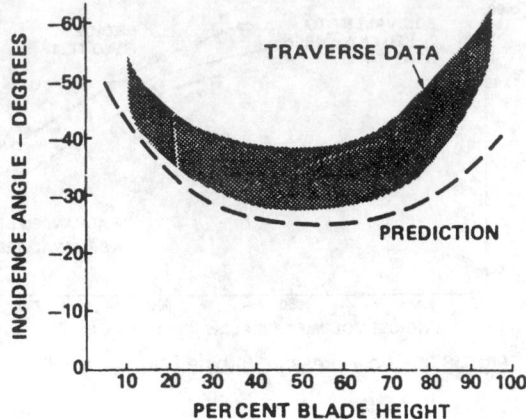

Fig. 58.37 Last stationary blade incidence angle at low end loading—6,000 lb/hr-ft².

By calculation, utilizing the rotor blade downstream traverse data, it is possible to determine the operating conditions relative to the rotor blade. That is, the conditions as seen by a traveler on the rotor blade can be determined. Figure 58.38 presents the Mach number leaving the last rotor blade (i.e., relative to it) as a function of blade height compared to the expected variation. Very high Mach numbers are experienced in this design and are a natural consequence of the high rotor blade tip speed, which approximates 2010 ft/sec, which converts to a wheel speed Mach number of 1.61.

In a manner similar to that applied at the inlet of the last stationary blade, the characteristics at the exit of this blade can be determined. Results of this traverse are shown in Fig. 58.39, where the effects of the stator wakes can be identified. The high Mach numbers experienced at the stationary blade hub make this particular measurement extremely difficult. The comparison shows good agreement, with expectations for this difficult measurement location.

The combination of these stationary blade exit traverses with velocity triangle calculations enables the flow conditions at the inlet of the rotor blade to be determined. Flow incidence as a function of blade height (diameter) is presented in Fig. 58.40 for the high-end load condition of Fig. 58.36. Significant flow losses can be incurred by off-design operation at high incidence levels. The hub and tip are particularly sensitive, and care must be taken in the design process to avoid this situation.

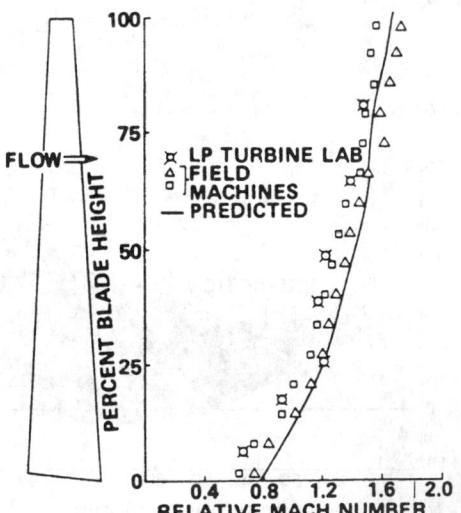

Fig. 58.38 Mach number relative to last rotating blade.

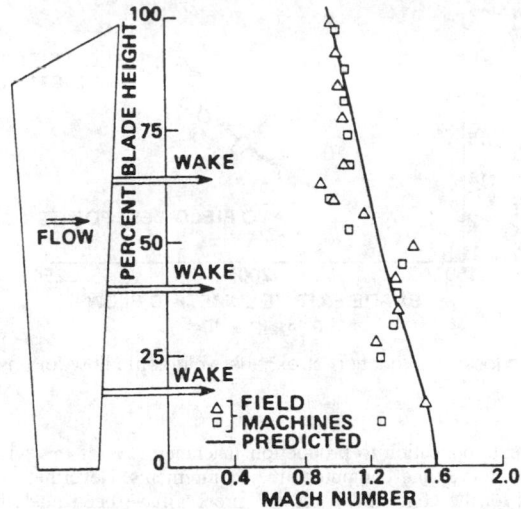

Fig. 58.39   Mach number at exit of last stationary blade.

Further confirmation of the design process can be achieved by evaluation of the kinetic energy leaving the last stage. This energy is lost to the turbine and represents a significant power output if it were possible to convert it to useful work. This kinetic energy, or *leaving loss,* is defined as

$$\overline{V^2} = \int_{\text{Hub}}^{\text{Tip}} (\rho V)\, \frac{V^2\, dA}{\dot{M}} \tag{58.33}$$

and plotted in Fig. 58.41 as a function of exhaust volumetric flow. Excellent agreement has been achieved between the test data and prediction.

The foregoing represents conventional practice in the design of low-pressure blading as well as the design and analysis of upstream blading located in the HP and IP cylinders. The capability of solving the Navier–Stokes equations is now available. Drawbacks to this approach are the complexities of constructing a three-dimensional model of the subject of interest and the computational time required. The current philosophy regarding the use of Navier–Stokes solvers is that their practical application is to new design concept developments in order to determine valid and optimum resolutions of these concepts. In this process, design guidelines and cause-and-effect relationships are

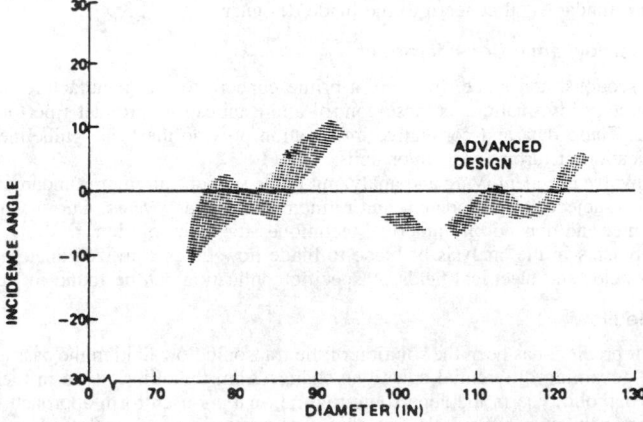

Fig. 58.40   Last rotating blade incidence angle at high-end loading.

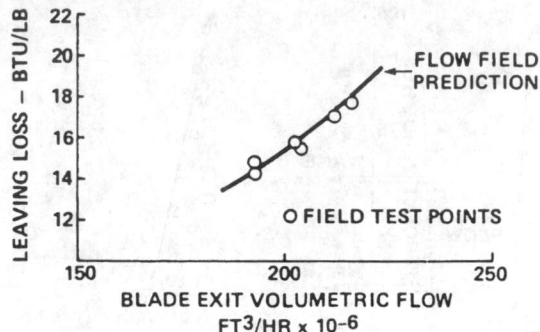

Fig. 58.41 Leaving loss as a function of exhaust volumetric flow for an advanced design.

developed for subsequent application to production machinery. As of this writing, their day-to-day use is prohibited by manpower and computer time requirements. Nevertheless, significant improvements in, and changes to, the conventional design process have been made by the Navier–Stokes solvers. For the steam turbine aerodynamicist, the most influential approaches are those developed by Denton,[11] a program *not* a Navier–Stokes solver but having three-dimensional inviscid capability, and that by Dawes,[12] a truly three-dimensional Navier–Stokes solver.

58.4.7 Blade-to-Blade Flow Analysis

The blade-to-blade flow field problem confronts the turbine designer with a variety of situations, some design, and some analysis of existing designs. The flow regime can vary from low subsonic (incompressible) to transonic with exit Mach numbers approaching 1.8. The questions to be resolved in blade design are what section will satisfy the flow field design, and how efficiently it will operate.

From the mechanical viewpoint, will the blade section, or combination of sections, be strong enough to withstand the steady forces required of it, centrifugal and steam loading, and will it be able to withstand the unknown unsteady forces impressed upon it? The structural strength of the blade can be readily evaluated as a function of its geometry, material characteristics, and steady stresses induced by steady loads. The response of the blade to unsteady forces, generally of an unknown nature, is a much more formidable problem. Tuning of the long, low pressure blade is a common occurrence, but as blades become shorter, in the high and intermediate pressure elements, and in the initial stages of the low-pressure element, tuning becomes a difficult task. Blades must be designed with sufficient strength, or margin, to withstand these unsteady loads and resonance conditions at high harmonics of running speed.

The aerodynamic design of the blade depends on its duty requirements, that is, what work is the blade expected to produce? The fluid turning is defined by the flow field design, which determines the radial distribution of inlet and exit angles. Inherent in the flow field design is the expectation of certain levels of efficiency (or losses) that have defined the angles themselves. The detailed blade design must then be accomplished in such a manner as to satisfy, or better, these design expectations. Losses then, are a fundamental concern to the blade designer.

54.4.8 Blade Aerodynamic Considerations

Losses in blade sections, and blade rows, are a prime concern to the manufacturer and are usually considered proprietary information, as these control and establish its product's performance level in the market place. These data and correlative information provide the basic guidelines affecting the design and application of turbine–generator units.

Techniques involved in identifying and analyzing blade sections are more fundamentally scientific and have been the subject of development and refinement for many years. Current processes utilize the digital computer and depend on numerical techniques for their solution.

Early developments in the analysis of blade-to-blade flow fields[13] utilized analogies existing between fluid flow fields and electrical fields. A specific application can be found in Ref. 11.

Transonic Blade Flow

The most difficult problem has been the solution of the transonic flow field in the passageway between two blades. The governing differential equations change from the elliptic type in the subsonic flow regime to the hyperbolic type in the supersonic regime, making a uniform approach to the problem solution mathematically quite difficult. In lieu of, and also in support of, analytic procedures to define the flow conditions within a transonic passage, experimental data have been heavily relied upon. These test programs also define the blade section losses, which is invaluable information in itself.

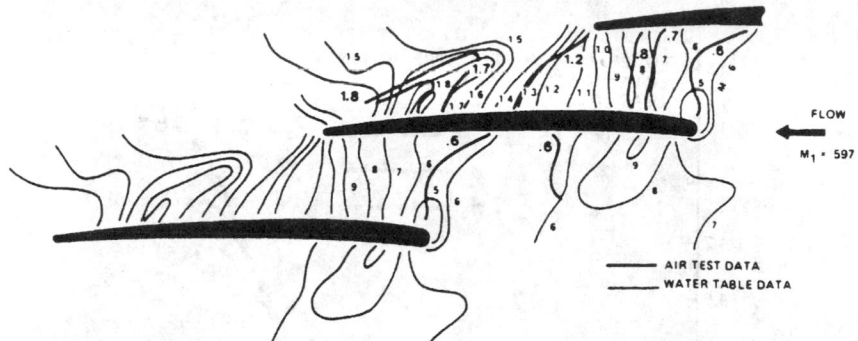

Fig. 58.42 Comparison of water table test results with interferometry air test data for Mach number distribution in transonic blade.

Testing and evaluating blade section performance is an art (and a science) unto itself. Of the several approaches to this type of testing, the air cascade is by far the most common and manageable. Pressure distributions (and hence Mach number) can be determined as a function of operating conditions such as inlet flow direction (incidence), Reynolds number, and overall pressure ratio. Losses also are determinable. Optical techniques are also useful in the evaluation of flow in transonic passages. Constant density parameters from interferometric photos can be translated into pressure and Mach number distributions.

A less common testing technique is the use of a free surface water table wherein local water depths are measured and converted to local Mach number.[14] The analogy between the water table and gas flow is valid for a gas with a ratio of specific heats equal to 2, an approximation, to be sure. A direct comparison of test results for the same blade cascade is presented in Fig. 58.42. The air test data were taken from interferometric photos and converted to parameters of constant Mach number shown in heavy solid lines. Superimposed on this figure are test data from a free-surface water table in lighter lines. A very good comparison is noted for these two sets of data from two completely different blade-testing concepts. The air test is more useful in that the section losses can be determined at the same time the optical studies are being done. The water table, however, produces reasonable results, comparable to the air cascade as far as blade surface distributions are concerned, and is an inexpensive and rapid means of obtaining good qualitative information and, with sufficient care in the experiment, good quantitative data.

Analytical Techniques

Various analytic means have been employed for the determination of the velocity and pressure distributions in transonic blade passages. A particular technique, termed *time marching*, solves the

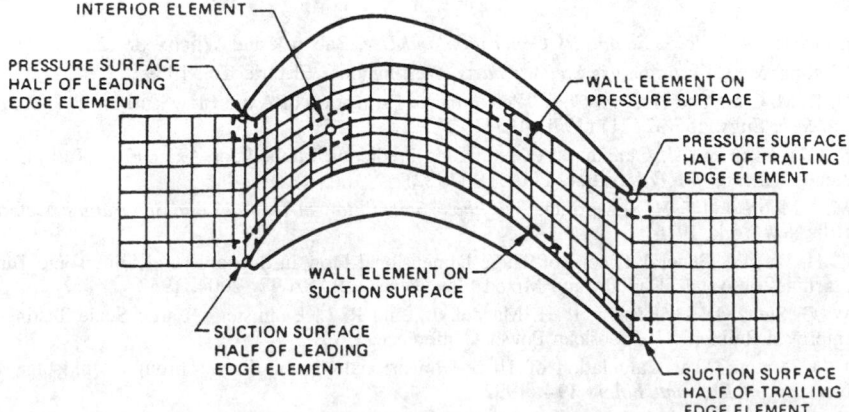

Fig. 58.43 Blade-to-blade calculational region.

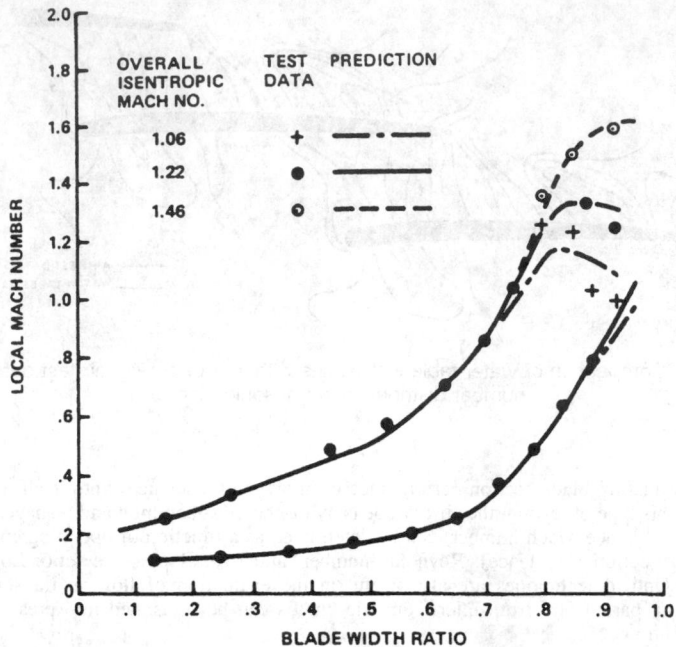

Fig. 58.44   Local Mach number distribution on transonic blade surfaces.

unsteady compressible flow equations by means of successive calculations in time utilizing a flow field comprised of finite area, or volume, elements. Figure 58.43 presents the general calculation region describing the flow passage between two blades. A result of the time marching method is shown in Fig. 58.44, which presents a plot of local Mach number along both the suction and pressure sides of a transonic blade. Parameters of exit isentropic Mach number show a significant variation in local surface conditions as the blade's overall total to static pressure ratio is varied. These data were obtained in air from a transonic cascade facility. In summary, sophisticated numerical processes are available and in common use for the determination of blade surface pressure and velocity distributions. These predictions have been verified by experimental programs.

REFERENCES

1. A. I. Kalina, "Recent Improvements in Kalina Cycles: Rationale and Methodology," *American Power Conf.* **55** (1993).
2. ASME, *The 1967 ASME Steam Tables,* ASME, New York.
3. W. G. Steltz, "The Critical and Two Phase Flow of Steam," *J. Engrg. for Power* **83**, Series A, No. 2 (1961).
4. Babcock and Wilcox, *Steam, Its Generation and Use,* Babcock and Wilcox, 1992.
5. Westinghouse, *Westinghouse Steam Charts,* Westinghouse Electric Co., 1969.
6. H. R. M. Craig and H. J. A. Cox, "Performance Estimation of Axial Flow Turbines," *Proc. Inst. of Mech. Engs.* **185**(32/71) (1970–1971).
7. S. C. Kacker and U. Okapuu, "A Mean Line Prediction Method for Axial Flow Turbine Efficiency," *J. Eng. for Power* **104**, 111–119 (1982).
8. M. J. Moore and C. H. Sieverding, *Two Phase Steam Flow in Turbines and Separators,* McGraw-Hill, New York, 1976.
9. C. H. Wu, "A General Theory of Three-Dimensional Flow in Subsonic and Supersonic Turbomachines of Axial-, Radial-, and Mixed-Flow Types," NASA TN 2604, 1952.
10. W. G. Steltz, D. D. Rosard, P. H. Maedel, Jr., and R. L. Bannister, "Large Scale Testing for Improved Reliability," American Power Conference, 1977.
11. J. D. Denton, "The Calculation of Three-Dimensional Viscous Flow Through Multistage Turbines," *J. of Turbomachinery* **114** (1992).
12. W. N. Dawes, "Toward Improved Throughflow Capability: The Use of Three-Dimensional Viscous Flow Solvers in a Multistage Environment," *J. of Turbomachinery* **114** (1992).

13. R. P. Benedict and C. A. Meyer, "Electrolytic Tank Analog for Studying Fluid Flow Fields within Turbomachinery," ASME Paper 57-A-120, 1957.

14. R. P. Benedict, "Analog Simulation," *Electro-Technology* (December 1963).

CHAPTER 59

INTERNAL COMBUSTION ENGINES

Ronald Douglas Matthews
**General Motors Foundation Combustion Sciences
and Automotive Research Laboratories
The University of Texas at Austin
Austin, Texas**

59.1	TYPES AND PRINCIPLES OF OPERATION	**1801**		59.3.1	Experimental Measurements	1814
	59.1.1 Spark Ignition Engines	1802		59.3.2	Theoretical Considerations	
	59.1.2 Compression Ignition (Diesel) Engines	1808			and Modeling	1816
				59.3.3	Engine Comparisons	1820
59.2	FUELS AND KNOCK	**1808**				
	59.2.1 Knock in Spark Ignition Engines	1808	59.4	EMISSIONS AND FUEL ECONOMY REGULATIONS		**1822**
	59.2.2 Knock in the Diesel Engine	1810		59.4.1	Light-Duty Vehicles	1822
	59.2.3 Characteristics of Fuels	1810		59.4.2	Heavy-Duty Vehicles	1825
				59.4.3	Nonhighway Heavy-Duty Standards	1826
59.3	PERFORMANCE AND EFFICIENCY	**1814**		**SYMBOLS**		**1826**

An internal combustion engine is a device that operates on an open thermodynamic cycle and is used to convert the chemical energy of a fuel to rotational mechanical energy. This rotational mechanical energy is most often used directly to provide motive power through an appropriate drive train, such as for an automotive application. The rotational mechanical energy may also be used directly to drive a propeller for marine or aircraft applications. Alternatively, the internal combustion engine may be coupled to a generator to provide electric power or may be coupled to hydraulic pump or a gas compressor. It may be noted that the favorable power-to-weight ratio of the internal combustion engine makes it ideally suited to mobile applications and therefore most internal combustion engines are manufactured for the motor vehicle, rail, marine, and aircraft industries. The high power-to-weight ratio of the internal combustion engine is also responsible for its use in other applications where a lightweight power source is needed, such as for chain saws and lawn mowers.

This chapter is devoted to discussion of the internal combustion engine, including types, principles of operation, fuels, theory, performance, efficiency, and emissions.

59.1 TYPES AND PRINCIPLES OF OPERATION

This chapter discusses internal combustion engines that have an intermittent combustion process. Gas turbines, which are internal combustion engines that incorporate a continuous combustion system, are discussed in a separate chapter.

Internal combustion (IC) engines may be most generally classified by the method used to initiate combustion as either spark ignition (SI) or compression ignition (CI or diesel) engines. Another general classification scheme involves whether the rotational mechanical energy is obtained via reciprocating piston motion, as is more common, or directly via the rotational motion of a rotor in a rotary (Wankel) engine (see Fig. 59.1). The physical principles of a rotary engine are equivalent to those of a piston engine if the geometric considerations are properly accounted for, so that the following discussion will focus on the piston engine and the rotary engine will be discussed only briefly. All of these IC engines include five general processes:

Mechanical Engineers' Handbook, 2nd ed., Edited by Myer Kutz.
ISBN 0-471-13007-9 © 1998 John Wiley & Sons, Inc.

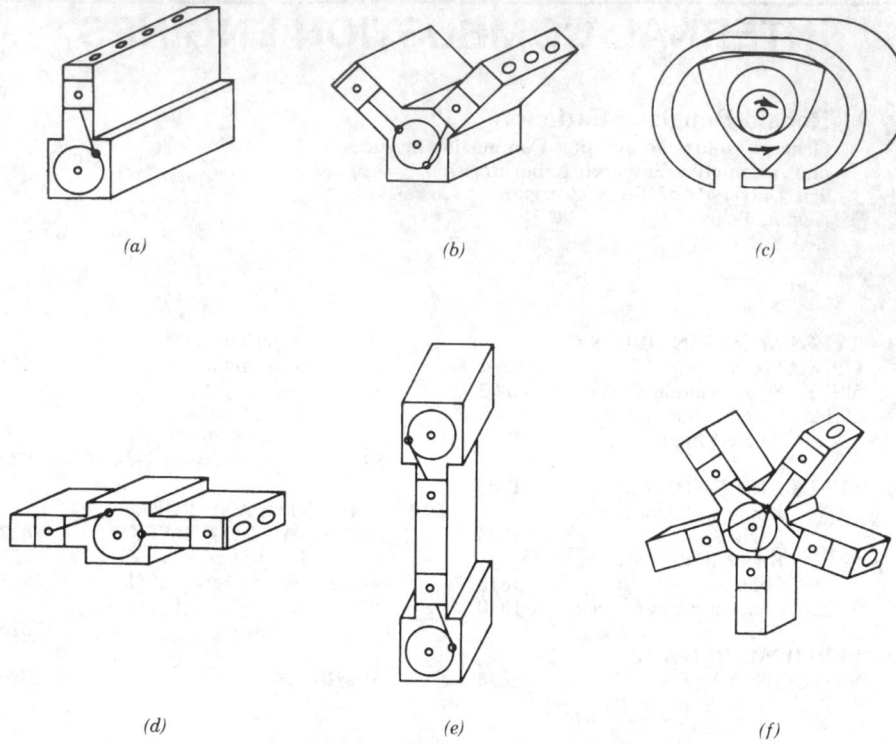

Fig. 59.1   IC engine configurations: (a) inline 4; (b) V6; (c) rotary (Wankel); (d) horizontal, flat, or opposed cylinder; (e) opposed piston; (f) radial.

1. An intake process, during which air or a fuel–air mixture is inducted into the combustion chamber

2. A compression process, during which the air or fuel–air mixture is compressed to higher temperature, pressure, and density

3. A combustion process, during which the chemical energy of the fuel is converted to thermal energy of the products of combustion

4. An expansion process, during which a portion of the thermal energy of the working fluid is converted to mechanical energy

5. An exhaust process, during which most of the products of combustion are expelled from the combustion chamber

The mechanics of how these five general processes are incorporated in an engine may be used to more specifically classify different types of internal combustion engines.

59.1.1   Spark Ignition Engines

In SI engines, the combustion process is initiated by a precisely timed discharge of a spark across an electrode gap in the combustion chamber. Before ignition, the combustible mixture may be either homogeneous (i.e., the fuel–air mixture ratio may be approximately uniform throughout the combustion chamber) or stratified (i.e., the fuel–air mixture ratio may be more fuel-lean in some regions of the combustion chamber than in other portions). In all SI engines, except the direct injection stratified charge (DISC) SI engine, the power output is controlled by controlling the air flow rate (and thus the volumetric efficiency) through the engine and the fuel–air ratio is approximately constant (and approximately stoichiometric) for almost all operating conditions. The power output of the DISC engine is controlled by varying the fuel flow rate, and thus the fuel–air ratio is variable while the volumetric efficiency is approximately constant. The fuel and air are premixed before entering the combustion chamber in all SI engines except the direct injection SI engine. These various categories of SI engines are discussed below.

Homogeneous Charge SI Engines

In the homogeneous charge SI engine, a mixture of fuel and air is inducted during the intake process. Traditionally, the fuel was mixed with the air in the venturi section of a carburetor. More recently, as more precise control of the fuel–air ratio became desirable, throttle body fuel injection took the place of carburetors for most automotive applications. Even more recently, intake port fuel injection has almost entirely replaced throttle body injection. The five processes mentioned above may be combined in the homogeneous charge SI engine to produce an engine that operates on either a 4-stroke cycle or on a 2-stroke cycle.

4-Stroke Homogeneous Charge SI Engines. In the more common 4-stroke cycle (see Fig. 59.2), the first stroke is the movement of the piston from top dead center (TDC—the closest approach of the piston to the cylinder head, yielding the minimum combustion chamber volume) to bottom dead center (BDC—when the piston is farthest from the cylinder head, yielding the maximum combustion chamber volume), during which the intake valve is open and the fresh fuel–air charge is inducted into the combustion chamber. The second stroke is the compression process, during which the intake and exhaust valves are both in the closed position and the piston moves from BDC back to TDC. The compression process is followed by combustion of the fuel–air mixture. Combustion is a rapid hydrocarbon oxidation process (not an explosion) of finite duration. Because the combustion process requires a finite, though very short, period of time, the spark is timed to initiate combustion slightly before the piston reaches TDC to allow the maximum pressure to occur slightly after TDC (peak pressure should, optimally, occur after TDC to provide a torque arm for the force caused by the high cylinder pressure). The combustion process is essentially complete shortly after the piston has receded away from TDC. However, for the purposes of a simple analysis and because combustion is very rapid, to aid explanation it may be approximated as being instantaneous and occurring while the piston is motionless at TDC. The third stroke is the expansion process or power stroke, during which the piston returns to BDC. The fourth stroke is the exhaust process, during which the exhaust valve is open and the piston proceeds from BDC to TDC and expels the products of combustion. The exhaust process for a 4-stroke engine is actually composed of two parts, the first of which is blowdown. When the exhaust valve opens, the cylinder pressure is much higher than the pressure in the exhaust manifold and this large pressure difference forces much of the exhaust out during what is called "blowdown" while the piston is almost motionless. Most of the remaining products of combustion are forced out during the exhaust stroke, but an "exhaust residual" is always left in the combustion chamber and mixes with the fresh charge that is inducted during the subsequent intake stroke. Once the piston reaches TDC, the intake valve opens and the exhaust valve closes and the cycle repeats, starting with a new intake stroke.

This explanation of the 4-stroke SI engine processes implied that the valves open or close instantaneously when the piston is either at TDC or BDC, when in fact the valves open and close relatively slowly. To afford the maximum open area at the appropriate time in each process, the exhaust valve opens before BDC during expansion, the intake valve closes after BDC during the compression stroke, and both the intake and exhaust valves are open during the valve overlap period since the intake valve opens before TDC during the exhaust stroke while the exhaust valve closes after TDC during the intake stroke. Considerations of valve timing are not necessary for this simple explanation of the 4-stroke cycle but do have significant effects on performance and efficiency. Similarly, spark timing will not be discussed in detail but does have significant effects on performance, fuel economy, and emissions.

The rotary (Wankel) engine is sometimes perceived to operate on the 2-stroke cycle because it shares several features with 2-stroke SI engines: a complete thermodynamic cycle within a single revolution of the output shaft (which is called an eccentric shaft rather than a crank shaft) and lack of intake and exhaust valves and associated valve train. However, unlike a 2-stroke, the rotary has a true exhaust "stroke" and a true intake "stroke" and operates quite well without boosting the pressure of the fresh charge above that of the exhaust manifold. That is, the rotary operates on the 4-stroke cycle.

2-Stroke Homogeneous Charge SI Engines. Alternatively, these five processes may be incorporated into a homogeneous charge SI engine that requires only two strokes per cycle (see Fig. 59.3). All commercially available 2-stroke SI engines are of the homogeneous charge type. That is, any nonuniformity of the fuel–air ratio within the combustion chamber is unintentional in current 2-stroke SI engines. The 2-stroke SI engine does not have valves, but rather has intake "transfer" and exhaust ports that are normally located across from each other near the position of the crown of the piston when the piston is at BDC. When the piston moves toward TDC, it covers the ports and the compression process begins. As previously discussed, for the ideal SI cycle, combustion may be perceived to occur instantaneously while the piston is motionless at TDC. The expansion process then occurs as the high pressure resulting from combustion pushes the piston back toward BDC. As the piston approaches BDC, the exhaust port is generally uncovered first, followed shortly thereafter by uncovering of the intake transfer port. The high pressure in the combustion chamber relative to that of the

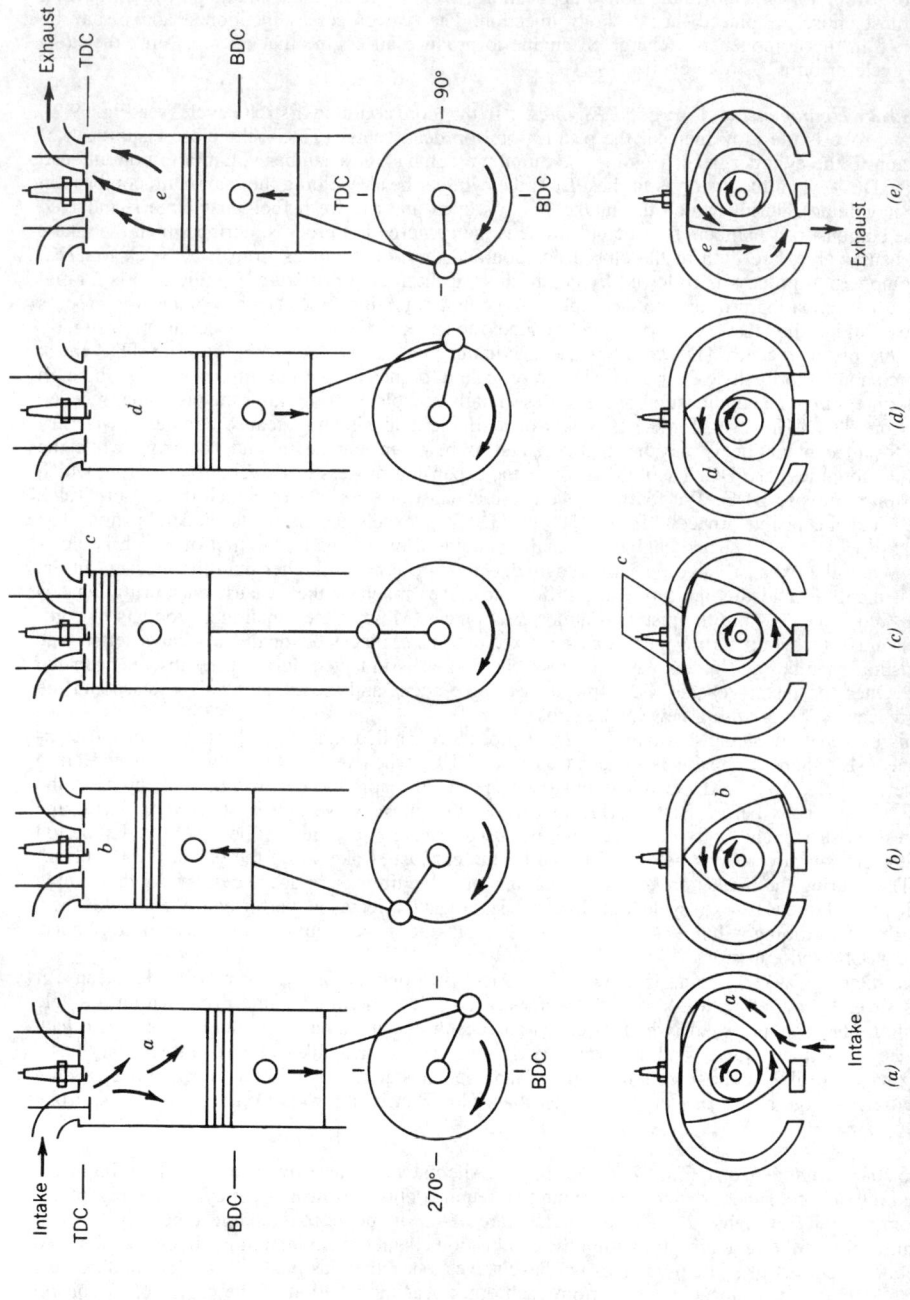

Fig. 59.2 Schematic of processes for 4-stroke SI piston and rotary engines (for 4-stroke CI, replace spark plug with fuel injector): (a) intake, (b) compression, (c) spark ignition and combustion (for CI, fuel injection, and autoignition), (d) expansion or power stroke, (e) exhaust.

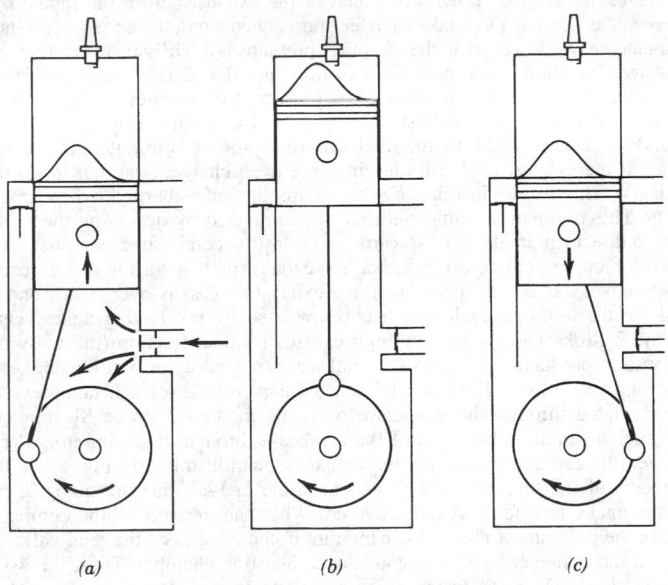

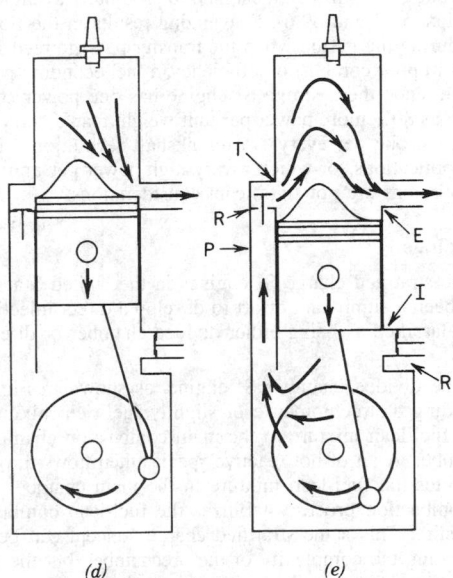

Fig. 59.3 Processes for 2-stroke crankcase compression SI engine (for CI engine, replace spark plug with fuel injector): (a) compression of trapped working fluid and simultaneous intake to crankcase, (b) spark ignition and combustion (for CI, fuel injection, and autoignition), (c) expansion or power, (d) beginning of exhaust, (e) intake and "loop" scavenging. E, exhaust port; I, intake port; P, transfer passage; R, read valve; T, transfer port.

exhaust manifold results in "blowdown" of much of the exhaust before the intake transfer port is uncovered. However, as soon as the intake transfer port is uncovered, the exhaust and intake processes can occur simultaneously. However, if the chamber pressure is high with respect to the pressure in the transfer passage, the combustion products can flow into the transfer passage. To prevent this, a reed valve can be located within the intake transfer passage, as illustrated in Fig. 59.3. Alternatively, a disc valve that is attached to the crankshaft can be used to control timing of the intake transfer process. Independent of when and how the intake transfer process is initiated, the momentum of the exhaust flowing out the exhaust port will entrain some fresh charge, resulting in short-circuiting of fuel out the exhaust. This results in relatively high emissions of unburned hydrocarbons and a fuel economy penalty. This problem is minimized but not eliminated by designing the port shapes and/ or piston crown to direct the intake flow toward the top of the combustion chamber so that the fresh charge must travel a longer path before reaching the exhaust port. After the piston reaches BDC and moves back up to cover the exhaust port again, the exhaust process is over. Thus, one of the strokes that is required for the 4-stroke cycle has been eliminated by not having an exhaust stroke. The penalty is that the 2-stroke has a relatively high exhaust residual fraction (the mass fraction of the remaining combustion products relative to the total mass trapped upon port closing).

As the piston proceeds from TDC to BDC on the expansion stroke, it compresses the fuel–air mixture which is routed through the crankcase on many modern 2-stroke SI engines. To prevent backflow of the fuel–air mixture back out of the crankcase through the carburetor, a reed valve may be located between the carburetor exit and the crankcase, as illustrated in Fig. 59.3. This crankcase compression process of the fuel–air mixture results in the fuel–air mixture being at relatively high pressure when the intake transfer port is uncovered. When the pressure in the combustion chamber becomes less than the pressure of the fuel–air mixture in the crankcase, the reed valve in the transfer passage opens and the intake charge flows into the combustion chamber. Thus, the 4-stroke's intake stroke is eliminated in the 2-stroke design by having both sides of the piston do work.

Because it is important to fill the combustion chamber as completely as possible with fresh fuel–air charge and thus important to purge the combustion chamber as completely as possible of combustion products, 2-stroke SI engines are designed to promote scavenging of the exhaust products via fluid dynamics (see Figs. 59.3 and 59.6). Scavenging results in the flow of some unburned fuel through the exhaust port during the period when the transfer passage reed valve and the exhaust port are both open. This results in poor combustion efficiency, a fuel economy penalty, and high emissions of hydrocarbons. However, since the 2-stroke SI engine has one power stroke per crankshaft revolution, it develops as much as 80% more power per unit weight than a comparable 4-stroke SI engine, which has only one power stroke per every two crankshaft revolutions. Therefore, the 2-stroke SI engine is best suited for applications for which a very high power per unit weight is needed and fuel economy and pollutant emissions are not significant considerations.

Stratified Charge SI Engines

All commercially available stratified charge SI engines in the United States operate on the 4-stroke cycle, although there has been a significant effort to develop a direct injection (stratified) 2-stroke SI engine. They may be subclassified as being either divided chamber or direct injection SI engines.

Divided Chamber. The divided chamber SI engine, as shown in Fig. 59.4, generally has two intake systems: one providing a stoichiometric or slightly fuel-rich mixture to a small prechamber and the other providing a fuel-lean mixture to the main combustion chamber. A spark plug initiates combustion in the prechamber. A jet of hot reactive species then flows through the orifice separating the two chambers and ignites the fuel-lean mixture in the main chamber. In this manner, the stoichiometric or fuel-rich combustion process stabilizes the fuel-lean combustion process that would otherwise be prone to misfire. This same stratified charge concept can be attained solely via fluid mechanics, thereby eliminating the complexity of the prechamber, but the motivation is the same as for the divided chamber engine. This overall fuel-lean system is desired since it can result in decreased emissions of the regulated pollutants in comparison to the usual, approximately stoichiometric, combustion process. Furthermore, lean operation produces a thermal efficiency benefit. For these reasons, there have been many attempts to develop a lean-burn homogeneous charge SI engine. However, the emissions of the oxides of nitrogen (NO_x) peak for a slightly lean mixture before decreasing to very low values when the mixture is extremely lean. Unfortunately, most lean-burn homogeneous charge SI engines cannot operate sufficiently lean—before encountering ignition problems—that they produce a significant NO_x benefit. The overall lean-burn stratified charge SI engine avoids these ignition limits by producing an ignitable mixture in the vicinity of the spark plug but a very lean mixture far from the spark plug. Unfortunately, the flame zone itself is nearly stoichiometric, resulting in much higher emissions of NO_x than would be expected from the overall extremely lean fuel–air ratio. For this reason, divided chamber SI engines are becoming rare. However, the direct injection process offers promise of overcoming this obstacle, as discussed below.

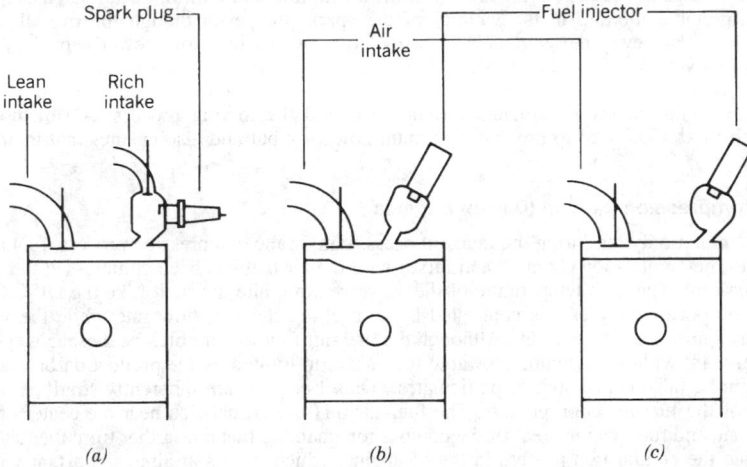

Fig. 59.4 Schematic cross sections of divided chamber engines: (a) prechamber SI, (b) prechamber IDI diesel, (c) swirl chamber IDI diesel.

Direct Injection.   In the direct injection engine, only air is inducted during the intake stroke. The direct injection engine can be divided into two categories: early and late injection.

The first 40 years of development of the direct injection SI engine focussed upon late injection. This version is commonly known as the *direct injection stratified charge* (DISC) engine. As shown in Fig. 59.5, fuel is injected late in the compression stroke near the center of the combustion chamber and ignited by a spark plug. The DISC engine has three primary advantages:

1. A wide fuel tolerance, that is, the ability to burn fuels with a relatively low octane rating without knock (see Section 59.2).

2. This decreased tendency to knock allows use of a higher compression ratio, which in turn results in higher power per unit displacement and higher efficiency (see Section 59.3).

3. Since the power output is controlled by the amount of fuel injected instead of the amount of air inducted, the DISC engine is not throttled (except at idle), resulting in higher volumetric efficiency and higher power per unit displacement for part load conditions (see Section 59.3).

Unfortunately, the DISC engine is also prone to high emissions of unburned hydrocarbons.

However, more recent developments in the DISC engine aim the fuel spray at the top of the piston to avoid wetting the cylinder liner with liquid fuel to minimize emissions of unburned hydro-

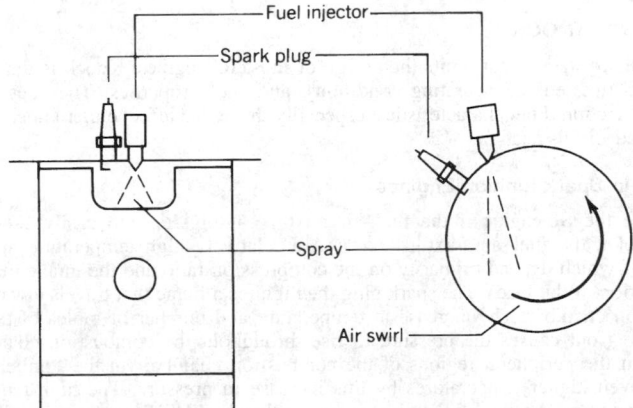

Fig. 59.5 Schematic of DISC SI engine combustion chambers.

carbons. The shape of the piston, together with the air motion and ignition location, ensure that there is still an ignitable mixture in the vicinity of the spark plug even though the overall mixture is extremely lean. However, the extremely lean operation results in a low power capability. Thus, at high loads this version of the direct injection engine uses early injection timing, as discussed below.

Early injection results in sufficient time available for the mixture to become essentially completely mixed before ignition, given sufficient turbulence to aid the mixing process. A stoichiometric or slightly rich mixture is used to provide maximum power output and also ensures that ignition is not a difficulty.

59.1.2 Compression Ignition (Diesel) Engines

CI engines induct only air during the intake process. Late in the compression process, fuel is injected directly into the combustion chamber and mixes with the air that has been compressed to a relatively high temperature. The high temperature of the air serves to ignite the fuel. Like the DISC SI engine, the power output of the diesel is controlled by controlling the fuel flow rate while the volumetric efficiency is approximately constant. Although the fuel–air ratio is variable, the diesel always operates overall fuel-lean, with a maximum allowable fuel–air ratio limited by the production of unacceptable levels of smoke (also called soot or particulates). Diesel engines are inherently stratified because of the nature of the fuel-injection process. The fuel–air mixture is fuel-rich near the center of the fuel-injection cone and fuel-lean in areas of the combustion chamber that are farther from the fuel injection cone. Unlike the combustion process in the SI engine, which occurs at almost constant volume, the combustion process in the diesel engine ideally occurs at constant pressure. That is, the combustion process in the CI engine is relatively slow, and the fuel–air mixture continues to burn during a significant portion of the expansion stroke (fuel continues to be injected during this portion of the expansion stroke) and the high pressure that would normally result from combustion is relieved as the piston recedes. After the combustion process is completed, the expansion process continues until the piston reaches BDC. The diesel may complete the five general engine processes through either a 2-stroke cycle or a 4-stroke cycle. Furthermore, the diesel may be subclassified as either an indirect injection diesel or a direct injection diesel.

Indirect injection (IDI) or divided chamber diesels are geometrically similar to divided chamber stratified charge SI engines. All IDI diesels operate on a 4-stroke cycle. Fuel is injected into the prechamber and combustion is initiated by autoignition. A glow plug is also located in the prechamber, but is only used to alleviate cold start difficulties. As shown in Fig. 59.4, the IDI may be designed so that the jet of hot gases issuing into the main chamber promotes swirl of the reactants in the main chamber. This configuration is called the *swirl chamber IDI diesel*. If the system is not designed to promote swirl, it is called the *prechamber IDI diesel*. The divided chamber design allows a relatively inexpensive pintle-type fuel injector to be used on the IDI diesel.

Direct injection (DI) or "open" chamber diesels are similar to DISC SI engines. There is no prechamber, and fuel is injected directly into the main chamber. Therefore, the characteristics of the fuel-injection cone have to be tailored carefully for proper combustion, avoidance of knock, and minimum smoke emissions. This requires the use of a high-pressure close-tolerance fuel injection system that is relatively expensive. The DI diesel may operate on either a 4-stroke or a 2-stroke cycle. Unlike the 2-stroke SI engine, the 2-stroke diesel often uses a mechanically driven blower for supercharging rather than crankcase compression and also may use multiple inlet ports in each cylinder, as shown in Fig. 59.6. Also, one or more exhaust valves in the top of the cylinder may be used instead of exhaust ports near the bottom of the cylinder, resulting in "through" or "uniflow" scavenging rather than "loop" or "cross" scavenging.

59.2 FUELS AND KNOCK

Knock is the primary factor that limits the design of most IC engines. Knock is the result of engine design characteristics, engine operating conditions, and fuel properties. The causes of knock are discussed in this section. Fuel characteristics, especially those that affect either knock or performance, are also discussed in this section.

59.2.1 Knock in Spark Ignition Engines

Knock occurs in the SI engine if the fuel–air mixture autoignites too easily. At the end of the compression stroke, the fuel–air mixture exists at a relatively high temperature and pressure, the specific values of which depend primarily on the compression ratio and the intake manifold pressure (which is a function of the load). The spark plug then ignites a flame that travels toward the periphery of the combustion chamber. The increase in temperature and number of moles of the burned gases behind the flame front causes the pressure to rise throughout the combustion chamber. The "end gases" located in the peripheral regions of the combustion chamber (in the "unburned zone") are compressed to even higher temperatures by this increase in pressure. The high temperature of the end gases can lead to a sequence of chemical reactions that are called *autoignition*. If the autoignition

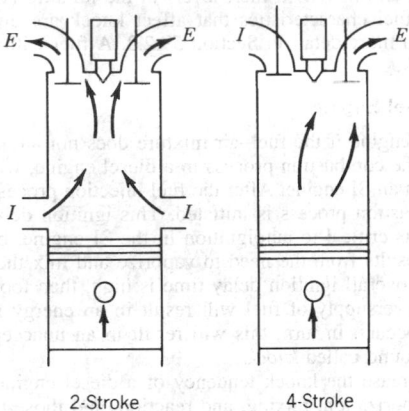

2-Stroke   4-Stroke

Fig. 59.6 Schematic of 2-stroke and 4-stroke DI diesels. The 2-stroke incorporates "uniflow" scavenging.

reactions have sufficient time at a sufficiently high temperature, the reaction sequence can produce strongly exothermic reactions such that the temperature in the unburned zone may increase at a rate of several million K/sec, which results in knock. That is, if the reactive end gases remain at a high temperature for a sufficient period of time (i.e., longer than the "ignition delay time"), then the autoignition reactions will produce knock. Normal combustion occurs if the flame front passes through the end gases before the autoignition reactions reach a strongly exothermic stage.

For most fuels, autoignition is characterized by three stages that are dictated by the unburned mixture (or end gas) temperature. Here, it is important to note that the temperature varies with crank angle due to compression by the piston motion and, after ignition, due to compression by the expanding flame front, and the entire temperature history shifts up or down due to the effects of load, ambient air temperature, etc. At "low" temperatures, the reactivity of the end gases increases with increasing temperature. As the temperature increases further, the rate of increase of the reactivity either slows markedly or even decreases (the so-called "negative temperature coefficient" regime). When the temperature increases to even higher values (typically, above ~900 K), the reactivity begins to increase extremely strongly, the autoignition reactions reach an energy liberating stage, enough energy may be released during this stage to initiate a "high" temperature (>1000 K) chemical mechanism,[1,2] and a runaway reaction occurs. If the rate of energy release is greater than the rate of expansion, then a strong pressure gradient will result. The steep pressure wave thus established will travel throughout the combustion chamber, reflect off the walls, and oscillate at the natural frequency characteristic of the combustion chamber geometry. This acoustic vibration results in an audible sound called *knock*. It should be noted that the flame speeds associated with knock are generally considered to be lower than the flame speeds associated with detonation (or explosion).[3,4] Nevertheless, the terms *knock* and *detonation* are often used interchangeably in reference to end gas autoignition.

The tendency of the SI engine to knock will be affected by any factors that affect the temperature and pressure of the end gases, the ignition delay time, the end gas residence time (before the normal flame passes through the end gases), and the reactivity of the mixture. The flame speed is a function of the turbulence intensity in the combustion chamber, and the turbulence intensity increases with increasing engine speed. Thus, the end gases will have a shorter residence time at high engine speed and there will be a decreased tendency to knock. As the load on the engine increases, the throttle plate is opened wider and the pressure in the intake manifold increases, thereby increasing the end gas pressure (and, thereby, temperature), resulting in a greater tendency to knock. Thus, knock is most likely to be observed for SI engines used in motor vehicles at conditions of high load and low engine speed, such as acceleration from a standing start.

Other factors that increase the knock tendency of an SI engine[1-5] include increased compression ratio, increased inlet air temperature, increased distance between the spark plug and the end gases, location of the hot exhaust valve near the region of the end gases that is farthest from the spark plug, and increased intake manifold temperature and pressure due to pressure boosting (supercharging or turbocharging). Factors that decrease the knock tendency of an SI engine[1-5] include retarding the spark timing, operation with either rich or lean mixtures (and thus the ability to operate the DISC SI engine at a higher compression ratio, since the end gases for this engine are extremely lean and therefore not very reactive), and increased inert levels in the mixture (via exhaust gas recirculation, water injection, etc.). The fuel characteristics that affect knock are quantified using octane rating tests, which are discussed in more detail in Section 59.2.3. A fuel with higher octane number has a decreased tendency to knock.

59.2.2 Knock in the Diesel Engine

Knock occurs in the diesel engine if the fuel–air mixture does not autoignite easily enough. Knock occurs at the beginning of the combustion process in a diesel engine, whereas it occurs near the end of the combustion process in an SI engine. After the fuel injection process begins, there is an ignition delay time before the combustion process is initiated. This ignition delay time is not caused solely by the chemical delay that is critical to autoignition in the SI engine, but is also due to a physical delay. The physical delay results from the need to vaporize and mix the fuel with the air to form a combustible mixture. If the overall ignition delay time is high, then too much fuel may be injected prior to autoignition. This oversupply of fuel will result in an energy release rate that is too high immediately after ignition occurs. In turn, this will result in an unacceptably high rate of pressure rise and cause the audible sound called *knock*.

The factors that will increase the knock tendency of a diesel engine[1,3,5] are those that decrease the rates of atomization, vaporization, mixing, and reaction, and those that increase the rate of fuel injection. The diesel engine is most prone to knock under cold start conditions because

1. The fuel, air, and combustion chamber walls are initially cold, resulting in high fuel viscosity (poor mixing and therefore a longer physical delay), poor vaporization (longer physical delay), and low initial reaction rates (longer chemical delay).
2. The low engine speed results in low turbulence intensity (poor mixing, yielding a longer physical delay) and may result in low fuel-injection pressures (poor atomization and longer physical delay).
3. The low starting load will lead to low combustion temperatures and thus low reaction rates (longer chemical delay).

After a diesel engine has attained normal operating temperatures, knock will be most liable to occur at high speed and low load (exactly the opposite of the SI engine). The low load results in low combustion temperatures and thus low reaction rates and a longer chemical delay. Since most diesel engines have a gear-driven fuel-injection pump, the increased rate of injection at high speed will more than offset the improved atomization and mixing (shorter physical delay).

Because the diesel knocks for essentially the opposite reasons than the SI engine, the factors that increase the knock tendency of an SI engine will decrease the knock tendency of a diesel engine: increased compression ratio, increased inlet air temperature, increased intake manifold temperature and pressure due to supercharging or turbocharging, and decreased concentrations of inert species. The knock tendency of the diesel engine will be increased if the injection timing is advanced or retarded from the optimum value and if the fuel has a low volatility, a high viscosity, and/or a low "cetane number." The cetane rating test and other fuel characteristics are discussed in more detail in the following section.

59.2.3 Characteristics of Fuels

Several properties are of interest for both SI engine fuels and diesel fuels. Many of these properties are presented in Table 59.1 for the primary reference fuels, for various types of gasolines and diesel fuels, and for the alternative fuels that are of current interest.

The stoichiometry, or relative amount of air and fuel, in the combustion chamber is usually specified by the air–fuel mass ratio (AF), the fuel–air mass ratio (FA = 1/AF), the equivalence ratio (ϕ), or the excess air ratio (λ). Measuring instruments may be used to determine the mass flow rates of air and fuel into an engine so that AF and FA may be easily determined. Alternatively, AF and FA may be calculated if the exhaust product composition is known, using any of several available techniques.[5] The equivalence ratio normalizes the actual fuel–air ratio by the stoichiometric fuel–air ratio (FA$_s$), where "stoichiometric" refers to the chemically-correct mixture with no excess air and no excess fuel. Recognizing that the stoichiometric mixture contains 100% "theoretical air" allows the equivalence ratio to be related to the actual percentage of theoretical air (TA, percentage by volume or mole):

$$\phi = FA/FA_s = AF_s/AF = 100/TA = 1/\lambda \tag{59.1}$$

The equivalence ratio is a convenient parameter because $\phi < 1$ refers to a fuel-lean mixture, $\phi > 1$ to a fuel-rich mixture, and $\phi = 1$ to a stoichiometric mixture.

The stoichiometric fuel–air and air–fuel ratios can be easily calculated from a reaction balance by assuming "complete combustion" [only water vapor (H_2O) and carbon dioxide (CO_2) are formed

Table 59.1 Properties for Various Fuels

Name	Formula	MW	AF_s	LHV_p	h_f^a	h_v^*	sg^b	RON	MCN	Ref.
Primary Reference Fuels										
Iso-octane	C_8H_{18}	114	15.1	44.6	-224.3^c	35.1^c	0.69	100	100	5
Normal heptane	C_7H_{16}	100	15.1	44.9	-187.9^c	36.6^c	0.68	0	0	5
Normal hexadecane	$C_{16}H_{34}$	226	14.9	44.1	-418.3^d	50.9^e	0.77	—	0^f	5
Alternative Fuels										
Average CNG	$CH_{3.88}$	17.4	16.3	47.9	-79.6	—	0.60^b	> 120	> 120	26, 27
LPG as propane[1]	C_3H_8	44	15.6	46.3	-103.9	15.1	0.50	112	97	5, 22
Methanol	CH_3OH	32	6.4	21.2	-201.3^g	37.5	0.79	112	91	5, 21
Ethanol	C_2H_5OH	46	9.0	27.8	-235.5^g	42.4	0.78	111	92	5, 21
*Gasolines***										
1988 U.S. avg.	$C_8H_{14.53}$	111	14.5	42.6	-189.9^h	NA^j	0.75	92.0	82.6	28
Certification[2]	$C_8H_{14.69}$	111	14.5	42.6	-202.4^h	NA	0.74	96.7	87.5	28
Cal. Phase 2 RFG[3]	$C_8H_{16.28}O_{0.24}$	116	14.2	41.6	-273.3^h	NA	0.74	NA	NA	29
Aviation	C_8H_{17}	113	14.9	41.9–43.1	-390.3^i	NA	0.72	NA	NA	3
Diesel Fuels										
Automotive	$C_{12}H_{23.7}$	168	14.7	40.6–44.4	-445.0^i	90.1–131.7	0.81–0.85	—	—	3
No. 1D	$C_{12}H_{26}$	170	15.0	42.4	-596.0	45.4	0.88	—	—	1
No. 2D	$C_{13}H_{28}$	184	15.0	41.8	-747.6	44.9	0.92	—	—	1
No. 4D	$C_{14}H_{30}$	198	15.0	41.3	-894.6	46.0	0.96	—	—	1

[a] Of vapor phase fuel at 298 K in MJ/kmole, except when noted otherwise.

[b] sg is ρ_F at 20°C/ρ_w at 4°C (1000 kg/m³), except values from Ref. 1 (reference temp. is 15°C) and CNG (referenced to air).

[c] From Ref. 23.

[d] Calculated.

[e] At 1 atm and boiling temperature.

[f] Estimate from Ref. 1, p. 147.

[g] Reference 22.

[h] Enthalpy of formation is for the liquid fuel (rather than the gaseous fuel), as calculated from fuel properties.

[i] Enthalpy of formation is for the liquid fuel as calculated from the average heating value.

[j] Typically 35–40.

* At 298 K and corresponding saturation pressure, except when noted otherwise.

** As C8.

[1] Liquefied petroleum gas has a variable composition, normally dominated by propane.

[2] The properties of emissions certification gasoline vary somewhat.

[3] Properties of California Phase 2 Reformulated Gasoline from a sample of Arco EC-X.

during the combustion process], even though the actual combustion process will almost never be complete. The reaction balance for the complete combustion of a stoichiometric mixture of air with a fuel of the atomic composition C_xH_y is

$$C_xH_y + (x + .25y)O_2 + 3.764(x + .25y)N_2 = xCO_2 + 0.5yH_2O + 3.764(x + .25y)N_2 \quad (59.2)$$

where air is taken to be 79% by volume "effective nitrogen" (N_2 plus the minor components in air) and 21% by volume oxygen (O_2) and thus the nitrogen-to-oxygen ratio of air is $0.79/0.21 = 3.764$. Given that the molecular weight (MW) of air is 28.967, the MW of carbon (C) is 12.011, and the MW of hydrogen (H) is 1.008, then AF_s and FA_s for any hydrocarbon fuel may be calculated from

$$AF_s = 1/FA_s = (x + 0.25y) \times 4.764 \times 28.967/(12.011x + 1.008y) \quad (59.3)$$

The stoichiometric air–fuel ratios for a number of fuels of interest are presented in Table 59.1.

The energy content of the fuel is most often specified using the constant-pressure lower heating value (LHV_p). The lower heating value is the maximum energy that can be released during combustion of the fuel if (1) the water in the products remains in the vapor phase, (2) the products are returned to the initial reference temperature of the reactants (298 K), and (3) the combustion process is carried out such that essentially complete combustion is attained. If the water in the products is condensed, then the higher heating value (HHV) is obtained. If the combustion system is a flow calorimeter, then the constant-pressure heating value is measured (and, most usually, this is HHV_p). If the combustion system is a bomb calorimeter, then the constant-volume heating value is measured (usually HHV_v). The constant-pressure heating value is the negative of the standard enthalpy of reaction (ΔH_R^{298}, also known as the heat of combustion) and ΔH_R^{298} is a function of the standard enthalpies of formation h_f^{298} of the reactant and product species. For a fuel of composition C_xH_y, Eq. (59.4) may be used to calculate the constant-pressure heating value (HV_p), given the enthalpy of formation of the fuel, or may be used to calculate h_f^{298} of the fuel, given HV_p:

$$HV_p = -\Delta H_R^{298} = \frac{(h_{f,C_xH_y}^{298} - \beta h_{v,C_xH_y}) + 393.418x + 0.5y(241.763 + 43.998\alpha)}{12.011x + 1.008y} \quad (59.4)$$

In Eq. (59.4): (1) $\alpha = 0$ if the water in the products is not condensed (yielding LHV_p) and $\alpha = 1$ if the water is condensed (yielding HHV_p); (2) $\beta = 0$ if the fuel is initially a vapor and $\beta = 1$ if the fuel is initially a liquid; (3) h_{v,C_xH_y} is the enthalpy of vaporization per kmole* of fuel at 298 K; (4) the standard enthalpies of formation are CO_2: -393.418 MJ/kmole, H_2O: -241.763 MJ/kmole, O_2: 0 MJ/kmole, N_2: 0 MJ/kmole; (5) the enthalpy of vaporization of H_2O at 298 K is 43.998 MJ/kmole; and (6) the denominator is simply the molecular weight of the fuel yielding the heating value in MJ/kg of fuel. Also, the relationship between the constant volume heating value (HV_v) and HV_p for a fuel C_xH_y is

$$HV_v = HV_p + \frac{2.478(0.25y - 1)}{12.011x + 1.008y} \quad (59.5)$$

Of the several heating values that may be defined, LHV_p is preferred for engine calculations since condensation of water in the combustion chamber is definitely to be avoided and since an engine is essentially a steady-flow device and thus the enthalpy is the relevant thermodynamic property (rather than the internal energy). For diesel fuels, HHV_v may be estimated from nomographs, given the density and the "mid-boiling point temperature" or given the "aniline point," the density, and the sulfur content of the fuel.[6]

Values for LHV_p, h_f^{298}, and h_v^{298} for various fuels of interest are presented in Table 59.1.

The specific gravity of a liquid fuel (sg_F) is the ratio of its density (ρ_F, usually at either 20°C or 60°F) to the density of water (ρ_w, usually at 4°C):

$$sg_F = \rho_F/\rho_w \qquad \rho_F = sg_F \cdot \rho_w \quad (59.6)$$

For gaseous fuels, such as natural gas, the specific gravity is referenced to air at standard conditions rather than to water. The specific gravity of a liquid fuel can be easily calculated from a simple measurement of the American Petroleum Institute gravity (API):

$$sg_F = 141.5/(API + 131.5) \quad (59.7)$$

*A kmole is a mole based on a kg, also referred to as a kg-mole.

Tables are available (SAE Standard J1082 SEP80) to correct for the effects of temperature if the fuel is not at the prescribed temperature when the measurement is performed. Values of sg_F for various fuels are presented in Table 59.1.

The knock tendency of SI engine fuels is rated using an octane number (ON) scale. A higher octane number indicates a higher resistance to knock. Two different octane-rating tests are currently used. Both use a single-cylinder variable-compression-ratio SI engine for which all operating conditions are specified (see Table 59.2). The fuel to be tested is run in the engine and the compression ratio is increased until knock of a specified intensity (standard knock) is obtained. Blends of two primary reference fuels are then tested at the same compression ratio until the mixture is found that produces standard knock. The two primary reference fuels are 2,2,4-trimethyl pentane (also called iso-octane), which is arbitrarily assigned an ON of 100, and n-heptane, which is arbitrarily assigned an ON of 0. The ON of the test fuel is then simply equal to the percentage of iso-octane in the blend that produced the same knock intensity at the same compression ratio. However, if the test fuel has an ON above 100, then iso-octane is blended with tetraethyl lead instead of n-heptane. After the knock tests are completed, the ON is then computed from

$$ON = 100 + \frac{28.28T}{1 + 0.736T + (1 + 1.472T - 0.035216T^2)^{1/2}} \tag{59.8}$$

where T is the number of milliliters of tetraethyl lead per U.S. gallon of iso-octane. The two different octane rating tests are called the Motor method (American Society of Testing and Materials, ASTM Standard D2700-82) and the Research method (ASTM D2600-82), and thus a given fuel (except these two primary reference fuels) will have two different octane numbers: a Motor octane number (MON) and a Research octane number (RON). The Motor method produces the lowest octane numbers, primarily because of the high intake manifold temperature for this technique, and thus the Motor method is said to be a more severe test for knock. The "sensitivity" of a fuel is defined as the RON minus the MON of that fuel. The "antiknock index" is the octane rating posted on gasoline pumps at service stations in the United States and is simply the average of RON and MON. Octane numbers for various fuels of interest are presented in Table 59.1.

The standard rating test for the knock tendency of diesel fuels (ASTM D613-82) produces the cetane number (CN). Because SI and diesel engines knock for essentially opposite reasons, a fuel with a high ON will have a low CN and therefore would be a poor diesel fuel. A single-cylinder variable-compression-ratio CI engine is used to measure the CN, and all engine operating conditions are specified. The compression ratio is increased until the test fuel exhibits an ignition delay of 13°. Here, it should be noted that ignition delay rather than knock intensity is measured for the CN technique. A blend of two primary reference fuels (n-hexadecane, which is also called n-cetane: $CN = 100$; and heptamethyl nonane, or i-cetane: $CN = 15$) are then run in the engine and the compression ratio is varied until a 13° ignition delay is obtained. The CN of this blend is given by

$$CN = \% \; n\text{-cetane} + 0.15 \times (\% \; \text{heptamethyl nonane}) \tag{59.9}$$

Various blends are tried until compression ratios are found that bracket the compression ratio of the test fuel. The CN is then obtained from a standard chart. General specifications for diesel fuels are presented in Table 59.3 along with characteristics of "average" diesel fuels for light duty vehicles.

Many other thermochemical properties of fuels may be of interest, such as vapor pressure, volatility, viscosity, cloud point, aniline point, mid-boiling-point temperature, and additives. Discussion of these characteristics is beyond the scope of this chapter but is available in the literature.[1,4–7]

Table 59.2   Test Specifications That Differ for Research and Motor Method Octane Tests—ASTM D2699-82 and D2700-82

Operating Condition	RON	MON
Engine speed (rpm)	600	900
Inlet air temperature (°C)	[a]	38°C
FA mixture temperature (°C)	[b]	149°C
Spark advance	13°BTDC	[c]

[a] Varies with barometric pressure.

[b] No control of fuel–air mixture temperature.

[c] Varies with compression ratio.

Table 59.3   Diesel Fuel Oil Specifications—ASTM D975-81

Property	Units	Fuel Type		
		1D[b]	2D[c]	4D[d]
Minimum flash point	°C	38	52	55
Maximum H_2O and sediment	Vol. %	0.05	0.05	0.50
Maximum carbon residue	%	0.15	0.35	—
Maximum ash	Wt. %	0.01	0.01	0.10
90% distillation temperature, min/max	°C	—/288	282/338	—/—
Kinematic viscosity,[a] min/max	mm²/sec	1.3/2.4	1.9/4.1	5.5/24.0
Maximum sulfur	Wt. %	0.5	0.5	2.0
Maximum Cu strip corrosion	—	No. 3	No. 3	—
Minimum cetane number	—	40	40	30

[a] At 40°C.

[b] Preferred for high-speed diesels, especially for winter use, rarely available. 1976 U.S. average properties[17,24]: API, 42.2; 220°C midboiling point; 0.081 wt. % sulfur; and cetane index[e] of 49.5.

[c] For high-speed diesels (passenger cars and trucks) 1972 U.S. average properties[17,24]: API = 35.7; MBP = 261°C; 0.253 wt. % sulfur; cetane index[e] = 48.4.

[d] Low- and medium-speed diesels.

[e] The cetane index is an approximation of the CN, calculated from ASTM D976-80 given the API and the midpoint temperature, and accurate within ± 2 CN for $30 \le CN \le 60$ for 75% of distillate fuels tested.

59.3   PERFORMANCE AND EFFICIENCY

The performance of an engine is generally specified through the brake power (bp), the torque (τ), or the brake mean effective pressure (bmep), while the efficiency of an engine is usually specified through the brake specific fuel consumption (bsfc) or the overall efficiency (η_e). Experimental and theoretical determination of important engine parameters is discussed in the following sections.

59.3.1   Experimental Measurements

Engine dynamometer (dyno) measurements can be used to obtain the various engine parameters using the relationships[5,8,9]

$$\text{bp} = LRN/9549.3 = LN/K \tag{59.10}$$

$$\tau = LR = 9549.3 \, \text{bp}/N \tag{59.11}$$

$$\text{bmep} = 60{,}000 \, \text{bp} \, X/DN \tag{59.12}$$

$$\text{bsfc} = \dot{m}_F/\text{bp} \tag{59.13}$$

$$\eta_e = \frac{3600 \, \text{bp}}{\dot{m}_F \text{LHV}_p} = \frac{3600}{\text{bsfc} \, \text{LHV}_p} \tag{59.14}$$

Definitions and standard units* for the variables in the above equations are presented in the symbols list. The constants in the above equations are simply unit conversion factors. The brake power is the useful power measured at the engine output shaft. Some power is used to overcome frictional losses in the engine and this power (the friction power, fp) is not available at the output shaft. The total rate of energy production within the engine is called the *indicated power* (ip)

$$\text{ip} = \text{bp} + \text{fp} \tag{59.15}$$

where the friction power can be determined from dyno measurements using:

$$\text{fp} = FRN/9549.3 = FN/K \tag{59.16}$$

*Standard units are not in strict compliance with the International System of units, in order to produce numbers of convenient magnitude.

The efficiency of overcoming frictional losses in the engine is called the *mechanical efficiency* (η_M), which is defined as

$$\eta_M = bp/ip = 1 - fp/ip \qquad (59.17)$$

The definitions of ip and η_M allow determination of the indicated mean effective pressure (imep) and the indicated specific fuel consumption (isfc):

$$imep = bmep/\eta_M = 60,000 \text{ ip } X/DN \qquad (59.18)$$

$$isfc = \eta_M bsfc = \dot{m}_F/ip \qquad (59.19)$$

Three additional efficiencies of interest are the volumetric efficiency (η_v), the combustion efficiency (η_c), and the indicated thermal efficiency (η_{ti}).

The volumetric efficiency is the effectiveness of inducting air into the engine[5,10–13] and is defined as the actual mass flow rate of air ($\dot{m}_A$) divided by the theoretical maximum air mass flow rate ($\rho_A DN/X$):

$$\eta_v = \frac{\dot{m}_A}{\rho_A DN/X} \qquad (59.20)$$

The combustion efficiency is the efficiency of converting the chemical energy of the fuel to thermal energy (enthalpy) of the products of combustion.[12–14] Thus,

$$\eta_c = -\Delta H^{298}_{R,\text{act}}/\text{LHV}_p \qquad (59.21)$$

The actual enthalpy of reaction ($\Delta H^{298}_{R,\text{act}}$) may be determined by measuring the mole fractions in the exhaust of CO_2, CO, O_2, and unburned hydrocarbons (expressed as "equivalent propane" in the following) and calculating the mole fractions of H_2O and H_2 from atom balances. For a fuel of composition C_xH_y,

$$-\Delta H^{298}_{R,\text{act}} = \frac{h^{298}_{f,C_xH_y} + x(Y_{CO_2}393.418 + Y_{H_2O}241.763 + Y_{CO}110.600 + Y_{C_3H_8}103.900)}{(Y_{CO_2} + Y_{CO} + 3Y_{C_3H_8})(12.011x + 1.008y)} \qquad (59.22)$$

where Y_i is the mole fraction of species i in the "wet" exhaust. In Eq. (59.22), a carbon balance was used to convert moles of species i per mole of product mixture to moles of species i per mole of fuel burned and the molecular weight of the fuel appears in the denominator to produce the enthalpy of reaction in units of MJ per kg of fuel burned. If a significant amount of soot is present in the exhaust (e.g., a diesel under high load), then the carbon balance becomes inaccurate and an oxygen balance would have to be substituted.

The indicated thermal efficiency is the efficiency of the actual thermodynamic cycle. This parameter is difficult to measure directly, but may be calculated from

$$\eta_{ti} = \frac{3600 \text{ ip}}{\eta_c \dot{m}_F \text{LHV}_p} = \frac{3600}{\text{isfc } \eta_c \text{LHV}_p} \qquad (59.23)$$

Because the engine performance depends on the air flow rate through it, the ambient temperature, barometric pressure, and relative humidity can affect the performance parameters and efficiencies. It is often desirable to correct the measured values to standard atmospheric conditions. The use of correction factors is discussed in the literature,[5,8,9] but is beyond the scope of this chapter.

The four fundamental efficiencies η_{ti}, η_c, η_v, and η_M are related to the global performance and efficiency parameters in the following section. Methods for modeling these efficiencies are also discussed in the following section.

59.3.2 Theoretical Considerations and Modeling

A set of *exact* equations relating the fundamental efficiencies to the global engine parameters is[12,13]

$$bp = \eta_{ti}\eta_c\eta_v\eta_M \, \rho_A \, D \, N \, LHV_p FA/(60X) \tag{59.24}$$

$$ip = \eta_{ti}\eta_c\eta_v \, \rho_A \, D \, N \, LHV_p FA/(60X) \tag{59.25}$$

$$\tau = 1000 \, \eta_{ti}\eta_c\eta_v\eta_M \, \rho_A \, D \, LHV_p FA/(2\pi X) \tag{59.26}$$

$$bmep = 1000 \, \eta_{ti}\eta_c\eta_v\eta_M \, \rho_A \, LHV_p FA \tag{59.27}$$

$$imep = 1000 \, \eta_{ti}\eta_c\eta_v \, \rho_A \, LHV_p FA \tag{59.28}$$

$$bsfc = 3600/(\eta_{ti}\eta_c\eta_M \, LHV_p) \tag{59.29}$$

$$isfc = 3600/(\eta_{ti}\eta_c \, LHV_p) \tag{59.30}$$

$$\eta_e = \eta_{ti}\eta_c\eta_M \tag{59.31}$$

where, again, the constants are simply units conversion factors. Equations (59.24)–(59.31) are of interest because they (1) can be derived solely from physical and thermodynamic considerations,[12] (2) can be used to explain observed engine characteristics,[12] and (3) can be used as a base for modeling engine performance.[13] For example, Eqs. (59.27) and (59.28) demonstrate that the mean effective pressure is useful for comparing different engines because it is a measure of performance that is essentially independent of displacement (D), engine speed (N), and whether the engine is a 2-stroke ($X = 1$) or a 4-stroke ($X = 2$). Similarly, Eqs. (59.24), (59.27), and (59.29) show that a diesel should have less power, lower bmep, and better bsfc than a comparable SI engine because the diesel generally has about the same LHV_p and η_c, higher η_{ti} and η_v, but lower η_M and much lower FA.

The performance of an engine may be theoretically predicted by modeling each of the fundamental efficiencies (η_{ti}, η_c, η_v, and η_M) and then combining these models using Eqs. (59.2)–(59.31) to yield the performance parameters. Simplified models for each of these fundamental efficiencies are discussed below. More detailed engine models are available with varying degrees of sophistication and accuracy,[2,4,11,13,15,16,25] but, because of their length and complexity, are beyond the scope of this chapter.

The combustion efficiency may be most simply modeled by assuming complete combustion. It can be shown[13] that for complete combustion

$$\eta_c = 1.0 \qquad \phi \leq 1 \tag{59.32a}$$

$$\eta_c = 1.0/\phi \qquad \phi \geq 1 \tag{59.32b}$$

Equation (59.32) implies that η_c is only dependent on FA. It has been shown that for homogeneous charge 4-stroke SI engines using fuels with a carbon-to-hydrogen ratio similar to that of gasoline, η_c is approximately independent of compression ratio, engine speed, ignition timing, and load. Although no data are available, it is expected that this is also true of 4-stroke stratified charge SI engines and diesel engines (at least up to the point of production of appreciable smoke). Such relationships will be less accurate for 2-stroke SI engines due to fuel short-circuiting. As shown in Fig. 59.7, Eq. (59.32) is accurate within 5–10% for fuel-lean combustion with accuracy decreasing to about 20% for the very fuel-rich equivalence ratio of 1.5 for the 4-stroke SI engine. Also shown in Fig. 59.7 is a quasi equilibrium equation that is slightly more accurate for fuel-lean systems and much more accurate for fuel-rich combustion:

$$\eta_c = 0.959 + 0.129\phi - 0.121\phi^2 \qquad 0.5 \leq \phi \leq 1.0 \tag{59.33a}$$

$$\eta_c = 2.594 - 2.173\phi + 0.546\phi^2 \qquad 1.0 \leq \phi \leq 1.5 \tag{59.33b}$$

The indicated thermal efficiency may be most easily modeled using air standard cycles. The values of η_{ti} predicted in this manner will be too high by a factor of about 2 or more, but the trends predicted will be qualitatively correct.

The SI engine ideally operates on the air standard Otto cycle, for which[2,4,5,10,11,14]

$$\eta_{ti} = 1 - (1/CR)^{k-1} \tag{59.34}$$

where CR is the compression ratio and k is the ratio of specific heats of the working fluid. The assumptions of the air standard Otto cycle are (1) no intake or exhaust processes and thus no exhaust residual and no pumping loss, (2) isentropic compression and expansion, (3) constant-volume heat addition and therefore instantaneous combustion, (4) constant-volume heat rejection replacing the exhaust blowdown process, and (5) air is the sole working fluid and is assumed to have a constant value of k. The errors in this model primarily result from failure to account for (1) a working fluid with variable composition and variable specific heats, (2) the finite duration of combustion, (3) heat

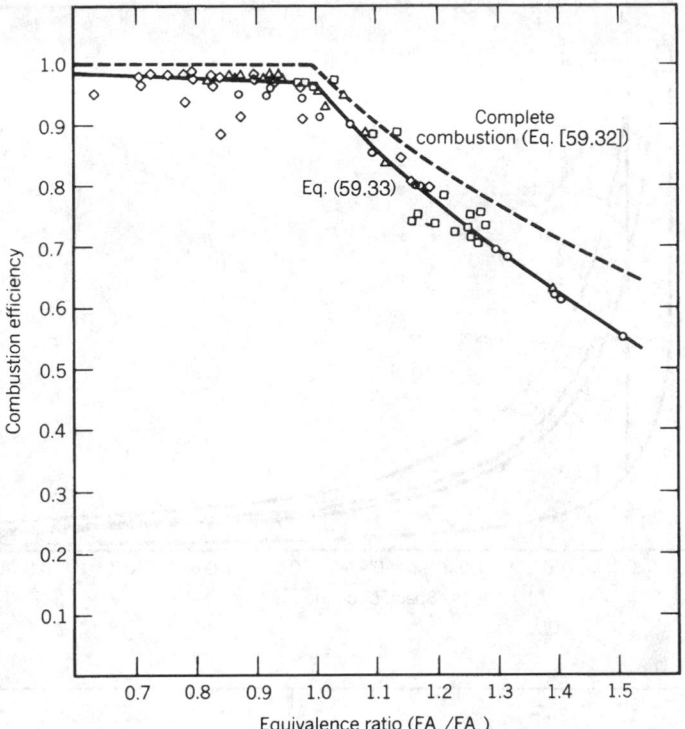

Fig. 59.7 Effect of equivalence ratio (normalized fuel–air ratio) on combustion efficiency for several different 4-stroke SI engines operating on indolene or iso-octane.[13] Model predictions of Eqs. (59.32) and (59.33) also shown.

losses, (4) fluid mechanics (especially as affecting flow past the intake and exhaust valves and the effects of these on the exhaust residual fraction), and (5) pumping losses. The air standard Otto cycle P–v and T–s diagrams are shown in Fig. 59.8. Equation (59.34) indicates that η_{ti} for the SI engine is only a function of CR. While this is not strictly correct, it has been shown[13] that, for the homogeneous charge 4-stroke SI engine, η_{ti} is not as strongly dependent on equivalence ratio, speed, and load as on CR. The predicted effect of CR on η_{ti} is compared with engine data in Fig. 59.9, showing that the theoretical trend is qualitatively correct. Thus, dividing the result of Eq. (59.34) by 2 will yield a reasonable estimate of the indicated thermal efficiency but will not reflect the effects of speed, load, spark timing, valve timing, or other engine design and operating conditions on η_{ti}.

The traditional simplified model for η_{ti} of the CI engine is the air standard Diesel cycle, but the air standard dual cycle is more representative of most modern diesel engines. Figure 59.8 compares the P–v and T–s diagrams for the air standard diesel, dual, and Otto cycles. For the air standard diesel cycle, it can be shown that [2,4,5,11,14]

$$\eta_{ti} = 1 - (1/CR)^{k-1}\,\frac{r_T^k - 1}{k(r_T - 1)} \tag{59.35}$$

where r_T is the ratio of the temperature at the end of combustion to the temperature at the beginning of combustion, and is thus a measure of the load. Assumptions for this cycle are (1) no intake or exhaust processes, (2) isentropic compression and expansion, (3) constant-pressure heat addition (combustion), (4) constant-volume heat rejection, and (5) air is the sole working fluid and has constant k. For the air standard dual cycle, it can be shown that[2,4,5,10,14]

$$\eta_{ti} = 1 - (1/CR)^{k-1}\,\frac{r_P r_v^k - 1}{(r_P - 1) + kr_P(r_v - 1)} \tag{59.36}$$

where r_P is the ratio of the maximum pressure to the pressure at the beginning of the combustion process and r_v is the ratio of the volume at the end of the combustion process to the volume at the

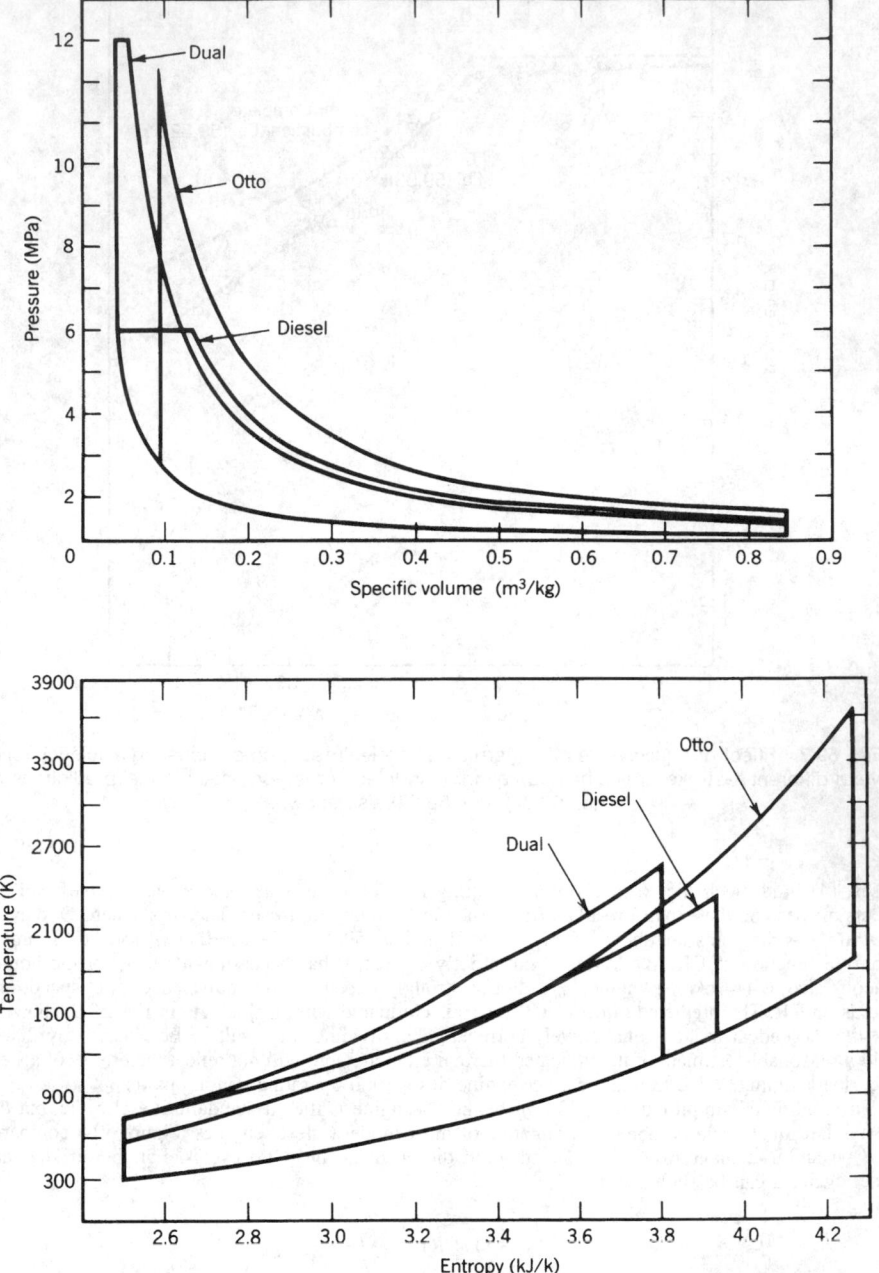

Fig. 59.8 Comparison of air standard Otto, diesel, and dual cycle P–v and T–s diagrams with $k = 1.3$. Otto with CR = 9:1, $\phi = 1.0$, C_8H_{18}. Diesel with CR = 20:1, $\phi = 0.7$, $C_{12}H_{26}$. Dual at same conditions as diesel, but with 50% of heat added at constant volume.

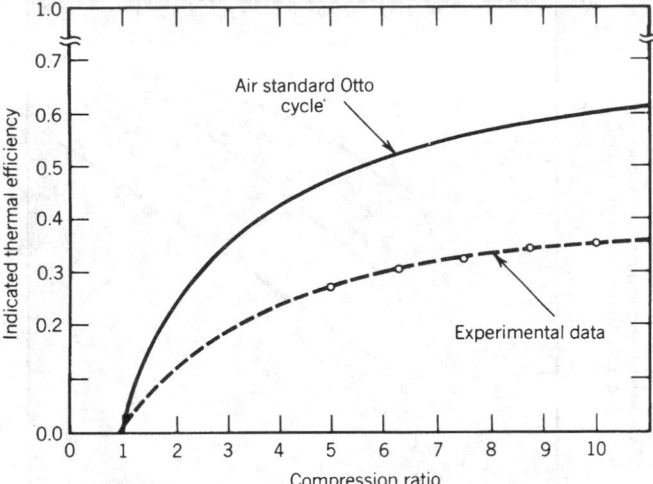

Fig. 59.9 Effect of compression ratio on indicated thermal efficiency of SI engine. Dashed line—experimental data for single cylinder 4-stroke SI engine.[13] Solid line—prediction of air standard Otto cycle [Eq. (59.34)].

beginning of the combustion process (TDC). The assumptions are the same as those for the air standard diesel cycle except that combustion is assumed to occur initially at constant volume with the remainder of the combustion process occurring at constant pressure. Equations (59.35) and (59.36) indicate that the η_{ti} of the diesel is a function of both compression ratio and load, predictions that are qualitatively correct.

The volumetric efficiency is the sole efficiency for which a simplified model cannot be developed from thermophysical principles without the need for iterative calculations. Factors affecting η_v include heat transfer, fluid mechanics (including intake and/or exhaust tuning and valve timing), and exhaust residual fraction. In fact, a 4-stroke engine with tuned intake (exhaust tuning is more important for the 2-stroke) and/or pressure boosting may have greater than 100% volumetric efficiency since Eq. (59.20) is referenced to the air inlet density rather than to the density of the air in the intake manifold. However, for unboosted engines, observed engine characteristics allow values to be estimated for η_v (for subsequent use in Eqs. (59.24)–(59.31)). For the unthrottled DISC SI engine, the diesel, and the SI engine at wide open throttle (full load), η_v is approximately independent of operating conditions other than engine speed (which is important due to valve timing and tuning effects). A peak value for η_v of 0.7–1.0 may be assumed for 4-stroke engines with untuned intake systems, recognizing that there is no justification for choosing any particular number unless engine data for that specific engine are available. For 4-stroke engines with tuned intake systems, a peak value of ~1.15–1.2 might be assumed. Because most SI engines control power output by varying η_v, the effect of load on η_v must be taken into account for this type of engine. Fortunately, other engine operating conditions have much less effect on part load η_v than does the load,[13] so that only load need be considered for this simplified approach. As shown in Fig. 59.10 for the 4-stroke SI engine, η_v is linearly related to load (imep). Because choked flow is attained at no load, a value for η_v at zero load at roughly one-half that assumed at full load may be used and a linear relationship between η_v and load (or intake manifold pressure) may then be used (the intake manifold pressure may drop well below that for choked flow during idle and deceleration, but the flow is still choked).

The mechanical efficiency may be most simply modeled by first determining η_{ti}, η_c, and η_v and then calculating the indicated power using Eq. (59.25). The friction power may then be calculated from an empirical relationship,[4] which has been shown to be reasonably accurate for a variety of production multicylinder SI engines (and it may apply reasonably well to diesels):

$$\text{fp} = 1.975 \times 10^{-9} \, D \, S \, \text{CR}^{1/2} N^2 \tag{59.37}$$

where S is the stroke in mm. It should be noted that oil viscosity (and therefore oil and coolant temperatures) can significantly affect fp, and that Eq. (59.37) applies for normal operating temperatures. Given fp and ip, η_M may be calculated using Eq. (59.17). Since load strongly affects ip and speed strongly affects fp, then η_M is more strongly dependent on speed and load than on other

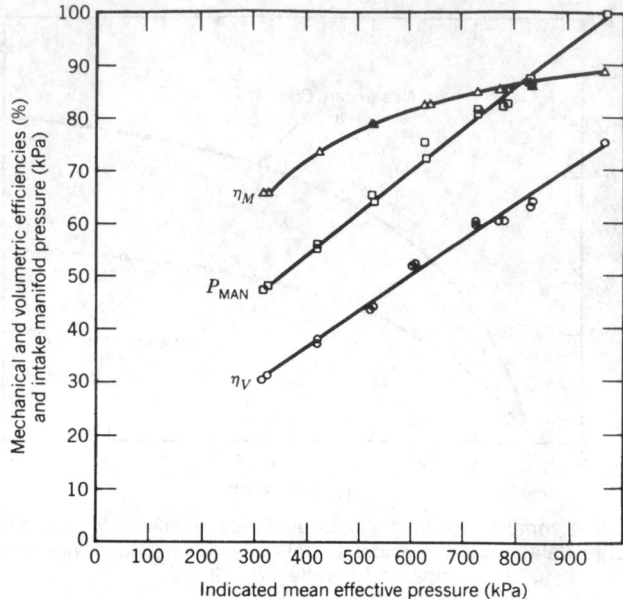

Fig. 59.10 Effect of load (imep) on volumetric efficiency, mechanical efficiency, and intake manifold absolute pressure for 4-stroke V6 SI engine at 2000 rpm.[13]

operating conditions. Figures 59.10 and 59.11 demonstrate the effects of speed and load on η_M for a 4-stroke SI engine. Similar trends would be observed for other types of engines.

The accuracy of the performance predictions discussed above depends on the accuracies of the prediction of η_{ti}, η_c, η_v, and fp. Of these, the models for η_c and fp are generally acceptable. Thus, the primary sources for error are the models of η_{ti} and η_v. As stated earlier, highly accurate models are available if more than qualitative performance predictions are needed.

59.3.3 Engine Comparisons

Table 59.4 presents comparative data for various types of engines.

One of the primary advantages of the SI engine is the wide speed range attainable owing to the short ignition delay, high flame speed, and low rotational mass. This results in high specific weight (bp/W) and high power per unit displacement (bp/D). Additionally, the energy content (LHV$_p$) of gasoline is about 3% greater than that of diesel fuel (on a mass basis; diesel fuel has ~13% more energy per gallon than gasoline). This aids both bmep and bp. The low CR results in relatively high η_M at full load and low engine weight. The low weight, combined with the relative mechanical simplicity (especially of the fueling system), results in low initial cost. The high full load η_M and the high FA (relative to the diesel) result in high bmep and bp. However, the low CR results in low η_{ti} (especially at part load), which in turn causes low η_e, high part load bsfc, and poor low speed τ. Because η_{ti} increases with load, reasonable bsfc is attained near full load. Also, since the product of η_{ti} and N initially increases faster with engine speed than η_M decreases, then good medium speed torque is attained and can be significantly augmented by intake tuning such that η_v peaks at medium engine speed. Another disadvantage of the SI engine is that the near stoichiometric FA used results in high engine-out emissions of the gaseous regulated exhaust pollutants: NO$_x$ (oxides of nitrogen), CO (carbon monoxide), and HCs (unburned hydrocarbons). On the other hand, the nature of the premixed combustion process results in particulate emission levels that are almost too low to measure.

One of the primary advantages of the diesel is that the high CR results in high η_{ti} and thus good full load bsfc and, because of the relationship between η_{ti} and load, much higher part load bsfc than the SI engine. The higher η_{ti} and the high η_v produce high η_e and good low speed τ. The low volatility of diesel fuel results in lower evaporative emissions and a lower risk of accidental fire. The high CR also results in relatively low η_M and high engine weight. The high W and the mechanical sophistication (especially of the fuel-injection system) lead to high initial cost. The low η_M, coupled with the somewhat lower LHV$_p$ and the much lower FA in comparison to the SI engine, yield lower bmep. The low bmep and limited range of engine speeds yield low bp and low bp/D. The low bp and high W produce low bp/W. The diesel is ideally suited to supercharging, since

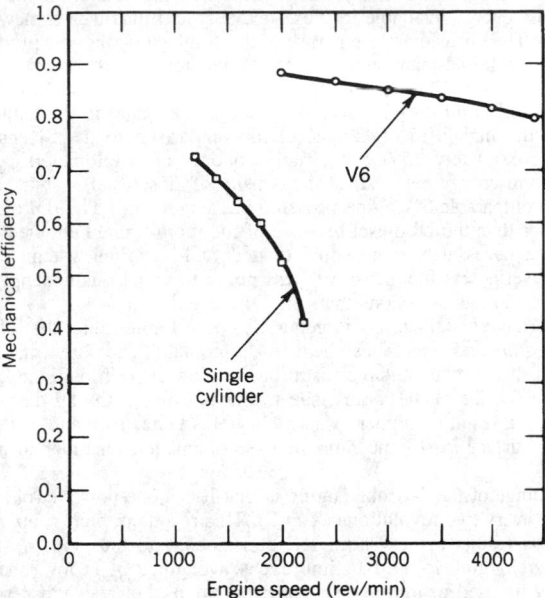

Fig. 59.11 Effect of engine speed on mechanical efficiency of 4-stroke SI V6 engine[13] and of 4-stroke SI single-cylinder research engine.

Table 59.4 Comparative Data for Motor Vehicle Engines[g]

Engine Type	CR	Peak N (rpm)	bmep (kPa)	bp/D (kW/liter)	bp/W (kW/kg)	bsfc (g/kWhr)	η_e (%)
Spark Ignition							
Motorcycles							
2-stroke	6.5–11	4500–7000[e]	400–550	20–50	0.17–0.40	600–400	14–18
4-stroke	6–10	5000–7500[e]	700–1000	30–60	0.18–0.40	340–270	25–31
Passenger cars							
2-stroke	6–8	4500–5400	450–600	30–45	0.18–0.40	480–340	17–18
4-stroke	7–11[c]	4000–7500[f]	700–1000	20–50	0.25–0.50	380–300	20–28
Rotary	8–9	6000–8000	950–1050	35–45	0.62–1.1	380–300	22–27
Trucks (4-stroke)	7–8	3600–5000	650–700	25–30	0.15–0.40	400–300	16–27
Diesel							
Passenger cars[a]	12–23[d]	4000–5000	500–750	18–22	0.20–0.40	340–240	23–28
Trucks—NA[b]	16–22	2100–4000	600–900	15–22	0.14–0.25	245–220	23–33
Trucks—TC[b]	15–22	2100–3200	1200–1800	18–26	0.14–0.29	245–220	—

[a] Exclusively IDI in United States. 2-stroke and 4-stroke DI may be used in other countries.
[b] NA: naturally aspirated, TC: turbocharged. Trucks are primarily DI.
[c] U.S. average about 9.
[d] U.S. average about 22.
[e] Many modern motorcycle engines have peak speeds exceeding 10,000 rpm.
[f] Production engines capable of exceeding 6000 rpm are rare.
[g] Adopted with permission from Adler and Bazlen.[3] Design trends to be read left to right.

supercharging inhibits knock. However, the lower exhaust temperatures of the diesel (500–600°C)[3] means that there is less energy available in diesel exhaust and thus it is somewhat more difficult to turbocharge the diesel. The diffusion-flame nature of the combustion process produces high particulate and NO_x levels. The overall fuel-lean nature of the combustion process produces relatively low levels of HCs and very low levels of CO. Diesel engines are much noisier than SI engines.

The IDI diesel is used in all diesel-powered passenger cars sold in the United States, which are attractive because of the high bsfc of the diesel in comparison to the SI engine. The IDI diesel passenger car has approximately 35% better fuel economy on a kilometers per liter of fuel basis (19% better on a per unit energy basis, since the density of diesel fuel is about 16% higher than that of gasoline) than the comparable SI-engine-powered passenger car.[17] The IDI diesel is currently used in passenger cars rather than the DI diesel because of the more limited engine speed range of the DI diesel and because the necessarily more sophisticated fuel-injection system leads to higher initial cost. Also, the IDI diesel is less fuel sensitive (less prone to knock), is generally smaller, runs more quietly, and has fewer odorous emissions than the DI diesel.[17]

The engine speed range of DI diesels is generally more limited than that of the IDI diesel. The DI diesel is easier to start and rejects less heat to the coolant. The somewhat higher exhaust temperature makes the DI diesel more suitable for turbocharging. At full load, the DI emits more smoke and is noisier than the IDI diesel. In comparison to the IDI diesel, the DI diesel has 15–20% better fuel economy.[17] This is a result of higher η_{ti} (which yields better bsfc and η_e) due to less heat loss because of the lower surface-to-volume ratio of the combustion chamber in absence of the IDI's prechamber.

The primary advantage of the 2-stroke engine is that it has one power stroke per revolution ($X = 1$) rather than one per every two revolutions ($X = 2$). This results in high τ, bp, and bp/W. However, the scavenging process results in very high HC emissions, and thus low η_c, bmep, bsfc, and η_e. When crankcase supercharging is used to improve scavenging, η_v is low,[3] being in the range of 0.3–0.7. Blowers may be used to improve scavenging and η_v, but result in decreased η_M. The mechanical simplicity of the valveless crankcase compression design results in high η_M, low weight, and low initial cost. Air-cooled designs have an even lower weight and initial cost, but have limited CR because of engine-cooling considerations. In fact, both air-cooled and water-cooled engines have high thermal loads because of the lack of no-load strokes. For the 2-stroke SI engine, the CR is also limited by the need to inhibit knock.

The primary advantages of the rotary SI engine are the resulting low vibration, high engine speed, and relatively high η_v. The high η_v produces high bmep and good low-speed τ. The high bmep and high attainable engine speed produce high bp. The valveless design results in decreased W. The high bp and low W result in very high bp/W. Because η_e and bsfc are independent of η_v, these parameters are approximately equal for the rotary SI and the 4-stroke piston SI engines.

59.4 EMISSIONS AND FUEL ECONOMY REGULATIONS*

Light-duty vehicles sold in the United States are subject to federal and state regulations regarding exhaust emissions, evaporative emissions, and fuel economy. Heavy-duty vehicles do not have to comply with any fuel economy standards, but are required to comply with standards for exhaust and evaporative emissions. Similar regulations are in effect in many foreign countries.

59.4.1 Light-Duty Vehicles

Light-duty-vehicle (LDV) emissions regulations are divided into those applicable to passenger cars and those applicable to light-duty trucks [defined as trucks of less than 6000 lb gross vehicle weight (GVW) rating prior to 1979 and less than 8500 lb after, except those with more than 45 ft² frontal area]. In turn, the light-duty-trucks (LDTs) are divided into four categories: LDT1s are "light light-duty trucks" (GVW < 6000 lb) that have a loaded vehicle weight (LVW) of <3750 lb and must meet the same emissions standards as passenger cars; LDT2s are also light light-duty trucks and have 3751 < LVW < 5750 lb; LDT3s are "heavy light-duty trucks" (GVW > 6000 lb) that have an adjusted loaded vehicle weight (ALVW) of 3751 < ALVW < 5750 lb; LDT4s are heavy light-duty trucks with 5751 < LVW < 8550 lb. Standards for the light-duty trucks in the LDT2–LDT4 categories are not presented or discussed for brevity, but the test procedure described below is used for these vehicles as well as for passenger cars.

U.S. federal and state of California exhaust emissions regulations for passenger cars are presented in Table 59.5. Emissions levels are for operation over the federal test procedure (FTP) transient driving cycle. The regulated emissions levels must be met at the end of the "useful life" of the vehicle. For passenger cars, the useful life is currently defined as five years or 50,000 miles, but the

---

*The regulations are specified in mixed English and metric units and these specified units are used in this section to avoid confusion.

Table 59.5 Federal and California Emissions Standards for Passenger Cars

	Useful Life (Miles)	THC (gm/mi)	NMHC[a] (gm/mi)	NMOG (gm/mi)	CO (gm/mi)	NO$_x$ (gm/mi)	Particulates (gm/mi)	HCHO (mg/mi)
Precontrol avg.		10.6			84.0	4.0		
Fed. 1975–77	50K	1.5			15.0	3.1		
Cal. 1975–77	50K	0.9			9.0	2.0		
Fed. 1977–79	50K	1.5			15.0	2.0		
Cal. 1977–79	50K	0.41			9.0	1.5		
Fed. 1980	50K	0.41			7.0	2.0		
Cal. 1980	50K	—	0.39		9.0	1.0		
Fed. 1981	50K	0.41			3.4	1.0		
Cal. 1981	50K		0.39		7.0	0.7		
Fed. 1982–84	50K	0.41			3.4	1.0	0.6	
Cal. 1982	50K		0.39		7.0	0.7	0.6	
Cal. 1983–84	50K		0.39		7.0	0.4	0.6	
Fed. 1985–90	50K	0.41			3.4	1.0	0.2	
Cal. 1985–92	50K		0.39		7.0	0.4	0.2	
Fed. Tier 0 (1991–94)	50K[b]	0.41	—	—	3.4	1.0		—
	100K[c]	0.80	—	—	10.0	1.2		—
Cal. Tier 1	50K	0.25	—		3.4	0.4		15[d]
	100K	0.31	—		4.2	0.6		15
Fed. Tier 1[e] (1994–)	50K	0.25	—		3.4	0.4		—
	100K	0.31	—		4.2	0.6		—
Fed. Tier 2 (2003)	100K	0.125	—		1.7	0.2		—
TLEV—alt. fuel	50K		—	0.125	3.4	0.4		15
	100K		—	0.156	4.2	0.6		18
TLEV—bi-fuel[f]	50K		—	0.250	3.4	0.4		15
	100K		—	0.310	4.2	0.6		18
LEV—alt. fuel	50K		—	0.075	3.4	0.2		15
	100K		—	0.090	4.2	0.3		18
LEV—bi-fuel[f]	50K		—	0.125	3.4	0.2		15
	100K		—	0.156	4.2	0.3		18
ULEV—alt. fuel	50K		—	0.040	1.7	0.2		8
	100K		—	0.055	2.1	0.3		11
ULEV—bi-fuel[f]	50K		—	0.075	1.7	0.2		8
	100K		—	0.090	2.1	0.3		11

[a] Organic material nonmethane hydrocarbon equivalent (OMNHCE) for methanol vehicles.

[b] Passenger cars can elect Tier 0 with the lower useful life, but cannot elect Tier 0 at the full useful life.

[c] LDTs that elect Tier 0 must use the full useful life.

[d] This formaldehyde standard must be complied with at 50,000 miles rather than at the end of the full useful life.

[e] Federal Tier 1 standards also require that all light-duty vehicles meet a 0.5% idle CO standard and a cold (20°F FTP) CO standard of 10 gm/mi.

[f] Flexible fuel and bi-fuel vehicles must certify both to the dedicated (alt. fuel) standard (while using the alternative fuel) *and* the bi-fuel standard when using the conventional fuel (e.g., certification gasoline).

Clean Air Act Amendments (CAAA) of 1990 redefined the useful life of passenger cars and LDT1s as 10 years/100,000 miles and 120,000 miles for LDT2–LDT4 light duty trucks. The longer useful life requirement is being phased-in and all vehicles must meet the longer useful life emissions standards in 2003. The first two phases of the FTP are illustrated in Fig. 59.12. Following these two phases (together, these are the LA-4 cycle), the engine is turned off, the vehicle is allowed to hot soak for 10 minutes, the engine is restarted, and the first 505 seconds are repeated as phase 3. The LA-4 cycle is intended to represent average urban driving habits and therefore has an average vehicle

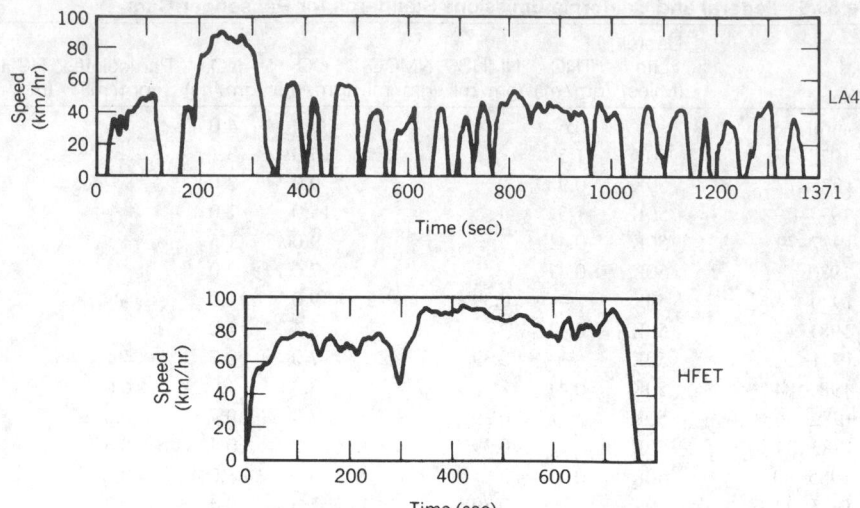

Fig. 59.12 FTP and HFET transient driving cycles.

speed of only 19.5 mi/hr (31.46 km/hr) and includes 2.29 stops per mile (1.42 per kilometer). The LA-4 test length is 7.5 miles (11.98 km) and the duration is 1372 seconds with 19% of this time spent with the vehicle idling. The motor vehicle manufacturer will commit at least two vehicles of each type to the certification procedure. The first vehicle will be used to accumulate 50,000 miles (or 100,000 or 120,000, depending upon the vehicle type and the useful life to which the vehicle is certified) so that a deterioration factor (DF) can be determined for each of the regulated species. The DF is determined by periodic FTP testing over the useful life and is intended to reveal how the emissions levels change with mileage accumulation. The second vehicle is subjected to the FTP test after 4000 miles of accumulation. The emission level of each regulated species is then multiplied by the DF for that species and the product is used to determine compliance with the regulatory standard. All data are submitted to the U.S. Environmental Protection Agency (EPA), which may perform random confirmatory testing, especially of the 4000-mile vehicles.

The 1981 standards represented a 96% reduction in HC and CO and 76% reduction in NO_x from precontrol levels. Beginning in 1980, the California HC standard was specified as either 0.41 grams of total hydrocarbons (THCs) per mile or 0.39 grams of nonmethane hydrocarbons (NMHCs) per mile, in recognition that methane is not considered to be photochemically reactive and therefore does not contribute to smog formation.[17,18] The federal CAAA of 1990 also recognized that methane does not contribute to smog formation and thus instituted a NMHC standard, rather than a THC standard, as part of the new emissions standards phased in from 1994 to 1996. Similarly, because the goal of the hydrocarbon standards is control of ambient ozone, the current California rules regulate Nonmethane Organic Gases (NMOG) rather than NMHCs. Furthermore, in recognition that each of the organic gases in vehicle exhaust has a different efficiency in ozone-formation chemistry, the California rulemaking does not regulate measured NMOG, but rather "reactivity-adjusted" NMOG. The measured NMOG is multiplied by a reactivity adjustment factor (RAF) and the resulting product is then judged against the NMOG standard. The RAF is both fuel-specific and emissions level-specific. The California emissions levels are TLEV (transitional low-emission vehicle), LEV (low-emission vehicle), ULEV (ultra-low-emission vehicle), and ZEV (zero-emission vehicle, or electric vehicle). For LEV vehicles, CARB has established the following RAFs: 1.0 for gasoline, 0.94 for California Phase 2 reformulated gasoline, 0.43 for CNG, 0.50 for LPG, and 0.41 for M85. The particulate emissions standards shown in Table 59.5 apply only to light-duty diesel (LDD) vehicles; light-duty gasoline (LDG) vehicles emit essentially immeasurable quantities of particulates, and therefore are not required to demonstrate compliance with the particulate standards.

LDVs are also required to meet evaporative emissions standards, as determined using a SHED (sealed housing for evaporative determination) test. The precontrol average for evaporative emissions was 50.6 grams per test, which is approximately equivalent to 4.3 gm HC per mile.[19] The federal regulations are 6 gm/test for 1978–80 and 2 gm/test thereafter. The California standard is also 2 gm/test. Evaporative emissions from the LDD are essentially immeasurable and thus are not required to demonstrate compliance with this regulation. California allows LDDs a 0.16 gm/mi credit on exhaust HCs in recognition of the negligible evaporative HC emissions of these vehicles. In the early

1990s, although retaining the standard of 2 gm/test, California instituted new rules requiring use of a new SHED test design, requiring that the test period increase from one day to four days, requiring that the maximum temperature increase to 104°F from 84°F and compliance for 120,000 miles/10 years instead of 50,000 miles/5 years. A similar federal SHED test will be implemented in the near future.

Beginning in 1978, the motor vehicle manufacturers were required to meet the corporate average fuel economy (CAFE) standards shown in Table 59.6 (a different, and lower, set of standards is in effect for "low-volume" manufacturers). The CAFE standard is the mandated average fuel economy, as determined by averaging over all vehicles sold by a given manufacturer. The CAFE requirement is based upon the "composite" fuel economy (mpg_c, in mi/gal), which is a weighted calculation of the fuel economy measured over the FTP cycle (mpg_u, "urban" fuel economy) plus the fuel economy measured over the highway fuel economy test (HFET, as shown in Fig. 59.12) transient driving cycle (mpg_h):

$$mpg_c = \frac{1.0}{\dfrac{0.55}{mpg_u} + \dfrac{0.45}{mpg_h}} \qquad (59.38)$$

The manufacturer is assessed a fine of $5 per vehicle produced for every 0.1 mpg below the CAFE standard. Beginning in 1985, an adjustment factor was added to the measured CAFE to account for the effects of changes in the test procedure. Additionally, the motor vehicle manufacturer is subjected to a "gas guzzler" tax for each vehicle sold that is below a minimum composite fuel economy, with the minimum level being: 1983, 19.0 mpg; 1984, 19.5 mpg; 1985, 21.0 mpg; 1986 on, 22.5 mpg. In 1986, the tax ranged from $500 for each vehicle with 21.5–22.4 mpg up to the maximum of $3850 for each vehicle with less than 12.5 mpg.

The federal and state regulations are intended to require compliance based on the best available technology. Therefore, future standards may be subjected to postponement, modification, or waiver, depending on available technology. For more detailed information regarding the standards and measurement techniques, refer to the U.S. Code of Federal Regulations, Title 40, Parts 86 and 88, and to the Society of Automotive Engineers (SAE) Recommended Practices.[20]

59.4.2 Heavy-Duty Vehicles

The test procedure for heavy-duty vehicles (HDVs) is different from that for the LDV. Because many manufacturers of heavy-duty vehicles offer the customer a choice of engines from various vendors, the engines rather than the vehicles must be certified. Thus, the emissions standards are based on grams of pollutant emitted per unit of energy produced (g/bhp-hr, mixed metric and English units, with bhp meaning brake horsepower) during engine operation over a prescribed transient test cycle. The federal standards for 1991 and newer heavy-duty engines are divided into two gross vehicle weight categories for spark ignition engines (less than and greater than 14,000 lb), but a single set of standards is used for diesel engines. The standards also depend upon the fuel used (diesel, gasoline, natural gas, etc.), as presented in Table 59.7. Beginning in 1985, the heavy-duty gasoline (HDG) vehicle was required to meet an idle CO standard of 0.47% and an evaporative emissions SHED Test standard of 3.0 g/test for vehicles with GVW less than 14,000 lb and 4.0 g/test for heavier vehicles.

Table 59.6 CAFE Standards for Passenger Cars Required by the Energy Policy and Conservation Act of 1975

Model Year	Average mpg
1978	18.0
1979	19.0
1980	20.0
1981	22.0[a]
1982	24.0[a]
1983	26.0[a]
1984	27.0[a]
1985	27.5[a]

[a] Set by the Secretary of the U.S. Department of Transportation.

Table 59.7 Emissions Standards[a] for 1991–2000 Model Year Heavy-Duty Engines (in g/bhp-hr)

GVW Category	Engine	Fuel	THCs[b]	NMHCs	CO	NO$_x$[c]	PM
≤ 14,000 lb	SI	gasoline	1.1	NA	14.4	5.0	NA
≤ 14,000 lb	SI	methanol	1.1	NA	14.4	5.0	NA
≤ 14,000 lb	SI	CNG	NA	0.9	14.4	5.0	NA
≤ 14,000 lb	SI	LPG	1.1	NA	14.4	5.0	NA
> 14,000 lb	SI	gasoline	1.9	NA	37.1	5.0	NA
> 14,000 lb	SI	methanol	1.9	NA	37.1	5.0	NA
> 14,000 lb	SI	CNG	NA	1.7	37.1	5.0	NA
> 14,000 lb	SI	LPG	1.9	NA	37.1	5.0	NA
any (> 8501 lb)	CI	diesel	1.3	NA	15.5	5.0	0.25[d]
any (> 8501 lb)	CI	methanol	1.3	NA	15.5	5.0	0.25[d]
any (> 8501 lb)	CI	CNG	NA	1.2	15.5	5.0	0.10
any (> 8501 lb)	CI	LPG	1.3	NA	15.5	5.0	0.10

[a] Other standards apply to clean fuel vehicles.
[b] Organic material hydrocarbon equivalent (OMHCE) for methanol.
[c] NO$_x$ standard decreases to 4.0 g/bhp-hr beginning in 1998.
[d] Particulate standard decreased to 0.10 g/bhp-hr beginning in 1994.

Prior to 1988, diesels were required only to meet smoke opacity standards (rather than particulate mass standards) of 20% opacity during acceleration, 15% opacity during lugging, and 50% peak opacity. Particulate mass standards for heavy duty diesels came into effect in 1988.

59.4.3 Nonhighway Heavy-Duty Standards

Stationary engines may be subjected to federal, state, and local regulations, and these regulations vary throughout the United States.[19] Diesel engines used in off-highway vehicles and in railroad applications may be subjected to state and local visible smoke limits, which are typically about 20% opacity.[19]

SYMBOLS

AF	air–fuel mass ratio [-]*
AF$_s$	stoichiometric air–fuel mass ratio [-]
ALVW	adjusted loaded vehicle weight = (curb weight + GVW)/2 [lb]
API	American Petroleum Institute gravity [°]
ASTM	American Society of Testing and Materials
BDC	bottom dead center
bmep	break mean effective pressure [kPa]
bp	brake power [kW]
bsfc	brake specific fuel consumption [g/kW-hr]
CARB	California Air Resources Board
CI	compression ignition (diesel) engine
CN	cetane number [-]
CO	carbon monoxide
CO$_2$	carbon dioxide
CR	compression ratio (by volume) [-]
C$_x$H$_y$	average fuel molecule, with x atoms of carbon and y atoms of hydrogen
D	engine displacement [liters]
DI	direct injection diesel engine
DISC	direct injection stratified charge SI engine

*Standard units, which do not strictly conform to metric practice, in order to produce numbers of convenient magnitude.

DF	emissions deteriorator factor [-]
F	force on dyno torque arm with engine being motored [N]
FA	fuel–air mass ratio [-]
FA_s	stoichiometric fuel–air mass ratio [-]
fp	friction power [kW]
FTP	federal test procedure transient driving cycle
GVW	gross vehicle weight rating
HCs	hydrocarbons
$h_{f,i}^{298}$	standard enthalpy of formation of gaseous species i at 298 K [MJ/kmole of species i]*
$h_{v,i}^{298}$	enthalpy of vaporization of species i at 298 K [MJ/kmole of species i]
ΔH_{RX}^{298}	enthalpy of reaction (heat of combustion) at 298 K [MJ/kg of fuel]
HDD	heavy-duty diesel vehicle
HDG	heavy-duty gasoline vehicle
HDV	heavy-duty vehicle
HFET	highway fuel economy test transient driving cycle
HV	heating value [MJ/kg of fuel]
HHV_p	constant-pressure higher heating value [MJ/kg of fuel]
HHV_v	constant-volume higher heating value [MJ/kg of fuel]
HV_p	constant-pressure heating value [MJ/kg of fuel]
HV_v	constant-volume heating value [MJ/kg of fuel]
IDI	indirect injection diesel engine
ILEV	inherently low emission vehicle
imep	indicated mean effective pressure [kPa]
ip	indicated power [kW]
isfc	indicated specific fuel consumption [g/kW-hr]
k	ratio of specific heats
K	dyno constant [N/kW-min]
L	force on dyno torque arm with engine running [N]
LDD	light-duty diesel vehicle
LDG	light-duty gasoline vehicle
LDT	light-duty truck
LDV	light-duty vehicle
LEV	low-emission vehicle
LHV_p	constant-pressure lower heating value [MJ/kg of fuel]
LHV_v	constant-volume lower heating value [MJ/kg of fuel]
LVW	loaded vehicle weight (curb weight plus 300 lb)
$\dot{m}_A$	air mass flow rate into engine [g/hr]
$\dot{m}_F$	fuel mass flow rate into engine [g/hr]
MON	octane number measured using Motor method [-]
mpg_c	composite fuel economy [mi/gal]
mpg_h	highway (HFET) fuel economy [mi/gal]
mpg_u	urban (FTP) fuel economy [mi/gal]
MW	molecular weight [kg/kmole]
N	engine rotational speed [rpm]
N_2	molecular nitrogen
NMHC	nonmethane hydrocarbons
NMOG	nonmethane organic gases (NMHCs plus carbonyls and alcohols)
NO_x	oxides of nitrogen
O_2	molecular oxygen

*A kmole is a mole based upon a kg; also called a kg-mole.

ON	octane number of fuel [-]
P	pressure [kPa]
R	dyno torque arm length [m]
RAF	reactivity adjustment factor
RON	octane number measured using Research method [-]
r_P	pressure ratio (end of combustion/end of compression) [-]
r_T	temperature ratio (end of combustion/end of compression) [-]
r_v	volume ratio (end of combustion/end of compression) [-]
s	entropy [kJ/kg-K]
S	piston stroke [nm]
SAE	Society of Automotive Engineers
sg_F	specific gravity of fuel
SHED	sealed housing for evaporative determination
SI	spark ignition engine
T	milliliters of tetraethyl lead per gallon of iso-octane [ml/gal]
T	temperature [K]
TA	percent theoretical air
TDC	top dead center
THCs	total hydrocarbons
TLEV	transitional low emission vehicle
ULEV	ultra low emission vehicle
v	specific volume [m^3/kg]
W	engine mass [kg]
X	crankshaft revolutions per power stroke
x	atoms of carbon per molecule of fuel
y	atoms of hydrogen per molecule of fuel
Y_i	mole fraction of species i in exhaust products [-]
ZEV	zero emission vehicle
α	1.0 for HHV, 0.0 for LHV [-]
β	1.0 for liquid fuel, 0.0 for gaseous fuel [-]
η_c	combustion efficiency [-]
η_e	overall engine efficiency [-]
η_M	mechanical efficiency [-]
η_{ti}	indicated thermal efficiency [-]
η_v	volumetric efficiency [-]
ϕ	equivalence ratio, FA/FA$_s$ [-]
ρ_A	density of air at engine inlet [kg/m^3]
ρ_F	density of fuel [kg/m^3]
ρ_w	density of water [kg/m^3]
λ	excess air ratio, AF/AF$_s$ [-]
τ	engine output torque [N-m]

REFERENCES

1. C. F. Taylor, *The Internal Combustion Engine in Theory and Practice,* Vol. 2, MIT Press, Cambridge, MA, 1979.
2. R. S. Benson and N. D. Whitehouse, *Internal Combustion Engines,* Pergamon Press, New York, 1979.
3. U. Adler and W. Baslen (eds.), *Automotive Handbook,* Bosch, Stuttgart, 1978.
4. L. C. Lichty, *Combustion Engine Processes,* McGraw-Hill, New York, 1967.
5. E. F. Obert, *Internal Combustion Engines and Air Pollution,* Harper and Row, New York, 1973.
6. Fuels and Lubricants Committee, "Diesel Fuels—SAE J312 APR82," in *SAE Handbook,* Vol. 3, pp. 23.34–23.39, 1983.
7. Fuels and Lubricants Committee, "Automotive Gasolines—SAE J312 JUN82," in *SAE Handbook,* Vol. 3, pp. 23.25–23.34, 1983.

8. Engine Committee, "Engine Power Test Code—Spark Ignition and Diesel—SAE J1349 DEC80," in *SAE Handbook,* Vol. 3, pp. 24.09–24.10, 1983.

9. Engine Committee, "Small Spark Ignition Engine Test Code—SAE J607a," in *SAE Handbook,* Vol. 3, pp. 24.14–24.15, 1983.

10. C. F. Taylor, *The Internal Combustion Engine in Theory and Practice,* Vol. 1, MIT Press, Cambridge, MA, 1979.

11. A. S. Campbell, *Thermodynamic Analysis of Combustion Engines,* Wiley, New York, 1979.

12. R. D. Matthews, "Relationship of Brake Power to Various Efficiencies and Other Engine Parameters: The Efficiency Rule," *International Journal of Vehicle Design* 4(5), 491–500 (1983).

13. R. D. Matthews, S. A. Beckel, S. Z. Miao, and J. E. Peters, "A New Technique for Thermodynamic Engine Modeling," *Journal of Energy* 7(6), 667–675 (1983).

14. B. D. Wood, *Applications of Thermodynamics,* Addison-Wesley, Reading, MA, 1982.

15. P. N. Blumberg, G. A. Lavoie, and R. J. Tabaczynski, "Phenomenological Model for Reciprocating Internal Combustion Engines," *Progress in Energy and Combustion Science* 5, 123–167 (1979).

16. J. N. Mattavi and C. A. Amann (eds.), *Combustion Modeling in Reciprocating Engines,* Plenum, New York, 1980.

17. Diesel Impacts Study Committee, *Diesel Technology,* National Academy Press, Washington, DC, 1982.

18. R. D. Matthews, "Emission of Unregulated Pollutants from Light Duty Vehicles," *International Journal of Vehicle Design* 5(4), 475–489 (1983).

19. Environmental Activities Staff, *Pocket Reference,* General Motors, Warren, MI, 1983.

20. Automotive Emissions Committee, "Emissions," in *SAE Handbook,* Vol. 3, Section 25, 1983.

21. Fuels and Lubricants Committee, "Alternative Automotive Fuels—SAE J1297 APR82," in *SAE Handbook,* Vol. 3, pp. 23.39–23.41, 1983.

22. R. C. Weast (ed.), *CRC Handbook of Chemistry and Physics,* 55th ed., CRC Press, Cleveland, OH, 1974.

23. ASTM Committee D-2, *Physical Constants of Hydrocarbons C_1 to C_{10},* American Society for Testing and Materials, Philadelphia, PA, 1963.

24. E. M. Shelton, *Diesel Fuel Oils, 1976,* U.S. Energy Research and Development Administration, Publication No. BERC/PPS-76/5, 1976.

25. C. M. Wu, C. E. Roberts, R. D. Matthews, and M. J. Hall, "Effects of Engine Speed on Combustion in SI Engines: Comparisons of Predictions of a Fractal Burning Model with Experimental Data," SAE Paper 932714, also in SAE Transactions: *Journal of Engines* 102(3), 2277–2291, 1994.

26. R. D. Matthews, J. Chiu, and D. Hilden, "CNG Compositions in Texas and the Effects of Composition on Emissions, Fuel Economy, and Driveability of NGVs," SAE Paper 962097, 1996.

27. W. E. Liss and W. H. Thrasher, *Variability of Natural Gas Composition in Select Major Metropolitan Areas of the United States,* American Gas Association Laboratories Report to the Gas Research Institute, GRI Report No. GRI-92/0123, 1991; also see SAE Paper 9123464, 1991.

28. R. H. Pahl and M. J. McNally, Fuel Blending and Analysis for the Auto/Oil Air Quality Improvement Research Program, SAE Paper 902098, 1990.

29. S. C. Mayotte, V. Rao, C. E. Lindhjem, and M. S. Sklar, *Reformulated Gasoline Effects on Exhaust Emissions: Phase II: Continued Investigation of Fuel Oxygenate Content, Oxygenate Type, Volatility, Sulfur, Olefins, and Distillation Parameters,* SAE Paper 941974, 1994.

30. J. Kubesh, S. R. King, and W. E. Liss, *Effect of Gas Composition on Octane Number of Natural Gas Fuels,* SAE Paper 922359, 1992.

31. K. Owen and T. Coley, *Automotive Fuels Handbook,* Society of Automotive Engineers, Warrendale, PA, 1990.

BIBLIOGRAPHY

Combustion and Flame, The Combustion Institute, Pittsburgh, PA, published monthly.

Combustion Science and Technology, Gordon and Breach, New York, published monthly.

International Journal of Vehicle Design, Interscience, Jersey, C.I., UK, published quarterly.

Proceedings of the Institute of Mechanical Engineers (Automobile Division), I Mech E, London.

Progress in Energy and Combustion Science, Pergamon, Oxford, UK, published quarterly.

SAE Technical Paper Series, Society of Automotive Engineers, Warrendale, PA.

Symposia (International) on Combustion, The Combustion Institute, Pittsburgh, PA, published biennially.

Transactions of the Society of Automotive Engineers, SAE, Warrendale, PA, published annually.

CHAPTER 60

HYDRAULIC SYSTEMS

Hugh R. Martin
University of Waterloo
Waterloo, Ontario, Canada

60.1	HYDRAULIC FLUIDS	1831	60.8	SYSTEM CLASSIFICATIONS	1847
60.2	CONTAMINATION CONTROL	1832	60.9	PUMP SETS AND ACCUMULATORS	1847
60.3	POSITIVE ASPECTS OF CONTAMINATION	1833	60.10	HYDROSTATIC TRANSMISSIONS	1851
60.4	DESIGN EQUATIONS— ORIFICES AND VALVES	1834	60.11	CONCEPT OF FEEDBACK CONTROL IN HYDRAULICS	1852
60.5	DESIGN EQUATIONS—PIPES AND FITTINGS	1835	60.12	IMPROVED MODEL	1854
60.6	HYDROSTATIC PUMPS AND MOTORS	1838	60.13	ELECTROHYDRAULIC SYSTEMS—ANALOG	1856
60.7	STIFFNESS IN HYDRAULIC SYSTEMS	1843	60.14	ELECTROHYDRAULIC SYSTEMS—DIGITAL	1860

60.1 HYDRAULIC FLUIDS

One of the results of the study of fluid mechanics has been the development of the use of hydraulic oil, a so-called incompressible fluid, for performing useful work. Fluids have been used to transmit power for many centuries, the most available fluid being water. While water is cheap and usually readily available, it does have the distinct disadvantages of promoting rusting, of freezing to a solid, and of having relatively poor lubrication properties.

Mineral oils have provided superior properties. Much of the success of modern hydraulic oils is due to the relative ease with which their properties can be altered by the use of additives, such as rust and foam inhibitors, without significantly changing fluid characteristics.

Although hydraulic oil is used mainly to transmit fluid power, it must also 1) provide lubrication for moving parts, such as spool valves, 2) absorb and transfer heat generated within the system, and 3) remain stable, both in storage and in use, over a wide range of possible physical and chemical changes.

It is estimated that 75% of all hydraulic equipment problems are directly related to the improper use of oil in the system. Contamination control in the system is a very important aspect of circuit design.

In certain industries, such as mining and nuclear power, it is critically important to control the potential for fire hazards. Hence, fire-resistant fluids have been playing an ever-increasing role in these types of industry. The higher pressure levels in modern fluid power circuits have made fire hazards more serious when petroleum oil is used, since a fractured component or line will result in a fine mist of oil that can travel as far as 40 ft and is readily ignited. The term *fire-resistant fluid*

Reprinted with permission from J. A. Schetz and A. E. Fuhs (eds.), *Handbook of Fluid Dynamics and Fluid Machinery.* © 1996 John Wiley & Sons, Inc.

Mechanical Engineers' Handbook, 2nd ed., Edited by Myer Kutz.
ISBN 0-471-13007-9   © 1998 John Wiley & Sons, Inc.

(FRF) generally relates to those liquids that fall into two broad classes: a) those where water provides the fire resistance, and b) those where a fire retardant is inherent in the chemical structure.[1-4] Fluids in the first group are water/glycol mixtures, water in oil emulsions (40–50% water), and oil in water emulsions (5–15% water). The second group are synthetic materials, in particular chlorinated hydrocarbons and phosphate esters.

A disadvantage with water-based fluids is that they are limited to approximately 50–60°C operating temperature because of evaporation. The high vapor pressure indicates this group is more prone to cavitation than mineral oils. Synthetic fluids such as the phosphate esters do not have this problem and also have far superior lubrication properties. Some typical characteristics of these various types of fluids are shown in Table 60.1.

Of all the physical properties that can be listed for hydraulic fluids, the essential characteristics of immediate interest to a designer are 1) bulk modulus, to assess system rigidity and natural frequency, 2) viscosity, to assess pipe work and component pressure losses, 3) density, to measure flow and pressure drop calculations, and 4) lubricity, to determine threshold and control accuracy assessments. The first three items are discussed in separate sections, as they relate directly to circuit design. *Lubricity,* the final item, is difficult to define, as it is very much a qualitative judgment. Lubricity affects the performance of a system, since it is a major factor in determining the level of damping in the system, that is, *viscous* or velocity-dependent *damping.* It also affects the accuracy of operation of a system because of its influence on the other type of friction, *coulomb friction,* which is velocity-independent.

Oil film strength is often referred to as the *anti-wear value* of a lubricant, which is the ability of the fluid to maintain a film between moving parts and thus prevent metal-to-metal contact. These characteristics are important for the moving parts in valves, cylinders, and pumps.[5]

60.2 CONTAMINATION CONTROL

There is little doubt that component failure or damage due to fluid contamination is an area of major concern to both the designer and user of fluid power equipment. Sources of contamination in fluid power equipment are many. Although oil is refined and blended under relatively clean conditions, it does accumulate small particles of debris during storage and transportation. It is not unusual for hydraulic oil circulating in a well maintained hydraulic circuit to be cleaner than that from a newly purchased drum. New components and equipment invariably have a certain amount of debris left from the manufacturing process, in spite of rigorous post-production flushing of the unit.

The contaminant level in a system can be increased internally due to local burning (oxidation) of oil to create sludges. This can be a result of running the oil temperature too high (normally 40–60°C is recommended) or due to local cavitation in the fluid.

The trend towards the use of higher system pressures in hydraulics generally results in narrower clearances between mating components. Under such design conditions, quite small particles in the range of 2–20 microns can block moving surfaces.

Extensive work on contamination classification has been carried out by Fitch and his co-workers.[6]

To take a specific example, consider the piston pump shown in Fig. 60.1. Component parts of the pump are loaded towards each other by forces generated by the pressure, and this same pressure always tends to force oil through the adjacent clearance. The life of the pump is related to the rate at which a relatively small amount of material is being worn away from a few critical surfaces. It is logical to assume, therefore, if the fluid in a clearance is contaminated with particles, rapid degradation and eventual failure can occur.

Although the geometric clearances are fixed, the actual clearances vary with eccentricity due to load and viscosity variations. Some typical clearances between moving parts are shown in Table 60.2.

Contamination control is the job of filtration. System reliability and life are related not only to the contamination level but also to contaminant size ranges. To maintain contaminant levels at a magnitude compatible with component reliability requires both the correct filter specification and suitable placement in the circuit. Filters can be placed in the suction line, pressure line, return line,

Table 60.1 Comparison of Some Hydraulic Fluids

Property	Units	FRF (Ester)	Mineral Oil	Water in Oil
Density (38 C)	kg m^{-3}	1136.0	858.2	980.0
Viscosity (38 C)	m^2 s^{-1}	4.6×10^{-6}	4.0×10^{-5}	0.15×10^{-5}
(99 C)		4.9×10^{-6}	5.8×10^{-6}	
Bulk modulus (38 C and 34.5 MPa)	N m^{-2}	2.25×10^9	1.38×10^9	2.18×10^9
Vapor pressure	kPa (abs)	6×10^{-5}	6×10^{-5}	1.0

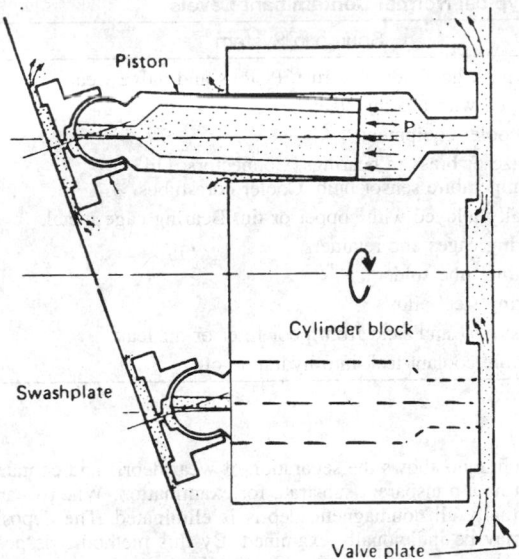

Fig. 60.1 Piston pump clearances.

or in a partial flow mode. To use a broad approach of just inserting a filter with a very low rating is unsatisfactory from the aspects of both cost and high pressure loss. The optimization of choice can be approached using simple computer modeling, as described by Foord.[7]

Dirt in hydraulic systems consists of many different types of material, ranging in size from less than 1 micron to greater than 100 microns. Since most general industrial hydraulics operating below 14 MPa are able to tolerate particles up to 25 microns, a 25-micron-rated filter is satisfactory. Equipment operating at pressures in the 14–21 MPa range should have 10–15-micron-rated filters, while high pressure pumps and precision servo valves need 5 micron-rated filtration. A good practical reference for filter selection has been written by Spencer.[8]

The size distribution of particles is of course random, and, generally speaking, the smaller the size range the greater the number of particles per 100 ml of fluid. Filters are not capable of removing all the contaminants, but for example, a 10-micron filter is one capable of removing about 98% of all particles exceeding 10 microns of a standard contaminant in a given concentration of prepared solution.

60.3 POSITIVE ASPECTS OF CONTAMINATION

Contamination buildup in a system can be used as a diagnostic tool. Regular sampling of the oil and examination of the particles can often give a clue to potential failure of components. In other words, this is a preventive maintenance tool. Many methods can be used for this type of examination, such as spectrochemical[9] or Ferrographic[10] methods. Sampling of the oil can be taken at any time and does not interfere with the operation of the equipment.

Table 60.3 shows the normally expected contaminant levels in parts per million (ppm); levels rising above these values and particularly rates of change of levels are indicative of potential failures.

Table 60.2 Typical Clearances in Pumps

Component	Clearance Range (micron)
Spool to sleeve in valve	1–10 diametrical
Gear pump tip to casing	0.5–5
Piston to bore	5–40
Valve plate to body of pump	0.5–5

Table 60.3 Some Typical Normal Contaminant Levels

Material	Source in System	Max Level (ppm)
Iron	Bearings, gears, or pipe rust. Pistons and valve wear.	20
Chromium	Alloyed with bearing steel	4
Aluminum	Air cooler equipment	10
Copper	Bronze or brass in bearings. Connectors. Oil temperature sensor bulb. Cooler core tubes.	30
Lead	Usually alloyed with copper or tin. Bearing cage metal.	20
Tin	Bearing cages and retainers	15
Silver	Cooling tube solder	3
Nickel	Bearing steel alloy	4
Silicon	Seals; dust and sand from poor filter or air leak	9
Sodium	Possible coolant leak into hydraulic oil	50

The Ferrographic technique allows the separation of wear debris and contaminants from the fluid and allows arrangement as a transparent substrate for examination. When wear particles are precipitated magnetically, virtually all nonmagnetic debris is eliminated. The deposited particles deposit according to size and may be individually examined. By this method it is possible to differentiate cutting wear, rubbing wear, erosion, and scuffing by the size and geometry of the particles. However, the Ferrographic method is expensive compared to other methods of analysis.[11]

60.4 DESIGN EQUATIONS—ORIFICES AND VALVES

The main controlling element in any hydraulic circuit is the orifice. The fluid equivalent of the electrical resistance, it can be fixed in size or can be variable, in the case of a spool valve. The orifice in its various configurations is also the main source of heat generation, resulting in the need for cooling techniques and a major source of noise.

The orifice equation is developed from Bernoulli's energy balance approach, which results in the following relationship:[12]

$$Q = \frac{C_c C_v A_o}{\sqrt{1 - \left(\dfrac{C_v A_o}{A_u}\right)^2}} \sqrt{\frac{2(p_u - p_{vc})}{\rho}} \tag{60.1}$$

where Q = volume flow rate, m³/s
 A_o = orifice area, m²
 A_u = upstream area, m²
 p_u = upstream static pressure, Pa
 p_{vc} = static pressure at *Vena contracta*, Pa
 C_c = contraction coefficient
 C_v = velocity coefficient
 ρ = mass density of hydraulic fluid, kg/m³

These parameters are shown in Fig. 60.2, together with the static pressure distribution on either side of a sharp-edged orifice. Experimental measurements show that the actual flow is about 60% of that given by Bernoulli's equation. Hence, the need for the contraction and velocity coefficients. This results in the practical form of Eq. (60.1) for typical industrial hydraulic oil

$$Q = 3.12 \times 10^{-2} A_o \sqrt{p_u - p_d} \ \text{m}^3\text{s}^{-1} \tag{60.2}$$

The symbols have the same definition as those for Eq. (60.1). The adequacy of Eq. (60.2) is demonstrated in Table 60.4.

In the case of a variable orifice, such as that found in a spool-type valve, the orifice area is a variable. In fact, it can be seen from Fig. 60.3 that the exposed area available for oil flow is part of a circle. If the orifice, in this case called a *control orifice* or *port,* is of radius r and the spool displacement from the closed position is x, then the uncovered area is

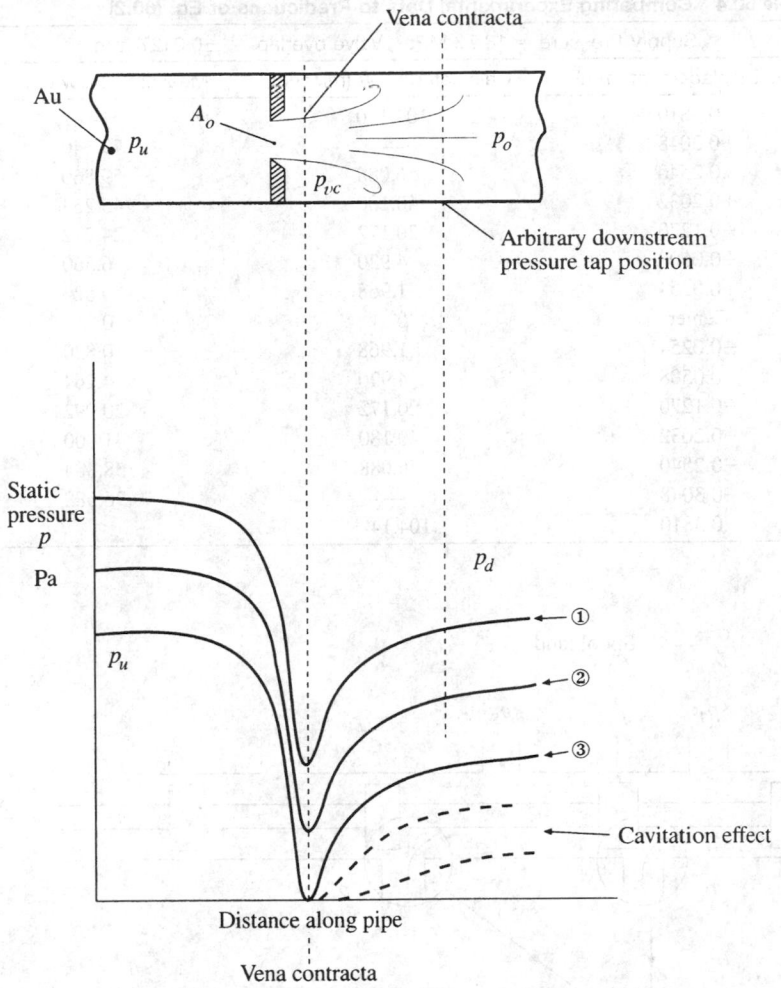

Fig. 60.2 Static pressure distribution.

$$A_o = [\theta - \cos (\theta/2)] \left(\frac{r^2}{2}\right) \qquad (60.3)$$

$$\theta = 2 \cos^{-1} [1 - (x/r)] \qquad (60.4)$$

The area displacement characteristic plotted in Fig. 60.3 shows the nonlinear nature of the curve.

One of the significant differences between the theoretical valve and the practical valve is the *lap*. It is not economical to produce zero lapped valves, so that only at the center position is the flow through the valve zero. Normally, the valve is either overlapped or underlapped, as shown in Fig. 60.4. An overlapped valve saves fluid loss when the spool is central. This is fine for directional control valves, but it produces both accuracy and stability problems if the valve is a precision control valve within a closed-loop configuration.

An underlapped valve gives much better control and stability, at the expense of a higher leakage rate (power loss). Many more details of valve design can be found in Martin and McCloy.[12]

60.5 DESIGN EQUATIONS—PIPES AND FITTINGS

While orifices serve the important function of controlling flow in the system, pipes and fittings are necessary to transmit fluid power from the input (usually a pump) to the output (usually a ram or motor). It is important to minimize losses through these conductors as well as through other com-

Table 60.4   Comparing Experimental Data to Predictions of Eq. (60.2)

Supply Pressure = 13.78 MPa Valve overlap = ± 0.0127 mm		
Valve Displacement (mm)	Calculated Flow (ml/sec)	Measured Flow (ml/sec)
+0.3810	104.140	—
+0.3048	—	79.540
+0.2540	56.088	59.860
+0.2032	40.180	45.264
+0.1270	20.172	24.272
+0.0508	4.920	6.560
+0.0254	1.968	0.820
Center	0	0
−0.0254	1.968	0.820
−0.0508	4.920	4.264
−0.1270	20.172	20.992
−0.2032	40.180	41.000
−0.2540	56.088	58.384
−0.3048	—	76.588
−0.3810	104.14	—

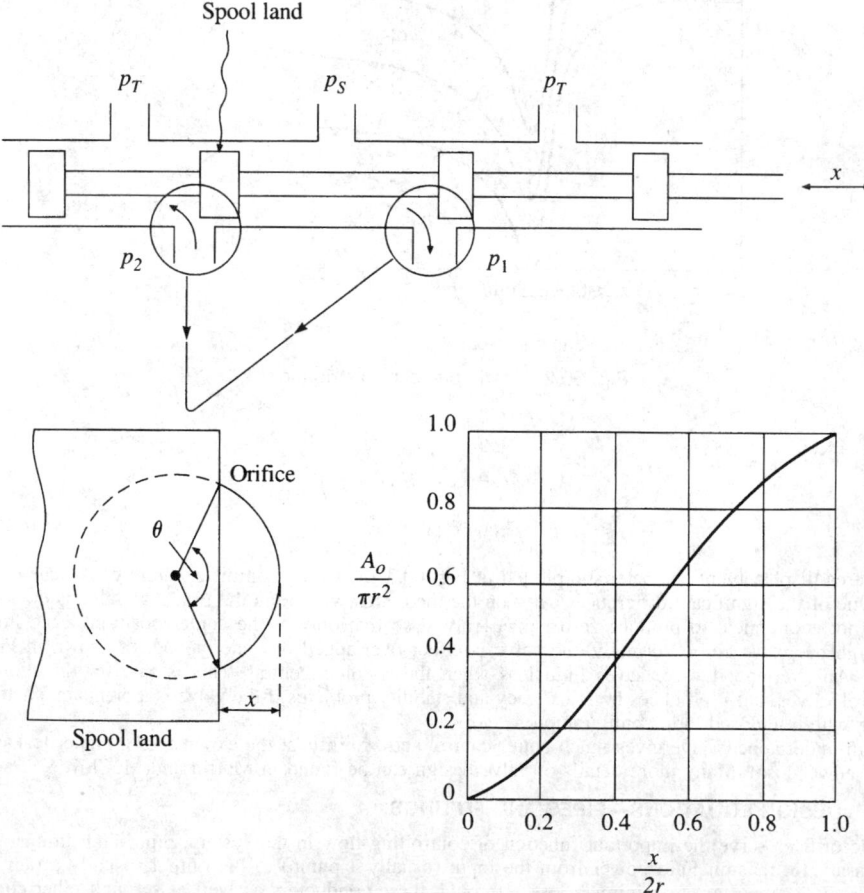

Fig. 60.3   Effective exposed orifice area for a spool-type valve.

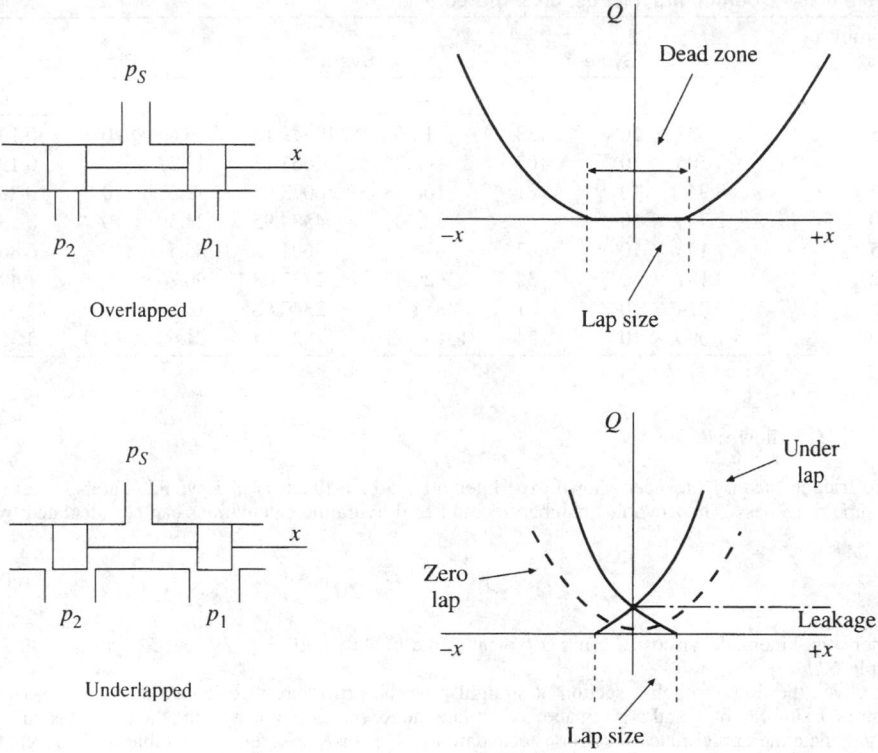

Fig. 60.4   Characteristics of valve lap.

ponents so that the maximum power is available for useful work at the circuit output. It is equally important to minimize component and piping cost. In some applications, it is also important to minimize weight and bulk size.

Pipe sizes are specified by nominal diameters, and the wall thickness by schedule number. The three schedules (or wall thicknesses) used in hydraulic piping are 40, 80, and 160, corresponding to *standard* pipe, *extra heavy,* and a little less than *double extra heavy.* The metric system of units has helped to complicate things for the designer during this transition period, more details can be found in Martin and McCloy.[12]

In the selection of piping for hydraulic circuits, the following are suggested:

- Suction lines to pumps should not carry fluid at velocities in excess of 1.5 m/sec in order to reduce the possibility of cavitation at the pump inlet.
- Delivery lines should not carry fluid at velocities in excess of 4.5 m/sec in order to prevent excessive shock loads in the pipework due to valve closure. Pressure loss due to friction in pipes should be limited to approximately 5% of the supply pressure and the recommendation also keeps heat generation to a reasonable level.
- Return lines should be of larger diameter than delivery lines to avoid back pressure buildup.

For typical industrial hydraulic oil, we can write

$$\Delta p = K_L / K_1 Q^2 \qquad (60.5)$$

where Δp = pressure drop along a straight pipe (kPa)
$\quad K_L$ = loss coefficient = $f \ell / d$
$\quad f$ = friction factor
$\quad \ell$ = pipe length (m)
$\quad d$ = internal pipe diameter (m)
$\quad K_1$ = see Table 60.5

Table 60.5   Coefficients for Eqs. 60.5 and 60.6

Nominal Bore		S.I. System		Old System		Pipe Area	
mm	in.	K_1	K_2	K_1	K_2	m²	in.²
8	¼	1.027×10^{-11}	138	12.64	10379.12	6.64×10^{-5}	0.1041
10	⅜	3.506×10^{-11}	102	42.26	7663.28	12.27×10^{-5}	0.1909
15	½	8.949×10^{-11}	81	108.08	6073.93	19.60×10^{-5}	0.3039
20	¾	2.740×10^{-10}	61	333.0	4584.95	34.30×10^{-5}	0.5333
25	1	7.195×10^{-10}	47	874.66	3601.52	55.57×10^{-5}	0.8643
32	1¼	2.181×10^{-9}	36	2620.49	2737.68	96.76×10^{-5}	1.496
40	1½	4.014×10^{-9}	31	4854.14	2340.66	13.13×10^{-4}	2.036
50	2	1.090×10^{-8}	24	13180.81	1823.64	21.65×10^{-4}	3.355

Q = flow rate (m³/sec)

The friction factor f has been shown experimentally to be a function of Reynolds number (Re) and of pipe roughness. The Reynolds number for industrial hydraulic calculations can be calculated from

$$\text{Re} = \frac{K_2}{v} Q \tag{60.6}$$

where the kinematic viscosity v has a typical value of 4.0×10^{-5} m²/s, and K_2 can be found in Table 60.5.

Given the flow through a section of straight pipe the procedure to calculate the pressure loss is simple. Using Eq. 60.6 and v given above calculate the Reynolds number. Using the Reynolds number to calculate the appropriate value for f, calculate a value for K_L. Referring to Table 60.5 for K_1, the pressure drop can be calculated using Eq. 60.5.

Unfortunately, not all piping is in straight runs so when a bend occurs the loss of pressure will be greater. The effective bend loss can be estimated from Eq. 60.7 and Figs. 60.5 and 60.6. These results are from Ref. 13.

$$\Delta p = (K + cK_B)Q^2/K_1 \tag{60.7}$$

where c = correction factor for bend angle (Fig. 60.5)
K_B = resistance coefficient for 90° bends (Fig. 60.6)

Further useful information about circuit design can be found in Keller.[14]

60.6   HYDROSTATIC PUMPS AND MOTORS

The source of power in a hydraulic circuit is the result of *hydrostatic* flow under pressure with the energy being transmitted by static pressure. In another type of fluid power, termed *hydrokinetic,* the transmission of energy is related to the change in velocity of the hydraulic fluid. While hydrostatic systems use positive displacement pumps, hydrokinetic systems use centrifugal pumps.[15]

Positive displacement machines have been in existence for many years. The concept is simply a variable displacement volume which can take the form of a piston in a cylinder, gear teeth engaging, or the sweeping action of a vane with eccentric axis placement. All these configurations are positive displacement in the sense that for each revolution of the pump shaft, a nearly constant quantity of fluid is delivered. In addition, there is some form of valving which either takes the form of nonreturn valves or a porting arrangement on a valve plate.

Examples of different types of pumps are shown in Figs. 60.7, 60.8, and 60.9. While torque and speed are the input variables to a pump, the output variables are pressure and flow. The product of these variables will give the input and output power. The difference between these values is a measure of the fluid and mechanical losses through the machine. These factors should be taken into account even for a simple analysis.

The torque required to drive the pump at constant speed can be divided into five components:

$$T_p = T_i + T_v + T_f + T_c \tag{60.8}$$

where T_p = actual required input torque (Nm)
T_i = ideal torque due to pressure differential and physical dimensions only

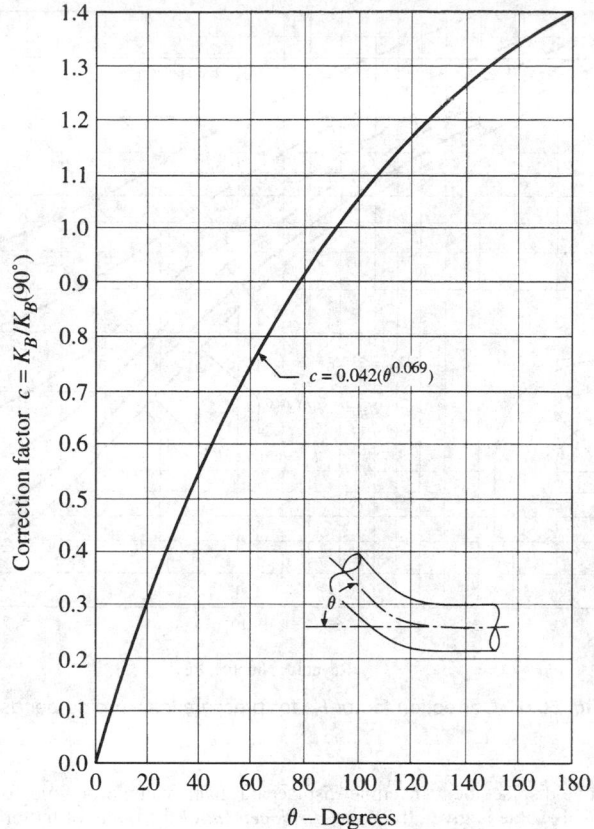

Fig. 60.5   Correction factor c. (Reproduced from AF Rocket Propulsion Lab., 1964.)

T_v = resisting torque due to viscous shearing of the fluid between stationary and moving parts of the pump, that is, viscous friction

T_f = resisting torque due to pressure and speed-dependent friction sources such as bearings and seals

T_c = remaining dry friction effects due to rubbing

The delivery from the pump can be expressed in a similar manner:

$$Q_p = Q_i - Q_l - Q_r \qquad (60.9)$$

where Q_p = actual pump delivery (ml/sec)
Q_i = ideal delivery of a pump due to geometric shape only
Q_l = viscous leakage flow
Q_r = loss in delivery due to inlet restriction[16]

If the pump is well designed and operating under its working specification, the loss represented by Q_r should not occur.

For a hydraulic motor, the procedure is reversed in the sense that flow and pressure are the input variables, and torque with angular velocity appears at the output. The corresponding equations are therefore

$$T_p = T_i - T_v - T_f - T_c \qquad (60.10)$$
$$Q_p = Q_i + Q_l \qquad (60.11)$$

Q_r is not a factor in motor performance.

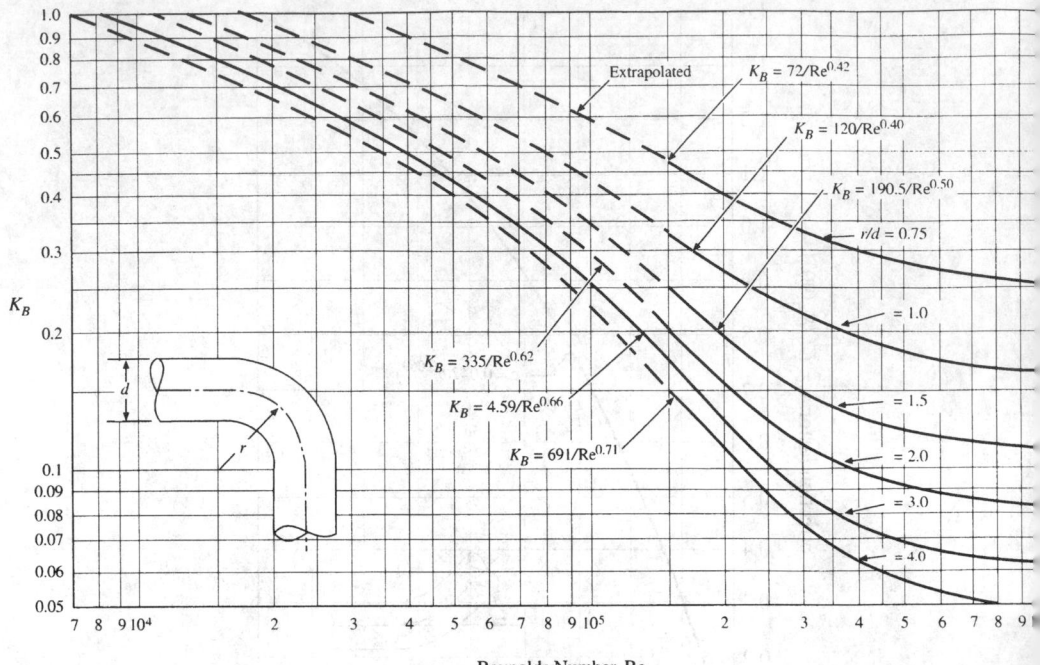

Fig. 60.6 Correction factor K_B for pressure loss in pipe bends.

The ideal positive-displacement machine displaces a given volume of fluid for every revolution of the input shaft. This value is given the name *displacement* of the pump or motor and is extensively used by manufacturers to label the pump size. Some typical characteristics for a hydraulic radial piston motor are shown in Fig. 60.10 and Table 60.6.

If the pump or motor rotates at N rpm, then

$$Q_i = D_p N \tag{60.12}$$

where D_p = swept volume per revolution = nV
V = swept volume per cylinder per revolution

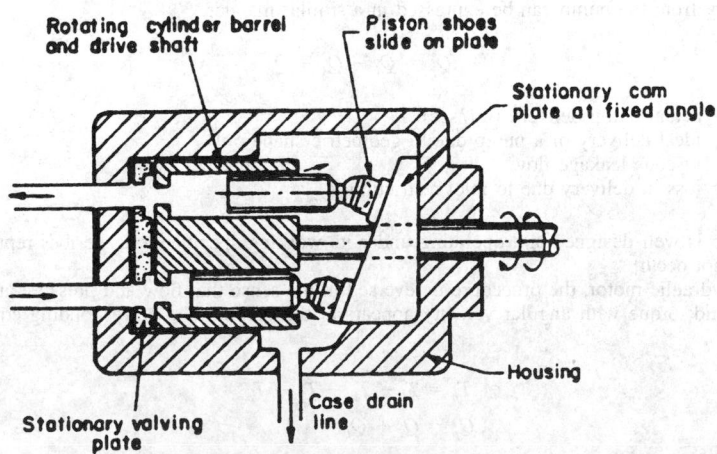

Fig. 60.7 Axial piston pump.

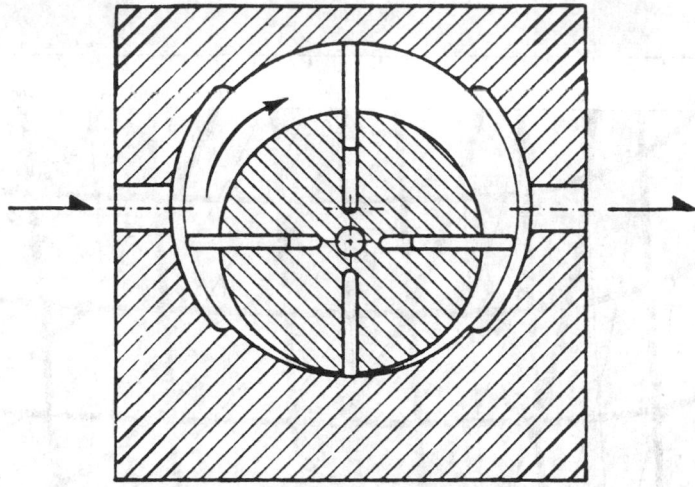

Fig. 60.8 Schematic cross section through a vane pump. (From J. Thoma, *Modern Oil Hydraulic Engineering.* © 1970 Technical and Trade Press. Reprinted with permission.)

n = number of cylinders in the pump or motor

The leakage term Q_l can be expressed in terms of a leakage coefficient C_s which is sometimes called the slip coefficient:

$$Q_l = \frac{\Delta p D_p C_s}{\mu} \tag{60.13}$$

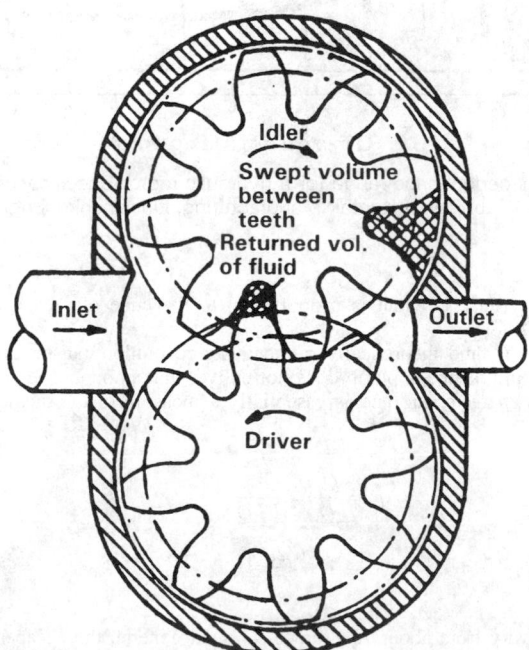

Fig. 60.9 Gear pump construction.

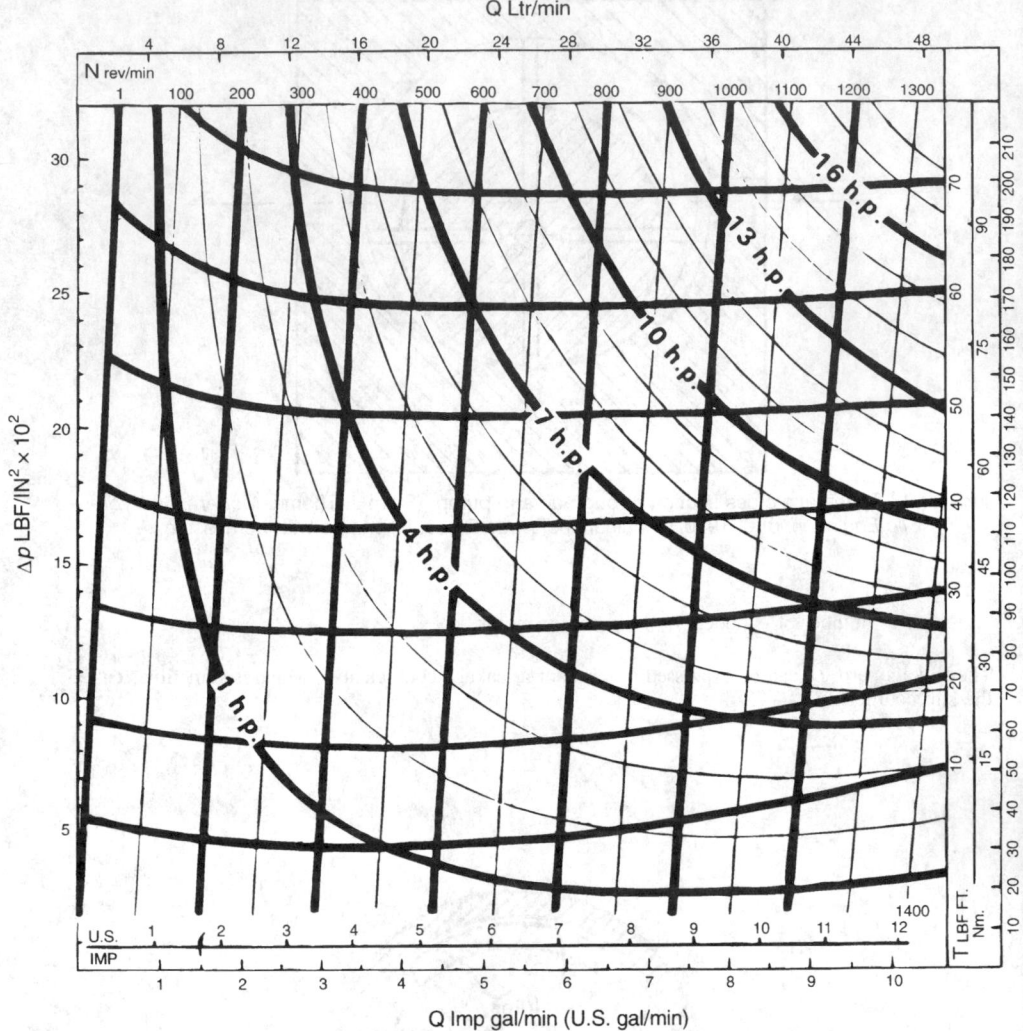

Fig. 60.10 Typical performance range for a hydraulic motor, specifications appear in Table 60.6. (Courtesy of Kontak Manufacturing, Lincolnshire, England.)

For most designs, the slip coefficient is proportional to the cube of typical clearances within the machine.[17]

While the volume of fluid theoretically pumped per revolution can be calculated from the geometry of the design, in practice, a pump does not deliver that amount. The *volumetric efficiency* η_V is used to assess this characteristic and is essentially a measure of the quality of machining or of wear in a pump.

$$\eta_V = \frac{Q_p}{Q_i} = \frac{Q_i - Q_l}{Q_i}$$

$$= 1 - \frac{Q_l}{Q_i} = 1 - \frac{\Delta p}{\mu N} C_s \tag{60.14}$$

The value of η_V can vary from about 75% up to 97%. In general, the cheaper the pump, the lower the volumetric efficiency.

The theoretical applied torque to a pump is given by

Table 60.6 Typical Performance Specification for a Radial Piston Motor[a]

Displacement	in.³/rev (cm³/rev)	1.99 (32.6)
Max. torque	lbf · ft (Nm)	74 (100)
Max. recommend speed	rev/min	1360
Max. output	hp	18
Max. recommended flow rate		10 imp.gal/min (45 ltr/min)
		(12 U.S.gal/min)
Max. pressure		3000 lbf/in.² (207 bar)
Approx. overall efficiency		85%
Max. back pressure (reversible)		3000 lbf/in.² (207 bar)
Max. drain line pressure (reversible)		50 lbf/in.² (3.5 bar)
Max. back pressure (uni-directional)		50 lbf/in.² (3.5 bar)
Weight		15 lb (6.8 kg)
Max. permissible shaft end load		2000 lbf (907 kgf)
Max. permissible shaft side load (¾ in. (19 mm) from shaft end)		2000 lbf (907 kgf)

[a]Acknowledgments to Kontak Manufacturing, Lincolnshire, England.

$$T_i = \Delta p D_p / 2\pi \tag{60.15}$$

In this case, the losses are assessed by the viscous drag coefficient C_d which is inversely proportional to the typical pump clearances, and by the drag coefficient C_f, which is proportional to the size of the pump. Referring to Eq. 60.10, T_c in a well-designed pump is normally small enough to ignore.

$$T_v = C_d D_p \mu N \tag{60.16}$$

$$T_f = C_f \frac{\Delta_p D_p}{2\pi} \tag{60.17}$$

Wilson[15] gives guidance as to the magnitude of these coefficients. His figures are given in Table 60.7.

The *mechanical efficiency* η_m of the device is a measure of the power wasted in friction. A reduction in the mechanical efficiency could, for example, be an indicator of bearing failure due to lack of lubrication.

$$\eta_m = \frac{T_i}{T_p} = \frac{T_i}{T_i + T_v + T_f}$$

$$\eta_m = \frac{1}{1 + \dfrac{2\pi C_d \mu N}{\Delta p} + C_f} \tag{60.18}$$

Finally, the *overall efficiency* of a pump or motor is the product of the volumetric and mechanical efficiencies. In general, gear pumps are suitable for pressures up to 17 MPa and have overall efficiencies of approximately 80%. A good-quality piston pump has an overall efficiency of 95% and is capable of operating with pressures up to 68.9 MPa.

60.7 STIFFNESS IN HYDRAULIC SYSTEMS

One important and often neglected aspect of hydraulic circuit design is the fact that in practice hydraulic oil is compressible. So far, only steady flow through the circuit has been discussed. How-

Table 60.7 Typical Hydraulic Pump Coefficients

Pump Type	D_p, in.³/rev	C_d	C_s	C_f
Piston	3.600	16.8×10^4	0.15×10^{-7}	0.045
Vane	2.865	7.3×10^4	0.477×10^{-7}	0.212
Spur gear	2.965	10.25×10^4	0.48×10^{-7}	0.179
Internal gear	2.965	9.77×10^4	1.02×10^{-7}	0.045

ever, when a demand for flow is changed or a valve is shut, flow and pressure in the system become subject to the rates of change. Under these conditions, natural modes of resonance can be excited, which can result in seemingly endless problems, ranging from excessive noise to fatigue failures.

Referring to Fig. 60.11, an increase in the applied force F to the piston will cause the volume of trapped oil to compress according to the relationship

$$-\Delta V = \frac{(p_2 - p_1)V_0}{\beta} \tag{60.19}$$

where $p_1 = F1/A$ = steady initial pressure
$p_2 = F2/A$ = steady final pressure
V_0 = original volume
β = bulk modulus of oil

The negative sign is to indicate that the oil volume reduces as the applied pressure increases. It is assumed that the walls of the container are rigid.

Although the change in volume is small, with a value of about 0.5% per 7 MPa applied pressure, it does result in high transient flow rates. As a comparison, air compresses about 50% for a pressure change of 0.1 MPa (1 atmosphere). The transient flow rates due to oil compressibility effects can be estimated from the first derivative of Eq. 60.19:

$$Q_c = \frac{V_0}{\beta} \frac{d\Delta p}{dt} \tag{60.20}$$

The actual value of oil bulk modulus is strongly dependent on the amount of air present in the form of bubbles. In practice, it is impossible not to have some level of air entrainment. The effective bulk modulus can then be estimated using

$$\beta_e = \frac{1}{\left[\dfrac{1}{\beta_0} + \dfrac{\alpha}{p}\right]} \tag{60.21}$$

where β_0 = oil bulk modulus with no air present
α = ratio of air volume to oil volume (typically 0.5%)
p = operating oil pressure

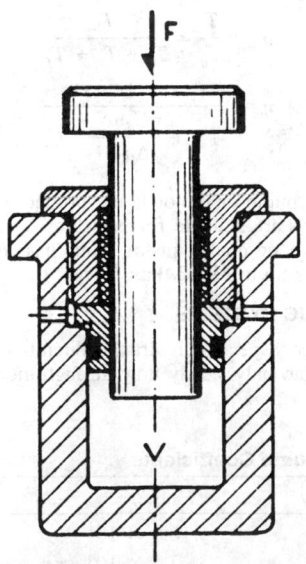

Fig. 60.11 Pressure chamber for measuring compressibility. (From J. Thoma, *Modern Oil Hydraulic Engineering.* © 1970 Technical and Trade Press. Reprinted with permission.)

These effects are illustrated in Fig. 60.12. The bulk modulus of the oil and the entrained air contribute to the effective spring a hydraulic system exhibits. For example, the hydraulic braking system of a vehicle feels spongy if there is air in the brake fluid, as a result of the circuit not being *bled* correctly.

The third factor in the system *stiffness* is the contribution from the containment vessel, which in this case is the steel pipework or reinforced rubber hose.[18] For a thin-walled metal pipe, the effective bulk modulus is estimated from[19]

$$\beta_c = \frac{TE}{D} \qquad (60.22)$$

where T = wall thickness, m
E = modulus of elasticity, Pa
D = pipe diameter, m

When the pipeline is a hydraulic hose, there is some difficulty in obtaining design information. Values for β_c in the range of 6.8×10^7 to 7.7×10^8 Pa have been quoted. Some further guidance is given in Ref. 18.

The total effective system bulk modulus taking all these effects into account can be calculated from

$$\beta_e = \left[\frac{1}{\beta_0} + \frac{\alpha}{p} + \frac{D}{TE} \right]^{-1} \qquad (60.23)$$

The effective stiffness of a system is important when the designer is concerned with reverse loading. For example, consider a hydraulic ram controlling a metal cutting tool. The loads on the tool can vary as it cuts through metal. If the hydraulic system is not very stiff, the tool will move about, giving a poor finish. Obviously there will be some movement, as it is not possible to design an infinitely stiff system. However, a high stiffness will make any such tendency to move very little.

The second problem due to system stiffness is related to dynamic behavior. Since hydraulic machines have moving parts, there will be masses and inertias to accelerate. The interaction of mass with stiffness results in natural resonant modes. These natural frequencies are normally passive, but, if excited by a power source of comparable frequency, the result can be significant noise and vibration or, in the extreme, structural failure. It is very important, therefore, for the designer to estimate these passive modes at the design stage.

Consider the case of a simple ram shown in Fig. 60.13, which is used to position a mass M. It can be shown in Martin and McCloy[12] that the flow into the ram is

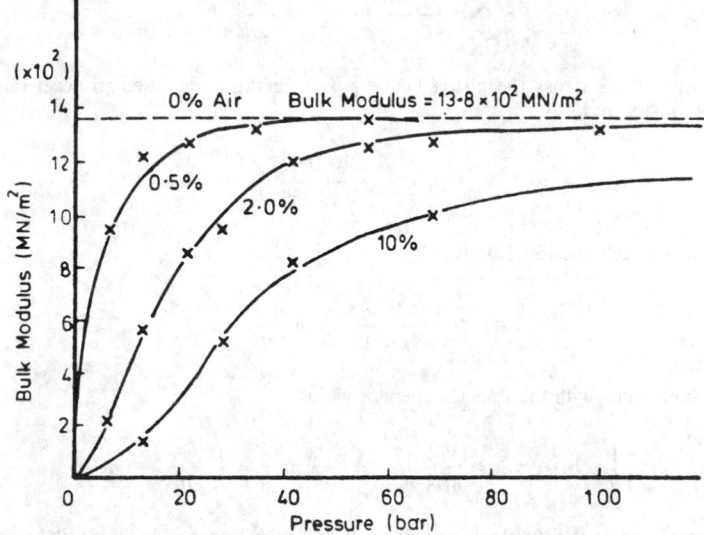

Fig. 60.12 Bulk modulus for a typical hydraulic oil including the effect of free air.

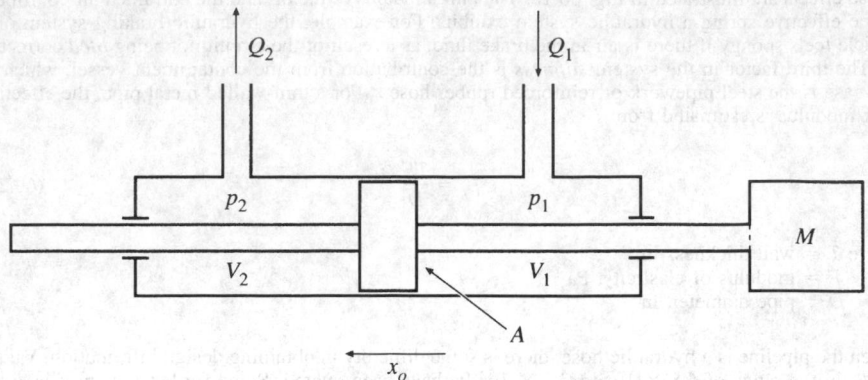

Fig. 60.13 Compressibility effects in a cylinder.

$$Q_1 = \frac{V_1}{\beta_1} \frac{dp_1}{dt} + A \frac{dx_0}{dt} \tag{60.24}$$

In other words, the first term on the right-hand side of Eq. (60.24) represents the contribution of compressibility to the total oil flow. If the flow is steady, this term disappears. The second term is the more commonly recognized flow into the ram as the piston bore volume geometry changes.

A similar argument can be applied to the left-hand side of the ram where the oil is being pushed out:

$$Q_2 = -\frac{V_2}{\beta_2} \frac{dp_2}{dt} + A \frac{dx_0}{dt} \tag{60.25}$$

The sign change is to differentiate between oil that is being compressed and oil that is expanding. The average flow through the ram can now be estimated by combining Eqs. (60.24) and (60.25). In practice, it is unlikely that there are two different fluids in the ram unless it is an air–oil system. Therefore, $\beta_1 = \beta_2 = \beta$. If V is the swept volume of the ram, then $V_1 = V_2 = V/2$ for the piston control. This results in the load flow equation

$$Q = \frac{V}{4\beta} \frac{d(p_1 - p_2)}{dt} + A \frac{dx_0}{dt} \tag{60.26}$$

Now, the pressure drop across the piston $(p_1 - p_2)$ is, in this case, used to accelerate the mass attached to this piston rod.

$$p_1 - p_2 = \frac{M}{A} \frac{d^2 x_0}{dt^2} \tag{60.27}$$

Combining Eqs. (60.26) and (60.27) gives

$$Q = \frac{VM}{4\beta A} \frac{d^3 x_0}{dt^3} + A \frac{dx_0}{dt}$$

Operating on both sides with the Laplace operator yields

$$\int_0^\infty Q(t) e^{-st} \, dt = \frac{VM}{4\beta A} \int_0^\infty e^{-st} \frac{d^3 x_0}{dt} \, dt + A \int_0^\infty \frac{dx_0}{dt} e^{-st} \, dt$$

With the appropriate initial conditions, namely $x_0 = dx_0/dt = 0$ at $t = 0$, the result is

$$\overline{Q}(s) = A \left[\frac{VM}{4\beta A^2} s^2 + 1 \right] s\overline{x}_0(s) \tag{60.28}$$

The bar over the symbol denotes the Laplace transform of the function, and s is the Laplace transform independent variable. The term inside the square brackets of Eq. (60.28) can be compared directly with the general equation for a second order system with zero damping ratio. Hence, the term $VM/4\beta A^2$ is the reciprocal of the system natural frequency squared.

$$f_n = \frac{1}{2\pi} \sqrt{\frac{\text{Stiffness}}{\text{mass}}} = \frac{1}{2\pi} \sqrt{\frac{4\beta A^2}{V} \times \frac{1}{m}} \tag{60.29}$$

If this system is excited, say by an impact on the mass, it will oscillate at the frequency defined by Eq. (60.29). Since there is no damping term, the mass would oscillate continuously. In practice, there will always be some damping, however small, from seal rubbing and oil film shear. The difficulty for the designer is to make some meaningful guesses as to what value to use. For example, one might use 15% of the stall load on the ram divided by the maximum velocity of the piston as a first guess.

60.8 SYSTEM CLASSIFICATIONS

Most hydraulic circuits, regardless of the application, fall into one or two general classifications.[20] The two major divisions are constant flow and constant pressure, depending on whether the output is mainly a function of flow (i.e., velocity, displacement, or acceleration) or mainly related to pressure (i.e., force or torque).

The simplest hydraulic circuits fall into the constant flow classification with open center valving. A simple example of this is shown in Fig. 60.14a. It is open center in the sense that when the control valve is centered the fluid is circulated directly back to tank. This method ensures minimum power loss and fluid heating in the quiescent periods. Compare this with Fig. 60.14b, which is a simple constant flow circuit in which oil is dumped through the relief valve at the end of the ram stroke. The restriction of the orifice in the relief valve generates high levels of heat and noise, thereby wasting power.

Figure 60.14a introduces the use of standard symbols for circuit design. The pressure relief valve is represented by a square with an offset arrow indicating that this valve is normally closed until sufficient pressure is developed in the pilot line (dotted) to push the valve open against the mechanical spring. This symbol is quite different from the physical drawing of a relief valve shown in Fig. 60.15, but it certainly conveys to the reader how the device is expected to operate.[21]

In the case of a directional control valve, it is always shown in its normally closed position. The reader is expected to visualize what happens when the valve is moved into its other two operating positions, as shown in Fig. 60.14c. In Europe, these symbols are standardized by the International Organization of Standards (ISO)[22] and also by the British Standards Institution (BSI).[23] In the United States, the American National Standards Institute develops the standards for the fluid power industry.[24]

The simplest form of flow control uses an adjustable orifice. These circuit configurations are shown in Fig. 60.16. When the orifice control is placed in the supply line upstream of the hydraulic cylinder, the system is said to be under *meter-in* flow control. Flow control is operative when flow is directed to the large-area side of the piston in Fig. 60.16a, while in the reverse direction, flow is dumped freely through the check valve. This type of control is best suited for resistive-type loads that the piston rod pushes against, and not for overrunning type loads.

When the orifice control is placed in the return line (Fig. 60.16b), the system is said to be under *meter-out* flow control. This type of control is best suited over running loads that are moving in the same direction as the motion of the actuator.

In *bleed-off* control (Fig. 60.16c), the flow control parallels the cylinder feed line. This approach can be used to adjust the cylinder speed over a range that is less than the maximum speed available. It has the advantage of using a small control valve, only large enough to handle the bleed flow and not the total flow. It also does not introduce a pressure drop in the main delivery line to the ram.

There are many pitfalls in designing fluid power circuits for the inexperienced, some of which are due to lack of design information. Some of these problems are reviewed in an excellent article by Achariga.[25]

60.9 PUMP SETS AND ACCUMULATORS

Any hydraulic circuit is useless unless there is a unit to provide the fluid power. Hence, the pump set is an important component of the system. Depending on the application, this can be designed and constructed by the user or the decision may be to purchase one of the many commercially available complete packages.

A typical design[12,26] is shown in Fig. 60.17. The motor driving the pump is usually electric for industrial application and runs at 1740 rpm. Other prime movers, such as diesel engines, could also

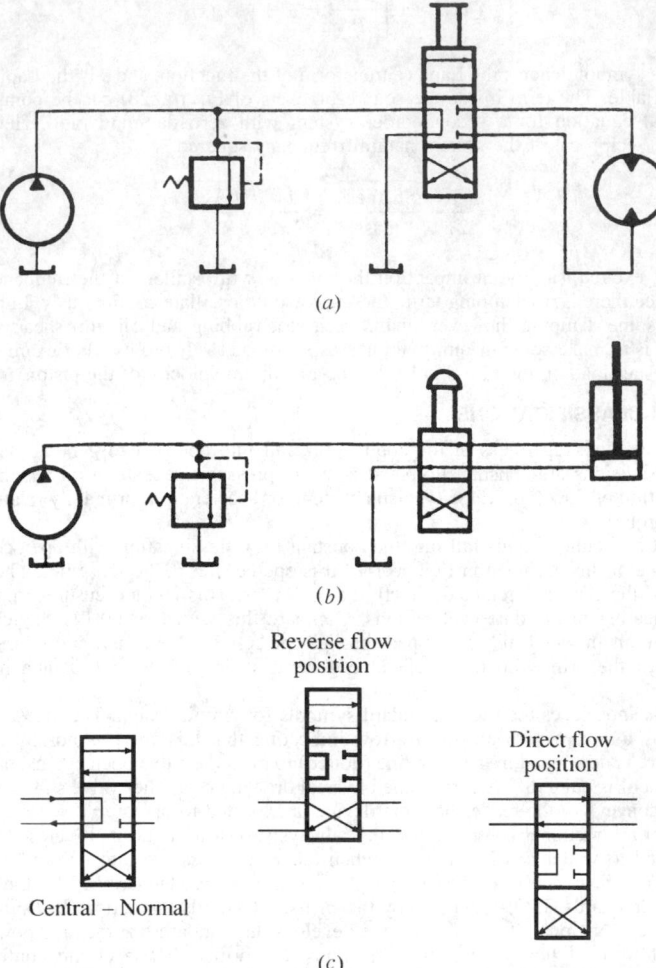

Fig. 60.14 Simple hydraulic circuits and valve symbols. (a) Open-center valve control; (b) simple constant flow circuit; (c) valve symbol meanings.

be used. The outlet from the fixed displacement pump is piped to a relief valve. This allows the working pressure to be selected. The nonreturn valve N prevents flow being forced back into the pump and also helps to stiffen the hydraulic circuit. An accumulator is included to smooth the pressure pulses developed in the pump. It will also provide additional flow for short-time high transient demands.

It is wise to include a shutoff cock, C1, to prevent oil spillage when the delivery line is disconnected and also to include a shutoff cock, C2, so that the accumulator can be discharged safely.

Good reservoir design, as in Fig. 60.17 is probably the most important aspect of preventive maintenance. It fulfills many functions besides containing sufficient oil to meet the demands of the complete system. It acts as a cooler and allows time for contaminants such as foam and dirt to settle out. The tank capacity is usually at least three times the maximum delivery of the pump in one minute and may be as large as six times if there are numerous valves in the circuit to generate heat. The inlet strainer removes larger debris, but care must be taken to ensure that it or any other component does not create significant pressure loss in the inlet (suction) side of the pump.

In starting a new hydraulic circuit, the following steps should be followed:

1. Make certain the pump is being driven in the correct direction for its design.
2. Make sure all nonreturn valves are located in the correct flow direction.

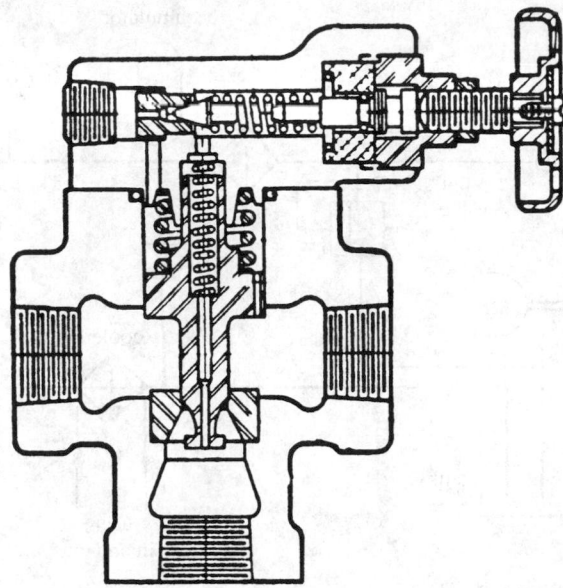

Fig. 60.15 Cross section through a pilot-operated pressure relief valve. (Courtesy of Vickers.)

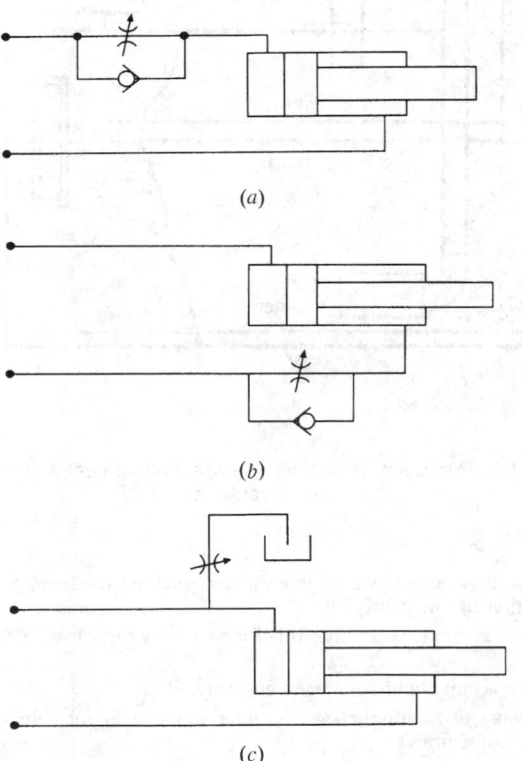

Fig. 60.16 Means of speed control. (a) Meter-in flow control; (b) meter-out flow control; (c) bleed-off.

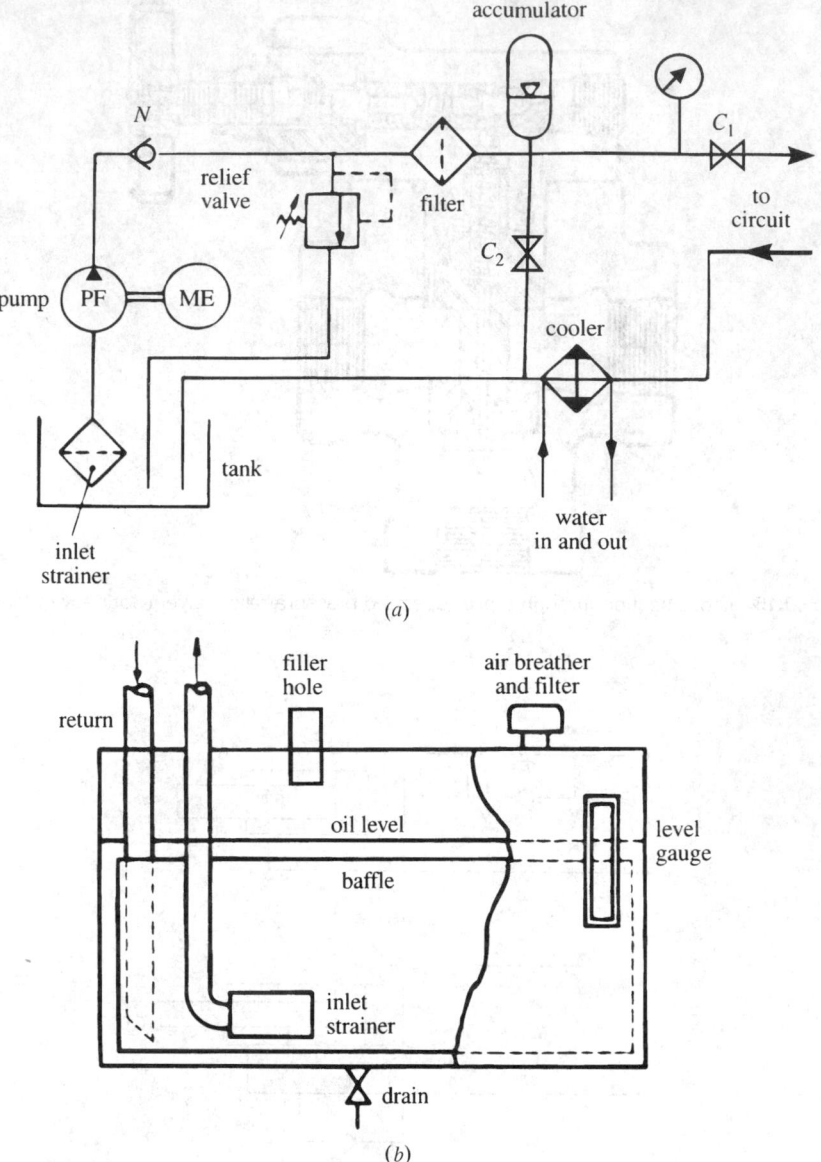

Fig. 60.17   Pump set design. (*a*) The basic pump circuit; (*b*) layout of a typical hydraulic reservoir.

3. Jog the pump drive motor two or three times watching for leaks and note system pressure for an indication of pump priming.
4. Check oil level in the tank and top off if a new circuit has used a significant amount of oil in filling.
5. Check that the accumulator is charged correctly.
6. After 5–10 hours of running, check the filter and strainer for debris left in the new system from parts manufacture.

The accumulator is an energy storage element into which hydraulic oil is pumped by the system to compress the contained gas. The accumulator can be used as a low pass filter to smooth out pump delivery fluctuations, or it can be used to supply small amounts of additional power for transient demands. It is sometimes cost-effective to use a small pump in a circuit where the duty cycle calls for low power requirements most of the time and large demands for relatively short periods. In this case, a larger accumulator can be used to store energy during the low-level part of the duty cycle and release it when high demand is required. A third application is to provide a stand-by short-term power source in case of failure of the pump set. This is particularly important in aerospace applications.

Accumulation size estimates are normally based on Boyle's law and assume that the charging gas temperature remains constant. The gas precharge in the accumulator is selected at about $\frac{1}{3}$ of the final maximum pressure required in the oil. It is advisable, of course, to use nitrogen as the gas. The usual stages in the operation of an accumulator are shown in Fig. 60.18, where

p_1, V_1 = gas precharge pressure and volume

p_2, V_2 = gas charge pressure when the pump is turned on and will correspond
to the system maximum pressure in the oil

p_3, V_3 = minimum pressure required in the circuit

If the volume of oil delivered from the accumulator is

$$V_0 = V_3 - V_2 \tag{60.30}$$

then for gas compressed isothermally,

$$V_1 = \frac{p_2/p_1 V_0}{(p_2/p_3 - 1)} \tag{60.31}$$

If there is shock loading present, the process is closer to adiabatic, and Eq. 60.31 is modified to

$$V_1 = \frac{(p_2/p_1)^{1/\gamma} V_0}{[(p_2/p_3)^{1/\gamma} - 1]} \tag{60.32}$$

where γ is the ratio of specific heats for the gas, which normally has a value $\gamma = 1.4$.

60.10 HYDROSTATIC TRANSMISSIONS

The hydrostatic transmission is usually a closed-circuit system in which the pump output flow is sent directly to a hydraulic motor. This package finds extensive use in many industries, such as steel and paper manufacturing. The electromechanical version is the Ward–Leonard drive, which consists of a dc generator supplying current to a dc motor, often used in locomotives. However, this unit tends to be very bulky in comparison to the hydraulic transmission.

Because of the ease of control of speed, torque, and direction of rotation, the hydrostatic transmission has become a popular choice in industrial equipment design. As can be seen from Fig. 60.19,

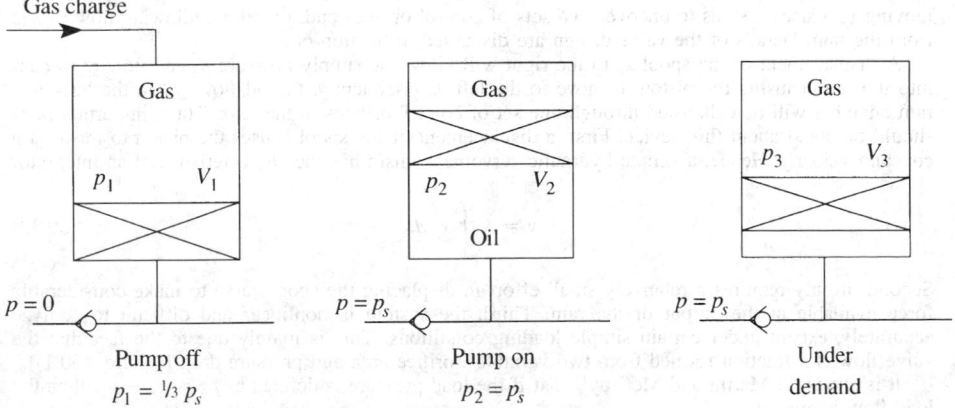

Fig. 60.18 Accumulator operation.

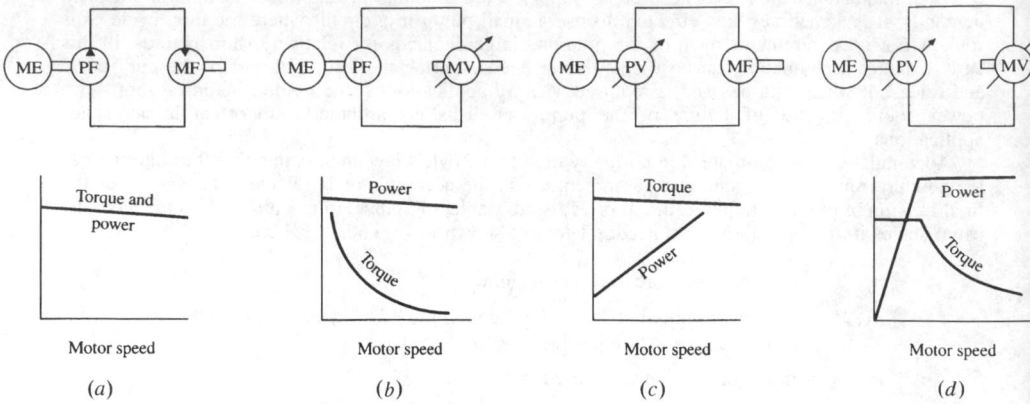

Fig. 60.19 Pump and motor combinations. (*a*) Fixed displacement pump and fixed displacement motor; (*b*) variable displacement motor and fixed displacement pump; (*c*) variable displacement pump and fixed displacement motor; (*d*) variable displacement pump and variable displacement motor.

there are several combinations whereby a positive displacement pump can be used to drive a motor. The choice will depend on the application.

The combination of a fixed displacement motor with a fixed displacement pump, Fig. 60.19*a* gives a fixed drive ratio. This is the simplest configuration where the output speed can only be controlled by altering the speed of the prime mover. For speed control, an alternative approach would be to include a bleed flow control valve from the main delivery line.[27,28]

When the motor is variable displacement (Fig. 60.19*b*), a fixed ratio drive at any given motor setting is obtained. The torque decreases with speed increase, making the characteristics compatible with winding machines. As a roll diameter increases, the rotational speed must decrease to hold a constant linear velocity of the material being wound while at the same time mass is increasing, requiring more drive torque.

By making the pump variable displacement instead of the motor, as in Fig. 60.19*c*, the torque remains constant over the speed range. With the pump at zero output, an idle condition is produced that is similar to a disengaged clutch.

The final configuration, shown in Fig. 60.19*d* introduces variability for both the pump and the motor. This combination can produce either a constant power or a constant torque drive. The combination has great flexibility—at a cost, of course—and can have both pump and motor adjusted together or separately. For example, where two separated parts of the same machine are to be driven at different speeds, a variable pump with two variable displacement motors can be used.

60.11 CONCEPT OF FEEDBACK CONTROL IN HYDRAULICS

A simple open-loop hydraulic servomechanism is shown in Fig. 60.20*a*. It consists of a spool valve, moving in a sleeve, so as to uncover two sets of control orifices and, therefore, allowing flow to and from the ram. Details of the valve design are discussed in Section 60.4.

A displacement of the spool x_v to the right will allow the supply pressure p_s and flow Q to pass into the ram, causing the piston to move to the left. Consequently, the oil flow $Q2$ in the left-hand ram chamber will be exhausted through one set of control orifices to the tank. Three important facts should be noted about this device. First, a displacement of the spool causes the piston to move at a constant velocity. Hence, a simple hydraulic servomechanism has the characteristics of an integrator

$$y = K \int x_v \, dt \tag{60.33}$$

Second, it only requires a relatively small effort in displacing the spool valve to make considerable force available at the output of the ram. Third, the system is nonlinear and difficult to analyze accurately, except under certain simple loading conditions. This is mainly due to the fact that the valve flow is a fraction formed from two variables, orifice area and pressure drop [see Eq. (60.1)].

It is shown in Martin and McCloy[12] that if the load pressure is defined as $p_L = p_1 - p_2$, then the load flow is given by

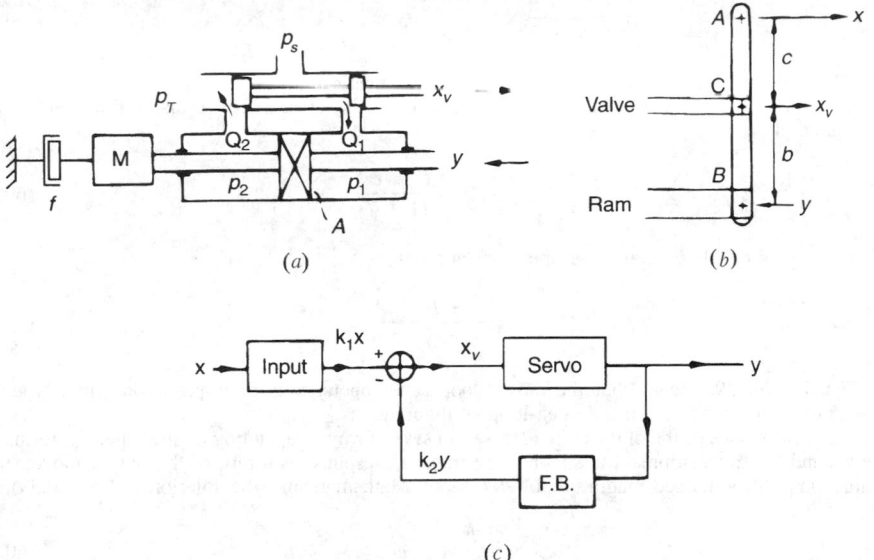

Fig. 60.20 Simple open loop hydraulic servomechanism. (a) Open loop configuration; (b) feedback linkage; (c) closed loop block diagram.

$$Q_L = C_d A_0 \sqrt{\frac{2}{\rho}} \sqrt{\frac{p_s - p_L}{2}} \tag{60.34}$$

If the mass M and damping f are very small (see Fig. 60.20a), then $p_1 = p_2 \approx \frac{1}{2}p_s$, and Eq. (60.34) becomes

$$Q_L = K_1 x_v \sqrt{\frac{p_s}{2}} = A \frac{dy}{dt} \tag{60.35}$$

where

$$K_1 x_v = C_d A_0 \sqrt{\frac{2}{\rho}}$$

and finally,

$$y = \frac{K_1}{A} \sqrt{\frac{p_s}{2}} \int x_v \, dt \tag{60.36}$$

Consider now the lever system shown in Fig. 60.20b. When this follow-up mechanism is attached to the valve and ram rod, a closed loop configuration results. A displacement of the input x causes a movement of the valve x_v, also to the right from a central closed position, since initially the lever $(c + b)$ pivots about B. The contribution to the valve displacement is now initially

$$x_v = \frac{b}{c + b} x = k_1 x \tag{60.37}$$

Once the spool valve opens, flow passes to the ram and the pivot B starts to move to the left. The top pivot point A is now held fixed by the input, hence the original valve displacement, Eq. 60.37, is now closed. The control equation is

$$x_v = \frac{b}{c + b} \times x - \frac{c}{c + b} \times y = k_1 x - k_2 y \qquad (60.38)$$

The block diagram in Fig. 60.20c should clarify the arrangement.

Combining Eq. (60.35) with Eq. (60.38) results in the closed loop transfer function for this configuration

$$y(s)/x = \frac{b/c}{(1 + Ts)} \qquad (60.39)$$

where the gain equals b/c, and the time constant T is

$$T = \frac{c + b}{c} \frac{A}{K_1 \sqrt{p_s/2}} \qquad (60.40)$$

Equation (60.39) means that the closed-loop servo operates as a simple exponential type lag, instead of an integrator, as in the open-loop configuration.

The performance of the unit can be assessed in several ways, depending on the type of information needed and the test equipment available. The transient response is a plot of the output movement y against time, for a defined magnitude of step input. Mathematically, the solution to Eq. (60.39) is

$$y = \frac{b}{c} (1 - e^{-t/T})x \qquad (60.41)$$

The plot of this equation is shown in Fig. 60.21a, where the time constant can be found from 63.2% of the final steady-state point. In theory, the steady state given by

$$y = bx/c \qquad (60.42)$$

will only be reached when t reaches infinity. However, it is normal in practice to use $4T = 98\%$ of final steady state as the practical steady-state value. Transient response testing is less costly and is easy to perform. However, the information is limited in that spectral information is difficult to interpret especially phase shift between the input and output. Some nonlinearities, such as dead zone, clipping, and small amplitude parasitic oscillations, can be identified.

A simple production test for assessing the level of friction in the moving parts of the servo-mechanism is to apply the ramp test depicted in Fig. 60.21b. In this case, the input is suddenly subjected to a constant velocity of ω_i rad/sec. The output tries to follow but lags by a steady state error of $T\omega_i$ where T is the system time constant. If the system were modeled by Eq. (60.39), this error could be reduced by changing the lever ratio $(c + b)/c$, for example.

60.12   IMPROVED MODEL

Experimental testing of an actual hydraulic servo system immediately reveals that the first order model discussed in the previous section does not adequately represent the actual performance. This is especially true at the prototype testing phase, where the system response has not been optimized. Instead of the smooth exponential type behavior shown in Fig. 60.21a, the response is likely to be quite oscillatory. The main reason for this is the fact that compressibility of the oil and the mass of the moving parts were considered negligible in the simple first-order model.

In addition, if mass is to be included, the load pressure p_L cannot be ignored as it was in Eq. (60.35). This means that the valve equation becomes a function of two variables: pressure drop and valve displacement.

Early attempts to solve this problem are recorded in a classic paper by Harpur[29] using a small perturbation method. This had the disadvantage that the valve characteristics at any instant in the motion of the servo were defined by the instantaneous values of the slopes of two nonlinear curves. In order to improve this situation, an alternative approach was suggested[30] that minimizes the average error and to a large extent overcomes this problem. Referring once more to Fig. 60.20, the flow into the actuator is given by

$$Q_1 = KA_{01}\sqrt{p_s - p_1} \qquad (60.43)$$

and the flow out is given by

$$Q_2 = KA_{02}\sqrt{p_2 - p_T} \qquad (60.44)$$

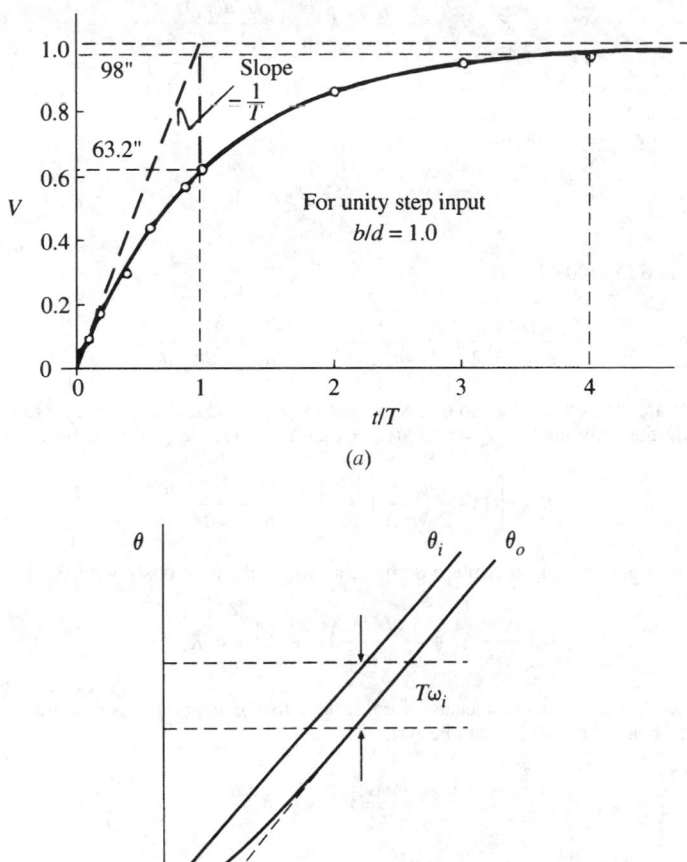

Fig. 60.21 First order performance–time domain. (a) Step response; (b) ramp input; $\theta = \omega_i t$ where t is time.

where $K = C_d \sqrt{2\phi}$ and the valve is symmetrical $A_{01} = A_{02} = A_0$. It can be shown (see Ref. 12) that the load flow through the valve is

$$Q_L = KA_0 \sqrt{\frac{P_s - P_L}{2}} \tag{60.45}$$

If compressibility of the oil is now included [see Eq. (60.26) of Section 60.7], then the load flow can be equated to the flow through the actuator so that

$Q_L =$ (flow due to piston movement) + (flow due to compressibility)

$$= A \frac{dy}{dt} + \frac{V}{4\beta} \frac{d}{dt} (p_L) \tag{60.46}$$

By equating Eqs. (60.45) and (60.46), the relationship between valve displacement x_v and output movement y can be obtained, and it is much more complex than in the previous model shown in Eq. (60.36).

$$K_v x_v \sqrt{\frac{p_s - p_L}{2}} = A \frac{dy}{dt} + \frac{V}{4\beta} \frac{d(p_L)}{dt} \qquad (60.47)$$

where a linear relationship between valve displacement and uncovered orifice area has been assumed.

The worst case for loading a system is when the output load is pure mass. It is the load pressure p_L that is used to accelerate the mass.

$$p_L = \frac{M}{A} \frac{d^2 y}{dt^2} \qquad (60.48)$$

Introduce this into Eq. (60.47):

$$\frac{K_v x_v}{A} \sqrt{\frac{p_s}{2} \left(1 - \frac{M}{p_s A} \frac{d^2 y}{dt^2}\right)} = \frac{dy}{dt} + \frac{VM}{4\beta A^2} \frac{d^3 y}{dt^3} \qquad (60.49)$$

If the constants are lumped together so that $K_r = K_v / A\sqrt{p_s/2}$, and if the square root term is expanded by the binomial theorem, neglecting terms greater than first order, Eq. (60.49) becomes

$$K_r x_v \left[1 - \frac{1}{2}\left(\frac{M}{p_s A}\right)\frac{d^2 y}{dt^2}\right] = \frac{dy}{dt} + \frac{VM}{4\beta A^2} \frac{d^3 y}{dt^3} \qquad (60.50)$$

Rearranging and transforming to Laplace domain, as was done previously with Eq. (60.28)

$$\left[\frac{VM}{4\beta A^2} s^2 + \frac{x_v}{2}\left(\frac{MK_r}{p_s A}\right) s + 1\right] s = K_r x_v \qquad (60.51)$$

The equation within the square brackets is equivalent to the general equation for a spring-mass-damper system, hence Eq. (60.51) can be written

$$\frac{1}{\omega_n^2} s^2 + \frac{2\xi}{\omega_n} s + s = K_r x_v \qquad (60.52)$$

It can now be seen that the damping contributed from the valve is partially determined by the valve displacement; the symbol ξ is the damping ratio. This explains why, in practical test results, the frequency response curves of a hydraulic servo change depending on the size of the input amplitude. Keating and Martin[30] discuss a further refinement to this model that results in an even better estimate of the dynamic response of this type of system.

60.13 ELECTROHYDRAULIC SYSTEMS—ANALOG

In the arrangements discussed in the previous section, signal transfer for feedback was done using mechanical linkages. It is much more convenient to employ electrical means to achieve these loops. However, this does require the use of transducers to convert mechanical and fluid signals into an electrical form.

Typical electrohydraulic servo valves use an electrical torque motor to move the spool arrangement. The most famous of these types of valves is the Moog 1500 series two-stage valve shown in Fig. 60.22a. It is also possible to have a single stage spool, as shown in Fig. 60.22b. In the example shown, the first stage is a double nozzle pilot valve controlling a second-stage spool valve. The torque motor is really a limited movement electric motor, arranged so that the flapper extends between two nozzles. This allows differential pressure to be applied across the sliding valve, whose movement in term meters fluid out of the valve.

While the flexibility of combining electrics with hydraulics is a major advantage, it should be noted that the torque motor does introduce an additional transfer function into the system. This in itself need not be a problem, provided that care is taken in the design process that the overall phase shift is not increased. As with any series type system, the overall performance is determined by the component with the poorest dynamic characteristics. It is important, therefore, to ensure that the torque motor valve assembly is not the weak link in the chain.

Feedback can now be achieved with the use of electrical transducers such as precision potentiometer, pressure devices, and accelerometers. A typical arrangement is shown in Fig. 60.23. For this particular arrangement, the minor loop is closed with position of the second-stage spool valve and the major loop is closed with position of the driving ram which controls the inertia type load. All

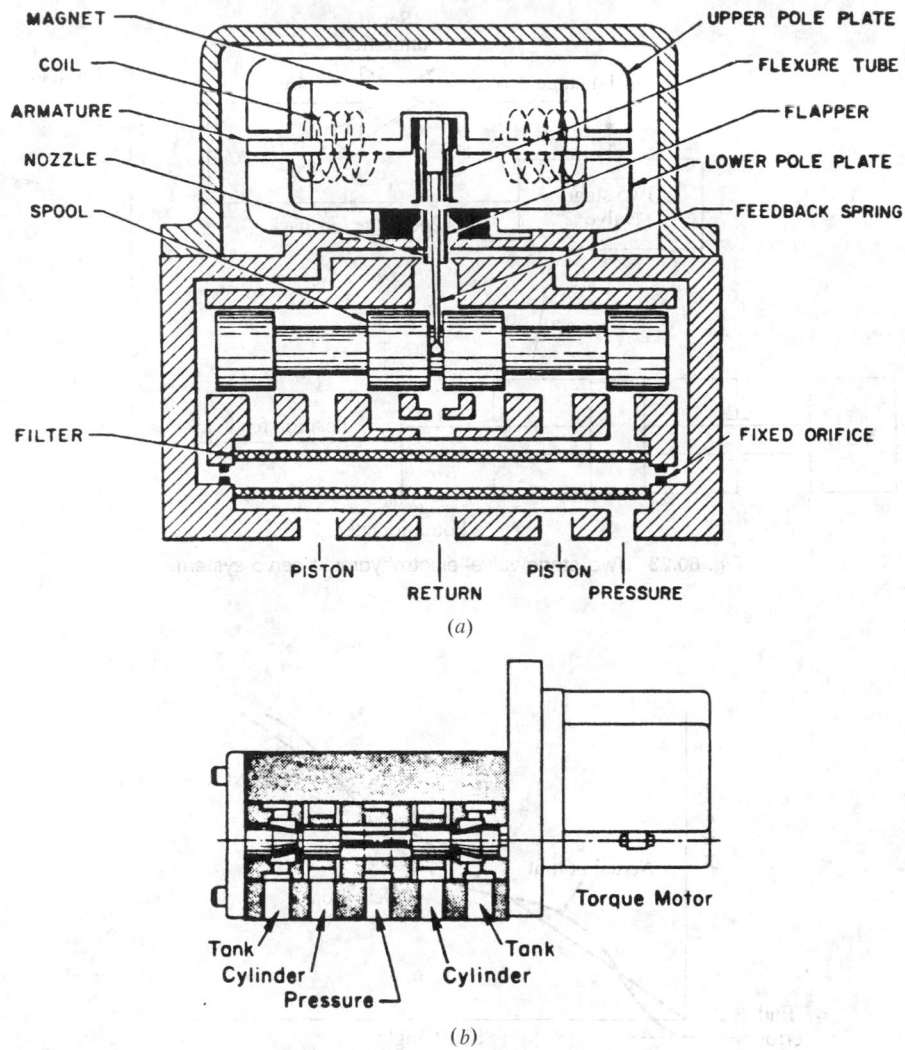

MAGNET

COIL

ARMATURE

NOZZLE

SPOOL

UPPER POLE PLATE

FLEXURE TUBE

FLAPPER

LOWER POLE PLATE

FEEDBACK SPRING

FILTER

FIXED ORIFICE

PISTON   PISTON
RETURN   PRESSURE

(a)

Torque Motor

Tank   Tank
Cylinder   Cylinder
   Pressure

(b)

Fig. 60.22 Typical electrohydraulic valves. (a) Two-stage valve (Moog Inc., East Aurora, NY); (b) single-stage valve.

the signals around the servo are now electrical and easy to adjust. The amplifier is used both as an easy method of gain adjustment and as a summing junction.

Potentiometers are the cheapest and simplest devices for converting linear and angular displacement information to electrical signals, but unless their performance, especially when coupled to other parts of the circuit, is understood, there can be real practical problems in getting an electrohydraulic servo to function correctly. The selection of a suitable instrument potentiometer requires the following specification items to be addressed.[31-33] The most commonly occurring terms are

1. *Linearity.* The deviation of the output voltage from a potentiometer from a linear law related to shaft rotation, Fig. 60.24a.
2. *Conformity.* Similar to linearity, but used in relation to potentiometers designed to follow a nonlinear law, Fig. 60.24b.
3. *Deviation.* Some suppliers of potentiometers quote deviation instead of linearity. The deviation is defined as the maximum permissible offset from the best straight line that can be

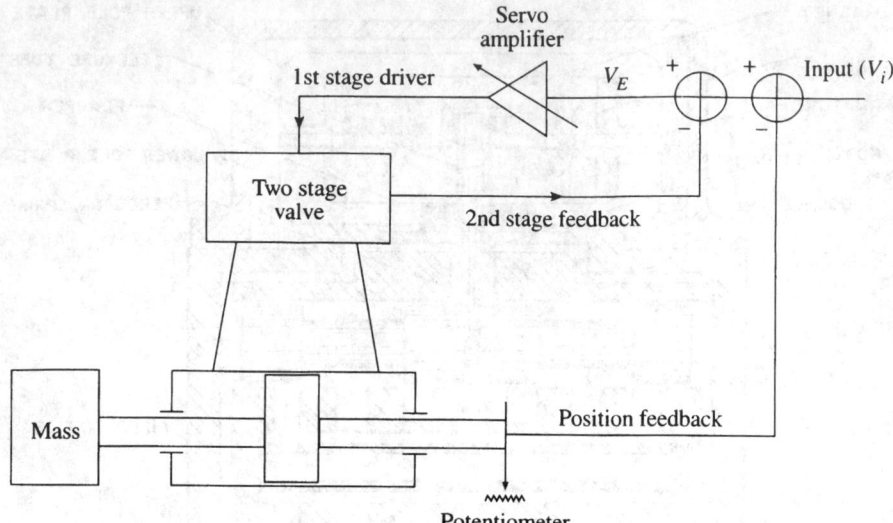

Fig. 60.23 Two-stage valve–electrohydraulic servo system.

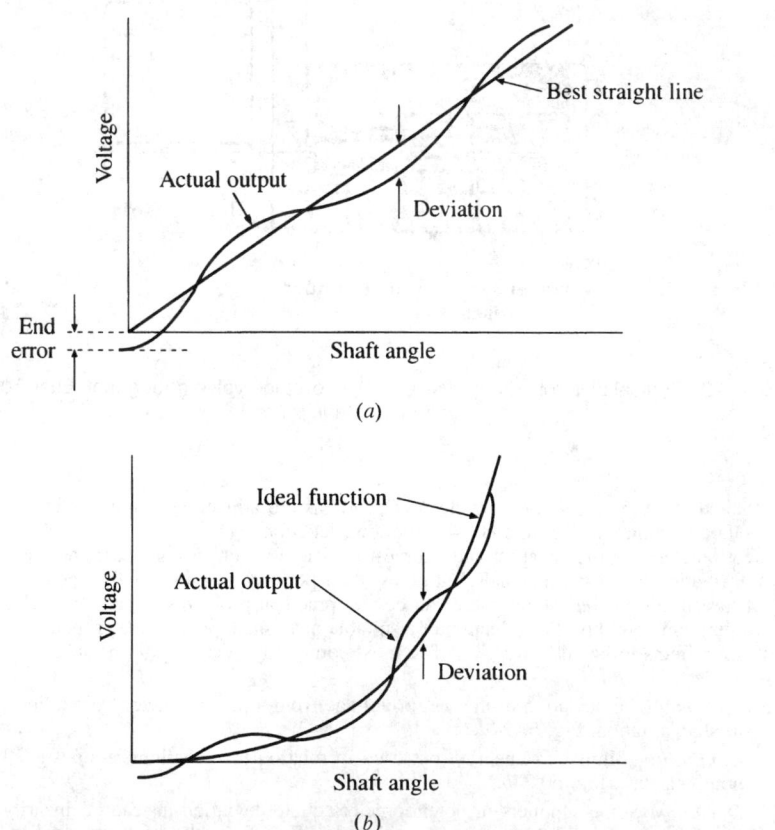

Fig. 60.24 Potentiometer linearity and conformity. (a) Linearity; (b) conformity.

drawn through the experimental points of measured resistance against rotation of the potentiometer. This tolerance is expressed as a percentage of the total resistance.

4. *Resolution.* The incremental rotation of the shaft necessary to produce the smallest incremental change of output voltage.

5. *Power rating.* The maximum continuous power that can be safely dissipated by the resistance at a specific temperature.

6. *Operating torque.* The torque required to start the wiper moving from rest. For small, general-purpose wire-wound potentiometers, this is in the range of 1–5 oz/in., while special-purpose units can be obtained having torques as low as 0.005 oz in.

7. *Electrical and mechanical angle.* The mechanical angle is the angle through which the potentiometer shaft can be rotated freely, while the electrical angle is that angle of rotation from which a voltage can be recorded.

A wide range of potentiometers are available, such as those using wire-wound cores, carbon tracks, and film deposits. Their mode of operation can be linear or nonlinear. Multiturn potentiometers are formed using a helix configuration and give much better resolution and accuracy than single-turn units.

Wire-wound potentiometers have the disadvantage that, when used in high-gain systems, the spacing between the wires produces a staircase effect in the output voltage. This results in a distinct roughness of motion in the servo output shaft. This problem can be resolved by using the more expensive film potentiometer. Carbon track versions tend to leave a deposit of carbon particles as the potentiometer wears in use. This also makes the servo output motion rough. These units need to be cleaned regularly. Operating torques, especially the starting torque for a potentiometer, become important when the unit is being driven from a low power source, such as the first or second stage of a spool valve. Another critical problem in electrohydraulic applications is to minimize loading errors on potentiometers, or in fact, any other transducer associated with the circuit.

Consider the arrangement shown in Fig. 60.25. In this circuit, a potentiometer is loaded by a circuit of resistance R_L, which represents the input impedance of the next stage. For the input circuit (Fig. 60.25a), $V_i = RI$, while for the output circuit, $V_o = kRI$, assuming the ideal case of $R_L = \infty$.

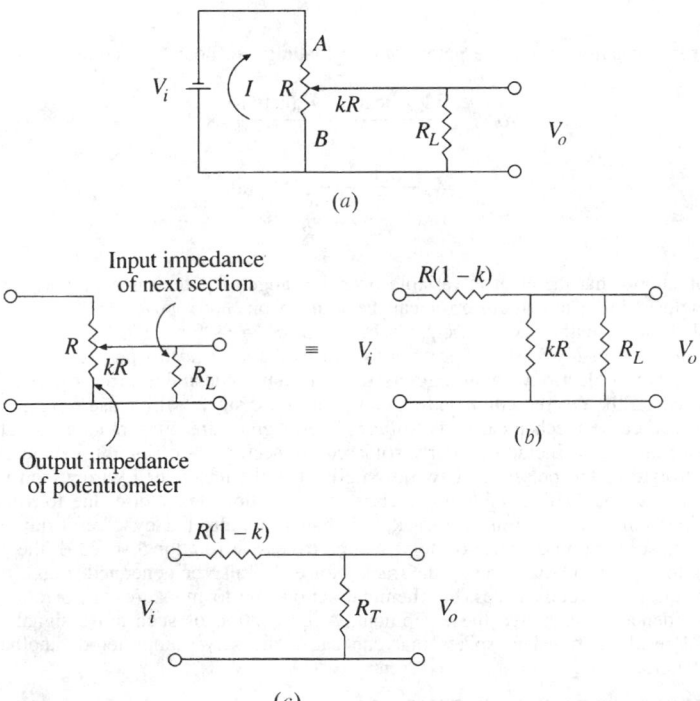

Fig. 60.25   Potentiometer loading. (a) Actual circuit; (b) equivalent circuit; (c) least complex equivalent circuit.

Since the current I in the circuit is common $V_i/V_o = 1/k$, which satisfies the conditions: $V_o = 0$ when wiper is at B, $k = 0$ and $V_o = V_i$ when wiper is at A, $k = 1$.

If now the practical case is considered, where R_L is finite in value, the circuit can be interpreted as shown in Fig. 60.25b. The parallel resistances can be replaced by

$$\frac{1}{R_T} = \frac{1}{kR} + \frac{1}{R_L}$$

$$R_T = \frac{kRR_L}{R_L + kR}$$

(60.53)

as shown in Fig. 60.25c. The total resistance as seen by V_i is now

$$R_i = R(1 - k) + \frac{kRR_L}{R_L + kR}$$

The current in the circuit must now be

$$I = \left(\frac{R_L + kR}{kR^2(1 - k) + RR_L} \right) V_i$$

But

$$V_o = R_T I = \frac{kRR_L}{R_L + kR} I$$

hence

$$\frac{V_o}{V_i} = \frac{k}{k(1 - k)\dfrac{R}{R_L} + 1}$$

(60.54)

The error in measurement due to potentiometer loading can be expressed as

$$\text{error } \% = \frac{V_o(\text{ideal}) - V_o(\text{actual})}{V_i} \times 100$$

$$= \frac{k^2(1 - k)}{k(1 - k) + \dfrac{R_L}{R}} \times 100$$

(60.55)

This equation shows that the error is variable over the range of potentiometer shaft angle positions. The shaft angle yielding maximum error can be found from $\partial(\text{error})/\partial k = 0$ and is a function of R_L/R. The shaft angle with maximum error occurs near $k^* = \frac{2}{3}$ for $0.2 < R_L/R$. As R_L/R becomes large, k^* becomes precisely $\frac{2}{3}$. If the error is to be less than 2% at this position, then $R_L/R \geq 7.2$.

The term *noise* in electrohydraulic circuits refers to any undesirable electrical circuit signal that is superimposed on the desired command signals. The noise signal will cause roughness in actuator movement or can cause mechanical parts to *buzz*. Such signals are often random in nature. Sources of circuit noise are many, including poorly soldered connections (dry joints), voltaic effects arising from an electrolyte in the presence of two dissimilar metals, unshielded wiring, and resistances of high impedance. Probably the major offender is the potentiometer. Noise due to vibration occurs when the wiper jumps away from the track; this can be controlled by careful adjustment of the contact arm pressure. Another cause of noise can be excessive rotational speed of the potentiometer shaft causing the wiper to bounce along the track. Noise will also be generated from dirt on the track and, in more unusual circumstances, by chemical action due to moisture, oil, or other liquids that may have accidentally penetrated the equipment. Amplification of such noise signals in the servo amplifier will result in transient spikes that can cause the servo amplifier or another instrument amplifier to saturate.

60.14   ELECTROHYDRAULIC SYSTEMS—DIGITAL

The electrohydraulic stepping motor is used as a high-torque, high-speed drive whose output motion is precise and repeatable. Some of the unique advantages are summarized in Benson[34] as: 1) position

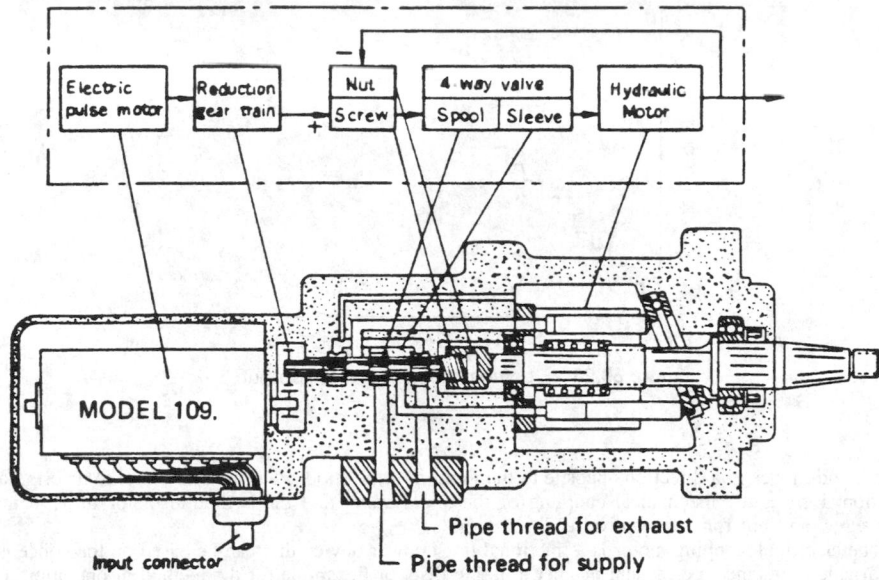

Fig. 60.26 Typical electrohydraulic stepping motor. (Courtesy of Fujitsu Ltd.)

feedback is not needed for positional control, unlike the electrohydraulic servo, 2) the operation of the unit is such that a microprocessor can often be interfaced directly (hence A/D and D/A converters are not needed), and 3) electrical tuning is not required, other than input of the command profile.

The electrohydraulic stepping motor has three main components: an electric stepping motor, a servo valve, and a rotary or linear actuator. A typical arrangement is shown in Fig. 60.26. The block diagram describes the function of the unit. The electric pulse motor controls the rate of flow and direction of flow of oil to the hydraulic motor.[35]

Command pulses are directed to the drive circuit amplifier. The pulses are fed in a phased sequence to the electric pulse motor, which, in turn, is connected through gears to a four-way spool valve. The gear on the stepping motor shaft has wide teeth so that the gear on the spool valve can move axially without disengaging. The other end of the spool has a lead screw, which is engaged with a nut on the hydraulic motor shaft. Thus, rotary motion of the stepping motor is transformed into axial motion of the spool, which in turn opens the four-way valve. This allows pressurized oil to flow to the hydraulic motor and cause it to rotate. As the hydraulic motor rotates, the nut on the motor shaft

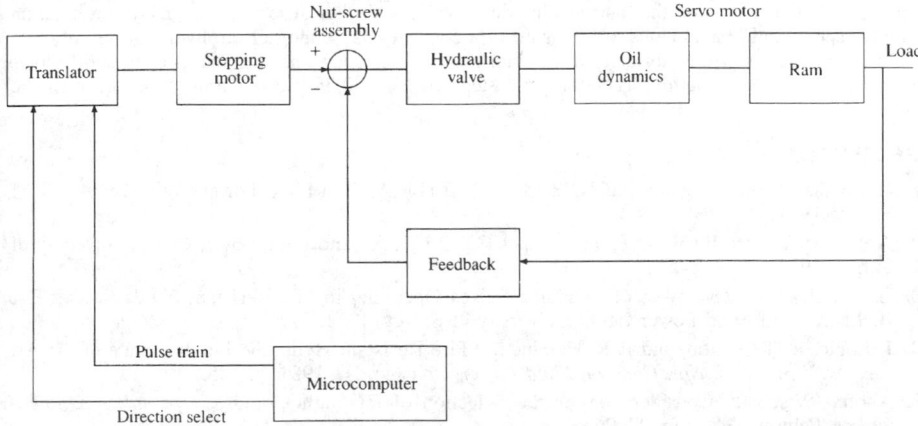

Fig. 60.27 Digital control system.

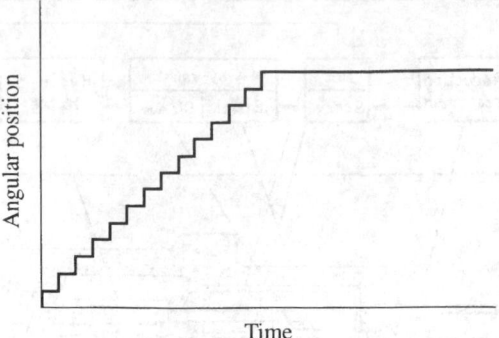

Fig. 60.28 Electrical stepping motor output.

rotates and moves in a direction opposite to the motion of the spool, returning the spool to its original position. Hence, the mechanical coupling of the spool valve and the hydraulic motor through the lead screw and nut form a negative feedback loop.

The electrical stepping motor is a fractional horsepower device that can be based on the concepts of variable reluctance, permanent magnet, rotating disc, or flexspline for the method of operation. In the case of the permanent magnet type, the magnetic rotor aligns itself with the magnetic orientation of the stator. When the stepping motor windings are correctly energized, the rotor will rotate one unit of angular displacement, typically $2\pi/200$ degrees/revolution, and then stop. Thus, any magnitude of motor rotation can be equated into the summation of some number of steps, as a result of a string of pulses from a microcomputer, applied at the input. Since the output position is not verified, the accuracy is solely a function of the ability of the motor to step through the exact number of steps commanded by the input. Hence, one important characteristic of a stepping motor is the maximum rate of the input pulses that can be followed.

The quantization of the motion into discrete steps is particularly well suited for a digital control device such as a microprocessor. The stepping motor drive (translator) accepts position and velocity profile commands in the form of a variable frequency pulse train and directional signal from the microprocessor. A complete block diagram is shown in Fig. 60.27.

The position of the electrical stepping motor shaft will be a sharply defined *staircase*, as shown in Fig. 60.28. The velocity of the shaft is, therefore, a function of the input pulse rate. The size of each step, both time- and position-wise, is determined by the pulse train and may also in practice have oscillatory overshoot if not well damped.

At frequencies exceeding 100 pulses/sec, the hydraulic valve in Fig. 60.26 does not have time to close. However, the number of pulses that the hydraulic motor lags behind the electric stepping motor is the amount of oil that should have been admitted into the hydraulic motor. This is remembered by the net and lead screw summing point at the valve shaft. Each stored pulse threads the screw into the nut a specific distance. When each stored pulse is used, the screw threads out the nut a specified distance, closing the four-way hydraulic valve a similar distance. This delay is a function of motor speed and is analogous to loop gain in a conventional analog electrohydraulic servo.

Since 100 pulses/sec is normally faster than the valve can respond, it remains open and follows the hydraulic motor smoothly. Therefore, the steps in Fig. 60.28 are not transmitted through the system.

REFERENCES

1. C. Stanley, "Fire Resistant Fluids," Paper F.2, B.H.R.A., 2nd Fluid Power Conf., University of Surrey, 1971.
2. A. Dalibert, "Fire Resistant Fluids," Paper F.3, B.H.R.A., 2nd Fluid Power Conf., University of Surrey, 1971.
3. E. S. Kelly, "Erosive Wear of Hydraulic Valves Operating in Fire Resistant Fluids," Paper F.4, B.H.R.A., 2nd Fluid Power Conf., University of Surrey, 1971.
4. J. Louie, R. T. Burton, and P. R. Ukrainetz, "Fire Resistant Hydraulic Fluids—State of the Art Review," *Proc. 37th Nat. Conf. on Fluid Power,* Chicago, IL, 1981, p. 285.
5. Anon., "Viscosity Considerations in the Selection of Hydraulic Fluids," *Hydraulics & Pneumatics,* Penton, 1980, pp. 23–26.
6. E. C. Fitch, L. E. Bensch, and R. K. Tessmann, *Contamination Control for the Fluid Power Industry,* Pacific Scientific, Montclair, CA, 1978.

7. B. A. Foord, "Specifying Fluid Filtration through Computer Simulation," in *Proc. 34th Nat. Conf. on Fluid Power,* Philadelphia, PA, 1978, p. 197.

8. J. Spencer, *Effective Contamination Control in Fluid Power Systems,* Sperry Vickers Publications, Troy, MI, 1980.

9. E. J. Forgeron and N. McCormack, "Controlled Maintenance System for Hydraulics through Used Oil Analysis," in *Proc. 34th Nat. Conf. on Fluid Power,* Philadelphia, PA, 1978, p. 187.

10. R. K. Tessman, "Ferrographic Measurements of Contaminant Wear in Gear Pumps," in *Proc. 34th Nat. Conf. on Fluid Power,* Philadelphia, PA, 1978, p. 179.

11. R. A. Collacott, *Mechanical Fault Diagnosis,* Chapman-Hall, 1977.

12. H. R. Martin and D. McCloy, *Control of Fluid Power,* 2nd ed., Ellis Horwood (John Wiley & Sons), New York, 1980.

13. *Aerospace Fluid Component Designer's Handbook,* RPL-TDR-64-25. AF Rocket Propulsion Lab., Edwards, CA, 1964.

14. G. Keller, "Hydraulic System Analysis," in *Hydraulics and Pneumatics,* Penton/IPC, Cleveland, OH, 1974.

15. J. Thoma, *Modern Oil Hydraulic Engineering,* Technical and Trade Press, 1970.

16. W. E. Wilson, *Positive Displacement Pumps and Fluid Motors,* Pitman Publications (and ASME Paper 48-SA-14), 1950.

17. D. E. Turnball, *Fluid Power Engineering,* Newnes–Butterworths, London, 1976.

18. H. R. Martin, "Effects of Pipes and Hoses on Hydraulic Circuit Noise and Performance," in *37th Nat. Conf. Fluid Power,* Chicago, IL, 1981, pp. 71–76.

19. H. E. Merritt, *Hydraulic Control Systems,* Wiley, New York, 1967.

20. *Fluid Power Designers Handbook,* Parker-Hannifin Corp., Cleveland, OH.

21. A. Esposito, *Fluid Power with Applications,* Prentice-Hall, Englewood Cliffs, NJ, 1980.

22. ISO 1219: 1976 Fluid Power Systems and Components—Graphical Symbols, International Standards Organization, 1976.

23. *BS 2917: 1977 Specification for Graphical Symbols Used on Diagrams for Fluid Power Systems and Components,* British Standards Institute, London, 1977.

24. USA SY32-10-1967, *Standard Symbols for Fluid Power Diagrams,* American National Standards Institute, New York, 1967.

25. R. Achariga, "Pitfalls the Inexperienced Fluid Power System Designer Should Avoid," *Hydraulic and Pneumatics,* Penton, 1982, pp. 152–168.

26. H. R. Martin, *The Design of Hydraulic Components and Systems,* Ellis Horwood (John Wiley & Sons), New York, 1995.

27. F. D. Yeaples, *Hydraulic and Pneumatic Power and Control,* McGraw-Hill, New York, 1966.

28. J. Kern, *Hydrostatic Transmission Systems,* Intertext Books, London, 1969.

29. N. F. Harpur, "Some Design Considerations of Hydraulic Servos of the Jack Type," in *Proc. Conf. on Hydraulic Seromechanisms, Int. Mech. Eng.,* London, 1953.

30. T. Keating and H. R. Martin, "Mathematical Models for the Design of Hydraulic Actuators," *ISA Transactions* **12**(2) (1973).

31. G. W. Dummer, *Variable Resistors and Potentiometers,* Pitman, 1963.

32. S. A. Davis and B. K. Ledgerwood, *Electromechanical Components for Servomechanisms,* McGraw-Hill, New York, 1961.

33. H. Neubert, *Instrument Transducers,* Oxford University Press, 1963.

34. B. Benson, "Rotary and Linear Electrohydraulic Stepping Actuators," in *Proc. 38th Nat. Conf. on Fluid Power,* Houston, TX, 1982.

35. T. Samar, "Electrohydraulic Stepping Motor Drives Machine Tools Without Feedback," in *Hydraulics & Pneumatics,* Penton, 1970.

CHAPTER 61

AIR COMPRESSORS

Joseph L. Foszcz
Senior Editor, Plant Engineering Magazine
Des Plaines, Illinois

61.1	INTRODUCTION	1865	61.4	SELECTION	1876
61.2	TYPES	1865	61.5	COST OF AIR LEAKS	1876
61.3	SIZING	1875			

61.1 INTRODUCTION

Compressed air provides power for many manufacturing operations. Energy stored in compressed air is directly convertible to work. Conversion from another form of energy, such as heat, is not involved. Compressed air can be supplied by several different types of compressors (Fig. 61.1). The choice depends on the amount, pressure, and quality of air a plant system requires.

The reciprocating compressor is manufactured in a broad range of configurations. Its pressure range is the broadest in the compressor family extending from vacuum to 40,000 psig. It declined in popularity from the late 1950s through the mid-1970s. Higher maintenance costs and lower capacity, when compared to the centrifugal compressor, contributed to this decline. The sudden rise in energy cost and the downsizing of new process plants have given the higher-efficiency, though lower-capacity, reciprocating compressor a more prominent role in new plant design.

Rotary compressors as a group make up the balance of positive displacement machines. This group of compressors has several features in common despite differences in construction. Probably the most important feature is the lack of valves as used in reciprocating compressors. The rotary is lighter in weight than the reciprocator and does not exhibit the shaking forces of the reciprocating compressor, making foundation requirements less rigorous. Though rotary compressors are relatively simple in construction, their physical design can very widely. Rotor design, both multiple and single, is one of the main items that distinguishes different types.

For certain applications, compression chamber lubricant oils cannot be tolerated in compressed air. The demand for oil-free air in processes where compressed air comes in direct contact with sensitive products, such as electronic components, instruments, food, and drugs, has increased the need for non-lubricated or oil-free air compressors.

Compressors are normally lubricated for a variety of reasons: to reduce wear, provide internal cooling, and effect a seal between moving parts. In reciprocating compressors, lubricant is distributed by a pressure or splash system to connecting rods, crank and piston pins, and main bearings. Rotary screw compressors inject oil into the screw to seal and cool the compressing air. Centrifugal and liquid ring compressors are, by design, oil-free.

Reciprocating, non-lubricated air compressors substitute low friction or self-lubricating materials such as carbon or Teflon for piston and packing rings. Oil-free screw and lobe type compressors are available with a design that does not require lubrication in the compression chamber for sealing and lubrication. Centrifugal air compressors are inherently nonlubricated.

Generally, nonlubricated compressors have a higher initial cost due to special designs and materials. Nonlubricated, reciprocating compressors have higher operating costs due to the increased maintenance of valves and rings, which tend to have short lives.

61.2 TYPES

Reciprocating single-acting compressors resemble automotive engines, are generally of one- or two-stage design, and are constant-capacity, variable-pressure units. They are very popular because of

Mechanical Engineers' Handbook, 2nd ed., Edited by Myer Kutz.
ISBN 0-471-13007-9 © 1998 John Wiley & Sons, Inc.

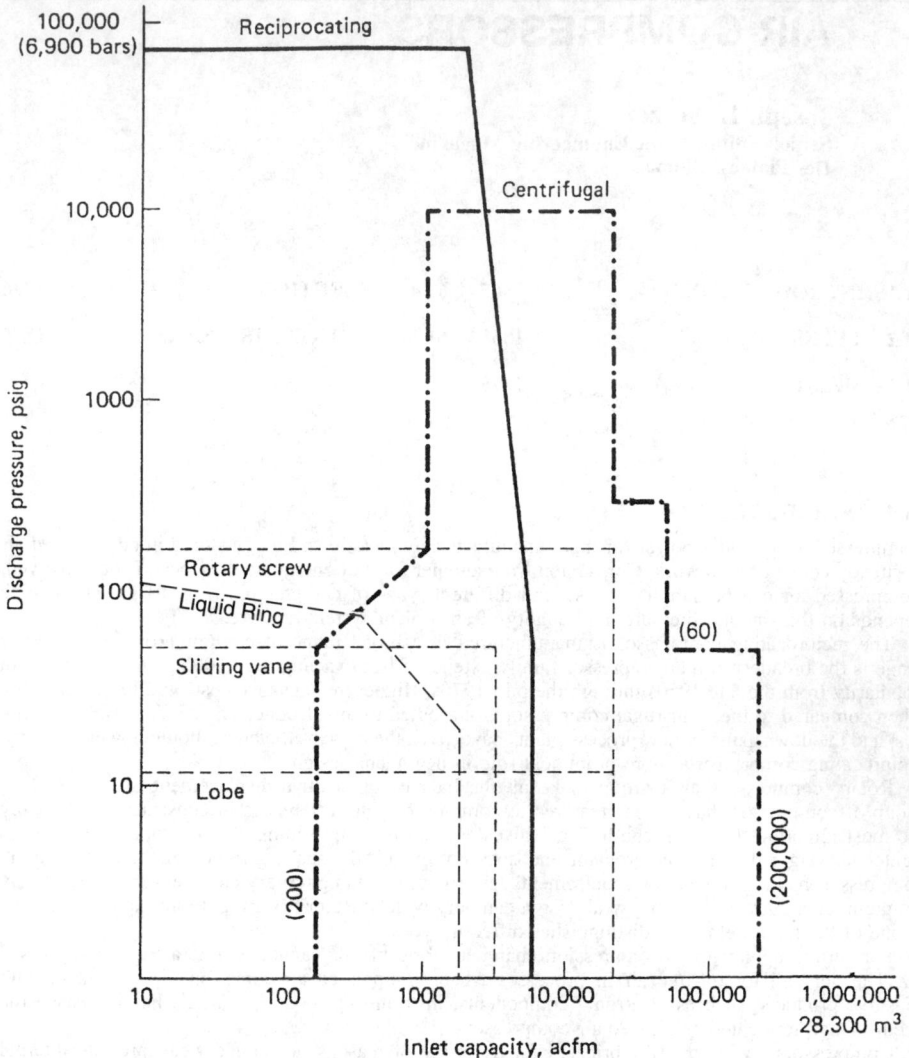

Fig. 61.1 Pressure-capacity chart showing the effective ranges of most compressors.[1]

their simplicity, efficiency, compactness, ease of maintenance, and relatively low price. In a single-stage compressor, air is compressed to the final pressure in a single stroke. This design is generally used for pressures from 25–100 psig. Units can be air- or liquid-cooled.

The two-stage design compresses air to an intermediate pressure in the first stage (Fig. 61.2). Most of the heat of compression is removed as the air passes through an intercooler, which is air or liquid cooled, to the second stage, where it is compressed to the final pressure. Two-stage compressors are generally used for pressures from 100–250 psig.

The reciprocating compressor is a positive displacement, intermittent-flow machine and operates at a fixed volume. One method of volume control is speed modulation. Another, more common, method is to use clearance pockets with or without valve unloading. With clearance pockets, cylinder performance is modified. With valve unloading, one or more inlet valves are physically opened. Capacity may be regulated in a single- or double-acting cylinder with single- or multiple-valve configurations.

Lubrication of compressor cylinders can be tailored to the application. The cylinders may be designed for normal hydrocarbon lubricants or can be modified for synthetic lubricants. The cylinder may also be designed for self-lubrication, in which case it is generally referred to as *nonlubed*. A

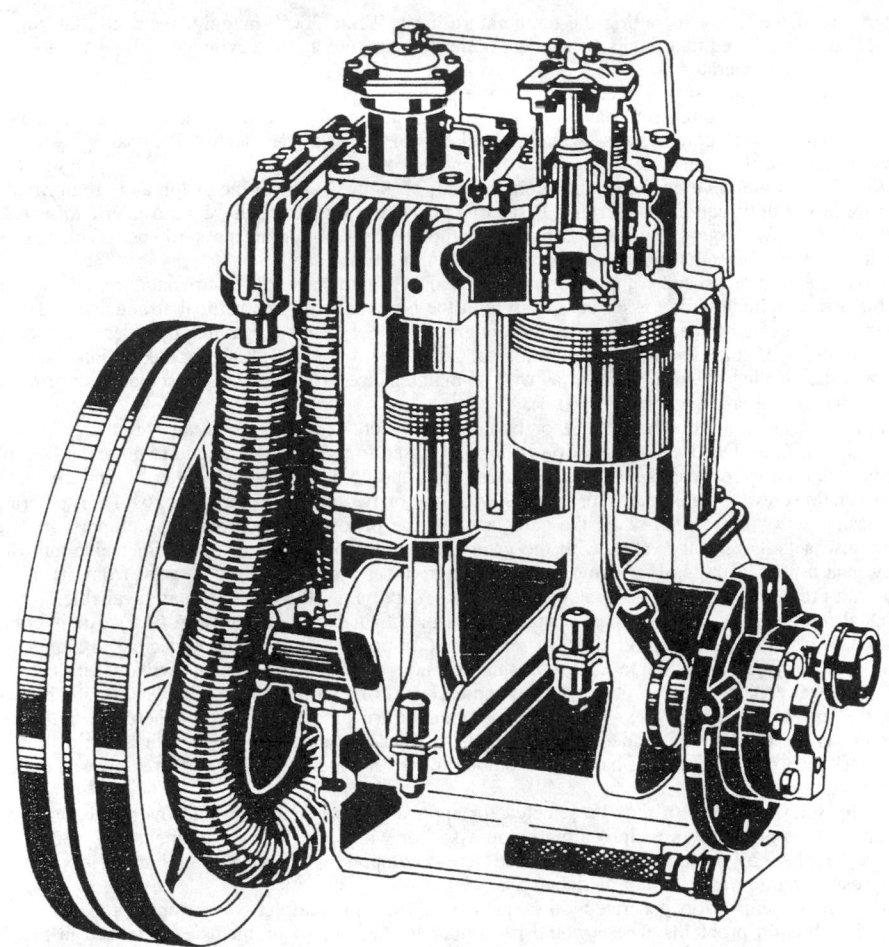

Fig. 61.2 Single-acting, two-stage compressor.[1]

compromise lubrication method that uses the nonlubed design but requires a small amount of lubricant is referred to as the *mini-lube* system.

Another feature necessary to the reciprocating compressor is cylinder cooling. Many compressors are furnished with water jackets as an integral part of the cylinder. Alternatively, particularly in smaller-size compressors, the cylinder can be designed for air cooling.

Reciprocating compressors can be classified into several types. One is the *automotive piston* type. The piston is connected to a connecting rod which is in turn connected directly to the crankshaft. This type of compressor has a single-acting cylinder and is limited to smaller air compressors.

Another common type of compressor for nonlube service is the *crosshead* type. The piston is driven by a fixed piston rod that passes through a stuffing or packing box and is connected to a crosshead. The crosshead, in turn, is connected to the crankshaft by a connecting rod. In this design, the cylinder is isolated from the crankcase by a distance piece. A variable-length or double-distance piece is used to keep crankcase lubricant from being exposed to the compressed air.

Reciprocating compressors usually will not tolerate liquids of any sort, particularly when delivered with the inlet air stream. A suction strainer or filter is mandatory for keeping ambient dirt and pipe scale out of the compressor. Fines from pipe scale and rust make short work of the internal bore of a cylinder and are not good for other components. The strainer should be removable for cleaning, particularly when it is intended for permanent installation. Under all circumstances, provision must be made to monitor the condition of the strainer.

Discharge temperatures should be limited to 300°F, as recommended by API 618. Higher temperatures cause problems with lubricant coking and valve deterioration. In nonlube service, the ring material is also a factor in setting the temperature limit. While 300°F may not seem all that hot, it should be remembered that this is an average outlet temperature and the cylinder will have hot spots exceeding this temperature.

Lubricated compressors use either a full-pressure or splash-lubricating system with oil in the crankcase. Oil-free compressors have a crosshead or distance piece between the crankcase and cylinders. Nonlubricated compressors use nonmetallic piston rings, guides, and sealed bearings with no lubricating oil in the crankcase.

Reciprocating double-acting designs compress air on both strokes of the piston and are normally used for heavy-duty, continuous service. Discharge pressures range from above atmospheric to several thousand psig. The largest single application is continuous-duty, supplying air at 100 psig. This design is available with the same modifications as single-acting compressors.

Double-acting crosshead compressors, when used as single-stage, have horizontal cylinders. The double-acting cylinder compressor is built in both the horizontal and the vertical arrangement. There is generally a design tradeoff to be made in this group of compressors regarding cylinder orientation. From a ring-wear consideration, the more logical orientation is vertical; however, taking into account size and the ensuing physical location as well as maintenance problems, most installations normally favor a horizontal arrangement (Fig. 61.3).

Rotary screw compressors use one or two rotors or screws and are constant-volume, variable-pressure machines. Oil or water injection is normally used to seal clearances and remove the heat of compression. Oil-free designs have reduced clearances and do not require any other sealing medium.

In single-screw designs, the rotor meshes with one or two pairs of gates (Fig. 61.4). The screw and casing act as a cylinder, while the gates act like the piston in a reciprocating compressor. The screw also acts as a rotary valve, with the gates and screw cooperating as a suction valve and the screw and a port in the casing acting as a discharge valve. Single-stage sizes range from 10–1200 cfm with pressures up to 150 psig. 250-psig designs, supplying 700–1200 cfm, are available.

Dual rotor designs use two intermeshing rotors in a twin-bore housing (Fig. 61.5). Air is compressed between the convex and concave rotors. The trapped volume of air is decreased along the rotor, increasing pressure. Single- and multistage versions are available with and without lubrication.

The power consumption of rotary screw compressors during unloaded operation is normally higher than that of reciprocating types. Recent developments have produced systems where the unloaded horsepower is 15–25% of loaded power. These systems are normally used with electric motor, constant-speed drives. Use as a base load compressor is recommended to avoid excessive unloaded power costs.

A dry screw compressor may be selected for applications where a high air-flow rate is required but space does not allow a reciprocating compressor, or where the flow requirement is greater than can be supplied by a single-unit, oil-flooded screw compressor. Packaged versions of dry screw compressors require a minimum of floor space.

Dry screw compressors generate high frequency pulsations that affect system piping and can cause acoustic vibration problems. These would be similar to the type of problems experienced in reciprocating compressor applications, except that the frequency is higher. While volume bottles work with the reciprocator, dry-type screw compressors require a manufacturer-supplied proprietary silencer to take care of the problem.

There is one problem this compressor can handle quite well: unlike most other compressors, it will tolerate a moderate amount of liquid. Injection for auxiliary cooling can be used, normally at a lower level than would be used in a flooded compressor. The compressor also works well in fouling service, if the material is not abrasive. The foulant tends to help seal the compressor and, in time, may improve performance.

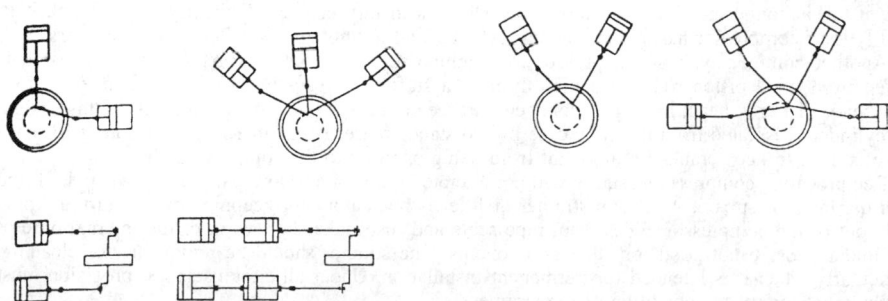

Fig. 61.3 Various cylinder arrangements used in displacement compressors. Some are suitable for single-acting compressors, while others are double-acting and require a crosshead and guide.[1]

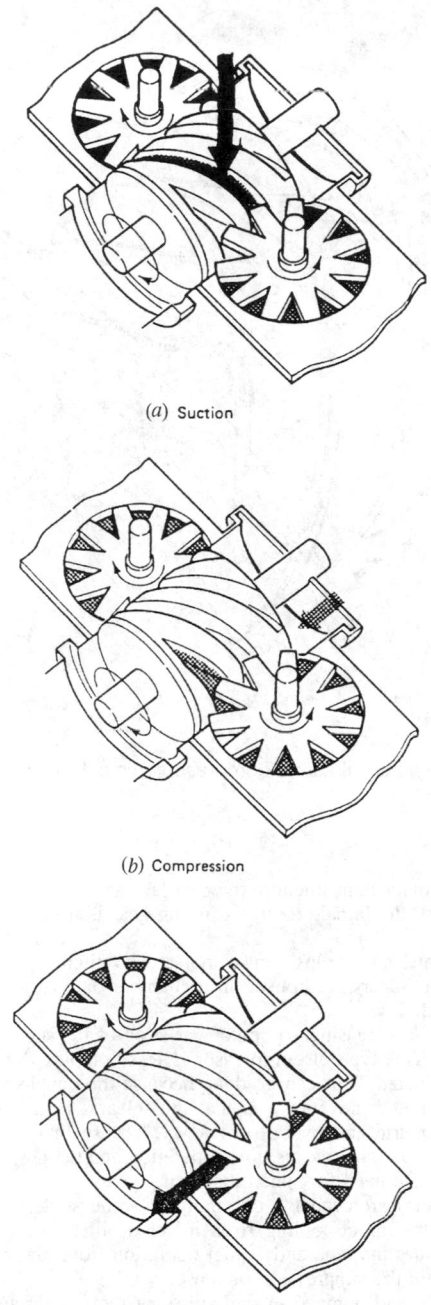

(*a*) Suction

(*b*) Compression

(*c*) Discharge

Fig. 61.4 Diagram showing the operation of the rotary, single-screw compressor.[1]

In dry screw compressors, the rotors are synchronized by timing gears. Because the male rotor, with a conventional profile, absorbs about 90% of the power transmitted to the compressor, only 10% of the power is transmitted through the gears. The gears have to be of good quality both to maintain the timing of the rotors and to minimize noise. Because the compressor will turn in reverse on gas backflow, keeping gear backlash to a minimum is important. A check valve should be included because compressors are sometimes subjected to reverse flow. To control backlash in the gears, a

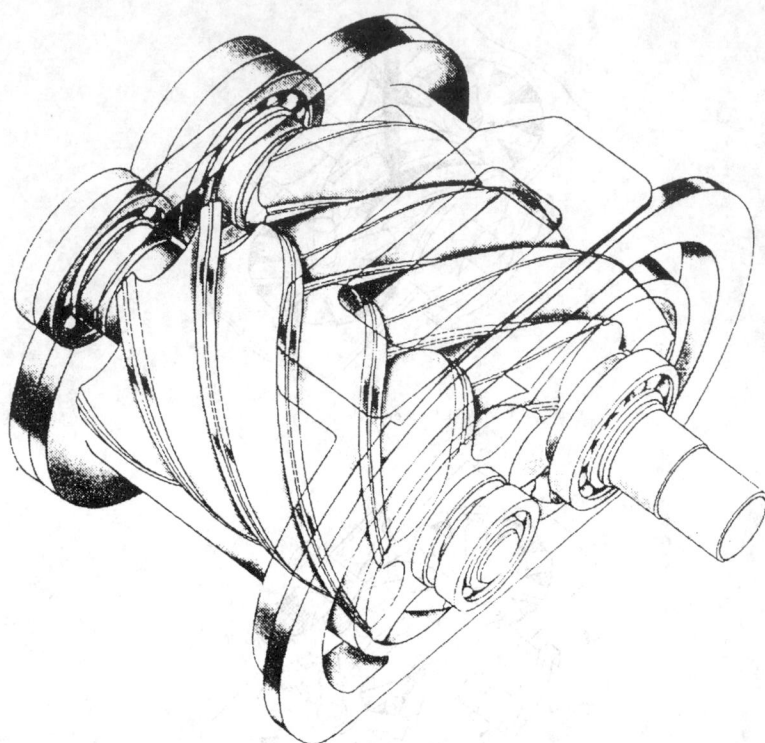

Fig. 61.5   Rotary, helical-screw compressor, typical single-stage design.[1]

split-driven gear is used to provide adjustment to the gear lash and maintain timing on reverse rotation. To provide timing adjustment, the female rotator's timing gear is made to be movable relative to its hub.

The gears are helical, which also helps control noise. The pitch line runout must be minimized to control torsional excitation. Gears are housed in a chamber outboard from the drive end and are isolated from the compressed air.

Oil-flooded versions are an increasingly popular variation of the screw compressor and are used in a variety of applications. This type of compressor is less complex than the dry version because timing gears have been eliminated. This can be done because the female rotor is driven by the male rotor through an oil film. Another advantage is that the oil acts as a seal for internal clearances, which means a higher volumetric and overall efficiency. The sealing improvement also results in higher efficiency at lower speeds. This means quiet operation and the possibility of direct connection to motor drivers, eliminating the need for speed increasing gears.

When gears are needed, they are available internally on some models. Higher pressure ratios can also be realized because of the direct cooling from injected oil. Pressure ratios as high as 21:1 in one casing are possible. Besides the inherently quiet operation from lower speed, oil dampens some of the internal pulses aiding in the suppression of noise.

The injected oil is sheared and pumped in the course of moving through the compressor. These power losses can be minimized by taking advantage of slower speed performance. There is an optimum speed where improvement in operation from oil offsets potential energy losses.

The points of injection are quite important for efficient operation. Oil should be injected in the casing wall at or near the intersection of the rotor bores on the discharge side of the machine.

Flooded compressors use a symmetric profile rotor extensively because of the rotor's efficiency. Flooded compressor size has recently been increased. The upper range is in the 7000 cfm range. While most applications are in air and refrigeration, certain modifications can make it applicable for process gas service. One consideration is the liquid used for flooding.

The fluid in a compressor is normally a petroleum-based lubricating oil, but not always. Factors to consider when selecting the lubricant include:

- Oxidation
- Condensation
- Viscosity
- Outgassing in the inlet
- Foaming
- Separation performance
- Chemical reaction

Some problems can be solved with specially selected oil grades. Another solution is synthetic oils, but cost is a problem, particularly with silicone oils. Alternatives need to be reviewed to match service life of the lubricant with lubrication requirements in the compressor.

One consideration for flooded compressors is the recovery of liquid. In conventional arrangements, the lubricating oil is separated at the compressor outlet, cooled, filtered, and returned to the compressor. This is fine for air service, where oil in the stream is not a major problem, but when oil-free air is needed, the separation problem becomes more complex. Because the machine is flooded and the discharge temperature is not high, separation is much easier relative to compressors that send small amounts of fluid at high temperature down stream. Usually part of the lubricant is in a vaporized form and is difficult to condense except where it is not wanted. To achieve quality oil-free air, such as that suitable for a desiccant-type dryer, separators that operate at the tertiary level should be considered. Here, the operator must be dedicated to separator maintenance, because these units require more than casual attention. Separation by refrigeration is not as critical if direct expansion chillers are used. In these applications, the oil moves through the tubes with the refrigerant and comes back to the compressor with no problem, if the temperature is not too low for the lubricant.

Advantages of helical screw compressors include smooth and pulse-free air output, compact size, high output volume, low vibration levels, and long life.

Centrifugal compressors are second only to reciprocating compressors in numbers of machines in service. Where capacity or horsepower rather than numbers is considered as a measure, the centrifugal, without a doubt, heads the compressor field. During the past 30 years, the centrifugal compressor, because of its smaller relative size and weight compared to the reciprocating machine, became much more popular for use in process plants, which were growing in size. The centrifugal compressor does not exhibit the inertially induced shaking forces of the reciprocator and therefore does not need the same massive foundation. Initially, the efficiency of the centrifugal was not as good as that of a well maintained reciprocating compressor. However, the centrifugal established its hold on the market in an era of cheap energy when power cost was rarely, if ever, evaluated.

The smaller compressor design was able to penetrate the general-process plant market, which had historically belonged to the reciprocating compressor. As the compressor grew in popularity, developments were begun to improve reliability, performance, and efficiency. With the increase in energy cost in the mid-1970s, efficiency improvements became a high priority. Initially, most development had concentrated on making the machine reliable, a goal that was reasonably well achieved. Run time between overhauls currently is three years or more, with six-year run times not unusual. As plant size increased, the pressure to maintain or improve reliability was very high because of the large economic impact of a nonscheduled shutdown.

Centrifugal compressors are dynamic types with rotating impellers that impart velocity and pressure to air (Fig. 61.6). Their design is simple and straightforward, consisting of one or more high-speed impellers with cooling sections. The only lubrication required is in the drive system, which is sealed off from the air system.

Integral gear-type centrifugal air compressors are generally used in central plant air applications requiring volumes ranging from 1000–30,000 cfm and discharge pressures from 100–125 psig.

Centrifugal air compressors are normally specified on the basis of required air-flow volume. However, there are several ways to calculate volume and serious problems can result unless both user and manufacturer use the same method. At the very least, the user can have problems comparing bids from competing manufacturers. At worst, he may choose the wrong compressor.

These problems can be avoided by specifying capacity in terms of actual inlet conditions and by understanding how compressor capacity is affected by variable ambient conditions such as inlet pressure, temperature, and relative humidity. Factors such as cooling water temperature and motor load must be considered before a compressor and its drive motor can be sized.

A multistage arrangement for integral gear-type compressors is shown in Fig. 61.7. The flow path is straight through the compressor, moving through each impeller and cooler in turn. This type of centrifugal compressor is probably the most common of any found in process service, with applications ranging from air to gas.

Sliding-vane compressors consist of a vane-type rotor mounted eccentrically in a housing (Fig. 61.8). As the rotor turns, the vanes slide out against the stator or housing. Air compression occurs when the volume of the space between the sliding vanes is reduced as the rotor turns. Single- and multistage versions are available.

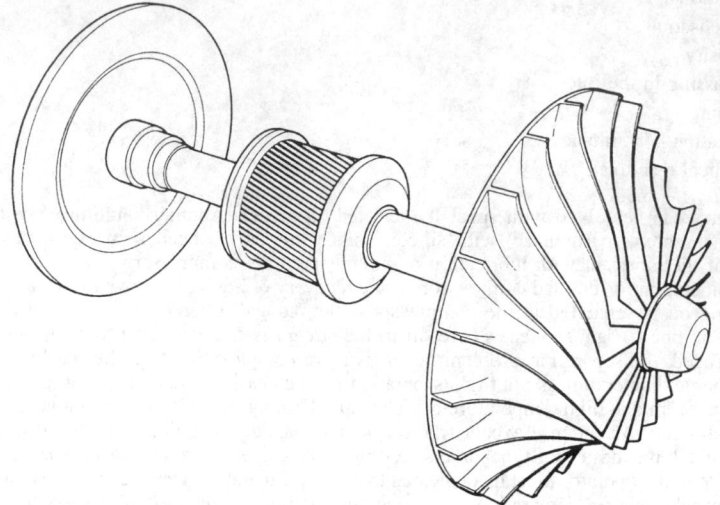

Fig. 61.6   Pinion of an intergral-gear unit having open, backward-curve-bladed impellers.[1]

The sliding-vane compressor consists of a single rotor mounted eccentrically in a cylinder slightly larger than the rotor. The rotor has a series of radial slots holding a set of vanes. The vanes are free to move radially within the rotor slots. They maintain contact with the cylinder wall by centrifugal force generated as the rotor turns.

The space between a pair of vanes, the rotor, and the cylinder wall forms crescent-shaped cells. As the rotor turns and a pair of vanes approach the inlet, air begins to fill the cell. The rotation and subsequent filling continue until the suction port edge has been passed by both vanes. Simultaneously,

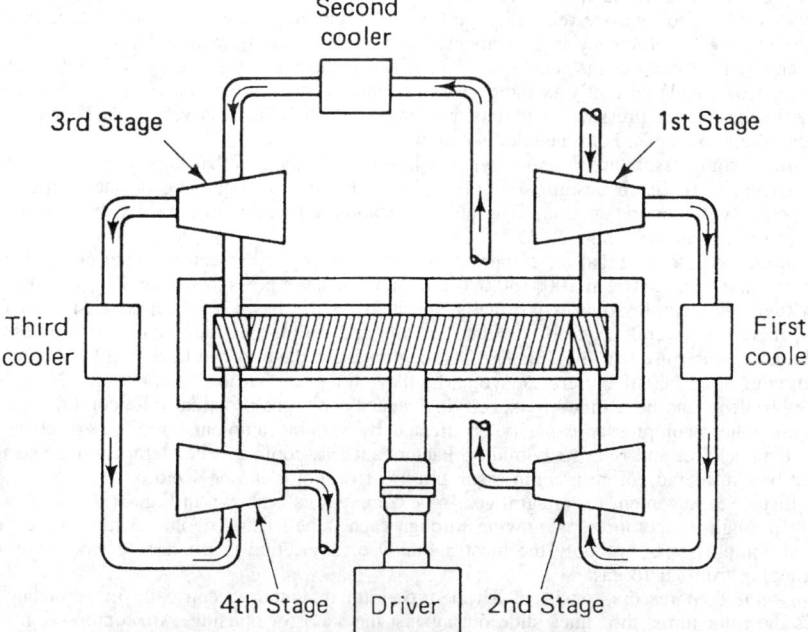

Fig. 61.7   Flow diagram of an integral-gear-type compressor showing stages of compression and including the cooling arrangement.[1]

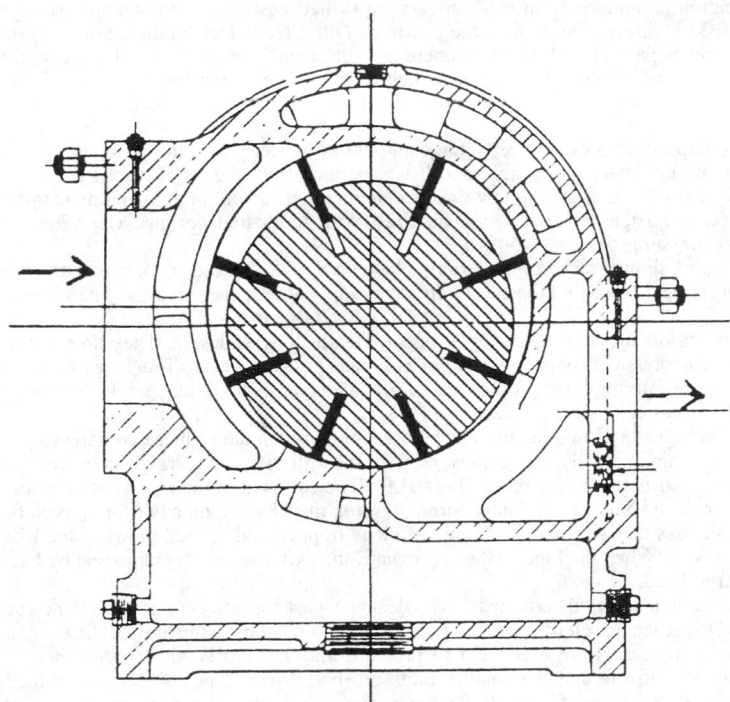

Fig. 61.8 Cross section of a sliding vane compressor (courtesy of A-C Compressor Corporation, Milwaukee, Wisconsin).[2]

the vanes have passed their maximum extension and begin to be pushed back into the rotor by the eccentricity of the cylinder wall. As the space becomes smaller, the air is compressed. The compression continues until the leading vane crosses the edge of the discharge port, and compressed air is discharged.

The sliding-vane compressor can be used to 50 psig in single-stage form and when staged can be used to 125 psig. An often overlooked application for the sliding-vane machine is that of vacuum service, where, in single-stage form, it can be used to 28 in. Hg. Volumes in vacuum service are in the 5000-cfm range. For pressure service, at the lower pressures, volumes are just under 4000 cfm and decrease to around 2000 cfm as the discharge pressure exceeds 30 psig.

The sliding-vane compressor efficiency is not as good as that of the reciprocating compressor, but the machine is rugged and light and lacks the foundation or skid weight requirement of the reciprocator.

Vane wear must be monitored in order to schedule replacement before the vanes become too short and wear the rotor slots. If the vanes are permitted to become too worn on the sides or too short, the vane may break and wedge between the rotor and the cylinder wall at the point of eccentricity, possibly breaking the cylinder. Shear pin couplings or equivalent torque-limiting couplings are sometimes used to prevent damage from a broken vane under sudden stall conditions.

As in most jacket-cooled compressors, the coolant acts as a heat sink to stabilize the cylinder dimensionally. The jacket outlet temperature should be around 115°F and be controlled by an automatic temperature regulator if the load or the water inlet temperature is prone to change.

Most of the drivers used with the sliding-vane compressor are electric motors. Variable-speed operation is possible within the limits of vane-speed requirements. The vanes must travel fast enough to seal against the cylinder wall but not so fast that they cause excessive wear. For smaller units, under 100 hp, V-belts are widely used. Direct connection to a motor, however, is possible for most compressors and is used throughout the size range.

For lubricated machines, vanes are made of a laminated asbestos impregnated with phenolic resin. For a nonlubricated design, carbon is used. The number of vanes influences the differential pressure between adjacent vane cells. The influence becomes less as the number of vanes increases.

Antifriction bearings are widely used, generally a roller type. Seals are either a packing or mechanical contact type. Packing and bearings are lubricated by a pressurized system. For nonflooded, lubricated compressors, a multiplunger pump, similar to one used with reciprocating compressors, is used. Lubrication is directed from the lubricator to drilled passages in the compressor cylinder and heads. One feed is directed to each of the bearings. Other feeds meter lubrication onto the cylinder wall. As the vanes pass the oil-injection openings, lubricant is spread around the cylinder walls to lubricate vane tips and eventually the vanes themselves. Oil entering the gas stream is separated in the discharge line. Because of high local heat, the lubricant may break down and not be suitable for recycling.

Flooded compressors pressure-feed a large amount of lubricant into the compressor, where it both cools the air and lubricates the compressor. It is separated from the air at discharge and recycled.

Oil-less designs are restricted to low-pressure applications due to high operating temperatures and sealing difficulties. Higher pressures are obtained with lubricated designs. Capacities range from 5–600 cfm at pressures from 80–150 psig.

Advantages of sliding-vane compressors include cool, clean, pulse-free air output, compact size, low noise levels, and low vibration levels. In some applications, there may not be a need for an air receiver.

Lobe compressors are a positive displacement, clearance-type design. They do not require lubrication in the compression chamber, only for the bearings and gears. The lobes do not drive one another and have intermeshing profiles that form a decreasing volume while rotating. Units are relatively vibration-free.

Lobe compressors are low-pressure machines. A feature unique to these compressors is that they do not compress air internally, as do most of the other rotaries. The straight-lobe compressor uses two rotors that intermesh as they rotate (Fig. 61.9). The rotors are timed by a set of gears. The lobe shape is either an involute or cycloidal form. A rotor may have either two or three lobes. As the rotors turn and pass the inlet port, a volume of air is trapped and carried between the lobes and the outer cylinder wall. When the lobe pushes air toward the exit, the air is compressed by back pressure in the discharge line.

Volumetric efficiency is determined by tip leakage past the rotors, not unlike the rotary screw compressor. This leakage, referred to as *slip*, is a function of rotor diameter and differential pressure.

Lobe-type compressors are used both in pressure and vacuum service. Larger units are direct-connected to their drivers and the smaller units are belt-driven. The drivers are normally electric motors. The main limitation of this rotary compressor is differential pressure on longer rotors, where deflection can be large. For a two-lobe machine, caution should be used when the rotor length is more than 1.5 times the rotor diameter at pressures in excess of 8 psi differential. Three-lobe compressors inherently have stiffer rotors and can sustain a higher pressure differential. A practical upper limit is about 10 psi differential for units above 3000 cfm and 12 psi differential for smaller units.

This type of compressor has a constant leakage rate for a fixed set of clearances, pressure, and temperature. Capacities range from 200–1500 cfm at 125 psig.

Liquid ring compressors employ a rotor to drive a captive ring of liquid within a cylindrical housing. The inner surface of the liquid ring serves as the face of a liquid piston operating within

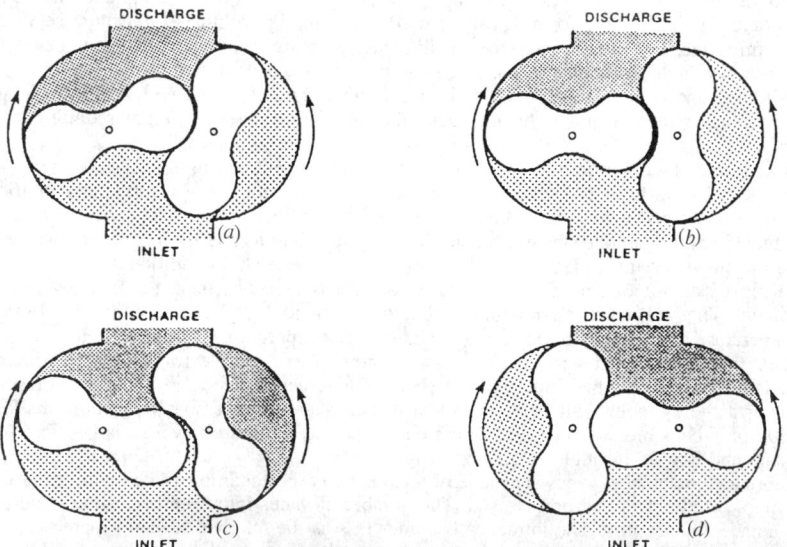

Fig. 61.9 Operating cycle of a straight-lobe rotary compressor (modified; courtesy of Ingersoll-Rand).[2]

each rotor chamber. At the inner diameter, these rotor chambers have openings that are sealed by, and revolve about, a stationary central plug or cone. This plug has permanently open ports that permit air to be taken into, and discharged from, the revolving rotor chambers.

As with the sliding-vane compressor, the single rotor is located eccentrically inside a cylinder or stator. The rotor has, extending from it, a series of vanes in a purely radial profile, or radial with forward-curved tips. Air inlet and outlet passages are located on the rotor. A liquid compressant partially fills the rotor and cylinder and orients itself in a ring-like manner as the rotor turns. Because of eccentricity, the ring moves in an oscillatory motion. The center of the ring connects with the inlet and outlet ports and forms an air pocket. As the rotor turns and the pocket moves away from the rotor, air enters through the inlet and fills the pocket. As the rotor turns, it carries the air pocket with it. Further turning takes the liquid ring from the maximum clearance area toward the minimum side. The liquid ring seals off the inlet port and traps the pocket of air. As the liquid ring is moved into the minimum clearance area, the pocket is compressed. When the ring uncovers the discharge port, the compressed pocket of air is discharged.

Efficiency of the liquid piston is about 50%, which is not very good compared to other rotary compressors. But because liquid is integral to the liquid piston compressor, taking in liquid with the air stream does not affect its operation as it would in other types of compressors. The liquid ring compressor is most often used in vacuum service, although it can also act as a positive pressure compressor. The compressor can be staged when the application requires more differential pressure than can be generated by a single stage. Liquid piston compressors can be used to compress air, in single-stage units of 35 psig and two-stage units of 125 psig. Vacuums of 26 in. Hg are possible. Flow capacity ranges from 2–16,000 cfm.

These compressors have only one solid moving part, the rotor. There is no metallic contact between the rotating and stationary elements. This design provides a continuous source of pressure without pulsation.

Delivered air is oil-free because the liquid ring is the piston and requires no lubrication. The liquid scrubs the air and removes solid particulates down to micron sizes. Many solids can pass through the compressor without doing damage. However, abrasive solids can shorten compressor life and should be removed with an inlet filter.

61.3 SIZING

Two conflicting factors influence the determination of total compressor capacity needed to supply a system: compressor efficiency and system demand. Constant-speed air compressors are most efficient when operated at full load or maximum capacity. The most efficient compressor is sized to handle the average load and would operate normally at full load. Undersized compressor capability results in reduced system-operating pressures. The inability to meet peak demands could result in decreased production and much greater overall plant operating cost.

Multiple compressors with sequential controls offer one solution to the dilemma of variable system demand by providing a better match of load and compressor capacity. Multiple compressors also permit compressor backup for maintenance and repairs. For example, three compressors, each with a capacity of 50% of peak load, is a configuration that offers these advantages. Disadvantages of multiple compressors are that full-load efficiency of smaller compressors is generally less than that of larger ones and that multiple units are more costly, per unit of capacity, to purchase and install.

If a new compressed-air system is being designed, system capacity is determined by analysis, where all known air users are identified and their expected consumption calculated. Air-consumption

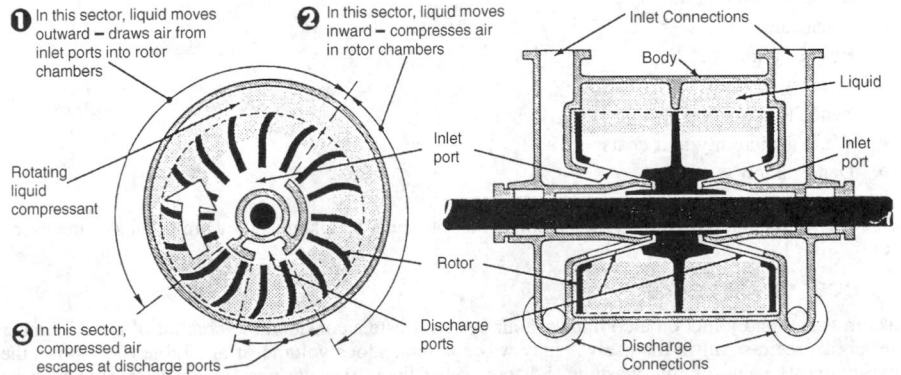

❶ In this sector, liquid moves outward – draws air from inlet ports into rotor chambers

❷ In this sector, liquid moves inward – compresses air in rotor chambers

Inlet Connections

Body

Liquid

Rotating liquid compressant

Inlet port

Inlet port

Rotor

Discharge ports

Discharge Connections

❸ In this sector, compressed air escapes at discharge ports

Fig. 61.10 A sectional and end view of a liquid ring compressor (courtesy of Nash Engineering Co.).[2]

rates of tools are available from manufacturers. A load factor is used to modify consumption by estimating the percentage of time that a pneumatic device is operating. Additional allowances must be made for leakage, typically no more than 10%, and for future plant growth.

The volumetric efficiency of a compressor is the ratio of the actual air delivered to total displacement. This efficiency is from 4–4.2 scfm/hp. The 4-scfm/hp figure can be used in calculations that will be sufficiently accurate for all practical applications.

It is normal practice to size water-cooled compressors 30% over system requirements and air-cooled compressors 40% over system requirements. These margins can be cut back if load estimates are based on specific plant experience rather than estimates.

If an existing system is being enlarged, load factors and required additional capacity are more easily and accurately measured and determined from operating experience. The proportion of the load handled adequately by the existing compressor system to that of the enlarged system can provide guidance for estimating additional capacity required. This is done by monitoring pressures at various locations throughout the plant during peak operating times.

61.4 SELECTION

The compressed air system is frequently a key utility in which reliability is absolutely essential. In turn, the air compressor is the heart of the compressed air system, and the proper compressor for the application is of paramount importance. Compressors vary widely in design or type, each with a fixed set of operating characteristics. It is the task of the air system designer to match the compressor type to system requirements.

Air compressor selection must take into account a wide variety of factors besides the type of machine. Topics that must be considered include

- Air requirements
- Driver
- Location
- Number of compressors
- Regulation
- Distribution
- Storage
- Piping
- Aftercoolers
- Separators
- Dryers
- Maintenance
- Noise limitations
- Subsoil or potential foundation problems
- Power rates or costs
- Hours per day of operation
- Percentage of time loaded
- First cost
- Lubricating oil costs
- Outdoor installation
- Attendance
- Resale value
- Installation time
- Ventilation
- Water availability and costs
- Depreciation

It is suggested that an individual assessment of the foregoing be taken as they are related to the user's needs.

61.5 COST OF AIR LEAKS

Leaks in valves and joints on a compressed-air system waste a considerable amount of air. A number of leaks that seem small in themselves may waste a tremendous volume of air. Table 61.1 shows the dollar-and-cents value of this wastage. For cost other than 10 cents per 100 cu ft, a ratio may be applied.

Table 61.1 Cost of Air Leaks

Size of Opening, in.	Cu Ft Air Wasted per Month at 100 psi, Based on an Orifice Coefficient of 0.65	Cost of Air Wasted per Month, Based on $0.10 per 1000 ft^3
3/8	6,671,890	$667.19
1/4	2,920,840	292.09
1/8	740,210	74.01
1/16	182,272	18.21
1/32	45,508	4.56

Often it is possible to determine the exact extent of air losses in a plant by finding what portion of the compressor capacity is required to keep pressure in the air lines when no equipment is being operated. Careful maintenance of air lines will more than pay for itself and may in some cases make unnecessary the replacement of the present compressor with one of larger capacity.

REFERENCES

1. J. P. Rollins, *Compressed Air and Gas Handbook,* 5th ed., Compressed Air and Gas Institute, Cleveland, OH, 1989.
2. R. N. Brown, *Compressors—Selection and Sizing,* Gulf, Houston, TX, 1986.
3. J. L. Foszcz, "A Guide to Air Compressors," Plant Engineering, 1995 (December, 1995).

CHAPTER 62

REFRIGERATION

Dennis L. O'Neal
Texas A & M University
College Station, Texas

K. W. Cooper
K. E. Hickman
Borg Warner Corporation
York, Pennsylvania

62.1	**INTRODUCTION**	1879		**62.6**	**STEAM JET REFRIGERATION**	**1894**
62.2	**BASIC PRINCIPLES**	1880		**62.7**	**INDIRECT REFRIGERATION**	**1894**
62.3	**REFRIGERATION CYCLES AND SYSTEM OVERVIEW**	**1881**			62.7.1 Use of Ice	1897
	62.3.1 Closed-Cycle Operation	1881		**62.8**	**SYSTEM COMPONENTS**	**1897**
	62.3.2 Open-Cycle Operation	1882			62.8.1 Compressors	1897
					62.8.2 Condensers	1901
62.4	**REFRIGERANTS**	**1883**			62.8.3 Evaporators	1903
	62.4.1 Regulations on the Production and Use of Refrigerants	1888			62.8.4 Expansion Devices	1905
	62.4.2 Refrigerant Selection for the Closed Cycle	1888		**62.9**	**DEFROST METHODS**	**1909**
					62.9.1 Hot Refrigerant Gas Defrost	1909
	62.4.3 Refrigerant Selection for the Open Cycle	1891			62.9.2 Air and Water Defrost	1910
62.5	**ABSORPTION SYSTEMS**	**1891**		**62.10**	**SYSTEM DESIGN CONSIDERATIONS**	**1910**
	62.5.1 Water–Lithium Bromide Absorption Chillers	1892		**62.11**	**REFRIGERATION SYSTEM SPECIFICATIONS**	**1910**
	62.5.2 Ammonia–Water Absorption Systems	1893				

62.1 INTRODUCTION

Refrigeration is the use of mechanical or heat-activated machinery for cooling purposes. The use of refrigeration equipment to produce temperatures below $-150°C$ is known as *cryogenics*.[1] When refrigeration equipment is used to provide human comfort, it is called *air conditioning*. This chapter focuses primarily on refrigeration applications, covering such diverse uses as food processing and storage, supermarket display cases, skating rinks, ice manufacture, and biomedical applications, such as blood and tissue storage or hypothermia used in surgery.

The first patent on a mechanically driven refrigeration system was issued to Jacob Perkins in 1834 in London.[2] The system used ether as the refrigerant. The first viable commercial system was produced in 1857 by James Harrison and D. E. Siebe and used ethyl ether as the refrigerant.[2]

Revised from *Kirk–Othmer Encyclopedia of Chemical Technology,* 3rd ed., Volume 20, Wiley, New York, 1982, by permission of the publisher.

Mechanical Engineers' Handbook, 2nd ed., Edited by Myer Kutz.
ISBN 0-471-13007-9 © 1998 John Wiley & Sons, Inc.

Refrigeration is used in installations covering a broad range of cooling capacities and temperatures. While the variety of applications results in a diversity of mechanical specifications and equipment requirements, the methods for producing refrigeration are well standardized.

62.2 BASIC PRINCIPLES

Most refrigeration systems utilize the *vapor-compression cycle* to produce the desired refrigeration effect. Besides vapor compression, two other, less common methods to produce refrigeration are the *absorption cycle* and *steam jet refrigeration.* These are described later in this chapter. With the vapor-compression cycle, a working fluid, called the *refrigerant,* evaporates and condenses at suitable pressures for practical equipment designs. The ideal (no pressure or frictional losses) vapor-compression refrigeration cycle is illustrated in Fig. 62.1 on a pressure–enthalpy diagram.

There are four basic components in every vapor-compression refrigeration system: (1) compressor, (2) condenser, (3) expansion device, and (4) evaporator. The compressor raises the pressure of the refrigerant vapor so that the refrigerant saturation temperature is slightly above the temperature of the cooling medium used in the condenser. The condenser is a heat exchanger used to reject heat from the refrigerant to a cooling medium. The refrigerant enters the condenser and usually leaves as a subcooled liquid. Typical cooling mediums used in condensers are air and water. After leaving the condenser, the liquid refrigerant expands to a lower pressure in the expansion valve. The expansion valve can be a passive device, such as a capillary tube or short-tube orifice, or an active device, such as a thermal expansion valve or electronic expansion valve. At the exit of the expansion valve, the refrigerant is at a temperature below that of the product to be cooled. As the refrigerant travels through the evaporator, it absorbs energy and is converted from a low-quality two-phase fluid to a superheated vapor under normal operating conditions. The vapor formed must be removed by the

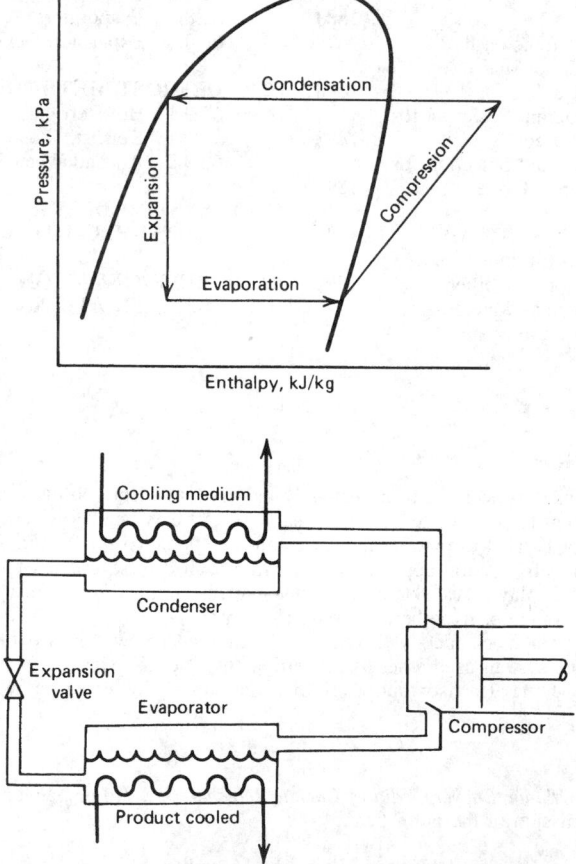

Fig. 62.1 Simple vapor-compression refrigeration cycle.[3]

compressor at a sufficient rate to maintain the low pressure in the evaporator and keep the cycle operating.

Pumped recirculation of refrigerant rather than direct evaporation of refrigerant is often used to service remotely located or specially designed heat exchangers. This technique provides the user with wide flexibility in applying refrigeration to complex processes and greatly simplifies operation. Secondary refrigerants or brines are also commonly used for simple control and operation. Direct application of ice and brine storage tanks may be used to level off batch cooling loads and reduce equipment size. This approach provides stored refrigeration where temperature control is vital as a safety consideration to prevent runaway reactions or pressure buildup.

All mechanical cooling results in the production of a greater amount of heat energy. In many instances, this heat energy is rejected to the environment directly to the air in the condenser or indirectly to water, where it is rejected in a cooling tower. Under some specialized applications, it may be possible to utilize this heat energy in another process at the refrigeration facility. This may require special modifications to the condenser. Recovery of this waste heat at temperatures up to 65°C can be used to achieve improved operating economy.

Historically, capacities of mechanical refrigeration systems have been stated in tons of refrigeration, a unit of measure related to the ability of an ice plant to freeze one short ton (907 kg) of ice in 24 hr. Its value is 3.51 kW_t (12,000 Btu/hr). Often a kilowatt of refrigeration capacity is identified as kW_t to distinguish it from the amount of electricity (kW_e) required to produce the refrigeration.

62.3 REFRIGERATION CYCLES AND SYSTEM OVERVIEW

Refrigeration can be accomplished in either *closed-cycle* or *open-cycle* systems. In a closed cycle, the refrigerant fluid is confined within the system and recirculates through the components (compressor, heat exchangers, and expansion valve) in the cycle. The system shown at the bottom of Fig. 62.1 is a closed cycle. In an open cycle, the fluid used as the refrigerant passes through the system once on its way to be used as a product or feedstock outside the refrigeration process. An example is the cooling of natural gas to separate and condense heavier components.

In addition to the distinction between open- and closed-cycle systems, refrigeration processes are also described as *simple cycles, compound cycles,* or *cascade cycles.* Simple cycles employ one set of components and a single refrigeration cycle, as shown in Fig. 62.1. Compound and cascade cycles use multiple sets of components and two or more refrigeration cycles. The cycles interact to accomplish cooling at several temperatures or to allow a greater span between the lowest and highest temperatures in the system than can be achieved with the simple cycle.

62.3.1 Closed-Cycle Operation

For a simple cycle, the lowest evaporator temperature that is practical in a closed-cycle system (Fig. 62.1) is set by the pressure-ratio capability of the compressor and by the properties of the refrigerant. Most high-speed reciprocating compressors are limited to a pressure ratio of 9:1, so that the simple cycle is used for evaporator temperatures of 2 to −50°C. Below these temperatures, the application limits of a single reciprocating compressor are reached. Beyond that limit, there is a risk of excessive heat, which may break down lubricants, high bearing loads, excessive oil foaming at startup, and inefficient operation because of reduced volumetric efficiency.

Centrifugal compressors with multiple stages can generate a pressure ratio up to 18:1, but their high discharge temperatures limit the efficiency of the simple cycle at these high pressure ratios. As a result, they operate with evaporator temperatures in the same range as reciprocating compressors.

The compound cycle (Fig. 62.2) achieves temperatures of approximately −100°C by using two or three compressors in series and a common refrigerant. This keeps the individual machines within their application limits. A refrigerant gas cooler is normally used between compressors to keep the final discharge temperature at a satisfactory level.

Below −100°C, most refrigerants with suitable evaporator pressures have excessively high condensing pressures. For some refrigerants, the refrigerant specific volume at low temperatures may be so great as to require compressors and other equipment of uneconomical size. With other refrigerants, the refrigerant specific volume may be satisfactory at low temperature but the specific volume may become too small at the condensing condition. In some circumstances, although none of the above limitations is encountered and a single refrigerant is practical, the compound cycle is not used because of oil-return problems or difficulties of operation.

To satisfy these conditions, the cascade cycle is used (Fig. 62.3). This consists of two or more separate refrigerants, each in its own closed cycle. The cascade condenser–evaporator rejects heat to the evaporator of the high-temperature cycle, which condenses the refrigerant of the low-temperature cycle. Refrigerants are selected for each cycle with pressure–temperature characteristics that are well suited for application at either the higher or lower portion of the cycle. For extremely low temperatures, more than two refrigerants may be cascaded to produce evaporator temperatures at cryogenic conditions (below −150°C). Expansion tanks, sized to handle the low-temperature refrigerant as a gas at ambient temperatures, are used during standby to hold pressure at levels suitable for economical equipment design.

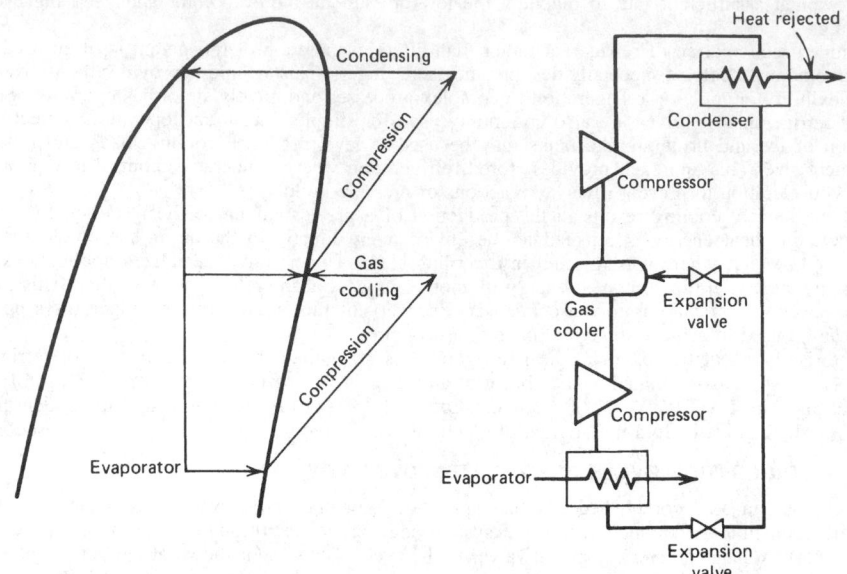

Fig. 62.2 Ideal compound refrigeration cycle.[3]

Compound cycles using reciprocating compressors, or any cycle using a multistage centrifugal compressor, allow the use of economizers or intercoolers between compression stages. Economizers reduce the discharge gas temperature from the preceding stage by mixing relatively cool gas with discharge gas before entering the subsequent stage. Either flash-type economizers, which cool refrigerant by reducing its pressure to the intermediate level, or surface-type economizers, which subcool refrigerant at condensing pressure, may be used to provide the cooler gas for mixing. This keeps the final discharge gas temperature low enough to avoid overheating of the compressor and improves compression efficiency.

Compound compression with economizers also affords the opportunity to provide refrigeration at an intermediate temperature. This provides a further thermodynamic efficiency gain because some of the refrigeration is accomplished at a higher temperature and less refrigerant must be handled by the lower-temperature stages. This reduces the power consumption and the size of the lower stages of compression.

Figure 62.4 shows a typical system schematic with flash-type economizers. Process loads at several different temperature levels can be handled by taking suction to an intermediate compression stage as shown. The pressure–enthalpy diagram illustrates the thermodynamic cycle.

Flooded refrigeration systems are a version of the closed cycle that may reduce design problems in some applications. In flooded systems, the refrigerant is circulated to heat exchangers or evaporators by a pump. Figure 62.5 shows the flooded cycle, which can use any of the simple or compound closed-refrigeration cycles.

The refrigerant-recirculating pump pressurizes the refrigerant liquid and moves it to one or more evaporators or heat exchangers, which may be remote from the receiver. The low-pressure refrigerant may be used as a single-phase heat-transfer fluid as in (A) of Fig. 62.5, which eliminates the extra heat-exchange step and increased temperature difference encountered in a conventional system that uses a secondary refrigerant or brine. This approach may simplify the design of process heat exchangers, where the large specific volumes of evaporating refrigerant vapor would be troublesome. Alternatively, the pumped refrigerant in the flooded system may be routed through conventional evaporators as in (B) and (C), or special heat exchangers as in (D).

The flooded refrigeration system is helpful when special heat exchangers are necessary for process reasons, or where multiple or remote exchangers are required.

62.3.2 Open-Cycle Operation

In many chemical processes, the product to be cooled can itself be used as the refrigerating liquid. An important example of this is in the gathering plants for natural gas. Gas from the wells is cooled, usually after compression and after some of the heavier components are removed as liquid. This

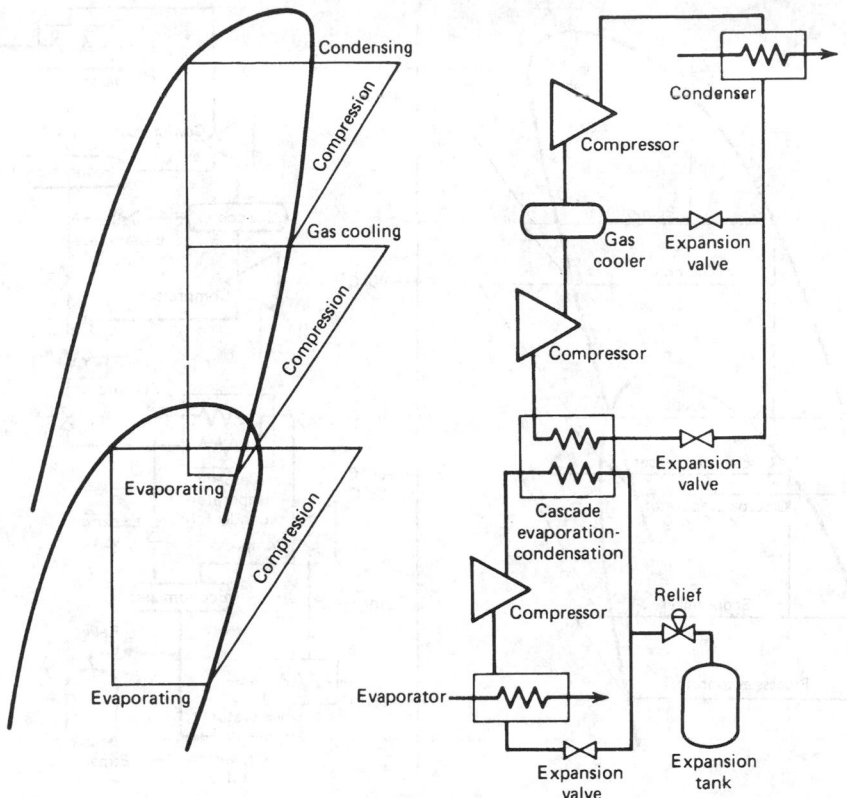

Fig. 62.3 Ideal cascade refrigeration cycle.[3]

liquid may be expanded in a refrigeration cycle to further cool the compressed gas, which causes more of the heavier components to condense. Excess liquid not used for refrigeration is drawn off as product. In the transportation of liquefied petroleum gas and of ammonia in ships and barges, the LPG or ammonia is compressed, cooled, and expanded. The liquid portion after expansion is passed on as product until the ship is loaded.

Open-cycle operation is similar to closed-cycle operation, except that one or more parts of the closed cycle may be omitted. For example, the compressor suction may be taken directly from gas wells, rather than from an evaporator. A condenser may be used and the liquefied gas may be drained to storage tanks.

Compressors may be installed in series or parallel for operating flexibility or for partial standby protection. With multiple reciprocating compressors, or with a centrifugal compressor, gas streams may be picked up or discharged at several pressures if there is refrigerating duty to be performed at intermediate temperatures. It always is more economical to refrigerate at the highest temperature possible.

Principal concerns in the open cycle involve dirt and contaminants, wet gas, compatibility of materials and lubrication circuits, and piping to and from the compressor. The possibility of gas condensing under various ambient temperatures either during operation or during standby must be considered. Beyond these considerations, the open-cycle design and its operation are governed primarily by the process requirements. The open system can use standard refrigeration hardware.

62.4 REFRIGERANTS

No one refrigerant is capable of providing cost-effective cooling over the wide range of temperatures and the multitude of applications found in modern refrigeration systems. Ammonia accounts for approximately half of the total worldwide refrigeration capacity.[4] Both chlorofluorocarbons (CFCs) and hydrochlorofluorocarbon (HCFC) refrigerants have historically been used in many supermarket and food storage applications. Most of these refrigerants are generally nontoxic and nonflammable.

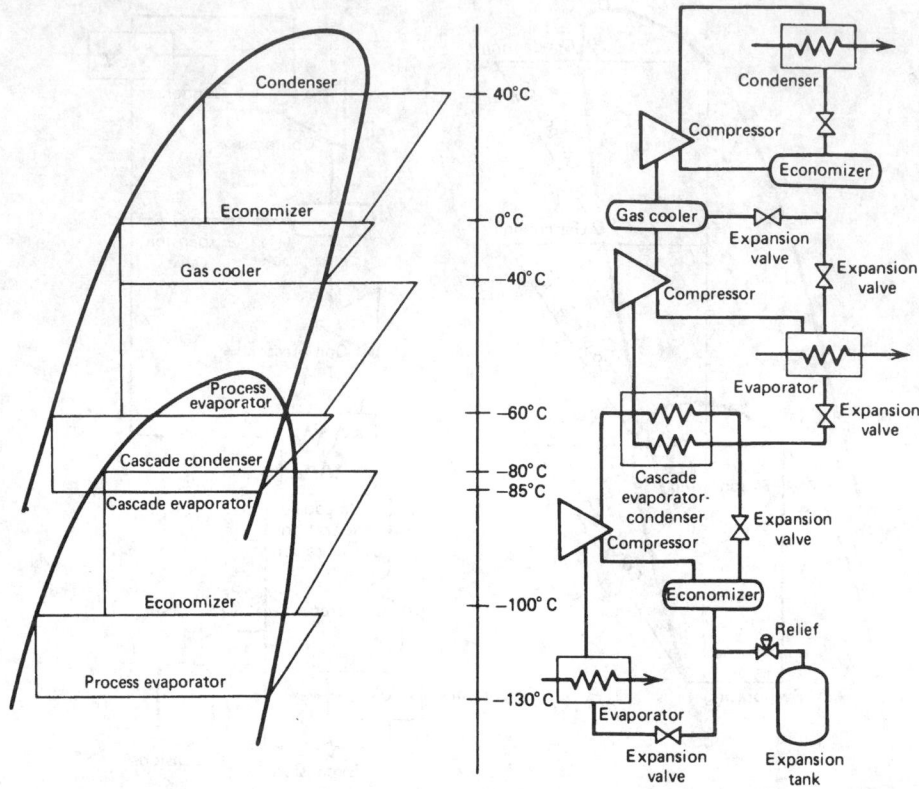

Fig. 62.4 Refrigeration cycle with flash economizers.[3]

However, recent U.S. federal and international regulations[5,6,7] have placed restrictions on the production and use of CFCs. Restrictions are also pending on HCFCs. Hydrofluorocarbons (HFCs) are now being used in some applications where CFCs were used. Regulations affecting refrigerants are discussed in the next section.

The chemical industry uses low-cost fluids such as propane and butane whenever they are available in the process. These hydrocarbon refrigerants, often thought of as too hazardous because of flammability, are suitable for use in modern compressors, and frequently add no more hazard than already exists in an oil refinery or petrochemical plant. These low-cost refrigerants are used in simple, compound, and cascade systems, depending on operating temperatures.

A standard numbering system, shown in Table 62.1, has been devised to identify refrigerants without the use of the cumbersome chemical name. There are many popular refrigerants in the methane and ethane series. These refrigerants are called *halocarbons* or *halogenated hydrocarbons* because of the presence of halogen elements such as fluorine or chlorine.[8] Halocarbons include CFCs, HCFCs, and HFCs.

Numbers assigned to the hydrocarbons and halohydrocarbons of the methane, ethane, propane, and cyclobutane series are such that the number uniquely specifies the refrigerant compound. The American National Standards Institute (ANSI) and American Society of Heating, Refrigerating, and Air Conditioning Engineers (ASHRAE) Standard 34-1992 describes the method of coding.[9]

Zeotropes and *azeotropes* are mixtures of two or more different refrigerants. A zeotropic mixture changes saturation temperatures as it evaporates (or condenses) at constant pressure. This phenomenon is called *temperature glide*. For example, R-407C has a boiling (bubble) point of −44°C and a condensation (dew) point of −37°C, which means it has a temperature glide of 7°C. An azeotropic mixture behaves much like a single-component refrigerant in that it does not change saturation temperatures appreciably as it evaporates or condenses at constant pressure. Some zeotropic mixtures, such as R-410A, actually have a small enough temperature glide (less than 0.5°C) that they are considered a near-azeotropic refrigerant mixture (nearm).

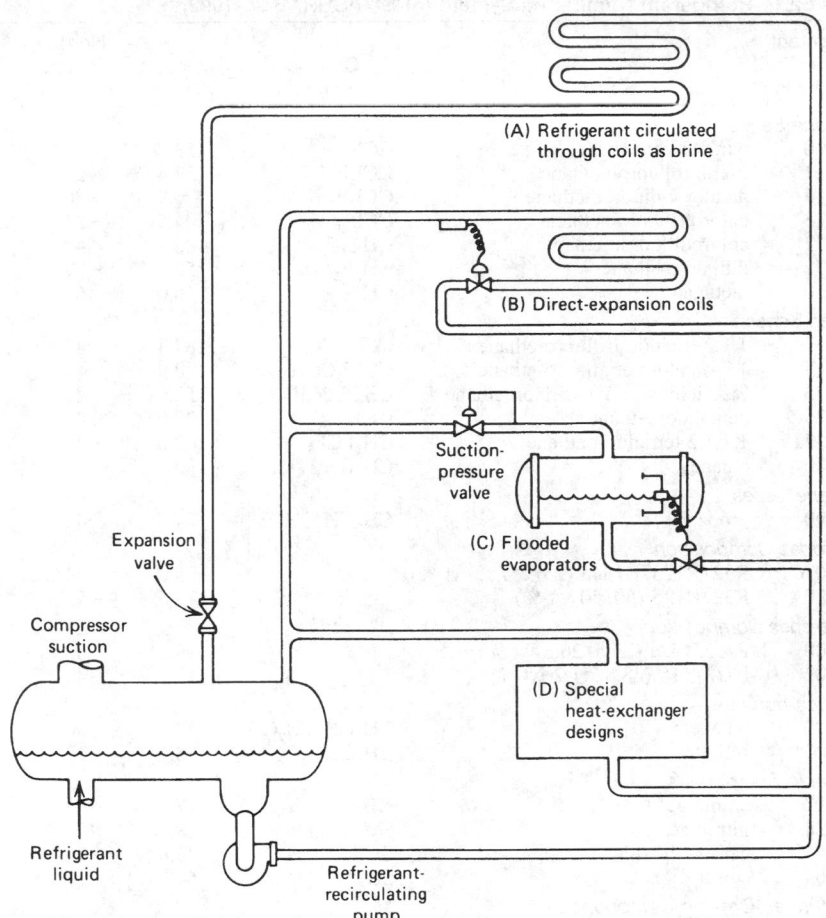

Fig. 62.5 Liquid recirculator.[3]

Because the bubble point and dew point temperatures are not the same for a given pressure, some zeotropic mixtures have been used to help control the temperature differences in low-temperature evaporators. These mixtures have been used in the lowest stage of some LNG plants.[10]

Refrigerants are grouped by their toxicity and flammability (Table 62.2).[9,11] Group A1 are non-flammable and least toxic, while Group B3 is flammable and most toxic. Toxicity is quantified by the threshold limit value–time-weighted average (TLV–TWA), which is the upper safety limit for airborne exposure to the refrigerant. If the refrigerant is non-toxic in quantities less than 400 parts per million, then it is a Class A refrigerant. If exposure to less than 400 parts per million is toxic, then the substance is given the B designation. The numerical designations refers to the flammability of the refrigerant. The last column of Table 62.1 shows the toxicity and flammability rating of many of the common refrigerants.

The A1 group of refrigerants generally fulfill the basic requirements for an ideal refrigerant with considerable flexibility as to refrigeration capacity. Many are used for comfort air conditioning since they are nontoxic and nonflammable. These refrigerants are also used extensively in refrigeration applications. Many CFCs are in the A1 group. With regulations banning the production and restricting the sale of all CFCs, the CFCs will eventually cease to be used. Common refrigerants in the A1 group include R-11, R-12, R-13, R-22, R-114, R-134a, and R-502.

Refrigerant 11, trichlorofluoromethane, is a CFC. It has a low-pressure and high-volume characteristic suitable for use in close-coupled centrifugal compressor systems for water or brine cooling. Its temperature range extends no lower than −7°C.

Table 62.1   Refrigerant Numbering System (ANSI/ASHRAE 34-1992)[a]

Refrigerant Number Designation	Chemical Name	Chemical Formula	Molecular Mass	Normal Boiling Point, °C	Safety Group
Methane Series					
10	tetrachloramethane	CCl_4	153.8	77	B1
11	trichlorofluoromethane	CCl_3F	137.4	24	A1
12	dichlorodifluoromethane	CCl_2F_2	120.9	−30	A1
13	chlorotrifluoromethane	$CClF_3$	104.5	−81	A1
22	chlorodifluoromethane	$CHClF_2$	86.5	−41	A1
32	difluoromethane	CH_2F_2	52.0	−52	A2
50	methane	CH_4	16.0	−161	A3
Ethane Series					
113	1,1,2-trichlorotrifluoro-ethane	CCl_2FCClF_2	187.4	48	A1
114	1,2-dichlorotetrafluoro-ethane	$CClF_2CClF_2$	170.9	4	A1
123	2,2-dichloro-1,1,1-trifluoroethane	$CHCL2CF3$	152.9	27	B1
125	pentafluoroethane	CHF_2CF_3	120.0	−49	A1
134a	1,1,1,2-tetrafluoroethane	CH_2FCF_3	102.0	−26	A1
170	ethane	CH_3CH_3	30	−89	A3
Propane Series					
290	propane	$CH_3CH_2CH_3$	44	−42	A3
Zeotropes Composition					
407C	R32/R125/R134a (23/25/52 wt %)		95.0	−44	A1
410A	R32/R125 (50/50 wt %)		72.6	−53	A1
Azeotropes Composition					
500	R-12/152a (73.8/26.2 wt %)		99.31	−33	A1
502	R-22/115 (48.8/51.2 wt %)		112	−45	A1
Hydrocarbons					
600	butane	$CH_3CH_2CH_2CH_3$	58.1	0	A3
600a	isobutane	$CH(CH_3)_3$	58.1	−12	A3
Inorganic Compounds					
717	ammonia	NH_3	17.0	−33	B2
728	nitrogen	N_2	28.0	−196	A1
744	carbon dioxide	CO_2	44.0	−78	A1
764	sulfur dioxide	SO_2	64.1	−10	B1
Unsaturated Organic Compounds					
1140	vinyl chloride	$CH_2=CHCl$	62.5	−14	B3
1150	ethylene	$CH_2=CH_2$	28.1	−104	A3
1270	propylene	$CH_3CH=CH_2$	42.1	−48	A3

[a]Reference 9. Reprinted with permission of American Society of Heating, Refrigerating and Air Conditioning Engineers from ANSI/ASHRAE Standard 34-1992.

Table 62.2   ANSI/ASHRAE Toxicity and Flammability Rating System[a]

Flammability	Group	Group
Highly	A3	B3
Moderate	A2	B2
Non	A1	B1
Threshold Limit Value (parts per million)	< 400	> 400

[a]Reference 9. Reprinted with permission of American Society of Heating, Refrigerating and Air Conditioning Engineers from ANSI/ASHRAE Standard 34-1992.

Refrigerant 12, dichlorodifluoromethane, is a CFC. It was the most widely known and used refrigerant for U.S. domestic refrigeration and automotive air conditioning applications until the early 1990s. It is ideal for close-coupled or remote systems ranging from small reciprocating to large centrifugal units. It has been used for temperatures as low as −90°C, although −85°C is a more practical lower limit because of the high gas volumes necessary for attaining these temperatures. It is suited for single-stage or compound cycles using reciprocating and centrifugal compressors.

Refrigerant 13, chlorotrifluoromethane, is a CFC. It is used in low-temperature applications to approximately −126°C. Because of its low volume, high condensing pressure, or both, and because of its low critical pressure and temperature, R-13 is usually cascaded with other refrigerants at a discharge pressure corresponding to a condensing temperature in the range of −56 to −23°C.

Refrigerant 22, chlorodifluoromethane, is an HCFC. It is used in many of the same applications, as R-12, but its lower boiling point and higher latent heat permit the use of smaller compressors and refrigerant lines than R-12. The higher-pressure characteristics also extend its use to lower temperatures in the range of −100°C.

Refrigerant 114, dichlorotetrafluoroethane, is a CFC. It is similar to R-11 but its slightly higher pressure and lower volume characteristic than R-11 extend its use to −17°C and higher capacities.

Refrigerant 134a, 1,1,1,2-tetrafluoroethane, is a hydrofluorocarbon (HFC). It is a replacement refrigerant for R-12 in both refrigeration and air conditioning applications. It has operating characteristics very similar to those of R-12.

Refrigerants 407C and 410A are both mixtures of HFCs. Both are considered replacements for R-22.

Refrigerant 502 is an azeotropic mixture of R-22 and R-115. Its pressure characteristics are similar to those of R-22 but it has a lower discharge temperature.

The B1 refrigerants are nonflammable, but have lower toxicity limits than those in the A1 group. Refrigerant 123, an HCFC, is used in many new low-pressure centrifugal chiller applications. Industry standards, such as ANSI/ASHRAE Standard 15-1994, provide detailed guidelines for safety precautions when using R-123 or any other refrigerant that is toxic or flammable.[11]

One of the most widely used industrial refrigerants is ammonia, even though it is moderately flammable and has a Class B toxicity rating. Ammonia liquid has a high specific heat, an acceptable density and viscosity, and high conductivity. Its enthalpy of vaporization is typically six to eight times higher than that of the commonly used halocarbons. These properties make it an ideal heat-transfer fluid with reasonable pumping costs, pressure drop, and flow rates. As a refrigerant, ammonia provides high heat transfer except when affected by oil at temperatures below approximately −29°C, where oil films become viscous. To limit the ammonia-discharge-gas temperature to safe values, its normal maximum condensing temperature is 38°C. Generally, ammonia is used with reciprocating compressors, although relatively large centrifugal compressors (≥ 3.5 MW, or 1.2 × 10⁶ Btu/hr) with 8 to 12 impeller stages required by its low molecular weight are in use today. Systems using ammonia should contain no copper (with the exception of Monel metal).

The flammable refrigerants (Groups A3 and B3) are generally applicable where a flammability or explosion hazard is already present and their use does not add to the hazard. These refrigerants have the advantage of low cost. Although they have fairly low molecular weight, they are suitable for centrifugal compressors of larger sizes. Because of the high acoustic velocity in these refrigerants, centrifugal compressors may be operated at high impeller tip speeds, which partly compensates for the higher head requirements than some of the nonflammable refrigerants.

These refrigerants should be used at pressures greater than atmospheric to avoid increasing the explosion hazard by the admission of air in case of leaks. In designing the system, it also must be recognized that these refrigerants are likely to be impure in refrigerant applications. For example, commercial propane liquid may contain about 2% (by mass) ethane, which in the vapor phase might represent as much as 16 to 20% (by volume). Thus, ethane may appear as a noncondensable. Either this gas must be purged or the compressor displacement must be increased about 20% if it is recycled from the condenser; otherwise, the condensing pressure will be higher than required for pure propane and the power requirement will be increased.

Refrigerant 290, propane, is the most commonly used flammable refrigerant. It is well suited for use with reciprocating and centrifugal compressors in close-coupled or remote systems. Its operating temperature range extends to −40°C.

Refrigerant 600, butane, occasionally is used for close-coupled systems in the medium temperature range of 2°C. It has a low-pressure and high-volume characteristic suitable for centrifugal compressors where the capacity is too small for propane and the temperature is within range.

Refrigerant 170, ethane, normally is used for close-coupled or remote systems at −87 to −7°C. It must be used in a cascade cycle because of its high-pressure characteristics.

Refrigerant 1150, ethylene, is similar to ethane but has a slightly higher-pressure, lower-volume characteristic that extends its use to −104 to −29°C. Like ethane, it must be used in the cascade cycle.

Refrigerant 50, methane, is used in an ultralow range of -160 to $-110°C$. It is limited to cascade cycles. Methane condensed by ethylene, which is in turn condensed by propane, is a cascade cycle commonly employed to liquefy natural gas.

Table 62.3 shows the comparative performance of different refrigerants at conditions more typical of some freezer applications. The data show the relatively large refrigerating effect that can be obtained with ammonia. Note also that for these conditions, both R-11 and R-123 would operate with evaporator pressures below atmospheric pressure.

62.4.1 Regulations on the Production and Use of Refrigerants

In 1974, Rowland and Molina put forth the hypothesis that CFCs destroyed the ozone layer.[13] By the late 1970s, the United States and Canada had banned the use of CFCs in aerosols. In 1985, Farmer noted a depletion in the ozone layer of approximately 40% over what had been measured in earlier years.[4] This depletion in the ozone layer became known as the ozone hole. In September 1987, 43 countries signed an agreement called the Montreal Protocol,[7] in which the participants agreed to freeze CFC production levels by 1990, then to decrease production by 20% by 1994 and 50% by 1999. The protocol, ratified by the United States in 1988, subjected the refrigeration industry, for the first time, to major CFC restrictions.

Regulations imposed restrictions on refrigerants.[4,6,14] Production of CFCs was to cease by January 1, 1996.[14] A schedule was also imposed on the phaseout of the production HCFCs by 2030. Refrigerants were divided into two classes. Class I were CFCs, halons, and other major ozone-depleting chemicals. Class II were HCFCs.

Two ratings are used to classify the harmful effects of a refrigerant on the environment.[15] The first, the *ozone depletion potential* (ODP), quantifies the potential damage that the refrigerant molecule has in destroying ozone in the stratosphere. When a CFC molecule is struck by ultraviolet light in the stratosphere, a chlorine atom breaks off and reacts with ozone to form oxygen and a chlorine/oxygen molecule. This molecule can then react with a free oxygen atom to form an oxygen molecule and a free chlorine. The chlorine can then react with another ozone molecule to repeat the process. The estimated atmospheric life of a given CFC or HCFC is an important factor in determining the value of the ODP.

The second rating is known as the *halocarbon global warming potential* (HGWP). It relates the potential for a refrigerant in the atmosphere to contribute to greenhouse effect. Like CO_2, refrigerants such as CFCs, HCFCs, and HFCs can block energy from the earth from radiating back into space. One molecule of R-12 can absorb as much energy as almost 5000 molecules of CO_2. Both the ODP and HGWP are normalized to the value of Refrigerant 11.

Table 62.4 shows the ODP and HGWP for a variety of refrigerants. As a class of refrigerants, the CFCs have the highest ODP and HGWP. Because HCFCs tend to be more unstable compounds and therefore have much shorter atmospheric lifetimes, their ODP and HGWP values are much smaller than those of the CFCs. All HFCs and their mixtures have zero ODP because fluorine does not react with ozone. However, some of the HFCs, such as R-125, R-134a, and R-143a do have HGWP values as large or larger than those of some of the HCFCs. From the standpoint of ozone depletion and global warming, hydrocarbons provide zero ODP and HGWP. However, hydrocarbons are also flammable, which makes them unsuitable in many applications.

62.4.2 Refrigerant Selection for the Closed Cycle

In any closed cycle, the choice of the operating fluid is based on the refrigerant whose properties are best suited to the operating conditions. The choice depends on a variety of factors, some of which may not be directly related to the refrigerant's ability to remove heat. For example, flammability, toxicity, density, viscosity, availability, and similar characteristics are often deciding factors. The suitability of a refrigerant also depends on factors such as the kind of compressor to be used (i.e., centrifugal, rotary, or reciprocating), safety in application, heat-exchanger design, application of codes, size of the job, and temperature ranges. The factors below should be taken into account when selecting a refrigerant.

Discharge (condensing) pressure should be low enough to suit the design pressure of commercially available pressure vessels, compressor casings, and so on. However, discharge pressure, that is, condenser liquid pressure, should be high enough to feed liquid refrigerant to all the parts of the system that require it.

Suction (evaporating) pressure should be above approximately 3.45 kPa (0.5 psia) for a practical compressor selection. When possible, it is preferable to have the suction pressure above atmospheric to prevent leakage of air and moisture into the system. Positive pressure normally is considered a necessity when dealing with hydrocarbons because of the explosion hazard presented by any air leakage into the system.

Standby pressure (saturation at ambient temperature) should be low enough to suit equipment design pressure unless there are other provisions in the system for handling the refrigerant during shutdown, such as inclusion of expansion tanks.

Table 62.3 Comparative Refrigeration Performance of Different Refrigerants at −23°C Evaporating Temperature and +37°C Condensing Temperature[a]

Refrigerant Number	Refrigerant Name	Evaporator Pressure (MPa)	Condenser Pressure (MPa)	Net Refrigerating Effect (kJ/kg)	Refrigerant Circulated (kg/h)	Compressor Displacement (L/s)	Power Input (kW)
11	Trichlorofluoromethane	0.013	0.159	145.8	24.7	7.65	0.297
12	Dichlorodifluoromethane	0.134	0.891	105.8	34.0	1.15	0.330
22	Chlorodifluoromethane	0.218	1.390	150.1	24.0	0.69	0.326
123	Dichlorotrifluoroethane	0.010	0.139	130.4	27.6	10.16	0.306
125	Pentafluoroethane	0.301	1.867	73.7	48.9	0.71	0.444
134a	Tetrafluoroethane	0.116	0.933	135.5	26.6	1.25	0.345
502	R-22/R-115 Azeotrope	0.260	1.563	91.9	39.2	0.72	0.391
717	Ammonia	0.166	1.426	1057.4	3.42	0.67	0.310

[a]Reference 12. Reprinted with permission of American Society of Heating, Refrigerating and Air Conditioning Engineers from *ASHRAE Handbook of Fundamentals*.

Table 62.4   Ozone-Depletion Potential and Halocarbon Global Warming Potential of Popular Refrigerants and Mixtures[a]

Refrigerant Number	Chemical Formula	Ozone Depletion Potential (ODP)	Halogen Global Warming Potential (HGWP)
Chlorofluorocarbons			
11	CCl_3F	1.0	1.0
12	CCl_2F_2	1.0	3.05
113	CCl_2FCClF_2	0.87	1.3
114	$CClF_2CClF_2$	0.74	4.15
115	$CClF_2CF_3$	1.43	9.6
Hydrochlorofluorocarbons			
22	$CHClF_2$	0.051	0.37
123	$CHCl_2CF_3$	0.016	0.019
124	$CHClFCF_3$	0.018	0.095
141b	CH_3CCl_2F	0.08	0.092
142b	CH_3CClF_2	0.056	0.37
Hydrofluorocarbons			
32	CH_2F_2	0	0.13
125	CHF_2CF_3	0	0.58
134a	CH_2FCF_3	0	0.285
143a	CH_3CF_3	0	0.75
152a	CH_3CHF_2	0	0.029
Hydrocarbons			
50	CH_4	0	0
290	$CH_3CH_2CH_3$	0	0
Zeotropes			
401A	R-22/R-152a/R-124 (53/13/34%wt)	0.03	0.22
407C	R-32/125/134a (23/25/52%wt)	0	0.22
410A	R-32/125 (50/50%wt)	0	0.44
Azeotropes			
500	R-12/152a (73.8/26.2%wt)	0.74	2.4
502	R-22/115 (48.8/51.2%wt)	0.23	5.1

[a]Compiled from References 4, 15, and 16.

Critical temperature and pressure should be well above the operating level. As the critical pressure is approached, less heat is rejected as latent heat compared to the sensible heat from desuperheating the compressor discharge gas, and cycle efficiency is reduced. Methane (R-50) and chlorotrifluoromethane (R-13) usually are cascaded with other refrigerants because of their low critical points.

Suction volume sets the size of the compressor. High suction volumes require centrifugal or screw compressors and low suction volumes dictate the use of reciprocating compressors. Suction volumes also may influence evaporator design, particularly at low temperatures, since they must include adequate space for gas–liquid separation.

Freezing point should be lower than minimum operating temperature. This generally is no problem unless the refrigerant is used as a brine.

Theoretical power required for adiabatic compression of the gas is slightly less with some refrigerants than others. However, this is usually a secondary consideration offset by the effects of particular equipment selections, such as line-pressure drops, on system power consumption.

Vapor density (or molecular weight) is an important characteristic when the compressor is centrifugal because the lighter gases require more impellers for a given pressure rise, that is, head, or temperature lift. On the other hand, centrifugal compressors have a limitation connected with the acoustic velocity in the gas, and this velocity decreases with the increasing molecular weight. Low vapor densities are desirable to minimize pressure drop in long suction and discharge lines.

Liquid density should be taken into account. Liquid velocities are comparatively low, so that pressure drop is usually no problem. However, static head may affect evaporator temperatures, and should be considered when liquid must be fed to elevated parts of the system.

Latent heat should be high because it reduces the quantity of refrigerant that needs to be circulated. However, large flow quantities are more easily controlled because they allow use of larger, less sensitive throttling devices and apertures.

Refrigerant cost depends on the size of the installation and must be considered both from the standpoint of initial charge and of composition owing to losses during service. Although a domestic refrigerator contains only a few dollars worth of refrigerant, the charge for a typical chemical plant may cost thousands of dollars.

Other desirable properties. Refrigerants should be stable and noncorrosive. For heat transfer considerations, a refrigerant should have low viscosity, high thermal conductivity, and high specific heat. For safety to life or property, a refrigerant should be nontoxic and nonflammable, should not contaminate products in case of a leak, and should have a low leakage tendency through normal materials of construction.

With a flammable refrigerant, extra precautions have to be taken in the engineering design if it is required to meet the explosion-proof classification. It may be more economical to use a higher-cost, but nonflammable, refrigerant.

62.4.3 Refrigerant Selection for the Open Cycle

Process gases used in the open cycle include chlorine, ammonia, and mixed hydrocarbons. These create a wide variety of operating conditions and corrosion problems. Gas characteristics affect both heat exchangers and compressors, but their impact is far more critical on compressor operation. All gas properties and conditions should be clearly specified to obtain the most economical and reliable compressor design. If the installation is greatly overspecified, design features result that not only add significant cost but also complicate the operation of the system and are difficult to maintain. Specifications should consider the following.

Composition. Molecular weight, enthalpy–entropy relationship, compressibility factor, and operating pressures and temperatures influence the selection and performance of compressors. If process streams are subject to periodic or gradual changes in composition, the range of variations must be indicated.

Corrosion. Special materials of construction and types of shaft seals may be necessary for some gases. Gases that are not compatible with lubricating oils or that must remain oil-free may necessitate reciprocating compressors designed with carbon rings or otherwise made oilless, or the use of centrifugal compressors designed with isolation seals. However, these features are unnecessary on most installations. Standard designs usually can be used to provide savings in first cost, simpler operation, and reduced maintenance.

Dirt and liquid carryover. Generally, the carryover of dirt and liquids can be controlled more effectively by suction scrubbers than by costly compressor design features. Where this is not possible, all anticipated operating conditions should be stated clearly so that suitable materials and shaft seals can be provided.

Polymerization. Gases that tend to polymerize may require cooling to keep the gas temperature low throughout compression. This can be handled by liquid injection or by providing external cooling between stages of compression. Provision may be necessary for internal cleaning with steam.

These factors are typical of those encountered in open-cycle gas compression. Each job should be thoroughly reviewed to avoid unnecessary cost and obtain the simplest possible compressor design for ease of operation and maintenance. Direct coordination between the design engineer and manufacturer during final stages of system design is strongly recommended.

62.5 ABSORPTION SYSTEMS

Ferdinand Carré patented the first absorption machine in 1859.[2] His design, which employed an ammonia/water solution, was soon produced in France, England, and Germany. By 1876, over 600 absorption systems had been sold in the United States. One of the primary uses for these machines was in the production of ice. During the late 1800s and early 1900s, different combinations of fluids were tested in absorption machines. These included such diverse combinations as ammonia with copper sulfate, camphor and naphthol with SO_2, and water with lithium chloride. The modern solution of lithium bromide and water was not used industrially until 1940.[2]

Absorption systems offer two distinct advantages over conventional vapor compression refrigeration. First, they do not use CFC or HCFC refrigerants. Second, absorption system can utilize a variety of heat sources, including natural gas, steam, solar-heated hot water, and waste heat from a turbine or industrial process. If the source of energy is from waste heat, absorption systems may provide the lowest-cost alternative for providing chilled water or refrigeration applications.

Two different systems are currently in use, a water–lithium bromide system where water is the refrigerant and lithium bromide is the absorbent, and a water–ammonia system where the ammonia is the refrigerant and the water is the absorbent.

Evaporator temperatures ranging from $-75°F$ to $50°F$ are achievable with absorption systems.[1] For water-chilling service, absorption systems generally use water as the refrigerant and lithium bromide as the absorbent solution. For process applications requiring chilled fluid below 7°C, the ammonia–water pair is used, with ammonia serving as the refrigerant.

62.5.1 Water–Lithium Bromide Absorption Chillers

Water–lithium bromide absorption machines can be classified by the method of heat input. *Indirect fired* chillers use steam or hot liquids as a heat source. *Direct fired* chillers use the heat from the firing of fossil fuels. *Heat-recovery* chillers use waste gases as the heat source.

A typical arrangement for a single-stage water–lithium bromide absorption system is shown schematically in Fig. 62.6. The absorbent, lithium bromide, may be thought of as a carrier fluid bringing spent refrigerant from the low-pressure side of the cycle (the absorber) to the high-pressure side (the generator). There, the waste heat, steam, or hot water that drives the system separates the water from the absorbent by a distillation process. The regenerated absorbent returns to the absorber, where it is cooled so it will absorb the refrigerant (water) vapor produced in the evaporator and thereby establish the low-pressure level that controls the evaporator temperature. Thermal energy released during the absorption process is transferred to the cooling water flowing through tubes in the absorber shell.

The external heat exchanger shown saves energy by heating the strong liquid flowing to the generator as it cools the hot absorbent flowing from the generator to the absorber. If the weak solution that passes through the regenerator to the absorber does not contain enough refrigerant and is cooled too much, crystallization can occur. Leaks or process upsets that cause the generator to overconcentrate the solution are indicated when this occurs. The slushy mixture formed does not harm the machine, but it interferes with continued operation. External heat and added water may be required to redissolve the mixture.

Single-stage absorption systems are most common where generator heat input is less than 95°C. The coefficient of performance (COP) of a system is the cooling achieved in the evaporator divided by the heat input to the generator. The COP of a single-stage lithium bromide machine generally is 0.65–0.70 for water-chilling duty. The heat rejected by the cooling tower from both the condenser

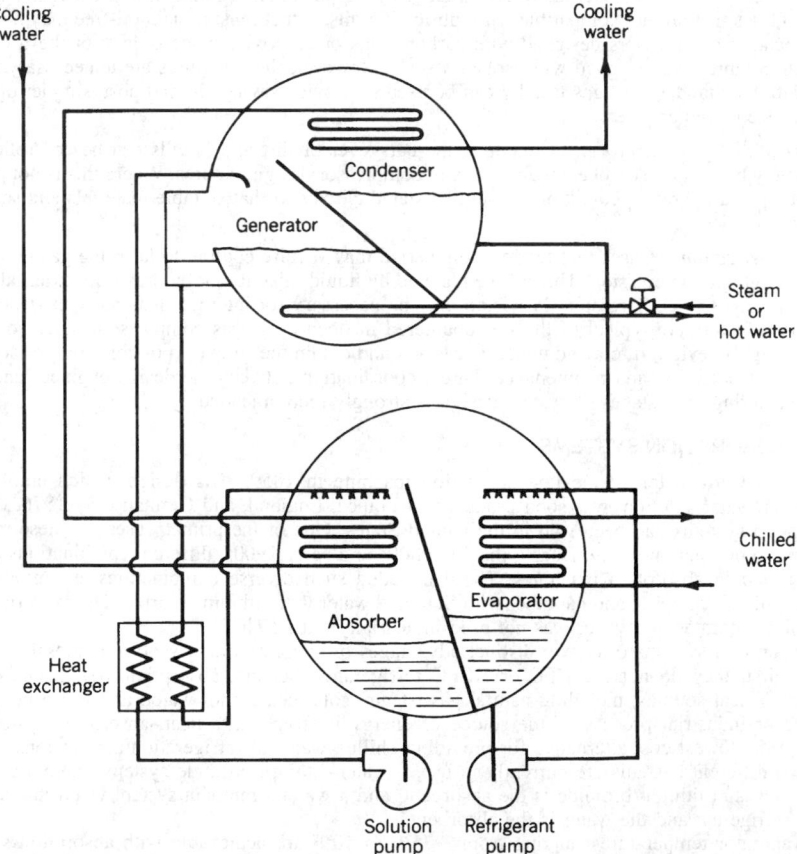

Fig. 62.6 Single-stage water–lithium bromide absorption system.[3]

and the absorber is the sum of the waste heat supplied plus the cooling produced, requiring larger cooling towers and cooling water flows than for vapor compression systems.

Absorption machines can be built with a two-stage generator (Fig. 62.7) with heat input temperatures greater than 150°C. Such machines are called dual-effect machines. The operation of the dual-effect machine is the same as the single-effect machine except that an additional generator, condenser, and heat exchanger are used. Energy from an external heat source is used to boil the dilute lithium bromide (absorbent) solution. The vapor from the primary generator flows in tubes to the second-effect generator. It is hot enough to boil and concentrate absorbent, which creates more refrigerant vapor without any extra energy input. Dual-effect machines typically use steam or hot liquids as input. Coefficients of performance above 1.0 can be obtained with these machines.

62.5.2 Ammonia–Water Absorption Systems

Ammonia–water absorption technology is used primarily in smaller chillers and small refrigerators found in recreational vehicles.[1] Refrigerators use a variation of the ammonia absorption cycle with ammonia, water, and hydrogen as the working fluids. They can be fired with both gas and electric heat. The units are hermetically sealed. A description of this technology can be found in Ref. 62.1.

Ammonia–water chillers have three major differences from water–lithium bromide systems. First, because the water is volatile, the regeneration of the weak absorbent to strong absorbent requires a

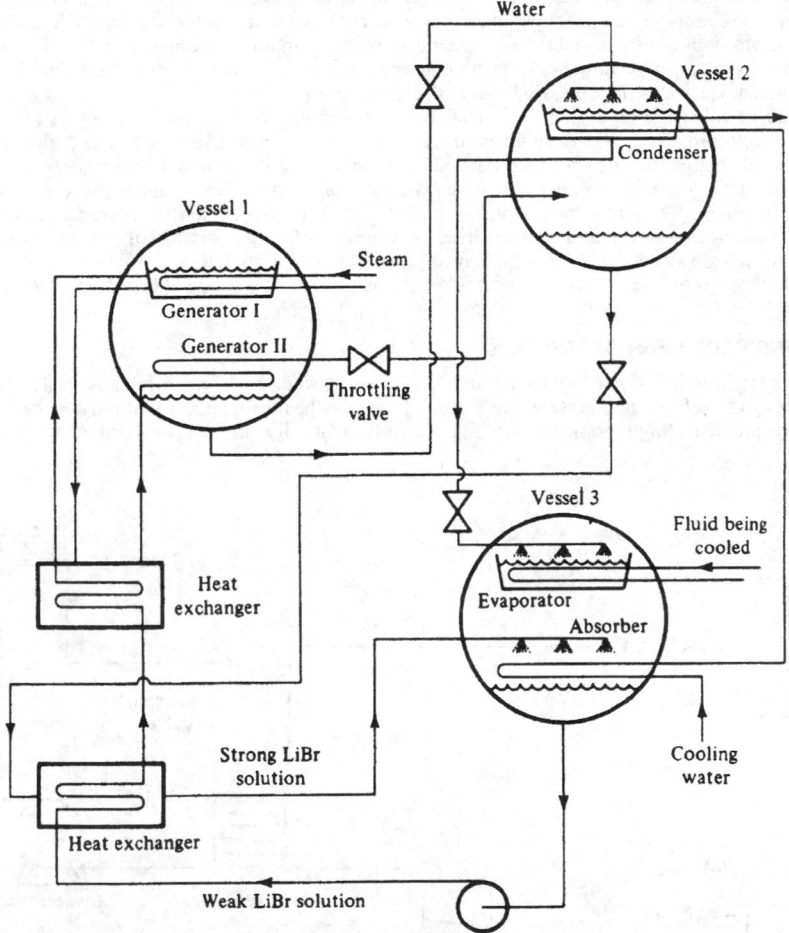

Fig. 62.7 Two-stage water–lithium bromide absorption system.[17] From W. F. Stoecker and J. W. Jones, *Refrigeration and Air Conditioning*, 2nd ed. © 1982 McGraw-Hill, Inc. Reprinted by permission.

distillation process. In a water–lithium bromide system, the generator is able to provide adequate distillation because the absorbent material (lithium bromide) is nonvolatile. In ammonia absorption systems, the absorbent (water) is volatile and tends to carry over into the evaporator, where it interferes with vaporization. This problem is overcome by adding a rectifier to purify the ammonia vapor flowing from the generator to the condenser.

A second difference between ammonia–water and water–lithium bromide systems are the operating pressures. In a water–lithium bromide system, evaporating pressures as low as 4–8 kPa are not unusual for the production of chilled water at 5–7°C. In contrast, an ammonia-absorption system would run evaporator pressures of between 400 and 500 kPa.

A third difference focuses on the type of heat transfer medium used in the condenser and absorber. Most lithium–bromide systems utilize water cooling in the condenser and absorber, while commercial ammonia systems use air cooling.

62.6 STEAM JET REFRIGERATION

Steam jet refrigeration represents yet another variation of the standard vapor compression cycle. Water is the refrigerant, so very large volumes of low-pressure (~1 kPa absolute) vapor must be compressed. A steam jet ejector offers a simple, inexpensive, but inefficient alternative to large centrifugal compressors required for systems of even moderate cooling capacity: 54 liter/sec of water vapor must be handled per kW of refrigeration at evaporator temperatures of 7°C.

The evaporator vessel should have a large surface area to enhance evaporative cooling. Sprays or cascades of water in sheets may be used. Because condenser pressure is subatmospheric (~7.6 kPa absolute), leakage of air into the system can cause poor condenser performance, so a small two-stage ejector is commonly used to remove the noncondensable vapors from the condenser. The condenser must condense not only the water vapor generated by the evaporator cooling load, but also the steam from the ejector primary flow nozzle. The condenser rejects two to three times the amount of heat that a mechanical vapor compression cycle would require.

Steam jet refrigeration systems are available in 35–3500 kW$_t$ capacities. Steam jet refrigeration can be used in process applications where direct vaporization can be used for concentration or drying of foods and chemicals. The cooling produced by the vaporization reduces the processing temperature and helps to preserve the product. No heat exchanger or indirect heat-transfer process is required, making the steam jet system more suitable than mechanical refrigeration for these applications. Examples are concentration of fruit juices, freeze-drying of foods, dehydration of pharmaceuticals, and chilling of leafy vegetables. When applied to process or batch applications such as these, the noncondensables ejector for the condenser must be large enough to obtain the system evacuation rate desired.

62.7 INDIRECT REFRIGERATION

The process fluid is cooled by an intermediate liquid, water or brine, which is itself cooled by evaporating the refrigerant, as shown in Fig. 62.8. Process heat exchangers that must be designed for corrosive products, high pressures, or high viscosities usually are not well suited for refrigerant

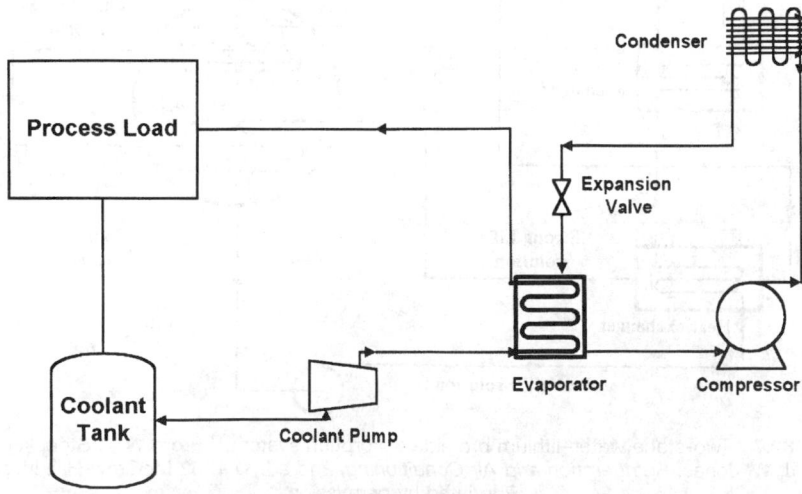

Fig. 62.8 Secondary coolant refrigeration system.

evaporators. Other problems preventing direct use of refrigerant are remote location, lack of sufficient pressures for the refrigerant liquid feed, difficulties with oil return, or inability to provide traps in the suction line to hold liquid refrigerant. Use of indirect refrigeration simplifies the piping system; it becomes a conventional hydraulic system.

The secondary coolant (brine) is cooled in the refrigeration evaporator and then is pumped to the process load. The brine system may include a tank, either open or closed but maintained at atmospheric pressure through a small vent pipe at the top, or may be a closed system pressurized by an inert, dry gas.

Secondary coolants can be broken into four categories:

1. *Coolants with a salt base.* These are water solutions of various concentrations and include the most common brines, that is, calcium chloride and sodium chloride.
2. *Coolants with a glycol base.* These are water solutions of various concentrations, most commonly ethylene glycol or propylene glycol.
3. *Coolants with an alcohol base.* Where low temperatures are not required, the alcohols are occasionally used in alcohol–water solutions.
4. *Coolants for low-temperature heat transfer.* These usually are pure substances such as methylene chloride, trichloroethylene, R-11, acetone, and methanol.

Coolants containing a mixture of calcium and sodium chloride are the most common refrigeration brines. These are applied primarily in industrial refrigeration and skating rinks. Glycols are used to lower the freezing point of water and used extensively as heat-transfer media in cooling systems. Low-temperature coolants include some common refrigerants (R-11, R-30, and R-1120). Alcohols and other secondary refrigerants, such as d-limonene ($C_{10}H_{16}$), are primarily used by the chemical processing and pharmaceutical industries.

A coolant needs to be compatible with other materials in the system where it is applied. It should have a minimum freezing point approximately 8°C below the lowest temperature to which it is exposed.[1] Table 62.5 shows a performance comparison of different types of coolants. Some coolants, such as the salts, glycols, and alcohols, are mixed with water to lower the freezing point of water. Different concentrations than listed in Table 62.5 will result in different freezing temperatures. The flow rate divided by capacity gives a way to compare the amount of flow (L/s) that will be needed

Table 62.5 Secondary Coolant Performance Comparisons[a]

Secondary Coolant	Concentration (by Weight), %	Freezing Point (°F)	Flow Rate/ Capacity (L/(s · kW))[b]	Heat Transfer Factor[c]	Energy Factor[d]
Salts					
calcium chloride	22	−22.1	0.0500	2.761	1.447
sodium chloride	23	−20.6	0.0459	2.722	1.295
Glycols					
propylene glycol	39	−20.6	0.0459	1.000	1.142
ethylene glycol	38	−21.6	0.0495	1.981	1.250
Alcohols					
methanol	26	−20.7	0.0468	2.307	1.078
Low-Temperature Fluids					
methylene chloride (R-30)	100	−96.7	0.1146	2.854	3.735
trichlorethylene (R-1120)	100	−86.1	0.1334	2.107	4.787
trichlorofluoromethane (R-11)	100	−111.1	0.1364	2.088	5.022
d-limonene	100	−96.7	0.1160	1.566	2.406

[a]Reference 18. Reprinted with permission of American Society of Heating, Refrigerating and Air Conditioning Engineers from *ASHRAE Handbook of HVAC Systems and Equipment.*
[b]Based on inlet secondary coolant temperature at the pmp of 25°F.
[c]Based on a curve fit of the Sieder & Tate heat transfer equation values using a 27-mm ID tube 4.9 m long and a film temperature of 2.8°C lower than the average bulk temperature with a 2.134 m/s velocity. The actual ID and length vary according to the specific loading and refrigerant applied with each secondary coolant, tube material, and surface augmentation.
[d]Based on the same pump head, refrigeration load, 20°F average temperature, 10°F range, and the freezing point (for water-based secondary coolants) 20–23°F below the lowest secondary coolant temperature.

to produce a kilowatt of cooling. The low-temperature coolants have the highest flow requirements of the four types of coolants. The heat transfer factor is a value normalized to propylene glycol. It is based on calculations inside a smooth tube. The salt mixtures and R-30 provide the highest heat-transfer factors of the fluids listed. The energy factor is a measure of the pumping requirements that will be needed for each of the coolants. The low temperature fluids require the largest pumping requirements.

Table 62.6 shows the general areas of application for the commonly used brines. Criteria for selection are discussed in the following paragraphs. The order of importance depends on the specific application.

Corrosion problems with sodium chloride and calcium chloride brines limit their use. When properly maintained in a neutral condition and protected with inhibitors, they will give 20 to 30 years of service without corrosive destruction of a closed system. Glycol solutions and alcohol–water solutions are generally less corrosive than salt brines, but they require inhibitors to suit the specific application for maximum corrosion protection. Methylene chloride, trichloroethylene, and trichloro-fluoromethane do not show general corrosive tendencies unless they become contaminated with im-purities such as moisture. However, methylene chloride and trichloroethylene must not be used with aluminum or zinc; they also attack most rubber compounds and plastics. Alcohol in high concentra-tions will attack aluminum. Reaction with aluminum is of concern because, in the event of leakage into the refrigeration compressor system, aluminum compressor parts will be attacked.

Toxicity is an important consideration in connection with exposure to some products and to op-erating personnel. Where brine liquid, droplets, or vapor may contact food products, as in an open spray-type system, sodium chloride and propylene glycol solutions are acceptable because of low toxicity. All other secondary coolants are toxic to some extent or produce odors that require that they be used only inside of pipe coils or a similar pressure-tight barrier.

Flash-point and explosive-mixture properties of some coolants require precautions against fire or explosion. Acetone, methanol, and ethanol are in this category but are less dangerous when used in closed systems.

Specific heat of a coolant determines the mass rate of flow that must be pumped to handle the cooling load for a given temperature rise. The low-temperature coolants, such as trichloroethylene, methylene chloride, and trichlorofluoromethane, have specific heats approximately one-third to one-fourth those of the water-soluble brines. Consequently, a significantly greater mass of the low-temperature brines must be pumped to achieve the same temperature change.

Stability at high temperatures is important where a brine may be heated as well as cooled. Above 60°C, methylene chloride may break down to form acid products. Trichloroethylene can reach 120°C before breakdown begins.

Viscosities of brines vary greatly. The viscosity of propylene gycol solutions, for example, makes them impractical for use below −7°C because of the high pumping costs and the low heat-transfer coefficient at the concentration required to prevent freezing. Mixtures of ethanol and water can become highly viscous at temperatures near their freezing points, but 190-proof ethyl alcohol has a low viscosity at all temperatures down to near the freezing point. Similarly, methylene chloride and R-11 have low viscosities down to −73°C. In this region, the viscosity of acetone is even more favorable.

Table 62.6 Application Information for Common Secondary Coolants[3,18]

Secondary Coolant	Toxic	Explosive	Corrosive
Salts			
calcium chloride	no	no	yes
sodium chloride	no	no	yes
Glycols			
propylene	no	no	some
ethanol	yes	no	some
Alcohols			
methanol	yes	yes	some
ethanol	yes	yes	some
Low-temperature fluids			
methylene chloride (R-30)	no	no	no
trichloroethylene (R-1120)	no	no	no
trichlorofluoromethane (R-11)	no	no	no
d-limonene	yes	yes	yes

Since a secondary coolant cannot be used below its freezing point, certain ones are not applicable at the lower temperatures. Sodium chloride's eutectic freezing point of −20°C limits its use to approximately −12°C. The eutectic freezing point of calcium chloride is −53°C, but achieving this limit requires such an accuracy of mixture that −40°C is a practical low limit of usage.

Water solubility in any open or semi-open system can be important. The dilution of a salt or glycol brine, or of alcohol by entering moisture, merely necessitates strengthening of the brine. But for a brine that is not water-soluble, such as trichloroethylene or methylene chloride, precautions must be taken to prevent free water from freezing on the surfaces of the heat exchanger. This may require provision for dehydration or periodic mechanical removal of ice, perhaps accompanied by replacement with fresh brine.

Vapor pressure is an important consideration for coolants that will be used in open systems, especially where it may be allowed to warm to room temperature between periods of operation. It may be necessary to pressurize such systems during periods of moderate temperature operation. For example, at 0°C the vapor pressure of R-11 is 39.9 kPa (299 mm Hg); that of a 22% solution of calcium chloride is only 0.49 kPa (3.7 mm Hg). The cost of vapor losses, the toxicity of the escaping vapors, and their flammability should be carefully considered in the design of the semiclosed or open system.

Environmental effects are important in the consideration of trichlorofluoromethane (R-11) and other chlorofluorocarbons. This is a refrigerant with a high ozone-depletion potential and halocarbon global warming potential. The environmental effect of each of the coolants should be reviewed before the use of it in a system is seriously considered.

Energy requirements of brine systems may be greater because of the power required to circulate the brine and because of the extra heat-transfer process, which necessitates the maintenance of a lower evaporator temperature.

62.7.1 Use of Ice

Where water is not harmful to a product or process, ice may be used to provide refrigeration. Direct application of ice or of ice and water is a rapid way to control a chemical reaction or remove heat from a process. The rapid melting of ice furnishes large amounts of refrigeration in a short time and allows leveling out of the refrigeration capacity required for batch processes. This stored refrigeration also is desirable in some processes where cooling is critical from the standpoint of safety or serious product spoilage.

Large ice plants, such as the block-ice plants built during the 1930s, are not being built today. However, ice still is used extensively, and equipment to make flake or cube ice at the point of use is commonly employed. This method avoids the loss of crushing and minimizes transportation costs.

62.8 SYSTEM COMPONENTS

There are four major components in any refrigeration system: compressor, condenser, evaporator, and expansion device. Each is discussed below.

62.8.1 Compressors

Both positive-displacement and centrifugal compressors are used in refrigeration applications. With positive-displacement compressors, the pressure of the vapor entering the compressor is increased by decreasing the volume of the compression chamber. Reciprocating, rotary, scroll, and screw compressors are examples of positive displacement compressors. Centrifugal compressors utilize centrifugal forces to increase the pressure of the refrigerant vapor. Refrigeration compressors can be used alone, in parallel, or in series combinations. Features of different compressors are described in this section.

Reciprocating Compressors

Modern high-speed reciprocating compressors with displacements up to 0.283–0.472 M³/sec (600–1000 cfm) generally are limited to a pressure ratio of about 9. The reciprocating compressor is basically a constant-volume variable-head machine. It handles various discharge pressures with relatively small changes in inlet volume flow rate, as shown by the heavy line in Fig. 62.9.

Open systems and many processes require nearly fixed compressor suction and discharge pressure levels. This load characteristic is represented by the horizontal typical open-system line in Fig. 62.9. In contrast, condenser operation in many closed systems is related to ambient conditions. For example, through cooling towers, the condenser pressure can be reduced as the outdoor temperature decreases. When the refrigeration load is lower, less refrigerant circulation is required. The resulting load characteristic is represented by the typical closed-system line in Fig. 62.9.

The compressor must be capable of matching the pressure and flow requirements imposed upon it by the system in which it operates. The reciprocating compressor matches the imposed discharge pressure at any level up to its limiting pressure ratio. Varying capacity requirements can be met by providing devices that unload individual or multiple cylinders. This unloading is accomplished by

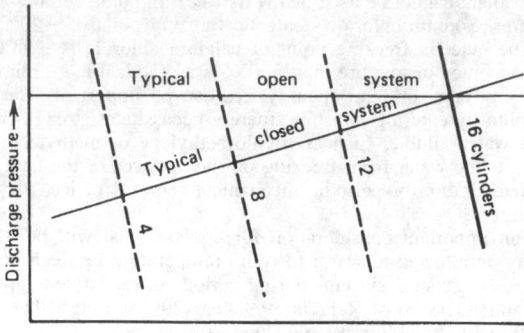

Fig. 62.9   Volume–pressure relationships for a reciprocating compressor.

blocking the suction or discharge valves that open either manually or automatically. Capacity can also be controlled through the use of variable speed or multispeed motors. When capacity control is implemented on a compressor, other factors at part-load conditions need to considered, such as effect on compressor vibration and sound when unloaders are used, the need for good oil return because of lower refrigerant velocities, and proper functioning of expansion devices at the lower capacities.

Most reciprocating compressors have a lubricated design. Oil is pumped into the refrigeration system during operation. Systems must be designed carefully to return oil to the compressor crankcase to provide for continuous lubrication and also to avoid contaminating heat-exchanger surfaces. At very low temperatures ($\sim -50°C$ or lower, depending on refrigerant used), oil becomes too viscous to return, and provision must be made for periodic plant shutdown and warmup to allow manual transfer of the oil.

Compressors usually are arranged to start unloaded so that normal torque motors are adequate for starting. When gas engines are used for reciprocating compressor drives, careful torsional analysis is essential.

Rotary Compressors

Rotary compressors include both rolling-piston and rotary-vane compressors. Rotary-vane compressors are primarily used in transportation air conditioning applications, while rolling-piston compressors are usually found in household refrigerators and small air conditioners up to inputs of 2 kW. Figure 62.10 shows the operation of a fixed-vane, rolling-piston rotary compressor.[8] The shaft is located in the center of the housing, while the roller is mounted on an eccentric. At position 1 of Fig. 62.10, the volume in chamber A is at its maximum. Suction gas enters directly into the suction port. As the roller rotates, the refrigerant vapor is compressed and is discharged into the compressor housing through the discharge valve.

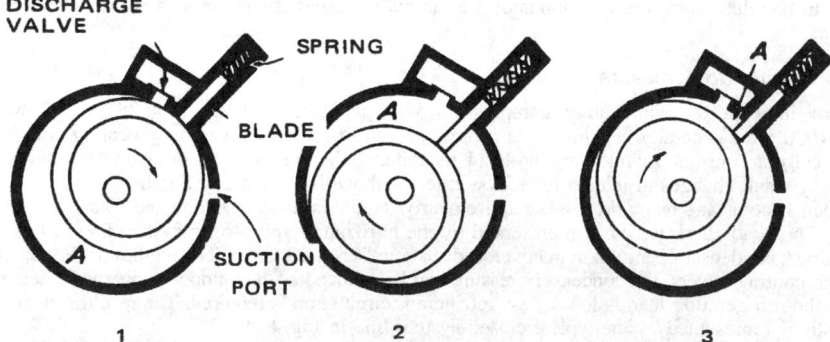

Fig. 62.10   Operation of a fixed-vane, rolling-piston rotary compressor.[8] (Courtesy of Business News Publishing Co.)

One difference between a rotary and reciprocating compressor is that the rotary is able to obtain a better vacuum during suction.[18] It has low re-expansion losses because there is no high-pressure discharge vapor present during suction, as with a reciprocating compressor.

Scroll Compressor

The principle of the scroll compressor was first patented in 1905.[19] However, the first commercial units were not built until the early 1980s.[20] Scroll compressors are primarily used in air conditioning and heat pump applications and some limited refrigeration applications. They range in capacity from 3–50 kW$_r$. Scroll compressors have two spiral-shaped scroll members that are assembled 180° out of phase (Fig. 62.11). One scroll is fixed while the other "orbits" the first. Vapor is compressed by sealing vapor off at the edge of the scrolls and reducing the volume of the gas as it moves inward toward the discharge port. Figure 62.11a shows the two scrolls at the instant that vapor has entered the compressor and compression begins. The orbiting motion of the second scroll forces the pocket of vapor toward the discharge port while decreasing its volume (Figs. 62.11b–62.11h). In Figs. 62.11c and 62.11f, the two scrolls open at the ends and allow new pockets of vapor to be admitted into the scrolls for compression. Compression is a nearly continuous process in a scroll compressor.

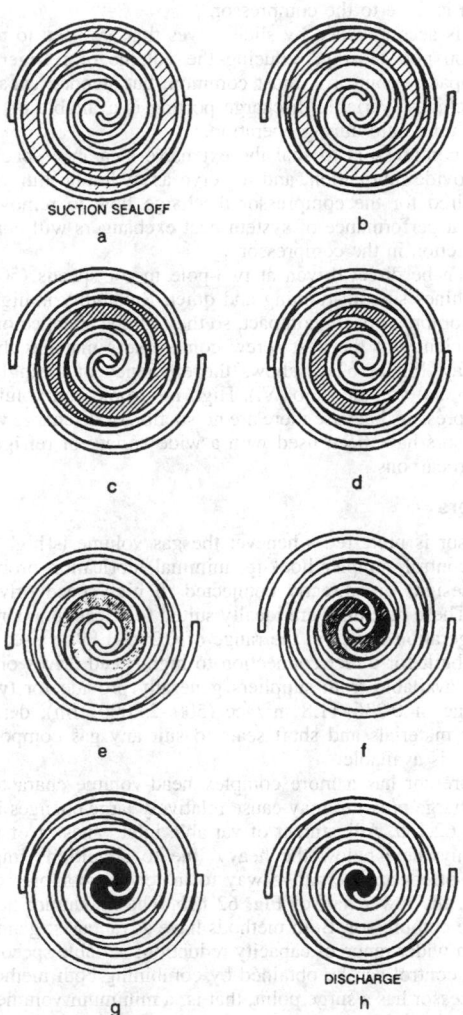

Fig. 62.11 Operation of the fixed and orbiting scrolls in a scroll compressor.[18]
See Table 62.5.

Scroll compressors offer several advantages over reciprocating compressors. First, relatively large suction and discharge ports can be used to reduce pressure losses. Second, the separation of the suction and discharge processes reduces the heat transfer between the discharge and suction processes. Third, with no valves and re-expansion losses, they have higher volumetric efficiencies. Capacities of systems with scroll compressors can be varied by the use of a variable-speed motors or of multiple suction ports at different locations within the two spiral members.

Screw Compressors

Screw compressors were first introduced in 1958.[2] These are positive displacement machines available in the capacity range from 15–1100 kW,, overlapping reciprocating compressors for lower capacities and centrifugal compressors for higher capacities. Both twin-screw and single-screw compressors are used in refrigeration duty.

Fixed suction and discharge ports, used instead of valves in reciprocating compressors, set the "built-in volume ratio" of the screw compressor. This is the ratio of the volume of fluid space in the meshing rotors at the beginning of the compression process to the volume in the rotors as the discharge port is first exposed. Associated with the built-in volume ratio is a pressure ratio that depends on the properties of the refrigerant being compressed. Peak efficiency is obtained if the discharge pressure imposed by the system matches the pressure developed by the rotors when the discharge port is exposed. If the interlobe pressure is greater or less than discharge pressure, energy losses occur but no harm is done to the compressor.

Capacity modulation is accomplished by slide valves that are used to provide a variable suction bypass or delayed suction port closing, reducing the volume of refrigerant actually compressed. Continuously variable capacity control is most common, but stepped capacity control is offered in some manufacturers' machines. Variable discharge porting is available on a few machines to allow control of the built-in volume ratio during operation.

Oil is used in screw compressors to seal the extensive clearance spaces between the rotors, to cool the machines, to provide lubrication, and to serve as hydraulic fluid for the capacity controls. An oil separator is required for the compressor discharge flow to remove the oil from the high-pressure refrigerant so that performance of system heat exchangers will not be penalized and the oil can be returned for reinjection in the compressor.

Screw compressors can be direct driven at two-pole motor speeds (50 or 60 Hz). Their rotary motion makes these machines smooth-running and quiet. Reliability is high when the machines are applied properly. Screw compressors are compact, so they can be changed out readily for replacement or maintenance. The efficiency of the best screw compressors matches that of reciprocating compressors at full load today. Figure 62.12 shows the efficiency of a single-screw compressor as a function of pressure ratio and volume ratio (Vi). High isentropic and volumetric efficiencies can be achieved with screw compressors because there are no suction or discharge valves and small clearance volumes. Screw compressors have been used with a wide variety of refrigerants, including halocarbons, ammonia, and hydrocarbons.

Centrifugal Compressors

The centrifugal compressor is preferred whenever the gas volume is high enough to allow its use, because it offers better control, simpler hookup, minimal lubrication problems, and lower maintenance. Single-impeller designs are directly connected to high-speed drives or driven through an internal speed increaser. These machines are ideally suited for clean, noncorrosive gases in moderate-pressure process or refrigeration cycles in the range of 0.236–1.89 m^3/sec (5 cfm). Multistage centrifugal compressors are built for direct connection to high-speed drives or for use with an external speed increaser. Designs available from suppliers generally provide for two to eight impellers per casing, covering the range of 0.236–11.8 m^3/sec (500–25,000 cfm), depending on the operating speed. A wide choice of materials and shaft seals to suit any gas composition, including dirty or corrosive process streams, is available.

The centrifugal compressor has a more complex head-volume characteristic than reciprocating machines. Changing discharge pressure may cause relatively large changes in inlet volume, as shown by the heavy line in Fig. 62.13a. Adjustment of variable inlet vanes or of a diffuser ring allows the compressor to operate anywhere below the heavy line to conditions imposed by the system. A variable-speed controller offers an alternative way to match the compressor's characteristics to the system load, as shown in the lower half of Fig. 62.13b. The maximum head capability is fixed by the operating speed of the compressor. Both methods have advantages: generally, variable inlet vanes or diffuser rings provide a wider range of capacity reduction; variable speed usually is more efficient. Maximum efficiency and control can be obtained by combining both methods of control.

The centrifugal compressor has a surge point, that is, a minimum-volume flow below which stable operation cannot be maintained. The percentage of load at which the surge point occurs depends on the number of impellers, design-pressure ratio, operating speed, and variable inlet-vane setting. The system design and controls must keep the inlet volume above this point by artificial loading, if necessary. This is accomplished with a bypass-valve-and-gas recirculation. Combined with a variable

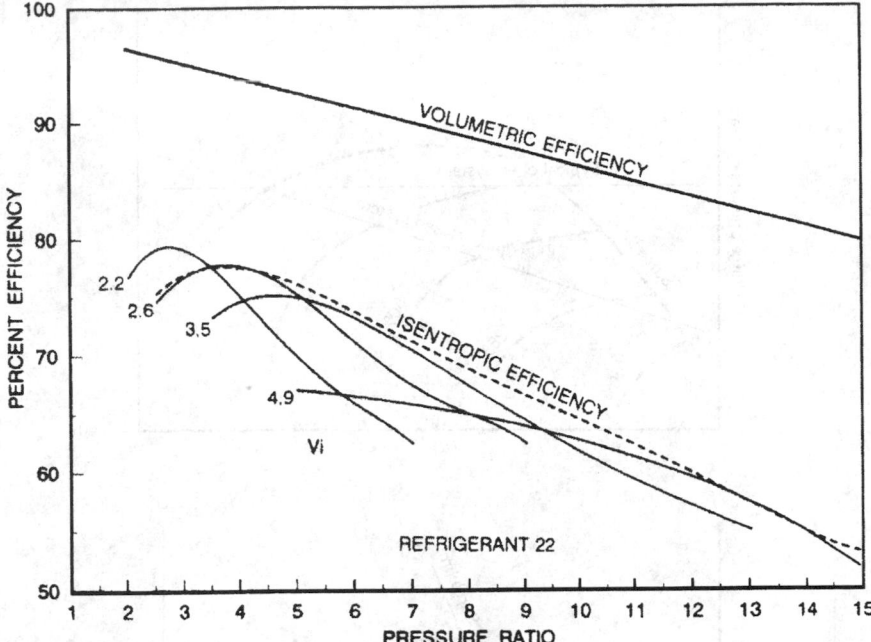

Fig. 62.12 Typical performance of a single-screw compressor.[18] See Table 62.5.

inlet-vane setting, variable diffuser ring, or variable speed control, the gas bypass allows stable operation down to zero load.

Compressor Operation

Provision for minimum load operation is strongly recommended for all installations because there will be fluctuations in plant load. For chemical plants, this permits the refrigeration system to be started up and thoroughly checked out independently of the chemical process.

Contrast between the operating characteristics of the positive displacement compressor and the centrifugal compressor is an important consideration in plant design to achieve satisfactory performance. Unlike positive displacement compressors, the centrifugal compressor will not rebalance abnormally high system heads. The drive arrangement for the centrifugal compressor must be selected with sufficient speed to meet the maximum head anticipated. The relatively flat head characteristics of the centrifugal compressor necessitates different control approaches than for positive displacement machines, particularly when parallel compressors are utilized. These differences, which account for most of the troubles experienced in centrifugal-compressor systems, cannot be overlooked in the design of a refrigeration system.

A system that uses centrifugal compressors designed for high pressure ratios and that requires the compressors to start with high suction density existing during standby will have high starting torque. If the driver does not have sufficient starting torque, the system must have provisions to reduce the suction pressure at startup. This problem is particularly important when using single-shaft gas turbine engines, or reduced-voltage starters on electric drives. Split-shaft gas turbines are preferred for this reason.

Drive ratings that are affected by ambient temperatures, altitudes, and so on, must be evaluated at the actual operating conditions. Refrigeration installations normally require maximum output at high ambient temperatures, a factor that must be considered when using drives such as gas turbines and gas engines.

62.8.2 Condensers

The refrigerant condenser is used to reject the heat of compression and the heat load picked up in the evaporator. This heat can be rejected to cooling water or air, both of which are commonly used.

The heat of compression depends on the compressor horsepower and becomes a significant part of the load on low-temperature systems affecting the size of condensers. Water-cooled shell-and-tube condensers designed with finned tubes and fixed tube sheets generally provide the most economical

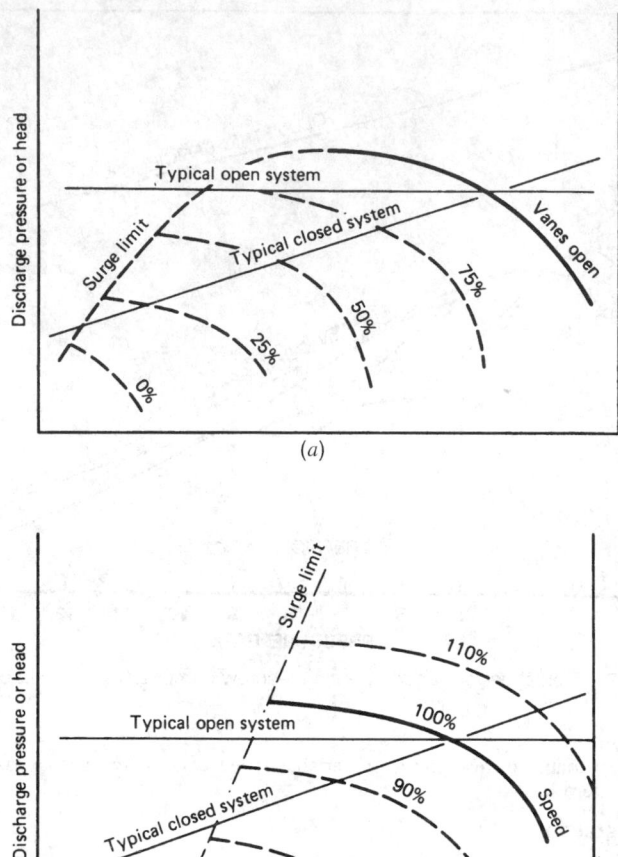

Fig. 62.13 Volume–pressure relationships in a centrifugal compressor: (a) with variable inlet-vane control at constant rotational speed; (b) with variable speed control at a constant inlet-vane opening.

exchanger design for refrigerant use. Figure 62.14 shows a typical refrigerant condenser. Commercially available condensers conforming to ASME Boiler and Pressure Vessel Code[21] construction adequately meet both construction and safety requirements for this duty.

Cooling towers and spray ponds are frequently used for water-cooling systems. These generally are sized to provide 29°C supply water at design load conditions. Circulation rates typically are specified so that design cooling loads are handled with a 5.6°C cooling-water temperature rise. Pump power, tower fans, makeup water (about 3% of the flow rate), and water treatment should be taken into account in operating cost studies. Water temperatures, which control condensing pressure, may have to be maintained above a minimum value to ensure proper refrigerant liquid feeding to all parts of the system.

River or well water, when available, provides an economical cooling medium. Quantities circulated will depend on initial supply temperatures and pumping cost, but are generally selected to handle the cooling load with 8.3–16.6°C water-temperature range. Water treatment and special exchanger materials frequently are necessary because of the corrosive and scale-forming characteristics of the water. Well water, in particular, must be analyzed for corrosive properties, with special attention

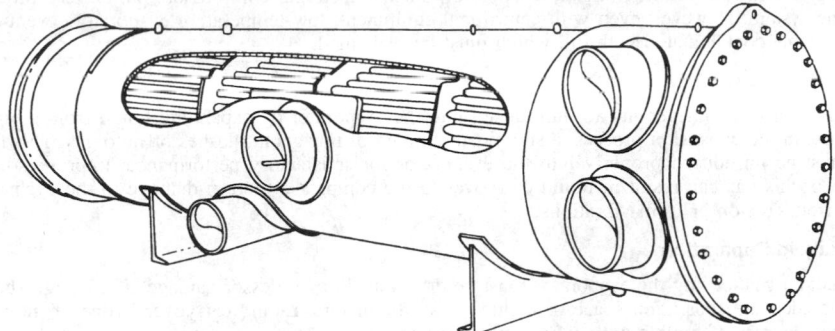

Fig. 62.14 Typical shell-in-tube refrigerant condenser.[3]

given to the presence of dissolved gases, such as H_2S and CO_2. These are extremely corrosive to many exchanger materials, yet difficult to detect in sampling. Pump power, water treatment, and special condenser material should be evaluated when considering costs.

Allowances must be made in heat-transfer calculations for fouling or scaling of exchanger surfaces during operation. This ensures sufficient surface to maintain rated performance over a reasonable interval of time between cleanings. Scale-factor allowances are expressed in $m^2 \cdot K/kW$ as additional thermal resistance.

Commercial practice normally includes a scale-factor allowance of 0.088. However, the long hours of operation usually associated with chemical-plant service and the type of cooling water frequently encountered generally justify a greater allowance to minimize the frequency of downtime for cleaning. Depending on these conditions, an allowance of 0.18 or 0.35 is recommended for chemical-plant service. Scale allowance can be reflected in system designs in two ways—as more heat-exchanger surface or as higher design condensing temperatures with attendant increase in compressor power. Generally, a compromise between these two approaches is most economical. For extremely bad water, parallel condensers, each with 60–100% capacity, may provide a more economical selection and permit cleaning one exchanger while the system is operating.

Use of air-cooled condensing equipment is on the increase. With tighter restrictions on the use of water, air-cooled equipment is used even on larger centrifugal-type refrigeration plants, although it requires more physical space than cooling towers. A battery of air-cooled with propeller fans located at the top, pull air over the condensing coil. Circulating fans and exchanger surface are usually selected to provide design condensing temperatures of 49–60°C with 35–38°C ambient dry bulb temperature.

The design dry bulb temperature should be carefully considered, since most weather data reflect an average or mean maximum temperature. If full load operation must be maintained at all times, care should be taken to provide sufficient condenser capacity for the maximum recorded temperature. This is particularly important when the compressor is centrifugal because of its flat-head character-istics and the need for adequate speed. Multiple-circuit or parallel air-cooled condensers must be provided with traps to prevent liquid backup into the idle circuit at light load. Pressure drop through the condenser coil must also be considered in establishing the compressor discharge pressure.

In comparing water-cooled and air-cooled condensers, the compression horsepower at design conditions is invariably higher with air-cooled condensing. However, ambient air temperatures are considerably below the design temperature most of the time, and operating costs frequently compare favorably over a full year. In addition, air-cooled condensers usually require less maintenance, al-though dirty or dusty atmospheres may affect this.

62.8.3 Evaporators

There are special requirements for evaporators in refrigeration service that are not always present in other types of heat-exchanger design. These include problems of oil return, flash-gas distribution, gas liquid separation, and submergence effects.

Oil Return

When the evaporator is used with reciprocating-compression equipment, it is necessary to ensure adequate oil return from the evaporator. If oil will not return in the refrigerant flow, it is necessary to provide an oil reservoir for the compression equipment and to remove oil mechanically from the low side of the system on a regular basis. Evaporators used with centrifugal compressors do not normally require oil return from the evaporator, since centrifugal compressors pump very little oil into the system. However, even with centrifugal equipment, low-temperature evaporators eventually may become contaminated with oil, which must be reclaimed.

Flash-Gas Distribution

As a general rule, refrigerants are introduced into the evaporator by expanding liquid from a higher pressure. In the expansion process, a significant amount of refrigerant flashes off into gas. This flash gas must be introduced properly into the evaporator for satisfactory performance. Improper distribution of this gas can result in liquid carryover to the compressor and in damage to the exchanger tubes from erosion or from vibrations.

Gas–Liquid Separation

The suction gas leaving the evaporator must be dry to avoid compressor damage. The design should provide adequate separation space or include mist eliminators. Liquid carryover is one of the most common sources of trouble with refrigeration systems.

Submergence Effect

In flooded evaporators, the evaporating pressure and temperature at the bottom of the exchanger surface are higher than at the top of the exchanger surface, owing to the liquid head. This static head or submergence effect significantly affects the performance of refrigeration evaporators operating at extremely low temperatures and low suction pressures.

Beyond these basic refrigeration-design requirements, the chemical industry imposes many special conditions. Exchangers frequently are applied to cool highly corrosive process streams. Consequently, special materials for evaporator tubes and channels of particularly heavy wall thickness are dictated. Corrosion allowances, that is, added material thicknesses, in evaporator design may be necessary in chemical service.

High-pressure and high-temperature design, particularly on the process side of refrigerant evaporators, is frequently encountered in chemical-plant service. Process-side construction may have to be suitable for pressures seldom encountered in commercial service, and differences between process inlet and leaving temperatures greater than 55°C are not uncommon. In such cases, special consideration must be given to thermal stresses within the refrigerant evaporator. U-tube construction or floating-tube-sheet construction may be necessary. Minor process-side modifications may permit use of less expensive standard commercial fixed-tube-sheet designs. However, coordination between the equipment supplier and chemical-plant designer is necessary to tailor the evaporator to the intended duty. Relief devices and safety precautions common to the refrigeration field normally meet chemical-plant needs but should be reviewed against individual plant standards. It must be the mutual responsibility of the refrigeration equipment supplier and the chemical-plant designer to evaluate what special features, if any, must be applied to modify commercial equipment for chemical-plant service.

Refrigeration evaporators are usually designed to meet the ASME Boiler and Pressure Vessel Code,[21] which provides for a safe reliable exchanger at economical cost. In refrigeration systems, these exchangers generally operate with relatively small temperature differentials, for which fixed-tube-sheet construction is preferred. Refrigerant evaporators also operate with simultaneous reduction in pressure as temperatures are reduced. This relationship results in extremely high factors of safety on pressure stresses, eliminating the need for expensive nickel steels from −59 to −29°C. Most designs are readily modified to provide suitable materials for corrosion problems on the process side.

The basic shell-and-tube heat exchanger with fixed tube sheets (Fig. 62.15) is most widely used for refrigeration evaporators. Most designs are suitable for fluids up to 2170 kPa (300 psig) and for operation with up to 38°C temperature differences. Above these limits, specialized heat exchangers generally are used to suit individual requirements.

With the fluid on the tube side, the shell side is flooded with refrigerant for efficient wetting of the tubes (see Fig. 62.16). Designs must provide for distribution of flash gas and liquid refrigerant entering the shell and for separation of liquid from the gas leaving the shell before it reaches the compressor.

In low-temperature applications and large evaporators, the exchanger surface may be sprayed rather than flooded. This eliminates the submergence effect or static-head penalty, which can be significant in large exchangers, particularly at low temperatures. The spray cooler (Fig. 62.17) is recommended for some large coolers to offset the cost of refrigerant inventory or charge that would be necessary for flooding.

Where the Reynolds number in the process fluid is low, as for a viscous or heavy brine, it may be desirable to handle the fluid on the shell side to obtain better heat transfer. In these cases, the refrigerant must be evaporated in the tubes. On small exchangers, commonly referred to as *direct-expansion coolers*, refrigerant feeding is generally handled with a thermal-expansion valve.

On large exchangers, this can best be handled by a small circulating pump to ensure adequate wetting of all tubes (Fig. 62.18). An oversize channel box on one end provides space for a liquid reservoir and for effective liquid–gas separation.

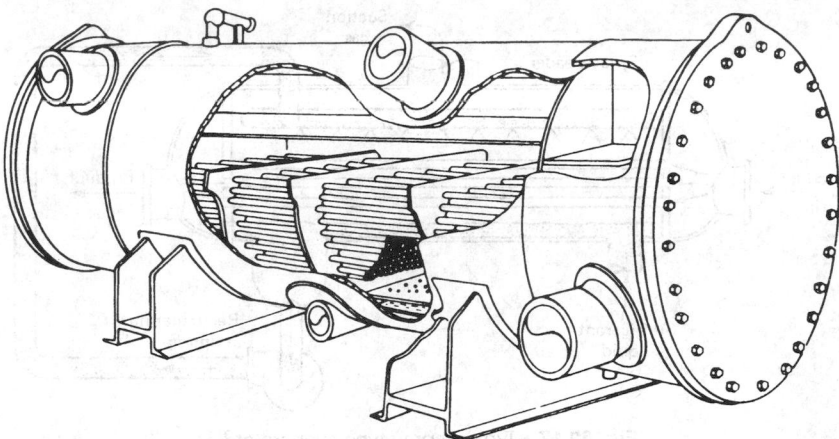

Fig. 62.15 Typical fixed-tube-sheet evaporator.[3]

62.8.4 Expansion Devices

The primary purpose of an expansion device is to control the amount of refrigerant entering the evaporator. In the process, the refrigerant entering the valve expands from a relatively high-pressure subcooled liquid to a saturated low-pressure mixture. Other types of flow-control devices, such as pressure regulators and float valves, can also be found in some refrigeration systems. Discussion of these can be found in Ref. 1. Five types of expansion devices can be found in refrigeration systems: (1) thermostatic expansion valves, (2) electronic expansion valves, (3) constant pressure expansion valves, (4) capillary tubes, and (5) short-tube restrictors. Each is discussed briefly below.

Thermostatic Expansion Valve

The thermostatic expansion valve (TXV) uses the superheat of the gas leaving the evaporator to control the refrigerant flow into the evaporator. Its primary function is to provide superheated vapor to the suction of the compressor. A TXV is mounted near the entrance to the evaporator and has a capillary tube extending from its top that is connected to a small bulb (Fig. 62.19). The bulb is mounted on the refrigerant tubing near the evaporator outlet. The capillary tube and bulb is filled with a substance called the *thermostatic charge*.[1] This charge often consists of a vapor or liquid that is the same substance as the refrigerant used in the system. The response of the TXV and the superheat setting can be adjusted by varying the type of charge in the capillary tube and bulb.

The operation of a TXV is straightforward. Liquid enters the TXV and expands to a mixture of liquid and vapor at pressure P_2. The refrigerant evaporates as it travels through the evaporator and reaches the outlet, where it is superheated. If the load on the evaporator is increased, the superheat leaving the evaporator will increase. This increase in superheat will increase the temperature and pressure (P_1) of the charge within the bulb and capillary tube. Within the top of the TXV is a

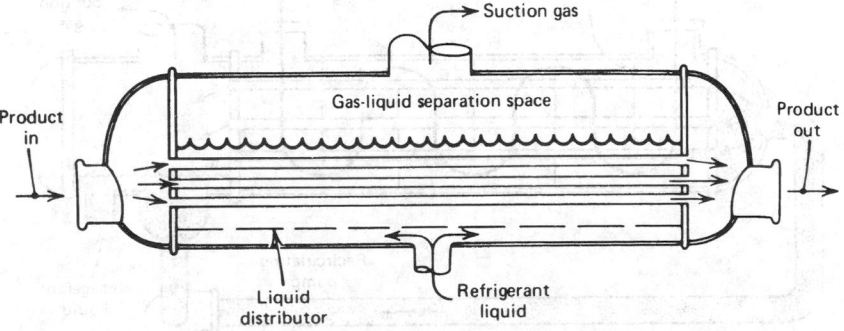

Fig. 62.16 Typical flooded shell-and-tube evaporator.[3]

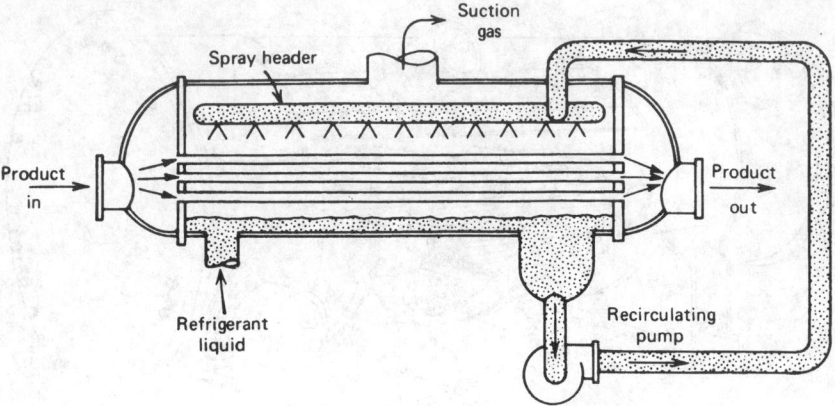

Fig. 62.17 Typical spray-type evaporator.[3]

diaphragm. With an increase in pressure of the thermostatic charge, a greater force is exerted on the diaphragm, which forces the valve port to open and allow more refrigerant into the evaporator. The larger refrigerant flow reduces the evaporator superheat back to the desired level.

The capacity of a TXV is determined on the basis of opening superheat values. TXV capacities are published for a range in evaporator temperatures and valve pressure drops. TXV ratings are based on liquid only entering the valve. The presence of flash gas will reduce the capacity substantially.

Electronic Expansion Valve

The electronic expansion valve (EEV) has become popular in recent years on larger or more expensive systems, where its cost can be justified. EEVs can be heat motor-activated, magnetically modulated, pulse width-modulated, and step motor-driven.[1] They can be used with digital control systems to provide control of the refrigeration system based on input variables from throughout the system.

Constant Pressure Expansion Valves

A constant pressure expansion valve controls the mass flow of the refrigerant entering the evaporator by maintaining a constant pressure in the evaporator. Its primary use is for applications where the refrigerant load is relatively constant. It is usually not applied where the refrigeration load may vary widely. Under these conditions, this expansion valve will provide too little flow to the evaporator at high loads and too much flow at low loads.

Capillary Tubes

Capillary tubes are used extensively in household refrigerators, freezers, and smaller air conditioners. The capillary tube consists of one or more small diameter tubes connecting the high-pressure liquid

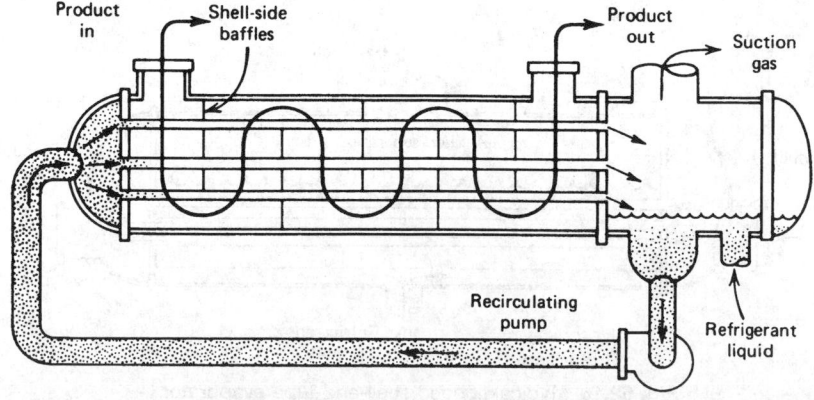

Fig. 62.18 Typical baffled-shell evaporator.[3]

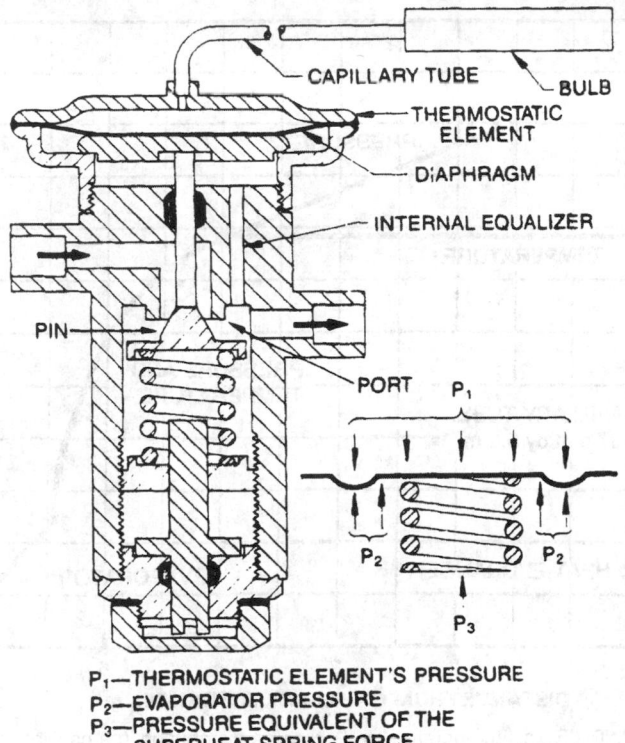

P₁—THERMOSTATIC ELEMENT'S PRESSURE
P₂—EVAPORATOR PRESSURE
P₃—PRESSURE EQUIVALENT OF THE
    SUPERHEAT SPRING FORCE

Fig. 62.19 Cross section of a thermal expansion valve.[1] Reprinted with permission of American Society of Heating, Refrigerating and Air Conditioning Engineers from *ASHRAE Handbook of Refrigeration Systems and Applications.*

line from the condenser to the inlet of the evaporator. Capillary tubes range in length from 1–6 m and diameters from 0.5–2 mm.[17]

After entering a capillary tube, the refrigerant remains a liquid for some length of the tube (Fig. 62.20). While the refrigerant is a liquid, the pressure drops, but the temperature remains relatively constant (from point 1 to 2 in Fig. 62.20). At point 2, the refrigerant enters into the saturation region, where a portion of the refrigerant begins to flash to vapor. The phase change accelerates the refrigerant and the pressure drops more rapidly. Because the mixture is saturated, its temperature drops with the pressure from 2 to 3. In many applications, the flow through a capillary tube is choked, which means that the mass flow through the tube is independent of downstream pressure.[17]

Because there are no moving parts to a capillary tube, it is not capable of making direct adjustments to variations in suction pressure or load. Thus, the capillary tube does not provide performance as good as TXVs when applied in systems that will experience a wide range in loads.

Even though the capillary tube is insensitive to changes in evaporator pressure, its flow rate will adjust to changes in the amount of refrigerant subcooling and condenser pressure. If the load in the condenser suddenly changes so that subcooled conditions are no longer maintained at the capillary tube inlet, the flow rate through the capillary tube will decrease. The decreased flow will produce an increase in condenser pressure and the refrigerant will achieve higher subcooling. The higher pressure and subcooling will increase the flow through the capillary tube.

The size of the compressor, evaporator, and condenser as well as the application (refrigerator or air conditioner) must all be considered when specifying the length and diameter of capillary tubes. Systems using capillary tubes are more sensitive to the amount of refrigerant charge than systems using TXVs or EEVs. Design charts for capillary tubes can be found in Ref. 1 for R-12 and R-22.

Short-Tube Restrictors

Short-tube restrictors are applied in many systems that formerly used capillary tubes. Figure 62.21 illustrates a short-tube restrictor and its housing. The restrictors are inexpensive, reliable, and easy to replace. In addition, for systems such as heat pumps that reverse cycle, short-tube restrictors eliminate the need for a check valve. Short tubes vary in length from 10–13 mm with a length-to-

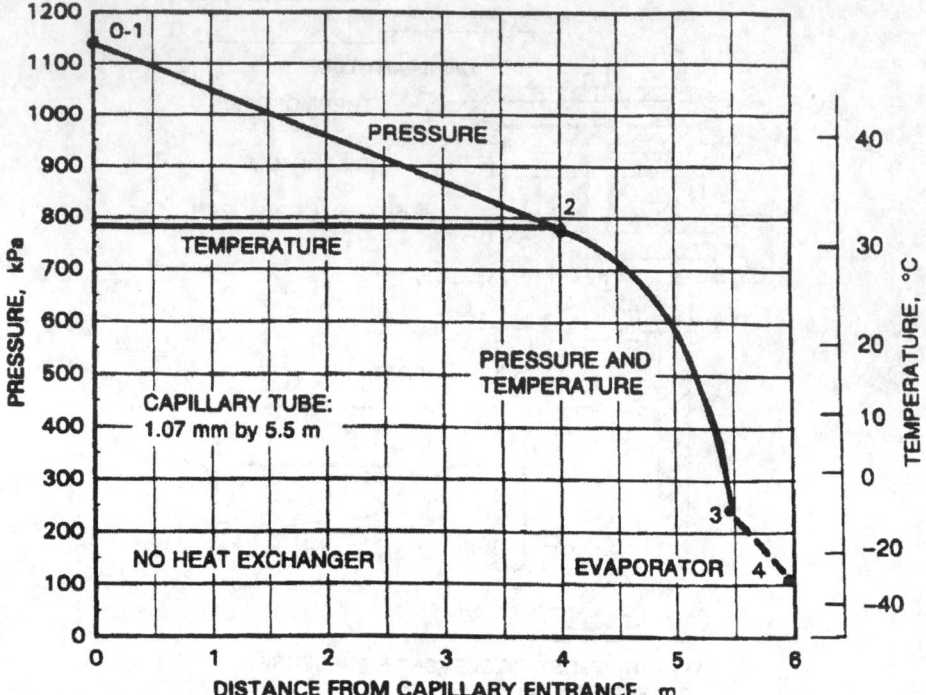

Fig. 62.20 Typical temperature and pressure distribution in a capillary tube.[1]
See Fig. 62.19.

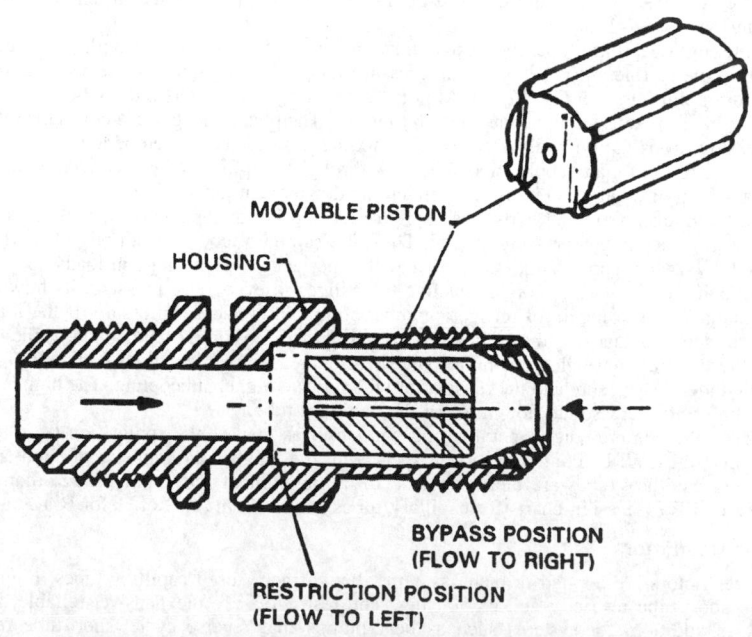

Fig. 62.21 Schematic of a short-tube restrictor.[1] See Fig. 62.19.

diameter ratio from 3 to 20.[1] Current applications for short-tube restrictors are primarily in air conditioners and heat pumps.

Like a capillary tube, short-tube restrictors operate with choked or near-choked flow in most applications.[22] The mass flow through the orifice is nearly independent of conditions downstream of the orifice. The flowrate does vary with changes in the condenser subcooling and pressure.

In applying short-tube restrictors, there are many similarities to capillary tubes. The size of the system components and type of system must be considered when sizing this expansion device. Sizing charts for the application of short-tube restrictors with R-22 can be found in Ref. 23.

62.9 DEFROST METHODS

When refrigeration systems operate below 0°C, frost can form on the heat-transfer surfaces of the evaporator. As frost grows, it begins to block the airflow passages and insulates the cold refrigerant from the warm, moist air that is being cooled by the refrigeration system. With increasing blockage of the airflow passages, the evaporator fan(s) are unable to maintain the design airflow through the evaporator. As airflow drops, the capacity of the system decreases and eventually degrades enough that the frost must be removed. This is accomplished with a defrost cycle.

Several defrost methods are used with refrigeration systems: hot refrigerant gas, air, and water. Each method can be used individually or in combination with the other.

62.9.1 Hot Refrigerant Gas Defrost

This method is the most common technique for defrosting commercial and industrial refrigeration systems. When the evaporator needs defrosting, hot gas from the discharge of the compressor can be diverted from the condenser to the evaporator by closing control valve number 2 and opening control valve number 1 in Fig. 62.22. The hot gas heats the evaporator and melts the frost. Some of the hot vapor condenses to liquid during the process. A special tank, such as an accumulator, can be used to protect the compressor from any liquid returning to the compressor.

During defrost operation, the evaporator fans are turned off. This allows the coil to reach higher temperatures faster, allows the liquid water to drain from the coil, and helps minimize the thermal load to the refrigerated space during defrost.

Defrost initiation is usually accomplished via demand defrost and time-initiated defrost. Demand defrost systems utilize a variable, such as pressure drop across the air side of the evaporator or a set

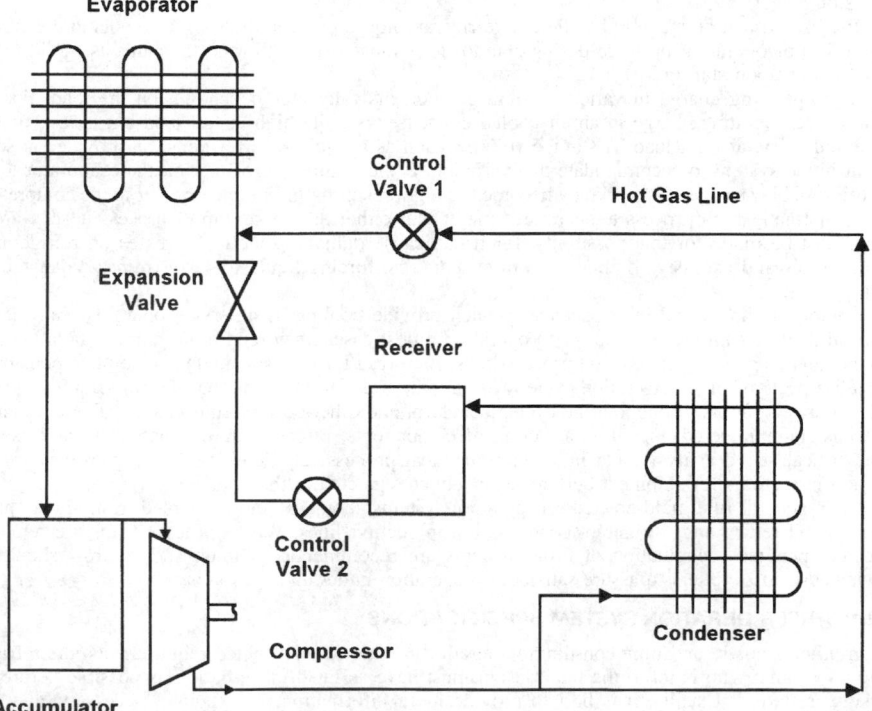

Fig. 62.22 Simplified diagram of a hot refrigerant gas defrost system.

of temperature inputs, to determine if frost has built up enough on the coil to require a defrost. Time-initiated defrost relies on a preset number of defrosts per day. The number of defrosts and length of time of each defrost can be adjusted. Ideally, the demand defrost system provides the most efficient defrost controls on a system because a defrost is initiated only if the evaporator needs it.

62.9.2 Air and Water Defrost

If the refrigerated space operates above 0°C, then the air in the space can be used directly to defrost the evaporator. Defrost is accomplished by continuing to operate the evaporator blower while the compressor is off. As the frost melts, some of it is evaporated into the airstream while the rest drains away from the coil as liquid water. The evaporated moisture adds an additional load to the evaporator when the compressor starts again.

Water can also be used to defrost the evaporator. The compressor and fans are shut off while water is sprayed over the evaporator. If the velocity of the water is kept high, it washes or melts the frost off the coil.

62.10 SYSTEM DESIGN CONSIDERATIONS

Associated with continuous operation are refrigeration startup and shutdown conditions that invariably differ, sometimes widely, from those of the process itself. These conditions, although they occupy very little time in the life of the installation, must be properly accommodated in the design of the refrigeration system. Consideration must be given to the amount of time required to achieve design operating conditions, the need for standby equipment, and so on.

In batch processing, operating conditions are expected to change with time, usually in a repetitive pattern. The refrigeration system must be designed for all extremes. Use of brine storage or ice banks can reduce equipment sizes for batch processes.

Closed-cycle operation involves both liquid and gas phases. System designs must take into account liquid-flow problems in addition to gas-flow requirements and must provide for effective separation of the liquid and gas phases in different parts of the system. These factors require careful design of all components and influence the arrangement or elevation of certain components in the cycle.

Liquid pressures must be high enough to feed liquid to the evaporators at all times, especially when evaporators are elevated or remotely located. In some cases, a pump must be used to suit the process requirements. The possibility of operation with reduced pressures caused by colder condensing temperatures than the specified design conditions must also be considered. Depending on the types of liquid valves and relative elevation of various parts of the system, it may be necessary to maintain condensing pressures above some minimum level, even if doing so increases the compression power.

Provision must be made to handle any refrigerant liquid that can drain to low spots in the system upon loss of operating pressure during shutdown. It must not be allowed to return as liquid to the compressor upon startup.

The operating charge in various system components fluctuates depending on the load. For example, the operating charge in an air-cooled condenser is quite high at full load but is low, that is, essentially dry, at light load. A storage volume, such as a liquid receiver, must be provided at some point in the system to accommodate this variation. If the liquid controls permit the evaporator to act as the variable storage, the level may become too high, resulting in liquid carryover to the compressor.

Abnormally high process temperatures may occur either during startup or process upsets. Provision must be made for this possibility, for it can cause damaging thermal stresses on refrigeration components and excessive boiling rates in evaporators, forcing liquid to carry over and damage the compressor.

Factory-designed and built packages, which provide cooling as a service or utility, can require several thousand kilowatts of power to operate. In most cases, they require no more installation than connection of power, utilities, and process lines. As a result, there is a single source of responsibility for all aspects of the refrigeration cycle involving the transfer and handling of both saturated liquids and saturated vapors throughout the cycle, oil return, and other design requirements. These packages are custom-engineered, including selection of components, piping, controls, base designs torsional and critical speed analysis, and individual chemical process requirements. Large packages are designed in sections for shipment but are readily interconnected in the field.

As a general rule, field-erected refrigeration systems should be close-coupled to minimize problems of oil return and refrigerant condensation in suction lines. Where process loads are remotely located, pumped recirculation or brine systems are recommended. Piping and controls should be reviewed with suppliers to assure satisfactory operation under all conditions.

62.11 REFRIGERATION SYSTEM SPECIFICATIONS

To minimize costly and time-consuming alterations owing to unexpected requirements, the refrigeration specialist who is to do the final design must have as much information as possible before the design is started. Usually, it is best to provide more information than thought necessary, and it is always wise to note where information may be sketchy, missing, or uncertain. Carefully spelling out

Table 62.7   Information Needed for the Design of a Refrigeration System

Process Flow sheets and Thermal Specifications	Basic Specifications	Instrumentation and Control Requirements	Off-Design Operation

Type of process
  batch
  continuous
Normal heat balances
Normal material balances
Normal material composition
Design operating pressure and temperatures
Design refrigeration loads
Energy recovery possibilities
Manner of supplying refrigeration (primary or secondary)
	Mechanical system details
	  construction standards
	    industry
	    company
	    local plant
	  insulation requirements
	  special corrosion-prevention requirements
	  special sealing requirements
	  process streams to the environment
	  process stream to refrigerant
	Operating environment
	  indoor or outdoor location
	  extremes
	  special requirements
	Special safety considerations
	  known hazards of process
	  toxicity and flammability constraints
	  maintenance limitations
	Reliability requirements
	  effect of loss of cooling on process safety
	  maintenance intervals and types that may be performed
	  Redundancy requirement
	  Acceptance test requirements
		Safety interlocks
		Process interlocks
		Special control requirements
		  at equipment
		  central control room
		Special or plant standard instruments
		Degree of automation: interface requirements
		Industry and company control standards
			Process startup sequence
			  degree of automation
			  refrigeration loads vs. time
			  time needed to bring process onstream
			  frequency of startup
			  process pressure, temperature, and composition changes during startup
			  special safety requirements
			Minimum load
			Need for standby capability
			Peak-load pressures and temperatures
			Composition extremes
			Process shutdown sequence
			  degree of automation
			  refrigeration load vs. time
			  shutdown timespan
			  process pressure, temperature, and composition changes
			  special safety requirements

the allowable margins in the most critical process variables and pointing out portions of the refrigeration cycle that are of least concern is always helpful to the designer.

A checklist of minimum information (Table 62.7) needed by a refrigeration specialist to design a cooling system for a particular application may be helpful.

Process flow sheets. For chemical process designs, seeing the process flow sheets is the best overall means for the refrigeration engineer to become familiar with the chemical process for which the refrigeration equipment is to be designed. In addition to providing all of the information shown in Table 62.7, they give the engineer a feeling for how the chemical plant will operate as a system and how the refrigeration equipment fits into the process.

Basic specifications. This portion of Table 62.7 fills in the detailed mechanical information that tells the refrigeration engineer how the equipment should be built, where it will be located, and specific safety requirements. This determines which standard equipment can be used and what special modifications need to be made.

Instrumentation and control requirements. These tell the refrigeration engineer how the system will be controlled by the plant operators. Particular controller types, as well as control sequencing and operation, must be spelled out to avoid misunderstandings and costly redesign. The refrigeration engineer needs to be aware of the degree of control required for the refrigeration system. For example, the process may require remote starting and stopping of the refrigeration system from the central control room. This could influence the way in which the refrigeration safeties and interlocks are designed.

Off-design operation. It is likely that the most severe operation of the refrigeration system will occur during startup or shutdown. The rapidly changing pressures, temperatures, and loads experienced by the refrigeration equipment can cause motor overloads, compressor surging, or loss of control if they are not anticipated during design.

REFERENCES

1. *ASHRAE Handbook of Refrigeration Systems and Applications*, American Society of Heating, Refrigerating and Air Conditioning Engineers, Atlanta, GA, 1994.

2. R. Thevenot, *A History of Refrigeration Throughout the World*, International Institute of Refrigeration, Paris, France, 1979, pp. 39–46.

3. K. W. Cooper and K. E. Hickman, "Refrigeration," in *Encyclopedia of Chemical Technology*, Vol. 20, 3rd ed., Wiley, New York, pp. 78–107.

4. C. E. Salas and M. Salas, *Guide to Refrigeration CFC's*, Fairmont Press, Liburn, GA, 1992.

5. U.S. Environmental Protection Agency, "The Accelerated Phaseout of Ozone-Depleting Substances," *Federal Register* **58**(236), 65018–65082 (December 10, 1993).

6. U.S. Environmental Protection Agency, "Class I Nonessential Products Ban, Section 610 of the Clean Air Act Amendments of 1990," *Federal Register* **58**(10) 4768–4799 (January 15, 1993).

7. United Nations Environmental Program (UNEP), *Montreal Protocol on Substances That Deplete the Ozone Layer—Final Act*, 1987.

8. G. King, *Basic Refrigeration*, Business News, Troy, MI, 1986.

9. ANSI/ASHRAE Standard 34-1992, *Number Designation and Safety Classification of Refrigerants*, American Society of Heating, Refrigerating, and Air Conditioning Engineers, Atlanta, GA, 1992.

10. G. G. Haselden, *Mech. Eng.* **44** (March 1981).

11. ANSI/ASHRAE 15-1994, *Safety Code for Mechanical Refrigeration*, American Society of Heating, Refrigerating and Air Conditioning Engineers, Atlanta, GA, 1994.

12. *ASHRAE Handbook of Fundamentals*, American Society of Heating, Refrigerating and Air Conditioning Engineers, Atlanta, GA, 1993, Chap. 16.

13. M. J. Molina and F. S. Rowland, "Stratospheric Sink for Chlorofluoromethanes: Chlorine Atoms Catalyzed Destruction of Ozone," *Nature* **249**, 810–812.

14. C. D. MacCracken, "The Greenhouse Effect on ASHRAE," *ASHRAE Journal* **31**(6), 52–55 (June 1996).

15. *Refrigerant Reference Guide*, National Refrigerants, Philadelphia, PA, 1995.

16. D. Didion, "Practical Considerations in the Use of Refrigerant Mixtures," Presented at the ASHRAE Winter Meeting, Atlanta, GA, February 1996.

17. W. F. Stoecker and J.W. Jones, *Refrigeration and Air Conditioning*, 2nd ed., McGraw-Hill, New York, 1982.

18. *ASHRAE Handbook of HVAC Systems and Equipment*, American Society of Heating, Refrigerating and Air Conditioning Engineers, Atlanta, GA, 1992, Chaps. 35 and 36.

19. K. Matsubara, K. Suefuji, and H. Kuno, "The Latest Compressor Technologies for Heat Pumps in Japan," in *Heat Pumps*, K. Zimmerman and R. H. Powell, Jr. (eds.), Lewis, Chelsea, MI, 1987.

20. T. Senshu, A. Araik, K. Oguni, and F. Harada, "Annual Energy-Saving Effect of Capacity-Modulated Air Conditioner Equipped with Inverter-Driven Scroll Compressor," *ASHRAE Transactions* **91**(2) (1985).

21. *ASME Boiler and Pressure Vessel Code*, Sect. VIII, Div. 1, American Society of Mechanical Engineers, New York, 1980.

22. Y. Kim and D. L. O'Neal, "A Comparison of Critical Flow Models for Estimating Two-Phase Flow of HCFC 22 and HFC 134a through Short Tube Orifices," *International Journal of Refrigeration* **18**(6) (December 1995).

23. Y. Kim and D. L. O'Neal, "Two-Phase Flow of Refrigerant-22 through Short-Tube Orifices," *ASHRAE Transactions* **100**(1) (1994).

CHAPTER 63

CRYOGENIC SYSTEMS

Leonard A. Wenzel
Lehigh University
Bethlehem, Pennsylvania

63.1	**CRYOGENICS AND CRYOFLUID PROPERTIES**	**1915**
63.2	**CRYOGENIC REFRIGERATION AND LIQUEFACTION CYCLES**	**1921**
	63.2.1 Cascade Refrigeration	1921
	63.2.2 The Linde or Joule–Thomson Cycle	1923
	63.2.3 The Claude or Expander Cycle	1924
	63.2.4 Low-Temperature Engine Cycles	1928
63.3	**CRYOGENIC HEAT-TRANSFER METHODS**	**1930**
	63.3.1 Coiled-Tube-in-Shell Exchangers	1931
	63.3.2 Plate-Fin Heat Exchangers	1933
	63.3.3 Regenerators	1933
63.4	**INSULATION SYSTEMS**	**1939**
	63.4.1 Vacuum Insulation	1940
	63.4.2 Superinsulation	1941
	63.4.3 Insulating Powders and Fibers	1943
63.5	**MATERIALS FOR CRYOGENIC SERVICE**	**1943**
	63.5.1 Materials of Construction	1943
	63.5.2 Seals and Gaskets	1953
	63.5.3 Lubricants	1953
63.6	**SPECIAL PROBLEMS IN LOW-TEMPERATURE INSTRUMENTATION**	**1953**
	63.6.1 Temperature Measurement	1953
	63.6.2 Flow Measurement	1955
	63.6.3 Tank Inventory Measurement	1955
63.7	**EXAMPLES OF CRYOGENIC PROCESSING**	**1955**
	63.7.1 Air Separation	1956
	63.7.2 Liquefaction of Natural Gas	1958
	63.7.3 Helium Recovery and Liquefaction	1962
63.8	**SUPERCONDUCTIVITY AND ITS APPLICATIONS**	**1963**
	63.8.1 Superconductivity	1963
	63.8.2 Applications of Superconductivity	1966
63.9	**CRYOBIOLOGY AND CRYOSURGERY**	**1969**

63.1 CRYOGENICS AND CRYOFLUID PROPERTIES

The science and technology of deep refrigeration processing occurring at temperatures lower than about 150 K is the field of cryogenics (from the Greek *kryos,* icy cold). This area has developed as a special discipline because it is characterized by special techniques, requirements imposed by physical limitations, and economic needs, and unique phenomena associated with low-thermal-energy levels.

Compounds that are processed within the cryogenic temperature region are sometimes called cryogens. There are only a few of these materials; they are generally small, relatively simple molecules, and they seldom react chemically within the cryogenic region. Table 63.1 lists the major cryogens along with their major properties, and with a reference giving more complete thermodynamic data.

Mechanical Engineers' Handbook, 2nd ed., Edited by Myer Kutz.
ISBN 0-471-13007-9 © 1998 John Wiley & Sons, Inc.

Table 63.1 Properties of Principal Cryogens

Name	Normal Boiling Point			Critical Point		Triple Point		Reference
	T (K)	Liquid Density (kg/m³)	Latent Heat (J/kg · mole)	T (K)	P (kPa)	T (K)	P (kPa)	
Helium	4.22	123.9	91,860	5.28	227			1
Hydrogen	20.39	70.40	902,300	33.28	1296	14.00	7.20	2, 3
Deuterium	23.56	170.0	1,253,000	38.28	1648	18.72	17.10	4
Neon	27.22	1188.7	1,737,000	44.44	2723	26.28	43.23	5
Nitrogen	77.33	800.9	5,579,000	126.17	3385	63.22	12.55	6
Air	78.78	867.7	5,929,000					7, 8
Carbon monoxide	82.11	783.5	6,024,000	132.9	3502	68.11	15.38	9
Fluorine	85.06	1490.6	6,530,000	144.2	5571			10
Argon	87.28	1390.5	6,504,000	151.2	4861	83.78		11, 12, 13
Oxygen	90.22	1131.5	6,801,000	154.8	5081	54.39	0.14	6
Methane	111.72	421.1	8,163,000	190.61	4619	90.67	11.65	14
Krypton	119.83	2145.4	9,009,000	209.4	5488	116.00	73.22	15
Nitric oxide	121.50	1260.2	13,809,000	179.2	6516	108.94		
Nitrogen trifluoride	144.72	1525.6	11,561,000	233.9	4530			
Refrigerant-14	145.11	1945.1	11,969,000	227.7	3737	89.17	0.12	16
Ozone	161.28	1617.8	14,321,000	261.1	5454			
Xenon	164.83	3035.3	12,609,000	289.8	5840	161.39	81.50	17
Ethylene	169.39	559.4	13,514,000	282.7	5068	104.00	0.12	18

All of the cryogens except hydrogen and helium have conventional thermodynamic and transport properties. If specific data are unavailable, the reduced properties correlation can be used with all the cryogens and their mixtures with at least as much confidence as the correlations generally allow. Qualitatively $T–S$ and $P–H$ diagrams such as those of Figs. 63.1 and 63.2 differ among cryogens only by the location of the critical point and freezing point relative to ambient conditions.

Air, ammonia synthesis gas, and some inert atmospheres are considered as single materials although they are actually gas mixtures. The composition of air is shown in Table 63.12. If a thermodynamic diagram for air has the lines drawn between liquid and vapor boundaries where the pressures are equal for the two phases, these lines will not be at constant temperature, as would be the case for a pure component. Moreover, these liquid and vapor states are not at equilibrium, for the equilibrium states have equal Ts and Ps, but differ in composition. That being so, one or both of these equilibrium mixtures is not air. Except for this difference the properties of air are also conventional.

Hydrogen and helium differ in that their molecular mass is small in relation to zero-point-energy levels. Thus quantum differences are large enough to produce measurable changes in gross thermodynamic properties.

Hydrogen and its isotopes behave abnormally because the small molecular weight allows quantum differences stemming from different molecular configurations to affect total thermodynamic properties. The hydrogen molecule consists of two atoms, each containing a single proton and a single electron. The electrons rotate in opposite directions as required by molecular theory. The protons, however, may rotate in opposed or parallel directions. Figure 63.3 shows a sketch of the two possi-

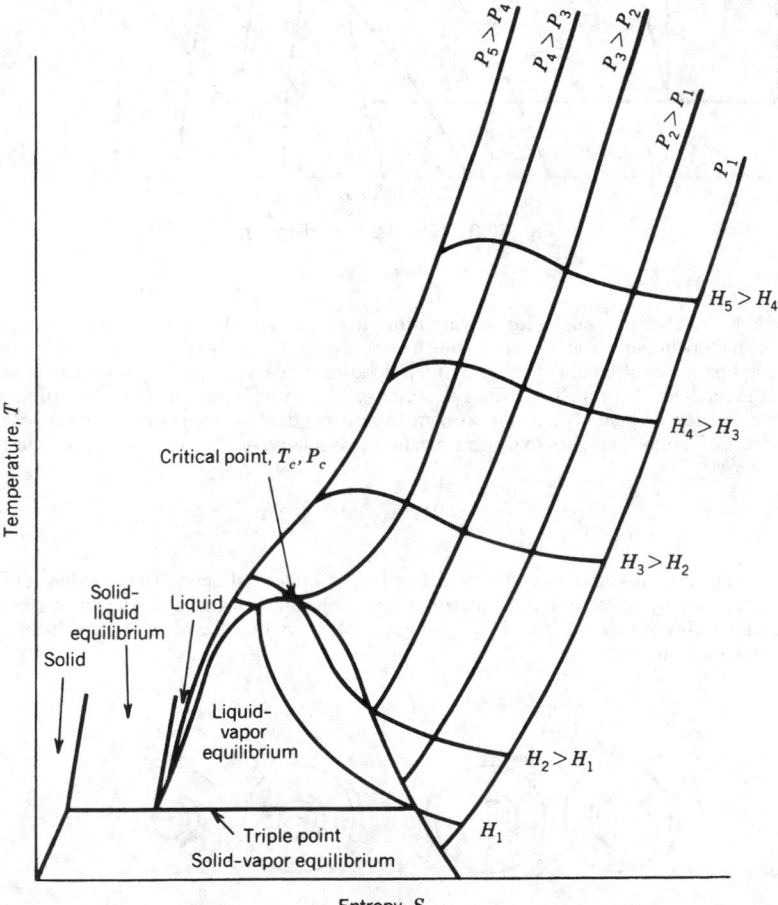

Fig. 63.1 Skeletal $T–S$ diagram.

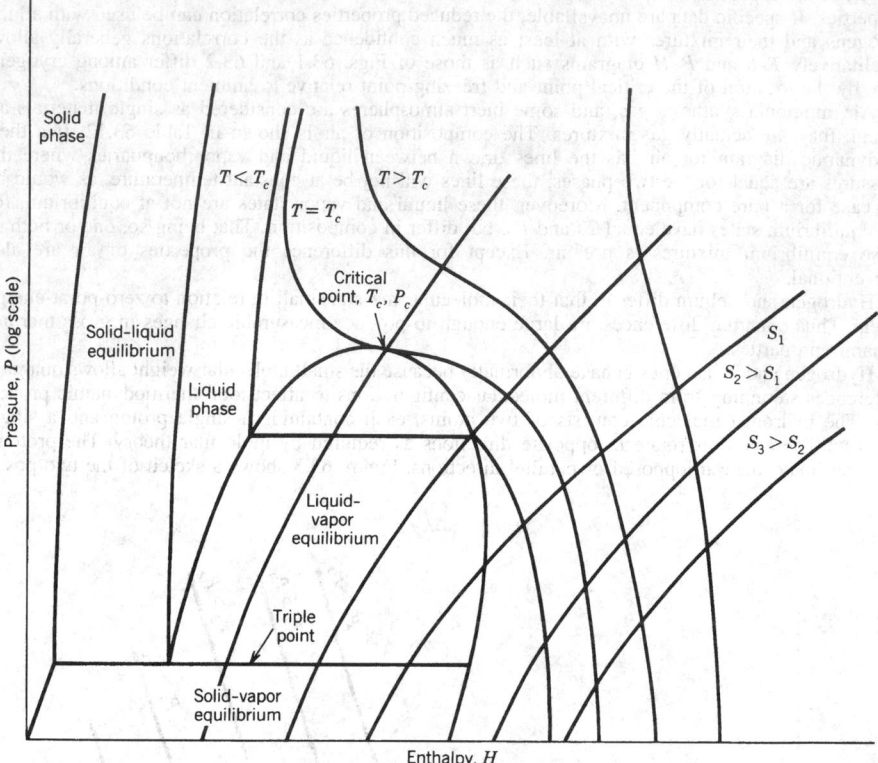

Fig. 63.2 Skeletal *P–H* diagram.

bilities, the parallel rotating nuclei identifying ortho-hydrogen and the opposite rotating nuclei identifying the parahydrogen. The quantum mechanics exhibited by these two molecule forms are different, and produce different thermodynamic properties. Ortho- and para-hydrogen each have conventional thermodynamic properties. However, ortho- and para-hydrogen are interconvertible with the equilibrium fraction of pure H_2 existing in para form dependent on temperature, as shown in Table 63.2. The natural ortho- and para-hydrogen reaction is a relatively slow one and of second order:[19]

$$\frac{dx}{d\theta} = 0.0114x^2 \quad \text{at} \quad 20 \text{ K}$$

where θ is time in hours and x is the mole fraction of ortho-hydrogen. The reaction rate can be greatly accelerated by a catalyst that interrupts the molecular magnetic field and possesses high surface area. Catalysts such as NiO_2/SiO_2 have been able to yield some of the highest heterogeneous reaction rates measured.[20]

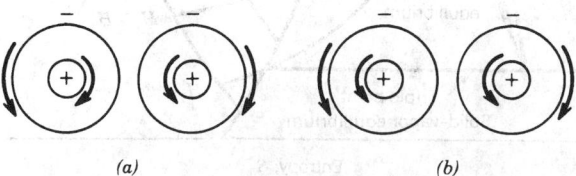

(a)   *(b)*

Fig. 63.3 Molecular configurations of (a) para- and (b) ortho-hydrogen.

Table 63.2 Equilibrium Para-Hydrogen Concentration as a Function of T (K)

T (K)	Equilibrium Percentage of Para-Hydrogen
20	99.82
30	96.98
40	88.61
60	65.39
80	48.39
100	38.51
150	28.54
273	25.13
500	25.00

Normally hydrogen exists as a 25 mole % p-H_2, 75 mole % o-H_2 mix. Upon liquefaction the hydrogen liquid changes to nearly 100% p-H_2. If this is done as the liquid stands in an insulated flask, the heat of conversion will suffice to evaporate the liquid, even if the insulation is perfect. For this reason the hydrogen is usually converted to para form during refrigeration by the catalyzed reaction, with the energy released added to the refrigeration load.

Conversely, liquid para-hydrogen has an enhanced refrigeration capacity if it is converted to the equilibrium state as it is vaporized and warmed to atmospheric condition. In certain applications recovery of this refrigeration is economically justifiable.

Helium, though twice the molecular weight of hydrogen, also shows the effects of flow molecular weight upon gross properties. The helium molecule is single-atomed and thus free from ortho–para-type complexities. Helium was liquefied conventionally first in 1908 by Onnes of Leiden, and the liquid phase showed conventional behavior at atmospheric pressure.

As temperature is lowered, however, a second-order phase change occurs at 2.18 K (0.05 atm) to produce a liquid called HeII. At no point does solidification occur just by evacuating the liquid. This results from the fact that the relationship between molecular volume, thermal energy (especially zero-point energy), and van der Waals attractive forces is such that the atoms cannot be trapped into a close-knit array by temperature reduction alone. Eventually, it was found that helium could be solidified if an adequate pressure is applied, but that the normal liquid helium (HeI)–HeII phase transition occurs at all pressures up to that of solidification. The phase diagram for helium is shown in Fig. 63.4. The HeI–HeII phase change has been called the lambda curve from the shape of the heat capacity curve for saturated liquid He, as shown in Fig. 63.5. The peculiar shape of the heat capacity curve produces a break in the curve for enthalpy of saturated liquid He as shown in Fig. 63.6.

HeII is a unique liquid exhibiting properties that were not well explained until after 1945. As liquid helium is evacuated to increasingly lower pressures, the temperature also drops along the vapor-pressure curve. If this is done in a glass vacuum-insulated flask, heat leaks into the liquid He causing boiling and bubble formation. As the temperature approaches 2.18 K, boiling gets more violent, but then suddenly stops. The liquid He is completely quiescent. This has been found to occur because the thermal conductivity of HeII is extremely large. Thus the temperature is basically constant and all boiling occurs from the surface where the hydrostatic head is least, producing the lowest boiling point.

Not only does HeII have very large thermal conductivity, but it also has near zero viscosity. This can be seen by holding liquid He in a glass vessel with a fine porous bottom such that normal He does not flow through. If the temperature is lowered into the HeII region, the helium will flow rapidly through the porous bottom. Flow does not seem to be enhanced or hindered by the size of the frit. Conversely, a propeller operated in liquid HeII will produce a secondary movement in a parallel propeller separated from the first by a layer of liquid HeII. Thus HeII has properties of finite and of infinitesimal viscosity.

These peculiar flow properties are also shown by the so-called thermal-gravimetric effect. There are two common demonstrations. If a tube with a finely fritted bottom is put into liquid HeII and the helium in the tube is heated, liquid flows from the main vessel into the fritted tube until the liquid level in the tube is much higher than that in the main vessel. A second, related, experiment uses a U-tube, larger on one leg than on the other with the two sections separated by a fine frit. If this tube is immersed, except for the end of the narrow leg, into liquid HeII and a strong light is

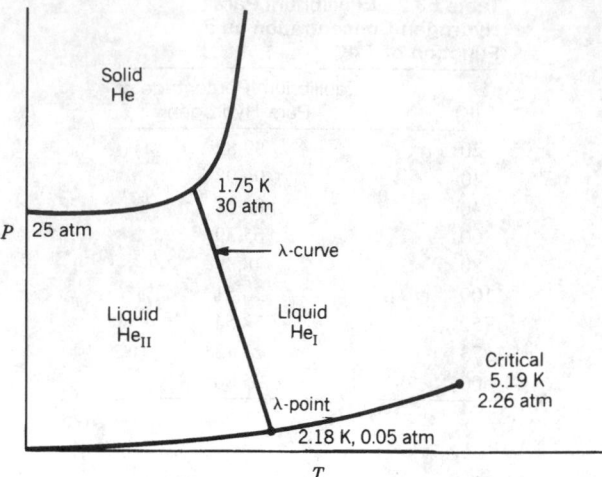

Fig. 63.4 Phase diagram for helium.

focused on the liquid He above the frit, liquid He will flow through the frit and out the small tube opening producing a fountain of liquid He several feet high.

These and other experiments[21] can be explained through the quantum mechanics of HeII. The pertinent relationships, the Bose–Einstein equations, indicate that HeII has a dual nature: it is both a "superfluid" which has zero viscosity and infinite thermal conductivity among other special properties, and a fluid of normal properties. The further the temperature drops below the lambda point the greater the apparent fraction of superfluid in the liquid phase. However, very little superfluid is required. In the flow through the porous frit the superfluid flows, the normal fluid is retained. However, if the temperature does not rise, some of the apparently normal fluid will apparently become superfluid. Although the superfluid flows through the frit, there is no depletion of superfluid in the liquid He left behind. In the thermogravimetric experiments the superfluid flows through the frit but is then changed to normal He. Thus there is no tendency for reverse flow.

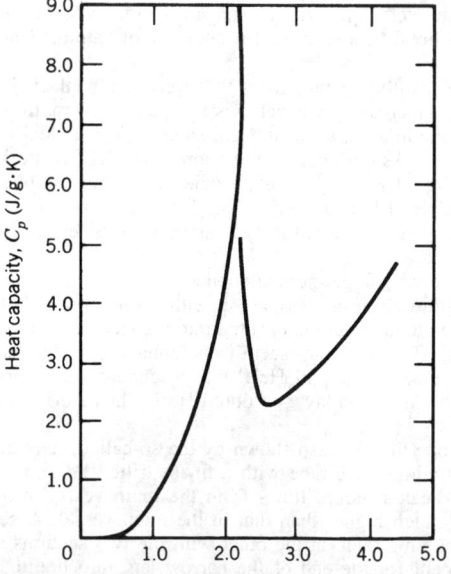

Fig. 63.5 Heat capacity of saturated liquid ⁴He.

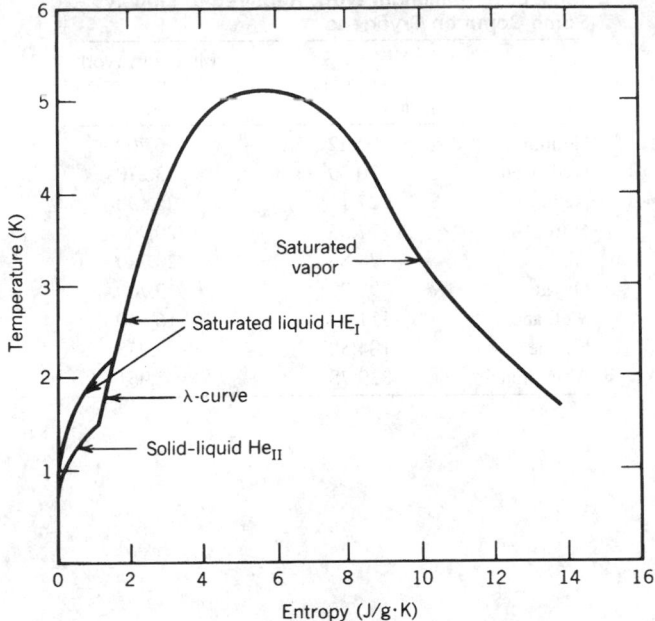

Fig. 63.6 Temperature–entropy diagram for saturation region of ^{4}He.

At this point applications have not developed for HeII. Still, the peculiar phase relationships and energy effects may influence the design of helium processes, and do affect the shape of thermodynamic diagrams for helium.

63.2 CRYOGENIC REFRIGERATION AND LIQUEFACTION CYCLES

One characteristic aspect of cryogenic processing has been its early and continued emphasis on process efficiency, that is, on energy conservation. This has been forced on the field by the very high cost of deep refrigeration. For any process the minimum work required to produce the process goal is

$$W_{min} = T_0 \Delta S - \Delta H \tag{63.2}$$

where W_{min} is the minimum work required to produce the process goal, ΔS and ΔH are the difference between product and feed entropy and enthalpy, respectively, and T_0 is the ambient temperature. Table 63.3 lists the minimum work required to liquefy 1 kg-mole of several common cryogens. Obviously, the lower the temperature level the greater the cost for unit result. The evident conflict in H_2 and He arises from their different molecular weights and properties. However, the temperature differences from ambient to liquid H_2 temperature and from ambient to liquid He temperatures are similar.

A refrigeration cycle that would approach the minimum work calculated as above would include ideal process steps as, for instance, in a Carnot refrigeration cycle. The cryogenic engineer aims for this goal while satisfying practical processing and capital cost limitations.

63.2.1 Cascade Refrigeration

The cascade refrigeration cycle was the first means used to liquefy air in the United States.[22] It uses conveniently chosen refrigeration cycles, each using the evaporator of the previous fluid cycle as condenser, which will produce the desired temperature. Figures 63.7 and 63.8 show a schematic T–S diagram of such a cycle and the required arrangement of equipment.

Obviously, this cycle is mechanically complex. After its early use it was largely replaced by other cryogenic cycles because of its mechanical unreliability, seal leaks, and poor mechanical efficiency. However, the improved reliability and efficiency of modern compressors has fostered a revival in the cascade cycle. Cascade cycles are used today in some base-load natural gas liquefaction (LNG) plants[23] and in the some peak-shaving LNG plants. They are also used in a variety of intermediate refrigeration processes. The cascade cycle is potentially the most efficient of cryogenic processes

**Table 63.3 Minimum Work Required to Liquefy
Some Common Cryogens**

Gas	Normal Boiling Point (K)	Minimum Work of Liquefaction (J/mole)
Helium	4.22	26,700
Hydrogen	20.39	23,270
Neon	27.11	26,190
Nitrogen	77.33	20,900
Air	78.8	20,740
Oxygen	90.22	19,700
Methane	111.67	16,840
Ethane	184.50	9,935
Ammonia	239.78	3,961

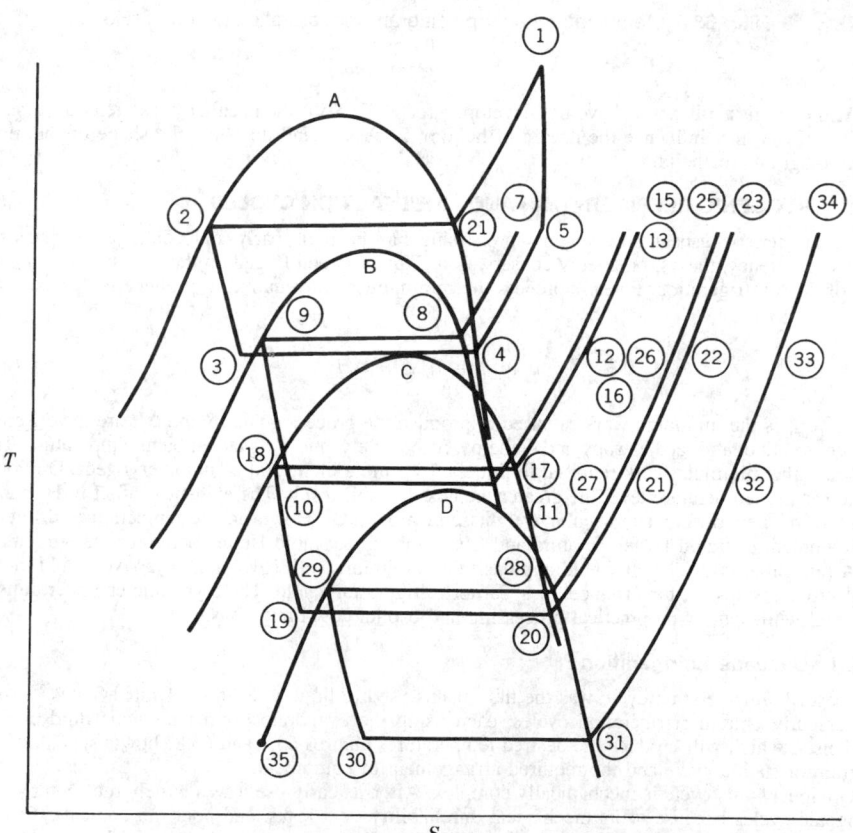

Fig. 63.7 Cascade refrigeration system on *T–S* coordinates. Note that *T–S* diagram for fluids A, B, C, and D are here superimposed. Numbers here refer to Fig. 63.8 flow points.

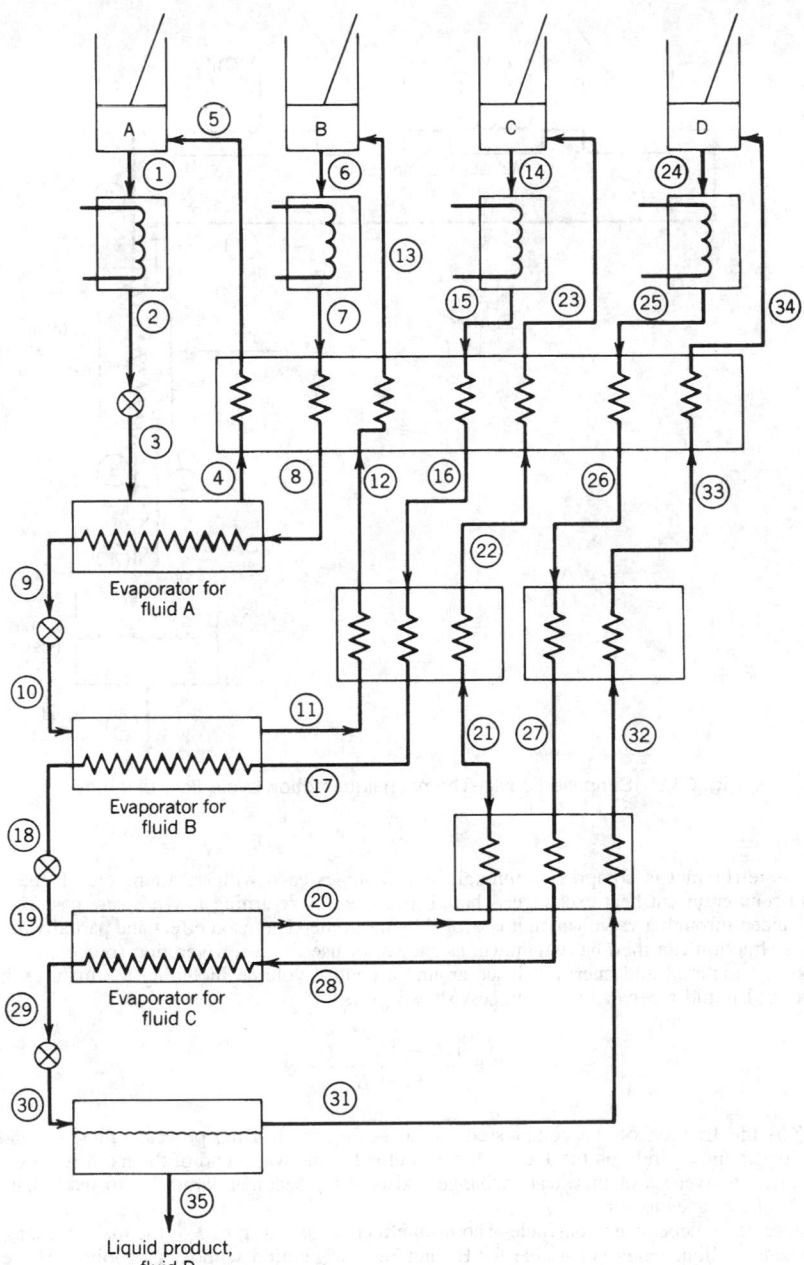

Fig. 63.8   Cascade liquefaction cycle—simplified flow diagram.

because the major heat-transfer steps are liquefaction–vaporization exchanges with each stream at a constant temperature. Thus heat transfer coefficients are high and ΔTs can be kept very small.

63.2.2   The Linde or Joule–Thomson Cycle

The Linde cycle was used in the earliest European efforts at gas liquefaction and is conceptually the simplest of cryogenic cycles. A simple flow sheet is shown in Fig. 63.9. Here the gas to be liquefied

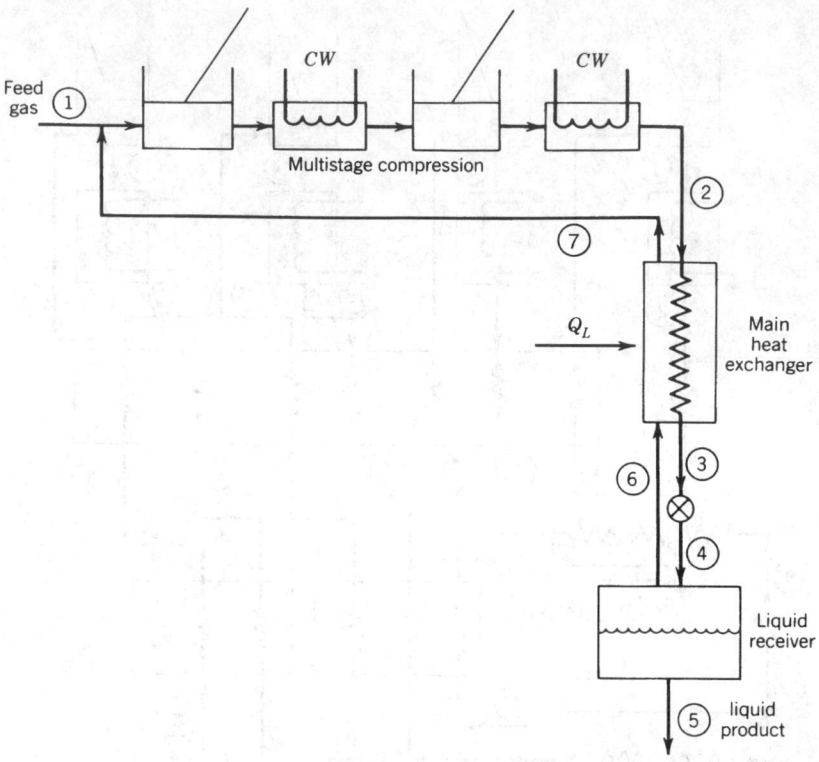

Fig. 63.9 Simplified Joule–Thomson liquefaction cycle flow diagram.

or used as refrigerant is compressed through several stages each with its aftercooler. It then enters the main countercurrent heat exchanger where it is cooled by returning low-pressure gas. The gas is then expanded through a valve where it is cooled by the Joule–Thomson effect and partially liquefied. The liquid fraction can then be withdrawn, as shown, or used as a refrigeration source.

Making a material and energy balance around a control volume including the main exchanger, JT valve, and liquid receiver for the process shown gives

$$X = \frac{(H_7 - H_2) - Q_L}{H_7 - H_5} \qquad (63.3)$$

where X is the fraction of the compressed gas to be liquefied. Thus process efficiency and even operability depend entirely on the Joule–Thomson effect at the warm end of the main heat exchanger and on the effectiveness of that heat exchanger. Also, if Q_L becomes large due to inadequate insulation, X quickly goes to zero.

Because of its dependence on Joule–Thomson effect at the warm end of the main exchanger, the Joule–Thomson liquefier is not usable for H_2 and He refrigeration without precooling. However, if H_2 is cooled to liquid N_2 temperature before it enters the JT cycle main heat exchanger, or if He is cooled to liquid H_2 temperature before entering the JT cycle main heat exchanger, further cooling to liquefaction can be done with this cycle. Even with fluids such as N_2 and CH_4 it is often advantageous to precool the gas before it enters the JT heat exchanger in order to take advantage of the greater Joule–Thomson effect at the lower temperature.

63.2.3 The Claude or Expander Cycle

Expander cycles have become workhorses of the cryogenic engineer. A simplified flow sheet is shown in Fig. 63.11. Here part of the compressed gas is removed from the main exchanger before being fully cooled, and is cooled in an expansion engine in which mechanical work is done. Otherwise, the system is the same as the Joule–Thomson cycle. Figure 63.12 shows a T–S diagram for this process. The numbers on the diagram refer to those on the process flow sheet.

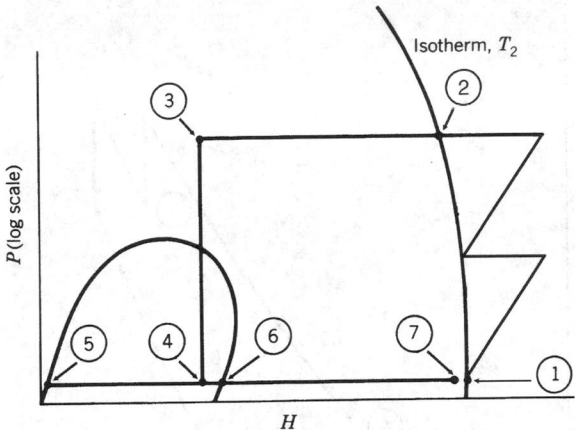

Fig. 63.10 Representation of the Joule–Thomson liquefaction cycle on a *P–H* diagram.

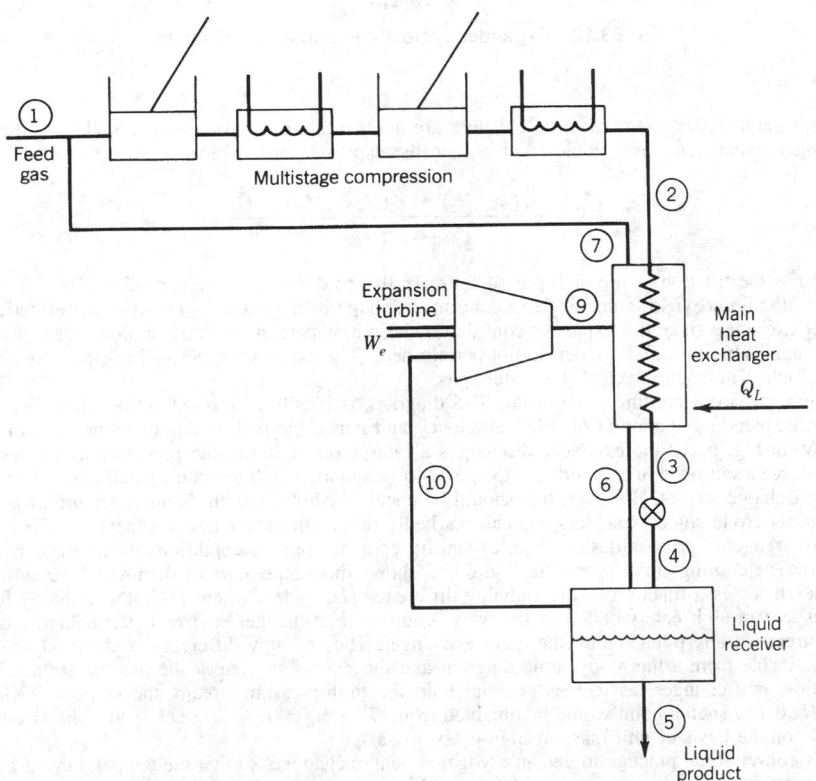

Fig. 63.11 Expander cycle simplified flow diagram.

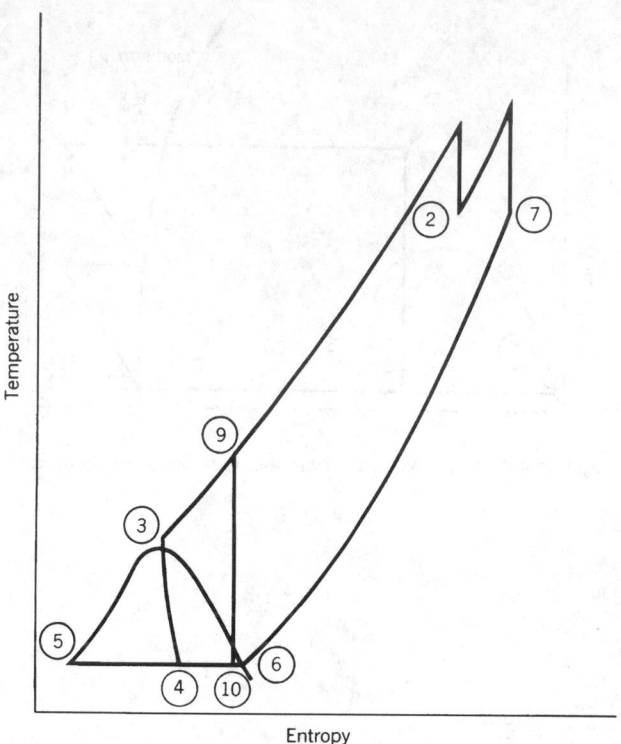

Fig. 63.12 Expander cycle shown on a *T–S* diagram.

If, as before, energy and material balances are made around a control volume including the main exchanger, expansion valve, liquid receiver, and the expander, one obtains

$$X = \frac{(H_7 - H_2) + Y(H_9 - H_{10}) - Q_L}{H_7 - H_5} \qquad (63.4)$$

where Y is the fraction of the high-pressure stream that is diverted to the expander.

Here the liquid yield is not so dependent on the shape of the warm isotherm or the effectiveness of heat exchange since the expander contributes the major part of the refrigeration. Also, the limitations applicable to a JT liquefier do not pertain here. The expander cycle will operate independent of the Joule–Thomson effect of the system gas.

The expansion step, line 9–10 on the *T–S* diagram, is ideally a constant entropy path. However, practical expanders operate at 60–90% efficiency and hence the path is one of modestly increasing entropy. In Fig. 63.12 the expander discharges a two-phase mixture. The process may be designed to discharge a saturated or a superheated vapor. Most expanders will tolerate a small amount of liquid in the discharge stream. However, this should be checked carefully with the manufacturer, for liquid can rapidly erode some expanders and can markedly reduce the efficiency of others.

Any cryogenic process design requires careful consideration of conditions in the main heat exchanger. The cooling curve plotted in Fig. 63.13 shows the temperature of the process stream being considered, T_i, as a function of the enthalpy difference $(H_o - H_i)$, where H_o is the enthalpy for the process stream as it enters or leaves the warm end of the exchanger, and H_i is the enthalpy of that same stream at any point within the main exchanger. The enthalpy difference is the product of the ΔH obtainable from a thermodynamic diagram and the mass flow rate of the process stream. If the mass flow rate changes, as it does at point 9 in the high-pressure stream, the slope will change. $H_o - H_i$, below such a point would be obtained from $H_o - H_i = (H_o - H_i) \cdot (1 - y)$ if the calculation is made on the basis of unit mass of high-pressure gas.

It is conventional practice to design cryogenic heat exchangers so that the temperature of a given process stream will be the same in each of the multiple passages of the exchanger at a given exchanger cross section. The temperature difference between the high- and low-pressure streams $(T_h - T_c)$ at

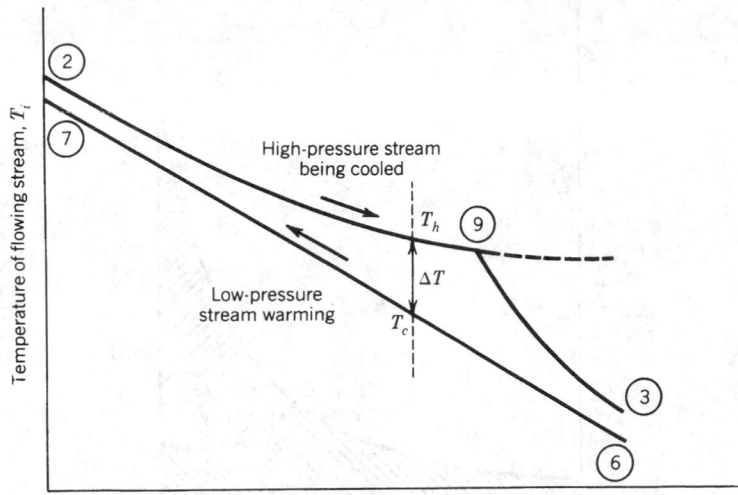

Fig. 63.13 Cooling curves showing temperatures throughout the main exchanger for the expander cycle.

that point is the ΔT available for heat transfer. Obviously, the simple ΔT_{lm} approach to calculation of heat-exchanger area will not be satisfactory here, for that method depends on linear cooling curves. The usual approach here is to divide the exchanger into segments of ΔH such that the cooling curves are linear for the section chosen and to calculate the exchanger area for each section. It is especially important to examine cryogenic heat exchangers in this detail because temperature ranges are likely to be large, thus producing heat-transfer coefficients that vary over the length of the exchanger, and because the curvature of the cooling curves well may produce regions of very small ΔT. In extreme cases the designer may even find that ΔT at some point in the exchanger reaches zero, or becomes negative, thus violating the second law. No exchanger designed in ignorance of this situation would operate as designed.

Minimization of cryogenic process power requirements, and hence operating costs, can be done using classical considerations of entropy gain. For any process

$$W = W_{\min} + \Sigma T_0 \Delta S_T \tag{63.5}$$

where W is the actual work required by the process, $W_{\min}$ is the minimum work [see Eq. (63.1)], and the last term represents the work lost in each process step. In that term T_0 is the ambient temperature, and ΔS_T is the entropy gain of the universe as a result of each process step.

In a heat exchanger

$$T_0 \Delta S_T = W_L = T_0 \int \frac{T_h - T_c}{T_h T_c} \, dH_i \tag{63.6}$$

where T_h and T_c represent temperatures of the hot and cold streams and the integration is carried out from end to end of the heat exchanger.

A comparison of the Claude cycle (so named because Georges Claude first developed a practical expander cycle for air liquefaction in 1902) with the Joule–Thomson cycle can thus be made by considering the W_L in the comparable process steps. In the cooling curve diagram, Fig. 63.13, the dotted line represents the high-pressure stream cooling curve of a Joule–Thomson cycle operating at the same pressure as does the Claude cycle. In comparison, the Claude cycle produces much smaller ΔTs at the cold end of the heat exchanger. If this is translated into lost work as done in Fig. 63.14, there is considerable reduction. The Claude cycle also reduces lost work by passing only a part of the high-pressure gas through a valve, which is a completely irreversible pressure reduction step. The rest of the high-pressure gas is expanded in a machine where most of the pressure lost produces usable work.

There are other ways to reduce the ΔT, and hence the W_L, in cryogenic heat exchangers. These methods can be used by the engineer as process conditions warrant. Figure 63.15 shows the effect

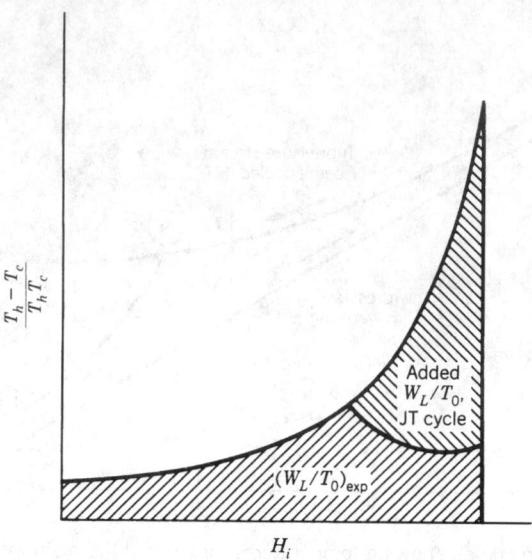

Fig. 63.14 Calculation of W_L in the main heat exchanger using Eq. (63.6) and showing the comparison between JT and Claude cycles.

of (*a*) intermediate refrigeration, (*b*) varying the amount of low-pressure gas in the exchanger, and (*c*) adding a third cold stream to the exchanger.

63.2.4 Low-Temperature Engine Cycles

The possibility that Carnot cycle efficiency could be approached by a refrigeration cycle in which all pressure change occurs by compression and expansion has encouraged the development of several cycles that essentially occur within an engine. In general, these have proven useful for small-scale cryogenic refrigeration under unusual conditions as in space vehicles. However, the Stirling cycle, discussed below, has been used for industrial-scale production situations.

The Stirling Cycle

In this cycle a noncondensable gas, usually helium, is compressed, cooled both by heat transfer to cooling water and by heat transfer to a cold solid matrix in a regenerator (see Section 63.3.3), and expanded to produce minimum temperature. The cold gas is warmed by heat transfer to the fluid or space being refrigerated and to the now-warm matrix in the regenerator and returned to the compressor. Figure 63.16 shows the process path on a *T–S* diagram. The process efficiency of this idealized cycle is identical to a Carnot cycle efficiency.

In application the Stirling cycle is usually operated in an engine where compression and expansion both occur rapidly with compression being nearly adiabatic. Figure 63.17 shows such a machine. The compressor piston (1) and a displacer piston (16 with cap 17) operate off the same crankshaft. Their motion is displaced by 90°. The result is that the compressor position is near the top or bottom of its cycle as the displacer is in most rapid vertical movement. Thus the cycle can be traced as follows:

1. With the displacer near the top of its stroke the compressor moves up compressing the gas in space 4.
2. The displacer moves down, and the gas moves through the annular water-cooled heat exchanger (13) and the annular regenerator (14) reaching the upper space (5) in a cooled, compressed state. The regenerator packing, fine copper wool, is warmed by this flow.
3. Displacer and compressor pistons move down together. Thus the gas in (5) is expanded and cooled further. This cold gas receives heat from the chamber walls (18) and interior fins (15) thus refrigerating these solid parts and their external finning.
4. The displacer moves up, thus moving the gas from space (5) to space (4). In flowing through the annular passages the gas recools the regenerator packing.

The device shown in Fig. 63.17 is arranged for air liquefaction. Room air enters at (23), passes through the finned structure where water and then CO_2 freeze out, and is then liquefied as it is further

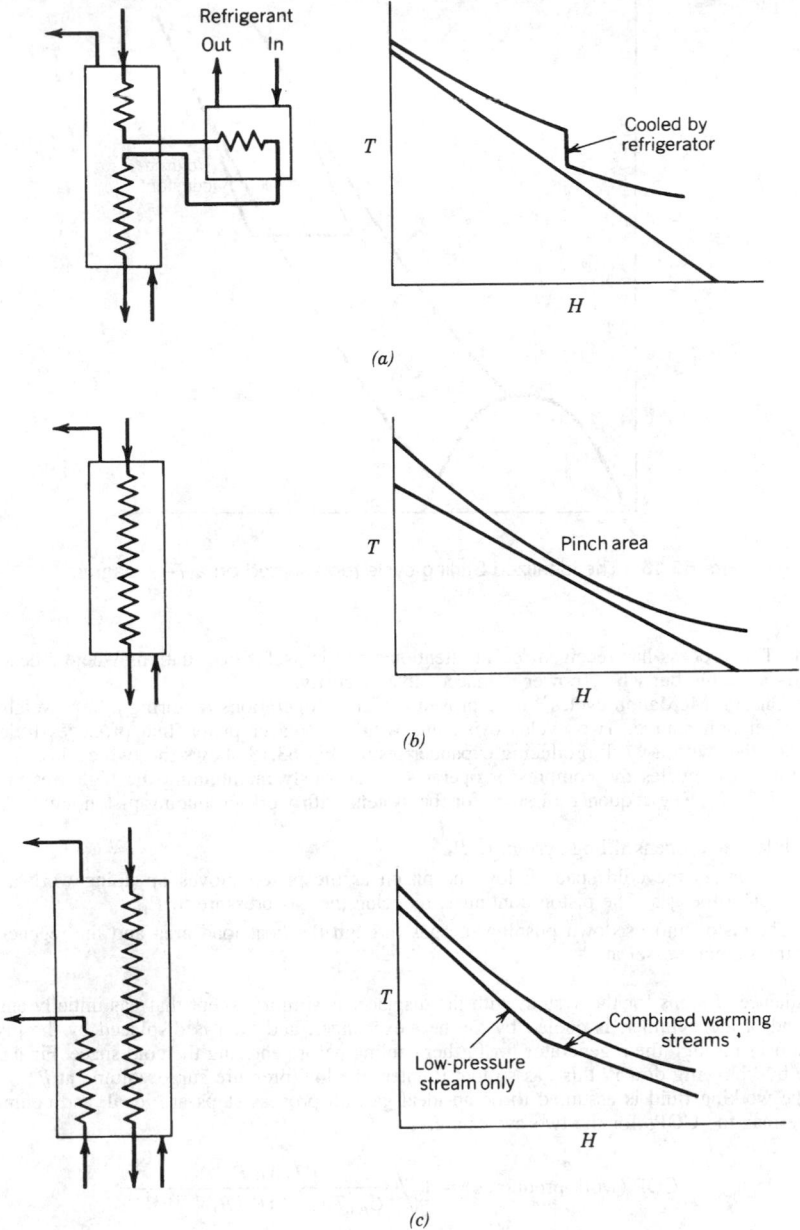

Fig. 63.15 Various cooling curve configurations to reduce W_L: (a) Cooling curve for intermediate refrigerator case. (b) Use of reduced warming stream to control ΔTs. (c) Use of an additional warming stream.

cooled as it flows over the finned surface (18) of the cylinder. The working fluid, usually He, is supplied as needed from the tank (27).

Other Engine Cycles

The Stirling cycle can be operated as a heat engine instead of as a refrigerator and, in fact, that was the original intent. In 1918 Vuilleumier patented a device that combines these two forms of Stirling cycle to produce a refrigerator that operated on a high-temperature heat source rather than on a source

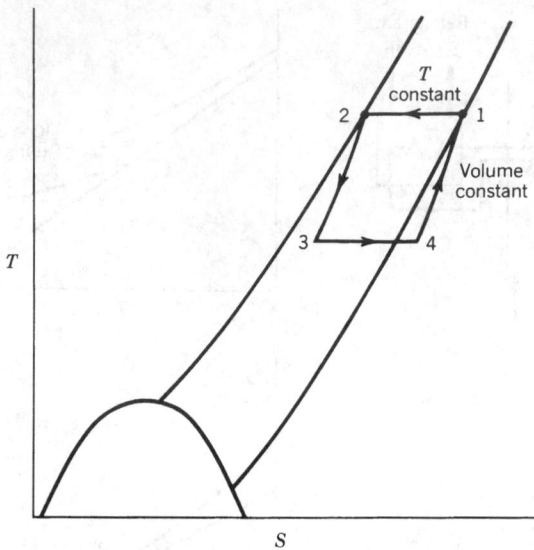

Fig. 63.16 The idealized Stirling cycle represented on a *T–S* diagram.

of work. This process has received recent attention[24] and is useful in situations where a heat source is readily available but where power is inaccessible or costly.

The Gifford–McMahon cycles[25] have proven useful for operations requiring a light-weight, compact refrigeration source. Two cycles exist: one with a displacer piston that produces little or no work; the other with a work-producing expander piston. Fig 63.18 shows the two cycles.

In both these cycles the compressor operates continuously maintaining the high-pressure surge volume of P_1, T_1. The sequence of steps for the system with work-producing piston are:

1. Inlet valve opens filling system to P_2.
2. Gas enters the cold space below the piston as the piston moves up doing work and thus cooling the gas. The piston continues, reducing the gas pressure to P_1.
3. The piston moves down pushing the gas through the heat load area and the regenerator to the storage vessel at P_1.

The sequence of steps for the system with the displacer is similar except that gas initially enters the warm end of the cylinder, is cooled by the heat exchanger, and then is displaced by the piston so that it moves through the regenerator for further cooling before entering the cold space. Final cooling is done by "blowing down" this gas so that it enters the low-pressure surge volume at P_1.

If the working fluid is assumed to be an ideal gas, all process steps are ideal, and compression is isothermal, the COPs for the two cycles are:

$$\text{COP (work producing)} = 1 \left/ \frac{RT_1 \ln P_2/P_1}{C_{P\,0}T_3[1 - (P_4/P_3)^{(k-1)/k}]^{-1}} \right.$$

$$\text{COP (displacer)} = \frac{P_3 - P_1}{P_3(T_1/T_3 - P_1/P_3) \ln P_2/P_1}$$

In these equations states 1 and 2 are those immediately before and after the compressor. State 3 is after the cooling step but before expansion, and state 4 is after the expansion at the lowest temperature.

63.3 CRYOGENIC HEAT-TRANSFER METHODS

In dealing with heat-transfer requirements the cryogenic engineer must effect large quantities of heat transfer over small ΔTs through wide temperature ranges. Commonly heat capacities and/or mass flows change along the length of the heat-transfer path, and often condensation or evaporation takes place. To minimize heat leak these complexities must be handled using exchangers with as large a heat-transfer surface area per exchanger volume as possible. Compact heat-exchanger designs of many sorts have been used, but only the most common types will be discussed here.

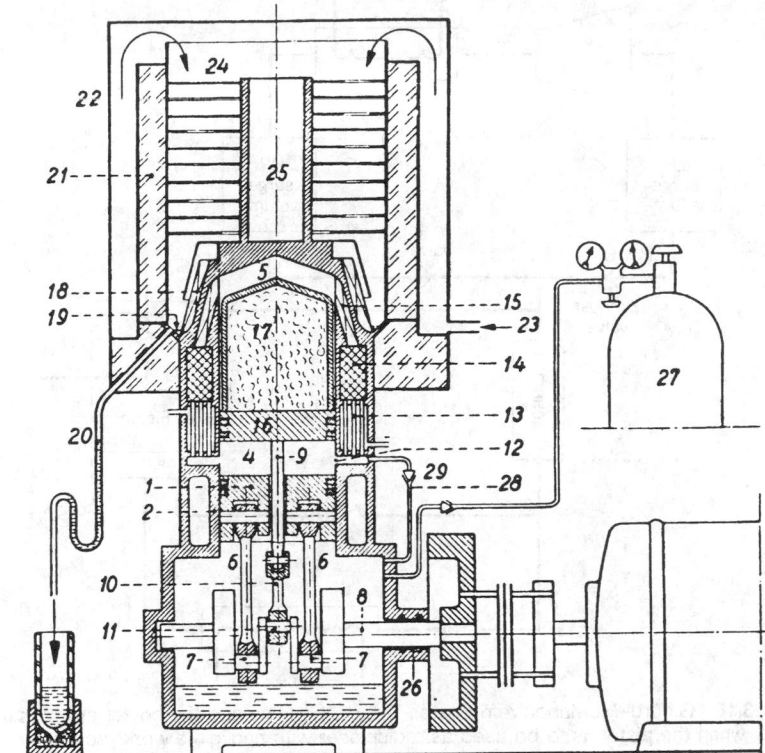

Fig. 63.17   Stirling cycle arranged for air liquefaction reference points have the following meanings: 1, compressor; 2, compression cylinder; 4, working fluid in space between compressor
and displacer; 5, working fluid in the cold head region of the machine; 6, two parallel connecting rods with cranks, 7, of the main piston; 8, crankshaft; 9, displacer rod, linked to connecting
rod, 10, and crank, 11, of the displacer; 12, ports; 13, cooler; 14, regenerator; 15, freezer; 16,
displacer piston, and 17, cap; 18, condenser for the air to be liquefied, with annular channel,
19, tapping pipe (gooseneck) 20, insulating screening cover, 21, and mantel 22; 23, aperture for
entry of air; 24, plates of the ice separator, joined by the tubular structure, 25, to the freezer
(15); 26, gas-tight shaft seal; 27, gas cylinder supplying refrigerant; 28, supply pipe with one-
way valve, 29. (Courtesy U.S. Philips Corp.)

63.3.1   Coiled-Tube-in-Shell Exchangers

The traditional heat exchanger for cryogenic service is the Hampson or coiled-tube-in-shell exchanger
as shown in Fig. 63.19. The exchanger is built by turning a mandrel on a specially built lathe, and
wrapping it with successive layers of tubing and spacer wires. Since longitudinal heat transfer down
the mandrel is not desired, the mandrel is usually made of a poorly conducting material such as
stainless steel, and its interior is packed with an insulating material to prevent gas circulation. Copper
or aluminum tubing is generally used. To prevent uneven flow distribution from tube to tube, tube
winding is planned so that the total length of each tube is constant independent of the layer on which
the tube is wound. This results in a constant winding angle, as shown in Fig. 63.20. For example,
the tube layer next to the mandrel might have five parallel tubes, whereas the layer next to the shell
might have 20 parallel tubes. Spacer wires may be laid longitudinally on each layer of tubes, or they
may be wound counter to the tube winding direction, or omitted. Their presence and size depends
on the flow requirements for fluid in the exchanger shell. Successive tube layers may be wound in
opposite or in the same direction.

After the tubes are wound on the mandrel they are fed into manifolds at each end of the tube
bundle. The mandrel itself may be used for this purpose, or hook-shaped manifolds of large diameter
tubing can be looped around the mandrel and connected to each tube in the bundle. Finally, the
exchanger is closed by wrapping a shell, usually thin-walled stainless steel, over the bundle and
welding on the required heads and nozzles.

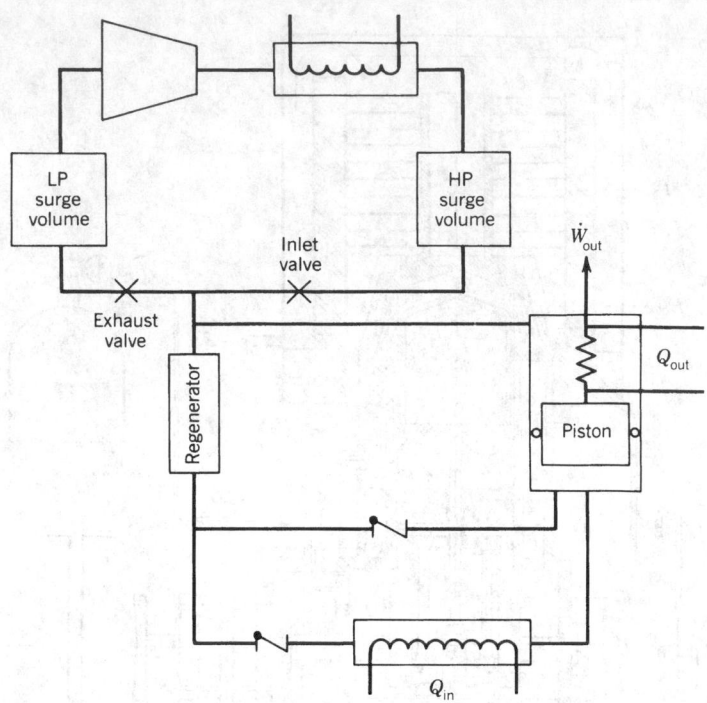

Fig. 63.18 Gifford–McMahon refrigerator. The dashed line and the cooler are present only when the piston is to be used as a displacer with negligible work production.

In application the low-pressure fluid flows through the exchanger shell, and high-pressure fluids flow through the tubes. This exchanger is easily adapted for use by three or more fluids by putting in a pair of manifolds for each tube-side fluid to be carried. However, tube arrangement must be carefully engineered so that the temperatures of all the cooling streams (or all the warming streams) will be identical at any given exchanger cross section. The exchanger is typically mounted vertically so that condensation and gravity effects will not result in uneven flow distribution. Most often the cold end is located at the top so that any liquids not carried by the process stream will move toward warmer temperatures and be evaporated.

Heat-transfer coefficients in these exchangers will usually vary from end to end of the exchangers because of the wide temperature range experienced. For this reason, and because of the nonlinear ΔT

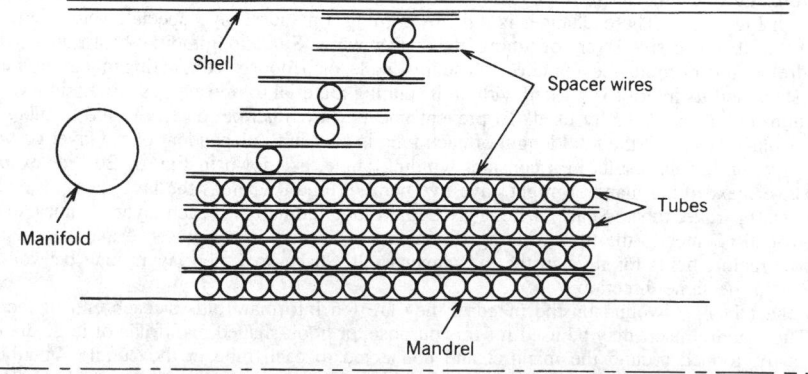

Fig. 63.19 Section of a coiled-tube-in-shell heat exchanger.

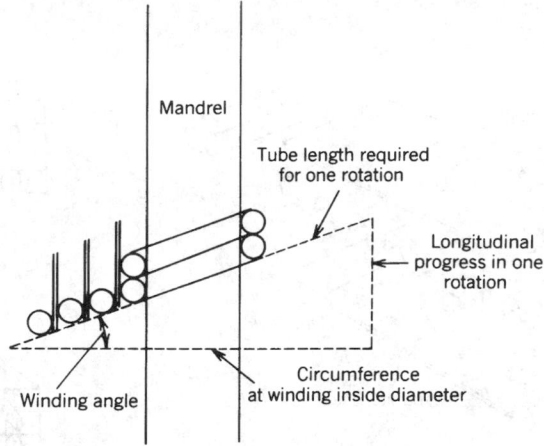

Fig. 63.20 Winding relationships for a coiled-tube-in-shell exchanger.

variations, the exchanger area must be determined by sections, the section lengths chosen so that linear ΔTs can be used and so that temperature ranges are not excessive. For inside tube heat-transfer coefficients with single-phase flow the Dittus–Boelter equation is used altered to account for the spiral flow:

$$\frac{hD}{k} = 0.023 \, N_{\text{Re}}^{0.8} N_{\text{Pn}}^{0.32} \left(1 + 3.5 \frac{d}{D} \right) \tag{63.7}$$

where D is the diameter of the helix and d the inside diameter.

For outside heat-transfer coefficients the standard design methods for heat transfer for flow across tube banks with in-line tubes are used. Usually the metal wall resistance is negligible. In some cases adjacent tubes are brazed or soldered together to promote heat transfer from one to the other. Even here wall resistance is usually a very small part of the total heat-transfer resistance.

Pressure drop calculations are made using equivalent design tools. Usually the low-pressure-side ΔP is critical in designing a usable exchanger.

The coiled-tube-in-shell exchanger is expensive, requiring a large amount of hand labor. Its advantages are that it can be operated at any pressure the tube can withstand, and that it can be built over very wide size ranges and with great flexibility of design. Currently these exchangers are little used in standard industrial cryogenic applications. However, in very large sizes (14 ft diameter × 120 ft length) they are used in base-load natural gas liquefaction plants, and in very small size (finger sized) they are used in cooling sensors for space and military applications.

63.3.2 Plate-Fin Heat Exchangers

The plate-fin exchanger has become the most common type used for cryogenic service. This results from its relatively low cost and high concentration of surface area per cubic foot of exchanger volume. It is made by clamping together a stack of alternate flat plates and corrugated sheets of aluminum coated with brazing flux. This assembly is then immersed in molten salt where the aluminum brazes together at points of contact. After removal from the bath the salt and flux are washed from the exchanger paths, and the assembly is enclosed in end plates and nozzles designed to give the desired flow arrangement. Usually the exchanger is roughly cubic, and is limited in size by the size of the available salt bath and the ability to make good braze seals in the center of the core. The core can be arranged for countercurrent flow or for cross flow. Figure 63.21 shows the construction of a typical plate-fin exchanger.

Procedures for calculating heat-transfer and pressure loss characteristics for plate-fin exchangers have been developed and published by the exchanger manufacturers. Table 63.4 and Fig. 63.22 present one set of these.

63.3.3 Regenerators

A regenerator is essentially a storage vessel filled with particulate solids through which hot and cold fluid flow alternately in opposite directions. The solids absorb energy as the hot fluid flows through, and then transfer this energy to the cold fluid. Thus this solid acts as a short-term energy-storage

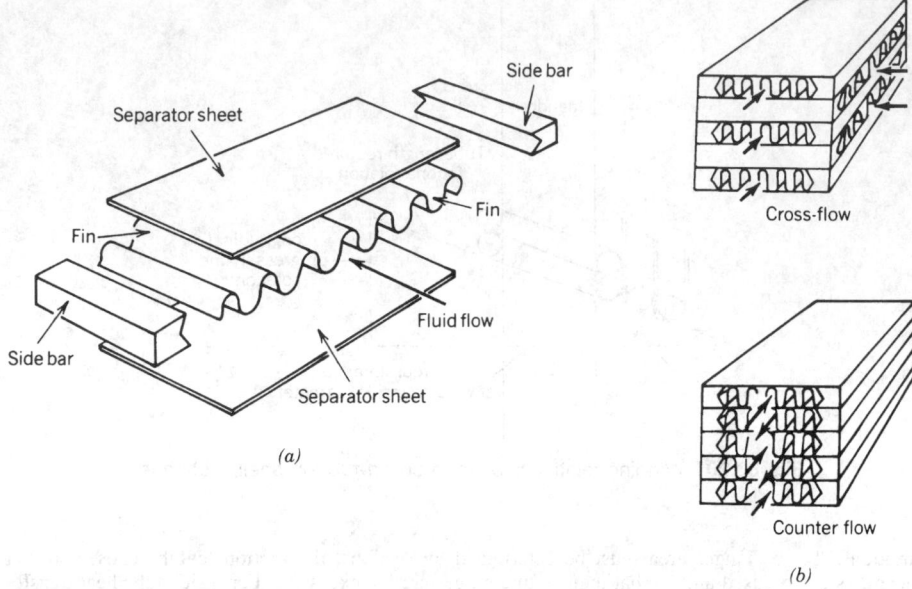

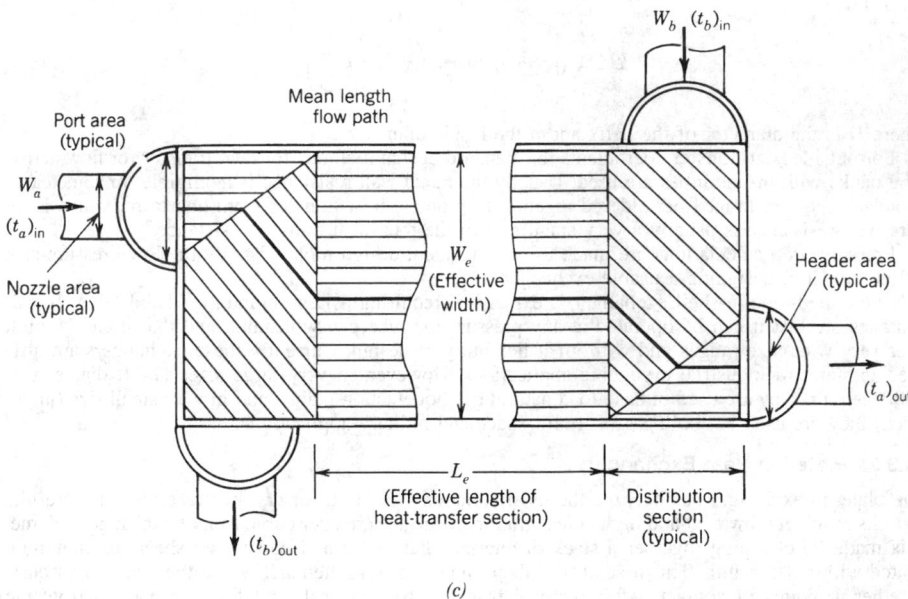

Fig. 63.21 Construction features of a plate-fin heat exchanger. (a) Detail of plate and fin. (b) Flow arrangements. (c) Total assembly arrangement.

medium. It should have high heat capacity and a large surface area, but should be designed as to avoid excessive flow pressure drop.

In cryogenic service regenerators have been used in two very different applications. In engine liquefiers very small regenerators packed with, for example, fine copper wire have been used. In these situations the alternating flow direction has been produced by the intake and exhaust strokes of the engine. In air separation plants very large regenerators in the form of tanks filled with pebbles have been used. In this application the regenerators have been used in pairs with one regenerator

Table 63.4 Computation of Fin Surface Geometrics[a]

Fin Height (in.)	Type of Surface	Fin Spacing (FPI)	Fin Thickness (in.)	A_c'	A_{ht}''	B	r_h	A_r/A_{ht}
0.200	Plain or perforated	14	0.008	0.001185	0.596	437	0.001986	0.751
0.200	Plain or perforated	14	0.012	0.001086	0.577	415	0.001884	0.760
0.250	Plain or perforated	10	0.025	0.001172	0.500	288	0.00234	0.750
0.375	Plain or perforated	8	0.025	0.001944	0.600	230	0.003240	0.778
0.375	Plain or perforated	15	0.008	0.00224	1.064	409	0.00211	0.862
0.250	⅛ lanced	15	0.012	0.001355	0.732	420	0.001855	0.813
0.250	⅛ lanced	14	0.020	0.001150	0.655	378	0.001751	0.817
0.375	⅛ lanced	15	0.008	0.00224	1.064	409	0.002108	0.862
0.455	Ruffled	16	0.005	0.002811	1.437	465	0.001956	0.893

[a] Definition and use of terms:

FPI = fins per inch

A_c' = free stream area factor, ft²/passage/in. of effective passage width

A_{ht}'' = heat-transfer area factor, ft²/passage/in./ft of effective length

B = heat-transfer area per unit volume between plates, ft²/ft³

r_h = hydraulic radius = cross section area/wetted perimeter, ft

A_r = effective heat-transfer area = $A_{ht} \cdot \eta_0$

A_{ht} = total heat-transfer area

η_0 = weighted surface effectiveness factor

   = $1 - (A_r/A_{ht})(1 - \eta_f)$

A_f = fin heat-transfer area

η_f = fin efficiency factor = [tanh (ml)]/ml

ml = fin geometry and material factor = $(b/s) \sqrt{2h/k}$

b = fin height, ft

h = film coefficient for heat transfer, Btu/hr · ft² · °F

k = thermal conductivity of the fin material, Btu/hr · ft · °F

s = fin thickness, ft

U = overal heat transfer coefficient = $1/(A/h_a A_a + A/h_b A_b)$

a,b = subscripts indicating the two fluids between which heat is being transferred

Courtesy Stewart-Warner Corp.

receiving hot fluid as cold fluid enters the other. Switch valves and check valves are used to alternate flow to the regenerator bodies, as shown in Fig. 63.23.

The regenerator operates in cyclical, unsteady-state conditions. Partial differential equations can be written to express temperatures of gas and of solid phase as a function of time and bed position under given conditions of flow rates, properties of gaseous and solid phases, and switch time. Usually these equations are solved assuming constant heat capacities, thermal conductivities, heat-transfer coefficients, and flow rates. It is generally assumed that flow is uniform throughout the bed cross section, that the bed has infinite conductivity in the radial direction but zero in the longitudinal direction, and that there is no condensation or vaporization occurring. Thermal gradients through the solid particles are usually ignored. These equations can then be solved by computer approximation. The results are often expressed graphically.[26]

An alternative approach compares the regenerator with a steady-state heat exchanger and uses exchanger design methods for calculating regenerator size.[27] Figure 63.24 shows the temperature–time relationship at several points in a regenerator body. In the central part of the regenerator ΔTs are nearly constant throughout the cycle. Folding the figure at the switch points superimposes the temperature data for this central section as shown in Fig. 63.25. It is clear that the solid plays only a time-delaying function as energy flows from the hot stream to the cold one. Temperature levels are set by the thermodynamics of the cooling curve such as Fig. 63.15 presents. Thus the $q = UA\Delta T$ equation can be used for small sections of the regenerator if a proper U can be determined.

During any half cycle the resistance to heat transfer from the gas to the solid packing will be just the gas-phase film coefficient. It can be calculated from empirical correlations for the packing material in use. For pebbles, the correlations for heat transfer to spheres in a packed bed[28] is normally used to obtain the film coefficient for heat transfer from gas to solid:

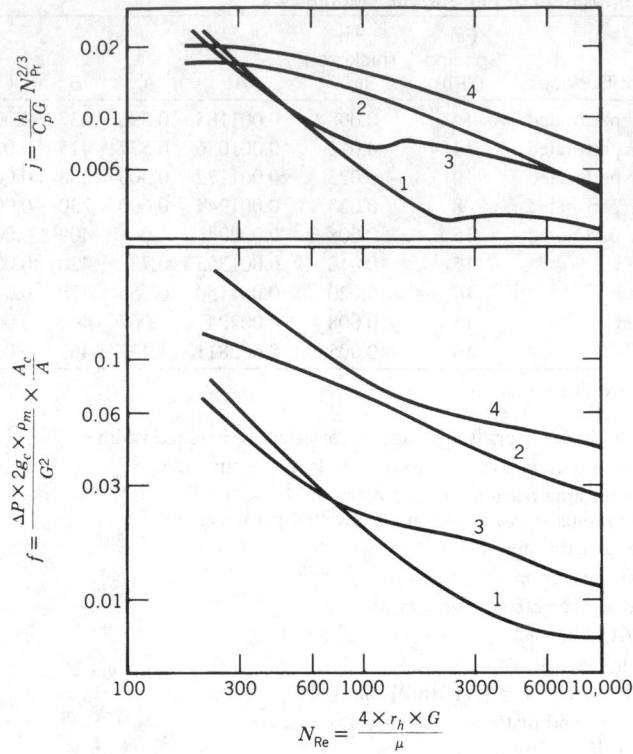

Fig. 63.22 Heat-transfer and flow friction factors in plate and fin heat exchangers. Curves 1: plain fin (0.200 in. height, 14 fins/in.—0.008 in. thick). Curves 2: ruffled fin (0.445 in. height, 16 fins/in.—0.005 in. thick). Curves 3: perforated fin (0.375 in. height, 8 fins/in.—0.025 in. thick). (Courtesy Stewart-Warner Corp.)

$$h_{gs} = 1.31(G/d)^{0.93} \qquad (63.8)$$

where h_{gs} = heat transfer from gas to regenerator packing or reverse, J/hr · m² · K
   G = mass flow of gas, kg/hr · m²
   d = particle diameter, m

The heat that flows to the packing surface diffuses into the packing by a conductive mode. Usually this transfer is fast relative to the transfer from the gas phase, but it may be necessary to calculate solid surface temperatures as a function of heat-transfer rate and adjust the overall ΔT accordingly. The heat-transfer mechanisms are typically symmetrical and hence the design equation becomes

$$A = \frac{q}{U\Delta T} = \frac{q}{h/2 \times \Delta T/2} \doteq \frac{4q}{h_{gs}\Delta T}$$

This calculation can be done for each section of the cooling curve until the entire regenerator area is calculated. However, at the ends of the regenerator temperatures are not symmetrical nor is the ΔT constant throughout the cycle. Figure 63.26 gives a correction factor that must be used to adjust the calculated area for these end effects. Usually a 10–20% increase in area results.

The cyclical nature of regenerator operation allows their use as trapping media for contaminants simultaneously with their heat-transfer function. If the contaminant is condensable, it will condense and solidify on the solid surfaces as the cooling phase flows through the regenerator. During the warming phase flow, this deposited condensed phase will evaporate flowing out with the return media.

Consider an air-separation process in which crude air at a moderate pressure is cooled by flow through a regenerator pair. The warmed regenerator is then used to warm up the returning nitrogen

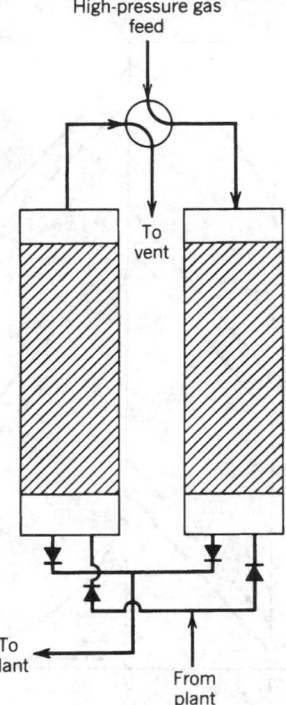

High-pressure gas
feed

To
vent

To
plant

From
plant

Fig. 63.23 Regenerator pair configuration.

at low pressure. The water and CO_2 in the air deposit on the regenerator surfaces and then reevaporate into the nitrogen. If deposition occurs at thermodynamic equilibrium, and assuming Raoult's law,

$$y_{H_2O} \text{ or } y_{CO_2} = \frac{P^\circ_{H_2O \text{ or } CO_2}}{P} \tag{63.9}$$

where y = mole fraction of H_2O or CO_2 in the gas phase
  P° = saturation vapor pressure of H_2O or CO_2
   P = total pressure of flowing stream

This equation can be applied to both the depositing, incoming situation and the reevaporating, outgoing situation. If the contaminant is completely removed in the regenerator, and the return gas is pure as it enters the regenerator, the moles of incoming gas times the mole fraction of contaminant must equal that same product for the outgoing stream if the contaminant does not accumulate in the regenerator. Since the vapor pressure is a function of temperature, and the returning stream pressure is lower than the incoming stream pressure, these relations can be combined to give the maximum stream-to-stream ΔT that may exist at any location in the regenerator. Figure 63.27 shows the results for one regenerator design condition. Also plotted on Fig. 63.27 is a cooling curve for these same design conditions. At the conditions given H_2O will be removed down to very low concentrations, but CO_2 solids may accumulate in the bottom of the regenerator. To prevent this it would be necessary to remove some of the air stream in the middle of the regenerator for further purification and cooling elsewhere.

  Cryogenic heat exchangers often are called on to condense or evaporate and two-phase heat-transfer commonly occurs, sometimes on both sides of a given heat exchanger. Heat-transfer coefficients and flow pressure losses are calculated using correlations taken from high-temperature data.[29] The distribution of multiphase processing streams into parallel channels is, however, a common and severe problem in cryogenic processing. In heat exchangers thousands of parallel paths may exist. Thus the designer must ensure that all possible paths offer the same flow resistance and that the two

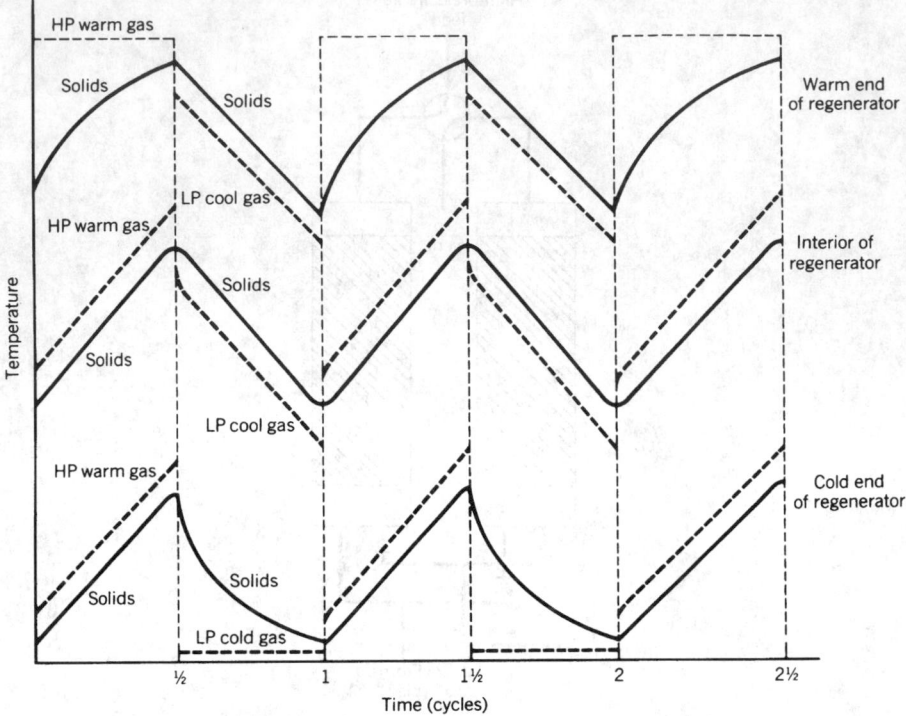

Fig. 63.24   Time–temperature histories in a regenerator.

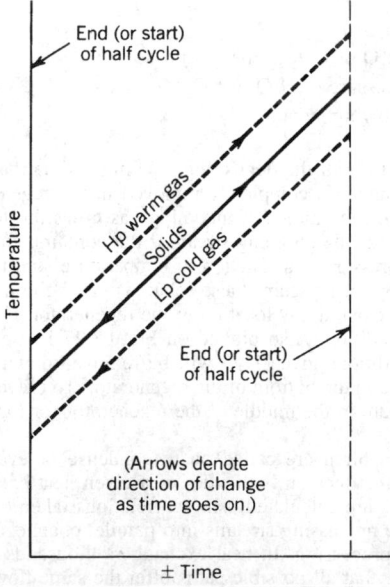

Fig. 63.25   Time–temperature history for a central slice through a regenerator.

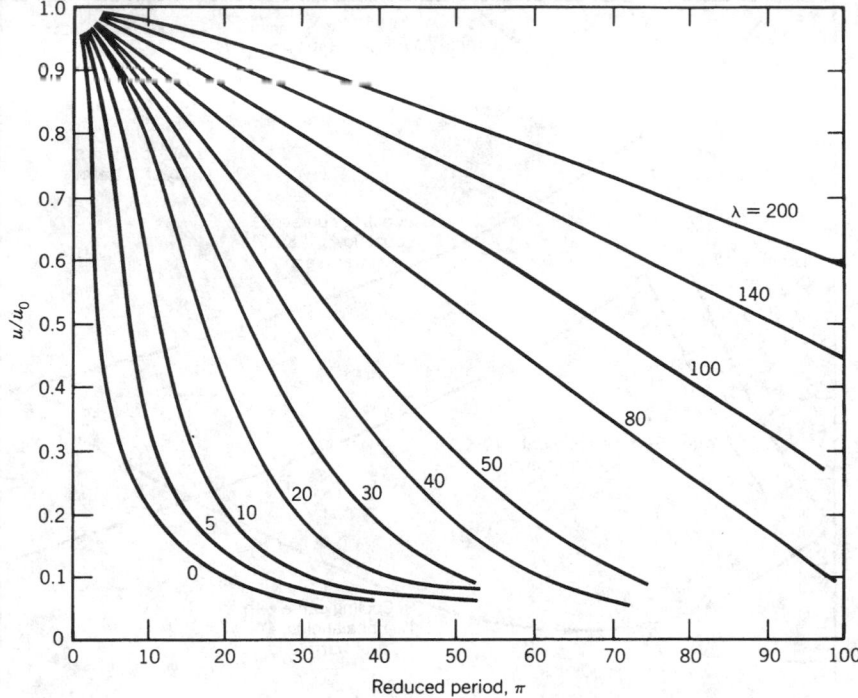

Fig. 63.26 End correction for regenerator heat transfer calculation using symmetrical cycle theory[27] (courtesy Plenum Press):

$$\lambda = \frac{4H_0 S(T_c + T_w)}{C_c T_c + C_w T_w} = \text{reduced length}$$

$$\pi = \frac{12 H_0(T_c + T_w)}{c\rho_s d} = \text{reduced period}$$

$$U_0 = \frac{1}{4}\left[\frac{1}{h} + \frac{0.1d}{k}\right]$$

where T_w, T_c = switching times of warm and cold streams, respectively, hr
S = regenerator surface area, m²
U_0 = overall heat transfer coefficient uncorrected for hysteresis, kcal/m² · hr · °C
U = overall heat transfer coefficient
C_w, C_c = heat capacity of warm and cold stream, respectively, kcal/hr · °C
c = specific heat of packing, kcal/kg · °C
d = particle diameter, m
ρ_s = density of solid, kg/m³

phases are well distributed in the flow stream approaching the distribution point. Streams that cool during passage through an exchanger are likely to be modestly self-compensating in that the viscosity of a cold gas is lower than that of a warmer gas. Thus a stream that is relatively high in temperature (as would be the case if that passage received more than its share of fluid) will have a greater flow resistance than a cooler system, so flow will be reduced. The opposite effect occurs for streams being warmed, so that these streams must be carefully balanced at the exchanger entrance.

63.4 INSULATION SYSTEMS

Successful cryogenic processing requires high-efficiency insulation. Sometimes this is a processing necessity, as in the Joule–Thomson liquefier, and sometimes it is primarily an economic requirement, as in the storage and transportation of cryogens. For large-scale cryogenic processes, especially those operating at liquid nitrogen temperatures and above, thick blankets of fiber or powder insulation, air

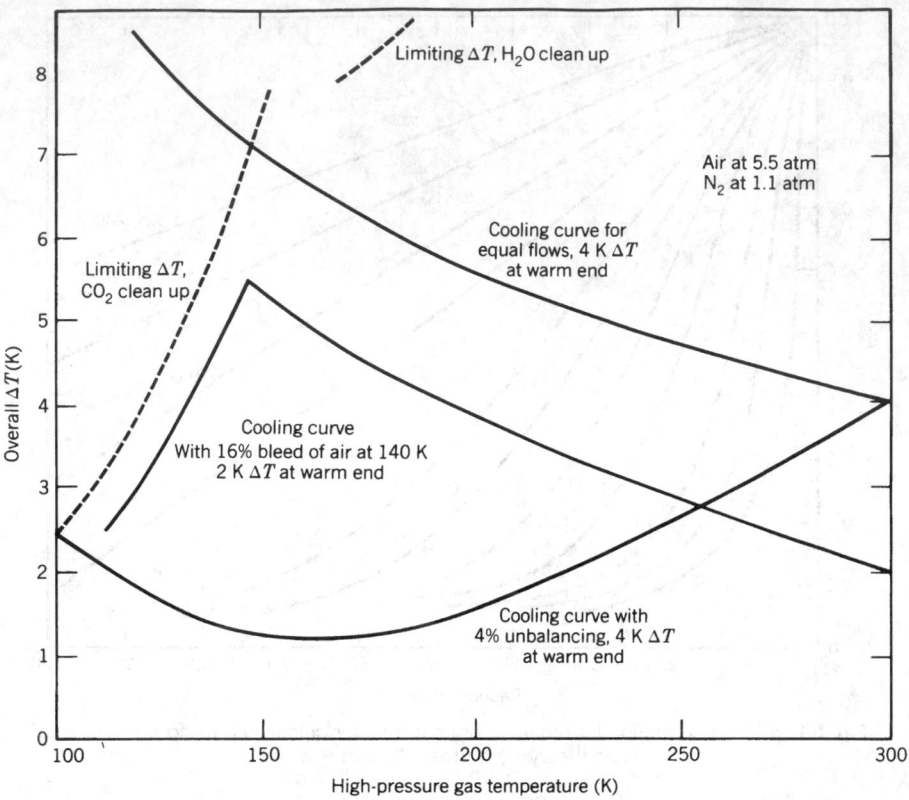

Fig. 63.27 ΔT limitation for contaminant cleanup in a regenerator.

or N_2 filled, have generally been used. For lower temperatures and for smaller units, vacuum insulation has been enhanced by adding one or many radiation shields, sometimes in the form of fibers or pellets, but often as reflective metal barriers. The use of many radiation barriers in the form of metal-coated plastic sheets wrapped around the processing vessel within the vacuum space has been used for most applications at temperatures approaching absolute zero.

63.4.1 Vacuum Insulation

Heat transfer occurs by convection, conduction, and radiation mechanisms. A vacuum space ideally eliminates convective and conductive heat transfer but does not interrupt radiative transfer. Thus heat transfer through a vacuum space can be calculated from the classic equation:

$$q = \sigma A F_{12}(T_1^4 - T_2^4) \qquad (63.10)$$

where q = rate of heat transfer, J/sec
   σ = Stefan-Boltzmann constant, 5.73×10^{-8} J/sec $\cdot$ m^2 $\cdot$ K
   F_{12} = combined emissivity and geometry factor
   T_1, T_2 = temperature (K) of radiating and receiving body, respectively

In this formulation of the Stefan–Boltzmann equation it is assumed that both radiator and receiver are gray bodies, that is, emissivity ϵ and absorptivity are equal and independent of temperature. It is also assumed that the radiating body loses energy to a totally uniform surroundings and receives energy from this same environment.

The form of the Stefan–Boltzmann equation shows that the rate of radiant energy transfer is controlled by the temperature of the hot surface. If the vacuum space is interrupted by a shielding surface, the temperature of that surface will become T_s, so that

$$q/A = F_{1s} (T_1^4 - T_s^4) = F_{s2} (T_s^4 - T_2^4) \qquad (63.11)$$

Since q/A will be the same through each region of this vacuum space, and assuming $F_{1s} = F_{s2} = F_{12}$

$$T_s = \sqrt[4]{\frac{T_1^4 + T_?^4}{2}} \tag{63.12}$$

For two infinite parallel plates or concentric cylinders or spheres with diffuse radiation transfer from one to the other,

$$F_{12} = 1 \left/ \frac{1}{\epsilon_1} + \frac{A_1}{A_2}\left(\frac{1}{\epsilon_2} - 1\right) \right. \tag{63.13}$$

If A_1 is a small body in a large enclosure, $F_{12} = \epsilon_1$. If radiator or receiver has an emissivity that varies with temperature, or if radiation is spectral, F_{12} must be found from a detailed statistical analysis of the various possible radiant beams.[30]

Table 63.5 lists emissivities for several surfaces of low emissivity that are useful in vacuum insulation.[31]

It is often desirable to control the temperature of the shield. This may be done by arranging for heat transfer between escaping vapors and the shield, or by using a double-walled shield in which is contained a boiling cryogen.

It is possible to use more than one radiation shield in an evacuated space. The temperature of intermediate streams can be determined as noted above, although the algebra becomes clumsy. However, mechanical complexities usually outweigh the insulating advantages.

63.4.2 Superinsulation

The advantages of radiation shields in an evacuated space have been extended to their logical conclusion in superinsulation, where a very large number of radiation shields are used. A thin, low emissivity material is wrapped around the cold surface so that the radiation train is interrupted often. The material is usually aluminum foil or aluminum-coated Mylar. Since the conductivity path must also be blocked, the individual layers must be separated. This may be done with glass fibers, perlite bits, or even with wrinkles in the insulating material; 25 surfaces/in. of thickness is quite common. Usually the wrapping does not fill in the insulating space. Table 63.6 gives properties of some available superinsulations.

Superinsulation has enormous advantages over other available insulation systems as can be seen from Table 63.6. In this table insulation performance is given in terms of effective thermal conductivity

$$k_e = \frac{q/A}{T/L} \tag{63.14}$$

where k_e = effective, or apparent, thermal conductivity
L = thickness of the insulation
$T = T_1 - T_2$

This insulating advantage translates into thin insulation space for a given rate of heat transfer, and into low weight. Hence designers have favored the use of superinsulation for most cryogen containers

Table 63.5 Emissivities of Materials Used for Cryogenic Radiation Shields

Material	Emissivity at		
	300 K	77.8 K	4.33 K
Aluminum plate	0.08	0.03	
Aluminum foil (bright finish)	0.03	0.018	0.011
Copper (commercial polish)	0.03	0.019	0.015
Monel	0.17	0.11	
304 stainless steel	0.15	0.061	
Silver	0.022		
Titanium	0.1		

Table 63.6   Properties of Various Multilayer Insulations (Warm Wall at 300 K)

Sample Thickness (cm)	Shields per Centimeter	Density (g/cm³)	Cold Wall T (K)	Conductivity (μW/cm · K)	Material[a]
3.7	26	0.12	76	0.7	1
3.7	26	0.12	20	0.5	1
2.5	24	0.09	76	2.3	2
1.5	76	0.76	76	5.2	3
4.5	6	0.03	76	3.9	4
2.2	6	0.03	76	3.0	5
3.2	24	0.045	76	0.85	5
1.3	47	0.09	76	1.8	5

[a] 1. Al foil with glass fiber mat separator.
2. Al foil with nylon net spacer.
3. Al foil with glass fabric spacer.
4. Al foil with glass fiber, unbonded spacer.
5. Aluminized Mylar, no spacer.

built for transport, especially where liquid H₂ or liquid He is involved, and for extraterrestrial space applications.

On the other hand, superinsulation must usually be installed in the field, and hence uniformity is difficult to achieve. Connections, tees in lines, and bends are especially difficult to wrap effectively. Present practice requires that layers of insulation be overlapped at a joint to ensure continuous coverage. Some configurations are shown in Fig. 63.28. Also, it has been found that the effectiveness

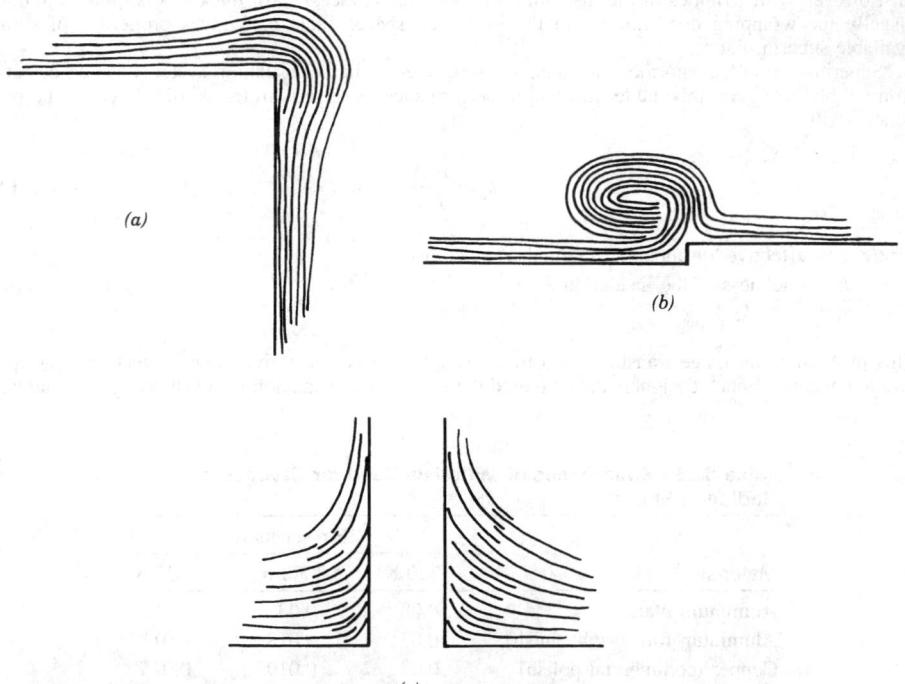

Fig. 63.28 Superinsulation coverage at joints and nozzles: (a) Lapped joint at corner. Also usable for nozzle or for pipe bend. (b) Rolled joint used at surface discontinuity, diameter change, or for jointure of insulation sections. (c) Multilayer insulation at a nozzle.

of superinsulation drops rapidly as the pressure increases. Pressures must be kept below 10^{-3} torr; evacuation is slow; a getter is required in the evacuated space; and all joints must be absolutely vacuum tight. Thus the total system cost is high.

63.4.3 Insulating Powders and Fibers

Fibers and powders have been used as insulating materials since the earliest of insulation needs. They retain the enormous advantage of ease of installation, especially when used in air, and low cost. Table 63.7 lists common insulating powders and fibers along with values of effective thermal conductivity.[32] Since the actual thermal conductivity is a function of temperature, these values may only be used for the temperature ranges shown.

For cryogenic processes of modest size and at temperatures down to liquid nitrogen temperature, it is usual practice to immerse the process equipment to be insulated in a cold box, a box filled with powder or fiber insulation. Insulation thickness must be large, and the coldest units must have the thickest insulation layer. This determines the placing of the process units within the cold box. Such a cold box may be assembled in the plant and shipped as a unit, or it can be constructed in the field. It is important to prevent moisture from migrating into the insulation and forming ice layers. Hence the box is usually operated at a positive gauge pressure using a dry gas, such as dry nitrogen. If rock wool or another such fiber is used, repairs can be made by tunneling through the insulation to the process unit. If an equivalent insulating powder, perlite, is used, the insulation will flow from the box through an opening into a retaining bag. After repairs are made, the insulation may be poured back into the box.

Polymer foams have also been used as cryogenic insulators. Foam-in-placed insulations have proven difficult to use because as the foaming takes place cavities are likely to develop behind process units. However, where the shape is simple and assembly can be done in the shop, good insulating characteristics can be obtained.

In some applications powders or fibers have been used in evacuated spaces. The absence of gas in the insulation pores reduces heat transfer by convection and conduction. Figure 63.29 shows the effect on a powder insulation of reducing pressure in the insulating space. Note that the pressures may be somewhat greater than that needed in a superinsulation system.

63.5 MATERIALS FOR CRYOGENIC SERVICE

Materials to be used in cryogenic service must operate satisfactorily in both ambient and cryogenic temperatures. The repeated temperature cycling that comes from starting up, operating, and shutting down this equipment is particularly destructive because of expansion and contraction that occur at every boundary and jointure.

63.5.1 Materials of Construction

Metals

Many of the normal metals used in equipment construction become brittle at low temperatures and fail with none of the prewarning of strain and deformation usually expected. Sometimes failure occurs at very low stress levels. The mechanism of brittle failure is still a topic for research. However, those metals that exhibit face-centered-cubic crystal lattice structure do not usually become brittle. The austenitic stainless steels, aluminum, copper, and nickel alloys are materials of this type. On the other hand, materials with body-centered-cubic crystal lattice forms or close-packed-hexagonal lattices are usually subject to a brittle transformation as the temperature is lowered. Such materials include the low-carbon steels and certain titanium and magensium alloys. Figure 63.30 shows these crystal forms and gives examples of notch toughness at room temperature and at liquid N_2 temperature for several example metals. In general carbon acts to raise the brittle transition temperature, and nickel lowers

Table 63.7 Effective Thermal Conductivity of Various Common Cryogenic Insulating Materials (300 to 76 K)

Material	Gas Pressure (mm Hg)	P (g/cm^2)	K (W/cm · K)
Silica aerogel (250A)	$<10^{-4}$	0.096	20.8×10^{-6}
	N_2 at 628	0.096	195.5×10^{-6}
Perlite (+30 mesh)	$<10^{-5}$	0.096	18.2×10^{-6}
	N_2 at 628	0.096	334×10^{-6}
Polystyrene foam	Air, 1 atm	0.046	259×10^{-6}
Polyurethane foam	Air, 1 atm	0.128	328×10^{-6}
Foamglas	Air, 1 atm	0.144	346×10^{-6}

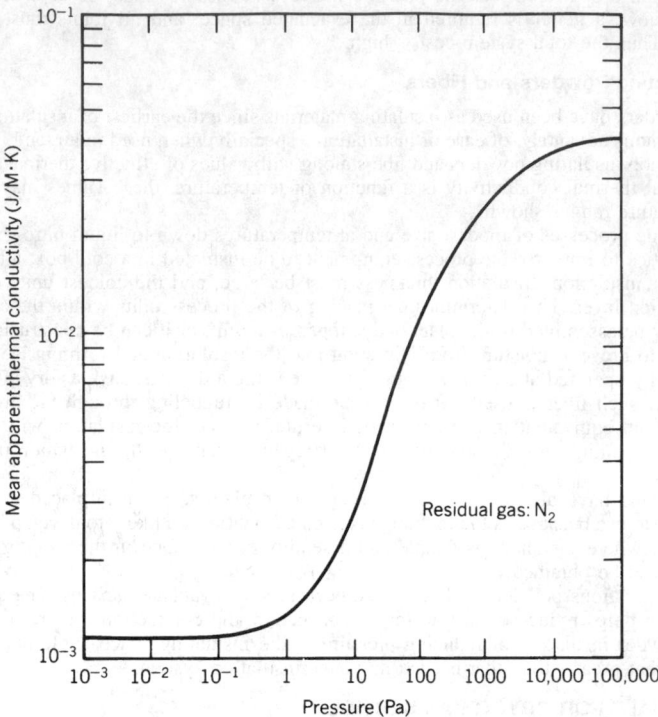

Fig. 63.29 Effect of residual gas pressure on the effective thermal conductivity of a powder insulation—perlite, 30–80 mesh, 300 to 78 K.

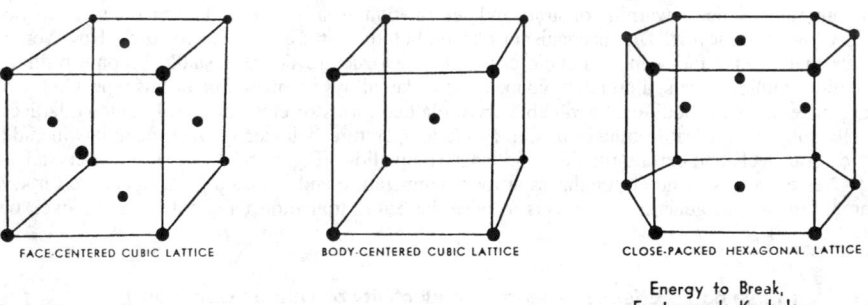

Metal	Crystal Lattice	Energy to Break, Foot-pounds Keyhole	
		Room Temperature	−320°F
Austenitic Stainless Steel	Face-centered Cubic	43	50
Aluminum	Face-centered Cubic	19	27
Copper	Face-centered Cubic	43	50
Nickel	Face-centered Cubic	89	99
Iron	Body-centered Cubic	78	1.5
Titanium	Close-packed Hexagonal	14.5	6.6
Magnesium	Close-packed Hexagonal	4	(3 at −105° F)

(Courtesy—American Society for Metals)

Fig. 63.30 Effect of crystal structure on brittle impact strengths of some metals. (Courtesy American Society for Metals.)

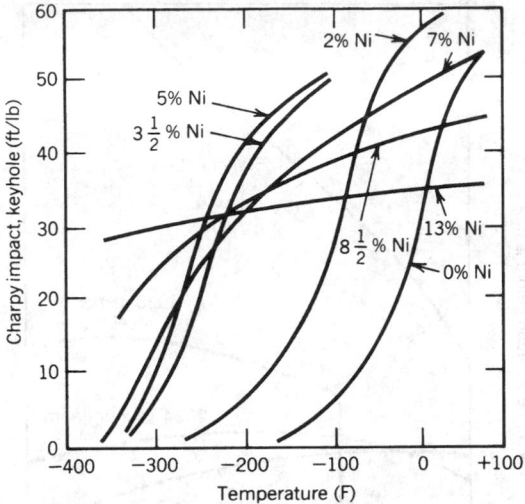

Fig. 63.31 Effect of nickel content in steels on Charpy impact values. (Courtesy American Iron and Steel Institute.)

it. Additional lowering can be obtained by fully killing steels by deoxidation with silicon and aluminum and by effecting a fine grain structure through normalizing by addition of selected elements.

In selecting a material for cryogenic service, several significant properties should be considered. The toughness or ductibility is of prime importance. Actually, these are distinctively different properties. A material that is ductile, as measured by elongation, may have poor toughness as measured by a notch impact test, particularly at cryogenic temperatures. Thus both these properties should be examined. Figures 63.31 and 63.32 show the effect of nickel content and heat treatment on Charpy impact values for steels. Figure 63.33 shows the tensile elongation before rupture of several materials used in cryogenic service.

Tensile and yield strength generally increase as temperature decreases. However, this is not always true, and the behavior of the particular material of interest should be examined. Obviously if the material becomes brittle, it is unusable regardless of tensile strength. Figure 63.34 shows the tensile and yield strength for several stainless steels.

Fatigue strength is especially important where temperature cycles from ambient to cryogenic are frequent, especially if stresses also vary. In cryogenic vessels maximum stress cycles for design are

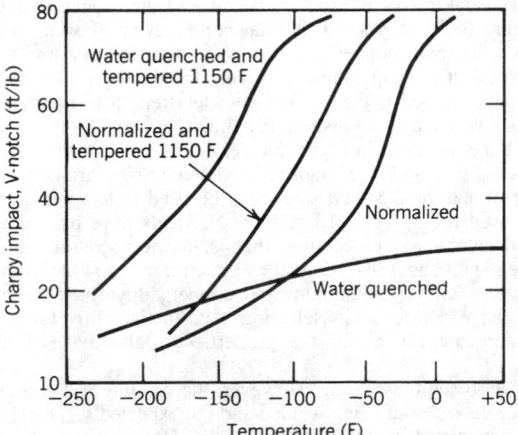

Fig. 63.32 Effect of heat treatment on Charpy impact values of steel. (Courtesy American Iron and Steel Institute.)

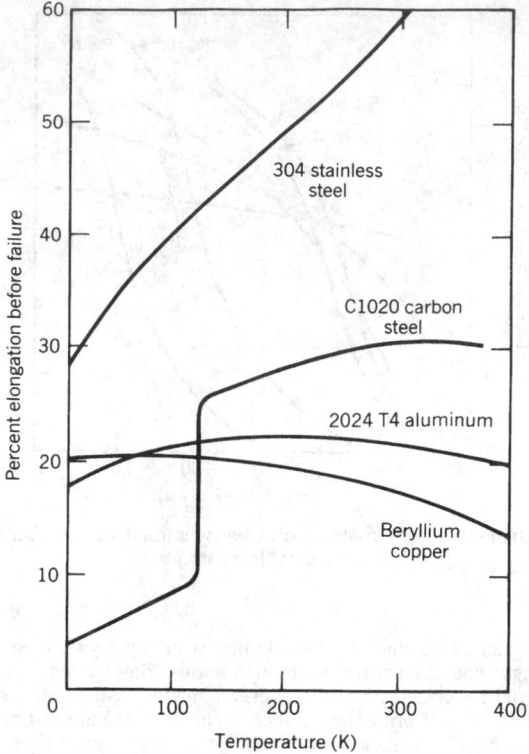

Fig. 63.33 Percent elongation before rupture of some materials used in cryogenic service.[33]

about 10,000–20,000 rather than the millions of cycles used for higher-temperature machinery design. Because fatigue strength data for low-temperature applications are scarce, steels used in cryogenic rotating equipment are commonly designed using standard room-temperature fatigue values. This allows a factor of safety because fatigue strength usually increases as temperature decreases.

Coefficient of expansion information is critical because of the stress that can be set up as temperatures are reduced to cryogenic or raised to ambient. This is particularly important where dissimilar materials are joined. For example, a 36-ft-long piece of 18-8 stainless will contract more than an inch in cooling from ambient to the boiling point of liquid H_2. And stainless steel has a coefficient of linear expansion much lower than that of copper or aluminum. This is seen in Fig. 63.35.

Thermal conductivity is an important property because of the economic impact of heat leaks into a cryogenic space. Figure 63.36 shows the thermal conductivity of some metals in the cryogenic temperature range. Note that pure copper shows a maximum at very low temperatures, but most alloys show only modest effect of temperature on thermal conductivity. One measure of the suitability of a material for cryogenic service is the ratio of tensile strength to thermal conductivity. On this basis stainless steel looks very attractive and copper much less so.

The most common materials used in cryogenic service have been the austenitic stainless steels, aluminum alloys, copper alloys, and aluminum-alloyed steels. Fine grained carbon-manganese steel and aluminum-killed steel and the 2.5% Ni steels can be used to temperatures as low as −50°C. A 3.5% Ni steel may be used roughly to −100°C; 5% Ni steels have been developed especially for applications in liquified natural gas processing, that is, for temperatures down to about −170°C. Austenitic stainless steels with about 9% Ni such as the common 304 and 316 types are usable well into the liquid H_2 range (−252°C). Aluminum and copper alloys have been used throughout the cryogenic temperature range. However, in selecting a particular alloy for a given application the engineer should consider carefully all of the properties of the material as they apply to that application.

Stainless steel may be joined by welding. However, the welding rod chosen and the joint design must both be selected for the material being welded and the expected service. For example, 9% nickel steel can be welded using nickel-based electrodes and a 60–80° single V joint design. Inert gas welding using Inconel-type electrodes is also acceptable. Where stress levels will not be high types

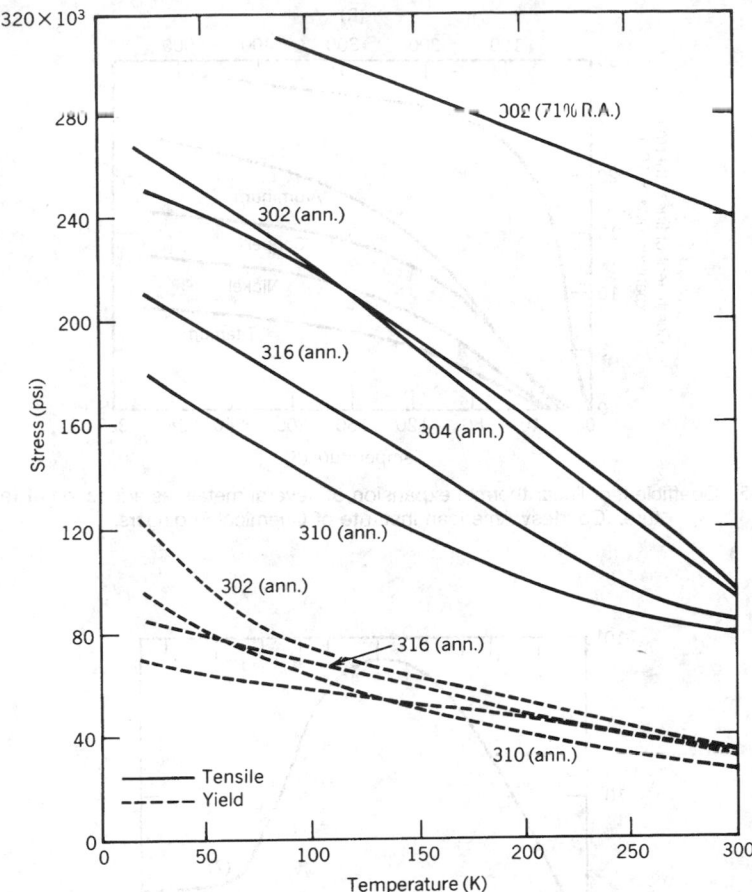

Fig. 63.34 Yield and tensile strength of several AISI 300 series stainless steels.[33] (Courtesy American Iron and Steel Institute.)

309 and 310 austenitic-stainless-steel electrodes can be used despite large differences in thermal expansion between the weld and the base metal.

Dissimilar metals can be joined for cryogenic service by soft soldering, silver brazing, or welding. For copper-to-copper joints a 50% tin/50% lead solder can be used. However, these joints have little ductility and so cannot stand high stress levels. Soft solder should not be used with aluminum, silicon-bronze, or stainless steel. Silver soldering is preferred for aluminum and silicon bronze and may also be used with copper and stainless steel.

Polymers

Polymers are frequently used as structural materials in research apparatus, as windows into cryogenic spaces, and for gaskets, O-rings, and other seals. Their suitability for the intended service should be as carefully considered as metals. At this point there is little accumulated, correlated data on polymer properties because of the wide variation in these materials from source to source. Hence properties should be obtained from the manufacturer and suitability for cryogenic service determined case by case.

Tables 63.8 and 63.9 list properties of some common polymeric materials. These are not all the available suitable polymers, but have been chosen especially for their compatibility with liquid O_2. For this service chemical inertness and resistance to flammability are particularly important. In addition to these, nylon is often used in cryogenic service because of its machinability and relative strength. Teflon and similar materials have the peculiar property of losing some of their dimensional stability at low temperatures; thus they should be used in confined spaces or at low stress levels.

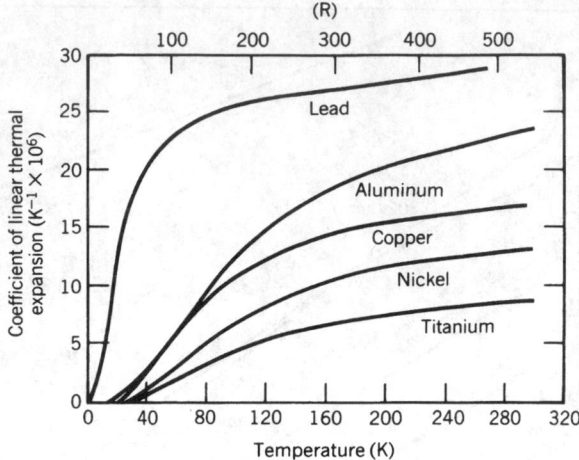

Fig. 63.35 Coefficient of linear thermal expansion of several metals as a function of temperature. (Courtesy American Institute of Chemical Engineers.)

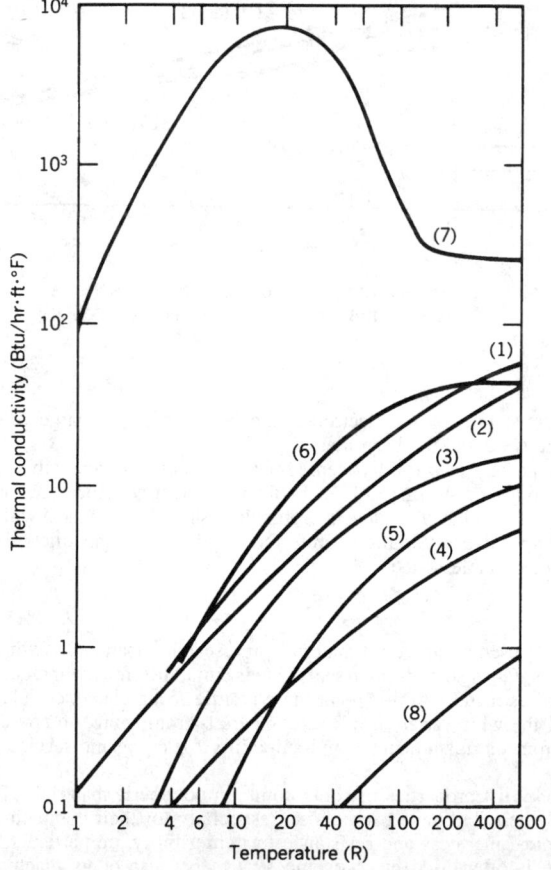

Fig. 63.36 Thermal conductivity of materials useful in low-temperature service. (1) 2024TA aluminum; (2) beryllium copper; (3) K-Monel; (4) titanium; (5) 304 stainless steel; (6) C1020 carbon steel; (7) pure copper; (8) Teflon.[35]

Table 63.8 Properties of Polymers Used in Cryogenic Service

Elastomer Type	Silicone Rubber	Vinylidene Fluoride Hexa-fluoropropylene	Fluorosilicone	Polytrifluoro-chloroethylene
Trade Name	Silastic[a] Silicone Rubber[e,f]	Viton[b] Fluore[c]	Silastic LS-53[a]	Kel-F[c,d]
Physical and Mechanical Properties				
Durometer range (shore A)	45–60	55–90	50–60	55–90
Specific gravity (base elastomer)	1.17–1.46	1.4–1.85	1.41–1.46	1.4–1.85
Density, lb/in.^{3} (base elastomer)	0.045	0.051–0.067	0.051	0.051–0.067
Tensile strength; psi:				
Pure gum	Under 400	>2000	1000	350–600
Reinforced	600–1500	—	—	—
Elongation, percent:				
Pure gum	Under 200	>350	200	500–800
Reinforced	200–800	—	—	—
Thermal conductivity, g, Btu/hr/ft^2/(°F/ft)	0.13	—	0.13	—
Coefficient of thermal expansion, cubical, in.^{3}/in.^{3}/°F	45×10^{-5}	27×10^{-5}	45×10^{-5}	—
Electrical insulation	Excellent	Excellent	Good	Excellent
Rebound				
Cold	Very good	Good	Very good	—
Hot	Very good	Excellent	Very good	—
Compression set	Good to excellent	Good to excellent	Good	Good to excellent
Resistance Properties				
Temperature:				
Tensile strength at 250°F, psi	850	300–800	—	300–800
Tensile strength at 400°F, psi	400	150–300	—	150–300
Elongation at 250°F, percent	350	100–350	—	100–350
Elongation at 400°F, percent	200	50–160	—	50–160
Low temperature brittle point, °F	−90 to −200	10 to −60	−90	10 to −60
Low temperature range of rapid stiffening, °F	−60 to −120	20 to −30	—	−20 to −30
Drift, room temperature	Poor to excellent	Good	—	Good
Drift, elevated temperature (158° to 212°F)	Excellent	Good to excellent	Excellent	Good to excellent

Table 63.8 (Continued)

Elastomer Type	Silicone Rubber	Vinylidene Fluoride Hexa-fluoropropylene	Fluorosilicone	Polytrifluoro-chloroethylene
Trade Name	Silastic[a] Silicone Rubber[e,f]	Viton[b] Fluore[c]	Silastic LS-53[a]	Kel-F[c,d]
Heat aging (212°F)	Excellent	Excellent	Excellent	Excellent
Maximum recommended continuous service temperature, °F	480	450	500	400
Minimum recommended service temperature, °F	−178	−50	−90	−60
Mechanical:				
Tear resistance	Poor	Poor to good	—	Poor to good
Abrasion resistance	Poor	Good	Poor	Good
Impact resistance (fatigue)	Poor	Poor to good	Poor	Poor to good
Chemical:				
Sunlight aging	Excellent	Excellent	Excellent	Excellent
Weather resistance	Excellent	Excellent	Excellent	$\longrightarrow$
Oxidation	Excellent	Excellent	Excellent	
Acids:				
Dilute	Very good	Excellent	Excellent	Poor to fair
Concentrated	Good	Good	Very good	Excellent
Alkali	Fair to excellent	Poor to fair	Fair to excellent	Good to excellent
Alcohol	Good	Excellent	Poor to good	Excellent
Petroleum products, resistance	Poor to fair	Good to excellent	Excellent	Good
Coal tar derivatives, resistance	Poor	Excellent	—	Excellent
Chlorinated solvents, resistance	Good	Good	—	Good
Hydraulic oils:				
Silicates	Poor	Good	—	Good
Phosphates	Good	Poor	—	Poor
Water swell resistance	Good	Good	Excellent	Good
Permeability to gases	Good	Excellent	Fair	Excellent

[a] Dow Corning Corp.
[b] E. I. duPont de Nemours.
[c] Minnesota Mining and Manufacturing Co.
[d] CTFE compounded with vinylidine fluoride.
[e] General Electric.
[f] Union Carbon and Carbide.

Table 63.9 Properties of Polymers Used in Cryogenic Service

Common Name	Fluorinated Ethylene Propylene	Polychlorotri-fluoroethylene	Polyvinylidene Fluoride	Polytetra-fluoro-ethylene	Polyimide
Trade Name	Teflon FEP[a]	Kel-F[b]	Kynar[c]	Fluorosint[c,e] Teflon TFE[a] Halon TFE[f]	Kapton H[a] Kapton F[a] Vespel[a] Polymer SP-1[a]
Physical and Mechanical Properties					
Specific gravity	2.14–2.17	2.1–2.2	1.76–1.77	2.13–2.22	1.42
Tensile strength, psi	2700–3100	4500–6000	7000	2000–4500	25,000[g]; 10,500
Elongation, percent	250–330	30–250	100–300	200–400	70[e]; 6–8
Tensile modulus, psi	0.5×10^5	$1.5 \times 10^5 - 3 \times 10^5$	1.2×10^5	0.58×10^5	4.3×10^5
Compressive strength, psi	2200	32,000–80,000	10,000	1700	24,400
Flexural strength, psi	—	7400–9300	—	—	14,000
Impact strength, ft-lb/in. of notch	No break	0.8–5.0	3.5	3.0	0.9
Rockwell hardness	R25	R110–R115	D80 (Shore)	D50–D65 (Shore)	H85–H95
Thermal conductivity, Btu/hr/ft²/(°F/in.)	1.75	0.9	0.9	1.75	2.2
Specific heat, Btu/lbm/°F	0.28	0.22	0.33	0.25	0.27
Coefficient of linear expansion, in./in./°F $\times 10^{-5}$	4.7×10^{-5} to 5.8×10^{-5}	5×10^{-5} to 15×10^{-5}	6.7×10^{-5}	5.5×10^{-5}	28×10^{-5} to 35×10^{-5}
Volume resistivity, ohm-cm	$>2 \times 10^{18}$	1.2×10^{18}	2×10^{14}	$>10^{18}$	10^{18}
Clarity	Transparent to translucent	transparent to translucent	Transparent to translucent	Opaque	Opaque
Processing Properties					
Molding qualities	Excellent	Excellent	Excellent	—	—
Injection molding temperature, °F	625–760	440–600	450–550	—	—
Mold shrinkage, in./in.	0.03–0.06	0.005–0.010	0.030	—	—
Machining qualities	Excellent	Excellent	Excellent	Excellent	—

Table 63.9 *(Continued)*

Common Name	Fluorinated Ethylene Propylene	Polychlorotrifluoroethylene	Polyvinylidene Fluoride	Polytetrafluoroethylene	Polyimide
Trade Name	Teflon FEP[a]	Kel-F[b]	Kynar[c]	Fluorosint[d,e] Teflon TFE[a] Halon TFE[f]	Kapton H[a] Kapton F[a] Vespel[a] Polymer SP-1[a]
Resistance Properties					
Mechanical abrasion and wear Tabor CS 17 wheel mg, loss/1000 cycles	—	0.01	17.6	—	—
Temperature:					
Flammability	None	None	Self-extinguishing	None	
Low temperature brittle point, °F	−420	−400	−80	−420	—
Resistance to heat, °F (continuous)	400	350–390	300	550	500
Deflection temperature under load, °F	—	258 (66 psi)	300 (66 psi), 195 (264 psi)	250 (66 psi)	—
Chemical:					
Effect of sunlight	None	None	Slight bleaching on long exposure	None	Degrades after prolonged exposure
Effect of weak acids	None →	→	None	None →	None
Effect of strong acids			Attacked by fuming sulfuric		None
Effect of weak alkalies			None		—
Effect of strong alkalies		Halogenated compounds cause slight swelling	None		Attacked
Effect of organic solvents			Resists most solvents		Resistant to most organic solvents

[a] E. I. duPont de Nemours.
[b] Minnesota Mining and Manufacturing Co.
[c] Pennsalt Chemicals Corp.
[d] Polymer Corp. of Pennsylvania.
[e] Polypenco, Inc.
[f] Allied Chemical Corp.
[g] Film.

Glass

Glasses, especially Pyrex and quartz, have proven satisfactory for cryogenic service because of their amorphous structure and very small coefficient of thermal expansion. They are commonly used in laboratory equipment, even down to the lowest cryogenic temperatures. They have also successfully been used as windows into devices such as hydrogen bubble chambers that are built primarily of metal.

63.5.2   Seals and Gaskets

In addition to careful selection of materials, seals must be specially designed for cryogenic service. Gaskets and O-rings are particularly subject to failure during thermal cycling. Thus they are best if confined and/or constructed of a metal-polymer combination. Such seals would be in the form of metal rings with C or wedge cross sections coated with a sealant such as Kel-F, Teflon, or soft metal. Various designs are available with complex cross sections for varying degrees of deflection. The surfaces against which these seal should be ground to specified finish. Elastomers such as neoprene and Viton-A have proven to be excellent sealants if captured in a space where they are subjected to 80% linear compression. This is true despite the fact that they are both extremely brittle at cryogenic temperatures without this stress.

Adhesive use at low temperatures is strictly done on an empirical basis. Still, adhesives have been used successfully to join insulating and vapor barrier blankets to metal surfaces. In every case the criteria are that the adhesive must not become crystalline at the operating temperature, must be resistant to aging, and must have a coefficient of contraction close to that of the base surface. Polyurethane, silicone, and various epoxy compounds have been used successfully in various cryogenic applications.

63.5.3   Lubricants

The lubrication of cryogenic machinery such as valves, pumps, and expanders is a problem that has generally been solved by avoidance. Valves usually have a long extension between the seat and the packing gland. This extension is gas filled so that the packing gland temperature stays close to ambient. For low-speed bearings babbitting is usually acceptable, as is graphite and molybdenum sulfide. For high-speed bearings, such as those in turboexpanders, gas bearings are generally used. In these devices some of the gas is leaked into the rotating bearing and forms a cushion for rotation. If out-leakage of the contained gas is undesirable, N_2 can be fed to the bearing and controlled so that leakage of N_2 goes to the room and not into the cryogenic system. Bearings of this sort have been operated at speeds up to 100,000 rpm.

63.6   SPECIAL PROBLEMS IN LOW-TEMPERATURE INSTRUMENTATION

Cryogenic systems usually are relatively clean and free flowing, and they often exist at a phase boundary where the degrees of freedom are reduced by one. Although these factors ease measurement problems, the fact that the system is immersed in insulation and therefore not easily accessible, the desire to limit thermal leaks to the system, and the likelihood that vaporization or condensation will occur in instrument lines all add difficulties.

Despite these differences all of the standard measurement techniques are used with low-temperature systems, often with ingenious changes to adapt the device to low-temperature use.

63.6.1   Temperature Measurement

Temperature may be measured using liquid-in-glass thermometers down to about $-40°C$, using thermocouples down to about liquid H_2 temperature, and using resistance thermometers and thermistors down to about 1 K. Although these are the usual devices of engineering measurement laboratory measurements have been done at all temperatures using gas thermometers and vapor pressure thermometers.

Table 63.10 lists the defining fixed points of the International Practical Temperature Scale of 1968. This scale does not define fixed points below the triple point of equilibrium He.[36] Below that range the NBS has defined a temperature scale to 1 K using gas thermometry.[37] At still lower temperatures measurement must be based on the fundamental theories of solids such as paramagnetic and superconducting phenomena.[38]

The usefulness of vapor pressure thermometry is limited by the properties of available fluids. This is evident from Table 63.11. For example, in the temperature range from 20.4 to 24.5 K there is no usable material. Despite this, vapor pressure thermometers are accurate and convenient. The major problem in their use is that the hydraulic head represented by the vapor line between point of measurement and the readout point must be taken into account. Also, the measurement point must be the coldest point experienced by the device. If not, pockets of liquid will form in the line between the point of measurement and the readout point greatly affecting the reading accuracy.

Standard thermocouples may be used through most of the cryogenic range, but, as shown in Fig. 63.37 for copper–constantan, the sensitivity with which they measure temperature drops as the tem-

Table 63.10   Defining Fixed Points of the International Practical Temperature Scale, 1968

Equilibrium Point	T (K)
Triple point of equilibrium H_2	13.81
Boiling point of equilibrium H_2 ($P = 33330.6$ N/m^2)	17.042
Boiling point of equilibrium H_2 ($P = 1$ atm)	20.28
Boiling point of neon ($P = 1$ atm)	27.102
Triple point of O_2	54.361
Boiling point of O_2 ($P = 1$ atm)	90.188
Triple point of H_2O ($P = 1$ atm)	273.16
Freezing point of Zn ($P = 1$ atm)	692.73
Freezing point of Ag ($P = 1$ atm)	1235.08
Freezing point of Au ($P = 1$ atm)	1337.58

perature decreases. At low temperatures heat transfer down the thermocouple wire may markedly affect the junction temperature. This is especially dangerous with copper wires, as can be seen from Fig. 63.36. Also, some thermocouple materials, for example, iron, become brittle as temperature decreases. To overcome these difficulties special thermocouple pairs have been used. These usually involve alloys of the noble metals. Figure 63.37 shows the thermoelectric power, and hence sensitivity of three of these thermocouple pairs.

Resistance thermometers are also very commonly used for cryogenic temperature measurement. Metal resistors, especially platinum, can be used from ambient to liquid He temperatures. They are extremely stable and can be read to high accuracy. However, expensive instrumentation is required because resistance differences are small requiring precise bridge circuitry. Resistance as a function of temperature for platinum is well known.[36]

At temperatures below 60 K, carbon resistors have been found to be convenient and sensitive temperature sensors. Since the change in resistance per given temperature difference is large (580 ohms/K would be typical at 4 K) the instrument range is small, and the resistor must be selected and calibrated for use in the narrow temperature range required.

Germanium resistors that are single crystals of germanium doped with minute quantities of impurities are also used throughout the cryogenic range. Their resistance varies approximately logarithmically with temperature, but the shape of this relation depends on the amount and type of dopant. Again, the germanium semiconductor must be selected and calibrated for the desired service.

Thermistors, that is, mixed, multicrystal semiconductors, like carbon and germanium resistors, give exponential resistance calibrations. They may be selected for order-of-magnitude resistance changes over very short temperature ranges or for service over wide temperature ranges. Calibration is necessary and may change with successive temperature cycling. For this reason they should be temperature-cycled several times before use. These sensors are cheap, extremely sensitive, easily read, and available in many forms. Thus they are excellent indicators of modest accuracy but of high sensitivity, such as sensors for control action. They do not, however, have the stability required for high accuracy.

Table 63.11   Properties of Cryogens Useful in Vapor Pressure Thermometers

Substance	Triple Point (K)	Boiling Point (K)	Critical Point (K)	dP/dT (mm/K)	Hydraulic Heat at Boiling Point (K/cm^2)
^{3}He	—	3.19	3.32	790	0.000054
^{4}He	—	4.215	5.20	715	0.00013
p-H_2 (20.4 K equilibrium)	13.80	20.27	32.98	224	0.00023
Ne	24.54	27.09	44.40	230	0.0039
N_2	63.15	77.36	126.26	89	0.0067
Ar	83.81	87.30	150.70	80	0.013
O_2	54.35	90.18	154.80	79	0.011

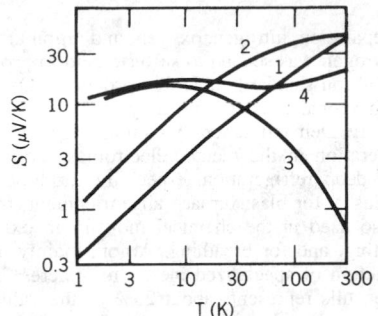

Fig. 63.37 Thermoelectric power of some thermocouples useful for cryogenic temperature measurement (courtesy Plenum Press): (1) Copper versus constantan; (2) Au + 2 at % Co versus silver normal (Ag + 0.37 at % Au); (3) Au + 0.03 at % Fe versus silver normal; (4) Au + 0.03 at % Fe versus Chromel.

63.6.2 Flow Measurement

Measurement of flow in cryogenic systems is often made difficult because of the need to deal with a liquid at its boiling point. Thus any significant pressure drop causes vaporization, which disrupts the measurement. This may be avoided by subcooling the liquid before measurement. Where this is possible, most measurement problems disappear, for cryogenic fluids are clean, low-viscosity liquids. Where subcooling is not possible, flow is most often measured using turbine flow meters or momentum meters.

A turbine meter has a rotor mounted axially in the flow stream and moved by the passing fluid. The rate of rotation, which is directly proportional to the volumetric flow rate, is sensed by an electronic counter that senses the passage of each rotor blade. There are two problems in the use of turbine meters in cryogenic fluids. First, these fluids are nonlubricating. Hence the meter rotor must be self-lubricated. Second, during cool-down or warm-up slugs of vapor are likely to flow past the rotor. These can flow rapidly enough to overspeed and damage the rotor. This can be avoided by locating a bypass around the turbine meter shutting off the meter during unsteady operation.

Momentum meters have a bob located in the flow stream to the support of which a strain gage is attached. The strain gage measures the force on the bob, which can be related through drag calculations or correlation to the rate of fluid flow past the bob. These meters are flexible and can be wide of range. They are sensitive to cavitation problems and to overstrain during upsets. Generally, each instrument must be calibrated.

63.6.3 Tank Inventory Measurement

The measurement of liquid level in a tank is made difficult by the cryogenic insulation requirements. This is true of stationary tanks, but even more so when the tank is in motion, as on a truck or spaceship, and the liquid is sloshing.

The simplest inventory measurement is by weight, either with conventional scales or by a strain gage applied to a support structure.

The sensing of level itself can be done using a succession of sensors that read differently when in liquid than they do in vapor. For instance, thermistors can be heated by a small electric current. Such devices cool quickly in liquid, and a resistance meter can "count" the number of thermistors in its circuit that are submerged.

A similar device that gives a continuous reading of liquid depth would be a vertical resistance wire, gently heated, while the total wire resistance is measured. The cold, submerged, fraction of the wire can be easily determined.

Other continuous reading devices include pressure gages, either with or without a vapor bleed, that read hydrostatic head, capacitance probes that indicate the fraction of their length that is submerged, ultrasonic systems that sense the time required for a wave to return from its reflectance off the liquid level, and light-reflecting devices.

63.7 EXAMPLES OF CRYOGENIC PROCESSING

Here three common, but greatly different, cryogenic technologies are described so that the interaction of the cryogenic techniques discussed above can be shown.

63.7.1 Air Separation

Among the products from air separation, nitrogen, oxygen, and argon are primary and are each major items of commerce. In 1994 nitrogen was second to sulfuric acid in production volume of industrial inorganic chemicals, with 932 billion standard cubic feet produced. Oxygen was third at 600 billion standard cubic feet produced. These materials are so widely used that their demand reflects the general trend in national industrial activity. Demand generally increases by 3 to 5%/year. Nitrogen is widely used for inert atmosphere generation in the metals, electronics and semiconductor, and chemical industries, and as a source of deep refrigeration, especially for food freezing and transporation. Oxygen is used in the steel industry for blast furnace air enrichment, for welding and scarfing, and for alloying operation. It is also used in the chemical industry in oxidation steps, for wastewater treatment, for welding and cutting, and for breathing. Argon, mainly used in welding, in stainless steel making, and in the production of specialized inert atmospheres, has a demand of only about 2% of that of oxygen. However, this represents about 25% of the value of oxygen shipments, and the argon demand is growing faster than that of oxygen or nitrogen.

Since all of the industrial gases are expensive to ship long distances, the industry was developed by locating a large number of plants close to markets and sized to meet nearby market demand. Maximum oxygen plant size has now grown into the 3000 ton/day range, but these plants are also located close to the consumer with the product delivered by pipe line. Use contracts are often long-term take-or-pay rental arrangements.

Air is a mixture of about the composition shown in Table 63.12. In an air separation plant O_2 is typically removed and distilled from liquified air. N_2 may also be recovered. In large plants argon may be recovered in a supplemental distillation operation. In such a plant the minor constituents (H_2–Xe) would have to be removed in bleed streams, but they are rarely collected. When this is done the Ne, Kr, Xe are usually adsorbed onto activated carbon at low temperature and separated by laboratory distillation.

Figure 63.38 is a simplified flow sheet of a typical small merchant oxygen plant meeting a variety of O_2 needs. Argon is not separated, and no use is made of the effluent N_2. Inlet air is filtered and compressed in the first of four compression stages. It is then sent to an air purifier where the CO_2 is removed by reaction with a recycling NaOH solution in a countercurrent packed tower. Usually the caustic solution inventory is changed daily. The CO_2-free gas is returned to the compressor for the final three stages after each of which the gas is cooled and water is separated from it. The compressed gas then goes to an adsorbent drier where the remaining water is removed onto silica gel or alumina. Driers are usually switched each shift and regenerated by using a slip stream of dry, hot N_2 and cooled to operating temperature with unheated N_2 flow.

The compressed, purified air is then cooled in the main exchanger (here a coiled tube type, but more usually of the plate-fin type) by transferring heat to both the returning N_2 and O_2. The process is basically a variation of that invented by Georges Claude where part of the high-pressure stream is withdrawn to the expansion engine (or turbine). The remainder of the air is further cooled in the main exchanger and expanded through a valve.

The combined air stream, nearly saturated or partly liquefied, enters the bottom of the high-pressure column. This distillation column condenses nearly pure N_2 at its top using boiling O_2 in the low-pressure column as heat sink. If the low-pressure column operates at about 140 kN/m^2 (20 psia), the high-pressure column must operate at about 690 kN/m^2 (100 psia). The bottom product, called crude O_2, is about 65 mole % N_2. The top product from the high-pressure column, nearly pure N_2, is used as N_2 reflux in the low-pressure column.

The crude O_2 is fed to an activated carbon bed where hydrocarbons are removed, is expanded to low-pressure column pressure, goes through a subcooler in which it supplies refrigeration to the

Table 63.12 Approximate Composition of Dry Air

Component	Composition (mole %)
N_2	78.03
O_2	20.99
Ar	0.93
CO_2	0.03
H_2	0.01
Ne	0.0015
He	0.0005
Kr	0.00011
Xe	0.000008

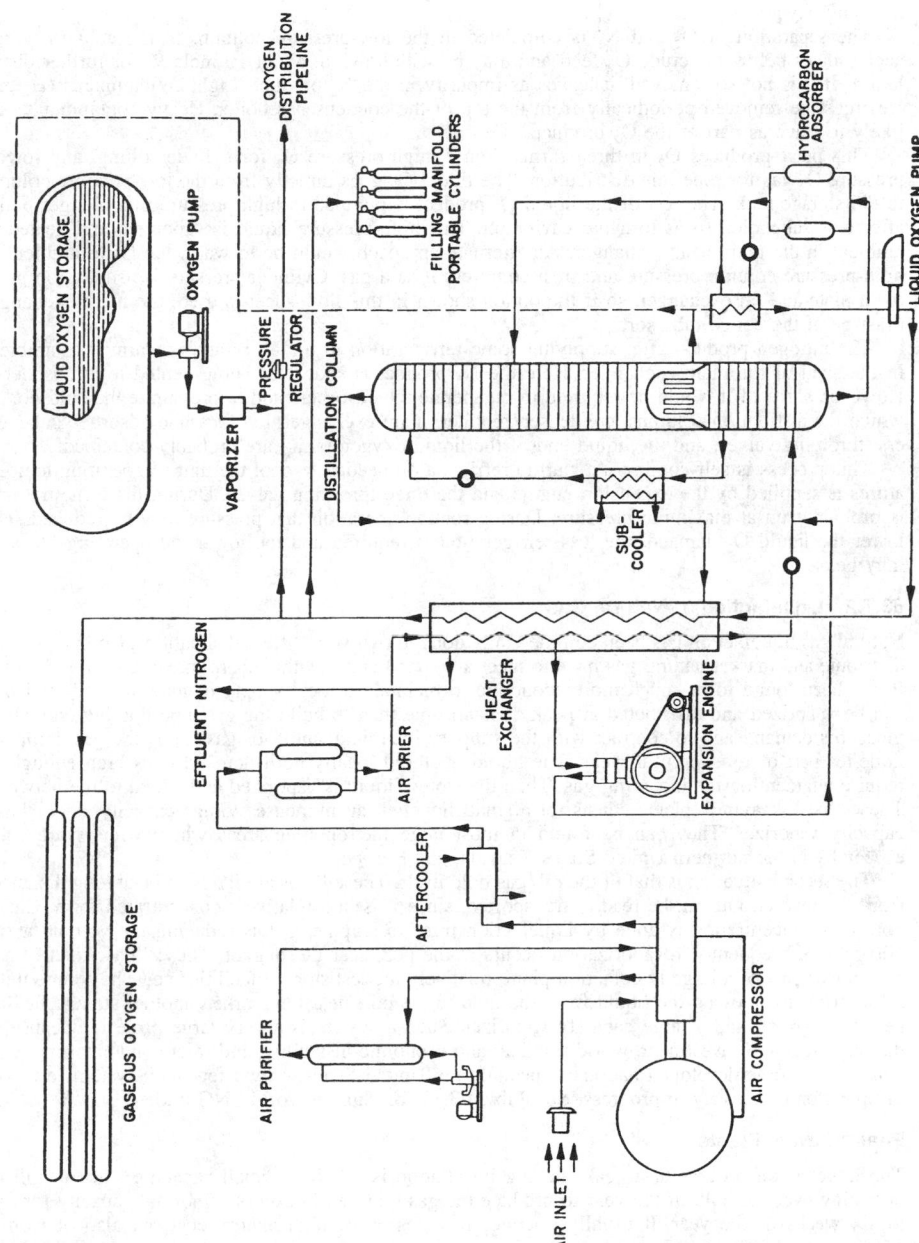

Fig. 63.38 Flow sheet of a merchant oxygen plant. (Courtesy Air Product and Chemicals, Inc.)

liquid O_2 product, and is fed to the low-pressure column. The hydrocarbons removed in the adsorber may come in as impurities in the feed or may be generated by decomposition of the compressor oil. If they are not fully removed, they are likely to precipitate in the liquid O_2 at the bottom of the low-pressure column. They accumulate there and can form an explosive mixture with oxygen whenever the plant is warmed up. Acetylene is especially dangerous in this regard because it is so little soluble in liquid oxygen.

The separation of O_2 and N_2 is completed in the low-pressure column. In the column, argon accumulates below the crude O_2 feed and may be withdrawn at about 10 mole % for further distillation. If it is not so removed, it leaves as impurity in the N_2 product. Light contaminants (H_2 and He) must be removed periodically from the top of the condenser/reboiler. Heavy contaminants are likely to leave as part of the O_2 product.

This plant produces O_2 in three forms: liquid, high-pressure O_2 for cylinder filling, and lower-pressure O_2 gas for pipe line distribution. The liquid O_2 goes directly from the low-pressure column to the storage tank. The rest of the liquid O_2 product is pumped to high pressure in a plunger pump after it is subcooled so as to avoid cavitation. This high-pressure liquid is vaporized and heated to ambient in the main heat exchanger. An alternate approach would be to warm the O_2 to ambient at high-pressure column pressure and then compress it as a gas. Cylinder pressure is usually too great for a plate-and-fin exchanger, so if the option shown in this flow sheet is used, the main exchanger must be of the coiled tube sort.

The nitrogen product, after supplying some refrigeration to the N_2 reflux, is warmed to ambient in the shell of the main exchanger. Here the N_2 product is shown as being vented to atmospheric. However, some of it would be required to regenerate the adsorbers and to pressurize the cold box in which the distillation columns, condenser/reboiler, main exchanger, hydrocarbon adsorber, subcoolers, throttling valves, and the liquid end of the liquid oxygen pump are probably contained.

This process is self-cooling. At startup refrigeration needed to cool the unit to operating temperatures is supplied by the expansion engine and the three throttling valves. During that time the unit is probably run at maximum pressure. During routine operation that pressure may be reduced. The lower the liquid O_2 demand, the less refrigeration is required and the lower the operating pressure may be.

63.7.2 Liquefaction of Natural Gas

Natural gas liquefaction has been commercially done in two very different situations. Companies that distribute and market natural gas have to meet a demand curve with a sharp maximum in midwinter. It has been found to be much more economic to maintain a local supply of natural gas liquid that can be vaporized and distributed at peak demand time than to build the gas pipe line big enough to meet this demand and to contract with the supplier for this quantity of gas. Thus the gas company liquefies part of its supply all year. The liquid is stored locally until demand rises high enough to require augmenting the incoming gas. Then the stored liquid is vaporized and added to the network. These "peak-shaving" plants consist of a small liquefier, an immense storage capacity, and a large capacity vaporizer. They can be found in most large metropolitan areas where winters are cold, especially in the northern United States, Canada, and Europe.

The second situation is that of the oil/gas field itself. These fields are likely to be at long distances from the market. Oil can be readily transported, since it is in a relatively concentrated form. Gas is not. This concentration is done by liquefaction prior to shipment, thus reducing the volume about 600-fold. Subsequently, revaporization occurs at the port near the market. These "base-load" LNG systems consist of a large liquefaction plant, relatively modest storage facilities near the source field, a train of ships moving the liquid from the field to the port near the market, another storage facility near the market, and a large capacity vaporizer. Such a system is a very large project. Because of the large required investment, world political and economic instability, and safety and environmental concerns in some developed nations, especially the United States, only a few such systems are now in operation or actively in progress. See Table 63.13 for data on world LNG trade.

Peak-Shaving Plants

The liquefaction process in a peak-shaving installation is relatively small capacity, since it will be operating over the bulk of the year to produce the gas required in excess of normal capacity for two to six weeks of the year. It usually operates in a region of high energy cost but also of readily available mechanical service and spare parts, and it liquefies relatively pure methane. Finally, operating reliability is not usually critical because the plant has capacity to liquefy the required gas in less time than in the maximum available.

For these reasons efficiency is more important than system reliability and simplicity. Cascade and various expander cycles are generally used, although a wide variety of processes have been used including the Stirling cycle.

Figure 63.39 shows a process in which an N_2 expander cycle is used for low-temperature refrigeration, whereas the methane itself is expanded to supply intermediate refrigeration. This is done because of the higher efficiency of N_2 expanders at low temperature and the reduced need for methane purification. The feed natural gas is purified and filtered and then split into two streams. The larger

Table 63.13   Data on World LNG Trade

Location	World's LNG Plants, 1994 Capacity, Million Metric Tons/yr	Parallel Liq. Trains	Country	World's LNG Imports, 1994 Quantity, Million Metric Tons/yr	Year	World's LNG Trade Amount, Million Metric Tons/yr
Kenai, Alaska	2.9	2	Japan	38.9	1980	22
Skikda, Algeria	6.2	8	S. Korea	4.4	1990	65
Arzew, Algeria	16.4	12	Taiwan	1.7	2000	90–95 (est)
Camel, Algeria	1.3	1	France	6.6	2010	130–160 (est)
Mersa, Libya	3.2	4	Other Europe	7.8		
Das Is., Abu Dhabi	4.3	2	U.S.A.	1.7		
Arun, Indonesia	9.0	5				
Bontang, Indonesia	13.2	7				
Lamut, Brunei	5.3	5				
Bintulu, Malaysia	7.5	3				
Barrup, Australia	6.0	3				

is cooled in part of the main exchanger, expanded in a turboexpander, and rewarmed to supply much of the warm end refrigeration, after which it is sent to the distribution system. The smaller fraction is cooled both by methane and by N_2 refrigeration until it is largely liquid, whereupon it goes to storage. Heavier liquids are removed by phase separation along the cooling path. Low-temperature refrigeration is supplied by a two-stage Claude cycle using N_2 as working fluid.

The LNG is stored in very large, insulated storage tanks. Typically such a tank might be 300 ft in diameter and 300 ft high. The height is made possible by the low density of LNG compared to other hydrocarbon liquids. LNG tanks have been built in ground as well as aboveground and of concrete as well as steel. However, the vast majority are aboveground steel tanks.

In designing and building LNG tanks the structural and thermal requirements added to the large size lead to many special design features. A strong foundation is necessary, and so the tank is often set on a concrete pad placed on piles. At the same time the earth underneath must be kept from freezing and later thawing and heaving. Thus electric cables or steam pipes are buried in the concrete to keep the soil above freezing. Over this pad a structurally sound layer of insulation, such as foam glass, is put to reduce heat leak to the LNG. The vertical tank walls are erected onto the concrete pad. The inner one is of stainless steel, the outer one is usually of carbon steel, and the interwall distance would be about 4 ft. The walls are field erected with welders carried in a tram attached to the top of the wall and lifted as the wall proceeds. The wall thickness is, of course, greater at the bottom than it is higher up.

The floor of the tank is steel laid over the foam glass and attached to the inner wall with a flexible joint. This is necessary because the tank walls will shrink upon cooling and expand when reheated. The dish roof is usually built within the walls over the floor. When the walls are completed, a flexible insulating blanket is put on the inside wall and the rest of the interwall space is filled with perlite. The blanket is necessary to counter the wall movement and prevent settling and crushing the perlite. At the end of construction the roof is lifted into position with slight air pressure. Usually this roof has hanging from it an insulated subroof that also rises and protects the LNG from heat leak to the roof. When this structure is in place, it is welded in and cover plates are put over the insulated wall spaces.

For safety considerations these tanks are usually surrounded by a berm designed to confine any LNG that escapes. LNG fire studies have shown such a fire to be less dangerous than a fire in an equivalent volume of gasoline. Still, the mass of LNG is so large that opportunities for disaster are seen as equally large. The fire danger will be reduced if the spill is more closely confined, and hence these berms tend to be high rather than large in diameter. In fact, a concrete tank berm built by the Philadelphia Gas Works is integral with the outside tank wall. That berm is of prestressed concrete thick enough to withstand the impact of a major commercial airliner crash.

Revaporization of LNG is done in large heat exchangers using air or water as heat sink. Shell and tube exchangers, radiators with fan-driven air for warming, and cascading liquid exchangers have all been used successfully, although the air-blown radiators tend to be noisy and subject to icing.

Base-Load LNG Plant

Table 63.13 lists the base-load LNG plants in operation in 1994. Products from these plants produce much of the natural gas used in Europe and in Japan, but United States use has been low, primarily because of the availability of large domestic gas fields.

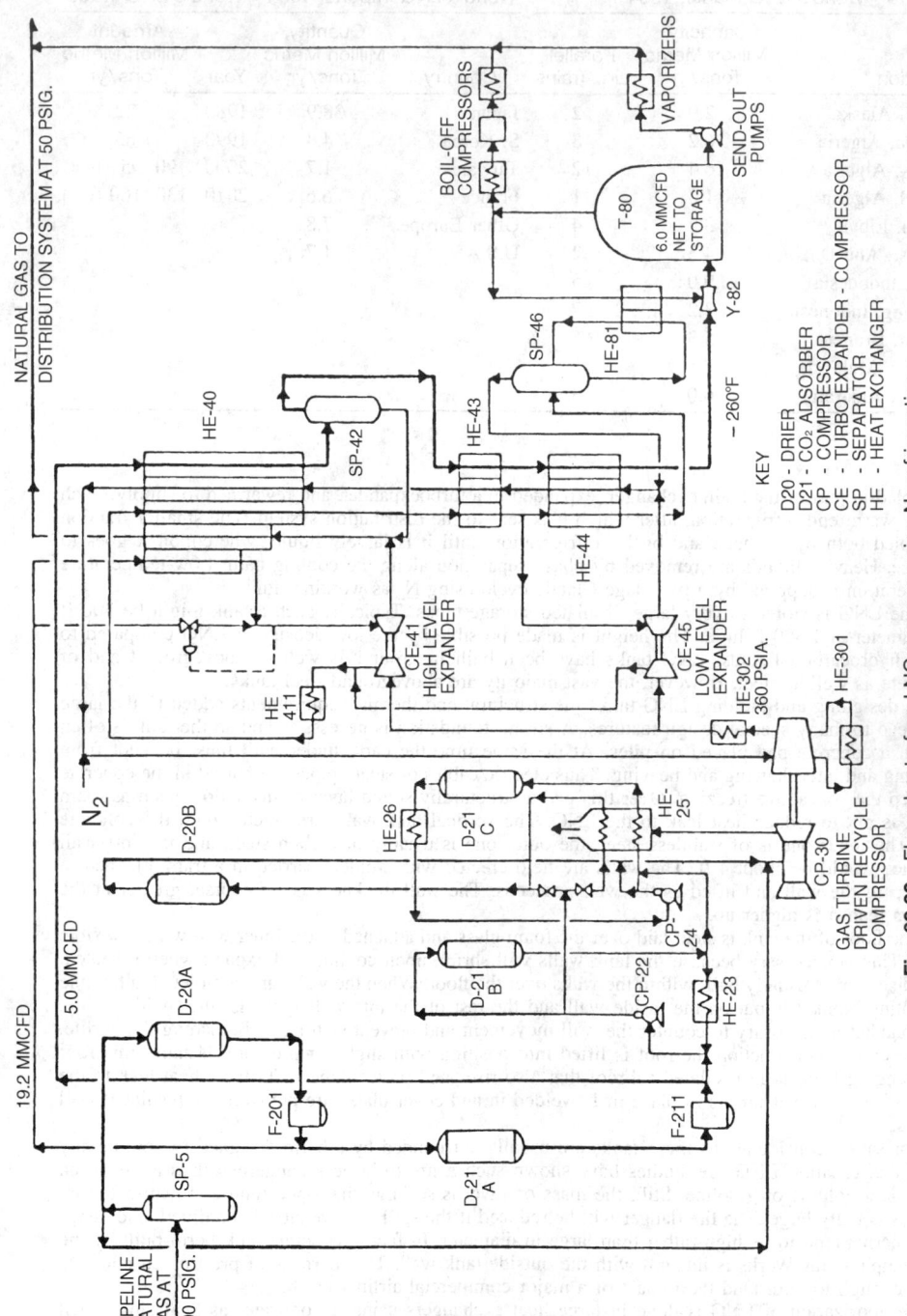

Fig. 63.39 Flow sheet of an LNG process using N_2 refrigeration.

KEY

D20 - DRIER
D21 - CO_2 ADSORBER
CP - COMPRESSOR
CE - TURBO EXPANDER - COMPRESSOR
SP - SEPARATOR
HE - HEAT EXCHANGER

In contrast to peak-shaving plants, liquefiers for these projects are large, primarily limited by the size of compressors and heat exchangers available in international trade. Also, these plants are located in remote areas where energy is cheap but repair facilities expensive or nonexistent. Thus, only two types of processes have been used: the classic cascade cycle and the mixed refrigerant cascade. Of these the mixed refrigerant cascade has gradually become dominant because of its mechanical simplicity and reliability.

Figure 63.40 shows a simplified process flow sheet of a mixed refrigerant cascade liquefier for natural gas. Here the natural gas passes through a succession of heat exchangers, or of bundles in a single heat exchanger, until liquified. The necessary refrigeration is supplied by a multicomponent refrigeration loop, which is essentially a Joule-Thomson cycle with successive phase separators to remove liquids as they are formed. These liquid streams are subcooled, expanded to low pressure, and used to supply the refrigeration required both by the natural gas and by the refrigerant mixture.

The success of this process depends on a selection of refrigerant composition that gives a cooling curve with shape closely matching the shape of the natural gas cooling curve. Thus all heat transfer will be across small ΔTs. This is shown in Fig. 63.41, a cooling curve for a mixed refrigerant cycle. The need to deal with a mixed refrigerant and to control the composition of the refrigerant mixture are the major difficulties with these processes. They complicate design, control, and general operation. For instance, a second process plant, nearly as large as the LNG plant, must be at hand to separate refrigerant components and supply makeup as needed by the liquefier.

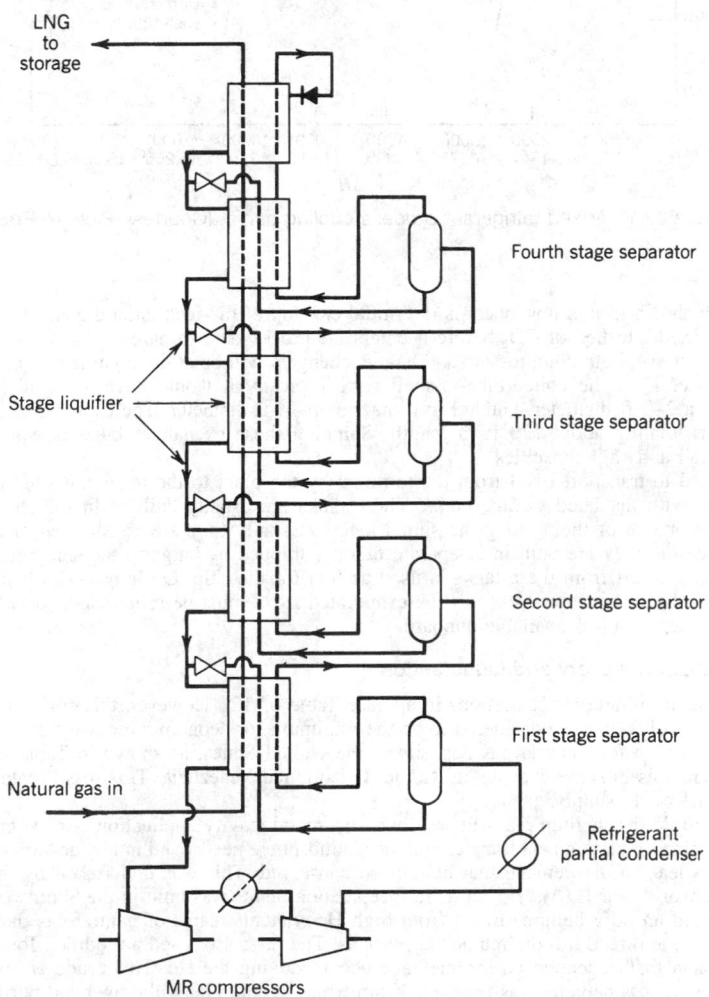

Fig. 63.40 Mixed refrigerant LNG process flow sheet. (Courtesy Plenum Press.)

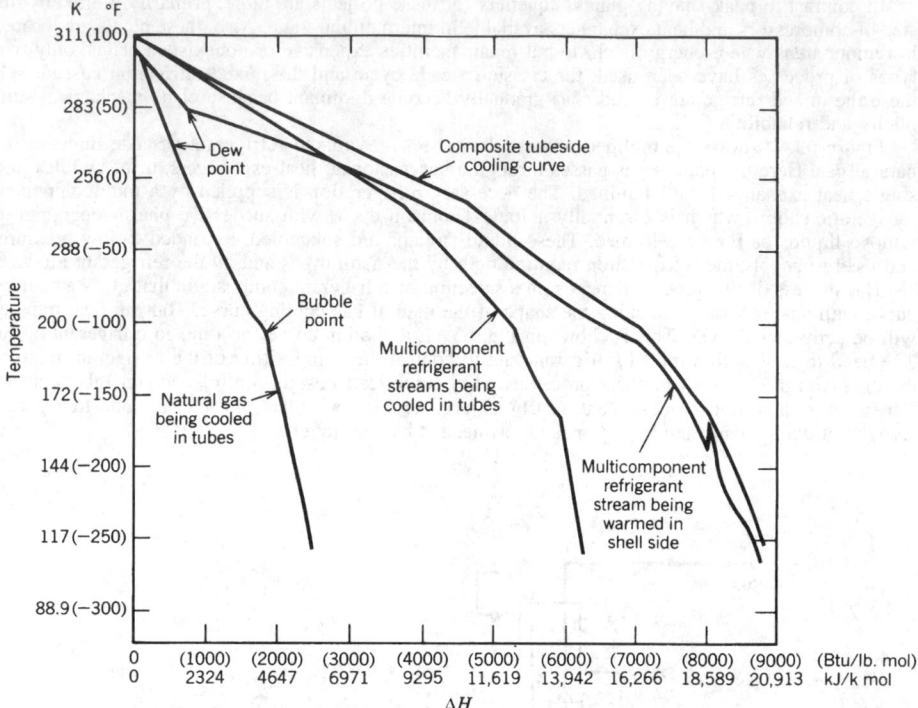

Fig. 63.41 Mixed refrigerant process cooling curve. (Courtesy Plenum Press.)

Also not shown in this flow sheet is the initial cleanup of the feed natural gas. This stream must be filtered, dried, purified of CO_2 before it enters the process shown here.

As noted above, both compressors and heat exchangers will be at the commercial maximum. The heat exchanger is of the coiled-tube-in-shell sort. Typically it would have ¾-in. aluminum tubes wrapped on a 2–3 ft diameter mandrel to a maximum 14-ft diameter. The exchanger is probably in two sections totaling about 120 ft in length. Shipping these exchanger bundles across the world challenges rail and ship capacities.

Ships used to transport LNG from the terminal by the plant to the receiving site are essentially supertankers with insulated storage tanks. These tanks are usually built to fit the ship hull. There may be four or five of them along the ship's length. Usually they are constructed at the shipyard, but in one design they are built in a separate facility, shipped by barge to the shipyard, and hoisted into position. Boiloff from these tanks is used as fuel for the ship. On long ocean hauls 6–10% of the LNG will be so consumed. In port the evaporated LNG must be reliquefied, for which purpose a small liquefier circuit is available onboard.

63.7.3 Helium Recovery and Liquefaction

Helium exists in minute concentrations in air (see Table 63.12). However, this concentration is well below the 0.3 vol % that is considered to be the minimum for economic recovery. It exists at higher concentrations in a few natural gas deposits in the United States, as shown in Table 63.14, and in like concentrations in some deposits in Russia, Poland, and Venezuela. This fossil material is apparently the total world supply.

The vital role that helium plays in welding, superconductivity applications, space program operations, medicine, in certain heat transfer and inert atmosphere needs, and in a wide variety of research requirements lead to the demand that helium be conserved. This was undertaken by the Bureau of Mines after World War II. A series of helium-separation plants was built in the Southwest. Generally these produced an 80% helium stream from high He-content steams of natural gas that would otherwise have gone directly to the municipal markets. The processes used a modified Joule-Thompson cooling system that depended on the methane accompanying the He. This crude He was stored in the Cliffside Field, a depleted gas reservoir, from which it could be withdrawn and purified. Most of

Table 63.14 Helium in Natural Gases in the United States

A. Composition of Some He-Rich Natural Gases in the United States

Location	Typical Composition (vol %)					
	CH_4	C_2H_6	N_2	CO_2	O_2	He
Colorado (Las Animas Co.)	0		77.6	14.7	0.3	7.4
Kansas (Waubaunsee, Elk, McPherson Cos.)	30	30	66.4	0.2	0	3.4
Michigan (Isabella Co.)	57.9	25.5	14.3	0	0.3	2.0
Montana (Musselshell)			54	30		16
Utah (Grand)	17		1.0	3.5		7.1

B. Estimated Helium Reserves (1994)

Location	Estimated Reserve (SCF)
Rocky Mountain area (Arizona, Colorado, Montana, New Mexico, Utah, Wyoming)	25×10^9
Midcontinent area (Kansas, Oklahoma, Texas)	169×10^9
He stored in the cliffside structure	30×10^9
Total	224×10^9

these plants shut down during the 1970s because of shifting government policies and budgetary limitations. In 1995 the last of these plants was closed down, as was the Bureau of Mines itself. The fate of the stored crude helium is being debated now (1996).

There are now about 30 billion standard cubic feet of crude He stored in the Cliffside reservoir, more than enough to supply the U.S. government needs estimated, at 10 Bcf through 2015. Total demand for U.S. helium is nearly constant at about 3 Bcf/yr (in 1994). Private industry supplies about 89% of this market, the rest coming from the stored government supply. The estimated He resources in helium-rich natural gas in the United States is about 240 Bcf as of 1994. With the stored He, this makes a total supply of about 270 Bcf, probably enough to supply the demand until the middle of the 21st century. Eventually technology will be needed to economically recover He from more dilute sources.

The liquefaction of He, or the production of refrigeration at temperatures in the liquid He range, requires special techniques. He, and also H_2, have negative Joule–Thomson coefficients at room temperature. Thus cooling must first be done with a modified Claude process to a temperature level of 30K or less. Often expanders are used in series to obtain temperatures close to the final temperature desired. An expansion valve may then be used to effect the actual liquefaction. Such a process is shown in Fig. 63.42. The goal of this process is the maintenance of a temperature low enough to sustain superconductivity (see below) using a conventional low-temperature superconductor. Since such processes are usually small, and since entropy gains at very low temperature are especially damaging to process efficiency, these processes must use very small ΔT's for heat transfer, require high-efficiency expanders, and must be insulated nearly perfectly. Note that in heat exchanger X4 the ΔT at the cold end is 0.55K.

63.8 SUPERCONDUCTIVITY AND ITS APPLICATIONS

For normal electrical conductors the resistance decreases sharply as temperature decreases, as shown in Fig. 63.42. For pure materials this decrease tends to level off at very low temperatures. This results from the fact that the resistance to electron flow results from two factors: the collision of electrons with crystal lattice imperfections and electron collisions with the lattice atoms themselves. The former effect is not temperature dependent, but the latter is. This relationship has, itself, proven of interest to engineers, and much thought and development has gone toward the building of power transmission lines operating at cryogenic temperatures and taking advantage of the reduced resistance.

63.8.1 Superconductivity

In 1911 Dr. Onnes of Leiden was investigating the electrical properties of metals at very low temperatures, helium having just been discovered and liquefied. He was measuring the resistance of frozen mercury as the temperature was reduced into the liquid He range. Suddenly the sample showed

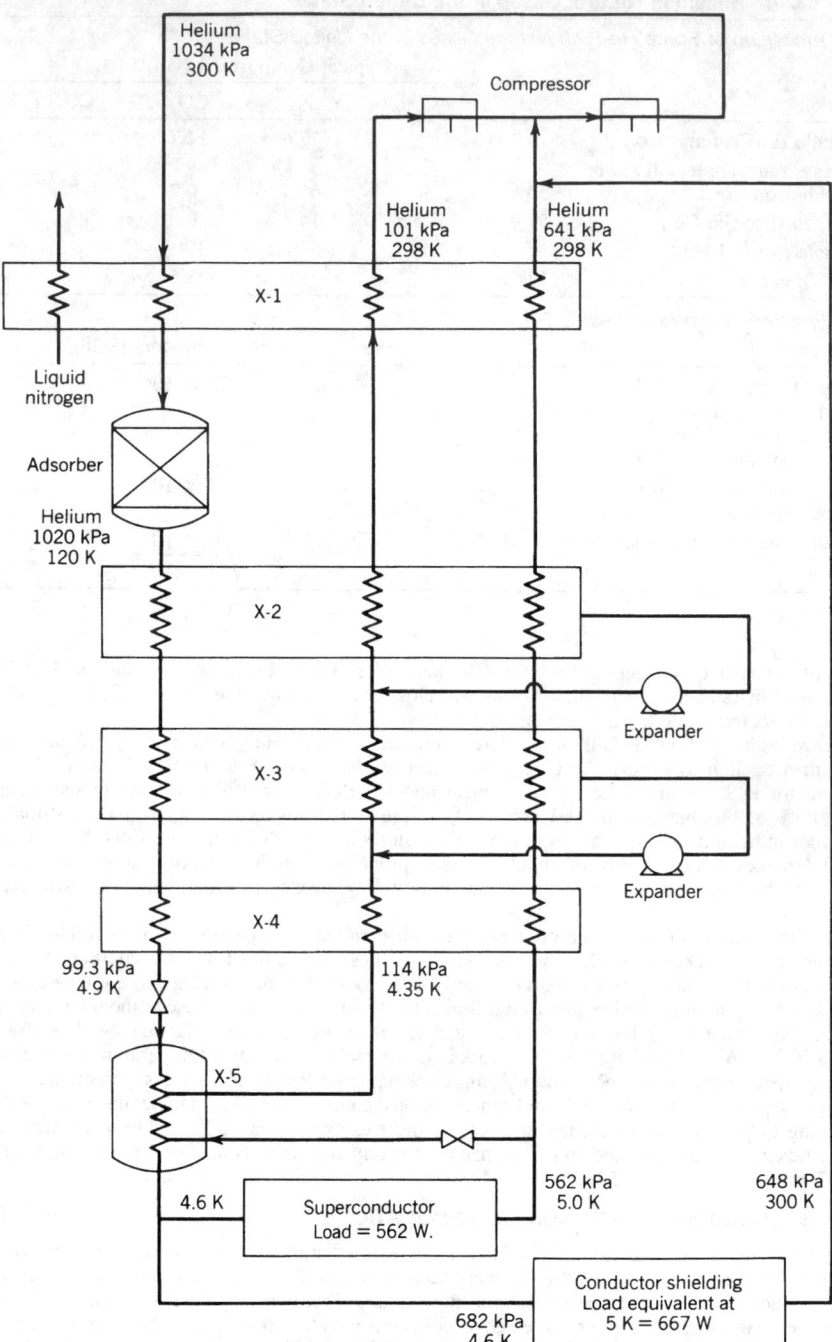

Fig. 63.42 Helium liquefier flow sheet.

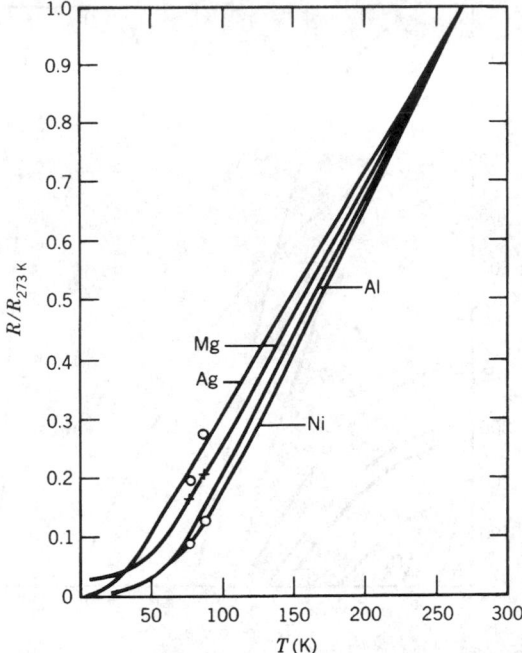

Fig. 63.43 Variation of resistance of metals with temperature.

zero resistance. At first a short circuit was suspected. However, very careful experiments showed that the electrical conductivity of the sample had dropped discontinuously to a very low value. The phenomenon of superconductivity has since been found to occur in a wide range of metals and alloys. The resistance of a superconductor has been found to be smaller than can be measured by the best instrumentation available. Possibly it is zero. Early on this was demonstrated by initiating a current in a superconducting ring which could then be maintained, undiminished, for months.

The phenomenon of superconductivity has been studied ever since in attempts to learn the extent of the phenomena, to develop a theory that will explain the basic mechanism and predict superconductive properties, and to use superconductivity in practical ways.

On an empirical basis it has been found that superconductors are diamagnetic, that is, they exclude a magnetic field, and that they exist within a region bounded by temperature and magnetic field strength. This is shown in Fig. 63.44. In becoming superconductive a material also changes in specific heat and in thermal conductivity.

The theory of supercoductivity developed after the discovery in 1933 by Meissner of the magnetic field exclusion. This led to a qualitative "two-fluid" model analogous to the theory underlying HeII. Since then this theory has been recast in quantum mechanical theory terms, most completely and successfully by J. Bardeen, L. N. Cooper, and R. Schrieffer of the University of Illinois in 1957 (BCS theory).

The BCS theory accounts for the Meissner effect and for other physical behavior phenomena of superconductors. It does not yet allow the prediction of superconductive transition points for new materials. The theory predicts an energy gap between normal and superconductive states existing simultaneously and visualizes the flow of paired electrons through the crystal lattice and the quantization of the magnetic flux. The quantized flux lines are fundamental to the explanation of the difference between type I superconductors that exhibit perfect Meissner effects, and which have relatively low transition temperatures and field tolerance, and type II superconductors that have imperfect Meissner effects, higher transition temperatures, and greater tolerance of magnetic fields. For example, Nb_3Sn, which is a type II superconductor, can be used for generation of magnetic fields of 100,000 gauss. These materials allow the penetration of magnetic field above a lower critical field strength, H_{c1}, but remain superconductive to a much greater field strength, H_{c2}. At fields above H_{c1}, flux enters the material in the form of quantized bundles that are pinned by dislocations so that the flux does not move easily and lead to normalization of the material.

Thus both HeII and superconductors may be considered examples of superfluids. Each exhibits a nondissipative process for mass transfer. In HeII the mass transferred is the fluid itself in inviscid

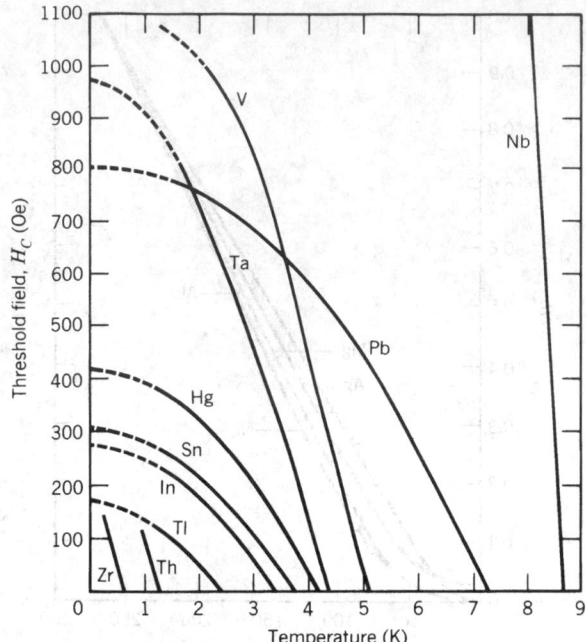

Fig. 63.44　Limits to superconductive behavior of some elements.

flow; with a superconductor it is the electrons flowing without encountering resistance. In both cases a flow velocity greater than a critical value restores normality. In superconductors circulating currents are set up in the penetration layer at the surface to cancel the applied magnetic field. When this field is increased to a critical value, the critical current is reached and the material reverts to the normal state.

63.8.2　Applications of Superconductivity

Development of applications of superconductivity has proceeded very rapidly over the past decade. However, by far the widest use of superconductivity has been in magnet applications. Superconductive magnets were constructed and tested soon after such practical superconductors as Nb_3Sn and rhodium–zirconium were discovered about 1960. Field strengths as high as 7 T were predicted then. Since then magnets in a wide range of sizes and shapes have been made, tested, and used with field strengths approaching 20 T.

The first three Nb alloys listed in Table 63.16 are ductile to the point that they are readily fabricated by conventional wire-drawing techniques. These form the cheapest finished product, but cannot be used at high field strengths. The Nb_3Sn, which is the most widely used superconductive

Table 63.15　Superconductive Elements with Zero-Field Transition Temperatures

Mg											Al 1.20^0	Si	P
Co	Sc	Ti 0.40^0	V 4.9^0	Cr	Mn	Fe	Co	Ni	Cu	Zn 0.88^0	Ga 1.10^0	Ge	As
Sr	Y	Zr 0.75^0	Nb 9.5^0	Mo 0.95^0	Tc 11.2^0	Ru 0.47^0	Rh	Pd	Ag	Cd 0.56^0	In 3.40^0	Sn 3.73^0	Sb
Ba	La 5.4^0	Hf 0.35^0	Ta 4.5^0	W	Re 1.7^0	Os 0.71^0	Ir 0.15^0	Pt	Au	Hg 4.15^0	Tl 2.39^0	Pb 7.22^0	Bi
Ra	Ac	Th 1.37^0	Pa	U-β 1.8^0									

Table 63.16 Some Commercially Available Superconductive Materials[a]

Material	T_c (K)	$H_c 2$ (T) at 4.2 K	J_c (10^5 A/cm^2 at 4.2 K)				Fabrication
			2.5 T	5 T	10 T	15 T	
Nb–25 wt% Zr	11	7.0	1.1	0.8	0	0	Fairly ductile
Nb–33 wt% Zr	11.5	8.0	0.9	0.8	0	0	Fairly ductile
Nb–48 wt% Ti	9.5	12.0	2.5	1.5	0.3	0	Ductile
Nb$_3$Sn	18.0	22.0	17.0	10.0	4.0	0.5	CVD diffusion bronze
V$_3$Ga	15.0	23.0	5.0	2.5	1.4	0.9	Diffusion bronze

[a] Courtesy Plenum Press.

material, and V$_3$Ga are formed into tape by chemical vapor deposition. The tape is clad with copper for stability and stainless steel for strength. Materials for multifilament conductor formation are produced by the bronze process. In this process filaments of Nb or V are drawn down in a matrix of Sn–Cu or Sn–Ga alloy, the bronze. Heat treatment then produces the Nb$_3$Sn or V$_3$Ga. The residual matrix is too resistive for satisfactory stabilization. Hence copper filaments are incorporated in the final conductors.

Multifilament conductors are then made by assembling superconductive filaments in a stabilizing matrix. For example, in one such conductor groups of 241 Ni–Ti filaments are sheathed in copper and cupronickel and packed in a copper matrix to make a 13,255-filament conductor. Such a conductor can be wound into an electromagnet or other large-scale electrical device.

Superconductive magnets have been used, or are planned to be used for particle acceleration in linear accelerators, for producing the magnetic fields in the plasma step of magnetohydrodynamics, for hydrogen bubble chambers, for producing magnetic "bottles" for nuclear fusion reactors such as the Tokomak, for both levitation and propulsion of ultra high speed trains, for research in solid-state physics, for field windings in motors, and for a host of small uses usually centered on research studies. In fact superconductive magnetics with field strength approaching 10 T are an item of commerce. They are usable where liquid helium temperatures are available and produce magnetic fields more conveniently and cheaply than can be done with a conventional electromagnet. Table 63.18 lists the superconductive magnets in use for various energy related applications.

Perhaps the most interesting of these applications is in high-speed railroads. Studies in Japan, Germany, Canada, and the United States are aimed at developing passenger trains that will operate at 300 mph and above. The trains would be levitated over the track by superconductive magnets, sinking to track level only at start and stop. Propulsion systems vary but are generally motors often with superconductive field windings. Such railroads are proposed for travel from Osaka to Tokyo and from San Diego to Los Angeles. Design criteria for the Japanese train are given in Table 63.19.

Superconductive electrical power transmission has been seriously considered for areas of high density use. Superconductors make it possible to bring the capacity of a single line up to 10,000–30,000 MW at a current density two orders of magnitude greater than conventional practice. The resulting small size and reduced energy losses reduce operating costs of transmission substantially.

The economic attraction of a superconductive transmission line depends on the cost of construction and the demand for power, but also on the cost of refrigeration. Thus a shield is built in and kept at liquid N$_2$ temperature to conserve on helium. Also, superinsulation is used around the liquid N$_2$ shield.

Other applications of superconductivity have been found in the microelectronics field. Superconductive switches have been proposed as high-speed, high-density memory devices and switches for computers and other electronic circuits. The ability of the superconductor to revert to normal and

Table 63.17 Composition on Critical Temperature of Some HTS Materials

Formula	a–b–b–d Values Reported	Critical Temp, T_c, K
Y$_a$ Ba$_b$ Cu$_c$ O$_n$	1–2–3, 1–2–4	80–92
Bi$_a$ Sr$_b$ Ca$_c$ Cu$_d$ O$_n$	2–2–1–2, 2–2–2–3	80–110
Tl$_a$ Ba$_b$ Ca$_c$ Cu$_d$ O$_n$	Several	–125
Hg$_a$ Ba$_b$ Ca$_c$ Cu$_d$ O$_n$	1–2–0–1, 1–2–2–3	95–155*

* High temperature obtained while subjecting the sample to external pressure.

Table 63.18 General Characteristics of Superconductive Magnets for Energy Conversion and Storage Systems[40,a]

Application	Magnet Type	Typical Stored Energy in the Winding (MJ)	Operated			Largest Prototype So Far
			dc	Pulsed	Transients	
MHD generators	Dipole magnet with warm aperature, possibly tapered	500–5000	Yes	No	Yes, from the MHD fluid	60-MJ magnet
Homopolar machines	Solenoid	10–100	Yes	No	No	3-MW generator
Synchronous machines	Rotating dipole or quadrupole winding	Power plant machines, 50–100; airborne systems, 0.5–1	Yes	No	Yes, in case of unbalanced load	5-MVA generator
Fusion magnets Tokamak or similar low-β confinement	Toroidal field coils	$\geq 10^5$	Yes	Yes, in case of fast voltage control	Yes, pulsed field harmonic components of poloidal field	—
	Poloidal field coils	$\geq 10^3$	No	Yes, with rise times of seconds	Yes, dc field components from the toroidal field	—
Mirror confinement	Baseball coils	$\geq 10^5$	Yes	No	No	Baseball coils with 9 MJ
Energy storage; operation of pulsed fusion magnets						
Theta pinch	No optimal shape defined yet	≥ 100 per unit	No	Yes, transfer time about 30 msec	—	300 kJ
Tokamak	No optimal shape defined yet	$\geq 10^4$	No	Yes, transfer time about seconds	—	—
Load leveling in the grid	No optimal shape defined yet	$\geq 10^8$	No	Yes, transfer time about hours	—	—

[a] Courtesy Plenum Press.

Table 63.19 Design Criteria for Japanese High Speed Train[41,a]

Maximum number of coaches/train	16
Maximum operation speed	550 km/hr
Maximum acceleration and deceleration:	
Acceleration	3 km/hr/sec
Deceleration, normal brake	5 km/hr/sec
Deceleration, emergency brake	10 km/hr/sec
Starting speed of levitation	100 km/hr
Effective levitation height (between coil centers)	250 mm
Accuracy of the track	± 10 mm/10 m
Hours of operation	From 6 AM to 12 PM at 15-min intervals
Period of operation without maintenance service	18 hr
Number of superconduction magnets	
Levitation	4×2 rows/coach
Guiding and drive	4×2 rows/coach
Carriage weight	30 tons
dimensions	25 m $\times$ 3.4 m $\times$ 3.4 m
Propulsion	Linear synchronous motor

[a] Courtesy Plenum Press.

again to superconductive in the presence or absence of a magnetic field makes an electric gate or a record of the presence of an electric current. However, these devices have been at least temporarily overshadowed by the rapid development of the electronic chip. Ultimately, of course, these chips will be immersed in a cryogen to reduce resistance and dissipate resistive heat.

63.9 CRYOBIOLOGY AND CRYOSURGERY

Cryogenics has found applications in medicine, food storage and transportation, and agriculture. In these areas the low temperature can be used to produce rapid tissue freezing and to maintain biological materials free of decay over long periods.

The freezing of food with liquid N_2 has become commonplace. Typically the loose, prepared food material is fed through an insulated chamber on a conveyor belt. Liquid N_2 is sprayed onto the food, and the evaporated N_2 flows countercurrent to the food movement to escape the chamber at the end in which the food enters. The required time of exposure depends on the size of individual food pieces and the characteristics of the food itself. For example, hamburger patties freeze relatively quickly because there is little resistance to nitrogen penetration. Conversely, whole fish may freeze rapidly on the surface, but the enclosing membranes prevent nitrogen penetration, so internal freezing occurs by conductive transfer of heat through the flesh. Usually a refrigerated holding period is required after the liquid N_2 spray chambers to complete the freezing process.

The advantages of liquid N_2 food freezing relative to more conventional refrigeration lie in the speed of freezing that produces less tissue damage and less chance for spoilage, and the inert nature of nitrogen, which causes no health hazard for the freezer plant worker or the consumer.

Liquid N_2 freezing and storage has also been used with parts of living beings such as red blood cells, bull semen, bones, and various other cells. Here the concern is for the survival of the cells upon thawing, for in the freezing process ice crystals form which may rupture cell walls upon freezing and thawing. The rate of survival has been found to depend on the rate of cooling and heating, with each class of material showing individual optima. Figure 63.45 shows the survival fractions of several cell types as a function of cooling velocity. Better than half the red blood cells survive at cooling rates of about 3000 K/min. Such a cooling rate would kill all of the yeast cells.

The mechanism of cell death is not clearly understood, and may result from any of several effects. The cell-wall rupture by crystals is the most obvious possibility. Another is the dehydration of the cell by water migration during the freezing process. In any case the use of additives such as glycerol, dimethyl sulfoxide, pyridine n-oxide, and methyl and dimethyl acetamide has greatly reduced cell mortality in various specific cases. The amount and type of additive that is most effective depends upon the specific cell being treated.

Controlled freezing has proven useful in several surgical procedures. In each of these the destruction of carefully selected cells and/or their removal has been the goal of the operation.

In treating Parkinson's disease destruction of some cells in the thalmus can lead to sharp reduction in tremors and muscular rigidity. The operation is done under local anesthetic using a very fine probe consisting of three concentric tubes. Liquid N_2 flows in through the center tube, returning as vapor through the central annulus. The outer annulus is evacuated and insulates all but the probe tip. The

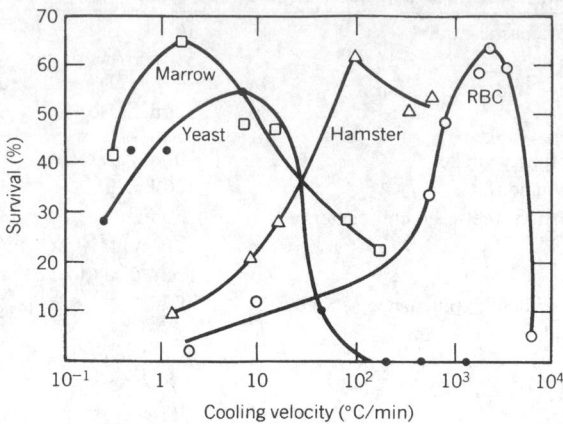

Fig. 63.45   Survival rate for various cells frozen to liquid N₂ temperature.[42]
(Courtesy Plenum Press.)

surgeon inserts the probe using X-ray pictures for guidance. He or she gently cools that probe tip using temperatures just below freezing. If the patient's tremors subside without other side effects, the right location has been found. Freezing of a quarter inch sphere around the probe tip can proceed.

In ophthalmic surgery cryogenic probes are used to lift cataracts from the lens of the eye. Here the cataract is frozen to the cryo-probe tip and carefully separated from the eye. Liquid N₂ is not needed and Freons or Joule–Thomson cooling is sufficient.

Malignant or surface tumors can also be removed cryogenically. The freezing of such a cell mass helps to prevent the escape of some of the cells into the blood stream or the body cavity.

REFERENCES

1. R. D. McCarty, "Thermodynamic Properties of Helium-4 from 2 to 1500 K at Pressures to 10^8 Pa," *J. Chem. Phys. Ref. Data*, **2**(4), 923 (1973); D. B. Mann, "Thermodynamic Properties of Helium from 3 to 300°K Between 0.5 and 100 Atmospheres," NBS Tech. Note 154 (Note 154A for British Units), Jan. 1962.

2. H. M. Roder and R. D. McCarty, "A Modified Benedict–Webb-Rubin Equation of State for Parahydrogen-2," NBS Report NBS1R 75-814, June 1975.

3. J. G. Hurst and R. B. Stewart, "A Compilation of the Property Differences of Ortho- and Para-Hydrogen or Mixtures of Ortho- and Para-Hydrogen," NBS Report 8812, May 1965.

4. R. Prydz, K. D. Timmerhaus, and R. B. Stewart, "The Thermodynamic Properties of Deuterium," *Adv. Cryo. Eng.* **13**, 384 (1968).

5. R. D. McCarty and R. B. Stewart, "Thermodynamic Properties of Neon from 25 to 300 K between 0.1 and 200 Atmospheres," Third Symposium on Thermophysical Properties, ASME, 1965, p. 84.

6. R. T. Jacobsen, R. B. Stewart, and A. F. Myers, "An Equation of State of Oxygen and Nitrogen," *Adv. Cryo. Eng.* **18**, 248 (1972).

7. A. Michels, T. Wassenaar, and G. Wolkers, *Appl. Sci. Res.* **A5**, 121 (1955).

8. T. R. Strobridge, "The Thermodynamic Properties of Nitrogen from 64 to 300 K between 0.1 and 200 Atmospheres," NBS Tech. Note 129 (Note 129A for British Units), Jan. 1962 and Feb. 1963.

9. J. G. Hust and R̄. B. Stewart, "Thermodynamic Properties Valves for Gaseous and Liquid Carbon Monoxide from 70 to 300 K with Pressures to 300 Atmospheres," NBS Tech. Note 202, Nov. 30, 1963.

10. R. Prydz, G. C. Straty, and K. D. Timmerhaus, "The Thermodynamic Properties of Fluorine," *Adv. Cryo. Eng.* **16**, 64 (1971).

11. E. Bendu, "Equations of State Exactly Represently the Phase Behavior of Pure Substances," *Proceedings of the 5th Symposium on Thermosphysical Properties*, ASME, 1970, p. 227.

12. A. L. Gosman, R. D. McCarty, and J. G. Hust, "Thermodynamic Properties of Argon from the Triple Point to 300 K at Pressures to 1000 Atmospheres," NBS Reference Data Series (NSRDS-NSB 27), Mar. 1969.

13. L. A. Weber, "Thermodynamic and Related Properties of Oxygen from the Triple Point to 300 K at Pressures to 330 Atmospheres," NBS Rpt. 9710 (Rpt. 9710A for British Units), June and Aug. 1968; L. A. Weber, *NSB J. Res.* **74A**(1), 93 (1970).

14. R. D. McCarty, "A Modified Benedict–Webb-Rubin Equation of State for Methane Using Recent Experimental Data," *Cryogenics* **14**, 276 (1974).

15. W. T. Ziegle, J. C. Mullins, B. S. Kirk, D. W. Yarborough, and A. R. Berquist, "Calculation of the Vapor Pressure and Heats of Vaporization and Sublimation of Liquids and Solids, Especially Below One Atmosphere Pressure: VI, Krypton," Tech. Rpt. No. 1, Proj. A-764, Georgia Inst. of Tech. Engrg. Expt. Sta., Atlanta, 1964.

16. R. C. Downing, "Refrigerant Equations," ASHRAE Paper 2313, *Trans. ASHRAE*, **80**, Part III, 1974, p. 158.

17. See Ref. 15, "VIII, Xenon," Tech. Rept. No. 3, Projs. A-764 and E-115, 1966.

18. E. Bendu, "Equations of State for Ethylene and Propylene," *Cryogenics* **15**, 667 (1975).

19. R. B. Scott, F. G. Brickwedde, H. C. Urey, and M. H. Wahl, *J. Chem. Phys.* **2**, 454 (1934).

20. A. H. Singleton, A. Lapin, and L. A. Wenzel, "Rate Model for Ortho-Para Hydrogen Reaction on a Highly Active Catalyst," *Adv. Cry. Eng.* **13**, 409–427 (1967).

21. C. T. Lane, *Superfluid Physics*, McGraw-Hill, New York, 1962, pp. 161–177.

22. L. S. Twomey (personal communication).

23. O. M. Bourguet, "Cryogenic Technology and Scaleup Problems of Very Large LNG Plants," *Adv. Crys. Prg.* **18**, K. Timmerhaus (ed.), Plenum Press, New York (1972), pp. 9–26.

24. T. T. Rule and E. B. Quale, "Steady State Operation of the Idealized Vuillurmier Refrigerator," *Adv. Cryo. Eng.* **14**, 1968.

25. W. E. Gifford, "The Gifford–McMahon Cycle," *Adv. Cryo. Eng.* **11**, 1965.

26. H. Hausen, "Warmeubutragung in Gegenstrom, Gluchstrom, und Kiezstrom," Springer-Verlag, Berlin, 1950.

27. D. E. Ward, "Some Aspects of the Design and Operation of Low Temperature Regenerator," *Adv. Cryo. Eng.* **6**, 525 (1960).

28. G. O. G. Lof and R. W. Hawley, *Ind. Ing. Clem.* **40**, 1061 (1948).

29. G. E. O'Connor and T. W. Russell, "Heat Transfer in Tubular Fluid-Fluid Systems," *Advances in Chemical Engineering*. T. B. Drew et al. (eds.), Academic Press, New York, 1978, Vol. 10, pp. 1–56.

30. M. Jacob, *Heat Transfer*, Wiley, New York, 1957, Vol. 2, pp. 1–199.

31. W. T. Ziegh and H. Cheung, *Adv. Cryo. Eng.* **2**, 100 (1960).

32. R. H. Kropschot, "Cryogenic Insulation," *ASHRAE Journal* (1958).

33. T. F. Darham, R. M. McClintock, and R. P. Reed, "Cryogenic Materials Data Book," Office of Tech. Services, Washington, DC, 1962.

34. R. J. Coruccini, *Chem. Eng. Prog.* 342 (July 1957).

35. R. B. Stewart and V. J. Johnson (eds.), "A Compedium of Materials at Low Temperatures, Phases I and II," WADD Tech. Rept. 60-56, NBS, Boulder, CO, 1961.

36. C. R. Barber et al., "The International Practical Temperature Scale," *Metrologia* **5**, 35 (1969).

37. F. G. Brickwedde, H. van Diyk, M. Durieux, J. M. Clement, and J. K. Logan, *J. Res. Natl. Bur. Stds.* **64A**, 1 (1960).

38. R. P. Reis and D. E. Mapother, *Temperature, Its Measurement in Science and Industry*, H. H. Plumb (ed.), 4, 885–895 (1972).

39. L. L. Sperikr, R. L. Powell, and W. J. Hall, "Progress in Cryogenic Thermocouples," *Adv. Cryo. Eng.* **14**, 316 (1968).

40. P. Komacek, "Applications of Superconductive Magnets to Energy with Particular Emphasis on Fusion Power," *Adv. Cryo. Eng.* **21**, 115 (1975).

41. K. Oshima and Y. Kyotani, "High Speed Transportation Levitated by Superconducting Magnet," *Adv. Cryo. Eng.* **19**, 154 (1974).

42. E. G. Cravalho, "The Application of Cryogenics to the Reversible Storage of Biomaterials," *Adv. Cryo. Eng.* **21**, 399 (1975).

43. K. D. Timmerhaus, "Cryogenics and Its Applications: Recent Developments and Outlook," *Bulletin of the Int. Inst. of Refrigeration* **66**(5), 3 (1994).

CHAPTER 64

INDOOR ENVIRONMENTAL CONTROL

Jerald D. Parker
F. C. McQuiston
Professors Emeritus
Oklahoma State University
Stillwater, Oklahoma

64.1	**MOIST AIR PROPERTIES AND CONDITIONING PROCESSES**	**1973**
	64.1.1 Properties of Moist Air	1973
	64.1.2 The Psychrometric Chart	1974
	64.1.3 Space Conditioning Processes	1975
	64.1.4 Human Comfort	1979
64.2	**SPACE HEATING**	**1982**
	64.2.1 Heat Transmission in Structures	1982
	64.2.2 Design Conditions	1985
	64.2.3 Calculation of Heat Losses	1986
	64.2.4 Air Requirements	1987
	64.2.5 Fuel Requirements	1987
64.3	**SPACE COOLING**	**1988**
	64.3.1 Heat Gain, Cooling Load, and Heat Extraction Rate	1988
	64.3.2 Design Conditions	1989
	64.3.3 Calculation of Heat Gains	1989
	64.3.4 Air Requirements	1990
	64.3.5 Fuel Requirements	1991
64.4	**AIR-CONDITIONING EQUIPMENT**	**1991**
	64.4.1 Central Systems	1991
	64.4.2 Unitary Systems	1995
	64.4.3 Heat Pump Systems	1996
64.5	**ROOM AIR DISTRIBUTION**	**1997**
	64.5.1 Basic Considerations	1997
	64.5.2 Jet and Diffuser Behavior	1998
64.6	**BUILDING AIR DISTRIBUTION**	**2000**
	64.6.1 Fans	2000
	64.6.2 Variable-Volume Systems	2003

64.1 MOIST AIR PROPERTIES AND CONDITIONING PROCESSES

64.1.1 Properties of Moist Air

Atmospheric air is a mixture of many gases plus water vapor and countless pollutants. Aside from the pollutants, which may vary considerably from place to place, the composition of the dry air alone is relatively constant, varying slightly with time, location, and altitude. In 1949 a standard composition of dry air was fixed by the International Joint Committee on Psychrometric Data, as shown in Table 64.1.[1]

Table 64.1   Composition of Dry Air[7]

Constituent	Molecular Mass	Volume Fraction
Oxygen	32.000	0.2095
Nitrogen	28.016	0.7809
Argon	39.944	0.0093
Carbon dioxide	44.010	0.0003

Mechanical Engineers' Handbook, 2nd ed., Edited by Myer Kutz.
ISBN 0-471-13007-9   © 1998 John Wiley & Sons, Inc.

The molecular mass M of dry air is 28.965, and the gas constant R is 53.353 ft $\cdot$ lbf/lbm $\cdot$ R or 287 J/kg $\cdot$ K.

The basic medium in air-conditioning practice is a mixture of dry air and water vapor. The amount of water vapor may vary from zero to a maximum determined by the temperature and pressure of the mixture. The latter case is called saturated air, a state of neutral equilibrium between the moist air and the liquid or solid phases of water.

Moist air up to about 3 atm pressure obeys the perfect gas law with sufficient accuracy for engineering calculations. The Gibb's–Dalton law for a mixture of perfect gases states that the mixture pressure is equal to the sum of the partial pressures of the constituents. Because the various constituents of the dry air may be considered to be one gas, it follows that the total pressure P of moist air is the sum of the partial pressures of the dry air p_a and the water vapor p_v:

$$P = p_a + p_v$$

Humidity ratio W (sometimes called the specific humidity) is the ratio of the mass of the water vapor m_v to the mass of the dry air m_a in the mixture:

$$W = \frac{m_v}{m_a}$$

Relative humidity ϕ is the ratio of the mole fraction of the water vapor x_v in a mixture to the mole fraction x_s of the water vapor in a saturated mixture at the same temperature and pressure:

$$\phi = \left(\frac{x_v}{x_s}\right)_{t,P}$$

For a mixture of perfect gases the mole fraction is equal to the partial pressure ratio of each constituent. The mole fraction of the water vapor is

$$x_v = \frac{p_v}{P}$$

Thus

$$\phi = \frac{p_v/P}{p_s/P} = \frac{p_v}{p_s}$$

Dew point temperature t_d is the temperature of saturated moist air at the same pressure and humidity ratio as the given mixture. It can be shown that

$$\phi = \frac{Wp_a}{0.6219 \, p_s}$$

where p_s is the saturation pressure of the water vapor at the mixture temperature.

The enthalpy i of a mixture of perfect gases is equal to the sum of the enthalpies of each constituent and is usually referenced to a unit mass of dry air:

$$i = i_a + Wi_v$$

Each term has the units of energy per unit mass of dry air. With the assumption of perfect-gas behavior the enthalpy is a function of temperature only. If zero Fahrenheit or Celsius is selected as the reference state where the enthalpy of dry air is zero, and if the specific heats c_{pa} and c_{pv} are assumed to be constant, simple relations result:

$$i_a = c_{pa}t$$
$$i_v = i_g + c_{pv}t$$

where the enthalpy of saturated water vapor i_g at 0°F is 1061.2 Btu/lbm and 2501.3 kJ/kg at 0°C.

64.1.2 The Psychrometric Chart

At a given pressure and temperature of an air–water vapor mixture one additional property is required to completely specify the state, except at saturation.

A practical device used to determine the third property is the psychrometer. This apparatus consists of two thermometers, or other temperature-sensing elements, one of which has a wetted cotton wick covering the bulb. The temperatures indicated by the psychrometer are called the wet bulb and the dry bulb temperatures. The wet bulb temperature is the additional property needed to determine the state of moist air.

To facilitate engineering computations, a graphical representation of the properties of moist air has been developed and is known as a psychrometric chart, Fig. 64.1.[2]

In Fig. 64.1 dry bulb temperature is plotted along the horizontal axis in degrees Fahrenheit or Celsius. The dry bulb temperature lines are straight but not exactly parallel and incline slightly to the left. Humidity ratio is plotted along the vertical axis on the right-hand side of the chart in lbm_v/lbm_a or kg_v/kg_a. The scale is uniform with horizontal lines. The saturation curve with values of the wet bulb temperature curves upward from left to right. Dry bulb, wet bulb, and dew point temperatures all coincide on the saturation curve. Relative humidity lines with a shape similar to the saturation curve appear at regular intervals. The enthalpy scale is drawn obliquely on the left of the chart with parallel enthalpy lines inclined downward to the right. Although the wet bulb temperature lines appear to coincide with the enthalpy lines, they diverge gradually in the body of the chart and are not parallel to one another. The spacing of the wet bulb lines is not uniform. Specific volume lines appear inclined from the upper left to the lower right and are not parallel. A protractor with two scales appears at the upper left of the chart. One scale gives the sensible heat ratio and the other the ratio of enthalpy difference to humidity ratio difference. The enthalpy, specific volume, and humidity ratio scales are all based on a unit mass of dry air.

64.1.3 Space Conditioning Processes

When air is heated or cooled without the loss or gain of moisture, the process is a straight horizontal line on the psychrometric chart because the humidity ratio is constant. Such processes can occur when moist air flows through a heat exchanger. In cooling, if the surface temperature is below the dew point temperature of the moist air, dehumidification will occur. This process will be considered later. Figure 64.2 shows a schematic of a device used to heat or cool air. Under steady-flow–steady-state conditions the energy balance becomes

$$\dot{m}_a i_2 + \dot{q} = \dot{m}_a i_1$$

The direction of the heat transfer is implied by the terms heating and cooling, and i_1 and i_2 may be obtained from the psychrometric chart. The convenience of the chart is evident. Figure 64.3 shows heating and cooling processes. The relative humidity decreases when the moist air is heated. The reverse process of cooling results in an increase in relative humidity.

When moist air is cooled to a temperature below its dew point, some of the water vapor will condense and leave the air stream. Figure 64.4 shows a schematic of a cooling and dehumidifying device and Fig. 64.5 shows the process on the psychrometric chart. Although the actual process path will vary considerably depending on the type surface, surface temperature, and flow conditions, the heat and mass transfer can be expressed in terms of the initial and final states. The total amount of heat transfer from the moist air is

$$\dot{q} = \dot{m}_a(i_1 - i_2) - \dot{m}_a(W_1 - W_2)i_w$$

The last term on the right-hand side is usually small compared to the others and is often neglected.

The cooling and dehumidifying process involves both sensible heat transfer, associated with the decrease in dry bulb temperature, and latent heat transfer, associated with the decrease in humidity ratio. We may also express the latent heat transfer as

$$\dot{q}_l = \dot{m}_a(i_1 - i_a)$$

and the sensible heat transfer is given by

$$\dot{q}_s = \dot{m}_a(i_a - i_2)$$

The energy of the condensate has been neglected. Obviously

$$\dot{q} = \dot{q}_s + \dot{q}_l$$

The sensible heat factor (SHF) is defined as $\dot{q}_s/\dot{q}$. This parameter is shown on the semicircular scale of Fig. 64.1.

A device to heat and humidify moist air is shown schematically in Fig. 64.6. An energy balance on the device and a mass balance on the water yields

$$\frac{i_2 - i_1}{W_2 - W_1} = \frac{\dot{q}}{\dot{m}_w} + i_w$$

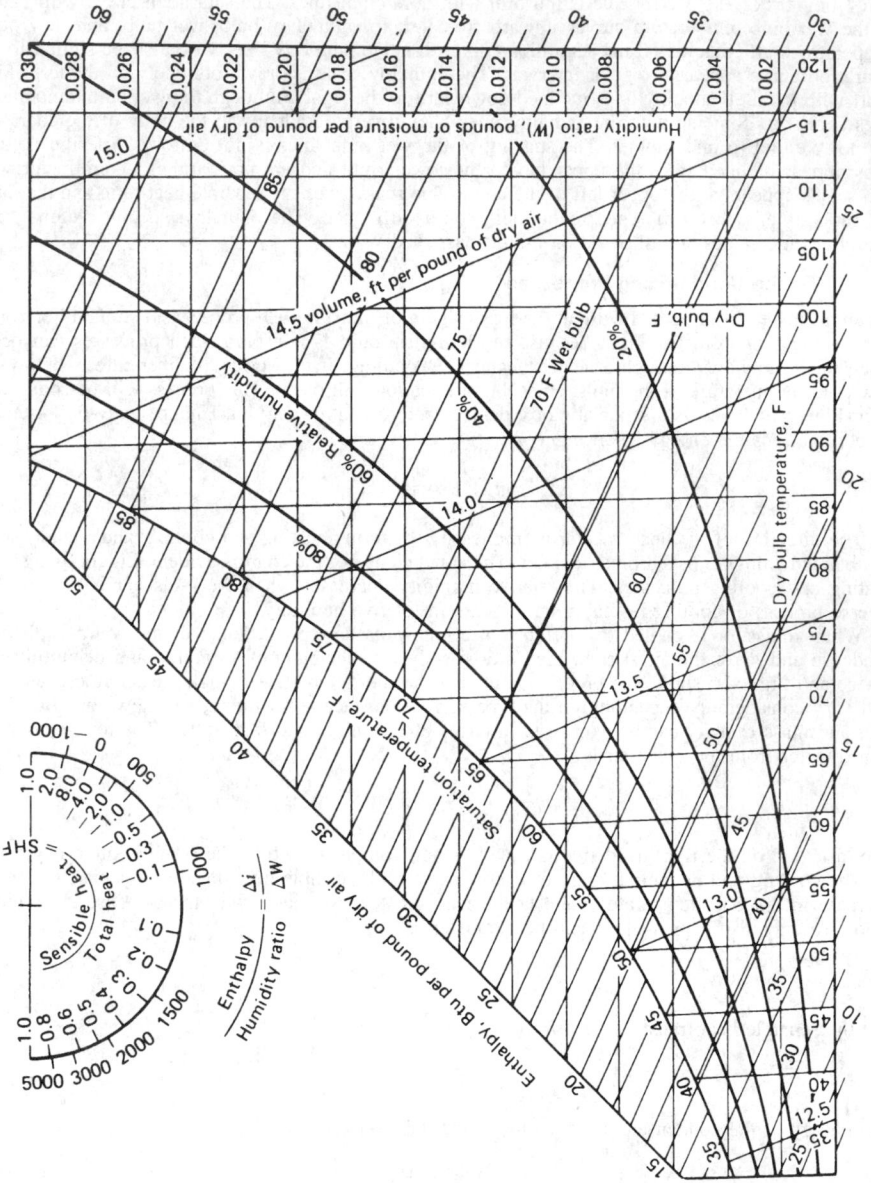

Fig. 64.1 Abridgment of ASHRAE psychrometric chart. (Reprinted by permission from *ASHRAE*.)

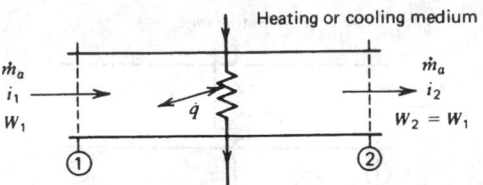

Fig. 64.2 Schematic of a heating or cooling device.[7]

This gives the direction of a straight line that connects the initial and final states on the psychrometric chart. Figure 64.7 shows a typical combined heating and humidifying process.

A graphical procedure makes use of the circular scale in Fig. 64.1 to solve for state 2. The ratio of enthalpy to humidity ratio $\Delta i / \Delta w$ is defined as

$$\frac{\Delta i}{\Delta W} = \frac{i_2 - i_1}{W_2 - W_1} = \frac{\dot{q}}{\dot{m}_w} + i_w$$

Figure 64.7 shows the procedure where a straight line is laid out parallel to the line on the protractor through state point 1. The intersection of this line with the computed value of w_2 determines the final state.

Moisture is frequently added without the addition of heat. In such cases, $q = 0$ and

$$\frac{\Delta i}{\Delta W} = \frac{i_2 - i_1}{W_2 - W_1} = i_w$$

The direction of the process on the psychrometric chart can therefore vary considerably. If the injected water is saturated vapor at the dry bulb temperature, the process will proceed at a constant dry bulb temperature. If the water enthalpy is greater than saturation, the air will be cooled and humidified. Figure 64.8 shows these processes. When liquid water at the wet bulb temperature is injected, the process follows a line of constant wet bulb temperature.

The mixing of air streams is quite common in air-condition systems, usually under adiabatic conditions and with steady flow. Figure 64.9 illustrates the mixing of two air streams. Combined energy and mass balances give

$$\frac{i_2 - i_3}{i_3 - i_1} = \frac{W_2 - W_3}{W_3 - W_1} = \frac{\dot{m}_{a1}}{\dot{m}_{a2}}$$

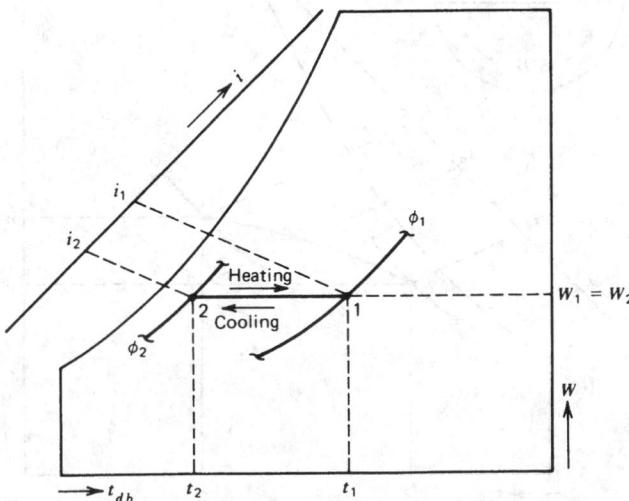

Fig. 64.3 Sensible heating and cooling process.[7]

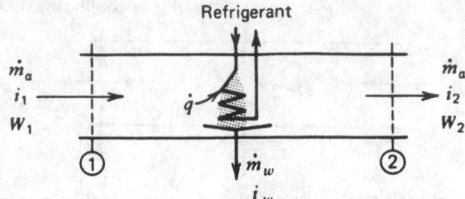

Fig. 64.4 Schematic of a cooling and dehumidifying device.[7]

This shows that the state of the mixed streams must lie on a straight line between states 1 and 2. This is shown in Fig. 64.10. The length of the various line segments are proportional to the masses of dry air mixed. This fact provides a very convenient graphical procedure for solving mixing problems.

The complete air-conditioning system may involve two or more of the processes just considered. In the air conditioning of a space during the summer the air supplied must have a sufficiently low temperature and moisture content to absorb the total heat gain of the space. Therefore, as the air flows through the space, it is heated and humidified. If the system is a closed loop, the air is then returned to the conditioning equipment where it is cooled and dehumidified and supplied to the space again. If fresh air is required in the space, outdoor air may be mixed with the return air before it goes to the cooling and dehumidifying equipment. During the winter months the same general processes occur but in reverse. During the summer months the heating and humidifying elements are inactive, and during the winter the cooling and dehumidifying coil is inactive. With appropriate controls, however, all of the elements may be continuously active to maintain precise conditions in the space.

The previous section treated the common-space air-conditioning problem assuming that the system was operating steadily at the design condition. Actually the space requires only a part of the designed capacity of the conditioning equipment most of the time. A control system functions to match the required cooling or heating of the space to the conditioning equipment by varying one or more system parameters. For example, the quantity of air circulated through the coil and to the space may be varied in proportion to the space load. This approach is known as variable air volume (VAV). Another approach is to circulate a constant amount of air to the space, but some of the return air is diverted around the coil and mixed with air coming off the coil to obtain a supply air temperature that is proportional to the space load. This is known as face and bypass control, because face and bypass dampers are used to divert the flow. Another possibility is to vary the coil surface temperature

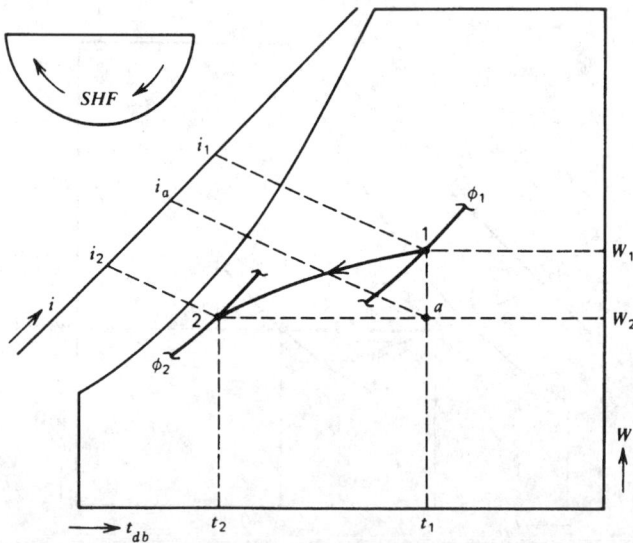

Fig. 64.5 Cooling and dehumidifying process.[7]

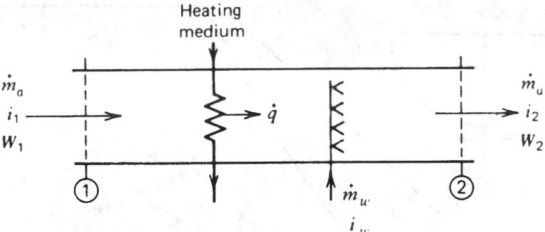

Fig. 64.6   Schematic of a heating and humidifying device.[7]

with respect to the required load by changing the temperature or the amount of heating or cooling fluid entering the coil. This technique is usually used in conjunction with VAV and face and bypass systems. However, control of the coolant temperature or quantity may be the only variable in some systems.

64.1.4   Human Comfort

Air conditioning is the simultaneous control of temperature, humidity, cleanliness, odor, and air circulation as required by the occupants of the space. We are concerned with the conditions that actually provide a comfortable and healthful environment. Not everyone within a given space can be made completely comfortable by one set of conditions, owing to a number of factors, many of which cannot be completely explained. However, clothing, age, sex, and the level of activity of each person are considerations. The factors that influence comfort, in their order of importance, are temperature, radiation, humidity, and air motion, and the quality of the air with regard to odor, dust, and bacteria. With a complete air-conditioning system all of these factors may be controlled simultaneously. In most cases a comfortable environment can be maintained when two or three of these factors are controlled. The *ASHRAE Handbook of Fundamentals* is probably the most up-to-date and complete source of information relating to the physiological aspects of thermal comfort.[3] ASHRAE Comfort Standard 55 defines acceptable thermal comfort as an environment that at least 80% of the occupants will find thermally acceptable.[4]

A complex regulating system in the body acts to maintain the deep body temperature at approximately 98.6°F or 36.9°C. If the environment is maintained at suitable conditions so that the body can easily maintain an energy balance, a feeling of comfort will result.

Two basic mechanisms within the body control the body temperature. The first is a decrease or increase in the internal energy production as the body temperature rises or falls, a process called metabolism. The metabolic rate depends on the level of activity such as rest, work, or exercise. The

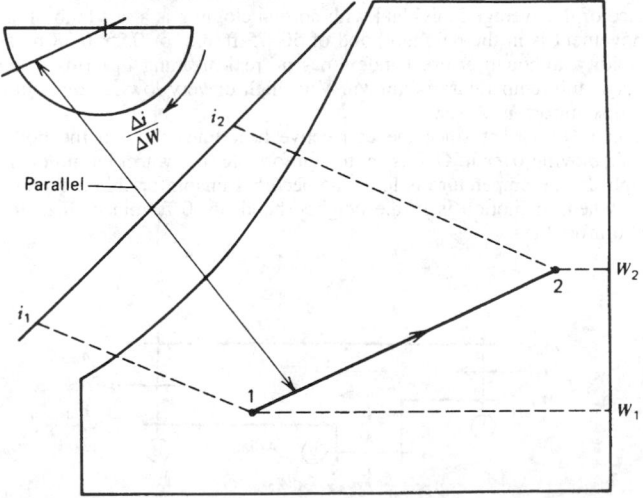

Fig. 64.7   Typical heating and humidifying process.[7]

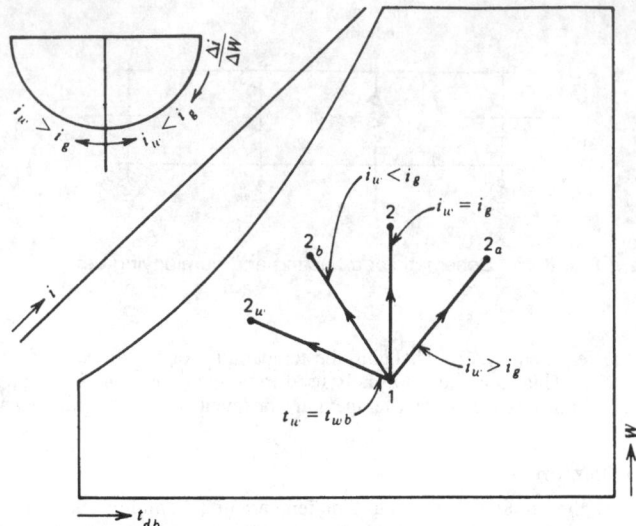

Fig. 64.8 Humidification processes without heat transfer.[7]

second is the control of the rate of heat dissipation by changing the rate of cutaneous blood circulation (the blood circulation near the surface of the skin). In this way heat transfer from the body can be increased or decreased.

Heat transfer to or from the body is principally by convection and conduction and, therefore, the air motion in the immediate vicinity of the body is a very important factor. Radiation exchange between the body and surrounding surfaces, however, can be important if the surfaces surrounding the body are at different temperatures than the air.

Another very important regulatory function of the body is sweating. Under very warm conditions great quantities of moisture can be released by the body to help cool itself.

There are many parameters to describe the environment in term of comfort. The dry bulb temperature is the single most important index of comfort. This is especially true when the relative humidity is between 40% and 60%. The dry bulb temperature is especially important for comfort in the colder regions. When humidity is high, the significance of the dry bulb temperature is less.

The dew point temperature is a good single measure of the humidity of the environment. The usefulness of the dew point temperature in specifying comfort conditions is, however, limited.

The wet bulb temperature is useful in describing comfort conditions in the regions of high temperature and high humidity where dry bulb temperature has less significance. For example, the upper limit for tolerance of the average individual with normal clothing is a wet bulb of about 86°F or 30°C when the air movement is in the neighborhood of 50–75 ft/mm or 0.25–0.38 m/sec.

Relative humidity, although a direct index, has no real meaning in terms of comfort unless the accompanying dry bulb temperature is known. Very high or very low relative humidity is generally associated with discomfort, however.

Air movement is important since the convective heat transfer from the body depends on the velocity of the air moving over it. One is more comfortable in a warm humid environment if the air movement is high. If the temperature is low, one becomes uncomfortable if the air movement is too high. Generally, when air motion is in the neighborhood of 50 ft/min or 0.25 m/sec, the average person will be comfortable.

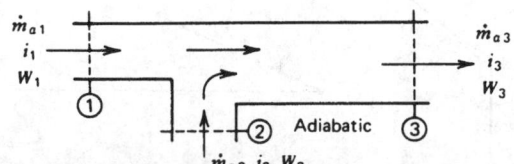

Fig. 64.9 Schematic adiabatic mixing of two air streams.[7]

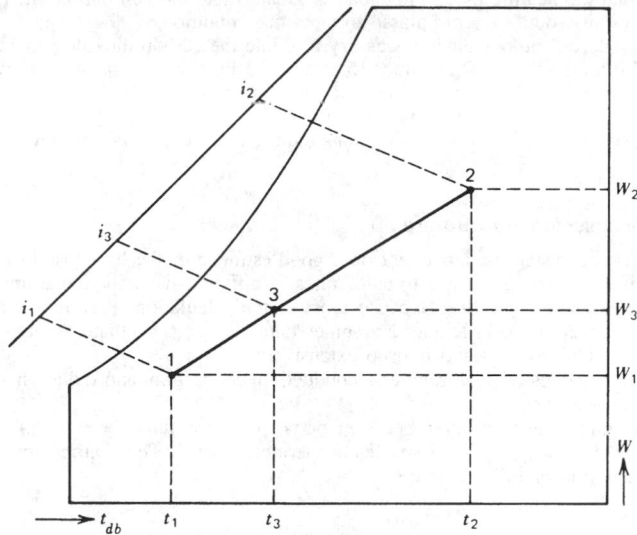

Fig. 64.10   Adiabatic mixing process.[7]

Clothing, through its insulation properties, is an important modifier of body heat loss and comfort. Clothing insulation can be described in terms of its clo value [1 clo = 0.88 ft² · hr · °F/Btu = 0.155 m² · C/W]. A heavy two-piece business suit and accessories has an insulation value of about 1 clo, whereas a pair of shorts is about 0.05 clo.

Ventilation.   The dominating function of outdoor air is to control air quality, and spaces that are more or less continuously occupied require some outdoor air. The required outdoor air is dependent on the rate of contaminant generation and the maximum acceptable contaminant level. In most cases more outdoor air than necessary is supplied. However, some overzealous attempts to save energy through reduction of outdoor air have caused poor-quality indoor air. Table 64.2, from ASHRAE Standard 62-89 (1989), prescribes the requirements for acceptable air quality.[4] Ventilation air is the combination of outdoor air, of acceptable quality, and of recirculated air from the conditioned space which after passing through the air-conditioning unit becomes supply air. The ventilation air may be 100% outdoor air. The term makeup air may be used synonymously with outdoor air, and the terms return and recirculated air are often used interchangeably. A situation could exist where the supply

Table 64.2   National Primary Ambient-Air Quality Standards for Outdoor Air as Set by the U.S. Environmental Protection Agency

	Long Term			Short Term		
	Concentration Averaging			Concentration Averaging		
Contaminant	µg/m³	ppm		µg/m³	ppm	
Sulfur dioxide	80	0.03	1 year	365[a]	0.14[a]	24 hours
Particles (PM 10)	50[b]	—	1 year	150[a]	—	24 hours
Carbon monoxide				40,000[a]	35[a]	1 hour
Carbon monoxide				10,000[a]	9[a]	8 hours
Oxidants (ozone)				235[c]	0.12[c]	1 hour
Nitrogen dioxide	100	0.055	1 year			
Lead	1.5	—	3 months[d]			

[a]Not to be exceeded more than once per year.

[b]Arithmetic mean

[c]Standard is attained when expected number of days per calendar year with maximal average concentrations above 0.12 ppm (235 µg/m³) is equal to or less than 1.

[d]Three-month period is a calendar quarter.

Source: Reprinted by permission from ANSI/ASHRAE Standard 62-89, 1989 (1).

air required to match the heating or cooling load is greater than the ventilation air. In that case an increased amount of air would be recirculated to meet this condition.

A minimum supply of outdoor air is necessary to dilute the carbon dioxide produced by metabolism and expired from the lungs. This value, 15 cfm or 7.5 liter/sec per person, allows an adequate factor of safety to account for health variations and some increased activity levels. Therefore, outdoor air requirements should never be less than 15 cfm or 7.5 liter/sec per person regardless of the treatment of the recirculated air. Some applications require more than this minimum.[4]

64.2 SPACE HEATING

64.2.1 Heat Transmission in Structures

The design of a heating system is dependent on a good estimate of the heat loss in the space to be conditioned. Precise calculation of heat-transfer rates is difficult, but experience and experimental data make reliable estimates possible. Because most of the calculations require a great deal of repetitive work, tables that list coefficients and other data for typical situations are used. Thermal resistance is a very useful concept and is used extensively.

Generally all three modes of heat transfer—conduction, convection, and radiation—are important in building heat gain and loss.

Thermal conduction is heat transfer between parts of a continuum because of the transfer of energy between particles or groups of particles at the atomic level. The Fourier equation expresses steady-state conduction in one dimension:

$$\dot{q} = -kA\,\frac{dt}{dx}$$

where q = heat transfer rate, Btu/hr or W
   k = thermal conductivity, Btu/hr · ft · °F or W/m · °C
   A = area normal to heat flow, ft or m
dt/dx = temperature gradient, °F/ft or °C/m

A negative sign appears because $\dot{q}$ flows in the positive direction of x when dt/dx is negative.

Consider the flat wall of Fig. 64.11a, where uniform temperatures t_1 and t_2 are assumed to exist on each surface. If the thermal conductivity, the heat-transfer rate, and the area are constant, integration gives

$$\dot{q} = \frac{-kA(t_2 - t_1)}{(x_2 - x_1)}$$

Another very useful form is

$$\dot{q} = \frac{-(t_2 - t_1)}{R'}$$

where R' is the thermal resistance defined by

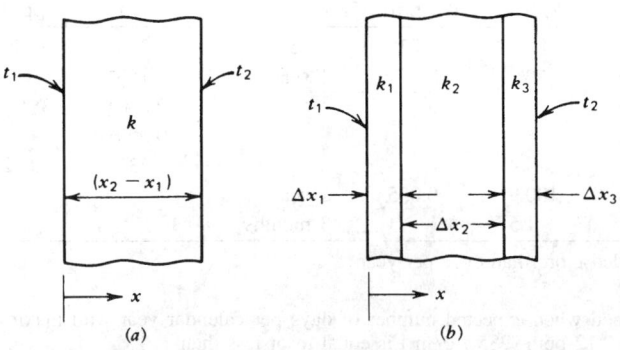

Fig. 64.11 Nomenclature for conduction in plane walls.[7]

$$R' = \frac{x_2 - x_1}{kA} = \frac{\Delta x}{kA}$$

The thermal resistance for a unit area of material is very commonly used. This quantity, sometimes called the R-factor, is referred to as the unit thermal resistance or simply the unit resistance, R. For a plane wall the unit resistance is

$$R = \frac{\Delta x}{k}$$

Note that thermal resistance R' is analogous to electrical resistance and q and $t_2 - t_1$ are analogous to current and potential difference in Ohm's law. This analogy provides a very convenient method of analyzing a wall or slab made up of two or more layers of dissimilar material. Figure 64.11b shows a wall constructed of three different materials. The heat transferred by conduction is given by

$$R' = R'_1 + R'_2 + R'_3 = \frac{\Delta x_1}{k_1 A} + \frac{\Delta x_2}{k_2 A} + \frac{\Delta x_3}{k_3 A}$$

Thermal convection is the transport of energy by mixing in addition to conduction. Convection is associated with fluids in motion, generally through a pipe or duct or along a surface. In the very thin layer of fluid next to the surface the transfer of energy is by conduction. In the main body of the fluid mixing is the dominant energy-transfer mechanism. A combination of conduction and mixing exists between these two regions. The transfer mechanism is complex and highly dependent on whether the flow is laminar or turbulent.

The usual, simplified approach in convection is to express the heat-transfer rate as

$$\dot{q} = hA(t - t_w)$$

where $\dot{q}$ = heat transfer rate from fluid to wall, Btu/hr or W
$\quad h$ = film coefficient, Btu/hr · ft^2 · °F or W/m^2 sec
$\quad t$ = bulk temperature of the fluid, °F or °C
$\quad t_w$ = wall temperature, °F or °C

The film coefficient h, sometimes called the unit surface conductance or alternatively the convective heat transfer coefficient, may also be expressed in terms of thermal resistance:

$$\dot{q} = \frac{t - t_w}{R'}$$

where

$$R' = \frac{1}{hA} \quad \text{(hr · °F/Btu or °C/W)}$$

or

$$R = \frac{1}{h} = \frac{1}{C}$$

where C is the unit thermal conductance. The thermal resistance for convection may be summed with the thermal resistances arising from pure conduction.

The film coefficient h depends on the fluid, the fluid velocity, the flow channel, and the degree of development of the flow field. Many correlations exist for predicting the film coefficient under various conditions. Correlations for forced convection are given in Chapter 3 of the *ASHRAE Handbook*.[2,5]

When the bulk of the fluid is moving relative to the heat-transfer surface, the mechanism is called forced convection, because such motion is usually caused by a blower, fan, or pump, which is forcing the flow. In forced convection buoyancy forces are negligible. In free convection, on the other hand, the motion of the fluid is due entirely to buoyancy forces, usually confined to a layer near the heated or cooled surface. Free convection is often referred to as natural convection.

Natural or free convection is an important part of HVAC applications. Various empirical relations for natural convection film coefficients can be found in the *ASHRAE Handbook of Fundamentals* (1997).[2]

Most building structures have forced convection along outer walls or roofs, and natural convection in inside air spaces and on the inner walls. There is considerable variation in surface conditions, and both the direction and magnitude of the air motion on outdoor surfaces are very unpredictable. The film coefficient for these situations usually ranges from about 1.0 Btu/hr · ft^2 · °F or 6 W/m^2 · °C for free convection up to about 6 Btu/hr · ft^2 · °F or 35 W/m^2 · °C for forced convection with an air velocity of about 15 miles per hour, 20 ft/sec, or 6 m/sec. Because of the low film coefficients the amount of heat transferred by thermal radiation may be equal to or larger than that transferred by free convection.

Thermal radiation, the transfer of thermal energy by electromagnetic waves, can occur in a perfect vacuum and is actually impeded by an intervening medium. The direct net transfer of energy by radiation between two surfaces which see only each other and which are separated by a nonabsorbing medium is given by

$$\dot{q}_{1-2} = \frac{\sigma(T_1^4 - T_2^4)}{\dfrac{1 - \epsilon_1}{A_1\epsilon_1} + \dfrac{1}{A_1F_{12}} + \dfrac{1 - \epsilon_2}{A_2\epsilon_2}}$$

where σ = Boltzmann constant, 0.1713×10^{-8} Btu/hr · ft^2 · °R^4 or 5.673×10^{-8} W/m · K^4
$\quad T$ = absolute temperature, °R or K
$\quad \epsilon$ = emittance
$\quad A$ = surface area, ft^2 or m^2
$\quad F$ = configuration factor, a function of geometry only

It has been assumed that both surfaces are "gray" (where the emittance ϵ equals the absorptance α).[6] Figure 64.12 shows situations where radiation may be a significant factor. For the wall,

$$\dot{q}_i = \dot{q}_w = \dot{q}_r + \dot{q}_o$$

and for the air space,

$$\dot{q}_i = \dot{q}_r + \dot{q}_c = \dot{q}_o$$

The resistances can be combined to obtain an equivalent overall resistance R' with which the heat-transfer rate can be computed using

$$\dot{q} = \frac{-(t_o - t_i)}{R'}$$

The thermal resistance for radiation is not easily computed, however, because of the fourth power temperature relationship.

Tables are available that give conductances and resistances for air spaces as a function of position, direction of heat flow, air temperature, and the effective emittance of the space.[5] The effective emittance E is given by

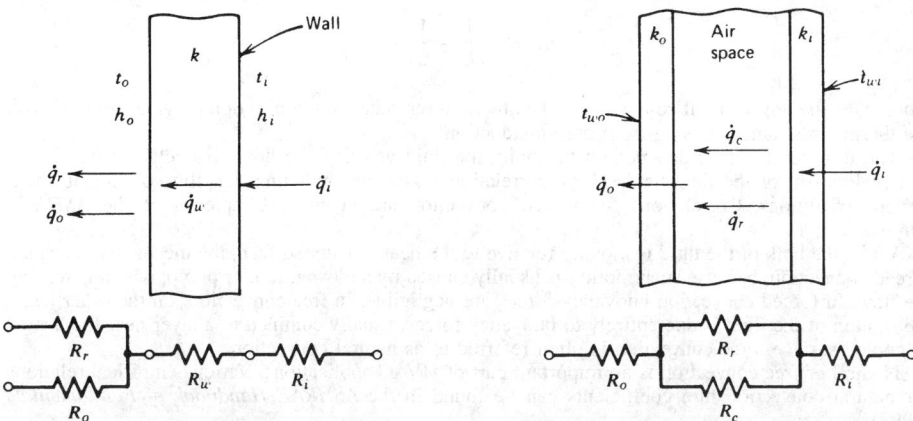

Fig. 64.12 Wall and air space illustrating thermal radiation effects.[7]

$$\frac{1}{E} = \frac{1}{\epsilon_1} + \frac{1}{\epsilon_2} - 1$$

where ϵ_1 and ϵ_2 are for each surface of the air space. Resistors connected in series may be replaced by an equivalent resistor equal to the sum of the series resistors; it will have an equivalent effect on the circuit:

$$R'_e = R'_1 + R'_2 + R'_3 + \cdots + R'_n$$

Figure 64.13 is an example of a wall being heated or cooled by a combination of convection and radiation on each surface and having five different resistances through which the heat must be conducted. The equivalent thermal resistance R'_e for the wall is given by

$$R'_e = R'_i + R'_1 + R'_2 + R'_3 + R'_o$$

Each of the resistances may be expressed in terms of fundamental variables giving

$$R'_e = \frac{1}{h_i A_i} + \frac{\Delta x_1}{k_1 A_1} + \frac{\Delta x_2}{k_2 A_2} + \frac{\Delta x_3}{k_3 A_3} + \frac{1}{h_o A_o}$$

The film coefficients and the thermal conductivities may be obtained from tables. For a plane wall, the areas are all equal and cancel.

The concept of thermal resistance is very useful and convenient in the analysis of complex arrangements of building materials. After the equivalent thermal resistance has been determined for a configuration, however, the overall unit thermal conductance, usually called the overall heat transfer coefficient U, is frequently used:

$$U = \frac{1}{R'A} = \frac{1}{R} \quad (\text{Btu/hr} \cdot \text{ft}^2 \cdot {}^\circ\text{F or W/m}^2 \cdot {}^\circ\text{C})$$

The heat transfer rate is then given by

$$\dot{q} = UA \, \Delta t$$

where UA = conductance, Btu/hr $\cdot$ °F or W/°C
     A = surface area, ft² or m²
     Δt = overall temperature difference, °F or °C

   Tabulated Overall Heat Transfer Coefficients.   For convenience of the designer, tables have been constructed that give overall coefficients for many common building sections including walls and floors, doors, windows, and skylights. The tables in the *ASHRAE Handbook* have a great deal of flexibility and are widely used.[2]

64.2.2   Design Conditions

Prior to the design of the heating system an estimate must be made of the maximum probable heat loss of each room or space to be heated. During the coldest months, sustained periods of very cold, cloudy, and stormy weather with relatively small variation in outdoor temperature may occur. In this situation heat loss from the space will be relatively constant and in the absence of internal heat gains will peak during the early morning hours. Therefore, for design purposes the heat loss is usually

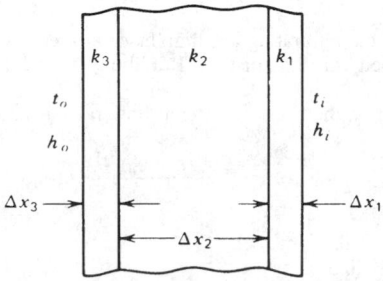

Fig. 64.13   Wall with thermal resistances in series.[7]

estimated for steady-state heat transfer for some reasonable design temperature. Transient analyses are often used to study the actual energy requirements of a structure in simulation studies. In such cases solar effects and internal heat gains are taken into account.

Here is the general procedure for calculation of design heat losses of a structure[7]:

1. Select the outdoor design conditions: temperature, humidity, and wind direction and speed.
2. Select the indoor design conditions to be maintained.
3. Estimate the temperature in any adjacent unheated spaces.
4. Select the transmission coefficients and compute the heat losses for walls, floors, ceilings, windows, doors, and floor slabs.
5. Compute the heat load due to infiltration.
6. Compute the heat load due to outdoor ventilation air. This may be done as part of the air quantity calculation.
7. Sum the losses due to transmission and infiltration.

The ideal heating system would provide enough heat to match the heat loss from the structure. However, weather conditions vary considerably from year to year, and heating systems designed for the worst weather conditions on record would have a great excess of capacity most of the time. The failure of a system to maintain design conditions during brief periods of severe weather is usually not critical. However, close regulation of indoor temperature may be critical for some industrial processes.

The outdoor design temperature should generally be the 97½% value. The 97½% value is the temperature equaled or exceeded 97½% of the total hours (2160) in December, January, and February. During a normal winter there would be about 54 hr at or below the 97½% value. For the Canadian stations the 97½% value pertains only to hours in January. If the structure is of lightweight construction, or poorly insulated, has considerable glass, and space temperature control is critical, however, 99% values should be considered. Should the outdoor temperature fall below the design value for some extended period, the indoor temperature may do likewise. The performance expected by the owner is a very important factor, and the designer should make clear to the owner the various factors considered in the design.

The indoor design temperature should be kept relatively low so that the heating equipment will not be oversized. Even properly sized equipment operates under partial load, at reduced efficiency, most of the time; therefore, any oversizing aggravates this condition and lowers the overall system efficiency. The indoor design value of relative humidity should be compatible with a healthful environment and the thermal and moisture integrity of the building envelope.

64.2.3 Calculation of Heat Losses

The heat transferred through walls, ceiling, roof, window glass, floors, and doors is all sensible heat transfer, referred to as transmission heat loss and computed from

$$\dot{q} = UA \, (t_i - t_o)$$

A separate calculation is made for each different surface in each room of the structure. To ensure a thorough job in estimating the heat losses, a worksheet should be used to provide a convenient and orderly way of recording all the coefficients and areas. Summations are conveniently made by room and for the complete structure.

All structures have some air leakage or infiltration. This means a heat loss because the cold dry outdoor air must be heated to the inside design temperature and moisture must be added to increase the humidity to the design value. The heat required is given by

$$\dot{q}_s = \dot{m}_o c_p (t_i - t_o)$$

where $\dot{m}_o$ = mass flow rate of the infiltrating air, lbm/hr or kg/sec
c_p = specific heat capacity of the moist air, Btu/lbm · °F or J/kg · °C

Infiltration is usually estimated on the basis of volume flow rate at outdoor conditions:

$$\dot{q}_s = \frac{\dot{Q} c_p \, (t_i - t_o)}{\nu_o}$$

where $\dot{Q}$ = volume flow rate, ft³/hr or m³/sec
ν_o = specific volume, ft³/lbm or m³/sec

The latent heat required to humidify the air is given by

$$\dot{q}_l = \dot{m}(W_i - W_o)i_{fg}$$

where $W_i - W_o$ = difference in design humidity ratio, lbm_v/lbm_a or kg_v/kg_a
$\quad\quad i_{fg}$ = latent heat of vaporization at indoor conditions, Btu/lbm_v or J/kg_v

In terms of volume flow rate,

$$\dot{q}_l = \frac{Q}{\nu_o}(W_i - W_o)i_{fg}$$

Infiltration can account for a large portion of the heating load.

Two methods are used in estimating air infiltration in building structures. In one method the estimate is based on the characteristics of the windows and doors and the pressure difference between inside and outside. This is known as the crack method, because of the cracks around window sash and doors. The other approach is the air-change method, which is based on an assumed number of air changes per hour for each room depending on the number of windows and doors. The crack method is generally considered to be the most accurate when the window and pressure characteristics can be properly evaluated. However, the accuracy of predicting air infiltration is restricted by the limited information on the air-leakage characteristics of the many components that make up a structure. The pressure differences are also difficult to predict because of variable wind conditions and stack effect in tall buildings.

64.2.4 Air Requirements

There are many cases, especially in residential and light commercial applications, when the latent heat loss is quite small and may be neglected. The air quantity is then computed from

$$\dot{q} = \dot{m}c_p\,(t_s - t_r)$$

or

$$\dot{q} = \frac{\dot{Q}c_p}{\nu_s}(t_s - t_r)$$

where ν_s = specific volume of supplied air, ft^3/lbm or m^3/kg
$\quad\quad t_s$ = temperature of supplied air, °F or °C
$\quad\quad t_r$ = room temperature, °F or °C

Residential and light commercial equipment operates with a temperature rise of 60–80°F or 33–44°C, whereas commercial applications will allow higher temperatures. The temperature of the air to be supplied must not be high enough to cause discomfort to occupants before it becomes mixed with room air.

In the unit-type equipment typically used for residences and small commercial buildings each size is able to circulate a relatively fixed quantity of air. Therefore, the air quantity is fixed within a narrow range when the heating equipment is selected. A slightly oversized unit is usually selected with the capacity to circulate a larger quantity of air than theoretically needed. Another condition that leads to greater quantities of circulated air for heating than needed is the greater air quantity sometimes required for cooling and dehumidifying. The same fan is used throughout the year and must therefore be large enough for the maximum air quantity required. Some units have different fan speeds for heating and for cooling.

After the total air flow rate required for the complete structure has been determined, the next step is to allocate the correct portion of the air to each room or space. This is necessary for design of the duct system. Obviously, the air quantity for each room should be apportioned according to the heating load for that space; therefore,

$$\dot{Q}_{rn} = \dot{Q}(\dot{q}_{rn}/\dot{q})$$

where Q_{rn} = volume flow rate of air supplied to room n, ft^3/min or m^3/sec
$\quad\quad \dot{q}_{rn}$ = total heat loss rate of room n, Btu/hr or W

The worksheet should have provisions for recording the air quantity for the structure and for each room.

64.2.5 Fuel Requirements

It is often desirable to estimate the quantity of energy necessary to heat the structure under typical weather conditions and with typical imputs from internal heat sources. This is a distinct procedure from design heat load calculations, which are usually made for one set of design conditions neglecting solar effects and internal heat sources. Simulation usually requires a digital computer.

In some cases where computer simulation is not possible or cannot be justified, such as residential buildings, reasonable results can be obtained using hand calculation methods such as the degree-day or bin method.

The degree-day procedure for computing fuel requirements is based on the assumption that, on a long-term basis, solar and internal gains will offset heat loss when the mean daily outdoor temperature is 65°F or 18°C. It is further assumed that fuel consumption will be proportional to the difference between the mean daily temperature and 65°F or 18°C. Degree days are defined by the relationship

$$DD = \frac{(t - t_a)N}{24}$$

where N is the number of hours for which the average temperature t_a is computed and t is 65°F or 18°C. The general relation for fuel calculations using this procedure is

$$F = \frac{24 \, DD \dot{q} C_D}{\eta(t_i - t_o)H}$$

where F = the quantity of fuel required for the period desired; the units depend on H
   DD = the degree days for period desired, °F-day or °C-day
   $\dot{q}$ = the total calculated heat loss based on design condition, t_i and t_o, Btu/hr or W
   η = an efficiency factor, which includes the effects of rated full load efficiency, part load performance, oversizing, and energy conservation devices
   H = the heating value of fuel, Btu or kWhr per unit volume or mass
   C_D = the interim correction factor for degree days based on 65°F or 18°C, Fig. 64.14

64.3 SPACE COOLING

64.3.1 Heat Gain, Cooling Load, and Heat Extraction Rate

A larger number of variables are considered in making cooling load calculations than in heating load calculations. In design for cooling, transient analysis must be used if satisfactory results are to be obtained. This is because the instantaneous heat gain into a conditioned space is quite variable with time primarily because of the strong transient effect created by the hourly variation in solar radiation. There may be an appreciable difference between the heat gain of the structure and the heat removed by the cooling equipment at a particular time. This difference is caused by the storage and subsequent transfer of energy from the structure and contents to the circulated air. If this is not taken into account, the cooling and dehumidifying equipment will usually be grossly oversized and estimates of energy requirements meaningless.

Heat gain is the rate at which energy is transferred to or generated within a space. It has two components, sensible heat and latent heat, which must be computed and tabulated separately. Heat gains usually occur in the following forms:

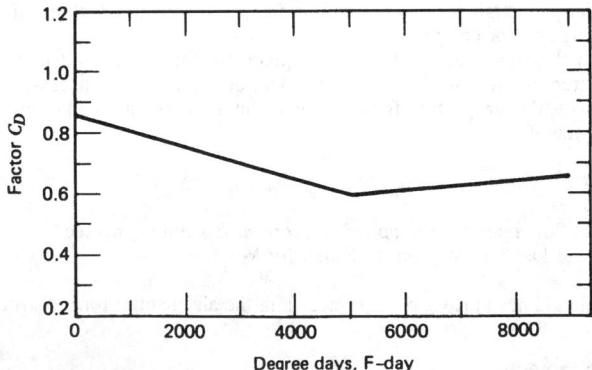

Fig. 64.14 Correction factor. (Reprinted by permission from *ASHRAE.*)

1. Solar radiation through openings.
2. Heat conduction through boundaries with convection and radiation from the inner surface into the space.
3. Sensible heat convection and radiation from internal objects.
4. Ventilation (outside) and infiltration air.
5. Latent heat gains generated within the space.

The cooling load is the rate at which energy must be removed from a space to maintain the temperature and humidity at the design values. The cooling load will generally differ from the heat gain at any instant of time, because the radiation from the inside surface of walls and interior objects as well as the solar radiation coming directly into the space through openings does not heat the air within the space directly. This radiant energy is mostly absorbed by floors, interior walls, and furniture, which are then cooled primarily by convection as they attain temperatures higher than that of the room air. Only when the room air receives the energy by convection does this energy become part of the cooling load. The heat-storage characteristics of the structure and interior objects determine the thermal lag and therefore the relationship between heat gain and cooling load. For this reason the thermal mass (product of mass and specific heat) of the structure and its contents must be considered in such cases. The reduction in peak cooling load because of the thermal lag can be quite important in sizing the cooling equipment.

The heat-extraction rate is the rate at which energy is removed from the space by the cooling and dehumidifying equipment. This rate may be equal to the cooling load. However, this is rarely true and some fluctuation in room temperature occurs. Because the cooling load is below the peak or design value most of the time, intermittent or variable operation of the cooling equipment is required.

64.3.2 Design Conditions

The problem of selecting outdoor design conditions for calculation of heat gain is similar to that for heat loss. Again it is not reasonable to design for the worst conditions on record because a great excess of capacity will result. The heat-storage capacity of the structure also plays an important role in this regard. A massive structure will reduce the effect of overload from short intervals of outdoor temperature above the design value. The *ASHRAE Handbook of Fundamentals* gives extensive outdoor design data.[2] Tabulation of dry bulb and mean coincident wet bulb temperatures that are equaled or exceeded 1%, 2½%, and 5% of the total hours during June through September (2928 hr) are given. The 2½% values are recommended for design purposes by ASHRAE.[2,5] The daily range of temperature is the difference between the average maximum and average minimum for the warmest month. The daily range is usually larger for the higher elevations where temperatures may be quite low late at night and during the early morning hours. The daily range has an effect on the energy stored by the structure. The variation in dry bulb temperature for a typical design day may be computed using the peak outdoor dry bulb temperature and the daily range, assuming a cosine relation with a maximum temperature at 3 PM and a minimum at 5 AM.

The local wind velocity for summer conditions is usually taken to be about one-half the winter design value but not less than about 7½ mph or 3.4 m/sec.

The indoor design conditions for the average job in the United States and Canada is 75°F or 24°C dry bulb and a relative humidity of 50%, when activity and dress of the occupants are light. The designer should be alert for unusual circumstances that may lead to uncomfortable conditions. Certain activities may require occupants to engage in active work or require heavy protective clothing, both of which would require lower design temperatures.

64.3.3 Calculation of Heat Gains

Design cooling loads for one day as well as long-term energy calculations may be done using the transfer-function approach. Details of this method are discussed in the *ASHRAE Handbook of Fundamentals.*[2]

It is not always practical to compute the cooling load using the transfer-function method; therefore, a hand calculation method has been developed from the transfer-function procedure and is referred to as the cooling-load-temperature-difference (CLTD) method. The method involves extensive use of tables and charts and various factors to express the dynamic nature of the problem and predicts cooling loads within about 5% of the values given by the transfer-function method.[5]

The CLTD method makes use of a temperature difference in the case of walls and roofs and cooling load factors (CLF) in the case of solar gain through windows and internal heat sources. The CLTD and CLF vary with time and are a function of environmental conditions and building parameters. They have been derived from computer solutions using the transfer-function procedure. A great deal of care has been taken to sample a wide variety of conditions in order to obtain reasonable accuracy. These factors have been derived for a fixed set of surface and environmental conditions; therefore, correction factors must often be applied. In general, calculations proceed as follows.

For walls and roofs,

$$\dot{q}_\theta = (U)(A)(CLTD)_\theta$$

where U = overall heat transfer coefficient, Btu/hr $\cdot$ ft^2 $\cdot$ °F or W/m^2 $\cdot$ °C
A = area, ft^2 or m^2
$(CLTD)_\theta$ = temperature difference which gives the cooling load at time θ, °F or °C

The CLTD accounts for the thermal response (lag) in the heat transfer through the wall or roof, as well as the response (lag) due to radiation of part of the energy from the interior surface of the wall to objects within the space.

For solar gain through glass

$$\dot{q}_\theta = (A)(SC)(SHGF)(CLF)_\theta$$

where A = area, ft^2 or m^2
   SC = shading coefficient (internal shade)
 $SHGF$ = solar heat gain factor, Btu/hr $\cdot$ ft^2 or W/m^2
$(CLF)_\theta$ = cooling load factor for time θ

The SHGF is the maximum for a particular month, orientation, and latitude. The CLF accounts for the variation of the SHGF with time, the massiveness of the structure, and internal shade. Again the CLF accounts for the thermal response (lag) of the radiant part of the solar input.

For internal heat sources

$$\dot{q}_\theta = (\dot{q}_i)(CLF)_\theta$$

where $\dot{q}_i$ = instantaneous heat gain from lights, people, and equipment, Btu/hr or W
$(CLF)_\theta$ = cooling load factor for time θ

The CLF accounts for the thermal response of the space to the various internal heat gains and is slightly different for each.

The time of day when the peak cooling load will occur must be estimated. In fact, two different types of peaks need to be determined. First, the time of the peak load for each room is needed in order to compute the air quantity for that room. Second, the time of the peak load for a zone served by a central unit is required to size the unit. It is at these peak times that cooling load calculations should be made. The estimated times when the peak load will occur are determined from the tables of CLTD and CLF values together with the orientation and physical characteristics of the room or space. The times of the peak cooling load for walls, roofs, windows, and so on, is obvious in the tables and the most dominant cooling load components will then determine the peak time for the entire room or zone. For example, rooms facing west with no exposed roofs will experience a peak load in the late afternoon or early evening. East-facing rooms tend to peak during the morning hours. A zone made up of east and west rooms with no exposed roofs will tend to peak when the west rooms peak. If there is a roof, the zone will tend to peak when the roof peaks. High internal loads may dominate the cooling load in some cases and cause an almost uniform load throughout the day.

The details of computing the various cooling load components are discussed in ASHRAE Cooling and Heating Load Calculation Manual.[5]

It is emphasized that the total space cooling load does not generally equal the load imposed on the central cooling unit or cooling coil. The outdoor ventilation air is usually mixed with return air and conditioned before it is supplied to the space. The air circulating fan may be upstream of the coil, in which case the fan power input is a load on the coil. In the case of vented light fixtures, the heat absorbed by the return air is imposed on the coil and not the room.

The next steps are to determine the air quantities and to select the equipment. These steps may be reversed depending on the type of equipment to be used.

64.3.4 Air Requirements

Computing air quantity for cooling and dehumidification requires the use of psychrometric charts. The cooling and dehumidifying coil is designed to match the sensible and latent heat requirements of a particular job and the fan is sized to handle the required volume of air. The fan, cooling coil, control dampers, and the enclosure for these components, referred to as an air handler, are assembled at the factory in a wide variety of coil and fan models to suit almost any requirement. The design engineer usually specifies the entering and leaving moist air conditions, the volume flow rate of the air, and the total pressure the fan must produce.

Specifically constructed equipment cannot be justified for small commercial and residential applications. Furthermore, these applications generally have a higher sensible heat factor, and dehumidification is not as critical as it is in large commercial buildings. Therefore, the equipment is manufactured to operate at or near one particular set of conditions. For example, typical residential and light commercial cooling equipment operates with a coil SHF of 0.75–0.8 with the air entering the coil at about 80°F or 27°C dry bulb and 67°F or 19°C wet bulb temperature. This equipment usually has a capacity of less than 10 tons or 35 kW. When the peak cooling load and latent heat requirements are appropriate, this less expensive type of equipment is used. In this case the air quantity is determined in a different way. The peak cooling load is first computed as 1.3 times the peak sensible cooling load for the structure to match the coil SHF. The equipment is then selected to match the peak cooling load as closely as possible. The air quantity is specified by the manufacturer for each unit and is about 400 cfm/ton or 0.0537 m³/sec · kW. The total air quantity is then divided among the various rooms according to the cooling load of each room.

64.3.5 Fuel Requirements

The only reliable methods available for estimating cooling equipment energy requirements require hour by hour predictions of the cooling load and must be done using a computer and representative weather data. This is mainly because of the great importance of thermal energy storage in the structure and the complexity of the equipment used. This approach is becoming much easier due to the development of personal computers. This complex problem is discussed in Ref. 3.

There has been recent work related to residential and light commercial applications that is adaptable to hand calculations. The analysis assumes a correctly sized system. Figure 64.15 summarizes the results of the study of compressor operating time for all locations inside the contiguous 48 states. With the compressor operating time it is possible to make an estimate of the energy consumed by the equipment for an average cooling season. The Air-Conditioning and Refrigeration Institute (ARI) publishes data concerning the power requirements of cooling and dehumidifying equipment and most manufacturers can furnish the same data. For residential systems it is generally best to cycle the circulating fan with the compressor. In this case fans and compressors operate at the same time. However, for light commercial applications the circulating fan will probably operate continuously, and this should be taken into account.

64.4 AIR-CONDITIONING EQUIPMENT

64.4.1 Central Systems

When the requirements of the system have been determined, the designer can select and arrange the various components. It is important that equipment be adequate, accessible for easy maintenance, and no more complex in arrangement and control than necessary to produce the conditions required.

Figure 64.16 shows the air-handling components of a central system for year-round conditioning. It is a built-up system, but most of the components are available in subassembled sections ready for bolting together in the field or completely assembled by the manufacturer. Other components not shown are the water heater or boiler, the chiller, condensing unit or cooling tower, pumps, piping, and controls.

All-Air Systems

An all-air system provides complete sensible heating and cooling and latent cooling by supplying only air to the conditioned space. In such systems there may be piping between the refrigerating and heat-producing devices and the air-handling device. In some applications heating is accomplished by a separate air, water, steam, or electric heating system. The term zone implies a provision or the need for separate thermostatic control, whereas the term room implies a partitioned area that may or may not require separate control.

All-air systems may be classified as (1) single-path systems and (2) dual-path systems. Single-path systems contain the main heating and cooling coils in a series flow air path using a common duct distribution system at a common air temperature to feed all terminal apparatus. Dual-path systems contain the main heating and cooling coils in a parallel flow or series-parallel flow air path using either (1) a separate cold and warm air duct distribution system that is blended at the terminal apparatus (dual-duct system), or (2) a single supply duct to each zone with a blending of warm and cold air at the main supply fan.

The all-air system is applied in buildings requiring individual control of conditions and having a multiplicity of zones such as office buildings, schools and universities, laboratories, hospitals, stores, hotels, and ships. Air systems are also used for many special applications where a need exists for close control of temperature and humidity.

The reheat system is to permit zone or space control for areas of unequal loading, or to provide heating or cooling of perimeter areas with different exposures, or for process or comfort applications where close control of space conditions is desired. The application of heat is a secondary process, being applied to either preconditioned primary air or recirculated room air. The medium for heating may be hot water, steam, or electricity.

Conditioned air is supplied from a central unit at a fixed temperature designed to offset the maximum cooling load in the space. The control thermostat activates the reheat unit when the temperature falls below the upper limit of the controlling instrument's setting. A schematic arrangement

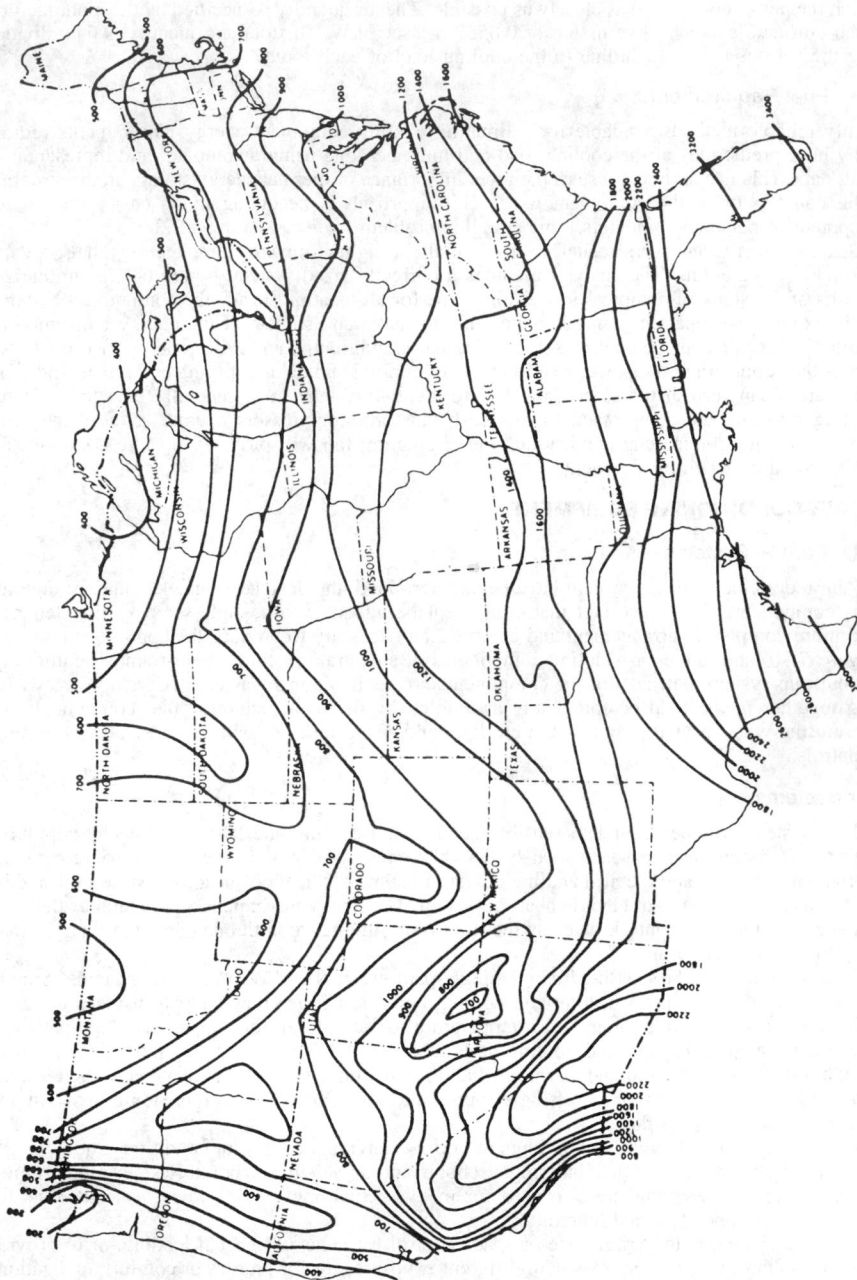

Fig. 64.15 Hours of compressor operation for residential systems. (Reprinted by permission from *ASHRAE*.)

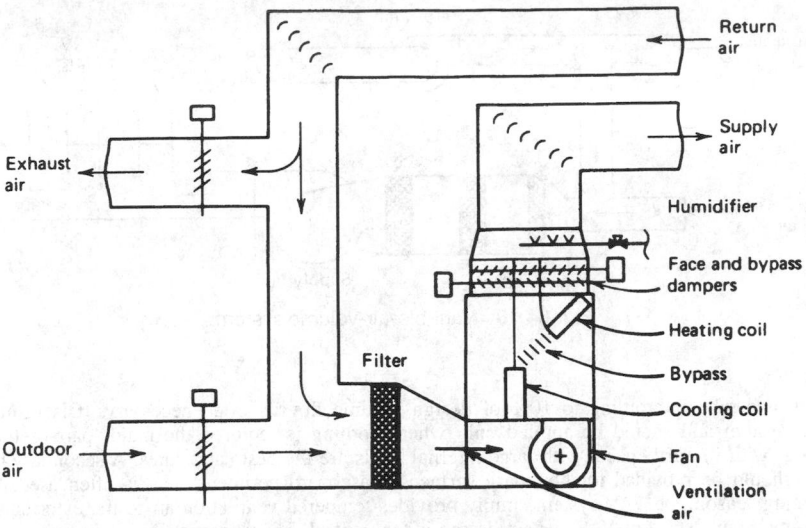

Fig. 64.16 Typical central air system.

of the components for a typical reheat system is shown in Fig. 64.17. To conserve energy reheat should not be used unless absolutely necessary. At the very least, reset control should be provided to maintain the cold air at the highest possible temperature to satisfy the space cooling requirement.

The variable-volume system compensates for varying load by regulating the volume of air supplied through a single duct. Special zoning is not required because each space supplied by a controlled outlet is a separate zone. Figure 64.18 is a schematic of a true variable-air-volume (VAV) system.

Significant advantages are low initial cost and low operating costs. The first cost of the system is low because it requires only single runs of duct and a simple control at the air terminal. Where diversity of loading occurs, smaller equipment can be used and operating costs are generally the lowest among all the air systems. Because the volume of air is reduced with a reduction in load, the refrigeration and fan horsepower follow closely the actual air-conditioning load of the building. During intermediate and cold seasons, outdoor air can be used for economy in cooling. In addition, the system is virtually self-balancing.

Until recently there were two reasons why variable-volume systems were not recommended for applications with loads varying more than 20%. First, throttling of conventional outlets down to 50–60% of their maximum design volume flow might result in the loss of control of room air motion with noticeable drafts resulting. Second, the use of mechanical throttling dampers produces noise, which increases proportionally with the amount of throttling.

With improvements in volume-throttling devices and aerodynamically designed outlets, this system can now handle interior areas as well as building perimeter areas where load variations are

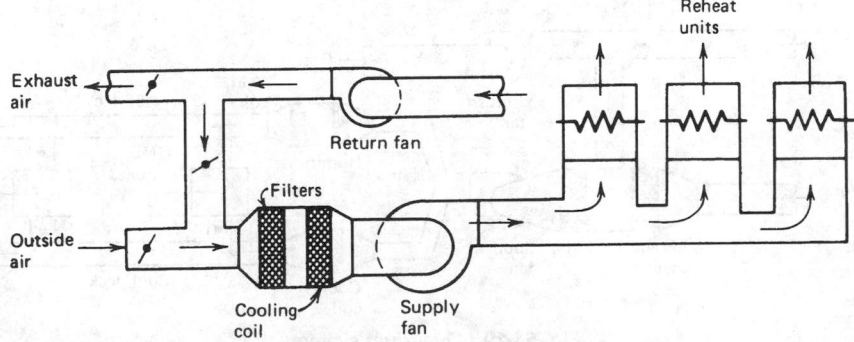

Fig. 64.17 Arrangement of components for a reheat system.

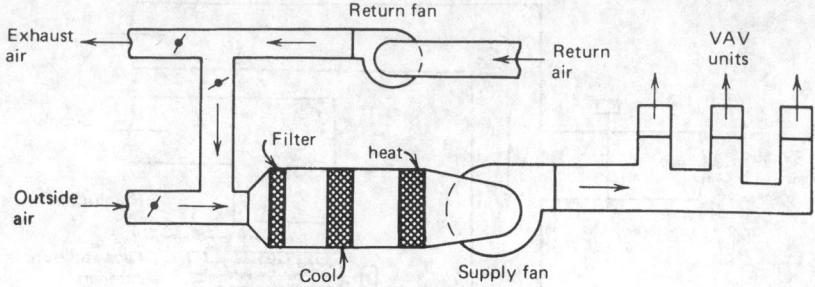

Fig. 64.18   Variable-air-volume system.

greatest, and where throttling to 10% of design volume flow is often necessary. It is primarily a cooling system and should be applied only where cooling is required the major part of the year. Buildings with internal spaces with large internal loads are the best candidates. A secondary heating system should be provided for boundary surfaces. Baseboard perimeter heat is often used. During the heating season, the VAV system simply provides tempered ventilation air to the exterior spaces.

An important aspect of VAV system design is fan control. There are significant fan power savings where fan speed is reduced in relation to the volume of air being circulated.

In the dual-duct system the central station equipment supplies warm air through one duct run and cold air through the other. The temperature in an individual space is controlled by a thermostat that mixes the warm and cool air in proper proportions. One form is shown in Fig. 64.19.

From the energy-conservation viewpoint the dual-duct system has the same disadvantage as reheat. Although many of these systems are in operation, few are now being designed and installed.

The multizone central station units provide a single supply duct for each zone, and obtain zone control by mixing hot and cold air at the central unit in response to room or zone thermostats. For a comparable number of zones this system provides greater flexibility than the single-duct and involves lower cost than the dual-duct system, but it is physically limited by the number of zones that may be provided at each central unit.

The multizone, blow-through system is applicable to locations and areas having high sensible heat loads and limited ventilation requirements. The use of many duct runs and control systems can make initial costs of this system high compared to other all-air systems. To obtain very fine control this system might require larger refrigeration and air-handling equipment.

The use of these systems with simultaneous heating and cooling is now discouraged for energy conservation.

Air and Water Systems

In an air and water system both air and water are distributed to each space to perform the cooling function. In virtually all air–water systems both cooling and heating functions are carried out by changing the air or water temperatures (or both) to permit control of space temperature during all seasons of the year.

The quantity of air supplied can be low compared to an all-air system, and less building space need be allocated for the cooling distribution system.

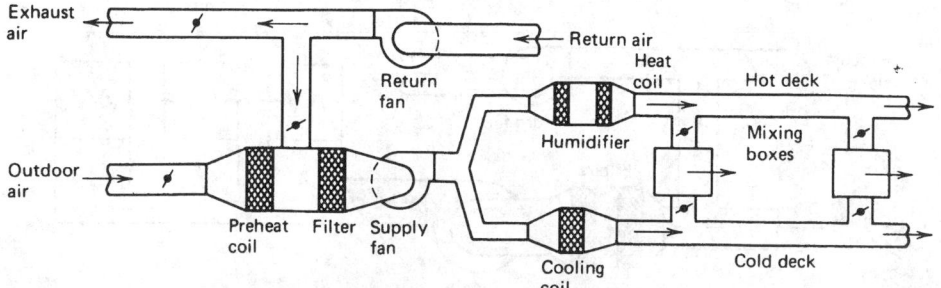

Fig. 64.19   Dual-duct system.

The reduced quantity of air is usually combined with a high-velocity method of air distribution to minimize the space required. If the system is designed so that the air supply is equal to the air needed to meet outside air requirements or that required to balance exhaust (including exfiltration) or both, the return air system can be eliminated for the areas conditioned in this manner.

The pumping power necessary to circulate the water throughout the building is usually significantly less than the fan power to deliver and return the air. Thus not only space but also operating cost savings can be realized.

Systems of this type have been commonly applied to office buildings, hospitals, hotels, schools, better apartment houses, research laboratories, and other buildings. Space saving has made these systems beneficial in high-rise structures.

Air and water systems are categorized as two-pipe, three-pipe, and four-pipe systems. They are basically similar in function, and all incorporate both cooling and heating capabilities for all-season air conditioning. However, arrangements of the secondary water circuits and control systems differ greatly.

All-Water Systems

All-water systems are those with fan-coil, unit ventilator, or valance-type room terminals. with unconditioned ventilation air supplied by an opening through the wall or by infiltration. Cooling and dehumidification are provided by circulating chilled water or brine through a finned coil in the unit. Heating is provided by supplying hot water through the same or a separate coil using two-, three-, or four-pipe water distribution from central equipment. Electric heating or a separate steam coil may also be used. Humidification is not practical in all-water systems unless a separate package humidifier is provided in each room.

The greatest advantage of the all-water system is its flexibility for adaptation to many building module requirements.

64.4.2 Unitary Systems

Unitary Air Conditioners

Unitary air-conditioning equipment consists of factory-matched refrigerant cycle components for inclusion in air-conditioning systems that are field designed to meet the needs of the user. They may vary in:

1. Arrangement: single or split (evaporator connected in the field)
2. Heat rejection: air cooled, evaporative condenser, water cooled
3. Unit exterior: decorative for in-space application, functional for equipment room and ducts, weatherproofed for outdoors
4. Placement: floor standing, wall mounted, ceiling suspended
5. Indoor air: vertical upflow, counterflow, horizontal, 90° and 180° turns, with fan, or for use with forced air furnace
6. Locations: indoor—exposed with plenums or furred in ductwork, concealed in closets, attics, crawl spaces, basements, garages, utility rooms, or equipment rooms; wall—built in, window, transom; outdoor—rooftop, wall mounted, or on ground
7. Heat: intended for use with upflow, horizontal, or counterflow forced air furnace, combined with furnace, combined with electrical heat, combined with hot water or steam coil

Unitary air conditioners as contrasted to room air conditioners are designed with fan capability for ductwork, although some units may be applied with plenums.

Heat pumps are also offered in many of the same types and capacities as unitary air conditioners.

Packaged reciprocating and centrifugal water chillers can be considered as unitary air conditioners particularly when applied with unitary-type chilled water blower coil units. Consequently, a higher level of design ingenuity and performance is required to develop superior system performance using unitary equipment than for central systems, since only a finite number of unitary models is available. Unitary equipment tends to fall automatically into a zoned system with each zone served by its own unit.

For large single spaces where central systems work best, the use of multiple units is often an advantage because of the movement of load sources within the larger space, giving flexibility to many smaller independent systems instead of one large central system.

A room air conditioner is an encased assembly designed as a unit primarily for mounting in a window, through a wall, or as a console. The basic function of a room air conditioner is to provide comfort by cooling, dehumidifying, filtering or cleaning, and circulating the room air. It may also provide ventilation by introducing outdoor air into the room, and by exhausting the room air to the outside. The conditioner may also be designed to provide heating by reverse cycle (heat pump) operation or by electric resistance elements.

64.4.3 Heat Pump Systems

The heat pump is a system in which refrigeration equipment is used such that heat is taken from a heat source and given up to the conditioned space when heating service is wanted and is removed from the space and discharged to a heat sink when cooling and dehumidification are desired. The thermal cycle is identical with that of ordinary refrigeration, but the application is equally concerned with the cooling effect produced at the evaporator and the heating effect produced at the condenser. In some applications both the heating and cooling effects obtained in the cycle are utilized.

Unitary heat pumps are shipped from the factory as a complete preassembled unit including internal wiring, controls, and piping. Only the ductwork, external power wiring, and condensate piping are required to complete the installation. For the split unit it is also necessary to connect the refrigerant piping between the indoor and outdoor sections. In appearance and dimensions, casings of unitary heat pumps closely resemble those of conventional air-conditioning units having equal capacity.

Heat Pump Types

The air-to-air heat pump is the most common type. It is particularly suitable for factory-built unitary heat pumps and has been widely used for residential and commercial applications. Outdoor air offers a universal heat-source, heat-sink medium for the heat pump. Extended-surface, forced-convection heat-transfer coils are normally used to transfer the heat between the air and the refrigerant.

Figure 64.20 shows typical curves of heat pump capacity versus outdoor dry bulb temperature. Imposed on the figure are approximate heating and cooling load curves for a building. In the heating mode it can be seen that the heat pump capacity decreases and the building load increases as the temperature drops. In the cooling mode the opposite trends are apparent. If the cooling load and heat pump capacity are matched at the cooling design temperature, then the balance point, where heating load and capacity match, is then fixed. This balance point will quite often be above the heating design temperature. In such cases supplemental heat must be furnished to maintain the desired indoor condition.

The most common type of supplemental heat for heat pumps in the United States is electrical-resistance heat. This is usually installed in the air-handler unit and is designed to turn on automatically, sometimes in stages, as the indoor temperature drops. In some systems the supplemental heat is turned on when the outdoor temperature drops below some preset value. Heat pumps which have fossil-fuel-fired supplemental heat are referred to as hybrid or bivalent heat pumps.

If the heat pump capacity is sized to match the heating load, care must be taken that there is not excessive cooling capacity for summer operation, which could lead to poor summer performance, particularly in dehumidification of the air.

Air-to-water heat pumps are commonly used in large buildings where zone control is necessary and are also sometimes used for the production of hot or cold water in industrial applications as well as heat reclaiming. Heat pumps for hot water heating are commercially available in residential sizes.

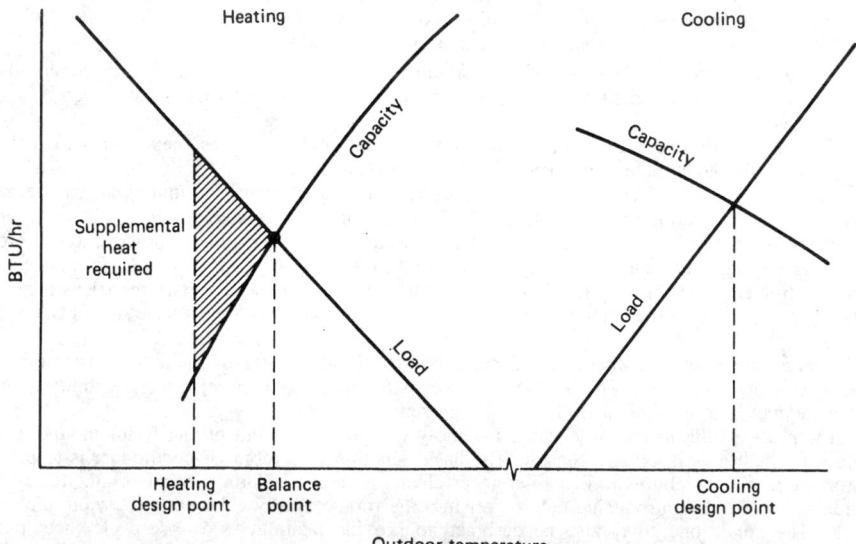

Fig. 64.20 Comparison of building heat loads with heat pump capacities.

A water-to-air heat pump uses water as a heat source and sink and uses air to transmit heat to or from the conditioned space.

A water-to-water heat pump uses water as the heat source and sink for both cooling and heating operation. Heating-cooling changeover may be accomplished in the refrigerant circuit, but in many cases it is more convenient to perform the switching in the water circuits.

Water may represent a satisfactory and in many cases an ideal heat source. Well water is particularly attractive because of its relatively high and nearly constant temperature, generally about 50°F or 10°C in northern areas and 60°F or 16°C and higher in the south. However, abundant sources of suitable water are not always available, and the application of this type of system is limited. Frequently, sufficient water may be available from wells, but the condition of the water may cause corrosion in heat exchangers or it may induce scale formation. Other considerations to be made are the costs of drilling, piping, and pumping, and the means for disposing of used water.

Surface or stream water may be used, but under reduced winter temperatures the cooling spread between inlet and outlet must be limited to prevent freeze-up in the water chiller, which is absorbing the heat.

Under certain industrial circumstances waste process water such as spent warm water in laundries and warm condenser water may be a source for specialized heat pump operations.

A building may require cooling in interior zones while needing heat in exterior zones. The needs of the north zones of a building may also be different from those of the south. In many cases a closed-loop heat pump system is a good choice. Closed-loop systems may be solar assisted. A closed-loop system is shown in Fig. 64.21.

Individual water-to-air heat pumps in each room or zone accept energy from or reject energy to a common water loop, depending on whether that area has a call for heating or for cooling. In the ideal case the loads will balance, and there will be no surplus or deficiency of energy in the loop. If cooling demand is such that more energy is rejected to the loop than is required for heating, the surplus is rejected to the atmosphere by a cooling tower. In the other case, an auxiliary furnace furnishes any deficiency.

The ground has been used successfully as a source–sink for heat pumps with both vertical and horizontal pipe installation. Water from the heat pump is pumped through plastic pipe and exchanges heat with the surrounding earth before being returned back to the heat pump, Fig. 64.22. Tests and analyses have shown rapid recovery in earth temperature around the pipe after the heat pump cycles off. Proper sizing depends on the nature of the earth surrounding the pipe, the water table level, and the efficiency of the heat pump.

Although still largely in the research stage, the use of solar energy as a heat source either on a primary basis or in combination with other sources is attracting increasing interest. Heat pumps may be used with solar systems in either a series or a parallel arrangement, or a combination of both.

64.5   ROOM AIR DISTRIBUTION

64.5.1   Basic Considerations

The object of air distribution in warm air heating, ventilating, and air-conditioning systems is to create the proper combination of temperature, humidity, and air motion in the occupied portion of the conditioned room. To obtain comfort conditions with this space, standard limits for an acceptable

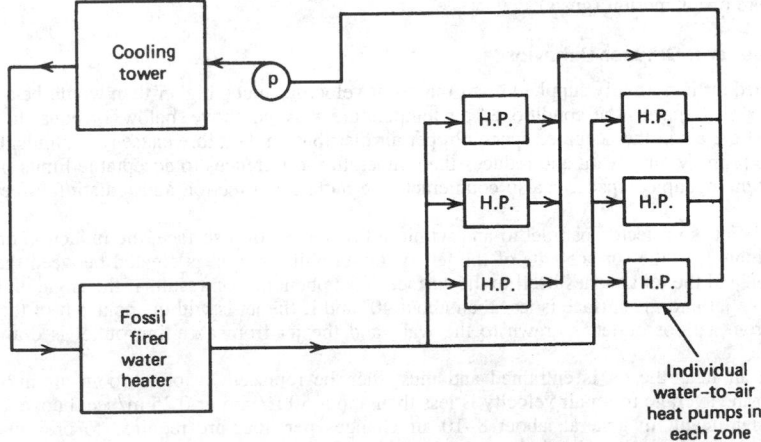

Fig. 64.21   Schematic of a closed-loop heat pump system.[7]

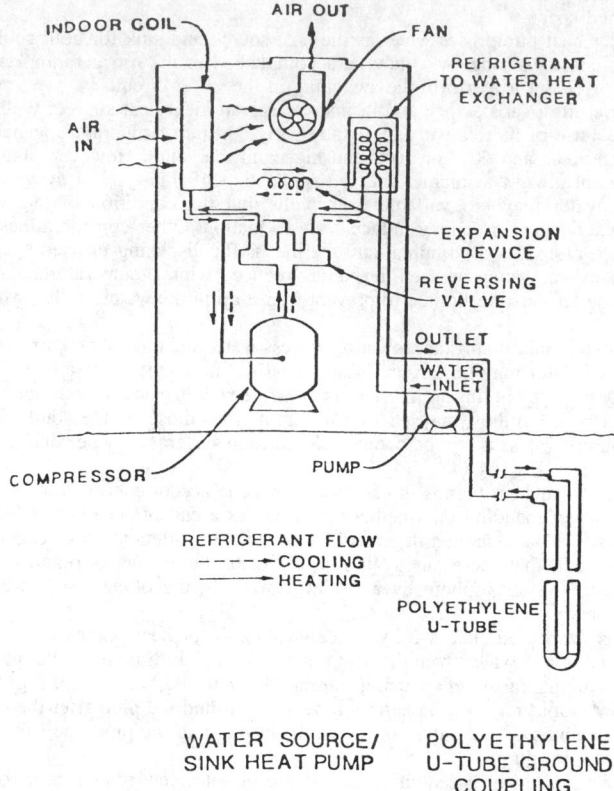

Fig. 64.22   Schematic of a ground-coupled heat pump system.

effective draft temperature have been established. This term comprises air temperature, air motion, relative humidity, and their physiological effect on the human body, any variation from accepted standards of one of these elements may result in discomfort to the occupants. Discomfort also may be caused by lack of uniform conditions within the space or by excessive fluctuation of conditions in the same part of the space. Such discomfort may arise owing to excessive room air temperature variations (horizontally, vertically, or both), excessive air motion (draft), failure to deliver or distribute the air according to the load requirements at the different locations, or rapid fluctuation of room temperature or air motion (gusts).

64.5.2   Jet and Diffuser Behavior

Conditioned air is normally supplied to air outlets at velocities much higher than would be acceptable in the occupied space. The conditioned air temperature may be above, below, or equal to the temperature of the air in the occupied space. Proper air distribution therefore causes entrainment of room air by the primary air stream and reduces the temperature differences to acceptable limits before the air enters the occupied space. It also counteracts the natural convection and radiation effects within the room.

When a jet is projected parallel to and within a few inches of a surface, the induction or entrainment is limited on the surface side of the jet. A low-pressure region is created between the surface and the jet, and the jet attaches itself to the surface. This phenomenon results if the angle of discharge between the jet and the surface is less than about 40° and if the jet is within about 1 ft of the surface. The jet from a floor outlet is drawn to the wall, and the jet from a ceiling outlet is drawn to the ceiling.

Room air near the jet is entrained and must then be replaced by other room air into motion. Whenever the average room air velocity is less than about 50 ft/min or 0.25 m/sec, buoyancy effects may be significant. In general, about 8–10 air changes per hour are required to prevent stagnant regions (velocity less than 15 ft/min or 0.08 m/sec). However, stagnant regions are not necessarily a serious condition. The general approach is to supply air in such a way that the high-velocity air from the outlet does not enter the occupied space. The region within 1 ft of the wall and above about 6 ft from the floor is out of the occupied space for practical purposes.[7]

Perimeter-type outlets are generally regarded as superior for heating applications. This is particularly true when the floor is over an unheated space or a slab and where considerable glass area exists in the wall. Diffusers with a wide spread are usually best for heating because buoyancy tends to increase the throw. For the same reason the spreading jet is not as good for cooling applications because the throw may not be adequate to mix the room air thoroughly. However, the perimeter outlet with a nonspreading jet is quite satisfactory for cooling. Diffusers are available that may be changed from the spreading to nonspreading type according to the season.

The high sidewall type of register is often used in mild climates and on the second and succeeding floors of multistory buildings. This type of outlet is not recommended for cold climates or with unheated floors. A considerable temperature gradient may exist between floor and ceiling when heating; however, this type outlet gives good air motion and uniform temperatures in the occupied zone for cooling application. These registers are generally selected to project air from about three-fourths to full room width.

The ceiling diffuser is very popular in commercial applications and many variations of it are available. Because the primary air is projected radially in all directions, the rate of entrainment is large, causing the high momentum jet to diffuse quickly. This feature enables the ceiling diffuser to handle larger quantities of air at higher velocities than most other types. The ceiling diffuser is quite effective for cooling applications but generally poor for heating. However, satisfactory results may be obtained in commercial structures when the floor is heated.

The return air intake generally has very little effect on the room air motion. But the location may have a considerable effect on the performance of the heating and cooling equipment. Because it is desirable to return the coolest air to the furnace and the warmest air to the cooling coil, the return air intake should be located in a stagnant region.

Noise produced by the air diffuser and air can be annoying to the occupants of the conditioned space. Noise criteria (NC) curves are used to describe the noise in HVAC systems.[5]

The selection and placement of the air outlets is ideally done purely on the basis of comfort. However, the architectural design and the functional requirements of the building often override comfort. When the designer is free to select the type of air-distribution system based on comfort, the perimeter type of system with vertical discharge of the supply air is to be preferred for exterior spaces when the heating requirements exceed 2000 degree (F) days. This type system is excellent for heating and satisfactory for cooling when adequate throw is provided. When the floors are warmed and the degree (F) day requirement is between about 3500 and 2000, the high sidewall outlet with horizontal discharge toward the exterior wall is acceptable for heating and quite effective for cooling. When the heating requirement falls below about 2000 degree (F) days, the overhead ceiling outlet or high sidewall diffuser is recommended because cooling is the predominant mode. Interior spaces in commercial structures are usually provided with overhead systems because cooling is required most of the time.

Commercial structures often are constructed in such a way that ducts cannot be installed to serve the desired air-distribution system. Floor space is very valuable and the floor area required for outlets may be covered by shelving or other fixtures, making a perimeter system impractical. In this case an overhead system must be used. In some cases the system may be a mixture of the perimeter and overhead type.

The Air Distribution Performance Index (ADPI) is defined as the percentage of measurements taken at many locations in the occupied zone of a space which meet a -3 to $2°F$ effective draft temperature criteria. The objective is to select and place the air diffusers so that an ADPI approaching 100% is achieved. ADPI is based only on air velocity and effective draft temperature, a local temperature difference from the room average, and is not directly related to the level of dry bulb temperature or relative humidity. These effects and other factors such as mean radiant temperature must be accounted for. The ADPI provides a means of selecting air diffusers in a rational way. There are no specific criteria for selection of a particular type of diffuser except as discussed above, but within a given type the ADPI is the basis for selecting the throw. The space cooling load per unit area is an important consideration. Heavy loading tends to lower the ADPI. However, loading does not influence design of the diffuser system significantly. Each type of diffuser has a characteristic room length. Table 64.3, the ADPI selection guide, gives the recommended ratio of throw to characteristic length that should maximize the ADPI. A range of throw-to-length ratios are also shown that should give a minimum ADPI. Note that the throw is based on a terminal velocity of 50 ft/min for all diffusers except the ceiling slot type. The general procedure for use of Table 64.3 is as follows:

1. Determine the airflow requirements and the room size.
2. Select the type of diffuser to be used.
3. Determine the room characteristic length.
4. Select the recommended throw-to-length ratio from Table 64.3.
5. Calculate the throw.

Table 64.3 ADPI Selection Guide[a]

Terminal Device	Room Load W/m²	Room Load Btu/hr · ft²	$T_{0.25}/L(T_{50}/L)$ for Max. ADPI	Maximum ADPI	For ADPI Greater Than	Range of $T_{0.25}/L(T_{50}/L)$
High sidewall grilles	250	80	1.8	68	—	—
	190	60	1.8	72	70	1.5–2.2
	125	40	1.6	78	70	1.2–2.3
	65	20	1.5	85	80	1.0–1.9
Circulr ceiling diffusers	250	80	0.8	76	70	0.7–1.3
	190	60	0.8	83	80	0.7–1.2
	125	40	0.8	88	80	0.5–1.5
	65	20	0.8	93	90	0.7–1.3
Sill grille straight vanes	250	80	1.7	61	60	1.5–1.7
	190	60	1.7	72	70	1.4–1.7
	125	40	1.3	86	80	1.2–1.8
	65	20	0.9	95	90	0.8–1.3
Sill grille spread vanes	250	80	0.7	94	90	0.8–1.5
	190	60	0.7	94	80	0.6–1.7
	125	40	0.7	94	—	—
	65	20	0.7	94	—	—
Ceiling slot diffusers[b]	250	80	0.3[b]	85	80	0.3–0.7
	190	60	0.3[b]	88	80	0.3–0.8
	125	40	0.3[b]	91	80	0.3–1.1
	65	20	0.3[b]	92	80	0.3–1.5
Light troffer diffusers	190	60	2.5	86	80	<3.8
	125	40	1.0	92	90	<3.0
	65	20	1.0	95	90	<4.5
Perforated and louvered ceiling diffusers	35–160	11–51	2.0	96	90	1.4–2.7
					80	1.0–3.4

Characteristic Room Length for Several Diffuser Types

Diffuser Type	Characteristic Length, L
High sidewall grille	Distance to wall perpendicular to jet
Circular ceiling diffuser	Distance to closest wall or intersecting air jet
Sill grille	Length of room in the direction of the jet flow
Ceiling slot diffuser	Distance to wall or midplane between outlets
Light troffer diffusers	Distance to midplane between outlets plus distance from ceiling to top of occupied zone
Perforated, louvered ceiling diffusers	Distance to wall or midplane between outlets

[a] Reprinted by permission from *ASHRAE Handbook of Fundamentals*, 1997.
[b] Given for $T_{0.50}/L(T_{100}/L)$.

6. Select the appropriate diffuser from catalog data.
7. Make sure any other specifications are met (noise, total pressure, etc.).

64.6 BUILDING AIR DISTRIBUTION

This section discusses the details of distributing the air to the various spaces in the structure. Proper design of the duct system and the selection of appropriate fans and accessories are essential. A poorly designed system may be noisy, inefficient, and lead to discomfort of occupants. Correction of faulty design is expensive and sometimes practically impossible.

64.6.1 Fans

The fan is an essential component of almost all heating and air-conditioning systems. Except in those cases where free convection creates air motion, a fan is used to move air through ducts and to induce air motion in the space. An understanding of the fan and its performance is necessary if one is to design a satisfactory duct system.

The centrifugal fan is the most widely used because it can effectively move large or small quantities of air over a wide range of pressures. The principle of operation is similar to the centrifugal pump in that rotating impeller mounted inside a scroll type of housing imparts energy to the air or gas being moved.

The vaneaxial fan is mounted on the centerline of the duct and produces an axial flow of the air. Guide vanes are provided before and after the wheel to reduce rotation of the air stream.

The tubeaxial fan is quite similar to the vaneaxial fan but does not have the guide vanes.

Axial flow fans are not capable of producing pressures as high as those of the centrifugal fan but can move large quantities of air at low pressure. Axial flow fans generally produce higher noise levels than centrifugal fans.

Fan efficiency may be expressed in two ways. The total fan efficiency is the ratio of total air power to the shaft power input:

$$\eta_t = \frac{\dot{W}_t}{\dot{W}_{sh}}$$

where Q = volume flow rate, ft³/min or m³/sec
$P_{01} - P_{02}$ = change in total pressure, lbf/ft² or Pa
$\dot{W}_{sh}$ = shaft power, ft · lbf/min or W

It has been common practice in the United States for Q to be in ft³/min, $P_{01} - P_{02}$ to be in inches of water, and for $\dot{W}_{sh}$ to be in horsepower. In this special case,

$$\eta_t = \frac{\dot{Q}(P_{01} - P_{02})}{6.350\dot{W}_{sh}}$$

The static fan efficiency is

$$\eta_s = \frac{\dot{Q}(P_1 - P_2)}{6350\dot{W}_{sh}}$$

Figure 64.23 illustrates typical performance curves for centrifugal fans. Note the differences in the pressure characteristics and in the point of maximum efficiency with respect to the point of maximum pressure.

Table 64.4 compares some of the more important characteristics of centrifugal fans.

The noise emitted by a fan is important in many applications. For a given pressure the noise level is proportional to the tip speed of the impeller and to the air velocity leaving the wheel. Fan noise is roughly proportional to the pressure developed regardless of the blade type; however, backward-curved fan blades generally have the better (lower) noise characteristics.

The pressure developed by a fan is limited by the maximum allowable speed. If noise is not a factor, the straight radial blade is superior. Fans may be operated in series to develop higher pressures, and multistage fans are also constructed. When fans are used in parallel, surging back and forth between fans may develop, particularly if the system demand is changing. Forward-curved blades are particularly unstable when operated at the point of maximum efficiency.

Combining both the system and fan characteristics on a plot is very useful in matching a fan to a system and to ensure fan operation at the desired conditions. Figure 64.24 illustrates the desired operating range for a forward-curved blade fan. The range is to the right of the point of maximum efficiency. The backward-curved blade fan has a selection range that brackets the range of maximum efficiency and is not so critical to the point of operation; however, this type should always be operated to the right of the point of maximum pressure. For a given system the efficiency does not change with speed; however capacity, total pressure, and power all depend on the speed. Changing the fan speed will not change the relative point of intersection between the system and fan characteristics. This can only be done by changing fans.

There are several simple relationships between fan capacity, pressure, speed, and power, which are referred to as the fan laws. The most useful fan laws are:

1. Capacity is directly proportional to fan speed.
2. Pressure (static, total, or velocity) is proportional to the square of the fan speed.
3. Power required is proportional to the cube of fan speed.

Three other fan laws are useful.

4. Pressure and power are proportional to the density of the air at constant speed and capacity.

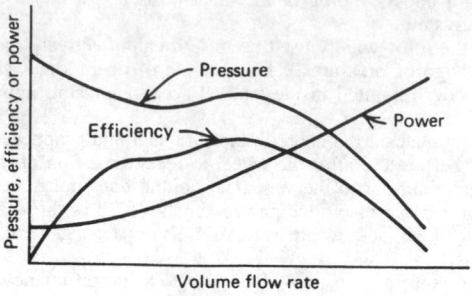

Forward-tip fan characteristics.

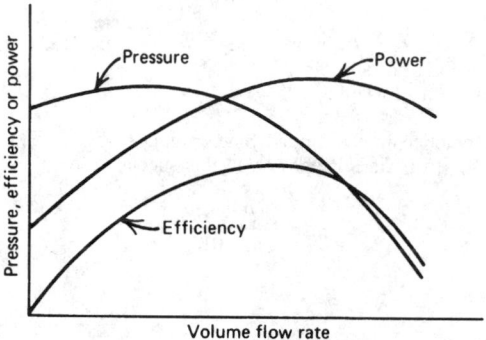

Backward-tip fan characteristics.

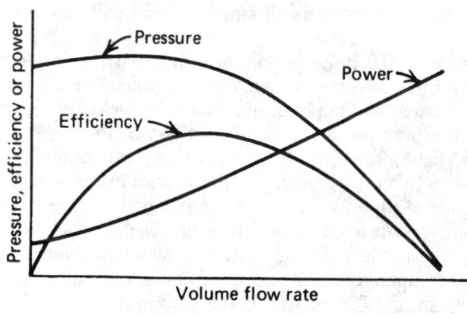

Radial-tip fan characteristics.

Fig. 64.23 Performance curves for centrifugal fans.[7]

Table 64.4 Comparison of Centrifugal Fan Types[7]

Item	Forward-Curved Blades	Radial Blades	Backward-Curved Blades
Efficiency	Medium	Medium	High
Space required	Small	Medium	Medium
Speed for given pressure rise	Low	Medium	High
Noise	Poor	Fair	Good

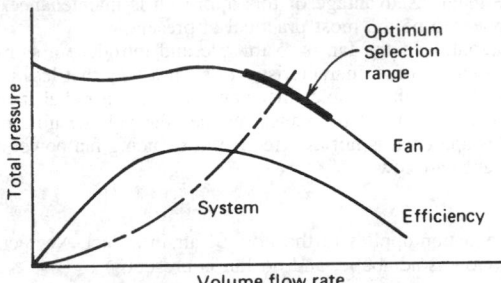

Fig. 64.24   Optimum match between system and forward-curved blade fan.[7]

5. Speed, capacity, and power are inversely proportional to the square root of the density at constant pressure.

6. Capacity, speed, and pressure are inversely proportional to the density, and the power is inversely proportional to the square of the density at a constant mass flow rate.

In a variable-air-volume system it is desirable to reduce fan speed as air-volume flow rate is reduced under part load conditions to reduce the fan power.

Fan Selection

To select a fan it is necessary to know the capacity and total pressure requirement of the system. The type and arrangement of the prime mover, the possibility of fans in parallel or series, nature of the load (variable or steady), and the noise constraints must also be considered. After the system characteristics have been determined, the main considerations in the actual fan selection are efficiency, reliability, size and weight, speed, noise, and cost.

To assist in the actual fan selection, manufacturers furnish graphs with the areas of preferred operation shown. In many cases manufacturers present their fan performance data in the form of tables. The static pressure is often given but not the total pressure. The total pressure may be computed from the capacity and the fan outlet dimensions. Data pertaining to noise are also available from most manufacturers.

It is important that the fan be efficient and quiet. Generally, a fan will generate the least noise when operated near the peak efficiency. Operation considerably beyond the point of maximum efficiency will be noisy. Forward-curved blades operated at high speeds will be noisy and straight blades are generally noisy, especially at high speed. Backward-curved blades may be operated on both sides of the peak efficiency at relatively high speeds with less noise than the other types of fans.

Fan Installation

The performance of a fan can be drastically reduced by improper connection to the duct system. In general, the duct connections should be such that the air may enter and leave the fan as uniformly as possible with no abrupt changes in direction or velocity. The designer must rely on good judgment and ingenuity in laying out the system. Space is often limited for the fan installation, and a less than optimum connection may have to be used. In this case the designer must be aware of the penalties (loss in total pressure and efficiency). Some manufacturers furnish application factors from which a modified fan curve can be computed.

The Air Movement and Control Association, Inc. (AMCA) has published system effect factors in their *Fan Applications Manual* that express the effect of various fan connections on system performance.[8]

64.6.2   Variable-Volume Systems

In variable-air-volume systems the total amount of circulated air may vary between some minimum and the full load design quantity. Normally, the minimum is about 20–25% of the maximum. The volume flow rate of the air is controlled independent of the fan by the terminal boxes, and the fan must respond to the system. The fan speed should be decreased as volume flow rate decreases. Variable speed electric motors have very low efficiency that offsets the benefit of lowering fan speed. Fan drives that make use of magnetic couplings have been developed and are referred to as eddy current drives. These are excellent devices with almost infinite adjustment of fan speed. Their only disadvantage is high cost. A change may be made in the fan speed by changing the diameter of the V-belt drive pulley by adjusting the pulley shives. This requires a mechanism that will operate while the drive is turning. The main disadvantage of this approach is maintenance. The eddy current and variable pulley drives appear to be the most practical at present.

Another approach to control of the fan is to throttle and introduce a swirling component to the air entering the fan that alters the fan characteristic in such a way that less power is required at the lower flow rates. This is done with variable inlet vanes that are a radial damper system located at the inlet to the fan. Gradual closing of the vanes reduces the volume flow rate of air and changes the fan characteristic. This approach is not as effective in reducing fan power as fan speed reduction, but the cost and maintenance are low.

Airflow in Ducts

The steady-flow energy equation applies to the flow of air in a duct. Neglecting the elevation head terms, assuming that the flow is adiabatic, and no fan is present,

$$\frac{g_c}{g}\frac{P_1}{\rho} + \frac{V_1^2}{2g} = \frac{g_c}{g}\frac{P_2}{\rho} + \frac{V_2^2}{2g} + l_f$$

and in terms of the total head

$$\frac{g_c}{g}\frac{P_{01}}{\rho} = \frac{g_c}{g}\frac{P_{02}}{\rho} + l_f$$

where V = average air velocity at a duct cross section, ft/min or m/sec
l_f = lost head due to friction, ft or m

The static and velocity head terms are interchangeable and may increase or decrease in the direction of flow depending on the duct cross-sectional area. Because the lost head must be positive, the total pressure always decreases in the direction of flow, as in Fig. 64.25.

For duct flow the units of each term are usually inches of water because of their small size. The equations may be written

$$H_{s1} + H_{v1} = H_{s2} + H_{v2} + l_f$$

and

$$H_{01} = H_{02} + l_f$$

For air at standard conditions

$$H_v = \left(\frac{V}{4005}\right)^2 \text{ in. } H_2O$$

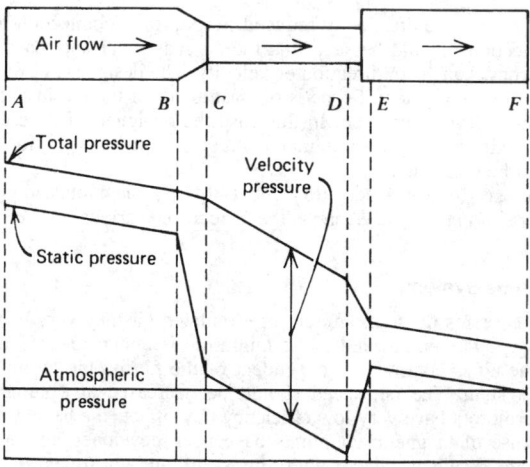

Fig. 64.25 Pressure changes during flow in ducts.[7]

where V is in ft/min,

$$P_v = \left(\frac{V}{1.29}\right)^2 \text{ Pa}$$

where V is in m/sec.

The lost head due to friction, l_f, in a straight, constant area duct may be determined by use of a friction factor. Because this approach becomes tedious when designing ducts, special charts have been prepared. Figure 64.26 is such a chart for air flowing in ducts. The chart is based on standard air and fully developed flow. For the temperature range of 50°F or 10°C to about 100°F or 38°C there is no need to correct for viscosity and density changes. Above 100°F or 38°C, however, a correction should be made. The density correction is also small for moderate pressure changes. For elevations below about 2000 ft or 610 m the correction is small. The correction for density and viscosity will normally be less than 1. The effect of roughness is an important consideration and difficult to assess.

A common problem to designers is determination of the roughness effect of fibrous glass duct liners and fibrous ducts. This material is manufactured in several grades with various degrees of absolute roughness. The usual approach to account for this roughness effect is to use a correction factor that is applied to the pressure loss obtained for galvanized metal duct.

The head loss due to friction is greater for a rectangular duct than for a circular duct of the same cross-sectional area and capacity. For most practical purposes ducts of aspect ratio not exceeding 8:1 will have the same lost head for equal length and mean velocity of flow as a circular duct of the same hydraulic diameter. When the duct sizes are expressed in terms of hydraulic diameter D and when the equations for friction loss in round and rectangular ducts are equated for equal length and capacity, an equation for the circular equivalent of a rectangular duct is obtained:

$$D_e = 1.3 \frac{(ab)^{5/8}}{(a + b)^{1/4}}$$

where a and b are the rectangular duct dimensions in any consistent units and D_e is the equivalent diameter. A table of equivalent diameters is given in the *ASHRAE Handbook*.[2]

Air Flow in Fittings

Whenever a change in area or direction occurs in a duct or when the flow is divided and diverted into a branch, substantial losses in total pressure may occur. These losses are usually of greater magnitude than the losses in the straight pipe and are referred to as dynamic losses.

Dynamic losses vary as the square of the velocity and are conveniently represented by

$$H_0 = (C)(H_v)$$

where the loss coefficient C is a constant. When different upstream and downstream areas are involved as in an expansion or contraction, either the upstream or downstream value of H_v may be used but C will be different in each case.

Fittings are classified as either constant flow, such as an elbow or transition, or as divided flow, such as a wye or tee. Tables give loss coefficients for many different types of constant flow fittings.[2] It should be kept in mind that the quality and type of construction may vary considerably for a particular type of fitting. Some manufacturers provide data for their own products.

Duct Design—General Considerations

The purpose of the duct system is to deliver a specified amount of air to each diffuser in the conditioned space at a specified total pressure. This is to ensure that the space load will be absorbed and the proper air motion within the space will be realized. The method used to lay out and size the duct system must result in a reasonably quiet system and must not require unusual adjustments to achieve the proper distribution of air to each space. A low noise level is achieved by limiting the air velocity, by using sound-absorbing duct materials or liners, and by avoiding drastic restrictions in the duct such as nearly closed dampers. Figure 64.26 gives recommended duct velocities for low- and high-velocity systems. A low-velocity duct system will generally have a pressure loss of less than 0.15 in. H₂0 per 100 ft (1.23 Pa/m), whereas high-velocity systems may have pressure losses up to about 0.7 in. H₂0 per 100 ft (5.7 Pa/m). Fibrous glass duct materials are very effective for noise control. The duct, insulation, and reflective vapor barrier are all the same piece of material. Metal ducts are usually lined with fibrous glass material in the vicinity of the air-distribution equip-

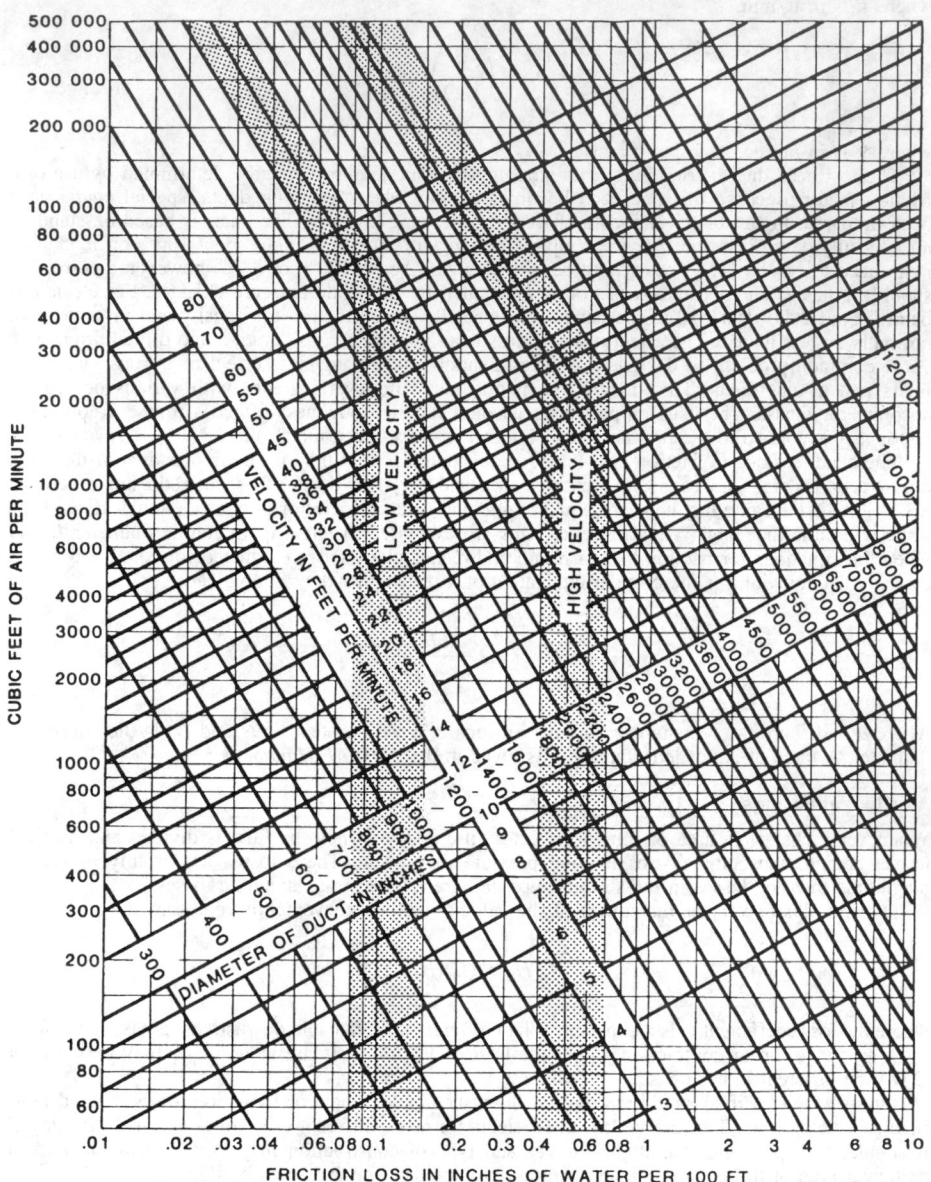

Fig. 64.26 Lost head due to friction for air flowing in ducts. (Reprinted by permission from *ASHRAE Handbook of Fundamentals*.)

ment. The remainder of the metal duct is then wrapped or covered with insulation and a vapor barrier. Insulation on the outside of the duct also reduces noise. The duct system should be relatively free of leaks, especially when the ducts are outside the conditioned space.

Generally, the location of the air diffusers and air-moving equipment is first selected with some attention given to how a duct system may be installed. The ducts are then laid out with attention given to space and ease of construction. It is very iinportant to design a duct system that can be constructed and installed in the allocated space. If this is not done, the installer may make changes in the field that lead to unsatisfactory operation.

From the standpoint of first cost, the ducts should be small; however, small ducts tend to give high air velocities, high noise levels, and large losses in total pressure. Therefore, a reasonable compromise between first cost, operating cost, and practice must be reached. A number of computer programs are available for this purpose.

For residential and light commercial applications all of the heating, cooling, and air-moving equipment is determined by the heating and/or cooling load. Therefore, the fan characteristics are known before the duct design is begun. The pressure losses in all other elements of the system except the supply and return ducts are known. The total pressure available for the ducts is then the difference between the total pressure characteristic of the fan and the sum of the total pressure losses of all of the other elements in the system excluding the ducts. Figure 64.27 shows a typical total pressure profile for a residential or light commercial system. In this case the fan is capable of developing 0.6 in. H_2O at the rated capacity. The return grille, filter, coils, and diffusers have a combined loss in total pressure of 0.38 in. H_2O. Therefore, the available total pressure for which the ducts must be designed is 0.22 in. H_2O. This is usually divided for low-velocity systems so that the supply duct system has about twice the total pressure loss of the return ducts.

Large duct systems are usually designed using velocity as a criterion, and the fan requirements are determined after the design is complete. For these systems the fan characteristics are specified and the correct fan is installed in the air handler.

Element	Total pressure loss, in. of water
Return grille	0.04
Return duct	0.08
Filter	0.08
Heat and cool coils	0.23
Supply ducts	0.14
Diffusers	0.03
Fan total pressure in. of water	0.60

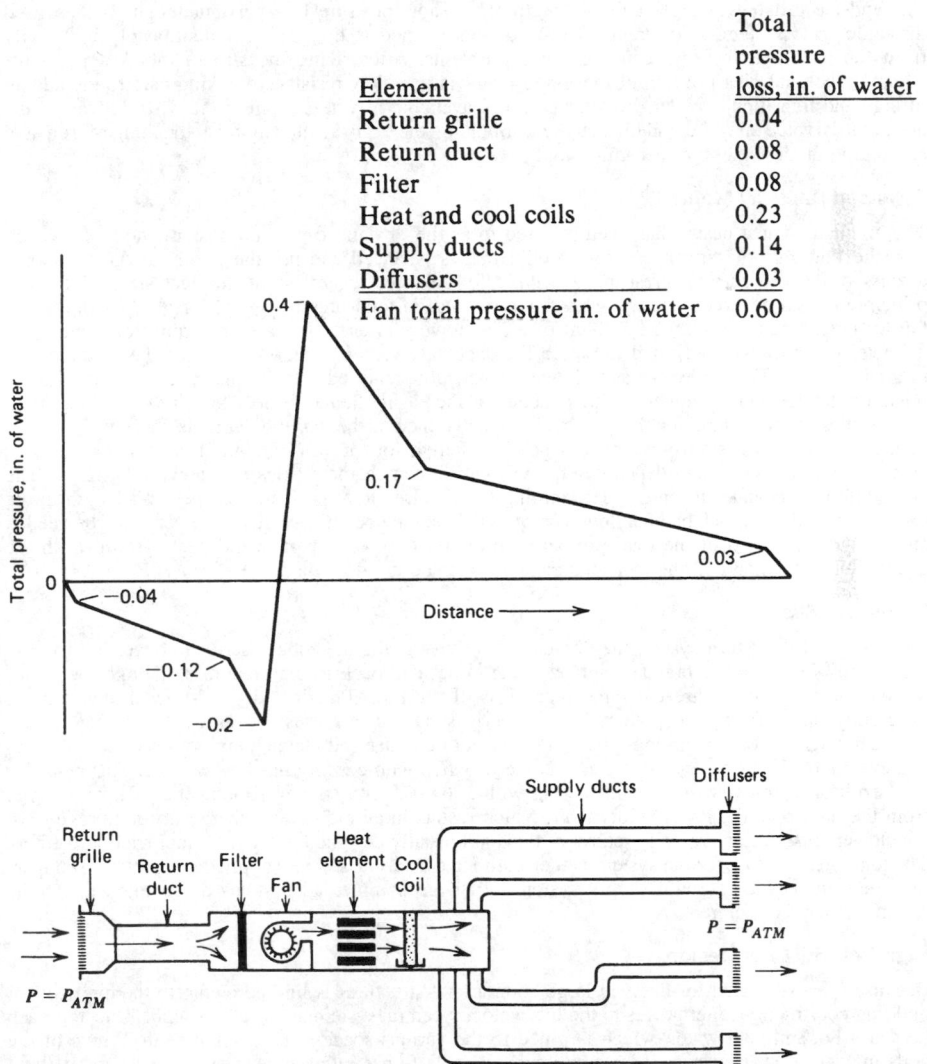

Fig. 64.27 Total pressure profile for a typical residential or light commercial system.[7]

Some designers neglect velocity pressure when dealing with low-velocity systems. This does not simplify the duct design procedure and is unnecessary. When the air velocities are high, the velocity pressure must be considered to achieve reasonable accuracy. If static and velocity pressure are computed separately, the problem becomes very complex. It is best to use total pressure in duct design because it is simpler and accounts for all of the flow energy.

Design of Low-Velocity Duct Systems

The methods described in this section pertain to low-velocity systems where the average velocity is less than about 1000 ft/min or S m/sec. These methods can be used for high-velocity system design, but the results will not be satisfactory in most cases.

Equal Friction Method

This method makes the pressure loss per foot of length the same for the entire system. If all runs from fan to diffuser are about the same length, this method will produce a well balanced design. However, most duct systems have duct runs ranging from long to short. The short runs will have to be dampered, which can cause considerable noise.

The usual procedure is to select the velocity in the main duct adjacent to the fan and to provide a satisfactory noise level for the particular application. The known flow rate then establishes the duct size and the lost pressure per unit of length. This same pressure loss per unit length is then used throughout the system. A desirable feature of this method is the gradual reduction of air velocity from fan to outlet, thereby reducing noise problems. After sizing the system the designer must compute the total pressure loss of the longest run (largest flow resistance), taking care to include all fittings and transitions. When the total pressure available for the system is known in advance, the design loss value may be established by estimating the equivalent length of the longest run and computing the low pressure per unit length.

Balanced Capacity Method

This method of duct design has been referred to as the "balanced pressure loss method." However, it is the flow rate or capacity of each outlet that is balanced and not the pressure. As previously discussed, the loss in total pressure automatically balances regardless of the duct sizes. The basic principle of this method of design is to make the loss in total pressure equal for all duct runs from fan to outlet when the required amount of air is flowing in each. For a given equivalent length the diameter can always be adjusted to obtain the necessary velocity that will produce the required loss in total pressure. There may be cases, however, when the required velocity may be too high to satisfy noise limitations and a damper or other means of increasing the equivalent length will be required.

The design procedure for the balanced capacity method begins the same as the equal friction method in that the design pressure loss per unit length for the run of longest equivalent length is determined in the same way depending on whether the fan characteristics are known in advance. The procedure then changes to one of determining the required total pressure loss per unit length in the remaining sections to balance the flow as required. The method shows where dampers may be needed and provides a record of the total pressure requirements of each part of the duct system. Both the equal friction method and the balanced capacity method are described in Ref. 7.

Return Air Systems

The design of the return system may be carried out using the methods described above. In this case the air flows through the branches into the main duct and back to the plenum. Although the losses in constant-flow fittings are the same regardless of the flow direction, divided-flow fittings behave differently and different equivalent lengths or loss coefficients must be used. Reference 2 gives considerable data for converging-type fittings of both circular and rectangular cross section. For low-velocity ratios the loss coefficient can become negative with converging-flow streams. This behavior is a result of a high-velocity stream mixing with a low-velocity stream. Kinetic energy is transferred from the higher- to the lower-velocity air, which results in an increase in energy or total pressure of the slower stream. Low-velocity return systems are usually designed using the equal friction method. The total pressure loss for the system is then estimated as discussed for supply duct systems. Dampers may be required just as with supply systems. In large commercial systems a separate fan for the return air may be required.

High-Velocity Duct Design

Because space allocated for ducts in large commercial structures is limited owing to the high cost of building construction, alternatives to the low-velocity central system are usually sought. One approach is to use hot and cold water, which is piped to the various spaces where small central units or fan coils may be used. However, it is sometimes desirable to use rather extensive duct systems without taking up too much space. The only way this can be done is to move the air at much higher velocities. High-velocity systems may use velocities as high as 6000 ft/min or about 30 m/sec. The use of high velocities reduces the duct sizes dramatically, but introduces some new problems.

Noise is probably the most serious consequence of high-velocity air movement. Special attention must be given to the design and installation of sound-attenuating equipment in the system. Because the air cannot be introduced to the conditioned space at a high velocity, a device called a terminal box is used to throttle the air to a low velocity, control the flow rate, and attenuate the noise. The terminal box is located in the general vicinity of the space it serves and may distribute air to several outlets.

The energy required to move the air through the duct system at high velocity is also an important consideration. The total pressure requirement is typically on the order of several inches of water. To partially offset the high fan power requirements, variable-speed fans are sometimes used.

Because the higher static and total pressures required by high-velocity systems aggravate the duct leakage problem, an improved duct fabrication system has been developed for high-velocity systems. The duct is generally referred to as spiral duct and has either a round or oval cross section. The fittings are machine formed and are especially designed to have low pressure losses and close fitting joints to prevent leakage.

The criterion for designing the high-velocity duct systems is somewhat different from that used for low-velocity systems. Emphasis is shifted from a self-balancing system to one that has minimum losses in total pressure.

REFERENCES

1. J. A. Goff, "Standardization of Thermodynamic Properties of Moist Air," *Transactions ASHVE,* **55** (1949).

2. *ASHRAE Handbook of Fundamentals,* American Society of Heating, Refrigerating and Air-Conditioning Engineers, New York, 1997.

3. ASHRAE Standard 55-92, *Thermal Environmental Condition for Human Occupancy,* American Society of Heating, Refrigerating and Air-Conditioning Engineers, New York, 1992.

4. ASHRAE Standard 62-89, *Standards for Ventilation Required for Minimum Acceptable Indoor Air Quality,* American Society of Heating, Refrigerating and Air-Conditioning Engineers, New York, 1981.

5. *Cooling and Heating Load Calculation Manual,* 2nd ed., American Society of Heating, Refrigerating and Air-Conditioning Engineers, New York, 1992.

6. J. D. Parker, J. H. Boggs, and E. F. Blick, *Introduction to Fluid Mechanics and Heat Transfer,* Addison-Wesley, Reading, MA, 1959.

7. F. C. McQuiston and J. D. Parker, *Heating, Ventilating, and Air Conditioning,* 4th ed., Wiley, New York, 1994.

8. *AMCA Fan Applications Manual,* Parts 1, 2, and 3, Air Movement and Control Association, Inc., 30 West University Drive, Arlington Heights, IL.

CHAPTER 65

AIR POLLUTION-CONTROL TECHNOLOGIES

C. A. Miller
United States Environmental Protection Agency
Research Triangle Park, North Carolina

65.1	**SULFUR DIOXIDE CONTROL**	**2012**
	65.1.1 Control Technologies	2013
	65.1.2 Alternative Control Strategies	2015
	65.1.3 Residue Disposal and Utilization	2015
	65.1.4 Costs of Control	2015
65.2	**OXIDES OF NITROGEN—FORMATION AND CONTROL**	**2015**
	65.2.1 NO_x Formation Chemistry	2015
	65.2.2 Combustion Modification NO_x Controls	2016
	65.2.3 Postcombustion NO_x Controls	2018
65.3	**CONTROL OF PARTICULATE MATTER**	**2020**
65.4	**CARBON MONOXIDE**	**2022**
65.5	**VOLATILE ORGANIC COMPOUNDS AND ORGANIC HAZARDOUS AIR POLLUTANTS**	**2022**
	65.5.1 Conventional Control Technologies	2023
	65.5.2 Alternative VOC Control Technologies	2024
65.6	**METAL HAZARDOUS AIR POLLUTANTS**	**2024**
65.7	**INCINERATION**	**2025**
65.8	**ALTERNATIVE POLLUTION-CONTROL APPROACHES**	**2025**
65.9	**GLOBAL CLIMATE CHANGE**	**2026**
	65.9.1 CO_2	2026
	65.9.2 Other Global Warming Gases	2027
	65.9.3 Ozone-Depleting Substances	2028

Air pollution and its control have played an increasingly important role in modern activities since the advent of the Industrial Revolution, particularly relative to industrial activities. Most industries engage in one or more activities that result in the release of pollutants into the atmosphere, and in the last 40 years, steps have been taken to reduce these emissions through the application of process modifications or the installation and use of air pollution control technologies. Air pollution sources are typically divided into two major categories, *mobile* and *stationary*. This chapter will discuss the use of technologies for reducing air pollution emissions from stationary sources, with an emphasis on the control of combustion-generated air pollution. Major stationary sources include utility power boilers, industrial boilers and heaters, metal smelting and processing plants, and chemical and other manufacturing plants.

Pollutants that are of primary concern are those that, in sufficient ambient concentrations, adversely impact human health and/or the quality of the environment. Those pollutants for which health criteria define specific acceptable levels of ambient concentrations are known in the United States as "criteria pollutants." The major criteria pollutants are carbon monoxide (CO), nitrogen dioxide (NO_2), ozone, particulate matter less than 10 μm in diameter (PM_{10}), sulfur dioxide (SO_2), and lead (Pb). Ambient concentrations of NO_2 are usually controlled by limiting emissions of both nitrogen oxide (NO) and NO_2, which combined are referred to as oxides of nitrogen (NO_x). NO_x and SO_2 are

Mechanical Engineers' Handbook, 2nd ed., Edited by Myer Kutz.
ISBN 0-471-13007-9 © 1998 John Wiley & Sons, Inc.

important in the formation of acid precipitation, and NO_x and volatile organic compounds (VOCs) can react in the lower atmosphere to form ozone, which can cause damage to lungs as well as to property.

Other compounds, such as benzene, polycyclic aromatic hydrocarbons (PAHs), other trace organics, and mercury and other metals, are emitted in much smaller quantities, but are more toxic and in some cases accumulate in biological tissue over time. These compounds have been grouped together as hazardous air pollutants (HAPs) or "air toxics," and have recently been the subject of increased regulatory control.[1] Also of increasing interest are emissions of compounds such as carbon dioxide (CO_2), methane (CH_4), or nitrous oxide (N_2O) that have the potential to affect the global climate by increasing the level of solar radiation trapped in the Earth's atmosphere, and compounds such as chlorofluorocarbons (CFCs) that react with and destroy ozone in the stratosphere, reducing the atmosphere's ability to screen out harmful ultraviolet radiation from the sun.

The primary method of addressing emissions of air pollutants in the United States has been the enactment of laws requiring sources of those pollutants to reduce emission rates to acceptable levels determined by the U.S. Environmental Protection Agency (EPA) and air pollution regulatory agencies at the state, regional, and local levels. Current standards vary between states and localities, depending upon the need to reduce ambient levels of pollutants. EPA typically sets "national ambient air quality standards" (NAAQS) for the criteria pollutants, and the states and localities then determine the appropriate methods to achieve and maintain those standards. EPA also sets minimum pollution performance requirements for new pollution sources, known as the "new source performance standards" or NSPS. For some pollutants (such as HAPs), EPA is required to set limits on the annual mass of emissions to reduce the total health risk associated with exposure to these pollutants. Other approaches include the limiting of the total national mass emissions of pollutants such as SO_2; this allows emissions trading to occur between different plants and between different regions of the country while ensuring that a limited level of SO_2 is available in the atmosphere for the formation of acid precipitation.

Combustion processes are a major anthropogenic source of air pollution in the United States, responsible for 24% of the total emissions of CO, NO_x, SO_2, VOCs, and particulates.[2] In 1992, 146 million tonnes (161 million tons) of these pollutants were emitted in the United States. Of these pollutants, stationary combustion processes emit 91% of the total U.S. SO_2 emissions, and 50% of the total U.S. NO_x emissions. The major combustion-generated pollutants (not including CO_2) by tonnage are CO, NO_x, PM, SO_2, and VOCs. Table 65.1 presents total estimated anthropogenic and combustion-generated emissions of selected air pollutants in the United States.

Combustion-generated air pollution can be viewed as originating through two major methods, although some overlap occurs between the two. The first of these methods is origination of pollution primarily from constituents in the fuel. Examples of these "fuel-borne" pollutants are SO_2 and trace metals. The second is the origination of pollutants through modification or reaction of constituents that are normally nonpolluting. CO, NO_x, and volatile organics are examples of "process-derived" pollutants. In the case of NO_x, fuel-borne nitrogen such as that in coal plays a major role in the formation of the pollutant; however, even such clean fuels as natural gas (which contains no appreciable nitrogen) can emit NO_x when combusted in nitrogen-containing air.

Major stationary sources of combustion-generated air pollution include steam electric generating stations, metal processing facilities, industrial boilers, and refinery and other process heaters. Table 65.2 shows the total U.S. emissions of criteria pollutants from these and other sources.

Given the wide variety of sources and pollutants, it is no surprise that there is a correspondingly wide variety of approaches to air pollution control. The three primary approaches are preprocess control, process modification, and postprocess control. Preprocess control usually involves cleaning of the fuel prior to introducing it into the combustion process, as in the case of coal cleaning. Process modifications are applied when the pollutant of interest is "process-derived," and such modifications do not adversely alter the product. Low-NO_x burners fall into this category. In the postprocess control approach, the pollution-forming process itself is not altered, and a completely separate pollution-cleaning process is added to clean up the pollutant after it has been formed. Flue gas desulfurization systems are an example of this approach.

Early work in the field of air pollution control technology focused on SO_2, NO_x, and particulates. Control technologies for these pollutants have been refined and tested extensively in service, and have in most cases reached the status of mature technologies. Nevertheless, work continues to improve performance, as measured by pollutant-reduction efficiency and operating and maintenance cost. These mature technologies are also being evaluated for their performance as control devices for HAPs, and as the bases for new hybrid technologies that seek to achieve pollutant emission reductions of 90% or more with minimal increase in capital or operating costs.

65.1 SULFUR DIOXIDE CONTROL

SO_2 emissions are controlled to a large degree by the use of flue gas desulfurization (FGD) systems. Although furnace sorbent injection has been demonstrated to provide some degree of SO_2 emission

Table 65.1 Anthropogenic Emissions of Selected Air Pollutants[a]

Pollutant	Anthropogenic Emissions, Tons/Year	Combustion Emissions, Tons/Year	Principal Source(s)	Reference
NO_x	2.3×10^7	2.2×10^7	Electric utilities/highway vehicles	16
N_2O	7.8×10^6	2.6×10^6	Biomass, mobile	17
SO_2	2.2×10^7	2.0×10^7	Electric utilities, industrial combustion	16
Total PM	1.1×10^7	2.1×10^6	Residential wood, off-highway vehicles	16
Metal PM	1.4×10^5	7.0×10^2	Metals same as listed below	16
Hg	3.3×10^2	2.1×10^2	MedWI, MWC, utility boilers	18
CO_2	5.5×10^9	5.4×10^9	Steam boilers, space heat, highway vehicles	16
CO	9.7×10^7	9.2×10^7	Highway and off-highway vehicles	16
PAH	3.6×10^4	1.8×10^4	Residential wood, open burning (16 PAHs)	16
CH_4	3.0×10^7	7.0×10^5	Stationary combustion	16
VOC	2.3×10^7	1.2×10^7	Highway and off-highway vehicles, wildfires	16
Organic HAPs	9.4×10^6	N/A		19
Pb	4.9×10^3	2.6×10^3	Highway vehicles, waste disposal	18

[a]Emission figures are for 1993 (CH_4 and CO_2 emission figures are for 1992 and HAPs are for 1991). Non-combustion HAPs reported through Toxic Release Inventory and do not include hydrogen chloride.

reductions, by far the most common FGD systems are wet or dry scrubbers. Other methods of reducing emissions of SO_2 include fuel desulfurization to remove at least a portion of the sulfur prior to burning, or switching to a lower-sulfur fuel.

65.1.1 Control Technologies

Wet scrubbers use a variety of means to ensure adequate mixing of the scrubber liquor and the flue gas. A venturi scrubber uses a narrowing of the flue gas flow path to confine the gas path. At the narrowest point, the scrubber liquor is sprayed into the flue gas, allowing the spray to cover as great a volume of gas as possible. Packed tower scrubbers utilize chemical reactor packing to create porous beds through which the flue gas and scrubber liquor pass, ensuring good contact between the two phases. The packing material is often plastic, but may be other materials as well. The primary

Table 65.2 Annual Combustion-Generated Emissions of Selected Pollutants by Stationary Source Category[a]

Pollutant	Stationary Fuel Combustion Emissions			% of Total
	Utility	Industrial	Other	
CO	311	714	5,154	6.4
NO_x	7,468	3,523	734	50.7
Total particulate	454	1,030	493	18.0
SO_2	15,841	3,090	589	88.7
VOC	32	279	394	3.1

[a]In thousand tons/year. Emissions values are for 1992.

requirements for the packing are to evenly distribute the gas and liquid across the tower cross section, provide adequate surface area for the reactions to occur, and allow the gas to pass through the bed without excessive pressure drop.

Perforated plate scrubbers usually are designed with the gas flowing upward and the liquid flowing in the opposite direction. The flow of the gas through the perforations is sufficiently high to retard the counterflow of the liquid, creating a liquid layer on the plate through which the gas must pass. This ensures good contact between the liquid and gas phases. Bubble cap designs also rely on a layer of liquid on the plate, but create the contact of the two phases through the design of the caps. Gas passes up into the cap and back down through narrow openings into the liquid. The liquid level is regulated by overflow weirs, through which the liquid passes to the next lower level. The gas pressure drop in this type of system increases with the height of the liquid and the gas flow rate.

Wet scrubbers for utility applications typically use either lime (CaO) or limestone (primarily calcium carbonate, $CaCO_3$) in an aqueous slurry, which is then sprayed into the flue gas flow in such a way as to maximize the contact between the SO_2-containing flue gas and the slurry. The reaction of the slurry and the SO_2 creates calcium sulfite ($CaSO_3$) or calcium sulfate ($CaSO_4$) in an aqueous solution. Because both these compounds have low water solubility, they may precipitate out of solution and create scale in the system piping and other components. Care must be taken during operation to minimize scale deposition by keeping the concentrations of $CaSO_3$ and $CaSO_4$ below the saturation point during operation. Wet scrubbers typically have high SO_2 removal efficiencies (90% or greater) and require relatively low flue gas energy requirements. In some cases, however, capital and operating costs may be higher than for dry scrubbers (see below) due to higher fan power requirements or increased maintenance due to excessive scaling.

Smaller industrial scrubbers typically use a clean liquor reagent, such as sodium carbonate or sodium hydroxide. Alkali compounds other than lime or limestone can also be used. Magnesium oxide (MgO) is used to form a slurry of magnesium hydroxide [$Mg(OH)_2$] to absorb the SO_2 and form magnesium sulfite or sulfate. The solid can be separated from the slurry, allowing the regeneration of the MgO and producing a relatively high concentration (10–15%) stream of SO_2. The SO_2 stream is then used to produce sulfuric acid.

Dual alkali scrubbing systems use two chemicals in a two-loop arrangement. A typical arrangement uses a more expensive sodium oxide or sodium hydroxide scrubbing liquor, which forms principally sodium sulfite (Na_2SO_3) when sprayed into the SO_2-containing flue gases. The spent liquor is then sent to the secondary loop, where a less expensive alkali, such as lime, is added. The calcium sulfate or sulfite precipitates out of the liquor, and the sodium-based liquor is regenerated for reuse in the scrubber. The calcium sulfate/sulfite is separated from the liquor and dried, and the solids are usually sent to a landfill for disposal. SO_2 removal efficiencies for such systems are typically 75% and higher, with many systems capable of reductions greater than 90%.

Dry scrubbers, or spray dryer absorbers (SDAs), also use an aqueous slurry of lime to capture the SO_2 in the flue gas. However, SDAs create a much finer spray, resulting in rapid evaporation of the water droplets and leaving the lime particles suspended in the flue gas flow. As SO_2 contacts these particles, reactions occur to create $CaSO_4$. The suspended particulate is then captured by a particle removal system, often a fabric filter (see below). An advantage of the dry scrubber is its lower capital and operating cost compared to the wet scrubber, and the production of a dry, rather than wet, waste material for disposal. In some cases, the dry slurry solids can be recycled and reused. Dry systems are typically less efficient than wet scrubbers, providing removal efficiencies of 70–90%.

Furnace sorbent injection is the direct injection of a solid calcium-based material, such as hydrated lime, limestone, or dolomite, into the furnace for the purpose of SO_2 capture. Depending upon the amount of SO_2 removal required, furnace sorbent injection can remove the need for FGD. SO_2 removal efficiencies of up to 70% have been demonstrated,[3] although 50% reductions are more typical. The effectiveness of furnace sorbent injection is dependent upon the calcium to sulfur ratio (Ca/S), furnace temperature, and humidity in the flue gas. A Ca/S of 2 is typically used. Furnace sorbent injection effectiveness decreases with increasing furnace temperature and increases as flue gas humidity levels decrease.

While the need for an SO_2 scrubbing system is eliminated, systems that use furnace sorbent injection require adequate capacity in their particulate removal equipment to remove the additional solid material injected into the furnace. In addition, increased soot blowing is also required to maintain clean heat transfer surfaces and prevent reduced heat-transfer efficiencies when furnace sorbent injection is used.

Fluid bed combustion (FBC) is another technology that allows the removal of SO_2 in a similar manner to furnace sorbent injection. In such systems, the fluidized bed contains a calcium-based solid that removes the sulfur as the coal is burned in the bed. FBC is limited to new plant designs, since it is an alternative design significantly different from conventional steam generation systems, and is not a retrofit technology. FBC systems typically remove 70–90% of the SO_2 generated in the combustion reactions.

65.1.2 Alternative Control Strategies

Coal cleaning (or fuel desulfurization) is also an option for removing a portion of the sulfur in the as-mined fuel. A significant portion of Eastern and Midwestern bituminous coals are currently cleaned to some degree to remove both sulfur and mineral matter. Cleaning may be done by crushing and screening the coal or by washing with water or a dense medium consisting of a slurry of water and magnetite. Washing is typically done by taking advantage of the different specific gravities of the different coal constituents. The sulfur in coal is typically in the form of iron pyrite (pyritic sulfur) or organic sulfur contained in the carbon structure of the coal.* The sulfur-reduction potential (or washability) of a coal depends on the relative amount and distribution of pyritic sulfur. The washability of U.S. coals varies from region to region and ranges from less than 10% to greater than 50%. For most Eastern U.S. high-sulfur coals, the sulfur reduction potential normally does not exceed 30%, limiting the use of physical coal cleaning for compliance coal production. Cleaning usually results in the generation of a solid or liquid waste that must be either disposed of or recycled.

Fuel switching is a further option for the reduction of SO_2 emissions. Fuel switching most often involves the change from a high-sulfur fuel to a lower-sulfur fuel of the same type. For coal, this change most commonly involves a change from a higher-sulfur Eastern coal to low-sulfur Western coals. In some instances, the change of coals may also result in restrictions to plant operability, usually due to changes in the slagging and fouling characteristics of the coal. However, many plants have found that the costs of compliance using a fuel-switching approach outweigh the operational changes. In some instances, fuel switching can also involve a change from high-sulfur coal to natural gas. In this case, not only are SO_2 emissions reduced, but the lack of nitrogen in natural gas also yields a reduction in NO_x emissions. Particulate emissions are also significantly reduced, as are emissions of trace metals.

65.1.3 Residue Disposal and Utilization

Flue gas desulfurization results in significant quantities of solid and/or liquid material that must be removed from the plant process. In some cases, the residues can be used as is, or processed to produce higher-quality materials for a number of applications. The cost of residue disposal can account for a significant portion of the total cost of SO_2 removal, particularly where landfilling costs are high. Early waste management approaches focussed on landfilling and, as such costs increased, more attention was given to utilization options.

For sludges from wet scrubbers, the use of forced oxidation of the spent scrubbing slurry produces $CaSO_4$ from the $CaSO_3$ in the slurry, which can then be processed to form a salable gypsum product. Some impurities can be removed by means of filtration and removal of the smaller particles, followed by the hydration of the $CaSO_4$ to form gypsum ($CaSO_4 \cdot 2H_2O$) and dewatering of the final solids. Depending upon the quality of the final product, the resulting solids can be used in building materials, soil stabilization and road base, aggregate products, or in agricultural applications. Spray dryer by-products have a higher free lime content, making them less acceptable as a building material. The most likely end use of these residues is as a road-base stabilization material.

65.1.4 Costs of Control

Many factors are involved in the costs of applying SO_2 control technologies, including the amount of sulfur in the coal, the level of control required, and the plant size and configuration (particularly for retrofit applications). However, there have been several studies conducted to compare the costs of SO_2 controls in terms of capital cost per kilowatt of plant capacity, annual cost in mills per kilowatt-hour, and dollars per ton of SO_2 removed. Table 65.3 shows ranges of estimated costs[4,5] and indicates that, although the capital and annual costs can vary significantly between the different approaches, the costs in dollars per ton of SO_2 removed are much more comparable. This is due in large part to the fact that the lower-cost SO_2 control strategies tend to result in lower SO_2 reductions compared to the more expensive control options.

65.2 OXIDES OF NITROGEN—FORMATION AND CONTROL

65.2.1 NO_x Formation Chemistry

NO_x formed by the combustion of fuel in air is typically composed of greater than 90% NO, with NO_2 making up the remainder. Unfortunately, NO is not amenable to flue gas scrubbing processes, as SO_2 is. An understanding of the chemistry of NO_x formation and destruction is helpful in understanding emission-control technologies for NO_x.

---

*Sulfur in coal may also be in the form of sulfates, particularly in weathered coal. Pyritic and organic sulfur are the two most common forms of sulfur in coal.

Table 65.3   Emission Reductions and Costs of Different SO$_2$ Control Technologies

Control Technology	SO$_2$ Reduction, %	Capital Cost, $/kW	Annual Cost, mils/ kW-hr	Cost, $/tonne SO$_2$ ($/ton SO$_2$) Ref. 5	Ref. 6
Wet scrubber	75–90+	150–180	16	385–660 (350–600)	1200 (1100)
Lime spray dryer	70–90	110–210	10	395–595 (360–540)	990 (900)
Furnace sorbent injection	50–70	50–120	6	460–825 (420–750)	825 (750)
Coal switching	60–70	27	4	NA	880 (800)

There are three major pathways to formation of NO in combustion systems: thermal NO$_x$, fuel NO$_x$, and prompt NO$_x$. Thermal NO$_x$ is created when the oxygen (O$_2$) and nitrogen (N$_2$) present in the air are exposed to the high temperatures of a flame, leading to a dissociation of O$_2$ and N$_2$ molecules and their recombination into NO. The rate of this reaction is highly temperature-dependent; therefore, a reduction in peak flame temperature can significantly reduce the level of NO$_x$ emissions. Thermal NO$_x$ is important in all combustion processes that rely on air as the oxidizer.

Fuel NO$_x$ is due to the presence of nitrogen in the fuel and is the greatest contributor to total NO$_x$ emissions in uncontrolled coal flames. By limiting the presence of O$_2$ in the region where the nitrogen devolatilizes from the solid fuel, the formation of fuel NO$_x$ can be greatly diminished. NO formation reactions depend upon the presence of hydrocarbon radicals and O$_2$, and since the hydrocarbon–oxygen reactions are much faster than the nitrogen–oxygen reactions, a controlled introduction of air into the devolatilization zone leads to the oxygen preferentially reacting with the hydrocarbon radicals (rather than with the nitrogen) to form water and CO. Finally, the combustion of CO is completed, and since this reaction does not promote NO production, the total rate of NO$_x$ production is reduced in comparison with uncontrolled flames. This staged combustion can be designed to take place within a single burner flame or within the entire furnace, depending on the type of control applied (see below). Fuel NO$_x$ is important primarily in coal combustion systems, although it is important in systems that use heavy oils, since both fuels contain significant amounts of fuel nitrogen.

Prompt NO$_x$ forms at a rate faster than equilibrium would predict for thermal NO$_x$ formation. Prompt NO$_x$ forms from nonequilibrium levels of oxide (O) and hydroxide (OH) radicals, through reactions initiated by hydrocarbon radicals with molecular nitrogen, and the reactions of O atoms with N$_2$ to form N$_2$O and finally the subsequent reaction of N$_2$O with O to form NO. Prompt NO$_x$ can account for more than 50% of NO$_x$ formed in fuel-rich hydrocarbon flames;[6] however, prompt NO does not typically account for a significant portion of the total NO emissions from combustion sources.

65.2.2   Combustion Modification NO$_x$ Controls

Because the rate of NO$_x$ formation is so highly dependent upon temperature as well as local chemistry within the combustion environment, NO$_x$ is ideally suited to control by means of modifying the combustion conditions. There are several methods of applying these combustion modification NO$_x$ controls, ranging from reducing the overall excess air levels in the combustor to burners specifically designed for low NO$_x$ emissions.

Low excess air (LEA) operation is the simplest form of NO$_x$ control, and relies on reducing the amount of combustion air fed into the furnace. LEA can also improve combustion efficiency where excess air levels are much too high. The drawbacks to this method are the relatively low NO$_x$ reduction and the potential for increased emissions of CO and unburned hydrocarbons if excess air levels are dropped too far. NO$_x$ emission reductions using LEA range between 5 and 20%, at relatively minimal cost if the reduction of combustion air does not also lead to incomplete combustion of fuel. Incomplete combustion significantly reduces combustion efficiency, increasing operating costs, and may result in high levels of CO or even carbonaceous soot emissions.[7]

Overfire air (OFA) is a simple method of staged combustion in which the burners are operated with very low excess air or at substoichiometric (fuel-rich) conditions, and the remaining combustion air is introduced above the primary flame zone to complete the combustion process and achieve the required overall stoichiometric ratio. The LEA or fuel-rich conditions result in lower peak flame temperatures and reduced levels of oxygen in the regions where the fuel-bound nitrogen devolatilizes from the solid fuel. These two effects result in lower NO$_x$ formation in the flame zone, and therefore lower emissions. Recent field studies showed approximately 20% reductions of NO_x emissions using advanced OFA in a coal-fired boiler.[8] OFA can be used for coal, oil, and natural gas, and to some degree for solid fuels such as municipal solid waste and biomass when combusted on stoker-grate units.

OFA typically requires special air-injection ports above the burners, as well as the associated combustion air ducting to the ports. In some cases, additional fan capability is required in order to ensure that the OFA is injected with enough momentum to penetrate the flue gases. Emissions of CO are usually not adversely affected by operation with OFA. Use of OFA can result in higher levels of carbon in fly ash when used in coal-fired applications, but proper design and operating may minimize this disadvantage. Another disadvantage to the application of OFA is the often corrosive nature of the flue gases in the fuel rich zone. If adequate precautions are not taken, this can lead to increased corrosion of boiler tubes.

Flue gas recirculation is a combustion-modification technique used to reduce the peak flame temperature by mixing some of the combustion gases back into the flame zone. This method is especially effective for fuels with little or no nitrogen, such as natural gas combustion systems. However, in many instances, the recirculation system requires a separate fan to compress the hot gases, and the fan capital and operating costs can be substantial. The resulting NO_x reductions can be significant, however, and emission reductions as high as 50% have been achieved.[9]

In the past 15 years, burners for both natural gas and coal have undergone major design improvements intended to incorporate the principles of staging and flue gas recirculation into the flow patterns of the fuel and air injected by the burner. These burners are generically referred to as *low NO_x burners* (LNBs), and are the most widely used NO_x control technology. Staging of fuel and air that is the basis for combustion modification NO_x control is achieved in LNBs by creating separate flow paths for the air and fuel. This is in contrast to earlier burner designs, in which the fuel and air flows were designed to mix as quickly and as turbulently as possible. While these highly turbulent flames were very successful in achieving rapid and complete combustion, they also resulted in very high peak flame temperatures and high levels of oxygen in the fuel devolatilization region, with correspondingly high levels of NO_x emissions. The controlled mixing of fuel and air flows typical of LNBs significantly reduced the rates of fuel and air mixing, leading to lower flame temperatures and considerable reductions of oxygen in the devolatilization regions of the flame, thereby reducing the production and emission of NO_x.

Low-NO_x burners may further reduce the formation of NO_x by inducing flue gases into the flame zone through recirculation. Careful design of the fluid dynamics of the air and fuel flows acts to recirculate the partially burned fuel and products of combustion back into the flame zone, further reducing the peak flame temperature and thus the rate of NO_x production. In some burner designs, this use of recirculated flue gas is taken a step further by using flue gas that has been extracted from the furnace, compressed, and fed back into the burner along with the fuel and air. These burners are typically used in natural gas-fired applications, and are among the "ultra-low NO_x burners" that can achieve emission levels as low as 5 ppm.

LNBs are standard on most new facilities. Some difficulties may be encountered during retrofit applications if the furnace dimensions do not allow for the longer flame lengths typical of these burners. The flame lengths can increase considerably due to the more controlled mixing of the fuel and air and, if adequate furnace lengths are not available, impingement of the flame on the opposite wall can lead to rapid cooling of the flame and therefore increased emissions of CO and organic compounds, as well as reduced heat transfer efficiency from the flame zone to the heat transfer fluid.

More precise control of air and fuel flows is often required for LNBs compared to conventional burners due to the reliance of many LNB designs on fluid dynamics to stage the air and fuel flows in particular patterns. Slight changes in the flow patterns can lead to significant drops in burner and boiler efficiencies, higher CO and organic compound emissions, and even damage to the burner from excessive coking of the fuel on the burner. In addition, the more strict operating conditions may impact the burners' ability to properly operate using fuels with different properties, primarily for coal-fired units. Coals with lower volatility or higher fuel nitrogen content may hamper NO_x reduction, and changes in the coals' slagging properties may lead to fouling of the burner ports. Further, improper air distribution within the burner may result in high levels of erosion within the burner, degrading performance and reducing operating life. An example of a typical pulverized coal LNB design is shown in Fig. 65.1

A further, relatively new method of controlling NO_x emissions by means of combustion modification is the application of *reburning*. Reburning is applied by injecting a portion of the fuel downstream of the primary burner zone, thereby creating a fuel-rich reburn zone in which high levels of hydrocarbon radicals react with the NO formed in the primary combustion zone to create H_2O, CO, and N_2. This is quite different from the other combustion modification techniques, which reduce NO_x emissions by preventing its formation. Reburning can use coal, oil, or natural gas as the reburn fuel, regardless of the fuel used in the main burners. Natural gas is an ideal reburn fuel, as it does not contain any fuel-bound nitrogen. Coals that exhibit rapid devolatilization and char burnout are also suitable for use as reburn fuels.

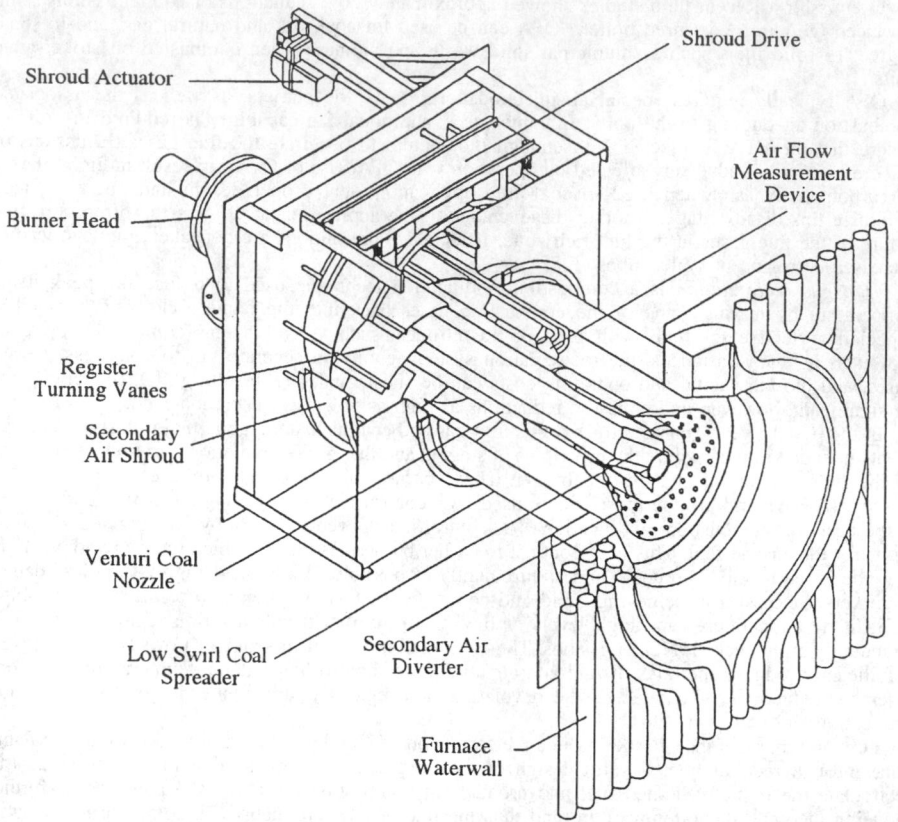

Fig. 65.1 Low-NO$_x$ burner.

In most applications, between 10 and 20% of the total heat input to the furnace is introduced in the reburn zone in the form of reburn fuel. The main burners are operated at slightly fuel-lean stoichiometries. This usually results in lower NO$_x$ levels leaving the primary zone, since the low excess air and lower flame temperatures produce lower NO$_x$. Above the primary zone, but far enough to allow for the combustion process to be nearly completed, the reburn fuel is introduced, and a reburn zone stoichiometry of 0.8 to 0.9 is created. Finally, sufficient air is injected downstream to burn out the remaining combustible materials (primarily CO) and reach the desired overall furnace stoichiometry (normally near 1.2). Reburning requires adequate furnace volume to allow the injection of the reburn fuel and the overfire air, as well as time for the combustion reactions to be completed.[10] Advanced reburning systems may utilize the injection of chemical reagents in addition to the reburn fuel to provide additional NO$_x$ reductions or to reduce the amounts of reburn fuel required for a given NO$_x$ reduction level. Reburning applied to full scale utility boilers has resulted in NO$_x$ emissions ranging from 50 to 65%.

65.2.3 Postcombustion NO$_x$ Controls

In some cases, either it is not possible to modify the combustion process or the levels of NO$_x$ reduction are beyond the capabilities of combustion modifications alone. In these instances, postcombustion controls must be used. There are two primary postcombustion NO$_x$ control technologies, *selective noncatalytic reduction* (SNCR) and *selective catalytic reduction* (SCR). Several systems have also been developed for scrubbing NO$_x$; however, since these remove only NO$_2$, they are not in broad commercial operation.

SNCR systems inject a nitrogen-based reagent into a relatively high temperature zone of the furnace, and rely on the chemical reaction of the reagent with the NO to produce N$_2$, N$_2$O, and H$_2$O. Removal efficiencies of up to 75% can be achieved with SNCR systems, but lower removal rates are typical. The SNCR reaction is highly temperature-dependent and, if not conducted properly, can result in either increased NO$_x$ emissions or considerable emissions of ammonia. The reagents most commonly used are ammonia (NH_3) and urea (NH_2CONH_2), although other chemicals have also been used, including cyanuric acid, ammonium sulfate, ammonium carbamate, and hydrazine hydrate. A number of proprietary reagents are also offered by several vendors, but all rely on similar chemical reaction processes. Proprietary reagents are used to vary the location and width of the temperature window, and to reduce the amount of ammonia slip to acceptable levels (typically less than 10–20 ppm).

The optimum temperature for SNCR systems will vary depending upon the reagent used, but ranges between 870 and 1150°C (1600 and 2100°F). Increased NO_x reductions can be obtained by using increasing amounts of reagent, although excessive use of reagent can lead to emissions of ammonia or, in some cases, conversion of the nitrogen in the ammonia to NO. Reduction efficiencies increase as the base NO level increases and, for systems with a low baseline NO level, removal efficiencies of less than 30% are not unusual. Adequate mixing of the reagent into the flue gases is also important in maximizing the performance of the SNCR process, and can be accomplished by the use of a grid of small nozzles across the gas path, adjusting the spray atomization to control droplet trajectories, or of an agent such as steam or air to transport the reagent into the flue gas.

Where the reagent is injected in larger amounts than the available NO, or where it is injected into a temperature too low to permit rapid reaction, the ammonia will pass through to the stack in the form of "ammonia slip." Where chlorine is present, a detached visible plume of ammonium chloride (NH_4Cl) may be formed if it is present in high enough levels. As plume temperature drops as it mixes in the atmosphere, the NH_4Cl changes from a liquid to a solid, resulting in a visible white plume. While these plumes may not indicate excessive NO_x or particulate emissions, they can result in perceptions of uncontrolled pollutant emissions.

SNCR systems typically have low capital cost, but much higher operating cost compared to low-NO_x burners due to the use of reagents. In some applications that have wide variations in load, additional injection locations may be required to ensure that the reagent is being injected into the proper temperature zone. In this case, more complex control and piping arrangements are also required.

SCR systems similarly rely on the use of an injected reagent (usually ammonia or urea) to convert the NO to N_2 and H_2O in the presence of a catalyst, and at lower temperatures (usually around 315–370°C [600–700°F]) than SNCR systems. Catalysts are typically titanium- and/or vanadium-based, and are installed in the flue gas streams at various locations in the gas path, depending upon the available volume, desired temperatures, and potential for solid particle plugging of the catalyst. SCR systems have not been installed in U.S. pulverized-coal-fired systems due to difficulties associated with plugging and fouling of the catalyst by the fly ash, poisoning of the catalyst by arsenic, and similar difficulties. However, recent tests have indicated the ability of SCR catalysts to maintain their performance over an extended period in U.S. pulverized-coal-fired applications.

Parameters of importance to SCR systems include the space velocity (volumetric gas flow per hour, per volume of catalyst), linear gas flow velocity, operating temperature, and baseline NO_x level. System designs must balance the increasing NO_x reductions with operating considerations such as catalyst cost, pressure drops across the catalyst bed, increased rate of catalyst deactivation, and increased NH_3 requirements. As NO_x reductions increase, the life of the catalyst decreases and the required amount of NH_3 injected increases. NO_x emissions can be reduced by over 90% if adequate catalyst and reagent are present and injection temperatures are optimized. For such reduction levels, catalysts may require replacement in as little as two years. It is possible to increase catalyst life where lower reductions are suitable.

Operational problems such as catalyst plugging and fouling can significantly reduce the effectiveness of SCR systems. Plugging can be a problem where the fuel used (e.g., coal) has a high particulate content. Interactions between sulfur and the injected reagent can lead to ammonium sulfate or bisulfate formation, which can result in fouling of the catalyst. In addition, the catalyst can convert SO_2 into sulfur trioxide (SO_3), which has a much higher dewpoint and can condense onto equipment and lead to excessive corrosion.

SCR systems are often more expensive to install than other NO_x removal systems due to the relatively high catalyst cost (10,600–14,100 $/m³ [300–400 $/ft³]). However, SCR systems can also remove higher levels of NO_x, resulting in costs in terms of $/ton of NO_x removed that are often competitive with other methods. Where very low NO_x emissions are required, SCR systems may be the only method of achieving the emission standard. SCR capital costs can be significant, particularly if large NO_x reductions are desired. In most cases, the largest portion of the cost is for the catalyst, which must be replaced periodically (approximately every three to four years). Costs for NH_3 must also be considered, but these costs are typically lower than for SNCR systems.

Hybrid systems combine SCR and SNCR by injecting a reagent into the furnace sections as the appropriate temperatures to take advantage of the SNCR NO_x reduction reactions, then passing the flue gases through a catalyst section to further reduce NO_x and provide some control of ammonia slip. Emissions of over 80% have been demonstrated on small-scale boilers using the hybrid approach.

Typical NO_x control performance and costs are shown in Table 65.4.

Table 65.4   Emission Reductions from Different NO$_x$ Control Technologies

Control Technology	Application	NO$_x$ Emission Reduction, %	Cost, $/tonne NO$_x$ ($/ton NO$_x$) Removed
Low excess air (LEA)	Boilers and furnaces	5–20	
Overfire air (OFA)	Pulverized-coal-fired-boilers Stoker-fired coal boilers	5–20	
Flue gas recirculation (FGR)	Natural-gas-fired boilers	20–50	
Low-NO$_x$ burners (LNB)	Natural-gas-fired boilers	40–60	
	Oil-fired boilers	20–40	
	Pulverized-coal, tangentially fired boilers	35–45	$130–$1300
	Pulverized-coal, wall fired boilers	40–65	($120–$1200)
LNB + FGR	Natural-gas-fired boilers	75–90	
LNB + OFA	Natural-gas-fired boilers	40–60	
	Oil-fired boilers	40–60	
	Pulverized-coal-fired boilers	45–65	$140–$1400 ($130–$1300)
Reburning	Natural gas reburn fuel with pulverized-coal main fuel	50–60	$420–$800 ($380–$730)
	Coal reburn fuel with pulverized-coal main fuel	40–60	$330–$990 ($300–$900)
Selective noncatalytic reduction (SNCR)	Combustion sources	30–75	$385–$1500 ($350–$1400)
Selective catalytic reduction (SCR)	Combustion sources	80–90	$420–$990 ($380–$900)

65.3 CONTROL OF PARTICULATE MATTER

Particulate matter (PM) control technologies can employ one or more of several techniques for removing particles from the gas stream in which they are suspended. These techniques are mechanical collection, wet scrubbing, electrostatic precipitation, and filtration. Large industrial and utility sources generally use electrostatic precipitators or fabric filters to remove fine particles from high-volume gas streams. Particulate removal efficiencies are shown in Table 65.5 for multiclones, electrostatic precipitators, fabric filters, and wet scrubbers.

Mechanical collection systems rely on the difference in inertial forces between the particles and the gas to separate the two. Examples of mechanical collection systems include cyclones and multiclones, rotary fan collectors, and settling chambers. Settling chambers use gravity to force the particles to "fall" out of the gas. Cyclones and multicyclones induce a spinning motion in the gas, forcing the heavier particles to the outside of the gas stream and against the inner cyclone wall. As the gas passes up through the cyclone, the particles strike the wall and fall to the bottom of the cylinder, where they are collected. Mechanical collection systems are primarily useful only in applications in which the particulate matter is relatively large (> 10 μm in diameter). Other applications include the initial stage of a multiprocess cleaning, where they remove the larger particles before the gas enters a higher-efficiency control device.

Table 65.5   Emission Reductions from Different PM Control Technologies[20]

Control Technology	Total Mass Emission Reduction, %	Mass Emission Reduction for Particles ≤ 0.3 μm, %
Multicyclone	50–70	0–15
Wet scrubber	95–99	30–85
Electrostatic precipitator	90–99.7	80–95
Fabric filter	99–99.9	99–99.8

Electrostatic precipitators (ESPs) operate by inducing an electrical charge onto the particles and then passing them through an electric field. This exerts a force on the charged particles, forcing them toward an electrode, where they are collected. The basic configuration of an ESP consists of one or more high-voltage electrodes that produce an ion-generating corona, in combination with a grounded collecting electrode. The generated ions charge the particles, and the high voltage between the electrodes results in an electric field that forces the particles toward the collecting electrode (see Fig. 65.2). The particles are removed from the collecting electrode by periodic rapping of the electrode, causing the particles to fall into a collection hopper below.

ESP performance can be significantly affected by the resistivity of the incoming particles. Particles of high resistivity are less easily charged, reducing the performance of the unit. Chemical additives can be introduced to the flue gas to reduce the effect of the high resistivity in a practice referred to as *gas conditioning*. Pulsing the electrodes with intense periodic high-voltage direct current can also improve performance with high-resistivity particles.

Performance can be improved by using separate charging and collection stages, which allows optimization of each process. Flushing the collected particles with continuous application of water can also provide collection of some of the gaseous pollutants in the flue gas, although this approach is likely to require the treatment of the resulting waste water.

Industrial-scale ESPs may have collecting electrodes 7.6–13.7 m (25–45 ft) high and 1.5–4.6 m (5–15 ft) long, with 60 or more gas flow lanes per section. Large units may have eight consecutive

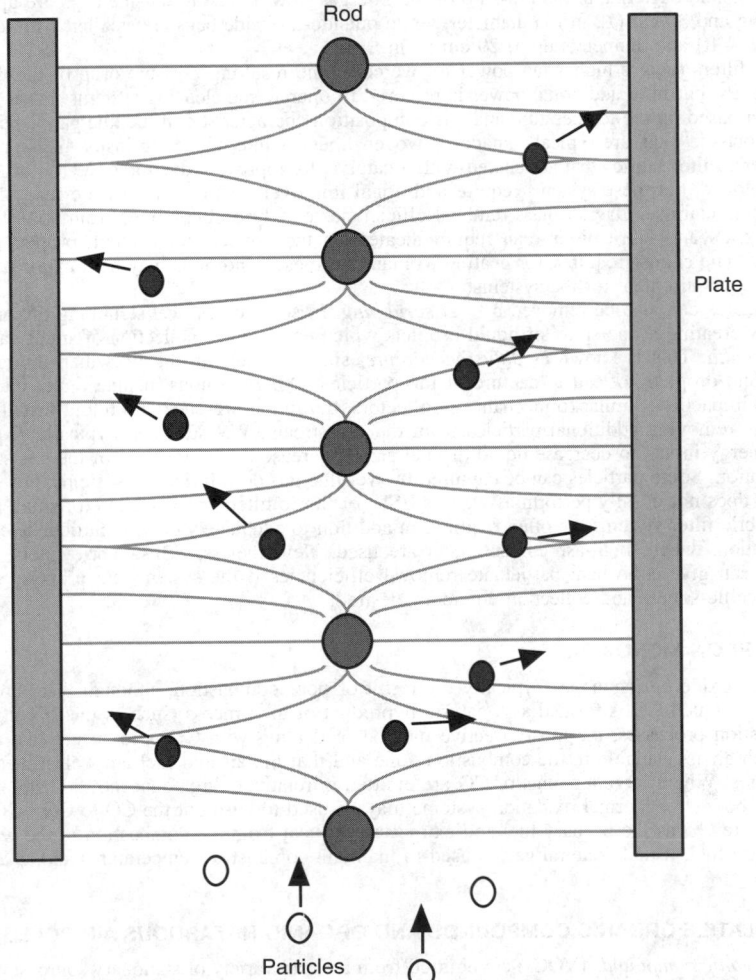

Fig. 65.2 Schematic of electrostatic precipitator operation.

sections, and gas flows of up to 85,000 m³/min (3,000,000 ft³/min). Voltages applied to the corona-generating electrode in industrial scale ESPs may range from 30 to 80 kV.

Pressure drops in ESPs are less than those of other particulate control devices, but they require significant amounts of electrical energy input to the electrodes. ESPs can remove 90–99.5% of the total particulate mass in the gas stream. However, particles are collected at lower efficiencies as their size decreases. In some applications, the combination of temperature and gas composition can lead to formation of trace organics, including polychlorinated dibenzodioxins (PCDDs) and polychlorinated dibenzofurans (PCDFs) in the ESP. Proper control of temperatures can significantly reduce or eliminate this problem.

In addition to large-scale industrial applications, ESPs have also been used for control of particulates in indoor air handling systems.

Fabric filters can also be used for industrial-scale removal of particulates. There are several filter configurations, including bags, envelopes, and cartridges. Bags are most common for large-volume gas streams, and are distributed in a baghouse. Industrial-scale baghouses may contain from less than 100 to several thousand individual bags, depending upon the gas flow rate, particle concentration, and allowable pressure drop.

Filter bags can be operated with gas flowing from inside to outside the bags, or with the gas flowing from outside to inside. In the inside-to-outside configuration, the particles are collected on the inner bag surface and the bags are cleaned periodically by reversing the gas flow intermittently for a short time, mechanically shaking the bags, or a combination of the two. For outside-to-inside bag configurations, a short pulse of air is injected down the inside of the bag to remove the particles that have collected on the outer bag surface. An internal cage is used to prevent the collapse of the outside-to-inside bags due to the pressure of the flue gas flow. Inside-to-outside bags are up to 11 m (36 ft) long and 30 cm (12 in.) in diameter, while outside-to-inside bags are smaller, with lengths up to 7.3 m (24 ft) and diameters up to 20 cm (8 in.).

Fabric filters require higher fan power to overcome the resulting pressure drop of the flue gases than do ESPs, but little additional power is required to operate the cleaning systems. Filter materials are chosen based on the temperature and gas composition characteristic of the flue gas to be cleaned. Inside-to-outside bags are typically made of woven fibers, while outside-to-inside bags are usually felted fibers. Filter fabrics that are coated with a catalyst to improve reduction of VOCs or NO_x have been developed, but these systems require additional improvements before they are ready for large-scale commercial use. Overall mass removal efficiencies of fabric filters range from 95–99.9%.

Both ESPs and fabric filters can require heaters in the hoppers to maintain proper collection conditions. The energy required for heating to maintain these conditions may be a major part of the total energy requirement for the systems.

Particulates can also be removed by *wet scrubbing.* In some cases, wet scrubbing of particulates is done by creating a fine spray of liquid droplets which enhance the collection of small particles in the gas stream. This is known as *diffusion capture,* since it relies on the Brownian motion of the particles and droplets to lead to capture of the particles. Other methods include direct interception or inertial impaction, similar to mechanical collectors, but in the presence of a liquid (usually water) to assist in removing additional particles from the gas stream. Wet scrubbers typically require significant energy inputs to decrease liquid droplet sizes, increase the momentum of the gas stream, or a combination. Some particles can be captured by wet flue gas desulfurization systems; however, wet scrubbing does not usually perform as well as ESPs or fabric filters in the capture of small particles. ESP or fabric filter systems are often required in addition to adequately control particulate emissions in applications where high-ash-content fuels are used. Nevertheless, ESPs, fabric filters, and wet scrubbers can give equivalent particulate removal efficiencies if the systems are properly designed for the specific source and collection efficiency desired.

65.4   CARBON MONOXIDE

Carbon monoxide emissions are typically the result of poor combustion, although there are several processes in which CO is formed as a natural byproduct of the process (such as the refining of oil). In combustion processes, the most effective method of dealing with CO is to ensure that adequate combustion air is available in the combustion zone and that the air and fuel are well mixed at high temperatures. Where large amounts of CO are emitted in relatively high concentration streams, dedicated CO boilers or thermal oxidation systems may be used to burn out the CO to CO_2. CO boilers use the waste CO as the primary fuel and extract useful heat from the combustion of the waste gas. An auxiliary fuel, usually natural gas, is used to maintain combustion temperatures and as a start-up fuel.

65.5   VOLATILE ORGANIC COMPOUNDS AND ORGANIC HAZARDOUS AIR POLLUTANTS

Volatile organic compounds (VOCs) are emitted from a broad variety of stationary sources, primarily manufacturing processes, and are of concern for two primary reasons. First, VOCs react in the atmosphere in the presence of sunlight to form photochemical oxidants (including ozone) that are harmful to human health. Second, many of these compounds are harmful to human health at relatively low concentrations. This second group of VOCs is referred to as *hazardous air pollutants* (HAPs) and is included for potential regulation under Title III of the Clean Air Act Amendments of 1990.[1]

Total VOC emissions in the U.S. have been declining over the past 10 years, primarily due to significant improvements in vehicle emission levels. During the same period, VOC emissions from industrial sources, solvent utilization, and chemical manufacturing have increased slightly, making these sources more important from a control perspective. In addition to VOCs, heavier organic compounds, such as polycyclic aromatic hydrocarbons (PAHs), nitrogenated PAHs, polychlorinated biphenyls (PCBs), and polychlorinated dibenzodioxins (PCDDs), are also important HAPs that may be emitted from a variety of sources. Combustion processes in general can form PAHs; however, proper equipment operation and maintenance typically results in PAH emissions from combustion sources on the order of parts per billion or less. Chlorinated organics emissions are characteristic of incineration processes in which chlorine is present; these compounds are discussed further below.

Control of VOCs and organic HAPs is less straightforward than for criteria pollutants due to the wide range of sources and the large number of different compounds that fall into this category. Much of the emissions of these compounds are due to fugitive emissions from process equipment such as valves, pumps, and transport systems, and emissions can be reduced considerably by proper maintenance and operation of existing equipment. Pump and valve seals and transfer equipment specially designed to reduce fugitive emissions can also provide significant reductions in such emissions. In other instances, alternative solvents or process modifications can eliminate the use of VOCs in manufacturing processes, thereby eliminating emissions. Often, these approaches can also reduce operating expenses by improving process efficiencies. In some cases, where the emission stream is relatively concentrated and characterized by a fixed pollutant or mixture of pollutants, several technologies are available that may allow recovery of the compound(s).

For process streams which cannot take advantage of these approaches, the emission stream often contains very dilute concentrations of the pollutant or pollutants, or the characteristics of the stream change significantly in composition and/or concentration. This makes generic prediction of control efficiencies and economics impossible for the broad category of VOCs and organic HAPs.

65.5.1 Conventional Control Technologies

Thermal oxidizers destroy organic compounds by passing them through high-temperature environments in the presence of oxygen. In practice, thermal oxidizers or incinerators typically operate by directing the pollutant stream into the combustion air stream, which is then mixed with a supplementary fuel (usually natural gas or fuel oil) and burned. Where the organic concentration is high enough to support combustion without a supplementary fuel, the organics are used as the fuel for incineration. Thermal incinerators are usually applied to emission streams containing dilute (less than 1000 ppm) of VOCs and organic HAPs. Destruction efficiencies can exceed 99%, but the effectiveness of the incinerator is a function of the temperature of the combustion chamber, the level of oxygen, and the degree of mixing of the air, supplementary fuel, and emission stream.

Boilers or industrial furnaces that are already present on a plant site can also be used as thermal incineration systems for appropriate streams of VOCs and organic HAPs. If the emission streams are of relatively low concentration, they can be added to the combustion air of the boiler or furnace and fed into the combustion environment for destruction. Where there is a very large emissions stream of relatively high concentration, the emission stream may be suitable as the primary fuel source for the boiler or furnace, with some supplemental fuel to maintain stable operation. Because these units are used to provide power or steam for plant processes, it is essential to maintain proper operation of these systems. The dual function of these units makes adequate monitoring and control essential to maintaining stable operation for both pollution control and process quality.

Flares are a simple form of thermal oxidation that do not use a confined combustion chamber. As with other forms of thermal oxidation, flares often require supplemental fuel. Flares are often used when the emission stream is intermittent or uncertain, such as the result of a process upset or emergency.

Catalytic oxidizers use a catalyst to promote the reaction of the organic compounds with oxygen, thereby requiring lower operating temperatures and reducing the need for supplemental fuel. Destruction efficiencies are typically near 95%, but can be increased by using additional catalyst or higher temperatures (and thus more supplemental fuel). The catalyst may be either fixed or mobile (fluid bed). Because catalysts may be poisoned by contacting improper compounds, catalytic oxidizers are neither as flexible nor as widely applied as thermal oxidation systems. Periodic replacement of the catalyst is necessary, even with proper usage.

Adsorption systems rely on a packed bed containing an adsorbent material to capture the VOC or organic HAP compound(s). Activated carbon is the most common adsorbent material for these systems, but alumina, silica gel, and polymers are also used. Adsorbers can achieve removal efficiencies of up to 99%, and in many cases allow for the recovery of the emitted compound. Organic compounds such as benzene, methyl ethyl ketone, and toluene are examples of compounds that are effectively captured by carbon bed adsorption systems. Adsorption beds must be regenerated or replaced periodically to maintain the bed's effectiveness. If absorbers are exposed to high-temperature gases (over 130°C), high humidity, or excessive organic concentrations, the organic compound will not be captured and "breakthrough" of the bed will occur. Monitoring of process conditions is therefore important to maintain the effectiveness of the adsorber performance.

Absorbers are similar to wet scrubbers in that they expose the emission stream to a solvent which removes the VOC or organic HAP. The solvent is selected to remove one or more particular compounds. Periodically, the solvent must be regenerated or replaced as it becomes saturated with the pollutant(s). Replacement of the solvent results in the need for disposal of the used solvent, often increasing potential for contamination of ground or surface water. Absorbers are therefore often used in conjunction with thermal oxidation systems in which the waste solvent can be destroyed.

Condensers are used to reduce the concentrations of VOCs and organic HAPs by lowering the temperature of the emission stream, thereby condensing these compounds. Condensers are most often used to reduce pollutant concentrations before the emission stream passes into other emission-reduction systems such as thermal or catalytic oxidizers, adsorbers, or absorbers.

65.5.2 Alternative VOC Control Technologies

Other technologies have been developed for the removal of VOCs and organic HAPs from emission streams that are not as widely applied as the control technologies noted above. In some cases, these alternatives are still under development, and hold promise for improving the capabilities of organic compound removal beyond conventional systems.

Biofilters rely on microorganisms to feed on and thus destroy the VOCs and organic HAPs. In these systems, the emission stream must come into direct contact with a filter containing the microorganism for sufficient time for the bioreaction to occur. Although biofilters can have lower overall costs than other technologies, technical problems, such as proper matching of the emission stream and the microorganisms, long-term operational stability, and disposal of the resulting solid wastes, may prevent their use in particular situations.

Corona destruction units use high-energy electric fields to destroy VOCs and organic HAPS as they pass through the corona. Two types of corona destruction units have been developed: packed bed corona and pulsed corona. In the packed bed system, the bed is packed with a high dielectric material and high voltage (15,000–20,000 V) is applied to electrodes at both ends of the bed. The resulting high-energy electric field fills the spaces between the packing material. As the emission stream passes through the bed, the organic compounds in the emission stream are destroyed. A pulsed corona system uses a single wire as one electrode and the walls of the reactor as the other. High voltages are intermittently applied across the electrodes at nanosecond intervals. The organic compounds passing through this electric field are destroyed. Disadvantages of the corona discharge systems include their high energy consumption and their potential for creating high levels of NO_x in the corona.

Plasma destruction systems rely on high temperatures generated by streams of ions to destroy the organic compounds. Plasma are incineration systems typically use an inert gas electrically heated to such high temperatures that the gas dissociates into a stream of ions. This high-temperature stream is then used to thermally break down organic materials into their simpler atomic constituents, which then recombine at lower temperatures into nontoxic products. The high temperatures result in high organic destruction efficiency; disadvantages include the high electrical energy requirement and the need to ensure that the entire waste gas stream adequately contacts the plasma.

Separation systems include membranes, hydrophobic molecular sieves, superactivated carbons, and improved absorption technologies. *Membranes* are used to separate and recover organics from an emission stream, particularly if the stream is consistent in composition. These systems can be used as control devices if high enough removals are achieved. High pressure drops in the emission stream and high sensitivity to contaminants are disadvantages of membrane systems.

Ultraviolet (UV) oxidizers use UV radiation as the basis for destruction of VOCs and organic HAPs. Absorption of UV light activates organic compounds, leading to photodissociation (or decomposition) or increased reactivity with other compounds. Reaction rates are sometimes increased by using a catalyst (typically titanium dioxide) in conjunction with UV radiation.

65.6 METAL HAZARDOUS AIR POLLUTANTS

Along with organic compounds, metals have risen in importance as pollutants of concern. Although the total mass of metal emissions is small compared to SO_2, NO_x, and total particulates, many metals have been shown to be toxic when people are exposed to even relatively small quantities over a long period of time.[11] In addition, some metals bioaccumulate, and people can be exposed to these metals through ingestion of contaminated food.

For the most part, metal emissions are in the form of particulate, although more volatile metals such as mercury can be emitted in the form of a vapor. In combustion systems, metal emissions are the result of the metals' being introduced into the combustion environment via the fuel. This is true for coal and heavy oils in particular. In other systems, such as metal-processing facilities, metal emissions are usually significantly higher in concentration, and are also more difficult to collect due to the lack of a single exit stream, as with coal-fired boilers.

If the metal particulate or vapor exits the plant via a single gas stream, most metal emissions can be reduced by installation of particulate removal equipment (see Section 65.3, above). For more volatile metals such as mercury, additional steps may be necessary to control emissions. One approach to mercury control has been injecting activated carbon into the flue gas downstream of the furnace and subsequently collecting the carbon with either a fabric filter or ESP. For systems with high uncontrolled mercury emissions, reductions of up to 90% can be achieved when using carbon injection in combination with fabric filters or ESPs. The level of reduction is dependent upon several factors, including the mercury species being removed, the temperature of the flue gases, and the amount and type of carbon being injected.

For applications such as metal-smelting facilities, reduction of metal-bearing emissions is much less straightforward. In some cases, a complete redesign of the process may be necessary to have a significant impact on emissions. In other cases, enclosure of a large area may be required to allow the channelling of the gases to a single exit point, where existing control technologies, such as fabric filters or ESPs, can then be applied.

Recent work has focused on the injection of sorbent materials into the flame to capture the metal. Sorbent injection can also influence the size of metal-bearing particulate by capturing many of the smaller particles, leading to a shift in particle-size distribution toward larger, more easily collected particles. In addition, if the sorbent reacts chemically with the metal, the leachability of the collected solid is reduced, making disposal of the collected particulate much less expensive. Sorbents have not yet been found that are universally applicable to the range of metals present in many combustion systems.

65.7 INCINERATION

Waste incineration presents unique problems due to the characteristics of the fuels employed. Uncontrolled emissions from hazardous waste incineration (HWI), municipal waste combustion (MWC), or medical waste incineration (MWI) are typically much higher in metals and halogenated compounds than those from combustion of fossil or biomass fuels. This is due to the high contents of metals and halogens, particularly chlorine, in the wastes being burned. Metal emissions can be reduced using the methods outlined above, the HCl emissions are usually reduced using a spray dryer, dry sorbent injection, or wet scrubber.

In some instances, waste incineration has resulted in the emissions of products of incomplete combustion (PICs) that are either not destroyed in the incineration process or are formed during some phase of the incineration and/or gas cleaning process. Trace quantities of organic compounds are typically produced during the combustion of hydrocarbon fuels, although, when proper temperatures and mixing of the fuel and oxygen are maintained, these emissions are near the detection limits of modern measurement methods, which are often in the range of several parts per billion by volume (although some compounds, such as dioxins, can be measured in the parts per trillion range).

Of particular concern are PCDDs and PCDFs, which are highly toxic. Studies have shown that flue gases containing HCl and fly ash at temperatures between 200 and 600°C can form PCDDs/PCDFs in the presence of a catalytic metal, such as copper. Due to the use of copper in ESPs and the need to maintain the flue gases at temperatures above the acid gas dew point, these conditions were often present in the air pollution-control systems of MWC units. Modifying the operating conditions of these units eliminated or greatly reduced the PCDD/PCDF formation potential.

65.8 ALTERNATIVE POLLUTION-CONTROL APPROACHES

In recent years, there has been a growing emphasis on the prevention of pollution rather than the removal of pollutants following their formation. Pollution prevention is in many cases a much more cost-effective approach than the installation and operation of traditional control technologies. Not only can prevention lead to reduced emissions of one or more pollutants, it can also improve the efficiency of the overall process, leading to significant improvements in energy efficiency and/or material use. Such approaches are frequently very process- and/or site-specific, and depend upon the equipment currently in use and process parameter limitations. However, periodic inspection and replacement of items such as valve and pump seals and gaskets, and reduction of atmospheric venting, can have significant impacts on the annual emissions of pollutants.

Where new equipment is being considered, prevention approaches can be very attractive, since the initial specifications can be set with emission minimization as a key parameter in combination with the process requirements. Investment in pollution-minimizing systems and processes can in many cases significantly reduce the requirements for additional equipment specifically for pollution control. Pollution prevention is most cost-effective when the entire production process can be designed with pollution control in mind. For instance, the use of paints that do not rely on the evaporation of an organic solvent can eliminate fugitive emissions of the organic solvent without the need for VOC control equipment. While these approaches can be very cost-effective, they often are applicable only when new equipment or processes are being installed.

One method of pollution prevention that can be applied without extensive replacement of existing equipment is the use of computer-based controls. Many current processes are poorly instrumented and are often controlled manually based on operational "rules of thumb." The rapidly increasing capabilities of desktop computers and programmable controllers provide many opportunities to improve process efficiency and reduce emissions of many pollutants. Many of these process controls can be applied in a straightforward manner, using simple feedback systems that maintain a given process setpoint. In more complex systems where feedstocks may be rapidly and unpredictably changing, the use of artificial intelligence (AI) methods may provide considerable improvements in control. Some examples of AI-based systems are expert systems, fuzzy logic controls, and artificial neural networks.

Expert systems are based on expert rules developed by system designers and operators, and in their simplest forms use a series of yes–no questions to arrive at an operational diagnosis of the system's performance and recommendations of process changes to improve that performance. These systems can act either as an operational advisor, taking process information and providing control guidance to operators, or in a closed loop capacity making the appropriate process changes automatically. Expert systems can be applied to pollution-control problems by signalling operators when plant conditions require adjustment to maintain proper pollution-control performance.

Fuzzy logic is an AI-based method that uses similar expert rules as the basis for automatic controls. Fuzzy-logic controls have been successfully applied to household appliances, automobiles, and robotic mechanisms, and are likely to be used in an increasing number of applications in the future. Their advantage over traditional feedback control systems is their simple design and ease of modification. Fuzzy logic is likely to become increasingly common in automatic controls for combustors and pollution control equipment.

Artificial neural networks (ANNs) are a pattern-recognition technique based on the neurons of the brain. ANNs have been used to identify patterns in the stock market, in computer vision systems and in radar signal-processing equipment. In pollution-control applications, ANNs may allow computerized control systems to identify automatically modes of operation that will maintain production efficiency while minimizing pollutant emissions, or may be used to identify conditions that indicate incipient failures of equipment, thereby minimizing or eliminating transient emissions due to equipment failure.

65.9 GLOBAL CLIMATE CHANGE

In recent years, emissions of certain gases from industrial and other human activities have been the focus of concern over their impact on global climate. In particular, emissions of CO_2, CH_4, and N_2O have been identified as having the potential to increase the concentrations of these compounds in the atmosphere to such a degree that the average global temperature could increase. Climatologists have predicted that slight changes in the average global temperature could have broad detrimental effects on local climates, including large and extended drought, flooding of coastlines, and decreased production of food. Because of the far-reaching impacts of such climate changes, considerable attention has been directed toward reducing the emissions of these global warming gases.

In addition, other compounds have been identified as having severe impacts on the stratospheric ozone layer that protects the Earth's surface from harmful ultraviolet radiation generated by the sun. Some of the ozone-depletion substances (ODSs) that have been identified include the broad category of chlorofluorocarbons (CFCs), which have been widely used as refrigerants, solvents, aerosol propellants, and foam-blowing agents.

65.9.1 CO_2

A number of commercially available processes exist for the removal of CO_2 from flue gases, or *CO_2 scrubbing*. These processes have been developed primarily for use in cleaning impurities, including CO_2, from natural gas, and are based on the use of an absorption solvent to scrub out the CO_2. In most cases, the solvent also removes other compounds, including H_2S, SO_2, HCN, COS, CS_2, and NH_3.[12] To date, no large-scale demonstration of a CO_2 scrubbing system has been conducted on a fossil fuel boiler or furnace, although engineering studies have been made to assess the costs and performance associated with CO_2 scrubbing. Removal efficiencies have been estimated at up to 95% for commercial CO_2 scrubbing processes.[13] However, applications of CO_2 scrubbers will reduce the overall efficiency of a plant, requiring additional CO_2 to be emitted in order to overcome the losses, bringing the overall process efficiency down. A schematic of a CO_2 scrubbing system is shown in Fig. 65.3.

Additional control technologies include cryogenic separation, which produces CO_2 in liquid form, or membrane separation, which also results in liquid CO_2 as the end product. Both these technologies are more energy-intensive than the absorption-based technologies, and are therefore likely to be suitable only in special instances unless significant breakthroughs are achieved.

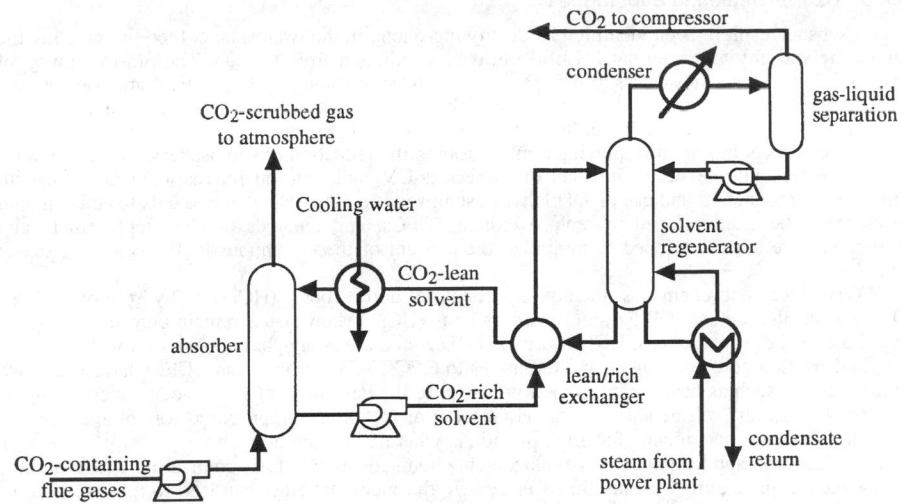

Fig. 65.3 Schematic of CO_2 scrubbing system.

Even if CO_2 control technologies can be shown to be cost-effective and technically feasible, CO_2 disposal is a critical issue that is as yet unresolved. Suggested options include injection of CO_2 into oil reservoirs to improve recovery, or injection into deep oceans. However, the technical uncertainties associated with these options are significant in terms of number and magnitude. Technical feasibility, environmental acceptability, cost, and safety of each of the proposed options are not yet understood to the degree that large-scale implementation is likely in the near term.

More feasible in the short term are CO_2 mitigation measures, such as increased efficiency and demand-side management to reduce the use of fossil fuels. These strategies are particularly important for controlling emissions from mobile sources, since removal technologies are not available for automotive or other sources in the transportation sector. A wider use of electrically powered vehicles will transfer the CO_2 emissions from the transportation to the utility sector, and may allow scrubbing technologies to be applied indirectly to the transportation sector. Biomass-derived or other non-fossil liquid fuels for transportation are another approach to mitigation of mobile source emissions. Ethanol, methanol, and hydrogen are three liquid fuels that have the potential to be used in mobile applications. However, supply, cost, and safety issues remain to be resolved before widespread use of these fuels is possible.

Other CO_2 mitigation approaches include increased use of biomass fuels, which are part of the global carbon cycle and do not add to the amounts of carbon that must be absorbed into the cycle, increased planting of forests, reduced destruction of existing forests, and increased efficiencies of energy use in developing countries.

65.9.2 Other Global Warming Gases

Sources of CH_4 emissions include the petroleum and natural gas production and processing industries, release of coal mine gases, escape of gases produced in solid waste landfills, and the raising of livestock. These emissions are often most efficiently reduced through stricter management of process releases and equipment leaks. In the cases of coal mine gas, landfill gas, and livestock releases, other approaches are necessary. Extraction and use of CH_4 from landfill gas as the feedstock for the generation of electricity from fuel cells has been demonstrated, and techniques for capturing emissions from cattle feedlots have been proposed.[14] Potential controls include improved nutrition, recovery of methane from covered waste lagoons, and the use of digesters. For coal mine CH_4 emissions, pre-mining degasification wells and capture and treatment of ventilation air are two possible control approaches.

Adipic acid plants produce an intermediary product in the production of nylon, and are major emitters of N_2O. Control technologies include thermal destruction in boilers, conversion to NO for recovery, and catalytic dissociation to molecular nitrogen and oxygen.[14] N_2O is also emitted by some combustion processes, particularly low-temperature combustion such as fluidized-bed combustors, but is not emitted in significant levels from conventional high-temperature combustion processes. Catalytic reduction can be applied to these systems with considerable success.

65.9.3 Ozone-Depleting Substances

Several substances have been identified as destroying ozone in the stratosphere, thereby reducing the atmosphere's ability to screen out harmful ultraviolet radiation from the sun. The most common of these ozone-depleting substances (ODSs) are chlorofluorocarbons (CFCs). CFCs are used as the working fluids in vapor-compression cycles, as solvents and leak-checking systems in industry, and in production of foams. Through venting and leakage, these compounds have been emitted into the atmosphere, where they are now playing a major role in the reduction of stratospheric ozone. Because of the potential adverse health impacts of increased UV radiation, an international agreement to eliminate the production and use of ODSs was established in 1990. New compounds have been, and are currently being, developed to replace existing CFCs, and methods of CFC replacement and destruction have been developed to minimize the amount of these compounds that reach the ozone layer.

CFC replacement chemicals include hydrochlorofluorocarbons (HCFCs), hydrofluorocarbons (HFCs), hydrofluoroethers (HFEs), and hydrocarbons (HCs). Many issues remain unresolved regarding the use of these compounds. HCFCs are also ODSs and are being phased out over the next 20–25 years and are therefore not long-term alternatives to CFCs. In addition, some HCFCs have their own unique markets such as heat pumps, which will require the development of replacement compounds. The use of "natural" refrigerants such as H_2O, CO_2, or NH_3 is being proposed for some cases, but these chemicals are not suitable for all applications where CFCs are now used.

In refrigeration applications, many replacement chemicals are not fully compatible with materials of construction in existing systems. In severe cases, the incompatibility can lead to a breakdown of the lubricant or other material, thus resulting in failure of the refrigeration unit. In such cases, changes in either the lubricant or the entire system are required for the replacement refrigerant to be used in a particular application. Additionally, steps are being taken with commercial refrigeration and vehicle air-conditioning systems to minimize CFC emissions by control of system leaks. Title VI of the 1990 Clean Air Act Amendments[1] now requires that motor vehicle air-conditioning system maintenance personnel be certified as having adequate training in the recovery and storage of CFCs. These substances must now be recycled and reused rather than being vented as in earlier practice.

Where CFCs are used as solvents, a dual approach is also being taken to reduce the emission of CFCs to the atmosphere. As with refrigeration applications, replacement chemicals are being used where possible. And in all processes where CFCs are being used, vapor-capture and recovery systems are being used to recycle and reuse as much of the chemical as possible. Vapor-capture and recovery systems are also being employed in the production of foams.

Destruction of CFCs once they have been recovered can be a difficult process, since many CFCs are also used as fire retardants and suppressants. However, studies have shown that CFCs can be incinerated, although incineration can produce high levels of hydrochloric and hydrofluoric acid (HCl and HF) gases in the exhaust. Due to the high levels of chlorine in the incineration flue gases, the production of PCDDs/PCDFs is also a concern, although sufficient flame temperatures and adequate gas-cleaning systems are usually sufficient to destroy any measurable levels of these compounds.[15]

REFERENCES

1. Public Law 101-549, Clean Air Act Amendments of 1990, November 15, 1990.
2. *National Air Pollutant Emission Trends Report,* 1990–1993, EPA-454/R-94-027 (NTIS PB95-171989), Office of Air Quality Planning and Standards, October 1994.
3. P. S. Nolan et al., "Results of the EPA LIMB Demonstration at Edgewater," in *Proceedings: 1990 SO₂ Control Symposium,* Vol. 1, EPA-600/9-91-015a (NTIS PB91-197210), Research Triangle Park, NC, May 1991.
4. R. J. Keeth, P. A. Ireland, and P. T. Radcliffe, "1990 Updated of FGD Economic Evaluations," in *Proceedings: 1990 SO₂ Control Symposium,* Vol. 1, EPA-600/9-91-015a (NTIS PB91-197210), Research Triangle Park, NC, May 1991.
5. T. E. Emmel, M. Maibodi, and N. Kaplan, "Retrofit Costs of SO₂ Controls in the United States and the Federal Republic of Germany," in *Proceedings: 1990 SO₂ Control Symposium,* Vol. 1, EPA-600/9-91-015a (NTIS PB91-197210), Research Triangle Park, NC, May 1991.
6. C. T. Bowman, "Control of Combustion-Generated Nitrogen Oxide Emissions: Technology Driven by Regulation," in XXIV Symposium (International) on Combustion, The Combustion Institute, Pittsburgh, PA, 1992, pp. 859–876.
7. K. J. Lim et al., *Environmental Assessment of Utility Boiler Combustion Modification NOₓ Controls: Volume 1. Technical Results,* EPA-600/7-80-075a (NTIS PB80-220957), Research Triangle Park, NC, April 1980.
8. S. M. Wilson, J. N. Sorge, L. L. Smith, and L. L. Larsen, "Demonstration of Low NOₓ Combustion Control Technologies on a 500 MWe Coal-Fired Utility Boiler," in *Proceedings: 1991 Joint Symposium on Stationary Combustion NOₓ Control,* Vol. 1, EPA-600/R-92-093a (NTIS PB93-212843), Research Triangle Park, NC, July 1992.

9. B. de Volo et al., "NO_x Reduction and Operational Performance of Two Full-Scale Utility Gas/Oil Burner Retrofit Installations," in *Proceedings: 1991 Joint Symposium on Stationary Combustion NO_x Control,* Vol. 3, EPA-600/R-92-093c (NTIS PB93-212868), Research Triangle Park, NC, July 1992.

10. S. L. Chen et al., "Bench and Pilot Scale Process Evaluation of Reburning for In-Furnace NO_x Reduction," in XXI Symposium (International) on Combustion, The Combustion Institute, Pittsburgh, PA, 1986, pp. 1159–1169.

11. W. P. Linak and J. O. L. Wendt, "Toxic Metal Emissions from Incineration: Mechanisms and Control," *Progress in Energy and Combustion Science* **19**, 145–185 (1993).

12. I. M. Smith and K. V. Thambimuthu, *Greenhouse Gases, Abatement and Control: The Role of Coal,* IEACR/39, London UK, IEA Coal Research, June 1991.

13. I. M. Smith, C. Nilsson, and D. M. B. Adams, *Greenhouse Gases—Perspectives on Coal,* IEAPER/12, London, UK, IEA Coal Research, August 1994.

14. P. G. Finlay, P. H. Pinault, and R. P. Hangebrauck, "Climate Change Prevention Technologies: Current and Future Trends," presented at the First North American Conference and Exhibition on Emerging Clean Air Technologies, Toronto, Canada, September 26–29, 1994.

15. United Nations Environment Programme, "Ad-Hoc Technical Advisory Committee on ODS Destruction Technologies," May 1992.

16. *National Air Pollutant Emission Trends, 1900–1993,* EPA-454/R-94-027 (NTIS PB95-171989), October 1994.

17. M. A. K. Khalil and R. A. Rasmussen, "The Global Sources of Nitrous Oxide," *J. Geophys. Res.* **97**, 14,651 (1992).

18. *Mercury Study Report to Congress,* Book 2 of 2, EPA-600/P-94-002ab (NTIS PB95-167342), Cincinnati, OH, 1995.

19. *National Air Pollutant Emission Trends, 1900–1992,* EPA-454/R-93-032 (NTIS PB94-152097), October 1993.

20. *Control Techniques for Particulate Emissions from Stationary Sources—Volume 1,* EPA-450/3-81-005a (NTIS PB83-127498), Research Triangle Park, NC, 1982.

CHAPTER 66

WATER POLLUTION CONTROL TECHNOLOGY

Carl A. Brunner, Ph.D.
Retired from U.S. Environmental Protection Agency

66.1	**INTRODUCTION**	**2031**	
66.2	**MUNICIPAL WASTEWATER TREATMENT**	**2032**	
66.3	**INDUSTRIAL WASTEWATER AND HAZARDOUS WASTE TREATMENT**	**2035**	
	66.3.1 Chemical Precipitation and Clarification	2036	

66.3.2	Activated Carbon Adsorption	2037
66.3.3	Air Stripping	2038
66.3.4	Steam Stripping	2039
66.3.5	Membrane Technologies	2039
66.3.6	Ion Exchange	2041
66.3.7	Other Methods	2043

66.1 INTRODUCTION

Various degrees of pollution management have been carried out for centuries. Roman sewers are well known. For much of this time, emphasis was on reducing esthetic problems in cities. When a connection was made between disease and microorganisms harbored in human wastes, the urgency to remove these wastes from human contact increased. Although this connection was suspected earlier, it was not proved until the 19th century. The early pollution-management methods did not include treatment and did not have a significant beneficial impact on the environment. Concerns about the impacts of wastes created by mankind upon the environment are relatively recent. In the 19th century, people began to notice that waterways were showing serious signs of pollution and a few states began to consider remedial measures.[1] Widespread use of treatment of wastewater did not occur prior to the early 20th century. The most commonly employed method of solid waste disposal well into the 20th century was simply dumping on the land or sometimes in local water bodies. Municipal incinerators began to be used by the end of the 19th century. The more acceptable sanitary landfill did not become a common substitute for the open dump until after World War II. Similarly, significant efforts to curb air pollution were not initiated until after World War II. Strong concern for pollution control, not only for health reasons, but also to improve the environment, originated in the 1960's with the development of very active environmental organizations. The federal government and some state governments responded by passing much stronger legislation than had been in force up to that time. The result of this legislation and strong public interest in the environment has been a greatly increased level of treatment of solid, liquid, and gaseous discharges.

This chapter will familiarize the reader with the most important treatment methods for treating waste materials, primarily liquid wastes. To understand why these methods are used, it is useful to know something about the regulatory framework that has developed, primarily as a result of federal legislation. Wastewater discharges have been regulated under a long series of legislative acts, but the key legislation that made very great changes to the degree of control required was the Federal Water Pollution Control Act Amendments of 1972. This act has been amended a number of times since and the overall legislation is commonly called the *Clean Water Act*. Under this legislation, specific requirements have been made for the direct discharges of both municipal and industrial wastewaters to a receiving water body and for the discharge of industrial wastewater to the public sewer system, or indirect discharge. For municipal wastewater, the minimum requirement is either for a reduction of biochemical oxygen demand (BOD) and suspended solids (SS) of at least 85% or effluent values

Mechanical Engineers' Handbook, 2nd ed., Edited by Myer Kutz.
ISBN 0-471-13007-9 © 1998 John Wiley & Sons, Inc.

of both BOD and SS not exceeding 30 mg/l, whichever is more stringent. For discharges to streams where the minimum requirement is viewed as having too great an impact, more stringent standards can be required, including parameters in addition to BOD and SS. Industries have been grouped into categories, of which there are 55 at this time. For each category, there are both discharge standards for direct discharge and pretreatment guidelines for indirect discharge.

For specific information in a community, the first point of contact should be local wastewater officials. Above the local level would be the state environmental agency and finally the U.S. Environmental Protection Agency (EPA), which maintains ten Regional Offices across the country. Changes in clean water legislation will continue to be made, possibly with significant impact on discharges.

Another law that has had a very great impact on control of wastes is the Resource Conservation and Recovery Act (RCRA). RCRA had its origin in 1976 as amendments to the Solid Waste Act. Regulations arising from RCRA and amendments are the primary method of control for hazardous wastes. For a waste to be hazardous, it must exhibit one or more of the following characteristics: ignitability, reactivity, corrosivity, or failure to pass a defined extraction procedure. The regulations are quite complex. One part deals with the land-disposal ban for listed substances. Hundreds of materials are included along with the best demonstrated available technology (BDAT) for controlling each. Questions regarding whether wastes are hazardous and their control should be referred to the state environmental agency.

A third law that has had a great impact on environmental control is the Comprehensive Environmental Response, Compensation and Liability Act of 1980 (CERCLA). Unlike most environmental laws, CERCLA deals with problems that are the result of past practice. Of the various aspects of CERCLA, the best known is the Superfund Program, which provides for remediation of wastes on the ground and involves a number of approaches for remediating contaminated soil and groundwater. Funds are provided by the federal government unless a party or parties responsible for the pollution can be found. Future legislation may modify this program significantly.

Although this chapter deals with waste treatment, it is important to point out that a widespread attitude has emerged that favors prevention of pollution over pollution control. There are many facets to a pollution-prevention program, including changes in raw materials to less hazardous materials and changes in manufacturing methods. From the standpoint of waste treatment, there should eventually be a measurable reduction in discharges or at least a reduction in the strengths of wastes being discharged. To some degree, treatment methods such as wastewater treatment should change to in-plant reuse and recovery systems with polluting materials and clean water being returned to the manufacturing operation.

66.2 MUNICIPAL WASTEWATER TREATMENT

Every day, billions of gallons of municipal wastewaters are discharged from plants ranging in size from those that handle the discharge from a few houses to more than one billion gallons per day. There are differences in the composition of municipal wastewater from city to city because of differences in the industrial and commercial contributions, and differences in overall contaminant concentration because of varying degrees of stormwater or groundwater entrance into the sanitary sewer system. Even with these differences, however, there is enough similarity that a small group of technologies is able to handle the treatment of most of the sources. Municipal wastewater contains an enormous number of chemical compounds including a wide range of metals and organic materials. The exact composition of all contaminants is never known. Two very common chemical elements, phosphorus and nitrogen, must be emphasized because they act as fertilizer in the receiving water body and can lead to excessive growth of nuisance plants such as algae. In addition, the wastewater contains large numbers of microorganisms, including human pathogens. Ideally, wastewater treatment would reduce all of the contaminants to low levels. In reality, the methods that are used have the primary function of reducing the oxygen demand of those materials that would biologically degrade in the receiving water body. When degradation occurs in the receiving water, oxygen is consumed. Because the stream biota require oxygen, its concentration should not be depressed significantly below the ambient level. As already indicated, requirements have been placed on the BOD and SS in the discharge because these parameters give a measure of the potential for decreasing dissolved oxygen. By settling to the bottom of the receiving water body, SS can also have a direct negative effect on sediment quality and bottom-dwelling aquatic life.

Nearly all municipal wastewater is treated by biological methods. The more common technologies are well described in the textbook literature.[2] These methods take advantage of the ability of a large group of natural microorganisms to utilize the organic materials in the wastewater for food. These organics include soluble materials and many of the insoluble materials that are solubilized by the complicated biochemical activity taking place. The method used by nearly all large cities and treating the greatest volume of wastewater is the activated sludge process. A second method, the trickling filter, was formerly very common, but has been displaced in many communities by activated sludge. There are a number of variations of these two methods, including combinations. Although these two methods account for a large fraction of total treatment capacity, there are a variety of other, usually less complicated, methods used predominantly by small communities.

Figure 66.1 is a diagram of a sequence of treatment operations and processes that are found in many treatment plants. Incoming wastewater passes through preliminary treatment, which might include screens for removal of very large objects, grinding for reducing the size of large objects and a short-term settling operation for removal of small to moderate but heavy particles (grit). The wastewater is usually pumped to a high enough level that it can flow through the remainder of the plant by gravity. Pumping may precede preliminary treatment or may follow at least some of the preliminary treatment to reduce wear on the pumps.

Following preliminary treatment is primary settling, which is intended to remove large organic particles not removed in preliminary treatment, and results in about 50–60% SS and about 30% BOD removal. This operation is carried out in tanks with circular or rectangular configurations. Residence time is up to two hours. Although the principal action is settling, there is also some biological activity. If the settled sludge is not removed quickly enough, biological activity can result in unpleasant odors and floating sludge. In large plants, there may be many settlers operating in parallel. A sludge-removal mechanism moves the sludge in the bottom of each settler to an opening for discharge to the sludge-processing system. Primary treatment is not required for all plants, but is almost always used in large plants.

The partially treated wastewater or primary effluent flows next to secondary treatment, where the major biological activity occurs. In the commonly used activated sludge process, treatment is carried out in tanks supplied with aeration devices. These may be various kinds of diffusers, mounted in or near the bottom of the tanks, that bubble air through the water or surface-mounted mixers or mechanical aerators. In addition to providing oxygen to the water, the aeration devices keep solids in suspension throughout the volume of the aeration tanks, which may be very large. In the aeration tanks, aerobic degradation of the organic matter in the wastewater takes place. Commonly a level of SS of about 2000 mg/l or higher is maintained. The very large number of microorganisms in these solids, commonly referred to as activated sludge, accomplish the biological degradation. The residence time in the aeration tanks is often about four to six hours, but there can be great variation from high rate treatment to extended aeration. In the latter, the residence time might be 24 hours. Extended aeration plants may not require primary treatment. Water leaving the aeration tanks flows to the secondary or final settling tanks. Although these settling tanks are used with technologies other than activated sludge, they play a unique role with activated sludge. As indicated in Fig. 66.1, part of the sludge removed from these tanks is returned to the aeration tanks. The amount of sludge is adjusted to maintain the desired SS concentration in these tanks.

The frequently used trickling filter form of secondary treatment consists of one or more beds of rock or other packing. The wastewater is dispersed over the upper surface of the filter bed and is collected at the bottom in an underdrain system. Biological slime accumulates on the packing. Oxygen is usually provided by the natural flow of air through the packing, although forced air may be employed. From time to time, some of the biomass sluffs from the filter bed and is carried, along with the water leaving the filter, to the secondary or final settler. Many trickling filters operate with return of some of the water from the underdrain system back to the top of the filter, where it is mixed with incoming feed water. Only very rarely is there any return of material from the final settler.

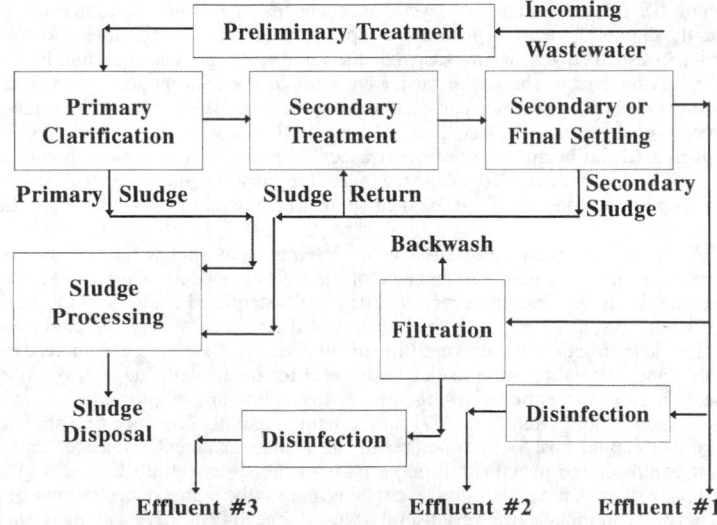

Fig. 66.1 Municipal wastewater treatment.

There are a number of variations and combinations of activated sludge and trickling filter or fixed film devices that are being operated. The rotating biological contractor (RBC) combines the two concepts. RBC systems include basins which contain some biologically active solids, but they also are equipped with banks of partially submerged vertical disks that rotate. Biomass accumulates on the disks. Oxygen is supplied to the system primarily through exposure of the rotating disks to the air. Adequate flocculation of the biomass from a trickling filter to provide effective settling in the final settler can be a problem. Adding a short-residence-time aeration tank following the trickling filter along with returning some sludge from the final settling tank has been found to improve flocculating and settling. These systems are referred to as trickling filter-solids contact.

The primary function of the final settling tanks is the removal of suspended matter from the water to a level that meets discharge requirements. For the standards of 30 mg/l BOD and 30 mg/l SS, this degree of treatment should be adequate, assuming there is no upset in the biological system. The settling tanks are of circular and rectangular configuration and typically have an overflow rate of 400 to 600 gallons/day per square foot. The important function of providing sludge to the aeration tanks in the activated sludge process has already been mentioned.

Figure 66.1 indicates three possible routes for the secondary effluent leaving the final settlers. Direct discharge to a receiving water is allowed in some locations. Disinfection is frequently required. To meet discharge requirements significantly less than the 30 mg/l requirements would necessitate a minimum of additional SS removal. Additional SS removal may also be a practical approach for meeting a phosphorus requirement.

Disinfection is frequently carried out using either elemental chlorine or a chlorine source, such as sodium hypochlorite. The chlorine is mixed with the water in a chlorine contact chamber, which often has a residence time of about 30 minutes. Because the addition of chlorine to waters containing organic materials has the potential for producing toxic chlorinated species, there is increasing concern about the health and ecological impacts of chlorination. Possible chemical substitutes are chlorine dioxide, which does not result in significant formation of chlorinated organics, and ozone. Use of these chemicals is more costly and at this time is rare. A non-chemical method, ultraviolet radiation (UV), is being used increasingly as a substitute for the traditional chlorination. This is a rapidly evolving technology with further improvements in the radiation sources being likely. In a typical installation, the water is passed through a chamber containing banks of tubular lamps. The lower the SS concentration, the more effective is the disinfection, which is dependent upon sufficient radiation reaching all of the water. There is also the obvious question of the ability of UV (and to a lesser degree with chemical methods) to disinfect within particles constituting the suspended matter. Parker and Darby reviewed the literature on this question[3] and carried out tests using rapid stirring to break up solids particles in secondary effluent that had been UV disinfected with low pressure UV lamps. They found some shielding of coliforms as measured by increases in these organisms in the water phase after stirring and stated the need for greater emphasis on pretreatment of the wastewater before disinfection, including particulate removal. Higher pressure UV sources are being developed which may largely overcome the problem.

The usual method in large plants for removal of SS below the concentration in secondary effluent is in-depth filtration. An in-depth filter contains two feet or more of sand, or a combination of sand covered with larger-particle-size-coal (dual media), or sand and coal over a fine, high-density material such as garnet (multimedia). Water is introduced at the top of the filter bed. These filters strain much of the remaining SS from the water by several mechanisms, including incorporation in biological slime. Eventually the solids deposits increase the pressure drop across the filters to the point that they must be backwashed. Backwashing suspends the filter media particles and removes most of the deposited solids. After backwashing, the media particles in the dual-media and multimedia filters naturally grade themselves with larger particles (coal) at the top. The reason for considering a combination of media such as sand and coal of a larger particle size is the increased solids holding capacity and increased run length between backwashes. The multimedia system should result in the highest degree of solids removal. To avoid the need for disposal, the backwash stream must be returned to an appropriate point in the treatment sequence, possibly before secondary treatment or the final settler.

In Fig. 66.1, the concentrated solids streams are referred to as sludge and are shown leading to sludge processing. To improve public acceptance of the treated sludge for use as a soil amendment, there is a tendency in the profession to refer to adequately stabilized sludge as biosolids. Although the volume of sludge produced is only a few percent of the volume of water treated, its processing constitutes a significant fraction of the overall treatment activity. A variety of methods exist that vary somewhat with plant size and possible modes of disposal for the treated sludge or biosolids. Metcalf and Eddy mention eight categories of sludge-processing operations or processes, under which are included 34 specific methods (Ref. 2, p. 767). In the liquid handling sequence of a sewage treatment plant, much of the capital cost is in concrete. In the sludge treatment sequence, there is greater opportunity for companies to manufacture major parts of the system. Both U.S. and foreign manufacturers are responding. Unless the sludge contains unusually high concentrations of hazardous materials, such as certain metals, it is very useful as a soil amendment. EPA and many environmental groups favor this beneficial use of sludge. Where application to the soil is not practical, incineration or landfilling are alternatives. For any of these methods, substantial processing of the primary and secondary sludge streams is necessary. Because their solids concentrations are only a few percent or less, these streams are usually first thickened. The simplest method is gravity thickening. If the sludge is to be incinerated, to reduce the fuel requirement to the incinerator, it must be further dewatered by mechanical means, such as filtration using belt filters or plate and frame filters, or centrifugation. Solids concentrations of up to 30% are desirable. For other than incineration, a very common method of sludge treatment is digestion. The objective of this biological process is to stabilize or further reduce the biodegradable content of the sludge and to reduce the concentration of pathogens. Digestion is commonly carried out under anaerobic conditions in large plants, but is also frequently conducted aerobically. Usually, the temperature is elevated, if possible, by using the heat generated by biological activity or, in the case of anaerobic digestion, by burning of methane produced by the digestion process. Another method to stabilize sludge is by adding chemicals which raise the pH and form a cement-like residue. In small plants, it is common to stabilize sludge by long-term storage in lagoons. Recent sludge disposal regulations determine to a considerable extent the condition of the sludge for landfilling, beneficial use on land and production of marketable products. For large-scale spreading on agricultural land, digestion may be sufficient. Composting has become common for producing a marketable product. The process has encountered odor problems that necessitate careful control of the operation.

Although very large plants will consist of a system included in Fig. 66.1, occasionally plants of 10–20 million gallons per day and many smaller plants may have somewhat simpler systems such as treatment ponds, or lagoons, wetland systems, and land application systems. Most of these systems will include some form of preliminary treatment. They may or may not include primary sedimentation. Biological degradation occurs in the ponds, on the surface of the land, or underground. Final settlers would not ordinarily be required. In some cases, such as land application with percolation, there is no surface discharge. These methods are land-intensive and are most appropriate in locations where large expanses of inexpensive land are available.

Although removal of organic materials is the usual objective of wastewater treatment, there is also a need in some locations to remove phosphorus and to control nitrogen. Phosphorus contributes to nuisance plant growths, such as algae blooms. Nitrogen exists in raw wastewater largely either as ammonia or in organic compounds that result in ammonia as they are degraded. Because ammonia exerts an oxygen demand from bacteria that convert it to nitrate, nitrification, there is sometimes a need to carry out this oxidation before discharge. In a few locations even the nitrate is undesirable because it, usually along with phosphorus, contributes to nuisance plant growths. In these cases, there is a need to remove the nitrogen entirely. Both phosphorus and nitrogen control can be attained biologically through manipulation of the dissolved oxygen, through recycle within the system, and sometimes by adding treatment stages to the system. Total nitrogen removal or denitrification is the most difficult, requiring oxidation followed by reduction to elemental nitrogen. For a very high degree of nitrogen removal, an organic supplement such as methanol is added to the last biological stage. Phosphorus can also be removed chemically by precipitating with iron or aluminum salts.

In the 1960s, there was much enthusiasm for advanced-waste-treatment systems that would produce very-high-quality water from municipal wastewater. To justify the cost of these systems, the water had to be reused for a high-quality purpose, including potable water. Although a pilot system operated at Denver, Colorado, indicated that, by all practical measures, water of potable quality could be produced, no full-scale plant has been constructed in the United States for the direct reuse of treated wastewater. There have been a small number of full-scale plants constructed, however, that do produce very-high-quality water. The systems include, among other technologies, chemical clarification, activated carbon treatment, and in at least one case reverse osmosis. These technologies are discussed in section 66.3.

Table 66.1 summarizes the municipal wastewater treatment technologies included above.

Capital and operating costs have not been included because of the wide variability, depending on site-specific conditions. As expected, there is an economy of scale for both capital and operating costs. At the time of writing, federal funds are available through each state to aid in payment for construction of municipal sewage-treatment plants.

66.3 INDUSTRIAL WASTEWATER AND HAZARDOUS WASTE TREATMENT

Although nonhazardous industrial wastewater and hazardous wastes are regulated under different legislation, methods for their control may be similar or even identical. In contrast to municipal wastewater, there is a very great diversity in the types of waste to be treated and an equally great variety in the kinds of treatment that might be used. Because it is impossible to include discussion of all of these methods, a few commonly used methods have been selected. It must be emphasized that biological treatment is used for treatment of many organic industrial wastewaters. Systems described for municipal wastewater treatment are applicable for this purpose.

Table 66.1 Municipal Wastewater Treatment Technologies

Technology	Usual Function
Preliminary treatment	Removal of large objects or reduction in size
	Removal of grit
Primary settling	Removal of large organic particles not removed in preliminary treatment
	Results in about 30% BOD reduction
Secondary treatment	Most common use is for further BOD and SS reduction to levels of 30 mg/l or less
	Can be designed to remove phosphorous, nitrify ammonia and denitrify nitrate
Disinfection	Reduction in levels of pathogenic microorganisms
Filtration (sand, dual media, multimedia)	Reduction of SS and BOD where effluent requirements are less than 30 mg/l
	Can be used to improve effectiveness of disinfection
Sludge processing	Preparation of sludge for beneficial use or disposal
	Can involve a wide variety of technologies for thickening, dewatering, biologically or chemically stabilizing and thermally treating, including drying and incineration

66.3.1 Chemical Precipitation and Clarification

Chemical clarification can be very effective for removal of contaminants from aqueous streams where the contaminants are particulate in nature. Figure 66.2 is a diagram of a typical chemical clarification system. To obtain a high degree of solids removal requires the formation of flocculent particles that will trap other particles and that will settle well. Flocculating chemicals such as aluminum sulfate (usually referred to as *alum*) or various iron salts are rapidly mixed with the feed under pH conditions that will cause the insoluble hydroxides of these metals to form. The water then flows to the flocculator, where it is slowly agitated. Under these conditions, large gelatinous particles form that settle to the bottom of the settler and are removed as sludge. The overflow from the settler may be of satisfactory clarity. If solids removal to very low levels is needed, a filter can be added to the system. In large installations, a commonly used type of filter would be a sand, a dual-media, or a multimedia filter. Similarly to the situation with municipal wastewater treatment, a large clarification system would consist largely of concrete tanks. For small systems, a number of manufacturers provide package systems which may have unique flocculator–settler combinations and filters.

As already indicated, clarification may be the only treatment needed for some industrial wastewaters where the intent is to reduce BOD and SS—for example, discharge to a public sewer. By carrying out precipitation of soluble materials, it may be possible to extend use of clarification to removing some hazardous materials, such as toxic metals. Lime can be used for this purpose where the hydroxide form of the metal is insoluble. In some situations, the precipitate that forms from lime addition settles well without adding other chemicals. If not, inorganic flocculants or a wide range of organic flocculants called polyelectrolytes can be added. Other materials that can be used for metals precipitation are compounds containing sulfide and carbonate, most likely sodium sulfide and sodium carbonate. Sulfides of many metals are very insoluble; some carbonates are insoluble. Because very toxic hydrogen sulfide can be formed from sulfides at low pH, care must be taken to prevent pH

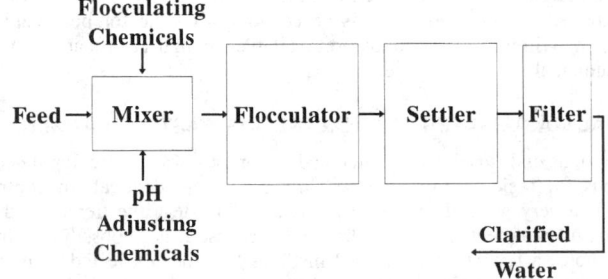

Fig. 66.2 Chemical clarification system.

reduction after precipitation has occurred. Sulfide dose must be controlled closely to prevent excess sulfide in the water leaving the system, which is a potential source of hydrogen sulfide. Also, sulfide is a reducing agent that exerts an oxygen demand until it is converted to sulfate.

Clarification and chemical precipitation may function as pretreatment prior to discharge to a public sewer system or prior to other treatment technologies where particulate matter causes problems. Two examples would be granular activated carbon treatment and most membrane processes.

66.3.2   Activated Carbon Adsorption

A very commonly used method for removing many organic materials from both gas and liquid streams is activated carbon adsorption. This discussion will involve only liquid (water) treatment, although much of what is said pertains also to treatment of gas streams. Activated carbon is made either from coal or organic materials that are carbonized and activated, usually by a thermal treatment that creates a pore structure within each carbon particle. This pore structure creates a large internal surface area that allows, under some circumstances, percentage amounts of contaminants, based on the mass of carbon, to be adsorbed from solution. The amount adsorbed varies from solute to solute, depending on the attraction between the carbon surface and the adsorbed material. The amount of adsorption for any solute is a function of the concentration of that solute in the aqueous phase. For a simple system with a single solute, adsorption capacity data can be obtained as a function of liquid concentration to produce a relationship called an isotherm. Very often the Freundlich isotherm is used to correlate the data. The form of this equation is as follows:

$$q = K_1 C^{1/n}$$

where q = amount of solute adsorbed, mass/mass carbon
  K_1 = a constant for the specific solute.
  C = the solute concentration in solution, mass/volume solution
  n = a constant for the specific solute

For low concentrations, it can often be assumed that there is a linear relationship between q and C, as follows:

$$q = K_2 C$$

where K_2 = a constant for the specific solute.

Observing either equation, it can be seen that the amount adsorbed decreases with decrease in solution concentration, approaching zero as the solution concentration approaches zero. For multicomponent systems, the isotherm for one component depends upon the overall composition. It is important to remember, however, that in general the amount of adsorption, q, increases or decreases with increase or decrease of that component in solution. To devise contactor systems that will make efficient use of the carbon, this principle must be kept in mind.

Activated carbon is available as a very fine powder or in a granular form with an average particle size of about one millimeter. Aqueous wastes are usually contacted with powdered carbon in a mixing and settling operation, conducted batchwise or as a continuous operation. A single tank can be used for batch operation by first mixing and then settling. Often a flocculating chemical must be added to attain adequate separation of the carbon. Continuous operation requires a mixer and a settler. Filtration of the effluent might be necessary for good removal of the carbon. Ultimately, in batch operation, and clearly in continuous operation, carbon is in contact with water of effluent concentration. Recalling the isotherm discussion, the loading of organic on the carbon will be low. More effective use of the carbon could be made using a multistage operation, but the cost of the system would be substantially increased. Use of powdered carbon does not lend itself easily to carbon reactivation.

Treatment with granular activated carbon is usually carried out by passing water through columns of the adsorbent. Figure 66.3 is a diagram of a two-column system using downflow operation. Operation can also be carried out in an upflow mode, either at a rate that allows carbon particles to remain as a packed bed or at a high enough rate to fluidize the bed. Conducting carbon treatment as shown in Fig. 66.3 results in much higher loading of organics on the carbon than is usually possible with powdered carbon treatment. In this case, carbon initially is in contact with water of feed concentration and, in terms of an isotherm, can adsorb an amount equivalent to that concentration. As the water passes downward, the concentration of contaminants in the water decreases, producing an S-shaped concentration curve (solution concentration versus column length) designated as the adsorption zone. Assuming the length of the carbon bed is somewhat longer than the adsorption zone, effluent concentrations of adsorbable materials can approach zero. As the carbon becomes loaded, the adsorption zone moves down the column and eventually breakthrough of contaminants occurs. At that point, it becomes necessary to replace the carbon with new or reactivated carbon. If the adsorption zone is very short, a single column may be sufficient. If the adsorption zone is long,

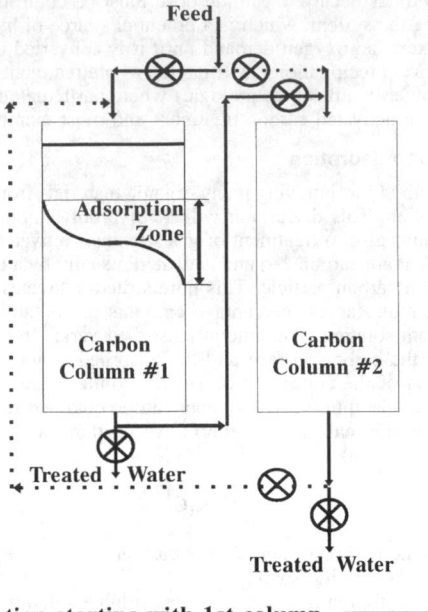

Fig. 66.3 Granular activated carbon treatment.

multiple columns in series can be used to increase the effectiveness of carbon utilization. With multiple columns, breakthrough is delayed until the lower end of the adsorption zone reaches the bottom of the last column. At that point, only the carbon in the first column needs to be replaced and the flow needs to be changed, as shown by the dotted lines in Fig. 66.3. With this mode of operation, high loading of most of the carbon in the system can be attained. If treatment cannot be interrupted, a standby column or multiple systems in parallel must be provided. Granular carbon is frequently reactivated because of the high cost of replacement carbon. Also, reactivation eliminates the need for and the cost of disposal of the contaminated carbon. Reactivation is a thermal process that is similar in function to the original activation. Reactivation can be done on-site if the size of the operation is large. Some companies provide a reactivation service if on-site reactivation is not practical. When dealing with carbon loaded with hazardous materials, careful attention must be given during reactivation to assure that the hazardous materials are either destroyed during reactivation or are controlled by an adequate air pollution-control system. Carbon loss during reactivation may be about 10%.

Operation of granular activated carbon systems has a number of problems, two of which can occur in essentially all cases. Activated carbon and ordinary steel form an electrolytic cell that corrodes the steel. To protect the metal, carbon columns are lined with rubber or coated with a nonporous organic coating. If the coating sustains damage, rapid corrosion of the tank wall can occur. Biological activity usually occurs in carbon systems when they have operated for a period of time. Surprisingly, this activity can be found even in systems treating toxic materials. If the activity remains at a low enough level that serious clogging does not occur, the result can be beneficial because biological degradation substitutes for loading of the carbon. If biological activity does cause clogging, backwash of the carbon columns is necessary. Not only does backwashing create an added operating cost, but it results in a biomass residual that must be further handled. Mixing of heavily loaded and lightly loaded carbon during backwashing reduces the overall time to breakthrough. Biological activity can create anaerobic conditions resulting in odor problems.

66.3.3   Air Stripping

Air stripping is useful for removal of volatile organic materials from a variety of aqueous streams, including industrial wastewaters, drinking water sources, contaminated groundwater, and contaminated soil itself. The treatment method is practical for dilute solutions with concentrations less than about 100 mg/l[4] and for hydrophobic compounds, such as chlorinated solvents, that exhibit large

Henry's law constants. Where the solutes being stripped are hazardous materials, air stripping may result in an exhaust air stream that is in violation of air pollution requirements, necessitating some form of air pollution control.

For an aqueous system with one contaminant at low concentrations, the equilibrium concentration of contaminant in air contacting the aqueous phase bears a linear relationship to the liquid concentration. This line passes through the origin. The slope of this curve is given by the Henry's law constant. For multicomponent systems, the Henry's law constant of any one component will almost certainly be affected by the presence of the other components, but generally, for each component, the concentration in the gas phase in equilibrium with the liquid will increase as aqueous phase concentration increases.

Air stripping can be carried out by simply bubbling air through water in a tank. In a batch operation, assuming one contaminant, the effect on a plot of gas concentration versus liquid concentration would be to move down that curve toward the origin. In a single-tank continuous operation, concentration of the liquid in the air stripping vessel is essentially that of the treated water and, therefore, the maximum concentration in the air must be very low. In either batch or continuous operation with this simple system, the amount of air required can be very large. This mode of operation may be acceptable if the air can be discharged without a need for pollutant removal. Much more efficient operation results from using a countercurrent contacting device. A commonly used example is a packed tower with downflow of the contaminated water countercurrent to the upward flow of air. The packing can be wood slats, rings of ceramic or other materials, or plastic packing of various geometries. The packing provides a large surface for mass transfer of contaminants from the aqueous phase to the air to take place. In a countercurrent contactor, treated water leaving from the bottom contacts clean air just as it would in a simple single-stage device. As the air travels upward, however, it contacts water of increasing contaminant concentrations. The result is a continuing transfer of the contaminant to the air. If the contactor were very long, representing a large number of equilibrium contacting stages, the air concentration would approach the value predicted by Henry's law for the feed, resulting in a greatly reduced air volume compared to the equivalent situation with a single-stage device. In an actual case, a compromise would be made that resulted in minimizing the cost of the overall operation, including air pollution control.

Although the packing in air strippers would ordinarily be substantially larger than particles of granular activated carbon, there is still the opportunity for clogging by suspended matter in the feed, biological growth, and possibly precipitation of inorganic material as a result of chemical oxidation. Chemical clarification or other methods of solids removal might be necessary.[4] If a significant precipitating oxidation reaction is expected, preoxidation with removal of the precipitate would be necessary.

66.3.4 Steam Stripping

Steam stripping can be used to remove volatile organic materials from aqueous streams at higher concentrations than is practical for air stripping. In addition, the method can remove lower-volatility materials than can reasonably be removed by air stripping. The process is more complicated, however, than air stripping and has higher capital and operating costs.

In its simplest form, steam stripping can be carried out by sparging steam into a vessel of water, either batchwise or as a continuous process, resulting in heating the water to boiling followed by conversion of some of the water to water vapor. The resulting vapor is condensed, producing a more concentrated aqueous solution of the volatile materials. Operating in this manner requires a large amount of energy and produces a condensate that is still quite dilute. The treatment method can be made much more energy efficient by utilizing a countercurrent stripping tower, as described for air stripping. The tower might be packed, but it might also contain multiple trays, typical of distillation columns. Steam stripping is actually a distillation operation. The water to be treated would be fed continuously to the top of the stripping tower typically after heat exchange with the hot treated water leaving the bottom of the tower. Steam would be injected at the bottom of the tower and would flow upward countercurrent to the downward flowing water. Just as in the case of air stripping, the water vapor rising through the tower would contact increasingly concentrated liquid water until it exited at the top of the tower approaching equilibrium with the feed water. The contaminated water vapor would then be condensed. This system not only makes efficient use of energy, but produces the highest possible concentration of contaminants in the condensate. Depending on solubility of the contaminants, it is possible for a separate contaminant phase to be formed that might be reusable in the manufacturing operation. Unlike air stripping, this process produces only a small amount of uncondensable gas to be discharged to the atmosphere.

66.3.5 Membrane Technologies

Natural membranes of various types have been used for centuries to remove solid materials from liquids. With the advent of synthetic polymers, mostly developed since World War II, the usefulness of membranes has been greatly expanded. It is now possible to produce membranes with removal capabilities ranging from what is usually considered ordinary filtration or removal of solid particles through ultrafiltration and hyperfiltration, which can remove large molecules, to reverse osmosis (some authors do not differentiate this from hyperfiltration), which can effectively remove even inorganic ions such as sodium and chloride to very low levels. A very extensive discussion of membrane technologies is given by Kirk and Othmer.[5] Much of the early development of these membranes was done for the purpose of producing potable water from mineralized sources such as brackish water and seawater. Potable water production continues to provide a significant market. Reverse osmosis, because it has such good rejection, has widespread potential in industrial wastewater treatment and in industrial processing for separating contaminated water into reusable water and a concentrate stream that also may be reusable. An example of the latter is in treatment of plating rinsewaters.

In addition to membrane systems such as reverse osmosis, where water passes through the membrane, leaving contaminants behind, there are systems in which the contaminants pass through the membrane. One of these that has been in use for many years is electrodialysis and a much newer method is pervaporation. In the latter method, volatile contaminants are vaporized through a membrane for which a vacuum or a stream of a carrier gas is maintained on the downstream side.

Reverse Osmosis

Ultrafiltration, hyperfiltration, and reverse osmosis have similar system configurations. Reverse osmosis will be described, but much of what is stated applies also to the other two operations. Reverse osmosis utilizes membranes that can exclude a large fraction of almost all solutes. The earliest membranes were made of cellulose acetate formulated in such a way that a very thin rejecting layer formed on the surface, with a much more porous layer beneath. Other materials are now available including thin-film composites. During operation, water under pressure is forced through the membrane. This water is called the permeate. To treat at practical flow rates, a large membrane area is needed. This need has been accommodated by development of several high area-to-volume configurations. One of these is the spiral wound module, which contains sandwiches of membrane and spacer material that are wrapped around an inner permeate collection tube and placed in a cylindrical pressure vessel. The permeate-carrying compartment is sealed around the edges. Feed entering one end of the cylinder is forced through the appropriate spaces in the membrane roll. Concentrate not passing through the membrane exits at the opposite end of the pressure vessel. Another high-surface configuration is the hollow fiber module. These modules contain bundles of reverse osmosis membrane in the form of fine hollow fibers. The ends of the fibers are mounted in a material such as epoxy. There are a number of specific configurations. These modules operate by having the water pass from the outside of the fibers into the hollow center, where it flows to one or both ends of the bundle. Two other configurations include tubular modules and plate-and-frame modules. Tubular modules contain much larger-diameter tubes with the reverse osmosis membrane in the interior of the tubes. Plate-and-frame modules contain stacks of circular membranes on supports with spacers between the membranes. The stacks are connected so that feed water flows upward through them and permeate is collected from each support. Neither the tubular nor plate-and-frame configurations have the surface-to-volume ratio that is possible with the other two configurations. The tubular modules have the advantage of being the easiest to clean.

To operate reverse osmosis, the feed water is forced under pressures of up to several hundred pounds per square inch through the modules. The minimum theoretical pressure to force water through the membrane is that which just exceeds the difference in osmotic pressure of the water on the upstream and downstream sides of the membrane. The osmotic pressure increases throughout the system as the contaminants in that stream are concentrated. In actuality, the operating pressure is usually well above the osmotic pressure. The water flows through the system until it emerges as a residual concentrate. The volume of this residual is an important consideration if it contains hazardous materials that interfere with simple methods of disposal. There are limits to which the water can reasonably be concentrated. Usually there is a minimum rate at which the water should flow along the membrane surface to minimize fouling and concentration gradients which can lead to unexpected precipitation of scaling materials on the membrane surface. As the feed stream is concentrated and reduced in volume by water passing through the membranes, the flow rate will decrease. In practice, a combination of staging of the membrane modules with decreasing membrane area in each stage, and some degree of recycling of the residual flow from a module back to that module to be mixed with new feed, could be used to maintain proper hydraulic conditions. As concentration on the upstream side increases, the concentration in the permeate also increases and may exceed either a discharge requirement or a reuse quality requirement. Finally, however, the concentrated stream reaches the point at which solubility limits are exceeded and precipitates begin to form. High turbulence can minimize precipitation or scaling on the membrane, but cannot totally prevent it. Calcium carbonate is a scale that commonly forms from treatment of natural waters because these waters contain calcium and a pH-dependent mixture of carbonate and bicarbonate alkalinity. To prevent this scale, feed water is acidified to convert the alkalinity to CO_2.

Fouling of the membranes is almost always found during operation of reverse osmosis and is also a problem with other membrane processes. Fouling is sometimes considered the same as scaling, but it differs from scale formation in not being the result of precipitation. The two can be found to occur together. Fouling results primarily from deposition on the membrane of suspended matter from the water. There is also the possibility of slimes forming from biological activity. One way to minimize fouling is to pretreat the feed water for removal of suspended matter. There are enzyme and chemical rinses that have been used to reduce fouling.

Electrodialysis

This treatment method is used to remove part of the ionic materials from water. It has been used for many years to partially demineralize heavily mineralized groundwater for production of potable water. The equipment consists of one or more stacks of membranes that have the ability to transmit either cations or anions, but not a significant amount of water. In each stack, the cation and anion membranes alternate and are separated by spacers that form a path for water passage. When a direct electric current is imposed on the stack of membranes, cations will tend to move toward the negative electrode and anions to the positive electrode. The ions can only pass through one membrane, however, before being blocked. The result is that the alternating spacer compartments become diluting compartments and concentrating compartments. All of the diluting streams are connected and produce the treated water. All of the concentrating streams are connected and produce a concentrate that must be disposed or possibly reused in the manufacturing operation.

Use of electrodialysis for pollution control is less flexible than reverse osmosis and some other treatment methods because it is not able to reduce the ionic content of the treated water to very low levels that might be required for discharge. Uses for industrial wastewater treatment would most likely occur where useful materials can be recovered from the concentrate stream and where the product water is acceptable for in-plant uses. Electrodialysis is subject to fouling and scale formation and operates most effectively on feed waters with very low concentration of suspended matter.

Pervaporation

As already indicated, this treatment method functions by the passage of volatile materials from an aqueous solution through a membrane to a gas phase consisting of a carrier gas or produced by a vacuum. Without considering the details of what occurs in the membrane, this method can be looked upon as a membrane-assisted evaporation that increases significantly the ratio of volatile materials to water in the gas phase compared to ordinary evaporation without a membrane. The treatment method accomplishes results similar to air stripping. It has the advantage over air stripping when using vacuum operation of capturing the volatiles in a highly concentrated form that may in some situations allow recovery and reuse. This technology is just beginning to reach the stage of commercial use, but could see rapid expansion in its application to process and waste streams.

66.3.6 Ion Exchange

Ion exchange has been used for many years for the softening of water and for production of deionized water. The technology is also being used for recovery of useful materials from in-plant streams and for treatment of some industrial wastewaters. With increased regulation of metals in wastewaters, increase in use can be expected. Traditionally, ion exchange has been used to remove only inorganic ionic materials by direct exchange with other ions on a solid matrix. Resins are now being produced by ion exchange resin manufacturers that remove some organic materials and selectively remove inorganic materials by more complicated mechanisms than simple ion exchange.

Ion exchange is carried out using granular, usually resinous, materials. The traditional ion exchange resins are cation or acid resins and anion or base resins. Furthermore, there are strong and weak acid resins and strong and weak base resins. The following equations describe two typical ion exchange reactions:

$$R_1H + M^+ \rightarrow R_1M + H^+$$
$$R_2OH + A^- \rightarrow R_2A + OH^-$$

where R_1 = a cation or acid resin
$\quad R_2$ = an anion or base resin
$\quad M^+$ = a metal ion
$\quad A^-$ = an anion

Both of these reactions are easily reversible. If the resin in the form R_1M were treated with acid, M^+ would be liberated and the resin would revert to the acid form R_1H. If the base resin were treated with a base, A^- would be liberated and the resin would revert to the base form R_2OH. If water containing the neutral compound MA were passed through a column of R_1H and then a column of R_2OH, the water could be demineralized. R_1H could be a strong acid or a weak acid cation resin. The difference is that the strong acid resin will function over a wide range of pH. Weak acid resins have a stronger affinity for H^+ than the strong acid resins and do not function well at low pH (high

H^+ concentration). If there were a large amount of M^+ in solution, the large amount of H^+ potentially entering the water phase might result in only partial M^+ removal. A strong acid resin would be a safer choice. Weak base resins, on the other hand, function well under acid conditions. A weak base resin could be used to complete the demineralization. If the weak base resin were placed before the cation exchange resin, it would not remove A^- significantly from a solution of a neutral salt.

As the ion exchange materials are exposed to more and more water, their ion exchange capacity is finally exhausted. The active sites in this case would have either M or A attached. At this point, the resin columns must be regenerated, the cation exchange resin with an acid such as sulfuric acid and the anion exchange resin with a base such as sodium hydroxide. The primary reason for using the weak acid and weak base resins whenever possible is their greater ease of regeneration. The most effective regeneration, utilizing only a slight excess of regenerant, would be accomplished by passing the regenerant solutions countercurrent to the direction of flow of the water being treated. In many systems, however, the resins are first backwashed before regeneration. Backwashing mixes the resin particles and eliminates the advantage of countercurrent regeneration. In these systems, both treatment and regeneration would most likely be downflow. Backwashing is done to flush from the beds particulate matter that might cause fouling and loss of ion exchange capacity. For most effective operation, the feed to an ion exchange system should have a very low concentration of particulate matter. Both the backwash and the used regenerant streams represent a waste. The volume of this waste would usually be only a small percentage of the volume of water treated. The objective of ion exchange is to concentrate the contaminants in a small volume. For industrial wastes containing hazardous materials, disposal becomes a problem. In these cases, recovery for reuse of the hazardous material has obvious advantages. There is also an obvious advantage to reusing the clean product water.

The above discussion was aimed at giving a general understanding of ion exchange. For industrial waste treatment, deionization would not be a likely objective. Possibly only a cation exchanger would be involved, or, as in the case of one chromium recovery system,[6] a cation exchanger for removing metallic ion impurities followed by an anion exchanger for removal of chromate. In addition, there are some organic removing resins that may use an acid or base for regenerant or may use a solvent.

Table 66.2 summarizes the industrial wastewater treatment technologies included above.

Table 66.2   Industrial Wastewater and Hazardous Waste Treatment Technologies

Technology	Contaminants Removed	Typical Applications
Chemical precipitation and clarification	Particulates BOD associated with particulates Trace metals when appropriate precipitants are used	Treatment prior to discharge to a public sewer Treatment prior to particulate sensitive technologies such as granular activated carbon and membrane processes
Activated carbon adsorption	A very broad range of soluble organic materials	Many industrial wastes containing soluble organic materials
Air stripping	Highly volatile materials	Dilute solutions of volatile materials, including industrial wastewaters, drinking water sources, contaminated groundwater and contaminated soil
Steam stripping	Volatile materials of lower volatility than appropriate for air stripping and at higher concentrations	Dilute to moderately concentrated solutions of a wide range of volatile materials
Reverse osmosis	A high proportion of organic and inorganic contaminants	To produce very-high-quality water
Electrodialysis	Inorganic ions	To remove a significant fraction of minerals from water
Pervaporation	Highly volatile materials	Not yet widely used Uses similar to air stripping
Ion exchange	Primarily inorganic ions	Deionization For recovery of ionic materials from industrial aqueous streams and industrial wastewaters

Capital and operating costs have not been included because of wide variability likely for different applications. Demineralization technologies can be expected to be somewhat more expensive than the other technologies.

66.3.7 Other Methods

There are a great many other technologies that could be included. In some cases, the methods are more closely associated with chemical processing than pollution control. In some cases, they have very limited or very specific uses. Some technologies might show promise, but are not far enough along in their stage of development to be considered for full-scale use. Treatment of contaminated soils and groundwaters has become increasingly important as the public has been informed by EPA and other environmental groups of the large number of contaminated sites that exist. Incinerators of various kinds have been used to destroy both concentrated hazardous organic wastes and soils contaminated with hazardous organic materials. At the time of this writing, incinerators are receiving a large amount of criticism because of the air pollution they can create if not properly operated. Thermal desorption, which vaporizes organic materials from contaminated soils, has been used at a number of sites. The air pollution risk from this process is substantially less than from incineration. Solvent extraction is usually thought of in the chemical industry as a method for concentrating solutes from a liquid processing stream into another liquid stream. The technology has been applied to contaminated soils for extraction of contaminants into a liquid stream. Solidification has been used to greatly reduce the pollution threat from contaminated soils. This technology utilizes addition of chemicals to the soil to create a solid mass with a very low rate of contaminants leaching. It is generally more effective with inorganic contaminants than organic. One form of solidification involves electrical heating of the soil to form a vitreous mass.

REFERENCES

1. E. L. Armstrong (ed.), *History of Public Works in the United States 1776–1976*, American Public Works Association, Chicago, IL, 1976, Chaps. 12 and 13.
2. Metcalf & Eddy, Inc., revised by G. Tchobanoglous and F. L. Burton, *Wastewater Engineering Treatment, Disposal and Reuse*, 3rd ed., McGraw-Hill, New York, 1991.
3. J. A. Parker and J. L. Darby, "Particle-Associated Coliform in Secondary Effluents: Shielding from Ultraviolet Light Disinfection," *Water Environment Research* **67**, 1065–1075 (1995).
4. J. V. Boegel, "Air Stripping and Steam Stripping," in *Standard Handbook of Hazardous Waste Treatment and Disposal*, H. M. Freeman (ed.), McGraw-Hill, New York, 1989, p. 6.108.
5. M. Grayson (ed.), *Kirk–Othmer Encyclopedia of Chemical Technology*, 3rd ed., Wiley, New York, 1982.
6. C. J. Brown, "Ion Exchange," in H. M. Freeman (ed.), *Standard Handbook of Hazardous Waste Treatment and Disposal*, McGraw-Hill, New York, 1989, p. 6.66.

PART 5

MANAGEMENT, FINANCE, QUALITY, LAW, AND RESEARCH

CHAPTER 67

MANAGEMENT CONTROL OF PROJECTS

Joseph A. Maciariello
Horton Professor of Management
Peter F. Drucker Graduate Management Center
Claremont Graduate School, and Claremont McKenna College
Claremont, California

Calvin J. Kirby
Vice-President Hughes Electronics and
Chief Executive Officer Hughes Avicom International

67.1	**GENERAL MODELS FOR THE MANAGEMENT CONTROL OF PROJECTS**	**2047**
	67.1.1 The Macro Cybernetic Model	2047
	67.1.2 Mutually Supportive Management Model for Complex Projects	2048
	67.1.3 The Cybernetic Model and Its Failure Modes	2050
67.2	**SYSTEMS DYNAMIC MODELS AND CONTROLLING THE WORK OF PROJECT TEAMS**	**2053**
	67.2.1 The Dynamics of Controlling a Project Team	2054
67.3	**SPECIFIC ISSUES IN THE PROJECT-CONTROL STRUCTURE**	**2056**
	67.3.1 Organizing for Complex Projects: Matrix Structure and Teams	2056
	67.3.2 Project Teams: A Case Study	2060
67.4	**SPECIFIC ISSUES IN THE PROJECT-CONTROL PROCESS**	**2063**
	67.4.1 Project-Planning and Control Process: Overview	2063
	67.4.2 The WBS	2064
	67.4.3 Network Plans—Time	2067
	67.4.4 Financial-Expenditure Planning: TV Transmission System Project	2070
	67.4.5 Scheduling Resources	2072
	67.4.6 The Budget Process	2077
	67.4.7 Systems of Reporting for Project Control	2080
67.5	**A SURVEY OF COMPUTER SOFTWARE FOR THE MANAGEMENT CONTROL OF PROJECTS**	**2083**

Projects are a very common feature of organizational work. They are prominent in aerospace and defense; construction; product development; public sector water, transportation and urban development; strategic thrusts; and in all kinds of team-related activity, including continuous improvement and reengineering activity.

67.1 GENERAL MODELS FOR THE MANAGEMENT CONTROL OF PROJECTS

67.1.1 The Macro Cybernetic Model

Figure 67.1 is a macro framework that places the entire task of the project control system design within a cybernetic framework. The framework can be understood best by viewing it from left to

Mechanical Engineers' Handbook, 2nd ed., Edited by Myer Kutz.
ISBN 0-471-13007-9   © 1998 John Wiley & Sons, Inc.

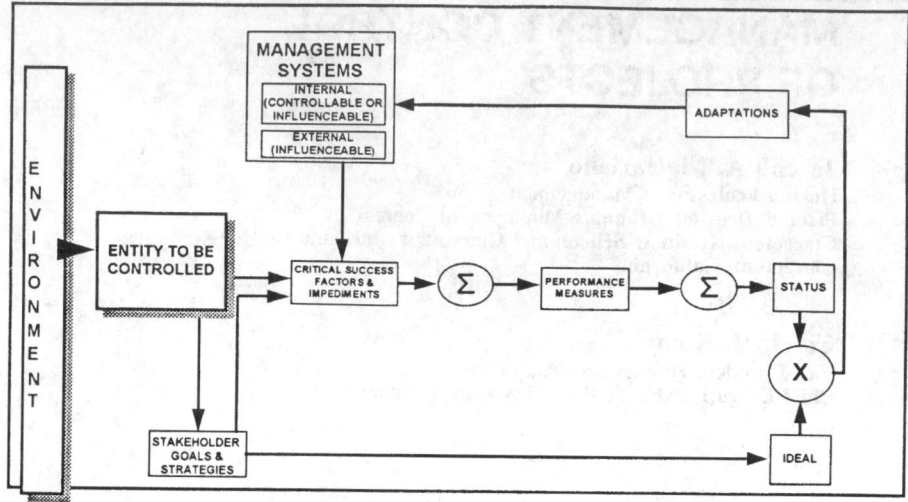

Fig. 67.1 Macro cybernetic control systems framework. (The design framework contained in Maciariello and Kirby, *Management Control Systems: Using Adaptive Systems to Attain Control,* 2nd ed., Prentice-Hall, Englewood Cliffs, NJ, 1994.)

right. The environment facing a project includes the customer, the competition, the technology, and the conditions of the various markets associated with the stakeholders of the project.

The development of effective and efficient project controls begins with the determination of the goals of each of the stakeholders of a project. The project manager should consider *who are the stakeholders* and *what is it that they seek from the project in order to continue to contribute to its success.* Next, strategies should be developed to meet the "inducements" necessary to satisfy the stakeholders, especially the critical stakeholders. Once these strategies are developed for each stakeholder, critical success factors (CSF's) for the attainment of these strategies as well as impediments to the attainment of these strategies should be identified. For each CSF, a performance measure should be developed to allow assessment of how well the project is performing in respect to stakeholder strategies.

Status reports are prepared periodically comparing actual performance against ideal performance. Gaps are a signal that changes must be made and improvements sought. Once performance is assessed and compared to the ideal, changes may be introduced in the management systems of the project in order to make the necessary improvements or adaptations.

The management systems of the project are themselves designed to exert control over the factors that can be controlled, to predict the values of uncontrollable factors and to influence the values of these uncontrollable factors.

67.1.2 Mutually Supportive Management Model for Complex Projects

Figure 67.2 identifies the key elements of the management systems that are required to control complex projects.

More recently, projects and subprojects are being managed by *project teams,* which resemble the matrix organization, although oftentimes the team or project leader has somewhat more formal authority over the functional resources assigned to the project than the project manager has under the matrix structure.

The *management style* of the project manager or team leader has to be predominately *participative,* since the manager often lacks full direct authority over functional personnel. The project leader is deeply involved in the integration of the work of a project. As a result, the project manager must be intimately familiar with the work, the technology, and the people involved on the project. As a result there are situations in which the project manager must be more directive and authoritative in order to accomplish the integration needed.

A project manager should seek to maintain an *open and candid culture.* This is required because there are numerous problems to be solved on a complex project and free and open communications are essential to cope with the dynamics of any complex project. The cross-disciplinary nature of many of the problems on projects require a *team orientation* throughout the duration of the entire

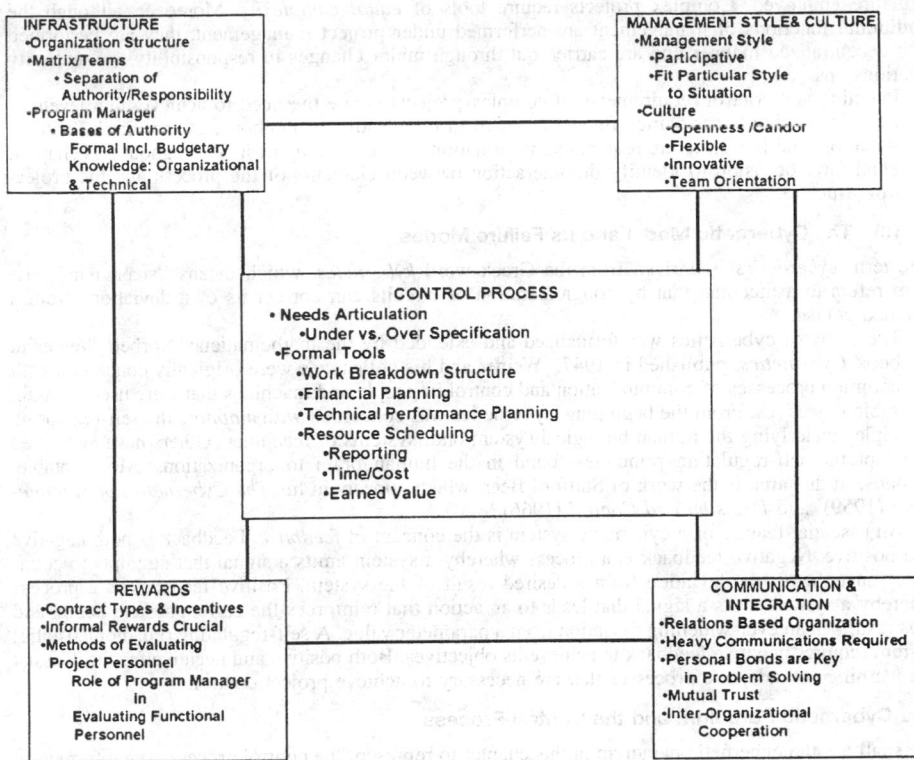

Fig. 67.2 Mutually supportive systems model (MSSM) for complex projects.

project. The high-technology characteristics of complex projects create the need for a culture that is innovative and flexible.

The *communication and integration* subsystem reflects the heavy communications requirement for the successful management of complex projects. *Personal relationships* that are related to flow of work and to problem-solving are key elements in the management of complex projects. The culture of openness should go a long way towards creating *mutual trust* among the various parties who must contribute to a project.

One of the key issues involving *rewards* is a decision about how the project and functional managers interrelate in the task of *evaluating performance of functional personnel* employed in project work. The most effective systems involve contributions from both functional and project managers in performance evaluation, but this has to be worked out in each organization.

Contracting methods on complex projects range from *fixed price* to *cost-plus-profit contracts*. There are various incentive arrangements negotiated between the project organization and the customer. Incentive arrangements are varied but are usually negotiated based upon *performance characteristics* of project deliverables, *schedule or cycle time, functionality,* and *cost performance.*

Informal rewards bestowed among project and functional personnel are especially important in the *relation-based organizations* that work on projects. Informal rewards involve various types of recognition of functional personnel by project managers and by peers.

Many of the new dimensions of control systems that are required in the control of complex projects are found in the control process. A whole new set of tools and concepts is required to more effectively facilitate the added coordination and integration required for project activity.

The *project-control process* is a procedure for the management of a project that operates through the project-control structure (style and culture, infrastructure, rewards, and coordination and integration subsystems) to achieve project goals. The process supports the formal and informal relationships embodied within the matrix or team structures in that it provides information to project and functional personnel upon which their decisions are based. The project-control structure and process must be mutually supportive if project goals are to be achieved.

Differences between processes required for project control and traditional management control occur because of the complexity of project activity and because of the difference in the organization structure employed. Complex projects require tools of *equal complexity.* Moreover, although the traditional functions of management are performed under project management, they are performed in a decentralized manner and are carried out through major changes in responsibility and authority relationships.

Planning and control requirements of complex projects create the need to achieve high levels of coordination without sacrificing efficiency, which in turn leads to the choice of the matrix or team organization, and this structure requires an information system to support it. Throughout this chapter, we shall have occasion to identify the interaction between elements of the process and the project control structure.

67.1.3 The Cybernetic Model and Its Failure Modes

The term *cybernetics* is derived from the Greek word *kybernetes,* which means "steersman." The term refers to a machine that by conglomeration of circuits can correct its own deviations from a planned course.

The study of cybernetics was formalized and extended by the mathematician Norbert Weiner in his book *Cybernetics,* published in 1947.[1] Weiner and his colleagues were originally concerned with the common processes of communication and control in people and machines that were used to attain desirable objectives. From the beginning, cybernetics was concerned with *mapping* the self-regulating principles underlying the human biological system onto systems of machines. Others have attempted to adapt the self-regulating principles found in the human brain to organizations. Most notable, perhaps, in this area is the work of Stafford Beer, which appears in his *The Cybernetics of Management* (1959)[2] and *Decision and Control* (1966).[3]

An essential feature of a cybernetic system is the concept of *feedback.* Feedback is both negative and positive. Negative feedback is a process whereby a system emits a signal that attempts to counteract an unfavorable deviation from a desired result of the system. Positive feedback is a process whereby a system emits a signal that leads to an action that reinforces the current system action and thus results in an ever-widening deviation from a parameter value. A self-regulating (i.e., homeostatic) system requires negative feedback to achieve its objectives. Both positive and negative feedback assist the learning and adaptive processes that are necessary to achieve project control.

The Cybernetic Paradigm and the Control Process

We shall use the cybernetic paradigm in this chapter to represent the control process. The information systems which support the control process are those in the middle box of Fig. 67.2. Figure 67.3

The Cybernetic Paradigm of the Control Process

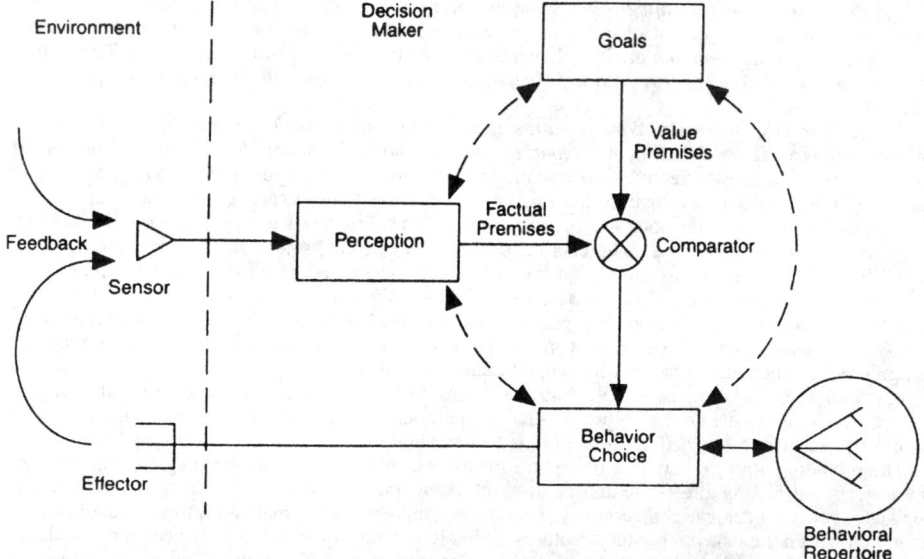

Fig. 67.3   The cybernetic paradigm (from Ref. 4; reproduced by permission of Donald W. Griesinger. A subsequent version of the paradigm appears in Ref. 5).

represents the cybernetic paradigm of the project control process. This particular version of the paradigm has been devised by Griesinger.[4,5]

The cybernetic paradigm represented in Fig. 67.3 allows us to capture the essential elements of the repetitive control process, which may be enumerated as follows:

1. Set goals and performance measures.
2. Measure achievement.
3. Compare achievement with goals.
4. Compute the variances as the result of the preceding comparison.
5. Report the variances.
6. Determine cause(s) of the variances.
7. Take action to eliminate the variances.
8. Follow up to ensure that goals are met.

These eight elements of the control process are captured in the cybernetic paradigm.

The paradigm begins with the assumption that decisions are explained as the result of the interaction between the manager/decision-maker and the environment faced by the decision-maker. Each project manager operates within an environment. The environment includes the "outside world" (i.e., the external environment of customers, suppliers, etc.) as well as other organizational units internal to the firm (i.e., the internal environment). A project manager must be responsive to changes in the external environment of the project as well as to changes within the internal environment.

A project or team manager scans the environment, either formally or informally, so as to absorb information or feedback pertaining to its condition. The manager comes into contact with the environment through the sensors of the project. Sensors are mechanisms used by managers to collect data. The mechanisms include reports that are reproduced as a result of formal attempts to scan the environment as well as "informal reports" that come to the attention of the manager through his or her senses of hearing and sight.

The manager constructs from these data certain beliefs concerning performance and the state of the external environment. These beliefs are referred to as *factual premises*. Factual premises are formed by passing these data through a cognitive process referred to as *perception,* which broadly refers to the psychological processes of extracting information from data and of interpreting the meaning of that information. Cognitive limitations prohibit decision-makers from assimilating all data in the environment, so the decision-maker uses past experiences, organizational goals, and personal and organizational aspirations to arrive at these beliefs about the actual state of the environment.

The manager uses these factual premises in a comparison process with organizational goals and performance measures. *Goals* are themselves a result of past learning concerning performance and accomplishments and represent the desired state for the manager. When a difference is determined to exist between what decision-makers desire (i.e., *value premises*) and their beliefs about the environment (i.e., *factual premises*), they are motivated to seek to close the gap. The *comparator* represents the comparison process that takes place between performance measures and performance information.

When a performance gap exists, decision-makers are motivated to search for courses of action that will move them closer to their goals. This choice, referred to as *behavioral choice,* is made by evoking from experience a limited set of alternatives that have been successful in solving similar problems in the past. The content of the set of alternatives evoked from the decision-maker's *behavioral repertoire* is itself a function of goals, past experience, and the decision-maker's perception as to the state of the environment. Search procedures are also included in the behavioral repertoire.

Alternative solutions are evoked from the behavioral repertoire according to established or learned search procedures. The first alternative found during the search that is believed to solve the problem is normally selected, so long as it meets project requirements. In the event that two or more alternatives are generated by the search procedure as potential solutions to the problem, the feasible alternative with the highest *subjective expected utility* that closes the gap will be chosen.

An alternative will be chosen only if it is expected to meet the goals of the decision-maker. If no alternative is expected to reduce or close the gap, the decision-maker will expand the search process. The search process is motivated by the presence of a gap and will stop when a feasible alternative is found that will close the gap.

Decisions require implementation. The *effector,* a manager, activates the decision, thus serving as a change agent. Control is brought about by action taken by the manager who next seeks to determine the effects of the action. This new information is referred to as *feedback*. If the new behavior leads to a reduction or elimination of the gap, the behavior is likely to be repeated in the future under similar circumstances. If goals are being met routinely, it is likely that the organization will eventually seek higher levels of performance.

In the event that goals are not achieved, the manager will repeat the process. If after repeated attempts the goals are not achieved, the manager will either alter the performance measures that are attended to and thereby distort his or her perceptions of reality, or reduce his or her goals. In either case, the performance gap is ultimately closed.

A certain amount of interaction takes place during the control process among the variables in the cybernetic paradigm. *Goals* direct the part of the environment that is *perceived* by managers. Perceptions about past performance influence current goals. We perceive that part of the environment that pertains to our goals. If decisions cannot be found that meet our goals, we change goals. Additional information may be introduced during the search process that can alter goals and alternatives considered, as well as the part of the environment that is attended to (perceptions).

Potential Failures in Project Steerability*

Potential failures in the task of successfully steering a project towards its objectives include those *external* to the decision-maker and those *internal* to the decision-maker and the organization. Figure 67.4 is the cybernetic model with potential failure modes associated with each external and internal variable of the model.

Steerability may be impeded as a result of four potential *environmental* failures, as defined in Fig. 67.4. *Lack of data* implies that a manager lacks information about the project environment that is necessary to achieve goals. A variant of the lack of data is the *lack of predictability* regarding environmental disturbances that impact performance towards a goal. *Overwhelming events* are environmental disturbances that overwhelm the manager's ability to cope. *Interference* occurs when factors or persons in the environment constrain the behavior of the manager in a way that prevents goal attainment.

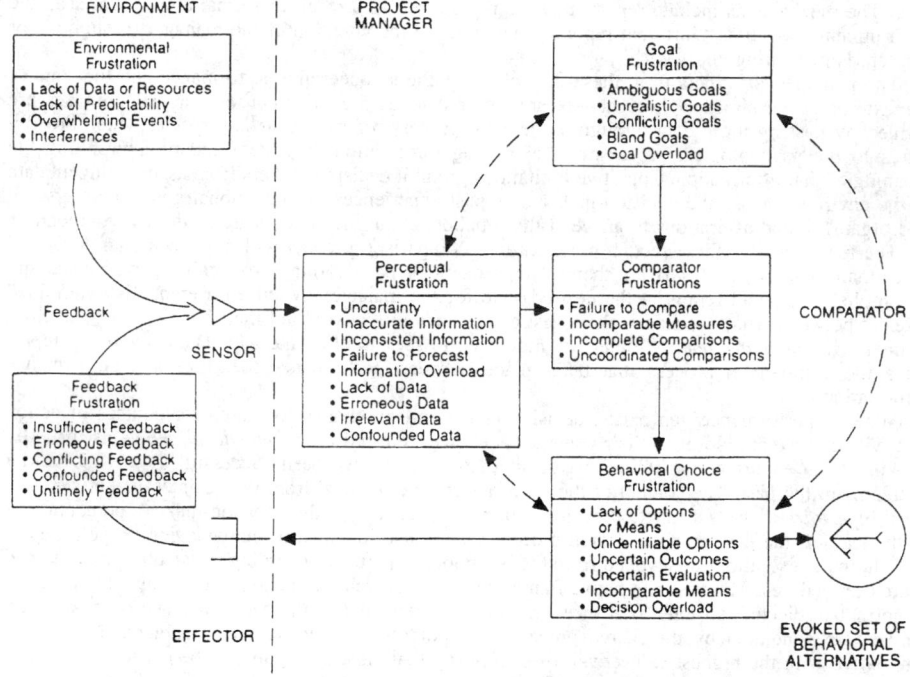

Fig. 67.4 Project management cybernetic failure modes.

*In general, this section is based on the cybernetic failure model as originally formulated by Graham (Ref. 6, pp. 32–47) and by Griesinger.[4,5] The cybernetic failure model was subsequently adapted by Edie Levenison, Joseph Maciariello, and Peter Zalkind for the purpose of analyzing the failure modes of an actual software project. This project and its failure modes are described in Maciariello and Kirby (Ref. 7, pp. 581–588).

Feedback is critical to steering an organization towards its goal. *Feedback frustrations* occur when data for evaluating the effectiveness of past decisions is insufficient. If feedback is insufficient, a manager does not know which actions to repeat and which to delete in the future. That is, there can be no organizational learning. *Erroneous feedback* as to the success of various actions can lead to decision error and goal failure. *Conflicting feedback* leaves the decision-maker confused as to the state of goal achievement. *Confounded feedback* occurs when the results of an action are mixed up with the results of other actions and with environmental changes so the decision-maker is confused as to the ultimate effects of a past decision. Finally, feedback may be *untimely* so far as necessary corrective decisions are concerned.

Perceptual frustration includes many of the same frustrations that are external to the decision-makers, except that these involve perceptual processes. *Uncertainty* occurs when a manager doesn't understand that a goal that is being pursued is in danger of being missed. *Inaccurate perceptions* are concerned with incorrect interpretation of the available data. *Inconsistent information* involves different interpretations of the same event or conflicting interpretations of multiple events. *Failure to forecast* is a failure to forecast the implications of trends that are at least partially visible. *Information overload* is a condition where accurate perception breaks down because of the inability to process environmental information effectively. *Lack of data* is the same frustration as discussed above for environmental variables, except this one pertains to perceptions that data are inadequate for making the necessary inferences. Similarly, *erroneous data* are data perceived to contain errors. *Irrelevant data* are those perceived as being inapplicable for necessary inferences. *Confounded data* may lead to spurious perceptions.

Goal frustrations are among the most serious impediments to steerability. *Ambiguous goals* are those for which criteria for achievement are not clear, thus frustrating the measurement process. *Unrealistic goals* are those that are simply beyond the individual's ability to achieve them. *Conflicting goals* are those individual and organizational goals that are incompatible and cannot be attained simultaneously because of the tradeoffs required for the accomplishment of each goal. *Bland* goals are those that are simply not highly valued, thus providing low motivation for their achievement. *Goal overload* occurs when the complexity of goals overwhelms the decision-maker's ability to sequence or prioritize them. Goal frustrations are so serious that the project manager must take extraordinary steps throughout the project life to ensure that there is continual congruence among crucial stakeholders regarding the goals of the project.

Comparator frustrations include the *failure to compare,* which is a case in which relevant perceptions are not compared to goals to determine if a gap exists. *Incomparable measures* is a case where goals and measurements of progress toward goals are conceptualized differently and incorrect surrogates for goals are measured. *Incomplete measures* is a case where the measure is a valid one for the goal but is incomplete as an assessment of performance towards the goal. *Uncoordinated comparisons* is a failure to compare perceptions and goals at the same point in time. This commonly occurs when there are long processing delays in preparing relevant information.

Behavioral choice frustrations are those involving the decision-making process itself. *Lack of options* is the frustration that occurs when, because of lack of ability, experience or free will, the decision-maker is unable to solve a problem and steer the organization towards its goal. Related, *unidentifiable options* are frustrations produced when appropriate behaviors, although knowable, are simply not accessed by the decision-maker as a result of inappropriate search procedures. *Uncertain evaluations and outcomes* occurs when the decision-maker is uncertain about predictions of the impact of alternatives upon the goal, thus making it difficult to choose effective remedies. *Incomparable means* involve two or more alternatives that are believed to make a contribution toward the goal but whose impacts upon the goal are not strictly comparable, thus frustrating rational choice. Finally, *decisional overload* occurs when too many decisions must be made in a given period of time, thus not allowing enough time for analysis of each decision.

67.2 SYSTEMS DYNAMIC MODELS AND CONTROLLING THE WORK OF PROJECT TEAMS

It is possible to examine the dynamics of the project-control system itself. These dynamics have a significant influence on the ability of the project manager to achieve control. As the project progresses in time, the various aspects of the management control systems interact with one another. These interactions can be described as various patterns of cause-and-effect relationships. When the various subsystems of the project-management-control system are appropriately aligned, they produce mutually supportive interactions that contribute to the efforts to achieve control. In contrast, when they are out of alignment, they frustrate attempts to achieve control.

Patterns of cause and effect in systems often are circular or "linking back" to the first variable. We call these circles of causality *causal loops.* Figure 67.5 is an example of a causal loop. Activity A influences B, which in turn influences C, which then influences A.

Let's assume that a member of the project expresses trust in another member (Action A). The second member, influenced by this action, might take on expanded responsibilities to ensure that an

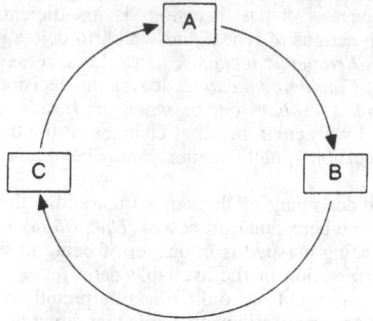

Fig. 67.5 Reinforcing causal diagram.

expected outcome is achieved (Action B). The improved outcome might lead a third member to comment to the first that the second member can be counted on to perform, thus increasing A's trust (Action C). This then is an example of a *reinforcing* causal loop in a positive direction. The opposite can also occur. Reducing trust might cause a member to reduce effort, thus further reducing trust. This, too, is a reinforcing spiral, but in a downward direction.

A reinforcing loop does not occur forever. Limiting forces materialize. Systems thinking recognizes that a change in one variable can cause changes in secondary variables. These secondary changes, not so obvious at first, can begin to feedback influences over time that limit the reinforcement process. For example, if a third party overheard the lack of trust member A has in member B, he or she might begin expressing trust in B, thus *balancing* the downward spiral. The second causal loop thus balances the first.

The second causal loop took some time to take effect. The *delay* is the third building block of systems thinking, along with *reinforcing* and *balancing* loops. Many of the dynamics we observe are due to unforeseen delays. From a management control perspective, the designer can use this kind of dynamic systems thinking to enhance the mutually supportive and adaptive dimensions of the control system.

A general principle to note based on the preceding discussion is:

> When a reinforcing process is set into motion in order to achieve a desired result, it also sets into motion secondary effects which usually slow down the primary effect.

Senge[8] and Kirby[9], and Severino (1992) using Kirby's results, have shown that these dynamic control problems can be minimized by creating a *learning* (*i.e., adaptive*) *organization*. Drawing from these studies, a learning organization requires the development of a *shared vision* of what the organization wants to accomplish, an environment that is continuously open to new ideas, one that encourages individual learning and mastery, and leadership by example.

We turn now to examine an example of an application of system dynamics to the management of project teams.*

67.2.1 The Dynamics of Controlling a Project Team

The first step a newly formed team must take is to develop a shared vision for its goal or objective. Then they must assess the current situation in terms of the vision. To do this, they must gather information to refine their understanding of the current situation and then determine appropriate action. As a result of the process of implementing the actions and gathering more feedback, members emerge in different roles more suited to the needs of the goal and vision. The team essentially learns to be effective. Each of these steps requires support from the control system.

Using findings from Kirby's study of successful and unsuccessful project teams, Fig. 67.6 shows the reinforcing system of informal activities that allowed the most successful teams to achieve their goals. The key environmental issue found in most successful teams was a culture of *trust and openness*. The leader of these teams had few preconceived assumptions or beliefs about the "best way"

---

*An expanded version of the models represented in this section may be accessed in "Team Dynamics in Adaptive Control Systems," by R. A. Severino, C. J. Kirby, J. A. Maciariello and N. N. Kelly, The Agility Forum, Bethlehem, PA, (https://www.agilityforum.org:445), 1996.

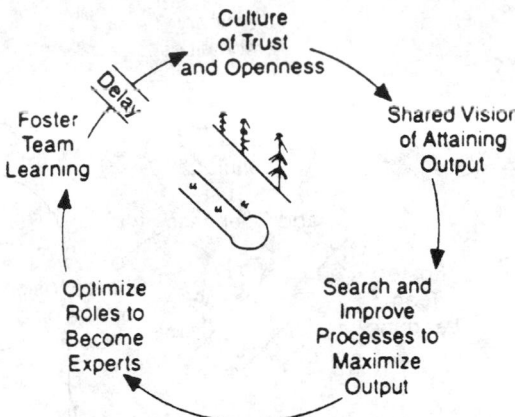

Fig. 67.6 Reinforcing system of informal activities.

to perform any given action. The groups used a free interchange of dialogue in their search procedures to weigh the benefits of any suggested actions. This culture led to the development of a shared vision of the desired objective and a search for the processes that optimize output. These activities represent informal planning activities. Once the process of making improvements was underway, team members assumed roles that better supported the process of further improvement. This led to an environment that fostered team or staff learning. After some time operating in this environment, during which time the teams developed and refined these skills, the process began to provide reinforcing feedback for increasing the level of trust and openness. This further reinforced the other activities, thus accelerating their efficiency.

Unfortunately, neither management staffs nor improvement teams will continue to improve their output endlessly. Figure 67.7 introduces the key limiting or balancing factors to this learning engine. There seem to be five such limiting factors. They should be expected in poorly performing teams and to some extent, eventually, even in successful project teams.

Teams that begin to fail or fail often have a leader or *dominant member* who carries strong preconceived beliefs about how the management team should act. This situation seems to block the discovery and ownership attributes present in the open dialogue of goal-seeking groups and *balances* the culture of openness and trust.

Proceeding clockwise along the reinforcing loop, another limiting factor is a gradual *erosion of the commitment* to the goals of the team that erodes the common vision. There are various degrees of commitment, a minimum being apathy and the maximum being total commitment. As levels of commitment fall, the amount of energy devoted to the goal falls.

Even if a staff remains committed to a goal and retains a culture of openness and trust, it can still be unsuccessful if it lacks an *adequate model of cause and effect*. These models are necessary to understand the meaning of data and in order to facilitate specific actions. The models available include many of the techniques associated with total quality management (TQM).

As team members seek to optimize their roles and become experts, a source of motivating energy propels them to close the gap between current performance and their goal. Senge[8] calls this source *creative tension* and its opposite *emotional tension*. Emotional tension distracts members from pursuit of their goal by forcing them to spend increasingly larger amounts of time in ambiguous roles. The matrix structure is particularly prone to role ambiguity because of the competing and often ambiguous instructions given to project participants by project and functional managers. Emotional tension tends to balance the positive forces that encourage mutually supportive emergent roles. Similarly, when staffs or teams exhibit *defensive routines* in reaction to team conflicts, team learning is curtailed. How teams respond to conflict frequently separates the excellent from the mediocre teams.

In summary, our analysis indicates that team learning is facilitated by the informal subsystems of control. But the formal elements of the control system also interact with the informal elements. Figure 67.8 shows the interaction of the informal with the formal elements of the project-control process.

Figure 67.8 illustrate the cybernetic behavior of both the informal and formal planing activities for a team working on an improvement project. The two activities shown on the left come from the informal systems and involve searching for data, seeking new directions, and formulating plans. These activities are most prevalent during times when teams are searching for solutions to pressing problems. The balancing feedback on the right illustrates the relationships of formal and informal processes. Formal planning and control processes are seen as the formal aspects of attaining output goals through

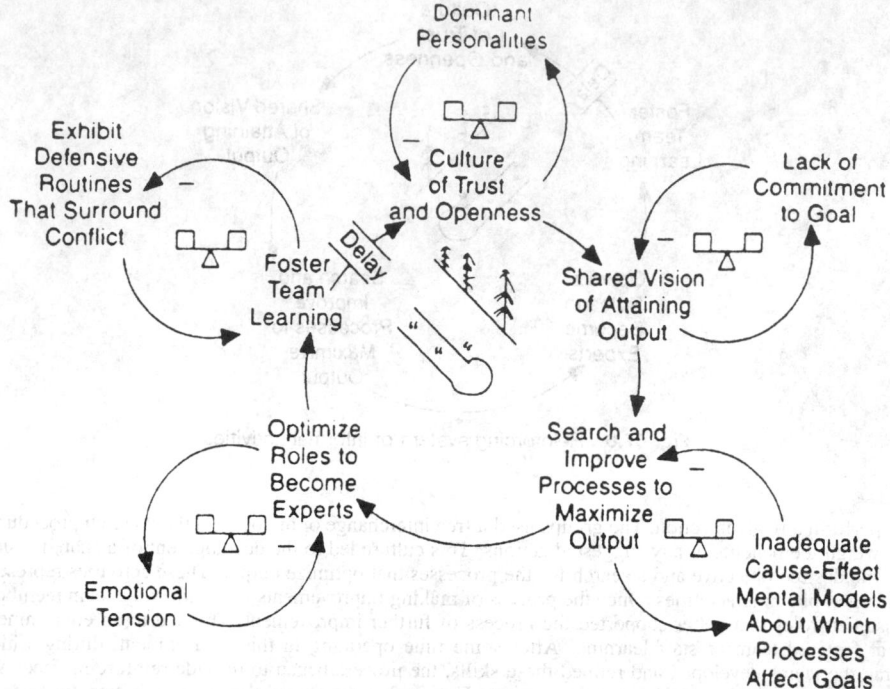

Fig. 67.7   Reinforcing system of informal activities with balancing items.

shared vision, assessing the gap between performance and vision, and taking steps in the planning and budgeting process to close the gap. The formal interacts with the informal to allow the team to achieve its goals. The formal gap measuring activities allows the team to steer its efforts towards goals.

Showing the formal and informal control activities within the context of the entire control system identifies a mutually supportive reinforcing learning system. Figure 67.9 displays a reinforcing loop that we might call the *adaptive control or learning engine*. The reinforcing loop shows the additional influence of formal and informal rewards (as a result of measurement) on team learning, thus linking the structural aspects of the control system to the process aspects.

Finally, we are in a position to view the entire dynamics of the control system as it affects team performance. Figure 67.10 is such a view. Two elements of the control structure not previously discussed are included at the top left. Both set the initial conditions for teams. Prior training or indoctrination for the team is one element and infrastructure or formal chartering of the team is the other.

The reinforcing loop of culture, vision, search, and so on is the upper reinforcing loop along with the five potentially balancing items shown along the outside. The formal control balancing activities support the informal planning processes. The outer loop of measurement and rewards reinforces team learning at the bottom of the structure.

The progress of the project team is determined largely by how close the leadership of the team can come to creating the mutually supporting reinforcing loops by using appropriate elements of the mutually supportive subsystems in a dynamic manner. At any given moment, some loops are rapidly cycling while others are sitting idle.

67.3   SPECIFIC ISSUES IN THE PROJECT-CONTROL STRUCTURE

67.3.1   Organizing for Complex Projects: Matrix Structure and Teams

Complex projects, although requiring close coordination, may not be large enough or of long enough duration to justify a separate project organization form, yet they often require far more coordination than is possible under a functional organization. In other words, neither extreme, a pure functional or a pure project organization, is ideally suited for such complex projects, and the tradeoff implied in the choice between the two is often unacceptable.

As the projects of an organization have become more complex and numerous, the functional organization has been forced to recognize the limitations of its structure. Often this recognition has

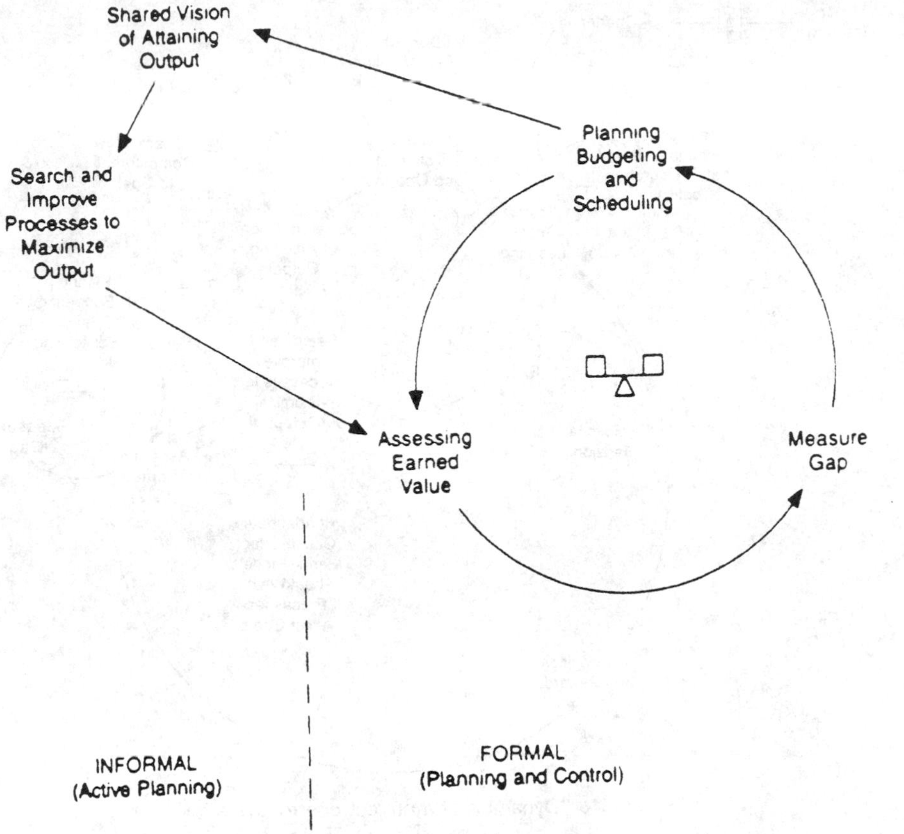

Shared Vision
of Attaining
Output

Planning
Budgeting
and
Scheduling

Search and
Improve
Processes to
Maximize
Output

Assessing
Earned
Value

Measure
Gap

INFORMAL
(Active Planning)

FORMAL
(Planning and Control)

Fig. 67.8 Cybernetic behavior of formal and informal activities.

led to formation of ad hoc teams of functional personnel to handle the "unique" coordination problems created by the high degree of interdependence required among functional disciplines on a given project. As these types of projects are recognized to be the very nature of the organization, team relationships often become less ad hoc and more formal.

Members of teams are selected based on functional expertise from the relevant functional groups. Leadership of a team is often assigned from the functional group that can make the largest contri-

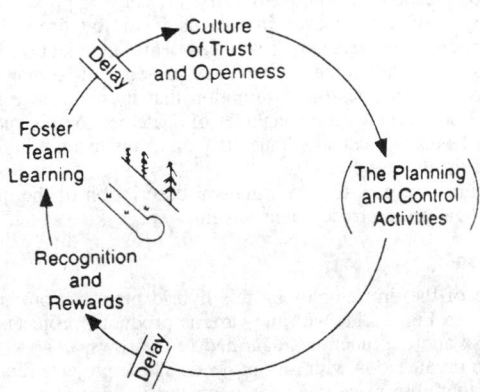

Culture
of Trust
and Openness

Foster
Team
Learning

The Planning
and Control
Activities

Recognition
and
Rewards

Delay

Delay

Fig. 67.9 Adaptive control engine.

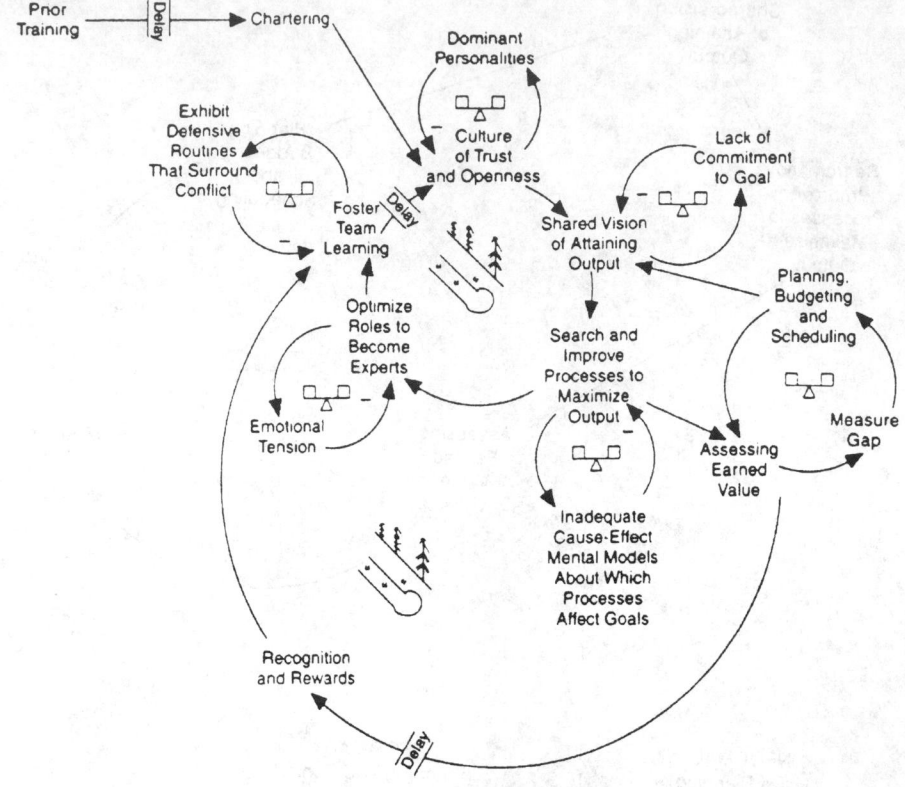

Fig. 67.10 Dynamics of a project control system.

bution toward solving the problem for which the team was formed. Once formed, it provides a device to achieve coordination across specialized functions and is an expedient means of getting complex tasks accomplished.

Although teams are disbanded upon completion of a project, the number of projects requiring such an arrangement should be expected to increase for any organization engaged in complex projects. In these organizations, the need for a more permanent structure that makes relationships among team members more formal soon becomes apparent. The new structure should allow the organization to coordinate these complex activities while retaining functional specialization and resulting scale economies. Neither of the two structures described allows an organization to achieve total coordination of complex activities and full-scale economies of specialization *simultaneously.*

A solution to this organizational dilemma was first provided approximately 40 years ago in the aerospace industry and has since been used in various forms by firms who are in businesses as divergent as food and produce and accounting services. This organization form has become known as the *matrix form* and is used today in many variants of project and product management. Regardless of the type of application, it embodies the assumption that it is not necessary to make an explicit tradeoff in organizational design between advantages of scale economies and coordination. Rather, it attempts to achieve high levels of each simultaneously and thus reach a higher level of effectiveness than either the functional or decentralized form.

With this as background, we now turn to a general description of the matrix structure, followed by illustration of its application to project management.

The Matrix Organization

Regardless of the nature of the firm employing this hybrid organization form, it may be described using a matrix, as shown in Fig. 67.11, with the various products, projects, or services identified in the columns of the matrix and the functions identified in the rows.

The matrix depicts an organization with four projects. These projects represent the primary output of the organization and therefore place demands upon the various functions. The functions, repre-

Projects/ Functions	Project 1	Project 2	Project 3	Project 4	Total Functional Output
Engineering					
Procurement					
Quality assurance					
Logistics support					
Manufacturing					
Project control					
Project management					
Overhead					
Total project requirements					

Fig. 67.11 The structure of matrix organization.

sented in the rows of the matrix, supply resources to various projects within the organization. The total output of a function for any particular period of time is found in the last column of the matrix and consists of the sum of all the contributions of a particular function to various projects. Each row of the last column represents the demand placed upon one function by multiple projects. This demand may be measured in physical units of input, such as resource hours, or in monetary units.

Functional contributions that are indirect, such as research and development, contract administration, personnel, business planning, public relations, and finance are included in the overhead row of the matrix.

The matrix itself, however does not uniquely define the distinguishing characteristics of the organization form used in project control, although the term is widely used to describe the organization form. Any organization may be described as a matrix; it need not be a hybrid form. All organizations produce outputs that may be identified in the columns of a matrix and use inputs that may be identified in the rows of the matrix. That is, all organizations have a purpose and use inputs or processes to fulfill the purpose. The truly distinguishing characteristics of the matrix organization structure, in all its variations, lie in the dual dimensions of management embodied within it and the allocation of responsibility and authority resulting from the dual management dimensions.

The matrix organization uses two overlapping dimensions of management, each of which may be identified with the matrix. The dual dimensions of management may be identified by referring to Fig. 67.11. Under the matrix organization structure, full responsibility for the goals of a project is given to project managers, and this responsibility is identified by the column dimension of Fig. 67.11. However, functional personnel who perform the work on the projects of an organization receive direction from functional management under the matrix structure, thus providing the second and overlapping dimension of management for the projects of the organization. The functional dimension of management is identified by the rows of the matrix.

Therefore, although the project manager assumes full responsibility for delivery of a product that meets performance specifications on a timely basis and in accordance with a contractual resource limits, he or she does not have full direct authority over the functional organizations that actually perform the work. If he or she did have such authority, the organization would not be a hybrid at all; it would be simply a decentralized project form. The distinguishing feature, therefore, of the matrix organization is the separation of the responsibility for the goals of a project from the authority to direct the work necessary to achieve those goals.

Furthermore, although there is a separation of responsibility and authority under the matrix structure, functional personnel actually do operate subject to dual sources of authority: the *knowledge-based authority* within the functional organization and the *resource-based authority* of the project manager. Unity of command is thus broken. This is rectified, to some degree, in organizations that use teams within the matrix to manage projects.

Even though the hallmark of the matrix is the separation of responsibility and authority, in practice we find that rather than a clear separation of responsibility and authority, we have formal and informal relationships among project and functional personnel that lead to a distribution of responsibility and authority.

Project-Matrix Organization. Project organizations are concerned with planning, coordination, and control of complex projects of an organization; projects require many activities proceeding both serially and simultaneously toward an *ultimate goal,* and continuous and intricate interaction among many different functional personnel of an organization. The goals for each project ordinarily include profit, either short term (i.e., the project only) or long term (i.e., future business). The goals are met by achieving agreed upon performance with respect to cost, quality, functionality, and time variables. Therefore, cost, quality, functionality, and schedule are ordinarily among the key success variables for a project. Upon completion of all activities the life of the project ceases and the organization is *dissolved.*

Figure 67.12 provides an example of a typical organization for the management of complex projects with the matrix structure superimposed. The formal organization chart, however, does not illustrate the dual dimensions of management embodied within the matrix. These dual dimensions were illustrated in Fig. 67.11.

The role of the project manager and his or her staff is to carry out the planning and control process of a project *without* getting involved in the actual direction of functional work. Under the matrix structure, the project manager plays the role of a planner, coordinator, and controller whose chief concern and responsibility is to produce a project on time, within cost constraints, and in accordance with quality and performance specifications. The project manager of a medium- to large-scale project generally has a small staff assigned directly to him or her. The staff is charged with responsibility for the planning, coordination, and control of subdivisions of the project.

Authority for directing functional work and accomplishing the technical tasks of a project lies with the managers and individual contributors of the various functions. The individual contributors to a project do not ordinarily report to the project manager, but rather to their respective functional managers. Often, in the matrix organization structure, a connection is made between the project office and functional groups by the appointment of assistant project managers, project engineers, or project leaders for each of the major functions, and these assistants report either directly or indirectly to the project manager.

67.3.2 Project Teams: A Case Study

Improving productivity and quality can follow many courses of action, but all improvement methods are forms of continuous adaptation. The focus of management in this example is upon the acceleration of significant quality and productivity improvements through the use of informal, ad hoc teams within a complex project.

The Project

The AN/BSY-1 Combat and Control System is produced for the Los Angeles class nuclear submarines used by the U.S. Navy. As a major subcontractor, Hughes Aircraft was responsible for a number of complex subsystems produced by other major contractors and integrated at their facilities.

As a major subcontractor to the AN/BSY-1 Project, Division 1E of the Ground Systems Group of Hughes Aircraft Company supplied a large suite of command and control electronics. The system, when installed on the Navy vessel, provided a complex stream of data to the operators in charge. The electronic systems consist of multiple electronic control units, modules, appropriate interfaces, and housings. The control units are built from an array of electronic modules that are contained in drawers. Without timely completion of the appropriate mix and quantity of modules, the system could not be assembled and the total project would have suffered a delay.

The Project-Control System. The project was organized in a formal matrix structure. A project manager, along with manufacturing, quality, engineering, and material managers, all reported to the division manager. Formal project reviews were held with each functional area, reporting on its specific activities. Formal cost information, material requisitions, and quality systems were used to control operational status. Project plans, including cost, schedule, and quality, had been developed early in the project and were tracked at project reviews.

The project struggled to meet objectives over a period of months. During this time, each manager had a good rationale for his particular problems. In general, the individual managers felt they were performing their specific functions adequately. A growing amount of internal pressure was being placed upon each functional group at project reviews to meet current goals. Interpersonal communications were sporadic, but largely directed at identifying causes for schedule slippage.

The essential formal control system was in place, but informal "teamwork" was not being utilized by management to support the project purpose or to adapt to current needs. Productivity and quality did not meet management expectations.

After a lengthy and complicated design phase, production was initiated in early 1986; product flow progressed slowly. In July 1987, module production stopped due to yield and productivity issues. Multiple internal inspections were implemented and 7700 complex module assemblies were being reviewed at various stages of assembly. The project was in danger of falling behind schedule. These problems were all internal issues. Management was determined to ship excellent hardware on time and on schedule to the customer.

The challenge facing management was to improve quality, increase productivity, and regain schedule integrity concurrently and quickly. The situation was not unlike many that are found in U.S. industry today. The customer desired strict compliance to contract specification as well as improved timeliness of delivery and increased productivity. The challenge was to dramatically improve all three parameters.

The Project-Improvement Effort. During the following 18-month period, no major capital investments were made, management personnel remained basically stable, customer pressure was intense but steady, and no major changes in the functional scope of the product were initiated. The

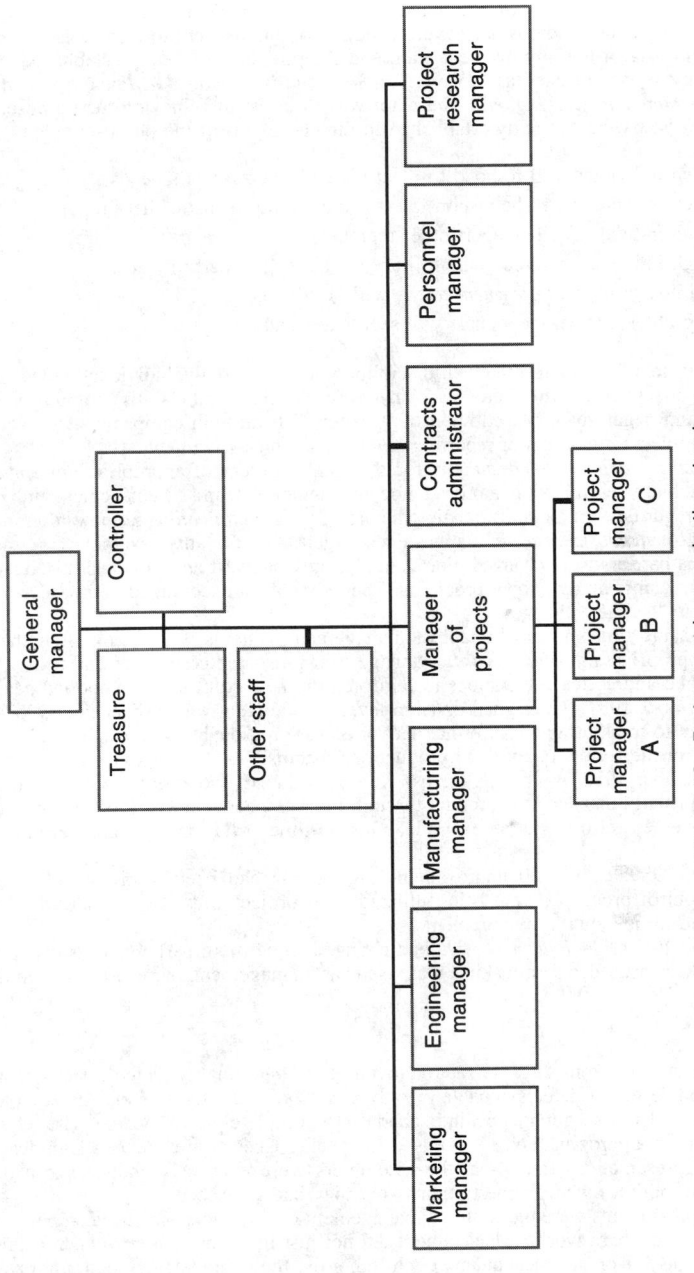

Fig. 67.12 Project matrix organization structure.

major change was an extensive effort to improve productivity and quality, which led to the development of appropriate informal management systems.

The improvement effort really began on July 10, 1987. A multidisciplinary ad hoc team chaired by the division manager was formed. Membership consisted of various individuals from all levels of both organizations. The team concept was supported both by the customer and by Hughes management, which had agreed to "work together" to meet the challenge that faced both organizations. As a result, both Hughes and customer were called upon, as needed, to resolve issues immediately. At first, the *purpose* of the improvement team and the project was clearly established. Specifically, the project purpose was restated as "to *produce high quality products at low cost, on time to meet the customer's complete satisfaction.*" The team was to assist in achieving the purpose. The following operative *values* were explicitly established in direct support of the purpose:

1. Each individual will respond to meet his or her customer's needs.
2. Each individual will be responsible for the quality of his or her output.
3. Each individual will support other members.
4. Each individual will continue to improve his or her performance.
5. Ad hoc *project improvement teams* will be utilized.
6. If problems arise, individuals will seek expert advice.

Over the next few months, the multidisciplinary team met daily. Individuals began surfacing needs, quality reports were clarified, process plans were clarified, and "skills" training was instituted. A great demand began to be placed upon the "experts" from both companies who were actually able to solve problems regardless of reporting level or subunit assignment.

Management demonstrated support for the establishment of appropriate *interpersonal relationships* by actually establishing more ad hoc management teams. Management utilized appropriate personnel regardless of organizational structure. By *communicating* the new dominant values, the functioning of *networks at all levels* was encouraged. Additionally, past values and norms that were perceived as barriers were changed. For example, previously, if an individual raised a concern, it was considered "complaining" about problems. The new values encouraged everyone to not only "complain" but to "suggest changes."

At the same time, informal adaptive management systems were being strengthened by implementing appropriate *informal rewards*. Management presented certificates and team awards and held informal cake-and-coffee celebrations to reinforce the new culture that supported adaptive controls.

In December 1987, the original division level ad hoc team was disbanded and local management teams began to track progress. Informal networks were reasonably well established, and changes had been made in the formal quality and production systems.

In September 1988, work cells (clans) were introduced. Workers were cross-training each other as they informally managed the product through their respective areas. Members began searching for new improvement efforts. As an example, "just-in-time" (JIT) production was introduced into the work flow.

By July 1989, over 1000 improvement suggestions had been received and implemented. The informal control processes were being utilized. The project was able to adapt quickly to schedule changes and to new quality requirements.

Additionally, the *management style* had changed to be more participative in support of the team concept. Consensus decision-making was used, as management more often assumed a *facilitative role*.

Results

Within the period from July 1987 to November 1989, quality defects were reduced from 23 defects/module to .04 defects/module or only *two thousandth* the previous defect total. The project schedule went from six months behind schedule to completely on schedule. The module cycle time decreased from approximately 39 weeks to 5 weeks. Costs were reduced significantly. The local customer representatives stated that the assemblers were extremely enthusiastic and conscientious and that the modules were the best quality that they had *ever* seen.

The improvements were the result of management cooperation, *increased teamwork,* a clear sense of purpose, and "hard work." Management did not just implement improvement teams or new techniques, such as JIT or statistical quality control, nor did they implement a set of improvement projects. Instead, management established the appropriate informal subsystems addressing issues in all five subsystems represented in Fig. 67.2 plus defining a clear purpose, complete with supporting values and beliefs. The result was tremendous improvements in both quality and productivity. Virtually no changes were made in the formal management systems, including the matrix organization structure.

Figure 67.13 displays the changes made in the informal control system of the division to foster the project improvement effort represented in this case.

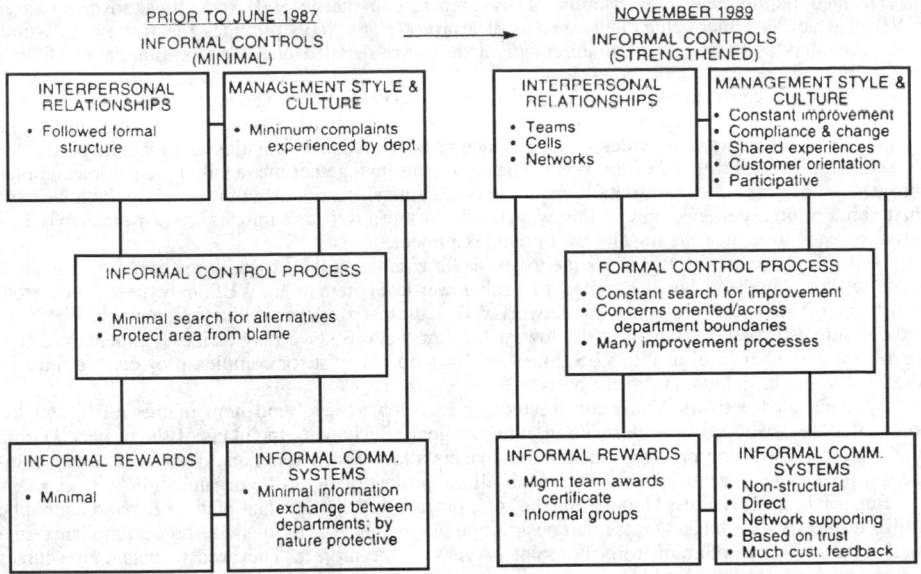

Fig. 67.13   BSY-1 Informal systems: before and during improvement effort.

67.4   SPECIFIC ISSUES IN THE PROJECT-CONTROL PROCESS

67.4.1   Project-Planning and Control Process: Overview

Figure 67.14 summarizes the project-planning and control process. The process provides for planning according to goals and requirements and control by exception. The process is initiated by establishing detailed project requirements, and in meeting them, we simultaneously achieve the goals of a project.

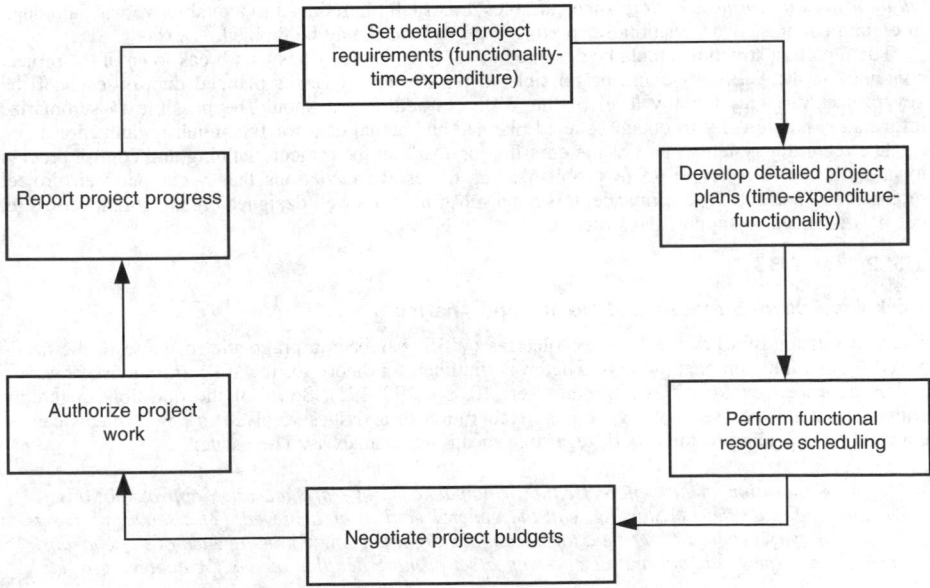

Fig. 67.14   Summary of project-control process.

Detailed requirements are established by preparing a means–end work breakdown structure (WBS), which is a hierarchical subdivision of a project. The WBS provides the framework within which we may establish project requirements and prepare detailed plans for the time, expenditures, and performance variables of the project.

Once all end items (purposes and subpurposes) of the project have been established, the next step of the process requires logical, consistent, and coordinated plans to achieve the end items of the project. Network analysis provides a tool for identifying functional activities that must be performed to achieve a lowest-level end of the WBS. That is, in putting together the detailed plans for a complex project, we begin at a level of detail where we can identify functional activities with which we have had some prior experience and in this way break up the novel task into its known elements. This process tends to reduce the novelty of a complex project.

Network plans and the WBS provide the basis for estimating the expenditures of a project. Labor, material, and overhead costs assigned to each lower-level item of a WBS may be derived from estimates of activities contained on networks. By summarizing vertically (i.e., up the WBS) all expenditure estimates beginning at the lowest level of the WBS, we may arrive at expenditure estimates for any other level of the WBS. Standard costs do not exist for complex projects and must be established on an individual project basis.

From detailed network plans, constructed for each lowest-level end item of the WBS, detailed schedules are developed in *each* function, with the goal of achieving the plans of the project. During the resource-scheduling process, functional managers allocate their functional resources among competing projects to maximize compliance with all the project plans of the organization.

Financial planning must be done for the total project, yet the financial plan so derived cannot be utilized directly as a budget, since its construction assumes that activities will be accomplished in a manner considered optimum from the point of view of the project. The need to balance resources among projects, observe institutional rules, and react to unexpected project change often requires us to accept less than optimum resource allocations. This means that *actual* resource allocation or scheduling decisions can be made only for those activities to be accomplished in a relatively short period of time, since long-range schedules would depend on long-range demands placed on a function by all projects, and these demands cannot be predicted with accuracy.

Once these allocation decisions are made, a block of work represented on the network, derived from the WBS, is authorized. The project-control process then turns to activities of control. Project office personnel are concerned with controlling actual performance to achieve a balance among expenditure, time, functionality, and quality variables of a project. Since we are required to achieve balance among these variables, our project-control system must contain and process progress information on each of these variables.

It is necessary, therefore, both to calculate variances for expenditure, time, and performance goals and to derive measures of combined variable performance whenever possible. Techniques of variance analysis are available for combining the time and cost variables into planned and actual measures of *value of work performed.* Performance variables are usually introduced in a qualitative way, although in certain circumstances, quantitative performance variances may be defined.

The reporting structure should be designed to conform to the means–end breakdown of the project contained in the WBS. It should be possible to retrieve actual versus planned data on each of the key project variables for any level of the WBS. In addition, it should be possible to summarize information horizontally to obtain detailed planned and actual data for functional organizations.

The reporting system is part of the contribution made in the project-planning and control process toward directing project effort to problem areas to resolve deviations that occur between project requirements and actual performance. It is not a substitute for a well-designed organizational structure, but it is intended to support the structure.

67.4.2 The WBS

Work Breakdown Structure and Means–End Analysis

The construction of work breakdown structures (WBS) has been a pragmatic response to the needs posed by new and complex projects. The broad outline of a theory for the WBS does exist, however, and is described by March and Simon (Ref. 10, pp. 190–191). Some of the questions regarding construction and the use of WBS's for the elaboration of activities involved in new projects can be clarified by appealing to their work regarding means–ends analysis. They state:

> In the elaboration of new projects, the principal technique of successive approximations is means–end analysis: (1) starting with the general goal to be achieved, (2) discovering a set of means, very generally specified, for accomplishing this goal, (3) taking each of these means, in turn, as a new subgoal and discovering a set of more detailed means for achieving it, etc.

How much detail should the WBS contain? Again referring to March and Simon (Ref. 10, p. 191):

It proceeds until it reaches a level of concreteness where known, existing projects can be employed to carry out the remaining detail. Hence the process connects new general purposes with an appropriate subset of existing repertory of generalized means. When the new goal lies in a relatively novel area, this process may have to go quite far before it comes into contact with that which is already known and programmed; when the goal is of a familiar kind, only a few levels need to be constructed of the hierarchy before it can be fitted into available programmed sequences.

The objective of the WBS, therefore, is to take innovative output requirements of a complex project and proceed through a hierarchical subdivision of the project down to a level of detail at which groups of familiar activities can be identified. Familiar activities are those for which the functional organizations have had some experience. What is familiar to one organization may not be familiar to another, depending on experience.

Project complexity is an organization-dependent variable, and the same project may require different levels of detail from different organizations. The *primary determinant* of complexity is organization-relevant technology. A project that is of relatively high technology for an organization requires more detailed analysis via the WBS than a project that is of relatively low technology. A project can be complex, however, even if the technology is low relative to what the organization is accustomed to; that is, it may be ill-structured, with many design options available, organizationally or interorganizationally interdependent, with many interactions required among functional disciplines, or very large. Therefore, the degree of detail found in a WBS for a given project depends on the relative level of technology required, the number of design options available, the interdependence of functional activities, and its size.

WBS and Project Management

Figure 67.15 provides an example of a WBS for a construction project. The objective of the project is to construct a television transmission tower and an associated building for housing television transmission equipment.* As a contractor for the project, we are given specifications for both the tower and building by our customer. We set out to prepare a proposal for this task that will be evaluated by the management of the television station.

As we see from the WBS, the main purpose or end item of the project (i.e., level 0 of the WBS) is provision of the TV transmission system. The primary means for providing this system are shown in level 1 of the WBS. That is, to complete the system we must provide the TV tower, the equipment building, the cable connecting the two, and a service road between the building and tower. These level 1 items are means for constructing the TV transmission system, but are also ends unto themselves for the level 2 items. For example, in order to construct the tower, we must prepare the site, erect the structure, and install the electrical system. These level 2 items are means for accomplishing level 1 ends, which themselves were means for achieving the level 0 end.

Similarly, to provide an equipment building, we must prepare the site, provide a structure, and install a fuel tank. These level 2 WBS ends are also means for constructing the structure of the equipment building. Furthermore, to provide a structure for the equipment building, we must provide a basement, main floor, roof, and interior. These level 3 WBS items are means for accomplishing the building, but also ends unto themselves.

For each level 1 WBS item, we proceed to elaborate means and ends until we arrive at means that are very familiar tasks, at which point we cease factoring the project into more detailed means. The amount of factoring done on a given end item and project therefore depends on the relative novelty associated with the project. Note that for the service road, we proceed *immediately* to final means (i.e., lay the base and grade) to achieve that end. Those two means are familiar activities to the organization and the factoring thus stops for that end item at level 1. Likewise, for the level 1 WBS item "underground cable," we simply insert one activity ("install the cable") and that ends the means–end chain for the cable.

Once we reach familiar means, we identify these as activities rather than ends, simply because they are final means, and, although our detailed planning may separate each of these activities into two or more tasks, there is no utility in identifying more detailed means. All other WBS elements, except at level 0, serve as both means and ends. Our detailed network planning begins at the level

---

*This example is based upon the case study "Peterson General Contractors," reproduced in R. A. Johnson, F. E. Kast, and J. E. Rosenweig, *The Theory and Management of Systems,* 3rd ed., McGraw-Hill, New York, pp. 268–273, 1973 and is included here by permission of the publisher. The case was written by Albert N. Schreiber and first appeared in A. N. Schreiber et al., *Cases in Manufacturing Management,* McGraw-Hill, New York, 1965, pp. 262–268.

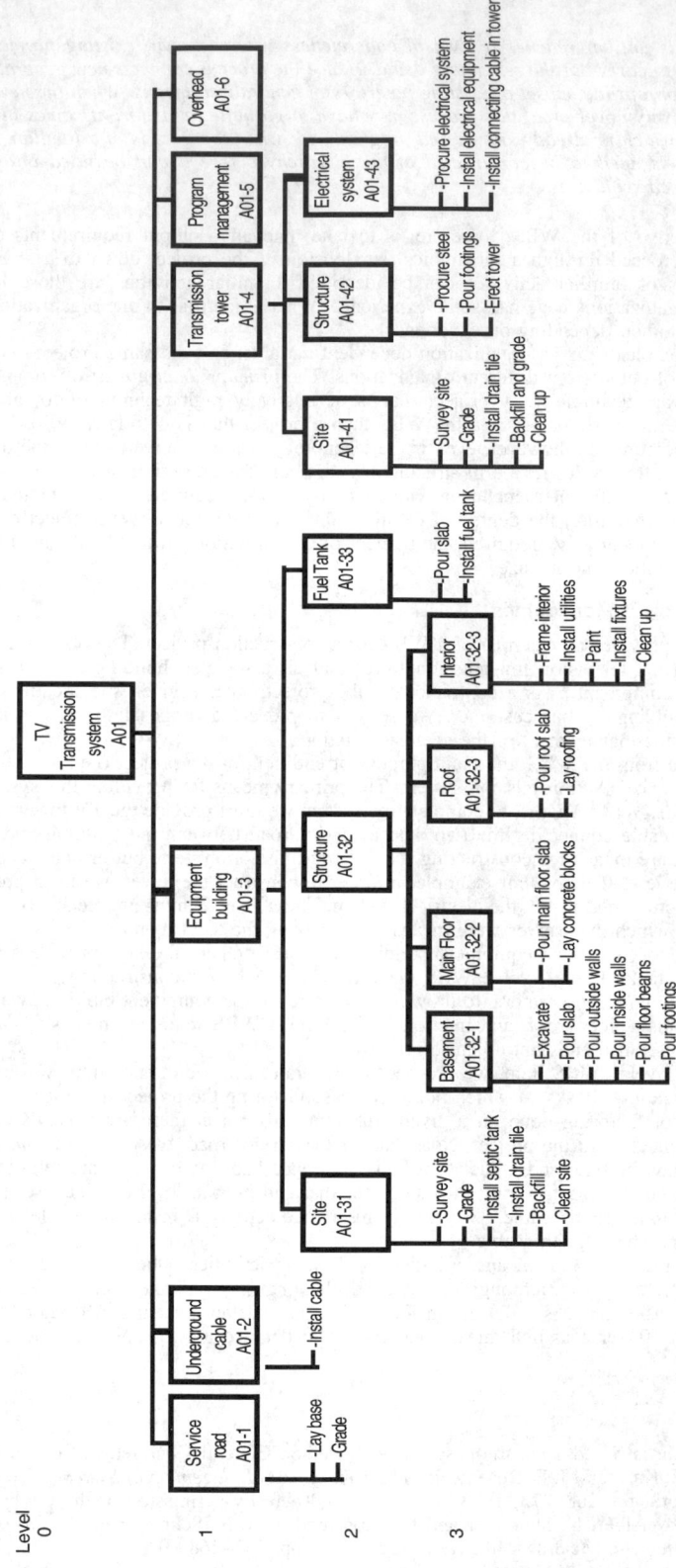

Fig. 67.15 WBS for a TV transmission system.

of the WBS, where these final means or activities are identified. Network planning thus begins at different levels of the WBS for various level 1 ends. For example, network planning will begin at level 1 for the service road, but at level 2 for the equipment building.

The elements of Fig. 67.15 that remain to be explained are the level 1 ends *project management* and *overhead*. Strictly speaking, we define our projects in terms of identifiable ends or outputs until we get down to the very last level, at which point we identify functions or activities; these latter activities are inputs rather than identifiable outputs. Because the input of project management is primarily that of planning, decision-making, and control, it cannot be traced directly to any one WBS item, but rather must be assigned directly to the project itself. We accomplish this by making it a level 1 item so as to include within the WBS framework all the resource costs associated with the project. Similarly, when deriving the WBS, we initially trace only those means that are directly related to each end item. Yet we also want the WBS to provide an accounting framework for accumulating total project costs. Therefore, we assign all indirect resources to the level 1 item called *overhead*.

Once the WBS is defined, we can assign an account-code structure to it. The purpose of the account code is to provide unique identification for each end item of the WBS to serve as the basis for the cost accumulation and reporting system of the project.

Any combination of alphabetical and numerical characters may be used; the only real requirement for the identification system is that each end item contain in its identification the account letter and number of its parent. For example, the identification assigned to the TV transmission system is A01 at level 0 of the WBS. The equipment building is identified as A01-3, indicating that its parent is the TV transmission system and that it is the third level 1 end item. The building structure is identified as A01-32, indicating that it is part of the equipment building (A01-3), which itself is a part of the TV transmission system (A01). The account-code structure proceeds down to the last end item of the WBS. Functional activities below lowest level ends of the WBS are assigned resource code numbers or letters for purposes of estimating and reporting financial expenditures by function.

67.4.3 Network Plans—Time

An Application of Network Analysis: TV Transmission System Project

We illustrate the process of network planning by constructing a network for the TV transmission system whose work breakdown structure was illustrated in Fig. 67.15. We use that WBS as the basis for network construction. From the WBS, we observe that most of the tasks to be performed are associated with either the equipment building or the transmission tower. Therefore, we shall draw one network for the building, one for the tower, and one for the service road.

The network for the building is given in Fig. 67.16. To draw the network plan for the building, it is necessary to identify the interrelationships among the lowest-level means on the WBS. We have assumed a set of interrelationships and have drawn the network accordingly. In this simple example, it is quite easy, although by no means trivial, to define optimum relationships among activities from the WBS. On more complex projects, the interrelationships must be ascertained by the planner from specialists in each functional discipline.

Notice the dashed lines that appear on this network for activities 5–6, 10–11, and 12–13. These dashed lines are called *dummy activities* and have two purposes in network analysis. First, the dummy activity is used to achieve unique numbering between parallel activities that originate at a common burst point and end at a common node. Dummy activity 5–6 is inserted for that reason. If it were

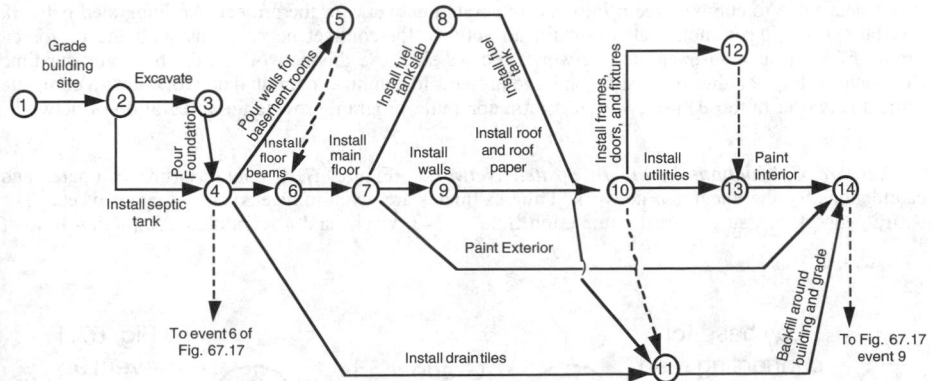

Fig. 67.16 Network for TV transmission building.

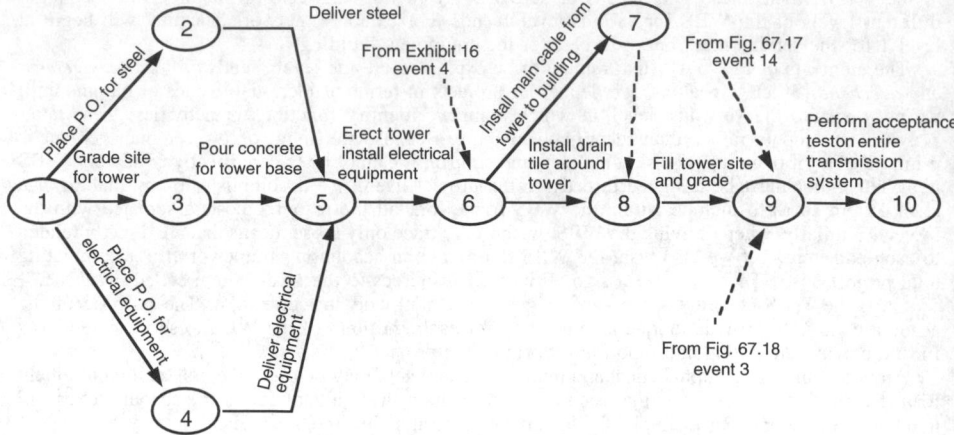

Fig. 67.17 Network for TV transmission tower.

not present, we would have two activities numbered identically (i.e., 4–6), thereby violating the uniqueness requirement. Second, the dummy activity is used to show a dependent relationship between activities where this dependency *does not consume resources*. For example, before we can fill in the foundation and grade it, the roof must be on the building and the drain tiles must be installed. Activity 10–11 depicts the dependent relationship between the fill work on the building (activity 11–14) and the installation of the roof (activity 9–10). Yet no resources are consumed by this dummy relationship. Dummy activities should be kept to a minimum in network construction, but often they are essential.

We turn now to the network for the transmission tower shown in Fig. 67.17. The transmission network is quite straightforward, with three notable exceptions. First, the *dashed lines* that flow into events 6 and 9 depict the interrelationships among the *individual networks* of the TV transmission system. Installation of the connecting cable between the tower and the building (activity 6–7) cannot begin until the tower is up (activity 5–6 of Fig. 67.17) and until the foundation of the *building* is poured (activity 3–4 of Fig. 67.16). Therefore, the dummy activity flowing into event 6 of Fig. 67.17 starts from event 4 of Fig. 67.16 and shows this physical dependency.

Second, final acceptance testing of the entire transmission system is shown on Fig. 67.17. The start of acceptance testing not only requires the tower to be complete, it also requires the completion of the building and the service road. Therefore, we have two dummy activities showing these dependencies, one from Fig. 67.16 and the other from Fig. 67.18. Figure 67.18 contains the two serial activities involved in laying the service road.

To summarize, we have constructed networks for each of the level 1 ends of the WBS. Because the connecting cable is a single simple activity, we have included it on Fig. 67.17 along with the tower. Moreover, the connecting road is a simple serial task, as shown in Fig. 67.18.

The networks constructed for this project are very simple, but realistic. Since they are quite simple, it is manageable to combine them into one integrated network for the project. An integrated network for the entire project should also contain an activity for contract negotiations with the customer. Figure 67.19 is such an integrated network, and we shall use this network as the basis for our time calculations. It is not always possible on large projects to combine individual networks into a complete project network. In those cases, we must let a computer program provide the integration of networks for us.

Network Calculations in the Integrated Network.   Figure 67.19 contains time estimates and calculations for the integrated network. Time estimates are given in weeks and tenths of weeks. The entire network has an expected completion time of 24.0 weeks and a scheduled completion time of

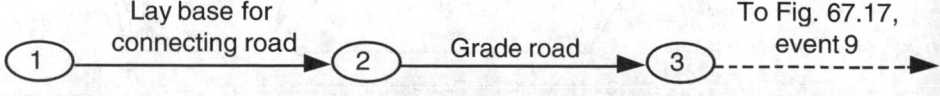

Fig. 67.18 Activities for connecting road.

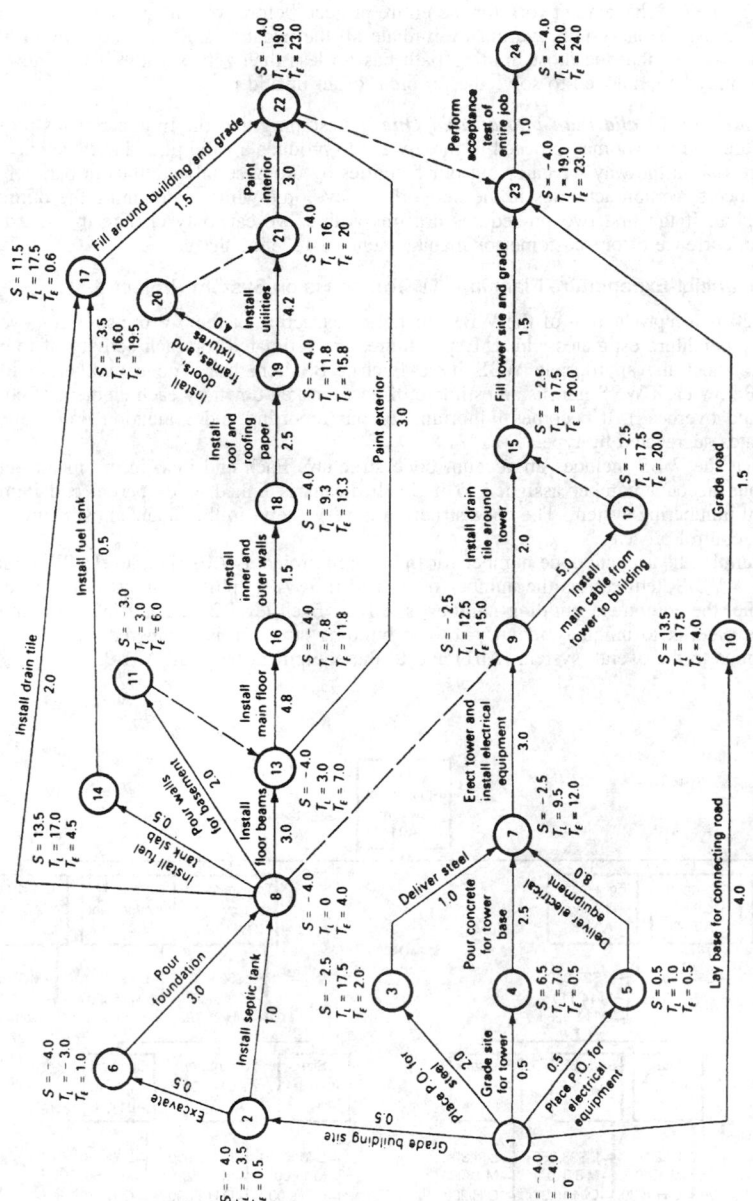

Fig. 67.19 Integrated network for TV transmission system.

20.0 weeks. The critical path has slack of −4.0 weeks and consists of activities 1–2, 2–6, 6–8, 8–13, 13–16, 16–18, 18–19, 19–21, 21–22, 22–23, and 23–24. Essentially, the critical path contains activities pertaining to the equipment building. Activity 8–11, another activity pertaining to the equipment building, has slack of −3.0 weeks and is therefore the second-most critical path.

The electrical tower is not in much better shape, either. It contains the third-most critical path of −2.5 weeks and includes activities 1–3, 3–7, 7–9, 9–12, and 12–15. You should trace through all other slack paths on the network before proceeding further.

Although we now have a network for the entire project, before we may consider this a valid plan for the TV transmission system, we must eliminate all the negative slack on the network in a non-arbitrary manner, so that the most limiting path has no less than zero slack. We turn now to alternatives that may be employed to solve the problem of an invalid plan.

Translating an Invalid Plan into a Valid One. Assuming that the time estimates provided on a network are correct, we may proceed in three ways to produce a valid plan. First, we may consider taking more risk in the way we carry out our activities by doing serial activities in parallel. Second, we may expedite certain activities in the network to save time while maintaining the optimum performance plan. If the first two procedures are impossible, we can only change the schedule date, with the concurrence of our customer or management, or redefine the project.

67.4.4 Financial-Expenditure Planning: TV Transmission System Project

Figure 67.20 is a reproduction of the WBS for the construction of the TV transmission system, but now with expenditure estimates added. Expenditures are estimated for each activity of the network and placed under the appropriate WBS item. Each WBS item has a code number to identify it uniquely. Below each WBS item is an estimate of cost, broken down by each element of cost (labor, material, and overhead). It becomes important for our reporting and evaluation procedure to have cost estimates segregated by type.

Note that the WBS includes an account code structure. Each end item in the means–end chain has a unique account number assigned to it. Each means is linked to its parent end item by this hierarchical numbering system. The code structure is very useful in the financial estimation phase of the project control process.

For example, the account code number for the overall project is A01 (i.e., level "0" of the WBS). Each level 1 WBS item carries the number of its end (i.e., A01) plus a unique suffix to identify it. For example, the equipment building number is A01-3. Each level 2 item carries the number of its parent plus a suffix to uniquely identify it. The building structure is numbered A01-3-2 to signify that it belongs to the overall system (A01) and to the equipment building A01-3.

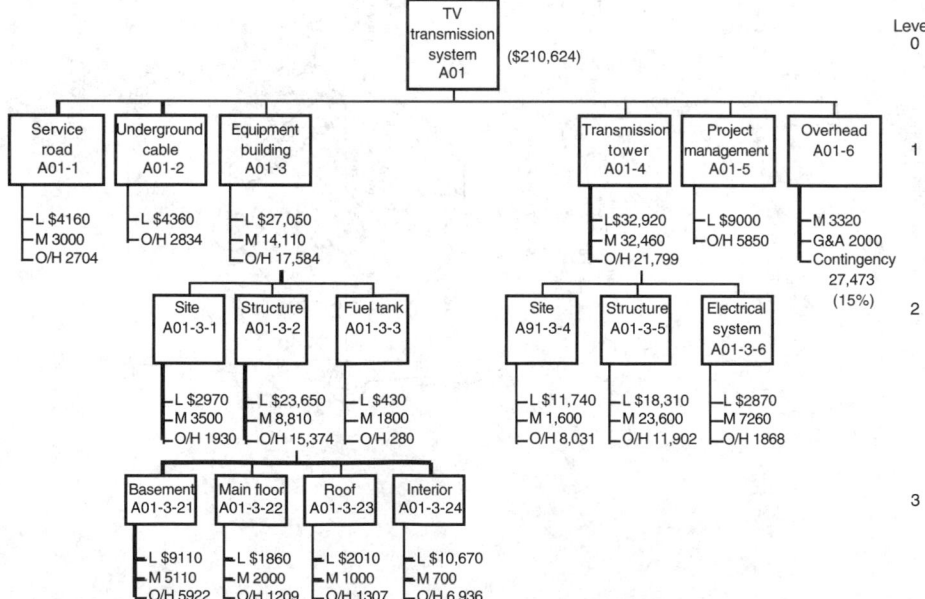

Fig. 67.20   WBS for TV transmission system.

The WBS of Exhibit 20 has three levels. To estimate standard costs for each end item of the WBS, we estimate each of the lowest-level end items and accumulate the standard costs up the WBS.

From the network of the TV system, we estimate direct labor and material cost for each of the lowest-level end items of the WBS. Since this is a relatively small project, each lowest-level end item is equal to one work package, and we develop *planned value of work* estimates for each of the lowest-level end items. We identify direct labor and direct material costs separately under each end item.

The standard overhead rate for this organization is 65% of labor costs. Labor cost is therefore the activity criterion. The organization has determined that indirect expenditures vary more directly with labor costs than with any other input or output variable. The overhead rate is thus computed by estimating overhead expenditures over the accounting period (normally a year) and dividing these expenditures by the expected or normal volume of labor costs for that same period.

Once we have arrived at the overhead rate, we simply apply it at each lowest-level end item to the standard assigned to the variable that serves as the activity criterion. This gives us the standard overhead charge for that end item. We then sum the three elements of cost to arrive at standard costs for an end item.

Since we can relate a lowest-level end item to the network, we shall be in a position in the reporting phase to collect actual costs for work performed and compare them to the planned value of work performed. Finally, we sum standard costs for each end item to its parent to find successively higher levels of project costs until we arrive at the standard cost for the entire system (i.e., A01 on the WBS).

Note that there are costs for project management and certain other overhead items that we choose not to allocate to project end items, instead identifying these separately at level 1 of the WBS. Of course, they too become part of our total estimated costs for the project. The estimated costs for the project may also be displayed by month, as in Fig. 67.21. Figure 67.21 becomes a control document. It does not contain profit or contingency, thus displaying a total cost $44,579 lower than the costs appearing on the WBS in Fig. 67.20.

The work package, which is a series of related activities, because it connects the WBS, the network, and the cost-accounting system for a meaningful segment of work, is the basic instrument for integrating the time and cost variables of a project. It is the lowest level of detail at which it is feasible to devise a combined measure of performance for time and cost.

The combined measure of performance is ordinarily called *the planned value of work* and it is arrived at simply by estimating the *budgeted value* of work represented on the network for each work package. Each work package thus contains estimates of its planned value, so that any major part of the work package is accorded a corresponding planned value.

Once work progresses, we collect data on actual expenditures and progress and assign *actual cost for work actually accomplished* for each work package. We then compare the *planned value for work actually accomplished* with the *actual cost for work accomplished* and compute the variance. The variance thus represents a measure of cost performance versus plan for the work actually accomplished. It integrates expenditures with schedule performance, thus achieving the joint measure of performance we seek. We shall discuss this integrated reporting measure further later in this chapter.

Months (Days) Worked

Element of Cost	1 (1-22)	2 (22-44)	3 (44-66)	4 (66-88)	5 (88-110)	6 (110-132)	7 (132-154)	8 (154-176)	Total
Labor	$17,200	$ 8,570	$5,220	$2,660	$15,100	$ 3,600	$ 14,110	$ 2,300	$ 68,760
Material Expenditures	30,860	21,730							$52,590
Applied O/H (65% of labor)	11,180	5,571	3,393	1,729	9,815	2,340	9,172	1,495	$44,695
Project Total Cost	59,240	35,871	8,613	4,389	24,915	5,940	23,282	3,795	$166,045
Cumulative Total Cost	$59,240	$95,111	$103,724	$108,113	$133,028	$138,968	$162,250	$166,045	

Fig. 67.21 Financial expenditure plan according to expected completion dates.

67.4.5 Scheduling Resources

Project plans represented by networks and financial plans provide functional management with the requirements, resources, and priorities for their function on each of the organization's projects. Although network plans provide a possible schedule for accomplishing the work, this schedule is not always practical or feasible when all other requirements placed on the function are considered. There are six specific requirements excluded during the planning process that must be considered during the resource allocation process. They are as follows:

1. Sufficient resources to perform each activity in an optimum manner is assumed to be available when formulating and optimizing plans. Limited availability of resources and the competition among projects for the same resources must be taken into account during the resource-allocation process.

2. The pattern of resource demands from all of the project plans must be considered not only in the light of resources available but also in terms of the distribution of demand placed on resources over time. Functional management cannot be expected to increase and reduce functional resource continuously in light of the fluctuating demands of each project. Functional resources levels are determined based on long-term organizational demands and their use must be relatively even from one period to the next.

3. Common facilities (e.g., computer time and testing equipment) are often required simultaneously by activities of the same project or by activities of different projects. The allocation process must resolve these conflicts.

4. Cash flow requirements of the projects are not always feasible for the organization, and these limitations enter into the allocation function.

5. State work laws and regulations must be observed in allocation decisions when overtime is being considered.

6. The nature of the contract negotiated between contractor and customer with regard to the relative value of various projects to the organization, as well as the long-term objectives of the organization, affect the relative priority that should be accorded various projects by the organization. This is another consideration of the resource-allocation process.

Not only must we recognize scheduling as a distinct activity in the project-control process separate from, yet related to, planning, but we must also establish different time horizons for these two activities. Project planning must be carried out for the entire duration of the project. Scheduling, on the other hand, ordinarily may be done profitably only on a short-term time horizon.

Scheduling requires commitment of resources on the part of functional management to specific tasks of the many projects of the organization. As the network relationships indicate, however, activities of one functional organization are dependent on the completion of activities of other functional organizations. Because of the dynamic, constantly changing nature of complex projects, we cannot expect network relationships and time estimates to be very precise. Expected start and completion times of activities become more tenuous the longer the elapsed time from the present. Therefore, functional organizations cannot establish realistic long-term schedules for carrying out the work of multiple projects. It is usually futile to allocate resources to specific jobs unless they are to be performed in the near term. More accurate scheduling can be done for these near-term activities, since most of the activities that limit their start are either in progress or complete.

Start dates for activities that are scheduled by the functional organizations must find their way back to appropriate project plans. Scheduled start dates are *superimposed* on network calculations, and they supersede expected start dates in calculation of the network so long as they are equal to or greater than expected start dates. Scheduled start dates that are earlier than expected start dates are invalid. Project office personnel must check the consistency of functional schedules and approve their implications. The portion of a project plan that has been scheduled is called a *scheduled plan.*

Although distant activities cannot be scheduled, it is important to preserve a valid plan for distant work, since the time estimates and interrelationships of the entire plan determine the time requirements (required dates) of work that can be scheduled.

To summarize this section, we may say that resource allocation or scheduling is a function with different purposes than planning. A network plan cannot ordinarily be used as a schedule for a project, yet it must serve as the basis for the schedule. Moreover, once activities are scheduled, these data must be incorporated into network plans. Thus, there is communication between these two important functions. If the plan alone is used as a schedule for performing the work, with slack used without considering other activities and competing projects, the ability to optimize performance in the organization is restricted and the value of the project-control system is lessened.

The resource-allocation process consists of three distinct but interrelated tasks: *resource loading, resource leveling,* and *constrained resource scheduling.* Resource loading is concerned with deriving the total demands of all projects placed on the resources of a function during a specified period of

Projects/ Functions	1	2	3	4	5	Total Functional Hours
A	20	40	8	5	0	73
B	30	30	15	0	0	75
C	25	25	20	8	20	98
D	10	15	20	14	30	89
E	40	10	12	10	0	72
Total Project Hours	125	120	75	37	50	407

Fig. 67.22 Resource loading in matrix format.

time. Resource leveling attempts to "smooth out" the demands to eliminate major peaks and troughs. Constrained resource scheduling is concerned with achieving all demands of the projects of an organization within the resource constraints of the function at minimum disruption to the plans of each of the organization's projects.

Resource Loading

To understand the resource-loading process, it is convenient to view the problem in matrix form. The various projects of an organization place demands on resources during a particular period of time, and the functional organizations supply these resources. A matrix illustrating this process appears in Fig. 67.22. The matrix represents the total demands placed on each of five functions by each project for a 10-week period of time. These demands, however, are not time-phased in this Figure. The resource demands in the matrix are taken from the work packages that are expected to be performed during the scheduling period.

Information on demands placed on each functional group during the scheduling horizon is only part of the information required in the resource-loading process. Slack information from each of the project plans is also required.

Figure 67.23 is an example of a report for one function, engineering, for one project for a 10-week period. This information is derived from project plans. The activities represented on the report have start dates, expected completion dates, required dates, and slack calculations.

Loading information from work packages is combined with calculations of slack from project plans into the resource-loading report for one functional organization. Figure 67.24 presents an example of a time-phased loading plan based on expected start and completion times for each of the activities. Where positive or negative slack exists, it is indicated by an extension of each bar to its right (for positive slack) or left (for negative slack—none shown on Fig. 67.24). Within each bar we have placed the number of persons per week required to achieve each task and have summed the total demands placed on the function vertically by week. The row on the bottom of the chart therefore contains an estimate of the total demands in terms of person-weeks of effort placed on the function of design engineering by all projects. Fig. 67.25 presents the loading plan graphically.

From Fig. 67.25, we note that there is an uneven distribution of demand for design resources over the 10-week period, with very high demands occurring in weeks 6–7 and 7–8. Even if resources are in good supply in the design organization, it is usually undesirable to have these large variations in

Function: Engineering Project: XXX Responsibility: John Smith-1997						
Preceding/Succeeding Event Numbers	Activity	Time Estimate	Start Date	Expected Completion Date	Required Completion Date	Slack
001-002	Prepare Detail Design—Component 101	2.0	01/01	01/15	01/29	2.0
011-012	Prepare Detail Design— Component 105	4.0	01/15	02/15	02/15	0.0
021-022	Prepare Detail Design— Component 208	3.0	02/01	02/22	02/15	-1.0
031-032	Prepare Detail Design— Component 304	1.5	02/01	02/11	02/04	-1.0
051-052	Prepare Detail Design— Component 508	2.5	02/15	03/03	03/01	-0.4

Fig. 67.23 Planned activities of functional organization.

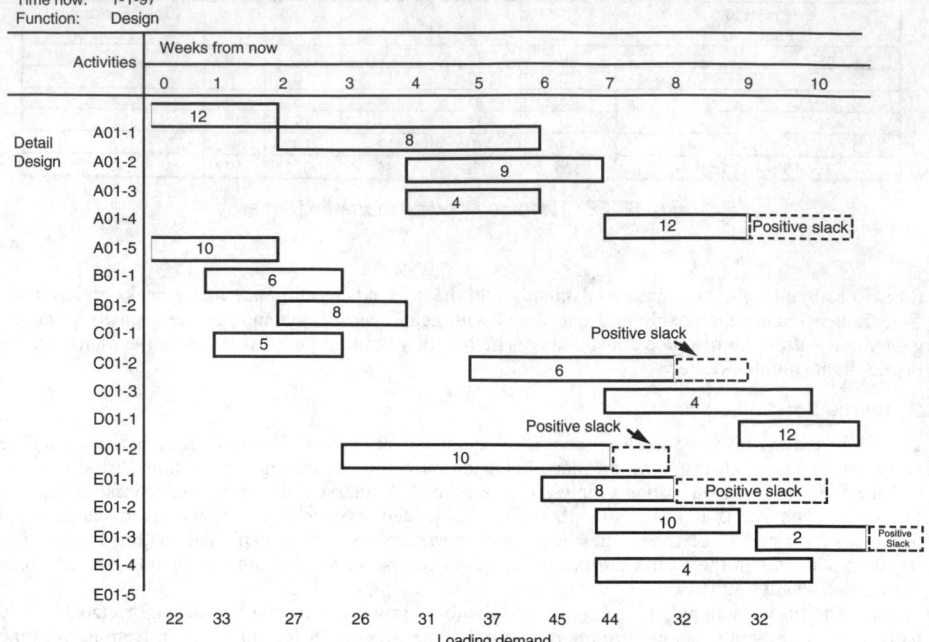

Fig. 67.24 Time-phased human resource loading plan.

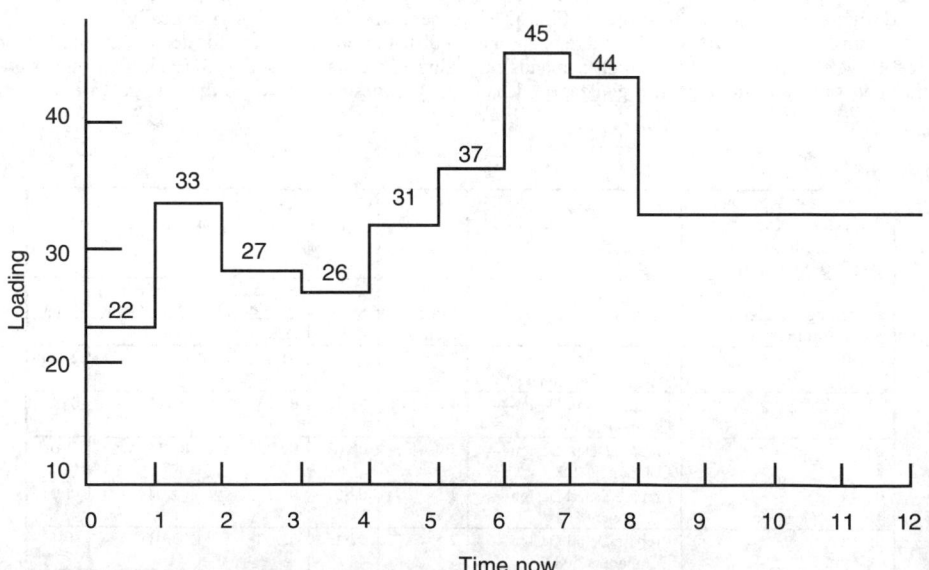

Fig. 67.25 Profile of demands for resources.

the demand for resources. The *resource-leveling process* attempts to remedy this situation by leveling or smoothing resource demands within the constraints of required dates for the various activities.

Resource Leveling

The resource-leveling process begins with the resource-loading and slack calculations of Fig. 67.24 and the resource profile of Fig. 67.25. It proceeds to *level* the demands for resources *without* exceeding the required dates of projects. The process is constrained by its leveling objectives and not by available resources.

By the use of slack calculations, some start times of activities may be adjusted to begin later than their earliest expected start date, thus shifting the demand for resources to a later point in time, *without* exceeding the original expected completion date of the network. Therefore, the resource-leveling process requires us to adjust performance times of activities according to slack calculations to produce a pattern of demand for resources that is as stable as possible over the scheduling horizon.

The resource requirements of the valid plan are taken as a beginning point for resource leveling. Required completion dates are treated as constraints so that we maintain valid plans. Adjustments are made by each functional manager; these adjustments are coordinated with personnel assigned to the various project-management offices to ensure that each functional organization does not frustrate the schedules of the other. That is, the use of slack by the various functions must be coordinated by the various project offices.

Free slack (or float), which we define to be that part of activity slack that, if used, does not affect slack calculations forward of the activity in the network, may be used immediately without approval of the project office, since its use cannot affect any other activity in the project. Normal slack, however, is identified with a path and although we may use it during resource leveling, its use must be coordinated with project office personnel whose project is affected. Coordination is necessary since only one activity on a path may use its positive slack; if all functions represented by activities on a single path use the slack, the combined network calculations would produce a negative slack path!

The resource-leveling function may be carried out manually unless the networks are large and involve multiple resources. Computer programs are available to assist in carrying out the resource-leveling process for complex networks.

Resource leveling for our sample projects results in the revised loading plan of Fig. 67.26 and the resource profile of Fig. 67.27. Note that in the leveling process we were able to reduce peak

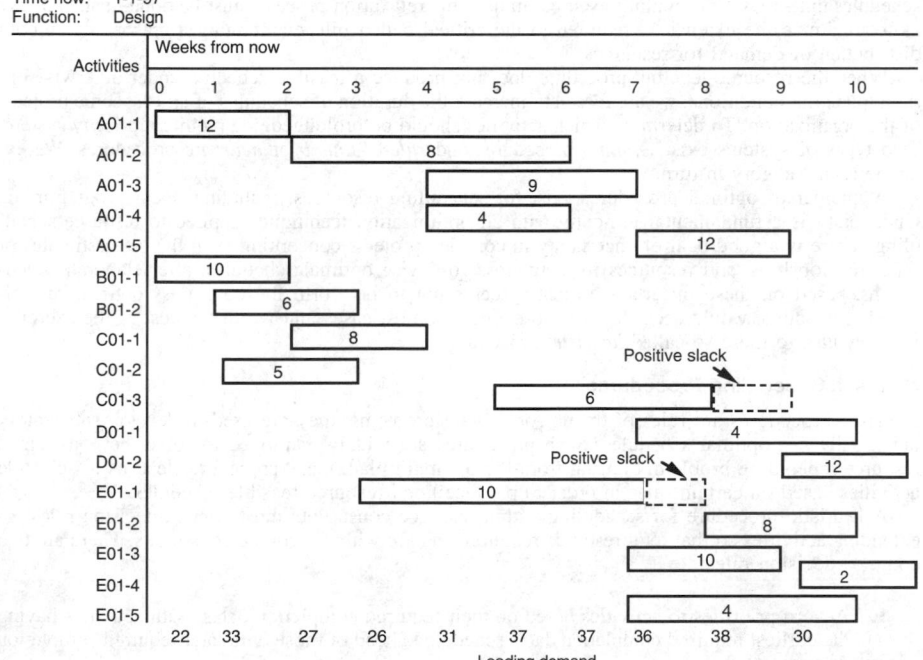

Fig. 67.26 Resource-leveled plan.

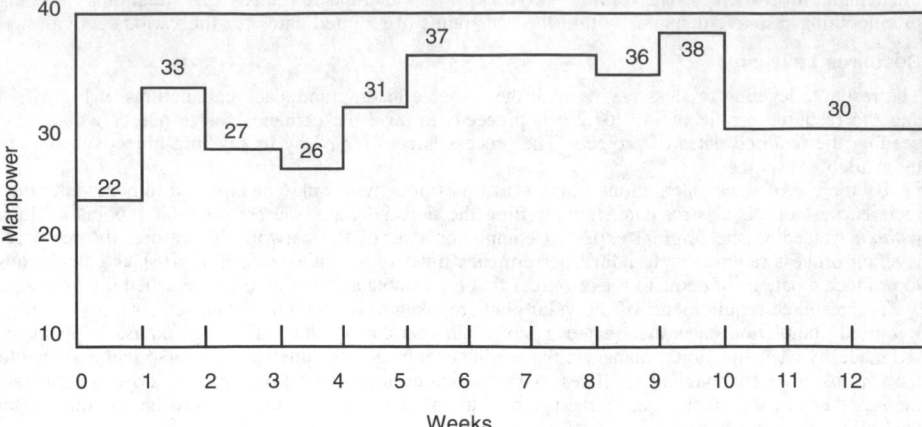

Fig. 67.27   Revised profile of demands for resources.

demands of weeks 6–7 and 7–8 significantly. This was accomplished by the selective use of positive slack that was available on the project. Assuming the use of slack meets the approval of the proper project office personnel, the leveling procedure results in a definite improvement in the distribution of resources of the design organization. The resource profile produced by the resource-leveling procedure, however, may not be feasible in light of known resource levels. The next problem, therefore, involves scheduling all these design tasks within the limits of available engineers. This is the task of scheduling subject to resource constraints to which we turn now.

Resource Scheduling Subject to Resource Constraints

If the resource-leveling process produces a loading profile within the resource limitations of the function, all is well. If, however, the loading demands of the leveling process produce resource requirements that exceed resource availability, including the use of overtime, then we must relax schedules until they "fit" within resource limits. This relaxation process must be done in such a way as to minimize the extensions required to the critical path while maintaining a reasonably smooth distribution of demand for resources.

When the resource-leveling procedure does not produce a feasible schedule under otherwise optimal planning conditions, we are forced to increase the duration of schedules of at least some projects of the organization. To determine which activities should be prolonged, we require a priority system. Two types of systems exist: *optimal* procedures and *rule-of-thumb* or *heuristic* procedures. We examine each category in turn.

A number of optimal procedures exist for scheduling resources, including linear programming. Uncertainty is a fundamental difficulty with all optimization techniques applied to resource scheduling. There is a good deal of uncertainty in complex projects concerning even the best estimates of time, relationships, and resources for activities. To devise optimal schedules, after elaborate calculations, based on these uncertain estimates seems not to be worth the cost. Less optimal rule-of-thumb procedures would seem to be good enough in most cases and worth the cost of the exercise. We now turn to these so-called *heuristic* procedures.

Heuristic Scheduling Procedures

Heuristic procedures are rules of thumb for solving problems; they are used to develop satisfactory but usually not optimal schedules. Such procedures are widely employed to solve the constrained resource scheduling problem. Starting from the optimum plan, these procedures lead us to schedule activities based on certain rules in order to produce good resource-feasible schedules.

A heuristic procedure for scheduling within resource constraints must contain decision rules for extending activities so that total resource requirements are within resource constraints. There are two common decision rules:

1. Accord priorities to activities based on their required completion dates, with activities having the earliest required completion dates scheduled ahead of those with later required completion dates.
2. Rank activities in order of duration and perform activities with the shortest duration first.

These two rules of thumb are given as examples of procedures that may be used to solve the constrained scheduling problem, given a leveled loading plan. All heuristic procedures proceed to extend activities that cannot be accommodated by available levels of resources through the use of one of these rules. A heuristic procedure must also have secondary rules for breaking ties. For example, if two activities have identical required dates yet cannot be performed simultaneously because of resource constraints, we might decide to perform the one with the shortest duration first.

It is important to realize that these rules of thumb are not likely to produce optimal schedules. They are designed to produce satisfactory feasible schedules. When placed in the context of the uncertainties found in organizations engaged in complex projects, however, rules of thumb such as these are operational and flexible enough to respond to the inevitable changes brought about by these uncertainties.

We should note that although we have described the resource allocation process as three distinct but interrelated tasks, in practice, they are often performed informally and simultaneously, depending on the magnitude of the task and the sophistication of the project-control process.

67.4.6   The Budget Process

A close parallel exists in the relationship between expenditure planning and budgeting to the relationship we described between network planning and resource scheduling. The resource-scheduling process begins with the activities, time requirements, and calculations of the network and proceeds to load resources, smooth resources, and construct schedules that are resource-feasible for a short period of time into the future. The portion of a network plan that has been scheduled for performance by functional groups is called a *scheduled plan*.

Similarly, the budgeting process begins with plans established during the financial planning process and proceeds to authorize expenditure limits within which, *on balance,* budgets are expected to adhere. The budget for a work package, however, is likely to differ in some important aspects from the financial plan, since the authorized work package must reflect decisions made in the resource-allocation process. The portion of the financial expenditure plan for which we have a budget is called a *budgeted plan*.

Work Package and Operating Budgets

Projects require an operating budget. For functional organizations engaged in complex projects, however, it is almost impossible to prepare an annual operating budget with any degree of confidence that it will be followed closely. Yet each organizational unit must perform resource and expenditure planning over a longer horizon than that which it can forecast perfectly.

This apparent budgeting dilemma is resolved by requiring both work package and operating budgets. Work package budgets, covering a short period of time, serve as work-authorization documents, whereas approved operating budgets serve to guide decisions regarding resource levels in each of the functional departments.

Financial data on work packages prepared during the expenditure planning process are far from ready to serve as budgets for the project. These financial plans were derived from estimates of network activities. As we saw previously, network plans ordinarily cannot be converted directly into schedules, but rather must be considered in light of available resources and other competing demands. Therefore, the financial plans of a given work package cannot be converted into budgets until the activities included in the work package have been scheduled, for only then do we know precisely when activities will be done and by whom and what resources will be used. Only a small portion of the financial plan is eligible to serve as a budget, for only a small portion of the network plan upon which the expenditure plan is based has been scheduled. The budget for the portion of the network plan that has been scheduled is negotiated between project and functional personnel. The approved budget then serves as the document that authorizes functional work.

The Budget as an Authorization Document.   We have seen that to achieve scale economies and coordination, the matrix structure causes us to violate the classical principle of *unity of command*. The stresses produced by the dual sources of command to which functional personnel must respond are nowhere potentially more divisive than in the budgeting and authorization process. This process, however, if performed correctly also possess opportunities to enhance identification with project goals, to improve performance, and to reduce or eliminate these natural tensions caused by dual lines of command.

Since the management function of directing project work formally may lie with functional managers under the matrix structure, the project manager should use the budget and related authorizing documents to exert "purse-string" authority over functional performance.

Once the scheduling is prepared for functional work, the budget implications may be derived by applying rates for each cost element as established in the project cost-accounting system. The schedule for functional work and its supporting budget serve as authorizing documents for functional work. By reviewing and approving the schedules and budgets of these work packages, the project manager begins to assert control over his or her project. Thus, a major portion of a project manager's time is spent negotiating budgets with functional managers in light of original financial planning for the work, current schedules, past performance by the various functions, and overall project status.

Project office personnel should ask the following questions regarding proposed work package budgets:

- Does the schedule as presented by the functional groups validate our project plans?
- Will the schedule work meet performance and quality specifications?
- Is the budget for the work consistent with planning estimates?

If each of these questions is answered in the affirmative, the project manager simply approves the budget and authorizes performance. He or she may authorize performance for the entire scheduling horizon (10 weeks in our examples) or for the duration of the work package. The authorization period is controlled by specifying the time period during which the project manager will accept charges against the account number assigned to a given work package.

Planned Value of Work.   The characteristics of complex projects require that we integrate time and expenditure plans into a measure of *planned value of work performed.* Once a schedule and budget are prepared and approved, *planned value* may be established for each work package, or it may be decided to integrate expenditure and time plans at a level somewhat higher on the WBS. The tightest control is achieved when planned values are established at the work package level.

If the integration is done at the work package level, the planned value of work becomes the approved budget for the work package. Later in the reporting process, actual costs for work performed on a work package can be compared with planned value for work to monitor in an integrated way time and cost performance.

The authorization document is a work package approved by the project manager or by his or her representative. It should contain detailed information on time, cost, planned value, and expected performance on that segment of work. When approved by both project and functional management, it becomes the agreement, or performance contract, between project and functional personnel. Figure 67.28 provides an illustration of an authorizing document.

Figure 67.28 contains a summary description of the work embodied in the work package, a milestone chart indicating scheduled completion dates for work package activities, a time-phased financial expenditure plan, and a time-phased work plan. Finally, it contains planned value of work calculations for major milestones.

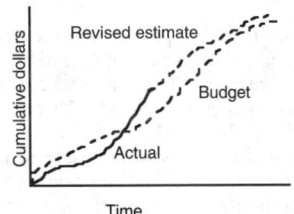
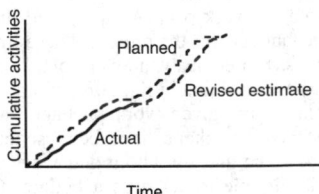

Function	Schedule		
Design	1▼	2▼	
Drafting	3 ▼	4▼	
Reliability		5▼	7▼
Manufacturing		6▼	
Test		8▼	
	Calendar weeks		

Planning value of work scheduled

Milestones:
1. Complete preliminary design
2. Complete final design
3. Prepare preliminary drawings
4. Prepare final detail drawings

5. Reliability summary
6. Manufacture
7. Reliability summary
8. Test

Work package description: The purpose of this task is to design and fabricate air inlet value according to specification xxx. WBS No. xxx-xxx-xxx

Fig. 67.28   Sample work package authorization document.

If the process leading to the issuance of the authorizing document is properly organized, the standards established should induce sufficient motivation on the part of functional personnel to perform the task and achieve or exceed the standards established.

The Budget Procedure. Operating budgets are ordinarily prepared for each project and functional organization on an annual basis. Each project manager prepares an operating budget for each period, and the budget normally represents a portion of the overall budget for the entire project. Each project budget contains an estimate of revenue and expenses. Expenses are ordinarily derived from a combination of approved work package budgets and project expenditure plans for the period, and normally include some funds for contingency. Estimating annual revenue for projects that are expected to extend over many budgeting periods is a difficult problem and tied to specific contract provisions.

Operating budgets are also prepared for each functional group for the budget period. If these functional budgets are tight and challenging, we must expect many functions to require some contingency funds for inevitable overexpenditures. These funds may be provided either through an appeal process to project managers, and a corresponding revision to the operating budget, or through a partial allocation of contingency funds from project budgets to functional organizations during the expenditure planning process. The latter procedure is likely to lead to goal-congruent behavior if functional organizations are treated as profits centers. The former procedure is likely to be most effective if functional organizations are treated as cost centers, which is the typical responsibility center designation for these organizational units.

Operating budgets for functional disciplines are prepared by summing approved work package budgets pertaining to a function with planned expenditures derived from approved project expenditure plans for the portion of the budget period not covered by approved work packages.

Contingency funds are required in these operating budgets to allow for some overexpenditures that are expected with tight budgets as well as unanticipated but inevitable differences that occur between financial expenditure plans and approved work package plans. That is, annual operating budgets for each function must be approved based largely on expenditure plans of the organization. Later, approved costs of work packages usually will deviate from the costs included in the original operating budget because of changes that occur between the planning process and the resource-allocation process.

If differences between approved work packages and operating budgets are large, revisions are called for in operating budgets; if differences are small, they should be absorbed into a contingency account, which itself may be treated as an overhead account, along with all other nonproject work. The functional department overhead account should also include idle time, company-sponsored research and development, proposal effort, indirect functional supplies, and functional supervision.

Functional Overhead Budgets. Functional overhead budgets normally contain estimates for all functional expenditures that cannot be traced directly to a funded project. Functional organizations are normally held responsible for performance regarding these overhead budgets. Under these circumstances, functional managers will attempt to keep their personnel employed on either a contractual project or an approved company-funded project, such as a proposal or a research project. Otherwise, the time of functional personnel must be charged to a special account called "idle time," and although some charges are expected to this account because of resource-allocation problems and because of normal transitions from one project to another that often involve delays, these charges must be kept to a minimum for the functional manager to achieve good performance regarding his or her overhead budget.

Project managers, on the other hand, seek to remove functional resources from their project as quickly as possible, since the performance of these managers is often evaluated based on profit. Functional managers must be able to use these personnel on other projects *for the organization as a whole* to achieve the cost reduction that is attributed to the project office. If the organization cannot use personnel so released, their time must be charged to idle time, which may cause functional managers to overexpend their overhead budgets! If functional overhead budgets are in jeopardy, the temptation is always present to mischarge functional time to contracts that appear to be able to absorb such charges and to prolong existing problem work longer than necessary.

Because many organizations treat their level of functional resources as essentially fixed during the short term, an unplanned underexpenditure on a project ordinary shifts an equivalent amount of costs somewhere else in the organization and does not result in a comparable organizational saving *unless* these resources may be absorbed profitably on another contract or on approved internally funded project.

These facts of organizational life leads us to place a premium upon planning and flexibility on the part of functional managers. Functional planning must always include provision for contingencies, whether this takes the form of preplanned effort on internally approved projects or plans to shift resources to new projects if these resources are released prematurely.

If this kind of contingency planning is not done in functional disciplines, pressures will build to mischarge contracts, overrun overhead budgets, and adjust the level of personnel in the organization at an undesirable rate.

Moreover, it is a mistake to place too much emphasis on performance regarding overhead budgets in the evaluation of functional managers to the exclusion of their performance regarding cost, schedule, and quality of all projects that are supported by the function. In addition, the quality of planning for the use of functional resources should play an important role in their performance evaluation.

67.4.7 Systems of Reporting for Project Control

To put the general requirement of a reporting system for complex projects into perspective, it is necessary to remember that we are describing a system that replaces the cost-accounting system in the management control process. The project cost-accounting system, however, *does have similarities* with conventional cost-accounting systems (e.g., account code structure, standards, variances, and overhead allocations), as we have seen. Therefore, when it comes to designing project-reporting systems, it is useful to begin by reviewing the reporting system established in conventional cost accounting for each element of cost. It turns out that each of the variances used in conventional cost accounting may be used in project cost accounting, but they must be supplemented with combined cost and schedule variances.

The labor variances of conventional cost accounting are subdivided into time and rate variances. The time variance is found for a task as follows:

$$(\text{standard hours} - \text{actual hours}) \times \text{standard rate} = \text{time variance} \tag{67.1}$$

The rate variance for labor is found by

$$(\text{standard rate} - \text{actual rate}) \times \text{actual hours} = \text{rate variance} \tag{67.2}$$

The total labor variance for a task is

$$\text{standard labor cost} - \text{actual labor cost} = \text{total labor variance} \tag{67.3}$$

Material variances are similarly subdivided into quantity and price variances. The quantity variance is computed as follows:

$$(\text{standard quantity} - \text{actual quantity}) \times \text{standard price} = \text{quantity variance} \tag{67.4}$$

The price variance is given by

$$(\text{standard price} - \text{actual price}) \times \text{actual quantity} = \text{price variance} \tag{67.5}$$

The total material variance is

$$\text{standard material cost} - \text{actual material cost} = \text{total material variance} \tag{67.6}$$

We omit the overhead variances, since our project-reporting system ought to focus primarily upon controllable project costs.

Now the difference between the requirements for project reporting and conventional cost reporting systems develops because the labor and material variances are only *cost* or *spending variances.* They essentially assume that the scheduled work was completed in the process of spending funds for labor and materials. This is a realistic assumption in most manufacturing operations. Not so, however, in the management of complex projects.

For any WBS item, we are interested in the relationship between *planned value of work* for a given time period and *actual cost of work* for the same time period. This will tell us our total variance for the task and we, by the computation of more detailed variances, seek to trace its causes. There are five potential causes for any total variance:

1. We did more or less work than scheduled.
2. We used more or less labor than planned for the actual work we did.
3. We paid more or less than planned for the actual labor used.
4. We used more material than planned for the work we accomplished.
5. We paid more or less than planned for the material we actually used.

The portion of the total variance attributable to number 1 is called the *schedule variance,* and the portion attributable to numbers 2–5 is called the *spending variance.* Therefore, the total variance for a given task or end item of a project is budgeted value of work planned − actual cost of work accomplished = total variance   (67.7)

The schedule variance is given by

budgeted value of work planned

$\qquad$ − budgeted value of work accomplished = schedule variance   (67.8)

The spending variance is

budgeted value of work accomplished

$\qquad$ − actual cost of work accomplished = spending variance   (67.9)

The spending variance is then subdivided into labor and material variances according to Eqs. (67.3) and (67.6). Labor and material variances may be subdivided further into rate and quantity variances according to Eqs. (67.1), (67.2), (67.4), and (67.5).

These nine variances, 67.1–67.9 may be computed for each level of the WBS and for each functional organization at regular intervals throughout the life of a project. The total variances for a WBS end item tell the responsible manager whether there is a problem or not regarding cost and schedule performance. If a problem exists, he or she can request more detailed reporting information for the next level of the WBS and find exactly where the problem is and whether the problem is concerned with schedule slippage or with a labor or material spending variance.

Note the only variance that is new is the schedule variance. Each of the other variances appears in conventional cost-accounting systems. The schedule variance requires for its computation data on planned and actual schedule performance together with normal cost data for each level of the WBS. The fact that there is only one new variance required should dispel any mystery surrounding the schedule and cost-reporting requirements for complex projects.

Conceptually, the time-and-cost reporting system for complex projects may be represented by Fig. 67.29. We should be capable of calculating a schedule variance and a cost variance for any level of the WBS. We should be able to divide the cost variance into its labor and material elements. We should be able to trace the schedule variance to a scheduled plan.

The reporting system should also contain the capability to provide information for each functional organization that is performing work on the project by WBS end item. This information should follow the same format as that for the project.

These, then, are the broad outlines that the formal reporting system should take with the recognition that the system should be flexible and adaptable to each organization.

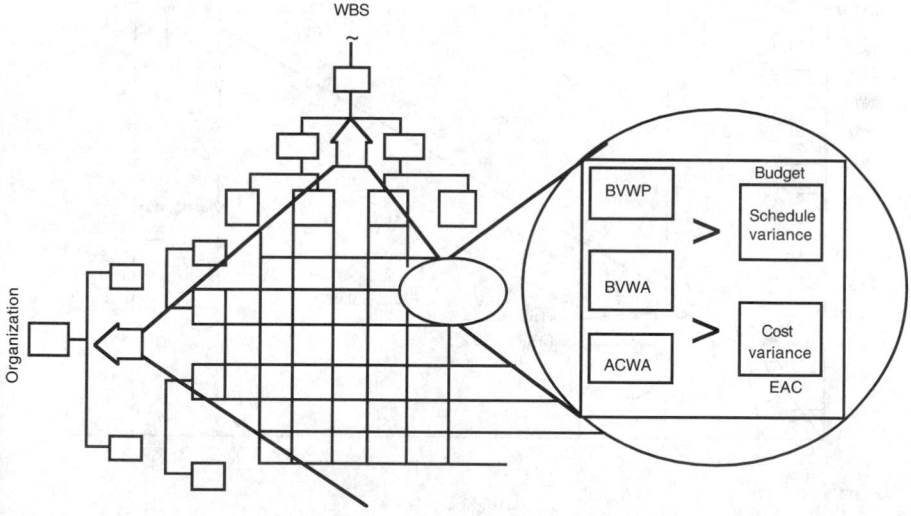

The variance terms in the diagram are defined as follows:
BVWP---budget value of work planned; BVWA---budget value of work accomplished;
ACWA---actual cost of work accomplished; EAC---estimate of cost at completion.

Fig. 67.29   Integrated time-cost reporting system.

Let us look at an example of an application of our variance system. Let us assume that we are considering a work package for the structure (A01-3-2) of our TV transmission building. Moreover, let us assume we placed the material orders at the estimated price and that we have chosen not to include overhead in our project control reports, since the project manager has little or no control over it. Therefore, our primary control variable is the estimated $23,650 of labor costs for this work package.

Approximately six weeks into the project, we have an integrated progress chart drawn up for us, as shown in Fig. 67.30. The chart shows that work on the structure is currently three weeks ahead of schedule, yet that is not the whole story. The schedule variance is positive, we have done more in the first six weeks than originally planned (i.e., BVWA > BVWP). Yet we have spent more than budgeted for the work accomplished (i.e., ACWA > BVWA). Projecting these trends to completion, we will spend approximately $2,000 more than estimated but we will finish three weeks early. Our conclusion at first look might be that the schedule gain was accomplished by spending more labor resources, and further investigation might show that to be true. Nevertheless, unless there are some changes made, the work package will overrun by approximately $2,000 at completion. The integrated report gives us a rather complete picture. It is much clearer than independent budget versus actual expenditure and schedule progress reports.

Reporting Delays and Bias

There always must be some delay between actual project progress and problems and their reporting, since the reporting process consumes time. All project status must be ascertained from the performing organizations. These data then must be processed. After processing, these reports must be analyzed to ensure that the processing was done correctly and to assess progress and problems. It is not unusual for this processing and analysis work to consume two weeks or more on a complex project although it could occur daily.

Moreover, the subsequent meetings and recommendations for action may take still another week or two. When action is finally taken, it may be to remedy a problem that existed a month ago! To further complicate the reporting problem, bias may creep into the reports.

If functional supervision is evaluated based on its rate of progress alone, then we should expect bias to enter into the reporting system. Bias can be prevented to some extent by explicit definition of activities that then become standard for the organization.

Often standardization of activity descriptions manifests itself in the preparation of a dictionary of terms and activities. The dictionary serves to accurately identify completed activities and improves communication within the project group while providing the basis for a historical file of time and cost date. This file becomes useful for estimating future projects containing similar activities.

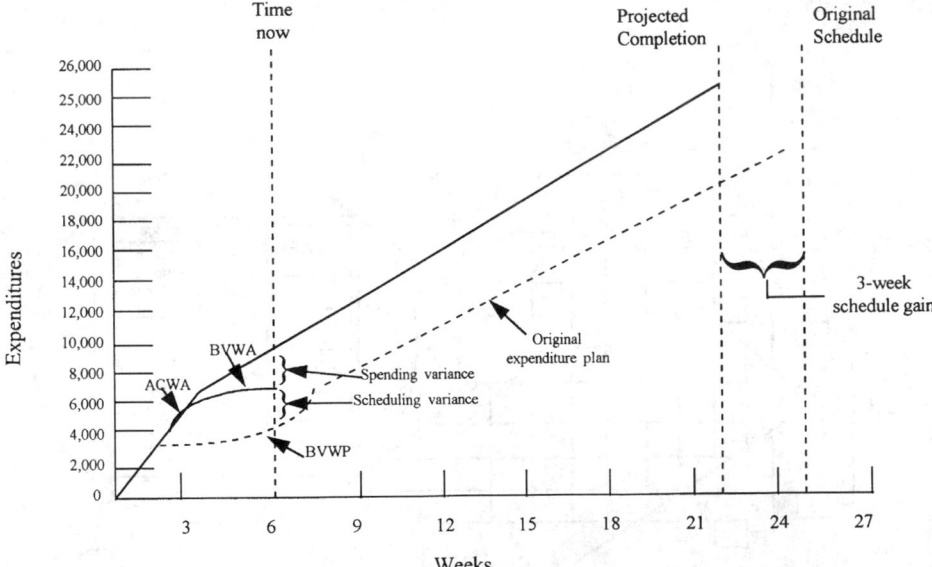

Fig. 67.30 Integrated progress chart for the structure of the TV transmission building A01-3-2. BVWA = budget value of work accomplished; ACWA, actual cost of work accomplished; BVWP = budget value of work planned. (Adapted by permission from G. Schillinglaw, *Managerial Cost Accounting*, 4th ed., Richard D. Irwin, Homewood, IL, p. 763.)

Finally, the frequency of reporting should vary from one project to the next, depending on the nature and complexity of the contract, the importance of the project to the organization, and customer reporting requirements.

67.5 A SURVEY OF COMPUTER SOFTWARE FOR THE MANAGEMENT CONTROL OF PROJECTS

There are currently more than 350 software packages available to assist in the management control of complex projects. These packages are designed to operate in mainframe, server, and personal computer environments. Many of these packages support group interactions and some even link project personnel working in different organizations. They range in price from $40 to $350,000.

Software packages are available for the purposes of constructing work breakdown structures; planning and scheduling; resource loading, leveling and constrained resource scheduling; financial planning and costing; performance analysis and reporting; multiple project planning, scheduling, and costing; and project graphing, including the drawing of project networks, schedules, and reports.

A complete list of these 350 packages is found in the 1995 Project Management Software Survey, published by PMI Communications, 323 West Main Street, Sylva, North Carolina, 28779. This survey describes the principal features of each software package, the hardware requirements, the price, and the address of vendors.

ACKNOWLEDGMENT

The author wishes to acknowledge the invaluable assistance of his secretary, Mrs. Elizabeth Rowe, Peter F. Drucker Graduate School of Management, Claremont Graduate University, in the preparation of this chapter. Her skill and patience are greatly appreciated.

REFERENCES

1. N. Weiner, *Cybernetics,* 2nd ed., MIT Press, Cambridge, MA, 1961.
2. S. Beer, *The Cybernetics of Management,* Wiley, New York, 1966.
3. S. Beer, *Decision and Control,* Wiley, New York, 1966.
4. D. W. Griesinger, *The Cybernetic Organization of Behavior,* Administration and Engineering Systems Monograph, Institution of Administration and Management, Union College, Schenectady, NY, 1979.
5. D. W. Griesinger, *Toward a Cybernetic Theory of the Firm,* Peter F. Drucker Management Center, Claremont Graduate School, Claremont, CA, 1992.
6. J. K. Graham, Jr., *The Cybernetic Design of Jobs with Application to Case Management in Community Mental Health,* Ph.D. dissertation, Union College and University, Schenectady, NY, 1980.
7. J. Maciariello and C. Kirby, *Management Control Systems: Using Adaptive Systems to Attain Control,* Prentice-Hall, Englewood Cliffs, NJ, 1994.
8. P. M. Senge, *The Fifth Discipline,* Doubleday/Currency, New York, 1990.
9. C. J. Kirby, *Performance Improvement Teams: A Study from the Adaptive Controls Perspective,* Ph.D. dissertation, Claremont Graduate School, Claremont, CA, 1992.
10. J. G. March and H. A. Simon, *Organizations,* Wiley, New York, 1958.

BIBLIOGRAPHY

Anthony, R. N., *The Management Control Function,* Harvard University Press, Cambridge, MA, 1988.

Barnard, C. I., *The Functions of the Executive,* Harvard University Press, Cambridge, MA, 1938.

Boston Consulting Group, *Reengineering and Beyond: A Senior Management Perspective,* Boston Consulting Group, Boston, MA, 1993.

Bullen, C. V., and J. F. Rockart, "A Primer on Critical Success Factors," in *The Rise of Managerial Computing,* Richard D. Irwin, Homewood, IL, 1986.

Clark, K., and S. Wheelwright, *Managing the New Product and Process Development: Text and Cases,* The Free Press, New York, 1993.

Davenport, T., *Process Innovation: Reengineering Work through Information Technology,* Harvard Business School Press, Boston, MA, 1993.

Drucker, P. F., *Management: Tasks, Responsibilities, and Practices.* Harper and Row, New York, 1973.

Forrester, J. W., "Counterintuitive Behavior of Social Systems," in *Simulation* **16**(1), 61–67 (January 1971).

Frame, D. J., *Managing Projects in Organizations,* rev. ed., Jossey-Bass, San Francisco, CA, 1995.

————, *The New Project Management,* Jossey-Bass, San Francisco, CA, 1994.

Freeman, E. R., *Strategic Management: A Stakeholder Approach,* Pittman, Mansfield, MA, 1984.

Griesinger, D. W., *Innovation for Stakeholder Advantage: A Stakeholder Approach.* Graduate Management Center, GMC 8401, Claremont Graduate School, Claremont, CA, 1984.

Hammer, M., and J. Champy, *Reengineering the Corporation,* Harper Business, New York, 1993.

Kirby, C. J., and J. Maciariello, "Integrated Product Development and Adaptive Management Systems," *Drucker Management* (Fall 1994).

Mintzberg, *The Structuring of Organizations,* Prentice-Hall, Englewood Cliffs, NJ, 1979.

PMI Communications, Project Management Institute, *1995 Project Management Software Survey,* Sylva, NC.

Sathe, V., *Culture and Related Corporate Realities,* Richard D. Irwin, Homewood, IL, 1985.

Sayles, L. R., *The Working Leader,* Free Press, New York, 1993.

Severino, R. A., "Systems Dynamics, Team Learning, and Paradigm Shifts for Program Managers," Unpublished Paper, Executive Management Program, Claremont Graduate School, Claremont, CA, 1992.

CHAPTER 68

MANAGING PEOPLE

Hans J. Thamhain
Department of Management
Bentley College
Waltham, Massachusetts

68.1	**CRITICAL ISSUES OF MANAGING IN TECHNOLOGY-BASED ENVIRONMENTS**	**2085**
68.2	**MOTIVATION AND ENGINEERING PERFORMANCE**	**2086**
	68.2.1 Sixteen Professional Needs That Affect Engineering Performance	2086
68.3	**MANAGING WITHOUT FORMAL AUTHORITY**	**2087**
68.4	**AN INCREASED FOCUS ON TEAM PERFORMANCE**	**2088**
68.5	**CHARACTERISTICS OF HIGH-PERFORMING ENGINEERING TEAMS**	**2089**
68.6	**BARRIERS TO HIGH TEAM PERFORMANCE**	**2091**
	68.6.1 Different Points of View	2091
	68.6.2 Role Conflict	2091
	68.6.3 Power Struggles	2091
	68.6.4 Group Think	2092
68.7	**BUILDING SELF-DIRECTED TEAMS**	**2092**
68.8	**RECOMMENDATIONS**	**2094**
	68.8.1 Clear Assignment	2094
	68.8.2 Clear Goals and Objectives	2094
	68.8.3 Effective Planning	2094
	68.8.4 Image Building	2094
	68.8.5 Process Definition and Team Structure	2094
	68.8.6 Interesting Work	2094
	68.8.7 Senior Management Support	2095
	68.8.8 Clear Communication	2095
	68.8.9 Commitment	2095
	68.8.10 Leadership	2095
	68.8.11 Reward System	2095
	68.8.12 Problem Avoidance	2095
	68.8.13 Personal Drive and Leadership	2095
68.9	**A FINAL MESSAGE**	**2095**

68.1 CRITICAL ISSUES OF MANAGING IN TECHNOLOGY-BASED ENVIRONMENTS

In their quest to remain competitive in our changing business environment, managers must work effectively with people, who are our most valuable asset. They are the heart and soul of a company's core competency, critical to the successful implementation of any strategic plan, operational initiative, or specific project undertaking. The mandate for managers is clear: they must weave together the best practices of both traditional and contemporary schools for teaching and directing their people toward desired results in a rapidly changing world. However, even the best practices do not guarantee success. They must be carefully integrated with the business process, its culture and value system. These challenges are especially present in today's technology-based organizations, which have become highly complex and multifaceted, requiring effective planning, organizing, and integration of complicated, multidisciplinary activities across functional lines in an environment of rapidly changing technology, global markets, regulations, and socioeconomic factors. Because of these dynamics, engineering organizations seldom are structured along traditional functional lines. Rather, they operate as matrices or hybrid project organizations that overlay the functional structure. Their management must share resources and power and establish communication channels that work both vertically and horizontally to integrate the many activities involved in modern engineering operations.

Mechanical Engineers' Handbook, 2nd ed., Edited by Myer Kutz.
ISBN 0-471-13007-9 © 1998 John Wiley & Sons, Inc.

68.2 MOTIVATION AND ENGINEERING PERFORMANCE

Understanding people is important in any management situation, and motivation is especially critical in today's technology-based organizations.[1] Leaders who succeed within these often unstructured work environments must confront untried problems to manage their complex tasks. They have to learn how to move across various organizational lines to gain services from personnel not reporting directly to them. They must build multidisciplinary teams into cohesive groups and deal with a variety of networks, such as line departments, staff groups, team members, clients, and senior management, each having different cultures, interests, expectations, and charters. To get results, these engineering managers must relate socially as well as technically and must understand the culture and value system of the organization in which they work. The days of the manager who gets by with only technical expertise or pure administrative skills are gone.

What works best? Observations of best-in-class practices show consistently and measurably two important characteristics of high performers: (1) they enjoy work and are excited about the contributions they make to their company and society, and (2) they have esteem needs fulfilled, that is, they feel good about themselves professionally. Specifically, field research studies have identified several professional needs strongly associated with job performance.

68.2.1 Sixteen Professional Needs That Affect Engineering Performance

Research studies show that the fulfillment of certain professional needs can drive engineering personnel to higher performance; conversely, the inability to fulfill these needs may become a barrier to individual performance and teamwork.[2-5] The rationale for this important correlation is found in the complex interaction of organizational and behavioral elements. Effective team management involves three primary issues: (1) people skills, (2) organizational structure, and (3) management style. All three issues are influenced by the specific task to be performed and the surrounding environment. That is, the degree of satisfaction of any of the needs is a function of (1) having the right mix of people with appropriate skills and traits, (2) organizing the people and resources according to the tasks to be performed, and (3) adopting the right leadership style. The sixteen professional needs critical to engineering performance are:

1. *Interesting and challenging work,* as intrinsic motivation of the individual, satisfies professional esteem needs and helps to integrate personal goals with the objectives of the organization.

2. *Professionally stimulating work environment* leads to professional involvement, creativity, and interdisciplinary support. It also fosters team building and is conducive to effective communication, conflict resolution, and commitment toward organizational goals. The quality of this work environment is defined through its organizational structure, facilities, and management style.

3. *Professional growth* is measured by promotional opportunities, salary advances, the learning of new skills and techniques, and professional recognition. A particular challenge exists for management in limited-growth or zero-growth businesses to compensate for lack of promotional opportunities by offering more intrinsic professional growth in terms of job satisfaction and skill building.

4. *Overall leadership* involves dealing effectively with individual contributors, managers, and support personnel within a specific functional discipline as well as across organizational lines. It involves technical expertise, information-processing skills, effective communications, and decision-making skills. Taken together, leadership means satisfying the need for clear direction and unified guidance toward established objectives.

5. *Tangible reward* include salary increases, bonuses, and incentives, as well as promotions, recognition, better offices, and educational opportunities. Although extrinsic, these financial rewards are necessary to sustain strong long-term efforts and motivation. Furthermore, they validate the "softer" intrinsic rewards, such as recognition and praise, and reassure people that higher goals are attainable.

6. *Technical expertise* means that personnel need to have all necessary interdisciplinary skills and expertise available within the project team to perform the required tasks. Technical expertise includes understanding the technicalities of the work, the technology and underlying concepts, theories and principles, design methods and techniques, and functioning and interrelationship of the various components that make up the total system.

7. *Assisting in problem solving:* examples include facilitating solutions to technical, administrative, and personal problems. It is a very important need, which, if not satisfied, often leads to frustration, conflict, and poor-quality work.

8. *Clearly defined objectives.* Goals, objectives, and outcomes of an effort must be clearly communicated to all affected personnel. Conflict can develop over ambiguities or missing information.

9. *Management control* is important for effective team performance. Managers must understand the interaction of organizational and behavior variables in order to exert the direction, leadership, and control required to steer the project effort toward established organizational goals without stifling innovation and creativity.

10. *Job security* is one of the very fundamental needs that must be satisfied before people consider higher-order growth needs.

11. *Senior management support* should be provided in four major areas: (1) financial resources, (2) effective operating charter, (3) cooperation from support departments, and (4) provision of necessary facilities and equipment. It is particularly crucial to larger, more complex undertakings.

12. *Good interpersonal relations* are required especially for effective teamwork; they foster a stimulating work environment with low conflict, high productivity, and involved, motivated personnel.

13. *Proper planning* is absolutely essential for the successful management of multidisciplinary activities. It requires communications and information-processing skills to define the actual resource requirements and administrative support necessary. It also requires the ability to negotiate resources and commitment from key personnel in various support groups across organizational lines.

14. *Clear role definition* helps to minimize role conflict and power struggles among team members and/or supporting organizations. Clear charters, plans, and good management direction are some of the powerful tools used to facilitate clear role definition.

15. *Open communications* satisfy the need for a free flow of information both horizontally and vertically, keeping personnel informed and functioning as a pervasive integrator of the overall project effort.

16. *Minimizing changes.* Although engineering managers have to live with constant change, their team members often see change as an unnecessary condition that impedes their creativity and timely performance. Advanced planning and proper communications can help to minimize changes and lessen their negative impact.

The significance of assessing these motivational forces lies in several areas. First, the above listing provides insight into the broad needs that engineering oriented professionals seem to have. These needs must be satisfied *continuously* before engineering personnel can reach high levels of performance. This is consistent with findings from other studies, which show that in technical environments a significant correlation exists between professional satisfaction and organizational performance.[2,6–9] From the above listing, we now know more specifically on what areas we should focus our attention. In fact, the above listing provides a model for *benchmarking;* that is, it provides managers with a framework for monitoring, defining, and assessing the needs of their people in specific ways. With their awareness of professional needs, managers can direct their personnel and build a work environment that is responsive to these needs. As an example, top-down the work structure and organizational goals might be fixed and not negotiable; however, engineering managers have a great deal of control over the way the work is distributed and assigned. The same degree of operational control also exists in most other need areas. Finally, the above listing of needs provides a topology for measuring organizational effectiveness as a function of the degree at which these needs seem to be satisfied.

Taken together, fulfilling professional needs helps to build people and eventually teams characterized by

- High levels of energy
- High ability to handle conflict and open communications
- High levels of innovation and creativity
- Commitment and ownership
- Willingness to take risks
- Team-oriented behavior
- High tolerance for stress, conflict, and change
- Cooperation and cross-functional linkages

These are precisely the ingredients necessary to work effectively in an environment characterized by technical complexities and rapid changes regarding technology, markets, regulations, and socioeconomic factors. It is also a work environment where traditional methods of authority-based direction, performance measures, and control are virtually ineffective.

68.3 MANAGING WITHOUT FORMAL AUTHORITY

Managers in technology-based work environments must often cross functional lines to get the required support. This is especially true for managers who operate within a matrix structure. Almost invariably,

Table 68.1 Authority Patterns of Engineering Organizations

Within technology-based work environments, management of professional people is largely characterized by:	Authority patterns that are defined only in part by formal organization charts and plans
	Authority that is largely perceived by the members of the organization based on earned credibility, expertise, and perceived priorities
	Multiple accountability of most personnel, especially in project-oriented environments
	Power that is shared between resource managers and project/task managers
	Individual autonomy and participation that is greater than in traditional organizations
	Weak superior–subordinate relationships in favor of stronger peer relationships
	Subtle shifts of personnel loyalties from functional to project lines
	Project performance depending on teamwork, group decision-making, and favoring the strongest organizations
	Reward and punishment power flowing along both vertical and horizontal lines in a highly dynamic pattern
	Rewards (and punishment) are influenced by many organizations and individuals
	Multiproject involvement of support personnel and sharing of resources among many activities

the manager must build multidisciplinary teams into cohesive work groups and successfully deal with a variety of interfaces, such as functional departments, staff groups, other support groups, clients, and senior management. In the *traditional organization,* position power comes largely in form of legitimate authority, reward, and punishment and is provided by these organizations. In contrast, engineering managers and team leaders have to build most of their power bases on their own. As shown in Table 68.1, they have to earn their authority and influence from other sources, including trust, respect, credibility, the image of a sound decision-maker, and the facilitation of a professionally stimulating work environment.

Position power is a necessary prerequisite for effective engineering project/team leadership. Like many other components of the management system, leadership style has also undergone changes over time. With increasing task complexity, increasing dynamics of the organizational environment, and the evolution of new organizational systems, such as the matrix, core team structures, design/build organizations, and process-oriented team concepts, a more adaptive and skill-oriented management style has evolved. This style complements the organizationally derived power bases—such as authority, reward, and punishment—with bases developed by the individual manager. Examples of these individually derived components of influence are technical and managerial expertise, friendship, work challenge, promotional ability, fund allocations, charisma, personal favors, project goal indemnification, recognition, and visibility. This so-called Style II management evolved particularly with the matrix. Effective engineering management combines both the organizationally derived and individually derived styles of influence.

Various research studies[3–6,10,11] provide an insight into the power spectrum available to engineering managers. These studies show that technical and managerial expertise, work challenge, and influence over salary were the most important influences that project leaders seem to have, while penalty factors, fund allocations, and traditional position-based authority appeared least important in gaining support from support staff and project team members.

68.4 AN INCREASED FOCUS ON TEAM PERFORMANCE

More than any other process, teamwork affects organizational performance.* Because of its potential for producing an economic advantage, work teams have been studied by many producing a consid-

---

*In response to this challenge, many researchers have investigated teamwork and its relationship to the innovation process.[12–18] Often, such research is especially related to technology-oriented developments because these multidisciplinary team efforts rely on interaction among various organizational, managerial, and environmental subsystems. Team members come from different organizations with different needs, backgrounds, interests, and expertise. To be effective, they must be transformed into an integrated work group that is unified toward the project objectives.

Table 68.2   Varibles Characterizing Effective Engineering Team Performance

Task Variables	People Variables	Leadership Variables	Organizational Variables
Technical success	Good communication	Organizational ability	Collaborative culture
Quality results	High involvement	Direction and leadership	Common goals and objectives
On-time	Capacity to resolve conflict	Facilitating group decision-making	Stable goals, objectives
On-budget	Mutual trust	Motivation	Risk sharing
Innovation, creativity	High team spirit	Conflict resolution	Involved management
Adaptability to change	High commitment	Team unification	Long-range strategy
	Team self-development	Viability and accessibility	Stimulating work environment
	Ability to interface	Top management	
	Need for achievement	Linkage	
	Collaborative spirit		

erable body of knowledge on the characteristics and behavior* of teams in various work settings.[7,10,12,21–22]

It is interesting to note that, in spite of changing leadership styles and continuously emerging new management practices, this established body of knowledge has formed an important and solid basis for guiding managers in our contemporary, more demanding work environment.[23] It also forms the basis for new management research, theory development, and tools and techniques.

In fact, work teams have long been considered an effective device to enhance organizational effectiveness. Since the discovery of the importance of social phenomena in the classic Hawthorne studies, management theorists and practitioners have tried to enhance group identity and cohesion in the workplace. Indeed, much of the "human relations movement" that occurred in the decades following Hawthorne is based on a group concept. McGregor's Theory Y, for example, spells out the criteria for an effective work group, and Likert called his highest form of management the *participative group,* or System 4.

In today's more complex and technologically sophisticated environment, the group has re-emerged in importance in the form of project teams. The management practices apply principles of interpersonal and group dynamics to create and guide project teams successfully.

68.5 CHARACTERISTICS OF HIGH-PERFORMING ENGINEERING TEAMS

The characteristics of a project team and its ultimate performance depend on many factors related to both people and structural issues. Obviously, each organization has its own way to measure and express performance of a project team. However, in spite of the existing cultural and philosophical differences, there seems to be a general agreement among managers on certain factors that are included in the characteristics of a successful project team. A simple framework is suggested in Table 68.2 for organizing the complex array of performance-related variables into four specific categories. (This framework resulted from several field studies. See Refs. 3–5, 24.)

Task-related variables are direct measures of task performance, such as the ability to produce quality results on time and within budget, innovative performance, and ability to change.

People-related variables affect the inner workings of the team and include good communications, high involvement, capacity to resolve conflict, mutual trust, and commitment to project objectives.

Leadership variables are associated with the various leadership positions within the project team. These positions can be created formally, such as the appointment of project managers and task leaders, or emerge dynamically within the work process as a result of individually developed power bases, such as expertise, trust, respect, credibility, friendship, and empathy. Typical leadership characteristics include the ability to organize and direct the task, facilitate group decision-making, motivate, assist in conflict and problem resolutions, and foster a work environment that satisfies the professional and personal needs of the team members.

Organizational variables include overall organizational climate, command-control-authority structure, policies, procedures, regulations, and regional cultures, values, and economic conditions. All of these variables are likely to be interrelated in a complex, intricate form.

---

*The characteristics of a high-performing technical project team have been studied extensively by Thamhain and Wilemon.[19,20] The studies found a strong association among project success, innovative performance, and certain leadership criteria that include the ability to (1) provide clear directions, (2) unify the team toward a common project goal, (3) foster clear communication channels and interfaces with other work groups, (4) provide stimulating work, (5) provide professional growth potential, (6) facilitate mutual trust and good interpersonal relations, and (7) involve management.

It is interesting to note that managers, when describing the characteristics of an effective, high-performing project team, focus not only on task-related skills for producing technical results on time and on budget, but also on the people and leadership-related qualities, as shown in Table 68.3.

The significance of grouping and categorizing team performances variables is in three areas:

1. It provides a model for determining the factors critical to high team performance in a particular project environment.
2. It provides a framework for diagnosing and stimulating team-building activities.
3. The team performance variable might be useful in benchmarking the team's characteristics against the "norm" of high-performing teams, hence providing the basis for self-assessment and continuous improvement.

Taken together, within an integrated team, members enjoy their group association and derive much of their personal and professional satisfaction from the integration with their team members. Specifically, some of the more important characteristics of such a truly integrated team are

- Satisfaction of individual needs
- Shared interests
- Pride and enjoyment in group activity
- Commitment to team objectives
- High trust, low conflict
- Comfortable with interdependence and change
- High degree of group interaction
- Strong performance norms and result orientation

Creating a climate and culture that produces such team characteristics is conducive to high performance and involves multifaceted challenges that increase with the complexities of the project and its global dimensions.

Table 68.3   Self-Directed Teams Defined

A self-directed work team is a group of people who can manage themselves and their work with a minimum of direct supervision. Yet these teams work within the boundaries of established organizational objectives, business plans, and strategies, as well as overall managerial direction and leadership. Most of the directions come from the work team, rather than from management. Specifically, the characteristics of self-directed teams can be described as follows.

Characteristics:

Members are encouraged (empowered) to take ownership in the work and self-control

Leadership evolves within the team based on expertise, trust, and respect

Members are highly committed to established team objectives

Has ability to organize task teams and define project plans within given objectives

Self-reliant, less dependent upon policies, procedures, and formal control systems

Interested in work, high involvement, energy, need for achievement, and pride in accomplishments

Rewards are significantly derived from recognition, accomplishments, and work challenge

Capacity for self-development of team members

Good intrateam communications and cross-functional linkages

Shared goals and values

Self-control, accountability, and ownership

Strong ability to seek out, share, and process information; group decision-making

Ability to share risks, mutual trust, and support

High level of team member involvement toward continuous improvement of work processes regarding quality and resource effectiveness of winning the day rather than looking for what is best for the team. There is also the possibility that lower-status individuals are being ignored, thus eliminating a potentially valuable resource

68.6 BARRIERS TO HIGH TEAM PERFORMANCE

As functioning groups, project teams are subject to all of the phenomena known as *group dynamics*. As a highly visible and focused work group, the project team often takes on a special significance and is accorded high status with commensurate expectations of performance. Although these groups bring significant energy and perspective to a task, the possibilities of malfunctions are great. A myth is that the assembly of talented and committed individuals automatically results in synergy and renders such a team impervious to many of the barriers commonly found in a project team environment. These barriers, while natural and predictable, take on additional facets in technology-oriented project situations that are exposed to the various challenges shown in Table 68.1. Understanding these barriers, their potential causes, and influencing factors is an important prerequisite for managing them effectively and hence facilitating a work environment where team members can focus their energy on desired results. The most common barriers to effective team performance are discussed below in the context of technology-oriented work environments.

68.6.1 Different Points of View

The purpose of a project team is to harness divergent skills and talents to accomplish project objectives. Having drawn upon various departments or perhaps even different organizations, there is the strong likelihood that team members will naturally see the world from their own unique point of view. There is a tendency to stereotype and devalue "other" views. Such tendencies are heightened when the project team includes people from different countries with different "work cultures," norms, values, needs, and interests. Further, these barriers are often particularly strong in highly technical project situations where members speak their own codes and languages. In addition, there may be historical conflict among organizational units. In such a case, representatives from these units will more than likely carry their prejudices into the team and potentially subvert attempts to create common objectives. Often these judgments are not readily known until the team actually begins its work and conflicts start developing.

68.6.2 Role Conflict

Project and matrix organizations are not only the product of ambiguity; they create ambiguity as well. Team members are actually in multiple roles and often report to different leaders, possibly creating conflicting loyalties. As "boundary role persons," they often do not know which constituency to satisfy. The "home" group or department has a set of expectations that might be at variance with the project team organization's. For example, a department may be run in a very mechanistic, hierarchical fashion while the project team may be more democratic and participatory. Team members might also experience time conflicts due to multiple task assignments that overlay and compete with traditional job responsibilities. The pull from these conflicting forces can either by exhilarating or a source of considerable tension for individual team members.

68.6.3 Power Struggles

While role conflict often occurs in a horizontal dimension (i.e., across units within the same division or across geographic and culture regions), conflict can also occur vertically as different authority levels are represented on the team. Individuals who occupy powerful positions elsewhere can try to recreate—or be expected to exercise—that influence in the group. Often such attempts to impose ideas or to exert leadership over the group are met with resistance, especially from others in similar positions. There can be subtle attempts to undermine potentially productive ideas, with the implicit goal of winning the day rather than looking for what is best for the team. There is also the possibility that lower-status individuals are being ignored, thus eliminating a potentially valuable resource.* While some struggle for power is inevitable in a diverse group, it must be managed to minimize potentially destructive consequences.

---

*An example of such power struggles occurred in a quality of work *life project team* in an engineering organization.[25] The team was set up as a collaborative employee-management group designed to devise ways to improve the quality of work life in one division of a utility company. The membership of this representative group was changed halfway through the project to include more top managers. When the managers came aboard, they continued in the role of "manager" rather than "team member." Subsequently, the weekly meetings became more like typical staff meetings than creative problem-solving sessions. Although there was considerable resistance, the differences were pushed under the table, as the staff people did not wish to confront their supervisors. There was also considerable posturing among the top managers in an effort to demonstrate their influence, although none would directly attempt to take the leadership position.

68.6.4 Group Think

This phenomenon of groups was identified by Janis[26] as a detriment to the decision-making process. It refers to the tendency for a highly cohesive group to develop a sense of detachment and elitism. It can particularly afflict groups that work on special projects. In an effort to maintain cohesion, the group creates shared illusions of invulnerability and unanimity. There is a reluctance to examine different points of view, as these are seen as dangerous to the group's existence. As a result, group members may censor their opinions as the group rationalizes the inherent quality and morality of its decisions. Because many project teams typically are labelled as special and often work under time pressure, they are particularly prone to the dangers of *group* think.

68.7 BUILDING SELF-DIRECTED TEAMS

As the work environment is changing toward higher levels of effectiveness, speed and quality, we are also encountering higher technical complexities, interdependencies across functional lines and geographic boundaries, and a critical need for innovative performance. With the changing environment, self-directed work teams are gradually replacing the more traditional, hierarchically structured project team. These emerging team processes are seen as significant tools for orchestrating the multifunctional activities that come into play during the execution of modern technology-based developments. As summarized in Table 68.3, these processes rely strongly on group interaction, resource- and power-sharing group decision-making, accountability, self-direction, and control. They also rely to a considerable extent on member-generated performance norms and evaluations, rather than hierarchical guidelines, policies, and procedures. While leveraging human resources via self-directed teams can improve project performance, with better resource utilization, speed and higher levels of innovation, it often requires radical changes from traditional management philosophy regarding organizational structure, motivation, leadership, and control. Leading such self-directed teams also requires a great deal of team management skills and overall guidance by senior management.

The key to continuous team development and effective team leadership is keeping the team focused. Field studies on multidisciplinary work groups show consistently and measurably that to be effective, managers and project leaders must not only recognize the potential drivers and barriers of high team performance, but also must know when in the life cycle of the project they are most likely to occur.[24,27,28] They also observe early warning signs of problems. A keen sensitivity to these warning signs and their diagnostics can help in dealing with developing problems in their early stages. Table 68.4 summarizes the most common warning signs of potential team performance problems. The list can also be used as metrics for benchmarking team strength, health, and potential for further development. Team leaders can focus on preventive actions and foster a work environment that is conducive to team building as an ongoing process. A crucial component of such a process is the sense of ownership and commitment of the team members. Team members must become stakeholders in the project, buying into the goals and objectives of the project, and willing to focus their efforts on the desired results.

Specific management insight has been gained from studies by Gemmill,[27] Thamhain,[3-5] and Wilemon[11,29] into work group dynamics of project teams. These studies clearly show significant correlations and interdependencies among work-environment factors and team performance. They indicate that high team performance involves four primary factors: managerial leadership, job content, personal goals and objectives, and work environment and organizational support. The actual correlation of 60 influence factors to the project team characteristics and performance provided some interesting insight into the strength and effect of these factors. One of the important findings was that only 12 of the 60 influence factors that were examined were found to be statistically significant. Other factors seem to be much less important to high team performance. Listed below are the 12 factors, classified as drivers, associated most strongly project team performance:

1. Professionally interesting and stimulating work
2. Recognition of accomplishment
3. Clear project objectives and directions
4. Sufficient resources
5. Experienced management personnel
6. Proper technical direction and leadership
7. Mutual trust, respect, low conflict
8. Qualified project team personnel
9. Involved, supportive upper management
10. Professional growth potential
11. Job security
12. Stable goals and priorities

Table 68.4 Early Warning Signs of Team Performance Problems

Observations and/or team member perceptions	Project perceived as unimportant
	Unclear task/project goals and objectives
	Excessive conflict among team members
	Unclear mission and business objectives
	Unclear requirements
	Perceived technical uncertainty and risks
	Low motivation, apathy, low team spirit
	Little team involvement during project planning
	Low degree of mutual trust and respect
	Disinterested, uninvolved management
	Lack of leadership credibility
	Poor communications among team members and/or support groups
	Problems attracting and holding team members
	Unclear roles, role conflict, power struggle
	Indecisions
	No agreement on project plans
	Surprises, contingencies, subtle problems
	Lack of performance feedback
	Professional skill obsolescence
	Perception of inadequate rewards and incentives
	Poor recognition, visibility of accomplishments
	Work not interesting, no challenge
	Perceived problems
	Fear of failure, potential for penalties
	Fear of evaluation
	Mistrust, collusion, protectionism
	Excessive documentation
	Excessive requests for directions
	Complaints about insufficient resources
	Strong resistance to change

It is interesting to note that these factors not only correlated favorably with the direct measures of high project team performance, such as technical success and on-time/on-budget performance, but also were positively associated with other desired team characteristics, such as commitment, effective communications, creativity, quality, change orientation, and needs for achievement. These measures are especially important in multicultural, multinational environments where management control is weak through traditional chain-of-command channels but relies more on the norms and desires established by the team and its individual members. What we find consistently is that successful project leaders pay attention to the human side. They seem to be effective in fostering a work environment conductive to innovative creative work, where people find the assignments challenging, leading to recognition and professional growth. Such a professionally stimulating environment also seems to lower communication barriers and conflict and enhanced the desire of personnel to succeed. Further, this seems to increase organizational awareness as well as the ability to respond to changing project requirements.

In addition, effective teams have good leadership. Team managers understand the task, the people, the organization, and all the factors crucial to success. They are action-oriented, provide the needed resources, properly direct the implementation of the project plan, and help in the identification and resolution of problems in their early stages.

Management and team leaders can help a great deal in keeping the project team focused. They must communicate and update organizational objectives and relate them to the project and its specific activities in various functional areas and geographic regions. Management can help in developing priorities by communicating project parameters and objectives related to organizational needs. While operationally the project might have to be fine-tuned to changing environments and evolving solutions, the top-down mission and project objectives should remain stable. Project team members need this stability to plan and organize their work toward unified results. This focus is also necessary for establishing benchmarks and integrating innovative activities across all disciplines. Moreover, clear goal-focus stimulates interest in the project and unifies the team. Ultimately it helps to refuel the commitment to established project objectives in such critical areas as technical performance, timing, and budgets.

68.8 RECOMMENDATIONS

A number of recommendations should help managers in dealing with people effectively. Special focus is on technology-based situations that involve the integration of multidisciplinary task teams.

68.8.1 Clear Assignment

At the outset of any new assignment, project leaders should discuss with their team members the overall task, its scope, and objectives. Involvement of the people during the early phases of the assignment, such as bid proposals, project and product planning, can produce great benefits toward plan acceptance, realism, buy-in, personnel matching, and unification of the task team. A thorough understanding of the task requirements comes usually with intense personal involvement, which can be stimulated through participation in project planning, requirements analysis, interface definition, or a producibility study. In addition, any committee-type activity, presentation, or data gathering will help to involve especially new team members and facilitate integration. It also will enable people to better understand their specific tasks and roles in the overall team effort. Senior management can help develop a "priority image" and communicate the basic project parameter and management guidelines.

68.8.2 Clear Goals and Objectives

Management must communicate and update the organizational goals and project objectives. The relationship and contribution of individual work to overall business plans and their goals, as well as of individual project objectives and their importance to the organizational mission must be clear to all personnel.

68.8.3 Effective Planning

Effective planning early in the life cycle of a project or specific mission will have a favorable impact on the work environment and team effectiveness. Because engineering managers and the project leaders have to integrate various tasks across many functional lines, proper planning requires the participation of the entire project team, including support departments, subcontractors, and management. Phased project planning (PPP), stage-gate concepts (SGC), and modern project-management techniques provide the conceptional framework and tools for effective cross-functional planning and organizing the work toward effective execution.

68.8.4 Image Building

Building a favorable image for an ongoing project, in terms of high priority, interesting work, importance to the organization, high visibility, and potential for professional rewards, is crucial for attracting and holding high-quality people. Senior management can help develop a "priority image" and communicate the key parameters and management guidelines for specific projects. Moreover, establishing and communicating clear and stable top-down objectives helps in building an image of high visibility, importance, priority, and interesting work. Such a pervasive process fosters a climate of active participation at all levels, helps attract and hold quality people, unifies the team, and minimizes dysfunctional conflict.

68.8.5 Process Definition and Team Structure

The proper setup and communication of the operational transfer process, such as concurrent engineering, stage-gate process. CAD/CAE/CAM, and design-build, is important for establishing the cross-functional linkages necessary for innovative engineering performance. Management must also define the basic team structure for each project early in its life cycle. The project plan, task matrix, project charter, and operating procedure are the principal management tools for defining organizational structure and business process.

68.8.6 Interesting Work

Whenever possible, managers should try to accommodate the professional interests and desires of their personnel. Interesting and challenging work is a perception that can be enhanced by the visibility of the work, management attention and support, priority image and the overlap of personnel values and perceived benefits with organizational objectives. Making work more interesting leads to increased involvement, better communication, lower conflict, higher commitment, stronger work effort, and higher levels of creativity.

68.8.7 Senior Management Support

It is critically important that senior management provide the proper environment for an engineering team to function effectively. At the onset of a new development, the responsible manager needs to negotiate the needed resources with the sponsor organization, and obtain commitment from management that these resources will be available. An effective working relationship among resource managers, project leaders, and senior management critically affects the perceived credibility, visibility, and priority of the engineering team and their work.

68.8.8 Clear Communication

Poor communication is a major barrier to teamwork and effective engineering performance. Management can facilitate the free flow of information, both horizontally and vertically, by work space design, regular meetings, reviews and information sessions. In addition, modern technology, such as voice mail, e-mail, electronic bulletin boards and conferencing, can greatly enhance communications, especially in complex organizational settings.

68.8.9 Commitment

Managers should ensure team-member commitment to their project plans, specific objectives, and results. If such commitments appear weak, managers should determine the reason for such lack of commitment of a team member and attempt to modify possible negative views. Because insecurity is often a major reason for low commitment, managers should try to determine why insecurity exists, then work to reduce the team members' fears and anxieties. Conflict with other team members and lack of interest in the project may be other reasons for such lack of commitment.

68.8.10 Leadership

Leadership positions should be carefully defined and staffed for all projects and support functions. Especially critical is the credibility of project leaders among team members, with senior management and with the program sponsor, for the leader's ability to manage multidisciplinary activities effectively across functional lines.

68.8.11 Reward System

Personnel evaluation and reward systems should be designed to reflect the desired power equilibrium and authority/responsibility-sharing of an organization. A QFD-philosophy helps to focus efforts toward desired results on company internal and external customers to foster a work environment that is strong on self-direction and self-control.

68.8.12 Problem Avoidance

Engineering managers should focus their efforts on problem avoidance. That is, managers and team leaders, through experience, should recognize potential problems and conflicts at their onset and deal with them before they become big and their resolutions consume a large amount of time and effort.

68.8.13 Personal Drive and Leadership

Managers can influence the work environment by their own actions. Concern for the team members, the ability to integrate personal needs of their staff with the goals of the organization, and the ability to create personal enthusiasm for a particular project can foster a climate of high motivation, work involvement, open communication, and ultimately high engineering performance.

68.9 A FINAL MESSAGE

Sophisticated people skills are crucial to effective role performance in technology-based organizations. Managers have to cross organizational, national, and cultural boundaries and work with people over whom they have little or no formal control. Alliances and collaborative ventures have forced these managers to focus more on cross-boundary relationships, negotiations, delegation, and commitment than on establishing formal command- and control-systems. To be effective in such a team environment, the manager must understand the interaction of organizational and behavioral variables. This understanding will facilitate a climate of active participation, minimal dysfunctional conflict, and effective communication. It will also foster an ambience conducive to chance, commitment, and self-direction. No single set of broad guidelines exists that guarantees instant managerial success. However, by understanding the variables and the interrelationships that drive people toward high performance in a technology-oriented environment, managers can examine and fine-tune leadership styles, actions, and resource allocations toward continuing organizational improvement.

REFERENCES

1. H. J. Thamhain, "Managing Technology: The People Factors," *Technical & Skill Training*, 24–31 (August 1990).

2. H. J. Thamhain, "Managing Engineers Effectively," *IEEE Transactions on Engineering Management,* 231–237 (August 1983).

3. H. J. Thamhain, *Engineering Management: Managing Effectively in Technology-Based Organizations,* New York, Wiley, 1992.

4. H. J. Thamhain, "Effective Leadership Style for Managing Project Teams," in *Handbook of Program and Project Management,* P. C. Dinsmore (ed.), AMACOM, New York, 1992, Chap. 22.

5. H. J. Thamhain, "Developing Engineering Program Management Skills," in *Handbook: Management of R&D and Engineering,* D. Kocaoglu (ed.), Wiley, New York, 1992, Chap. 22.

6. G. R. Gemmill and H. J. Thamhain, "Influence Styles of Project Managers: Some Project Performance Correlates," *Academy of Management Journal* (June 1974).

7. R. Katz, "Managing High-Performance R&D Teams," *European Management Journal* 12(3), 243–252 (September 1994).

8. D. Hague, "The Development of Managers' Knowledge and Skills," *International Journal of Technology Management* 2, 699–710 (1987).

9. J. White, "Developing Leaders for the High Performance Workplace," *Human Resource Management* 33(1), 161–168 (Spring 1994).

10. A. Nurick and H. Thamhain, "Project Team Development in Multinational Environments," in *Global Project Management Handbook,* D. Cleland (ed.), McGraw-Hill, New York, 1993, Chap. 38.

11. A. Shenhar and H. Thamhain, "A New Mixture of Project Management Skills," *Human Systems Management Journal* 13(1), 27–40 (March 1994).

12. J. J. Aquilino, "Multi-skilled Work Teams: Productivity Benefits," *California Management Review* (Summer 1977).

13. R. L. Klien and H. B. Anderson, "Teambuilding Styles and Their Impact on Project Management Results," *Project Management Journal* 27(1), 41–50 (March 1996).

14. S. Oderwald, "Global Work Teams," *Training and Development* 5(2), 35–45 (February 1996).

15. H. J. Thamhain, "Managing Technologically Innovative Team Efforts toward New Product Success," *Journal of Product Innovation Management* 7(1), 5–18 (March 1990).

16. H. J. Thamhain, "Working with Project Teams," in *Project Management Strategic Design and Implementation,* D. Cleland (ed.), McGraw-Hill, New York, 1998, Chap. 38.

17. H. J. Thamhain and D. L. Wilemon, "Building High Performance Engineering Project Teams," in *The Human Side of Managing Technological Innovation,* R. Katz (ed.), Oxford, Oxford University Press, 1996, Chap. 12, pp. 125–136.

18. J. Shaw, C. Fisher, and A. Randolph, "From Maternalism to Accountability," *Academy of Management Executive* 5(1), 7–20 (February 1991).

19. H. J. Thamhain, "Managing Self-Directed Teams toward Innovative Results," *Engineering Management Journal* 8(3) (September 1996).

20. H. J. Thamhain and D. L. Wilemon, "Leadership Conflict and Project Management Effectiveness," *Executive Bookshelf on Generating Technological Innovations, Sloan Management Review,* 68–87 (Fall 1987).

21. P. R. Harris, "Building a High-Performance Team," *Training Development Journal* (April 1986).

22. D. A. Roming, *Breakthrough Teamwork,* New York: Irwin, New York, 1996.

23. H. R. Jessup, "New Roles of Leadership," *Training and Development Journal* 44(11) (November 1990).

24. H. J. Thamhain, "Managing Technologically Innovative Team Efforts towards New Product Success," *Journal of Product Innovation Management* 7(1), 5–18 (March 1990).

25. A. J. Nurick, "Facilitating Effective Work Teams," *SAM Advanced Management Journal* 57(2) (Spring 1992).

26. I. Janis, *Victims of Groupthink,* Boston: Houghton Mifflin, Boston, 1972.

27. G. M. Parker (ed.), *Handbook of Best Practices for Teams,* HRD Press, 1996.

28. H. J. Thamhain and D. L. Wilemon, "Building High-Performing Engineering Project Teams," in R. Katz (ed.), *The Human Side of Managing Technological Innovation,* New York: Oxford University Press, New York, 1997, Chap. 12.

29. D. L. Wilemon, "Cross-functional Teams and New Product Development," in *Proceedings, Portland International Conference on Management of Engineering and Technology (PICMET '97),* Portland, OR, July 1997.

CHAPTER 69

FINANCE AND THE ENGINEERING FUNCTION

William Brett
New York, New York

69.1	**INTRODUCTION AND OUTLINE**	**2097**
	69.1.1 Needs of Owners, Investors, and Lenders	2098
	69.1.2 Needs of Top Managers	2098
	69.1.3 Needs of Middle Managers of Line Functions	2098
	69.1.4 Needs of Staff Groups (Product Planners, Engineers, Market Researchers)	2099
	69.1.5 Needs of Accountants	2099
69.2	**A FINANCIAL MODEL**	**2100**
69.3	**BALANCE SHEET**	**2100**
	69.3.1 Current Assets	2102
	69.3.2 Current Liabilities	2102
	69.3.3 Accrual Accounting	2102
	69.3.4 Interest-Bearing Current Liabilities	2102
	69.3.5 Net Working Capital	2103
	69.3.6 Current Ratio	2103
	69.3.7 Fixed Assets	2103
	69.3.8 Total Capital	2104
	69.3.9 Second Year Comparison	2104
69.4	**PROFIT AND LOSS STATEMENT**	**2105**
	69.4.1 Financial Ratios	2105
69.5	**CASH FLOW OR SOURCE AND APPLICATION OF FUNDS**	**2107**
	69.5.1 Accelerated Depreciation	2108
69.6	**EVALUATING RESULTS AND TAKING ACTION**	**2111**
	69.6.1 Comparing Current Results with Budgets and Forecasts	2111
	69.6.2 Identifying Problems and Solutions	2113
	69.6.3 Initiating Action	2114
69.7	**FINANCIAL TOOLS FOR THE INDEPENDENT PROFESSIONAL ENGINEER**	**2114**
	69.7.1 Simple Record-Keeping	2116
	69.7.2 Getting the System Started	2116
	69.7.3 Operating the System	2116
69.8	**CONCLUSIONS**	**2116**

69.1 INTRODUCTION AND OUTLINE

Finance is fundamental; accounting is merely the set of procedures, techniques, and reports that make possible the effective execution of the finance function. Harold Geneen, the legendary chairman of International Telephone and Telegraph, included in his *Sayings of Chairman Hal*, "The worst thing a manager can do is run out of money." He meant it! The corporate function of Finance is that function which makes the decisions, or rather provides the recommendations to top management who really make the decisions, that prevent the enterprise from running out of money. Accounting gathers, organizes, and disseminates information that make it possible to make these decisions accurately and timely. In modern business, accounting performs many correlative functions, some in such detail and so esoteric as to appear to be an end in themselves.

The objectives of this chapter on finance and accounting are to describe:

- How accounting systems work to provide information for top managers and owners
- How financial management is carried out

Mechanical Engineers' Handbook, 2nd ed., Edited by Myer Kutz.
ISBN 0-471-13007-9   © 1998 John Wiley & Sons, Inc.

Additionally, this chapter provides a concise description of how an accounting system is constructed to provide for the needs of middle management and staff groups such as engineers and marketers.

The purposes and uses of accounting systems, data, and reports are quite different for different people and functions in the business community. The engineer needs to understand accounting principles and processes as they apply to his or her function and also to understand the way in which others of the enterprise view business and what their information needs are. The following are five major groups that have distinctly differing points of view and objectives:

- Owners, investors, lenders, and boards of directors
- Top managers
- Middle managers of line functions
- Staff groups such as product planners, engineers, and market researchers
- Accountants

69.1.1 Needs of Owners, Investors, and Lenders

The first group—owners, investors, and lenders—have as their primary concern the preservation and protection of the capital or the assets of the business. The Board of Directors represents the interest of the owners and can be considered to be the agents of the owners (stockholders). The board members provide continuing review of the performance of top management as well as approval or disapproval of policies and key investment decisions.

This entire group wants to be assured that the property of the business—fixed plant and equipment, inventories, etc.—is being conserved. Next, they want to be assured that there will be sufficient liquidity, which means only that there will be enough cash available to pay all the bills as they come due. Finally, they want to see evidence of some combination of regular payout or growth in value—a financial return such as regular dividends or indications that the enterprise is increasing in value. Increase in value may be evidenced by growth in sales and profits, by increases in the market value of the stock, or by increased value of the assets owned. If the dividend payout is small, the growth expectations will be large.

The information available to the owners is, at a minimum, that which is published for public companies—the balance sheet, cash flow, and profit and loss statement. Special reports and analyses are also provided when indicated.

69.1.2 Needs of Top Managers

The top managers must be sensitive to the needs and desires of the owners as expressed by the Board of Directors and of the bankers and other lenders so that all of the purposes and objectives of owners and lenders are also the objectives of top managers. Additionally, top management has the sole responsibility for:

Developing long-range strategic plans and objectives
Approving short-range operating and financial plans
Ensuring that results achieved are measuring up to plan
Initiating broad gage corrective programs when results are not in conformance with objectives

Reports of financial results to this group must be in considerable detail and identified by major program, product, or operating unit in order to give insight sufficient to correct problems in time to prevent disasters. The degree of detail is determined by the management style of the top executive. Usually such reports are set up so that trouble points are automatically brought to the top executives' attention, and the detail is provided in order to make it possible to delve into the problems.

In addition to the basic financial reports to the owners, directors and top managers need:

- Long-term projections
- One-year budgets
- Periodic comparison of budget to actual
- Unit or facility results
- Product line results
- Performance compared to standard cost

69.1.3 Needs of Middle Managers of Line Functions

For our present purposes we will consider only managers of the sales and the manufacturing groups and their needs for financial, sales, and cost information. The degree to which the chief executive shares information down the line varies greatly among companies, ranging from a highly secretive handling of all information to a belief that sharing all the facts of the business improves performance and involvement through greater participation. In the great bulk of publicly held, large corporations, with modern management, most of the financial information provided to top management is available to staff and middle management, either on routine basis or on request. There are additional data that are needed by lower-level line managers where adequate operational control calls for much greater detail than that which is routinely supplied to top executives.

The fundamental assignment of the line manager in manufacturing and sales is to execute the policies of top management. In order to do this effectively, the manager needs to monitor actions and evaluate results. In an accounting context this means the manufacturing manager, either by formal rules or by setting his or her personal rules of thumb, needs to:

- Set production goals
- Set worker and machine productivity standards
- Set raw material consumption standards
- Set overhead cost goals
- Establish product cost standards
- Compare actual performance against goals
- Develop remedial action plans to correct deficiencies
- Monitor progress in correcting variances

The major accounting and control tools needed to carry out this mission include:

- Production standards
- Departmental budgets
- Standard costs
- Sales and production projections
- Variance reports
- Special reports

It is important that the line manager understands the profit and loss picture in his or her area of control and that job performance is not merely measured against preset standards but that he or she is considered to be an important contributor to the entire organization. It is, then, important that managers understand the total commercial environment in which they are working, so that full disclosure of product profits is desirable. Such a philosophy requires that accounting records and reports be clear and straightforward, with the objective of exposing operating issues rather than being designed for a tax accountant or lawyer.

The top marketing executive must have a key role in the establishment of prices and the determination of the condition of the market so that he or she is a full partner in managing the enterprise for profits. He or she therefore needs to participate with the manufacturing executive in the development of budgets and longer-range financial plans. Thus the budget becomes a joint document of marketing and manufacturing, with both committed to its successful execution.

The marketing executive needs to be furnished with all of the information indicated above as appropriate for the manufacturing manager.

69.1.4 Needs of Staff Groups (Product Planners, Engineers, Market Researchers)

The major requirement of accounting information for staff is that it provide a way to measure the economic effect of proposed changes to the enterprise. For the engineer this may mean changes in equipment or tooling or redesign of the product as a most frequent kind of change that must be evaluated before funds can be committed.

Accounting records that show actual and standard costs by individual product and discrete operation are invaluable in determining the effect of change in design or process. If changes in product or process can result in changes in total unit sales or in price, the engineer needs to know those projected effects. His or her final projections of improved profits will then incorporate the total effect of engineering changes.

The accounting records need to be in sufficient detail that new financial projections can be made reliably, with different assumptions of product features, sales volume, cost, and price.

69.1.5 Needs of Accountants

The accounting system must satisfy the strategic, operational, and control requirements of the organization as outlined above, but it has other external demands that must be satisfied. The accountants have the obligation to maintain records and prepare reports to shareholders that are "in conformity with generally accepted accounting principles consistently applied." Therefore, traditional approaches are essential so that the outside auditor as well as the tax collector will understand the reports and find them acceptable. There seems to be little need to sacrifice the development of good, effective control information for operating executives in order to satisfy the requirements of the tax collector or the auditor. The needs are compatible.

The key financial reporting and accounting systems typically used by each group are explained next.

69.2 A FINANCIAL MODEL

A major concern of the owners or the Board of Directors and the lenders to the business must be to ensure the security of the assets of the business. The obvious way to do this in a small enterprise is occasionally to take a look. It is certainly appropriate for directors to visit facilities and places where inventories are housed to ensure that the assets really do exist, but this can only serve as a spot check and an activity comparable to a military inspection—everything looks very good when the troops know that the general is coming. The most useful and convenient way, as well as the most reliable way, to protect the assets is by careful study of financial records and a comparison with recent history to determine the trends in basic values within the business. A clear and consistent understanding of the condition of the assets of the business requires the existence of a uniform and acceptable system of accounting for them and for reporting their condition. The accounting balance sheet provides this.

In the remainder of this chapter, a set of examples based on the experience of one fictitious company is developed. The first element in the case study is the corporate balance sheet. From there the case moves back to the profit and loss and the cash flow statements. The case moves eventually back to the basic statements of expense and revenue to demonstrate how these records are used by the people managing the business—how these records enable them to make decisions concerning pricing, product mix, and investment in new plant and processes. The case will also show how these records help management to direct the business into growth patterns, a strengthened financial position, or increased payout to the owners.

The name of the fictitious company is the Commercial Construction Tool Company, Incorporated, and will be referred to as CCTCO throughout the remainder of this chapter. The company manufactures a precision hand tool, which is very useful in the positioning and nailing of various wooden structural members as well as sheathing in the construction of frame houses. The tool is a proprietary product on which the patents ran out some time ago; however, the company has had a reputation for quality and performance that has made it very difficult for competition to gain much headway. The tool has a reputation and prestige among users such that no apprentice carpenter would be without one. The product is sold through hardware distributors who supply lumber yards and independent retail hardware stores. About three years ago the company introduced a lighter weight and somewhat simplified model for use in the "do-it-yourself" market. Sales of the home-use model have been good and growing rapidly, and there is some concern that the HOMMODEL (home model) is cannibalizing sales of the COMMODEL (commercial model).

The company has one manufacturing facility and its general offices and sales offices are at the same location.

At the first directors' meeting affer the year-end closing of the books the board is presented with the financial statements starting with the balance sheets for the beginning and end of the year. The principle of the balance sheet is that the enterprise has a net value to the owners (net worth) equal to the value of what is owned (the assets) less the amount owed to others (the liabilities).

69.3 BALANCE SHEET

When any business starts, the first financial statement is the balance sheet. In the case of CCTCO, the company was started many years ago to exploit the newly patented product. The beginning balance sheet was the result of setting up the initial financing. To get the enterprise started the original owners determined that $1000 (represents one million dollars, since in all of the exhibits and tables the last three zeros are deleted) was needed. The inventor and friends and associates put up $600 as the owners share—600,000 shares of common stock at a par value of $1 per share. Others, familiar with the product and the originators of the business, provided $400 represented by notes to be paid in 20 years—long-term debt. The original balance sheet was as shown below:

Assets		Liabilities and Net Worth	
		Liabilities	
Cash	1000	Long-term debt	400
		Net worth	–0–
		Capital stock	600
Total assets	1000	Total liabilities and net worth	1000

The first financial steps of the company were to purchase equipment and machinery for $640 and raw materials for $120. The equipment was sent COD, but the raw material was to be paid for in 30 days. Immediately the balance sheet became more complex. There were now current assets—cash and inventory of raw materials—as well as fixed assets—machinery. Current liabilities showed up now in the form of accounts payable-the money owed for the raw material. All this before anything was produced. Now the balance sheet had become:

Assets		Liabilities and Net Worth	
		Liabilities	
Cash	360	Accounts payable	120
Inventories	120	Current liabilities	120
Current assets	480	Long-term debt	400
Fixed assets	640	Total liabilities	560
		Net worth	
		Capital stock	600
Total assets	1120	Total liabilities and net worth	1120

Affer a number of years of manufacturing and selling product the balance sheet became as shown below in Table 69.1. This important financial report requires explanation.

Assets are generally of three varieties:

- *Current.* Usually liquid and will probably be turned over at least once each year.
- *Fixed.* Usually real estate and the tools of production, frequently termed plant, property, and equipment.
- *Intangible.* Assets without an intrinsic value, such as good will or development costs which are not written off as a current expense but are declared an asset until the development has been commercialized.

Table 69.1 Commercial Construction Tool Co., Inc.

Balance Sheet	Beginning
Assets	
Current assets	
Cash	52
Accounts receivable	475
Inventories	941
Total current assets	1468
Fixed assets	
Gross plant and equipment	2021
Less reserve for depreciation	471
Net plant and equipment	1550
Total assets	3018
Liabilities	
Current liabilities	
Accounts payable	457
Short-term debt	565
Long-term debt becoming current	130
Total current liabilities	1152
Long-term liabilities	
Interest-bearing debt	843
Total liabilities	1995
Net worth	
Capital stock	100
Earned surplus	923
Total net worth	1023
Total liabilities and net worth	3018

69.3.1 Current Assets

In CCTCO's balance sheet the first item to occur is cash, which the company tries to keep relatively low, sufficient only to handle the flow of checks. Any excess over that amount the treasurer applies to pay off short-term debt, which has been arranged with local banks at one-half of one percent over the prime rate.

Accounts receivable are trade invoices not yet paid. The terms offered by CCTCO are typical—2% 10 days net 30, which means that if the bill is paid by the customer within 10 days after receipt, he or she can take a 2% discount, otherwise the total amount is due within 30 days. Distributors in the hardware field are usually hard pressed for cash and are frequently slow payers. As a result, receivables are the equivalent of two and a half month's sales, tying up a significant amount of the company's capital.

Inventories are the major element of current assets and consist of purchased raw materials, primarily steel, paint, and purchased parts; work in process, which includes all material that has left the raw material inventory point but has not yet reached the stage of completion where it is ready to be shipped; and finished goods. In order to provide quick delivery service to customers, CCTCO finds it necessary to maintain inventories at the equivalent of about three months' shipments—normally about 25% of the annual cost of goods sold.

69.3.2 Current Liabilities

Skipping to the liability section of the report, in order to look at all the elements of the liquid segment of the balance sheet, we next evaluate the condition of current liabilities. This section is composed of two parts: interest-bearing debt and debt that carries no interest charge. The noninterest-bearing part is primarily accounts payable, which is an account parallel but opposite to accounts receivable. It consists of the trade obligations not yet paid for steel, paint, and parts as well as office supplies and other material purchases. Sometimes included in this category are estimates of taxes that have been accrued during the period but not yet paid as well as other services used but not yet billed or paid for.

69.3.3 Accrual Accounting

At this point it is useful to define the term "accrued" or "accrual" as opposed to "cash" basis accounting. Almost all individual, personal accounting is done on a cash basis, that is, for individual tax accounting, no transaction exists until the money changes hands—by either writing a check or paying cash. In commercial and industrial accounting the accrual system is normally used, in which the transaction is deemed to occur at the time of some other overt act. For example, a sale takes place when the goods are shipped against a bona fide order, even though money will not change hands for another month. Taxes are charged based on the pro rata share for the year even though they may not be paid until the subsequent year. Thus costs and revenues are charged when it is clear that they are in fact obligated. This tends to anticipate and level out income and costs and to reduce the effect of fluctuations resulting only from the random effect of the time at which payments are made. Business managers wish to eliminate, as far as possible, wide swings in financial results and accrual accounting assists in this, as well as providing a more clearly cause-related set of financial statements. It also complicates the art of accounting quite considerably.

69.3.4 Interest-Bearing Current Liabilities

Interest-bearing current obligations are of two types: short-term bank borrowings and that portion of long-term debt that must be paid during the current year. Most businesses, and particularly those with a seasonal variation in sales, find it necessary to borrow from banks on a regular basis. The fashion clothing industry needs to produce three or four complete new lines each year and must borrow from the banks to provide the cash to pay for labor and materials to produce the fall, winter, and spring lines. When the shipments have been made to the distributors and large retail chains and their invoices have been paid, the manufacturer can "get out of the banks," only to come back to finance the next season's line. Because CCTCO's sales have a significant summer bulge at the retail level, they must have heavy inventories in the early spring, which drop to a fairly low level in the fall. Bank borrowings are usually required in February through May, but CCTCO is normally out of the banks by year end, so that the year-end balance sheet has a sounder look than it would have in April. The item "short-term borrowings" of $565 consists of bank loans that had not been paid back by the year's end.

The second part of interest-bearing current liabilities is that part of the long-term debt that matures within 12 months, and will have to be paid within the 12-month period. Such obligations are typically bonds or long-term notes. These current maturities represent an immediate drain on the cash of the business and are therefore classed as a current liability. As CCTCO has an important bond issue with maturities taking place uniformly over a long period, it has long-term debt maturing in practically every year.

69.3.5 Net Working Capital

The total of current assets less current liabilities is known as "net working capital." Although it is not usually defined in the balance sheet, it is important in the financial management of a business because it represents a large part of the capitalization of an enterprise and because, to some degree, it is controllable in the short run.

In times of high interest rates and cash shortages, companies tend to take immediate steps to collect their outstanding bills much more quickly. They will carefully "age" their receivables, which means that an analysis showing receivables ranked by the length of time they have been unpaid will be made and particular pressure will be brought to bear on those invoices that have been outstanding for a long time. On the other hand, steps will be taken to slow the payment of obligations; discounts may be passed up if the need for cash is sufficiently pressing and a general slowing of payments will occur.

Considerable pressure will be exerted to reduce inventories in the three major categories of raw material, work in process, and finished goods as well as stocks of supplies. Annual inventory turns can sometimes be significantly improved. There are, however, irreducible minimums for net working capital, and going beyond those points may result in real damage to the business through reducing service, increasing delivery times, damaging credit ratings, and otherwise upsetting customer and supplier relationships.

The effect of reducing net working capital, in a moderate and constructive way, spreads through the financial structure of the enterprise. The need for borrowing is reduced and interest expense is thereby reduced and profits are increased. Also, another effect on the balance sheet further improves the financial position. As the total debt level is reduced and the net worth is increased, the ratio of debt to equity is reduced, thus improving the financial community's assessment of strength. An improved rating for borrowing purposes may result, making the company eligible for lower interest rates. Other aspects of this factor will be covered in more detail in the discussion of net worth and long-term debt.

69.3.6 Current Ratio

The need to maintain the strength of another important analysis ratio puts additional resistance against the objective of holding net working capital to the minimum. Business owners feel the need to maintain a healthy "current ratio." In order to be in a position to pay current bills, the aggregate of cash, receivables, and inventories must be available in sufficient amount. One measure of the ability to pay current obligations is the ratio of current assets to current liabilities, the current ratio. In more conservative times and before the days of leverage, a ratio of 2.0 or even 3.0 was considered strong, an indication of financial stability. In times of high interest rates and with objectives of rapid growth, much lower ratios are acceptable and even desirable. CCTCO's ratio of 1.27 ($1468/$1152) is considered quite satisfactory.

69.3.7 Fixed Assets

Continuing the evaluation on the asset side of the balance sheet we find the three elements of fixed assets, that is, gross plant and equipment, reserve for depreciation, and net plant and equipment. Gross plant is the original cost of all the assets now owned and is a straightforward item. The concept of depreciation is one which is frequently misunderstood and partly because of the name "reserve for depreciation." The name seems to indicate that there is a reserve of cash, put away somewhere that can be used to replace the old equipment when necessary. This is not the case. Accountants have a very special meaning for the word reserve in this application. It means, to an accountant, the sum of the depreciation expense that has been applied over the life, up to now, of the asset.

When an asset, such as a machine, is purchased, it is assigned an estimated useful life in years. In a linear depreciation system, the value of the asset is reduced each year by the same percentage that one year is to its useful life. For example, an asset with a 12-year useful life would have an 8.33% annual depreciation rate (100 times the reciprocal of 12). The critical reason for reducing the value each year is to reduce the profit by an amount equivalent to the degree to which equipment is transformed into product. With high income taxes, the depreciation rate is critical to ensuring that taxes are held to the legal minimum. When the profit and loss statement is covered, the effect on profits and cash flow as a result of using nonlinear, accelerated depreciation rates will be covered. The important point to understand is that the reserve for depreciation does not represent a reserve of cash but only an accounting artifice to show how much depreciation expense has been taken (charged against profits) so far and, by difference, to show the amount of depreciation expense that may be taken in the future.

The difference between gross plant and reserve for depreciation, net plant and equipment, is not necessarily the remaining market value of the equipment at all, but is the amount of depreciation expense that may be charged against profits in future years. The understanding of this principle of depreciation is critical to the later understanding of profits and cash flow.

69.3.8 Total Capital

Together, the remaining items (long-term debt and net worth) on the liability side of the balance sheet make up the basic investment in the business. In the beginning, the entrepreneurs looked for money to get the business started. It came from two sources, equity investors and lenders. The equity investors were given an ownership share in the business, with the right to a portion of whatever profits might be made or a pro rata share of the proceeds of liquidation, if that became necessary. The lenders were given the right to regular and prescribed interest payments and were promised repayment of principal on a scheduled basis. They were not to share in the profits, if any. A third source of capital became available as the enterprise prospered. Profits not paid out in dividends were reinvested in plant and equipment and working capital. Each of these sources has an official name.

Lenders: Long-Term Debt
Equity Investors: Capital Stock
Profits Reinvested: Earned Surplus

In many cases the cash from equity investors is divided into two parts, the par value of the common shares issued, traditionally $1 each, and the difference between par and the actual proceeds from the sale of stock. For example, the sale of 1000 shares of par value $1 stock, for $8000 net of fees, would be expressed:

Capital Stock (1000 shares at $1 par): $1000
Paid in Surplus: $7000
Total Capitalization: $8000

The final item on the balance sheet, earned surplus or retained earnings, represents the accumulated profits generated by the business which have not been paid out, but were reinvested.

Net worth is the total of capital stock and earned surplus and can also be defined as the difference, at the end of an accounting period, between the value of the assets, as stated on the corporate books, and the obligations of the business.

All of this is a simplified view of the balance sheet. In actual practice there are a number of other elements that may exist and take on great importance. These include preferred stocks, treasury stock, deferred income taxes, and goodwill. When any of these special situations occur, a particular review of the specific case is needed in order to understand the implications to the business and their effect on the financial condition of the enterprise.

69.3.9 Second Year Comparison

The balance sheet in Table 69.1 is a statement of condition. It tells the financial position of the company at the beginning of the period. At the end of the year the Board of Directors is presented two balance sheets—the condition of the business at the beginning and at the end of the period, as shown in Table 69.2. The Board is interested in the trends represented by the change in the balance sheet over a 1-year period.

Total assets have increased by $395 over the period—probably a good sign. Net worth or owners' equity has increased by $27, which is $368 less than the increase in assets. The money for the increase in assets comes from creating substantially more liabilities or obligations as well as the very small increase in the net worth. A look at the liabilities shows the following (note the errors from rounding that result from the use of computer models for financial statements):

	Increase
Accounts payable	$ 46
Short-term debt	36
Long-term debt	286
	$368

Changes in net working capital are evaluated to determine the efficiency in the use of cash and the soundness of the short-term position. No large changes that would raise significant questions have taken place. Current assets increased $167 and current liabilities by $82. These increases result from the fact that sales had increased, which had required higher inventories and receivables. The current ratio (current assets over current liabilities) had strengthened to 1.33 from 1.27 at the beginning of the period, indicating an improved ability to pay bills and probably increased borrowing power.

A major change in the left-hand (asset) side was the increase in fixed assets. Gross plant was up $500, nearly 25%, indicating an aggressive expansion or improvement program.

Net worth and earned surplus were up by $27, an important fact, sure to receive attention from the board.

Table 69.2 Commercial Construction Tool Co., Inc.—Costs and Revenues, Bad Year—Actual

Balance Sheet	Beginning	Ending	Change
Assets			
Current assets			
Cash	52	62	10
Accounts receivable	475	573	98
Inventories	941	1000	59
Total current assets	1468	1635	167
Fixed assets			
Gross plant and equipment	2021	2521	500
Less reserve for depreciation	471	744	273
Net plant and equipment	1550	1777	227
Total assets	3018	3413	395
Liabilities			
Current liabilities			
Accounts payable	457	503	46
Short-term debt	565	600	36
Long-term debt becoming current	130	130	0
Total current liabilities	1152	1233	82
Long-term liabilities			
Interest-bearing debt	843	1129	286
Total liabilities	1995	2362	368
Net worth			
Capital stock	100	100	0
Earned surplus	923	950	27
Total net worth	1023	1050	27
Total liabilities and net worth	3018	3413	395

In order to understand why the balance sheet had changed and to further evaluate the year's results, the directors needed a profit and loss statement and a cash flow statement.

69.4 PROFIT AND LOSS STATEMENT

The profit and loss statement (P&L) is probably the best understood and most used statement provided by accountants: It summarizes most of the important annual operating data and it acts as a bridge from one balance sheet to the next. It is a summary of transactions for the year—where the money came from and where most of it went. Table 69.3 is the P&L for CCTCO for the year.

For the sake of simplicity, net sales are shown as Sales. In many statements, particularly internal reports, gross sales are shown followed by returns and discounts to give a net sales figure. Cost of sales is a little more complex. Sales may be made from inventory or off the production line on special order. Stocks of finished goods or inventories are carried on the books at their cost of production. The formula for determining the cost of product shipped to customers is:

$$\text{beginning inventory} + \text{cost of production} - \text{ending inventory} = \text{cost of sales}$$

Additionally, CCTCO uses a standard burden rate system of applying overhead costs to production. The difference between the overhead charged to production at standard burden rates and the actual overhead costs for the period, in this case $62, is called unabsorbed burden and is added to the cost of production for the year, or it may be charged off as a period cost. The procedures for developing burden rates will be treated in more detail in a subsequent section.

Gross margin is the difference between sales dollars and the cost of manufacture. After deducting the costs of administrative overhead and selling expense, operating profit remains. Interest expense is part of the total cost of capital of the business and is therefore separate from operations. The last item, income tax, only occurs when there is a profit.

69.4.1 Financial Ratios

The combination of the P&L and the balance sheet makes it possible to calculate certain ratios that have great significance to investors. The ratios are shown in Table 69.4. The first and most commonly

Table 69.3 Commercial Construction Tool Co., Inc.—Costs and Revenues, Bad Year—Actual

Profit and Loss Statement	($000)
Sales	4772
Cost of production	4097
Beginning inventory	941
Ending inventory	1000
Net change	59
Cost of sales	4038
Gross margin	734
Selling expense	177
Administrative	249
Operating profit	308
Interest	169
Profit before tax	138
Income tax	66
Net income	72

used as a measure of success is the return on sales. This is a valuable ratio to measure progress of a company from year to year, but is of less importance in comparing one company to another. A more useful ratio would be returns to value added. Value added is the difference between the cost of purchased raw materials and net sales, and represents the economic contribution of the enterprise. It is a concept used more extensively in Europe than the United States and is the basis of the Value Added Tax (VAT), quite common in Europe and at this writing being considered in the United States.

Return on assets begins to get closer to the real interest of the investor. It represents the degree to which assets are profitable, and would indicate, from an overall economic point of view, whether the enterprise was an economic and competitive application of production facilities.

A ratio even more interesting to the investor is the return on invested capital. Total assets, as was described earlier, are financed by three sources:

- Equity—made up of stock, that is, owners' investment and profits retained in the business
- Interest-bearing debt—composed of bonds, notes, and bank loans
- Current liabilities—composed of operating debts such as accounts payable and taxes payable, which do not require interest payments

Because the current liabilities are normally more than offset by current assets, the economic return is well described by the return on total or invested capital, which is net profit after taxes divided by the sum of equity plus interest-bearing debt.

A rate of return percentage of great interest to the owner is the return on equity. This rate of return compared to the return on total capital represents the degree to which the investment is or can be leveraged. It is to the interest of the investor to maximize the return on his or her dollars invested, so, to the degree that money can be borrowed at interest rates well below the capacity of the business to provide a return, the total profits to the owners will increase. Return on equity is a function of

Table 69.4 Commercial Construction Tool Co., Inc.—Costs and Revenues, Bad Year—Actual

Financial Ratios	
Return on sales	1.51
Return on assets	2.24
Return on invested capital	3.56
Return on equity	6.94
Asset to sales ratio	0.67
Debt percent to debt plus equity	69
Average cost of capital	20.67

the ratio of debt to debt plus equity (total capital) and is a measure of the leverage percentage in the business. It is to the advantage of the owners to increase this ratio in order to increase the return on equity up to the point that the investment community, including bankers, concludes that the company is excessively leveraged and is in unsound financial condition. At that point it becomes more difficult to borrow money and interest rates of willing lenders increase significantly. Fashions in leverage change depending on the business cycle. In boom times with low interest rates, highly leveraged enterprises are popular, but tend to fall into disfavor when times are tough.

A more direct measure of leverage is "debt percent to debt plus equity" or debt to total capital. The 69% for CCTCO indicates that lenders really "own" 69% of the company and investors only 31%.

Another ratio of interest to investors is the asset turnover or asset to sales ratio. If sales from a given asset base can be very high, the opportunity to achieve high profits appears enhanced. On the other hand, it is very difficult to change the asset to sales ratio very much without changing the basic business. Certain industries or businesses are characterized as being capital intensive, which means they have a high asset to sales ratio or a low asset turnover. It is fundamental to the integrated forest products industries that they have a high asset to sales ratio, typically one to one. The opposite extreme, for example, the bakery industry, may have a ratio of 0.3–0.35 and turn over assets about three times per year. Good management and very effective use of facilities coupled with low inventories can make the best industry performer 10% better than the average, but there is no conceivable way that the fundamental level can be dramatically and permanently changed.

The final figure in Table 69.4, that of average cost of capital, cannot be calculated from only the P&L and balance sheet. One component of the total cost of capital is the dividend payout, which is not included in either report. It was stated previously that the P&L shows where most of the money went—it does not include dividends and payments for new equipment and other capital goods. For this we need the cash flow, also known as the source and application of funds, shown in Table 69.5.

69.5 CASH FLOW OR SOURCE AND APPLICATION OF FUNDS

There are two sources of operating cash for any business: the net profits after tax and noncash expenses. In Table 69.5, the cash generated by the business is shown as $344, the sum of net profit and depreciation. This is actually the operating cash generated and does not include financing cash sources, which are also very important. These sources include loans, capital contributions, and the sale of stock and are included in the cash flow statement as well as in the balance sheet where they have already been reviewed in a previous section of this chapter.

It seems clear and not requiring further explanation that the net profit after tax represents money remaining at the end of the period, but the treatment of noncash expenses as a source of operating funds is less self-evident. Included in the cost of production and sales in the previous section were materials and labor and many indirect expenses such as rent and depreciation, which were included in the P&L in order to achieve two objectives:

- Do not overstate annual earnings.
- Do not pay more income taxes than the law requires.

Table 69.5 Commercial Construction Tool Co., Inc.—Costs and Revenues, Bad Year—Actual

Source and Application of Funds	($000)
Net profit after tax	72
Depreciation expense	273
Cash generated	344
Increase in net working capital	
Change in cash	10
Change in receivables	98
Change in inventories	59
Change in payables	−46
Net change	121
Capital expenditures	500
Operating cash requirements	621
Operating cash flow	−276
Dividends	45
Net cash needs	−321
Increase in debt	321

In the section on fixed assets, when discussing the balance sheet, it was pointed out that the reserve for depreciation is not an amount of money set aside and available for spending. It is the total of the depreciation expense charged so far against a still existing asset. The example was a piece of equipment with a useful life of 12 years, the total value of which was reduced by 8.33% (the reciprocal of 12 times 100) each year. This accounting action is taken to reduce profits to a level that takes into consideration the decreasing value of equipment over time, to reduce taxes, and to avoid overstating the value of assets. Depreciation expense is not a cash expense—no check is written—it is an accounting convention. The cash profit to the business is therefore understated in the P&L statement because less money was spent for expenses than indicated. The overstatement is the amount of depreciation and other noncash expenses included in costs for the year.

In the P&L in Table 69.4, included in the cost of sales of $4038, is $273 of depreciation expense. If this noncash item were not included as an expense of doing business, profit before tax would be increased from $138 to $411. Taxes were calculated at a 48% rate, so the revised net profit after tax would be $214. This new net profit would also be cash generated from operations instead of the $344 actually generated ($72 profit plus $273 depreciation) when noncash expenses are included as costs. The reduction in cash available to the business resulting from ignoring depreciation is exactly equal to the increase in taxes paid on profits. The anomaly is that the business has more money left at the end of the year when profits are lower!

69.5.1 Accelerated Depreciation

This is a logical place to examine various kinds of depreciation systems. So far, only a straight-line approach has been considered—the example used was a 12-year life resulting in an 8.33% annual expense or writedown rate. Philosophical arguments have been developed to support a larger writedown in the early years and reducing the depreciation rate in later years. Some of the reasons advanced include:

- A large loss in value is suffered when a machine becomes second hand.
- The usefulness and productivity of a machine is greater in the early years.
- Maintenance and repair costs of older machines are larger.
- The value of older machines does not change much from one year to the next.

The reason that accelerated systems have come into wide use is more practical than philosophical. With faster, early writedowns the business reduces its taxes now and defers them to a later date. Profits are reduced in the early years but cash flow is improved. There are two common methods of accelerating depreciation in the early years of a machine's life:

Sum of the digits
Double declining balance

Table 69.6 compares the annual depreciation expense for the two accelerated systems to the straightline approach. For these examples a salvage value of zero is assumed at the end of the period of useful life. At the time of asset retirement and sale, a capital gain or loss would be realized as compared to the residual, undepreciated value of the asset, or zero, if fully depreciated.

The methods of calculation are represented by the following equations and examples where:

$$N = \text{number of years of useful life}$$
$$A = \text{year for which depreciation is calculated}$$
$$P = \text{original price of the asset}$$
$$D_a = \text{depreciation in year } A$$
$$B = \text{book value at year end}$$

The equation for straight-line depreciation is

$$D_a = \frac{1}{n} \times P$$
$$B = P - (D_1 + D_2 + \cdots + D_a)$$

In the example with an asset costing $40,000 with an 8-year useful life:

$$D_a = \frac{1}{8} \times 40{,}000 = 0.125 \times 40{,}000 = 5000$$

To calculate depreciation by the sum of the years' digits method, use

Table 69.6 Accelerated Depreciation Methods[a]

Straight-Line Method[b]

Year	Rate	Depreciation Expense	Book Value, Year End
1	0.125	5000	35000
2	0.125	5000	30000
3	0.125	5000	25000
4	0.125	5000	20000
5	0.125	5000	15000
6	0.125	5000	10000
7	0.125	5000	5000
8	0.125	5000	0

Sum of the Years' Digits Method[c]

Year	Rate	Depreciation Expense	Book Value, Year End
1	0.2222222	8889	31111
2	0.1944444	7778	23333
3	0.1666667	6667	16667
4	0.1388889	5556	11111
5	0.1111111	4444	6667
6	0.0833333	3333	3333
7	0.0555556	2222	1111
8	0.0277778	1111	0

Double the Declining Balance Method[d]

Year	Rate Rate	Depreciation Expense	Book Value, Year End
1	0.25	10000	30000
2	0.1875	7500	22500
3	0.140625	5625	16875
4	0.1054688	4219	12656
5	0.0791016	3164	9492
6	0.0593262	2373	7119
7	0.0444946	1780	5339
8	0.1334839	5339	0

[a]Basic assumptions: equipment life, 8 years; original price, $40,000; estimated salvage value, $0.
[b]Annual rate equation: one divided by the number of years times the original price.
[c]Annual rate equation: sum of the number of years divided into the years of life remaining.
[d]Annual rate equation: twice the straight-line rate times the book value at the end of the preceding year.

$$D_a = [(N + 1 - A)/(N + N - 1 + N - 2 + \cdots + 1)] \times P$$

For the third year, for example,

$$D_3 = [(8 + 1 - 3)/(8 \times 7 + 6 + 5 + 4 + 3 + 2 + 1)] \times 40,000$$
$$D_3 = [(6)/(36)] \times 40,000 = 0.1667 \times 40,000 = 6667$$

The depreciation rates shown in Table 69.6 under the double declining balance method are calculated to show a comparison of write-off rates between systems. The actual calculations are done quite differently:

$$D_a = \frac{2}{n} \times B_{a-1}$$

$$B_a = P - (D_1 + D_2 + \cdots D_{a-1})$$

In the third year, then,

$$D_3 = \frac{2}{8} \times 22,500 = 5625$$

and

$$B_3 = 40,000 - (10,000 + 7500 + 5625) = 16,875$$

Note in Table 69.6 that the double declining balance method, as should be expected, if allowed to continue forever, never succeeds in writing off the entire value. The residue is completely written off in the final year of the asset's life. The sum of the years' digits is a straight line and provides for a full write-off at the end of the period.

Figure 69.1 depicts, graphically, the annual depreciation expense using the three methods.

In many cases, a company will succeed in attaining both the advantages to cash flow and tax minimization of accelerated depreciation as well as the maximizing of earnings by using straight-line depreciation. This is done by having one set of books for the tax collector and another for the shareholders and the investing public. This practice is an accepted approach and, where followed, is explained in the fine print of the annual report.

A number of special depreciation provisions and investment tax credit arrangements are available to companies from time to time. The provisions change as tax laws are revised either to encourage investment and growth or to plug tax loopholes, depending on which is politically popular at the time. The preceding explains the theory—applications vary considerably with changes in the law and differences in corporate objectives and philosophy.

The cash generated by the business has, as its first use, the satisfaction of the needs for working capital, that is, the needs for funds to finance increases in inventories, receivables, and cash in the bank. Each of these assets requires cash in order to provide them. Offsetting these uses of cash are the changes that may take place in the short-term debts of the enterprise and accounts payable. In Table 69.5, we see that $121 is required in increased net working capital, essentially all of which goes to provide for increased inventories and receivables needed to support sales increases.

The largest requirement for cash is the next item, that of capital expenditures, which has consumed $500 of the cash provided to the business. The total needs of the company for cash—the operating cash requirements—have risen to $621 compared to the cash generated of $344, and that is not the end of cash needs. The shareholders have become accustomed to a return on their investment—an annual cash dividend. The dividend is not considered part of operating cash flow nor is it a tax deductible expense as interest payments are. The dividend, added to the net operating cash flow of −$276, results in a borrowing requirement for the year of $321.

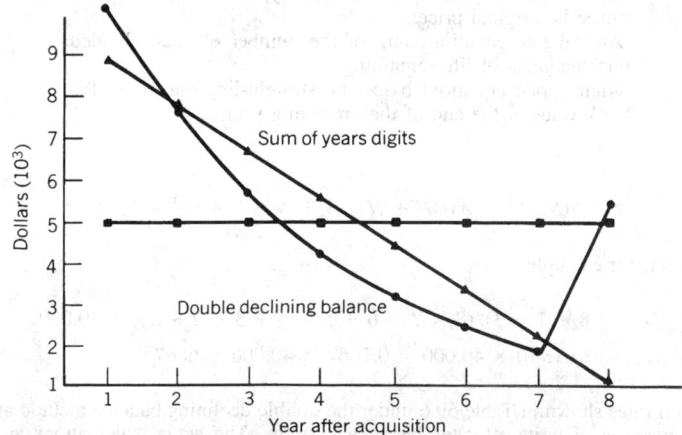

Fig. 69.1 Comparison of depreciation methods. ($40,000 original price; 8-year life; and no residual salvage value.)

Table 69.7 Commercial Construction Tool Co., Inc.—P&L Statement ($000)

	Budget	Actual	Variance	Percent
Sales	5261	4772	−489	−9.29
Cost of production	3972	4097	−1	−0.03
Beginning inventory	941	941	0	0.00
Ending inventory	1007	1000	−7	−0.68
Net change	66	59	−7	−10.36
Cost of sales	3906	4038	−132	−3.37
Gross margin	1355	734	−621	−45.81
Selling expense	160	177	−17	−10.90
Administrative	231	249	−18	−7.88
Operating profit	964	308	−656	−68.08
Interest	154	169	−15	−9.93
Profit before tax	810	138	−671	−82.92
Income tax	389	66	322	82.92
Net income	421	72	−349	−82.92

To summarize, the Board of Directors has been furnished a set of operating statements and financial ratios as shown in Tables 69.2–69.5. These ratios show a superficial picture of the economics of the enterprise from a financial viewpoint and present some issues and problems to the directors. The condition of the ratios and rates of return for CCTCO are of great concern to the directors and lead to some hard questions for management.

Why, when the total cost of capital, that is, interest plus dividends as a percentage of debt plus equity is 20.67%, is the return on total capital only 3.56%? Why does it take $0.67 worth of assets to provide $1 worth of sales in a year? Why is it that profit after tax is only 1.5 1% of sales? The board will not be pleased with performance and will want to know what can and will be done to improve. The banks will perhaps have concerns about further loans and shareholders or prospective shareholders will wonder about the price of the stock.

The answers to these questions require a level of cost and revenue information normally supplied to top management.

69.6 EVALUATING RESULTS AND TAKING ACTION

Corporate chief executives who allowed themselves to be as badly surprised by poor results at the end of the year as the chief executive of CCTCO would be unlikely to last long enough to take corrective action. However, the results at CCTCO can provide clear examples of the usefulness of accounting records in determining the cause of business problems and in pointing in the direction of practical solutions.

69.6.1 Comparing Current Results with Budgets and Forecasts

The first step of the chief executive at CCTCO was to compare actual results with those projected for the year. It had been the practice at CCTCO to prepare a comprehensive business plan and budget at the beginning of each year. Monthly and yearly, reports comparing actual with budget were made available to top officers of the company. Tables 69.7–69.10 show a comparison of the budgeted P&L,

Table 69.8 Commercial Construction Tool Co., Inc.—Financial Ratios

	Budget	Actual	Variance
Return on sales	8.00	1.51	−6.50
Return on assets	12.98	2.24	−10.74
Return on invested capital	20.47	3.56	−16.92
Return on equity	34.76	6.94	−27.82
Asset to sales ratio	0.62	0.67	−0.06
Debt percent to debt plus equity	60	69	−9.53
Average cost of capital	16.43	20.67	−4.24

Table 69.9   Commercial Construction Tool Co., Inc.—Variance Analysis, Balance Sheet ($000)

	Budget	Actual	Variance	Percent
Assets				
Current assets				
Cash	56	62	6	10.78
Accounts receivable	631	573	−59	−9.29
Inventories	1007	1000	−7	−0.68
Total current assets	1695	1635	−59	−3.51
Fixed assets				
Gross plant and equipment	2521	2521	0	0.00
Less reserve for depreciation	744	744	0	0.00
Net plant and equipment	1777	1777	0	0.00
Total assets	3472	3413	−59	−1.71
Liabilities				
Current liabilities				
Accounts payable	491	503	13	2.59
Short-term debt	604	600	−4	−0.68
Long-term debt becoming current	130	130	0	0.00
Total current liabilities	1225	1233	9	0.70
Long-term liabilities				
Interest-bearing debt	848	1129	281	33.17
Total liabilities	2073	2362	290	13.98
Net worth				
Capital stock	100	100	0	0.00
Earned surplus	1300	950	−349	−26.87
Total net worth	1400	1050	−349	−24.95
Total liabilities and net worth	3472	3413	−59	−1.71

Table 69.10   Commercial Construction Tool Co., Inc.—Variance Analysis, Source and Application of Funds ($000)

	Budget	Actual	Variance	Percent
Net profit after tax	419	72	−347	−82.84
Depreciation expense	273	273	0	0.00
Cash generated	692	344	−347	−50.20
Increase in net working capital				
Change in cash	4	10	6	150.90
Change in receivables	156	98	−59	−37.53
Change in inventories	66	59	−7	−10.36
Change in payables	−34	−46	−13	37.85
Net change	193	121	−72	−37.40
Capital expenditures	500	500	0	0.00
Operating cash requirements	693	621	−72	−10.42
Operating cash flow	−1	−276	−275	27500.00
Dividends	45	45	0	0.00
Net cash needs	−46	−321	−275	593.31
Increase in debt	46	321	275	593.31

performance ratios, balance sheet, and cash flow for the year compared to the actual performance already reviewed by the board.

An examination of the budget/actual comparisons revealed many serious deviations from plan. Net worth and long-term debt were trouble spots. Profits were far from expected results, and cash flow was far below plan.

The president searched the reports for the underlying causes in order to focus his attention and questions on those corporate functions and executives that appeared to be responsible for the failures. He concluded that there were seven critical variances from the budget, which when understood, should eventually lead to the underlying real causes. They included

Element	Variance	Percent
Sales	−489	−9.29
Cost of sales	−132	−3.37
Selling expense	−17	−10.90
Administrative expense	−18	−7.88
Interest	−15	−9.93
Net working capital	−68	−14.50

The president asked the VP Sales and the VP Manufacturing to report to him as to what had happened to cause these variances from plan and what corrective action could be taken. He instructed the Controller to provide all the cost and revenue analyses needed to arrive at answers.

In two weeks the three executives made a presentation to the president that provided a comprehensive understanding of the problems, recommended solutions to them, and a timetable to implement the program. The following is a summary of that report.

69.6.2 Identifying Problems and Solutions

Causes of Last Year's Results

The poor operating results of last year are caused almost entirely by a change in product mix from the previous year and not contemplated in the budget established 15 months ago. The introduction of the HOMMODEL nearly two years ago resulted in very few sales in the early months following its initial availability. However, early last year, sales accelerated dramatically, caught up with, and passed those of the COMMODEL. For a number of reasons this has had a poor effect on the financial structure of our company:

- Lack of experience on the new product has resulted in costs higher than standard.
- Standard margins are lower for the HOMMODEL.
- Travel and communications costs were high because of the new product introduction.
- Prices on the HOMMODEL were lower than standard because of special introductory dealer discounts and deals.
- Receivables increased because of providing initial stocking plans for new dealers handling the HOMMODEL.
- Higher interest expense resulted from higher debt—a direct result of cash flow shortfall.

The only significant variance unrelated to the new product was the fact that factory and office rents were raised during the year.

The following product mix table summarizes a number of accounting documents and shows the effect of product mix on profits.

Recommended Corrective Action

As the major problems are caused by the new product cannibalizing sales of the old COMMODEL, action is directed toward increasing margins on the HOMMODEL to nearly that of the COMMODEL and increasing the proportion of sales of the latter. This will be accomplished by simultaneously reducing unit cost and increasing selling price of the new product. The following program will be undertaken:

- Increase the unit price to 3.52 and eliminate deals and promotion pricing for a margin improvement of 0.34.
- Productivity improvements realized in the last two months of the year will reduce costs by 0.15 for the year.
- Proposed changes in material and finish will further reduce costs by 0.032.

Table 69.11　Product Line Comparison: Unit Volume, Price, and Costs

	Budget	Actual	Variance
Commodel			
Sales (1000s)	740	530	(210)
Unit price	4.203	4.280	0.077
Unit cost	3.0612	3.139	0.078
Unit margin	1.142	1.141	−0.001
Sales $	$3,110,220	$2,268,400	(841,820)
Cost $	$2,265,140	$1,663,670	($16,380)
Margin $	$845,080	$604,730	(240,350)
HOMMODEL			
Sales (1000s)	670	830	160
Unit price	3.210	3.016	(0.194)
Unit cost	2.449	2.932	(0.483)
Unit margin	0.761	0.084	(0.677)
Sales $	$2,150,700	$2,503,280	352,580
Cost $	$1,640,830	$2,433,560	792,730
Margin $	$509,870	$69,720	(440,150)
Total			
Sales $	$5,260,920	$4,771,680	(489,240)
Cost $	$3,905,970	$4,097,230	$776,350
Margin $	$1,354,950	$674,450	(680,500)
Selling expense	$160,000	$177,000	17,000
Administrative expense	$231,000	$249,000	18,000
Operating profit	$963,950	$248,450	($715,500)

These changes in price and cost will bring the standard margin of the HOMMODEL to 1.21, slightly more than that of the COMMODEL, thus eliminating any unfavorable effect of cannibalizing.

This report enabled the president to assure the board that the recommended steps would be taken and the year to come would provide better results.

69.6.3　Initiating Action

Following Board approval, the president asked the manufacturing manager, in conjunction with marketing, to prepare a five-year projection of operating results. The projection, as shown in Table 69.12, was prepared in a personal computer spreadsheet by the manufacturing manager and showed an increase in operating profit to just over $1,000,000 by the end of the five-year period.

The manufacturing manager was able to demonstrate the logic of his conclusions by showing the economic and operating assumptions on which the projections were based, as shown below:

Concerning the COMMODEL:

1. Unit sales will increase 1.5% annually.
2. Unit prices will increase at 2.5% annually, 0.5% less than the expected inflation rate of 3.0%.
3. Unit costs will increase at the same rate as prices.

Concerning the HOMMODEL:

1. Unit sales will increase at 4.0% annually.
2. Unit prices will increase at the same rate as for the COMMODEL, 2.5% annually.
3. Unit costs will increase at the same rate as prices.

Concerning expenses, both selling and administrative expenses will increase at 3.0% annually.

Using his model, the manufacturing manager was able to demonstrate to the board the reasonableness and the sensitivity of his projections. The cell formulae used in the spreadsheet are shown in Table 69.13.

69.7　FINANCIAL TOOLS FOR THE INDEPENDENT PROFESSIONAL ENGINEER

In the 1990s and for some years prior to that time, it became common for engineers to become independent consultants or "free lances." This was partly brought about by corporate downsizing and the tendency of companies to bring in part-time technical assistance for specific projects rather

Table 69.12 Product Line Comparison: Unit Volume, Price, and Costs

	Projections				
	Year 1	Year 2	Year 3	Year 4	Year 5
COMMODEL					
Sales (1000's)	700	717	735	754	773
Unit price	4.280	4.387	4.497	4.609	4.724
Unit cost	3.139	3.217	3.298	3.380	3.465
Unit margin	1.141	1.170	1.199	1.229	1.259
Sales $	$2,996,000	$3,070,900	$3,147,672	$3,226,364	$3,307,023
Cost $	$2,197,300	$2,252,233	$2,308,538	$2,366,252	$2,425,408
Margin $	$798,700	$818,668	$839,134	$860,113	$881,615
HOMMODEL					
Sales (1000's)	650	666	683	700	717
Unit price	3.520	3.608	3.698	3.791	3.885
Unit cost	2.750	2.819	2.889	2.961	3.035
Unit margin	0.770	0.789	0.809	0.829	0.850
Sales $	$2,288,000	$2,403,830	$2,525,524	$2,653,379	$2,787,706
Cost $	$1,787,500	$1,877,992	$1,973,066	$2,072,952	$2,177,895
Margin $	$500,500	$525,838	$552,458	$580,427	$609,811
Total					
Sales $	$5,284,000	$5,474,730	$5,673,196	$5,879,743	$6,094,729
Cost $	$3,984,800	$4,130,225	$4,281,604	$4,439,204	$4,603,303
Margin $	$1,299,200	$1,344,505	$1,391,593	$1,440,539	$1,491,426
Selling expense	$177,000	$182,310	$187,779	$193,413	$199,215
Administrative expense	$249,000	$256,470	$264,164	$272,089	$280,252
Operating profit	$873,200	$905,725	$939,649	$975,037	$1,011,959

Table 69.13 Cell Formulae Used for Projections

	Product Line Comparison: Unit Volume, Price, and Costs		
	Year 1	Year 2	Formula
COMMODEL			
Sales (1000's)	700	710	+C7*1.015
Unit price	4.280	4.387	+C8*1.025
Unit cost	3.139	3.217	+C9*1.025
Unit margin	1.141	1.170	+D8-D9
Sales $	$2,996,000	$3,116,963	+D7*D8*1000
Cost $	$2,197,300	$2,286,016	+D7*D9*1000
Margin $	$798,700	$830,948	+D11-D9*D7*1000
HOMMODEL			
Sales (1000's)	650	676	+C15*1.04
Unit price	3.520	3.608	+C16*1.025
Unit cost	2.750	2.819	+C17*1.025
Unit margin	0.770	0.789	+D16-D17
Sales $	$2,288,000	$2,439,008	+D15*D16*1000
Cost $	$1,787,500	$1,905,475	+D15*D17*1000
Margin $	$500,500	$533,533	+D19-D17*D15*1000
Total			
Sales $	$5,284,000	$5,555,971	+D19+D11
Cost $	$3,984,800	$4,191,491	+D20+D12
Margin $	$1,299,200	$1,364,481	+D21+D13
Selling expense	$177,000	$182,310	+C27*1.03
Administrative expense	$249,000	$256,470	+C28*1.03
Operating profit	$873,200	$925,701	+D25-D27-D28

In a like manner, relationship and cell formulae can be developed for year-by-year balance sheets and cash flows.

than to develop an in-house capability that was not needed at all times. One of the implications of this development is that the engineer needs to be able to account for his own expenses and income as a "business." This accounting must satisfy the requirement of the U.S. Internal Revenue Service and records need to be adequate to convince the IRS that tax submissions are accurate, that they satisfy the tax law, and that there is no fraud or indication of deception.

69.7.1 Simple Record-Keeping

With present home and business accounting software for the personal computer, the keeping of basic records can be made accurate, simple, and convincing to an IRS investigator and to the engineer's accountant.

The records of a private engineering practice should be *cash* rather than accrual and therefore can be based on bank and credit card transactions. Small cash transactions can be handled through a petty cash account that is replenished by check and that contains a journal of expenditures. A personal computer system can be set up that will automatically categorize each check written and even split a check into a number of categories, when necessary.

At the time of the publication of this edition the most popular program for personal finance was *QUICKEN,* but others are available and some banks will provide software and on-line access to a checking account. These systems make it possible to group and print out with full back-up and audit trail capability so that full quick disclosure is constantly available in a format that makes IRS audits become a matter solely of interpreting the law rather than tracking obscure expenditures or elements of income.

69.7.2 Getting the System Started

The first step should be to select an accountant. Although it is possible to maintain all needed records and prepare tax returns with computer software, the use of an accountant will probably save taxes through his knowledge of the law and is, for most engineers, essential. Following are some of the early decisions that should made with the accountant:

- Incorporation or not
- Computer needs
- Software needs
- Definition of categories or accounts
- Setting up bank accounts
- Level of accountant involvement

69.7.3 Operating the System

The basic approach to relatively painless small business accounting is that when a check is written or a deposit made, the transaction is entered in the computer at the time of the transaction and never again! As bills become due, the check is entered in and printed by the software or base; the funds are even transferred to the payees by the software. From that base, transaction lists, tabulations, and groupings are all done without writing or performing manual arithmetic. Cross columns *always* balance.

At the end of the fiscal year, the data can be transferred into a tax preparation program that will sort data and calculate the tax. At that time, the data can be transmitted to the accountant with a detailed, by category, listing of each transaction in hard copy or machine language or both. The accountant has very little number-crunching to do and accounting fees are minimal.

In the past, the problems of accounting for a business were a significant deterrent to freelancing. Sound, simple computer approaches eliminate that part of the terror of being on your own.

69.8 CONCLUSIONS

This chapter is intended to portray the principles of financial reporting without describing the underlying cost accounting systems needed to manage a business. These become so complex and are so varied that they are beyond the scope of this work.

The capacity to understand the meaning of financial reports and to make time projections based on historical reports coupled with sound assumptions for the future is frequently important to the engineer. Additionally, the ability to devise and administer a simple accounting system used to manage an engineering practice is, especially today, a useful skill.

The section is designed to provide a basis in these capabilities.

CHAPTER 70

DETAILED COST ESTIMATING

Rodney D. Stewart
Mobile Data Services
Huntsville, Alabama

70.1	**THE ANATOMY OF A DETAILED ESTIMATE**	**2118**	**70.7**	**FUNCTIONAL ELEMENTS DESCRIBED** 2125
	70.1.1 Time, Skills, and Labor-Hours Required to Prepare an Estimate	2119	**70.8**	**PHYSICAL ELEMENTS DESCRIBED** 2125
70.2	**DISCUSSION OF TYPES OF COSTS**	**2121**	**70.9**	**TREATMENT OF RECURRING AND NONRECURRING**
	70.2.1 Initial Acquisition Costs	2121		**ACTIVITIES** 2128
	70.2.2 Fixed and Variable Costs	2121		
	70.2.3 Recurring and Nonrecurring Costs	2121	**70.10**	**WORK BREAKDOWN STRUCTURE**
	70.2.4 Direct and Indirect Costs	2121		**INTERRELATIONSHIPS** 2128
				70.10.1 Skill Matrix in a Work Breakdown Structure 2128
70.3	**COLLECTING THE INGREDIENTS OF THE ESTIMATE**	**2121**		70.10.2 Organizational Relationships to a Work Breakdown Structure 2129
	70.3.1 Labor-Hours	2121		
	70.3.2 Materials and Subcontracts	2122	**70.11**	**METHODS USED WITHIN THE DETAILED**
	70.3.3 Labor Rates and Factors	2123		**ESTIMATING PROCESS** 2129
	70.3.4 Indirect Costs, Burden, and Overhead	2123		70.11.1 Detailed Resource Estimating 2129
	70.3.5 General and Administrative Costs	2123		70.11.2 Direct Estimating 2129
	70.3.6 Fee, Profit, or Earnings	2123		70.11.3 Estimating by Analogy (Rules of Thumb) 2129
	70.3.7 Assembly of the Ingredients	2123		70.11.4 Firm Quotes 2129
				70.11.5 Handbook Estimating 2130
70.4	**THE FIRST QUESTIONS TO ASK (AND WHY)**	**2124**		70.11.6 The Learning Curve 2130
	70.4.1 What Is It?	2124		70.11.7 Labor-Loading Methods 2131
	70.4.2 What Does It Look Like?	2124		70.11.8 Statistical and Parametric Estimating as Inputs to Detailed Estimating 2131
	70.4.3 When Is It to Be Available?	2124		
	70.4.4 Who Will Do It?	2124	**70.12**	**DEVELOPING A SCHEDULE** 2132
	70.4.5 Where Will It Be Done?	2124		
			70.13	**TECHNIQUES USED IN SCHEDULE PLANNING** 2134
70.5	**THE ESTIMATE SKELETON: THE WORK BREAKDOWN STRUCTURE**	**2125**		
			70.14	**ESTIMATING ENGINEERING ACTIVITIES** 2134
70.6	**THE HIERARCHICAL RELATIONSHIP OF A DETAILED WORK BREAKDOWN STRUCTURE**	**2125**		70.14.1 Engineering Skill Levels 2134

Adapted from Rodney D. Stewart, *Cost Estimating,* 2d ed., Wiley, 1991, by permission of the publisher.

Mechanical Engineers' Handbook, 2nd ed., Edited by Myer Kutz.
ISBN 0-471-13007-9 © 1998 John Wiley & Sons, Inc.

	70.14.2	Design	2134
	70.14.3	Analysis	2134
	70.14.4	Drafting	2134
70.15	**MANUFACTURING/ PRODUCTION ENGINEERING**		**2135**
	70.15.1	Engineering Documentation	2136
70.16	**ESTIMATING MANUFACTURING/ PRODUCTION AND ASSEMBLY ACTIVITIES**		**2136**
70.17	**MANUFACTURING ACTIVITIES**		**2137**
70.18	**IN-PROCESS INSPECTION**		**2137**
70.19	**TESTING**		**2137**
70.20	**COMPUTER SOFTWARE COST ESTIMATING**		**2139**
70.21	**LABOR ALLOWANCES**		**2140**

	70.21.1	Variance from Measured Labor-Hours	2140
	70.21.2	Personal, Fatigue, and Delay (PFD) Time	2140
	70.21.3	Tooling and Equipment Maintenance	2140
	70.21.4	Normal Rework and Repair	2141
	70.21.5	Engineering Change Allowance	2141
	70.21.6	Engineering Prototype Allowance	2141
	70.21.7	Design Growth Allowance	2141
	70.21.8	Cost Growth Allowance	2141
70.22	**ESTIMATING SUPERVISION, DIRECT MANAGEMENT, AND OTHER DIRECT CHARGES**		**2141**
70.23	**THE USE OF "FACTORS" IN DETAILED ESTIMATING**		**2142**
70.24	**CONCLUDING REMARKS**		**2142**

70.1 THE ANATOMY OF A DETAILED ESTIMATE

The detailed cost estimating process, like the manufacture of a product, is comprised of parallel and sequential steps that flow together and interact to culminate in a completed estimate. Figure 70.1 shows the anatomy of a detailed estimate. This figure depicts graphically how the various cost estimate ingredients are synthesized from the basic man-hour estimates and material quantity estimates. Man-hour estimates of each basic skill required to accomplish the job are combined with the labor rates for these basic skills to derive labor-dollar estimates. In the meantime, material quantities are estimated in terms of the units by which they are measured or purchased, and these material quantities

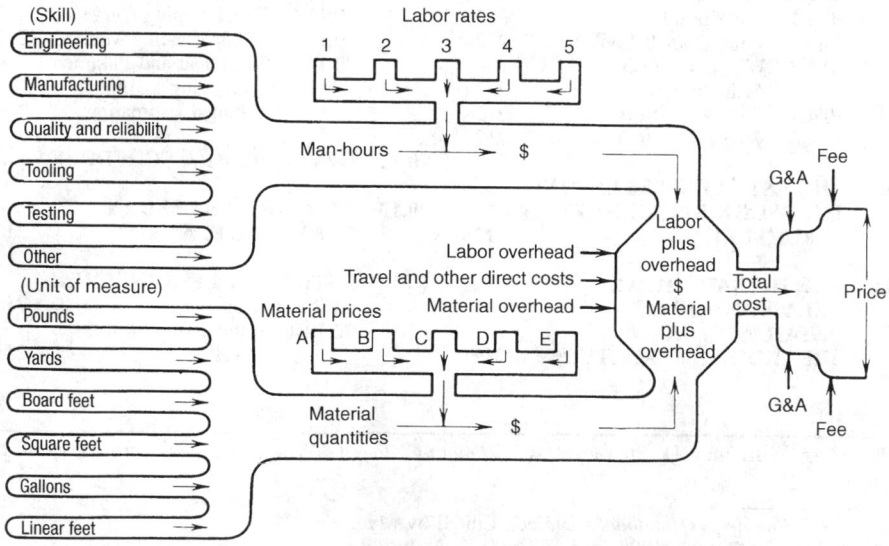

Fig. 70.1 Anatomy of an estimate.

are combined with their costs per unit to develop detailed direct material dollar estimates. Labor overhead or burden is applied to direct material costs. Then travel costs and other direct costs are added to produce total costs; general and administrative expenses and fee or profit are added to derive the "price" of the final estimate.

The labor rates applied to the basic man-hour estimates are usually "composite" labor rates; that is, they represent an average of the rates within a given skill category. For example, the engineering skill may include draftsmen, designers, engineering assistants, junior engineers, engineers, and senior engineers. The number and titles of engineering skills vary widely from company to company, but the use of a composite labor rate for the engineering skill category is common practice. The composite labor rate is derived by multiplying the labor rate for each skill by the percentage of man-hours of that skill required to do a given task and adding the results. For example, if each of the six skills have the following labor rates and percentages, the composite labor rate is computed as follows:

Skill	Labor Rate ($/h)	Percentage in the Task
Draftsman	12.00	7
Designer	16.00	3
Engineering assistant	20.00	10
Junior engineer	26.00	20
Engineer	30.00	50
Senior engineer	36.00	10
Total		100

Composite labor rate $- (0.07 \times \$12.00) + (0.03 \times \$16.00) + (0.10 \times \$20.00) + (0.20 \times \$26.00) + (0.50 \times \$30.00) + (0.10 \times \$36.00) = \$27.12$. Similar computations can be made to obtain the composite labor rate for skills within any of the other categories.

Another common practice is to establish separate overhead or burden pools for each skill category. These burden pools carry the peripheral costs that are related to and are a function of the labor-hours expended in that particular skill category. Assuming that the burden pool is established for each of the labor skills shown in Fig. 70.1, one can write an equation to depict the entire process. This equation is shown in Fig. 70.2. Thus far we have only considered a one-element cost estimate. The addition of multi-element work activities or work outputs will greatly increase the number of mathematical computations, and it becomes readily evident that the anatomy of an estimate is so complex that computer techniques for computation are essential for all but the simplest estimate.

70.1.1 Time, Skills, and Labor-Hours Required to Prepare an Estimate

The resources (skills, calendar time, and labor-hours) required to prepare a cost estimate depend on a number of factors. One factor is the estimating method utilized. Another is the level of technology or state of the art involved in the job or task being estimated. A rule of thumb can be utilized to develop a rough idea of the estimating time required. The calendar time required to develop an accurate and credible estimate is usually about 8% of the calendar time required to accomplish a task involving existing technology and 18% for a task involving a high technology (i.e., nuclear plant construction, aerospace projects). These percentages are divided approximately as shown in Table 70.1.

Note that the largest percentage of the required estimating time is for defining the output. This area is most important because it establishes a good basis for estimate credibility and accuracy, as well as making it easier for the estimator to develop supportable labor-hour and material estimates. These percentages also assume that the individuals who are going to perform the task or who have intimate working knowledge of the task are going to assist in estimate preparation. Hence the skill mix for estimating is very similar to the skill mix required for actually performing the task.

Labor-hours required for preparation of a cost estimate can be derived from these percentages by multiplying the task's calendar period in years by 2000 labor-hours per year, multiplying the result by the percentage in Table 70.1, and then multiplying the result by 0.1 and by the number of personnel on the estimating team. Estimating team size is a matter of judgment and depends on the complexity of the task, but it is generally proportional to the skills required to perform the task (as mentioned). Examples of the application of these rules of thumb for determining the resources required to prepare a cost estimate follow:

1. A three-year, high-technology project involving 10 basic skills or disciplines would require the following number of labor-hours to estimate:

$$3 \times 2000 \times 0.18 \times 100 = 1080 \text{ labor-hours}$$

$$T = \{[(E_H \times E_R) \times (1 + E_O)] + [(M_H \times M_R) \times (1 + M_O)] + [(TO_H \times TO_R)$$
$$\times (1 + TO_O)] + [(Q_H \times Q_R) \times (1 + Q_O)] + [(TE_H + TE_R) \times (1 + TE_O)]$$
$$+ [(O_H \times O_R) \times (1 + O_O)] + S_D + S_O + [M_D \times (1 + M_{OH})]$$
$$+ T_D + C_D + OD_D\} \times \{GA + 1.00\} \times \{F + 1.00\}$$

(a)

$$T = \{(L1_H \times L1_R) \times (1 + L1_O)] + [(L2_H \times L2_R) \times (1 + L2_O) \cdots$$
$$+ [(LN_H \times LN_R \times (1 + LN_O)] + S_D + S_O + [M_D \times (1 + M_{OH})]$$
$$+ T_D + CD + OD_D\} \times \{1 + GA\} \times \{1 \times F\}$$

where $L1, L2, \ldots LN$ are various labor rate categories

(b)

Symbols:
T = total cost
E_H = engineering labor hours
E_R = engineering composite labor rate in dollars per hour
E_O = engineering overhead rate in decimal form (i.e., 1.15 = 115%)
M_H = manufacturing labor hours
M_R = manufacturing composite labor rate in dollars per hour
M_O = manufacturing overhead rate in decimal form
TO_H = tooling labor hours
TO_R = tooling composite labor rate in dollars per hour
TO_O = tooling overhead in decimal form
Q_H = quality, reliability, and safety labor hours
Q_R = quality, reliability, and safety composite labor rate in dollars per hour
Q_O = quality, reliability, and safety overhead rate in decimal form
TE_H = testing labor hours
TE_R = testing composite labor rate in dollars per hour
TE_O = testing overhead rate in decimal form
O_H = other labor hours
O_R = labor rate for other hours category in dollars per hour
O_O = overhead rate for the hours category in decimal form
S_D = major subcontract dollar
S_O = other subcontract dollars
M_D = material dollars
M_{OH} = material overhead in decimal form (10% = 0.10)
T_D = travel dollars
C_D = computer dollars
OD_D = other direct dollars
GA = general and administrative expense in decimal form (25% = 0.25)
F = fee in decimal form (0.10 = 10%)

Fig. 70.2   Generalized equation for cost estimating.

Table 70.1   Estimating Time as a Percentage of Total Job Time

	Existing Technology (%)	High Technology (%)
Defining the output	4.6	14.6
Formulating the schedule and ground rules	1.2	1.2
Estimating materials and labor-hours	1.2	1.2
Estimating overhead, burden, and G&A	0.3	0.3
Estimating fee, profit, and earnings	0.3	0.3
Publishing the estimate	0.4	0.4
Total	8.0	18.0

2. A six-month "existing-technology" project requiring five skills or disciplines would require $0.6 \times 2000 \times 0.08 \times 0.1 \times 5 = 48$ labor-hours to develop an estimate.

These relationships are drawn from the author's experience in preparing and participating in cost estimates and can be relied on to give you a general guideline in preparing for the estimating process. But remember that these are "rules of thumb," and exercise caution and discretion in their application.

70.2 DISCUSSION OF TYPES OF COSTS

Detailed estimating requires the understanding of and the distinction between initial acquisition costs, fixed and variable costs, recurring and nonrecurring costs, and direct and indirect costs. These distinctions are described in the material that follows.

70.2.1 Initial Acquisition Costs

Businesspersons, consumers, and government officials are becoming increasingly aware of the need to estimate accurately and to justify the initial acquisition cost of an item to be purchased, manufactured, or built. Initial acquisition costs usually refer to the total costs to procure, install, and put into operation a piece of equipment, a product, or a structure. Initial acquisition costs do not consider costs associated with the use and possession of the item. Individuals or businesses who purchase products now give serious consideration to maintenance, operation, depreciation, energy, insurance, storage, and disposal costs before purchasing or fabricating an item, whether it be an automobile, home appliance, suit of clothes, or industrial equipment. Initial acquisition costs include planning, estimating, designing, and/or purchasing the components of the item; manufacturing, assembly, and inspection of the item; and installing and testing the item. Initial acquisition costs also include marketing, advertising, and markup of the price of the item as it flows through the distribution chain.

70.2.2 Fixed and Variable Costs

The costs of all four categories of productive outputs (processes, products, projects, and services) involve both fixed and variable costs. The relationship between fixed and variable costs depends on a number of factors, but it is principally related to the kind of output being estimated and the rate of output. Fixed cost is that group of costs involved in an ongoing activity whose total will remain relatively constant regardless of the quantity of output or the phase of the output cycle being estimated. Variable cost is the group of costs that vary in relationship to the rate of output. Therefore, where it is desirable to know the effect of output rate on costs, it is important to know the relationship between the two forms of cost as well as the magnitude of these costs. Fixed costs are meaningful only if they are considered at a given point in time, since inflation and escalation will provide a variable element to "fixed" costs. Fixed costs may only be truly fixed over a given range of outputs. Rental of floor space for a production machine is an example of a fixed cost, and its use of electrical power will be a variable cost.

70.2.3 Recurring and Nonrecurring Costs

Recurring costs are repetitive in nature and depend on continued output of a like kind. They are similar to variable costs because they depend on the quantity or magnitude of output. Nonrecurring costs are incurred to generate the very first item of output. It is important to separate recurring and nonrecurring costs if it is anticipated that the costs of continued or repeated production will be required at some future date.

70.2.4 Direct and Indirect Costs

As discussed earlier, direct costs are those that are attributable directly to the specific work activity or work output being estimated. Indirect costs are those that are spread across several projects and allocable on a percentage basis to each project. Table 70.2 is a matrix giving examples of these costs for various work outputs.

70.3 COLLECTING THE INGREDIENTS OF THE ESTIMATE

Before discussing the finer points of estimating, it is important to define the ingredients and to provide a preview of the techniques and methods utilized to collect these estimate ingredients.

70.3.1 Labor-Hours

Since the expenditure of labor-hours is the basic reason for the incurrence of costs, the estimating of labor-hours is the most important aspect of cost estimating. Labor-hours are estimated by four basic techniques: (1) use of methods, time, and measurement (MTM) techniques; (2) the labor-loading or staffing technique; (3) direct judgment of man-hours required; and (4) use of estimating handbooks. MTM methods are perhaps the most widespread methods of deriving labor-hour and skill estimates for industrial processes. These methods are available from and taught by the MTM Association for Standards and Research, located in Fair Lawn, New Jersey. The association is international in scope

Table 70.2 Examples of Costs for Various Outputs

	Process	Product	Project	Service
Initial acquisition costs	Plant construction costs	Manufacturing costs, marketing costs, and profit	Planning costs, design costs, manufacturing costs, test and checkout costs, and delivery costs	
Fixed costs	Plant maintenance costs	Plant maintenance costs	Planning costs and design costs	Building rental
Variable costs	Raw material costs	Labor costs	Manufacturing costs, test and checkout costs, and delivery costs	Labor costs
Recurring costs	Raw material costs	Labor and material costs	Manufacturing costs, test and checkout costs, and delivery costs	Labor costs
Nonrecurring costs	Plant construction costs	Plant construction costs	Planning costs and design costs	Initial capital equipment investment
Direct costs	Raw material	Manufacturing costs	Planning, design manufacturing, test and checkout and delivery costs	Labor and materials costs
Indirect costs	Energy costs	Marketing costs and profit	Energy costs	Energy costs

and has developed five generations of MTM systems for estimating all aspects of industrial, manu-facturing, or machining operations. The MTM method subdivides operator motions into small incre-ments that can be measured, and provides a means for combining the proper manual operations in a sequence to develop labor-hour requirements for accomplishing a job.

The labor-loading or staffing technique is perhaps the simplest and most widely used method for estimating the labor-hours required to accomplish a given job. In this method, the estimator envisions the job, the work location, and the equipment or machines required, and estimates the number of people and skills that would be needed to staff a particular operation. The estimate is usually ex-pressed in terms of a number of people for a given number of days, weeks, or months. From this staffing level, the estimated on-the-job labor-hours required to accomplish a given task can be computed.

Another method closely related to this second method is the use of direct judgment of the number of labor-hours required. This judgment is usually made by an individual who has had direct hands-on experience in either performing or supervising a like task.

Finally, the use of handbooks is a widely utilized and accepted method of developing labor-hour estimates. Handbooks usually provide larger time increments than the MTM method and require a specific knowledge of the work content and operation being performed.

70.3.2 Materials and Subcontracts

Materials and subcontract dollars are estimated in three ways: (1) drawing "takeoffs" and handbooks, (2) dollar-per-pound relationships, and (3) direct quotations or bids. The most accurate way to esti-mate material costs is to calculate material quantities directly from a drawing or specification of the completed product. Using the quantities required for the number of items to be produced, the appro-priate materials manufacturer's handbook, and an allowance for scrap or waste, one can accurately compute the material quantities and prices. Where detailed drawings of the item to be produced are not available, a dollar-per-pound relationship can be used to determine a rough order of magnitude cost. Firm quotations or bids for the materials or for the item to be subcontracted are better than any of the previously mentioned ways of developing a materials estimate because the supplier can be held to the bid.

70.3.3 Labor Rates and Factors

The labor rate, or number of dollars required per labor-hour, is the quantity that turns a labor-hour estimate into a cost estimate; therefore, the labor rate and any direct cost factors that are added to it are key elements of the cost estimate. Labor rates vary by skill, geographical location, calendar date, and the time of day or week applied. Overtime, shift premiums, and hazardous-duty pay are also added to hourly wages to develop the actual labor rate to be used in developing a cost estimate. Wage rate structures vary considerably, depending on union contract agreements. Once the labor rate is applied to the labor-hour estimate to develop a labor cost figure, other factors are commonly used to develop other direct cost allowances, such as travel costs and direct material costs.

70.3.4 Indirect Costs, Burden, and Overhead

Burden or overhead costs for engineering activities very often are as high as 100% of direct engineering labor costs, and manufacturing overheads go to 150% and beyond. A company that can keep its overhead from growing excessively, or a company that can successfully trim its overhead, can place itself in an advantageously competitive position. Since overhead more than doubles the cost of a work activity or work output, trimming the overhead has a significant effect on reducing overall costs.

70.3.5 General and Administrative Costs

General and administrative costs range up to 20% of total direct and indirect costs for large companies. General and administrative costs are added to direct and overhead costs and are recognized as a legitimate business expense.

70.3.6 Fee, Profit, or Earnings

The fee, profit, or earnings will depend on the amount of risk the company is taking in marketing the product, the market demand for the item, and the required return on the company's investment. This subject is one that deserves considerable attention by the cost estimator. Basically, the amount of profit depends on the astute business sense of the company's management. Few companies will settle for less than 10% profit, and many will not make an investment or enter into a venture unless they can see a 20 to 30% return on their investment.

70.3.7 Assembly of the Ingredients

Once resource estimates have been accumulated, the process of reviewing, compiling, organizing, and computing the estimate begins. This process is divided into two general subdivisions of work: (1) reviewing, compiling, and organizing the input resource data, and (2) computation of the costs based on desired or approved labor rates and factors. A common mistake made in developing cost estimates is the failure to perform properly the first of these work subdivisions. In the process of reviewing, compiling, and organizing the data, duplications in resource estimates are discovered and eliminated; omissions are located and remedied; overlapping or redundant effort is recognized and adjusted; and missing or improper rationale, backup data, or supporting data are identified, corrected, or supplied. A thorough review of the cost estimate input data by the estimator or estimating team, along with an adjustment and reconciliation process, will accomplish these objectives.

Computation of a cost estimate is mathematically simple since it involves only multiplication and addition. The number of computations can escalate rapidly, however, as the number of labor skills, fiscal years, and work breakdown structure elements are increased. One who works frequently in industrial engineering labor hour and material-based cost estimating will quickly come to the conclusion that some form of computer assistance is required.

With the basic ingredients and basic tools available, we are now ready to follow the steps required to develop a good detailed cost estimate. All steps are needed for any good cost estimate. The manner of accomplishing each step, and the depth of information needed and time expended on each step, will vary considerably, depending on what work activity or work output is being estimated. These steps are as follows:

1. Develop the work breakdown structure.
2. Schedule the work elements.
3. Retrieve and organize historical cost data.
4. Develop and use cost estimating relationships.
5. Develop and use production learning curves.

6. Identify skill categories, levels, and rates.
7. Develop labor-hour and material estimates.
8. Develop overhead and administrative costs.
9. Apply inflation and escalation factors.
10. Price (compute) the estimated costs.
11. Analyze, adjust, and support the estimate.
12. Publish, present, and use the estimate.

70.4 THE FIRST QUESTIONS TO ASK (AND WHY)

Whether you are estimating the cost of a process, product, or service, there are some basic questions you must ask to get started on a detailed cost estimate. These questions relate principally to the requirements, descriptions, location, and timing of the work.

70.4.1 What Is It?

A surprising number of detailed cost estimates fail to be accurate or credible because of a lack of specificity in describing the work that is being estimated. The objectives, ground rules, constraints, and requirements of the work must be spelled out in detail to form the basis for a good cost estimate. First, it is necessary to determine which of the four generic work outputs (process, product, project, or service) or combination of work outputs best describe the work being estimated. Then it is necessary to describe the work in as much detail as possible.

70.4.2 What Does It Look Like?

Work descriptions usually take the form of detailed specifications, sketches, drawings, materials lists, and parts lists. Weight, size, shape, material type, power, accuracy, resistance to environmental hazards, and quality are typical factors that are described in detail in a specification. Processes and services are usually defined by the required quality, accuracy, speed, consistency, or responsiveness of the work. Products and projects, on the other hand, usually require a preliminary or detailed design of the item or group of items being estimated. In general, more detailed designs will produce more accurate cost estimates. The principal reason for this is that as a design proceeds, better definitions and descriptions of all facets of this design unfold. The design process is an interactive one in which component or subsystem designs proceed in parallel; component or subsystem characteristics reflect on and affect one another to alter the configuration and perhaps even the performance of the end item. Another reason that a more detailed design results in a more accurate and credible cost estimate is that the amount of detail itself produces a greater awareness and visibility of potential inconsistencies, omissions, duplications, and overlaps.

70.4.3 When Is It to Be Available?

Production rate, production quantity, and timing of production initiation and completion are important ground rules to establish before starting a cost estimate. Factors such as raw material availability, labor skills required, and equipment utilization often force a work activity to conform to a specific time period. It is important to establish the optimum time schedule early in the estimating process, to establish key milestone dates, and to subdivide the overall work schedule into identifiable increments that can be placed on a calendar time scale. A work output schedule placed on a calendar time scale will provide the basic inputs needed to compute start-up costs, fiscal-year funding, and inflationary effects.

70.4.4 Who Will Do It?

The organization or organizations that are to perform an activity, as well as the skill categories and skill levels within these organizations, must be known or assumed to formulate a credible cost estimate. Given a competent organization with competent employees, another important aspect of developing a competitive cost estimate is the determination of the make or buy structure and the skill mix needs throughout the time period of a work activity. Judicious selection of the performers and wise time phasing of skill categories and skill levels can rapidly produce prosperity for any organization with a knowledge of its employees, its products, and its customers.

70.4.5 Where Will It Be Done?

Geographical factors have a strong influence on the credibility and competitive stature of a cost estimate. In addition to the wide variation in labor costs for various locations, material costs vary substantially from location to location, and transportation costs are entering even more heavily into the cost picture than in the past. The cost estimator must develop detailed ground rules and assumptions concerning location of the work, and then estimate costs accurately in keeping with all location-oriented factors.

70.5   THE ESTIMATE SKELETON: THE WORK BREAKDOWN STRUCTURE

The first step in developing a cost estimate of any type of work output is the development of a work breakdown structure. The work breakdown structure serves as a framework for collecting, accumulating, organizing, and computing the direct and directly related costs of a work activity or work output. It also can be and usually is utilized for managing and reporting resources and related costs throughout the lifetime of the work. There is considerable advantage in using the work breakdown structure and its accompanying task descriptions as the basis for scheduling, reporting, tracking, and organizing, as well as for initial costing. Hence it is important to devote considerable attention to this phase of the overall estimating process. A work breakdown structure is developed by subdividing a process, product, project, or service into its major work elements, then breaking the major work elements into subelements, and subelements into sub-subelements, and so on. There are usually 5 to 10 subelements under each major work element.

The purpose of developing the work breakdown structure is fivefold:

1. To provide a lower-level breakout of small tasks that are easy to identify, man-load, schedule, and estimate
2. To ensure that all required work elements are included in the work output
3. To reduce the possibility of overlap, duplication, or redundancy of tasks
4. To furnish a convenient hierarchical structure for the accumulation of resource estimates
5. To give greater overall visibility as well as depth of penetration into the makeup of any work activity

70.6   THE HIERARCHICAL RELATIONSHIP OF A DETAILED WORK BREAKDOWN STRUCTURE

A typical work breakdown structure is shown in Fig. 70.3. Note that the relationship resembles a hierarchy where each activity has a higher activity, parallel activities, and lower activities. A basic principle of work breakdown structures is that the resources or content of each work breakdown are made up of the sum of the resources or content of elements below it. No work element that has lower elements exceeds the sum of those lower elements in resource requirements. The bottommost elements are estimated at their own level and sum to higher levels. Many numbering systems are feasible and workable. The numbering system utilized here is one that has proved workable in a wide variety of situations.

One common mistake in using work breakdown structures is to try to input or allocate effort to every element, even those at a higher level. Keep in mind that this should not be done because each block or work element contains only that effort included in those elements *below* it. If there are no elements below it, then it can contain resources. If there is need to add work activities or resources not included in a higher-level block, add an additional block below it to include the desired effort. Level 1 of a work breakdown structure is usually the top level, with lower levels numbered sequentially as shown. The "level" is usually equal to the number of digits in the work element block. For example, the block numbered 1.1.3.2 is in level 4 because it contains four digits.

70.7   FUNCTIONAL ELEMENTS DESCRIBED

When subdividing a work activity or work output into its elements, the major subdivisions can be either functional or physical elements. The second level in a work breakdown structure usually consists of a combination of functional and physical elements if a product or project is being estimated. For a process or service, all second-level activities could be functional. Functional elements of a production or project activity can include activities such as planning, project management, systems engineering and integration, testing, logistics, and operations. A process or service can include any of hundreds of functional elements. Typical examples of the widely dispersed functional elements that can be found in a work breakdown structure for a service are advising, assembling, binding, cleaning, fabricating, inspecting, packaging, painting, programming, receiving, testing, and welding.

70.8   PHYSICAL ELEMENTS DESCRIBED

The physical elements of a work output are the physical structures, hardware, products, or end items that are supplied to the consumer. These physical elements represent resources because they require labor and materials to produce. Hence they can and should be a basis for the work breakdown structure.

Figure 70.4 shows a typical work breakdown structure of just the physical elements of a well-known consumer product, the automobile. The figure shows how just one automobile company chose to subdivide the components of an automobile. For any given product or project, the number of ways that a work breakdown structure can be constructed are virtually unlimited. For example, the company

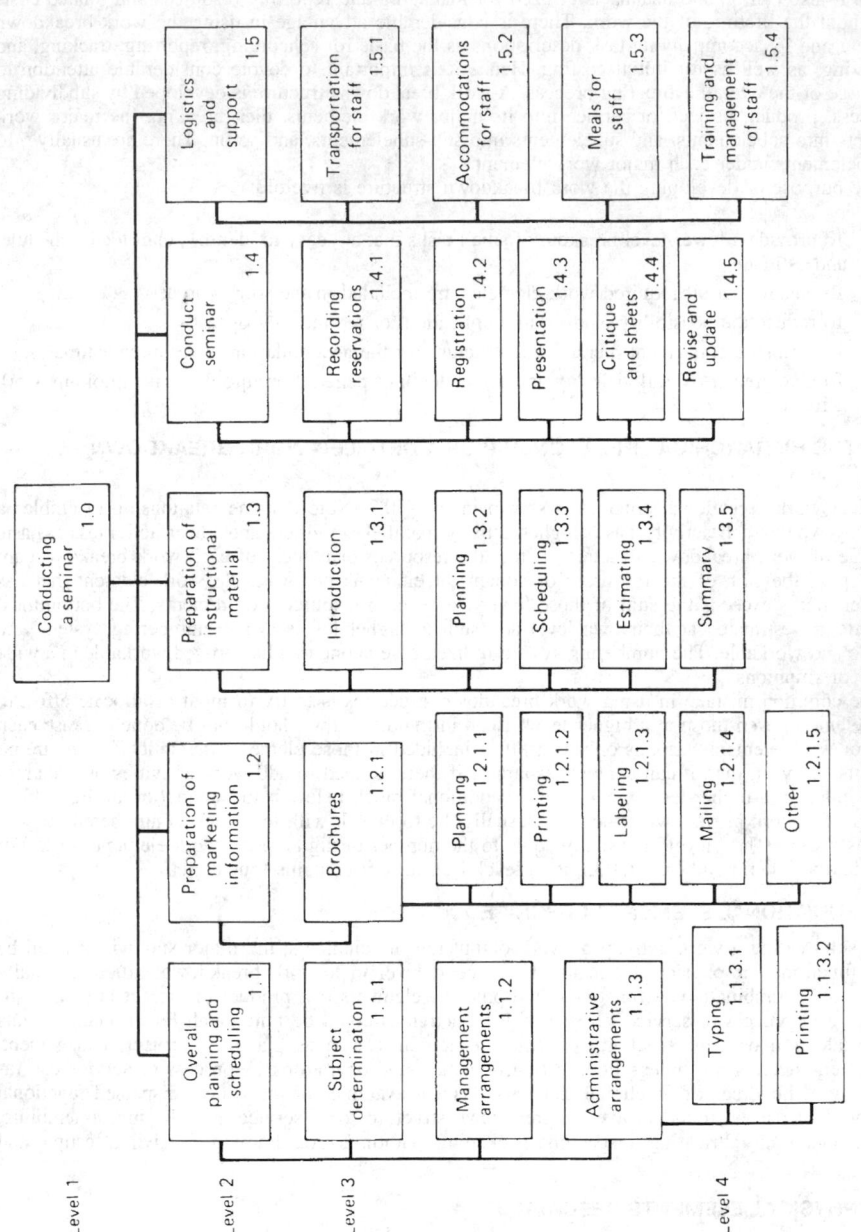

Fig. 70.3 Typical work breakdown structure.

Level 1

Level 2

Level 3

Level 4

Conducting a seminar 1.0

Overall planning and scheduling 1.1

Subject determination 1.1.1

Management arrangements 1.1.2

Administrative arrangements 1.1.3

Typing 1.1.3.1

Printing 1.1.3.2

Preparation of marketing information 1.2

Brochures 1.2.1

Planning 1.2.1.1

Printing 1.2.1.2

Labeling 1.2.1.3

Mailing 1.2.1.4

Other 1.2.1.5

Preparation of instructional material 1.3

Introduction 1.3.1

Planning 1.3.2

Scheduling 1.3.3

Estimating 1.3.4

Summary 1.3.5

Conduct seminar 1.4

Recording reservations 1.4.1

Registration 1.4.2

Presentation 1.4.3

Critique and sheets 1.4.4

Revise and update 1.4.5

Logistics and support 1.5

Transportation for staff 1.5.1

Accommodations for staff 1.5.2

Meals for staff 1.5.3

Training and management of staff 1.5.4

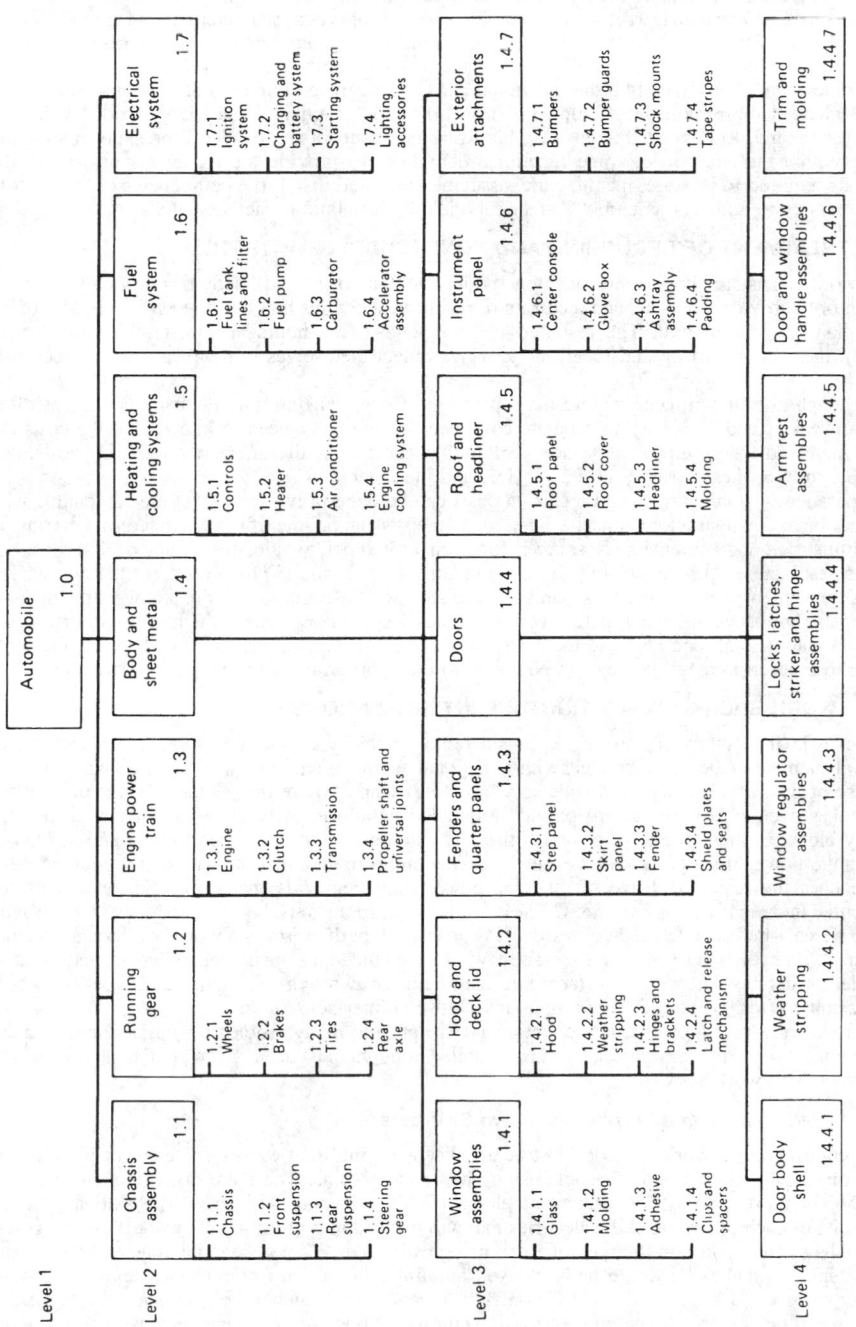

Fig. 70.4 Work breakdown structure of an automobile.

could have included the carburetor and engine cooling system as part of the engine assembly (this might have been a more logical and workable arrangement since it is used in costing a mass-production operation). Note that the structure shows a level-3 breakout of the body and sheet metal element, and the door (a level-3 element) is subdivided into its level-4 components.

This *physical element* breakout demonstrates several important characteristics of a work break-down structure. First, note that level 5 would be the individual component parts of each assembly or subassembly. It only took three subdivisions of the physical hardware to get down to a point where the next level breakout would be the individual parts. One can see rapidly that breaking down every level-2 element three more levels (down to level 5) would result in a very large work breakdown structure. Second, to convert this physical hardware breakout into a true work breakdown structure would require the addition of some functional activities. To provide the manpower as well as the materials required to procure, manufacture, assemble, test, and install the components of each block, it is necessary to add an "assembly," "fabrication," or "installation" activity block.

70.9 TREATMENT OF RECURRING AND NONRECURRING ACTIVITIES

Most work consists of both nonrecurring activities, or "one-of-a-kind" activities needed to produce an item or to provide a service, and recurring or repetitive activities that must be performed to provide more than one output unit. The resources requirements (labor-hours and materials) necessary to perform these nonrecurring and recurring activities reflect themselves in nonrecurring and recurring costs.

Although not all estimates require the separation of nonrecurring and recurring costs, it is often both convenient and necessary to separate costs because one may need to know what the costs are for an increased work output rate. Since work output rate principally affects the recurring costs, it is desirable to have these costs readily accessible and identifiable.

Separation of nonrecurring and recurring costs can be done in two ways that are compatible with the work breakdown structure concept. First, the two costs can be identified, separated, and accounted for within each work element. Resources for each task block would, then, include three sets of resource estimates: (1) nonrecurring costs, (2) recurring costs, and (3) total costs for that block. The second convenient method of cost separation is to start with identical work breakdown structures for both costs, and develop two separate cost estimates. A third estimate, which sums the two cost estimates into a total, can also use the same basic work breakdown structure. If there are elements unique to each cost category, they can be added to the appropriate work breakdown structure.

70.10 WORK BREAKDOWN STRUCTURE INTERRELATIONSHIPS

As shown in the automobile example, considerable flexibility exists concerning the placement of physical elements (the same is true with functional elements) in the work breakdown structure. Because of this, and because it is necessary to define clearly where one element leaves off and the other takes over, it is necessary to provide a detailed definition of what is included in each work activity block. In the automotive example, the rear axle unit could have been located and defined as part of the power train or as part of the chassis assembly rather than as part of the running gear. Where does the rear axle leave off and the power train begin? Is the differential or part of the differential included in the power train? These kinds of questions must be answered—and they usually are answered—before a detailed cost estimate is generated, in the form of a work breakdown structure dictionary. The dictionary describes exactly what is included in each work element and what is excluded; it defines where the interface is located between two work elements; and it defines where the assembly effort is located to assemble or install two interfacing units.

A good work breakdown structure dictionary will prevent many problems brought about by over-laps, duplications, and omissions, because detailed thought has been given to the interfaces and content of each work activity.

70.10.1 Skill Matrix in a Work Breakdown Structure

When constructing a work breakdown structure, keep in mind that each work element will be per-formed by a person or group of people using one or more skills. There are two important facets of the labor or work activity for each work element: skill mix and skill level. The skill mix is the proportion of each of several skill categories that will be used in performing the work. *Skill categories* vary widely and depend on the type of work being estimated. For a residential construction project, for example, typical skills would be bricklayer, building laborer, carpenter, electrician, painter, plas-terer, or plumber. Other typical construction skills are structural steelworker, cement finisher, glazier, roofer, sheet metal worker, pipefitter, excavation equipment operator, and general construction laborer. Professional skill categories such as lawyers, doctors, financial officers, administrators, project man-agers, engineers, printers, writers, and so forth are called on to do a wide variety of direct-labor activities. Occasionally, skills will be assembled into several broad categories (such as engineering, manufacturing, tooling, testing, and quality assurance) that correspond to overhead or burden pools.

Skill level, on the other hand, depicts the experience or salary level of an individual working within a given skill category. For example, engineers are often subdivided into various categories such as principal engineers, senior engineers, engineers, associate engineers, junior engineers, and engineering technicians. The skilled trades are often subdivided into skill levels and given names that depict their skill level; for example, carpenters could be identified as master carpenters, journeymen, apprentices, and helpers. Because skill categories and skill levels are designated for performing work within each work element, it is not necessary to establish separate work elements for performance of each skill. A work breakdown structure for home construction would not have an element designated *carpentry,* because carpentry is a skill needed to perform one or more of the work elements (i.e., roof construction, wall construction).

70.10.2 Organizational Relationships to a Work Breakdown Structure

Frequently all or part of a work breakdown structure will have a direct counterpart in the performing organization. Although it is not necessary for the work breakdown structure to be directly correlatable to the organizational structure, it is often convenient to assign the responsibility for estimating and for performing a specific work element to a specific organizational segment. This practice helps to motivate the performer, since it assigns responsibility for an identifiable task, and it provides the manager greater assurance that each part of the work will be accomplished. In the planning and estimating process, early assignment of work elements to those who are going to be responsible for performing the work will motivate them to do a better job of estimating and will provide greater assurance of completion of the work within performance, schedule, and cost constraints, because the functional organizations have set their own goals. Job performance and accounting for work accomplished versus funds spent can also be accomplished more easily if an organizational element is held responsible for a specific work element in the work breakdown structure.

70.11 METHODS USED WITHIN THE DETAILED ESTIMATING PROCESS

The principal methods used *within* the detailed estimating process are detailed resource estimating, direct estimating, estimating by analogy, firm quotes, handbook estimating, and the parametric estimating technique mentioned earlier. These methods are described briefly in the following sections.

70.11.1 Detailed Resource Estimating

Detailed resource estimating involves the synthesis of a cost estimate from resource estimates made at the lowest possible level in the work breakdown structure. Detailed estimating presumes that a detailed design of the product or project is available and that a detailed manufacturing, assembly, testing, and delivery schedule is available for the work. This type of estimating assumes that skills, labor-hours, and materials can be identified for each work element through one or more of the methods that follow. A detailed estimate is usually developed through a synthesis of work element estimates developed by various methods.

70.11.2 Direct Estimating

A direct estimate is a judgmental estimate made in a "direct" method by an estimator or performer who is familiar with the task being estimated. The estimator will observe and study the task to be performed and then forecast resources in terms of labor-hours, materials, and/or dollars. For example, a direct estimate could be quoted as "so many dollars." Many expert estimators can size up and estimate a job with just a little familiarization. One estimator I know can take a fairly complex drawing and, within just a few hours, develop a rough order-of-magnitude estimate of the resources required to build the item. Direct estimating is a skill borne of experience in both estimating and in actually performing the "hands-on" work.

70.11.3 Estimating by Analogy (Rules of Thumb)

This method is similar to the direct estimating method in that considerable judgment is required, but an additional feature is the comparison with some existing or past task of similar description. The estimator collects resource information on a similar or analogous task and compares the task to be estimated with the similar or analogous activity. The estimator would say that "this task should take about twice the time (man-hours, dollars, materials, etc.) as the one used as a reference." This judgmental factor (a factor of 2) would then be multiplied by the resources used for the reference task to develop the estimate for the new task. A significant pitfall in this method of estimating is the potential inability of the estimator to identify subtle differences in the two work activities and, hence, to be estimating the cost of a system based on one that is really not similar or analogous.

70.11.4 Firm Quotes

One of the best methods of estimating the resources required to complete a work element or to perform a work activity is the development of a firm quotation by the supplier or vendor. The two keys to the development of a realistic quotation are (1) the solicitation of bids from at least three sources, and (2) the development of a detailed and well-planned request for quotation. Years of experience by many organizations in the field of procurement have indicated that three bids are optimum from the standpoint of achieving the most realistic and reasonable price at a reasonable expenditure of effort. The solicitation of at least three bids provides sufficient check and balance and furnishes bid prices and conditions for comparison, evaluation, and selection. A good request for quotation (RFQ) is essential, however, to evaluate the bids effectively. The RFQ should contain ground rules, schedules, delivery locations and conditions, evaluation criteria, and specifications for the work. The RFQ should also state and specify the format required for cost information. A well-prepared RFQ will result in a quotation or proposal that will be easily evaluated, verified, and compared with independent estimates.

70.11.5 Handbook Estimating

Handbooks, catalogs, and reference books containing information on virtually every conceivable type of product, part, supplies, equipment, raw material, and finished material are available in libraries and bookstores and directly from publishers. Many of these handbooks provide labor estimates for installation or operation, as well as the purchase costs of the item. Some catalogs either do not provide price lists or provide price lists as a separate insert to permit periodic updates of prices without changing the basic catalog description. Information services provide microfilmed cassettes and on-line databases for access to the descriptions and costs of thousands and even tens of thousands of items.

If you produce a large number of estimates, it may pay to subscribe to a microfilm catalog and handbook data access system or, at least, to develop your own library of databases, handbooks, and catalogs.

70.11.6 The Learning Curve

The learning curve is a mathematical and graphical representation of the reduction in time, resources, or costs either actually or theoretically encountered in the conduct of a repetitive human activity. The theory behind the learning curve is that successive identical operations will take less time, use fewer resources, or cost less than preceding operations. The term *learning* is used because it relates primarily to the improvement of mental or manual skills observed when an operation is repeated, but *learning* can also be achieved by a shop or organization through the use of improved equipment, purchasing, production, or management techniques. When the learning curve is used in applications other than those involving the feedback loop that brings improvement of an individual's work activities, it is more properly named by one or more of the following terms:

Productivity improvement curve	Production improvement curve
Manufacturing progress function	Production acceleration curve
Experience curve	Time reduction curve
Progress curve	Cost improvement curve
Improvement curve	

Learning curve theory is based on the concept that as the total quantity of units produced doubles, the hours required to produce the last unit of this doubled quantity will be reduced by a constant percentage. This means that the hours required to produce unit 2 will be a certain percentage less than the hours required to produce unit 1; the hours required to produce unit 4 will be the same percentage less than the hours required to produce unit 2; the hours required to produce unit 8 will be the same percentage less than unit 4; and this constant percentage of reduction will continue for doubled quantities as long as uninterrupted production of the same item continues. The complement of this constant percentage of reduction is commonly referred to as the *slope*. This means that if the constant percentage of reduction is 10%, the slope would be 90%. Table 70.3 gives an example of a learning curve with 90% slope when the number of hours required to produce the first unit is 100.

Table 70.3 Learning Curve Values

Cumulative Units	Hours per Unit	Percent Reduction
1	100.00	
2	90.00	10
4	81.00	10
8	72.90	10
16	65.61	10
32	59.05	10

The reason for using the term *slope* in naming this reduction will be readily seen when the learning curve is plotted on coordinates with logarithmic scales on both the *x* and *y* axes (in this instance, the learning "curve" actually becomes a straight line). But first, let us plot the learning curve on conventional coordinates. You can see by the plot in Fig. 70.5 that it is truly a curve when plotted on conventional coordinates, and that the greater the production quantity, the smaller the incremental reduction in labor-hours required from unit to unit.

When the learning curve is plotted on log–log coordinates. as shown in Fig. 70.6, it becomes a straight line. The higher the slope, the flatter the line; the lower the slope, the steeper the line.

The effects of plotting curves on different slopes can be seen in Fig. 70.7, which shows the effects on labor-hour reductions of doubling the quantities produced 12 times. Formulas for the unit curve and the cumulative average curve are shown in Table 70.4.

Care should be taken in the use of the learning curve to avoid an overly optimistic (low) learning curve slope and to avoid using the curve for too few units in production. Most learning curve textbooks point out that this technique is credibly applicable only to operations that are done by hand (employ manual or physical operations) and that are highly repetitive.

70.11.7   Labor-Loading Methods

One of the most straightforward methods of estimating resources or labor-hours required to accomplish a task is the labor-loading or shop-loading method. This estimating technique is based on the fact that an experienced participant or manager of any activity can usually perceive, through judgment and knowledge of the activity being estimated, the number of individuals of various skills needed to accomplish a task. The shop-loading method is similar in that the estimator can usually predict what portion of an office or shop's capacity will be occupied by a given job. This percentage shop-loading factor can be used to compute labor-hours or resources if the total shop labor or total shop operation costs are known. Examples of the labor-loading and shop-loading methods based on 1896 labor-hours of on-the-job work per year are shown in Table 70.5.

70.11.8   Statistical and Parametric Estimating as Inputs to Detailed Estimating

Statistical and parametric estimating involves collecting and organizing historical information through mathematical techniques and relating this information to the work output that is being estimated. There are a number of methods that can be used to correlate historical cost and manpower information; the choice depends principally on mathematical skills, imagination, and access to data. These mathematical and statistical techniques provide some analytical relationship between the product, project, or service being estimated and its physical characteristics. The format most commonly used for statistical and parametric estimating is the *estimating relationship,* which relates some physical characteristic of the work output (weight, power requirements, size, or volume) with the cost or labor-hours required to produce it. The most widely used estimating relationship is linear. That is, the

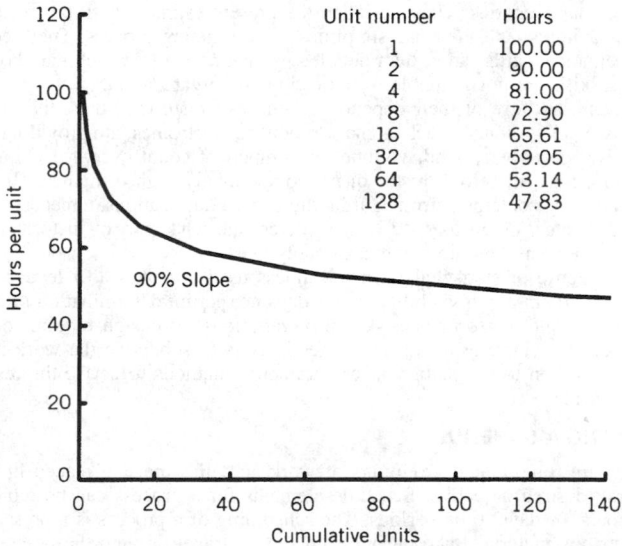

Fig. 70.5   Learning curve on a linear plot.

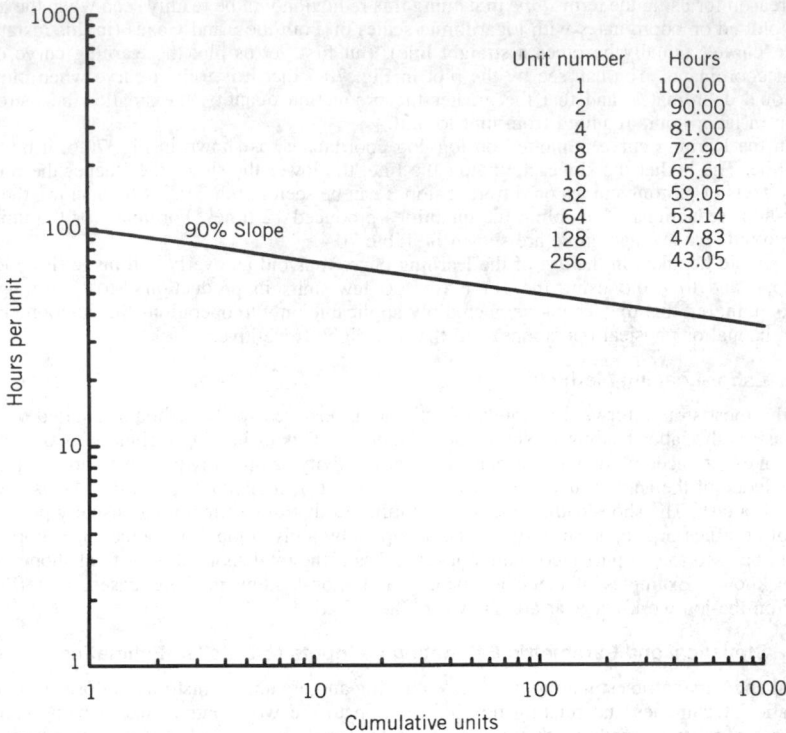

Fig. 70.6 Learning curve on a log–log plot.

mathematical equation representing the relationship is a linear equation, and the relationship can be depicted by a straight line when plotting on a graph with conventional linear coordinates for the x (horizontal) and y (vertical) axes. Other forms of estimating relationships can be derived based on curve-fitting techniques.

Estimating relationships have some advantages but certain distinct limitations. They have the advantage of providing a quick estimate even though very little is known about the work output except its physical characteristics. They correlate the present estimate with past history of resource utilization on similar items, and their use simplifies the estimating process. They require the use of statistical or mathematical skills rather than detailed estimating skills, which may be an advantage if detailed estimating skills are not available to the estimating organization.

On the other hand, because of their dependence on past (historical) data, they may erroneously indicate cost trends. Some products, such as mass-produced electronics, are providing more capability per pound, and lower costs per pound, volume, or component count every year. Basing electronics costs on past history may, therefore, result in noncompetitively high estimates. History should not be repeated if that history contains detrimental inefficiencies, duplications, unnecessary redundancies, rework, and overestimates. Often it is difficult to determine what part of historical data should be used to reflect future resource requirements accurately.

Finally, the parametric or statistical estimate, unless used at a very low level in the estimating process, does not provide in-depth visibility, and it does not permit determination of cost effects from subtle changes in schedule, performance, skill mix variations, or design requirements. The way to use the statistical or parametric estimate most effectively is to subdivide the work into the smallest possible elements and then to use statistical or parametric methods to derive the resources required for these small elements.

70.12 DEVELOPING A SCHEDULE

Schedule elements are time-related groupings of work activities that are placed in sequence to accomplish an overall desired objective. Schedule elements for a process can be represented by very small (minutes, hours, or days) time periods. The scheduling of a process is represented by the time the raw material or raw materials take during each step to travel through the process. The schedule for manufacturing a product or delivery of a service is, likewise, a time flow of the various components or actions into a completed item or activity.

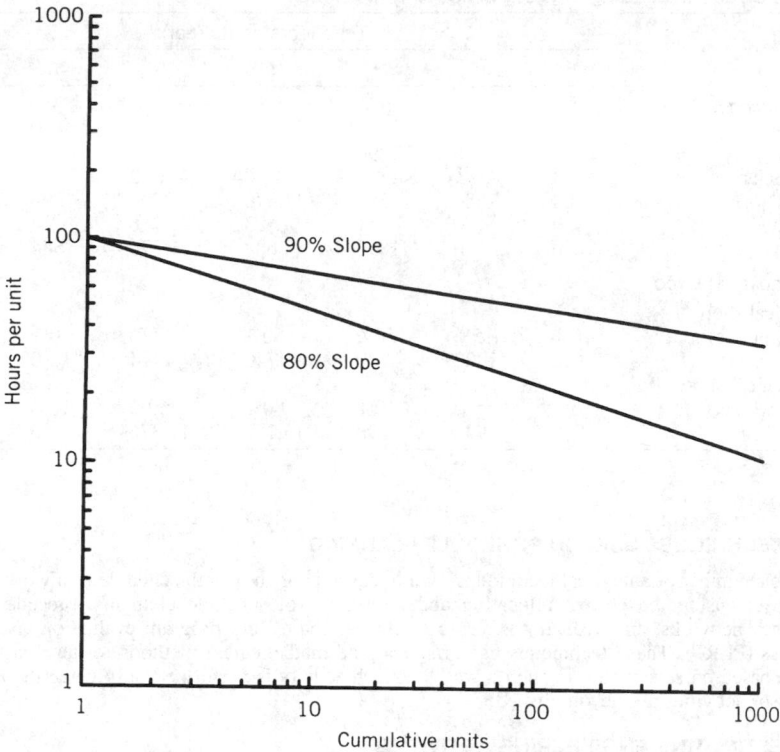

Fig. 70.7   Comparison between two learning curves.

A project (the construction or development of a fairly large, complex, or multidisciplinary tangible work output) contains distinct schedule elements called *milestones*. These milestones are encountered in one form or another in almost all projects:

1. Study and analysis
2. Design
3. Procurement of raw materials and purchased parts
4. Fabrication or manufacturing of components and subsystems
5. Assembly of the components and subsystems
6. Testing of the combined system to qualify the unit for operation in its intended environment
7. Acceptance testing, preparation, packaging, shipping, and delivery of the item
8. Operation of the item

Table 70.4   Learning Curve Formulas

Unit curve $\quad Y_x = KX^N$

where $\qquad Y_x$ = number of direct labor-hours required to produce the Xth unit

$\qquad K$ = number of direct labor-hours required to produce the first unit

$\qquad x$ = number of units produced

$\qquad n$ = slope of curve expressed in positive hundredths (e.g., $n = 0.80$ for an 80% curve)

$$N = \frac{\log_{10} n}{\log_{10} 2}$$

Cumulative average curve

$$V_x \approx \frac{K}{X(1 + N)} [(x + 0.5)^{(1+N)} - (0.5)^{(1-N)}]$$

where $\qquad V_x$ = the cumulative average number of direct labor-hours required to produce x units

Table 70.5   Labor-Loading and Shop-Loading Methods

	Time Increment (Year)						
	1	2	3	4	5	6	7
Labor-loading Method							
Engineers	1	1	1	2	1	0	0
Hours	1896	1896	1896	3792	1896	0	0
Technicians	3	4	4	6	2	1	0
Hours	5688	7584	7584	11,376	3792	1896	0
Draftsmen	0	0	1	3	6	4	2
Hours	0	0	1896	5688	11,376	7584	3792
Shop-loading Method							
Electrical shop							
(5 workers)	10%	15%	50%	50%	5%	0%	0%
Hours	948	1422	4740	4740	474	0	0
Mechanical shop							
(10 workers)	5%	5%	10%	80%	60%	10%	5%
Hours	948	948	1896	15,168	11,376	1896	948

70.13   TECHNIQUES USED IN SCHEDULE PLANNING

There are a number of analytical techniques used in developing an overall schedule of a work activity that help to ensure the correct allocation and sequencing of schedule elements: precedence and dependency networks, arrow diagrams, critical path bar charts, and program evaluation and review techniques (PERT). These techniques use graphical and mathematical methods to develop the best schedule based on sequencing in such a way that each activity is performed only when the required predecessor activities are accomplished.

70.14   ESTIMATING ENGINEERING ACTIVITIES

Engineering activities include the design, drafting, analysis, and redesign activities required to produce an end item. Costing of engineering activities is usually based on labor-loading and staffing resource estimates.

70.14.1   Engineering Skill Levels

The National Society of Professional Engineers has developed position descriptions and recommended annual salaries for nine levels of engineers. These skills levels are broad enough in description to cover a wide variety of engineering activities. The principal activities performed by engineers are described in the following paragraphs.

70.14.2   Design

The design activity for any enterprise includes conceptual design, preliminary design, final design, and design changes. The design engineer must design prototypes, components for development or preproduction testing, special test equipment used in development or preproduction testing, support equipment, and production hardware. Since design effort is highly dependent on the specific work output description, design hours must be estimated by a design professional experienced in the area being estimated.

70.14.3   Analysis

Analysis goes hand-in-hand with design and employs the same general skill level as design engineering. Categories of analysis that support, augment, or precede design are thermal, stress, failure, dynamics, manufacturing, safety, and maintainability. Analysis is estimated by professionals skilled in analytical techniques. Analysis usually includes computer time as well as labor-hours.

70.14.4   Drafting

Drafting, or engineering drawing, is one area in the engineering discipline where labor-hours can be correlated to a product: the completed engineering drawing. Labor-hour estimates must still be quoted in ranges, however, because the labor-hours required for an engineering drawing will vary considerably depending on the complexity of the item being drawn. The drafting times given in Table 70.6 are approximations for class-A drawings of nonelectronic (mechanical) parts where all the design information is available and where the numbers represent "board time," that is, actual time that the

Table 70.6 Engineering Draft Times

Drawing Letter Designation	Size	Approximate Board-Time Hours for Drafting of Class A Drawings (h)
A	8½ × 11	1–4
B	11 × 17	2–8
C	17 × 22	4–12
D	22 × 34	8–16
E and F	34 × 44 and 28 × 40	16–40
J	34 × 48 and larger	40–80

draftsman is working on the drawing. A class-A drawing is one that is fully dimensioned and has full supporting documentation. An additional eight hours per drawing is usually required to obtain approval and signoffs of stress, thermal, supervisors, and drawing release system personnel. If a "shop drawing" is all that is required (only sufficient information for manufacture of the part with some informal assistance from the designer and/or draftsman), the board time labor-hours required would be approximately 50% of that listed in Table 70.6.

70.15 MANUFACTURING/PRODUCTION ENGINEERING

The manufacturing/production engineering activity required to support a work activity is preproduction planning and operations analysis. This differs from the general type of production engineering wherein overall manufacturing techniques, facilities, and processes are developed. Excluded from this categorization is the design time of production engineers who redesign a prototype unit to conform to manufacturing or consumer requirements, as well as time for designing special tooling and special test equipment. A listing of some typical functions of manufacturing engineering follows:

1. Fabrication planning.
   a. Prepare operations sheets for each part.
   b. List operational sequence for materials, machines, and functions.
   c. Recommend standard and special tooling.
   d. Make up tool order for design and construction of special tooling.
   e. Develop standard time data for operations sheets.
   f. Conduct liaison with production and design engineers.
2. Assembly planning.
   a. Develop operations sheets for each part.
   b. Build first sample unit.
   c. Itemize assembly sequence and location of parts.
   d. Order design and construction of special jigs and fixtures.
   e. Develop exact component dimensions.
   f. Build any special manufacturing aids, such as wiring harness jig boards.
   g. Apply standard time data to operations sheet.
   h. Balance time cycles of final assembly line work stations.
   i. Effect liaison with production and design engineers.
   j. Set up material and layout of each work station in accordance with operations sheet.
   k. Instruct technicians in construction of the first unit.
3. Test planning.
   a. Determine overall test method to meet performance and acceptance specifications.
   b. Break total test effort into positions by function and desired time cycle.
   c. Prepare test equipment list and schematic for each position.
   d. Prepare test equipment design order for design and construction of special purpose test fixtures.
   e. Prepare a step-by-step procedure for each position.
   f. Effect liaison with production and design engineers.
   g. Set up test positions and check out.
   h. Instruct test operator on first unit.

Table 70.7 New Documentation

Function	Labor-Hours per Page
Research, liaison, technical writing, editing, and supervision	5.7
Typing and proofreading	0.6
Illustrations	4.3
Engineering	0.7
Coordination	0.2
Total[a]	11.5

[a]A range of 8 to 12 labor-hours per page can be used.

4. Sustaining manufacturing engineering.
    a. Debug, as required, engineering design data.
    b. Debug, as required, manufacturing methods and processes.
    c. Recommend more efficient manufacturing methods throughout the life of production.

The following statements may be helpful in deriving manufacturing engineering labor-hour estimates for high production rates:

1. Total fabrication and assembly labor-hours, divided by the number of units to be produced, multiplied by 20, gives manufacturing engineering start-up costs.
2. For sustaining manufacturing engineering, take the unit fabrication and assembly man-hours, multiply by 0.07. (These factors are suggested for quantities up to 100 units.)

70.15.1 Engineering Documentation

A large part of an engineer's time is spent in writing specifications, reports, manuals, handbooks, and engineering orders. The complexity of the engineering activity and the specific document requirements are important determining factors in estimating the engineering labor-hours required to prepare engineering documentation.

The hours required for engineering documentation (technical reports, specifications, and technical manuals) will vary considerably depending on the complexity of the work output; however, average labor-hours for origination and revision of engineering documentation have been derived based on experience, and these figures can be used as average labor-hours per page of documentation. (See Tables 70.7 and 70.8.)

70.16 ESTIMATING MANUFACTURING/PRODUCTION AND ASSEMBLY ACTIVITIES

A key to successful estimating of manufacturing activities is the process plan. A process plan is a listing of all operations that must be performed to manufacture a product or to complete a project, along with the labor-hours required to perform each operation. The process plan is usually prepared by an experienced foreman, engineer, or technician who knows the company's equipment, personnel, and capabilities, or by a process-planning department chartered to do all of the process estimating. The process planner envisions the equipment, work station, and environment; estimates the number

Table 70.8 Revised Documentation

Function	Labor-Hours per Page
Research, liaison, technical writing, editing, and supervision	4.00
Typing and proofreading	0.60
Illustrations	0.75
Engineering	0.60
Coordination	0.20
Total[a]	6.15

[a]A range of 4 to 8 labor-hours per page can be used.

of persons required; and estimates how long it will take to perform each step. From this information the labor-hours required are derived. Process steps are numbered, and space is left between operations listed to allow easy insertion of operations or activities as the process is modified.

A typical process plan for a welded cylinder assembly is given in Table 70.9. The process plan is used not only to plan and estimate a manufacturing or construction process, but often also as part of the manufacturing or construction work order itself. As such, it shows the shop or construction personnel each step to take in the completion of the work activity. Fabrication of items from metals, plastics, or other materials in a shop is usually called *manufacturing,* whereas fabrication of buildings, structures, bridges, dams, and public facilities on site is usually called *construction.* Different types of standards and estimating factors are used for each of these categories of work. Construction activities are covered in a subsequent chapter.

70.17 MANUFACTURING ACTIVITIES

Manufacturing activities are broken into various categories of effort, such as metal working and forming; welding, brazing, and soldering; application of fasteners; plating, printing, surface treating, heat treating; and manufacturing of electronic components (a special category). The most common method of estimating the time and cost required for manufacturing activities is the industrial engineering approach, whereby standards or target values are established for various operations. The term *standards* is used to indicate standard time data. All possible elements of work are measured, assigned a standard time for performance, and documented. When a particular job is to be estimated, all of the applicable standards for all related operations are added together to determine the total time.

The use of standards produces more accurate and more easily justifiable estimates. Standards also promote consistency between estimates as well as among estimators. Where standards are used, personal experience is desirable or beneficial, but not mandatory. Standards have been developed over a number of years through the use of time studies and synthesis of methods analysis. They are based on the level of efficiency that could be attained by a job shop producing up to 1000 units of any specific work output. Standards are actually synoptical values of more detailed times. They are adaptations, extracts, or benchmark time values for each type of operation. The loss of accuracy occasioned by summarization and/or averaging is acceptable when the total time for a system is being developed. If standard values are used with judgment and interpolations for varying stock sizes, reasonably accurate results can be obtained.

Machining operations make up a large part of the manufacturing costs of many products and projects. Machining operations are usually divided into setup times and run times. Setup time is the time required to establish and adjust the tooling, to set speeds and feeds on the metal-removal machine, and to program for the manufacture of one or more identical or similar parts. Run time is the time required to complete each part. It consists of certain fixed positioning times for each item being machined, as well as the actual metal-removal and cleanup time for each item. Values are listed for "soft" and "hard" materials. Soft values are for aluminum, magnesium, and plastics. Hard values are for stainless steel, tool steel, and beryllium. Between these two times would be standard values for brass, bronze, and medium steel.

70.18 IN-PROCESS INSPECTION

The amount of in-process inspection performed on any process, product, project, or service will depend on the cost of possible scrappage of the item as well as the degree of reliability required for the final work output. In high-rate production of relatively inexpensive items, it is often economically desirable to forgo in-process inspection entirely in favor of scrapping any parts that fail a simple go, no-go inspection at the end of the production line. On the other hand, expensive and sophisticated precision-manufactured parts may require nearly 100% inspection. A good rule of thumb is to add 10% of the manufacturing and assembly hours for in-process inspection. This in-process inspection does not include the in-process testing covered in the following paragraphs.

70.19 TESTING

Testing usually falls into three categories: (1) development testing, (2) qualification testing, and (3) production acceptance testing.

Rules of thumb are difficult to come by for estimating development testing, because testing varies with the complexity, uncertainty, and technological content of the work activity. The best way to estimate the cost of development testing is to produce a detailed test plan for the specific project and to cost each element of this test plan separately, being careful to consider all skills, facilities, equipment, and material needed in the development test program.

Qualification testing is required in most commercial products and on all military or space projects to demonstrate adequately that the article will operate or serve its intended purpose in environments far more severe than those intended for its actual use. Automobile crash tests are an example. Military products must often undergo severe and prolonged tests under high shock, thermal, and vibration loads as well as heat, humidity, cold, and salt spray environments. These tests must be meticulously planned and scheduled before a reasonable estimate of their costs can be generated.

Table 70.9   Process Plan

Operation Number	Labor-Hours	Description
010	—	Receive and inspect material (skins and forgings)
020	24	Roll form skin segments
030	60	Mask and chem-mill recessed pattern in skins
040	—	Inspect
050	36	Trim to design dimension and prepare in welding skin segments into cylinders (two)
060	16	Locate segments on automatic seam welder tooling fixture and weld per specification (longitudinal weld)
070	2	Remove from automatic welding fixture
080	18	Shave welds on inside diameter
090	16	Establish trim lines (surface plate)
100	18	Install in special fixture and trim to length
110	8	Remove from special fixture
120	56	Install center mandrel—center ring, forward and aft sections (cylinders)—forward and aft mandrel—forward and aft rings—and complete special feature setup
130	—	Inspect
140	24	Butt weld (4 places)
150	8	Remove from special feature and remove mandrels
160	59	Radiograph and dye penetrant inspect
170	—	Inspect dimensionally
180	6	Reinstall mandrels in preparation for final machining
190	14	Finish OD-aft
	10	Finish OD-center
	224	Finish OD-forward
200	40	Program for forward ring
220	30	Handwork (3 rings)
230	2	Reinstall cylinder assembly with mandrels still in place or on the special fixture
240	16	Clock and drill index holes
250	—	Inspect
260	8	Remove cylinder from special fixture—remove mandrel
270	1	Install in holding cradle
280	70	Locate drill jig on forward end and hand-drill leak check vein (drill and tap), and hand-drill hole pattern
290	64	Locate drill jig on aft ring and hand-drill hole pattern
300	—	Inspect forward and aft rings
310	8	Install protective covers on each end of cylinder
320	—	Transfer to surface treat
340	24	Remove covers and alodine
350	—	Inspect
360	8	Reinstall protective covers and return to assembly area

Table 70.10   Test Estimating Ratios

	Percent of Direct Labor		
	Simple	Average	Complex
Fabrication and Assembly Labor Base			
Receiving test	1	2	4
Production test	9	18	36
Total	10	20	40
Assembly Labor Base			
Receiving test	2	3	7
Production test	15	32	63
Total	17	35	70

Receiving inspection, production testing, and acceptance testing can be estimated using experience factors and ratios available from previous like-work activities. Receiving tests are tests performed on purchased components, parts, and/or subassemblies prior to acceptance by the receiving department. Production tests are tests of subassemblies, units, subsystems, and systems during and after assembly. Experience has shown, generally, that test labor varies directly with the amount of fabrication and assembly labor. The ratio of test labor to other production labor will depend on the complexity of the item being tested. Table 70.10 gives the test labor percentage of direct fabrication and assembly labor for simple, average, and complex items.

Special-purpose tooling and special-purpose test equipment are important items of cost because they are used only for a particular job; therefore, that job must bear the full cost of the tool or test fixture. In contrast to the special items, general-purpose tooling or test equipment is purchased as capital equipment, and costs are spread over many jobs. Estimates for tooling and test equipment are included in overall manufacturing start-up ratios shown in Table 70.11. Under "degree of precision and complexity," "high," means high-precision multidisciplinary systems, products, or subsystems; "medium" means moderately complex subsystems or components; and "low" means simple, straight-forward designs of components or individual parts. Manual and computer-aided design hours required for test equipment are shown in Table 70.12 CAD drawings take approximately 67.5% of the time required (on the average) to produce manual drawings.

70.20   COMPUTER SOFTWARE COST ESTIMATING

Detailed cost estimates must include the cost of computer software development and testing where necessary to provide deliverable source code or to run the analysis or testing programs needed to develop products or services.

Because of the increasing number and types of computers and computer languages, it is difficult to generate overall ground rules or rules of thumb for computer software cost estimating. Productivity in computer programming is greatly affected by the skill and competence of the computer analyst or programmer. The advent of computer-aided software engineering (CASE) tools has dramatically

Table 70.11   Manufacturing Startup Ratios

Cost Element	Degree of Precision and Complexity	Recurring Manufacturing Costs Lot Quantity (%)			
		10	100	1000	10,000
Production planning	High	20	6	1.7	0.5
	Medium	10	3	0.8	0.25
	Low	5	1.5	0.4	0.12
Special tooling	High	10	6	3.5	2
	Medium	5	3	2	1
	Low	3	1.5	1	—
Special test equipment	High	10	6	3.5	2
	Medium	6	3	2	1
	Low	3	1.5	1	0.5
Composite total	High	40	18	8.7	4.5
	Medium	21	9	4.8	2.25
	Low	11	4.5	2.4	1.12

Table 70.12   Design Hours for Test Equipment

Type Design	Manual Hours/ Square Foot	Standard Drawing Size	Square Feet/ Drawing	Manual Hours/ Drawing	CAD Hours/ Drawing
Original concept	15	C	2.5	38	26
		D	5.0	75	51
		H	9.0	135	91
		J	11.0	165	111
Layout	10	B	1.0	10	7
		C	2.5	25	17
		D	5.0	50	34
		H	9.0	90	61
		J	11.0	110	74
Detail or copy	3	A	0.7	2.1	1.4
		B	1.0	3.0	2.0
		C	2.5	7.5	5.1
		D	5.0	15.0	10.1
		H	9.0	27.0	18.2
		J	11.0	33.0	22.3

accelerated the process of software analysis, development, testing, and documentation. Productivity is highly dependent on which CASE tools, if any, are utilized.

Complicated flight software for aircraft and space systems is subjected to design review and testing in simulations and on the actual flight computer hardware. A software critical design review is usually conducted about 43% of the way through the program; an integrated systems test is performed at the 67% completion mark; prototype testing is done at 80% completion; installation with the hardware is started with about 7% of the time remaining (at the 93% completion point).

70.21   LABOR ALLOWANCES

"Standard times" assume that the workers are well trained and experienced in their jobs, that they apply themselves to the job 100% of the time, that they never make a mistake, take a break, lose efficiency, or deviate from the task for any reason. This, of course, is an unreasonable assumption because there are legitimate and numerous unplanned work interruptions that occur with regularity in any work activity. Therefore, labor allowances must be added to any estimate that is made up of an accumulation of standard times. These labor allowances can accumulate to a factor of 1.5 to 2.5. The total standard time for a given work activity, depending on the overall inherent efficiency of the shop, equipment, and personnel, will depend on the nature of the task. Labor allowances are made up of a number of factors that are described in the following sections.

70.21.1   Variance from Measured Labor-Hours

Standard hours vary from actual measured labor-hours because workers often deviate from the standard method or technique used or planned for a given operation. This deviation can be caused by a number of factors ranging from the training, motivation, or disposition of the operator to the use of faulty tools, fixtures, or machines. Sometimes shortages of materials or lack of adequate supervision are causes of deviations from standard values. These variances can add 5 to 20% to standard time values.

70.21.2   Personal, Fatigue, and Delay (PFD) Time

Personal times are for personal activities such as coffee breaks, trips to the restroom or water fountain, unforeseen interruptions, or emergency telephone calls. Fatigue time is allocated because of the inability of a worker to produce at the same pace all day. Operator efficiency decreases as the job time increases. Delays include unavoidable delays caused by the need for obtaining supervisory instructions, equipment breakdown, power outages, or operator illness. PFD time can add 10 to 20% to standard time values.

70.21.3   Tooling and Equipment Maintenance

Although normal or routine equipment maintenance can be done during times other than operating shifts, there is usually some operator-performed machine maintenance activity that must be performed during the machine duty cycle. These activities include adjusting tools, sharpening tools, and periodically cleaning and oiling machines. In electroplating and processing operations, the operator maintains solutions and compounds, and handles and maintains racks and fixtures. Tooling and equipment maintenance can account for 5 to 12% of standard time values.

70.21.4 Normal Rework and Repair

The overall direct labor-hours derived from the application of the preceding three allowance factors to standard times must be increased by additional amounts to account for normal rework and repair. Labor values must be allocated for rework of defective purchased materials, rework of in-process rejects, final test rejects, and addition of minor engineering changes. Units damaged on receipt or during handling must also be repaired. This factor can add 10 to 20% direct labor-hours to those previously estimated.

70.21.5 Engineering Change Allowance

For projects where design stability is poor, where production is initiated prior to final design release, and where field testing is being performed concurrently with production, an engineering change allowance should be added of up to 10% of direct labor-hours. Change allowances vary widely for different types of work activities. Even fairly well defined projects, however, should contain a change allowance.

70.21.6 Engineering Prototype Allowance

The labor-hours required to produce an engineering prototype are greater than those required to produce the first production model. Reworks are more frequent, and work is performed from sketches or unreleased drawings rather than from production drawings. An increase over first production unit labor of 15 to 25% should be included for each engineering prototype.

70.21.7 Design Growth Allowance

Where estimates are based on incomplete drawings, or where concepts or early breadboards only are available prior to the development of a cost estimate, a design growth allowance is added to all other direct labor costs. This design growth allowance is calculated by subtracting the percentage of design completion from 100%, as shown in the following tabulation:

Desirable Design Completion (%)	Design Completed (%)	Design Growth Allowance (%)
100	50	50
100	75	25
100	80	20
100	90	10
100	100	0

70.21.8 Cost Growth Allowance

Occasionally a cost estimate will warrant the addition of allowances for cost growth. Cost growth allowances are best added at the lowest level of a cost estimate rather than at the top levels. These allowances include reserves for possible misfortunes, natural disasters, strikes, and other unforeseen circumstances. Reserves should not be used to account for normal design growth. Care should be taken in using reserves in a cost estimate because they are usually the first cost elements that come under attack for removal from the cost estimate or budget. Remember, cost growth with an incomplete design is a certainty, not a reserve or contingency! Defend your cost growth allowance, but be prepared to relinquish your reserve if necessary.

70.22 ESTIMATING SUPERVISION, DIRECT MANAGEMENT, AND OTHER DIRECT CHARGES

Direct supervision costs will vary with the task and company organization. Management studies have shown that the span of control of a supervisor over a complex activity should not exceed 12 workers. For simple activities, the ratio of supervisors to employees can go down. But the 1:12 ratio (8.3%) will usually yield best results. Project management for a complex project can add an additional 10 to 14%. Other direct charges are those attributable to the project being accomplished but not included in direct labor or direct materials. Transportation, training, and reproduction costs, as well as special service or support contracts and consultants, are included in the category of "other direct costs."

Two cost elements of "other direct costs" that are becoming increasingly prominent are travel and transportation costs. A frequent check on public and private conveyance rates and costs is mandatory. Most companies provide a private vehicle mileage allowance for employees who use their own vehicles in the conduct of company business. Rates differ and depend on whether the private conveyance is being utilized principally for the benefit of the company or principally for the convenience of the traveler. Regardless of which rate is used, the mileage allowance must be periodically updated to keep pace with actual costs. Many companies purchase or lease vehicles to be used by their employees on official business.

Per diem travel allowances or reimbursement for lodging, meals, and miscellaneous expenses must also be included in overall travel budgets. These reimbursable expenses include costs of a motel or hotel room; food, tips, and taxes; local transportation and communication; and other costs such as laundry, mailing costs, and on-site clerical services. Transportation costs include the transport of equipment, supplies, and products, as well as personnel, and can include packaging, handling, shipping, postage, and insurance charges.

70.23 THE USE OF "FACTORS" IN DETAILED ESTIMATING

The practice of using factors is becoming increasingly common, particularly in high-technology work activities and work outputs. One company uses an "allocation factor," which allocates miscellaneous labor-oriented functions to specific functions such as fabrication or assembly. This company adds 14.4% to fabrication hours and 4.1% to assembly hours to cover miscellaneous labor-hour expenditures associated with these two functions. It is also common to estimate hours for planning, tooling, quality and inspection, production support, and sustaining engineering based on percentages of manufacturing and/or assembly hours. Tooling materials and computer supplies are sometimes estimated based on so much cost per tooling hour, and miscellaneous shop hardware (units, bolts, fasteners, cleaning supplies, etc.), otherwise known as *pan stock,* is estimated at a cost per manufacturing hour.

The disadvantage of the use of such factors is that inefficiencies can become embedded in the factored allowances and eventually cause cost growth. A much better method of estimating the labor-hours and materials required to accomplish these other direct activities is to determine the specific tasks and materials required to perform the job by laborloading, shoploading, or process-planning methods. When the materials, labor-hours, and other direct costs have been estimated, the basic direct resources required to do the job have been identified. The estimator can now move into the final steps of the detailed estimating process with the full assurance that all work elements and all cost elements have been included in the detailed estimate.

70.24 CONCLUDING REMARKS

In summary, detailed cost estimating involves meticulous penetration into the smallest feasible portions of a work output or work activity and the systematic and methodical assembly of the resources in all cost, work, and schedule elements. Detailed estimating requires detailed design, manufacturing, and test descriptions, and involves great time, effort, and penetration into the resources required to do the job. Wherever possible, detailed estimates should be used to establish a firm and credible cost estimate and to verify and substantiate higher-level parametric estimates.

CHAPTER 71

INVESTMENT ANALYSIS

Byron W. Jones
Kansas State University
Manhattan, Kansas

71.1	**ESSENTIALS OF FINANCIAL ANALYSIS**	**2143**		71.2.2	Classification of Alternatives	2149
	71.1.1 Sources of Funding for Capital Expenditures	2143		71.2.3	Analysis Period	2151
	71.1.2 The Time Value of Money	2144	**71.3**	**EVALUATION METHODS**		**2152**
	71.1.3 Discounted Cash Flow and Interest Calculations	2144		71.3.1	Establishing Cash Flows	2152
				71.3.2	Present Worth	2154
				71.3.3	Annual Cash Flow	2155
71.2	**INVESTMENT DECISIONS**	**2148**		71.3.4	Rate of Return	2155
	71.2.1 Allocation of Capital			71.3.5	Benefit—Cost Ratio	2157
	Funds	2148		71.3.6	Payback Period	2157

71.1 ESSENTIALS OF FINANCIAL ANALYSIS

71.1.1 Sources of Funding for Capital Expenditures

Engineering projects typically require the expenditure of funds for implementation and in return provide a savings or increased income to the firm. In this sense an engineering project is an investment for the firm and must be analyzed as an investment. This is true whether the project is a major new plant, a minor modification of some existing equipment, or anything in between. The extent of the analysis of course must be commensurate with the financial importance of the project. Financial analysis of an investment has two parts: funding of the investment and evaluation of the economics of the investment. Except for very large projects, such as a major plant expansion or addition, these two aspects can be analyzed independently. All projects generally draw from a common pool of capital funds rather than each project being financed separately. The engineering function may require an in-depth evaluation of the economics of a project, while the financing aspect generally is not dealt with in detail if at all. The primary reason for being concerned with the funding of projects is that the economic evaluation often requires at least an awareness if not an understanding of this function.

The funds used for capital expenditures come from two sources: debt financing and equity financing. Debt financing refers to funds that are borrowed from outside the company. The two common sources are bank loans and the sale of bonds. Bank loans are typically used for short-term financing, and bonds are used for long-term financing. Debt financing is characterized by a contractual arrangement specifying interest payments and repayment. The lender does not share in the profits of the investments for which the funds are used nor does it share the associated risks except through the possibility of the company defaulting. Equity financing refers to funds owned by the company. These funds may come from profits earned by the company or from funds set aside for depreciation allowances. Or, the funds may come from the sale of new stock. Equity financing does not require any specified repayment; however, the owners of the company (stockholders) do expect to make a reasonable return on their investment.

The decisions of how much funding to secure and the relative amounts to secure from debt and equity sources are very complicated and require considerable subjective judgment. The current stock market, interest rates, projections of future market conditions, etc., must be addressed. Generally, a company will try to maintain approximately a constant ratio of funding from the different sources. This mix will be selected to maximize earnings without jeopardizing the company's financial well

Mechanical Engineers' Handbook, 2nd ed., Edited by Myer Kutz.
ISBN 0-471-13007-9   © 1998 John Wiley & Sons, Inc.

being. However, the ratio of debt to equity financing does vary considerably from company to company reflecting different business philosophies.

71.1.2   The Time Value of Money

The time value of money is frequently referred to as interest or interest rate in economic analyses. Actually, the two are not exactly the same thing. Interest is a fee paid for borrowed funds and is established when the loan is made. The time value of money is related to interest rates, but it includes other factors also. The time value of money must reflect the cost of money. That is, it must reflect the interest that is paid on loans and bonds, and it must also reflect the dividends paid to the stockholders. The cost of money is usually determined as a weighted average of the interest rates and dividend rates paid for the different sources of funds used for capital expenditures. The time value of money must also reflect opportunity costs. The opportunity cost is the return that can be earned on available, but unused, projects.

In principle, the time value of money is the greater of the cost of money and the opportunity cost. The determination of the time value of money is difficult, and the reader is referred to advanced texts on this topic for a complete discussion (see, for example, Bussey[1]). The determination is usually made at the corporate level and is not the responsibility of engineers. The time value of money is frequently referred to as the interest rate for economic evaluations, and one should be aware that the terms interest rate and time value of money are used interchangeably. The time value of money is also referred to as the required rate of return.

Another factor that may or may not be reflected in a given time value of money is inflation. Inflation results in a decreased buying power of the dollar. Consequently, the cost of money generally is higher during periods of high inflation, since the funds are repaid in dollars less valuable than those in which the funds were obtained. Opportunity costs are not necessarily directly affected by inflation except that inflation affects the cash flows used in evaluating the returns for the projects on which the opportunity costs are based. It is usually up to the engineer to verify that inflation has been included in a specified time value of money, since this information is not normally given. In some applications it is beneficial to use an inflation-adjusted time value of money. The relationship is

$$1 + i_r = \frac{1 + i_a}{1 + f} \qquad (71.1)$$

where i_a is the time value of money, which reflects the higher cost of money and the higher opportunity cost due to inflation; f is the inflation rate; and i_r is the inflation-adjusted time value of money, which actually reflects the true cost of capital in terms of constant value. The variables i_r and i_a may be referred to as the real and apparent time values of money, respectively. i_r, i_a, and f are all expressed in fractional rather than percentage form in Eq. (71.1) and all must be expressed on the same time basis, generally an annual rate. See Section 71.3 for additional discussion on the use of i_r and i_a. Also see Jones[2] for a detailed discussion.

The time value of money may also reflect risks associated with a project. This is particularly true for projects where there is a significant probability of poor return or even failure (e.g., the development of a new product). In principle, risk can be evaluated in assessing the economics of a project by including the probabilities of various outcomes (see Riggs[3] for example). However, these calculations are complicated and are often dependent on subjective judgment. The more common approach is to simply use a time value of money which is greater for projects that are more risky. This is why some companies will use different values of i for different types of investments (e.g., expansion versus cost reduction versus diversification, etc.). Such adjustments for risk are usually made at the corporate level and are based on experience and other subjective inputs as much as they are on formal calculations. The engineer usually is not concerned with such adjustments, at least for routine economic analyses. If the risks of a project are included in an economic analysis, then it is important that the time value of money also not be adjusted for risks, since this would represent an overcompensation and would distort the true economic picture.

71.1.3   Discounted Cash Flow and Interest Calculations

For the purpose of economic analysis, a project is represented as a group of cash flows showing the expenditures and the income or savings attributable to the project. The object of economic analysis is normally to determine the profitability of the project based on these cash flows. However, the profitability cannot be assessed simply by summing up the cash flows, owing to the effect of the time value of money. The time value of money results in the value of a cash flow depending not only on its magnitude but also on when it occurs according to the equation

$$P = F \frac{1}{(1 + i)^n} \qquad (71.2)$$

Table 71.1   Discounted Cash Flow Calculation

Year	Estimated Cash Flows for Project	Discounted Cash Flows[a]
0	−$120,000	−$120,000
1	− 75,000	− 68,200
2	+ 50,000	+ 41,300
3	+ 60,000	+ 45,100
4	+ 70,000	+ 47,800
5	+ 30,000	+ 18,000
6	+ 20,000	+ 11,300

[a] Based on $i = 10\%$.

where F is a cash flow that occurs sometime in the future, P is the equivalent value of that cash flow now, i is the annual time value of money in fractional form, and n is the number of years from now when cash flow F occurs. Cash flow F is often referred to as the future value or future amount, while cash flow P is referred to as the present value or present amount. Equation (71.2) can be used to convert a set of cash flows for a project to a set of economically equivalent cash flows. These equivalent cash flows are referred to as discounted cash flows and reflect the reduced economic value of cash flows that occur in the future. Table 71.1 shows a set of cash flows that have been discounted.

Equation (71.2) is the basis for a more general principle referred to as economic equivalence. It shows the relative economic value of cash flows occurring at different points in time. It is not necessary that P refer to a cash flow that occurs at the present; rather, it simply refers to a cash flow that occurs n years before F. The equation works in either direction for computing equivalent cash flows. That is, it can be used to find a cash flow F that occurs n years after P and that is equivalent to P, or to find a cash flow P that is equivalent to F but which occurs n years before F. This principle of equivalence allows cash flows to be manipulated as needed to facilitate economic calculations.

The time value of money is usually specified as an annual rate. However, several other forms are sometimes encountered or may be required to solve a particular problem. An interest rate* as used in Eq. (71.2) is referred to as a discrete interest rate, since it specifies interest for a discrete time period of 1 year and allows calculations in multiples (n) of this time period. This time period is referred to as the compounding period. If it is necessary to change an interest rate stated for one compounding period to an equivalent interest rate for a different compounding period, it can be done by

$$i_1 = (1 + i_2)^{\Delta t_1 / \Delta t_2} - 1 \qquad (71.3)$$

where Δt_1 and Δt_2 are compounding periods and i_1 and i_2 are the corresponding interest rates, respectively. Interest rates i_1 and i_2 are in fractional form. If an interest rate with a compounding period different than 1 year is used in Eq. (71.2) then n in that equation refers to the number of those compounding periods and not the number of years.

Interest may also be expressed in a nominal form. Nominal interest rates are frequently used to describe the interest associated with borrowing but are not used to express the time value of money. A nominal interest rate is stated as an annual interest rate but with a compounding period different from 1 year. A nominal rate must be converted to an equivalent compound interest rate before being used in calculations. The relationship between a nominal interest rate (i_n) and a compound interest rate (i_c) is $i_c = i_n / m$, where m is the number of compounding periods per year. The compounding period (Δt) for i_c is 1 year/m. For example, a 10% nominal interest rate compounded quarterly translates to a compound interest rate of 2.5% with a compounding period of ¼ year. Equation (71.3) may be used to convert the resulting interest rate to an equivalent interest rate with annual compounding. This later interest rate is referred to as the effective annual interest rate. For the 10% nominal interest above, the effective annual interest rate is 10.38%.

Interest may also be defined in continuous rather than discrete form. With continuous interest (sometimes referred to as continuous compounding), interest accrues continuously. Equation (71.2) can be rewritten for continuous interest as

---

*The term *interest rate* is used here instead of the time value of money. The results apply to interest associated with borrowing and interest in the context of the time value of money.

$$P = F \times e^{-rt} \qquad\qquad (71.4)$$

where r is the continuous interest rate and has units of inverse time (but is normally expressed as a percentage per unit of time), t is the time between P and F, and e is the base of the natural logarithm. Note that the time units on r and t must be consistent with the year normally used. Discretely compounded interest may be converted to continuously compounded interest by

$$r = \frac{1}{\Delta t} \ln(1 + i)$$

and continuously compounded interest to discretely compounded interest by

$$i = e^{r\,\Delta t} - 1$$

Note that these are dimensional equations; the units for r and Δt must be consistent and the interest rates are in fractional form.

It is often desirable to manipulate groups of cash flows rather than just single cash flows. The principles used in Eq. (71.2) or (71.3) may be extended to multiple cash flows if they occur in a regular fashion or if they flow continuously over a period of time at some defined rate. The types of cash flows that can be readily manipulated are:

1. A uniform series in which cash flows occur in equal amounts on a regular periodic basis.
2. An exponentially increasing series in which cash flows occur on a regular periodic basis and increase by a constant percentage each year as compared to the previous year.
3. A gradient in which cash flows occur on a regular periodic basis and increase by a constant amount each year.
4. A uniform continuous cash flow where the cash flows at a constant rate over some period of time.
5. An exponentially increasing continuous cash flow where the cash flows continuously at an exponentially increasing rate.

These cash flows are illustrated in Figs. 71.1–71.5.

Any one of these groups of cash flows may be related to a single cash flow P as shown in these figures. The relationship between a group of cash flows and a single cash flow (or a single to single cash flow) may be reduced to an interest factor. The interest factors resulting in a single present amount are shown in Table 71.2. Derivations for most of these interest factors can be found in an introductory text on engineering economics (see, for example, Grant et al.[4]). The interest factor gives the relationship between the group of cash flows and the single present amount. For example,

$$P = A \cdot (P/A, i, n)$$

The term $(P/A, i, n)$ is referred to as the interest factor and gives the ratio of P to A and shows that it is a function of i and n. Other interest factors are used accordingly. The interest factors may be manipulated as if they are the mathematical ratio they represent. For example,

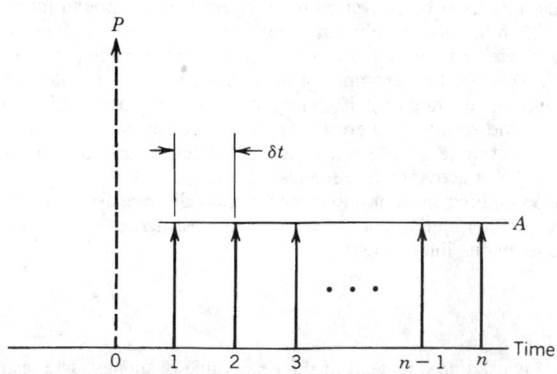

Fig. 71.1 Uniform series of cash flows.

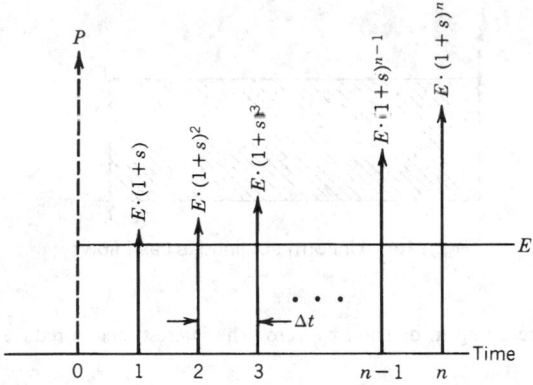

Fig. 71.2 Exponentially increasing series cash flows.

$$(A/P, i, n) = \frac{1}{(P/A, i, n)}$$

Thus for each interest factor represented in Table 71.2 a corresponding inverse interest factor may be generated as above. Two interest factors may be combined to generate a third interest factor in some cases. For example,

$$(F/A, i, n) = (P/A, i, n)(F/P, i, n)$$

or

$$(F/C, r, t) = (P/C, r, t)(F/P, r, t)$$

Again the interest factors are manipulated as the ratios they represent. In theory, any combination of interest factors may be used in this manner. However, it is wise not to mix interest factors using discrete interest with those using continuous interest, and it is usually best to proceed one step at a time to ensure that the end result is correct.

There are several important limitations when using the interest factors in Table 71.2. The time relationship between cash flows must be adhered to rigorously. Special attention must be paid to the time between P and the first cash flow. The time between the periodic cash flows must be equal to the compounding period when interest factors with discrete interest rates are used. If these do not match, Eq. (71.3) must be used to find an interest rate with the appropriate compounding period. It is also necessary to avoid dividing by zero. This is not usually a problem; however, it is possible

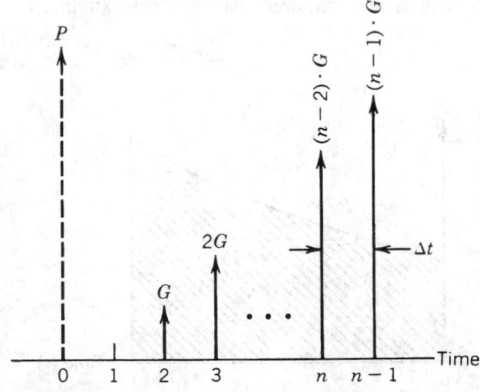

Fig. 71.3 Cash flow gradient.

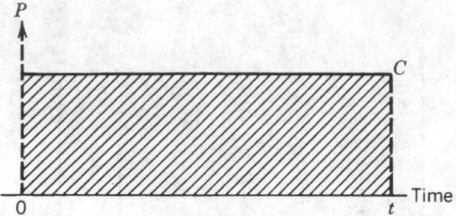

Fig. 71.4 Uniform continuous cash flow.

that $i = s$ or $r = a$, resulting in division by zero. The interest factors reduce to simpler forms in these special cases:

$$(P/F, i, s, n) = n, \qquad s = i$$
$$(P/C, r, a, t) = c \cdot t, \qquad a = r$$

It is sometimes necessary to deal with groups of cash flows that extend over very long periods of time ($n \to \infty$ or $t \to \infty$). The interest factors in this case reduce to simpler forms; but, limitations may exist if a finite value of P is to result. These reduced forms and limitations are presented in Table 71.3.

Several of the more common interest factors may be referred to by name rather than the notation used here. These interest factors and the corresponding names are presented in Table 71.4.

71.2 INVESTMENT DECISIONS

71.2.1 Allocation of Capital Funds

In most companies there are far more projects available than there are funds to implement them. It is necessary, then, to allocate these funds to the projects that provide the maximum return on the funds invested. The question of how to allocate capital funds will generally be handled at several different levels within the company, with the level at which the allocation is made depending on the size of the projects involved. At the top level, major projects such as plant additions or new product developments are considered. These may be multimillion or even multibillion dollar projects and have major impact on the future of the company. At this level both the decisions of which projects to fund and how much total capital to invest may be addressed simultaneously. At lower levels a capital budget may be established. Then, the projects that best utilize the funds available must be determined.

The basic principle utilized in capital rationing is illustrated in Fig. 71.6. Available projects are ranked in order of decreasing return. Those that are within the capital constraint are funded, those that are outside this constraint are not. However, it is not desirable to fund projects that have a rate of return less than the cost of money even if sufficient capital funds are available. If the size of individual projects is not small compared to the funds available, such as is the case with major investments, then it may be necessary to use linear programming techniques to determine the best set of projects to fund and the amount of funding to secure. As with most financial analyses, a fair amount of subjective judgment is also required. At the other extreme, the individual projects are

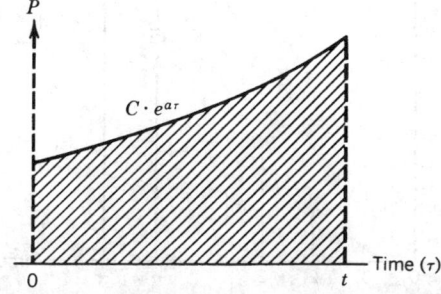

Fig. 71.5 Exponentially increasing continuous cash flow.

Table 71.2 Mathematical Expression of Interest Factors for Converting Cash Flows to Present Amounts[a]

Type of Cash Flows	Interest Factor	Mathematical Expression
Single	$(P/F, i, n)$	$= \dfrac{1}{(1 + i)^n}$ $n = \dfrac{t_2 - t_1}{\Delta t}$
Single	$(P/F, r, t)$	$= e^{-rt}$ $t = t_2 - t_1$
Uniform series	$(P/A, i, n)$	$= \dfrac{1}{i}\left[1 - \dfrac{1}{(1 + i)^n}\right]$ $n = $ number of cash flows
Uniform series	$(P/A, r, n, \Delta t)$	$= \dfrac{1}{e^{r\Delta t} - 1}(1 - e^{-rn\,\Delta t})$ $\Delta t = $ time between cash flows $n = $ number of cash flows
Exponentially increasing series	$(P/E, i, s, n)$	$= \dfrac{1 + s}{i - s}\left[1 - \left(\dfrac{1 + s}{1 + i}\right)^n\right]$ $n = $ number of cash flows $s = $ escalation rate
Exponentially increasing series	$(P/E, r, s, \Delta t, n)$	$= \dfrac{1 + s}{(e^{r\,\Delta t} - s - 1)}\left[1 - \left(\dfrac{1 + s}{e^{r\,\Delta t}}\right)^n\right]$ $n = $ number of cash flows $\Delta t = $ time between cash flows $s = $ escalation rate
Gradient	$(P/G, i, n)$	$= \dfrac{1}{i}\left[\dfrac{(1 + i)^n - 1}{i(1 + i)^n} - \dfrac{n}{(1 + i)^n}\right]$
Continuous	$(P/C, r, t)$	$= \dfrac{1 - e^{-rt}}{r}$ $t = $ duration of cash flow
Continuous increasing exponentially	$(P/C, r, a, t)$	$= \dfrac{1 - e^{-(r-a)t}}{r - a}$ $t = $ duration of cash flow $a = $ rate of increase of cash flow

[a] See Figs. 71.1–71.5 for definitions of variables n, Δt, s, a, E, G, and C. Variables i, r, a, and s are in fractional rather than percentage form.

relatively small compared to the capital available. It is not practical to use the same optimization techniques for such a large number of projects; and little benefit is likely to be gained anyway. Rather, a cutoff rate of return (also called required rate of return or minimum attractive rate of return) is established. Projects with a return greater than this amount are considered for funding; projects with a return less than this amount are not. If the cutoff rate of return is selected appropriately, then the total funding required for the projects considered will be approximately equal to the funds available. The use of a required rate of return allows the analysis of routine projects to be evaluated without using sophisticated techniques. The required rate of return may be thought of as an opportunity cost, since there are presumably unfunded projects available that earn approximately this return. The required rate of return may be used as the time value of money for routine economic analyses, assuming it is greater than the cost of money. For further discussion of the allocation of capital see Grant et al.[4] or other texts on engineering economics.

71.2.2 Classification of Alternatives

From the engineering point of view, economic analysis provides a means of selecting among alternatives. These alternatives can be divided into three categories for the purpose of economic analysis: independent, mutually exclusive, and dependent.

Independent alternatives do not affect one another. Any one or any combination may be implemented. The decision to implement one alternative has no effect on the economics of another alter-

Table 71.3   Interest Factors for $n \to \infty$ or $t \to \infty$

Interest Factor	Limitation
$(P/A, i, n) = \dfrac{1}{i}$	$i > 0$
$(P/A, r, n, \Delta t) = \dfrac{1}{e^{r\,\Delta t} - 1}$	$r > 0$
$(P/E, i, s, n) = \dfrac{1 + s}{i - s}$	$s < i$
$(P/E, r, s, \Delta t, n) = \dfrac{1 + s}{e^{r\,\Delta t} - s - 1}$	$s < e^{r\,\Delta t} - 1$
$(P/G, i, n) = \left(\dfrac{1}{i}\right)^2$	$i > 0$
$(P/C, r, t) = \dfrac{1}{r}$	$r > 0$
$(P/C, r, a, t) = \dfrac{1}{r - a}$	$a < r$

native that is independent. For example, a company may wish to reduce the delivered cost of some heavy equipment it manufactures. Two possible alternatives might be to (1) add facilities for rail shipment of the equipment directly from the plant and (2) add facilities to manufacture some of the subassemblies in-house. Since these alternatives have no effect on one another, they are independent. Each independent alternative may be evaluated on its own merits. For routine purposes the necessary criterion for implementation is a net profit based on discounted cash flows or a return greater than the required rate of return.

Mutually exclusive alternatives are the opposite extreme from independent alternatives. Only one of a group of mutually exclusive alternatives may be selected, since implementing one eliminates the possibility of implementing any of the others. For example, some particular equipment in the field may be powered by a diesel engine, a gasoline engine, or an electric motor. These alternatives are mutually exclusive since once one is selected there is no reason to implement any of the others. Mutually exclusive alternatives cannot be evaluated separately but must be compared to each other. The single most profitable, or least costly, alternative as determined by using discounted cash flows is the most desirable from an economic point of view. The alternative with the highest rate of return is not necessarily the most desirable. The possibility of not implementing any of the alternatives should also be considered if it is a feasible alternative.

Dependent alternatives are like independent alternatives in that any one or any combination of a group of independent alternatives may be implemented. Unlike independent alternatives, the decision to implement one alternative will affect the economics of another, dependent alternative. For example, the expense for fuel for a heat-treating furnace might be reduced by insulating the furnace and by modifying the furnace to use a less costly fuel. Either one or both of these alternatives could be implemented. However, insulating the furnace reduces the amount of energy required by the furnace thus reducing the savings of switching to a less costly fuel. Likewise, switching to a less costly fuel reduces the energy cost thus reducing the savings obtained by insulating the furnace. These alternatives are dependent, since implementing one affects the economics of the other one. Not recognizing dependence between alternatives is a very common mistake in economic analysis. The dependence can occur in many ways, sometimes very subtly. Different projects may share costs or cause each

Table 71.4   Interest Factor Names

Factor	Name
$(F/P, i, n)$	Compound amount factor
$(P/F, i, n)$	Present worth factor
$(F/A, i, n)$	Series compound amount factor
$(P/A, i, n)$	Series present worth factor
$(A/P, i, n)$	Capital recovery factor
$(A/F, i, n)$	Sinking fund factor

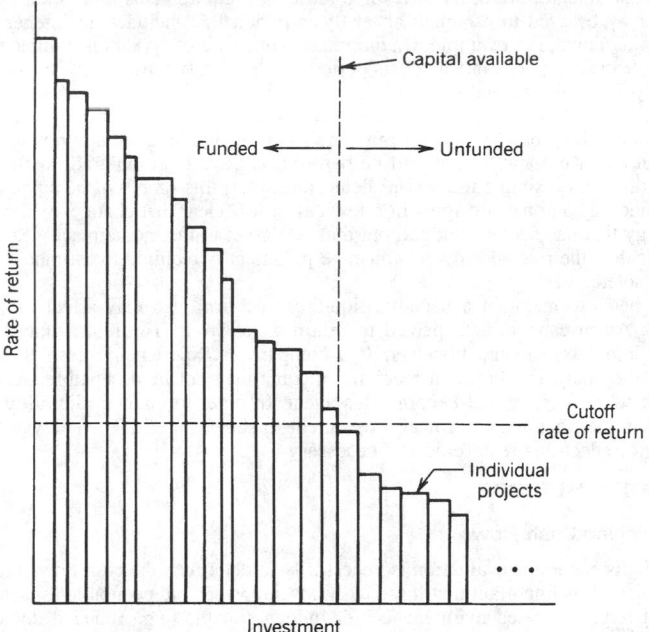

Fig. 71.6 Principle of capital allocation.

other to be more costly. They may interfere with each other or complement each other. Whenever a significant dependence between alternatives is identified, they should not be evaluated as being independent. The approach required to evaluate dependent alternatives is to evaluate all feasible combinations of the alternatives. The combination that provides the greatest profit, or least cost, when evaluated using discounted cash flows indicates which alternatives should be implemented and which should not. The number of possible combinations becomes very large quite quickly if very many dependent alternatives are to be evaluated. Therefore, an initial screening of combinations to eliminate obviously undesirable ones is useful.

Many engineering decisions do not deal with discrete alternatives but rather deal with one or more parameters that may vary over some range. Such situations result in continuous alternatives, and, in concept at least, an infinite number of possibilities exist but only one may be selected. For example, foam insulation may be sprayed on a storage tank. The thickness may vary from a few millimeters to several centimeters. The various thicknesses represent a continuous set of alternatives. The approach used to determine the most desirable value of the parameter for continuous alternatives is to evaluate the profit, or cost, using discounted cash flows for a number of different values to determine an optimum value. Graphical presentation of the results is often very helpful.

71.2.3 Analysis Period

An important part of an economic analysis is the determination of the appropriate analysis period. The concept of life-cycle analysis is used to establish the analysis period. Life-cycle analysis refers to an analysis period that extends over the life of the entire project including the implementation (e.g., development and construction). In many situations an engineering project addresses a particular need (e.g., transporting fluid from a storage tank to the processing plant). When it is clear that this need exists only for a specific period of time, then this period of need may establish the analysis period. Otherwise the life of the equipment involved will establish the project life.

Decisions regarding the selection, replacement, and modification of particular equipment and machinery are often a necessary part of many engineering functions. The same principles used for establishing an analysis period for the larger projects apply in this situation as well. If the lives of the equipment are greater than the period of need, then that period of need establishes the analysis period. If the lives of the equipment are short compared to the analysis period, then the equipment lives establish the analysis period. The lives of various equipment alternatives are often significantly different. The concept of life-cycle analysis requires that each alternative be evaluated over its full life. Fairness in the comparison requires that the same analysis period be applied to all alternatives that serve the same function. In order to resolve these two requirements more than one life cycle for the equipment may be used to establish an analysis period that includes an integer number of life cycles of each alternative. For example, if equipment with a life of 6 years is compared to equipment with a life of 4 years, an appropriate analysis period is 12 years, two life cycles of the first alternative and three life cycles of the second.

Obsolescence must also be considered when selecting an analysis period. Much equipment becomes uneconomical long before it wears out. The life of equipment, then, is not set by how long it can function but rather by how long it will be before it is desirable to upgrade with a newer design. Unfortunately, there is no simple method to determine when this time will be since obsolescence is due to new technology and new designs. In a few cases, it is clear that changes will soon occur (e.g., a new technology that has been developed but that is not yet in the marketplace). Such cases are the exception rather than the rule and much subjective judgment is required to estimate when something will become obsolete.

The requirements to maintain acceptable liquidity in a firm also may affect the selection of an analysis period. An investment is expected to return a net profit. The return may be a number of years after the initial investment, however. If a company is experiencing cash flow difficulties or anticipates that they may, this delay in receiving income may not be acceptable. A long-term profit is of little value to a company that becomes insolvent. In order to maintain liquidity, an upper limit may be placed on the time allowed for an investment to show a profit. The analysis period must be shortened then to reflect this requirement if necessary.

71.3 EVALUATION METHODS

71.3.1 Establishing Cash Flows

The first part of any economic evaluation is necessarily to determine the cash flows that appropriately describe the project. It is important that these cash flows represent all economic aspects of the project. All hidden cost (e.g., increased maintenance) or hidden benefits (e.g., reduced downtime) must be included as well as the obvious expenses, incomes, or savings associated with the project. Wherever possible, nonmonetary factors (e.g., reduced hazards) should be quantified and included in the analysis. Also, taxes associated with a project should not be ignored. (Some companies do allow a before-tax calculation for routine analysis.) Care should be taken that no factor be included twice. For example, high maintenance costs of an existing machine may be considered an expense in the alternative of keeping that machine or a savings in the alternative of replacing it with a new one, but it should not be considered both ways when comparing these two alternatives. Expenses or incomes that are irrelevant to the analysis should not be included. In particular, sunk costs, those expenses which have already been incurred, are not a factor in an economic analysis except for how they affect future cash flows (e.g., past equipment purchases may affect future taxes owing to depreciation allowances). The timing of cash flows over a project's life is also important, since cash flows in the near future will be discounted less than those that occur later. It is, however, customary to treat all of the cash flows that occur during a year as a single cash flow either at the end of the year (year-end convention) or at the middle of the year (mid-year convention).

Estimates of the cash flows for a project are generally determined by establishing what goods and services are going to be required to implement and sustain a project and by establishing the goods, services, benefits, savings, etc., that will result from the project. It is then necessary to estimate the associated prices, costs, or values. There are several sources of such information including historical data, projections, bids, or estimates by suppliers, etc. Care must be exercised when using any of these sources to be sure they accurately reflect the true price when the actual transaction will occur. Historical data are misleading, since they reflect past prices, not current prices, and may be badly in error owing to inflation that has occurred in recent years. Current prices may not accurately reflect future prices for the same reason. When historical data are used, they should be adjusted to reflect changes that have occurred in prices. This adjustment can be made by

$$p(t_0) = p(t_1) \frac{PI(t_0)}{PI(t_1)} \tag{71.5}$$

where $p(t)$ is the price, cost, or value of some item at time t; $PI(t)$ is the price index at time t, and t_0 is the present time. The price index reflects the change in prices for an item or group of items. Indexes for many categories of goods and services are available from the Bureau of Labor Statistics.[2] Current prices should also be adjusted when they refer to future transactions. Many companies have projections for prices for many of their more important products. Where such projections are not available, a relationship similar to Eq. (71.5) may be used except that price indexes for future years are not available. Estimates of future inflation rates may be substituted instead:

$$p(t_2) = p(t_0)(1 + f)^n \tag{71.6}$$

where f is the annual inflation rate, n is the number of years from the present until time t_2 ($t_2 - t_0$

in years), and t_0 is the present time. The inflation rate in Eq. (71.6) is the overall inflation rate unless it is expected that the particular item in question will increase in price much faster or slower than prices in general. In this case, an inflation rate pertaining to the particular item should be used.

Changing prices often distort the interpretation of cash flows. This distortion may be minimized by expressing all cash flows in a reference year's dollars (e.g., 1990 dollars). This representation is referred to as constant dollar cash flows. Historic data may be converted to constant dollar representation by

$$Y^c = Y^d \frac{\overline{PI(t_0)}}{\overline{PI(t)}} \tag{71.7}$$

where Y^c is the constant dollar representation of a dollar cash flow U^d, $\overline{PI(t)}$ is the value of the price index at time t, t_0 is the reference year, and t is the year in which Y^d occurred. An overall price index such as the Wholesale Price Index or the Gross National Product Implicit Price Deflator is used in this calculation, whereas a more specific price index is used in Eq. (71.5). Future cash flows may be expressed using constant dollar representation by

$$Y^c = Y^d \frac{1}{(1 + f)^n} \tag{71.8}$$

where f is the projected annual inflation rate and n is the number of years after t_0 that Y^d occurs. It is usually convenient to let t_0 be the present. Then n is equal to the number of years from the present and, also, present prices may be used to make most constant dollar cash flow estimates. The use of constant dollar representation simplifies the economic analysis in many situations. However, it is important that the time value of money be adjusted as indicated in Eq. (71.2). Additional discussion on this topic may be found in Ref. 2.

Mutually exclusive alternatives and dependent alternatives often yield cash flows that are either all negative or predominantly negative, that is, they only deal with expenses. It is not possible to view each alternative in terms of an investment (initial expense) and a return (income). However, two alternatives may be used to create a set of cash flows that represent an investment and return as shown in Fig. 71.7. Alternative B is more expensive to implement than A but costs less to operate or sustain. Cash flow C is the difference between B and A. It shows the extra investment required for B and the savings it produces. Cash flow C may then be analyzed as an investment to determine

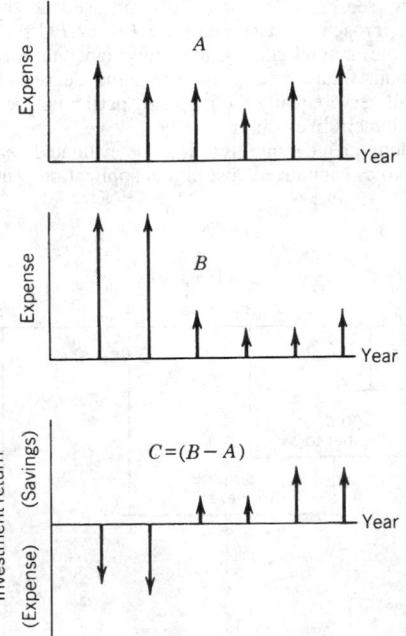

Fig. 71.7 Investment and return generated by comparing two alternatives.

if the extra investment required for C is worthwhile. This approach works well when only two alternatives are considered. If there are three or more alternatives, the comparison gets more complicated. Figure 71.8 shows the process required. Two alternatives are compared, the winner of that comparison is compared to a third, the winner of that comparison to a fourth, and so on until all alternatives have been considered and the single best alternative identified. When using this procedure, it is customary, but not necessary, to order the alternatives from least expensive to most expensive according to initial cost. This analysis of multiple alternatives may be referred to as incremental analysis, since only the difference, or incremental cash flow between alternatives, is considered. The same concept may be applied to continuous alternatives.

71.3.2 Present Worth

All of the cash flows for a project may be reduced to a single equivalent cash flow using the concepts of time value of money and cash flow equivalence. This single equivalent cash flow is usually calculated for the present time or the project initiation, hence the term present worth (also present value). However, this single cash flow can be calculated for any point in time if necessary. Occasionally, it is desired to calculate it for the end of a project rather than the beginning, the term future worth (also future value) is applied then. This single cash flow, either a present worth or future worth, is a measure of the net profit for the project and thus is an indication of whether or not the project is worthwhile economically. It may be calculated using interest factors and cash flow manipulations as described in preceding sections. Modern calculators and computers usually make it just as easy to calculate the present worth directly from the project cash flows. The present worth PW of a project is

$$PW = \sum_{j=0}^{n} Y_j \frac{1}{(1 + i)^j} \tag{71.9}$$

where Y_j is the project cash flow in year j, n is the length of the analysis period, and i is the time value of money. This equation uses the sign convention of income or savings being positive and expenses being negative.

In the case of independent alternatives, the PW of a project is a sufficient measure of the project's profitability. Thus, a positive present worth indicates an economically desirable project and a negative present worth indicates an economically undesirable project. In the case of mutually exclusive, dependent, or continuous alternatives the present worth of a given alternative means little. The alternative, or combination of dependent alternatives, that has the highest present worth is the most desirable economically. Often cost is predominant in these alternatives. It is customary then to reverse the sign convention in Eq. (71.9) and call the result the present cost. The alternative with the smallest present cost is then the most desirable economically. It is also valid to calculate the present worth of the incremental cash flow (see Figs. 71.7 and 71.8) and use it as the basis for choosing between alternatives. However, the approach of calculating the PW or PC for each alternative is generally much easier. Regardless of the method chosen, it is important that all alternatives be treated fairly; use similar assumptions about future prices, use the same degree of conservatism in estimating expenses, make sure they all serve equally well, etc. In particular, be sure that the proper analysis period is selected when equipment lives differ.

Projects that have very long or indefinite lives may be evaluated using the present worth method. The present cost is referred to as capitalized cost in this application. The capitalized cost can be used

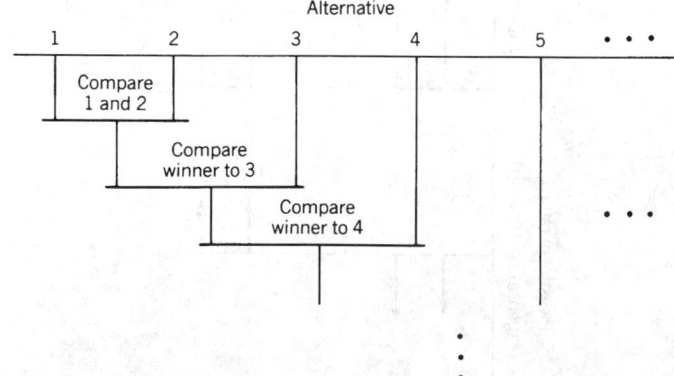

Fig. 71.8 Comparison of multiple alternatives using incremental cash flows.

for economic analysis in the same manner as the present cost; however, it cannot be calculated using Eq. (71.9), since the number of calculations required would be rather large as $n \rightarrow \infty$. The interest factors in Table 71.3 usually can be used to reduce the portion of the cash flows that continue indefinitely to a single equivalent amount, which can then be dealt with as any other single cash flow.

71.3.3 Annual Cash Flow

The annual cash flow method is very similar in concept to the present worth method and generally can be used whenever a present worth analysis can. The present worth or present cost of a project can be converted to an annual cash flow, ACF, by

$$ACF = PW \cdot (A/P, i, n) \qquad (71.10)$$

where $(A/P, i, n)$ is the capital recovery factor, i is the time value of money, and n is the number of years in the analysis period. Since ACF is proportional to PW, a positive ACF indicates a profitable investment and a negative ACF indicates an unprofitable investment. Similarly, the alternative with the largest ACF will also have the largest PW. The PW in Eq. (71.10) can be replaced with PC, and ACF can be used to represent a cost when that is more appropriate. The ACF is thus equally as useful for economic analysis as PW or PC. It also has the advantage of having more intuitive meaning. ACF represents the equivalent annual income or cost over the life of a project.

Annual cash flow is particularly useful for analyses involving equipment with unequal lives. The n in Eq. (71.10) refers to the length of the analysis period and for unequal lives that means integer multiples of the life cycles. The annual cash flow for the analysis period will be the same as the analysis period for a single life cycle, as shown in Fig. 71.9, as long as the cash flows for the equipment repeat from one life cycle to the next. The annual cash flow for each equipment alternative can then be calculated for its own life cycle rather than for a number of life cycles. Unfortunately, prices generally increase from one life cycle to the next due to inflation and the cash flows from one life cycle to the next will be more like those shown in Fig. 71.10. The errors caused by this change from life cycle to life cycle usually will be acceptable if inflation is moderate (e.g., less than 5%) and the lives of various alternatives do not differ greatly (e.g., 7 versus 9 years). If inflation is high or if alternatives have lives that differ greatly, significant errors may result. The problem can often be circumvented by converting the cash flows to constant dollars using Eq. (71.8). With the inflationary price increases removed, the cash flows will usually repeat from one life cycle to the next.

71.3.4 Rate of Return

The rate of return (also called the internal rate of return) method is the most frequently used technique for evaluating investments. The rate of return is based on Eq. (71.9) except that rather than solving for PW, the PW is set to zero and the equation is solved for i. The resulting interest rate is the rate of return of the investment:

$$O = \sum_{j=0}^{n} Y_j \frac{1}{(1 + i)^j} \qquad (71.11)$$

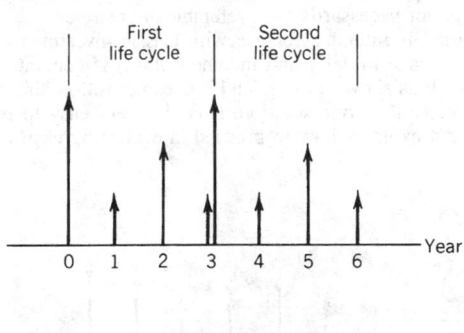

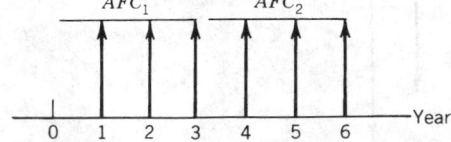

Fig. 71.9 Annual cash flow for equipment with repeating cash flows (3-year life).

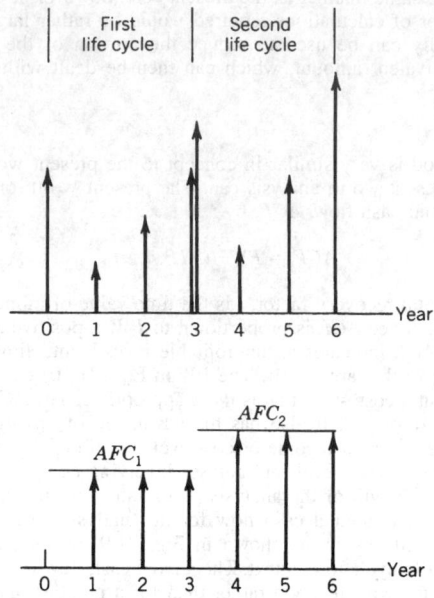

Fig. 71.10 Annual cash flow for equipment with nonrepeating cash flows (3-year life).

where Y_j is the cash flow for year j, i is the investment's rate of return (rather than the time value of money), and n is the length of the analysis period. It is usually necessary to solve Eq. (71.11) by trial and error, except for a few very simple situations. If constant dollar representation is used, the resulting rate of return is referred to as the real rate of return or the inflation-corrected rate of return. It may be converted to a dollar rate of return using Eq. (71.2).

A rate of return for an investment greater than the time value of money indicates that the investment is profitable and worthwhile economically. Likewise an investment with a rate of return less than the time value of money is unprofitable and is not worthwhile economically. The rate of return calculation is generally preferred over the present worth method or annual cash flow method by decision makers since it gives a readily understood economic measure. However, the rate of return method only allows a single investment to be evaluated or two projects compared using the incremental cash flow. When several mutually exclusive alternatives, dependent alternative combinations, or continuous alternatives exist, it is necessary to compare two investments at a time as shown in Fig. 71.8 using incremental cash flows. Present worth or annual cash flow methods are simpler to use in these instances. It is important to realize that with these types of decisions the alternative with the highest rate of return is not necessarily the preferable alternative.

The rate of return method is intended for use with classic investments as shown in Fig. 71.11. An expense (investment) is made initially and income (return) is generated in later years. If a particular set of cash flows, such as shown in Fig. 71.12, does not follow this pattern, it is possible that Eq. (71.11) will generate more than one solution. It is also very easy to misinterpret the results of such cash flows. Reference 4 explains how to proceed in evaluating cash flows of the nature shown in Fig. 71.12.

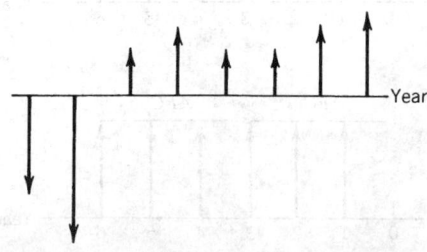

Fig. 71.11 Example of pure investment cash flows.

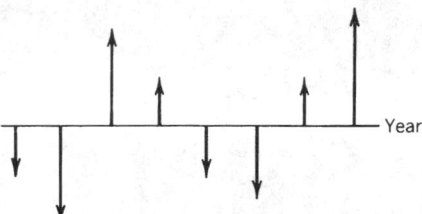

Fig. 71.12   Example of mixed cash flows.

71.3.5   Benefit–Cost Ratio

The benefit–cost ratio (B/C) calculation is a form of the present worth method. With B/C, Eq. (71.9) is used to calculate the present worth of the income or savings (benefits) of the investment and the expenses (costs) separately. These two quantities are then combined to form the benefit–cost ratio:

$$B/C = \frac{PW_B}{PC_E}$$ (71.12)

where PW_B is the present worth of the benefits and PC_E is the present cost of the expenses. A B/C greater than 1 indicates that the benefits outweigh the costs and a B/C less than 1 indicates the opposite. The B/C also gives some indication as to how good an investment is. A B/C of about 1 indicates a marginal investment, whereas a B/C of 3 or 4 indicates a very good one. The PC_E usually refers to the initial investment expense. The PW_B includes the income and savings less operating cost and other expenses. There is some leeway in deciding whether a particular expense should be included in PC_E or subtracted from PW_B. The placement will change B/C some, but will never make a B/C which is less than 1 become greater than 1 or vice versa.

The benefit–cost calculation can only be applied to a single investment or used to compare two investments using the incremental cash flow. When evaluating several mutually exclusive alternatives, dependent alternative combinations, or continuous alternatives, the alternatives must be compared two at a time as shown in Fig. 71.8 using incremental cash flows. The single alternative with the largest B/C is not necessarily the preferred alternative in this case.

71.3.6   Payback Period

The payback calculation is not a theoretically valid measure of the profitability of an investment and is frequently criticized for this reason. However, it is widely used and does provide useful information. The payback period is defined as the period of time required for the cumulative net cash flow to be equal to zero; that is, the time required for the income or savings to offset the initial costs and other expenses. The payback period does not measure the profitability of an investment but rather its liquidity. It shows how fast the money invested is recovered. It is a useful measure for a company experiencing cash flow difficulties and which cannot afford to tie up capital funds for a long period of time. A maximum allowed payback period may be specified in some cases. A short payback period is generally indicative of a very profitable investment, but that is not ensured since there is no accounting for the cash flows that occur after the payback period. Most engineering economists agree that the payback period should not be used as the means of selecting among alternatives.

The payback period is sometimes calculated using discounted cash flows rather than ordinary cash flows. This modification does not eliminate the criticisms of the payback calculation although it does usually result in only profitable investments having a finite payback period. A maximum allowed payback period may also be used with this form. This requirement is equivalent to arbitrarily shortening the analysis period to the allowed payback period to reflect liquidity requirements. Since there are different forms of the payback calculation and the method is not theoretically sound, extreme care should be exercised in using the payback period in decision making.

REFERENCES

1. L. E. Bussey, *The Economic Analysis of Industrial Projects,* Prentice-Hall, Englewood Cliffs, NJ, 1978.

2. B. W. Jones, *Inflation in Engineering Economic Analysis,* Wiley, New York, 1982.

3. J. L. Riggs, *Engineering Economics,* 2nd ed., McGraw-Hill, New York, 1982.

4. E. L. Grant, W. G. Ireson, and R. S. Leavenworth, *Principles of Engineering Economy,* 7th ed., Wiley, New York, 1982.

CHAPTER 72

TOTAL QUALITY MANAGEMENT AND THE MECHANICAL ENGINEER

R. Alan Kemerling
Staff Quality Systems Engineer—New Product Development
Ethicon Endo-Surgery, Inc.
Cincinnati, Ohio

Jack B. ReVelle
Hughes Missile Systems Company
Tucson, Arizona

72.1	**WHAT IS TOTAL QUALITY MANAGEMENT?**	**2159**	
	72.1.1 The Traditional Approach to Quality	2159	
	72.1.2 The New Paradigm of Total Quality Management	2160	
72.2	**DEFINITIONS OF *QUALITY***	**2160**	
72.3	**WHAT ARE THE BENEFITS FOR MY COMPANY?**	**2161**	
72.4	**HOW WILL IT CHANGE MY ROLE?**	**2162**	
	72.4.1 As a Mechanical Engineer	2162	
	72.4.2 As a Manager of Mechanical Engineers	2163	
72.5	**WHAT ARE THE TOOLS OF TOTAL QUALITY MANAGEMENT AND HOW DO I USE THEM?**	**2164**	
	72.5.1 Technical Tools—Quality Function Deployment (QFD)	2164	

	72.5.2 Technical Tools—Seven Management and Planning (7 MP) Tools	2166	
	72.5.3 Technical Tools—Design of Experiments (DOE)	2168	
	72.5.4 Technical Tools—SPC, SQC, and 7 QC	2171	
	72.5.5 Technical Tools—Process Capability or Validation Studies	2172	
	72.5.6 Technical Tools—Other TQM Tools	2173	
	72.5.7 Cultural/Social Tools—Concurrent Engineering	2173	
	72.5.8 Cultural/Social Tools—Teams	2175	
	72.5.9 Cultural/Social Tools—The Variability Reduction Process (VRP)	2175	
72.6	**SUMMARY**	**2176**	

72.1 WHAT IS TOTAL QUALITY MANAGEMENT?

72.1.1 The Traditional Approach to Quality

Before considering a definition of Total Quality Management, for contrast let's review the traditional approach to quality. During the Industrial Revolution, a major change that allowed manufacturing to achieve significant efficiency gains was a division of labor for all aspects of manufacturing work. This approach, led by Frederick W. Taylor, advocated management of factory work by dividing it into simple, repetitive tasks that could be executed quickly and easily with a minimum of skill.

Mechanical Engineers' Handbook, 2nd ed., Edited by Myer Kutz.
ISBN 0-471-13007-9 © 1998 John Wiley & Sons, Inc.

Generally, Taylor's approach worked well for the time, making durable consumer items affordable for many.

During World War II, the Department of Defense pressed for a similar specialization in the quality function as a means to assure the quality of war materials. The government's document for quality, MIL-Q-9858, specified a separate and independent quality department with the responsibility to plan, audit, and assure that required quality levels were met. Usually, outgoing quality levels were met by significant amounts of inspection and test of the final product. Goods or services that did not conform to requirements were made to conform (reworked) or scrapped. Other documents, such as MIL-STD-105, specified how to sample and what decisions to make, based on the results of inspections. Commercial firms have often followed this organizational approach, some even adopting government inspection standards.

The practical effect of this organizational approach, as shown in Fig. 72.1, was to make the *quality* of the finished goods or services the *responsibility* of the quality department. There was little incentive for any other operation in the company to be concerned with quality. After all, the quality department *was* the department paid to find and fix defective goods or services.

By Frederick Taylor's logic, this arrangement still made sense. Quality engineers could improve their ability to plan for quality, develop inspection and test plans, and direct inspection staff. However, this was one area where division of labor and separation of responsibilities did not prove to be the most efficient approach for the entire enterprise, especially as products and services became more and more complex. First of all, inspection, particularly visual inspection, is never 100% successful in catching defects. As a result, there were still dissatisfied customers and warranty costs, even with significant levels of inspection. Second, it became apparent to some far-sighted business leaders that inspection and test were not adding value, but businesses were in fact supporting an entire "hidden factory" of extra floor space, materials, labor, and machinery to take care of rework and scrapped material. Some organizations paid lip service to the concept that "quality cannot be inspected into" the product, but few made an attempt to change. Those that did began to grasp the fact that the quality of goods and services, as perceived by the customer, is a function of the entire enterprise. Hence, the entire enterprise must be engaged in planning for quality and delivering quality results. As suggested in Fig. 72.2, it will take a different organizational approach to answer the new quality requirements.

72.1.2 The New Paradigm of Total Quality Management

This insight leads to a review of Total Quality Management (TQM). First, here is a definition of TQM for discussion purposes: "Total Quality Management is an evolving management philosophy and methodology for guiding the continuous improvement of products, processes and services with the objective of realizing optimum customer value and satisfaction. It fosters the engagement of everyone in the enterprise toward this end."[1] As is evident from the definition, TQM departs from the division of labor theory of Taylorism to assert that what the customer perceives as quality is the responsibility of everyone in the organization. This doesn't mean that the assembler of the engine is responsible for the finish on the hood of the car. The tools of TQM include methods to deploy and measure appropriate quality characteristics for each operation in the organization.

72.2 DEFINITIONS OF *QUALITY*

Several definitions of *quality* have been used over the years. Following are some of the predominant ones.

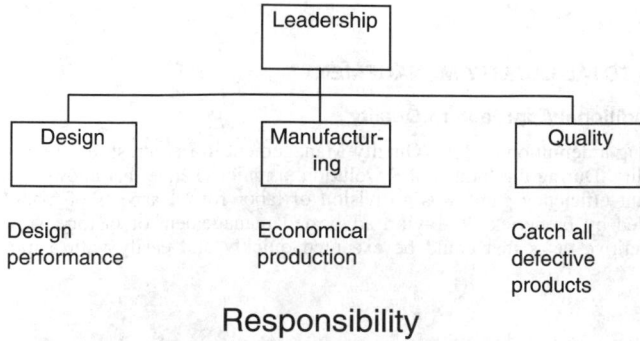

Responsibility

Fig. 72.1 Who has the responsibility for quality?

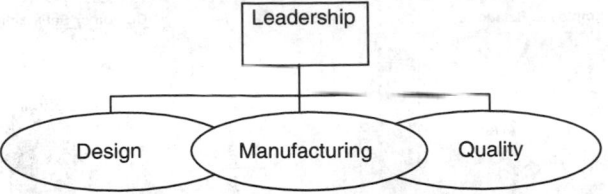

Design recognizes the responsbility to produce a
design that can be manufactured economically.
Manufacturing recognizes the responsibility to
develop stable processes and maintain control.
Quality audits products and systems to foster
continuous improvement.

Fig. 72.2   A unified approach is needed.

- Freedom from defects[2]
- Fitness for use[3]
- The totality of features and characteristics of a product or service that bear on its ability to satisfy given needs[4]
- The features and characteristics that delight the customer[5]

A review of these definitions will show a progression from a narrow consideration of the absence or presence of defects to a more holistic consideration of the ability of the product or service to *satisfy* the customer. This progression parallels the evolution of quality management from just the management of inspection to TQM.

72.3   WHAT ARE THE BENEFITS FOR MY COMPANY?

There are several benefits stemming from the adoption of an active and effective TQM program. These include:

- Improved customer satisfaction from better products and services
- Improved profit margins from reduced costs
- Easier introduction of new products and services
- Higher worker satisfaction due to involvement with improvement teams, integrated product and process development teams, and design for manufacture and assembly (DFMA) teams

These are strong claims, but they can easily be supported by data. The first study to address the effects of TQM application beyond the quality of products and services was conducted by the General Accounting Office (GAO) at the request of Congressman Donald Ritter (R—Pa).[6] This study looked at 20 companies that received a site visit for the Malcolm Baldrige National Quality Award (MBNQA) (see Chapter 73) in 1988 and 1989. To receive a site visit for the MBNQA indicates that the company is a "finalist" in this assessment of TQM applications.

The GAO study considered data (where available) in four broad areas with a number of specific elements in each: (1) employee relations, (2) operating procedures, (3) customer satisfaction, and (4) financial performance. In each case, the available companies' data were analyzed for trends from the time the company reported it started its TQM initiatives. In addition, the companies' data were compared with metrics available from their specific industry. The results are shown in Fig. 72.3. All charts are to the same scale, represent average annual percent improvement, and have the results stated so that a positive bar represents a favorable result for the company. The specific elements for each area are printed under the bar.

In the area of employee-related indicators, the survey looked at employee satisfaction (from surveys), attendance, turnover, safety/health (lost work days due to work-related injury and illness), and suggestions received. These measures show the degree of personnel engagement in TQM and staff response to the initiative.

The survey also looked at operating indicators. These are metrics of the quality and costs of products and services. The categories of measurements included (1) reliability, (2) timeliness of

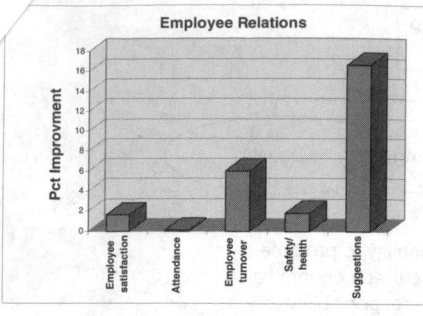
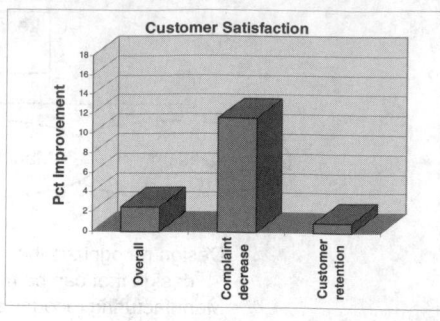
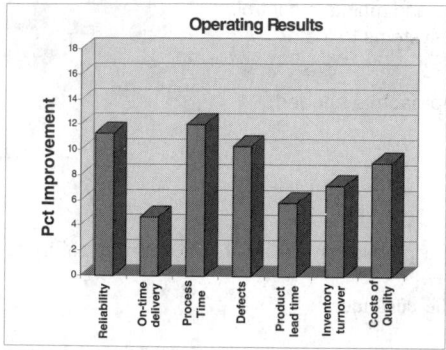
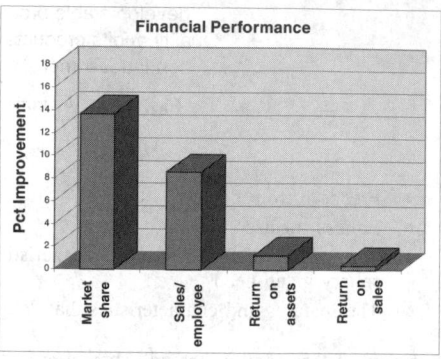

Fig. 72.3 Charts of results from the GAO TQM study.

delivery, (3) order-processing time, (4) errors or defects, (5) product lead time, (6) inventory turnover, (7) costs of quality, and (8) cost savings. These metrics are an expansion of "traditional" quality measures. They represent a measure of quality system effectiveness.

Customer satisfaction is a very important indicator for any business. If customers are not satisfied, the company's profitability will be affected at some point, usually sooner than later. This survey looked at three measures of customer satisfaction: (1) overall customer satisfaction, (2) customer complaints, and (3) customer retention.

The survey looked at the increased financial performance of the companies applying TQM. The metrics looked at were (1) market share, (2) sales per employee, (3) return on assets, and (4) return on sales. These measures put to rest the theory that TQM efforts do not offer an attractive return on investment. How much is a 14% annual increase in market share worth to your company?

72.4 HOW WILL IT CHANGE MY ROLE?

72.4.1 As a Mechanical Engineer

Traditionally, engineers become engineers because they have an aptitude for or prefer to deal with data and things. The typical mechanical engineer is most focused on one key responsibility, the performance of his or her design or process. This is still an important consideration, but as your organization adopts TQM, whether due to customer requirements or competitive pressures, some new dimensions will be added to your role. As shown in Fig. 72.4, TQM has many aspects that affect both the organization and the individuals. This section will include a brief discussion of some of them.

First of all, a mechanical engineer working in a TQM environment will probably be part of a multifunctional team, usually an integrated product and process development team (more on this will be found in a later section of this chapter). This will require what may be new skills, such as listening to other viewpoints on a design, reaching consensus on decisions, and achieving alignment on customer needs. To the mechanical engineer, teams may appear inefficient, slowing down "important" design work, but the performance of a well-developed team has often proven superior to other organizational forms.

Another change that a mechanical engineer may note in TQM is a focus on processes. In the past, engineers usually felt that the result was important, not necessarily the means. TQM focusses on the means (processes) as much as the results. This is one way to achieve minimum variation in

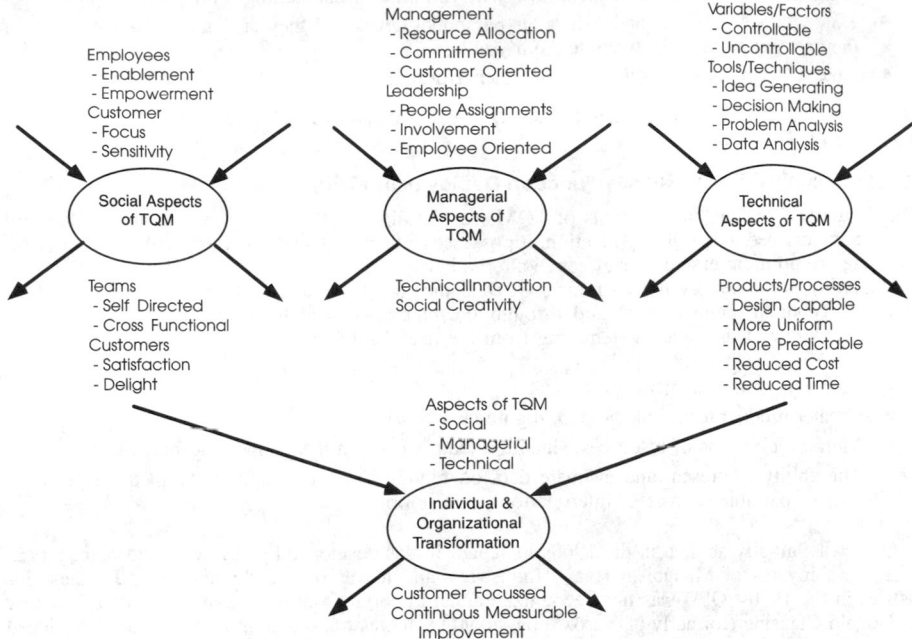

Employees
- Enablement
- Empowerment
Customer
- Focus
- Sensitivity

Management
- Resource Allocation
- Commitment
- Customer Oriented
Leadership
- People Assignments
- Involvement
- Employee Oriented

Variables/Factors
- Controllable
- Uncontrollable
Tools/Techniques
- Idea Generating
- Decision Making
- Problem Analysis
- Data Analysis

Social Aspects
of TQM

Managerial
Aspects of
TQM

Technical
Aspects of TQM

Teams
- Self Directed
- Cross Functional
Customers
- Satisfaction
- Delight

TechnicalInnovation
Social Creativity

Products/Processes
- Design Capable
- More Uniform
- More Predictable
- Reduced Cost
- Reduced Time

Aspects of TQM
- Social
- Manageriul
- Technical

Individual &
Organizational
Transformation

Customer Focussed
Continuous Measurable
Improvement

Fig. 72.4 The comprehensive model of TQM.

results, to consistently use the best process available. At first thought, this may appear restrictive, but it is not. TQM is serious about continuous improvement. This means that processes will not remain static, but when the current "best process" is discovered, all functions that can use it are expected to use it.

A final key change that a mechanical engineer might note in an organization adopting TQM involves the engineer's relationship with the management structure. To free up the creative capability in the organization and to make it more agile, management must move from a directive relationship to a coaching or guiding relationship. Of course, this will be a significant change for the manager and engineer and sometimes the transition is not smooth.

72.4.2 As a Manager of Mechanical Engineers

If you are a manager of mechanical engineers in an organization deploying TQM, you will be in for changes that may make you feel insecure in your position. You will see a drive to reduce your apparent authority, to place your staff on teams, and to turn your position into that of "coach." It's possible that you'll stop receiving funding to supply personnel for projects. Instead the funding will go directly to the team. Your personnel will most likely be located with their team, perhaps geographically removed from you.

We have emphasized this negative picture to draw attention to the focus on management in TQM. A significant part of the pressure to change and the pressure from change falls on management. If you think that TQM is something to assign to someone or something that staff can do without your involvement, you are on a path to a failed implementation.

In addition to the personal considerations, there are other concerns that you must consider for a TQM implementation.

- Does your organization have a plan for identifying what teams, how many are needed, and how you will task them?
- Do you have a way to assign team leaders and team members?
- How are you going to equip teams with the TQM tools and team skills to succeed?
- Do you have subject matter experts (SMEs) identified for TQM tools and team skills?
- Do you currently have data systems on your processes?
- Do you know what your customers expect?
- How will you fund the teams?

If the funding goes to the teams, how will you know what staffing levels to maintain?

- How will you evaluate and help your personnel develop if they are on a team, especially if they are geographically separate from you?
- How will you know when a team is not performing?

72.5 WHAT ARE THE TOOLS OF TOTAL QUALITY MANAGEMENT AND HOW DO I USE THEM?

72.5.1 Technical Tools—Quality Function Deployment (QFD)

QFD is the first of the "major" tools of TQM we will discuss. By "major" we mean that the tool fulfills a major need in a TQM application, it possesses a fairly extensive research and literature base, and there are no more efficient or effective alternatives.

If *quality* is defined by the customer, QFD is the tool to assure that the customers' *vision of quality* is captured, defined, deployed through the enterprise, and linked to the activities of the enterprise. A few of the benefits stemming from the use of QFD are:

- More satisfied customers
- Greater product team linkage and alignment
- More efficient use of resources, since the team works on the "important things first"
- The ability to present and evaluate data on requirements, alternatives, competitive position, targets, possible sources of interrelations, and priorities

QFD was initially applied in the 1960s in Japan. It was developed by engineers and managers in the Kobe shipyards of Mitsubishi Heavy Industries, and it was refined through other Japanese industries in the 1970s. QFD was first recognized as an important tool for use in the United States by Dr. Donald Clausing (formerly of Xerox, now at MIT). It was translated into English and introduced to the U.S. in the early 1980s. Following publication of the first book on the subject, *Better Designs in Half the Time*,[5] it has been applied in many diverse U.S. situations.

At the heart of applying QFD are one or more matrices. These matrices are the key to QFD's ability to link customer requirements (referred to as the *voice of the customer* or *customer WHATs* in QFD literature) with the organization's plans, product or service features, options, and analysis (referred to as *HOWs*). The first matrix used in a major application of QFD will usually be a form of the A-1 matrix (Ref. 5, pp. 2–6). This matrix often includes features not always applied in the other matrices. As a result, it often takes a characteristic form and is called the *House of Quality* (HOQ) in QFD literature. Figure 72.5 presents the basic form of the HOQ.

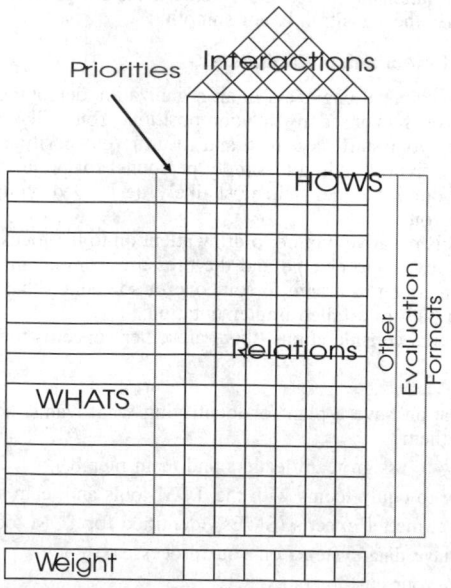

Fig. 72.5 The House of Quality (HOQ) and its major elements.

The A-1 matrix starts with either raw (verbatim) or restated customer WHATs and their priorities. The priorities are usually coded from 10 to 1, with 10 representing the most important item(s) and 1 representing the least. These WHATs and their priorities are listed as row headings down the left side of the matrix. Frequently we find that customer WHATs are qualitative requirements that are difficult to directly relate to design requirements, so the project team will develop a list of substitute quality characteristics and place these as column headings on this matrix. The column headings in QFD matrices are referred to as *HOWs* in QFD literature. Substitute quality characteristics are usually quantifiable measures that function as high-level product or process design targets and metrics. For example, a customer may want good gas mileage (a WHAT), but the design team needs to set a specific miles-per-gallon target (a HOW). Next the team develops a consensus on the correlation between the WHATs and the HOWs. Each correlation is marked in the row–column intersections using symbols having an associated numeric weight. The convention is 9 points for a high correlation between a WHAT and a HOW, with 3, 1, and 0 for medium, low, and no correlations, respectively. The assignment of points to the various correlation levels and the prioritization of customer WHATs are used to develop a weighted list of HOWs. The correlation values (9, 3, 1, and 0) are multiplied by the WHATs priority values and summed over each HOW column. These column summations indicate the relative importance of the substitute quality characteristics and their strength of linkage to the customer requirements.

The other major element of the A-1 matrix is the characteristic triangular *roof* (an isosceles triangle) which contains the interrelationship assessments of the HOWs. In many cases, improvement in one or more substitute quality characteristics may foster improvement in or be detrimental to others. These positive and negative interrelationships are noted in the column–column intersections of the roof. For example, if customer WHATs for a car include "good acceleration" and "economical fuel consumption," these may be translated into substitute quality characteristics (HOWs) such as the 0–60 mph time, time required to pass, and highway mileage (mpg). Subsequent design effort to improve the 0–60 mph time will likely improve the time to pass, but will also likely reduce the highway mileage. These would be reflected as positive and negative interrelationships, respectively.

Other features that may be added to the A-1 matrix include target values, competitive assessments, risk assessments, and others. These are typically entered as separate rows or columns on the bottom or right side of the A-1 matrix.

The key output of the A-1 matrix is a prioritized list of substitute quality characteristics. This list may be used as the inputs (WHATs) to other matrices. For example, in Fig. 72.6 we show the HOWs

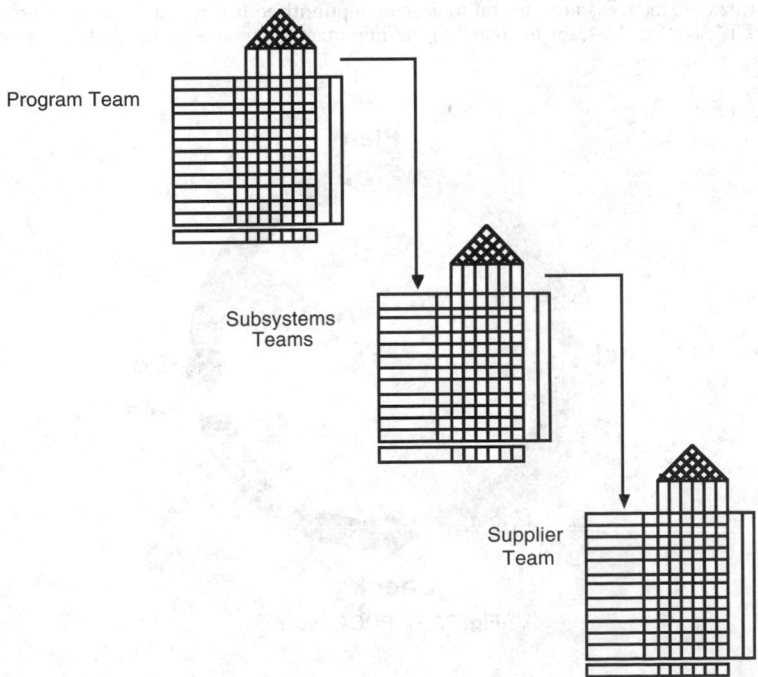

Program Team

Subsystems Teams

Supplier Team

Fig. 72.6 QFD matrices may be used to "flowdown" customer requirements.

of the project A-1 matrix flowing down to become WHATs for subsystem teams. Their HOWs may then be flowed down as inputs (WHATs) for their suppliers. Following the car mileage example, target mileage requirements may be flowed to the engine team and efficiency requirements flowed to the transmission team. They may then break their requirements out to fuel injection, piston, gear, and any other suppliers. This assures that the *voice of the customer* is deployed throughout the enterprise and that all activities are linked with customer requirements.

72.5.2 Technical Tools—Seven Management and Planning (7 MP) Tools

Dr. Deming proposed that TQM applications should follow what is now known as the PDCA (plan, do, check, act)* cycle, as pictured in Fig. 72.7. The PDCA cycle is a logical approach that parallels the scientific method of "observe, hypothesize, test hypothesis, modify hypothesis." Most early TQM tools addressed the "do, check, act" portion of the cycle. In later years, a suite of tools were developed to assist the planning efforts of TQM. These have become known as the *7 MP* tools:[7]

1. Affinity diagram
2. Tree diagram
3. Prioritization matrix
4. Interrelationship digraph
5. Matrix diagram
6. Activity network diagram
7. Process decision program chart

The first tool widely used in the 7 MP suite is the affinity diagram, which is excellent for generating and grouping ideas and concepts. Teams will find the affinity diagram useful for exploring issues in a new project or factors to consider during implementation. This tool often uses simple sticky papers or cards to generate and collect team ideas. These are then arranged into "affinity" groupings by the team and assigned a descriptive header. The affinity header descriptions represent the key issues or concepts identified by the team. The number of cards under each header indicates the breadth of team consensus on the issue.

The tree diagram, pictured in Fig. 72.8, is a good tool to break down a complex project into manageable tasks. The team starts with the overall project or goal description, which is broken down into the next logical division of effort. Each new element may be further divided (if it makes sense) until the team has a list of self-contained tasks that may be assigned to one or more subteams or individuals.

A prioritization matrix is most useful to develop a prioritized list from a large set of options. This tool makes it easy for the team to focus on the important items and avoid "hidden agendas" that

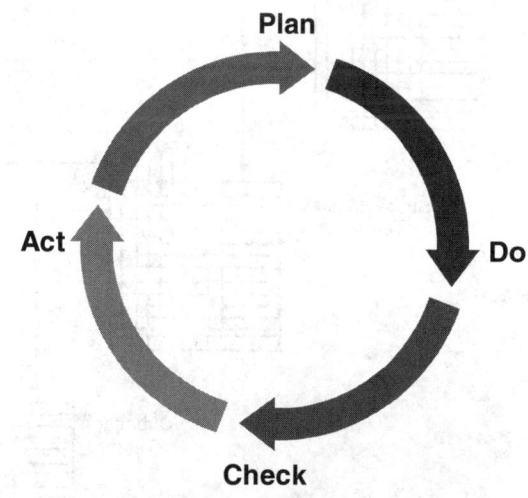

Fig. 72.7 PDCA cycle.

*Since early writings, Dr. Deming has modified this to PDSA—plan, do, study, act.

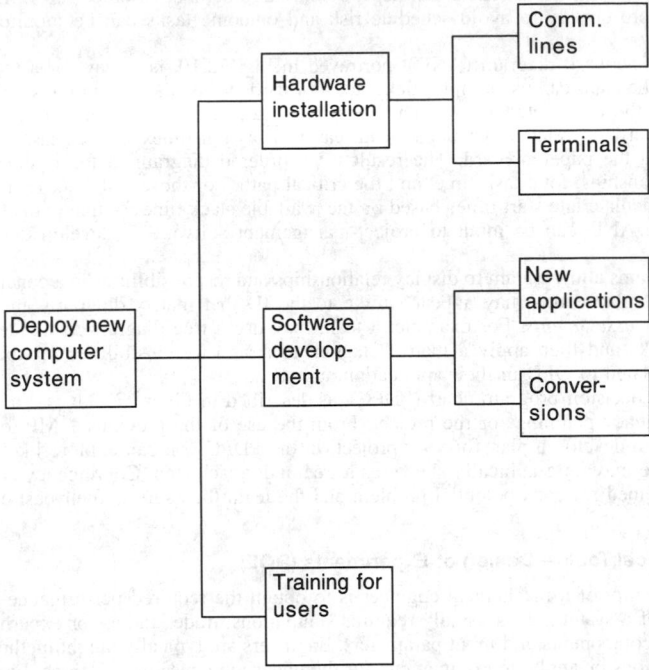

Fig. 72.8 Example tree diagram.

may drive the team. In this tool, the team uses pair-wise comparisons to determine the overall relationship of a large number of elements.

An interrelationship digraph (ID), as presented in Fig. 72.9, helps a team discover the relationships and dependencies between project activities. Using simple graphical techniques, the team indicates task relationships one by one. When all the pair-wise comparisons are completed, the team has the information necessary to identify the driver tasks (tasks that drive or precede a large number of other

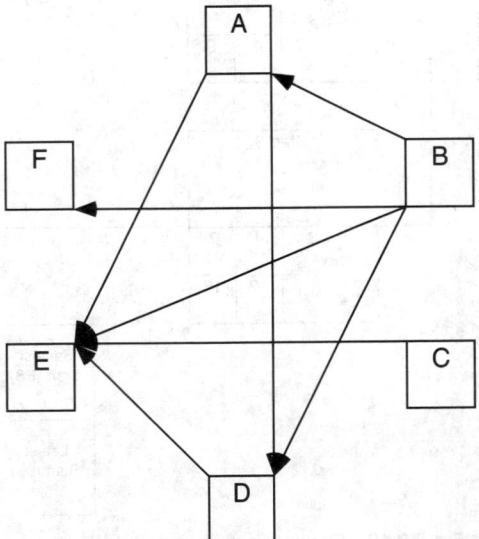

Fig. 72.9 Example ID (arrows represent influence or predecessor relations).

tasks) and the outcomes tasks (tasks that depend on a large number of other tasks). Driver tasks can be managed more closely to avoid schedule risk and outcome tasks can be monitored for project performance.

The activity network diagram (AND), portrayed in Fig. 72.10, is a way for a team to schedule project tasks. The team can use simple sticky notes or cards to list the program tasks. These can then be arranged in the anticipated flow order (sequential, parallel, or a combination) with directional arrows drawn between related tasks. The team can then assign times to each task placing the task process time on the paper or card. The result is an ordered diagram that can show predecessor/successor relationships, total task time, and the critical path. For those tasks not on the critical path, the team can calculate late start times based on the available slack time for that path. The information contained in an AND can be input to project-management software to develop the familiar Gantt chart.

Matrix diagrams allow a team to display relationships and responsibilities in a concise and efficient manner. At first glance this may appear similar to the ID, but matrix diagrams are most used for *assignments* not *assessments*. For example, a team may use a tree diagram to divide a project into manageable tasks and then apply a matrix diagram to assign responsibilities for the tasks. Matrix diagrams are related to QFD in their application approach.

The process decision program chart (PDPC), as described in Figure 72.11, is a tool that helps to develop contingency planning for the project. From the use of the previous 7 MP tools, your team should be able to develop a plan for your project. In the PDPC you can explore likely problems for each step. These may be graphically shown as a tree under each step. Contingency countermeasures can then be planned for each potential problem and the team then selects their best choice from the options.

72.5.3 Technical Tools—Design of Experiments (DOE)

A key responsibility of a mechanical engineer is to obtain the required performance from a system or component of a system. This usually requires simulations, trade studies, or experimentation with various system components and input parameters. Engineers are typically taught methods that require certain assumptions or apply approximations for the underlying system equations. For best performance, this may not be sufficient. Approximations may not be accurate enough and are singularly inadequate to guide variability reduction.

Design of experiments of DOE is the tool of choice for trade studies and system or component experimentation. A properly planned and conducted DOE will yield the most useful information possible from a series of experimental runs, giving the engineer not only the identity of key pa-

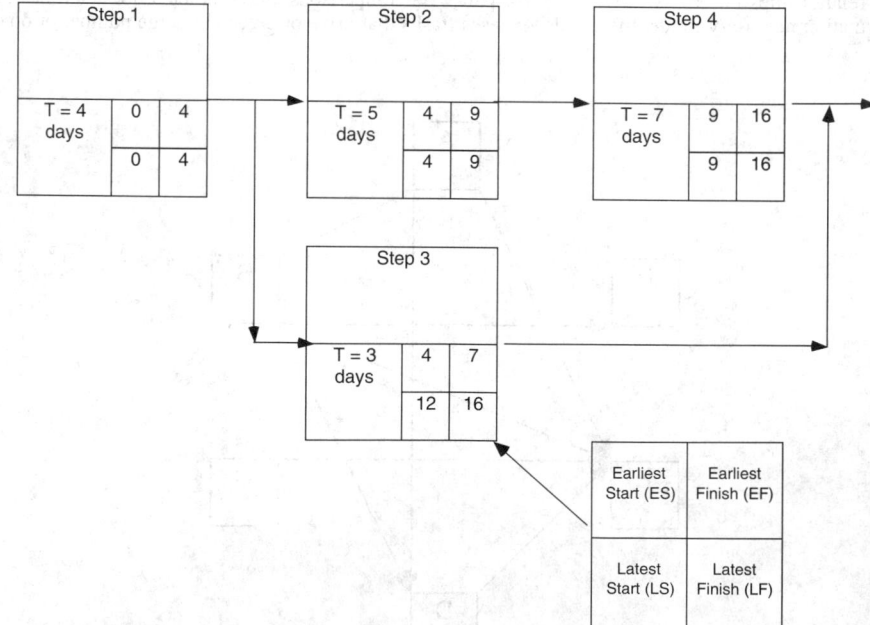

Fig. 72.10 Example activity network diagram.

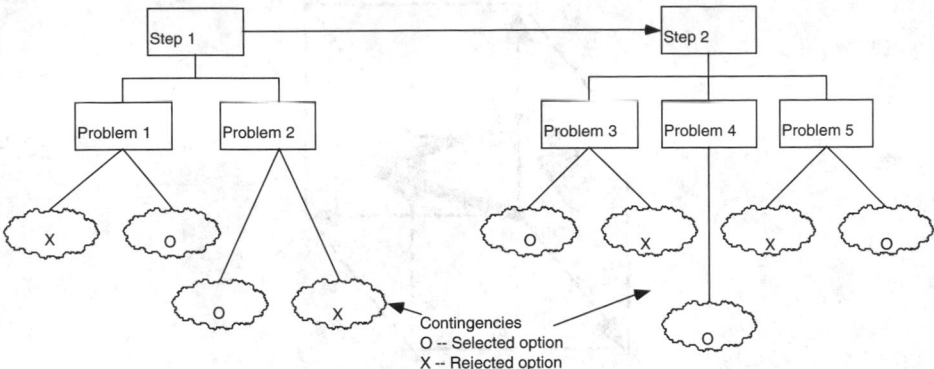

Fig. 72.11 Example process decision program chart.

rameters, but also an estimate of the underlying performance equation. This will allow the engineer to efficiently set the system up for optimum performance in nearly all cases.

The chief competitor to DOE is one-factor-at-a-time (OFAAT) experimentation, where an engineer holds all but one factor constant. That factor is varied on one or more experimental runs to see if it has an effect on the system response. This is repeated for the other factors. Unfortunately, OFAAT leads to only linear, and usually only first-order, information on each experimental factor. If there are significant system interactions or higher-order effects, OFAAT will not reveal them. In Fig. 72.12, a system space is shown for a system with three factors, each at two levels. Experimenting through OFAAT will only explore the four points (circled in Fig. 72.12) where first-order information is available. If there is a significant two-factor interaction in the system, it will show at the appropriate corner point where both factors are changed. If there is a three-factor interaction, it will require information from the corner where all three factors are changed.

Another competitor to OFAAT is "random" experimentation, as displayed in Fig. 72.13. In this approach, a number of process factors are changed each time the experiment is done. With this

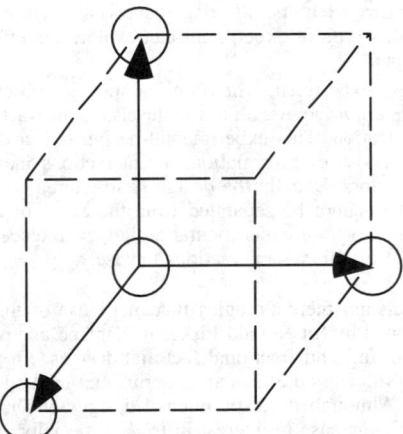

For a process with three factors at two levels, one factor at a time experiments explore only a limited part of the process domain. We will gain no knowledge of interactions with this approach.

Fig. 72.12 One-factor-at-a-time.

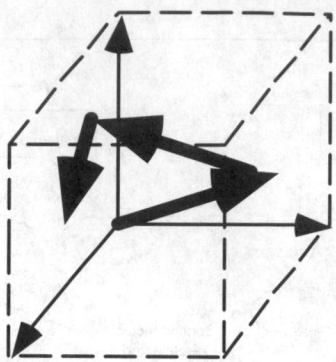

For a process with three factors,
random change to all factors
represents random movement in the
experimental domain.

Fig. 72.13   Random experimental movement.

approach, if the process improves or grows worse, the team will not know which factor or factors were the influence.

In contrast to OFAAT and random experimentation, DOE systematically measures the system response as multiple factors are changed. The orderly and planned change of system factors is the key to DOE. Prior to the experiment, the engineer, often using a multifunctional team, will determine which factors (system inputs or parameters) might affect system response. The experimental levels (factor settings) for each factor will also be determined. Finally, the team should decide how much experimentation the project can afford. This and other preferences will determine the type of experiment to conduct.

There are many types of experimental designs. Generally, an experiment with more than one factor falls into one of the following major classifications:

- *Full factorial.* An experiment where all possible combinations of factor level settings are run at least once. If there are n factors, all at two levels, this will result in 2^n experimental runs for one replication. This type of experiment can explore the effects of all factors and factor interaction combinations.
- Fractional factorial. An experiment where only a specific subset of the possible factor level settings is run. If there are n factors, all at two levels, a half-fractional experiment will require 2^{n-1} runs for one replication. This experimental design reduces the number of experimental runs, but the cost is a loss of information, as interactions may be confounded with other interactions or main factors. Usually the design is structured so that higher level interactions (three-factor or higher) cannot be separated from the effect of another factor or lower-level factor interaction. In this type of experimental design, experience and knowledge are essential to avoid an experiment that mixes interactions unwisely.

There are several experimental methodologies that make use of these key experimental design types. Classical DOE, developed by Sir Ronald Fisher in England and promoted in the U.S. by Box, Hunter, and Hunter, uses both full and fractional factorial designs.[8] In the early 1980s, Dr. Genichi Taguchi began to promote in the United States an experimental methodology that uses special set of fractional factorial designs.[9] Although the experimental designs of Dr. Taguchi are not unique, his approach generated a dramatic increase in interest in DOE, especially among engineers. Dr. Taguchi made three major contributions to DOE. First, he developed a DOE methodology that offered clearer guidance to engineers than earlier approaches. Secondly, he promoted the concept of "robust design" and showed how DOE could be used to obtain it. Finally, he promoted the application of a quality loss function, expressed in dollars, showing how the enterprise, and society in general, are affected by variation from a target value.[10]

Usually experiments are run with factors at two levels. Occasionally an experiment deals with attribute factors (qualitative factors such as material types) at more than two levels. Sometimes nonlinear effects are expected, so even continuous factors (factors with settings on some continuous scale, such as temperature) are run at three or more levels.

72.5.4 Technical Tools—SPC, SQC, and 7 QC

One technical tool for TQM that came to early public attention was SPC (statistical process control). After somewhat rocky first application attempts, many companies are finding SPC to be useful for reducing defects, lowering defect rates, and making key processes much more consistent and dependable. The key to successful SPC application is understanding what SPC does and doesn't do.

SPC is the application of statistical (often in graphical form) methods to identify when a process may have been influenced by a "special" cause of variation. Dr. Walter Shewhart, who developed the earliest concepts and applications of SPC, divided process variation into two types. One type of variation he described is often called "common cause" or "normal" process variation in the literature. Normal variation results from the myriad of factors inherent to the process interacting with each other. Examples of normal process variation sources in a simple drilling operation include drill splay, variation in bits, variation in material, and so on. These factors interact and create a resulting pattern of variation in hole size, location, and so on. The second form of variation described by Dr. Shewhart is often referred to as "special cause" variation. Examples of special causes in the previously mentioned drilling operation might include changes in personnel, excess bit wear, changes in material clamping technique, changes in material, and so on.

We make the distinction between these sources of variation to separate the manageable from the unmanageable. Special causes of variation can usually be identified and removed from the process. Normal causes of variation can only be removed or reduced by changing the process, which often requires management involvement and/or capital expenditure. Although process changes may be necessary, usually removing special causes variation sources is more cost-effective and should be addressed first.

How does SPC fit into this? Dr. Shewhart, working in an AT&T Western Electric plant, saw that their processes had a lot of variation and that operators were constantly adjusting. He suspected that they were often reacting to normal variations and that their additional adjustments were adding to the process variation. He proposed the use of SPC and SPC charts to signal when a process may have been influenced by a special cause of variation. Then the operators, engineers, or managers could pursue adjustments or investigations, as necessary.

SPC charts come in many forms, but in general all plot one or more statistics (a descriptive measure from a unit or sample) on a chart that contains control limits, such as the chart in Fig. 72.14. The control limits are derived from past stable process data and usually represent $\overline{X} \pm 3s$ for each statistic (note that some statistics do not have a lower limit) where $\overline{X}$ is the long-run average for the statistic and $3s$ is three times the standard deviation of the statistic. If the statistic follows the normal distribution (and nearly all will, due to the central limit theorem), a point outside the control limit would only occur 0.27% of the time. Thus, a point outside either limit most likely reflects the influence of a special cause of variation. In addition to watching for points beyond the control limit, SPC practitioners also apply tests for patterns in consecutive points. Such patterns, such as trends of seven points in a row increasing or decreasing, also reflect events that would not likely happen in a process operating only with normal causes of variation. In Fig. 72.14, we see an $\overline{X}$ and R chart. In this chart, we plot sample averages ($\overline{X}$) and the range (R) for each subgroup. A subgroup usually consists of 2 to 10 samples for this type of chart. This type of chart detects both a shift in the process average and a change in process variation. Following are some rules for abnormal patterns in SPC charts:[11]

- One point beyond a control limit
- A run of seven or more points either up or down or consecutive above or below the centerline
- Two of three consecutive points outside 2 sigma, but still inside the 3-sigma line
- Four of five consecutive points beyond 1 sigma

While SPC deals with in-process measures, often our only significant way to measure the process result is by measuring the performance of the finished product. For example, when we assemble an electronic circuit, there are in-process measures to be monitored, but the final performance can only be measured by final test. As with in-process measures, final performance variation is a function of the variation resulting from normal and special causes. SPC can be used in this case to identify when to investigate for a special cause and apply corrective action. Often this approach is called *statistical quality control* (SQC). The same charts and approaches are often used. We should note that SQC should not be used as a substitute for SPC. Since SPC is directed at process inputs, not later in the cycle, it offers faster detection and correction of problems.

SPC and SQC are powerful tools, but they essentially do only one thing: they identify when a process was probably influenced by a special cause of variation. When that occurs, the team must

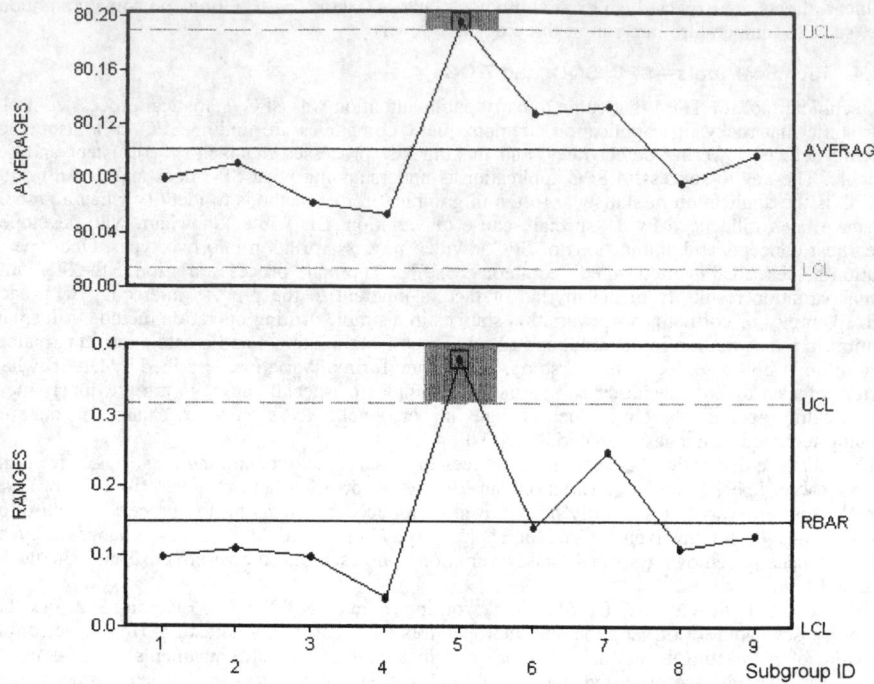

Fig. 72.14   Example X bar (X) and R chart (with one point out of control).

determine what happened and remove the cause to return the process to the normal state. Many of the tools for this job are grouped with SPC/SQC in what are called the *seven quality-control* (7 QC) tools:[11,12]

1. SPC/SQC
2. Histograms
3. Scatter plots
4. Pareto charts
5. Fishbone diagrams
6. Check sheets
7. Defect maps

Application of these tools with SPC will enable the team to maintain a stable process.

72.5.5 Technical Tools—Process Capability or Validation Studies

One of the more useful methodologies coming from TQM applications is the joining of manufacturing process capability assessment and the processes of developing design requirements. As was previously discussed, there have often been barriers between design and manufacturing. There was distrust, finger-pointing, and a general lack of teamwork.

For most companies, engineering design has been slow to recognize that they had a responsibility to work with manufacturing to develop a design package meeting customers' needs that was *manufacturable*. For their part, manufacturing has not been proactive in work to develop consistent processes with minimum variation. There is plenty of blame to go around, so how does an organization change? A key way to change without arguing is to look at facts and data. Characterize your processes according to what you expect of them (engineering requirements). Based on the results, you may decide that it is more cost-effective to change the design for some parameters if they appear to be controlled too tightly. If the design requires certain performance, but the current process can't reliably meet requirements, you must improve the process! Following are the steps for doing so. They are easy to follow.

1. Prioritize your processes and start working on the highest one(s), i.e., the vital few.
2. If the process doesn't have SPC, apply it!
3. Get the process under statistical control, i.e., predictable.
4. From the SPC chart, obtain estimates of the process average and standard deviation.
5. Assess the process C_{pk}.
6. Based on the C_{pk} and economic considerations, change the product specifications or improve the process to obtain C_{pk} goals.
7. Move on to the next process.

First of all, you should develop a strategy of work. Since you probably don't have resources to do everything, make sure you do the important things first. The next two steps are key. If you don't have SPC on the process, you can't determine if it's stable. If the process is not stable, all subsequent assessments will be worthless.

In steps 4 and 5, you obtain estimates of the process average and standard deviation and then apply them to an assessment of performance called the process performance index (C_{pk}). This measure (calculations and performance values are given in Fig. 72.15) shows how well three standard deviations fit between the process average and the closest specification limit. What value is appropriate? Many organizations use a C_{pk} of 1.33 as a minimum value. This means that four standard deviations fit in the distance between the process average and the closest specification. A few companies are using C_{pk} values of 1.50 as their target. Such higher values of C_{pk} allow more margin if the process shifts. You can see this in the values listed in Fig. 72.15 that show the effect of 1 and 1.5 standard deviation shifts.

The last two steps must not be ignored. If you find that the process capability is not acceptable, you must change the design requirements, improve the process, or *live with poor process performance for as long as you make the product*. The decision of which to address—design, process, or both—is an economic one. When you have completed this project, move on to the next one. One element of process assessment that should not be neglected is gage repeatability or reproducibility assessment. If the major source of process variation is in the measurement, it is usually the cheapest way to improve the process.

72.5.6 Technical Tools—Other TQM Tools

By some counts there are more than 100 TQM tools that may be applied for different aspects of TQM applications.[12] These range from simple graphical procedures for data exploration to complex tools like DOE. A partial list follows:

- Activity-based costing
- Bar chart
- Benchmarking
- Brainstorming
- Business process re-engineering
- Continuous improvement
- Cost of quality
- Critical path method (CPM)
- Cycle time management
- Data-collection strategy
- Defect map
- Delphi method
- Deployment chart
- Design for manufacture/assembly
- Events log
- Failure mode and effects analysis
- Fault tree analysis
- Five whys
- Gap analysis
- Imagineering
- Just-in-time
- Nominal group technique
- Policy deployment
- Problem solving
- Ranking
- Sampling
- Scatter analysis
- Spider chart
- Stratification
- Survey analysis
- Synchronous workshop
- Systems analysis
- Thematic content analysis
- Time study sheet
- Value engineering

72.5.7 Cultural/Social Tools—Concurrent Engineering

In the past, a new product-development effort followed a predictable path. Design engineers worked with marketing and customers on initial feasibility studies. If these studies looked favorable, one or more prototypes were then built, usually in a special prototype facility. An initial design was then formulated and a pilot production scheduled. During this time, manufacturing engineers were drawn

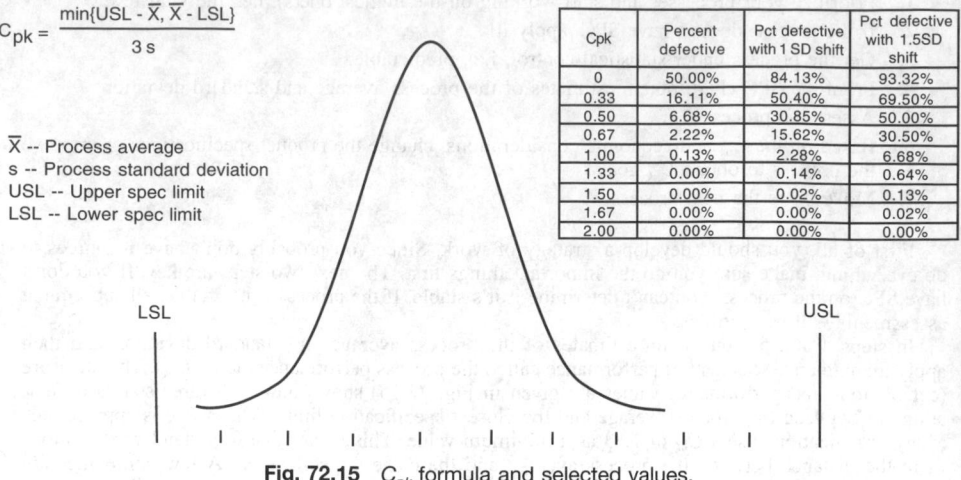

$$C_{pk} = \frac{\min\{USL - \overline{X}, \overline{X} - LSL\}}{3s}$$

$\overline{X}$ -- Process average
s -- Process standard deviation
USL -- Upper spec limit
LSL -- Lower spec limit

Cpk	Percent defective	Pct defective with 1 SD shift	Pct defective with 1.5SD shift
0	50.00%	84.13%	93.32%
0.33	16.11%	50.40%	69.50%
0.50	6.68%	30.85%	50.00%
0.67	2.22%	15.62%	30.50%
1.00	0.13%	2.28%	6.68%
1.33	0.00%	0.14%	0.64%
1.50	0.00%	0.02%	0.13%
1.67	0.00%	0.00%	0.02%
2.00	0.00%	0.00%	0.00%

Fig. 72.15 C_{pk} formula and selected values.

into the project. At the same time, marketing's involvement was reduced, since the design group had their input and the project became a production problem. At this point, engineering changes increased as producibility problems and cost issues emerged.

As full-scale production begins, after-market support's involvement increases. Additionally, marketing often gets involved again with new input from early customers and competitive comparisons. Since the whole process may take some time, this new marketing input can represent a significant customer change in tastes and reaction to competing products. This adds to the engineering change rate. In many projects, the change rate may continue at a high level well into full-scale production. This phenomenon, described as the *engineering version of rework,* can be very significant in cost.[13]

Besides the cost involved, this approach is very time-consuming. More agile competitors can beat the enterprise to market. Since a significant portion of profit from a new product or service comes early in the production cycle, it is important to the enterprise that it not be ceded to competitors.[14]

To combat the problem of long development cycles and to reduce the degree of late engineering change, concurrent engineering was proposed for especially complex design efforts. Concurrent engineering promised to remove the problems in a design cycle by concurrently developing the product design as well as the processes necessary for production, test, and after-market support.

The concept was quite simple and theoretically dealt with the problem. Unfortunately, except for a few isolated cases, concurrent engineering did not fulfill its promise. It fell short for two rather simple reasons. First of all, by its nature it still involved only *engineering.* There was still no drive to include marketing, finance, production operators, testers, and so on. These people bring significant

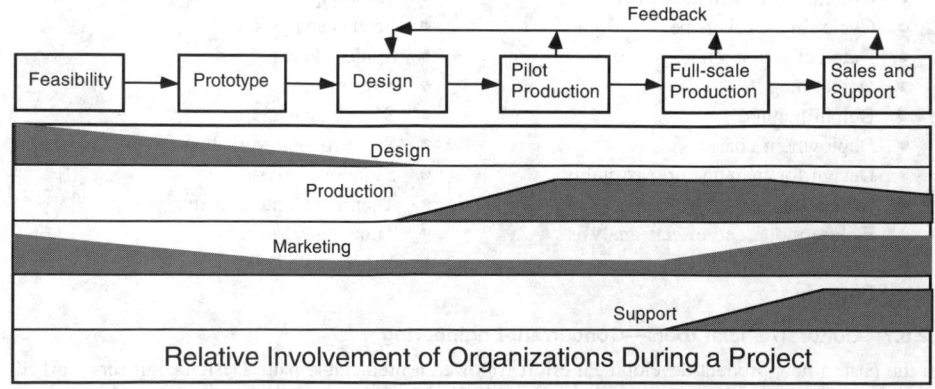

Fig. 72.16 In traditional design, involvement is often partitioned.

insight into issues that affect cost and reliability. The second reason for concurrent engineering's lack of success comes from the nature of organizations. As they currently exist for most companies, functional organizations do not communicate well. Since concurrent engineering did nothing to improve this problem, those outside product design still often had to design their processes in a vacuum, isolated from each other.

Obviously, concurrent engineering, by itself, was not the answer. It would take more to improve the design process.

72.5.8   Cultural / Social Tools—Teams

In the 1980s and before, some leaders started to picture a vision of a radically different organizational structure. One 1990 annual report pictured "a boundaryless company . . . where we knock down the walls that separate us from each other on the inside and our key constituencies on the outside" (Ref. 15, p. 63). Increasingly, business leaders saw teams as a way to solve the design cycle problem and make the enterprise more flexible and agile.

To see how this works, consider the traditional hierarchical organization. Individual elements of this organization are connected through their management chain. How does any department request support of another? Since the powers of budget and personnel evaluation flow from the manager, department staff respond to their manager. Requests for support must be made through the management chain and must often be accompanied with necessary funding. Such funding must be authorized by the giving department's manager and usually involves the two supporting finance organizations, one to prepare the document authorizing funding and one to receive the funding and set up charge-collection systems. A relatively simple request for support can easily involve six people and significant documentation. This is not conducive to a rapid response!

Now let's picture another approach. In this organization, a project team is formed with the responsibility to complete the project. This team may have total responsibility for the new product or service, or it may have responsibility for a subset of the project. The team is given the budget for the project. The team is staffed with representatives of all pertinent functional areas (a multifunctional team). Such a team has the capability to overcome the barriers of traditional organizations.

Teams have been successfully applied on many projects, but the most recent evolution of team applications finally fulfills the promises of concurrent engineering. Referred to as an *integrated product and process development* (IPPD or IPD) team, this approach uses multifunctional teams to develop concurrently the processes and the design of new products and services.

72.5.9   Cultural / Social Tools—The Variability Reduction Process (VRP)

One clear message has emerged from research and observation of various companies' attempts at TQM. Implemented correctly, TQM can be an important strategic weapon for the enterprise. Implemented poorly, it can not only fail to yield promised results, it can be a drag on the enterprise as time and resources are diverted to poorly planned exercises.

The way to avoid an ineffective TQM initiative is to insure that it drives toward goals that can really help the business. A way to achieve such an impact is to use the VRP to focus your TQM efforts. As can be seen in Fig. 72.17, any business has certain key core functions. No matter what the enterprise does, it must

1. Identify customer needs
2. Develop or deploy needed business functions

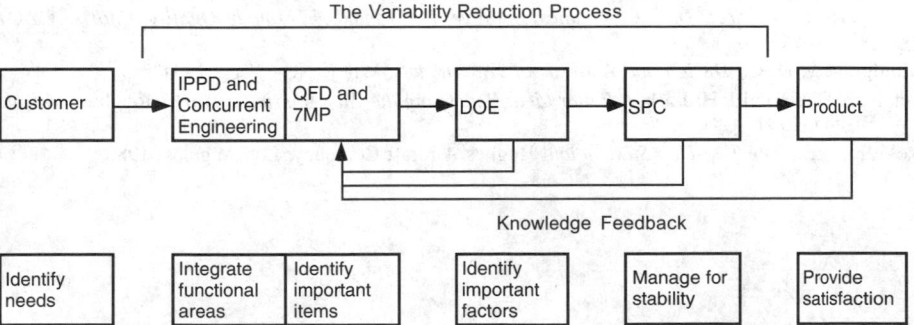

Fig. 72.17   The variability reduction process will guide TQM application.

3. Identify key processes
4. Set key process factors to deliver required performance
5. Manage the processes in a stable manner
6. Meet customer needs

The VRP organizes key TQM tools around the core business functions. These tools may be applied to improve each step. The effect is to engage the whole enterprise in continuous improvement of all processes with a focus on customer needs. Such an approach can significantly transform the enterprise.

72.6 SUMMARY

TQM is a strategic tool for many world-class companies today. It will be a part of the work life for most mechanical engineers and managers. Greater knowledge of the tools and methodologies will be beneficial to your career and to your employees.

REFERENCES

1. J. B. ReVelle, *Becoming the Total Quality Manager (TQM)*, A Workshop for the Institute of Industrial Engineers (IIE), Atlanta, 1995.
2. J. M. Juran (ed.), *Juran's Quality Control Handbook*. 4th ed., McGraw-Hill, New York, 1988.
3. J. M. Juran and F. M. Gryna, *Quality Planning and Analysis*, McGraw-Hill, New York, 1980.
4. R. C. Swanson, *Quality Improvement Handbook, Team Guide to Tools and Techniques*, St. Lucie Press, Delray Beach, FL, 1995.
5. B. King, *Better Designs in Half the Time: Implementing QFD in America*, GOAL/QPC, Methuen, MA, 1987.
6. General Accounting Office (GAO), *Management Practices, U.S. Companies Improve Performance Through Quality Efforts*, GAO/NSIAD-91-190, Washington, DC, May 1991.
7. M. Brassard and D. Ritter, *The Memory Jogger*, GOAL/QPC, Methuen, MA, 1985.
8. G. E. P. Box, W. Hunter, and J. S. Hunter, *Statistics for Experimenters*, Wiley, New York, 1978.
9. G. Taguchi, *System of Experimental Design*, UNIPUB/Kraus International Publications, New York, 1987.
10. G. S. Peace, *Taguchi Methods: A Hands-On Approach*, Addison-Wesley, Reading, MA, 1993.
11. D. C. Montgomery, *Introduction to Statistical Quality Control*, 2nd ed., Wiley, New York, 1991.
12. J. B. ReVelle, R. A. Kemerling, and H. K. Jackson Jr., *TQM ToolSchool*™, Quality America (software), Tucson, AZ, 1995.
13. J. P. Womack, D. T. Jones, and D. Roos, *The Machine That Changed the World*, Rawson Associates, New York, 1990.
14. J. B. ReVelle, N. L. Frigon Sr., and A. K. Jackson Jr., *From Concept to Customer: The Practical Guide to Integrated Product and Process Development and Business Process Reengineering*, Van Nostrand Reinhold, New York, 1995.
15. J. H. Boyett and J. T. Boyett, *Beyond Workplace 2000: Essential Strategies for the New American Corporation*, Dutton, New York, 1995.

BIBLIOGRAPHY

Brassard, M., and D. Ritter, *The Memory Jogger II: A Pocket Guide of Tools for Continuous Improvement & Effective Planning*, GOAL/QPC, Methuen, MA, 1994.

Management Practice: U.S. Companies Improve Performance Through Quality Efforts, GAO/NSIAD-91-190.

Montgomery, D. C., *Design and Analysis of Experiments*, Wiley, New York, 1991.

Peterson, D. E., and J. Hillkirk, *A Better Idea: Redefining the Way Americans Work*, Houghton Mifflin, Boston, 1991.

ReVelle, J. B., *The Two-Day Statistician*, Hughes Aircraft Company, Los Angeles, 1985.

CHAPTER 73

REGISTRATIONS, CERTIFICATIONS, AND AWARDS

Jack B. ReVelle
Cynthia M. Scribner
Hughes Missile Systems Co.
Tucson, Arizona

73.1	**INTRODUCTION**	**2177**	**73.3**	**QUALITY AWARDS**		**2180**
				73.3.1	Deming Prize	2180
73.2	**REGISTRATIONS AND**			73.3.2	Malcolm Baldrige	
	CERTIFICATIONS	**2177**			National Quality Award	2182
	73.2.1 ISO 9000	2178		73.3.3	European Quality Award	2184
	73.2.2 ISO 9000 Certification/			73.3.4	Shingo Prize for	
	Registration	2179			Excellence in American	
	73.2.3 QS 9000	2179			Manufacturing	2185
	73.2.4 TE 9000	2179		73.3.5	State Quality Awards	2191
	73.2.5 Other Quality System			73.3.6	How Do They Compare?	2191
	Standards	2180				
	73.2.6 ISO 14000	2180				

73.1 INTRODUCTION

As the concept and practice of Total Quality Management (TQM) has evolved over the past decade, a number of external influences have appeared. Of these, the most notable are registrations and certifications to international standards and quality awards offered by local, national, and international bodies. It is interesting to note that the United Kingdom and other European countries first accepted the ISO 9000 registration process wholeheartedly while only recently beginning to create national quality awards. In the United States, the Malcolm Baldrige National Quality Award was the first of these external influences to gain support. The ISO 9000 standard met with strong resistance in the United States and is only now gaining in acceptance.[1]

As companies engage in the process of achieving certifications, registrations and awards, mechanical engineers may be asked to participate, assisting their companies in preparations for a certification audit, or writing sections of an award application. This chapter provides a general overview of the most widely recognized programs. Keep in mind that standards are revised periodically, and award criteria may be updated annually. Use the contact information at the end of each section to obtain the latest information.

73.2 REGISTRATIONS AND CERTIFICATIONS

While the concept of certifying or registering quality systems to an industry or international standard is becoming accepted practice throughout the world, the terminology is often misunderstood. For all practical purposes, it does not matter whether the term *registration* or *certification* is used. When a company seeks validation of its ISO quality-management system by hiring a third-party registrar, the quality system is certified as meeting the ISO requirements, and the registrar issues a certificate.[2] The certification is then entered in a register of certified companies. Thus, companies meeting the requirements of a standard are both certified and registered. The term *certification* is most often used for this process in Europe. In the United States, it is more common to hear the process called *registration*.

Mechanical Engineers' Handbook, 2nd ed., Edited by Myer Kutz.
ISBN 0-471-13007-9 © 1998 John Wiley & Sons, Inc.

73.2.1 ISO 9000

As the European Trading Community began to take shape in the 1980s, there was a perceived need for a common quality standard for all nations. The International Organization for Standardization assigned this task to Technical Committee 176, and in 1987, the ISO 9000 Quality System Standards were issued. Since then, a 1994 revision has been released. The standards are published in the United States as ANSI/ASQC Q9000, a joint effort between the American National Standards Institute (ANSI) and the American Society for Quality (ASQ).

The ISO 9000-series of standards is composed of several guidelines and three separate conformance models: ISO 9001, 9002, and 9003. The appropriate model is determined by the scope of an organization's activities. ISO 9001 contains provisions for companies that perform design/development, production, installation, and servicing; ISO 9002 is appropriate when the organization does not design any products, but performs all other tasks; and ISO 9003 is limited to provisions for quality assurance in final inspection and test.[3] The ISO 9000 Standards contain 20 elements of a quality-management system, although some of these do not apply to ISO 9002 and 9003. See Fig. 73.1 for a list of the elements and the ISO models to which each pertains.

In addition to the quality system models, there are ISO guidelines to augment understanding of the requirements. Guidelines are not requirements and need not be followed to obtain ISO 9000 registration. Some of these additional documents, however, can enhance understanding of the basic requirements and provide assistance for companies creating or improving quality systems. These include:

- ISO 8402: quality terminology and concepts and a cross reference of common quality terms used in Europe and the United States
- ISO 9000: a set of guidelines to help the user select the appropriate quality system model (ISO 9001, ISO 9002, ISO 9003)
- ISO 9000-3: the guideline for software quality

9001	9002	9003	ISO ELEMENT
✗	✗	✗	4.1 Management Responsibility
✗	✗	✗	4.2 Quality System
✗	✗	✗	4.3 Contract Review
✗			4.4 Design Control
✗	✗	✗	4.5 Document and Data Control
✗	✗	✗	4.6 Purchasing
✗	✗	✗	4.7 Control of Customer Supplied Product
✗	✗	✗	4.8 Product Identification and Traceability
✗	✗		4.9 Process Control
✗	✗	✗	4.10 Inspection and Testing
✗	✗	✗	4.11 Control of Inspection, Measuring and Test Equipment
✗	✗	✗	4.12 Inspection and Test Status
✗	✗	✗	4.13 Control of Nonconforming Product
✗	✗	✗	4.14 Corrective and Preventive Action
✗	✗	✗	4.15 Handling, Storage, Packaging, Preservation, and Delivery
✗	✗	✗	4.16 Control of Quality Records
✗	✗	✗	4.17 Internal Quality Audits
✗	✗	✗	4.18 Training
✗	✗		4.19 Servicing
✗	✗	✗	4.20 Statistical Techniques

Fig. 73.1 ISO conformance models by element.

- ISO 9004-1: explanations and suggested implementation methods for the elements of ISO 9001
- ISO 10011: guideline for internal quality audits
- ISO 10013: suggested formats and contents for an ISO 9000 quality manual

73.2.2   ISO 9000 Certification / Registration

Separate from the ISO 9000 Standards per se is a certification/registration process that has become institutionalized in many countries. The process requires that a third-party *registrar* review a company's documented quality system and the implementation of that system through on-site audits. The third-party registrar certifies that the system meets all of the requirements of a specific ISO 9000 model. The registration of the quality system can then be publicized. The registrar also performs periodic recertification audits.

The American Society for Quality (ASQ) is a good source of information on registrars in the United States. The Registrar Accreditation Board (RAB) is the U.S. agency that accredits agencies to serve as registrars. The RAB is a wholly owned, not-for-profit subsidiary of ASQ.[4]

The effort to obtain ISO 9000 registration typically takes 12 to 18 months from the time a company makes the commitment to become registered until its quality system receives the certificate from its third-party registrar. The cost of registration varies depending on the size and complexity of the company, the number of locations to be included on the registration certificate, and the state of its existing quality system when the decision to obtain registration is made.

Third-party registrars are generally contracted for three years. In addition to the initial assessment for registration, the registrar may be asked to perform a pre-assessment audit. A registering agency cannot perform the duties of an ISO consultant to companies for which it will be conducting the third-party assessment. Many companies find it helpful to hire an outside consultant to help prepare for ISO registration. There are many texts available on the subject of ISO 9000 quality systems and the registration process.

To obtain copies of ANSI/ASQC Q9000 documents, contact ASQ at 1-800-248-1946.

73.2.3   QS 9000

QS 9000 is an enhanced version of ISO 9000 created by the Big Three U.S. auto makers (General Motors, Ford, and Chrysler) in conjunction with other car and truck manufacturers. Although not an international standard, QS 9000 includes all of the requirements of ISO 9001 plus industry-specific requirements and a section of requirements specific to either Chrysler, Ford, or General Motors. QS 9000 was first issued in 1994 by the Automotive Industry Advisory Group (AIAG).[5]

The goal of QS 9000 is to reduce defects and waste in the supply chain while continuously improving quality and productivity. It is seen as a benefit to suppliers because it reduces duplication of systems, reporting methods, and audits while enhancing communication throughout the industry. For most suppliers, having a single quality-management system required by all automakers represents an opportunity for significant savings.

QS 9000 includes seven documents, all of which must be referenced to create a compliant system. The auto industry standard is more prescriptive than ISO 9001. There is a continuing debate as to whether QS 9000 is more rigorous than its ISO counterpart. A comparison of the number of "shalls" in each reveals 137 in ISO 9000 as compared to 300 in QS 9000.[3] This may be reflective of complexity, rigor, or both.

In the United States, the RAB (Registrar Accreditation Board) performs accreditation of registrars to QS 9000, and there is a certification/registration process in place despite the fact that the document is not controlled by ANSI, the International Organization for Standardization, or any other recognized standards-issuing body. The Big Three automakers have announced that third-party registration to QS 9000 will be required of all first-tier suppliers by 1997. First-tier suppliers are internal and external suppliers of production materials, production or service parts, and heat treating, painting, plating, or other finish services supplied directly to General Motors, Ford, or Chrysler. This could include as many as 14,000 companies worldwide. As these first-tier suppliers begin requiring QS 9000 compliance or registration of their own suppliers, more than 40,000 second-tier suppliers could be affected.

The QS 9000 documents are copyrighted by AIAG, which is the sole source of the documents, thus they must be purchased from them. To order these documents, contact AIAG at 1-800-358-3570.

73.2.4   TE 9000

Another of the auto-industry standards, TE 9000, is expected to be released as a supplement to QS 9000. This standard will be applied to tooling and equipment manufacturers that supply the non-production parts used in automobile manufacturing processes. Similar to QS 9000, the TE quality system standard will include ISO 9001 in its entirety along with industry- and auto company-specific requirements. The Big Three are expected to require third-party registration of quality systems to TE

9000. These registrations will be performed by registrars already accredited to perform ISO 9000 registrations. Although a publication date for TE 9000 has not been announced, affected companies are being encouraged to seek ISO 9001 registration as well as to follow the guidelines in the auto industry's *Reliability and Maintainability Guideline for Manufacturing Machinery.*[3]

When released, TE 9000 standards will be available for purchase from AIAG at 1-800-358-3570.

73.2.5   Other Quality System Standards

Although the auto-industry standards have gained acceptance, other attempts to create specialized quality system requirements have not fared as well. The Japanese created JIS Z9901, a software quality standard modeled after ISO 9000. So far, the standard has not been released or made mandatory to companies selling products in Japan. There is a concern that such specialized requirements may be used as trade barriers, limiting entry into global markets.[6]

73.2.6   ISO 14000

The ISO 14000 series of environmental management standards was released in 1996. The standards represent the work of the International Organization for Standardization's Technical Committee 207, and provide requirements for managing compliance to environmental regulations.[7] It is expected to affect all aspects of a company's environmental operations, including:

- Environmental management systems
- Environmental auditing
- Labeling requirements and formats
- Environmental performance evaluation
- Life-cycle assessments

It is expected that the ISO 14001 registration process will be similar to that of the quality system standard, ISO 9001. At this writing, the exact registration process has not been finalized. The Registration Accreditation Board (RAB) will most likely serve as the U.S. accrediting body in association with the American National Standards Institute (ANSI). Registration will require:

- Procedures for implementing an environmental management system that maintains compliance with applicable government regulations
- Proof that procedures are being followed
- Commitment to continuous improvement
- Commitment to pollution reduction

Certification to the ISO 14001 standard may become requisite to doing business in Europe in much the same way that ISO 9000 is now required by many companies both in Europe and the United States. The environmental standard is expected to minimize trade barriers and synchronize national environmental laws, labeling requirements, and other procedures that can enhance entry into global markets. Certification to the standard may also provide companies with some degree of legal protection.[8]

The environmental performance reporting requirements at the core of ISO 14001 are causing concern for some U.S. companies. There is a perception that such reports could supply the Environmental Protection Agency (EPA) with incriminating evidence resulting in fines and other penalties. However, there is also a possibility that registration to ISO 14001 might become incorporated into EPA requirements.[9]

At this writing, ISO 14000 has not been released. Contact ANSI at (212) 642-4900 for status and ordering information.

73.3   QUALITY AWARDS

73.3.1   Deming Prize

The Deming Prize was created in 1951 by the Union of Japanese Scientists and Engineers (JUSE). It was named after Dr. W. Edwards Deming to recognize his contributions to Japanese quality control. Deming was invited to Japan in 1950 to present a series of lectures on quality control and statistical techniques. At the time, Japan was still occupied by Allied forces and the Japanese were beginning to rebuild their industries. Deming's approach to quality control was instituted throughout Japan. It was later broadened to include total quality management (TQM), although Deming disavowed any relationship to TQM.

There are two types of Deming Prizes: Individual Person and Application. The Application Prize is offered in four categories: Overall Organization, Overseas Company, Division, and Small Enterprise. In addition, there is a Quality Control for Factory Prize.

The criteria for the Application Prize is contained on a broad, 10-point Deming Prize Checklist (see Fig. 73.2). There is no weighting for these criteria as is found in the Malcolm Baldrige National Quality Award criteria. In addition, other, unwritten criteria are also used by the judges when considering an organization for the prize. These can include:

- Cost Controls
- Inspection
- Inventories
- Processes
- Research
- Training

- Equipment Maintenance
- Instrumentation
- Personnel
- Profits
- Safety

The Deming Prize Committee administers the prize process. The Committee is chaired by the chairman of the JUSE board of directors or a person selected by the board. The prize committee is made up of quality experts chosen by its chairman. These experts review applications, conduct site visits, and select the individuals and organizations to receive the Deming Prize.[10]

The Deming Application Prize involves a process that can take several years and cost a great deal. Implied in this process is the use of JUSE consultants for months or years to assist the applicant in putting the prescribed quality control systems into place. The consultants perform a quality-control diagnosis and recommend changes. The organization creates its application for the Deming Prize the year after the JUSE consultants have completed their work. The length of the application is set according to the size of the company, ranging from 50 pages for organizations with fewer than 100

The Deming Prize Checklist
1. POLICIES. How are policies determined and transmitted? What results have been achieved?
2. ORGANIZATION and its management. How are scopes of responsibility and authority defined? How is cooperation promoted and quality control managed?
3. EDUCATION and dissemination. How is quality control taught, and how is training delivered to employees? To what extent are QC and statistical techniques understood? How are QC circle activities utilized?
4. COLLECTION, dissemination, and use of information on quality. How is information collected and disseminated at various locations inside and outside the company? How well is it used? How quickly?
5. ANALYSIS. Are critical problems grasped and analyzed against overall quality and the production process? Are they interpreted appropriately, using the correct statistical methods?
6. STANDARDIZATION. How are standards used, controlled, and systematized? What is their role in enhancement of company technology?
7. CONTROL. Are quality procedures reviewed for maintenance and improvement? Are responsibility and authority scrutinized, control charts and statistical techniques checked?
8. QUALITY ASSURANCE. Are all elements of the production operation that are essential for quality and reliability (from product development to service) examined, along with the quality assurance management system?
9. EFFECTS (results). Are products of sufficiently good quality being sold? Have there been improvements in quality, quantity, and cost? Has the whole company been improved in quality, profit, scientific way of thinking, and will to work?
10. FUTURE PLANS. Are strong and weak points in the present situation recognized? Is promotion of quality control planned and likely to continue?

Fig. 73.2 Deming Prize criteria.

employees to 75 pages for 100–2,000 employees plus 5 pages for each additional 500 employees over 2,000. Applications are due in November and notification from the Committee on whether the application meets eligibility and technical requirements is made in December.[4]

Applications that pass the initial review must submit a Description of QC Practices and a company business prospectus in January. Both documents must be written in Japanese. If the Description is approved by the Committee, an on-site inspection is scheduled between March and September of that year.

In its first 38 years, the Deming Prize was awarded to a total of 139 companies. Only one prize was awarded in the category of Overseas Company, to Florida Power and Light in 1988. Two U.S. companies, Texas Instruments and Xerox, have been part-owners of Japanese companies that won the Deming Prize for Overall Organizations.[11]

For information on the Deming Prize for Overseas Companies, contact:

The Deming Prize Committee

Union of Japanese Scientists and Engineers

5-10-11 Sendagaya, Shibuya-ku

Tokyo 151

Japan

(011) 03-5379-1227, 1232, 03-3225-1813 Fax

73.3.2 Malcolm Baldrige National Quality Award

Although not the oldest quality award, the Malcolm Baldrige National Quality Award (MBNQA) has had the greatest influence on TQM in the United States. Named after the U.S. Secretary of Commerce who died in a tragic rodeo accident in 1987, this award was created by U.S. Public Law 100-107 on August 20, 1987.[12] It was designed to help U.S. companies enhance their competitiveness through focus on two results-oriented goals:

1. Delivery of ever-improving value to customers, resulting in marketplace success
2. Improvement of overall company performance and capabilities

The award is offered only to U.S. for-profit companies in one of three categories:

1. Manufacturing companies
2. Service companies
3. Small businesses with less than 500 employees

A maximum of two awards per year may be given in each category. There is no minimum number of awards that must be given.

The Department of Commerce is responsible for administering the MBNQA program. The National Institute of Standards and Technology (NIST), an agency of the Department of Commerce's Technology Administration, manages the award program. The American Society for Quality (ASQ) assists in administering the program under contract to NIST.

Applicants must complete an application of up to 70 pages describing their businesses in seven main categories (Fig. 73.3). Points are awarded on a weighted scale (Fig. 73.4) with a maximum of 1000 points possible. Typically, winners score in the 700s. (See Fig. 73.5 for list of winners and categories for each.)

The seven criteria Categories are broken into subcategories called *Items*. Each Item has points assigned and contains Areas to Address. There are 54 Areas to Address in the 1996 MBNQA criteria. Each Area to Address must be covered in the application unless the area does not apply to a company's business.[13]

The MBNQA criteria is results-oriented and focuses on a company's business, customer, and competitive results. The greatest changes to the criteria were made in 1995, when the word *quality* was almost entirely removed, broadening the scope of the award criteria to encompass the entire business operations and not just TQM. Quality-management systems must be fully integrated into a company's operations.

Applications for the MBNQA are evaluated by five to ten members of the Board of Examiners. The Board is composed of approximately 250 examiners, a volunteer group of recognized experts in the areas of quality and continuous improvement. Board members are selected annually through an application process. Applications are scored during the first stage of the award process.

Applicants that received high scores from the examiners (generally, over 600 points out of a possible 1,000) receive site visits. The findings from the site visits are summarized in a site visit report that is presented to a panel of judges for review. The judges can recommend up to two winners in each category. The judges' recommendations are given to NIST, which makes the final recom-

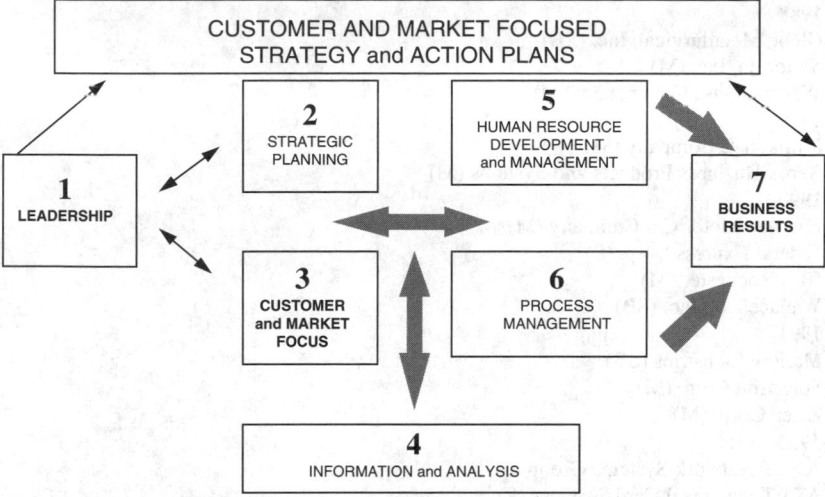

Fig. 73.3 MBNQA criteria framework.

mendations to the U.S. Secretary of Commerce. All applicants receive a detailed feedback report that itemizes strengths and areas for improvement.

The application fees for the MBNQA range from $1,200 for small businesses to $4,000 for large companies. In addition, expenses incurred during a site visit are reimbursed by the applicant. These fees are minimal when compared to the amount that would be charged by consultants for an analysis as detailed as the feedback report.[14]

Some of the past winners, however, have spent large sums to prepare their companies to apply for the award. The total cost of consultants, systems enhancements, and labor to create the application have ranged from several thousand to estimates in the millions. NIST has tracked the financial performance of past winners, however, and found stock performance many times better than the average Standard and Poors 500 performance (Fig. 73.6).

The number of applications for the MBNQA declined sharply in 1995, with only 47 applicants and 13 site visits (Fig. 73.7). This may not indicate a loss of interest in quality awards so much as a dramatic increase in state awards based on the Baldrige criteria. Many companies have developed self-assessment checklists and processes using the MBNQA criteria. The influence of the criteria may well be growing even as the applications decline.

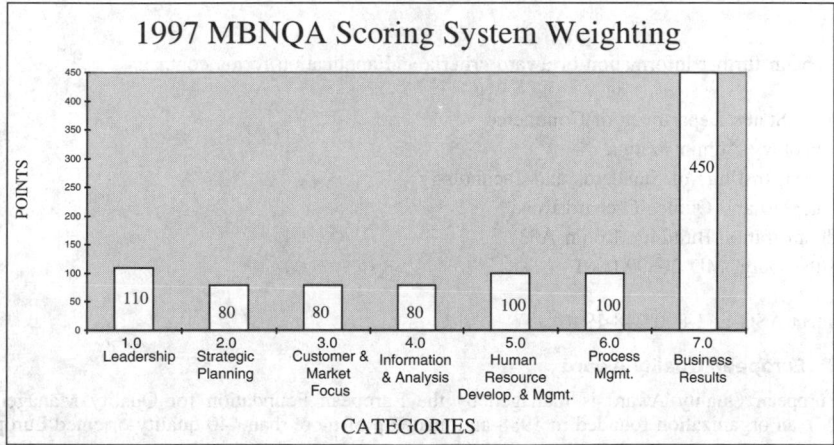

Fig. 73.4 Weights of the Malcolm Baldrige National Quality Award 1997 application criteria.

1988
Globe Metallurgical, Inc. (SB)
Motorola, Inc. (M)
Westinghouse Commercial (M)
1989
Milliken & Company (M)
Xerox Business Products and Systems (M)
1990
Cadillac Motor Car Company (M)
Federal Express Corp. (S)
IBM Rochester (M)
Wallace Co., Inc. (SB)
1991
Marlow Industries (SB)
Solectron Corp. (M)
Zytec Corp. (M)
1992
AT&T Network Systems Group (M)
AT&T Universal Card Services (S)
Granite Rock Company (SB)
Texas Instruments, Inc. (M)
1993
Ames Rubber Corp. (SB)
Eastman Chemical Co. (M)
1994
AT&T Consumer Communications (S)
GTE Directories (S)
Wainwright Industries (SB)
1995
Armstrong World Industries (M)
Corning Telecommunications (M)
1996
ADAC Laboratories (M)
Custom Research, Inc. (SB)
Dana Commercial Credit Corporation (S)
Trident Precision Manufacturing, Inc. (SB)

Fig. 73.5 MBNQA Award Winners—1988–1996. (From NIST's MBNQA homepage, located at http://www.nist.gov:8012/). (M) = Manufacturing, (S) = Service, (SB) = Small Business.

To obtain further information or award criteria and application forms contact:

United States Department of Commerce
Technology Administration
National Institute of Standards and Technology
Route 270 and Quince Orchard Road
Administration Building, Room A537
Gaithersburg, MD 20899-0001

Or contact ASQ at 1-800-248-1946.

73.3.3 European Quality Award

The European Quality Award is managed by the European Foundation for Quality Management (EFQM), an organization founded in 1988 and made up of more than 440 quality-oriented European businesses and organizations. It was created to enhance European competitiveness and effectiveness through the application of TQM principles in all aspects of organizations. EFQM headquarters is located in the Netherlands.

Date of Investment	Whole Company Winner or Parent (Subsidiary Winner)	Stock Purchases		Aug. 1, 1995 Close		
		Price	$ Invested	Price	$Value	%Change
4/4/88	Motorola	11.125**	$1,000	76 1/2	$6,876	587.6
4/4/88	Westinghouse (CNFD)	25.56*	17.78***	13 5/8	9	-46.7
4/3/89	Xerox (Business Products and Systems)	60.25	790***	119 3/8	1,565	98.1
4/2/90	General Motors (Cadillac Motor Car Division)	45.5	13.39***	48 3/4	14	7.1
4/2/90	Federal Express	55.38	1,000	67 1/2	1,219	21.9
4/2/90	IBM (IBM Rochester)	105.88	17.62***	108 7/8	18	2.8
4/1/91	Solectron	4.1875**	1,000	36 3/8	8,687	768.7
4/1/92	AT&T (Universal Card Services)	40.38	6.53***	52 3/4	9	30.7
4/1/92	AT&T (Transmission Sys. Bus. Unit)	40.38	37.54	52 3/4	49	30.7
4/1/92	Texas Instruments (Defense Sys. & Elec. Group)	32	246.61	156 1/4	1,204	388.3
11/11/93	Zytec	10.38	1,000	8 1/4	795	-20.5
4/1/94	Eastman Chemical	45.25	1,000	64	1,414	41.1
4/1/94	AT&T (Consumer Communications Serv.)	51.25	159.26	52 3/4	164	2.9
4/1/94	GTE (GTE Directories)	31	41.88	35 1/2	48	14.5
TOTALS:	**S&P 500** **Baldrige Award-Winning Companies**		6330.61 6330.61		10,033 22,072	58.5 248.7

```
*      Adjusted for 2 for 1 stock split after investment date
**     Adjusted for two separate stock splits of 2 for 1 after investment date
***    For subsidiaries, the sum invested is $1,000 × the % of the parent company's employee base that
       the subsidiary represents
```

Fig. 73.6   NIST stock study of Malcolm Baldrige National Quality Award winners, updated 21 March 1996.

The European Quality Award program was instituted in 1991, and the first prizes were awarded in 1992. The award system consists of several European Quality Prizes given to organizations that show their approach to TQM has contributed significantly over the years to satisfying the expectations of their customers, employees and other stakeholders. One of these prize winners is selected to receive the top award, the European Quality Award.[15]

This awards program is open to any European company or public service organization. European divisions of companies whose parent organizations are located outside Europe are also eligible. Xerox was the winner of the first European Quality Award in 1992, and Texas Instruments Europe received the award in 1995.[16]

The European Quality Award criteria is weighted and scored on a scale of 0 to 1,000 in a manner similar to the criteria for the Malcolm Baldrige National Quality Award. The criteria are divided into two main categories: Enabler Criteria and Results Criteria. (See Fig. 73.8 for details of the criteria and scoring system.)

73.3.4   Shingo Prize for Excellence in American Manufacturing

The Shingo Prize promotes world-class manufacturing in North America. It is administered by the College of Business, Utah State University, in partnership with the National Association of Manufacturers. The prize has been awarded to 17 companies since its inception in 1988.

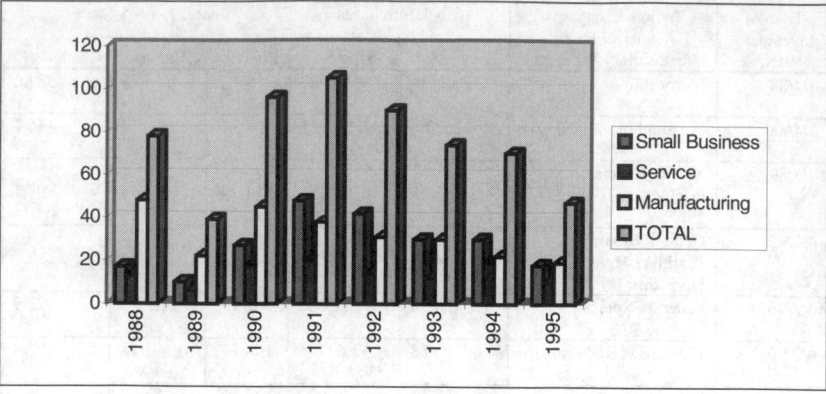

Fig. 73.7   MBNQA applications per year by type of organization.

ENABLER CRITERIA (How results are being achieved)	POINTS %
I. Leadership: How the executive team and all other managers inspire, drive, and reflect TQM as the organization's fundamental process for continuous improvement.	100 (10%)
II. Policy and Strategy: How the organization's policy and strategy reflect the concept of TQ, and how the principles of TQ are used in formulation, deployment, review, and improvement of policy and strategy.	80 (8%)
III. People Management: How the organization releases the full potential of its people to continuously improve its business.	90 (9%)
IV. Resources: How the organization's resources are effectively deployed in support of policy and strategy.	90 (9%)
V. Processes: How processes are identified, reviewed, and if necessary revised to ensure continuous improvement.	140 (14%)
RESULTS CRITERIA (What the organization has achieved and is achieving)	
VI. Customer Satisfaction: What the organization is achieving in relation to the satisfaction of its external customers.	200 (20%)
VII. People Satisfaction: What the organization is achieving in relation to the satisfaction of its people.	90 (9%)
VIII. Impact on Society: What the organization is achieving in satisfying the expectations of the community at large. This includes perceptions of the organization's approach to quality of life and the environment.	60 (6%)
IX. Business Results: What the organization is achieving in relation to its planned business objectives and in satisfying the needs and expectations of everyone with a financial interest or stake in the organization.	150 (15%)
TOTAL POINTS	1,000 (100%)

Fig. 73.8   European Quality Award criteria.

The Shingo Prize for Excellence in Manufacturing honors Dr. Shigeo Shingo, a leading expert on improving the manufacturing process. He created, with Taiichi Ohno, many of the facets of just-in-time manufacturing while working with Toyota Production Systems. Shingo is known for his books, including *Zero Quality Control: Source Inspection and the Poka-yoke System*; *Non-Stock Production: The Shingo System for Continuous Improvement*; and *The Shingo Production Management: Improving Process Functions*.

The philosophy of the Shingo Prize is that world-class status may be achieved through focused improvements in core manufacturing processes, implementing lean, just-in-time philosophies and systems, eliminating waste, and achieving zero defects, while continuously improving products and costs.

The mission of the Shingo Prize is to:

- Facilitate an increased awareness by the manufacturing community of lean, just-in-time manufacturing processes, systems, and methodologies that will maintain and enhance a company's competitive position in the world marketplace
- Foster enhanced understanding and subsequent sharing of successful core manufacturing-improvement methodologies
- Encourage research and study of manufacturing processes and production improvements in both the academic and business arenas

The Shingo Prize is awarded annually to:

- Manufacturing companies, divisions, and plants in the United States, Canada, and Mexico
- Research and writing that addresses innovative manufacturing, quality and productivity improvements, systems, and processes

The prize uses weighted criteria and requires a written application. See Fig. 73.9 for the criteria and weighting.[4]

For further information on the Shingo Prize, the application process, or the criteria, contact:

CRITERIA	POINTS
Total Quality and Productivity Management Culture and Infrastructure • Leading: 100 • Empowering: 100 • Partnering: 75	275 pts.
Manufacturing Strategy, Processes, and Systems • Manufacturing Vision and Strategy: 50 • Manufacturing Process Integration: 125 • Quality and Productivity Methods Integration: 125 • Manufacturing and Business Integration: 125	425 pts.
Measured Customer Service • Customer Satisfaction: 100	100 pts.
Measured Quality and Productivity • Quality Enhancement: 100 • Productivity Improvement: 100	200 pts.
TOTAL	1,000 pts.

Fig. 73.9 Shingo Prize criteria and weighting.

STATE	TYPE OF QUALITY AWARD	CRITERIA & ELIGIBILITY	CONTACT
AR	1) Interest 2) Commitment 3) Achievement 4) Governor's	Based on Baldrige Award. Open to public and private organizations in state of Arkansas.	Arkansas Quality Award, Inc. 1111 West Capitol, Room 1013 Little Rock, AR 72201 501/373-1300
AZ	1) Prospecting 2) Pioneer 3) Governor's	Based on last year's Baldrige Award criteria with some modifications. Open to public, private and non-profit in state.	Arizona Quality Alliance 1435 N. Hayden Rd. Scotsdale, AZ 85257 602/481-3454
CA-1	1) Management 2) Marketplace 3) Workplace 4) Community 5) Overall Excellence	Uses Baldrige Award concepts. Open to for-profit manufacturing, service and small business in state.	California Center for Quality Education and Development PO Box 2231 Sacramento, CA 95812-2231 916/322-3590
CA-2	Eureka Award	Based on Baldrige Award. Open to California-based service, non-profit, governmental and educational institutions.	California Council for Quality & Service PO Box 880774 San Diego, CA 92168 619/491-3050
CT-1	Connecticut Award for Excellence (CAFE)	Uses Baldrige criteria. Open to business, education, government, and health care.	Connecticut Award for Excellence PO Box 38 Rocky Hill, CT 06067 800/392-2122
CT-2	Connecticut Quality Improvement Award	Uses current year Baldrige criteria. Open to for-profit and not-for-profit organizations.	Connecticut Quality Improvement Award, Inc. PO Box 1396 Stamford, CT 06904-1396 203/322-9534
DE	Up to 10 awards per year.	Modified Baldrige criteria. Open to manufacturing, non-manufacturing and non-profit in large and small categories.	Delaware Quality Consortium, Inc. Delaware Economic Development Off. PO Box 1401 Dover, DE 19903 302/739-4271
FL	Governor's Sterling Award	Baldrige criteria. Open to private manufacturing, private service, education, health care, and public.	Florida Sterling Council Governor's Sterling Award Office Room 313, Carlton Building Tallahassee, FL 32399-0001 904/922-5316
HA	State Award for Excellence: Gold (highest) Red (significant) Purple (high)	Patterned after Baldrige Award. Open to any organization which provides products or services to people of Hawaii.	Pacific Region Institute for Service Excellence Chamber of Commerce of Hawaii 1132 Bishop Street, Suite 200 Honolulu, HI 96813 808/545-4355
IL	Lincoln Award	New, 1/96. Open to large and small industry and service in state.	Lincoln Award for Business Excellence 520 W. Jackson Blvd., #600 Chicago, IL 60607 312/258-4074
LA	Louisiana Quality Award	Patterned after Baldrige Award. Open to any size or type organization in state.	Louisiana Quality Foundation c/o LSU at Alexandria 8100 Highway 71 South Alexandria, LA 71302-9121 318/473-6453
MA	Massachusetts Quality Award	Uses Baldrige criteria. Open to manufacturing, service and non-profit organizations.	Massachusetts State Quality Award 3 Robinson Drive Bedford, MA 01730 617/275-1200
MD	Maryland Excellence Award	Criteria from Maryland Senate Productivity Award. Two categories: Education and small business.	Maryland Center for Quality & Productivity College of Business and Management University of Maryland College Park, MD 20742-7215 301/405-7099
ME	Maine Quality Award	Modeled after Baldrige Award. Open to large and small manufacturing and service companies and non-profits of any size.	Maine Quality Award Program Margaret Chase Smith Library PO Box 3152 Skowhegan, ME 04976 207/474-0513
MI	Michigan Quality Leadership Award	Based on Baldrige criteria. Manufacturing, Service, Health Care, Education, Public Sector, and Small Enterprise categories.	Michigan Quality Council Oakland University 525 O'Dowd Hall Rochester, MI 48309-4401 810/370-4552

Fig. 73.10 State quality awards (information supplied by National Institute of Standards and Technology).

MN	Minnesota Quality Award	Uses Baldrige categories and items but does not include areas to address. Categories: Manufacturing, Service, and Education	Minnesota Quality Award Minnesota Council for Quality 2850 Metro Drive, Suite 300 Bloomington, MN 55425 612/851-3181
MO	Missouri Quality Award	Patterned after Baldrige Award. Categories: Manufacturing, Service, Health Care, Education, Public Sector in 3 sizes.	Excellence in Missouri Foundation Harry S. Truman State Office Building Room 620, 301 W. High Street Jefferson City, MO 65102 314/526-1725
MS	Mississippi Quality Award. Four award levels: 1) Interest 2) Commitment 3) Award 4) Governor's	Patterned after Baldrige Award. Any MS public or private organization may apply.	Mississippi Quality Award Center for Quality and Productivity 3825 Ridgewood Rd. Jackson, MS 39211 601/982-6739
NC	North Carolina Quality Leadership Award	Uses previous year's Baldrige Award criteria. Seven categories: 1) Education, 2-4) Manufacturing (small, medium and large) and 5-7) Service (small, medium and large).	North Carolina Quality Leadership Award 4904 Professional Court, Suite 100 Raleigh, NC 27609 919/872-8198
NE	Edgerton Quality Awards: 1) Continuous Process Improvement 2) Adaptation of Technology	Patterned after Baldrige criteria (plus 8th category for "Sharing of Information") and Minnesota award program. Two categories: Manufacturing and Service.	The Edgerton Quality Award Program Nebraska Department of Economic Develop. Existing Business Assistance Division PO Box 94666, 301 Centennial Mall South Lincoln, NE 68509-4666 402/471-4167
NH	New Hampshire Quality Award	Patterned after Baldrige Award. Two categories for small (>200 employees) and large organizations 1) Manufacturing, 2) Service.	New Hampshire Quality Council PO Box 3128 Portsmouth, NH 03802 603/427-2280
NJ	New Jersey Quality Achievement Award	Uses Baldrige Award criteria. Categories: Manufacturing, Service, Small Business, Education, Government	New Jersey Quality Achievement Award Mary G. Roebling Building, CN 827 Trenton, NJ 08625-0827 609/777-0939
NM	1) Pinon-commitment 2) Roadrunner-progress 3) Zia-excellence 4) Quality Hero	Pinon requires written description of seven Baldrige categories. Roadrunner requires 28 Baldrige items, and Zia requires complete Baldrige criteria application. Quality Heros are individuals cited for outstanding service. Open to any public or privately-held organization of any size.	Quality New Mexico PO Box 25005 Albuquerque, NM 87125 505/242-7903
NY	Governor's Excelsior Award	Modeled on Baldrige Award. Open to any organization, any size.	The Excelsior Award, Inc. 152 Washington Avenue Albany, NY 12210-2289 518/465-1706
OK	Oklahoma Quality Award	Patterned after Baldrige Award. Categories: Manufacturing (large, medium, small) and Service (large, medium, small).Plans to expand to public sector.	Oklahoma State Quality Award Foundation 6601 N. Broadway Oklahoma City, OK 73116 405/841-5295
OR	Oregon Quality Award	Modified Baldrige criteria. Applicants complete self-assessment with areas for improvement. Categories: Manufacturing, Service, Education, Health Care, and Government.	Oregon Quality Award One World Trade Center 121 S.W. Salmon, Suite 1140 Portland, OR 97204 503/224-4606
PA	Pennsylvania Quality Leadership Award: 1) Cornerstone 2) Keystone 3) Governor's	Use Baldrige criteria. Open to any public or private organization. Categories: Manufacturing (large and small), and Service (large and small).	Pennsylvania Quality Leadership Foundation PO Box 4129 Harrisburg, PA 17111-0129 717/561-7180

Fig. 73.10   (Continued)

RI	1) RI Award for Competitiveness and Excellence 2) Quality Achievement Award 3) AT&T/URI Quality and Education Award	Uses modified Baldrige Award criteria. Open to any RI organization except U.S. Government, professional and trade organizations. AT&T/URI winner receives $10,000 (K-12).	Rhode Island Area Coalition for Excellence PO Box 6766 Providence, RI 02940 401/454-3030
SC	1) Achiever's Award 2) Governor's Award	Uses previous year's Baldrige Award criteria. Open to public and private organizations.	South Carolina Quality Forum c/o Quality Institute University of South Carolina at Spartanburg 800 University Way Spartanburg, SC 29303 803/599-2990
TN	1) Quality Interest 2) Commitment 3) Achievement 4) Governor's	Modeled after Baldrige Award with four levels. Open to any public or private organization.	Tennessee Quality Award Office 2233 Highway 75, Suite 1 Blountville, TN 37617-5840 615/279-0037
TX	Texas Quality Award	Patterned after Baldrige Award. Categories: small (< 100 employees) and large (> 100). Open to for-profit and not-for-profit organizations.	Quality Texas PO Box 684157 Austin, TX 78768-4157 512/477-8137
UT	1) Improvement 2) Progress 3) Governor's 4) Continuous (for past winners of Governor's Award)	Based on Baldrige Award with separate criteria for government and education. Four award categories: 1) manufacturing, 2) service, 3) Education, 4) Government.	Utah Quality Council 2120 State Office Building Salt Lake City, UT 84114 801/538-3067
WA		Categories: 1) manufacturing, 2) service, 3) government and education, 4) not-for-profit; all categories judged in large or small (< 200) division.	Department of Labor and Industries N. 901 Monroe, Suite 100 Spokane, WA 99201 509/324-2534

Fig. 73.10   (Continued)

	Deming	ISO 9000	Baldrige	Shingo	European	State
Year Created	1951	1987	1987	1988	1991	various
Form	Long-term prize	Certification	Annual Awards	Annual Prizes	Annual Prizes and Award	Annual Awards
Emphasis	Statistics; Quality Control	Documented Procedures and Compliance	Customer Satisfaction, Business Results, All Parts of the Organization	Just-in time Manufacturing, Process Improvements	Customer Satisfaction, Business Results, Processes	Similar to Baldrige
Missing from TQ Perspective	Customer Focus	Customer Focus, Business Results, Support Organizations	Complete	Business and Support Organizations, Customer Focus	Complete	Complete
Cost	Very High	Low-- Medium	Medium-- High	Low	Medium-- High	Very Low

Fig. 73.11   Comparison of TQM elements in ISO 9000 and the quality awards.

The Shingo Prize for Excellence in Manufacturing
College of Business, Utah State University
Logan, UT 84322-3521
(801) 797-2279, (801) 797-3440 Fax

73.3.5 State Quality Awards

In the United States, quality awards have been initiated in over 40 states. Many of the awards are based on the Baldrige Award criteria, although eligibility has been extended in most cases to not-for-profit and governmental organizations as well as manufacturing and service companies. While the number of applicants for the Malcolm Baldrige National Quality Award has declined in recent years, state applications have increased dramatically. In 1995, there were only 47 applications submitted for the MBNQA but the total for state awards was 450, a 33% increase from the previous year. Even though state awards were almost nonexistent in 1990, over 40 states had initiated award programs by 1997, with others expected to follow suit.

In most states, the award process adheres to the MBNQA with a written application, site visits, and an award ceremony in the fall. Application costs are often less than half the Baldrige application fee and, in most cases, the organization receives a feedback report. Figure 73.10 is a listing of state awards and contact points.

73.3.6 How Do They Compare?

There have been many attempts to compare the various award and registration initiatives looking for common and missing elements of a Total Quality (TQ) system. Figure 73.11 illustrates some of the differences among the more popular TQ initiatives.

The value of any of the registrations, certifications, or awards is not necessarily in achieving the certificate or plaque. The benefit is derived from the process itself, which serves to drive continuous improvement.

REFERENCES

1. R. Hilary, "Behind the Stars and Stripes: Quality in the USA," *Quality Progress*, 31–35 (January 1996).
2. L. A. Wilson, *Eight-Step Process to Successful ISO 9000 Implementation: A Quality Management System Approach*, ASQC Quality Press, 1995.
3. L. Streubing, "9000 Standards?" *Quality Progress*, 23–28 (January 1996).
4. F. X. Mahoney and C. G. Thor, *The TQM Trilogy*, American Management Association, 1994.
5. P. Stein, Untitled, Internet Web page, Quality.Org., File date: Dec. 5, 1995, amended Jan. 8, 1996.
6. A. Zuckerman, "Standards Battles Heating Up Worldwide," *Quality Progress*, 23–24 (September 1995).
7. C. G. Hemenway and G. J. Hale, "Are You Ready for ISO 14000?" *Quality*, 26–28 (November 1995).
8. T. Tibor, "Hurdles to Implementation," *Chemical Week*, 50 (November 8, 1995).
9. B. Rothery, "Why ISO 14000 Will Catch ISO 9000," *Manufacturing Engineering*, 128 (November 1995).
10. T. B. Kinni, "Total-Quality Bargains: Francis Mahoney on the *TQM Trilogy*," *Industry Week*, 65 (September 4, 1995).
11. D. Greising, "Selling a Bright Idea—Along with the Kilowatts," *Business Week*, 59 (August 8, 1994).
12. M. G. Brown, *Baldrige Award Winning Quality*, 5th ed., ASQC Quality Press, 1995.
13. *Malcolm Baldrige National Quality Award 1996 Application Forms and Instructions*, National Institute of Standards and Technology.
14. *Malcolm Baldrige National Quality Award 1996 Award Criteria*, National Institute of Standards and Technology.
15. P. Wendel, "The European Quality Award, and How Texas Instruments Europe Took the Trophy Home," *The Quality Observer* **5**, 39–48 (January 1996).
16. Anonymous, "Evaluating the Operation of the European Quality Award (EQA) for Self-Assessment," *Management Accounting—London*, 8 (April 1995).

CHAPTER 74

SAFETY ENGINEERING

Jack B. ReVelle
Hughes Missile Systems Company
Tucson, Arizona

74.1	**INTRODUCTION**	**2194**
	74.1.1 Background	2194
	74.1.2 Employee Needs and Expectations	2194
74.2	**GOVERNMENT REGULATORY REQUIREMENTS**	**2195**
	74.2.1 Environmental Protection Agency (EPA)	2195
	74.2.2 Occupational Safety and Health Administration (OSHA)	2196
	74.2.3 State-Operated Compliance Programs	2197
74.3	**SYSTEM SAFETY**	**2197**
	74.3.1 Methods of Analysis	2198
	74.3.2 Fault Tree Technique	2199
	74.3.3 Criteria for Preparation/Review of System Safety Procedures	2199
74.4	**HUMAN FACTORS ENGINEERING/ERGONOMICS**	**2202**
	74.4.1 Human–Machine Relationships	2202
	74.4.2 Human Factors Engineering Principles	2203
	74.4.3 General Population Expectations	2204
74.5	**ENGINEERING CONTROLS FOR MACHINE TOOLS**	**2205**
	74.5.1 Basic Concerns	2205
	74.5.2 General Requirements	2205
	74.5.3 Danger Sources	2207
74.6	**MACHINE SAFEGUARDING METHODS**	**2207**
	74.6.1 General Classifications	2207
	74.6.2 Guards, Devices, and Feeding and Ejection Methods	2208
74.7	**ALTERNATIVES TO ENGINEERING CONTROLS**	**2208**
	74.7.1 Substitution	2208
	74.7.2 Isolation	2213
	74.7.3 Ventilation	2213
74.8	**DESIGN AND REDESIGN**	**2213**
	74.8.1 Hardware	2213
	74.8.2 Process	2213
74.9	**PERSONAL PROTECTIVE EQUIPMENT**	**2214**
	74.9.1 Background	2214
	74.9.2 Planning and Implementing the Use of Protective Equipment	2215
	74.9.3 Adequacy, Maintenance, and Sanitation	2216
74.10	**MANAGING THE SAFETY FUNCTION**	**2217**
	74.10.1 Supervisor's Role	2217
	74.10.2 Elements of Accident Prevention	2217
	74.10.3 Management Principles	2218
	74.10.4 Eliminating Unsafe Conditions	2219
	74.10.5 Unsafe Conditions Involving Mechanical or Physical Facilities	2221
74.11	**SAFETY TRAINING**	**2223**
	74.11.1 Specialized Courses	2223
	74.11.2 Job Hazard Analysis Training	2224
	74.11.3 Management's Overview of Training	2224

Mechanical Engineers' Handbook, 2nd ed., Edited by Myer Kutz.
ISBN 0-471-13007-9 © 1998 John Wiley & Sons, Inc.

74.1 INTRODUCTION

74.1.1 Background

More than ever before, engineers are aware of and concerned with employee safety and health. The necessity for this involvement was accelerated with the passage of the OSHAct in 1970, but much of what has occurred since that time would have happened whether or not the OSHAct had become the law.

As workplace environments become more technologically complex, the necessity for protecting the work force from safety and health hazards continues to grow. Typical workplace operations from which workers should be protected are presented in Table 74.1. Whether they should be protected through the use of personal protective equipment, engineering controls, administrative controls, or a combination of these approaches, one fact is clear; it makes good sense to ensure that they receive the most cost-effective protection available. Arguments in support of engineering controls over personal protective equipment and vice versa are found everywhere in the current literature. Some of the most persuasive discussions are included in this chapter.

74.1.2 Employee Needs and Expectations

In 1981 ReVelle and Boulton asked the question, "Who cares about the safety of the worker on the job?" in their award-winning two-part article in *Professional Safety,* "Worker Attitudes and Perceptions of Safety." The purpose of their study was to learn about worker attitudes and perceptions of safety. To accomplish this objective, they established the following working definition:

WORKER ATTITUDES AND PERCEPTIONS   As a result of continuing observation, an awareness is developed, as is a tendency to behave in a particular way regarding safety.

To learn about these beliefs and behaviors, they inquired to find out:

1. Do workers think about safety?
2. What do they think about safety in regard to:
    (a) Government involvement in their workplace safety.
    (b) Company practices in training and hazard prevention.
    (c) Management attitudes as perceived by the workers.
    (d) Coworkers' concern for themselves and others.
    (e) Their own safety on the job.
3. What do workers think should be done, and by whom, to improve safety in their workplace?

Table 74.1 Operations Requiring Engineering Controls and/or Personal Protective Equipment

Acidic/basic process and treatments	Grinding
Biological agent processes and treatments	Hoisting
Blasting	Jointing
Boiler/pressure vessel usage	Machinery (mills, lathes, presses)
Burning	Mixing
Casting	Painting
Chemical agent processes and treatments	Radioactive source processes and treatments
Climbing	Sanding
Compressed air/gas usage	Sawing
Cutting	Shearing
Digging	Soldering
Drilling	Spraying
Electrical/electronic assembly and fabrication	Toxic vapor, gas, and mists and dust exposure
Electrical tool usage	Welding
Flammable/combustible/toxic liquid usage	Woodworking

The major findings of the ReVelle–Boulton study are summarized here.*

Half the workers think that government involvement in workplace safety is about right; almost one-fourth think more intervention is needed in such areas as more frequent inspections, stricter regulations, monitoring, and control.

Workers in large companies expect more from their employers in providing a safe workplace than workers in small companies. Specifically, they want better safety programs, more safety training, better equipment and maintenance of equipment, more safety inspections and enforcement of safety regulations, and provision of more personal protective equipment.

Supervisors who talk to their employees about safety and are perceived by them to be serious are also seen as being alert for safety hazards and representative of their company's attitude.

Coworkers are perceived by other employees to care for their own safety and for the safety of others.

Only 20% of the surveyed workers consider themselves to have received adequate safety training. But more than three-fourths of them feel comfortable with their knowledge to protect themselves on the job.

Men are almost twice as likely to wear needed personal protective equipment as women.

Half the individuals responding said they would correct a hazardous condition if they saw it.

Employees who have had no safety training experience almost twice as many on-the-job accidents as their fellow workers who have received such training.

Workers who experienced accidents were generally candid and analytical in accepting responsibility for their part in the accident; and 85% said their accidents could have been prevented.

The remainder of this chapter addresses those topics and provides that information which engineering practitioners require to professionally perform their responsibilities with respect to the safety of the work force.

74.2 GOVERNMENT REGULATORY REQUIREMENTS†

Two relatively new agencies of the federal government enforce three laws that impact many of the operational and financial decisions of American businesses, large and small. The Environmental Protection Agency (EPA) has responsibility for administering the Toxic Substances Control Act (TSCA) and the Resource Conservation and Recovery Act (RCRA), both initially enforced in 1976. The Occupational Safety and Heath Act (OSHAct) of 1970 is enforced by the Occupational Safety and Health Administration (OSHA), a part of the Department of Labor. This section addresses the regulatory demands of these federal statutes from the perspective of whether to install engineering controls that would enable companies to meet these standards or simply to discontinue certain operations altogether, that is, can they justify the associated costs of regulatory compliance.

74.2.1 Environmental Protection Agency (EPA)

Toxic Substances Control Act (TSCA)

Until the TSCA, the federal government was not empowered to prevent chemical hazards to health and the environment by banning or limiting chemical substances at a germinal, premarket stage. Through the TSCA of 1975, production workers, consumers, indeed every American, would be protected by an equitably administered early warning system controlled by the EPA. This broad law authorizes the EPA Administrator to issue rules to prohibit or limit the manufacturing, processing, or distribution of any chemical substance or mixture that "may present an unreasonable risk of injury to health or the environment." The EPA Administrator may require testing—at a manufacturer's or processor's expense—of a substance alter finding that:

- The substance may present an unreasonable risk to health or the environment.
- There may be a substantial human or environmental exposure to the substance.
- Insufficient data and experience exist for judging a substance's health and environmental effects.
- Testing is necessary to develop such data.

---

*Reprinted with permission from the January 1982 issue of *Professional Safety,* official publication of the American Society of Safety Engineers.

†"Engineering Controls: A Comprehensive Overview," by Jack B. ReVelle. Used by permission of The Merritt Company, Publisher, from T. S. Ferry, *Safety Management Planning,* copyright © 1982, The Merritt Company, Santa Monica, CA 90406.

This legislation is designed to cope with hazardous chemicals like kepone, vinyl chloride, asbestos, fluorocarbon compounds (Freons), and polychlorinated biphenyls (PCBs).

Resource Conservation and Recovery Act (RCRA)

Enacted in 1976 as an amendment to the Solid Waste Disposal Act, the RCRA sets up a "cradle-to-grave" regulatory mechanism, that is, a tracking system for such wastes from the moment they are generated to their final disposal in an environmentally safe manner. The act charges the EPA with the development of criteria for identifying hazardous wastes, creating a manifest system for tracking wastes through final disposal, and setting up a permit system based on performance and management standards for generators, transporters, owners, and operators of waste treatment, storage, and disposal facilities. It is expected that the RCRA will be a strong force for innovation and eventually lead to a broad rethinking of chemical processes, that is, to look at hazardous waste disposal *not* just in terms of immediate costs, but rather with respect to life-cycle costs.

74.2.2   Occupational Safety and Health Administration (OSHA)*

The Occupational Safety and Health Act (OSHAct), a federal law that became effective on April 28, 1971, is intended to pull together all federal and state occupational safety and health-enforcement efforts under a federal program designed to establish uniform codes, standards, and regulations. The expressed purpose of the act is "to assure, as far as possible, every working woman and man in the Nation safe and healthful working conditions, and to preserve our human resources." To accomplish this purpose, the promulgation and enforcement of safety and health standards is provided for, as well as research, information, education, and training in occupational safety and health.

Perhaps no single piece of federal legislation has been more praised and, conversely, more criticized than the OSHAct, which basically is a law requiring virtually all employers to ensure that their operations are free of hazards to workers.

Occupational Safety and Health Standards

When Congress passed the OSHAct of 1970, it authorized the promulgation, without further public comment or hearings, of groups of already codified standards. The initial set of standards of the act (Part 1910, published in the *Federal Register* on May 29, 1971) thus consisted in part of standards that already had the force of law, such as those issued by authority of the Walsh–Healey Act, the Construction Safety Act, and the 1958 amendments to the Longshoremen's and Harbor Workers' Compensation Act. A great number of the adopted standards, however, derived from voluntary national consensus standards previously prepared by groups such as the American National Standards Institute (ANSI) and the National Fire Protection Association (NFPA).

The OSHAct defines the term "occupational safety and health standard" as meaning "a standard which requires conditions or the adoption or use of one or more practices, means, methods, operations or processes, reasonably necessary or appropriate to provide safe or healthful employment and places of employment." Standards contained in Part 1910† are applicable to general industry. Those contained in Part 1926 are applicable to the construction industry; and standards applicable to ship repairing, shipbuilding, and longshoring are contained in Parts 1915–1918. These OSHA standards fall into the following four categories, with examples for each type:

1. *Specification Standards.*   Standards that give specific proportions, locations, and warning symbols for signs that must be displayed.

2. *Performance Standards.*   Standards that require achievement of, or within, specific minimum or maximum criteria.

3. *Particular Standards (Vertical).*   Standards that apply to particular industries, with specifications that relate to the individual operations.

4. *General Standards (Horizontal).*   Standards that can apply to any workplace and relate to broad areas (environmental control, walking surfaces, exits, illumination, etc.).

The Occupational Health and Safety Administration is authorized to promulgate, modify, or revoke occupational safety and health standards. It also has the authority to promulgate emergency temporary standards where it is found that employees are exposed to grave danger. Emergency temporary standards can take effect immediately on publication in the *Federal Register.* Such standards remain in

---

*R. De Reamer, *Modern Safety and Health Technology,* copyright © 1980. Reprinted by permission of Wiley, New York.

†The Occupation Safety and Health Standards, Title 29, CFR Chapter XVIII, Parts 1910, 1926, and 1915–1918 are available at all OSHA regional and area offices.

effect until superseded by a standard promulgated under the procedures prescribed by the OSHAct—notice of proposed rule in the *Federal Register,* invitation to interested persons to submit their views, and a public hearing if required.

Required Notices and Records

During an inspection the compliance officer will ascertain whether the employer has:

- Posted notice informing employees of their rights under the OSHAct (Job Safety and Health Protection, OSHAct poster).
- Maintained log of recordable injuries and illnesses (OSHA Form No. 200, Log and Summary of Occupational Injuries and Illnesses).
- Maintained the Supplementary Record of Occupational Injuries and Illnesses (OSHA Form No. 101).
- Annually posted the Summary of Occupational Injuries and Illnesses (OSHA Form No. 200). This form must be posted no later than February 1 and must remain in place until March 1.
- Made a copy of the OSHAct and OSHA safety and health standards available to employees on request.
- Posted boiler inspection certificates, boiler licenses, elevator inspection certificates, and so on.

74.2.3   State-Operated Compliance Programs

The OSHAct encourages each state to assume the fullest responsibility for the administration and enforcement of occupational safety and health programs. For example, federal law permits any state to assert jurisdiction, under state law, over any occupational or health standard not covered by a federal standard.

In addition, any state may assume responsibility for the development and enforcement of its own occupational safety and health standards for those areas now covered by federal standards. However, the state must first submit a plan for approval by the Labor Department's Occupational Safety and Health Administration. Many states have done so.

Certain states are now operating under an approved state plan. These states may have adopted the existing federal standards or may have developed their own standards. Some states also have changed the required poster. You need to know whether you are covered by an OSHA-approved state plan operation, or are subject to the federal program, in order to determine which set of standards and regulations (federal or state) apply to you. The easiest way to determine this is to call the nearest OSHA Area Office.

If you are subject to state enforcement, the OSHA Area Office will explain this, explain whether the state is using the federal standards, and provide you with information on the poster and on the OSHA recordkeeping requirements. After that, the OSHA Area Office will refer you to the appropriate state government office for further assistance.

This assistance also may include free on-site consultation visits. If you are subject to state enforcement, you should take advantage of this service.

For your information, the following are operating under OSHA-approved state plans, as of September 1, 1997

Alaska	New Mexico
Arizona	New York
California	Oregon
Connecticut	Puerto Rico
Guam	South Carolina
Hawaii	Tennessee
Indiana	Utah
Iowa	Vermont
Kentucky	Virginia
Maryland	Virgin Islands
Michigan	Washington
Minnesota	Wyoming
Nevada	

74.3   SYSTEM SAFETY*

System safety is when situations having accident potential are examined in a step-by-step cause–effect manner, tracing a logical progression of events from start to finish. System safety techniques can

---

*R. De Reamer, *Modern Safety and Health Technology,* copyright © 1980. Reprinted by permission of Wiley, New York.

provide meaningful predictions of the frequency and severity of accidents. However, their greatest asset is the ability to identify many accident situations in the system that would have been missed if less detailed methods had been used.

74.3.1 Methods of Analysis

A system cannot be understood simply in terms of its individual elements or component parts. If an operation of a system is to be effective, all parts must interact in a predictable and a measurable manner, within specific performance limits and operational design constraints.

In analyzing any system, three basic components must be considered: (1) the equipment (or machines); (2) the operators and supporting personnel (maintenance technicians, material handlers, inspectors, etc.); and (3) the environment in which both workers and machines are performing their assigned functions. Several analysis methods are available:

- *Gross-Hazard Analysis.* Performed early in design; considers overall system as well as individual components; it is called "gross" because it is the initial safety study undertaken.
- *Classification of Hazards.* Identifies types of hazards disclosed in the gross-hazard analysis, and classifies them according to potential severity (Would defect or failure be catastrophic?); indicates actions and/or precautions necessary to reduce hazards. May involve preparation of manuals and training procedures.
- *Failure Modes and Effects.* Considers kinds of failures that might occur and their effect on the overall product or system. Example: effect on system that will result from failure of single component (e.g., a resistor or hydraulic valve).
- *Hazard-Criticality Ranking.* Determines statistical, or quantitative, probability of hazard occurrence; ranking of hazards in the order of "most critical" to "least critical."
- *Fault-Tree Analysis.* Traces probable hazard progression. Example: If failure occurs in one component or part of the system, will fire result? Will it cause a failure in some other component?
- *Energy-Transfer Analysis.* Determines interchange of energy that occurs during a catastrophic accident or failure. Analysis is based on the various energy inputs to the product or system and how these inputs will react in event of failure or catastrophic accident.
- *Catastrophe Analysis.* Identifies failure modes that would create a catastrophic accident.
- *System-Subsystem Integration.* Involves detailed analysis of interfaces, primarily between systems.
- *Maintenance-Hazard Analysis.* Evaluates performance of the system from a maintenance standpoint. Will it be hazardous to service and maintain? Will maintenance procedures be apt to create new hazards in the system?
- *Human-Error Analysis.* Defines skills required for operation and maintenance. Considers failure modes initiated by human error and how they would affect the system. The question of whether special training is necessary should be a major consideration in each step.
- *Transportation-Hazard Analysis.* Determines hazards to shippers, handlers, and bystanders. Also considers what hazards may be "created" in the system during shipping and handling.

There are other quantitative methods that have successfully been used to recommend a decision to adopt engineering controls, personal protective equipment, or some combination. Some of these methods are:*

- *Expected Outcome Approach.* Since safety alternatives involve accident costs that occur more or less randomly according to probabilities which might be estimated, a valuable way to perform needed economic analyses for such alternatives is to calculate expected outcomes.
- *Decision Analysis Approach.* A recent extension of systems analysis, this approach provides useful techniques for transforming complex decision problems into a sequentially oriented series of smaller, simpler problems. This means that a decision-maker can select reasoned choices that will be consistent with his or her perceptions about the uncertainties involved in a particular problem together with his or her fundamental attitudes toward risk-taking.
- *Mathematical Modeling.* Usually identified as an "operations research" approach, there are numerous mathematical models that have demonstrated potential for providing powerful anal-

---

*J. B. ReVelle, *Engineering Controls: A Comprehensive Overview.* Used by permission of The Merritt Company, Publisher, from T. S. Ferry, *Safety Management Planning,* copyright © 1982, The Merritt Company, Santa Monica, CA 90406.

ysis insights into safety problems. These include dynamic programming, inventory-type modeling, linear programming, queue-type modeling, and Monte Carlo simulation.

There is a growing body of literature about these formal analytical methods and others not mentioned in this chapter, including failure mode and effect (FME), technique for human error prediction (THERP), system safety hazard analysis, and management oversight and risk tree (MORT).

All have their place. Each to a greater or lesser extent provides a means of overcoming the limitations of intuitive, trial-and-error analysis.

Regardless of the method or methods used, the systems concept of hazard recognition and analysis makes available a powerful tool of proven effectiveness for decision making about the acceptability of risks. To cope with the complex safety problems of today and the future, engineers must make greater use of system safety techniques.

74.3.2 Fault Tree Technique*

When a problem can be stated quantitatively, management can assess the risk and determine the trade-off requirements between risk and capital outlay. Structuring key safety problems or vital decision-making in the form of fault paths can greatly increase communication of data and subjective reasoning. This technique is called fault-tree analysis. The transferability of data among management, engineering staff, and safety personnel is a vital step forward.

Another important aspect of this system safety technique is a phenomenon that engineers have long been aware of in electrical networks. That is, an end system formed by connecting several subsystems is likely to have entirely different characteristics from any of the subsystems considered alone. To fully evaluate and understand the entire system's performance with key paths of potential failure, the engineer must look at the entire system—only then can he or she look meaningfully at each of the subsystems.

Figure 74.1 introduces the most commonly used symbols used in fault-tree analysis.

74.3.3 Criteria for Preparation/Review of System Safety Procedures†

Correlation Between Procedure and Hardware

1. Statement of hardware configuration to which it was written?
2. Background descriptive or explanatory information where needed?
3. Reflect or reference latest revisions of drawings, manuals, or other procedures?

Adequacy of the Procedure

1. The best way to do the job?
2. Procedure easy to understand?
3. Detail appropriate—not too much, not too little?
4. Clear, concise, and free from ambiguity that could lead to wrong decisions?
5. Calibration requirements clearly defined?
6. Critical red-line parameters identified and clearly defined? Required values specified?
7. Corrective controls of above parameters clearly defined?
8. All values, switches, and other controls identified and defined?
9. Pressure limits, caution notes, safety distances, or hazards peculiar to this operation clearly defined?
10. Hard-to-locate components adequately defined and located?
11. Jigs and arrangements provided to minimize error?
12. Job safety requirements defined, for example, power off, pressure down, and tools checked for sufficiency?
13. System operative at end of job?
14. Hardware evaluated for human factors and behavioral stereotype problems? If not corrected, are any such clearly identified?
15. Monitoring points and methods of verifying adherence specified?

---

*R. De Reamer, *Modern Safety and Health Technology.* Copyright © 1980. Reprinted by permission of Wiley, New York.

†Reprinted from *MORT* Safety Assurance Systems, pp. 278–283, by courtesy of Marcel Dekker, Inc., New York.

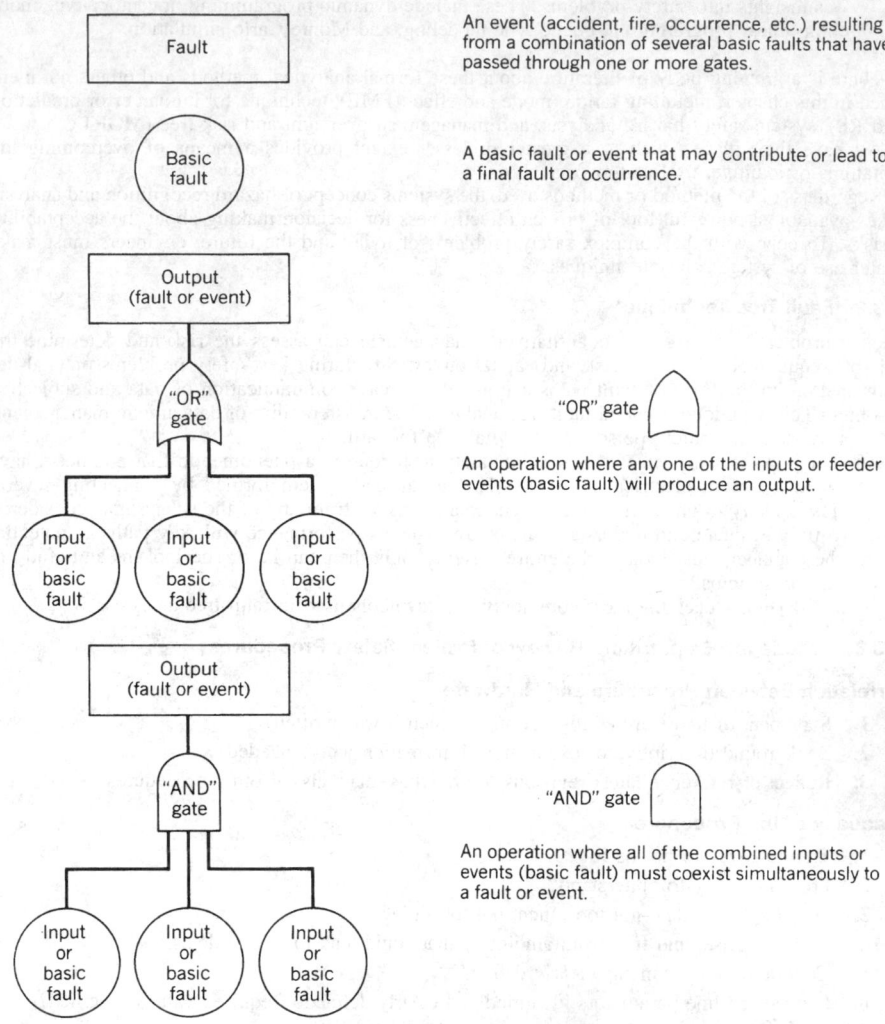

Fig. 74.1 Most common symbols used in fault-tree analysis.

16. Maintenance and/or inspection to be verified? If so, is a log provided?
17. Safe placement of process personnel or equipment specified?
18. Errors in previous, similar processes studied for cause? Does this procedure correct such causes?

Accuracy of the Procedure

1. Capacity to accomplish specified purpose verified by internal review?
2. All gauges, controls, valves, etc., called out, described, and labeled exactly as they actually are?
3. All setpoints or other critical controls, etc., compatible with values in control documents?
4. Safety limitations adequate for job to be performed?
5. All steps in the proper sequence?

Adequacy and Accuracy of Supporting Documentation

1. All necessary supporting drawings, manuals, data sheets, sketches, etc., either listed or attached?

2. All interfacing procedures listed?

Securing Provisions

1. Adequate instructions to return the facility or hardware to a safe operating or standby condition?
2. Securing instructions provide step-by-step operations?

Backout Provisions

1. Can procedure put any component or system in a condition which could be dangerous?
2. If so, does procedure contain emergency shutdown or backout procedures either in an appendix or as an integral part?
3. Backout procedure (or instructions for its use) included at proper place?

Emergency Measures

1. Procedures for action in case of emergency conditions?
2. Does procedure involve critical actions such that preperformance briefing on possible hazards is required?
3. Are adequate instructions either included or available for action to be taken under emergency conditions? Are they in the right place?
4. Are adequate shutdown procedures available? Cover all systems involved? Available for emergency reentry teams?
5. Specify requirements for emergency team for accident recovery, troubleshooting, or investigative purposes where necessary? Describe conditions under which emergency team will be used? Hazards they may encounter or must avoid?
6. Does procedure consider interfaces in shutdown procedures?
7. How will changes be handled? What are thresholds for changes requiring review?
8. Emergency procedures tested under range of conditions that may be encountered, for example, at night during power failure?

Caution and Warning Notes

1. Caution and warning notes included where appropriate?
2. Caution and warning notes precede operational steps containing potential hazards?
3. Adequate to describe the potential hazard?
4. Major cautions and warnings called out in general introduction, as well as prior to steps?
5. Separate entries with distinctive bold type or other emphatic display?
6. Do they include supporting safety control (health physics, safety engineer, etc.) if needed at specific required steps in procedure?

Requirements for Communications and Instrumentation

1. Adequate means of communication provided?
2. Will loss of communications create a hazard?
3. Course of action clearly defined for loss of required communications?
4. Verification of critical communication included prior to point of need?
5. Will loss of control or monitoring capability of critical functions create a hazard to people or hardware?
6. Alternate means, or a course of action to regain control or monitoring functions, clearly defined?
7. Above situations flagged by cautions and warnings?

Sequence-of-Events Considerations

1. Can any operation initiate an unscheduled or out-of-sequence event?
2. Could it induce a hazardous condition?
3. Identified by warnings or cautions?
4. Covered by emergency shutdown and backout procedures?
5. All steps sequenced properly? Sequence will not contribute to or create a hazard?
6. All steps which, if performed out-of-sequence, could cause a hazard identified and flagged?
7. Have all noncompatible simultaneous operations been identified and suitably restricted?
8. Have these been prohibited by positive callout or separation in step-by-step inclusion within the text of the procedure?

Environmental Considerations (Natural or Induced)

1. Environmental requirements specified that constrain the initiation of the procedure or require shutdown or evacuation, once in progress?
2. Induced environments (toxic or explosive atmospheres, etc.) considered?
3. All latent hazards (pressure, height, voltage, etc.) in adjacent environments considered?
4. Are there induced hazards from simultaneous performance of more than one procedure by personnel within a given space?

Personnel Qualification Statements

1. Requirement for certified personnel considered?
2. Required frequency of recheck of personnel qualifications considered?

Interfacing Hardware and Procedures Noted

1. All interfaces described by detailed callout?
2. Interfacing operating procedures identified, or written to provide ready equipment?
3. Where more than one organizational element is involved, are proper liaison and areas of responsibility established?

Procedure Sign-Off

1. Procedure to be used as an in-hand, literal checklist?
2. Step sign-off requirements considered and identified and appropriate spaces provided in the procedure?
3. Procedure completion sign-off requirements indicated (signature, authority, date, etc.)?
4. Supervisor verification of correct performance required?

General Requirements

1. Procedure discourages a shift change during performance or accommodates a shift change?
2. Where shift changes are necessary, include or reference shift overlap and briefing requirements?
3. Mandatory inspection, verification, and system validation required whenever procedure requires breaking into and reconnecting a system?
4. Safety prerequisites defined? All safety instructions spelled out in detail to all personnel?
5. Require prechecks of supporting equipment to ensure compatibility and availability?
6. Consideration for unique operations written in?
7. Procedures require walk-through or talk-through dry runs?
8. General supervision requirements, for example, what is protocol for transfer of supervisor responsibilities to a successor?
9. Responsibilities of higher supervision specified?

Reference Considerations

1. Applicable quality assurance and reliability standards considered?
2. Applicable codes, standards, and regulations considered?
3. Procedure complies with control documents?
4. Hazards and system safety degradations identified and considered against specific control manuals, standards, and procedures?
5. Specific prerequisite administrative and management approvals complied with?
6. Comments received from the people who will do the work?

Special Considerations

1. Has a documented safety analysis been considered for safety-related deviations from normal practices or for unusual or unpracticed maneuvers?
2. Have new restrictions or controls become effective that affect the procedure in such a manner that new safety analyses may be required?

74.4 HUMAN FACTORS ENGINEERING/ERGONOMICS*

74.4.1 Human–Machine Relationships

- Human factors engineering is defined as "the application of the principles, laws, and quantitative relationships which govern man's response to external stress to the analysis and design

---

*R. De Reamer, *Modern Safety and Health Technology*. Copyright © 1980. Reprinted by permission of Wiley, New York.

of machines and other engineering structures, so that the operator of such equipment will not be stressed beyond his/her proper limit or the machine forced to operate at less than its full capacity in order for the operator to stay within acceptable limits of human capabilities."*

- A principal objective of the supervisor and safety engineer in the development of safe working conditions is the elimination of bottlenecks, stresses and strains, and psychological booby traps that interfere with the free flow of work. It is an accepted concept that the less an operator has to fear from his or her job or machine, the more attention he or she can give to his or her work.

- In the development of safe working conditions, attention is given to many things, including machine design and machine guarding, personal protective equipment, plant layout, manufacturing methods, lighting, heating, ventilation, removal of air contaminants, and the reduction of noise. Adequate consideration of each of these areas will lead to a proper climate for accident prevention, increased productivity, and worker satisfaction.

- The human factors engineering approach to the solution of the accident problem is to build machines and working areas around the operator, rather than place him or her in a setting without regard to his or her requirements and capacities. Unless this is done, it is hardly fair to attribute so many accidents to human failure, as is usually the case.

- If this point of view is carried out in practice, fewer accidents should result, training costs should be reduced, and extensive redesign of equipment after it is put into use should be eliminated.

- All possible faults in equipment and in the working area, as well as the capacities of the operator, should be subjected to advance analysis. If defects are present, it is only a matter of time before some operator "fails" and has an accident.

- Obviously, the development of safe working conditions involves procedures that may go beyond the occasional safety appraisal or search for such obvious hazards as an oil spot on the floor, a pallet in the aisle, or an unguarded pinch point on a new lathe.

Human–machine relationships have improved considerably with increased mechanization and automation. Nevertheless, with the decrease in manual labor has come specialization, increased machine speeds, and monotonous repetition of a single task, which create work relationships involving several physiological and psychological stresses and strains. Unless this scheme of things is recognized and dealt with effectively, many real problems in the field of accident prevention may be ignored.

74.4.2 Human Factors Engineering Principles

- Human factors engineering or *ergonomics,*† as it is sometimes called, developed as a result of the experience in the use of highly sophisticated equipment in World War II. The ultimate potentialities of complex instruments of war could not be realized because the human operators lacked the necessary capabilities and endurance required to operate them. This discipline now has been extended to many areas. It is used extensively in the aircraft and aerospace industry and in many other industries to achieve more effective integration of humans and machines.

- The analysis should consider all possible faults in the equipment, in the work area, and in the worker—including a survey of the nature of the task, the work surroundings, the location of controls and instruments, and the way the operator performs his or her duties. The questions of importance in the analysis of machines, equipment, processes, plant layout, and the worker will vary with the type and purpose of the operation, but usually will include the following (pertaining to the worker):‡

  1. What sense organs are used by the operator to receive information? Does he or she move into action at the sound of a buzzer, blink of a light, reading of a dial, verbal order? Does the sound of a starting motor act as a cue?

---

*Theodore F. Hatch, Professor (retired), by permission.

†The term *ergonomics* was coined from the Greek roots *ergon* (work) and *nomos* (law, rule) and is now currently used to deal with the interactions between humans and such environmental elements as atmospheric contaminants, heat, light, sound, and all tools and equipment pertaining to the workplace.

‡R. A. McFarland, "Application of Human Factors Engineering to Safety Engineering Problems," *National Safety Congress Transactions,* 1967, Vol. 12. Permission granted by the National Safety Council.

2. What sort of discrimination is called for? Does the operator have to distinguish between lights of two different colors, tones of two different pitches, or compare two dial readings?

3. What physical response is he or she required to make: Pull a handle? Turn a wheel? Step on a pedal? Push a button?

4. What overall physical movements are required in the physical response? Do such movements interfere with his or her ability to continue receiving information through his or her sense organs? (For example, would pulling a handle obstruct his or her line of vision to a dial he or she is required to watch?) What forces are required (e.g., torque in turning a wheel)?

5. What are the speed and accuracy requirements of the machine? Is the operator required to watch two pointers to a hairline accuracy in a split second? Or is fairly close approximation sufficient? If a compromise is necessary, which is more essential: speed or accuracy?

6. What physiological and environmental conditions are likely to be encountered during normal operation of the machine? Are there any unusual temperatures, humidity conditions, crowded workspace, poor ventilation, high noise levels, toxic chemicals, and so on?

• Pertaining to the machine, equipment, and the surrounding area, these key questions should be asked:

1. Can the hazard be eliminated or isolated by a guard, ventilating equipment, or other device?

2. Should the hazard be identified by the use of color, warning signs, blinking lights, or alarms?

3. Should interlocks be used to protect the worker when he or she forgets or makes the wrong move?

4. Is it necessary to design the machine, the electrical circuit, or the pressure circuit so it will always be fail-safe?

5. Is there need for standardization?

6. Is there need for emergency controls, and are controls easily identified and accessible?

7. What unsafe conditions would be created if the proper operating sequence were not followed?

74.4.3 General Population Expectations*

• The importance of standardization and normal behavior patterns has been recognized in business and industry for many years. A standard tool will more likely be used properly than will a nonstandard one, and standard procedures will more likely be followed.

• People expect things to operate in a certain way and certain conditions to conform with established standards. These general population "expectations"—the way in which the ordinary person will react to a condition or stimulus—must not be ignored or workers will be literally trapped into making mistakes. A list of "General Population Expectations" follows:

1. Doors are expected to be at least 6 feet, 6 inches in height.

2. The level of the floor at each side of a door is expected to be the same.

3. Stair risers are expected to be of the same height.

4. It is a normal pattern for persons to pass to the left on motorways (some countries excluded).

5. People expect guardrails to be securely anchored.

6. People expect the hot-water faucet to be on the left side of the sink, the cold-water faucet on the right, and the faucet to turn to the left (counterclockwise) to let the water run and to the right to turn the water off.

7. People expect floors to be nonslippery.

8. Flammable solvents are expected to be found in labeled, red containers.

9. The force required to operate a lever, push a cart, or turn a crank is expected to go unchanged.

10. Knobs on electrical equipment are expected to turn clockwise for "on," to increase current, and counterclockwise for "off."

*R. De Reamer, *Modern Safety and Health Technology.* Copyright © 1980. Reprinted by permission of Wiley, New York.

11. For control of vehicles in which the operator is riding, the operator expects a control motion to the right or clockwise to result in a similar motion of his or her vehicle and vice versa.

12. Very large objects or dark objects imply "heaviness." Small objects or light-colored ones imply "lightness." Large heavy objects are expected to be "at the bottom." Small, light objects are expected to be "at the top."

13. Seat heights are expected to be at a certain level when a person sits down.

74.5   ENGINEERING CONTROLS FOR MACHINE TOOLS*

74.5.1   Basic Concerns

Machine tools (such as mills, lathes, shearers, punch presses, grinders, drills, and saws) provide an example of commonplace conditions where there is only a limited number of items of personal protective gear available for use. In such cases as these, the problem to be solved is not personal protective equipment versus engineering controls, but rather which engineering control(s) should be used to protect the machine operator. A summary of employee safeguards is contained in Table 74.2. The list of possible machinery-related injuries is presented in Section 74.10. There seem to be as many hazards created by moving machine parts as there are types of machines. Safeguards are essential for protecting workers from needless and preventable injuries.

A good rule to remember is: Any machine part, function, or process that may cause injury must be safeguarded. Where the operation of a machine or accidental contact with it can injure the operator or others in the vicinity, the hazard must be either controlled or eliminated.

Dangerous moving parts in these three basic areas need safeguarding:

- The point of operation: that point where work is performed on the material, such as cutting, shaping, boring, or forming of stock.
- Power transmission apparatus: all components of the mechanical system that transmit energy to the part of the machine performing the work. These components include flywheels, pulleys, belts, connecting rods, couplings, cams, spindles, chains, cranks, and gears.
- Other moving parts: all parts of the machine that move while the machine is working. These can include reciprocating, rotating, and transverse moving parts, as well as feed mechanisms and auxiliary parts of the machine.

A wide variety of mechanical motions and actions may present hazards to the worker. These can include the movement of rotating teeth, and any parts that impact or shear. These different types of hazardous mechanical motions and actions are basic to nearly all machines, and recognizing them is the first step toward protecting workers from the danger they present.

The basic types of hazardous mechanical motions and actions are:

Motions	Actions
Rotating (including in-running nip points)	Cutting
	Punching
Reciprocating	Shearing
Transverse	Bending

74.5.2   General Requirements

What must a safeguard do to protect workers against mechanical hazards? Engineering controls must meet these minimum general requirements:

- *Prevent Contact.* The safeguard must prevent hands, arms, or any other part of a worker's body from making contact with dangerous moving parts. A good safeguarding system eliminates the possibility of operators or workers placing their hands near hazardous moving parts.
- *Secure.* Workers should not be able to easily remove or tamper with the safeguard, because a safeguard that can easily be made ineffective is no safeguard at all. Guards and safety devices should be made of durable material that will withstand the conditions of normal use. They must be firmly secured to the machine.

---

*J. B. ReVelle, *Engineering Controls: A Comprehensive Overview.* Used by permission of the Merritt Company, Publisher, from T. S. Ferry, *Safety Management Planning,* copyright © 1982, The Merritt Company, Santa Monica, CA 90406.

Table 74.2   Summary of Employee Safeguards

To Protect	Personal Protective Equipment to Use	Engineering Controls to Use
Breathing	Self-contained breathing apparatus, gas masks, respirators, alarm systems	Ventilation, air-filtration systems, critical level warning systems, electrostatic precipitators
Eyes/face	Safety glasses, filtered lenses, safety goggles, face shield, welding goggles/helmets, hoods	Spark deflectors, machine guards
Feet/legs	Safety boots/shoes, leggings, shin guards	
Hands/arms/body	Gloves, finger cots, jackets, sleeves, aprons, barrier creams	Machine guards, lockout devices, feeding and ejection methods
Head/neck	Bump caps, hard hats, hair nets	Toe boards
Hearing	Ear muffs, ear plugs, ear valves	Noise reduction/isolation by equipment modification/substitution, equipment lubrication/maintenance programs, eliminate/dampen noise sources, reduce compressed air pressure, change operations[a]
Excessively high/low temperatures	Reflective clothing, temperature controlled clothing	Fans, air conditioning, heating, ventilation, screens, shields, curtains
Overall	Safety belts, lifelines, grounding mats, slap bars	Electrical circuit grounding, polarized plugs/outlets, safety nets

[a]Examples of the types of changes that should be considered include:
  • Grinding instead of chipping.
  • Electric tools in place of pneumatic tools.
  • Pressing instead of forging.
  • Welding instead of riveting.
  • Compression riveting over pneumatic riveting.
  • Mechanical ejection in place of air-blast ejection.
  • Wheels with rubber or composition tires on plant trucks and cars instead of all-metal wheels.
  • Wood or plastic tote boxes in place of metal tote boxes.
  • Use of an undercoating on machinery covers.
  • Wood in place of all-metal workbenches.

Machines often produce noise (unwanted sound), and this can result in a number of hazards to workers. Not only can it startle and disrupt concentration, but it can interfere with communications, thus hindering the worker's safe job performance. Research has linked noise to a whole range of harmful health effects, from hearing loss and aural pain to nausea, fatigue, reduced muscle control, and emotional disturbances. Engineering controls such as the use of sound-dampening materials, as well as less sophisticated hearing protection, such as ear plugs and muffs, have been suggested as ways of controlling the harmful effects of noise. Vibration, a related hazard that can cause noise and thus result in fatigue and illness for the worker, may be avoided if machines are properly aligned, supported, and, if necessary, anchored.

Because some machines require the use of cutting fluids, coolants, and other potentially harmful substances, operators, maintenance workers, and others in the vicinity may need protection. These substances can cause ailments ranging from dermatitis to serious illnesses and disease. Specially constructed safeguards, ventilation, and protective equipment and clothing are possible temporary solutions to the problem of machinery-related chemical hazards until these hazards can be better controlled or eliminated from the workplace.

- *Protect from Falling Objects.*   The safeguard should ensure that no objects can fall into moving parts. A small tool that is dropped into a cycling machine could easily become a projectile that could strike and injure someone.

- *Create No New Hazards.*   A safeguard defeats its own purpose if it creates a hazard of its own such as a shear point, a jagged edge, or an unfinished surface that can cause a laceration. The edges of guards, for instance, should be rolled or bolted in such a way that eliminates sharp edges.

- *Create No Interference.*   Any safeguard which impedes a worker from performing the job quickly and comfortably might soon be overridden or disregarded. Proper safeguarding can actually enhance efficiency, since it can relieve the worker's apprehensions about injury.

- *Allow Safe Lubrication.*   If possible, one should be able to lubricate the machine without removing the safeguards. Locating oil reservoirs outside the guard, with a line leading to the lubrication point, will reduce the need for the operator or maintenance worker to enter the hazardous area.

74.5.3   Danger Sources

All power sources for machinery are potential sources of danger. When using electrically powered or controlled machines, for instance, the equipment as well as the electrical system itself must be properly grounded. Replacing frayed, exposed, or old wiring will also help to protect the operator and others from electrical shocks or electrocution. High-pressure systems, too, need careful inspection and maintenance to prevent possible failure from pulsation, vibration, or leaks. Such a failure could cause explosions or flying objects.

74.6   MACHINE SAFEGUARDING METHODS*

74.6.1   General Classifications

There are many ways to safeguard machinery. The type of operation, the size or shape of stock, the method of handling the physical layout of the work area, the type of material, and production requirements or limitations all influence selection of the appropriate safeguarding method(s) for the individual machine.

As a general rule, power transmission apparatus is best protected by fixed guards that enclose the danger area. For hazards at the point of operation, where moving parts actually perform work on stock, several kinds of safeguarding are possible. One must always choose the most effective and practical means available.

1. Guards
    - (a) Fixed
    - (b) Interlocked
    - (c) Adjustable
    - (d) Self-adjusting
2. Devices
    - (a) Presence sensing
        - Photoelectrical (optical)
        - Radio frequency (capacitance)
        - Electromechanical
    - (b) Pullback
    - (c) Restraint
    - (d) Safety controls
        - Safety trip controls
            - Pressure-sensitive body bar
            - Safety triprod
            - Safety tripwire cable
        - Two-hand control
        - Two-hand trip

---

*J. B. ReVelle, *Engineering Controls: A Comprehensive Overview.* Used by permission of The Merritt Company, Publisher, from T. S. Ferry, *Safety Management Planning,* copyright © 1982, The Merritt Company, Santa Monica, CA 90406.

    (e)   Gates
           Interlocked
           Other
3. Location/distance
4. Potential feeding and ejection methods to improve safety for the operator
    (a) Automatic feed
    (b) Semiautomatic feed
    (c) Automatic ejection
    (d) Semiautomatic ejection
    (e) Robot
5. Miscellaneous aids
    (a) Awareness barriers
    (b) Miscellaneous protective shields
    (c) Hand-feeding tools and holding fixtures

74.6.2 Guards, Devices, and Feeding and Ejection Methods

Tables 74.3–74.5 provide the interested reader with specifics regarding machine safeguarding.

74.7 ALTERNATIVES TO ENGINEERING CONTROLS*

Engineering controls are an alternative to personal protective equipment, or is it the other way around? This chicken-and-egg situation has become an emotionally charged issue with exponents on both sides arguing their beliefs with little in the way of well-founded evidence to support their cases. The reason for this unfortunate situation is that there is no single solution to all the hazardous operations found in industry. The only realistic answer to the question concerning which of the two methods of abating personnel hazards is—it depends. Each and every situation requires an independent analysis considering all the known factors so that a truly unbiased decision can be reached.

This section presents material useful to engineers in the selection and application of solutions to industrial safety and health problems. Safety and health engineering control principles are deceptively few: substitution; isolation; and ventilation, both general and localized. In a technological sense, an appropriate combination of these strategic principles can be brought to bear on any industrial safety or hygiene control problem to achieve a satisfactory quality of the work environment. It may not be, and usually is not, necessary or appropriate to apply all these principles to any specific potential hazard. A thorough analysis of the control problem must be made to ensure that a proper choice from among these methods will produce the proper control in a manner that is most compatible with the technical process, is acceptable to the workers in terms of day-to-day operation, and can be accomplished with optimal balance of installation and operating expenses.

74.7.1 Substitution

Although frequently one of the most simple engineering principles to apply, substitution is often overlooked as an appropriate solution to occupational safety and health problems. There is a tendency to analyze a particular problem from the standpoint of correcting rather than eliminating it. For example, the first inclination in considering a vapor-exposure problem in a degreasing operation is to provide ventilation of the operation rather than consider substituting a solvent having a much lower degree of hazard associated with its use. However, substitution of less hazardous substances, changing from one type of process equipment to another, or, in some cases, even changing the process itself, may provide an effective control of a hazard at minimal expense.

This strategy is often used in conjunction with safety equipment: substituting safety glass for regular glass in some enclosures, replacing unguarded equipment with properly guarded machines, replacing safety gloves or aprons with garments made of material more impervious to the chemicals being handled. Since substitution of equipment frequently is done as an immediate response to an obvious problem, it is not always recognized as an engineering control, even though the end result is every bit as effective.

Substituting one process or operation for another may not be considered except in major modifications. In general, a change in any process from a batch to a continuous type of operation carries with it an inherent reduction in potential hazard. This is true primarily because the frequency and duration of potential contact of workers with the process materials are reduced when the overall

---

*J. B. ReVelle, *Engineering Controls: A Comprehensive Overview.* Used by permission of The Merritt Company, Publisher, from T. S. Ferry, *Safety Management Planning,* copyright © 1982, The Merritt Company, Santa Monica, CA 90406.

Table 74.3 Machine Safeguarding: Guards

Method	Safeguarding Action	Advantages	Limitations
Fixed	Provides a barrier	Can be constructed to suit many specific applications In-plant construction is often possible Can provide maximum protection Usually requires minimum maintenance Can be suitable to high production, repetitive operations	May interfere with visibility Can be limited to specific operations Machine adjustment and repair often requires its removal, thereby necessitating other means of protection for maintenance personnel
Interlocked	Shuts off or disengages power and prevents starting of machine when guard is open; should require the machine to be stopped before the worker can reach into the danger area	Can provide maximum protection Allows access to machine for removing jams without time-consuming removal of fixed guards	Requires careful adjustment and maintenance May be easy to disengage
Adjustable	Provides a barrier that may be adjusted to facilitate a variety of production operations	Can be constructed to suit many specific applications Can be adjusted to admit varying sizes of stock	Hands may enter danger area—protection may not be complete at all times May require frequent maintenance and/or adjustment The guard may be made ineffective by the operator May interfere with visibility
Self-adjusting	Provides a barrier that moves according to the size of the stock entering danger area	Off-the-shelf guards are often commercially available	Does not always provide maximum protection May interfere with visibility May require frequent maintenance and adjustment

Table 74.4 Machine Safeguarding: Devices

Method	Safeguarding Action	Advantages	Limitations
Photoelectric	Machine will not start cycling when the light field is interrupted When the light field is broken by any part of the operator's body during the cycling process, immediate machine braking is activated	Can allow freer movement for operator	Does not protect against mechanical failure May require frequent alignment and calibration Excessive vibration may cause lamp filament damage and premature burnout Limited to machines that can be stopped
Radio frequency (capacitance)	Machine cycling will not start when the capacitance field is interrupted When the capacitance field is disturbed by any part of the operator's body during the cycling process, immediate machine braking is activated	Can allow freer movement for operator	Does not protect against mechanical failure Antennae sensitivity must be properly adjusted Limited to machines that can be stopped
Electromechanical	Contact bar or probe travels a predetermined distance between the operator and the danger area Interruption of this movement prevents the starting of machine cycle	Can allow access at the point of operation	Contact bar or probe must be properly adjusted for each application; this adjustment must be maintained properly
Pullback	As the machine begins to cycle, the operator's hands are pulled out of the danger area	Eliminates the need for auxiliary barriers or other interference at the danger area	Limits movement of operator May obstruct workspace around operator Adjustments must be made for specific operations and for each individual Requires frequent inspections and regular maintenance Requires close supervision of the operator's use of the equipment
Restraint (holdback)	Prevents the operator from reaching into the danger area	Little risk of mechanical failure	Limits movements of operator May obstruct workspace Adjustments must be made for specific operations and each individual Requires close supervision of the operator's use of the equipment

Method	Safeguarding Action	Limitations
Safety trip controls Pressure-sensitive body bar Safety tripod Safety tripwire cable	Stops machine when tripped Simplicity of use	All controls must be manually activated May be difficult to activate controls because of their location Only protects the operator May require special fixtures to hold work May require a machine brake
Two-hand control	Concurrent use of both hands is required, preventing the operator from entering the danger area Operator's hands are at a predetermined location Operator's hands are free to pick up a new part after first half of cycle is completed	Requires a partial cycle machine with a brake Some two-hand controls can be rendered unsafe by holding with arm or blocking, thereby permitting one-hand operation Protects only the operator
Two-hand trip	Concurrent use of two hands on separate controls prevents hands from being in danger area when machine cycle starts Operator's hands are away from danger area Can be adapted to multiple operations No obstruction to hand feeding Does not require adjustment for each operation	Operator may try to reach into danger area after tripping machine Some trips can be rendered unsafe by holding with arm or blocking, thereby permitting one-hand operation Protects only the operator May require special fixtures
Gates Interlocked Other	Provides a barrier between danger area and operator or other personnel Can prevent reaching into or walking into the danger area	May require frequent inspection and regular maintenance May interfere with operator's ability to see the work

Table 74.5 Machine Safeguarding: Feeding and Ejection Methods

Method	Safeguarding Action	Advantages	Limitations
Automatic feed	Stock is fed from rolls, indexed by machine mechanism, etc.	Eliminates the need for operator involvement in the danger area	Other guards are also required for operator protection—usually fixed barrier guards Requires frequent maintenance May not be adaptable to stock variation
Semiautomatic feed	Stock is fed by chutes, movable dies, dial feed, plungers, or sliding bolster		
Automatic ejection	Workpieces are ejected by air or mechanical means		May create a hazard of blowing chips or debris Size of stock limits the use of this method Air ejection may present a noise hazard
Semiautomatic ejection	Workpieces are ejected by mechanical means, which are initiated by the operator	Operator does not have to enter danger area to remove finished work	Other guards are required for operator protection May not be adaptable to stock variation
Robots	Perform work usually done by operator	Operator does not have to enter danger area Are suitable for operations where high stress factors are present, such as heat and noise	Can create hazards themselves Require maximum maintenance Are suitable only to specific operations

process approach becomes one of continuous operation. The substitution of processes can be applied on a fundamental basis, for example, substitution of airless spray for conventional spray equipment can reduce the exposure of a painter to solvent vapors. Substitution of a paint dipping operation for the paint spray operation can reduce the potential hazard even further. In any of these cases, the automation of the process can further reduce the potential hazard (Table 74.5).

74.7.2 Isolation

Application of the principle of isolation is frequently envisioned as consisting of the installation of a physical barrier (such as a machine guard or device-refer to Tables 74.3 and 74.4) between a hazardous operation and the workers. Fundamentally, however, this isolation can be provided *without* a physical barrier through the appropriate use of distance and, in some situations, time.

Perhaps the most common example of isolation as a control strategy is associated with storage and use of flammable solvents. The large tank farms with dikes around the tanks, underground storage of some solvents, the detached solvent sheds, and fireproof solvent storage rooms within buildings are all commonplace in American industry. Frequently, the application of the principle of isolation maximizes the benefits of additional engineering concepts such as excessive noise control, remote control materials handling (as with radioactive substances), and local exhaust ventilation.

74.7.3 Ventilation

Workplace air quality is affected directly by the design and performance of the exhaust system. An improperly designed hood or a hood evacuated with an insufficient volumetric rate of air will contaminate the occupational environment and affect workers in the vicinity of the hazard source. This is a simple, but powerful, symbolic representation of one form of the close relationship between atmospheric emissions (as regulated by the Environmental Protection Agency) and occupational exposure (as regulated by the Occupational Safety and Health Administration). What is done with gases generated as a result of industrial operations/processes? These emissions can be exhausted directly to the atmosphere, indirectly to the atmosphere (from the workplace through the general ventilation system), or recirculated to the workplace. The effectiveness of the ventilation system design and operation impacts directly on the necessity and type of respiratory gear needed to protect the work force.

74.8 DESIGN AND REDESIGN*

74.8.1 Hardware

Designers of machines must consider the performance characteristics of machine operators as a major constraint in the creation or modification of both mechanical and electrical equipment. To do less would be tantamount to ignoring the limitations of human capabilities. Equipment designers especially concerned with engineering controls to be incorporated into machines, whether at the time of initial conceptualization or later when alterations are to be made, must also be cognizant of the principles of human factors (ergonomics). Equipment designers are aware that there are selected tasks that people can perform with greater skill and dependability than machines, and vice versa. Some of these positive performance characteristics are noted in Table 74.6. In addition, designers of equipment and engineering controls are knowledgeable of human performance limitations, both physically and psychologically. They know that the interaction of forces between people and their operating environment presents a never-ending challenge in assessing the complex interrelationships that provide the basis for that often fine line between safety versus hazard or health versus contaminant. Table 74.7 identifies the six pertinent sciences most closely involved in the design of machines and engineering controls.

It is both rational and reasonable to expect that, when engineering controls are being considered to eliminate or reduce hazards or contaminants, designers make full use of the principles established by specialists in these human performance sciences.

74.8.2 Process

A stress (or stressor) is some physical or psychological feature of the environment that requires an operator to be unduly exerted to continue performing. Such exertion is termed strain as in "stress and strain." Common physical stressors in industrial workplaces are poor illumination, excessive noise, vibration, heat, and the presence of excessive, harmful atmospheric contaminants.

Unfortunately, much less is known about their effects when they occur at the same time, in rapid sequence, or over extended periods of time. Research suggests that such effects are not simply

*J. B. ReVelle, *Engineering Controls: A Comprehensive Overview.* Used by permission of The Merritt Company, Publisher, from T. S. Ferry, *Safety Management Planning,* copyright © 1982, the Merritt Company, Santa Monica, CA 90406.

Table 74.6 Positive Performance Characteristics—Some Things Done Better by

People	Machines
Detect signals in high noise fields	Respond quickly to signals
Recognize objects under widely different conditions	Sense energies outside human range
Perceive patterns	Consistently perform precise, routine, repetitive operations
Sensitive to a wide variety of stimuli	
Long-term memory	Recall and process enormous amounts of data
Handle unexpected or low-probability events	Monitor people or other machines
Reason inductively	Reason deductively
Profit from experience	Exert enormous power
Exercise judgment	Relatively uniform performance
Flexibility, improvision, and creativity	Rapid transmission of signals
Select and perform under overload conditions	Perform several tasks simultaneously
Adapt to changing environment	Expendable
Appreciate and create beauty	Resistance to many environmental stresses
Perform fine manipulations	
Perform when partially impaired	
Relatively maintenance-free	

additive, but synergistic, thus compounding their detrimental effects. In addition, when physical work environments are unfavorable to equipment operators, two or more stressors are generally present: high temperature and excessive noise, for example. The solution to process design and redesign is relatively easy to specify, but costly to implement—design the physical environment so that all physical characteristics are within an acceptable range.

Marketed in the United States since the early 1960s, industrial robots offer both hardware and process designers a technology that can be used when hazardous or uncomfortable working conditions are expected or already exist. Where a job situation poses potential dangers or the workplace is hot or in some other way unpleasant, a robot should be considered as a substitute for human operators. Hot forging, die casting, and spray painting fall into this category. If workparts or tools are awkward or heavy, an industrial robot may fill the job. Some robots are capable of lifting items weighing several hundred pounds.

An industrial robot is a general purpose, programmable machine that possesses certain humanlike capabilities. The most obvious characteristic is the robot's arm, which, when combined with the robot's capacity to be programmed, makes it ideally suited to a variety of uncomfortable/undesirable production tasks. Hardware and process designers now possess an additional capability for potential inclusion in their future designs and redesigns.

74.9 PERSONAL PROTECTIVE EQUIPMENT*

74.9.1 Background

Engineering controls, which eliminate the hazard at the source and do not rely on the worker's behavior for their effectiveness, offer the best and most reliable means of safeguarding. Therefore, engineering controls must be first choice for eliminating machinery hazards. But whenever an extra measure of protection is necessary, operators must wear protective clothing or personal protective equipment.

If it is to provide adequate protection, the protective clothing and equipment selected must always be:

- Appropriate for the particular hazards
- Maintained in good condition
- Properly stored when not in use, to prevent damage or loss
- kept clean and sanitary

---

*J. B. ReVelle, *Engineering Controls: A Comprehensive Overview.* Used by permission of The Merritt Company, Publisher, from T. S. Ferry, *Safety Management Planning,* copyright © 1982, The Merritt Company, Santa Monica, CA 90406.

Table 74.7 People Performance Sciences

Anthropometry	Pertains to the measurement of physical features and characteristics of the static human body.
Biomechanics	A study of the range, strength, endurance, and accuracy of movements of the human body.
Ergonomics	Human factors engineering-especially biomechanics aspects.
Human factors engineering	Designing for human use.
Kinesiology	A study of the principles of mechanics and anatomy of human movement.
Systems safety engineering	Designing that considers the operator's qualities, the equipment, and the environment relative to successful task performance.

Protective clothing is available for every part of the human body. Hard hats can protect the head from falling objects when the worker is handling stock; caps and hair nets can help keep the worker's hair from being caught in machinery. If machine coolants could splash, or particles could fly into the operator's eyes or face, then face shields, safety goggles, glasses, or similar kinds of protection must be used. Hearing protection may be needed when workers operate noisy machinery. To guard the trunk of the body from cuts or impacts from heavy or rough-edged stock, there are certain protective coveralls, jackets, vests, aprons, and full-body suits. Workers can protect their hands and arms from the same kinds of injury with special sleeves and gloves. And safety shoes and boots, or other acceptable foot guards, can shield the feet against injury in case the worker needs to handle heavy stock which might drop.

It is important to note that protective clothing and equipment themselves can create hazards. A protective glove which can become caught between rotating parts, or a respirator facepiece which hinders the wearer's vision require alertness and careful supervision whenever they are used.

Other aspects of the worker's dress may present additional safety hazards. Loose-fitting clothing might possibly become entangled in rotating spindles or other kinds of moving machinery. Jewelry, such as bracelets and rings, can catch on machine parts or stock and lead to serious injury by pulling a hand into the danger area.

Naturally, each situation will vary. In some simple cases, respirators, chemical goggles, aprons, and gloves may be sufficient personal protective equipment to afford the necessary coverage. In more complicated situations, even the most sophisticated equipment may not be enough and engineering controls would become mandatory. Safety, industrial, and plant engineers should be expected to provide the necessary analyses to ascertain the extent of the hazard to employees whose work causes them to be exposed to the corrosive fumes.

74.9.2 Planning and Implementing the Use of Protective Equipment*

This section reviews ways to help plan, implement, and maintain personal protective equipment. This can be considered in terms of the following nine phases: (1) need analysis, (2) equipment selection, (3) program communication, (4) training, (5) fitting and adjustment, (6) target date setting, (7) break-in period, (8) enforcement, and (9) follow-through.

The first phase of promoting the use of personal protective equipment is called *need analysis*. Before selecting protective equipment, the hazards or conditions the equipment must protect the employee from must be determined. To accomplish this, questions such as the following must be asked:

- What standards does the law require for this type of work in this type of environment?
- What needs do our accident statistics point to?
- What hazards have we found in our safety and/or health inspections?
- What needs show up in our job analysis and job observation activities?
- Where is the potential for accidents, injuries, illnesses, and damage?
- Which hazards cannot be eliminated or segregated?

*J. B. ReVelle and Joe Stephenson, *Safety Training Methods,* Copyright © 1995. Reprinted by permission of Wiley, New York.

The second phase of promoting the use of protective equipment is *equipment selection.* Once a need has been established, proper equipment must be selected. Basic consideration should include the following:

- Conformity to the standards
- Degree of protection provided
- Relative cost
- Ease of use and maintenance
- Relative comfort

The third phase is *program communication.* It is not appropriate to simply announce a protective equipment program, put it into effect, and expect to get immediate cooperation. Employees tend to resist change unless they see it as necessary, comfortable, or reasonable. It is helpful to use various approaches to publicity and promotion to teach employees why the equipment is necessary. Various points can be covered in supervisor's meetings, in safety meetings, by posters, on bulletin boards, in special meetings, and in casual conversation. Gradually, employees will come to expect or to request protective equipment to be used on the job. The main points in program communication are to educate employees in why protective equipment is necessary and to encourage them to want it and to use it.

Training is an essential step in making sure protective equipment will be used properly. The employees should learn why the equipment is necessary, when it must be used, who must use it, where it is required, what the benefits are, and how to use it and take care of it. Do not forget that employee turnover will bring new employees into the work area. Therefore, you will continually need to train new employees in the use of the protective equipment they will handle.

After the training phase comes the *fitting and adjustment* phase. Unless the protective equipment fits the individual properly, it may not give the necessary protection. There are many ways to fit or to adjust protective equipment. For example, face masks have straps that hold them snug against the contours of the head and face and prevent leaks; rubberized garments have snaps or ties that can be drawn up snugly, to keep loose and floppy garments from getting caught in machinery.

The next phase is *target date setting.* After the other phases have been completed, set specific dates for completion of the various phases. For example, all employees shall be fitted with protective equipment before a certain date; all training shall be completed by a certain date; after a certain date, all employees must wear their protective equipment while in the production area.

After setting the target dates, expect a *break-in period.* There will usually be a period of psychological adjustment whenever a new personal protective program is established. Remember two things:

- Expect some gripes, grumbles, and problems.
- Appropriate consideration must be given to each individual problem; then strive toward a workable solution.

It might also be wise to post signs that indicate the type of equipment needed. For example, a sign might read, "Eye protection must be worn in this area."

After the break-in phase comes *enforcement.* If all the previous phases were successful, problems in terms of enforcement should be few. In case disciplinary action is required, sound judgment must be used and each case must be evaluated on an individual basis.

If employees fail to use protective equipment, they may be exposed to hazards. Do not forget, the employer can be penalized if employees do not use their protection.

The final phase is *follow-through.* Although disciplinary action may sometimes be necessary, positive motivation plays a more effective part in a successful protective equipment program. One type of positive motivation is a proper example set by management. Managers must wear their protective equipment, just as employees are expected to wear theirs.

74.9.3 Adequacy, Maintenance, and Sanitation

Before selling safety shoes and supplying safety goggles at a company store, the attendants must be guided by a well-structured program of equipment maintenance, preferably preventive maintenance.

Daily maintenance of different types of equipment might include: adjustment of the suspension system on a safety hat; cleaning of goggle lenses, glasses, or spectacles; scraping residue from the sole of a safety shoe; or proper adjustment of a face mask when donning an air-purifying respirator.

Performing these functions should be coupled with periodic inspections for weaknesses or defects in the equipment. How often this type of check is made, of course, depends on the particular type of equipment used. For example, sealed-canister gas masks should be weighed on receipt from the manufacturer, and the weight should be marked indelibly on each canister. Stored units should then be reweighed periodically, and those exceeding a recommended weight should be discarded even though the seal remains unbroken.

Sanitation, as spelled out in the OSHAct, is a key part of any operation, and it requires the use of personal protective equipment, not only to eliminate cross-infection among users of the same unit of equipment, but because unsanitary equipment is objectionable to the wearer.

Procedures and facilities that are necessary to sanitize or disinfect equipment can be an integral part of an equipment maintenance program. For example, the OSHAct says, "Respirators used routinely shall be inspected during cleaning." Without grime and dirt to hinder an inspection, gauges can be read better, rubber or elastomer parts can be checked for pliability and signs of deterioration, and valves can be checked.

74.10 MANAGING THE SAFETY FUNCTION

74.10.1 Supervisor's Role

The responsibilities of the first-line supervisor are many. Direction of the work force includes the following supervisory functions:

- Setting goals
- Improving present work methods
- Delegating work
- Allocating manpower
- Meeting deadlines
- Controlling expenditures
- Following progress of work
- Evaluating employee performance
- Forecasting manpower requirements
- Supervising on-the-job training
- Reviewing employee performance
- Handling employee complaints
- Enforcing rules
- Conducting meetings
- Increasing safety awareness*

Supervisory understanding of the interrelationships of these responsibilities is a learned attribute. Organizations that expect their supervisors to offer a high quality of leadership to their employees must provide appropriate training and experiential opportunities to current supervisors and supervisory trainees alike.

74.10.2 Elements of Accident Prevention†

- Safety policy must be clearly defined and communicated to all employees.
- The safety record of a company is a barometer of its efficiency. An American Engineering Council study revealed "maximum productivity is ordinarily secured only when the accident rate tends toward the unreducible minimum.
- Unless line supervisors are accountable for the safety of all employees, no safety program will be effective. Top management must let all supervisors and managers know what is expected of them in safety.
- Periodic progress reports are required to let managers and employees know what they have accomplished in safety.
- Meetings with supervisors and managers to review accident reports, compensation costs, accident-cause analysis, and accident-prevention procedures are important elements of the overall safety program.
- The idea of putting on a big safety campaign with posters, slogans, and safety contests is wrong. The Madison Avenue approach does not work over the long run.

*B. D. Lewis, Jr., *The Supervisor in 1975,* copyright © September 1973. Reprinted with permission of *Personnel Journal,* Costa Mesa, CA; all rights reserved.

†R. De Reamer, *Modern Safety and Health Technology.* Copyright © 1980. Reprinted by permission of Wiley, New York.

- Good housekeeping and the enforcement of safety rules show that management has a real concern for employee welfare. They are important elements in the development of good morale. (A U.S. Department of Labor study has revealed that workers are vitally concerned with safety and health conditions of the workplace. A surprisingly high percentage of workers ranked protection against work-related injuries and illness and pleasant working conditions as having a priority among their basic on-the-job needs. In fact, they rated safety higher than fringe benefits and steady employment.)
- The use of personal protective equipment (safety glasses, safety shoes, hard hats, etc.) must be a condition of employment in all sections of the plant where such protection is required.
- Safety files must be complete and up to date to satisfy internal information requirements as well as external inspections by OSHA Compliance Officers and similar officials (Table 74.8).

74.10.3 Management Principles*

- Regardless of the industry or the process, the role of supervisors and managers in any safety program takes precedence over any of the other elements. This is not to say that the managerial role is necessarily more important than the development of safe environments, but without

Table 74.8   Requirements for Safety Files[a]

The following items are presented for your convenience as you review your administrative storage index to determine the adequacy of your safety-related files.

Number	Action Required	Action Completed	
		Yes	No
1. Is there a separate section for safety-related files?			
2. Are the following subjects provided for in the safety section of the files:			
a. Blank OSHA forms?			
b. Completed OSHA forms?			
c. Blank company safety forms?			
d. Completed company safety forms?			
e. Blank safety checklists?			
f. Completed safety checklists?			
g. Agendas of company safety meetings?			
h. Minutes of company safety meetings?			
i. Records of safety equipment purchases?			
j. Records of safety equipment checkouts?			
k. Incoming correspondence related to safety?			
l. Outgoing correspondence related to safety?			
m. Record of safety projects assigned?			
n. Record of safety projects completed?			
o. Record of fire drills (if applicable)?			
p. Record of external assistance used to provide specialized safety expertise?			
q. Record of inspections by fire department, insurance companies, state and city inspectors, and OSHA compliance officers?			
r. National Safety Council catalogs and brochures for films, posters, and other safety-related materials?			
3. Are the files listed in item 2 reviewed periodically:			
a. To ensure that they are current?			
b. To retire material over five years old?			
4. Are safety-related files reviewed periodically to determine the need to eliminate selected files and to add new subjects?			
5. Is the index to the file current, so that an outsider could easily understand the system?			

[a] J. B. ReVelle and Joe Stephenson, *Safety Training Methods,* 2nd ed. Copyright © 1995. Reprinted by permission of Wiley, New York.

*R. De Reamer, *Modern Safety and Health Technology.* Copyright © 1980. Reprinted by permission of Wiley, New York.

manager and supervisor participation, the other elements have a lukewarm existence. There is a dynamic relationship between management and the development of safe working conditions, and management and the development of safety awareness, and the relationship must not be denied.

- Where responsibility for preventing accidents and providing a healthful work environment is sloughed off to the safety department or a safety committee, any reduction in the accident rate is minimal. To reduce the accident rate, and in particular, to make a good rate better, line managers must be held responsible and accountable for safety. Every member of the management team must have a role in the safety program. Admittedly, this idea is not new, but application of the concept still requires crystal-clear definition and vigorous promotion.

- Notwithstanding the many excellent examples of outstanding safety records that have been achieved because every member of management had assumed full responsibility for safety, there are still large numbers of companies, particularly the small establishments, using safety contests, posters, or safety committees as the focal point of their safety programs—but with disappointing results. Under such circumstances safety is perceived as an isolated aspect of the business operation with rather low ceiling possibilities at best. But there are some who feel that gimmicks must be used because foremen and the managers do not have time for safety.

- As an example of the case in point, a handbook on personnel contains the statement that "A major disadvantage of some company-sponsored safety programs is that the supervisor can't spare sufficient time from his regular duties for running the safety program." Significantly, this was not a casual comment in a chapter on safety. It was indented and in bold print for emphasis. Yet it is a firmly accepted fact that to achieve good results in safety, managers and supervisors *must* take the time to fulfill their safety responsibilities. Safety is one of their *regular* duties.

- The interrelationships of the many components of an effective industrial safety program are portrayed in Fig. 74.2.

74.10.4 Eliminating Unsafe Conditions*

The following steps should be taken to effectively and efficiently eliminate an unsafe condition:

- *Remove.* If at all possible, have the hazard eliminated.
- *Guard.* If danger point (i.e., high tension wires) cannot be removed, see to it that hazard is shielded by screens, enclosures, or other guarding devices.
- *Warn.* If guarding is impossible or impractical, warn of the unsafe condition. If a truck must back up across a sidewalk to a loading platform, the sidewalk cannot be removed or a fence built around the truck. All that can be done is to warn that an unsafe condition exists. This is done by posting a danger sign or making use of a bell, horn, whistle, signal light, painted striped lines, red flag, or other device.
- *Recommend.* If you cannot remove or guard an unsafe condition on your own, notify the proper authorities about it. Make specific recommendations as to how the unsafe condition can be eliminated.
- *Follow Up.* After a reasonable length of time, check to see whether the recommendation has been acted on, or whether the unsafe condition still exists. If it remains, the person or persons to whom the recommendations were made should be notified.

The following factors should be considered in organizing a plant that provides for maximum productivity and employee well-being:

- The general arrangement of the facility should be efficient, orderly, and neat.
- Workstations should be clearly identified so that employees can be assigned according to the most effective working arrangement.
- Material flow should be designed to prevent unnecessary employee movement for given work.
- Materials storage, distribution, and handling should be routinized for efficiency and safety.
- Decentralized tool storage should be used wherever possible. Where centralized storage is essential (e.g., general supply areas, locker areas, and project storage areas), care should be

---

*J. B. ReVelle and Joe Stephenson, *Safety Training Methods,* 2nd ed. Copyright © 1995. Reprinted by permission of Wiley, New York.

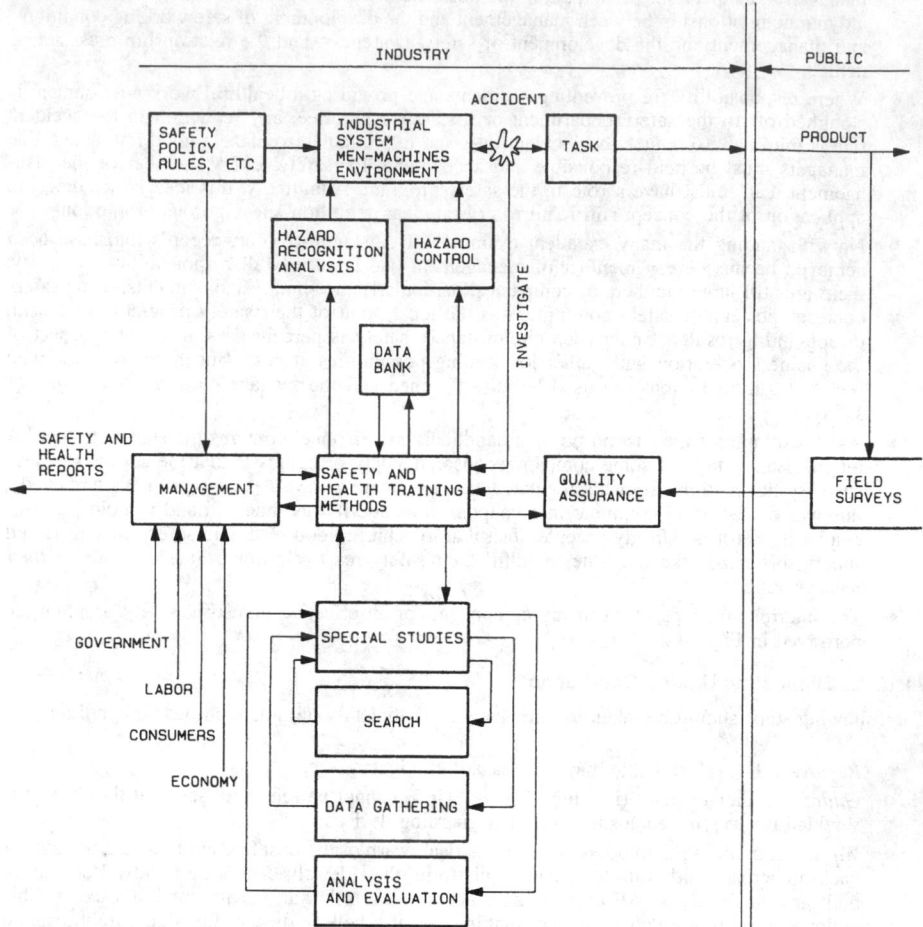

Fig. 74.2 Basic functions of an effective safety program. Reprinted with permission from *Industrial Engineering Magazine.* Copyright © 1979 American Institute of Industrial Engineers, Inc., 25 Technology Park/Atlanta, Norcross, GA 30092.

given to establish a management system that will avoid unnecessary crowding or congested traffic flow. (Certain procedures, such as time staggering, may reduce congestion.)

- Time-use plans should be established for frequently used facilities to avoid having workers wait for a particular apparatus.
- A warning system and communications network should be established for emergencies such as fire, explosion, storm, injuries, and other events that would affect the well-being of employees.

The following unsafe conditions checklist presents a variety of undesirable characteristics to which both employers and employees should be alert:

- *Unsafe Conditions—Mechanical Failure.* These are types of unsafe conditions that can lead to occupational accidents and injuries. *Note:* Keep in mind that unsafe conditions often come about as a result of unsafe acts.
- *Lack of Guards.* This applies to hazardous places like platforms, catwalks, or scaffolds where no guardrails are provided; power lines or explosive materials that are not fenced off or enclosed in some way; and machines or other equipment where moving parts or other danger points are not safeguarded.

- *Inadequate Guards.*   Often a hazard that is partially guarded is more dangerous than it would be if there were no guards. The employee, seeing some sort of guard, may feel secure and fail to take precautions that would ordinarily be taken if there were no guards at all.
- *Defects.*   Equipment or materials that are worn, torn, cracked, broken, rusty, bent, sharp, or splintered; buildings, machines, or tools that have been condemned or are in disrepair.
- *Hazardous Arrangement (Housekeeping).*   Cluttered floors and work areas; improper layout of machines and other production facilities; blocked aisle space or fire exits; unsafely stored or piled tools and material; overloaded platforms and vehicles; inadequate drainage and disposal facilities for waste products.
- *Improper Illumination.*   Insufficient light; too much light; lights of the wrong color; glare; arrangement of lighting systems that result in shadows and too much contrast.
- *Unsafe Ventilation.*   Concentration of vapors, dusts, gases, fumes; unsuitable capacity, location, or arrangement of ventilation system; insufficient air changes, impure air source used for air changes; abnormal temperatures and humidity.

In describing conditions for each item to be inspected, terms such as the following should be used:

Broken	Leaking
Corroded	Loose (or slipping)
Decomposed	Missing
Frayed	Rusted
Fuming	Spillage
Gaseous	Vibrating
Jagged	

An alphabetized listing of possible problems to be inspected is presented in Table 74.9.

Hazard Classification

It is important to differentiate the *degrees of severity* of different hazards. The commonly used standards are given below.

- *Class A Hazard.*   Any condition or practice with *potential* for causing *loss* of life or body part and/or extensive loss of structure, equipment, or material.
- *Class B Hazard.*   Any condition or practice with *potential* for causing serious injury, illness, or property damage, but less severe than Class A.
- *Class C Hazard.*   Any condition or practice with *probable potential* for causing *nondisabling* injury or illness, or *nondisruptive* property damage.

74.10.5 Unsafe Conditions Involving Mechanical or Physical Facilities*

The total working environment must be under constant scrutiny because of changing conditions, new employees, equipment additions and modifications, and so on. The following checklist is presented as a guide to identify potential problems:

1. Building
   Correct ceiling height
   Correct floor type; in acceptable condition
   Adequate illumination
   Adequate plumbing and heating pipes and equipment
   Windows with acceptable opening, closing, and holding devices; protection from breakage
   Acceptable size doors with correct swing and operational quality
   Adequate railing and nonslip treads on stairways and balconies
   Adequate ventilation
   Adequate storage facilities
   Adequate electrical distribution system in good condition
   Effective space allocation
   Adequate personal facilities (restrooms, drinking fountains, washup facilities, etc.)
   Efficient traffic flow

---

*J. B. ReVelle and Joe Stephenson, *Safety Training Methods,* 2nd ed. Copyright © 1995. Reprinted by permission of Wiley, New York.

Table 74.9 List of Possible Problems to Be Inspected[a]

Acids	Dusts	Railroad cars
Aisles	Electric motors	Ramps
Alarms	Elevators	Raw materials
Atmosphere	Explosives	Respirators
Automobiles	Extinguishers	Roads
Barrels	Flammables	Roofs
Bins	Floors	Safety devices
Blinker lights	Forklifts	Safety glasses
Boilers	Fumes	Safety shoes
Borers	Gas cylinders	Scaffolds
Buggies	Gas engines	Shafts
Buildings	Gases	Shapers
Cabinets	Hand tools	Shelves
Cables	Hard hats	Sirens
Carboys	Hoists	Slings
Catwalks	Hoses	Solvents
Caustics	Hydrants	Sprays
Chemicals	Ladders	Sprinkler systems
Claxons	Lathes	Stairs
Closets	Lights	Steam engines
Connectors	Mills	Sumps
Containers	Mists	Switches
Controls	Motorized carts	Tanks
Conveyors	Piping	Trucks
Cranes	Pits	Vats
Crossing lights	Platforms	Walkways
Cutters	Power tools	Walls
Docks	Presses	Warning devices
Doors	Racks	

[a]*Principles and Practices of Occupational Safety and Health: A Programmed Instruction Course,* OSHA 2213, Student Manual Booklet 1, U.S. Department of Labor, Washington, DC, p. 40.

    Adequate functional emergency exits
    Effective alarms and communications systems
    Adequate fire prevention and extinguishing devices
    Acceptable interior color scheme
    Acceptable noise absorption factor
    Adequate maintenance and cleanliness

2. Machinery and Equipment
    Acceptable placement, securing, and clearance
    Clearly marked safety zones
    Adequate nonskid flooring around machines
    Adequate guard devices on all pulleys
    Sharp, secure knives and cutting edges
    Properly maintained and lubricated machines, in good working condition
    Functional, guarded, magnetic-type switches on all major machines
    Properly wired and grounded machines
    Functional hand and portable power tools, in good condition and grounded
    Quality machines adequate to handle the expected work load
    Conspicuously posted safety precautions and rules near each machine
    Guards for all pinch points within 7 ft of the floor

74.11 SAFETY TRAINING

74.11.1 Specialized Courses*

First-Aid Training

First-aid courses pay big dividends in industry. This statement is based on clear evidence that people trained in first aid are more safety conscious and less likely to have an accident.

The importance of first-aid training from the safety standpoint is that it teaches much more than applying a bandage or a splint. According to the Red Cross, "The primary purpose of first aid training is the prevention of accidents." Each lesson teaches the student to analyze (1) how the accident happened, (2) how the accident could have been prevented, and (3) how to treat the injury. But the biggest dividend of first-aid training is the lives that have been saved because trainees were prepared to apply mouth-to-mouth resuscitation, to stop choking using the Heimlich maneuver (ejection of foreign object by forceful compression of diaphragm), or to stem the flow of blood.

Since the OSHAct, first-aid training has become a matter of federal law—the act stipulates that in absence of an infirmary, clinic, or hospital in proximity to the workplace, a person or persons shall be adequately trained to render first aid. The completion of the basic American National Red Cross first-aid course will be considered as having met this requirement. Just what constitutes *proximity* to a clinic or hospital? The OSH Review Commission recognizing that first aid must be given within 3 minutes of serious accidents concluded that an employer whose plant had no one trained in first aid present and was located 9 minutes from the nearest hospital violated the standard (1910.151, Medical and First Aid).

Driver Training

The number 1 accident killer of *employees* is the traffic accident. Each year more than 27,000 workers die in non-work-related motor-vehicle accidents, and an additional 3900 employees are killed in work-related accidents. The employer pays a heavy toll for these accidents. Those that are work related are compensable, but the others are, nonetheless, costly. The loss of a highly skilled worker, a key scientist, or a company executive could have a serious impact on the success of the business.

There is, fortunately, something constructive that employers can do to help protect their employees and their executives from the tragedy and waste of traffic accidents. Driver training for workers and executives can be provided either in-house or through community training agencies.

Companies that have conducted driver-training programs report that the benefits of such training were not limited to the area of improved traffic-accident performance. These companies also experienced lower on-the-job injury frequency rates (the training produced an increase in safety awareness) and improved employee–community relations.

Companies have taken several approaches to driver training:

- A course has been made available to employees on a volunteer basis, either on- or off-hours.
- Driver training has been made mandatory for employees who operate a motor vehicle on company business.
- The company has promoted employee attendance at community-agency-operated programs.
- Full-scale driver-training programs have been conducted for all employees and members of their families. This is done off-hours, and attendance is voluntary.

Fire Protection Training

All employees must know what to do when a fire alarm sounds. All employees must know something about the equipment provided for fire protection and what they can do toward preventing a fire. They must know:

- The plan established for evacuation of the building in case of emergency.
- How to use the first-aid fire appliances provided (extinguishers, hose, etc.).
- How to use other protective equipment. (Every employee should know that water to extinguish fires comes out of the pipes of the sprinkler systems and that stock must not be piled so close to sprinkler lines that it prevents good distribution of water from sprinkler heads on a fire in the piled material. They should know that fire doors must be kept operative and not obstructed by stock piles, tools, or other objects.)

---

*R. De Reamer, *Modern Safety and Health Technology*. Copyright © 1980. Reprinted by permission of Wiley, New York.

- How to give a fire alarm and how to operate plant fire alarm boxes and street boxes in the public alarm system.
- Where smoking in the plant is permitted and where, for fire-safety reasons, it is prohibited.
- The housekeeping routine (disposal of wiping rags and waste, handling of packing materials, and other measures for orderliness and cleanliness throughout the plant).
- Hazards of any special process in which the employee is engaged.

All these "what-to-do" items can appropriately be covered in training sessions and evacuation drills.

Other Specialized Courses

Some of the other specialized courses that can be given for safety training are:

- Accident investigation
- Accident report preparation
- Hazard inspection
- Personal protective equipment
- Powered equipment and vehicles
- Safety recordkeeping
- Specific disasters

74.11.2 Job Hazard Analysis Training*

Admittedly, the conventional mass approach to safety training takes little of the supervisor's time. Group training sessions, safety posters, films, and booklets are handled by the plant safety engineer or other staff people. On the other hand, where safety training is carried out on a personalized basis, the first-line supervisor must necessarily do the training. This will take more of his or her time and require more attention to detail, but this additional effort pays off because of the increased effectiveness of the training method.

In launching a personalized safety-training program, the first step is the preparation of a job-hazard analysis for each job in the plant. To make the job-hazard analysis in an organized manner, use of a form similar to the one shown in Table 74.10 is suggested. The key elements of the form are: (1) job description; (2) job location; (3) key job steps; (4) tool used; (5) potential health and injury hazards; and (6) safe practices, apparel, and equipment.

A review of the form will indicate the steps in making a job-hazard analysis. To start an analysis, the key steps of the job are listed in order in the first column of the form. Where pertinent, the tool used to perform the job step is listed in the second column. Then, in the third column opposite each job step, the hazards of the particular step are indicated. Finally, in the fourth column of the form are listed the safe practices the employee must be shown and have discussed. Here the supervisor lists the safe work habits that must be stressed and the safety equipment and clothing required for the job.

In making the analysis, an organized approach is required so the less obvious accident hazards will not be missed. This means going out on the floor and actually watching the job being performed and jotting down key steps and hazards as they are observed. Supervisors who make such a job-hazard analysis are often surprised to find hazards in the job cycle that they had missed seeing in the past. Their original negative reaction to the thought of additional paperwork soon disappears. In the long run, supervisors realize that proper hazard analysis will help them do a better training job.

As previously stated, a job-hazard analysis is made for each job. In most cases, each supervisor will have to make from 5 to 10 different analyses. Of course, in maintenance and construction work, the variety of jobs covers a much wider range. Fortunately, these jobs can be grouped by the type of work performed and a job-hazard analysis can be made for each category of work, rather than for each job. For example, repair, installation, and relocation of equipment; cleaning motors; and unloading cars might be a few of the various categories of maintenance work to be analyzed.

74.11.3 Management's Overview of Training†

An effective accident prevention program requires proper job performance from everyone in the workplace. All employees must know about the materials and equipment they are working with, what

---

*R. De Reamer, *Modern Safety and Health Technology.* Copyright © 1980. Reprinted by permission of Wiley, New York.

†J. B. ReVelle and Joe Stephenson, *Safety Training Methods,* 2nd ed. Copyright © 1995. Reprinted by permission of Wiley, New York.

Table 74.10 Job Hazard Analysis

Job Description: Three Spindle Drill Press—Impeller 34C6
Job Location: Bldg. 19-2, Pump Section

Key Job Steps	Tool Used	Potential Health and Injury Hazard	Safe Practices, Apparel, and Equipment
Get material from operation	Tote box	Dropping tote box on foot.	Wear safety shoes. Have firm grip on box.
		Back strain from lifting.	Stress proper lifting methods.
		Picking up overloaded boxes.	Tell employee to get help or lighten load.
Inspect and set up drill press	Drill press	Check for defective machines.	Do not operate if defective. Attach red or yellow "do not operate" tag.
		Chuck wrench not removed.	Always remove chuck wrench immediately after use.
		Making adjustments when machine is running.	Always stop spindle before making adjustments.
Drilling		Hair, clothing, or jewelry catching on spindle.	Wear head covering, snug-fitting clothing. No loose sleeves. Avoid wearing rings, bracelets, or wristwatches.
		Spinning work or fixture.	Use proper blocks or clamps to hold work and fixture securely.
		Injury to hands—cuts, etc.	Never wear gloves. Use hook, brush, or other tool to remove chips. Use compressed air only when instructed.
		Drill sticks in work.	Stop spindle, free drill by hand.
		Flying chips.	Wear proper eye protection.
		Pinch points at belts.	Always stop press before adjusting belts.
		Broken drills.	Do not attempt to force drill, apply pressure.

James Black
Signature

4/22/97	1 of 3
Date	Page

hazards are known in the operation, and how these hazards have been controlled or eliminated. Each individual employee needs to know and understand the following points (especially if they have been included in the company policy and in a "code of safe practices"):

No employee is expected to undertake a job until he or she has received instruction on how to do it properly and has been authorized to perform that job.

No employee should undertake a job that appears to be unsafe.

Mechanical safeguards are in place and must be kept in place.

Each employee is expected to report all unsafe conditions encountered during work.

Even slight injury or illness suffered by an employee must be reported at once.

In addition to the points above, any safety rules that are a condition of employment, such as the use of safety shoes or eye protection, should be explained clearly and enforced at once.

The first-line supervisors must know how to train employees in the proper way of doing their jobs. Encourage and consider providing for supervisory training for these supervisors. (Many colleges offer appropriate introductory management training courses.)

Some specific training requirements in the OSHA standards must be met, such as those that pertain to first aid and powered industrial trucks (including forklifts). In general, they deal with situations where the use of untrained or improperly trained operators on skill machinery could cause hazardous situations to develop, not only for the operator, but possibly for nearby workers, too.

Particular attention must be given to new employees. Immediately on arriving at work, new employees begin to learn things and to form attitudes about the company, the job, their boss, and their fellow employees. Learning and attitude formation occur regardless of whether the employer makes a training effort. If the new employees are trained during those first few hours and days to do things the right way, considerable losses may be avoided later.

At the same time, attention must be paid to regular employees, including the old-timers. Old habits can be wrong habits. An employee who continues to repeat an unsafe procedure is not working safely, even if an "accident" has not resulted from this behavior.

Although every employee's attitude should be one of determination that "accidents" can be prevented, one thing more may be needed. It should be stressed that the responsibility assigned to the person in charge of the job—as well as to all other supervisors—is to be sure that there is a concerted effort under way at all times to follow every safe work procedure and health practice applicable to that job. It should be clearly explained to these supervisors that they should never silently condone unsafe or unhealthful activity in or around any workplace.

BIBLIOGRAPHY

"Accident Prevention: Your Key to Controlling Surging Workers' Compensation Costs," *Occupational Hazards,* 35 (November 1979).

"Accident Related Losses Make Cost Soar," *Industrial Engineering,* 26 (May 1979).

"Analyzing a Plant Energy-Management Program: Part I—Measuring Performance," *Plant Engineering,* 59 (October 30, 1980).

"Analyzing a Plant Energy-Management Program: Part II—Forecasting Consumption" *Plant Engineering,* 149 (November 13, 1980).

"Anatomy of a Vigorous In-Plant Program," *Occupational Hazards,* 32 (July 1979).

"A Shift Toward Protective Gear," *Business Week,* 56H (April 13, 1981).

"A Win for OSHA," *Business Week,* 62 (June 29, 1981).

"Buyers Should Get Set for Tougher Safety Rules," *Purchasing,* 34 (May 25, 1976).

"Complying with Toxic and Hazardous Substances Regulations—Part I," *Plant Engineering,* 283 (March 6, 1980).

"Complying with Toxic and Hazardous Substances Regulations—Part II," *Plant Engineering,* 157 (April 17, 1980).

"Computers Help Pinpoint Worker Exposure," *Chemecology,* 11 (May 1981).

"Conserving Energy by Recirculating Air from Dust Collection Systems," *Plant Engineering,* 151 (April 17, 1980).

"Control Charts Help Set Firm's Energy Management Goals," *Industrial Engineering,* 56 (December 1980).

"Controlling Noise and Reverberation with Acoustical Baffles," *Plant Engineering,* 131 (April 17, 1980).

"Controlling Plant Noise Levels," *Plant Engineering,* 127 (June 24, 1976).

"Cost-Benefit Decision Jars OSHA Reform," *Industry Week,* 18 (June 29, 1981).

"Cost Factors for Justifying Projects," *Plant Engineering,* 145 (October 16, 1980).

"Costs, Benefits, Effectiveness, and Safety: Setting the Record Straight," *Professional Safety,* 28 (August 1975).

"Costs Can Be Cut Through Safety," *Professional Safety,* 34 (October 1976).

"Cutting Your Energy Costs," *Industry Week,* 43 (February 23, 1981).

R. De Reamer, *Modern Safety and Health Technology,* Wiley, New York, 1980.

"Elements of Effective Hearing Protection," *Plant Engineering,* 203 (January 22, 1981).

"Energy Constraints and Computer Power Will Greatly Impact Automated Factories in the Year 2000," *Industrial Engineering,* 34 (November 1980).

"Energy Managers Gain Power," *Industry Week,* 62 (March 17, 1980).

"Energy Perspective for the Future," *Industry Week,* 67 (May 26, 1980).

"Engineering and Economic Considerations for Baling Plant Refuse," 34 (April 30, 1981).

Engineering Control Technology Assessment for the Plastics and Resins Industry, NIOSH Research Report Publication No. 78-159.

"Engineering Project Planner, A Way to Engineer Out Unsafe Conditions," *Professional Safety,* 16 (November 1976).

"EPA Gears Up to Control Toxic Substances," *Occupational Hazards,* 68 (May 1977).

T. S. Ferry, *Safety Management Planning,* The Merritt Company, Santa Monica, CA, 1982.

"Fume Incinerators for Air Pollution Control," *Plant Engineering,* 108 (November 13, 1980).

"Groping for a Scientific Assessment of Risk," *Business Week,* 120J (October 20, 1980).

"Hand and Body Protection: Vital to Safety Success," *Occupational Hazards,* 31 (February 1979).

"Hazardous Wastes: Coping with a National Health Menace," *Occupational Hazards,* 56 (October 1979).

"Hearing Conservation—Implementing an Effective Program," *Professional Safety,* 21 (October 1978).

"How Do You Know Your Hazard Control Program Is Effective?," *Professional Safety,* 18 (June 1981).

"How to Control Noise," *Plant Engineering,* 90 (October 5, 1972).

"Human Factors Engineering—A Neglected Art," *Professional Safety,* 40 (March 1978).

"IE Practices Need Reevaluation Due to Energy Trends," *Industrial Engineering,* 52 (December 1980).

"Industrial Robots: A Primer on the Present Technology," *Industrial Engineering,* 54 (November 1980).

"Job-Safety Equipment Comes Under Fire, Are Hard Hats a Solution or a Problem?," *The Wall Street Journal,* 40 (November 18, 1977).

W. G. Johnson, *MORT Safety Assurance Systems,* Marcel Dekker, Inc., New York, 1979.

R. A. McFarland, "Application of Human Factors Engineering to Safety Engineering Problems," *National Safety Congress Transactions,* National Safety Council, Chicago, 1967, Vol. 12.

"New OSHA Focus Led to Noise-Rule Delay," *Industry Week,* 13 (June 15, 1981).

"OSHA Communique," *Occupational Hazards,* 27 (June 1981).

"OSHA Moves Health to Front Burner," *Purchasing,* 46 (September 26, 1979).

"OSHA to Analyze Costs, Benefits of Lead Standard," *Occupational Health & Safety,* 13 (June 1981).

Patty's Industrial Hygiene and Toxicology, 3rd revised ed., Wiley-Interscience, New York, 1978, Vol. 1.

"Practical Applications of Biomechanics in the Workplace," *Professional Safety,* 34 (July 1975).

"Private Sector Steps Up War on Welding Hazards," *Occupational Hazards,* 50 (June 1981).

"Putting Together a Cost Improvement Program," *Industrial Engineering,* 16 (December 1979).

"Reduce Waste Energy with Load Controls," *Industrial Engineering,* 23 (July 1979).

"Reducing Noise Protects Employee Hearing," *Chemecology,* 9 (May 1981).

"Regulatory Relief Has Its Pitfalls, Too," *Industry Week,* 31 (June 29, 1981).

J. B. ReVelle and Joe Stephenson, *Safety Training Methods,* 2nd ed. Wiley, New York, 1995.

"ROI Analysis for Cost-Reduction Projects," *Plant Engineering,* 109 (May 15, 1980).

"Safety & Profitability—Hand in Hand," *Professional Safety,* 36 (March 1978).

"Safety Managers Must Relate to Top Management on Their Terms," *Professional Safety,* 22 (November 1976).

"Superfund Law Spurs Cleanup of Abandoned Sites," *Occupational Hazards,* 67 (April 1981).

"Taming Coal Dust Emissions," *Plant Engineering,* 123 (May 15, 1980).

"The Cost–Benefit Argument—Is the Emphasis Shifting?," *Occupational Hazards,* 55 (February 1980).

"The Cost/Benefit Factor in Safety Decisions," *Professional Safety,* 17 (November 1978).

"The Design of Manual Handling Tasks," *Professional Safety,* 18 (March 1980).

"The Economics of Safety . . . A Review of the Literature and Perspective," *Professional Safety,* 31 (December 1977).

"The Hidden Cost of Accidents," *Professional Safety,* 36 (December 1975).

"The Human Element in Safe Man-Machine Systems," *Professional Safety,* 27 (March 1981).

"The Problem of Manual Materials Handling," *Professional Safety,* 28 (April 1976).

"Time for Decisions on Hazardous Waste," *Industry Week,* 51 (June 15, 1981).

"Tips for Gaining Acceptance of a Personal Protective Equipment Program," *Professional Safety,* 20 (March 1976).

"Toxic Substances Control Act," *Professional Safety,* 25 (December 1976).

"TSCA: Landmark Legislation for Control of Chemical Hazards," *Occupational Hazards,* 79 (May 1977).

"Were Engineering Controls 'Economically Feasible'?," *Occupational Hazards,* 27 (January 1981).

"Were Noise Controls 'Technologically Feasible'?," *Occupational Hazards,* 37 (January 1981).

"What Are Accidents Really Costing You?" *Occupational Hazards,* 41 (March 1979).

"What's Being Done About Hazardous Wastes?," *Occupational Hazards,* 63 (April 1981).

"Where OSHA Stands on Cost–Benefit Analysis," *Occupational Hazards,* 49 (November 1980).

"Worker Attitudes and Perceptions of Safety," *Professional Safety,* 28 (December 1981); 20 (January 1982).

CHAPTER 75

WHAT THE LAW REQUIRES OF THE ENGINEER

Alvin S. Weinstein, Ph.D., J.D., P.E.
Martin S. Chizek, M.S., J.D., A.S.P.
Weinstein Associates
Brunswick, Maine

75.1 THE ART OF THE ENGINEER	**2229**	
75.1.1 Modeling for the Real World	2229	
75.1.2 The Safety Factor	2230	
75.2 PROFESSIONAL LIABILITY	**2230**	
75.2.1 Liability of an Employee	2231	
75.2.2 Liability of a Business	2233	
75.3 THE LAWS OF PRODUCT LIABILITY	**2235**	
75.3.1 Definition	2235	
75.3.2 Negligence	2236	
75.3.3 Strict Liability	2236	
75.3.4 Express Warranty and Misrepresentation	2236	
75.4 THE NATURE OF PRODUCT DEFECTS	**2237**	
75.4.1 Production or Manufacturing Flaws	2237	
75.4.2 Design Flaws	2238	
75.4.3 Instructions and Warnings	2238	
75.5 UNCOVERING PRODUCT DEFECTS	**2239**	
75.5.1 Hazard Analysis	2239	
75.5.2 Hazard Index	2240	
75.5.3 Design Hierarchy	2240	
75.6 DEFENSES TO PRODUCT LIABILITY	**2241**	
75.6.1 State of the Art	2241	
75.6.2 Contributory/Comparative Negligence	2241	
75.6.3 Assumption of the Risk	2242	
75.7 RECALLS, RETROFITS, AND THE CONTINUING DUTY TO WARN	**2243**	
75.7.1 After-Market Hazard Recognition	2243	
75.7.2 Types of Corrective Action	2243	
75.8 DOCUMENTATION OF THE DESIGN PROCESS	**2244**	
75.9 A FINAL WORD	**2245**	

75.1 THE ART OF THE ENGINEER

75.1.1 Modeling for the Real World

Engineers believe that they practice their craft in a world of certainty. Nothing could be further from the truth!

Because this chapter deals with the interface between law and technology, and because products liability is likely to be the legal area of concern to the engineer, our principal focus will be on the engineering (design) of products, or components of products.

Think for a moment about the usual way an engineer proceeds from a product concept to the resulting device. The engineer generally begins the design process with some type of specifications for the eventual device to meet, such as performance parameters, functional capabilities, size, weight, cost, and so on.

Implicit, if not explicit, in the specifications are assumptions about the device's ultimate interaction with the real world. If the specification concerns, for example, loading or power needs that the device is either to produce or to withstand, someone has created boundaries within which the product is to

Mechanical Engineers' Handbook, 2nd ed., Edited by Myer Kutz.
ISBN 0-471-13007-9 © 1998 John Wiley & Sons, Inc.

function. Clearly there are bound to be some uncertainties, despite the specifying of precise values for the designers to meet.

Even assuming that a given loading for a certain component is known with precision and repeatability, the design of the component more than likely will involve various assumptions: *how* the loading acts (e.g., point-load or distributed); *when* it acts (e.g., static or dynamic); *where* it acts (e.g., two or three dimensions); and *what* it acts on (e.g., how sophisticated an analysis technique to use).

The point is that even with sophisticated and powerful computational tools and techniques, the real world is always modeled into one that can be analyzed and, as a result, is truly artificial. That is, a measure of uncertainty will always exist in any result, whatever the computational power. The question that is often unanswered or ignored in the design process is: How *much* uncertainty is there about the subtleties and exigencies of the true behavior of the environment (including people) on the product and the uncertainties in our, yes, artificial modeling technique?

75.1.2 The Safety Factor

To mask the uncertainties and, frankly, to admit that, despite our avowal, the world from which we derive our design is not real but artificial, we incorporate a "safety factor." Truly, it should be viewed as a factor of ignorance. We use it in an attempt to reestablish the real world from the one we have modified and simplified by our assumptions, and to make it tractable; that is, so we can meet the product specifications. The function of the safety factor, then, is to bridge the gap between the computational world and the one in which the product must actually function.

There are, in general, three considerations to be incorporated into the safety factor:

1. Uncertainties in material properties
2. Uncertainties in quality assurance
3. Uncertainties in the interaction of persons and the product—from the legal perspective, the most important of all

To illustrate:

Example: Truck-Mounted Crane

Consider a truck-mounted crane, whose design specification is that it is to be capable of lifting 30 tons. The intent is, of course, that only under certain specific conditions, i.e., the boom angle, boom extension, rotational location of boom, etc., will the crane be able to lift 30 tons.

Inherent in the design, however, must be a safety factor cushion, not only to account for, e.g., the uncertainties in the yield stress of the steel or the possibility of some welds not being full penetration during fabrication, but also for the uncertainties of the crane operator not knowing the precise weight of the load. In the real world, it is foreseeable that there will be times when no one on the job site knows, or has ready access to sufficient data to know, with reasonable certainty the weight of the load to be lifted.

The dilemma for the engineer–designer is how much latitude to allow in the load-lifting capability of the crane to accommodate uncertainty in the load weight. That is, the third component of the safety factor must reflect a realistic assessment of real world uncertainties. The difficulty, of course, is that there are serious competing tradeoffs to be considered in deciding upon this element of the safety factor. For each percent above the 30-ton load specification that the engineer builds into the safety factor, the crane is likely to be heavier, larger, perhaps less maneuverable, etc. That is, the utility of the crane is likely to be increasingly compromised in one or more ways as the safety factor is increased.

Yet the engineer's creed requires that the product must function in its true environment of use and do so with reasonable safety and reliability. The art of the engineer, then, is to balance competing tradeoffs in design decision-making to minimize the existence of hazards, while acknowledging and accounting for human frailties, reasonably foreseeable product uses and misuses, and the true environment of product use.

And that is what the law requires of the engineer as well. We will explore some of these considerations later in this chapter. But first, let's look at the issues of professional liability.

75.2 PROFESSIONAL LIABILITY

Whether engaged in research, development, manufacturing, engineering services, or technical consulting, today's engineer must be cognizant that the law imposes substantial accountability on both individual engineers and technology-related companies. The engineer can never expect to insulate himself entirely from legal liability. However, he can limit his liability by maintaining a fundamental understanding of the legal concepts he is likely to encounter in the course of his career, such as professional negligence, agency, employment agreements, intellectual property rights, contractual obligations, and liability insurance.

75.2.1   Liability of an Employee

Negligence and the Standard of Care

A lawsuit begins when a person (corporations, as well, are considered as "persons" for legal purposes) whose body or property is injured or damaged alleges that the injury was caused by the acts of another and files a complaint. The person asserting the complaint is the *plaintiff;* the person against whom the complaint is brought is the *defendant.*

In the complaint, the plaintiff must state a *cause of action* (a legal theory or principle) that would, if proven to the satisfaction of the jury, permit the plaintiff to recover damages. If the cause of action asserted is *negligence,* then the plaintiff must prove, first, that the defendant owed the plaintiff a *duty* (i.e., had a responsibility toward the plaintiff). Then the plaintiff must show that the defendant *breached* that duty and consequently, that the breach of duty by the defendant was the *cause* of the plaintiff's injury.

The doctrine of negligence rests on the duty of every person to exercise due care in his or her conduct toward others. A breach of this duty of care that results in injury to persons or property may result in a *tort* claim, which is a civil wrong (as opposed to a criminal wrong) for which the legal system compensates the successful plaintiff by awarding money damages. To make out a cause of action in negligence, it is not necessary for the plaintiff to establish that the defendant either intended harm or acted recklessly in bringing about the harm. Rather, the plaintiff must show that the defendant's actions fell below the *standard of care* established by law.

In general, the standard of care that must be exercised is that conduct that the *average reasonable person* of ordinary prudence would follow under the same or similar circumstances. The standard of care is an external and objective one and has nothing to do with individual subjective judgment, though higher duties may be imposed by specific statutory provisions or by reason of special knowledge.

Example: Negligent or Not?

Suppose a person is running down the street knocking people aside and causing injuries. Is this person breaching the duty to care to society and acting negligently? To determine this, we need to undertake a risk/utility analysis, i.e., does the utility of the action outweigh the harm caused?

If this person is running to catch the last bus to work, then the risk probably outweighs the utility. However, if the person has seen a knife-wielding assailant attacking someone and is trying to reach the policeman on the corner, then the utility (saving human life) is great. In such a case, perhaps society should allow the possible harm caused and thus not find the person negligent, even though other persons were injured in the attempt to reach the police officer.

No duty is imposed upon a person to take precautions against events that cannot reasonably be foreseen. However, the professional must utilize such superior judgment, skill, and knowledge as he actually possesses. Thus, the professional mechanical engineer might be held liable for miscalculating the load-lifting capability in the crane example, while a general engineering technician might not.

The duty to exercise reasonable care and avoid negligence does not mean that engineers guarantee the results of their professional efforts. Indeed, if an engineer can show that everything a reasonably prudent engineer might do was, in fact, done correctly, then liability cannot attach.

Example: Collapse of a Reasonably Designed Overpass

A highway overpass, when designed, utilized all of the acceptable analysis techniques and incorporated all of the features that were considered to be appropriate for earthquake resistance at that time. Years later, the overpass collapses when subjected to an earthquake of moderate intensity. At the time of the collapse, there are newer techniques and features that, in all likelihood, would have prevented the collapse had they been incorporated into the design.

It is unlikely that liability would attach to the engineers who created the original design and specifications as long as they utilized techniques that were reasonable at that time.

Additionally, liability depends on a showing that the negligence of the engineer was the direct and proximate cause of the damages. If it can be shown that there were other superseding causes responsible for the damages, the engineer may escape liability even though his actions deviated from professional standards.

Example: Collapse of a Negligently Designed Overpass

Suppose, instead, that after the collapse of the overpass in the preceding example, a review of the original analysis conducted by the engineers reveals several deficiencies in critical specifications that reasonably prudent engineers would not have overlooked. However, the intensity of the earthquake was of such a magnitude that, with reasonable certainty, the overpass would have collapsed even if it had been designed using the appropriate specifications. The engineers, in this scenario, are likely to escape liability.

However, the law does allow "joint and severable" liability against multiple parties who either act in concert or independently to cause injury to a plaintiff. Other defenses to an allegation of negligence include the "state of the art" argument, contributory/comparative negligence, and assumption of the risk. These are discussed in Section 75.6.

An employer is generally liable for the negligence, carelessness, errors, and omissions of its employees. However, as we will see in the next section, liability may attach to the engineer employee under the law of agency.

Agency and Authority

Agency is generally defined as the relationship that arises when one person (the principal) manifests an intention that another person (the agent) shall act on his behalf. A principal may appoint an agent to do any act except an act that, by its nature, or by contract, requires personal performance by the principal. An engineer employee may act as an agent of his employer, just as an engineering consultant may act as an agent of her client.

The agent, of course, has whatever duties are expressly stated in the contract with the principal. Additionally, in the absence of anything contrary in the agreement, the agent has three major duties implied by law:

1. The fiduciary duty of an agent to his principal is one of undivided loyalty, e.g., no self-dealing or obtaining secret profits;
2. An agent must obey all reasonable directions of the principal; and
3. An agent owes a duty to the principal to carry out his duties with reasonable care, in light of local community standards and taking into account any special skills of the agent.

Just as the agent has duties, the principal owes the agent a duty to compensate the agent reasonably for his services, indemnify the agent for all expenses or losses reasonably incurred in discharging any authorized duties, and, of course, to comply with the terms of any contract with the agent.

With regard to tort liability in the context of the employer–employee relationship, an employer can be liable only for those torts committed by a person who is considered an employee; he is not generally liable for torts committed by an agent functioning as an independent contractor. An example of an employee is one who works full-time for his employer, is compensated on a time basis, and is subject to the supervision of the principal in the details of his work. An example of an independent contractor is one who has a calling of her own, is hired to perform a particular job, is paid a given amount for that job, and followed her own discretion in carrying out the job. Engineering consultants are usually considered to be independent contractors.

Even when the employer–employee relationship is established, however, the employer is not liable for the torts of an employee unless the employee was acting within the scope of, or incidental to, the employer's business. Additionally, the employer is usually not liable for the intentional torts of an employee on the simple ground that an intentional tort (e.g., fraud) is clearly outside the scope of employment. However, where the employee intentionally chooses a wrongful means to promote the employer's business, such as fraud or misrepresentation, the employer may be held liable.

With regard to contractual liability under the law of agency, a principal will be bound on a contract that an agent enters into on his behalf if that agent has *actual authority*, i.e., authority expressly or implicitly contained within the agency agreement. The agent cannot be held liable to the principal for breach since he acted within the scope of his authority. To ensure knowledge of actual authority, the engineer should always obtain clear, written evidence of his job description, duties, responsibilities, "sign-off" authority, and so on.

Even where employment or agency actually exists, unless it is unequivocally clear that the individual engineer is acting on behalf of an employer or other disclosed principal, an injured third party has the right to proceed against either the engineer or the employer/principal or both under the rule that an agent for an undisclosed or partially disclosed principal is liable on the transaction together with her principal. Thus, engineers acting as employees or agents should always include their title, authority, and the name of the employer/principal when signing any contract or business document.

Even if the agent lacks actual authority, the principal can still be held liable on contracts entered into on his behalf if the agent had *apparent authority*, that is, where a third party reasonably believed, based upon the circumstances, that the agent possessed actual authority to perform the acts in question. In this case, however, the agent may be held liable for losses incurred by the principal for unauthorized acts conducted outside the scope of the agent's actual authority.

Employment Agreements

Rather than relying entirely on the law of agency to control the employer–employee relationship, most employers require engineers to sign a variety of employment agreements as a condition of employment. These agreements are generally valid and legally enforceable to the extent that they are reasonable in duration and scope.

A clause typically found in an engineer's employment contract is the agreement of the employee to transfer the entire right, title, and interest in and to all ideas, innovations, and creations to the company. These generally include designs, developments, inventions, improvements, trade secrets, discoveries, writings, and other works, including software, databases, and other computer-related products and processes. As long as the work is within the scope of the company's business, research, or investigation, or the work resulted from or is suggested by any of the work performed for the company, its ownership is required to be assigned to the company.

Another common employment agreement is a non-competition provision whereby the engineer agrees not to compete during his or her employment by the company and for some period after leaving the company's employ. These are also enforceable as long as the scope of the exclusion is reasonable in time and distance, when taking the nature of the product or service into account and the relative status of the employee. For example, courts would likely find invalid a two-year, nation-wide noncompetition agreement against a junior CAD/CAM engineer in a small company; however, this agreement might be found fully enforceable against the chief design engineer of a large aircraft manufacturer. In any case, engineers should inform new/prospective employers of any prior employment agreement that is still in effect.

As will be seen in the next section, however, even if an employment agreement was not executed, ex-employees are not free to disclose or utilize proprietary information gained from their previous employers.

Intellectual Property

A *patent* is a legally recognized and enforceable property right for the exclusive use, manufacture, or sale of an invention by its inventor (or heirs or assignees) for a limited period of time that is granted by the government. In the United States, exclusive control of the invention is granted for a period of 20 years from the date of filing the patent, and in consideration for which the right to free and unrestricted use passes to the general public. Patents may be granted to one or more individuals for *new* and *useful* processes, machines, manufacturing techniques, and materials, including improvements that are not obvious to one skilled in the particular art. The inventor, in turn, may license, sell, or assign patent rights to a third party. Remedies against patent infringers include monetary damages and injunctions against further infringement.

Engineers working with potentially patentable technology must follow certain formalities in the documentation and publication of information relating to the technology in order to preserve patent protection. Conversely, engineers or companies considering marketing a newly developed product or technology should have a patentability search conducted to ensure that they are not infringing existing patents.

Many companies rely on *trade secrets* to protect their technical processes and products. A trade secret is any information, design, device, process, composition, technique, or formula that is not known generally and that affords its owner a competitive business advantage. Advantages of trade secret protection include avoiding the cost and effort involved in patenting, and the possibility of perpetual protection. The main disadvantage of a trade secret is that protection vanishes when the public is able to discover the "secret," whether by inspection, analysis, or reverse engineering. Trade secret protection thus lends itself more readily to intangible "know-how" than to end products.

Trade secrets have legal status and are protected by state common law. In some states, the illegal disclosure of trade secrets is classified as fraud, and employees can be fined or even jailed for such activity. Customer lists, supplier's identities, equipment, and plant layouts cannot be patented, yet they can be important in the conduct of a business and therefore are candidates for protection as trade secrets.

75.2.2 Liability of a Business

Negligence for Services

Negligence (as defined in Section 75.2.1) and standards of care apply not only to individual engineers, but also to consulting and engineering firms. At least one State Supreme Court has defined the standard of care for engineering services as follows:

> *In performing professional services for a client, an engineer has the duty to have that degree of learning and skill ordinarily possessed by reputable engineers, practicing in the same or a similar locality and under similar circumstances. It is his further duty to use the care and skill ordinarily used in like cases by reputable members of his profession practicing in the*

same or a similar locality, under similar circumstances, and to use reasonable diligence and his best judgment in the exercise of his professional skills and in the application of his learning, in an effort to accomplish the purpose for which he was employed. *

Occasionally, an engineer's duty to the general public may supersede the duty to her client. For example, an engineer retained to investigate the integrity of a building, and who determined the building was at imminent risk of collapse, would have a duty to warn the occupants even if the owner requested that the engineer treat the results of the investigation as confidential.†

The engineer also has a duty to adhere to applicable state and federal safety requirements. For example, the U.S. Department of Labor Occupational Safety and Health Administration has established safety and health standards for subjects ranging from the required thickness of a worker's hardhat to the maximum decibel noise level in a plant. In many jurisdictions, the violation of a safety code, standard or statute that results in injury is "negligence per se," that is, a conclusive presumption of duty and breach of duty. Engineers should be aware, however, that the reverse of this rule does not hold true: compliance with required safety standards does not necessarily establish reasonable care.

Contractual Obligations

A viable contract, whether it be a simple purchase order to a vendor or a complex joint venture, requires the development of a working agreement that is mutually acceptable to both parties. An agreement (contract) binds each of the parties to do something or perhaps even refrain from doing something. As part of such an agreement, each of the parties acquires a legally enforceable right to the fulfillment of the promises made by the other. Breach of the contract may result in a court awarding damages for losses sustained by the non-breaching party, or requiring "specific performance" of the contract by the breaching party.

An oral contract can constitute just as binding a commitment as a written contract, although, by statute, some types of contracts are required to be in writing. As a practical matter, agreements of any importance should always be, and generally are, reduced to writing. However, a contract may also be created by implication based upon the conduct of one party toward another.

In general, a contract must embody certain key elements, including (a) mutual assent as consisting of an offer and its acceptance between competent parties based on (b) valid consideration for a (c) lawful purpose or object in (d) clear-cut terms. In the absence of any one of these elements, a contract will generally not exist and hence will not be enforceable in a court of law.

Mutual assent is often referred to as a "meeting of the minds." The process by which parties reach this meeting of the minds generally is some form of negotiation, during which, at some point, one party makes a proposal (offer) and the other agrees to it (acceptance). A counteroffer has the same effect as a rejection of the original offer.

In order to have a legally enforceable contract, there must generally be a bargained-for exchange of "consideration" between the parties, that is, a benefit received by the promisor or a detriment incurred by the promisee. The element of bargain assures that, at least when the contract is formed, both parties see an advantage in contracting for the anticipated performance.

If the subject matter of a contract (either the consideration or the object of a contract) is illegal, then the contract is void and unenforceable. Generally, illegal agreements are classified as such either because they are expressly prohibited by law (e.g., contracts in restraint of trade), or because they violate public policy (e.g., contracts to defraud others).

Problems with contracts can occur when the contract terms are incomplete, ambiguous, or susceptible to more than one interpretation, or where there are contemporaneous conflicting agreements. In these cases, courts may allow other oral or written evidence to vary the terms of the contract.

A party that breaches a contract may be liable to the nonbreaching party for "expectation" damages, that is, sufficient damages to buy substitute performance. The breaching party may also be liable for any reasonably foreseeable consequential damages resulting from the breach.

Contract law generally permits claims to be made under a contract only by those who are "in privity," that is, those parties among whom a contractual relationship actually exists. However, when a third party is an intended beneficiary of the contract or when contractual rights or duties have been transferred to a third party, then that third party may also have certain legally enforceable rights.

The same act can be, and very often is, both negligent and a breach of contract. In fact, negligence in the nature of malpractice alleged by a client against an engineering firm will almost invariably constitute a breach of contract as well as negligence, since the engineer, by contracting with the client, undertakes to comply with the standard of practice employed by average local engineers. If the condition is not expressed, it is generally implied by the courts.

Clark v. City of Seward, 659 P.2d 1227 (Alaska, 1983).

†California Attorney General's Opinion, Opinion No. 85-208 (1985).

Insurance for Engineers

It is customary for most businesses, and some individual engineers, to carry comprehensive liability insurance. The insurance industry recognizes that engineers, because of their occupation, are susceptible to special risks of liability. Therefore, when a carrier issues a comprehensive liability policy to an engineering consultant or firm, it may exclude from the insurance afforded by the policy the risk of professional negligence, malpractice, and "errors and omissions." The engineer should seek independent advice on the extent and type of the coverage being offered before accepting coverage. However, depending on the wording of the policy and the specific nature of the claim, the comprehensive liability carrier may be under a duty to defend an action against the insured and sometimes must also pay the loss. When a claim is made against an insured engineering consultant or firm, they should retain a competent attorney to review the policy prior to accepting the conclusions of the insurance agent as to the absence of coverage.

While the engineer employee of a well insured firm probably has limited liability exposure, the professional engineering consultant should be covered by professional liability (malpractice) insurance. However, many engineers decide to forgo malpractice insurance because of high premium rates. Claims may be infrequent, but can be economically devastating when incurred. The proper amount of coverage should be worked out with a competent underwriter, and will vary by engineering discipline and type of work. A policy should be chosen that not only pays damages, but also underwrites the costs of attorney's fees, expert witnesses, and so on.

Case Study

The following case serves to illustrate the importance of developing a fundamental understanding of the professional liability concepts discussed above.

S&W Engineering was retained by Chesapeake Paper Products to provide engineering services in connection with the expansion of Chesapeake's paper mill. S&W's vice president met with Chesapeake's project manager and provided him with a proposed engineering contract and price quotations. Several weeks later Chesapeake's project manager verbally authorized S&W to proceed with the work. S&W's engineering contract was never signed by Chesapeake; instead, Chesapeake sent S&W a Purchase Order (P.O.) that authorized engineering services "in accordance with the terms and conditions" of S&W's engineering contract. However, Chesapeake's P.O. also contained language in smaller print stating "This order may be accepted only upon the terms and conditions specified above and on the reverse side."

The drawings supplied by S&W to Chesapeake's general contractor subsequently contained errors and omissions, resulting in delays and increased costs to Chesapeake. Chesapeake sued S&W for breach of contract, arguing that the purchase order issued by Chesapeake constituted the parties' contract and that this P.O. contained a clause requiring S&W's standard of care to be "free from defects in workmanship." Additionally, another P.O. clause required indemnification of all expenses "which might incur as a result of the agreement."

S&W agreed that its engineering drawings had contained some inconsistencies, but denied that those errors constituted a breach of contract. S&W claimed that the parties' contract consisted of the terms in its proposed Engineering Contract it had delivered to Chesapeake at the outset of the Project. S&W's Engineering Contract provided that the "Engineer shall provide detail engineering services . . . conforming with good engineering practice." S&W's proposed contract also contained a clause precluding the recovery of any consequential damages.

At a jury trial, 14 witnesses testified and the parties introduced more than 1,000 exhibits. The jury found that the parties' "operative contract" was the P.O. and that S&W's services did not meet the contractually required standard of care. Chesapeake was awarded $4,665,642 in damages.*

75.3 THE LAWS OF PRODUCT LIABILITY

75.3.1 Definition

In Section 75.1, the art of engineering was characterized as a progression from real-world product specifications to the world modified by assumptions. This assumed world permits establishing precise component design parameters. Finally, the engineer must attempt to return to the real world by using a "safety factor" to bridge the gap between the ideal, but artificial, world of precise design calculations to the real world of uncertainties in who, how, and where the product will actually function.

The laws of product liability sharpen and intensify this focus on product behavior in the real world. *Product liability* is the descriptive term for a legal action brought by an injured person (the plaintiff) against another party (the defendant) alleging that a product sold (or manufactured or assembled) by the defendant was in a substandard condition and that this substandard condition was a principal factor in causing the harm of the plaintiff.

---

*Chesapeake Paper Products v. Stone & Webster Engineering, No. 94-1617 (4th Cir., 1995).

The key phrase for the engineer is *substandard condition*. In legal parlance, this means that the product is alleged to contain a *defect*. During litigation, the product is put on trial so that the jury can decide whether the product contained a defect and, if so, whether the defect caused the injury.

The laws of product liability take a retrospective look at the product and how it functioned as it interacted with the persons who used it within the environment surrounding the product and the persons. Three legal principles generally govern the considerations brought to this retrospective look at the engineer's art:

1. Negligence
2. Strict liability
3. Express warranty and misrepresentation

75.3.2 Negligence

This principle is based upon the conduct or fault of the parties, as discussed in Section 75.2.1. From the plaintiff's point of view, it asks two things: first, whether the defendant acted as a *reasonable person* (or company) in producing and selling the product in the condition in which it was sold, and second, if not, whether the condition of the product was a substantial factor in causing the plaintiff's injury.

The test of *reasonableness* is to ask what risks the defendant (i.e., designer, manufacturer, assembler, or seller) foresaw as reasonably occurring when the product was used by the expected population of users within the actual environment of use. Obviously, the plaintiff argues that if the defendant had acted reasonably, the product designer would have foreseen the risk actually faced by the plaintiff and would have eliminated it during the design phase and before the product was marketed. That is, the argument is that the defendant, in ignoring or not accounting for this risk in the design of the product, did not properly balance the risks to product users against the utility of the product to society.

It is the *reasonableness,* or lack thereof, of *the defendant's behavior* (in designing, manufacturing or marketing the product, or in communicating to the user through instructions and warnings) that is the question under the principle of negligence. These issues will be fully discussed in Section 75.5.

75.3.3 Strict Liability

In contrast to negligence, strict liability ignores the defendant's behavior. It is, at least in theory, of no consequence whether the manufacturer behaved reasonably in designing, manufacturing, and marketing the product. The only concern here is the quality of the product as it actually functions in society.

Essentially, the question to be resolved by the jury under strict liability is whether or not the risks associated with the real-world use of the product by the expected user population exceed the utility of the product and, if so, whether there was a reasonable alternative to the design that would have reduced the risks without seriously impairing the product's utility or making it unduly expensive.

If the jury decides that the risks outweighed the product's utility and a reasonable alternative to reducing the risk existed, then the product is judged to be in a *defective condition unreasonably dangerous.*

Under strict liability, a product is defective when it contains *unreasonable* dangers, and only unreasonable dangers in the product can trigger liability. While it is unlikely the marketing department will ever use the phrase in a promotion campaign, a product may contain *reasonable* dangers without liability. In the eyes of the law, a product whose only dangers are reasonable ones is *not* defective.

Stated positively, a product that does not contain unreasonable dangers is *reasonably safe*—and that is all the law requires. This means that any residual risks associated with the product have been transferred *appropriately* to the ultimate user of the product.

Section 75.5 discusses the methodology for uncovering unreasonable dangers associated with products.

75.3.4 Express Warranty and Misrepresentation

The third basic legal principle governing possible liability has nothing to do with either the manufacturer's conduct (negligence) or the quality of the product (strict liability). Express warranty and misrepresentation are concerned only with what is communicated to the potential buyer that becomes part of the "basis of the bargain."

An express warranty is created whenever any type of communication to the potential buyer describes some type of *objectively measurable* characteristic of the product.

Sample Express Warranties

- This truck will last 10 years.
- This glass is shatterproof.

- This automatic grinder will produce 10,000 cutter blades per hour.
- This transmission tower will withstand the maximum wind velocities and ice loads in your area.

If such a communication is, first, at least a part of the reason that the product was purchased and then, if reasonably foreseeable circumstances ultimately prove the communication invalid, there has been misrepresentation, and the buyer is entitled to recover damages consistent with the failed promise.

It doesn't matter one whit if the product cannot possibly live up to the promise. This is not the issue. It is the failure to keep a promise that becomes part of the basis of the bargain, and that the buyer did not have sufficient expertise for not believing the promise, that can trigger the liability.

Someone with a legal bent might argue, against the misrepresentation claim, that the back of the sales form clearly and unequivocally disclaims all liability arising from any warranties not contained in the sales document (i.e., the contract). The courts, when confronted with what appears to be a conflict between the express warranty communicated to the buyer and the fine print on the back of the document disclaiming everything, inevitably side with the buyer who believed the express warranty to the extent that it became a part of the "basis of the bargain."

The communications creating the express warranty can be in any form: verbal, written, visual, or any combination of these. In the old days, courts used to view advertising as mere puffing and rarely sided with the buyer arguing about exaggerated claims made about the product. In recent years, however, the courts have acknowledged that buying is engendered in large part by media representations. Now, when such representations can be readily construed as express warranties, the buyer's claim is likely to be upheld. It should also be noted that misrepresentation claims have been upheld when both the plaintiff and the defendant are sophisticated, have staffs of engineers and lawyers, and the dealings between the parties are characterized as "arm's length."

In precarious economic times, the exuberance of salespersons, in their quest to make the sale, may oversell the product and create express warranties that the engineer cannot meet. This can then trigger liability, despite the engineer's best efforts.

Because it is so easy to create, albeit unintentionally, an express warranty, all departments that deal in any with a product must recognize this potential problem and structure methods and procedures to minimize its occurrence. The means that engineering, manufacturing, sales, marketing customer service, and upper management must create a climate in which there is agreement among the appropriate entities that what is being promised to the buyer can actually be delivered.

75.4   THE NATURE OF PRODUCT DEFECTS

The law recognizes four areas that can create a "defective condition unreasonably dangerous to the user or consumer":

1. Production or Manufacturing
2. Design
3. Instructions
4. Warnings

75.4.1   Production or Manufacturing Flaws

A production or manufacturing defect can arise when the product fails to emerge as the manufacturer intended. The totalities of the specifications, tolerances, and so on, define the product and all of the bits and pieces that make it up, and collectively they prescribe the manufacturer's intent for exactly how the product is to emerge from the production line.

If there is a deviation from any of these defining characteristics of the product (e.g., specifications, tolerances, etc.), then there exists a production or manufacturing flaw. If this flaw or deviation can cause the product to fail or malfunction under reasonably foreseeable conditions of use, and these conditions are within the expected performance requirements for the product, then the product can be defective.

What is important to note here is that the deviation from the specifications must be *serious* enough to be able to precipitate the failure or malfunction of the product within the foreseeable uses and performance envelope of the product, hence creating unreasonable dangers. To illustrate, let's return to the crane described in the first section of this chapter.

Example: Truck-Crane—Flaw or Defect?

Suppose that a critical weld is specified to be 4 in. in length and to have full penetration. After a failure, the crane is examined and the weld is full-penetration but only 3½ in. long, which escaped the notice of the quality inspectors. There is a deviation or flaw. However, whether this flaw rises to the level of defect depends on several considerations:

First, what safety factor considerations entered into the design of the weld? It may be that the designer calculated the necessary weld length to be 3 in. and specified 4 in. to account for the uncertainties described in Section 75.1. Next, if it can be shown by the crane manufacturer that a 3½ in. weld was adequate for all reasonably foreseeable use conditions of the crane, than it could be argued that the failure was due to crane misuse and not due to the manufacturing flaw.

Alternatively, the plaintiff could argue that the engineer's assumptions as to the magnitude of the safety factor did not realistically assess the uncertainty of the weight loads to be lifted; if they had done so, the minimum acceptable length would have been the 4 in. actually specified.

While this is a hypothetical example, it illustrates the interplay of several important elements that must be considered when deciding if a production flaw can rise to the level of a defect. Foreseeable uses and misuses of the product, and its prescribed or implicit performance requirements, are two of the most important.

75.4.2 Design Flaws

The standard for measuring the existence of a production flaw is simple. One need only compare the product's attributes as it actually leaves the production line with what the manufacturer intended them to be, by examining the manufacturer's internal documents that prescribe the entire product.

To uncover a design flaw, however, requires comparing the correctly manufactured product with a standard that is not as readily prescribed as the manufacturer's own specifications and is significantly more complex. The standard is a societal one in which the risks of the product are balanced against its utility to establish whether the product contains unreasonable dangers. If there are unreasonable dangers, then the design flaw becomes a defect.

In the crane example, assume that there has been a boom failure and that the crane met all of the manufacturer's specifications, that is, no manufacturing defect is alleged. The plaintiff alleges, instead, that if the boom had been fabricated from a heavier gage as well as a stronger alloy steel, the collapse would have been avoided. The plaintiff's contention can be considered a design flaw. There is no question that the boom could have been fabricated using the plaintiff's proposed specifications and, for the sake of our discussion, we will also assume the boom would not have failed using the different material.

The critical question, however, is should the boom have been designed that way? The answer is, only if the original design created unreasonable dangers. The existence of unreasonable dangers, therefore a defective condition, can be deduced from a risk/utility analysis of the interaction of crane users, users, and the environments within which the crane is expected to function.

The analysis must consider, first, the foreseeability of crane loads of uncertain magnitude that could cause the original design to fail, but not the modified design. Balanced against that consideration will be a reduction in the utility of the crane because of its increased weight and/or size if the proposed design alterations are incorporated. There will be also an increased cost. It is this analysis of competing tradeoffs that the designer must consider before deciding on the proposed design specifications. Fundamentally, though, as in the discussion of a production defect, the consideration is that of the safety factor, bridging the gap between *assumed* product function and *actual* product function.

75.4.3 Instructions and Warnings

A product can be perfectly manufactured from a design that contains no unreasonable dangers and yet be defective because of inadequate instructions. Instructions are the communications between the manufacturer and the user that describe how the product is to be used to achieve the intended function of the product.

Warnings are to communicate any residual hazards, the consequent risks of injury, and the actions the user must take to avoid injury. If the warnings are inadequate, the product can be defective even if the design, manufacturing, and instructions meet the legal tests.

While the courts have not given clear or unequivocal guidelines for assessing the adequacy of instructions and warnings, there are several basic considerations that should underlie their development:

- They must be understood by the expected user population.
- They must be effective in a multilingual population.
- There must be some reasonable and objective evidence to prove that the warnings and instructions can be understood and are likely to be effective.

Simply put, writing instructions and warnings is deceptively easy. However, gathering evidence to support the contention that they are *adequate* can be extremely difficult, costly, and time-consuming. To do this means surveying the actual user population and describing those characteristics that are likely to govern comprehension, such as age, education, reading capability, sex, cultural and ethnic background, and so on. Then a statistically selected random sample of the identified user population must be chosen to test the communication for comprehension, using the method suggested in the American National Standards Institute standard ANSI Z535.3. Finally, the whole process must be documented. Then, and only then, can a manufacturer argue that the user communications, that is, instructions and warnings, are adequate.

75.5   UNCOVERING PRODUCT DEFECTS

75.5.1   Hazard Analysis

In the preceding section, a risk/utility analysis was described as a basis for assessing whether or not the product was in a defective condition unreasonably dangerous. Now we will consider the methodology and the process of the risk/utility analysis.

We begin with a disclaimer: Neither the process nor the methodology about to be discussed is readily quantifiable. However, this fact does not lessen their importance; it only emphasizes the care that must be exercised.

The process is one of scenario-building. The first step is to characterize, as accurately as possible, the users of the product, the ways in which they will use the product, and the environment in which they will use it. These elements must be quantified as much as possible.

Example: Foreseeable Users of a Hand-Held Tool

Will the user population be comprised of younger users, female users, elderly users? If so, these populations are likely to need special ergonomic or human factors considerations in the design of handgrips, operating controls, etc. Will the tool be found in the home? If so, inadvertent use by small children is likely to be a consideration in designing the controls. Certainly the ability to read and understand instructions and warnings must be a significant element of the characterization of the users.

If the best of all worlds, the only product uses the engineer would be concerned with are the *intended* uses. Unfortunately, the law requires that the product design acknowledge and account for *reasonably foreseeable misuses* of the product. Of all the concepts the engineer must deal with, this one is perhaps the hardest to analyze and the most difficult to accept. Part of the reason, of course, is the difficulty of distinguishing between uses that are *reasonably foreseeable* and those uses that the manufacturer can argue are truly *misuse* for which no account must be taken in design.

The concept of legal unforeseeability is a difficult one. Many people might think that if they have ever talked about the possibility of misusing a product in a certain way, then they have "foreseen" that misuse and therefore must account for it in their design. This is not the case. Legally, *unforeseeable misuse* means a use so egregious, or so bizarre, or so remote that it is termed *unforeseeable*, even when such a misuse has been a topic of discussion.

A simple illustration might help.

Example: How Many Ways Can You Use a Screwdriver?

There is no question that the intended purpose and function of a screwdriver is to insert and remove screws. This means that ideally, the shank of a screwdriver is subjected only to a twisting motion, or torque.

But how do most people open paint cans? With a screwdriver, of course. In that context, however, the shank is subjected to a bending moment, not a torque. Any manufacturer who produced and marketed a screwdriver with shank material able to withstand high torque, but without sufficient bending resistance to open a paint can without shattering, would have a difficult time avoiding liability for any injuries that occurred.

The reason, of course, is that using a screwdriver to open paint cans would be considered as a reasonably foreseeable misuse, and should be accounted for in the design. On the other hand, suppose someone uses a screwdriver as a cold chisel to loosen a rusted nut and the screwdriver shatters, causing injury. The manufacturer could argue that such a use was a misuse that the manufacturer had no duty to account for in the design.

Finding the line that separates the misuses the engineer must account for from the misuses that are legally unforeseeable is not easy, nor is the line a precise one. All that is required, however, is for the engineer to show the reasonableness of the process of how the line was ultimately decided, while attempting to meet competing tradeoffs in selecting the product's specifications. Unquestionably, we can always imagine all types of bizarre situations in which a product is misused and someone is injured. Does this mean that all such situations must somehow be accounted for in design? Of course not. But what is required is to make a reasonable attempt to separate user behavior into two categories: that which can reasonably be accounted for in design and that which is beyond reasonable considerations.

The third element in the risk/utility process is the environment within which the user and product interact. If it is cold, how cold? If it is hot, how hot? Will it be dark, making warnings and instructions difficult to read? Will the product be used near water. If so, both fresh and salt? How long will the product last? Will it be repainted, scraped, worn, and so on? These, too, would be considerations in warning adequately.

The scenario building must integrate the three elements of the hazard analysis: the users, the uses, and the environment. By asking "What if . . . ?," a series of hazards can be postulated from integrating the users with the uses within an environment.

Example: "What if an Air-Operated Sander . . . ?"

What if an air-operated sander is used in a marine environment? What if the user inadvertently drops it overboard and then continues to use it without having it disassembled and cleaned? What hazards could arise? Could corrosion ultimately freeze the control valve continually open, leading to loss of control at some future time, long after the event in question?

75.5.2 Hazard Index

After completion of the hazard analyses, the hazards should be rank-ordered from the most serious to the least serious. One way to do this is to assign a numerical probability of the event occurring and then to assess, also using a numerical scale, the seriousness of the harm. The product of these two numbers is the *Hazard Index* and permits a relative ranking of the hazards. The scales chosen to provide some measure of probability and seriousness should be limited; the scale may run, for example, from 0 to 4. A 0 implies that the event is so unlikely to occur, or that the resulting harm is so minimal, as to be negligible. Correspondingly, a 4 would mean that an event was almost certain to occur, or that the result would be death or serious irreparable injury. With this scale, the hazard index could range from 0 to 16.

Once this is done, attention is then focused on the most serious hazards, eventually working down to the least serious one.

75.5.3 Design Hierarchy

Ideally, for each such event, the objective would be, first, to "design out" the hazard. If a hazard can be designed out, it can never return to cause harm.

Failing the ability to design out the hazard, the next consideration must be guarding. Can an unobtrusive barrier be placed between the user and the hazard? It must be noted that if a guarding configuration greatly impairs the utility of the product, or greatly increases the time needed to carry out the product's intended function, it is likely to be removed. In such a case, the user is not protected from the hazard, nor is the manufacturer likely to be protected from liability if an injury results, because removing an obtrusive guard may be considered a foreseeable misuse.

If the hazard cannot be designed out, nor can an effective guard be devised, then *and only then* should the last element of the design hierarchy be considered: a warning.

A warning must be viewed as an element of the design process, not as an afterthought. To be perfectly candid, if the engineer has to resort to a warning to minimize or eliminate a risk of injury from that hazard, it may be an admission of a failure in the design process.

Yet there are innumerable instances where a warning must be given. Section 75.4 described the considerations necessary to develop an adequate warning, the legal standard. What was not described there are the three necessary elements that must be included before the process of establishing adequacy begins:

1. The nature of the hazard
2. The potential magnitude of the injury
3. The action to be taken to avoid the injury

A warning paraphrased from an aerosol can of hair spray provides an exercise for the reader:

⚠ **WARNING**

- **Harmful vapors**
- **Inhalation may cause death or blindness**
- **Use in a well-ventilated area**

The reader should analyze these three phrases carefully and critically, then describe the user populations to which the warning might apply, then answer the question of whether or not it is likely that injury could be avoided by that user population. Suppose that a foreseeable portion of the population using this aerosol can are people whose English reading ability is at the 3rd or 4th grade level. (It is estimated that about half of English-speaking Americans cannot read beyond the 4th grade level.) What can you conclude about comprehension and the ability to avoid injury?

Warnings are, in fact, the most difficult way to minimize or eliminate hazards to users.

75.6 DEFENSES TO PRODUCT LIABILITY

Up to now, we have only looked at the factors that permit an analysis of whether or not the product contains a defect, i.e., an unreasonable danger. Certainly the ultimate defense to an allegation that the product was defective is to show through a risk/utility analysis that, on balance, the product's utility outweighs its risks and, in addition, that there were no feasible alternatives to the present design.

It may be, however, that the plaintiff's suggested design alternative is, in fact, viable as of the time the incident occurred. Is there any analysis that could offer a defense? There may be, by considering a *state-of-the-art* argument.

75.6.1 State of the Art

Decades ago, the phrase *state of the art* meant, simply, what the custom and practice was of the particular industry in question. Because of the concern that an entire industry could delay introduction of newer, safer designs by relying on the "custom and practice" argument to defeat a claim of negligence, the courts have adopted a broader definition of the term.

The definition today is "what is both technologically and economically feasible." The time at which this analysis is performed is, in general, the date the product in question was manufactured.

Thus, while a plaintiff's suggested alternative design may have been technologically and economically feasible at the time the incident occurred, their argument may not be viable if the product was manufactured 10 years before the incident occurred.

To make that argument convincing, however, means that engineers must always be actively seeking new and emerging technology, looking to its potential applicability to their industry and products. It is expected, too, that technological advances are sought, not only in the engineer's own industry, but in related and allied fields as well. Keeping current has an added dimension, that of being alert to broader vistas of technological change outside one's own industry.

The second element of today's state-of-the-art principle is that innovative advances must be economically viable as well. It is generally, but incorrectly, assumed that the term *economic viability* is limited to the incremental cost of incorporating the technological advance into the product and how it will affect the direct cost of manufacturing and the subsequent profit margin.

The courts, however, are concerned with another cost in measuring economic viability, in addition to the direct cost of incorporating a safety improvement in the product: the cost to society and ultimately to the manufacturer if the technological advance is *not* incorporated into the product and injuries occur as a result. The technological advances we are concerned with here are those that are likely to enhance safety.

While it is more difficult and certainly cannot be predicted with a great deal of precision, an estimate of costs of the probable harm to product users is part of the equation. An approach to this analysis was described in Section 75.5. Estimating both the probability and seriousness of the harm from a realistic vantage point if the technological advance is *not* incorporated can form the basis for estimating the downside risk of not including the design feature.

75.6.2 Contributory/Comparative Negligence

We have not yet really considered what role, if any, the plaintiff's behavior plays in defending a product against an allegation of defect. We have earlier touched on misuse of the product, which is a use so egregious, and so bizarre, or so remote, that is it termed *legally unforeseeable.* You may recall the example discussing the hypothetical use of a screwdriver as a cold chisel to illustrate what could very likely be considered as misuse.

But what about the plaintiff's behavior that is not so extreme? Does that enter at all into the equation of how fault is apportioned? Yes, it does, in the form of contributory or comparative negligence, if the legal theory embracing the litigation is negligence. You will recall that under negligence, the defendant's behavior is measured by asking if that party was acting as a *reasonable* person (or manufacturer, or engineer) would have acted under the same or similar circumstances. And the reasonableness of the behavior is the result of having foreseen the risks of one's actions by having undertaken a risk/utility balancing prior to engaging in the action.

In a negligence action, the plaintiff's behavior is measured in exactly the same way. The defendant asks the jury to consider whether the plaintiff was behaving as a reasonable person would have under the same or similar circumstances. Did the plaintiff contribute to his or her harm by not acting reasonably? This is called *contributory negligence.*

While some states still retain the original concept that *any* contributory negligence on the part of the plaintiff totally bars his or her recovery of damages, most states have adopted some form of comparative negligence. Generally, the jury is asked to assess the behavior of both the plaintiff and the defendant and apportion the fault in causing the harm between them, making certain the percentages total 100%. The plaintiff's award, if any, is then reduced by the percentage of his or her comparative negligence.

The test of the defendant's negligence and the plaintiff's contributory negligence is termed an objective one. That is, the jury is asked to judge the actions of the parties relative to what a reasonable person would have done in the same or similar circumstances. The jury does not, as a rule, consider whether anything in that party's background, training, age, experience, education, and so on played any role in the actions that led to the injury.

75.6.3 Assumption of the Risk

There is another defense involving the plaintiff's behavior that does consider the plaintiff's characteristics in assessing his or her culpability. It is termed *assumption of the risk*. In essence, this defense argues that the plaintiff consented to being injured by the product. In one common form, used for analyzing this aspect of the plaintiff's behavior, the jury is asked if the plaintiff *voluntarily* and *unreasonably* assumed a *known* risk. To prevail, the defendant must present evidence on all three of these elements and must prevail on all three for a jury to conclude that the plaintiff's "assumed the risk."

The first element, asking whether the plaintiff voluntarily confronted the danger, and the third element, considering whether the risk was known, are both subjective elements. That is, the jury must determine the state of the mind of the plaintiff, assessing what he or she actually knew or believed or what can reasonably be inferred about his or her behavior at the instant prior to the event that led to injury. Thus, the plaintiff's background, education, training, experience, and so on become critical elements in this assessment.

A couple of points should be made here. First, in determining whether the plaintiff voluntarily confronted the hazard, the test is whether or not the plaintiff had *viable* alternatives.

Example: Work or Walk

In a workplace setting, a worker is given a choice of either using a now-unguarded press or being fired. The press had been properly guarded for all the time the plaintiff had used it in the past, but the employer has removed the guards to increase productivity and now tells the employee either to use the press as-is or be fired. The courts do not consider that the plaintiff had viable alternatives, since the choice between working on an unguarded press or being fired is no choice at all. The lesson to the engineer in this example is that the guarding slowed productivity and was removed, leaving the press-user in a no-win situation. The design should have incorporated, to the extent possible, guarding that did not *slow production.*

Second, the same in-depth consideration must also be given to knowledge of the risk by the plaintiff. The plaintiff's background, education, and so on must provide a reasonable appreciation of the actual nature of the harm that could befall him or her.

Example: The Truly Combustible Car

The driver of a car is confronted by a slight smell of smoke the first time the windshield wipers are used, and is trying to bring the car to the dealer in a rainstorm to see what the trouble is when the car literally bursts into flames, causing injury. Has the driver assumed the risk of injury by continuing to drive after smelling smoke? Can the car manufacturer successfully argue that the risks of injury were known to the driver? The question can only be answered by examining those elements in the driver's background that could, in any way, lead a jury to conclude that the driver should have recognized that smoke from electrically operated wipers could lead to a conflagration. The old adage of "where there's smoke, there's fire" is insufficient to charge the plaintiff with knowledge of the precise risk he or she faced without more knowledge of the driver's background.

The final element of assumption of the risk, the unreasonableness of the plaintiff's choice in voluntarily confronting a known risk, is an objective element, exactly the same as in negligence. That is, what would a reasonable person have done under the same or similar circumstances?

Example: The Truly Combustible Car Meets the Good Samaritan

A passerby observes the car from the previous example. It is on fire, and the driver is struggling to get out. The passerby rescues the driver, but is seriously burned and suffers smoke inhalation in the process. The driver files suit against the manufacturer alleging a defect that created unreasonable danger when the wipers were turned on. The passerby also files suit against the automobile manufacturer to recover for the injuries suffered as a result of the rescue, arguing that the rescue would not have been necessary if there had been no defect. Would this good Samaritan be found to have assumed the risk of injury? Clearly the choice to try to rescue the driver was voluntary and the risks of injury were from a fire were apparent to anyone, including the rescuer. But was the act of rescuing the car's occupant a reasonable or unreasonable one? If the jury concludes that it was a reasonable choice, the passerby would not have been found to have assumed the risk, despite having voluntarily exposed himself to a known risk.

The defendant must prevail in all three of the elements, not just two. Needless to say, raising and succeeding in the defense of assumption of the risk is not an easy one for the defendant.

One final word about these defenses: While the "assumption of the risk" defense applies both in a claim of negligence and strict liability, the contributory/comparative negligence defense does *not* apply in strict liability. The reason is that strict liability is a no-fault concept whereas negligence is a fault-based concept. It would be inconsistent to argue no-fault theory (strict liability) against the defendant and permit the defendant to argue a fault-based defense (contributory negligence) concerning the plaintiff's behavior.

75.7 RECALLS, RETROFITS, AND THE CONTINUING DUTY TO WARN

Manufacturers generally have a post-sale or continuing duty to warn of latent defects in their products that are revealed through consumer use.

Sometimes, however, even a post-sale warning may be inadequate to render a product reasonably safe. In those circumstances, it may be necessary for a manufacturer to retrofit the product by adding certain safety devices or guards. Moreover, there may be instances where it is not feasible to add guards or safety devices, or where the danger of the product is so great that the product simply must be removed from the market by being recalled.

75.7.1 After-Market Hazard Recognition

The manufacturer is responsible for establishing feedback mechanisms from customers, distributors, and sales personnel that will ensure that post-sale problems are discovered. Applicable data may include product performance and test data, orders for repair parts, complaint files, quality-control and inspection records, and instruction and warning modifications. Another source of hazard recognition information comes from previous accident investigations, claims, and lawsuits. The manufacturer should also have an ongoing program of compiling and evaluating risk data from historical, field and/or laboratory testing, and fault-tree, failure modes, and hazard analyses.

Once the manufacturer has determined that a previously sold product is defective (that is, contains unreasonable dangers) and is still in use, it must decide what response is appropriate. If the product is currently being produced, an initial assessment as to the seriousness of the problem must be made in order to decide whether production is to be halted immediately and inventories frozen in the warehouses and on dealers' shelves in order to limit distribution.

Following this assessment, the nature of the defect must be established. If the problem is safety-related, and depending upon the type of the product, appropriate regulatory agencies may have to be immediately notified. The manufacturer must then consider the magnitude of the hazards by estimating the probability of occurrence of events and the likely seriousness of injury or damage. The necessity for postulating such data is to provide some measure of the magnitude of the consequences if no action is taken, or to decide the extent of the action to be taken in light of the estimated consequences. Alternatively, if the consequences of even a low probability of occurrence could result in serious injury or death, or could seriously affect the marketability of the product or the corporate reputation, the decision to take action should be independent of such estimates.

Once the decision to take action is made, the origin, extent, and cause of the problem must be addressed in order to plan effective corrective measures. Is the origin of the defect in the raw material, fabrication, or quality control? If the problem is one of fabrication, did it occur in-house or from a purchased part? Where are the faulty products—that is, are the products in inventory, in shipment, in dealers' stock, or in the hands of the buyers? Does the defect arise from poor design, inadequate inspection, improper materials, fabrication procedure, ineffective or absent testing, or a combination of these events?

75.7.2 Types of Corrective Action

After the decision to take action has been made, and the origin, extent, and cause of the problem have been investigated, the appropriate corrective action must be determined. Possible options are to recall the product and replace it with another one; to develop a retrofit and either send personnel into the field to retrofit the product or have the customer return the product to the manufacturer for repair; to send out the parts and have the customer fix the product; or simply to send out a warning about the problem. This process should be fully documented to substantiate the reasons for the selection of a particular response. The urgency with which the corrective action is taken will be determined by the magnitude of the hazard.

Warnings

A manufacturer is not required to warn of every known danger, even with actual knowledge of that danger. A warning is required where a product can be dangerous for its intended and reasonably foreseeable uses and where the nature of the hazard is unlikely to be readily recognized by the expected user class. When a hazard associated with a product that was previously unknown to the manufacturer becomes apparent after the product has been in use, the manufacturer has a threshold duty to warn the existing user population.

Factors to consider in determining whether to issue a post-sale warning include the manufacturer's ability to warn (i.e., how readily and completely the product users can be identified and located), the product's life expectancy (the longer life expectancy, the greater risk of potential harm if post-sale warnings are not given), and the obviousness of the danger. Thus, the practicality, cost, and burden of providing an effective warning must be weighed against the potential harm of omitting the warning.

Recalls

Where the potential harm to the consumer is so great that a warning alone is not adequate to eliminate the danger, the proper remedy may be to institute a recall of the product either for repair or replacement. For some products, a recall may be mandated by statute or a governmental regulatory agency. Where a recall is not mandated, however, the decision to institute a product recall should be made using the analysis undertaken in Section 75.7.1.

Retrofits

A recall campaign may not be an appropriate solution, particularly if the equipment is large or cannot be easily removed from an installation. For equipment with potentially serious hazards or requiring complicated modification, the manufacturer should send its personnel to perform (and document) the retrofit. For equipment with relatively minor potential hazards for which there is a simple fix, the manufacturer may opt to send to the owners the parts necessary to solve the problem.

Regardless of the type of corrective action program selected, it is essential that all communications directed to the owners and/or users urging them to participate in the corrective action program be clear and concise. Most important, however, is the necessity for the communication to identify the nature of the risks and the potential seriousness of the harm that could befall the product user.

75.8 DOCUMENTATION OF THE DESIGN PROCESS

There are conflicting arguments by attorneys about what documentation, if any, the manufacturer should retain in the files (or on the floppies, the hard drive, or tape back-up). Since it would be well-nigh impossible to run a business without documentation or some sort, it only makes sense to preserve the type of documentation that can, if the product is challenged in court, demonstrate the care and concern that went into the design, manufacturing, marketing, and user communications of the product.

The first principle of documentation is to minimize or eliminate potential adverse use of the documentation by an adverse party. For example, words such as *defect* should not appear in the company's minutes, notes, and so on. There can be *deviations, flaws, departures,* and so on from specifications or tolerances. These are not defects unless they could create unreasonable dangers in the use of the product.

Also, all adverse criticism of the product, whether internally from employees or externally from customers, dealers, and so on must be considered and addressed in writing by the responsible corporate person having the appropriate authority.

Apart from these considerations, the company should make an effort to create a *documentation tree,* delineating what paper is needed, who should write it, where it should be kept, who should keep it, and for how long. The retention period for documents, for the most part, should be based on common sense. If a government or other agency requires or suggests the length of time certain documents be kept, obviously those rules should be followed. For the rest, the length of time should be based upon sound business practices. If the product has certain critical components that, if they fail before the end of the product's useful life, could result in a serious safety problem, the documentation supporting the efficacy of these parts should be retained for as long as the product is likely to be in service.

Because the law requires only that a product be reasonably safe, clearly the documentation to be preserved should be that which will support the argument that all of the critical engineering decisions that balanced competing tradeoffs were reasonable and were based on reducing the risks from all foreseeable hazards. The rationales underlying these decisions should be part of the record, for two reasons. First, because it will give those who will review the designs when the product is to be updated or modified in subsequent years, the bases for existing design decisions. If the prior assumptions and rationales are still valid, they need not be altered. Conversely, if some do not reflect current thinking, then only those aspects of the design need to be altered. Without these rationales, all the design parameters will have to be re-examined for efficacy.

Secondly, and just as importantly, having the rationales in writing for those safety-critical decisions can provide a solid legal defense if the design is ever challenged as defective.

Thus, the documentation categories that are appropriate both for subsequent design review and for creating strong legal defense positions are these:

- Hazard and risk data that formed the bases for the safety considerations
- Design safety formulations, including fault-tree and failure-models and effects analyses
- Warnings and instructions formulation, together with the methodology used for development and testing
- Standards used, including in-house standards, and the rationale for the requirements utilized in the design
- Quality assurance program, including the methodology and rationale for the processes and procedures
- Performance of the product in use, describing reporting procedures, follow-up data acquisition and analysis, and a written recall and retrofit policy

This type of documentation will permit recreating the process by which the reasonably safe product was designed, manufactured, and marketed.

75.9 A FINAL WORD

In the preceding pages, we have only touched on a few of the areas where the law can have a significant impact on engineers' discharge of their professional responsibilities. As part of the process of product design, the law asks the engineer to consider that for the product that emerges from the mind of the designer and the hand of the worker to play a role in enhancing society's well-being, it must

- Account for reasonably foreseeable product misuse
- Acknowledgment human frailties and the characteristics of the actual users
- Function in the true environment of product use
- Eliminate or guard against the hazards
- Not substitute warnings for effective design and guards

What has been discussed here and summarized above is, after all, just good engineering. Our objective is to help the engineer recognize those considerations that are necessary to bridge the gap between the preliminary product concept and the finished product that has to function in the real world, with real users and for real uses, for all of its useful life.

Apart from understanding and utilizing these considerations during the product design process, engineers have an obligation, both personally and professionally, to maintain competence in their chosen field so that there can be no question that all actions, decisions, and recommendations, in retrospect, were reasonable.

That is, after all, what the law requires of all of us.

CHAPTER 76

PATENTS

David A. Burge
David A. Burge Co., L.P.A.
Cleveland, Ohio

Benjamin D. Burge
Chips & Technologies, Inc.
San Jose, California

76.1 **WHAT DOES IT MEAN TO OBTAIN A PATENT** 2248
   76.1.1 Utility, Design, and Plant Patents 2248
   76.1.2 Patent Terms and Expiration 2248
   76.1.3 Four Types of Applications 2249
   76.1.4 Why File a Provisional Application 2249
   76.1.5 Understanding That a Patent Grants a "Negative Right" 2249

76.2 **WHAT CAN BE PATENTED AND BY WHOM** 2250
   76.2.1 Ideas, Inventions, and Patentable Inventions 2250
   76.2.2 The Requirement of Statutory Subject Matter 2250
   76.2.3 The Requirement of Originality of Inventorship 2251
   76.2.4 The Requirement of Novelty 2252
   76.2.5 The Requirement of Utility 2253
   76.2.6 The Requirement of Nonobviousness 2253
   76.2.7 Statutory Bar Requirements 2254

76.3 **PREPARING TO APPLY FOR A PATENT** 2255
   76.3.1 The Patentability Search 2255
   76.3.2 Putting the Invention in Proper Perspective 2255
   76.3.3 Preparing the Application 2256
   76.3.4 Enablement, Best Mode, Description, and Distinctness Requirements 2256

76.3.5 Functional Language in Claims 2257
76.3.6 Product-by-Process Claims 2257
76.3.7 Claim Format 2257
76.3.8 Executing the Application 2258
76.3.9 Patent and Trademark Office Fees 2258
76.3.10 Small Entity Status 2259
76.3.11 Express Mail Filing 2260

76.4 **PROSECUTING A PENDING PATENT APPLICATION** 2260
   76.4.1 Patent Pending 2260
   76.4.2 Secrecy of Pending Applications 2260
   76.4.3 Duty of Candor 2260
   76.4.4 Initial Review of an Application 2261
   76.4.5 Response to an Office Action 2261
   76.4.6 Reconsideration in View of the Filing of a Response 2262
   76.4.7 Interviewing the Examiner 2263
   76.4.8 Restriction and Election Requirements 2263
   76.4.9 Double-Patenting Rejections 2263
   76.4.10 Patent Issuance 2264
   76.4.11 Safeguarding the Original Patent Document 2264
   76.4.12 Continuation, Divisional and Continuation-in-Part Applications 2264
   76.4.13 Maintaining a Chain of Patent Applications 2264

Mechanical Engineers' Handbook, 2nd ed., Edited by Myer Kutz.
ISBN 0-471-13007-9 © 1998 John Wiley & Sons, Inc.

76.5 PATENT PROTECTIONS
AVAILABLE ABROAD **2265**
76.5.1 Canadian Filing 2265
76.5.2 Foreign Filing in Other
Countries 2265
76.5.3 Annual Maintenance Taxes
and Working Requirements 2265
76.5.4 Filing under International
Convention 2265
76.5.5 Filing on a Country-by-
Country Basis 2266

76.5.6 The Patent Cooperation
Treaty 2266
76.5.7 The European Patent
Convention 2266
76.5.8 Advantages and
Disadvantages of
International Filing 2266
76.5.9 Trends in International
Patent Protection 2267

76.1 WHAT DOES IT MEAN TO OBTAIN A PATENT

Before meaningfully discussing such topics as inventions that qualify for patent protection and procedures that are involved in obtaining a patent, it is necessary to know about such basics as the four different types of patent applications that can be filed, the three different types of patents that can be obtained, and what rights are associated with the grant of a patent.

76.1.1 Utility, Design, and Plant Patents

When one speaks of obtaining a patent, it is ordinarily assumed that what is intended is a utility patent. Unless stated otherwise, the discussion of patents presented in this chapter applies only to U.S. patents, and principally to utility patents.

Utility patents are granted to protect processes, machines, articles of manufacture, and compositions of matter that are new, useful, and unobvious.

Design patents are granted to protect the ornamental appearances of articles of manufacture—that is, shapes, configurations, ornamentation, and other appearance-defining characteristics that are new, unobvious, and not dictated primarily by functional considerations.

Plant patents are granted to protect new varieties of plants that have been asexually reproduced, with the exception of tuber-propagated plants and those found in an uncultivated state. New varieties of roses and shrubs often are protected by plant patents.

Both utility and design patent protections may be obtained on some inventions. A utility patent typically will have claims that define novel combinations of structural features, techniques of manufacture, and/or methods of use of a product. A design patent typically will cover outer configuration features that are not essential to the function of the product but rather give the product an esthetically pleasing appearance.

Genetically engineered products may qualify for plant patent protection, for utility patent protection, and/or for other protections provided for by statute that differ from patents. Computer software and other computer-related products may qualify for design and/or utility patent protections. These are developing areas of intellectual property law.

76.1.2 Patent Terms and Expiration

Plant patents have a normal term of 17 years, measured from the dates these patents were granted (their *issue dates*).

Design patents that currently are in force have normal terms of 14 years, measured from their issue dates. Prior to a change of law that took effect during 1982, it was impossible for design patent owners to elect shorter terms of $3\frac{1}{2}$ or 7 years.

Utility patents that expired prior to June 8, 1995, had a normal term of 17 years, measured from their issue dates. Utility patents that (1) were in force on June 8, 1995, or (2) issue from applications that were filed prior to June 8, 1995, have normal terms that expire either 17 years, measured from their issue dates, or 20 years, measured from the filing date of the applications from which these patents issued, whichever is later. Utility patents that issue from applicants that were filed on or after June 8, 1995, have normal terms that expire 20 years from filing.

The filing date from which the 20-year measurement is taken to calculate the normal expiration date of a utility patent is the earliest applicable filing date. If, for example, a patent issues from a continuation application, a divisional application or a continuation-in-part application that claims the benefit of the filing date of an earlier filed "parent" application, the 20-year measurement is taken from the filing date of the "parent" application.

The normal term of a patent may be shortened due to a variety of circumstances. If, for example, a court of competent jurisdiction should declare that a patent is "invalid," the normal term of the patent will have been brought to an early close. In some instances, a "terminal disclaimer" may have been filed by the owner of a patent to cause early termination. The filing of a terminal disclaimer is sometimes required by the Patent and Trademark Office during the examination of an application that is so closely subject-matter-related to an earlier-filled application that there may be a danger that two patents having different expiration dates will issue covering substantially the same invention.

76.1.3 Four Types of Applications

Three types of patent applications are well known. A *utility* application is what one files to obtain a utility patent. A *design* application is what one files to obtain a design patent. A *plant* application is what one files to obtain a plant patent.

Effective June 8, 1995, it became possible to file one or more *provisional* applications as a precursor to the filing of a utility application. The filing of a provisional application will *not* result in the issuance of any kind of patent. In fact, no examination will be made of the merits of the invention described in a provisional application. Examination "on the merits" takes place only if a utility application is filed; and, if examination takes place, it centers on the content of the utility application, not on the content of any provisional applications that are referred to in the utility application.

The filing of a provisional application that adequately described an invention will establish a filing date that can be relied on in a later-filed utility application relating to the same invention (1) if the utility application is filed within one year of the filing date of the provisional application and (2) if the utility application makes proper reference to the provisional application. While the filing date of a provisional application can be relied on to establish a reduction to practice of an invention, the filing date of a provisional application does *not* start the 20-year clock that determines the normal expiration date of a utility patent.

Absent the filing of a implant application within one year from the filing date of a provisional application, the Patent and Trademark Office will destroy the provisional application once it has been pending a full year.

76.1.4 Why File a Provisional Application

If a provisional application will not be examined and will not result in the issuance of any form of patent whatsoever, why would one want to file a provisional application? Actually there are several reasons why the filing of one or more provisional applications may be advantageous before a full-blown utility application is put on file.

If foreign filing rights are to be preserved, it often is necessary for a U.S. application to be filed before *any* public disclosure is made of an invention so that one or more foreign applications can be filed within one year of the filing date of the U.S. application, with the result that the foreign applications will be afforded the benefit of the filing date of the U.S. application (due to a treaty referred to as the *Paris Convention*), thereby ensuring that the foreign applications comply with "absolute novelty" requirements of foreign patent law.

If the one-year grace period provided by U.S. law (which permits applicants to file a U.S. application anytime within a full one-year period from the date of the first activity that starts the clock running on the one-year grace period) is about to expire, it may be desirable to file a provisional application rather than a utility application because (1) preparing and filing a provisional application usually can be done less expensively, (2) preparing and filing a provisional application usually can be done more quickly, inasmuch as it usually involves less effort, and (3) everything that one may want to include in a utility application may not have been discerned (i.e., invention development and testing may still be underway), hence it may be desirable to postpone for as much as a full year the drafting and filing of a utility application.

If development work is still underway when a first provisional application is filed, and if the result of the development program brings additional invention improvements to light, it may be desirable (during the permitted period of one year between the filing of the first provisional application and the filing of a full-blown utility application) to file one or more additional provisional applications, all of which can be referred to and can have their filing dates relied upon when a utility application is filed within one year, measured from the filing date of the earliest-filed provisional application.

76.1.5 Understanding That a Patent Grants a "Negative Right"

It is surprisingly common to find that even those who hold several patents fail to properly understand the "negative" nature of the rights that are embodied in the grant of a patent.

What a patent grants is the "negative right" to "exclude others" from making, using, or selling an invention that is covered by the patent. *Not* included in the grant of a patent is a "positive right" enabling the patent owner to actually make, use, or sell the invention. In fact, a patent owner may be precluded, by the existence of other patents, from making, using, and selling his or her patented invention. Illustrating this often misunderstood concept is the following example, referred to as *The Parable of the Chair.*

If inventor A invents a three-legged stool at an early time when such an invention is not known to others, A's invention may be viewed as being "basic" to the art of seats, probably will be held to be patentable, and the grant of a patent probably will have the practical effect of enabling A both (1) to prevent others from making, using, and selling three-legged stools and (2) to be the only entity who *can* legally make, use, and sell three-legged stools during the term of A's patent.

If, during the term of A's patent, inventor B improves upon A's stool by adding a fourth leg for stability and an upright back for enhanced support and comfort (whereby a chair is born), and if B obtains a patent on his chair invention, the grant of B's chair patent will enable B to prevent others from making, using, and selling four-legged back-carrying seats. However, B's patent will do nothing at all to permit B to make, use, or sell chairs—the problem being that a four-legged seat having a back *infringes* A's patent because each of B's chairs *includes* three legs that support a seat, which is what A can exclude others from making, using, or selling. To legally make, use, or sell chairs, B must obtain a license from A.

And if, during the terms of the patents granted to A and B, C invents and patents the improvement of providing curved rocking rails that connect with leg bottoms, and arms that connect with the back and seat, thereby bringing into existence a rocking chair, C can exclude others during the term of his patent from making, using, or selling rocking chairs, but must obtain licenses from both A and B in order to make, use, or sell rocking chairs, for a rocking chair includes three legs and a seat and includes four legs, a seat, and a back.

Invention improvements may represent very legitimate subject matter for the grant of a patent. However, patents that cover invention improvements may not give the owners of these patents any right at all to make, use, or sell their patented inventions unless licenses are obtained from those who obtained patents on inventions that are more basic in nature. Once the terms of the more basic patents expire, owners of improvement patents then may be able to practice and profit from their inventions on an exclusive basis during the remaining portions of the terms of their patents.

76.2 WHAT CAN BE PATENTED AND BY WHOM

For an invention to be patentable, it must meet several requirements set up to ensure that patents are not issued irresponsibly. Some of these standards are complex to understand and apply. Let us simplify and summarize the essence of these requirements.

76.2.1 Ideas, Inventions, and Patentable Inventions

Invention is a misleading term because it is used in so many different senses. In one, it refers to the act of inventing. In another, it refers to the product of the act of inventing. In still another, the term designates a patentable invention, the implication mistakenly being that if an invention is not patentable, it is not an invention.

In the context of modern patent law, invention is the conception of a novel and useful contribution followed by its reduction to practice. Conception is the beginning of an invention; it is the creation in the mind of an inventor of a useful means for solving a particular problem. Reduction to practice can be either actual, as when an embodiment of the invention is tested to prove its successful operation under typical conditions of service, or constructive, as when a patent application is filed containing a complete description of the invention.

Ideas, per se, are not inventions and are not patentable. They are the tools of inventors, used in the development of inventions. Inventions are patentable only insofar as they meet certain criteria established by law. For an invention to be protectable by the grant of a utility patent, it must satisfy the following conditions:

1. Fit within one of the statutorily recognized classes of patentable subject matter
2. Be the true and original product of the person seeking to patent the invention as its inventor
3. Be new at the time of its invention by the person seeking to patent it
4. Be useful in the sense of having some beneficial use in society
5. Be nonobvious to one of ordinary skill in the art to which the subject matter of the invention pertains at the time of its invention
6. Satisfy certain statutory bars that require the inventor to proceed with due diligence in pursuing efforts to file and prosecute a patent application

76.2.2 The Requirement of Statutory Subject Matter

As stated in the Supreme Court decision of *Kewanee Oil v. Bicron Corp.*, 416 U.S. 470, 181 U.S.P.Q. 673 (1974), no utility patent is available for any discovery, however, useful, novel, and nonobvious, unless it falls within one of the categories of patentable subject matter prescribed by Section 101 of Title 35 of the United States Code. Section 101 makes this process.

Whoever invents or discovers a new and useful process, machine, manufacture, or composition of matter, or any new and useful improvement thereof may obtain a patent therefore, subject to the conditions and requirements of this title.

The effect of establishing a series of statutory classes of eligible subject matter has been to limit the pursuit of patent protection to the useful arts. Patents directed to processes, machines, articles of manufacture, and compositions of matter have come to be referred to as utility patents, inasmuch as these statutorily recognized classes encompass the useful arts.

Three of the four statutorily recognized classes of eligible subject matter may be thought of as products, namely, machines, manufactures, and compositions of matter. *Machine* has been interpreted in a relatively broad manner to include a wide variety of mechanisms and mechanical elements. *Manufactures* is essentially a catch-all term covering products other than machines and compositions of matter. *Compositions of matter,* another broad term, embraces such elements as new molecules, chemical compounds, mixtures, alloys, and the like. *Manufactures* and *compositions of matter* arguably include such genetically engineered life forms as are not products of nature. The fourth class, *processes,* relates to procedures leading to useful results.

Subject matter held to be ineligible for patent protection includes printed matter, products of nature, ideas, and scientific principles. Alleged inventions of perpetual motion machines are refused patents. A mixture of ingredients such as foods and medicines cannot be patented unless there is more to the mixture than the mere cumulative effect of its components. So-called patent medicines are seldom patented.

While no patent can be issued on an old product despite the fact that it has been found to be derivable through a new process, the new process for producing the product may well be patentable. That a product has been reduced to a purer state than was previously available in the prior art does not render the product patentable, but the process of purification may be patentable. A new use for an old product does not entitle one to obtain product patent protection, but may entitle one to obtain process patent protection, assuming the process meets other statutory requirements.

A newly discovered law of nature, regardless of its importance, is not entitled to patent protection. Methods of conducting business and processes that either require a mental step to be performed or depend on aesthetic or emotional reactions have been held not to constitute statutory subject matter.

While the requirement of statutory subject matter fails principally within the bounds of 35 U.S.C. 101, other laws also operate to restrict the patenting of certain types of subject matter. For example, several statutes have been passed by Congress affecting patent rights in subject matter relating to atomic energy, aeronautics, and space. Still another statute empowers the Commissioner of Patents and Trademarks to issue secrecy orders regarding patent applications disclosing inventions that might be detrimental to the national security of the United States.

The foreign filing of patent applications on inventions made in the United States is prohibited for a brief period of time until a license has been granted by the Commissioner of Patents and Trademarks to permit foreign filing. This prohibition period enables the Patent and Trademark Office to review newly filed applications, locate any containing subject matter that may pose concerns to national security, and, after consulting with other appropriate agencies of government, issue secrecy orders preventing the contents of these applications from being publicly disclosed. If a secrecy order issues, an inventor may be barred from filing applications abroad on penalty of imprisonment for up to two years or a $10,000 fine or both. In the event a patent application is withheld under a secrecy order, the patent owner has a right to recover compensation from the government for damage caused by the secrecy order and/or for the use the government may have made of the invention.

Licenses permitting expedited foreign filing are almost always automatically granted by the Patent and Trademark Office at the time of issuing an official filing receipt, which advises the inventor of the filing date and serial number assigned to his or her application. Official filing receipts usually issue within a month of the date of filing and bear a statement attesting to the grant of a foreign filing license.

76.2.3 The Requirement of Originality of Inventorship

Under U.S. patent law, only the true and original inventor or inventors may apply to obtain patent protection. If the inventor has derived an invention from any other source or person, he or she is not entitled to apply for or obtain a patent.

The laws of our country are strict regarding the naming of the proper inventor or joint inventors in a patent application. When one person acting alone conceives an invention, he or she is the sole inventor and he or she alone must be named as the inventor in a patent application filed on that invention. When a plurality of people contribute to the conception of an invention, these persons must be named as joint inventors if they have contributed to the inventive features that are claimed in a patent application filed on the invention.

Joint inventorship occurs when two or more persons collaborate in some fashion, with each contributing to conception. It is not necessary that exactly the same idea should have occurred to each of the collaborators at the same time. Section 116 of Title 35 of the United States Code includes the following provision:

Inventors may apply for a patent jointly even though (1) they did not physically work together or at the same time, (2) each did not make the same type or amount of contribution, or (3) each did not make a contribution to the subject matter of every claim of the patent.

Those who may have assisted the inventor or inventors by providing funds or materials for development or by building prototypes are not deemed to be inventors unless they contributed to the conception of the invention. While inventors may have a contractual obligation to assign rights in an invention to their employers, this obligation, absent a contribution to conception, does not entitle a supervisor or an employer to be named as an inventor. When a substantial number of patentable features relating to a single overall development have occurred as the result of different combinations of sole inventors acting independently and/or joint inventors collaborating at different times, the patent law places a burden on the inventors to sort out "who invented what." Patent protection on the overall development must be pursued in the form of a number of separate patent applications, each directed to such patentable aspects of the development as originated with a different inventor or group of inventors. In this respect, U.S. patent practice is unlike that of many foreign countries, where the company for whom all the inventors work is often permitted to file a single patent application in its own name covering the overall development.

Misjoinder of inventors occurs when a person who is not a joint inventor has been named as such in a patent application. *Nonjoinder of inventors* occurs when there has been a failure to include a person who should have been named as a joint inventor. *Misdesignation of inventorship* occurs when none of the true inventors are named in an application. Only in recent years has correction of a misdesignation been permitted. If a problem of misjoinder, nonjoinder, or misdesignation has arisen without deceptive intent, provisions of the patent law permit correction of the error as long as such is pursued with diligence following the discovery.

76.2.4 The Requirement of Novelty

Section 101 of Title 35 of the United States Code requires that a patentable invention be new. What is meant by *new* is defined in Sections 102(a), 102(e), and 102(g). Section 102(a) bars the issuance of a patent on an invention "known or used by others in this country, or patented or described in a printed publication in this or a foreign country, before the invention thereof by the applicant for patent." Section 102(e) bars the issuance of a patent on an invention "described in a patent granted on an application for patent by another filed in the United States before the invention thereof by the applicant for patent, or in an international application by another." Section 102(g) bars the issuance of a patent on an invention that "before the applicant's invention thereof . . . was made in this country by another who had not abandoned, suppressed, or concealed it."

These novelty requirements amount to negative rules of invention, the effect of which is to prohibit the issuance of a patent on an invention if the invention is not new. The novelty requirements of 35 U.S.C. 102 should not be confused with the statutory bar requirements of 35 U.S.C. 102, which are discussed in Section 76.2.7. A comparison of the novelty and statutory bar requirements of 35 U.S.C. 102 is presented in Table 76.1. The statutory bar requirements are distinguishable from the novelty requirements in that they do not relate to the newness of the invention, but to ways an inventor, who would otherwise have been able to apply for patent protection, has lost that right by tardiness.

To understand the novelty requirements of 35 U.S.C. 102, one must understand the concept of anticipation. A claimed invention is anticipated if a single prior art reference contains all the essential elements of the claimed invention. If teachings from more than one reference must be combined to

---

Table 76.1   Summary of the Novelty and Statutory Bar Requirements of 35 U.S.C. 102

Novelty Requirements

One may not patent an invention if, prior to its date of invention, the invention was any of the following:

1. Known or used by others in this country.
2. Patented or described in a printed publication in this or a foreign country.
3. Described in a patent granted on an application for patent by another filed in the United States.
4. Made in this country by another who had not abandoned, suppressed, or concealed it.

Statutory Bar Requirements

One may not patent an invention he or she has previously abandoned. One may not patent an invention if, more than one year prior to the time his or her patent application is filed, the invention was any of the following:

1. Patented or described in a printed publication in this or a foreign country.
2. In public use or on sale in this country.
3. Made the subject of an inventor's certificate in a foregin country.
4. Made the subject of a foreign patent application, which results in the issuance of a foreign patent before an application is filed in this country.

show that the claimed combination of elements exists, there is no anticipation, and novelty exists. Combining references to render a claimed invention unpatentable brings into play the nonobviousness requirements of 35 U.S.C. 103, not the novelty requirement of 35 U.S.C. 102. Novelty hinges on anticipation and is a much easier concept to understand and apply than that of nonobviousness.

35 U.S.C. 102(a) Known or Used by Others in This Country Prior to the Applicant's Invention

In interpreting whether an invention has been known or used in this country, it has been held that the knowledge must consist of a complete and adequate description of the claimed invention and that this knowledge must be available, in some form, to the public. Prior use of an invention in this country by another will be disabling only if the invention in question has actually been reduced to practice and its use has been accessible to the public in some minimal sense. For a prior use to be disabling under Section 102(a), the use must have been of a complete and operable product or process that has been reduced to practice.

35 U.S.C. 102(a) Described in a Printed Publication in This or a Foreign Country Prior to the Applicant's Invention

For a printed publication to constitute a full anticipation of a claimed invention, the printed publication must adequately describe the claimed invention. The description must be such that it enables a person of ordinary skill in the art to which the invention pertains to understand and make the invention. The question of whether a publication has taken place is construed quite liberally by the courts to include almost any act that might legitimately constitute publication. The presence of a single thesis in a college library has been held to constitute publication. Similar liberality has been applied in construing the meaning of the term *printed*.

35 U.S.C. 102(a) Patented in This or a Foreign Country

An invention is not deemed to be novel if it was patented in this country or any foreign country prior to the applicant's date of invention. For a patent to constitute a full anticipation and thereby render an invention unpatentable for lack of novelty, the patent must provide an adequate, operable description of the invention. The standard to be applied under Section 102(a) is whether the patent "describes" a claimed invention. A pending patent application is treated as constituting a "patent" for purposes of applying Section 102(a) as of the date of its issuance.

35 U.S.C. 102(e) Described in a Patent Filed in This Country Prior to the Applicant's Invention

Section 102(e) prescribes that if another inventor has applied to protect an invention before you invent the same invention, you cannot patent the invention. The effective date of a U.S. patent, for purposes of a Section 102(e) determination, is the filing date of its application, rather than the date of patent issuance.

35 U.S.C. 102(g) Abandoned, Suppressed, or Concealed

For the prior invention of another person to stand as an obstacle to the novelty of one's invention under Section 102(g), the invention made by another must not have been abandoned, suppressed, or concealed. Abandonment, suppression, or concealment may be found when an inventor has been inactive for a significant period of time in pursuing reduction to practice of an invention. This is particularly true when the inventor's becoming active again has been spurred by knowledge of entry into the field of a second inventor.

76.2.5 The Requirement of Utility

To comply with the utility requirements of U.S. patent law, an invention must be capable of achieving some minimal useful purpose that is not illegal, immoral, or contrary to public policy. The invention must be operable and capable of being used for some beneficial purpose. The invention does not need to be a commercially successful product in order to satisfy the requirement of utility. While the requirement of utility is ordinarily a fairly easy one to meet, problems do occasionally arise with chemical compounds and processes, particularly in conjunction with various types of drugs. An invention incapable of being used to effect the proposed object of the invention may be held to fail the utility requirement.

76.2.6 The Requirement of Nonobviousness

The purpose of the novelty requirements of 35 U.S.C. 102 and the nonobviousness requirement of 35 U.S.C. 103 are the same—to limit the issuance of patents to those innovations that do, in fact, advance the state of the useful arts. While the requirements of novelty and nonobviousness may seem very much alike, the requirement of nonobviousness is a more sweeping one. This requirement maintains that if it would have been obvious (at the time an invention was made) to anyone ordinarily skilled in the art to produce the invention in the manner disclosed, then the invention does not rise to the dignity of a patentable invention and is therefore not entitled to patent protection.

The question of nonobviousness must be wrestled with by patent applicants in the event the Patent and Trademark Office rejects some or all their claims based on an assertion that the claimed invention is obvious in view of the teaching of one or a combination of two or more prior art references. When a combination of references is relied on in rejecting a claim, the argument the examiner is making is that it is obvious to combine the teachings of these references to produce the claimed invention. When such a rejection has been made, the burden is on the applicant to establish to the satisfaction of the examiner that the proposed combination of references would not have been obvious to one skilled in the art at the time the invention was made; and/or that, even if the proposed combination of references is appropriate, it still does not teach or suggest the claimed invention.

In an effort to ascertain whether a new development is nonobvious, the particular facts and circumstances surrounding the development must be considered and weighed as a whole. While the manner in which an invention was made must not be considered to negate the patentability of an invention, care must be taken to ensure that the question of nonobviousness is judged as of the time the invention was made and in light of the then existing knowledge and state of the art. This test of nonobviousness has been found to be an extremely difficult one for courts to apply.

The statutory language prescribing the nonobviousness requirement appears at Title 35, Section 103, stating:

> *A patent may not be obtained . . . if the differences between the subject matter sought to be patented and the prior art are such that the subject matter as a whole would have been obvious at the time the invention was made to a person having ordinary skill in the art to which said subject matter pertains.*

In the landmark decision of *Graham v. John Deere,* 383 U.S.1, 148 U.S.P.Q. 459 (1966), the U.S. Supreme Court held that several basic factual inquiries should be made in determining nonobviousness. These inquiries prescribe a four-step procedure or approach for judging nonobviousness. First, the scope and content of the prior art in the relevant field or fields must be ascertained. Second, the level of ordinary skill in the relevant field or fields must be ascertained. Second, the level of ordinary skill in the pertinent art is determined. Third, the differences between the prior art and the claims at issue are examined. Fourth and finally, a determination is made as to whether these differences would have been obvious to one of ordinary skill in the applicable art at the time the invention was made.

76.2.7 Statutory Bar Requirements

Despite the fact that an invention may be new, useful, and nonobvious and that it may satisfy the other requirements of the patent law, an inventor can still lose the right to pursue patent protection on the invention unless he or she complies with certain requirements of the law called *statutory bars.* The statutory bar requirements ensure that inventors will act with diligence in pursuing patent protection.

While 35 U.S.C. 102 includes both the novelty and the statutory bar requirements of the law, it intertwines these requirements in a complex way that is easily misinterpreted. The novelty requirements are basic to a determination of patentability in the same sense as are the requirements of statutory subject matter, originality, and nonobviousness. The statutory bar requirements are not basic to a determination of patentability, but rather operate to decline patent protection to an invention that may have been patentable at one time.

Section 102(b) bars the issuance of a patent if an invention was "in public use or on sale" in the United States more than one year prior to the date of the application for a patent. Section 102(c) bars the issuance of a patent if a patent applicant has previously abandoned the invention. Section 102(d) bars the issuance of a patent if the applicant has caused the invention to be first patented in a foreign country and has failed to file an application in the United States within one year after filing for a patent in a foreign country. Table 76.1 summarizes the statutory bar requirements of Section 102.

Once an invention has been made, the inventor is under no specific duty to file a patent application within any certain period of time. However, should one of the "triggering" events described in Section 102 occur, regardless of whether this occurrence may have been the result of action taken by the inventor or by actions of others, the inventor must apply for a patent within the prescribed period of time or be barred from obtaining a patent.

Some of the events that trigger statutory bar provisions are the patenting of an invention in this or a foreign country; the describing in a printed publication of the invention in this or a foreign country; the public use of the invention in this country; or putting the invention on sale in this country. Some public uses and putting an invention on sale in this country will not trigger statutory bars if these activities were incidental to experimentation. Whether a particular activity amounts to experimental use has been the subject of much judicial dissension. The doctrine of experimental use is a difficult one to apply because of the conflicting decisions issued on this subject.

Certainly, the safest approach to take is to file for patent protection well within one year of any event leading to the possibility of any statutory bar coming into play. If foreign patent protections are to be sought, the safest approach is to file an application in this country before any public disclosure is made of the invention.

76.3 PREPARING TO APPLY FOR A PATENT

Conducting a patentability search and preparing a patent application are two of the most important stages in efforts to pursue patent protection. This section points out pitfalls to avoid in both stages.

76.3.1 The Patentability Search

Conducting a patentability search prior to the preparation of a patent application can be extremely beneficial even when an inventor is convinced that no one has introduced a similar invention into the marketplace. A properly performed patentability study will guide not only the determination of the scope of patent protection to be sought, but also the claim-drafting approaches to be used. In almost every instance, a patent attorney who has at hand the results of a carefully conducted patentability study can do a better job of drafting a patent application, thereby helping to ensure that it will be prosecuted smoothly, at minimal expense, through the rigors of examination in the Patent and Trademark Office.

Occasionally, a patentability search will indicate that an invention is totally unpatentable. When this is the case, the search will have saved the inventor the cost of preparing and filing a patent application. At times a patentability search turns up one or more newly issued patents that pose infringement concerns. A patentability search is not, however, as extensive a search as is one conducted to locate possible infringement concerns when a great deal of money is being invested in a new product.

Some reasonable limitation is ordinarily imposed on the scope of a patentability search to keep search costs within a relatively small budget. The usual patentability search covers only U.S. patents and does not extend to foreign patents or to publications. Only the most pertinent Patent and Trademark Office subclasses are covered. However, despite the fact that patentability studies are not of exhaustive scope, a carefully conducted patentability search ordinarily can be relied on to give a decent indication of whether an invention is worthy of pursuing patent coverage to protect.

Searches do occasionally fail to turn up one or more pertinent references despite the best efforts of a competent searcher. Several reasons explain why a reference may be missed. One is that the files of the Public Search Room of the Patent and Trademark Office are incomplete. The Patent and Trademark Office estimates that as many as 7% of the Search Room references are missing or misfiled. Another reason is that the Public Search Room files do not contain some Patent Office subclasses. The searcher must review these missing subclasses in the "examiners' art," the files of patents used by Patent and Trademark Office examiners, where the examiners are free to remove references and take them to their offices as they see fit. Since most patents are cross-referenced in several subclasses, a careful searcher will try to ensure that the field encompassed by a search extends to enough subclasses that patents are located that should have been found in other subclasses, but were not.

76.3.2 Putting the Invention in Proper Perspective

It is vitally important that a client take whatever time is needed to make certain that his or her patent attorney fully understands the character of an invention before the attorney undertakes the preparation of a patent application. The patent attorney should be given an opportunity to talk with those involved in the development effort from which an invention has emerged. He or she should be told what features these people believe are important to protect. Moreover, the basic history of the art to which the invention relates should be described, together with a discussion of the efforts made by others to address the problems solved by the present invention.

The client should also convey to his or her patent attorney how the present invention fits into the client's overall scheme of developmental activities. Much can be done in drafting a patent application to lay the groundwork for protection of future developments. Additionally, one's patent attorney needs to know how product liability concerns may arise with regard to the present invention so that statements he or she makes in the patent application will not be used to the client's detriment in product liability litigation. Personal injury lawyers have been known to scrutinize the representations made in a manufacturer's patents to find language that will assist in obtaining recoveries for persons injured by patented as well as unpatented inventions of the manufacturer.

Before preparation of an application is begun, careful consideration should be given to the scope and type of claims that will be included. In many instances, it is possible to pursue both process and product claims. Also, in many instances, it is possible to present claims approaching the invention from varying viewpoints so different combinations of features can be covered. Frequently, it is possible to couch at least two of the broadest claims in different language so efforts of competitors to design around the claim language will be frustrated.

Careful considerations must be given to approaches competitors may take in efforts to design around the claimed invention. The full range of invention equivalents also needs to be taken into account so that claims of appropriate scope will be presented in the patent application.

76.3.3 Preparing the Application

A well drafted patent application is a work of art. It should be a readable and understandable teaching document. If it is not, insist that your patent attorney rework the document. A patent application that accurately describes an invention without setting forth the requisite information in a clear and convincing format may be legally sufficient, but it does not represent the quality of work a client has the right to expect.

A well drafted patent application should include an introductory section that explains accurately, yet interestingly, the background of the invention and the character of the problems that are overcome. It should discuss the closest prior art known to the applicant and should indicate how the invention patentably differs from prior art proposals. It should present a summary of the invention that brings out the major advantages of the invention and explains how prior-art drawbacks are overcome. These elements of a patent application may occupy several typed pages. They constitute an introduction to the remainder of the document.

Following this introductory section, the application should present a brief description of such drawings as may accompany the application. Then follows a detailed description of the best mode known to the inventor for carrying out the invention. In the detailed description, one or more preferred embodiments of the invention are described in sufficient detail to enable a person having ordinary skill in the art to which the invention pertains to practice the invention. While some engineering details, such as dimensions, materials of construction, circuit component values, and the like, may be omitted, all details critical to the practice of the invention must be included. If there is any question about the essential character of a detail, prudent practice would dictate its inclusion.

The written portion of the application concludes with a series of claims. The claims are the most difficult part of the application to prepare. While the claims tend to be the most confusing part of the application, the applicant should spend enough time wrestling with the claims and/or discussing this section with the patent attorney to make certain that the content of the claims is fully understood. Legal gibberish should be avoided, such as endless uses of the word *said*. Elements unessential to the practice of the invention should be omitted from the claims. Essential elements should be described in the broadest possible terms in at least some of the claims so the equivalents of the preferred embodiment of the invention will be covered.

The patent application will usually include one or more sheets of drawings and will be accompanied by a suitable declaration or oath to be signed by the inventor or inventors. The drawings of a patent application should illustrate each feature essential to the practice of the invention and show every feature to which reference is made in the claims. The drawings must comply in size and format with a lengthy set of technical rules promulgated and frequently updated by the Patent and Trademark Office. The preparation of patent drawings is ordinarily best left to an experienced patent draftsperson.

If a patent application is prepared properly, it should pave the way for smooth handling of the patent application during its prosecution. If a patent application properly tells the story of the invention, it should constitute a teaching document that will stand on its own and be capable of educating a court regarding the character of the art to which the invention pertains, as well as the import of this invention to that art. Since patent suits are tried before judges who rarely have technical backgrounds, it is important that a patent application make an effort to present the basic features of the invention in terms understandable by those having no technical training. It is unusual for an invention to be so impossibly complex that its basic thrust defies description in fairly simple terms. A patent application is suspect if it wholly fails to set forth, at some point, the pitch of the invention in terms a grade school student can grasp.

76.3.4 Enablement, Best Mode, Description, and Distinctness Requirements

Once a patent application has been prepared and is in the hands of the inventor for review, it is important that the inventor keep in mind the enablement, best mode, description, and distinctness requirements of the patent law.

The enablement requirement calls for the patent application to present sufficient information to enable a person skilled in the relevant art to make and use the invention. The disclosure presented in the application must be such that it does not require one skilled in the art to experiment to any appreciable degree to practice the invention.

The best-mode requirement mandates that an inventor disclose, at the time he or she files a patent application, the best mode he or she then knows about for carrying out or practicing the invention.

The description requirement also relates to the descriptive part of a patent application and the support it must provide for any claims that may need to be added after the application has been filed. Even though a patent application may adequately teach how to make and use the subject matter of the claimed invention, a problem can arise during the prosecution of a patent application where one determines it is desirable to add claims that differ in language from those filed originally. If the claim language one wants to add does not find full support in the originally filed application, the benefit of the original filing date will be lost with regard to the subject matter of the claims to be added—a problem referred to as *late claiming*, about which much has been written in court decisions of the past 40 years. Therefore, in reviewing a patent application prior to its being executed, an inventor should keep in mind that the description that forms a part of the application should include support for any language he or she may later want to incorporate in the claims of the application.

The distinctness requirement applies to the content of the claims. In reviewing the claims of a patent application, an inventor should endeavor to make certain the claims particularly point out and distinctly claim the subject matter that he or she regards as his or her invention. The claims must be *definite* in the sense that their language must clearly set forth the area over which an applicant seeks exclusive rights. The language used in the claims must find antecedent support in the descriptive portion of the application. The claims must not include within their scope of coverage any prior art known to the inventor, and yet should present the invention in the broadest possible terms that patentably distinguish the invention over the prior art.

76.3.5 Functional Language in Claims

While functional language in claims may tend to draw objection, there is statutory support for using a particular type of functional claim language. Section 112 of Title 35 of the United States Code includes this statement:

> *An element in a claim for a combination may be expressed as a means or step for performing a specified function without the recital of structure, material, or acts in support thereof, and such claim shall be construed to cover the corresponding structure, material, or acts described in the specification and equivalents thereof.*

Using a *means-plus-function* or a *step-plus-function* format to claim an invention can be one of the most effective avenues to take in an effort to achieve the broadest possible coverage of alternative approaches that competitors may explore. However, in drafting a claim in mean-plus-function format, care must be taken to ensure that what is being claimed amounts to more than a single means (i.e., a single element defined in means-plus-function format), since the requirement of the patent law that means-plus-function language be used only in a claim for a *combination* is not met by such a claim. Such a claim is deemed to be of undue breadth for, in essence, it claims every conceivable means for achieving a stated result.

During recent years, much has been written regarding how a means- or step-plus-function claim limitation should be interpreted. This continues to be a developing area of patent law.

76.3.6 Product-by-Process Claims

In some instances, it is possible to claim a product by describing the process or method of its manufacture.

Even though a *product-by-process* claim is limited and defined by the process it recites, a determination of patentability of the claimed product does not depend on its method of production. If the claimed product is the same as or obvious from a product of the prior art, the claimed product is deemed unpatentable even though the prior product was made by a different process. If the prior art discloses a product that is identical with or only slightly different than a claimed product, an alternative rejection based either on Section 102 or 103 may be given by the Patent and Trademark Office. Once the Patent and Trademark Office has rejected a product-by-process claim by showing that the claimed product appears to be the same or similar to a prior art product, although produced by a different process, the burden falls on the applicant to prove that there exists an unobvious difference between the claimed product and the prior art product.

76.3.7 Claim Format

A patent applicant has some freedom in selecting the terminology he uses to define and claim his invention, for it has long been held that "and applicant is his own lexicographer." However, the meanings that an applicant assigns to the terminology he or she uses must not be repugnant to the well known usages of such terminology. When an applicant does not define the terms he uses, such terms must be given their "plain meaning," namely the meanings given to such terms by those of ordinary skill in the relevant art.

Each claim is a complete sentence. In many instances the first part of the sentence of each claim appears at the beginning of the claims section and reads, "What is claimed is:." Each claim typically includes three parts: preamble, transition, and body. The preamble introduces the claim by summarizing the field of the invention, its relation to the prior art, and its intended use, or the like. The transition is a word or phrase connecting the preamble to the body. The terms *comprises* or *comprising* often perform this function. The body is the listing of elements and limitations that define the scope of what is being claimed.

Claims are either *independent* or *dependent*. An independent claim stands on its own and makes no reference to any other claim. A dependent claim refers to another claim that may be independent or dependent, and adds to the subject matter of the referenced claim. If a dependent claim depends from (makes reference to) more than one other claim, it is called a *multiple dependent* claim.

One type of claim format that can be used gained notoriety in a 1917 decision of the Commissioner of Patents, *Ex parte Jepson,* 1917 C.D. 62. In a claim of the *Jepson* format, the preamble recites all the elements deemed to be old, the body of the claim includes only such new elements as constitute improvements, and the transition separates the old from the new. The Patent and Trademark Office favors the use of *Jepson*-type claims since this type of claim is thought to assist in segregating what is old in the art from what the applicant claims as his or her invention.

In 1966, the Patent and Trademark Office sought to encourage the use of *Jepson*-type claims by prescribing the following rule 75(e):

> *Where the nature of the case admits, as in the case of an improvement, any independent claim should contain in the following order, (1) a preamble comprising a general description of the elements or steps of the claimed combination which are conventional or known, (2) a phrase such as "wherein the improvement comprises," and (3) those elements, steps and/or relationships which constitute that portion of the claimed combination which the applicant considers as the new or improved portion.*

Thankfully, the use of the term *should* in Rule 75(e) makes use of *Jepson*-type claims permissive rather than mandatory. Many instances occur when it is desirable to include several distinctly old elements in the body of the claim. The preamble in a *Jepson*-type claim has been held to constitute a limitation for purposes of determining patentability and infringement, while the preambles of claims presented in other types of format may not constitute limitations. A proper understanding of the consequences of presenting claims in various types of formats and the benefits thereby obtained will be taken into account by one's patent attorney.

76.3.8 Executing the Application

Once an inventor has satisfied himself or herself with the content of a proposed patent application, he or she should read carefully the oath or declaration accompanying the application. The required content of this formal document recently has been simplified. In it the inventor states that he or she

1. Has reviewed and understands the content of the application, including the claims, as amended by any amendment specifically referred to in the oath or declaration
2. Believes the named inventor or inventors to be the original and first inventor or inventors of the subject matter which is claimed and for which a patent is sought
3. Acknowledges the duty to disclose to the Patent and Trademark Office during examination of the application all information known to the person to be material to patentability

If the application is being filed as a division, continuation, or continuation-in-part of one or more co-pending parent applications, the parent case or cases are identified in the oath or declaration. Additionally, if a claim to the benefit of a foreign-filed application is being made, it is recited in the oath or declaration.

Absolutely no changes should be made in any part of a patent application once it has been executed. If some change, no matter how ridiculously minor, is found to be required after an application has been signed, the executed oath or declaration must be destroyed and a new one signed after the application has been corrected. If an application is executed without having been inspected by the applicant or is altered after having been executed, it may be stricken from the files of the Patent and Trademark Office.

76.3.9 Patent and Trademark Office Fees

The Office charges a fee to file an application, a fee to issue a patent, fees to maintain a patent if it is to be kept alive for its full available term, and a host of other fees for such things as obtaining an extension of time to respond to an Office Action. The schedule of fees charged by the Office is updated periodically, usually resulting in fee increases as the Office has increasingly become self-supporting. Such fee increases often take effect on or about October 1, when the government's new fiscal year begins. This has been known to result in increased numbers of September filings of applications during years when sizable fee increases have taken effect.

As of this writing, the basic fee required to file an application for a utility patent has increased more than 1500-fold to $790 since it was first set at 50 cents under the Patent Act of 1790.

In addition to the basic fee of $790, $82 is charged for each independent claim in excess of a total of three; $22 is charged for each claim of any kind in excess of a total of 20, and $270 is charged for any application that includes one or more multiple dependent claims. However, if the applicant is entitled to claim the benefits of small entity status, the entire filing fee (including each of the filing fee components just described) is halved, as are most other fees that are associated with the handling of a patent application.

Provisional applications require a $150 filing fee that may be halved for small entities. Applications for design patents require a filing fee that presently stands at $330 unless small-entity status is established, whereupon this fee also may be halved. Plant patent applications require a $540 filing fee that may also be halved for small entities.

New rules now permit the Office to assign a filing date before the filing fee and oath of declaration have been received. While the filing fee and an oath or declaration are still needed to complete an application, a filing date will now be assigned as of the date of receipt of the descriptive portion of an application (known as the specification), accompanied by at least one claim, any required drawings, and a statement of the names of the inventors.

The issue fee charged by the Office for issuing a utility patent on an allowed application stands at $1320. Establishing a right to the benefits of small entity status permits reduction of this fee to $660. The issue fee for a design application is $450, which also may be halved with the establishment of small-entity status. A plant patent requires an issue fee of $670, which may be halved for small entities. There is no issue fee associated with a provisional application since a provisional application does not issue as a patent unless it is supplemented within one year of its filing date by the filing of a complete utility application.

Maintenance fees must be paid to keep an issued utility patent in force during its term. No maintenance fees are charged on design or plant patents, or on utility patents that have issued from applications filed before December 12, 1980. As of this writing, maintenance fees of $1050, $2100 and $3160 are due no later than $3\frac{1}{2}$, $7\frac{1}{2}$, and $11\frac{1}{2}$ years, respectively, from a utility patent's issue date. Qualification for the benefits of small entity status allows these fees to be reduced to $525, $1050, and $1580, respectively. Failure to timely pay any maintenance fee, or to late-pay it during a six-month grace period following its due date accompanied by a late payment surcharge of $130 ($65 for small entities), will cause a patent to lapse permanently.

76.3.10   Small Entity Status

The practice of providing half-price fees to individual inventors, nonprofit organizations, and small businesses came into existence concurrently with the implementation of an October 1, 1982, fee increase.

Qualification for small-entity status requires only the filing of a verified statement prior to or with the first fee paid as a small entity. All entities having rights with respect to an application or patent must each be able to qualify for small entity status; otherwise, small entity status cannot be achieved. Statements as to qualification as a small entity must be filed by all entities having rights with respect to an application or patent in order to qualify. Once qualification has been achieved, there is a continuing duty to advise the Office before or at the time of paying the next fee if qualification for small-entity status has been lost.

Those who qualify for small-entity status include

1.   A sole inventor who has not transferred his or her rights and is under no obligation to transfer his or her rights to an entity that fails to qualify

2.   Joint inventors where no one among them has transferred his or her rights and is under no obligation to transfer his or her rights to an entity that fails to qualify

3.   A nonprofit organization such as an institution of higher education or an IRS-qualified and exempted nonprofit organization

4.   A small business that has no more than 500 employees after taking into account the average number of employees (including full-time, part-time, and temporary) during the fiscal year of the business entity in question, and of its affiliates, with the term *affiliate* being defined by a broad-reaching "control" test

Attempting to establish small-entity status fraudulently or establishing such status improperly or through gross negligence, is considered a fraud on the Office. An application could be disallowed for such an act. Failure to establish small-entity status on a timely basis forfeits the right to small-entity status benefits with respect to a fee being paid. However, if small-entity status is established within two months after a fee was paid, a refund of the excess amount paid may be obtained if a request for a refund is received by the Patent and Trademark Office within the two-month period. A good-faith error made in establishing small-entity status may be excused by paying any deficient fees. However, if the payment is made more than three months after the error occurred, a verified statement establishing good faith and explaining the error must be filed.

76.3.11   Express Mail Filing

During 1983, a new procedure was adopted by the Office that permits any paper or fee to be filed with the Office by using the "Express Mail Post Office to Addressee" service of the U.S. Postal Service. When this is done, the filing date of the paper or fee will be that shown on the "Express Mail" mailing label.

To qualify for the filed-when-mailed advantage, each paper must bear the number of the "Express Mail" mailing label, must be addressed to the Assistant Commissioner for Patents, Washington, DC 20231, and must comply with other requirements that are changed from time to time.

The practical and very important effect of this new procedure is to eliminate the hassle that has long been associated with the last-minute attempts to effect physical delivery of patent applications and other papers and fee payments to the Office in time to meet a bar date or comply with a convention filing date.

76.4   PROSECUTING A PENDING PATENT APPLICATION

Once an executed patent application has been received by the Patent and Trademark Office, the patent application is said to be pending. The prosecution period of a patent application is the time during which an application is pending, it begins when a patent application is filed in the Patent and Trademark Office and continues until either a patent is granted or the application is abandoned. The activities that take place during this time are referred to as *prosecution.*

76.4.1   Patent Pending

Once an application for a patent has been received by the Patent and Trademark Office, the applicant may mark products embodying the invention and literature or drawings relating to the invention with an indication of "Patent Pending" or "Patent Applied For." These expressions mean a patent application has been filed and has neither been abandoned nor issued as a application. The terms do not mean that the Patent and Trademark Office has taken up examination of the merits of an application, much less approved the application for issuance as a patent.

The fact that a patent application has been filed, or is pending or applied for, does not provide any protection against infringement by competitors. While pending patent applications are held in secrecy by the Patent and Trademark Office and therefore do not constitute a source of information available to competitors regarding the activities of an inventor, nothing prevents competitors from independently developing substantially the same invention and seeking to market it. Unless and until a patent actually issues, there is no legal basis for stopping a competitor from purchasing a product bearing a designation "Patent Pending" and copying the invention embodied in the purchased product. Infringement liability does not attach to infringements that may have occurred prior to the issue date of a patent.

As a practical matter, however, marking products with the designation "Patent Pending" often has the effect of discouraging competitors from copying an invention, whereby the term of the patent that eventually issues may effectively be extended to include the period during which the application is pending. In many instances, competitors will not risk a substantial investment in preparation for the manufacture and merchandising of a product marked with the designation "Patent Pending," for they know their efforts may be legally interrupted as soon as a patent issues.

76.4.2   Secrecy of Pending Applications

With the exception of applications filed to reissue and requests to re-examine existing patents, pending patent applications are maintained in strictest confidence by the Patent and Trademark Office. No information regarding a pending application will be given out by the Office without authority from the applicant or owner of the application. However, if an interested third party learns of the pendency of an application, he or she may file a protest to its issuance.

The file of a pending application can only be inspected as a matter of right by the named inventor, an assignee of record, an exclusive licensee, an attorney of record, or such persons as have received written authority from someone permitted by right to inspect the file. This provision of secrecy extends to abandoned applications as well as to pending applications. In the event an abandoned application is referred to in an issued patent, access to the file of the abandoned case will be granted to members of the public on request. Should a pending patent application be referred to in an issued patent, access may usually be obtained by petition. All reissue applications are open to inspection by the general public.

76.4.3   Duty of Candor

The Patent and Trademark Office has placed increased emphasis on the duty an applicant has to deal candidly with the Patent and Trademark Office.

In accordance with Patent Office guidelines, a patent applicant is urged to submit an Information Disclosure Statement either concurrently with the filing of an application or within three months of its filing. When these guidelines were imposed in 1977, what are now called *information disclosure statements* were referred to as *prior art statements.* An information disclosure statement may be either separate from or incorporated in a patent application. It should include a listing of patents, publications, or other information that is believed to be "material" and a concise explanation of the relevance of each listed item, and should be accompanied by copies of each listed patent or publication. Items are deemed to be "material" where there is a "substantial likelihood that a reasonable examiner would consider it important in deciding whether to allow the application to issue as a patent."

To ensure that the Patent and Trademark Office will give due consideration to an information disclosure statement, the information disclosure statement must be (1) filed within three months of the filing date of a normal U.S. application, or (2) within three months of entry of a U.S.-filed international application into its national stage, or (3) before the mailing date of a first Office communication treating the merits of the claimed invention (known as an "Office Action"), whatever occurs last. Consideration thereafter can be had only if other requirements are met, which typically include the certification of certain information, the filing of a petition, and/or the payment of a fee. Information disclosure statements filed before the grant of a patent that do not comply with the requirements of the Office are not considered by the Office but will be placed in the official file of the patent (which becomes public on the issue date of a patent).

The courts have held that those who participate in proceedings before the Office have the "highest duty of candor and good faith." While the courts differ in their holding of the consequences of misconduct, fraud on the Patent and Trademark Office has been found to be a proper basis for taking a wide variety of punitive actions, such as striking applications from the records of the Office, cancelling issued patents, denying enforcement of patents in infringement actions, awarding attorney's fees to defendants in infringement actions, and imposing criminal sanctions on those who were involved in fraudulently procuring patents. Inequitable conduct other than outright fraud has been recognized as a defense against enforcement of a patent, as a basis for awarding attorney's fees in an infringement action, and as a basis of antitrust liability.

In short, the duty of candor one has in dealings with the Office should be taken very seriously. Prudent practice would urge that if there is any question concerning whether a reference or other facts are "material," a citation should be made promptly to the Office so that the examiner can decide the issue.

76.4.4 Initial Review of an Application

Promptly after an application is filed, it is examined to make certain it is complete and satisfies formal requirements sufficiently to permit its being assigned a filing date and serial number. Once a patent application has been received by the Patent and Trademark Office and assigned a filing date and serial number, the classification of the subject matter of the claimed invention is determined and the application is assigned to the appropriate examining group. In the group, the application is assigned to a particular examiner. Each examiner is instructed to take up considerations of the applications assigned to him or her in the order of their filing.

Although more than 2000 examiners staff the Patent and Trademark Office, a backlog of several months of cases awaits action in most of the examining sections called *group art units*. This results in a delay of several months between the time an application is filed and when it receives its first thorough examination on the merits. At the time of this writing, the Office is granting about 115,000 patents per year.

Once an examiner reaches an application and begins the initial review, he or she checks the application still further for compliance with formal requirements and conducts a search of the prior art to determine the novelty and nonobviousness of the claimed invention. The examiner prepares an Office Action, in which he or she notifies the applicant of any objections to the application or requirements regarding election of certain claims for present prosecution, and/or any rejections he or she believes should be made of the claims.

In the event the examiner deems all the claims in the application to be patentable, he or she notifies the applicant of this fact and issues a notice of allowance. Applications that are allowed "lock, stock and barrel" on the first Office Action are sometimes regarded with suspicion—a nagging concern being that, had broader claims been sought, perhaps they too might have been allowed.

In some instances, the examiner will find it necessary to object to the form of the application. One hopes that these formal objections are not debilitating and can be corrected by relatively minor amendments made in response to the Office action.

In treating the merits of the claims, especially in a first Office Action, it is not uncommon for an examiner to reject a majority if not all of the claims. Some examiners feel strongly that they have a duty to cite the closest art they are able to find and to present rejections based on this art to encourage or force the inventor to put on record in the file of the application such arguments as are needed to illustrate to the public exactly how the claimed invention distinguishes patentably over the cited art.

76.4.5 Response to an Office Action

In the event the first Office Action issued by an examiner is adverse in any respect and/or leaves one or more issues unresolved, the applicant may reply in almost any form that constitutes a bona fide attempt to advance the prosecution of the application. The applicant is entitled to at least one reconsideration by the Office following the issuance of the first Office Action; however, as a minimum, a response must present at least some argument or other basis for requesting reconsideration.

Since the file of a patent application will become open to public inspection on the issuance of a patent and because an issued patent must be interpreted in view of the content of its file, the character of any arguments presented to the Patent and Trademark Office in support of a claimed invention are critical. In responding to an Office Action, it is essential that care be taken in the drafting of arguments to ensure that no misrepresentations are made and that the arguments will not result in an unfavorable interpretation of allowed claims being made during the years when the resulting patent is in force.

Years ago, it was not unusual for half a dozen or more Office Actions to issue during the course of pendency of a patent application. During recent years, however, the Office has placed emphasis on "compacting" the prosecution of patent applications, and insists that responses to Office Actions make a genuine, full-fledged effort to advance the prosecution of the application. Today it is not unusual for the prosecution of a patent application to be concluded on the issuance of the second or third Office Action. In an increasing number of cases, a final rejection is made as early as the second or third Office Action.

When an Office Action is mailed from the Patent and Trademark Office, a time period for filing a response begins. In the event a response is not filed within the time set by law, the application will automatically become abandoned, and rights to a patent may be lost forever. Ordinarily, a response must be filed within a three-month period from the mailing date of the Office Action. An extension of time of up to three months ordinarily can be had so long as the statutory requirement of filing a response within six months of the issuance of an Office Action is met. In previous years, it was necessary to present reasons to justify the grant of an extension of time, and the grant of an extension request was discretionary. Now, however, extensions are granted automatically upon receipt of a petition for extension accompanied by a proper response to the Office Action and the required fee.

The fee for obtaining an extension of time increases as the number of months covered by the extension is requested increases. The fees for one, two, and three-month extensions of time currently are $110, $400, and $950, respectively, unless small-entity status is established, in which case the fees are $55, $200, and $475, respectively.

In responding to an Office Action, each objection and rejection made by the examiner must be treated. If the inventor agrees that certain of his or her claims should not be allowed in view of art cited by the examiner, these claims may be cancelled or amended to better distinguish the claimed invention over the cited art.

Typical responses to Office Actions involve the addition, cancellation, substitution, or amendment of claims; the amendment of the descriptive portion of the application to correct typographical errors and the like (which can be done if no "new matter" is added); and the presentation of arguments regarding the allowability of the claimed invention in which explanations are provided that point out how the claims patentably distinguish over the cited references. A response may also include the submission of an affidavit to overcome a cited reference either by establishing a date of invention before the effective date of the reference or by presenting factual evidence supporting patentability over the reference.

If the Office has objected to the drawings, corrections must be made (at a time before the issue fee is paid) by providing substitute drawings that include the required corrections.

76.4.6 Reconsideration in View of the Filing of a Response

Once the applicant has responded, the examiner reexamines the case and issues a second Office Action apprising the applicant of his or her findings. If the examiner agrees to allow all of the claims that remain active in the application, prosecution on the merits is closed and the applicant may not present further amendments or add other claims as a matter of right. If the Office Action is adverse with regard to the merits of the claims, the prosecution of the case continues until such time as the examiner issues an Office Action that presents a final rejection.

The examiner makes a rejection final once a clear and unresolved issue has developed between the examiner and the applicant. After a final rejection has issued, the character of the responses that may be made by the applicant is limited. The applicant may appeal the final rejection to an intra-agency Board of Appeals, cancel the rejected claims, comply with all of the requirements for allowance if any have been laid down by the examiner, or file a continuation application whereby the examination procedure is begun again.

If an initial appeal taken to the Board of Patent Appeals should result in an unfavorable decision, a further appeal may be taken either to the U.S. District Court for the District of Columbia or to the U.S. Court of Appeals for the Federal Circuit (which, as of October 1, 1982, replaced what was previously known as the U.S. Court of Customs and Patent Appeals). In some instances, further appeals may be pursued to higher courts.

In the majority of instances during the period of prosecution, the application eventually reaches a form acceptable to the examiner handling the application, and the examiner will issue a notice of allowance. If it is impossible to reach accord with the examiner handling the application, the inventor can make use of the procedures for appeal.

If the record of examination of an application does not otherwise reveal the reasons for allowance, an examiner may put a comment in the file explaining his or her reasons for allowing the case. If the reason stated by an examiner for allowing a patent is shown during litigation to be faulty, this can cause the patent to be held invalid. Therefore, if a statement of reasons for allowance is provided by an examiner, it should be reviewed with care and commented upon, in writing, if deemed to be necessary.

76.4.7 Interviewing the Examiner

If, during the prosecution of a patent application, it appears that substantial differences of opinion or possible misunderstandings are being encountered in dealing with the examiner to whom the application has been assigned, it often is helpful for the attorney to conduct a personal interview with the examiner. While the applicant has a right to attend such a meeting, this right is best exercised sparingly and usually requires that the applicant spend time with the attorney to become better prepared to advance rather than to detract from his or her position.

Considering the relatively sterile and terse nature of many office actions, it may prove difficult to determine accurately what the examiner's opinion may be regarding how the application should be further prosecuted. While word-processing equipment acquired by the Office has made it easier for examiners to expound the reasons underlying their rejections, situations still arise where it is quite clear that an examiner and an attorney are not communicating in the full sense of the word. At times, a personal interview will be found to provide valuable guidance for bringing the prosecution of the application to a successful conclusion. In other instances, an interview will be beneficial in ascertaining the true character of any difference of opinion between the applicant and the examiner, thereby enabling the exact nature of this issue to be addressed thoroughly in the next response filed by the applicant.

76.4.8 Restriction and Election Requirements

If a patent examiner determines that an application contains claims to more than one independent and distinct invention, the examiner may impose what is called a *restriction requirement*. In the event the examiner finds that the application claims alternative modes or forms of an invention, he or she may require the applicant to elect one of these species for present prosecution. This is called a *species election requirement*. Once a restriction or election requirement has been imposed, the applicant must elect one of the designated inventions or species for present prosecution in the original application. The applicant may file divisional applications on the nonelected inventions or species any time during the pendency of the original application, which often results in a plurality of related patents issuing on different aspects of what the inventor regards as a single invention.

When responding to an Office Action that includes a restriction and/or election requirement, it often is desirable to present arguments in an effort to traverse the requirement and request its reconsideration. After traversing, the examiner is obliged to reconsider the requirement, but he or she may repeat it and make it final. Sometimes the examiner can be persuaded to modify or withdraw a restriction and/or election requirement, thereby permitting a larger number of claims to be considered during the prosecution of the pending application. As a practical matter, unless the examiner has set out a restriction and/or election requirement that is utterly and completely absurd, seeking reconsideration tends to be a waste of effort. Thankfully, it is seldom that an examiner decides that a simple application defines an excessive number of separate inventions, whereby the need to petition for reconsideration is a rarity.

76.4.9 Double-Patenting Rejections

Occasionally, one may receive a rejection based on the doctrine of double patenting. This doctrine precludes the issuance of a second patent on the same invention already claimed in a previously issued patent.

One approach to overcoming a double-patenting rejection is to establish a clear line of demarcation between the claimed subject matter of the second application and that of the earlier patent. If the line of demarcation is such that the claimed subject matter of the pending application is nonobvious in view of the invention claimed in the earlier patent, no double-patenting rejection is proper. If the claimed subject matter of the patenting application defines merely an obvious variation of the claimed invention of the earlier issued patent, the double-patenting rejection may be overcome by the filing of a terminal disclaimer. The terminal portion of any patent issuing on the pending application is disclaimed so that any patent issuing on the pending application will expire on the same day the already existing patent expires. If the claimed subject matter of the pending application is identical to the claimed subject matter in the earlier issued patent, it is not possible to establish a line of demarcation between the two cases and the pending application is not patentable even if a terminal disclaimer is filed.

The courts have held that double-patenting problems may occur when a utility patent application and a design patent application have been filed on the same invention. However, the fact that both a utility patent and a design patent may have issued on various features of a common invention does not necessarily mean a double-patenting problem exists. If it is possible to practice the ornamental appearance covered by the design patent without necessarily infringing any of the claims of the utility patent, and if it is possible to practice the claimed invention of the utility patent without infringing the ornamental appearance covered by the design patent, no double-patenting problem is present.

76.4.10 Patent Issuance

Once a notice of allowance has been mailed by the Office, the applicant has an inextensible period of three months to pay the issue fee. Payment of the issue fee is a prerequisite to the issuance of a patent. If payment of the issue fee is unavoidably or unintentionally late, the application becomes abandoned but usually can be revived within a year of the payment due date. Reviving an unintentionally abandoned application requires a much higher fee payment than does revival of an unavoidably abandoned application. A patent will not issue unless this fee is paid.

A few weeks before the patent issues, the Office mails a notice of issuance, which advises the applicant of the issue date and patent number.

Once a patent issues, its file history is no longer held secret. The several documents that form the complete file history of a patent are referred to collectively as the *official file* or the *file wrapper.*

Upon receipt of a newly issued patent, it should be reviewed with care to check for printing errors. If printing errors of misleading or otherwise significant nature are detected, it is desirable to petition for a certificate of correction. If errors of a clerical or typographical nature have been made by the applicant or by his or her attorney and if these errors are not the fault of the Patent and Trademark Office, a fee must be paid to obtain the issuance of a Certificate of Correction. If the errors are the fault of the Patent and Trademark Office, no such fee need be paid.

The issuance of a patent carries with it a presumption of validity. As was well stated by Judge Markey in *Roper Corp. v. Litton Systems, Inc.,* 757 F.2d 1266 (Fed. Cir., 1985), "A patent is born valid and remains valid until a challenger proves it was stillborn or had birth defects. . . ." If the validity of a patent is put in question, the challenger has the burden of establishing invalidity by evidence that is clear and convincing.

76.4.11 Safeguarding the Original Patent Document

The original patent document merits appropriate safeguarding. It is printed on heavy bond paper, its pages are fastened together by a blue ribbon, and it bears the Official Seal of the United States Patent and Trademark Office. The patent owner should preserve this original document in a safe place as evidence of his or her proprietary interest in the invention. If an infringer must be sued, the patent owner may be called on to produce the original letters patent document in court.

76.4.12 Continuation, Divisional, and Continuation-in-Part Applications

During the pendency of an application, it may be desirable to file either a continuation or a divisional application. A continuation application may be filed if the prosecution of a pending application has not proceeded as desired, whereby a further opportunity for reconsideration can be had before an appeal is taken. A divisional application may be filed when two or more inventions are disclosed in the original application and claims to only one of these inventions have been considered during examination of the originally filed case.

It frequently occurs during the pendency of a patent application that a continuing program of research and development program being conducted by the inventor results in the conception of improvements in the original invention. Because of a prohibition in the patent law against amending the content of a pending patent application to include "new matter," any improvements made in the invention after the time an application is filed cannot be incorporated into a pending application. When improvements are made that are deemed to merit patent protection, a continuation-in-part application is filed. Such an application can be filed only during the pendency of an earlier-filed application commonly called the *parent case.* The continuation-in-part case receives the benefit of the filing date of the parent case with regard to such subject matter as is common to the parent case. Any subject matter uncommon to the parent case is entitled only to the benefit of the filing date of the continuation-in-part case.

In some instances when a continuation-in-part application has been filed, the improvements that form the subject matter of the continuation-in-part case are closely associated with the subject matter of the earlier-filed application, and the earlier application may be deliberately abandoned in favor of the continuation-in-part case. In other instances, the new matter that is the subject of the continuation-in-part application clearly constitutes an invention in and of itself. In such a situation, it may be desirable to continue the prosecution of the original application to obtain one patent that covers the invention claimed in the original application and a second patent that covers the improvement features.

76.4.13 Maintaining a Chain of Pending Applications

If a continuing development program is under way that produces a series of improvements, it can be highly advantageous to maintain on file in the Patent and Trademark Office a continuing series of pending applications—an unbroken chain of related cases. If an original parent application is initially filed, and a series of continuation, division, and/or continuation-in-part applications are filed in such a manner that ensures the existence of an uninterrupted chain of pending cases, any patent or patents that may issue on the earlier cases cannot be used as references cited by the Office as obstacles in the path of allowance of later applications in the chain. This technique of maintaining a series or chain of pending applications is an especially important technique to use when the danger exists that the closest prior art the Office may be able to cite against the products of a continuing research and development effort is the patent protection that issued on early aspects of this effort.

76.5 PATENT PROTECTIONS AVAILABLE ABROAD

U.S. patents provide no protection abroad and can be asserted against a foreigner only in the event the foreigner's activities infringe within the geographical bounds of our country. This section briefly outlines some of the factors one should consider if patent protection outside the United States is desired.

76.5.1 Canadian Filing

Many U.S. inventors file in Canada. Filing an application in Canada tends to be somewhat less expensive than filing in other countries. With the exception of a stringently enforced unity requirement, which necessitates that all the claims in an application strictly define a single inventive concept, Canadian patent practice essentially parallels that of the United States. If one has success in prosecuting an application in the United States, it is not unusual for the Canadian Intellectual Property Office to agree to allow claims of substantially the same scope as those allowed in the United States.

76.5.2 Foreign Filing in Other Countries

Obtaining foreign patent protection on a country-by-country basis in countries other than Canada, particularly in non-English-speaking countries, has long been an expensive undertaking. In almost all foreign countries, local agents or attorneys must be employed, and the requirements of the laws of each country must be met. Some countries exempt large areas of subject matter, such as pharmaceuticals, from what may be patented.

Filing abroad often necessitates that one provide a certified copy of the United States case for filing in each foreign country selected. Translations are needed in most non-English-speaking countries. In such countries as Japan, even the retyping of a patent application to put it in proper form can be costly.

With the exception of a few English-speaking countries, it is not at all uncommon for the cost of filing an application in a single foreign country to equal, if not substantially exceed, the costs that have been incurred in filing the original U.S. application. These seemingly unreasonably high costs prevail even though the U.S. application from which a foreign application is prepared already provides a basic draft of the essential elements of the foreign case.

76.5.3 Annual Maintenance Taxes and Working Requirements

In many foreign countries, annual fees must be paid to maintain the active status of a patent. Some countries require annual maintenance fee payments even during the time that the application remains pending. In some countries, the fees escalate each year on the theory that the invention must be worth more as it is more extensively put into practice. These annual maintenance fees not only benefit foreign economies, but also become so overwhelming in magnitude as to cause many patent owners to dedicate their foreign invention rights to the public. Maintaining patents in force in several foreign countries is often unjustifiably expensive.

In many foreign countries, there are requirements that an invention be "worked" or practiced within these countries if patents within these countries are to remain active. Licensing of a citizen of or business entity domesticated within the country to practice an invention satisfies the working requirement in some countries.

76.5.4 Filing under International Convention

If applications are filed abroad within one year of the filing date of an earlier-filed U.S. case, the benefit of the filing date of the earlier-filed U.S. case usually can be attributed to the foreign applications. Filing within one year of the filing date of a U.S. case is known as filing under international convention. The convention referred to is the Paris Convention, which has been ratified by our country and by almost all other major countries. Taiwan is among the few countries that do not honor this treaty.

Most foreign countries do not provide the one-year grace period afforded by U.S. statute to file an application. Instead, certain foreign countries require that an invention be "absolutely novel" at the time of filing of a patent application in these countries. If the U.S. application has been filed prior to any public disclosure of an invention, the absolute novelty requirements of most foreign countries can be met by filing applications in these countries under international convention, whereby the effective filing date of the foreign cases is the same as that of the U.S. case.

76.5.5 Filing on a Country-by-Country Basis

If one decides to file abroad, one approach is to file separate applications in each selected country. Most U.S. patent attorneys have associates in foreign countries with whom they work in pursuing patent protections abroad. It is customary for the U.S. attorney to advise a foreign associate about how he or she believes the prosecution of an application should be handled, but to leave final decisions to the expertise of the foreign associate.

76.5.6 The Patent Cooperation Treaty

Since June 1978, U.S. applicants have been able to file an application in the United States Patent and Trademark Office in accordance with the terms of the Patent Cooperation Treaty (PCT), which has been ratified by the United States and by the vast majority of developed countries.

PCT member countries include such major countries as Australia, Austria, Belgium, Brazil, Canada, China, Denmark, Finland, France, Germany, Hungary, Japan, Mexico, Netherlands, Norway, Russia, Sweden, Switzerland, the United Kingdom, and the United States. In filing a PCT case, a U.S. applicant can designated the application for eventual filing in the national offices of such other countries as have ratified the treaty.

One advantage of PCT filing is that the applicant is afforded an additional eight months beyond the one-year period he or she would otherwise have had under the Paris Convention to decide whether he or she wants to complete filings in the countries he or she has designated. Under the Patent Cooperation Treaty, an applicant has 20 months from the filing date of his or her U.S. application to make the final foreign filing decision.

Another advantage of PCT filing is that it can be carried out literally at the last minute of the one-year convention period measured from the date of filing of U.S. application. Thus, in situations where a decision to file abroad to effect filings has been postponed until it is impractical if not impossible to effect filings of separate applications in individual countries, a single PCT case can be filed on a timely basis in the United States Patent and Trademark Office designating the desired countries.

Still another feature of PCT filing is that, by the time the applicant must decide on whether to complete filings in designated countries, he or she has the benefit of the preliminary search report (a first Office Action) on which to base his or her decision. If the applicant had elected instead to file applications on a country-by-country basis under international convention, it is possible that he or she might not have received a first Office Action from the Patent and Trademark Office within the one year permitted for filing under international convention.

76.5.7 The European Patent Convention

Another option available to U.S. citizens since June 1978 is to file a single patent application to obtain protection in one or more of the countries of Europe, most of which are parties to the so-called *European Patent Convention* (EPC).

Two routes are available to U.S. citizens to effect EPC filing. One is to act directly through a European patent agent or attorney. The other is to use PCT filing through the United States Patent and Trademark Office and to designate EPC filing as a *selected country.*

A European Patent Office (EPO) has been set up in Munich, Germany. Before applications are examined by the EPO in Munich, a Receiving Section located at The Hague inspects newly filed applications for form. A novelty search report on the state of the art is provided by the International Patent Institute at The Hague. Within 18 months of filing, The Hague will publish an application to seek views on patentability from interested parties. Once publication has been made and the examination fee paid by the applicant, examination moves to Munich, where a determination is made of patentability and prosecution is carried out with the applicant responding to objections received from the examiner. The EPO decides whether a patent will issue, after which time a copy of the patent application is transferred to the individual patent offices of the countries designated by the applicant. The effect of EPC filing is that, while only a single initial application need be filed and prosecuted, in the end, separate and distinct patents issue in the designated countries. Any resulting patents have terms of 20 years measured from the effective date of filing of the original application.

76.5.8 Advantages and Disadvantages of International Filing

An advantage of both PCT and EPC filing is that the required applications can be prepared in exactly the same format. Their form and content will be accepted in all countries that have adhered to the EPC and/or PCT programs. Therefore, the expense of producing applications in several different formats and in different languages is eliminated. The fact that both PCT and EPC applications can, in their initial stages, be prepared and prosecuted in the English language is another important advantage for U.S. citizens.

A principal disadvantages of both of these types of international patent filings is their cost. Before savings over the country-by-country approach are achieved, filing must be anticipated in several countries, perhaps as many as four to six, depending on which countries are selected. A disadvantage of EPC filing is that a single examination takes place for all the designated countries, and patent protection in all these countries is determined through this single examination procedure.

76.5.9   Trends in International Patent Protection

With the advent of the PCT and EPC programs, a significant step forward has been taken that may someday lead to the development of a multinational patent system. For the predictable future, however, it seems clear that the major countries of the world intend to maintain intact their own patent systems.

CHAPTER 77

ELECTRONIC INFORMATION RESOURCES: YOUR ON-LINE BOOKSHELF

Robert N. Schwarzwalder, Jr.
Michelle Kazmer
Ford Motor Company
Dearborn, Michigan

77.1	**BACKGROUND AND DEFINITIONS**	**2269**	**77.3**	**ACCESS OPTIONS FOR ELECTRONIC INFORMATION RESOURCES** 2278
77.2	**INTERNET RESOURCES**	**2271**	77.3.1	Internet Access Options 2278
	77.2.1 Approaches to Using the Internet	2271	77.3.2	Database and Commercial Services 2279
	77.2.2 The World Wide Web	2272	77.3.3	Databases of Importance for Mechanical Engineers 2282
	77.2.3 Telnetting, Listservs, Usenet, FTP	2276	77.3.4	CD-ROM and Desktop Databases 2283
	77.2.4 The Intranet: Information Resources within the Corporation	2277	77.3.5	Options for Using Electronic Information 2284
	77.2.5 Future of the Internet	2278		

77.1 BACKGROUND AND DEFINITIONS

Why read this chapter? Perhaps out of curiosity about the World Wide Web. What if we could offer you a way to get better results in your job, tackle new projects without extensive gear-up time, and avoid costly dead ends? The point is, there is a great deal of technical information available on most of the technical problems you will be involved with this year. By tapping into that information, you can avoid mistakes and get a fast start on new projects. It is this competitive edge that has corporations excited about Intranets.

In the last several years, there has been an explosion of interest in the Internet and on-line information resources. These resources have emerged from obscurity and come to occupy a place of prominence in the corporate world as companies have realized that intellectual capital is as valuable as financial capital. Due to the vast quantity of information available to the mechanical engineer, the old standby of having a bookshelf of the essential handbooks is no longer sufficient. The globalization of research and development and the decreasing cycle time of product development demand rapid access to the world-wide literature of engineering. All of this at a time when professional engineers are being asked to be more productive and more time-efficient! Therefore, effective access to data and ideas requires computer interfaces that can speed access to and delivery of the information necessary for the job.

Fortunately, today's mechanical engineer has a wide variety of tools to create an electronic bookshelf, offering rapid access to a global library of technical information. In this chapter, we will discuss a variety of tools available to you and offer suggestions as to how those systems can be used to provide quick, easy access to a wide variety of information. These tools consist of Internet resources

Communications should be addressed to Robert Schwarzwalder, Ford Motor Company, MD 1153 SRL, Dearborn, MI 48121, or by email at rschwar3@ford.com.

Mechanical Engineers' Handbook, 2nd ed., Edited by Myer Kutz.
ISBN 0-471-13007-9 © 1998 John Wiley & Sons, Inc.

and on-line databases. They share a common mode of access in that they are accessed by connecting to a remote computer, the host, by using a telephone or network connection from your microcomputer. In the corporate or academic workplace, this network access is typically provided by the organization. We have supplied a list of services that provide network access to information for individuals who are employed as consultants or are part of a small firm. We have also included a section about "desktop" databases available on CD-ROM, which are accessed locally on your own computer or on a local computer network.

The Internet is an interconnected series of computer networks. Begun as a Department of Defense-sponsored project, the Internet has grown into an increasingly commercial service. Because of the wide variety of services and systems available through this network of networks, it is difficult to describe. The following terms are related to the discussion of the Internet, database, and on-line information resources.

Agent. A software device that filters information before it reaches the end-user, or locates and sends information to the end-user.

ASCII. Common keyboard characters; refers to an interface that only allows the user to view and use common text and numerical characters.

Bulletin board system (BBS). Electronic discussion forums available to subscribers through networks. Messages posted to the bulletin board and the responses travel as e-mail.

CD-ROM. Compact Disk, Read Only Memory, a digital storage medium used for desktop databases.

Client. A software application mounted on your computer which extracts some service from a server somewhere else on the network. This relationship is often referred to as a *client–server* application.

Database. A computer-based search and retrieval system that allows a user to retrieve and display information based upon a series of command protocols.

Downloading. The transfer of electronic data from a larger system to a small system, such as from a mainframe computer to a desktop machine.

E-mail. Electronic mail permits an individual to post a message to the mailbox of another user. Each individual/mailbox has a unique address that can receive mail from anywhere on the Internet.

End-user. The person who ultimately uses the information. This term is typically used to distinguish situations where the person who searches an information system is the same as the person who will use the results of the search, as opposed to having an information professional search for results.

Firewall. A security system designed to keep unauthorized users out of a computer network.

FTP. File Transfer Protocol is a system for retrieving data or text files from a remote computer.

False drops. Unwanted information which is inadvertently retrieved in an Internet or on-line search.

FAQ (frequently asked questions). Often a list of frequently asked questions and their answers. Many USENET news groups, and increasingly World Wide Web sites, maintain FAQ lists in order to cut down the number of repetitive questions.

File server. A host machine that stores and provides access to files; remote users often use ftp to access a file server.

Gopher. A menu-based system for exploring Internet resources developed at the University of Minnesota in 1991. By choosing menu items, the gopher will link you to those sources.

GUI (graphical user interface). A system, such as the World Wide Web, that allows the user to view and use graphics, as opposed to an ASCII interface.

Host. A network computer that has resources that are shared with others.

Hostname. Identifies a computer, the host, by a name of the machine and the domain name. The domain name may describe a single computer or a group of computers. For example, *this.machine.com* and *that.machine.com* are two computers named *this* and *that* in the domain *machine.com.*

HTML. HyperText Markup Language, the language used to create World Wide Web documents.

Internet. A collection of interconnected networks that speak the Internet Protocol (IP) and related protocols. The Internet provides file transfer, remote login, electronic mail, news, and other services. As of April 1993, there were 10,000 networks connected to the Internet.

Knowbot. A term introduced by Vince Cerf meaning a "robotic librarian." These experimental information-retrieval agents are under development.

Listserv. A mailing list devoted to a specific topic. Any message posted to the topically oriented listserv is forwarded to all subscribers.

Node. Any computer on a network.

PDF. Portable Document Format, a computer-platform-independent electronic file format developed by Adobe Systems, Inc.

Remote login. The process of accessing a host mounted on another network, which is usually accomplished using telnet protocol.

Server. Used to mean either (1) software that allows a computer to offer a service to another computer (i.e., client), or (2) the computer on which the server software runs.

SLIP/PPP. Connection types that allow the user to access graphical interfaces, such as the Web. SLIP or PPP connections typically require client software such as WINSOCK for the PC or MacTCP or MacPPP for the Macintosh.

Telnet (TELetype NETwork). A software application utility for TCP-IP that provides terminal emulation, and thus remote login capability from a microcomputer to some remote host.

TCP-IP (Transmission Control Protocol/Internet Protocol). The most common communication protocol for regional and national networks.

URL (Universal Resource Locator). An electronic address on the World Wide Web.

USENET. A distributed bulletin board and discussion system that generally requires access to a UNIX host. Also known as *NEWS* or *NETNEWS.*

Veronica. A search system under development at the University of Nevada that creates a mini-gopher based on a one-word search across gophers.

VT100. A standard, commonly used protocol for terminal emulation, often used when logging into a remote host or network.

WAIS (Wide Area Information Servers). A method of searching the Internet, WAIS searches specially created indexes of databases, not the databases themselves.

WWW (the Web, or the World Wide Web). A system of interconnecting resources offering flexible multimedia coverage of a variety of topics. Netscape is the primary software package for accessing the WWW.

In addition to the systems resident on the Internet are a variety of services that can be accessed through the Internet or by direct modem connection. These services consist of databases that allow the engineer to search and retrieve materials including citations to articles, technical data, financial information, and the full text of articles. This information is available from thousands of different databases. Individual databases tend to be subject-specific, although the scope of coverage can vary greatly. Databases are created by commercial publishers, research institutions, or government agencies. Hundreds of different on-line vendors, or databanks, provide the search and retrieval software to allow remote users to manipulate these databases. Unlike the Internet, where only recently have the resources to maintain a high-quality product been available, these databases are typically well supported and can be depended upon to be consistent and reliable.

77.2 INTERNET RESOURCES

77.2.1 Approaches to Using the Internet

If you have no idea of what the Internet is about, there are some good ways to get your feet wet. One is simply to find someone well versed in various Internet systems to sit down with you as you experiment. You take the keyboard; that way you can maintain the pace of the demonstration. You can also take advantage of the flurry of Internet workshops being offered at conferences or through local organizations. *The Whole Internet: User's Guide and Catalog,* 2nd edition, by Ed Krol, is an excellent Internet user's manual; we recommend it highly. But the big thing is to get started.

Why use the Internet? Simply because it offers information unobtainable via the commercial vendors and free access to some information for which you would otherwise need to pay. Internet databases such as the EPA's provide information available through the commercial database NTIS, and some information that NTIS doesn't have. Fedworld, a bulletin board of federal information resources, hooks you in to scores of federal agencies and allows you to download some documents directly. The list could go on and on. Suffice it to say that, among all of the trash, there is a great deal of real substance to the Internet.

Once you are surfing along that electronic superhighway, you'll wonder where all of that useful information is hidden. The first adjective one would use to describe the Internet is *unorganized.* Be very careful which listservs you join and which usenet groups you subscribe to; you may find yourself waist-deep in information. We have included a list of some useful Internet resources later in this chapter. These systems offer a number of advantages. Not only do they provide a unique way of interacting with professionals in your specialization, but they also allow you to keep close tabs on the newest technical issues in your field.

If you are having difficulty finding information on a topic, you can post a question to one or more of these groups. While there is no guarantee that you will get an answer, you will be surprised by the number of information professionals and engineers who will take the time to help you. But beware; there are plenty of "lurkers" who observe but remain silent. Lurkers are interested in knowing who is doing what. It is remarkably easy to figure what people are doing by what they ask. If you represent a corporation or are working on cutting-edge research, you should either not post questions or else subdivide your query into questions that cannot be easily reconstructed by competitors.

Where can you find useful information on the Internet? Start with the list of resources provided in this chapter. Once you are actively involved, you will find it fairly easy to pick up new sources. Most Internet systems build new sources into their structure. For example, in the World Wide Web, links are rapidly being established to new sources. The nature of the network is such that many of these additions are done in a seamless manner; however, you need to be prepared for rapid and dramatic changes in the Internet. Sometimes links to valuable resources change without notice and seem to vanish. Sometimes new tools emerge that make older systems obsolete overnight. Think of it as the price we pay for innovation. You can also find out what's new on the Internet by subscribing to a listserv emphasizing your area of interest and by reading this chapter and other literature devoted to information resources in engineering and technology.

While the list below is certainly not comprehensive, it will give you a feel for the sorts of engineering resources available through the Internet. We have included brief comments on the nature and use of each type of resource. Any book on the Internet will provide in-depth discussions of these services.

77.2.2 The World Wide Web

There has been a firestorm of interest in the World Wide Web in the last few years. This system provides the user with connections to graphics, sound, and animation in a manner that allows easy browsing. The Web allows hypertextual connections. That is, by clicking your mouse button on a highlighted word or image, you can immediately connect to that remote host. So, as you browse through a homepage, you can select from a variety of options and be connected to them without needing to know where they are or what their addresses are. Ease of use is one of the Web's biggest selling factors; however, the ability to view graphics and use multimedia to present ideas is where it really excels.

Unlike a number of other resources on the Internet, the Web requires that you have client software on your computer. The most common client software is Netscape. Netscape is a Windows-type tool for navigating the Web. As the Web grows, more Web sites are requiring special software to view documents and run programs. Software packages such as Adobe Acrobat are readily available as downloadable files through the Web. Typically, Web sites requiring special software will provide instructions for obtaining the programs. While obtaining software from well established companies is fairly safe, use caution when downloading programs from unfamiliar sources. This is an excellent way to import a computer virus into your computer or network! Web addresses are referred to as *Universal Resource Locators* or *URLs*.

Alta Vista Web Search

URL: http://www.altavista.digital.com/
Alta Vista is one of many Web search engines. By entering terms of interest, you can retrieve lists of Web sites that may be relevant to your needs. The large number of Web sites, rather than flaws in the search engines, account for the false drops. Alta Vista is widely respected for its speed, ease of use, and strong coverage of the Internet.

ANSI Online (American National Standards Institute)

URL: http://www.ansi.org/docs/home.html
While it functions mostly as an on-line sales brochure, this site does help you identify ANSI and ISO standards and provides information for ordering the desired documents. The homepage provides some information on forthcoming standards and symposia, but its clear focus is on selling standards.

ASMENet (American Society of Mechanical Engineers)

URL: http://www.asme.org/
ASMENet is a one-stop-shop for any information emanating from the Society. It includes information about ASME conferences and publications, education, employment, professional development, and other information of interest to mechanical engineers.

American Society for Testing and Materials (ASTM)

URL: http://www.astm.org/
Besides producing standards, ASTM devises a large number of the test methods used to establish standards compliance. This homepage offers searching of the ASTM standards and test methods, a list of publications and services, directory information, and a full listing of the various ASTM committees.

Engineering Information Inc. Engineering Information Village (Fig. 77.1)

URL: http://www.ei.org/ (800/221-1044 or ei@ei.org)

This Web site deserves special attention due to its unique nature and relevance to mechanical engineering. Engineering Information, Inc. (Ei) has crafted a very strong service on the Web that provides access to Ei's Compendex database and about 180 other commercial databases, as well as thousands of Web sites, listservs, and news groups. The availability of document delivery and assistance from senior engineers nicely rounds out the service. What you end up with isn't really a Web homepage, but a unique service that takes advantage of Web, e-mail, and other Internet technologies. The main graphic of the Village is mapped so that a click on a subject icon launches you to that section of the homepage. By adding connections to business, government, travel services, commercial Web services, and news resources, Ei has created a tool that provides an electronic bookshelf for the engineer. Unlike most Web resources, this Web service has a charge for using it, due to the inclusion of commercial database access, document delivery, and certain consultation services.

Fedworld

URL: http://www.fedworld.gov

Fedworld is the Web clearinghouse for U.S. federal information through the Internet. Since the early 1990s, the federal government has shifted to electronic publication of information in order to save money. Through Fedworld, you can access report databases from the EPA, DOE and NASA, census and economic data, and a wide variety of federal information. This is one of the most valuable Web sites available today.

General Electric

URL: http://www.ge.com/

General Electric provides an example of the types of corporate information available through the Web. GE uses its Web site to help advertise its products and services as well as to provide business information to existing and potential stockholders. By browsing Web pages such as GE's, you can obtain a great deal of information about a company.

ICE (Internet Connections for Engineers)

URL: http://www.englib.cornell.edu/ice/ice-index.html

ICE provides one of the best compilations of Web resources for engineering and the hard sciences. Produced at Cornell University, ICE is also a center for the collection of Web-based educational tools for engineering. Cornell has been a leading university in developing the Internet and this site demonstrates some of that vision and hard work.

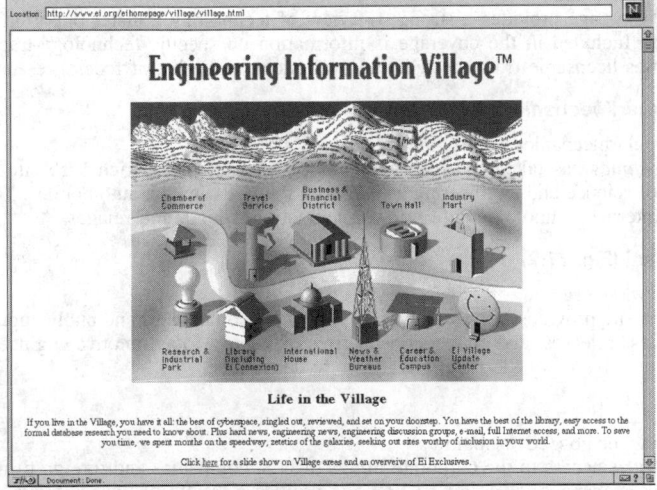

Fig. 77.1 The Engineering Information Village provides a wide range of engineering information and services through the Web.

Information Analysis Centers (IACs)

URL: http://www.dtic.dla.mil/iac/

Information Analysis Centers are agencies of the U.S. Department of Defense charged with the critical analysis of data. These centers analyze materials data for metals, ceramics, polymers, and so on for DOD efforts and have a secondary objective of providing information to U.S. industry. The IAC Web pages are one of the more innovative programs of the DOD in providing technical information to the public.

International Business Information on the WWW

URL: http://www.ciber.msu.edu/busres.htm

This Web site, developed by Michigan State University, provides links to on-line periodicals, U.S. Commerce Department information, and country-specific pages related to international business. Some of the information available from non-U.S. corporations through the Web is unobtainable from any other source.

International Organization for Standardization (ISO)

URL: http://www.iso.ch/welcome.html

The ISO Web site provides a catalog of ISO standards and information on ISO and its activities. Also included are news and background information on the ISO 9000 standard on quality manufacturing.

Journal of Mechanical Design

URL: http://www-jmd.engr.ucdavis.edu:80/jmd/index.html

As this handbook goes to press, there are scores of Web-based journals. This seems to be a trend that will accelerate in the coming years. The *Journal of Mechanical Design* offers an index of articles back to 1993, information for authors, and general information on the journal.

NTRS (NASA Technical Reports Server)

URL: http://techreports.larc.nasa.gov/cgi-bin/NTRS

NTRS is a clearinghouse search that allows you to search full-text or abstracts of technical reports from several NASA Research Centers, Flight Centers, and Laboratories.

National Center for Manufacturing Sciences

URL: http://www.ncms.org/

NCMS is an organization of companies and individuals dedicated to the study and advancement of manufacturing. Their Web site is an excellent source of information related to manufacturing. Members of NCMS have access to citation databases and others services, but there are a variety of resources open to the general public.

National Technology Transfer Center (NTTC)

URL: http://www.nttc.edu/

The NTTC Web page provides a strong overview of technology transfer opportunities from the U.S. government. Included in the coverage is information on specific technology-transfer opportunities, technologies licensable from the federal government, and a list of federal research facilities.

PM Zone (Popular Mechanics)

URL: http://popularmechanics.com/homepage2d.html

Popular Mechanics has taken the Internet by storm with a content-rich Web site that provides broad coverage of science and engineering as well as a focused view on automobiles. This homepage creatively uses Internet technologies to offer you animation, sound, and images.

SAE International (Fig. 77.2)

URL: http://www.sae.org/

The SAE Web site provides information on SAE's many conferences and publications, collections of automotive press releases, and connections to sites of interest to automotive engineers. Fee-based access to an automotive news file is also available to subscribers.

SEC Edgar

URL: http://www.sec.gov/edgarhp.htm

This site provides access to the corporate filings of U.S. companies and foreign firms represented on the U.S. stock exchanges. 10K and 10Q reports provide a wealth of financial data on companies as well as excerpts from annual and quarterly reports. This is an excellent free source of corporate financial information.

Fig. 77.2   SAE's Web site promotes the society's activities with lists of publications and conferences, along with automotive news. (Reprinted with permission, copyright Society of Automotive Engineers, Inc.)

Society of Manufacturing Engineers

URL: http://www.sme.org/

With over 70,000 members, SME is one of the largest societies serving the mechanical engineering profession. The SME homepage offers information on the society's conferences and publications, educational and accreditation opportunities, and a collection of information of interest to members.

Thomas Register (Fig. 77.3)

URL: http://www.thomasregister.com:8000/index-to.html

The *Thomas Register* is one of the most respected names in equipment catalogs. The Web version offers free on-line use of the catalog and an easy way to search for companies. Like many Web services, free access to this catalog may not last forever.

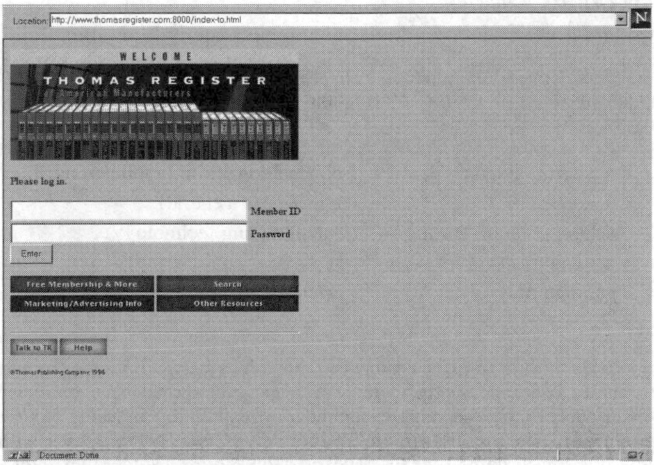

Fig. 77.3   The Thomas Register homepage provides searching of technical equipment retailers on the Web.

77.2.3   Telnetting, Listservs, Usenet, Gophers, FTP

Telnetting

Telnet is an Internet application that allows you to enter remote hosts to search databases, use software, access bulletin boards, or perform e-mail functions. One application of telnetting is to tie directly into remote library catalogues to determine if the system has a particular book or journal. Another application is to get into your own host system if you are away from your system, but have access to a foreign host. In addition, more and more governmental agency and specialized societal databases are becoming available on the Internet. The telnet protocol is very easy. Simply enter the command

telnet HOSTNAME

where HOSTNAME is the name of the host you wish to access. Once in the host, you must follow its commands and protocols. Your system may have security measures, such as a firewall, that require special passwords to enter. Ask your local systems office for details. As always, enter HELP for assistance or a list of commands. Most systems use a word such as *quit, logoff, stop, end,* or *bye* to end the connection. If that doesn't work, simply turn off your machine.

Listservs

To subscribe to a listserv, send an e-mail message to one of the addresses below with a single line as message:

SUBSCRIBE LISTSERV.NAME FIRST NAME LAST NAME

For example, to subscribe to MECH-L, John Doe would send a message to LISTSERV@ UTARLVM1.UTA.EDU:

SUBSCRIBE MECH-L John Doe

The first message you receive from the listserv will be a set of instructions for using that system. We advise that you save this message for further reference. For a list of listservs, send the following command to any LISTSERV address:
LIST GLOBAL

CAEDS-L (Computer-Aided Engineering Design)

Address: *listserv@listserv.syr.edu*

MATERIALS-L (Materials Science)

Address: *listserv@listserv.liv.qc.uk*

MECH-L (Mechanical Engineering)

Address: *listserv@listserv.utq.edu*

Usenet Groups

Usenet may be accessed from some gophers or directly using a variety of software packages. Consult your local computer systems office for information on your best option. The contents under each group will consist of series of comments on a variety of subtopics.

sci.comp-aided	(sci/tech computing applications)
sci.engr	(engineering, general)
sci.engr.manufacturing	(manufacturing technology)
sci.engr.mech	(mechanical engineering)
sci.materials	(materials science)

FTP

The ftp application uses the file transfer protocol to move files between your computer and a remote computer. For example, you can use ftp to retrieve a text file, spreadsheet, or image from another site. If you have an account on the remote computer, you will log in using your user name and password. There are also anonymous ftp sites, which allow users to take files without having an account. In general, the log-in name for these sites is *anonymous* or sometimes *guest.* The password is either identical to the login; your e-mail address; or there will be no password prompt.

There are two types of ftp applications that you might see. One is the command-based ftp, done from a command prompt. Type *ftp HOSTNAME*

The command to get a file is *get <filename>*; to list the files available in the directory on the remote machine in which you're working, the command is *ls*. To get out of ftp, type *quit* or *bye*.

There are also graphical (or Windows-based) ftp applications that work with the point-and-click technique rather than requiring you to enter commands at a prompt.

77.2.4 The Intranet: Information Resources within the Corporation

1995 saw the advent of the Intranet, the private Internet within an organization or corporation. Intranets experienced explosive growth in 1996 because they offered a flexible option to more expensive software solutions for communication, document management, and knowledge management. These internal webs provided security, speed, graphical communications, and low cost, a combination that proved irresistible to large and small organizations. For the information customer, finding information has become as simple as obtaining a Web browser and using the corporate homepage or search engine. While most Intranets are not well organized, the information universe is small enough to be searched in a reasonable amount of time.

The opportunity and challenge of the Intranet is that anyone and everyone can become a publisher and an information manager. Bringing internal information to the Intranet is the subject of a great deal of corporate scrutiny. Issues of converting text documents to Web documents (in formats such as HTML or PDF), Web security (through authentication or encryption systems), and data storage have been addressed at great length. However, questions of providing external information feeds to augment these internal resources have been all but ignored. External information feeds can take the form of a CD-ROM system at an individual's desktop or internationally accessible, locally mounted database tapes in a major corporation. Deciding which options to take can play a major part in determining competitive position and expenses. Figure 77.4 illustrates some of the options available for information access, depending upon the size of the user population and the need for information currency. The use of CD-ROMs in a networked environment should be limited to those applications where the number of concurrent users will be very small. Even in a wide area network (WAN) configuration, significant input–output problems occur when multiple users attempt to use the same CD-ROM. In a similar fashion, while individual on-line database accounts can work well on a small scale, the companies who offer these services are not generally efficient at offering corporate-wide services. The traditional answer to developing enterprise-wide information access has been to load large databases locally. This solution is capable of supporting large numbers of concurrent users, but is very expensive. In the last few years, a number of Web-based databases have offered companies universal access to information through the Web. The fact that these systems require less network support and training make them a very attractive option for Intranet architects.

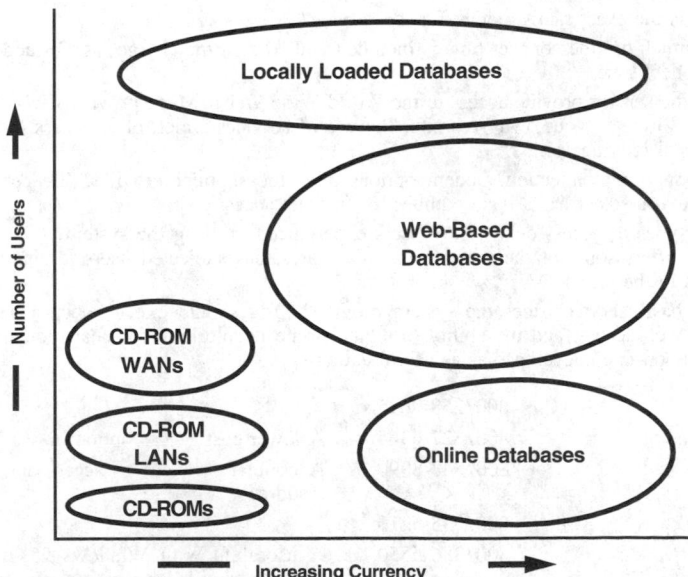

Fig. 77.4 Decision matrix represents corporate information options as a function of currency versus number of users.

77.2.5 Future of the Internet

In 1990, the Internet was primarily of interest to the federal government and academic institutions. The overwhelming concentration of use was electronic mail, much of that informal. The World Wide Web, a graphical, hypertextual medium, opened up almost limitless options for information-sharing. In the early 1990s, universities began to use the Web. They were quickly joined by businesses and individuals. By the middle of the decade, corporations had begun to exploit the flexibility of the Web to provide internal information systems. By 1997, as we write this, the growth in these Intranets has exceeded that of the publicly available Internet.

While the beginning of the Internet boom was publicly financed and free to most users, the trend in the last few years has been towards the provision of high-value, fee-based services through the Web. With the financing of commercial Web services and of corporations erecting private Web-based information networks, information value has replaced novelty as the benchmark of a Web site. It is clear from existing trends that the Web has matured from the plaything of 1990 to a fundamental tool of business and industry.

For the engineer in the corporate setting, Intranets will provide a means of easily acquiring internal corporate information and of sharing your information with others. The future of corporate information in this environment involves being able to convert graphics and documents on the fly to Web-enabled documents, thus eliminating the need for time-consuming records-retention practices. Information available through these internal Webs will be available in full format, text and graphics at the click of a mouse. The hypertextual nature of the Web will allow the combination of video and tutorials into documents, expanding the potential of electronic information from a static to a dynamic medium.

For the consulting engineer, the growth in Internet services will provide all of the advantages of a large corporate library without the space and overhead. By tapping into remote databases and document delivery options, you can scan through a global collection of literature in seconds. As these services expand, expect to see a greater abundance of materials in full text. As network services improve, we will begin to see electronic documents with accompanying graphics.

77.3 ACCESS OPTIONS FOR ELECTRONIC INFORMATION RESOURCES

77.3.1 Internet Access Options

One of the first things you will need to do to access Internet and commercial database services is to obtain a network connection. A variety of services provide Internet access to anyone with a microcomputer and a modem. Each of these services has its own unique billing structure with access fees that typically run from $11 to $25 per month. Which service is best for you depends upon which features you want to take advantage of and how frequently you want to use them. For example, some services allow unlimited use of electronic mail, while others include added charges for e-mail use beyond a specified limit. When negotiating with these services, you will want to determine

1. What is the exact rate structure for the service?
2. How much on-line connect time is included and what is the charge rate for additional hours of connect time?
3. Does the service provide access to the World Wide Web? (Many providers will offer you the option to create your own Web site. Beware of services which offer access through a non-graphical interface!)
4. What type of connection/modem options does the supplier provide? (Be certain that the service you select has a local number for Internet access.)
5. What types of access delays have users experienced in using the system? (America Online, which offers some of the most competitive rates, has subjected users to significant access delays in the past!)
6. What basic services does the system provide? (You should expect electronic mail, World Wide Web access, and the ability to telnet. Some people look for chat rooms and features such as on-line encyclopedias as useful extras.)

AlterNet	800/258-9695	
America Online	800/827-6364	A lower-cost access option
CompuServe	800/848-8990	A popular provider of access and information resources
InternetMCI	800/955-5210	
MicroSoft Network	800/386-5550	A useful fit with Windows 95 or subsequent systems from MicroSoft
Netcom On-Line Communication Services	800/353-6600	

Prodigy	800/776-3449	A popular on-line service, Prodigy does include advertisements
PSINet Pipeline	800/827-7482	
UUNet Technologies	800/265-2213	
The Well	415/332-9200	One of the first Internet service providers, the Well is well known for its extensive conferencing system

In addition, there are probably a number of local Internet service providers in your area. Local companies may provide better price options for local customers than the national services.

77.3.2 Database and Commercial Services

Commercial Database Services for Mechanical Engineers

There are several large companies who coordinate access to databases that you will find useful for your work. Through these services, you can gain access to retrieve technical, news, and business information. By using their services, you can purchase use of scores or hundreds of databases through a single supplier. With these systems, you can search millions of references in seconds and find references that would either have taken you days to locate or would have been impossible to find! By working on the same system, you need only learn one set of search and retrieval commands that may be used for hundreds of different databases.

Typically, when you search a database, you are retrieving citations to publications rather than the publications themselves. Increasingly, the full text of articles is available through these databases, especially for the business literature and news reports, but these still lack the graphs and figures often necessary to interpret the text. However, most items are available only as a reference. If you do not have access to a corporate, university, or public library to provide copies of the actual publications, you can use one of the document suppliers mentioned at the end of this section.

DIALOG

This system began in the 1970s and is, perhaps, the largest of the database vendors. DIALOG offers a variety of options for you to access their system through a telnet session or through the World Wide Web. If you are not interested in learning the system's command language, a menu-driven system is available. For DIALOG, and for all of the database providers, you sacrifice search power for ease of use by opting for a menu system. Such systems work well for simple requests or to retrieve a few references on a popular subject, but may fail you if you have a difficult search question. DIALOG's strength is its wide variety of databases, including many options for searching patents, trademarks, and business information, as well as references in the social sciences and humanities. While DIALOG is not very strong in its coverage of international literature, it offers a gateway to DATA-STAR, a European system that contains a number of unique databases, a few of which are not searchable in English. A useful feature of DIALOG and other database vendors is the ability to search several databases simultaneously and remove any records that appear in more than one database. In large systems such as DIALOG and STN, this feature allows you to search large collections of multidisciplinary literature quickly and economically.

DIALOG
2440 W. El Camino Real
Mountain View, CA 94040
800/334-2564

Sample DIALOG Record from Engineering Information Inc.'s Compendex Plus Database:

FN- DIALOG(R)File 8:Ei Compendex(R)|
CZ- (c) 1996 Engineering Info. Inc. All rts. reserv.|
AN- <DIALOG> 04510805|
AN- <EI NUMBER> EIP96103350983|
TI- Engineering in the global marketplace|
AU- Valenti, Michael|
SO- Mechanical Engineering v 118 n 9 Sept 1996. p 74–78|
SO- <PY> 1996|
CO- MEENAH|
SN- 0025-6501|
LA- English|

DT- JA^(Journal Article)|

TC- G^(General Review)|

JA- 9611W4|

AB- Today, mechanical engineering firms are reaping new profits by expanding into new international markets. Key to the development of a successful global strategy for many companies is the assistance provided by private consultants and government agencies. This aid ranges from using interpretive skills and explaining international tax codes to identifying reliable partners and smoothing out difficult negotiations.‖

DIALOG's search language is straightforward. The basic DIALOG commands are summarized in Table 77.1.

STN International

While STN is smaller than DIALOG, it is a specialized resource for engineering and scientific information. STN contains most of the same engineering resources as DIALOG and adds a variety of specialized chemical, energy, and environment databases. This system exists as a union of three database centers, one in the United States, one in Germany, and one in Japan. While it has strong coverage of the patent literature, STN is weaker on sources of news and lacks the coverage of nontechnical disciplines. It makes up for this weakness by its unique strength in materials and chemical information. Not only is STN the best source of international materials information, especially for metals, but it provides some very sophisticated ways to search for this information. Materials data, almost completely absent from DIALOG, is abundantly available in STN.

STN International

2540 Olentangy River Road

P.O. Box 3012

Columbus, OH 43210-0012

800/848-6538

Examination of the basic STN commands (Table 77.2) will demonstrate only a few differences between it and DIALOG. Indeed most of the major systems differ only slightly in the structure and syntax of their commands.

Knowledge Express

This service is much more limited than KR DIALOG or STN International, but it is designed for the end-user and has a collection of resources that is geared to the technical business user. Knowledge Express stresses business news, federal and university research, licensable technologies, patents, and

Table 77.1

Command	Short Form	Function	Example
Begin	B	To open access to a database (e.g., B 8 . . . to open Compendex*Plus)	B 8
Search	S	Retrieves references to the specified term or phrase	S aluminum
		Allows word adjacency	S metal( )forming
		Allows word proximity (e.g., terms within three words of each other)	S spark(3w)timing
		Wild card (e.g., all terms beginning with manufactur, including manufacture, manufacturing, etc.)	S manufactur?
Expand	E	Displays an index of terms alphabetically related to selected term (e.g., engineer, engineering, engineers)	E engineer
Help	?	To get information or help	? Limit
Limit	L	Restricts output (e.g., set 3 to English language only)	L s3/eng
Type	T	Displays record(s) in a variety of formats (e.g., set 3, full text of records 1–6)	T 3/9/1–6
Quit	Q	To end session	Q

Table 77.2 Basic STN Commands

Command	Short Form	Function	Example
File	FIL	To open access to a database (e.g., FIL CA, to open Chemical Abstracts)	FIL CA
Search	S	Retrieves references to the specified term or phrase	S aluminum
		Allows word adjacency	S metal(w)forming
		Allows word proximity (e.g., terms within 3 words of each other)	S spark(3w)timing
		Wild card (e.g., all terms beginning with manufactur, including manufacture, manufacturing, etc.)	S manufactur?
		Wild card for internal characters (e.g., for U.S. and British spellings)	S alumin!um
Expand	E	Displays an index of terms alphabetically related to selected term (e.g., engineer, engineering, engineers)	E engineer
Help	Help	To get information or help	Help Commands
Display	D	Displays record(s) in a variety of formats (e.g., set 3, citation and abstracts for records 1–6)	D L3 bib,ab 1–6
Logoff	Log	To quit system	Log

technology transfer. Because of its strength in these areas and its lack of coverage of the mainstream technical literature, Knowledge Express is best seen as a complement to one of the larger services, rather than a substitute for either of them.

> Knowledge Express
> One Westlakes
> 1235 Westlakes Drive, Suite 210
> Berwyn, PA 19312-9877
> 800/529-5337

Document Suppliers

In most cases, you will be obtaining citations from these databases rather than full text. There are a number of companies who will mail or fax documents to you upon request. In some cases, these services are directly accessible through the on-line database vendors. If you live near a major university, inquire with the library to see if fee-based document delivery is available. Often such university-based options cost less than the commercial services. The cost of a photocopy will include $12 to $20 for handling; the royalty fees (which are highly variable); plus added charges for rush delivery. Listed below are some of the major document delivery vendors who cover the technical literature.

Article Express	800/238-3458	For copies of journal articles and conference papers
Global Engineering Documents	800/854-7179 303/792-2181	Supplies articles, papers and industrial standards. Provides worldwide coverage through a number of "Global Information Centres"
ANSI	212/642-4900	ANSI can provide U.S. and international standards
IEEE	800/678-4333 908/981-0060	Source for worldwide IEEE standards
NTIS	800/553-6847	Call NTIS with requests for federal technical reports
Powells Technical Books	800/225-6911	One of the best sources of engineering books, even out of print or rare books (http://www.technical.powells.portland.or.us/)

77.3.3   Databases of Importance for Mechanical Engineers

Engineering Databases

These databases contain references to journal articles, conference papers, patents, books, and technical reports. A few databases, like those in the MDP Network, contain actual data as opposed to references to literature.

Compendex (Availability: DIALOG, STN, CD-ROM)

Compendex is produced by Engineering Information, Inc. This company has been providing access to the literature of engineering since the late 1800s, electronic access since 1970. The scope of the database is very broad, covering all subjects within engineering; however, it is best for more traditional areas like mechanical and civil engineering. While the service is not limited to U.S. literature, it is strongest for journals and conferences produced in this country. While you may not want to restrict your use to this database, it is a MUST for the mechanical engineer. (Compendex is also available through the World Wide Web. See Section 77.2.2 for details.)

ISMEC (Availability: DIALOG, STN, CD-ROM)

The database of the *Mechanical Engineering Abstracts* provides indexing of journal articles, conference papers, books, technical reports, and patents dealing with mechanical engineering, production engineering, and engineering management. ISMEC's coverage extends back to 1973. While it is significantly smaller and less frequently updated than Compendex, its strong focus on mechanical engineering makes it a valuable resource for anyone involved in this discipline.

METADEX (Availability: DIALOG, STN, CD-ROM)

This is the definitive metals database and indexes the commercial literatures on metals and alloys and metallurgy. This database references citations to articles, books, and conferences papers, as opposed to the MPD files (below), which contain data.

MPD (Materials Property Data) Network (Availability: STN)

The MPD Network is a collection of databases containing physical property data for a wide variety of materials. The network's coverage is strongest for metals and polymer data, but also includes ceramics. Of the metals, the coverage of steels and aluminum alloys is especially strong. Beside retrieving data points on particular materials, this system may be used to select a range of materials meeting specific property ranges. In this manner, you can use the MPD Network as a tool to assist you in determining materials options for a specific application.

1MOBILITY (SAE Literature) (Availability: STN, DIALOG, CD-ROM)

The Society of Automotive Engineers has produced an index of transportation engineering since the early 1900s. They produce this database, which contains citations to technical literature and news related to automotive (approximately 75%) and aviation (approximately 25%) engineering. One advantage of this database is its coverage of the reports literature of non-U.S. automotive societies.

INSPEC (Availability: DIALOG, STN, CD-ROM)

While Compendex is the best source for most engineering literature, it is no equal for INSPEC for computer science, electrical engineering, or advanced materials research. In addition, since INSPEC is produced in Britain, it has better coverage of the European literature. If your question involves mainstream mechanical engineering, this file is probably not of interest. However, it is an excellent source of computer controls and automation information.

NTIS (Availability: DIALOG, STN, CD-ROM)

The federal government spends billion of dollars each year on research and development. Unfortunately, almost none of this work is available through traditional indexes or databases. NTIS, the National Technical Information Service, provides the database of the same name to make this material accessible to the public. You can expect to find a variety of reports from agencies such as the DOE, DOD, NASA, and EPA here. Of special interest are results of Small Business Innovation Grants in cutting-edge technologies such as robotics and automation control that are given to start-up companies.

Chemical Abstracts (Availability: STN best, DIALOG, CD-ROM)

This database is the world's largest and covers anything even remotely involving chemistry or chemical engineering. If you need information on specific metal alloys or tribology, you may want to give this system a try. A companion database, the "Registry" file, contains chemical registration numbers that uniquely identify specific substances. On STN, these numbers may be used to find materials information in other databases on the network. The CD-ROM option lacks some of the search features of the on-line databases.

Aerospace Database (Availability: DIALOG, CD-ROM)

For mechanical engineers involved in aeronautics or astronautics, this database is a must. Providing a broad coverage of the aerospace literature since 1962, this database is one of the few that includes federal research in its scope.

University Technologies (Availability: Knowledge Express)

Knowledge Express's University Technologies file provides information on collaborative research and license opportunities from over 150 universities. The database covers all areas of research.

News and Business Databases

The news and business files can be of value for two reasons. First, they provide a way of locating information on companies that you may be benchmarking or are considering as suppliers. This kind of heads-up can save you time and expense. Second, there are times when the only information available on a new technology is the public statements of the company who has developed it. While this type of release is seldom a good source of technical information, it may be all there is.

ABI-Inform (Availability: DIALOG, STN, CD-ROM)

This is an excellent database of the mainstream business literature, indexing popular business periodicals along with major trade publications. It is a very good place to look for what the press is saying about a company or corporate trend.

AP NEWS (Availability: DIALOG)

A full-text database, AP News allows you to tap into the newswire service that serves the news industry. This database is updated constantly and provides up-to-the-minute news and financial information.

CORPTECH (Availability: Knowledge Express)

This directory provides detailed information on over 40,000 technology-related companies, including contact information, company descriptions, and financial information. The database is updated quarterly.

PAPERS (Availability: DIALOG)

PAPERS indexes scores of U.S. newspapers and provides a good source of background information. Often a local newspaper will get information that a major publication would not due to its everyday exposure to the company.

Patent Databases

Your first exposure to patents will leave no doubt that this is a literature written by and for lawyers. It is difficult to comprehend even if you have a firm understanding of the underlying technology. Nevertheless, this is an essential literature for engineers, since it is often the only source of technical information for certain technologies. Most of the R&D in applied areas is done by companies who have no desire to publish their results. Doing so would share millions of dollars of technology with their competitors. However, in order to protect their innovations, they need to reveal details of these technologies in the patents they file with the U.S. and foreign patent offices. By examining these patents, you can gain access to cutting-edge information long before it appears in the journal literature.

Derwent World Patents Index (Availability: DIALOG, CD-ROM)

This is an excellent database for the world's patent literature. By providing extra indexing and adding natural language keywords to each title, Derwent makes the patents easier to find. The fact that it takes 12 to 18 months for a patent application to show up in the U.S. literature, coupled with the fact that most large companies file simultaneous patent applications in other countries, suggests that searching the global literature is preferable even if you only want U.S. patents.

CLAIMS/U.S. Patents Abstracts (Availability: DIALOG)

If you are determined to limit your search to U.S. patents, this database provides strong U.S. coverage at a discount to the price of the Derwent database.

77.3.4 CD-ROM and Desktop Databases

CD-ROMs provide information that resides locally on the user's computer, rather than the information residing on a remote computer and the customer paying to access those data. CD-ROMs are sold individually or on a subscription basis. An individual CD-ROM will be a "snapshot" of information, whether it offers information traditionally found in book form, such as conference proceedings, or a database of indexing and perhaps full-text documents such as articles, papers, and standards. With the subscription system, you will receive a new CD-ROM at a specified interval (monthly, quarterly, yearly) that supersedes the old CD.

Another important difference between CD-ROMs and some of the on-line services is that the CD-ROM may include graphical images of figures, tables, and so on that aren't available through a text-based on-line service. Also, the CD-ROM can comprise a selection of data that are part of a larger database but have been extracted to provide coverage of a more specialized subject area.

Information contained on a CD-ROM requires special software in order to read it, and you must have access to a computer with a CD-ROM drive. The software should be included with the price of the CD, and will either be included on the disk itself or packaged separately. Once the software is installed, the CD may be inserted into the CD-ROM drive at any time; the installed software can be started like any other computer program; and the data will be retrieved from the CD.

There are several factors to consider in selecting the CD-ROM option. CD-ROM provides a "pay-once" alternative to the "pay-as-you-go" system you'll get from the on-line database providers. This can be a good option, especially if you're likely to spend a lot of time searching one database. You don't watch the clock ticking, metering up on-line time, when using a CD-ROM. If, however, you need access to a large variety of databases (perhaps simultaneously), the on-line route may be the way to go. Another factor is frequency of update; if a monthly or quarterly snapshot is current enough for your application, CD-ROM is a good option. Also, find out how retrospective the information contained on the CD-ROM is; if your need is for information from the recent few years, it may be contained on one CD.

The databases available on CD-ROM are generally the same databases available through the on-line services (see Section 77.3.3). One exception to this is the Society of Automotive Engineers (SAE), who have packaged information available on CD-ROM a bit differently than what is available in the 1MOBILITY (Global Mobility) database described above. (See below for further discussion.)

Three primary providers of CD-ROMs for engineering information are DIALOG, Silver Platter, and SAE. Also, several databases, such as *Chemical Abstracts* and Derwent patent products, are available on CD-ROM directly from the database producers.

DIALOG

The CD-ROM products from DIALOG, like their on-line system, offer both a command line search option (identical to the DIALOG search language described above) and a menu-driven system. Most of the databases indicated in the above section as being available through DIALOG are also available on CD-ROM. An example of a CD-ROM that includes part of a larger database geared towards a more specialized field is the Derwent Petroleum and Power Engineering database, extracted from the Derwent World Patents Index. Call 1-800-334-2564 for information.

Silver Platter

Silver Platter tends to have a lot of "general reference" databases, but they do have some of the basics for engineering literature, in particular, Ei Compendex, INSPEC, NTIS, and ABI/INFORM. Silver Platter has been in the CD-ROM business for many years (since 1983) and also has World Wide Web-based databases available. Call 617-769-2599 for information.

SAE

The databases available on CD-ROM through SAE are different from those available from the previous two vendors. SAE offers their technical papers, standards, and handbook on CD-ROM; and various repackaged segments of the Global Mobility Database covering specialized areas such as highway vehicles safety and automotive fuels and lubricants. Call 412-776-4841 for information.

77.3.5 Options for Using Electronic Information

In this chapter, we have surveyed a variety of options for you to obtain information directly through the Internet and various commercial services. At this point you have the tools necessary to develop your own unique suite of information services. Your corporate library or nearest university engineering library is an excellent place to visit for more information on these tools. Many technical libraries have become designers of Internet and Intranet systems. Their familiarity with the information systems and their experience with engineering information make them a valuable resource.

What brings many people to this point is the excitement over the Web and the Internet; what keeps them is the advantages they gain from having ready access to information. If you work for a large corporation, you are probably finding that your company is increasingly using the Intranet, LotusNotes, or e-mail as a means of corporate communication. As companies rely more heavily upon electronic communications, your ability to function effectively in this medium will affect your ability to function effectively within the organization. Those of you who work for a small firm or as consultants will find that information is your greatest tool. Technologies and business directions are changing so rapidly these days that without a means of staying current, you are at a severe competitive disadvantage.

What will information do for you? It can mean knowing which firms to benchmark and what questions to ask. It can mean getting a quick start on a new project. It can mean not missing technical options that save money or time. It can mean having better options in less time than your competition. Electronic information expands your collection of handbooks to include a global library of resources. When coupled with document delivery, the resources discussed in this chapter allow you to have greater access to information than any library in the world! For the mechanical engineer of the next century, these tools will be as common as the desktop computer is today.

BIBLIOGRAPHY

Krol, E., *The Whole Internet: User's Guide and Catalog,* 2nd ed., O'Reilly and Associates, Sebastopol, CA, 1994.

　　Krol's book is easily accessible to the novice and packed with practical, easy to follow information. Krol provides extensive treatment on how to use the Internet, what's allowed there, and how to deal with network problems. Although Krol did the whole book based on UNIX, he includes DOS and Mac directions also.

O'Leary, M., *The Online 100: ONLINE Magazine's Field Guide to the 100 Most Important Online Databases.* Pemberton Press Books, Wilton, CT, 1995.

　　This is a good overview of the major database options for all subject disciplines. While it is weak on engineering, it will be useful if your interests are more eclectic.

"Internet Watch" column in *Automotive Engineering* (for automotive engineers)

　　This monthly column provides tips on using the Internet along with updates on new Web sites.

CHAPTER 78

SOURCES OF MECHANICAL ENGINEERING INFORMATION

Fritz Dusold
Retired from Mid-Manhattan Library
Science and Business Department
New York, New York

Myer Kutz
Myer Kutz Associates, Inc.
New York, New York

78.1	INTRODUCTION	2287		78.5	CODES, SPECIFICATIONS, AND STANDARDS	2289
78.2	THE PRIMARY LITERATURE	2287				
	78.2.1 Periodicals	2288		78.6	GOVERNMENT PUBLICATIONS	2290
	78.2.2 Conference Proceedings	2288				
78.3	INDEXES AND ABSTRACTS	2288		78.7	ENGINEERING SOCIETIES	2290
	78.3.1 Manual Searching	2288				
	78.3.2 Online Searching	2288		78.8	LIBRARIES	2291
78.4	ENCYCLOPEDIAS AND HANDBOOKS	2289		78.9	INFORMATION BROKERS	2291

78.1 INTRODUCTION

This chapter is designed to enable the engineer to find information efficiently and to take advantage of all available information. The emphasis is placed on publications and services designed to identify and obtain information. Because of space limitations references to individual works, which contain the required information, are limited to a few outstanding or unusual items.

78.2 THE PRIMARY LITERATURE

The most important source of information is the primary literature. It consists mainly of the articles published in periodicals and of papers presented at conferences. New discoveries are first reported in the primary literature. It is, therefore, a major source of current information. Peer review and editorial scrutiny, prior to publication of an article, are imposed to ensure that the article passes a standard of quality. Most engineers are familiar with a few publications, but are not aware of the extent of the total production of primary literature. *Engineering Index*[1] (known as *Compendex* in its electronic version) alone abstracts material from thousands of periodicals and conferences.

Handbooks and encyclopedias are part of the secondary literature. They are derived from primary sources and make frequent references to periodicals. Handbooks and encyclopedias are arranged to present related materials in an organized fashion and provide quick access to information in a condensed form.

While monographs—books written for professionals—are either primary or secondary sources of knowledge and information, textbooks are part of the tertiary literature. They are derived from primary

Mechanical Engineers' Handbook, 2nd ed., Edited by Myer Kutz.
ISBN 0-471-13007-9 © 1998 John Wiley & Sons, Inc.

and secondary sources. Textbooks provide extensive explanations and proofs for the material covered to provide the student with an opportunity to understand a subject thoroughly.

78.2.1 Periodicals

In most periodicals published by societies and commercial publishers, articles are identified usually by issue, and/or volume, date, and page number. Bibliographic control is excellent, and it is usually a routing matter to obtain a copy of a desired article. But some problems do exist. The two most common are periodicals that are known by more than one name, and the use of nonstandard abbreviations. Both of these problems could be solved by using the International Standard Serial Number (ISSN), which accurately identifies each publication. The increasing size and use of automated databases should provide an impetus to increased use of ISSN or some other standard.

The first scientific periodical, *Le Journal des Scavans,* was published January 4, 1665. The second, *Philosophical Transactions* of the Royal Society (London), appeared on March 6, 1665. The number of scientific periodicals has been increasing steadily with some setbacks caused by wars and natural catastrophies. The accumulated body of knowledge is tremendous. Much of this information can be retrieved by consulting indexes, abstracts, and bibliographies.

78.2.2 Conference Proceedings

The bibliographic control for papers presented at conferences is not nearly as good as for periodicals. The responsibility for publishing the papers usually falls upon the sponsoring agent or host group. For major conferences the sponsoring agency is frequently a professional society or a department of a university. In these cases an individual with some experience in publishing is usually found to act as an editor of the *Proceedings.* In other instances the papers are issued prior to the conference as *Preprints.* In still other situations the papers will be published in a periodical as a special issue or distributed over several issues of one or more periodicals. An additional, unknown, percentage of papers are never published and are only available in manuscript form from the author.

78.3 INDEXES AND ABSTRACTS

Toward the end of the last century the periodical literature had reached a volume that made it impossible for the "educated man" to review all publications. In order to retrieve the desired or needed information, indexes and abstracts were prepared by individual libraries and professional societies. In the 1960s computers became available for storing and manipulating information. This lead to the creation and marketing of automated data banks.

78.3.1 Manual Searching

The major abstracts typically provide the name of the author, a brief abstract of the article, and the title of the article, and identify where the article was published. Alphabetical author and subject indexes are usually provided, and a unique number is assigned to refer to the abstract. Many abstracts are published monthly or more frequently. Annual cumulations are available in many cases. The most important abstracts for engineers are:

> *Engineering Index*[1]
> *Science Abstracts*[2]
>> Series A: Physics Abstracts
>> Series B: Electrical and Electronics Abstracts
>> Series V: Computer and Control Abstracts
> *Chemical Abstracts*[3]
> *Metals Abstracts*[4]

A comprehensive listing of abstracts and indexes can be found in *Ulrich's International Periodical Directory.*[5]

78.3.2 Online Searching

Most of the major indexes and abstracts are now available in machine-readable form. For a comprehensive list of databases and online vendors see *Information Industry Market Place.*[6] The names of online databases frequently differ from their paper counterparts. *Engineering Index,*[1] for example, offers COMPENDEX and EI Engineering Meetings online. Most of the professional societies producing online databases will undertake a literature search. A society member is frequently entitled to reduced charges for this service. In addition to indexes and abstracts, periodicals, encyclopedias, and handbooks are available online. There seems to be virtually no limit to the information that can be made available online or on CDROM's, which can be networked in large institutions with many potential users. The high demand for quick information retrieval ensures the expansion of this service.

In addition to the online indexes, several library networks and consortia, such as OCLC, the Online Computer Library Center, located in Columbus, Ohio, produce online databases. These are essentially equivalent to the catalogs of member libraries and can be used to determine which library owns a particular book or subscribes to a particular periodical.

78.4 ENCYCLOPEDIAS AND HANDBOOKS

There are well over 300 encyclopedias and handbooks covering science and technology. "amazon.com" and "barnesandnoble.com" are Internet sites with comprehensive catalogs of books. The date of publication should be checked before using any of these works if the required information is likely to have been affected by recent progress. The following list represents only a sampling of available works of outstanding value.

The *Kirk–Othmer Encyclopedia of Chemical Technology*[7] provides in 24 volumes, plus a separate index, a comprehensive and authoritative treatment of a wide range of subjects, with heaviest concentration on materials and processes. The basic set is updated by supplements.

The *Encyclopedia of Polymer Science and Engineering*[8] is one of the major works in this important area of materials.

Metals Handbook[9] provides an encyclopedic treatment of metallurgy and related subjects. Each of the volumes is devoted to a separate topic such as mechanical testing, powder metallurgy, and heat treating. Each of the articles is written by a committee of experts on that particular topic.

The *CRC Handbook of Chemistry and Physics,*[10] popularly known as the "Rubber Handbook," is probably the most widely available handbook. It is updated annually to include new materials and to provide more accurate information on previously published sections as soon as the information becomes available. The CRC Press is one of the major publishers of scientific and technical handbooks. A *Composite Index of CRC Handbooks*[11] provides access to the information covering the latest editions of 57 CRC handbooks, some of them multivolume sets, which were published prior to 1977.

The increasing concern with industrial health and safety has placed an additional responsibility on the engineer to see that materials are handled in a safe manner. Sax's *Dangerous Properties of Industrial Materials*[12] provides an authoritative treatment of this subject. This book also covers handling and shipping regulations for a large variety of materials.

Engineers have always been concerned with interaction between humans and machines. This area has become increasingly sophisticated and specialized. The *Human Factors Design Handbook*[13] is written for the design engineer rather than the human factor specialist. The book provides the engineer with guidelines for designing products for convenient use by people. The second edition offers new material drawn from the lessons learned, for example, in the computer and space industries. Another important title is the *Handbook of Human Factors and Ergonomics.*[14]

Engineering work frequently requires a variety of calculations. The *Standard Handbook of Engineering Calculations*[15] provides the answer to most problems. Although most of the information in this handbook is easily adaptable to computer programming, a new edition would probably take greater advantage of the increasing availability of computers.

Conservation of energy remains an important consideration, owing to energy's increasing cost. Two titles dealing with this subject are the *CRC Handbook of Energy Efficiency*[16] and the *Energy Management Handbook.*[17]

Composite materials frequently offer advantages in properties and economy over conventional materials. Information about composites can be found in several McGraw-Hill and CRC Press handbooks.

When England converted to the metric system the British Standards Institution published *Metric Standards for Engineers.*[18] With the increasing worldwide distribution of products, metric units will gain in importance regardless of the official position of the U.S. government. This handbook offers the engineer an authoritative and detailed treatment of metrification.

78.5 CODES, SPECIFICATIONS, AND STANDARDS

Codes, specifications, and standards are produced by government agencies, professional societies, businesses, and organizations devoted almost exclusively to the production of standards. In the United States the American National Standards Institute[19] (ANSI) acts as a clearinghouse for industrial standards. ANSI frequently represents the interests of U.S. industries at international meetings. Copies of standards from most industrial countries can be purchased from ANSI as well as from the issuer.

Copies of standards issued by government agencies are usually supplied by the agency along with the contract. They are also available from several centers maintained by the government for the distribution of publications. Most libraries do not collect government specifications.

Many of the major engineering societies issue specifications in areas related to their functions. These specifications are usually developed, and revised, by membership committees.

The American Society of Mechanical Engineers[20] (ASME) has been a pioneer in publishing codes concerned with areas in which mechanical engineers are active. In 1885 ASME formed a Standardization Committee on Pipe and Pipe Threads to provide for greater interchangeability. In 1911 the

Boiler Code Committee was formed to enhance the safety of boiler operation. The 1983 ASME Boiler and Pressure Vessel Code was published in a metric (SI) edition, in addition to the edition using U.S. customary units, to reflect its increasing worldwide acceptance. The boiler code covers the design, materials, manufacture, installation, operation, and inspection of boilers and pressure vessels. Revisions, additions, and deletions to the code are published twice yearly during the three-year cycle of the code.

A frequently used collection of specifications is the *Annual Book of Standards*[21] issued by the American Society for Testing and Materials (ASTM). These standards are prepared by committees drawn primarily from the industry most immediately concerned with the topic.

The standards written by individual companies are usually prepared by a member of the standards department. They are frequently almost identical to standards issued by societies and government agencies and make frequent references to these standards. The main reason for these "in-house" standards is to enable the company to revise a standard quickly in order to impose special requirements on a vendor.

The large number of standards issued by a variety of organizations has resulted in a number of identical or equivalent standards. Information Handling Services (IHS)[22] makes available virtually all standards on CD-ROM.

78.6 GOVERNMENT PUBLICATIONS

The U.S. government is probably the largest publisher in the world. Most of the publications are available from the Superintendent of Documents.[23] Publication catalogs are available on the Government Printing Office web site, GPO.gov. Increasingly, the GPO is relying on electronic dissemination rather than print. These publications are provided, free of charge, to depository libraries throughout the country. Depository libraries are obligated to keep these publications for a minimum of five years and to make them readily available to the public.

The government agencies most likely to publish information of interest to engineers are probably the National Institute of Science and Technology, the Geological Survey, the National Oceanic and Atmospheric Administration, and the National Technical Information Service.

78.7 ENGINEERING SOCIETIES

Engineering societies have exerted a strong influence on the development of the profession. The ASME publishes the following periodicals in order to keep individuals informed of new developments and to communicate other important information:

Applied Mechanics Reviews (monthly)
CIME (*Computers in Mechanical Engineering,* published by Springer-Verlag, New York)
Mechanical Engineering (monthly)
Transactions (quarterly)
The *Transactions* cover the following fields: power, turbomachinery, industry, heat transfer, applied mechanics, bioengineering, energy resources technology, solar energy engineering, dynamic systems, measurement and control, fluids engineering, engineering materials and technology, pressure vessel technology, and tribology.

Many engineering societies have prepared a code of ethics in order to guide and protect engineers. Societies frequently represent the interests of the profession at government hearings and keep the public informed on important issues. They also provide an opportunity for continuing education, particularly for preparing for professional engineers examinations. The major societies and trade associations in the United States are

American Concrete Institute
American Institute of Chemical Engineers
American Institute of Steel Construction
American Society of Civil Engineers
American Society of Heating, Refrigerating, and Air-Conditioning Engineers
American Society of Mechanical Engineers
Institute of Electrical and Electronics Engineers
Instrument Society of America
National Association of Corrosion Engineers
National Electrical Manufacturers Association
National Fire Protection Association
Society of Automotive Engineers

Technical Association for the Pulp and Paper Industry
Underwriters Laboratories

78.8 LIBRARIES

The most comprehensive collections of engineering information can be found at large research libraries. Four of the largest in the United States are:

John Crerar Library
35 West 33rd Street
Chicago, IL 60616

Library of Congress
Washington, DC 20540

Linda Hall Library
5109 Cherry Street
Kansas City, MO 64110

New York Public Library—Science, Industry and Business Library
188 Madison Avenue
New York, NY 10016

These libraries are accessible to the public. They do provide duplicating services and will answer telephoned or written reference questions.

Substantial collections also exist at universities and engineering schools. These libraries are intended for use by faculty and students, but outsiders can frequently obtain permission to use these libraries by appointment, upon payment of a library fee, or through a cooperative arrangement with a public library.

Special libraries in business and industry frequently have excellent collections on the subjects most directly related to their activity. They are usually only available for use by employees and the company.

Public libraries vary considerably in size, and the collection will usually reflect the special interests of the community. Central libraries, particularly in large cities, may have a considerable collection of engineering books and periodicals. Online searching is becoming an increasingly frequent service provided by public libraries.

Regardless of the size of a library, the reference librarian should prove helpful in obtaining materials not locally available. These services include interlibrary loans from networks, issuing of courtesy cards to provide access to nonpublic libraries, and providing the location of the nearest library that owns needed materials.

78.9 INFORMATION BROKERS

During the last decade a large number of information brokers have come into existence. For an international listing see *The Burwell World Directory of Information Brokers.*[24] Information brokers can be of considerable use in researching the literature and retrieving information, particularly in situations where the engineer does not have the time and resources to do the searching. The larger brokers have a staff of trained information specialists skilled in online and manual searching. Retrieval of needed items is usually accomplished by sending a messenger to make copies at a library. It is therefore not surprising that most information brokers are located near research libraries or are part of an information center.

The larger information brokers usually cover all subjects and offer additional services, such as translating foreign language materials. Smaller brokers, and those associated with a specialized agency, frequently offer searching in a limited number of subjects. The selection of the most appropriate information broker should receive considerable attention if a large amount of work is required or a continuing relationship is expected.

REFERENCES

1. *Engineering Index,* Engineering Information Inc. (monthly).
2. *Science Abstracts,* INSPEC, Institution of Electrical Engineers (semimonthly).
3. *Chemical Abstracts,* American Chemical Society (weekly).
4. *Metals Abstracts,* Metals Information (monthly).
5. *Ulrich's International Periodicals Directory,* R. R. Bowker, New York (annual).
6. *Information Industry Market Place,* An International Directory of Information Products and Services, R. R. Bowker, New York.
7. *Kirk–Othmer Encyclopedia of Chemical Technology,* 4th ed., Wiley-Interscience, New York, 1991–1997 (24 vols.).

8. *Encyclopedia of Polymer Science and Engineering,* 2nd ed., Wiley-Interscience, New York, 1985–1989 (19 vols.).

9. *Metals Handbook,* American Society for Metals, Metals Park, OH.

10. *CRC Handbook of Chemistry and Physics,* CRC Press Inc., Boca Raton, FL (annual).

11. *Composite Index of CRC Handbooks,* CRC Press Inc., Boca Raton, FL, 1977.

12. Sax's *Dangerous Properties of Industrial Materials,* 9th ed., Van Nostrand Reinhold, New York, 1996.

13. W. E. Woodson et al., *Human Factors Design Handbook,* 2nd ed., McGraw-Hill, New York, 1991.

14. G. Salvendy (ed.), *Handbook of Human Factors,* Wiley, New York, 1997.

15. T. G. Hicks et al., *Standard Handbook of Engineering Calculations,* McGraw-Hill, New York, 1994.

16. F. Kreith and R. West, *CRC Handbook of Energy Efficiency,* CRC Press, Boca Raton, FL, 1996.

17. W. C. Turner, *Energy Management Handbook,* Fairmont Press, Lisburn, GA, 1997.

18. British Standards Institution, *Metric Standards for Engineers,* BSI, London, 1967.

19. American National Standards Institute, 11 West 42nd Street, New York, NY 10036.

20. American Society of Mechanical Engineers, 345 East 47th Street, New York, NY 10017.

21. American Society for Testing and Materials, *Annual Book of Standards,* ASTM, 100 Barr Harbor Drive, West Conshohocken, PA 19428-2959.

22. Information Handling Services, 15 Inverness Way East, Englewood, CO 80150.

23. Superintendent of Documents, U.S. Government Printing Office, Washington, DC 20402.

24. Burwell Enterprises, *The Burwell World Directory of Information Brokers,* Houston, TX, 1996.

INDEX

ABI-Inform, 2283
Abrasive barrel finishing, 1129–1130
Abrasive belt finishing, 1130
Abrasive flow machining, 1079, 1083
Abrasive jet machining, 1079, 1083
Abrasive machining:
    grain size and spacing, 1073
    machines, 1078
    structure, 1073
    synthetic abrasives, 1074
Abrasives, synthetic, 1074
ABS, 118
Absorbers, 2024
Absorption:
    refrigeration, 1891–1894
Absorptivity, 1405–1406
    furnace, 1465
    mean, 1471
    solar incident radiation, 1407, 1409
Acceleration, 1295
Acceleration measurement, 681–692, 928
    advantages, 690
    attachment, 690
    basic design, 684–685
    cable, 686
    charge-sensing amplifiers, 685
    constant peak velocity, 690–691
    crude, 681–683
    crystal, 684–685
    difficulties in using, 689–690
    intermediate amplifiers, 686
    location at point of small motion, 690–691
    packaging with solid-state amplifier, 689
    piezoelectric, 683, 690
    piezoresistive, 683–684
    sensitivity, 681–682, 684, 689, 690
    servo, 682–684
    shock calibration and measurement, 693–695
    shock pulse response, 691
    static calibration, 683
    static measurements, 683
    strain-gage, 681–682
    useful range, 682
    voltage measurement, 685–686
    voltage-sensing amplifiers, 686–687
Acceptance sampling, 1183–1184
Accident prevention, 2217–2218
Accountants, financial concerns, 2099–2100, 2102
Accounting, 2099
Accounts payable, 2102
Accounts receivable, 2102
Accrual accounting, 2102
Accumulator, 1850–1851
    operation, 1851
    size, 1851
    see also Pump sets
Accuracy, 918

ACIS, 309
Acoustic enclosures, 720
Activated adsorption, 2037–2038
    air stripping, 2038–2039
    chemical precipitation and clarification, 2036–2037
Actuator:
    electromechanical, 875–876
    hydraulic, 876–878
    pneumatic, 878–880
Adiabatic, 1334
    flow, 1302, 1310
    quasi-static process, 1342
AISI alloy steels, 36
Adaptive control, 914
Adams-Bashforth predictor methods, 844
Adams-Moultron corrector methods, 844–845
Adhesive wear, 456, 458
Adhesives, 344–345
Adsorption, 618–619
    systems, 2023–2024
Advanced Turbine System, 1772
Aerospace Database, 2283
AFBMA method, 605
Agency, 2232
Aggregate planning, 1003–1006
    alternative strategies to meet demand fluctuations, 1003
    approaches, 1004–1005
        classification, 1004
    costs, 1004
    difficulties, 1003
    dilemma, 1005–1006
    levels, 1005
    purpose, 1003
Air:
    ambient, quality standards, 1981
    convection coefficient, 1481–1483
    distribution:
        building, see Building air distribution
        room, 1997–2000
    dry, composition, 1973
    dry, approximate composition, 1956
    gaseous, thermophysical properties, 1254–1256
    heat loss across tube banks, 1486
    liquid, thermodynamic properties, 1253
    makeup, 1981
    moist, properties, 1973–1974
    pressure drop, 1487–1489
    separation, cryogenic processing, 1956–1958
    space cooling requirements, 1990–1991
    space heating requirements, 1991
    thermodynamic properties, 1461–1463
Air change method, 1987
Air conditioning, 1976, 1991–1997
    central systems, 1991, 1993–1995
    economizers, 1601

Air conditioning (*Continued*)
  efficiency and chilled-water temperatures,
    1600
  heat pumps, 1996–1997
  unitary systems, 1995
  see also Conditioning process, HVAC; Space
    cooling
Air compressor, 1865–1877
  antifriction bearings, 1873–1874
  cost, 1865
    air leaks, 1877
  cylinder:
    arrangements, 1868
    cooling, 1867
    mini-lubed, 1867
    nonlubed, 1866
  efficiency, 1865, 1875
    liquid piston, 1875
    volumetric, 1874
  gears, 1869–1870
  lubricant selection, 1870–1871
  pressure-capacity chart, 1866
  selection, 1876
  sizing, 1875–1876
  system demand, 1875
  types, 1865–1875
    centrifugal, 1865, 1871
    crosshead, 1867
    double-acting crosshead, 1868
    dry screw, 1868–1869
    dual rotor, 1868
    flooded, 1870, 1874
    helical screw, 1871
    integral-gear type, 1871–1872
    jacket-cooled, 1873
    liquid piston, 1875
    liquid-ring, 1874
    lobe, 1874
    lubricating, double-acting, 1868
    oil-flooded, 1870
    oil-free screw, 1865, 1874
    reciprocating, single acting, 1865–1867
    reciprocating, double-acting, 1868
    rotary screw, 1865, 1868–1869
    single-acting, two-stage, 1866–1867
    sliding-vane, 1871–1873, 1875
  vane wear, 1874
Air-cooled condenser, 1629
Air-cooled heat exchanger, 1611, 1613, 1614,
    1625–1627
Aircraft gas turbines, 1755–1757
  auxiliary power unit, 1757
  by-pass ratio, 1759
  design progress, 1755
  engine performance, 1755
  jet propulsion, 1727, 1755–1756
  origins, 1727
  propulsion efficiency, 1756
  weight, 1755
  Whittle jet engine, 1727, 1728
Air Distribution Performance Index, 1999
Airfoil cooling, 1736–1737
Air heating, 1641–1647
Air pollution:
  annual combustion-generated emissions, 2013

anthropogenic emissions, 2013
carbon monoxide, 1973, 2022
combustible contaminants, 45–88
combustion-generated, control, 2011–2012
control, furnace, 1494–1496
corporate average fuel economy standards,
    1825
criteria pollutants, 2011
emission standards, 1823
federal test procedure transient driving cycle,
    1822, 1824
global warming gases, 2027
heavy duty vehicles, 1825–1826
highway fuel economy test transient driving
    cycle, 1825
incineration, 2025
lead, 2011
light duty vehicles, 1822–1825
metal hazardous air pollutants, 2024–2025
nitrogen dioxide, 2011
nonhighway heavy duty standards, 1826
ozone, 2011
ozone-depleting substances, 2028
particulate matter, 2011
sulfur dioxide, 2011
volatile organic compounds and organic
    hazardous air pollutants, 2022–2024
see also specific air pollutants
Air pollution, control technologies, 2011–2029
  artificial neural networks, 2026
  carbon dioxide scrubbing, 2026–2027
  carbon monoxide, 2022
  computer-based, 20316
  cryogenic separation, 2026
  enactment of laws, 2012
  expert systems, 2026
  fuzzy logic, 2026
  global climate change, 2026
  incineration, 2025
  metal hazardous air pollutants, 2024–2025
  nitrogen oxides control, 2016–2020
    combustion modification, 2016–2018
    emission reductions, 2020
    formation chemistry, 2015–2016
    post-combustion control, 2018–2020
  particulate matter, control, 2020–2022
    emission reductions, 2020
    schematic of electrostatic precipitator
      operation, 2021
  postprocess control, 2012
  preprocess control, 2012
  prevention of pollution, 2025
  process modification, 2012
  sulfur dioxide control, 2012–2015
    cost, 2015–2016
    residue disposal and utilization, 2015
  volatile organic compounds and organic
    hazardous air pollutants, 2022–2024
    alternative technologies, 2024
    conventional control technologies,
      2023–2024
Air and water defrost, 1910
Air stripping, 2038–2039

AISI 300 stainless steels, yield and tensile
    strength, 1947
AISI 1541, 1172
Algorithmic effectiveness of performance
    objectives achievement evaluation,
    790–791
Algorithms:
    block diagram, 906
    control design, 904
    Dahlin's algorithm, 909
    digital, 903–909
    direct design, 908
    feedforward command compensation, 904
    finite settling time, 908–909
Alkyd resins, 129
Allowance:
    bend, 1114–1115
    shear stress, 145
Alloy:
    composition, 13
Alloy steel, 36
    AISI, 36
    alloying element functions, 36
    austenitic stainless steels, 39
    composition, high temperature, 40
    electrical resistivities and conductivities, 761
    ferrite stainless steels, 39
    high-strength low-alloy, 36
    high-temperature service, heat-resisting,
    38–40
    maraging, 41
    martensitic stainless, 37
    material composition effects on S-N curve,
    401
    notch sensitivity index, 420
    quenched and tempered low-carbon
    constructional, 41
    silicon-steel electrical sheets, 41
    stainless, 37
    standard numerical designations, 32
    thermomechanical treatment, 36
    tool steels, 37
Alta Vista Web Search, 2272
Aluminum:
    influence of crack length on gross failure
    stress for center crack plate, 385
    plain strain fracture toughness, 389
    properties, 1494
    24ST Aluminum, notch sensitivity index, 420
    yield strength, 389
Aluminum alloys, 46
    applications, 52–53
    arc cutting, 54
    brazeability, 51–54
    cast, designation system, 47
    comparative characteristics, 51
    corrosion, 51, 54–56
    cutting tools, 53
    designation systems, 46–48
    electrolytically deposited coloring, 57
    finishing, 56
    galvanic corrosion, 56
    grain size effects on S-N curves, 403
    grinding, 54
    heat-treatable, mechanical properties, 50

influence of nonzero mean stress on fatigue
    behavior, 402, 405, 410, 424
    machining, 53–54
    material composition effects on S-N curve,
    403
    mechanical finishes, 56
    mechanical properties, 48–49
    milling, 54
    nonheat-treatable, mechanical properties, 49
    notch sensitivity index $vs.$ notch radius, 421
    oxide film properties 54–55
    pitting, 55
    routing, 54
    sawing, 54
    shearing, 54
    single-point tool operations, 53–54
    strain hardened, 48
    tempers, 48
    thermally, treated, 48–49
    thickness ranges, joining processes, 52
    weldability, 51
    workability, 52
    working stresses, 49
    wrought, 47, 50, 52
18S Aluminum alloy, grain size effects on
    S-N curves, 403
2014 T6 Aluminum alloy, influence on
    nonzero mean stress on fatigue behavior,
    411
2024 Aluminum alloy, combined creep and
    fatigue, 445
7075-T6 Aluminum alloy, S-N-P curves, 401
American National Standards Institute, 2178,
    2289
American Society of Mechanical Engineers,
    2289
American Society for Quality Control, 2178
American Society for Testing and Materials,
    2272
Amino resins, 129
Ammonia, as refrigerant, 1883, 1887, 1888
Ammonia-water absorption, 1892–1893
Amplifier:
    charge-sensing, 685–686
    differential, 872–873
    feedback compensation, 869, 886
    packaging with accelerometer, 689
    pneumatic nozzle-flapper amplifier, 878–880
    voltage-sensing, 686–687
Analog-to-digital, 902
Analysis:
    dynamic, 281–282
    experimental, 282
    finite element, 279–280
    kinetic, 281
    static, 281
AND gate, 494–495, 2200
Angle bending, 1115
Annealing, 29–31, 1494
Annual Book of Standards, 2290
Anodizing, 1132
ANSI/ASQC Q8980, 2178–2179
ANSI Online, 2272
Anthracite, 1539
AP NEWS, 2283

Application of material components, 163–184
  aerospace and defense applications, 164–166
  biomedical applications, 179–180
  corrosion applications, 176, 179
  dimensionally stable devices, 178, 181
  electronic packaging and thermal control,
      168
  high temperature applications, 176–177
  internal combustion engines, 168–170
  machine components, 166–167
  marine structure, 182–183
  miscellaneous applications, 182–184
  offshore and onshore oil exploration and
      production equipment, 178, 180–181
  process industries, 176
  sports and leisure equipment, 180, 182
  transportation, 170, 172–176
  wear- and erosion-resistant applications, 176,
      178
Archard adhesive wear constant, 458
Artificial neural networks, 2026
Asbestos, laminated, 1873
Ash:
  base: acid ratio, 1544
  characteristics, 1543–1545
  fouling, 1544
  fouling factor, 1545
  fusion temperatures, 1541
  iron/calcium ratio, 1544
  mineral analysis, 1540
  percent, 1540
  resistivity, 1545
  silica/alumina ratio, 1544
  silica percentage, 1545
  slagging factor, 1544
ASHRAE Handbook of Fundamentals, 1979,
    1989
ASHRAE Standard 55, 1979
ASHRAE Standard 15-1994, 1887
ASHRAE Standard 34-1992, 1884, 1886
ASHRAE Standard 34-78
ASHRAE Standard 93-77, 1568
ASHRAE Standard 96-80, 1568
ASME code, pressure vehicle, 1704
ASME properties of Steam, 1772
ASME Performance Test Code, 1792
ASMENet, 2272
Assembly:
  product design, 935–950
Assembly line balancing, 1023–1029
Assets, 2100–2103
ASTM A325, 1142, 1154, 1155
ASTM A490, 1142, 1154
ASTM D197, 1546
ASTM D388, 1538
ASTM D409, 1541
ASTM D613-82, 1813
ASTM D718, 1541
ASTM D975-81, 1540, 1814
ASTM D1855, 1541
ASTM D2013, 1545, 1546
ASTM D2015, 1540
ASTM D2234, 1545, 1546
ASTM D2361, 1540
ASTM D2492, 1540

ASTM D2600-82, 1813
ASTM D2697-82, 1813
ASTM D2698-82, 1813
ASTM D2793, 1540
ASTM D2797, 1541
ASTM D3172, 1540
ASTM D3173, 1540, 1546
ASTM D3174, 1540
ASTM D3175, 1540
ASTM D3176, 1540
ASTM D3177, 1540
ASTM D3286, 1540
ASTM D3302, 1540, 1546
Atomic Energy Act of 1944, 1700, 1707
Atomic Energy Commission, 1700, 1707, 1722
Atomization, vaporization by, 1441
Atmosphere, standard, 1293
  defining properties, 1294
  thermophysical properties, 1257
Austenite, 18, 20–21
Autodesk 3-D, 310
Autoignition temperature, liquid fuels, 1445
Automation:
  defined, 1188
  guided vehicle system, 1234, 1238
  reasons for, 1188–1189
  storage and retrieval system, 1234, 1236,
      1239
  see also Computer
Automobile:
  gas turbine applications, 1728, 1757
Available heat, 1435
  fuel oils, 1438
  natural gas, 1437
  ratios, 1456
Averaging decibels, 715
Avogadro's number, 1343
Awards, quality, 2180–2191
  comparison, 2190–2191
  Deming Prize, 2180–2182
  European Quality Award, 2184–2186
  Malcolm Baldrige National Quality Award,
      2182–2186
  Shingo Prize for Excellence in American
      Manufacturing, 2185, 2187, 2191
  state quality awards, 2188–2191
Axial flow turbine, 1734

Bags, bulk materials packaging, 1214,
    1216–1217
Bainite, 18, 23
Balance sheet, 2100–2105, 2112
  accrural accounting, 2102
  current assets, 2102
  current liabilities, 2102
  current ratio, 2103
  example, 2101
  fixed assets, 2103
  interest-bearing current liabilities, 2102
  net working capital, 2103
  second year comparison, 2104–2105
  total capital, 2104
Balanced capacity method, 2008
Balanced pressure loss method, 2008
Ball bearings, 576–578

angular-contact, 577, 579, 598
applied load, 595
characteristics, 576–579
elliptical contact, capacity formulas, 606
geometry, 582–588
radial, 584
radially loaded, 595–597
self-aligning, 577
split-ring, 578
thrust-load–axial defection curve, 601
thrust-loaded, 597–599
see also specific types of bearings
Band friction cutting blades, 1074
Band saw blades, 1074
Base-load LNG plant, 1959, 1961–1962
plants in operation, 1959
BASIC, 302
Basic reliability networks, 488
k-out-of-*n* network, 489–490
parallel network, 488
parallel network diagram, 489
series network, 488
block diagram, 486
standby system, 488–489
block diagram, 490
Bar bending stress, 361
Basic oxygen furnace, design 357–359
oxygen demand cycle, 357
oxygen production system, 357
BCS theory, 1965
Beam, 207–224
axially loaded, 222
bending moment, 209, 213–215
cantilever, 207–208, 222
constrained, 207
continuous, 207, 217–219
curved, 219–221
dangerous section, 209
deflection, 212–215
design, 212, 214–217
elements of sections, 210–211
equilibrium conditions, 208–209
flexure, 207, 209
impact stresses, 220–223
impulsive loading, 224
impulsive stresses, 224
reaction at supports, 207–208
rectangular, uniform strength, 218
restrained, 214
shear center, 215
simple, 207
safe load, 216
timber, 214
types, 207
vertical shear, 213–215
vibration load, 224
web shear, 215
welded, 357–360
Bearing:
asperities, 616–617
balancing machines, 672
elliptical, 528
exaggerated deformation, 1139
externally pressurized, 520
four-lobe, 528

gas-lubricated, 545–546
hydrodynamic film, 512
journal, *see* Journal bearing
load per unit area, 513
lubricated surfaces, 616–617
oil-impregnated porous metal, 514
recommended lubricant viscosities, 517
rolling-element, *see* Roller bearings
rubbing, 514
selection, 514
slider, 512, 520
squeeze film, 520
three-lobe, 528
thrust, *see* Thrust bearing
Bearing-type connection:
complex butt joint:
sample problem, 1139–1142
preliminary calculations, 1140–1142
simple lap joint:
strength, 1138–1139
Behavior, evaluation, 791
Benchmarking, 477–478, 483–484
competitive, 483
industrial, 483
internal, 483
shadow, 483
world-class, 483
Bending:
cold working, 1114–1116
of fasteners, 1159
and torsional stresses, 229
Bending moment, 213–215
Benefit-cost ratio, 2157
Bernoulli's equation, 1300, 1327, 1682, 1834
Binary cycle conversion, 1588–1590
Biofilters, 2024
Bituminous coal, 1455, 1539
suitability for gasification, 1752
Black-body radiation, 1464–1465
absorptivity, 1465
charts, 1465
coefficient for source temperature, 1467
electromagnetic spectrum, 1403
emissive power, 1402
emissivity, 1400, 1403, 1465–1466
exchange, shape factor, 1410
as function of load surface temperature, 1467
Planck's distribution law, 1402–1403
Stefan-Boltzmann equation, 1402, 1464
view factors, 1465–1466, 1468–1470
Wein's displacement law, 1403–1404
Blade aerodynamic considerations, 1796–1798
analytical techniques, 1797–1798
local Mach number distribution on transonic
blade surfaces, 1798
transonic blade flow, 1796–1799
water table test results with interferometry air
test data, comparison, 1797
Blade-to-blade:
calculation region, 1797
flow analysis, 1796
Blading:
gas turbine, 1724, 1736
Blanking, 1117
Blasius equation, 1317

Blast furnace gas, available heat ratios, 1458
Blasting, 1128–1129
Block diagram:
  armature-controlled dc motor, 875–876
  digital controller, 906
  feedback-control system, 868
  position control system, 871
  pressure regulator, 869
  thermostat system, 868
  velocity-control system, 882
Blow molding, 1126
Bode diagrams, 830–831
Bode plot, 896, 898
Boiler, 1419
  heat-recovery, see Heat-recovery boiler
  feedwater heaters, 1613
Boiling, 1420, 1422–1423
  film pool, 1422
  nucleate, 1420, 1422
  pool, 1420
  in water, simplified relations, 1422–1423
  wire in pool of water, 1420
Boiling-water reactor, 1704–1705
  containment, 1720
  control system, 1720
  power production, 1720
  fuel assembly, 1707
  vessel and core configuration, 1706
Bolt stretch, 1156–1158
  ultrasonic measurement, 1156
Bolted joint:
  behavior, 1143
  clamping force, 1144
  combined loads, 1145
  connectors, 1137
  connector rows, 1137
  critical external load, 1148–1149
  design-allowable stress, 1144
  efficiency, 1138
  elastic curves, 1147–1148
  flange rotation, 1145
  gasket crush, 1145
  load, 1145
  preload, 1145
  shear stress allowance, 1145
  stress-strain curve, 1143
  stress cracking, 1145
  thread stripping strength, 1133
  torsional stress factor, 1144
  ultimate strength, 1143
  very large external loads, 1149–1153
  yield strength, 1144, 1147
Bonding, 4
Boolean logic, used in graphics software,
    301–302, 304–305
Boring, 1048
Boundary layers, see Fluid mechanics,
    boundary layers
Boxes, bulk materials packaging, 1216, 1218
Boyle's law, 1851
Brabbee's design, 1325
Brainstorming, 773
Brainwriting, 773
Brayton gas turbine cycle, 1724, 1742,
    1749–1750

Brine, see Indirect refrigeration
Brinelling, 379
Brittle fracture, 379
Broaching, 1068–1070
Brush seals, 654–657
  applications, 657
  cross section, 654
  flow modeling, 656
  leakage performance comparisons, 655–656
    effects of speed, 655
    installed performance, 655–656
  on-going developments, 657
  materials, 656–657
  multiple, 655
  single-stage, 655
Bucket, air-cooled, gas turbine, 1735
Buckingham Pi theorem, 1304
Buckling, 313, 382
Budget process, 2077–2079
  budget as authorization document,
    2077–2078
  contingency funds, 2078
  functional overhead budgets, 2079–2080
  operating budget, 2077
  planned value of work, 2078
  procedure, 2079
  sample work process, 2078
  work package, 2077
Buffing, 1079
Building:
  commercial, energy consumption, 1593–1596
  materials, thermal properties, 1373
  unsafe conditions, 2221
Building air distribution, 2000–2009
  fans, 2000–2003
  variable-volume systems, 2003–2009
    airflow in ducts and fittings, 2004–2005
    balanced capacity method, 2008
    duct design, 2005–2008
    dynamic losses, 2005–2006
    equal friction method, 2008
    high-velocity duct design, 2008–2009
    lost head due to friction, 2006
    low-velocity duct systems, 2008
    pressure changes during flow in ducts,
      2004
    rectangular duct, 2005
    return air systems, 2008
    steady-flow energy equation, 2004
Bulk material handling:
  conveying, 1206–1212
  packaging:
    bags, 1214, 1216–1217
    boxes, 1216, 1218
    drums, 1219
  storage, 1212, 1214
  transportation, 1218–1219
Bulk modulus, 1832, 1844, 1845
Burner, 1439–1442
  delayed-mix, 1440
  draft-type, 1439
  dual-fuel, 1492, 1494
  fuel-directed, 1440–1441
  gaseous fuels, 1439–1440
  high-velocity, 1442

integral fan, 1440
liquid fuels, 1442
nozzle mix, 1440
oxy-fuel, 1447
packaged, 1440
ports, 1494
premix, 1440–1441
sealed-in, power, 1439
types, 1489, 1491–1494
    air inspirator, 1493
    dual fuel burner, 1494
    flame pattern, burner without premixing, 1491
    high-pressure gas inspirator, 1493
    one-way fixed furnaces, 1489
    venturi tube mixing, 1491
vaporization by automization, 1441
wall radiating, 1440, 1443
windbox, 1439–1440
Burnishing, 1114
Burnout point, 1420
Butt joints, 1137
By-product fuels, 1435

C., 300–301
    and BR Toolkits, 323
C++, 301
    and VR Toolkits, 323
CAD software:
    applications, 314–317
        compared with VR system, 320–326
        computer-aided manufacturing, 317
        optimization applications, 314–315
        rapid prototyping, 316–317
        virtual prototyping, 315–316
    common sweep methods, 305–306
    editing features, 304
    graphical representation of image data, 309–310
    graphics definition, 302
    solid modeling, 302–309
    SolidView, 940
    transformations, 307–309
CAD Standards and translators, 309–314
    ACIS, 310
    analysis software, 311
    DXF, 310
    IGES, 309–310
    STEP, 310
Cams, 513
Canada, patent filing, 2265
Capillary tube, 1906–1907
Capital:
    cost of, 2107
    fund allocation, 2148–2149
    net working, 2103
    stock, 2104
    total, 2104
Carbide, 1046
Carbon, 1873
Carbon/carbon matrix material (CCCs), 132, 143
    characteristics, 133
    design and analysis, 187
    manufacturing process, 163

mechanical properties, 153
physical properties, 153, 161
Carbon dioxide:
    condensed and saturated vapor, thermophysical properties, 1258
    emissions, 1731–1734
    emissivity, 1416, 1418
    enthalpy-log pressure diagram, 1260
    gaseous, thermophysical properties, 1259
Carbon monoxide, 2022
Carbon steels:
    cast, microstructure, 32
    cold working, 34
    effect of surface finish, 407
    heat treatment, 34
    hot working, 33
    microstructure and grain size, 32
    properties, 32
    residual elements, 35
    standard numerical designation, 32
    variations in mechanical properties, 33
Carnot cycle, 1928
Carnot efficiency, 1767, 1771
Carousel, 1236–1238, 1241
Cartesian coordinate system, 1295
Cascade algorithm, 1670–1671
Cascade control, 893
Cascade refrigeration, 1921–1923
Cash flow, 2108
    adjustment of historical data to reflect changes, 2152
    accelerated depreciation, 2108–2111
    annual, 2155–2156
    conversion to constant dollar representation, 2153
    discounted, 2144–2148
    establishing, 2152–2154
    exponentially increasing continuous, 2146
    exponentially increasing series, 2147
    gradient, 2147
    interest factors, conversion to present amounts, 2149
    mixed, 2157
    pure investment, 2156
    single equivalent, 2154
    types, 2146
    uniform continuous, 2148
    uniform series, 2146
Casting:
    centrifugal, 1121–1123
    ceramic process, 1125
    cost comparisons of systems, 1124
    full-mold, 1125–1126
    investment, 1125–1126
    methods, 1120
    permanent mold, 1123–1125
    plaster-mold, 1125
    sand, 1120–1122
Casual loop diagrams, 775
Catalytic oxidizers, 2023
Catastrophe analysis, 2198
Causal diagram, 2054
CD-ROM and Desktop database, 2283–2284
Central air systems, 1991, 1993–1995
    air-and-water, 1994–1995

Central air systems (*Continued*)
  all air, 1991, 1993–1994
  all-water, 1995
  dual-duct, 1994
  variable-air-volume, 1994
Centrifugal casting, 1121–1123
Centrifugal pumps and fans, 2001
  axial flow, 1687
  backward curved blades, 1687, 1689
  Bernoulli equation, 1682
  blade inlet angle, importance, 1689
  characteristics, 1486–1487, 1492
  comparison, 2002
  compared with hydrostatic pumps and
      motors, 1838
  D'Arcy equation, 1682
  determination of fluid heat required,
      1682–1684
  diffuser ring, 1687
  efficiency curve, fixed geometry, and
      constant angular frequency, 1690
  energy transfer, 1687
  engineering data for pressure loss, 1685
  equivalent lengths, 1683
  Euler's theory, velocity vector relationships,
      1688
  flow coefficient, 1688
  fluid system, 1682
  forward curved blades, 1689
  friction loss, 1685–1686
  head coefficient, 1688
  hydrokinetic system, 1838
  limit load, 1689
  mixed flow, 1689
  modeling, 1691–1692
  motor size, 1693
  nondimensional performance characteristics,
      1687–1689
  performance curves, 2002
  power coefficient, 1688
  power requirement, 1691
  pump and fan similarity, 1681
  scroll, 1687
  selection, 1692–1696
  specific speed, 1690–1691
  systems head curves, 1684, 1686–1687
  tip speed, 1687
  total developed head of fan, 1684
  variation of physical appearance and
      expected efficiency with specific speed,
      1690
  volute, 1687
  see also Fan; Pump
Ceramic matrix composites (CMCs), 132–133,
      138–139, 142
  characteristics, 132
  design and analysis, 187
  manufacturing process, 163
  mechanical properties, 150, 152–153
  physical properties, 161
Ceramics:
  class-ceramics, 15
  crystalline, 14
  for use in electronic packaging, 345
  noncrystalline, 14
  process, 1125
  structure, 14–15
Cetane number, 1527, 1810, 1813
Chain reaction, 1712, 1715–1718
  effect of delayed neutrons on reactor
      behavior, 1717–1718
  multiplication factor, 1716
  neutron generation time, 1716
  reactor behavior, 1715–1717
  utilization factor, 1716
Charette, 773
Charpe impact values, steel, 1945
Chart:
  activity relationship, 1229
  assembly, 1224
  process flow, 1224
  operations process, 1223
  see also specific charts
Chemical Abstracts, 2284
Chemical conversions, 1132
Chemical exergy, 1357–1359
Chemical machining, 1096, 1098
Chemical oxide coatings, 1132
Chemical precipitation and clarification,
      2036–2037
Chillers:
  direct fired, 1892
  heat-recovery, 1892
  indirect fired, 1892
  water-lithium bromide, 1892–1893
Chip module thermal resistance, 1656–1661
  definition, 1656
  external, 1658–1660
  flow, 1660
  internal, 1657–1658
  substrate or PCB conduction, 1658
  total, single chip packages, 1660–1661
Chromate coatings, 1132
Circuit design, 1847–1848
Circular saw, 1073
CLAIMS/U.S. Patents Abstracts, 2283
Clamping force limits, 1144–1146
  chart for design decisions, hypothetical joint,
      1146
  combined loads, 1145–1146
  design-allowable bolt stress and assembly
      stress limits, 1144
  flange rotation, 1145
  gasket crush, 1145
  shear stress allowance, 1145
  stress cracking, 1145
  thread stripping strength, 1144
  torsion stress factor, 1144–1145
  yield strength of bolt, 1144
Classification systems, history, 949. *See also*
      specific types of systems
Clauded cycle, *see* Expander cycle
Cleaning, 1128–1130
Clean Water Act, 2031
Closed-loop system, 868–869
Clothing, 2214–2215
Cluster algorithm, 1671
CNC system, 1198
Coal:
  agglomerating character, 1537

approximation formulas, 1537
ash, *see* Ash
burning characteristics, 1541–1543
caloric value, 1538
chemical and physical properties, 1540–1541
classification, 1537–1539
cleaning, 1546–1547, 2015
combustion, 1541
compared with gas, 1751–1753
cost comparison with gas, 1753–1755
cumulative washability data, 1547
fixed carbon, 1538, 1540
float and sink tests, 1546
free swelling index, 1538
froth floatation, 1546
gasification, 1455
gravity separation, 1547
grindability, 1541
heavy-media vessels, 1547
mineral-matter-free basis, 1537
nature, 1535
Parr formulas, 1537
percent ash, 1540
percent volatile matter, 1540
proximate analysis, 1540
pulverized output, 1541
reserves, 1535–1536
sampling, 1545–1546
sulfur content and form, 1540
types, 1539–1540
ultimate analysis, 1540
uses, 1539
washability data, 1547
Coatings, 15, 1130–1132
  organic, 1130
Coefficient of friction, 524
  as function of shear strength, 621
  atmosphere and, 623
  copper lubricated with hydrocarbons, 620
  fixed-incline pad thrust bearings, 534
  load and, 621
  pivoted pad thrust bearings, 537
  relationship of film thickness, 621–622
  speed and, 621, 623
  temperature and, 623
  type of lubrication, 616–618, 623–624
Coefficients of thermal expansion, 1462
Coiled-tube-in-shell exchanger, 1931–1933
  Dittus-Boelter equation, 1933
  section, 1932
  winding relationship, 1933
"Coke button" test, 1541
Coke oven gas, 1456, 1458
Cold plate, 1672–1674
Cold roll forming, 1116
Cold-working processes, 1112–1120
  advantages and disadvantages, 1112
  bending, 1114–1116
  classification, 1112–1113
  drawing, 1118–1120
  forming with rubber, 1119
  high energy rate forming, 1120
  shearing, 1116
  squeezing processes, 1113–1114

values of percent penetration and shear
    strength, 1117
  types, 1112
Colebrook-White equation, 1317
Collocation services, 262–264
  application sharing, 263
  audio technology, 264
  computer-supported meetings, 262–263
  desktop conferencing, 263
  electronic messaging, 262
  transmission technology, 264
  video technology, 264
Columns, 229–234
  definitions, 229–230
  eccentric loads, 231
  end designs, 229
  slenderness ratio, 230
  steel, 232, 234
  subjected to transverse or cross-bending
    loads, 231
  theory, 230–231
  wooden, 232–234
Combined creep and fatigue, 380, 443–449
  creep-limited static stresses, 442
  failure prediction, 444
  frequency modified strain-range method, 444
  governing life prediction equation, 448
  hysteresis loop, 448
  strain-range partitioning method, 447
  summation of damage fractions, 447
  total time to fracture, 447
Combined loads, 1145–1146
Combined stresses, 199–203
Combining decibels, 712–713
Combustion, 1431–1447
  air deficiency, 1433
  air-fuel ratio, 1431–1433
  autoignition temperature, 1445
  available heat, *see* Available heat
  catalytic, 1733–1734
  characteristics of fuels, 1433, 1435
  control equipment, 1494–1495
  convection, 1439
  equivalence ratio, 1433
  excess air, 1433
  explosion hazard, 1433
  flame monitoring devices, 1443
  flame radiation, 1436
  flash point, 1445
  gas turbine, 1730–1731
  heat transferred from, 1436
  lean ratio, 1433
  oxy-fuel firing, 1447
  oxygen, 1431
  oxygen enrichment, 1433, 1459–1460
  products, sensible enthalpies, 1287–1288
  purposes, 1435–1439
  requirements for, 1444
  safety, 1442–1447
  Sankey diagram, 1436
  stability, 1433
  stoichiometry, 1431, 1433
  vitiated air, 1433
  see also Burner; Fuels
Combustors, 1724

Combustors (*Continued*)
  Nox x combustors, 1732–1734
Compact heat exchangers, 1611
Compensation:
  capillary, 543–544
  constant-flow-valve, 545
  controller, 893–896
  digital feedforward, 904
  digital series, 907
  feedback, 869, 886
  feedforth, 894–895
  orifice, 544
  see also Controller, compensation
Completing square, 816
Complex butt joint, bearing-type connection,
    1139–1142
  preliminary calculations, 1140–1142
  rivet capacity solution, 1140–1141
  strength, 1139
  tearing capacity, 1141
Complex *s* plane, 824–825
Component mounting:
  discrete components, 341
  general, 341
  printed circuit board components, 341–343
  specific components, 341
Composite Index of CRC Handbooks, 2289
Composite materials:
  applications, 163–184
  characteristics, 132
  classes, 132
  comparative properties, 133–136
  design and analysis, 184–187
  mechanical properties, 144–153
  minimalizing stress, 134–135
  physical properties, 153–169
  pitch fibers, 133, 136
  properties, 142–161
  reinforcement forms, 133
  reinforcement and matrix materials, 136–143
  specific tensile strength, 134
  types, 132
Composite panel, 721
  transmission loss of a composite wall, 722
Composites:
  coatings, 15–16
  fiberglass, 15
Compounds:
  thermochemical properties, 1286
  thermophysical properties, 1251–1252
Comprehensive Environmental Response,
    Compensation and Liability Act of 1980,
  2032
Compressor:
  air, *see* Air compressor
  axial, 1728
  centrifugal, 1728–1729, 1865, 1871,
    1900–1901
  crosshead, 1867
  double-acting crosshead, 1868
  dry screw, 1868–1869
  dual rotor, 1868
  dynamic, 1728
  efficiency, 1728–1729
  flooded, 1870, 1874

  helical screw, 1871
  integral-gear type, 1871–1872
  jacket-cooled, 1873
  liquid piston, 1875
  liquid ring, 1865, 1874
  lobe, 1874
  lubricating, double-acting, 1868
  oil-flooded, 1870
  oil-free screw, 1865, 1874
  operating time, space cooling, 1991
  operation, 1901
  pressure ratio, 1724, 1748–1749, 1751
  reciprocating, single acting, 1865–1867,
    1897–1898
  reciprocating, double-acting, 1868
  rotary screw, 1865, 1868–1869, 1898
  screw, 1900–1901
  scroll, 1899–1900
  single-acting, two-stage, 1866–1867
  sliding-vane, 1871–1875
  rotary, 1865
  thermodynamics, 1347
Computed tomography, 744–746
  of a flashlight, 745
  of a pencil, 745
Computed torque method, 904
Computer:
  application to design process, 278
  automated storage and retrieval systems,
    1234, 1236, 1239
  categories, 284–285
  CISC, 285–286
  central processing unit (CPU), 285
  engineering PCs, 286
  engineering workstations, 286–287
  evolution, 284
  hardware, 282–283
  imput/output, CPU, 282–283
  mainframe, 284–285
  microcomputer, 284–285
  minicomputers, 284–285
  parallel processing, 287
  RISC, 285–286
  supercomputers, 284–285
  use in design of heat exchangers, 1631–1635
Computer-aided design, 275–318, 1187
  analytical methods, 279–283
  applications, 314–317
  applying computers to design, 278–282
  CAD software, 301–309
  CAD/CAM part programming, 1193–1194
  computer, 283–287
  design process, 276–278
  hardware, 282–283
  input devices, 290–293
  historical perspective, 276
  memory systems, 287–290
  modeling, 279–280
  output devices, 293–296
  software, 296–302
  standards and translators, 309–314
Computer-aided process planning, 1203
Computer-integrated manufacturing, 1188,
    1197–1199
  see also Group technology; Industrial robots

Computer graphics
    three dimensional vs. VR, 324–325
    immersive VR systems, 324
Concurrent engineering:
    adoption, 250
    and the individual, 252
    applicability, 252
    application of principles, 258
    barriers to, 251–252
    definition, 250
    essence, 250–251
    innovation, 253
    lessons, 253
    origin, 249–250
    team, 250–251, 261, 350
    teamwork leading to chaos, 252
    technologies, 253–258
        collocation services, 262–264
        communication, 253, 262
        coordination services, 264–267
        corporate history management services,
            271–274
        data-sharing, 256–257
        electronic design notebooks, 257
        information sharing, 267–270
        negotiation/tradeoff, 255
        process libraries, 257–258
        task coordination, 254–255
Condensation, 1423–1424
    dropwise, 1423
    effect of noncondensable gases, 1424
    film, 1423
Condenser, 1901–1903, 2024
    inadequate venting or drainage, 1630
Conditioning process, 1975, 1977–1979
    adiabatic mixing process, 1981
    air stream mixing, 1980
    cooling and dehumidifying process, 1978
    heating or cooling device schematic, 1977
    heating and humidifying process, 1979
    heat transfer, 1980
    sensible heat factor, 1975
    sensible heating and cooling process, 1977
    space, humidification without heat transfer,
        1980
Conductance, unit surface, 1983
Conduction:
    conduction shape factor, 1377, 1379–1380
    defined, 1369
    Fourier's law, 1369, 1649
    heat conduction with convection heat transfer
        on boundaries, 1381
    heat transfer, 1369–1385, 1649–1652
    interface/contact resistance, 1650–1652
    non-steady state, 1474–1477
    one-dimensional conduction, 1649–1650
    one-dimensional conduction with internal
        heat generation, 1650
    one-dimensional steady heat conduction,
        1375, 1378
    shape factor, 1379–1380
    spreading resistance, 1650
    steady state, 1472–1474
    thermal resistance for circular heat source,
        1651

transient conduction, see Transient heat
    conduction
two-dimensional steady heat conduction,
    1377, 1381
Conductivities, refractory and insulating
    materials, 1473
Conformal surfaces, 512
Conjection molding, 1126
Contact stresses, 242–244, 558
    areas of contact and pressure, 243
    bearing race, 598
    elastic solids, geometry, 569
    elliptical, 558–561, 563
    rectangular, 563–566
Contamination control, 1832–1833
    contaminant levels, 1834
    dirt, 1833
    filtration, 1832–1833
    positive aspects, 1833–1834
        clearances in pumps, 1833
        Ferrographic method, 1833–1834
        piston pump clearances, 1833
        spectrochemical method, 1833
Continuous versus noncontinuous fields,
    752–753
Contraction coefficients, potential flow theory,
    1303–1304
Control charts, 482
Control laws, 880–885
    derivative control, 884
    integral, 883–884
    PD control, 885
    performance specifications, 889–891
    PID control, 885
        algorithm, 905
        op-amp implementation, 889
        open-loop design, 898–899
    proportional control, 881–882
    proportional-plus-integral control, 884
Controller:
    compensation, 893–896
    design, 886, 904
    digital, 902, 906
    digital process, diagram, 911
    electronic, 886–887
    hardware, 886–887
    motion, 910
    performance and nonlinearities, 893
    process, 910
    programmable logic controllers, 909
    single-loop system, 886
    two-loop process, application, 912
Control systems design, 867–915
    adaptive control, 914
    block diagram, 863, 871
    closed loop system, 868–869
    dead-time elements, 898
    design, 899–901
    disturbance rejection, 868
    Fuzzy logic, 913
    gain margin, 897–898
    integral absolute-error, 890
    integral-of-time multiplied absolute-error, 890
    integral-of-time-multiplied squared-error, 890
    neural networks, 914

Control systems design (*Continued*)
  nonlinear control, 914
  nonlinearities and controller performance,
      893
  Nyquist stability theorem, 896–898
  optimal control, 891, 914
  performance indices, 889
  phase margin, 897
  reset windup, 893
  standard diagram, 870
  steady-state error, 872
  structure, 869–872
  system-type number and error coefficients,
      872
  transfer functions, 871
  trends, 912–914
  Ziegler-Nichols rules, 891–893
  see also Control laws
Convection, 1385, 1389, 1391–1401
  coefficient, 1381, 1652
  combustion products, 1439
  configuration factors, 1484
  forced, *see* Forced convection
  formula, 1481
  free, *see* Free convection
  heat transfer, 1481–1483, 1652–1655
  log-mean temperature difference, 1400
  natural, 1653–1654, *see also* Free convection
  see also Boiling; Condensation
Convergence loops, 1631–1632
Conveyors:
  belt, 1207–1212
  chain-driven, 1230–1232
  continuous-flow, 1208, 1214
  gravity, 1229, 1231
  pneumatic, 1208, 1215
  power-and-free, 1232–1233
  powdered, 1230, 1232, 1233
  roller, 1232
  screw, 1207–1209
  types and functions, 1207
  unit material handling, 1224
  vibrating, 1208, 1214
Cooling load, 1988–1991
Cooling-load-temperature-difference method,
      1989–1990
Coordinate system, Cartesian, 1192–1193
Coordination services, 264–267
  common visibility 265
  constraint management, 266–267
  managing workflow, 266
  tracking design progress, 266
Copper:
  composition, 59
  corrosion, 60
  fabrication, 60
  hardening, 60
  lubricated with hydrocarbons, coefficient of
      friction, 620
  properties, 59–60
  relationship of friction and thickness of
      films, 621–622
Copper alloys, 59–68
  application, 61–63
  bridge bearing plates, 68

  casting specifications, 65–67
  conductivity, 68
  fabrication, 62
  gear bronzes, 68
  piston rings, 68
  properties, 63–64
  sand cast, 60–62
  selection, 62
CORBA Standards, 268–269
Corona destruction unit, 2024
Corporate history management services:
  issues, 271–274
    creation and update, 273
    environment of a VTM, 274
    navigation, 273
    process modeling, 271
    representation, 272
    retrieval, 272–273
    security, 273
    user acceptance and validation, 274
    user interface, 273
CORPTECH, 2283
Correction for background noise, 715–716
Corrosion, 379–382, 462–468, 491
  alloy 600, 81
  alloy 625, 81
  alloy 798, 81
  alloy 823, 81
  alloy C-276, 81
  alloy G, 81
  alloy X-748, 81
  aluminum, 54–56
  biological, 380, 4673
  cavitation, 380, 466–467
  classification, 379–380, 463
  copper, 60
  copper nickel alloys, 80–82
  crevice, 378, 464–465
  decarburization, 467
  defined, 462
  direct chemical attack, 380, 463
  dry, 81
  erosion, 380, 466
  of fasteners, 1159
  fatigue, 382
  fretting, 381, 454, 457–458
  galvanic, 380, 463–464
  high-nickel, 81
  hydrogen damage, 380, 467
  hydrogen embrittlement, 380, 467
  intergranular, 380, 466–467
  nickel alloys, 72–80
  nickel content, effect on air oxidation, 82
  nickel-copper alloys, 81
  of materials selected for electronic
      packaging, 345
  pitting, 82, 85, 380, 466
  potentials in flowing seawater, 83
  prevention, 379–381, 463–467
  protection, 1132
  pure nickel, 81
  rate data, 464
  selecting leaching, 380
  S-N curves affected by, 408
  stress-corrosion cracking, 82, 380, 382

titanium alloys, 99–100
variables, 463
wet, 81
Corrosive wear, 382
Cost estimating, 2117–2142
  administrative, 2123
  by analogy, 2129
  anatomy, 2118–2119
  assembly, 2123–2124
  basic steps, 2123–2124
  detailed, 2129
  direct, 2129
  earnings, 2123
  engineering activities, 2134–2135
  factors, 2142
  fee, 2123
  firm quotes, 2129–2130
  generalized equation, 2119–2120
  geographical factors, 2124
  growth allowances, 2141
  handbook estimating, 2130
  indirect costs, 2123
  in-process inspection, 2137
  labor, 2121–2122, 2123, 2140
  labor-loading methods, 2131, 2134
  learning curve, 2130–2133
  management, 2141–2142
  manufacturing activities, 2136–2138
  manufacturing startup ratios, 2139
  materials, 2122–2123
  overhead, 2123
  parametric estimating, 2131–2132
  preparation, required resources, 2119–2121
  profit, 2123
  estimating relationships, 2131–2132
  scheduling, 2132, 2134
  shop-loading methods, 2134
  software, 2139–2140
  special tooling, 2140
  statistical estimating, 2132
  subcontracts, 2122–2123
  supervision, 2141–2142
  symbols, 2120
  test equipment, 2137
  work description, 2124
  work element structure, 2125–2129
Costs:
  air heating, 1643
  aggregate planning, 1004
    holding, 995
    inventory, 1004
    item, 995
    ordering, 995
    production, 1004
    shortage, 995
    workforce and production rate changes,
      1004
  analysis, 2134
  direct, 2121–2122
  drafting, 2134
  fixed, 2121–2122
  indirect, 2121–2122
  initial acquisition, 2121–2122
  nonrecurring, 2121–2122, 2128
  nuclear power plants, 1706, 1709, 1757

  overhead, 2123
  recurring, 2121–2122, 2128
  variable, 2121–2122
Cost-benefit analysis of ergonomics and design
  research, 337
Crack:
  cyclic-load range, effect, 432
  growth to critical size, 434
  growth rate, 432–435
  initiation period, 429–430
  rain-flow cycle counting method, 430
  safe-life-with-cracks requirement, 436
Crack propagation:
  coordinates, 388
  crack-opening mode, 386–387
  critical stress intensity, 389
  forward sliding mode, 386–387
  influence of crack length on gross failure
    stress, 385–386
  model for crack-tip stress, 386
  modes of crack displacement, 386–387
  plane strain, 389
  plane stress plastic zone adjustment factor,
    389–390
  plastic zone at tip, 390
  side sliding mode, 386–387
  stress intensity factors, *see* Stress intensity
    factor
  see also Fatigue
Crack-tip stress field, 386–387
CRC Handbook of Chemistry and Physics,
  2289
CRC Handbook of Energy Efficiency, 2289
Creep, 382, 437–443, 491
  buckling, 382
  constant rate, 440–442
  cumulative, 442–443
  curves plotted against on log-log coordinates,
    441
  definition, 195–196, 203–205
  determination rate, 205
  equations, 204–205
  failure, 381, 437
  log-log stress-time creep law, 442
  long-term behavior, 439–440
  mechanism of failure, 204
  melting temperatures, 437–438
  multiaxial state of stress, 442
  parabolic, 441
  Robinson hypothesis, 442–443
  rupture, 438
  strain, 442
  stress relaxation, 204
  uniaxial state of stress, 440–442
Critical point, 1346
Critical-state properties, 1346
Crossflow:
  coefficient, 1400
  cylinders, 1394, 1663
  liquid metals over banks of tubes, 1396
  noncircular cylinders, 1394, 1663
Cryobiology, 1969–1970
Cryogenic heat transfer, *see* Coiled-tube-in-
  Shell exchangers; Plate-fin heat
  exchangers; Regenerator

Cryogenic insulation, 1939–1943
  insulating powders and fibers, 1943–1944
  multilayer insulations, 1942
  superinsulation, 1941–1943
  vacuum insulation, 1940–1941
Cryogenic processing, 1955–1963
  air separation, 1956–1958
  helium purification and liquefaction,
    1962–1963
  merchant oxygen plant flow sheet, 1957
  natural gas liquefaction, *see* Liquid natural
    gas
Cryogenic refrigeration, 1921–1930
  cascade refrigeration, 1921–1923
  expander cycle, 1924–1930
  Gifford-McMahan cycles, 1930, 1932
  Joule-Thomson cycle, 1923–1924
  minimum work, 1921
  sequence of steps, 1930
  Stirling cycle, 1928–1931
  Vuillemeumer cycle, 1929
Cryogenics, 1360–1361, 1915–1921
Cryogenic separation, 2026
Cryogenic service:
  materials, 1943–1953
    glass, 1953
    lubricants, 1953
    metals, *see* Metal, cryogenic service
    polymers, 1947
    seals and gaskets, 1953
  see also Low-temperature instrumentation
Cryogens:
  air, 1917
  ammonia syrhesis gas, 1917
  helium, 1917–1921
  hydrogen, 19176–1919
  minimum work required to liquefy, 1922
  P-H diagram, 1918
  properties, 1916
  T-S diagram, 1917
  useful in vapor pressure thermometers, 1954
Cryosurgery, 1969–1970
Crystallography, 5–7
CTE, 133–134
  minimalizing stress, 134
  variation of, with particle volume fraction,
    135
Cu-Au system, 9
Cubic boron nitride, 1046
Cumulative damage theory, 410, 413–414, 430
Current liabilities, *see* Liability, current
Current ratio, 2103–2104
Customer needs mapping method, 482
Cutoff, 1118
Cutting-off, metal cutting, 1073–1074
Cybernetic:
  behavior of formal and informal activities,
    2057
  failure modes, 2050, 2052
  framework, 2048
  model, 2047–2048
  paradigm, 2050–2052
  potential failures in project steerability, 2052
Cyclic stress-strain behavior, 425
Cylinders:

press fit, 236
thick wall, 235–236
thin wall, 235

Dahlin's algorithm, 909
Damage tolerance, 436–437
Damper:
  system, single unit, 695, 697
Dangerous Properties of Industrial Materials,
    2289
D'Arcy equation, 1682
Darcy friction factor, 1315
Data:
  reduction, 364–365
DC motor:
  armature-controlled, 875–876
  field-controlled, 876
  position-controlled system, 869–870
  stepper motor, 876
Dead-time elements, 898
Decibels and levels, 712
Decision analysis approach, 2198
Defects, 8–11
  dislocations, 11
  grain boundaries, 12
  point, 10
  solid solution, 9
Deformation, 558
  impact, 380
  maximum, 562, 564
  resistance, 82–83
  strain hardening, 82, 84
  wear, 379, 460
Defrost methods, 1909–1910
  air and water defrost, 1910
  hot refrigerant gas defrost, 1909–1910
Deming's approach to TQM, 477
Deming Prize, 2180–2182
Depreciation, 2103
  accelerated, 2108–2111
  expense, 2108
  linear, 2103
  rate, 2103
Derwent World Patents Index, 2283
Density, lubricants, 521
Describing function method:
  backlash nonlinearity, 857
  dead zone nonlinearity, 854
  defined, 853–854
  saturation nonlinearity, 856
  three-position on-off device with hysteresis,
    858
Design:
  computer, heat exchangers, 1631–1635
  cyclical tension loads, 1158–1159
  ergonomic factors, 330–337
  hierarchy, 2240–2241
  product, 935–950
    documentation of process, 2244–2245
    practices, 935
  quality, 477–480
    guidelines for improving design quality,
      479–480
    Kumes' approach for process
      improvement, 480

process design review, 479–480
product, design review, 477–478
Taguchi's quality philosophy summary, 480
top quality management, 475–486
reliability-based, 491–492
selected documents developed by the U.S. Department of Defense, 492
Design for manufacturing and assembly, 935–950
analysis, 940
comparison of results, 946
components, 938
definition, 936–950
human factors test, 946
importance, 950
machining analysis summary report, 942
methodology, 946
metrics, 949–950
motor drive assembly, 938, 940
new products checklist, 949–950
principles, 940
redesign of motor assembly, 944
role of management, 942–946
team environment, 946–948, 950
Design life-cycle costing, 501
Design research methods, 337
competitive product analysis, 337
product performance analysis, 337
usability studies, 337
Design reliability tools, 492–501
failure modes and effects (FMEA), 492–494, 503
procedure for performing FMEA, 493–494
failure rate modeling and parts count method, 496–497
fault trees, 494–496, 503
Markov modeling, 498–500
network reduction method, 498
safety factors, 500–501
stress-strength interference theory approach, 497–498
Design solution sources:
computers, 244–245
testing, 245
Design techniques, 340
Development of aids for the systems design process, 786
detailed design, testing and implementation phase, 788
evaluation phase, 788–789
operational deployment phase, 789
preliminary conceptual design phase, 787
requirements specification phase, 786
Devices:
gyrating, 1214
machine safeguarding, 2210–2211
whirlpool, 1214
Dew point temperature, 1974, 1978
Diagonalized canonical form, 839–840
Diallyl Phthalate, 129
DIALOG, search example, 2279–2280, 2282–2283
Diamonds, metal cutting tools, 1046
Diathermal, 1334

Die casting, 1124
Diesel engine:
advantages, 1820–1822
applications, 1822
comparison with gas turbine and steam turbine, 1762–1763
descriptions, 1526, 1528
diesel index, 1527
direct injection, 1808
fuels, 1526–1528, 1810, 1813
indirect injection, 1808
knock, 1810, 1813
properties, 1530
requirements, 1529
Difference equations, 861
Diffuser:
behavior, 1998–2000
thermodynamics, 1352
Digital control, 901–903
controller structure, 902
forms of PID control, 902–903
hardware, 909–911
software, 911–912
structure, 902
Digital signal processors, 909
Digital simulation:
Adams-Bashforth predictor methods, 844
Adams-Moultron corrector methods, 844–845
continuous-system, 841–842
critical step size, 845
Euler method, 842–843
general state-variable equation, 841
languages, 846
numerical integration errors, 845
predictor-corrector methods, 844–845
Runge-Kutta methods, 843–844
selecting integration method, 846
time constants and steps, 845–846
Digital-to-analog, 902
Dimensionless numbers, 1304–1305
Dimensionless parameters, 1653
Direct steam conversion, 1587–1588
Disaggregation:
levels, 1005
Discrete and hybrid systems:
difference equations, 861
pulse transfer functions, 864–865
uniform sampling, 862–864
zero-order data hold, 863–864
z transform, 864–865
Disk:
abrasive, 1072
steel friction, 1073
Distributed-parameter models, 850–851
Distribution function, 850
Dittus-Boelter equation, 1392, 1933
DNC system, 1198
Document suppliers, 2281
Double sampling, 1184
Drag, 1323
Drag coefficient, 1309
infinite circular cylinders and spheres, 1323
three-dimensional shapes, 1324
two-dimensional shapes, 1324
Drawing, 1107–1110

Drawing (*Continued*)
 cold, 1118–1120
 hot, 1107–1109
 blank diameters, 1109–1110
 pressure estimation, 1110
Drilling machines, 1051–1060
 accuracy, 1057, 1060
 advancing rate, 1055
 cutting speed, 1054
 drill geometry, 1051, 1054
 metal removal rate, 1051–1053
 motor power, 1058–1059
 oversize diameters, 1060
 recommended feeds, 1056
 spade-drill blade elements, 1057, 1060
 thrust forces and torque, 1051, 1054
 torque and thrust constants, 1055
 work-material constants, 1055
Drinking, 1118
Double walls, 723
Duct:
 adiabatic flow, 1311, 1322
 airflow, 2004–2005
 design, 2005–2008
 drag coefficient, 1323–1324
 expansion factors, 1328
 expansion losses, 1319
 flow, 1290, 1305
 flow meters, 1327
 head losses, 1315, 1317
 high-velocity design, 2008–2009
 incompressible flow, 1311, 1313, 1315
 installation, 2007
 isentropic flow, 1310, 1328
 isothermal flow, 1320, 1322
 laminar flow, 1314–1316, 1320
 loss coefficient, 1321
 lost head due to friction, 2006
 low-velocity systems, 2008
 noise, 2006, 2009
 pipe fitting losses, 1319–1320
 pressure gradient, 1314–1316
 subsonic flow, 1320, 1322
 total pressure profile, 2007
 turbulent flow, 1316–1319
Ductile materials, fatigue failure, 424
Ductile rupture, 379
DXF, 310
Dynamic balancing, in-place, 673
Dynamic imbalance, 668–669
Dynamic response, 314
Dynamic seals, 638–657
 brush seals, 654
 emission concerns, 644–646
 honeycomb seals, 653–654
 initial seal selection, 638–641
 labyrinth seals, 650
 mechanical seals, 642–644
 non-contracting seals for high-speed/
  aerospace applications, 646–650

Economics:
 furnace, fuel-saving improvements,
  1502–1503
 metal cutting, 1043–1046

production, designing, *see* Designing for
  economic production
Eddy current inspection, 746–750
 absolute versus differential coil configuration,
  750
 impedance plane, 746–748
  complex impedance plane, 748
  encircling coil on a solid cylinder, 747
  for a thin-walled tube, 749
 lift off of inspection coil from the specimen,
  747–749
  effects of changes in conditions on local
   signal changes, 749
 skin effect, 746
Editing features, of CAD, 306–307
Efficacy evaluation, 791
 attributes, 791
Efficiency, first- and second-laws, 1336
Eigenstructure, 831, 838–840
Elastic constants, 196
Elastic deformation, 379, 380, 382
 axial direction, 382
 distortion energy theory, 383
 Hooke's law equation, 382
 maximum shearing stress theory, 383
 plastic true strain equations, 383
Elastic limit, defined, 194
Elastic strain, 382
Elastomers, 126–129
 properties, 126–127
Electric motors, efficiency, 1601–1602
Electrical discharge:
 grinding, 1093
 machining, 1093–1095
 sawing, 1094
 wire cutting, 1094–1095
Electrical resistivities and conductivities of
  metals and alloys, 761
Electromechanical actuators, 875–876
Electromechanical machining, 1084–1085
Electronic controllers, 886–887. *See also*
  Op-amp
Electronic displays:
 direct-view storage tubes, 294
 raster-scan terminals, 294
 vector refresh terminals, 293
Electronic equipment, 1649–16512
 chip module thermal resistance, 1656–1661
 cooling:
  cold plate, 1672–1674
  natural convection, 1661–1662
  thermoelectric coolers, 1674–1678
  see also Contact resistance; Forced
   convection
 extended surface, *see* Extended surface; Fin
 heat transfer:
  conduction, 1649–1652
  convective, 1652–1655
  radiative, 1655–1656
 shake tests, 705–709
Electronic information, 2269–2285
 definitions, 2269
 database and commercial services,
  2279–2284
  CD-ROM, 2283

commercial services for mechanical engineers, 2279
desktop, 2283
DIALOG, 2279–2280, 2282–2283
document suppliers, 2281
important databases for mechanical engineers, 2282–2283
Knowledge Express, 2281
STN International, 2280
internet:
  access options, 2278
  approaches, 2271–2272
  ftp, 2276
  future, 2278
  listservs, 2276
  telnetting, 2276–2277
  terms, 2270–2271
  usenet groups, 2276
  World Wide Web, 2272–2275
intranet, 2277
options for use, 2284
Electronic packaging:
  component mounting, 340–342
  conditions, 340
  design techniques, 339
  fastening and joining, 342–344
  interconnection, 344–345
  manufacturability, 350
  materials selection, 345
  protective packaging, 350–351
  scope, 339–340
  shock and vibration, 345
  structural design, 347–348
  thermal design, 348–349
Electric power:
  costs, nuclear power plants, 1706, 1709, 1757
Electricity generation, gas turbine applications, 1753–1753
Electrochemical, 1087–1092
  deburring, 1087
  discharge grinding, 1088
  grinding, 1088
  honing, 1089–1090
  machining, 1089–1090
  polishing, 1090–1091
  sharpening, 1090–1091
  turning, 1091–1092
Electrohydraulic systems:
  analog, 1856–1860
    advantage, 1856
    Moog 1500 valve, 1856–1857
    noise, 1860
    potentiometers, 1857–1860
    typical system, 1858
  digital, 1860–1862
    advantages, 1860–1861
    control system, 1861
    purpose, 1860
    stepping motor, 1861–1862
Electron-beam machining, 1092–1093
Electroplating, 1131
Electropolishing, 1098, 1130
Electrostatic precipitators, 2021–2022
Electro-stream, 1091–1092

Elements, thermophysical properties, 1249–1251
Elenbaas number, 1654
Elevators, bucket, 1208, 1212–1213
Embossing, 1119
Emission concerns, 644–646
  sealing approaches, 644–646
  application guide, 647
    double seals, 645–646
    properties of common barrier fluids for tandem or double seals, 646
    single seals, 644–645
    tandem seals, 645–646
Emissions, *see* Air pollution
Emissivity, 1404
  carbon dioxide, 1416
  correction factor, 1418–1419
  furnace, 1465
  mean, 1471
  metallic surfaces, 1407–1408
  nometallic materials, 1407, 1409
  water, 1418–1409
Employment agreement, 2233
Enamel, 1131
Encyclopedia of Polymer Science and Engineering, 2289
Energy:
  management, 1600–1601. *See also* Energy audit
  transfer analysis, 2234
Energy audit, 1591–1605
  auditor, 1591–1592
  capital-intensive conservation measures, 1600–1602
  defined, 1591
  electric motor efficiency, 1601–1602
  energy savings as result of implementing recommendations, 1592
  evaluating conservation opportunities, 1602–1604
  heat recovery, 1601
  HVAC, *see* HVAC
  identifying opportunities for saving energy, 1597–1602
  lighting, *see* Lighting
  low-cost conservation, 1598–1599
  presenting results, 1604
  procedure, 1592
  records of energy use, 1592–1597
Energy conversion, thermal to mechanical, 1776–1777, 1781–1782
  Euler's turbine equation, 1782
  general energy relationship, 1781
  process, 1776–1777
  turbine blade row steamtube, 1781
  turbomachinery blading, purpose, 1776
Energy Management Handbook, 2289
Engine:
  comparisons, 1820–1822
  diesel, *see* Diesel engine
  performance measurements, 1814–1816
  radiator fins, 1383
  spark ignition, 1802–1808
    advantages, 1820
    applications, 1806

Engine (*Continued*)
    direct injection stratified charge,
      1807–1808
    divided chamber, 1806–1807
    fuels, *see* Fuels
    knock, 1808–1810, 1813
    processes, 1802
    2-stroke, 1803, 1805–1806
    4-stroke, 1803–1804
    theoretical modeling, 1816–1820
    types, 1801–1802
    see also specific types of engines
Engineering activities, estimating, 2134–2135
Engineering Information Inc.'s Global
    Engineering Village, 2273
Engineering Index, 2287–2288
Engineering materials, 959–966
    energy requirements, 962
    shortages, 962
    taxonomy:
      availability, 966
      basis, 963
      chemical properties, 965
      condition, 965
      customizing, 963
      families, 964
      format, 964
      material code, 964–965
      mechanical properties, 965
      objectives, 962–963
      physical properties, 965
      processability, 966
    varieties, 962
Engineering organizations, authority patterns,
    2088
Engineering societies, as information sources,
    2290–2291
Engineering system components, 1347–1350
Engineers:
    contractual obligations, 2234
    financial concerns, 2099
    financial tools, 2114, 2116
      decision-making with accountant, 2116
      record-keeping, 2116
    insurance, 2235
    liability, *see* Liability; Product liability
    management tools, 2164–2176, *see also*
      Total quality management
    professional needs that affect performance,
      2086–2087
    professional responsibility, 2245
Enthalpy, 1339
    combustion products, 1287–1288
Entropy, 1337–1338
    generation, 1337–1338
      minimization, 1359–1360
    transfer, 1338
Environment:
    thermodynamic definition, 1331
Environmental pollution, *see specific types of*
    pollution
Environmental Protection Agency, 2195–2196
Epoxy resins, 128
Equal friction method, 2008
Equipment:

maintenance, 2140
    repair, 2141
    rework, 2141
    see also Electronic equipment; Fabrication,
      equipment classification
Ergonomic analyses:
    anatomical analysis, 334–335
    biochemical analysis, 333–334
      accelerometers, 334
      data glove, 334
      force plates, 334
      force sensors, 334
    cost-benefit analysis of ergonomics and
      design research, 337
    design research methods, 336
    link analysis, 335
    low-speed cine analysis, 336
    motion analysis, 335
    task analysis, 335
    thermographic imprint analysis, 336
Ergonomics:
    analyses, 332–336
    design process, 330–331
    design research, 331–332
      methods, 336
    generally, 329
    human performance, 329–330
      general interdisciplinary nature of human
        factors, 330
      perceptual and cognitive ergonomics, 330
      physical ergonomics, 330
        body dimensions, 331
Error analysis, 923–927
    external estimates, 925–927
    internal estimates, 923–924
    introduction, 923
    probability curve, 924
    use of normal distribution to calculate the
      probable error in X, 924
Error:
    coefficient, 872
    detectors, 874
    steady-state, 872
Eschka method, 1540
Estimating relationships, 2131–2132
Euler equation of motion, 1300
Euler method, 842–843
Euler's formula, 230
Euler's theory, 1688
Euler's turbine equation, 1782
European Foundation for Quality Management,
    2184
European Quality Award, 2184–2186
Eutectoid steel, 19–20, 22
Evaluation methodology and evaluation criteria,
    790
Evaluation test instruments, 791
Evaporators, 1903–1904
    baffled-shell, 1906
    fixed-tube sheet, 1905
    flash-gas distribution, 1904
    flooded-shell-and-tube, 1905
    gas-liquid separation, 1904
    oil return, 1903
    short-tube restrictors, 1907

spray type, 1906
submergence effect, 1904
Exergy, 1353–1359
  analysis, 1353
  chemical, 1357–1359
  physical, 1355–1357
  symbols and units, 1353–1355
Exhaust heat, gas turbines, 1749–1751
  efficiency of electric power, 1750
  LM500 aero-derivative gas turbine, 1754
  performance of combined cycle
    arrangements, comparison, 1753
  plant costs, 1750
  power plants, multi-unit, 1753
  regeneration, 1750–1751
  simple combined cycle power plant, 1752
  three-pressure, reheat cycle, 1752
Exhaust system, gas turbine, 1738
Expandable-Bead Molding, 1126
Expander, thermodynamics, 1347
Expander cycle:
  calculation of W l, 1928
  comparison with Joule-Thomson cycle,
    1927–1928
  cooling curves, 1927
  diagram, 1925
  N 2, 1956
  T-S diagram, 1926
Expansion devices, 1905–1909
  capillary tubes, 1906–1907
  constant pressure expansion valves, 1906
  electronic expansion valve, 1906
  thermostatic expansion valve, 1905
Ex parte Jepson, 2258
Expected outcome approach, system safety,
  2198
Explosion, 1433, 1442
Express Mail, patent filing, 2260
Extended surface, 1667–1671
  arrays, 1670–1671
  assumptions, 1668
  cascade algorithm, 1670
  cluster algorithm, 1671
  cylindrical spine, 1670
  parallel algorithm, 1671–1672
  see also Fin
Extruding, 1126
Extrusion:
  hot-working, 1107–1108
  cold-working, 1113–1114

Fabric filters, 2022
Fabrication:
  equipment classification, 974–980
    capital resources control, 974
    customizing, 977
    equipment code, 977–978
    equipment identification, 978
    equipment selection, 975
    family code, 978
    manufacturer code, 978
    manufacturing engineering services, 975
    model number, 978
    operation codes, 978–980
    process code, 978

quality assurance, 976
  rationale, 976
  specific sheets, 978–979
  standard and special equipment, 976
process taxonomy, 967–974
  basis, 970
  capabilities, 973–974
  classification rules and procedures, 970
  code, 973
  divisions, 969
  nonshaping processes, 969
  objectives, 968
  purpose, 967–968
  shaping processes, 969
strain hardening, 82, 84
tool, *see* Tools
Factors, use in estimating, 2142
Failure, 377–473
  analysis, 468
  bearing, 1139
  classification of structures by path load, 437
  creep, 437–439
  criteria, 377
  data, 504
    rates for selected mechanical parts, 505
    sources, 504
  defined, 377–378
  fatigue, 492, 1158–1159
  impact, 380
  load, 1139
  mode and effect analysis, 479
  modes and causes, 491, 2198
  modes and effect, 378–382
  modes and effects analysis (FMEA),
    492–494
  plane strain fracture toughness as prediction
    criteria, 389–390
  prediction:
    combined creep and fatigue, 444
    equations, 410
    maximum shearing stress multiaxial
      fatigue theory, 410
    maximum shearing stress theory, 383
    modified Goodman relationship, 413
    Palmgren-Miner hypothesis, 413–414
  retrospective design, 468
  shear, 1138
  shear behind connector, 1140
  shear of rivet, 1141
  static, 491
  stress corrosion cracking, 382, 467–468
  summary, 378–382
  synergistic, 376
  tear out, 1140
  tearing main plate of joint, 1138–1139
  see also specific types of failure
Failure modes and effects analysis (FMEA),
  492–494, 503
  procedure, 493–494
Failure rate modeling and parts count method,
  496–497
Fan, 1696–1698, 2000–2003
  centrifugal, *see* Centrifugal pumps and fans
  effectiveness of methods of controlling, 1697
  laws, 2001, 2003

Fan (*Continued*)
  multirating table, 1697
  noise, 2001
  optimum match between system and
      forward-curved blade fan, 2001, 2003
  power and efficiency, 1696
  power requirement, air-cooled heat
      exchangers, 1626–1627
  selection, 1696, 1698, 2003
  similarity to pumps, 1681
  static efficiency, 2001
  types, 1696, 2001
  variable volume service, 1698
Fanning friction factor, 1315
Fastening and joining:
  adhesives, 344
  general, 342
  mechanical fastening, 342–343
  welding and soldering, 343–344
Fatigue, 205–207, 379
  accumulated fatigue damage, 413
  combined with creep, *see* Combined creep
      and fatigue
  computer-controlled closed-loop fatigue
      testing machine, 400
  cyclic loading tests, 435
  cyclic strain-controlled, *see* Fatigue, low-
      cycle, 421
  damage, 413–414, 447
  definitions, 205
  design considerations, 397–398
  domains of cyclic stressing on straining, 397
  fretting, 379
  full-scale testing, 435–436
  high-cycle, 379, 397, 405
  impact, 380
  influence of nonzero mean stress, 411
  initiation, 397
  life prediction, *see* Fracture mechanics,
      methods for fatigue life prediction
  limit, 401
  loading, laboratory testing, 397–401
  low-cycle, 379, 397, 426
  modes of failure, 200–201
  properties, fretted steel and titanium
      specimens, 455
  properties, fretting under states of stress, 454
  quasirandom stress-time pattern, 399
  residual, subsequent to fretting, 450–451
  service loading simulation test, 435–436
  S-N-P curves, *see* S-N-P curves
  strain, nonzero mean, 402, 405, 410,
      424–425
  strength, 401
  stress-time pattern, 398–399
  surface, 379
  thermal, 379
  see also Stress
Fault tree, 494–496, 503
  analysis:
      symbols, 494, 2200. *See also* Logic gates
  for kitchen without hot water, 495
  system safety, 2198
  technique, 2199
Federal Energy Policy of 1992, 1772

Fedworld, 2273
Feedback:
  compensation, 886
      amplifier, 868
      cascade control, 893
  control system, 869
  properties, 868
  pseudoderivative, 896
  state-variable, 895
  see also Systems engineering
Feedback-control system, terminology and basic
      structure, 871
  first order performance-time domain test,
      1855
  in hydraulics, 1852–1854
  open-loop servomechanism, 1853
Feedforth compensation, 894–895
Feedforward command compensation, 904
Feeding methods, machine safeguarding, 2212
Ferrographic method, 1831–1834
Fiber-optic connections, 344–345
Fibers, 137–139
  aramid, 139
  based on alumina, 139
  based on silicon carbide, 138–139
  boron fibers, 138
  carbon (graphite fibers), 138
  glass fibers, 137–138
  high-density polyethylene fibers, 139
  properties, 137
Field test verification, flow field design,
      1791–1796
  aerodynamic loss, 1793
  ASME Performance Test Code, 1792
  flow incidence, 1793
  in-service evaluation, 1791
  kinetic energy, 1795
  last rotating blade incidence, 1795
  last stationary blade incidence angle,
      1793–1794
  low-pressure blade, Campbell diagram, 1792
  low-pressure turbine blade path, 1791
  low-pressure turbine field test results, 1793
  low-pressure turbine laboratory test results,
      1792
  Mach number relative to mast rotating blade,
      1795
  Navier-Stokes equation, 1795–1796
Film condensation, 1422–1424
Film temperature, 1394, 1396
Film-based radiography, 742–743
  density *vs.* relative exposure, 743
  density or darkness vs. relative exposure for
      common films, 742
Fillets, 1158
Film coefficient, 1985
  jet impingement, 1482–1483
Fin, 1383–1384
  air flow across, 1665
  combining into arrays, 1670
  effectiveness, 1383
  efficiency, 1383
  heat transfer rate, 1384
  radial, 1669
  rectangular profile, 1668–1670

singular, 1672
temperature distribution, 1384
Finance:
    needs of:
        accountants, 2099–2100
        middle managers of line functions,
            2098–2099
        owners, investors, and lenders, 2098
        staff groups, 2099
        top managers, 2098
Financial-expenditure planning, 2070–2071
Financial model, 2100
Financial ratios, 2105–2107, 2111
Financial results, evaluating, 2111–2114
    cell formulae used for projections, 2115
    comparison with budgets and forecasts, 2111,
        2113
    identifying problems and solutions,
        2113–2114
    initiating action, 2114
    product line comparison, 2114–2115
    variance analysis, 2112
Finishing:
    surface, 1078
Finite element analysis, 279–280, 311–313
Fire protection training, 2223–2224
Firing temperature, 1734, 1747–1748
First law of thermodynamics:
    closed systems, 1333–1335
    open system, 1338
    steam turbines, 1769
Fishbone diagram, 480
Fission, 1700, 1711–1712
    cross section, 1716
    emitted and recoverable energies, 1711
    products, 1707, 1711, 1719
    see also Chain reaction
Fittings, air flow, 2005
Flammability limits, 1445, 1512
Flange rotation, 1145
Flanging, 1116
Flare, 2023
Flash point:
    liquid fuels, 1445
Flashed steam conversion, 1588–1589
Floating body, 1294
Flow:
    coefficient, 1327
    control valve, pneumatic, 880
    maldistribution, heat exchangers, 1629
    transducer, 874
Flow control:
    bleed off, 1847, 1849
    meter-in, 1847, 1849
    meter-out, 1847, 1849
Flow diagram, 1225
Flow field:
    considerations, 1789–1790
    design, field test verification, 1791–1796
    solution techniques, 1790–1791
Flow shops, 1017–1018
Flue gas:
    convection coefficient, 1482–1485
    desulfurization system, 2012, 2015
    heat loss across tube banks, 1486

pressure drop, 1487–1490
    thermodynamic properties, 1463
Fluid:
    definition, 1290
    flow furnace, 1485–1488, 1494
    hydraulic, 1831–1832
    properties, 1290
Fluidized-bed:
    heat transfer, 1483
Fluid mechanics, 1289–1329
    boundary layers, 1307–1310
    contraction coefficients, 1303–1304
    drag, 1309, 1323
    dynamic similarity, 1304–1306
        dimensionless numbers, 1304–1305
        Froude number, 1304–1305
        Mach number, 1304–1305
        pressure coefficient, 1304
        Reynolds number, 1304–1305, 1327
        Weber number, 1304–1305
    dynamic similitude, 1305–1306
    energy, 1301–1303
    fluid statics, 1290–1294
    gases, see Gas dynamics
    kinematics, 1294–1298
        acceleration, 1295
        circulation, 1297–1298
        continuity equations, 1298–1299
        fluid element deformation, 1295, 1297
        streamlines, 1295
        velocity, 1295
        vorticity, 1297–1298
    lift, 1323–1324
    mass flow fluid measurements, 1326–1328
    moments of inertia, plane surface about
        center of gravity, 1292
    momentum, 1298–1300
    pressure measurements, 1324–1325
    sonic speed, 1290
    velocity measurements, 1325–1326
    volumetric flow rate, 1326
    volumetric measurements, 1326–1328
Fluid statics, 1290–1294
    aerostatics, 1293
    constant centrifugal acceleration, 1290
    constant linear acceleration, 1290
    liquid forces on submerged surfaces, 1291
    manometers, 1291
    stability, 1293–1294
Fluorinated ethylene-propylene, 125, 128
Fluorinated thermoplastics, 124–125, 128
Fluoropolymers, properties, 125, 128
Flux-cored arc welding, 1166–1167
Force field analysis, 481
Forced convection, 1654, 1662–1667
    air flow over electronic components,
        1665–1666
    boiling, 1422
    correlation factor, 1665
    crossflow, see Crossflow
    external flow, 1393–1397, 1662
        average Nusselt number, 1393–1394
        across banks of tubes, 1395
        circular cylinder in crossflow, 1394
        critical Reynolds number, 1393

Forced convection (*Continued*)
  high-speed, over flat plate, 1396
  high-speed gas, past cones, 1396
  laminar, on flat plate, 1393
  liquid metals crossflow over banks of
    tubes, 1396
  noncircular cylinders in crossflow, 1394
  past sphere, 1394
  stagnation point handling for gases, 1396
  turbulent, on flat plate, 1393
flow:
  across arrays of pin fins, 1665
  across spheres, 1664
  across tube banks, 1664
  disturbance effects, 1663
internal flow:
  bulk temperature, 1385
  Dittus-Boelter equation, 1392
  fully developed turbulent, of liquid metals
    in circular tubes, 1393
  hydraulic diameter, 1393
  laminar, for short tubes, 1391–1392
  laminar fully developed, 1389, 1391
  Nusselt number, 1389, 1391–1393
  Peclet number, 1393
  Sieder-Tate, equation, 1391
  turbulent, in circular tubes, 1392–1393
nucleate boiling, 1422
plane surface, 1662
Reynolds number 1663
tubes, pipes, ducts, and annuli, 1666–1667
Forecasting, 988–994
  conclusions, 994
  definitions, 988
  error analysis, 993–994
    mean absolute deviation, 993
  general concepts, 988
  qualitative, 988
  quantitative, 988–993
    basic regression analysis, 992
    casual methods, 992–993
    exponential smoothing, 991
    index numbers, 989
    methods of analysis of time series,
      989–990
    moving average, 990
    quadratic regression, 993
    weighted moving average, 990–991
Forced convection, 1481
  convection coefficient, 1482–1485
Forging:
  cold rolling, 1113
  hot-working, 1105–1106
Formation chemistry, nitrogen oxides,
  2015–2036
FORTRAN, 299–300
Fouling, 1544
  factor, 1545
  heat exchangers, 1627–1628
Fourier's law, 1649
Fracture:
  control, 436–437
  in compression, 196
  types, 196
Fracture mechanics, 383–384

analog of material at critical point in
    structure, 429
biaxial brittle fracture strength data compared
    to maximum normal stress theory, 384
fracture control, 384
maximum normal stress theory, 385
methods for fatigue life prediction, 429–435
monotonics fracture stength-ductility
    combinations, 428
prediction for cracks, 389
stress intensity factors, *see* Stress intensity
    factor
see also Crack growth
Free convection, 1397, 1399–1410, 1983
  average Nusselt number, 1397
  boiling, 1420
  cavities between walls, 1397–1400
  enclosed spaces, 1397
  flat plates and cylinders, 1397–1399
  Grashof number, 1397
  heat-transfer correlation, 1399
  Rayleigh number, 1397
  spheres, 1397
Frequency and wavelength, 712
Frequency response, 828–829
  performance measures, 829
  plots, 832–839
Fresnel-type concentrators, 1564–1565
Fretting, 380, 449–456
  abrasive pit-digging hypothesis, 450–451
  asperity-contact microcrack initiation
    mechanism, 451–452
  corrosion, 382, 454, 457–458
  damage, 450–451
  defined, 449
  delamination theory, 452
  fatigue, 381
  friction-generated cyclic stress fretting
    hypothesis, 451, 453
  impact, 380, 454
  minimizing or preventing, 456
  principal effect, 454
  residual fatigue properties subsequent to,
    452, 454
  S-N curves affected by, 410
  variables, 450–451
  wear, 380, 453–454
  weight loss, 453
Friction-type connections, 1142–1144
  tensile stress, 1142
Froude number, 1304–1305
FTP, 2276
Fuel oils, 1517–1533
  analysis, 1453, 1518, 1522
  API gravity, 1518
  ASTM grades, 1525
  available heat, 1435, 1457
  burners, 1441–1442
  classification, 1517–1518
  conditioning, 1442
  flammability data, 1445
  gravities, 1520
  heating requirements, 1521
  kerosene, 1519, 1525
  orimulsion, 1528–1529, 1532

properties, 1519
shale oils, 1528, 1531
specifications, 1523–1524
tar and oils, 1528, 1532
viscosity-temperature relations, 1444
see also specific fuel oils
Fuels, 1433–1435
air ratio furnaces, 1489
available heat ratios, 1457–1459
aviation turbine, 1525–1526
by-product, 1435
combustion characteristics, 1456
composition, 1813
demand, 1454
diesel, 1526–1528, 1810, 1813
cetane number, 1527, 1810, 1813
descriptions, 1528
diesel index, 1527
properties, 1530
requirements, 1529
diesel oil specifications, 1814
economy, federal standards, 1824
enthalpy of formation, 1812
enthalpy of vaporization, 1812
filters, 1812
flame stability, 1509
gaseous, 1433
analysis, 1508
burners, 1439–1441
caloric properties, 1512
combustion characteristics, 1510
properties, 1509
see also Liquified gases; Natural gas
heating values, 1812
heats of formation, 1456
higher heating value, 1435
hydrogen/carbon ratios, 1506
liquid, 1435
autoignition temperature, 1445
burners, 1441
flammability data, 1445
flash point, 1445
properties, 1519
uses, 1518
see also Fuel oils
lower heating value, 1435, 1507
octane ratings, 1810, 1813
properties, 1811
shut-off valve, 1446
space cooling requirements, 1991
space heating requirements, 1987–1988
spark ignition engines, 1808–1809, 1813
specific density, 1812
stoichiometry, 1810, 1812
test specifications, 1813
units, 1454
Fuel switching, 2015
Fuel system, gas turbine, 1739
Full-mold casting, 1125
Full-scale testing, 435
Funds, source and application, 2107–2111
Furnace, 1449–1503
air pollution control, 1496
burner, *see* Burner
capacity, 1501

cast or rammed refractories, 1453
classification by heat source, 1453
combustion control equipment, 1494–1496
complex thermal processes, 1499, 1501
construction, 1453–1454
continuous-types, 1452
conversion of metric to English units, 1452
economics, 1502–1503
fluid flow, 1485–1488
fuel-saving improvements, 1503
furnace wall losses, 1461–1462
heat content of materials, 1460
heating rates, 1501–1502
heat transfer, *see* Heat transfer, furnace;
      Black-body radiation
industrial, 2023
modules, selecting, 1502
operating schedule, 1503
oxygen enrichment of combustion air,
      1459–1460
probable errors in calculations, 1450
regenerative, diagram, 1459
selecting number of modules, 1502
self-emptying, 1452
standards and conditions, 1450
symbols and abbreviations, 1451
temperature profiles, 1501
thermal diffusivity, 1461
thermal expansion of structures, 1453–1454
thermal properties of materials, 1460–1462
types, 1450, 1452–1453
waste heat recovery systems, 1496–1497. *See
      also* Heat-recovery boiler
Furnace bed combustion, 2014
Fusion, 1701, 1710, 1712–1714
cross section, 1713
products, 1712
reactions, 1713
Fusion welding:
spectrum of heat intensities, 1157
Fuzzy logic, 913, 2026

Gain margin, 897–898
Galling, 381
Galvanic corrosion, 380, 463–464
aluminum, 56
titanium alloys in seawater, 101
Galling, 381
Gamma radiation, 1705, 1711, 1714
GAMS, 373
Gap analysis method, 484
Gas-cooled-graphite-moderated reactors, 1701
Gas-cooled reactor, 1701
Gas dynamics, 1310–1313
adiabatic flow, 1310
isentropic flow, 1310–1311
Mach number, 1310–1312
normal shocks, 1311–1313
nozzle, 1311
oblique shocks, 1313
Rankine-Hugoniot equation, 1312
stagnation pressure and temperature, 1310
supersonic flow, 1314
Gas gravity, 1509
Gas inspirator system, 1491, 1493

Gas metal arc welding, 1162–1166
Gases:
  at atmospheric pressure, ratio of principal
    specific heats, 1282
  dynamics, *see* Gas dynamics
  flow in pipes, 1320–1323
  global warming, 2027
  ideal, 1341–1344, 1363
    constant values, 1342
    mixtures, 1355–1356
    quasi-static process, 1342
    temperature, 1335–1336
    universal gas constant, 1344
  mean beam length, 1415, 1417
  noncondensable, effect on condensation,
    1424
  radiation, 1415–1416, 1466–1471
  specific heat, 1342
  stagnation point heating, 1406–1397
  thermal conductivity, 1370, 1377
  thermal radiation properties, 1415–1417
  thermophysical properties, 1254–1256
  see also Air pollution; Air pollution control
    technologies
Gasket crush, 1145
Gaskets, cryogenic service, 1953
Gasket/rope seal materials, 637
Gaskets, 629–634
  bolt pattern, 632–633
  bolting pattern indicating good *vs.* poor
    sealing areas, 633
  common gasket materials, gasket factors and
    minimum design seating stress table,
    630–631
  flange thickness, 632
  flange surfaces, 632
  metallic, 631
  non-metallic, 631
  original *vs.* redesigned gasket for improved
    sealing, 633
    ASME method, 631–632
    simplified method, 632
  thickness and compressibility, 633–634
Gas turbine, 1723–1764
  airfoil cooling, 1736
  applications, 1749–1759
    aircraft, *see* Aircraft gas turbines
    automobile, 1728
    combined cycle power plant, 1751, 1753
    industrial, 1753
    integrated gasification, 1751–1753, 1754,
      1755
    surface transportation, 1727–1728,
      1757–1759
    railroad, 1728
    use of coal, 1751
    use of exhaust heat, 1749–1751
  availability, 1761
  axial, 1734
  blading, 1724
  buckets, 1724
  combustors, 1724
  comparisons with diesel and steam engines,
    1762
  components, 1728–1735

  combustion system, 1730–1731
  compressors, 1728–1730
  cost, 1747
  emissions, 1731–1734
  controls, 1737
  cooling water and air systems, 1739
  efficiency, 1725–1728, 1755
  enclosures and lagging, 1738–1739
  engine types, 1724, 1728
  exhaust system, 1738
  firing temperature, 1724–1725
  first stage nozzle, 1735
  first-stage air-cooled bucket, 1735
  fossil fuel technologies, comparison,
    1762–1763
  fuel systems, 1739
  history, 1727–1728
  inlet systems, 1738
  lubricating systems, 1739
  maintenance, 1759, 1761
  nonaviation fuels, 1526–1527
  operating principles, 1723–1727
  operation, 1740
  performance, 1724–1725, 1740–1749
    configurations and cycle characteristics,
      1740–1746
    gas generator, 1741
    intercooling, 1742–1744
    multishaft turbine, 1741–1742
    power turbine, 1741
    regenerative, multishaft turbine, 1746
    reheat or refired turbine, 1745
    simple-cycle, single-shaft turbine, 1741
    single-shaft, intercooled turbine, 1744
    temperature-entropy diagrams, 1743–1744,
      1746, 1750
    trends, 1747–1749
    two-shaft turbine, 1742, 1761
  pressure ratio, 1724, 1748–1749, 1751
  radial in-flow, 1734
  regeneration 1749
  reliability, 1761
  temperature changes, 1725–1726
  thermodynamic cycle, *see* Thermodynamic
    cycle
  water-wash systems, 1739
  working fluids, 1724
Gear:
  finishing, 1067
  machining methods, 1063, 1067
  manufacturing, 1063, 1067
  teeth, 513
General Electric, 2273
Geometry data, 1634–1635
Geopressured resources, 1585–1587
Geothermal energy conversion, 1587–1590
  binary cycle, 1588–1590
  direct steam, 1587–1588
  flashed steam, 1588–1589
  hybrid fossil/geothermal plants, 1590
Geothermal resources:
  classification, 1584
  energy conversion, 1587–1590
    binary cycle conversion, 1588–1589
    direct steam conversion, 1587–1588

flashed steam conversion, 1588–1589
  hybrid fossil/geothermal plants, 1590
geopressured resources, 1585–1587
hot dry rock resources, 1585–1586
hot igneous resources, 1584
hydrothermal resources, 1584–1585
  liquid-dominated, 1585
  vapor-dominated, 1585
Geysers, The, 1584
Gibbs free energy, 1339
Gifford-McMahan cycles, 1930, 1932
Glass:
  cryogenic service, 1947, 1953
  for use in electronic packaging, 345
Global climate change, 2026
Goodman relationship diagram, modified, 413
Graham v. John Deere, 2254
Granular activated carbon treatment, 2038
Grashof number, 1653
Grease, lubrication, 518
Greenhouse, 1573–1574
Grinding, 1074
  electrical discharge, 1093
  electrochemical, 1088
  electrochemical discharge, 1088
  low-stress, 1079, 1082
  machines, 1078
  wheels, standard system, 1074
Gross hazard analysis, 2198
Group technology, 1199–1203
  computer-aided process planning, 1203
  machine cell designs, 1201
  part family formation, 1200
  parts, classification and coding, 1200–1202
  production flow analysis, 1201
Guards, 2209, 2220–2221
Guidelines for improving design quality,
  479–480

Haaland's equation, 1317
Handbook of Human Factors and Ergonomics,
  2289
Handbooks, as information sources, 2289
Hardenability, 25. *See also* Steel, hardenability
Hardgrove method, 1541
Hardware:
  VR hardware, 320–322
Hazard:
  analysis, 2239–2240
  classification, 2198, 2221
  critically ranking, 2198
  index, 2240
  see also Safety engineering
Hazardous waste treatment, 2035
Hearth-type pusher furnace, 1450
Heat:
  available, *see* Available heat
  loss:
    calculation, 1986
    flow of air or flue gas across tube banks,
      1486
  sinks, *see* Extended surface
Heat engine, 1336, 1767–1772
  bottoming unit, 1771
  Carnot efficiency, 1766, 1771

conventional steam turbine cycle, 1771
thermal efficiency, 1767–1768
heat rate, 1768
Kalina cycle, 1772
Mollier diagram, 1771–1772, 1775
simple turbine cycle, 1768
topping unit, 1771
typical fossil unit turbine cycle heat balance,
  1770
typical nuclear unit turbine cycle heat
  balance, 1770
Heat exchanger:
  air cooled, *see* Air-cooled heat exchanger
  baffle spacing, 1628
  boiler feedwater heaters, 1613
  ceramic, 1497
  design, *see* Heat exchanger design
  equations for required surface, 1614
  external, 1927–1928
  latent heat transfer, 1614
  mean temperature difference, 1614
  nomenclature, 1635
  operational problems, 1627–1630
    critical heat flux in vaporizers, 1630
    flow maldistribution, 1629
    fowling, 1627–1628
    inadequate venting, drainage or blowdown,
      1631
    instability, 1630
    shellside flow maldistribution, 1629
    temperature pinch, 1629–1630
    tubeside flow maldistribution, 1629
    vibration, 1628
  overall heat-transfer coefficient, 1615–1616
  plate-type, 1610
  pressure drop, 1616
  process, 1894
  program quality and selection, 1633
  rating methods, 1616–1627
  recommended minimum temperatures to
    avoid acid rain, 1644
  recuperator, 1613
  regenerator, 1613
  sensible heat transfer, 1614
  shell and tube, *see* Shell and tube heat
    exchanger
  size and cost estimation, 1613–1616
  spiral plate, 1610–1611, 1613
  surface, 1601
  temperature profiles, 1397, 1399
  thermodynamics, 1352
  use of computers in thermal design process,
    1631–1635
Heat exchanger design, using computers,
    1631–1635
  design, 1633
  determining and organizing imput data,
    1633–1635
    fouling, 1635
    geometry data, 1634–1635
    process data, 1634
  incrementation, 1631
  main convergence loops, 1631–1632
  program quality and selection, 1633
  rating, 1632
  simulation, 1633

Heat extraction rates, 1989
Heat flux:
    critical, 1630
    maximum, shell and tube reboilers and
        vaporizers, 1625
Heat of formation, 1456
Heat gain, 1988–1989
Heating, 1982–1988
    air requirements, 1987
    conduction in plane walls, 1982
    design conditions, 1985–1986
    film coefficient, 1985
    forced convection, 1983
    fuel requirements, 1987–1988
    heat loss calculation, 1986–1987
    heat transfer, 1982
    heat transmission in structures, 1982–1985
    indoor design temperature, 1986
    mechanical solar, 1569
    outdoor design temperature, 1986
    overall heat transfer coefficient, 1985
    overall resistance, 1985
    passive solar, 1569, 1571
    rates, furnace, 1501–1502
    thermal conduction, 1982–1983
    thermal convection, 1983
    thermal radiation, 1984
    thermal resistance, 1983–1985
    wall with thermal resistances in series, 1985
    see also HVAC
Heating, ventilating, and air conditioning, see
        HVAC
Heat pump:
    air-to-air, 1996
    air-to-water, 1996
    balance point, 1996
    closed loop, 1997
    comparison of building heat loads with pump
        capacities, 1996
    ground-coupled, 1998
    supplemental heat, 1996
    water-to-air, 1997
    water-to-water, 1997
Heat pipes, 1424–1429
    boiling limit, 1427
    capillary pressure, 1425
    construction, 1424
    limitation:
        transport, 1424
        sonic, 1426
        viscous, 1426
    Mach number, 1426
    operation, 1424
    pressure drop:
        liquid, 1425
        normal and axial hydrostatic, 1425
        vapor, 1426
    Reynolds number, 1426
    thermal resistance, 1427–1429
    wick structures:
        capillary radius, 1425
        thermal conductivity, 1429
Heat pump, 1337
Heat-recovery boiler, 1601

Heat transfer, 1334–1335, 1361–1362,
        1367–1430, 1462–1485
    to and from body, 1980
    combustion, 1439
    conduction, see Conduction
    convection, see Convection
    correlations, electronic equipment cooling,
        see Contact resistance; Forced
        convection; Free convection
    furnace, 1462–1463
        coefficients for gas and solid radiation,
            1473
        combined coefficients, 1483–1485
        combined emissivity and absorptivity
            factors, 1465
        combined radiation factors, 1472–1473
        from combustion gases to load, 1479
        convection, 1481–1483
        emissivity-absorptivity, 1465–1466
        equivalent temperature profiles, 1480–1481
        fluidized-bed, 1483
        gas radiation, 1466–1467, 1471
        heating time as function of loading
            pattern, square billets, 1478
        heating time as function of thickness,
            carbon steel slabs, 1479
        log mean temperature difference, 1464
        mean emissivity and absorptivity, 1471
        mean value of gas temperature, 1472
        non-steady-state conduction, 1472, 1474,
            1477
        overall ratio, 1481
        radiation charts, 1465–1469
        solid-state radiation, see Black-body
            radiation
        steady-state conduction, 1472–1474
        temperature profiles, 1479–1480
    latent, 1975
    minimum entropy generation, 1362
    modes:
        conduction, 346–349
        evaporation, 350
        forced convection, 349
        free convection, 349
        radiation, 349
    negligible load thermal resistance, 1477
    Newman method, 1477
    phase change, 1654
    radiative, 1655–1656
    sensible, 1975
Heat transfer coefficient, 1393, 1652–1655
    air-cooled heat exchangers, 1625–1627
    convective, 1983
    heat exchangers, 1615–1616
    hollow cylinder, 1378
    hollow sphere, 1378
    nucleate boiling, 1624–1625
    overall, 1381, 1985
    plane well, 1378
    shell and tube condensers, 1620–1622
    shell and tube reboilers and vaporizers,
        1623–1625
    shell and tube single-phase exchangers, 1618
    tube banks in cross flow, 1395–1396
Heat treatment:

nickel alloys, 84–86
Heavy-water-moderated natural-uranium-fueled
    reactor, 1701
Helium:
  cryogen, 1917
  cryogenic processing:
    liquifier flow sheet, 1964
    purification and liquification, 1962–1963
    recovery process, 1962
  HEII, 1919, 1965
  heat capacity, 1920
  in natural gases, 1963
  phase diagram, 1920
  temperature-entropy diagram, 1921
Helmholtz free energy, 1339
Herringbone groove, 555–561
Hertzian stress, 566
Heuristics/priority dispatching rules,
    1020–1023
Hexadecanol, physical adsorption, 618
Hierarchical computer control, 1197–1198
High-energy-rate forming, 1119–1120
High-strength bolts, installation, 1153–1155
  nut rotation, 1155
High-temperature gas-cooled reactors, 1700,
    1705
  vessel and core configuration, 1708
Hobbing, 1114
Honeycomb seals, 653–654
  annular seals, 654
  materials, 653–654
Hooke's law, 193, 200, 313–314
Hopper trucks, 1218–1219
Hoppers, 1212, 1214
Hoshin planning method, 484
Hot-dip plating, 1131
Hot dry rock resources, 1585–1586
Hot refrigerant gas defrost, 1909–1910
Hot-working processes, 1102–1112
  classification, 1103
  defined, 1102
  drawing, 1107–1110
    blank diameters, 1109–1110
    pressure estimation, 1110
  extrusion, 1107–1108
  forging, 1105
    types, 1105–1106
  rolling, 1103–1105
    isothermal, 1104–1105
  spinning, 1110–1111
    setup and dimensional relations, 1111
    tube, 1111
  piercing, 1111–1112, 1118
  pipe welding, 1111
Human comfort, indoor environmental control,
    1979–1982
Human error analysis, 2198
Human Factors Design Handbook, 2289
Human factors engineering, 2202–2205
  defined, 2202–2203
  evaluation, 789
  general population expectations, 2204–2205
  human-machine relationships, 2202–2203
  principles, 2203–2204
Human factors test, 946

Human performance and ergonomics, 329–330
Humidity ratio, 1974
HVAC:
  energy audit, 1593–1598
  energy consumption, 1593
  equipment control, *see* Building, automation
    systems
  heat loss through steam traps, 5–6
  low-cost conservation measures, 1598–1599
  outside-air flow recommendations, 1598
  preventative maintenance, 1596–1597
  system performance, 1596
  see also Air conditioning; Heating;
    Ventilation
Hybrid fossil/geothermal plants, 1590
Hybrid systems, *see* Discrete and hybrid
    systems
Hydraulics, 1831–1863
  accumulator, 1850–1851
  circuits, 1842, 1848
  classifications, 1847
    constant flow and constant pressure, 1847
    means of speed control, 1849
    pressure, 1847
  contamination:
    control, 1832–1833
    positive aspects, 1833–1834
  electrohydraulic systems:
    analog, 1856–1860
    digital, 1860–1862
  feedback control, 1852–1854, 1856
  first-order performance-time domain, 1855
  fluids, 1831–1832
    comparison, 1832
    fire-resistant, 1831–1832
    mineral oils, 1831
    use, 1831
  hydrostatic transmission, 1851–1852
  hydraulic radial piston motor, 1840,
    1842–1843
  improved model, 1854–1856
  orifices, design equations, 1833–1834, 1836
  pipes and fittings, design equations, 1835,
    1837–1838
  pump set, 1847
  pumps and motors, hydrostatic, 1838–1843
    compared with centrifugal pumps, 1838
    types, 1838–1840
  stiffness, 1843–1847
    bulk modulus, 1844–1845
    compressibility effects in cylinder, 1846
    dynamic behavior, 1845
    Laplace transform variable, 1847
    pressure chamber for measuring
      compressibility, 1844
    resulting problems, 1844
    reversed loading, 1845
    value of oil, 1844–1845
  valves:
    design equations, 1834–1835, 1837
    symbols, 1848
Hydrodynamic machining, 1079, 1084
Hydrogen:
  cryogen, 1917–1919
  ortho-hydrogen, 1918

Hydrogen (*Continued*)
  para-hydrogen, 1918–1919
Hydrogen/carbon ratios, 1506
Hydrokinetic system, 1838
  use of centrifugal pumps, 1838
Hydrostatic bearing, 536–545
  advantages, 537
  annular thrust bearing, 540–542
  bearing pad coefficient determination,
      539–542
  capillary compensation, 543–544
  circular step bearing pad, 540
  constant-flow-valve compensation, 545
  formation of fluid film, 537, 540
  vs. hydrodynamic, 535
  orifice compensation, 544
  pad coefficients, 538–539
Hydrostatic pumps and motors, 1838–1843
  axial piston pump, 1840
  coefficients, 1843
  compared with centrifugal pumps, 1838
  efficiency:
      mechanical efficiency, 1843
      overall efficiency, 1843
      slip coefficient, 1842
      volumetric efficiency, 1842
  gear pump construction, 1841
  imput variables, 1838–1839
  leakage, 1839, 1841
  output variables, 1838–1839
  radial piston motor, 1840, 1842–1843
  torque, 1838–1839, 1842
  types, 1838–1840
  vane pump, 1841
Hydrostatic transmission, 1851–1852
  pump and motor combinations, 1852
Hydrothermal resources:
  liquid-dominated, 1585
  major U.S., 1584
  vapor dominated, 1585
Hypereutectoid steel, changes on heating and
      cooling, 20
Hysteresis loop:
  cyclic loading, 422
  loading and unloading, 200

ICE, 2273
Ice:
  refrigeration use, 1897
Ice/water, thermodynamic properties, 1273
Ideal elements, 796–802
  amplifier, 802
  capacitive storage elements, 799
  inductive storage elements, 799
  lumped-element models, 799
  modulator, 802
  multiport elements, 799, 802–805
  one-port element, 798–801
      constitutive relationships, 798
      laws, 798–801
  physical variables, 796–797
  power, 797–798
  primary and secondary physical variables,
      796–798
  pure gyrator, 802

  pure transducer, 802
  pure transformer, 799
  pure transmitters, 802
  resistive elements, 799
  two-port element, 799
Ideas, patentability, 2250
IGES, 309
Ignition, 1512
Impedance concepts, 919–923
  application of Thevenin's theorem, 920
  idealized elastic structure, 921
  measuring the reactive force at the tip, 922
  measuring the tip deflection, 922
Impedance plane, 746–748
  complex impedance plane, 748
  encircling coil on a solid cylinder, 747
  for a thin-walled tube, 749
Impression-die drop forging, 1106
Impulse response, 818
Incineration, 2043, 2025
Incompressible substances, 1344
Indexes and abstracts, as information sources,
      2288–2289
Indoor environmental control, 1973–2009
  air conditioning, *see* Air conditioning
  clothing, 1981
  human comfort, 1979–1982
  psychrometric chart, 1974–1976
  space conditioning processes, 1975,
      1977–1978
  space heating, *see* Heating
  ventilation, 1924
Indirect refrigeration, 1894–1897
  use of ice, 1897
Industrial robots, 1195–1197
  applications, 1197
  configurations, 1196
  control and programming, 1197
  defined, 1195–1196
Industrial wastewater and hazardous waste
      treatment, 2035–2043
  see also Water pollution, control technology
Industries, types, 1189
Information Industry Market Place, 2288
Information sources, 2287–2292
  codes, 2289–2290
  conference proceedings, 2288
  electronic, *see* Electronic information
  encyclopedias, 2289
  engineering societies, 2290–2291
  government publications, 2290
  handbooks, 2289
  indexes and abstracts, 2288–2289
  information brokers, 2291
  libraries, 2291
  library network, 2291
  online searching, 2288–2289, 2291
  periodicals, 2288
  specifications, 2289
  standards, 2289
  see also Online databases; Databases
Information Analysis Centers, 2274
Information in systems engineering, 783–784
  types of information, 783

Information processing by humans and
organizations, 777–782
biases, 780–781
cognitive illusions, 782
prescriptions to biases, 781
Information sharing, 267–270
Carnot, 270
CORBA Standards, 268–269
ISS, 270
mediators, 269–270
scripting languages, 269
Web, 269–270
World Wide Web, 268
Infrared cameras, 750–751
Initial seal selection, 638
mechanical seals, 642
radial lip, 639–640
seal selection chart, 642
turbine engine seals, 641
types of rotary seal, 640
Initial-value theorem, 817
Injection molding, 1126
Inlet systems, 1738
Input devices:
digitizer, 292
digitizing tablet and cursor, 292
for virtual reality hardware, 320–322
keyboard, 290
light pen, 291
mouse, 291
scanners, 293
touch pad, 291
trackball, 291
Input-output model, manufacturing process,
1191
Inspection process, 753
Insulation:
critical radius for cylinders, 1381
materials, thermal properties, 1373–1374
see also Cryogenic insulation
Integer programming, 366
Integrated gasification combined cycle,
1751–1755
comparison with coal, 1751–1752
fuels suitable, 1752
gasifiers, 1753–1754
Intellectual property, 2233
Interaction matrices, 773
Interconnection:
board level, 344
discrete wiring, 344
fiber-optic connections, 344–345
general, 344
interequipment, 344
intermodule, 344
intramodule, 344
laws, 806–807
Intercooling, 1742–1743
Interest, 2144
calculations, 2144
compounded, 2145–2146
continuous, 2146
factors, 2146–2148
nominal rate, 2145

International Business Information on the
WWW, 2274
International Organization for Standardization,
2178, 2274
International Practical Temperature Scale, 1953
Internet:
access options, 2278
approaches, 2271–2272
ftp, 2276
future, 2278
listservs, 2276
telnetting, 2276–2277
terms, 2270–2271
usenet groups, 2276
World Wide Web, 2272–2275
Intranet, 2277
Invention:
conception, 2250
defined, 2250
nonobvious, 2253–2254
patentable, 2250
putting into perspective, 2255–2256
Inventor, 2250
Inventorship, requirement of originality, 2250
Inventory models, 994–1003
approach, 996
general, 996
demand, 994
definitions, 996
general discussion, 994
holding cost, 995
item cost, 995
maximum shortage, 995
modeling approach, 996–1003
conclusions, 1003
general, 996
models of situations, 997–1003
ordering cost, 995
procurement quantity, 994
shortage cost, 995
types, 995–996
constant demand, 995
deterministic, 995
lumpy demand, 995
probabilistic, 996
Investment analysis, 2143–2157
allocation of capital funds, 2148–2149
alternatives, comparing, 2149–2151
analysis period, 2151–2152
annual cash flow, 2155
benefit-cash ratio, 2157
capital expenditure funding sources,
2143–2144
classification of alternatives, 2149–2151
discounted cash flow, 2144–2148
establishing cash flows, 2152–2154
interest calculation, 2149
payback period, 2157
present worth, 2154–2155
rate of return, 2155
time value of money, 2144
Investment casting, 1125–1126
Investors:
equity, 2104
financial concerns, 2098

Ion exchange, 2041–2043
Iron:
  crystalline structure, 19
  pure, changes on heating and cooling, 19
Ironing, 1119
Iron/calcium ratio, 1544
Iron-iron carbide phase diagram, 19–25
  austenite, 20–21
  bainite, 23
  continuous cooling, 24
  effect of alloys, 20
  eutectoid steel heating and cooling, 19–20
  fine and coarse grain steels, 21
  hypereutectoid steel heating and cooling, 20
  hypoeutectoid steel heating and cooling, 20
  isothermal transformal diagram, 21–23
  martensite, 23
  microscopic grain size determination, 21
  pearlite, 23
  transformation rates, 23
  trends in heat-treated products, 21
Isentropic flow, 1310
ISO conformance models, 2178
ISO 8382, 2178
ISO 8980, 2178–2180
ISO 8980-3, 2178
ISO 8984-1, 2179
ISO 8981, 2178, 2180
ISO 10011, 2179
ISO 10013, 2179
ISO 14000, 2180
ISO 14001, 2180
Isothermal:
  creep and fatigue data, 445
  rolling, 1104
  transformation diagram, steel, 21–22
Issue analysis, 775–779
  continuous-time dynamic simulation, 776
  discrete event digital simulation, 776
  econometrics, 776
  expert opinion methods, 775
  mathematical programming, 777
  microeconomic models, 776
  model credibility, 777
  modeling methods, 777
  model verification, 776
  objective, 776
  optimization, 776
  optimum systems control, 777
  parameter estimation, 776
  policy or planning models, 777
  simulation model, steps in construction, 776
  structural modeling, 775–776
  trend extrapolation/time series forecasting,
      776
  uses for models, 777
Issue analysis, systems engineering, 775–779
  characteristics, 779
  interpretive structural modeling, 775–779
  forecasting methods, 778
  mathematical programming, 779
  optimum systems control, 779
Issue formulation, 771–775
  asking approach, 771–772
  brainstorming, 773

  casual loop diagrams, 774–775
  complexity, 772
  conflicting concerns, 773
  DELPHI, 773
  descriptive identification, 773
  experimental discovery, 772
  interaction matrices, 773
  normative synthesis, 772
  questionnaires, 773
  scenario writing, 774
  surveys, 773
  system definition matrix, 774
  system synthesis, 772
  systems engineering, 771–775
    attributes of objectives, 774
    collective inquiry methods, 775
    modeling aids, 775–776
      prototype, 774
    trees, 774

James Watt's flyball governor for speed control,
    877–878
Japanese manufacturing philosophy, 1029–1030
  Kanban, 1030
Jepson-type claims, 2258
JIS Z9899, 2180
Job sequencing and scheduling, 1014–1029
  assembly line balancing, 1024–1029
    automobile back-up light assembly, 1025
    definitions, 1023–1024
    design, 1026
    goal chasing method, 1028
    goals-coordinating method, 1028
    life balancing techniques, 1026
    mixed model, 1028–1029
      problem areas, 1028–1029
    precedence diagram, 1026
    structure of balancing problem, 1024–1026
    two-card Kanban system, 1027
  differences, 1014
  flow shops, 1017–1018
    more than three machines, 1018
    product flow, 1017
    three machines/n jobs, 1018
    two machines/n jobs, 1017
  heuristics/priority dispatching rules,
      1020–1023
    assembly line configuration, 1024
    example problem, 1022–1024
    global, 1020
    local, 1020
    types, 1020
  Japanese manufacturing philosophy,
      1029–1030
    Kanban, 1030
  job shops, 1018–1020
    product flow, 1019
    two machines/n jobs, 1018–1020
    m Machines/n jobs, 1020
  single-machine problem, 1015
    maximum lateness/maximum tardiness,
        1016
    minimize number of tardy jobs,
        1016–1017
    mean flow time, 1015–1016

mean lateness, 1016
  weighted mean flow time, 1016
structure, 1014–1015
time-based competition, 1030–1031
  first to customer for existing products, 1031
  first to market for new products, 1031
Job shops, 1019–1020
Johnson's Apparent Elastic Limit, 194
Joints:
  behavior under tensile loads, 1146–1153
  bolted, 1136–1138
  butt, 1137, 1139–1142
  elastic curves, 1147–1148
  lap, 1137–1139
  preload diagram, 1147
  short-term relaxation, 1146
  spring rate, 1151
  riveted, 1136–1138
  summary diagram, 1149
    critical external load, 1148–1149, 1151
    very large external loads, 1149–1153
  welded, 1159–1170
  see also specific types of joints
Jordan canonical form, 840
Joule-Thomson coefficient, 1340
Joule-Thomson cycle, 1923–1924
  comparison with expander cycle, 1927–1928
  diagram, 1924
  liquefaction cycle diagram, 1925
Joule-Thomson liquefier, 1939
Journal bearings, 512, 514, 524
  attitude angle, 524, 527
  bearing number, 554
  determining whirl frequency ratio, 528–530
  diagram, 524
  full, 524–525
  gas-lubricated, 546–548
  guide to, 514–515
  herringbone grove, 548, 554–560
  liquid lubricated, 524–530
  load co-efficient determination, 548–549, 552–553
  maximum load, 524, 548, 552
  nonplain, 525
  optimal film thickness determination, 556
  optimal groove, 557
  120 degree partial, 524–527
  180 degree partial, 524–527
  pivoted pad, 546–548
  pivoted thickness determination, 550
  plain, 524
  shoe stiffness coefficient determination, 554
  stability parameter, 546
  trailing-edge film thickness determination, 551
  whirl instability, 546
Journal of Mechanical Design, 2274
Just in time concept, 1029

Kaizen method, 481–482
Kalina cycle, 1772
Kanban, 1030
Kerosene, 1519, 1525
Kewanee Oil v. Bicron Corp., 2250

Kinetic analysis, 281
Kirchhoff's law of radiation, 1407
Kirchhoff voltage, 1675
Kirk-Othmer Encyclopedia of Chemical Technology, 2289
Knowledge Express, 2281
Kume's approach for process improvement, 480

Labor:
  estimates, 2140–2141
  factors, 2123
  hours, 2121–2122
  rates, 2123
Labor-hour estimates, 2119–2121, 2131
Labyrinth seals, 650–653
  applications, 652
  clearance factor, 653
  computer analysis tools: labyrinth seals, 653
  leakage flow modeling, 651–652
  seal configurations, 650–651
Laminar flow, 1307
  boundary layer parameters, 1307–1310, 1487–1488
  ducts, 1315–1316
  entrance effects, 1320
  flat plate, 1393
  friction flow, 1316
  fully developed, 1389, 1391
  growth of boundary layers, 1315
  pressure drop, 1319
  short tubes, 1391–1392
  transition from laminar to turbulent boundary layer, 1307
Lancing, 1118
Lap joints, 1137–1140
  bearing failure, 1139
  heat failure, 1138
  shear failure, 1139
  tear of plate, 1139–1140
Laplace transform, 813
  definition, 813
  pairs, 813–814
  properties, 815
  variable, 1847
Larson-Miller theory, 440
Laser-beam machining, 1095–1096
Laser-beam torch, 1096–1097
Lathe:
  cutting speeds, 1050
  operations, 1048
  size, 1051
  tool angles and cutting speeds, 1050
  tool geometry, 1049
  types, 1050
Least squares, 364
Le Journal des Scavans, 2288
Lenders, financial concerns, 2098
Liability, 2101
  business:
    negligence for services, 2233–2234
    case study, 2235
    contractual obligations, 2234
    insurance for engineers, 2235
  current, 2101–2102, 2105–2106
  employee:

Liability (*Continued*)
  agency and authority, 2232
  employment agreements, 2233
  intellectual property, 2233
  negligence and standard of care,
    2231–2232
  product liability defenses:
    assumption of risk, 2242
    contributory/comparative negligence,
      2241–2242
    state of the art, 2241
  product liability, laws:
    definition, 2235–2236
    express warranty and misrepresentation,
      2236–2237
    negligence, 2236
    principles, 2236
    strict liability, 2236
    substandard condition, 2236
  professional, 2230
Libraries, 2291
Life cycle, 765–770
  phases, 771
Lift coefficient, 1323
Lift off of inspection coil from the specimen,
    747–750
  effects of changes in conditions on local
    signal changes, 749
Lighting:
  conversions, 1603
  efficiency, 1601, 1602
  energy audit, 1593–1597
  energy consumption, 1593
  illuminance for activities, 1595
  improper illumination, 2221
  low-cost conservation measures, 1598–1599
  output reduction as function of fixture
    cleaning frequency, 1594
  weighting factors, 1596
Light-water-moderated enriched-uranium-fueled
    reactor, 1701
Light-water reactor, 1700
Lignites, 1539
  suitability for gasification, 1752
Linde cycle, *see* Joule-Thomson cycle
Linearizing approximations, 853
Linear models:
  I/O form, 808–810
  standard forms, 807–813
  state-variable form, 810–813
Linear statistic analysis, 313
Linearity, 919
  independent linearity, 919
Liquid-dominated resources, 1584–1585
Liquid-fueled reactors, 1701
Liquid fuels, *see* Fuels, liquid
Liquid-metal-cooled fast breeder reactor, 1701
Liquid-metal solid-fueled fast spectrum
    reactors, 1701
Liquid natural gas:
  cryogenic processing, 1955–1963
    base-load plant, 1959–1962
    data on world trade, 1959
    mixed refrigerant cascade liquifier, 1962

mixed refrigerant process cooling curve,
    1961
  peak-shaving plants, 1958–1959
  storage, 1958
  using N-2 refrigeration, flow sheet, 1960
  revaporization, 1958
Liquid penetrants, 730–732
  categories, 730
  limitations of penetrant inspections, 730, 732
  process, 730
    cleaning, 731
    penetrant flaw indication in turbine blade,
      731
  reference standards, 730
Liquids:
  at atmospheric pressure, ratio of principal
    heats, 1282
  saturated:
    specific heat at constant pressure, 1281
    thermal conductivity, 1284
    viscosity, 1285
  surface tension, 1283
  thermal conductivity, 1369, 1375
Liquified petroleum gases, 1514–1515
Listserv, 2276
Log mean temperature difference, 1400
Logic gate, 494–495, 2200
  AND, 494–495, 2200
  OR, 494–495
Loss coefficient, 1321
Lot-sizing techniques, 1010, 1014
  lot-for-lot, 1014
Low cycle fatigue, 420–429
Low-stress grinding, 1079–1082
Low-temperature instrumentation, 1953–1955
  flow measurement, 1955
  tank inventory measurement, 1955
  temperature measurement, 1953–1954
Lubricant, 516–518
  coefficient of friction, 623
  cryogenic service, 1953
  density, 521
  elastic deformation, 522
  extreme-pressure, 623–624
  film shape, 521
  flow:
    fixed incline pad thrust bearings, 530–534
    pivoted pad thrust bearings, 532–533,
      536–538
  grease, 518
  hydraulic fluid, 1832
  manufacturer and designation, 522
  oil, 517
    vs. grease, 516
  pressure-viscosity coefficients, 523
  reactivity, relationship to wear, 622
  viscosity, 516–517
Lubrication, 507–627
  boundary, 616–627
  coefficient of friction, 616–617
  elastohydrodynamic, 508, 513, 520, 556–558
    contact stresses, *see* Contact stresses, 558
    contour plots, 575
    defined, 556–558
    deformation, *see* Deformation, 558

dimensionless grouping, 566–568
elasticity parameter, 566, 573
film parameter, 573
hard, 520, 556, 568
isoviscous-elastic regime, 574
isoviscous-rigid regime, 574
map of regimes, 575–577
piezoviscous-elastic regime, 575
piezoviscous-rigid regime, 574
soft, 520, 557–558, 568–570
viscosity parameter, 573
ellipticity parameter, 566
equations, 520–523
factor, 605–609
film thickness, 566, 573
grease, 518
hydrodynamic, 508, 513, 523
load parameter, 566
load support mechanisms, 520
material parameter, 567
oil, 517–518
regimes, 518–520
boundary, 520
elastohydrodynamic, 520–521
film conditions, 520
hydrodynamic, 520
map, 575–577
wear rate determination, 616–617
speed parameter, 566
symbols, 508–512
Lumped-element model:
automobile suspension system, 805
milling machine, 809
uniform heat transfer through wall, 850
Lumped-parameter model, uniform heat transfer
through wall, 851
Lundberg-Palmer theory, 604–605

Mach number, 1756–1757, 1773, 1789, 1791,
1793–1795, 1797–1798
Machinery noise control, 719
Machines in semireverberant locations,
716–717
Machinability, 1048
data, prediction, 1195
Machine cell, design types, 1201
Machinery:
design and redesign for engineering control,
2213
engineering controls, 2205–2207
alternatives, 2208
isolation, 2213
substitutions, 2208, 2213
ventilation, 2213
hazardous mechanical motions and actions,
2205
measurement of noise, 716
safeguarding methods, 2207–2212
unsafe conditions, 2221–2222
Machining:
abrasive flow, 1079, 1083
abrasive jet, 1079, 1083
allowances, sand casting, 1122
chemical, 1096, 1098
electrical discharge, 1093–1095

electrochemical, 1087–1092
deburring, 1087
discharge grinding, 1088
grinding, 1088
honing, 1089–1090
polishing, 1090–1091
sharpening, 1090–1091
turning, 1091–1092
electromechanical, 1084–1085
electron-beam, 1092–1093
electropolishing, 1098
electro-stream, 1091–1092
hydrodynamic, 1079, 1084
laser-beam, 1095–1096
laser-beam torch, 1096–1097
low-stress grinding, 1079, 1084
magnesium alloys, 110
non-traditional, 1079–1081
current processes, 1080
surface roughness and tolerances, 1081
see also specific types of machining
nickel alloys, 86
photochemical, 1098–1099
plasma-beam, 1096–1097
power, 1037–1040
shaped-tube electrolytic, 1091–1093
temperature, 1078
thermally assisted, 1084–1085
thermochemical, 1099
total form, 1085
ultrasonic machining, 1078, 1086
water-jet, 1086–1087
Magnesium:
applications, 109–110
physical properties, 110, 112
Magnesium alloys, 109–110
applications, 109–110
castings, mechanical properties, 110–111
in common use, 111
corrosion, 113
fabrication, 109–110
finishing, 113
physical properties, 113
wrought products, mechanical properties,
110, 112
Magnetic disks, 289
floppy disks, 289
hard disks, 289
Magnetic lines of flux in a ferromagnetic metal
near a flaw, 751
Magnetic particle method, 751–753
continuous versus noncontinuous fields,
752–753
demagnetizing the part, 753
inspection process, 753
magnetic lines of flux in a ferromagnetic
metal near a flaw, 751
the magnetizing field, 751–752
flux intensity versus density, 751
Magnetizing field, 751–752
flux intensity versus density, 751
Mainframes, 284–285
Malcolm Baldrige National Quality Award,
2182–2186
Management:

Management (*Continued*)
  costs, 2141–2142
  financial concerns, 2098
  line manager, financial concerns, 2098–2099
  tools, 2162–2176
  see also Total quality management
Managing people:
  engineering performance, 2086–2087
  without formal authority, 2087–2088
  motivation, 2086
  recommendations, 2094–2095
    clear assignment, 2094
    clear communication, 2095
    clear goals and objectives, 2094
    commitment, 2095
    effective planning, 2094
    interesting work, 2094
    image building, 2094
    leadership, 2095
    personal drive and leadership, 2095
    problem avoidance, 2095
    process definition, 2094
    reward system, 2095
    senior management support, 2095
    team structure, 2094
  in technology-based environments, issues,
    2085
  style, 2048
  see also Project management; Teamwork
Manometers, 1291
Manson-Coffin-Morrow equation, 444
Manson-Coffin-Morrow expression, 448
Manson-Haferd theory, 440
Manufacturability:
  assembly considerations, 350
  concurrent engineering, 350
  design to process, 350
Manufacturing:
  activities, 2137
  cell, 1198–1199
  computer-integrated, *see* Computer-integrated
    manufacturing
  computer-aided planning, 1203
  cycle, 1189
  engineering:
    estimating, 2136–2142
    sustaining, 2136
  lead time, 1192
  process:
    input-output model, 1191
    see also Fabrication
  product design, 935–950
  startup ratios, 2139
Manufacturing system:
  computer-integrated, 1187
  flexible, 1198–1199
  numerical-control, 1192–1195
    adaptive-control, 1194
    CAD/CAM part programming, 1193–1194
    coordinated system, 1192–1193
    machinability data prediction, 1195
    parts selection, 1193
    programming by scanning and digitizing,
      1194
Maraging steels, 41

Market researchers, financial concerns, 2099
Markov modeling, 498–501
  transition diagram for a two-state system,
    500
Martensite, 19
  phase properties, 23
  tempered, 23
  transformation to, 23–24
Mass absorption coefficient *vs.* atomic number,
  741
Master diagram, 402, 412
Material handling, 1205–1242
  bulk, *see* Bulk material handling
  defined, 1205–1206
  equipment considerations and examples,
    1225–1239
    automated guided vehicle systems, 1234,
      1236, 1239
    automated storage and retrieval systems,
      1234, 1236, 1239
    carousel systems, 1236–1238, 1241
    conveyors, 1226–1233
    hoists, cranes and monorail, 1233, 1235
    industrial trucks, 1234–1237
    plan development, 1225–1226
    shelving, bin, drawer, and rack storage,
      1238–1239, 1241–1242
  implementing the solution, 1239–1240
  storage, 1212–1219
  twenty principles, 1222, 1224
  unit, 1219–1225
    analysis of systems, 1220
    data collection, 1220–1223
    problem identification and definition, 1220
    utilizing loads, 1223–1225
Materials requirements planning, 1006–1014
  calculations, 1010
  conclusions, 1010
  definitions, 1006
  diagram, 1009
  lot-sizing techniques, 1010, 1014
  master production schedule, 1006–1007
    format, 1008
  planned order release schedule, 1014
  procedures and required inputs, 1006–1010
  product structure tree, 1009
  purpose, 1006
Materials selection:
  ceramics and glasses, 345
  corrosion, 345
  general, 345
  materials, 345
  metals, 345
  plastics and adhesives, 345
Mathematical models, 795–866
  analogs, 807
  choice of, 796
  classifications, 846–865
    discrete and hybrid systems, *see* Discrete
      and hybrid systems
    distributed-parameter models, 850–851
    nonlinear systems, *see* Nonlinear systems
    stochastic systems, 846–850
    summary, 848
  compatibility, 806–807

continuity, 806
duals, 807
ideal elements, *see* Ideal elements
iterative approach to control system design,
    796–797
interconnection laws, 806–807
linear models, standard forms, 807–813
lumped-parameter, 796
output equations, 810–811
rationale, 795–796
simple *vs.* complex, 794
state equations, 810–811
state-variable methods, *see* State-variable
    methods
superposition, 852
system safety, 2198
time-varying systems, 851–852
transform methods, *see* Transform methods
see also Simulation
Matra Data's Euliked CAD system, 939
Matrix materials, 136–143
    carbon, 142
    ceramic, 142
    metals, 142
    polymer, 139–142
    properties, 140–142
Matrix solution technique, 1790
Mean absolute deviation, 993
Measurements, 917–932
    appendix, 928–932
        acceleration measurement, 928
        shock measurement, 928
        sound measurement, 928–932
        vibration measurement, 928
    error analysis, 923–927
        external estimates, 925–927
        internal estimates, 923–924
        introduction, 923
        use of normal distribution to calculate the
            probable error in X, 924–925
    impedance concepts, 919
    machine noise, 716
    standards and accuracy, 917–919
        accuracy and precision, 918
        linearity, 919
        sensitivity or resolution, 918–919
        standards, 917–918
Mechanical design:
    reliability, 487–507
    top quality management, 475–486
Mechanical failure modes, 491, 503
    corrosion, 491
    creep, 491
    failure fatigue, 491
    static failure, 491
    types, 491
    wear, 491
Mechanical fastening, 342–343
Mechanically fastened joints:
    clamping force, 1144–1145
    fatigue failure and design for tension loads,
        1158–1159
    differences, 1136–1137
    efficiency, 1138
    evaluation of slip characteristics, 1153

exaggerated bearing deformation, 1139
installation of high-strength bolts, 1153–1155
purpose, 1136
sample problem of complex butt joint,
    1139–1142
shear failure, 1138
solutions:
    bearing-type connections, 1138–1142
    friction-type connections, 1142–1144
theoretical behavior under tensile loads,
    1146–1153
torque and turn, 1155
ultrasonic measurement of bolt strength or
    tension, 1135
see also Bolted joint; Riveted joints;
    Complex butt joint; Welded joints
Mechanical seal leakage, 643
    gas flow, 643
    liquid flow, 643
        for coned faces, 643
        for parallel faces, 643
Mechanical seals, 642–644
    balance, 642–643
    face seal materials, 644
    illustration of face seal balance conditions,
        642
    leakage, 643
    seal face flatness, 643
    types of rotary seal, 640
Meissner effect, 1965
Melting point:
    surface film, 619–620
Membrane technologies, 2039–2040
    electrodialysis, 2041
    pervaporation, 2041
    reverse osmosis, 2040–2041
Memory systems, 287–290
    external memory, 289
    internal memory and related techniques, 288
        cache memory, 288
        memory addressing, 288
        metal oxide semiconductor (MOS) RAM,
            288
        registers, 288
        virtual memory, 288
    magnetic disks, 289
    magnetic tape, 290
    optical data storage, 290
    organizational methods, 287–288
Mercury, saturated, thermodynamic properties,
    1261
    enthalpy-log pressure diagram, 1262
Metal:
    cryogenic service, 1943–1947
    crystal structure, effect on brittle impact
        strengths, 1944
    liquid:
        cross flow over banks of tubes, 1396
        fully turbulent flow in circular tubes, 1393
        thermal properties, 1376
    thermal conductivity, 1370
Metal cutting:
    abrasive machining, 1074
    broaching, 1068–1070
    buffing, 1079

Metal cutting (*Continued*)
  conventional machining processes, 1037
  cutting fluids, 1047
  cutting forces, 1039–1041
  cutting speeds, 1043–1044, 1048, 1050, 1062
  cutting velocities, 1036, 1039
  depth of cut, 1075
  drilling machines, *see* Drilling machines
  economics, 1043–1046
  factors, 1036
  feeds, 1041, 1075
  finishing, 1067
  gear manufacturing, 1063–1067
  grain size, 1075
  grain spacing, 1075
  grinding, 1074, 1077–1078
  grinding fluids, 1078
  machinability, 1048
  machining plastics, 1074
  machining time, 1048
  mechanics, 1036, 1038
  metal removal rate, 1052–1053, 1064–1065
  oblique cutting, 1036, 1038
  relationship of cost factors, 1045
  sawing, shearing, and cutting off, 1073–1074
  shaping, planning, and slotting, 1070–1073
  shear angle, 1039
  surface finishing, 1078
  synthetic abrasives, 1074
  thread cutting and forming, 1067–1068
  tool life, 1041–1043
  tools, 1046–1048
  turning machines, 1048–1051
  ultrasonic machining, 1078, 1086
Metallizing, 1131
Metal matrix composites (MMCs):
  characteristics, 132
  comparative properties, 133
  design and analysis, 187
  manufacturing process, 163
  mechanical properties, 148–150
  physical properties, 153–154, 156, 158,
      161–162
  thermal conductivity, 136
Metals:
  electrical resistivity, 136, 761
  for use in electronic packaging, 345
  for use in MMC matric materials, 142
  structure, 5–8, 12–13
  thermal conductivity, 136
Metals Handbook, 2289
Methane:
  at atmospheric pressure, thermophysical
      properties, 1264
  available heat, 1457, 1459
  combustion with 110% air, 1459, 1471
  saturated, thermodynamic properties, 1263
  used for gas fuel, 1739
Metric Standards for Engineers, 2289
Microcomputers, 284–285
Milling:
  allowances, 1063
  cutter, 1060
  cutting action, 1063
  cutting speeds, 1061, 1062

  feed, 1060, 1062
  horizontal, 1061
  horsepower, 1062
  machining time, 1062
  metal removal rate, 1062
  vertical, 1061
Minicomputers, 284–285
Minimum selection, sand castings, 1122
Miscellaneous substances at atmospheric
      pressure, thermophysical properties,
      1276–1277
Mixed model assembly lines, 1028–1029
  problem areas, 1028–1029
Modal matrix, 840
Models:
  for purchase discounts, 999–1003
    fixed holding cost, 999
    total cost function, 1001
    variable holding cost, 1001
  inventory, 994–1003. *See also inventory
      models*
  manufacturing model with shortage
      permitted, 999–1000
  manufacturing model with shortage
      prohibited, 998–999
  purchase model with shortage permitted, 998
  purchase model with shortage prohibited,
      997–998
Moist air, properties, 1973–1974
Modeling, 279
  hybrid solid, 279
  Markov, 498–501
  solid, 279, 302–309
  surface, 279
  wireframe, 279–280
Modular cabinet drawer storage, 1240–1241
Modulator, 800
Mohr's Circle, 199, 201
Molding:
  plastic-molding, 1126–1127
  powder metallurgy, 1127–1128
Momentum equation, 1314
Momentum, theorem, 1299–1300
Monorail, 1233, 1235
Motor drive assembly, 938–940
Mufflers, 725, 727
Multiattribute utility theory, 780
Multiaxial fatigue failure theory, 410
Multiaxial principal shearing stresses, 405
Multiple and sequential sampling, 1184
Multiport elements, 799, 802–805
Municipal wastewater treatment, 2032–2035
  biological methods, 2032
  disinfection, 2034
  gravity thickening, 2035
  in-depth filtration, 2034
  sequence diagram, 2033
  trickling filter, 2032
Mutually supportive systems model, 2049

National Center for Manufacturing Sciences,
      2274
National Technology Transfer Center, 2274
Natural convection, 1653–1654
  in confined spaces, 1661–1662

Natural gas, 1505–1515
  analysis, 1507
  available heat, 1437, 1437
  caloric value, 1507, 1512
  combustion characteristics, 1510–1511
  distribution, 1505
  environmental impact, 1505
  flame, 1509, 1512–1513
  flammability limits, 1512
  gas gravity, 1509
  minimum ignition temperature, 1512
  net heating value, 1507
  properties, 1507
  sources, 1507
  storage, 1507
  types, 1507
  uses, 1505
  Wobbe Index, 1512
  see also Fuels, gaseous; Liquid natural gas
Navier-Stokes equations, 507, 1300, 1304, 1795–1796
NC machining, parts selection, 1193
Net profit, 2107
Netscape, 2272
Network reduction method, 498
  block diagram, 498
Newt positive suction head, 1693
Neuman Method, 1477–1479
Neural networks, 914
Neutron radiography, 741–742
  showing bondline flaws, 741
News and Business Databases, 2283
Newton's second law, 1300
Nibbling, 1118
Nickel, 71
Nickel 200, 72
Nickel 201, 72
Nickel alloys, 72–89
  chemical composition, 73–74
  classification, 72
  cold work, effect on hardness, 87
  duranickel alloy, 72
  fabrication, 82
    shear load, 87
    shear strength, 87
  heat treatment, 84
  machining, 86
  mechanical properties, 75
  modifications to product special properties, 78
  registered trademarks, 88
  rupture stress, 76
  welding, 86
Nickel-chromium-iron alloys, 77
Nickel-chromium molybdenum alloys, 79–80
Nickel-copper alloys, 76
Nickel-iron alloys, 79
Nickel-iron chromium alloys, 78–79
  Incoloy alloy 798, 79
  Incoloy alloy 798H, 79–80
  Incoloy alloy 823, 79
  Incoloy alloy 79
  Pyromet, 79
  Refractaloy 26, 79
Nickel powder alloys, 80

Nitrogen oxides, control, 2016–2020
  combustion modification, 2016–2018
    flue gas recirculation, 2017
    low excess air, 2016
    low nitrogen burner, 2017–2018
    overfire air, 2016–2017
    reburning, 2017
  emission reductions, 2020
  formation chemistry, 2015–2016
  post-combustion control, 2018–2020
    selective catalytic reduction, 2018–2020
    selective noncatalytic reduction, 2018–2020
Noise measurement and control, 711–728
  acoustic enclosures, 722
  averaging decibels, 715
  combining decibels, 712–713
  composite panel, 721
    transmission loss of a composite wall, 722
  correction for background noise, 715–716
  decibels and levels, 712
  double walls, 723
  frequency and wavelength, 712
  machinery noise control, 719
  machines in semireverberant locations, 716–717
  measurement of machine noise, 716
  mufflers, 725, 727
  noise reduction due to increased absorption in room, 720
  single panel, 721
  sound absorption, 719–720
    absorption coefficients, 719
  sound analyzers, 715
  sound characteristics, 711–712
  sound control recommendations, 727–728
  sound isolation, 720–721
    transmission loss of building materials, 721
  sound-level meter, 715
  sound power and pressure, 712
  sound produced by several machines of the same type, 713–714
  two-surface method, 717–718
    area ratio, 718
    microphone locations, 718
  velocity of sound, 712
    in gases, 714
    in liquids, 714
    in solids, 713
  vibration damping, 725
  vibration isolation, 723–726
    transmissibility of flexible mountings, 726
Noise reduction due to increased absorption in room, 720
Nonconformal contacts, 507
Nonconformal surfaces, 512
Noncontacting seals for high speed/aerospace applications, 646–650
  comparison of brush, labyrinth and self-acting film-riding face seal leakage rates, 650
  computer analysis tools: face/annular seals, 647, 649–650

Noncontacting seals for high speed/aerospace applications (*Continued*)
  hydrodynamic noncontacting mechanical seals, 649
  self-energized hydrostatic noncontacting mechanical seals, 648
Nondestructive testing, 729–760
  capabilities of common methods, 731
  eddy current inspection, 746–750
  electrical resistivities and conductivities of commercial metals and alloys, 761
  introduction, 729
  liquid penetrants, 730–732
  magnetic particle method, 751–753
  radiography, 739–740
  thermal methods, 750–751
  ultrasonic methods, 732–733
  ultrasonic properties of common materials, 754
Nonferrous alloys, cast, 1046
Nonlinear control, 914
Nonlinear structural analysis, 316
Nonlinear systems, 852–860
  backlash nonlinearities, 857
  dead zone nonlinearity, 854
  describing functions, 853–855
  Jacobian matrices, 853
  limit cycles, 852
  linearizing approximation, 853
  linear *vs.* nonlinear behaviors, 852
  phase-plane method, 855–857, 859–860
  response to sinusoidal inputs, 852, 856
  saturation nonlinearity, 856
Nonmetals, emissivity, 1407
Normal modes analysis, 315–316
Notch sensitivity index, 417, 420–421
Notching, 1118
Nozzle:
  converging-diverging, gas flow, 1311
  gas turbines, 1735
  thermodynamics, 1347
NTRS, 2274
Nuclear binding energies, 1711
Nuclear power, 1699–1722
  chain reaction, *see* Chain reaction
  history, 1699–1700
  probalistic risk analysis, 1720
  radiation, 1712–1715
  U.S. development, 1700
  U.S. policy, 1707–1710
  see also Fission; Fusion
Nuclear power plant:
  capacity factor, 1706
  environmental consideration, 1709
  in operation, 1702
  proliferation, 1709–1710
  waste material, 1710
  see also Reactor
Nuclear Regulatory Commission, 1707, 1715, 1720–1722
Nuclear weapons, 1700, 1705
  proliferation, 1709–1710
Numerical control:
  defined, 1192

manufacturing systems, *see* Manufacturing system, numerical-control
Numerical integration:
  Adams-Bashforth predictor methods, 844
  Adams-Moultron corrector methods, 844–845
  errors, 845
  Euler method, 842–843
  method selection, 846
  predictor-corrector methods, 844–845
  Runge-Kutta methods, 843–844
Nusselt number, 1389, 1391–1393, 1397, 1653, 1666
  average, 1393–1394
Nyquist stability theorem, 896–898

Occupational Safety and Health Administration, 2196
  safety file requirements, 2218
  training requirements, 2226
Occupational Safety and Health Standards, 2196
Octane ratings, 1810, 1813
Offset factor, 528
Oil:
  fuel, *see* Fuel oils
  lubrication, 517
  whirl, 528
Oil-water emulsions, 1528, 1531–1532
One-port element, *see* Ideal elements, one-port element
Online databases, 2279–2285
  CD-ROM, 2283
  commercial database services for mechanical engineers, 2279
  Compendex, 2282
  DIALOG, 2279–2280, 2282–2283
  document suppliers, 2281
  engineering, 2282
  important databases for mechanical engineers, 2282–2283
  INSPEC, 2282
  ISMEC, 2282
  Knowledge Express, 2281
  METADEX, 2282
  MPD Network, 2282
  1MOBILITY, 2282
  NTIS, 2282
  options for use, 2284
  patent, 2283
  STN International, 2280
  searching, 2291
  see also CD-ROM and Desktop database
Op-amp:
  adder circuit, 887
  block diagram, 886–887
  PD control, implementation of, 886, 899
  PI control, implementation of, 886, 888
  PID control, implementation of, 886, 889
  proportional control, implementation of, 886–887
Open-die hammer forging, 1105–1106
Open-loop design for PID control, 898–899
Operation characteristic curve, 1180
Opitz system, 1202
Optimal control, 914

Optimization methods, 368–373
 computer programs, 372–373
 constrained, 369–372
 descent direction, 369
 multiple unconstrained, 369
 single variable, 368–369
 unconstrained, 368
Optimization problem:
 classification, 367
 constrained, 366
 gradient vector, 367
 integer programming, 366, 372–373
 linear fractional programming, 368
 linear programming, 366, 372–373
 mixed integer nonlinear programming, 368,
   372–373
 programming, 366
 quadratic programming, 366
 structure, 366–368
 unconstrained, 368
Optimization theory, 353
 applications, 356–365
  data reduction, 364–365
  economic machining problem, 362–364
  nonlinear curve fitting, 365
  oxygen supply system, 357–362
  welded beam, 359–360
 formulation, 354
 gradient vector, 367
 independent variables, 355
 model, 355
 operations and planning applications, 362
 optimal design process, 356
 performance measure, 354
OR gate, 494–495, 2200
Orifice, design equations, 1833–1834, 1836
 Bernouilli's equation, 1834
 effective exposed, for spool-type valve, 1836
 static pressure distribution, 1835
 velocity coefficient, 1834
 see also Valve
Orimulsion fuel, characteristics, 1532
O-Rings, 634–637
 basic sealing mechanism, 634
 effect of percent compression and material
   shore hardness on seal compression, 635
 material selection/chemical compatibility,
   635–636
  rotary applications, 635
 preload, 634–635
 thermal effects, 634
OSHA, see Occupational Safety and Health
   Administration
Output devices:
 electronic displays, 293–294
 for virtual reality hardware, 322–323
 hard copy devices, 294–296
Overhead:
 costs, 2123
 functional overhead budgets, 2079–2080
 work breakdown structure, 2067
Overlapping stress concentrations, 1158
Owners, financial concerns, 2098
Oxidizers:
 catalytic oxidizers, 2023

thermal oxidizers, 2023
 ultraviolet, 2024
Oxygen:
 for combustion, 1431
 enrichment, 1433
 merchant plant, 1957
Oxy-Fuel firing, 1447
Ozone-depleting substances, 2028

Packings and braided rope seals, 637–638
 effect of temperature, pressure and
   representative compression on seal flow,
   639
 flow vs. pressure data for 4 temperatures,
   639
 for high temperatures, 637–638
 gasket/rope seal materials, 637
 materials, 635
 most important elastomers and their
   properties, 636
 schematic of turbine vane seal, 639
Paints, 1131
Pallet-loading patterns, 1229
Pallet rack storage, 1242
Palmgren-Miner hypothesis, 413–414
PAPERS, 2283
Parallel algorithm, 1671–1672
Parametric estimating, 2131–2132
Pareto diagram, 481
Paris Convention, patent filing, 2265
Parkinson's disease, cryosurgery, 1970
Parr formulas, 1537
Part family:
 classification, 949–950, 958, 1200–1202
  application, 952–953
  attributable selection matrix production
    estimation, 953
  basic premises, 954
  basic shape taxonomy, 950
  bistring representation, 955
  design retrieval, 950
  E-tree concept, 954
  generative process planning, 952–953
  history, 949–950
  keywords, 955
  N-tree concept, 955
  parametric and generative design, 953
  parametric part programming, 954
  production estimating, 953
  tool design standardization, 954
 code, 955–958
  basic shape, 956
  complexity code, 957
  format and design, 956
  material code, 958–961
  name or function code, 956
  precision class code, 957, 958
  purpose, 955
  size code, 957
  special features, 957
 formation, 1200
Particulate matter, control, 2020–2022
 diffusion capture,. 2022
 electrostatic precipitators, 2021–2022
 fabric filters, 2022

Particulate matter, control (*Continued*)
  wet scrubbing, 2022
Parts:
  production flow analysis, 1201
Patent, 2247–2267
  annual maintenance, 2265
  application types, 2249
  best mode requirement, 2256
  Canadian filing, 2265
  chain of pending cases, 2264–2265
  claims, 2257–2258
  composition of matter, 2251
  continuation, 2264
  continuation-in-part, 2264
  description requirement, 2256–2257
  design, 2248
  distinctiveness requirement, 2257
  division, 2264
  double patenting, 2263–2264
  drawings, 2256
  election requirement, 2263
  enablement requirement, 2256
  European Patent Convention, 2266
  Express Mail filing, 2260
  file wrapper, 2264
  final rejection, 2262
  foreign filing, 2251, 2265
  functional language in claims, 2257
  ideas, 2250
  international filing, 2265–2267
    advantages and disadvantages, 2266–2267
  interviewing examiner, 2263
  invention, 2250
  issuance, 2264
  Jepson claims, 2258
  Kewanee Oil v. Bicron Corp., 2250
  machine, 2251
  manufacturers, 2251
  means-plus-function language, 2257
  negative right, 2249–2250
  new matter, 2264
  nonobvious requirement, 2253–2254
  Notice of Allowance, 2262
  novelty requirement, 2252
  Office Action, 2261–2262
  originality of inventorship, 2251–2252
  patentability, 2255
  Patent and Trademark Office fees,
    2258–2259
  Patent Cooperation Treaty, 2266
  pending, 2260
  plant, 2248
  preparing application, 2256, 2258
  processes, 2251
  product-by-process claims, 2257
  prosecuting, 2260–2265
  putting invention in perspective, 2255–2256
  reasons for allowance, 2263
  restriction requirement, 2263
  safeguarding original document, 2264
  secrecy of pending applications, 2260–2261
  small entity status, 2259
  statutory bar requirement, 2254–2255
  statutory subject matter, 2250–2251
  terminal disclaimer, 2263
  terms and expiration, 2248
  utility, 2248, 2253
  working requirements, 2265
  see also 35 U.S.C. 102
Patent and Trademark Office:
  applications received, 2260
  Board of Patent Appeals, 2262
  duty of candor, 2260–2261
  examiners, 2261–2262
  fees, 2258–2259
  interviewing examiner, 2263
  Office Action, 2261–2262
  Public Search Room, 2255
Patent Cooperation Treaty, 2266
Patent databases, 2283
Pattern shrinkage allowance, sand casting, 1121
Payback period, 2157
PCB conduction, 1658
Peak-shaving plants, 1958–1959
Pearlite 18, 23
Peat, 1542
Peclet number, 1393
Peening, 1114
Peltier voltage, 1675
Penetrameter, 743–744
  radiography of penetrameter mentioned
    below, 744
  schematic of typical film penetrameter, 743
Perforating, 1118
Performance criterion, 354
Performance indices, 889–891
Periodicals, as information sources, 2288
Permanent-mold casting, 1123–1125
Perpendicularity, 1158
Personal protective equipment, 2206,
    2214–2217
  adequacy, maintenance, and sanitation, 2216
  background, 2214–2215
  break-in period, 2216
  creation of additional hazard, 2215
  enforcement, 2216
  equipment selection, 2216
  fitting and adjustment, 2216
  follow through, 2216
  need analysis, 2215
  operations requiring engineering controls, *see*
    Machinery, engineering controls:
  planning and implementing use, 2215–2216
  program communication, 2216
  target date setting, 2216
  training, 2216
Petroleum:
  liquified gases, 1514–1515
  see also specific products
Phase:
  diagram, Fe-C, 9
Phenolic resins, 128
Philosophical Transactions, 2288
Photovoltaic conversions, sunlight to electricity,
    1577–1581
PID control:
  digital forms, 902–903
Pipes and fittings:
  correction factor, loss of pressure from bend,
    1838–1840

design equations, 1835, 1837–1838
  kinetic viscosity, 1838
  Reynolds number, 1838
  selection, 1837
  sizes, 1837
Phosphate coatings, 1132
Pitting:
  aluminum, 55
Phase:
  variables, 812–813
Phase-plane method, 855–857, 869–870
  phase portrait, 856
Photochemical machining, 1098–1099
Physical exergy, 1355–1357
Physical systems:
  example, 802, 804
  mathematical models, *see* Mathematical
    models
  pattern of interconnections, 804
  system graph, 804–806
Pickling, 1128
Piercing:
  cold, 1118
  hot, 1111–1112
Pipe welding, 1111
Pitot tube, 1325
Planck's distribution law, 1402–1403
Planing, metal cutting, 1070, 1073
Planning:
  aggregate, 1003–1006
  financial-expenditure, 2070–2071
  job sequencing and scheduling, 1014–1029
  materials requirements, 1006–1014
  network, 2067–2070
  production, 987–1033
    forecasting, 988–994
    inventory models, 994–1003
  timeliness, 987
Plasma destruction system, 2024
Plasma-beam machining, 1096–1097
Plaster-mold casting, 1125
Plastic deformation, *see* Elastic deformation
Plasticity, effect on stress and strain
    concentration, 419
Plastics, 115–121
  for use in electronic packaging, 345
  machining, 1074
  molding processes, 1126–1127
  see also specific plastics
Plate-fin heat exchangers, 1933
  computation of fin surface geometrics, 1935
  construction features, 1934
  heat transfer and flow friction factors, 1936
Plate-type heat exchangers, 1610, 1612
Plates, 237
  formulas, 238–241
Plotters, 294–296
  raster, 295–296
  vector, 294
PM Zone, 2274
Poisson's Ratio, 193, 312
Poisson strains, 380
Poka-Yoke method, 482
Polar diagram, 1325
Polishing, 1130

Polyacetals, 121–122
Polyamides, 120–121, 123–124
Poly(butylene terephthalate), properties, 120
Polycarbonates, 122–123
Poly(chlorotrifluoroethylene), 124
Polyesters:
  thermoplastic, 120
  unsaturated, 128
Poly(ethylene, chlorotrifluoroethylene), 128
Poly(ethylene terephthalate, 119
Polyimides, 123–125
Polymer matrix composites (PMCs), 132–142
  design and analysis, 185–186
  manufacturing process, 163
  maximum stress, 135
  mechanical properties, 144–148
  physical properties, 156, 159–160
  thermoplastic resins, 142
  thermoplastics, 139, 141–142
  thermosetting resins, 139, 142
  thermosets, 139, 141–142
Polymers:
  cryogenic service, 1947–1953
  foams, cryogenic insulation, 1943
  fibre-reinforced, 133
  structure, 15
Poly(methyl methacrylate) properties, 119
Polymorphism, 8
Polyphenylene ether, modified, 123–124
Polyphenylene sulfide, 121–122
Polypropylene, 116–117
Polystyrene, 117
Polysulfone, 122–124
Poly(tetrafluoroethylene), 124
Polyvinyl chloride, 118–119
Poly(vinyl fluoride), 128
Poly(vinylidene chloride) 119
Polyvinylidene fluoride, 125
Position-control system, 870–871, 881
Positive displacement pump, 1692–1693
Potentiometer, 873
  linearity and conformity, 1858
  loading, 1859
  terms of selection, 1857, 1859
  types, 1859
Powders:
  metallurgy, 1127–1128
Power, 797–798
  coefficient, 1688
  fluid energy, 1302
  loss:
    fixed-incline pad thrust bearings, 534
    pivoted pad thrust bearings, 536
  net flow, 799
  spectrum, 850
Power plants, 1364–1367
  central receiver, 1575
  comparisons, 1753
  driven by hot-single-phase liquid, 1365
  refrigerator model, 1366
  solar, 1364, 1575
  with bypass heat leak, 1365
Prandtl law, pipe friction, 1317
Prandtl number, 1482, 1653
Prandtl's equation, 1316

Prandtl-Schlichtling equation, 1308
Precision, 918
Predictor-corrector methods, 844–845
Preload factor, 528–529
Present worth, 2154–2155
Press forging, 1106
Pressure, 1339
  balance loops, 1631–1632
  capillary, 1425
  coefficient, 1304
  drop:
    factors for flow of air or flue gas through
      tube banks, 1489
    function of Reynolds number, 1490
    liquid, 1425
    normal and axial hydrostatic, 1425
    vapor, 1426
    heat exchangers, 1616
    shell and tube condensers, 1622
    shell and tube single-phase exchangers,
      1618–1619
    velocity heads for flow of air or flue gas,
      1487–1488
  measurements, 1324–1325
  ratio, 1724, 1748–1749
  regulator, block diagram, 868
  function of Reynolds number, 1490
  velocity convergence loop, 1632
  velocity head for flow of air or flue gas,
    1487–1488
Pressurized-water reactor, 1701, 1703–1704
  arrangement of primary system, 1718
  containment, 1719
  emergency cooling system, 1719
  power production, 1719
  steam generator, 1704
  vessel and core configurations, 1703
Probabilistic risk analysis, 1722
Process:
  annealing, steel, 29–30
  data, 1634
  design and redesign, 2213–2214
Process flow chart, 1224
Process flow sheets, 1912
Process industries:
  solar energy applications, 1575
Process-reaction method, 891
Product:
  planners, financial concerns, 2099
Product defects:
  design flaws, 2238
  design hierarchy, 2240–2241
  hazard analysis, 2239–2240
  hazard index, 2240
  instructions and warnings, 2238–2239
  legal definition, 2237
  production or manufacturing flaws, 2237
  uncovering defects, 2239–2241
Product design, 935–950
  practices, 935
  role of management, 943–944
  teamwork, 946
  unit manufacturing cost, 935–936
Product design review, 477–479
  benchmarking, 477–478, 483

quality function deployment, 477–478
quality loss function, 477–478
Product flow, 1017, 1019
Product, legal duty to warn:
  after-market hazard recognition, 2242
    post-sale warning, 2242
  corrective action, 2242–2244
    recalls, 2244
    retrofits, 2244
    warnings, 2244
Production:
  engineering, estimating, 2135–2136
  flow analysis, 1201
  operations, 1189–1192
  planning, *see* Planning, production;
    Aggregate planning; Forecasting;
    Scheduling
  plants, 1190
Product structure tree, 1009
Profit and loss statement, 2105–2107
  financial ratios, 2105–2107, 2111
Programmable logic controllers, 909
Programming:
  by scanning and digitizing, 1194
  CAD/CAM part, 1193–1194
  cost estimating, 2139–2140
  industrial robots, 1195–1197
Programming languages:
  high-level, 299–301
  low-level, 299
  VRML, 323
Project management, 2047–2084
  adaptive control engine, 2057
  communication, 2049
  computer software available, 2083
  cybernetic behavior of formal and informal
    activities, 2057
  dominant member, 2055
  dynamics, 2053–2056
    causal loops, 2053–2054
  models, 2047–2049
    macro cybernetic model, 2047–2048,
      2050–2053
    mutually supportive systems model, 2049
  process, 2049, 2063–2082
    budgeting, 2077–2077
    control, 2064
    plan coordination, 2064
    financial-expenditure planning, 2064,
      2070–2071
    means-end analysis, 2064
    network analysis, 2064
    network calculations, 2068
    network planning, 2067–2070
    profile of demands for resources, 2074
    performance of personnel, 2064
    project planning, 2063–2064
    reporting structure, 2064
    resource leveling, 2075
    resource loading, 2073–2075
    scheduling, 2064, 2072–2077
    summary, 2063
    time-phased human resource loading plan,
      2074
    variance analysis, 2064

work breakdown structure, 2064–2067
overhead, 2067
project teams, 2048, 2060–2062
reinforcing system of informal activities, 2055
reinforcing system of informal activities with balancing items, 2055
reporting systems, 2080–2082
reporting delays, 2082
structure, 2056–2060
ad hoc teams, 2057
case study, 2060–2062
chart, 2061
dynamics, 2058
project control system, 2060
project improvement effort, 2060, 2062
project-matrix organization, 2059
matrix organization, 2058–2060
organizing complex projects, 20563
trust and openness, 2054
style, 2048
see also Management; Teamwork
Propane:
available heat, 1457
Protective packaging:
general, 350
shipping environment protection, 351
storage environment protection, 350
Psychrometric chart, 1289, 1974–1976
Pulse echo instruments, 1156
Pulse transfer functions, 864–865
Pump:
centrifugal, see Centrifugal pumps and fans
model, performance characteristics, 1692
positive displacement, 1692–1693
selection, 1692–1696
characteristics of pumps, 1692
net positive suction head, 1693
operating performance in system, 1694–1695
throttling vs. variable speed drive, 1695–1696
similarity to fans, 1681
specific speed, 1690–1691
thermodynamics, 1347
Pump sets, 1847–1848, 1850–1851
good reservoir design, 1848
inlet strainer, 1848
starting new hydraulic circuit, steps, 1848, 1850
tank capacity, 1848
typical design, 1850
see also Accumulator
Punches and dies, cutting force, 1117–1118

QS 8980, 2179
Quality, definitions, 2160–2161
Quality control, 1175–1179
acceptance sampling, 1183–1184
control charts, 1176–1179
control charts for attributes, 1180–1183
defence department acceptance sampling by variables, 1184–1185
dimension, 1175
interrelationship of tolerances of assembled products, 1179
measurements, 1175
operation characteristic curve, 1180
tolerance, 1175
Quality function deployment, 2164–2166
Quality tools and methods, 480–485
benchmarking, 483–484
control charts, 482
customer needs mapping method, 482
fishbone diagram, 480
force field analysis, 481
fishbone diagram layout, 481
gap analysis method, 484
Hoshin planning method, 484
Kaizen method, 481–482
Pareto diagram, 481
Poka-Yoke method, 482
Quattro-Pro, 373

Radial in-flow turbine, 1734
Radiation, 1400, 1402, 1417
absorptivity, 1405–1407, 1409
biological effects, 1714–1715
black enclosure of uniform temperature, 1416
configuration factor, 1407–1368, 1410
damage, 381, 1713–1714
diffuse-gray surface, 1410, 1413–1414
diffuse surface, 1407
emissivity, see Emissivity
exposure, cancer probability, 1714
flame, 1436
factors, 1472
function, 1406
gasses, 1415–1417
gray enclosure, 1417
gray surface, 1407
heat transfer coefficient, 1415, 1477
irradiation, 1405
Kirchhoff's law, 1407
moon beam length, 1417
participating and nonparticipating gases, 1415
properties, 1404–1407
radiosity, 1410
reflectivity, 1406
refractory, 1437
shield, 1414
shields, cryogenic, emissivities of materials used, 1941
types, nuclear power, 1714
see also Black-box radiation
Radioactive wastes, disposal, 1709
Radiography, 738–746
attenuation of X-radiation, 741–742
computed tomography, 744–746
of a flashlight, 745
of a pencil, 745
crack in end of an aluminum tubing, 739
film-based radiography, 742–743
density vs. relative exposure, 743
density or darkness vs. relative exposure for common films, 742
generation and absorption of X-radiation, 739–740

Radiography (*Continued*)
   equivalence factors, 740
     X-ray tube voltage *vs.* thickness of
       industrial materials, 740
   mass absorption coefficient *vs.* atomic
     number, 741
   neutron radiography, 741–742
     showing bondline flaws, 741
   penetrameter, 743–744
     radiography of penetramer mentioned
      below, 744
     schematic of typical film penetrameter,
      743
   real-time radiography, 744
   typical composite with typical flaws, 738
Railroad:
   gas turbine applications, 1727–1728, 1759
   hopper cars, 1218
Rain-flow cycle counting method, 431
Rankine thermodynamic cycle, 1723,
     1749–1750
Rankine-Hugoniot curve, 1312
Rankine's Theory, 200
Rate action, 884
Rate of return, 2106, 2155–2156
Rating, 1632
Rational functions, 814
   proper and improper, 816
Rayleigh number, 1653
Reactivity, negative temperature coefficient,
     1705
Reactor:
   behavior, effect of delayed neutrons,
     1717–1718
   boiling water, 1704–1707, 1720
   constraints, 1705–1706
   critical, 1715
   design basis accidents, 1721
   gas-cooled, 1701
   gas-cooled graphite-moderated, 1700
   heavy-water-moderated natural-uranium-
     fueled, 1701
   high-temperature gas-cooled, 1700, 1705,
     1708
   light water, 1700
   light-water-moderated enriched-uranium-
     fueled, 1701
   liquid-fueled, 1700
   liquid-metal-cooled fast breeder, 1701
   liquid-metal solid-fueled fast spectrum, 1700
   liquid-sodium-coolcd-graphite-moderated,
     1700
   nonleakage probability, 1716
   pressurized-water, *see* Pressurized-water
     reactor
   safety analysis, 1720–1722
Reactor Safety Study, 1722
Real-time radiography, 744
Recalls, 2244
Reciprocity relations, 1410
Recuperative furnaces, 1497–1498
Recuperator, 1613
   combinations, 1498–1499
   concenric tube, 1500
   cross-flow-type, 1499

   radiant-tube, 1500
   sack-type recuperator, 1498
Reduction to practice, 2250
Reflection and transmission of sound, 733–735
   data collection and display in the A-scan
     mode, 734
Refractories, 1453–1454
Refraction of sound, 735–737
   amplitude and phase of reflected coefficient and
     transmitted amplitude *vs.* angle of
     incidence for longitudinal wave incident
     on water-steel interface, 736
     Snell's law and direction of propagation of
      acoustic waves, 735
Refrigerant, 1883–1891
   ammonia, 1883, 1887, 1888
   ANSI classification, 1884, 1886
   azeotropes, 1884
   chlorofluorocarbons, 1883
   comparative performance of different
     refrigerants, 1889
   composition, 1891
   corrosion, 1891
   cost, 1888
   critical temperature and pressure, 1890
   dirt and liquid carryover, 1891
   discharge pressure, 1888
   flammable, 1887
   freezing point, 1890
   halocarbons, 1884
   hydrochlorofluorocarbons, 1883
   latent heat, 1890
   liquid density, 1890
   mixtures, 1890
   numbered, physical properties, 1278–1280
   numbering system, 1886
   ozone-depletion potential and halocarbon
     global warming potential, 1890
   polymerization, 1891
   regulations on use, 1888
   selection:
     closed cycle, 1888
     open-cycle, 1891
   standby pressure, 1888
   suction pressure, 1888
   suction volume, 1890
   theoretical power, 1890
   toxicity and flammability rating system, 1886
   vapor density, 1890
   zeotropes, 1884
Refrigerant 11, 1885
Refrigerant 12, 1887
Refrigerant 13, 1887
Refrigerant 22, 1887
   at atmospheric pressure, thermophysical
     properties, 1266
   enthalpy-log diagram, 1266
   saturated, thermophysical properties, 1265
Refrigerant 50, 1888
Refrigerant 114, 1887
Refrigerant 123, 1887
Refrigerant 134a, 1887
   interim thermophysical properties,
     1268–1269
   compressibility factor, 1270

enthalpy-log pressure diagram, 1271
saturated, thermodynamic properties, 1267
Refrigerant 170, 1887
Refrigerant 290, 1887
Refrigerant 407C, 1887
Refrigerant 410A, 1887
Refrigerant 502, 1887
Refrigerant 600, 1887
Refrigerant 1150, 1887
Refrigeration, 1879–1913
  absorption systems, 1891–1894
    ammonia-water, 1891
    fired chillers, 1892
    heat recovery chillers, 1892
    indirect fired chillers, 1892
    water-lithium bromide absorption chillers,
      1892–1893
  basic principles, 1880–1881
  closed-cycle operation, 1881–1882
    cascade cycle, 1881, 1883
    compound cycle, 1881–1882
    flooded cycle, 1919–1920
    refrigerant numbering system, 1886
    refrigerant selection, 1882
    system with flash economizers, 1882, 1884
  compressors, 1897–1901
    centrifugal, 1900–1901
    operation, 1901
    rotary, 1898
    reciprocating, 1897–1898
    screw, 1900
    scroll, 1899–1900
  condensers, 1901–1903
  defrost methods, 1909–1910
    air and water defrost, 1910
    hot refrigerant gas defrost, 1909–1910
  evaporators, 1903–1904
    baffled-shell, 1906
    fixed-tube-sheet, 1905
    flash-gas distribution, 1904
    flooded shell-and-tube, 1905
    gas-liquid separation, 1904
    oil return, 1903
    short-tube restrictors, 1907
    spray-type, 1906
    submergence effect, 1904
  expansion devices, 1905–1909
    capillary tubes, 1906–1907
    constant pressure expansion valves, 1906
    electronic expansion valve, 1906
    thermostatic expansion valve, 1905
  heat recovery, 1601
  ice, 1897
  indirect, 1894–1897
    application of brine coolants, 1896
    brine properties, 1896–1897
    efficiency, 1895
    secondary brine system, 1895
  instrumentation and control requirements,
    1912
  off-design operation, 1912
  open-cycle operation, 1882–1883
  process flow sheets, 1912
  steam jet, 1894
  system design, 1910

system specifications, 1910–1912
vapor cycle, 1880
Regenerative air preheating, 1496–1497
Regenerator, 1613, 1933–1939
  applications, 1934–1935
  cyclical, unsteady-state conditions, 1935
  design equation, 1936
  end correction for heat transfer calculation,
    1939
  film coefficient for heat transfer from gas to
    solid, 1935
  gas turbines, 1750
  pair configurations, 1937
  Raoult's law, 1937
  stream-to-stream temperature difference
    limitation for containment cleanup, 1940
  time-temperature history, 1938
  trapping of contaminants, 1936
Registration/certification, quality systems, 2177
  ISO conformance models, 2178
  ISO 8382, 2178
  ISO 8980, 2178–2179
  ISO 8980-3, 2178
  ISO 8981, 2180
  ISO 8984-1, 2179
  ISO 10011, 2179
  ISO 10013, 2179
  ISO 14000, 2180
  ISO 14001, 2180
  JIS Z9899, 2180
  QS 8980, 2179
  TE 8980, 2179–2180
Reinforced-plastic molding, 1127
Reinforcements, 136–143
  continuous fibers, 137
  cycles to failure, 137
  discontinuous fibers, 137
  particles, 137
  whiskers, 137
Reliability in mechanical design:
  basic reliability networks, 488–491
  design life-cycle costing, 501
  design-reliability tools, 492–501
  failure data, 504–506
  history, 487
  mechanical failure modes and causes, 491
  reliability-based design, 491–492
  risk assessment, 501–504
Repair costs, 2141
Reset windup, 893
Resilience, defined, 197
Resource:
  leveling, 2075–2076
  loading, 2073–2075
  scheduling, 2072–2077
Resources Conservation and Recovery Act,
  2032, 2196
Retrofits, 2244
ReVelle-Boulton study, 2194
Rework costs, 2141
Reynolds equation, 521
Reynolds number, 545, 1290, 1300, 1327
  boundary layer, 1307–1309
  condensate flow, 1424
  critical, 1393, 1487

Reynolds number (*Continued*)
    flow of air around single cylinders, 1663
    flow of air or flue gas through or across tube
        banks, 1490
    flow coefficient and, 1327
    flow momentum in viscous dissipation, 1653
    forced convection, 1481, 1482
    friction factor dependence, 1304
    function of friction factor, 1838
    heat transfer, 1666
    pressure drop of one velocity head as
        function of, 1490
Reynolds transport theorem, 1301
Risk assessment, 501–504
    process and application benefits, 502
    techniques, 502–504
        consequence analysis, 503
        event tree analysis (ETA), 503
        failure modes and effects analysis
            (FMEA), 503
        fault tree analysis (FTA), 503
        frequency analysis, 504
        hazard and operability study, 503
        hazard identification, 503
        risk estimation, 503
Riveted joints, 1137, 1137–1138
Riveting, 1114
Roll bending, 1116
Roll forging, 1106
Rolled threads, 1158
Roller bearings, 512–515, 520, 578–582
    AFBMA method, 605
    applications, 609–616
    applied load, 596
    capacity formulas, mixed rectangular and
        elliptical contacts, 606–607
    contact fatigue, 600–601
    cylindrical, 578
        applications, 609–612
        characteristics, 580
        film thickness, 611–612
        operating conditions, 609
    distribution of fatigue failures, 602
    fatigue spell, 601
    geometry, 582–589
        crowning, 588
        curvature sum and difference, 590
        free endplay and contact angle, 589
        race conformity, 589
        spherical, 589
    group fatigue life determination, 608
    kinematics, 591–594
        angular velocities, 591–592
        ball spin axis orientations, 592
        contact angles, 591
        race control, 591
        surface velocities, 593
    life adjustment factors:
        lubrication, 608–609
        materials, 606–608
    load-deflection, 594–595
    lubrication-life correction factor
        determination, 609
    Lundberg-Palmer theory, 604–605
    material factor, 607

    needle, 581–582, 584
    preloading, 600
    radial, 604
        application, 612–616
        features of geometry, 612–614
        maximum load elastic compression,
            614–615
        minimum film thickness, 615–616
        operating conditions, 612
    radially loaded, 595–597
    rectangular contact, capacity formulas,
        606–607
    spherical, 578, 590
        characteristics, 581–582
        standardized double-row, 581
    static load, 594
    tapered, 578
        characteristics, 583
        simplified geometry, 594
    types, 576
    Weibull distribution, 601–604
    Weibull plot of fatigue failures, 604
Rolling:
    cold, 1113
    hot, 1103–1105
    isothermal, 1104–1105
    mills, furnace modules, 1502
Room air distribution, 1997–2000
    Air Distribution Performance Index, 1999
    ceiling diffuser, 1999
    characteristic room length for several diffuser
        types, 2000
    diffuser behavior, 1998–2000
    high sidewall type of register, 1999
    jet behavior, 1998–2000
    perimeter-type outlets, 1999
Root-locus, 899–901
    PD control, 899–900
    PI control, 900–901
    series lead and lag compensators, 901
Rotating elements:
    blades, 244
    disks, 244
    shafts, 244
Rotating fluid machines, *see* Centrifugal pumps
    and fans
Rotational imbalance, 668–673
    balancing machines, 672–673
    dynamic imbalance, 671–673
    in-place balancing, 673
    static imbalance, 668, 673
Rotomolding, 1126
Runge-Kutta methods, 843–844

SAE International, 2274–2275, 2284
Safety:
    combustion, 1442–1447
    factor, 500–501, 2230
    furnace control system, 1458
    nuclear power, 1707–1708
Safety engineering, 2193–2228
    alternatives to engineering controls, 2208,
        2213
    basic functions of safety program, 2220
    controls for machinery, 2205–2207

basic concerns, 2205
danger sources, 2207
employee safeguards, 2206
general requirements, 2205–2207
design and redesign, 2213–2214
elements of accident prevention, 2217–2218
employee needs and expectations, 2194–2195
Environmental Protection Agency, 2195–2196
hazard classification, 2221
human positive performance characteristics, 2214
machine positive performance characteristics, 2214
machine safeguarding methods, *see* Machinery, safeguarding methods
management principles, 2218–2219
Occupational Safety and Health Administration, 2196
operations from which workers should be protected, 2196
people performance sciences, 2215
personal protective equipment, 2214–2217
possible problems to be inspected, 2222
required notices and records, 2197
requirement for safety files, 2218
Resources Conservation and Recovery Act, 2196
state-operated compliance programs, 2197
supervisor's role, 2217
system safety, *see* System safety
Toxic Substances Control Act, 2195–2196
training, *see* Training, safety
unsafe conditions:
    checklist, 2220–2221
    eliminating, 2219–2221
    involving mechanical or physical facilities, 2221–2222
    see also Human factors engineering
Salt gradient ponds, 1571
SAN, *see* Styrene/acrylonitrile copolymer
Sand casting, 1120–1121
Sankey diagram, 1436
Saturated liquid, 1346
Saturated vapor, 1346
Sawing, metal cutting, 1073–1074
Scenario-writing, 774
Scheduling, 2132–2134
    heuristic, 2076–2077
    resources, 2072–2077
    See also Aggregate planning, 1003–1006
Job sequencing, 1014–1029
Scrubbing systems, 2013–2014, 2022
    carbon dioxide, 2026–2027
Seals:
    cryogenic service, 1953
    dynamic, 657
        brush seals, 654–657
        emission concerns, 644–646
        honeycomb seals, 653
        initial seal selection, 638–641
        labyrinth seals, 650
        mechanical seals, 642–644
        noncontacting seals for high-speed/aerospace applications, 646–650

generally, 629
static, 629–638
    gaskets, 629–634
    o-rings, 634–637
    packings and braided rope seals, 637–638
Seaming, 1116
SEC Edgar, 2274
Secant formula, 231
Second law of thermodynamics:
    closed systems, 1335–1337
    open system, 1338
    steam turbines, 1771–1772
Seebeck coefficient, 1675
Seizure, 381
Selective catalytic reduction, 2018–2020
Selective noncatalytic reduction, 2018–2020
Self-fulfilling prophecies, 778
Sensitivity, 918–919
Separation systems, 2024
Shaft:
    round, torsion formula, 225
    torsion, 224–229
        angle of twist, 226
        definitions, 224–225
        formulas, 227–228
        noncircular, 226
        ultimate strength, 228
        transmitted horsepower, 226
    unbalanced, 669–670
Shaker:
    electromagnetic, 706–707
    imput signals, 709
    shock testing, 701, 704
Shake tests, electronic assemblies, 705–709
    sine-wave test, 707
    stress screening, 707
    see also Vibration, random
Shale oils, 1528, 1531
Shaped-tube electrolytic machining, 1091–1093
Shape factor, 1408–1414
    additive property, 1410
    aligned parallel rectangles, 1412
    black-body radiation exchange, 1410
    coaxial parallel circular disks, 1412
    radiation exchange between surfaces, 1414
    reciprocity relations, 1410
    rectangles with common edge, 1413
    relation in enclosure, 1410
    simple geometries, 1411
Shaping, metal cutting, 1070, 1073
Shaving, 1118
Shear:
    horizontal, 208–209, 213–214
    resisting, 208
    vertical, 208, 213
Shear strength, 620–621
Shearing:
    cold-working, 1116–1118
    angle, metal cutting, 1073–1074
Shielded metal arc welding, 1167, 1169
Shell and tube condenser, 1619–16416
    correction for mixture effects, 1622
    heat-transfer coefficients, 1620–1622
    pressure drop, 1622
    selection, 1619

Shell and tube condenser (*Continued*)
temperature profiles, 1619–1620
Shell and tube heat exchanger, 1607–1610
baffle types, 1610, 1612
E-type, 1608–1609
fouling, 1635
F-type shell, 1609
G-type, 1609–1610
H-type, 1609, 1611
J-type, 1609
K-type, 1610–1611
overall heat-transfer coefficient, 1615–1616
X-type, 1609–1610
Shell and tube reboiler, 1622–1625
heat-transfer coefficients, 1623–1625
maximum heat flux, 1625
mean temperature difference, 1623
selection, 1622–1623
temperature pinch, 1629–1630
temperature profiles, 1623
Shell and tube single-phase exchanger,
1616–1619
baffle spacing and cut, 1617
bypass coefficient for heat transfer, 1618
cross-sectional flow areas and flow velocities,
1617–1618
heat-transfer coefficients, 1618
pressure drop, 1618–1619
tube length and shell diameter, 1617
Shelving, 1238–1239
Shingo Prize for Excellence in American
Manufacturing, 2185, 2187, 2191
Shipping environment protection, 351
Shock, 692–695
calibration, 693, 695
damper system, 695, 697
defined, 692
drop-ball calibrator, 693–694
filtering, 693
normal, gas dynamics, 1311–1313
oblique, gas dynamics, 1313–1314
pulse:
conversion from time to frequency domain,
695–696
half-sine, 693
oscilloscope views, 692–693
response spectrum, 695, 697
spectra, 695–696
Shock testing, 695–705
big-bang tests, 695, 697–698
carriages that rattle, 692
complex pulses, 700–701, 704–705
energy transmission paths, underwater
explosion to hull of submarine, 697, 699
equipment needed, 700, 703
flow, spectrum testing to required spectrum,
701, 704
flow charts, 701, 704
free-flow gravity drop tester, 700, 701
hammer tests, 697, 699
high-frequency hash, 693
in-flight shock analysis, 700–701, 704
machine using dry nitrogen, 698, 703
machine using shop compressed air, 698, 702
Navy test machines, 698

parallel filters method, 704
shaker, 700
simple pulse shapes, 700, 705
spectra in surface ship, 697, 700
test requirement envelope, 700
Shock and vibration, 345–347
environmental loads, 345
general, 345–346
life, 346
testing, 347
Shock measurement, 928
Short-tube restrictors, 1907
Shot peening:
effect on S-N curves, 408
Shunting capacitance, accelerometer, 685–686
Sieder-Tate, equation, 1391
Silica, coal ash, 1545
Silica/alumina ratio, coal ash, 1546
Silos, 1212, 1214
Silver Platter, 2284
Simulation:
analog, *see* Analog simulation
digital, *see* Digital simulation
experimental analysis of model behavior,
840–841
model, steps in constructing, 776
performance prediction, heat exchanger, 1633
Single panel, 721
Sizing, 1113
Skill levels, engineering, 2134
Skill matrix, 2128–2129
Skill mix, 2119, 2121
Skin effect, 746
Slagging, 1544
Slip-resistant joint, 1153
summary of coefficients, 1154
Slotting, metal cutting, 1070, 1073
S-N-P curves, 401–413
4340 steel, 412
2014-T6 aluminum alloy, 411
7075-T6 aluminum alloy, 401
constant stress level testing program, 398
development, 401
factors affecting, 402–411
accumulated fatigue damage, 411
corrosion, 408
fretting, 410
geometrical discontinuities, 406
grain flow direction, 404
grain size, 403
heat treatment, 404
material composition, 402
operating speed, 410
operating temperature, 408
shot peening, 407
size, 406
surface finish, 407
ultimate strength, 410
welding detail, 405
failure prediction, 405, 410
low-cycle fatigue, 422
multiaxial fatigue stresses, 402, 405, 410
notched and unnotched specimens, 421
nonzero mean cyclic stresses, 402, 405
4130 alloy steel, 412

4340 steel, 412
S-N curve:
  best-fit for notched 4130 alloy steel sheet, 412
  fatigue testing of new alloy, 400–401
  standard fatigue testing method, 402
spectrum loading, 410, 413–414
types of material response to cyclic loading, 402
Snell's law, 735
Social judgment theory, 780
Society of Manufacturing Engineers, 2275
Socket action, 237, 242
Sodium, saturated, thermodynamic properties, 1272
Software:
  computer languages, 299–301
  graphical user interface, 298–299
  operating systems, 296–297
  project management, 2083
  virtual reality software, 322–323
Solar constant, 1407, 1554
Solar energy, 1549–1581
  availability, 1549–1560
  central receiver, 1575
  conversion, 1364
    coupling, 1364
  defined, 1549
  extraterrestrial solar flux, 1554–1555
  hourly solar flux conversions, 1559, 1559
  industrial solar space heating systems, 1575
  mechanical solar space heating systems, 1569
  monthly averaged, daily solar flux
    conversions, 1559–1560
  nonthermal applications, 1577–1581
  passive:
    direct gain, 1571
    greenhouse, 1573–1574
    solar space heating systems, 1569, 1571
    thermal storage wall, 1571
  performance prediction, 1576–1577, 1577
  photovoltaic conversions, 1577–1579
  ratio of monthly averaged diffuse to total
    flux, 1559–1560
  solar constant, 1407, 1554
  solar fraction, 1577–1579
  solar geometry, 1549–1552
    azimuth angle, 1552
    celestial sphere, 1551
    declination, 1549–1551
    hour angle, 1549–1551
    solar-altitude, 1552
    solar position, 1551–1552
  solar pond, 1571–1574
  solar-powered cooling, 1575
  solar water heating, 1569
  sunrise and sunset 1552–1554
    equations for tracking collectors, 1556
    hour angles, 1553
    incidence angle, 1552
    solar incidence angle, 1552–1555
    sun paths, 1554
  terrestrial solar flux, 1555
  thermal power production, 1575
  utilizability, 1576–1579

see also Solar thermal collectors
Solar flux:
  daily absorbed, 1576
  extraterrestrial, 1554–1555
  hourly conversions, 1559–1560
  monthly averaged data, 1559–1560
    daily conversions, 1559–1560
  terrestrial, 1555
  useful, 1577
Solar fraction, 1577–1579
Solar One, 1575
Solar ponds, 1571–1574
Solar thermal collectors, 1560–1569
  central receiver, 1566
  closed-loop testing configuration, 1568
  collector efficiency, 1564
  compound curvature concentrators, 1566–1568
  concentrating collectors, 1564
  cross-sections, 1562
  flat-plate, 1560
  Fresnel-type concentrators, 1564–1565
  geometric concentration ratio, 1565
  with one cover, diagram, 1562
  optical efficiency, 1564
  overall thermal conductance, 1560
  paraboloidal concentrator, 1564–1565
  performance, 1564
  segmented mirror approximation, 1566
  selected surfaces, 1564
  single-curvature concentrators, 1565
  solar incidence angle equations, 1556
  testing, 1568–1569
  trough, 1564–1566
  useful heat, 1560
Solid modeling, 302–309
  dimension-driven design, 303
  feature-based modeling, 304–305
  parametric modeling, 303–304
  variational modeling, 303
Solid-state radiation, see Black-body radiation
SolidView, 940
Solidification, 2043
Solids:
  crystalline, 7
  noncrystalline, 8
Solvents, 1128
Sommerfeld number, 525–530
Sorbent injection, 2025
Sound control recommendations, 727–728
Sound:
  characteristics, 709–710
  reflection and transmission, 733–735
  refraction, 735–737
Sound measurement, 929–932
  acoustic intensity-diagnosis of vibrations, 930
    intensity and power measurements, 931
    intensity measuring probe, 932
  frequency characteristics for weighing
    functions of sound level meters, 929
  introduction, 928
  measurement in enclosed spaces, 930
  measurements in open spaces, 929
  sound absorption, 719–720
    absorption coefficients, 719

Sound measurement (*Continued*)
  sound analyzers, 717
  sound characteristics, 711–712
  sound isolation, 720–721
    transmission loss of building materials, 721
  sound-level meter, 715, 928
Sound power and pressure, 712
Sound produced by several machines of the same type, 713–714
Sound waves, 733
Space cooling, 1988–1991
  air requirements, 1990–1991
  compressor operating time, 1991
  design conditions, 1989
  fuel requirements, 1991
  heat gain:
    calculations, 1989–1990
  cooling load, and heat extraction rate, 1989
  peak load, 1991
Specific heat:
  ratio, liquids and gases, 1282
  saturated liquids, 1281
Spectrum loading, 410, 413–414
Spheres:
  thick wall, 236–237
  thin wall, 235
Spinning, hot, 1110–1111
  setup and dimensional relations, 1111
  tube, 1111
Spiral-type heat exchangers, 1610–1611, 1613
Squeezing processes, 1113–1114
Stage:
  design, turbine, 1782–1784
    blade reaction required to prevent negative base reaction, 1784
    blade section losses, 1784
    effect on aerodynamic efficiency, 1783
    impulse and reaction stage geometries, comparison, 1783
    Rateau stage, 1782
    velocity ratio, 1782
  performance characteristics, steam turbine, 1784–1788
    expansion process, 1785
    flow number, 1786
    flow relationships, 1787
    gas dynamic flow relationships, 1787
    maps, 1788
    referred flow rate, 1786
    stage velocity triangles, 1785
    thermodynamic conditions, 1786
    velocity ratio, 1785
Staking, 1114
Standard Handbook of Engineering Calculations, 2289
Standards and accuracy, 917–918
  accuracy and precision, 918
  linearity, 919
  manufacturing activity estimates, 2137
  occupational safety and health standards, 2196–2197
  sensitivity or resolution, 918–919
Start-up ratios, manufacturing, 2139
State, 7–8

Fe-C phase diagram, 9
State equation, 829
State-variable methods, 829–831, 839–840
  diagonalized canonical variable, 839–840
  Jordan canonical form, 840
  modal matrix, 840
  state equation solution, 829
  state transition matrix, 838
State quality awards, 2188–2191
State-variable model, 810–811
Static imbalance, 668–669
Static seals:
  gaskets, 629–634
  o-rings, 634–637
  packings and braided rope seals, 637–638
  see also listings under various seals
Statutory bar requirements, 2252, 2254–2255
Steady-state radiation view factors:
  parallel planes connected by reradiating sidewalls, 1470
  ribbon-type electric heating elements, 1470
  spaced cylinders, 1469
Steam jet refrigeration, 1894
Steam turbines:
  blade path design, 1772–1798
    blade aerodynamic considerations, 1796–1798
    blade-to-blade analysis, 1796
    field test verification of flow field design, 1791–1796
    flow field solution techniques, 1790–1791
    low-pressure turbine design, 1788–1790
    stage performance characteristics, 1784–1787
    thermal to mechanical energy conversion, 1776–1777, 1781–1782
    turbine stage designs, 1782–1784
  Carnot cycle, 1775
  Carnot efficiency, 1767, 1771
  design, low-pressure, 1788–1789
    advances, 1789
    aerothermodynamic equations, 1789–1790
    meridional plane, 1790
  Eddystone No. 1 unit, 1766–1767
  field test verification of flow field design, 1791–1796
  heat engine and energy conversion processes, 1767–1772
  heat rate improvement, 1766
  history, 1765–1767
  process, 1765
  steam cycle development, 1766
  thermal efficiency, 1767–1768, 1771
  thermodynamic properties, 1772–1775
  turbine cycle, 1768–1771
  see also Energy conversion, thermal to mechanical
Steam/water, thermophysical properties, 1274–1275
Steel, 17–43
  alloy, see Alloy steels
  carbon, see Carbon steels
  cast, microstructure, 33
  dual-phase sheet, 35
  hardenability, 25–31
    annealing, 29

austempering, 28
austenization, 26
carburizing, 31
cooling rate, 26
isothermal annealing, 29–30
martempering, 28
measurement, 26
nitriding, 31
normalizing, 29
process annealing, 31
quenching, 27–28
severity of quench, 26
spheroidization annealing, 31
tempering, 27–28
in weldments, 1171
heat-resisting products, 38–39
heat treatment, 18–19
high carbon, 1046
high speed, 1046
iron-iron carbide phase diagram, see Iron-iron carbide phase diagram
metallography, 18–19
plane strain fracture toughness, 389
shot peening, effect on S-N curves, 407
size effects on S-N curves, 404
yield strength, 405
Stefan-Boltzmann equation, 1464, 1940
Stefan-Boltzmann Law, 1402
Stefan-Boltzmann law for solids, 1436
Ships, gas turbine applications, 1759
STEP, 310
Stirling cycle, 1928–1931
arranged for air liquefaction, 1931
T-S diagram, 1930
STN International, 2280
Stochastic systems, 846–850
autocorrelation, 850
correlation coefficient, 849
cross-correlation function, 850
density function, 847
distribution function, 847
processes, 849–850
random variable, 846–847, 849
state-variable formulation, 846
Storage bins, 1212, 1214, 1216, 1238–1239
Storage environment protection, 350
Storage piles, 1212
Storage systems, 1363
entropy generation during sensible-heat storage, 1363
Straightening, 1116
Strain:
amplitude:
cyclic, 422, 424
vs. fatigue life, 424–427
at fracture, 446
concentration, effect of plasticity, 419
creep, 440–442
cyclic strain hardening and softening phenomena, 423
elastic, see Elastic deformation
equivalent total strain range, 428
hardening, 82–84
nonzero mean, 424–425
tensile inelastic, 448

Strain-range partitioning method, 447
Streamline curvature technique, 1790–1791
Steam stripping, 2039
Strength:
reduction factor, 418
short columns, 234
thread stripping, 1144
yield strength of bolt, 1144
Stress:
assembly, 1144
combined, 199–203
compressive, 192
concentration, 199, 415–420
concept, 415
due to circular hole, 417
elastic stress concentration factor, 414–416
examples, 416
fatigue stress concentration factor, 199–200, 417–419
function of material, 417
highly local, 416
notch sensitivity index, 417, 420–421
plasticity, effect of, 419
widely distributed, 416
contact, see Contact stresses
corrosion, 380, 462, 467–468
cracking, 1145
definitions, 191–197
design-allowable bolt, 1144
dynamic, 197
equivalent, 428
failure, 1143
Goodman's linear relationship, 402, 405, 410
live loads, 222
master diagram, 402, 412
maximum shear, 462
maximum shearing stress, multiaxial fatigue theory, 405
multiaxial principal shearing, 410
multiaxial state, creep, 442
Mohr's Circle, 199, 201
nonzero mean, 402, 425
failure prediction equations, 410
multiaxial, 410
uniaxial, 405
normal, defined, 192–193
overlapping concentrations, 1158
principal, 199
relaxation, 204
relative orientation, 199, 201
residual, 1147
rupture, 381, 437–438, 1143
screening, 707
shear, 192–193
static, 191–197
static, creep limited, 443
tensile, defined, 192
thread, distribution, 1158–1159
torsional, 229, 1144–1145
uniaxial state, creep, 440–442
unit compressive, 231
Stress-concentration factor, 199–200, 419
curved beams, 220–222
shaft with a fillet, 418
Stress intensity factors, 388–392

Stress intensity factors (*Continued*)
  center cracked specimen, 392
  critical magnitude, 389
  part-through thumbnail surface cracks, 391, 396
  range and crack growth rate, 430
  single edge notch specimen, 393
  single through-the-thickness edge crack, 391, 392, 394
  strain energy release rate, relation to, 390
  through-the-thickness crack from circular hole, 391, 395
  various types of cracks, 391
Stress-strain relationship, 200–202
  comparison of theories of failure, 202
  defined, 193
  true, 195
Stress-strength interference theory approach, 497–498
Stretch, forming, 1118–1119
Strict liability, 2236
Structural modeling, 775–776
Structural design:
  complexity, 347
  degree of enclosure, 347
  general, 347
  strength, 347
  thermal expansion and stresses, 348
Structure:
  atomic, 3–4
  bcc, 6, 13
  bonding, see Bonding
  bravais lattices, 7
  ceramics, 14–15
  close packed layer, 5
  crystallography, 5
  defects, see Defects
  fcc, 5–6
  hcp, 5–6, 12
  polymorphism, 8
  project control, 2056–2062
    ad hoc teams, 2057
    case study, 2060–2062
    chart, 2061
    dynamics, 2058
    project control system, 2060
    project improvement effort, 2060, 2062
    project-matrix organization, 2059
    matrix organization, 2058–2060
    organizing complex projects, 2056
  simple, 4–5
  tertahedral and ochtahedral sites, 6
  work breakdown, 2064–2068, 2070–2071
Styrene/acrylonitrile copolymer, 117–118
Subbituminous coals, 1539
Subcontract costs, 2122–2123
Submerged arc welding, 1160–1162
Substrate conduction, 1658
Submerged body, 1294
Sulfur:
  content in coal, 1540
Sulfur dioxide, control, 2012
  coal cleaning, 2015
  cost of control, 2015
  financial cost, 2015–2016

flue gas desulfurization system, 2012, 2015
  fluid bed combustion, 2014
  fuel switching, 2015
  furnace sorbent injection, 2014
  residue disposal and utilization, 2015
  scrubbers, 2013–2014
    dry, 2014
    dual alkali systems, 2014
    industrial, 2014
    perforated plate, 2014
    wet, 2013–2014
Sunspace, 1571, 1574
Superconductivity, 1963–1969
  applications, 1966–1969
  BCS theory, 1965
  commercially available materials, 1967
  electrical power transmission, 1967
  high-speed railroads, 1967
  limits to superconductive behavior, 1966
  magnets, 1967–1969
  Meissner effect, 1965
  superconductive elements, 16452
  variation of resistance of metals with temperature, 1965
Superinsulation, cryogenic insulation, 1941–1943
Supervision costs, 2141–2142
Surface conditions, fasteners, 1159
Surface fatigue wear, 379
Surface tension, liquid, 1283
Surface treatment, chemical conversions, 1128–1132. *See also specific types of treatment*
Surface waves, generation and propagation, 735
Swagging, 1106, 1113
Synchro, 872–874
System:
  boundaries, defining, 354
  definition matrix, 774
  design, 784–792
    algorithmic effectiveness of performance objectives achievement evaluation, 790–791
    behavioral or human factors evaluation, 791
    benefits, 793
    characteristics, 785, 792–793
    considerations, refrigeration, 1910
    decision situation model, 785–786
    detailed design, testing and implementation phase, 788
    effectiveness aspects, 792
    efficacy evaluation, 791
    evaluation, 790–791
    leadership and training requirements, 789–790
    operational deployment phase, 789
    operational environments and decision situation models, 785–786
    phases, 788–791
    preliminary conceptual design phase, 787
    purposes, 782–783
    requirements specification phase, 786
    systems evaluation, 790–791

systemic process aids and systems, 786–789
graph, 803–806
model, 355
parameters, 354
purposes, 784–785
relations, 806–807
safety, 2197–2202
  accuracy of procedure, 2200
  adequacy and accuracy of supporting documentation, 2200–2201
  adequacy of procedure, 2199
  analysis methods, 2198–2199
  backout provisions, 2201
  caution and warning notes, 2201
  communications and instrumentation requirements, 2201
  correlation between procedure and hardware, 2199
  emergency measures, 2201
  environmental considerations, 2202
  general requirements, 2202
  interfacing hardware and procedures noted, 2202
  personnel qualification statements, 2202
  procedure sign-off, 2202
  reference considerations, 2202
  securing provisions, 2201
  sequence-of-events considerations, 2201
  special considerations, 2202
subsystem integration, 2198
thermodynamic, 1333
variables, 354
System life cycle, functional elements:
analysis, 765
conceptual illustration of three levels of systems engineering, 769
formulation, 765
interpretation, 766
logic structure, 768
methodology, 768
phases and steps in a 49-element two-dimensional framework, 769
representation of structure systems engineering and management functional efforts, 769
representation of three systems engineering steps, 769
system-level architecture, 769
Systems engineering, 763–794
conditions for use, 766–767
definition, 765–767
design, 786–794
functional elements, 765–770
life cycle, see System life cycle
logic dimension, 766
objectives, 770
management, 769
methodology and methods, 771–784
  central role of information in systems engineering, 783–784
  information processing biases, 780–781
  information processing by humans and organizations, 779–782
  issue analysis, see Issue analysis

issue formulation, see Issue formulation
types of information, 783
morphology, 766
multidisciplines, 769
objectives, 770–771
potential results, 766
process structure, 767
systems, science and operations research, 765–766
see also System design
Systems engineering objectives, 770–771

Taguchi's quality philosophy summary, 480
Tank inventory measurement, 1955
Tar sands, 1528, 1532
TE 8980, 2179–2180
Team environment, 946–948, 950
Teamwork:
barriers to high performance, 2091–2092
  different points of view, 2091
  group think, 2092
  power struggles, 2091
  role conflict, 2091
building self-directed teams, 2092–2094
characteristics of high-performance, 2089–2092
cultural/social tools, 2175
dominant member, 2055
dynamics, 2054–2056
early-warning signs of performance problems, 2093
management recommendations, 2094–2095
performance, 2088–2089
shared vision, 2054
see also Managing people; Project management
Tear out, 1140
Telnetting, 2276–2277
Temperature, 1339
absolute, 1337
bulk, 1385
ideal gas, 1341
loops, 1631–1632
measurement, low-temperature, 1953–1955
pinch, heat exchangers, 1639–1630
profiles:
  shell and tube condensers, 1619–1620
  shell and tube reboilers and vaporizers, 1623
relative, 1334
stagnation, 1396
Tensile loads, 1146–1153
Tensile strength, AISI 300 stainless steels, 1947
Tension loads, design, 1158–1159
Terminals:
direct-view storage tube, 294
raster-scan terminals, 294
vector refresh terminals, 293–294
Test estimating ratios, 2139
Testing:
cost estimating, 2137, 2139
equipment:
  cost estimating, 2139
  design hours, 2140
  estimating ratios, 2139

Testing (*Continued*)
  imput signals, 819
  planning, 2138
Thermal comfort, 1979
Thermal conductance, 1375
  overall solar collectors, 1560
Thermal conduction, 1982
  solid materials, 1472
Thermal conductivity, 1370–1374, 1460
  gases, 1370–1371, 1377
  insulating materials, 1373–1374
  liquids, 1376
  liquid-saturated wick structures, 1429
  materials useful in low temperature service,
    1948
  metals, 1370, 1372
  nonmetals, 1370–1371, 1373
Thermal design:
  general, 348
  heat transfer modes, 348–350
Thermal desorption, 2043
Thermal diffusivity, 1461
Thermal expansion coefficient, linear, 1397,
    1948
Thermal expansion and stresses, 348
Thermally assisted machining, 1084–1085
Thermal methods, 750–751
  infrared cameras, 750–751
  paints, 751
  testing, 751
Thermal modeling, 1649–1661
Thermal oxidizers, 2023
Thermal processes, complex, furnace, 1499,
    1501
Thermal properties:
  air and flue gas, 1461–1463
  materials heated in furnaces, 1460–1461
Thermal radiation, 1984
Thermal relaxation, 381
Thermal resistance, 1375, 1427–1429, 1381,
    1655–1656, 1983–1984
  chip module, 1656–1661
  heat pipe, 1427–1429
  negligible load, 1477
Thermal shock, 381
Thermal storage wall, 1571
Thermistors, 1954
Thermochemical machining, 1099
Thermochemical properties, compounds,
    1286–1287
Thermocouple, 1494–1495, 1953
  thermoelectric power, 1955
Thermodynamic cycle, 1723
  firing temperature, 1724–1725
  intercooler, 1743
  pressure ratio, 1724, 1749
  pressures and temperature, effects of, 1725
  regeneration, 1745
  temperature changes, ideal cycle, 1725
  temperature ratio, 1725
Thermodynamics, 1331–1350
  applicability canonical relation, 1339
  conditions, steam turbine, 1786
  coupled behavior, 1334
  engineering system components, 1347–1350

heat engine, 1767
incompressible substances, 1344
laws, *see specific laws*
P, *v*, T surface, 1340
quality, 1346
reversible cycle, 1335
thermodynamic properties, relations among,
    1339–1340
thermodynamic properties, *see*
    Thermodynamic properties, steam,
    1772–1775
two-phase states, 1344
see also Exergy; Gases, ideal
Thermodynamic properties, steam, 1772–1775
  chocking (sonic) velocity as function of
    temperature and pressure, 1777
  chocking velocity for water/steam mixture,
    1778
  critical choking mass flow rate, 1776
  flow number, 1774
  isentropic expansion exponent, 1775
  Mollier diagram, 1773
  product for high-temperature steam, 1780
  product for low temperature steam, 1779
Thermoelectric coolers, 1674–1678
  analysis, 1677
  design equations, 1675–1676
  optimization, 1676–1677
  schematic, 1676
Thermoelectric effects, equations, 1674–1678
Thermoforming, 1127
Thermometer:
  resistance, 1954
  vapor pressure, 1954
Thermometry, vapor pressure, 1954
Thermophysical properties:
  conversion factors, 1249
  phase transition:
    compounds, 1251–1252
    elements, 1249–1251
Thermoplastics:
  engineering, 121–124
  fluorinated, 124–125, 128
  properties, 141–142
Thermosets, 128–129, 139
  properties, 141–142
Thermosiphon, 1569
Thevenin's theorem, 920
*The Burwell World Directory of Information
    Brokers,* 2291
The Whole Internet: User's Guide and Catalog,
    2271
Thomas Register, 2275
Thomson voltage, 1675
Thread:
  cutting and forming, 1067–1068
  cutting speeds, 1068
  internal, 1067–1068
  rolling, 1068
  run-out, 1158
  single-thread cutter, 1068
  stress distribution, 1158–1159
Thrust bearing, 605
  annular, 540–542
  bearing number, 552

characteristics, 579
dimensionless load capacity and stiffness determination, 563
fixed-incline pad, 530–532
  allowable minimum outlet film thickness, 532–533
  coefficient of friction determination, 532, 534
  configuration, 530
  lubricant flow determination, 532, 535
  power loss determination, 532–533
gas-lubricated, 548–556
guide to, 514, 515
liquid-lubricated, 530–535
maximum-load-capacity, 552, 563
maximum-stiffness, 552, 563
optimal step parameter determination, 563
pivoted pad, 532–533
  coefficient of friction determination, 532, 534
  configuration, 532–536
  design, 532–533
  film thickness ratio determination, 532–533
  lubricant flow determination, 533, 535
  outlet film thickness determination, 533, 537
  pivot position determination, 532–533
  power loss determination, 532–533
Rayleigh step bearing, 548–553, 562
spiral groove, 553–556
  configuration, 565
  flow determination, 567
  groove factor determination, 564
  groove length fraction determination, 569
  load determination, 565
  optimal groove geometry determination, 568
  stiffness determination, 567
  torque determination, 568
step sector, 533, 539
Tilt factors, 1555–1560
Time:
  constant, 819
  estimate preparation, 2119–2121
  personal, fatigue, and delay, 2140
  standards, 2137
  learning curve, 2130–2131
Time-based competition, 1030–1031
  first to customer for existing products, 1031
  first to market for new products, 1031
Titanium:
  Corrosion data: ASTM grade 2, 99–100
  crystalline forms, 87
  heat-transfer qualities, 97
  history, 91–92
  physical properties, 96–97
  plane strain fracture toughness, 389
  yield strength, 405
Titanium alloys:
  aerospace alloys, 92–107
  alpha-beta transformation temperature, 92
  alpha stabilizers, 92–93
  ASTM grade requirements, 96
  bending, 104

beta-eutectoid elements, 93
beta stabilizers, 92–94
boiler code, 98
company names, 106
corrosion resistance, 97 98
cutting, 104
design stress intensity values in tension, 103
drawing, 100, 104
fabrication, 98
galvanic series in flowing seawater, 101
grinding, 104
health and safety factors, 107
hydrogen-storage alloys, 96
material composition effects on *S-N* curve, 403
maximum allowable stress values in tension, 102
NaCl solution, corrosion, 101
O, N, and C, effects on ultimate tensile strength, 93
nonaerospace alloys, 95–96
phase diagrams, 92
properties, 94–95
quality control, 105, 107
shape-memory alloys, 96
specifications, 94, 105
standards, 105
superconducting alloys, 96
tensile properties, comparison on weld and parent metal, 105
uses, 107–108
welding, 104–105
wrought applications, 94–95
Tools:
  for biochemical analysis, 334
  ceramic, 1046
  classification, 981
    control, 981
    customizing, 983
    investment, 981
    rationale, 982
    standard and special tooling, 982
    taxonomy charts, 982
  coated, 1046
  coding, 982
  control, 981
  cost estimating, 2139
  cutting fluids, 1047
  design standardization, 954
  geometry, 1046–1047
  life, 1041–1043
    cutting speeds and, 1042
    exponent *n* values, 1042
    maximum production and cost, 1046
  maintenance, 2140
  materials, 1046
  oxide, 1046
  special:
    defined, 982
  specification sheets, 985
    acquisition information, 984
    identification, 984
    parameters, 984
  standard, 982
    defined, 982

Tools (*Continued*)
  wear:
    factor, 1042
    types, 1042
Top quality management:
  Deming's approach, 477
  history, 475
  quality, 476
  quality tools, 480–485
  management, 476–477
  quality in the design phase, 477–478
  total involvement, 476
Torque:
  and turn, 1155
    turn-of-nut techniques, 1155
  dc motor, 875–876
  drilling machines, 1051, 1054
    constants, 1055
Total form machining, 1085
Total quality management, 2159–2176
  awards, *see* Awards
  activity network diagram, 2168
  approach:
    organizational, 2160
    unified, 2161
  benefits for company, 2161–2162
  chart results, 2162
  cultural/social tools, 2173
    concurrent engineering, 2173–2175
    teams, 2175
    variability reduction process, 2175–2176
  definitions, 2160–2161
  house of quality, 2164
  interrelationship digraph, 2167–2168
  matrices, 2164
  model, 2163
  one-factor-at-a-time, 2169
  PDCA cycle, 2166
  process decision program chart, 2168, 2169
  random experimentation, 2169–2170
  role change:
    as manager of mechanical engineers, 2163
    as mechanical engineer, 2162–2163
  new paradigm, 2160
  technical tools:
    design of experiments, 2168–2171
    other types, 2173
    process capability, validation studies,
      2172–2173
    quality function deployment, 2164–2166
    statistical process control, 2171–2172
    statistical quality control, 2171–2172
    seven management and planning tools,
      2166–2168
  traditional approach, 2159–2160
  tree diagram, 2167
  see also Teamwork
Toughness:
  comparison, 195
  defined, 194–195
  relative, 198
Toxic Substances Control Act, 2195–2196
Training:
  safety, 2223–2226
    driver, 2223

  fire protection, 2223–2224
  first-aid, 2223
  job hazard analysis, 2224–2225
  manager's overview, 2224–2226
  personalized program, 2224
Transducer:
  defined, 872
  displacement, 872–874
  dynamic response, 875
  flow, 874
  potentiometer, 873
  temperature, 874
  velocity, 872–874
Transfer function, 876
  armature-controlled dc motor, 875–876
  cascaded elements, 893–894
  closed-loop, 869
  command, 870
  disturbance, 871
  field-controlled dc motor, 876
  hydraulic servometer, 877
  pneumatic actuator, 878–880
  pulse, 864–865
  see also Ziegler-Nichols rules
Transform methods, 813–838
  block diagrams, 818–820
  complex poles, 816
  final-value theorem, 817
  impulse response, 818
  initial-value theorem, 817
  inversion by partial-fraction expansion,
    815–816
  Laplace transforms, 813–815
  poles, 814–815
  rational functions, 814, 816
  repeated poles, 816
  response to periodic inputs, 829
    frequency response, 828
    frequency response performance measures,
      829
    frequency response plots, 829
    general periodic inputs, 829
    transfer function plots, 832–839
  transfer functions, 817
  transient analysis, 813–827
    complex s-plane, 825
    damping ratio, 824
    effect of zeros, 827
    first-order response, 819–823
    high-order systems, 825
    overdamped response, 824
    parts of complete response, 818–819
    performance measures, 827
    response to unit step input, 819–822
    second-order response, 823–825
    singularity functions, 819, 821
    test inputs, 819, 821
    transfer function, 823
  zeros, 814–815
Transformation ratio, 802
Transformer:
  linear variable differential, 872–873
  pure, 799
  synchro transmitter-control, 872–874
Transformations, of graphical data, 309–311

Transportation:
 hazard analysis, 2198
Transient heat conduction:
 centerline temperature function of time,
  infinite cylinder, 1388
 center temperature as function of time,
  sphere, 1389
 internal energy change as function of time
  infinite cylinder, 1389
  plan wall, 1387
  sphere, 1391
 midplane temperature as function of time,
  plane wall, 1386
 temperature distribution:
  fin base, 1384
  infinite cylinder, 1389
  plane wall, 1387
  sphere, 1391
Transit time instruments, 1156
Transmissivity, 1406
Transportation:
 bulk materials, 1218–1219
Trees, as graphical aids, 774
Trunnion, 237, 241
Trimming, 1118
Tube spinning, 1111
Turbine:
 aviation fuels, 1525–1526
 thermodynamics, 1347
Turbulent flow, 1294, 1307, 1487–1488
 boundary layer parameters, 1309
 circular tubes, 1392–1393
 ducts, 1316–1319
 flat plate, 1393
 flow rate, 1317
 friction factors for commercial pipes,
  1318
 growth of boundary layers, 1315
 pipe size, 1317, 1320
 transition from laminar to turbulent boundary
  layer, 1307
 turbulence level, 1307
 velocity fluctuations, 1307
Turning machines, 1048, 1050–1051
 break-even conditions, 1051
 lathe size, 1051
 tool angles and cutting speeds, 1048,
  1050
 see also Boring; Lathe
Two-card Kanban system, 1027
Two machines/n jobs, 1017–1018, 1020
Two-phase states:
 Clapeyron relation, 1346
 critical point, 1346
 critical-state properties, 1346
 locus, 1345
 moderately compressed liquid, 1347
 saturated solid, 1347
 slightly superheated vapor, 1347
Two-surface method, 717–718
 area ratio, 718
 microphone locations, 718
Ulrich's International Periodical Directory,
 2288

Ultrasonic methods, 732–738
 generation and propagation of surface waves,
  735
 inspection process, 737–738
  C-scan image of composite specimen, 738
  representation of ultrasonic data collection,
   737
 reflection and transmission of sound,
  733–735
  data collection and display in the A-scan
   mode, 734
 refraction of sound, 735–737
 amplitude and phase of reflected coefficient and
  transmitted amplitude vs. angle of
  incidence for longitudinal wave incident
  on water-steel interface, 736
 Snell's law and direction of propagation of
  acoustic waves, 735
 sound waves, 733
Ultrasonic machining, 1078, 1086
Ultrasonic properties of common materials,
 754–758
 calculated properties of composites, 758
 of liquids, 754
 of solids:
 ceramics, 756–757
 metals, 754–756
 polymers, 757–758
Ultraviolet oxidizers, 2024
Uniform sampling, 862, 864
Unit load design, 1228
Unit manufacturing cost, 935–936
Unitary air conditioners, 1995
University Technologies, 2283
Upset forging, 1106
UNIX, 297, 320, 373
U.S. Environmental Protection Agency, 2012,
 2032, 2180
Usenet groups, 2276
User interface issues, model, 334
35 U.S.C. 101, 2252
35 U.S.C. 102, 2252–2253
 novelty and statutory bar requirements, 2252
 statutory bar requirements, 2252
35 U.S.C. 102(a):
 described in printed publication, 2252–2253
 known or used by others, 2253
 patented in this or a foreign country, 2253
35 U.S.C. 102(e), described in patent,
 2252–2253
35 U.S.C. 102(g), abandoned, suppressed, or
 concealed, 2252–2253
35 U.S.C. 103, 2253
 nonobvious requirement, 2253
35 U.S.C. 112, 2257

Vacuum insulation, cryogenic insulation,
 1940–1941
 emissivities of materials for radiation shields,
  1941
 Stefan-Boltzmann equation, 1940
Vacuum metallizing, 1131
Valve:
 comparing experimental data to predictions,
  1836